HAZARDOUS CHEMICALS DATA BOOK

Hazardous Chemicals Data Book

Edited by G. Weiss

NOYES DATA CORPORATION
Park Ridge, New Jersey, U.S.A.
1980

Library of Congress Catalog Card Number: 80-21634
ISBN: 0-8155-0831-X
Printed in the United States

Published in the United States of America by
Noyes Data Corporation
Noyes Building, Park Ridge, New Jersey 07656

Library of Congress Cataloging in Publication Data

Main entry under title:

Hazardous chemicals data book.

Bibliography: p.
Includes index.
1. Hazardous substances--Handbooks, manuals,
etc. I. Weiss, G., 1940- II. Series.
T55.3.H3H396 604.7 80-21634
ISBN 0-8155-0831-X

FOREWORD

Instant information for decision-making in emergency situations by personnel involved with chemical accidents is the prime purpose of this compilation of about 1,350 hazardous chemicals. The book, prepared in clear, concise, easy-to-locate format, should become a valuable source on any library or laboratory shelf. It is intended for use by scientists, engineers, managers, transportation personnel, or anyone who may have contact with or require data on a particular chemical.

There is a large amount of pertinent data provided for each chemical, and examples of a few of the headings are:

Toxicity and Health Hazards	Flammability and Explosion Hazards
Hazards Relating to Water Pollution	Chemical Reactivity
Spill and Leak Procedures	Extinguishing Media
Special Protective Equipment	Special Precautions

The volume is set up in two parts, each alphabetically arranged. The first part is based on the Department of Transportation's Chemical Hazard Response Information System. The second part is based on Material Safety Data Sheets obtained from the Oak Ridge National Laboratory. It should be noted that Part 2 includes only those chemicals which were not listed in Part 1. The use of a convenient all-inclusive chemical index furnishes the responsible individual with the desired information in a minimum of time.

The two information sources used to prepare this book are:

CHRIS—Hazardous Chemical Data (GPO 050-012-00147-2), issued by the U.S. Department of Transportation Coast Guard, October 1978.

Material Safety Data Sheets—The Basis for Control of Toxic Chemicals, Volumes 1-3 (ORNL/TM-6981/V1, V2, V3), by E.E. Ketchen and W.E. Porter of Oak Ridge National Laboratory for the U.S. Department of Energy, September 1979.

In order to keep the price of this large and extensive book to a reasonable level, it has been reproduced by photo-offset directly from the original reports and the cost savings passed on to the reader. Due to this method of publishing, certain portions of the reports may be less legible than desired.

TABLE OF CONTENTS

PART 1: DEPARTMENT OF TRANSPORTATION CHEMICAL HAZARD RESPONSE INFORMATION SYSTEM (CHRIS) CHEMICALS

PART 2: OAK RIDGE NATIONAL LABORATORY CHEMICALS

Part 1

Department of Transportation Chemical Hazard Response Information System (CHRIS) Chemicals

This part was prepared from *CHRIS–Hazardous Chemical Data* (GPO 050-012-00147-2), issued by the U.S. Department of Transportation Coast Guard, October 1978.

INTRODUCTION

The Chemical Hazard Response Information System (CHRIS) was designed to provide information needed for decision-making by responsible Coast Guard personnel during emergencies that occur during the water transport of hazardous chemicals. The system consists of four handbooks—*A Condensed Guide to Chemical Hazards, Hazardous Chemical Data, Hazard Assessment Handbook* (CG-446-3), and *Response Methods Handbook* (CG-446-4); a hazard assessment computer system (HACS), and technical support personnel located at Coast Guard headquarters.

Only the hazardous chemical data sheets are presented here, as it was thought that this information would be of value to industry. The publication of the data sheets alone will add a significant body of instant information for use by the scientist, engineer, manager, or transportation personnel in either the laboratory or the library.

EXPLANATION OF TERMS

This section explains the special terms used in the data sheets which follow, gives the sources of specific items, and includes other information that will be useful to the reader in interpreting the data. The paragraphs below are keyed to the relevant portions of the sheets by the subheading and number used in the data sheets.

The expression "Not pertinent" means that the data item either has no real meaning (such as the flash point of a nonflammable chemical) or is not required for assessing a hazardous situation. The expression "Data not available" means that the information sought was not found in the general or specialized data sources listed in the "Data Sources" section. In a few cases where important data were not available, values were estimated by usually reliable procedures; all such values are labeled "(est.)".

The *name* used for each of the chemicals included in the CHRIS manuals is either (1) that specified in the Code of Federal Regulations, Title 46, Table 151.01-10(b) (Compounds Regulated by Subchapter 0) and Table 151.01-10(d) (Compounds to be Regulated by Subchapter D) or (2) a common name of those chemicals not now regulated by Subchapters O and D but known to be hazardous during shipment. The data sheets are arranged in alphabetical order by chemical name, not by the 3-letter code.

The *3-letter code* is designed to facilitate correct identification of chemicals in oral or written communication. The code should be used only *in addition* to the compound name; it should not be used alone.

1. RESPONSE TO DISCHARGE

In every case of a discharge or leak, it is obvious that an effort should be made to reduce, stop, or contain the flow of material at its source if this can be done safely. The purpose of the terms used in this section is to describe in a general way the cautionary and corrective responses that are described in greater detail in the Response Methods Handbook.

- *"Issue warning"* is used when the chemical is a *poison,* has a *high flammability,* is a *water contaminant,* is an *air contaminant* (so as to be hazardous to life), is an *oxidizing material,* or is *corrosive.* In the Response Methods Handbook the italicized properties are the bases for cautionary responses to restrict ignition, and to restrict contaminated water for human use, farm use, and industrial use.
- *"Restrict access"* is used only for those chemicals that are unusually and immediately hazardous to personnel unless they are protected properly by gas masks, eye goggles, protective clothing, etc. In the Response Methods Handbook this cautionary response is sometimes used in a broader sense to ensure exclusion of spectators and others who might ignite flammable compounds.
- *"Evacuate area"* is used primarily for unusually poisonous chemicals or those that ignite easily. The same expression is used in the Response Methods Handbook for this cautionary response.
- *"Mechanical containment"* is used for water-insoluble chemicals that float and do not evaporate readily. In the Response Methods Handbook the corresponding corrective response is "Contain".
- *"Should be removed"* is used for chemicals that cannot be allowed to disperse because of their harmful effect on humans or on the ecological system in general. The term is not used unless there is a reasonable chance of preventing dispersal, after a discharge or leak, by chemical and physical treatment.
- *"Chemical and physical treatment"* is recommended for chemicals that can, be removed by skimming, pumping, dredging, burning, neutralization, absorption, coagulation, or precipitation. The corrective response may also include the use of dipersing agents, sinking agents, and biological treatment. Specific procedures to be followed are found in the Response Methods Handbook. Cleaning of the shoreline is also specified there for some compounds.
- *"Disperse and flush"* is used for chemicals that can be made nonhazardous to humans by simple dilution with water; a corresponding corrective response is used in the Response Methods Handbook. In a few cases the response is indicated even when the compound reacts with water because, when proper care is taken, dilution is still the most effective way of removing the primary hazard.

2. LABELS

The labels illustrated in Section 2 are those specified by the United Nations Labelling System (see CFR, Title 49, Part 172). Labels are shown for all chemicals that are listed by name in the CFR and also for most flammable chemicals that fall under listed generic categories, such as "Solvents," or "Alcohols."

3. CHEMICAL DESIGNATIONS

3.1 Synonyms —Alternative systematic chemical names and commonly used trivial names are given. Commercial or trade names are shown in a few cases where they are in common use.

3.2 Coast Guard Compatibility Classification – An entry is made when the chemical has been assigned to one of the 43 cargo groups listed in Navigation and Vessel Inspection Circular No. 4-75, "Guide to Compatibility of Chemicals." Appropriate parts of the Guide are included in this manual. Some chemicals not now included in the Guide were assigned to a group by the Cargo and Hazardous Materials Division, Coast Guard Headquarters. If the chemical is not a liquid carried in bulk in ships' tanks, this data item is "Not applicable."

3.3 Chemical Formula – This has been limited to a commonly used one-line formula. In the case of some organic compounds it has not been possible to represent chemical structure within such a limitation.

3.4 IMCO/United Nations Numerical Designation – The designation is that of the "International Maritime Dangerous Goods Code" published by the Inter-Governmental Maritime Consultative Organization (IMCO), London, 1972.

4. OBSERVABLE CHARACTERISTICS

4.1 Physical State (as shipped) – All chemicals that are listed in Code of Federal Regulations, Title 46, Tables 151.01-10(b) and 151.01-10(d) are shipped as liquids. Other designations include liquefied gas, liquefied compressed gas, and solid. Where a compound may be shipped either as a liquid or solid, both designations are given.

4.2 Color – All color descriptions found in the common reference sources are included. The color description is that for pure material. Occasionally the color of a chemical changes when it dissolves in water or becomes a gas.

4.3 Odor – All odor descriptions found in the common reference sources are included. The expression "characteristic" is used only when no other reasonable description was found. The odor description is that for pure material.

5. HEALTH HAZARDS

5.1 Personal Protective Equipment – The items listed are those recommended by (a) manufacturers, either in technical bulletins or in Material Safety Data Sheets, (b) the Manufacturing Chemists Association, or (c) the National Safety Council, for use by personnel while responding to fire or accidental discharge of the chemical. They are intended to protect the lungs, eyes, and skin. Safety showers and eyewash fountains are considered to be important protective equipment for the handling of almost all chemicals; they are not usually listed.

5.2 Symptoms Following Exposure – These are brief descriptions of the effects observed in humans when the vapor (gas) is inhaled, when the liquid or solid is ingested (swallowed), and when the liquid or solid comes in contact with the eyes or skin.

5.3 Treatment for Exposure – "First-aid" procedures are recommended. They deal with exposure to the vapor (gas), liquid, or solid and include inhalation, ingestion (swallowing) and contact with eyes or skin. The instruction "Do NOT induce vomiting" is given if an unusual hazard is associated with the chemical being sucked into the lungs (aspiration) while the patient is vomiting. "Seek medical attention" or "Call a doctor" is recommended in those cases where only competent medical personnel can treat the injury properly. In all cases of human exposure, seek medical assistance as soon as possible.

5.4 Toxicity by Inhalation (Threshold Limit Value) – The threshold limit value (TLV) is usually expressed in units of parts per million (ppm) – i.e., the parts of vapor (gas) per million parts of contaminated air by volume at 25°C (77°F) and atmospheric pressure. For a chemical that forms a fine mist or dust, the concentration is given in milligrams per cubic meter (mg/m^3). The TLV is defined as the concentration of the substance in air that can be breathed for five consecutive eight-hour workdays (40-hour work week) by most people without adverse effect.* As some people become ill after exposure to concentrations lower than the TLV, this value cannot be used to define exactly what is a "safe" or "dangerous" concentration.

No entry appears when the chemical is a mixture; it is possible to calculate the TLV for a mixture only when the TLV for each component of the mixture is known and the composition of the mixture by weight is also known.

---

*American Conference of Governmental Industrial Hygienists, "Threshold Limit Values for Substance in Workroom Air, Adopted by ACGIH for 1972"

5.5 Short-Term Inhalation Limits – The parts of vapor (gas) per million parts of contaminated air by volume at 25°C (77°F) and atmospheric pressure is given. The limits are given in milligrams per cubic meter for chemicals that can form a fine mist or dust. The values given are the maximum permissible average exposures for the time periods specified. Most of the data came from the Commonwealth of Pennsylvania, Department of Environmental Resources, Title 25, Article IV, Chapter 201 (1971). Some of the data (called "Emergency Exposure Limits") were provided by the Committee on Toxicology, National Academy of Science – National Academy of Engineering, National Research Council.

In some instances the values disagree, or the short-term limits overlap the TLV given in 5.4. These are not errors; the values were supplied by several laboratories, each of which used its own experimental techniques and methods of calculation.

5.6 Toxicity by Ingestion – The Grade and corresponding LD_{50} value are those defined by the National Academy of Sciences, Committee on Hazardous Materials, "Evaluation of the Hazard of Bulk Water Transportation of Industrial Chemicals, A Tentative Guide," Washington, D.C., 1972. Data were also collected from other sources and converted to the appropriate Grade before entry in this manual. The term LD_{50} signifies that about 50% of the animals given the specified dose by mouth will die. Thus, for a Grade 4 chemical (below 50 mg/kg) the toxic dose for 50% of animals weighing 70 kg (150 lb) is 70 x 50 = 3500 mg = 3.5 g, or less than 1 teaspoonful; it might be as little as a few drops. For a Grade 1 chemical (5 to 15g/kg), the LD_{50} would be between a pint and a quart for a 150-lb man. All LD_{50} values have been obtained using small laboratory animals such as rodents, cats, and dogs. The substantial risks taken in using these values for estimating human toxicity are the same as those taken when new drugs are administered to humans for the first time.

5.7 Late Toxicity – Where there is evidence that the chemical can cause cancer, mutagenic effects, teratogenic effects, or a delayed injury to vital organs such as the liver or kidney, a qualitative description of the effect is given.

5.8 Vapor (Gas) Irritant Characteristics – The most appropriate of five statements listed below is given. (Source: National Academy of Sciences, Committee on Hazardous Materials, "Evaluation of the Hazard of Bulk Water Transportation of Industrial Chemicals, A Tentative Guide," Washington, D.C., 1970.)

(1) Vapors are nonirritating to eyes and throat.

(2) Vapors cause a slight smarting of the eyes or respiratory system if present in high concentrations. The effect is temporary.

(3) Vapors cause moderate irritation such that personnel will find high concentrations unpleasant. The effect is temporary.

(4) Vapors are moderately irritating such that personnel will not usually tolerate moderate or high concentrations.

(5) Vapors cause severe irritation of eyes and throat and can cause eye and lung injury. They cannot be tolerated even at low concentrations.

5.9 Liquid or Solid Irritant Characteristics – The most appropriate of the following five statements is given (same source as 5.8 above):

(1) No appreciable hazard. Practically harmless to the skin.

(2) Minimum hazard. If spilled on clothing and allowed to remain, may cause smarting and reddening of skin.

(3) Causes smarting of the skin and first-degree burns on short exposure; may cause second-degree burns on long exposure.

(4) Fairly severe skin irritant. May cause pain and second-degree burns after a few minutes' contact.

(5) Severe skin irritant. Causes second- and third-degree burns on short contact and is very injurious to the eyes.

5.10 Odor Threshold – This is the lowest concentration in air that most humans can detect by smell. The value cannot be relied on to prevent over-exposure, because human sensitivity to odors varies over wide limits, some chemicals cannot be smelled at toxic concentrations, odors can be masked by other odors, and some compounds rapidly deaden the sense of smell.

6. FIRE HAZARDS

6.1 Flash Point – This is defined as the lowest temperature at which vapors above a volatile combustible substance will ignite in air when exposed to a flame. Depending on the test method used, the values given are either Tag closed cup (C.C.) (ASTM D56) or Cleveland open cup (O.C.) (ASTM D93). The values, along with those in 6.2 and 6.7 below, give an indication of the relative flammability of the chemical. In general, the open cup value is about 10° to 15°F higher than the closed cup value.

6.2 Flammable Limits in Air – The percent concentration in air (by volume) is given for the lower (LFL) and upper (UFL) limit. The values, along with those in 6.1 and 6.7, give an indication of the relative flammability of the chemical. The limits are sometimes referred to as "lower explosive limit" (LEL) and "upper explosive limit" (UEL).

6.3 Fire Extinguishing Agents – The agents are listed in decreasing order of importance. The general capabilities of all agents are described by G.H. Tryon, "Fire Protection Handbook," 13th ed., National Fire Protection Association, Boston, Mass., 1969.

6.4 Fire Extinguishing Agents Not to be Used – The agents listed must not be used because they react with the chemical and create an additional hazard. In some cases they are listed because they are ineffective in putting out the fire.

6.5 Special Hazards of Combustion Products – Some chemicals decompose or burn to give off toxic and irritating gases. Such gases may also be given off by chemicals that vaporize in the heat of a fire without either decomposing or burning. If no entry appears, the combustion products are thought to be similar to those formed by the burning of oil, gasoline, or alcohol; they include carbon monoxide (poisonous), carbon dioxide, and water vapor. The specific combustion products are usually not well known over the wide variety of conditions existing in fires; some may be hazardous.

6.6 Behavior in Fire – Any characteristic behavior that might increase significantly the hazard involved in a fire is described. The formation of dense smoke or flammable vapor clouds, and the possibility of polymerization and explosions is stated. Unusual difficulty in extinguishing the fire is also noted.

6.7 Ignition Temperature – This is the minimum temperature at which the material will ignite without a spark or flame being present. Along with the values in 6.1 and 6.2 above, it gives an indication of the relative flammability of the chemical. It is sometimes called the "autoignition temperature." The method of measurement is given in ASTM D-2155.

6.8 Electrical Hazard – The ease with which the chemical is ignited by electrical equipment is indicated by the *Group* and *Class* assignment made in "Fire Codes," Vol. 5, National Fire Protection Association, Boston, Mass., 1972, pp. 70-289. This information is available for relatively few chemicals, so an absence of data does not necessarily mean that the substance is not hazardous in the presence of electrical equipment.

6.9 Burning Rate – The value is the rate (in millimeters per minute) at which the depth of a pool of liquid decreases as the liquid burns. Details of measurement are given by D.S. Burgess, A. Strasser, and J. Grumer, "Diffusive Burning of Liquid Fuels in Open Trays," Fire Research Abstracts and Reviews, 3, 177 (1961).

7. CHEMICAL REACTIVITY

7.1 Reactivity with Water – The term "No reaction" means that no hazard results when the chemical reacts or mixes with water. Where a hazard does result, it is described.

7.2 Reactivity with Common Materials – This is limited to hazardous reactions with fuels and with common materials of construction such as metal, wood, plastics, cement, and glass. The nature of the hazard, such as severe corrosion or formation of a flammable gas, is described.

7.3 Stability During Transport – The term "Stable" means that the chemical will not decompose in a hazardous manner under the conditions of temperature, pressure, and mechanical shock that are normally encountered during shipment; the term does *not* apply to fire situations. Where there is a possibility of hazardous decomposition, an indication of the conditions and the nature of the hazard is given.

7.4 Neutralizing Agents for Acids and Caustics – In all cases involving accidental discharge, dilution with water may be followed by use of the agent specified, particularly if the material cannot be flushed away; the agent specified need not necessarily be used. More detailed information for this kind of response is given in the Response Methods Handbook, CG-446-4.

7.5 Polymerization – A few chemicals can undergo rapid polymerization to form sticky, resinous materials, with the liberation of much heat. The containers may explode. For these chemicals the conditions under which the reaction can occur are given. See Section 13.16 for quantitative data.

7.6 Inhibitor of Polymerization – The chemical names and concentrations of inhibitors added by the manufacturer to prevent polymerization are given.

8. WATER POLLUTION

8.1 Aquatic Toxicity – The form of data presentation used by the Environmental Protection Agency's "Oil and Hazardous Material-Technical Assistance Data System (OHM-TADS)" is used here. Reading from left to right and separated by slashes (/) are the following data:

- Concentration in parts per million by weight (or milligrams per liter) at which the chemical was tested;

- Time of exposure in hours;

- Name of the aquatic species studied;

- Effect observed; LC_{50} means that approximately 50% of the fish will die under the conditions of concentration and time given. TL_m (Median Tolerance Limit) means that approximately 50% of the fish will show abnormal behavior (including death) under the conditions of concentration and time given; the term EC_{50} (Effective $Concentration_{50}$) is used sometimes instead of TL_m;

- The kind of water used in the test (fresh or salt).

Some chemicals have been tested with many species of fish. Where the data were available, the data sheet cites one illustrative test in fresh water and one in salt water.

8.2 Waterfowl Toxicity –Very little information is available. In a few cases there is entered the LD_{50} value, which indicates the dose (in milligrams per kilogram of body weight) that is lethal to about half the waterfowl tested.

8.3 Biological Oxygen Demand (BOD) – Also called "biochemical oxygen demand," this is a standard way of describing how much oxygen dissolved in water is consumed by biological oxidation of the chemical during the stated period of time. The unit *lb/lb* indicates the pounds of oxygen consumed by each pound of chemical during the time stated. When given in *percent,* the values indicate the pounds of oxygen consumed by each 100 pounds of chemical during the time stated. If the percentage is followed by "(theor.)", it indicates the pounds of oxygen theoretically required to completely oxidize 100 pounds of the chemical.

8.4 Food Chain Concentration Potential – If the chemical is consumed by fish, marine plants, waterfowl, etc., that are in turn eaten by other species, the substance may accumulate and ultimately be consumed by humans. Where this occurs, an indication of the potential hazard and its significance is given.

9. SELECTED MANUFACTURERS

Three manufacturers are named wherever possible. When production data for the chemical were available, the manufacturers were selected from among the major producers. The "1972 Directory of Chemical Producers, United States of

America" (Chemical Information Services, Stanford Research Institute, Menlo Park, California) and "Buyers Guide Issue," *Chemical Week,* **111**, No. 7, Part Two (October 25, 1972) were the sources used.

10. SHIPPING INFORMATION

10.1 Grades or Purity – The grades USP (United States Pharmacopoeia) and CP (chemically pure) are quite pure. Where "Technical" or "Commercial" grades are given, the percent by weight of the pure chemical present is usually indicated. In a few cases the identity of the major impurities is given. If the properties of the less pure grades differ significantly from those of the pure substance, the differences in properties are described in general terms.

10.2 Storage Temperature – The range of temperatures at which the chemical is normally shipped in bulk by water transport is given. "Ambient" means the temperature of the surroundings.

10.3 Inert Atmosphere – The terms used are "inerted," "padded," "ventilated (forced)," "ventilated (natural)," and "no requirement." They are given when found in the Code of Federal Regulations, Title 46, beginning in Part 151.05.

10.4 Venting – The terms used are "open," "pressure-vacuum," and "safety relief" (same source as 10.3 above).

11. HAZARD ASSESSMENT CODE

The code designates the assessment calculation route to be used in the Hazard Assessment Handbook when evaluating the hazard.

12. HAZARD CLASSIFICATIONS

12.1 Code of Federal Regulations – This is the hazard class specified in the Code of Federal Regulations, Title 49, Part 172, December 31, 1976. Chemicals not specifically listed therein have been classified as "Flammable" if their flash point (closed cup) is below 100°F.

12.2 NAS Hazard Rating for Bulk Water Transportation – The indicated ratings are given in "Evaluation of the Hazard of Bulk Water Transportation of Industrial Chemicals, A Tentative Guide," National Academy of Sciences, Committee on Hazardous Materials, Washington, D.C., 1972. An outline of the rating system is given in Table 1.

TABLE 1

EXPLANATION OF NAS HAZARD RATINGS

Rating	Fire	Health: Vapor Irritant	Health: Liquid or Solid Irritant	Health: Poisons
0	No hazard	No effect	No effect	No effect
1	Flash point (C.C.) above 140°F	Slight effect	Causes skin smarting	Slightly toxic
2	Flash point (C.C.) 100 to 140°F	Moderate irritation; temporary effect	First-degree burns, short exposure	Intermediate toxicity
3	Flash point (C.C.) below 100°F; boiling point above 100°F	Irritating; cannot be tolerated	Second-degree burns, few minutes' exposure	Moderately toxic
4	Flash point (C.C.) below 100°F; boiling point below 100°F	Severe effect; may do permanent injury	Second-degree and third-degree burns	Severely toxic

Rating	Water Pollution: Human Toxicity	Water Pollution: Aquatic Toxicity	Water Pollution: Aesthetic Effect
0	Nontoxic; LD_{50}> 15 g/kg	Acute threshold limits above 10,000 ppm	No significant pollution; gases and odorless liquids
1	Practically nontoxic; LD_{50} 5 to 15 g/kg	Threshold limits 1,000 to 10,000 ppm	Mild-odored light oils and soluble chemicals
2	Slightly toxic; LD_{50} 0.5 to 5 g/kg	Threshold limits 100 to 1,000 ppm	Mild-odored, colorless, water-insoluble oils; boiling point 150-450°F
3	Moderately toxic; LD_{50} 50 to 500 mg/kg	Threshold limits 1 to 100 ppm	Light-colored high-boiling oils; odorous water-soluble compounds
4	Toxic; LD_{50}<50 mg/kg	Threshold limits below 1 ppm	Heavy oils, colored or bad odors

TABLE 1 (Continued)

Rating	Reactivity: Other Chemicals	Reactivity: Water	Reactivity: Self-reaction
0	Inactive; may be attacked by materials rated 4	No reaction	No reaction
1	React only with materials rated 4	Mild reaction; unlikely to be hazardous	Mild self-reaction under some conditions
2	React with materials rated 3 or 4	Moderate reaction	Will undergo self-reaction if contaminated; do not require stabilizer
3	React with each other and with materials rated 2 or 4	More vigorous reaction; may be hazardous	Vigorous self-reaction; require stabilizer
4	React with each other and materials rated 0-3	Vigorous reaction; likely to be hazardous	Self-oxidizing chemical; capable of explosion or detonation

12.3 NFPA Hazard Classifications — The indicated ratings are given in "Fire Protection Guide on Hazardous Materials," Code 704M, 4th ed., National Fire Protection Association, Boston, Mass., 1972. Changes in the rating that are described in "1973 NFPA Technical Committee Reports, Volume A" are included. The classifications are defined in Table 2 below. This system uses the diamond-shaped warning symbol, shown below. The top, left, and right boxes refer to flammability, health, and reactivity hazards respectively and contain a number from 0 to 4. The bottom box is used for special hazards; the most common of these is a warning against the use of water, indicated by the symbol ~~W~~.

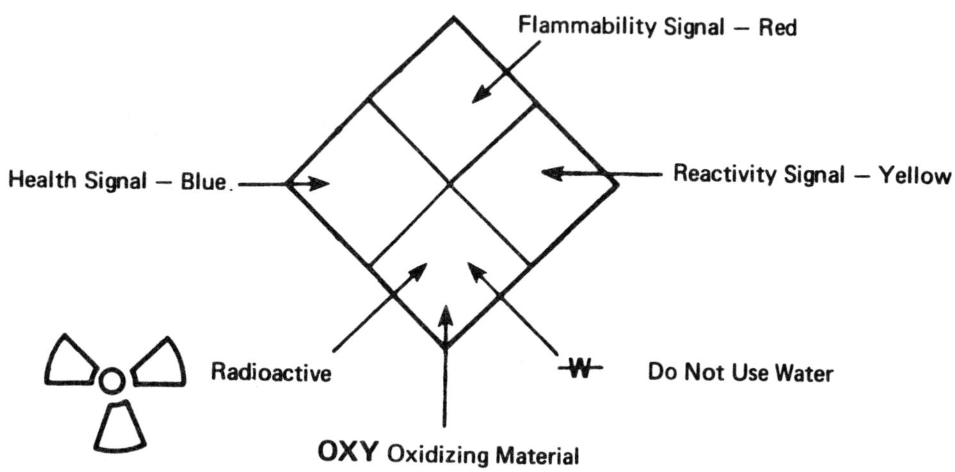

TABLE 2

EXPLANATION OF NFPA HAZARD CLASSIFICATIONS

Classification Health Hazard (blue)	Definition
4	Materials which on very short exposure could cause death or major residual injury even though prompt medical treatment were given.
3	Materials which on short exposure could cause serious temporary or residual injury even though prompt medical treatment were given.
2	Materials which on intense or continued exposure could cause temporary incapacitation or possible residual injury unless prompt medical treatment is given.
1	Materials which on exposure would cause irritation but only minor residual injury even if no treatment is given.
0	Materials which on exposure under fire conditions would offer no hazard beyond that of ordinary combustible material.
Flammability (red)	
4	Materials which will rapidly or completely vaporize at atmospheric pressure and normal ambient temperature, or which are readily dispersed in air and which will burn readily.
3	Liquids and solids that can be ignited under almost all ambient temperature conditions.
2	Materials that must be moderately heated or exposed to relatively high ambient temperatures before ignition can occur.

TABLE 2 (Continued)

Classification	Definition
Flammability (red) cont'd	
1	Materials that must be preheated before ignition can occur.
0	Materials that will not burn.
Reactivity (yellow)	
4	Materials which in themselves are readily capable of detonation or of explosive decomposition or reaction at normal temperatures and pressures.
3	Materials which in themselves are capable of detonation or explosive reaction but require a strong initiating source or which must be heated under confinement before initiation or which react explosively with water.
2	Materials which in themselves are normally unstable and readily undergo violent chemical change but do not detonate. Also materials which may react violently with water or which may form potentially explosive mixtures with water.
1	Materials which in themselves are normally stable, but which can become unstable at elevated temperatures and pressures or which may react with water with some release of energy but not violently.
0	Materials which in themselves are normally stable, even under fire exposure conditions, and which are not reactive with water.
Other (white)	
~~W~~	Materials which react so violently with water that a possible hazard results when they come in contact with water, as in a fire situation. Similar to Reactivity Classification 2.
Oxy	Oxidizing material; any solid or liquid that readily yields oxygen or other oxidizing gas, or that readily reacts to oxidize combustible materials.

13. PHYSICAL AND CHEMICAL PROPERTIES

13.1 Physical State at 15°C and 1 atm – The statement indicates whether the chemical is a solid, liquid, or gas after it has reached equilibrium with its surroundings at "ordinary" conditions of temperature and pressure.

13.2 Molecular Weight – The value given is the weight of a molecule of the chemical relative to a value of 12 for one atom of carbon.

The molecular weight is useful in converting from molecular units to weight units, and in calculating the pressure, volume and temperature relationships for gaseous materials. The ratio of the densities of any two gases is approximately equal to the ratio of their molecular weights (see 13.10).

The molecular weights of mixtures can be calculated if both the identity and quantity of each component of the mixture are known. Because the composition of mixtures described in this manual is not known exactly, or because it varies from one shipment to another, no molecular weights are given for such mixtures.

13.3 Boiling Point at 1 atm – The value is the temperature of a liquid when its vapor pressure is 1 atm. For example, when water is heated to 100°C (212°F) its vapor pressure rises to 1 atm and the liquid boils.

The boiling point at 1 atm indicates whether a liquid will boil and become a gas at any particular temperature and sea-level atmospheric pressure.

13.4 Freezing Point – The freezing point is the temperature at which a liquid changes to a solid. For example, liquid water changes to solid ice at 0°C (32°F). Some liquids solidify very slowly even when cooled below their freezing point. When liquids are not pure (for example, salt water) their freezing points are lowered slightly.

13.5 Critical Temperature – The maximum temperature at which a liquid can exist, no matter what the pressure on it, is called the critical temperature. For example, the critical temperature of water is 372°C (705°F). The value can be used to estimate many properties whose values are not immediately available.

13.6 Critical Pressure – The vapor pressure of a chemical at the critical temperature (see 13.5) is called the critical pressure. For example, the critical pressure of water is 218 atm. Values are given in pounds per square inch absolute, atmospheres, and meganewtons per square meter. The value can be used for estimating many property values that are not immediately available.

13.7 Specific Gravity – The specific gravity of a chemical is the ratio of the weight of the solid or liquid to the weight of an equal volume of water at 4°C (or at some other specified temperature).

If the specific gravity is less than 1.0 (or less than 1.03 in seawater) the chemical will float; if higher, it will sink.

13.8 Liquid Surface Tension – This property is a measure of the tensile force at the surface of a liquid that tends to shape liquid fragments into spherical drops. Values are expressed in dynes per centimeter and newtons per meter. Liquids with high surface tensions show less tendency to spread. Water has a surface tension of about 73 dynes/cm; seawater has a slightly higher value.

13.9 Liquid-Water Interfacial Tension – The value is a measure of the tensile forces existing at the interface between a liquid and water. Approximately, it is the difference between the individual surface tension of the liquid and that of water. Low values of the interfacial tension indicate that the chemical spreads readily on a water surface. The units are the same as in 13.8.

13.10 Vapor (Gas) Specific Gravity – The value is the ratio of the weight of vapor to the weight of an equal volume of dry air at the same conditions of temperature and pressure. Buoyant vapors have a vapor specific gravity less than one. The value may be approximated by the ratio M/29, where M is the molecular weight of the chemical (see 13.2).

In some cases the vapor may be at a temperature different from that of the surrounding air. For example, the vapor from a container of boiling methane at -172°F sinks in warm air, even though the vapor specific gravity of methane at 60°F is about 0.6.

13.11 Ratio of Specific Heats of Vapor (Gas) – This property is the ratio of the specific heat at constant pressure (C_p) to the specific heat at constant volume (C_v); its value is always greater than one. In most cases it was calculated by use of the expression

$$\frac{C_p}{C_v} = \frac{C_p}{(C_p - R)}$$

where R is the Universal Gas Constant.

The ratio varies slightly with temperature; the value given is at 20°C (68°F). The ratio is often of value in estimating temperature changes when gases are compressed or expanded. Higher values of the ratio lead to larger temperature changes for a given pressure change.

13.12 Latent Heat of Vaporization – The value is the heat that must be added to the specified weight of a liquid before it can change to vapor (gas). It varies with temperature; the value given is that at the boiling point at 1 atm (see 13.3). The units used are Btu per pound, calories per gram, and joules per kilogram.

No value is given for chemicals with very high boiling points at 1 atm, because such substances are considered essentially nonvolatile.

13.13 Heat of Combustion – The value is the amount of heat liberated when the specified weight is burned in oxygen at 25°C. The products of combustion, including water, are assumed to remain as gases; the value given is usually referred to as the "lower heat value." The negative sign before the value indicates that heat is given off when the chemical burns. Units are the same as in 13.12.

13.14 Heat of Decomposition – The value is the amount of heat liberated when the specified weight decomposes to more stable substances. The value is given for very few chemicals, because most are stable and do not decompose under the conditions of temperature and pressure encountered during shipment. The negative sign before the value simply indicates that heat is given off during the decomposition. The value does *not* include heat given off when the chemical burns. Units are the same as in 13.13.

13.15 Heat of Solution – The value represents the heat liberated when the specified weight of chemical is dissolved in a relatively large amount of water at 25°C ("infinite dilution"). A negative sign before the value indicates that heat is given off, causing a rise in temperature. (A few chemicals absorb heat when they dissolve, causing the temperature to fall.) Units are the same as in 13.12.

In those few cases where the chemical reacts with water and the reaction products dissolve, the heat given off during the reaction is included in the heat of solution.

13.16 Heat of Polymerization – The value is the heat liberated when the specified weight of the compound (usually called the monomer) polymerizes to form the polymer. In some cases the heat liberated is so great that the temperature rises significantly, and the material may burst its container or catch fire. The negative sign before the value indicates that heat is given off during the polymerization reaction. Units are the same as in 13.12.

GUIDE TO COMPATIBILITY OF CHEMICALS

The Guide is based in part upon information provided to the Coast Guard by the National Academy of Sciences – U.S. Coast Guard Advisory Committee on Hazardous Materials and represents the latest information available to the Coast Guard on chemical compatibility.

The accidental mixing of one chemical cargo with another can in some cases be expected to result in a vigorous and hazardous chemical reaction. The generation of toxic gases, the heating, overflow, and rupture of cargo tanks, and fire and explosion are possible consequences of such reactions.

The purpose of the Compatibility Chart is to show chemical combinations believed to be dangerously reactive in the case of accidental mixing. It should be recognized, however, that the Chart provides a broad grouping of chemicals with an extensive variety of possible binary combinations. Although one group, generally speaking, can be considered dangerously reactive with another group where an "X" appears on the Chart, there may exist between the groups some combinations which would not dangerously react. The Chart should therefore not be used as an infallible guide. It is offered as an aid in the safe loading of bulk chemical cargoes, with the recommendation that proper safeguards be taken to avoid accidental mixing of binary mixtures for which an "X" appears on the Chart. Proper safeguards would include consideration of such factors as avoidance of the use of common cargo and vent lines and carriage in adjacent tanks having a common bulkhead.

The following procedure explains how the Guide should be used in determining compatibility information:

(1) Determine the reactivity group of a particular product by referring to the alphabetical list in Table 3.

(2) Enter the Chart with the reactivity group. Proceed across the page. An "X" indicates a reactivity group that forms an unsafe combination with the product in question.

For example, crotonaldehyde is listed in Table 3 as belonging in Group 19 (Aldehydes). The Chart shows that chemicals in this group should be segregated from sulfuric and nitric acids, caustics, ammonia, and all types of amines (aliphatic, alkanol, and aromatic). According to note A, crotonaldehyde is also incompatible with non-oxidizing mineral acids.

It is recognized there are wide variations in the reaction rates of individual chemicals within the broad groupings shown reactive by the Compatibility Chart. Some individual materials in one group will react violently with some of the materials in another group and cause great hazard; others will react slowly, or not at all. Accordingly, a useful addition to the Guide would be the identification of specific binary combinations which are found *not* to be dangerously reactive, even though an "X" appears on the chart for those two chemicals. A few such combinations are listed in Table 5.

COMPATIBILITY CHART

CARGO GROUPS / REACTIVE GROUPS	1. NON-OXIDIZING MINERAL ACIDS	2. SULFURIC ACID	3. NITRIC ACID	4. ORGANIC ACIDS	5. CAUSTICS	6. AMMONIA	7. ALIPHATIC AMINES	8. ALKANOLAMINES	9. AROMATIC AMINES	10. AMIDES	11. ORGANIC ANHYDRIDES	12. ISOCYANATES	13. VINYL ACETATE	14. ACRYLATES	15. SUBSTITUTED ALLYLS	16. ALKYLENE OXIDES	17. EPICHLOROHYDRIN	18. KETONES	19. ALDEHYDES	20. ALCOHOLS, GLYCOLS	21. PHENOLS, CRESOLS	22. CAPROLACTAM SOLUTION	
1. NON-OXIDIZING MINERAL ACIDS		X			X	X	X	X	X	X	X	X	X			X	X		A	E			1
2. SULFURIC ACID	X		X	X	X	X	X	X	X	X	X	X	X	X	X	X	X	X	X	X	X	X	2
3. NITRIC ACID		X			X	X	X	X	X	X	X	X	X	X	X	X	X	X	X	X	X		3
4. ORGANIC ACIDS		X			X	X	X	X	C			X				X	X			F			4
5. CAUSTICS	X	X	X	X							X	X				X	X		X	X	X	X	5
6. AMMONIA	X	X	X	X						X	X	X	X			X	X		X				6
7. ALIPHATIC AMINES	X	X	X	X							X	X	X	X	X	X	X	X	X	X	X	X	7
8. ALKANOLAMINES	X	X	X	X							X	X	X	X	X	X	X	B	X				8
9. AROMATIC AMINES	X	X	X	C							X	X							X				9
10. AMIDES	X	X	X			X						X									X		10
11. ORGANIC ANHYDRIDES	X	X	X		X	X	X	X	X														11
12. ISOCYANATES	X	X	X	X	X	X	X	X	X	X					D					X		X	12
13. VINYL ACETATE	X	X	X			X	X	X															13
14. ACRYLATES		X	X				X	X															14
15. SUBSTITUTED ALLYLS		X	X				X	X				D											15
16. ALKYLENE OXIDES	X	X	X	X	X	X	X	X															16
17. EPICHLOROHYDRIN	X	X	X	X	X	X	X	X															17
18. KETONES		X	X				X	B															18
19. ALDEHYDES	A	X	X		X	X	X	X	X														19
20. ALCOHOLS, GLYCOLS	E	X	X	F	X		X					X											20
21. PHENOLS, CRESOLS		X	X		X		X			X													21
22. CAPROLACTAM SOLUTION		X			X		X					X											22
30. OLEFINS		X	X																				30
31. PARAFFINS																							31
32. AROMATIC HYDROCARBONS			X																				32
33. MISCELLANEOUS HYDROCARBON MIXTURES			X																				33
34. ESTERS		X	X																				34
35. VINYL HALIDES			X																			X	35
36. HALOGENATED HYDROCARBONS		G			H		I																36
37. NITRILES		X																					37
38. CARBON DISULFIDE							X	X															38
39. SULFOLANE																							39
40. GLYCOL ETHERS		X										X											40
41. ETHERS		X	X																				41
42. NITROCOMPOUNDS					X	X	X	X	X														42
43. MISCELLANEOUS WATER SOLUTIONS		X										X											43
	1	2	3	4	5	6	7	8	9	10	11	12	13	14	15	16	17	18	19	20	21	22	

(Letters refer to notes on following page)

NOTES TO COMPATIBILITY CHART: REACTIVITY DIFFERENCES (DEVIATIONS) WITHIN CHEMICAL GROUPS

A Acrolein (19), Crotonaldehyde (19), and 2-Ethyl-3-propyl acrolein (19) are not compatible with Group 1, Non-Oxidizing Mineral Acids.

B Isophorone (18), and Mesityl Oxide (18) are not compatible with Group 8, Alkanolamines.

C Acrylic Acid (4) is not compatible with Group 9, Aromatic Amines.

D Allyl Alcohol (15) is not compatible with Group 12, Isocyanates.

E Furfuryl Alcohol (20) is not compatible with Group 1, Non-oxidizing Mineral Acids.

F Furfuryl Alcohol (20) is not compatible with Group 4, Organic Acids.

G Dichloroethyl Ether (36) is not compatible with Group 2, Sulfuric Acid.

H Trichloroethylene (36) is not compatible with Group 5, Caustics.

I Ethylenediamine (7) is not compatible with Ethylene Dichloride (36).

TABLE 3

ALPHABETICAL LISTING OF COMPOUNDS

Name	Group No.
Acetaldehyde	19
Acetic Acid	4
Acetic Anhydride	11
Acetone	18
Acetonitrile	37
Acrolein (inhibited)	19
Acrylic Acid (inhibited)	4
Acrylonitrile (inhibited)	15
Adiponitrile	37
Allyl Alcohol	15
Allyl Chloride	15
Aminoethylethanolamine	8
Ammonia, Anhydrous	6
Ammonium Hydroxide (28% or less)	6
Ammonium Nitrate, Urea, Water Solutions (containing Ammonia)	6
Ammonium Nitrate, Urea, Water Solutions (not containing Ammonia)	43
Amyl Acetate	34
Amyl Alcohol	20
Amyl Tallate	34
Aniline	9
Asphalt	33
Asphalt Blending Stocks:	
Roofers Flux	33
Straight Run Residue	33
Benzene	32
Benzene, Toluene Xylene (crude)	32
Butadiene (inhibited)	30
Butane	31
Butyl Acrylate (inhibited)	14
Butyl Acetate	34
Butyl Alcohol	20
Butylamine	7
Butyl Benzyl Phthalate	34
Butylene	30
1,3-Butylene Glycol	20
Butylene Oxide	16
Butyl Ether	41
Butyl Methacrylate (inhibited)	14
Butyraldehyde	19
Butyric Acid	4
Camphor Oil (light)	18
Caprolactam Solution	22
Carbolic Oil	21
Carbon Disulfide	38
Carbon Tetrachloride	36
Caustic Potash Solution	5
Caustic Soda Solution	5
Chlorine	*
Chlorobenzene	36
Chloroform	36
Chlorosulfonic Acid	*
Corn Syrup	43
Creosote, Coal Tar	21
Cresols	21
Cresylate Spent Caustic Solution	5
Crotonaldehyde	19
Cumene	32
Cycloaliphatic Resins	31
Cyclohexane	31
Cyclohexanol	20
Cyclohexanone	18
Cyclohexylamine	7
Cymene	32
Decaldehyde	19
Decane	31
Decene	30
Decyl Alcohol	20
Decyl Acrylate (inhibited)	14

ALPHABETICAL LISTING OF COMPOUNDS (Continued)

ALPHABETICAL LISTING OF COMPOUNDS (Continued)

ALPHABETICAL LISTING OF COMPOUNDS (Continued)

ALPHABETICAL LISTING OF COMPOUNDS (Continued)

Compound	Group
Polybutene	30
Polyethylene Glycols	40
Polymethylene Polyphenyl-isocyanate	12
Polypropylene	30
Polypropylene Glycol Methyl Ether	40
Polypropylene Glycols	40
Propane	31
Propanolamine	8
Propionaldehyde	19
Propionic Acid	4
Propionic Anhydride	11
Propyl Acetate	34
Propyl Alcohol	20
Propylamine	7
Propylene	30
Propylene Butylene Polymer	30
Propylene Glycol	20
Propylene Oxide	16
Propylene Tetramer	30
Propyl Ether	41
Pyridine	9
Sodium Hydrosulfide Solution (45% or less)	5
Sorbitol	20
Styrene (inhibited)	30
Sulfolane	39
Sulfur (molten)	*
Sulfuric Acid	2
Sulfuric Acid, Spent	2
Tallow	34
Tallow Fatty Alcohol	20
1,1,2,2-Tetrachloro-ethane	36
Tetradecanol	20
Tetradecene	30
Tetradecylbenzene	32
Tetraethylene Glycol	40
Tetraethylenepentamine	7
Tetrahydrofuran	41
Tetrahydronaphthalene	32
Tetrasodium Salt of EDTA Solution	43
Toluene	32
Toluene Diisocyanate	12
1,2,4-Trichlorobenzene	36
Trichloroethylene	36
Tridecanol	20
Tridecene	30
Tridecylbenzene	32
Triethanolamine	8
Triethylamine	7
Triethyl Benzene	32
Triethylene Glycol	40
Triethylenetetramine	7
Tripropylene Glycol	40
Turpentine	30
Undecanol	20
Undecene	30
Undecylbenzene	32
Valeraldehyde	19
Vinyl Acetate (inhibited)	13
Vinyl Chloride (inhibited)	35
Vinylidene Chloride (inhibited)	35
Vinyl Toluene (inhibited)	30
Xylene	32

* Because of very high reactivity or unusual conditions of carriage, this product is not included in the Compatibility Chart. If compatibility information is needed for a shipment, contact Commandant (G-MHM-1/83), U.S. Coast Guard, 400 Seventh Street, S.W., Washington, D. C. 20590.

TABLE 4

REACTIVITY GROUPS

1. Non-Oxidizing Mineral Acids

Hydrochloric Acid
Hydrofluoric Acid
Phosphoric Acid

2. Sulfuric Acids

Spent Sulfuric Acid
Sulfuric Acid (98% or less)

3. Nitric Acid

Nitric Acid (70% or less)

4. Organic Acids

Acetic Acid
Butyric Acid
Formic Acid
Propionic Acid
Acrylic Acid (inhibited)

5. Caustics

Caustic Potash Solution
Caustic Soda Solution
Cresylate Spent Caustic Solution
Sodium Hydrosulfide Solution
(45% or less)

6. Ammonia

Ammonia, Anhydrous
Ammonium Hydroxide (28% or less)
Ammonium Nitrate, Urea, Water Solutions (containing Ammonia)

7. Aliphatic Amines

Butylamine
Cyclohexylamine
Dibutylamine
Diethylamine
Diethylenetriamine
Diisopropylamine
Dimethylamine
Di-n-propylamine
Ethylamine
Ethylenediamine
Hexamethyleneimine
Methylamine
Morpholine
Propylamine
Tetraethylenepentamine
Triethylamine
Triethylenetetramine

8. Alkanolamines

Aminoethylethanolamine
Diethanolamine
Diethylethanolamine
Diisopropanolamine
Dimethylethanolamine
Ethanolamine
Propanolamine
Triethanolamine

9. Aromatic Amines

Aniline
Pyridine
2-Methyl-5-Ethylpyridine

10. Amides

Dimethylformamide

11. Organic Anhydrides

Acetic Anhydride
Phthalic Anhydride
Propionic Anhydride

12. Isocyanates

Diphenylmethane Diisocyanate
Polyphenyl Polymethylene-isocyanate
Toluene Diisocyanate

13. Vinyl Acetate

Vinyl Acetate (inhibited)

14. Acrylates

Butyl Acrylate (inhibited)
Butyl Methacrylate (inhibited)
Decyl Acrylate (inhibited)
Ethyl Acrylate (inhibited)
2-Ethylhexyl Acrylate (inhibited)
Ethyl Methacrylate (inhibited)
Methyl Acrylate (inhibited)
Methyl Methacrylate (inhibited)

15. Substituted Allyls

Acrylonitrile (inhibited)
Allyl Alcohol
Allyl Chloride
1,3-Dichloropropene

16. Alkylene Oxides

Propylene Oxide
Butylene Oxide

17. Epichlorohydrin

Epichlorohydrin

18. Ketones

Acetone
Camphor Oil
Cyclohexanone
Diisobutyl Ketone
Isophorone
Mesityl Oxide
Methyl Ethyl Ketone
Methyl Isobutyl Ketone

19. Aldehydes

Acetaldehyde
Acrolein (inhibited)
Butyraldehyde
Decaldehyde
Ethylhexaldehyde
Formaldehyde
Glutaraldehyde Solution
Glyoxal Solution
Methylbutyraldehyde
Octyl Aldehyde
Pentyl Aldehyde
Propionaldehyde
Valeraldehyde

20. Alcohols, Glycols

Amyl Alcohol
Butyl Alcohol
1,3-Butylene Glycol
Cyclohexanol
Decyl Alcohol
Diacetone Alcohol
Diisobutyl Carbinol
Dodecanol
Ethanol
Ethoxylated Alcohols C_{11}-C_{15}

Ethyl Alcohol
Ethylbutanol
Ethylene Chlorohydrin
Ethylene Cyanohydrin
Ethylene Glycol
2-Ethyl Hexanol
Furfuryl Alcohol
Glycerin
Hexanol
Hexylene Glycol
Methanol
Methyl Alcohol
Methylamyl Alcohol
Methylisobutyl Carbinol
Octyl Alcohol
Nonyl Alcohol
Pentadecanol
Propyl Alcohol
Propylene Glycol
Sorbitol
Tallow Fatty Alcohol
Tetradecanol
Tridecanol
Undecanol

21. Phenols and Cresols

Carbolic Oil
Creosote, Coal Tar
Cresols
Nonyl Phenol
Phenol

22. Caprolactam Solution

Caprolactam Solution

23 - 29. Unassigned

30. Olefins

Butadiene (inhibited)
Butene
Butylene
Decene
Dicyclopentadiene
Diisobutylene
Dodecene
Ethylene
Hexene
Isoprene (inhibited)
Methyl Acetylene, Propadiene Mixture (stabilized)
(alpha-) Methyl Styrene (inhibited)
Nonene
Octene
Pentene
Polybutene
Polypropylene
Propylene
Propylene Butylene Polymer
Propylene Tetramer
Styrene (inhibited)
Vinyl Toluene (inhibited)
Tetradecene
Tridecene
Turpentine
Undecene

31. Paraffins

Butane
Cycloaliphatic Resins
Cyclohexane
Decane
Dodecane
Ethane
Heptane
Hexane
Methane
Nonane
Octane
Pentane
Propane

32. Aromatic Hydrocarbons

Benzene
Benzene, Toluene, Xylene (crude)
Cumene
Cymene
Decylbenzene
Diethylbenzene
Dodecylbenzene
Ethylbenzene
Naphthalene
Tetradecylbenzene
Tetrahydronaphthalene
Toluene
Tridecylbenzene
Triethylbenzene
Undecylbenzene
Xylene

33. Misc. Hydrocarbon Mixtures

Asphalt
Asphalt Blending Stocks
Diphenyl - Diphenyl Oxide
Distillates
Gas Oil, Cracked
Gasoline Blending Stocks
Gasolines
Jet Fuels
Kerosene
Mineral Spirits
Naphtha
Oils, Crude
Oils, Diesel
Oils, Coal
Oils, Fuel (No. 1 thru No. 6)
Oils, Residual
Oils, Road
Oils, Transformer
Petrolatum
Petroleum Naphtha

34. Esters

Amyl Acetate
Amyl Tallate
Butyl Acetate
Butyl Benzyl Phthalate
Castor Oil
Coconut Oil
Cottonseed Oil
Dibutyl Phthalate
Diethylene Glycol Monobutyl Ether Acetate
Diheptyl Phthalate
Diisodecyl Phthalate
Dinonyl Phthalate
Dioctyl Phthalate
Diundecyl Phthalate
Ethyl Acetate
Ethylene Glycol Monobutyl Ether Acetate
Ethylene Glycol Monoethyl Ether Acetate
Ethylhexyl Tallate
Fish Oil
Glycol Diacetate
Lard
Methyl Acetate
Methyl Amyl Acetate
Octyl Epoxy Tallate
Olive Oil
Palm Oil
Peanut Oil
Propyl Acetate
Safflower Oil
Soybean Oil
Tallow
Tucum Oil
Vegetable Oil

35. Vinyl Halides

Vinyl Chloride (inhibited)
Vinylidene Chloride (inhibited)

36. Halogenated Hydrocarbons

Carbon Tetrachloride
Chlorobenzene

Chloroform
Dichlorobenzene
1,1-Dichloroethane
Dichloroethyl Ether
Dichloromethane
1,1-Dichloropropane
1,2-Dichloropropane
Ethyl Chloride
Ethylene Dibromide
Ethylene Dichloride
Methyl Bromide
Methyl Chloride
Pentachloroethane
Perchloroethylene
1,1,2,2-Tetrachloroethane
1,2,4-Trichlorobenzene
Trichloroethylene

37. Nitriles

Acetonitrile
Adiponitrile

38. Carbon Disulfide

39. Sulfolane

40. Glycol Ethers

Diethylene Glycol
Diethylene Glycol Monobutyl Ether
Diethylene Glycol Monoethyl Ether
Diethylene Glycol Monomethyl Ether
Dipropylene Glycol
Ethoxy Triglycol
Ethylene Glycol Monobutyl Ether
Ethylene Glycol Monethyl Ether
Ethylene Glycol Monomethyl Ether
Nonylphenol, Ethoxylated
Polyethylene Glycols
Polypropylene Glycols
Polypropylene Glycol Methyl Ether
Soybean Oil, Epoxidized
Tetraethylene Glycol
Triethylene Glycol
Tripropylene Glycol

41. Ethers

Butyl Ether
1,4-Dioxane
Ethyl Ether
Methyl Formal (Dimethyl Formal)
Propyl Ether
Tetrahydrofuran

42. Nitrocompounds

(mono-) Nitrobenzene
1- or 2-Nitropropane
Nitrotoluene

43. Miscellaneous Water Solutions

Ammonium Nitrate, Urea, Water Solutions (not containing Ammonia)
Corn Syrup
Dextrose Solution
Latex Solutions
Tetrasodium Salt of EDTA Solution

TABLE 5

COMBINATIONS NOT DANGEROUSLY REACTIVE
(as tested in accordance with procedure outlined in NVC 5-70)

Chemical	Chemical
Acetone (8)	Caustic soda solution (3)
Acrylonitrile (inhibited) (14)	Methyl alcohol (6)
Acrylonitrile (inhibited) (14)	Niax polyol (6)*
Acrylonitrile (inhibited) (14)	Polyol 3030 (6)*
Acrylonitrile (inhibited) (14)	Propylene glycol (6)
Acrylonitrile (inhibited) (14)	Voranol CP 4100 (6)*
Benzene (10)	Phosphoric acid (1)
Butyl acetate (n-, iso-*) (13)	Caustic soda solution (3)
Butyl acrylate (inhibited) (14)	Methyl alcohol (6)
Butyl acrylate (inhibited) (14)	Voranol CP 4100 (6)*
n-Butyl alcohol (6)	Styrene (inhibited) (14)
n-Butyl alcohol (6)	Vinyl acetate (inhibited) (14)
Carbon tetrachloride (5)	Caustic soda solution (3)
Caustic soda solution (3)	Acetone (8)
Caustic soda solution (3)	Butyl acetate (iso-*, n-) (13)
Caustic soda solution (3)	Carbon tetrachloride (5)
Caustic soda solution (3)	Oils, edible: coconut (13)*
Caustic soda solution (3)	Oils, edible: cottonseed (13)
Caustic soda solution (3)	Dichloropropane (5)
Caustic soda solution (3)	Dichloropropene (5)
Caustic soda solution (3)	Diisodecyl phthalate (13)*
Caustic soda solution (3)	Di-normal-alkyl phthalate (13)*
Caustic soda solution (3)	Dioctyl phthalate (13)
Caustic soda solution (3)	Ethyl acetate (13)
Caustic soda solution (3)	Ethylene dichloride (5)
Caustic soda solution (3)	Oils, edible: fish (13)
Caustic soda solution (3)	Grease (inedible, yellow) (13)*
Caustic soda solution (3)	Lard (edible) (13)*
Caustic soda solution (3)	Linseed oil (raw) (13)*
Caustic soda solution (3)	Methylene chloride (5)*
Caustic soda solution (3)	Methyl ethyl ketone (8)
Caustic soda solution (3)	Methyl isobutyl ketone (8)
Caustic soda solution (3)	Palm oil (13)*
Caustic soda solution (3)	Perchloroethylene (5)*
Caustic soda solution (3)	Propyl acetate (iso-*, n-) (13)
Caustic soda solution (3)	Oils, edible: soya bean (13)
Caustic soda solution (3)	Oils, miscellaneous: sperm
Caustic soda solution (3)	Styrene (inhibited) (14)
Caustic soda solution (3)	Tallow (13)
Caustic soda solution (3)	Trichloroethane (5)
Dichloropropane (5)	Caustic soda solution (3)
Dichloropropene (5)	Caustic soda solution (3)
Diisodecyl phthalate (13)*	Caustic soda solution (3)
Di-normal-alkyl phthalate (13)*	Caustic soda solution (3)
Dimethylformamide (4)	Furfural (7)
Dimethylformamide (4)	Phenol (15)
Dioctyl phthalate (13)	Caustic soda solution (3)
Dioctyl phthalate (13)	Ethylenediamine (4)
Diphenylmethanediisocyanate	Ethylene dichloride (5)
Ethyl acetate (13)	Caustic soda solution (3)
Ethyl acrylate (inhibited) (14)	Ethylene glycol (6)
Ethyl acrylate (inhibited) (14)	2-Ethyl hexanol (6)
Ethyl acrylate (inhibited) (14)	Voranol CP 4100 (6)*
Ethyl alcohol (6)	Methyl methacrylate (inhibited) (14)
Ethylenediamine (4)	Dioctyl phthalate (13)
Ethylene dichloride (5)	Caustic soda solution (3)
Ethylene dichloride (5)	Diphenylmethanediisocyanate
Ethylene glycol (6)	Ethyl acrylate (inhibited) (14)
Ethylene glycol (6)	Styrene (inhibited) (14)
Ethylene glycol (6)	Vinyl acetate (inhibited) (14)

*Not presently included in CHRIS.

COMBINATIONS NOT DANGEROUSLY REACTIVE (Continued)

2-Ethyl hexanol (6)	Ethyl acrylate (inhibited) (14)
2-Ethyl hexanol (6)	Styrene (inhibited) (14)
Furfural (7)	Dimethylformamide (4)
Furfural (7)	Isopropyl alcohol (6)
Furfural (7)	Methyl ethyl ketone (8)
Grease (inedible, yellow) (13)*	Caustic soda solution (3)
Isobutyl alcohol (6)	Styrene (inhibited) (14)
Isobutyl alcohol (6)	Vinyl acetate (inhibited) (14)
Isodecyl alcohol (6)	Vinyl acetate (inhibited) (14)
Isooctyl alcohol (6)	Methyl methacrylate (inhibited) (14)
Isooctyl alcohol (6)	Styrene (inhibited) (14)
Isooctyl alcohol (6)	Vinyl acetate (inhibited) (14)
Isopropyl alcohol (6)	Furfural (7)
Isopropyl alcohol (6)	Styrene (inhibited) (14)
Isopropyl alcohol (6)	Vinyl acetate (inhibited) (14)
Lard (edible) (13)*	Caustic soda solution (3)
Linseed oil (raw) (13)*	Caustic soda solution (3)
Methyl alcohol (6)	Acrylonitrile (inhibited) (14)
Methyl alcohol (6)	Butyl acrylate (inhibited) (14)
Methyl alcohol (6)	Styrene (inhibited) (14)
Methyl alcohol (6)	Vinyl acetate (inhibited) (14)
Methylene chloride (5)*	Caustic soda solution (3)
Methyl ethyl ketone (8)	Caustic soda solution (3)
Methyl ethyl ketone (8)	Furfural (7)
Methyl isobutyl ketone (8)	Caustic soda solution (3)
Methyl methacrylate (inhibited) (14)	Ethyl alcohol (6)
Methyl methacrylate (inhibited) (14)	Isooctyl alcohol (6)
Niax polyol (6)*	Acrylonitrile (inhibited) (14)
Niax polyol (6)*	Vinyl acetate (inhibited) (14)
Oils, edible: coconut (13)*	Caustic soda solution (3)

Oils, edible: cottonseed (13)	Caustic soda solution (3)
Oils, edible: fish (13)	Caustic soda solution (3)
Oils, edible: soya bean (13)	Caustic soda solution (3)
Oils, miscellaneous: sperm	Caustic soda solution (3)
Palm oil (13)*	Caustic soda solution (3)
Perchloroethylene (5)*	Caustic soda solution (3)
Phenol (15)	Dimethyl formamide (4)
Phosphoric acid (1)	Benzene (10)
Phosphoric acid (1)	Toluene (10)
Phosphoric acid (1)	Xylene (10)
Polyol 3030 (6)*	Acrylonitrile (inhibited) (14)
Propyl acetate (iso-*, n-) (13)	Caustic soda solution (3)
Propylene glycol (6)	Acrylonitrile (inhibited) (14)
Propylene glycol (6)	Styrene (inhibited) (14)
Styrene (inhibited) (14)	n-Butyl alcohol (6)
Styrene (inhibited) (14)	Caustic soda solution (3)
Styrene (inhibited) (14)	Ethylene glycol (6)
Styrene (inhibited) (14)	2-Ethyl hexanol (6)
Styrene (inhibited) (14)	Isobutyl alcohol (6)
Styrene (inhibited) (14)	Isooctyl alcohol (6)
Styrene (inhibited) (14)	Isopropyl alcohol (6)
Styrene (inhibited) (14)	Methyl alcohol (6)
Styrene (inhibited) (14)	Propylene glycol (6)
Styrene (inhibited) (14)	Trichloroethylene (5)
Tallow (13)	Caustic soda solution (3)
Toluene (10)	Phosphoric acid (1)
Trichloroethane (5)	Caustic soda solution (3)
Trichloroethylene (5)	Styrene (inhibited) (14)
Vinyl acetate (inhibited) (14)	n-Butyl alcohol (6)
Vinyl acetate (inhibited) (14)	Ethylene glycol (6)

*Not presently included in CHRIS.

COMBINATIONS NOT DANGEROUSLY REACTIVE (Continued)

Vinyl acetate (inhibited) (14)	Isobutyl alcohol (6)
Vinyl acetate (inhibited) (14)	Isodecyl alcohol (6)
Vinyl acetate (inhibited) (14)	Isooctyl alcohol (6)
Vinyl acetate (inhibited) (14)	Isopropyl alcohol (6)
Vinyl acetate (inhibited) (14)	Methyl alcohol (6)
Vinyl acetate (inhibited) (14)	Niax polyol (6)*
Vinyl acetate (inhibited) (14)	Voranol CP 4100 (6)*
Voranol CP 4100 (6)*	Acrylonitrile (inhibited) (14)
Voranol CP 4100 (6)*	Butyl acrylate (inhibited) (14)
Voranol CP 4100 (6)*	Ethyl acrylate (inhibited) (14)
Voranol CP 4100 (6)*	Vinyl acetate (inhibited) (14)
Xylene (10)	Phosphoric acid (1)

Toluene 2,4-diisocyanate (TDI), diphenylmethanediisocyanate (MDI), and polymethylene polyphenyl isocyanate (PAPI)* are considered compatible with reactivity groups 9, 10, 11, 12, 18, and 21.

*Not presently included in CHRIS.

DATA SOURCES

The source of every item of data contained in the data sheets is recorded in master data files and is available on request. The principal sources are listed below. Many other sources were consulted, but most of them provided only a few items and are not given here. In a few cases the information given is based on an analogy with that for a closely related chemical; the analogy was drawn by an expert in the field, whose identity appears in the master data file.

Where a source was used for a single category of data, the source is given in the "Explanation of Terms" and is not repeated here.

GENERAL SOURCES

The following sources contained data for many of the 13 data categories used:

1. Manufacturers' Technical Bulletins – This is usually the best single source of general information about the chemical. All bulletins were solicited in late 1972 and contain the most recent data. Bulletins were not available for a few chemicals that are not items of commerce but are intermediates shipped from one manufacturing site to another.

2. Material Safety Data Sheets – These were provided by the manufacturer using the U.S. Department of Labor Form OSHA-20 or an approved modification.

3. Code of Federal Regulations – Office of the Federal Register, Archives and Record Service, Washington, D.C., 1972. Titles 46 (Shipping) and 49 (Transportation) were used in the most recent revision available, in all cases since January 1, 1971.

4. Chemical Safety Data Sheets – Manufacturing Chemists Association, Washington, D.C.

5. Industrial Safety Data Sheets – National Safety Council, Chicago, Illinois.

6. International Maritime Dangerous Goods Code – Inter-Governmental Maritime Consultative Organization (IMCO), London, 1972.

7. Petroleum Products Handbook – V.B. Guthrie (ed.), McGraw-Hill, New York, 1960.

8. Glossary of Terms Used in Petroleum Refining – 2nd edition, American Petroleum Institute, New York, 1962.

9. The Handling and Storage of Liquid Propellants – Office of Defense Research and Engineering, U.S. Government Printing Office, Washington, D.C., 1963.

10. Industrial Chemicals – W.L. Faith, D.B. Keyes, and R.L. Clark, 3rd edition, Wiley, New York, 1965.

11. Chemical Technology of Petroleum – W.A. Gruse and D.R. Stevens, 3rd edition, McGraw-Hill, New York, 1960.

12. Chemical Rocket/Propellant Hazards – CPIA Publication No. 194, Vol. III, 1970.

13. Organic Solvents – J.A. Riddick and W.B. Bunger, 3rd edition, Wiley-Interscience, New York, 1970.

14. Transport of Dangerous Goods – (4 vols) United Nations, New York, 1970.

15. Kirk-Othmer Encyclopedia of Chemical Technology – 1st edition (1947-1960) and 2nd edition (1963-1970), Interscience-Wiley, New York.

16. Evaluation of the Hazard of Bulk Water Transportation of Industrial Chemicals, A Tentative Guide – National Academy of Sciences, Washington, D.C., 1970; includes supplement with additions to March 1972.

17. Matheson Gas Data Book – 5th edition, Matheson Gas Products, East Rutherford, New Jersey, 1971.

18. Explosive and Toxic Hazardous Materials – J. H. Meidl, Glencoe Press, Beverly Hills, California, 1969.

20. Organic Phosphorus Compounds – G. M. Kosolapoff and L. Maier (6 Vols.), Wiley-Interscience, New York.

21. The Chemistry of Organo-Phosphorus Pesticides – C. Fest and K. J. Schmidt, Springer-Verlag, New York, 1973.

CHEMICAL DESIGNATIONS

1. Commercial Organic Chemical Names – Compiled by the Synthetic Organic Chemical Manufacturers Association (SOCMA), Chemical Abstracts Service, Columbus, Ohio, 1965.

2. Chemical Synonyms and Trade Names – W. Gardner and E.I. Cooke, 7th edition, CRC press, Cleveland, Ohio, 1971.

3. The Merck Index of Chemicals and Drugs – 8th edition, Merck and Co., Rahway, New Jersey, 1968.

HEALTH HAZARDS

1. Industrial Hygiene and Toxicology – F.A. Patty, 2nd edition, Vol. II, Interscience, New York, 1963.

2. Toxicity and Metabolism of Industrial Solvents – E. Browning, Elsevier, New York, 1965.

3. Practical Toxicology of Plastics – R. Lefaux, CRC Press, Cleveland, Ohio, 1968.

4. Industrial Toxicology – L.T. Fairhall, Williams and Wilkins, 2nd edition, Baltimore, Maryland, 1957.

5. Toxicology of Drugs and Chemicals – W.B. Deichmann and H.W. Girarde, Academic Press, New York, 1969.

6. Clinical Toxicology of Commercial Products – M.N. Gleason, et al., 3rd edition, Williams and Wilkins, Baltimore, Maryland, 1969.

7. Handbook of Toxicology: Acute Toxicities of Solids, Liquids and Gases to Laboratory Animals – W.S. Spector, Saunders, Philadelphia, Pa., 1956.

8. Occupational Diseases: A Guide to their Recognition – U.S. Department of Health, Education, and Welfare. Public Health Service Publication No. 1097. Superintendent of Documents, Washington, D.C., 1964.

9. First Aid Textbook – American National Red Cross, Washington, D.C., 1972.

10. Electrical Safety Practice: Odor Warning for Safety – Monograph 113 Instrument Society of America (ISA), Pittsburgh, Pa., 1972.

11. Toxic Substances – Annual List 1971 – H.E. Christensen, U.S. Department of Health, Education, and Welfare, Superintendent of Documents, Washington, D.C., 1971.

12. Hygienic Guide Series – American Industrial Hygiene Association, Detroit, Michigan, 48227.

13. Toxicity of Industrial Metals – E. Browning, 2nd Edition, Appleton-Century-Crofts, New York, 1969.

FIRE HAZARDS

1. The Fire and Explosion Hazards of Commercial Oils – W. Vlachos and C.A. Vlachos, Vlachos and Co., Philadelphia, Pa., 1921.

2. 1972 Annual Book of ASTM Standards – American Society for Testing and Materials, Philadelphia, Pa., 1972.

3. Fire Protection Guide on Hazardous Materials – 4th edition, Nos. 325A, 325M, 49, 491M, and 704M, National Fire Protection Association (NFPA), Boston, Mass., 1972.

4. Fire Protection Handbook – G.H. Tryon (ed.), 13th edition, National Fire Protection Association (NFPA), Boston, Mass., 1969.

5. Handbook of Industrial Loss Prevention – 2nd edition, Factory Mutual Engineering Corp., McGraw-Hill, New York, 1967.

WATER POLLUTION

1. Water Quality Criteria Data Book – Vol. 1 – Organic Chemicals (1970) and Vol. 2 – Inorganic Chemicals (1971), United States Environmental Protection Administration, Superintendent of Documents, Washington, D.C.

2. Engineering Management of Water Quality – P.H. McGauhey, McGraw-Hill, New York, 1968.

3. The BOD of Textile Chemicals – Proceedings of the American Association of Textile Chemists and Colorists, *American Dyestuff Reporter,* August 29, 1966, p. 39.

4. Biodegradable Surfactants for the Textile Industry – *American Dyestuff Reporter,* January 30, 1967.

5. Water Quality Criteria – J.E. McKee and M.W. Wolf, 2nd edition, California State Water Quality Control Board, Sacramento, California, 1963.

6. Water Quality Criteria – National Technical Advisory Committee, Federal Water Pollution Control Administration, Washington, D.C., 1968.

7. Water Quality Characteristics of Hazardous Materials – R. W. Hann, Jr., and P. A. Jensen, Environmental Engineering Division, Texas A and M University, College Station, Texas, 1974.

PHYSICAL AND CHEMICAL PROPERTIES

1. Solubilities of Inorganic and Organic Compounds – H. Stephen and T. Stephen, Macmillan, New York, 1963, Vol. 1, Part 1.

2. The Critical Constants of Organic Compounds – A.P. Kudchadker, G.H. Alani and B.J. Zwolinski, *Chemical Reviews,* **68**,659 (1968).

3. Physical Properties of Hydrocarbons – Vol. 1 (1968) and Vol. 2 (1970), R.W. Gallant, Gulf Publishing Co., Houston, Texas.

4. International Critical Tables – McGraw-Hill, New York, 1926.

5. Handbook of Chemistry and Physics – R.C. Weast (ed.), 53rd edition, CRC Publishing Co., Cleveland, Ohio, 1972.

6. The Properties of Gases and Liquids – R.C. Reid and T.K. Sherwood, 2nd edition, McGraw-Hill, New York, 1966.

7. Thermal Conductivity of Gases and Liquids – N.V. Tsederberg, MIT Press, Cambridge, Mass., 1965.

8. Lange's Handbook of Chemistry – N.A. Lange, 10th edition, McGraw-Hill, New York, 1969.

9. The Chemical Thermodynamics of Organic Compounds – D.R. Stull, et al., Wiley, New York, 1969.

10. Matheson Gas Data Book – 4th edition, Matheson Co., Inc., 1966.

11. Physical Properties of Chemical Compounds – Vol. 1 (1955), Vol. 2 (1959), and Vol. 3 (1961), R.R. Dreisbach, American Chemical Society, Washington, D.C.

12. Beilsteins Handbuch der Organischen Chemie – Springer, Berlin, Germany.

13. Gmelins Handbuch der Anorganischen Chemie – Verlag Chemie, Weinheim, Germany.

14. Solubilities of Inorganic and Organic Compounds – 3rd edition and supplement, A. Seidell and W.F. Linke, Van Nostrand, New York, 1941-1952.

15. Selected Values of Physical and Thermodynamic Properties of Hydrocarbons and Related Compounds – F.D. Rossini, et al., American Petroleum Institute Project 44, American Petroleum Institute, Pittsburgh, Pa., 1953.

16. Heat of Combustion and Formation of Organic Compounds – E.S. Domalski, *Journal of Physical and Chemical Reference Data,* **1**, 221 (1972).

17. Surface Tension of Pure Liquid Compounds – J.L. Jasper, *J. Phys. Chem. Ref. Data,* **1**, 841 (1972).

18. JANAF Thermochemical Tables – NSRDS-NBS-37 (1970); 1974 Supplement and complete index, *J. Phys. Chem. Ref. Data,* **3**, 311 (1974).

19. Physical and Thermodynamic Properties of Aliphatic Alcohols – R.C. Wilhoit and B. J. Zwolinski, *J. Phys. Chem. Ref. Data,* **2** (1973), Supplement 1.

20. Critical Constants of Hydrocarbons – C. A. Passut and R. P. Danner, *Ind. Eng. Chem., Proc. Des. Devel.,* **12**, 365 (1973).

AAD — ACETALDEHYDE

Common Synonyms	Watery liquid	Colorless	Sharp, fruity odor
Acetic aldehyde Ethanal	Floats and mixes with water. Flammable, irritating vapor is produced. Boiling point is 69° F.		

Avoid contact with liquid and vapor. Keep people away.
Wear goggles and self-contained breathing apparatus.
Shut off ignition sources and call fire department.
Stay upwind and use water spray to "knock down" vapor.
Stop discharge if possible.
Isolate and remove discharged material.
Notify local health and pollution control agencies.

Fire

FLAMMABLE.
Flashback along vapor trail may occur.
Vapor may explode if ignited in an enclosed area.

Wear goggles and self-contained breathing apparatus.
Combat fires from safe distance or protected location.
Extinguish with dry chemical, alcohol foam, or carbon dioxide.
Water may be ineffective on fire.
Cool exposed containers with water.

Exposure

CALL FOR MEDICAL AID.

VAPOR
Irritating to eyes, nose and throat.
If inhaled, will cause nausea, vomiting, headache, difficult breathing, or loss of consciousness.

Move to fresh air.
If breathing has stopped, give artificial respiration.
If breathing is difficult, give oxygen.

LIQUID
Will burn skin and eyes.
Harmful if swallowed.

Remove contaminated clothing and shoes.
Flush affected areas with plenty of water.
IF IN EYES, hold eyelids open and flush with plenty of water.
IF SWALLOWED and victim is CONSCIOUS, have victim drink water or milk.

Water Pollution

HARMFUL TO AQUATIC LIFE IN VERY LOW CONCENTRATIONS.
May be dangerous if it enters water intakes.

Notify local health and wildlife officials.
Notify operators of nearby water intakes.

1. RESPONSE TO DISCHARGE

(See Response Methods Handbook, CG 446-4)

Issue warning—high flammability
Disperse and flush

2. LABEL

Red

3. CHEMICAL DESIGNATIONS

3.1 **Synonyms:** Acetic aldehyde
Ethanal
Ethyl aldehyde

3.2 **Coast Guard Compatibility Classification:** Aldehyde

3.3 **Chemical Formula:** CH_3CHO

3.4 **IMCO/United Nations Numerical Designation:** 3.1/1089

4. OBSERVABLE CHARACTERISTICS

4.1 **Physical State (as shipped):** Liquid

4.2 **Color:** Colorless

4.3 **Odor:** Penetrating, fruity; sharp pungent

5. HEALTH HAZARDS

5.1 **Personal Protective Equipment:** Rubber gloves, eye goggles, and other equipment to prevent any contact with body. Organic canister or air pack as required.

5.2 **Symptoms Following Exposure:** Breathing vapors will be irritating and may cause nausea, vomiting, headache, and unconsciousness. Contact with eyes may cause burns. Skin contact from clothing wet with the chemical causes burns or severe irritation.

5.3 **Treatment for Exposure:** INHALATION: remove victim to fresh air; if breathing has stopped, give artificial respiration; if breathing is difficult, give oxygen; call a physician at once. SKIN: wash with soap and water. EYES: flush with water.

5.4 **Toxicity by Inhalation (Threshold Limit Value):** 100 ppm

5.5 **Short-Term Inhalation Limits:** 50 ppm for 60 min

5.6 **Toxicity by Ingestion:** Grade 2; LD_{50} 0.5 to 5 g/kg (rat)

5.7 **Late Toxicity:** Data not available

5.8 **Vapor (Gas) Irritant Characteristics:** Vapor is moderately irritating such that personnel will not usually tolerate moderate or high concentrations.

5.9 **Liquid or Solid Irritant Characteristics:** Minimum hazard. If spilled on clothing and allowed to remain, may cause smarting and reddening of the skin.

5.10 **Odor Threshold:** 0.21 ppm

6. FIRE HAZARDS

6.1 **Flash Point:** −36°F C.C.; −58°F O.C.

6.2 **Flammable Limits in Air:** 4%–60%

6.3 **Fire Extinguishing Agents:** Dry chemical, alcohol foam, carbon dioxide

6.4 **Fire Extinguishing Agents Not to be Used:** Water may be ineffective

6.5 **Special Hazards of Combustion Products:** Produces irritating vapor when heated

6.6 **Behavior in Fire:** Vapors are heavier than air and may travel a considerable distance to a source of ignition and flash back.

6.7 **Ignition Temperature:** 365°F

6.8 **Electrical Hazard:** Class I, Group C

6.9 **Burning Rate:** 3.3 mm/min.

7. CHEMICAL REACTIVITY

7.1 **Reactivity with Water:** No reaction

7.2 **Reactivity with Common Materials:** No reaction

7.3 **Stability During Transport:** Stable

7.4 **Neutralizing Agents for Acids and Caustics:** Not pertinent

7.5 **Polymerization:** May occur. Avoid heat, dust, strong oxidizing or reducing substances, strong acids and bases

7.6 **Inhibitor of Polymerization:** None

8. WATER POLLUTION

8.1 **Aquatic Toxicity:**
53 ppm/96 hr/sunfish/TL_m/fresh water
70 ppm/24 hr/pin perch/TL_m/salt water

8.2 **Waterfowl Toxicity:** Data not available

8.3 **Biological Oxygen Demand (BOD):** 93–127%, 5 days

8.4 **Food Chain Concentration Potential:** None

9. SELECTED MANUFACTURERS

1. Celanese Corp.
Celanese Chemical Co. Division
245 Park Ave.
New York, N. Y. 10017
2. Eastman Kodak Co.
Texas Eastman Co. Division
Longview, Texas 75601
3. Union Carbide Corp.
Chemicals & Plastics Division
270 Park Ave.
New York, N. Y. 10017

10. SHIPPING INFORMATION

10.1 **Grades or Purity:** >99%

10.2 **Storage Temperature:** Ambient

10.3 **Inert Atmosphere:** Inerted

10.4 **Venting:** Safety relief

11. HAZARD ASSESSMENT CODE

(See Hazard Assessment Handbook, CG 446-3)

A-B-C-K-L-M-N-Z

12. HAZARD CLASSIFICATIONS

12.1 **Code of Federal Regulations:** Flammable liquid

12.2 **NAS Hazard Rating for Bulk Water Transportation:**

Category	Rating
Fire	4
Health	
Vapor Irritant	3
Liquid or Solid Irritant	1
Poisons	2
Water Pollution	
Human Toxicity	2
Aquatic Toxicity	3
Aesthetic Effect	2
Reactivity	
Other Chemicals	2
Water	0
Self-Reaction	1

12.3 **NFPA Hazard Classifications:**

Category	Classification
Health Hazard (Blue)	2
Flammability (Red)	4
Reactivity (Yellow)	2

13. PHYSICAL AND CHEMICAL PROPERTIES

13.1 **Physical State at 15°C and 1 atm:** Liquid

13.2 **Molecular Weight:** 44.05

13.3 **Boiling Point at 1 atm:** 68.7°F = 20.4°C = 293.6°K

13.4 **Freezing Point:** −189°F = −123°C = 150°K

13.5 **Critical Temperature:** 370°F = 188°C = 461°K

13.6 **Critical Pressure:** 820 psia = 56 atm = 5.7 MN/m^2

13.7 **Specific Gravity:** 0.780 at 20°C (liquid)

13.8 **Liquid Surface Tension:** Not pertinent

13.9 **Liquid-Water Interfacial Tension:** Not pertinent

13.10 **Vapor (Gas) Specific Gravity:** 1.5

13.11 **Ratio of Specific Heats of Vapor (Gas):** 1.182

13.12 **Latent Heat of Vaporization:** 245 Btu/lb = 136 cal/g = 5.69×10^5 J/kg

13.13 **Heat of Combustion:** −10,600 Btu/lb = −5890 cal/g = -246.4×10^5 J/kg

13.14 **Heat of Decomposition:** Not pertinent

13.15 **Heat of Solution:** Not pertinent

13.16 **Heat of Polymerization:** Not pertinent

NOTES

REVISED 1978

ACETIC ACID

Common Synonyms			
Glacial acetic acid Ethanoic acid Vinegar acid	Watery liquid	Colorless	Strong vinegar odor
	Sinks and mixes with water. Irritating vapor is produced. Freezing point is 62° F.		

AVOID CONTACT WITH LIQUID AND VAPOR. Keep people away.
Wear goggles, self-contained breathing apparatus, and rubber overclothing (including gloves).
Stop discharge if possible.
Call fire department.
Isolate and remove discharged material.
Notify local health and pollution control agencies.

Fire

Combustible.
Vapor may explode if ignited in an enclosed area.
Wear goggles, self-contained breathing apparatus, and rubber overclothing (including gloves).
Extinguish with water, dry chemical, alcohol foam, or carbon dioxide.
Cool exposed containers with water.

Exposure

CALL FOR MEDICAL AID.

VAPOR
Irritating to nose and throat.
If inhaled, will cause coughing, nausea, vomiting, or difficult breathing.
Move to fresh air.
If breathing has stopped, give artificial respiration.
If breathing is difficult, give oxygen.

LIQUID OR SOLID
Will burn skin and eyes.
Harmful if swallowed.
Remove contaminated clothing and shoes.
Flush affected areas with plenty of water.
IF IN EYES, hold eyelids open and flush with plenty of water.
IF SWALLOWED and victim is CONSCIOUS, have victim drink water or milk.
DO NOT INDUCE VOMITING.

Water Pollution

HARMFUL TO AQUATIC LIFE IN VERY LOW CONCENTRATIONS.
May be dangerous if it enters water intakes.
Notify local health and wildlife officials.
Notify operators of nearby water intakes.

1. RESPONSE TO DISCHARGE

(See Response Methods Handbook, CG 446-4)

Issue warning – corrosive
Disperse and flush

2. LABEL

3. CHEMICAL DESIGNATIONS

3.1 **Synonyms:** Ethanoic acid; Glacial acetic acid; Vinegar acid
3.2 **Coast Guard Compatibility Classification:** Organic acid
3.3 **Chemical Formula:** CH_3COOH
3.4 **IMCO/United Nations Numerical Designation:** 3.3/1842

4. OBSERVABLE CHARACTERISTICS

4.1 **Physical State (as shipped):** Liquid
4.2 **Color:** Colorless
4.3 **Odor:** Characteristic vinegar; pungent; vinegar-like; sharp

5. HEALTH HAZARDS

5.1 **Personal Protective Equipment:** Protective clothing should be worn when skin contact might occur. Respiratory protection necessary when exposed to vapor. Complete eye protection.
5.2 **Symptoms Following Exposure:** Breathing of vapors causes coughing, chest pain, and irritation of nose and throat; may cause nausea and vomiting. Contact with skin and eye causes burns.
5.3 **Treatment for Exposure:** INHALATION: move victim at once to fresh air; if breathing becomes difficult, give oxygen; get medical care quickly. INGESTION: if victim is conscious, have him drink water or milk; do NOT induce vomiting. SKIN OR EYE CONTACT: flush immediately with plenty of clean running water; wash eyes for at least 15 min. and get medical care as quickly as possible; remove contaminated clothing and launder before wearing again.
5.4 **Toxicity by Inhalation (Threshold Limit Value):** 10 ppm
5.5 **Short-Term Inhalation Limits:** 40 ppm for 5 min.
5.6 **Toxicity by Ingestion:** Grade 2; LD_{50} 0.5 to 5.0 g/kg (rat)
5.7 **Late Toxicity:** Data not available
5.8 **Vapor (Gas) Irritant Characteristics:** Vapors cause moderate irritation such that personnel will find high concentrations unpleasant. The effect is temporary.
5.9 **Liquid or Solid Irritant Characteristics:** Fairly severe skin irritant; may cause pain and second-degree burns after a few minutes of contact.
5.10 **Odor Threshold:** 1.0 ppm

6. FIRE HAZARDS

6.1 **Flash Point:** 112°F O.C.; 104°F C.C.
6.2 **Flammable Limits in Air:** 5.4% – 16.0%
6.3 **Fire Extinguishing Agents:** Water, alcohol foam, dry chemical or carbon dioxide
6.4 **Fire Extinguishing Agents Not to be Used:** None
6.5 **Special Hazards of Combustion Products:** Irritating vapor generated when heated.
6.6 **Behavior in Fire:** Not pertinent
6.7 **Ignition Temperature:** 800°F
6.8 **Electrical Hazard:** Not pertinent
6.9 **Burning Rate:** 1.6 mm/min.

7. CHEMICAL REACTIVITY

7.1 **Reactivity with Water:** No reaction
7.2 **Reactivity with Common Materials:** Corrosive, particularly when diluted. Attacks most common metals, including most stainless steels. Excellent solvent for many synthetic resins or rubber.
7.3 **Stability During Transport:** Stable
7.4 **Neutralizing Agents for Acids and Caustics:** Dilute with water, rinse with sodium bicarbonate solution.
7.5 **Polymerization:** Not pertinent
7.6 **Inhibitor of Polymerization:** Not pertinent

8. WATER POLLUTION

8.1 **Aquatic Toxicity:**
75 ppm/96 hr/bluegill/TL_m/fresh water
100 ppm/96 hr/goldfish/TL_m/fresh water
100-330 ppm/48 hr/shrimp/LC_{50}/aerated water
8.2 **Waterfowl Toxicity:** Not pertinent
8.3 **Biological Oxygen Demand (BOD):** 52 – 62%, 5 days
8.4 **Food Chain Concentration Potential:** None noted

9. SELECTED MANUFACTURERS

1. Celanese Corp.
Celanese Chemical Co. Div.
245 Park Ave.
New York, N. Y. 10017
2. Eastman Kodak Co.
Tennessee Eastman Co. Div.
Kingsport, Tenn. 37662
3. Union Carbide Corp.
Chemicals & Plastics Div.
270 Park Ave.
New York, N. Y. 10017

10. SHIPPING INFORMATION

10.1 **Grades or Purity:** Commercial; USP;CP
10.2 **Storage Temperature:** Ambient
10.3 **Inert Atmosphere:** No requirement
10.4 **Venting:** Open

11. HAZARD ASSESSMENT CODE

(See Hazard Assessment Handbook, CG 446-3)

A-P-Q

12. HAZARD CLASSIFICATIONS

12.1 **Code of Federal Regulations:** Corrosive Material

12.2 **NAS Hazard Rating for Bulk Water Transportation:**

Category	Rating
Fire	2
Health	
Vapor Irritant	2
Liquid or Solid Irritant	3
Poisons	2
Water Pollution	
Human Toxicity	1
Aquatic Toxicity	2
Aesthetic Effect	2
Reactivity	
Other Chemicals	2
Water	0
Self-Reaction	0

12.3 **NFPA Hazard Classifications:**

Category	Classification
Health Hazard (Blue)	2
Flammability (Red)	2
Reactivity (Yellow)	1

13. PHYSICAL AND CHEMICAL PROPERTIES

13.1 **Physical State at 15°C and 1 atm:** Liquid
13.2 **Molecular Weight:** 60.05
13.3 **Boiling Point at 1 atm:** 244°F = 117.9°C = 391.1°K
13.4 **Freezing Point:** 62.1°F = 16.7°C = 290°K
13.5 **Critical Temperature:** 611°F = 321.6°C = 594.8°K
13.6 **Critical Pressure:** 839 psia = 57.1 atm = 5.78 MN/m^2
13.7 **Specific Gravity:** 1.051 at 20°C (liquid)
13.8 **Liquid Surface Tension:** Not pertinent
13.9 **Liquid-Water Interfacial Tension:** Not pertinent
13.10 **Vapor (Gas) Specific Gravity:** Not pertinent
13.11 **Ratio of Specific Heats of Vapor (Gas):** 1.145
13.12 **Latent Heat of Vaporization:** 17.1 Btu/lb = 96.7 cal/g = 4.05×10^5 J/kg
13.13 **Heat of Combustion:** −5645 Btu/lb = −3136 cal/g = -131.3×10^5 J/kg
13.14 **Heat of Decomposition:** Not pertinent
13.15 **Heat of Solution:** Not pertinent
13.16 **Heat of Polymerization:** Not pertinent

NOTES

ACA — ACETIC ANHYDRIDE

Common Synonyms			
Ethanoic anhydride	Watery liquid	Colorless	Strong vinegar odor
	Sinks and reacts slowly with water. Irritating vapor is produced.		

Avoid contact with liquid and vapor. Keep people away.
Wear goggles, self-contained breathing apparatus, and rubber overclothing (including gloves).
Stop discharge if possible.
Call fire department.
Isolate and remove discharged material.
Notify local health and pollution control agencies.

Fire

Combustible.
Vapor may explode if ignited in an enclosed area.

Wear goggles, self-contained breathing apparatus, and rubber overclothing (including gloves).
Combat fires from safe distance or from behind barrier.
Extinguish with water, dry chemical, alcohol foam, or carbon dioxide.
Cool exposed containers with water.

Exposure

CALL FOR MEDICAL AID.
VAPOR
Will burn eyes.
Irritating to nose and throat.
If inhaled, will cause nausea, vomiting, or difficult breathing.

Move to fresh air.
If in eyes, hold eyelids open and flush with plenty of water.
If breathing has stopped, give artificial respiration.
If breathing is difficult, give oxygen.
LIQUID
Will burn skin and eyes.
Harmful if swallowed.

Remove contaminated clothing and shoes.
Flush affected areas with plenty of water.
IF IN EYES, hold eyelids open and flush with plenty of water.
IF SWALLOWED and victim is CONSCIOUS, have victim drink water or milk.
DO NOT INDUCE VOMITING.

Water Pollution

HARMFUL TO AQUATIC LIFE IN VERY LOW CONCENTRATIONS.
May be dangerous if it enters water intakes.

Notify local health and wildlife officials.
Notify operators of nearby water intakes.

1. RESPONSE TO DISCHARGE

(See Response Methods Handbook, CG 446-4)

Issue warning – corrosive
Disperse and flush

2. LABEL

3. CHEMICAL DESIGNATIONS

3.1 **Synonyms:** Ethanoic anhydride
3.2 **Coast Guard Compatibility Classification:** Organic anhydride
3.3 **Chemical Formula:** $CH_3CO\text{-}O\text{-}COCH_3$
3.4 **IMCO/United Nations Numerical Designation:** 8.0/1715

4. OBSERVABLE CHARACTERISTICS

4.1 **Physical State (as shipped):** Liquid
4.2 **Color:** Colorless
4.3 **Odor:** Very strong; pungent; vinegar-like; characteristic

5. HEALTH HAZARDS

5.1 **Personal Protective Equipment:** Protective clothing when skin contact might occur; respiratory protection necessary for all exposures; complete eye protection.
5.2 **Symptoms Following Exposure:** Liquid is volatile and causes little irritation on uncovered skin. However, causes severe burns when clothing is wet with the chemical or if it enters gloves or shoes. Causes skin and eye burns and irritation of respiratory tract. Nausea and vomiting may develop after exposure.
5.3 **Treatment for Exposure:** INHALATION: move victim at once to fresh air; if breathing becomes difficult, give oxygen; get medical care quickly. INGESTION: do NOT induce vomiting. SKIN OR EYE CONTACT WITH LIQUID OR VAPOR: flush immediately with plenty of clean running water; wash eyes for at least 15 min. and get medical care quickly.
5.4 **Toxicity by Inhalation (Threshold Limit Value):** 5 ppm
5.5 **Short-Term Inhalation Limits:** Data not available
5.6 **Toxicity by Ingestion:** Grade 2; 0.5 to 5 g/kg (rat)
5.7 **Late Toxicity:** Not pertinent
5.8 **Vapor (Gas) Irritant Characteristics:** Vapor is moderately irritating such that personnel will not usually tolerate moderate or high vapor concentrations.
5.9 **Liquid or Solid Irritant Characteristics:** Fairly severe skin irritant, may cause pain and second-degree burns after a few minutes of contact.
5.10 **Odor Threshold:** 0.14 ppm

6. FIRE HAZARDS

6.1 **Flash Point:** 136°F O.C.; 120°F C.C.
6.2 **Flammable Limits in Air:** 2.7% – 10.0%
6.3 **Fire Extinguishing Agents:** Water spray, dry chemical, alcohol foam, or carbon dioxide
6.4 **Fire Extinguishing Agents Not to be Used:** Water and foam react, but heat liberated is not enough to create a hazard. Dry chemical forced below the surface can cause foaming and boiling.
6.5 **Special Hazards of Combustion Products:** Irritating vapors are generated when heated.
6.6 **Behavior in Fire:** Not pertinent
6.7 **Ignition Temperature:** 600°F
6.8 **Electrical Hazard:** Not pertinent
6.9 **Burning Rate:** 3.3 mm/min.

7. CHEMICAL REACTIVITY

7.1 **Reactivity with Water:** Reacts slowly with water; considerable heat liberated when water spray is used.
7.2 **Reactivity with Common Materials:** Corrodes iron, steel, and other metals.
7.3 **Stability During Transport:** Stable
7.4 **Neutralizing Agents for Acids and Caustics:** Dilute with water; use sodium bicarbonate solution to rinse.
7.5 **Polymerization:** Not pertinent
7.6 **Inhibitor of Polymerization:** Not pertinent

8. WATER POLLUTION

8.1 **Aquatic Toxicity:**
75 ppm/96 hr/bluegill/TL_m/fresh water
100-300 ppm/48 hr/shrimp/LC_{50}/salt water
8.2 **Waterfowl Toxicity:** Not pertinent
8.3 **Biological Oxygen Demand (BOD):** 53%, 1-5 days
8.4 **Food Chain Concentration Potential:** None noted

9. SELECTED MANUFACTURERS

1. Celanese Corp.
Celanese Chemical Co. Div.
245 Park Ave.
New York, N. Y. 10017
2. Eastman Kodak Co.
Tennessee Eastman Co. Div.
Kingsport, Tenn. 37662
3. Union Carbide Corp.
Chemicals and Plastics Div.
270 Park Ave.
New York, N. Y. 10017

10. SHIPPING INFORMATION

10.1 **Grades or Purity:**
Pure: 99% min.
Technical: 75-98.5% min.
10.2 **Storage Temperature:** Ambient
10.3 **Inert Atmosphere:** No requirement
10.4 **Venting:** Pressure-vacuum

11. HAZARD ASSESSMENT CODE

(See Hazard Assessment Handbook, CG 446-3)

A-P-Q

12. HAZARD CLASSIFICATIONS

12.1 **Code of Federal Regulations:** Corrosive Material
12.2 **NAS Hazard Rating for Bulk Water Transportation:**

Category	Rating
Fire	2
Health	
Vapor Irritant	3
Liquid or Solid Irritant	3
Poisons	3
Water Pollution	
Human Toxicity	1
Aquatic Toxicity	2
Aesthetic Effect	2
Reactivity	
Other Chemicals	3
Water	2
Self-Reaction	0

12.3 **NFPA Hazard Classifications:**

Category	Classification
Health Hazard (Blue)	2
Flammability (Red)	2
Reactivity (Yellow)	1
	W̶

13. PHYSICAL AND CHEMICAL PROPERTIES

13.1 **Physical State at 15°C and 1 atm:** Liquid
13.2 **Molecular Weight:** 102.09
13.3 **Boiling Point at 1 atm:** 282°F = 139°C = 412°K
13.4 **Freezing Point:** −101°F = −74.1°C = 199.1°K
13.5 **Critical Temperature:** 565°F = 296°C = 569°K
13.6 **Critical Pressure:** 679 psia = 46.2 atm = 4.68 MN/m^2
13.7 **Specific Gravity:** 1.08 at 20°C (liquid)
13.8 **Liquid Surface Tension:** Not pertinent
13.9 **Liquid-Water Interfacial Tension:** Not pertinent
13.10 **Vapor (Gas) Specific Gravity:** Not pertinent
13.11 **Ratio of Specific Heats of Vapor (Gas):** 1.093
13.12 **Latent Heat of Vaporization:** 119Btu/lb = 66.2 cal/g = 2.77×10^5 J/kg
13.13 **Heat of Combustion:** −7058 Btu/lb = −3921 cal/g = -164.2×10^5 J/kg
13.14 **Heat of Decomposition:** Not pertinent
13.15 **Heat of Solution:** Not pertinent
13.16 **Heat of Polymerization:** Not pertinent

NOTES

REVISED 1978

ACT	**ACETONE**

Common Synonyms Dimethyl ketone Propanone	Watery liquid Colorless Sweet odor Floats and mixes with water. Flammable, irritating vapor is produced.

Stay upwind and use water spray to "knock down" vapor.
Shut off ignition sources and call fire department. Keep people away.
Stop discharge if possible.
Isolate and remove discharged material.
Avoid contact with liquid and vapor.
Notify local health and pollution control agencies.

Fire	FLAMMABLE. Flashback along vapor trail may occur. Vapor may explode if ignited in an enclosed area. Extinguish with dry chemical, alcohol foam, or carbon dioxide. Water may be ineffective on fire. Cool exposed containers with water.
Exposure	CALL FOR MEDICAL AID. VAPOR Irritating to eyes, nose and throat. If inhaled, may cause difficult breathing or loss of consciousness. Move to fresh air. If breathing has stopped, give artificial respiration. If breathing is difficult, give oxygen. LIQUID Irritating to eyes. Not irritating to skin. IF IN EYES, hold eyelids open and flush with plenty of water.
Water Pollution	Dangerous to aquatic life in high concentrations. May be dangerous if it enters water intakes. Notify local health and pollution control officials. Notify operators of nearby water intakes.

1. RESPONSE TO DISCHARGE

(See Response Methods Handbook, CG 446-4)

Issue warning—high flammability
Disperse and flush

2. LABEL

Red

3. CHEMICAL DESIGNATIONS

3.1 **Synonyms:** Dimethyl ketone
2-Propanone
3.2 **Coast Guard Compatibility Classification:** Ketone
3.3 **Chemical Formula:** CH_3COCH_3
3.4 **IMCO/United Nations Numerical Designation:** 3.1/1090

4. OBSERVABLE CHARACTERISTICS

4.1 **Physical State (as shipped):** Liquid
4.2 **Color:** Colorless
4.3 **Odor:** Sweetish; pleasant, resembling that of mint or fruit; pungent; sharp, penetrating residual; ketonic; pleasant, non-residual

5. HEALTH HAZARDS

5.1 **Personal Protective Equipment:** Organic vapor canister or air-supplied mask; synthetic rubber gloves; chemical safety goggles or face splash shield.
5.2 **Symptoms Following Exposure:** INHALATION: vapor irritating to eyes and mucous membranes; acts as an anesthetic in very high concentrations. INGESTION: low order of toxicity but very irritating to mucous membranes. SKIN: prolonged excessive contact causes defatting of the skin, possibly leading to dermatitis.
5.3 **Treatment for Exposure:** INHALATION: if victim is overcome, remove to fresh air and call a physician; administer artificial respiration if breathing is irregular or stopped. INGESTION: if victim has swallowed large amounts and is conscious and not having convulsions, induce vomiting and get medical help promptly; no specific antidote known. SKIN: wash well with water. EYES: flush with water immediately for at least 15 min. Consult a physician.
5.4 **Toxicity by Inhalation (Threshold Limit Value):** 1000 ppm
5.5 **Short-Term Inhalation Limits:** 1000 ppm for 30 min.
5.6 **Toxicity by Ingestion:** Grade 1; LD_{50} 5 to 15 g/kg (dog)
5.7 **Late Toxicity:** Not pertinent
5.8 **Vapor (Gas) Irritant Characteristics:** If present in high concentrations, vapors cause moderate irritation of the eyes or respiratory system. Effect is temporary.
5.9 **Liquid or Solid Irritant Characteristics:** No appreciable hazard. Practically harmless to the skin because it is very volatile and evaporates quickly from the skin.
5.10 **Odor Threshold:** 100 ppm

6. FIRE HAZARDS

6.1 **Flash Point:** 4°F O.C.; 0°F C.C.
6.2 **Flammable Limits in Air:** 2.6%–12.8%
6.3 **Fire Extinguishing Agents:** Alcohol foam, dry chemical, carbon dioxide
6.4 **Fire Extinguishing Agents Not to be Used:** Water in straight hose stream will scatter and spread fire and should not be used.
6.5 **Special Hazards of Combustion Products:** Not pertinent
6.6 **Behavior in Fire:** Not pertinent
6.7 **Ignition Temperature:** 869°F
6.8 **Electrical Hazard:** Class I, Group D
6.9 **Burning Rate:** 3.9 mm/min.

7. CHEMICAL REACTIVITY

7.1 **Reactivity with Water:** No reaction
7.2 **Reactivity with Common Materials:** No reaction
7.3 **Stability During Transport:** Stable
7.4 **Neutralizing Agents for Acids and Caustics:** Not pertinent
7.5 **Polymerization:** Not pertinent
7.6 **Inhibitor of Polymerization:** Not pertinent

8. WATER POLLUTION

8.1 **Aquatic Toxicity:**
14,250 ppm/24 hr/sunfish/killed/tap water
13,000 ppm/48 hr/mosquito fish/TL_m/turbid water
8.2 **Waterfowl Toxicity:** Not pertinent
8.3 **Biological Oxygen Demand (BOD):** (Theor) 122%, 5 days
8.4 **Food Chain Concentration Potential:** None noted

9. SELECTED MANUFACTURERS

1. Allied Chemical Corp.
Specialty Chemicals Div.
Wilmington Turnpike
Marcus Hook, Pa. 19061
2. Shell Chemical Co.
Industrial Chemicals Div.
Houston, Texas 77001
3. Union Carbide Corp.
Chemicals and Plastics Div.
270 Park Ave.
New York, N. Y. 10017

10. SHIPPING INFORMATION

10.1 **Grades or Purity:**
Technical: 99.5% plus 0.5% water
Reagent: 99.5% plus 0.5% water
10.2 **Storage Temperature:** Ambient
10.3 **Inert Atmosphere:** No requirement
10.4 **Venting:** Open (flame arrester) or pressure-vacuum

11. HAZARD ASSESSMENT CODE

(See Hazard Assessment Handbook, CG 446-3)

A-P-Q-R-S

12. HAZARD CLASSIFICATIONS

12.1 **Code of Federal Regulations:** Flammable liquid
12.2 **NAS Hazard Rating for Bulk Water Transportation:**

Category	Rating
Fire	3
Health	
Vapor Irritant	1
Liquid or Solid Irritant	0
Poisons	0
Water Pollution	
Human Toxicity	1
Aquatic Toxicity	1
Aesthetic Effect	1
Reactivity	
Other Chemicals	2
Water	0
Self-Reaction	1

12.3 **NFPA Hazard Classifications:**

Category	Classification
Health Hazard (Blue)	1
Flammability (Red)	3
Reactivity (Yellow)	0

13. PHYSICAL AND CHEMICAL PROPERTIES

13.1 **Physical State at 15°C and 1 atm:** Liquid
13.2 **Molecular Weight:** 58.08
13.3 **Boiling Point at 1 atm:** 133°F = 56.1°C = 329.3°K
13.4 **Freezing Point:** −138°F = −94.7°C = 178.5°K
13.5 **Critical Temperature:** 455°F = 235°C = 508°K
13.6 **Critical Pressure:** 682 psia = 46.4 atm = 4.70 MN/m^2
13.7 **Specific Gravity:** 0.791 at 20°C (liquid)
13.8 **Liquid Surface Tension:** Not pertinent
13.9 **Liquid-Water Interfacial Tension:** Not pertinent
13.10 **Vapor (Gas) Specific Gravity:** 2.0
13.11 **Ratio of Specific Heats of Vapor (Gas):** 1.127
13.12 **Latent Heat of Vaporization:** 220 Btu/lb = 122 cal/g = 5.11 × 10^5 J/kg
13.13 **Heat of Combustion:** −12,250 Btu/lb = −6808 cal/g = −285.0 × 10^5 J/kg
13.14 **Heat of Decomposition:** Not pertinent
13.15 **Heat of Solution:** Not pertinent
13.16 **Heat of Polymerization:** Not pertinent

NOTES

REVISED 1978

ACY | ACETONE CYANOHYDRIN

Common Synonyms			
2-Methyl lactonitrile Alpha-hydroxy isobutronitrile	Watery liquid	Colorless	Mild, almond odor.
	Floats and mixes with water. Poisonous vapor is produced.		

AVOID CONTACT WITH LIQUID AND VAPOR. KEEP PEOPLE AWAY.
Wear chemical protective suit with self-contained breathing apparatus.
Stop discharge if possible.
Stay upwind and use water spray to "knock-down" vapor.
Call fire department.
Isolate and remove discharged material.
Notify local health and pollution control agencies.

Fire

Combustible.
POISONOUS GASES ARE PRODUCED WHEN HEATED.
Vapor may explode if ignited in an enclosed area.
WEAR CHEMICAL PROTECTIVE SUIT WITH SELF-CONTAINED BREATHING APPARATUS.
Combat fires from safe distance or from protected location.
Extinguish with water, dry chemical, alcohol foam, or carbon dioxide.
Cool exposed containers with water.

Exposure

CALL FOR MEDICAL AID.

VAPOR
POISONOUS IF INHALED.
Irritating to eyes, nose and throat.
Move to fresh air.
If breathing has stopped, give artificial respiration.
If breathing is difficult, give oxygen.

LIQUID
POISONOUS IF SWALLOWED.
Will burn skin and eyes.
Remove contaminated clothing and shoes.
Flush affected areas with plenty of water.
IF SWALLOWED and victim is CONSCIOUS, have victim drink water or milk and have victim induce vomiting.
IF SWALLOWED and victim is UNCONSCIOUS OR HAVING CONVULSIONS, do nothing except keep victim warm.

Water Pollution

HARMFUL TO AQUATIC LIFE IN VERY LOW CONCENTRATIONS.
May be dangerous if it enters water intakes.
Notify local health and wildlife officials.
Notify operators of nearby water intakes.

1. RESPONSE TO DISCHARGE

(See Response Methods Handbook, CG 446-4)

Issue warning—poison
Restrict access
Should be removed—human toxicity

2. LABEL

3. CHEMICAL DESIGNATIONS

3.1 **Synonyms:** alpha-Hydroxyisobutyronitrile; 2-Methyllactonitrile
3.2 **Coast Guard Compatibility Classification:** Cyanohydrin
3.3 **Chemical Formula:** $(CH_3)_2C(OH)CN$
3.4 **IMCO/United Nations Numerical Designation:** 6.1/1541

4. OBSERVABLE CHARACTERISTICS

4.1 **Physical State (as shipped):** Liquid
4.2 **Color:** Colorless
4.3 **Odor:** Characteristic; distinct, strong cyanide

5. HEALTH HAZARDS

5.1 **Personal Protective Equipment:** Air-supplied mask with canister approved for use with acrylonitrile in less than 2% concentrations; rubber or plastic gloves; cover goggles or face mask; rubber boots; slicker suit; safety helmet.

5.2 **Symptoms Following Exposure:** At low dosages the earliest symptoms may be weakness, headaches, confusion and occasionally nausea and vomiting. Respiratory rate and depth will usually be increased at the beginning and at later stages become slow and gasping.

5.3 **Treatment for Exposure:** Call a physician for all cases of exposure. INHALATION: remove victim to fresh air. (Rescuer should wear suitable mask.) INGESTION: if victim is conscious, induce vomiting by having him drink strong salt water. SKIN: remove contaminated clothing and wash affected skin thoroughly with soap and water. EYES: hold eyelids apart and wash with continuous, gentle stream of water for at least 15 min.
If breathing has stopped, give artificial respiration until physician arrives. If victim is unconscious, administer amyl nitrite by crushing an ampule in a cloth and holding it under his nose for 15 seconds in every minute. Do not interrupt artificial respiration during this procedure. Replace ampule when its strength is spent; continue treatment until victim's condition improves or physician arrives.

5.4 **Toxicity by Inhalation (Threshold Limit Value):** Data not available
5.5 **Short-Term Inhalation Limits:** Data not available
5.6 **Toxicity by Ingestion:** Grade 4; below 50 mg/kg (mice)
5.7 **Late Toxicity:** Causes liver damage in rats
5.8 **Vapor (Gas) Irritant Characteristics:** Vapors irritate the eyes and respiratory system if present in high concentrations. The effect is temporary.

6. FIRE HAZARDS

6.1 **Flash Point:** 165°F C.C.
6.2 **Flammable Limits in Air:** 2.2%–12%
6.3 **Fire Extinguishing Agents:** Water spray, dry chemical, alcohol foam, carbon dioxide
6.4 **Fire Extinguishing Agents Not to be Used:** Not pertinent
6.5 **Special Hazards of Combustion Products:** Toxic hydrogen cyanide is generated when heated
6.6 **Behavior in Fire:** Not pertinent
6.7 **Ignition Temperature:** 1270°F
6.8 **Electrical Hazard:** Not pertinent
6.9 **Burning Rate:** Data not available

7. CHEMICAL REACTIVITY

7.1 **Reactivity with Water:** No reaction
7.2 **Reactivity with Common Materials:** No reaction
7.3 **Stability During Transport:** Stable
7.4 **Neutralizing Agents for Acids and Caustics:** Not pertinent
7.5 **Polymerization:** Not pertinent
7.6 **Inhibitor of Polymerization:** Not pertinent

8. WATER POLLUTION

8.1 **Aquatic Toxicity:** Data not available
8.2 **Waterfowl Toxicity:** Not pertinent
8.3 **Biological Oxygen Demand (BOD):** Data not available
8.4 **Food Chain Concentration Potential:** Data not available

9. SELECTED MANUFACTURERS

1. Aldrich Chemical Co.
   940 West St. Paul Ave.
   Milwaukee, Wis. 53233
2. Eastman Kodak Co.
   Rochester, New York 14650
3. Rohm and Haas Co.
   Independence Mall West
   Philadelphia, Pa. 19105

10. SHIPPING INFORMATION

10.1 **Grades or Purity:** 98–99%
10.2 **Storage Temperature:** Ambient
10.3 **Inert Atmosphere:** No requirement
10.4 **Venting:** Pressure-vacuum

11. HAZARD ASSESSMENT CODE

(See Hazard Assessment Handbook, CG 446-3)

A-P-Q

12. HAZARD CLASSIFICATIONS

12.1 **Code of Federal Regulations:** Poisonous liquid or solid, Class B

12.2 **NAS Hazard Rating for Bulk Water Transportation:**

Category	Rating
Fire	1
Health	
Vapor Irritant	1
Liquid or Solid Irritant	2
Poisons	4
Water Pollution	
Human Toxicity	4
Aquatic Toxicity	3
Aesthetic Effect	3
Reactivity	
Other Chemicals	2
Water	3
Self-Reaction	0

12.3 **NFPA Hazard Classifications:**

Category	Classification
Health Hazard (Blue)	4
Flammability (Red)	1
Reactivity (Yellow)	2

13. PHYSICAL AND CHEMICAL PROPERTIES

13.1 **Physical State at 15°C and 1 atm:** Liquid
13.2 **Molecular Weight:** 85.11
13.3 **Boiling Point at 1 atm:** Decomposes
13.4 **Freezing Point:** $-5.8°F = -21°C = 252°K$
13.5 **Critical Temperature:** Not pertinent
13.6 **Critical Pressure:** Not pertinent
13.7 **Specific Gravity:** 0.925 at 25°C (liquid)
13.8 **Liquid Surface Tension:** Not pertinent
13.9 **Liquid-Water Interfacial Tension:** Not pertinent
13.10 **Vapor (Gas) Specific Gravity:** Not pertinent
13.11 **Ratio of Specific Heats of Vapor (Gas):** (est.) 1.074
13.12 **Latent Heat of Vaporization:** Data not available
13.13 **Heat of Combustion:** Data not available
13.14 **Heat of Decomposition:** Not pertinent
13.15 **Heat of Solution:** Not pertinent
13.16 **Heat of Polymerization:** Not pertinent

5. HEALTH HAZARDS (Cont'd.)

5.9 **Liquid or Solid Irritant Characteristics:** Causes smarting of the skin and first-degree burns on short exposure and may cause secondary burns on long exposure.
5.10 **Odor Threshold:** Data not available

REVISED 1978

ATN

ACETONITRILE

Common Synonyms			
Methyl cyanide Cyanomethane Ethanenitrile	Watery liquid	Colorless	Sweet odor
	Floats and mixes with water. Flammable, irritating vapor is produced.		

Wear goggles and self-contained breathing apparatus.
Shut off ignition sources and call fire department.
Stay upwind and use water spray to "knock down" vapor.
Stop discharge if possible.
Avoid contact with liquid and vapor. Keep people away.
Isolate and remove discharged material.
Notify local health and pollution control agencies.

Fire

FLAMMABLE
POISONOUS GASES MAY BE PRODUCED WHEN HEATED.
Flashback along vapor trail may occur.
Vapor may explode if ignited in an enclosed area.
Wear goggles and self-contained breathing apparatus.
Extinguish with dry chemical, alcohol foam, or carbon dioxide.
Water may be ineffective on fire.
Cool exposed containers with water.

Exposure

CALL FOR MEDICAL AID.

VAPOR
Irritating to eyes, nose and throat.
If inhaled, will cause difficult breathing.
Move to fresh air.
If breathing has stopped, give artificial respiration.
If breathing is difficult, give oxygen.

LIQUID
Irritating to skin and eyes.
Harmful if swallowed.
Remove contaminated clothing and shoes.
Flush affected areas with plenty of water.
IF IN EYES, hold eyelids open and flush with plenty of water.
IF SWALLOWED and victim is CONSCIOUS, have victim drink water or milk.

Water Pollution

Dangerous to aquatic life in high concentrations.
May be dangerous if it enters water intakes.
Notify local health and pollution control officials.
Notify operators of nearby water intakes.

1. RESPONSE TO DISCHARGE
(See Response Methods Handbook, CG 446-4)

Issue warning—high flammability
Disperse and flush

2. LABEL

Red

3. CHEMICAL DESIGNATIONS

3.1 **Synonyms:** Ethanenitrile; Ethyl nitrile; Cyanomethane; Methyl cyanide
3.2 **Coast Guard Compatibility Classification:** Nitrile
3.3 **Chemical Formula:** CH_3CN
3.4 **IMCO/United Nations Numerical Designation:** 6.1/1648

4. OBSERVABLE CHARACTERISTICS

4.1 **Physical State (as shipped):** Liquid
4.2 **Color:** Colorless
4.3 **Odor:** Sweet; ethereal

5. HEALTH HAZARDS

5.1 **Personal Protective Equipment:** Wear self-contained breathing apparatus
5.2 **Symptoms Following Exposure:** Exposure to 160 ppm for 4 hours causes flushing of the face and a feeling of constriction in the chest; 500 ppm for brief periods is irritating to the nose and throat. Severe exposures cause irritability, skin eruptions, confusion, delirium, convulsions, paralysis, and death due to central nervous system depression.
5.3 **Treatment for Exposure:** Remove victim from contaminated atmosphere. Give artificial respiration and oxygen if respiration is impaired.
5.4 **Toxicity by Inhalation (Threshold Limit Value):** 40 ppm
5.5 **Short-Term Inhalation Limits:** 40 ppm for 60 min.
5.6 **Toxicity by Ingestion:** Grade 3; 50–500 mg/kg (guinea pig)
5.7 **Late Toxicity:** Not pertinent
5.8 **Vapor (Gas) Irritant Characteristics:** Vapors cause a slight smarting of the eyes or respiratory system if present in high concentrations. Effect is temporary.
5.9 **Liquid or Solid Irritant Characteristics:** Minimum hazard. If spilled on clothing and allowed to remain, may cause smarting and reddening of the skin.
5.10 **Odor Threshold:** 40 ppm

6. FIRE HAZARDS

6.1 **Flash Point:** 42°F O.C.
6.2 **Flammable Limits in Air:** 4.4%–16%
6.3 **Fire Extinguishing Agents:** Alcohol foam, dry chemical, carbon dioxide
6.4 **Fire Extinguishing Agents Not to be Used:** Water may be ineffective
6.5 **Special Hazards of Combustion Products:** Toxic vapors are generated when heated
6.6 **Behavior in Fire:** Vapor heavier than air and may travel a considerable distance to a source of ignition and flash back.
6.7 **Ignition Temperature:** 975°F
6.8 **Electrical Hazard:** Not pertinent
6.9 **Burning Rate:** 2.7 mm/min.

7. CHEMICAL REACTIVITY

7.1 **Reactivity with Water:** No reaction
7.2 **Reactivity with Common Materials:** No reaction
7.3 **Stability During Transport:** Stable
7.4 **Neutralizing Agents for Acids and Caustics:** Not pertinent
7.5 **Polymerization:** Not pertinent
7.6 **Inhibitor of Polymerization:** Not pertinent

8. WATER POLLUTION

8.1 **Aquatic Toxicity:** 1150 ppm/24 hr/fathead minnow/TL_m/hard water
8.2 **Waterfowl Toxicity:** Not pertinent
8.3 **Biological Oxygen Demand (BOD):** 17%, 5 days
8.4 **Food Chain Concentration Potential:** None noted

9. SELECTED MANUFACTURERS

1. American Cyanamid Co.
   Industrial Chemicals and Plastics Division
   New Orleans, La. 70121
2. Eastman Kodak Co.
   Texas Eastman Co. Division
   Longview, Texas 75601
3. Vistron Corp.
   Midland Building
   Cleveland, Ohio 44115

10. SHIPPING INFORMATION

10.1 **Grades or Purity:** Data not available
10.2 **Storage Temperature:** Ambient
10.3 **Inert Atmosphere:** No requirement
10.4 **Venting:** Pressure-vacuum

11. HAZARD ASSESSMENT CODE
(See Hazard Assessment Handbook, CG 446-3)

A-P-Q-R-S

12. HAZARD CLASSIFICATIONS

12.1 **Code of Federal Regulations:** Flammable liquid
12.2 **NAS Hazard Rating for Bulk Water Transportation:**

Category	Rating
Fire	3
Health	
Vapor Irritant	1
Liquid or Solid Irritant	1
Poisons	3
Water Pollution	
Human Toxicity	2
Aquatic Toxicity	1
Aesthetic Effect	1
Reactivity	
Other Chemicals	2
Water	0
Self-Reaction	0

12.3 **NFPA Hazard Classifications:**

Category	Classification
Health Hazard (Blue)	2
Flammability (Red)	3
Reactivity (Yellow)	0

13. PHYSICAL AND CHEMICAL PROPERTIES

13.1 **Physical State at 15°C and 1 atm:** Liquid
13.2 **Molecular Weight:** 41.05
13.3 **Boiling Point at 1 atm:** 179°F = 81.6°C = 354.8°K
13.4 **Freezing Point:** −50.3°F = −45.7°C = 227.5°K
13.5 **Critical Temperature:** 526.5°F = 274.7°C = 547.9°K
13.6 **Critical Pressure:** 701 psia = 47.7 atm = 4.83 MN/m^2
13.7 **Specific Gravity:** 0.787 at 20°C (liquid)
13.8 **Liquid Surface Tension:** Not pertinent
13.9 **Liquid-Water Interfacial Tension:** Not pertinent
13.10 **Vapor (Gas) Specific Gravity:** 1.4
13.11 **Ratio of Specific Heats of Vapor (Gas):** 1.192
13.12 **Latent Heat of Vaporization:** 313 Btu/lb = 174 cal/g = 7.29×10^5 J/kg
13.13 **Heat of Combustion:** −13,360 Btu/lb = −7420 cal/g = -310.7×10^5 J/kg
13.14 **Heat of Decomposition:** Not pertinent
13.15 **Heat of Solution:** Not pertinent
13.16 **Heat of Polymerization:** Not pertinent

NOTES

REVISED 1978

ACP	ACETOPHENONE

Common Synonyms Acetylbenzene Methyl phenyl ketone	Liquid Colorless Flowery, sweet odor Sinks slowly in water. Freezing point is 68° F.

Stop discharge if possible.
Call fire department.
Avoid contact with liquid.
Isolate and remove discharged material.
Notify local health and pollution control agencies.

Fire	Combustible. Extinguish with water, dry chemical, foam, or carbon dioxide. Cool exposed containers with water.
Exposure	CALL FOR MEDICAL AID. LIQUID OR SOLID Irritating to skin and eyes. Harmful if swallowed. Remove contaminated clothing and shoes. Flush affected areas with plenty of water. IF IN EYES, hold eyelids open and flush with plenty of water. IF SWALLOWED and victim is CONSCIOUS, have victim drink water or milk.
Water Pollution	Effect of low concentrations on aquatic life is unknown. Fouling to shoreline. May be dangerous if it enters water intakes. Notify local health and pollution control officials. Notify operators of nearby water intakes.

1. RESPONSE TO DISCHARGE

(See Response Methods Handbook, CG 446-4)

Issue warning—water contaminant
Disperse and flush

2. LABELS

No hazard label required by Code of Federal Regulations

3. CHEMICAL DESIGNATIONS

3.1 **Synonyms:** Acetylbenzene
Methyl phenyl ketone
3.2 **Coast Guard Compatibility Classification:** Ketone
3.3 **Chemical Formula:** $C_6H_5COCH_3$
3.4 **IMCO/United Nations Numerical Designation:** Not listed

4. OBSERVABLE CHARACTERISTICS

4.1 **Physical State (as shipped):** Liquid
4.2 **Color:** Colorless
4.3 **Odor:** Persistent; like orange blossoms and jasmine

5. HEALTH HAZARDS

5.1 **Personal Protective Equipment:** Protect eyes and skin from direct contact.
5.2 **Symptoms Following Exposure:** No toxicity expected from inhalation or ingestion except slight narcotic effect. Liquid can cause eye and skin irritation on contact.
5.3 **Treatment for Exposure:** SKIN OR EYE CONTACT: irrigate affected area with water for 15 min.
5.4 **Toxicity by Inhalation (Threshold Limit Value):** Data not available
5.5 **Short-Term Inhalation Limits:** Data not available
5.6 **Toxicity by Ingestion:** Grade 2; LD_{50} 0.5 to 5 g/kg
5.7 **Late Toxicity:** Data not available
5.8 **Vapor (Gas) Irritant Characteristics:** Nonirritating to the eyes and throat
5.9 **Liquid or Solid Irritant Characteristics:** Minimum hazard. If spilled on clothing and allowed to remain, may cause smarting and reddening of the skin.
5.10 **Odor Threshold:** 0.01 - 0.025 mg/m^3

6. FIRE HAZARDS

6.1 **Flash Point:** 180°F O.C.
6.2 **Flammable Limits in Air:** Data not available
6.3 **Fire Extinguishing Agents:** Water, foam, dry chemical, carbon dioxide
6.4 **Fire Extinguishing Agents Not to be Used:** None
6.5 **Special Hazards of Combustion Products:** Not pertinent
6.6 **Behavior in Fire:** Not pertinent
6.7 **Ignition Temperature:** 1058°F
6.8 **Electrical Hazard:** Not pertinent
6.9 **Burning Rate:** Data not available

7. CHEMICAL REACTIVITY

7.1 **Reactivity with Water:** No reaction
7.2 **Reactivity with Common Materials:** No reaction
7.3 **Stability During Transport:** Stable
7.4 **Neutralizing Agents for Acids and Caustics:** Not pertinent
7.5 **Polymerization:** Not pertinent
7.6 **Inhibitor of Polymerization:** Not pertinent

8. WATER POLLUTION

8.1 **Aquatic Toxicity:** Data not available
8.2 **Waterfowl Toxicity:** Data not available
8.3 **Biological Oxygen Demand (BOD):** Data not available
8.4 **Food Chain Concentration Potential:** None noted

9. SELECTED MANUFACTURERS

1. Givaudan Corp.
Aroma Chemicals Div.
125 Delawanna Ave.
Clifton, N. J. 07014
2. Orbis Products Corp.
475 Tenth Ave.
New York, N. Y. 10018
3. Union Carbide Corp.
Chemicals & Plastics Div.
270 Park Ave.
New York, N. Y. 10017

10. SHIPPING INFORMATION

10.1 **Grades or Purity:** Technical: 99+% acetophenone
10.2 **Storage Temperature:** Data not available
10.3 **Inert Atmosphere:** Data not available
10.4 **Venting:** Data not available

11. HAZARD ASSESSMENT CODE

(See Hazard Assessment Handbook, CG 446-3)

A-T-U-X-Y

12. HAZARD CLASSIFICATIONS

12.1 **Code of Federal Regulations:** Not listed
Combustible Liquid
12.2 **NAS Hazard Rating for Bulk Water Transportation:** Not listed
12.3 **NFPA Hazard Classifications:**

Category	Classification
Health Hazard (Blue)	1
Flammability (Red)	2
Reactivity (Yellow)	0

13. PHYSICAL AND CHEMICAL PROPERTIES

13.1 **Physical State at 15°C and 1 atm:** Liquid
13.2 **Molecular Weight:** 120.15
13.3 **Boiling Point at 1 atm:** 395.1°F = 201.7°C = 474.9°K
13.4 **Freezing Point:** 67.5°F = 19.7°C = 292.9°K
13.5 **Critical Temperature:** 802°F = 428°C = 701°K
13.6 **Critical Pressure:** 560 psia = 38 atm = 3.8 MN/m^2
13.7 **Specific Gravity:** 1.028 at 20°C (liquid)
13.8 **Liquid Surface Tension:** 12 dymes/cm = 0.012 N/m at 30°C
13.9 **Liquid-Water Interfacial Tension:** (est.) 40 dynes/cm = 0.04 N/m at 27°C
13.10 **Vapor (Gas) Specific Gravity:** Not pertinent
13.11 **Ratio of Specific Heats of Vapor (Gas):** (est.) 1.071
13.12 **Latent Heat of Vaporization:** 150 Btu/lb = 83.6 cal/g = 3.5×10^5 J/kg
13.13 **Heat of Combustion:** −14,850 Btu/lb = −8250 cal/g = -345.4×10^5 J/kg
13.14 **Heat of Decomposition:** Not pertinent
13.15 **Heat of Solution:** Not pertinent
13.16 **Heat of Polymerization:** Not pertinent

NOTES

REVISED 1978

ATA | ACETYLACETONE

Common Synonyms
2,4-Pentanedione
Diacetylmethane

Liquid — Colorless — Unpleasant Odor

Floats and mixes slowly with water. Flammable, irritating vapor is produced.

Shut off ignition sources. Call fire department.
Stop discharge if possible. Keep people away.
Avoid contact with liquid and vapor.
Isolate and remove discharged material.
Notify local health and pollution control agencies.

Fire

Combustible.
Flashback along vapor trail may occur.
Vapor may explode if ignited in an enclosed area.
Wear goggles and self-contained breathing apparatus.
Extinguish with dry chemicals, alcohol foam, or carbon dioxide.
Water may be ineffective on fire.
Cool exposed containers with water.

Exposure

CALL FOR MEDICAL AID

VAPOR
Irritating to eyes.
If inhaled will cause dizziness, coughing, headache, or loss of consciousness.
If in eyes, hold eyelids open and flush with plenty of water.
If breathing has stopped, give artificial respiration.
If breathing is difficult, give oxygen.

LIQUID
Irritating to skin and eyes.
Harmful if swallowed.
Remove contaminated clothing and shoes.
Flush affected areas with plenty of water.
IF IN EYES, hold eyelids open and flush with plenty of water.
IF SWALLOWED and victim is UNCONSCIOUS OR HAVING CONVULSIONS, do nothing except keep victim warm.

Water Pollution

Effect of low concentrations on aquatic life is unknown.
May be dangerous if it enters water intakes.
Notify local health and wildlife officials.
Notify operators of nearby water intakes.

1. RESPONSE TO DISCHARGE
(See Response Methods Handbook, CG 446-4)

Issue warning — water contaminant
Disperse and flush

2. LABELS

No hazard label required by Code of Federal Regulations

3. CHEMICAL DESIGNATIONS

3.1 **Synonyms:** Diacetylmethane; 2,4-Pentanedione
3.2 **Coast Guard Compatibility Classification:** Not listed
3.3 **Chemical Formula:** $CH_3COCH_2COCH_3$
3.4 **IMCO/United Nations Numerical Designation:** Not listed

4. OBSERVABLE CHARACTERISTICS

4.1 **Physical State (as shipped):** Liquid
4.2 **Color:** Water-white
4.3 **Odor:** Unpleasant

5. HEALTH HAZARDS

5.1 **Personal Protective Equipment:** Safety glasses; eye bath and safety shower; air-supplied mask for concentrations above 2%
5.2 **Symptoms Following Exposure:** Inhalation causes dizziness, headache, nausea, vomiting, and loss of consciousness. Contact with liquid irritates eyes.
5.3 **Treatment for Exposure:** INHALATION: remove to fresh air; if victim is not breathing, give artificial respiration and then oxygen; call a physician. EYES or SKIN: flush with water.
5.4 **Toxicity by Inhalation (Threshold Limit Value):** Data not available
5.5 **Short-Term Inhalation Limits:** Data not available
5.6 **Toxicity by Ingestion:** Grade 2; oral LD_{50} = 1,000 mg/kg (rat)
5.7 **Late Toxicity:** Data not available
5.8 **Vapor (Gas) Irritant Characteristics:** Data not available
5.9 **Liquid or Solid Irritant Characteristics:** Data not available
5.10 **Odor Threshold:** Data not available

6. FIRE HAZARDS

6.1 **Flash Point:** 105°F O.C.; 93°F C.C.
6.2 **Flammable Limits in Air:** 2.4% – 11.6%
6.3 **Fire Extinguishing Agents:** Dry chemical, "alcohol" foam, carbon dioxide
6.4 **Fire Extinguishing Agents Not to be Used:** Water may be ineffective on fire.
6.5 **Special Hazards of Combustion Products:**
6.6 **Behavior in Fire:** Vapor is heavier than air and may travel to a source of ignition and flash back.
6.7 **Ignition Temperature:** 644°F
6.8 **Electrical Hazard:** Data not available
6.9 **Burning Rate:** 3.6 mm/min.

7. CHEMICAL REACTIVITY

7.1 **Reactivity with Water:** No reaction
7.2 **Reactivity with Common Materials:** May dissolve plastics
7.3 **Stability During Transport:** Stable
7.4 **Neutralizing Agents for Acids and Caustics:** Not pertinent
7.5 **Polymerization:** Not pertinent
7.6 **Inhibitor of Polymerization:** Not pertinent

8. WATER POLLUTION

8.1 **Aquatic Toxicity:** Data not available
8.2 **Waterfowl Toxicity:** Data not available
8.3 **Biological Oxygen Demand (BOD):** Data not available
8.4 **Food Chain Concentration Potential:** None

9. SELECTED MANUFACTURERS

1. Union Carbide Corp.
   Chemicals and Plastics Div.
   270 Park Avenue
   New York, N. Y. 10017
2. Aldrich Chemical Co.
   940 W. Saint Paul Ave.
   Milwaukee, Wis. 53233
3. Eastman Organic Chemicals
   343 State Street
   Rochester, N. Y. 14650

10. SHIPPING INFORMATION

10.1 **Grades or Purity:** Commercial
10.2 **Storage Temperature:** Ambient
10.3 **Inert Atmosphere:** No requirement
10.4 **Venting:** Open (flame arrester)

11. HAZARD ASSESSMENT CODE
(See Hazard Assessment Handbook, CG 446-3)

A-P-Q

12. HAZARD CLASSIFICATIONS

12.1 **Code of Federal Regulations:** Not listed
12.2 **NAS Hazard Rating for Bulk Water Transportation:** Not listed
12.3 **NFPA Hazard Classifications:**

Category	Classification
Health Hazard (Blue)	2
Flammability (Red)	2
Reactivity (Yellow)	0

13. PHYSICAL AND CHEMICAL PROPERTIES

13.1 **Physical State at 15°C and 1 atm:** Liquid
13.2 **Molecular Weight:** 100.12
13.3 **Boiling Point at 1 atm:** 284.7°F = 140.4°C = 413.6°K
13.4 **Freezing Point:** −10.3°F = −23.5°C = 249.7°K
13.5 **Critical Temperature:** Not pertinent
13.6 **Critical Pressure:** Not pertinent
13.7 **Specific Gravity:** 0.975 at 20°C
13.8 **Liquid Surface Tension:** 31.2 dynes/cm = 0.0312 N/m at 20°C
13.9 **Liquid-Water Interfacial Tension:** Not pertinent
13.10 **Vapor (Gas) Specific Gravity:** 3.45
13.11 **Ratio of Specific Heats of Vapor (Gas):** (est.) 1.072
13.12 **Latent Heat of Vaporization:** 194 Btu/lb = 108 cal/g = 4.52×10^5 J/kg
13.13 **Heat of Combustion:** −11,070 Btu/lb = −6,150 cal/g = -257×10^5 J/kg
13.14 **Heat of Decomposition:** Not pertinent
13.15 **Heat of Solution:** −11.5 Btu/lb = −6.4 cal/g = -0.27×10^5 J/kg
13.16 **Heat of Polymerization:** Not pertinent

NOTES

ABM — ACETYL BROMIDE

Common Synonyms	Liquid	Colorless	Sharp unpleasant odor
	Flammable, irritating vapor is produced.		

AVOID CONTACT WITH LIQUID AND VAPOR. KEEP PEOPLE AWAY.
Wear goggles, self-contained breathing apparatus, and rubber overclothing (including gloves).
Shut off ignition sources. Call fire department.
Stop discharge if possible.
Isolate and remove discharged material.
Notify local health and pollution control agencies.

Fire

FLAMMABLE.
Irritating gases are produced when heated.
Flashback along vapor trail may occur.
Vapor may explode if ignited in an enclosed area.
Wear goggles, self-contained breathing apparatus and rubber overclothing (including gloves).
Extinguish with dry chemicals or carbon dioxide.
DO NOT USE WATER ON FIRE.

Exposure

CALL FOR MEDICAL AID.

VAPOR
Irritating to eyes, nose and throat.
If inhaled will cause difficult breathing.
If in eyes, hold eyelids open and flush with plenty of water.
If breathing has stopped, give artificial respiration.
If breathing is difficult, give oxygen.

LIQUID
Will burn skin and eyes.
Harmful if swallowed.
Remove contaminated clothing and shoes.
Flush affected areas with plenty of water.
IF IN EYES, hold eyelids open and flush with plenty of water.
IF SWALLOWED and victim is CONSCIOUS, have victim drink water or milk.
IF SWALLOWED and victim is UNCONSCIOUS OR HAVING CONVULSIONS, do nothing except keep victim warm.

Water Pollution

Effect of low concentrations on aquatic life is unknown.
May be dangerous if it enters water intakes.
Notify local health and wildlife officials.
Notify operators of nearby water intakes.

1. RESPONSE TO DISCHARGE

(See Response Methods Handbook, CG 446-4)

Issue warning corrosive
Restrict access
Evacuate area
Disperse and flush

2. LABEL

3. CHEMICAL DESIGNATIONS

3.1 **Synonyms:** No common synonyms
3.2 **Coast Guard Compatibility Classification:** Not listed
3.3 **Chemical Formula:** CH_3COBr
3.4 **IMCO/United Nations Numerical Designation:** 8/1716

4. OBSERVABLE CHARACTERISTICS

4.1 **Physical State (as shipped):** Liquid
4.2 **Color:** Colorless
4.3 **Odor:** Acrid and sharp

5. HEALTH HAZARDS

5.1 **Personal Protective Equipment:** Safety goggles; gloves; adequate ventilation; provisions for flushing eyes or skin with water
5.2 **Symptoms Following Exposure:** Inhalation produces primary irritation of the respiratory tract; symptoms of lung damage may be delayed. Contact with liquid produces primary irritation of eyes and severe skin damage; delayed blistering is not uncommon.
5.3 **Treatment for Exposure:** INHALATION: remove victim from exposure; if breathing has stopped, give artificial respiration; if breathing is difficult, give oxygen; watch for delayed lung damage. EYES: flush with water for at least 15 min.; get medical attention. SKIN: flush with water; treat burns as needed.
5.4 **Toxicity by Inhalation (Threshold Limit Value):** Data not available
5.5 **Short-Term Inhalation Limits:** Data not available
5.6 **Toxicity by Ingestion:** Grade 2; oral rat LD_{50} = 3,310 mg/kg (acetic acid). Decomposes violently in water, forming bromic acid and acetic acid)
5.7 **Late Toxicity:** Data not available
5.8 **Vapor (Gas) Irritant Characteristics:** Data not available
5.9 **Liquid or Solid Irritant Characteristics:** Data not available
5.10 **Odor Threshold:** 5.0×10^{-4} ppm

6. FIRE HAZARDS

6.1 **Flash Point:** Data not available
6.2 **Flammable Limits in Air:** Data not available
6.3 **Fire Extinguishing Agents:** Carbon dioxide, dry chemical
6.4 **Fire Extinguishing Agents Not to be Used:** Water
6.5 **Special Hazards of Combustion Products:** Toxic and irritating hydrogen bromide fumes may form in fires.
6.6 **Behavior in Fire:** Do not apply water to adjacent fires. Reacts with water to produce toxic and irritating gases.
6.7 **Ignition Temperature:** Data not available
6.8 **Electrical Hazard:** Data not available
6.9 **Burning Rate:** Data not available

7. CHEMICAL REACTIVITY

7.1 **Reactivity with Water:** Reacts violently, forming corrosive and toxic fumes of hydrogen bromide
7.2 **Reactivity with Common Materials:** Attacks and corrodes wood and most metals in the presence of moisture. Flammable hydrogen gas may collect in enclosed spaces.
7.3 **Stability During Transport:** Stable if protected from moisture
7.4 **Neutralizing Agents for Acids and Caustics:** Flood with water, rinse with dilute sodium bicarbonate or soda ash solution.
7.5 **Polymerization:** Not pertinent
7.6 **Inhibitor of Polymerization:** Not pertinent

8. WATER POLLUTION

8.1 **Aquatic Toxicity:** Data not available
8.2 **Waterfowl Toxicity:** Data not available
8.3 **Biological Oxygen Demand (BOD):** Data not available
8.4 **Food Chain Concentration Potential:** None

9. SELECTED MANUFACTURERS

1. Eastman Organic Chemicals
343 State Street
Rochester, N. Y. 14650
2. Aldrich Chemical Co.
940 W. Saint Paul Ave.
Milwaukee, Wis. 53233
3. Gallard-Schlesinger Chemical Mfg. Co.
584 Mineola Ave.
Carle Place, N. Y. 11514

10. SHIPPING INFORMATION

10.1 **Grades or Purity:** Analytical; Commercial
10.2 **Storage Temperature:** Ambient
10.3 **Inert Atmosphere:** Padded
10.4 **Venting:** Pressure-vacuum

11. HAZARD ASSESSMENT CODE

(See Hazard Assessment Handbook, CG 446-3)

A-O

12. HAZARD CLASSIFICATIONS

12.1 **Code of Federal Regulations:** Corrosive liquid
12.2 **NAS Hazard Rating for Bulk Water Transportation:** Not listed
12.3 **NFPA Hazard Classifications:** Not listed

13. PHYSICAL AND CHEMICAL PROPERTIES

13.1 **Physical State at 15°C and 1 atm:** Liquid
13.2 **Molecular Weight:** 122.95
13.3 **Boiling Point at 1 atm:** 169°F = 76°C = 349°K
13.4 **Freezing Point:** −141.7°F = −96.5°C = 176.7°K
13.5 **Critical Temperature:** Not pertinent
13.6 **Critical Pressure:** Not pertinent
13.7 **Specific Gravity:** 1.66 at 16°C (liquid)
13.8 **Liquid Surface Tension:** Not pertinent
13.9 **Liquid-Water Interfacial Tension:** Not pertinent
13.10 **Vapor (Gas) Specific Gravity:** 4.24
13.11 **Ratio of Specific Heats of Vapor (Gas):** (est.) 1.144
13.12 **Latent Heat of Vaporization:** 106 Btu/lb = 59 cal/g = 2.5×10^5 J/kg
13.13 **Heat of Combustion:** Data not available
13.14 **Heat of Decomposition:** Not pertinent
13.15 **Heat of Solution:** Data not available
13.16 **Heat of Polymerization:** Not pertinent

NOTES

ACC — ACETYL CHLORIDE

Common Synonyms	Liquid Colorless Sharp odor Reacts violently with water. Irritating visible vapor cloud is produced.

Shut off ignition sources. Call fire department.
Avoid contact with liquid and vapor. Keep people away.
Wear goggles, self-contained breathing apparatus and rubber overclothing (including gloves).
Stop discharge if possible.
Isolate and remove discharged material.
Notify local health and pollution control agencies.

Fire	FLAMMABLE. POISONOUS GASES ARE PRODUCED IN FIRE. Flashback along vapor trail may occur. Vapor may explode if ignited in an enclosed area. Wear chemical protective suit with self-contained breathing apparatus. Extinguish with dry chemicals or carbon dioxide. DO NOT USE WATER OR FOAM ON FIRE. Evacuate surrounding area.
Exposure	Call for medical aid. VAPOR Irritating to eyes, nose and throat. If inhaled will cause difficult breathing. Move victim to fresh air. If breathing is difficult, give oxygen. LIQUID Will burn skin and eyes. Harmful if swallowed. Remove contaminated clothing and shoes. Flush affected areas with plenty of water. IF IN EYES, hold eyelids open and flush with plenty of water. IF SWALLOWED, and victim is CONSCIOUS, have victim drink water or milk. DO NOT INDUCE VOMITING.
Water Pollution	Effect of low concentrations on aquatic life is unknown. May be dangerous if it enters water intakes. Notify local health and wildlife officials. Notify operators of nearby water intakes.

1. RESPONSE TO DISCHARGE
(See Response Methods Handbook, CG 446-4)

Issue warning—high flammability, corrosive
Restrict access
Disperse and flush

2. LABEL

Red

3. CHEMICAL DESIGNATIONS

3.1 **Synonyms:** No common synonyms
3.2 **Coast Guard Compatibility Classification:** Not applicable
3.3 **Chemical Formula:** CH_3COCl
3.4 **IMCO/United Nations Numerical Designation:** 8/1717

4. OBSERVABLE CHARACTERISTICS

4.1 **Physical State (as shipped):** Liquid
4.2 **Color:** Colorless
4.3 **Odor:** Pungent

5. HEALTH HAZARDS

5.1 **Personal Protective Equipment:** Safety goggles; rubber or plastic gloves; self-contained breathing apparatus.
5.2 **Symptoms Following Exposure:** Vapor irritates mucous membranes. Ingestion of liquid or contact with eyes or skin causes severe irritation.
5.3 **Treatment for Exposure:** INHALATION: remove from exposure; seek medical attention. EYES: flush with copious amounts of water. SKIN: flush with copious amounts of water. INGESTION: give plenty of water; do NOT induce vomiting.
5.4 **Toxicity by Inhalation (Threshold Limit Value):** Data not available
5.5 **Short-Term Inhalation Limits:** Data not available
5.6 **Toxicity by Ingestion:** Readily hydrolyzes to form hydrochloric and acetic acids. Oral human LD_{LO} = 1470 mg/kg (acetic acid). Grade 2; oral rat LD_{50} = 3310 mg/kg (acetic acid).
5.7 **Late Toxicity:** None
5.8 **Vapor (Gas) Irritant Characteristics:** Vapors cause severe irritation of eyes and throat and can cause eye and lung injury. They cannot be tolerated even at low concentrations.
5.9 **Liquid or Solid Irritant Characteristics:** Severe skin irritant. Causes second- and third-degree burns on short contact and is very injurious to the eyes.
5.10 **Odor Threshold:** Acetic acid – 1 ppm; hydrochloric acid – 1 ppm

6. FIRE HAZARDS

6.1 **Flash Point:** 40°F C.C.
6.2 **Flammable Limits in Air:** Data not available
6.3 **Fire Extinguishing Agents:** Carbon dioxide, dry chemical
6.4 **Fire Extinguishing Agents Not to be Used:** Water, foam
6.5 **Special Hazards of Combustion Products:** When heated to decomposition, hydrogen chloride and phosgene, extremely poisonous gases, are evolved.
6.6 **Behavior in Fire:** Vapor is heavier than air and may travel a considerable distance to a source of ignition and flash back.
6.7 **Ignition Temperature:** 734°F
6.8 **Electrical Hazard:** Data not available
6.9 **Burning Rate:** 2.6 mm/min.

7. CHEMICAL REACTIVITY

7.1 **Reactivity with Water:** Reacts vigorously with water, evolving hydrogen chloride fumes (hydrochloric acid).
7.2 **Reactivity with Common Materials:** Is highly corrosive to most metals in the presence of moisture.
7.3 **Stability During Transport:** Stable
7.4 **Neutralizing Agents for Acids and Caustics:** Following dilution with water, limestone or sodium bicarbonate can be used.
7.5 **Polymerization:** Not pertinent
7.6 **Inhibitor of Polymerization:** Not pertinent

8. WATER POLLUTION

8.1 **Aquatic Toxicity:** Data not available
8.2 **Waterfowl Toxicity:** Data not available
8.3 **Biological Oxygen Demand (BOD):** Data not available
8.4 **Food Chain Concentration Potential:** None

9. SELECTED MANUFACTURERS

1. Eastman Kodak Co.
   Eastman Organic Chemicals
   Rochester, N. Y. 14650
2. Allied Chemical Corp.
   Specialty Chemicals Division
   P. O. Box 1087R
   Morristown, N. J. 07960
3. White Chemical Corp.
   P. O. Box 278
   Bayonne, N. J. 07002

10. SHIPPING INFORMATION

10.1 **Grades or Purity:** Commercial
10.2 **Storage Temperature:** Ambient
10.3 **Inert Atmosphere:** No requirement
10.4 **Venting:** Pressure-vacuum

11. HAZARD ASSESSMENT CODE
(See Hazard Assessment Handbook, CG 446-3)

A-O

12. HAZARD CLASSIFICATIONS

12.1 **Code of Federal Regulations:** Corrosive material
12.2 **NAS Hazard Rating for Bulk Water Transportation:**

Category	Rating
Fire	3
Health	
Vapor Irritant	4
Liquid or Solid Irritant	4
Poisons	3
Water Pollution	
Human Toxicity	2
Aquatic Toxicity	3
Aesthetic Effect	2
Reactivity	
Other Chemicals	3
Water	4
Self-Reaction	0

12.3 **NFPA Hazard Classifications:**

Category	Classification
Health Hazard (Blue)	3
Flammability (Red)	3
Reactivity (Yellow)	2 W

13. PHYSICAL AND CHEMICAL PROPERTIES

13.1 **Physical State at 15°C and 1 atm:** Liquid
13.2 **Molecular Weight:** 78.5
13.3 **Boiling Point at 1 atm:** 124°F = 51°C = 324°K
13.4 **Freezing Point:** −170°F = −112°C = 161°K
13.5 **Critical Temperature:** (est.) 475°F = 246°C = 519°K
13.6 **Critical Pressure:** (est.) 845 psia = 57.5 atm = 5.83 MN/m²
13.7 **Specific Gravity:** 1.1039 at 21°C (liquid)
13.8 **Liquid Surface Tension:** 26 dynes/cm = 0.026 N/m at 20°C
13.9 **Liquid-Water Interfacial Tension:** Not pertinent
13.10 **Vapor (Gas) Specific Gravity:** 3
13.11 **Ratio of Specific Heats of Vapor (Gas):** 1.1467
13.12 **Latent Heat of Vaporization:** 160 Btu/lb = 88 cal/g = 3.7×10^5 J/kg
13.13 **Heat of Combustion:** −6,000 Btu/lb = −3,300 cal/g = -140×10^5 J/kg
13.14 **Heat of Decomposition:** Not pertinent
13.15 **Heat of Solution:** (est.) −54 Btu/lb = −30 cal/g = -1.3×10^5 J/kg
13.16 **Heat of Polymerization:** Not pertinent

NOTES

ACE	ACETYLENE

Common Synonyms			
Ethyne Ethine	Compressed gas	Colorless	Mild garlic odor
	This flammable gas is slightly lighter than air, and will disperse slowly unless confined.		

Shut off ignition sources and call fire department.
Stop discharge if possible. Keep people away.
Stay upwind and use water spray to "knock down" vapor.
Notify local health and pollution control agencies.

Fire	FLAMMABLE. Containers may explode in fire. Flashback along gas trail may occur. May explode if ignited in an enclosed area. Stop flow of gas if possible. Cool exposed containers and protect men effecting the shutoff with water. Let fire burn.
Exposure	CALL FOR MEDICAL AID. GAS Not irritating to eyes, nose or throat. If inhaled will cause headache, difficult breathing, or loss of consciousness. Move to fresh air. If breathing has stopped, give artificial respiration. If breathing is difficult, give oxygen.
Water Pollution	Not harmful to aquatic life.

1. RESPONSE TO DISCHARGE

(See Response Methods Handbook, CG 446-4)

Issue warning—high flammability

2. LABEL

Red

3. CHEMICAL DESIGNATIONS

3.1 **Synonyms:** Ethine
Ethyne

3.2 **Coast Guard Compatibility Classification:** Not applicable

3.3 **Chemical Formula:** C_2H_2

3.4 **IMCO/United Nations Numerical Designation:** 2.0/1001

4. OBSERVABLE CHARACTERISTICS

4.1 **Physical State (as shipped):** Compressed gas

4.2 **Color:** Colorless

4.3 **Odor:** Distinctive; garlic-like

5. HEALTH HAZARDS

5.1 **Personal Protective Equipment:** Air supply respirator in areas of high concentration. Avoid all sources of ignition.

5.2 **Symptoms Following Exposure:** Headache, dizziness and loss of consciousness may occur. Death from "smothering" may occur if oxygen content of the air is severely reduced by dilution with acetylene.

5.3 **Treatment for Exposure:** INHALATION: no specific antidote known; remove victim to fresh air, keep him warm and quiet, and call a doctor; recovery is usually rapid. If patient is unconscious, administer oxygen; if breathing has stopped, give artificial respiration.

5.4 **Toxicity by Inhalation (Threshold Limit Value):** 5000 ppm (suggested)

5.5 **Short-Term Inhalation Limits:** 100,000 ppm for 30 - 60 min.

5.6 **Toxicity by Ingestion:** Not pertinent

5.7 **Late Toxicity:** Not pertinent

5.8 **Vapor (Gas) Irritant Characteristics:** None

5.9 **Liquid or Solid Irritant Characteristics:** Not pertinent

5.10 **Odor Threshold:** Data not available

6. FIRE HAZARDS

6.1 **Flash Point:** Gas

6.2 **Flammable Limits in Air:** 2.5%—80%

6.3 **Fire Extinguishing Agents:** Stop flow of gas

6.4 **Fire Extinguishing Agents Not to be Used:** Carbon dioxide, dry chemical and water spray are not generally recommended because the discharged gas or volatile liquid may create a more serious explosion hazard.

6.5 **Special Hazards of Combustion Products:** Not pertinent

6.6 **Behavior in Fire:** May explode in fire

6.7 **Ignition Temperature:** 581°F

6.8 **Electrical Hazard:** Class I, Group A

6.9 **Burning Rate:** Not pertinent

7. CHEMICAL REACTIVITY

7.1 **Reactivity with Water:** No reaction

7.2 **Reactivity with Common Materials:** Under certain conditions forms spontaneously explosive compounds with copper

7.3 **Stability During Transport:** Stable as shipped

7.4 **Neutralizing Agents for Acids and Caustics:** Not pertinent

7.5 **Polymerization:** Not pertinent

7.6 **Inhibitor of Polymerization:** Not pertinent

8. WATER POLLUTION

8.1 **Aquatic Toxicity:** 1000 cc/l /1 hr/sunfish/not killed/fresh water

8.2 **Waterfowl Toxicity:** Not pertinent

8.3 **Biological Oxygen Demand (BOD):** Not pertinent

8.4 **Food Chain Concentration Potential:** Not pertinent

9. SELECTED MANUFACTURERS

1. Monochem Inc.
P. O. Box 488
Geismar, Louisiana 70734
2. Monsanto Co.
Monsanto Polymers & Petrochemicals Co.
800 North Lindbergh Blvd.
St. Louis, Mo. 63166
3. Union Carbide Corp.
Chemicals & Plastics Div.
270 Park Ave.
New York, N. Y. 10017

10. SHIPPING INFORMATION

10.1 **Grades or Purity:** Commercial grade acetylene is supplied dissolved in acetone under pressure in cylinders.

10.2 **Storage Temperature:** Data not available

10.3 **Inert Atmosphere:** Data not available

10.4 **Venting:** Data not available

11. HAZARD ASSESSMENT CODE

(See Hazard Assessment Handbook, CG 446-3)

A-B-C

12. HAZARD CLASSIFICATIONS

12.1 **Code of Federal Regulations:** Flammable compressed gas

12.2 **NAS Hazard Rating for Bulk Water Transportation:** Not listed

12.3 **NFPA Hazard Classifications:**

Category	Classification
Health Hazard (Blue)	1
Flammability (Red)	4
Reactivity (Yellow)	2

13. PHYSICAL AND CHEMICAL PROPERTIES

13.1 **Physical State at 15°C and 1 atm:** Compressed gas

13.2 **Molecular Weight:** 26.04

13.3 **Boiling Point at 1 atm:** $-119°F = -84.0°C = 189.2°K$

13.4 **Freezing Point:** Not pertinent

13.5 **Critical Temperature:** $95.3°F = 35.2°C = 308.4°K$

13.6 **Critical Pressure:** 890.7 psia $= 60.59$ atm $= 6.138$ MN/m^2

13.7 **Specific Gravity:** 0.613 at −80°C (liquid)

13.8 **Liquid Surface Tension:** Not pertinent

13.9 **Liquid-Water Interfacial Tension:** Not pertinent

13.10 **Vapor (Gas) Specific Gravity:** 0.9

13.11 **Ratio of Specific Heats of Vapor (Gas):** 1.235

13.12 **Latent Heat of Vaporization:** Not pertinent

13.13 **Heat of Combustion:** $-20{,}747$ Btu/lb $= -11{,}526$ cal/g $= -482.57 \times 10^5$ J/kg

13.14 **Heat of Decomposition:** Not pertinent

13.15 **Heat of Solution:** Not pertinent

13.16 **Heat of Polymerization:** Not pertinent

NOTES

REVISED 1978

APS	ACETYL PEROXIDE SOLUTION

ACETYL PEROXIDE SOLUTION

Common Synonyms			
Diacetyl peroxide solution	Liquid	Colorless	Sharp odor
	Sinks in water.		

Stop discharge if possible. Keep people away.
Call fire department.
Avoid contact with liquid.
Isolate and remove discharged material.
Notify local health and pollution control agencies.

Fire

Combustible.
May explode on contact with combustibles.
Containers may explode in fire.
Combat fires from safe distance or protected location.
Flood discharge area with water.
Cool exposed containers with water.

Exposure

Call for medical aid.

VAPOR
Irritating to eyes, nose and throat.
Move victim to fresh air.

LIQUID
Irritating to skin and eyes.
Harmful if swallowed.
Remove contaminated clothing and shoes.
Flush affected areas with plenty of water.
IF IN EYES, hold eyelids open and flush with plenty of water.
IF SWALLOWED, and victim is CONSCIOUS, have victim drink water or milk and have victim induce vomiting.
IF SWALLOWED and victim is UNCONSCIOUS OR HAVING CONVULSIONS, do nothing except keep victim warm.

Water Pollution

Effect of low concentrations on aquatic life is unknown.
May be dangerous if it enters water intakes.
Notify local health and wildlife officials.
Notify operators of nearby water intakes.

1. RESPONSE TO DISCHARGE

(See Response Methods Handbook, CG 446-4)

Issue warning – oxidizing material, water contaminant
Should be removed
Chemical and physical treatment

2. LABEL

Yellow

3. CHEMICAL DESIGNATIONS

3.1 **Synonyms:** Diacetyl peroxide solution
3.2 **Coast Guard Compatibility Classification:** Not applicable
3.3 **Chemical Formula:** $CH_3CO(O_2)OCCH_3$ in dimethyl phthalate
3.4 **IMCO/United Nations Numerical Designation:** 5.2/1519

4. OBSERVABLE CHARACTERISTICS

4.1 **Physical State (as shipped):** Liquid
4.2 **Color:** Colorless
4.3 **Odor:** Pungent

5. HEALTH HAZARDS

5.1 **Personal Protective Equipment:** Protective goggles; rubber apron and gloves.
5.2 **Symptoms Following Exposure:** Contact with liquid causes irritation of eyes and skin. If ingested, irritates mouth and stomach.
5.3 **Treatment for Exposure:** EYES: wash with plenty of water and get medical attention. SKIN: wash with plenty of soap and water. INGESTION: induce vomiting and call a physician.
5.4 **Toxicity by Inhalation (Threshold Limit Value):** Data not available
5.5 **Short-Term Inhalation Limits:** Data not available
5.6 **Toxicity by Ingestion:** Data not available
5.7 **Late Toxicity:** Data not available
5.8 **Vapor (Gas) Irritant Characteristics:** Data not available
5.9 **Liquid or Solid Irritant Characteristics:** Data not available
5.10 **Odor Threshold:** Data not available

6. FIRE HAZARDS

6.1 **Flash Point:** 113°F O.C.
6.2 **Flammable Limits in Air:** Not pertinent
6.3 **Fire Extinguishing Agents:** Water, dry chemical, carbon dioxide
6.4 **Fire Extinguishing Agents Not to be Used:** Not pertinent
6.5 **Special Hazards of Combustion Products:** Not pertinent
6.6 **Behavior in Fire:** May explode. Burns with accelerating intensity.
6.7 **Ignition Temperature:** Explodes
6.8 **Electrical Hazard:** Data not available
6.9 **Burning Rate:** Data not available

7. CHEMICAL REACTIVITY

7.1 **Reactivity with Water:** No reaction
7.2 **Reactivity with Common Materials:** May ignite combustible materials such as wood.
7.3 **Stability During Transport:** Heat-and shock-sensitive crystals may separate at very low temperatures during transport.
7.4 **Neutralizing Agents for Acids and Caustics:** Not pertinent
7.5 **Polymerization:** Not pertinent
7.6 **Inhibitor of Polymerization:** Not pertinent

8. WATER POLLUTION

8.1 **Aquatic Toxicity:** Data not available
8.2 **Waterfowl Toxicity:** Data not available
8.3 **Biological Oxygen Demand (BOD):** Data not available
8.4 **Food Chain Concentration Potential:** None

9. SELECTED MANUFACTURERS

1. Pennwalt Corporation
Lucidol Division
1740 Military Road
Buffalo, N. Y. 14240
2. Aztec Chemicals Division
Dart Industries, Inc.
555 Garden Street
Elyria, Ohio 44035

10. SHIPPING INFORMATION

10.1 **Grades or Purity:** 25% acetyl peroxide; 75% dimethyl phthalate
10.2 **Storage Temperature:** 32°F – 100°F (32-41°F optimum)
10.3 **Inert Atmosphere:** No requirement
10.4 **Venting:** Pressure-vacuum

11. HAZARD ASSESSMENT CODE

(See Hazard Assessment Handbook, CG 446-3)

A-X-Y

12. HAZARD CLASSIFICATIONS

12.1 **Code of Federal Regulations:** Organic Peroxide
12.2 **NAS Hazard Rating for Bulk Water Transportation:** Not listed
12.3 **NFPA Hazard Classifications:**

Category	Classification
Health Hazard (Blue)	1
Flammability (Red)	2
Reactivity (Yellow)	4

13. PHYSICAL AND CHEMICAL PROPERTIES

13.1 **Physical State at 15°C and 1 atm:** Liquid
13.2 **Molecular Weight:** Mixture
13.3 **Boiling Point at 1 atm:** Decomposes
13.4 **Freezing Point:** 17°F = –8°C = 265°K
13.5 **Critical Temperature:** Not pertinent
13.6 **Critical Pressure:** Not pertinent
13.7 **Specific Gravity:** 1.2 at 20°C (liquid)
13.8 **Liquid Surface Tension:** (est.) 30 dynes/cm = 0.030 N/m at 20°C
13.9 **Liquid-Water Interfacial Tension:** (est.) 30 dynes/cm = 0.030 N/m at 20°C
13.10 **Vapor (Gas) Specific Gravity:** Not pertinent
13.11 **Ratio of Specific Heats of Vapor (Gas):** Not pertinent
13.12 **Latent Heat of Vaporization:** Not pertinent
13.13 **Heat of Combustion:** (est.) –15,700 Btu/lb = –8,750 cal/g = -366×10^5 J/kg
13.14 **Heat of Decomposition:** (est.) –50 Btu/lb = –28 cal/g = -1.2×10^5 J/kg
13.15 **Heat of Solution:** Not pertinent
13.16 **Heat of Polymerization:** Not pertinent

NOTES

ACD ACRIDINE

Common Synonyms	Solid	Yellow	Weak irritating odor
Dibenzo[b,e]pyridine 10-Azaanthracene Benzo(b)quinoline	Sinks in water.		

Avoid contact with solid and dust. Keep people away.
Shut off ignition sources. Call fire department.
Stop discharge if possible.
Isolate and remove discharged material.
Notify local health and pollution control agencies.

Fire

Combustible.
POISONOUS GASES MAY BE PRODUCED IN FIRE.
Wear goggles, self-contained breathing apparatus and rubber overclothing (including gloves).
Extinguish with water or foam.
Carbon dioxide and dry chemicals may be ineffective on fire.

Exposure

CALL FOR MEDICAL AID.

DUST
Irritating to eyes, nose and throat.
If inhaled will cause coughing or difficult breathing.
If in eyes, hold eyelids open and flush with plenty of water.
If breathing has stopped, give artificial respiration.
If breathing is difficult, give oxygen.

SOLID
Irritating to skin and eyes.
Harmful if swallowed.
Remove contaminated clothing and shoes.
Flush affected areas with plenty of water.
IF IN EYES, hold eyelids open and flush with plenty of water.
IF SWALLOWED and victim is UNCONSCIOUS OR HAVING CONVULSIONS, do nothing except keep victim warm.

Water Pollution

HARMFUL TO AQUATIC LIFE IN VERY LOW CONCENTRATIONS.
May be dangerous if it enters water intakes.
Notify local health and wildlife officials.
Notify operators of nearby water intakes.

1. RESPONSE TO DISCHARGE
(See Response Methods Handbook, CG 446-4)

Issue warning—water contaminant
Mechanical containment
Should be removed
Chemical and physical treatment

2. LABELS

No hazard label required by Code of Federal Regulations

3. CHEMICAL DESIGNATIONS

3.1 **Synonyms:** 10-Azaanthracene
Benzo (b) quinoline
Dibenzo [b, e] pyridine
3.2 **Coast Guard Compatibility Classification:** Not listed
3.3 **Chemical Formula:** $C_{13}H_9N$
3.4 **IMCO/United Nations Numerical Designation:** Not listed

4. OBSERVABLE CHARACTERISTICS

4.1 **Physical State (as shipped):** Solid
4.2 **Color:** Yellow
4.3 **Odor:** Weak, somewhat irritating

5. HEALTH HAZARDS

5.1 **Personal Protective Equipment:** Dust respirator; chemical goggles; rubber gloves
5.2 **Symptoms Following Exposure:** Inhalation irritates respiratory system and causes sneezing, crying, and vomiting. Contact with liquid irritates eyes, skin, and mucous membranes. At high temperature and during sun exposure, damage to the cornea, skin, and mucous membranes may occur following the liberation of acridine vapor.
5.3 **Treatment for Exposure:** INHALATION: remove victim to fresh air; if breathing has stopped, give artificial respiration; if breathing is difficult, give oxygen. EYES: wash with copious amounts of water for 20 min.; seek medical attention. SKIN: wash with large amounts of water for 20 min.
5.4 **Toxicity by Inhalation (Threshold Limit Value):** Data not available
5.5 **Short-Term Inhalation Limits:** Data not available
5.6 **Toxicity by Ingestion:** Grade 2; oral rat LD_{50} = 2,000 mg/kg
5.7 **Late Toxicity:** Data not available
5.8 **Vapor (Gas) Irritant Characteristics:** Data not available
5.9 **Liquid or Solid Irritant Characteristics:** Data not available
5.10 **Odor Threshold:** Data not available

6. FIRE HAZARDS

6.1 **Flash Point:** Not pertinent (combustible solid)
6.2 **Flammable Limits in Air:** Not pertinent
6.3 **Fire Extinguishing Agents:** Water, foam, monoammonium phosphate, dry chemical
6.4 **Fire Extinguishing Agents Not to be Used:** Carbon dioxide and other dry chemicals may not be effective.
6.5 **Special Hazards of Combustion Products:** Toxic oxides of nitrogen may form in fire.
6.6 **Behavior in Fire:** Sublimes before melting
6.7 **Ignition Temperature:** Data not available
6.8 **Electrical Hazard:** Not pertinent
6.9 **Burning Rate:** Not pertinent

7. CHEMICAL REACTIVITY

7.1 **Reactivity with Water:** No reaction
7.2 **Reactivity with Common Materials:**
7.3 **Stability During Transport:** Stable
7.4 **Neutralizing Agents for Acids and Caustics:** Not pertinent
7.5 **Polymerization:** Not pertinent
7.6 **Inhibitor of Polymerization:** Not pertinent

8. WATER POLLUTION

8.1 **Aquatic Toxicity:**
0.7 ppm/*/perch/kill/fresh water
*Time period not specified
8.2 **Waterfowl Toxicity:** Data not available
8.3 **Biological Oxygen Demand (BOD):** Data not available
8.4 **Food Chain Concentration Potential:** None

9. SELECTED MANUFACTURERS

1. Aldrich Chemical Co.
940 W. Saint Paul Avenue
Milwaukee, Wis. 53233
2. Pfaltz and Bauer, Inc.
126-04 Northern Boulevard
Flushing, N. Y. 11368
3. J. T. Baker Chemical Co.
Phillipsburg, N. J. 08865

10. SHIPPING INFORMATION

10.1 **Grades or Purity:** Commercial
10.2 **Storage Temperature:** Ambient
10.3 **Inert Atmosphere:** No requirement
10.4 **Venting:** Open

11. HAZARD ASSESSMENT CODE
(See Hazard Assessment Handbook, CG 446-3)

II

12. HAZARD CLASSIFICATIONS

12.1 **Code of Federal Regulations:** Not listed
12.2 **NAS Hazard Rating for Bulk Water Transportation:** Not listed
12.3 **NFPA Hazard Classifications:** Not listed

13. PHYSICAL AND CHEMICAL PROPERTIES

13.1 **Physical State at 15°C and 1 atm:** Solid
13.2 **Molecular Weight:** 179.08
13.3 **Boiling Point at 1 atm:** 655°F = 346°C = 619°K
13.4 **Freezing Point:** 230°F = 110°C = 383°K
13.5 **Critical Temperature:** Not pertinent
13.6 **Critical Pressure:** Not pertinent
13.7 **Specific Gravity:** (approx.) 1.2 at 20°C (solid)
13.8 **Liquid Surface Tension:** Not pertinent
13.9 **Liquid-Water Interfacial Tension:** Not pertinent
13.10 **Vapor (Gas) Specific Gravity:** Not pertinent
13.11 **Ratio of Specific Heats of Vapor (Gas):** Not pertinent
13.12 **Latent Heat of Vaporization:** Not pertinent
13.13 **Heat of Combustion:** $-15,800$ Btu/lb $= -8790$ cal/g $= -368 \times 10^5$ J/kg
13.14 **Heat of Decomposition:** Not pertinent
13.15 **Heat of Solution:** Not pertinent
13.16 **Heat of Polymerization:** Not pertinent

NOTES

ARL ACROLEIN

Common Synonyms			
Acraldehyde Acrylic aldehyde Allyl aldehyde Ethylene aldehyde 2-propenal Acrylaldehyde	Watery liquid	Colorless to light yellow	Sharp, irritating odor
	Floats and mixes with water. Poisonous, flammable vapor is produced.		

AVOID CONTACT WITH LIQUID AND VAPOR. KEEP PEOPLE AWAY.
Wear goggles, self-contained breathing apparatus, and rubber overclothing (including gloves).
Shut off ignition sources. Call fire department.
Stop discharge if possible.
Evacuate area in case of large discharge.
Isolate and remove discharged material.
Notify local health and pollution control agencies.

Fire

FLAMMABLE.
POISONOUS GASES ARE PRODUCED WHEN HEATED.
Containers may explode in fire.
Flashback along vapor trail may occur.
Vapor may explode if ignited in an enclosed area.
Combat fires from safe distance or protected location.
Extinguish with dry chemicals, alcohol foam, or carbon dioxide.
Water may be ineffective on fire.
Cool exposed containers with water.

Exposure

CALL FOR MEDICAL AID.

VAPOR
POISONOUS IF INHALED.
Irritating to eyes, nose and throat.
Move victim to fresh air.
If in eyes, hold eyelids open and flush with plenty of water.
If breathing has stopped, give artificial respiration.
If breathing is difficult, give oxygen.

LIQUID
POISONOUS IF SWALLOWED.
Will burn eyes.
Irritating to skin.
Remove contaminated clothing and shoes.
Flush affected areas with plenty of water.
IF IN EYES, hold eyelids open and flush with plenty of water.
IF SWALLOWED and victim is CONSCIOUS, have victim drink water or milk and have victim induce vomiting.
IF SWALLOWED and victim is UNCONSCIOUS OR HAVING CONVULSIONS, do nothing except keep victim warm.

Water Pollution

HARMFUL TO AQUATIC LIFE IN VERY LOW CONCENTRATIONS.
May be dangerous if it enters water intakes.
Notify local health and wildlife officials.
Notify operators of nearby water intakes.

1. RESPONSE TO DISCHARGE

(See Response Methods Handbook, CG 446-4)

Issue warning—high flammability, air contaminant
Restrict access
Evacuate area
Disperse and flush

2. LABELS

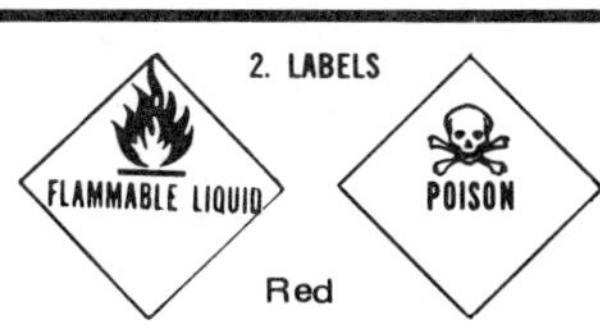

Red

3. CHEMICAL DESIGNATIONS

3.1 **Synonyms:** Acraldehyde, Acrylic aldehyde, 2-Propenal, Acrylaldehyde
3.2 **Coast Guard Compatibility Classification:** Aldehyde
3.3 **Chemical Formula:** $CH_2{=}CH\cdot CHO$
3.4 **IMCO/United Nations Numerical Designation:** 3.1/1092

4. OBSERVABLE CHARACTERISTICS

4.1 **Physical State (as shipped):** Liquid
4.2 **Color:** Colorless to slightly yellow
4.3 **Odor:** Extremely sharp—lachrymator; piercing disagreeable; extremely irritating, acrid, pungent

5. HEALTH HAZARDS

5.1 **Personal Protective Equipment:** Chemical safety goggles and face shield; self-contained breathing apparatus, positive-pressure hose mask, airline mask or industrial canister-type gas mask; rubber safety shoes; clothing made of rubber or other impervious material.

5.2 **Symptoms Following Exposure:** Inhalation causes irritation of nose and throat, a feeling of pressure in the chest, and shortness of breath. Nausea and vomiting commonly occur. Loss of consciousness if exposure has been sufficiently great. Congestion in the chest may be present in varying amounts, and fluid may collect in the lungs (pulmonary edema) of severely exposed persons. Vapor also causes severe eye irritation (redness, weeping, and swelling of lids); liquid burns eyes, contact with skin causes reddening or blistering. Ingestion causes severe irritation of mouth and stomach.

5.3 **Treatment for Exposure:** Keep patient warm; if he is conscious, give coffee; *call physician after all exposures to this compound.* INHALATION: remove patient to fresh air; if breathing becomes difficult, give oxygen. If breathing has stopped, start artificial respiration. EYES: immediately flush with plenty of water for at least 15 min. If medical attention is not immediately available, continue eye irrigation for another 15-min. period. Upon completion of the first 15 min. of irrigation, it is permissible to instill 2 or 3 drops of an effective aqueous local eye anesthetic for relief of pain. No oils or ointments should be used unless ordered by the physician. SKIN: flush at once with large volumes of water. Wash thoroughly with soap and large quantities of running water. INGESTION: have victim drink large amounts of water. Induce vomiting by sticking a finger down the throat or by giving salt water (one tablespoon of table salt to a glass of water). Keep patient warm and quiet.

5.4 **Toxicity by Inhalation (Threshold Limit Value):** 0.1 ppm

6. FIRE HAZARDS

6.1 **Flash Point:** < 0°F O.C.; −13°F C.C.
6.2 **Flammable Limits in Air:** 2.8%–31%
6.3 **Fire Extinguishing Agents:** Foam, dry chemical, carbon dioxide
6.4 **Fire Extinguishing Agents Not to be Used:** Water may be ineffective.
6.5 **Special Hazards of Combustion Products:** Poisonous vapor of acrolein is formed from hot liquid.
6.6 **Behavior in Fire:** Vapor is heavier than air and may travel a considerable distance to a source of ignition and flash back. Polymerization may take place, and containers may explode in fire.
6.7 **Ignition Temperature:** 453°F
6.8 **Electrical Hazard:** Data not available
6.9 **Burning Rate:** 3.8 mm/min.

7. CHEMICAL REACTIVITY

7.1 **Reactivity with Water:** No reaction
7.2 **Reactivity with Common Materials:** No reaction
7.3 **Stability During Transport:** Stable when inhibited
7.4 **Neutralizing Agents for Acids and Caustics:** Not pertinent
7.5 **Polymerization:** Undergoes uncatalyzed polymerization reaction around 200°C. Light promotes polymerization.
7.6 **Inhibitor of Polymerization:** Hydroquinone: 0.10 to 0.25%

8. WATER POLLUTION

8.1 **Aquatic Toxicity:**
0.08 ppm/24hr/salmon/TL_m/fresh water
0.055 ppm/96 hr/oyster/EC_{50}/salt water
8.2 **Waterfowl Toxicity:** Data not available
8.3 **Biological Oxygen Demand (BOD):** 33%, 10 days
8.4 **Food Chain Concentration Potential:** None

9. SELECTED MANUFACTURERS

1. Union Carbide Corporation
Chemicals and Plastics Division
270 Park Avenue
New York, N. Y. 10017
2. Shell Chemical Company
Industrial Chemicals Division
P. O. Box 2463
Houston, Texas 77001

10. SHIPPING INFORMATION

10.1 **Grades or Purity:** Industrial, 92+%
10.2 **Storage Temperature:** Ambient
10.3 **Inert Atmosphere:** No requirement
10.4 **Venting:** Pressure-vacuum

11. HAZARD ASSESSMENT CODE

(See Hazard Assessment Handbook, CG 446-3)

A-P-Q-R-S-Z

12. HAZARD CLASSIFICATIONS

12.1 **Code of Federal Regulations:** Flammable liquid

12.2 **NAS Hazard Rating for Bulk Water Transportation:**

Category	Rating
Fire	3
Health	
Vapor Irritant	4
Liquid or Solid Irritant	3
Poisons	4
Water Pollution	
Human Toxicity	4
Aquatic Toxicity	3
Aesthetic Effect	3
Reactivity	
Other Chemicals	3
Water	0
Self-Reaction	3

12.3 **NFPA Hazard Classifications:**

Category	Classification
Health Hazard (Blue)	3
Flammability (Red)	3
Reactivity (Yellow)	2

13. PHYSICAL AND CHEMICAL PROPERTIES

13.1 **Physical State at 15°C and 1 atm:** Liquid
13.2 **Molecular Weight:** 56.1
13.3 **Boiling Point at 1 atm:** 127°F = 53°C = 326°K
13.4 **Freezing Point:** −125°F = −87°C = 186°K
13.5 **Critical Temperature:** (est.) 489°F = 254°C = 527°K
13.6 **Critical Pressure:** (est.) 737 psia = 50.0 atm = 5.08 MN/m^2
13.7 **Specific Gravity:** 0.843 at 20°C (liquid)
13.8 **Liquid Surface Tension:** 24 dynes/cm = 0.024 N/m at 20°C
13.9 **Liquid-Water Interfacial Tension:** (est.) 35 dynes/cm = 0.035 N/m at 20°C
13.10 **Vapor (Gas) Specific Gravity:** 1.94
13.11 **Ratio of Specific Heats of Vapor (Gas):** 1.1487
13.12 **Latent Heat of Vaporization:** 216 Btu/lb = 120 cal/g = 5.02×10^5 J/kg
13.13 **Heat of Combustion:** −12,500 Btu/lb = −6,950 cal/g = -290×10^5 J/kg
13.14 **Heat of Decomposition:** Not pertinent
13.15 **Heat of Solution:** Not pertinent
13.16 **Heat of Polymerization:** (est.) −50 Btu/lb = −28 cal/g = -1.2×10^5 J/kg

5. HEALTH HAZARDS (Cont'd.)

5.5 **Short-Term Inhalation Limits:** 0.5 ppm—5 min; 0.2 ppm—60 min.
5.6 **Toxicity by Ingestion:** Grade 4; LD_{50} below 50 mg/kg
5.7 **Late Toxicity:** Grade 4; oral rat LD_{50} = 46 mg/kg. Grade 4; oral rabbit LD_{50} = 7 mg/kg
5.8 **Vapor (Gas) Irritant Characteristics:** Vapors cause severe irritation of eyes and throat and can cause eye and lung injury. They cannot be tolerated even at low concentrations.
5.9 **Liquid or Solid Irritant Characteristics:** Causes smarting of the skin and first-degree burns on short exposure; may cause second-degree burns on long exposure.
5.10 **Odor Threshold:** 0.21 ppm

AAM — ACRYLAMIDE

Common Synonyms
Propenamide (50%)
Acrylic acid amide (50%)

Liquid — Colorless — Odorless

Sinks and mixes with water.

AVOID CONTACT WITH LIQUID. KEEP PEOPLE AWAY.
Wear rubber overclothing (including gloves).
Stop discharge if possible.
Isolate and remove discharged material.
Notify local health and pollution control agencies.

Fire

Not flammable.

Exposure

CALL FOR MEDICAL AID.

VAPOR
Irritating to eyes, nose and throat.
Harmful if inhaled.
If in eyes, hold eyelids open and flush with plenty of water.
If breathing has stopped, give artificial respiration.
If breathing is difficult, give oxygen.

LIQUID
Will burn skin and eyes.
Harmful if swallowed.
Remove contaminated clothing and shoes.
Flush affected areas with plenty of water.
IF IN EYES, hold eyelids open and flush with plenty of water.
IF SWALLOWED and victim is CONSCIOUS, have victim drink water or milk and have victim induce vomiting.
IF SWALLOWED and victim is UNCONSCIOUS OR HAVING CONVULSIONS, do nothing except keep victim warm.

Water Pollution

Effect of low concentrations on aquatic life is unknown.
May be dangerous if it enters water intakes.
Notify local health and wildlife officials.
Notify operators of nearby water intakes.

1. RESPONSE TO DISCHARGE

(See Response Methods Handbook, CG 446-4)

Issue warning—water contaminant
Disperse and flush

2. LABELS

No hazard label required by Code of Federal Regulations

3. CHEMICAL DESIGNATIONS

3.1 **Synonyms:** Acrylic amide 50%; Propenamide 50%
3.2 **Coast Guard Compatibility Classification:** Not listed
3.3 **Chemical Formula:** $CH_2{=}CHCONH_2 - H_2O$
3.4 **IMCO/United Nations Numerical Designation:** Not listed

4. OBSERVABLE CHARACTERISTICS

4.1 **Physical State (as shipped):** Liquid
4.2 **Color:** Clear
4.3 **Odor:** None

5. HEALTH HAZARDS

5.1 **Personal Protective Equipment:** Safety glasses with side shields; clean body-covering clothing; rubber gloves, boots, apron as dictated by circumstances; in absence of proper environmental control, use approved dust respirator.
5.2 **Symptoms Following Exposure:** Has produced central nervous system damage, which is partly reversible. Effects can be produced by oral or skin contact as well as by injection. Chronic acrylamide poisoning can cause midbrain disturbance and peripheral neuropathy. Contact with liquid can cause moderate irritation of eyes and skin and may cause moderate transient corneal injury.
5.3 **Treatment for Exposure:** INHALATION: if ill effects occur, immediately get patient to fresh air, keep him quiet and warm, and get medical help. INGESTION: if ingested, immediately give large amounts of water (or milk if immediately available), then induce vomiting and get medical help. EYES: immediately flush with plenty of water for at least 15 min. and get medical help promptly. SKIN: immediate, continuous, and thorough washing in flowing water is imperative, preferably deluge shower with abundant soap; if burns are present, get medical help; discard all contaminated clothing and wearing accessories.
5.4 **Toxicity by Inhalation (Threshold Limit Value):** 0.3 mg/m³
5.5 **Short-Term Inhalation Limits:** Data not available
5.6 **Toxicity by Ingestion:** Grade 3; oral rat LD_{50} = 170 mg/kg
5.7 **Late Toxicity:** Repeated exposure to small amounts may cause essentially reversible neurological effects

6. FIRE HAZARDS

6.1 **Flash Point:** Not flammable
6.2 **Flammable Limits in Air:** Not flammable
6.3 **Fire Extinguishing Agents:** Not pertinent
6.4 **Fire Extinguishing Agents Not to be Used:** Not pertinent
6.5 **Special Hazards of Combustion Products:** Toxic oxides of nitrogen may form in fire.
6.6 **Behavior in Fire:** Sealed containers may burst as a result of polymerization.
6.7 **Ignition Temperature:** Not pertinent
6.8 **Electrical Hazard:** Not pertinent
6.9 **Burning Rate:** Not pertinent

7. CHEMICAL REACTIVITY

7.1 **Reactivity with Water:** No reaction
7.2 **Reactivity with Common Materials:**
7.3 **Stability During Transport:** Stable
7.4 **Neutralizing Agents for Acids and Caustics:** Not pertinent
7.5 **Polymerization:** May occur at temperatures above 50°C (120°F).
7.6 **Inhibitor of Polymerization:** Oxygen (air) plus 50 ppm of copper as copper sulfate

8. WATER POLLUTION

8.1 **Aquatic Toxicity:** 130 ppm/96 hr/ harlequin fish/LC_{50}
8.2 **Waterfowl Toxicity:** Data not available
8.3 **Biological Oxygen Demand (BOD):** Data not available
8.4 **Food Chain Concentration Potential:** None

9. SELECTED MANUFACTURERS

1. Dow Chemical Co.
Midland, Mich. 48640
2. Vistron Corp.
Midland Building
Cleveland, Ohio 44115

10. SHIPPING INFORMATION

10.1 **Grades or Purity:** 15%—50% solution in water
10.2 **Storage Temperature:** Below 50°C (122°F)
10.3 **Inert Atmosphere:** Ventilated (natural)
10.4 **Venting:** Open

11. HAZARD ASSESSMENT CODE

(See Hazard Assessment Handbook, CG 446-3)

A-P-Z

12. HAZARD CLASSIFICATIONS

12.1 **Code of Federal Regulations:** Not listed
12.2 **NAS Hazard Rating for Bulk Water Transportation:**

Category	Rating
Fire	0
Health	
Vapor Irritant	0
Liquid or Solid Irritant	1
Poisons	3
Water Pollution	
Human Toxicity	3
Aquatic Toxicity	3
Aesthetic Effect	0
Reactivity	
Other Chemicals	2
Water	0
Self-Reaction	3

12.3 **NFPA Hazard Classifications:** Data not available

13. PHYSICAL AND CHEMICAL PROPERTIES

13.1 **Physical State at 15°C and 1 atm:** Liquid
13.2 **Molecular Weight:** 71 (solute only)
13.3 **Boiling Point at 1 atm:** Data not available (Vapor Pressure = 0.033 atm at 125°C)
13.4 **Freezing Point:** 183°F = 84°C = 357°K
13.5 **Critical Temperature:** Not pertinent
13.6 **Critical Pressure:** Not pertinent
13.7 **Specific Gravity:** 1.05 at 25°C (liquid)
13.8 **Liquid Surface Tension:** Data not available
13.9 **Liquid-Water Interfacial Tension:** Not pertinent
13.10 **Vapor (Gas) Specific Gravity:** Not pertinent
13.11 **Ratio of Specific Heats of Vapor (Gas):** Not pertinent
13.12 **Latent Heat of Vaporization:** Not pertinent
13.13 **Heat of Combustion:** Not pertinent
13.14 **Heat of Decomposition:** Not pertinent
13.15 **Heat of Solution:** Not pertinent
13.16 **Heat of Polymerization:** Data not available

5. HEALTH HAZARDS (Cont'd)

5.8 **Vapor (Gas) Irritant Characteristics:** Data not available
5.9 **Liquid or Solid Irritant Characteristics:** Data not available
5.10 **Odor Threshold:** Not pertinent

ACR	ACRYLIC ACID

Common Synonyms			
Propenoic Acid	Watery liquid	Colorless	Irritating odor
	Sinks and mixes with water. Irritating vapor is produced. Freezing point is 54° F.		

Avoid contact with liquid and vapor. Keep people away.
Wear goggles, self-contained breathing apparatus, and rubber overclothing (including gloves).
Stop discharge if possible.
Call fire department.
Isolate and remove discharged material.
Notify local health and pollution control agencies.

Fire

Combustible.
POISONOUS GAS MAY BE PRODUCED IN FIRE.
Containers may explode when heated.
Vapor may explode if ignited in an enclosed area.

Wear goggles, self-contained breathing apparatus, and rubber overclothing (including gloves).
Combat fires from safe distance or protected location.
Extinguish with dry chemicals, alcohol foam, or carbon dioxide.
Cool exposed containers with water.

Exposure

CALL FOR MEDICAL AID.

VAPOR
Irritating to eyes, nose and throat.

Move to fresh air.
If breathing has stopped, give artificial respiration.
If breathing is difficult, give oxygen.

LIQUID OR SOLID
Will burn skin and eyes.
Harmful if swallowed.

Remove contaminated clothing and shoes.
Flush affected areas with plenty of water.
IF IN EYES, hold eyelids open and flush with plenty of water.
IF SWALLOWED and victim is CONSCIOUS, have victim drink water or milk.
DO NOT INDUCE VOMITING.

Water Pollution

Effect of low concentrations on aquatic life is unknown.
May be dangerous if it enters water intakes.

Notify local health and pollution control officials.
Notify operators of nearby water intakes.

1. RESPONSE TO DISCHARGE

(See Response Methods Handbook, CG 446-4)

Issue warning—corrosive
Restrict access
Disperse and flush

2. LABEL

3. CHEMICAL DESIGNATIONS

3.1 **Synonyms:** Propenoic acid
3.2 **Coast Guard Compatibility Classification:** Organic acid
3.3 **Chemical Formula:** $CH_2=CHCOOH$
3.4 **IMCO/United Nations Numerical Designation:** Not listed

4. OBSERVABLE CHARACTERISTICS

4.1 **Physical State (as shipped):** Liquid
4.2 **Color:** Colorless
4.3 **Odor:** Acrid

5. HEALTH HAZARDS

5.1 **Personal Protective Equipment:** Chemical respirator at ambient temperatures to avoid inhalation of noxious fumes; rubber gloves if exposed to wet material; acid goggles or face shield for splash exposure; safety shower and/or eye fountain may be required.
5.2 **Symptoms Following Exposure:** May burn skin or eyes upon short contact. INHALATION: eye and nasal irritation and lacrimation. INGESTION: may cause severe damage to the gastrointestinal tract.
5.3 **Treatment for Exposure:** Get medical attention promptly for all exposures. INHALATION: remove victim to fresh air. INGESTION: do NOT induce vomiting. SKIN OR EYES: flush with water for at least 15 min.
5.4 **Toxicity by Inhalation (Threshold Limit Value):** Data not available
5.5 **Short-Term Inhalation Limits:** Data not available
5.6 **Toxicity by Ingestion:** Grade 2; LD_{50} 0.5 to 5 g/kg (rat)
5.7 **Late Toxicity:** Not pertinent
5.8 **Vapor (Gas) Irritant Characteristics:** Vapor is moderately irritating such that personnel will not usually tolerate moderate or high vapor concentrations.
5.9 **Liquid or Solid Irritant Characteristics:** Fairly severe skin irritant; may cause pain and second-degree burns after a few minutes of contact.
5.10 **Odor Threshold:** Data not available

6. FIRE HAZARDS

6.1 **Flash Point:** (Glacial) 118°F O.C.
6.2 **Flammable Limits in Air:** (Tech.) 2.4% LEL
6.3 **Fire Extinguishing Agents:** Water spray, alcohol foam, dry chemical, or carbon dioxide
6.4 **Fire Extinguishing Agents Not to be Used:** Not pertinent
6.5 **Special Hazards of Combustion Products:** Toxic vapors are generated when heated
6.6 **Behavior in Fire:** May polymerize and explode
6.7 **Ignition Temperature:** 374°F
6.8 **Electrical Hazard:** Not pertinent
6.9 **Burning Rate:** 1.6 mm/min.

7. CHEMICAL REACTIVITY

7.1 **Reactivity with Water:** No reaction
7.2 **Reactivity with Common Materials:** No reaction
7.3 **Stability During Transport:** Normally unstable but will not detonate.
7.4 **Neutralizing Agents for Acids and Caustics:** Wash with water, rinse with sodium bicarbonate solution.
7.5 **Polymerization:** May occur on contact with acids, iron salts, or at elevated temperatures and release high energy rapidly; may cause explosion under confinement.
7.6 **Inhibitor of Polymerization:** Monomethyl ether of hydroquinone 180-200 ppm; phenothiazine (for tech. grades) 1000 ppm; hydroquinone (0.1%); methylene blue (0.5-1%); N, N'-diphenyl-p-phenylenediamine (0.05%)

8. WATER POLLUTION

8.1 **Aquatic Toxicity:** Data not available
8.2 **Waterfowl Toxicity:** Data not available
8.3 **Biological Oxygen Demand (BOD):** 35%, 10 days
8.4 **Food Chain Concentration Potential:** None listed

9. SELECTED MANUFACTURERS

1. Celanese Corporation
   Celanese Chemical Co. Division
   245 Park Ave.
   New York, N. Y. 10017
2. Rohm and Haas Co.
   Independence Mall West
   Philadelphia, Pa. 19105
3. Union Carbide Corp.
   Chemicals and Plastics Division
   270 Park Ave.
   New York, N. Y. 10017

10. SHIPPING INFORMATION

10.1 **Grades or Purity:** Technical: 94.0%; glacial: 98.0-99.5%
10.2 **Storage Temperature:** 60°-75°F
10.3 **Inert Atmosphere:** Data not available
10.4 **Venting:** Data not available

11. HAZARD ASSESSMENT CODE

(See Hazard Assessment Handbook, CG 446-3)

A-P-Q-Z

12. HAZARD CLASSIFICATIONS

12.1 **Code of Federal Regulations:** Corrosive Material

12.2 **NAS Hazard Rating for Bulk Water Transportation:**

Category	Rating
Fire	2
Health	
Vapor Irritant	3
Liquid or Solid Irritant	3
Poisons	2
Water Pollution	
Human Toxicity	1
Aquatic Toxicity	3
Aesthetic Effect	2
Reactivity	
Other Chemicals	2
Water	0
Self-Reaction	3

12.3 **NFPA Hazard Classifications:**

Category	Classification
Health Hazard (Blue)	3
Flammability (Red)	2
Reactivity (Yellow)	2

13. PHYSICAL AND CHEMICAL PROPERTIES

13.1 **Physical State at 15°C and 1 atm:** Liquid
13.2 **Molecular Weight:** 72.06
13.3 **Boiling Point at 1 atm:** 286.3°F = 141.3°C = 414.5°K
13.4 **Freezing Point:** 54.1°F = 12.3°C = 285.5°K
13.5 **Critical Temperature:** 648°F = 342°C = 615°K
13.6 **Critical Pressure:** 840 psia = 57 atm = 5.8 MN/m^2
13.7 **Specific Gravity:** 1.0497 at 20°C (liquid)
13.8 **Liquid Surface Tension:** Not pertinent
13.9 **Liquid-Water Interfacial Tension:** Not pertinent
13.10 **Vapor (Gas) Specific Gravity:** Not pertinent
13.11 **Ratio of Specific Heats of Vapor (Gas):** 1.121
13.12 **Latent Heat of Vaporization:** 272.7 Btu/lb = 151.5 cal/g = 6.343×10^5 J/kg
13.13 **Heat of Combustion:** −8100 But/lb ≈ −4500 cal/g = -188.4×10^5 J/kg
13.14 **Heat of Decomposition:** Not pertinent
13.15 **Heat of Solution:** Not pertinent
13.16 **Heat of Polymerization:** −463 Btu/lb = −257 cal/g = -10.8×10^5 J/kg

NOTES

REVISED 1978

ACRYLONITRILE

Common Synonyms	Watery liquid	Colorless to light yellow	Irritating odor
	Floats on water. Poisonous, flammable vapor is produced.		

AVOID CONTACT WITH LIQUID AND VAPOR. KEEP PEOPLE AWAY.
Wear goggles, self-contained breathing apparatus and rubber overclothing (including gloves).
Shut off ignition sources and call fire department.
Stop discharge if possible.
Stay upwind and use water spray to "knock down" vapor.
Evacuate area in case of large discharge.
Isolate and remove discharged material.
Notify local health and pollution control agencies.

Fire

FLAMMABLE.
POISONOUS GASES MAY BE PRODUCED IN FIRE.
Flashback along vapor trail may occur.
Vapor may explode if ignited in an enclosed area.
Wear goggles, self-contained breathing apparatus, and rubber overclothing (including gloves).
Combat fires from a safe distance or protected location.
Extinguish with dry chemical, alcohol foam, or carbon dioxide.
Water may be ineffective on fire.
Cool exposed containers with water.

Exposure

CALL FOR MEDICAL AID.

VAPOR
POISONOUS IF INHALED.
Irritating to eyes.
Move to fresh air.
If breathing has stopped, give artificial respiration.
If breathing is difficult, give oxygen.

LIQUID
POISONOUS IF SWALLOWED
Irritating to skin and eyes.
Remove contaminated clothing and shoes.
Flush affected areas with plenty of water.
IF IN EYES, hold eyelids open and flush with plenty of water.
IF SWALLOWED and victim is CONSCIOUS, have victim drink water or milk and have victim induce vomiting.
IF SWALLOWED and victim is UNCONSCIOUS OR HAVING CONVULSIONS, do nothing except keep victim warm.

Water Pollution

HARMFUL TO AQUATIC LIFE IN VERY LOW CONCENTRATIONS.
Fouling to shoreline.
May be dangerous if it enters water intakes.
Notify local health and wildlife officials.
Notify operators of nearby water intakes.

1. RESPONSE TO DISCHARGE

(See Response Methods Handbook, CG 446-4)

Issue warning—poison; highly flammable
Restrict access
Disperse and flush

2. LABELS

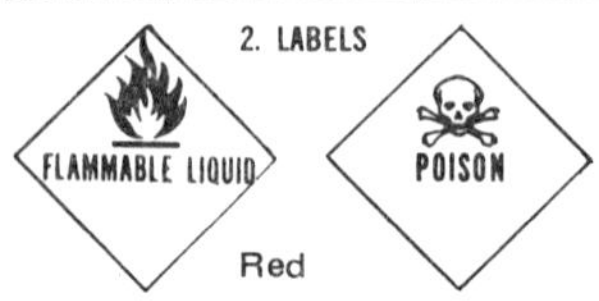

3. CHEMICAL DESIGNATIONS

3.1 **Synonyms:** Cyanoethylene
Fumigrain
Ventox
Vinyl cyanide

3.2 **Coast Guard Compatibility Classification:** Substituted allyl

3.3 **Chemical Formula:** $CH_2{=}CHCN$

3.4 **IMCO/United Nations Numerical Designation:** 3.1/1093

4. OBSERVABLE CHARACTERISTICS

4.1 **Physical State (as shipped):** Liquid

4.2 **Color:** Colorless

4.3 **Odor:** Mild; pungent, resembling that of peach seed kernels

5. HEALTH HAZARDS

5.1 **Personal Protective Equipment:** Air-supplied mask, industrial chemical type, with approved canister for acrylonitrile in low (less than 2%) concentrations; rubber or plastic gloves; cover goggles or face mask; rubber boots; slicker suit; safety helmet.

5.2 **Symptoms Following Exposure:** Similar to those of hydrogen cyanide. Vapor inhalation may cause weakness, headache, sneezing, abdominal pain, and vomiting. Similar symptoms shown if large amounts of liquid are absorbed through the skin; lesser amounts cause stinging and sometimes blisters; contact with eyes causes severe irritation. Ingestion produces nausea, vomiting and abdominal pain.

5.3 **Treatment for Exposure:** Skilled medical treatment is necessary; call physician for all cases of exposure. INHALATION: remove victim to fresh air. (Wear an oxygen or fresh-air-supplied mask when entering contaminated area.) INGESTION: induce vomiting by administering strong solution of salt water, but only if victim is conscious. SKIN: remove contaminated clothing and wash affected area thoroughly with soap and water. EYES: hold eyelids apart and wash with continuous gentle stream of water for at least 15 min.
If victim is not breathing, give artificial respiration until physician arrives. If he is unconscious, crush an amyl nitrite ampule in a cloth and hold it under his nose for 15 seconds in every minute. Do not interrupt artificial respiration while doing this. Replace ampule when its strength is spent and continue treatment until condition improves or physician arrives.

5.4 **Toxicity by Inhalation (Threshold Limit Value):** 20 ppm

5.5 **Short-Term Inhalation Limits:** 40 ppm for 30 minutes

5.6 **Toxicity by Ingestion:** Grade 3; LD_{50} 50 to 500 mg/kg (rat, guinea pig)

5.7 **Late Toxicity:** Data not available

6. FIRE HAZARDS

6.1 **Flash Point:** 30°F C.C.; 31°F O.C.

6.2 **Flammable Limits in Air:** 3.05% – 17.0%

6.3 **Fire Extinguishing Agents:** Dry chemical, alcohol foam, carbon dioxide

6.4 **Fire Extinguishing Agents Not to be Used:** Water or foam may cause frothing

6.5 **Special Hazards of Combustion Products:** When heated or burned, ACN may evolve toxic hydrogen cyanide gas and oxides of nitrogen.

6.6 **Behavior in Fire:** Vapor is heavier than air and may travel a considerable distance to a source of ignition and flash back. May polymerize and explode.

6.7 **Ignition Temperature:** 898°F

6.8 **Electrical Hazard:** Class I, Group D

6.9 **Burning Rate:** Data not available

7. CHEMICAL REACTIVITY

7.1 **Reactivity with Water:** No reaction

7.2 **Reactivity with Common Materials:** Attacks copper and copper alloys; these metals should not be used. Penetrates leather, so contaminated leather shoes and gloves should be destroyed. Attacks aluminum in high concentrations.

7.3 **Stability During Transport:** Stable

7.4 **Neutralizing Agents for Acids and Caustics:** Not pertinent

7.5 **Polymerization:** May occur spontaneously in absence of oxygen or on exposure to visible light or excessive heat, violently in the presence of alkali.
Pure ACN is subject to self-polymerization with rapid pressure development. The commercial product is inhibited and not subject to this reaction.

7.6 **Inhibitor of Polymerization:** Methylhydroquinone (35-45 ppm)

8. WATER POLLUTION

8.1 **Aquatic Toxicity:**
100 ppm/24 hr/all fish/100% killed/fresh water
0.05-1 ppm/24 hr/bluegill/lethal/salt water

8.2 **Waterfowl Toxicity:** Not pertinent

8.3 **Biological Oxygen Demand (BOD):** 70%, 5 days

8.4 **Food Chain Concentration Potential:** None noted

9. SELECTED MANUFACTURERS

1. E. I. duPont de Nemours & Co., Inc.
Electrochemicals Dept.
Wilmington, Del. 19898
2. Monsanto Co.
Monsanto Polymers & Petrochemicals Co.
800 North Lindbergh Blvd.
St. Louis, Mo. 63166
3. Vistron Corp.
Cleveland, Ohio 44115

10. SHIPPING INFORMATION

10.1 **Grades or Purity:** Technical: 98-100%

10.2 **Storage Temperature:** Ambient

10.3 **Inert Atmosphere:** No requirement

10.4 **Venting:** Pressure-vacuum

11. HAZARD ASSESSMENT CODE

(See Hazard Assessment Handbook, CG 446-3)

A-P-Q-R-S-Z

12. HAZARD CLASSIFICATIONS

12.1 **Code of Federal Regulations:** Flammable liquid

12.2 **NAS Hazard Rating for Bulk Water Transportation:**

Category	Rating
Fire	3
Health	
Vapor Irritant	3
Liquid or Solid Irritant	1
Poisons	3
Water Pollution	
Human Toxicity	4
Aquatic Toxicity	3
Aesthetic Effect	2
Reactivity	
Other Chemicals	3
Water	0
Self-Reaction	3

12.3 **NFPA Hazard Classifications:**

Category	Classification
Health Hazard (Blue)	4
Flammability (Red)	3
Reactivity (Yellow)	2

13. PHYSICAL AND CHEMICAL PROPERTIES

13.1 **Physical State at 15°C and 1 atm:** Liquid

13.2 **Molecular Weight:** 53.06

13.3 **Boiling Point at 1 atm:** 171°F = 77.4°C = 350.6°K

13.4 **Freezing Point:** −118°F = −83.6°C = 189.6°K

13.5 **Critical Temperature:** 505°F = 263°C = 536°K

13.6 **Critical Pressure:** 660 psia = 45 atm = 4.6 MN/m^2

13.7 **Specific Gravity:** 0.8075 at 20°C (liquid)

13.8 **Liquid Surface Tension:** Not pertinent

13.9 **Liquid-Water Interfacial Tension:** Not pertinent

13.10 **Vapor (Gas) Specific Gravity:** 1.8

13.11 **Ratio of Specific Heats of Vapor (Gas):** 1.151

13.12 **Latent Heat of Vaporization:** 265 Btu/lb = 147 cal/g = 6.16×10^5 J/kg

13.13 **Heat of Combustion:** -14,300 Btu/lb = -7930 cal/g = -332×10^5 J/kg

13.14 **Heat of Decomposition:** Not pertinent

13.15 **Heat of Solution:** Not pertinent

13.16 **Heat of Polymerization:** Not pertinent

5. HEALTH HAZARDS (Cont'd.)

5.8 **Vapor (Gas) Irritant Characteristics:** Vapor is moderately irritating such that personnel will not usually tolerate moderate or high vapor concentrations.

5.9 **Liquid or Solid Irritant Characteristics:** If spilled on clothing and allowed to remain, may cause smarting and reddening of the skin. Large amounts may be absorbed through the skin and cause poisoning.

5.10 **Odor Threshold:** 21.4 ppm (Sense of smell fatigues rapidly).

REVISED 1978

ADA — ADIPIC ACID

Common Synonyms			
Adipinic acid 1,4-Butanedicarboxylic acid Hexanedioic acid	Solid crystals	White	Odorless
	Sinks and mixes slowly with water.		

Stop discharge if possible. Keep people away.
Shut off ignition sources. Call fire department.
Avoid contact with solid and dust.
Isolate and remove discharged material.
Notify local health and pollution control agencies.

Fire

Combustible.
Dust cloud may explode if ignited in an enclosed area.
Extinguish with water, dry chemicals, foam, or carbon dioxide.

Exposure

CALL FOR MEDICAL AID.

DUST
Irritating to eyes, nose and throat.
If inhaled will cause coughing or difficult breathing.
If in eyes, hold eyelids open and flush with plenty of water.
If breathing has stopped, give artificial respiration.
If breathing is difficult, give oxygen.

SOLID
Irritating to skin and eyes.
Harmful if swallowed.
Remove contaminated clothing and shoes.
Flush affected areas with plenty of water.
IF IN EYES, hold eyelids open and flush with plenty of water.
IF SWALLOWED and victim is UNCONSCIOUS OR HAVING CONVULSIONS, do nothing except keep victim warm.

Water Pollution

Dangerous to aquatic life in high concentrations.
May be dangerous if it enters water intakes.
Notify local health and wildlife officials.
Notify operators of nearby water intakes.

1. RESPONSE TO DISCHARGE
(See Response Methods Handbook, CG 446-4)

Should be removed
Chemical and physical treatment

2. LABELS

No hazard label required by Code of Federal Regulations

3. CHEMICAL DESIGNATIONS

3.1 **Synonyms:** Adipinic acid; 1,4-Butanedicarboxylic acid; Hexanedioic acid
3.2 **Coast Guard Compatibility Classification:** Data not available
3.3 **Chemical Formula:** $HOOC(CH_2)_4COOH$
3.4 **IMCO/United Nations Numerical Designation:** Data not available

4. OBSERVABLE CHARACTERISTICS

4.1 **Physical State (as shipped):** Crystalline
4.2 **Color:** White
4.3 **Odor:** None

5. HEALTH HAZARDS

5.1 **Personal Protective Equipment:** Normal protection against exposure to finely divided organic solids (rubber gloves, plastic goggles)
5.2 **Symptoms Following Exposure:** Inhalation of vapor irritates mucous membranes of the nose and lungs, causing coughing and sneezing. Contact with liquid irritates eyes and has a pronounced drying effect on the skin; may produce dermatitis.
5.3 **Treatment for Exposure:** INHALATION: remove victim to fresh air; get medical attention if irritation persists. EYES: flush with water for at least 15 min. SKIN: flush with water.
5.4 **Toxicity by Inhalation (Threshold Limit Value):** Data not available
5.5 **Short-Term Inhalation Limits:** Data not available
5.6 **Toxicity by Ingestion:** Grade 2; oral mouse LD_{50} = 1,900 mg/kg
5.7 **Late Toxicity:** Data not available
5.8 **Vapor (Gas) Irritant Characteristics:** Data not available
5.9 **Liquid or Solid Irritant Characteristics:** Data not available
5.10 **Odor Threshold:** Data not available

6. FIRE HAZARDS

6.1 **Flash Point:** Combustible solid; 300°F O.C.; 376°F C.C.
6.2 **Flammable Limits in Air:** (dust) 10-15 mg/l
6.3 **Fire Extinguishing Agents:** Foam, water fog, carbon dioxide, or dry chemical.
6.4 **Fire Extinguishing Agents Not to be Used:**
6.5 **Special Hazards of Combustion Products:**
6.6 **Behavior in Fire:** Melts and may decompose to give volatile acidic vapors of valeric acid and other substances. Dust may form explosive mixture with air.
6.7 **Ignition Temperature:** 788°F
6.8 **Electrical Hazard:** Not pertinent
6.9 **Burning Rate:** Not pertinent

7. CHEMICAL REACTIVITY

7.1 **Reactivity with Water:** No reaction
7.2 **Reactivity with Common Materials:**
7.3 **Stability During Transport:** Stable
7.4 **Neutralizing Agents for Acids and Caustics:** Rinse with dilute sodium bicarbonate or soda ash solution.
7.5 **Polymerization:** Not pertinent
7.6 **Inhibitor of Polymerization:** Not pertinent

8. WATER POLLUTION

8.1 **Aquatic Toxicity:** <330 ppm/24 hr/bluegill/TL_m/fresh water
8.2 **Waterfowl Toxicity:** Data not available
8.3 **Biological Oxygen Demand (BOD):** (theoretical) 1.3%, 0.5 days
8.4 **Food Chain Concentration Potential:** None

9. SELECTED MANUFACTURERS

1. Celanese Chemical Corp.
   A Division of Celanese Corp.
   245 Park Avenue
   New York, N. Y. 10017
2. E. I. duPont de Nemours & Co., Inc.
   1007 Market St.
   Wilmington, Del. 19898
3. Monsanto Company
   800 North Lindbergh Blvd.
   St. Louis, Mo. 63166

10. SHIPPING INFORMATION

10.1 **Grades or Purity:** Commercial, 99.8%
10.2 **Storage Temperature:** Ambient
10.3 **Inert Atmosphere:** No requirement
10.4 **Venting:** Open

11. HAZARD ASSESSMENT CODE
(See Hazard Assessment Handbook, CG 446-3)

SS-11

12. HAZARD CLASSIFICATIONS

12.1 **Code of Federal Regulations:** Not listed
12.2 **NAS Hazard Rating for Bulk Water Transportation:** Not listed
12.3 **NFPA Hazard Classifications:**

Category	Classification
Health Hazard (Blue)	1
Flammability (Red)	1
Reactivity (Yellow)	0

13. PHYSICAL AND CHEMICAL PROPERTIES

13.1 **Physical State at 15°C and 1 atm:** Solid
13.2 **Molecular Weight:** 146.1
13.3 **Boiling Point at 1 atm:** Not pertinent (decomposes)
13.4 **Freezing Point:** 304°F = 151°C = 424°K
13.5 **Critical Temperature:** Not pertinent
13.6 **Critical Pressure:** Not pertinent
13.7 **Specific Gravity:** 1.36 at 20°C (solid)
13.8 **Liquid Surface Tension:** Not pertinent
13.9 **Liquid-Water Interfacial Tension:** Not pertinent
13.10 **Vapor (Gas) Specific Gravity:** Not pertinent
13.11 **Ratio of Specific Heats of Vapor (Gas):** Not pertinent
13.12 **Latent Heat of Vaporization:** Not pertinent
13.13 **Heat of Combustion:** $-8,242$ Btu/lb $= -4,579$ cal/g $= -191.6 \times 10^5$ J/kg
13.14 **Heat of Decomposition:** Not pertinent
13.15 **Heat of Solution:** Not pertinent
13.16 **Heat of Polymerization:** Not pertinent

NOTES

ADN

ADIPONITRILE

Common Synonyms			
1,4-Dicyanobutane	Liquid	Colorless to light yellow	Odorless
	Floats on water. Freezing point is 36° F.		

Stop discharge if possible.
Call fire department.
Avoid contact with liquid.
Isolate and remove discharged material.
Notify local health and pollution control agencies.

Fire

Combustible.
POISONOUS GASES MAY BE PRODUCED IN FIRE.
Vapor may explode if ignited in an enclosed area.
Wear chemical protective suit with self-contained breathing apparatus.
Extinguish with water, dry chemicals, foam or carbon dioxide.
Cool exposed containers with water.

Exposure

CALL FOR MEDICAL AID.

LIQUID OR SOLID
Irritating to skin and eyes.
If swallowed, will cause nausea or vomiting.
Remove contaminated clothing and shoes.
Flush affected areas with plenty of water.
IF IN EYES, hold eyelids open and flush with plenty of water.
IF SWALLOWED and victim is CONSCIOUS, have victim drink water or milk.

Water Pollution

Dangerous to aquatic life in high concentrations.
Fouling to shoreline.
May be dangerous if it enters water intakes.
Notify local health and pollution control officials.
Notify operators of nearby water intakes.

1. RESPONSE TO DISCHARGE

(See Response Methods Handbook, CG 446-4)

Issue warning—poison, water contaminant
Mechanical containment
Should be removed

2. LABELS

No hazard label required by Code of Federal Regulations

3. CHEMICAL DESIGNATIONS

3.1 **Synonyms:** 1,4-Dicyanobutane
3.2 **Coast Guard Compatibility Classification:** Nitrile
3.3 **Chemical Formula:** $CN(CH_2)_4CN$
3.4 **IMCO/United Nations Numerical Designation:** Not listed

4. OBSERVABLE CHARACTERISTICS

4.1 **Physical State (as shipped):** Liquid
4.2 **Color:** Water-white to light yellow
4.3 **Odor:** Practically odorless

5. HEALTH HAZARDS

5.1 **Personal Protective Equipment:** Rubber gloves and clothing giving full body and face protection to avoid contact with skin. Air or oxygen mask.
5.2 **Symptoms Following Exposure:** Ingestion of a few ml. may cause weakness, mental confusion, vomiting, rapid respiration, and tachycardia and convulsions. Headache and convulsions can result from exposure to vapor.
5.3 **Treatment for Exposure:** Symptomatic treatment. Call physician. Thiosulfate should be considered. Administer vapor of amyl nitrite if patient is unconscious.
5.4 **Toxicity by Inhalation (Threshold Limit Value):** 50 ppm
5.5 **Short-Term Inhalation Limits:** Data not available
5.6 **Toxicity by Ingestion:** Data not available
5.7 **Late Toxicity:** Data not available
5.8 **Vapor (Gas) Irritant Characteristics:** If present in high concentrations, vapors cause a slight smarting of the eyes or respiratory system and may also cause more severe symptoms such as headache and convulsions.
5.9 **Liquid or Solid Irritant Characteristics:** If spilled on clothing and allowed to remain, may cause smarting and reddening of the skin. If absorbed by skin may cause more severe symptoms such as headache and convulsions.
5.10 **Odor Threshold:** Data not available

6. FIRE HAZARDS

6.1 **Flash Point:** 199°F O.C.
6.2 **Flammable Limits in Air:** LFL = 1.0% at 200°C
6.3 **Fire Extinguishing Agents:** Water spray, dry chemical, foam, or carbon dioxide
6.4 **Fire Extinguishing Agents Not to be Used:** Not pertinent
6.5 **Special Hazards of Combustion Products:** Toxic gases are generated in fires
6.6 **Behavior in Fire:** Not pertinent
6.7 **Ignition Temperature:** Data not available
6.8 **Electrical Hazard:** Not pertinent
6.9 **Burning Rate:** 2.7 mm/min.

7. CHEMICAL REACTIVITY

7.1 **Reactivity with Water:** No reaction
7.2 **Reactivity with Common Materials:** No reaction
7.3 **Stability During Transport:** Stable
7.4 **Neutralizing Agents for Acids and Caustics:** Not pertinent
7.5 **Polymerization:** Not pertinent
7.6 **Inhibitor of Polymerization:** Not pertinent

8. WATER POLLUTION

8.1 **Aquatic Toxicity:** 1250 ppm/24 hr/sunfish/TL_m/fresh water
8.2 **Waterfowl Toxicity:** Data not available
8.3 **Biological Oxygen Demand (BOD):** 40%, 5 days
8.4 **Food Chain Concentration Potential:** Data not available

9. SELECTED MANUFACTURERS

1. Celanese Corporation
   Celanese Chemical Co. Division
   245 Park Ave.
   New York, N. Y. 10017
2. E. I. duPont de Nemours, Inc.
   Plastics Department
   DuPont Building
   Wilmington, Del. 19898
3. Monsanto Company
   Monsanto Textiles Co.
   800 N. Lindbergh Blvd.
   St. Louis, Mo. 63166

10. SHIPPING INFORMATION

10.1 **Grades or Purity:** Data not available
10.2 **Storage Temperature:** Ambient
10.3 **Inert Atmosphere:** No requirement
10.4 **Venting:** Pressure vacuum

11. HAZARD ASSESSMENT CODE

(See Hazard Assessment Handbook, CG 446-3)

A-T-U

12. HAZARD CLASSIFICATIONS

12.1 **Code of Federal Regulations:** Combustible Liquid
12.2 **NAS Hazard Rating for Bulk Water Transportation:**

Category	Rating
Fire	1
Health	
Vapor Irritant	1
Liquid or Solid Irritant	1
Poisons	3
Water Pollution	
Human Toxicity	3
Aquatic Toxicity	2
Aesthetic Effect	3
Reactivity	
Other Chemicals	2
Water	0
Self-Reaction	0

12.3 **NFPA Hazard Classifications:**

Category	Classification
Health Hazard (Blue)	4
Flammability (Red)	2
Reactivity (Yellow)	0

13. PHYSICAL AND CHEMICAL PROPERTIES

13.1 **Physical State at 15°C and 1 atm:** Liquid
13.2 **Molecular Weight:** 108
13.3 **Boiling Point at 1 atm:** 554°F = 290°C = 563°K
13.4 **Freezing Point:** 36°F = 2.4°C = 275.5°K
13.5 **Critical Temperature:** Not pertinent
13.6 **Critical Pressure:** Not pertinent
13.7 **Specific Gravity:** 0.9611 at 25°C (liquid)
13.8 **Liquid Surface Tension:** Not pertinent
13.9 **Liquid-Water Interfacial Tension:** Not pertinent
13.10 **Vapor (Gas) Specific Gravity:** Not pertinent
13.11 **Ratio of Specific Heats of Vapor (Gas):** Not pertinent
13.12 **Latent Heat of Vaporization:** (est.) 240 Btu/lb = 134 cal/g = 5.59×10^5 J/kg
13.13 **Heat of Combustion:** -14,230 Btu/lb = -7910 cal/g = -331×10^5 J/kg
13.14 **Heat of Decomposition:** Not pertinent
13.15 **Heat of Solution:** Not pertinent
13.16 **Heat of Polymerization:** Not pertinent

NOTES

REVISED 1978

ALD	ALDRIN

Common Synonyms			
1,2,3,4,10,10-hexachloro-1,4,4a,5,8,8a-hexahydro-1,4-endo-exo-5,8-dimethanonaphthalene.	Solid crystals, or solution	Light to dark brown	Mild chemical odor
	Solid sinks in water; solution floats on water.		

AVOID CONTACT WITH LIQUID OR SOLID. KEEP PEOPLE AWAY.
Wear goggles, self-contained breathing apparatus, and rubber overclothing (including gloves).
Stop discharge if possible.
Call fire department if solution is discharged.
Isolate and remove discharged material.
Notify local health and pollution control agencies.

Fire

Solid is not flammable but usually is dissolved in a combustible liquid.
POISONOUS GASES ARE PRODUCED WHEN HEATED.
Wear goggles, self-contained breathing apparatus, and rubber overclothing (including gloves).
Extinguish with water, dry chemical, foam or carbon dioxide.
Cool exposed containers with water.

Exposure

CALL FOR MEDICAL AID.

SOLID OR SOLUTION
POISONOUS IF SWALLOWED OR IF SKIN IS EXPOSED.
Irritating to skin, eyes.
Remove contaminated clothing and shoes.
Flush affected areas with plenty of water.
IF IN EYES, hold eyelids open and flush with plenty of water.
IF SWALLOWED and victim is CONSCIOUS, have victim drink water or milk and have victim induce vomiting.
IF SWALLOWED and victim is UNCONSCIOUS OR HAVING CONVULSIONS, do nothing except keep victim warm.

Water Pollution

HARMFUL TO AQUATIC LIFE IN VERY LOW CONCENTRATIONS.
May be dangerous if it enters water intakes.
Notify local health and wildlife officials.
Notify operators of nearby water intakes.

1. RESPONSE TO DISCHARGE

(See Response Methods Handbook, CG 446-4)

Issue warning – poison, water contaminant; liquid forms are flammable
Mechanical containment (of liquid form)
Should be removed

2. LABEL

3. CHEMICAL DESIGNATIONS

3.1 **Synonyms:** endo-, exo-1, 2, 3, 4, 10, 10-hexachloro-1, 4, 4a, 5, 8, 8a-hexahydro-1,4:5,8-dimethanonaphthalene
HHDN

3.2 **Coast Guard Compatibility Classification:** Not applicable

3.3 **Chemical Formula:** $C_{12}H_8Cl_6$

3.4 **IMCO/United Nations Numerical Designation:** 6.1/1542

4. OBSERVABLE CHARACTERISTICS

4.1 **Physical State (as shipped):** Solid

4.2 **Color:** Tan to dark brown

4.3 **Odor:** Mild chemical

5. HEALTH HAZARDS

5.1 **Personal Protective Equipment:** During prolonged exposure to mixing and loading operations, wear clean synthetic rubber gloves and mask or respirator of the type passed by the U. S. Bureau of Mines for aldrin protection.

5.2 **Symptoms Following Exposure:** Ingestion, inhalation, or skin absorption of a toxic dose will induce nausea, vomiting, hyperexcitability, tremors, epileptiform convulsions, and ventricular fibrillation. Aldrin may cause temporary reversible kidney and liver injury. Symptoms may be seen after ingestion of less than 1 gram in an adult; ingestion of 25 mg has caused death in children.

5.3 **Treatment for Exposure:** SKIN CONTACT: wash with soap and running water. If material gets into eyes, wash immediately with running water for at least 15 minutes; get medical attention. INGESTION: call physician immediately; induce vomiting immediately. Repeat until vomit fluid is clear. Never give anything by mouth to an unconscious person. Keep patient prone and quiet. PHYSICIAN: administer barbituates as anti-convulsant therapy. Observe patient carefully because repeated treatment may be necessary.

5.4 **Toxicity by Inhalation (Threshold Limit Value):** 0.25 mg/m^3

5.5 **Short-Term Inhalation Limits:** 1 mg/m^3 for 30 min.

5.6 **Toxicity by Ingestion:** Grade 3; LD 50 to 500 mg/kg (rat)

5.7 **Late Toxicity:** Chronic exposure produces benign tumors in mice.

5.8 **Vapor (Gas) Irritant Characteristics:** Vapors cause slight smarting of the eyes or respiratory system if present in high concentrations. Effect is temporary.

5.9 **Liquid or Solid Irritant Characteristics:** Minimum hazard. If spilled on clothing and allowed to remain, may cause smarting and reddening if the skin.

5.10 **Odor Threshold:** Data not available

6. FIRE HAZARDS

6.1 **Flash Point:** Not flammable

6.2 **Flammable Limits in Air:** Not pertinent

6.3 **Fire Extinguishing Agents:** Water spray, dry chemical, foam or carbon dioxide for fires involving solutions of aldrin in hydrocarbon solvents.

6.4 **Fire Extinguishing Agents Not to be Used:** Not pertinent

6.5 **Special Hazards of Combustion Products:** Irritating fumes of hydrochloric acid and chlorinated decomposition products are given off

6.6 **Behavior in Fire:** Not pertinent

6.7 **Ignition Temperature:** Not pertinent

6.8 **Electrical Hazard:** Not pertinent

6.9 **Burning Rate:** Not pertinent

7. CHEMICAL REACTIVITY

7.1 **Reactivity with Water:** No reaction

7.2 **Reactivity with Common Materials:** No reaction

7.3 **Stability During Transport:** Stable

7.4 **Neutralizing Agents for Acids and Caustics:** Not pertinent

7.5 **Polymerization:** Not pertinent

7.6 **Inhibitor of Polymerization:** Not pertinent

8. WATER POLLUTION

8.1 **Aquatic Toxicity:**
0.130 ppm/24 hr/bluegill/LC_{50}/fresh water
0.05 ppm/24 hr/goldfish/LC_{50}/fresh water
0.01 ppm/24 hr/oyster/sublethal effect/salt water

8.2 **Waterfowl Toxicity:** 520 mg/kg

8.3 **Biological Oxygen Demand (BOD):** Not pertinent

8.4 **Food Chain Concentration Potential:** High

9. SELECTED MANUFACTURERS

Shell Chemical Co.
Agricultural Div.
2401 Crow Canyon Rd.
San Ramon, Calif. 94583

10. SHIPPING INFORMATION

10.1 **Grades or Purity:** 20-95% aldrin, 5-80% inert ingredients. Several solutions in hydrocarbon solvents.

10.2 **Storage Temperature:** Data not available

10.3 **Inert Atmosphere:** Data not available

10.4 **Venting:** Data not available

11. HAZARD ASSESSMENT CODE

(See Hazard Assessment Handbook, CG 446-3)

I I

12. HAZARD CLASSIFICATIONS

12.1 **Code of Federal Regulations:** Poisonous liquid or solid, Class B

12.2 **NAS Hazard Rating for Bulk Water Transportation:**

Category	Rating
Fire	0
Health	
Vapor Irritant	1
Liquid or Solid Irritant	1
Poisons	3
Water Pollution	
Human Toxicity	4
Aquatic Toxicity	4
Aesthetic Effect	2
Reactivity	
Other Chemicals	2
Water	0
Self-Reaction	0

12.3 **NFPA Hazard Classifications:**

Category	Classification
Health Hazard (Blue)	2
Flammability (Red)	0
Reactivity (Yellow)	0

13. PHYSICAL AND CHEMICAL PROPERTIES

13.1 **Physical State at 15°C and 1 atm:** Solid

13.2 **Molecular Weight:** 364.93

13.3 **Boiling Point at 1 atm:** Not pertinent

13.4 **Freezing Point:** 219°F = 104°C = 377°K

13.5 **Critical Temperature:** Not pertinent

13.6 **Critical Pressure:** Not pertinent

13.7 **Specific Gravity:** 1.6 at 20°C (solid)

13.8 **Liquid Surface Tension:** Not pertinent

13.9 **Liquid-Water Interfacial Tension:** Not pertinent

13.10 **Vapor (Gas) Specific Gravity:** Not pertinent

13.11 **Ratio of Specific Heats of Vapor (Gas):** Not pertinent

13.12 **Latent Heat of Vaporization:** Not pertinent

13.13 **Heat of Combustion:** Not pertinent

13.14 **Heat of Decomposition:** Not pertinent

13.15 **Heat of Solution:** Not pertinent

13.16 **Heat of Polymerization:** Not pertinent

NOTES

REVISED 1978

ABS	ALKYLBENZENESULFONIC ACIDS

Common Synonyms			
Decylbenzenesulfonic acid Un / Do / Tri / Tetra / Penta / Hexa } benzenesulfonic acid	Liquid	White to yellow	Odorless
	Mixes with water.		

Stop discharge if possible. Keep people away.
Call fire department.
Avoid contact with liquid.
Isolate and remove discharged material.
Notify local health and pollution control agencies.

Fire	Combustible. Flammable gas may be produced on contact with metals. Extinguish with dry chemicals, alcohol foam, or carbon dioxide. Water may be ineffective on fire.
Exposure	CALL FOR MEDICAL AID. LIQUID Irritating to skin and eyes. If swallowed will cause nausea. Remove contaminated clothing and shoes. Flush affected areas with plenty of water. IF IN EYES, hold eyelids open and flush with plenty of water. IF SWALLOWED and victim is CONSCIOUS, have victim drink water or milk. IF SWALLOWED and victim is UNCONSCIOUS OR HAVING CONVULSIONS, do nothing except keep victim warm.
Water Pollution	Effect of low concentrations on aquatic life is unknown. May be dangerous if it enters water intakes. Notify local health and wildlife officials. Notify operators of nearby water intakes.

1. RESPONSE TO DISCHARGE
(See Response Methods Handbook, CG 446-4)

Issue warning—corrosive
Disperse and flush

2. LABELS

No hazard label required by Code of Federal Regulations

3. CHEMICAL DESIGNATIONS

3.1 **Synonyms:** Un / Do / Tri / Tetra / Penta / Hexa } decylbenzenesulfonic acid

3.2 **Coast Guard Compatibility Classification:** Not listed

4. OBSERVABLE CHARACTERISTICS

4.1 **Physical State (as shipped):** Liquid

4.2 **Color:** Colorless to yellow

4.3 **Odor:** None

5. HEALTH HAZARDS

5.1 **Personal Protective Equipment:** Goggles or face shield; rubber gloves

5.2 **Symptoms Following Exposure:** Ingestion causes irritation of mouth and stomach. Contact with liquid irritates eyes and skin.

5.3 **Treatment for Exposure:** INGESTION: give large amount of water. EYES: flush with water for at least 15 min. SKIN: flush with water; wash with soap and water.

5.4 **Toxicity by Inhalation (Threshold Limit Value):** Data not available

5.5 **Short-Term Inhalation Limits:** Not pertinent

5.6 **Toxicity by Ingestion:** Grade 2; LD_{50} 0.5-5 g/kg (rat)

5.7 **Late Toxicity:** Data not available

5.8 **Vapor (Gas) Irritant Characteristics:** Data not available

5.9 **Liquid or Solid Irritant Characteristics:** Data not available

5.10 **Odor Threshold:** Data not available

6. FIRE HAZARDS

6.1 **Flash Point:** 395°F O.C.

6.2 **Flammable Limits in Air:** Data not available

6.3 **Fire Extinguishing Agents:** Alcohol foam, dry chemical, carbon dioxide

6.4 **Fire Extinguishing Agents Not to be Used:** Water may be ineffective.

6.5 **Special Hazards of Combustion Products:** Irritating sulfuric acid mist may form in fire.

6.6 **Behavior in Fire:**

6.7 **Ignition Temperature:** Data not available

6.8 **Electrical Hazard:** Data not available

6.9 **Burning Rate:** Data not available

7. CHEMICAL REACTIVITY

7.1 **Reactivity with Water:** No reaction

7.2 **Reactivity with Common Materials:** May attack metals, causing accumulation of flammable hydrogen gas in enclosed spaces.

7.3 **Stability During Transport:** Stable

7.4 **Neutralizing Agents for Acids and Caustics:** Flush with water, rinse with dilute sodium bicarbonate or soda ash solution

7.5 **Polymerization:** Not pertinent

7.6 **Inhibitor of Polymerization:** Not pertinent

8. WATER POLLUTION

8.1 **Aquatic Toxicity:** Data not available

8.2 **Waterfowl Toxicity:** Data not available

8.3 **Biological Oxygen Demand (BOD):** Data not available

8.4 **Food Chain Concentration Potential:** None

9. SELECTED MANUFACTURERS

1. Continental Oil Co.
   Petrochemical Dept.
   Park 80 Plaza East
   Saddle Brook, N. J. 07662
2. Witco Chemical Co.
   Organics Div.
   277 Park Avenue
   New York, N. Y. 10017

10. SHIPPING INFORMATION

10.1 **Grades or Purity:** Commercial, 88%—97%

10.2 **Storage Temperature:** Ambient

10.3 **Inert Atmosphere:** No requirement

10.4 **Venting:** Open

11. HAZARD ASSESSMENT CODE
(See Hazard Assessment Handbook, CG 446-3)

A-P

12. HAZARD CLASSIFICATIONS

12.1 **Code of Federal Regulations:** Not listed

12.2 **NAS Hazard Rating for Bulk Water Transportation:** Not listed

12.3 **NFPA Hazard Classifications:** Not listed

13. PHYSICAL AND CHEMICAL PROPERTIES

13.1 **Physical State at 15°C and 1 atm:** Liquid

13.2 **Molecular Weight:** 310—394

13.3 **Boiling Point at 1 atm:** Not pertinent (decomposes)

13.4 **Freezing Point:** Data not available

13.5 **Critical Temperature:** Not pertinent

13.6 **Critical Pressure:** Not pertinent

13.7 **Specific Gravity:** 1.0 to 1.4 at 20°C (liquid)

13.8 **Liquid Surface Tension:** Data not available

13.9 **Liquid-Water Interfacial Tension:** Not pertinent

13.10 **Vapor (Gas) Specific Gravity:** Not pertinent

13.11 **Ratio of Specific Heats of Vapor (Gas):** Not pertinent

13.12 **Latent Heat of Vaporization:** Not pertinent

13.13 **Heat of Combustion:** Data not available

13.14 **Heat of Decomposition:** Not pertinent

13.15 **Heat of Solution:** Data not available

13.16 **Heat of Polymerization:** Not pertinent

5. HEALTH HAZARDS (Cont'd)

3.3 **Chemical Formula:** $C_nH_{2n+1}C_6H_4SO_3H$ (n = 10 − 16)

3.4 **IMCO/United Nations Numerical Designation:** Not listed

ALA

ALLYL ALCOHOL

Common Synonyms			
Vinyl carbinol 2-Propen-1-ol	Watery liquid	Colorless	Sharp mustard odor
	Floats and mixes with water. Poisonous, flammable vapor is produced.		

AVOID CONTACT WITH LIQUID AND VAPOR. KEEP PEOPLE AWAY.
Shut off ignition sources and call fire department.
Wear goggles, self-contained breathing apparatus, and rubber overclothing (including gloves).
Stop discharge if possible.
Stay upwind and use water spray to "knock down" vapor.
Isolate and remove discharged material.
Notify local health and pollution control agencies.

Fire

FLAMMABLE.
POISONOUS GASES MAY BE PRODUCED IN FIRE.
Flashback along vapor trail may occur.
Vapor may explode if ignited in an enclosed area.

Wear goggles, self-contained breathing apparatus, and rubber overclothing (including gloves).
Extinguish with dry chemical, alcohol foam, or carbon dioxide.
Water may be ineffective on fire.
Cool exposed containers with water.

Exposure

CALL FOR MEDICAL AID.

VAPOR
POISONOUS IF INHALED OR IF SKIN IS EXPOSED.
Irritating to eyes, nose and throat.
Move to fresh air.
If in eyes, hold eyelids open and flush with plenty of water.
If breathing has stopped, give artificial respiration.
If breathing is difficult, give oxygen.

LIQUID
POISONOUS IF SWALLOWED OR IF SKIN IS EXPOSED.
Will burn eyes.
Remove contaminated clothing and shoes.
Flush affected areas with plenty of water.
IF IN EYES, hold eyelids open and flush with plenty of water.
IF SWALLOWED and victim is CONSCIOUS, have victim drink water or milk.

Water Pollution

HARMFUL TO AQUATIC LIFE IN VERY LOW CONCENTRATIONS.
May be dangerous if it enters water intakes.
Notify local health and wildlife officials.
Notify operators of nearby water intakes.

1. RESPONSE TO DISCHARGE
(See Response Methods Handbook, CG 446-4)

Issue warning—high flammability, poison
Restrict access
Disperse and flush

2. LABELS

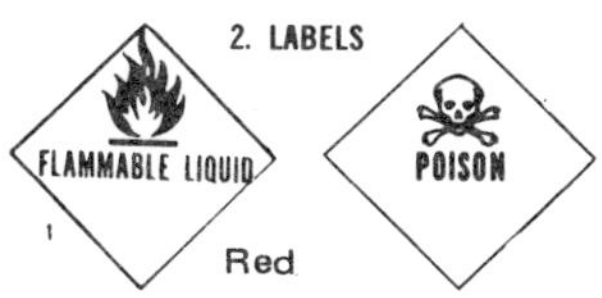

3. CHEMICAL DESIGNATIONS

3.1 **Synonyms:** 2-Propen-1-ol; Vinylcarbinol
3.2 **Coast Guard Compatibility Classification:** Substituted allyl
3.3 **Chemical Formula:** $CH_2{=}CHCH_2OH$
3.4 **IMCO/United Nations Numerical Designation:** 3.2/1098

4. OBSERVABLE CHARACTERISTICS

4.1 **Physical State (as shipped):** Liquid
4.2 **Color:** Colorless
4.3 **Odor:** Characteristic, pungent; sharp; causes tears.

5. HEALTH HAZARDS

5.1 **Personal Protective Equipment:** Organic canister or air pack; rubber gloves, goggles; other protective equipment as required to prevent all body contact.
5.2 **Symptoms Following Exposure:** Vapors are quite irritating to eyes, nose, and throat. Eye irritation may be accompanied by complaints of photophobia and pain in the eyeball; pain may not begin until 6 hours after exposure. Liquid may cause first- and second-degree burns of the skin, with blister formation; underlying part will become swollen and painful, and local muscle spasms may occur.
5.3 **Treatment for Exposure:** INHALATION: remove victim from contaminated area and administer oxygen; get medical attention immediately. SKIN: remove liquid with soap and water. EYES: flush with continuous stream of water for 15 min.
5.4 **Toxicity by Inhalation (Threshold Limit Value):** 2 ppm
5.5 **Short-Term Inhalation Limits:** 5 ppm for 30 min.
5.6 **Toxicity by Ingestion:** Grade 3; LD_{50} 50 to 500 mg/kg (mouse, rat)
5.7 **Late Toxicity:** Data not available
5.8 **Vapor (Gas) Irritant Characteristics:** Vapor is moderately irritating such that personnel will not usually tolerate moderate or high vapor concentrations.
5.9 **Liquid or Solid Irritant Characteristics:** Causes smarting of the skin and first-degree burns on short exposure; may cause secondary burns on long exposure.
5.10 **Odor Threshold:** 0.78 ppm

6. FIRE HAZARDS

6.1 **Flash Point:** 72°F C.C.; 90°F O.C.
6.2 **Flammable Limits in Air:** 2.5%–18%
6.3 **Fire Extinguishing Agents:** Dry chemical, alcohol foam, carbon dioxide
6.4 **Fire Extinguishing Agents Not to be Used:** Water may be ineffective
6.5 **Special Hazards of Combustion Products:** Toxic vapors are generated when heated
6.6 **Behavior in Fire:** Vapor heavier than air and may travel a considerable distance to a source of ignition and flash back
6.7 **Ignition Temperature:** 829°F
6.8 **Electrical Hazard:** Not pertinent
6.9 **Burning Rate:** 2.7 mm/min.

7. CHEMICAL REACTIVITY

7.1 **Reactivity with Water:** No reaction
7.2 **Reactivity with Common Materials:** No reaction
7.3 **Stability During Transport:** Stable at ordinary temperatures and pressures
7.4 **Neutralizing Agents for Acids and Caustics:** Not pertinent
7.5 **Polymerization:** Not pertinent
7.6 **Inhibitor of Polymerization:** Not pertinent

8. WATER POLLUTION

8.1 **Aquatic Toxicity:**
10 ppm/*/threshold/fresh water
2.5 ppm/*/bivalve larvae/lethal/salt water
*Time period not specified.
8.2 **Waterfowl Toxicity:** Data not available
8.3 **Biological Oxygen Demand (BOD):** 57%, 10 days; 20%, 5 days
8.4 **Food Chain Concentration Potential:** None noted

9. SELECTED MANUFACTURERS

1. FMC Corp.
Organic Chemicals Div.
633 Third Ave.
New York, N. Y. 10012
2. Olin Corp.
Chemicals Div.
Brandenburg, Ky. 40108
3. Shell Chemical Co.
Industrial Chemicals Div.
Houston, Texas 77001

10. SHIPPING INFORMATION

10.1 **Grades or Purity:** 98%
10.2 **Storage Temperature:** Ambient
10.3 **Inert Atmosphere:** No requirement
10.4 **Venting:** Pressure-vacuum

11. HAZARD ASSESSMENT CODE
(See Hazard Assessment Handbook, CG 446-3)

A-P-Q

12. HAZARD CLASSIFICATIONS

12.1 **Code of Federal Regulations:** Flammable Liquid
12.2 **NAS Hazard Rating for Bulk Water Transportation:**

Category	Rating
Fire	3
Health	
Vapor Irritant	3
Liquid or Solid Irritant	2
Poisons	3
Water Pollution	
Human Toxicity	2
Aquatic Toxicity	3
Aesthetic Effect	2
Reactivity	
Other Chemicals	2
Water	0
Self-Reaction	1

12.3 **NFPA Hazard Classifications:**

Category	Classification
Health Hazard (Blue)	3
Flammability (Red)	3
Reactivity (Yellow)	0

13. PHYSICAL AND CHEMICAL PROPERTIES

13.1 **Physical State at 15°C and 1 atm:** Liquid
13.2 **Molecular Weight:** 58.08
13.3 **Boiling Point at 1 atm:** 206°F = 96.9°C = 370.1°K
13.4 **Freezing Point:** −200°F = −129°C = 144°K
13.5 **Critical Temperature:** 521.4°F = 271.9°C = 545.1°K
13.6 **Critical Pressure:** 840 psia = 57 atm = 5.8 MN/m^2
13.7 **Specific Gravity:** 0.852 at 20°C (liquid)
13.8 **Liquid Surface Tension:** Not pertinent
13.9 **Liquid-Water Interfacial Tension:** Not pertinent
13.10 **Vapor (Gas) Specific Gravity:** 2.0
13.11 **Ratio of Specific Heats of Vapor (Gas):** 1.12
13.12 **Latent Heat of Vaporization:** 295 Btu/lb = 164 cal/g = 6.87×10^5 J/kg
13.13 **Heat of Combustion:** −13,720 Btu/lb = −7620 cal/g = -319.0×10^5 J/kg
13.14 **Heat of Decomposition:** Not pertinent
13.15 **Heat of Solution:** (est.) Negligible
13.16 **Heat of Polymerization:** Not pertinent

NOTES

REVISED 1978

ABR	ALLYL BROMIDE

Common Synonyms			
3-bromopropene Bromallylene 3-bromopropylene	Liquid	Colorless to light yellow	Irritating odor
	Sinks in water. Flammable, irritating vapor is produced.		

Shut off ignition sources. Call fire department.
Evacuate area in case of large discharge.
Avoid contact with liquid and vapor.
Isolate and remove discharged material.
Notify local health and pollution control agencies.

Fire

FLAMMABLE.
POISONOUS GASES MAY BE PRODUCED IN FIRE.
Flashback along vapor trail may occur.
Vapor may explode if ignited in an enclosed area.
Wear goggles, self-contained breathing apparatus and rubber overclothing (including gloves).
Extinguish with dry chemicals, alcohol foam, or carbon dioxide.
Water may be ineffective on fire.
Cool exposed containers with water.

Exposure

Call for medical aid.

VAPOR
Irritating to eyes, nose and throat.
If inhaled will cause headache, dizziness, coughing or difficult breathing.
Move victim to fresh air.
If breathing is difficult, give oxygen.

LIQUID
Irritating to skin and eyes.
Remove contaminated clothing and shoes.
Flush affected areas with plenty of water.
IF IN EYES, hold eyelids open and flush with plenty of water.
IF SWALLOWED and victim is CONSCIOUS, have victim drink water or milk.

Water Pollution

Effect of low concentrations on aquatic life is unknown.
May be dangerous if it enters water intakes.
Notify local health and wildlife officials.
Notify operators of nearby water intakes.

1. RESPONSE TO DISCHARGE

(See Response Methods Handbook, CG 446-4)

Issue warning—high flammability
Restrict access
Should be removed
Chemical and physical treatment

2. LABEL

Red

3. CHEMICAL DESIGNATIONS

3.1 **Synonyms:** Bromallylene; 3-Bromopropene; 3-Bromopropylene
3.2 **Coast Guard Compatibility Classification:** Not applicable
3.3 **Chemical Formula:** $CH_2 = CH \cdot CH_2Br$
3.4 **IMCO/United Nations Numerical Designation:** 3.2/1099

4. OBSERVABLE CHARACTERISTICS

4.1 **Physical State (as shipped):** Liquid
4.2 **Color:** Colorless to light yellow
4.3 **Odor:** Irritating, unpleasant; pungent

5. HEALTH HAZARDS

5.1 **Personal Protective Equipment:** Goggles and face shield; protective clothing; self-contained breathing apparatus for high vapor concentrations.
5.2 **Symptoms Following Exposure:** Inhalation of vapor irritates mucous membranes and causes dizziness, headache, and lung irritation. Contact with liquid irritates eyes and skin. Ingestion causes irritation of mouth and stomach.
5.3 **Treatment for Exposure:** INHALATION: remove from exposure; if breathing is difficult, give oxygen; call physician. EYES: flush with water for at least 15 min. and call physician. SKIN: flush with water; get medical attention for skin irritation. INGESTION: do NOT induce vomiting; get medical attention.
5.4 **Toxicity by Inhalation (Threshold Limit Value):** Data not available
5.5 **Short-Term Inhalation Limits:** Data not available
5.6 **Toxicity by Ingestion:** Grade 4; oral LD_{50} = 30 mg/kg (guinea pig)
5.7 **Late Toxicity:** Data not available
5.8 **Vapor (Gas) Irritant Characteristics:** Vapors are moderately irritating such that personnel will not usually tolerate moderate or high vapor concentrations.
5.9 **Liquid or Solid Irritant Characteristics:** Fairly severe skin irritant. May cause pain and second-degree burns after a few minutes' contact.
5.10 **Odor Threshold:** Data not available

6. FIRE HAZARDS

6.1 **Flash Point:** 30°F C.C.
6.2 **Flammable Limits in Air:** 4.4%–7.3%
6.3 **Fire Extinguishing Agents:** Dry chemical, alcohol foam, or carbon dioxide
6.4 **Fire Extinguishing Agents Not to be Used:** Water may be ineffective.
6.5 **Special Hazards of Combustion Products:** Toxic hydrogen bromide gas formed in fire.
6.6 **Behavior in Fire:** Vapor is heavier than air and may travel a considerable distance to a source of ignition and flash back.
6.7 **Ignition Temperature:** 563°F
6.8 **Electrical Hazard:** Data not available
6.9 **Burning Rate:** 3.5 mm/min.

7. CHEMICAL REACTIVITY

7.1 **Reactivity with Water:** No reaction
7.2 **Reactivity with Common Materials:** No reaction
7.3 **Stability During Transport:** Stable
7.4 **Neutralizing Agents for Acids and Caustics:** Not pertinent
7.5 **Polymerization:** Not pertinent
7.6 **Inhibitor of Polymerization:** Not pertinent

8. WATER POLLUTION

8.1 **Aquatic Toxicity:** Data not available
8.2 **Waterfowl Toxicity:** Data not available
8.3 **Biological Oxygen Demand (BOD):** Data not available
8.4 **Food Chain Concentration Potential:** None

9. SELECTED MANUFACTURERS

1. White Chemical Corporation
East 22nd Street
Box 278
Bayonne, N. J. 07002
2. American Hoechst Corporation
Chemicals and Plastics Division
Somerville, N. J. 08876
3. Columbia Organic Chemical Co.
912 Drake St.
Columbia, S. C. 29205

10. SHIPPING INFORMATION

10.1 **Grades or Purity:** Commercial
10.2 **Storage Temperature:** Ambient
10.3 **Inert Atmosphere:** No requirement
10.4 **Venting:** Pressure-vacuum

11. HAZARD ASSESSMENT CODE

(See Hazard Assessment Handbook, CG 446-3)

A-X-Y

12. HAZARD CLASSIFICATIONS

12.1 **Code of Federal Regulations:** Flammable liquid
12.2 **NAS Hazard Rating for Bulk Water Transportation:**

Category	Rating
Fire	3
Health	
Vapor Irritant	3
Liquid or Solid Irritant	3
Poisons	2
Water Pollution	
Human Toxicity	
Aquatic Toxicity	2
Aesthetic Effect	2
Reactivity	
Other Chemicals	2
Water	0
Self-Reaction	1

12.3 **NFPA Hazard Classifications:**

Category	Classification
Health Hazard (Blue)	3
Flammability (Red)	3
Reactivity (Yellow)	1

13. PHYSICAL AND CHEMICAL PROPERTIES

13.1 **Physical State at 15°C and 1 atm:** Liquid
13.2 **Molecular Weight:** 121
13.3 **Boiling Point at 1 atm:** 158°F = 70°C = 343°K
13.4 **Freezing Point:** −182°F = −119°C = 154°K
13.5 **Critical Temperature:** Not pertinent
13.6 **Critical Pressure:** Not pertinent
13.7 **Specific Gravity:** 1.4161 at 20°C (liquid)
13.8 **Liquid Surface Tension:** 26.9 dynes/cm = 0.0269 N/m at 20°C
13.9 **Liquid-Water Interfacial Tension:** (est.) 40 dynes/cm = 0.040 N/m at 20°C
13.10 **Vapor (Gas) Specific Gravity:** 4.2
13.11 **Ratio of Specific Heats of Vapor (Gas):** 1.1210
13.12 **Latent Heat of Vaporization:** (est.) 110 Btu/lb = 59 cal/g = 2.5×10^5 J/kg
13.13 **Heat of Combustion:** (est.) 6,700 Btu/lb = 3,700 cal/g = 150×10^5 J/kg
13.14 **Heat of Decomposition:** Not pertinent
13.15 **Heat of Solution:** Not pertinent
13.16 **Heat of Polymerization:** Not pertinent

NOTES

ALC

ALLYL CHLORIDE

Common Synonyms			
3-Chloropropene 3-Chloropropylene	Liquid	Colorless to yellowish brown or red	Sharp, irritating odor
	Floats on water. Flammable, irritating vapor is produced.		

Avoid contact with liquid and vapor. Keep people away.
Shut off ignition sources and call fire department.
Wear goggles, self-contained breathing apparatus, and rubber overclothing (including gloves).
Stop discharge if possible.
Stay upwind and use water spray to "knock down" vapor.
Isolate and remove discharged material.
Notify local health and pollution control agencies.

Fire

FLAMMABLE.
POISONOUS GASES ARE PRODUCED IN FIRE.
Flashback along vapor trail may occur.
Vapor may explode if ignited in an enclosed area.
Wear goggles, self-contained breathing apparatus, and rubber overclothing (including gloves).
Extinguish with dry chemical, alcohol foam, or carbon dioxide.
Cool exposed containers with water.

Exposure

CALL FOR MEDICAL AID.

VAPOR
POISONOUS IF INHALED OR IF SKIN IS EXPOSED.
Irritating to eyes, nose and throat.
Move to fresh air.
If breathing has stopped, give artificial respiration.
If breathing is difficult, give oxygen.

LIQUID
POISONOUS IF SWALLOWED OR IF SKIN IS EXPOSED.
Will burn eyes.
Remove contaminated clothing and shoes.
Flush affected areas with plenty of water.
IF IN EYES, hold eyelids open and flush with plenty of water.
IF SWALLOWED and victim is CONSCIOUS, have victim drink water or milk and have victim induce vomiting.
IF SWALLOWED and victim is UNCONSCIOUS OR HAVING CONVULSIONS, do nothing except keep victim warm.

Water Pollution

HARMFUL TO AQUATIC LIFE IN VERY LOW CONCENTRATIONS.
May be dangerous if it enters water intakes.
Notify local health and wildlife officials.
Notify operators of nearby water intakes.

1. RESPONSE TO DISCHARGE
(See Response Methods Handbook, CG 446-4)

Issue warning—high flammability
Restrict access

2. LABEL

Red

3. CHEMICAL DESIGNATIONS

3.1 **Synonyms:** 3-Chloropropene; 3-Chloropropylene
3.2 **Coast Guard Compatibility Classification:** Substituted allyl
3.3 **Chemical Formula:** $CH_2{=}CHCH_2Cl$
3.4 **IMCO/United Nations Numerical Designation:** 3.1/1100

4. OBSERVABLE CHARACTERISTICS

4.1 **Physical State (as shipped):** Liquid
4.2 **Color:** Colorless
4.3 **Odor:** Sharp; penetrating; irritating; characteristic pungent

5. HEALTH HAZARDS

5.1 **Personal Protective Equipment:** Respiratory protection: (1 ppm to 2% for 1/2 hr or less) full face mask and canister; (greater concentration) self-contained breathing apparatus or its equivalent. Rubber or neoprene gloves, apron, boots; clean body-covering clothes; chemical goggles, gas-tight goggles, or equivalent; full face shield.
5.2 **Symptoms Following Exposure:** Causes marked irritation of skin and may burn. Burns the eyes; effect may be delayed.
5.3 **Treatment for Exposure:** INHALATION: if ill effects develop, move person to fresh air, keep him warm and quiet. Get medical attention immediately. Start artificial respiration if breathing stops. INGESTION: promptly induce vomiting. Get medical attention immediately; no known antidote; treat symptomatically. EYES: immediately flush with plenty of water for at least 15 min.; get medical attention promptly. SKIN: remove clothing and flush affected area thoroughly.
5.4 **Toxicity by Inhalation (Threshold Limit Value):** 1 ppm
5.5 **Short-Term Inhalation Limits:** Data not available
5.6 **Toxicity by Ingestion:** Grade 2; LD_{50} 0.5 to 5 g/kg (rat)
5.7 **Late Toxicity:** Lung, liver and kidney damage in experimental animals.
5.8 **Vapor (Gas) Irritant Characteristics:** Vapor is moderately irritating such that personnel will not usually tolerate moderate or high vapor concentrations.
5.9 **Liquid or Solid Irritant Characteristics:** Causes smarting of the skin and first-degree burns on short exposure and may cause secondary burns on long exposure.
5.10 **Odor Threshold:** 0.47 ppm

6. FIRE HAZARDS

6.1 **Flash Point:** −20°F C.C.
6.2 **Flammable Limits in Air:** 3.3% 11.1%
6.3 **Fire Extinguishing Agents:** Foam, dry chemical, carbon dioxide
6.4 **Fire Extinguishing Agents Not to be Used:** Water fog
6.5 **Special Hazards of Combustion Products:** Releases irritating hydrogen chloride gas on combustion
6.6 **Behavior in Fire:** Not pertinent
6.7 **Ignition Temperature:** 737°F
6.8 **Electrical Hazard:** Not pertinent
6.9 **Burning Rate:** Data not available

7. CHEMICAL REACTIVITY

7.1 **Reactivity with Water:** No reaction
7.2 **Reactivity with Common Materials:** No reaction
7.3 **Stability During Transport:** Stable
7.4 **Neutralizing Agents for Acids and Caustics:** Not pertinent
7.5 **Polymerization:** Not pertinent
7.6 **Inhibitor of Polymerization:** Not pertinent

8. WATER POLLUTION

8.1 **Aquatic Toxicity:** 48 ppm/96 hr/guppy/TL_m/fresh water
8.2 **Waterfowl Toxicity:** Data not available
8.3 **Biological Oxygen Demand (BOD):** Data not available
8.4 **Food Chain Concentration Potential:** None noted

9. SELECTED MANUFACTURERS

1. Aldrich Chemical Co.
940 West St. Paul Ave.
Milwaukee, Wis. 53233
2. Dow Chemical Co.
Midland, Michigan 48640
3. Eastman Kodak Co.
Rochester, N. Y. 14650

10. SHIPPING INFORMATION

10.1 **Grades or Purity:** 97%
10.2 **Storage Temperature:** Ambient
10.3 **Inert Atmosphere:** No requirement
10.4 **Venting:** Pressure-vacuum

11. HAZARD ASSESSMENT CODE
(See Hazard Assessment Handbook, CG 446-3)

A-T-U-V-W

12. HAZARD CLASSIFICATIONS

12.1 **Code of Federal Regulations:** Flammable liquid
12.2 **NAS Hazard Rating for Bulk Water Transportation:**

Category	Rating
Fire	3
Health	
Vapor Irritant	3
Liquid or Solid Irritant	2
Poisons	3
Water Pollution	
Human Toxicity	2
Aquatic Toxicity	1
Aesthetic Effect	2
Reactivity	
Other Chemicals	2
Water	0
Self-Reaction	1

12.3 **NFPA Hazard Classifications:**

Category	Classification
Health Hazard (Blue)	3
Flammability (Red)	3
Reactivity (Yellow)	1

13. PHYSICAL AND CHEMICAL PROPERTIES

13.1 **Physical State at 15°C and 1 atm:** Liquid
13.2 **Molecular Weight:** 76.53
13.3 **Boiling Point at 1 atm:** 113°F = 45°C = 318°K
13.4 **Freezing Point:** −210.1°F = −134.5°C = 138.7°K
13.5 **Critical Temperature:** 466°F = 241°C = 514°K
13.6 **Critical Pressure:** 690 psia = 47 atm = 4.8 MN/m^2
13.7 **Specific Gravity:** 0.938 at 20°C (liquid)
13.8 **Liquid Surface Tension:** 28.9 dynes/cm = 0.0289 N/m at 15°C
13.9 **Liquid-Water Interfacial Tension:** Data not available
13.10 **Vapor (Gas) Specific Gravity:** 2.6
13.11 **Ratio of Specific Heats of Vapor (Gas):** 1.124
13.12 **Latent Heat of Vaporization:** Data not available
13.13 **Heat of Combustion:** −9749 Btu/lb = −5416 cal/g = -226.8×10^5 J/kg
13.15 **Heat of Solution:** Data not available
13.16 **Heat of Polymerization:** Not pertinent

NOTES

REVISED 1978

ACF	ALLYL CHLOROFORMATE

Common Synonyms			
Allyl chlorocarbonate	Watery liquid	Colorless	Extremely irritating odor
	Sinks in water. Flammable, irritating vapor is produced.		

Stop discharge if possible. Keep people away.
Shut off ignition sources. Call fire department.
Isolate and remove discharged material.
Notify local health and pollution control agencies.

Fire

FLAMMABLE
Flashback along vapor trail may occur.
Vapor may explode if ignited in an enclosed area.

Wear goggles, self-contained breathing apparatus and rubber overclothing (including gloves).
Extinguish with dry chemicals, foam or carbon dioxide.
Water may be ineffective on fire.
Cool exposed containers with water.

Exposure

Call for medical aid.

VAPOR
Irritating to eyes, nose and throat.
If inhaled will cause difficult breathing.
Move victim to fresh air.
If breathing is difficult, give oxygen.

LIQUID
Irritating to skin and eyes.
Harmful if swallowed.
Remove contaminated clothing and shoes.
Flush affected areas with plenty of water.
IF IN EYES, hold eyelids open and flush with plenty of water.
IF SWALLOWED and victim is CONSCIOUS, have victim drink water or milk.

Water Pollution

Effect of low concentrations on aquatic life is unknown.
May be dangerous if it enters water intakes.
Notify local health and wildlife officials.
Notify operators of nearby water intakes.

1. RESPONSE TO DISCHARGE

(See Response Methods Handbook, CG 446-4)

Issue warning—corrosive
Restrict access
Disperse and flush

2. LABEL

Red

3. CHEMICAL DESIGNATIONS

3.1 **Synonyms:** Allyl chlorocarbonate
3.2 **Coast Guard Compatibility Classification:** Not applicable
3.3 **Chemical Formula:** $CH_2{=}CH\cdot CH_2\cdot O\cdot COCl$
3.4 **IMCO/United Nations Numerical Designation:** 8/1722

4. OBSERVABLE CHARACTERISTICS

4.1 **Physical State (as shipped):** Liquid
4.2 **Color:** Colorless
4.3 **Odor:** Extremely irritating, causes tears; pungent

5. HEALTH HAZARDS

5.1 **Personal Protective Equipment:** Vapor-proof protective goggles and face shield; plastic or rubber gloves, shoes and clothing; gas mask or self-contained breathing apparatus.
5.2 **Symptoms Following Exposure:** Vapor irritates eyes and respiratory tract. Contact with liquid causes eye and skin irritation, and ingestion irritates mouth and stomach.
5.3 **Treatment for Exposure:** INHALATION: remove from exposure; support respiration if necessary; call physician. EYES: if irritated by either vapor or liquid, flush with water for at least 15 min. SKIN: wash with large amounts of water for at least 15 min. INGESTION: do NOT induce vomiting; give water; call physician.
5.4 **Toxicity by Inhalation (Threshold Limit Value):** Data not available
5.5 **Short-Term Inhalation Limits:** Data not available
5.6 **Toxicity by Ingestion:** Grade 3; LD_{50} 50 to 500 mg/kg
5.7 **Late Toxicity:** Data not available
5.8 **Vapor (Gas) Irritant Characteristics:** Vapors are moderately irritating such that personnel will not usually tolerate moderate or high vapor concentrations.
5.9 **Liquid or Solid Irritant Characteristics:** Fairly severe skin irritant. May cause pain and second-degree burns after a few minutes' contact.
5.10 **Odor Threshold:** 1.4 ppm

6. FIRE HAZARDS

6.1 **Flash Point:** 92°F O.C.; 88°F C.C.
6.2 **Flammable Limits in Air:** Data not available
6.3 **Fire Extinguishing Agents:** Dry chemical, foam, carbon dioxide
6.4 **Fire Extinguishing Agents Not to be Used:** Water may be ineffective
6.5 **Special Hazards of Combustion Products:** When heated to decomposition, emits highly toxic phosgene gas.
6.6 **Behavior in Fire:** Vapor is heavier than air and may travel considerable distance to a source of ignition and flash back.
6.7 **Ignition Temperature:** Data not available
6.8 **Electrical Hazard:** Data not available
6.9 **Burning Rate:** 4.9 mm/min.

7. CHEMICAL REACTIVITY

7.1 **Reactivity with Water:** Reacts slowly, generating hydrogen chloride.
7.2 **Reactivity with Common Materials:** Corrodes metals
7.3 **Stability During Transport:** Stable
7.4 **Neutralizing Agents for Acids and Caustics:** Flush with water, rinse with lime or sodium bicarbonate solution.
7.5 **Polymerization:** Not pertinent
7.6 **Inhibitor of Polymerization:** Not pertinent

8. WATER POLLUTION

8.1 **Aquatic Toxicity:** Data not available
8.2 **Waterfowl Toxicity:** Data not available
8.3 **Biological Oxygen Demand (BOD):** Data not available
8.4 **Food Chain Concentration Potential:** None

9. SELECTED MANUFACTURERS

1. Chemetron Corporation
   Organic Chemicals Division
   393 Seventh Avenue
   New York, N. Y. 10001
2. Ott Chemical Co.
   500 Agard Rd.
   Muskegon, Mich. 49445
3. Polysciences, Inc.
   Paul Valley Industrial Park
   Warrington, Pa. 18976

10. SHIPPING INFORMATION

10.1 **Grades or Purity:** Commercial, 97+%
10.2 **Storage Temperature:** Keep cool
10.3 **Inert Atmosphere:** No requirement
10.4 **Venting:** Pressure-vacuum

11. HAZARD ASSESSMENT CODE

(See Hazard Assessment Handbook, CG 446-3)

A-O-X-Y

12. HAZARD CLASSIFICATIONS

12.1 **Code of Federal Regulations:** Corrosive Flammable Liquid
12.2 **NAS Hazard Rating for Bulk Water Transportation:**

Category	Rating
Fire	3
Health	
Vapor Irritant	3
Liquid or Solid Irritant	3
Poisons	3
Water Pollution	
Human Toxicity	3
Aquatic Toxicity	3
Aesthetic Effect	3
Reactivity	
Other Chemicals	3
Water	0
Self-Reaction	1

12.3 **NFPA Hazard Classifications:**

Category	Classification
Health Hazard (Blue)	3
Flammability (Red)	3
Reactivity (Yellow)	1

13. PHYSICAL AND CHEMICAL PROPERTIES

13.1 **Physical State at 15°C and 1 atm:** Liquid
13.2 **Molecular Weight:** 120.5
13.3 **Boiling Point at 1 atm:** 113°F = 45°C = 318°K
13.4 **Freezing Point:** −112°F = −80°C = 193°K
13.5 **Critical Temperature:** Not pertinent
13.6 **Critical Pressure:** Not pertinent
13.7 **Specific Gravity:** 1.139 at 20°C (liquid)
13.8 **Liquid Surface Tension:** (est.) 25 dynes/cm = 0.025 N/m at 20°C
13.9 **Liquid-Water Interfacial Tension:** Not pertinent
13.10 **Vapor (Gas) Specific Gravity:** 4.15
13.11 **Ratio of Specific Heats of Vapor (Gas):** 1.0804
13.12 **Latent Heat of Vaporization:** (est.) 100 Btu/lb = 56 cal/g = 2.3×10^5 J/kg
13.13 **Heat of Combustion:** (est.) −7,800 Btu/lb = −4,300 cal/g = -180×10^5 J/kg
13.14 **Heat of Decomposition:** Not pertinent
13.15 **Heat of Solution:** Not pertinent
13.16 **Heat of Polymerization:** Not pertinent

NOTES

ATC	ALLYLTRICHLOROSILANE

Common Synonyms			
Allylsilicone trichloride	Liquid	Colorless	Sharp, irritating odor
	Reacts violently with water. Irritating visible vapor cloud is produced.		

Avoid contact with liquid. Keep people away.
Stop discharge if possible.
Call fire department.
Isolate and remove discharged material.
Notify local health and pollution control agencies.

Fire

Combustible.
POISONOUS GASES MAY BE PRODUCED IN FIRE.
Wear goggles, self-contained breathing apparatus and rubber overclothing (including gloves).
Extinguish with dry chemicals or carbon dioxide.
DO NOT USE WATER OR FOAM ON FIRE.

Exposure

Call for medical aid.

VAPOR
Irritating to eyes, nose and throat.
If inhaled will cause coughing or difficult breathing.
Move victim to fresh air.
If breathing is difficult, give oxygen.

LIQUID
Will burn skin and eyes.
Harmful if swallowed.
Remove contaminated clothing and shoes.
Flush affected areas with plenty of water.
IF IN EYES, hold eyelids open and flush with plenty of water.
IF SWALLOWED and victim is CONSCIOUS, have victim drink water or milk.
DO NOT INDUCE VOMITING.

Water Pollution

Effect of low concentrations on aquatic life is unknown.
May be dangerous if it enters water intakes.
Notify local health and wildlife officials.
Notify operators of nearby water intakes.

1. RESPONSE TO DISCHARGE

(See Response Methods Handbook, CG 446-4)

Issue warning—corrosive
Restrict access
Disperse and flush

2. LABEL

3. CHEMICAL DESIGNATIONS

3.1 **Synonyms:** Allylsilicone trichloride
3.2 **Coast Guard Compatibility Classification:** Not applicable
3.3 **Chemical Formula:** $CH_2{=}CH \cdot CH_2 \cdot SiCl_3$
3.4 **IMCO/United Nations Numerical Designation:** 8/1724

4. OBSERVABLE CHARACTERISTICS

4.1 **Physical State (as shipped):** Liquid
4.2 **Color:** Colorless
4.3 **Odor:** Sharp; pungent, irritating, like hydrochloric acid

5. HEALTH HAZARDS

5.1 **Personal Protective Equipment:** Acid-vapor-type respiratory protection; rubber gloves; chemical goggles; other equipment necessary to protect skin and eyes.
5.2 **Symptoms Following Exposure:** Inhalation of vapor irritates mucous membranes. Liquid causes severe burns of eyes and skin and severe internal burns if ingested.
5.3 **Treatment for Exposure:** Get medical attention after all exposures to this compound. INHALATION: remove from exposure; support respiration. EYES: flush with water 15 min. SKIN: flush with water. INGESTION: do NOT induce vomiting; give water.
5.4 **Toxicity by Inhalation (Threshold Limit Value):** Data not available
5.5 **Short-Term Inhalation Limits:** Data not available
5.6 **Toxicity by Ingestion:** Grade 3; LD_{50} 50 to 500 mg/kg
5.7 **Late Toxicity:** Data not available
5.8 **Vapor (Gas) Irritant Characteristics:** Vapors cause severe irritation of eyes and throat and can cause eye or lung injury. They cannot be tolerated even at low concentrations.
5.9 **Liquid or Solid Irritant Characteristics:** Severe skin irritant. Causes second- and third-degree burns on short contact and is very injurious to the eyes.
5.10 **Odor Threshold:** Data not available

6. FIRE HAZARDS

6.1 **Flash Point:** 100°F O.C.; 95°F C.C.
6.2 **Flammable Limits in Air:** Data not available
6.3 **Fire Extinguishing Agents:** Dry chemical, carbon dioxide
6.4 **Fire Extinguishing Agents Not to be Used:** Water
6.5 **Special Hazards of Combustion Products:** Irritating vapors of hydrogen chloride and phosgene may form.
6.6 **Behavior in Fire:** Difficult to extinguish. Re-ignition may occur.
6.7 **Ignition Temperature:** Data not available
6.8 **Electrical Hazard:** Data not available
6.9 **Burning Rate:** 2.2 mm/min.

7. CHEMICAL REACTIVITY

7.1 **Reactivity with Water:** Reacts vigorously, generating hydrogen chloride (hydrochloric acid).
7.2 **Reactivity with Common Materials:** Corrodes metal because of hydrochloric acid formed.
7.3 **Stability During Transport:** Stable
7.4 **Neutralizing Agents for Acids and Caustics:** Flush with water, rinse with sodium bicarbonate
7.5 **Polymerization:** Not pertinent
7.6 **Inhibitor of Polymerization:** Not pertinent

8. WATER POLLUTION

8.1 **Aquatic Toxicity:** Data not available
8.2 **Waterfowl Toxicity:** Data not available
8.3 **Biological Oxygen Demand (BOD):** Data not available
8.4 **Food Chain Concentration Potential:** None

9. SELECTED MANUFACTURERS

1. Dow Corning Corporation
   P. O. Box 592
   Midland, Mich. 48640
2. Polysciences, Inc.
   Paul Valley Industrial Park
   Warrington, Pa. 18976
3. Pfaltz and Bauer, Inc.
   126-04 Northern Boulevard
   Flushing, N. Y. 11368

10. SHIPPING INFORMATION

10.1 **Grades or Purity:** Commercial
10.2 **Storage Temperature:** Ambient
10.3 **Inert Atmosphere:** No requirement
10.4 **Venting:** Pressure-vacuum

11. HAZARD ASSESSMENT CODE

(See Hazard Assessment Handbook, CG 446-3)

A-O

12. HAZARD CLASSIFICATIONS

12.1 **Code of Federal Regulations:** Corrosive
12.2 **NAS Hazard Rating for Bulk Water Transportation:**

Category	Rating
Fire	3
Health	
Vapor Irritant	4
Liquid or Solid Irritant	4
Poisons	3
Water Pollution	
Human Toxicity	3
Aquatic Toxicity	3
Aesthetic Effect	2
Reactivity	
Other Chemicals	3
Water	4
Self-Reaction	4

12.3 **NFPA Hazard Classifications:** Not listed

13. PHYSICAL AND CHEMICAL PROPERTIES

13.1 **Physical State at 15°C and 1 atm:** Liquid
13.2 **Molecular Weight:** 175.5
13.3 **Boiling Point at 1 atm:** 241°F = 116°C = 389°K
13.4 **Freezing Point:** Not pertinent
13.5 **Critical Temperature:** Not pertinent
13.6 **Critical Pressure:** Not pertinent
13.7 **Specific Gravity:** 1.215 at 20°C (liquid)
13.8 **Liquid Surface Tension:** (est.) 20 dynes/cm = 0.020 N/m at 20°C
13.9 **Liquid-Water Interfacial Tension:** Not pertinent
13.10 **Vapor (Gas) Specific Gravity:** 6
13.11 **Ratio of Specific Heats of Vapor (Gas):** 1.0863
13.12 **Latent Heat of Vaporization:** 97 Btu/lb = 54 cal/g = 2.3×10^5 J/kg
13.13 **Heat of Combustion:** (est.) −5,200 Btu/lb = −2,900 cal/g = -120×10^5 J/kg
13.14 **Heat of Decomposition:** Not pertinent
13.15 **Heat of Solution:** Data not available
13.16 **Heat of Polymerization:** Not pertinent

NOTES

ACL	# ALUMINUM CHLORIDE

Common Synonyms			
Anhydrous aluminum chloride	Solid crystals or powder	Yellow-orange to grayish-white	Irritating odor
	Sinks in water. Poisonous gas is produced on contact with water.		

Keep people away. Evacuate area in case of large discharge.
Avoid contact with solid or dust.
Wear goggles, self-contained breathing apparatus, and rubber overclothing (including gloves).
Isolate and remove discharged material.
Notify local health and pollution control agencies.

Fire

Not flammable.
Wear goggles, self-contained breathing apparatus, and rubber overclothing (including gloves).
Do not use water on adjacent fires.
Extinguish adjacent fires with dry chemical or foam.

Exposure

CALL FOR MEDICAL AID.

DUST
Irritating to eyes, nose and throat.
Harmful if inhaled.
Move to fresh air.
If breathing has stopped, give artificial respiration.
If breathing is difficult, give oxygen.

SOLID
Will burn skin and eyes.
Harmful if swallowed.
Remove contaminated clothing and shoes.
Flush affected areas with plenty of water.
IF IN EYES, hold eyelids open and flush with plenty of water.
IF SWALLOWED and victim is CONSCIOUS, have victim drink water or milk.
DO NOT INDUCE VOMITING.

Water Pollution

HARMFUL TO AQUATIC LIFE IN VERY LOW CONCENTRATIONS.
May be dangerous if it enters water intakes.
Notify local health and wildlife officials.
Notify operators of nearby water intakes.

1. RESPONSE TO DISCHARGE
(See Response Methods Handbook, CG 446-4)

Disperse and flush with care
Issue warning—corrosive

2. LABELS

No hazard label required by Code of Federal Regulations

3. CHEMICAL DESIGNATIONS

3.1 **Synonyms:** Anhydrous aluminum chloride
3.2 **Coast Guard Compatibility Classification:** Not applicable
3.3 **Chemical Formula:** $AlCl_3$
3.4 **IMCO/United Nations Numerical Designation:** 8.0/1726

4. OBSERVABLE CHARACTERISTICS

4.1 **Physical State (as shipped):** Solid
4.2 **Color:** Orange to yellow through gray to white
4.3 **Odor:** Like hydrogen chloride; like hydrochloric acid

5. HEALTH HAZARDS

5.1 **Personal Protective Equipment:** All personnel in the area should wear safety clothing, including fully closed goggles, rubber or plastic-coated gloves, rubber shoes, and coveralls of acid-resistant material. An acid-vapor canister mask should be carried in case of emergency. In certain applications, it may be advisable to wear this equipment on a routine basis.
5.2 **Symptoms Following Exposure:** Contact with the skin or eyes in the presence of moisture causes thermal and acid burns.
5.3 **Treatment for Exposure:** INGESTION: if victim is conscious have him drink water or milk. Do NOT induce vomiting. SKIN: flush immediately with plenty of water. For eye contact, flush with water for at least 15 mins. and get medical attention immediately.
5.4 **Toxicity by Inhalation (Threshold Limit Value):** 5 ppm (hydrogen chloride)
5.5 **Short-Term Inhalation Limits:** 5 ppm for 5 min.; 30 ppm for 10 min.; 20 ppm for 20 min.; 10 ppm for 60 min. (all for hydrogen chloride.)
5.6 **Toxicity by Ingestion:** No systemic effects, but severe burns of mouth.
5.7 **Late Toxicity:** None recognized
5.8 **Vapor (Gas) Irritant Characteristics:** Vapor (of hydrogen chloride) is moderately irritating such that personnel will not usually tolerate moderate or high vapor concentrations.
5.9 **Liquid or Solid Irritant Characteristics:** Fairly severe skin irritant; may cause pain and second-degree burns after a few minutes' contact.
5.10 **Odor Threshold:** 1-5 ppm (hydrogen chloride)

6. FIRE HAZARDS

6.1 **Flash Point:** Not flammable
6.2 **Flammable Limits in Air:** Not flammable
6.3 **Fire Extinguishing Agents:** Not pertinent
6.4 **Fire Extinguishing Agents Not to be Used:** Do not use water on adjacent fires
6.5 **Special Hazards of Combustion Products:** Not pertinent
6.6 **Behavior in Fire:** Reacts violently with water used in extinguishing adjacent fires
6.7 **Ignition Temperature:** Not flammable
6.8 **Electrical Hazard:** Not pertinent
6.9 **Burning Rate:** Not flammable

7. CHEMICAL REACTIVITY

7.1 **Reactivity with Water:** Reacts violently with water, liberating hydrogen chloride gas and heat.
7.2 **Reactivity with Common Materials:** None if dry. If wet it attacks metals because of hydrochloric acid formed; flammable hydrogen is formed.
7.3 **Stability During Transport:** Stable if kept dry and protected from atmospheric moisture.
7.4 **Neutralizing Agents for Acids and Caustics:** Hydrochloric acid formed by reaction with water can be flushed away with water. Rinse with sodium bicarbonate or lime solution.
7.5 **Polymerization:** Not pertinent
7.6 **Inhibitor of Polymerization:** Not pertinent

8. WATER POLLUTION

8.1 **Aquatic Toxicity:** Not pertinent
8.2 **Waterfowl Toxicity:** Not pertinent
8.3 **Biological Oxygen Demand (BOD):** Not pertinent
8.4 **Food Chain Concentration Potential:** Not pertinent

9. SELECTED MANUFACTURERS

1. Allied Chemical Corp.
   Industrial Chemicals Division
   Morristown, N. J. 07960
2. Pearsall Corp.
   Pearsall Chemical Co. Div.
   4635 Southwest Freeway
   Houston, Texas 77027
3. Stauffer Chemical Co.
   Industrial Chemical Div.
   Baton Rouge, La. 70821

10. SHIPPING INFORMATION

10.1 **Grades or Purity:** Pure: 99.7%; technical: 98.5%
10.2 **Storage Temperature:** Data not available
10.3 **Inert Atmosphere:** Data not available
10.4 **Venting:** Data not available

11. HAZARD ASSESSMENT CODE
(See Hazard Assessment Handbook, CG 446-3)

RR-C

12. HAZARD CLASSIFICATIONS

12.1 **Code of Federal Regulations:** Not listed
12.2 **NAS Hazard Rating for Bulk Water Transportation:** Not listed
12.3 **NFPA Hazard Classifications:**

Category	Classification
Health Hazard (Blue)	3
Flammability (Red)	0
Reactivity (Yellow)	2 W (struck through)

13. PHYSICAL AND CHEMICAL PROPERTIES

13.1 **Physical State at 15°C and 1 atm:** Solid
13.2 **Molecular Weight:** 133.34
13.3 **Boiling Point at 1 atm:** Not pertinent
13.4 **Freezing Point:** Not pertinent
13.5 **Critical Temperature:** Not pertinent
13.6 **Critical Pressure:** Not pertinent
13.7 **Specific Gravity:** 2.44 at 25°C (solid)
13.8 **Liquid Surface Tension:** Not pertinent
13.9 **Liquid-Water Interfacial Tension:** Not pertinent
13.10 **Vapor (Gas) Specific Gravity:** Not pertinent
13.11 **Ratio of Specific Heats of Vapor (Gas):** Not pertinent
13.12 **Latent Heat of Vaporization:** Not pertinent
13.13 **Heat of Combustion:** Not pertinent
13.14 **Heat of Decomposition:** Not pertinent
13.15 **Heat of Solution:** Not pertinent
13.16 **Heat of Polymerization:** Not pertinent

NOTES

ALF	ALUMINUM FLUORIDE

Common Synonyms	Solid powder or granules White Odorless Sinks in water.

Avoid contact with dust.
Isolate and remove discharged material.
Notify local health and pollution control agencies.

Fire	Not flammable. POISONOUS GASES MAY BE PRODUCED WHEN HEATED. Wear goggles and self-contained breathing apparatus.
Exposure	DUST If inhaled, irritating to nose and throat. Move to fresh air.
Water Pollution	HARMFUL TO AQUATIC LIFE IN VERY LOW CONCENTRATIONS. May be dangerous if it enters water intakes. Notify local health and wildlife officials. Notify operators of nearby water intakes.

1. RESPONSE TO DISCHARGE

(See Response Methods Handbook, CG 446-4)

Disperse and flush

2. LABELS

No hazard label required by Code of Federal Regulations

3. CHEMICAL DESIGNATIONS

3.1 **Synonyms:** No common synonyms
3.2 **Coast Guard Compatibility Classification:** Not applicable
3.3 **Chemical Formula:** $AlF_3 \cdot 3H_2O$
3.4 **IMCO/United Nations Numerical Designation:** Not listed

4. OBSERVABLE CHARACTERISTICS

4.1 **Physical State (as shipped):** Solid
4.2 **Color:** White
4.3 **Odor:** None

5. HEALTH HAZARDS

5.1 **Personal Protective Equipment:** Goggles to protect against airborne particles and possibly respirator for intermittent heavy dust exposures.
5.2 **Symptoms Following Exposure:** ACUTE: respiratory irritation; possible nose bleeding or vomiting; CHRONIC: aggravates bronchitis/asthma; increased bone density.
5.3 **Treatment for Exposure:** For acute poisoning, oral administration of limewater, intravenous infusion of glucose, and intravenous injections of calcium gluconates.
5.4 **Toxicity by Inhalation (Threshold Limit Value):** Not pertinent
5.5 **Short-Term Inhalation Limits:** Not pertinent
5.6 **Toxicity by Ingestion:** LDL_O = 600 mg/kg (guinea pig)
5.7 **Late Toxicity:** Skeletal fluorosis (bone abnormalities) in humans, working in aluminum plant for 12 years.
5.8 **Vapor (Gas) Irritant Characteristics:** Not pertinent
5.9 **Liquid or Solid Irritant Characteristics:** No appreciable hazard. Practically harmless to the skin.
5.10 **Odor Threshold:** Not pertinent

6. FIRE HAZARDS

6.1 **Flash Point:** Not flammable
6.2 **Flammable Limits in Air:** Not flammable
6.3 **Fire Extinguishing Agents:** Not pertinent
6.4 **Fire Extinguishing Agents Not to be Used:** Not pertinent
6.5 **Special Hazards of Combustion Products:** When heated to sublimation condition, emits toxic fumes of fluoride
6.6 **Behavior in Fire:** Not pertinent
6.7 **Ignition Temperature:** Not flammable
6.8 **Electrical Hazard:** Not pertinent
6.9 **Burning Rate:** Not flammable

7. CHEMICAL REACTIVITY

7.1 **Reactivity with Water:** No reaction
7.2 **Reactivity with Common Materials:** No reaction
7.3 **Stability During Transport:** Stable
7.4 **Neutralizing Agents for Acids and Caustics:** Not pertinent
7.5 **Polymerization:** Not pertinent
7.6 **Inhibitor of Polymerization:** Not pertinent

8. WATER POLLUTION

8.1 **Aquatic Toxicity:** 60 ppm/*/fish/lethal/fresh water
*Time period not specified.
8.2 **Waterfowl Toxicity:** Data not available
8.3 **Biological Oxygen Demand (BOD):** Not pertinent
8.4 **Food Chain Concentration Potential:** None noted

9. SELECTED MANUFACTURERS

1. Allied Chemical Corp.
Industrial Chemicals Division
Morristown, N. J. 07960
2. Olin Corporation
Chemicals Division
120 Long Ridge Rd.
Stamford, Conn. 06904
3. Stauffer Chemical Co.
Industrial Chemicals Division
Bastrop, La. 71220

10. SHIPPING INFORMATION

10.1 **Grades or Purity:** 90.7%
10.2 **Storage Temperature:** Data not available
10.3 **Inert Atmosphere:** Data not available
10.4 **Venting:** Data not available

11. HAZARD ASSESSMENT CODE

(See Hazard Assessment Handbook, CG 446-3)

II

12. HAZARD CLASSIFICATIONS

12.1 **Code of Federal Regulations:** Not listed
12.2 **NAS Hazard Rating for Bulk Water Transportation:** Not listed
12.3 **NFPA Hazard Classifications:**

Category	Classification
Health Hazard (Blue)	1
Flammability (Red)	0
Reactivity (Yellow)	0

13. PHYSICAL AND CHEMICAL PROPERTIES

13.1 **Physical State at 15°C and 1 atm:** Solid
13.2 **Molecular Weight:** 83.98
13.3 **Boiling Point at 1 atm:** Not pertinent
13.4 **Freezing Point:** Not pertinent
13.5 **Critical Temperature:** Not pertinent
13.6 **Critical Pressure:** Not pertinent
13.7 **Specific Gravity:** 2.88 at 25°C (solid)
13.8 **Liquid Surface Tension:** Not pertinent
13.9 **Liquid-Water Interfacial Tension:** Not pertinent
13.10 **Vapor (Gas) Specific Gravity:** Not pertinent
13.11 **Ratio of Specific Heats of Vapor (Gas):** Not pertinent
13.12 **Latent Heat of Vaporization:** Not pertinent
13.13 **Heat of Combustion:** Not pertinent
13.14 **Heat of Decomposition:** Not pertinent
13.15 **Heat of Solution:** Not pertinent
13.16 **Heat of Polymerization:** Not pertinent

NOTES

REVISED 1978

ALN — ALUMINUM NITRATE

Common Synonyms			
Aluminum nitrate nonahydrate Nitric acid, aluminum salt	Solid	White	Odorless
	Sinks and mixes slowly with water.		

Stop discharge if possible. Keep people away.
Avoid contact with solid and dust.
Isolate and remove discharged material.
Notify local health and pollution control agencies.

Fire

Not flammable.
POISONOUS GASES MAY BE PRODUCED IN FIRE.
Wear goggles and self-contained breathing apparatus.

Exposure

CALL FOR MEDICAL AID.

DUST
Irritating to eyes, nose and throat.
Harmful if inhaled.
If in eyes, hold eyelids open and flush with plenty of water.
If breathing has stopped, give artificial respiration.
If breathing is difficult, give oxygen.

SOLID
Irritating to skin and eyes.
If swallowed will cause nausea or vomiting.
Remove contaminated clothing and shoes.
Flush affected areas with plenty of water.
IF IN EYES, hold eyelids open and flush with plenty of water.
IF SWALLOWED and victim is CONSCIOUS, have victim drink water or milk.
IF SWALLOWED and victim is UNCONSCIOUS OR HAVING CONVULSIONS, do nothing except keep victim warm.

Water Pollution

HARMFUL TO AQUATIC LIFE IN VERY LOW CONCENTRATIONS.
May be dangerous if it enters water intakes.
Notify local health and wildlife officials.
Notify operators of nearby water intakes.

1. RESPONSE TO DISCHARGE

(See Response Methods Handbook, CG 446-4)

Issue warning—water contaminant
Disperse and flush

2. LABEL

Yellow

3. CHEMICAL DESIGNATIONS

3.1 **Synonyms:** Aluminum nitrate nonahydrate; Nitric acid, aluminum salt
3.2 **Coast Guard Compatibility Classification:** Not listed
3.3 **Chemical Formula:** $Al(NO_3)_3 \cdot 9H_2O$
3.4 **IMCO/United Nations Numerical Designation:** 5.1/1438

4. OBSERVABLE CHARACTERISTICS

4.1 **Physical State (as shipped):** Solid
4.2 **Color:** White
4.3 **Odor:** None

5. HEALTH HAZARDS

5.1 **Personal Protective Equipment:** Goggles or face shield; dust respirator; rubber gloves
5.2 **Symptoms Following Exposure:** Ingestion of large doses causes gastric irritation, nausea, vomiting, and purging. Contact with dust irritates eyes and skin.
5.3 **Treatment for Exposure:** EYES: flush with water for at least 15 min. SKIN: flush with water; wash with soap and water.
5.4 **Toxicity by Inhalation (Threshold Limit Value):** Data not available
5.5 **Short-Term Inhalation Limits:** Data not available
5.6 **Toxicity by Ingestion:** Grade 3; oral rat LD_{50} = 264 mg/kg (nonahydrate)
5.7 **Late Toxicity:** Data not available
5.8 **Vapor (Gas) Irritant Characteristics:** Data not available
5.9 **Liquid or Solid Irritant Characteristics:** Data not available
5.10 **Odor Threshold:** Odorless

6. FIRE HAZARDS

6.1 **Flash Point:** Not flammable
6.2 **Flammable Limits in Air:** Not flammable
6.3 **Fire Extinguishing Agents:** Not pertinent
6.4 **Fire Extinguishing Agents Not to be Used:** Not pertinent
6.5 **Special Hazards of Combustion Products:** Toxic oxides of nitrogen may form in fire.
6.6 **Behavior in Fire:** May increase the intensity of fire when in contact with combustible material
6.7 **Ignition Temperature:** Not pertinent
6.8 **Electrical Hazard:** Not pertinent
6.9 **Burning Rate:** Not pertinent

7. CHEMICAL REACTIVITY

7.1 **Reactivity with Water:** Dissolves and forms a weak solution of nitric acid. The reaction is not hazardous.
7.2 **Reactivity with Common Materials:** May corrode metals in presence of moisture
7.3 **Stability During Transport:** Stable
7.4 **Neutralizing Agents for Acids and Caustics:** Flush with water.
7.5 **Polymerization:** Not pertinent
7.6 **Inhibitor of Polymerization:** Not pertinent

8. WATER POLLUTION

8.1 **Aquatic Toxicity:** 0.07 ppm/10 days/stickleback/killed/fresh water
8.2 **Waterfowl Toxicity:** Data not available
8.3 **Biological Oxygen Demand (BOD):** None
8.4 **Food Chain Concentration Potential:** None

9. SELECTED MANUFACTURERS

1. Mallinckrodt Chemical Works
223 Westside Ave.
P. O. Box 384
Jersey City, N. J. 07303
2. Allied Chemical Corp.
Specialty Chemicals Div.
P. O. Box 1087R
Morristown, N. J. 07960
3. J. T. Baker Chemical Co.
Phillipsburg, N. J. 08865

10. SHIPPING INFORMATION

10.1 **Grades or Purity:** Reagent, 99+%; Technical
10.2 **Storage Temperature:** Ambient
10.3 **Inert Atmosphere:** No requirement
10.4 **Venting:** Open

11. HAZARD ASSESSMENT CODE

(See Hazard Assessment Handbook, CG 446-3)

SS

12. HAZARD CLASSIFICATIONS

12.1 **Code of Federal Regulations:** Oxidizer
12.2 **NAS Hazard Rating for Bulk Water Transportation:** Not listed
12.3 **NFPA Hazard Classifications:** Not listed

13. PHYSICAL AND CHEMICAL PROPERTIES

13.1 **Physical State at 15°C and 1 atm:** Solid
13.2 **Molecular Weight:** 375.13
13.3 **Boiling Point at 1 atm:** Not pertinent (decomposes)
13.4 **Freezing Point:** 163°F = 73°C = 346°K
13.5 **Critical Temperature:** Not pertinent
13.6 **Critical Pressure:** Not pertinent
13.7 **Specific Gravity:** >1 at 20°C (solid)
13.8 **Liquid Surface Tension:** Not pertinent
13.9 **Liquid-Water Interfacial Tension:** Not pertinent
13.10 **Vapor (Gas) Specific Gravity:** Not pertinent
13.11 **Ratio of Specific Heats of Vapor (Gas):** Not pertinent
13.12 **Latent Heat of Vaporization:** Not pertinent
13.13 **Heat of Combustion:** Not pertinent
13.14 **Heat of Decomposition:** Not pertinent
13.15 **Heat of Solution:** Not pertinent
13.16 **Heat of Polymerization:** Not pertinent

NOTES

ALM

ALUMINUM SULFATE

Common Synonyms
Cake aluminum
Patent aluminum

Solid | Gray-white | Odorless

Sinks and mixes slowly with water.

AVOID CONTACT WITH LIQUID AND VAPOR. KEEP PEOPLE AWAY.
Wear goggles, self-contained breathing apparatus, and rubber overclothing (including gloves).
Shut off ignition sources. Call fire department.
Stop discharge if possible.
Isolate and remove discharged material.
Notify local health and pollution control agencies.

Fire

Not flammable.

Wear goggles, self-contained breathing apparatus and rubber overclothing (including gloves).
Extinguish with dry chemicals or carbon dioxide.
DO NOT USE WATER ON FIRE.

Exposure

CALL FOR MEDICAL AID.

DUST
Irritating to eyes, nose and throat.
If inhaled will cause difficult breathing.
If in eyes, hold eyelids open and flush with plenty of water.
If breathing has stopped, give artificial respiration.
If breathing is difficult, give oxygen.

SOLID
Irritating to skin and eyes.
If swallowed will cause nausea or vomiting.
Remove contaminated clothing and shoes.
Flush affected areas with plenty of water.
IF IN EYES, hold eyelids open and flush with plenty of water.
IF SWALLOWED and victim is CONSCIOUS, have victim drink water or milk.
IF SWALLOWED and victim is UNCONSCIOUS OR HAVING CONVULSIONS, do nothing except keep victim warm.

HARMFUL TO AQUATIC LIFE IN VERY LOW CONCENTRATIONS.
May be dangerous if it enters water intakes.
Notify local health and wildlife officials.
Notify operators of nearby water intakes.

1. RESPONSE TO DISCHARGE

(See Response Methods Handbook, CG 446-4)

Issue warning—water contaminant
Should be removed
Chemical and physical treatment

2. LABELS

No hazard label required by Code of Federal Regulations

3. CHEMICAL DESIGNATIONS

3.1 **Synonyms:** Cake alum
Patent alum
3.2 **Coast Guard Compatibility Classification:** Not listed
3.3 **Chemical Formula:** $Al_2(SO_4)_3 \cdot 18H_2O$
3.4 **IMCO/United Nations Numerical Designation:** Not listed

4. OBSERVABLE CHARACTERISTICS

4.1 **Physical State (as shipped):** Solid
4.2 **Color:** Gray-white
4.3 **Odor:** None

5. HEALTH HAZARDS

5.1 **Personal Protective Equipment:** Dust respirator; goggles or face shield; rubber gloves
5.2 **Symptoms Following Exposure:** Inhalation of dust irritates nose and mouth. Ingestion of large doses causes gastric irritation, nausea, vomiting, and purging. Dust irritates eyes and skin.
5.3 **Treatment for Exposure:** INHALATION: rinse nose and mouth with water. INGESTION: give large amounts of water. EYES: flush with water for at least 15 min. SKIN: flush with water, wash with soap and water.
5.4 **Toxicity by Inhalation (Threshold Limit Value):** Data not available
5.5 **Short-Term Inhalation Limits:** Data not available
5.6 **Toxicity by Ingestion:** Grade 2; oral mouse LD_{50} = 770 mg/kg
5.7 **Late Toxicity:** Data not available
5.8 **Vapor (Gas) Irritant Characteristics:** Data not available
5.9 **Liquid or Solid Irritant Characteristics:** Data not available
5.10 **Odor Threshold:** Data not available

6. FIRE HAZARDS

6.1 **Flash Point:** Not flammable
6.2 **Flammable Limits in Air:** Not flammable
6.3 **Fire Extinguishing Agents:** Not pertinent
6.4 **Fire Extinguishing Agents Not to be Used:** Not pertinent
6.5 **Special Hazards of Combustion Products:** Not pertinent
6.6 **Behavior in Fire:**
6.7 **Ignition Temperature:** Not pertinent
6.8 **Electrical Hazard:** Not pertinent
6.9 **Burning Rate:** Not pertinent

7. CHEMICAL REACTIVITY

7.1 **Reactivity with Water:** No reaction
7.2 **Reactivity with Common Materials:** May corrode metals in presence of moisture
7.3 **Stability During Transport:** Stable
7.4 **Neutralizing Agents for Acids and Caustics:** Flush with water.
7.5 **Polymerization:** Not pertinent
7.6 **Inhibitor of Polymerization:** Not pertinent

8. WATER POLLUTION

8.1 **Aquatic Toxicity:**
14ppm/36 hr/fundulus/fatal/fresh water
240ppm/48 hr/mosquitofish/TL_m/*
*Water type not specified.
8.2 **Waterfowl Toxicity:** Data not available
8.3 **Biological Oxygen Demand (BOD):** None
8.4 **Food Chain Concentration Potential:** None

9. SELECTED MANUFACTURERS

1. Allied Chemical Corp.
Industrial Chemicals Div.
Morristown, N. J. 07960
2. American Cyanamid Co.
Industrial Chemicals and Plastics Div.
Wayne, N. J. 07470
3. W. R. Grace & Co.
Davison Chemicals Div.
Baltimore, Md. 21203

10. SHIPPING INFORMATION

10.1 **Grades or Purity:** Technical
10.2 **Storage Temperature:** Ambient
10.3 **Inert Atmosphere:** No requirement
10.4 **Venting:** Open

11. HAZARD ASSESSMENT CODE

(See Hazard Assessment Handbook, CG 446-3)

SS

12. HAZARD CLASSIFICATIONS

12.1 **Code of Federal Regulations:** Not listed
12.2 **NAS Hazard Rating for Bulk Water Transportation:** Not listed
12.3 **NFPA Hazard Classifications:** Not listed

13. PHYSICAL AND CHEMICAL PROPERTIES

13.1 **Physical State at 15°C and 1 atm:** Solid
13.2 **Molecular Weight:** 666.4
13.3 **Boiling Point at 1 atm:** Not pertinent
13.4 **Freezing Point:** Not pertinent
13.5 **Critical Temperature:** Not pertinent
13.6 **Critical Pressure:** Not pertinent
13.7 **Specific Gravity:** 1.7 at 20°C (solid)
13.8 **Liquid Surface Tension:** Not pertinent
13.9 **Liquid-Water Interfacial Tension:** Not pertinent
13.10 **Vapor (Gas) Specific Gravity:** Not pertinent
13.11 **Ratio of Specific Heats of Vapor (Gas):** Not pertinent
13.12 **Latent Heat of Vaporization:** Not pertinent
13.13 **Heat of Combustion:** Not pertinent
13.14 **Heat of Decomposition:** Not pertinent
13.15 **Heat of Solution:** −22.1 Btu/lb = −12.3 cal/g = -0.515×10^5 J/kg
13.16 **Heat of Polymerization:** Not pertinent

NOTES

AEA — AMINOETHYLETHANOLAMINE

Common Synonyms			
N-(2-Aminoethyl) ethanolamine 2-[(2-Aminoethyl) amino] ethanol N-β-Hydroxyethylethylenediamine N-Hydroxethyl-1, 2-ethanediamine	Liquid	Colorless	Mild ammonia odor
	Sinks and mixes with water.		

Avoid contact with liquid. Keep people away.
Wear rubber overclothing.
Stop discharge if possible.
Call fire department.
Isolate and remove discharged material.
Notify local health and pollution control agencies.

Fire

Combustible.
Extinguish with dry chemical, alcohol foam, or carbon dioxide.
Water may be ineffective on fire.

Exposure

CALL FOR MEDICAL AID.

LIQUID
Will burn skin and eyes.
Remove contaminated clothing and shoes.
Flush affected areas with plenty of water.
IF IN EYES, hold eyelids open and flush with plenty of water.
IF SWALLOWED and victim is CONSCIOUS, have victim drink water or milk.
DO NOT INDUCE VOMITING.

Water Pollution

Effect of low concentrations on aquatic life is unknown.
May be dangerous if it enters water intakes.
Notify local health and pollution control officials.
Notify operators of nearby water intakes.

1. RESPONSE TO DISCHARGE
(See Response Methods Handbook, CG 446-4)

Disperse and flush

2. LABELS

No hazard label required by Code of Federal Regulations

3. CHEMICAL DESIGNATIONS

3.1 **Synonyms:** 2-[(2-Aminoethyl) amino] ethanol; N-(2-Aminoethyl) ethanolamine; N-beta-Hydroxyethylethylenediamine; N-Hydroxyethyl-1, 2-ethanediamine

3.2 **Coast Guard Compatibility Classification:** Alkanolamine

3.3 **Chemical Formula:** $HOCH_2CH_2NHCH_2CH_2NH_2$

3.4 **IMCO/United Nations Numerical Designation:** Not listed

4. OBSERVABLE CHARACTERISTICS

4.1 **Physical State (as shipped):** Liquid

4.2 **Color:** Colorless

4.3 **Odor:** Mild ammonia-like

5. HEALTH HAZARDS

5.1 **Personal Protective Equipment:** Goggles or face shield, protective clothing

5.2 **Symptoms Following Exposure:** Skin contact will cause mild irritation; eye contact will cause more severe irritation.

5.3 **Treatment for Exposure:** INGESTION: do NOT induce vomiting; call physician immediately. SKIN: wash area with plenty of water. EYES: flush thoroughly with running water, preferably for 15 min.

5.4 **Toxicity by Inhalation (Threshold Limit Value):** Data not available

5.5 **Short-Term Inhalation Limits:** Data not available

5.6 **Toxicity by Ingestion:** Grade 2; LD_{50} 0.5 to 5 g/kg

5.7 **Late Toxicity:** Data not available

5.8 **Vapor (Gas) Irritant Characteristics:** High concentrations of vapor may cause a slight smarting of the eyes or respiratory system. The effect is temporary.

5.9 **Liquid or Solid Irritant Characteristics:** Fairly severe skin irritant; may cause pain and second-degree burns after a few minutes' contact.

5.10 **Odor Threshold:** Data not available

6. FIRE HAZARDS

6.1 **Flash Point:** 265°F O.C.

6.2 **Flammable Limits in Air:** 1%–8% (calc.)

6.3 **Fire Extinguishing Agents:** Alcohol foam, dry chemical or carbon dioxide

6.4 **Fire Extinguishing Agents Not to be Used:** Water or foam may cause frothing

6.5 **Special Hazards of Combustion Products:** Not pertinent

6.6 **Behavior in Fire:** Not pertinent

6.7 **Ignition Temperature:** 695°F

6.8 **Electrical Hazard:** Not pertinent

6.9 **Burning Rate:** Data not available

7. CHEMICAL REACTIVITY

7.1 **Reactivity with Water:** No reaction

7.2 **Reactivity with Common Materials:** No reaction

7.3 **Stability During Transport:** Stable

7.4 **Neutralizing Agents for Acids and Caustics:** Dilute with water

7.5 **Polymerization:** Not pertinent

7.6 **Inhibitor of Polymerization:** Not pertinent

8. WATER POLLUTION

8.1 **Aquatic Toxicity:** Data not available

8.2 **Waterfowl Toxicity:** Data not available

8.3 **Biological Oxygen Demand (BOD):** Data not available

8.4 **Food Chain Concentration Potential:** Data not available

9. SELECTED MANUFACTURERS

1. Dow Chemical Co.
   Midland, Mich. 48640
2. Jefferson Chemical Co.
   3336 Richmond Ave.
   Houston, Tex. 77052
3. Union Carbide Corp.
   Chemicals and Plastics Division
   270 Park Ave.
   New York, N. Y. 10017

10. SHIPPING INFORMATION

10.1 **Grades or Purity:** Data not available

10.2 **Storage Temperature:** Ambient

10.3 **Inert Atmosphere:** No requirement

10.4 **Venting:** Open

11. HAZARD ASSESSMENT CODE
(See Hazard Assessment Handbook, CG 446-3)

A-P-Q

12. HAZARD CLASSIFICATIONS

12.1 **Code of Federal Regulations:** Not listed

12.2 **NAS Hazard Rating for Bulk Water Transportation:**

Category	Rating
Fire	1
Health	
Vapor Irritant	1
Liquid or Solid Irritant	3
Poisons	1
Water Pollution	
Human Toxicity	1
Aquatic Toxicity	1
Aesthetic Effect	3
Reactivity	
Other Chemicals	3
Water	0
Self-Reaction	0

12.3 **NFPA Hazard Classifications:** Not listed

13. PHYSICAL AND CHEMICAL PROPERTIES

13.1 **Physical State at 15°C and 1 atm:** Liquid

13.2 **Molecular Weight:** 104.15

13.3 **Boiling Point at 1 atm:** 469°F = 243°C = 516°K

13.4 **Freezing Point:** Data not available

13.5 **Critical Temperature:** Not pertinent

13.6 **Critical Pressure:** Not pertinent

13.7 **Specific Gravity:** 1.028 at 25°C (liquid)

13.8 **Liquid Surface Tension:** Not pertinent

13.9 **Liquid-Water Interfacial Tension:** Not pertinent

13.10 **Vapor (Gas) Specific Gravity:** Not pertinent

13.11 **Ratio of Specific Heats of Vapor (Gas):** (est.) 1.053

13.12 **Latent Heat of Vaporization:** (est.) 209 Btu/lb = 116 cal/g = 4.85×10^5 J/kg

13.13 **Heat of Combustion:** (est.) −12,300 Btu/lb = −6860 cal/g = -287×10^5 J/kg

13.14 **Heat of Decomposition:** Not pertinent

13.15 **Heat of Solution:** (est.) −4 Btu/lb = −2 cal/g = -0.1×10^5 J/kg

13.16 **Heat of Polymerization:** Not pertinent

NOTES

AMMONIA, ANHYDROUS

Common Synonyms			
Liquid Ammonia	Liquefied compressed gas	Colorless	Ammonia odor
	Floats and boils on water. Poisonous, visible vapor cloud is produced.		

Avoid contact with liquid and vapor. Keep people away.
Wear goggles, self-contained breathing apparatus, and rubber overclothing (including gloves).
Stop discharge if possible.
Stay upwind and use water spray to "knock down" vapor.
Call fire department.
Isolate and remove discharged material.
Notify local health and pollution control agencies.

Fire

Combustible.
Wear goggles, self-contained breathing apparatus, and rubber overclothing (including gloves).
Stop flow of gas or liquid if possible.
Cool exposed containers and protect men effecting shutoff with water.
Let fire burn.

Exposure

CALL FOR MEDICAL AID.

VAPOR
POISONOUS IF INHALED.
Irritating to eyes, nose and throat.
Move to fresh air.
If in eyes, hold eyelids open and flush with plenty of water.
If breathing has stopped, give artificial respiration.
If breathing is difficult, give oxygen.

LIQUID
Will burn skin and eyes.
Harmful if swallowed.
Will cause frostbite.
Remove contaminated clothing and shoes.
Flush affected areas with plenty of water.
DO NOT RUB AFFECTED AREAS.
IF IN EYES, hold eyelids open and flush with plenty of water.
IF SWALLOWED and victim is CONSCIOUS, have victim drink water or milk.

Water Pollution

HARMFUL TO AQUATIC LIFE IN VERY LOW CONCENTRATIONS.
May be dangerous if it enters water intakes.
Notify local health and wildlife officials.
Notify operators of nearby water intakes.

1. RESPONSE TO DISCHARGE

(See Response Methods Handbook, CG 446-4)

Issue warning—poison
Restrict access
Evacuate area
Disperse and flush

2. LABEL

Green

3. CHEMICAL DESIGNATIONS

3.1 **Synonyms:** Liquid ammonia
3.2 **Coast Guard Compatibility Classification:** Ammonia
3.3 **Chemical Formula:** NH_3
3.4 **IMCO/United Nations Numerical Designation:** 2.0/1005

4. OBSERVABLE CHARACTERISTICS

4.1 **Physical State (as shipped):** Compressed liquefied gas
4.2 **Color:** Colorless
4.3 **Odor:** Pungent; extremely pungent

5. HEALTH HAZARDS

5.1 **Personal Protective Equipment:** Gas-tight chemical goggles, self-contained breathing apparatus, rubber boots, rubber gloves, emergency shower and eye bath.
5.2 **Symptoms Following Exposure:** 700 ppm causes eye irritation, and permanent injury may result if prompt remedial measures are not taken; 5000 ppm can cause immediate death from spasm, inflammation, or edema of the larynx. Contact of the liquid with skin freezes the tissue and then produces a caustic burn.
5.3 **Treatment for Exposure:** INHALATION: move victim to fresh air and give artificial repiration if necessary. Oxygen may be useful. Observe for laryngeal spasm and perform tracheostomy if indicated. SKIN OR EYES: flood immediately with running water for 15 min. Treat subsequently as thermal burn.
5.4 **Toxicity by Inhalation (Threshold Limit Value):** 25 ppm
5.5 **Short-Term Inhalation Limits:** 50 ppm for 5 min.
5.6 **Toxicity by Ingestion:** Not pertinent
5.7 **Late Toxicity:** Not pertinent
5.8 **Vapor (Gas) Irritant Characteristics:** Vapors cause severe eye or throat irritation and may cause eye or lung injury; vapors cannot be tolerated even at low concentrations.
5.9 **Liquid or Solid Irritant Characteristics:** Causes smarting of the skin and first-degree burns on short exposure; may cause secondary burns on long exposure.
5.10 **Odor Threshold:** 46.8 ppm

6. FIRE HAZARDS

6.1 **Flash Point:** Not flammable under conditions likely to be encountered
6.2 **Flammable Limits in Air:** 15.50%–27.00%
6.3 **Fire Extinguishing Agents:** Stop flow of gas or liquid. Let fire burn.
6.4 **Fire Extinguishing Agents Not to be Used:** None
6.5 **Special Hazards of Combustion Products:** Not pertinent
6.6 **Behavior in Fire:** Not pertinent
6.7 **Ignition Temperature:** 1204°F
6.8 **Electrical Hazard:** Class I, Group D
6.9 **Burning Rate:** 1 mm/min.

7. CHEMICAL REACTIVITY

7.1 **Reactivity with Water:** Dissolves with mild heat effect
7.2 **Reactivity with Common Materials:** Corrosive to copper and galvanized surfaces.
7.3 **Stability During Transport:** Stable
7.4 **Neutralizing Agents for Acids and Caustics:** Dilute with water
7.5 **Polymerization:** Not pertinent
7.6 **Inhibitor of Polymerization:** Not pertinent

8. WATER POLLUTION

8.1 **Aquatic Toxicity:**
2.0 - 2.5 ppm/1-4 days/goldfish and yellow perch/LC
60 - 80 ppm/3 days/ crayfish/LC_{100}
8.2 ppm/96 hr/fathead minnow/TL_m
*Time period not specified.
8.2 **Waterfowl Toxicity:** 120 ppm
8.3 **Biological Oxygen Demand (BOD):** Not pertinent
8.4 **Food Chain Concentration Potential:** None

9. SELECTED MANUFACTURERS

1. American Oil Co.
910 South Michigan Ave.
Chicago, Illinois 60605
2. CF Industries Inc.
100 S. Wacker Drive
Chicago, Illinois 60606
3. The Williams Companies
Agrico Chemicals Co., Inc.
5050 Poplar Ave.
Memphis, Tenn. 38101

10. SHIPPING INFORMATION

10.1 **Grades or Purity:** Commercial, industrial, refrigeration, electronic, and metallurgical grades all have purity greater than 99.5%
10.2 **Storage Temperature:** Ambient for pressurized ammonia; low temperature for ammonia at atmospheric pressure
10.3 **Inert Atmosphere:** No requirement
10.4 **Venting:** Safety relief 250 psi for ammonia under pressure. Pressure-vacuum for ammonia at atmospheric pressure.

11. HAZARD ASSESSMENT CODE

(See Hazard Assessment Handbook, CG 446-3)

A-B-C-K-L-M-N-O

12. HAZARD CLASSIFICATIONS

12.1 **Code of Federal Regulations:** Nonflammable compressed gas
12.2 **NAS Hazard Rating for Bulk Water Transportation:**

Category	Rating
Fire	1
Health	
Vapor Irritant	4
Liquid or Solid Irritant	2
Poisons	2
Water Pollution	
Human Toxicity	2
Aquatic Toxicity	2
Aesthetic Effect	2
Reactivity	
Other Chemicals	3
Water	2
Self-Reaction	0

12.3 **NFPA Hazard Classifications:**

Category	Classification gas	liquefied
Health Hazard (Blue)	2	3
Flammability (Red)	1	1
Reactivity (Yellow)	0	0

13. PHYSICAL AND CHEMICAL PROPERTIES

13.1 **Physical State at 15°C and 1 atm:** Gas
13.2 **Molecular Weight:** 17.03
13.3 **Boiling Point at 1 atm:** $-28.1°F = -33.4°C = 239.8°K$
13.4 **Freezing Point:** $-108°F = -77.7°C = 265.5°K$
13.5 **Critical Temperature:** $271°F = 133°C = 406°K$
13.6 **Critical Pressure:** $1636 \text{ psia} = 111.3 \text{ atm} = 11.27 \text{ MN/m}^2$
13.7 **Specific Gravity:** 0.682 at −33.4°C (liquid)
13.8 **Liquid Surface Tension:** Not pertinent
13.9 **Liquid-Water Interfacial Tension:** Not pertinent
13.10 **Vapor (Gas) Specific Gravity:** 0.6
13.11 **Ratio of Specific Heats of Vapor (Gas):** 1.3 at 20°C
13.12 **Latent Heat of Vaporization:** $589 \text{ Btu/lb} = 327 \text{ cal/g} = 13.7 \times 10^5 \text{ J/kg}$
13.13 **Heat of Combustion:** $-7992 \text{ Btu/lb} = -4440 \text{ cal/g} = -185.9 \times 10^5 \text{ J/kg}$
13.14 **Heat of Decomposition:** Not pertinent
13.15 **Heat of Solution:** $-232 \text{ Btu/lb} = -129 \text{ cal/g} = -5.40 \times 10^5 \text{ J/kg}$
13.16 **Heat of Polymerization:** Not pertinent

NOTES

REVISED 1978

AAT

AMMONIUM ACETATE

Common Synonyms			
Acetic acid, ammonium salt	Solid	White	Weak ammonia odor
	Sinks and mixes with water.		

Stop discharge if possible. Keep people away.
Avoid contact with solid and dust.
Isolate and remove discharged material.
Notify local health and pollution control agencies.

Fire	Not flammable. Irritating gases may be produced when heated. Wear goggles and self-contained breathing apparatus.
Exposure	CALL FOR MEDICAL AID. DUST Irritating to eyes, nose and throat. If inhaled will cause difficult breathing. If in eyes, hold eyelids open and flush with plenty of water. If breathing has stopped, give artificial respiration. If breathing is difficult, give oxygen. SOLID Irritating to skin and eyes. If swallowed will cause nausea. Remove contaminated clothing and shoes. Flush affected areas with plenty of water. IF IN EYES, hold eyelids open and flush with plenty of water. IF SWALLOWED and victim is CONSCIOUS, have victim drink water or milk. IF SWALLOWED and victim is UNCONSCIOUS OR HAVING CONVULSIONS, do nothing except keep victim warm.
Water Pollution	Dangerous to aquatic life in high concentrations. May be dangerous if it enters water intakes. Notify local health and wildlife officials. Notify operators of nearby water intakes.

1. RESPONSE TO DISCHARGE

(See Response Methods Handbook, CG 446-4)

Disperse and flush

2. LABELS

No hazard label required by Code of Federal Regulations

3. CHEMICAL DESIGNATIONS

3.1 **Synonyms:** Acetic acid, ammonium salt

3.2 **Coast Guard Compatibility Classification:** Not listed

3.3 **Chemical Formula:** $NH_4C_2H_3O_2$

3.4 **IMCO/United Nations Numerical Designation:** Not listed

4. OBSERVABLE CHARACTERISTICS

4.1 **Physical State (as shipped):** Solid

4.2 **Color:** White

4.3 **Odor:** Weak ammonia

5. HEALTH HAZARDS

5.1 **Personal Protective Equipment:** Dust mask; goggles or face shield; rubber gloves

5.2 **Symptoms Following Exposure:** Inhalation of dust irritates nose and mouth. Ingestion irritates mouth and stomach. Contact with dust causes irritation of eyes and mild irritation of skin.

5.3 **Treatment for Exposure:** INHALATION: remove victim to fresh air; rinse nose and mouth with water. INGESTION: give large amounts of water. EYES: flush with water for at least 15 min. SKIN: flush with water.

5.4 **Toxicity by Inhalation (Threshold Limit Value):** Data not available

5.5 **Short-Term Inhalation Limits:** Data not available

5.6 **Toxicity by Ingestion:** Data not available

5.7 **Late Toxicity:** Data not available

5.8 **Vapor (Gas) Irritant Characteristics:** Data not available

5.9 **Liquid or Solid Irritant Characteristics:** Data not available

5.10 **Odor Threshold:** Data not available

6. FIRE HAZARDS

6.1 **Flash Point:** Not flammable

6.2 **Flammable Limits in Air:** Not flammable

6.3 **Fire Extinguishing Agents:** Not pertinent

6.4 **Fire Extinguishing Agents Not to be Used:** Not pertinent

6.5 **Special Hazards of Combustion Products:** Irritating vapors of ammonia and acetic acid may form in fires.

6.6 **Behavior in Fire:**

6.7 **Ignition Temperature:** Not pertinent

6.8 **Electrical Hazard:** Not pertinent

6.9 **Burning Rate:** Not pertinent

7. CHEMICAL REACTIVITY

7.1 **Reactivity with Water:** No reaction

7.2 **Reactivity with Common Materials:**

7.3 **Stability During Transport:** Stable

7.4 **Neutralizing Agents for Acids and Caustics:** Not pertinent

7.5 **Polymerization:** Not pertinent

7.6 **Inhibitor of Polymerization:** Not pertinent

8. WATER POLLUTION

8.1 **Aquatic Toxicity:** 238 ppm/24 hr/mosquito fish/TL_m/fresh water

8.2 **Waterfowl Toxicity:** Data not available

8.3 **Biological Oxygen Demand (BOD):** 79%, 1-5 days

8.4 **Food Chain Concentration Potential:** None

9. SELECTED MANUFACTURERS

1. Allied Chemical Corp.
   Specialty Chemicals Div.
   P. O. Box 1087R
   Morristown, N. J. 07960
2. Mallinckrodt Chemical Works
   223 Westside Ave.
   P. O. Box 384
   Jersey City, N. J. 07303
3. J. T. Baker Chemical Co.
   Phillipsburg, N. J. 08865

10. SHIPPING INFORMATION

10.1 **Grades or Purity:** Reagent; CP; Technical, 97+%

10.2 **Storage Temperature:** Ambient

10.3 **Inert Atmosphere:** No requirement

10.4 **Venting:** Open

11. HAZARD ASSESSMENT CODE

(See Hazard Assessment Handbook, CG 446-3)

SS

12. HAZARD CLASSIFICATIONS

12.1 **Code of Federal Regulations:** Not listed

12.2 **NAS Hazard Rating for Bulk Water Transportation:** Not listed

12.3 **NFPA Hazard Classifications:** Not listed

13. PHYSICAL AND CHEMICAL PROPERTIES

13.1 **Physical State at 15°C and 1 atm:** Solid

13.2 **Molecular Weight:** 77.08

13.3 **Boiling Point at 1 atm:** Not pertinent (decomposes)

13.4 **Freezing Point:** Not pertinent

13.5 **Critical Temperature:** Not pertinent

13.6 **Critical Pressure:** Not pertinent

13.7 **Specific Gravity:** 1.17 at 20°C (solid)

13.8 **Liquid Surface Tension:** Not pertinent

13.9 **Liquid-Water Interfacial Tension:** Not pertinent

13.10 **Vapor (Gas) Specific Gravity:** Not pertinent

13.11 **Ratio of Specific Heats of Vapor (Gas):** Not pertinent

13.12 **Latent Heat of Vaporization:** Not pertinent

13.13 **Heat of Combustion:** Not pertinent

13.14 **Heat of Decomposition:** Not pertinent

13.15 **Heat of Solution:** -5.8 Btu/lb $= -3.2$ cal/g $= -0.13 \times 10^5$ J/kg

13.16 **Heat of Polymerization:** Not pertinent

NOTES

ABZ AMMONIUM BENZOATE

Common Synonyms			
Benzoic acid, ammonium salt	Solid	White	Odorless
	Sinks and mixes slowly with water.		

Stop discharge if possible. Keep people away.
Avoid contact with solid and dust.
Isolate and remove discharged material.
Notify local health and pollution control agencies.

Fire

Combustible.
Dust cloud may explode if ignited in an enclosed area.
POISONOUS GASES MAY BE PRODUCED IN FIRE.
Irritating gases may be produced when heated.
Wear goggles and self-contained breathing apparatus.
Extinguish with water, dry chemicals, foam, or carbon dioxide.

Exposure

CALL FOR MEDICAL AID.

DUST
Irritating to eyes, nose and throat.
Harmful if inhaled.
If in eyes, hold eyelids open and flush with plenty of water.
If breathing has stopped, give artificial respiration.
If breathing is difficult, give oxygen.

SOLID
Irritating to skin and eyes.
Remove contaminated clothing and shoes.
Flush affected areas with plenty of water.
IF IN EYES, hold eyelids open and flush with plenty of water.
IF SWALLOWED and victim is CONSCIOUS, have victim drink water or milk.
IF SWALLOWED and victim is UNCONSCIOUS OR HAVING CONVULSIONS, do nothing except keep victim warm.

Water Pollution

Effect of low concentrations on aquatic life is unknown.
May be dangerous if it enters water intakes.
Notify local health and wildlife officials.
Notify operators of nearby water intakes.

1. RESPONSE TO DISCHARGE
(See Response Methods Handbook, CG 446-4)

Disperse and flush

2. LABELS

No hazard label required by Code of Federal Regulations

3. CHEMICAL DESIGNATIONS

3.1 **Synonyms:** Benzoic acid, ammonium salt
3.2 **Coast Guard Compatibility Classification:** Not listed
3.3 **Chemical Formula:** $C_6H_5COONH_4$
3.4 **IMCO/United Nations Numerical Designation:** Not listed

4. OBSERVABLE CHARACTERISTICS

4.1 **Physical State (as shipped):** Solid
4.2 **Color:** White
4.3 **Odor:** None

5. HEALTH HAZARDS

5.1 **Personal Protective Equipment:** Safety glasses, gloves
5.2 **Symptoms Following Exposure:** Inhalation of dust may irritate nose. Contact with eyes causes irritation.
5.3 **Treatment for Exposure:** INHALATION: move to uncontaminated atmosphere. EYES: flush with water for at least 15 min. SKIN: flush with water.
5.4 **Toxicity by Inhalation (Threshold Limit Value):** Data not available
5.5 **Short-Term Inhalation Limits:** Data not available
5.6 **Toxicity by Ingestion:** Data not available
5.7 **Late Toxicity:** Data not available
5.8 **Vapor (Gas) Irritant Characteristics:** Data not available
5.9 **Liquid or Solid Irritant Characteristics:** Data not available
5.10 **Odor Threshold:** Data not available

6. FIRE HAZARDS

6.1 **Flash Point:** Not pertinent (combustible solid)
6.2 **Flammable Limits in Air:** Not pertinent
6.3 **Fire Extinguishing Agents:** Water, foam, dry chemical, carbon dioxide
6.4 **Fire Extinguishing Agents Not to be Used:**
6.5 **Special Hazards of Combustion Products:** Irritating and toxic ammonia gas may form in fires.
6.6 **Behavior in Fire:**
6.7 **Ignition Temperature:** Not pertinent
6.8 **Electrical Hazard:** Not pertinent
6.9 **Burning Rate:** Not pertinent

7. CHEMICAL REACTIVITY

7.1 **Reactivity with Water:** No reaction
7.2 **Reactivity with Common Materials:**
7.3 **Stability During Transport:** Slowly releases ammonia gas, which may collect in closed container
7.4 **Neutralizing Agents for Acids and Caustics:** Not pertinent
7.5 **Polymerization:** Not pertinent
7.6 **Inhibitor of Polymerization:** Not pertinent

8. WATER POLLUTION

8.1 **Aquatic Toxicity:** Data not available
8.2 **Waterfowl Toxicity:** Data not available
8.3 **Biological Oxygen Demand (BOD):** Data not available
8.4 **Food Chain Concentration Potential:** None

9. SELECTED MANUFACTURERS

1. Ashland Oil, Inc.
   P. O. Box 2219
   Columbus, Ohio 43216
2. Heico, Inc.
   Delaware Water Gap, Pa. 18327
3. Gallard-Schlesinger Chemical Mfg. Co.
   584 Mineola Ave.
   Carle Place, N. Y. 11514

10. SHIPPING INFORMATION

10.1 **Grades or Purity:** Technical
10.2 **Storage Temperature:** Ambient
10.3 **Inert Atmosphere:** No requirement
10.4 **Venting:** Open

11. HAZARD ASSESSMENT CODE
(See Hazard Assessment Handbook, CG 446-3)

SS

12. HAZARD CLASSIFICATIONS

12.1 **Code of Federal Regulations:** Not listed
12.2 **NAS Hazard Rating for Bulk Water Transportation:** Not listed
12.3 **NFPA Hazard Classifications:** Not listed

13. PHYSICAL AND CHEMICAL PROPERTIES

13.1 **Physical State at 15°C and 1 atm:** Solid
13.2 **Molecular Weight:** 139.15
13.3 **Boiling Point at 1 atm:** Not pertinent (decomposes)
13.4 **Freezing Point:** 388°F = 198°C = 471°K
13.5 **Critical Temperature:** Not pertinent
13.6 **Critical Pressure:** Not pertinent
13.7 **Specific Gravity:** 1.26 at 25°C (solid)
13.8 **Liquid Surface Tension:** Not pertinent
13.9 **Liquid-Water Interfacial Tension:** Not pertinent
13.10 **Vapor (Gas) Specific Gravity:** Not pertinent
13.11 **Ratio of Specific Heats of Vapor (Gas):** Not pertinent
13.12 **Latent Heat of Vaporization:** Not pertinent
13.13 **Heat of Combustion:** Data not available
13.14 **Heat of Decomposition:** Not pertinent
13.15 **Heat of Solution:** 34 Btu/lb = 19 cal/g = 0.80×10^5 J/kg
13.16 **Heat of Polymerization:** Not pertinent

NOTES

ABC AMMONIUM BICARBONATE

Common Synonyms			
Acid ammonium carbonate Ammonium hydrogen carbonate Carbonic acid, monoammonium salt	Solid	White	Weak ammonia odor
	Sinks and mixes slowly with water. Freezing point is 95°F.		

Stop discharge if possible. Keep people away.
Avoid contact with solid and dust.
Isolate and remove discharged material.
Notify local health and pollution control agencies.

Fire

Not flammable.
POISONOUS GASES MAY BE PRODUCED IN FIRE.
Irritating gases may be produced when heated.
Wear goggles and self-contained breathing apparatus.

Exposure

CALL FOR MEDICAL AID.

DUST
Irritating to eyes, nose and throat.
If inhaled will cause difficult breathing.
If in eyes, hold eyelids open and flush with plenty of water.
If breathing has stopped, give artificial respiration.
If breathing is difficult, give oxygen.

SOLID
Irritating to skin and eyes.
Harmful if swallowed.
Remove contaminated clothing and shoes.
Flush affected areas with plenty of water.
IF IN EYES, hold eyelids open and flush with plenty of water.
IF SWALLOWED and victim is CONSCIOUS, have victim drink water or milk.
IF SWALLOWED and victim is UNCONSCIOUS OR HAVING CONVULSIONS, do nothing except keep victim warm.

Water Pollution

Effect of low concentrations on aquatic life is unknown.
May be dangerous if it enters water intakes.
Notify local health and wildlife officials.
Notify operators of nearby water intakes.

1. RESPONSE TO DISCHARGE

(See Response Methods Handbook, CG 446-4)

Disperse and flush

2. LABELS

No hazard label required by Code of Federal Regulations

3. CHEMICAL DESIGNATIONS

3.1 **Synonyms:**
Acid ammonium carbonate
Ammonium hydrogen carbonate
Carbonic acid, monoammoniun salt

3.2 **Coast Guard Compatibility Classification:** Not listed

3.3 **Chemical Formula:** NH_4HCO_3

3.4 **IMCO/United Nations Numerical Designation:** Not listed

4. OBSERVABLE CHARACTERISTICS

4.1 **Physical State (as shipped):** Solid

4.2 **Color:** White

4.3 **Odor:** Slight ammonia

5. HEALTH HAZARDS

5.1 **Personal Protective Equipment:** Work gloves; dust respirator; safety glasses or chemical safety goggles if dusty

5.2 **Symptoms Following Exposure:** Inhalation may cause respiratory irritation. Ingestion could be harmful. Contact with eyes or skin causes irritation.

5.3 **Treatment for Exposure:** Consult physician in case of ingestion or severe irritation. INHALATION: move to uncontaminated atmosphere. EYES OR SKIN: wash with large amounts of water.

5.4 **Toxicity by Inhalation (Threshold Limit Value):** Data not available

5.5 **Short-Term Inhalation Limits:** Data not available

5.6 **Toxicity by Ingestion:** Data not available

5.7 **Late Toxicity:** Data not available

5.8 **Vapor (Gas) Irritant Characteristics:** Data not available

5.9 **Liquid or Solid Irritant Characteristics:** Data not available

5.10 **Odor Threshold:** Data not available

6. FIRE HAZARDS

6.1 **Flash Point:** Not flammable

6.2 **Flammable Limits in Air:** Not flammable

6.3 **Fire Extinguishing Agents:** Not pertinent

6.4 **Fire Extinguishing Agents Not to be Used:** Not pertinent

6.5 **Special Hazards of Combustion Products:** Irritating and toxic ammonia gas may form in fires.

6.6 **Behavior in Fire:** Decomposes, but reaction is not explosive. Ammonia gas is formed.

6.7 **Ignition Temperature:** Not pertinent

6.8 **Electrical Hazard:** Not pertinent

6.9 **Burning Rate:** Not pertinent

7. CHEMICAL REACTIVITY

7.1 **Reactivity with Water:** No reaction

7.2 **Reactivity with Common Materials:** May attack copper, nickel, and zinc

7.3 **Stability During Transport:** Decomposes above 34°C (91°F) with formation of ammonia gas, which may collect in closed containers

7.4 **Neutralizing Agents for Acids and Caustics:** Not pertinent

7.5 **Polymerization:** Not pertinent

7.6 **Inhibitor of Polymerization:** Not pertinent

8. WATER POLLUTION

8.1 **Aquatic Toxicity:** Data not available

8.2 **Waterfowl Toxicity:** Data not available

8.3 **Biological Oxygen Demand (BOD):** Data not available

8.4 **Food Chain Concentration Potential:** None

9. SELECTED MANUFACTURERS

1. Diamond Shamrock Chemical Co.
300 Union Commerce Building
Cleveland, Ohio 44115
2. Allied Chemical Corp.
Industrial Chemicals Division
P. O. Box 1139R
Morristown, N. J. 07960
3. J. T. Baker Chemical Co.
Phillipsburg, N. J. 08865

10. SHIPPING INFORMATION

10.1 **Grades or Purity:** Food; Reagent

10.2 **Storage Temperature:** Below 33°C (91°F)

10.3 **Inert Atmosphere:** No requirement

10.4 **Venting:** Open

11. HAZARD ASSESSMENT CODE

(See Hazard Assessment Handbook, CG 446-3)

SS

12. HAZARD CLASSIFICATIONS

12.1 **Code of Federal Regulations:** Not listed

12.2 **NAS Hazard Rating for Bulk Water Transportation:** Not listed

12.3 **NFPA Hazard Classifications:** Not listed

13. PHYSICAL AND CHEMICAL PROPERTIES

13.1 **Physical State at 15°C and 1 atm:** Solid

13.2 **Molecular Weight:** 79.06

13.3 **Boiling Point at 1 atm:** Not pertinent (decomposes)

13.4 **Freezing Point:** Not pertinent (decomposes at 35°C)

13.5 **Critical Temperature:** Not pertinent

13.6 **Critical Pressure:** Not pertinent

13.7 **Specific Gravity:** 1.57 at 20°C (solid)

13.8 **Liquid Surface Tension:** Not pertinent

13.9 **Liquid-Water Interfacial Tension:** Not pertinent

13.10 **Vapor (Gas) Specific Gravity:** Not pertinent

13.11 **Ratio of Specific Heats of Vapor (Gas):** Not pertinent

13.12 **Latent Heat of Vaporization:** Not pertinent

13.13 **Heat of Combustion:** Not pertinent

13.14 **Heat of Decomposition:** Not pertinent

13.15 **Heat of Solution:** 140 Btu/lb = 80 cal/g = 3.3×10^5 J/kg

13.16 **Heat of Polymerization:** Not pertinent

NOTES

ABF	AMMONIUM BIFLUORIDE

Common Synonyms			
Ammonium hydrogen fluoride Acid ammonium fluoride Ammonium acid fluoride	Solid	White	Odorless
	Sinks and mixes with water.		

Avoid contact with solid and dust. Keep people away.
Wear rubber overclothing (including gloves).
Stop discharge if possible.
Isolate and remove discharged material.
Notify local health and pollution control agencies.

Fire	Not flammable. POISONOUS GASES MAY BE PRODUCED IN FIRE. Wear goggles and self-contained breathing apparatus.
Exposure	CALL FOR MEDICAL AID. DUST Irritating to eyes, nose and throat. If inhaled will cause coughing or difficult breathing. If in eyes, hold eyelids open and flush with plenty of water. If breathing has stopped, give artificial respiration. If breathing is difficult, give oxygen. SOLID Will burn skin and eyes. If swallowed will cause nausea or vomiting. Remove contaminated clothing and shoes. Flush affected areas with plenty of water. IF IN EYES, hold eyelids open and flush with plenty of water. IF SWALLOWED and victim is CONSCIOUS, have victim drink water or milk. IF SWALLOWED and victim is UNCONSCIOUS OR HAVING CONVULSIONS, do nothing except keep victim warm.
Water Pollution	Effect of low concentrations on aquatic life is unknown. May be dangerous if it enters water intakes. Notify local health and wildlife officials. Notify operators of nearby water intakes.

1. RESPONSE TO DISCHARGE

(See Response Methods Handbook, CG 446-4)

Issue warning—corrosive, water contaminant
Restrict access
Disperse and flush

2. LABEL

3. CHEMICAL DESIGNATIONS

3.1 **Synonyms:**
Acid ammonium fluoride
Ammonium acid fluoride
Ammoniun hydrogen fluoride

3.2 **Coast Guard Compatibility Classification:** Not listed

3.3 **Chemical Formula:** NH_4HF_2

3.4 **IMCO/United Nations Numerical Designation:** 8/1727

4. OBSERVABLE CHARACTERISTICS

4.1 **Physical State (as shipped):** Solid

4.2 **Color:** White

4.3 **Odor:** None

5. HEALTH HAZARDS

5.1 **Personal Protective Equipment:** Bu. Mines approved respirator; rubber gloves; safety goggles.

5.2 **Symptoms Following Exposure:** Inhalation of dust may cause irritation of the respiratory system. Ingestion causes irritation of mouth and stomach, vomiting, abdominal pain, convulsions, collapse, acute toxic nephritis. Contact with dust irritates eyes and may cause burns or rash on skin. High concs. of fluorine in the urine have been reported following skin contact.

5.3 **Treatment for Exposure:** *Begin first aid as quickly as possible.* INHALATION: remove victim to fresh air. INGESTION: perform gastric lavage with lime water or 1% calcium chloride solution; support respiration; call a physician. EYES: flush with water for at least 15 min.; consult physician. SKIN: flush with water; treat burns. OTHER: remove all contaminated clothing in the shower at once.

5.4 **Toxicity by Inhalation (Threshold Limit Value):** 2.5 mg/m³ (as fluorine)

5.5 **Short-Term Inhalation Limits:** Data not available

5.6 **Toxicity by Ingestion:** Grade 3; LD_{50} = 50 mg/kg (guinea pig), 60 mg/kg (rat)

5.7 **Late Toxicity:** Data not available

5.8 **Vapor (Gas) Irritant Characteristics:** Data not available

5.9 **Liquid or Solid Irritant Characteristics:** Data not available

5.10 **Odor Threshold:** Data not available

6. FIRE HAZARDS

6.1 **Flash Point:** Not flammable

6.2 **Flammable Limits in Air:** Not flammable

6.3 **Fire Extinguishing Agents:** Not pertinent

6.4 **Fire Extinguishing Agents Not to be Used:** Do not apply water to adjacent fires.

6.5 **Special Hazards of Combustion Products:** Toxic ammonia and hydrogen fluoride gases may form in fire.

6.6 **Behavior in Fire:**

6.7 **Ignition Temperature:** Not pertinent

6.8 **Electrical Hazard:** Not pertinent

6.9 **Burning Rate:** Not pertinent

7. CHEMICAL REACTIVITY

7.1 **Reactivity with Water:** Dissolves and forms a weak solution of hydrofluoric acid.

7.2 **Reactivity with Common Materials:** In presence of moisture will corrode glass, cement, and most metals. Flammable hydrogen gas may collect in enclosed spaces.

7.3 **Stability During Transport:** Stable

7.4 **Neutralizing Agents for Acids and Caustics:** Flush with water, rinse with dilute solution of sodium bicarbonate or soda ash.

7.5 **Polymerization:** Not pertinent

7.6 **Inhibitor of Polymerization:** Not pertinent

8. WATER POLLUTION

8.1 **Aquatic Toxicity:** Data not available

8.2 **Waterfowl Toxicity:** Data not available

8.3 **Biological Oxygen Demand (BOD):** Data not available

8.4 **Food Chain Concentration Potential:** None.

9. SELECTED MANUFACTURERS

1. American Hoechst Corp.
Chemicals and Plastics Division
Route 202-206 North
Somerville, N. J. 08876

2. The Harshaw Chemical Co.
1945 E. 97th Street
Cleveland, Ohio 44106

3. Allied Chemical Corp.
Specialty Chemicals Division
P. O. Box 1087R
Morristown, N. J. 07960

10. SHIPPING INFORMATION

10.1 **Grades or Purity:** Pure, 99+%; Technical, 97—98.5%

10.2 **Storage Temperature:** Ambient

10.3 **Inert Atmosphere:** No requirement

10.4 **Venting:** Open

11. HAZARD ASSESSMENT CODE

(See Hazard Assessment Handbook, CG 446-3)

SS

12. HAZARD CLASSIFICATIONS

12.1 **Code of Federal Regulations:** Corrosive solid

12.2 **NAS Hazard Rating for Bulk Water Transportation:** Not listed

12.3 **NFPA Hazard Classifications:**

Category	Classification
Health Hazard (Blue)	3
Flammability (Red)	0
Reactivity (Yellow)	1

13. PHYSICAL AND CHEMICAL PROPERTIES

13.1 **Physical State at 15°C and 1 atm:** Solid

13.2 **Molecular Weight:** 57.04

13.3 **Boiling Point at 1 atm:** 463.1°F = 239.5°C = 512.7°K

13.4 **Freezing Point:** 258.0°F = 125.6°C = 398.8°K

13.5 **Critical Temperature:** Not pertinent

13.6 **Critical Pressure:** Not pertinent

13.7 **Specific Gravity:** 1.5 at 20°C (solid)

13.8 **Liquid Surface Tension:** Not pertinent

13.9 **Liquid-Water Interfacial Tension:** Not pertinent

13.10 **Vapor (Gas) Specific Gravity:** Not pertinent

13.11 **Ratio of Specific Heats of Vapor (Gas):** Not pertinent

13.12 **Latent Heat of Vaporization:** Not pertinent

13.13 **Heat of Combustion:** Not pertinent

13.14 **Heat of Decomposition:** Not pertinent

13.15 **Heat of Solution:** 154 Btu/lb = 85.7 cal/g = 3.59×10^5 J/kg

13.16 **Heat of Polymerization:** Not pertinent

NOTES

ACB AMMONIUM CARBONATE

Common Synonyms	Solid	White	Strong Ammonia Odor
	Sinks and mixes with water.		

Stop discharge if possible. Keep people away.
Avoid contact with solid and dust.
Isolate and remove discharged material.
Notify local health and pollution control agencies.

Fire

Not flammable.
POISONOUS GASES MAY BE PRODUCED IN FIRE.
Wear goggles and self-contained breathing apparatus.

Exposure

CALL FOR MEDICAL AID.

DUST
Irritating to eyes, nose and throat.
If inhaled will cause difficult breathing.

If in eyes, hold eyelids open and flush with plenty of water.
If breathing has stopped, give artificial respiration.
If breathing is difficult, give oxygen.

SOLID
Irritating to skin and eyes.
If swallowed will cause nausea.

Remove contaminated clothing and shoes.
Flush affected areas with plenty of water.
IF IN EYES, hold eyelids open and flush with plenty of water.
IF SWALLOWED and victim is CONSCIOUS, have victim drink water or milk.
IF SWALLOWED and victim is UNCONSCIOUS OR HAVING CONVULSIONS, do nothing except keep victim warm.

Water Pollution

Effect of low concentrations on aquatic life is unknown.
May be dangerous if it enters water intakes.

Notify local health and wildlife officials.
Notify operators of nearby water intakes.

1. RESPONSE TO DISCHARGE
(See Response Methods Handbook, CG 446-4)

Disperse and flush

2. LABELS

No hazard label required by Code of Federal Regulations

3. CHEMICAL DESIGNATIONS

3.1 **Synonyms:** Hartshorn; Sal volatile
3.2 **Coast Guard Compatibility Classification:** Not listed
3.3 **Chemical Formula:** $(NH_4)_2CO_3$
3.4 **IMCO/United Nations Numerical Designation:** Not listed

4. OBSERVABLE CHARACTERISTICS

4.1 **Physical State (as shipped):** Solid
4.2 **Color:** White
4.3 **Odor:** Strong ammonia

5. HEALTH HAZARDS

5.1 **Personal Protective Equipment:** Dust respirator; protection against ammonia vapors
5.2 **Symptoms Following Exposure:** Inhalation causes irritation of nose and throat. Ingestion may cause gastric irritation. Contact with eyes or skin causes irritation.
5.3 **Treatment for Exposure:** INHALATION: leave contaminated area. INGESTION: give large amount of water. EYES: flush with copious amounts of water. SKIN: flush with water.
5.4 **Toxicity by Inhalation (Threshold Limit Value):** Data not available
5.5 **Short-Term Inhalation Limits:** Data not available
5.6 **Toxicity by Ingestion:** Data not available
5.7 **Late Toxicity:** Data not available
5.8 **Vapor (Gas) Irritant Characteristics:** Data not available
5.9 **Liquid or Solid Irritant Characteristics:** Data not available
5.10 **Odor Threshold:** < 5 ppm (as ammonia gas)

6. FIRE HAZARDS

6.1 **Flash Point:** Not pertinent
6.2 **Flammable Limits in Air:** Not pertinent
6.3 **Fire Extinguishing Agents:** Water
6.4 **Fire Extinguishing Agents Not to be Used:** Not pertinent
6.5 **Special Hazards of Combustion Products:** Toxic ammonia gas will form in fires.
6.6 **Behavior in Fire:** Decomposes, but reaction is not explosive. Ammonia gas is formed.
6.7 **Ignition Temperature:** Not pertinent
6.8 **Electrical Hazard:** Not pertinent
6.9 **Burning Rate:** Not pertinent

7. CHEMICAL REACTIVITY

7.1 **Reactivity with Water:** No reaction
7.2 **Reactivity with Common Materials:**
7.3 **Stability During Transport:** Stable
7.4 **Neutralizing Agents for Acids and Caustics:** Not pertinent
7.5 **Polymerization:** Not pertinent
7.6 **Inhibitor of Polymerization:** Not pertinent

8. WATER POLLUTION

8.1 **Aquatic Toxicity:**
24 ppm/3.5 hr/goldfish/killed/fresh water
10 ppm/> 100 hr/goldfish/tolerated/fresh water
8.2 **Waterfowl Toxicity:** Data not available
8.3 **Biological Oxygen Demand (BOD):** Data not available
8.4 **Food Chain Concentration Potential:** None

9. SELECTED MANUFACTURERS

1. Agrico Chemical Co.
National Bank of Tulsa Building
Tulsa, Okla. 74103
2. Allied Chemical Corp.
Specialty Chemicals Division
P. O. Box 1087R
Morristown, N. J. 07960
3. Mallinckrodt Chemical Works
223 Westside Avenue
P. O. Box 384
Jersey City, N. J. 07303

10. SHIPPING INFORMATION

10.1 **Grades or Purity:** CP; NF; USP; Reagent; Technical
10.2 **Storage Temperature:** Ambient
10.3 **Inert Atmosphere:** No requirement
10.4 **Venting:** Open

11. HAZARD ASSESSMENT CODE
(See Hazard Assessment Handbook, CG 446-3)

SS

12. HAZARD CLASSIFICATIONS

12.1 **Code of Federal Regulations:** Not listed
12.2 **NAS Hazard Rating for Bulk Water Transportation:** Not listed
12.3 **NFPA Hazard Classifications:** Not listed

13. PHYSICAL AND CHEMICAL PROPERTIES

13.1 **Physical State at 15°C and 1 atm:** Solid
13.2 **Molecular Weight:** 157.1
13.3 **Boiling Point at 1 atm:** Not pertinent (decomposes)
13.4 **Freezing Point:** Not pertinent
13.5 **Critical Temperature:** Not pertinent
13.6 **Critical Pressure:** Not pertinent
13.7 **Specific Gravity:** 1.5 at 20°C (solid)
13.8 **Liquid Surface Tension:** Not pertinent
13.9 **Liquid-Water Interfacial Tension:** Not pertinent
13.10 **Vapor (Gas) Specific Gravity:** Not pertinent
13.11 **Ratio of Specific Heats of Vapor (Gas):** Not pertinent
13.12 **Latent Heat of Vaporization:** Not pertinent
13.13 **Heat of Combustion:** Not pertinent
13.14 **Heat of Decomposition:** Not pertinent
13.15 **Heat of Solution:** Not pertinent
13.16 **Heat of Polymerization:** Not pertinent

NOTES

AMC	# AMMONIUM CHLORIDE

Common Synonyms			
Ammonium muriate Sal ammoniac Salmiac Amchloride Ammoneric	Solid	White	Odorless
	Sinks and mixes slowly with water.		

Stop discharge if possible. Keep people away.
Avoid contact with solid and dust.
Isolate and remove discharged material.
Notify local health and pollution control agencies.

Fire	Not flammable. POISONOUS GASES MAY BE PRODUCED IN FIRE. Irritating gases may be produced when heated. Wear goggles and self-contained breathing apparatus.
Exposure	CALL FOR MEDICAL AID. DUST Irritating to eyes, nose and throat. If inhaled will cause coughing or difficult breathing. If in eyes, hold eyelids open and flush with plenty of water. If breathing has stopped, give artificial respiration. If breathing is difficult, give oxygen. SOLID Irritating to skin and eyes. If swallowed will cause nausea. Remove contaminated clothing and shoes. Flush affected areas with plenty of water. IF IN EYES, hold eyelids open and flush with plenty of water. IF SWALLOWED and victim is CONSCIOUS, have victim drink water or milk. IF SWALLOWED and victim is UNCONSCIOUS OR HAVING CONVULSIONS, do nothing except keep victim warm.
Water Pollution	HARMFUL TO AQUATIC LIFE IN VERY LOW CONCENTRATIONS. May be dangerous if it enters water intakes. Notify local health and wildlife officials. Notify operators of nearby water intakes.

1. RESPONSE TO DISCHARGE
(See Response Methods Handbook, CG 446-4)

Disperse and flush

2. LABELS

No hazard label required by Code of Federal Regulations

3. CHEMICAL DESIGNATIONS

3.1 **Synonyms:** Amchlor; Ammoneric; Ammonium muriate; Sal ammoniac; Salmiac
3.2 **Coast Guard Compatibility Classification:** Not listed
3.3 **Chemical Formula:** NH_4Cl
3.4 **IMCO/United Nations Numerical Designation:** Not listed

4. OBSERVABLE CHARACTERISTICS

4.1 **Physical State (as shipped):** Solid
4.2 **Color:** White
4.3 **Odor:** None

5. HEALTH HAZARDS

5.1 **Personal Protective Equipment:** Gloves of any material; safety glasses or chemical safety goggles; dust mask or respirator as necessary
5.2 **Symptoms Following Exposure:** Inhalation of fumes irritates respiratory passages. Ingestion irritates mouth and stomach. Fumes are irritating to eyes. Contact with skin may cause irritation.
5.3 **Treatment for Exposure:** INHALATION: remove to fresh air. INGESTION: give large amount of water; get medical attention if irritation persists. EYES OR SKIN: flush with plenty of water.
5.4 **Toxicity by Inhalation (Threshold Limit Value):** 10 mg/m^3
5.5 **Short-Term Inhalation Limits:** Data not available
5.6 **Toxicity by Ingestion:** Grade 2; oral rat LD_{50} = 1,650 mg/kg
5.7 **Late Toxicity:** Data not available
5.8 **Vapor (Gas) Irritant Characteristics:** Data not available
5.9 **Liquid or Solid Irritant Characteristics:** Data not available
5.10 **Odor Threshold:** Odorless

6. FIRE HAZARDS

6.1 **Flash Point:** Not flammable
6.2 **Flammable Limits in Air:** Not flammable
6.3 **Fire Extinguishing Agents:** Not pertinent
6.4 **Fire Extinguishing Agents Not to be Used:** Not pertinent
6.5 **Special Hazards of Combustion Products:** Toxic and irritating ammonia and hydrogen chloride gases may form in fire.
6.6 **Behavior in Fire:** May volatilize and condense on cool surfaces.
6.7 **Ignition Temperature:** Not pertinent
6.8 **Electrical Hazard:** Not pertinent
6.9 **Burning Rate:** Not pertinent

7. CHEMICAL REACTIVITY

7.1 **Reactivity with Water:** No reaction
7.2 **Reactivity with Common Materials:**
7.3 **Stability During Transport:** Stable
7.4 **Neutralizing Agents for Acids and Caustics:** Not pertinent
7.5 **Polymerization:** Not pertinent
7.6 **Inhibitor of Polymerization:** Not pertinent

8. WATER POLLUTION

8.1 **Aquatic Toxicity:** 6 ppm/96 hr/sunfish TL_m/fresh water
8.2 **Waterfowl Toxicity:** Data not available
8.3 **Biological Oxygen Demand (BOD):** Data not available
8.4 **Food Chain Concentration Potential:** None

9. SELECTED MANUFACTURERS

1. Pennwalt Corp.
   Three Parkway
   Philadelphia, Pa. 19102
2. Allied Chemical Corp.
   Industrial Chemicals Division
   P. O. Box 1139 R
   Morristown, N. J. 07960
3. Mallinckrodt Chemical Works
   223 Westside Avenue
   P. O. Box 384
   Jersey City, N. J. 07303

10. SHIPPING INFORMATION

10.1 **Grades or Purity:** USP; Reagent; Technical, 99+%
10.2 **Storage Temperature:** Ambient
10.3 **Inert Atmosphere:** No requirement
10.4 **Venting:** Open

11. HAZARD ASSESSMENT CODE
(See Hazard Assessment Handbook, CG 446-3)

SS

12. HAZARD CLASSIFICATIONS

12.1 **Code of Federal Regulations:** Not listed
12.2 **NAS Hazard Rating for Bulk Water Transportation:** Not listed
12.3 **NFPA Hazard Classifications:** Not listed

13. PHYSICAL AND CHEMICAL PROPERTIES

13.1 **Physical State at 15°C and 1 atm:** Solid
13.2 **Molecular Weight:** 53.5
13.3 **Boiling Point at 1 atm:** Not pertinent (decomposes)
13.4 **Freezing Point:** Not pertinent
13.5 **Critical Temperature:** Not pertinent
13.6 **Critical Pressure:** Not pertinent
13.7 **Specific Gravity:** 1.53 at 20°C (solid)
13.8 **Liquid Surface Tension:** Not pertinent
13.9 **Liquid-Water Interfacial Tension:** Not pertinent
13.10 **Vapor (Gas) Specific Gravity:** Not pertinent
13.11 **Ratio of Specific Heats of Vapor (Gas):** Not pertinent
13.12 **Latent Heat of Vaporization:** Not pertinent
13.13 **Heat of Combustion:** Not pertinent
13.14 **Heat of Decomposition:** Not pertinent
13.15 **Heat of Solution:** 130 Btu/lb = 72 cal/g = 3.0×10^5 J/kg
13.16 **Heat of Polymerization:** Not pertinent

NOTES

ACI	AMMONIUM CITRATE

Common Synonyms			
Diammonium citrate Ammonium citrate Citric acid, Diammonium salt	Solid	White	Weak Ammonia Odor
	Sinks and mixes with water.		

Stop discharge if possible. Keep people away.
Avoid contact with solid and dust.
Isolate and remove discharged material.
Notify local health and pollution control agencies.

Fire	Combustible. POISONOUS GASES MAY BE PRODUCED IN FIRE. Wear goggles and self-contained breathing apparatus. Extinguish with water.
Exposure	DUST Irritating to eyes, nose and throat. If inhaled will cause coughing or difficult breathing. If in eyes, hold eyelids open and flush with plenty of water. If breathing has stopped, give artificial respiration. If breathing is difficult, give oxygen. SOLID Irritating to skin and eyes. If swallowed will cause nausea. Remove contaminated clothing and shoes. Flush affected areas with plenty of water. IF IN EYES, hold eyelids open and flush with plenty of water. IF SWALLOWED and victim is CONSCIOUS, have victim drink water or milk and have victim induce vomiting. IF SWALLOWED and victim is UNCONSCIOUS OR HAVING CONVULSIONS, do nothing except keep victim warm.
Water Pollution	Effect of low concentrations on aquatic life is unknown. May be dangerous if it enters water intakes. Notify local health and wildlife officials. Notify operators of nearby water intakes.

1. RESPONSE TO DISCHARGE
(See Response Methods Handbook, CG 446-4)

Disperse and flush

2. LABELS
No hazard label required by Code of Federal Regulations

3. CHEMICAL DESIGNATIONS
3.1 **Synonyms:** Ammonium citrate, dibasic; Citric acid, diammonium salt; Diammonium citrate
3.2 **Coast Guard Compatibility Classification:** Not listed
3.3 **Chemical Formula:** $(NH_4)_2HC_6H_5O_7$
3.4 **IMCO/United Nations Numerical Designation:** Not listed

4. OBSERVABLE CHARACTERISTICS
4.1 **Physical State (as shipped):** Solid
4.2 **Color:** White
4.3 **Odor:** Slight ammoniacal

5. HEALTH HAZARDS
5.1 **Personal Protective Equipment:** Approved dust respirator; eye goggles
5.2 **Symptoms Following Exposure:** Inhalation causes respiratory irritation. Ingestion causes diarrhea. Contact with eyes causes mild irritation.
5.3 **Treatment for Exposure:** INHALATION: move to fresh air. EYES: flush with water or physiological saline; get medical care if irritation persists. SKIN: flush with water.
5.4 **Toxicity by Inhalation (Threshold Limit Value):** Data not available
5.5 **Short-Term Inhalation Limits:** Data not available
5.6 **Toxicity by Ingestion:** Data not available
5.7 **Late Toxicity:** Data not available
5.8 **Vapor (Gas) Irritant Characteristics:** Data not available
5.9 **Liquid or Solid Irritant Characteristics:** Data not available
5.10 **Odor Threshold:** Data not available

6. FIRE HAZARDS
6.1 **Flash Point:** Not pertinent (combustible solid)
6.2 **Flammable Limits in Air:** Not pertinent
6.3 **Fire Extinguishing Agents:** Water
6.4 **Fire Extinguishing Agents Not to be Used:** Not pertinent
6.5 **Special Hazards of Combustion Products:** Toxic ammonia gas may form in fire.
6.6 **Behavior in Fire:**
6.7 **Ignition Temperature:** Not pertinent
6.8 **Electrical Hazard:** Not pertinent
6.9 **Burning Rate:** Not pertinent

7. CHEMICAL REACTIVITY
7.1 **Reactivity with Water:** No reaction
7.2 **Reactivity with Common Materials:**
7.3 **Stability During Transport:** Stable
7.4 **Neutralizing Agents for Acids and Caustics:** Not pertinent
7.5 **Polymerization:** Not pertinent
7.6 **Inhibitor of Polymerization:** Not pertinent

8. WATER POLLUTION
8.1 **Aquatic Toxicity:** Data not available
8.2 **Waterfowl Toxicity:** Data not available
8.3 **Biological Oxygen Demand (BOD):** Data not available
8.4 **Food Chain Concentration Potential:** None

9. SELECTED MANUFACTURERS
1. Pfizer Chemicals Div.
   235 E. 42nd St.
   New York, N.Y. 10017
2. Allied Chemical Corp.
   Specialty Chemicals Div.
   P. O. Box 1087R
   Morristown, N. J. 07960
3. J. T. Baker Chemical Co.
   Phillipsburg, N. J. 08865

10. SHIPPING INFORMATION
10.1 **Grades or Purity:** Reagent
10.2 **Storage Temperature:** Ambient
10.3 **Inert Atmosphere:** No requirement
10.4 **Venting:** Open

11. HAZARD ASSESSMENT CODE
(See Hazard Assessment Handbook, CG 446-3)

SS

12. HAZARD CLASSIFICATIONS
12.1 **Code of Federal Regulations:** Not listed
12.2 **NAS Hazard Rating for Bulk Water Transportation:** Not listed
12.3 **NFPA Hazard Classifications:** Not listed

13. PHYSICAL AND CHEMICAL PROPERTIES
13.1 **Physical State at 15°C and 1 atm:** Solid
13.2 **Molecular Weight:** 226
13.3 **Boiling Point at 1 atm:** Not pertinent (decomposes)
13.4 **Freezing Point:** Not pertinent
13.5 **Critical Temperature:** Not pertinent
13.6 **Critical Pressure:** Not pertinent
13.7 **Specific Gravity:** 1.48 at 20°C (solid)
13.8 **Liquid Surface Tension:** Not pertinent
13.9 **Liquid-Water Interfacial Tension:** Not pertinent
13.10 **Vapor (Gas) Specific Gravity:** Not pertinent
13.11 **Ratio of Specific Heats of Vapor (Gas):** Not pertinent
13.12 **Latent Heat of Vaporization:** Not pertinent
13.13 **Heat of Combustion:** Data not available
13.14 **Heat of Decomposition:** Not pertinent
13.15 **Heat of Solution:** Not pertinent
13.16 **Heat of Polymerization:** Not pertinent

NOTES

AMD	AMMONIUM DICHROMATE

Common Synonyms			
Ammonium bichromate	Solid crystals or powder	Orange to red	Odorless
	Sinks and mixes with water.		

Stop discharge if possible. Keep people away.
Call fire department.
Avoid contact with solid.
Isolate and remove discharged material.
Notify local health and pollution control agencies.

Fire

FLAMMABLE.
May cause fire on contact with combustibles.
Containers may explode in fire.
Combat fires from safe distance or protected location with unmanned hose holder or monitor nozzle.
Flood discharge area with water.
Cool exposed containers with water.

Exposure

Call for medical aid.

DUST
Irritating to eyes, nose and throat.
If inhaled will cause coughing or difficult breathing.
Move victim to fresh air.
If in eyes, hold eyelids open and flush with plenty of water.
If breathing is difficult, give oxygen.

SOLID
Irritating to skin and eyes.
Harmful if swallowed.
Remove contaminated clothing and shoes.
Flush affected areas with plenty of water.
IF IN EYES, hold eyelids open and flush with plenty of water.
IF SWALLOWED and victim is CONSCIOUS, have victim drink water or milk and have victim induce vomiting.
IF SWALLOWED and victim is UNCONSCIOUS OR HAVING CONVULSIONS, do nothing except keep victim warm.

Water Pollution

Dangerous to aquatic life in high concentrations.
May be dangerous if it enters water intakes.
Notify local health and wildlife officials.
Notify operators of nearby water intakes.

1. RESPONSE TO DISCHARGE
(See Response Methods Handbook, CG 446-4)

Issue warning—high flammability, oxidizing material
Disperse and flush

2. LABEL

Yellow

3. CHEMICAL DESIGNATIONS

3.1 **Synonyms:** Ammonium bichromate
3.2 **Coast Guard Compatibility Classification:** Not applicable
3.3 **Chemical Formula:** $(NH_4)_2Cr_2O_7$
3.4 **IMCO/United Nations Numerical Designation:** 5.1/1439

4. OBSERVABLE CHARACTERISTICS

4.1 **Physical State (as shipped):** Solid
4.2 **Color:** Orange-yellow; bright red-orange; orange to red
4.3 **Odor:** None

5. HEALTH HAZARDS

5.1 **Personal Protective Equipment:** Dust respirator; protective goggles, gloves, clothing.
5.2 **Symptoms Following Exposure:** Inhalation causes irritation or ulceration of the mucous membranes of the nose, throat or respiratory tract. Respiratory irritation can produce symptoms resembling those of asthma. Continuing irritation of the nose may lead to perforation of the nasal septum. External contact can cause eye irritation and conjunctivitis, irritation and ulceration of skin wounds, and rash or external ulcers. If ingested, irritates mucous membrane and causes vomiting.
5.3 **Treatment for Exposure:** INHALATION: remove to clean air and summon medical attention. EYES: immediately flush with water for at least 15 min. and consult a physician. SKIN: flush with water; if skin irritation develops, get medical attention. INGESTION: vomiting should occur; follow with an emetic of soapy water; give large amounts of water.
5.4 **Toxicity by Inhalation (Threshold Limit Value):** 0.1 mg/m^3
5.5 **Short-Term Inhalation Limits:** Not pertinent
5.6 **Toxicity by Ingestion:** Data not available
5.7 **Late Toxicity:** Data not available
5.8 **Vapor (Gas) Irritant Characteristics:** Not pertinent
5.9 **Liquid or Solid Irritant Characteristics:** Data not available
5.10 **Odor Threshold:** Not pertinent

6. FIRE HAZARDS

6.1 **Flash Point:** Flammable solid
6.2 **Flammable Limits in Air:** Not pertinent
6.3 **Fire Extinguishing Agents:** Water
6.4 **Fire Extinguishing Agents Not to be Used:** Not pertinent
6.5 **Special Hazards of Combustion Products:** Greenish chromic oxide smoke may cause irritation of lungs and mucous membranes.
6.6 **Behavior in Fire:** Decomposes at about 180°C. Decomposition self-sustaining at about 225°C with spectacular swelling and evolution of heat and nitrogen, leaving chromic oxide residue. Pressure of confined gases can burst closed containers explosively.
6.7 **Ignition Temperature:** 437°F
6.8 **Electrical Hazard:** Data not available
6.9 **Burning Rate:** Not pertinent

7. CHEMICAL REACTIVITY

7.1 **Reactivity with Water:** No reaction
7.2 **Reactivity with Common Materials:** Can ignite combustible material such as wood shavings.
7.3 **Stability During Transport:** Stable
7.4 **Neutralizing Agents for Acids and Caustics:** Not pertinent
7.5 **Polymerization:** Not pertinent
7.6 **Inhibitor of Polymerization:** Not pertinent

8. WATER POLLUTION

8.1 **Aquatic Toxicity:** 136 ppm/96 hr/mosquito fish/TL_m/fresh water
8.2 **Waterfowl Toxicity:** Data not available
8.3 **Biological Oxygen Demand (BOD):** None
8.4 **Food Chain Concentration Potential:** None

9. SELECTED MANUFACTURERS

1. Allied Chemical Corporation
Specialty Chemicals Division
P. O. Box 1087R
Morristown, N. J. 07960
2. J. T. Baker Chemical Co.
222 Red School Lane
Phillipsburg, N. J. 08865
3. Mallinckrodt Chemical Works
223 West Side Avenue
P. O. Box 384
Jersey City, N. J. 07303

10. SHIPPING INFORMATION

10.1 **Grades or Purity:** Analytical reagent grade: 99.0%; technical (photolitho) grade; technical granular grade: 99.7%; C. P. granular grade: 99.8%.
10.2 **Storage Temperature:** Ambient
10.3 **Inert Atmosphere:** No requirement
10.4 **Venting:** Open

11. HAZARD ASSESSMENT CODE
(See Hazard Assessment Handbook, CG 446-3)

SS-Z

12. HAZARD CLASSIFICATIONS

12.1 **Code of Federal Regulations:** Oxidizer
12.2 **NAS Hazard Rating for Bulk Water Transportation:** Not listed
12.3 **NFPA Hazard Classifications:**

Category	Classification
Health Hazard (Blue)	
Flammability (Red)	
Reactivity (Yellow)	
	oxy

13. PHYSICAL AND CHEMICAL PROPERTIES

13.1 **Physical State at 15°C and 1 atm:** Solid
13.2 **Molecular Weight:** 252.06
13.3 **Boiling Point at 1 atm:** Not pertinent
13.4 **Freezing Point:** Not pertinent
13.5 **Critical Temperature:** Not pertinent
13.6 **Critical Pressure:** Not pertinent
13.7 **Specific Gravity:** 2.15 at 25°C (solid)
13.8 **Liquid Surface Tension:** Not pertinent
13.9 **Liquid-Water Interfacial Tension:** Not pertinent
13.10 **Vapor (Gas) Specific Gravity:** Not pertinent
13.11 **Ratio of Specific Heats of Vapor (Gas):** Not pertinent
13.12 **Latent Heat of Vaporization:** Not pertinent
13.13 **Heat of Combustion:** Not pertinent
13.14 **Heat of Decomposition:** Not pertinent
13.15 **Heat of Solution:** 41 Btu/lb = 23 cal/g = 0.96×10^5 J/kg
13.16 **Heat of Polymerization:** Not pertinent

NOTES

AFR — AMMONIUM FLUORIDE

Common Synonyms			
Neutral ammonium fluoride	Solid	White	Odorless
	Sinks and mixes with water.		

AVOID CONTACT WITH SOLID AND DUST. KEEP PEOPLE AWAY.
Wear dust respirator and rubber overclothing (including gloves).
Stop discharge if possible.
Isolate and remove discharged material.
Notify local health and pollution control agencies.

Fire

Not flammable.
POISONOUS GASES MAY BE PRODUCED IN FIRE.
Wear goggles and self-contained breathing apparatus.

Exposure

CALL FOR MEDICAL AID.

DUST
Irritating to eyes, nose and throat.
If inhaled will cause coughing or difficult breathing.
If in eyes, hold eyelids open and flush with plenty of water.
If breathing has stopped, give artificial respiration.
If breathing is difficult, give oxygen.

SOLID
POISONOUS IF SWALLOWED.
Will burn skin and eyes.
Remove contaminated clothing and shoes.
Flush affected areas with plenty of water.
IF IN EYES, hold eyelids open and flush with plenty of water.
IF SWALLOWED and victim is CONSCIOUS, have victim drink water or milk.
IF SWALLOWED and victim is UNCONSCIOUS OR HAVING CONVULSIONS, do nothing except keep victim warm.

Water Pollution

Effect of low concentrations on aquatic life is unknown.
May be dangerous if it enters water intakes.
Notify local health and wildlife officials.
Notify operators of nearby water intakes.

1. RESPONSE TO DISCHARGE
(See Response Methods Handbook, CG 446-4)

Issue warning—water contaminant
Disperse and flush

2. LABELS

No hazard label required by Code of Federal Regulations

3. CHEMICAL DESIGNATIONS

3.1 **Synonyms:** Neutral ammonium fluoride
3.2 **Coast Guard Compatibility Classification:** Not listed
3.3 **Chemical Formula:** NH_4F
3.4 **IMCO/United Nations Numerical Designation:** Not listed

4. OBSERVABLE CHARACTERISTICS

4.1 **Physical State (as shipped):** Solid
4.2 **Color:** White
4.3 **Odor:** None

5. HEALTH HAZARDS

5.1 **Personal Protective Equipment:** Dust mask; goggles or face shield; rubber gloves
5.2 **Symptoms Following Exposure:** Inhalation of dust may cause irritation of respiratory system. Ingestion is harmful; readily soluble fluorides may be fatal if relatively small quantities are swallowed. Contact with eyes causes local irritation of the mucous membrane. Contact with skin may cause burns. High concs. of fluorine in the urine have been reported following skin contact.
5.3 **Treatment for Exposure:** *Begin first aid as quickly as possible.* INHALATION: remove to fresh air. INGESTION: perform gastric lavage with limewater or 1% calcium chloride solution; support respiration; call a physician. EYES: flush with water for 15 min.; consult physician. SKIN: shower immediately with large quantities of water; remove all contaminated clothing in the shower at once; consult physician.
5.4 **Toxicity by Inhalation (Threshold Limit Value):** 2.5 mg/m^3 (as fluorine)
5.5 **Short-Term Inhalation Limits:** Data not available
5.6 **Toxicity by Ingestion:** Data not available
5.7 **Late Toxicity:** Data not available
5.8 **Vapor (Gas) Irritant Characteristics:** Data not available
5.9 **Liquid or Solid Irritant Characteristics:** Data not available
5.10 **Odor Threshold:** Data not available

6. FIRE HAZARDS

6.1 **Flash Point:** Not flammable
6.2 **Flammable Limits in Air:** Not flammable
6.3 **Fire Extinguishing Agents:** Not pertinent
6.4 **Fire Extinguishing Agents Not to be Used:** Not pertinent
6.5 **Special Hazards of Combustion Products:** Toxic ammonia and hydrogen fluoride gases are formed in fires
6.6 **Behavior in Fire:** May sublime when hot and condense on cool surfaces
6.7 **Ignition Temperature:** Not pertinent
6.8 **Electrical Hazard:** Not pertinent
6.9 **Burning Rate:** Not pertinent

7. CHEMICAL REACTIVITY

7.1 **Reactivity with Water:** Dissolves and forms dilute solution of hydrofluoric acid
7.2 **Reactivity with Common Materials:** May corrode glass, cement, and most metals
7.3 **Stability During Transport:** Stable
7.4 **Neutralizing Agents for Acids and Caustics:** Not pertinent
7.5 **Polymerization:** Not pertinent
7.6 **Inhibitor of Polymerization:** Not pertinent

8. WATER POLLUTION

8.1 **Aquatic Toxicity:** Data not available
8.2 **Waterfowl Toxicity:** Data not available
8.3 **Biological Oxygen Demand (BOD):** Data not available
8.4 **Food Chain Concentration Potential:** None

9. SELECTED MANUFACTURERS

1. Allied Chemical Corp.
   General Chemical Div.
   40 Rector St.
   New York, N. Y. 10006
2. J. T. Baker Chemical Co.
   Phillipsburg, N. J. 08865
3. Gallard-Schlesinger Chemical Mfg. Co.
   584 Mineola Ave.
   Carle Place, N. Y. 11514

10. SHIPPING INFORMATION

10.1 **Grades or Purity:** Technical, 96.0%; Reagent; Electronic; Low sodium MOS
10.2 **Storage Temperature:** Ambient
10.3 **Inert Atmosphere:** No requirement
10.4 **Venting:** Open

11. HAZARD ASSESSMENT CODE
(See Hazard Assessment Handbook, CG 446-3)

SS

12. HAZARD CLASSIFICATIONS

12.1 **Code of Federal Regulations:** ORM-B
12.2 **NAS Hazard Rating for Bulk Water Transportation:** Not listed
12.3 **NFPA Hazard Classifications:** Not listed

13. PHYSICAL AND CHEMICAL PROPERTIES

13.1 **Physical State at 15°C and 1 atm:** Solid
13.2 **Molecular Weight:** 37.04
13.3 **Boiling Point at 1 atm:** Not pertinent (decomposes)
13.4 **Freezing Point:** Not pertinent
13.5 **Critical Temperature:** Not pertinent
13.6 **Critical Pressure:** Not pertinent
13.7 **Specific Gravity:** 1.32 at 25°C (solid)
13.8 **Liquid Surface Tension:** Not pertinent
13.9 **Liquid-Water Interfacial Tension:** Not pertinent
13.10 **Vapor (Gas) Specific Gravity:** Not pertinent
13.11 **Ratio of Specific Heats of Vapor (Gas):** Not pertinent
13.12 **Latent Heat of Vaporization:** Not pertinent
13.13 **Heat of Combustion:** Not pertinent
13.14 **Heat of Decomposition:** Not pertinent
13.15 **Heat of Solution:** 72 Btu/lb = 40 cal/g = 1.7×10^5 J/kg
13.16 **Heat of Polymerization:** Not pertinent

NOTES

AFM — AMMONIUM FORMATE

Common Synonyms			
Formic acid, ammonium salt	Solid	White	Weak ammonia odor
	Sinks and mixes slowly with water.		

Stop discharge if possible. Keep people away.
Avoid contact with solid and dust.
Isolate and remove discharged material.
Notify local health and pollution control agencies.

Fire

Combustible
POISONOUS GASES MAY BE PRODUCED IN FIRE.
Irritating gases may be produced when heated.
Wear goggles, self-contained breathing apparatus and rubber overclothing (including gloves).
Extinguish with water or foam.

Exposure

CALL FOR MEDICAL AID.

DUST
Irritating to eyes, nose and throat.
If inhaled will cause coughing or difficult breathing.
If in eyes, hold eyelids open and flush with plenty of water.
If breathing has stopped, give artificial respiration.
If breathing is difficult, give oxygen.

SOLID
Irritating to skin and eyes.
If swallowed will cause nausea.
Remove contaminated clothing and shoes.
Flush affected areas with plenty of water.
IF IN EYES, hold eyelids open and flush with plenty of water.
IF SWALLOWED and victim is CONSCIOUS, have victim drink water or milk.
IF SWALLOWED and victim is UNCONSCIOUS OR HAVING CONVULSIONS, do nothing except keep victim warm.

Water Pollution

Effect of low concentrations on aquatic life is unknown.
May be dangerous if it enters water intakes.
Notify local health and wildlife officials.
Notify operators of nearby water intakes.

1. RESPONSE TO DISCHARGE

(See Response Methods Handbook, CG 446-4)

Disperse and flush

2. LABELS

No hazard label required by Code of Federal Regulations

3. CHEMICAL DESIGNATIONS

3.1 **Synonyms:** Formic acid, ammonium salt
3.2 **Coast Guard Compatibility Classification:** Not listed
3.3 **Chemical Formula:** $HCOONH_4$
3.4 **IMCO/United Nations Numerical Designation:** Not listed

4. OBSERVABLE CHARACTERISTICS

4.1 **Physical State (as shipped):** Solid
4.2 **Color:** White
4.3 **Odor:** Weak ammonia

5. HEALTH HAZARDS

5.1 **Personal Protective Equipment:** Dust mask; goggles or face shield; rubber gloves
5.2 **Symptoms Following Exposure:** Inhalation causes irritation of nose and throat. Ingestion irritates mouth and stomach. Contact with eyes or skin causes irritation.
5.3 **Treatment for Exposure:** INHALATION: remove to fresh air. INGESTION: give large amounts of water; get medical attention. EYES: flush with water for at least 15 min. SKIN: flush with water.
5.4 **Toxicity by Inhalation (Threshold Limit Value):** Data not available
5.5 **Short-Term Inhalation Limits:** Data not available
5.6 **Toxicity by Ingestion:** Grade 2; oral LD_{50} = 2,250 mg/kg (mouse)
5.7 **Late Toxicity:** Data not available
5.8 **Vapor (Gas) Irritant Characteristics:** Data not available
5.9 **Liquid or Solid Irritant Characteristics:** Data not available
5.10 **Odor Threshold:** Data not available

6. FIRE HAZARDS

6.1 **Flash Point:** Not pertinent
6.2 **Flammable Limits in Air:** Not pertinent
6.3 **Fire Extinguishing Agents:** Water, foam
6.4 **Fire Extinguishing Agents Not to be Used:**
6.5 **Special Hazards of Combustion Products:** Toxic and irritating ammonia and formic acid gases may form in fire.
6.6 **Behavior in Fire:**
6.7 **Ignition Temperature:** Not pertinent
6.8 **Electrical Hazard:** Not pertinent
6.9 **Burning Rate:** Not pertinent

7. CHEMICAL REACTIVITY

7.1 **Reactivity with Water:** No reaction
7.2 **Reactivity with Common Materials:**
7.3 **Stability During Transport:** Stable
7.4 **Neutralizing Agents for Acids and Caustics:** Not pertinent
7.5 **Polymerization:** Not pertinent
7.6 **Inhibitor of Polymerization:** Not pertinent

8. WATER POLLUTION

8.1 **Aquatic Toxicity:** Data not available
8.2 **Waterfowl Toxicity:** Data not available
8.3 **Biological Oxygen Demand (BOD):** Data not available
8.4 **Food Chain Concentration Potential:** None

9. SELECTED MANUFACTURERS

1. R.S.A. Corporation
   690 Saw Mill River Rd.
   Ardsley, N. Y. 10502
2. Mallinckrodt Chemical Works
   223 Westside Avenue
   P. O. Box 384
   Jersey City, N. Y. 07303
3. Gallard-Schlesinger Chemical Mfg. Co.
   584 Mineola Avenue
   Carle Place, N. Y. 11514

10. SHIPPING INFORMATION

10.1 **Grades or Purity:** Analytical grade; Organic chemical grade
10.2 **Storage Temperature:** Ambient
10.3 **Inert Atmosphere:** No requirement
10.4 **Venting:** Open

11. HAZARD ASSESSMENT CODE

(See Hazard Assessment Handbook, CG 446-3)

SS

12. HAZARD CLASSIFICATIONS

12.1 **Code of Federal Regulations:** Not listed
12.2 **NAS Hazard Rating for Bulk Water Transportation:** Not listed
12.3 **NFPA Hazard Classifications:** Not listed

13. PHYSICAL AND CHEMICAL PROPERTIES

13.1 **Physical State at 15°C and 1 atm:** Solid
13.2 **Molecular Weight:** 63.06
13.3 **Boiling Point at 1 atm:** Not pertinent (decomposes)
13.4 **Freezing Point:** 241°F = 116°C = 389°K
13.5 **Critical Temperature:** Not pertinent
13.6 **Critical Pressure:** Not pertinent
13.7 **Specific Gravity:** 1.28 at 25°C (solid)
13.8 **Liquid Surface Tension:** Not pertinent
13.9 **Liquid-Water Interfacial Tension:** Not pertinent
13.10 **Vapor (Gas) Specific Gravity:** Not pertinent
13.11 **Ratio of Specific Heats of Vapor (Gas):** Not pertinent
13.12 **Latent Heat of Vaporization:** Not pertinent
13.13 **Heat of Combustion:** Not pertinent
13.14 **Heat of Decomposition:** Not pertinent
13.15 **Heat of Solution:** 84.7 Btu/lb = 47.1 cal/g = 1.97×10^5 J/kg
13.16 **Heat of Polymerization:** Not pertinent

NOTES

AGC	AMMONIUM GLUCONATE

Common Synonyms	Solid White Weak ammonia odor Sinks and mixes with water.

Stop discharge if possible. Keep people away.
Avoid contact with solid and dust.
Isolate and remove discharged material.
Notify local health and pollution control agencies.

Fire	Combustible POISONOUS GASES MAY BE PRODUCED IN FIRE. Wear goggles, self-contained breathing apparatus and rubber overclothing (including gloves). Extinguish with water.
Exposure	CALL FOR MEDICAL AID. DUST Irritating to eyes, nose and throat. If inhaled will cause coughing or difficult breathing. If in eyes, hold eyelids open and flush with plenty of water. If breathing has stopped, give artificial respiration. If breathing is difficult, give oxygen. SOLID Irritating to skin and eyes. Harmful if swallowed. Remove contaminated clothing and shoes. Flush affected areas with plenty of water. IF IN EYES, hold eyelids open and flush with plenty of water. IF SWALLOWED and victim is CONSCIOUS, have victim drink water or milk. IF SWALLOWED and victim is UNCONSCIOUS OR HAVING CONVULSIONS, do nothing except keep victim warm.
Water Pollution	Effect of low concentrations on aquatic life is unknown. May be dangerous if it enters water intakes. Notify local health and wildlife officials. Notify operators of nearby water intakes.

1. RESPONSE TO DISCHARGE
(See Response Methods Handbook, CG 446-4)

Disperse and flush

2. LABELS

No hazard label required by Code of Federal Regulations

3. CHEMICAL DESIGNATIONS

3.1 **Synonyms:** No common synonyms
3.2 **Coast Guard Compatibility Classification:** Not listed
3.3 **Chemical Formula:** $NH_4C_6H_{11}O_7$
3.4 **IMCO/United Nations Numerical Designation:** Not listed

4. OBSERVABLE CHARACTERISTICS

4.1 **Physical State (as shipped):** Solid
4.2 **Color:** White
4.3 **Odor:** Weak ammonia

5. HEALTH HAZARDS

5.1 **Personal Protective Equipment:** Respirator for nuisance dust
5.2 **Symptoms Following Exposure:** Inhalation of dust irritates nose and throat. Contact with eyes causes mild irritation.
5.3 **Treatment for Exposure:** INHALATION: remove to fresh air. EYES or SKIN: flush with water.
5.4 **Toxicity by Inhalation (Threshold Limit Value):** Data not available
5.5 **Short-Term Inhalation Limits:** Data not available
5.6 **Toxicity by Ingestion:** Data not available
5.7 **Late Toxicity:** Data not available
5.8 **Vapor (Gas) Irritant Characteristics:** Data not available
5.9 **Liquid or Solid Irritant Characteristics:** Data not available
5.10 **Odor Threshold:** Data not available

11. HAZARD ASSESSMENT CODE
(See Hazard Assessment Handbook, CG 446-3)

SS

12. HAZARD CLASSIFICATIONS

12.1 **Code of Federal Regulations:** Not listed
12.2 **NAS Hazard Rating for Bulk Water Transportation:** Not listed
12.3 **NFPA Hazard Classifications:** Not listed

13. PHYSICAL AND CHEMICAL PROPERTIES

13.1 **Physical State at 15°C and 1 atm:** Solid
13.2 **Molecular Weight:** 213
13.3 **Boiling Point at 1 atm:** Not pertinent
13.4 **Freezing Point:** Not pertinent
13.5 **Critical Temperature:** Not pertinent
13.6 **Critical Pressure:** Not pertinent
13.7 **Specific Gravity:** > 1 at 20°C (solid)
13.8 **Liquid Surface Tension:** Not pertinent
13.9 **Liquid-Water Interfacial Tension:** Not pertinent
13.10 **Vapor (Gas) Specific Gravity:** Not pertinent
13.11 **Ratio of Specific Heats of Vapor (Gas):** Not pertinent
13.12 **Latent Heat of Vaporization:** Not pertinent
13.13 **Heat of Combustion:** Data not available
13.14 **Heat of Decomposition:** Not pertinent
13.15 **Heat of Solution:** Data not available
13.16 **Heat of Polymerization:** Not pertinent

NOTES

6. FIRE HAZARDS

6.1 **Flash Point:** Not pertinent
6.2 **Flammable Limits in Air:** Not pertinent
6.3 **Fire Extinguishing Agents:** Water
6.4 **Fire Extinguishing Agents Not to be Used:** Not pertinent
6.5 **Special Hazards of Combustion Products:** Toxic oxides of nitrogen may form in fires.
6.6 **Behavior in Fire:**
6.7 **Ignition Temperature:** Not pertinent
6.8 **Electrical Hazard:** Not pertinent
6.9 **Burning Rate:** Not pertinent

7. CHEMICAL REACTIVITY

7.1 **Reactivity with Water:** No reaction
7.2 **Reactivity with Common Materials:**
7.3 **Stability During Transport:** Stable
7.4 **Neutralizing Agents for Acids and Caustics:** Not pertinent
7.5 **Polymerization:** Not pertinent
7.6 **Inhibitor of Polymerization:** Not pertinent

8. WATER POLLUTION

8.1 **Aquatic Toxicity:** Data not available
8.2 **Waterfowl Toxicity:** Data not available
8.3 **Biological Oxygen Demand (BOD):** Data not available
8.4 **Food Chain Concentration Potential:** None

9. SELECTED MANUFACTURERS

1. Pfizer, Incorporated
Chemicals Div.
235 E 42nd St.
New York, N. Y. 10017
2. Pfaltz and Bauer, Inc.
126-04 Northern Blvd.
Flushing, N. Y. 11368
3. Schwartz/Mann
Orangeburg, N. Y. 10962

10. SHIPPING INFORMATION

10.1 **Grades or Purity:** Pure
10.2 **Storage Temperature:** Ambient
10.3 **Inert Atmosphere:** No requirement
10.4 **Venting:** Open

AMH	**AMMONIUM HYDROXIDE** (< 28% AQUEOUS AMMONIA)

Common Synonyms			
Ammonia water Aqueous ammonia	Watery liquid	Colorless	Ammonia odor
	Floats and mixes with water. Irritating vapor is produced.		

Avoid contact with liquid and vapor. Keep people away.
Wear goggles, self-contained breathing apparatus, and rubber overclothing (including gloves).
Stop discharge if possible.
Stay upwind and use water spray to "knock down" vapor.
Isolate and remove discharged material.
Notify local health and pollution control agencies.

Fire

Not flammable.

Exposure

CALL FOR MEDICAL AID.

VAPOR
Irritating to skin, eyes, nose and throat.
If inhaled, will cause nausea, vomiting, difficult breathing, or loss of consciousness.
Move to fresh air.
IF IN EYES, hold eyelids open and flush with plenty of water.
If breathing has stopped, give artificial respiration.
If breathing is difficult, give oxygen.

LIQUID
Will burn skin and eyes.
Harmful if swallowed.
Remove contaminated clothing and shoes.
Flush affected areas with plenty of water.
IF IN EYES, hold eyelids open and flush with plenty of water.
IF SWALLOWED and victim is CONSCIOUS, have victim drink water or milk.

Water Pollution

HARMFUL TO AQUATIC LIFE IN VERY LOW CONCENTRATIONS.
May be dangerous if it enters water intakes.
Notify local health and wildlife officials.
Notify operators of nearby water intakes.

1. RESPONSE TO DISCHARGE

(See Response Methods Handbook, CG 446-4)

Issue warning—air contaminant
Restrict access
Disperse and flush

2. LABEL

3. CHEMICAL DESIGNATIONS

3.1 **Synonyms:** Ammonia water; Aqueous ammonia; Household ammonia
3.2 **Coast Guard Compatibility Classification:** Ammonia
3.3 **Chemical Formula:** $NH_4OH \cdot H_2O$
3.4 **IMCO/United Nations Numerical Designation:** 2/1005

4. OBSERVABLE CHARACTERISTICS

4.1 **Physical State (as shipped):** Liquid
4.2 **Color:** Colorless
4.3 **Odor:** Pungent

5. HEALTH HAZARDS

5.1 **Personal Protective Equipment:** Rubber boots, gloves, apron, and coat; broad-brimmed rubber or felt hat; safety goggles. Use of protective oil will reduce skin irritation from ammonia.
5.2 **Symptoms Following Exposure:** Contact of liquid or vapor with skin, mucous membranes, lungs, or gastroenteric tract causes marked local irritation. Ingestion causes burning pain in mouth, throat, stomach, and thorax, constriction of throat, and coughing. This is soon followed by vomiting of blood or by passage of loose stools containing blood. Breathing difficulty, convulsions, and shock may result. Brief exposure to 5000 ppm or ingestion of 3-4 ml may be fatal.
5.3 **Treatment for Exposure:** INHALATION: give artificial respiration and oxygen if needed; enforce rest. INGESTION: do NOT induce vomiting; lavage stomach with water or lemon juice, milk, or demulcents; delay may cause perforation of esophagus or stomach; swelling of glottis may necessitate tracheostomy. EYES OR SKIN: wash with plenty of water.
5.4 **Toxicity by Inhalation (Threshold Limit Value):** 1 ppm
5.5 **Short-Term Inhalation Limits:** (ammonia gas) 100 ppm for 30 min.; 500 ppm for 10 min.
5.6 **Toxicity by Ingestion:** Grade 3; oral rat, LD_{50} = 350 mg/kg
5.7 **Late Toxicity:** Data not available
5.8 **Vapor (Gas) Irritant Characteristics:** Vapors cause moderate irritation such that personnel will find high concentrations intolerable. The effect is temporary.
5.9 **Liquid or Solid Irritant Characteristics:** Causes smarting of the skin and first-degree burns on short exposure; may cause second-degree burns on long exposure.
5.10 **Odor Threshold:** 50 ppm

6. FIRE HAZARDS

6.1 **Flash Point:** Not flammable
6.2 **Flammable Limits in Air:** Not flammable
6.3 **Fire Extinguishing Agents:** Not pertinent
6.4 **Fire Extinguishing Agents Not to be Used:** Not pertinent
6.5 **Special Hazards of Combustion Products:** Not pertinent
6.6 **Behavior in Fire:** Not pertinent
6.7 **Ignition Temperature:** Not flammable
6.8 **Electrical Hazard:** Data not available
6.9 **Burning Rate:** Not flammable

7. CHEMICAL REACTIVITY

7.1 **Reactivity with Water:** Mild liberation of heat
7.2 **Reactivity with Common Materials:** Corrosive to copper, copper alloys, aluminum alloys, galvanized surfaces
7.3 **Stability During Transport:** Stable
7.4 **Neutralizing Agents for Acids and Caustics:** Dilute with water
7.5 **Polymerization:** Not pertinent
7.6 **Inhibitor of Polymerization:** Not pertinent

8. WATER POLLUTION

8.1 **Aquatic Toxicity:**
6.25 ppm/24 hr/trout/lethal/fresh water
15 ppm/48 hr/sunfish/TL_m/Phila. tap water
8.2 **Waterfowl Toxicity:** Data not available
8.3 **Biological Oxygen Demand (BOD):** Data not available
8.4 **Food Chain Concentration Potential:** None

9. SELECTED MANUFACTURERS

1. American Oil Co.
   910 S. Michigan Ave.
   Chicago, Ill. 60605
2. CF Industries, Inc.
   100 S. Wacker Drive
   Chicago, Ill. 60606
3. Williams Companies
   Agrico Chemicals Division
   5050 Poplar Ave.
   Memphis, Tenn. 38101

10. SHIPPING INFORMATION

10.1 **Grades or Purity:** Grade A: 29.4% NH_3; B: 25%; C: 15%. USP: 27 to 29%. CP: 28%
10.2 **Storage Temperature:** Ambient
10.3 **Inert Atmosphere:** No requirement
10.4 **Venting:** Pressure-vacuum

11. HAZARD ASSESSMENT CODE

(See Hazard Assessment Handbook, CG 446-3)

A-P-R-S

12. HAZARD CLASSIFICATIONS

12.1 **Code of Federal Regulations:** Corrosive Material
12.2 **NAS Hazard Rating for Bulk Water Transportation:** Not listed
12.3 **NFPA Hazard Classifications:** Not listed

13. PHYSICAL AND CHEMICAL PROPERTIES

13.1 **Physical State at 15°C and 1 atm:** Liquid
13.2 **Molecular Weight:** Not pertinent
13.3 **Boiling Point at 1 atm:** Not pertinent
13.4 **Freezing Point:** Not pertinent
13.5 **Critical Temperature:** Not pertinent
13.6 **Critical Pressure:** Not pertinent
13.7 **Specific Gravity:** 0.89 at 20°C (liquid)
13.8 **Liquid Surface Tension:** Not pertinent
13.9 **Liquid-Water Interfacial Tension:** Not pertinent
13.10 **Vapor (Gas) Specific Gravity:** Not pertinent
13.11 **Ratio of Specific Heats of Vapor (Gas):** Not pertinent
13.12 **Latent Heat of Vaporization:** Not pertinent
13.13 **Heat of Combustion:** Not pertinent
13.14 **Heat of Decomposition:** Not pertinent
13.15 **Heat of Solution:** Not pertinent
13.16 **Heat of Polymerization:** Not pertinent

NOTES

REVISED 1978

AID — AMMONIUM IODIDE

Common Synonyms	Solid	White	Odorless
	Sinks and mixes with water.		

Stop discharge if possible. Keep people away.
Avoid contact with solid and dust.
Isolate and remove discharged material.
Notify local health and pollution control agencies.

Fire

Not flammable.
POISONOUS GASES MAY BE PRODUCED IN FIRE.
Irritating gases may be produced when heated.
Wear goggles, self-contained breathing apparatus and rubber overclothing (including gloves).

Exposure

CALL FOR MEDICAL AID.

DUST
Irritating to eyes, nose and throat.
If inhaled will cause coughing or difficult breathing.
If in eyes, hold eyelids open and flush with plenty of water.
If breathing has stopped, give artificial respiration.
If breathing is difficult, give oxygen.

SOLID
Irritating to skin and eyes.
Harmful if swallowed.
Remove contaminated clothing and shoes.
Flush affected areas with plenty of water.
IF IN EYES, hold eyelids open and flush with plenty of water.
IF SWALLOWED and victim is CONSCIOUS, have victim drink water or milk.
IF SWALLOWED and victim is UNCONSCIOUS OR HAVING CONVULSIONS, do nothing except keep victim warm.

Water Pollution

Effect of low concentrations on aquatic life is unknown.
May be dangerous if it enters water intakes.
Notify local health and wildlife officials.
Notify operators of nearby water intakes.

1. RESPONSE TO DISCHARGE

(See Response Methods Handbook, CG 446-4)

Issue warning—water contaminant
Disperse and flush

2. LABELS

No hazard label required by Code of Federal Regulations

3. CHEMICAL DESIGNATIONS

3.1 **Synonyms:** No common synonyms
3.2 **Coast Guard Compatibility Classification:** Not listed
3.3 **Chemical Formula:** NH_4I
3.4 **IMCO/United Nations Numerical Designation:** Not listed

4. OBSERVABLE CHARACTERISTICS

4.1 **Physical State (as shipped):** Solid
4.2 **Color:** White
4.3 **Odor:** None

5. HEALTH HAZARDS

5.1 **Personal Protective Equipment:** Dust mask; goggles or face shield; rubber gloves
5.2 **Symptoms Following Exposure:** Inhalation causes irritation of nose and throat. Contact with eyes causes irritation.
5.3 **Treatment for Exposure:** INHALATION: remove to fresh air. INGESTION: if significant amount has been ingested, get medical attention. EYES: flush with water.
5.4 **Toxicity by Inhalation (Threshold Limit Value):** Data not available
5.5 **Short-Term Inhalation Limits:** Data not available
5.6 **Toxicity by Ingestion:** Data not available
5.7 **Late Toxicity:** Data not available
5.8 **Vapor (Gas) Irritant Characteristics:** Data not available
5.9 **Liquid or Solid Irritant Characteristics:** Data not available
5.10 **Odor Threshold:** Odorless

6. FIRE HAZARDS

6.1 **Flash Point:** Not flammable
6.2 **Flammable Limits in Air:** Not flammable
6.3 **Fire Extinguishing Agents:** Not pertinent
6.4 **Fire Extinguishing Agents Not to be Used:** Not pertinent
6.5 **Special Hazards of Combustion Products:** Toxic and irritating fumes of hydrogen iodide, iodine, and oxides of nitrogen may form in fire.
6.6 **Behavior in Fire:** Compound may sublime in fire and condense on cold surfaces.
6.7 **Ignition Temperature:** Not pertinent
6.8 **Electrical Hazard:** Not pertinent
6.9 **Burning Rate:** Not pertinent

7. CHEMICAL REACTIVITY

7.1 **Reactivity with Water:** No reaction
7.2 **Reactivity with Common Materials:**
7.3 **Stability During Transport:** Stable
7.4 **Neutralizing Agents for Acids and Caustics:** Not pertinent
7.5 **Polymerization:** Not pertinent
7.6 **Inhibitor of Polymerization:** Not pertinent

8. WATER POLLUTION

8.1 **Aquatic Toxicity:** Data not available
8.2 **Waterfowl Toxicity:** Data not available
8.3 **Biological Oxygen Demand (BOD):** Data not available
8.4 **Food Chain Concentration Potential:** None

9. SELECTED MANUFACTURERS

1. American Hoechst Corp.
   Chemicals and Plastics Division
   Somerville, N. J. 08876
2. Mallinckrodt Chemical Works
   223 Westside Avenue
   P. O. Box 384
   Jersey City, N. J. 07303
3. J. T. Baker Chemical Co.
   Phillipsburg, N. J. 08865

10. SHIPPING INFORMATION

10.1 **Grades or Purity:** NF; Reagent 99%
10.2 **Storage Temperature:** Ambient
10.3 **Inert Atmosphere:** No requirement
10.4 **Venting:** Open

11. HAZARD ASSESSMENT CODE

(See Hazard Assessment Handbook, CG 446-3)

SS

12. HAZARD CLASSIFICATIONS

12.1 **Code of Federal Regulations:** Not listed
12.2 **NAS Hazard Rating for Bulk Water Transportation:** Not listed
12.3 **NFPA Hazard Classifications:** Not listed

13. PHYSICAL AND CHEMICAL PROPERTIES

13.1 **Physical State at 15°C and 1 atm:** Solid
13.2 **Molecular Weight:** 144.94
13.3 **Boiling Point at 1 atm:** Not pertinent (sublimes)
13.4 **Freezing Point:** Not pertinent
13.5 **Critical Temperature:** Not pertinent
13.6 **Critical Pressure:** Not pertinent
13.7 **Specific Gravity:** 2.56 at 20°C (solid)
13.8 **Liquid Surface Tension:** Not pertinent
13.9 **Liquid-Water Interfacial Tension:** Not pertinent
13.10 **Vapor (Gas) Specific Gravity:** Not pertinent
13.11 **Ratio of Specific Heats of Vapor (Gas):** Not pertinent
13.12 **Latent Heat of Vaporization:** Not pertinent
13.13 **Heat of Combustion:** Not pertinent
13.14 **Heat of Decomposition:** Not pertinent
13.15 **Heat of Solution:** 43 Btu/lb = 24 cal/g = 1.0×10^5 J/kg
13.16 **Heat of Polymerization:** Not pertinent

NOTES

ALT

AMMONIUM LACTATE

Common Synonyms			
DL-Lactic acid, ammonium salt Ammonium lactate syrup	Solid or liquid	White	Odorless
	Sinks and mixes with water.		

Stop discharge if possible. Keep people away.
Avoid contact with solid and dust.
Isolate and remove discharged material.
Notify local health and pollution control agencies.

Fire

Combustible.
POISONOUS GASES MAY BE PRODUCED IN FIRE.
Wear goggles and self-contained breathing apparatus.
Extinguish with water or foam.

Exposure

CALL FOR MEDICAL AID.

DUST
Irritating to eyes, nose and throat.
If inhaled will cause coughing or difficult breathing.
If in eyes, hold eyelids open and flush with plenty of water.
If breathing has stopped, give artificial respiration.
If breathing is difficult, give oxygen.

LIQUID OR SOLID
Irritating to skin and eyes.
Harmful if swallowed.
Remove contaminated clothing and shoes.
Flush affected areas with plenty of water.
IF IN EYES, hold eyelids open and flush with plenty of water.
IF SWALLOWED and victim is CONSCIOUS, have victim drink water or milk
IF SWALLOWED and victim is UNCONSCIOUS OR HAVING CONVULSIONS, do nothing except keep victim warm.

Water Pollution

Effect of low concentrations on aquatic life is unknown.
May be dangerous if it enters water intakes.
Notify local health and wildlife officials.
Notify operators of nearby water intakes.

1. RESPONSE TO DISCHARGE

(See Response Methods Handbook, CG 446-4)

Disperse and flush

2. LABELS

No hazard label required by Code of Federal Regulations

3. CHEMICAL DESIGNATIONS

3.1 **Synonyms:** Ammonium lactate syrup; *dl*—Lactic acid; ammonium salt

3.2 **Coast Guard Compatibility Classification:** Not listed

3.3 **Chemical Formula:** $CH_3CH(OH)COONH_4$

3.4 **IMCO/United Nations Numerical Designation:** Not listed

4. OBSERVABLE CHARACTERISTICS

4.1 **Physical State (as shipped):** Solid or liquid

4.2 **Color:** White

4.3 **Odor:** None

5. HEALTH HAZARDS

5.1 **Personal Protective Equipment:** Dust mask; goggles or face shield; rubber gloves

5.2 **Symptoms Following Exposure:** Inhalation causes irritation of nose and throat. Contact with eyes causes irritation.

5.3 **Treatment for Exposure:** INHALATION: remove to fresh air. EYES: flush with water.

5.4 **Toxicity by Inhalation (Threshold Limit Value):** Data not available

5.5 **Short-Term Inhalation Limits:** Data not available

5.6 **Toxicity by Ingestion:** Data not available

5.7 **Late Toxicity:** Data not available

5.8 **Vapor (Gas) Irritant Characteristics:** Data not available

5.9 **Liquid or Solid Irritant Characteristics:** Data not available

5.10 **Odor Threshold:** Data not available

6. FIRE HAZARDS

6.1 **Flash Point:** Not pertinent (Combustible solid)

6.2 **Flammable Limits in Air:** Not pertinent

6.3 **Fire Extinguishing Agents:** Water, foam

6.4 **Fire Extinguishing Agents Not to be Used:** Not pertinent

6.5 **Special Hazards of Combustion Products:** Toxic oxides of nitrogen may be formed in a fire.

6.6 **Behavior in Fire:**

6.7 **Ignition Temperature:** Not pertinent

6.8 **Electrical Hazard:** Not pertinent

6.9 **Burning Rate:** Not pertinent

7. CHEMICAL REACTIVITY

7.1 **Reactivity with Water:** No reaction

7.2 **Reactivity with Common Materials:**

7.3 **Stability During Transport:** Stable

7.4 **Neutralizing Agents for Acids and Caustics:** Not pertinent

7.5 **Polymerization:** Not pertinent

7.6 **Inhibitor of Polymerization:** Not pertinent

8. WATER POLLUTION

8.1 **Aquatic Toxicity:** Data not available

8.2 **Waterfowl Toxicity:** Data not available

8.3 **Biological Oxygen Demand (BOD):** Data not available

8.4 **Food Chain Concentration Potential:** None

9. SELECTED MANUFACTURERS

1. Tanatex Chemical Corp.
   P. O. Box 388
   Lyndhurst, N. J. 07071
2. Pfaltz and Bauer, Inc.
   126-04 Northern Boulevard
   Flushing, N. Y. 11368
3. Gallard-Schlesinger Chemical Mfg. Co.
   584 Mineola Avenue
   Carle Place, N. Y. 11514

10. SHIPPING INFORMATION

10.1 **Grades or Purity:** Pure; 60% syrup in water

10.2 **Storage Temperature:** Ambient

10.3 **Inert Atmosphere:** No requirement

10.4 **Venting:** Open

11. HAZARD ASSESSMENT CODE

(See Hazard Assessment Handbook, CG 446-3)

SS

12. HAZARD CLASSIFICATIONS

12.1 **Code of Federal Regulations:** Not listed

12.2 **NAS Hazard Rating for Bulk Water Transportation:** Not listed

12.3 **NFPA Hazard Classifications:** Not listed

13. PHYSICAL AND CHEMICAL PROPERTIES

13.1 **Physical State at 15°C and 1 atm:** Solid

13.2 **Molecular Weight:** 107.11

13.3 **Boiling Point at 1 atm:** Not pertinent

13.4 **Freezing Point:** Not pertinent

13.5 **Critical Temperature:** Not pertinent

13.6 **Critical Pressure:** Not pertinent

13.7 **Specific Gravity:** 1.2 at 15°C (solid)

13.8 **Liquid Surface Tension:** Not pertinent

13.9 **Liquid-Water Interfacial Tension:** Not pertinent

13.10 **Vapor (Gas) Specific Gravity:** Not pertinent

13.11 **Ratio of Specific Heats of Vapor (Gas):** Not pertinent

13.12 **Latent Heat of Vaporization:** Not pertinent

13.13 **Heat of Combustion:** Not pertinent

13.14 **Heat of Decomposition:** Not pertinent

13.15 **Heat of Solution:** Data not available

13.16 **Heat of Polymerization:** Not pertinent

NOTES

ALS — AMMONIUM LAURYL SULFATE

Common Synonyms		
Dodecyl sulfate, ammonium salt Lauryl ammonium sulfate	Liquid	Light yellow
	May float or sink and mix with water.	

Stop discharge if possible. Keep people away.
Avoid contact with liquid.
Isolate and remove discharged material.
Notify local health and pollution control agencies.

Fire

Not flammable.
POISONOUS GASES MAY BE PRODUCED IN FIRE.
Wear goggles and self-contained breathing apparatus.

Exposure

CALL FOR MEDICAL AID.

LIQUID
Irritating to skin and eyes.
Harmful if swallowed.
Remove contaminated clothing and shoes.
Flush affected areas with plenty of water.
IF IN EYES, hold eyelids open and flush with plenty of water.
IF SWALLOWED and victim is CONSCIOUS, have victim drink water or milk.
IF SWALLOWED and victim is UNCONSCIOUS OR HAVING CONVULSIONS, do nothing except keep victim warm.

Water Pollution

Effect of low concentrations on aquatic life is unknown.
May be dangerous if it enters water intakes.
Notify local health and wildlife officials.
Notify operators of nearby water intakes.

1. RESPONSE TO DISCHARGE

(See Response Methods Handbook, CG 446-4)

Disperse and flush

2. LABELS

No hazard label required by Code of Federal Regulations

3. CHEMICAL DESIGNATIONS

3.1 **Synonyms:** Dodecyl sulfate, ammonium salt; Lauryl ammonium sulfate
3.2 **Coast Guard Compatibility Classification:** Not listed
3.3 **Chemical Formula:** $C_{12}H_{25}OSO_2ONH_4-H_2O$
3.4 **IMCO/United Nations Numerical Designation:** Not listed

4. OBSERVABLE CHARACTERISTICS

4.1 **Physical State (as shipped):** Liquid
4.2 **Color:** Light yellow
4.3 **Odor:** Data not available

5. HEALTH HAZARDS

5.1 **Personal Protective Equipment:** Rubber gloves; goggles or face shield
5.2 **Symptoms Following Exposure:** Contact with liquid irritates eyes and may have drying effect on the skin. Prolonged contact will cause skin irritation.
5.3 **Treatment for Exposure:** EYES OR SKIN: flush with water.
5.4 **Toxicity by Inhalation (Threshold Limit Value):** Data not available
5.5 **Short-Term Inhalation Limits:** Data not available
5.6 **Toxicity by Ingestion:** Data not available
5.7 **Late Toxicity:** Data not available
5.8 **Vapor (Gas) Irritant Characteristics:** Data not available
5.9 **Liquid or Solid Irritant Characteristics:** Data not available
5.10 **Odor Threshold:** Data not available

6. FIRE HAZARDS

6.1 **Flash Point:** Not pertinent (non-combustible water solution)
6.2 **Flammable Limits in Air:** Not pertinent
6.3 **Fire Extinguishing Agents:** Not pertinent
6.4 **Fire Extinguishing Agents Not to be Used:** Not pertinent
6.5 **Special Hazards of Combustion Products:** Toxic oxides of nitrogen and sulfur may form in fires.
6.6 **Behavior in Fire:**
6.7 **Ignition Temperature:** Not pertinent
6.8 **Electrical Hazard:** Not pertinent
6.9 **Burning Rate:** Not pertinent

7. CHEMICAL REACTIVITY

7.1 **Reactivity with Water:** No reaction
7.2 **Reactivity with Common Materials:**
7.3 **Stability During Transport:** Stable
7.4 **Neutralizing Agents for Acids and Caustics:** Not pertinent
7.5 **Polymerization:** Not pertinent
7.6 **Inhibitor of Polymerization:** Not pertinent

8. WATER POLLUTION

8.1 **Aquatic Toxicity:** Data not available
8.2 **Waterfowl Toxicity:** Data not available
8.3 **Biological Oxygen Demand (BOD):** Data not available
8.4 **Food Chain Concentration Potential:** None

9. SELECTED MANUFACTURERS

1. Richardson Company
   Organic Chemicals Div.
   100 New Street
   Paterson, N.J. 07501
2. Stepan Chemical Co.
   Edens and Winnetka
   Northfield, Ill. 60093

10. SHIPPING INFORMATION

10.1 **Grades or Purity:** 28—30% solution in water
10.2 **Storage Temperature:** Ambient
10.3 **Inert Atmosphere:** No requirement
10.4 **Venting:** Open

11. HAZARD ASSESSMENT CODE

(See Hazard Assessment Handbook, CG 446-3)

A—P

12. HAZARD CLASSIFICATIONS

12.1 **Code of Federal Regulations:** Not listed
12.2 **NAS Hazard Rating for Bulk Water Transportation:** Not listed
12.3 **NFPA Hazard Classifications:** Not listed

13. PHYSICAL AND CHEMICAL PROPERTIES

13.1 **Physical State at 15°C and 1 atm:** Liquid
13.2 **Molecular Weight:** 283 (solute only)
13.3 **Boiling Point at 1 atm:** Not pertinent
13.4 **Freezing Point:** Not pertinent
13.5 **Critical Temperature:** Not pertinent
13.6 **Critical Pressure:** Not pertinent
13.7 **Specific Gravity:** 1.03 at 20°C (liquid)
13.8 **Liquid Surface Tension:** Data not available
13.9 **Liquid-Water Interfacial Tension:** Not pertinent
13.10 **Vapor (Gas) Specific Gravity:** Not pertinent
13.11 **Ratio of Specific Heats of Vapor (Gas):** Not pertinent
13.12 **Latent Heat of Vaporization:** Not pertinent
13.13 **Heat of Combustion:** Not pertinent
13.14 **Heat of Decomposition:** Not pertinent
13.15 **Heat of Solution:** Not pertinent
13.16 **Heat of Polymerization:** Not pertinent

NOTES

AMB	AMMONIUM MOLYBDATE

Common Synonyms			
Molybdic acid (58%)	Solid	Colorless to greenish-yellow or white	Odorless
	Sinks and mixes with water.		

Stop discharge if possible. Keep people away.
Avoid contact with solid and dust.
Isolate and remove discharged material.
Notify local health and pollution control agencies.

Fire

Not flammable.
POISONOUS GASES MAY BE PRODUCED IN FIRE.
Wear goggles and self-contained breathing apparatus.

Exposure

CALL FOR MEDICAL AID.

DUST
Irritating to eyes, nose and throat.
If inhaled will cause coughing, or difficult breathing.
If in eyes, hold eyelids open and flush with plenty of water.
If breathing has stopped, give artificial respiration.
If breathing is difficult, give oxygen.

SOLID
Irritating to skin and eyes.
Harmful if swallowed.
Remove contaminated clothing and shoes.
Flush affected areas with plenty of water.
IF IN EYES, hold eyelids open and flush with plenty of water.
IF SWALLOWED and victim is CONSCIOUS, have victim drink water or milk.
IF SWALLOWED and victim is UNCONSCIOUS OR HAVING CONVULSIONS, do nothing except keep victim warm.

Water Pollution

Effect of low concentrations on aquatic life is unknown.
May be dangerous if it enters water intakes.
Notify local health and wildlife officials.
Notify operators of nearby water intakes.

1. RESPONSE TO DISCHARGE
(See Response Methods Handbook, CG 446-4)

Issue warning—water contaminant
Should be removed
Chemical and physical treatment

2. LABELS

No hazard label required by Code of Federal Regulations

3. CHEMICAL DESIGNATIONS

3.1 **Synonyms:** Molybdic acid, 85%
3.2 **Coast Guard Compatibility Classification:** Not listed
3.3 **Chemical Formula:** $(NH_4)_6Mo_7O_{24}\cdot 4H_2O$ (for CP and Reagent grades only)
3.4 **IMCO/United Nations Numerical Designation:** Not listed

4. OBSERVABLE CHARACTERISTICS

4.1 **Physical State (as shipped):** Solid
4.2 **Color:** White or greenish-yellow; colorless to yellow
4.3 **Odor:** None

5. HEALTH HAZARDS

5.1 **Personal Protective Equipment:** Dust mask; goggles or face shield; rubber gloves
5.2 **Symptoms Following Exposure:** Inhalation causes irritation of nose and throat. Contact with eyes causes irritation.
5.3 **Treatment for Exposure:** INHALATION: move to fresh air. EYES: flush with water for at least 15 min.
5.4 **Toxicity by Inhalation (Threshold Limit Value):** 5 mg/m^3 (as molybdenum)
5.5 **Short-Term Inhalation Limits:** Data not available
5.6 **Toxicity by Ingestion:** Grade 3; oral rat LD_{50} = 333 mg/kg
5.7 **Late Toxicity:** Data not available
5.8 **Vapor (Gas) Irritant Characteristics:** Data not available
5.9 **Liquid or Solid Irritant Characteristics:** Data not available
5.10 **Odor Threshold:** Data not available

6. FIRE HAZARDS

6.1 **Flash Point:** Not flammable
6.2 **Flammable Limits in Air:** Not flammable
6.3 **Fire Extinguishing Agents:** Not pertinent
6.4 **Fire Extinguishing Agents Not to be Used:** Not pertinent
6.5 **Special Hazards of Combustion Products:** Toxic oxides of nitrogen may form in fire.
6.6 **Behavior in Fire:**
6.7 **Ignition Temperature:** Not pertinent
6.8 **Electrical Hazard:** Not pertinent
6.9 **Burning Rate:** Not pertinent

7. CHEMICAL REACTIVITY

7.1 **Reactivity with Water:** No reaction
7.2 **Reactivity with Common Materials:**
7.3 **Stability During Transport:** Stable
7.4 **Neutralizing Agents for Acids and Caustics:** Not pertinent
7.5 **Polymerization:** Not pertinent
7.6 **Inhibitor of Polymerization:** Not pertinent

8. WATER POLLUTION

8.1 **Aquatic Toxicity:** Data not available
8.2 **Waterfowl Toxicity:** Data not available
8.3 **Biological Oxygen Demand (BOD):** Data not available
8.4 **Food Chain Concentration Potential:** Data not available

9. SELECTED MANUFACTURERS

1. J. T. Baker Chemical Co.
   Phillipsburg, N. J. 08865
2. Climax Molybdenum Company
   One Greenwich Plaza
   Greenwich, Conn. 06830
3. Var-lac-oid Chemical Co.
   666 South Front St.
   Elizabeth, N. J. 07202

10. SHIPPING INFORMATION

10.1 **Grades or Purity:** Reagent; CP. A closely related substance is known as "molybdic acid 85%".
10.2 **Storage Temperature:** Ambient
10.3 **Inert Atmosphere:** No requirement
10.4 **Venting:** Open

11. HAZARD ASSESSMENT CODE
(See Hazard Assessment Handbook, CG 446-3)

SS

12. HAZARD CLASSIFICATIONS

12.1 **Code of Federal Regulations:** Not listed
12.2 **NAS Hazard Rating for Bulk Water Transportation:** Not listed
12.3 **NFPA Hazard Classifications:** Not listed

13. PHYSICAL AND CHEMICAL PROPERTIES

13.1 **Physical State at 15°C and 1 atm:** Solid
13.2 **Molecular Weight:** 123.6
13.3 **Boiling Point at 1 atm:** Not pertinent
13.4 **Freezing Point:** Not pertinent
13.5 **Critical Temperature:** Not pertinent
13.6 **Critical Pressure:** Not pertinent
13.7 **Specific Gravity:** 1.4 at 20°C (solid)
13.8 **Liquid Surface Tension:** Not pertinent
13.9 **Liquid-Water Interfacial Tension:** Not pertinent
13.10 **Vapor (Gas) Specific Gravity:** Not pertinent
13.11 **Ratio of Specific Heats of Vapor (Gas):** Not pertinent
13.12 **Latent Heat of Vaporization:** Not pertinent
13.13 **Heat of Combustion:** Not pertinent
13.14 **Heat of Decomposition:** Not pertinent
13.15 **Heat of Solution:** Data not available
13.16 **Heat of Polymerization:** Not pertinent

NOTES

AMN — AMMONIUM NITRATE

Common Synonyms			
Nitram	Solid pellets or flakes	White to light gray or brown	Odorless
	Sinks and mixes with water.		

Call Fire department. Keep people away.
Evacuate area in case of large discharge.
Isolate and remove discharged material.
Notify local health and pollution control agencies.

Fire

May cause fire and explode on contact with combustibles.
CONTAINERS MAY EXPLODE IN FIRE.
POISONOUS GASES MAY BE PRODUCED WHEN HEATED.
Wear self-contained breathing apparatus.
Evacuate surrounding area.
Combat fires from protected location with unmanned hose holder or monitor nozzle.
Flood discharge area with water.
Cool exposed containers with water.
Continue cooling after fire has been extinguished.

Exposure

CALL FOR MEDICAL AID.

DUST
Irritating to eyes, nose, and throat.
If inhaled, may cause coughing or difficult breathing.
Move to fresh air.
If in eyes, hold eyelids open and flush with plenty of water.
If breathing has stopped, give artificial respiration.
If breathing is difficult, give oxygen.

Water Pollution

Effect of low concentrations on aquatic life is unknown.
May be dangerous if it enters water intakes.
Notify local health and wildlife officials.
Notify operators of nearby water intakes.

1. RESPONSE TO DISCHARGE

(See Response Methods Handbook, CG 446-4)

Disperse and flush

2. LABEL

Yellow

3. CHEMICAL DESIGNATIONS

3.1 **Synonyms:** Nitram
3.2 **Coast Guard Compatibility Classification:** Ammonia
3.3 **Chemical Formula:** NH_4NO_3
3.4 **IMCO/United Nations Numerical Designation:** 5.1/1943

4. OBSERVABLE CHARACTERISTICS

4.1 **Physical State (as shipped):** Solid
4.2 **Color:** Colorless (pure) to gray or brown (fertilizer grades)
4.3 **Odor:** None

5. HEALTH HAZARDS

5.1 **Personal Protective Equipment:** Wear self-contained breathing apparatus
5.2 **Symptoms Following Exposure:** Irritation of eyes and mucous membranes. Absorption via ingestion or inhalation causes urination and acid urine. Large amount causes systemic acidosis and metheglobinemia (abnormal hemoglobin).
5.3 **Treatment for Exposure:** Remove ıιom exposure — symptoms reversible.
5.4 **Toxicity by Inhalation (Threshold Limit Value):** Not pertinent
5.5 **Short-Term Inhalation Limits:** Not pertinent
5.6 **Toxicity by Ingestion:** Data not available
5.7 **Late Toxicity:** Data not available
5.8 **Vapor (Gas) Irritant Characteristics:** Not pertinent
5.9 **Liquid or Solid Irritant Characteristics:** None
5.10 **Odor Threshold:** Not pertinent

6. FIRE HAZARDS

6.1 **Flash Point:** Not flammable
6.2 **Flammable Limits in Air:** Not flammable
6.3 **Fire Extinguishing Agents:** Use flooding amounts of water in early stages of fire. When large quantities are involved in massive fires, control efforts should be confined to protecting from explosion.
6.4 **Fire Extinguishing Agents Not to be Used:** Not pertinent
6.5 **Special Hazards of Combustion Products:** Decomposes, giving off extremely toxic oxides of nitrogen.
6.6 **Behavior in Fire:** May explode in fires. Supports combustion of common organic fuels.
6.7 **Ignition Temperature:** Not flammable
6.8 **Electrical Hazard:** Not pertinent
6.9 **Burning Rate:** Not flammable

7. CHEMICAL REACTIVITY

7.1 **Reactivity with Water:** No reaction
7.2 **Reactivity with Common Materials:** No reaction
7.3 **Stability During Transport:** If heated strongly, decomposes, giving off toxic gases and gases which support combustion. Undergoes detonation if heated under confinement.
7.4 **Neutralizing Agents for Acids and Caustics:** Not pertinent
7.5 **Polymerization:** Not pertinent
7.6 **Inhibitor of Polymerization:** Not pertinent

8. WATER POLLUTION

8.1 **Aquatic Toxicity:** Data not available
8.2 **Waterfowl Toxicity:** Data not available
8.3 **Biological Oxygen Demand (BOD):** Data not available
8.4 **Food Chain Concentration Potential:** None

9. SELECTED MANUFACTURERS

1. Allied Chemical Corp.
   Agricultural Div.
   Morristown, N. J. 07960
2. Hercules Inc.
   Explosives & Chemical Propulsion Dept.
   Bessemer, Ala. 35020
3. Monsanto Co.
   Monsanto Commercial Products Co.
   Agricultural Div.
   800 North Lindbergh Blvd.
   St. Louis, Mo. 63166

10. SHIPPING INFORMATION

10.1 **Grades or Purity:** Pure grade; fertilizer grade (33.5% nitrogen)
10.2 **Storage Temperature:** Data not available
10.3 **Inert Atmosphere:** Data not available
10.4 **Venting:** Data not available

11. HAZARD ASSESSMENT CODE

(See Hazard Assessment Handbook, CG 446-3)

SS

12. HAZARD CLASSIFICATIONS

12.1 **Code of Federal Regulations:** Oxidizing material
12.2 **NAS Hazard Rating for Bulk Water Transportation:** Not listed
12.3 **NFPA Hazard Classifications:**

Category	Classification*	
Health Hazard (Blue)	0	2
Flammability (Red)	1	1
Reactivity (Yellow)	3	3
	oxy	oxy

*First column refers to non-fire situation.

13. PHYSICAL AND CHEMICAL PROPERTIES

13.1 **Physical State at 15°C and 1 atm:** Solid
13.2 **Molecular Weight:** 80.05
13.3 **Boiling Point at 1 atm:** Not pertinent
13.4 **Freezing Point:** 337.8°F = 169.9°C = 443.1°K
13.5 **Critical Temperature:** Not pertinent
13.6 **Critical Pressure:** Not pertinent
13.7 **Specific Gravity:** 1.72 at 20°C (solid)
13.8 **Liquid Surface Tension:** Not pertinent
13.9 **Liquid-Water Interfacial Tension:** Not pertinent
13.10 **Vapor (Gas) Specific Gravity:** Not pertinent
13.11 **Ratio of Specific Heats of Vapor (Gas):** Not pertinent
13.12 **Latent Heat of Vaporization:** Not pertinent
13.13 **Heat of Combustion:** Not pertinent
13.14 **Heat of Decomposition:** Not pertinent
13.15 **Heat of Solution:** Not pertinent
13.16 **Heat of Polymerization:** Not pertinent

NOTES

REVISED 1978

ANP	# AMMONIUM NITRATE-PHOSPHATE MIXTURE

Common Synonyms	Solid	Grayish-white	Odorless
	Sinks and mixes with water.		

Stop discharge if possible. Keep people away.
Call fire department.
Avoid contact with solid and dust.
Isolate and remove discharged material.

Fire	Not flammable. Will increase the intensity of a fire. POISONOUS GASES MAY BE PRODUCED IN FIRE. Irritating gases may be produced when heated. Containers may explode in fire. Wear goggles and self-contained breathing apparatus. Flood discharge area with water.
Exposure	CALL FOR MEDICAL AID. DUST Irritating to eyes, nose and throat. If inhaled will cause coughing or difficult breathing. If in eyes, hold eyelids open and flush with plenty of water. If breathing has stopped, give artificial respiration. If breathing is difficult, give oxygen. SOLID Irritating to skin and eyes. Harmful if swallowed. Remove contaminated clothing and shoes. Flush affected areas with plenty of water. IF IN EYES, hold eyelids open and flush with plenty of water. IF SWALLOWED and victim is CONSCIOUS, have victim drink water or milk. IF SWALLOWED and victim is UNCONSCIOUS OR HAVING CONVULSIONS, do nothing except keep victim warm.
Water Pollution	Effect of low concentrations on aquatic life is unknown. May be dangerous if it enters water intakes. Notify local health and wildlife officials. Notify operators of nearby water intakes.

1. RESPONSE TO DISCHARGE

(See Response Methods Handbook, CG 446-4)

Issue warning—oxidizing material
Disperse and flush

2. LABEL

OXIDIZER

Yellow

3. CHEMICAL DESIGNATIONS

3.1 **Synonyms:** No common synonyms
3.2 **Coast Guard Compatibility Classification:** Not listed
3.3 **Chemical Formula:** NH_4NO_3—$(NH_4)_2HPO_4$—$NH_4H_2PO_4$—$CaHPO_4$—KCl—K_2SO_4
3.4 **IMCO/United Nations Numerical Designation:** 9/—

4. OBSERVABLE CHARACTERISTICS

4.1 **Physical State (as shipped):** Solid
4.2 **Color:** Grayish-white
4.3 **Odor:** Characteristic

5. HEALTH HAZARDS

5.1 **Personal Protective Equipment:** Self-contained breathing apparatus must be used when fighting fires. At other times a dust mask is adequate.
5.2 **Symptoms Following Exposure:** Inhalation causes irritation of nose and throat. Contact with eyes causes irritation.
5.3 **Treatment for Exposure:** INHALATION: move to fresh air. EYES: flush with water for 15 min.
5.4 **Toxicity by Inhalation (Threshold Limit Value):** Data not available
5.5 **Short-Term Inhalation Limits:** Data not available
5.6 **Toxicity by Ingestion:** Data not available
5.7 **Late Toxicity:** Data not available
5.8 **Vapor (Gas) Irritant Characteristics:** Data not available
5.9 **Liquid or Solid Irritant Characteristics:** Data not available
5.10 **Odor Threshold:** Odorless

6. FIRE HAZARDS

6.1 **Flash Point:** Not pertinent
6.2 **Flammable Limits in Air:** Not pertinent
6.3 **Fire Extinguishing Agents:** Water
6.4 **Fire Extinguishing Agents Not to be Used:** Steam, inert gases, foam, dry chemical
6.5 **Special Hazards of Combustion Products:** Toxic and irritating oxides of nitrogen may form in fires.
6.6 **Behavior in Fire:** Will increase intensity of fire when in contact with combustible material. Containers may explode.
6.7 **Ignition Temperature:** Not pertinent
6.8 **Electrical Hazard:** Not pertinent
6.9 **Burning Rate:** Not pertinent

7. CHEMICAL REACTIVITY

7.1 **Reactivity with Water:** No reaction
7.2 **Reactivity with Common Materials:** Corrodes metals to same degree as ordinary fertilizer; the reaction is not hazardous.
7.3 **Stability During Transport:** Stable
7.4 **Neutralizing Agents for Acids and Caustics:** Not pertinent
.5 **Polymerization:** Not pertinent
7.6 **Inhibitor of Polymerization:** Not pertinent

8. WATER POLLUTION

8.1 **Aquatic Toxicity:** Data not available
8.2 **Waterfowl Toxicity:** Data not available
8.3 **Biological Oxygen Demand (BOD):** Data not available
8.4 **Food Chain Concentration Potential:** None

9. SELECTED MANUFACTURERS

Chevron Chemical Co.
940 Hensley Street
Richmond, Calif. 94804

10. SHIPPING INFORMATION

10.1 **Grades or Purity:** Several grades of varying composition, all of which contain less than 70% of ammonium nitrate, the only hazardous ingredient
10.2 **Storage Temperature:** Ambient
10.3 **Inert Atmosphere:** Ventilated (natural)
10.4 **Venting:** Open

11. HAZARD ASSESSMENT CODE

(See Hazard Assessment Handbook, CG 446-3)

SS

12. HAZARD CLASSIFICATIONS

12.1 **Code of Federal Regulations:** Oxidizing material
12.2 **NAS Hazard Rating for Bulk Water Transportation:** Not listed
12.3 **NFPA Hazard Classifications:** Not listed

13. PHYSICAL AND CHEMICAL PROPERTIES

13.1 **Physical State at 15°C and 1 atm:** Solid
13.2 **Molecular Weight:** Not pertinent (mixture)
13.3 **Boiling Point at 1 atm:** Not pertinent
13.4 **Freezing Point:** Not pertinent
13.5 **Critical Temperature:** Not pertinent
13.6 **Critical Pressure:** Not pertinent
13.7 **Specific Gravity:** 1.8 at 20°C (solid)
13.8 **Liquid Surface Tension:** Not pertinent
13.9 **Liquid-Water Interfacial Tension:** Not pertinent
13.10 **Vapor (Gas) Specific Gravity:** Not pertinent
13.11 **Ratio of Specific Heats of Vapor (Gas):** Not pertinent
13.12 **Latent Heat of Vaporization:** Not pertinent
13.13 **Heat of Combustion:** Not pertinent
13.14 **Heat of Decomposition:** Data not available
13.15 **Heat of Solution:** Data not available
13.16 **Heat of Polymerization:** Not pertinent

NOTES

ANS	AMMONIUM NITRATE-SULFATE MIXTURE

Common Synonyms	Solid	Grayish-white	Odorless
	Sinks and mixes with water.		

Stop discharge if possible. Keep people away.
Call fire department.
Avoid contact with solid and dust.
Isolate and remove discharged material.
Notify local health and pollution control agencies.

Fire

Not flammable
Will increase the intensity of a fire.
POISONOUS GASES MAY BE PRODUCED IN FIRE.
Irritating gases may be produced when heated.
Containers may explode in fire.
Wear goggles and self-contained breathing apparatus.
Flood discharge area with water.

Exposure

CALL FOR MEDICAL AID.

DUST
Irritating to eyes, nose and throat.
If inhaled will cause coughing, or difficult breathing.
If in eyes, hold eyelids open and flush with plenty of water.
If breathing has stopped, give artificial respiration.
If breathing is difficult, give oxygen.

SOLID
Irritating to skin and eyes.
Harmful if swallowed.
Remove contaminated clothing and shoes.
Flush affected areas with plenty of water.
IF IN EYES, hold eyelids open and flush with plenty of water.
IF SWALLOWED and victim is CONSCIOUS, have victim drink water or milk.
IF SWALLOWED and victim is UNCONSCIOUS OR HAVING CONVULSIONS, do nothing except keep victim warm.

Water Pollution

Effect of low concentrations on aquatic life is unknown.
May be dangerous if it enters water intakes.
Notify local health and wildlife officials.
Notify operators of nearby water intakes.

1. RESPONSE TO DISCHARGE

(See Response Methods Handbook, CG 446-4)

Issue warning—oxidizing material
Disperse and flush

2. LABELS

No hazard label required by Code of Federal Regulations

3. CHEMICAL DESIGNATIONS

3.1 **Synonyms:** No common synonyms
3.2 **Coast Guard Compatibility Classification:** Not listed
3.3 **Chemical Formula:** NH_4NO_3—$(NH_4)_2SO_4$
3.4 **IMCO/United Nations Numerical Designation:** 9/—

4. OBSERVABLE CHARACTERISTICS

4.1 **Physical State (as shipped):** Solid
4.2 **Color:** Grayish-white
4.3 **Odor:** Characteristic

5. HEALTH HAZARDS

5.1 **Personal Protective Equipment:** Self-contained breathing apparatus must be used when fighting fires. At other times a dust mask is adequate.
5.2 **Symptoms Following Exposure:** Inhalation causes irritation of nose and throat. Contact with eyes causes irritation.
5.3 **Treatment for Exposure:** INHALATION: move to fresh air. EYES: flush with water for 15 min.
5.4 **Toxicity by Inhalation (Threshold Limit Value):** Data not available
5.5 **Short-Term Inhalation Limits:** Data not available
5.6 **Toxicity by Ingestion:** Grade 3; oral rat LD_{50} = 58 mg/kg (ammonium sulfate)
5.7 **Late Toxicity:** Data not available
5.8 **Vapor (Gas) Irritant Characteristics:** Not pertinent
5.9 **Liquid or Solid Irritant Characteristics:** Data not available
5.10 **Odor Threshold:** Odorless

6. FIRE HAZARDS

6.1 **Flash Point:** Not pertinent
6.2 **Flammable Limits in Air:** Not pertinent
6.3 **Fire Extinguishing Agents:** Water
6.4 **Fire Extinguishing Agents Not to be Used:** Steam, inert gases, foam, dry chemical
6.5 **Special Hazards of Combustion Products:** Toxic and irritating oxides of nitrogen may form in fires.
6.6 **Behavior in Fire:** Will increase intensity of fire when in contact with combustible material. Containers may explode.
6.7 **Ignition Temperature:** Not pertinent
6.8 **Electrical Hazard:** Not pertinent
6.9 **Burning Rate:** Not pertinent

7. CHEMICAL REACTIVITY

7.1 **Reactivity with Water:** No reaction
7.2 **Reactivity with Common Materials:** Corrodes metals to same degree as ordinary fertilizer; the reaction is not hazardous.
7.3 **Stability During Transport:** Stable
7.4 **Neutralizing Agents for Acids and Caustics:** Not pertinent
7.5 **Polymerization:** Not pertinent
7.6 **Inhibitor of Polymerization:** Not pertinent

8. WATER POLLUTION

8.1 **Aquatic Toxicity:** Data not available
8.2 **Waterfowl Toxicity:** Data not available
8.3 **Biological Oxygen Demand (BOD):** Data not available
8.4 **Food Chain Concentration Potential:** None

9. SELECTED MANUFACTURERS

Chevron Chemical Co.
940 Hensley Street
Richmond, Calif. 94804

10. SHIPPING INFORMATION

10.1 **Grades or Purity:** Several grades of varying composition, all of which contain less than 70% of ammonium nitrate, the only hazardous ingredient.
10.2 **Storage Temperature:** Ambient
10.3 **Inert Atmosphere:** Ventilated (natural)
10.4 **Venting:** Open

11. HAZARD ASSESSMENT CODE

(See Hazard Assessment Handbook, CG 446-3)

SS

12. HAZARD CLASSIFICATIONS

12.1 **Code of Federal Regulations:** ORM-C
12.2 **NAS Hazard Rating for Bulk Water Transportation:** Not listed
12.3 **NFPA Hazard Classifications:** Not listed

13. PHYSICAL AND CHEMICAL PROPERTIES

13.1 **Physical State at 15°C and 1 atm:** Solid
13.2 **Molecular Weight:** Not pertinent
13.3 **Boiling Point at 1 atm:** Not pertinent
13.4 **Freezing Point:** Not pertinent
13.5 **Critical Temperature:** Not pertinent
13.6 **Critical Pressure:** Not pertinent
13.7 **Specific Gravity:** 1.8 at 20°C (solid)
13.8 **Liquid Surface Tension:** Not pertinent
13.9 **Liquid-Water Interfacial Tension:** Not pertinent
13.10 **Vapor (Gas) Specific Gravity:** Not pertinent
13.11 **Ratio of Specific Heats of Vapor (Gas):** Not pertinent
13.12 **Latent Heat of Vaporization:** Not pertinent
13.13 **Heat of Combustion:** Not pertinent
13.14 **Heat of Decomposition:** Data not available
13.15 **Heat of Solution:** Data not available
13.16 **Heat of Polymerization:** Not pertinent

NOTES

ANU

AMMONIUM NITRATE-UREA SOLUTION

Common Synonyms

Liquid | Colorless | Slight ammonia odor

Sinks and mixes with water.

Stop discharge if possible. Keep people away.
Isolate and remove discharged material.
Notify local health and pollution control agencies.

Fire

Not flammable.

Exposure

Call for medical aid.

LIQUID
Irritating to skin and eyes.
Harmful if swallowed.
Remove contaminated clothing and shoes.
Flush affected areas with plenty of water.
IF IN EYES, hold eyelids open and flush with plenty of water.
IF SWALLOWED and victim is CONSCIOUS, have victim drink water or milk.

Water Pollution

Effect of low concentrations on aquatic life is unknown.
May be dangerous if it enters water intakes.
Notify local health and wildlife officials.
Notify operators of nearby water intakes.

1. RESPONSE TO DISCHARGE

(See Response Methods Handbook, CG 446-4)

Disperse and flush

2. LABELS

No label required by Code of Federal Regulations

3. CHEMICAL DESIGNATIONS

3.1 **Synonyms:** Solar nitrogen solutions
Nitrex nitrogen solutions
(non-pressure)

3.2 **Coast Guard Compatibility Classification:** Misc. water solutions

3.3 **Chemical Formula:** $NH_4NO_3 - NH_2CONH_2 - H_2O$

3.4 **IMCO/United Nations Numerical Designation:** Not listed

4. OBSERVABLE CHARACTERISTICS

4.1 **Physical State (as shipped):** Liquid

4.2 **Color:** Colorless

4.3 **Odor:** Weak ammonia

5. HEALTH HAZARDS

5.1 **Personal Protective Equipment:** Goggles or face shield; rubber gloves.

5.2 **Symptoms Following Exposure:** Liquid irritates eyes, particularly those grades that contain a little free ammonia.

5.3 **Treatment for Exposure:** EYES: wash with plenty of water for 10-15 min. and rinse with a 5% boric acid solution; call a doctor. SKIN: rinse well with water.

5.4 **Toxicity by Inhalation (Threshold Limit Value):** Data not available

5.5 **Short-Term Inhalation Limits:** Data not available

5.6 **Toxicity by Ingestion:** Grade 2; LD_{50} 0.5 to 5 g/kg

5.7 **Late Toxicity:** None

5.8 **Vapor (Gas) Irritant Characteristics:** Vapors are nonirritating to the eyes and throat.

5.9 **Liquid or Solid Irritant Characteristics:** Minimum hazard. If spilled on clothing and allowed to remain, may cause smarting and reddening of the skin.

5.10 **Odor Threshold:** Not pertinent

6. FIRE HAZARDS

6.1 **Flash Point:** Not flammable

6.2 **Flammable Limits in Air:** Not flammable

6.3 **Fire Extinguishing Agents:** Not pertinent

6.4 **Fire Extinguishing Agents Not to be Used:** Not pertinent

6.5 **Special Hazards of Combustion Products:** Not pertinent

6.6 **Behavior in Fire:** Water of solution may evaporate, and remaining solids may then explode.

6.7 **Ignition Temperature:** Not pertinent

6.8 **Electrical Hazard:** Not pertinent

6.9 **Burning Rate:** Not pertinent

7. CHEMICAL REACTIVITY

7.1 **Reactivity with Water:** No reaction

7.2 **Reactivity with Common Materials:** No reaction

7.3 **Stability During Transport:** Stable

7.4 **Neutralizing Agents for Acids and Caustics:** Not pertinent

7.5 **Polymerization:** Not pertinent

7.6 **Inhibitor of Polymerization:** Not pertinent

8. WATER POLLUTION

8.1 **Aquatic Toxicity:** Data not available

8.2 **Waterfowl Toxicity:** Data not available

8.3 **Biological Oxygen Demand (BOD):** Data not available

8.4 **Food Chain Concentration Potential:** None

9. SELECTED MANUFACTURERS

1. Solar Nitrogen Chemicals
Lima, Ohio
2. W. R. Grace & Co.
Agricultural Chemicals Group
P. O. Box 277
Memphis, Tenn. 38101

10. SHIPPING INFORMATION

10.1 **Grades or Purity:** Data are for non-pressure solutions containing 44.3% ammonium nitrate, 35.4% urea, and 20.3% water. Other grades contain 0-70% ammonium nitrate, 0-43% urea, 0-36.8% free ammonia, and water. Those containing more than 2% free ammonia are stored under pressure (0-30 psig at 104°F); for hazards of these, see Ammonium Hydroxide.

10.2 **Storage Temperature:** Ambient

11. HAZARD ASSESSMENT CODE

(See Hazard Assessment Handbook, CG 446-3)

A-P

12. HAZARD CLASSIFICATIONS

12.1 **Code of Federal Regulations:** Not listed

12.2 **NAS Hazard Rating for Bulk Water Transportation:**

Category	Rating
Fire	0
Health	
Vapor Irritant	0
Liquid or Solid Irritant	1
Poisons	1
Water Pollution	
Human Toxicity	2
Aquatic Toxicity	2
Aesthetic Effect	0
Reactivity	
Other Chemicals	2
Water	0
Self-Reaction	0

12.3 **NFPA Hazard Classifications:** Not listed

13. PHYSICAL AND CHEMICAL PROPERTIES

13.1 **Physical State at 15°C and 1 atm:** Liquid

13.2 **Molecular Weight:** Not pertinent

13.3 **Boiling Point at 1 atm:** > 212°F = > 100°C = > 373°K

13.4 **Freezing Point:** 32°F = 0°C = 273°K

13.5 **Critical Temperature:** Not pertinent

13.6 **Critical Pressure:** Not pertinent

13.7 **Specific Gravity:** 1.327 at 20°C (liquid)

13.8 **Liquid Surface Tension:** Not pertinent

13.9 **Liquid-Water Interfacial Tension:** Not pertinent

13.10 **Vapor (Gas) Specific Gravity:** Not pertinent

13.11 **Ratio of Specific Heats of Vapor (Gas):** Not pertinent

13.12 **Latent Heat of Vaporization:** Not pertinent

13.13 **Heat of Combustion:** Not pertinent

13.14 **Heat of Decomposition:** Not pertinent

13.15 **Heat of Solution:** 0 Btu/lb = 0 cal/g = 0 J/kg

13.16 **Heat of Polymerization:** Not pertinent

10. SHIPPING INFORMATION (Cont'd.)

10.3 **Inert Atmosphere:** No requirement

10.4 **Venting:** Open; if > 2% free ammonia, then pressure-vacuum.

AOL	AMMONIUM OLEATE

Common Synonyms
Oleic acid, ammonium salt
Ammonia soap

Solid paste — Yellow-brown — Weak ammonia odor

Sinks and mixes with water.

Stop discharge if possible. Keep people away.
Avoid contact with solid.
Isolate and remove discharged material.
Notify local health and pollution control agencies.

Fire

Combustible.
POISONOUS GASES MAY BE PRODUCED IN FIRE.
Wear goggles and self-contained breathing apparatus.
Extinguish with dry chemicals or water.

Exposure

CALL FOR MEDICAL AID.

SOLID
Irritating to skin and eyes.
Harmful if swallowed.
Remove contaminated clothing and shoes.
Flush affected areas with plenty of water.
IF IN EYES, hold eyelids open and flush with plenty of water.
IF SWALLOWED and victim is CONSCIOUS, have victim drink water or milk.
IF SWALLOWED and victim is UNCONSCIOUS OR HAVING CONVULSIONS, do nothing except keep victim warm.

Water Pollution

Effect of low concentrations on aquatic life is unknown.
May be dangerous if it enters water intakes.
Notify local health and wildlife officials.
Notify operators of nearby water intakes.

1. RESPONSE TO DISCHARGE
(See Response Methods Handbook, CG 446-4)

Disperse and flush

2. LABELS

No hazard label required by Code of Federal Regulations

3. CHEMICAL DESIGNATIONS

3.1 **Synonyms:** Ammonia soap; Oleic acid, ammonium salt
3.2 **Coast Guard Compatibility Classification:** Not listed
3.3 **Chemical Formula:** $C_{17}H_{33}COONH_4—H_2O$
3.4 **IMCO/United Nations Numerical Designation:** Not listed

4. OBSERVABLE CHARACTERISTICS

4.1 **Physical State (as shipped):** Pasty solid
4.2 **Color:** Yellow-brown
4.3 **Odor:** Weak ammonia

5. HEALTH HAZARDS

5.1 **Personal Protective Equipment:** Goggles or face shield; rubber gloves
5.2 **Symptoms Following Exposure:** Contact with eyes causes irritation. Prolonged contact may cause skin irritation.
5.3 **Treatment for Exposure:** EYES OR SKIN: flush with water.
5.4 **Toxicity by Inhalation (Threshold Limit Value):** Not pertinent
5.5 **Short-Term Inhalation Limits:** Not pertinent
5.6 **Toxicity by Ingestion:** Data not available
5.7 **Late Toxicity:** Data not available
5.8 **Vapor (Gas) Irritant Characteristics:** Not pertinent
5.9 **Liquid or Solid Irritant Characteristics:** Data not available
5.10 **Odor Threshold:** Data not available

6. FIRE HAZARDS

6.1 **Flash Point:** Not pertinent
6.2 **Flammable Limits in Air:** Not pertinent
6.3 **Fire Extinguishing Agents:** Water, dry chemical
6.4 **Fire Extinguishing Agents Not to be Used:** Not pertinent
6.5 **Special Hazards of Combustion Products:** Toxic oxides of nitrogen may form in fire.
6.6 **Behavior in Fire:**
6.7 **Ignition Temperature:** Not pertinent
6.8 **Electrical Hazard:** Not pertinent
6.9 **Burning Rate:** Not pertinent

7. CHEMICAL REACTIVITY

7.1 **Reactivity with Water:** No reaction
7.2 **Reactivity with Common Materials:**
7.3 **Stability During Transport:** Stable
7.4 **Neutralizing Agents for Acids and Caustics:** Not pertinent
7.5 **Polymerization:** Not pertinent
7.6 **Inhibitor of Polymerization:** Not pertinent

8. WATER POLLUTION

8.1 **Aquatic Toxicity:** Data not available
8.2 **Waterfowl Toxicity:** Data not available
8.3 **Biological Oxygen Demand (BOD):** Data not available
8.4 **Food Chain Concentration Potential:** None

9. SELECTED MANUFACTURERS

1. Emkay Chemical Co.
319 Second Street
Elizabeth, N. J. 07206
2. Pfaltz and Bauer, Inc.
126-04 Northern Blvd.
Flushing, N. Y. 11368

10. SHIPPING INFORMATION

10.1 **Grades or Purity:** 70% in water; Technical
10.2 **Storage Temperature:** Ambient
10.3 **Inert Atmosphere:** Ventilated (natural)
10.4 **Venting:** Open

11. HAZARD ASSESSMENT CODE
(See Hazard Assessment Handbook, CG 446-3)

SS

12. HAZARD CLASSIFICATIONS

12.1 **Code of Federal Regulations:** Not listed
12.2 **NAS Hazard Rating for Bulk Water Transportation:** Data not available
12.3 **NFPA Hazard Classifications:** Data not available

13. PHYSICAL AND CHEMICAL PROPERTIES

13.1 **Physical State at 15°C and 1 atm:** Solid or liquid
13.2 **Molecular Weight:** 299.5 (solute)
13.3 **Boiling Point at 1 atm:** Not pertinent
13.4 **Freezing Point:** Not pertinent
13.5 **Critical Temperature:** Not pertinent
13.6 **Critical Pressure:** Not pertinent
13.7 **Specific Gravity:** >1 at 20°C (liquid or solid)
13.8 **Liquid Surface Tension:** Not pertinent
13.9 **Liquid-Water Interfacial Tension:** Not pertinent
13.10 **Vapor (Gas) Specific Gravity:** Not pertinent
13.11 **Ratio of Specific Heats of Vapor (Gas):** Not pertinent
13.12 **Latent Heat of Vaporization:** Not pertinent
13.13 **Heat of Combustion:** Data not available
13.14 **Heat of Decomposition:** Not pertinent
13.15 **Heat of Solution:** Not pertinent
13.16 **Heat of Polymerization:** Not pertinent

NOTES

AOX — AMMONIUM OXALATE

Common Synonyms			
Oxalic acid, diammonium salt Diammonium oxalate Ammonium oxalate hydrate	Solid	White	Odorless
	Sinks and mixes slowly with water.		

AVOID CONTACT WITH SOLID AND DUST. KEEP PEOPLE AWAY.
Wear dust respirator and rubber overclothing (including gloves).
Stop discharge if possible.
Call fire department.
Isolate and remove discharged material.
Notify local health and pollution control agencies.

Fire

Combustible.
POISONOUS GASES MAY BE PRODUCED IN FIRE.
Wear goggles and self-contained breathing apparatus.
Extinguish with water or foam.

Exposure

CALL FOR MEDICAL AID.

DUST
POISONOUS IF INHALED OR IF SKIN IS EXPOSED.
If inhaled will cause coughing or difficult breathing.
If in eyes, hold eyelids open and flush with plenty of water.
If breathing has stopped, give artificial respiration.
If breathing is difficult, give oxygen.

SOLID
POISONOUS IF SWALLOWED OR IF SKIN IS EXPOSED.
Will burn skin and eyes.
Remove contaminated clothing and shoes.
Flush affected areas with plenty of water.
IF IN EYES, hold eyelids open and flush with plenty of water.
IF SWALLOWED and victim is CONSCIOUS, have victim drink water or milk and have victim induce vomiting.
IF SWALLOWED and victim is UNCONSCIOUS OR HAVING CONVULSIONS, do nothing except keep victim warm.

Water Pollution

Effect of low concentrations on aquatic life is unknown.
May be dangerous if it enters water intakes.
Notify local health and wildlife officials.
Notify operators of nearby water intakes.

1. RESPONSE TO DISCHARGE

(See Response Methods Handbook, CG 446-4)

Issue warning—poison, water contaminant
Restrict access
Should be removed
Chemical and physical treatment

2. LABELS

No hazard label required by Code of Federal Regulations

3. CHEMICAL DESIGNATIONS

3.1 **Synonyms:** Ammonium oxalate hydrate; Diammonium oxalate; Oxalic acid, diammonium salt
3.2 **Coast Guard Compatibility Classification:** Not listed
3.3 **Chemical Formula:** $(NH_4)_2C_2O_4 \cdot H_2O$
3.4 **IMCO/United Nations Numerical Designation:** Not listed

4. OBSERVABLE CHARACTERISTICS

4.1 **Physical State (as shipped):** Solid
4.2 **Color:** White
4.3 **Odor:** None

5. HEALTH HAZARDS

5.1 **Personal Protective Equipment:** Approved dust respirator; rubber or plastic-coated gloves; chemical goggles
5.2 **Symptoms Following Exposure:** Ingestion or excessive inhalation of dust causes systemic poisoning; possible symptoms include pain in throat, esophagus, and stomach; mucous membranes turn white; vomiting, severe purging, weak pulse, cardiovascular collapse, neuromuscular symptoms. Contact with eyes causes irritation. Contact with skin causes irritation or severe burns.
5.3 **Treatment for Exposure:** *Speed is essential.* INHALATION: remove to fresh air. INGESTION: call physician immediately; induce vomiting. EYES: flush with water and seek medical attention. SKIN: flush with water. OTHER: watch for swelling of the glottis and delayed constriction of the esophagus.
5.4 **Toxicity by Inhalation (Threshold Limit Value):** Data not available
5.5 **Short-Term Inhalation Limits:** Data not available
5.6 **Toxicity by Ingestion:** Data not available
5.7 **Late Toxicity:** Kidney damage
5.8 **Vapor (Gas) Irritant Characteristics:** Data not available
5.9 **Liquid or Solid Irritant Characteristics:** Data not available
5.10 **Odor Threshold:** Odorless

6. FIRE HAZARDS

6.1 **Flash Point:** Not pertinent (combustible solid)
6.2 **Flammable Limits in Air:** Not pertinent
6.3 **Fire Extinguishing Agents:** Water, foam
6.4 **Fire Extinguishing Agents Not to be Used:** Not pertinent
6.5 **Special Hazards of Combustion Products:** Toxic oxides of nitrogen may form in fire.
6.6 **Behavior in Fire:**
6.7 **Ignition Temperature:** Not pertinent
6.8 **Electrical Hazard:** Not pertinent
6.9 **Burning Rate:** Not pertinent

7. CHEMICAL REACTIVITY

7.1 **Reactivity with Water:** No reaction
7.2 **Reactivity with Common Materials:**
7.3 **Stability During Transport:** Stable
7.4 **Neutralizing Agents for Acids and Caustics:** Not pertinent
7.5 **Polymerization:** Not pertinent
7.6 **Inhibitor of Polymerization:** Not pertinent

8. WATER POLLUTION

8.1 **Aquatic Toxicity:** Data not available
8.2 **Waterfowl Toxicity:** Data not available
8.3 **Biological Oxygen Demand (BOD):** Data not available
8.4 **Food Chain Concentration Potential:** None

9. SELECTED MANUFACTURERS

1. Pfizer Chemical Div.
   235 E. 42nd Street
   New York, N. Y. 10017
2. Heico, Inc.
   Delaware Water Gap, Pa. 18327
3. American Hoechst Corp.
   Chemicals and Plastics Div.
   Rt. 202-206 North
   Somerville, N. J. 08876

10. SHIPPING INFORMATION

10.1 **Grades or Purity:** Pure, 98—100.5%
10.2 **Storage Temperature:** Ambient
10.3 **Inert Atmosphere:** No requirement
10.4 **Venting:** Open

11. HAZARD ASSESSMENT CODE

(See Hazard Assessment Handbook, CG 446-3)

SS—II

12. HAZARD CLASSIFICATIONS

12.1 **Code of Federal Regulations:** Not listed
12.2 **NAS Hazard Rating for Bulk Water Transportation:** Not listed
12.3 **NFPA Hazard Classifications:** Not listed

13. PHYSICAL AND CHEMICAL PROPERTIES

13.1 **Physical State at 15°C and 1 atm:** Solid
13.2 **Molecular Weight:** 142.11
13.3 **Boiling Point at 1 atm:** Not pertinent (decomposes at 70°C)
13.4 **Freezing Point:** Not pertinent (decomposes at 70°C)
13.5 **Critical Temperature:** Not pertinent
13.6 **Critical Pressure:** Not pertinent
13.7 **Specific Gravity:** 1.50 at 18.5°C (solid)
13.8 **Liquid Surface Tension:** Not pertinent
13.9 **Liquid-Water Interfacial Tension:** Not pertinent
13.10 **Vapor (Gas) Specific Gravity:** Not pertinent
13.11 **Ratio of Specific Heats of Vapor (Gas):** Not pertinent
13.12 **Latent Heat of Vaporization:** Not pertinent
13.13 **Heat of Combustion:** Data not available
13.14 **Heat of Decomposition:** Not pertinent
13.15 **Heat of Solution:** 101 Btu/lb = 56 cal/g = 2.3×10^5 J/kg
13.16 **Heat of Polymerization:** Not pertinent

NOTES

APB

AMMONIUM PENTABORATE

Common Synonyms
Ammonium pentaborate tetrahydrate
Ammonium decaborate octahydrate

Solid | White | Odorless

Sinks and mixes slowly with water.

Stop discharge if possible. Keep people away.
Avoid contact with solid and dust.
Isolate and remove discharged material.
Notify local health and pollution control agencies.

Fire

Not flammable.
POISONOUS GASES MAY BE PRODUCED IN FIRE.
Wear goggles and self-contained breathing apparatus.

Exposure

CALL FOR MEDICAL AID.

DUST
Irritating to eyes, nose and throat.
If inhaled will cause coughing or difficult breathing.
If in eyes, hold eyelids open and flush with plenty of water.
If breathing has stopped, give artificial respiration.
If breathing is difficult, give oxygen.

SOLID
Irritating to skin and eyes.
If swallowed will cause nausea and vomiting.
Remove contaminated clothing and shoes.
Flush affected areas with plenty of water.
IF IN EYES, hold eyelids open and flush with plenty of water.
IF SWALLOWED and victim is CONSCIOUS, have victim drink water or milk and have victim induce vomiting.
IF SWALLOWED and victim is UNCONSCIOUS OR HAVING CONVULSIONS, do nothing except keep victim warm.

Water Pollution

Effect of low concentrations on aquatic life is unknown.
May be dangerous if it enters water intakes.
Notify local health and wildlife officials.
Notify operators of nearby water intakes.

1. RESPONSE TO DISCHARGE

(See Response Methods Handbook, CG 446-4)

Disperse and flush

2. LABELS

No hazard label required by Code of Federal Regulations

3. CHEMICAL DESIGNATIONS

3.1 **Synonyms:** Ammonium decaborate octahydrate; Ammonium pentaborate tetrahydrate
3.2 **Coast Guard Compatibility Classification:** Not listed
3.3 **Chemical Formula:** $NH_4B_5O_8 \cdot 4H_2O$ or $(NH_4)_2B_{10}O_{16} \cdot 8H_2O$
3.4 **IMCO/United Nations Numerical Designation:** Not listed

4. OBSERVABLE CHARACTERISTICS

4.1 **Physical State (as shipped):** Solid
4.2 **Color:** White
4.3 **Odor:** None

5. HEALTH HAZARDS

5.1 **Personal Protective Equipment:** Disposable-type respirator; side-shield safety spectacles; full-brimmed hard hat; goggles; ammonia gas mask
5.2 **Symptoms Following Exposure:** Inhalation of dust may result in non-specific irritation of upper respiratory tract. Ingestion of large quantities may produce symptoms of non-specific irritation of the gastrointestinal tract, nausea, vomiting, cramps, diarrhea. Contact with dust causes moderate eye irritation and minor skin irritation.
5.3 **Treatment for Exposure:** INHALATION: move from contaminated atmosphere; if respiratory discomfort persists, see a physician. INGESTION: give large amounts of water or warm salty water to induce vomiting and continue until vomitus is clear; obtain medical attention if abdominal discomfort persists. EYES: flush with large quantities of running water for a minimum of 15 min.; obtain medical help if irritation persists. SKIN: immediately flush affected areas with water; obtain medical help if irritation persists.
5.4 **Toxicity by Inhalation (Threshold Limit Value):** 10 mg/m^3 (as boric oxide)
5.5 **Short-Term Inhalation Limits:** Data not available
5.6 **Toxicity by Ingestion:** Data not available
5.7 **Late Toxicity:** Data not available
5.8 **Vapor (Gas) Irritant Characteristics:** Data not available
5.9 **Liquid or Solid Irritant Characteristics:** Data not available
5.10 **Odor Threshold:** Data not available

6. FIRE HAZARDS

6.1 **Flash Point:** Not flammable
6.2 **Flammable Limits in Air:** Not flammable
6.3 **Fire Extinguishing Agents:** Not pertinent
6.4 **Fire Extinguishing Agents Not to be Used:** Not pertinent
6.5 **Special Hazards of Combustion Products:** Toxic oxides of nitrogen may form in fire.
6.6 **Behavior in Fire:**
6.7 **Ignition Temperature:** Not pertinent
6.8 **Electrical Hazard:** Not pertinent
6.9 **Burning Rate:** Not pertinent

7. CHEMICAL REACTIVITY

7.1 **Reactivity with Water:** No reaction
7.2 **Reactivity with Common Materials:**
7.3 **Stability During Transport:** Stable
7.4 **Neutralizing Agents for Acids and Caustics:** Not pertinent
7.5 **Polymerization:** Not pertinent
7.6 **Inhibitor of Polymerization:** Not pertinent

8. WATER POLLUTION

8.1 **Aquatic Toxicity:** Data not available
8.2 **Waterfowl Toxicity:** Data not available
8.3 **Biological Oxygen Demand (BOD):** Data not available
8.4 **Food Chain Concentration Potential:** None

9. SELECTED MANUFACTURERS

1. U. S. Borax and Chemical Corp.
   3075 Wilshire Blvd.
   Los Angeles, Calif. 90010
2. American Hoechst Corp.
   Chemicals and Plastics Div.
   Rt. 206-206 North
   Somerville, N. J. 08876
3. Stauffer Chemical Co.
   Industrial Chemical Div.
   Westport, Conn. 06880

10. SHIPPING INFORMATION

10.1 **Grades or Purity:** Pure, 99%; Radio, 99.98%; Technical, 99.8%
10.2 **Storage Temperature:** Ambient
10.3 **Inert Atmosphere:** No requirement
10.4 **Venting:** Open

11. HAZARD ASSESSMENT CODE

(See Hazard Assessment Handbook, CG 446-3)

SS

12. HAZARD CLASSIFICATIONS

12.1 **Code of Federal Regulations:** Not listed
12.2 **NAS Hazard Rating for Bulk Water Transportation:** Not pertinent
12.3 **NFPA Hazard Classifications:** Not pertinent

13. PHYSICAL AND CHEMICAL PROPERTIES

13.1 **Physical State at 15°C and 1 atm:** Solid
13.2 **Molecular Weight:** 272.20
13.3 **Boiling Point at 1 atm:** Not pertinent
13.4 **Freezing Point:** Not pertinent
13.5 **Critical Temperature:** Not pertinent
13.6 **Critical Pressure:** Not pertinent
13.7 **Specific Gravity:** 1.58 at 15°C (solid)
13.8 **Liquid Surface Tension:** Not pertinent
13.9 **Liquid-Water Interfacial Tension:** Not pertinent
13.10 **Vapor (Gas) Specific Gravity:** Not pertinent
13.11 **Ratio of Specific Heats of Vapor (Gas):** Not pertinent
13.12 **Latent Heat of Vaporization:** Not pertinent
13.13 **Heat of Combustion:** Not pertinent
13.14 **Heat of Decomposition:** Not pertinent
13.15 **Heat of Solution:** Data not available
13.16 **Heat of Polymerization:** Not pertinent

NOTES

AMP

AMMONIUM PERCHLORATE

Common Synonyms	Solid crystals — White — Odorless Sinks and mixes with water.

Wear goggles and self-contained breathing apparatus.
Isolate and remove discharged material.
Notify local health and pollution control agencies.

Fire	May cause fire and explode on contact with combustibles. CONTAINERS MAY EXPLODE IN FIRE. POISONOUS GASES MAY BE PRODUCED IN FIRE. Wear goggles and self-contained breathing apparatus. Evacuate surrounding area. Combat large fires from protected location with unmanned hose holder or monitor nozzle.
Exposure	CALL FOR MEDICAL AID. SOLID Irritating to eyes and skin. Harmful if swallowed. Flush affected areas with plenty of water. IF IN EYES, hold eyelids open and flush with plenty of water. IF SWALLOWED and victim is CONSCIOUS, have victim drink water or milk.
Water Pollution	Effect of low concentrations on aquatic life is unknown. May be dangerous if it enters water intakes. Notify local health and wildlife officials. Notify operators of nearby water intakes.

1. RESPONSE TO DISCHARGE

(See Response Methods Handbook, CG 446-4)

Issue warning—oxidizing material
Disperse and flush

2. LABEL

Yellow

3. CHEMICAL DESIGNATIONS

3.1 **Synonyms:** No common synonyms
3.2 **Coast Guard Compatibility Classification:** Not applicable
3.3 **Chemical Formula:** NH_4ClO_4
3.4 **IMCO/United Nations Numerical Designation:** 5.1/1442

4. OBSERVABLE CHARACTERISTICS

4.1 **Physical State (as shipped):** Solid
4.2 **Color:** White
4.3 **Odor:** None

5. HEALTH HAZARDS

5.1 **Personal Protective Equipment:** Data not available
5.2 **Symptoms Following Exposure:** Irritating to skin and mucous membranes.
5.3 **Treatment for Exposure:** Data not available
5.4 **Toxicity by Inhalation (Threshold Limit Value):** Not pertinent
5.5 **Short-Term Inhalation Limits:** Not pertinent
5.6 **Toxicity by Ingestion:** Grade 2; oral rat LD_{50} = 3500 mg/kg
5.7 **Late Toxicity:** Not pertinent
5.8 **Vapor (Gas) Irritant Characteristics:** Not pertinent
5.9 **Liquid or Solid Irritant Characteristics:** Data not available
5.10 **Odor Threshold:** Not pertinent

6. FIRE HAZARDS

6.1 **Flash Point:** Not flammable
6.2 **Flammable Limits in Air:** Not flammable
6.3 **Fire Extinguishing Agents:** Water (from protected location)
6.4 **Fire Extinguishing Agents Not to be Used:** Not pertinent
6.5 **Special Hazards of Combustion Products:** Toxic gases are produced in a fire.
6.6 **Behavior in Fire:** May explode when involved in fire or exposed to shock or friction.
6.7 **Ignition Temperature:** Not flammable
6.8 **Electrical Hazard:** Not pertinent
6.9 **Burning Rate:** Not flammable

7. CHEMICAL REACTIVITY

7.1 **Reactivity with Water:** No reaction
7.2 **Reactivity with Common Materials:** No reaction
7.3 **Stability During Transport:** If contaminated with carbonaceous materials, can become an explosive which is sensitive to shock and friction. Readily detonates or explodes.
7.4 **Neutralizing Agents for Acids and Caustics:** Not pertinent
7.5 **Polymerization:** Not pertinent
7.6 **Inhibitor of Polymerization:** Not pertinent

8. WATER POLLUTION

8.1 **Aquatic Toxicity:** Data not available
8.2 **Waterfowl Toxicity:** Data not available
8.3 **Biological Oxygen Demand (BOD):** None
8.4 **Food Chain Concentration Potential:** None

9. SELECTED MANUFACTURERS

1. Kerr McGee Chemical Corp.
   Kerr McGee Bldg.
   Oklahoma City, Okla. 73102
2. Pacific Engineering & Production Co. of Nevada
   Henderson, Nevada 89015
3. G. Frederick Smith Chemical Co.
   P. O. Box 23344
   Columbus, Ohio 43223

10. SHIPPING INFORMATION

10.1 **Grades or Purity:** Data not available
10.2 **Storage Temperature:** Data not available
10.3 **Inert Atmosphere:** Data not available
10.4 **Venting:** Data not available

11. HAZARD ASSESSMENT CODE

(See Hazard Assessment Handbook, CG 446-3)

SS

12. HAZARD CLASSIFICATIONS

12.1 **Code of Federal Regulations:** Oxidizing material
12.2 **NAS Hazard Rating for Bulk Water Transportation:** Not listed
12.3 **NFPA Hazard Classifications:**

Category	Classification*	
Health Hazard (Blue)	0	2
Flammability (Red)	1	1
Reactivity (Yellow)	4	4
	oxy	oxy

*First column refers to non-fire situation.

13. PHYSICAL AND CHEMICAL PROPERTIES

13.1 **Physical State at 15°C and 1 atm:** Solid
13.2 **Molecular Weight:** 117.49
13.3 **Boiling Point at 1 atm:** Not pertinent
13.4 **Freezing Point:** Not pertinent
13.5 **Critical Temperature:** Not pertinent
13.6 **Critical Pressure:** Not pertinent
13.7 **Specific Gravity:** 1.95 at 15°C (solid)
13.8 **Liquid Surface Tension:** Not pertinent
13.9 **Liquid-Water Interfacial Tension:** Not pertinent
13.10 **Vapor (Gas) Specific Gravity:** Not pertinent
13.11 **Ratio of Specific Heats of Vapor (Gas):** Not pertinent
13.12 **Latent Heat of Vaporization:** Not pertinent
13.13 **Heat of Combustion:** Not pertinent
13.14 **Heat of Decomposition:** Not pertinent
13.15 **Heat of Solution:** Not pertinent
13.16 **Heat of Polymerization:** Not pertinent

NOTES

REVISED 1978

APE AMMONIUM PERSULFATE

Common Synonyms			
Ammonium peroxydisulfate Peroxydisulfuric acid, diammonium salt	Solid	Light straw to colorless	Mild unpleasant odor
	Sinks and mixes with water.		

Stop discharge if possible. Keep people away.
Call fire department.
Avoid contact with solid and dust.
Isolate and remove discharged material.
Notify local health and pollution control agencies.

Fire	Not flammable. Will increase the intensity of a fire. POISONOUS GASES MAY BE PRODUCED IN FIRE. Wear goggles and self-contained breathing apparatus. Flood discharge area with water.
Exposure	CALL FOR MEDICAL AID. DUST Irritating to eyes, nose and throat. Harmful if inhaled. If in eyes, hold eyelids open and flush with plenty of water. If breathing has stopped, give artificial respiration. If breathing is difficult, give oxygen. SOLID Irritating to skin and eyes. Harmful if swallowed. Remove contaminated clothing and shoes. Flush affected areas with plenty of water. IF IN EYES, hold eyelids open and flush with plenty of water. IF SWALLOWED and victim is CONSCIOUS, have victim drink water or milk. IF SWALLOWED and victim is UNCONSCIOUS OR HAVING CONVULSIONS, do nothing except keep victim warm.
Water Pollution	Dangerous to aquatic life in high concentrations. May be dangerous if it enters water intakes. Notify local health and wildlife officials. Notify operators of nearby water intakes.

1. RESPONSE TO DISCHARGE

(See Response Methods Handbook, CG 446-4)

Issue warning—oxidizing material
Disperse and flush

2. LABEL

Yellow

3. CHEMICAL DESIGNATIONS

3.1 **Synonyms:** Ammonium peroxydisulfate; Peroxydisulfuric acid, diammonium salt
3.2 **Coast Guard Compatibility Classification:** Not listed
3.3 **Chemical Formula:** $(NH_4)_2S_2O_8$
3.4 **IMCO/United Nations Numerical Designation:** 5.1/1444

4. OBSERVABLE CHARACTERISTICS

4.1 **Physical State (as shipped):** Solid
4.2 **Color:** Light straw to colorless
4.3 **Odor:** Slight acrid

5. HEALTH HAZARDS

5.1 **Personal Protective Equipment:** U. S. Bu. Mines approved toxic dust mask; chemical goggles; rubber gloves; neoprene-coated shoes
5.2 **Symptoms Following Exposure:** Inhalation produces slight toxic effects. Contact with dust irritates eyes and causes skin rash.
5.3 **Treatment for Exposure:** INHALATION: remove to fresh air. EYES: wash with water for 20 min.; call a physician. SKIN: wash with water.
5.4 **Toxicity By Inhalation (Threshold Limit Value):** Data not available
5.5 **Short-Term Inhalation Limits:** Data not available
5.6 **Toxicity by Ingestion:** Grade 2; oral rat LD_{50} = 820 mg/kg
5.7 **Late Toxicity:** Data not available
5.8 **Vapor (Gas) Irritant Characteristics:** Data not available
5.9 **Liquid or Solid Irritant Characteristics:** Data not available
5.10 **Odor Threshold:** Data not available

6. FIRE HAZARDS

6.1 **Flash Point:** Not pertinent
6.2 **Flammable Limits in Air:** Not pertinent
6.3 **Fire Extinguishing Agents:** Water
6.4 **Fire Extinguishing Agents Not to be Used:**
6.5 **Special Hazards of Combustion Products:** Toxic oxides of nitrogen and sulfuric acid fumes may form in fire.
6.6 **Behavior in Fire:** Decomposes with loss of oxygen that increases intensity of fire
6.7 **Ignition Temperature:** Not pertinent
6.8 **Electrical Hazard:** Not pertinent
6.9 **Burning Rate:** Not pertinent

7. CHEMICAL REACTIVITY

7.1 **Reactivity with Water:** No reaction
7.2 **Reactivity with Common Materials:** Contact with grease, wood, and othe combustibles may result in a fire.
7.3 **Stability During Transport:** Stable
7.4 **Neutralizing Agents for Acids and Caustics:** Not pertinent
7.5 **Polymerization:** Not pertinent
7.6 **Inhibitor of Polymerization:** Not pertinent

8. WATER POLLUTION

8.1 **Aquatic Toxicity:** 120 ppm/48 hr/daphnia/TL_m/fresh water
8.2 **Waterfowl Toxicity:** Data not available
8.3 **Biological Oxygen Demand (BOD):** None
8.4 **Food Chain Concentration Potential:** None

9. SELECTED MANUFACTURERS

1. FMC Corporation
6810 River Rd. and Sawyer Ave.
Tonawanda, N. Y. 14150
2. Allied Chemical Corp.
Specialty Chemicals Div.
P. O. Box 1087R
Morristown, N. J. 07960
3. J. T. Baker Chemical Co.
Phillipsburg, N. J. 08865

10. SHIPPING INFORMATION

10.1 **Grades or Purity:** Reagent; pure
10.2 **Storage Temperature:** Ambient
10.3 **Inert Atmosphere:** No requirement
10.4 **Venting:** Open

11. HAZARD ASSESSMENT CODE

(See Hazard Assessment Handbook, CG 446-3)

SS

12. HAZARD CLASSIFICATIONS

12.1 **Code of Federal Regulations:** Oxidizing solid
12.2 **NAS Hazard Rating for Bulk Water Transportation:** Not listed
12.3 **NFPA Hazard Classifications:** Not listed

13. PHYSICAL AND CHEMICAL PROPERTIES

13.1 **Physical State at 15°C and 1 atm:** Solid
13.2 **Molecular Weight:** 228.20
13.3 **Boiling Point at 1 atm:** Not pertinent (decomposes at 120°C)
13.4 **Freezing Point:** Not pertinent (decomposes at 120°C)
13.5 **Critical Temperature:** Not pertinent
13.6 **Critical Pressure:** Not pertinent
13.7 **Specific Gravity:** 1.98 at 20°C (solid)
13.8 **Liquid Surface Tension:** Not pertinent
13.9 **Liquid-Water Interfacial Tension:** Not pertinent
13.10 **Vapor (Gas) Specific Gravity:** Not pertinent
13.11 **Ratio of Specific Heats of Vapor (Gas):** Not pertinent
13.12 **Latent Heat of Vaporization:** Not pertinent
13.13 **Heat of Combustion:** Not pertinent
13.14 **Heat of Decomposition:** Data not available
13.15 **Heat of Solution:** 77 Btu/lb = 43 cal/g = 1.8×10^5 J/kg
13.16 **Heat of Polymerization:** Not pertinent

NOTES

APP — AMMONIUM PHOSPHATE

Common Synonyms			
Monoammonium orthophosphate Ammonium phosphate, dibasic Secondary ammonium phosphate Diammonium hydrogen phosphate Diammonium orthophosphate	Solid	White	Weak ammonia odor
	Sinks and mixes with water.		

Stop discharge if possible. Keep people away.
Avoid contact with solid and dust.
Isolate and remove discharged material.
Notify local health and pollution control agencies.

Fire

Not flammable.
Irritating gases may be produced when heated.

Exposure

CALL FOR MEDICAL AID.

DUST
Irritating to eyes, nose and throat.
If inhaled will cause coughing or difficult breathing.
If in eyes, hold eyelids open and flush with plenty of water.
If breathing has stopped, give artificial respiration.
If breathing is difficult, give oxygen.

SOLID
Irritating to skin and eyes.
Harmful if swallowed.
Remove contaminated clothing and shoes.
Flush affected areas with plenty of water.
IF IN EYES, hold eyelids open and flush with plenty of water.
IF SWALLOWED and victim is CONSCIOUS, have victim drink water or milk.
IF SWALLOWED and victim is UNCONSCIOUS OR HAVING CONVULSIONS, do nothing except keep victim warm.

Water Pollution

Effect of low concentrations on aquatic life is unknown.
May be dangerous if it enters water intakes.
Notify local health and wildlife officials.
Notify operators of nearby water intakes.

1. RESPONSE TO DISCHARGE

(See Response Methods Handbook, CG 446-4)

Disperse and flush

2. LABELS

No hazard label required by Code of Federal Regulations

3. CHEMICAL DESIGNATIONS

3.1 **Synonyms:** Ammonium phosphate, dibasic
Diammonium hydrogen phosphate
Diammonium orthophosphate
Monoammonium orthophosphate
Secondary ammonium phosphate

3.2 **Coast Guard Compatibility Classification:** Not listed

3.3 **Chemical Formula:** $NH_4H_2PO_4$ and $(NH_4)_2HPO_4$

(Continued on page 4)

4. OBSERVABLE CHARACTERISTICS

4.1 **Physical State (as shipped):** Solid

4.2 **Color:** White

4.3 **Odor:** Diammonium – faint ammonia
Monoammonium – faint acid

5. HEALTH HAZARDS

5.1 **Personal Protective Equipment:** Dust mask, protective gloves, and goggles. When diammonium phosphate is stored in closed area, self-contained breathing apparatus is required to protect against ammonia fumes.

5.2 **Symptoms Following Exposure:** Inhalation of monoammonium form causes irritation of mucous membranes; with diammonium form, ammonia vapors in closed area can cause pulmonary edema and asphyxia. Contact with solid or with ammonia gas causes irritation of eyes and skin.

5.3 **Treatment for Exposure:** INHALATION: if exposed to ammonia fumes from diammonium phosphate, give artificial respiration and oxygen if needed; enforce rest. EYES: flush with water for at least 15 min.; if irritation persists, get medical attention. SKIN: flush with water.

5.4 **Toxicity by Inhalation (Threshold Limit Value):** Data not available

5.5 **Short-Term Inhalation Limits:** Data not available

5.6 **Toxicity by Ingestion:** Data not available

5.7 **Late Toxicity:** Data not available

5.8 **Vapor (Gas) Irritant Characteristics:** Data not available

5.9 **Liquid or Solid Irritant Characteristics:** Data not available

5.10 **Odor Threshold:** Odorless

6. FIRE HAZARDS

6.1 **Flash Point:** Not flammable

6.2 **Flammable Limits in Air:** Not flammable

6.3 **Fire Extinguishing Agents:** Not pertinent

6.4 **Fire Extinguishing Agents Not to be Used:** Not pertinent

6.5 **Special Hazards of Combustion Products:** Toxic and irritating fumes of ammonia and oxides of nitrogen may form in fires.

6.6 **Behavior in Fire:**

6.7 **Ignition Temperature:** Not pertinent

6.8 **Electrical Hazard:** Not pertinent

6.9 **Burning Rate:** Not pertinent

7. CHEMICAL REACTIVITY

7.1 **Reactivity with Water:** No reaction

7.2 **Reactivity with Common Materials:**

7.3 **Stability During Transport:** Stable

7.4 **Neutralizing Agents for Acids and Caustics:** Not pertinent

7.5 **Polymerization:** Not pertinent

7.6 **Inhibitor of Polymerization:** Not pertinent

8. WATER POLLUTION

8.1 **Aquatic Toxicity:** 155 ppm/96 hr/fathead minnow/LC_{50}

8.2 **Waterfowl Toxicity:** Data not available

8.3 **Biological Oxygen Demand (BOD):** Data not available

8.4 **Food Chain Concentration Potential:** None

9. SELECTED MANUFACTURERS

1. J. T. Baker Chemical Co.
Phillipsburg, N. J. 08865
2. Mobil Oil Corporation
150 E. 42nd Street
New York, N. Y. 10017
3. Occidental Chemical Co.
P. O. Box 198
Lathrop, Calif. 95330

10. SHIPPING INFORMATION

10.1 **Grades or Purity:** Reagent; Technical

10.2 **Storage Temperature:** Ambient

10.3 **Inert Atmosphere:** Ventilated (forced)

10.4 **Venting:** Open

11. HAZARD ASSESSMENT CODE

(See Hazard Assessment Handbook, CG 446-3)

SS

12. HAZARD CLASSIFICATIONS

12.1 **Code of Federal Regulations:** Not listed

12.2 **NAS Hazard Rating for Bulk Water Transportation:** Not listed

12.3 **NFPA Hazard Classifications:** Not listed

13. PHYSICAL AND CHEMICAL PROPERTIES

13.1 **Physical State at 15°C and 1 atm:** Solid

13.2 **Molecular Weight:**
Monoammonium: 115
Diammonium: 132

13.3 **Boiling Point at 1 atm:** Not pertinent (begins to decompose at 100°C)

13.4 **Freezing Point:** Not pertinent (begins to decompose at 100°C)

13.5 **Critical Temperature:** Not pertinent

13.6 **Critical Pressure:** Not pertinent

13.7 **Specific Gravity:**
Diammonium: 1.8 at 20°C (solid)
Monoammonium: 1.6 at 20°C (solid)

13.8 **Liquid Surface Tension:** Not pertinent

13.9 **Liquid-Water Interfacial Tension:** Not pertinent

13.10 **Vapor (Gas) Specific Gravity:** Not pertinent

13.11 **Ratio of Specific Heats of Vapor (Gas):** Not pertinent

13.12 **Latent Heat of Vaporization:** Not pertinent

13.13 **Heat of Combustion:** Not pertinent

13.14 **Heat of Decomposition:** Not pertinent

13.15 **Heat of Solution:** 42 Btu/lb = 23 cal/g = 0.97×10^5 J/kg

13.16 **Heat of Polymerization:** Not pertinent

3. CHEMICAL DESIGNATIONS (Cont'd)

3.4 **IMCO/United Nations Numerical Designation:** Not listed

ASL	AMMONIUM SILICOFLUORIDE

Common Synonyms Ammonium fluosilicate	Solid White Odorless Sinks and mixes slowly with water.

AVOID CONTACT WITH SOLID AND DUST. KEEP PEOPLE AWAY.
Wear dust respirator and rubber overclothing (including gloves).
Stop discharge if possible.
Isolate and remove discharged material.
Notify local health and pollution control agencies.

Fire	Not flammable. POISONOUS GASES MAY BE PRODUCED IN FIRE. Irritating gases may be produced when heated. Wear goggles and self-contained breathing apparatus.
Exposure	CALL FOR MEDICAL AID. DUST POISONOUS IF INHALED OR IF SKIN IS EXPOSED. If inhaled will cause coughing or difficult breathing. If in eyes, hold eyelids open and flush with plenty of water. If breathing has stopped, give artificial respiration. If breathing is difficult, give oxygen. SOLID POISONOUS IF SWALLOWED OR IF SKIN IS EXPOSED. Will burn skin and eyes. Harmful if swallowed. Remove contaminated clothing and shoes. Flush affected areas with plenty of water. IF IN EYES, hold eyelids open and flush with plenty of water. IF SWALLOWED and victim is CONSCIOUS, have victim drink water or milk and have victim induce vomiting. IF SWALLOWED and victim is UNCONSCIOUS OR HAVING CONVULSIONS, do nothing except keep victim warm.
Water Pollution	Effect of low concentrations on aquatic life is unknown. May be dangerous if it enters water intakes. Notify local health and wildlife officials. Notify operators of nearby water intakes.

1. RESPONSE TO DISCHARGE
(See Response Methods Handbook, CG 446-4)

Issue warning—water contaminant
Restrict access
Disperse and flush

2. LABELS

No hazard label required by Code of Federal Regulations

3. CHEMICAL DESIGNATIONS

3.1 **Synonyms:** Ammonium fluosilicate
3.2 **Coast Guard Compatibility Classification:** Not listed
3.3 **Chemical Formula:** $(NH_4)_2SiF_6$
3.4 **IMCO/United Nations Numerical Designation:** Not listed

4. OBSERVABLE CHARACTERISTICS

4.1 **Physical State (as shipped):** Solid
4.2 **Color:** White
4.3 **Odor:** None

5. HEALTH HAZARDS

5.1 **Personal Protective Equipment:** Dust respirator; acid-resistant clothing and hat; rubber gloves; goggles and safety shoes
5.2 **Symptoms Following Exposure:** Inhalation of dust can cause pulmonary irritation and can be fatal in some cases. Ingestion may be fatal. Contact with dust causes irritation of eyes and irritation or ulceration of skin.
5.3 **Treatment for Exposure:** INHALATION: remove patient to fresh air. INGESTION: cause vomiting by giving soapy water or mustard water; have patient drink large quantities of lime water; if necessary, give stimulant such as strong coffee; keep patient warm. EYES: flush with water for 20 min., holding eyelids open. SKIN: wash with soap and water.
5.4 **Toxicity by Inhalation (Threshold Limit Value):** 2.5 mg/m^3
5.5 **Short-Term Inhalation Limits:** Data not available
5.6 **Toxicity by Ingestion:** Grade 3; LD_{50} 100 mg/kg (rat)
5.7 **Late Toxicity:** Data not available
5.8 **Vapor (Gas) Irritant Characteristics:** Data not available
5.9 **Liquid or Solid Irritant Characteristics:** Data not available
5.10 **Odor Threshold:** Data not available

6. FIRE HAZARDS

6.1 **Flash Point:** Not flammable
6.2 **Flammable Limits in Air:** Not flammable
6.3 **Fire Extinguishing Agents:** Not pertinent
6.4 **Fire Extinguishing Agents Not to be Used:** Not pertinent
6.5 **Special Hazards of Combustion Products:** Toxic and irritating hydrogen fluoride, silicon tetrafluoride, and oxides of nitrogen may form in fires.
6.6 **Behavior in Fire:**
6.7 **Ignition Temperature:** Not pertinent
6.8 **Electrical Hazard:** Not pertinent
6.9 **Burning Rate:** Not pertinent

7. CHEMICAL REACTIVITY

7.1 **Reactivity with Water:** No reaction
7.2 **Reactivity with Common Materials:**
7.3 **Stability During Transport:** Stable
7.4 **Neutralizing Agents for Acids and Caustics:** Not pertinent
7.5 **Polymerization:** Not pertinent
7.6 **Inhibitor of Polymerization:** Not pertinent

8. WATER POLLUTION

8.1 **Aquatic Toxicity:** Data not available
8.2 **Waterfowl Toxicity:** Data not available
8.3 **Biological Oxygen Demand (BOD):** Data not available
8.4 **Food Chain Concentration Potential:** None

9. SELECTED MANUFACTURERS

1. Agrico Chemical Co.
P. O. Box 67
Pierce (Polk County) Fla. 33867
2. Essex Chemical Co.
Chemicals Division
1401 Broad Street
Clifton, N. J. 07015
3. W. R. Grace & Co.
Agricultural Chemicals Group
Baltimore, Md. 21203

10. SHIPPING INFORMATION

10.1 **Grades or Purity:** Pure, 99+%; Commercial, 98+%
10.2 **Storage Temperature:** Ambient
10.3 **Inert Atmosphere:** No requirement
10.4 **Venting:** Open

11. HAZARD ASSESSMENT CODE
(See Hazard Assessment Handbook, CG 446-3)

SS

12. HAZARD CLASSIFICATIONS

12.1 **Code of Federal Regulations:** Not listed
12.2 **NAS Hazard Rating for Bulk Water Transportation:** Not listed
12.3 **NFPA Hazard Classifications:** Not listed

13. PHYSICAL AND CHEMICAL PROPERTIES

13.1 **Physical State at 15°C and 1 atm:** Solid
13.2 **Molecular Weight:** 178.14
13.3 **Boiling Point at 1 atm:** Not pertinent (decomposes)
13.4 **Freezing Point:** Not pertinent
13.5 **Critical Temperature:** Not pertinent
13.6 **Critical Pressure:** Not pertinent
13.7 **Specific Gravity:** 2.0 at 20°C (solid)
13.8 **Liquid Surface Tension:** Not pertinent
13.9 **Liquid-Water Interfacial Tension:** Not pertinent
13.10 **Vapor (Gas) Specific Gravity:** Not pertinent
13.11 **Ratio of Specific Heats of Vapor (Gas):** Not pertinent
13.12 **Latent Heat of Vaporization:** Not pertinent
13.13 **Heat of Combustion:** Not pertinent
13.14 **Heat of Decomposition:** Not pertinent
13.15 **Heat of Solution:** 85 Btu/lb = 47 cal/g = 2.0×10^5 J/kg
13.16 **Heat of Polymerization:** Not pertinent

NOTES

AMMONIUM STEARATE

Common Synonyms			
Ammonium stearate dispersion Stearic acid, ammonium salt	Solid paste or liquid	White to yellow	Weak ammonia odor
	May float or sink in water.		

Stop discharge if possible. Keep people away.
Call fire department.
Avoid contact with liquid and solid.
Isolate and remove discharged material.
Notify local health and pollution control agencies.

Fire

Combustible solid.
Solution not flammable.
POISONOUS GASES MAY BE PRODUCED IN FIRE.
Wear goggles and self-contained breathing apparatus.
Extinguish with water, dry chemicals, foam, or carbon dioxide.

Exposure

CALL FOR MEDICAL AID.

LIQUID OR SOLID
Irritating to skin and eyes.
If swallowed will cause nausea.
Remove contaminated clothing and shoes.
Flush affected areas with plenty of water.
IF IN EYES, hold eyelids open and flush with plenty of water.
IF SWALLOWED and victim is CONSCIOUS, have victim drink water or milk.
IF SWALLOWED and victim is UNCONSCIOUS OR HAVING CONVULSIONS, do nothing except keep victim warm.

Water Pollution

Effect of low concentrations on aquatic life is unknown.
May be dangerous if it enters water intakes.
Notify local health and wildlife officials.
Notify operators of nearby water intakes.

1. RESPONSE TO DISCHARGE

(See Response Methods Handbook, CG 446-4)

Disperse and flush

2. LABELS

No hazard label required by Code of Federal Regulations

3. CHEMICAL DESIGNATIONS

3.1 **Synonyms:** Ammonium stearate dispersion
Stearic acid, ammonium salt
3.2 **Coast Guard Compatibility Classification:** Not listed
3.3 **Chemical Formula:** $C_{18}H_{35}COONH_4 - H_2O$
3.4 **IMCO/United Nations Numerical Designation:** Not listed

4. OBSERVABLE CHARACTERISTICS

4.1 **Physical State (as shipped):** Pasty solid or liquid
4.2 **Color:** White to yellow
4.3 **Odor:** Slightly ammoniacal

5. HEALTH HAZARDS

5.1 **Personal Protective Equipment:** Gloves and goggles
5.2 **Symptoms Following Exposure:** None experienced under normal conditions of use. Ingestion causes irritation of mouth and stomach. Contact with eyes causes irritation.
5.3 **Treatment for Exposure:** INHALATION or INGESTION: no treatment required. EYES: flush with copious quantities of tap water. SKIN: flush with water.
5.4 **Toxicity by Inhalation (Threshold Limit Value):** Data not available
5.5 **Short-Term Inhalation Limits:** Data not available
5.6 **Toxicity by Ingestion:** Data not available
5.7 **Late Toxicity:** Data not available
5.8 **Vapor (Gas) Irritant Characteristics:** Not pertinent
5.9 **Liquid or Solid Irritant Characteristics:** Data not available
5.10 **Odor Threshold:** Data not available

6. FIRE HAZARDS

6.1 **Flash Point:** > 140°F C.C. (pure material only; solution not flammable)
6.2 **Flammable Limits in Air:** Not pertinent
6.3 **Fire Extinguishing Agents:** Water, foam, dry chemical, carbon dioxide
6.4 **Fire Extinguishing Agents Not to be Used:**
6.5 **Special Hazards of Combustion Products:** Toxic ammonia and oxides of nitrogen may form in fires.
6.6 **Behavior in Fire:**
6.7 **Ignition Temperature:** Data not available
6.8 **Electrical Hazard:** Not pertinent
6.9 **Burning Rate:** Not pertinent

7. CHEMICAL REACTIVITY

7.1 **Reactivity with Water:** No reaction
7.2 **Reactivity with Common Materials:**
7.3 **Stability During Transport:** Stable
7.4 **Neutralizing Agents for Acids and Caustics:** Not pertinent
7.5 **Polymerization:** Not pertinent
7.6 **Inhibitor of Polymerization:** Not pertinent

8. WATER POLLUTION

8.1 **Aquatic Toxicity:** Data not available
8.2 **Waterfowl Toxicity:** Data not available
8.3 **Biological Oxygen Demand (BOD):** Data not available
8.4 **Food Chain Concentration Potential:** None

9. SELECTED MANUFACTURERS

1. Diamond Shamrock Chemical Co.
1100 Superior Avenue
Cleveland, Ohio 44114
2. Emkay Chemical Co.
319 Second Street
Elizabeth, N. J. 07206

10. SHIPPING INFORMATION

10.1 **Grades or Purity:** 33% dispersion in water; technical paste
10.2 **Storage Temperature:** Ambient
10.3 **Inert Atmosphere:** No requirement
10.4 **Venting:** Open

11. HAZARD ASSESSMENT CODE

(See Hazard Assessment Handbook, CG 446-3)

A-P-SS

12. HAZARD CLASSIFICATIONS

12.1 **Code of Federal Regulations:** Not listed
12.2 **NAS Hazard Rating for Bulk Water Transportation:** Not listed
12.3 **NFPA Hazard Classifications:** Not listed

13. PHYSICAL AND CHEMICAL PROPERTIES

13.1 **Physical State at 15°C and 1 atm:** Solid or liquid
13.2 **Molecular Weight:** 301.5
13.3 **Boiling Point at 1 atm:** Not pertinent (decomposes)
13.4 **Freezing Point:** Not pertinent
13.5 **Critical Temperature:** Not pertinent
13.6 **Critical Pressure:** Not pertinent
13.7 **Specific Gravity:** 1.01 at 20°C (liquid)
13.8 **Liquid Surface Tension:** Not pertinent
13.9 **Liquid-Water Interfacial Tension:** Not pertinent
13.10 **Vapor (Gas) Specific Gravity:** Not pertinent
13.11 **Ratio of Specific Heats of Vapor (Gas):** Not pertinent
13.12 **Latent Heat of Vaporization:** Not pertinent
13.13 **Heat of Combustion:** Data not available
13.14 **Heat of Decomposition:** Not pertinent
13.15 **Heat of Solution:** Not pertinent
13.16 **Heat of Polymerization:** Not pertinent

NOTES

ASM

AMMONIUM SULFAMATE

Common Synonyms
Ammate
Sulfamic acid, monoammonium salt
Ammonium amidosulfonate
Ammonium amidosulphate
AMS

Solid | White or brownish-gray | Odorless

Sinks and mixes with water.

Stop discharge if possible. Keep people away.
Avoid contact with solid and dust.
Isolate and remove discharged material.
Notify local health and pollution control agencies.

Fire

Not flammable.
POISONOUS GASES MAY BE PRODUCED IN FIRE.
Wear goggles and self-contained breathing apparatus.

Exposure

CALL FOR MEDICAL AID.

DUST
Irritating to eyes, nose and throat.
If inhaled will cause coughing or difficult breathing.

If in eyes, hold eyelids open and flush with plenty of water.
If breathing has stopped, give artificial respiration.
If breathing is difficult, give oxygen.

SOLID
Irritating to skin and eyes.
If swallowed will cause nausea or vomiting.

Remove contaminated clothing and shoes.
Flush affected areas with plenty of water.
IF IN EYES, hold eyelids open and flush with plenty of water.
IF SWALLOWED and victim is CONSCIOUS, have victim drink water or milk.
IF SWALLOWED and victim is UNCONSCIOUS OR HAVING CONVULSIONS, do nothing except keep victim warm.

Water Pollution

Dangerous to aquatic life in high concentrations.
May be dangerous if it enters water intakes.

Notify local health and wildlife officials.
Notify operators of nearby water intakes.

1. RESPONSE TO DISCHARGE

(See Response Methods Handbook, CG 446-4)

Disperse and flush

2. LABELS

No hazard label required by Code of Federal Regulations

3. CHEMICAL DESIGNATIONS

3.1 **Synonyms:** Ammate; Ammonium amidosulfonate; Ammonium amidosulphate; AMS; Sulfamic acid, monoammonium salt

3.2 **Coast Guard Compatibility Classification:** Not listed

3.3 **Chemical Formula:** $NH_2SO_3NH_4$

3.4 **IMCO/United Nations Numerical Designation:** Not listed

4. OBSERVABLE CHARACTERISTICS

4.1 **Physical State (as shipped):** Solid

4.2 **Color:** White or brownish-gray

4.3 **Odor:** None

5. HEALTH HAZARDS

5.1 **Personal Protective Equipment:** Dust mask; goggles or face shield; rubber gloves.

5.2 **Symptoms Following Exposure:** Inhalation causes irritation of nose and throat. Ingestion causes gastrointestinal disturbances. Dust irritates eyes.

5.3 **Treatment for Exposure:** INHALATION: remove to fresh air. INGESTION: give large amount of water; get medical attention. EYES: flush with water for 15 min. SKIN: flush with water.

5.4 **Toxicity by Inhalation (Threshold Limit Value):** 10 mg/m³

5.5 **Short-Term Inhalation Limits:** Data not available

5.6 **Toxicity by Ingestion:** Grade 2; oral rat LD_{50} = 1,600 mg/kg

5.7 **Late Toxicity:** Data not available

5.8 **Vapor (Gas) Irritant Characteristics:** Data not available

5.9 **Liquid or Solid Irritant Characteristics:** Data not available

5.10 **Odor Threshold:** Data not available

6. FIRE HAZARDS

6.1 **Flash Point:** Not flammable

6.2 **Flammable Limits in Air:** Not flammable

6.3 **Fire Extinguishing Agents:** Not pertinent

6.4 **Fire Extinguishing Agents Not to be Used:** Not pertinent

6.5 **Special Hazards of Combustion Products:** Toxic oxides of nitrogen may form in fires.

6.6 **Behavior in Fire:**

6.7 **Ignition Temperature:** Not pertinent

6.8 **Electrical Hazard:** Not pertinent

6.9 **Burning Rate:** Not pertinent

7. CHEMICAL REACTIVITY

7.1 **Reactivity with Water:** No reaction

7.2 **Reactivity with Common Materials:**

7.3 **Stability During Transport:** Stable

7.4 **Neutralizing Agents for Acids and Caustics:** Not pertinent

7.5 **Polymerization:** Not pertinent

7.6 **Inhibitor of Polymerization:** Not pertinent

8. WATER POLLUTION

8.1 **Aquatic Toxicity:** 259 ppm/24 hr/catfish/50% kill/fresh water

8.2 **Waterfowl Toxicity:** Data not available

8.3 **Biological Oxygen Demand (BOD):** Data not available

8.4 **Food Chain Concentration Potential:** None

9. SELECTED MANUFACTURERS

1. Heico, Inc.
Delaware Water Gap, Pa. 18327

2. Gallard-Schlesinger Chemical Mfg. Co.
584 Mineola Ave.
Carle Place, N. Y. 11514

3. Pfaltz and Bauer, Inc.
126-04 Northern Blvd.
Flushing, N. Y. 11368

10. SHIPPING INFORMATION

10.1 **Grades or Purity:** Reagent, 99.0%; Commercial, 80%

10.2 **Storage Temperature:** Ambient

10.3 **Inert Atmosphere:** No requirement

10.4 **Venting:** Open

11. HAZARD ASSESSMENT CODE

(See Hazard Assessment Handbook, CG 446-3)

SS

12. HAZARD CLASSIFICATIONS

12.1 **Code of Federal Regulations:** Not listed

12.2 **NAS Hazard Rating for Bulk Water Transportation:** Not listed

12.3 **NFPA Hazard Classifications:** Not listed

13. PHYSICAL AND CHEMICAL PROPERTIES

13.1 **Physical State at 15°C and 1 atm:** Solid

13.2 **Molecular Weight:** 114.13

13.3 **Boiling Point at 1 atm:** Decomposes above 200°C

13.4 **Freezing Point:** 268°F = 131°C = 404°K

13.5 **Critical Temperature:** Not pertinent

13.6 **Critical Pressure:** Not pertinent

13.7 **Specific Gravity:** > 1 at 20°C (solid)

13.8 **Liquid Surface Tension:** Not pertinent

13.9 **Liquid-Water Interfacial Tension:** Not pertinent

13.10 **Vapor (Gas) Specific Gravity:** Not pertinent

13.11 **Ratio of Specific Heats of Vapor (Gas):** Not pertinent

13.12 **Latent Heat of Vaporization:** Not pertinent

13.13 **Heat of Combustion:** Not pertinent

13.14 **Heat of Decomposition:** Not pertinent

13.15 **Heat of Solution:** Data not available

13.16 **Heat of Polymerization:** Not pertinent

NOTES

AMS — AMMONIUM SULFATE

Common Synonyms	Solid	White	Odorless
	Sinks and mixes with water.		

Isolate and remove discharged material.
Notify local health and pollution control agencies.

Fire	Not flammable.
Exposure	Not harmful.
Water Pollution	Effect of low concentrations on aquatic life is unknown. May be dangerous if it enters water intakes. Notify local health and wildlife officials. Notify operators of nearby water intakes.

1. RESPONSE TO DISCHARGE
(See Response Methods Handbook, CG 446-4)

Disperse and flush

2. LABELS

No hazard label required by Code of Federal Regulations

3. CHEMICAL DESIGNATIONS

3.1 **Synonyms:** No common synonyms
3.2 **Coast Guard Compatibility Classification:** Not applicable
3.3 **Chemical Formula:** $(NH_4)_2SO_4$
3.4 **IMCO/United Nations Numerical Designation:** Not listed

4. OBSERVABLE CHARACTERISTICS

4.1 **Physical State (as shipped):** Solid
4.2 **Color:** White
4.3 **Odor:** None

5. HEALTH HAZARDS

5.1 **Personal Protective Equipment:** Data not available
5.2 **Symptoms Following Exposure:** Data not available
5.3 **Treatment for Exposure:** Data not available
5.4 **Toxicity by Inhalation (Threshold Limit Value):** Not pertinent
5.5 **Short-Term Inhalation Limits:** Not pertinent
5.6 **Toxicity by Ingestion:** Grade 3; LD_{50} = 58 mg/kg (rat)
5.7 **Late Toxicity:** Not pertinent
5.8 **Vapor (Gas) Irritant Characteristics:** Not pertinent
5.9 **Liquid or Solid Irritant Characteristics:** Not pertinent
5.10 **Odor Threshold:** Not pertinent

6. FIRE HAZARDS

6.1 **Flash Point:** Not flammable
6.2 **Flammable Limits in Air:** Not flammable
6.3 **Fire Extinguishing Agents:** Not pertinent
6.4 **Fire Extinguishing Agents Not to be Used:** Not pertinent
6.5 **Special Hazards of Combustion Products:** Not pertinent
6.6 **Behavior in Fire:** Not pertinent
6.7 **Ignition Temperature:** Not flammable
6.8 **Electrical Hazard:** Not pertinent
6.9 **Burning Rate:** Not flammable

7. CHEMICAL REACTIVITY

7.1 **Reactivity with Water:** No reaction
7.2 **Reactivity with Common Materials:** No reaction
7.3 **Stability During Transport:** Stable
7.4 **Neutralizing Agents for Acids and Caustics:** Not pertinent
7.5 **Polymerization:** Not pertinent
7.6 **Inhibitor of Polymerization:** Not pertinent

8. WATER POLLUTION

8.1 **Aquatic Toxicity:**
1290 ppm/96 hr/mosquitofish/fresh-water/TL_m
292 ppm/96 hr/Daphnia magna/fresh-water/TL_m
8.2 **Waterfowl Toxicity:** Data not available
8.3 **Biological Oxygen Demand (BOD):** None
8.4 **Food Chain Concentration Potential:** None

9. SELECTED MANUFACTURERS

1. Allied Chemical Corp.
   Agricultural Div.
   Morristown, N. J. 07960
2. Occidental Petroleum Corp.
   Occidental Chemical Co.
   Western Division
   Lathrop, Calif. 95330
3. Valley Nitrogen Producers, Inc.
   Fresno, Calif. 93717

10. SHIPPING INFORMATION

10.1 **Grades or Purity:** Data not available
10.2 **Storage Temperature:** Data not available
10.3 **Inert Atmosphere:** Data not available
10.4 **Venting:** Data not available

11. HAZARD ASSESSMENT CODE
(See Hazard Assessment Handbook, CG 446-3)

SS

12. HAZARD CLASSIFICATIONS

12.1 **Code of Federal Regulations:** Not listed
12.2 **NAS Hazard Rating for Bulk Water Transportation:** Not listed
12.3 **NFPA Hazard Classifications:** Not listed

13. PHYSICAL AND CHEMICAL PROPERTIES

13.1 **Physical State at 15°C and 1 atm:** Solid
13.2 **Molecular Weight:** 132.14
13.3 **Boiling Point at 1 atm:** Not pertinent
13.4 **Freezing Point:** Not pertinent
13.5 **Critical Temperature:** Not pertinent
13.6 **Critical Pressure:** Not pertinent
13.7 **Specific Gravity:** 1.78 at 15°C (solid)
13.8 **Liquid Surface Tension:** Not pertinent
13.9 **Liquid-Water Interfacial Tension:** Not pertinent
13.10 **Vapor (Gas) Specific Gravity:** Not pertinent
13.11 **Ratio of Specific Heats of Vapor (Gas):** Not pertinent
13.12 **Latent Heat of Vaporization:** Not pertinent
13.13 **Heat of Combustion:** Not pertinent
13.14 **Heat of Decomposition:** Not pertinent
13.15 **Heat of Solution:** Not pertinent
13.16 **Heat of Polymerization:** Not pertinent

NOTES

REVISED 1978

ASF	# AMMONIUM SULFIDE

Common Synonyms			
Ammonium sulfide solution Ammonium hydrogen sulfide solution Ammonium sulfhydrate solution	Liquid	Colorless to yellow	Strong odor of rotten eggs and ammonia
	Mixes with water. Irritating vapor is produced. Boiling point is 104°F.		

Avoid contact with liquid and vapor. Keep people away.
Wear rubber overclothing (including gloves).
Shut off ignition sources. Call fire department.
Stop discharge if possible.
Stay upwind. Use water spray to "knock down" vapor.
Isolate and remove discharged material.
Notify local health and pollution control agencies.

Fire

FLAMMABLE.
POISONOUS GASES MAY BE PRODUCED IN FIRE.
Wear goggles and self-contained breathing apparatus.
Extinguish with water, dry chemicals, or carbon dioxide.

Exposure

CALL FOR MEDICAL AID.

VAPOR
Irritating to eyes, nose and throat.
If inhaled will cause dizziness, headache, coughing, or difficult breathing.
If in eyes, hold eyelids open and flush with plenty of water.
If breathing has stopped, give artificial respiration.
If breathing is difficult, give oxygen.

LIQUID
Will burn skin and eyes.
If swallowed will cause nausea.
Remove contaminated clothing and shoes.
Flush affected areas with plenty of water.
IF IN EYES, hold eyelids open and flush with plenty of water.
IF SWALLOWED and victim is CONSCIOUS, have victim drink water or milk.
IF SWALLOWED and victim is UNCONSCIOUS OR HAVING CONVULSIONS, do nothing except keep victim warm.
DO NOT INDUCE VOMITING.

Water Pollution

Dangerous to aquatic life in high concentrations.
May be dangerous if it enters water intakes.
Notify local health and wildlife officials.
Notify operators of nearby water intakes.

1. RESPONSE TO DISCHARGE
(See Response Methods Handbook, CG 446-4)

Issue warning—air contaminant, water contaminant
Restrict access
Should be removed
Chemical and physical treatment

2. LABEL

Red

3. CHEMICAL DESIGNATIONS

3.1 **Synonyms:** Ammonium hydrogen sulfide solution; Ammonium sulfhydrate solution; Ammonium sulfide solution
3.2 **Coast Guard Compatibility Classification:** Not listed
3.3 **Chemical Formula:** $(NH_4)_2S$—NH_4SH—H_2O
3.4 **IMCO/United Nations Numerical Designation:** Not listed

4. OBSERVABLE CHARACTERISTICS

4.1 **Physical State (as shipped):** Liquid
4.2 **Color:** Yellow to orange (fuming solution); colorless to yellow
4.3 **Odor:** Strong odor of sulfide and ammonia

5. HEALTH HAZARDS

5.1 **Personal Protective Equipment:** Self-contained breathing apparatus; rubber or plastic gloves; splash goggles; rubber shoes
5.2 **Symptoms Following Exposure:** Inhalation of 500 ppm for 30 min. produces headaches, dizziness, bronchial pneumonia; 600 ppm for 30 min. can cause death. Ingestion causes severe irritation of mucous membranes and stomach. Contact with liquid causes severe burns of eyes and severe skin irritation. May be absorbed through skin and cause hydrogen sulfide poisoning.
5.3 **Treatment for Exposure:** *Get medical attention following all overexposures to this compound.* INHALATION: move victim to fresh air; give artificial respiration, oxygen; consult physician. INGESTION: give large amount of water. EYES OR SKIN: wash with large quantities of water for 15 min.; consult physician.
5.4 **Toxicity by Inhalation (Threshold Limit Value):** Data not available
5.5 **Short-Term Inhalation Limits:** Data not available
5.6 **Toxicity by Ingestion:** Data not available
5.7 **Late Toxicity:** Data not available
5.8 **Vapor (Gas) Irritant Characteristics:** Data not available
5.9 **Liquid or Solid Irritant Characteristics:** Data not available
5.10 **Odor Threshold:** Data not available

6. FIRE HAZARDS

6.1 **Flash Point:** 72°F C.C.
6.2 **Flammable Limits in Air:** 4%—46% (hydrogen sulfide)
6.3 **Fire Extinguishing Agents:** Water, dry chemical, carbon dioxide
6.4 **Fire Extinguishing Agents Not to be Used:**
6.5 **Special Hazards of Combustion Products:** Toxic hydrogen sulfide gas is released when solution is heated. If ignited, this will form irritating sulfur dioxide gas.
6.6 **Behavior in Fire:**
6.7 **Ignition Temperature:** Data not available
6.8 **Electrical Hazard:** Not pertinent
6.9 **Burning Rate:** Not pertinent

7. CHEMICAL REACTIVITY

7.1 **Reactivity with Water:** No reaction
7.2 **Reactivity with Common Materials:** Severely corrodes copper, zinc, and their alloys
7.3 **Stability During Transport:** Stable, but toxic hydrogen sulfide and ammonia gases may collect in enclosed spaces.
7.4 **Neutralizing Agents for Acids and Caustics:** Dilute with water. Do not attempt to neutralize with acid.
7.5 **Polymerization:** Not pertinent
7.6 **Inhibitor of Polymerization:** Not pertinent

8. WATER POLLUTION

8.1 **Aquatic Toxicity:**
100 ppm/72 hr/goldfish/killed/fresh water
248 ppm/48 hr/mosquitofish/TL_m/fresh water
8.2 **Waterfowl Toxicity:** Data not available
8.3 **Biological Oxygen Demand (BOD):** Data not available
8.4 **Food Chain Concentration Potential:** None

9. SELECTED MANUFACTURERS

1. Argus Chemical Corp.
   Halby Division
   600 Terminal Avenue
   New Castle, Del. 19720
2. J. T. Baker Chemical Co.
   Phillipsburg, N. J. 08865
3. Chemical Products Corporation
   Cartersville, Ga. 30120

10. SHIPPING INFORMATION

10.1 **Grades or Purity:** Technical, 45—50% in water; Reagent, 52—60% in water
10.2 **Storage Temperature:** Ambient
10.3 **Inert Atmosphere:** Ventilated (natural)
10.4 **Venting:** Open (flame arrester)

11. HAZARD ASSESSMENT CODE
(See Hazard Assessment Handbook, CG 446-3)

A-P

12. HAZARD CLASSIFICATIONS

12.1 **Code of Federal Regulations:** Flammable Liquid
12.2 **NAS Hazard Rating for Bulk Water Transportation:** Not listed
12.3 **NFPA Hazard Classifications:** Not listed

13. PHYSICAL AND CHEMICAL PROPERTIES

13.1 **Physical State at 15°C and 1 atm:** Solid or liquid
13.2 **Molecular Weight:** 68.14 (solute)
13.3 **Boiling Point at 1 atm:** 104°F = 40°C = 313°K
13.4 **Freezing Point:** Not pertinent
13.5 **Critical Temperature:** Not pertinent
13.6 **Critical Pressure:** Not pertinent
13.7 **Specific Gravity:** 0.99—1.01 at 20°C (liquid)
13.8 **Liquid Surface Tension:** Data not available
13.9 **Liquid-Water Interfacial Tension:** Not pertinent
13.10 **Vapor (Gas) Specific Gravity:** Not pertinent
13.11 **Ratio of Specific Heats of Vapor (Gas):** Not pertinent
13.12 **Latent Heat of Vaporization:** Not pertinent
13.13 **Heat of Combustion:** Not pertinent
13.14 **Heat of Decomposition:** Not pertinent
13.15 **Heat of Solution:** 95.0 Btu/lb = 52.8 cal/g = 2.21×10^5 J/kg
13.16 **Heat of Polymerization:** Not pertinent

NOTES

AMF	AMMONIUM SULFITE

Common Synonyms	Solid	White	Odorless
	Sinks and mixes slowly with water.		

Stop discharge if possible. Keep people away.
Avoid contact with solid and dust.
Isolate and remove discharged material.
Notify local health and pollution control agencies.

Fire	Not flammable. POISONOUS GASES MAY BE PRODUCED IN FIRE. Wear goggles and self-contained breathing apparatus.
Exposure	CALL FOR MEDICAL AID. DUST Irritating to eyes, nose and throat. If inhaled will cause coughing or difficult breathing. If in eyes, hold eyelids open and flush with plenty of water. If breathing has stopped, give artificial respiration. If breathing is difficult, give oxygen. SOLID Irritating to skin and eyes. If swallowed will cause nausea. Remove contaminated clothing and shoes. Flush affected areas with plenty of water. IF IN EYES, hold eyelids open and flush with plenty of water. IF SWALLOWED and victim is CONSCIOUS, have victim drink water or milk. IF SWALLOWED and victim is UNCONSCIOUS OR HAVING CONVULSIONS, do nothing except keep victim warm.
Water Pollution	Dangerous to aquatic life in high concentrations May be dangerous if it enters water intakes. Notify local health and wildlife officials. Notify operators of nearby water intakes.

1. RESPONSE TO DISCHARGE
(See Response Methods Handbook, CG 446-4)

Disperse and flush

2. LABELS

No hazard label required by Code of Federal Regulations

3. CHEMICAL DESIGNATIONS

3.1 **Synonyms:** No common synonyms
3.2 **Coast Guard Compatibility Classification:** Not listed
3.3 **Chemical Formula:** $(NH_4)_2SO_3$
3.4 **IMCO/United Nations Numerical Designation:** Not listed

4. OBSERVABLE CHARACTERISTICS

4.1 **Physical State (as shipped):** Solid
4.2 **Color:** White
4.3 **Odor:** None

5. HEALTH HAZARDS

5.1 **Personal Protective Equipment:** Dust mask; goggles or face shield; rubber gloves
5.2 **Symptoms Following Exposure:** Inhalation of dust causes irritation of nose and throat. Ingestion causes irritation of mouth and stomach. Dust irritates eyes.
5.3 **Treatment for Exposure:** INHALATION: move to fresh air. INGESTION: give large amount of water; get medical attention. EYES: flush with water for 15 min. SKIN: flush with water.
5.4 **Toxicity by Inhalation (Threshold Limit Value):** Data not available
5.5 **Short-Term Inhalation Limits:** Data not available
5.6 **Toxicity by Ingestion:** Data not available
5.7 **Late Toxicity:** Data not available
5.8 **Vapor (Gas) Irritant Characteristics:** Data not available
5.9 **Liquid or Solid Irritant Characteristics:** Data not available
5.10 **Odor Threshold:** Data not available

6. FIRE HAZARDS

6.1 **Flash Point:** Not flammable
6.2 **Flammable Limits in Air:** Not flammable
6.3 **Fire Extinguishing Agents:** Not pertinent
6.4 **Fire Extinguishing Agents Not to be Used:** Not pertinent
6.5 **Special Hazards of Combustion Products:** Toxic sulfur dioxide and oxides of nitrogen may form in fires.
6.6 **Behavior in Fire:**
6.7 **Ignition Temperature:** Not pertinent
6.8 **Electrical Hazard:** Not pertinent
6.9 **Burning Rate:** Not pertinent

7. CHEMICAL REACTIVITY

7.1 **Reactivity with Water:** No reaction
7.2 **Reactivity with Common Materials:**
7.3 **Stability During Transport:** Stable
7.4 **Neutralizing Agents for Acids and Caustics:** Dilute with water. Do not attempt to neutralize with acids.
7.5 **Polymerization:** Not pertinent
7.6 **Inhibitor of Polymerization:** Not pertinent

8. WATER POLLUTION

8.1 **Aquatic Toxicity:** 240 ppm/48 hr/mosquitofish/TL_m/fresh water
8.2 **Waterfowl Toxicity:** Data not available
8.3 **Biological Oxygen Demand (BOD):** Data not available
8.4 **Food Chain Concentration Potential:** None

9. SELECTED MANUFACTURERS

1. J. T. Baker Chemical Co.
   Phillipsburg, N. J. 08865
2. Pfaltz and Bauer, Inc.
   126-04 Northern Blvd.
   Flushing, N. Y. 11368

10. SHIPPING INFORMATION

10.1 **Grades or Purity:** Anhydrous; Purified monohydrate
10.2 **Storage Temperature:** Ambient
10.3 **Inert Atmosphere:** No requirement
10.4 **Venting:** Open

11. HAZARD ASSESSMENT CODE
(See Hazard Assessment Handbook, CG 446-3)

SS

12. HAZARD CLASSIFICATIONS

12.1 **Code of Federal Regulations:** Not listed
12.2 **NAS Hazard Rating for Bulk Water Transportation:** Data not available
12.3 **NFPA Hazard Classifications:** Data not available

13. PHYSICAL AND CHEMICAL PROPERTIES

13.1 **Physical State at 15°C and 1 atm:** Solid
13.2 **Molecular Weight:** 134.2
13.3 **Boiling Point at 1 atm:** Not pertinent
13.4 **Freezing Point:** Not pertinent
13.5 **Critical Temperature:** Not pertinent
13.6 **Critical Pressure:** Not pertinent
13.7 **Specific Gravity:** > 1.1 at 20°C (solid)
13.8 **Liquid Surface Tension:** Not pertinent
13.9 **Liquid-Water Interfacial Tension:** Not pertinent
13.10 **Vapor (Gas) Specific Gravity:** Not pertinent
13.11 **Ratio of Specific Heats of Vapor (Gas):** Not pertinent
13.12 **Latent Heat of Vaporization:** Not pertinent
13.13 **Heat of Combustion:** Not pertinent
13.14 **Heat of Decomposition:** Not pertinent
13.15 **Heat of Solution:** 20.5 Btu/lb $= 11.4$ cal/g $= 0.477 \times 10^5$ J/kg
13.16 **Heat of Polymerization:** Not pertinent

NOTES

ATR	AMMONIUM TARTRATE

Common Synonyms			
l-Tartaric acid, ammonium salt	Solid	White	Odorless
	Sinks and mixes with water.		

Stop discharge if possible. Keep people away.
Call fire department.
Avoid contact with solid and dust.
Isolate and remove discharged material.
Notify local health and pollution control agencies.

Fire	Combustible. POISONOUS GASES MAY BE PRODUCED IN FIRE. Wear goggles and self-contained breathing apparatus. Extinguish with water.
Exposure	CALL FOR MEDICAL AID. DUST Irritating to eyes, nose and throat. Harmful if inhaled. If in eyes, hold eyelids open and flush with plenty of water. If breathing has stopped, give artificial respiration. If breathing is difficult, give oxygen. SOLID Irritating to skin and eyes. Harmful if swallowed. Remove contaminated clothing and shoes. Flush affected areas with plenty of water. IF IN EYES, hold eyelids open and flush with plenty of water. IF SWALLOWED and victim is CONSCIOUS, have victim drink water or milk. IF SWALLOWED and victim is UNCONSCIOUS OR HAVING CONVULSIONS, do nothing except keep victim warm.
Water Pollution	Effect of low concentrations on aquatic life is unknown. May be dangerous if it enters water intakes. Notify local health and wildlife officials. Notify operators of nearby water intakes.

1. RESPONSE TO DISCHARGE

(See Response Methods Handbook, CG 446-4)

Disperse and flush

2. LABELS

No hazard label required by Code of Federal Regulations

3. CHEMICAL DESIGNATIONS

3.1 **Synonyms:** *l*-Tartaric acid, ammonium salt
3.2 **Coast Guard Compatibility Classification:** Not listed
3.3 **Chemical Formula:** $C_4H_{12}N_2O_6$
3.4 **IMCO/United Nations Numerical Designation:** Not listed

4. OBSERVABLE CHARACTERISTICS

4.1 **Physical State (as shipped):** Solid
4.2 **Color:** White
4.3 **Odor:** None

5. HEALTH HAZARDS

5.1 **Personal Protective Equipment:** Dust mask; goggles or face shield; rubber gloves
5.2 **Symptoms Following Exposure:** Contact with solid may irritate eyes.
5.3 **Treatment for Exposure:** EYES: flush with water.
5.4 **Toxicity by Inhalation (Threshold Limit Value):** Data not available
5.5 **Short-Term Inhalation Limits:** Data not available
5.6 **Toxicity by Ingestion:** Data not available
5.7 **Late Toxicity:** Data not available
5.8 **Vapor (Gas) Irritant Characteristics:** Data not available
5.9 **Liquid or Solid Irritant Characteristics:** Data not available
5.10 **Odor Threshold:** Data not available

6. FIRE HAZARDS

6.1 **Flash Point:** Not pertinent (combustible solid)
6.2 **Flammable Limits in Air:** Not pertinent
6.3 **Fire Extinguishing Agents:** Water
6.4 **Fire Extinguishing Agents Not to be Used:** Not pertinent
6.5 **Special Hazards of Combustion Products:** Toxic oxides of nitrogen or ammonia gas may form in fires.
6.6 **Behavior in Fire:**
6.7 **Ignition Temperature:** Not pertinent
6.8 **Electrical Hazard:** Not pertinent
6.9 **Burning Rate:** Not pertinent

7. CHEMICAL REACTIVITY

7.1 **Reactivity with Water:** No reaction
7.2 **Reactivity with Common Materials:**
7.3 **Stability During Transport:** Stable
7.4 **Neutralizing Agents for Acids and Caustics:** Not pertinent
7.5 **Polymerization:** Not pertinent
7.6 **Inhibitor of Polymerization:** Not pertinent

8. WATER POLLUTION

8.1 **Aquatic Toxicity:** Data not available
8.2 **Waterfowl Toxicity:** Data not available
8.3 **Biological Oxygen Demand (BOD):** Data not available
8.4 **Food Chain Concentration Potential:** None

9. SELECTED MANUFACTURERS

1. Ganes Chemical Works, Inc.
   611 Broad Street
   Carlstadt, N. J. 07072
2. American Hoechst Corp.
   Chemicals and Plastics Division
   Rt. 202-206 North
   Somerville, N. J. 08876
3. Gallard-Schlesinger Chemical Mfg. Co.
   584 Mineola Ave.
   Carle Place, N. Y. 11514

10. SHIPPING INFORMATION

10.1 **Grades or Purity:** Analytical; Pure
10.2 **Storage Temperature:** Ambient
10.3 **Inert Atmosphere:** No requirement
10.4 **Venting:** Open

11. HAZARD ASSESSMENT CODE

(See Hazard Assessment Handbook, CG 446-3)

SS

12. HAZARD CLASSIFICATIONS

12.1 **Code of Federal Regulations:** Not listed
12.2 **NAS Hazard Rating for Bulk Water Transportation:** Data not available
12.3 **NFPA Hazard Classifications:** Data not available

13. PHYSICAL AND CHEMICAL PROPERTIES

13.1 **Physical State at 15°C and 1 atm:** Solid
13.2 **Molecular Weight:** 184
13.3 **Boiling Point at 1 atm:** Not pertinent (decomposes)
13.4 **Freezing Point:** Not pertinent
13.5 **Critical Temperature:** Not pertinent
13.6 **Critical Pressure:** Not pertinent
13.7 **Specific Gravity:** 1.60 at 25°C (solid)
13.8 **Liquid Surface Tension:** Not pertinent
13.9 **Liquid-Water Interfacial Tension:** Not pertinent
13.10 **Vapor (Gas) Specific Gravity:** Not pertinent
13.11 **Ratio of Specific Heats of Vapor (Gas):** Not pertinent
13.12 **Latent Heat of Vaporization:** Not pertinent
13.13 **Heat of Combustion:** Data not available
13.14 **Heat of Decomposition:** Not pertinent
13.15 **Heat of Solution:** Data not available
13.16 **Heat of Polymerization:** Not pertinent

NOTES

AMT — AMMONIUM THIOCYANATE

Common Synonyms	Solid or solution (in water)	White	Odorless
Ammonium sulfocyanate Ammonium rhodanate Ammonium rhodanide Ammonium sulfocyanide Thiocyanic acid, ammonium salt	Sinks and mixes with water.		

Stop discharge if possible. Keep people away.
Call fire department.
Avoid contact with liquid and solid.
Avoid contact with solid and dust.
Isolate and remove discharged material.
Notify local health and pollution control agencies.

Fire

Combustible Solid. Solution not flammable.
POISONOUS GASES MAY BE PRODUCED IN FIRE.
Wear goggles and self-contained breathing apparatus.
Extinguish with water.

Exposure

CALL FOR MEDICAL AID.

VAPOR OR DUST
Irritating to eyes, nose and throat.
If inhaled will cause coughing or difficult breathing.
If in eyes, hold eyelids open and flush with plenty of water.
If breathing has stopped, give artificial respiration.
If breathing is difficult, give oxygen.

LIQUID OR SOLID
Irritating to skin and eyes.
If swallowed will cause nausea or vomiting.
Remove contaminated clothing and shoes.
Flush affected areas with plenty of water.
IF IN EYES, hold eyelids open and flush with plenty of water.
IF SWALLOWED and victim is CONSCIOUS, have victim drink water or milk.
IF SWALLOWED and victim is UNCONSCIOUS OR HAVING CONVULSIONS, do nothing except keep victim warm.

Water Pollution

Dangerous to aquatic life in high concentrations.
May be dangerous if it enters water intakes.
Notify local health and wildlife officials.
Notify operators of nearby water intakes.

1. RESPONSE TO DISCHARGE
(See Response Methods Handbook, CG 446-4)

Issue warning—water contaminant
Disperse and flush

2. LABELS

No hazard label required by Code of Federal Regulations

3. CHEMICAL DESIGNATIONS

3.1 **Synonyms:** Ammonium rhodanate; Ammonium rhodanide; Ammonium sulfocyanate; Ammonium sulfocyanide; Thiocyanic acid, ammonium salt
3.2 **Coast Guard Compatibility Classification:** Not listed
3.3 **Chemical Formula:** NH_4SCN
3.4 **IMCO/United Nations Numerical Designation:** Not listed

4. OBSERVABLE CHARACTERISTICS

4.1 **Physical State (as shipped):** Solid, or solution in water
4.2 **Color:** White
4.3 **Odor:** None

5. HEALTH HAZARDS

5.1 **Personal Protective Equipment:** Rubber or plastic gloves; rubber or plastic apron; standard goggles
5.2 **Symptoms Following Exposure:** Inhalation of dust causes irritation of nose and throat. Ingestion causes dizziness, cramps, nervous disturbances. Dust irritates eyes. Can be absorbed through skin; prolonged contact may produce various skin eruptions, dizziness, cramps, nausea, and mild to severe disturbance of the nervous system.
5.3 **Treatment for Exposure:** INHALATION: move to fresh air. INGESTION: give large amount of water; get medical attention. EYES OR SKIN: wash with water; consult physician.
5.4 **Toxicity by Inhalation (Threshold Limit Value):** Data not available
5.5 **Short-Term Inhalation Limits:** Data not available
5.6 **Toxicity by Ingestion:** Grade 2; oral rat LD_{50} = 854 mg/kg
5.7 **Late Toxicity:** Data not available
5.8 **Vapor (Gas) Irritant Characteristics:** Data not available
5.9 **Liquid or Solid Irritant Characteristics:** Data not available
5.10 **Odor Threshold:** Data not available

6. FIRE HAZARDS

6.1 **Flash Point:** Solid may be combustible; solution is not flammable.
6.2 **Flammable Limits in Air:** Not pertinent
6.3 **Fire Extinguishing Agents:** Water
6.4 **Fire Extinguishing Agents Not to be Used:**
6.5 **Special Hazards of Combustion Products:** Decomposes to form ammonia, hydrogen sulfide, and hydrogen cyanide. Oxides of nitrogen may also form. All of these products are toxic.
6.6 **Behavior in Fire:**
6.7 **Ignition Temperature:** Not pertinent
6.8 **Electrical Hazard:** Not pertinent
6.9 **Burning Rate:** Not pertinent

7. CHEMICAL REACTIVITY

7.1 **Reactivity with Water:** No reaction
7.2 **Reactivity with Common Materials:**
7.3 **Stability During Transport:** Stable
7.4 **Neutralizing Agents for Acids and Caustics:** Not pertinent
7.5 **Polymerization:** Not pertinent
7.6 **Inhibitor of Polymerization:** Not pertinent

8. WATER POLLUTION

8.1 **Aquatic Toxicity:**
280—300 ppm/1 hr/orange-spotted sunfish/killed/fresh water
420 ppm/48 hr/mosquitofish/TL_m/fresh water
8.2 **Waterfowl Toxicity:** Data not available
8.3 **Biological Oxygen Demand (BOD):** $<$ 0.010 lb/lb, 5 days
8.4 **Food Chain Concentration Potential:** None

9. SELECTED MANUFACTURERS

1. Argus Chemical Corp.
Halby Division
600 Terminal Ave.
New Castle, Del. 19720
2. Mallinckrodt Chemical Works
223 Westside Avenue
P. O. Box 384
Jersey City, N. J. 07303
3. J. T. Baker Chemical Co.
Phillipsburg, N. J. 08865

10. SHIPPING INFORMATION

10.1 **Grades or Purity:** Reagent; Technical, 50—65% solution in water
10.2 **Storage Temperature:** Ambient
10.3 **Inert Atmosphere:** No requirement
10.4 **Venting:** Open

11. HAZARD ASSESSMENT CODE
(See Hazard Assessment Handbook, CG 446-3)

A-P-SS

12. HAZARD CLASSIFICATIONS

12.1 **Code of Federal Regulations:** Not listed
12.2 **NAS Hazard Rating for Bulk Water Transportation:** Not listed
12.3 **NFPA Hazard Classifications:** Not listed

13. PHYSICAL AND CHEMICAL PROPERTIES

13.1 **Physical State at 15°C and 1 atm:** Solid or liquid
13.2 **Molecular Weight:** 76.12
13.3 **Boiling Point at 1 atm:** (solution; solid decomposes) 239°F = 115°C = 388°K
13.4 **Freezing Point:** (solid) 320°F = 160°C = 433°K
13.5 **Critical Temperature:** Not pertinent
13.6 **Critical Pressure:** Not pertinent
13.7 **Specific Gravity:**
$>$ 1.1 at 20°C (solid)
1.1—1.15 at 20°C (solution)
13.8 **Liquid Surface Tension:** Not pertinent
13.9 **Liquid-Water Interfacial Tension:** Not pertinent
13.10 **Vapor (Gas) Specific Gravity:** Not pertinent
13.11 **Ratio of Specific Heats of Vapor (Gas):** Not pertinent
13.12 **Latent Heat of Vaporization:** Not pertinent
13.13 **Heat of Combustion:** Not pertinent
13.14 **Heat of Decomposition:** Not pertinent
13.15 **Heat of Solution:** 133 Btu/lb = 74 cal/g = 3.1×10^5 J/kg
13.16 **Heat of Polymerization:** Not pertinent

NOTES

ATF AMMONIUM THIOSULFATE

Common Synonyms Ammonium hyposulfite	Solid or solution (in water)	Colorless liquid or white solid	Ammonia odor
	Sinks and mixes with water.		

Stop discharge if possible. Keep people away.
Avoid contact with liquid and solid.
Avoid contact with solid and dust.
Isolate and remove discharged material.
Notify local health and pollution control agencies.

Fire	Not flammable. POISONOUS GASES MAY BE PRODUCED IN FIRE. Wear goggles and self-contained breathing apparatus.
Exposure	CALL FOR MEDICAL AID. DUST Irritating to eyes, nose and throat. If inhaled will cause difficult breathing. If in eyes, hold eyelids open and flush with plenty of water. If breathing has stopped, give artificial respiration. If breathing is difficult, give oxygen. SOLUTION OR SOLID Irritating to skin and eyes. Harmful if swallowed. Remove contaminated clothing and shoes. Flush affected areas with plenty of water. IF IN EYES, hold eyelids open and flush with plenty of water. IF SWALLOWED and victim is CONSCIOUS, have victim drink water or milk. IF SWALLOWED and victim is UNCONSCIOUS OR HAVING CONVULSIONS, do nothing except keep victim warm.
Water Pollution	Effect of low concentrations on aquatic life is unknown. May be dangerous if it enters water intakes. Notify local health and wildlife officials. Notify operators of nearby water intakes.

1. RESPONSE TO DISCHARGE
(See Response Methods Handbook, CG 446-4)

Issue warning—water contaminant
Disperse and flush

2. LABELS

No hazard label required by Code of Federal Regulations

3. CHEMICAL DESIGNATIONS

3.1 **Synonyms:** Ammonium hypo; Ammonium hyposulfite
3.2 **Coast Guard Compatibility Classification:** Not listed
3.3 **Chemical Formula:** $(NH_4)_2S_2O_3$
3.4 **IMCO/United Nations Numerical Designation:** Not listed

4. OBSERVABLE CHARACTERISTICS

4.1 **Physical State (as shipped):** Solid, or solution in water
4.2 **Color:** Clear, colorless
4.3 **Odor:** Ammonia

5. HEALTH HAZARDS

5.1 **Personal Protective Equipment:** Respirator; rubber gloves; safety glasses; face shield; rubber apron, shield, or suit as necessary to prevent skin contact
5.2 **Symptoms Following Exposure:** Inhalation of dust may irritate respiratory system. Ingestion could be harmful. Contact with eyes or skin may cause irritation.
5.3 **Treatment for Exposure:** INHALATION: move to fresh air. INGESTION: get medical attention at once. EYES: flush with plenty of water for at least 15 min. and get immediate medical attention. SKIN: wash with water.
5.4 **Toxicity by Inhalation (Threshold Limit Value):** Data not available
5.5 **Short-Term Inhalation Limits:** Data not available
5.6 **Toxicity by Ingestion:** Data not available
5.7 **Late Toxicity:** Data not available
5.8 **Vapor (Gas) Irritant Characteristics:** Data not available
5.9 **Liquid or Solid Irritant Characteristics:** Data not available
5.10 **Odor Threshold:** Data not available

6. FIRE HAZARDS

6.1 **Flash Point:** Not flammable
6.2 **Flammable Limits in Air:** Not flammable
6.3 **Fire Extinguishing Agents:** Not pertinent
6.4 **Fire Extinguishing Agents Not to be Used:** Not pertinent
6.5 **Special Hazards of Combustion Products:** Toxic ammonia, hydrogen sulfide, and oxides of nitrogen and sulfur may form in fires.
6.6 **Behavior in Fire:**
6.7 **Ignition Temperature:** Not pertinent
6.8 **Electrical Hazard:** Not pertinent
6.9 **Burning Rate:** Not pertinent

7. CHEMICAL REACTIVITY

7.1 **Reactivity with Water:** No reaction
7.2 **Reactivity with Common Materials:**
7.3 **Stability During Transport:** Stable, but toxic ammonia gas may collect in enclosed spaces.
7.4 **Neutralizing Agents for Acids and Caustics:** Not pertinent
7.5 **Polymerization:** Not pertinent
7.6 **Inhibitor of Polymerization:** Not pertinent

8. WATER POLLUTION

8.1 **Aquatic Toxicity:** Data not available
8.2 **Waterfowl Toxicity:** Data not available
8.3 **Biological Oxygen Demand (BOD):** 0.62 lb/lb, 5 days
8.4 **Food Chain Concentration Potential:** None

9. SELECTED MANUFACTURERS

1. Allied Chemical Corp.
Industrial Chemicals Div.
P. O. Box 1139R
Morristown, N. J. 07960
2. Narco Corporation
12520 Cerise Avenue
Hawthorne, Calif. 90250
3. Gallard-Schlesinger Chemical Mfg. Co.
584 Mineola Avenue
Carle Place, N. Y. 11514

10. SHIPPING INFORMATION

10.1 **Grades or Purity:** Technical; Analytical; Technical, 60% solution in water
10.2 **Storage Temperature:** Ambient
10.3 **Inert Atmosphere:** Ventilated (natural)
10.4 **Venting:** Open

11. HAZARD ASSESSMENT CODE
(See Hazard Assessment Handbook, CG 446-3)

A-P-SS

12. HAZARD CLASSIFICATIONS

12.1 **Code of Federal Regulations:** Not listed
12.2 **NAS Hazard Rating for Bulk Water Transportation:** Not listed
12.3 **NFPA Hazard Classifications:** Not listed

13. PHYSICAL AND CHEMICAL PROPERTIES

13.1 **Physical State at 15°C and 1 atm:** Solid or liquid
13.2 **Molecular Weight:** 148.2
13.3 **Boiling Point at 1 atm:** Not pertinent (decomposes)
13.4 **Freezing Point:** Not pertinent
13.5 **Critical Temperature:** Not pertinent
13.6 **Critical Pressure:** Not pertinent
13.7 **Specific Gravity:** 2 at 20°C (solid)
13.8 **Liquid Surface Tension:** Not pertinent
13.9 **Liquid-Water Interfacial Tension:** Not pertinent
13.10 **Vapor (Gas) Specific Gravity:** Not pertinent
13.11 **Ratio of Specific Heats of Vapor (Gas):** Not pertinent
13.12 **Latent Heat of Vaporization:** Not pertinent
13.13 **Heat of Combustion:** Not pertinent
13.14 **Heat of Decomposition:** Not pertinent
13.15 **Heat of Solution:** Data not available
13.16 **Heat of Polymerization:** Not pertinent

NOTES

AML	AMYL ACETATE

Common Synonyms			
Acetic acid n-amyl ester	Watery liquid	Colorless to yellow	Banana odor
	Floats on water. Flammable, irritating vapor is produced.		

Shut off ignition sources and call fire department.
Stop discharge if possible. Keep people away.
Avoid contact with liquid and vapor.
Stay upwind and use water spray to "knock down" vapor.
Isolate and remove discharged material.
Notify local health and pollution control agencies.

Fire

FLAMMABLE.
Flashback along vapor trail may occur.
Vapor may explode if ignited in an enclosed area.
Extinguish with dry chemical, alcohol foam, or carbon dioxide.
Water may be ineffective on fire.
Cool exposed containers with water.

Exposure

CALL FOR MEDICAL AID.

VAPOR
Irritating to eyes, nose and throat.
If inhaled, will cause nausea, headache or dizziness.
Move to fresh air.
If breathing has stopped, give artificial respiration.
If breathing is difficult, give oxygen.

LIQUID
Irritating to skin and eyes.
Remove contaminated clothing and shoes.
Flush affected areas with plenty of water.
IF IN EYES, hold eyelids open and flush with plenty of water.

Water Pollution

Effect of low concentrations on aquatic life is unknown.
Fouling to shoreline.
May be dangerous if it enters water intakes.
Notify local health and pollution control officials.
Notify operators of nearby water intakes.

1. RESPONSE TO DISCHARGE

(See Response Methods Handbook, CG 446-4)

Issue warning – high flammability
Mechanical containment
Chemical and physical treatment

2. LABEL

Red

3. CHEMICAL DESIGNATIONS

3.1 **Synonyms:** Amyl acetate, mixed isomers
Pentyl acetates

3.2 **Coast Guard Compatibility Classification:** Ester

3.3 **Chemical Formula:** $CH_3COOC_5H_{11}$

3.4 **IMCO/United Nations Numerical Designation:** 3.2/1104

4. OBSERVABLE CHARACTERISTICS

4.1 **Physical State (as shipped):** Liquid

4.2 **Color:** Colorless to yellow

4.3 **Odor:** Pleasant banana-like; mild; characteristic banana- or pear-like odor

5. HEALTH HAZARDS

5.1 **Personal Protective Equipment:** Air-supplied mask or chemical cartridge respirator; protective gloves, goggles, safety shower, and eye bath.

5.2 **Symptoms Following Exposure:** Irritation of eyes, nose and throat. Dizziness, nausea, headache.

5.3 **Treatment for Exposure:** INHALATION: move victim to fresh air; call physician; administer oxygen. SKIN OR EYES: flush with water.

5.4 **Toxicity by Inhalation (Threshold Limit Value):** 100 ppm

5.5 **Short-Term Inhalation Limits:** 200 ppm for 30 min.

5.6 **Toxicity by Ingestion:** Grade 1; LD_{50} = 6.5 g/kg (rat)

5.7 **Late Toxicity:** None

5.8 **Vapor (Gas) Irritant Characteristics:** Vapors cause a slight smarting of the eyes or respiratory system if present in high concentration. The effect is temporary.

5.9 **Liquid or Solid Irritant Characteristics:** No appreciable hazard. Practically harmless to the skin. If spilled on clothing and allowed to remain, may cause smarting and reddening of the skin.

5.10 **Odor Threshold:** 0.067 ppm

6. FIRE HAZARDS

6.1 **Flash Point:** (iso-): 69°F C.C.
(n-): 91° F C.C.

6.2 **Flammable Limits in Air:** 1.1%–7.5% (n)

6.3 **Fire Extinguishing Agents:** Alcohol foam, dry chemical, carbon dioxide

6.4 **Fire Extinguishing Agents Not to be Used:** Water in straight hose stream will scatter and spread fire and should not be used

6.5 **Special Hazards of Combustion Products:** Not pertinent

6.6 **Behavior in Fire:** Not pertinent

6.7 **Ignition Temperature:** 572°F (n)

6.8 **Electrical Hazard:** Not pertinent

6.9 **Burning Rate:** 4.1 mm/min.

7. CHEMICAL REACTIVITY

7.1 **Reactivity with Water:** No reaction

7.2 **Reactivity with Common Materials:** No reaction

7.3 **Stability During Transport:** Stable

7.4 **Neutralizing Agents for Acids and Caustics:** Not pertinent

7.5 **Polymerization:** Not pertinent

7.6 **Inhibitor of Polymerization:** Not pertinent

8. WATER POLLUTION

8.1 **Aquatic Toxicity:**
120 ppm/48 hr/daphnia/TL_m/turbid water
180 ppm/96 hr/scenedesmus/TL_m/fresh water

8.2 **Waterfowl Toxicity:** Data not available

8.3 **Biological Oxygen Demand (BOD):** 0.3–0.8 lb/lb, 5 days

8.4 **Food Chain Concentration Potential:** None

9. SELECTED MANUFACTURERS

1. Commercial Solvents Corp.
245 Park Avenue
New York, N. Y. 10017

2. Publicker Industries, Inc.
1429 Walnut St.
Philadelphia, Pa. 19102

3. Union Carbide Corp.
Chemicals and Plastics Division
270 Park Ave.
New York, N. Y. 10017

10. SHIPPING INFORMATION

10.1 **Grades or Purity:** 85–96% (technical, commercial)

10.2 **Storage Temperature:** Ambient (cool)

10.3 **Inert Atmosphere:** No requirement

10.4 **Venting:** Data not available

11. HAZARD ASSESSMENT CODE

(See Hazard Assessment Handbook, CG 446-3)

A-T-U

12. HAZARD CLASSIFICATIONS

12.1 **Code of Federal Regulations:** Flammable liquid

12.2 **NAS Hazard Rating for Bulk Water Transportation:** Not listed

12.3 **NFPA Hazard Classifications:**

Category	Classification
Health Hazard (Blue)	1
Flammability (Red)	3
Reactivity (Yellow)	0

13. PHYSICAL AND CHEMICAL PROPERTIES

13.1 **Physical State at 15°C and 1 atm:** Liquid

13.2 **Molecular Weight:** 130.19

13.3 **Boiling Point at 1 atm:** 295°F = 146°C = 419°K

13.4 **Freezing Point:** <−148°F = <−100°C = <173°K

13.5 **Critical Temperature:** Not pertinent

13.6 **Critical Pressure:** Not pertinent

13.7 **Specific Gravity:** 0.876 at 20°C (liquid)

13.8 **Liquid Surface Tension:** 12 dynes/cm = 0.012 N/m at 30°C

13.9 **Liquid-Water Interfacial Tension:** (est.) 50 dynes/cm = 0.05 N/m at 17°C

13.10 **Vapor (Gas) Specific Gravity:** Not pertinent

13.11 **Ratio of Specific Heats of Vapor (Gas):** 1.1

13.12 **Latent Heat of Vaporization:** 140 Btu/lb = 75 cal/g = 3.1×10^5 J/kg

13.13 **Heat of Combustion:** −13,360 Btu/lb = −7423 cal/g = -310.8×10^5 J/kg

13.14 **Heat of Decomposition:** Not pertinent

13.15 **Heat of Solution:** Not pertinent

13.16 **Heat of Polymerization:** Not pertinent

NOTES

REVISED 1978

AAN — n-AMYL ALCOHOL

Common Synonyms
Pentyl alcohol
l-Pentanol
n-Butylcarbinol

Liquid — Colorless — Mild, sweet odor

Floats on water. Flammable, irritating vapor is produced.

Shut off ignition sources and call fire department.
Stop discharge if possible. Keep people away.
Avoid contact with liquid and vapor.
Stay upwind and use water spray to "knock down" vapor.
Isolate and remove discharged material.
Notify local health and pollution control agencies.

Fire

FLAMMABLE.
Flashback along vapor trail may occur.
Vapor may explode if ignited in an enclosed area.
Extinguish with dry chemical, alcohol foam, or carbon dioxide.
Water may be ineffective on fire.
Cool exposed containers with water.

Exposure

CALL FOR MEDICAL AID.

VAPOR
Irritating to eyes, nose and throat.
If inhaled, will cause coughing, nausea, headache, or difficult breathing.
Move to fresh air.
If breathing has stopped, give artificial respiration.
If breathing is difficult, give oxygen.

LIQUID
Irritating to eyes.
Harmful if swallowed.
Not irritating to skin.
Flush affected areas with plenty of water.
IF IN EYES, hold eyelids open and flush with plenty of water.
IF SWALLOWED and victim is CONSCIOUS, have victim drink water or milk.

Water Pollution

Effect of low concentrations on aquatic life is unknown.
Fouling to shoreline.
May be dangerous if it enters water intakes.
Notify local health and pollution control officials.
Notify operators of nearby water intakes.

1. RESPONSE TO DISCHARGE

(See Response Methods Handbook, CG 446-4)

Disperse and flush

2. LABEL

Red

3. CHEMICAL DESIGNATIONS

3.1 **Synonyms:** l-Amyl alcohol; n-Butylcarbinol; l-Pentanol; Pentyl alcohol

3.2 **Coast Guard Compatibility Classification:** Alcohol

3.3 **Chemical Formula:** $CH_3(CH_2)_3CH_2OH$

3.4 **IMCO/United Nations Numerical Designation:** 3.2/1105

4. OBSERVABLE CHARACTERISTICS

4.1 **Physical State (as shipped):** Liquid

4.2 **Color:** Colorless

4.3 **Odor:** Like alcohol; causes coughing

5. HEALTH HAZARDS

5.1 **Personal Protective Equipment:** Face splash shield, goggles, protective clothing, and cartridge respirator.

5.2 **Symptoms Following Exposure:** Irritation of skin, eyes, and respiratory tract; headache and vertigo; dyspnea and cough; nausea, vomiting, and diarrhea. Double vision, deafness, delirium, and occasionally fatal poisoning, preceded by severe nervous symptoms, have been reported. Coma, glycosuria, and methemoglobinemia can occur.

5.3 **Treatment for Exposure:** SKIN: remove chemical by thorough washing with soap and water. EYES: wash promptly with large quantities of water for at least 15 mins. Call a doctor.

5.4 **Toxicity by Inhalation (Threshold Limit Value):** Data not available.

5.5 **Short-Term Inhalation Limits:** 150 ppm for 30 min.

5.6 **Toxicity by Ingestion:** Grade 2; LD_{50} 0.5 to 5 g/kg

5.7 **Late Toxicity:** None

5.8 **Vapor (Gas) Irritant Characteristics:** Vapors cause a slight smarting of the eyes or respiratory system if present in high concentrations. The effect is temporary.

5.9 **Liquid or Solid Irritant Characteristics:** No appreciable hazard. Practically harmless to the skin.

5.10 **Odor Threshold:** 0.12 ppm

6. FIRE HAZARDS

6.1 **Flash Point:** 77°F C.C.

6.2 **Flammable Limits in Air:** 1.1% 7.5%

6.3 **Fire Extinguishing Agents:** Alcohol foam, dry chemical or carbon dioxide

6.4 **Fire Extinguishing Agents Not to be Used:** Water may be ineffective

6.5 **Special Hazards of Combustion Products:** Not pertinent

6.6 **Behavior in Fire:** Not pertinent

6.7 **Ignition Temperature:** 680°F

6.8 **Electrical Hazard:** Not pertinent

6.9 **Burning Rate:** 3.6 mm/min.

7. CHEMICAL REACTIVITY

7.1 **Reactivity with Water:** No reaction

7.2 **Reactivity with Common Materials:** No reaction

7.3 **Stability During Transport:** Stable

7.4 **Neutralizing Agents for Acids and Caustics:** Not pertinent

7.5 **Polymerization:** Not pertinent

7.6 **Inhibitor of Polymerization:** Not pertinent

8. WATER POLLUTION

8.1 **Aquatic Toxicity:** Data not available

8.2 **Waterfowl Toxicity:** Data not available

8.3 **Biological Oxygen Demand (BOD):** 155%, 5 days

8.4 **Food Chain Concentration Potential:** None

9. SELECTED MANUFACTURERS

1. Allied Chemical Corp.
   Specialty Chemicals Division
   Marcus Hook, Pa. 19061
2. Exxon Chemical Company
   Baton Rouge, La. 70821
3. Union Carbide Corp.
   Chemicals and Plastics Division
   270 Park Ave.
   New York, N. Y. 10017

10. SHIPPING INFORMATION

10.1 **Grades or Purity:** 98%; 74% plus 25% 2-methyl-1-butanol

10.2 **Storage Temperature:** Ambient

10.3 **Inert Atmosphere:** No requirement

10.4 **Venting:** Open (flame arrester)

11. HAZARD ASSESSMENT CODE

(See Hazard Assessment Handbook, CG 446-3)

A-P-Q-T-U

12. HAZARD CLASSIFICATIONS

12.1 **Code of Federal Regulations:** Flammable Liquid

12.2 **NAS Hazard Rating for Bulk Water Transportation:**

Category	Rating
Fire	3
Health	
Vapor Irritant	1
Liquid or Solid Irritant	0
Poisons	2
Water Pollution	
Human Toxicity	2
Aquatic Toxicity	2
Aesthetic Effect	2
Reactivity	
Other Chemicals	2
Water	0
Self-Reaction	0

12.3 **NFPA Hazard Classifications:**

Category	Classification
Health Hazard (Blue)	1
Flammability (Red)	3
Reactivity (Yellow)	0

13. PHYSICAL AND CHEMICAL PROPERTIES

13.1 **Physical State at 15°C and 1 atm:** Liquid

13.2 **Molecular Weight:** 88.15

13.3 **Boiling Point at 1 atm:** 280.2°F = 137.9°C = 411.1°K

13.4 **Freezing Point:** −110°F = −79°C = 194°K

13.5 **Critical Temperature:** 595°F = 313°C = 586°K

13.6 **Critical Pressure:** Not pertinent

13.7 **Specific Gravity:** 0.818 at 15°C (liquid)

13.8 **Liquid Surface Tension:** 25.60 dynes/cm = .02560 N/m at 20°C

13.9 **Liquid-Water Interfacial Tension:** 5 dynes/cm = 0.005 N/m at 20°C

13.10 **Vapor (Gas) Specific Gravity:** Not pertinent

13.11 **Ratio of Specific Heats of Vapor (Gas):** 1.06

13.12 **Latent Heat of Vaporization:** 217.1 Btu/lb = 120.6 cal/g = 5.049×10^5 J/kg

13.13 **Heat of Combustion:** −16,200 Btu/lb = −9000 cal/g = -376.8×10^5 J/kg

13.14 **Heat of Decomposition:** Not pertinent

13.15 **Heat of Solution:** Not pertinent

13.16 **Heat of Polymerization:** Not pertinent

NOTES

AMY — n-AMYL CHLORIDE

Common Synonyms			
Amyl chloride 1-Chlorpentane n-Butylcarbinyl chloride 1-Pentyl chloride Chloride of amyl	Liquid	Colorless to purple	Pleasant odor
	Floats on water. Flammable vapor is produced.		

Shut off ignition sources. Call fire department.
Stop discharge if possible. Keep people away.
Evacuate area in case of large discharge.
Isolate and remove discharged material.
Notify local health and pollution control agencies.

Fire

FLAMMABLE.
Flashback along vapor trail may occur.
Vapor may explode if ignited in an enclosed area.
Irritating gases may be produced when heated.
Extinguish with dry chemicals, foam or carbon dioxide.
Water may be ineffective on fire.

Exposure

Call for medical aid.

VAPOR
Irritating to eyes, nose and throat.
Move victim to fresh air.

LIQUID
Irritating to skin and eyes.
Remove contaminated clothing and shoes.
Flush affected areas with plenty of water.
IF IN EYES, hold eyelids open and flush with plenty of water.
IF SWALLOWED, and victim is CONSCIOUS, have victim drink water or milk.

Water Pollution

Effect of low concentrations on aquatic life is unknown.
Fouling to shoreline.
May be dangerous if it enters water intakes.
Notify local health and wildlife officials.
Notify operators of nearby water intakes.

1. RESPONSE TO DISCHARGE

(See Response Methods Handbook, CG 446-4)

Mechanical containment
Chemical and physical treatment

2. LABEL

Red

3. CHEMICAL DESIGNATIONS

3.1 **Synonyms:** Amyl chloride; 1-Chloropentane; n-Butylcarbinyl chloride; 1-Pentyl chloride; Chloride of amyl

3.2 **Coast Guard Compatibility Classification:** Halogenated compounds (5)

3.3 **Chemical Formula:** $CH_3CH_2CH_2CH_2CH_2Cl$

3.4 **IMCO/United Nations Numerical Designation:** 3.2/1107

4. OBSERVABLE CHARACTERISTICS

4.1 **Physical State (as shipped):** Liquid

4.2 **Color:** Colorless, straw; deep purple

4.3 **Odor:** Aromatic

5. HEALTH HAZARDS

5.1 **Personal Protective Equipment:** Protective goggles or face shield; rubber gloves.

5.2 **Symptoms Following Exposure:** Inhalation causes mild irritation of mucous membranes. Ingestion of liquid or contact with skin or eyes causes mild irritation.

5.3 **Treatment for Exposure:** INHALATION: remove to fresh air; apply artificial respiration if required. EYES: flush with water. SKIN: wash well with soap and water. INGESTION: induce vomiting; give water.

5.4 **Toxicity by Inhalation (Threshold Limit Value):** Data not available

5.5 **Short-Term Inhalation Limits:** Data not available

5.6 **Toxicity by Ingestion:** Grade 1; LD_{50} 5 to 15 g/kg

5.7 **Late Toxicity:** Data not available

5.8 **Vapor (Gas) Irritant Characteristics:** Vapors cause a slight smarting of the eyes or respiratory system if present in high concentrations. The effect is temporary.

5.9 **Liquid or Solid Irritant Characteristics:** Minimum hazard. If spilled on clothing and allowed to remain, may cause smarting and reddening of the skin.

5.10 **Odor Threshold:** Data not available

6. FIRE HAZARDS

6.1 **Flash Point:** 55°F O.C.; 34°F C.C.

6.2 **Flammable Limits in Air:** 1.4% 8.6%

6.3 **Fire Extinguishing Agents:** Dry chemical, foam, carbon dioxide

6.4 **Fire Extinguishing Agents Not to be Used:** Water may be ineffective.

6.5 **Special Hazards of Combustion Products:** Irritating hydrogen chloride and toxic phosgene may be formed in fires.

6.6 **Behavior in Fire:** Not pertinent

6.7 **Ignition Temperature:** 500°F

6.8 **Electrical Hazard:** Data not available

6.9 **Burning Rate:** 4.9 mm/min.

7. CHEMICAL REACTIVITY

7.1 **Reactivity with Water:** No reaction

7.2 **Reactivity with Common Materials:** No reaction

7.3 **Stability During Transport:** Stable

7.4 **Neutralizing Agents for Acids and Caustics:** Not pertinent

7.5 **Polymerization:** Not pertinent

7.6 **Inhibitor of Polymerization:** Not pertinent

8. WATER POLLUTION

8.1 **Aquatic Toxicity:** Data not available

8.2 **Waterfowl Toxicity:** Data not available

8.3 **Biological Oxygen Demand (BOD):** Data not available

8.4 **Food Chain Concentration Potential:** None

9. SELECTED MANUFACTURERS

1. Eastman Kodak Co.
Eastman Organic Chemicals
Rochester, N. Y. 14650
2. Lachat Chemicals, Inc.
20200 Ashland Avenue
Chicago Heights, Ill. 60411
3. Pfaltz and Bauer, Inc.
126-04 Northern Boulevard
Flushing, N. Y. 11368

10. SHIPPING INFORMATION

10.1 **Grades or Purity:** Commercial

10.2 **Storage Temperature:** Ambient

10.3 **Inert Atmosphere:** No requirement

10.4 **Venting:** Open (flame arrester)

11. HAZARD ASSESSMENT CODE

(See Hazard Assessment Handbook, CG 446-3)

A-T-U

12. HAZARD CLASSIFICATIONS

12.1 **Code of Federal Regulations:** Flammable liquid

12.2 **NAS Hazard Rating for Bulk Water Transportation:**

Category	Rating
Fire	3
Health	
Vapor Irritant	1
Liquid or Solid Irritant	1
Poisons	2
Water Pollution	
Human Toxicity	1
Aquatic Toxicity	2
Aesthetic Effect	2
Reactivity	
Other Chemicals	1
Water	0
Self-Reaction	0

12.3 **NFPA Hazard Classifications:**

Category	Classification
Health Hazard (Blue)	1
Flammability (Red)	3
Reactivity (Yellow)	0

13. PHYSICAL AND CHEMICAL PROPERTIES

13.1 **Physical State at 15°C and 1 atm:** Liquid

13.2 **Molecular Weight:** 106.6

13.3 **Boiling Point at 1 atm:** 226°F = 108°C = 381°K

13.4 **Freezing Point:** −146°F = −99°C = 174°K

13.5 **Critical Temperature:** Not pertinent

13.6 **Critical Pressure:** Not pertinent

13.7 **Specific Gravity:** 0.8834 at 20°C (liquid)

13.8 **Liquid Surface Tension:** 24.9 dynes/cm = 0.0249 N/m at 20°C

13.9 **Liquid-Water Interfacial Tension:** (est.) 35 dynes/cm = 0.035 N/m at 20°C

13.10 **Vapor (Gas) Specific Gravity:** 3.7

13.11 **Ratio of Specific Heats of Vapor (Gas):** 1.0650

13.12 **Latent Heat of Vaporization:** 132.1 Btu/lb = 73.40 cal/g = 3.073×10^5 J/kg

13.13 **Heat of Combustion:** −13,500 Btu/lb = −7,500 cal/g = -314×10^5 J/kg

13.14 **Heat of Decomposition:** Not pertinent

13.15 **Heat of Solution:** Not pertinent

13.16 **Heat of Polymerization:** Not pertinent

NOTES

AMM	n-AMYL MERCAPTAN

Common Synonyms			
1-Pentanethiol Amyl hydrosulfide Amyl sulfhydrate Amyl thioalcohol	Liquid	Colorless to yellow	Strong garlic odor
	Floats on water. Flammable vapor is produced.		

Shut off ignition sources. Call fire department.
Stop discharge if possible. Keep people away.
Isolate and remove discharged material.
Notify local health and pollution control agencies.

Fire	FLAMMABLE. POISONOUS GASES ARE PRODUCED IN FIRE. Flashback along vapor trail may occur. Vapor may explode if ignited in an enclosed area. Containers may explode in fire. Wear goggles and self-contained breathing apparatus. Extinguish with dry chemicals, foam or carbon dioxide. Water may be ineffective on fire. Cool exposed containers with water.
Exposure	Call for medical aid. VAPOR Move victim to fresh air. LIQUID Irritating to skin and eyes. Remove contaminated clothing and shoes. Flush affected areas with plenty of water. IF IN EYES, hold eyelids open and flush with plenty of water. IF SWALLOWED and victim is CONSCIOUS, have victim drink water or milk.
Water Pollution	Effect of low concentrations on aquatic life is unknown. Fouling to shoreline. May be dangerous if it enters water intakes. Notify local health and wildlife officials. Notify operators of nearby water intakes.

1. RESPONSE TO DISCHARGE

(See Response Methods Handbook, CG 446-4)

Issue warning—high flammability
Mechanical containment
Should be removed
Chemical and physical treatment

2. LABEL

Red

3. CHEMICAL DESIGNATIONS

3.1 **Synonyms:** 1-Pentanethiol; Amyl hydrosulfide; Amyl sulfhydrate; Amyl thioalcohol
3.2 **Coast Guard Compatibility Classification:** Not listed
3.3 **Chemical Formula:** $CH_3CH_2CH_2CH_2CH_2SH$
3.4 **IMCO/United Nations Numerical Designation:** 3.2/1111

4. OBSERVABLE CHARACTERISTICS

4.1 **Physical State (as shipped):** Liquid
4.2 **Color:** Colorless; water-white to yellow
4.3 **Odor:** Strong, offensive; garlic

5. HEALTH HAZARDS

5.1 **Personal Protective Equipment:** Plastic gloves; goggles.
5.2 **Symptoms Following Exposure:** Inhalation may cause nausea because of offensive odor. Contact with eyes or skin causes slight irritation. Ingestion may cause vomiting.
5.3 **Treatment for Exposure:** INHALATION: remove to fresh air; apply artificial respiration if required. EYES: wash with water; see a physician. SKIN: wash with soap and water. INGESTION: induce vomiting if it does not occur spontaneously.
5.4 **Toxicity by Inhalation (Threshold Limit Value):** Data not available
5.5 **Short-Term Inhalation Limits:** Data not available
5.6 **Toxicity by Ingestion:** Data not available
5.7 **Late Toxicity:** None
5.8 **Vapor (Gas) Irritant Characteristics:** Data not available
5.9 **Liquid or Solid Irritant Characteristics:** Data not available
5.10 **Odor Threshold:** 0.3 mg/m³

6. FIRE HAZARDS

6.1 **Flash Point:** 65°F O.C.
6.2 **Flammable Limits in Air:** Data not available
6.3 **Fire Extinguishing Agents:** Dry chemical, foam, carbon dioxide
6.4 **Fire Extinguishing Agents Not to be Used:** Water may be ineffective
6.5 **Special Hazards of Combustion Products:** Sulfur dioxide gas is formed
6.6 **Behavior in Fire:** Vapor is heavier than air and may travel considerable distance to a source of ignition and flash back.
6.7 **Ignition Temperature:** Data not available
6.8 **Electrical Hazard:** Data not available
6.9 **Burning Rate:** 4.7 mm/min.

7. CHEMICAL REACTIVITY

7.1 **Reactivity with Water:** No reaction
7.2 **Reactivity with Common Materials:** No reaction
7.3 **Stability During Transport:** Stable
7.4 **Neutralizing Agents for Acids and Caustics:** Not pertinent
7.5 **Polymerization:** Not pertinent
7.6 **Inhibitor of Polymerization:** Not pertinent

8. WATER POLLUTION

8.1 **Aquatic Toxicity:** Data not available
8.2 **Waterfowl Toxicity:** Data not available
8.3 **Biological Oxygen Demand (BOD):** Data not available
8.4 **Food Chain Concentration Potential:** None

9. SELECTED MANUFACTURERS

1. Pennwalt Corporation
   Industrial Chemicals Department
   Three Parkway
   Philadelphia, Pa. 19102
2. Eastman Kodak Co.
   Eastman Organic Chemicals
   Rochester, N. Y. 14650
3. Aldrich Chemical Co.
   940 West St. Paul Ave.
   Milwaukee, Wis. 53233

10. SHIPPING INFORMATION

10.1 **Grades or Purity:** 96.0+%
10.2 **Storage Temperature:** Ambient
10.3 **Inert Atmosphere:** No requirement
10.4 **Venting:** Pressure-vacuum

11. HAZARD ASSESSMENT CODE

(See Hazard Assessment Handbook, CG 446-3)

A-T-U

12. HAZARD CLASSIFICATIONS

12.1 **Code of Federal Regulations:** Flammable liquid
12.2 **NAS Hazard Rating for Bulk Water Transportation:** Not listed
12.3 **NFPA Hazard Classifications:**

Category	Classification
Health Hazard (Blue)	2
Flammability (Red)	3
Reactivity (Yellow)	0

13. PHYSICAL AND CHEMICAL PROPERTIES

13.1 **Physical State at 15°C and 1 atm:** Liquid
13.2 **Molecular Weight:** 104.2
13.3 **Boiling Point at 1 atm:** 248°F = 120°C = 393°K
13.4 **Freezing Point:** −105°F = −76°C = 197°K
13.5 **Critical Temperature:** 610°F = 321°C = 594°K
13.6 **Critical Pressure:** 508 psia = 34.5 atm = 3.50 MN/m²
13.7 **Specific Gravity:** 0.8392 at 25°C (liquid)
13.8 **Liquid Surface Tension:** 26.8 dynes/cm = 0.0268 N/m at 20°C
13.9 **Liquid-Water Interfacial Tension:** (est.) 35 dynes/cm = 0.035 N/m at 20°C
13.10 **Vapor (Gas) Specific Gravity:** 3.59
13.11 **Ratio of Specific Heats of Vapor (Gas):** 1.0622
13.12 **Latent Heat of Vaporization:** 171 Btu/lb = 94.9 cal/g = 3.97 × 10⁵ J/kg
13.13 **Heat of Combustion:** −17,070 Btu/lb = −9,480 cal/g = −397 × 10⁵ J/kg
13.14 **Heat of Decomposition:** Not pertinent
13.15 **Heat of Solution:** Not pertinent
13.16 **Heat of Polymerization:** Not pertinent

NOTES

AMK — n-AMYL METHYL KETONE

Common Synonyms			
Methyl amyl ketone 2-Heptanone 2-Ketoheptane Methyl pentyl ketone Pentyl methyl ketone	Liquid	White	Penetrating fruity odor
	Floats and mixes slowly with water.		

Stop discharge if possible. Keep people away.
Shut off ignition sources. Call fire department.
Stay upwind. Use water spray to "knock down" vapor.
Avoid contact with liquid and vapor.
Isolate and remove discharged material.
Notify local health and pollution control agencies.

Fire

Combustible.

Extinguish with dry chemicals, alcohol foam, or carbon dioxide.
Water may be ineffective on fire.
Cool exposed containers with water.

Exposure

CALL FOR MEDICAL AID.

VAPOR
Irritating to eyes, nose and throat.
If inhaled will cause dizziness, headache, or difficult breathing.

If in eyes, hold eyelids open and flush with plenty of water.
If breathing has stopped, give artificial respiration.
If breathing is difficult, give oxygen.

LIQUID
Irritating to skin and eyes.
If swallowed will cause nausea or vomiting.

Remove contaminated clothing and shoes.
Flush affected areas with plenty of water.
IF IN EYES, hold eyelids open and flush with plenty of water.
IF SWALLOWED and victim is CONSCIOUS, have victim drink water or milk.
IF SWALLOWED and victim is UNCONSCIOUS OR HAVING CONVULSIONS, do nothing except keep victim warm.

Water Pollution

Effect of low concentrations on aquatic life is unknown.
Fouling to shoreline.
May be dangerous if it enters water intakes.

Notify local health and wildlife officials.
Notify operators of nearby water intakes.

1. RESPONSE TO DISCHARGE

(See Response Methods Handbook, CG 446-4)

Issue warning—water contaminant
Mechanical containment
Should be removed
Chemical and physical treatment

2. LABELS

No hazard label required by Code of Federal Regulations

3. CHEMICAL DESIGNATIONS

3.1 **Synonyms:** 2-Heptanone; 2-Ketoheptane; Methyl amyl ketone; Methyl pentyl ketone; Pentyl methyl ketone

3.2 **Coast Guard Compatibility Classification:** Not listed

3.3 **Chemical Formula:** $CH_3CH_2CH_2CH_2CH_2COCH_3$

3.4 **IMCO/United Nations Numerical Designation:** 3.3/1110

4. OBSERVABLE CHARACTERISTICS

4.1 **Physical State (as shipped):** Liquid

4.2 **Color:** Colorless

4.3 **Odor:** Characteristic

5. HEALTH HAZARDS

5.1 **Personal Protective Equipment:** Gloves and goggles

5.2 **Symptoms Following Exposure:** Inhalation of concentrated vapor may have narcotic effect. Ingestion causes gastrointestinal disturbances. Contact with eyes causes irritation. Prolonged and repeated contact with skin may cause defatting with resultant irritation.

5.3 **Treatment for Exposure:** INHALATION: move to fresh air. INGESTION: get medical attention. EYES: flush with water for 15 to 20 min. SKIN: flush affected areas with water.

5.4 **Toxicity by Inhalation (Threshold Limit Value):** 100 ppm (suggested)

5.5 **Short-Term Inhalation Limits:** Data not available

5.6 **Toxicity by Ingestion:** Grade 2; oral LD_{50} = 1,670 mg/kg (rat)

5.7 **Late Toxicity:** Data not available

5.8 **Vapor (Gas) Irritant Characteristics:** Data not available

5.9 **Liquid or Solid Irritant Characteristics:** Data not available

5.10 **Odor Threshold:** 0.897 ppb

6. FIRE HAZARDS

6.1 **Flash Point:** 117°F O.C.; 102°F C.C.

6.2 **Flammable Limits in Air:** 1.11%–7.9%

6.3 **Fire Extinguishing Agents:** Dry chemical, foam, carbon dioxide

6.4 **Fire Extinguishing Agents Not to be Used:** Water may be ineffective.

6.5 **Special Hazards of Combustion Products:**

6.6 **Behavior in Fire:**

6.7 **Ignition Temperature:** Data not available

6.8 **Electrical Hazard:** Data not available

6.9 **Burning Rate:** Data not available

7. CHEMICAL REACTIVITY

7.1 **Reactivity with Water:** No reaction

7.2 **Reactivity with Common Materials:** Will attack some forms of plastic.

7.3 **Stability During Transport:** Stable

7.4 **Neutralizing Agents for Acids and Caustics:** Not pertinent

7.5 **Polymerization:** Not pertinent

7.6 **Inhibitor of Polymerization:** Not pertinent

8. WATER POLLUTION

8.1 **Aquatic Toxicity:** Data not available

8.2 **Waterfowl Toxicity:** Data not available

8.3 **Biological Oxygen Demand (BOD):** 0.5 lb/lb, 10 days

8.4 **Food Chain Concentration Potential:** None

9. SELECTED MANUFACTURERS

1. Eastman Chemical Products, Inc.
   Industrial Chemicals Div.
   Kingsport, Tenn. 37662
2. Aldrich Chemical Co.
   940 W. Saint Paul Ave.
   Milwaukee, Wis. 53233
3. Pfaltz and Bauer, Inc.
   126–04 Northern Blvd.
   Flushing, N. Y. 11368

10. SHIPPING INFORMATION

10.1 **Grades or Purity:** Technical; Pure

10.2 **Storage Temperature:** Ambient

10.3 **Inert Atmosphere:** Ventilated (natural)

10.4 **Venting:** Open (flame arrester)

11. HAZARD ASSESSMENT CODE

(See Hazard Assessment Handbook, CG 446-3)

A-T-U

12. HAZARD CLASSIFICATIONS

12.1 **Code of Federal Regulations:** Not listed

12.2 **NAS Hazard Rating for Bulk Water Transportation:** Not listed

12.3 **NFPA Hazard Classifications:** Not listed

13. PHYSICAL AND CHEMICAL PROPERTIES

13.1 **Physical State at 15°C and 1 atm:** Liquid

13.2 **Molecular Weight:** 114.19

13.3 **Boiling Point at 1 atm:** 304.7°F = 151.5°C = 424.7°K

13.4 **Freezing Point:** −31°F = −35°C = 238°K

13.5 **Critical Temperature:** Data not available

13.6 **Critical Pressure:** Data not available

13.7 **Specific Gravity:** 0.8204 at 15°C (liquid)

13.8 **Liquid Surface Tension:** 26.17 dynes/cm = 0.02617 N/m at 25°C

13.9 **Liquid-Water Interfacial Tension:** 12.4 dynes/cm = 0.0124 N/m at 25°C

13.10 **Vapor (Gas) Specific Gravity:** 3.94

13.11 **Ratio of Specific Heats of Vapor (Gas):** (est.) 1.051

13.12 **Latent Heat of Vaporization:** 148.9 Btu/lb = 82.7 cal/g = 3.46×10^5 J/kg

13.13 **Heat of Combustion:** Data not available

13.14 **Heat of Decomposition:** Not pertinent

13.15 **Heat of Solution:** Not pertinent

13.16 **Heat of Polymerization:** Not pertinent

NOTES

ANT	n-AMYL NITRATE

Common Synonyms			
Mixed primary amyl nitrates Diesel ignition improver	Liquid	Colorless to light straw	Ether-like odor
	May float or sink in water.		

Stop discharge if possible. Keep people away.
Avoid contact with liquid and vapor.
Shut off ignition sources. Call fire department.
Stay upwind. Use water spray to "knock down" vapor.
Isolate and remove discharged material.
Notify local health and pollution control agencies.

Fire

Combustible.
POISONOUS GASES MAY BE PRODUCED IN FIRE.
May explode if exposed to heat or flames.
Wear goggles and self-contained breathing apparatus.
Extinguish with dry chemicals, foam, or carbon dioxide.
Water may be ineffective on fire.
Cool exposed containers with water.

Exposure

CALL FOR MEDICAL AID.

VAPOR
Irritating to eyes, nose and throat.
If inhaled will cause headache
If in eyes, hold eyelids open and flush with plenty of water.
If breathing has stopped, give artificial respiration.
If breathing is difficult, give oxygen.

LIQUID
Irritating to skin and eyes.
If swallowed will cause nausea or headache.
Remove contaminated clothing and shoes.
Flush affected areas with plenty of water.
IF IN EYES, hold eyelids open and flush with plenty of water.
IF SWALLOWED and victim is CONSCIOUS, have victim drink water or milk and have victim induce vomiting.
IF SWALLOWED and victim is UNCONSCIOUS OR HAVING CONVULSIONS, do nothing except keep victim warm.

Water Pollution

Effect of low concentrations on aquatic life is unknown.
Fouling to shoreline.
May be dangerous if it enters water intakes.
Notify local health and wildlife officials.
Notify operators or nearby water intakes.

1. RESPONSE TO DISCHARGE

(See Response Methods Handbook, CG 446-4)

Issue warning—air contaminant
Restrict access
Mechanical containment
Should be removed
Chemical and physical treatment

2. LABELS

No hazard label required by Code of Federal Regulations

3. CHEMICAL DESIGNATIONS

3.1 **Synonyms:** Diesel ignition improver; Mixed primary amyl nitrates
3.2 **Coast Guard Compatibility Classification:** Not listed
3.3 **Chemical Formula:** $C_5H_{11}ONO_2$
3.4 **IMCO/United Nations Numerical Designation:** 3.3/1112

4. OBSERVABLE CHARACTERISTICS

4.1 **Physical State (as shipped):** Liquid
4.2 **Color:** Light straw; water white
4.3 **Odor:** Ethereal

5. HEALTH HAZARDS

5.1 **Personal Protective Equipment:** Respirator with canister for vapors at high concentrations
5.2 **Symptoms Following Exposure:** Inhalation or ingestion may cause headache, methemoglobin, and nausea. Liquid and vapor irritate eyes. Contact with skin may cause slight irritation.
5.3 **Treatment for Exposure:** INHALATION: move to fresh air; support respiration; get medical attention. INGESTION: induce vomiting; get medical attention. EYES: irrigate thoroughly with water. SKIN: wash with soap and water.
5.4 **Toxicity by Inhalation (Threshold Limit Value):** Data not available
5.5 **Short-Term Inhalation Limits:** Data not available
5.6 **Toxicity by Ingestion:** Data not available
5.7 **Late Toxicity:** Data not available
5.8 **Vapor (Gas) Irritant Characteristics:** Data not available
5.9 **Liquid or Solid Irritant Characteristics:** Data not available
5.10 **Odor Threshold:** Data not available

6. FIRE HAZARDS

6.1 **Flash Point:** 120°F O.C.
6.2 **Flammable Limits in Air:** Data not available
6.3 **Fire Extinguishing Agents:** Foam, dry chemical, carbon dioxide
6.4 **Fire Extinguishing Agents Not to be Used:**
6.5 **Special Hazards of Combustion Products:** Toxic oxides of nitrogen may form in a fire.
6.6 **Behavior in Fire:** Overheated material may detonate.
6.7 **Ignition Temperature:** Data not available
6.8 **Electrical Hazard:** Data not available
6.9 **Burning Rate:** Data not available

7. CHEMICAL REACTIVITY

7.1 **Reactivity with Water:** No reaction
7.2 **Reactivity with Common Materials:** May form combustible mixture with wood or other combustibles. Liquid will attack some plastics.
7.3 **Stability During Transport:** Stable
7.4 **Neutralizing Agents for Acids and Caustics:** Not pertinent
7.5 **Polymerization:** Not pertinent
7.6 **Inhibitor of Polymerization:** Not pertinent

8. WATER POLLUTION

8.1 **Aquatic Toxicity:** Data not available
8.2 **Waterfowl Toxicity:** Data not available
8.3 **Biological Oxygen Demand (BOD):** Data not available
8.4 **Food Chain Concentration Potential:** None

9. SELECTED MANUFACTURERS

Ethyl Corporation
Industrial Chemicals Div.
451 Florida St.
Baton Rouge, La. 70801

10. SHIPPING INFORMATION

10.1 **Grades or Purity:** Mixture containing n-amyl nitrate, 60%; iso-amyl nitrate, 5%; 2 methylbutyl nitrate, 35%
10.2 **Storage Temperature:** Ambient
10.3 **Inert Atmosphere:** No requirement
10.4 **Venting:** Open

11. HAZARD ASSESSMENT CODE

(See Hazard Assessment Handbook, CG 446-3)

A-T-U-X-Y

12. HAZARD CLASSIFICATIONS

12.1 **Code of Federal Regulations:** Combustible Liquid
12.2 **NAS Hazard Rating for Bulk Water Transportation:** Not listed
12.3 **NFPA Hazard Classifications:**

Category	Classification*	
Health Hazard (Blue)	1	2
Flammability (Red)	2	2
Reactivity (Yellow)	0	0
	oxy	oxy

*First column refers to nonfire situation

13. PHYSICAL AND CHEMICAL PROPERTIES

13.1 **Physical State at 15°C and 1 atm:** Liquid
13.2 **Molecular Weight:** 133
13.3 **Boiling Point at 1 atm:** 292—314°F = 144—156°C = 417—429°K
13.4 **Freezing Point:** −190°F = −123°C = 150°K
13.5 **Critical Temperature:** Not pertinent
13.6 **Critical Pressure:** Not pertinent
13.7 **Specific Gravity:** 1.0 at 20°C (liquid)
13.8 **Liquid Surface Tension:** Data not available
13.9 **Liquid-Water Interfacial Tension:** Data not available
13.10 **Vapor (Gas) Specific Gravity:** 4.59
13.11 **Ratio of Specific Heats of Vapor (Gas):** Not pertinent
13.12 **Latent Heat of Vaporization:** Data not available
13.13 **Heat of Combustion:** Data not available
13.14 **Heat of Decomposition:** Data not available
13.15 **Heat of Solution:** Not pertinent
13.16 **Heat of Polymerization:** Not pertinent

NOTES

ANI	iso-AMYL NITRITE

iso-AMYL NITRITE

Common Synonyms	Liquid	Colorless to light yellow	Pleasant fruity odor
Isopentyl nitrite Amyl nitrite 3-Methylbutyl nitrite	Floats on water. Poisonous gas is produced on contact with water.		

Shut off ignition sources. Call fire department.
Wear goggles, self-contained breathing apparatus and rubber overclothing (including gloves).
Avoid contact with liquid. Keep people away.
Stop discharge if possible.
Isolate and remove discharged material.
Notify local health and pollution control agencies.

Fire

FLAMMABLE.

POISONOUS GASES ARE PRODUCED IN FIRE.
Flashback along vapor trail may occur.
Vapor may explode if ignited in an enclosed area.
DO NOT USE WATER ON FIRE.
Cool exposed containers with water.

Exposure

CALL FOR MEDICAL AID.

VAPOR
POISONOUS IF INHALED.
Irritating to eyes, nose and throat.
Move to fresh air.
If breathing has stopped, give artificial respiration.
If breathing is difficult, give oxygen.

LIQUID
Will burn skin and eyes.
If swallowed will cause dizziness, headache or loss of consciousness.
Remove contaminated clothing and shoes.
Flush affected areas with plenty of water.
IF IN EYES, hold eyelids open and flush with plenty of water.
IF SWALLOWED and victim is CONSCIOUS, have victim drink water or milk.
DO NOT INDUCE VOMITING.

Water Pollution

Effect of low concentrations on aquatic life is unknown.
Fouling to shoreline.
May be dangerous if it enters water intakes.
Notify local health and wildlife officials.
Notify operators of nearby water intakes.

1. RESPONSE TO DISCHARGE
(See Response Methods Handbook, CG 446-4)

Issue warning—high flammability
Restrict access
Disperse and flush

2. LABEL

Red

3. CHEMICAL DESIGNATIONS

3.1 **Synonyms:** Isopentyl nitrite; Amyl nitrite; 3-Methylbutyl nitrite
3.2 **Coast Guard Compatibility Classification:** Not applicable
3.3 **Chemical Formula:** $(CH_3)_2CHCH_2CH_2ONO$
3.4 **IMCO/United Nations Numerical Designation:** 3.1/1113

4. OBSERVABLE CHARACTERISTICS

4.1 **Physical State (as shipped):** Liquid
4.2 **Color:** Light yellow, transparent
4.3 **Odor:** Pleasant, fragrant, fruity

5. HEALTH HAZARDS

5.1 **Personal Protective Equipment:** Protective goggles or face shield; self-contained breathing apparatus; protective gloves and clothing.
5.2 **Symptoms Following Exposure:** Inhalation or ingestion causes flushing of the face, pulsatile headache, disturbing tachycardia, cyanosis (methemoglobinemia), weakness, confusion, restlessness, faintness, and collapse. Contact with eyes or skin causes irritation.
5.3 **Treatment for Exposure:** INHALATION or INGESTION: place patient in recumbent position; if necessary, administer oxygen. For treatment of severe methemoglobinemia, transfuse with whole blood or give I.V. or I.M. a dose of 1-2 mg/kg methylene blue or an oral dose of 3-5 mg/kg. EYES: after contact with liquid, irrigate with large quantities of water for 15 min.; call physician. SKIN: after contact with liquid, wash with large amounts water. Call physician.
5.4 **Toxicity by Inhalation (Threshold Limit Value):** Data not available
5.5 **Short-Term Inhalation Limits:** Data not available
5.6 **Toxicity by Ingestion:** Grade 1; LD_{50} 5 to 15 g/kg
5.7 **Late Toxicity:** Methemoglobinemia may occur.
5.8 **Vapor (Gas) Irritant Characteristics:** Vapors are nonirritating to the eyes and throat.
5.9 **Liquid or Solid Irritant Characteristics:** Severe skin irritant. Causes second- and third-degree burns on short contact and is very injurious to the eyes.
5.10 **Odor Threshold:** Data not available

6. FIRE HAZARDS

6.1 **Flash Point:** 0°F O.C.
6.2 **Flammable Limits in Air:** Data not available
6.3 **Fire Extinguishing Agents:** Dry chemical, alcohol foam, carbon dioxide
6.4 **Fire Extinguishing Agents Not to be Used:** Water
6.5 **Special Hazards of Combustion Products:** Toxic oxides of nitrogen are formed.
6.6 **Behavior in Fire:** Containers may explode.
6.7 **Ignition Temperature:** 410°F
6.8 **Electrical Hazard:** Data not available
6.9 **Burning Rate:** 3.4 mm/min.

7. CHEMICAL REACTIVITY

7.1 **Reactivity with Water:** Decomposes on exposure to air, light, or water, evolving toxic oxides of nitrogen which are orange in color.
7.2 **Reactivity with Common Materials:** May corrode metals if wet.
7.3 **Stability During Transport:** Stable if kept sealed and not exposed to light.
7.4 **Neutralizing Agents for Acids and Caustics:** Not pertinent
7.5 **Polymerization:** Not pertinent
7.6 **Inhibitor of Polymerization:** Not pertinent

8. WATER POLLUTION

8.1 **Aquatic Toxicity:** Data not available
8.2 **Waterfowl Toxicity:** Data not available
8.3 **Biological Oxygen Demand (BOD):** Data not available
8.4 **Food Chain Concentration Potential:** None

9. SELECTED MANUFACTURERS

1. Eastman Kodak Co.
   Eastman Organic Chemicals
   Rochester, N. Y. 14650
2. Aldrich Chemical Co.
   940 West St. Paul Avenue
   Milwaukee, Wis. 53233
3. Pfaltz and Bauer, Inc.
   126-04 Northern Boulevard
   Flushing, N. Y. 11368

10. SHIPPING INFORMATION

10.1 **Grades or Purity:** Commercial; USP
10.2 **Storage Temperature:** Ambient
10.3 **Inert Atmosphere:** No requirement
10.4 **Venting:** Pressure-vacuum

11. HAZARD ASSESSMENT CODE
(See Hazard Assessment Handbook, CG 446-3)

A-T-U-V-W

12. HAZARD CLASSIFICATIONS

12.1 **Code of Federal Regulations:** Flammable liquid
12.2 **NAS Hazard Rating for Bulk Water Transportation:**

Category	Rating
Fire	3
Health	
Vapor Irritant	0
Liquid or Solid Irritant	4
Poisons	1
Water Pollution	
Human Toxicity	1
Aquatic Toxicity	1
Aesthetic Effect	2
Reactivity	
Other Chemicals	3
Water	0
Self-Reaction	4

12.3 **NFPA Hazard Classifications:**

Category	Classification
Health Hazard (Blue)	1
Flammability (Red)	—
Reactivity (Yellow)	2

13. PHYSICAL AND CHEMICAL PROPERTIES

13.1 **Physical State at 15°C and 1 atm:** Liquid
13.2 **Molecular Weight:** 117.1
13.3 **Boiling Point at 1 atm:** 210°F = 99°C = 372°K
13.4 **Freezing Point:** Not pertinent
13.5 **Critical Temperature:** Not pertinent
13.6 **Critical Pressure:** Not pertinent
13.7 **Specific Gravity:** 0.8758 at 20°C (liquid)
13.8 **Liquid Surface Tension:** (est.) 20 dynes/cm = 0.020 N/m at 20°C
13.9 **Liquid-Water Interfacial Tension:** (est.) 40 dynes/cm = 0.040 N/m at 20°C
13.10 **Vapor (Gas) Specific Gravity:** 4
13.11 **Ratio of Specific Heats of Vapor (Gas):** 1.0709
13.12 **Latent Heat of Vaporization:** 212 Btu/lb = 118 cal/g = 4.94×10^5 J/kg
13.13 **Heat of Combustion:** −12,500 Btu/lb = −6,930 cal/g = -290×10^5 J/kg
13.14 **Heat of Decomposition:** Not pertinent
13.15 **Heat of Solution:** Not pertinent
13.16 **Heat of Polymerization:** Not pertinent

NOTES

ATS	n-AMYLTRICHLOROSILANE

Common Synonyms			
Trichloropentylsilane Pentylsilicon trichloride Trichloroamylsilane	Liquid	Colorless	Sharp irritating odor
	Reacts violently with water. Irritating visible vapor cloud is produced.		

Avoid contact with liquid and vapor. Keep people away.
Wear goggles, self-contained breathing apparatus, and rubber overclothing (including gloves).
Stop discharge if possible.
Call fire department.
Isolate and remove discharged material.
Notify local health and pollution control agencies.

Fire	Combustible. POISONOUS GASES ARE PRODUCED IN FIRE. Wear goggles and self-contained breathing apparatus. Extinguish with dry chemicals or carbon dioxide. DO NOT USE WATER OR FOAM ON FIRE.
Exposure	Call for medical aid. VAPOR Irritating to eyes, nose and throat. If inhaled will cause difficult breathing. Move victim to fresh air. If breathing has stopped, give artificial respiration. If breathing is difficult, give oxygen. LIQUID Will burn skin and eyes. Harmful if swallowed. Remove contaminated clothing and shoes. Flush affected areas with plenty of water. IF IN EYES, hold eyelids open and flush with plenty of water. IF SWALLOWED, and victim is CONSCIOUS, have victim drink water or milk. DO NOT INDUCE VOMITING.
Water Pollution	Effect of low concentrations on aquatic life is unknown. May be dangerous if it enters water intakes. Notify local health and wildlife officials. Notify operators of nearby water intakes.

1. RESPONSE TO DISCHARGE

(See Response Methods Handbook, CG 446-4)

Issue warning—corrosive
Restrict access
Disperse and flush

2. LABEL

3. CHEMICAL DESIGNATIONS

3.1 **Synonyms:** Trichloropentylsilane; Pentylsilicon trichloride; Trichloroamylsilane
3.2 **Coast Guard Compatibility Classification:** Not applicable
3.3 **Chemical Formula:** $CH_3CH_2CH_2CH_2CH_2SiCl_3$
3.4 **IMCO/United Nations Numerical Designation:** 8/1728

4. OBSERVABLE CHARACTERISTICS

4.1 **Physical State (as shipped):** Liquid
4.2 **Color:** Colorless
4.3 **Odor:** Sharp, like hydrochloric acid; pungent; acrid

5. HEALTH HAZARDS

5.1 **Personal Protective Equipment:** Acid-vapor-type respiratory protection; rubber gloves; chemical worker's goggles.
5.2 **Symptoms Following Exposure:** Inhalation causes irritation of mucous membrane. Contact of liquid with eyes or skin causes severe burns, and ingestion causes severe burns of mouth and stomach.
5.3 **Treatment for Exposure:** Get medical attention immediately after exposure to this compound. INHALATION: remove from exposure; support respiration. EYES: flush with water for 15 min. SKIN: flush with water. INGESTION: give large amounts of water.
5.4 **Toxicity by Inhalation (Threshold Limit Value):** Data not available
5.5 **Short-Term Inhalation Limits:** Data not available
5.6 **Toxicity by Ingestion:** Grade 2; oral rat LD_{50} = 2,340 mg/kg
5.7 **Late Toxicity:** Data not available
5.8 **Vapor (Gas) Irritant Characteristics:** Vapors are moderately irritating such that personnel will not usually tolerate moderate or high vapor concentrations.
5.9 **Liquid or Solid Irritant Characteristics:** Severe skin irritant. Causes second- and third-degree burns on short contact and is very injurious to the eyes.
5.10 **Odor Threshold:** Data not available

6. FIRE HAZARDS

6.1 **Flash Point:** 145°F O.C.
6.2 **Flammable Limits in Air:** Data not available
6.3 **Fire Extinguishing Agents:** Dry chemical, carbon dioxide
6.4 **Fire Extinguishing Agents Not to be Used:** Water, foam
6.5 **Special Hazards of Combustion Products:** Irritating hydrogen chloride and toxic phosgene may be formed.
6.6 **Behavior in Fire:** Difficult to extinguish. Re-ignition may occur.
6.7 **Ignition Temperature:** Data not available
6.8 **Electrical Hazard:** Data not available
6.9 **Burning Rate:** 2.5 mm/min.

7. CHEMICAL REACTIVITY

7.1 **Reactivity with Water:** Reacts vigorously to generate toxic hydrogen chloride gas (hydrochloric acid).
7.2 **Reactivity with Common Materials:** Corrodes metal
7.3 **Stability During Transport:** Stable
7.4 **Neutralizing Agents for Acids and Caustics:** After flushing with water, rinse with sodium bicarbonate solution or lime water.
7.5 **Polymerization:** Not pertinent
7.6 **Inhibitor of Polymerization:** Not pertinent

8. WATER POLLUTION

8.1 **Aquatic Toxicity:** Data not available
8.2 **Waterfowl Toxicity:** Data not available
8.3 **Biological Oxygen Demand (BOD):** Data not available
8.4 **Food Chain Concentration Potential:** None

9. SELECTED MANUFACTURERS

1. Union Carbide Corporation, Chemicals and Plastics Division, 270 Park Avenue, New York, N. Y. 10017
2. Dow Corning Corporation, P. O. Box 592, Midland, Mich. 48640
3. PCR, Inc., P. O. Box 1466, Gainesville, Fla. 32602

10. SHIPPING INFORMATION

10.1 **Grades or Purity:** Commercial
10.2 **Storage Temperature:** Ambient
10.3 **Inert Atmosphere:** No requirement
10.4 **Venting:** Pressure-vacuum

11. HAZARD ASSESSMENT CODE

(See Hazard Assessment Handbook, CG 446-3)

A-O

12. HAZARD CLASSIFICATIONS

12.1 **Code of Federal Regulations:** Corrosive
12.2 **NAS Hazard Rating for Bulk Water Transportation:**

Category	Rating
Fire	2
Health	
Vapor Irritant	3
Liquid or Solid Irritant	4
Poisons	3
Water Pollution	
Human Toxicity	3
Aquatic Toxicity	3
Aesthetic Effect	2
Reactivity	
Other Chemicals	3
Water	4
Self-Reaction	1

12.3 **NFPA Hazard Classifications:** Not listed

13. PHYSICAL AND CHEMICAL PROPERTIES

13.1 **Physical State at 15°C and 1 atm:** Liquid
13.2 **Molecular Weight:** 205.6
13.3 **Boiling Point at 1 atm:** 320°F = 160°C = 433°K
13.4 **Freezing Point:** Not pertinent
13.5 **Critical Temperature:** Not pertinent
13.6 **Critical Pressure:** Not pertinent
13.7 **Specific Gravity:** 1.137 at 25°C (liquid)
13.8 **Liquid Surface Tension:** (est.) 20 dynes/cm = 0.020 N/m at 20°C
13.9 **Liquid-Water Interfacial Tension:** Not pertinent
13.10 **Vapor (Gas) Specific Gravity:** 7.1
13.11 **Ratio of Specific Heats of Vapor (Gas):** Data not available
13.12 **Latent Heat of Vaporization:** (est.) 86.8 Btu/lb = 48.2 cal/g = 2.02×10^5 J/kg
13.13 **Heat of Combustion:** (est.) −6,630 Btu/lb = −3,680 cal/g = -154×10^5 J/kg
13.14 **Heat of Decomposition:** Not pertinent
13.15 **Heat of Solution:** (est.) 180 Btu/lb = 100 cal/g = 4.0×10^5 J/kg
13.16 **Heat of Polymerization:** Not pertinent

NOTES

ANL — ANILINE

Common Synonyms			
Aminobenzene Aniline oil Phenylamine Blue oil	Oily liquid Sinks slowly in water.	Colorless to yellowish brown	Weak odor

AVOID CONTACT WITH LIQUID AND VAPOR. KEEP PEOPLE AWAY.
Wear chemical protective suit with self-contained breathing apparatus.
Stop discharge if possible.
Stay upwind and use water spray to "knock down" vapor.
Call fire department.
Isolate and remove discharged material.
Notify local health and pollution control agencies.

Fire

Combustible.
POISONOUS GAS IS PRODUCED WHEN HEATED.
Vapor may explode if ignited in an enclosed area.
Wear chemical protective suit with self-contained breathing apparatus.
Extinguish with water, dry chemical, foam, or carbon dioxide.
Cool exposed containers with water.

Exposure

CALL FOR MEDICAL AID.

LIQUID
POISONOUS IF SWALLOWED OR IF SKIN IS EXPOSED.
Irritating to eyes.
Remove contaminated clothing and shoes.
Flush affected areas with plenty of water.
IF IN EYES, hold eyelids open and flush with plenty of water.
IF SWALLOWED and victim is CONSCIOUS, have victim drink water or milk.

Water Pollution

Dangerous to aquatic life in high concentrations.
May be dangerous if it enters water intakes.
Notify local health and wildlife officials.
Notify operators of nearby water intakes.

1. RESPONSE TO DISCHARGE

(See Response Methods Handbook, CG 446-4)

Issue warning—poison, water contaminant
Restrict access
Should be removed
Chemical and physical treatment

2. LABEL

3. CHEMICAL DESIGNATIONS

3.1 **Synonyms:** Aminobenzene; Aniline oil; Blue oil; Phenylamine
3.2 **Coast Guard Compatibility Classification:** Aromatic amine
3.3 **Chemical Formula:** $C_6H_5NH_2$
3.4 **IMCO/United Nations Numerical Designation:** 6.1/1547

4. OBSERVABLE CHARACTERISTICS

4.1 **Physical State (as shipped):** Liquid
4.2 **Color:** Colorless to pale brown
4.3 **Odor:** Aromatic amine-like; characteristic, peculiar; strongly amine-like

5. HEALTH HAZARDS

5.1 **Personal Protective Equipment:** Respirator for organic vapors, splashproof goggles, rubber gloves, boots.
5.2 **Symptoms Following Exposure:** ACUTE EXPOSURE: Blue discoloration of finger-tips, cheeks, lips and nose; nausea, vomiting, headache and drowsiness followed by delirium, coma and shock. CHRONIC EXPOSURE: Loss of appetite, loss of weight, headaches, visual disturbances; skin lesions.
5.3 **Treatment for Exposure:** Remove victim to fresh air and call a physician at once. SKIN, EYE CONTACT: immediately flush skin or eyes with plenty of water for at least 15 min. If cyanosis is present, shower with soap and warm water, with special attention to scalp and fingernails. Administer oxygen until physician arrives.
5.4 **Toxicity by Inhalation (Threshold Limit Value):** 5 ppm
5.5 **Short-Term Inhalation Limits:** 50 ppm for 30 min.; 5 ppm for 8 hr.
5.6 **Toxicity by Ingestion:** Grade 3; LD_{50} 50 to 500 mg/kg
5.7 **Late Toxicity:** None recognized
5.8 **Vapor (Gas) Irritant Characteristics:** Vapors cause a slight smarting of the eyes or respiratory system if present in high concentrations. The effect is temporary.
5.9 **Liquid or Solid Irritant Characteristics:** If spilled on clothing and allowed to remain, may cause smarting and reddening of the skin.
5.10 **Odor Threshold:** 0.5 ppm

6. FIRE HAZARDS

6.1 **Flash Point:** 168°F O.C.; 158°F C.C.
6.2 **Flammable Limits in Air:** 1.3% (LEL)
6.3 **Fire Extinguishing Agents:** Water, foam, dry chemical, or carbon dioxide
6.4 **Fire Extinguishing Agents Not to be Used:** Not pertinent
6.5 **Special Hazards of Combustion Products:** Toxic vapors are generated when heated
6.6 **Behavior in Fire:** Not pertinent
6.7 **Ignition Temperature:** 1418°F
6.8 **Electrical Hazard:** Not pertinent
6.9 **Burning Rate:** 3.0 mm/min.

7. CHEMICAL REACTIVITY

7.1 **Reactivity with Water:** No reaction
7.2 **Reactivity with Common Materials:** No reaction
7.3 **Stability During Transport:** Stable
7.4 **Neutralizing Agents for Acids and Caustics:** Flush with water and rinse with dilute acetic acid
7.5 **Polymerization:** Not pertinent
7.6 **Inhibitor of Polymerization:** Not pertinent

8. WATER POLLUTION

8.1 **Aquatic Toxicity:** 1020 ppm/1 hr/sunfish/killed/fresh water; 10 ppm/96 hr/scenedesmus/TL_m/fresh water
8.2 **Waterfowl Toxicity:** Data not available
8.3 **Biological Oxygen Demand (BOD):** 150%, 5 days
8.4 **Food Chain Concentration Potential:** None

9. SELECTED MANUFACTURERS

1. Baychem Corp.
   Mobay Chemical Co. Division
   Penn Lincoln Pkwy. West
   Pittsburgh, Pa. 15205
2. E. I. duPont de Nemours and Co.
   Explosives Department
   Wilmington, Del. 19898
3. First Mississippi Corp.
   First Chemical Corp.
   Pascagoula, Miss. 39567

10. SHIPPING INFORMATION

10.1 **Grades or Purity:** Commercial: 99.5%
10.2 **Storage Temperature:** Ambient
10.3 **Inert Atmosphere:** No requirement
10.4 **Venting:** Pressure-vacuum

11. HAZARD ASSESSMENT CODE

(See Hazard Assessment Handbook, CG 446-3)

A-P-Q-T-U-X-Y

12. HAZARD CLASSIFICATIONS

12.1 **Code of Federal Regulations:** Poisonous liquid or solid, Class B
12.2 **NAS Hazard Rating for Bulk Water Transportation:**

Category	Rating
Fire	1
Health	
Vapor Irritant	1
Liquid or Solid Irritant	1
Poisons	3
Water Pollution	
Human Toxicity	3
Aquatic Toxicity	2
Aesthetic Effect	4
Reactivity	
Other Chemicals	3
Water	0
Self-Reaction	0

12.3 **NFPA Hazard Classifications:**

Category	Classification
Health Hazard (Blue)	3
Flammability (Red)	2
Reactivity (Yellow)	0

13. PHYSICAL AND CHEMICAL PROPERTIES

13.1 **Physical State at 15°C and 1 atm:** Liquid
13.2 **Molecular Weight:** 93.13
13.3 **Boiling Point at 1 atm:** 363.6°F = 184.2°C = 457.4°K
13.4 **Freezing Point:** 21°F = −6.1°C = 267.1°K
13.5 **Critical Temperature:** 798.1°F = 425.6°C = 698.8°K
13.6 **Critical Pressure:** 770 psia = 52.4 atm = 5.31 MN/m^2
13.7 **Specific Gravity:** 1.022 at 20°C (liquid)
13.8 **Liquid Surface Tension:** 45.5 dynes/cm = .0455 N/m at 20°C
13.9 **Liquid-Water Interfacial Tension:** 5.8 dynes/cm = 0.0058 N/m at 20°C
13.10 **Vapor (Gas) Specific Gravity:** Not pertinent
13.11 **Ratio of Specific Heats of Vapor (Gas):** 1.1
13.12 **Latent Heat of Vaporization:** 198 Btu/lb = 110 cal/g = 4.61×10^5 J/kg
13.13 **Heat of Combustion:** −14,980 Btu/lb = −8320 cal/g = -348.3×10^5 J/kg
13.14 **Heat of Decomposition:** Not pertinent
13.15 **Heat of Solution:** Not pertinent
13.16 **Heat of Polymerization:** Not pertinent

NOTES

REVISED 1978

ASC ANISOYL CHLORIDE

Common Synonyms		
p-Anisoyl chloride	Liquid	Yellow to brown Sharp, penetrating odor
		Reacts with water. Irritating vapor is produced. Freezing point is 72° F.

Wear goggles and self-contained breathing apparatus.
Stop discharge if possible. Keep people away.
Isolate and remove discharged material.
Notify local health and pollution control agencies.

Fire

Combustible.
POISONOUS GASES MAY BE PRODUCED WHEN HEATED.
Extinguish with dry chemicals or carbon dioxide.
DO NOT USE WATER ON ADJACENT FIRES.

Exposure

Call for medical aid.

VAPOR
Irritating to eyes, nose and throat.
Move victim to fresh air.

LIQUID
Irritating to skin and eyes.
Harmful if swallowed.
Remove contaminated clothing and shoes.
Flush affected areas with plenty of water.
IF IN EYES, hold eyelids open and flush with plenty of water.
IF SWALLOWED, and victim is CONSCIOUS, have victim drink water or milk.

Water Pollution

Effect of low concentrations on aquatic life is unknown.
May be dangerous if it enters water intakes.
Notify local health and wildlife officials.
Notify operators of nearby water intakes.

1. RESPONSE TO DISCHARGE

(See Response Methods Handbook, CG 446-4)

Issue warning—corrosive
Restrict access
Disperse and flush

2. LABEL

3. CHEMICAL DESIGNATIONS

3.1 **Synonyms:** p-Anisoyl chloride
3.2 **Coast Guard Compatibility Classification:** Not applicable
3.3 **Chemical Formula:** p-$CH_3OC_6H_4COCl$
3.4 **IMCO/United Nations Numerical Designation:** 8/1729

4. OBSERVABLE CHARACTERISTICS

4.1 **Physical State (as shipped):** Liquid
4.2 **Color:** Yellow; brown
4.3 **Odor:** Sharp, penetrating

5. HEALTH HAZARDS

5.1 **Personal Protective Equipment:** Goggles and face shield; plastic gloves; protective clothing.
5.2 **Symptoms Following Exposure:** Vapor irritates mucous membranes. Contact of liquid with eyes or skin causes severe irritation. Ingestion causes severe irritation of mouth and stomach.
5.3 **Treatment for Exposure:** INHALATION: remove to fresh air. EYES: flush with water for at least 15 min.; get medical attention. SKIN: flush with water; wash well with soap and water. INGESTION: do NOT induce vomiting; give large amounts of water.
5.4 **Toxicity by Inhalation (Threshold Limit Value):** Data not available
5.5 **Short-Term Inhalation Limits:** Data not available
5.6 **Toxicity by Ingestion:** Data not available
5.7 **Late Toxicity:** Data not available
5.8 **Vapor (Gas) Irritant Characteristics:** Data not available
5.9 **Liquid or Solid Irritant Characteristics:** Data not available
5.10 **Odor Threshold:** Data not available

6. FIRE HAZARDS

6.1 **Flash Point:** Data not available
6.2 **Flammable Limits in Air:** Not pertinent
6.3 **Fire Extinguishing Agents:** Carbon dioxide, dry chemical
6.4 **Fire Extinguishing Agents Not to be Used:** Water, foam
6.5 **Special Hazards of Combustion Products:** Irritating hydrogen chloride fumes may be formed.
6.6 **Behavior in Fire:** Not pertinent
6.7 **Ignition Temperature:** Data not available
6.8 **Electrical Hazard:** Data not available
6.9 **Burning Rate:** Data not available

7. CHEMICAL REACTIVITY

7.1 **Reactivity with Water:** Reacts slowly to generate hydrogen chloride (hydrochloric acid). The reaction is not hazardous.
7.2 **Reactivity with Common Materials:** Corrodes metal slowly
7.3 **Stability During Transport:** Stable
7.4 **Neutralizing Agents for Acids and Caustics:** Flush with water, rinse with sodium bicarbonate or lime solution.
7.5 **Polymerization:** Not pertinent
7.6 **Inhibitor of Polymerization:** Not pertinent

8. WATER POLLUTION

8.1 **Aquatic Toxicity:** Data not available
8.2 **Waterfowl Toxicity:** Data not available
8.3 **Biological Oxygen Demand (BOD):** Data not available
8.4 **Food Chain Concentration Potential:** None

9. SELECTED MANUFACTURERS

1. Eastman Kodak Co.
   Eastman Organic Chemicals
   Rochester, N. Y. 14650
2. Aldrich Chemical Co.
   940 West St. Paul Avenue
   Milwaukee, Wis. 53233
3. Pfaltz and Bauer, Inc.
   126-04 Northern Boulevard
   Flushing, N. Y. 11368

10. SHIPPING INFORMATION

10.1 **Grades or Purity:** Commercial
10.2 **Storage Temperature:** Ambient
10.3 **Inert Atmosphere:** No requirement
10.4 **Venting:** Pressure-vacuum

11. HAZARD ASSESSMENT CODE

(See Hazard Assessment Handbook, CG 446-3)

A-O-X

12. HAZARD CLASSIFICATIONS

12.1 **Code of Federal Regulations:** Corrosive
12.2 **NAS Hazard Rating for Bulk Water Transportation:** Not listed
12.3 **NFPA Hazard Classifications:** Not listed

13. PHYSICAL AND CHEMICAL PROPERTIES

13.1 **Physical State at 15°C and 1 atm:** Liquid
13.2 **Molecular Weight:** 171.6
13.3 **Boiling Point at 1 atm:** 504°F = 262°C = 535°K
13.4 **Freezing Point:** 72°F = 22°C = 295°K
13.5 **Critical Temperature:** Not pertinent
13.6 **Critical Pressure:** Not pertinent
13.7 **Specific Gravity:** 1.26 at 20°C (liquid)
13.8 **Liquid Surface Tension:** (est.) 25 dynes/cm = 0.025 N/m at 20°C
13.9 **Liquid-Water Interfacial Tension:** Not pertinent
13.10 **Vapor (Gas) Specific Gravity:** Not pertinent
13.11 **Ratio of Specific Heats of Vapor (Gas):** Not pertinent
13.12 **Latent Heat of Vaporization:** Not pertinent
13.13 **Heat of Combustion:** (est.) −10,500 Btu/lb = −5,830 cal/g = -244×10^5 J/kg
13.14 **Heat of Decomposition:** Not pertinent
13.15 **Heat of Solution:** (est.) 90 Btu/lb = 50 cal/g = 2.1×10^5 J/kg
13.16 **Heat of Polymerization:** Not pertinent

NOTES

ATH	ANTHRACENE

Common Synonyms
Anthracin
Paranaphthalene
Green oil

Solid | White to yellow | Weak aromatic odor

Sinks in water.

Stop discharge if possible. Keep people away.
Avoid contact with solid and dust.
Isolate and remove discharged material.
Notify local health and pollution control agencies.

Fire

Combustible.
Dust cloud may explode if ignited in an enclosed area.
Extinguish with water, dry chemicals, foam, or carbon dioxide.

Exposure

CALL FOR MEDICAL AID.

DUST
Irritating to eyes, nose and throat.
If inhaled will cause coughing or difficult breathing.
If in eyes, hold eyelids open and flush with plenty of water.
If breathing has stopped, give artificial respiration.
If breathing is difficult, give oxygen.

SOLID
Irritating to skin and eyes.
Harmful if swallowed.
Remove contaminated clothing and shoes.
Flush affected areas with plenty of water.
IF IN EYES, hold eyelids open and flush with plenty of water.
IF SWALLOWED and victim is CONSCIOUS, have victim drink water or milk.
IF SWALLOWED and victim is UNCONSCIOUS OR HAVING CONVULSIONS, do nothing except keep victim warm.

Water Pollution

Effect of low concentrations on aquatic life is unknown.
May be dangerous if it enters water intakes.
Notify local health and wildlife officials.
Notify operators of nearby water intakes.

1. RESPONSE TO DISCHARGE
(See Response Methods Handbook, CG 446-4)

Should be removed
Chemical and physical treatment

2. LABELS

No hazard label required by Code of Federal Regulations

3. CHEMICAL DESIGNATIONS

3.1 **Synonyms:** Anthracin; Green oil; Paranaphthalene
3.2 **Coast Guard Compatibility Classification:** Not listed
3.3 **Chemical Formula:** $C_{14}H_{10}$
3.4 **IMCO/United Nations Numerical Designation:** Not listed

4. OBSERVABLE CHARACTERISTICS

4.1 **Physical State (as shipped):** Solid
4.2 **Color:** White to yellow
4.3 **Odor:** Weak aromatic

5. HEALTH HAZARDS

5.1 **Personal Protective Equipment:** Dust mask; goggles or face shield; rubber gloves
5.2 **Symptoms Following Exposure:** Inhalation of dust irritates nose and throat. Contact with eyes causes irritation.
5.3 **Treatment for Exposure:** INHALATION: move to fresh air. EYES: flush with water for 15 min.
5.4 **Toxicity by Inhalation (Threshold Limit Value):** Data not available
5.5 **Short-Term Inhalation Limits:** Data not available
5.6 **Toxicity by Ingestion:** Data not available
5.7 **Late Toxicity:** Data not available
5.8 **Vapor (Gas) Irritant Characteristics:** Data not available
5.9 **Liquid or Solid Irritant Characteristics:** Data not available
5.10 **Odor Threshold:** Data not available

6. FIRE HAZARDS

6.1 **Flash Point:** Not pertinent (combustible solid)
6.2 **Flammable Limits in Air:** Not pertinent
6.3 **Fire Extinguishing Agents:** Water, foam, dry chemical, carbon dioxide
6.4 **Fire Extinguishing Agents Not to be Used:**
6.5 **Special Hazards of Combustion Products:**
6.6 **Behavior in Fire:**
6.7 **Ignition Temperature:** 1004°F
6.8 **Electrical Hazard:** Not pertinent
6.9 **Burning Rate:** Not pertinent

7. CHEMICAL REACTIVITY

7.1 **Reactivity with Water:** No reaction
7.2 **Reactivity with Common Materials:**
7.3 **Stability During Transport:** Stable
7.4 **Neutralizing Agents for Acids and Caustics:** Not pertinent
7.5 **Polymerization:** Not pertinent
7.6 **Inhibitor of Polymerization:** Not pertinent

8. WATER POLLUTION

8.1 **Aquatic Toxicity:** 5 ppm/24 hr/trout & bluegill/no effect
8.2 **Waterfowl Toxicity:** Data not available
8.3 **Biological Oxygen Demand (BOD):** Data not available
8.4 **Food Chain Concentration Potential:** None

9. SELECTED MANUFACTURERS

1. Gallard-Schlesinger Chemical Mfg. Co.
   584 Mineola Avenue
   Carle Place, N. Y. 11514
2. Aldrich Chemical Co.
   940 W. Saint Paul Ave.
   Milwaukee, Wis. 53233
3. Eastman Organic Chemicals
   Rochester, N. Y. 14650

10. SHIPPING INFORMATION

10.1 **Grades or Purity:** Various fluorescence grades; Scintillation grade; Technical grade, 90—98%
10.2 **Storage Temperature:** Ambient
10.3 **Inert Atmosphere:** No requirement
10.4 **Venting:** Open

11. HAZARD ASSESSMENT CODE
(See Hazard Assessment Handbook, CG 446-3)

II

12. HAZARD CLASSIFICATIONS

12.1 **Code of Federal Regulations:** Not listed
12.2 **NAS Hazard Rating for Bulk Water Transportation:** Not listed
12.3 **NFPA Hazard Classifications:**

Category	Classification
Health Hazard (Blue)	0
Flammability (Red)	1
Reactivity (Yellow)	—

13. PHYSICAL AND CHEMICAL PROPERTIES

13.1 **Physical State at 15°C and 1 atm:** Solid
13.2 **Molecular Weight:** 178.23
13.3 **Boiling Point at 1 atm:** 646.2°F = 341.2°C = 614.4°K
13.4 **Freezing Point:** 421.7°F = 216.5°C = 489.7°K
13.5 **Critical Temperature:** Not pertinent
13.6 **Critical Pressure:** Not pertinent
13.7 **Specific Gravity:** 1.24 at 20°C (solid)
13.8 **Liquid Surface Tension:** Not pertinent
13.9 **Liquid-Water Interfacial Tension:** Not pertinent
13.10 **Vapor (Gas) Specific Gravity:** Not pertinent
13.11 **Ratio of Specific Heats of Vapor (Gas):** Not pertinent
13.12 **Latent Heat of Vaporization:** Not pertinent
13.13 **Heat of Combustion:** $-17{,}100$ Btu/lb $= -9{,}510$ cal/g $= -398 \times 10^5$ J/kg
13.14 **Heat of Decomposition:** Not pertinent
13.15 **Heat of Solution:** Not pertinent
13.16 **Heat of Polymerization:** Not pertinent

NOTES

ANTIMONY PENTACHLORIDE

Common Synonyms	Liquid	Colorless to brown	Unpleasant odor
Antimony (V) chloride Antimony perchloride	Sinks in water. Irritating vapor is produced. Freezing point is 37° F.		

Avoid contact with liquid. Keep people away.
Wear rubber overclothing (including gloves).
Stop discharge if possible.
Isolate and remove discharged material.
Notify local health and pollution control agencies.

Fire

Not flammable.
POISONOUS GASES ARE PRODUCED WHEN HEATED.
DO NOT USE WATER ON ADJACENT FIRES.

Exposure

Call for medical aid.

VAPOR
Irritating to eyes, nose and throat.
If inhaled will cause coughing or difficult breathing.
Move victim to fresh air.
If breathing has stopped, give artificial respiration.
If breathing is difficult, give oxygen.

LIQUID
Will burn skin and eyes.
If swallowed will cause nausea, vomiting or loss of consciousness.
Remove contaminated clothing and shoes.
Flush affected areas with plenty of water.
IF IN EYES, hold eyelids open and flush with plenty of water.
IF SWALLOWED and victim is CONSCIOUS, have victim drink water or milk.
DO NOT INDUCE VOMITING.

Water Pollution

Effect of low concentrations on aquatic life is unknown.
May be dangerous if it enters water intakes.
Notify local health and wildlife officials.
Notify operators of nearby water intakes.

1. RESPONSE TO DISCHARGE

(See Response Methods Handbook, CG 446-4)

Issue warning—corrosive
Restrict access
Disperse and flush

2. LABEL

3. CHEMICAL DESIGNATIONS

3.1 **Synonyms:** Antimony (V) chloride; Antimony perchloride
3.2 **Coast Guard Compatibility Classification:** Not applicable
3.3 **Chemical Formula:** $SbCl_5$
3.4 **IMCO/United Nations Numerical Designation:** 8/1730

4. OBSERVABLE CHARACTERISTICS

4.1 **Physical State (as shipped):** Liquid
4.2 **Color:** Colorless to medium brown; yellow; red-brown
4.3 **Odor:** Pungent; offensive

5. HEALTH HAZARDS

5.1 **Personal Protective Equipment:** Organic vapor-acid gas type canister mask; rubber, neoprene, vinyl, etc. gloves; chemical safety goggles, plus face shield where appropriate; acid-resistant clothing, plus apron for splash protection; rubber safety shoes or boots; hard hat.
5.2 **Symptoms Following Exposure:** Inhalation causes irritation of nose and throat. Contact of liquid with eyes or skin causes severe burns. Ingestion causes vomiting and severe burns of mouth and stomach. Overexposure by any route can cause bloody stools, slow pulse, low blood pressure, coma, convulsions, cardiac arrest.
5.3 **Treatment for Exposure:** INHALATION: remove to clean air; rinse mouth and gargle with water; if overexposure is serious, get prompt medical attention. EYES: flush eyes and eye-lids thoroughly with large amounts of water; get prompt medical attention. SKIN: flush thoroughly with water; remove contaminated clothing; wash affected area with soap and water; if overexposure is serious, get prompt medical attention. INGESTION: dilute by drinking water; if vomiting occurs, administer more water. If overexposure is serious, get prompt medical attention.
5.4 **Toxicity by Inhalation (Threshold Limit Value):** 0.5 mg/m^3 as antimony
5.5 **Short-Term Inhalation Limits:** Data not available
5.6 **Toxicity by Ingestion:** Grade 2; oral LD_{50} = 1,115 mg/kg (rat), 900 mg/kg (guinea pig)
5.7 **Late Toxicity:** Antimony poisoning may result.
5.8 **Vapor (Gas) Irritant Characteristics:** Vapors are moderately irritating such that personnel will not usually tolerate moderate or high vapor concentrations.

6. FIRE HAZARDS

6.1 **Flash Point:** Not flammable
6.2 **Flammable Limits in Air:** Not flammable
6.3 **Fire Extinguishing Agents:** Not pertinent
6.4 **Fire Extinguishing Agents Not to be Used:** Do not use water or foam on adjacent fires.
6.5 **Special Hazards of Combustion Products:** Not pertinent
6.6 **Behavior in Fire:** Irritating fumes of hydrogen chloride given off when water or foam is used to extinguish adjacent fire.
6.7 **Ignition Temperature:** Not pertinent
6.8 **Electrical Hazard:** Not pertinent
6.9 **Burning Rate:** Not pertinent

7. CHEMICAL REACTIVITY

7.1 **Reactivity with Water:** Reacts to form hydrogen chloride gas (hydrochloric acid).
7.2 **Reactivity with Common Materials:** Causes corrosion of metal.
7.3 **Stability During Transport:** Stable
7.4 **Neutralizing Agents for Acids and Caustics:** Soda ash or soda ash-lime mixture
7.5 **Polymerization:** Not pertinent
7.6 **Inhibitor of Polymerization:** Not pertinent

8. WATER POLLUTION

8.1 **Aquatic Toxicity:** Data not available
8.2 **Waterfowl Toxicity:** Data not available
8.3 **Biological Oxygen Demand (BOD):** None
8.4 **Food Chain Concentration Potential:** None

9. SELECTED MANUFACTURERS

1. Hooker Chemical Corporation
   Specialty Chemicals Division
   4700 Buffalo Avenue
   Niagara Falls, N. Y. 14302
2. J. T. Baker Chemical Co.
   Phillipsburg, N. J. 08865
3. Mallinckrodt Chemical Works
   223 West Side Avenue
   Jersey City, N. J. 07303

10. SHIPPING INFORMATION

10.1 **Grades or Purity:** 99+%
10.2 **Storage Temperature:** Ambient
10.3 **Inert Atmosphere:** No requirement
10.4 **Venting:** Pressure-vacuum

11. HAZARD ASSESSMENT CODE

(See Hazard Assessment Handbook, CG 446-3)

A-O

12. HAZARD CLASSIFICATIONS

12.1 **Code of Federal Regulations:** Corrosive
12.2 **NAS Hazard Rating for Bulk Water Transportation:**

Category	Rating
Fire	0
Health	
Vapor Irritant	3
Liquid or Solid Irritant	4
Poisons	3
Water Pollution	
Human Toxicity	–
Aquatic Toxicity	3
Aesthetic Effect	2
Reactivity	
Other Chemicals	3
Water	3
Self-Reaction	0

12.3 **NFPA Hazard Classifications:** Not listed

13. PHYSICAL AND CHEMICAL PROPERTIES

13.1 **Physical State at 15°C and 1 atm:** Liquid
13.2 **Molecular Weight:** 299.05
13.3 **Boiling Point at 1 atm:** (est.) 347°F = 175°C = 448°K
13.4 **Freezing Point:** 37°F = 3°C = 276°K
13.5 **Critical Temperature:** Not pertinent
13.6 **Critical Pressure:** Not pertinent
13.7 **Specific Gravity:** 2.354 at 20°C (liquid)
13.8 **Liquid Surface Tension:** (est.) 15 dynes/cm = 0.015 N/m at 20°C
13.9 **Liquid-Water Interfacial Tension:** Not pertinent
13.10 **Vapor (Gas) Specific Gravity:** Not pertinent
13.11 **Ratio of Specific Heats of Vapor (Gas):** Not pertinent
13.12 **Latent Heat of Vaporization:** 68.9 Btu/lb = 38.3 cal/g = 1.60×10^5 J/kg
13.13 **Heat of Combustion:** Not pertinent
13.14 **Heat of Decomposition:** Not pertinent
13.15 **Heat of Solution:** −211.9 Btu/lb = −117.7 cal/g = -4.925×10^5 J/kg
13.16 **Heat of Polymerization:** Not pertinent

5. HEALTH HAZARDS (Cont'd.)

5.9 **Liquid or Solid Irritant Characteristics:** Severe skin irritant; causes second- and third-degree burns on short contact and is very injurious to the eyes.
5.10 **Odor Threshold:** Data not available

APF

ANTIMONY PENTAFLUORIDE

Common Synonyms

Liquid — Colorless — Sharp odor

Reacts violently with water. Poisonous gas is produced on contact with water. Freezing point is 45° F.

AVOID CONTACT WITH LIQUID AND VAPOR. KEEP PEOPLE AWAY.
Wear goggles and self-contained breathing apparatus.
Stop discharge if possible.
Call fire department.
Isolate and remove discharged material.
Notify local health and pollution control agencies.

Fire

Not flammable.
May cause fire on contact with combustibles.
POISONOUS GASES ARE PRODUCED WHEN HEATED.
DO NOT USE WATER OR FOAM ON FIRE OR ON ADJACENT FIRES.
Extinguish with dry chemicals or carbon dioxide.

Exposure

CALL FOR MEDICAL AID.

VAPOR
POISONOUS IF INHALED.
Irritating to eyes, nose and throat.
Move victim to fresh air.
If breathing has stopped, give artificial respiration.
If breathing is difficult, give oxygen.

LIQUID
POISONOUS IF SWALLOWED.
Will burn skin and eyes.
Remove contaminated clothing and shoes.
Flush affected areas with plenty of water.
IF IN EYES, hold eyelids open and flush with plenty of water.
IF SWALLOWED and victim is CONSCIOUS, have victim drink water or milk.
DO NOT INDUCE VOMITING.

Water Pollution

Effect of low concentrations on aquatic life is unknown.
May be dangerous if it enters water intakes.
Notify local health and wildlife officials.
Notify operators of nearby water intakes.

1. RESPONSE TO DISCHARGE

(See Response Methods Handbook, CG 446-4)

Issue warning—corrosive, air contaminant
Restrict access
Disperse and flush

2. LABEL

3. CHEMICAL DESIGNATIONS

3.1 **Synonyms:** No common synonyms
3.2 **Coast Guard Compatibility Classification:** Not applicable
3.3 **Chemical Formula:** SbF_5
3.4 **IMCO/United Nations Numerical Designation:** 8/1732

4. OBSERVABLE CHARACTERISTICS

4.1 **Physical State (as shipped):** Liquid
4.2 **Color:** Colorless
4.3 **Odor:** Pungent

5. HEALTH HAZARDS

5.1 **Personal Protective Equipment:** Acid-gas-type canister mask; rubber gloves, protective clothing; safety goggles and face shield.
5.2 **Symptoms Following Exposure:** Inhalation causes irritation of nose and throat. Contact of liquid with eyes or skin causes severe burns. Ingestion causes vomiting and severe burns of mouth and throat. Overexposure by any route can cause bloody stools, slow pulse, low blood pressure, coma, convulsions, cardiac arrest.
5.3 **Treatment for Exposure:** INHALATION: remove to fresh air; rinse mouth with water; give oxygen if necessary to assist breathing; get medical attention. EYES: irrigate with copious amounts of water for at least 15 min.; get medical attention. SKIN: flush with copious amounts of water; wash well with soap and water. INGESTION: dilute by drinking water; if vomiting occurs, drink more water; get medical attention promptly.
5.4 **Toxicity by Inhalation (Threshold Limit Value):** 0.5 mg/m^3 as antimony
5.5 **Short-Term Inhalation Limits:** Data not available
5.6 **Toxicity by Ingestion:** Data not available
5.7 **Late Toxicity:** Antimony poisoning may result.
5.8 **Vapor (Gas) Irritant Characteristics:** Data not available
5.9 **Liquid or Solid Irritant Characteristics:** Data not available
5.10 **Odor Threshold:** Data not available

6. FIRE HAZARDS

6.1 **Flash Point:** Not flammable (but see 7.2)
6.2 **Flammable Limits in Air:** Not flammable
6.3 **Fire Extinguishing Agents:** Not pertinent
6.4 **Fire Extinguishing Agents Not to be Used:** Do not use water or foam on adjacent fires.
6.5 **Special Hazards of Combustion Products:** Not pertinent
6.6 **Behavior in Fire:** Gives off toxic hydrogen fluoride fumes when water is used to extinguish adjacent fire.
6.7 **Ignition Temperature:** Not pertinent
6.8 **Electrical Hazard:** Not pertinent
6.9 **Burning Rate:** Not pertinent

7. CHEMICAL REACTIVITY

7.1 **Reactivity with Water:** Reacts vigorously to form toxic hydrogen fluoride (hydrofluoric acid).
7.2 **Reactivity with Common Materials:** When moisture is present, causes severe corrosion of metals (except steel) and glass. If confined and wet can cause explosion. May cause fire in contact with combustible material.
7.3 **Stability During Transport:** Stable
7.4 **Neutralizing Agents for Acids and Caustics:** Flush with water, rinse with sodium bicarbonate or lime solution.
7.5 **Polymerization:** Not pertinent
7.6 **Inhibitor of Polymerization:** Not pertinent

8. WATER POLLUTION

8.1 **Aquatic Toxicity:** Data not available
8.2 **Waterfowl Toxicity:** Data not available
8.3 **Biological Oxygen Demand (BOD):** None
8.4 **Food Chain Concentration Potential:** None

9. SELECTED MANUFACTURERS

1. Ozark-Mahoning Company
   Special Chemicals Division
   1870 South Boulder
   Tulsa, Okla. 74119
2. Ventron Corp.
   Alfa Products Division
   P. O. Box 159
   Beverly, Mass. 01915
3. PCR, Inc.
   P. O. Box 1466
   Gainesville, Fla. 32602

10. SHIPPING INFORMATION

10.1 **Grades or Purity:** Commercial
10.2 **Storage Temperature:** Ambient
10.3 **Inert Atmosphere:** No requirement
10.4 **Venting:** Pressure-vacuum

11. HAZARD ASSESSMENT CODE

(See Hazard Assessment Handbook, CG 446-3)

A-O

12. HAZARD CLASSIFICATIONS

12.1 **Code of Federal Regulations:** Corrosive
12.2 **NAS Hazard Rating for Bulk Water Transportation:** Not listed
12.3 **NFPA Hazard Classifications:** Not listed

13. PHYSICAL AND CHEMICAL PROPERTIES

13.1 **Physical State at 15°C and 1 atm:** Liquid
13.2 **Molecular Weight:** 216.7
13.3 **Boiling Point at 1 atm:** 289°F = 143°C = 416°K
13.4 **Freezing Point:** 45°F = 7°C = 280°K
13.5 **Critical Temperature:** Not pertinent
13.6 **Critical Pressure:** Not pertinent
13.7 **Specific Gravity:** 2.340 at 30°C (liquid)
13.8 **Liquid Surface Tension:** (est.) 20 dynes/cm = 0.020 N/m at 20°C
13.9 **Liquid-Water Interfacial Tension:** Not pertinent
13.10 **Vapor (Gas) Specific Gravity:** Not pertinent
13.11 **Ratio of Specific Heats of Vapor (Gas):** Not pertinent
13.12 **Latent Heat of Vaporization:** (est.) 79 Btu/lb = 44 cal/g = 1.8 × 10^5 J/kg
13.13 **Heat of Combustion:** Not pertinent
13.14 **Heat of Decomposition:** Not pertinent
13.15 **Heat of Solution:** Data not available
13.16 **Heat of Polymerization:** Not pertinent

NOTES

APT	ANTIMONY POTASSIUM TARTRATE

Common Synonyms			
Tartar emetic Tartrated antimony Tartarized antimony Potassium antimonyl tartrate	Solid	White	Odorless
	Sinks in water.		

AVOID CONTACT WITH SOLID AND DUST. KEEP PEOPLE AWAY.
Wear dust respirator and rubber overclothing (including gloves).
Stop discharge if possible.
Isolate and remove discharged material.
Notify local health and pollution control agencies.

Fire

Not flammable.

Exposure

CALL FOR MEDICAL AID.

DUST
POISONOUS IF INHALED OR IF SKIN IS EXPOSED.
If inhaled will cause dizziness, headache, coughing, or difficult breathing.
If in eyes, hold eyelids open and flush with plenty of water.
If breathing has stopped, give artificial respiration.
If breathing is difficult, give oxygen.

SOLID
Will burn skin and eyes.
If swallowed will cause nausea, dizziness, loss of consciousness.
Remove contaminated clothing and shoes.
Flush affected areas with plenty of water.
IF IN EYES, hold eyelids open and flush with plenty of water.
IF SWALLOWED and victim is CONSCIOUS, have victim drink water or milk and have victim induce vomiting.
IF SWALLOWED and victim is UNCONSCIOUS OR HAVING CONVULSIONS, do nothing except keep victim warm.

Water Pollution

HARMFUL TO AQUATIC LIFE IN VERY LOW CONCENTRATIONS.
May be dangerous if it enters water intakes.
Notify local health and wildlife officials.
Notify operators or nearby water intakes.

1. RESPONSE TO DISCHARGE

(See Response Methods Handbook, CG 446-4)

Issue warning—poison, water contaminant
Restrict access
Disperse and flush

2. LABELS

No hazard label required by Code of Federal Regulations

3. CHEMICAL DESIGNATIONS

3.1 **Synonyms:** Potassium antimonyl tartrate; Tartar emetic; Tartarized antimony; Tartrated antimony

3.2 **Coast Guard Compatibility Classification:** Not listed

3.3 **Chemical Formula:** KOOC·CHOH·CHOH·COO(SbO)·½H_2O

3.4 **IMCO/United Nations Numerical Designation:** 6.1/1551

4. OBSERVABLE CHARACTERISTICS

4.1 **Physical State (as shipped):** Solid

4.2 **Color:** White

4.3 **Odor:** None

5. HEALTH HAZARDS

5.1 **Personal Protective Equipment:** Dust respirator; rubber or plastic-coated gloves; chemical goggles; tightly woven, close fitting clothes; Bu. Mines approved respirator

5.2 **Symptoms Following Exposure:** Inhalation causes inflammation of membranes of nose and throat, upper respiratory irritation, headache, dizziness. Ingestion causes gastrointestinal upset, strong irritation, vomiting. Contact with eyes or skin causes irritation. Further symptoms of exposure include nervous complaints (i.e., irritability, dizziness, muscular and neurological pain).

5.3 **Treatment for Exposure:** INHALATION: move to fresh air. INGESTION: call physician immediately; use water (plain, soapy, or salty) or milk (3-4 glasses) to provoke vomiting. EYES: flush with water for 15 min.; consult a physician. SKIN: flush with water; wash well with soap and water.

5.4 **Toxicity by Inhalation (Threshold Limit Value):** 0.5 mg/m³ (as antimony)

5.5 **Short-Term Inhalation Limits:** Data not available

5.6 **Toxicity by Ingestion:** Grade 3; oral rat LD_{50} = 115 mg/kg

5.7 **Late Toxicity:** Data not available

5.8 **Vapor (Gas) Irritant Characteristics:** Data not available

5.9 **Liquid or Solid Irritant Characteristics:** Data not available

5.10 **Odor Threshold:** Odorless

6. FIRE HAZARDS

6.1 **Flash Point:** Not flammable

6.2 **Flammable Limits in Air:** Not flammable

6.3 **Fire Extinguishing Agents:** Not pertinent

6.4 **Fire Extinguishing Agents Not to be Used:** Not pertinent

6.5 **Special Hazards of Combustion Products:**

6.6 **Behavior in Fire:**

6.7 **Ignition Temperature:** Not pertinent

6.8 **Electrical Hazard:** Not pertinent

6.9 **Burning Rate:** Not pertinent

7. CHEMICAL REACTIVITY

7.1 **Reactivity with Water:** No reaction

7.2 **Reactivity with Common Materials:**

7.3 **Stability During Transport:** Stable

7.4 **Neutralizing Agents for Acids and Caustics:** Not pertinent

7.5 **Polymerization:** Not pertinent

7.6 **Inhibitor of Polymerization:** Not pertinent

8. WATER POLLUTION

8.1 **Aquatic Toxicity:**
20 ppm*/96 hr/fathead minnow/TL_m/fresh (soft) water
12 ppm*/96 hr/fathead minnow/TL_m/fresh (hard) water
*as antimony

8.2 **Waterfowl Toxicity:** Data not available

8.3 **Biological Oxygen Demand (BOD):** Data not available

8.4 **Food Chain Concentration Potential:** High

9. SELECTED MANUFACTURERS

1. Pfizer Chemicals Div.
   235 E. 42nd St.
   New York, N. Y. 10017
2. The Harshaw Chemical Co.
   1945 E. 97th St.
   Cleveland, Ohio 44106
3. Mallinckrodt Chemical Works
   223 Westside Ave.
   P. O. Box 384
   Jersey City, N. J. 07303

10. SHIPPING INFORMATION

10.1 **Grades or Purity:** Pure, 99–103%

10.2 **Storage Temperature:** Ambient

10.3 **Inert Atmosphere:** No requirement

10.4 **Venting:** Open

11. HAZARD ASSESSMENT CODE

(See Hazard Assessment Handbook, CG 446-3)

SS

12. HAZARD CLASSIFICATIONS

12.1 **Code of Federal Regulations:** ORM-A

12.2 **NAS Hazard Rating for Bulk Water Transportation:** Not listed

12.3 **NFPA Hazard Classifications:** Not listed

13. PHYSICAL AND CHEMICAL PROPERTIES

13.1 **Physical State at 15°C and 1 atm:** Solid

13.2 **Molecular Weight:** 334

13.3 **Boiling Point at 1 atm:** Not pertinent

13.4 **Freezing Point:** Not pertinent

13.5 **Critical Temperature:** Not pertinent

13.6 **Critical Pressure:** Not pertinent

13.7 **Specific Gravity:** 2.60 at 20°C (solid)

13.8 **Liquid Surface Tension:** Not pertinent

13.9 **Liquid-Water Interfacial Tension:** Not pertinent

13.10 **Vapor (Gas) Specific Gravity:** Not pertinent

13.11 **Ratio of Specific Heats of Vapor (Gas):** Not pertinent

13.12 **Latent Heat of Vaporization:** Not pertinent

13.13 **Heat of Combustion:** Not pertinent

13.14 **Heat of Decomposition:** Not pertinent

13.15 **Heat of Solution:** Not pertinent

13.16 **Heat of Polymerization:** Not pertinent

NOTES

ATM

ANTIMONY TRICHLORIDE

Common Synonyms			
Butter of antimony Antimony (III) Chloride	Solid	White to pale yellow	Sharp unpleasant odor
	Sinks and mixes violently with water.		

AVOID CONTACT WITH SOLID AND DUST. KEEP PEOPLE AWAY.
Wear dust respirator and rubber overclothing (including gloves).
Stop discharge if possible.
Isolate and remove discharged material.
Notify local health and pollution control agencies.

Fire

Not flammable.
POISONOUS GASES MAY BE PRODUCED IN FIRE.
Irritating gases may be produced when heated.
Wear goggles and self-contained breathing apparatus.
DO NOT USE WATER ON ADJACENT FIRES.

Exposure

CALL FOR MEDICAL AID.

DUST
POISONOUS IF INHALED OR IF SKIN IS EXPOSED.
If inhaled will cause coughing, or difficult breathing.
If in eyes, hold eyelids open and flush with plenty of water.
If breathing has stopped, give artificial respiration.
If breathing is difficult, give oxygen.

SOLID
Will burn skin and eyes.
If swallowed will cause nausea, vomiting or loss of consciousness.
Remove contaminated clothing and shoes.
Flush affected areas with plenty of water.
IF IN EYES, hold eyelids open and flush with plenty of water.
IF SWALLOWED and victim is CONSCIOUS, have victim drink water or milk.
IF SWALLOWED and victim is UNCONSCIOUS OR HAVING CONVULSIONS, do nothing except keep victim warm.

Water Pollution

HARMFUL TO AQUATIC LIFE IN VERY LOW CONCENTRATIONS.
May be dangerous if it enters water intakes.
Notify local health and wildlife officials.
Notify operators of nearby water intakes.

1. RESPONSE TO DISCHARGE
(See Response Methods Handbook, CG 446-4)

Issue warning—water contaminant, corrosive
Disperse and flush

2. LABEL

3. CHEMICAL DESIGNATIONS

3.1 **Synonyms:** Antimony butter; Antimony (III) chloride; Butter of antimony
3.2 **Coast Guard Compatibility Classification:** Not listed
3.3 **Chemical Formula:** $SbCl_3$
3.4 **IMCO/United Nations Numerical Designation:** 8/1733

4. OBSERVABLE CHARACTERISTICS

4.1 **Physical State (as shipped):** Solid
4.2 **Color:** White to pale yellow
4.3 **Odor:** Sharp, acrid

5. HEALTH HAZARDS

Note: The respiratory system is the chief avenue of entrance of antimony and its compounds into the body.

5.1 **Personal Protective Equipment:** Bu. Mines approved respirator; chemical safety goggles; face shield; leather or rubber safety shoes; rubber apron; rubber gloves
5.2 **Symptoms Following Exposure:** Inhalation of small amounts may cause only irritation of the nose, throat and air passages; large exposures result in severe air-passage irritation. Ingestion causes vomiting, purging with bloody stools, slow pulse and low blood pressure; slow, shallow breathing; coma and convulsions sometimes followed by death. Contact with eyes causes severe eye burns or at least severe eye irritation. Contact of dry chemical with skin may result in deep chemical burns.
5.3 **Treatment for Exposure:** INHALATION: move victim at once to fresh air and keep him warm, but not hot; call a physician immediately; nasal passages may be irrigated from a gently flowing hose. INGESTION: induce vomiting by giving large quantities of warm salt water; have a physician see the patient at once. EYES: flush with copious amounts of water for at least 15 min.; call a physician at once. SKIN: flush with large quantities of flowing water followed by washing of skin surfaces with soap and water; remove all contaminated clothing promptly.
5.4 **Toxicity by Inhalation (Threshold Limit Value):** 0.5 mg/m^3 (as antimony)
5.5 **Short-Term Inhalation Limits:** Data not available
5.6 **Toxicity by Ingestion:** Grade 2; oral rat LD_{50} = 675 mg/kg

6. FIRE HAZARDS

6.1 **Flash Point:** Not flammable
6.2 **Flammable Limits in Air:** Not flammable
6.3 **Fire Extinguishing Agents:** Not pertinent
6.4 **Fire Extinguishing Agents Not to be Used:** Do not use water on adjacent fires.
6.5 **Special Hazards of Combustion Products:** Toxic and irritating antimony oxide and hydrogen chloride may form in fires.
6.6 **Behavior in Fire:**
6.7 **Ignition Temperature:** Not pertinent
6.8 **Electrical Hazard:** Not pertinent
6.9 **Burning Rate:** Not pertinent

7. CHEMICAL REACTIVITY

7.1 **Reactivity with Water:** Reacts vigorously to form a strong solution of hydrochloric acid.
7.2 **Reactivity with Common Materials:** Corrodes most metals in presence of moisture. Flammable hydrogen gas may collect in enclosed spaces.
7.3 **Stability During Transport:** Stable
7.4 **Neutralizing Agents for Acids and Caustics:** Large amounts of water followed by sodium bicarbonate or soda ash solution
7.5 **Polymerization:** Not pertinent
7.6 **Inhibitor of Polymerization:** Not pertinent

8. WATER POLLUTION

8.1 **Aquatic Toxicity:**
17 ppm*/96 hr/fathead minnow/TL_m/fresh (hard) water
9 ppm*/96 hr/fathead minnow/TL_m/fresh (soft) water
*as antimony
8.2 **Waterfowl Toxicity:** Data not available
8.3 **Biological Oxygen Demand (BOD):** Data not available
8.4 **Food Chain Concentration Potential:** High

9. SELECTED MANUFACTURERS

1. J. T. Baker Chemical Co.
Phillipsburg, N. J. 08865
2. Stauffer Chemical Company
Specialty Chemical Division
Westport, Conn. 06880
3. Gallard-Schlesinger Chemical Mfg. Co.
584 Mineola Ave.
Carle Place, N. Y. 11514

10. SHIPPING INFORMATION

10.1 **Grades or Purity:** Commercial, 99+%; Analytical; Anhydrous
10.2 **Storage Temperature:** Ambient
10.3 **Inert Atmosphere:** Padded
10.4 **Venting:** Pressure-vacuum

11. HAZARD ASSESSMENT CODE
(See Hazard Assessment Handbook, CG 446-3)

RR

12. HAZARD CLASSIFICATIONS

12.1 **Code of Federal Regulations:** Corrosive solid
12.2 **NAS Hazard Rating for Bulk Water Transportation:** Not listed
12.3 **NFPA Hazard Classifications:** Not listed

13. PHYSICAL AND CHEMICAL PROPERTIES

13.1 **Physical State at 15°C and 1 atm:** Solid
13.2 **Molecular Weight:** 228
13.3 **Boiling Point at 1 atm:** 433°F = 223°C = 496°K
13.4 **Freezing Point:** 163°F = 73°C = 346°K
13.5 **Critical Temperature:** Not pertinent
13.6 **Critical Pressure:** Not pertinent
13.7 **Specific Gravity:** 3.14 at 20°C (solid)
13.8 **Liquid Surface Tension:** Not pertinent
13.9 **Liquid-Water Interfacial Tension:** Not pertinent
13.10 **Vapor (Gas) Specific Gravity:** Not pertinent
13.11 **Ratio of Specific Heats of Vapor (Gas):** Not pertinent
13.12 **Latent Heat of Vaporization:** Not pertinent
13.13 **Heat of Combustion:** Not pertinent
13.14 **Heat of Decomposition:** Not pertinent
13.15 **Heat of Solution:** −70 Btu/lb = −39 cal/g = -1.6×10^5 J/kg
13.16 **Heat of Polymerization:** Not pertinent

5. HEALTH HAZARDS (Cont'd)

5.7 **Late Toxicity:** Data not available
5.8 **Vapor (Gas) Irritant Characteristics:** Data not available
5.9 **Liquid or Solid Irritant Characteristics:** Data not available
5.10 **Odor Threshold:** Data not available

ATT	ANTIMONY TRIFLUORIDE

Common Synonyms	Solid crystals	White	Odorless
	Sinks in water.		

Avoid contact with solid.
Isolate and remove discharged material.
Notify local health and pollution control agencies.

Fire

Not flammable.

Exposure

CALL FOR MEDICAL AID.

SOLID
Irritating to skin and eyes.
Harmful if swallowed.
Remove contaminated clothing and shoes.
Flush affected areas with plenty of water.
IF IN EYES, hold eyelids open and flush with plenty of water.
IF SWALLOWED and victim is CONSCIOUS, have victim drink water or milk.

Water Pollution

HARMFUL TO AQUATIC LIFE IN VERY LOW CONCENTRATIONS.
Fouling to shoreline.
May be dangerous if it enters water intakes.
Notify local health and wildlife officials.
Notify operators of nearby water intakes.

1. RESPONSE TO DISCHARGE

(See Response Methods Handbook, CG 446-4)

Issue warning—water contaminant
Disperse and flush

2. LABELS

No hazard label required by Code of Federal Regulations

3. CHEMICAL DESIGNATIONS

3.1 **Synonyms:** No common synonyms
3.2 **Coast Guard Compatibility Classification:** Not applicable
3.3 **Chemical Formula:** SbF_3
3.4 **IMCO/United Nations Numerical Designation:** 6.1/1549

4. OBSERVABLE CHARACTERISTICS

4.1 **Physical State (as shipped):** Solid
4.2 **Color:** White
4.3 **Odor:** None

5. HEALTH HAZARDS

5.1 **Personal Protective Equipment:** Approved respirator, rubber gloves.
5.2 **Symptoms Following Exposure:** Resemble those of lead and arsenic poisoning. ACUTE POISONING: irritation of the mouth, nose, stomach and intestines; vomiting, purging with bloody stools; slow pulse and low blood pressure; slow, shallow breathing; coma and convulsions sometimes followed by death from cardiac and respiratory exhaustion. CHRONIC POISONING: dryness of throat; pain on swallowing; occasional vomiting and persistent nausea; susceptibility to fainting; diarrhea, loss of appetite and weight; giddiness; dermatitis, either pustular or ulcerative; anemia
5.3 **Treatment for Exposure:** If any symptoms, however slight, are noticed, the affected individual should be removed from contact with the chemical and placed under the care of a physician who is versed in the treatment necessary.
5.4 **Toxicity by Inhalation (Threshold Limit Value):** 0.5 mg/m^3
5.5 **Short-Term Inhalation Limits:** Not pertinent
5.6 **Toxicity by Ingestion:** Grade 3; LD_{50} 50 to 500 mg/kg (guinea pig)
5.7 **Late Toxicity:** Data not available
5.8 **Vapor (Gas) Irritant Characteristics:** Not pertinent
5.9 **Liquid or Solid Irritant Characteristics:** Fairly severe skin irritant. May cause pain and second-degree burns after a few minutes' contact.
5.10 **Odor Threshold:** Not pertinent

6. FIRE HAZARDS

6.1 **Flash Point:** Not flammable
6.2 **Flammable Limits in Air:** Not flammable
6.3 **Fire Extinguishing Agents:** Not pertinent
6.4 **Fire Extinguishing Agents Not to be Used:** Not pertinent
6.5 **Special Hazards of Combustion Products:** Not pertinent
6.6 **Behavior in Fire:** Not pertinent
6.7 **Ignition Temperature:** Not flammable
6.8 **Electrical Hazard:** Not pertinent
6.9 **Burning Rate:** Not flammable

7. CHEMICAL REACTIVITY

7.1 **Reactivity with Water:** No reaction
7.2 **Reactivity with Common Materials:** No reaction
7.3 **Stability During Transport:** Stable
7.4 **Neutralizing Agents for Acids and Caustics:** Not pertinent
7.5 **Polymerization:** Not pertinent
7.6 **Inhibitor of Polymerization:** Not pertinent

8. WATER POLLUTION

8.1 **Aquatic Toxicity:** 200 ppm/<24 hr/tinca vulgaris (fish)/killed/fresh water
8.2 **Waterfowl Toxicity:** Data not available
8.3 **Biological Oxygen Demand (BOD):** Data not available
8.4 **Food Chain Concentration Potential:** Data not available

9. SELECTED MANUFACTURERS

1. Allied Chemical Corp.
Specialty Chemicals Div.
Wilmington Turnpike
Marcus Hook, Pa. 19061
2. Ozark-Mahoning Co.
1870 South Boulder Ave.
Tulsa, Oklahoma 74119

10. SHIPPING INFORMATION

10.1 **Grades or Purity:** 98%
10.2 **Storage Temperature:** Data not available
10.3 **Inert Atmosphere:** Data not available
10.4 **Venting:** Data not available

11. HAZARD ASSESSMENT CODE

(See Hazard Assessment Handbook, CG 446-3)

SS

12. HAZARD CLASSIFICATIONS

12.1 **Code of Federal Regulations:** Not listed
12.2 **NAS Hazard Rating for Bulk Water Transportation:** Not listed
12.3 **NFPA Hazard Classifications:** Not listed

13. PHYSICAL AND CHEMICAL PROPERTIES

13.1 **Physical State at 15°C and 1 atm:** Solid
13.2 **Molecular Weight:** 178.75
13.3 **Boiling Point at 1 atm:** Not pertinent
13.4 **Freezing Point:** 558°F = 292°C = 565°K
13.5 **Critical Temperature:** Not pertinent
13.6 **Critical Pressure:** Not pertinent
13.7 **Specific Gravity:** 4.38 at 21°C (solid)
13.8 **Liquid Surface Tension:** Not pertinent
13.9 **Liquid-Water Interfacial Tension:** Not pertinent
13.10 **Vapor (Gas) Specific Gravity:** Not pertinent
13.11 **Ratio of Specific Heats of Vapor (Gas):** Not pertinent
13.12 **Latent Heat of Vaporization:** Not pertinent
13.13 **Heat of Combustion:** Not pertinent
13.14 **Heat of Decomposition:** Not pertinent
13.15 **Heat of Solution:** Not pertinent
13.16 **Heat of Polymerization:** Not pertinent

NOTES

REVISED 1978

ATX	# ANTIMONY TRIOXIDE

Common Synonyms Diantimony trioxide Senarmontite Valentinite Exitelite Weisspiessglanz	Solid	White	Odorless
	Sinks in water.		

AVOID CONTACT WITH SOLID AND DUST. KEEP PEOPLE AWAY.
Wear dust respirator and rubber overclothing (including gloves).
Stop discharge if possible.
Isolate and remove discharged material.
Notify local health and pollution control agencies.

Fire

Not flammable.

Exposure

CALL FOR MEDICAL AID.

DUST
POISONOUS IF INHALED OR IF SKIN IS EXPOSED.
If inhaled will cause coughing, difficult breathing or loss of consciousness.
If in eyes, hold eyelids open and flush with plenty of water.
If breathing has stopped, give artificial respiration.
If breathing is difficult, give oxygen.

SOLID
POISONOUS IF SWALLOWED OR IF SKIN IS EXPOSED.
If swallowed will cause dizziness, nausea, vomiting or loss of consciousness.
Remove contaminated clothing and shoes.
Flush affected areas with plenty of water.
IF IN EYES, hold eyelids open and flush with plenty of water.
IF SWALLOWED and victim is CONSCIOUS, have victim drink water or milk and have victim induce vomiting.
IF SWALLOWED and victim is UNCONSCIOUS OR HAVING CONVULSIONS, do nothing except keep victim warm.

Water Pollution

HARMFUL TO AQUATIC LIFE IN VERY LOW CONCENTRATIONS.
May be dangerous if it enters water intakes.
Notify local health and wildlife officials.
Notify operators of nearby water intakes.

1. RESPONSE TO DISCHARGE
(See Response Methods Handbook, CG 446-4)

Issue warning—poison
Restrict access
Should be removed
Chemical and physical treatment

2. LABELS

No hazard label required by Code of Federal Regulations

3. CHEMICAL DESIGNATIONS

3.1 **Synonyms:** Diantimony trioxide; Exitelite; Flowers of antimony; Senarmontite; Valentinite; Weisspiessglanz
3.2 **Coast Guard Compatibility Classification:** Not listed
3.3 **Chemical Formula:** Sb_2O_3
3.4 **IMCO/United Nations Numerical Designation:** 6.1/1549

4. OBSERVABLE CHARACTERISTICS

4.1 **Physical State (as shipped):** Solid
4.2 **Color:** White
4.3 **Odor:** None

5. HEALTH HAZARDS

5.1 **Personal Protective Equipment:** Rubber gloves; safety goggles; dust mask
5.2 **Symptoms Following Exposure:** Inhalation causes inflammation of upper and lower respiratory tract, including pneumonitis. Ingestion causes irritation of the mouth, nose, stomach and intestines; vomiting, purging with bloody stools; slow pulse and low blood pressure; slow, shallow breathing; coma and convulsions sometimes followed by death. Contact with eyes causes conjunctivitis. Contact with skin causes dermatitis and rhinitis.
5.3 **Treatment for Exposure:** *If any of the symptoms of poisoning, even slight, are noticed, the affected individual should be removed from contact with the chemical and placed under care of a physician.* INGESTION: induce vomiting. EYES: flush with water for at least 15 min. SKIN: wash well with soap and water.
5.4 **Toxicity by Inhalation (Threshold Limit Value):** 0.5 mg/m^3 (as antimony)
5.5 **Short-Term Inhalation Limits:** Data not available
5.6 **Toxicity by Ingestion:** Grade 0; oral rat LD_{50} = 20,000 mg/kg
5.7 **Late Toxicity:** Data not available
5.8 **Vapor (Gas) Irritant Characteristics:** Data not available
5.9 **Liquid or Solid Irritant Characteristics:** Data not available
5.10 **Odor Threshold:** Data not available

6. FIRE HAZARDS

6.1 **Flash Point:** Not flammable
6.2 **Flammable Limits in Air:** Not flammable
6.3 **Fire Extinguishing Agents:** Not pertinent
6.4 **Fire Extinguishing Agents Not to be Used:** Not pertinent
6.5 **Special Hazards of Combustion Products:**
6.6 **Behavior in Fire:**
6.7 **Ignition Temperature:** Not pertinent
6.8 **Electrical Hazard:** Not pertinent
6.9 **Burning Rate:** Not pertinent

7. CHEMICAL REACTIVITY

7.1 **Reactivity with Water:** No reaction
7.2 **Reactivity with Common Materials:**
7.3 **Stability During Transport:** Stable
7.4 **Neutralizing Agents for Acids and Caustics:** Not pertinent
7.5 **Polymerization:** Not pertinent
7.6 **Inhibitor of Polymerization:** Not pertinent

8. WATER POLLUTION

8.1 **Aquatic Toxicity:**
>80 ppm*/96 hr/fathead minnow/TL_m/hard or soft fresh water
*as antimony
8.2 **Waterfowl Toxicity:** Data not available
8.3 **Biological Oxygen Demand (BOD):** None
8.4 **Food Chain Concentration Potential:** High

9. SELECTED MANUFACTURERS

1. J. T. Baker Chemical Co.
Phillipsburg, N. J. 08865
2. McGean Chemical Co.
1250 Terminal Tower
Cleveland, Ohio 44113
3. Gallard-Schlesinger Chemical Mfg. Co.
584 Mineola Ave.
Carle Place, N. Y. 11514

10. SHIPPING INFORMATION

10.1 **Grades or Purity:** Reagent, 99.9+%; Optical grade
10.2 **Storage Temperature:** Ambient
10.3 **Inert Atmosphere:** No requirement
10.4 **Venting:** Open

11. HAZARD ASSESSMENT CODE
(See Hazard Assessment Handbook, CG 446-3)

II

12. HAZARD CLASSIFICATIONS

12.1 **Code of Federal Regulations:** Not listed
12.2 **NAS Hazard Rating for Bulk Water Transportation:** Not listed
12.3 **NFPA Hazard Classifications:** Not listed

13. PHYSICAL AND CHEMICAL PROPERTIES

13.1 **Physical State at 15°C and 1 atm:** Solid
13.2 **Molecular Weight:** 291.50
13.3 **Boiling Point at 1 atm:** Not pertinent
13.4 **Freezing Point:** Not pertinent
13.5 **Critical Temperature:** Not pertinent
13.6 **Critical Pressure:** Not pertinent
13.7 **Specific Gravity:** 5.2 at 25°C (solid)
13.8 **Liquid Surface Tension:** Not pertinent
13.9 **Liquid-Water Interfacial Tension:** Not pertinent
13.10 **Vapor (Gas) Specific Gravity:** Not pertinent
13.11 **Ratio of Specific Heats of Vapor (Gas):** Not pertinent
13.12 **Latent Heat of Vaporization:** Not pertinent
13.13 **Heat of Combustion:** Not pertinent
13.14 **Heat of Decomposition:** Not pertinent
13.15 **Heat of Solution:** Not pertinent
13.16 **Heat of Polymerization:** Not pertinent

NOTES

ASA | ARSENIC ACID

Common Synonyms			
Orthoarsenic acid Arsenic pentoxide	Solid crystals or solution	White or colorless	Odorless
	Sinks and mixes with water. Freezing point is 95° F.		

AVOID CONTACT WITH SOLUTION, SOLID AND DUST.
KEEP PEOPLE AWAY.
Wear rubber overclothing (including gloves).
Stop discharge if possible.
Isolate and remove discharged material.
Notify local health and pollution control agencies.

Fire

Not flammable.

Exposure

CALL FOR MEDICAL AID.

DUST
POISONOUS IF INHALED.
Irritating to eyes, nose and throat.
Move victim to fresh air.
If in eyes, hold eyelids open and flush with plenty of water.
If breathing is difficult, give oxygen.

SOLUTION OR SOLID
POISONOUS IF SWALLOWED.
Irritating to skin and eyes.
Remove contaminated clothing and shoes.
Flush affected areas with plenty of water.
IF IN EYES, hold eyelids open and flush with plenty of water.
IF SWALLOWED and victim is CONSCIOUS, have victim drink water or milk and have victim induce vomiting.
IF SWALLOWED and victim is UNCONSCIOUS OR HAVING CONVULSIONS, do nothing except keep victim warm.

Water Pollution

Effect of low concentrations on aquatic life is unknown.
May be dangerous if it enters water intakes.
Notify local health and wildlife officials.
Notify operators of nearby water intakes.

1. RESPONSE TO DISCHARGE

(See Response Methods Handbook, CG 446-4)

Issue warning—poison, water contaminant
Restrict access
Should be removed
Chemical and physical treatment

2. LABEL

3. CHEMICAL DESIGNATIONS

3.1 **Synonyms:** Arsenic pentoxide
Orthoarsenic acid
3.2 **Coast Guard Compatibility Classification:** Not applicable
3.3 **Chemical Formula:** As_2O_5 or $H_3AsO_4 \cdot \frac{1}{2}H_2O$
3.4 **IMCO/United Nations Numerical Designation:** 6.1/1553

4. OBSERVABLE CHARACTERISTICS

4.1 **Physical State (as shipped):** Solid; a concentrated water solution is sometimes shipped.
4.2 **Color:** White
4.3 **Odor:** None

5. HEALTH HAZARDS

5.1 **Personal Protective Equipment:** Calamine lotion and zinc oxide powder on hands and other skin areas; rubber gloves; U. S. Bureau of Mines approved dust respirator.
5.2 **Symptoms Following Exposure:** Ingestion causes irritation of stomach, weakness, other gastrointestinal symptoms. Overdose can cause arsenic poisoning, but symptoms are delayed.
5.3 **Treatment for Exposure:** Get medical attention after all exposures to this compound. Be alert for arsenic poisoning symptoms. SKIN: wash well with soap and water. INGESTION: induce vomiting; drink freely lime water, milk, or raw egg; give a cathartic.
5.4 **Toxicity by Inhalation (Threshold Limit Value):** 0.5 mg/m^3 as arsenic
5.5 **Short-Term Inhalation Limits:** Data not available
5.6 **Toxicity by Ingestion:** Grade 4; oral LD_{50} = 48 mg/kg (young rats)
5.7 **Late Toxicity:** Arsenic compounds may be carcinogenic.
5.8 **Vapor (Gas) Irritant Characteristics:** Vapors are nonirritating to eyes and throat.
5.9 **Liquid or Solid Irritant Characteristics:** Minimum hazard. If spilled on clothing and allowed to remain, may cause smarting and reddening of the skin.
5.10 **Odor Threshold:** Odorless

6. FIRE HAZARDS

6.1 **Flash Point:** Not flammable
6.2 **Flammable Limits in Air:** Not flammable
6.3 **Fire Extinguishing Agents:** Not pertinent
6.4 **Fire Extinguishing Agents Not to be Used:** Not pertinent
6.5 **Special Hazards of Combustion Products:** Not pertinent
6.6 **Behavior in Fire:** Not pertinent
6.7 **Ignition Temperature:** Not pertinent
6.8 **Electrical Hazard:** Not pertinent
6.9 **Burning Rate:** Not pertinent

7. CHEMICAL REACTIVITY

7.1 **Reactivity with Water:** No reaction
7.2 **Reactivity with Common Materials:** Will corrode metal and may give off toxic arsine gas.
7.3 **Stability During Transport:** Stable
7.4 **Neutralizing Agents for Acids and Caustics:** Flush with water, rinse with sodium bicarbonate or lime solution.
7.5 **Polymerization:** Not pertinent
7.6 **Inhibitor of Polymerization:** Not pertinent

8. WATER POLLUTION

8.1 **Aquatic Toxicity:** Data not available
8.2 **Waterfowl Toxicity:** Data not available
8.3 **Biological Oxygen Demand (BOD):** None
8.4 **Food Chain Concentration Potential:** None

9. SELECTED MANUFACTURERS

1. J. T. Baker Chemical Co.
Phillipsburg, N. J. 08865
2. Gallard-Schlesinger Chemical Manufacturing Corp.
584 Mineola Avenue
Carle Place, N. Y. 11514
3. Cerac, Inc.
13460 W. Silver Spring Rd.
Menomenee Falls, Wis. 53051

10. SHIPPING INFORMATION

10.1 **Grades or Purity:** Commercial
10.2 **Storage Temperature:** Ambient
10.3 **Inert Atmosphere:** No requirement
10.4 **Venting:** Pressure-vacuum

11. HAZARD ASSESSMENT CODE

(See Hazard Assessment Handbook, CG 446-3)

SS

12. HAZARD CLASSIFICATIONS

12.1 **Code of Federal Regulations:** Poisonous, Class B
12.2 **NAS Hazard Rating for Bulk Water Transportation:**

Category	Rating
Fire	0
Health	
Vapor Irritant	0
Liquid or Solid Irritant	1
Poisons	4
Water Pollution	
Human Toxicity	4
Aquatic Toxicity	3
Aesthetic Effect	1
Reactivity	
Other Chemicals	2
Water	0
Self-Reaction	0

12.3 **NFPA Hazard Classifications:** Not listed

13. PHYSICAL AND CHEMICAL PROPERTIES

13.1 **Physical State at 15°C and 1 atm:** Solid
13.2 **Molecular Weight:** 229.8
13.3 **Boiling Point at 1 atm:** Not pertinent
13.4 **Freezing Point:** Not pertinent
13.5 **Critical Temperature:** Not pertinent
13.6 **Critical Pressure:** Not pertinent
13.7 **Specific Gravity:** 2.2 at 20°C (solid)
13.8 **Liquid Surface Tension:** Not pertinent
13.9 **Liquid-Water Interfacial Tension:** Not pertinent
13.10 **Vapor (Gas) Specific Gravity:** Not pertinent
13.11 **Ratio of Specific Heats of Vapor (Gas):** Not pertinent
13.12 **Latent Heat of Vaporization:** Not pertinent
13.13 **Heat of Combustion:** Not pertinent
13.14 **Heat of Decomposition:** Not pertinent
13.15 **Heat of Solution:** 3.1 Btu/lb = 1.7 cal/g = 0.071 × 10^5 J/kg
13.16 **Heat of Polymerization:** Not pertinent

NOTES

ARD ARSENIC DISULFIDE

Common Synonyms			
Realgar Red arsenic glass Red arsenic sulfide Red orpiment Ruby arsenic	Solid Sinks in water.	Red-brown	Odorless

AVOID CONTACT WITH SOLID AND DUST. Keep people away.
Wear chemical protective suit with self-contained breathing apparatus.
Isolate and remove discharged material.
Notify local health and pollution control agencies.

Fire	Not Flammable. Wear chemical protective suit with self-contained breathing apparatus.
Exposure	CALL FOR MEDICAL AID. DUST POISONOUS IF INHALED. Harmful to skin. Move to fresh air. If breathing has stopped, give artificial respiration. If breathing is difficult, give oxygen. SOLID POISONOUS IF SWALLOWED. Will burn eyes and skin. Remove contaminated clothing and shoes. Flush affected areas with plenty of water. IF IN EYES, hold eyelids open and flush with plenty of water. IF SWALLOWED and victim is CONSCIOUS, have victim drink water or milk and have victim induce vomiting. IF SWALLOWED and victim is UNCONSCIOUS OR HAVING CONVULSIONS, do nothing except keep victim warm.
Water Pollution	HARMFUL TO AQUATIC LIFE IN VERY LOW CONCENTRATIONS. May be dangerous if it enters water intakes. Notify local health and wildlife officials. Notify operators of nearby water intakes.

1. RESPONSE TO DISCHARGE

(See Response Methods Handbook, CG 446-4)

Issue warning—poison, water contaminant
Restrict access
Should be removed
Chemical and physical treatment

2. LABEL

3. CHEMICAL DESIGNATIONS

3.1 **Synonyms:** Realgar; Red arsenic glass; Red arsenic sulfide; Red orpiment; Ruby arsenic
3.2 **Coast Guard Compatibility Classificaton:** Not listed
3.3 **Chemical Formula:** As_2S_2
3.4 **IMCO/United Nations Numerical Designation:** 6.1/1557

4. OBSERVABLE CHARACTERISTICS

4.1 **Physical State (as shipped):** Solid
4.2 **Color:** Red-brown
4.3 **Odor:** None

5. HEALTH HAZARDS

5.1 **Personal Protective Equipment:** Approved respirator; goggles; rubber gloves; clean protective clothing
5.2 **Symptoms Following Exposure:** (Acute and sub-acute poisoning are not common.) Repeated inhalation causes irritation of nose, laryngitis, mild bronchitis. Ingestion causes weakness, loss of appetite, gastrointestinal disturbances, peripheral neuritis, occasional hepatitis. Contact with eyes causes irritation. Irritates skin, especially where moist; if not treated, may cause ulceration.
5.3 **Treatment for Exposure:** *Consult physician after all overexposure to this compound.* INHALATION: move to fresh air. INGESTION: induce vomiting by giving warm salt water; repeat until vomit is clear. EYES: flush with water for at least 15 min. SKIN: wash well with water.
5.4 **Toxicity by Inhalation (Threshold Limit Value):** 0.5 mg/m^3 (as arsenic)
5.5 **Short-Term Inhalation Limits:** Data not available
5.6 **Toxicity by Ingestion:** Grade 4; LD_{50} <50 mg/kg
5.7 **Late Toxicity:** Possible skin and lung cancer.
5.8 **Vapor (Gas) Irritant Characteristics:** Data not available
5.9 **Liquid or Solid Irritant Characteristics:** Data not available
5.10 **Odor Threshold:** Odorless

6. FIRE HAZARDS

6.1 **Flash Point:** Not pertinent
6.2 **Flammable Limits in Air:** Not pertinent
6.3 **Fire Extinguishing Agents:** Water
6.4 **Fire Extinguishing Agents Not to be Used:** Not pertinent
6.5 **Special Hazards of Combustion Products:** Poisonous fumes of the compound may be formed in fires. If ignited, will form sulfur dioxide gas.
6.6 **Behavior in Fire:** May ignite at very high temperatures
6.7 **Ignition Temperature:** Not pertinent
6.8 **Electrical Hazard:** Not pertinent
6.9 **Burning Rate:** Not pertinent

7. CHEMICAL REACTIVITY

7.1 **Reactivity with Water:** No reaction
7.2 **Reactivity with Common Materials:**
7.3 **Stability During Transport:** Stable
7.4 **Neutralizing Agents for Acids and Caustics:** Not pertinent
7.5 **Polymerization:** Not pertinent
7.6 **Inhibitor of Polymerization:** Not pertinent

8. WATER POLLUTION

8.1 **Aquatic Toxicity:** Data not available
8.2 **Waterfowl Toxicity:** Data not available
8.3 **Biological Oxygen Demand (BOD):** Data not available
8.4 **Food Chain Concentration Potential:** Data not available

9. SELECTED MANUFACTURERS

1. Gallard-Schlesinger Chemical Mfg. Co.
584 Mineola Ave.
Carle Place, N. Y. 11514
2. Electronic Space Products, Inc.
854 S. Robertson Blvd.
Los Angeles, Calif. 90035
3. Research Organic/Inorganic Chemical Corp.
Sun Valley, Calif. 91352

10. SHIPPING INFORMATION

10.1 **Grades or Purity:** 99+%; Technical
10.2 **Storage Temperature:** Ambient
10.3 **Inert Atmosphere:** No requirement
10.4 **Venting:** Open

11. HAZARD ASSESSMENT CODE

(See Hazard Assessment Handbook, CG 446-3)

II

12. HAZARD CLASSIFICATIONS

12.1 **Code of Federal Regulations:** Poisonous solid
12.2 **NAS Hazard Rating for Bulk Water Transportation:** Not listed
12.3 **NFPA Hazard Classifications:** Not listed

13. PHYSICAL AND CHEMICAL PROPERTIES

13.1 **Physical State at 15°C and 1 atm:** Solid
13.2 **Molecular Weight:** 214
13.3 **Boiling Point at 1 atm:** 1,049°F = 565°C = 838°K
13.4 **Freezing Point:** 585°F = 307°C = 580°K
13.5 **Critical Temperature:** Not pertinent
13.6 **Critical Pressure:** Not pertinent
13.7 **Specific Gravity:** 3.5 at 20°C (solid)
13.8 **Liquid Surface Tension:** Not pertinent
13.9 **Liquid-Water Interfacial Tension:** Not pertinent
13.10 **Vapor (Gas) Specific Gravity:** Not pertinent
13.11 **Ratio of Specific Heats of Vapor (Gas):** Not pertinent
13.12 **Latent Heat of Vaporization:** Not pertinent
13.13 **Heat of Combustion:** Not pertinent
13.14 **Heat of Decomposition:** Not pertinent
13.15 **Heat of Solution:** Not pertinent
13.16 **Heat of Polymerization:** Not pertinent

NOTES

AST	# ARSENIC TRICHLORIDE

Common Synonyms			
Fuming liquid arsenic Arsenic (III) trichloride Arsenic chloride Arsenous chloride Caustic arsenic chloride	Liquid	Colorless	Unpleasant odor
	Sinks and reacts in water. Poisonous visible vapor cloud is produced.		

AVOID CONTACT WITH LIQUID AND VAPOR. KEEP PEOPLE AWAY.
Wear goggles and self-contained breathing apparatus.
Stop discharge if possible.
Isolate and remove discharged material.
Notify local health and pollution control agencies.

Fire	Not Flammable. POISONOUS GASES ARE PRODUCED WHEN HEATED.
Exposure	CALL FOR MEDICAL AID. VAPOR POISONOUS IF INHALED. Move victim to fresh air. If breathing is difficult, give oxygen. LIQUID POISONOUS IF SWALLOWED. Irritating to skin and eyes. Remove contaminated clothing and shoes. Flush affected areas with plenty of water. IF IN EYES, hold eyelids open and flush with plenty of water. IF SWALLOWED and victim is CONSCIOUS, have victim drink water or milk and have victim induce vomiting. IF SWALLOWED and victim is UNCONSCIOUS OR HAVING CONVULSIONS, do nothing except keep victim warm.
Water Pollution	Effect of low concentrations on aquatic life is unknown. May be dangerous if it enters water intakes. Notify local health and wildlife officials. Notify operators of nearby water intakes.

1. RESPONSE TO DISCHARGE
(See Response Methods Handbook, CG 446-4)

Issue warning—poison, water contaminant, corrosive
Restrict access
Disperse and flush

2. LABEL

3. CHEMICAL DESIGNATIONS

3.1 **Synonyms:** Arsenic (III) trichloride; Arsenic chloride; Arsenous chloride; Arsenious chloride; Butter of arsenic; Caustic arsenic chloride; Caustic oil of arsenic; Fuming liquid arsenic
3.2 **Coast Guard Compatibility Classification:** Not applicable
3.3 **Chemical Formula:** $AsCl_3$
3.4 **IMCO/United Nations Numerical Designation:** 6.1/1560

4. OBSERVABLE CHARACTERISTICS

4.1 **Physical State (as shipped):** Liquid
4.2 **Color:** Colorless
4.3 **Odor:** Acrid

5. HEALTH HAZARDS

5.1 **Personal Protective Equipment:** Safety goggles and face shield; acid-type canister gas mask; rubber gloves; protective clothing.
5.2 **Symptoms Following Exposure:** Inhalation causes irritation of nose and throat. Contact of liquid with eyes or skin causes severe irritation. Ingestion causes weakness and severe irritation of mouth and stomach. Overdose can cause arsenic poisoning, but symptoms are delayed.
5.3 **Treatment for Exposure:** Get medical attention after all exposures to the compound. Be alert for arsenic poisoning symptoms. INHALATION: remove to fresh air; give artificial respiration if needed. EYES: flush with water for at least 15 min. SKIN: flush with water. INGESTION: give large amounts of water, then induce vomiting; give lime water, milk, or raw egg; give a cathartic.
5.4 **Toxicity by Inhalation (Threshold Limit Value):** 0.5 mg/m^3 as arsenic
5.5 **Short-Term Inhalation Limits:** Data not available
5.6 **Toxicity by Ingestion:** Grade 3; oral rat LD_{50} = 138 mg/kg; fatal human dose 70-180 mg, depending on weight.
5.7 **Late Toxicity:** Arsenic compounds may be carcinogenic.
5.8 **Vapor (Gas) Irritant Characteristics:** Data not available
5.9 **Liquid or Solid Irritant Characteristics:** Data not available
5.10 **Odor Threshold:** Data not available

6. FIRE HAZARDS

6.1 **Flash Point:** Not flammable
6.2 **Flammable Limits in Air:** Not flammable
6.3 **Fire Extinguishing Agents:** Not pertinent
6.4 **Fire Extinguishing Agents Not to be Used:** Avoid water on adjacent fires.
6.5 **Special Hazards of Combustion Products:** Irritating and toxic hydrogen chloride formed when involved in fire.
6.6 **Behavior in Fire:** Becomes gaseous and causes irritation. Forms hydrogen chloride (hydrochloric acid) by reaction with water used on adjacent fires.
6.7 **Ignition Temperature:** Not pertinent
6.8 **Electrical Hazard:** Not pertinent
6.9 **Burning Rate:** Not pertinent

7. CHEMICAL REACTIVITY

7.1 **Reactivity with Water:** Reacts with water to generate hydrogen chloride (hydrochloric acid).
7.2 **Reactivity with Common Materials:** Corrodes metal.
7.3 **Stability During Transport:** Stable
7.4 **Neutralizing Agents for Acids and Caustics:** Flush with water, rinse with sodium bicarbonate or lime solution.
7.5 **Polymerization:** Not pertinent
7.6 **Inhibitor of Polymerization:** Not pertinent

8. WATER POLLUTION

8.1 **Aquatic Toxicity:** Data not available
8.2 **Waterfowl Toxicity:** Data not available
8.3 **Biological Oxygen Demand (BOD):** Data not available
8.4 **Food Chain Concentration Potential:** None

9. SELECTED MANUFACTURERS

1. Cerac, Inc.
13460 W. Silver Spring Rd.
Menomenee Falls, Wis. 53051
2. Ventron Corporation
Alfa Products
P. O. Box 159
Beverly, Mass. 01915
3. Gallard Schlesinger Chemical Manufacturing Co.
584 Mineola Avenue
Carle Place, N. Y. 11514

10. SHIPPING INFORMATION

10.1 **Grades or Purity:** Commercial
10.2 **Storage Temperature:** Ambient
10.3 **Inert Atmosphere:** No requirement
10.4 **Venting:** Pressure-vacuum

11. HAZARD ASSESSMENT CODE
(See Hazard Assessment Handbook, CG 446-3)

A-O

12. HAZARD CLASSIFICATIONS

12.1 **Code of Federal Regulations:** Poisonous, Class B
12.2 **NAS Hazard Rating for Bulk Water Transportation:** Not listed
12.3 **NFPA Hazard Classifications:** Not listed

13. PHYSICAL AND CHEMICAL PROPERTIES

13.1 **Physical State at 15°C and 1 atm:** Liquid
13.2 **Molecular Weight:** 181.3
13.3 **Boiling Point at 1 atm:** 266.4°F = 130.2°C = 403.4°K
13.4 **Freezing Point:** 9°F = −13°C = 260°K
13.5 **Critical Temperature:** Not pertinent
13.6 **Critical Pressure:** Not pertinent
13.7 **Specific Gravity:** 2.156 at 25°C (liquid)
13.8 **Liquid Surface Tension:** (est.) 20 dynes/cm = 0.020 N/m at 20°C
13.9 **Liquid-Water Interfacial Tension:** Not pertinent
13.10 **Vapor (Gas) Specific Gravity:** Not pertinent
13.11 **Ratio of Specific Heats of Vapor (Gas):** Not pertinent
13.12 **Latent Heat of Vaporization:** 88.31 Btu/lb = 49.06 cal/g = 2.054×10^5 J/kg
13.13 **Heat of Combustion:** Not pertinent
13.14 **Heat of Decomposition:** Not pertinent
13.15 **Heat of Solution:** (est.) −18 Btu/lb = −10 cal/g = -0.42×10^5 J/kg
13.16 **Heat of Polymerization:** Not pertinent

NOTES

ATO — ARSENIC TRIOXIDE

Common Synonyms			
Arsenous acid Arsenous acid anhydride Arsenous oxide Arsenic sesquioxide White arsenic	Solid crystals or powder	White	Odorless
	Sinks and mixes slowly with water.		

AVOID CONTACT WITH SOLID AND DUST. KEEP PEOPLE AWAY.
Stay upwind. Use water spray to "knock down" dust.
Isolate and remove discharged material.
Notify local health and pollution control agencies.

Fire

Not flammable.
POISONOUS GASES MAY BE PRODUCED WHEN HEATED.

Exposure

CALL FOR MEDICAL AID.

DUST
POISONOUS IF INHALED.
Irritating to eyes, nose and throat.
Move victim to fresh air.
If in eyes, hold eyelids open and flush with plenty of water.
If breathing is difficult, give oxygen.

SOLID
POISONOUS IF SWALLOWED.
Irritating to skin and eyes.
Remove contaminated clothing and shoes.
Flush affected areas with plenty of water.
IF IN EYES, hold eyelids open and flush with plenty of water.
IF SWALLOWED and victim is CONSCIOUS, have victim drink water or milk and have victim induce vomiting.
IF SWALLOWED and victim is UNCONSCIOUS OR HAVING CONVULSIONS, do nothing except keep victim warm.

Water Pollution

HARMFUL TO AQUATIC LIFE IN VERY LOW CONCENTRATIONS.
May be dangerous if it enters water intakes.
Notify local health and wildlife officials.
Notify operators of nearby water intakes.

1. RESPONSE TO DISCHARGE

(See Response Methods Handbook, CG 446-4)

Issue warning—poison, water contaminant
Restrict access
Should be removed
Chemical and physical treatment

2. LABEL

3. CHEMICAL DESIGNATIONS

3.1 **Synonyms:** Arsenous acid; Arsenous acid anhydride; Arsenous oxide; Arsenic sesquioxide; White arsenic

3.2 **Coast Guard Compatibility Classification:** Not applicable

3.3 **Chemical Formula:** As_2O_3

3.4 **IMCO/United Nations Numerical Designation:** 6.1/1561

4. OBSERVABLE CHARACTERISTICS

4.1 **Physical State (as shipped):** Solid

4.2 **Color:** White

4.3 **Odor:** Like garlic; none

5. HEALTH HAZARDS

5.1 **Personal Protective Equipment:** Bureau of Mines approved respirator; protective gloves, eye protection; full protective coveralls.

5.2 **Symptoms Following Exposure:** Ingestion causes irritation of mucous membrane, weakness, loss of appetite, gastrointestinal disturbances. Overdose can cause arsenic poisoning, but symptoms are delayed.

5.3 **Treatment for Exposure:** Get medical attention after all exposures to this compound. Be alert for arsenic poisoning symptoms. SKIN: wash thoroughly with soap and water; remove contaminated clothing and shower with soap and water; irritations, except for milder cases which disappear in a day or two, should have medical attention. INGESTION: vomiting should be induced and a physician should be called at once; drink freely of lime water, sweet milk, or raw eggs, followed by castor oil or any brisk cathartic.

5.4 **Toxicity by Inhalation (Threshold Limit Value):** 0.5 mg/m^3 as arsenic

5.5 **Short-Term Inhalation Limits:** Data not available

5.6 **Toxicity by Ingestion:** Grade 4; oral mouse LD_{50} = 45 mg/kg

5.7 **Late Toxicity:** Arsenic compounds may be carcinogenic.

5.8 **Vapor (Gas) Irritant Characteristics:** Data not available

5.9 **Liquid or Solid Irritant Characteristics:** Data not available

5.10 **Odor Threshold:** Odorless

6. FIRE HAZARDS

6.1 **Flash Point:** Not flammable

6.2 **Flammable Limits in Air:** Not flammable

6.3 **Fire Extinguishing Agents:** Not pertinent

6.4 **Fire Extinguishing Agents Not to be Used:** Not pertinent

6.5 **Special Hazards of Combustion Products:** Toxic fumes of arsenic trioxide and arsine may be formed in fire situations.

6.6 **Behavior in Fire:** May volatilize and form toxic fumes of arsenic trioxide.

6.7 **Ignition Temperature:** Not pertinent

6.8 **Electrical Hazard:** Not pertinent

6.9 **Burning Rate:** Not pertinent

7. CHEMICAL REACTIVITY

7.1 **Reactivity with Water:** No reaction

7.2 **Reactivity with Common Materials:** No reaction

7.3 **Stability During Transport:** Stable

7.4 **Neutralizing Agents for Acids and Caustics:** Flush with water

7.5 **Polymerization:** Not pertinent

7.6 **Inhibitor of Polymerization:** Not pertinent

8. WATER POLLUTION

8.1 **Aquatic Toxicity:**
5.3 mg/l /8 days/salmon/harmful/*
*Type of water not specified

8.2 **Waterfowl Toxicity:** Data not available

8.3 **Biological Oxygen Demand (BOD):** Data not available

8.4 **Food Chain Concentration Potential:** None

9. SELECTED MANUFACTURERS

1. American Smelting and Refining Company
120 Broadway
New York, N. Y. 10005

2. Gallard-Schlesinger Chemical Manufacturing Co.
584 Mineola Avenue
Carle Place, N. Y. 11514

3. Ventron Corp.
Alfa Products
P. O. Box 159
Beverly, Mass. 01915

10. SHIPPING INFORMATION

10.1 **Grades or Purity:** Refined: 99%; Crude: 95%

10.2 **Storage Temperature:** Ambient

10.3 **Inert Atmosphere:** No requirement

10.4 **Venting:** Pressure-vacuum

11. HAZARD ASSESSMENT CODE

(See Hazard Assessment Handbook, CG 446-3)

II-SS

12. HAZARD CLASSIFICATIONS

12.1 **Code of Federal Regulations:** Poisonous, Class B

12.2 **NAS Hazard Rating for Bulk Water Transportation:** Not listed

12.3 **NFPA Hazard Classifications:** Not listed

13. PHYSICAL AND CHEMICAL PROPERTIES

13.1 **Physical State at 15°C and 1 atm:** Solid

13.2 **Molecular Weight:** 197.8

13.3 **Boiling Point at 1 atm:** 855°F = 457°C = 730°K

13.4 **Freezing Point:** 599°F = 315°C = 588°K

13.5 **Critical Temperature:** Not pertinent

13.6 **Critical Pressure:** Not pertinent

13.7 **Specific Gravity:** 3.7 at 20°C (solid)

13.8 **Liquid Surface Tension:** Not pertinent

13.9 **Liquid-Water Interfacial Tension:** Not pertinent

13.10 **Vapor (Gas) Specific Gravity:** Not pertinent

13.11 **Ratio of Specific Heats of Vapor (Gas):** Not pertinent

13.12 **Latent Heat of Vaporization:** Not pertinent

13.13 **Heat of Combustion:** Not pertinent

13.14 **Heat of Decomposition:** Not pertinent

13.15 **Heat of Solution:** Data not available

13.16 **Heat of Polymerization:** Not pertinent

NOTES

ART	ARSENIC TRISULFIDE

Common Synonyms			
Arsenic yellow King's gold King's yellow Orpiment Yellow arsenic sulfide	Solid Sinks in water.	Yellow-orange	Odorless

AVOID CONTACT WITH SOLID AND DUST. Keep people away.
Wear chemical protective suit with self-contained breathing apparatus.
Isolate and remove discharged material.
Notify local health and pollution control agencies.

Fire

Not flammable.
Wear chemical protective suit with self-contained breathing apparatus.

Exposure

CALL FOR MEDICAL AID.

DUST
POISONOUS IF INHALED.
Harmful to skin.
Move to fresh air.
If breathing has stopped, give artificial respiration.
If breathing is difficult, give oxygen.

SOLID
POISONOUS IF SWALLOWED.
Will burn eyes and skin.
Remove contaminated clothing and shoes.
Flush affected areas with plenty of water.
IF IN EYES, hold eyelids open and flush with plenty of water.
IF SWALLOWED and victim is CONSCIOUS, have victim drink water or milk and have victim induce vomiting.
IF SWALLOWED and victim is UNCONSCIOUS OR HAVING CONVULSIONS, do nothing except keep victim warm.

Water Pollution

HARMFUL TO AQUATIC LIFE IN VERY LOW CONCENTRATIONS.
May be dangerous if it enters water intakes.
Notify local health and wildlife officials.
Notify operators of nearby water intakes.

1. RESPONSE TO DISCHARGE

(See Response Methods Handbook, CG 446-4)

Issue warning—poison, water contaminant
Restrict access
Should be removed
Chemical and physical treatment

2. LABEL

3. CHEMICAL DESIGNATIONS

3.1 **Synonyms:** Arsenic yellow; King's gold; King's yellow; Orpiment; Yellow arsenic sulfide
3.2 **Coast Guard Compatibility Classification:** Not listed
3.3 **Chemical Formula:** As_2S_3
3.4 **IMCO/United Nations Numerical Designation:** 6.1/1557

4. OBSERVABLE CHARACTERISTICS

4.1 **Physical State (as shipped):** Solid
4.2 **Color:** Yellow-orange
4.3 **Odor:** None

5. HEALTH HAZARDS

5.1 **Personal Protective Equipment:** Self-contained breathing apparatus; goggles; rubber gloves; clean protective clothing
5.2 **Symptoms Following Exposure:** (Acute and sub-acute poisoning are not common.) Repeated inhalation causes irritation of nose, laryngitis, mild bronchitis. Ingestion causes weakness, loss of appetite, gastrointestinal disturbances, peripheral neuritis, occasional hepatitis. Contact with eyes causes irritation. Irritates skin, especially where moist; if not treated, may cause ulceration
5.3 **Treatment for Exposure:** *Consult physician after all overexposures to this compound.* INHALATION: move to fresh air. INGESTION: induce vomiting by giving warm salt water; repeat until vomit is clear. EYES: flush with water for at least 15 min. SKIN: wash well with water.
5.4 **Toxicity by Inhalation (Threshold Limit Value):** 0.5 mg/m^3 (as arsenic)
5.5 **Short-Term Inhalation Limits:** Data not available
5.6 **Toxicity by Ingestion:** Grade 4; LD_{50} <50 mg/kg
5.7 **Late Toxicity:** Possible skin and lung cancer
5.8 **Vapor (Gas) Irritant Characteristics:** Data not available
5.9 **Liquid or Solid Irritant Characteristics:** Data not available
5.10 **Odor Threshold:** Odorless

6. FIRE HAZARDS

6.1 **Flash Point:** Not pertinent
6.2 **Flammable Limits in Air:** Not pertinent
6.3 **Fire Extinguishing Agents:** Water
6.4 **Fire Extinguishing Agents Not to be Used:** Not pertinent
6.5 **Special Hazards of Combustion Products:** Poisonous fumes of the compound may be formed in fires.
6.6 **Behavior in Fire:** May ignite at very high temperatures.
6.7 **Ignition Temperature:** Not pertinent
6.8 **Electrical Hazard:** Not pertinent
6.9 **Burning Rate:** Not pertinent

7. CHEMICAL REACTIVITY

7.1 **Reactivity with Water:** No reaction
7.2 **Reactivity with Common Materials:**
7.3 **Stability During Transport:** Stable
7.4 **Neutralizing Agents for Acids and Caustics:** Not pertinent
7.5 **Polymerization:** Not pertinent
7.6 **Inhibitor of Polymerization:** Not pertinent

8. WATER POLLUTION

8.1 **Aquatic Toxicity:** Data not available
8.2 **Waterfowl Toxicity:** Data not available
8.3 **Biological Oxygen Demand (BOD):** Data not available
8.4 **Food Chain Concentration Potential:** Data not available

9. SELECTED MANUFACTURERS

1. Gallard-Schlesinger Chemical Mfg. Co.
   584 Mineola Ave.
   Carle Place, N. Y. 11514
2. Cerac, Inc.
   13460 W. Silver Spring Rd.
   Menonomee Falls, Wis. 53051
3. Var-lac-oid Chemical Co.
   666 S. Front Street
   Elizabeth, N. J. 07202

10. SHIPPING INFORMATION

10.1 **Grades or Purity:** Technical; Pure, 99+%; Optical grade, 99.99+%
10.2 **Storage Temperature:** Ambient
10.3 **Inert Atmosphere:** No requirement
10.4 **Venting:** Open

11. HAZARD ASSESSMENT CODE

(See Hazard Assessment Handbook, CG 446-3)

II

12. HAZARD CLASSIFICATIONS

12.1 **Code of Federal Regulations:** Poisonous solid
12.2 **NAS Hazard Rating for Bulk Water Transportation:** Not listed
12.3 **NFPA Hazard Classifications:** Not listed

13. PHYSICAL AND CHEMICAL PROPERTIES

13.1 **Physical State at 15°C and 1 atm:** Solid
13.2 **Molecular Weight:** 246
13.3 **Boiling Point at 1 atm:** Not pertinent
13.4 **Freezing Point:** 572°F = 300°C = 573°K
13.5 **Critical Temperature:** Not pertinent
13.6 **Critical Pressure:** Not pertinent
13.7 **Specific Gravity:** 3.43 at 20°C (solid)
13.8 **Liquid Surface Tension:** Not pertinent
13.9 **Liquid-Water Interfacial Tension:** Not pertinent
13.10 **Vapor (Gas) Specific Gravity:** Not pertinent
13.11 **Ratio of Specific Heats of Vapor (Gas):** Not pertinent
13.12 **Latent Heat of Vaporization:** Not pertinent
13.13 **Heat of Combustion:** Not pertinent
13.14 **Heat of Decomposition:** Not pertinent
13.15 **Heat of Solution:** Not pertinent
13.16 **Heat of Polymerization:** Not pertinent

NOTES

ASP	# ASPHALT

Common Synonyms
Asphaltic bitumen
Bitumen
Petroleum asphalt

Thick liquid (generally heated) Dark Brown to black Tar odor

May float or sink in water. Rubbery solid is produced when cooled.

Stop discharge if possible.
Avoid contact with liquid.
Call fire department.
Isolate and remove discharged material.
Notify local health and pollution control agencies.

Fire

Combustible.

Extinguish with water, dry chemical, foam, or carbon dioxide.
Cool exposed containers with water.

Exposure

LIQUID
Will burn skin and eyes.

Flush affected areas with plenty of water.

Water Pollution

Effect of low concentrations on aquatic life is unknown.
FOULING TO SHORELINE.
May be dangerous if it enters water intakes.

Notify local health and pollution control officials.
Notify operators of nearby water intakes.

1. RESPONSE TO DISCHARGE

(See Response Methods Handbook, CG 446-4)

Mechanical containment
Chemical and physical treatment

2. LABELS

No hazard label required by Code of Federal Regulations

3. CHEMICAL DESIGNATIONS

3.1 **Synonyms:**
Asphalt cements
Asphaltic bitumen
Bitumen
Petroleum asphalt

3.2 **Coast Guard Compatibility Classification:** Miscellaneous hydrocarbon mixtures

3.3 **Chemical Formula:** Not pertinent

3.4 **IMCO/United Nations Numerical Designation:** 3.2/1999; 3.3/1999

4. OBSERVABLE CHARACTERISTICS

4.1 **Physical State (as shipped):** Liquid

4.2 **Color:** Dark brown to black

4.3 **Odor:** Tarry

5. HEALTH HAZARDS

5.1 **Personal Protective Equipment:** Protective clothing; face and eye protection when handling hot material.

5.2 **Symptoms Following Exposure:** Contact with skin may cause dermatitis. Inhalation of vapors may cause moderate irritation of nose and throat. Hot liquid burns skin.

5.3 **Treatment for Exposure:** Severe burns may result from contact with hot asphalt. If molten asphalt strikes the exposed skin, cool the skin immediately by quenching with cold water. A burn should be covered with a sterile dressing, and the patient should be taken immediately to a hospital.

5.4 **Toxicity by Inhalation (Threshold Limit Value):** 5 mg/m^3

5.5 **Short-Term Inhalation Limits:** Data not available

5.6 **Toxicity by Ingestion:** Grade 1; LD_{50} 5 to 15 g/kg

5.7 **Late Toxicity:** None observed

5.8 **Vapor (Gas) Irritant Characteristics:** Vapors cause a slight smarting of the eyes or respiratory system if present in high concentrations. The effect is temporary.

5.9 **Liquid or Solid Irritant Characteristics:** Causes smarting of the skin and first-degree burns on short exposure; may cause secondary burns on long exposure.

5.10 **Odor Threshold:** Data not available

6. FIRE HAZARDS

6.1 **Flash Point:** 300°F—550°F O.C.

6.2 **Flammable Limits in Air:** Not pertinent

6.3 **Fire Extinguishing Agents:** Water spray, dry chemical, foam or carbon dioxide

6.4 **Fire Extinguishing Agents Not to be Used:** Water or foam may cause frothing

6.5 **Special Hazards of Combustion Products:** Not pertinent

6.6 **Behavior in Fire:** Not pertinent

6.7 **Ignition Temperature:** 400°F—700°F

6.8 **Electrical Hazard:** Not pertinent

6.9 **Burning Rate:** Data not available

7. CHEMICAL REACTIVITY

7.1 **Reactivity with Water:** No reaction

7.2 **Reactivity with Common Materials:** No reaction

7.3 **Stability During Transport:** Stable

7.4 **Neutralizing Agents for Acids and Caustics:** Not pertinent

7.5 **Polymerization:** Not pertinent

7.6 **Inhibitor of Polymerization:** Not pertinent

8. WATER POLLUTION

8.1 **Aquatic Toxicity:** Data not available

8.2 **Waterfowl Toxicity:** Data not available

8.3 **Biological Oxygen Demand (BOD):** Data not available

8.4 **Food Chain Concentration Potential:** Data not available

9. SELECTED MANUFACTURERS

1. Johns-Manville Co.
Greenwood Plaza
Denver, Colo. 80217

2. Sun Oil Co.
1608 Walnut St.
Philadelphia, Pa. 19103

3. Witco Chemical Corp.
Pioneer Division
277 Park Ave.
New York, N. Y. 10017

10. SHIPPING INFORMATION

10.1 **Grades or Purity:** Each of the following is available in several grades: asphalt cement, rapid-curing liquid asphalt, medium-curing liquid asphalt, slow-curing liquid asphalt (road oil), emulsified asphalt, inverted asphaltic emulsion, oxidized (air-blown) asphalt.

10.2 **Storage Temperature:** Elevated

10.3 **Inert Atmosphere:** No requirement

10.4 **Venting:** Open (flame arrester)

11. HAZARD ASSESSMENT CODE

(See Hazard Assessment Handbook, CG 446-3)

A-T-U-X-Y

12. HAZARD CLASSIFICATIONS

12.1 **Code of Federal Regulations:** Combustible liquid

12.2 **NAS Hazard Rating for Bulk Water Transportation:**

Category	Rating
Fire	1
Health	
Vapor Irritant	1
Liquid or Solid Irritant	2
Poisons	1
Water Pollution	
Human Toxicity	0
Aquatic Toxicity	1
Aesthetic Effect	4
Reactivity	
Other Chemicals	0
Water	0
Self-Reaction	0

12.3 **NFPA Hazard Classifications:**

Category	Classification
Health Hazard (Blue)	0
Flammability (Red)	1
Reactivity (Yellow)	0

13. PHYSICAL AND CHEMICAL PROPERTIES

13.1 **Physical State at 15°C and 1 atm:** Liquid

13.2 **Molecular Weight:** Not pertinent

13.3 **Boiling Point at 1 atm:** Not pertinent

13.4 **Freezing Point:** Not pertinent

13.5 **Critical Temperature:** Not pertinent

13.6 **Critical Pressure:** Not pertinent

13.7 **Specific Gravity:** (est.) 1.00 at 20°C (liquid)

13.8 **Liquid Surface Tension:** Data not available

13.9 **Liquid-Water Interfacial Tension:** 70 dynes/cm = 0.07 N/m at 77 °C

13.10 **Vapor (Gas) Specific Gravity:** Not pertinent

13.11 **Ratio of Specific Heats of Vapor (Gas):** Not pertinent

13.12 **Latent Heat of Vaporization:** Not pertinent

13.13 **Heat of Combustion:** Data not available

13.14 **Heat of Decomposition:** Not pertinent

13.15 **Heat of Solution:** Not pertinent

13.16 **Heat of Polymerization:** Not pertinent

NOTES

REVISED 1978

ARF	ASPHALT BLENDING STOCKS: ROOFERS FLUX

Common Synonyms	
Liquid asphalt Asphaltum Fluxing oil Road oil Petroleum tailings Dust-laying oil	Oily liquid (generally heated) Dark brown to black Tar odor May float or sink in water. Rubbery solid is produced when cooled.

Stop discharge if possible.
Call fire department.
Avoid contact with liquid.
Isolate and remove discharged material.
Notify local health and pollution control agencies.

Fire	Combustible. Extinguish with water, dry chemical, foam, or carbon dioxide. Cool exposed containers with water.
Exposure	CALL FOR MEDICAL AID. LIQUID Will burn skin and eyes. Harmful if swallowed. Flush affected areas with plenty of water. IF IN EYES, hold eyelids open and flush with plenty of water. IF SWALLOWED and victim is CONSCIOUS, have victim drink water or milk. DO NOT INDUCE VOMITING.
Water Pollution	Effect of low concentrations on aquatic life is unknown. FOULING TO SHORELINE. May be dangerous if it enters water intakes. Notify local health and pollution control officials. Notify operators of nearby water intakes.

1. RESPONSE TO DISCHARGE

(See Response Methods Handbook, CG 446-4)

Mechanical containment
Chemical and physical treatment

2. LABELS

No hazard label required by Code of Federal Regulations

3. CHEMICAL DESIGNATIONS

3.1 **Synonyms:**
Asphaltum oil
Flux oil
Petroleum tailings
Residual oil
Road oil

3.2 **Coast Guard Compatibility Classification:** Miscellaneous hydrocarbon mixtures

3.3 **Chemical Formula:** Not pertinent

3.4 **IMCO/United Nations Numerical Designation:** 3.2/1999; 3.3/1999

4. OBSERVABLE CHARACTERISTICS

4.1 **Physical State (as shipped):** Liquid

4.2 **Color:** Brown to black

4.3 **Odor:** Tarry

5. HEALTH HAZARDS

5.1 **Personal Protective Equipment:** Protective clothing; face and eye protection

5.2 **Symptoms Following Exposure:** Inhalation of vapors from semi-solid materials causes moderate irritation of nasal and upper respiratory tract passages. Aspiration causes slow onset and low degree of chemical pneumonitis with clinical symptoms of lower respiratory tract irritation. Ingestion produces irritation of gastrointestinal tract.

5.3 **Treatment for Exposure:** INHALATION OR ASPIRATION: treatment usually unnecessary. INGESTION: do NOT induce vomiting; do NOT lavage; administer 2-4 oz of olive oil and 1-2 oz of activated charcoal. EYES: wash with plenty of water. SKIN: wipe off material and wash with soap and water.

5.4 **Toxicity by Inhalation (Threshold Limit Value):** Data not available

5.5 **Short-Term Inhalation Limits:** Data not available

5.6 **Toxicity by Ingestion:** Grade 2; LD_{50} 0.5 to 5 g/kg

5.7 **Late Toxicity:** None observed

5.8 **Vapor (Gas) Irritant Characteristics:** Vapors cause irritation of eyes or respiratory system if present in high concentrations. The effect is temporary.

5.9 **Liquid or Solid Irritant Characteristics:** Causes smarting of the skin and first-degree burns on short exposure; may cause secondary burns on long exposure.

5.10 **Odor Threshold:** Data not available

6. FIRE HAZARDS

6.1 **Flash Point:** 300-550°F C.C.

6.2 **Flammable Limits in Air:** Not pertinent

6.3 **Fire Extinguishing Agents:** Water, foam, dry chemical or carbon dioxide

6.4 **Fire Extinguishing Agents Not to be Used:** Water or foam may cause frothing

6.5 **Special Hazards of Combustion Products:** Not pertinent

6.6 **Behavior in Fire:** Not pertinent

6.7 **Ignition Temperature:** 400-700°F

6.8 **Electrical Hazard:** Not pertinent

6.9 **Burning Rate:** Data not available

7. CHEMICAL REACTIVITY

7.1 **Reactivity with Water:** No reaction

7.2 **Reactivity with Common Materials:** No reaction

7.3 **Stability During Transport:** Stable

7.4 **Neutralizing Agents for Acids and Caustics:** Not pertinent

7.5 **Polymerization:** Not pertinent

7.6 **Inhibitor of Polymerization:** Not pertinent

8. WATER POLLUTION

8.1 **Aquatic Toxicity:** Data not available

8.2 **Waterfowl Toxicity:** Data not availalbe

8.3 **Biological Oxygen Demand (BOD):** Data not available

8.4 **Food Chain Concentration Potential:** Data not available

9. SELECTED MANUFACTURERS

1. Johns-Manville Co.
   Greenwood Plaza
   Denver, Colo. 80217
2. Sun Oil Co.
   1608 Walnut St.
   Philadelphia, Pa. 19103
3. Witco Chemical Corp.
   Pioneer Division
   277 Park Ave.
   New York, N. Y. 10017

10. SHIPPING INFORMATION

10.1 **Grades or Purity:** Data not available

10.2 **Storage Temperature:** Elevated

10.3 **Inert Atmosphere:** No requirement

10.4 **Venting:** Open (flame arrester)

11. HAZARD ASSESSMENT CODE

(See Hazard Assessment Handbook, CG 446-3)

A-T-U

12. HAZARD CLASSIFICATIONS

12.1 **Code of Federal Regulations:** Not Listed

12.2 **NAS Hazard Rating for Bulk Water Transportation:**

Category	Rating
Fire	1
Health	
Vapor Irritant	1
Liquid or Solid Irritant	2
Poisons	1
Water Pollution	
Human Toxicity	0
Aquatic Toxicity	1
Aesthetic Effect	4
Reactivity	
Other Chemicals	0
Water	0
Self-Reaction	0

12.3 **NFPA Hazard Classifications:**

Category	Classification
Health Hazard (Blue)	0
Flammability (Red)	1
Reactivity (Yellow)	0

13. PHYSICAL AND CHEMICAL PROPERTIES

13.1 **Physical State at 15°C and 1 atm:** Liquid

13.2 **Molecular Weight:** Not pertinent

13.3 **Boiling Point at 1 atm:** Not pertinent

13.4 **Freezing Point:** 20 to 110°F = −7 to 43°C = 266 to 316°K

13.5 **Critical Temperature:** Not pertinent

13.6 **Critical Pressure:** Not pertinent

13.7 **Specific Gravity:** (est.) 1.11 at 50°C (liquid)

13.8 **Liquid Surface Tension:** Data not available

13.9 **Liquid-Water Interfacial Tension:** Data not available

13.10 **Vapor (Gas) Specific Gravity:** Not pertinent

13.11 **Ratio of Specific Heats of Vapor (Gas):** Not pertinent

13.12 **Latent Heat of Vaporization:** Not pertinent

13.13 **Heat of Combustion:** Data not available

13.14 **Heat of Decomposition:** Not pertinent

13.15 **Heat of Solution:** Not pertinent

13.16 **Heat of Polymerization:** Not pertinent

NOTES

REVISED 1978

ASR — ASPHALT BLENDING STOCKS: STRAIGHT RUN RESIDUE

Common Synonyms			
Petroleum residue Road binder Seal-coating material Residual asphalt Petroleum asphalt Petroleum pitch	Oily liquid (generally heated)	Black	Tar odor
	May float or sink in water. Rubbery solid is produced when cooled.		

Avoid contact with liquid.
Stop discharge if possible.
Call fire department.
Isolate and remove discharged material.
Notify local health and pollution control agencies.

Fire

Combustible.

Extinguish with water, dry chemical, foam or carbon dioxide.
Cool exposed containers with water.

Exposure

CALL FOR MEDICAL AID.
LIQUID
Will burn skin and eyes.

Flush affected areas with plenty of water.
IF IN EYES, hold eyelids open and flush with plenty of water.
IF SWALLOWED and victim is CONSCIOUS, have victim drink water or milk.
DO NOT INDUCE VOMITING.

Water Pollution

Effect of low concentrations on aquatic life is unknown.
FOULING TO SHORELINE.
May be dangerous if it enters water intakes.

Notify local health and pollution control officials.
Notify operators of nearby water intakes.

1. RESPONSE TO DISCHARGE

(See Response Methods Handbook, CG 446-4)

Mechanical containment
Chemical and physical treatment

2. LABELS

No hazard label required by Code of Federal Regulations

3. CHEMICAL DESIGNATIONS

3.1 **Synonyms:** Carpeting medium; Petroleum asphalt; Petroleum pitch; Petroleum residue; Residual asphalt; Road binder; Seal-coating material
3.2 **Coast Guard Compatibility Classification:** Miscellaneous hydrocarbon mixtures
3.3 **Chemical Formula:** Not pertinent
3.4 **IMCO/United Nations Numerical Designation:** 3.2/1999; 3.3/1999

4. OBSERVABLE CHARACTERISTICS

4.1 **Physical State (as shipped):** Liquid
4.2 **Color:** Black
4.3 **Odor:** Tarry

5. HEALTH HAZARDS

5.1 **Personal Protective Equipment:** Protective clothing; eye and face protection
5.2 **Symptoms Following Exposure:** Inhalation may cause moderate irritation of nose and throat. Skin contact may cause dermatitis.
5.3 **Treatment for Exposure:** Severe burns may result from contact with hot asphalt. If molten asphalt strikes the exposed skin, cool the skin immediately by quenching with cold water. A burn should be covered with a sterile dressing, and the patient should be taken immediately to a hospital.
5.4 **Toxicity by Inhalation (Threshold Limit Value):** Data not available
5.5 **Short-Term Inhalation Limits:** Data not available
5.6 **Toxicity by Ingestion:** Grade 1; LD_{50} 5 to 15 g/kg
5.7 **Late Toxicity:** None observed
5.8 **Vapor (Gas) Irritant Characteristics:** Vapors cause a slight smarting of the eyes or respiratory system if present in high concentrations. The effect is temporary.
5.9 **Liquid or Solid Irritant Characteristics:** Causes smarting of the skin and first-degree burns on short exposure; may cause secondary burns on long exposure.
5.10 **Odor Threshold:** Data not available

6. FIRE HAZARDS

6.1 **Flash Point:** 400 – 600°F O.C.
6.2 **Flammable Limits in Air:** Not pertinent
6.3 **Fire Extinguishing Agents:** Water, foam, carbon dioxide or dry chemical
6.4 **Fire Extinguishing Agents Not to be Used:** Water or foam may cause frothing.
6.5 **Special Hazards of Combustion Products:** Not pertinent
6.6 **Behavior in Fire:** Not pertinent
6.7 **Ignition Temperature:** 450 – 700°F
6.8 **Electrical Hazard:** Not pertinent
6.9 **Burning Rate:** Data not available

7. CHEMICAL REACTIVITY

7.1 **Reactivity with Water:** No reaction
7.2 **Reactivity with Common Materials:** No reaction
7.3 **Stability During Transport:** Stable
7.4 **Neutralizing Agents for Acids and Caustics:** Not pertinent
7.5 **Polymerization:** Not pertinent
7.6 **Inhibitor of Polymerization:** Not pertinent

8. WATER POLLUTION

8.1 **Aquatic Toxicity:** Data not available
8.2 **Waterfowl Toxicity:** Data not available
8.3 **Biological Oxygen Demand (BOD):** Data not available
8.4 **Food Chain Concentration Potential:** Data not available

9. SELECTED MANUFACTURERS

1. Johns-Manville Co.
   Greenwood Plaza
   Denver, Colo. 80217
2. Sun Oil Co.
   1608 Walnut St.
   Philadelphia, Pa. 19103
3. Witco Chemical Corp.
   Pioneer Division
   277 Park Ave.
   New York, N. Y. 10017

10. SHIPPING INFORMATION

10.1 **Grades or Purity:** Data not available
10.2 **Storage Temperature:** Elevated
10.3 **Inert Atmosphere:** No requirement
10.4 **Venting:** Open (flame arrester)

11. HAZARD ASSESSMENT CODE

(See Hazard Assessment Handbook, CG 446-3)

A-T-U

12. HAZARD CLASSIFICATIONS

12.1 **Code of Federal Regulations:** Not Listed
12.2 **NAS Hazard Rating for Bulk Water Transportation:**

Category	Rating
Fire	1
Health	
Vapor Irritant	1
Liquid or Solid Irritant	2
Poisons	1
Water Pollution	
Human Toxicity	0
Aquatic Toxicity	1
Aesthetic Effect	4
Reactivity	
Other Chemicals	0
Water	0
Self-Reaction	0

12.3 **NFPA Hazard Classifications:**

Category	Classification
Health Hazard (Blue)	0
Flammability (Red)	1
Reactivity (Yellow)	0

13. PHYSICAL AND CHEMICAL PROPERTIES

13.1 **Physical State at 15°C and 1 atm:** Solid
13.2 **Molecular Weight:** Not pertinent
13.3 **Boiling Point at 1 atm:** Not pertinent
13.4 **Freezing Point:** 80 to 225°F = 26 to 107°C = 299 to 380°K
13.5 **Critical Temperature:** Not pertinent
13.6 **Critical Pressure:** Not pertinent
13.7 **Specific Gravity:** Data not available
13.8 **Liquid Surface Tension:** Data not available
13.9 **Liquid-Water Interfacial Tension:** Data not available
13.10 **Vapor (Gas) Specific Gravity:** Not pertinent
13.11 **Ratio of Specific Heats of Vapor (Gas):** Not pertinent
13.12 **Latent Heat of Vaporization:** Not pertinent
13.13 **Heat of Combustion:** Data not available
13.14 **Heat of Decomposition:** Not pertinent
13.15 **Heat of Solution:** Not pertinent
13.16 **Heat of Polymerization:** Not pertinent

NOTES

REVISED 1978

ATZ | ATRAZINE

Common Synonyms		
2-Chloro-4-ethylamino-6-isopropylamino-s-triazine Aatrex Herbicide	Solid crystals Sinks in water.	White

AVOID CONTACT WITH SOLID AND DUST. KEEP PEOPLE AWAY.
Wear goggles and self-contained breathing apparatus.
Stay upwind. Use water spray to "knock down" dust.
Isolate and remove discharged material.
Notify local health and pollution control agencies.

Fire

Not flammable.
POISONOUS GASES MAY BE PRODUCED WHEN HEATED.

Exposure

CALL FOR MEDICAL AID.

DUST
POISONOUS IF INHALED.
Irritating to eyes, nose and throat.
Move victim to fresh air.
If in eyes, hold eyelids open and flush with plenty of water.
If breathing is difficult, give oxygen.

SOLID
POISONOUS IF SWALLOWED.
Irritating to skin and eyes.
Remove contaminated clothing and shoes.
Flush affected areas with plenty of water.
IF IN EYES, hold eyelids open and flush with plenty of water.
IF SWALLOWED and victim is CONSCIOUS, have victim drink water or milk and have victim induce vomiting.
IF SWALLOWED and victim is UNCONSCIOUS OR HAVING CONVULSIONS, do nothing except keep victim warm.

Water Pollution

HARMFUL TO AQUATIC LIFE IN VERY LOW CONCENTRATIONS.
May be dangerous if it enters water intakes.
Notify local health and wildlife officials.
Notify operators of nearby water intakes.

1. RESPONSE TO DISCHARGE

(See Response Methods Handbook, CG 446-4)

Issue warning—water contaminant
Should be removed
Chemical and physical treatment

2. LABELS

No hazard label required by Code of Federal Regulations

3. CHEMICAL DESIGNATIONS

3.1 **Synonyms:** 2-Chloro-4-ethylamino-6-isopropylamino-S-triazine; Aatrex herbicide
3.2 **Coast Guard Compatibility Classification:** To be developed
3.3 **Chemical Formula:** $C_8H_{14}N_5Cl$
3.4 **IMCO/United Nations Numerical Designation:** 6.1/1609

4. OBSERVABLE CHARACTERISTICS

4.1 **Physical State (as shipped):** Solid
4.2 **Color:** White
4.3 **Odor:** Data not available

5. HEALTH HAZARDS

5.1 **Personal Protective Equipment:** Dust mask; goggles; rubber gloves.
5.2 **Symptoms Following Exposure:** Irritates eyes and skin. If ingested, irritates mouth and stomach.
5.3 **Treatment for Exposure:** EYES: flush with copious amounts of water for 15 min. SKIN: wash with large amounts of water. INGESTION: no specific antidote; induce vomiting and give a saline laxative and supportive therapy.
5.4 **Toxicity by Inhalation (Threshold Limit Value):** Data not available
5.5 **Short-Term Inhalation Limits:** Data not available
5.6 **Toxicity by Ingestion:** Grade 2; oral rat LD_{50} = 3080 mg/kg
5.7 **Late Toxicity:** Data not available
5.8 **Vapor (Gas) Irritant Characteristics:** Vapors are nonirritating to eyes and throat.
5.9 **Liquid or Solid Irritant Characteristics:** Causes smarting of the skin and first-degree burns on short exposure and may cause second-degree burns on long exposure.
5.10 **Odor Threshold:** Data not available

6. FIRE HAZARDS

6.1 **Flash Point:** Not flammable
6.2 **Flammable Limits in Air:** Not flammable
6.3 **Fire Extinguishing Agents:** Not pertinent
6.4 **Fire Extinguishing Agents Not to be Used:** Not pertinent
6.5 **Special Hazards of Combustion Products:** Irritating hydrogen chloride and toxic oxides of nitrogen may be formed.
6.6 **Behavior in Fire:** Not pertinent
6.7 **Ignition Temperature:** Not pertinent
6.8 **Electrical Hazard:** Not pertinent
6.9 **Burning Rate:** Not pertinent

7. CHEMICAL REACTIVITY

7.1 **Reactivity with Water:** No reaction
7.2 **Reactivity with Common Materials:** No reaction
7.3 **Stability During Transport:** Stable
7.4 **Neutralizing Agents for Acids and Caustics:** Not pertinent
7.5 **Polymerization:** Not pertinent
7.6 **Inhibitor of Polymerization:** Not pertinent

8. WATER POLLUTION

8.1 **Aquatic Toxicity:** 12.6 mg/l /48 hr/rainbow trout/TL_m/fresh water
8.2 **Waterfowl Toxicity:** > 2000 mg/kg LD_{50}
8.3 **Biological Oxygen Demand (BOD):** Data not available
8.4 **Food Chain Concentration Potential:** None

9. SELECTED MANUFACTURERS

Ciba-Geigy Corporation
Agricultural Division
P. O. Box 11422
Greensboro, N. C. 27409

10. SHIPPING INFORMATION

10.1 **Grades or Purity:** Various grades, 70-80%. Mixtures with sodium chlorate and sodium metaborate.
10.2 **Storage Temperature:** Ambient
10.3 **Inert Atmosphere:** No requirement
10.4 **Venting:** Open

11. HAZARD ASSESSMENT CODE

(See Hazard Assessment Handbook, CG 446-3)

II

12. HAZARD CLASSIFICATIONS

12.1 **Code of Federal Regulations:** Not Listed
12.2 **NAS Hazard Rating for Bulk Water Transportation:**

Category	Rating
Fire	0
Health	
Vapor Irritant	0
Liquid or Solid Irritant	2
Poisons	0
Water Pollution	
Human Toxicity	2
Aquatic Toxicity	3
Aesthetic Effect	1
Reactivity	
Other Chemicals	1
Water	0
Self-Reaction	0

12.3 **NFPA Hazard Classifications:** Not listed

13. PHYSICAL AND CHEMICAL PROPERTIES

13.1 **Physical State at 15°C and 1 atm:** Solid
13.2 **Molecular Weight:** 215.7
13.3 **Boiling Point at 1 atm:** Decomposes
13.4 **Freezing Point:** 347°F = 175°C = 348°K
13.5 **Critical Temperature:** Not pertinent
13.6 **Critical Pressure:** Not pertinent
13.7 **Specific Gravity:** (est.) 1.2 at 20°C (solid)
13.8 **Liquid Surface Tension:** Not pertinent
13.9 **Liquid-Water Interfacial Tension:** Not pertinent
13.10 **Vapor (Gas) Specific Gravity:** Not pertinent
13.11 **Ratio of Specific Heats of Vapor (Gas):** Not pertinent
13.12 **Latent Heat of Vaporization:** Not pertinent
13.13 **Heat of Combustion:** (est.) −9,500 Btu/lb = −5,300 cal/g = −220 × 10^5 J/kg
13.14 **Heat of Decomposition:** Not pertinent
13.15 **Heat of Solution:** Not pertinent
13.16 **Heat of Polymerization:** Not pertinent

NOTES

AZM — AZINPHOSMETHYL

Common Synonyms		
0, 0-Dimethyl S-[(4-oxo-1,2, 3-benzotriazine-3-(4H)-yl) methyl] phosphorodithioate Guthion insecticide Gusathion insecticide	Solid Sinks in water.	Brown

AVOID CONTACT WITH SOLID AND DUST. KEEP PEOPLE AWAY.
Wear goggles and self-contained breathing apparatus.
Stay upwind. Use water spray to "knock down" dust.
Isolate and remove discharged material.
Notify local health and pollution control agencies.

Fire

Not flammable.
POISONOUS GASES MAY BE PRODUCED WHEN HEATED.

Exposure

CALL FOR MEDICAL AID.

DUST
POISONOUS IF INHALED.
Move victim to fresh air.
If in eyes, hold eyelids open and flush with plenty of water.
If breathing is difficult, give oxygen.

SOLID
POISONOUS IF SWALLOWED.
IF SWALLOWED and victim is CONSCIOUS, have victim drink water or milk and have victim induce vomiting.
IF SWALLOWED and victim is UNCONSCIOUS OR HAVING CONVULSIONS, do nothing except keep victim warm.

Water Pollution

HARMFUL TO AQUATIC LIFE IN VERY LOW CONCENTRATIONS.
May be dangerous if it enters water intakes.
Notify local health and wildlife officials.
Notify operators of nearby water intakes.

1. RESPONSE TO DISCHARGE

(See Response Methods Handbook, CG 446-4)

Issue warning—water contaminant
Should be removed
Chemical and physical treatment

2. LABEL

3. CHEMICAL DESIGNATIONS

3.1 **Synonyms:** O, O-Dimethyl S-[(4-oxo-1,2,3-benzotriazine-3(4H)-yl)methyl] phosphorodithioate; Guthion insecticide; Gusathion insecticide

3.2 **Coast Guard Compatibility Classification:** Not applicable

3.3 **Chemical Formula:** $C_{10}H_{12}N_3O_3PS_2$

3.4 **IMCO/United Nations Numerical Designation:** 6.1/1615

4. OBSERVABLE CHARACTERISTICS

4.1 **Physical State (as shipped):** Solid

4.2 **Color:** Brown

4.3 **Odor:** Data not available

5. HEALTH HAZARDS

5.1 **Personal Protective Equipment:** Dust mask; protective goggles; rubber gloves.

5.2 **Symptoms Following Exposure:** Dust irritates eyes. Inhalation or ingestion causes sweating, constriction of pupils of eyes, asthmatic symptoms, cramps, weakness, convulsions, collapse.

5.3 **Treatment for Exposure:** INHALATION: remove to fresh air; keep warm; call doctor. EYES: flush with water for at least 15 min. SKIN: flush with water, wash with soap and water. INGESTION: get medical attention at once; give water slurry of charcoal; do NOT give milk or alcohol.

5.4 **Toxicity by Inhalation (Threshold Limit Value):** 0.2 mg/m^3

5.5 **Short-Term Inhalation Limits:** Data not available

5.6 **Toxicity by Ingestion:** Grade 4; oral rat LD_{50} = 11-18.5 mg/kg

5.7 **Late Toxicity:** Data not available

5.8 **Vapor (Gas) Irritant Characteristics:** Data not available

5.9 **Liquid or Solid Irritant Characteristics:** Data not available

5.10 **Odor Threshold:** Data not available

6. FIRE HAZARDS

6.1 **Flash Point:** Not flammable

6.2 **Flammable Limits in Air:** Not flammable

6.3 **Fire Extinguishing Agents:** Not pertinent

6.4 **Fire Extinguishing Agents Not to be Used:** Not pertinent

6.5 **Special Hazards of Combustion Products:** Oxides of sulfur and phosphorus may be formed when in fire situation.

6.6 **Behavior in Fire:** Data not available

6.7 **Ignition Temperature:** Not pertinent

6.8 **Electrical Hazard:** Not pertinent

6.9 **Burning Rate:** Not pertinent

7. CHEMICAL REACTIVITY

7.1 **Reactivity with Water:** No reaction

7.2 **Reactivity with Common Materials:** No reaction

7.3 **Stability During Transport:** Stable

7.4 **Neutralizing Agents for Acids and Caustics:** Not pertinent

7.5 **Polymerization:** Not pertinent

7.6 **Inhibitor of Polymerization:** Not pertinent

8. WATER POLLUTION

8.1 **Aquatic Toxicity:** 0.005 mg/l /96 hr/bluegill/TL_m/fresh water

8.2 **Waterfowl Toxicity:** 136 mg/kg LD_{50}

8.3 **Biological Oxygen Demand (BOD):** Data not available

8.4 **Food Chain Concentration Potential:** None

9. SELECTED MANUFACTURERS

Baychem Corporation
Chemagro Division
Box 4913
Kansas City, Mo. 64120

10. SHIPPING INFORMATION

10.1 **Grades or Purity:** Technical; 50% wettable powder; water emulsions

10.2 **Storage Temperature:** Ambient

10.3 **Inert Atmosphere:** No requirement

10.4 **Venting:** Open

11. HAZARD ASSESSMENT CODE

(See Hazard Assessment Handbook, CG 446-3)

II

12. HAZARD CLASSIFICATIONS

12.1 **Code of Federal Regulations:** Poisonous, Class B

12.2 **NAS Hazard Rating for Bulk Water Transportation:** Not listed

12.3 **NFPA Hazard Classifications:** Not listed

13. PHYSICAL AND CHEMICAL PROPERTIES

13.1 **Physical State at 15°C and 1 atm:** Solid

13.2 **Molecular Weight:** 317

13.3 **Boiling Point at 1 atm:** Decomposes

13.4 **Freezing Point:** 163°F = 73°C = 346°K

13.5 **Critical Temperature:** Not pertinent

13.6 **Critical Pressure:** Not pertinent

13.7 **Specific Gravity:** 1.4 at 20°C (solid)

13.8 **Liquid Surface Tension:** Not pertinent

13.9 **Liquid-Water Interfacial Tension:** Not pertinent

13.10 **Vapor (Gas) Specific Gravity:** Not pertinent

13.11 **Ratio of Specific Heats of Vapor (Gas):** Not pertinent

13.12 **Latent Heat of Vaporization:** Not pertinent

13.13 **Heat of Combustion:** (est.) −8,600 Btu/lb = −4,800 cal/g = −200 × 10^5 J/kg

13.14 **Heat of Decomposition:** Not pertinent

13.15 **Heat of Solution:** Not pertinent

13.16 **Heat of Polymerization:** Not pertinent

NOTES

BRC BARIUM CARBONATE

Common Synonyms

Solid crystals or powder — White — Odorless

Sinks in water.

Isolate and remove discharged material.
Notify local health and pollution control agencies.

Fire

Not flammable.

Exposure

CALL FOR MEDICAL AID.
SOLID
If swallowed, will cause nausea and vomiting.

IF SWALLOWED and victim is CONSCIOUS, have victim drink water or milk.

Water Pollution

HARMFUL TO AQUATIC LIFE IN VERY LOW CONCENTRATIONS.
Fouling to shoreline.
May be dangerous if it enters water intakes.
Notify local health and wildlife officials.
Notify operators of nearby water intakes.

1. RESPONSE TO DISCHARGE

(See Response Methods Handbook, CG 446-4)

Should be removed
Chemical and physical treatment

2. LABELS

No hazard label required by Code of Federal Regulations

3. CHEMICAL DESIGNATIONS

3.1 **Synonyms:** No common synonyms
3.2 **Coast Guard Compatibility Classification:** Not applicable
3.3 **Chemical Formula:** $BaCO_3$
3.4 **IMCO/United Nations Numerical Designation:** 6.1/1564

4. OBSERVABLE CHARACTERISTICS

4.1 **Physical State (as shipped):** Solid
4.2 **Color:** White
4.3 **Odor:** None

5. HEALTH HAZARDS

5.1 **Personal Protective Equipment:** Dust respirator
5.2 **Symptoms Following Exposure:** (INGESTION ONLY): excessive salivation, vomiting, severe abdominal pain, and violent purging with watery and bloody stools; a slow and often irregular pulse and a transient elevation in arterial blood pressure; tinnitus, giddiness and vertigo; muscle twitchings, progressing to convulsions and/or paralysis; dilated pupils with impaired accommodation; confusion and increasing somnolence, without coma; collapse and death from respiratory failure and cardiac arrest.
5.3 **Treatment for Exposure:** Rapid oral administration of a soluble sulfate in water, such as magnesium or sodium sulfate (2 oz), alum (4 gm), or very dilute sulfuric acid (30 ml of a 10% solution diluted to 1 qt). These agents precipitate barium as the insoluble sulfate. Gastric lavage or induced emesis. Seek medical attention.
5.4 **Toxicity by Inhalation (Threshold Limit Value):** Not pertinent
5.5 **Short-Term Inhalation Limits:** Not pertinent
5.6 **Toxicity by Ingestion:** Grade 2; LD_{50} 0.5 to 5 g/kg (rabbit, rat, guinea pig)
5.7 **Late Toxicity:** None observed
5.8 **Vapor (Gas) Irritant Characteristics:** Not pertinent
5.9 **Liquid or Solid Irritant Characteristics:** None
5.10 **Odor Threshold:** Not pertinent

6. FIRE HAZARDS

6.1 **Flash Point:** Not flammable
6.2 **Flammable Limits in Air:** Not flammable
6.3 **Fire Extinguishing Agents:** Not pertinent
6.4 **Fire Extinguishing Agents Not to be Used:** Not pertinent
6.5 **Special Hazards of Combustion Products:** Not pertinent
6.6 **Behavior in Fire:** Not pertinent
6.7 **Ignition Temperature:** Not flammable
6.8 **Electrical Hazard:** Not pertinent
6.9 **Burning Rate:** Not flammable

7. CHEMICAL REACTIVITY

7.1 **Reactivity with Water:** No reaction
7.2 **Reactivity with Common Materials:** No reaction
7.3 **Stability During Transport:** Stable
7.4 **Neutralizing Agents for Acids and Caustics:** Not pertinent
7.5 **Polymerization:** Not pertinent
7.6 **Inhibitor of Polymerization:** Not pertinent

8. WATER POLLUTION

8.1 **Aquatic Toxicity:** Data not available
8.2 **Waterfowl Toxicity:** Data not available
8.3 **Biological Oxygen Demand (BOD):** None
8.4 **Food Chain Concentration Potential:** Data not available

9. SELECTED MANUFACTURERS

1. Chemical Products Corp.
   Cartersville, Georgia 30120
2. FMC Corp.
   Inorganic Chemicals Div.
   633 Third Ave.
   New York, N. Y. 10017
3. Sherwin Williams Co.
   Sherwin Williams Chemicals Div.
   P. O. Box 5638
   Cleveland, Ohio 44101

10. SHIPPING INFORMATION

10.1 **Grades or Purity:** Reagent grade 99.0%; Ceramic grade 99.1%; Ceramic and chemical grade 99.3%; Glass grade 98.8%; Electronic ceramic grade 99.6%
10.2 **Storage Temperature:** Data not available
10.3 **Inert Atmosphere:** Data not available
10.4 **Venting:** Data not available

11. HAZARD ASSESSMENT CODE

(See Hazard Assessment Handbook, CG 446-3)

I I

12. HAZARD CLASSIFICATIONS

12.1 **Code of Federal Regulations:** Not listed
12.2 **NAS Hazard Rating for Bulk Water Transportation:** Not listed
12.3 **NFPA Hazard Classifications:** Not listed

13. PHYSICAL AND CHEMICAL PROPERTIES

13.1 **Physical State at 15°C and 1 atm:** Solid
13.2 **Molecular Weight:** 197.35
13.3 **Boiling Point at 1 atm:** Not pertinent
13.4 **Freezing Point:** Not pertinent
13.5 **Critical Temperature:** Not pertinent
13.6 **Critical Pressure:** Not pertinent
13.7 **Specific Gravity:** 4.3 at 20°C (solid)
13.8 **Liquid Surface Tension:** Not pertinent
13.9 **Liquid-Water Interfacial Tension:** Not pertinent
13.10 **Vapor (Gas) Specific Gravity:** Not pertinent
13.11 **Ratio of Specific Heats of Vapor (Gas):** Not pertinent
13.12 **Latent Heat of Vaporization:** Not pertinent
13.13 **Heat of Combustion:** Not pertinent
13.14 **Heat of Decomposition:** Not pertinent
13.15 **Heat of Solution:** Not pertinent
13.16 **Heat of Polymerization:** Not pertinent

NOTES

BCR	# BARIUM CHLORATE

Common Synonyms	Solid	White	Odorless
	Sinks and mixes with water.		

AVOID CONTACT WITH SOLID AND DUST. KEEP PEOPLE AWAY.
Wear goggles, self-contained breathing apparatus, and rubber overclothing (including gloves).
Call fire department.
Isolate and remove discharged material.
Notify local health and pollution control agencies.

Fire

Not flammable.
May cause fire on contact with combustibles.
POISONOUS GASES MAY BE PRODUCED IN FIRE.
Containers may explode in fire.
Combat fires from safe distance or protected location.
Flood discharge area with water.
Cool exposed containers with water.

Exposure

CALL FOR MEDICAL AID.

DUST
POISONOUS IF INHALED.
Irritating to eyes, nose and throat.
Move victim to fresh air.
If in eyes, hold eyelids open and flush with plenty of water.
If breathing is difficult, give oxygen.

SOLID
POISONOUS IF SWALLOWED.
Irritating to skin and eyes.
Remove contaminated clothing and shoes.
Flush affected areas with plenty of water.
IF IN EYES, hold eyelids open and flush with plenty of water.
IF SWALLOWED and victim is CONSCIOUS, have victim drink water or milk and have victim induce vomiting.
IF SWALLOWED and victim is UNCONSCIOUS OR HAVING CONVULSIONS, do nothing except keep victim warm.

Water Pollution

Effect of low concentrations on aquatic life is unknown.
May be dangerous if it enters water intakes.
Notify local health and wildlife officials.
Notify operators of nearby water intakes.

1. RESPONSE TO DISCHARGE

(See Response Methods Handbook, CG 446-4)

Issue warning—oxidizing material, water contaminant
Should be removed
Chemical and physical treatment
Disperse and flush

2. LABEL

Yellow

3. CHEMICAL DESIGNATIONS

3.1 **Synonyms:** Barium chlorate monohydrate
3.2 **Coast Guard Compatibility Classification:** Not applicable
3.3 **Chemical Formula:** $Ba(ClO_3)_2 \cdot H_2O$
3.4 **IMCO/United Nations Numerical Designation:** 5.1/1445

4. OBSERVABLE CHARACTERISTICS

4.1 **Physical State (as shipped):** Solid
4.2 **Color:** White
4.3 **Odor:** None

5. HEALTH HAZARDS

5.1 **Personal Protective Equipment:** Goggles or face shield; dust respirator (U. S. Bureau of Mines approved); rubberized shoes and gloves; coveralls or other suitable outer clothing.
5.2 **Symptoms Following Exposure:** Inhalation causes irritation of upper respiratory system. Contact with eyes or skin causes irritation. Ingestion causes abdominal pain, nausea and vomiting, diarrhea, pallor, blueness, shortness of breath, excessive salivation, convulsive tremors, slow, hard pulse, elevated blood pressure, unconsciousness. Hemorrhages may occur in the stomach, intestines, and kidneys. Muscular paralysis may follow.
5.3 **Treatment for Exposure:** Get medical attention. Alert doctor to possibility of barium poisoning, particularly if compound was swallowed. INHALATION: remove to fresh air. EYES: flush with copious quantities of water for at least 15 min.; get medical attention. SKIN: flush with water. INGESTION: induce vomiting and call a physician; have victim drink aqueous 10% solution of magnesium or sodium sulfate; for severe intoxication, calcium or a magnesium salt may have to be given I.V. with caution; treatment otherwise is supportive and symptomatic.
5.4 **Toxicity by Inhalation (Threshold Limit Value):** 0.5 mg/m^3
5.5 **Short-Term Inhalation Limits:** Data not available
5.6 **Toxicity by Ingestion:** Data not available
5.7 **Late Toxicity:** Barium poisoning
5.8 **Vapor (Gas) Irritant Characteristics:** Not pertinent
5.9 **Liquid or Solid Irritant Characteristics:** Data not available
5.10 **Odor Threshold:** Not pertinent

6. FIRE HAZARDS

6.1 **Flash Point:** Not flammable (but see 7.2)
6.2 **Flammable Limits in Air:** Not pertinent
6.3 **Fire Extinguishing Agents:** Not pertinent
6.4 **Fire Extinguishing Agents Not to be Used:** Data not available
6.5 **Special Hazards of Combustion Products:** Yields toxic fumes when involved in fire.
6.6 **Behavior in Fire:** May cause an explosion when involved in a fire.
6.7 **Ignition Temperature:** Not pertinent
6.8 **Electrical Hazard:** Not pertinent
6.9 **Burning Rate:** Not pertinent

7. CHEMICAL REACTIVITY

7.1 **Reactivity with Water:** No reaction
7.2 **Reactivity with Common Materials:** Can form explosive mixtures with combustible materials such as oil and wood; these can be ignited by friction.
7.3 **Stability During Transport:** Stable
7.4 **Neutralizing Agents for Acids and Caustics:** Not pertinent
7.5 **Polymerization:** Not pertinent
7.6 **Inhibitor of Polymerization:** Not pertinent

8. WATER POLLUTION

8.1 **Aquatic Toxicity:** Data not available
8.2 **Waterfowl Toxicity:** Data not available
8.3 **Biological Oxygen Demand (BOD):** None
8.4 **Food Chain Concentration Potential:** None

9. SELECTED MANUFACTURERS

1. Barium and Chemicals, Inc.
   P. O. Box 218
   Steubenville, Ohio 43952
2. Cerac, Inc.
   13460 W. Silver Spring Road
   Menomenee Falls, Wis. 53051
3. Gallard-Schlesinger Chemical Manufacturing Co.
   584 Mineola Avenue
   Carle Place, N. Y. 11514

10. SHIPPING INFORMATION

10.1 **Grades or Purity:** Technical; Reagent
10.2 **Storage Temperature:** Ambient
10.3 **Inert Atmosphere:** No requirement
10.4 **Venting:** Open

11. HAZARD ASSESSMENT CODE

(See Hazard Assessment Handbook, CG 446-3)

SS

12. HAZARD CLASSIFICATIONS

12.1 **Code of Federal Regulations:** Oxidizing material
12.2 **NAS Hazard Rating for Bulk Water Transportation:** Not listed
12.3 **NFPA Hazard Classifications:**

Category	Classification*	
Health Hazard (Blue)	0	1
Flammability (Red)	0	0
Reactivity (Yellow)	2	2
	oxy	oxy

*First column refers to nonfire situation.

13. PHYSICAL AND CHEMICAL PROPERTIES

13.1 **Physical State at 15°C and 1 atm:** Solid
13.2 **Molecular Weight:** 332 (monohydrate)
13.3 **Boiling Point at 1 atm:** Not pertinent
13.4 **Freezing Point:** 777°F = 414°C = 687°K
13.5 **Critical Temperature:** Not pertinent
13.6 **Critical Pressure:** Not pertinent
13.7 **Specific Gravity:** 3.18 at 20°C (solid)
13.8 **Liquid Surface Tension:** Not pertinent
13.9 **Liquid-Water Interfacial Tension:** Not pertinent
13.10 **Vapor (Gas) Specific Gravity:** Not pertinent
13.11 **Ratio of Specific Heats of Vapor (Gas):** Not pertinent
13.12 **Latent Heat of Vaporization:** Not pertinent
13.13 **Heat of Combustion:** Not pertinent
13.14 **Heat of Decomposition:** Not pertinent
13.15 **Heat of Solution:** 36 Btu/lb = 20 cal/g = 0.84×10^5 J/kg
13.16 **Heat of Polymerization:** Not pertinent

NOTES

BNT

BARIUM NITRATE

Common Synonyms	Solid	White	Odorless
	Sinks and mixes with water.		

AVOID CONTACT WITH SOLID AND DUST. KEEP PEOPLE AWAY.
Wear goggles and self-contained breathing apparatus.
Call fire department.
Isolate and remove discharged material.
Notify local health and pollution control agencies.

Fire

Not flammable.
May cause fire on contact with combustibles.
POISONOUS GASES MAY BE PRODUCED IN FIRE.
Flood discharge area with water.

Exposure

CALL FOR MEDICAL AID.

DUST
POISONOUS IF INHALED.
Irritating to eyes, nose and throat.
Move victim to fresh air.
If in eyes, hold eyelids open and flush with plenty of water.
If breathing is difficult, give oxygen.

SOLID
POISONOUS IF SWALLOWED.
Irritating to skin and eyes.
Remove contaminated clothing and shoes.
Flush affected areas with plenty of water.
IF IN EYES, hold eyelids open and flush with plenty of water.
IF SWALLOWED and victim is CONSCIOUS, have victim drink water or milk and have victim induce vomiting.
IF SWALLOWED and victim is UNCONSCIOUS OR HAVING CONVULSIONS, do nothing except keep victim warm.

Water Pollution

Dangerous to aquatic life in high concentrations.
May be dangerous if it enters water intakes.
Notify local health and wildlife officials.
Notify operators of nearby water intakes.

1. RESPONSE TO DISCHARGE
(See Response Methods Handbook, CG 446-4)

Issue warning—oxidizing material, water contaminant
Should be removed
Chemical and physical treatment
Disperse and flush

2. LABEL

Yellow

3. CHEMICAL DESIGNATIONS

3.1 **Synonyms:** No common synonyms
3.2 **Coast Guard Compatibility Classification:** Not applicable
3.3 **Chemical Formula:** $Ba(NO_3)_2$
3.4 **IMCO/United Nations Numerical Designation:** 5.1/1446

4. OBSERVABLE CHARACTERISTICS

4.1 **Physical State (as shipped):** Solid
4.2 **Color:** White
4.3 **Odor:** None

5. HEALTH HAZARDS

5.1 **Personal Protective Equipment:** Goggles or face shield; dust respirator; rubber gloves and shoes; suitable coveralls.
5.2 **Symptoms Following Exposure:** Inhalation or contact with eyes or skin causes irritation. Ingestion causes excessive salivation, vomiting, colic, diarrhea, convulsive tremors, slow, hard pulse, elevated blood pressure. Hemorrhages may occur in the stomach, intestines, and kidneys. Muscular paralysis may follow.
5.3 **Treatment for Exposure:** Get medical attention. Alert doctor to possibility of barium poisoning, particularly if compound was swallowed. INHALATION: remove to fresh air. EYES: flush with water for at least 15 min. SKIN: flush with water. INGESTION: oral administration of an aqueous 10% solution of magnesium or sodium sulfate; in severe intoxication, calcium or a magnesium salt may have to be given I.V. with caution; treatment otherwise is supportive and symptomatic.
5.4 **Toxicity by Inhalation (Threshold Limit Value):** 0.5 mg/m^3
5.5 **Short-Term Inhalation Limits:** Data not available
5.6 **Toxicity by Ingestion:** Grade 3; oral rat LD_{50} = 355 mg/kg
5.7 **Late Toxicity:** Barium poisoning
5.8 **Vapor (Gas) Irritant Characteristics:** Data not available
5.9 **Liquid or Solid Irritant Characteristics:** Data not available
5.10 **Odor Threshold:** Not pertinent

6. FIRE HAZARDS

6.1 **Flash Point:** Not flammable (but see 7.2)
6.2 **Flammable Limits in Air:** Not flammable
6.3 **Fire Extinguishing Agents:** Not pertinent
6.4 **Fire Extinguishing Agents Not to be Used:** Not pertinent
6.5 **Special Hazards of Combustion Products:** Yields toxic gaseous oxides of nitrogen when involved in fire.
6.6 **Behavior in Fire:** Mixtures with combustible materials are readily ignited and may burn fiercely. Containers may explode.
6.7 **Ignition Temperature:** Not pertinent
6.8 **Electrical Hazard:** Not pertinent
6.9 **Burning Rate:** Not pertinent

7. CHEMICAL REACTIVITY

7.1 **Reactivity with Water:** No reaction
7.2 **Reactivity with Common Materials:** Contact with combustible material may cause fire.
7.3 **Stability During Transport:** Stable
7.4 **Neutralizing Agents for Acids and Caustics:** Not pertinent
7.5 **Polymerization:** Not pertinent
7.6 **Inhibitor of Polymerization:** Not pertinent

8. WATER POLLUTION

8.1 **Aquatic Toxicity:** 500 ppm/168 hr/stickleback/average survival/fresh water
8.2 **Waterfowl Toxicity:** Data not available
8.3 **Biological Oxygen Demand (BOD):** None
8.4 **Food Chain Concentration Potential:** None

9. SELECTED MANUFACTURERS

1. J. T. Baker Chemical Co.
   Phillipsburg, N. J. 08865
2. Barium and Chemicals, Inc.
   P. O. Box 218
   Steubenville, Ohio 43952
3. Ventron Corp.
   Alfa Products
   P. O. Box 159
   Beverly, Mass. 01915

10. SHIPPING INFORMATION

10.1 **Grades or Purity:** Technical; Reagent
10.2 **Storage Temperature:** Ambient
10.3 **Inert Atmosphere:** No requirement
10.4 **Venting:** Open

11. HAZARD ASSESSMENT CODE
(See Hazard Assessment Handbook, CG 446-3)

SS

12. HAZARD CLASSIFICATIONS

12.1 **Code of Federal Regulations:** Oxidizing material
12.2 **NAS Hazard Rating for Bulk Water Transportation:** Not listed
12.3 **NFPA Hazard Classifications:**

Category	Classification
Health Hazard (Blue)	1
Flammability (Red)	0
Reactivity (Yellow)	0
	oxy

13. PHYSICAL AND CHEMICAL PROPERTIES

13.1 **Physical State at 15°C and 1 atm:** Solid
13.2 **Molecular Weight:** 261.35
13.3 **Boiling Point at 1 atm:** Decomposes
13.4 **Freezing Point:** 1,098°F = 592°C = 865°K
13.5 **Critical Temperature:** Not pertinent
13.6 **Critical Pressure:** Not pertinent
13.7 **Specific Gravity:** 3.24 at 23°C (solid)
13.8 **Liquid Surface Tension:** Not pertinent
13.9 **Liquid-Water Interfacial Tension:** Not pertinent
13.10 **Vapor (Gas) Specific Gravity:** Not pertinent
13.11 **Ratio of Specific Heats of Vapor (Gas):** Not pertinent
13.12 **Latent Heat of Vaporization:** Not pertinent
13.13 **Heat of Combustion:** Not pertinent
13.14 **Heat of Decomposition:** Not pertinent
13.15 **Heat of Solution:** 36 Btu/lb = 20 cal/g = 0.84×10^5 J/kg
13.16 **Heat of Polymerization:** Not pertinent

NOTES

BPC — BARIUM PERCHLORATE

Common Synonyms			
	Solid	White	Odorless
	Sinks and mixes with water.		

AVOID CONTACT WITH SOLID AND DUST. KEEP PEOPLE AWAY.
Wear rubber overclothing (including gloves).
Isolate and remove discharged material.
Notify local health and pollution control agencies.

Fire

Not flammable.
May explode on contact with combustibles.
POISONOUS GASES MAY BE PRODUCED WHEN HEATED.
Containers may explode in fire.
Combat fires from safe distance or protected location.
Flood discharge area with water.
Cool exposed containers with water.

Exposure

CALL FOR MEDICAL AID.

DUST
POISONOUS IF INHALED.
Irritating to eyes, nose and throat.
Move victim to fresh air.
If in eyes, hold eyelids open and flush with plenty of water.
If breathing is difficult, give oxygen.

SOLID
POISONOUS IF SWALLOWED.
Irritating to skin and eyes.
Remove contaminated clothing and shoes.
Flush affected areas with plenty of water.
IF IN EYES, hold eyelids open and flush with plenty of water.
IF SWALLOWED and victim is CONSCIOUS, have victim drink water or milk.

Water Pollution

Effect of low concentrations on aquatic life is unknown.
May be dangerous if it enters water intakes.
Notify local health and wildlife officials.
Notify operators of nearby water intakes.

1. RESPONSE TO DISCHARGE

(See Response Methods Handbook, CG 446-4)

Issue warning—oxidizing material, water contaminant
Should be removed
Chemical and physical treatment
Disperse and flush

2. LABEL

Yellow

3. CHEMICAL DESIGNATIONS

3.1 **Synonyms:** Barium perchlorate trihydrate
3.2 **Coast Guard Compatibility Classification:** Not applicable
3.3 **Chemical Formula:** $Ba(ClO_4)_2 \cdot 3H_2O$
3.4 **IMCO/United Nations Numerical Designation:** 5.1/1447

4. OBSERVABLE CHARACTERISTICS

4.1 **Physical State (as shipped):** Solid
4.2 **Color:** White
4.3 **Odor:** None

5. HEALTH HAZARDS

5.1 **Personal Protective Equipment:** Goggles or face shield; dust respirator; rubber gloves and shoes; suitable coveralls.
5.2 **Symptoms Following Exposure:** Inhalation or contact with eyes or skin causes irritation. Ingestion causes excessive salivation, vomiting, colic, diarrhea, convulsive tremors, slow, hard pulse, and elevated blood pressure; hemorrhages may occur in the stomach, intestines, and kidneys; muscular paralysis may follow.
5.3 **Treatment for Exposure:** Get medical attention. Alert doctor to possibility of barium poisoning, particularly if compound was swallowed. INHALATION: remove to fresh air. EYES: flush with water for at least 15 min. SKIN: flush with water. INGESTION: oral administration of an aqueous 10% solution of magnesium or sodium sulfate; for severe intoxication, calcium or a magnesium salt may have to be given I.V. with caution; treatment otherwise is supportive and symptomatic.
5.4 **Toxicity by Inhalation (Threshold Limit Value):** 0.5 mg/m^3
5.5 **Short-Term Inhalation Limits:** Data not available
5.6 **Toxicity by Ingestion:** Data not available
5.7 **Late Toxicity:** Barium poisoning
5.8 **Vapor (Gas) Irritant Characteristics:** Data not available
5.9 **Liquid or Solid Irritant Characteristics:** Data not available
5.10 **Odor Threshold:** Not pertinent

6. FIRE HAZARDS

6.1 **Flash Point:** Not flammable (but see 7.2)
6.2 **Flammable Limits in Air:** Not pertinent
6.3 **Fire Extinguishing Agents:** Not pertinent
6.4 **Fire Extinguishing Agents Not to be Used:** Not pertinent
6.5 **Special Hazards of Combustion Products:** Not pertinent
6.6 **Behavior in Fire:** Increases the intensity of fire. Containers may explode.
6.7 **Ignition Temperature:** Not pertinent
6.8 **Electrical Hazard:** Not pertinent
6.9 **Burning Rate:** Not pertinent

7. CHEMICAL REACTIVITY

7.1 **Reactivity with Water:** No reaction
7.2 **Reactivity with Common Materials:** When mixed with combustible material or finely divided metals, can cause explosions.
7.3 **Stability During Transport:** Stable
7.4 **Neutralizing Agents for Acids and Caustics:** Not pertinent
7.5 **Polymerization:** Not pertinent
7.6 **Inhibitor of Polymerization:** Not pertinent

8. WATER POLLUTION

8.1 **Aquatic Toxicity:** Data not available
8.2 **Waterfowl Toxicity:** Data not available
8.3 **Biological Oxygen Demand (BOD):** None
8.4 **Food Chain Concentration Potential:** None

9. SELECTED MANUFACTURERS

1. Barium and Chemicals, Inc.
P. O. Box 218
Steubenville, Ohio 43952
2. Cerac, Inc.
13460 W. Silver Spring Rd.
Menomenee Falls, Wis. 53051
3. Gallard-Schlesinger Chemical Manufacturing Co.
584 Mineola Ave.
Carle Place, N. Y. 11514

10. SHIPPING INFORMATION

10.1 **Grades or Purity:** Technical; Reagent
10.2 **Storage Temperature:** Ambient
10.3 **Inert Atmosphere:** No requirement
10.4 **Venting:** Open

11. HAZARD ASSESSMENT CODE

(See Hazard Assessment Handbook, CG 446-3)

SS

12. HAZARD CLASSIFICATIONS

12.1 **Code of Federal Regulations:** Oxidizing material
12.2 **NAS Hazard Rating for Bulk Water Transportation:** Not listed
12.3 **NFPA Hazard Classifications:** Not listed

13. PHYSICAL AND CHEMICAL PROPERTIES

13.1 **Physical State at 15°C and 1 atm:** Solid
13.2 **Molecular Weight:** 390.35
13.3 **Boiling Point at 1 atm:** Decomposes
13.4 **Freezing Point:** 941°F = 505°C = 778°K
13.5 **Critical Temperature:** Not pertinent
13.6 **Critical Pressure:** Not pertinent
13.7 **Specific Gravity:** 3.2 at 20°C (solid)
13.8 **Liquid Surface Tension:** Not pertinent
13.9 **Liquid-Water Interfacial Tension:** Not pertinent
13.10 **Vapor (Gas) Specific Gravity:** Not pertinent
13.11 **Ratio of Specific Heats of Vapor (Gas):** Not pertinent
13.12 **Latent Heat of Vaporization:** Not pertinent
13.13 **Heat of Combustion:** Not pertinent
13.14 **Heat of Decomposition:** Not pertinent
13.15 **Heat of Solution:** 9 Btu/lb = 5 cal/g = 0.2×10^5 J/kg
13.16 **Heat of Polymerization:** Not pertinent

NOTES

BPM	# BARIUM PERMANGANATE

Common Synonyms	Solid	Dark purple to black	Odorless
	Sinks and mixes with water.		

AVOID CONTACT WITH SOLID AND DUST. KEEP PEOPLE AWAY.
Wear rubber overclothing (including gloves).
Isolate and remove discharged material.
Notify local health and pollution control agencies.

Fire

Not flammable.
May cause fire on contact with combustibles.
Containers may explode in fire.
Combat fires from safe distance or protected location.
Flood discharge area with water.
Cool exposed containers with water.

Exposure

CALL FOR MEDICAL AID.

DUST
POISONOUS IF INHALED.
Irritating to eyes, nose and throat.
Move victim to fresh air.
If in eyes, hold eyelids open and flush with plenty of water.
If breathing is difficult, give oxygen.

SOLID
POISONOUS IF SWALLOWED.
Irritating to skin and eyes.
Remove contaminated clothing and shoes.
Flush affected areas with plenty of water.
IF IN EYES, hold eyelids open and flush with plenty of water.
IF SWALLOWED and victim is CONSCIOUS, have victim drink water or milk and have victim induce vomiting.
IF SWALLOWED and victim is UNCONSCIOUS OR HAVING CONVULSIONS, do nothing except keep victim warm.

Water Pollution

HARMFUL TO AQUATIC LIFE IN VERY LOW CONCENTRATIONS.
May be dangerous if it enters water intakes.
Notify local health and wildlife officials.
Notify operators of nearby water intakes.

1. RESPONSE TO DISCHARGE

(See Response Methods Handbook, CG 446-4)

Issue warning—oxidizing material, water contaminant
Should be removed
Chemical and physical treatment
Disperse and flush

2. LABEL

Yellow

3. CHEMICAL DESIGNATIONS

3.1 **Synonyms:** No common synonyms
3.2 **Coast Guard Compatibility Classification:** Not applicable
3.3 **Chemical Formula:** $Ba(MnO_4)_2$
3.4 **IMCO/United Nations Numerical Designation:** 5.1/1448

4. OBSERVABLE CHARACTERISTICS

4.1 **Physical State (as shipped):** Solid
4.2 **Color:** Dark purple to black
4.3 **Odor:** None

5. HEALTH HAZARDS

5.1 **Personal Protective Equipment:** Goggles or face shield; dust respirator; rubber gloves and shoes.
5.2 **Symptoms Following Exposure:** Inhalation or contact with eyes or skin causes irritation. Ingestion causes abdominal pain, nausea, vomiting, pallor, shortness of breath.
5.3 **Treatment for Exposure:** Get medical attention. Alert doctor to possibility of barium poisoning, particularly if compound was swallowed. INHALATION: remove to fresh air. EYES: flush with copious amounts of water for 15 min. SKIN: wash with copious amounts of water. INGESTION: induce vomiting; give a 10% water solution of Epsom salt.
5.4 **Toxicity by Inhalation (Threshold Limit Value):** 0.5 mg/m^3
5.5 **Short-Term Inhalation Limits:** Data not available
5.6 **Toxicity by Ingestion:** Data not available
5.7 **Late Toxicity:** Barium poisoning
5.8 **Vapor (Gas) Irritant Characteristics:** Data not available
5.9 **Liquid or Solid Irritant Characteristics:** Data not available
5.10 **Odor Threshold:** Not pertinent

6. FIRE HAZARDS

6.1 **Flash Point:** Not flammable
6.2 **Flammable Limits in Air:** Not flammable
6.3 **Fire Extinguishing Agents:** Not pertinent
6.4 **Fire Extinguishing Agents Not to be Used:** Not pertinent
6.5 **Special Hazards of Combustion Products:** Not pertinent
6.6 **Behavior in Fire:** Can increase the intensity of fire.
6.7 **Ignition Temperature:** Not pertinent
6.8 **Electrical Hazard:** Not pertinent
6.9 **Burning Rate:** Not pertinent

7. CHEMICAL REACTIVITY

7.1 **Reactivity with Water:** No reaction
7.2 **Reactivity with Common Materials:** When mixed with combustible material, can ignite by friction or acids; may be spontaneously combustible.
7.3 **Stability During Transport:** Stable
7.4 **Neutralizing Agents for Acids and Caustics:** Not pertinent
7.5 **Polymerization:** Not pertinent
7.6 **Inhibitor of Polymerization:** Not pertinent

8. WATER POLLUTION

8.1 **Aquatic Toxicity:**
2.2 to 4.1 mg/l of Mn/8 to 18 hr/fish/killed/*
*Type of water not specified.
8.2 **Waterfowl Toxicity:** Data not available
8.3 **Biological Oxygen Demand (BOD):** None
8.4 **Food Chain Concentration Potential:** None

9. SELECTED MANUFACTURERS

1. Cerac, Inc.
13460 W. Silver Spring Rd.
Menomenee Falls, Wis. 53051
2. Ventron Corp.
Alfa Products
P. O. Box 159
Beverly, Mass. 01915

10. SHIPPING INFORMATION

10.1 **Grades or Purity:** Commercial
10.2 **Storage Temperature:** Ambient
10.3 **Inert Atmosphere:** No requirement
10.4 **Venting:** Open

11. HAZARD ASSESSMENT CODE

(See Hazard Assessment Handbook, CG 446-3)

SS

12. HAZARD CLASSIFICATIONS

12.1 **Code of Federal Regulations:** Oxidizing material
12.2 **NAS Hazard Rating for Bulk Water Transportation:** Not listed
12.3 **NFPA Hazard Classifications:** Not listed

13. PHYSICAL AND CHEMICAL PROPERTIES

13.1 **Physical State at 15°C and 1 atm:** Solid
13.2 **Molecular Weight:** 375
13.3 **Boiling Point at 1 atm:** Decomposes
13.4 **Freezing Point:** Not pertinent
13.5 **Critical Temperature:** Not pertinent
13.6 **Critical Pressure:** Not pertinent
13.7 **Specific Gravity:** 3.77 at 20°C (solid)
13.8 **Liquid Surface Tension:** Not pertinent
13.9 **Liquid-Water Interfacial Tension:** Not pertinent
13.10 **Vapor (Gas) Specific Gravity:** Not pertinent
13.11 **Ratio of Specific Heats of Vapor (Gas):** Not pertinent
13.12 **Latent Heat of Vaporization:** Not pertinent
13.13 **Heat of Combustion:** Not pertinent
13.14 **Heat of Decomposition:** Not pertinent
13.15 **Heat of Solution:** Data not available
13.16 **Heat of Polymerization:** Not pertinent

NOTES

BPO	BARIUM PEROXIDE

Common Synonyms			
Barium dioxide Barium superoxide Barium binoxide	Solid	Light gray to tan	Odorless
	Sinks in water.		

AVOID CONTACT WITH SOLID AND DUST. KEEP PEOPLE AWAY.
Wear rubber overclothing (including gloves).
Isolate and remove discharged material.
Notify local health and pollution control agencies.

Fire

Not flammable.
May cause fire on contact with combustibles.
Containers may explode in fire.
Combat fires from safe distance or protected location.
Flood discharge area with water.
Cool exposed containers with water.

Exposure

CALL FOR MEDICAL AID.

DUST
POISONOUS IF INHALED.
Irritating to eyes, nose and throat.
Move victim to fresh air.
If in eyes, hold eyelids open and flush with plenty of water.
If breathing is difficult, give oxygen.

SOLID
POISONOUS IF SWALLOWED.
Will burn skin and eyes.
Remove contaminated clothing and shoes.
Flush affected areas with plenty of water.
IF IN EYES, hold eyelids open and flush with plenty of water.
IF SWALLOWED and victim is CONSCIOUS, have victim drink water or milk.
DO NOT INDUCE VOMITING.

Water Pollution

Effect of low concentrations on aquatic life is unknown.
May be dangerous if it enters water intakes.
Notify local health and wildlife officials.
Notify operators of nearby water intakes.

1. RESPONSE TO DISCHARGE

(See Response Methods Handbook, CG 446-4)

Issue warning—oxidizing material, water contaminant
Should be removed
Chemical and physical treatment

2. LABEL

Yellow

3. CHEMICAL DESIGNATIONS

3.1 **Synonyms:** Barium dioxide
Barium superoxide
Barium binoxide

3.2 **Coast Guard Compatibility Classification:** Not applicable

3.3 **Chemical Formula:** BaO_2

3.4 **IMCO/United Nations Numerical Designation:** 5.1/1449

4. OBSERVABLE CHARACTERISTICS

4.1 **Physical State (as shipped):** Solid

4.2 **Color:** Light grayish-tan; grayish-white

4.3 **Odor:** None

5. HEALTH HAZARDS

5.1 **Personal Protective Equipment:** U. S. B. M.-approved toxic dust respirator; liquid-proof PVC gloves; chemical safety goggles; full cover clothing.

5.2 **Symptoms Following Exposure:** Inhalation causes irritation of mucous membranes, throat, and nose. Contact with eyes or skin causes severe burns. Ingestion causes excessive salivation, vomiting, colic, diarrhea, convulsive tremors, slow, hard pulse, and elevated blood pressure; hemorrhages may occur in the stomach, intestines, and kidneys; muscular paralysis may follow.

5.3 **Treatment for Exposure:** Get medical attention. Alert doctor to possibility of barium poisoning, particularly if compound was swallowed. INHALATION: remove to fresh air. EYES: flush with water for 15 min. SKIN: flush with water. INGESTION: oral administration of an aqueous 10% solution of magnesium or sodium sulfate; in a severe intoxication, calcium or a magnesium salt may have to be given I.V. with caution; treatment otherwise is supportive and symptomatic.

5.4 **Toxicity by Inhalation (Threshold Limit Value):** 0.5 mg/m^3

5.5 **Short-Term Inhalation Limits:** Data not available

5.6 **Toxicity by Ingestion:** Data not available

5.7 **Late Toxicity:** Barium poisoning

5.8 **Vapor (Gas) Irritant Characteristics:** Data not available

5.9 **Liquid or Solid Irritant Characteristics:** Data not available

5.10 **Odor Threshold:** Not pertinent

6. FIRE HAZARDS

6.1 **Flash Point:** Not flammable but may cause fire on contact with combustibles.

6.2 **Flammable Limits in Air:** Not pertinent

6.3 **Fire Extinguishing Agents:** Flood with water, dry powder (e.g. graphite, powdered limestone).

6.4 **Fire Extinguishing Agents Not to be Used:** Not pertinent

6.5 **Special Hazards of Combustion Products:** Not pertinent

6.6 **Behavior in Fire:** Can increase intensity of fire.

6.7 **Ignition Temperature:** Not pertinent

6.8 **Electrical Hazard:** Not pertinent

6.9 **Burning Rate:** Not pertinent

7. CHEMICAL REACTIVITY

7.1 **Reactivity with Water:** Decomposes slowly. The reaction is not hazardous.

7.2 **Reactivity with Common Materials:** Corrodes metal slowly. If mixed with combustible material or finely divided metals, can ignite spontaneously or by friction.

7.3 **Stability During Transport:** Stable

7.4 **Neutralizing Agents for Acids and Caustics:** Not pertinent

7.5 **Polymerization:** Not pertinent

7.6 **Inhibitor of Polymerization:** Not pertinent

8. WATER POLLUTION

8.1 **Aquatic Toxicity:** Data not available

8.2 **Waterfowl Toxicity:** Data not available

8.3 **Biological Oxygen Demand (BOD):** None

8.4 **Food Chain Concentration Potential:** None

9. SELECTED MANUFACTURERS

1. FMC Corporation
Industrial Chemicals Division
633 Third Avenue
New York, N. Y. 10017
2. J. T. Baker Chemical Co.
Phillipsburg, N. J. 08865
3. Barium and Chemicals, Inc.
P. O. Box 218
Steubenville, Ohio 43952

10. SHIPPING INFORMATION

10.1 **Grades or Purity:** Technical: 91-92.5%; high-purity reagent

10.2 **Storage Temperature:** Ambient

10.3 **Inert Atmosphere:** No requirement

10.4 **Venting:** Pressure-vacuum

11. HAZARD ASSESSMENT CODE

(See Hazard Assessment Handbook, CG 446-3)

II

12. HAZARD CLASSIFICATIONS

12.1 **Code of Federal Regulations:** Oxidizing material

12.2 **NAS Hazard Rating for Bulk Water Transportation:** Not listed

12.3 **NFPA Hazard Classifications:**

Category	Classification
Health Hazard (Blue)	1
Flammability (Red)	0
Reactivity (Yellow)	0
	oxy

13. PHYSICAL AND CHEMICAL PROPERTIES

13.1 **Physical State at 15°C and 1 atm:** Solid

13.2 **Molecular Weight:** 169.4

13.3 **Boiling Point at 1 atm:** Decomposes

13.4 **Freezing Point:** 842°F = 450°C = 723°K

13.5 **Critical Temperature:** Not pertinent

13.6 **Critical Pressure:** Not pertinent

13.7 **Specific Gravity:** 4.96 at 20°C (solid)

13.8 **Liquid Surface Tension:** Not pertinent

13.9 **Liquid-Water Interfacial Tension:** Not pertinent

13.10 **Vapor (Gas) Specific Gravity:** Not pertinent

13.11 **Ratio of Specific Heats of Vapor (Gas):** Not pertinent

13.12 **Latent Heat of Vaporization:** Not pertinent

13.13 **Heat of Combustion:** Not pertinent

13.14 **Heat of Decomposition:** -194 Btu/lb $= -108$ cal/g $= -4.52 \times 10^5$ J/kg

13.15 **Heat of Solution:** Not pertinent

13.16 **Heat of Polymerization:** Not pertinent

NOTES

BZD BENZALDEHYDE

Common Synonyms		
Oil of bitter almond Benzoic aldehyde	Watery liquid Colorless to pale yellow Bitter almond odor May float or sink in water.	

Stop discharge if possible. Keep people away.
Avoid contact with liquid.
Call fire department.
Isolate and remove discharged material.
Notify local health and pollution control agencies.

Fire	Combustible. Extinguish with water, dry chemical, foam, or carbon dioxide. Cool exposed containers with water.
Exposure	CALL FOR MEDICAL AID. LIQUID Irritating to skin and eyes. Harmful if swallowed. Remove contaminated clothing and shoes. Flush affected areas with plenty of water. IF IN EYES, hold eyelids open and flush with plenty of water. IF SWALLOWED and victim is CONSCIOUS, have victim drink water or milk and have victim induce vomiting. IF SWALLOWED and victim is UNCONSCIOUS OR HAVING CONVULSIONS, do nothing except keep victim warm.
Water Pollution	Effect of low concentrations on aquatic life is unknown. Fouling to shoreline. May be dangerous if it enters water intakes. Notify local health and pollution control officials. Notify operators of nearby water intakes.

1. RESPONSE TO DISCHARGE

(See Response Methods Handbook, CG 446-4)

Issue warning—water contaminant
Should be removed

2. LABELS

No hazard label required by Code of Federal Regulations

3. CHEMICAL DESIGNATIONS

3.1 **Synonyms:** Benzoic aldehyde
Oil of bitter almond

3.2 **Coast Guard Compatibility Classification:** Not applicable

3.3 **Chemical Formula:** C_6H_5CHO

3.4 **IMCO/United Nations Numerical Designation:** 3.3/1990

4. OBSERVABLE CHARACTERISTICS

4.1 **Physical State (as shipped):** Liquid

4.2 **Color:** Colorless to pale yellow

4.3 **Odor:** Like almonds

5. HEALTH HAZARDS

5.1 **Personal Protective Equipment:** Chemical goggles and protective clothing.

5.2 **Symptoms Following Exposure:** Inhalation of concentrated vapor may irritate eyes, nose and throat. Liquid is irritating to the eyes. Prolonged contact with the skin may cause irritation.

5.3 **Treatment for Exposure:** SKIN, EYE CONTACT: move victim to fresh air. Call physician immediately. Wash contaminated skin area with water. Flush eyes with plenty of water for at least 15 min. INGESTION: induce vomiting. Call a physician.

5.4 **Toxicity by Inhalation (Threshold Limit Value):** Data not available

5.5 **Short-Term Inhalation Limits:** Data not available

5.6 **Toxicity by Ingestion:** Grade 2; LD_{50} 0.5 to 5 g/kg

5.7 **Late Toxicity:** Data not available

5.8 **Vapor (Gas) Irritant Characteristics:** Vapors cause severe irritation of eyes and throat and can cause eye and lung injury. They cannot be tolerated even at low concentrations.

5.9 **Liquid or Solid Irritant Characteristics:** Minimum hazard. If spilled on clothing and allowed to remain, may cause smarting and reddening of the skin.

5.10 **Odor Threshold:** 0.042 ppm.

6. FIRE HAZARDS

6.1 **Flash Point:** 148°F C.C.; 165°F O.C.

6.2 **Flammable Limits in Air:** Data not available

6.3 **Fire Extinguishing Agents:** Water spray, foam, carbon dioxide or dry chemical.

6.4 **Fire Extinguishing Agents Not to be Used:** Not pertinent

6.5 **Special Hazards of Combustion Products:** Not pertinent

6.6 **Behavior in Fire:** Not pertinent

6.7 **Ignition Temperature:** 378°F

6.8 **Electrical Hazard:** Not pertinent

6.9 **Burning Rate:** 3.8 mm/min.

7. CHEMICAL REACTIVITY

7.1 **Reactivity with Water:** No reaction

7.2 **Reactivity with Common Materials:** No reaction

7.3 **Stability During Transport:** Stable

7.4 **Neutralizing Agents for Acids and Caustics:** Not pertinent

7.5 **Polymerization:** Not pertinent

7.6 **Inhibitor of Polymerization:** Not pertinent

8. WATER POLLUTION

8.1 **Aquatic Toxicity:** Data not available

8.2 **Waterfowl Toxicity:** Data not available

8.3 **Biological Oxygen Demand (BOD):**
50%, 10 days
150%, 5 days

8.4 **Food Chain Concentration Potential:** None

9. SELECTED MANUFACTURERS

1. Fritzsche, Dodge and Olcott, Inc.
Clifton, N. J. 07015

2. Stauffer Chemical Co.
Specialty Chemicals Division
Edison, N. J. 08817

3. Tenneco, Inc.
Tenneco Chemicals, Inc.
280 Park Ave.
New York, N. Y. 10017

10. SHIPPING INFORMATION

10.1 **Grades or Purity:** Technical grade 98.0%
NF (FCC) grade 98.0%

10.2 **Storage Temperature:** Ambient

10.3 **Inert Atmosphere:** Inerted

10.4 **Venting:** Data not available

11. HAZARD ASSESSMENT CODE

(See Hazard Assessment Handbook, CG 446-3)

A-T-U-X-Y

12. HAZARD CLASSIFICATIONS

12.1 **Code of Federal Regulations:** Combustible liquid

12.2 **NAS Hazard Rating for Bulk Water Transportation:** Not listed

12.3 **NFPA Hazard Classifications:**

Category	Classification
Health Hazard (Blue)	2
Flammability (Red)	2
Reactivity (Yellow)	0

13. PHYSICAL AND CHEMICAL PROPERTIES

13.1 **Physical State at 15°C and 1 atm:** Liquid

13.2 **Molecular Weight:** 106.12

13.3 **Boiling Point at 1 atm:** 354°F = 179°C = 452°K

13.4 **Freezing Point:** Not pertinent

13.5 **Critical Temperature:** 666°F = 352°C = 625°K

13.6 **Critical Pressure:** 316 psia = 21.5 atm = 2.18 MN/m^2

13.7 **Specific Gravity:** 1.046 at 20°C (liquid)

13.8 **Liquid Surface Tension:** 40.0 dynes/cm = 0.040 N/m at 20°C

13.9 **Liquid-Water Interfacial Tension:** 15.5 dynes/cm = 0.0155 N/m at 20°C

13.10 **Vapor (Gas) Specific Gravity:** Not pertinent

13.11 **Ratio of Specific Heats of Vapor (Gas):** 1.1

13.12 **Latent Heat of Vaporization:** 156 Btu/lb = 86.5 cal/g = 3.62×10^5 J/kg

13.13 **Heat of Combustion:** −13,730 Btu/lb = −7630 cal/g = -319.5×10^5 J/kg

13.14 **Heat of Decomposition:** Not pertinent

13.15 **Heat of Solution:** Not pertinent

13.16 **Heat of Polymerization:** Not pertinent

NOTES

REVISED 1978

BNZ	# BENZENE

Common Synonyms Benzol Benzole	Watery liquid Colorless Gasoline-like odor Floats on water. Flammable, irritating vapor is produced. Freezing point is 42° F.

Avoid contact with liquid and vapor. Keep people away.
Wear goggles and self-contained breathing apparatus.
Shut off ignition sources and call fire department.
Stop discharge if possible.
Stay upwind and use water spray to "knock down" vapor.
Isolate and remove discharged material.
Notify local health and pollution control agencies.

Fire	FLAMMABLE. Flashback along vapor trail may occur. Vapor may explode if ignited in an enclosed area. Wear goggles and self-contained breathing apparatus. Extinguish with dry chemical, foam, or carbon dioxide. Water may be ineffective on fire. Cool exposed containers with water.
Exposure	CALL FOR MEDICAL AID. VAPOR Irritating to eyes, nose and throat. If inhaled, will cause headache, difficult breathing, or loss of consciousness. Move to fresh air. If breathing has stopped, give artificial respiration. If breathing is difficult, give oxygen. LIQUID Irritating to skin and eyes. Harmful if swallowed. Remove contaminated clothing and shoes. Flush affected areas with plenty of water. IF IN EYES, hold eyelids open and flush with plenty of water. IF SWALLOWED and victim is CONSCIOUS, have victim drink water or milk.
Water Pollution	HARMFUL TO AQUATIC LIFE IN VERY LOW CONCENTRATIONS. May be dangerous if it enters water intakes. Notify local health and wildlife officials. Notify operators of nearby water intakes.

1. RESPONSE TO DISCHARGE
(See Response Methods Handbook, CG 446-4)

Issue warning—high flammability
Restrict access

2. LABEL

Red

3. CHEMICAL DESIGNATIONS

3.1 **Synonyms:** Benzol
Benzole
3.2 **Coast Guard Compatibility Classification:** Aromatic hydrocarbon
3.3 **Chemical Formula:** C_6H_6
3.4 **IMCO/United Nations Numerical Designation:** 3.2/1114

4. OBSERVABLE CHARACTERISTICS

4.1 **Physical State (as shipped):** Liquid
4.2 **Color:** Colorless
4.3 **Odor:** Aromatic; rather pleasant aromatic odor; characteristic odor

5. HEALTH HAZARDS

5.1 **Personal Protective Equipment:** Hydrocarbon vapor canister, supplied air or a hose mask; hydrocarbon-insoluble rubber or plastic gloves; chemical goggles or face splash shield; hydrocarbon-insoluble apron such as neoprene.
5.2 **Symptoms Following Exposure:** Dizziness, excitation, pallor, followed by flushing, weakness, headache, breathlessness, chest constriction. Coma and possible death.
5.3 **Treatment for Exposure:** SKIN: flush with water followed by soap and water; remove contaminated clothing and wash skin. EYES: flush with plenty of water until irritation subsides. INHALATION: remove from exposure immediately. Call a physician. If breathing is irregular or stopped, start resuscitation, administer oxygen.
5.4 **Toxicity by Inhalation (Threshold Limit Value):** 25 ppm
5.5 **Short-Term Inhalation Limits:** 75 ppm for 30 min.
5.6 **Toxicity by Ingestion:** Grade 3; LD_{50} 50 to 500 mg/kg
5.7 **Late Toxicity:** Leukemia
5.8 **Vapor (Gas) Irritant Characteristics:** If present in high concentrations, vapors may cause irritation of eyes or respiratory system. The effect is temporary.
5.9 **Liquid or Solid Irritant Characteristics:** Minimum hazard. If spilled on clothing and allowed to remain, may cause smarting and reddening of the skin.
5.10 **Odor Threshold:** 4.68 ppm

6. FIRE HAZARDS

6.1 **Flash Point:** 12°F C.C.
6.2 **Flammable Limits in Air:** 1.3%–7.9%
6.3 **Fire Extinguishing Agents:** Dry chemical, foam, or carbon dioxide
6.4 **Fire Extinguishing Agents Not to be Used:** Water may be ineffective
6.5 **Special Hazards of Combustion Products:** Not pertinent
6.6 **Behavior in Fire:** Vapor is heavier than air and may travel considerable distance to a source of ignition and flash back
6.7 **Ignition Temperature:** 1097°F
6.8 **Electrical Hazard:** Class I, Group D
6.9 **Burning Rate:** 6.0 mm/min.

7. CHEMICAL REACTIVITY

7.1 **Reactivity with Water:** No reaction
7.2 **Reactivity with Common Materials:** No reaction
7.3 **Stability During Transport:** Stable
7.4 **Neutralizing Agents for Acids and Caustics:** Not pertinent
7.5 **Polymerization:** Not pertinent
7.6 **Inhibitor of Polymerization:** Not pertinent

8. WATER POLLUTION

8.1 **Aquatic Toxicity:**
5 ppm/6 hr/minnow/lethal/distilled water
20 ppm/24 hr/sunfish/TL_m/tap water
8.2 **Waterfowl Toxicity:** Data not available
8.3 **Biological Oxygen Demand (BOD):** 1.2 lb/lb, 10 days
8.4 **Food Chain Concentration Potential:** None

9. SELECTED MANUFACTURERS

1. Commonwealth Oil Refining Co., Inc.
Commonwealth Petrochemicals Co.
Penuelas, Puerto Rico 00724
2. Phillips Petroleum Co.
Phillips Puerto Rico Core, Inc.
Banco Popular Center
Hato Rey, P. R. 00936
3. Shell Chemical Co.
Petrochemicals Div.
P. O. Box 2463
Houston, Texas 77001

10. SHIPPING INFORMATION

10.1 **Grades or Purity:**

Industrial pure	99+%
Thiophene-free	99+%
Nitration	99+%
Industrial 90%	85+%
Reagent	99+%

10.2 **Storage Temperature:** Ambient
10.3 **Inert Atmosphere:** No requirement
10.4 **Venting:** Pressure-vacuum

11. HAZARD ASSESSMENT CODE
(See Hazard Assessment Handbook, CG 446-3)

A-T-U-V-W

12. HAZARD CLASSIFICATIONS

12.1 **Code of Federal Regulations:** Flammable liquid
12.2 **NAS Hazard Rating for Bulk Water Transportation:**

Category	Rating
Fire	3
Health	
Vapor Irritant	1
Liquid or Solid Irritant	1
Poisons	3
Water Pollution	
Human Toxicity	1
Aquatic Toxicity	3
Aesthetic Effect	2
Reactivity	
Other Chemicals	1
Water	0
Self-Reaction	0

12.3 **NFPA Hazard Classifications:**

Category	Classification
Health Hazard (Blue)	2
Flammability (Red)	3
Reactivity (Yellow)	0

13. PHYSICAL AND CHEMICAL PROPERTIES

13.1 **Physical State at 15°C and 1 atm:** Liquid
13.2 **Molecular Weight:** 78.11
13.3 **Boiling Point at 1 atm:** 176°F = 80.1°C = 353.3°K
13.4 **Freezing Point:** 42.0°F = 5.5°C = 278.7°K
13.5 **Critical Temperature:** 552.0°F = 288.9°C = 562.1°K
13.6 **Critical Pressure:** 710 psia = 48.3 atm = 4.89 MN/m^2
13.7 **Specific Gravity:** 0.879 at 20°C (liquid)
13.8 **Liquid Surface Tension:** 28.9 dynes/cm = 0.0289 N/m at 20°C
13.9 **Liquid-Water Interfacial Tension:** 35.0 dynes/cm = 0.035 N/m at 20°C
13.10 **Vapor (Gas) Specific Gravity:** 2.7
13.11 **Ratio of Specific Heats of Vapor (Gas):** 1.061
13.12 **Latent Heat of Vaporization:** 169 Btu/lb = 94.1 cal/g = 3.94×10^5 J/kg
13.13 **Heat of Combustion:** −17,460 Btu/lb = −9698 cal/g = -406.0×10^5 J/kg
13.14 **Heat of Decomposition:** Not pertinent
13.15 **Heat of Solution:** Not pertinent
13.16 **Heat of Polymerization:** Not pertinent

NOTES

REVISED 1978

BENZENE HEXACHLORIDE

Common Synonyms
1,2,3,4,5,6-Hexachlorocyclohexane
Lindane
BHC
Gammexane
Tri-6

Solid crystals or powder, or solution | Solid is light to dark brown | Musty odor

Solid sinks in water. Solution generally floats on water.

Avoid contact with liquid and solid. Keep people away.
Wear goggles and self-contained breathing apparatus.
Stop discharge if possible.
Call fire department.
Stay upwind and use water spray to "knock down" vapor.
Isolate and remove discharged material.
Notify local health and pollution control agencies.

Fire

Solid not flammable, but usually dissolved in combustible liquid.
POISONOUS GASES ARE PRODUCED WHEN SOLID IS HEATED OR WHEN SOLUTION BURNS.
Wear goggles and self-contained breathing apparatus.
Extinguish with foam, dry chemical, or carbon dioxide.

Exposure

CALL FOR MEDICAL AID.

SOLID OR SOLUTION
POISONOUS IF SWALLOWED.
Irritating to skin and eyes.
Remove contaminated clothing and shoes.
Flush affected areas with plenty of water.
IF IN EYES, hold eyelids open and flush with plenty of water.
IF SWALLOWED and victim is CONSCIOUS, have victim drink water or milk.

Water Pollution

HARMFUL TO AQUATIC LIFE IN VERY LOW CONCENTRATIONS.
Fouling to shoreline.
May be dangerous if it enters water intakes.
Notify local health and pollution control officials.
Notify operators of nearby water intakes.

1. RESPONSE TO DISCHARGE

(See Response Methods Handbook, CG 446-4)

Issue warning—water contaminant, poison
Should be removed
Chemical and physical treatment

2. LABEL

3. CHEMICAL DESIGNATIONS

3.1 **Synonyms:**
BHC
Gammexane
1, 2, 3, 4, 5, 6-Hexachlorocyclohexane
Lindane

3.2 **Coast Guard Compatibility Classification:** Not applicable

3.3 **Chemical Formula:** $C_6H_6Cl_6$

3.4 **IMCO/United Nations Numerical Designation:** 6.1/1615

4. OBSERVABLE CHARACTERISTICS

4.1 **Physical State (as shipped):** Solid

4.2 **Color:** Light tan to dark brown

4.3 **Odor:** Characteristic

5. HEALTH HAZARDS

5.1 **Personal Protective Equipment:** Data not available

5.2 **Symptoms Following Exposure:** Hyperirritability and central nervous excitation: notably vomiting, restlessness, muscle spasms, ataxia, and clonic and tonic convulsions. Subsequent central nervous depression leading to respiratory failure. Occasional dermatitis and urticaria.

5.3 **Treatment for Exposure:** Gastric lavage and saline cathartics (not oil laxatives because they promote absorption). Sedatives: pentobarbital or phenobarbital in amounts adequate to control convulsions. Calcium gluconate intravenously may be used in conjunction with sedatives to control convulsions. Rest and quiet. Do NOT use epinephrine because ventricular fibrillation may result.

5.4 **Toxicity by Inhalation (Threshold Limit Value):** 0.5 mg/m^3

5.5 **Short-Term Inhalation Limits:** 1 mg/m^3 for 30 min.

5.6 **Toxicity by Ingestion:** Gamma isomer (Lindane): Grade 3; LD_{50} 50 to 500 mg/kg (rat)
Technical mixture: Grade 2; LD_{50} 0.5 to 5 g/kg

5.7 **Late Toxicity:** Mutagen to human lymphocytes.

5.8 **Vapor (Gas) Irritant Characteristics:** Moderately irritating. Personnel will not usually tolerate moderate or high concentrations.

5.9 **Liquid or Solid Irritant Characteristics:** Minimum hazard. If spilled on clothing and allowed to remain, may cause smarting and reddening of the skin.

5.10 **Odor Threshold:** Data not available

6. FIRE HAZARDS

6.1 **Flash Point:** Not flammable

6.2 **Flammable Limits in Air:** Not flammable

6.3 **Fire Extinguishing Agents:** Not pertinent

6.4 **Fire Extinguishing Agents Not to be Used:** Not pertinent

6.5 **Special Hazards of Combustion Products:** Toxic gases are generated when solid is heated or when solution burns.

6.6 **Behavior in Fire:** Not pertinent

6.7 **Ignition Temperature:** Not flammable

6.8 **Electrical Hazard:** Not pertinent

6.9 **Burning Rate:** Not flammable

7. CHEMICAL REACTIVITY

7.1 **Reactivity with Water:** No reaction

7.2 **Reactivity with Common Materials:** No reaction

7.3 **Stability During Transport:** Stable

7.4 **Neutralizing Agents for Acids and Caustics:** Not pertinent

7.5 **Polymerization:** Not pertinent

7.6 **Inhibitor of Polymerization:** Not pertinent

8. WATER POLLUTION

8.1 **Aquatic Toxicity:**
0.77 ppm/96 hr/bluegill/TL_m/fresh water
0.138 ppm/96 hr/guppy/TL_m/fresh water
0.034 ppm/10 hr/coho fry/100% lethal/salt water

8.2 **Waterfowl Toxicity:** 2000 mg/kg

8.3 **Biological Oxygen Demand (BOD):** Data not available

8.4 **Food Chain Concentration Potential:** High

9. SELECTED MANUFACTURERS

Hooker Chemical Corp.
Industrial Chemicals Div.
Niagara Falls, N. Y. 14302

10. SHIPPING INFORMATION

10.1 **Grades or Purity:**
Fortified grade: 40–45% gamma isomer
Lindane: pure gamma isomer

10.2 **Storage Temperature:** Data not available

10.3 **Inert Atmosphere:** Data not available

10.4 **Venting:** Data not available

11. HAZARD ASSESSMENT CODE

(See Hazard Assessment Handbook, CG 446-3)

II

12. HAZARD CLASSIFICATIONS

12.1 **Code of Federal Regulations:** Poisonous liquid or solid, Class B

12.2 **NAS Hazard Rating for Bulk Water Transportation:** Not listed

12.3 **NFPA Hazard Classifications:**

Category	Classification*	
Health Hazard (Blue)	2	2
Flammability (Red)	0	1
Reactivity (Yellow)	0	0

*First column refers to non-fire situation

13. PHYSICAL AND CHEMICAL PROPERTIES

13.1 **Physical State at 15°C and 1 atm:** Solid

13.2 **Molecular Weight:** 290.83

13.3 **Boiling Point at 1 atm:** Not pertinent

13.4 **Freezing Point:** Not pertinent

13.5 **Critical Temperature:** Not pertinent

13.6 **Critical Pressure:** Not pertinent

13.7 **Specific Gravity:** 1.891 at 19°C (solid)

13.8 **Liquid Surface Tension:** Not pertinent

13.9 **Liquid-Water Interfacial Tension:** Not pertinent

13.10 **Vapor (Gas) Specific Gravity:** Not pertinent

13.11 **Ratio of Specific Heats of Vapor (Gas):** Not pertinent

13.12 **Latent Heat of Vaporization:** Not pertinent

13.13 **Heat of Combustion:** Not pertinent

13.14 **Heat of Decomposition:** Not pertinent

13.15 **Heat of Solution:** Not pertinent

13.16 **Heat of Polymerization:** Not pertinent

NOTES

REVISED 1978

BPD	BENZENE PHOSPHORUS DICHLORIDE

Common Synonyms			
Phenylphosphorus dichloride	Liquid	Colorless	Unpleasant odor
	Reacts violently with water. Poisonous visible vapor cloud is produced.		

Stop discharge if possible. Keep people away.
Avoid contact with liquid and vapor.
Isolate and remove discharged material.
Notify local health and pollution control agencies.

Fire

Fire data not available.

Exposure

CALL FOR MEDICAL AID.

GAS PRODUCED IN REACTION WITH WATER.
POISONOUS IF INHALED.
Irritating to eyes, nose and throat.
Move to fresh air.
If breathing has stopped, give artificial respiration.
If breathing is difficult, give oxygen.

LIQUID
Irritating to skin and eyes.
Harmful if swallowed.
Remove contaminated clothing and shoes.
Flush affected areas with plenty of water.
IF IN EYES, hold eyelids open and flush with plenty of water.
IF SWALLOWED and victim is CONSCIOUS, have victim drink water or milk.

Water Pollution

Effect of low concentrations on aquatic life is unknown.
May be dangerous if it enters water intakes.
Notify local health and wildlife officials.
Notify operators of nearby water intakes.

1. RESPONSE TO DISCHARGE
(See Response Methods Handbook, CG 446-4)

Issue warning—corrosive
Restrict access
Disperse and flush

2. LABEL

3. CHEMICAL DESIGNATIONS

3.1 **Synonyms:** Phenyl phosphonous dichloride; Phenylphosphine dichloride; Dichlorophenylphosphine

3.2 **Coast Guard Compatibility Classification:** Not listed

3.3 **Chemical Formula:** $C_6H_5PCl_2$

3.4 **IMCO/United Nations Numerical Designation:** Not listed

4. OBSERVABLE CHARACTERISTICS

4.1 **Physical State (as shipped):** Liquid

4.2 **Color:** Colorless

4.3 **Odor:** Acrid; pungent

5. HEALTH HAZARDS

5.1 **Personal Protective Equipment:** Self-contained breathing apparatus; acid-type canister mask; goggles and face shield; rubber gloves; protective clothing.

5.2 **Symptoms Following Exposure:** Inhalation causes irritation of nose and throat; pulmonary edema may develop following severe exposures. Contact with skin or eyes causes severe burns. Ingestion causes severe burns of mouth and stomach.

5.3 **Treatment for Exposure:** Get medical attention following all exposures to this compound. INHALATION: remove to fresh air; if breathing has stopped, start mouth-to-mouth resuscitation. EYES: flush with water for at least 15 min.; do not use oils or ointments. SKIN: flush with water; wash with soap and water. INGESTION: give large amounts of milk or water; do NOT induce vomiting; if vomiting does occur, give milk or beaten eggs at one-hour intervals.

5.4 **Toxicity by Inhalation (Threshold Limit Value):** Data not available

5.5 **Short-Term Inhalation Limits:** Data not available

5.6 **Toxicity by Ingestion:** Data not available

5.7 **Late Toxicity:** Data not available

5.8 **Vapor (Gas) Irritant Characteristics:** Data not available

5.9 **Liquid or Solid Irritant Characteristics:** Data not available

5.10 **Odor Threshold:** Data not available

6. FIRE HAZARDS

6.1 **Flash Point:** 215°F O.C. May be lower because of presence of dissolved phosphorus.

6.2 **Flammable Limits in Air:** Not pertinent

6.3 **Fire Extinguishing Agents:** Large amounts of water.

6.4 **Fire Extinguishing Agents Not to be Used:** Not pertinent

6.5 **Special Hazards of Combustion Products:** Toxic fumes include oxides of phosphorus and hydrogen chloride.

6.6 **Behavior in Fire:** Containers may rupture. Hot liquid is spontaneously flammable because of presence of dissolved phosphorus.

6.7 **Ignition Temperature:** 319°F

6.8 **Electrical Hazard:** Data not available

6.9 **Burning Rate:** Data not available

7. CHEMICAL REACTIVITY

7.1 **Reactivity with Water:** Reacts vigorously to form hydrogen chloride (hydrochloric acid).

7.2 **Reactivity with Common Materials:** Corrodes metal, except 316 stainless steel, nickel, or Hastelloy.

7.3 **Stability During Transport:** Stable

7.4 **Neutralizing Agents for Acids and Caustics:** Flush with water, rinse with sodium bicarbonate or lime solution.

7.5 **Polymerization:** Not pertinent

7.6 **Inhibitor of Polymerization:** Not pertinent

8. WATER POLLUTION

8.1 **Aquatic Toxicity:** Data not available

8.2 **Waterfowl Toxicity:** Data not available

8.3 **Biological Oxygen Demand (BOD):** Data not available

8.4 **Food Chain Concentration Potential:** None

9. SELECTED MANUFACTURERS

1. PCR, Inc.
   P. O. Box 1466
   Gainesville, Fla. 32602
2. Aldrich Chemical Co.
   940 West St. Paul Ave.
   Milwaukee, Wis. 53233
3. Stauffer Chemical Co.
   Agricultural Chemical Division
   Westport, Conn. 06880

10. SHIPPING INFORMATION

10.1 **Grades or Purity:** Commercial

10.2 **Storage Temperature:** Ambient

10.3 **Inert Atmosphere:** No requirement

10.4 **Venting:** Pressure-vacuum

11. HAZARD ASSESSMENT CODE
(See Hazard Assessment Handbook, CG 446-3)

A-O

12. HAZARD CLASSIFICATIONS

12.1 **Code of Federal Regulations:** Corrosive

12.2 **NAS Hazard Rating for Bulk Water Transportation:** Not listed

12.3 **NFPA Hazard Classifications:** Not listed

13. PHYSICAL AND CHEMICAL PROPERTIES

13.1 **Physical State at 15°C and 1 atm:** Liquid

13.2 **Molecular Weight:** 179.0

13.3 **Boiling Point at 1 atm:** 430°F = 221°C = 494°K

13.4 **Freezing Point:** −60°F = −51°C = 222°K

13.5 **Critical Temperature:** Not pertinent

13.6 **Critical Pressure:** Not pertinent

13.7 **Specific Gravity:** 1.140 at 25°C (liquid)

13.8 **Liquid Surface Tension:** (est.) 25 dynes/cm = 0.025 N/m at 20°C

13.9 **Liquid-Water Interfacial Tension:** Not pertinent

13.10 **Vapor (Gas) Specific Gravity:** Not pertinent

13.11 **Ratio of Specific Heats of Vapor (Gas):** Not pertinent

13.12 **Latent Heat of Vaporization:** Not pertinent

13.13 **Heat of Combustion:** (est.) −8,200 Btu/lb = −4,500 cal/g = -190×10^5 J/kg

13.14 **Heat of Decomposition:** Not pertinent

13.15 **Heat of Solution:** −72 Btu/lb = −40 cal/g = -1.7×10^5 J/kg

13.16 **Heat of Polymerization:** Not pertinent

NOTES

BPT	BENZENE PHOSPHORUS THIODICHLORIDE

Common Synonyms			
Benzenethiophosphonyl chloride Phenylphosphonothioic dichloride Phenylphosphine thiodichloride	Liquid	Colorless	Unpleasant odor
	Sinks and reacts in water. Poisonous visible vapor cloud is produced.		

Stop discharge if possible. Keep people away.
Avoid contact with liquid and vapor.
Isolate and remove discharged material.
Notify local health and pollution control agencies.

Fire

Fire data not available.

Exposure

CALL FOR MEDICAL AID.

VAPOR PRODUCED IN REACTION WITH WATER.

POISONOUS IF INHALED.
Irritating to eyes, nose and throat.
Move to fresh air.
If breathing has stopped, give artificial respiration.
If breathing is difficult, give oxygen.

LIQUID
Irritating to skin and eyes.
Harmful if swallowed.
Remove contaminated clothing and shoes.
Flush affected areas with plenty of water.
IF IN EYES, hold eyelids open and flush with plenty of water.
IF SWALLOWED and victim is CONSCIOUS, have victim drink water or milk.

Water Pollution

Effect of low concentrations on aquatic life is unknown.
May be dangerous if it enters water intakes.
Notify local health and wildlife officials.
Notify operators of nearby water intakes.

1. RESPONSE TO DISCHARGE

(See Response Methods Handbook, CG 446-4)

Issue warning—corrosive
Restrict access
Disperse and flush

2. LABEL

3. CHEMICAL DESIGNATIONS

3.1 **Synonyms:** Benzenethiophosphonyl chloride; Phenylphosphonothioic dichloride; Phenylphosphine thiodichloride

3.2 **Coast Guard Compatibility Classification:** Not listed

3.3 **Chemical Formula:** $C_6H_5PSCl_2$

3.4 **IMCO/United Nations Numerical Designation:** Not listed

4. OBSERVABLE CHARACTERISTICS

4.1 **Physical State (as shipped):** Liquid

4.2 **Color:** Colorless to light yellow

4.3 **Odor:** Acrid; pungent

5. HEALTH HAZARDS

5.1 **Personal Protective Equipment:** Self-contained breathing apparatus; acid-type canister mask; goggles and face shield; rubber gloves; protective clothing.

5.2 **Symptoms Following Exposure:** Inhalation of vapor irritates nose and throat; pulmonary edema may result. Contact with eyes or skin causes severe irritation. Ingestion causes severe irritation of mouth and stomach.

5.3 **Treatment for Exposure:** Get medical attention following all exposures to this compound. INHALATION: remove to fresh air. EYES: flush with water for at least 15 min.; do not use oils or ointments. SKIN: flush with water; wash with soap and water. INGESTION: give large amounts of water or milk; induce vomiting; give milk, eggs, or olive oil.

5.4 **Toxicity by Inhalation (Threshold Limit Value):** Data not available

5.5 **Short-Term Inhalation Limits:** Data not available

5.6 **Toxicity by Ingestion:** Data not available

5.7 **Late Toxicity:** Data not available

5.8 **Vapor (Gas) Irritant Characteristics:** Data not available

5.9 **Liquid or Solid Irritant Characteristics:** Data not available

5.10 **Odor Threshold:** Data not available

6. FIRE HAZARDS

6.1 **Flash Point:** 252°F O.C.

6.2 **Flammable Limits in Air:** Not pertinent

6.3 **Fire Extinguishing Agents:** Water

6.4 **Fire Extinguishing Agents Not to be Used:** Not pertinent

6.5 **Special Hazards of Combustion Products:** Toxic fumes include oxides of phosphorus and sulfur and hydrogen chloride.

6.6 **Behavior in Fire:** Containers may rupture.

6.7 **Ignition Temperature:** 338°F

6.8 **Electrical Hazard:** Data not available

6.9 **Burning Rate:** Data not available

7. CHEMICAL REACTIVITY

7.1 **Reactivity with Water:** Forms hydrogen chloride fumes (hydrochloric acid). The reaction is slow unless water is hot.

7.2 **Reactivity with Common Materials:** Corrodes metal slowly.

7.3 **Stability During Transport:** Stable

7.4 **Neutralizing Agents for Acids and Caustics:** Flush with water, rinse with sodium bicarbonate or lime solution.

7.5 **Polymerization:** Not pertinent

7.6 **Inhibitor of Polymerization:** Not pertinent

8. WATER POLLUTION

8.1 **Aquatic Toxicity:** Data not available

8.2 **Waterfowl Toxicity:** Data not available

8.3 **Biological Oxygen Demand (BOD):** Data not available

8.4 **Food Chain Concentration Potential:** None

9. SELECTED MANUFACTURERS

1. Aldrich Chemical Co.
   940 West St. Paul Ave.
   Milwaukee, Wis. 53233
2. PCR, Inc.
   P. O. Box 1466
   Gainesville, Fla. 32602
3. Stauffer Chemical Co.
   Agricultural Chemical Division
   Westport, Conn. 06880

10. SHIPPING INFORMATION

10.1 **Grades or Purity:** Commercial

10.2 **Storage Temperature:** Ambient

10.3 **Inert Atmosphere:** No requirement

10.4 **Venting:** Pressure-vacuum

11. HAZARD ASSESSMENT CODE

(See Hazard Assessment Handbook, CG 446-3)

A-O

12. HAZARD CLASSIFICATIONS

12.1 **Code of Federal Regulations:** Corrosive

12.2 **NAS Hazard Rating for Bulk Water Transportation:** Not listed

12.3 **NFPA Hazard Classifications:** Not listed

13. PHYSICAL AND CHEMICAL PROPERTIES

13.1 **Physical State at 15°C and 1 atm:** Liquid

13.2 **Molecular Weight:** 211

13.3 **Boiling Point at 1 atm:** 518°F = 270°C = 543°K

13.4 **Freezing Point:** −11.2°F = −24.0°C = 249.2°K

13.5 **Critical Temperature:** Not pertinent

13.6 **Critical Pressure:** Not pertinent

13.7 **Specific Gravity:** 1.378 at 20°C (liquid)

13.8 **Liquid Surface Tension:** (est.) 25 dynes/cm = 0.025 N/m at 20°C

13.9 **Liquid-Water Interfacial Tension:** Not pertinent

13.10 **Vapor (Gas) Specific Gravity:** Not pertinent

13.11 **Ratio of Specific Heats of Vapor (Gas):** Not pertinent

13.12 **Latent Heat of Vaporization:** Not pertinent

13.13 **Heat of Combustion:** (est.) −7,700 Btu/lb = −4,300 cal/g = -180×10^5 J/kg

13.14 **Heat of Decomposition:** Not pertinent

13.15 **Heat of Solution:** (est.) −9 Btu/lb = −5 cal/g = -0.2×10^5 J/kg

13.16 **Heat of Polymerization:** Not pertinent

NOTES

BZA BENZOIC ACID

Common Synonyms Benzenecarboxylic acid Carboxylbenzene	Solid crystals or powder	White	Faint pleasant odor
	Sinks in water.		

Avoid contact with solid and dust. Keep people away.
Wear goggles and self-contained breathing apparatus.
Stay upwind and use water spray to "knock down" dust.
Call fire department.
Isolate and remove discharged material.
Notify local health and pollution control agencies.

Fire

Combustible.
Vapor may explode if ignited in an enclosed area.
Dust may form explosive mixture with air.
Extinguish with water, dry chemical, chemical foam, or carbon dioxide.

Exposure

CALL FOR MEDICAL AID.

DUST
Irritating to nose and throat if inhaled.
Move to fresh air.

SOLID
Irritating to skin and eyes.
Flush affected areas with plenty of water.
IF IN EYES, hold eyelids open and flush with plenty of water.

Water Pollution

HARMFUL TO AQUATIC LIFE IN VERY LOW CONCENTRATIONS.
May be dangerous if it enters water intakes.
Notify local health and wildlife officials.
Notify operators of nearby water intakes.

1. RESPONSE TO DISCHARGE

(See Response Methods Handbook, CG 446-4)

Disperse and flush

2. LABELS

No hazard label required by Code of Federal Regulations

3. CHEMICAL DESIGNATIONS

3.1 **Synonyms:** Benzenecarboxylic acid; Carboxybenzene; Dracyclic acid

3.2 **Coast Guard Compatibility Classification:** Not applicable

3.3 **Chemical Formula:** C_6H_5COOH

3.4 **IMCO/United Nations Numerical Designation:** Not listed

4. OBSERVABLE CHARACTERISTICS

4.1 **Physical State (as shipped):** Solid

4.2 **Color:** White

4.3 **Odor:** Faint, pleasant; slight aromatic

5. HEALTH HAZARDS

5.1 **Personal Protective Equipment:** Bureau of Mines dust respirator; when melted material present, use eye protection and organic respirator for fumes.

5.2 **Symptoms Following Exposure:** Dust may be irritating to nose and eyes. At elevated temperatures, fumes may cause irritation of eyes, respiratory system, and skin.

5.3 **Treatment for Exposure:** Remove patient to fresh air. EYE CONTACT: flush eyes with water.

5.4 **Toxicity by Inhalation (Threshold Limit Value):** Not pertinent

5.5 **Short-Term Inhalation Limits:** Not pertinent

5.6 **Toxicity by Ingestion:** Grade 2; LD_{50} 0.5 to 5 g/kg

5.7 **Late Toxicity:** None

5.8 **Vapor (Gas) Irritant Characteristics:** Not pertinent

5.9 **Liquid or Solid Irritant Characteristics:** Minimum hazard. If spilled on clothing and allowed to remain, may cause smarting and reddening of the skin. Dust may irritate nose and eyes.

5.10 **Odor Threshold:** Not pertinent

6. FIRE HAZARDS

6.1 **Flash Point:** 250°F C.C.

6.2 **Flammable Limits in Air:** Not pertinent

6.3 **Fire Extinguishing Agents:** Dry powder, chemical foam, water fog, carbon dioxide

6.4 **Fire Extinguishing Agents Not to be Used:** None

6.5 **Special Hazards of Combustion Products:** Not pertinent

6.6 **Behavior in Fire:** Vapor from molten benzoic acid may form explosive mixture with air. Concentrated dust may form explosive mixture.

6.7 **Ignition Temperature:** 1063°F

6.8 **Electrical Hazard:** Not pertinent

6.9 **Burning Rate:** Not pertinent

7. CHEMICAL REACTIVITY

7.1 **Reactivity with Water:** No reaction

7.2 **Reactivity with Common Materials:** No reaction

7.3 **Stability During Transport:** Stable

7.4 **Neutralizing Agents for Acids and Caustics:** Not pertinent

7.5 **Polymerization:** Not pertinent

7.6 **Inhibitor of Polymerization:** Not pertinent

8. WATER POLLUTION

8.1 **Aquatic Toxicity:**
200 ppm/7 hr/goldfish/lethal/fresh water
500 ppm/1 hr/sunfish/lethal/fresh water

8.2 **Waterfowl Toxicity:** Data not available

8.3 **Biological Oxygen Demand (BOD):** 165%, 5 days

8.4 **Food Chain Concentration Potential:** None

9. SELECTED MANUFACTURERS

1. Monsanto Co.
Monsanto Industrial Chemicals Co.
800 North Lindbergh Blvd.
St. Louis, Mo. 63166

2. Northwest Industries Inc.
Velsicol Chemical Corp.
341 E. Ohio St.
Chicago, Ill. 60611

3. Tenneco Chemicals Inc.
Tenneco Intermediates Div.
5 Turner Place
Piscataway, N. J. 08854

10. SHIPPING INFORMATION

10.1 **Grades or Purity:** USP, FCC grade: 99.5% – 100.5%

10.2 **Storage Temperature:** Data not available

10.3 **Inert Atmosphere:** Data not available

10.4 **Venting:** Data not available

11. HAZARD ASSESSMENT CODE

(See Hazard Assessment Handbook, CG 446-3)

II

12. HAZARD CLASSIFICATIONS

12.1 **Code of Federal Regulations:** Not listed

12.2 **NAS Hazard Rating for Bulk Water Transportation:** Not listed

12.3 **NFPA Hazard Classifications:** Not listed

13. PHYSICAL AND CHEMICAL PROPERTIES

13.1 **Physical State at 15°C and 1 atm:** Solid

13.2 **Molecular Weight:** 122.12

13.3 **Boiling Point at 1 atm:** 480.6°F = 249.2°C = 522.4°K

13.4 **Freezing Point:** 252.1°F = 122.3°C = 395.5°K

13.5 **Critical Temperature:** 894°F = 479°C = 752°K

13.6 **Critical Pressure:** 660 psia = 45 atm = 4.6 MN/m^2

13.7 **Specific Gravity:** 1.316 at 28°C (solid)

13.8 **Liquid Surface Tension:** Not pertinent

13.9 **Liquid-Water Interfacial Tension:** Not pertinent

13.10 **Vapor (Gas) Specific Gravity:** Not pertinent

13.11 **Ratio of Specific Heats of Vapor (Gas):** Not pertinent

13.12 **Latent Heat of Vaporization:** Not pertinent

13.13 **Heat of Combustion:** Not pertinent

13.14 **Heat of Decomposition:** Not pertinent

13.15 **Heat of Solution:** Not pertinent

13.16 **Heat of Polymerization:** Not pertinent

NOTES

BZN	# BENZONITRILE

Common Synonyms			
Phenylcyanide Cyanobenzene Benzoic acid nitrile	Liquid	Colorless	Almond-like odor
	May float or sink in water.		

AVOID CONTACT WITH LIQUID AND VAPOR. KEEP PEOPLE AWAY.
Stop discharge if possible.
Call fire department.
Isolate and remove discharged material.
Notify local health and pollution control agencies.

Fire	Combustible. POISONOUS GASES MAY BE PRODUCED IN FIRE. Wear goggles and self-contained breathing apparatus. Extinguish with dry chemicals, foam, or carbon dioxide. Water may be ineffective on fire. Cool exposed containers with water.
Exposure	CALL FOR MEDICAL AID. VAPOR Irritating to eyes, nose and throat. If inhaled will cause headache, difficult breathing or loss of consciousness. If in eyes, hold eyelids open and flush with plenty of water. If breathing has stopped, give artificial respiration. If breathing is difficult, give oxygen. LIQUID Irritating to skin and eyes. If swallowed will cause headache, nausea, vomiting or loss of consciousness. Remove contaminated clothing and shoes. Flush affected areas with plenty of water. IF IN EYES, hold eyelids open and flush with plenty of water. IF SWALLOWED and victim is CONSCIOUS, have victim drink water or milk. IF SWALLOWED and victim is UNCONSCIOUS OR HAVING CONVULSIONS, do nothing except keep victim warm.
Water Pollution	Effect of low concentrations on aquatic life is unknown. May be dangerous if it enters water intakes. Notify local health and wildlife officials. Notify operators of nearby water intakes.

1. RESPONSE TO DISCHARGE

(See Response Methods Handbook, CG 446-4)

Issue warning—water contaminant
Mechanical containment
Should be removed
Chemical and physical treatment

2. LABELS

No hazard label required by Code of Federal Regulations

3. CHEMICAL DESIGNATIONS

3.1 **Synonyms:** Benzoic acid nitrile; Cyanobenzene; Phenylcyanide
3.2 **Coast Guard Compatibility Classification:** Not listed
3.3 **Chemical Formula:** C_6H_5CN
3.4 **IMCO/United Nations Numerical Designation:** Not listed

4. OBSERVABLE CHARACTERISTICS

4.1 **Physical State (as shipped):** Liquid
4.2 **Color:** Colorless
4.3 **Odor:** Almond-like

5. HEALTH HAZARDS

5.1 **Personal Protective Equipment:** Rubber gloves; splash-proof goggles; rubber boots or rubber overshoes; impervious clothing for splash protection; cartridge-type mask or other protection against vapor must be worn for working in poorly ventilated area or where poisoning by inhalation may be possible.

5.2 **Symptoms Following Exposure:** Benzonitrile may enter the human body by ingestion, absorption through the skin, or inhalation. The earliest symptoms of cyano compound intoxication may be weakness, headaches, confusion, and occasionally nausea and vomiting. The respiratory rate and depth will usually be increased at the beginning and at later stages become slow and gasping. Blood pressure is usually normal, especially in the mild or moderately severe cases, although the pulse rate is usually more rapid than normal.

5.3 **Treatment for Exposure:** INHALATION: remove patient to fresh air; get immediate medical attention. INGESTION: *Call physician immediately*. Until physician arrives, take the following steps: a. Provide for inhalation of amyl nitrite vapor from ampules crushed in a handkerchief and held to the nose of the victim. b. Induce vomiting unless patient is unconscious. (Gastric lavage should be employed by, or under the supervision of, a physician.) c. Keep patient warm and quiet until medical attention arrives. EYES: immediately flush with large volumes of water for at least 15 min. SKIN: wash thoroughly at once, without scrubbing, with large amounts of soap and water. OTHER: exposed personnel should be checked periodically for chronic toxic effects.

5.4 **Toxicity by Inhalation (Threshold Limit Value):** Data not available
5.5 **Short-Term Inhalation Limits:** Data not available
5.6 **Toxicity by Ingestion:** Grade 2; oral rat LD_{50} = 800 mg/kg

6. FIRE HAZARDS

(Material is combustible but burns with difficulty)

6.1 **Flash Point:** 167°F C.C.
6.2 **Flammable Limits in Air:** Data not available
6.3 **Fire Extinguishing Agents:** Foam, dry chemical, carbon dioxide
6.4 **Fire Extinguishing Agents Not to be Used:** Water may be ineffective.
6.5 **Special Hazards of Combustion Products:** Toxic hydrogen cyanide and oxides of nitrogen may form in fire.
6.6 **Behavior in Fire:**
6.7 **Ignition Temperature:** Data not available
6.8 **Electrical Hazard:** Data not available
6.9 **Burning Rate:** Difficult to burn

7. CHEMICAL REACTIVITY

7.1 **Reactivity with Water:** No reaction
7.2 **Reactivity with Common Materials:** Will attack some plastics
7.3 **Stability During Transport:** Stable
7.4 **Neutralizing Agents for Acids and Caustics:** Not pertinent
7.5 **Polymerization:** Not pertinent
7.6 **Inhibitor of Polymerization:** Not pertinent

8. WATER POLLUTION

8.1 **Aquatic Toxicity:**
5 ppm/24 hr/rainbow trout/no effect/fresh water
135 ppm/96 hr/fathead minnow/TL_m/soft fresh water
78 ppm/96 hr/fathead minnow/TL_m/hard fresh water
8.2 **Waterfowl Toxicity:** Data not available
8.3 **Biological Oxygen Demand (BOD):** 60% (theo.), 18 days
8.4 **Food Chain Concentration Potential:** None

9. SELECTED MANUFACTURERS

1. Velsicol Chemical Corp.
341 E. Ohio Street
Chicago, Ill. 60611
2. Eastman Organic Chemicals
Rochester, N. Y. 14650
3. Aldrich Chemical Co.
940 W. Saint Paul Ave.
Milwaukee, Wis. 53233

10. SHIPPING INFORMATION

10.1 **Grades or Purity:** Pure, 99+%
10.2 **Storage Temperature:** Ambient
10.3 **Inert Atmosphere:** Ventilated (natural)
10.4 **Venting:** Open (flame arrester)

11. HAZARD ASSESSMENT CODE

(See Hazard Assessment Handbook, CG 446-3)

A-T-X

12. HAZARD CLASSIFICATIONS

12.1 **Code of Federal Regulations:** Not listed
12.2 **NAS Hazard Rating for Bulk Water Transportation:** Not listed
12.3 **NFPA Hazard Classifications:** Not listed

13. PHYSICAL AND CHEMICAL PROPERTIES

13.1 **Physical State at 15°C and 1 atm:** Liquid
13.2 **Molecular Weight:** 103.12
13.3 **Boiling Point at 1 atm:** 376°F = 191°C = 464°K
13.4 **Freezing Point:** 9.0°F = −12.8°C = 260.4°K
13.5 **Critical Temperature:** 799.2°F = 426.2°C = 699.4°K
13.6 **Critical Pressure:** 611 psia = 41.6 atm = 4.22 MN/m²
13.7 **Specific Gravity:** 1.01 at 25°C (liquid)
13.8 **Liquid Surface Tension:** 34.7 dynes/cm = 0.0347 N/m at 25°C
13.9 **Liquid-Water Interfacial Tension:** Data not available
13.10 **Vapor (Gas) Specific Gravity:** 3.6
13.11 **Ratio of Specific Heats of Vapor (Gas):** 1.091
13.12 **Latent Heat of Vaporization:** 157.7 Btu/lb = 87.6 cal/g = 3.67 × 10⁵ J/kg
13.13 **Heat of Combustion:** −15,100 Btu/lb = −8,400 cal/g = −351 × 10⁵ J/kg
13.14 **Heat of Decomposition:** Not pertinent
13.15 **Heat of Solution:** Not pertinent
13.16 **Heat of Polymerization:** Not pertinent

5. HEALTH HAZARDS (Cont'd.)

5.7 **Late Toxicity:** Data not available
5.8 **Vapor (Gas) Irritant Characteristics:** Data not available
5.9 **Liquid or Solid Irritant Characteristics:** Data not available
5.10 **Odor Threshold:** Data not available

BZP — BENZOPHENONE

Common Synonyms
Benzoyl benzene
Diphenyl ketone
Diphenyl methanone
Alpha-oxodiphenylmethane
Alpha-oxoditane

Liquid or solid — White — Flowery odor

May float or sink in water.

Stop discharge if possible. Keep people away.
Avoid contact with liquid and solid.
Isolate and remove discharged material.
Notify local health and pollution control agencies.

Fire

Combustible.
Extinguish with dry chemicals, alcohol foam, or carbon dioxide.
Water may be ineffective on fire.

Exposure

CALL FOR MEDICAL AID.

LIQUID OR SOLID
Irritating to skin and eyes.
If swallowed will cause nausea, or vomiting.
Remove contaminated clothing and shoes.
Flush affected areas with plenty of water.
IF IN EYES, hold eyelids open and flush with plenty of water.
IF SWALLOWED and victim is CONSCIOUS, have victim drink water or milk.
IF SWALLOWED and victim is UNCONSCIOUS OR HAVING CONVULSIONS, do nothing except keep victim warm.

Water Pollution

Effect of low concentrations on aquatic life is unknown.
Fouling to shoreline.
May be dangerous if it enters water intakes.
Notify local health and wildlife officials.
Notify operators of nearby water intakes.

1. RESPONSE TO DISCHARGE
(See Response Methods Handbook, CG 446-4)

Issue warning—water contaminant
Mechanical containment
Should be removed
Chemical and physical treatment

2. LABELS

No hazard label required by Code of Federal Regulations

3. CHEMICAL DESIGNATIONS

3.1 **Synonyms:** Benzoylbenzene; Diphenyl ketone; Diphenyl methanone; *alpha*-Oxodiphenylmethane; *alpha*-Oxoditane
3.2 **Coast Guard Compatibility Classification:** Not listed
3.3 **Chemical Formula:** $C_6H_5COC_6H_5$
3.4 **IMCO/United Nations Numerical Designation:** Not listed

4. OBSERVABLE CHARACTERISTICS

4.1 **Physical State (as shipped):** Liquid or solid
4.2 **Color:** White
4.3 **Odor:** Characteristic

5. HEALTH HAZARDS

5.1 **Personal Protective Equipment:** Goggles or face shield; rubber gloves
5.2 **Symptoms Following Exposure:** Ingestion causes gastrointestinal disturbances. Contact causes eye irritation and, if prolonged, irritation of skin.
5.3 **Treatment for Exposure:** INHALATION: remove to fresh air. INGESTION: get medical attention. EYES: flush with water for at least 15 min.; get medical attention if irritation persists. SKIN: flush with water, wash with soap and water.
5.4 **Toxicity by Inhalation (Threshold Limit Value):** Data not available
5.5 **Short-Term Inhalation Limits:** Data not available
5.6 **Toxicity by Ingestion:** Grade 1; acute oral rat $LD_{50} \geq$ 10,000 mg/kg
5.7 **Late Toxicity:** Data not available
5.8 **Vapor (Gas) Irritant Characteristics:** Data not available
5.9 **Liquid or Solid Irritant Characteristics:** Data not available
5.10 **Odor Threshold:** Data not available

6. FIRE HAZARDS
(Material is combustible)

6.1 **Flash Point:** Data not available
6.2 **Flammable Limits in Air:** Data not available
6.3 **Fire Extinguishing Agents:** Foam, dry chemical, carbon dioxide
6.4 **Fire Extinguishing Agents Not to be Used:** Water may be ineffective.
6.5 **Special Hazards of Combustion Products:**
6.6 **Behavior in Fire:**
6.7 **Ignition Temperature:** Data not available
6.8 **Electrical Hazard:** Data not available
6.9 **Burning Rate:** Not pertinent

7. CHEMICAL REACTIVITY

7.1 **Reactivity with Water:** No reaction
7.2 **Reactivity with Common Materials:** Will attack some plastics
7.3 **Stability During Transport:** Stable
7.4 **Neutralizing Agents for Acids and Caustics:** Not pertinent
7.5 **Polymerization:** Not pertinent
7.6 **Inhibitor of Polymerization:** Not pertinent

8. WATER POLLUTION

8.1 **Aquatic Toxicity:** Data not available
8.2 **Waterfowl Toxicity:** Data not available
8.3 **Biological Oxygen Demand (BOD):** Data not available
8.4 **Food Chain Concentration Potential:** None

9. SELECTED MANUFACTURERS

1. Eastman Organic Chemicals
Rochester, N. Y. 14650
2. Aldrich Chemical Co.
940 W. Saint Paul Ave.
Milwaukee, Wis. 53233
3. Gallard-Schlesinger Chemical Mfg. Co.
584 Mineola Avenue
Carle Place, N. Y. 11514

10. SHIPPING INFORMATION

10.1 **Grades or Purity:** 99+%
10.2 **Storage Temperature:** Ambient
10.3 **Inert Atmosphere:** No requirement
10.4 **Venting:** Open

11. HAZARD ASSESSMENT CODE
(See Hazard Assessment Handbook, CG 446-3)

A-T-X or II

12. HAZARD CLASSIFICATIONS

12.1 **Code of Federal Regulations:** Not listed
12.2 **NAS Hazard Rating for Bulk Water Transportation:** Not listed
12.3 **NFPA Hazard Classifications:** Not listed

13. PHYSICAL AND CHEMICAL PROPERTIES

13.1 **Physical State at 15°C and 1 atm:** Solid
13.2 **Molecular Weight:** 182
13.3 **Boiling Point at 1 atm:** 581.9°F = 305.5°C = 578.7°K
13.4 **Freezing Point:** 118.2°F = 47.9°C = 321.1°K
13.5 **Critical Temperature:** Not pertinent
13.6 **Critical Pressure:** Not pertinent
13.7 **Specific Gravity:** 1.085 at 50°C (liquid)
13.8 **Liquid Surface Tension:** 42 dynes/cm = 0.042 N/m at 50°C
13.9 **Liquid-Water Interfacial Tension:** Data not available
13.10 **Vapor (Gas) Specific Gravity:** Not pertinent
13.11 **Ratio of Specific Heats of Vapor (Gas):** Not pertinent
13.12 **Latent Heat of Vaporization:** 126.0 Btu/lb = 70.0 cal/g = 2.93×10^5 J/kg
13.13 **Heat of Combustion:** −15,400 Btu/lb = −8,550 cal/g = -358×10^5 J/kg
13.14 **Heat of Decomposition:** Not pertinent
13.15 **Heat of Solution:** Not pertinent
13.16 **Heat of Polymerization:** Not pertinent

NOTES

BZC	BENZOYL CHLORIDE

Common Synonyms	
Benzenecarbonyl chloride	Watery liquid Colorless to slightly brown Sinks and reacts slowly with water producing a poisonous gas.

Avoid contact with liquid and vapor.
Wear goggles and self-contained breathing apparatus.
Keep people away. Evacuate area in case of large discharge.
Stop discharge if possible.
Call fire department.
Isolate and remove discharged material.
Notify local health and pollution control agencies.

Fire	Combustible. POISONOUS GASES ARE PRODUCED IN FIRE AND WHEN HEATED. Wear goggles and self-contained breathing apparatus. DO NOT USE WATER. Extinguish with foam, dry chemical, or carbon dioxide.
Exposure	CALL FOR MEDICAL AID. LIQUID Will burn skin and eyes. Harmful if swallowed. Remove contaminated clothing and shoes. Flush affected areas with plenty of water. IF IN EYES, hold eyelids open and flush with plenty of water. IF SWALLOWED and victim is CONSCIOUS, have victim drink water or milk. DO NOT INDUCE VOMITING.
Water Pollution	HARMFUL TO AQUATIC LIFE IN VERY LOW CONCENTRATIONS. May be dangerous if it enters water intakes. Notify local health and wildlife officials. Notify operators of nearby water intakes.

1. RESPONSE TO DISCHARGE

(See Response Methods Handbook, CG 446-4)

Issue warning—corrosive
Restrict access
Chemical and physical treatment

2. LABEL

3. CHEMICAL DESIGNATIONS

3.1 **Synonyms:** Benzenecarbonyl chloride
3.2 **Coast Guard Compatibility Classification:** Not applicable
3.3 **Chemical Formula:** C_6H_5COCl
3.4 **IMCO/United Nations Numerical Designation:** 8/1736

4. OBSERVABLE CHARACTERISTICS

4.1 **Physical State (as shipped):** Liquid
4.2 **Color:** Colorless; may become slightly brownish on standing
4.3 **Odor:** Pungent characteristic

5. HEALTH HAZARDS

5.1 **Personal Protective Equipment:** Full protective clothing, including full-face respirator for acid gases and organic vapors (yellow GMC canister), close-fitting goggles, nonslip rubber gloves, plastic apron, face shield.
5.2 **Symptoms Following Exposure:** INHALATION: may irritate eyes, nose and throat. INGESTION: causes acute discomfort. SKIN: causes irritation and burning.
5.3 **Treatment for Exposure:** INHALATION: remove to fresh air; administer oxygen with patient in sitting position. INGESTION: give water; call physician at once; give milk. EYES: flush with water for 15 min.; get medical attention. SKIN: wash with plenty of soap and water.
5.4 **Toxicity by Inhalation (Threshold Limit Value):** Data not available
5.5 **Short-Term Inhalation Limits:** Data not available
5.6 **Toxicity by Ingestion:** Data not available
5.7 **Late Toxicity:** Data not available
5.8 **Vapor (Gas) Irritant Characteristics:** Vapors cause severe irritation of eyes and throat and can cause eye and lung injury. They cannot be tolerated even at low concentrations.
5.9 **Liquid or Solid Irritant Characteristics:** Severe skin irritant. Causes second- and third-degree burns on short contact and is very injurious to the eyes.
5.10 **Odor Threshold:** Data not available

6. FIRE HAZARDS

6.1 **Flash Point:** 162°F O.C.
6.2 **Flammable Limits in Air:** 1.2%–4.9%
6.3 **Fire Extinguishing Agents:** Foam, carbon dioxide, dry chemical, water fog
6.4 **Fire Extinguishing Agents Not to be Used:** Water spray. Do not allow water to enter containers.
6.5 **Special Hazards of Combustion Products:** Highly poisonous phosgene gas may be formed in fires.
6.6 **Behavior in Fire:** At fire temperatures the compound may react violently with water or steam.
6.7 **Ignition Temperature:** 185°F
6.8 **Electrical Hazard:** Not pertinent
6.9 **Burning Rate:** Data not available

7. CHEMICAL REACTIVITY

7.1 **Reactivity with Water:** Slow reaction with water to produce hydrochloric acid fumes. Reaction much faster with steam.
7.2 **Reactivity with Common Materials:** Slow corrosion of metals, but no immediate hazard.
7.3 **Stability During Transport:** Not pertinent
7.4 **Neutralizing Agents for Acids and Caustics:** Soda ash and water; lime
7.5 **Polymerization:** Does not occur
7.6 **Inhibitor of Polymerization:** Not pertinent

8. WATER POLLUTION

8.1 **Aquatic Toxicity:**
200 ppm/7 hr/goldfish/lethal/fresh water
500 ppm/1 hr/sunfish/lethal/fresh water
8.2 **Waterfowl Toxicity:** Data not available
8.3 **Biological Oxygen Demand (BOD):** 165%, 5 days
8.4 **Food Chain Concentration Potential:** None

9. SELECTED MANUFACTURERS

1. Northwest Industries, Inc.
Velsicol Chemical Corp.
341 E. Ohio St.
Chicago, Ill. 60611
2. Occidental Petroleum Corp.
Hooker Chemical Corp.
Industrial Chemicals Division
Niagara Falls, N. Y. 14302
3. Stauffer Chemical Co.
Specialty Chemicals Division
Green Farms Rd.
Westport, Conn. 06880

10. SHIPPING INFORMATION

10.1 **Grades or Purity:** 99+%; special grade
10.2 **Storage Temperature:** Store in cool, dry area
10.3 **Inert Atmosphere:** Data not available
10.4 **Venting:** Pressure-vacuum

11. HAZARD ASSESSMENT CODE

(See Hazard Assessment Handbook, CG 446-3)

A-X-Y-O

12. HAZARD CLASSIFICATIONS

12.1 **Code of Federal Regulations:** Corrosive material
12.2 **NAS Hazard Rating for Bulk Water Transportation:**

Category	Rating
Fire	1
Health	
Vapor Irritant	4
Liquid or Solid Irritant	4
Poisons	2
Water Pollution	
Human Toxicity	2
Aquatic Toxicity	2
Aesthetic Effect	2
Reactivity	
Other Chemicals	3
Water	4
Self-Reaction	0

12.3 **NFPA Hazard Classifications:**

Category	Classification
Health Hazard (Blue)	3
Flammability (Red)	2
Reactivity (Yellow)	1 W̶

13. PHYSICAL AND CHEMICAL PROPERTIES

13.1 **Physical State at 15°C and 1 atm:** Liquid
13.2 **Molecular Weight:** 140.57
13.3 **Boiling Point at 1 atm:** 387°F = 197.3°C = 470.5°K
13.4 **Freezing Point:** 30.9°F = −0.6°C = 272.6°K
13.5 **Critical Temperature:** Not pertinent
13.6 **Critical Pressure:** Not pertinent
13.7 **Specific Gravity:** 1.211 at 25°C (liquid)
13.8 **Liquid Surface Tension:** 36.3 dynes/cm = 0.0363 N/m at 20°C
13.9 **Liquid-Water Interfacial Tension:** Not pertinent
13.10 **Vapor (Gas) Specific Gravity:** Not pertinent
13.11 **Ratio of Specific Heats of Vapor (Gas):** Not pertinent
13.12 **Latent Heat of Vaporization:** Data not available
13.13 **Heat of Combustion:** −10,030 Btu/lb = −5570 cal/g = -233.2×10^5 J/kg
13.14 **Heat of Decomposition:** Not pertinent
13.15 **Heat of Solution:** Not pertinent
13.16 **Heat of Polymerization:** Not pertinent

NOTES

REVISED 1978

BAL — BENZYL ALCOHOL

Common Synonyms			
Phenylcarbinol Benzenecarbinol Alpha-hydroxytoluene Phenylmethanol Phenylmethyl alcohol	Liquid	Colorless	Mild, pleasant odor
	May float or sink in water.		

AVOID CONTACT WITH LIQUID AND VAPOR. KEEP PEOPLE AWAY.
Wear rubber overclothing (including gloves).
Stop discharge if possible.
Call fire department.
Isolate and remove discharged material.
Notify local health and pollution control agencies.

Fire

Combustible.
Extinguish with dry chemicals or carbon dioxide.
Water or foam may be ineffective on fire.
Cool exposed containers with water.

Exposure

CALL FOR MEDICAL AID.

VAPOR
Irritating to eyes, nose and throat.
If inhaled will cause headache, coughing, or difficult breathing.
If in eyes, hold eyelids open and flush with plenty of water.
If breathing has stopped, give artificial respiration.
If breathing is difficult, give oxygen.

LIQUID
Irritating to skin and eyes.
If swallowed will cause nausea or vomiting.
MAY BE READILY ABSORBED THROUGH THE SKIN.
Remove contaminated clothing and shoes.
Flush affected areas with plenty of water.
IF IN EYES, hold eyelids open and flush with plenty of water.
IF SWALLOWED and victim is CONSCIOUS, have victim drink water or milk and have victim induce vomiting.
IF SWALLOWED and victim is UNCONSCIOUS OR HAVING CONVULSIONS, do nothing except keep victim warm.

Water Pollution

Effect of low concentrations on aquatic life is unknown.
May be dangerous if it enters water intakes.

Notify local health and wildlife officials.
Notify operators of nearby water intakes.

1. RESPONSE TO DISCHARGE

(See Response Methods Handbook, CG 446-4)

Issue warning—water contaminant
Disperse and flush

2. LABELS

No hazard label required by Code of Federal Regulations

3. CHEMICAL DESIGNATIONS

3.1 **Synonyms:** Benzenecarbinol; *alpha*-Hydroxytoluene; Phenylcarbinol; Phenylmethanol; Phenylmethyl alcohol
3.2 **Coast Guard Compatibility Classification:** Not listed
3.3 **Chemical Formula:** $C_6H_5CH_2OH$
3.4 **IMCO/United Nations Numerical Designation:** Not listed

4. OBSERVABLE CHARACTERISTICS

4.1 **Physical State (as shipped):** Liquid
4.2 **Color:** Colorless
4.3 **Odor:** Mild, pleasant

5. HEALTH HAZARDS

5.1 **Personal Protective Equipment:** Rubber gloves; chemical safety goggles
5.2 **Symptoms Following Exposure:** Inhalation of vapor may cause irritation of upper respiratory tract. Prolonged or excessive inhalation may result in headache, nausea, vomiting, and diarrhea. In severe cases, respiratory stimulation followed by respiratory and muscular paralysis, convulsions, narcosis and death may result. Ingestion may produce severe irritation of the gastrointestinal tract, followed by nausea, vomiting, cramps and diarrhea; tissue ulceration may result. Contact with eyes causes local irritation. Material can be absorbed through skin with anesthetic or irritant effect.
5.3 **Treatment for Exposure:** INHALATION: remove victim from contaminated atmosphere; call physician immediately. INGESTION: induce vomiting and contact a physician. EYES: flush with plenty of water for 15 min. and contact a physician. SKIN: flush with water, wash with soap and water; obtain medical attention in case of irritation or central nervous system depression.
5.4 **Toxicity by Inhalation (Threshold Limit Value):** Data not available
5.5 **Short-Term Inhalation Limits:** Data not available
5.6 **Toxicity by Ingestion:** Grade 2; oral rat LD_{50} = 1,230 mg/kg
5.7 **Late Toxicity:** Data not available
5.8 **Vapor (Gas) Irritant Characteristics:** Data not available
5.9 **Liquid or Solid Irritant Characteristics:** Data not available
5.10 **Odor Threshold:** 5.5 ppm

6. FIRE HAZARDS

6.1 **Flash Point:** 220°F O.C.; 213°F C.C.
6.2 **Flammable Limits in Air:** Data not available
6.3 **Fire Extinguishing Agents:** "Alcohol" foam, dry chemical, carbon dioxide
6.4 **Fire Extinguishing Agents Not to be Used:** Water or foam may cause frothing.
6.5 **Special Hazards of Combustion Products:**
6.6 **Behavior in Fire:**
6.7 **Ignition Temperature:** 817°F
6.8 **Electrical Hazard:** Data not available
6.9 **Burning Rate:** 3.74 mm/min.

7. CHEMICAL REACTIVITY

7.1 **Reactivity with Water:** No reaction
7.2 **Reactivity with Common Materials:** Will attack some plastics
7.3 **Stability During Transport:** Stable
7.4 **Neutralizing Agents for Acids and Caustics:** Not pertinent
7.5 **Polymerization:** Not pertinent
7.6 **Inhibitor of Polymerization:** Not pertinent

8. WATER POLLUTION

8.1 **Aquatic Toxicity:** 360 ppm/48 hr/daphnia magna/TL_m/*
*Water type not specified.
8.2 **Waterfowl Toxicity:** Data not available
8.3 **Biological Oxygen Demand (BOD):** 155%, 5 days
8.4 **Food Chain Concentration Potential:** None

9. SELECTED MANUFACTURERS

1. Velsicol Chemical Corp.
   341 E. Ohio St.
   Chicago, Ill. 60611
2. Stauffer Chemical Company
   Specialty Chemical Div.
   Meadow Road
   Edison, N. J. 08817
3. Givaudan Corporation
   125 Delawanna Avenue
   Clifton, N. J. 07014

10. SHIPPING INFORMATION

10.1 **Grades or Purity:** NF; Photographic; Technical; Textile
10.2 **Storage Temperature:** Ambient
10.3 **Inert Atmosphere:** No requirement
10.4 **Venting:** Open

11. HAZARD ASSESSMENT CODE

(See Hazard Assessment Handbook, CG 446-3)

A-P-T-X

12. HAZARD CLASSIFICATIONS

12.1 **Code of Federal Regulations:** Not listed
12.2 **NAS Hazard Rating for Bulk Water Transportation:** Not listed
12.3 **NFPA Hazard Classifications:**

Category	Classification
Health Hazard (Blue)	2
Flammability (Red)	1
Reactivity (Yellow)	0

13. PHYSICAL AND CHEMICAL PROPERTIES

13.1 **Physical State at 15°C and 1 atm:** Liquid
13.2 **Molecular Weight:** 108.13
13.3 **Boiling Point at 1 atm:** 401°F = 205°C = 478°K
13.4 **Freezing Point:** 4.5°F = −15.3°C = 257.9°K
13.5 **Critical Temperature:** 757°F = 403°C = 676°K
13.6 **Critical Pressure:** 663 psia = 45.0 atm = 4.57 MN/m^2
13.7 **Specific Gravity:** 1.050 at 15/15°C (liquid)
13.8 **Liquid Surface Tension:** 39.0 dynes/cm = 0.0390 N/m at 20°C
13.9 **Liquid-Water Interfacial Tension:** Not pertinent
13.10 **Vapor (Gas) Specific Gravity:** 3.73
13.11 **Ratio of Specific Heats of Vapor (Gas):** 1.070
13.12 **Latent Heat of Vaporization:** 193 Btu/lb = 107 cal/g = 4.48×10^5 J/kg
13.13 **Heat of Combustion:** −14,850 Btu/lb = −8,260 cal/g = -345×10^5 J/kg
13.14 **Heat of Decomposition:** Not pertinent
13.15 **Heat of Solution:** Not pertinent
13.16 **Heat of Polymerization:** Not pertinent

NOTES

BZM	BENZYLAMINE

Common Synonyms Alpha-aminotoluene Phenylmethyl amine	Liquid Colorless to light yellow Strong ammonia odor Floats and mixes with water.

Avoid contact with liquid and vapor. Keep people away.
Wear rubber overclothing (including gloves).
Stop discharge if possible.
Call fire department.
Isolate and remove discharged material.
Notify local health and pollution control agencies.

Fire	Combustible. POISONOUS GASES MAY BE PRODUCED IN FIRE. Wear goggles and self-contained breathing apparatus. Extinguish with dry chemicals, alcohol foam, or carbon dioxide. Water may be ineffective on fire. Cool exposed containers with water.
Exposure	CALL FOR MEDICAL AID. VAPOR Irritating to eyes, nose and throat. If inhaled will cause dizziness, headache, or difficult breathing. If in eyes, hold eyelids open and flush with plenty of water. If breathing has stopped, give artificial respiration. If breathing is difficult, give oxygen. LIQUID Will burn skin and eyes. Harmful if Swallowed. Remove contaminated clothing and shoes. Flush affected areas with plenty of water. IF IN EYES, hold eyelids open and flush with plenty of water. IF SWALLOWED and victim is CONSCIOUS, have victim drink water or milk. IF SWALLOWED and victim is UNCONSCIOUS OR HAVING CONVULSIONS, do nothing except keep victim warm.
Water Pollution	HARMFUL TO AQUATIC LIFE IN VERY LOW CONCENTRATIONS. May be dangerous if it enters water intakes. Notify local health and wildlife officials. Notify operators of nearby water intakes.

1. RESPONSE TO DISCHARGE

(See Response Methods Handbook, CG 446-4)

Issue warning—air contaminant, water contaminant
Restrict access
Disperse and flush

2. LABELS

No hazard label required by Code of Federal Regulations

3. CHEMICAL DESIGNATIONS

3.1 **Synonyms:** *alpha*-Aminotoluene; Phenylmethyl amine
3.2 **Coast Guard Compatibility Classification:** Not listed
3.3 **Chemical Formula:** $C_6H_5CH_2NH_2$
3.4 **IMCO/United Nations Numerical Designation:** Not listed

4. OBSERVABLE CHARACTERISTICS

4.1 **Physical State (as shipped):** Liquid
4.2 **Color:** Colorless to light yellow
4.3 **Odor:** Strong ammonia

5. HEALTH HAZARDS

5.1 **Personal Protective Equipment:** Self-contained breathing apparatus; goggles or face shield; rubber gloves
5.2 **Symptoms Following Exposure:** Inhalation of vapor causes irritation of the mucous membranes of the nose and throat, and lung irritation with respiratory distress and cough. Headache, nausea, faintness, and anxiety can occur. Exposure to vapor produces eye irritation with lachrymation, conjunctivitis, and corneal edema resulting in halos around lights. Direct local contact with liquid is known to produce severe and sometimes permanent eye damage and skin burns. Vapors may also produce primary skin irritation and dermatitis.
5.3 **Treatment for Exposure:** INHALATION: remove victim from exposure; if breathing is difficult, administer oxygen; if breathing has stopped, begin artificial respiration. EYES or SKIN: wash with copious amounts of water for 15 min.
5.4 **Toxicity by Inhalation (Threshold Limit Value):** Data not available
5.5 **Short-Term Inhalation Limits:** Data not available
5.6 **Toxicity by Ingestion:** Data not available
5.7 **Late Toxicity:** Data not available
5.8 **Vapor (Gas) Irritant Characteristics:** Data not available
5.9 **Liquid or Solid Irritant Characteristics:** Data not available
5.10 **Odor Threshold:** Data not available

6. FIRE HAZARDS

6.1 **Flash Point:** 168°F O.C.
6.2 **Flammable Limits in Air:** Data not available
6.3 **Fire Extinguishing Agents:** Alcohol foam, dry chemical, carbon dioxide
6.4 **Fire Extinguishing Agents Not to be Used:** Water may be ineffective.
6.5 **Special Hazards of Combustion Products:** Toxic nitrogen oxides may form in a fire.
6.6 **Behavior in Fire:**
6.7 **Ignition Temperature:** Data not available
6.8 **Electrical Hazard:** Data not available
6.9 **Burning Rate:** 4.13 mm/min.

7. CHEMICAL REACTIVITY

7.1 **Reactivity with Water:** No reaction
7.2 **Reactivity with Common Materials:** In presence of moisture may weakly corrode some metals. Liquid will attack some plastics.
7.3 **Stability During Transport:** Stable
7.4 **Neutralizing Agents for Acids and Caustics:** Flush with water.
7.5 **Polymerization:** Not pertinent
7.6 **Inhibitor of Polymerization:** Not pertinent

8. WATER POLLUTION

8.1 **Aquatic Toxicity:** 60 ppm/48 hr/D. magna/TL_m/fresh water
8.2 **Waterfowl Toxicity:** Data not available
8.3 **Biological Oxygen Demand (BOD):** Data not available
8.4 **Food Chain Concentration Potential:** None

9. SELECTED MANUFACTURERS

1. Miles Laboratories, Inc. Sumner Division Elkhart, Ind. 46514
2. The Ames Laboratories, Inc. 200 Rock Lane Milford, Conn. 06460
3. Eastman Organic Chemicals Rochester, N. Y. 14650

10. SHIPPING INFORMATION

10.1 **Grades or Purity:** Commercial, 98.5+%
10.2 **Storage Temperature:** Ambient
10.3 **Inert Atmosphere:** Ventilated (natural)
10.4 **Venting:** Open (flame arrester)

11. HAZARD ASSESSMENT CODE

(See Hazard Assessment Handbook, CG 446-3)

A-P-Q

12. HAZARD CLASSIFICATIONS

12.1 **Code of Federal Regulations:** Not listed
12.2 **NAS Hazard Rating for Bulk Water Transportation:** Not listed
12.3 **NFPA Hazard Classifications:** Not listed

13. PHYSICAL AND CHEMICAL PROPERTIES

13.1 **Physical State at 15°C and 1 atm:** Liquid
13.2 **Molecular Weight:** 107.16
13.3 **Boiling Point at 1 atm:** 364.1°F = 184.5°C = 457.7°K
13.4 **Freezing Point:** (approx.) −51°F = −46°C = 227°K
13.5 **Critical Temperature:** Not pertinent
13.6 **Critical Pressure:** Not pertinent
13.7 **Specific Gravity:** 0.98 at 20°C (liquid)
13.8 **Liquid Surface Tension:** 39.5 dynes/cm = 0.0395 N/m at 20°C
13.9 **Liquid-Water Interfacial Tension:** Not pertinent
13.10 **Vapor (Gas) Specific Gravity:** 3.70
13.11 **Ratio of Specific Heats of Vapor (Gas):** (est.) 1.070
13.12 **Latent Heat of Vaporization:** (est.) 164 Btu/lb = 91 cal/g = 3.8×10^5 J/kg
13.13 **Heat of Combustion:** −16,260 Btu/lb = −9,040 cal/g = -378×10^5 J/kg
13.14 **Heat of Decomposition:** Not pertinent
13.15 **Heat of Solution:** −43 Btu/lb = −24 cal/g = -1.0×10^5 J/kg
13.16 **Heat of Polymerization:** Not pertinent

NOTES

BBR | BENZYL BROMIDE

Common Synonyms			
alpha-Bromotoluene omega-Bromotoluene Bromotoluene, alpha	Liquid	Colorless to yellow	Sharp irritating odor
	Sinks in water.		

Avoid contact with liquid and vapor. Keep people away.
Stop discharge if possible.
Call fire department.
Isolate and remove discharged material.
Notify local health and pollution control agencies.

Fire	Combustible. Irritating gases are produced when heated. Wear goggles and self-contained breathing apparatus. Extinguish with water, dry chemicals, foam, or carbon dioxide. Cool exposed containers with water.
Exposure	Call for medical aid. LIQUID Irritating to skin and eyes. Harmful if swallowed. Remove contaminated clothing and shoes. Flush affected areas with plenty of water. IF IN EYES, hold eyelids open and flush with plenty of water. IF SWALLOWED and victim is CONSCIOUS, have victim drink water or milk.
Water Pollution	HARMFUL TO AQUATIC LIFE IN VERY LOW CONCENTRATIONS. May be dangerous if it enters water intakes. Notify local health and wildlife officials. Notify operators of nearby water intakes.

1. RESPONSE TO DISCHARGE
(See Response Methods Handbook, CG 446-4)

Issue warning—corrosive
Restrict access
Should be removed
Chemical and physical treatment

2. LABEL

3. CHEMICAL DESIGNATIONS

3.1 **Synonyms:** alpha-Bromotoluene
omega-Bromotoluene
Bromotoluene, alpha

3.2 **Coast Guard Compatibility Classification:** Not listed

3.3 **Chemical Formula:** $C_6H_5CH_2Br$

3.4 **IMCO/United Nations Numerical Designation:** 8/1737

4. OBSERVABLE CHARACTERISTICS

4.1 **Physical State (as shipped):** Liquid

4.2 **Color:** Colorless to yellow

4.3 **Odor:** Very sharp; pungent; like tear gas

5. HEALTH HAZARDS

5.1 **Personal Protective Equipment:** Self-contained breathing apparatus; goggles; rubber gloves; protective clothing.

5.2 **Symptoms Following Exposure:** Inhalation causes irritation of nose and throat; severe exposure may cause pulmonary edema. Vapors cause severe eye irritation; liquid can burn eyes. Skin contact causes irritation. Ingestion causes irritation of mouth and stomach.

5.3 **Treatment for Exposure:** INHALATION: remove to fresh air. EYES: irrigate with copious amounts of water for 15 min. SKIN: flush with water, wash with soap and water. INGESTION: do NOT induce vomiting; give large amounts of water.

5.4 **Toxicity by Inhalation (Threshold Limit Value):** Data not available

5.5 **Short-Term Inhalation Limits:** Data not available

5.6 **Toxicity by Ingestion:** Data not available

5.7 **Late Toxicity:** Data not available

5.8 **Vapor (Gas) Irritant Characteristics:** Data not available

5.9 **Liquid or Solid Irritant Characteristics:** Data not available

5.10 **Odor Threshold:** Data not available

6. FIRE HAZARDS

6.1 **Flash Point:** 174°F C.C.

6.2 **Flammable Limits in Air:** Not pertinent

6.3 **Fire Extinguishing Agents:** Water, dry chemical, foam, carbon dioxide

6.4 **Fire Extinguishing Agents Not to be Used:** Not pertinent

6.5 **Special Hazards of Combustion Products:** Irritating hydrogen bromide gas may be formed.

6.6 **Behavior in Fire:** Forms vapor that is a powerful tear gas.

6.7 **Ignition Temperature:** Data not available

6.8 **Electrical Hazard:** Data not available

6.9 **Burning Rate:** 2.6 mm/min.

7. CHEMICAL REACTIVITY

7.1 **Reactivity with Water:** Reacts slowly to generate hydrogen bromide (hydrobromic acid).

7.2 **Reactivity with Common Materials:** Decomposes rapidly in the presence of all common metals except nickel and lead, liberating heat and hydrogen bromide.

7.3 **Stability During Transport:** Stable

7.4 **Neutralizing Agents for Acids and Caustics:** Rinse with sodium bicarbonate or lime solution.

7.5 **Polymerization:** Polymerizes with evolution of heat and hydrogen bromide when in contact with all common metals except nickel and lead.

7.6 **Inhibitor of Polymerization:** None used.

8. WATER POLLUTION

8.1 **Aquatic Toxicity:**
0.05 mg/l/*/marine fish/no irritant response/salt water
0.1 mg/l/*/marine fish/violent irritant activity/salt water
*Time period not specified.

8.2 **Waterfowl Toxicity:** Data not available

8.3 **Biological Oxygen Demand (BOD):** Data not available

8.4 **Food Chain Concentration Potential:** None

9. SELECTED MANUFACTURERS

1. White Chemical Corporation
East 22nd Street
P. O. Box 278
Bayonne, N. J. 07002
2. Eastman Kodak Co.
Eastman Organic Chemicals
Rochester, N. Y. 14650
3. Aldrich Chemical Co.
940 West St. Paul Avenue
Milwaukee, Wis. 53233

10. SHIPPING INFORMATION

10.1 **Grades or Purity:** Commercial

10.2 **Storage Temperature:** Ambient

10.3 **Inert Atmosphere:** No requirement

10.4 **Venting:** Pressure-vacuum

11. HAZARD ASSESSMENT CODE
(See Hazard Assessment Handbook, CG 446-3)

A-O-X-Y

12. HAZARD CLASSIFICATIONS

12.1 **Code of Federal Regulations:** Corrosive

12.2 **NAS Hazard Rating for Bulk Water Transportation:** Not listed

12.3 **NFPA Hazard Classifications:**

Category	Classification
Health Hazard (Blue)	2
Flammability (Red)	2
Reactivity (Yellow)	0

13. PHYSICAL AND CHEMICAL PROPERTIES

13.1 **Physical State at 15°C and 1 atm:** Liquid

13.2 **Molecular Weight:** 171.0

13.3 **Boiling Point at 1 atm:** 388°F = 198°C = 471°K

13.4 **Freezing Point:** 25.0°F = −3.9°C = 269.3°K

13.5 **Critical Temperature:** Not pertinent

13.6 **Critical Pressure:** Not pertinent

13.7 **Specific Gravity:** 1.441 at 22°C (liquid)

13.8 **Liquid Surface Tension:** 32.3 dynes/cm = 0.0323 N/m at 20°C

13.9 **Liquid-Water Interfacial Tension:** (est.) 35 dynes/cm = 0.035 N/m at 20°C

13.10 **Vapor (Gas) Specific Gravity:** 5.9

13.11 **Ratio of Specific Heats of Vapor (Gas):** Data not available

13.12 **Latent Heat of Vaporization:** 120 Btu/lb = 66.4 cal/g = 2.78×10^5 J/kg

13.13 **Heat of Combustion:** (est.) −9,000 Btu/lb = −5,000 cal/g = -210×10^5 J/kg

13.14 **Heat of Decomposition:** Not pertinent

13.15 **Heat of Solution:** Not pertinent

13.16 **Heat of Polymerization:** Not pertinent

NOTES

BBP — BENZYL n-BUTYL PHTHALATE

Common Synonyms: Butyl benzyl phthalate; Phthalic acid, benzyl butyl ether

Liquid — Colorless — Slight odor

Sinks in water.

Stop discharge if possible.
Avoid contact with liquid.
Isolate and remove discharged material.
Notify local health and pollution control agencies.

Fire

Combustible.
Irritating gases may be produced when heated.

Extinguish with dry chemicals, alcohol foam, or carbon dioxide.
Water may be ineffective on fire.
Cool exposed containers with water.

Exposure

CALL FOR MEDICAL AID.

LIQUID
Irritating to skin and eyes.

Remove contaminated clothing and shoes.
Flush affected areas with plenty of water.

Water Pollution

Effect of low concentrations on aquatic life is unknown.
May be dangerous if it enters water intakes.

Notify local health and wildlife officials.
Notify operators of nearby water intakes.

1. RESPONSE TO DISCHARGE

(See Response Methods Handbook, CG 446-4)

Should be removed
Chemical and physical treatment

2. LABELS

No hazard label required by Code of Federal Regulations

3. CHEMICAL DESIGNATIONS

3.1 **Synonyms:** Butyl benzyl phthalate; Phthalic acid, benzyl butyl ester
3.2 **Coast Guard Compatibility Classification:** Ester
3.3 **Chemical Formula:** $C_{19}H_{21}O_4$
3.4 **IMCO/United Nations Numerical Designation:** Not listed

4. OBSERVABLE CHARACTERISTICS

4.1 **Physical State (as shipped):** Liquid
4.2 **Color:** Colorless
4.3 **Odor:** Slight characteristic

5. HEALTH HAZARDS

5.1 **Personal Protective Equipment:** Protective gloves and goggles
5.2 **Symptoms Following Exposure:** Prolonged contact with liquid causes some irritation of eyes and skin.
5.3 **Treatment for Exposure:** EYES: flush with water for 15 min. SKIN: wash well with soap and water.
5.4 **Toxicity by Inhalation (Threshold Limit Value):** Data not available
5.5 **Short-Term Inhalation Limits:** Data not available
5.6 **Toxicity by Ingestion:** Grade 1; oral rat LD_{50} = 13,500 mg/kg
5.7 **Late Toxicity:** Data not available
5.8 **Vapor (Gas) Irritant Characteristics:** Vapors are nonirritating to eyes and throat.
5.9 **Liquid or Solid Irritant Characteristics:** No appreciable hazard; practically harmless to skin.
5.10 **Odor Threshold:** Data not available

6. FIRE HAZARDS

6.1 **Flash Point:** 390°F O.C.
6.2 **Flammable Limits in Air:** Data not available
6.3 **Fire Extinguishing Agents:** Dry chemical, carbon dioxide, foam
6.4 **Fire Extinguishing Agents Not to be Used:** Water or foam may cause frothing.
6.5 **Special Hazards of Combustion Products:** Irritating vapors of unburned chemical may form in fires.
6.6 **Behavior in Fire:**
6.7 **Ignition Temperature:** Data not available
6.8 **Electrical Hazard:** Data not available
6.9 **Burning Rate:** Data not available

7. CHEMICAL REACTIVITY

7.1 **Reactivity with Water:** No reaction
7.2 **Reactivity with Common Materials:**
7.3 **Stability During Transport:** Stable
7.4 **Neutralizing Agents for Acids and Caustics:** Not pertinent
7.5 **Polymerization:** Not pertinent
7.6 **Inhibitor of Polymerization:** Not pertinent

8. WATER POLLUTION

8.1 **Aquatic Toxicity:** Data not available
8.2 **Waterfowl Toxicity:** Data not available
8.3 **Biological Oxygen Demand (BOD):** Data not available
8.4 **Food Chain Concentration Potential:** None

9. SELECTED MANUFACTURERS

1. Monsanto Co.
800 North Lindbergh Blvd.
St. Louis, Mo. 63166
2. Pfaltz and Bauer, Inc.
126-04 Northern Blvd.
Flushing, N. Y. 11368

10. SHIPPING INFORMATION

10.1 **Grades or Purity:** Commercial
10.2 **Storage Temperature:** Ambient
10.3 **Inert Atmosphere:** No requirement
10.4 **Venting:** Open

11. HAZARD ASSESSMENT CODE

(See Hazard Assessment Handbook, CG 446-3)

A-X

12. HAZARD CLASSIFICATIONS

12.1 **Code of Federal Regulations:** Not listed
12.2 **NAS Hazard Rating for Bulk Water Transportation:**

Category	Rating
Fire	1
Health	
Vapor Irritant	0
Liquid or Solid Irritant	0
Poisons	0
Water Pollution	
Human Toxicity	0
Aquatic Toxicity	1
Aesthetic Effect	3
Reactivity	
Other Chemicals	1
Water	0
Self-Reaction	0

12.3 **NFPA Hazard Classifications:**

Category	Classification
Health Hazard (Blue)	1
Flammability (Red)	1
Reactivity (Yellow)	0

13. PHYSICAL AND CHEMICAL PROPERTIES

13.1 **Physical State at 15°C and 1 atm:** Liquid
13.2 **Molecular Weight:** 313
13.3 **Boiling Point at 1 atm:** 698°F = 380°C = 643°K
13.4 **Freezing Point:** Not pertinent
13.5 **Critical Temperature:** Not pertinent
13.6 **Critical Pressure:** Not pertinent
13.7 **Specific Gravity:** 1.12 at 20°C (liquid)
13.8 **Liquid Surface Tension:** Data not available
13.9 **Liquid-Water Interfacial Tension:** Data not available
13.10 **Vapor (Gas) Specific Gravity:** Not pertinent
13.11 **Ratio of Specific Heats of Vapor (Gas):** Not pertinent
13.12 **Latent Heat of Vaporization:** Data not available
13.13 **Heat of Combustion:** −14,550 Btu/lb = −8,090 cal/g = -338×10^5 J/kg
13.14 **Heat of Decomposition:** Not pertinent
13.15 **Heat of Solution:** Not pertinent
13.16 **Heat of Polymerization:** Not pertinent

NOTES

BCL	# BENZYL CHLORIDE

Common Synonyms			
alpha-Chlorotoluene omega-Chlorotoluene	Watery liquid	Colorless to yellow	Sharp irritating odor
	Sinks in water.		

Avoid contact with liquid. Keep people away.
Stop discharge if possible.
Call fire department.
Isolate and remove discharged material.
Notify local health and pollution control agencies.

Fire	Combustible. Irritating gases are produced when heated. Wear goggles and self-contained breathing apparatus. Extinguish with water, dry chemicals, foam, or carbon dioxide. Cool exposed containers with water.
Exposure	Call for medical aid. LIQUID Will burn skin and eyes. If swallowed will cause nausea, and vomiting. Remove contaminated clothing and shoes. Flush affected areas with plenty of water. IF IN EYES, hold eyelids open and flush with plenty of water. IF SWALLOWED and victim is CONSCIOUS, have victim drink water or milk. DO NOT INDUCE VOMITING.
Water Pollution	HARMFUL TO AQUATIC LIFE IN VERY LOW CONCENTRATIONS. May be dangerous if it enters water intakes. Notify local health and wildlife officials. Notify operators of nearby water intakes.

1. RESPONSE TO DISCHARGE

(See Response Methods Handbook, CG 446-4)

Issue warning - corrosive
Restrict access
Should be removed
Chemical and physical treatment

2. LABEL

3. CHEMICAL DESIGNATIONS

3.1 **Synonyms:** alpha-Chlorotoluene
omega-Chlorotoluene
Chlorotoluene, alpha

3.2 **Coast Guard Compatibility Classification:** Not listed

3.3 **Chemical Formula:** $C_6H_5CH_2Cl$

3.4 **IMCO/United Nations Numerical Designation:** 8/1738

4. OBSERVABLE CHARACTERISTICS

4.1 **Physical State (as shipped):** Liquid

4.2 **Color:** Colorless to pale yellow

4.3 **Odor:** Pungent, irritating

5. HEALTH HAZARDS

5.1 **Personal Protective Equipment:** Chemical safety goggles or face shield; self-contained breathing apparatus, positive-pressure hose mask, air-line mask, industrial canister-type gas mask, or chemical cartridge respirator; rubber gloves; protective clothing.

5.2 **Symptoms Following Exposure:** Inhalation causes severe irritation of upper respiratory tract with coughing, burning of the throat, headache, dizziness, and weakness; lung damage and pulmonary edema may occur after severe exposure; chronic irritation of the upper respiratory tract may occur after prolonged and repeated exposure to vapors. Immediate and severe eye irritation may result from contact with the liquid or vapors; prolonged or permanent eye damage may result. Vapors irritate skin, and liquid may cause severe burns. Ingestion may cause immediate and severe burns of the mouth, throat, and gastrointestinal tract; nausea, vomiting, cramps, and diarrhea may follow; gastrointestinal damage and systemic effects may result.

5.3 **Treatment for Exposure:** INHALATION: remove from contaminated atmosphere; if breathing has ceased, start mouth-to-mouth resuscitation; oxygen, if available, should be administered only by an experienced person when authorized by a physician; keep patient warm and comfortable; call a physician immediately. EYES: immediately flush with large quantities of running water for a minimum of 15 min.; hold eyelids apart during irrigation to insure flushing of the entire surface of the eye and lids with water; do not attempt to neutralize with chemical agents; obtain medical attention as soon as possible; oils or ointments should not be used unless directed by a physician; continue irrigation for an additional 15 min. if physician is not available. SKIN: immediately flush affected areas with water; remove contaminated clothing under shower; continue washing with water; do not attempt to neutralize with chemical agents; obtain medical attention if irritation persists. INGESTION: give large amounts of water; do NOT induce vomiting.

6. FIRE HAZARDS

6.1 **Flash Point:** 165°F O.C.; 140°F C.C.

6.2 **Flammable Limits in Air:** 1.1% (LFL)

6.3 **Fire Extinguishing Agents:** Water, dry chemical, foam, and carbon dioxide

6.4 **Fire Extinguishing Agents Not to be Used:** Not pertinent

6.5 **Special Hazards of Combustion Products:** Irritating hydrogen chloride gas may form.

6.6 **Behavior in Fire:** Forms vapor that is a powerful tear gas.

6.7 **Ignition Temperature:** 1,161°F

6.8 **Electrical Hazard:** Data not available

6.9 **Burning Rate:** 4.2 mm/min.

7. CHEMICAL REACTIVITY

7.1 **Reactivity with Water:** Undergoes slow hydrolysis, liberating hydrogen chloride (hydrochloric acid).

7.2 **Reactivity with Common Materials:** Decomposes rapidly in the presence of all common metals (with the exception of nickel and lead), liberating heat and hydrogen chloride.

7.3 **Stability During Transport:** Stable

7.4 **Neutralizing Agents for Acids and Caustics:** Rinse with sodium bicarbonate or lime solution.

7.5 **Polymerization:** Polymerizes with evolution of heat and hydrogen chloride when in contact with all common metals except nickel and lead.

7.6 **Inhibitor of Polymerization:** Triethylamine, propylene oxide, or sodium carbonate.

8. WATER POLLUTION

8.1 **Aquatic Toxicity:**
0.05 mg/l /*/marine fish/no irritant response/salt water
0.1 mg/l /*/marine fish/violent irritant activity/salt water
*Time period not specified.

8.2 **Waterfowl Toxicity:** Data not available

8.3 **Biological Oxygen Demand (BOD):** Data not available

8.4 **Food Chain Concentration Potential:** None

9. SELECTED MANUFACTURERS

1. Stauffer Chemical Company
Specialty Chemical Division
Westport, Conn. 06880
2. Velsicol Chemical Corporation
341 East Ohio Street
Chicago, Ill. 60611
3. Monsanto Company
Monsanto Industrial Chemicals
800 North Lindbergh Boulevard
St. Louis, Mo. 63166

10. SHIPPING INFORMATION

10.1 **Grades or Purity:** 98.5+%, either anhydrous or stabilized

10.2 **Storage Temperature:** Ambient

10.3 **Inert Atmosphere:** No requirement

10.4 **Venting:** Pressure-vacuum

11. HAZARD ASSESSMENT CODE

(See Hazard Assessment Handbook, CG 446-3)

A-O-X-Y

12. HAZARD CLASSIFICATIONS

12.1 **Code of Federal Regulations:** Corrosive

12.2 **NAS Hazard Rating for Bulk Water Transportation:**

Category	Rating
Fire	1
Health	
Vapor Irritant	4
Liquid or Solid Irritant	4
Poisons	2
Water Pollution	
Human Toxicity	2
Aquatic Toxicity	3
Aesthetic Effect	3
Reactivity	
Other Chemicals	3
Water	2
Self-Reaction	3

12.3 **NFPA Hazard Classifications:**

Category	Classification
Health Hazard (Blue)	2
Flammability (Red)	2
Reactivity (Yellow)	1

13. PHYSICAL AND CHEMICAL PROPERTIES

13.1 **Physical State at 15°C and 1 atm:** Liquid

13.2 **Molecular Weight:** 126.6

13.3 **Boiling Point at 1 atm:** 354.9°F = 179.4°C = 452.6°K

13.4 **Freezing Point:** −38.6°F = −39.2°C = 234.0°K

13.5 **Critical Temperature:** (est.) 772°F = 411°C = 684°K

13.6 **Critical Pressure:** (est.) 567 psia = 38.5 atm = 3.91 MN/m^2

13.7 **Specific Gravity:** 1.10 at 25°C (liquid)

13.8 **Liquid Surface Tension:** 37.5 dynes/cm = 0.0375 N/m at 20°C

13.9 **Liquid-Water Interfacial Tension:** (est.) 30 dynes/cm = 0.030 N/m at 20°C

13.10 **Vapor (Gas) Specific Gravity:** 4.36

13.11 **Ratio of Specific Heats of Vapor (Gas):** 1.0689

13.12 **Latent Heat of Vaporization:** 130 Btu/lb = 70 cal/g = 2.9 × 10^5 J/kg

13.13 **Heat of Combustion:** −12,000 Btu/lb = −6,700 cal/g = −280 × 10^5 J/kg

13.14 **Heat of Decomposition:** Not pertinent

13.15 **Heat of Solution:** Not pertinent

13.16 **Heat of Polymerization:** Not pertinent

5. HEALTH HAZARDS (Cont'd.)

5.4 **Toxicity by Inhalation (Threshold Limit Value):** 1 ppm

5.5 **Short-Term Inhalation Limits:** Data not available

5.6 **Toxicity by Ingestion:** Grade 2; oral rat LD_{50} = 1,231 mg/kg

5.7 **Late Toxicity:** Data not available

5.8 **Vapor (Gas) Irritant Characteristics:** Vapors cause severe irritation of eyes and throat and can cause eye and lung injury. They cannot be tolerated even at low concentrations.

5.9 **Liquid or Solid Irritant Characteristics:** Severe skin irritant. Causes second- and third-degree burns on short contact and is very injurious to the eyes.

5.10 **Odor Threshold:** 0.047 ppm

BCF	# BENZYL CHLOROFORMATE

Common Synonyms			
Carbobenzoxy chloride Chloroformic acid, benzyl ester Benzylcarbonyl chloride Benzyl chlorocarbonate	Liquid	Colorless	Sharp, irritating odor
	Sinks in water. Reacts with water.		

Call fire department.
Avoid contact with liquid. Keep people away.
Stop discharge if possible.
Isolate and remove discharged material.
Notify local health and pollution control agencies.

Fire	Combustible. POISONOUS GASES ARE PRODUCED IN FIRE. Containers may explode in fire. Wear chemical protective suit with self-contained breathing apparatus. Extinguish with dry chemicals, foam or carbon dioxide.
Exposure	Call for medical aid. LIQUID Irritating to skin and eyes. Harmful if swallowed. Remove contaminated clothing and shoes. Flush affected areas with plenty of water. IF IN EYES, hold eyelids open and flush with plenty of water. IF SWALLOWED and victim is CONSCIOUS, have victim drink water or milk.
Water Pollution	Effect of low concentrations on aquatic life is unknown. May be dangerous if it enters water intakes. Notify local health and wildlife officials. Notify operators of nearby water intakes.

1. RESPONSE TO DISCHARGE

(See Response Methods Handbook, CG 446-4)

Issue warning—corrosive
Restrict access
Should be removed
Chemical and physical treatment

2. LABEL

3. CHEMICAL DESIGNATIONS

3.1 **Synonyms:** Carbobenzoxy chloride
Chloroformic acid, benzyl ester
Benzylcarbonyl chloride
Benzyl chlorocarbonate

3.2 **Coast Guard Compatibility Classification:** Not applicable

3.3 **Chemical Formula:** $C_6H_5CH_2OCOCl$

3.4 **IMCO/United Nations Numerical Designation:** 8/1739

4. OBSERVABLE CHARACTERISTICS

4.1 **Physical State (as shipped):** Liquid

4.2 **Color:** Colorless

4.3 **Odor:** Irritating; sharp, penetrating

5. HEALTH HAZARDS

5.1 **Personal Protective Equipment:** Self-contained breathing apparatus or acid-type canister mask; goggles or face shield; rubber gloves; protective clothing.

5.2 **Symptoms Following Exposure:** Inhalation causes mucous membrane irritation. Eyes are irritated by excessive exposure to vapor. Liquid causes severe irritation of eyes and irritates skin. Ingestion causes irritation of mouth and stomach.

5.3 **Treatment for Exposure:** INHALATION: remove from exposure; support respiration; call physician. EYES: irrigate with copious amounts of water for 15 min. SKIN: flush with large quantities of water; wash with soap and water. INGESTION: give large amounts of water; do NOT induce vomiting.

5.4 **Toxicity by Inhalation (Threshold Limit Value):** Data not available

5.5 **Short-Term Inhalation Limits:** Data not available

5.6 **Toxicity by Ingestion:** Grade 3; LD_{50} 50 to 500 mg/kg

5.7 **Late Toxicity:** Data not available

5.8 **Vapor (Gas) Irritant Characteristics:** Vapors cause moderate irritation such that personnel will find high concentrations unpleasant. The effect is temporary.

5.9 **Liquid or Solid Irritant Characteristics:** Causes smarting of the skin and first-degree burns on short exposure and may cause second-degree burns on long exposure.

5.10 **Odor Threshold:** Data not available

6. FIRE HAZARDS

6.1 **Flash Point:** 176°F O.C.; 227°F C.C. Vigorous decomposition occurs at these temperatures; thus these values are anomalous due to the effect of the decomp. products (benzyl chloride and CO_2).

6.2 **Flammable Limits in Air:** Not pertinent

6.3 **Fire Extinguishing Agents:** Dry chemical, foam and carbon dioxide.

6.4 **Fire Extinguishing Agents Not to be Used:** Data not available

6.5 **Special Hazards of Combustion Products:** Toxic phosgene, hydrogen chloride, and benzyl chloride vapors may form.

6.6 **Behavior in Fire:** Containers may explode.

6.7 **Ignition Temperature:** Data not available

6.8 **Electrical Hazard:** Data not available

6.9 **Burning Rate:** 4.0 mm/min.

7. CHEMICAL REACTIVITY

7.1 **Reactivity with Water:** Forms hydrogen chloride (hydrochloric acid). The reaction is not very vigorous in cold water.

7.2 **Reactivity with Common Materials:** Slow corrosion of metal.

7.3 **Stability During Transport:** Stable

7.4 **Neutralizing Agents for Acids and Caustics:** Flush with water, rinse with sodium bicarbonate or lime solution.

7.5 **Polymerization:** Not pertinent

7.6 **Inhibitor of Polymerization:** Not pertinent

8. WATER POLLUTION

8.1 **Aquatic Toxicity:** Data not available

8.2 **Waterfowl Toxicity:** Data not available

8.3 **Biological Oxygen Demand (BOD):** Data not available

8.4 **Food Chain Concentration Potential:** None

9. SELECTED MANUFACTURERS

1. Chemetron Corporation
Organic Chemicals Division
393 Seventh Avenue
New York, N. Y. 10001
2. Aldrich Chemical Co.
940 West St. Paul Avenue
Milwaukee, Wis. 53233
3. PCR, Inc.
P. O. Box 1466
Gainesville, Fla. 32602

10. SHIPPING INFORMATION

10.1 **Grades or Purity:** 97+%

10.2 **Storage Temperature:** Ambient, in cool place

10.3 **Inert Atmosphere:** No requirement

10.4 **Venting:** Pressure-vacuum

11. HAZARD ASSESSMENT CODE

(See Hazard Assessment Handbook, CG 446-3)

A-O-X-Y

12. HAZARD CLASSIFICATIONS

12.1 **Code of Federal Regulations:** Corrosive

12.2 **NAS Hazard Rating for Bulk Water Transportation:**

Category	Rating
Fire	1
Health	
Vapor Irritant	2
Liquid or Solid Irritant	2
Poisons	3
Water Pollution	
Human Toxicity	3
Aquatic Toxicity	3
Aesthetic Effect	3
Reactivity	
Other Chemicals	2
Water	3
Self-Reaction	3

12.3 **NFPA Hazard Classifications:** Not listed

13. PHYSICAL AND CHEMICAL PROPERTIES

13.1 **Physical State at 15°C and 1 atm:** Liquid

13.2 **Molecular Weight:** 170.6

13.3 **Boiling Point at 1 atm:** (decomposes) 306°F = 152°C = 425°K

13.4 **Freezing Point:** Not pertinent

13.5 **Critical Temperature:** Not pertinent

13.6 **Critical Pressure:** Not pertinent

13.7 **Specific Gravity:** 1.22 at 20°C (liquid)

13.8 **Liquid Surface Tension:** (est.) 25 dynes/cm = 0.025 N/m at 20°C

13.9 **Liquid-Water Interfacial Tension:** Not pertinent

13.10 **Vapor (Gas) Specific Gravity:** Not pertinent

13.11 **Ratio of Specific Heats of Vapor (Gas):** Not pertinent

13.12 **Latent Heat of Vaporization:** (est.) 90 Btu/lb = 50 cal/g = 2.1×10^5 J/kg

13.13 **Heat of Combustion:** (est.) −10,000 Btu/lb = −5,700 cal/g = -240×10^5 J/kg

13.14 **Heat of Decomposition:** Not pertinent

13.15 **Heat of Solution:** Not pertinent

13.16 **Heat of Polymerization:** Not pertinent

NOTES

BZO BENZYLDIMETHYLOCTADECYLAMMONIUM CHLORIDE

Common Synonyms			
Stearyldimethylbenzyl-ammonium chloride Tallow benzyl dimethyl ammonium chloride Benzyldimethylstearyl-ammonium chloride	Solid or thick liquid	White	Mild odor
	Sinks and mixes with water.		

Stop discharge if possible. Keep people away.
Avoid contact with liquid and solid.
Isolate and remove discharged material.
Notify local health and pollution control agencies.

Fire

Combustible.
POISONOUS GASES MAY BE PRODUCED IN FIRE.
Wear goggles and self-contained breathing apparatus.
Extinguish with water.

Exposure

CALL FOR MEDICAL AID.

LIQUID OR SOLID
Irritating to skin and eyes.
If swallowed will cause nausea or vomiting.

Remove contaminated clothing and shoes.
Flush affected areas with plenty of water.
IF IN EYES, hold eyelids open and flush with plenty of water.
IF SWALLOWED and victim is CONSCIOUS, have victim drink water or milk.
IF SWALLOWED and victim is UNCONSCIOUS OR HAVING CONVULSIONS, do nothing except keep victim warm.
DO NOT INDUCE VOMITTING.

Water Pollution

Effect of low concentrations on aquatic life is unknown.
May be dangerous if it enters water intakes.

Notify local health and wildlife officials.
Notify operators of nearby water intakes.

1. RESPONSE TO DISCHARGE

(See Response Methods Handbook, CG 446-4)

Issue warning—water contaminant
Disperse and flush

2. LABELS

No hazard label required by Code of Federal Regulations

3. CHEMICAL DESIGNATIONS

3.1 **Synonyms:** Benzyldimethylstearylammonium chloride; Dimethylbenzyloctadecylammonium chloride; Dimethyloctadecylbenzyl-ammonium chloride; Octadecyl-dimethylbenzylammonium chloride

3.2 **Coast Guard Compatibility Classification:** Not listed

(Continued on page 4)

4. OBSERVABLE CHARACTERISTICS

4.1 **Physical State (as shipped):** Solid or viscous liquid

4.2 **Color:** White

4.3 **Odor:** Mild

5. HEALTH HAZARDS

5.1 **Personal Protective Equipment:** Goggles or face shield; rubber gloves

5.2 **Symptoms Following Exposure:** Ingestion causes gastrointestinal disturbances. Contact with chemical irritates eyes and skin and may damage eyes.

5.3 **Treatment for Exposure:** INGESTION: do NOT induce vomiting; give large quantities of fluid and call physician immediately. EYES: flush with plenty of water for at least 15 min.; call a physician. SKIN: remove contaminated clothing; flush skin with plenty of water.

5.4 **Toxicity by Inhalation (Threshold Limit Value):** Data not available

5.5 **Short-Term Inhalation Limits:** Data not available

5.6 **Toxicity by Ingestion:** Grade 2; oral rat LD_{50} = 4,000 mg/kg

5.7 **Late Toxicity:** Data not available

5.8 **Vapor (Gas) Irritant Characteristics:** Data not available

5.9 **Liquid or Solid Irritant Characteristics:** Data not available

5.10 **Odor Threshold:** Data not available

6. FIRE HAZARDS

6.1 **Flash Point:** Not pertinent

6.2 **Flammable Limits in Air:** Not pertinent

6.3 **Fire Extinguishing Agents:** Not pertinent

6.4 **Fire Extinguishing Agents Not to be Used:** Not pertinent

6.5 **Special Hazards of Combustion Products:** Toxic oxides of nitrogen and hydrochloric acid fumes may form in fires.

6.6 **Behavior in Fire:**

6.7 **Ignition Temperature:** Not pertinent

6.8 **Electrical Hazard:** Not pertinent

6.9 **Burning Rate:** Not pertinent

7. CHEMICAL REACTIVITY

7.1 **Reactivity with Water:** No reaction

7.2 **Reactivity with Common Materials:**

7.3 **Stability During Transport:** Stable

7.4 **Neutralizing Agents for Acids and Caustics:** Not pertinent

7.5 **Polymerization:** Not pertinent

7.6 **Inhibitor of Polymerization:** Not pertinent

8. WATER POLLUTION

8.1 **Aquatic Toxicity:** 5 ppm/48 hr/australoboris snails/80% kill/fresh water

8.2 **Waterfowl Toxicity:** Data not available

8.3 **Biological Oxygen Demand (BOD):** Data not available

8.4 **Food Chain Concentration Potential:** None

9. SELECTED MANUFACTURERS

1. Fine Organics Inc.
   205 Main Street
   Lodi, N. J. 07644
2. Lonza, Inc.
   Fair Lawn, N. J. 07410
3. Ashland Chemical Company
   P. O. Box 2219
   Columbus, Ohio 43216

10. SHIPPING INFORMATION

10.1 **Grades or Purity:** Pure, 95+%; 24—26% solution in water.

10.2 **Storage Temperature:** Ambient

10.3 **Inert Atmosphere:** No requirement

10.4 **Venting:** Open

11. HAZARD ASSESSMENT CODE

(See Hazard Assessment Handbook, CG 446-3)

SS

12. HAZARD CLASSIFICATIONS

12.1 **Code of Federal Regulations:** Not listed

12.2 **NAS Hazard Rating for Bulk Water Transportation:** Not listed

12.3 **NFPA Hazard Classifications:** Not listed

13. PHYSICAL AND CHEMICAL PROPERTIES

13.1 **Physical State at 15°C and 1 atm:** Solid

13.2 **Molecular Weight:** 411

13.3 **Boiling Point at 1 atm:** Not pertinent (decomposes at 120°C)

13.4 **Freezing Point:** Not pertinent

13.5 **Critical Temperature:** Not pertinent

13.6 **Critical Pressure:** Not pertinent

13.7 **Specific Gravity:** >1.1 at 20°C (solid)

13.8 **Liquid Surface Tension:** Not pertinent

13.9 **Liquid-Water Interfacial Tension:** Not pertinent

13.10 **Vapor (Gas) Specific Gravity:** Not pertinent

13.11 **Ratio of Specific Heats of Vapor (Gas):** Not pertinent

13.12 **Latent Heat of Vaporization:** Not pertinent

13.13 **Heat of Combustion:** Not pertinent

13.14 **Heat of Decomposition:** Not pertinent

13.15 **Heat of Solution:** Data not available

13.16 **Heat of Polymerization:** Not pertinent

3. CHEMICAL DESIGNATIONS (Cont'd.)

3.3 **Chemical Formula:** $(C_6H_5CH_2)(CH_3)_2(C_{18}H_{37})NCl$

3.4 **IMCO/United Nations Numerical Designation:** Not listed

BMA	BENZYLTRIMETHYLAMMONIUM CHLORIDE

Common Synonyms			
BTMAC Trimethylbenzylammonium chloride	Liquid	Light yellow	Mild almond odor
	May float or sink in water.		

Stop discharge if possible. Keep people away.
Avoid contact with liquid.
Isolate and remove discharged material.
Notify local health and pollution control agencies.

Fire	Combustible. POISONOUS GASES MAY BE PRODUCED IN FIRE. Wear goggles and self-contained breathing apparatus. Extinguish with dry chemicals, foam or carbon dioxide. Water may be ineffective on fire.
Exposure	CALL FOR MEDICAL AID. LIQUID Irritating to skin and eyes. If swallowed will cause nausea or vomiting. Remove contaminated clothing and shoes. Flush affected areas with plenty of water. IF IN EYES, hold eyelids open and flush with plenty of water. IF SWALLOWED and victim is CONSCIOUS, have victim drink water or milk. IF SWALLOWED and victim is UNCONSCIOUS OR HAVING CONVULSIONS, do nothing except keep victim warm.
Water Pollution	Effect of low concentrations on aquatic life is unknown. May be dangerous if it enters water intakes. Notify local health and wildlife officials. Notify operators of nearby water intakes.

1. RESPONSE TO DISCHARGE

(See Response Methods Handbook, CG 446-4)

Issue warning—water contaminant
Disperse and flush

2. LABELS

No hazard label required by Code of Federal Regulations

3. CHEMICAL DESIGNATIONS

3.1 **Synonyms:** BTMAC; Trimethylbenzylammonium chloride
3.2 **Coast Guard Compatibility Classification:** Not listed
3.3 **Chemical Formula:** $C_6H_5CH_2N(CH_3)_3Cl\text{-}H_2O$
3.4 **IMCO/United Nations Numerical Designation:** Not listed

4. OBSERVABLE CHARACTERISTICS

4.1 **Physical State (as shipped):** Liquid
4.2 **Color:** Light yellow
4.3 **Odor:** Mild almond

5. HEALTH HAZARDS

5.1 **Personal Protective Equipment:** Goggles and rubber gloves
5.2 **Symptoms Following Exposure:** Ingestion causes gastrointestinal disturbances. Contact with liquid irritates eyes and may irritate skin.
5.3 **Treatment for Exposure:** INGESTION: give large amount of water; get medical attention. EYES: flush with water for at least 15 min.; if irritation continues, get medical attention. SKIN: flush well with water.
5.4 **Toxicity by Inhalation (Threshold Limit Value):** Data not available
5.5 **Short-Term Inhalation Limits:** Data not available
5.6 **Toxicity by Ingestion:** Grade 3; LD_{50} 50-500 mg/kg
5.7 **Late Toxicity:** Data not available
5.8 **Vapor (Gas) Irritant Characteristics:** Data not available
5.9 **Liquid or Solid Irritant Characteristics:** Data not available
5.10 **Odor Threshold:** Data not available

6. FIRE HAZARDS

6.1 **Flash Point:** Not pertinent
6.2 **Flammable Limits in Air:** Not pertinent
6.3 **Fire Extinguishing Agents:** Not pertinent
6.4 **Fire Extinguishing Agents Not to be Used:** Not pertinent
6.5 **Special Hazards of Combustion Products:** Toxic oxides of nitrogen and hydrochloric acid fumes may form in fire.
6.6 **Behavior in Fire:**
6.7 **Ignition Temperature:** Not pertinent
6.8 **Electrical Hazard:** Not pertinent
6.9 **Burning Rate:** Not pertinent

7. CHEMICAL REACTIVITY

7.1 **Reactivity with Water:** No reaction
7.2 **Reactivity with Common Materials:**
7.3 **Stability During Transport:** Stable
7.4 **Neutralizing Agents for Acids and Caustics:** Not pertinent
7.5 **Polymerization:** Not pertinent
7.6 **Inhibitor of Polymerization:** Not pertinent

8. WATER POLLUTION

8.1 **Aquatic Toxicity:** Data not available
8.2 **Waterfowl Toxicity:** Data not available
8.3 **Biological Oxygen Demand (BOD):** Data not available
8.4 **Food Chain Concentration Potential:** None

9. SELECTED MANUFACTURERS

1. Tanatex Chemical Company
   Page and Schuyler Avenues
   Lyndhurst, N. J. 07071
2. Sun Chemical Co.
   P. O. Box 70
   Chester, S. C. 29706
3. Ashland Chemical Company
   P. O. Box 2219
   Columbus, Ohio 43216

10. SHIPPING INFORMATION

10.1 **Grades or Purity:** 50—60% solution in water
10.2 **Storage Temperature:** Ambient
10.3 **Inert Atmosphere:** No requirement
10.4 **Venting:** Open

11. HAZARD ASSESSMENT CODE

(See Hazard Assessment Handbook, CG 446-3)

A-P

12. HAZARD CLASSIFICATIONS

12.1 **Code of Federal Regulations:** Not listed
12.2 **NAS Hazard Rating for Bulk Water Transportation:** Not listed
12.3 **NFPA Hazard Classifications:** Not listed

13. PHYSICAL AND CHEMICAL PROPERTIES

13.1 **Physical State at 15°C and 1 atm:** Liquid
13.2 **Molecular Weight:** 172 (solute)
13.3 **Boiling Point at 1 atm:** Not pertinent (decomposes)
13.4 **Freezing Point:** Not pertinent
13.5 **Critical Temperature:** Not pertinent
13.6 **Critical Pressure:** Not pertinent
13.7 **Specific Gravity:** 1.07 at 20°C (liquid)
13.8 **Liquid Surface Tension:** Data not available
13.9 **Liquid-Water Interfacial Tension:** Not pertinent
13.10 **Vapor (Gas) Specific Gravity:** Not pertinent
13.11 **Ratio of Specific Heats of Vapor (Gas):** Not pertinent
13.12 **Latent Heat of Vaporization:** Not pertinent
13.13 **Heat of Combustion:** Not pertinent
13.14 **Heat of Decomposition:** Not pertinent
13.15 **Heat of Solution:** Not pertinent
13.16 **Heat of Polymerization:** Not pertinent

NOTES

BEC

BERYLLIUM CHLORIDE

Common Synonyms	Solid	White to green	Sharp odor
	Sinks and mixes violently with water.		

AVOID CONTACT WITH SOLID AND DUST. KEEP PEOPLE AWAY.
Wear dust respirator and rubber overclothing (including gloves).
Stop discharge if possible.
Isolate and remove discharged material.
Notify local health and pollution control agencies.

Fire

Not flammable.
Irritating gases may be produced when heated.

Wear goggles and self-contained breathing apparatus.
DO NOT USE WATER ON ADJACENT FIRES.

Exposure

CALL FOR MEDICAL AID.
DUST
POISONOUS IF INHALED OR IF SKIN IS EXPOSED.
If inhaled will cause coughing, difficult breathing, or loss of consciousness.

If in eyes, hold eyelids open and flush with plenty of water.
If breathing has stopped, give artificial respiration.
If breathing is difficult, give oxygen.

SOLID
POISONOUS IF SWALLOWED OR IF SKIN IS EXPOSED.
Will burn skin and eyes.
If swallowed will cause nausea, coughing, or loss of consciousness.
Remove contaminated clothing and shoes.
Flush affected areas with plenty of water.
IF IN EYES, hold eyelids open and flush with plenty of water.
IF SWALLOWED and victim is CONSCIOUS, have victim drink water or milk and have victim induce vomiting.
IF SWALLOWED and victim is UNCONSCIOUS OR HAVING CONVULSIONS, do nothing except keep victim warm.

Water Pollution

HARMFUL TO AQUATIC LIFE IN VERY LOW CONCENTRATIONS.
May be dangerous if it enters water intakes.

Notify local health and wildlife officials.
Notify operators of nearby water intakes.

1. RESPONSE TO DISCHARGE

(See Response Methods Handbook, CG 446-4)

Issue warning—poison, corrosive
Restrict access
Should be removed
Chemical and physical treatment

2. LABEL

3. CHEMICAL DESIGNATIONS

3.1 **Synonyms:** No common synonyms
3.2 **Coast Guard Compatibility Classification:** Not listed
3.3 **Chemical Formula:** $BeCl_2$
3.4 **IMCO/United Nations Numerical Designation:** 6.1/1566

4. OBSERVABLE CHARACTERISTICS

4.1 **Physical State (as shipped):** Solid
4.2 **Color:** White to green
4.3 **Odor:** Sharp, acrid

5. HEALTH HAZARDS

5.1 **Personal Protective Equipment:** Respiratory protection; gloves; freshly laundered clothing; chemical safety goggles
5.2 **Symptoms Following Exposure:** Inhalation causes pneumonitis, nasopharyngitis, tracheobronchitis, dyspnea, chronic cough. Ingestion causes irritation of mouth and stomach. Contact with dust causes conjunctival inflammation of eyes and irritation of skin. Any dramatic, unexplained weight loss should be considered as a possible first indication of beryllium disease.
5.3 **Treatment for Exposure:** INHALATION: chest x-ray should be taken immediately for evidence of pneumonitis. EYES: flush with water for at least 15 min.; if irritation persists, get medical attention. SKIN: cuts or puncture wounds in which beryllium may be embedded under the skin should be thoroughly cleansed immediately by a physician.
5.4 **Toxicity by Inhalation (Threshold Limit Value):** 0.002 mg/m^3 (as beryllium)
5.5 **Short-Term Inhalation Limits:** 0.025 mg/m^3 less than 30 min.
5.6 **Toxicity by Ingestion:** Grade 3; oral rat LD_{50} = 86 mg/kg
5.7 **Late Toxicity:** Be produces a chronic systemic disease that primarily affects the lung but also can involve other organs such as lymph nodes, liver, bones, and kidney.
5.8 **Vapor (Gas) Irritant Characteristics:** Data not available
5.9 **Liquid or Solid Irritant Characteristics:** Data not available
5.10 **Odor Threshold:** Data not available

6. FIRE HAZARDS

6.1 **Flash Point:** Not flammable
6.2 **Flammable Limits in Air:** Not flammable
6.3 **Fire Extinguishing Agents:** Not pertinent
6.4 **Fire Extinguishing Agents Not to be Used:** Do not use water on adjacent fires.
6.5 **Special Hazards of Combustion Products:** Toxic and irritating beryllium oxide fumes and hydrogen chloride may form in fires.
6.6 **Behavior in Fire:**
6.7 **Ignition Temperature:** Not pertinent
6.8 **Electrical Hazard:** Not pertinent
6.9 **Burning Rate:** Not pertinent

7. CHEMICAL REACTIVITY

7.1 **Reactivity with Water:** Reacts vigorously with evolution of heat. Forms beryllium oxide and hydrochloric acid solution.
7.2 **Reactivity with Common Materials:** Corrodes most metals in presence of moisture. Flammable and explosive hydrogen gas may collect in enclosed spaces.
7.3 **Stability During Transport:** Stable
7.4 **Neutralizing Agents for Acids and Caustics:** Flush with water, rinse with dilute solution of sodium bicarbonate or soda ash.
7.5 **Polymerization:** Not pertinent
7.6 **Inhibitor of Polymerization:** Not pertinent

8. WATER POLLUTION

8.1 **Aquatic Toxicity:**
0.15 ppm*/96 hr/fathead minnow/TL_m/soft fresh water
15 ppm*/96 hr/fathead minnow/TL_m/hard fresh water
*as beryllium
8.2 **Waterfowl Toxicity:** Data not available
8.3 **Biological Oxygen Demand (BOD):** None

9. SELECTED MANUFACTURERS

1. Kawecki Berylco Industries, Inc.
Hazleton, Pa. 18201
2. Var-lac-oid Chemical Co.
666 South Front St.
Elizabeth, N. J. 07202
3. Research Organic/Inorganic Chemical Corp.
11686 Sheldon
Sun Valley, Calif. 91352

10. SHIPPING INFORMATION

10.1 **Grades or Purity:** Commercial, 99+%
10.2 **Storage Temperature:** Ambient
10.3 **Inert Atmosphere:** No requirement
10.4 **Venting:** Open

11. HAZARD ASSESSMENT CODE

(See Hazard Assessment Handbook, CG 446-3)

RR

12. HAZARD CLASSIFICATIONS

12.1 **Code of Federal Regulations:** Poisonous solid, Class B
12.2 **NAS Hazard Rating for Bulk Water Transportation:** Not listed
12.3 **NFPA Hazard Classifications:** Not listed

13. PHYSICAL AND CHEMICAL PROPERTIES

13.1 **Physical State at 15°C and 1 atm:** Solid
13.2 **Molecular Weight:** 79.9
13.3 **Boiling Point at 1 atm:** (sublimes) 968°F = 520°C = 793°K
13.4 **Freezing Point:** 824°F = 440°C = 713°K
13.5 **Critical Temperature:** Not pertinent
13.6 **Critical Pressure:** Not pertinent
13.7 **Specific Gravity:** 1.90 at 25°C (solid)
13.8 **Liquid Surface Tension:** Not pertinent
13.9 **Liquid-Water Interfacial Tension:** Not pertinent
13.10 **Vapor (Gas) Specific Gravity:** Not pertinent
13.11 **Ratio of Specific Heats of Vapor (Gas):** Not pertinent
13.12 **Latent Heat of Vaporization:** Not pertinent
13.13 **Heat of Combustion:** Not pertinent
13.14 **Heat of Decomposition:** Not pertinent
13.15 **Heat of Solution:** −1,000 Btu/lb = −557 cal/g = -23.3×10^5 J/kg
13.16 **Heat of Polymerization:** Not pertinent

8. WATER POLLUTION (Cont'd.)

8.4 **Food Chain Concentration Potential:** Bioconcentration of 100-fold can occur under constant exposure. Not significant in spill conditions.

BEF	BERYLLIUM FLUORIDE

Common Synonyms	Solid	White	Odorless
	Sinks and mixes with water.		

AVOID CONTACT WITH SOLID AND DUST. KEEP PEOPLE AWAY.
Wear chemical protective suit with self-contained breathing apparatus.
Isolate and remove discharged material.
Notify local health and pollution control agencies.

Fire

Not flammable.

Exposure

CALL FOR MEDICAL AID.

DUST
POISONOUS IF INHALED, IF SWALLOWED, OR IF SKIN IS EXPOSED.
Will burn eyes.
Move to fresh air.
If breathing has stopped, give artificial respiration.
If breathing is difficult, give oxygen.

SOLID
POISONOUS IF SWALLOWED OR IF SKIN IS EXPOSED.
Will burn eyes.
Remove contaminated clothing and shoes.
Flush affected areas with plenty of water.
IF IN EYES, hold eyelids open and flush with plenty of water.
IF SWALLOWED and victim is CONSCIOUS, have victim drink water or milk and have victim induce vomiting.
IF SWALLOWED and victim is UNCONSCIOUS OR HAVING CONVULSIONS, do nothing except keep victim warm.

Water Pollution

HARMFUL TO AQUATIC LIFE IN VERY LOW CONCENTRATIONS.
May be dangerous if it enters water intakes.
Notify local health and wildlife officials.
Notify operators of nearby water intakes.

1. RESPONSE TO DISCHARGE

(See Response Methods Handbook, CG 446-4)

Issue warning—poison, water contaminant
Restrict access
Should be removed
Chemical and physical treatment

2. LABEL

3. CHEMICAL DESIGNATIONS

3.1 **Synonyms:** No common synonyms
3.2 **Coast Guard Compatibility Classification:** Not listed
3.3 **Chemical Formula:** BeF_2
3.4 **IMCO/United Nations Numerical Designation:** 6.1/1566

4. OBSERVABLE CHARACTERISTICS

4.1 **Physical State (as shipped):** Solid
4.2 **Color:** White
4.3 **Odor:** None

5. HEALTH HAZARDS

5.1 **Personal Protective Equipment:** Respiratory protection; gloves; goggles
5.2 **Symptoms Following Exposure:** Any dramatic weight loss should be considered as possible first indication of beryllium disease. Inhalation causes irritation of nose, throat, and lungs, severe pneumonitis, and/or pulmonary edema. Ingestion causes fatigue, weakness, loss of appetite. Contact with eyes causes severe irritation and burns. Contact with skin causes dermatitis and non-healing ulcers.
5.3 **Treatment for Exposure:** INHALATION: move to fresh air; chest x-ray should be taken immediately to detect pneumonitis, if exposure has been severe. INGESTION: induce vomiting; get medical attention. EYES: flush with water for at least 15 min.; get medical attention. SKIN: flush with water; get medical attention if skin has been broken.
5.4 **Toxicity by Inhalation (Threshold Limit Value):** 0.002 mg/m^3 (as beryllium)
5.5 **Short-Term Inhalation Limits:** 0.025 mg/m^3, less than 30 min.
5.6 **Toxicity by Ingestion:** Grade 3; oral LD_{50} = 100 mg/kg (mouse)
5.7 **Late Toxicity:** Berylliosis of lungs may occur from 3 months to 15 years after exposure. Chronic systemic diseases of the liver, spleen, lymph nodes, bone, kidney, and other organs may also occur.
5.8 **Vapor (Gas) Irritant Characteristics:** Data not available
5.9 **Liquid or Solid Irritant Characteristics:** Data not available
5.10 **Odor Threshold:** Data not available

6. FIRE HAZARDS

6.1 **Flash Point:** Not flammable
6.2 **Flammable Limits in Air:** Not flammable
6.3 **Fire Extinguishing Agents:** Not pertinent
6.4 **Fire Extinguishing Agents Not to be Used:** Not pertinent
6.5 **Special Hazards of Combustion Products:** Toxic and irritating vapor of unburned material may form in fire.
6.6 **Behavior in Fire:**
6.7 **Ignition Temperature:** Not pertinent
6.8 **Electrical Hazard:** Not pertinent
6.9 **Burning Rate:** Not pertinent

7. CHEMICAL REACTIVITY

7.1 **Reactivity with Water:** No reaction
7.2 **Reactivity with Common Materials:**
7.3 **Stability During Transport:** Stable
7.4 **Neutralizing Agents for Acids and Caustics:** Not pertinent
7.5 **Polymerization:** Not pertinent
7.6 **Inhibitor of Polymerization:** Not pertinent

8. WATER POLLUTION

8.1 **Aquatic Toxicity:**
0.15 ppm*/96 hr/fathead minnow/TL_m/fresh (soft) water
15 ppm*/96 hr/fathead minnow/TL_m/fresh (hard) water
*As beryllium
8.2 **Waterfowl Toxicity:** Data not available
8.3 **Biological Oxygen Demand (BOD):** None
8.4 **Food Chain Concentration Potential:** Bioconcentration of 100-fold under constant exposure only. Not significant under spill conditions.

9. SELECTED MANUFACTURERS

1. Kawecki Berylco Industries, Inc.
P. O. Box 429
Hazleton, Pa. 18201
2. Brush Wellman, Inc.
South River Road
Elmore, Ohio 43416
3. Gallard-Schlesinger Chemical Mfg. Co.
584 Mineola Ave.
Carle Place, N. Y. 11514

10. SHIPPING INFORMATION

10.1 **Grades or Purity:** Purified, 99.99%; Chemically pure, 99+%
10.2 **Storage Temperature:** Ambient
10.3 **Inert Atmosphere:** No requirement
10.4 **Venting:** Open

11. HAZARD ASSESSMENT CODE

(See Hazard Assessment Handbook, CG 446-3)

SS

12. HAZARD CLASSIFICATIONS

12.1 **Code of Federal Regulations:** Poisonous solid, Class B
12.2 **NAS Hazard Rating for Bulk Water Transportation:** Not listed
12.3 **NFPA Hazard Classifications:** Not listed

13. PHYSICAL AND CHEMICAL PROPERTIES

13.1 **Physical State at 15°C and 1 atm:** Solid
13.2 **Molecular Weight:** 47
13.3 **Boiling Point at 1 atm:** Not pertinent (sublimes at 800°C)
13.4 **Freezing Point:** Not pertinent
13.5 **Critical Temperature:** Not pertinent
13.6 **Critical Pressure:** Not pertinent
13.7 **Specific Gravity:** 1.99 at 20°C (solid)
13.8 **Liquid Surface Tension:** Not pertinent
13.9 **Liquid-Water Interfacial Tension:** Not pertinent
13.10 **Vapor (Gas) Specific Gravity:** Not pertinent
13.11 **Ratio of Specific Heats of Vapor (Gas):** Not pertinent
13.12 **Latent Heat of Vaporization:** Not pertinent
13.13 **Heat of Combustion:** Not pertinent
13.14 **Heat of Decomposition:** Not pertinent
13.15 **Heat of Solution:** −180 Btu/lb = −98 cal/g = -4.1×10^5 J/kg
13.16 **Heat of Polymerization:** Not pertinent

NOTES

BEM — BERYLLIUM METALLIC

Common Synonyms	Solid	Silver color	Odorless
	Sinks in water.		

AVOID CONTACT WITH SOLID AND DUST. KEEP PEOPLE AWAY.
Wear dust respirator and rubber overclothing (including gloves).
Stop discharge if possible.
Call fire department.
Isolate and remove discharged material.
Notify local health and pollution control agencies.

Fire

Combustible.
POISONOUS GASES MAY BE PRODUCED IN FIRE.
Dust cloud may explode if ignited in an enclosed area.
Wear goggles and self-contained breathing apparatus.
Extinguish with dry graphite, soda ash, or other inert powder.
DO NOT USE WATER ON FIRE.

Exposure

CALL FOR MEDICAL AID.

DUST
POISONOUS IF INHALED OR IF SKIN IS EXPOSED.
If inhaled will cause coughing or difficult breathing.
If in eyes, hold eyelids open and flush with plenty of water.
If breathing has stopped, give artificial respiration.
If breathing is difficult, give oxygen.

SOLID
POISONOUS IF SWALLOWED OR IF SKIN IS EXPOSED.
Remove contaminated clothing and shoes.
Flush affected areas with plenty of water.
IF IN EYES, hold eyelids open and flush with plenty of water.
IF SWALLOWED and victim is CONSCIOUS, have victim drink water or milk and have victim induce vomiting.
IF SWALLOWED and victim is UNCONSCIOUS OR HAVING CONVULSIONS, do nothing except keep victim warm.

Water Pollution

Effect of low concentrations on aquatic life is unknown.
May be dangerous if it enters water intakes.
Notify local health and wildlife officials.
Notify operators of nearby water intakes.

1. RESPONSE TO DISCHARGE
(See Response Methods Handbook, CG 446-4)

Issue warning—poison, water contaminant
Restrict access
Should be removed
Chemical and physical treatment

2. LABELS

No hazard label required by Code of Federal Regulations

3. CHEMICAL DESIGNATIONS

3.1 **Synonyms:** No common synonyms
3.2 **Coast Guard Compatibility Classification:** Not listed
3.3 **Chemical Formula:** Be
3.4 **IMCO/United Nations Numerical Designation:** 6.1/1567

4. OBSERVABLE CHARACTERISTICS

4.1 **Physical State (as shipped):** Solid
4.2 **Color:** White
4.3 **Odor:** None

5. HEALTH HAZARDS

5.1 **Personal Protective Equipment:** Bu. Mines approved respirator; clean work clothes daily; gloves; eye protection
5.2 **Symptoms Following Exposure:** Any dramatic, unexplained weight loss should be considered as possible first indication of beryllium disease. Dust is extremely toxic when inhaled; symptoms include coughing, shortness of breath, and acute or chronic lung disease. There is no record of illness from ingestion of beryllium. Contact with dust causes conjunctival inflammation of eyes and dermatitis.
5.3 **Treatment for Exposure:** INHALATION: acute disease may require hospitalization with administration of oxygen; chest x-ray should be taken immediately. EYES: flush with water for at least 15 min. SKIN: flush with water; wash with soap and water; all cuts, scratches or other injuries should receive prompt medical attention.
5.4 **Toxicity by Inhalation (Threshold Limit Value):** 0.002 mg/m^3
5.5 **Short-Term Inhalation Limits:** 0.025 mg/m^3, 5 min.
5.6 **Toxicity by Ingestion:** Grade 3; LD_{50} 50-500 mg/kg
5.7 **Late Toxicity:** Beryllium disease may occur in lung, lymph nodes, liver, spleen, kidney, and other organs.
5.8 **Vapor (Gas) Irritant Characteristics:** Data not available
5.9 **Liquid or Solid Irritant Characteristics:** Data not available
5.10 **Odor Threshold:** Data not available

6. FIRE HAZARDS

6.1 **Flash Point:** Not pertinent (combustible solid)
6.2 **Flammable Limits in Air:** Not pertinent
6.3 **Fire Extinguishing Agents:** Graphite, sand, or any other inert dry powder
6.4 **Fire Extinguishing Agents Not to be Used:** Water
6.5 **Special Hazards of Combustion Products:** Combustion yields beryllium oxide fume, which is toxic if inhaled.
6.6 **Behavior in Fire:** Powder may form explosive mixture with air.
6.7 **Ignition Temperature:** Not pertinent
6.8 **Electrical Hazard:** Not pertinent
6.9 **Burning Rate:** Not pertinent

7. CHEMICAL REACTIVITY

7.1 **Reactivity with Water:** No reaction
7.2 **Reactivity with Common Materials:**
7.3 **Stability During Transport:** Stable
7.4 **Neutralizing Agents for Acids and Caustics:** Not pertinent
7.5 **Polymerization:** Not pertinent
7.6 **Inhibitor of Polymerization:** Not pertinent

8. WATER POLLUTION

8.1 **Aquatic Toxicity:** Data not available
8.2 **Waterfowl Toxicity:** Data not available
8.3 **Biological Oxygen Demand (BOD):** Data not available
8.4 **Food Chain Concentration Potential:** Data not available

9. SELECTED MANUFACTURERS

1. Var-lac-oid Chemical Co.
   666 South Front St.
   Elizabeth, N. J. 07202
2. Gallard-Schlesinger Chemical Mfg. Co.
   584 Mineola Avenue
   Carle Place, N. Y. 11514
3. Pfaltz and Bauer, Inc.
   126-04 Northern Blvd.
   Flushing, N. Y. 11368

10. SHIPPING INFORMATION

10.1 **Grades or Purity:** Grade AA, 99.96+%; Grade A, 99.87+%; Nuclear grade
10.2 **Storage Temperature:** Ambient
10.3 **Inert Atmosphere:** No requirement
10.4 **Venting:** Open

11. HAZARD ASSESSMENT CODE
(See Hazard Assessment Handbook, CG 446-3)

II

12. HAZARD CLASSIFICATIONS

12.1 **Code of Federal Regulations:** Not listed
12.2 **NAS Hazard Rating for Bulk Water Transportation:** Not listed
12.3 **NFPA Hazard Classifications:**

Category	Classification*
Health Hazard (Blue)	4
Flammability (Red)	1
Reactivity (Yellow)	1

*Applies to dust or powder.

13. PHYSICAL AND CHEMICAL PROPERTIES

13.1 **Physical State at 15°C and 1 atm:** Solid
13.2 **Molecular Weight:** 9.01
13.3 **Boiling Point at 1 atm:** Not pertinent
13.4 **Freezing Point:** Not pertinent
13.5 **Critical Temperature:** Not pertinent
13.6 **Critical Pressure:** Not pertinent
13.7 **Specific Gravity:** 1.85 at 20°C (solid)
13.8 **Liquid Surface Tension:** Not pertinent
13.9 **Liquid-Water Interfacial Tension:** Not pertinent
13.10 **Vapor (Gas) Specific Gravity:** Not pertinent
13.11 **Ratio of Specific Heats of Vapor (Gas):** Not pertinent
13.12 **Latent Heat of Vaporization:** Not pertinent
13.13 **Heat of Combustion:** −28,000 Btu/lb = −15,560 cal/g = −652 × 10^5 J/kg
13.14 **Heat of Decomposition:** Not pertinent
13.15 **Heat of Solution:** Not pertinent
13.16 **Heat of Polymerization:** Not pertinent

NOTES

BEN — BERYLLIUM NITRATE

Common Synonyms			
Beryllium nitrate trihydrate	Solid	White	Odorless
	Sinks and mixes with water.		

AVOID CONTACT WITH SOLID AND DUST. KEEP PEOPLE AWAY.
Wear dust respirator and rubber overclothing (including gloves).
Stop discharge if possible.
Isolate and remove discharged material.
Notify local health and pollution control agencies.

Fire

Not flammable.
Will increase the intensity of a fire.
POISONOUS GASES MAY BE PRODUCED IN FIRE.
Wear goggles and self-contained breathing apparatus.
Flood discharge area with water.

Exposure

CALL FOR MEDICAL AID.

DUST
POISONOUS IF INHALED OR IF SKIN IS EXPOSED.
If inhaled will cause coughing or difficult breathing.
If in eyes, hold eyelids open and flush with plenty of water.
If breathing has stopped, give artificial respiration.
If breathing is difficult, give oxygen.

SOLID
POISONOUS IF SWALLOWED OR IF SKIN IS EXPOSED.
If swallowed will cause nausea, vomiting or loss of consciousness.
Remove contaminated clothing and shoes.
Flush affected areas with plenty of water.
IF IN EYES, hold eyelids open and flush with plenty of water.
IF SWALLOWED and victim is CONSCIOUS, have victim drink water or milk and have victim induce vomiting.
IF SWALLOWED and victim is UNCONSCIOUS OR HAVING CONVULSIONS, do nothing except keep victim warm.

Water Pollution

HARMFUL TO AQUATIC LIFE IN VERY LOW CONCENTRATIONS.
May be dangerous if it enters water intakes.
Notify local health and wildlife officials.
Notify operators of nearby water intakes.

1. RESPONSE TO DISCHARGE
(See Response Methods Handbook, CG 446-4)

Issue warning—poison, water contaminant
Restrict access
Should be removed
Chemical and physical treatment

2. LABEL

3. CHEMICAL DESIGNATIONS
3.1 **Synonyms:** Beryllium nitrate trihydrate
3.2 **Coast Guard Compatibility Classification:** Not listed
3.3 **Chemical Formula:** $Be(NO_3)_2 \cdot 3H_2O$
3.4 **IMCO/United Nations Numerical Designation:** 6.1/1566

4. OBSERVABLE CHARACTERISTICS
4.1 **Physical State (as shipped):** Solid
4.2 **Color:** White
4.3 **Odor:** None

5. HEALTH HAZARDS
5.1 **Personal Protective Equipment:** Respiratory protection; gloves; freshly laundered clothing; chemical safety goggles
5.2 **Symptoms Following Exposure:** Any dramatic, unexplained weight loss should be considered as possible first indication of beryllium disease. Inhalation causes pneumonitis, nasopharyngitis, tracheobronchitis, dyspnea, chronic cough. Ingestion causes anorexia, fatigue, weakness, malaise. Contact with eyes causes conjunctival inflammation. Contact with skin causes dermatitis and non-healing ulcers.
5.3 **Treatment for Exposure:** INHALATION: remove to fresh air; take chest x-ray immediately to check for pneumonitis. INGESTION: induce vomiting; get medical attention. EYES: flush with water for at least 15 min.; get medical attention. SKIN: cuts or puncture wounds in which beryllium may be embedded under the skin should be thoroughly cleansed immediately by a physician.
5.4 **Toxicity by Inhalation (Threshold Limit Value):** 0.002 mg/m^3 (as beryllium)
5.5 **Short-Term Inhalation Limits:** 0.025 mg/m^3 less than 30 min.
5.6 **Toxicity by Ingestion:** Data not available
5.7 **Late Toxicity:** May cause chronic systemic disease of the lung as well as other organs such as liver, spleen, lymph nodes, bone, and kidney.
5.8 **Vapor (Gas) Irritant Characteristics:** Data not available
5.9 **Liquid or Solid Irritant Characteristics:** Data not available
5.10 **Odor Threshold:** Data not available

6. FIRE HAZARDS
6.1 **Flash Point:** Not combustible
6.2 **Flammable Limits in Air:** Not combustible
6.3 **Fire Extinguishing Agents:** Water
6.4 **Fire Extinguishing Agents Not to be Used:** Not pertinent
6.5 **Special Hazards of Combustion Products:** Toxic and irritating beryllium oxide and oxides of nitrogen may form in fire.
6.6 **Behavior in Fire:** May increase intensity of fire when in contact with combustible material
6.7 **Ignition Temperature:** Not pertinent
6.8 **Electrical Hazard:** Not pertinent
6.9 **Burning Rate:** Not pertinent

7. CHEMICAL REACTIVITY
7.1 **Reactivity with Water:** Reacts to form weak solution of nitric acid; the reaction is not hazardous.
7.2 **Reactivity with Common Materials:** In presence of moisture will damage wood and corrode most metals.
7.3 **Stability During Transport:** Stable
7.4 **Neutralizing Agents for Acids and Caustics:** Not pertinent
7.5 **Polymerization:** Not pertinent
7.6 **Inhibitor of Polymerization:** Not pertinent

8. WATER POLLUTION
8.1 **Aquatic Toxicity:**
0.15 ppm*/96 hr/fathead minnow/TL_m/soft water
20 ppm*/96 hr/fathead minnow/TL_m/hard water
*as beryllium
8.2 **Waterfowl Toxicity:** Data not available
8.3 **Biological Oxygen Demand (BOD):** None
8.4 **Food Chain Concentration Potential:** Bioconcentration of 100-fold can occur under constant exposure. Not significant in spill conditions.

9. SELECTED MANUFACTURERS
1. Kawecki Berylco Industries, Inc.
Hazleton, Pa. 18201
2. Var-lac-oid Chemical Co.
666 South Front St.
Elizabeth, N. J. 07202
3. Gallard-Schlesinger Chemical Mfg. Co.
584 Mineola Avenue
Carle Place, N. Y. 11514

10. SHIPPING INFORMATION
10.1 **Grades or Purity:** Purified
10.2 **Storage Temperature:** Ambient
10.3 **Inert Atmosphere:** No requirement
10.4 **Venting:** Open

11. HAZARD ASSESSMENT CODE
(See Hazard Assessment Handbook, CG 446-3)

SS

12. HAZARD CLASSIFICATIONS
12.1 **Code of Federal Regulations:** Poisonous solid, Class B
12.2 **NAS Hazard Rating for Bulk Water Transportation:** Not listed
12.3 **NFPA Hazard Classifications:** Not listed

13. PHYSICAL AND CHEMICAL PROPERTIES
13.1 **Physical State at 15°C and 1 atm:** Solid
13.2 **Molecular Weight:** 205.1
13.3 **Boiling Point at 1 atm:** Not pertinent (decomposes)
13.4 **Freezing Point:** Not pertinent
13.5 **Critical Temperature:** Not pertinent
13.6 **Critical Pressure:** Not pertinent
13.7 **Specific Gravity:** 1.56 at 20°C (solid)
13.8 **Liquid Surface Tension:** Not pertinent
13.9 **Liquid-Water Interfacial Tension:** Not pertinent
13.10 **Vapor (Gas) Specific Gravity:** Not pertinent
13.11 **Ratio of Specific Heats of Vapor (Gas):** Not pertinent
13.12 **Latent Heat of Vaporization:** Not pertinent
13.13 **Heat of Combustion:** Not pertinent
13.14 **Heat of Decomposition:** Not pertinent
13.15 **Heat of Solution:** Data not available
13.16 **Heat of Polymerization:** Not pertinent

NOTES

BEO	# BERYLLIUM OXIDE

Common Synonyms			
Beryllia Bromellite	Solid	White	Odorless
	Sinks in water.		

AVOID CONTACT WITH SOLID AND DUST. KEEP PEOPLE AWAY.
Wear Dust respirator and rubber overclothing (including gloves).
Stop discharge if possible.
Isolate and remove discharged material.
Notify local health and pollution control agencies.

Fire

Not flammable.
POISONOUS GASES MAY BE PRODUCED IN FIRE.
Wear goggles and self-contained breathing apparatus.

Exposure

CALL FOR MEDICAL AID.

DUST
POISONOUS IF INHALED.
If inhaled will cause coughing and difficult breathing.
If in eyes, hold eyelids open and flush with plenty of water.
If breathing has stopped, give artificial respiration.
If breathing is difficult, give oxygen.

SOLID
POISONOUS IF SWALLOWED.
Irritating to skin and eyes.
Remove contaminated clothing and shoes.
Flush affected areas with plenty of water.
IF IN EYES, hold eyelids open and flush with plenty of water.
IF SWALLOWED and victim is CONSCIOUS, have victim drink water or milk and have victim induce vomiting.
IF SWALLOWED and victim is UNCONSCIOUS OR HAVING CONVULSIONS, do nothing except keep victim warm.

Water Pollution

Effect of low concentrations on aquatic life is unknown.
May be dangerous if it enters water intakes.
Notify local health and wildlife officials.
Notify operators of nearby water intakes.

1. RESPONSE TO DISCHARGE

(See Response Methods Handbook, CG 446-4)

Issue warning—poison, water contaminant
Restrict access
Should be removed
Chemical and physical treatment

2. LABEL

3. CHEMICAL DESIGNATIONS

3.1 **Synonyms:** Beryllia; Bromellite
3.2 **Coast Guard Compatibility Classification:** Not listed
3.3 **Chemical Formula:** BeO
3.4 **IMCO/United Nations Numerical Designation:** 6.1/1566

4. OBSERVABLE CHARACTERISTICS

4.1 **Physical State (as shipped):** Solid
4.2 **Color:** White
4.3 **Odor:** None

5. HEALTH HAZARDS

5.1 **Personal Protective Equipment:** Respiratory protection; gloves; freshly laundered clothing; chemical safety goggles
5.2 **Symptoms Following Exposure:** Any dramatic, unexplained weight loss should be considered as possible first indication of beryllium disease. Other symptoms include anorexia, fatigue, weakness, malaise. Inhalation causes pneumonitis, nasopharyngitis, tracheobronchitis, dyspnea, chronic cough. Contact with dust causes conjunctival inflammation of eyes and irritation of skin.
5.3 **Treatment for Exposure:** INHALATION: take chest x-ray immediately to check for pneumonitis. INGESTION: induce vomiting; get medical attention. EYES: flush with water for at least 15 min.; get medical attention. SKIN: cuts or puncture wounds in which beryllium may be embedded under the skin should be thoroughly cleansed immediately by a physician.
5.4 **Toxicity by Inhalation (Threshold Limit Value):** 0.002 mg/m^3 (as beryllium)
5.5 **Short-Term Inhalation Limits:** 0.025 mg/m^3 less than 30 min.
5.6 **Toxicity by Ingestion:** Data not available
5.7 **Late Toxicity:** Beryllium disease may occur in lymph nodes, liver, spleen, kidney, etc. as well as lung.
5.8 **Vapor (Gas) Irritant Characteristics:** Data not available
5.9 **Liquid or Solid Irritant Characteristics:** Data not available
5.10 **Odor Threshold:** Data not available

6. FIRE HAZARDS

6.1 **Flash Point:** Not flammable
6.2 **Flammable Limits in Air:** Not flammable
6.3 **Fire Extinguishing Agents:** Not pertinent
6.4 **Fire Extinguishing Agents Not to be Used:** Not pertinent
6.5 **Special Hazards of Combustion Products:** Toxic beryllium oxide fume may form in fire.
6.6 **Behavior in Fire:**
6.7 **Ignition Temperature:** Not pertinent
6.8 **Electrical Hazard:** Not pertinent
6.9 **Burning Rate:** Not pertinent

7. CHEMICAL REACTIVITY

7.1 **Reactivity with Water:** No reaction
7.2 **Reactivity with Common Materials:**
7.3 **Stability During Transport:** Stable
7.4 **Neutralizing Agents for Acids and Caustics:** Not pertinent
7.5 **Polymerization:** Not pertinent
7.6 **Inhibitor of Polymerization:** Not pertinent

8. WATER POLLUTION

8.1 **Aquatic Toxicity:** Data not available
8.2 **Waterfowl Toxicity:** Data not available
8.3 **Biological Oxygen Demand (BOD):** None
8.4 **Food Chain Concentration Potential:** Bioconcentration of 100-fold can occur under constant exposure. Not significant in spill conditions.

9. SELECTED MANUFACTURERS

1. Var-lac-oid Chemical Co.
   666 South Front St.
   Elizabeth, N. J. 07202
2. Gallard-Schlesinger Chemical Mfg. Co.
   584 Mineola Avenue
   Carle Place, N. Y. 11514
3. Ventron Corporation
   P. O. Box 159
   Beverly, Mass. 01915

10. SHIPPING INFORMATION

10.1 **Grades or Purity:** Technical; Nuclear
10.2 **Storage Temperature:** Ambient
10.3 **Inert Atmosphere:** No requirement
10.4 **Venting:** Open

11. HAZARD ASSESSMENT CODE

(See Hazard Assessment Handbook, CG 446-3)

II

12. HAZARD CLASSIFICATIONS

12.1 **Code of Federal Regulations:** Poisonous solid, Class B
12.2 **NAS Hazard Rating for Bulk Water Transportation:** Not listed
12.3 **NFPA Hazard Classifications:** Not listed

13. PHYSICAL AND CHEMICAL PROPERTIES

13.1 **Physical State at 15°C and 1 atm:** Solid
13.2 **Molecular Weight:** 25
13.3 **Boiling Point at 1 atm:** Not pertinent
13.4 **Freezing Point:** Not pertinent
13.5 **Critical Temperature:** Not pertinent
13.6 **Critical Pressure:** Not pertinent
13.7 **Specific Gravity:** 3.0 at 20°C (solid)
13.8 **Liquid Surface Tension:** Not pertinent
13.9 **Liquid-Water Interfacial Tension:** Not pertinent
13.10 **Vapor (Gas) Specific Gravity:** Not pertinent
13.11 **Ratio of Specific Heats of Vapor (Gas):** Not pertinent
13.12 **Latent Heat of Vaporization:** Not pertinent
13.13 **Heat of Combustion:** Not pertinent
13.14 **Heat of Decomposition:** Not pertinent
13.15 **Heat of Solution:** Not pertinent
13.16 **Heat of Polymerization:** Not pertinent

NOTES

BES	# BERYLLIUM SULFATE

Common Synonyms			
Beryllium sulfate tetrahydrate	Solid	White	Odorless
	Sinks and mixes with water		

AVOID CONTACT WITH SOLID AND DUST. KEEP PEOPLE AWAY.
Wear dust respirator and rubber overclothing (including gloves).
Stop discharge if possible.
Isolate and remove discharged material.
Notify local health and pollution control agencies.

Fire

Not flammable.
POISONOUS GASES MAY BE PRODUCED IN FIRE.
Wear goggles and self-contained breathing apparatus.

Exposure

CALL FOR MEDICAL AID.

DUST
POISONOUS IF INHALED OR IF SKIN IS EXPOSED.
If inhaled will cause coughing or difficult breathing.
If in eyes, hold eyelids open and flush with plenty of water.
If breathing has stopped, give artificial respiration.
If breathing is difficult, give oxygen.

SOLID
Irritating to skin and eyes.
Harmful if swallowed.
Remove contaminated clothing and shoes.
Flush affected areas with plenty of water.
IF IN EYES, hold eyelids open and flush with plenty of water.
IF SWALLOWED and victim is CONSCIOUS, have victim drink water or milk and have victim induce vomiting.
IF SWALLOWED and victim is UNCONSCIOUS OR HAVING CONVULSIONS, do nothing except keep victim warm.

Water Pollution

HARMFUL TO AQUATIC LIFE IN VERY LOW CONCENTRATIONS.
May be dangerous if it enters water intakes.
Notify local health and wildlife officials.
Notify operators of nearby water intakes.

1. RESPONSE TO DISCHARGE

(See Response Methods Handbook, CG 446-4)

Issue warning—poison, water contaminant
Restrict access
Should be removed
Chemical and physical treatment

2. LABEL

3. CHEMICAL DESIGNATIONS

3.1 **Synonyms:** Beryllium sulfate tetrahydrate
3.2 **Coast Guard Compatibility Classification:** Not listed
3.3 **Chemical Formula:** $BeSO_4 \cdot 4H_2O$
3.4 **IMCO/United Nations Numerical Designation:** 6.1/1566

4. OBSERVABLE CHARACTERISTICS

4.1 **Physical State (as shipped):** Solid
4.2 **Color:** White
4.3 **Odor:** None

5. HEALTH HAZARDS

5.1 **Personal Protective Equipment:** Respiratory protection; gloves; freshly laundered clothing; chemical safety goggles
5.2 **Symptoms Following Exposure:** Any dramatic, unexplained weight loss should be considered as possible first indication of beryllium disease. Other symptoms include anorexia, fatigue, weakness, malaise. Inhalation causes pneumonitis, nasopharyngitis, tracheobronchitis, dyspnea, chronic cough. Contact with eyes causes conjunctival inflammation. Contact with skin causes dermatitis of primary irritant or sensitization type; causes ulcer formation when in contact with cuts.
5.3 **Treatment for Exposure:** INHALATION: take chest x-ray immediately to check for evidence of pneumonitis. INGESTION: induce vomiting; get medical attention. EYES: flush with water for at least 15 min.; get medical attention. SKIN: cuts or puncture wounds in which beryllium may be embedded under the skin should be thoroughly cleansed immediately by a physician.
5.4 **Toxicity by Inhalation (Threshold Limit Value):** 0.002 mg/m^3 as beryllium
5.5 **Short-Term Inhalation Limits:** 0.025 mg/m^3 less than 30 min.
5.6 **Toxicity by Ingestion:** Grade 3; oral rat LD_{50} = 82 mg/kg
5.7 **Late Toxicity:** Beryllium disease may occur in the lymph nodes, liver, spleen, kidney, etc., as well as lung.
5.8 **Vapor (Gas) Irritant Characteristics:** Data not available
5.9 **Liquid or Solid Irritant Characteristics:** Data not available
5.10 **Odor Threshold:** Data not available

6. FIRE HAZARDS

6.1 **Flash Point:** Not flammable
6.2 **Flammable Limits in Air:** Not flammable
6.3 **Fire Extinguishing Agents:** Not pertinent
6.4 **Fire Extinguishing Agents Not to be Used:** Not pertinent
6.5 **Special Hazards of Combustion Products:** Toxic beryllium oxide and sulfuric acid fumes may form in fire.
6.6 **Behavior in Fire:**
6.7 **Ignition Temperature:** Not pertinent
6.8 **Electrical Hazard:** Not pertinent
6.9 **Burning Rate:** Not pertinent

7. CHEMICAL REACTIVITY

7.1 **Reactivity with Water:** No reaction
7.2 **Reactivity with Common Materials:**
7.3 **Stability During Transport:** Stable
7.4 **Neutralizing Agents for Acids and Caustics:** Not pertinent
7.5 **Polymerization:** Not pertinent
7.6 **Inhibitor of Polymerization:** Not pertinent

8. WATER POLLUTION

8.1 **Aquatic Toxicity:**
11 ppm/96 hr/fathead minnow/TL_m/hard fresh water
0.2 ppm/96 hr/fathead minnow/TL_m/soft fresh water
8.2 **Waterfowl Toxicity:** Data not available
8.3 **Biological Oxygen Demand (BOD):** None
8.4 **Food Chain Concentration Potential:** Bioconcentration of 100-fold can occur under constant exposure. Not significant in spill conditions.

9. SELECTED MANUFACTURERS

1. Brush Wellman Inc.
   Elmore, Ohio 43416
2. Kawecki, Berylco Industries, Inc.
   Hazleton, Pa. 18201
3. Var-lac-oid Chemical Co.
   666 South Front St.
   Elizabeth, N. J. 07202

10. SHIPPING INFORMATION

10.1 **Grades or Purity:** High purity; Analytical grade
10.2 **Storage Temperature:** Ambient
10.3 **Inert Atmosphere:** No requirement
10.4 **Venting:** Open

11. HAZARD ASSESSMENT CODE

(See Hazard Assessment Handbook, CG 446-3)

SS

12. HAZARD CLASSIFICATIONS

12.1 **Code of Federal Regulations:** Poisonous solid, Class B
12.2 **NAS Hazard Rating for Bulk Water Transportation:** Not listed
12.3 **NFPA Hazard Classifications:** Not listed

13. PHYSICAL AND CHEMICAL PROPERTIES

13.1 **Physical State at 15°C and 1 atm:** Solid
13.2 **Molecular Weight:** 177.14
13.3 **Boiling Point at 1 atm:** Not pertinent (decomposes)
13.4 **Freezing Point:** Not pertinent
13.5 **Critical Temperature:** Not pertinent
13.6 **Critical Pressure:** Not pertinent
13.7 **Specific Gravity:** 1.71 at 11°C (solid)
13.8 **Liquid Surface Tension:** Not pertinent
13.9 **Liquid-Water Interfacial Tension:** Not pertinent
13.10 **Vapor (Gas) Specific Gravity:** Not pertinent
13.11 **Ratio of Specific Heats of Vapor (Gas):** Not pertinent
13.12 **Latent Heat of Vaporization:** Not pertinent
13.13 **Heat of Combustion:** Not pertinent
13.14 **Heat of Decomposition:** Not pertinent
13.15 **Heat of Solution:** −11 Btu/lb = −6 cal/g = -0.3×10^5 J/kg
13.16 **Heat of Polymerization:** Not pertinent

NOTES

BOC	BISMUTH OXYCHLORIDE

Common Synonyms Bismuth chloride oxide Basic bismuth choride Bismuthyl chloride Bismuth subchloride	Solid White Odorless Sinks in water.

Stop discharge if possible. Keep people away.
Avoid contact with solid and dust.
Isolate and remove discharged material.
Notify local health and pollution control agencies.

Fire	Not flammable. Irritating gases may be produced when heated. Wear goggles and self-contained breathing apparatus.
Exposure	CALL FOR MEDICAL AID. DUST Irritating to eyes, nose and throat. Harmful if inhaled. If in eyes, hold eyelids open and flush with plenty of water. If breathing has stopped, give artificial respiration. If breathing is difficult, give oxygen. SOLID Irritating to skin and eyes. Harmful if swallowed. Remove contaminated clothing and shoes. Flush affected areas with plenty of water. IF IN EYES, hold eyelids open and flush with plenty of water. IF SWALLOWED and victim is CONSCIOUS, have victim drink water or milk. IF SWALLOWED and victim is UNCONSCIOUS OR HAVING CONVULSIONS, do nothing except keep victim warm.
Water Pollution	Effect of low concentrations on aquatic life is unknown. May be dangerous if it enters water intakes. Notify local health and wildlife officials. Notify operators of nearby water intakes.

1. RESPONSE TO DISCHARGE

(See Response Methods Handbook, CG 446-4)

Should be removed
Chemical and physical treatment

2. LABELS

No hazard label required by Code of Federal Regulations

3. CHEMICAL DESIGNATIONS

3.1 **Synonyms:** Basic bismuth chloride; Bismuth chloride oxide; Bismuth subchloride; Bismuthyl chloride; Pearl white
3.2 **Coast Guard Compatibility Classification:** Not listed
3.3 **Chemical Formula:** BiOCl
3.4 **IMCO/United Nations Numerical Designation:** Not listed

4. OBSERVABLE CHARACTERISTICS

4.1 **Physical State (as shipped):** Solid
4.2 **Color:** White
4.3 **Odor:** None

5. HEALTH HAZARDS

5.1 **Personal Protective Equipment:** Goggles or face shield; protective gloves; dust mask
5.2 **Symptoms Following Exposure:** Contact with dust causes mild eye irritation and can cause skin rashes.
5.3 **Treatment for Exposure:** EYES: flush with water.
5.4 **Toxicity by Inhalation (Threshold Limit Value):** Data not available
5.5 **Short-Term Inhalation Limits:** Data not available
5.6 **Toxicity by Ingestion:** Grade 0; LD_{50} > 21.5 g/kg (rat)
5.7 **Late Toxicity:** Data not available
5.8 **Vapor (Gas) Irritant Characteristics:** Data not available
5.9 **Liquid or Solid Irritant Characteristics:** Data not available
5.10 **Odor Threshold:** Data not available

6. FIRE HAZARDS

6.1 **Flash Point:** Not flammable
6.2 **Flammable Limits in Air:** Not flammable
6.3 **Fire Extinguishing Agents:** Not pertinent
6.4 **Fire Extinguishing Agents Not to be Used:** Not pertinent
6.5 **Special Hazards of Combustion Products:** Irritating hydrogen chloride gas may form in fire.
6.6 **Behavior in Fire:**
6.7 **Ignition Temperature:** Not pertinent
6.8 **Electrical Hazard:** Not pertinent
6.9 **Burning Rate:** Not pertinent

7. CHEMICAL REACTIVITY

7.1 **Reactivity with Water:** No reaction
7.2 **Reactivity with Common Materials:**
7.3 **Stability During Transport:** Stable
7.4 **Neutralizing Agents for Acids and Caustics:** Not pertinent
7.5 **Polymerization:** Not pertinent
7.6 **Inhibitor of Polymerization:** Not pertinent

8. WATER POLLUTION

8.1 **Aquatic Toxicity:** Data not available
8.2 **Waterfowl Toxicity:** Data not available
8.3 **Biological Oxygen Demand (BOD):** None
8.4 **Food Chain Concentration Potential:** Data not available

9. SELECTED MANUFACTURERS

1. The Mearl Corp.
41 East 42nd St.
New York, N. Y. 10017
2. Mallinckrodt Chemical Works
223 Westside Ave.
P. O. Box 384
Jersey City, N. J. 07303
3. Whittaker Corporation
Rona Pearl Company Division
East 21st Street
Bayonne, N. J. 07002

10. SHIPPING INFORMATION

10.1 **Grades or Purity:** Dry powder, 100%; aqueous concentrates; dispersions of solid in mineral oil or castor oil.
10.2 **Storage Temperature:** Ambient
10.3 **Inert Atmosphere:** No requirement
10.4 **Venting:** Open

11. HAZARD ASSESSMENT CODE

(See Hazard Assessment Handbook, CG 446-3)

II

12. HAZARD CLASSIFICATIONS

12.1 **Code of Federal Regulations:** Not listed
12.2 **NAS Hazard Rating for Bulk Water Transportation:** Not listed
12.3 **NFPA Hazard Classifications:** Not listed

13. PHYSICAL AND CHEMICAL PROPERTIES

13.1 **Physical State at 15°C and 1 atm:** Solid
13.2 **Molecular Weight:** 260.4
13.3 **Boiling Point at 1 atm:** Not pertinent (decomposes)
13.4 **Freezing Point:** Not pertinent
13.5 **Critical Temperature:** Not pertinent
13.6 **Critical Pressure:** Not pertinent
13.7 **Specific Gravity:** 7.7 at 20°C (solid)
13.8 **Liquid Surface Tension:** Not pertinent
13.9 **Liquid-Water Interfacial Tension:** Not pertinent
13.10 **Vapor (Gas) Specific Gravity:** Not pertinent
13.11 **Ratio of Specific Heats of Vapor (Gas):** Not pertinent
13.12 **Latent Heat of Vaporization:** Not pertinent
13.13 **Heat of Combustion:** Not pertinent
13.14 **Heat of Decomposition:** Not pertinent
13.15 **Heat of Solution:** Not pertinent
13.16 **Heat of Polymerization:** Not pertinent

NOTES

BPA

BISPHENOL A

Common Synonyms 4,4'-Isopropylidendiphenol p,p'-Dihydroxydiphenyl-dimethylmethane 2,2-Bis (4-hydroxyphenyl)-propane	Solid flakes or powder Sinks in water.	White to light brown	Weak medicine odor

Avoid contact with solid and dust. Keep people away.
Call fire department.
Isolate and remove discharged material.
Notify local health and pollution control agencies.

Fire

Combustible.
Extinguish with dry chemical, foam, or carbon dioxide.
Cool exposed containers with water.

Exposure

CALL FOR MEDICAL AID.

DUST
Irritating to nose and throat if inhaled.
Move to fresh air.

SOLID
Irritating to skin and eyes.
Harmful if swallowed.
Flush affected areas with plenty of water.
IF IN EYES, hold eyelids open and flush with plenty of water.
IF SWALLOWED and victim is CONSCIOUS, have victim drink water or milk and have victim induce vomiting.
IF SWALLOWED and victim is UNCONSCIOUS OR HAVING CONVULSIONS, do nothing except keep victim warm.

Water Pollution

Effect of low concentrations on aquatic life is unknown.
May be dangerous if it enters water intakes.
Notify local health and pollution control officials.
Notify operators of nearby water intakes.

1. RESPONSE TO DISCHARGE
(See Response Methods Handbook, CG 446-4)

Should be removed

2. LABELS

No hazard label required by Code of Federal Regulations

3. CHEMICAL DESIGNATIONS

3.1 **Synonyms:**
2, 2-Bis (4-hydroxyphenyl) propane
p, p'-Dihydroxydiphenyldimethylmethane
4, 4'-Isopropylidenediphenol
Ucar bisphenol HP

3.2 **Coast Guard Compatibility Classification:** Not applicable

3.3 **Chemical Formula:** p-$HOC_6H_4C(CH_3)_2C_6H_4OH$-p

3.4 **IMCO/United Nations Numerical Designation:** Not listed

4. OBSERVABLE CHARACTERISTICS

4.1 **Physical State (as shipped):** Solid

4.2 **Color:** White to cream

4.3 **Odor:** Very weak phenolic

5. HEALTH HAZARDS

5.1 **Personal Protective Equipment:** Approved dust mask and clean, body-covering clothing sufficient to prevent excessive or repeated exposure to dust, fumes, or solutions. Safety glasses with side shields.

5.2 **Symptoms Following Exposure:** Dusts irritating to upper respiratory passages; may cause sneezing.

5.3 **Treatment for Exposure:** SKIN: Wash with soap and plenty of water. Avoid wearing contaminated clothing; EYES: Promptly flush with plenty of water for at least 15 minutes and get medical help. INGESTION: If large amounts are swallowed, induce vomiting promptly and get medical help promptly. No known antidote.

5.4 **Toxicity by Inhalation (Threshold Limit Value):** Not pertinent

5.5 **Short-Term Inhalation Limits:** Not pertinent

5.6 **Toxicity by Ingestion:** Grade 2; LD_{50} 0.5 to 5 g/kg (rat)

5.7 **Late Toxicity:** Lowered hemoglobin and erythrocyte (red blood cell) counts below normal in rats.

5.8 **Vapor (Gas) Irritant Characteristics:** Not pertinent

5.9 **Liquid or Solid Irritant Characteristics:** Minimum hazard. If spilled on clothing and allowed to remain, may cause smarting and reddening of the skin.

5.10 **Odor Threshold:** Not pertinent

6. FIRE HAZARDS

6.1 **Flash Point:** 415°F O.C.

6.2 **Flammable Limits in Air:** Not pertinent

6.3 **Fire Extinguishing Agents:** Foam, dry chemical, carbon dioxide

6.4 **Fire Extinguishing Agents Not to be Used:** Data not available

6.5 **Special Hazards of Combustion Products:** Not pertinent

6.6 **Behavior in Fire:** Not pertinent

6.7 **Ignition Temperature:** Data not available

6.8 **Electrical Hazard:** Not pertinent

6.9 **Burning Rate:** Not pertinent

7. CHEMICAL REACTIVITY

7.1 **Reactivity with Water:** No reaction

7.2 **Reactivity with Common Materials:** No reaction

7.3 **Stability During Transport:** Stable

7.4 **Neutralizing Agents for Acids and Caustics:** Not pertinent

7.5 **Polymerization:** Not pertinent

7.6 **Inhibitor of Polymerization:** Not pertinent

8. WATER POLLUTION

8.1 **Aquatic Toxicity:** Data not available

8.2 **Waterfowl Toxicity:** Data not available

8.3 **Biological Oxygen Demand (BOD):** Data not available

8.4 **Food Chain Concentration Potential:** Data not available

9. SELECTED MANUFACTURERS

1. Dow Chemical Co.
   Midland, Michigan 48640
2. General Electric Co.
   1 Plastics Ave.
   Pittsfield, Mass. 01201
3. Shell Chemical Co.
   Polymers Div.
   P. O. Box 2463
   Houston, Texas 77001

10. SHIPPING INFORMATION

10.1 **Grades or Purity:** Commercial; high purity

10.2 **Storage Temperature:** Data not available

10.3 **Inert Atmosphere:** Data not available

10.4 **Venting:** Data not available

11. HAZARD ASSESSMENT CODE
(See Hazard Assessment Handbook, CG 446-3)

II

12. HAZARD CLASSIFICATIONS

12.1 **Code of Federal Regulations:** Not listed

12.2 **NAS Hazard Rating for Bulk Water Transportation:** Not listed

12.3 **NFPA Hazard Classifications:** Not listed

13. PHYSICAL AND CHEMICAL PROPERTIES

13.1 **Physical State at 15°C and 1 atm:** Solid

13.2 **Molecular Weight:** 228.28

13.3 **Boiling Point at 1 atm:** Not pertinent

13.4 **Freezing Point:** 315°F = 157°C = 430°K

13.5 **Critical Temperature:** Not pertinent

13.6 **Critical Pressure:** Not pertinent

13.7 **Specific Gravity:** 1.195 at 25°C (solid)

13.8 **Liquid Surface Tension:** Not pertinent

13.9 **Liquid-Water Interfacial Tension:** Not pertinent

13.10 **Vapor (Gas) Specific Gravity:** Not pertinent

13.11 **Ratio of Specific Heats of Vapor (Gas):** Not pertinent

13.12 **Latent Heat of Vaporization:** Not pertinent

13.13 **Heat of Combustion:** Not pertinent

13.14 **Heat of Decomposition:** Not pertinent

13.15 **Heat of Solution:** Not pertinent

13.16 **Heat of Polymerization:** Not pertinent

NOTES

REVISED 1978

BDE	BISPHENOL A DIGLYCIDYL ETHER

Common Synonyms		
4,4' Isopropylidenediphenol - epichlorohydrin resin Bisphenol A - epichlorohydrin condensate 2,2-Bis(p-hydroxyphenyl) propane, diglycidyl ether	Liquid Sinks in water.	Yellowish brown

Stop discharge if possible. Keep people away.
Call fire department.
Avoid contact with liquid.
Isolate and remove discharged material.
Notify local health and pollution control agencies.

Fire

Combustible.
Extinguish with water, dry chemicals, foam, or carbon dioxide.

Exposure

CALL FOR MEDICAL AID.

LIQUID
Irritating to skin and eyes.
Harmful if swallowed.

Remove contaminated clothing and shoes.
Flush affected areas with plenty of water.
IF IN EYES, hold eyelids open and flush with plenty of water.
IF SWALLOWED and victim is CONSCIOUS, have victim drink water or milk.
IF SWALLOWED and victim is UNCONSCIOUS OR HAVING CONVULSIONS, do nothing except keep victim warm.

Water Pollution

Effect of low concentrations on aquatic life is unknown.
May be dangerous if it enters water intakes.

Notify local health and wildlife officials.
Notify operators of nearby water intakes.

1. RESPONSE TO DISCHARGE

(See Response Methods Handbook, CG 446-4)

Should be removed
Chemical and physical treatment

2. LABELS

No hazard label required by Code of Federal Regulations

3. CHEMICAL DESIGNATIONS

3.1 **Synonyms:** 4,4' Isopropylidenediphenol-epichlorohydrin resin; Bisphenol A-epichlorohydrin condensate; 2,2-Bis (p-hydroxyphenyl) propane, diglycidyl ether
3.2 **Coast Guard Compatibility Classification:** Not listed
3.3 **Chemical Formula:** $C_{21}H_{24}O_4$
3.4 **IMCO/United Nations Numerical Designation:** Not listed

4. OBSERVABLE CHARACTERISTICS

4.1 **Physical State (as shipped):** Liquid
4.2 **Color:** Amber
4.3 **Odor:** Data not available

5. HEALTH HAZARDS

5.1 **Personal Protective Equipment:** Rubber gloves; goggles; protective clothing and protective creams. Good personal hygiene is necessary, with instruction of personnel and adequate cleaning facilities.
5.2 **Symptoms Following Exposure:** Contact with liquid irritates eyes. Prolonged or repeated contact with skin causes irritation and dermatitis.
5.3 **Treatment for Exposure:** EYES: flush with water for at least 15 min. SKIN: remove chemical with water or waterless skin cleaner.
5.4 **Toxicity by Inhalation (Threshold Limit Value):** Data not available
5.5 **Short-Term Inhalation Limits:** Data not available
5.6 **Toxicity by Ingestion:** Grade 1; LD_{50} 5 – 15 g/kg
5.7 **Late Toxicity:** Data not available
5.8 **Vapor (Gas) Irritant Characteristics:** Vapors are nonirritating to eyes and throat.
5.9 **Liquid or Solid Irritant Characteristics:** Minimum hazard. If spilled on clothing and allowed to remain, may cause smarting and reddening of skin.
5.10 **Odor Threshold:** Odorless

6. FIRE HAZARDS

6.1 **Flash Point:** 175°F O.C.
6.2 **Flammable Limits in Air:** Data not available
6.3 **Fire Extinguishing Agents:** Water, foam, dry chemical, carbon dioxide
6.4 **Fire Extinguishing Agents Not to be Used:**
6.5 **Special Hazards of Combustion Products:**
6.6 **Behavior in Fire:**
6.7 **Ignition Temperature:** Data not available
6.8 **Electrical Hazard:** Data not available
6.9 **Burning Rate:** Data not available

7. CHEMICAL REACTIVITY

7.1 **Reactivity with Water:** No reaction
7.2 **Reactivity with Common Materials:**
7.3 **Stability During Transport:** Stable
7.4 **Neutralizing Agents for Acids and Caustics:** Not pertinent
7.5 **Polymerization:** Not pertinent
7.6 **Inhibitor of Polymerization:** Not pertinent

8. WATER POLLUTION

8.1 **Aquatic Toxicity:** Data not available
8.2 **Waterfowl Toxicity:** Data not available
8.3 **Biological Oxygen Demand (BOD):** Data not available
8.4 **Food Chain Concentration Potential:** None

9. SELECTED MANUFACTURERS

Shell Chemical Co.
1 Shell Plaza
Houston, Texas 77002

10. SHIPPING INFORMATION

10.1 **Grades or Purity:** Commercial
10.2 **Storage Temperature:** Ambient
10.3 **Inert Atmosphere:** No requirement
10.4 **Venting:** Open

11. HAZARD ASSESSMENT CODE

(See Hazard Assessment Handbook, CG 446-3)

A–X

12. HAZARD CLASSIFICATIONS

12.1 **Code of Federal Regulations:** Not listed
12.2 **NAS Hazard Rating for Bulk Water Transportation:**

Category	Rating
Fire	1
Health	
Vapor Irritant	0
Liquid or Solid Irritant	1
Poisons	1
Water Pollution	
Human Toxicity	1
Aquatic Toxicity	0
Aesthetic Effect	0
Reactivity	
Other Chemicals	2
Water	0
Self-Reaction	1

12.3 **NFPA Hazard Classifications:** Not listed

13. PHYSICAL AND CHEMICAL PROPERTIES

13.1 **Physical State at 15°C and 1 atm:** Liquid
13.2 **Molecular Weight:** 340
13.3 **Boiling Point at 1 atm:** Not pertinent (decomposes)
13.4 **Freezing Point:** Not pertinent
13.5 **Critical Temperature:** Not pertinent
13.6 **Critical Pressure:** Not pertinent
13.7 **Specific Gravity:** 1.16 at 20°C (liquid)
13.8 **Liquid Surface Tension:** Data not available
13.9 **Liquid-Water Interfacial Tension:** Data not available
13.10 **Vapor (Gas) Specific Gravity:** Not pertinent
13.11 **Ratio of Specific Heats of Vapor (Gas):** Not pertinent
13.12 **Latent Heat of Vaporization:** Data not available
13.13 **Heat of Combustion:** (est.) $-14{,}900$ Btu/lb $= -8{,}300$ cal/g $= -350 \times 10^5$ J/kg
13.14 **Heat of Decomposition:** Not pertinent
13.15 **Heat of Solution:** Not pertinent
13.16 **Heat of Polymerization:** Data not available

NOTES

BCP

BOILER COMPOUND, LIQUID

Common Synonyms

Liquid | Colorless to brown | Odorless or mild odor

Sinks and mixes with water.

Avoid contact with liquid. Keep people away.
Stop discharge if possible.
Isolate and remove discharged material.
Notify local health and pollution control agencies.

Fire

Not flammable.
Flammable gas may be produced on contact with metals.
Cool exposed containers with water.

Exposure

Call for medical aid.

LIQUID
Will burn skin and eyes.
Harmful if swallowed.

Remove contaminated clothing and shoes.
Flush affected areas with plenty of water.
IF IN EYES, hold eyelids open and flush with plenty of water.
IF SWALLOWED, and victim is CONSCIOUS, have victim drink water or milk.
DO NOT INDUCE VOMITING.

Water Pollution

Effect of low concentrations on aquatic life is unknown.
May be dangerous if it enters water intakes.
Notify local health and wildlife officials.
Notify operators of nearby water intakes.

1. RESPONSE TO DISCHARGE

(See Response Methods Handbook, CG 446-4)

Issue warning—corrosive, water contaminant
Restrict access
Disperse and flush

2. LABEL

3. CHEMICAL DESIGNATIONS

3.1 **Synonyms:** Alkaway liquid alkaline deruster
3.2 **Coast Guard Compatibility Classification:** Not listed
3.3 **Chemical Formula:** Not pertinent
3.4 **IMCO/United Nations Numerical Designation:** Not listed

4. OBSERVABLE CHARACTERISTICS

4.1 **Physical State (as shipped):** Liquid
4.2 **Color:** Colorless to brown
4.3 **Odor:** None to mild

5. HEALTH HAZARDS

5.1 **Personal Protective Equipment:** Goggles or face shield; rubber gloves; protective clothing.
5.2 **Symptoms Following Exposure:** Contact of liquid with eyes causes severe caustic burns. Also causes caustic burns of skin if not washed off immediately. Ingestion causes caustic burns of mouth and stomach.
5.3 **Treatment for Exposure:** Basic treatment is identical with that for caustic soda or caustic potash solutions. EYES: flush with water for at least 15 min.; call a doctor. SKIN: flush with water; wash with soap and water. INGESTION: give large amounts of water.
5.4 **Toxicity by Inhalation (Threshold Limit Value):** 5 mg/m^3
5.5 **Short-Term Inhalation Limits:** Data not available
5.6 **Toxicity by Ingestion:** Data not available
5.7 **Late Toxicity:** Not pertinent
5.8 **Vapor (Gas) Irritant Characteristics:** Not pertinent
5.9 **Liquid or Solid Irritant Characteristics:** Data not available
5.10 **Odor Threshold:** Not pertinent

6. FIRE HAZARDS

6.1 **Flash Point:** Not flammable
6.2 **Flammable Limits in Air:** Not flammable
6.3 **Fire Extinguishing Agents:** Not pertinent
6.4 **Fire Extinguishing Agents Not to be Used:** Not pertinent
6.5 **Special Hazards of Combustion Products:** Not pertinent
6.6 **Behavior in Fire:** May burst container
6.7 **Ignition Temperature:** Not pertinent
6.8 **Electrical Hazard:** Not pertinent
6.9 **Burning Rate:** Not pertinent

7. CHEMICAL REACTIVITY

7.1 **Reactivity with Water:** No reaction
7.2 **Reactivity with Common Materials:** Attacks aluminum and zinc; the reaction may form flammable hydrogen gas.
7.3 **Stability During Transport:** Stable
7.4 **Neutralizing Agents for Acids and Caustics:** Flush with water
7.5 **Polymerization:** Not pertinent
7.6 **Inhibitor of Polymerization:** Not pertinent

8. WATER POLLUTION

8.1 **Aquatic Toxicity:** Data not available
8.2 **Waterfowl Toxicity:** Data not available
8.3 **Biological Oxygen Demand (BOD):** None
8.4 **Food Chain Concentration Potential:** None

9. SELECTED MANUFACTURERS

1. BASF Wyandotte Corp.
   Chemical Specialties Division
   Wyandotte, Mich. 48192
2. Nalco Chemical Co.
   180 N. Michigan Ave.
   Chicago, Ill. 60601
3. Olin Water Services
   3155 Fiberglas Road
   Kansas City, Kan. 66115

10. SHIPPING INFORMATION

10.1 **Grades or Purity:** Various commercial grades, some of which contain chelating and complexing agents for metals.
10.2 **Storage Temperature:** Ambient, preferably 40-100°F
10.3 **Inert Atmosphere:** No requirement
10.4 **Venting:** Open

11. HAZARD ASSESSMENT CODE

(See Hazard Assessment Handbook, CG 446-3)

A-P

12. HAZARD CLASSIFICATIONS

12.1 **Code of Federal Regulations:** Corrosive
12.2 **NAS Hazard Rating for Bulk Water Transportation:** Not listed
12.3 **NFPA Hazard Classifications:** Not listed

13. PHYSICAL AND CHEMICAL PROPERTIES

13.1 **Physical State at 15°C and 1 atm:** Liquid
13.2 **Molecular Weight:** Not pertinent
13.3 **Boiling Point at 1 atm:** >220°F = >104°C = >377°K
13.4 **Freezing Point:** Not pertinent
13.5 **Critical Temperature:** Not pertinent
13.6 **Critical Pressure:** Not pertinent
13.7 **Specific Gravity:** 1.48 at 20°C (liquid)
13.8 **Liquid Surface Tension:** Not pertinent
13.9 **Liquid-Water Interfacial Tension:** Not pertinent
13.10 **Vapor (Gas) Specific Gravity:** Not pertinent
13.11 **Ratio of Specific Heats of Vapor (Gas):** Not pertinent
13.12 **Latent Heat of Vaporization:** Not pertinent
13.13 **Heat of Combustion:** Not pertinent
13.14 **Heat of Decomposition:** Not pertinent
13.15 **Heat of Solution:** Not pertinent
13.16 **Heat of Polymerization:** Not pertinent

NOTES

BAC	BORIC ACID

Common Synonyms Orthoboric acid Boracic acid	Solid White Odorless Sinks and mixes with water.

Stop discharge if possible. Keep people away.
Avoid contact with solid.
Isolate and remove discharged material.
Notify local health and pollution control agencies.

Fire	Not flammable.
Exposure	CALL FOR MEDICAL AID. SOLID Irritating to skin and eyes. If swallowed will cause nausea or vomiting. Remove contaminated clothing and shoes. Flush affected areas with plenty of water. IF IN EYES, hold eyelids open and flush with plenty of water. IF SWALLOWED and victim is CONSCIOUS, have victim drink water or milk and have victim induce vomiting. IF SWALLOWED and victim is UNCONSCIOUS OR HAVING CONVULSIONS, do nothing except keep victim warm.
Water Pollution	Dangerous to aquatic life in high concentrations. May be dangerous if it enters water intakes. Notify local health and wildlife officials. Notify operators of nearby water intakes.

1. RESPONSE TO DISCHARGE

(See Response Methods Handbook, CG 446-4)

Issue warning—water contaminant
Should be removed
Chemical and physical treatment

2. LABELS

No hazard label required by Code of Federal Regulations

3. CHEMICAL DESIGNATIONS

3.1 **Synonyms:** Boracic acid; Orthoboric acid
3.2 **Coast Guard Compatibility Classification:** Not listed
3.3 **Chemical Formula:** H_3BO_3
3.4 **IMCO/United Nations Numerical Designation:** Not listed

4. OBSERVABLE CHARACTERISTICS

4.1 **Physical State (as shipped):** Solid
4.2 **Color:** White
4.3 **Odor:** None

5. HEALTH HAZARDS

5.1 **Personal Protective Equipment:** Chemical goggles
5.2 **Symptoms Following Exposure:** Although no adverse effects have been reported from inhaling boric acid dust, it is absorbed through mucous membranes. Ingestion of 5 grams or more may irritate gastrointestinal tract and affect central nervous system. Contact with dust or aqueous solutions may irritate eyes; no chronic effects have been recognized, but continued contact should be avoided. Dust and solutions are absorbed through burns and open wounds but not through unbroken skin.
5.3 **Treatment for Exposure:** INHALATION: remove from contaminated atmosphere. INGESTION: obtain medical attention as soon as possible; if the patient is conscious, induce vomiting by giving warm salty water (2 tablespoons of table salt to a pint of water) or warm soapy water; if this measure is unsuccessful, vomiting may be induced by tickling the back of the patient's throat with a finger; vomiting should be encouraged about three times or until the vomitus is clear; additional water may be given to wash out the stomach. EYES: immediately flush the eyes with large quantities of running water for a minimum of 15 min.; hold the eyelids apart during the irrigation to ensure flushing of the entire surface of the eye and lids with water; obtain medical attention as soon as possible; continue the irrigation for an additional 15 min. if the physician is not available. SKIN: immediately flush affected area with water; remove contaminated clothing under the shower; continue washing with water—do not attempt to neutralize with chemical agents; obtain medical attention unless burn is minor.
5.4 **Toxicity by Inhalation (Threshold Limit Value):** 10 mg/m^3 (as boric oxide)
5.5 **Short-Term Inhalation Limits:** Data not available

6. FIRE HAZARDS

6.1 **Flash Point:** Not flammable
6.2 **Flammable Limits in Air:** Not flammable
6.3 **Fire Extinguishing Agents:** Not pertinent
6.4 **Fire Extinguishing Agents Not to be Used:** Not pertinent
6.5 **Special Hazards of Combustion Products:**
6.6 **Behavior in Fire:**
6.7 **Ignition Temperature:** Not pertinent
6.8 **Electrical Hazard:** Not pertinent
6.9 **Burning Rate:** Not pertinent

7. CHEMICAL REACTIVITY

7.1 **Reactivity with Water:** No reaction
7.2 **Reactivity with Common Materials:**
7.3 **Stability During Transport:** Stable
7.4 **Neutralizing Agents for Acids and Caustics:** Not pertinent
7.5 **Polymerization:** Not pertinent
7.6 **Inhibitor of Polymerization:** Not pertinent

8. WATER POLLUTION

8.1 **Aquatic Toxicity:** 1800 ppm/24 hr/mosquito fish/TL_m/fresh water
8.2 **Waterfowl Toxicity:** Data not available
8.3 **Biological Oxygen Demand (BOD):** None
8.4 **Food Chain Concentration Potential:** None

9. SELECTED MANUFACTURERS

1. Kerr-McGee Chemical Corp.
   Kerr-McGee Bldg.
   Oklahoma City, Okla. 73102
2. U. S. Borax and Chemical Co.
   3075 Wilshire Blvd.
   Los Angeles, Calif. 90010
3. Stauffer Chemical Co.
   Industrial Chemicals Div.
   Westport, Conn. 06880

10. SHIPPING INFORMATION

10.1 **Grades or Purity:** Radio, 99.98%; Technical, 99.9%; N.F., 99.5%
10.2 **Storage Temperature:** Ambient
10.3 **Inert Atmosphere:** No requirement
10.4 **Venting:** Open

11. HAZARD ASSESSMENT CODE

(See Hazard Assessment Handbook, CG 446-3)

SS-II

12. HAZARD CLASSIFICATIONS

12.1 **Code of Federal Regulations:** Not listed
12.2 **NAS Hazard Rating for Bulk Water Transportation:** Not listed
12.3 **NFPA Hazard Classifications:** Not listed

13. PHYSICAL AND CHEMICAL PROPERTIES

13.1 **Physical State at 15°C and 1 atm:** Solid
13.2 **Molecular Weight:** 61.83
13.3 **Boiling Point at 1 atm:** Not pertinent (decomposes)
13.4 **Freezing Point:** Not pertinent
13.5 **Critical Temperature:** Not pertinent
13.6 **Critical Pressure:** Not pertinent
13.7 **Specific Gravity:** 1.51 at 14°C (solid)
13.8 **Liquid Surface Tension:** Not pertinent
13.9 **Liquid-Water Interfacial Tension:** Not pertinent
13.10 **Vapor (Gas) Specific Gravity:** Not pertinent
13.11 **Ratio of Specific Heats of Vapor (Gas):** Not pertinent
13.12 **Latent Heat of Vaporization:** Not pertinent
13.13 **Heat of Combustion:** Not pertinent
13.14 **Heat of Decomposition:** Not pertinent
13.15 **Heat of Solution:** -157 Btu/lb $= -87$ cal/g $= -3.7 \times 10^5$ J/kg
13.16 **Heat of Polymerization:** Not pertinent

5. HEALTH HAZARDS (Cont'd)

5.6 **Toxicity by Ingestion:** Grade 2; oral rat LD_{50} = 2,660 mg/kg
5.7 **Late Toxicity:** Data not available
5.8 **Vapor (Gas) Irritant Characteristics:** Data not available
5.9 **Liquid or Solid Irritant Characteristics:** Data not available
5.10 **Odor Threshold:** Odorless

BTB	# BORON TRIBROMIDE

Common Synonyms	Liquid	Colorless	Sharp odor
	Reacts violently with water. Poisonous vapor is produced.		

Avoid contact with liquid and vapor. Keep people away.
Stop discharge if possible.
Isolate and remove discharged material.
Notify local health and pollution control agencies.

Fire

Not flammable.
POISONOUS GASES MAY BE PRODUCED IN FIRE.
Wear goggles and self-contained breathing apparatus.
DO NOT USE WATER ON ADJACENT FIRES.

Exposure

CALL FOR MEDICAL AID.

VAPOR
Irritating to eyes, nose and throat.
If inhaled will cause coughing or difficult breathing.
If in eyes, hold eyelids open and flush with plenty of water.
If breathing has stopped, give artificial respiration.
If breathing is difficult, give oxygen.

LIQUID
Will burn skin and eyes.
If swallowed will cause nausea and vomiting.
Remove contaminated clothing and shoes.
Flush affected areas with plenty of water.
IF IN EYES, hold eyelids open and flush with plenty of water.
IF SWALLOWED and victim is CONSCIOUS, have victim drink water or milk.
IF SWALLOWED and victim is UNCONSCIOUS OR HAVING CONVULSIONS, do nothing except keep victim warm.

Water Pollution

Effect of low concentrations on aquatic life is unknown.
May be dangerous if it enters water intakes.
Notify local health and wildlife officials.
Notify operators of nearby water intakes.

1. RESPONSE TO DISCHARGE

(See Response Methods Handbook, CG 446-4)

Issue warning: corrosive, air contaminant
Restrict access
Disperse and flush with care

2. LABEL

3. CHEMICAL DESIGNATIONS

3.1 **Synonyms:** No common synonyms
3.2 **Coast Guard Compatibility Classification:** Not listed
3.3 **Chemical Formula:** BBr_3
3.4 **IMCO/United Nations Numerical Designation:** Not listed

4. OBSERVABLE CHARACTERISTICS

4.1 **Physical State (as shipped):** Liquid
4.2 **Color:** Colorless
4.3 **Odor:** Sharp, irritating

5. HEALTH HAZARDS

5.1 **Personal Protective Equipment:** Chemical safety glasses or face mask, rubber gloves, and respiratory protection
5.2 **Symptoms Following Exposure:** Inhalation causes severe irritation of mucous membranes. Ingestion causes burns of mouth and stomach. Contact with eyes or skin causes severe burns.
5.3 **Treatment for Exposure:** *Get medical attention for all exposures to this chemical.* INHALATION: remove from exposure; support respiration. INGESTION: give large amounts of water. EYES: flush with water for at least 15 min. SKIN: wash thoroughly with water; launder clothing before reuse.
5.4 **Toxicity by Inhalation (Threshold Limit Value):** 1 ppm
5.5 **Short-Term Inhalation Limits:** Data not available
5.6 **Toxicity by Ingestion:** Data not available
5.7 **Late Toxicity:** Data not available
5.8 **Vapor (Gas) Irritant Characteristics:** Data not available
5.9 **Liquid or Solid Irritant Characteristics:** Data not available
5.10 **Odor Threshold:** Data not available

6. FIRE HAZARDS

6.1 **Flash Point:** Not flammable
6.2 **Flammable Limits in Air:** Not flammable
6.3 **Fire Extinguishing Agents:** Not pertinent
6.4 **Fire Extinguishing Agents Not to be Used:** Do not use water or foam on adjacent fires.
6.5 **Special Hazards of Combustion Products:** Toxic fumes of the chemical or hydrogen bromide may form in fires.
6.6 **Behavior in Fire:**
6.7 **Ignition Temperature:** Not pertinent
6.8 **Electrical Hazard:** Not pertinent
6.9 **Burning Rate:** Not pertinent

7. CHEMICAL REACTIVITY

7.1 **Reactivity with Water:** Reacts violently to form hydrobromic acid solution and fumes.
7.2 **Reactivity with Common Materials:** Strongly attacks metals and wood. Flammable hydrogen gas may collect in enclosed spaces.
7.3 **Stability During Transport:** Stable
7.4 **Neutralizing Agents for Acids and Caustics:** Flush with water, rinse with dilute solution of sodium bicarbonate or soda ash.
7.5 **Polymerization:** Not pertinent
7.6 **Inhibitor of Polymerization:** Not pertinent

8. WATER POLLUTION

8.1 **Aquatic Toxicity:** Data not available
8.2 **Waterfowl Toxicity:** Data not available
8.3 **Biological Oxygen Demand (BOD):** None
8.4 **Food Chain Concentration Potential:** None

9. SELECTED MANUFACTURERS

1. Eagle-Picher Industries, Inc.
   Miami Research Laboratories
   Miami, Okla. 74354
2. Gallard-Schlesinger Chemical Mfg. Co.
   584 Mineola Avenue
   Carle Place, N.Y. 11514
3. Ventron Corporation
   P.O. Box 159
   Beverly, Mass. 01915

10. SHIPPING INFORMATION

10.1 **Grades or Purity:** Epitaxial, 99.999+%; Pure, 99.99+%; Technical
10.2 **Storage Temperature:** Ambient
10.3 **Inert Atmosphere:** Padded
10.4 **Venting:** Pressure-vacuum

11. HAZARD ASSESSMENT CODE

(See Hazard Assessment Handbook, CG 446-3)

A-O

12. HAZARD CLASSIFICATIONS

12.1 **Code of Federal Regulations:** Corrosive liquid
12.2 **NAS Hazard Rating for Bulk Water Transportation:** Not listed
12.3 **NFPA Hazard Classifications:** Not listed

13. PHYSICAL AND CHEMICAL PROPERTIES

13.1 **Physical State at 15°C and 1 atm:** Liquid
13.2 **Molecular Weight:** 250.5
13.3 **Boiling Point at 1 atm:** 196°F = 91°C = 364°K
13.4 **Freezing Point:** −51°F = −46°C = 227°K
13.5 **Critical Temperature:** Not pertinent
13.6 **Critical Pressure:** Not pertinent
13.7 **Specific Gravity:** 2.645 at 20°C (liquid)
13.8 **Liquid Surface Tension:** 29.1 dynes/cm = 0.0291 N/m at 22°C
13.9 **Liquid-Water Interfacial Tension:** Not pertinent
13.10 **Vapor (Gas) Specific Gravity:** 8.64
13.11 **Ratio of Specific Heats of Vapor (Gas):** 1.140
13.12 **Latent Heat of Vaporization:** 52 Btu/lb = 29 cal/g = 1.2×10^5 J/kg
13.13 **Heat of Combustion:** Not pertinent
13.14 **Heat of Decomposition:** Not pertinent
13.15 **Heat of Solution:** Data not available
13.16 **Heat of Polymerization:** Not pertinent

NOTES

BRT BORON TRICHLORIDE

Common Synonyms

Liquid	Colorless	Irritating odor

Reacts violently with water. Irritating visible vapor cloud is produced. Boiling point is 54° F.

AVOID CONTACT WITH LIQUID AND VAPOR. KEEP PEOPLE AWAY.
Wear goggles, self-contained breathing apparatus, and rubber overclothing (including gloves).
Stop discharge if possible.
Isolate and remove discharged material.
Notify local health and pollution control agencies.

Fire

Not flammable.
POISONOUS GASES MAY BE PRODUCED WHEN HEATED.
Cool exposed containers with water.

Exposure

CALL FOR MEDICAL AID.

VAPOR
Irritating to eyes, nose and throat.
If inhaled will cause coughing or difficult breathing.
Move victim to fresh air.
If breathing has stopped, give artificial respiration.
If breathing is difficult, give oxygen.

LIQUID
POISONOUS IF SWALLOWED.
Will burn skin and eyes.
Remove contaminated clothing and shoes.
Flush affected areas with plenty of water.
IF IN EYES, hold eyelids open and flush with plenty of water.
IF SWALLOWED and victim is CONSCIOUS, have victim drink water or milk.
DO NOT INDUCE VOMITING.

Water Pollution

Effect of low concentrations on aquatic life is unknown.
May be dangerous if it enters water intakes.
Notify local health and wildlife officials.
Notify operators of nearby water intakes.

1. RESPONSE TO DISCHARGE

(See Response Methods Handbook, CG 446-4)

Issue warning—corrosive, air contaminant
Restrict access
Evacuate area
Disperse and flush

2. LABEL

3. CHEMICAL DESIGNATIONS

3.1 **Synonyms:** Boron chloride
3.2 **Coast Guard Compatibility Classification:** Not applicable
3.3 **Chemical Formula:** BCl_3
3.4 **IMCO/United Nations Numerical Designation:** 2/1741

4. OBSERVABLE CHARACTERISTICS

4.1 **Physical State (as shipped):** Liquid
4.2 **Color:** Colorless
4.3 **Odor:** Acrid and irritating

5. HEALTH HAZARDS

5.1 **Personal Protective Equipment:** Chemical goggles; rubber protective clothing and gloves; self-contained breathing apparatus.
5.2 **Symptoms Following Exposure:** Inhalation causes edema and severe irritation of the upper respiratory system. Contact with liquid causes acid burns of eyes and severe burns of skin. Ingestion causes severe burns of mouth and stomach.
5.3 **Treatment for Exposure:** INHALATION: remove to fresh air; give oxygen or apply artificial respiration; keep warm; call a doctor at once; observe for pulmonary edema. EYES: wash with plenty of water for 15 min.; consult an eye specialist. SKIN: wash off with plenty of water. INGESTION: do NOT induce vomiting; give large amount of water.
5.4 **Toxicity by Inhalation (Threshold Limit Value):** Data not available
5.5 **Short-Term Inhalation Limits:** Data not available
5.6 **Toxicity by Ingestion:** Grade 2; LD_{50} 0.5 to 5 g/kg
5.7 **Late Toxicity:** Data not available
5.8 **Vapor (Gas) Irritant Characteristics:** Vapors cause severe irritation of eyes and throat and can cause eye or lung injury. They cannot be tolerated even at low concentrations.
5.9 **Liquid or Solid Irritant Characteristics:** Severe skin irritant. Causes second- and third-degree burns on short contact and is very injurious to the eyes.
5.10 **Odor Threshold:** Decomposes in moist air, releasing hydrochloric acid and decomposition products. Hydrochloric acid – 1 ppm

6. FIRE HAZARDS

6.1 **Flash Point:** Not flammable
6.2 **Flammable Limits in Air:** Not flammable
6.3 **Fire Extinguishing Agents:** Not pertinent
6.4 **Fire Extinguishing Agents Not to be Used:** Not pertinent
6.5 **Special Hazards of Combustion Products:** Not pertinent
6.6 **Behavior in Fire:** Toxic fumes of hydrogen chloride are given off upon contact with water applied to adjacent fires.
6.7 **Ignition Temperature:** Not pertinent
6.8 **Electrical Hazard:** Not pertinent
6.9 **Burning Rate:** Not pertinent

7. CHEMICAL REACTIVITY

7.1 **Reactivity with Water:** Reacts vigorously to liberate heat and forms hydrogen chloride fumes (hydrochloric acid) and boric acid.
7.2 **Reactivity with Common Materials:** Vigorously attacks elastomers and packing materials. Viton, Tygon, Saran, or silastic elastomers and natural and synthetic rubbers are not recommended for service. Lead and graphite-impregnated asbestos are to be avoided. In the presence of moisture, highly corrosive to most metals.
7.3 **Stability During Transport:** Stable
7.4 **Neutralizing Agents for Acids and Caustics:** Flush with water, rinse with sodium bicarbonate or lime solution.
7.5 **Polymerization:** Not pertinent
7.6 **Inhibitor of Polymerization:** Not pertinent

8. WATER POLLUTION

8.1 **Aquatic Toxicity:** Data not available
8.2 **Waterfowl Toxicity:** Data not available
8.3 **Biological Oxygen Demand (BOD):** None
8.4 **Food Chain Concentration Potential:** None

9. SELECTED MANUFACTURERS

1. Kerr-McGee Chemical Corp.
   Kerr-McGee Building
   Oklahoma City, Okla. 73102
2. Matheson Gas Products Co.
   East Rutherford, N. J. 07073
3. Union Carbide Corp.
   Linde Division
   270 Park Avenue
   New York, N. Y. 10017

10. SHIPPING INFORMATION

10.1 **Grades or Purity:** C.P. (99.9+%)
10.2 **Storage Temperature:** Ambient
10.3 **Inert Atmosphere:** No requirement
10.4 **Venting:** Pressure-vacuum

11. HAZARD ASSESSMENT CODE

(See Hazard Assessment Handbook, CG 446-3)

A-C-O

12. HAZARD CLASSIFICATIONS

12.1 **Code of Federal Regulations:** Corrosive
12.2 **NAS Hazard Rating for Bulk Water Transportation:**

Category	Rating
Fire	0
Health	
Vapor Irritant	4
Liquid or Solid Irritant	4
Poisons	3
Water Pollution	
Human Toxicity	2
Aquatic Toxicity	1
Aesthetic Effect	2
Reactivity	
Other Chemicals	4
Water	4
Self-Reaction	0

12.3 **NFPA Hazard Classifications:** Not listed

13. PHYSICAL AND CHEMICAL PROPERTIES

13.1 **Physical State at 15°C and 1 atm:** Gas
13.2 **Molecular Weight:** 117.2
13.3 **Boiling Point at 1 atm:** 54.3°F = 12.4°C = 285.6°K
13.4 **Freezing Point:** −161°F = −107°C = 166°K
13.5 **Critical Temperature:** 352°F = 178°C = 451°K
13.6 **Critical Pressure:** 566 psia = 38.5 atm = 3.90 MN/m^2
13.7 **Specific Gravity:** 1.35 at 11°C (liquid)
13.8 **Liquid Surface Tension:** 16.7 dynes/cm = 0.0167 N/m at 20°C
13.9 **Liquid-Water Interfacial Tension:** Not pertinent
13.10 **Vapor (Gas) Specific Gravity:** 4
13.11 **Ratio of Specific Heats of Vapor (Gas):** 1.1470
13.12 **Latent Heat of Vaporization:** 68.8 Btu/lb = 38.2 cal/g = 1.60×10^5 J/kg
13.13 **Heat of Combustion:** Not pertinent
13.14 **Heat of Decomposition:** Not pertinent
13.15 **Heat of Solution:** −13,000 Btu/lb = −7,200 cal/g = -300×10^5 J/kg
13.16 **Heat of Polymerization:** Not pertinent

NOTES

BRX	BROMINE

Common Synonyms	Watery liquid — Reddish-brown — Sharp irritating odor Sinks in water. Irritating brown vapor is produced.

AVOID CONTACT WITH LIQUID AND VAPOR. Keep people away.
Wear goggles, self-contained breathing apparatus, and rubber overclothing (including gloves).
Stop discharge if possible.
Notify local health and pollution control agencies.

Fire	Not flammable. May cause fire on contact with combustibles. POISONOUS GASES ARE PRODUCED IN FIRE. Cool exposed containers with water. Wear goggles, self-contained breathing apparatus, and rubber overclothing (including gloves).
Exposure	CALL FOR MEDICAL AID. VAPOR Irritating to eyes, nose and throat. If inhaled, will cause coughing, difficult breathing, or loss of consciousness. Move to fresh air. If breathing has stopped, give artificial respiration (but NOT mouth to mouth). If breathing is difficult, give oxygen. LIQUID Will burn skin and eyes. Harmful if swallowed. Remove contaminated clothing and shoes. Flush affected areas with plenty of water. IF IN EYES, hold eyelids open and flush with plenty of water. IF SWALLOWED and victim is CONSCIOUS, have victim drink water or milk.
Water Pollution	HARMFUL TO AQUATIC LIFE IN VERY LOW CONCENTRATIONS. May be dangerous if it enters water intakes. Notify local health and wildlife officials. Notify operators of nearby water intakes.

1. RESPONSE TO DISCHARGE
(See Response Methods Handbook, CG 446-4)

Issue warning—poison, air contaminant
Restrict access
Evacuate area
Disperse and flush

2. LABEL

3. CHEMICAL DESIGNATIONS

3.1 **Synonyms:** No common synonyms
3.2 **Coast Guard Compatibility Classification:** Halogen
3.3 **Chemical Formula:** Br_2
3.4 **IMCO/United Nations Numerical Designation:** 8.0/1744

4. OBSERVABLE CHARACTERISTICS

4.1 **Physical State (as shipped):** Liquid
4.2 **Color:** Dark red; red-brown
4.3 **Odor:** Sharp, harsh, penetrating

5. HEALTH HAZARDS

5.1 **Personal Protective Equipment:** Chemical safety goggles, face shield; self-contained air-line canister mask; rubber suit.
5.2 **Symptoms Following Exposure:** SKIN: contact with liquid or vapor may cause acne and slow-healing ulcers. INHALATION: induces severe irritation of the respiratory passages and pulmonary edema. Probable lethal oral dose for an adult is 1 ml. A brief exposure to 1000 ppm may be fatal.
5.3 **Treatment for Exposure:** SKIN AND EYES: wash well with water and sodium bicarbonate solution. RESPIRATORY SYSTEM: if there is obstruction to breathing establish airway by pulling tongue forward, inserting an airway tube, or doing a tracheostomy; begin artificial respiration; if difficulty in breathing is a result of pulmonary edema, treatment should be carried out with the patient in the sitting position. Administration of oxygen is most important; INGESTION: do not induce vomiting. Have victim drink water and milk.
5.4 **Toxicity by Inhalation (Threshold Limit Value):** 0.1 ppm
5.5 **Short-Term Inhalation Limits:** 0.4 ppm for 30 min.
5.6 **Toxicity by Ingestion:** Not pertinent
5.7 **Late Toxicity:** None
5.8 **Vapor (Gas) Irritant Characteristics:** Causes severe eye or throat irritations which can cause eye or lung injury; cannot be tolerated even at low concentrations.
5.9 **Liquid or Solid Irritant Characteristics:** Severe skin irritant. Causes second- and third-degree burns on short contact; very injurious to the eyes.
5.10 **Odor Threshold:** 3.5 ppm

6. FIRE HAZARDS

6.1 **Flash Point:** Not flammable
6.2 **Flammable Limits in Air:** Not flammable
6.3 **Fire Extinguishing Agents:** Use water spray to cool exposed containers and to wash away spills.
6.4 **Fire Extinguishing Agents Not to be Used:** Not pertinent
6.5 **Special Hazards of Combustion Products:** Toxic and irritating gases are generated when heated or in fires
6.6 **Behavior in Fire:** Not pertinent
6.7 **Ignition Temperature:** Not flammable
6.8 **Electrical Hazard:** Not pertinent
6.9 **Burning Rate:** Not flammable

7. CHEMICAL REACTIVITY

7.1 **Reactivity with Water:** No reaction
7.2 **Reactivity with Common Materials:** Reacts violently with aluminum. May cause fire in contact with wood, cotton, straw. Iron, steel, stainless steel, and copper are corroded by bromine and are especially subject to attack by wet bromine. Of the plastics, only those which are highly fluorinated resist bromine attack.
7.3 **Stability During Transport:** Stable
7.4 **Neutralizing Agents for Acids and Caustics:** Not pertinent
7.5 **Polymerization:** Not pertinent
7.6 **Inhibitor of Polymerization:** Not pertinent

8. WATER POLLUTION

8.1 **Aquatic Toxicity:**
10 ppm/10 hr/cladophora/killed/fresh water
10 ppm/*/fish/irritant/salt water
*Time period not specified.
8.2 **Waterfowl Toxicity:** Data not available
8.3 **Biological Oxygen Demand (BOD):** Not pertinent
8.4 **Food Chain Concentration Potential:** None

9. SELECTED MANUFACTURERS

1. Arkansas Chemicals Inc.
El Dorado, Arkansas 71730
2. Bromet Co.
Magnolia, Arkansas 71753
3. Dow Chemical Co.
Midland, Michigan 48640

10. SHIPPING INFORMATION

10.1 **Grades or Purity:** Commercial, technical
10.2 **Storage Temperature:** Cool but above 20°F to prevent freezing
10.3 **Inert Atmosphere:** Data not available
10.4 **Venting:** Data not available

11. HAZARD ASSESSMENT CODE
(See Hazard Assessment Handbook, CG 446-3)

A-P-X

12. HAZARD CLASSIFICATIONS

12.1 **Code of Federal Regulations:** Corrosive material
12.2 **NAS Hazard Rating for Bulk Water Transportation:**

Category	Rating
Fire	0
Health	
Vapor Irritant	4
Liquid or Solid Irritant	4
Poisons	4
Water Pollution	
Human Toxicity	4
Aquatic Toxicity	3
Aesthetic Effect	3
Reactivity	
Other Chemicals	4
Water	1
Self-Reaction	0

12.3 **NFPA Hazard Classifications:**

Category	Classification
Health Hazard (Blue)	4
Flammability (Red)	0
Reactivity (Yellow)	0
	oxy

13. PHYSICAL AND CHEMICAL PROPERTIES

13.1 **Physical State at 15°C and 1 atm:** Liquid
13.2 **Molecular Weight:** 159.81
13.3 **Boiling Point at 1 atm:** 138°F = 58.8°C = 332°K
13.4 **Freezing Point:** 19°F = −7.2°C = 266°K
13.5 **Critical Temperature:** Not pertinent
13.6 **Critical Pressure:** Not pertinent
13.7 **Specific Gravity:** 3.12 at 20°C (Liquid)
13.8 **Liquid Surface Tension:** 41 dynes/cm = .041 N/m at 20°C
13.9 **Liquid-Water Interfacial Tension:** Data not available
13.10 **Vapor (Gas) Specific Gravity:** 5.5 at 20°C
13.11 **Ratio of Specific Heats of Vapor (Gas):** 1.3
13.12 **Latent Heat of Vaporization:** 80.6 Btu/lb = 44.8 cal/g = 1.88×10^5 J/kg
13.13 **Heat of Combustion:** Not pertinent
13.14 **Heat of Decomposition:** Not pertinent
13.15 **Heat of Solution:** Not pertinent
13.16 **Heat of Polymerization:** Not pertinent

NOTES

BPF	# BROMINE PENTAFLUORIDE

Common Synonyms	Liquefied gas	Colorless	Irritating odor
	Reacts violently with water. Poisonous vapor is produced.		

Stop discharge if possible. Keep people away.
Avoid contact with liquid and vapor.
Isolate and remove discharged material.
Notify local health and pollution control agencies.

Fire	Not flammable. May cause fire on contact with combustibles. Will increase the intensity of a fire. POISONOUS GASES MAY BE PRODUCED IN FIRE. Containers may explode in fire. Wear goggles and self-contained breathing apparatus. DO NOT USE WATER OR FOAM ON FIRE.
Exposure	CALL FOR MEDICAL AID. VAPOR Irritating to eyes, nose and throat. If inhaled will cause coughing or difficult breathing. If in eyes, hold eyelids open and flush with plenty of water. If breathing has stopped, give artificial respiration. If breathing is difficult, give oxygen. LIQUID Will burn skin and eyes. Harmful if swallowed. Remove contaminated clothing and shoes. Flush affected areas with plenty of water. IF IN EYES, hold eyelids open and flush with plenty of water. IF SWALLOWED and victim is CONSCIOUS, have victim drink water or milk. IF SWALLOWED and victim is UNCONSCIOUS OR HAVING CONVULSIONS, do nothing except keep victim warm.
Water Pollution	Effect of low concentrations on aquatic life is unknown. May be dangerous if it enters water intakes. Notify local health and wildlife officials. Notify operators of nearby water intakes.

1. RESPONSE TO DISCHARGE
(See Response Methods Handbook, CG 446-4)

Issue warning—corrosive, air contaminant
Restrict access
Evacuate area
Disperse and flush

2. LABEL

Yellow

3. CHEMICAL DESIGNATIONS

3.1 **Synonyms:** No common synonyms
3.2 **Coast Guard Compatibility Classification:** Not listed
3.3 **Chemical Formula:** BrF_5
3.4 **IMCO/United Nations Numerical Designation:** 8/1745

4. OBSERVABLE CHARACTERISTICS

4.1 **Physical State (as shipped):** Liquefied gas
4.2 **Color:** Colorless
4.3 **Odor:** Highly irritating

5. HEALTH HAZARDS

5.1 **Personal Protective Equipment:** Self-contained breathing apparatus, acid suit, and gloves
5.2 **Symptoms Following Exposure:** Chemical is highly corrosive and toxic. Inhalation causes severe burns of mucous membrane. Ingestion causes severe burns of mouth. Contact with eyes or skin causes severe burns.
5.3 **Treatment for Exposure:** *Get medical attention IMMEDIATELY for any exposure to this chemical, even if no adverse effects are evident.* INHALATION: remove victim from area; apply artificial respiration if breathing has ceased. INGESTION: give large amount of water. EYES: wash with copious amounts of water for 15 min. SKIN: wash with large amounts of water and follow with lime water; remove contaminated clothing.
5.4 **Toxicity by Inhalation (Threshold Limit Value):** 0.1 ppm
5.5 **Short-Term Inhalation Limits:** Data not available
5.6 **Toxicity by Ingestion:** Data not available
5.7 **Late Toxicity:** Data not available
5.8 **Vapor (Gas) Irritant Characteristics:** Data not available
5.9 **Liquid or Solid Irritant Characteristics:** Data not available
5.10 **Odor Threshold:** Data not available

6. FIRE HAZARDS

Chemical is strong oxidizer and may cause fire in contact with organic materials such as wood, cotton, or straw.

6.1 **Flash Point:** Not flammable
6.2 **Flammable Limits in Air:** Not flammable
6.3 **Fire Extinguishing Agents:** Not pertinent
6.4 **Fire Extinguishing Agents Not to be Used:** Do not use water or foam on adjacent fires.
6.5 **Special Hazards of Combustion Products:** Toxic and irritating fumes of hydrogen fluoride and bromine may form in fires.
6.6 **Behavior in Fire:** Containers may burst when exposed to heat of fire.
6.7 **Ignition Temperature:** Not pertinent
6.8 **Electrical Hazard:** Not pertinent
6.9 **Burning Rate:** Not pertinent

7. CHEMICAL REACTIVITY

7.1 **Reactivity with Water:** Reacts violently with water, evolving hydrogen fluoride, an extremely irritating and corrosive gas.
7.2 **Reactivity with Common Materials:** Reacts violently with many metals and materials of construction such as wood, glass, some plastics.
7.3 **Stability During Transport:** Stable
7.4 **Neutralizing Agents for Acids and Caustics:** Flush with water.
7.5 **Polymerization:** Not pertinent
7.6 **Inhibitor of Polymerization:** Not pertinent

8. WATER POLLUTION

8.1 **Aquatic Toxicity:** Data not available
8.2 **Waterfowl Toxicity:** Data not available
8.3 **Biological Oxygen Demand (BOD):** None
8.4 **Food Chain Concentration Potential:** None

9. SELECTED MANUFACTURERS

1. Union Carbide Corp.
   Linde Division
   270 Park Avenue
   New York, N. Y. 10017
2. Gallard-Schlesinger Chemical Mfg. Co.
   584 Mineola Avenue
   Carle Place, N. Y. 11514
3. Pfaltz and Bauer, Inc.
   126–04 Northern Blvd.
   Flushing, N. Y. 11368

10. SHIPPING INFORMATION

10.1 **Grades or Purity:** Technical, 98.0+%; Pure, 99.9%
10.2 **Storage Temperature:** Ambient
10.3 **Inert Atmosphere:** Padded
10.4 **Venting:** Safety relief

11. HAZARD ASSESSMENT CODE
(See Hazard Assessment Handbook, CG 446-3)

A-O-Z

12. HAZARD CLASSIFICATIONS

12.1 **Code of Federal Regulations:** Oxidizer
12.2 **NAS Hazard Rating for Bulk Water Transportation:** Not listed
12.3 **NFPA Hazard Classifications:**

Category	Classification
Health Hazard (Blue)	4
Flammability (Red)	0
Reactivity (Yellow)	3
	₩, oxy

13. PHYSICAL AND CHEMICAL PROPERTIES

13.1 **Physical State at 15°C and 1 atm:** Liquid
13.2 **Molecular Weight:** 174.9
13.3 **Boiling Point at 1 atm:** 106°F = 41°C = 314°K
13.4 **Freezing Point:** −76°F = −60°C = 213°K
13.5 **Critical Temperature:** 387°F = 197°C = 470°K
13.6 **Critical Pressure:** Data not available
13.7 **Specific Gravity:** 2.48 at 20°C (liquid)
13.8 **Liquid Surface Tension:** Data not available
13.9 **Liquid-Water Interfacial Tension:** Not pertinent
13.10 **Vapor (Gas) Specific Gravity:** 6.03
13.11 **Ratio of Specific Heats of Vapor (Gas):** 1.089 at 25°C
13.12 **Latent Heat of Vaporization:** 76.8 Btu/lb = 42.7 cal/g = 1.79×10^5 J/kg
13.13 **Heat of Combustion:** Not pertinent
13.14 **Heat of Decomposition:** Not pertinent
13.15 **Heat of Solution:** Data not available
13.16 **Heat of Polymerization:** Not pertinent

NOTES

BTF — BROMINE TRIFLUORIDE

Common Synonyms

Liquid — Colorless — Extremely irritating odor

Reacts violently with water. Poisonous gas is produced on contact with water. Freezing point is 28° F.

AVOID CONTACT WITH LIQUID AND VAPOR. KEEP PEOPLE AWAY.
Stop discharge if possible.
Call fire department.
Isolate and remove discharged material.
Notify local health and pollution control agencies.

Fire

Not flammable.
May cause fire on contact with combustibles.
POISONOUS GASES MAY BE PRODUCED IN FIRE.
Containers may explode in fire.
Wear goggles and self-contained breathing apparatus.
Extinguish with dry chemicals or carbon dioxide.
DO NOT USE WATER OR FOAM ON FIRE.

Exposure

CALL FOR MEDICAL AID.

VAPOR
POISONOUS IF INHALED.
Irritating to eyes, nose and throat.
Move victim to fresh air.
If in eyes, hold eyelids open and flush with plenty of water.
If breathing has stopped, give artificial respiration.
If breathing is difficult, give oxygen.

LIQUID
POISONOUS IF SWALLOWED.
Will burn skin and eyes.
Remove contaminated clothing and shoes.
Flush affected areas with plenty of water.
IF IN EYES, hold eyelids open and flush with plenty of water.
IF SWALLOWED and victim is CONSCIOUS, have victim drink water or milk.
DO NOT INDUCE VOMITING.

Water Pollution

Effect of low concentrations on aquatic life is unknown.
May be dangerous if it enters water intakes.
Notify local health and wildlife officials.
Notify operators of nearby water intakes.

1. RESPONSE TO DISCHARGE

(See Response Methods Handbook, CG 446-4)

Issue warning—corrosive, air contaminant
Restrict access
Evacuate area
Disperse and flush

2. LABELS

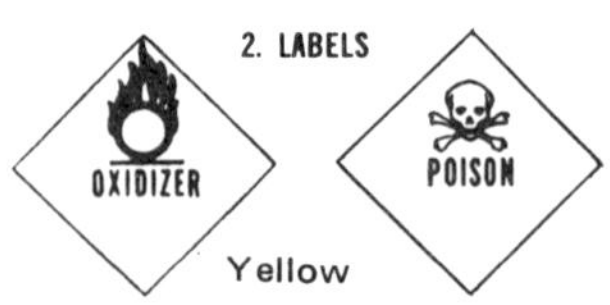

3. CHEMICAL DESIGNATIONS

3.1 **Synonyms:** No common synonyms
3.2 **Coast Guard Compatibility Classification:** Not applicable
3.3 **Chemical Formula:** BrF_3
3.4 **IMCO/United Nations Numerical Designation:** 8/1746

4. OBSERVABLE CHARACTERISTICS

4.1 **Physical State (as shipped):** Liquid
4.2 **Color:** Colorless to gray-yellow
4.3 **Odor:** Extremely irritating

5. HEALTH HAZARDS

5.1 **Personal Protective Equipment:** Self-contained breathing apparatus; complete protective clothing; safety glasses; face shield.
5.2 **Symptoms Following Exposure:** Inhalation causes severe irritation of upper respiratory system. Contact with liquid or vapor causes severe burns of eyes and can cause ulcers and blindness. Contact with skin causes severe burns. Ingestion causes severe burns of mucous membranes.
5.3 **Treatment for Exposure:** Get immediate medical attention for all exposures. INHALATION: remove from exposure; support respiration. EYES: irrigate with copious amounts of water for at least 15 min. SKIN: wash with large amounts of water for at least 15 min., then rinse with sodium bicarbonate or lime solution.
5.4 **Toxicity by Inhalation (Threshold Limit Value):** 0.1 ppm (suggested)
5.5 **Short-Term Inhalation Limits:** 50 ppm/30 min.; 100 ppm/3 min.
5.6 **Toxicity by Ingestion:** Data not available
5.7 **Late Toxicity:** Data not available
5.8 **Vapor (Gas) Irritant Characteristics:** Data not available
5.9 **Liquid or Solid Irritant Characteristics:** Data not available
5.10 **Odor Threshold:** Data not available

6. FIRE HAZARDS

6.1 **Flash Point:** Not flammable but may cause fire on contact with combustibles.
6.2 **Flammable Limits in Air:** Not flammable
6.3 **Fire Extinguishing Agents:** Dry chemical, carbon dioxide
6.4 **Fire Extinguishing Agents Not to be Used:** Water, foam
6.5 **Special Hazards of Combustion Products:** Data not available
6.6 **Behavior in Fire:** Forms very toxic and irritating fumes.
6.7 **Ignition Temperature:** Not pertinent
6.8 **Electrical Hazard:** Not pertinent
6.9 **Burning Rate:** Not pertinent

7. CHEMICAL REACTIVITY

7.1 **Reactivity with Water:** Reacts vigorously to generate toxic hydrogen fluoride gas (hydrofluoric acid).
7.2 **Reactivity with Common Materials:** Will cause severe corrosion of common metals and glass. May cause fire in contact with organic materials such as wood, cotton, or straw.
7.3 **Stability During Transport:** Stable
7.4 **Neutralizing Agents for Acids and Caustics:** Flush with water, rinse with sodium bicarbonate or lime solution.
7.5 **Polymerization:** Not pertinent
7.6 **Inhibitor of Polymerization:** Not pertinent

8. WATER POLLUTION

8.1 **Aquatic Toxicity:** Data not available
8.2 **Waterfowl Toxicity:** Data not available
8.3 **Biological Oxygen Demand (BOD):** None
8.4 **Food Chain Concentration Potential:** None

9. SELECTED MANUFACTURERS

1. Matheson Gas Products Co.
   East Rutherford, N. J. 07073
2. Union Carbide Corp.
   Linde Division
   270 Park Avenue
   New York, N. Y. 10017
3. Air Products and Chemicals, Inc.
   Specialty Gas Department
   P. O. Box 538
   Allentown, Pa. 18105

10. SHIPPING INFORMATION

10.1 **Grades or Purity:** 98+%
10.2 **Storage Temperature:** Ambient
10.3 **Inert Atmosphere:** No requirement
10.4 **Venting:** Pressure-vacuum

11. HAZARD ASSESSMENT CODE

(See Hazard Assessment Handbook, CG 446-3)

A-O

12. HAZARD CLASSIFICATIONS

12.1 **Code of Federal Regulations:** Oxidizer
12.2 **NAS Hazard Rating for Bulk Water Transportation:** Not listed
12.3 **NFPA Hazard Classifications:** Not listed

13. PHYSICAL AND CHEMICAL PROPERTIES

13.1 **Physical State at 15°C and 1 atm:** Liquid
13.2 **Molecular Weight:** 136.9
13.3 **Boiling Point at 1 atm:** 258.4°F = 125.8°C = 399.0°K
13.4 **Freezing Point:** 47.8°F = 8.8°C = 282.0°K
13.5 **Critical Temperature:** (est.) 621°F = 327°C = 600°K
13.6 **Critical Pressure:** Not pertinent
13.7 **Specific Gravity:** 2.81 at 20°C (liquid)
13.8 **Liquid Surface Tension:** 36.3 dynes/cm = 0.0363 N/m at 20°C
13.9 **Liquid-Water Interfacial Tension:** Not pertinent
13.10 **Vapor (Gas) Specific Gravity:** 4.7
13.11 **Ratio of Specific Heats of Vapor (Gas):** 1.1428
13.12 **Latent Heat of Vaporization:** 130 Btu/lb = 74 cal/g = 3.1×10^5 J/kg
13.13 **Heat of Combustion:** Not pertinent
13.14 **Heat of Decomposition:** Not pertinent
13.15 **Heat of Solution:** Data not available
13.16 **Heat of Polymerization:** Not pertinent

NOTES

BBZ BROMOBENZENE

Common Synonyms			
Monobromobenzene Phenyl bromide Bromobenzol	Liquid Sinks in water.	Colorless	Pleasant odor

Stop discharge if possible. Keep people away.
Call fire department.
Isolate and remove discharged material.
Notify local health and pollution control agencies.

Fire

Combustible.
Extinguish with water, dry chemicals, foam, or carbon dioxide.
Cool exposed containers with water.

Exposure

Call for medical aid.

LIQUID

Irritating to skin and eyes.
Harmful if swallowed.

Remove contaminated clothing and shoes.
Flush affected areas with plenty of water.
IF IN EYES, hold eyelids open and flush with plenty of water.
IF SWALLOWED and victim is CONSCIOUS, have victim drink water or milk.

Water Pollution

Effect of low concentrations on aquatic life is unknown.
May be dangerous if it enters water intakes.
Notify local health and wildlife officials.
Notify operators of nearby water intakes.

1. RESPONSE TO DISCHARGE

(See Response Methods Handbook, CG 446-4)

Issue warning—water contaminant
Should be removed
Chemical and physical treatment

2. LABELS

No label required by Code of Federal Regulations

3. CHEMICAL DESIGNATIONS

3.1 **Synonyms:** Monobromobenzene
Phenyl bromide
Bromobenzol

3.2 **Coast Guard Compatibility Classification:** Not listed

3.3 **Chemical Formula:** C_6H_5Br

3.4 **IMCO/United Nations Numerical Designation:** Not listed

4. OBSERVABLE CHARACTERISTICS

4.1 **Physical State (as shipped):** Liquid

4.2 **Color:** Colorless

4.3 **Odor:** Pleasant

5. HEALTH HAZARDS

5.1 **Personal Protective Equipment:** Goggles or face shield; rubber gloves and apron.

5.2 **Symptoms Following Exposure:** Contact with liquid causes irritation of eyes and mild irritation of skin. Ingestion causes mild irritation of mouth and stomach.

5.3 **Treatment for Exposure:** EYES: flush with water for at least 15 min. SKIN: wipe off, wash with soap and water. INGESTION: induce vomiting; consult a doctor.

5.4 **Toxicity by Inhalation (Threshold Limit Value):** Data not available

5.5 **Short-Term Inhalation Limits:** Data not available

5.6 **Toxicity by Ingestion:** Data not available

5.7 **Late Toxicity:** Data not available

5.8 **Vapor (Gas) Irritant Characteristics:** Data not available

5.9 **Liquid or Solid Irritant Characteristics:** Data not available

5.10 **Odor Threshold:** Data not available

6. FIRE HAZARDS

6.1 **Flash Point:** 124°F C.C.

6.2 **Flammable Limits in Air:** Not pertinent

6.3 **Fire Extinguishing Agents:** Water, dry chemical, foam, carbon dioxide

6.4 **Fire Extinguishing Agents Not to be Used:** Not pertinent

6.5 **Special Hazards of Combustion Products:** Irritating hydrogen bromide and other gases may be produced in fire.

6.6 **Behavior in Fire:** Not pertinent

6.7 **Ignition Temperature:** 1,049°F

6.8 **Electrical Hazard:** Data not available

6.9 **Burning Rate:** 3.8 mm/min.

7. CHEMICAL REACTIVITY

7.1 **Reactivity with Water:** No reaction

7.2 **Reactivity with Common Materials:** No reaction

7.3 **Stability During Transport:** Stable

7.4 **Neutralizing Agents for Acids and Caustics:** Not pertinent

7.5 **Polymerization:** Not pertinent

7.6 **Inhibitor of Polymerization:** Not pertinent

8. WATER POLLUTION

8.1 **Aquatic Toxicity:** Data not available

8.2 **Waterfowl Toxicity:** Data not available

8.3 **Biological Oxygen Demand (BOD):** Data not available

8.4 **Food Chain Concentration Potential:** Data not available

9. SELECTED MANUFACTURERS

1. Dow Chemical Co.
Midland, Mich. 48640
2. Great Lakes Chemical Corp.
P. O. Box 2200
West Lafayette, Ind. 47906
3. Michigan Chemical Corp.
351 East Ohio Street
Chicago, Ill. 60611

10. SHIPPING INFORMATION

10.1 **Grades or Purity:** Commercial

10.2 **Storage Temperature:** Ambient

10.3 **Inert Atmosphere:** No requirement

10.4 **Venting:** Open (flame arrester)

11. HAZARD ASSESSMENT CODE

(See Hazard Assessment Handbook, CG 446-3)

A-X-Y

12. HAZARD CLASSIFICATIONS

12.1 **Code of Federal Regulations:** Combustible Liquid

12.2 **NAS Hazard Rating for Bulk Water Transportation:** Not listed

12.3 **NFPA Hazard Classifications:**

Category	Classification
Health Hazard (Blue)	2
Flammability (Red)	2
Reactivity (Yellow)	0

13. PHYSICAL AND CHEMICAL PROPERTIES

13.1 **Physical State at 15°C and 1 atm:** Liquid

13.2 **Molecular Weight:** 157

13.3 **Boiling Point at 1 atm:** 313°F = 156°C = 429°K

13.4 **Freezing Point:** −23.1°F = −30.6°C = 242.6°K

13.5 **Critical Temperature:** 747°F = 397°C = 670°K

13.6 **Critical Pressure:** 655 psia = 44.6 atm = 4.52 MN/m²

13.7 **Specific Gravity:** 1.49 at 25°C (liquid)

13.8 **Liquid Surface Tension:** 36 dynes/cm = 0.036 N/m at 20°C

13.9 **Liquid-Water Interfacial Tension:** (est.) 30 dynes/cm = 0.030 N/m at 20°C

13.10 **Vapor (Gas) Specific Gravity:** 5.4

13.11 **Ratio of Specific Heats of Vapor (Gas):** 1.0931

13.12 **Latent Heat of Vaporization:** 104 Btu/lb = 58 cal/g = 2.4×10^5 J/kg

13.13 **Heat of Combustion:** −8,510 Btu/lb = −4,730 cal/g = -198×10^5 J/kg

13.14 **Heat of Decomposition:** Not pertinent

13.15 **Heat of Solution:** Not pertinent

13.16 **Heat of Polymerization:** Not pertinent

NOTES

BRU	# BRUCINE

Common Synonyms			
10,11-Dimethoxystrychnine (-) Brucine dihydrate	Solid	White	Odorless
	Sinks in water.		

AVOID CONTACT WITH SOLID AND DUST. KEEP PEOPLE AWAY.
Wear dust respirator and rubber overclothing (including gloves).
Stop discharge if possible.
Isolate and remove discharged material.
Notify local health and pollution control agencies.

Fire

Combustile.
POISONOUS GASES MAY BE PRODUCED IN FIRE.
Wear goggles and self-contained breathing apparatus.
Extinguish with water, dry chemicals, foam, or carbon dioxide.

Exposure

CALL FOR MEDICAL AID.

DUST
POISONOUS IF INHALED.
Irritating to eyes, nose and throat.
If in eyes, hold eyelids open and flush with plenty of water.
If breathing has stopped, give artificial respiration.
If breathing is difficult, give oxygen.

SOLID
POISONOUS IF SWALLOWED OR IF SKIN IS EXPOSED.
If swallowed will cause nausea and vomiting.
Remove contaminated clothing and shoes.
Flush affected areas with plenty of water.
IF IN EYES, hold eyelids open and flush with plenty of water.
IF SWALLOWED and victim is CONSCIOUS, have victim drink water or milk and have victim induce vomiting.
IF SWALLOWED and victim is UNCONSCIOUS OR HAVING CONVULSIONS, do nothing except keep victim warm.

Water Pollution

Effect of low concentrations on aquatic life is unknown.
May be dangerous if it enters water intakes.
Notify local health and wildlife officials.
Notify operators of nearby water intakes.

1. RESPONSE TO DISCHARGE
(See Response Methods Handbook, CG 446-4)

Issue warning—poison, water contaminant
Restrict access
Should be removed
Chemical and physical treatment

2. LABEL

3. CHEMICAL DESIGNATIONS

3.1 **Synonyms:** Brucine dihydrate; 10,11-Dimethoxystrychnine
3.2 **Coast Guard Compatibility Classification:** Not listed
3.3 **Chemical Formula:** $C_{23}H_{26}N_2O_4$ or $C_{23}H_{26}N_2O_4 \cdot 2H_2O$
3.4 **IMCO/United Nations Numerical Designation:** 6.1/1570

4. OBSERVABLE CHARACTERISTICS

4.1 **Physical State (as shipped):** Solid
4.2 **Color:** White
4.3 **Odor:** None

5. HEALTH HAZARDS

5.1 **Personal Protective Equipment:** Dust mask; goggles or face shield; rubber gloves
5.2 **Symptoms Following Exposure:** Chemical is toxic if inhaled, swallowed, or absorbed through skin. Inhalation produces intense bitter taste. Ingestion causes nausea, vomiting, restlessness, excitement, twitching, and (rarely) convulsions. Contact with dust irritates eyes.
5.3 **Treatment for Exposure:** INHALATION: remove victim from exposure. INGESTION: induce vomiting; get medical attention at once. EYES: flush with water for 15 min.
5.4 **Toxicity by Inhalation (Threshold Limit Value):** Data not available
5.5 **Short-Term Inhalation Limits:** Data not available
5.6 **Toxicity by Ingestion:** Grade 4; oral rat LD_{50} = 1 mg/kg
5.7 **Late Toxicity:** Data not available
5.8 **Vapor (Gas) Irritant Characteristics:** Data not available
5.9 **Liquid or Solid Irritant Characteristics:** Data not available
5.10 **Odor Threshold:** Odorless

6. FIRE HAZARDS

6.1 **Flash Point:** Not pertinent (combustible solid)
6.2 **Flammable Limits in Air:** Not pertinent
6.3 **Fire Extinguishing Agents:** Water, foam, dry chemical, carbon dioxide
6.4 **Fire Extinguishing Agents Not to be Used:**
6.5 **Special Hazards of Combustion Products:** Toxic oxides of nitrogen may form in fires.
6.6 **Behavior in Fire:**
6.7 **Ignition Temperature:** Not pertinent
6.8 **Electrical Hazard:** Not pertinent
6.9 **Burning Rate:** Not pertinent

7. CHEMICAL REACTIVITY

7.1 **Reactivity with Water:** No reaction
7.2 **Reactivity with Common Materials:**
7.3 **Stability During Transport:** Stable
7.4 **Neutralizing Agents for Acids and Caustics:** Not pertinent
7.5 **Polymerization:** Not pertinent
7.6 **Inhibitor of Polymerization:** Not pertinent

8. WATER POLLUTION

8.1 **Aquatic Toxicity:** Data not available
8.2 **Waterfowl Toxicity:** Data not available
8.3 **Biological Oxygen Demand (BOD):** Data not available
8.4 **Food Chain Concentration Potential:** Possible bioaccumulation problem for the duration of 1 week.

9. SELECTED MANUFACTURERS

1. Gallard-Schlesinger Chemical Mfg. Co.
584 Mineola Ave.
Carle Place, N. Y. 11514
2. Aldrich Chemical Co.
940 W. Saint Paul Ave.
Milwaukee, Wis. 53233
3. Norse Laboratories, Inc.
2038 Alameda Padre Serra
Santa Barbara, Calif. 93103

10. SHIPPING INFORMATION

10.1 **Grades or Purity:** Pure
10.2 **Storage Temperature:** Ambient
10.3 **Inert Atmosphere:** No requirement
10.4 **Venting:** Open

11. HAZARD ASSESSMENT CODE
(See Hazard Assessment Handbook, CG 446-3)

II

12. HAZARD CLASSIFICATIONS

12.1 **Code of Federal Regulations:** Poisonous solid, Class B
12.2 **NAS Hazard Rating for Bulk Water Transportation:** Not listed
12.3 **NFPA Hazard Classifications:** Not listed

13. PHYSICAL AND CHEMICAL PROPERTIES

13.1 **Physical State at 15°C and 1 atm:** Solid
13.2 **Molecular Weight:** 394.4
13.3 **Boiling Point at 1 atm:** Not pertinent (decomposes)
13.4 **Freezing Point:** 352°F = 178°C = 451°K
13.5 **Critical Temperature:** Not pertinent
13.6 **Critical Pressure:** Not pertinent
13.7 **Specific Gravity:** > 1 at 20°C (solid)
13.8 **Liquid Surface Tension:** Not pertinent
13.9 **Liquid-Water Interfacial Tension:** Not pertinent
13.10 **Vapor (Gas) Specific Gravity:** Not pertinent
13.11 **Ratio of Specific Heats of Vapor (Gas):** Not pertinent
13.12 **Latent Heat of Vaporization:** Not pertinent
13.13 **Heat of Combustion:** −13,400 Btu/lb = −7,440 cal/g = -311×10^5 J/kg
13.14 **Heat of Decomposition:** Not pertinent
13.15 **Heat of Solution:** Not pertinent
13.16 **Heat of Polymerization:** Not pertinent

NOTES

BDI — BUTADIENE, INHIBITED

Common Synonyms Divinyl Vinylethylene 1,3-Butadiene	Liquefied compressed gas Colorless Gasoline-like odor Floats and boils on water. Flammable visible vapor cloud is produced.

Avoid contact with liquid and gas. Keep people away.
Wear goggles, self-contained breathing apparatus, and rubber overclothing (including gloves).
Stop discharge if possible.
Shut off ignition sources and call fire department.
Evacuate area in case of large discharge.
Stay upwind and use water spray to "knock down" vapor.
Isolate and remove discharged material.
Notify local health and pollution control agencies.

Fire	FLAMMABLE. Containers may explode in fire. Flashback along vapor trail may occur. Vapor may explode if ignited in an enclosed area. Wear goggles, self-contained breathing apparatus, and rubber overclothing (including gloves). Stop flow of gas if possible. Cool exposed containers and protect men effecting shut-off with water. Let fire burn.
Exposure	CALL FOR MEDICAL AID. VAPOR Irritating to eyes, nose and throat. Move to fresh air. LIQUID Irritating to skin and eyes. Remove contaminated clothing and shoes. Flush affected areas with plenty of water. IF IN EYES, hold eyelids open and flush with plenty of water.
Water Pollution	Not harmful to aquatic life. May be dangerous if it enters water intakes. Notify operators of nearby water intakes.

1. RESPONSE TO DISCHARGE

(See Response Methods Handbook, CG 446-4)

Issue warning - high flammability
Evacuate area

2. LABEL

Red

3. CHEMICAL DESIGNATIONS

3.1 **Synonyms:** Biethylene
1,3-Butadiene
Bivinyl
Divinyl
Vinylethylene

3.2 **Coast Guard Compatibility Classification:** Olefin

3.3 **Chemical Formula:** $CH_2{=}CHCH{=}CH_2$

3.4 **IMCO/United Nations Numerical Designation:** 2.0/1010

4. OBSERVABLE CHARACTERISTICS

4.1 **Physical State (as shipped):** Gas

4.2 **Color:** Colorless

4.3 **Odor:** Mildly aromatic

5. HEALTH HAZARDS

5.1 **Personal Protective Equipment:** Chemical-type safety goggles; rescue harness and life line for those entering a tank or enclosed storage space; hose mask with hose inlet in a vapor-free atmosphere; self-contained breathing apparatus; rubber suit.

5.2 **Symptoms Following Exposure:** Slight anesthetic effect at high concentrations; causes "frostbite" from skin contact; slight irritation to eyes and nose at high concentrations.

5.3 **Treatment for Exposure:** Remove from exposure immediately. Call a physician. INHALATION: if breathing is irregular or stopped, start resuscitation, administer oxygen. SKIN CONTACT: remove contaminated clothing and wash affected skin area. EYE CONTACT: irrigate with water for 15 min.

5.4 **Toxicity by Inhalation (Threshold Limit Value):** 1,000 ppm

5.5 **Short-Term Inhalation Limits:** Data not available

5.6 **Toxicity by Ingestion:** Data not available

5.7 **Late Toxicity:** None

5.8 **Vapor (Gas) Irritant Characteristics:** Vapors cause a slight smarting of the eyes or respiratory system if present in high concentrations. The effect is temporary.

5.9 **Liquid or Solid Irritant Characteristics:** Minimum hazard. If spilled on clothing and allowed to remain, may cause smarting and reddening of the skin because of frostbite.

5.10 **Odor Threshold:** 4 mg/m^3

6. FIRE HAZARDS

6.1 **Flash Point:** −105°F (est.)

6.2 **Flammable Limits in Air:** 2.0%–11.5%

6.3 **Fire Extinguishing Agents:** Stop flow of gas

6.4 **Fire Extinguishing Agents Not to be Used:** Not pertinent

6.5 **Special Hazards of Combustion Products:** Not pertinent

6.6 **Behavior in Fire:** Vapors heavier than air and may travel a considerable distance to a source of ignition and flash back. Containers may explode in a fire due to polymerization.

6.7 **Ignition Temperature:** 788°F

6.8 **Electrical Hazard:** Class I, Group B

6.9 **Burning Rate:** 8.0 mm/min.

7. CHEMICAL REACTIVITY

7.1 **Reactivity with Water:** No reaction

7.2 **Reactivity with Common Materials:** No reaction

7.3 **Stability During Transport:** Explosive decomposition when contaminated with peroxides formed by reaction with air.

7.4 **Neutralizing Agents for Acids and Caustics:** Not pertinent

7.5 **Polymerization:** Stable when inhibitors present

7.6 **Inhibitor of Polymerization:** tert-Butylcatechol (0.01–0.02%)

8. WATER POLLUTION

8.1 **Aquatic Toxicity:** Not pertinent

8.2 **Waterfowl Toxicity:** Not pertinent

8.3 **Biological Oxygen Demand (BOD):** Not pertinent

8.4 **Food Chain Concentration Potential:** Not pertinent

9. SELECTED MANUFACTURERS

1. Neches Butane Products Co.
Port Neches, Texas 77651
2. Petro-Tex Chemical Corp.
Houston, Texas 77017
3. Phillips Petroleum Co.
Bartlesville, Oklahoma 74004

10. SHIPPING INFORMATION

10.1 **Grades or Purity:**

Research grade:	99.86 mole%
Special purity:	99.5 mole%
Rubber grade:	99.0 mole%
Commercial:	98%

10.2 **Storage Temperature:** Ambient

10.3 **Inert Atmosphere:** No requirement

10.4 **Venting:** Safety relief

11. HAZARD ASSESSMENT CODE

(See Hazard Assessment Handbook, CG 446-3)

A-B-C-D-E-F-G-Z

12. HAZARD CLASSIFICATIONS

12.1 **Code of Federal Regulations:** Flammable compressed gas

12.2 **NAS Hazard Rating for Bulk Water Transportation:**

Category	Rating
Fire	4
Health	
Vapor Irritant	1
Liquid or Solid Irritant	1
Poisons	1
Water Pollution	
Human Toxicity	0
Aquatic Toxicity	1
Aesthetic Effect	1
Reactivity	
Other Chemicals	2
Water	0
Self-Reaction	3

12.3 **NFPA Hazard Classifications:**

Category	Classification
Health Hazard (Blue)	2
Flammability (Red)	4
Reactivity (Yellow)	2

13. PHYSICAL AND CHEMICAL PROPERTIES

13.1 **Physical State at 15°C and 1 atm:** Gas

13.2 **Molecular Weight:** 54.09

13.3 **Boiling Point at 1 atm:** 24.1°F = −4.4°C = 268.8°K

13.4 **Freezing Point:** −164°F = −108.9°C = 164.3°K

13.5 **Critical Temperature:** 306°F = 152°C = 425°K

13.6 **Critical Pressure:** 628 psia = 42.7 atm = 4.32 MN/m^2

13.7 **Specific Gravity:** 0.621 at 20°C (liquid)

13.8 **Liquid Surface Tension:** 13.4 dynes/cm = 0.0134 N/m at 20°C

13.9 **Liquid-Water Interfacial Tension:** (est.) 67 dynes/cm = 0.067 N/m at 22°C

13.10 **Vapor (Gas) Specific Gravity:** 1.9 at 20°C

13.11 **Ratio of Specific Heats of Vapor (Gas):** 1.1

13.12 **Latent Heat of Vaporization:** 180 Btu/lb = 100 cal/g = 4.19 × 10^5 J/kg

13.13 **Heat of Combustion:** −19,008 Btu/lb = −10,560 cal/g = −442.13 × 10^5 J/kg

13.14 **Heat of Decomposition:** Not pertinent

13.15 **Heat of Solution:** Not pertinent

13.16 **Heat of Polymerization:** −549 Btu/lb = −305 cal/g = −12.8 × 10^5 J/kg

NOTES

REVISED 1978

BUTANE

BUT

Common Synonyms	
n-Butane	Liquefied compressed gas Colorless Gasoline-like odor Floats and boils on water. Flammable, visible vapor cloud is formed.

Stop discharge if possible. Keep people away.
Shut off ignition sources and call fire department.
Stay upwind and use water spray to "knock down" vapor.
Avoid contact with liquid and vapor.
Notify local health and pollution control agencies.

Fire

FLAMMABLE.
Flashback along vapor trail may occur.
Vapor may explode if ignited in an enclosed area.
Stop flow of gas if possible.
Cool exposed containers and protect men effecting shut-off with water.
Let fire burn.

Exposure

CALL FOR MEDICAL AID.

VAPOR
If inhaled, will cause dizziness or difficult breathing.
Not irritating to eyes, nose or throat.
Move to fresh air.
If breathing has stopped, give artificial respiration.
If breathing is difficult, give oxygen.

LIQUID
Will cause frostbite.
Flush affected areas with plenty of water.
DO NOT RUB AFFECTED AREAS.

Water Pollution

Not harmful to aquatic life.
May be dangerous if it enters water intakes.
Notify operators of nearby water intakes.

1. RESPONSE TO DISCHARGE

(See Response Methods Handbook, CG 446-4)

Issue warning—high flammability
Restrict access
Evacuate area

2. LABEL

Red

3. CHEMICAL DESIGNATIONS

3.1 **Synonyms:** n-Butane
3.2 **Coast Guard Compatibility Classification:** Paraffin
3.3 **Chemical Formula:** n-C_4H_{10}
3.4 **IMCO/United Nations Numerical Designation:** 2.0/1011

4. OBSERVABLE CHARACTERISTICS

4.1 **Physical State (as shipped):** Compressed gas
4.2 **Color:** Colorless
4.3 **Odor:** Like gasoline

5. HEALTH HAZARDS

5.1 **Personal Protective Equipment:** Self-contained breathing apparatus and safety goggles.
5.2 **Symptoms Following Exposure:** High exposure produces drowsiness but no other evidence of systemic effect.
5.3 **Treatment for Exposure:** ORAL AND ASPIRATION: No treatment required. INHALATION: Guard against self-injury if stuporous, confused, or anesthetized. Apply artificial respiration if not breathing. Avoid administration of epinephrine or other sympathomimetic amines. Prevent aspirations of vomitus by proper positioning of the head. Give symptomatic and supportive treatment.
5.4 **Toxicity by Inhalation (Threshold Limit Value):** 500 ppm
5.5 **Short-Term Inhalation Limits:** Data not available
5.6 **Toxicity by Ingestion:** Not pertinent
5.7 **Late Toxicity:** None
5.8 **Vapor (Gas) Irritant Characteristics:** None
5.9 **Liquid or Solid Irritant Characteristics:** No appreciable hazard. Practically harmless to the skin because it is very volatile and evaporates quickly from the skin. Some frostbite possible.
5.10 **Odor Threshold:** 6.16 ppm

6. FIRE HAZARDS

6.1 **Flash Point:** −100°F (est.)
6.2 **Flammable Limits in Air:** 1.8%–8.4%
6.3 **Fire Extinguishing Agents:** Stop flow of gas
6.4 **Fire Extinguishing Agents Not to be Used:** Not pertinent
6.5 **Special Hazards of Combustion Products:** Not pertinent
6.6 **Behavior in Fire:** Not pertinent
6.7 **Ignition Temperature:** 807°F
6.8 **Electrical Hazard:** Class I, Group D
6.9 **Burning Rate:** 7.9 mm/min.

7. CHEMICAL REACTIVITY

7.1 **Reactivity with Water:** No reaction
7.2 **Reactivity with Common Materials:** No reaction
7.3 **Stability During Transport:** Stable
7.4 **Neutralizing Agents for Acids and Caustics:** Not pertinent
7.5 **Polymerization:** Not pertinent
7.6 **Inhibitor of Polymerization:** Not pertinent

8. WATER POLLUTION

8.1 **Aquatic Toxicity:** None
8.2 **Waterfowl Toxicity:** None
8.3 **Biological Oxygen Demand (BOD):** None
8.4 **Food Chain Concentration Potential:** None

9. SELECTED MANUFACTURERS

1. Neches Butane Products Co.
   Port Neches, Texas 77651
2. Petro-Tex Chemical Corp.
   Houston, Texas 77017
3. Phillips Petroleum Co.
   Bartlesville, Oklahoma 74004

10. SHIPPING INFORMATION

10.1 **Grades or Purity:** Research: 99.95%; Pure: 99.4%; Technical: 97.6%
10.2 **Storage Temperature:** Ambient
10.3 **Inert Atmosphere:** No requirement
10.4 **Venting:** Safety relief

11. HAZARD ASSESSMENT CODE

(See Hazard Assessment Handbook, CG 446-3)

A-B-C-D-E-F-G

12. HAZARD CLASSIFICATIONS

12.1 **Code of Federal Regulations:** Flammable compressed gas

12.2 **NAS Hazard Rating for Bulk Water Transportation:**

Category	Rating
Fire	4
Health	
Vapor Irritant	0
Liquid or Solid Irritant	0
Poisons	0
Water Pollution	
Human Toxicity	0
Aquatic Toxicity	0
Aesthetic Effect	0
Reactivity	
Other Chemicals	0
Water	0
Self-Reaction	0

12.3 **NFPA Hazard Classifications:**

Category	Classification
Health Hazard (Blue)	1
Flammability (Red)	4
Reactivity (Yellow)	0

13. PHYSICAL AND CHEMICAL PROPERTIES

13.1 **Physical State at 15°C and 1 atm:** Gas
13.2 **Molecular Weight:** 58.12
13.3 **Boiling Point at 1 atm:** 31.1°F = −0.48°C = 272.72°K
13.4 **Freezing Point:** −216°F = −138°C = 135°K
13.5 **Critical Temperature:** 306°F = 152°C = 425°K
13.6 **Critical Pressure:** 550.8 psia = 37.47 atm = 3.796 MN/m²
13.7 **Specific Gravity:** 0.60 at 0°C (liquid)
13.8 **Liquid Surface Tension:** 14.7 dynes/cm = .0147 N/m at 0°C
13.9 **Liquid-Water Interfacial Tension:** (est.) 65 dynes/cm = 0.065 N/m at 22°C
13.10 **Vapor (Gas) Specific Gravity:** 20 at 20°C
13.11 **Ratio of Specific Heats of Vapor (Gas):** 1.092
13.12 **Latent Heat of Vaporization:** 170 Btu/lb = 92 cal/g = 3.9 × 10⁵ J/kg
13.13 **Heat of Combustion:** −19,512 Btu/lb = −10,840 cal/g = −453.85 × 10⁵ J/kg
13.14 **Heat of Decomposition:** Not pertinent
13.15 **Heat of Solution:** Not pertinent
13.16 **Heat of Polymerization:** Not pertinent

NOTES

REVISED 1978

BDO	# 1,4-BUTANEDIOL

Common Synonyms 1,4-dihydroxybutane Tetramethylene glycol	Thick liquid Colorless Odorless Sinks and mixes with water. Freezing point is 68° F.

Stop discharge if possible.
Call fire department.
Isolate and remove discharged material.
Notify local health and pollution control agencies.

Fire	Combustible. Extinguish with dry chemical, alcohol foam, or carbon dioxide. Water may be ineffective on fire.
Exposure	CALL FOR MEDICAL AID. LIQUID OR SOLID Irritating to skin or eyes. Harmful if swallowed. Flush affected areas with plenty of water. IF SWALLOWED and victim is CONSCIOUS, have victim drink water or milk.
Water Pollution	Effect of low concentrations on aquatic life is unknown. May be dangerous if it enters water intakes. Notify local health and pollution control officials. Notify operators of nearby water intakes.

1. RESPONSE TO DISCHARGE

(See Response Methods Handbook, CG 446-4)

Disperse and flush

2. LABELS

No hazard label required by Code of Federal Regulations

3. CHEMICAL DESIGNATIONS

3.1 **Synonyms:** 1, 4-Dihydroxylbutane; Tetramethyleneglycol

3.2 **Coast Guard Compatibility Classification:** Alcohols

3.3 **Chemical Formula:** $HOCH_2(CH_2)_2CH_2OH$

3.4 **IMCO/United Nations Numerical Designation:** 3.3/1987

4. OBSERVABLE CHARACTERISTICS

4.1 **Physical State (as shipped):** Liquid

4.2 **Color:** Colorless

4.3 **Odor:** Odorless

5. HEALTH HAZARDS

5.1 **Personal Protective Equipment:** Eye protection

5.2 **Symptoms Following Exposure:** Ingestion of large amounts needed to produce any symptoms.

5.3 **Treatment for Exposure:** SKIN OR EYES: wash off immediately with plenty of water.

5.4 **Toxicity by Inhalation (Threshold Limit Value):** Not pertinent

5.5 **Short-Term Inhalation Limits:** Not pertinent

5.6 **Toxicity by Ingestion:** Grade 2; LD_{50} 0.5 to 5 g/kg (rat)

5.7 **Late Toxicity:** None

5.8 **Vapor (Gas) Irritant Characteristics:** None

5.9 **Liquid or Solid Irritant Characteristics:** None

5.10 **Odor Threshold:** Not pertinent

6. FIRE HAZARDS

6.1 **Flash Point:** >250°F O.C.

6.2 **Flammable Limits in Air:** Data not available

6.3 **Fire Extinguishing Agents:** Alcohol foam, dry chemical or carbon dioxide

6.4 **Fire Extinguishing Agents Not to be Used:** Water or foam may cause frothing

6.5 **Special Hazards of Combustion Products:** Not pertinent

6.6 **Behavior in Fire:** Not pertinent

6.7 **Ignition Temperature:** Data not available

6.8 **Electrical Hazard:** Not pertinent

6.9 **Burning Rate:** Data not available

7. CHEMICAL REACTIVITY

7.1 **Reactivity with Water:** No reaction

7.2 **Reactivity with Common Materials:** No reaction

7.3 **Stability During Transport:** Stable

7.4 **Neutralizing Agents for Acids and Caustics:** Not pertinent

7.5 **Polymerization:** Not pertinent

7.6 **Inhibitor of Polymerization:** Not pertinent

8. WATER POLLUTION

8.1 **Aquatic Toxicity:** Data not available

8.2 **Waterfowl Toxicity:** Data not available

8.3 **Biological Oxygen Demand (BOD):** Data not available

8.4 **Food Chain Concentration Potential:** None

9. SELECTED MANUFACTURERS

GAF Corporation
Chemical Division
Calvert City, Kentucky 42029

10. SHIPPING INFORMATION

10.1 **Grades or Purity:** Regular grade: 99%; Anhydrous grade: 99.3%

10.2 **Storage Temperature:** 75°–100°F

10.3 **Inert Atmosphere:** Data not available

10.4 **Venting:** Data not available

11. HAZARD ASSESSMENT CODE

(See Hazard Assessment Handbook, CG 446-3)

A-P-Q

12. HAZARD CLASSIFICATIONS

12.1 **Code of Federal Regulations:** Not listed

12.2 **NAS Hazard Rating for Bulk Water Transportation:** Not listed

12.3 **NFPA Hazard Classifications:** Not listed

13. PHYSICAL AND CHEMICAL PROPERTIES

13.1 **Physical State at 15°C and 1 atm:** Liquid

13.2 **Molecular Weight:** 90.12

13.3 **Boiling Point at 1 atm:** 442°F = 228°C = 501°K

13.4 **Freezing Point:** 68.2°F = 20.1°C = 293.3°K

13.5 **Critical Temperature:** 716°F = 380°C = 653°K

13.6 **Critical Pressure:** 720 psia = 49 atm = 5.0 MN/m^2

13.7 **Specific Gravity:** 1.017 at 20°C (liquid)

13.8 **Liquid Surface Tension:** Not pertinent

13.9 **Liquid-Water Interfacial Tension:** Not pertinent

13.10 **Vapor (Gas) Specific Gravity:** Not pertinent

13.11 **Ratio of Specific Heats of Vapor (Gas):** Not pertinent

13.12 **Latent Heat of Vaporization:** Not pertinent

13.13 **Heat of Combustion:** (est.) −11,900 Btu/lb = −6630 cal/g = -277×10^5 J/kg

13.14 **Heat of Decomposition:** Not pertinent

13.15 **Heat of Solution:** Not pertinent

13.16 **Heat of Polymerization:** Not pertinent

NOTES

BUD	1,4-BUTENEDIOL

Common Synonyms 2-Butene-1, 4-diol 1,4-Dihydroxy-2-butene	Thick liquid	Light yellow	Odorless
	Sinks and mixes with water. Freezing point is 45° F.		

Stop discharge if possible.
Call fire department.
Avoid contact with liquid.
Isolate and remove discharged material.
Notify local health and pollution control agencies.

Fire	Combustible. Extinguish with dry chemical, alcohol foam, or carbon dioxide. Water may be ineffective on fire.
Exposure	CALL FOR MEDICAL AID. LIQUID Irritating to skin and eyes. Harmful if swallowed. Remove contaminated clothing and shoes. Flush affected areas with plenty of water. IF IN EYES, hold eyelids open and flush with plenty of water. IF SWALLOWED and victim is CONSCIOUS, have victim drink water or milk.
Water Pollution	Effect of low concentrations on aquatic life is unknown. May be dangerous if it enters water intakes. Notify local health and pollution control officials. Notify operators of nearby water intakes.

1. RESPONSE TO DISCHARGE
(See Response Methods Handbook, CG 446-4)

Disperse and flush

2. LABELS

No hazard label required by Code of Federal Regulations

3. CHEMICAL DESIGNATIONS

3.1 **Synonyms:** cis-2-Butene-1, 4-diol; 1, 4-Dihydroxy-2-butene

3.2 **Coast Guard Compatibility Classification:** Not applicable

3.3 **Chemical Formula:** $HOCH_2CH{=}CHCH_2OH$

3.4 **IMCO/United Nations Numerical Designation:** 3.3/1987

4. OBSERVABLE CHARACTERISTICS

4.1 **Physical State (as shipped):** Liquid

4.2 **Color:** Pale yellow

4.3 **Odor:** Odorless

5. HEALTH HAZARDS

5.1 **Personal Protective Equipment:** Eye protection

5.2 **Symptoms Following Exposure:** Data not available

5.3 **Treatment for Exposure:** SKIN OR EYE CONTACT: flush well with water. Consult physician in cases of skin irritation, eye contact, or accidental ingestion.

5.4 **Toxicity by Inhalation (Threshold Limit Value):** Not pertinent

5.5 **Short-Term Inhalation Limits:** Not pertinent

5.6 **Toxicity by Ingestion:** Grade 2; LD_{50} 0.5 to 5 g/kg

5.7 **Late Toxicity:** Data not available

5.8 **Vapor (Gas) Irritant Characteristics:** Not pertinent

5.9 **Liquid or Solid Irritant Characteristics:** Minimum hazard. If spilled on clothing and allowed to remain, may cause smarting and reddening of the skin.

5.10 **Odor Threshold:** Not pertinent

6. FIRE HAZARDS

6.1 **Flash Point:** 263°F O.C.

6.2 **Flammable Limits in Air:** Not pertinent

6.3 **Fire Extinguishing Agents:** Alcohol foam, dry chemical or carbon dioxide

6.4 **Fire Extinguishing Agents Not to be Used:** Foam or water may cause frothing

6.5 **Special Hazards of Combustion Products:** Not pertinent

6.6 **Behavior in Fire:** Not pertinent

6.7 **Ignition Temperature:** Data not available

6.8 **Electrical Hazard:** Not pertinent

6.9 **Burning Rate:** Data not available

7. CHEMICAL REACTIVITY

7.1 **Reactivity with Water:** No reaction

7.2 **Reactivity with Common Materials:** No reaction

7.3 **Stability During Transport:** Stable

7.4 **Neutralizing Agents for Acids and Caustics:** Not pertinent

7.5 **Polymerization:** Stable

7.6 **Inhibitor of Polymerization:** Not pertinent

8. WATER POLLUTION

8.1 **Aquatic Toxicity:** Data not available

8.2 **Waterfowl Toxicity:** Data not available

8.3 **Biological Oxygen Demand (BOD):** Data not available

8.4 **Food Chain Concentration Potential:** None

9. SELECTED MANUFACTURERS

GAF Corporation
Chemical Division
Calvert City, Kentucky 42029

10. SHIPPING INFORMATION

10.1 **Grades or Purity:** 95%

10.2 **Storage Temperature:** Above 45°F

10.3 **Inert Atmosphere:** Inerted

10.4 **Venting:** Data not available

11. HAZARD ASSESSMENT CODE
(See Hazard Assessment Handbook, CG 446-3)

A-P-Q

12. HAZARD CLASSIFICATIONS

12.1 **Code of Federal Regulations:** Not listed

12.2 **NAS Hazard Rating for Bulk Water Transportation:** Not listed

12.3 **NFPA Hazard Classifications:** Not listed

13. PHYSICAL AND CHEMICAL PROPERTIES

13.1 **Physical State at 15°C and 1 atm:** Liquid

13.2 **Molecular Weight:** 88.11

13.3 **Boiling Point at 1 atm:** 453°F = 234°C = 507°K

13.4 **Freezing Point:** 45°F = 7°C = 280°K

13.5 **Critical Temperature:** Not pertinent

13.6 **Critical Pressure:** Not pertinent

13.7 **Specific Gravity:** 1.07 at 25°C (liquid)

13.8 **Liquid Surface Tension:** Not pertinent

13.9 **Liquid-Water Interfacial Tension:** Not pertinent

13.10 **Vapor (Gas) Specific Gravity:** Not pertinent

13.11 **Ratio of Specific Heats of Vapor (Gas):** Not pertinent

13.12 **Latent Heat of Vaporization:** Not pertinent

13.13 **Heat of Combustion:** (est.) −10,800 Btu/lb = −5980 cal/g = -250×10^5 J/kg

13.14 **Heat of Decomposition:** Not pertinent

13.15 **Heat of Solution:** (est.) 9 Btu/lb = 5 cal/g = 0.2×10^5 J/kg

13.16 **Heat of Polymerization:** Not pertinent

NOTES

BCN	# n-BUTYL ACETATE

Common Synonyms Acetic acid, n-butyl ester Butyl acetate Butyl ethanoate	Watery liquid Colorless Pleasant fruity odor Floats on water. Flammable irritating vapor is produced.

Stop discharge if possible. Keep people away.
Shut off ignition sources and call fire department.
Avoid contact with liquid and vapor.
Stay upwind and use water spray to "knock down" vapor.
Isolate and remove discharged material.
Notify local health and pollution control agencies.

Fire

FLAMMABLE.
Flashback along vapor trail may occur.
Vapor may explode if ignited in an enclosed area.

Extinguish with dry chemical, foam or carbon dioxide.
Water may be ineffective on fire.
Cool exposed containers with water.

Exposure

CALL FOR MEDICAL AID.

VAPOR
Irritating to eyes, nose and throat.
If inhaled, will cause nausea, dizziness, headache or difficult breathing.

Move to fresh air.
If breathing has stopped, give artificial respiration.
If breathing is difficult, give oxygen.

LIQUID
Irritating to skin and eyes.
Harmful if swallowed.

Remove contaminated clothing and shoes.
Flush affected areas with plenty of water.
IF IN EYES, hold eyelids open and flush with plenty of water.
IF SWALLOWED and victim is CONSCIOUS, have victim drink water or milk, and have victim induce vomiting.
IF SWALLOWED and victim is UNCONSCIOUS OR HAVING CONVULSIONS, do nothing except keep victim warm.

Water Pollution

Effect of low concentrations on aquatic life is unknown.
Fouling to shoreline.
May be dangerous if it enters water intakes.

Notify local health and pollution control officials.
Notify operators of nearby water intakes.

1. RESPONSE TO DISCHARGE

(See Response Methods Handbook, CG 446-4)

Issue warning—flammability
Mechanical containment

2. LABEL

Red

3. CHEMICAL DESIGNATIONS

3.1 **Synonyms:** Acetic acid, butyl ester; Butyl acetate; Butyl ethanoate

3.2 **Coast Guard Compatibility Classification:** Ester

3.3 **Chemical Formula:** $CH_3COO(CH_2)_3CH_3$

3.4 **IMCO/United Nations Numerical Designation:** 3.2/1123

4. OBSERVABLE CHARACTERISTICS

4.1 **Physical State (as shipped):** Liquid

4.2 **Color:** Colorless

4.3 **Odor:** Characteristic; agreeable fruity (in low conc.), disagreeable (in higher conc.); nonresidual

5. HEALTH HAZARDS

5.1 **Personal Protective Equipment:** All-purpose canister mask, chemical safety goggles, rubber gloves.

5.2 **Symptoms Following Exposure:** SKIN: prolonged or frequently repeated exposures may lead to drying. INHALATION: headaches, dizziness, nausea, irritation of respiratory passages and eyes.

5.3 **Treatment for Exposure:** EYES: in case of contact, flush with water for at least 15 min. INHALATION: remove from exposure immediately. Call a physician. If breathing is irregular or stopped, start resuscitation, administer oxygen. INGESTION: induce vomiting and call a physician.

5.4 **Toxicity by Inhalation (Threshold Limit Value):** 150-200 ppm

5.5 **Short-Term Inhalation Limits:** 300 ppm for 30 min.

5.6 **Toxicity by Ingestion:** Grade 2; LD_{50} 0.5 to 5 g/kg

5.7 **Late Toxicity:** None

5.8 **Vapor (Gas) Irritant Characteristics:** Vapors cause a slight smarting of the eyes or respiratory system if present in high concentrations. The effect is temporary.

5.9 **Liquid or Solid Irritant Characteristics:** Minimum hazard. If spilled on clothing and allowed to remain, may cause smarting and reddening of the skin.

5.10 **Odor Threshold:** 10 ppm

6. FIRE HAZARDS

6.1 **Flash Point:** 99°F O.C.; 75°F C.C.

6.2 **Flammable Limits in Air:** 1.7%–7.6%

6.3 **Fire Extinguishing Agents:** Foam, dry chemical, carbon dioxide

6.4 **Fire Extinguishing Agents Not to be Used:** Water in straight hose stream will scatter and spread fire and should not be used.

6.5 **Special Hazards of Combustion Products:** Not pertinent

6.6 **Behavior in Fire:** Not pertinent

6.7 **Ignition Temperature:** 760°F

6.8 **Electrical Hazard:** Class I, Group D

6.9 **Burning Rate:** 4.4 mm/min.

7. CHEMICAL REACTIVITY

7.1 **Reactivity with Water:** No reaction

7.2 **Reactivity with Common Materials:** No reaction

7.3 **Stability During Transport:** Stable

7.4 **Neutralizing Agents for Acids and Caustics:** Not pertinent

7.5 **Polymerization:** Not pertinent

7.6 **Inhibitor of Polymerization:** Not pertinent

8. WATER POLLUTION

8.1 **Aquatic Toxicity:** 44 ppm/48 hr/daphnia/TL_m/fresh water

8.2 **Waterfowl Toxicity:** Data not available

8.3 **Biological Oxygen Demand (BOD):** 0.15 to 0.5 lb/lb, 5 days (theor.) 52%, 5 days

8.4 **Food Chain Concentration Potential:** None

9. SELECTED MANUFACTURERS

1. Celanese Corp.
   Celanese Chemical Co. Div.
   245 Park Ave.
   New York, N. Y. 10017
2. Eastman Kodak Co.
   Tennessee Eastman Co. Div.
   P. O. Box 511
   Kingsport, Tenn. 37662
3. Union Carbide Corp.
   Chemicals and Plastics Div.
   270 Park Ave.
   New York, N. Y. 10017

10. SHIPPING INFORMATION

10.1 **Grades or Purity:** Urethane: 99.5%; pure: 98%; commercial: 90-92%

10.2 **Storage Temperature:** Ambient

10.3 **Inert Atmosphere:** No requirement

10.4 **Venting:** Open (flame arrester)

11. HAZARD ASSESSMENT CODE

(See Hazard Assessment Handbook, CG 446-3)

A-T-U

12. HAZARD CLASSIFICATIONS

12.1 **Code of Federal Regulations:** Flammable liquid

12.2 **NAS Hazard Rating for Bulk Water Transportation:**

Category	Rating
Fire	3
Health	
Vapor Irritant	1
Liquid or Solid Irritant	1
Poisons	2
Water Pollution	
Human Toxicity	0
Aquatic Toxicity	2
Aesthetic Effect	2
Reactivity	
Other Chemicals	1
Water	0
Self-Reaction	0

12.3 **NFPA Hazard Classifications:**

Category	Classification
Health Hazard (Blue)	1
Flammability (Red)	3
Reactivity (Yellow)	0

13. PHYSICAL AND CHEMICAL PROPERTIES

13.1 **Physical State at 15°C and 1 atm:** Liquid

13.2 **Molecular Weight:** 116.16

13.3 **Boiling Point at 1 atm:** 259°F = 126°C = 399°K

13.4 **Freezing Point:** −100°F = −73.5°C = 199.7°K

13.5 **Critical Temperature:** 582.6°F = 305.9°C = 579.1°K

13.6 **Critical Pressure:** 455 psia = 31 atm = 3.1 MN/m²

13.7 **Specific Gravity:** 0.875 at 20°C (liquid)

13.8 **Liquid Surface Tension:** 14.5 dynes/cm = 0.0145 N/m at 25°C

13.9 **Liquid-Water Interfacial Tension:** (est.) 57 dynes/cm = 0.057 N/m at 22°C

13.10 **Vapor (Gas) Specific Gravity:** Not pertinent

13.11 **Ratio of Specific Heats of Vapor (Gas):** 1.058

13.12 **Latent Heat of Vaporization:** 133 Btu/lb = 73.9 cal/g = 3.09×10^5 J/kg

13.13 **Heat of Combustion:** −13,130 Btu/lb = −7294 cal/g = -305.4×10^5 J/kg

13.14 **Heat of Decomposition:** Not pertinent

13.15 **Heat of Solution:** Not pertinent

13.16 **Heat of Polymerization:** Not pertinent

NOTES

REVISED 1978

BTA	# sec-BUTYL ACETATE

Common Synonyms
Acetic acid,
Sec-butyl ester

Watery liquid | Colorless | Pleasant, fruity odor

Floats on water. Flammable, irritating vapor is produced.

Stop discharge if possible. Keep people away.
Shut off ignition sources and call fire department.
Avoid contact with liquid and vapor.
Stay upwind and use water spray to "knock down" vapor.
Isolate and remove discharged material.
Notify local health and pollution control agencies.

Fire

FLAMMABLE.
Flashback along vapor trail may occur.
Vapor may explode if ignited in an enclosed area.
Extinguish with dry chemical, foam or carbon dioxide.
Water may be ineffective on fire.
Cool exposed containers with water.

Exposure

CALL FOR MEDICAL AID.

VAPOR
Irritating to eyes, nose and throat.
If inhaled, will cause nausea, headache or difficult breathing.
Move to fresh air.
If breathing has stopped, give artificial respiration.
If breathing is difficult, give oxygen.

LIQUID
Irritating to skin and eyes.
Remove contaminated clothing and shoes.
Flush affected areas with plenty of water.
IF IN EYES, hold eyelids open and flush with plenty of water.

Water Pollution

Effect of low concentrations on aquatic life is unknown.
Fouling to shoreline.
May be dangerous if it enters water intakes.
Notify local health and pollution control officials.
Notify operators of nearby water intakes.

1. RESPONSE TO DISCHARGE
(See Response Methods Handbook, CG 446-4)

Issue warning: high flammability
Mechanical containment

2. LABEL

Red

3. CHEMICAL DESIGNATIONS

3.1 **Synonyms:** Acetic acid, sec-butyl ester
3.2 **Coast Guard Compatibility Classification:** Ester
3.3 **Chemical Formula:** $CH_3COOCH(CH_3)CH_2CH_3$
3.4 **IMCO/United Nations Numerical Designation:** 3.2/1124

4. OBSERVABLE CHARACTERISTICS

4.1 **Physical State (as shipped):** Liquid
4.2 **Color:** Colorless
4.3 **Odor:** Pleasant

5. HEALTH HAZARDS

5.1 **Personal Protective Equipment:** Organic vapor canister or air-supplied mask; chemical goggles or face splash shield.
5.2 **Symptoms Following Exposure:** Headaches, dizziness, nausea, irritation of respiratory passage and eyes.
5.3 **Treatment for Exposure:** INHALATION: if victim is overcome by vapors, remove from exposure immediately; call a physician; if breathing is irregular or stopped, start resuscitation and administer oxygen. EYES: flush with water for at least 15 min.
5.4 **Toxicity by Inhalation (Threshold Limit Value):** 200 ppm
5.5 **Short-Term Inhalation Limits:** Data not available
5.6 **Toxicity by Ingestion:** Data not available
5.7 **Late Toxicity:** None
5.8 **Vapor (Gas) Irritant Characteristics:** Vapors cause a slight smarting of the eyes or respiratory system if present in high concentrations. The effect is temporary.
5.9 **Liquid or Solid Irritant Characteristics:** Minimum hazard. If spilled on clothing and allowed to remain, may cause smarting and reddening of the skin.
5.10 **Odor Threshold:** Data not available

6. FIRE HAZARDS

6.1 **Flash Point:** 62°F C.C.; 88°F O.C.
6.2 **Flammable Limits in Air:** 1.7% 9.8%
6.3 **Fire Extinguishing Agents:** Foam, carbon dioxide, or dry chemical
6.4 **Fire Extinguishing Agents Not to be Used:** Water may be ineffective
6.5 **Special Hazards of Combustion Products:** Not pertinent
6.6 **Behavior in Fire:** Not pertinent
6.7 **Ignition Temperature:** Data not available
6.8 **Electrical Hazard:** Not pertinent
6.9 **Burning Rate:** 4.4 mm/min. (approx.)

7. CHEMICAL REACTIVITY

7.1 **Reactivity with Water:** No reaction
7.2 **Reactivity with Common Materials:** No reaction
7.3 **Stability During Transport:** Stable
7.4 **Neutralizing Agents for Acids and Caustics:** Not pertinent
7.5 **Polymerization:** Not pertinent
7.6 **Inhibitor of Polymerization:** Not pertinent

8. WATER POLLUTION

8.1 **Aquatic Toxicity:** Data not available
8.2 **Waterfowl Toxicity:** Data not available
8.3 **Biological Oxygen Demand (BOD):** 0.15 to 0.5 lb/lb, 5 days
8.4 **Food Chain Concentration Potential:** None

9. SELECTED MANUFACTURERS

1. Exxon Chemical Co.
Houston, Texas 77001
2. Hercules Inc.
Polymers Dept.
Parlin, N. J. 08859
3. Publicker Industries Inc.
1429 Walnut St.
Philadelphia, Pa. 19102

10. SHIPPING INFORMATION

10.1 **Grades or Purity:** Technical and Pure
10.2 **Storage Temperature:** Ambient
10.3 **Inert Atmosphere:** No requirement
10.4 **Venting:** Open (flame arrester) or pressure-vacuum

11. HAZARD ASSESSMENT CODE
(See Hazard Assessment Handbook, CG 446-3)

A-T-U

12. HAZARD CLASSIFICATIONS

12.1 **Code of Federal Regulations:** Flammable liquid
12.2 **NAS Hazard Rating for Bulk Water Transportation:**

Category	Rating
Fire	3
Health	
Vapor Irritant	1
Liquid or Solid Irritant	1
Poisons	2
Water Pollution	
Human Toxicity	1
Aquatic Toxicity	2
Aesthetic Effect	2
Reactivity	
Other Chemicals	1
Water	0
Self-Reaction	0

12.3 **NFPA Hazard Classifications:**

Category	Classification
Health Hazard (Blue)	1
Flammability (Red)	3
Reactivity (Yellow)	0

13. PHYSICAL AND CHEMICAL PROPERTIES

13.1 **Physical State at 15°C and 1 atm:** Liquid
13.2 **Molecular Weight:** 116.16
13.3 **Boiling Point at 1 atm:** 234°F = 112°C = 385°K
13.4 **Freezing Point:** −100°F = −73.5°C = 199.7°K
13.5 **Critical Temperature:** 550°F = 288°C = 561°K
13.6 **Critical Pressure:** 469 psia = 32 atm = 3.2 MN/m^2
13.7 **Specific Gravity:** 0.872 at 20°C (liquid)
13.8 **Liquid Surface Tension:** 23.3 dynes/cm = 0.0233 N/m at 21°C
13.9 **Liquid-Water Interfacial Tension:** (est.) 58 dynes/cm = 0.058 N/m at 17°C
13.10 **Vapor (Gas) Specific Gravity:** Not pertinent
13.11 **Ratio of Specific Heats of Vapor (Gas):** 1.061
13.12 **Latent Heat of Vaporization:** (est.) 130 Btu/lb = 74 cal/g = 3.1×10^5 J/kg
13.13 **Heat of Combustion:** (est.) −13,100 Btu/lb = −7300 cal/g = -305×10^5 J/kg
13.14 **Heat of Decomposition:** Not pertinent
13.15 **Heat of Solution:** Not pertinent
13.16 **Heat of Polymerization:** Not pertinent

NOTES

REVISED 1978

BAI

iso-BUTYL ACRYLATE

Common Synonyms Isobutyl 2-propenoate Acrylic acid, isobutyl ester	Watery liquid	Colorless	Sharp fragrant odor
	Floats on water. Irritating vapor is produced.		

Stop discharge if possible.
Call fire department.
Avoid contact with liquid and vapor.
Isolate and remove discharged material.
Notify local health and pollution control agencies.

Fire

FLAMMABLE
Flashback along vapor trail may occur.
Vapor may explode if ignited in an enclosed area.
Extinguish with dry chemical, foam, or carbon dioxide.
Cool exposed containers with water.

Exposure

CALL FOR MEDICAL AID.

VAPOR
Irritating to eyes, nose and throat.
Move to fresh air.
If breathing has stopped, give artificial respiration.
If breathing is difficult, give oxygen.

LIQUID
Irritating to skin and eyes.
Harmful if swallowed.
Remove contaminated clothing and shoes.
Flush affected areas with plenty of water.
IF IN EYES, hold eyelids open and flush with plenty of water.
IF SWALLOWED and victim is CONSCIOUS, have victim induce vomiting.
IF SWALLOWED and victim is UNCONSCIOUS OR HAVING CONVULSIONS, do nothing except keep victim warm.

Water Pollution

Effect of low concentrations on aquatic life is unknown.
Fouling to shoreline.
May be dangerous if it enters water intakes.
Notify local health and pollution control officials.
Notify operators of nearby water intakes.

1. RESPONSE TO DISCHARGE

(See Response Methods Handbook, CG 446-4)

Mechanical containment
Chemical and physical treatment

2. LABEL

Red

3. CHEMICAL DESIGNATIONS

3.1 **Synonyms:** Acrylic acid, isobutyl ester
3.2 **Coast Guard Compatibility Classification:** Acrylate
3.3 **Chemical Formula:** $CH_2{=}CHCOOCH_2CH(CH_3)_2$
3.4 **IMCO/United Nations Numerical Designation:** Data not available

4. OBSERVABLE CHARACTERISTICS

4.1 **Physical State (as shipped):** Liquid
4.2 **Color:** Colorless
4.3 **Odor:** Sharp, fragrant

5. HEALTH HAZARDS

5.1 **Personal Protective Equipment:** Self-contained breathing apparatus, rubber gloves, chemical goggles.
5.2 **Symptoms Following Exposure:** Moderate toxicity when swallowed. Contact with the eyes causes minor irritation no worse than that caused by hand soap.
5.3 **Treatment for Exposure:** INHALATION: move victim to fresh air at once; give oxygen if breathing is difficult or artificial respiration if breathing has stopped; call a doctor. INGESTION: make victim vomit by sticking a finger down the throat or by giving strong, warm salt water to drink; get medical attention. SKIN AND EYES: remove chemical by flushing with plenty of clean, running water; remove contaminated clothing and wash exposed skin with soap and water.
5.4 **Toxicity by Inhalation (Threshold Limit Value):** Data not available
5.5 **Short-Term Inhalation Limits:** Data not available
5.6 **Toxicity by Ingestion:** Data not available
5.7 **Late Toxicity:** Not pertinent
5.8 **Vapor (Gas) Irritant Characteristics:** Vapors cause a slight smarting of the eyes or respiratory system if present in high concentrations. The effect is temporary.
5.9 **Liquid or Solid Irritant Characteristics:** Minimum hazard. If spilled on clothing and allowed to remain, may cause smarting and reddening of the skin.
5.10 **Odor Threshold:** Data not available

6. FIRE HAZARDS

6.1 **Flash Point:** 94°F O.C.
6.2 **Flammable Limits in Air:** 1.9%–8.0%
6.3 **Fire Extinguishing Agents:** Dry chemical, foam, or carbon dioxide
6.4 **Fire Extinguishing Agents Not to be Used:** Not pertinent
6.5 **Special Hazards of Combustion Products:** Not pertinent
6.6 **Behavior in Fire:** Not pertinent
6.7 **Ignition Temperature:** 644°F
6.8 **Electrical Hazard:** Not pertinent
6.9 **Burning Rate:** 4.8 mm/min.

7. CHEMICAL REACTIVITY

7.1 **Reactivity with Water:** No reaction
7.2 **Reactivity with Common Materials:** No reaction
7.3 **Stability During Transport:** Stable
7.4 **Neutralizing Agents for Acids and Caustics:** Not pertinent
7.5 **Polymerization:** Will polymerize when hot. Uncontrolled bulk polymerization can be explosive.
7.6 **Inhibitor of Polymerization:** Methyl ether of hydroquinone: 10-60 ppm; Hydroquinone: 5 ppm

8. WATER POLLUTION

8.1 **Aquatic Toxicity:** Data not available
8.2 **Waterfowl Toxicity:** Data not available
8.3 **Biological Oxygen Demand (BOD):** Data not available
8.4 **Food Chain Concentration Potential:** Data not available

9. SELECTED MANUFACTURERS

1. Celanese Corp.
   Celanese Chemical Co. Division
   245 Park Ave.
   New York, N. Y. 10017
2. Rohm and Haas Co.
   Independence Mall West
   Philadelphia, Pa. 19105
3. Union Carbide Corp.
   Chemicals and Plastics Division
   270 Park Ave.
   New York, N. Y. 10017

10. SHIPPING INFORMATION

10.1 **Grades or Purity:** 99.0%
10.2 **Storage Temperature:** Ambient
10.3 **Inert Atmosphere:** No requirement
10.4 **Venting:** Pressure-vacuum

11. HAZARD ASSESSMENT CODE

(See Hazard Assessment Handbook, CG 446-3)

A-T-U-Z

12. HAZARD CLASSIFICATIONS

12.1 **Code of Federal Regulations:** Flammable Liquid
12.2 **NAS Hazard Rating for Bulk Water Transportation:**

Category	Rating
Fire	2
Health	
Vapor Irritant	1
Liquid or Solid Irritant	1
Poisons	1
Water Pollution	
Human Toxicity	1
Aquatic Toxicity	2
Aesthetic Effect	2
Reactivity	
Other Chemicals	2
Water	0
Self-Reaction	3

12.3 **NFPA Hazard Classifications:** Not listed

13. PHYSICAL AND CHEMICAL PROPERTIES

13.1 **Physical State at 15°C and 1 atm:** Liquid
13.2 **Molecular Weight:** 128.17
13.3 **Boiling Point at 1 atm:** 280.2°F = 137.9°C = 411.1°K
13.4 **Freezing Point:** −78.0°F = −61.1°C = 212.1°K
13.5 **Critical Temperature:** 599°F = 315°C = 588°K
13.6 **Critical Pressure:** 440 psia = 30 atm = 3.0 MN/m²
13.7 **Specific Gravity:** 0.889 at 20°C (liquid)
13.8 **Liquid Surface Tension:** 2.47 dynes/cm = 0.0247 N/m at 25°C
13.9 **Liquid-Water Interfacial Tension:** (est.) 35 dynes/cm = 0.035 N/m at 27°C
13.10 **Vapor (Gas) Specific Gravity:** Not pertinent
13.11 **Ratio of Specific Heats of Vapor (Gas):** 1.044
13.12 **Latent Heat of Vaporization:** 130 Btu/lb = 71 cal/g = 3.0×10^5 J/kg
13.13 **Heat of Combustion:** −13,500 Btu/lb = −7500 cal/g = -314×10^5 J/kg
13.14 **Heat of Decomposition:** Not pertinent
13.15 **Heat of Solution:** Not pertinent
13.16 **Heat of Polymerization:** −229 Btu/lb = −127 cal/g = -5.32×10^5 J/kg

NOTES

REVISED 1978

BTC

n-BUTYL ACRYLATE

Common Synonyms
Butyl acrylate
n-Butyl 2-propenoate
Acrylic acid, n-butyl ester

Watery liquid | Colorless | Sharp, fragrant odor

Floats on water.

Stop discharge if possible.
Call fire department.
Avoid contact with liquid.
Isolate and remove discharged material.
Notify local health and pollution control agencies.

Fire

Combustible.
Containers may explode in fire.
Extinguish with dry chemical, foam, or carbon dioxide.
Cool exposed containers with water.

Exposure

CALL FOR MEDICAL AID.

LIQUID
Irritating to skin and eyes.
Harmful if swallowed.
Remove contaminated clothing and shoes.
Flush affected areas with plenty of water.
IF IN EYES, hold eyelids open and flush with plenty of water.
IF SWALLOWED, and victim is CONSCIOUS have victim drink water or milk.
IF SWALLOWED and victim is UNCONSCIOUS OR HAVING CONVULSIONS, do nothing except keep victim warm.

Water Pollution

Effect of low concentrations on aquatic life is unknown.
Fouling to shoreline.
May be dangerous if it enters water intakes.
Notify local health and pollution control officials.
Notify operators of nearby water intakes.

1. RESPONSE TO DISCHARGE

(See Response Methods Handbook, CG 446-4)

Mechanical containment
Chemical and physical treatment

2. LABELS

No hazard label required by Code of Federal Regulations

3. CHEMICAL DESIGNATIONS

3.1 **Synonyms:**
Acrylic acid, butyl ester
Butyl acrylate
Butyl 2-propenoate

3.2 **Coast Guard Compatibility Classification:** Acrylate

3.3 **Chemical Formula:** $CH_2{=}CHCOO(CH_2)_3CH_3$

3.4 **IMCO/United Nations Numerical Designation:** Not listed

4. OBSERVABLE CHARACTERISTICS

4.1 **Physical State (as shipped):** Liquid

4.2 **Color:** Colorless

4.3 **Odor:** Characteristic acrylic

5. HEALTH HAZARDS

5.1 **Personal Protective Equipment:** Self-contained breathing apparatus, rubber gloves, acid goggles.

5.2 **Symptoms Following Exposure:** Vapor is irritating when breathed at high concentrations. Contact with liquid causes irritation of skin and burning of eyes.

5.3 **Treatment for Exposure:** INHALATION: remove to fresh air; administer artificial respiration or oxygen if indicated; call a physician. SKIN AND EYES: wash with plenty of water.

5.4 **Toxicity by Inhalation (Threshold Limit Value):** Data not available

5.5 **Short-Term Inhalation Limits:** LD_{50}: 1000ppm, 4 hr

5.6 **Toxicity by Ingestion:** Grade 2; 0.5 to 5 g/kg (rat)

5.7 **Late Toxicity:** Data not available

5.8 **Vapor (Gas) Irritant Characteristics:** Vapors cause a slight smarting of the eyes or respiratory system if present in high concentrations. The effect is temporary.

5.9 **Liquid or Solid Irritant Characteristics:** Minimum hazard. If spilled on clothing and allowed to remain, may cause smarting and reddening of the skin.

5.10 **Odor Threshold:** Data not available

6. FIRE HAZARDS

6.1 **Flash Point:** 118°F O.C.

6.2 **Flammable Limits in Air:** 1.4%–9.4%

6.3 **Fire Extinguishing Agents:** Dry chemical, foam or carbon dioxide

6.4 **Fire Extinguishing Agents Not to be Used:** Not pertinent

6.5 **Special Hazards of Combustion Products:** Not pertinent

6.6 **Behavior in Fire:** Not pertinent

6.7 **Ignition Temperature:** 534°F

6.8 **Electrical Hazard:** Not pertinent

6.9 **Burning Rate:** 4.7 mm/min.

7. CHEMICAL REACTIVITY

7.1 **Reactivity with Water:** No reaction

7.2 **Reactivity with Common Materials:** No reaction

7.3 **Stability During Transport:** Stable

7.4 **Neutralizing Agents for Acids and Caustics:** Not pertinent

7.5 **Polymerization:** Will polymerize on application of heat; uncontrolled bulk polymerization can be explosive.

7.6 **Inhibitor of Polymerization:** Methyl ether of hydroquinone: 15–100 ppm. Store in contact with air.

8. WATER POLLUTION

8.1 **Aquatic Toxicity:** Data not available

8.2 **Waterfowl Toxicity:** Data not available

8.3 **Biological Oxygen Demand (BOD):** Data not available

8.4 **Food Chain Concentration Potential:** None

9. SELECTED MANUFACTURERS

1. Celanese Corp.
Celanese Chemical Co. Division
245 Park Ave.
New York, N. Y. 10017

2. Rohm and Haas Co.
Independence Mall West
Philadelphia, Pa. 19105

3. Union Carbide Corp.
Chemicals and Plastic Div.
270 Park Ave.
New York, N. Y. 10017

10. SHIPPING INFORMATION

10.1 **Grades or Purity:** 99+%

10.2 **Storage Temperature:** Ambient

10.3 **Inert Atmosphere:** No requirement

10.4 **Venting:** Pressure-vacuum

11. HAZARD ASSESSMENT CODE

(See Hazard Assessment Handbook, CG 446-3)

A-T-U-Z

12. HAZARD CLASSIFICATIONS

12.1 **Code of Federal Regulations:** Combustible Liquid

12.2 **NAS Hazard Rating for Bulk Water Transportation:**

Category	Rating
Fire	2
Health	
Vapor Irritant	1
Liquid or Solid Irritant	1
Poisons	1
Water Pollution	
Human Toxicity	1
Aquatic Toxicity	2
Aesthetic Effect	2
Reactivity	
Other Chemicals	2
Water	0
Self-Reaction	3

12.3 **NFPA Hazard Classifications:**

Category	Classification
Health Hazard (Blue)	2
Flammability (Red)	2
Reactivity (Yellow)	2

13. PHYSICAL AND CHEMICAL PROPERTIES

13.1 **Physical State at 15°C and 1 atm:** Liquid

13.2 **Molecular Weight:** 128.17

13.3 **Boiling Point at 1 atm:** 299.8°F = 148.8°C = 422.0°K

13.4 **Freezing Point:** −83°F = −64°C = 209°K

13.5 **Critical Temperature:** 621°F = 327°C = 600°K

13.6 **Critical Pressure:** 426 psia = 29 atm = 2.9 MN/m^2

13.7 **Specific Gravity:** 0.899 at 20°C (liquid)

13.8 **Liquid Surface Tension:** (est.) 20 dynes/cm = 0.020 N/m at 27°C

13.9 **Liquid-Water Interfacial Tension:** (est.) 60 dynes/cm = 0.060 N/m at 27°C

13.10 **Vapor (Gas) Specific Gravity:** Not pertinent

13.11 **Ratio of Specific Heats of Vapor (Gas):** 1.080

13.12 **Latent Heat of Vaporization:** 120 Btu/lb = 66.4 cal/g = 2.78×10^5 J/kg

13.13 **Heat of Combustion:** −13,860 Btu/lb = −7700 cal/g = -322.4×10^5 J/kg

13.14 **Heat of Decomposition:** Not pertinent

13.15 **Heat of Solution:** Not pertinent

13.16 **Heat of Polymerization:** −25.9 Btu/lb = −144 cal/g = -6.03×10^5 J/kg

NOTES

REVISED 1978

BAN n-BUTYL ALCOHOL

Common Synonyms			
Butanol Butyl alcohol 1-Butanol 1-Hydroxybutane n-Propylcarbinol	Watery liquid	Colorless	Alcohol odor
	Floats and mixes slowly with water. Flammable, irritating vapor is produced.		

Stop discharge if possible. Keep people away.
Call fire department.
Avoid contact with liquid and vapor.
Stay upwind and use water spray to "knock down" vapor.
Isolate and remove discharged material.
Notify local health and pollution control agencies.

Fire

FLAMMABLE
Flashback along vapor trail may occur.
Vapor may explode if ignited in an enclosed area.
Extinguish with dry chemical or carbon dioxide.
Water and alcohol foam may be ineffective on fire.
Cool exposed containers with water.

Exposure

CALL FOR MEDICAL AID.

VAPOR
Irritating to eyes, nose and throat.
If inhaled, will cause nausea, headache, dizziness.
Move to fresh air.
If breathing has stopped, give artificial respiration.
If breathing is difficult, give oxygen.

LIQUID
Irritating to skin and eyes.
Harmful if swallowed.
Remove contaminated clothing and shoes.
Flush affected areas with plenty of water.
IF IN EYES, hold eyelids open and flush with plenty of water.
IF SWALLOWED, and victim is CONSCIOUS, have victim drink water or milk.

Water Pollution

Dangerous to aquatic life in high concentrations.
May be dangerous if it enters water intakes.
Notify local health and pollution control officials.
Notify operators of nearby water intakes.

1. RESPONSE TO DISCHARGE

(See Response Methods Handbook, CG 446-4)

Restrict access
Disperse and flush

2. LABEL

Red

3. CHEMICAL DESIGNATIONS

3.1 **Synonyms:** Butanol
Butyl alcohol
1-Butanol
1-Hydroxybutane
n-Propylcarbinol

3.2 **Coast Guard Compatibility Classification:** Alcohol

3.3 **Chemical Formula:** $CH_3(CH_2)_2CH_2OH$

3.4 **IMCO/United Nations Numerical Designation:** 3.3/1120

4. OBSERVABLE CHARACTERISTICS

4.1 **Physical State (as shipped):** Liquid

4.2 **Color:** Colorless

4.3 **Odor:** Alcohol-like; pungent; strong; characteristic; mildly alcoholic, non residual

5. HEALTH HAZARDS

5.1 **Personal Protective Equipment:** Organic vapor canister or air-supplied mask; chemical goggles or face splash shield.

5.2 **Symptoms Following Exposure:** Anesthesia, nausea, headache, dizziness, irritation of respiratory passages. Mildly irritating to the skin and eyes.

5.3 **Treatment for Exposure:** INHALATION: remove from exposure immediately; call a physician; if breathing is irregular or has stopped, start resuscitation and administer oxygen. INGESTION: induce vomiting and call a physician. EYES: flush with water for at least 15 minutes.

5.4 **Toxicity by Inhalation (Threshold Limit Value):** 100 ppm

5.5 **Short-Term Inhalation Limits:** 150 ppm for 30 min.

5.6 **Toxicity by Ingestion:** Grade 2; LD_{50} 0.5 to 5 g/kg (rat)

5.7 **Late Toxicity:** None

5.8 **Vapor (Gas) Irritant Characteristics:** Vapors cause a slight smarting of the eyes or respiratory system if present in high concentrations. The effect is temporary.

5.9 **Liquid or Solid Irritant Characteristics:** Minimum hazard. If spilled on clothing and allowed to remain, may cause smarting and reddening of the skin.

5.10 **Odor Threshold:** 2.5 ppm

6. FIRE HAZARDS

6.1 **Flash Point:** 84°F C.C.; 97°F O.C.

6.2 **Flammable Limits in Air:** 1.4% 11.2%

6.3 **Fire Extinguishing Agents:** Carbon dioxide, dry chemicals

6.4 **Fire Extinguishing Agents Not to be Used:** Not pertinent

6.5 **Special Hazards of Combustion Products:** Not pertinent

6.6 **Behavior in Fire:** Not pertinent

6.7 **Ignition Temperature:** 650°F

6.8 **Electrical Hazard:** Class I, Group D

6.9 **Burning Rate:** 3.2 mm/min.

7. CHEMICAL REACTIVITY

7.1 **Reactivity with Water:** No reaction

7.2 **Reactivity with Common Materials:** No reaction

7.3 **Stability During Transport:** Stable

7.4 **Neutralizing Agents for Acids and Caustics:** Not pertinent

7.5 **Polymerization:** Not pertinent

7.6 **Inhibitor of Polymerization:** Not pertinent

8. WATER POLLUTION

8.1 **Aquatic Toxicity:** 1000 ppm/24 hr/goldfish/died/fresh water

8.2 **Waterfowl Toxicity:** Data not available

8.3 **Biological Oxygen Demand (BOD):** 1.1 1.92 lb/lb, 5 days; 157%, 5 days

8.4 **Food Chain Concentration Potential:** None

9. SELECTED MANUFACTURERS

1. Celanese Corp.
Celanese Chemical Co. Div.
245 Park Ave.
New York, N. Y. 10017
2. Dow Badische Co.
Williamsburg, Virginia 23185
3. Union Carbide Corp.
Chemicals & Plastics Div.
270 Park Ave.
New York, N. Y. 10017

10. SHIPPING INFORMATION

10.1 **Grades or Purity:** 99+%

10.2 **Storage Temperature:** Ambient

10.3 **Inert Atmosphere:** No requirement

10.4 **Venting:** Open (flame arrester)

11. HAZARD ASSESSMENT CODE

(See Hazard Assessment Handbook, CG 446-3)

A-P-Q

12. HAZARD CLASSIFICATIONS

12.1 **Code of Federal Regulations:** Flammable Liquid

12.2 **NAS Hazard Rating for Bulk Water Transportation:**

Category	Rating
Fire	3
Health	
Vapor Irritant	1
Liquid or Solid Irritant	1
Poisons	2
Water Pollution	
Human Toxicity	2
Aquatic Toxicity	2
Aesthetic Effect	2
Reactivity	
Other Chemicals	2
Water	0
Self-Reaction	0

12.3 **NFPA Hazard Classifications:**

Category	Classification
Health Hazard (Blue)	1
Flammability (Red)	3
Reactivity (Yellow)	0

13. PHYSICAL AND CHEMICAL PROPERTIES

13.1 **Physical State at 15°C and 1 atm:** Liquid

13.2 **Molecular Weight:** 74.12

13.3 **Boiling Point at 1 atm:** 243.9°F = 117.7°C = 390.9°K

13.4 **Freezing Point:** −129°F = −89.3°C = 183.9°K

13.5 **Critical Temperature:** 553.6°F = 289.8°C = 563.0°K

13.6 **Critical Pressure:** 640.2 psia = 43.55 atm = 4.412 MN/m^2

13.7 **Specific Gravity:** 0.810 at 20°C (liquid)

13.8 **Liquid Surface Tension:** 24.6 dynes/cm = .0246 N/m at 20°C

13.9 **Liquid-Water Interfacial Tension:** (est.) 56 dynes/cm = 0.056 N/m at 27°C

13.10 **Vapor (Gas) Specific Gravity:** Not pertinent

13.11 **Ratio of Specific Heats of Vapor (Gas):** 1.083

13.12 **Latent Heat of Vaporization:** 256 Btu/lb = 142 cal/g = 5.95×10^5 J/kg

13.13 **Heat of Combustion:** −14,230 Btu/lb = −7906 cal/g = -331.0×10^5 J/kg

13.14 **Heat of Decomposition:** Not pertinent

13.15 **Heat of Solution:** Data not available

13.16 **Heat of Polymerization:** Not pertinent

NOTES

REVISED 1978

BAS | sec-BUTYL ALCOHOL

Common Synonyms			
2-Butanol 2-Hydroxybutane Methylethylcarbinol	Watery liquid	Colorless	Alcohol odor
	Floats and mixes slowly with water. Flammable, irritating vapor is produced.		

Stop discharge if possible. Keep people away.
Shut off ignition sources and call fire department.
Avoid contact with liquid and vapor.
Stay upwind and use water spray to "knock down" vapor.
Isolate and remove discharged material.
Notify local health and pollution control agencies.

Fire

FLAMMABLE.
Flashback along vapor trail may occur.
Vapor may explode if ignited in an enclosed area.
Extinguish with dry chemical or carbon dioxide.
Water and alcohol foam may be ineffective on fire.
Cool exposed containers with water.

Exposure

CALL FOR MEDICAL AID.

VAPOR
Irritating to eyes, nose and throat.
If inhaled, will cause headache, dizziness or difficult breathing.
Move to fresh air.
If breathing has stopped, give artificial respiration.
If breathing is difficult, give oxygen.

LIQUID
Irritating to eyes.
If in eyes, hold eyelids open and flush with plenty of water.

Water Pollution

Effect of low concentrations on aquatic life is unknown.
May be dangerous if it enters water intakes.
Notify local health and pollution control officials.
Notify operators of nearby water intakes.

1. RESPONSE TO DISCHARGE
(See Response Methods Handbook, CG 446-4)

Restrict access
Disperse and flush

2. LABEL

Red

3. CHEMICAL DESIGNATIONS

3.1 **Synonyms:**
2-Butanol
Butylene hydrate
2-Hydroxybutane
Methylethylcarbinol

3.2 **Coast Guard Compatibility Classification:** Alcohol

3.3 **Chemical Formula:** $CH_3CH_2CH(OH)CH_3$

3.4 **IMCO/United Nations Numerical Designation:** 3.3/1121

4. OBSERVABLE CHARACTERISTICS

4.1 **Physical State (as shipped):** Liquid

4.2 **Color:** Colorless

4.3 **Odor:** Strong, pleasant

5. HEALTH HAZARDS

5.1 **Personal Protective Equipment:** Organic vapor canister or air-supplied mask; chemical goggles or face splash shield.

5.2 **Symptoms Following Exposure:** Headache, dizziness, and respiratory irritation. Liquid is severely irritating to the eyes and may cause eyeburn.

5.3 **Treatment for Exposure:** INHALATION: remove from exposure immediately; call a physician; if breathing is irregular or has stopped, start resuscitation and administer oxygen. INGESTION: induce vomiting and call a physician. EYES: flush with water for at least 15 minutes.

5.4 **Toxicity by Inhalation (Threshold Limit Value):** 150 ppm

5.5 **Short-Term Inhalation Limits:** 200 ppm for 60 min.

5.6 **Toxicity by Ingestion:** Grade 1; 5-15 g/kg (rat-single oral dose)

5.7 **Late Toxicity:** None

5.8 **Vapor (Gas) Irritant Characteristics:** Vapors cause a slight smarting of the eyes or respiratory system if present in high concentrations. The effect is temporary.

5.9 **Liquid or Solid Irritant Characteristics:** No appreciable hazard. Practically harmless to the skin.

5.10 **Odor Threshold:** Data not available

6. FIRE HAZARDS

6.1 **Flash Point:** 75°F C.C.

6.2 **Flammable Limits in Air:** 1.7% – 9.0%

6.3 **Fire Extinguishing Agents:** Carbon dioxide, dry chemicals

6.4 **Fire Extinguishing Agents Not to be Used:** Not pertinent

6.5 **Special Hazards of Combustion Products:** Not pertinent

6.6 **Behavior in Fire:** Not pertinent

6.7 **Ignition Temperature:** 763°F

6.8 **Electrical Hazard:** Class I, Group D

6.9 **Burning Rate:** 3.1 mm/min.

7. CHEMICAL REACTIVITY

7.1 **Reactivity with Water:** No reaction

7.2 **Reactivity with Common Materials:** No reaction

7.3 **Stability During Transport:** Stable

7.4 **Neutralizing Agents for Acids and Caustics:** Not pertinent

7.5 **Polymerization:** Not pertinent

7.6 **Inhibitor of Polymerization:** Not pertinent

8. WATER POLLUTION

8.1 **Aquatic Toxicity:** Data not available

8.2 **Waterfowl Toxicity:** Data not available

8.3 **Biological Oxygen Demand (BOD):** Data not available

8.4 **Food Chain Concentration Potential:** None

9. SELECTED MANUFACTURERS

1. Exxon Chemical Co.
Houston, Texas 77001
2. Shell Chemical Co.
Industrial Chemicals Div.
Houston, Texas 77001
3. Union Carbide Corp.
Chemicals and Plastics Division
270 Park Ave.
New York, N. Y. 10017

10. SHIPPING INFORMATION

10.1 **Grades or Purity:** 99+%

10.2 **Storage Temperature:** Ambient

10.3 **Inert Atmosphere:** No requirement

10.4 **Venting:** Open (flame arrester) or pressure-vacuum

11. HAZARD ASSESSMENT CODE
(See Hazard Assessment Handbook, CG 446-3)

A-P-Q

12. HAZARD CLASSIFICATIONS

12.1 **Code of Federal Regulations:** Flammable Liquid

12.2 **NAS Hazard Rating for Bulk Water Transportation:**

Category	Rating
Fire	3
Health	
Vapor Irritant	1
Liquid or Solid Irritant	0
Poisons	1
Water Pollution	
Human Toxicity	1
Aquatic Toxicity	1
Aesthetic Effect	2
Reactivity	
Other Chemicals	2
Water	0
Self-Reaction	0

12.3 **NFPA Hazard Classifications:** Not listed

13. PHYSICAL AND CHEMICAL PROPERTIES

13.1 **Physical State at 15°C and 1 atm:** Liquid

13.2 **Molecular Weight:** 74.12

13.3 **Boiling Point at 1 atm:** 211°F = 99.5°C = 372.7°K

13.4 **Freezing Point:** −174.5°F = −114.7°C = 158.5°K

13.5 **Critical Temperature:** 505.0°F = 262.8°C = 536.0°K

13.6 **Critical Pressure:** 608.4 psia = 41.39 atm = 4.193 MN/m²

13.7 **Specific Gravity:** 0.807 at 20°C (liquid)

13.8 **Liquid Surface Tension:** 23.0 dynes/cm = .023 N/m at 20°C

13.9 **Liquid-Water Interfacial Tension:** Data not available

13.10 **Vapor (Gas) Specific Gravity:** Not pertinent

13.11 **Ratio of Specific Heats of Vapor (Gas):** 1.080

13.12 **Latent Heat of Vaporization:** 243 Btu/lb = 135 cal/g = 5.65×10^5 J/kg

13.13 **Heat of Combustion:** -15,500 Btu/lb = -8600 cal/g = -360×10^5 J/kg

13.14 **Heat of Decomposition:** Not pertinent

13.15 **Heat of Solution:** Data not available

13.16 **Heat of Polymerization:** Not pertinent

NOTES

REVISED 1978

BAT	tert-BUTYL ALCOHOL

Common Synonyms	
2-Methyl-2-propanol Trimethylcarbinol	Oily liquid Colorless Sharp alcohol odor Floats and mixes with water. Flammable, irritating vapor is produced. Freezing point is 78° F.

Stop discharge if possible. Keep people away.
Shut off ignition sources and call fire department.
Avoid contact with liquid and vapor.
Stay upwind and use water spray to "knock down" vapor.
Isolate and remove discharged material.
Notify local health and pollution control agencies.

Fire	FLAMMABLE. Flashback along vapor trail may occur. Vapor may explode if ignited in an enclosed area. Extinguish with dry chemical or carbon dioxide. Water and alcohol foam may be ineffective on fire. Cool exposed containers with water.
Exposure	CALL FOR MEDICAL AID. VAPOR Irritating to eyes, nose, throat. If inhaled, will cause dizziness, difficult breathing. Move to fresh air. If breathing has stopped, give artificial respiration. If breathing is difficult, give oxygen. LIQUID Irritating to skin and eyes. Harmful if swallowed. Remove contaminated clothing and shoes. Flush affected areas with plenty of water. IF IN EYES, hold eyelids open and flush with plenty of water. IF SWALLOWED and victim is CONSCIOUS, have victim drink water or milk.
Water Pollution	Effect of low concentrations on aquatic life is unknown. May be dangerous if it enters water intakes. Notify local health and pollution control officials. Notify operators of nearby water intakes.

1. RESPONSE TO DISCHARGE

(See Response Methods Handbook, CG 446-4)

Issue warning—high flammability
Restrict access
Disperse and flush

2. LABEL

Red

3. CHEMICAL DESIGNATIONS

3.1 **Synonyms:** 2-Methyl-2-propanol; Trimethylcarbinol
3.2 **Coast Guard Compatibility Classification:** Alcohol
3.3 **Chemical Formula:** $(CH_3)_3COH$
3.4 **IMCO/United Nations Numerical Designation:** 3.3/1122

4. OBSERVABLE CHARACTERISTICS

4.1 **Physical State (as shipped):** Liquid. Sometimes freezes below 75°F
4.2 **Color:** Colorless
4.3 **Odor:** Characteristic; camphor-like; pungent

5. HEALTH HAZARDS

5.1 **Personal Protective Equipment:** Air pack or organic canister mask, rubber gloves, and goggles
5.2 **Symptoms Following Exposure:** Vapor is narcotic in action and irritating to respiratory passages. Liquid is irritating to skin and eyes.
5.3 **Treatment for Exposure:** INHALATION: remove victim from exposure and restore breathing. SKIN, EYE CONTACT: remove liquid from skin with water. Flush eyes with water.
5.4 **Toxicity by Inhalation (Threshold Limit Value):** 100 ppm
5.5 **Short-Term Inhalation Limits:** 150 ppm for 30 min.
5.6 **Toxicity by Ingestion:** Grade 2; 0.5 to 5.0 g/kg (rat)
5.7 **Late Toxicity:** Data not available
5.8 **Vapor (Gas) Irritant Characteristics:** Vapors cause a slight smarting of the eyes or respiratory system if present in high concentrations. The effect is temporary.
5.9 **Liquid or Solid Irritant Characteristics:** No appreciable hazard. Practically harmless to the skin.
5.10 **Odor Threshold:** Data not available

6. FIRE HAZARDS

6.1 **Flash Point:** 52°F C.C.; 61°F O.C.
6.2 **Flammable Limits in Air:** 2.35%—8.00%
6.3 **Fire Extinguishing Agents:** Dry chemical, carbon dioxide
6.4 **Fire Extinguishing Agents Not to be Used:** Not pertinent
6.5 **Special Hazards of Combustion Products:** Not pertinent
6.6 **Behavior in Fire:** Not pertinent
6.7 **Ignition Temperature:** 896°F
6.8 **Electrical Hazard:** Class I, Group D
6.9 **Burning Rate:** 3.4 mm/min.

7. CHEMICAL REACTIVITY

7.1 **Reactivity with Water:** No reaction
7.2 **Reactivity with Common Materials:** No reaction
7.3 **Stability During Transport:** Stable
7.4 **Neutralizing Agents for Acids and Caustics:** Not pertinent
7.5 **Polymerization:** Not pertinent
7.6 **Inhibitor of Polymerization:** Not pertinent

8. WATER POLLUTION

8.1 **Aquatic Toxicity:** Data not available
8.2 **Waterfowl Toxicity:** Data not available
8.3 **Biological Oxygen Demand (BOD):** 0%, 5 days
8.4 **Food Chain Concentration Potential:** None

9. SELECTED MANUFACTURERS

1. Eastman Kodak Co.
   Rochester, N. Y. 14650
2. Shell Chemical Co.
   Industrial Chemicals Div.
   Houston, Texas 77001

10. SHIPPING INFORMATION

10.1 **Grades or Purity:** 99+%
10.2 **Storage Temperature:** Ambient
10.3 **Inert Atmosphere:** No requirement
10.4 **Venting:** Open (flame arrester) or pressure-vacuum

11. HAZARD ASSESSMENT CODE

(See Hazard Assessment Handbook, CG 446-3)

A-P-Q-R-S

12. HAZARD CLASSIFICATIONS

12.1 **Code of Federal Regulations:** Flammable liquid
12.2 **NAS Hazard Rating for Bulk Water Transportation:**

Category	Rating
Fire	3
Health	
Vapor Irritant	1
Liquid or Solid Irritant	0
Poisons	1
Water Pollution	
Human Toxicity	2
Aquatic Toxicity	1
Aesthetic Effect	2
Reactivity	
Other Chemicals	2
Water	0
Self-Reaction	0

12.3 **NFPA Hazard Classifications:**

Category	Classification
Health Hazard (Blue)	1
Flammability (Red)	3
Reactivity (Yellow)	0

13. PHYSICAL AND CHEMICAL PROPERTIES

13.1 **Physical State at 15°C and 1 atm:** Liquid
13.2 **Molecular Weight:** 74.12
13.3 **Boiling Point at 1 atm:** 181°F = 82.6°C = 355.8°K
13.4 **Freezing Point:** 78.3°F = 25.7°C = 298.9°K
13.5 **Critical Temperature:** 451°F = 233°C = 506°K
13.6 **Critical Pressure:** 576 psia = 39.2 atm = 3.97 MN/m²
13.7 **Specific Gravity:** 0.78 at 26°C (liquid)
13.8 **Liquid Surface Tension:** 20.7 dynes/cm = 0.0207 N/m at 25°C
13.9 **Liquid-Water Interfacial Tension:** Not pertinent
13.10 **Vapor (Gas) Specific Gravity:** 2.6
13.11 **Ratio of Specific Heats of Vapor (Gas):** 1.080
13.12 **Latent Heat of Vaporization:** 234 Btu/lb = 130 cal/g = 5.44×10^5 J/kg
13.13 **Heat of Combustion:** −14,000 Btu/lb = −7780 cal/g = -325.7×10^5 J/kg
13.14 **Heat of Decomposition:** Not pertinent
13.15 **Heat of Solution:** Not pertinent
13.16 **Heat of Polymerization:** Not pertinent

NOTES

REVISED 1978

BAM

n-BUTYLAMINE

Common Synonyms		
1-Aminobutane Butylamine Mono-n-butylamine Norvalamine	Liquid Colorless Fishy, ammonia-like odor Mixes with water.	

AVOID CONTACT WITH LIQUID AND VAPOR. KEEP PEOPLE AWAY.
Wear goggles, self-contained breathing apparatus, and rubber overclothing (including gloves).
Shut off ignition sources. Call fire department.
Stop discharge if possible.
Stay upwind. Use water spray to "knock down" vapor.
Isolate and remove discharged material.
Notify local health and pollution control agencies.

Fire

Flammable.
POISONOUS GASES MAY BE PRODUCED IN FIRE.
Flashback along vapor trail may occur.
Vapor may explode if ignited in an enclosed area.
Wear goggles and self-contained breathing apparatus.
Extinguish with dry chemicals, foam, or carbon dioxide.
Water may be ineffective on fire.
Cool exposed containers with water.

Exposure

CALL FOR MEDICAL AID.

VAPOR
POISONOUS IF INHALED OR IF SKIN IS EXPOSED.
If inhaled will cause dizziness, headache, coughing, or difficult breathing.
If in eyes, hold eyelids open and flush with plenty of water.
If breathing has stopped, give artificial respiration.
If breathing is difficult, give oxygen.

LIQUID
Will burn skin and eyes.
If swallowed will cause nausea and vomiting.
Remove contaminated clothing and shoes.
Flush affected areas with plenty of water.
IF IN EYES, hold eyelids open and flush with plenty of water.
IF SWALLOWED and victim is CONSCIOUS, have victim drink water or milk.
IF SWALLOWED and victim is UNCONSCIOUS OR HAVING CONVULSIONS do nothing except keep victim warm.

Water Pollution

HARMFUL TO AQUATIC LIFE IN VERY LOW CONCENTRATIONS.
May be dangerous if it enters water intakes.
Notify local health and wildlife officials.
Notify operators of nearby water intakes.

1. RESPONSE TO DISCHARGE

(See Response Methods Handbook, CG 446-4)

Issue warning—high flammability, air contaminant, water contaminant
Restrict access
Disperse and flush

2. LABEL

Red

3. CHEMICAL DESIGNATIONS

3.1 **Synonyms:** 1-Aminobutane; Butylamine; Mono-n-butylamine; Norvalamine
3.2 **Coast Guard Compatibility Classification:** Aliphatic amine
3.3 **Chemical Formula:** $CH_3(CH_2)_3NH_2$
3.4 **IMCO/United Nations Numerical Designation:** 3.2/1125

4. OBSERVABLE CHARACTERISTICS

4.1 **Physical State (as shipped):** Liquid
4.2 **Color:** Colorless
4.3 **Odor:** Fish-like; ammonia-like

5. HEALTH HAZARDS

5.1 **Personal Protective Equipment:** Air-supplied mask; rubber gloves; coverall goggles; face shield; butyl rubber apron
5.2 **Symptoms Following Exposure:** Inhalation causes irritation, nausea, vomiting, headache, faintness, severe coughing and chest pains; can cause lung edema. Ingestion causes severe irritation of mouth and stomach. Contact with eyes causes severe irritation and edema of the cornea. Contact with skin causes burns; absorption through skin may cause nausea, vomiting and shock.
5.3 **Treatment for Exposure:** INHALATION: remove victim to fresh air; call a physician; give oxygen if breathing is difficult; if not breathing, give artificial respiration. INGESTION: give large amounts of water; get medical attention. EYES: flush with water at least 15 min.; get medical care. SKIN: remove contaminated clothing; flush skin with plenty of water at least 15 min.
5.4 **Toxicity by Inhalation (Threshold Limit Value):** 5 ppm
5.5 **Short-Term Inhalation Limits:** 5 ppm, 5 min.
5.6 **Toxicity by Ingestion:** Grade 3; oral LD_{50} = 500 mg/kg (rat)
5.7 **Late Toxicity:** Data not available
5.8 **Vapor (Gas) Irritant Characteristics:** Data not available
5.9 **Liquid or Solid Irritant Characteristics:** Data not available
5.10 **Odor Threshold:** Data not available

6. FIRE HAZARDS

6.1 **Flash Point:** 30°F O.C.; 10°F C.C.
6.2 **Flammable Limits in Air:** 1.7% 9.8%
6.3 **Fire Extinguishing Agents:** "Alcohol" foam, dry chemical, carbon dioxide
6.4 **Fire Extinguishing Agents Not to be Used:** Water may be ineffective.
6.5 **Special Hazards of Combustion Products:** Toxic oxides of nitrogen may form in fire.
6.6 **Behavior in Fire:** Vapor is heavier than air and may travel to a source of ignition and flash back. Containers may explode in fire.
6.7 **Ignition Temperature:** 594°F
6.8 **Electrical Hazard:** Data not available
6.9 **Burning Rate:** 5.79 mm/min.

7. CHEMICAL REACTIVITY

7.1 **Reactivity with Water:** No reaction
7.2 **Reactivity with Common Materials:** May corrode some metals in presence of water.
7.3 **Stability During Transport:** Stable
7.4 **Neutralizing Agents for Acids and Caustics:** Flush with water.
7.5 **Polymerization:** Not pertinent
7.6 **Inhibitor of Polymerization:** Not pertinent

8. WATER POLLUTION

8.1 **Aquatic Toxicity:** 30–70 ppm/24 hr/creek chub/critical range/fresh water
8.2 **Waterfowl Toxicity:** Data not available
8.3 **Biological Oxygen Demand (BOD):** 26.5% theo., 5 days
8.4 **Food Chain Concentration Potential:** None

9. SELECTED MANUFACTURERS

1. Union Carbide Corp. Chemicals and Plastics Div. 270 Park Avenue New York, N. Y. 10017
2. Virginia Chemicals, Inc. 3340 West Norfolk Rd. Portsmouth, Va. 23703
3. Air Products and Chemicals, Inc. U. S. Highway 90 Pace, Fla. 32502

10. SHIPPING INFORMATION

10.1 **Grades or Purity:** Pure, 100%
10.2 **Storage Temperature:** Ambient
10.3 **Inert Atmosphere:** No requirement
10.4 **Venting:** Open (flame arrester)

11. HAZARD ASSESSMENT CODE

(See Hazard Assessment Handbook, CG 446-3)

A-P-Q-R-S

12. HAZARD CLASSIFICATIONS

12.1 **Code of Federal Regulations:** Flammable liquid
12.2 **NAS Hazard Rating for Bulk Water Transportation:** Not listed
12.3 **NFPA Hazard Classifications:**

Category	Classification
Health Hazard (Blue)	2
Flammability (Red)	3
Reactivity (Yellow)	0

13. PHYSICAL AND CHEMICAL PROPERTIES

13.1 **Physical State at 15°C and 1 atm:** Liquid
13.2 **Molecular Weight:** 73.14
13.3 **Boiling Point at 1 atm:** 171.3°F = 77.4°C = 350.6°K
13.4 **Freezing Point:** −56°F = −49°C = 224°K
13.5 **Critical Temperature:** 484°F = 251°C = 524°K
13.6 **Critical Pressure:** 603 psia = 41 atm = 4.16 MN/m²
13.7 **Specific Gravity:** 0.741 at 20°C (liquid)
13.8 **Liquid Surface Tension:** 53.11 dynes/cm = 0.05311 N/m at 20°C
13.9 **Liquid-Water Interfacial Tension:** Not pertinent
13.10 **Vapor (Gas) Specific Gravity:** 2.5
13.11 **Ratio of Specific Heats of Vapor (Gas):** (est.) 1.071
13.12 **Latent Heat of Vaporization:** 180 Btu/lb = 100 cal/g = 4.2×10^5 J/kg
13.13 **Heat of Combustion:** −17,595 Btu/lb = −9,775 cal/g = -409.0×10^5 J/kg
13.14 **Heat of Decomposition:** Not pertinent
13.15 **Heat of Solution:** −137 Btu/lb = −76.2 cal/g = -3.19×10^5 J/kg
13.16 **Heat of Polymerization:** Not pertinent

NOTES

BTL	sec-BUTYLAMINE

Common Synonyms	Liquid	White	Ammonia-like odor
	Mixes with water.		

Shut off ignition sources. Call fire department.
Avoid contact with liquid and vapor. Keep people away.
Wear rubber overclothing (including gloves).
Stop discharge if possible.
Stay upwind. Use water spray to "knock down" vapor.
Isolate and remove discharged material.
Notify local health and pollution control agencies.

Fire

FLAMMABLE.
POISONOUS GASES MAY BE PRODUCED IN FIRE.
Flashback along vapor trail may occur.
Vapor may explode if ignited in an enclosed area.
Wear goggles and self-contained breathing apparatus.
Extinguish with dry chemicals, alcohol foam, or carbon dioxide.
Water may be ineffective on fire.
Cool exposed containers with water.

Exposure

CALL FOR MEDICAL AID.

VAPOR
Irritating to eyes, nose and throat.
If inhaled will cause coughing or difficult breathing.
If in eyes, hold eyelids open and flush with plenty of water.
If breathing has stopped, give artificial respiration.
If breathing is difficult, give oxygen.

LIQUID
Will burn skin and eyes.
If swallowed will cause nausea and vomiting
Remove contaminated clothing and shoes.
Flush affected areas with plenty of water.
IF IN EYES, hold eyelids open and flush with plenty of water.
IF SWALLOWED and victim is CONSCIOUS, have victim drink water or milk and have victim induce vomiting.
IF SWALLOWED and victim is UNCONSCIOUS OR HAVING CONVULSIONS, do nothing except keep victim warm.

Water Pollution

Effect of low concentrations on aquatic life is unknown.
May be dangerous if it enters water intakes.
Notify local health and wildlife officials.
Notify operators of nearby water intakes.

1. RESPONSE TO DISCHARGE
(See Response Methods Handbook, CG 446-4)

Issue warning—high flammability, air contaminant, water contaminant
Restrict access
Disperse and flush

2. LABEL

Red

3. CHEMICAL DESIGNATIONS

3.1 **Synonyms:** No common synonyms
3.2 **Coast Guard Compatibility Classification:** Aliphatic amine
3.3 **Chemical Formula:** $CH_3CH_2CH(CH_3)NH_2$
3.4 **IMCO/United Nations Numerical Designation:** Not listed

4. OBSERVABLE CHARACTERISTICS

4.1 **Physical State (as shipped):** Liquid
4.2 **Color:** Colorless
4.3 **Odor:** Ammonia-like

5. HEALTH HAZARDS

5.1 **Personal Protective Equipment:** Chemical safety goggles; rubber gloves and apron; respiratory protective equipment; non-sparking shoes
5.2 **Symptoms Following Exposure:** Inhalation causes irritation or burns of the respiratory system; exposure to concentrated vapors can cause asphyxiation. Ingestion causes burns of mouth and stomach. Contact with eyes causes lachrymation, conjunctivitis, burns, corneal edema. Contact with skin causes irritation or burns, dermatitis.
5.3 **Treatment for Exposure:** INHALATION: remove patient from exposure; keep him quiet; contact physician. INGESTION: give large amount of water; induce vomiting; consult a physician. EYES: flush thoroughly with water for 15 min.; call physician immediately. SKIN: remove all contaminated clothing; flood affected area with large quantities of water; consult a physician.
5.4 **Toxicity by Inhalation (Threshold Limit Value):** Data not available
5.5 **Short-Term Inhalation Limits:** Data not available
5.6 **Toxicity by Ingestion:** Grade 3; oral LD_{50} = 380 mg/kg (rat)
5.7 **Late Toxicity:** Data not available
5.8 **Vapor (Gas) Irritant Characteristics:** Vapors are moderately irritating such that personnel will not usually tolerate moderate or high concentrations.
5.9 **Liquid or Solid Irritant Characteristics:** Severe skin irritant. Causes second- and third-degree burns on short contact and is very injurious to the eyes.
5.10 **Odor Threshold:** Data not available

6. FIRE HAZARDS

6.1 **Flash Point:** 16°F C.C.
6.2 **Flammable Limits in Air:** Data not available
6.3 **Fire Extinguishing Agents:** "Alcohol" foam, dry chemical, carbon dioxide
6.4 **Fire Extinguishing Agents Not to be Used:** Water may be ineffective.
6.5 **Special Hazards of Combustion Products:** Toxic oxides of nitrogen may be formed in fire.
6.6 **Behavior in Fire:** Vapor is heavier than air and may travel to a source of ignition and flash back. Containers may explode in fire.
6.7 **Ignition Temperature:** 712°F
6.8 **Electrical Hazard:** Data not available
6.9 **Burning Rate:** 6.18 mm/min.

7. CHEMICAL REACTIVITY

7.1 **Reactivity with Water:** No reaction
7.2 **Reactivity with Common Materials:** May corrode some metals in presence of water.
7.3 **Stability During Transport:** Stable
7.4 **Neutralizing Agents for Acids and Caustics:** Flush with water.
7.5 **Polymerization:** Not pertinent
7.6 **Inhibitor of Polymerization:** Not pertinent

8. WATER POLLUTION

8.1 **Aquatic Toxicity:** Data not available
8.2 **Waterfowl Toxicity:** Data not available
8.3 **Biological Oxygen Demand (BOD):** Data not available
8.4 **Food Chain Concentration Potential:** None

9. SELECTED MANUFACTURERS

1. Pennwalt Corporation
   Three Parkway
   Philadelphia, Pa. 19102
2. Virginia Chemicals, Inc.
   3340 West Norfolk Rd.
   Portsmouth, Va. 23703
3. BASF Wyandotte Corp.
   1609 Biddle Avenue
   Wyandotte, Mich. 48192

10. SHIPPING INFORMATION

10.1 **Grades or Purity:** Pure
10.2 **Storage Temperature:** Ambient
10.3 **Inert Atmosphere:** No requirement
10.4 **Venting:** Open (flame arrester)

11. HAZARD ASSESSMENT CODE
(See Hazard Assessment Handbook, CG 446-3)

A-P-Q-R-S

12. HAZARD CLASSIFICATIONS

12.1 **Code of Federal Regulations:** Flammable liquid
12.2 **NAS Hazard Rating for Bulk Water Transportation:**

Category	Rating
Fire	3
Health	
Vapor Irritant	3
Liquid or Solid Irritant	4
Poisons	4
Water Pollution	
Human Toxicity	2
Aquatic Toxicity	2
Aesthetic Effect	1
Reactivity	
Other Chemicals	3
Water	0
Self-Reaction	0

12.3 **NFPA Hazard Classifications:**

Category	Classification
Health Hazard (Blue)	3
Flammability (Red)	3
Reactivity (Yellow)	—

13. PHYSICAL AND CHEMICAL PROPERTIES

13.1 **Physical State at 15°C and 1 atm:** Liquid
13.2 **Molecular Weight:** 73.1
13.3 **Boiling Point at 1 atm:** 145°F = 63°C = 336°K
13.4 **Freezing Point:** −155°F = −104°C = 169°K
13.5 **Critical Temperature:** Data not available
13.6 **Critical Pressure:** Data not available
13.7 **Specific Gravity:** 0.721 at 20°C (liquid)
13.8 **Liquid Surface Tension:** 22.42 dynes/cm = 0.02242 N/m at 20°C
13.9 **Liquid-Water Interfacial Tension:** Not pertinent
13.10 **Vapor (Gas) Specific Gravity:** 2.52
13.11 **Ratio of Specific Heats of Vapor (Gas):** (est.) 1.073 at 20°C
13.12 **Latent Heat of Vaporization:** 178.09 Btu/lb = 98.94 cal/g = 4.160×10^5 J/kg
13.13 **Heat of Combustion:** −17,600 Btu/lb = −9,780 cal/g = -409×10^5 J/kg
13.14 **Heat of Decomposition:** Not pertinent
13.15 **Heat of Solution:** −170 Btu/lb = −93 cal/g = -3.9×10^5 J/kg
13.16 **Heat of Polymerization:** Not pertinent

NOTES

BUA — tert-BUTYLAMINE

Common Synonyms			
2-Aminoisobutane 2-Amino-2-methylpropane 1,1-Dimethylethylamine TBA Trimethylaminomethane	Liquid	Colorless	Ammonia-like odor
	Floats and mixes with water. Flammable, irritating vapor is produced.		

Shut off ignition sources and call fire department.
Avoid contact with liquid and vapor. Keep people away.
Wear goggles, self-contained breathing apparatus, and rubber overclothing (including gloves).
Stop discharge if possible.
Stay upwind and use water spray to "knock down" vapor.
Isolate and remove discharged material.
Notify local health and pollution control agencies.

Fire	FLAMMABLE. Flashback along vapor trail may occur. Vapor may explode if ignited in an enclosed area. Extinguish with dry chemical, alcohol foam, or carbon dioxide. Water may be ineffective on fire. Cool exposed containers with water.
Exposure	CALL FOR MEDICAL AID. VAPOR Irritating to eyes, nose and throat. If inhaled, will cause difficult breathing. Move to fresh air. If breathing has stopped, give artificial respiration. If breathing is difficult, give oxygen. LIQUID Irritating to skin and eyes. Harmful if swallowed. Remove contaminated clothing and shoes. Flush affected areas with plenty of water. IF IN EYES, hold eyelids open and flush with plenty of water. IF SWALLOWED and victim is CONSCIOUS, have victim drink water or milk.
Water Pollution	Effect of low concentrations on aquatic life is unknown. May be dangerous if it enters water intakes. Notify local health and wildlife officials. Notify operators of nearby water intakes.

1. RESPONSE TO DISCHARGE

(See Response Methods Handbook, CG 446-4)

Issue warning—high flammability, air contaminant
Restrict access
Evacuate area
Disperse and flush

2. LABEL

Red

3. CHEMICAL DESIGNATIONS

3.1 **Synonyms:** 2-Aminoisobutane; 2-Amino-2-methylpropane; 1,1-Dimethylethylamine; TBA; Trimethylaminomethane
3.2 **Coast Guard Compatibility Classification:** Aliphatic amine
3.3 **Chemical Formula:** $(CH_3)_3CNH_2$
3.4 **IMCO/United Nations Numerical Designation:** 3.2/1993

4. OBSERVABLE CHARACTERISTICS

4.1 **Physical State (as shipped):** Liquid
4.2 **Color:** Colorless
4.3 **Odor:** Like ammonia

5. HEALTH HAZARDS

5.1 **Personal Protective Equipment:** Self-contained breathing apparatus; goggles or face shield; rubber gloves
5.2 **Symptoms Following Exposure:** Inhalation causes irritation of nose, mouth, and lungs. Ingestion causes irritation of mouth and stomach. Contact with liquid causes severe irritation of eyes and moderate irritation of skin.
5.3 **Treatment for Exposure:** INHALATION: move to fresh air; give artificial respiration if breathing has stopped. INGESTION: give large amounts of water and induce vomiting. EYES: immediately flush with water for at least 15 min.; get medical attention. SKIN: flush with water; wash with soap and water.
5.4 **Toxicity by Inhalation (Threshold Limit Value):** Data not available
5.5 **Short-Term Inhalation Limits:** Data not available
5.6 **Toxicity by Ingestion:** Grade 3; oral LD_{50} = 180 mg/kg (rat)
5.7 **Late Toxicity:** Data not available
5.8 **Vapor (Gas) Irritant Characteristics:** Data not available
5.9 **Liquid or Solid Irritant Characteristics:** Data not available
5.10 **Odor Threshold:** Data not available

6. FIRE HAZARDS

6.1 **Flash Point:** 16°F C.C.
6.2 **Flammable Limits in Air:** 1.7%–8.9% (at 212°F)
6.3 **Fire Extinguishing Agents:** Dry chemical, alcohol foam, carbon dioxide
6.4 **Fire Extinguishing Agents Not to be Used:** Water may be ineffective.
6.5 **Special Hazards of Combustion Products:** Toxic oxides of nitrogen may form in fire.
6.6 **Behavior in Fire:**
6.7 **Ignition Temperature:** 716°F
6.8 **Electrical Hazard:** Data not available
6.9 **Burning Rate:** ~7 mm/min.

7. CHEMICAL REACTIVITY

7.1 **Reactivity with Water:** No reaction
7.2 **Reactivity with Common Materials:** Liquid will attack some forms of plastics.
7.3 **Stability During Transport:** Stable
7.4 **Neutralizing Agents for Acids and Caustics:** Flush with water
7.5 **Polymerization:** Not pertinent
7.6 **Inhibitor of Polymerization:** Not pertinent

8. WATER POLLUTION

8.1 **Aquatic Toxicity:** Data not available
8.2 **Waterfowl Toxicity:** Data not available
8.3 **Biological Oxygen Demand (BOD):** Data not available
8.4 **Food Chain Concentration Potential:** None

9. SELECTED MANUFACTURERS

1. Monsanto Company
800 N. Lindbergh Blvd.
St. Louis, Mo. 63166
2. Eastman Organic Chemicals
Rochester, N. Y. 14650
3. Aldrich Chemical Co.
940 W. Saint Paul Avenue
Milwaukee, Wis. 53233

10. SHIPPING INFORMATION

10.1 **Grades or Purity:** 99+%
10.2 **Storage Temperature:** Ambient
10.3 **Inert Atmosphere:** No requirement
10.4 **Venting:** Open

11. HAZARD ASSESSMENT CODE

(See Hazard Assessment Handbook, CG 446-3)

A-P-Q-R-S

12. HAZARD CLASSIFICATIONS

12.1 **Code of Federal Regulations:** Flammable liquid
12.2 **NAS Hazard Rating for Bulk Water Transportation:** Not listed
12.3 **NFPA Hazard Classifications:** Not listed

13. PHYSICAL AND CHEMICAL PROPERTIES

13.1 **Physical State at 15°C and 1 atm:** Liquid
13.2 **Molecular Weight:** 73.14
13.3 **Boiling Point at 1 atm:** 113°F = 45°C = 318°K
13.4 **Freezing Point:** Not pertinent
13.5 **Critical Temperature:** Data not available
13.6 **Critical Pressure:** Data not available
13.7 **Specific Gravity:** 0.696 at 20°C (liquid)
13.8 **Liquid Surface Tension:** 19 dynes/cm = 0.019 N/m at 20°C
13.9 **Liquid-Water Interfacial Tension:** Not pertinent
13.10 **Vapor (Gas) Specific Gravity:** 8.13
13.11 **Ratio of Specific Heats of Vapor (Gas):** Data not available
13.12 **Latent Heat of Vaporization:** 167.0 Btu/lb = 92.8 cal/g = 3.88×10^5 J/kg
13.13 **Heat of Combustion:** −17,600 Btu/lb = −9,790 cal/g = -410×10^5 J/kg
13.14 **Heat of Decomposition:**
13.15 **Heat of Solution:** −170 Btu/lb = −96 cal/g = -4.0×10^5 J/kg
13.16 **Heat of Polymerization:** Not pertinent

NOTES

BTN	BUTYLENE

Common Synonyms 1-Butene	Liquefied compressed gas Colorless Fragrant gasoline-like odor Floats and boils on water. Flammable, visible vapor cloud is produced.

Stop discharge if possible. Keep people away.
Shut off ignition sources and call fire department.
Stay upwind and use water spray to "knock down" vapor.
Avoid contact with liquid and vapor.
Notify local health and pollution control agencies.

Fire	FLAMMABLE. Flashback along vapor trail may occur. Vapor may explode if ignited in an enclosed area. Stop flow of gas if possible. Cool exposed containers and protect men effecting the shutoff with water. Let fire burn.
Exposure	CALL FOR MEDICAL AID. VAPOR If inhaled, will cause dizziness and difficult breathing. Move to fresh air. If breathing has stopped, give artificial respiration. If breathing is difficult, give oxygen. LIQUID Will cause frostbite. Flush affected areas with plenty of water. DO NOT RUB AFFECTED AREAS.
Water Pollution	Not harmful to aquatic life. May be dangerous if it enters water intakes. Notify operators of nearby water intakes.

1. RESPONSE TO DISCHARGE
(See Response Methods Handbook, CG 446-4)

Issue warning—high flammability
Restrict access
Evacuate area

2. LABEL

Red

3. CHEMICAL DESIGNATIONS

3.1 **Synonyms:** 1-Butene
3.2 **Coast Guard Compatibility Classification:** Olefin
3.3 **Chemical Formula:** $CH_3CH_2CH{=}CH_2$
3.4 **IMCO/United Nations Numerical Designation:** 2.0/1012

4. OBSERVABLE CHARACTERISTICS

4.1 **Physical State (as shipped):** Compressed gas
4.2 **Color:** Colorless
4.3 **Odor:** Sweetish

5. HEALTH HAZARDS

5.1 **Personal Protective Equipment:** Chemical goggles, gloves, self-contained breathing apparatus or organic canister.
5.2 **Symptoms Following Exposure:** May act as an asphyxiant or slight anesthetic at high vapor concentrations. Vapor concentrations are not usually a hazard at room temperature except in enclosed spaces.
5.3 **Treatment for Exposure:** INHALATION: remove victim to fresh air and apply resuscitation. Call a doctor. EYES AND SKIN: flush with water for at least 15 minutes.
5.4 **Toxicity by Inhalation (Threshold Limit Value):** Data not available
5.5 **Short-Term Inhalation Limits:** Data not available
5.6 **Toxicity by Ingestion:** Not pertinent
5.7 **Late Toxicity:** None
5.8 **Vapor (Gas) Irritant Characteristics:** Vapors are non-irritating to the eyes and throat.
5.9 **Liquid or Solid Irritant Characteristics:** No appreciable hazard. Practically harmless to the skin because it is very volatile and evaporates quickly. May cause frostbite.
5.10 **Odor Threshold:** Data not available

6. FIRE HAZARDS

6.1 **Flash Point:** Not pertinent
6.2 **Flammable Limits in Air:** 1.6%–10%
6.3 **Fire Extinguishing Agents:** Stop flow of gas
6.4 **Fire Extinguishing Agents Not to be Used:** Not pertinent
6.5 **Special Hazards of Combustion Products:** Not pertinent
6.6 **Behavior in Fire:** Containers may explode in fire. Vapor is heavier than air and may travel long distance to a source of ignition and flash back.
6.7 **Ignition Temperature:** 725°F
6.8 **Electrical Hazard:** Not pertinent
6.9 **Burning Rate:** 8.8 mm/min.

7. CHEMICAL REACTIVITY

7.1 **Reactivity with Water:** No reaction
7.2 **Reactivity with Common Materials:** No reaction
7.3 **Stability During Transport:** Stable
7.4 **Neutralizing Agents for Acids and Caustics:** Not pertinent
7.5 **Polymerization:** Not pertinent
7.6 **Inhibitor of Polymerization:** Not pertinent

8. WATER POLLUTION

8.1 **Aquatic Toxicity:** None
8.2 **Waterfowl Toxicity:** None
8.3 **Biological Oxygen Demand (BOD):** None
8.4 **Food Chain Concentration Potential:** None

9. SELECTED MANUFACTURERS

1. Air Products and Chemicals, Inc.
   Allentown, Pa. 18105
2. Gulf Oil Corp.
   Gulf Oil Chemicals Co. Division
   Pittsburgh, Pa. 15230
3. Petro-Tex Chemical Corp.
   Houston, Texas 77017

10. SHIPPING INFORMATION

10.1 **Grades or Purity:** Data not available
10.2 **Storage Temperature:** Ambient
10.3 **Inert Atmosphere:** No requirement
10.4 **Venting:** Safety relief

11. HAZARD ASSESSMENT CODE
(See Hazard Assessment Handbook, CG 446-3)

A-B-C-D-E-F-G

12. HAZARD CLASSIFICATIONS

12.1 **Code of Federal Regulations:** Flammable compressed gas
12.2 **NAS Hazard Rating for Bulk Water Transportation:** Not listed
12.3 **NFPA Hazard Classifications:** Not listed

13. PHYSICAL AND CHEMICAL PROPERTIES

13.1 **Physical State at 15°C and 1 atm:** Gas
13.2 **Molecular Weight:** 56.10
13.3 **Boiling Point at 1 atm:** 20.7°F = −6.3°C = 266.9°K
13.4 **Freezing Point:** −297°F = −183°C = 90°K
13.5 **Critical Temperature:** 295.5°F = 146.4°C = 419.6°K
13.6 **Critical Pressure:** 584 psia = 39.7 atm = 4.02 MN/m²
13.7 **Specific Gravity:** 0.595 at 20°C (liquid)
13.8 **Liquid Surface Tension:** 12.5 dynes/cm = 0.0125 N/m at 20°C
13.9 **Liquid-Water Interfacial Tension:** (est.) 68 dynes/cm = 0.068 N/m at 0°C
13.10 **Vapor (Gas) Specific Gravity:** 1.9
13.11 **Ratio of Specific Heats of Vapor (Gas):** 1.104
13.12 **Latent Heat of Vaporization:** 168 Btu/lb = 93.4 cal/g = 3.91×10^5 J/kg
13.13 **Heat of Combustion:** −19,487 Btu/lb = −10,826 cal/g = -453.26×10^5 J/kg
13.14 **Heat of Decomposition:** Not pertinent
13.15 **Heat of Solution:** Not pertinent
13.16 **Heat of Polymerization:** Not pertinent

NOTES

REVISED 1978

BTO	# BUTYLENE OXIDE

Common Synonyms 1,2-Butylene oxide 1,2-Epoxybutane 1-Butene oxide Alpha-butylene oxide	Liquid Colorless Sharp odor Mixes with water.

Shut off ignition sources. Call fire department.
Avoid contact with liquid and vapor. Keep people away.
Wear rubber overclothing (including gloves).
Stop discharge if possible.
Stay upwind. Use water spray to "knock down" vapor.
Isolate and remove discharged material.
Notify local health and pollution control agencies.

Fire	FLAMMABLE. Flashback along vapor trail may occur. Vapor may explode if ignited in an enclosed area. **Extinguish with dry chemicals, foam, or carbon dioxide.** **Water may be ineffective on fire.** **Cool exposed containers with water.**
Exposure	CALL FOR MEDICAL AID. VAPOR Irritating to eyes, nose and throat. If inhaled will cause coughing or difficult breathing. **If in eyes, hold eyelids open and flush with plenty of water.** **If breathing has stopped, give artificial respiration.** **If breathing is difficult, give oxygen.** LIQUID Will burn skin and eyes. If swallowed will cause nausea and vomiting. **Remove contaminated clothing and shoes.** **Flush affected areas with plenty of water.** **IF IN EYES, hold eyelids open and flush with plenty of water.** **IF SWALLOWED and victim is CONSCIOUS, have victim drink water or milk and have victim induce vomiting.** **IF SWALLOWED and victim is UNCONSCIOUS OR HAVING CONVULSIONS, do nothing except keep victim warm.**
Water Pollution	Effect of low concentrations on aquatic life is unknown. May be dangerous if it enters water intakes. **Notify local health and wildlife officials.** **Notify operators of nearby water intakes.**

1. RESPONSE TO DISCHARGE
(See Response Methods Handbook, CG 446-4)

Issue warning - high flammability, air contaminant
Restrict access
Disperse and flush

2. LABEL

Red

3. CHEMICAL DESIGNATIONS

3.1 **Synonyms:** 1-Butene oxide; 1,2-Butylene oxide; *alpha*-Butylene oxide; 1,2-Epoxybutane
3.2 **Coast Guard Compatibility Classification:** Alkylene oxide
3.3 **Chemical Formula:** $C_2H_5CHCH_2O$
3.4 **IMCO/United Nations Numerical Designation:** Not listed

4. OBSERVABLE CHARACTERISTICS

4.1 **Physical State (as shipped):** Liquid
4.2 **Color:** Colorless
4.3 **Odor:** Pungent

5. HEALTH HAZARDS

5.1 **Personal Protective Equipment:** Clean protective clothing; rubber gloves; chemical worker's goggles; self-contained breathing apparatus
5.2 **Symptoms Following Exposure:** Inhalation: intolerable odor and irritation; respiratory injury may occur at higher levels. Ingestion causes irritation of mouth and stomach. Contact with either liquid or vapor may cause burns of eyes. Liquid produces frostbite-type of skin burn if free to evaporate; if confined to skin, burn may cause skin sensitization; not readily absorbed in toxic amounts.
5.3 **Treatment for Exposure:** INHALATION: if any ill effects occur, immediately remove person to fresh air and get medical help; if breathing stops, start artificial respiration. INGESTION: induce vomiting promptly and get medical help. EYES: promptly flush with plenty of water for at least 15 min. and get medical help. SKIN: promptly flush with plenty of water; remove all contaminated clothing and wash before reuse.
5.4 **Toxicity by Inhalation (Threshold Limit Value):** Data not available
5.5 **Short-Term Inhalation Limits:** Data not available
5.6 **Toxicity by Ingestion:** Grade 2; oral LD_{50} = 1,410 mg/kg (rat)
5.7 **Late Toxicity:** Data not available
5.8 **Vapor (Gas) Irritant Characteristics:** Data not available
5.9 **Liquid or Solid Irritant Characteristcs:** Data not available
5.10 **Odor Threshold:** Data not available

6. FIRE HAZARDS

6.1 **Flash Point:** $<-20°F$ O.C.
6.2 **Flammable Limits in Air:** 1.5% - 18.3%
6.3 **Fire Extinguishing Agents:** Dry chemical, alcohol foam, carbon dioxide
6.4 **Fire Extinguishing Agents Not to be Used:** Water may be ineffective.
6.5 **Special Hazards of Combustion Products:**
6.6 **Behavior in Fire:** Containers may explode in fire. Use water to cool container from safe distance.
6.7 **Ignition Temperature:** 959°F
6.8 **Electrical Hazard:** Data not available
6.9 **Burning Rate:** Data not available

7. CHEMICAL REACTIVITY

7.1 **Reactivity with Water:** No reaction
7.2 **Reactivity with Common Materials:**
7.3 **Stability During Transport:** Stable
7.4 **Neutralizing Agents for Acids and Caustics:** Not pertinent
7.5 **Polymerization:** May occur when in contact with strong acids or bases
7.6 **Inhibitor of Polymerization:** Data not available

8. WATER POLLUTION

8.1 **Aquatic Toxicity:** Data not available
8.2 **Waterfowl Toxicity:** Data not available
8.3 **Biological Oxygen Demand (BOD):** Data not available
8.4 **Food Chain Concentration Potential:** None

9. SELECTED MANUFACTURERS

1. Dow Chemical U.S.A.
   Midland, Mich. 48640
2. Aldrich Chemical Co.
   940 W. Saint Paul Ave.
   Milwaukee, Wis. 53233

10. SHIPPING INFORMATION

10.1 **Grades or Purity:** Technical, 99%
10.2 **Storage Temperature:** Ambient
10.3 **Inert Atmosphere:** No requirement
10.4 **Venting:** Pressure-vacuum

11. HAZARD ASSESSMENT CODE
(See Hazard Assessment Handbook, CG 446-3)

A-P-Q-R-S

12. HAZARD CLASSIFICATIONS

12.1 **Code of Federal Regulations:** Flammable liquid
12.2 **NAS Hazard Rating for Bulk Water Transportation:** Not listed
12.3 **NFPA Hazard Classifications:**

Category	Classification
Health Hazard (Blue)	2
Flammability (Red)	3
Reactivity (Yellow)	2

13. PHYSICAL AND CHEMICAL PROPERTIES

13.1 **Physical State at 15°C and 1 atm:** Liquid
13.2 **Molecular Weight:** 72
13.3 **Boiling Point at 1 atm:** 145°F = 63°C = 336°K
13.4 **Freezing Point:** $<-58°F = <-50°C = <223°K$
13.5 **Critical Temperature:** Data not available
13.6 **Critical Pressure:** Data not available
13.7 **Specific Gravity:** 0.826 at 25°C (liquid)
13.8 **Liquid Surface Tension:** Data not available
13.9 **Liquid-Water Interfacial Tension:** Not pertinent
13.10 **Vapor (Gas) Specific Gravity:** 2.49
13.11 **Ratio of Specific Heats of Vapor (Gas):** Data not available
13.12 **Latent Heat of Vaporization:** (est.) 180 Btu/lb = 100 cal/g = 4.2×10^5 J/kg
13.13 **Heat of Combustion:** −15,200 Btu/lb = −8,470 cal/g = -354×10^5 J/kg
13.14 **Heat of Decomposition:** Not pertinent
13.15 **Heat of Solution:** Not pertinent
13.16 **Heat of Polymerization:** Data not available

(Continued on pages 5 and 6)

NOTES

BHP | tert-BUTYL HYDROPEROXIDE

Common Synonyms
Cadox TBH

Watery liquid | Colorless | Odorless

Floats and mixes slowly with water.

Stop discharge if possible.
Keep people away. Evacuate area in case of large discharge.
Shut off ignition sources and call fire department.
Stay upwind and flood spill area with water.
Avoid contact with liquid.
Notify local health and pollution control agencies.

Fire

FLAMMABLE.
May explode if subjected to heat, flame, or shock.
May cause fire and explode on contact with combustibles.
Vapor may explode if ignited in an enclosed area.
Evacuate surrounding area.
Combat fires from safe distance or protected location.
Flood discharge area with water.
Extinguish with dry chemical, foam or carbon dioxide.
Cool exposed containers with water.

Exposure

CALL FOR MEDICAL AID.

VAPOR
Irritating to eyes, nose and throat.
Move to fresh air.
If breathing has stopped, give artificial respiration.
If breathing is difficult, give oxygen.

LIQUID
Irritating to skin and eyes.
Harmful if swallowed.
Remove contaminated clothing and shoes.
Flush affected areas with plenty of water.
IF IN EYES, hold eyelids open and flush with plenty of water.
IF SWALLOWED and victim is CONSCIOUS, have victim drink water or milk, and have victim induce vomiting.
IF SWALLOWED and victim is UNCONSCIOUS OR HAVING CONVULSIONS, do nothing except keep victim warm.

Water Pollution

Effect of low concentrations on aquatic life is unknown.
May be dangerous if it enters water intakes.
Notify local health and pollution control officials.
Notify operators of nearby water intakes.

1. RESPONSE TO DISCHARGE

(See Response Methods Handbook, CG 446-4)

Issue warning—high flammability
Restrict access
Mechanical containment
Chemical and physical treatment

2. LABELS

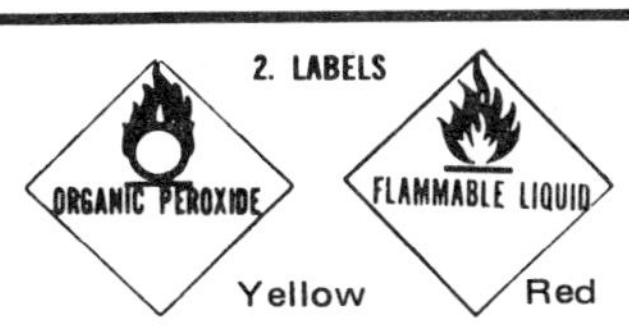

3. CHEMICAL DESIGNATIONS

3.1 **Synonyms:** Cadox TBH
3.2 **Coast Guard Compatibility Classification:** Not applicable
3.3 **Chemical Formula:** $(CH_3)_3C{\cdot}O{\cdot}OH$
3.4 **IMCO/United Nations Numerical Designation:** 5.2/1949

4. OBSERVABLE CHARACTERISTICS

4.1 **Physical State (as shipped):** Liquid
4.2 **Color:** Colorless to pale yellow
4.3 **Odor:** Data not available

5. HEALTH HAZARDS

5.1 **Personal Protective Equipment:** Goggles, well-fitting gloves, barrier creams
5.2 **Symptoms Following Exposure:** Liquid causes severe burns of skin and eyes.
5.3 **Treatment for Exposure:** INGESTION: induce vomiting and follow with gastric lavage. INHALATION: remove individual from contaminated atmosphere; give artificial respiration and oxygen if needed. SKIN, EYE, AND MUCOUS MEMBRANE CONTACT: flood affected tissues with water.
5.4 **Toxicity by Inhalation (Threshold Limit Value):** Data not available
5.5 **Short-Term Inhalation Limits:** Data not available
5.6 **Toxicity by Ingestion:** Grade 3; 50 to 500 mg/kg
5.7 **Late Toxicity:** None
5.8 **Vapor (Gas) Irritant Characteristics:** Vapors cause a slight smarting of the eyes or respiratory system if present in high concentrations. The effect is temporary.
5.9 **Liquid or Solid Irritant Characteristics:** Powerful irritant of skin and eyes.
5.10 **Odor Threshold:** Data not available

6. FIRE HAZARDS

6.1 **Flash Point:** 100°F O.C.
6.2 **Flammable Limits in Air:** Not pertinent
6.3 **Fire Extinguishing Agents:** Dry chemical, foam or carbon dioxide.
6.4 **Fire Extinguishing Agents Not to be Used:** Not pertinent
6.5 **Special Hazards of Combustion Products:** Not pertinent
6.6 **Behavior in Fire:** May explode in fire
6.7 **Ignition Temperature:** Not pertinent
6.8 **Electrical Hazard:** Not pertinent
6.9 **Burning Rate:** Not pertinent

7. CHEMICAL REACTIVITY

7.1 **Reactivity with Water:** No reaction
7.2 **Reactivity with Common Materials:** Reacts vigorously with easily oxidized materials, including wood and some metals.
7.3 **Stability During Transport:** Shock- and heat-sensitive; self-accelerating decomposition at 200°F.
7.4 **Neutralizing Agents for Acids and Caustics:** Not pertinent
7.5 **Polymerization:** Not pertinent
7.6 **Inhibitor of Polymerization:** Not pertinent

8. WATER POLLUTION

8.1 **Aquatic Toxicity:** Data not available
8.2 **Waterfowl Toxicity:** Data not available
8.3 **Biological Oxygen Demand (BOD):** Data not available
8.4 **Food Chain Concentration Potential:** None

9. SELECTED MANUFACTURERS

1. Oxirane Corp.
   120 Alexander St.
   Princeton, N. J. 08540
2. Pennwalt Corp.
   Lucidol Division
   Buffalo, N. Y. 14240
3. Witco Chemical Corp.
   U. S. Peroxygen Division
   Marshall, Texas 75670

10. SHIPPING INFORMATION

10.1 **Grades or Purity:** 70—90%
10.2 **Storage Temperature:** 65—85°F
10.3 **Inert Atmosphere:** Data not available
10.4 **Venting:** Data not available

11. HAZARD ASSESSMENT CODE

(See Hazard Assessment Handbook, CG 446-3)

A-P-Q-Z

12. HAZARD CLASSIFICATIONS

12.1 **Code of Federal Regulations:** Flammable Liquid
12.2 **NAS Hazard Rating for Bulk Water Transportation:**

Category	Rating
Fire	2
Health	
Vapor Irritant	1
Liquid or Solid Irritant	1
Poisons	2
Water Pollution	
Human Toxicity	3
Aquatic Toxicity	
Aesthetic Effect	1
Reactivity	
Other Chemicals	4
Water	0
Self-Reaction	4

12.3 **NFPA Hazard Classifications:**

Category	Classification
Health Hazard (Blue)	1
Flammability (Red)	4
Reactivity (Yellow)	4 oxy

13. PHYSICAL AND CHEMICAL PROPERTIES

13.1 **Physical State at 15°C and 1 atm:** Liquid
13.2 **Molecular Weight:** 90.12
13.3 **Boiling Point at 1 atm:** Decomposes
13.4 **Freezing Point:** $-31°F = -35°C = 238°K$
13.5 **Critical Temperature:** Not pertinent
13.6 **Critical Pressure:** Not pertinent
13.7 **Specific Gravity:** 0.880 at 25°C (liquid)
13.8 **Liquid Surface Tension:** Data not available
13.9 **Liquid-Water Interfacial Tension:** Data not available
13.10 **Vapor (Gas) Specific Gravity:** Not pertinent
13.11 **Ratio of Specific Heats of Vapor (Gas):** Not pertinent
13.12 **Latent Heat of Vaporization:** Not pertinent
13.13 **Heat of Combustion:** (est.) $-13{,}000$ Btu/lb $= -7200$ cal/g $= -300 \times 10^5$ J/kg
13.14 **Heat of Decomposition:** -675 Btu/lb $= -375$ cal/g $= -15.7 \times 10^5$ J/kg
13.15 **Heat of Solution:** Not pertinent
13.16 **Heat of Polymerization:** Not pertinent

(Continued on pages 5 and 6)

NOTES

REVISED 1978

n-BUTYL MERCAPTAN

BTM

Common Synonyms			
1-Butanethiol Thiobutyl alcohol	Liquid	Colorless to yellow	Skunk like odor
	Floats on water. Poisonous vapor is produced.		

AVOID CONTACT WITH LIQUID AND VAPOR. KEEP PEOPLE AWAY.
Wear goggles and self-contained breathing apparatus.
Shut off ignition sources. Call fire department.
Stop discharge if possible.
Evacuate area in case of large discharge.
Isolate and remove discharged material.
Notify local health and pollution control agencies.

Fire

FLAMMABLE.
Irritating gases may be produced when heated.
Containers may explode in fire.
Flashback along vapor trail may occur.
Vapor may explode if ignited in an enclosed area.

Extinguish with dry chemicals, alcohol foam, or carbon dioxide.
Water may be ineffective on fire.
Cool exposed containers with water.

Exposure

CALL FOR MEDICAL AID.

VAPOR
POISONOUS IF INHALED.
Irritating to eyes.

Move victim to fresh air.
If breathing has stopped, give artificial respiration.
If breathing is difficult, give oxygen.

LIQUID
Irritating to skin and eyes.
Harmful if swallowed.

Remove contaminated clothing and shoes.
Flush affected areas with plenty of water.
IF IN EYES, hold eyelids open and flush with plenty of water.
IF SWALLOWED and victim is CONSCIOUS, have victim drink water or milk and have victim induce vomiting.
IF SWALLOWED and victim is UNCONSCIOUS OR HAVING CONVULSIONS, do nothing except keep victim warm.

Water Pollution

HARMFUL TO AQUATIC LIFE IN VERY LOW CONCENTRATIONS.
Fouling to shoreline.
May be dangerous if it enters water intakes.

Notify local health and wildlife officials.
Notify operators of nearby water intakes.

1. RESPONSE TO DISCHARGE

(See Response Methods Handbook, CG 446-4)

Issue warning—high flammability
Restrict access
Mechanical containment
Should be removed
Chemical and physical treatment

2. LABEL

Red

3. CHEMICAL DESIGNATIONS

3.1 **Synonyms:** 1-Butanethiol
Thiobutyl alcohol

3.2 **Coast Guard Compatibility Classification:** Not applicable

3.3 **Chemical Formula:** $CH_3CH_2CH_2CH_2SH$

3.4 **IMCO/United Nations Numerical Designation:** 3.1/1228

4. OBSERVABLE CHARACTERISTICS

4.1 **Physical State (as shipped):** Liquid

4.2 **Color:** Colorless to yellow

4.3 **Odor:** Strong skunk-like

5. HEALTH HAZARDS

5.1 **Personal Protective Equipment:** Plastic gloves, goggles; self-contained breathing apparatus.

5.2 **Symptoms Following Exposure:** Inhalation causes loss of sense of smell; muscular weakness, convulsions, and respiratory paralysis may follow prolonged exposure. Contact of liquid with eyes or skin causes slight irritation. Ingestion causes nausea.

5.3 **Treatment for Exposure:** INHALATION: remove victim from contaminated atmosphere; give artificial respiration and oxygen if needed; observe for signs of pulmonary edema. EYES: wash with plenty of water; see a physician. SKIN: wash with soap and water. INGESTION: induce vomiting and follow with gastric lavage.

5.4 **Toxicity by Inhalation (Threshold Limit Value):** 0.5 ppm

5.5 **Short-Term Inhalation Limits:** Data not available

5.6 **Toxicity by Ingestion:** Grade 2; oral LD_{50} = 1,500 mg/kg (rat)

5.7 **Late Toxicity:** Data not available

5.8 **Vapor (Gas) Irritant Characteristics:** Data not available

5.9 **Liquid or Solid Irritant Characteristics:** Data not available

5.10 **Odor Threshold:** 0.001 ppm

6. FIRE HAZARDS

6.1 **Flash Point:** 55°F O.C.

6.2 **Flammable Limits in Air:** Data not available

6.3 **Fire Extinguishing Agents:** Dry chemical, alcohol foam, carbon dioxide

6.4 **Fire Extinguishing Agents Not to be Used:** Water

6.5 **Special Hazards of Combustion Products:** Irritating sulfur dioxide may form.

6.6 **Behavior in Fire:** Vapors are heavier than air and may travel long distance to a source of ignition and flash back.

6.7 **Ignition Temperature:** Data not available

6.8 **Electrical Hazard:** Data not available

6.9 **Burning Rate:** 7.4 mm/min.

7. CHEMICAL REACTIVITY

7.1 **Reactivity with Water:** No reaction

7.2 **Reactivity with Common Materials:** No reaction

7.3 **Stability During Transport:** Stable

7.4 **Neutralizing Agents for Acids and Caustics:** Not pertinent

7.5 **Polymerization:** Not pertinent

7.6 **Inhibitor of Polymerization:** Not pertinent

8. WATER POLLUTION

8.1 **Aquatic Toxicity:**
7.3 mg/l /24 hr/fish/TL_m/fresh water

8.2 **Waterfowl Toxicity:** Data not available

8.3 **Biological Oxygen Demand (BOD):** Data not available

8.4 **Food Chain Concentration Potential:** None

9. SELECTED MANUFACTURERS

1. Pennwalt Corporation
Industrial Chemicals Department
Three Parkway
Philadelphia, Pa. 19102

2. Phillips Petroleum Company
Chemical Department
Special Products Division
Phillips Building
Bartlesville, Okla. 74004

3. Eastman Kodak Co.
Eastman Organic Chemicals
Rochester, N. Y. 14650

10. SHIPPING INFORMATION

10.1 **Grades or Purity:** 98+%

10.2 **Storage Temperature:** Ambient

10.3 **Inert Atmosphere:** No requirement

10.4 **Venting:** Pressure-vacuum

11. HAZARD ASSESSMENT CODE

(See Hazard Assessment Handbook, CG 446-3)

A-T-U-V-W

12. HAZARD CLASSIFICATIONS

12.1 **Code of Federal Regulations:** Flammable liquid

12.2 **NAS Hazard Rating for Bulk Water Transportation:** Not listed

12.3 **NFPA Hazard Classifications:**

Category	Classification
Health Hazard (Blue)	2
Flammability (Red)	3
Reactivity (Yellow)	0

13. PHYSICAL AND CHEMICAL PROPERTIES

13.1 **Physical State at 15°C and 1 atm:** Liquid

13.2 **Molecular Weight:** 90.2

13.3 **Boiling Point at 1 atm:**
209.3°F = 98.5°C = 371.7°K

13.4 **Freezing Point:**
−176.2°F = −115.7°C = 157.5°K

13.5 **Critical Temperature:**
554°F = 290°C = 563°K

13.6 **Critical Pressure:**
572 psia = 38.9 atm = 3.94 MN/m^2

13.7 **Specific Gravity:** 0.841 at 20°C (liquid)

13.8 **Liquid Surface Tension:**
26.1 dynes/cm = 0.0261 N/m at 20°C

13.9 **Liquid-Water Interfacial Tension:** (est.)
30 dynes/cm = 0.030 N/m at 20°C

13.10 **Vapor (Gas) Specific Gravity:** 3.1

13.11 **Ratio of Specific Heats of Vapor (Gas):**
1.0770 at 16°C

13.12 **Latent Heat of Vaporization:** 154.0 Btu/lb
= 85.58 cal/g = 3.583 × 10^5 J/kg

13.13 **Heat of Combustion:** −16,601 Btu/lb
= −9,223 cal/g = −386 × 10^5 J/kg

13.14 **Heat of Decomposition:** Not pertinent

13.15 **Heat of Solution:** Not pertinent

13.16 **Heat of Polymerization:** Not pertinent

NOTES

BMN — n-BUTYL METHACRYLATE

Common Synonyms			
Methacrylic acid, butyl ester Butyl methacrylate Butyl 2-methacrylate n-Butyl alpha-methylacrylate	Liquid Floats on water.	Colorless	Mild odor

Stop discharge if possible. Keep people away.
Call fire department.
Avoid contact with liquid.
Isolate and remove discharged material.
Notify local health and pollution control agencies.

Fire	Combustible. Containers may explode in fire. **Extinguish with dry chemicals, foam or carbon dioxide.** **Water may be ineffective on fire.** **Cool exposed containers with water.**
Exposure	**Call for medical aid.** LIQUID Irritating to skin and eyes. Harmful if swallowed. **Remove contaminated clothing and shoes.** **Flush affected areas with plenty of water.** **IF IN EYES, hold eyelids open and flush with plenty of water.** **IF SWALLOWED and victim is CONSCIOUS, have victim drink water or milk and have victim induce vomiting.** **IF SWALLOWED and victim is UNCONSCIOUS OR HAVING CONVULSIONS, do nothing except keep victim warm.**
Water Pollution	Effect of low concentrations on aquatic life is unknown. Fouling to shoreline. May be dangerous if it enters water intakes. **Notify local health and wildlife officials.** **Notify operators of nearby water intakes.**

1. RESPONSE TO DISCHARGE
(See Response Methods Handbook, CG 446-4)

Mechanical containment
Should be removed
Chemical and physical treatment

2. LABELS

No label required by Code of Federal Regulations

3. CHEMICAL DESIGNATIONS

3.1 **Synonyms:** Methacrylic acid, butyl ester; Butyl methacrylate; Butyl 2-methacrylate; n-Butyl alpha-methyl acrylate; Butyl 2-methyl-2-propenoate
3.2 **Coast Guard Compatibility Classification:** Acrylate
3.3 **Chemical Formula:** $CH_2{=}C(CH_3)COOCH_2CH_2CH_2CH_3$
3.4 **IMCO/United Nations Numerical Designation:** Not listed

4. OBSERVABLE CHARACTERISTICS

4.1 **Physical State (as shipped):** Liquid
4.2 **Color:** Colorless
4.3 **Odor:** Moderate acrylate

5. HEALTH HAZARDS

5.1 **Personal Protective Equipment:** Self-contained respirator; impervious gloves; chemical splash goggles.
5.2 **Symptoms Following Exposure:** Inhalation may cause nausea because of offensive odor. Contact with liquid causes irritation of eyes and mild irritation of skin. Ingestion causes irritation of mouth and stomach.
5.3 **Treatment for Exposure:** INHALATION: remove to fresh air; give oxygen or artificial respiration as required. EYES: flush with copious amounts of water for 15 min. and consult physician. SKIN: wash with soap and water. INGESTION: induce vomiting; call a physician.
5.4 **Toxicity by Inhalation (Threshold Limit Value):** Data not available
5.5 **Short-Term Inhalation Limits:** Data not available
5.6 **Toxicity by Ingestion:** Grade 0; LD_{50}> 15 g/kg
5.7 **Late Toxicity:** Birth defects in rats (gross and skeletal abnormalities)
5.8 **Vapor (Gas) Irritant Characteristics:** Data not available
5.9 **Liquid or Solid Irritant Characteristics:** Data not available
5.10 **Odor Threshold:** Data not available

6. FIRE HAZARDS

6.1 **Flash Point:** 150°F O.C.
6.2 **Flammable Limits in Air:** (est.) 2%–8%
6.3 **Fire Extinguishing Agents:** Dry chemical, foam, carbon dioxide
6.4 **Fire Extinguishing Agents Not to be Used:** Water may be ineffective.
6.5 **Special Hazards of Combustion Products:** Not pertinent
6.6 **Behavior in Fire:** Containers may explode.
6.7 **Ignition Temperature:** 562°F
6.8 **Electrical Hazard:** Data not available
6.9 **Burning Rate:** 4.8 mm/min.

7. CHEMICAL REACTIVITY

7.1 **Reactivity with Water:** No reaction
7.2 **Reactivity with Common Materials:** No reaction
7.3 **Stability During Transport:** Stable
7.4 **Neutralizing Agents for Acids and Caustics:** Not pertinent
7.5 **Polymerization:** May occur when heated.
7.6 **Inhibitor of Polymerization:** 9-15 ppm monomethyl ether of hydroquinone; 90-120 ppm hydroquinone

8. WATER POLLUTION

8.1 **Aquatic Toxicity:** Data not available
8.2 **Waterfowl Toxicity:** Data not available
8.3 **Biological Oxygen Demand (BOD):** Data not available
8.4 **Food Chain Concentration Potential:** None

9. SELECTED MANUFACTURERS

1. Rohm and Haas Co.
   Independence Mall West
   Philadelphia, Pa. 19105
2. E. I. du Pont de Nemours & Co.
   Plastics Department
   1007 Market St.
   Wilmington, Del. 19898

10. SHIPPING INFORMATION

10.1 **Grades or Purity:** 98.5+%
10.2 **Storage Temperature:** Ambient
10.3 **Inert Atmosphere:** No requirement
10.4 **Venting:** Pressure-vacuum

11. HAZARD ASSESSMENT CODE
(See Hazard Assessment Handbook, CG 446-3)

A-T-U

12. HAZARD CLASSIFICATIONS

12.1 **Code of Federal Regulations:** Not listed
12.2 **NAS Hazard Rating for Bulk Water Transportation:** Not listed
12.3 **NFPA Hazard Classifications:**

Category	Classification
Health Hazard (Blue)	2
Flammability (Red)	2
Reactivity (Yellow)	0

13. PHYSICAL AND CHEMICAL PROPERTIES

13.1 **Physical State at 15°C and 1 atm:** Liquid
13.2 **Molecular Weight:** 142.2
13.3 **Boiling Point at 1 atm:** 325°F = 163°C = 436°K
13.4 **Freezing Point:** <32°F = <0°C = <273°F
13.5 **Critical Temperature:** Not pertinent
13.6 **Critical Pressure:** Not pertinent
13.7 **Specific Gravity:** 0.8975 at 20°C (liquid)
13.8 **Liquid Surface Tension:** (est.) 30 dynes/cm = 0.030 N/m at 20°C
13.9 **Liquid-Water Interfacial Tension:** (est.) 35 dynes/cm = 0.035 N/m at 20°C
13.10 **Vapor (Gas) Specific Gravity:** Not pertinent
13.11 **Ratio of Specific Heats of Vapor (Gas):** Not pertinent
13.12 **Latent Heat of Vaporization:** Not pertinent
13.13 **Heat of Combustion:** (est.) −14,800 Btu/lb = −8,230 cal/g = -344×10^5 J/kg
13.14 **Heat of Decomposition:** Not pertinent
13.15 **Heat of Solution:** Not pertinent
13.16 **Heat of Polymerization:** −180 Btu/lb = −100 cal/g = -4.2×10^5 J/kg

NOTES

BTP p-tert-BUTYLPHENOL

Common Synonyms	Solid · White · Disinfectant-like odor May float or sink in water.

Avoid contact with solid and dust. Keep people away.
Wear rubber overclothing (including gloves).
Stop discharge if possible.
Call fire department.
Isolate and remove discharged material.
Notify local health and pollution control agencies.

Fire	Combustible. Extinguish with dry chemicals, foam, or carbon dioxide. Water may be ineffective on fire.
Exposure	CALL FOR MEDICAL AID. DUST Irritating to eyes, nose and throat. If inhaled will cause difficult breathing. If in eyes, hold eyelids open and flush with plenty of water. If breathing has stopped, give artificial respiration. If breathing is difficult, give oxygen. SOLID Will burn skin and eyes. If swallowed will cause nausea and vomiting. Remove contaminated clothing and shoes. Flush affected areas with plenty of water. IF IN EYES, hold eyelids open and flush with plenty of water. IF SWALLOWED and victim is CONSCIOUS, have victim drink water or milk and have victim induce vomiting. IF SWALLOWED and victim is UNCONSCIOUS OR HAVING CONVULSIONS, do nothing except keep victim warm.
Water Pollution	Effect of low concentrations on aquatic life is unknown. May be dangerous if it enters water intakes. Notify local health and wildlife officials. Notify operators of nearby water intakes.

1. RESPONSE TO DISCHARGE
(See Response Methods Handbook, CG 446-4)

Mechanical containment
Should be removed
Chemical and physical treatment

2. LABELS

No hazard label required by Code of Federal Regulations

3. CHEMICAL DESIGNATIONS

3.1 **Synonyms:** No common synonyms
3.2 **Coast Guard Compatibility Classification:** Not listed
3.3 **Chemical Formula:** 1,4$-(CH_3)_3CC_6H_4OH$
3.4 **IMCO/United Nations Numerical Designation:** Not listed

4. OBSERVABLE CHARACTERISTICS

4.1 **Physical State (as shipped):** Solid
4.2 **Color:** White
4.3 **Odor:** Aromatic

5. HEALTH HAZARDS

5.1 **Personal Protective Equipment:** Chemical workers' goggles; clean, body-protecting clothing
5.2 **Symptoms Following Exposure:** Inhalation of vapors causes irritation of respiratory system. Ingestion causes irritation of mouth and stomach. Contact with eyes causes burns. Contact with dry skin produces no significant irritation, but wet skin is subject to moderate irritation, even a mild burn.
5.3 **Treatment for Exposure:** INHALATION: move to fresh air; begin artificial respiration if breathing has ceased. INGESTION: force milk or water and then immediately induce vomiting; treat symptomatically. EYES: immediately flush eyes with plenty of water for at least 15 min., get medical attention promptly. SKIN: flush with plenty of water; remove grossly contaminated clothing.
5.4 **Toxicity by Inhalation (Threshold Limit Value):** Data not available
5.5 **Short-Term Inhalation Limits:** Data not available
5.6 **Toxicity by Ingestion:** Grade 2; oral LD_{50} = 3,250 mg/kg (rat)
5.7 **Late Toxicity:** Data not available
5.8 **Vapor (Gas) Irritant Characteristics:** Data not available
5.9 **Liquid or Solid Irritant Characteristics:** Data not available
5.10 **Odor Threshold:** Data not available

6. FIRE HAZARDS

6.1 **Flash Point:** 235°F C.C. (liquid)
6.2 **Flammable Limits in Air:** Not pertinent
6.3 **Fire Extinguishing Agents:** Dry chemical, foam, carbon dioxide
6.4 **Fire Extinguishing Agents Not to be Used:** Water may be ineffective
6.5 **Special Hazards of Combustion Products:**
6.6 **Behavior in Fire:**
6.7 **Ignition Temperature:** Not pertinent
6.8 **Electrical Hazard:** Not pertinent
6.9 **Burning Rate:** Not pertinent

7. CHEMICAL REACTIVITY

7.1 **Reactivity with Water:** No reaction
7.2 **Reactivity with Common Materials:**
7.3 **Stability During Transport:** Stable
7.4 **Neutralizing Agents for Acids and Caustics:** Not pertinent
7.5 **Polymerization:** Not pertinent
7.6 **Inhibitor of Polymerization:** Not pertinent

8. WATER POLLUTION

8.1 **Aquatic Toxicity:** Data not available
8.2 **Waterfowl Toxicity:** Data not available
8.3 **Biological Oxygen Demand (BOD):** Data not available
8.4 **Food Chain Concentration Potential:** None

9. SELECTED MANUFACTURERS

1. The Dow Chemical Co.
2030 Dow Center
Midland, Mich. 48640
2. Aldrich Chemical Co.
940 W. Saint Paul Ave.
Milwaukee, Wis. 53233
3. Eastman Organic Chemicals
Rochester, N.Y. 14650

10. SHIPPING INFORMATION

10.1 **Grades or Purity:** Technical, 98.5%
10.2 **Storage Temperature:** Ambient
10.3 **Inert Atmosphere:** No requirement
10.4 **Venting:** Open

11. HAZARD ASSESSMENT CODE
(See Hazard Assessment Handbook, CG 446-3)

II

12. HAZARD CLASSIFICATIONS

12.1 **Code of Federal Regulations:** Not listed
12.2 **NAS Hazard Rating for Bulk Water Transportation:** Not listed
12.3 **NFPA Hazard Classifications:** Not listed

13. PHYSICAL AND CHEMICAL PROPERTIES

13.1 **Physical State at 15°C and 1 atm:** Solid
13.2 **Molecular Weight:** 150
13.3 **Boiling Point at 1 atm:** 463.1°F = 239.5°C = 512.7°K
13.4 **Freezing Point:** 210°F = 99°C = 372°K
13.5 **Critical Temperature:** Not pertinent
13.6 **Critical Pressure:** Not pertinent
13.7 **Specific Gravity:** 1.037 at 25°C (solid)
13.8 **Liquid Surface Tension:** Not pertinent
13.9 **Liquid-Water Interfacial Tension:** Not pertinent
13.10 **Vapor (Gas) Specific Gravity:** Not pertinent
13.11 **Ratio of Specific Heats of Vapor (Gas):** Not pertinent
13.12 **Latent Heat of Vaporization:** Not pertinent
13.13 **Heat of Combustion:** (est.) $-16,900$ Btu/lb = $-9,410$ cal/g = -394×10^5 J/kg
13.14 **Heat of Decomposition:** Not pertinent
13.15 **Heat of Solution:** Not pertinent
13.16 **Heat of Polymerization:** Not pertinent

NOTES

BCS — BUTYLTRICHLOROSILANE

Common Synonyms			
n-Butyltrichlorosilane	Liquid	Colorless	Sharp, irritating odor
	Reacts violently with water. Irritating visible vapor cloud is produced.		

AVOID CONTACT WITH LIQUID AND VAPOR. KEEP PEOPLE AWAY.
Wear goggles, self-contained breathing apparatus and rubber overclothing (including gloves).
Stop discharge if possible.
Call fire department.
Isolate and remove discharged material.
Notify local health and pollution control agencies.

Fire

Combustible.
POISONOUS GASES MAY BE PRODUCED IN FIRE.
Extinguish with dry chemicals or carbon dioxide.
DO NOT USE WATER OR FOAM ON FIRE.

Exposure

Call for medical aid.

VAPOR
Irritating to eyes, nose and throat.
Harmful if inhaled.
Move victim to fresh air.
If breathing is difficult, give oxygen.

LIQUID
Will burn skin and eyes.
Harmful if swallowed.
Remove contaminated clothing and shoes.
Flush affected areas with plenty of water.
IF IN EYES, hold eyelids open and flush with plenty of water.
IF SWALLOWED and victim is CONSCIOUS, have victim drink water or milk.
DO NOT INDUCE VOMITING.

Water Pollution

Effect of low concentrations on aquatic life is unknown.
May be dangerous if it enters water intakes.
Notify local health and wildlife officials.
Notify operators of nearby water intakes.

1. RESPONSE TO DISCHARGE
(See Response Methods Handbook, CG 446-4)

Issue warning—corrosive
Restrict access
Disperse and flush

2. LABEL

3. CHEMICAL DESIGNATIONS
3.1 **Synonyms:** n-Butyltrichlorosilane
3.2 **Coast Guard Compatibility Classification:** Not applicable
3.3 **Chemical Formula:** $CH_3CH_2CH_2CH_2SiCl_3$
3.4 **IMCO/United Nations Numerical Designation:** 8/1747

4. OBSERVABLE CHARACTERISTICS
4.1 **Physical State (as shipped):** Liquid
4.2 **Color:** Colorless
4.3 **Odor:** Sharp, hydrochloric acid-like; pungent and irritating

5. HEALTH HAZARDS
5.1 **Personal Protective Equipment:** Acid-vapor-type respiratory protection; rubber gloves, chemical worker's goggles, and other protective equipment as necessary to protect skin and eyes.
5.2 **Symptoms Following Exposure:** Inhalation of vapor irritates upper respiratory system. Contact of liquid with eyes or skin causes severe burns. Ingestion causes burns of mouth and stomach.
5.3 **Treatment for Exposure:** Seek medical attention after all exposures to this compound. INHALATION: remove to fresh air; apply artificial respiration if required. EYES: flush with water for 15 min. SKIN: flush with water. INGESTION: do NOT induce vomiting; give large amounts of water.
5.4 **Toxicity by Inhalation (Threshold Limit Value):** Data not available
5.5 **Short-Term Inhalation Limits:** Data not available
5.6 **Toxicity by Ingestion:** Data not available
5.7 **Late Toxicity:** Data not available
5.8 **Vapor (Gas) Irritant Characteristics:** Data not available
5.9 **Liquid or Solid Irritant Characteristics:** Data not available
5.10 **Odor Threshold:** Data not available

6. FIRE HAZARDS
6.1 **Flash Point:** 130°F O.C.; 126°F C.C.
6.2 **Flammable Limits in Air:** Data not available
6.3 **Fire Extinguishing Agents:** Dry chemical, carbon dioxide
6.4 **Fire Extinguishing Agents Not to be Used:** Water, foam
6.5 **Special Hazards of Combustion Products:** Hydrogen chloride, chlorine, or phosgene may be formed.
6.6 **Behavior in Fire:** Difficult to extinguish. Re-ignition may occur.
6.7 **Ignition Temperature:** Data not available
6.8 **Electrical Hazard:** Data not available
6.9 **Burning Rate:** 2.2 mm/min.

7. CHEMICAL REACTIVITY
7.1 **Reactivity with Water:** Reacts vigorously to generate hydrogen chloride (hydrochloric acid).
7.2 **Reactivity with Common Materials:** Will react with common metals to evolve hydrogen chloride and cause severe corrosion.
7.3 **Stability During Transport:** Stable
7.4 **Neutralizing Agents for Acids and Caustics:** Flush with water, rinse with sodium bicarbonate or lime solution.
7.5 **Polymerization:** Not pertinent
7.6 **Inhibitor of Polymerization:** Not pertinent

8. WATER POLLUTION
8.1 **Aquatic Toxicity:** Data not available
8.2 **Waterfowl Toxicity:** Data not available
8.3 **Biological Oxygen Demand (BOD):** Data not available
8.4 **Food Chain Concentration Potential:** None

9. SELECTED MANUFACTURERS
1. Dow Corning Corporation
P. O. Box 592
Midland, Mich. 48640
2. Dynamit-Nobel of America, Inc.
105 Stonehurst Court
Northvale, N. J. 07647
3. PCR, Inc.
P. O. Box 1466
Gainesville, Fla. 32602

10. SHIPPING INFORMATION
10.1 **Grades or Purity:** 95+%
10.2 **Storage Temperature:** Ambient
10.3 **Inert Atmosphere:** No requirement
10.4 **Venting:** Pressure-vacuum

11. HAZARD ASSESSMENT CODE
(See Hazard Assessment Handbook, CG 446-3)

A-O

12. HAZARD CLASSIFICATIONS
12.1 **Code of Federal Regulations:** Corrosive
12.2 **NAS Hazard Rating for Bulk Water Transportation:** Not listed
12.3 **NFPA Hazard Classifications:**

Category	Classification
Health Hazard (Blue)	2
Flammability (Red)	2
Reactivity (Yellow)	0

13. PHYSICAL AND CHEMICAL PROPERTIES
13.1 **Physical State at 15°C and 1 atm:** Liquid
13.2 **Molecular Weight:** 191.5
13.3 **Boiling Point at 1 atm:** 288°F = 142°C = 415°K
13.4 **Freezing Point:** Not pertinent
13.5 **Critical Temperature:** Not pertinent
13.6 **Critical Pressure:** Not pertinent
13.7 **Specific Gravity:** 1.16 at 20°C (liquid)
13.8 **Liquid Surface Tension:** (est.) 25 dynes/cm = 0.025 N/m at 20°C
13.9 **Liquid-Water Interfacial Tension:** Not pertinent
13.10 **Vapor (Gas) Specific Gravity:** 6.5
13.11 **Ratio of Specific Heats of Vapor (Gas):** Not pertinent
13.12 **Latent Heat of Vaporization:** (est.) 81 Btu/lb = 45 cal/g = 1.9×10^5 J/kg
13.13 **Heat of Combustion:** (est.) −4,300 Btu/lb = −2,400 cal/g = -100×10^5 J/kg
13.14 **Heat of Decomposition:** Not pertinent
13.15 **Heat of Solution:** Data not available
13.16 **Heat of Polymerization:** Not pertinent

NOTES

BTD	1,4-BUTYNEDIOL

Common Synonyms		
2-Butyne-1,4-diol 1,4-Dihydroxy-2-butyne	Solid crystals, or watery solution	Solid is white to light yellow Solution is brownish yellow
	Sinks and mixes with water.	

Stop discharge if possible.
Call fire department.
Avoid contact with liquid and solid.
Isolate and remove discharged material.
Notify local health and pollution control agencies.

Fire	Combustible Extinguish with water, alcohol foam, dry chemical or carbon dioxide.
Exposure	CALL FOR MEDICAL AID. LIQUID OR SOLID Irritating to skin and eyes. Harmful if swallowed. Remove contaminated clothing and shoes. Flush affected areas with plenty of water. IF IN EYES, hold eyelids open and flush with plenty of water. IF SWALLOWED and victim is CONSCIOUS, have victim drink water or milk.
Water Pollution	Effect of low concentrations on aquatic life is unknown. May be dangerous if it enters water intakes. Notify local health and pollution control officials. Notify operators of nearby water intakes.

1. RESPONSE TO DISCHARGE

(See Response Methods Handbook, CG 446-4)

Disperse and flush

2. LABELS

No hazard label required by Code of Federal Regulations

3. CHEMICAL DESIGNATIONS

3.1 **Synonyms:** 2-Butyne-1, 4-diol; 1, 4-Dihydroxy-2-butyne

3.2 **Coast Guard Compatibility Classification:** Not applicable

3.3 **Chemical Formula:** $HOCH_2C{\equiv}CCH_2OH$

3.4 **IMCO/United Nations Numerical Designation:** Not listed

4. OBSERVABLE CHARACTERISTICS

4.1 **Physical State (as shipped):** Solid or 35% aqueous solution.

4.2 **Color:** Solid: colorless to pale yellow. Solution:straw to amber color

4.3 **Odor:** Data not available

5. HEALTH HAZARDS

5.1 **Personal Protective Equipment:** Neoprene rubber gloves and safety goggles or face shield.

5.2 **Symptoms Following Exposure:** May cause dermatitis.

5.3 **Treatment for Exposure:** SKIN CONTACT: wash affected skin area thoroughly with water. EYE CONTACT: immediately wash with water for at least 15 minutes and get medical attention.

5.4 **Toxicity by Inhalation (Threshold Limit Value):** Not pertinent

5.5 **Short-Term Inhalation Limits:** Not pertinent

5.6 **Toxicity by Ingestion:** Grade 3; LD_{50} 50 to 500 mg/kg

5.7 **Late Toxicity:** Data not available

5.8 **Vapor (Gas) Irritant Characteristics:** Not pertinent

5.9 **Liquid or Solid Irritant Characteristics:** Minimum hazard. If spilled on clothing and allowed to remain, may cause smarting and reddening of the skin.

5.10 **Odor Threshold:** Not pertinent

6. FIRE HAZARDS

6.1 **Flash Point:** 263° F O.C. (pure butynediol)

6.2 **Flammable Limits in Air:** Not pertinent

6.3 **Fire Extinguishing Agents:** Water, alcohol foam, dry chemical, or carbon dioxide

6.4 **Fire Extinguishing Agents Not to be Used:** Not pertinent

6.5 **Special Hazards of Combustion Products:** Not pertinent

6.6 **Behavior in Fire:** Not pertinent

6.7 **Ignition Temperature:** Data not available

6.8 **Electrical Hazard:** Not pertinent

6.9 **Burning Rate:** Data not available

7. CHEMICAL REACTIVITY

7.1 **Reactivity with Water:** No reaction

7.2 **Reactivity with Common Materials:** No reaction

7.3 **Stability During Transport:** Stable

7.4 **Neutralizing Agents for Acids and Caustics:** Not pertinent

7.5 **Polymerization:** Not pertinent

7.6 **Inhibitor of Polymerization:** Not pertinent

8. WATER POLLUTION

8.1 **Aquatic Toxicity:** Data not available

8.2 **Waterfowl Toxicity:** Data not available

8.3 **Biological Oxygen Demand (BOD):** Data not available

8.4 **Food Chain Concentration Potential:** None

9. SELECTED MANUFACTURERS

GAF Corporation
Chemical Division
Calvert City, Kentucky 42029

10. SHIPPING INFORMATION

10.1 **Grades or Purity:** Tech. Flake: 96%; 35% solution

10.2 **Storage Temperature:** Data not available

10.3 **Inert Atmosphere:** Data not available

10.4 **Venting:** Data not available

11. HAZARD ASSESSMENT CODE

(See Hazard Assessment Handbook, CG 446-3)

SS

12. HAZARD CLASSIFICATIONS

12.1 **Code of Federal Regulations:** Not listed

12.2 **NAS Hazard Rating for Bulk Water Transportation:** Not listed

12.3 **NFPA Hazard Classifications:** Not listed

13. PHYSICAL AND CHEMICAL PROPERTIES

13.1 **Physical State at 15°C and 1 atm:** Solid

13.2 **Molecular Weight:** 36.09

13.3 **Boiling Point at 1 atm:** 460°F = 238°C = 511°K

13.4 **Freezing Point:** 140°F = 58°C = 331°K

13.5 **Critical Temperature:** Not pertinent

13.6 **Critical Pressure:** Not pertinent

13.7 **Specific Gravity:** 1.07 at 20°C (solid)

13.8 **Liquid Surface Tension:** Not pertinent

13.9 **Liquid-Water Interfacial Tension:** Not pertinent

13.10 **Vapor (Gas) Specific Gravity:** Not pertinent

13.11 **Ratio of Specific Heats of Vapor (Gas):** Not pertinent

13.12 **Latent Heat of Vaporization:** Not pertinent

13.13 **Heat of Combustion:** −11,020 Btu/lb = −6120 cal/g = -256.2×10^5 J/kg

13.14 **Heat of Decomposition:** Not pertinent

13.15 **Heat of Solution:** Not pertinent

13.16 **Heat of Polymerization:** Not pertinent

NOTES

BAD	iso-BUTYRALDEHYDE

Common Synonyms	
2-Methylpropanal Isobutyric aldehyde Isobutyraldehyde Isobutylaldehyde	Watery liquid Colorless Pleasant gasoline-like odor Floats and mixes slowly with water. Flammable, irritating vapor is produced.

Avoid contact with liquid and vapor. Keep people away.
Wear goggles and self-contained breathing apparatus.
Shut off ignition sources and call fire department.
Stop discharge if possible.
Stay upwind and use water spray to "knock down" vapor.
Isolate and remove discharged material.
Notify local health and pollution control agencies.

Fire	FLAMMABLE. Flashback along vapor trail may occur. Vapor may explode if ignited in an enclosed area. **Wear goggles and self-contained breathing apparatus.** **Extinguish with dry chemical, foam or carbon dioxide.** **Water may be ineffective on fire.** **Cool exposed containers with water.**
Exposure	CALL FOR MEDICAL AID. VAPOR Irritating to eyes, nose and throat. If inhaled, will cause difficult breathing. **Move to fresh air.** **If breathing has stopped, give artificial respiration.** **If breathing is difficult, give oxygen.** LIQUID Irritating to skin. Will burn eyes. Harmful if swallowed. **Remove contaminated clothing and shoes.** **Flush affected areas with plenty of water.** **IF IN EYES, hold eyelids open and flush with plenty of water.** **IF SWALLOWED and victim is CONSCIOUS, have victim drink water or milk.**
Water Pollution	Effect of low concentrations on aquatic life is unknown. May be dangerous if it enters water intakes. **Notify local health and pollution control officials.** **Notify operators of nearby water intakes.**

1. RESPONSE TO DISCHARGE

(See Response Methods Handbook, CG 446-4)

Issue warning – high flammability
Restrict access
Disperse and flush

2. LABEL

Red

3. CHEMICAL DESIGNATIONS

3.1 **Synonyms:**
Isobutyric aldehyde
Isobutyraldehyde
Isobutylaldehyde
2-Methylpropanal

3.2 **Coast Guard Compatibility Classification:** Aldehyde

3.3 **Chemical Formula:** $(CH_3)_2CHCHO$

3.4 **IMCO/United Nations Numerical Designation:** 3.2/2045

4. OBSERVABLE CHARACTERISTICS

4.1 **Physical State (as shipped):** Liquid

4.2 **Color:** Colorless

4.3 **Odor:** Pungent

5. HEALTH HAZARDS

5.1 **Personal Protective Equipment:** Appropriate protective clothing, including rubber gloves, rubber shoes and protective eyewear.

5.2 **Symptoms Following Exposure:** Vapor is irritating to the eyes and mucous membranes.

5.3 **Treatment for Exposure:** EYES: immediately flush with plenty of water for at least 15 min.

5.4 **Toxicity by Inhalation (Threshold Limit Value):** Data not available

5.5 **Short-Term Inhalation Limits:** Data not available

5.6 **Toxicity by Ingestion:** Grade 2; 0.5 to 5 g/kg (rat)

5.7 **Late Toxicity:** Data not available

5.8 **Vapor (Gas) Irritant Characteristics:** Vapors cause a moderate irritation such that personnel will find high concentrations unpleasant. The effect is temporary.

5.9 **Liquid or Solid Irritant Characteristics:** Minimum hazard. If spilled on clothing and allowed to remain, may cause smarting and reddening of the skin.

5.10 **Odor Threshold:** 0.047 ppm

6. FIRE HAZARDS

6.1 **Flash Point:** 13°F O.C.; –40°F C.C.

6.2 **Flammable Limits in Air:** 2.0% – 10.0%

6.3 **Fire Extinguishing Agents:** Foam, dry chemical or carbon dioxide

6.4 **Fire Extinguishing Agents Not to be Used:** Data not available

6.5 **Special Hazards of Combustion Products:** Not pertinent

6.6 **Behavior in Fire:** Vapors are heavier than air and may travel considerable distance to a source of ignition and flash back. Fires are difficult to control due to ease of reignition.

6.7 **Ignition Temperature:** 385°F

6.8 **Electrical Hazard:** Not pertinent

6.9 **Burning Rate:** 4.8 mm/min.

7. CHEMICAL REACTIVITY

7.1 **Reactivity with Water:** No reaction

7.2 **Reactivity with Common Materials:** No reaction

7.3 **Stability During Transport:** Stable

7.4 **Neutralizing Agents for Acids and Caustics:** Not pertinent

7.5 **Polymerization:** Not pertinent

7.6 **Inhibitor of Polymerization:** Not pertinent

8. WATER POLLUTION

8.1 **Aquatic Toxicity:** Data not available

8.2 **Waterfowl Toxicity:** Data not available

8.3 **Biological Oxygen Demand (BOD):** 162 lb/lb in 5 days

8.4 **Food Chain Concentration Potential:** None

9. SELECTED MANUFACTURERS

1. Dow Badische Co.
   Williamsburg, Va. 23185
2. Eastman Kodak Co.
   Texas Eastman Co. Division
   Longview, Texas 75601
3. Union Carbide Corp.
   Chemicals and Plastics Division
   270 Park Ave.
   New York, N. Y. 10017

10. SHIPPING INFORMATION

10.1 **Grades or Purity:** Dry grade: 98.0%; wet grade: 96.0%; commercial: 97%

10.2 **Storage Temperature:** Ambient

10.3 **Inert Atmosphere:** No requirement

10.4 **Venting:** Pressure-vacuum

11. HAZARD ASSESSMENT CODE

(See Hazard Assessment Handbook, CG 446-3)

A-P-Q

12. HAZARD CLASSIFICATIONS

12.1 **Code of Federal Regulations:** Flammable liquid

12.2 **NAS Hazard Rating for Bulk Water Transportation:**

Category	Rating
Fire	3
Health	
Vapor Irritant	2
Liquid or Solid Irritant	1
Poisons	2
Water Pollution	
Human Toxicity	2
Aquatic Toxicity	2
Aesthetic Effect	3
Reactivity	
Other Chemicals	2
Water	0
Self-Reaction	1

12.3 **NFPA Hazard Classifications:**

Category	Classification
Health Hazard (Blue)	2
Flammability (Red)	3
Reactivity (Yellow)	0

13. PHYSICAL AND CHEMICAL PROPERTIES

13.1 **Physical State at 15°C and 1 atm:** Liquid

13.2 **Molecular Weight:** 72.11

13.3 **Boiling Point at 1 atm:** 147°F = 64.1°C = 337.3°K

13.4 **Freezing Point:** –112°F = –80°C = 193°K

13.5 **Critical Temperature:** 464°F = 240°C = 513°K

13.6 **Critical Pressure:** 600 psia = 41 atm = 4.2 MN/m²

13.7 **Specific Gravity:** 0.791 at 20°C (liquid)

13.8 **Liquid Surface Tension:** 22.0 dynes/cm = 0.0220 N/m at 24°C

13.9 **Liquid-Water Interfacial Tension:** 7.2 dynes/cm = 0.0072 N/m at 22.7°C

13.10 **Vapor (Gas) Specific Gravity:** 2.5

13.11 **Ratio of Specific Heats of Vapor (Gas):** 1.093

13.12 **Latent Heat of Vaporization:** 180 Btu/lb = 98 cal/g = 4.1×10^5 J/kg

13.13 **Heat of Combustion:** –13,850 Btu/lb = –7693 cal/g = -322.1×10^5 J/kg

13.14 **Heat of Decomposition:** Not pertinent

13.15 **Heat of Solution:** Not pertinent

13.16 **Heat of Polymerization:** Not pertinent

NOTES

REVISED 1978

BTR

n-BUTYRALDEHYDE

Common Synonyms	Watery liquid	Colorless	Pungent odor
Butyraldehyde Butaldehyde Butyric aldehyde Butanal Butylaldehyde	Floats and mixes slowly with water. Flammable, irritating vapor is produced.		

Avoid contact with liquid and vapor. Keep people away.
Wear goggles and self-contained breathing apparatus.
Shut off ignition sources and call fire department.
Stop discharge if possible.
Stay upwind and use water spray to "knock down" vapor.
Isolate and remove discharged material.
Notify local health and pollution control agencies.

Fire	FLAMMABLE. Flashback along vapor trail may occur. Vapor may explode if ignited in an enclosed area. **Wear goggles and self-contained breathing apparatus.** **Extinguish with dry chemical, foam or carbon dioxide.** **Water may be ineffective on fire.** **Cool exposed containers with water.**
Exposure	**CALL FOR MEDICAL AID.** VAPOR Irritating to eyes, nose and throat. If inhaled, will cause nausea, vomiting, headache or loss of consciousness. **Move to fresh air.** **If breathing has stopped, give artificial respiration.** **If breathing is difficult, give oxygen.** LIQUID Irritating to skin. Will burn eyes. Harmful if swallowed. **Remove contaminated clothing and shoes.** **Flush affected areas with plenty of water.** **IF IN EYES, hold eyelids open and flush with plenty of water.** **IF SWALLOWED and victim is CONSCIOUS, have victim drink water or milk.**
Water Pollution	Effect of low concentrations on aquatic life is unknown. May be dangerous if it enters water intakes. **Notify local health and pollution control officials.** **Notify operators of nearby water intakes.**

1. RESPONSE TO DISCHARGE

(See Response Methods Handbook, CG 446-4)

Issue warning—highly flammable
Restrict access
Disperse and flush

2. LABEL

Red

3. CHEMICAL DESIGNATIONS

3.1 **Synonyms:** Butanal; Butyraldehyde; Butyric aldehyde; Butyl aldehyde

3.2 **Coast Guard Compatibility Classification:** Aldehyde

3.3 **Chemical Formula:** $CH_3CH_2CH_2CHO$

3.4 **IMCO/United Nations Numerical Designation:** 3.2/1129

4. OBSERVABLE CHARACTERISTICS

4.1 **Physical State (as shipped):** Liquid

4.2 **Color:** Colorless

4.3 **Odor:** Pungent aldehyde; pungent and intense

5. HEALTH HAZARDS

5.1 **Personal Protective Equipment:** Protective goggles, gloves, and organic canister gas mask.

5.2 **Symptoms Following Exposure:** Inhalation will cause irritation and possibly nausea, vomiting, headache, and loss of consciousness. Contact with eyes causes burns. Skin contact may be irritating.

5.3 **Treatment for Exposure:** INHALATION: remove victim to fresh air; if breathing has stopped, give artificial respiration; if breathing is difficult, give oxygen; call a doctor at once. SKIN AND EYES: immediately flush with water for at least 15 min.; get medical care for eyes; remove contaminated clothing and wash underlying skin.

5.4 **Toxicity by Inhalation (Threshold Limit Value):** Data not available

5.5 **Short-Term Inhalation Limits:** Data not available

5.6 **Toxicity by Ingestion:** Grade 1; 5-15 g/kg (rat)

5.7 **Late Toxicity:** Data not available

5.8 **Vapor (Gas) Irritant Characteristics:** Vapors cause moderate irritation such that personnel will find high concentrations unpleasant. The effect is temporary.

5.9 **Liquid or Solid Irritant Characteristics:** Minimum hazard. If spilled on clothing and allowed to remain, may cause smarting and reddening of the skin.

5.10 **Odor Threshold:** 0.0046 ppm

6. FIRE HAZARDS

6.1 **Flash Point:** 15°F O.C.; 20°F C.C.

6.2 **Flammable Limits in Air:** 2.5%—10.6%

6.3 **Fire Extinguishing Agents:** Dry chemical, carbon dioxide, foam

6.4 **Fire Extinguishing Agents Not to be Used:** Not pertinent

6.5 **Special Hazards of Combustion Products:** Not pertinent

6.6 **Behavior in Fire:** Vapors are heavier than air and may travel considerable distance to a source of ignition and flash back. Fires are difficult to control due to ease of reignition.

6.7 **Ignition Temperature:** 446°F

6.8 **Electrical Hazard:** Not pertinent

6.9 **Burning Rate:** 4.4 mm/min.

7. CHEMICAL REACTIVITY

7.1 **Reactivity with Water:** None

7.2 **Reactivity with Common Materials:** None

7.3 **Stability During Transport:** Stable

7.4 **Neutralizing Agents for Acids and Caustics:** Not pertinent

7.5 **Polymerization:** May occur in presence of heat, acids, or alkalis

7.6 **Inhibitor of Polymerization:** Not pertinent

8. WATER POLLUTION

8.1 **Aquatic Toxicity:** Data not available

8.2 **Waterfowl Toxicity:** Data not available

8.3 **Biological Oxygen Demand (BOD):** 1.62 lb/lb, 5 days; 106%, 5 days (theor.)

8.4 **Food Chain Concentration Potential:** Data not available

9. SELECTED MANUFACTURERS

1. Dow Badische Co.
   Williamsburg, Va. 23185
2. Eastman Kodak Co.
   Texas Eastman Division
   Longview, Texas 75601
3. Union Carbide Corp.
   Chemicals and Plastics Division
   270 Park Ave.
   New York, N. Y. 10017

10. SHIPPING INFORMATION

10.1 **Grades or Purity:** Water saturated: 97%; dry: 99.5%

10.2 **Storage Temperature:** Ambient

10.3 **Inert Atmosphere:** No requirement

10.4 **Venting:** Pressure-vacuum

11. HAZARD ASSESSMENT CODE

(See Hazard Assessment Handbook, CG 446-3)

A-P-Q

12. HAZARD CLASSIFICATIONS

12.1 **Code of Federal Regulations:** Flammable liquid

12.2 **NAS Hazard Rating for Bulk Water Transportation:**

Category	Rating
Fire	3
Health	
Vapor Irritant	2
Liquid or Solid Irritant	1
Poisons	2
Water Pollution	
Human Toxicity	1
Aquatic Toxicity	3
Aesthetic Effect	3
Reactivity	
Other Chemicals	2
Water	0
Self-Reaction	1

12.3 **NFPA Hazard Classifications:**

Category	Classification
Health Hazard (Blue)	2
Flammability (Red)	3
Reactivity (Yellow)	0

13. PHYSICAL AND CHEMICAL PROPERTIES

13.1 **Physical State at 15°C and 1 atm:** Liquid

13.2 **Molecular Weight:** 72.11

13.3 **Boiling Point at 1 atm:** 167°F = 74.8°C = 348.0°K

13.4 **Freezing Point:** −142°F = −96.4°C = 176.8°K

13.5 **Critical Temperature:** 484°F = 251°C = 524°K

13.6 **Critical Pressure:** 590 psia = 40 atm = 4.1 MN/m²

13.7 **Specific Gravity:** 0.803 at 20°C (liquid)

13.8 **Liquid Surface Tension:** 24.6 dynes/cm = 0.0246 N/m at 20°C

13.9 **Liquid-Water Interfacial Tension:** 5.7 dynes/cm = 0.0057 N/m at 22.3°C

13.10 **Vapor (Gas) Specific Gravity:** 2.5

13.11 **Ratio of Specific Heats of Vapor (Gas):** 1.089

13.12 **Latent Heat of Vaporization:** 184 Btu/lb = 102 cal/g = 4.27×10^5 J/kg

13.13 **Heat of Combustion:** −15,210 Btu/lb = −8450 cal/g = -353.8×10^5 J/kg

13.14 **Heat of Decomposition:** Not pertinent

13.15 **Heat of Solution:** Not pertinent

13.16 **Heat of Polymerization:** Not pertinent

NOTES

REVISED 1978

BRA	# n-BUTYRIC ACID

Common Synonyms			
Butanic acid Butanoic acid Ethylacetic acid Propanecarboxylic acid Butyric acid	Liquid	Colorless	Rancid butter odor
	Floats and mixes with water. Freezing point is 23°F.		

Avoid contact with liquid and vapor. Keep people away.
Wear rubber overclothing (including gloves).
Stop discharge if possible.
Call fire department.
Isolate and remove discharged material.
Notify local health and pollution control agencies.

Fire	Combustible. Extinguish with dry chemicals, alcohol foam, or carbon dioxide. Water may be ineffective on fire.
Exposure	CALL FOR MEDICAL AID. VAPOR Irritating to eyes, nose and throat. If inhaled will cause coughing or difficult breathing. If in eyes, hold eyelids open and flush with plenty of water. If breathing has stopped, give artificial respiration. If breathing is difficult, give oxygen. LIQUID Will burn skin and eyes. If swallowed will cause nausea and vomiting. Remove contaminated clothing and shoes. Flush affected areas with plenty of water. IF IN EYES, hold eyelids open and flush with plenty of water. IF SWALLOWED and victim is CONSCIOUS, have victim drink water or milk and have victim induce vomiting. IF SWALLOWED and victim is UNCONSCIOUS OR HAVING CONVULSIONS, do nothing except keep victim warm.
Water Pollution	Dangerous to aquatic life in high concentrations. May be dangerous if it enters water intakes. Notify local health and wildlife officials. Notify operators of nearby water intakes.

1. RESPONSE TO DISCHARGE
(See Response Methods Handbook, CG 446-4)

Issue warning—water contaminant
Disperse and flush

2. LABEL

3. CHEMICAL DESIGNATIONS

3.1 **Synonyms:** Butanic acid; Butanoic acid; Butyric acid; Ethylacetic acid; Propanecarboxylic acid
3.2 **Coast Guard Compatibility Classification:** Organic acid
3.3 **Chemical Formula:** $CH_3CH_2CH_2COOH$
3.4 **IMCO/United Nations Numerical Designation:** Not listed

4. OBSERVABLE CHARACTERISTICS

4.1 **Physical State (as shipped):** Liquid
4.2 **Color:** Clear
4.3 **Odor:** Rancid, disagreeable; strong, penetrating, like rancid butter

5. HEALTH HAZARDS

5.1 **Personal Protective Equipment:** Self-contained breathing apparatus; rubber gloves; vapor-proof plastic goggles; impervious apron and boots
5.2 **Symptoms Following Exposure:** Inhalation causes irritation of mucous membrane and respiratory tract; may cause nausea and vomiting. Ingestion causes irritation of mouth and stomach. Contact with eyes may cause serious injury. Contact with skin may cause burns; chemical is readily absorbed through the skin and may cause damage by this route.
5.3 **Treatment for Exposure:** INHALATION: remove victim to fresh air; give oxygen if breathing is difficult; call a physician. INGESTION: give large amount of water and induce vomiting. EYES: irrigate with water for 15 min. and get medical attention. SKIN: flush affected areas immediately and thoroughly with water.
5.4 **Toxicity by Inhalation (Threshold Limit Value):** Data not available
5.5 **Short-Term Inhalation Limits:** Data not available
5.6 **Toxicity by Ingestion:** Grade 2; oral LD_{50} = 2,940 mg/kg (rat)
5.7 **Late Toxicity:** Data not available
5.8 **Vapor (Gas) Irritant Characteristics:** Vapors cause moderate irritation such that personnel will find high concentrations unpleasant. The effect is temporary.
5.9 **Liquid or Solid Irritant Characteristics:** Fairly severe skin irritant. May cause pain and second-degree burns after a few minutes' contact.
5.10 **Odor Threshold:** 0.001 ppm

6. FIRE HAZARDS

6.1 **Flash Point:** 166°F O.C.; 160°F C.C.
6.2 **Flammable Limits in Air:** 2.19% – 13.4%
6.3 **Fire Extinguishing Agents:** Dry chemical, "alcohol" foam, carbon dioxide
6.4 **Fire Extinguishing Agents Not to be Used:** Water may be ineffective.
6.5 **Special Hazards of Combustion Products:**
6.6 **Behavior in Fire:**
6.7 **Ignition Temperature:** 842°F
6.8 **Electrical Hazard:** Data not available
6.9 **Burning Rate:** 2.7 mm/min.

7. CHEMICAL REACTIVITY

7.1 **Reactivity with Water:** No reaction
7.2 **Reactivity with Common Materials:** May attack aluminum or other light metals with formation of flammable hydrogen gas.
7.3 **Stability During Transport:** Stable
7.4 **Neutralizing Agents for Acids and Caustics:** Flush with water
7.5 **Polymerization:** Not pertinent
7.6 **Inhibitor of Polymerization:** Not pertinent

8. WATER POLLUTION

8.1 **Aquatic Toxicity:**
400 ppm/*/trout/lethal/fresh water
200 ppm/24 hr/bluegill/TL_m/fresh water
*Time period not specified.
8.2 **Waterfowl Toxicity:** Data not available
8.3 **Biological Oxygen Demand (BOD):** 1.150 lb/lb, 5 days; 1.450 lb/lb, 20 days
8.4 **Food Chain Concentration Potential:** Seafood may be tainted following a spill but chemical does not concentrate in food chain.

9. SELECTED MANUFACTURERS

1. Celanese Chemical Co.
   522 Fifth Avenue
   New York, N. Y. 10036
2. Eastman Chemical Products, Inc.
   Kingsport, Tenn. 37662
3. Mallinckrodt Chemical Works
   223 Westside Avenue
   P. O. Box 384
   Jersey City, N. J. 07303

10. SHIPPING INFORMATION

10.1 **Grades or Purity:** Commercial, 99.5+%
10.2 **Storage Temperature:** Ambient
10.3 **Inert Atmosphere:** No requirement
10.4 **Venting:** Open

11. HAZARD ASSESSMENT CODE
(See Hazard Assessment Handbook, CG 446-3)

A-P-Q

12. HAZARD CLASSIFICATIONS

12.1 **Code of Federal Regulations:** Corrosive Material
12.2 **NAS Hazard Rating for Bulk Water Transportation:**

Category	Rating
Fire	1
Health	
Vapor Irritant	2
Liquid or Solid Irritant	3
Poisons	0
Water Pollution	
Human Toxicity	1
Aquatic Toxicity	2
Aesthetic Effect	3
Reactivity	
Other Chemicals	2
Water	0
Self-Reaction	0

12.3 **NFPA Hazard Classifications:**

Category	Classification
Health Hazard (Blue)	2
Flammability (Red)	2
Reactivity (Yellow)	0

13. PHYSICAL AND CHEMICAL PROPERTIES

13.1 **Physical State at 15°C and 1 atm:** Liquid
13.2 **Molecular Weight:** 88.1
13.3 **Boiling Point at 1 atm:** 327°F = 164°C = 437°K
13.4 **Freezing Point:** 23°F = −5°C = 268°K
13.5 **Critical Temperature:** 671°F = 355°C = 628°K
13.6 **Critical Pressure:** 764 psia = 52 atm = 5.3 MN/m²
13.7 **Specific Gravity:** 0.958 at 20°C
13.8 **Liquid Surface Tension:** 26.74 dynes/cm = 0.02674 N/m at 20°C
13.9 **Liquid-Water Interfacial Tension:** Not pertinent
13.10 **Vapor (Gas) Specific Gravity:** 3.0
13.11 **Ratio of Specific Heats of Vapor (Gas):** 1.079 at 20°C
13.12 **Latent Heat of Vaporization:** 167 Btu/lb = 92.7 cal/g = 3.88×10^5 J/kg
13.13 **Heat of Combustion:** −10,620 Btu/lb = −5,900 cal/g = -247×10^5 J/kg
13.14 **Heat of Decomposition:** Not pertinent
13.15 **Heat of Solution:** −82 Btu/lb = −45 cal/g = -1.9×10^5 J/kg
13.16 **Heat of Polymerization:** Not pertinent

NOTES

CDA	# CACODYLIC ACID

Common Synonyms		
Hydroxydimethylarsine oxide Dimethylarsinic acid ANSAR Silvisar 510	Solid Sinks and mixes with water.	Colorless or dyed blue

AVOID CONTACT WITH SOLID AND DUST. KEEP PEOPLE AWAY.
Wear goggles, self-contained breathing apparatus, and rubber overclothing (including gloves).
Isolate and remove discharged material.
Notify local health and pollution control agencies.

Fire	Not flammable. POISONOUS GASES MAY BE PRODUCED WHEN HEATED.
Exposure	CALL FOR MEDICAL AID. DUST POISONOUS IF INHALED OR IF SKIN IS EXPOSED. Move victim to fresh air. If in eyes, hold eyelids open and flush with plenty of water. SOLID POISONOUS IF SWALLOWED OR IF SKIN IS EXPOSED. Remove contaminated clothing and shoes. Flush affected areas with plenty of water. IF IN EYES, hold eyelids open and flush with plenty of water. IF SWALLOWED and victim is CONSCIOUS, have victim drink water or milk and have victim induce vomiting. IF SWALLOWED and victim is UNCONSCIOUS OR HAVING CONVULSIONS, do nothing except keep victim warm.
Water Pollution	Effect of low concentrations on aquatic life is unknown. May be dangerous if it enters water intakes. Notify local health and wildlife officials. Notify operators of nearby water intakes.

1. RESPONSE TO DISCHARGE

(See Response Methods Handbook, CG 446-4)

Issue warning - poison, water contaminant
Should be removed
Chemical and physical treatment

2. LABEL

3. CHEMICAL DESIGNATIONS

3.1 **Synonyms:** Hydroxydimethylarsine oxide; Dimethylarsinic acid; Ansar; Silvisar 510
3.2 **Coast Guard Compatibility Classification:** Not applicable
3.3 **Chemical Formula:** $(CH_3)_2AsOOH$
3.4 **IMCO/United Nations Numerical Designation:** 6.1/1572

4. OBSERVABLE CHARACTERISTICS

4.1 **Physical State (as shipped):** Solid
4.2 **Color:** White; water solutions may be dyed blue.
4.3 **Odor:** None

5. HEALTH HAZARDS

5.1 **Personal Protective Equipment:** Dust respirator; goggles; protective clothing.
5.2 **Symptoms Following Exposure:** Chemical is essentially non-irritating in contact with skin or eyes. Ingestion causes arsenic poisoning, but symptoms are delayed.
5.3 **Treatment for Exposure:** Be alert for delayed arsenic poisoning symptoms. EYES or SKIN: flush with water. INGESTION: induce vomiting and call physician at once.
5.4 **Toxicity by Inhalation (Threshold Limit Value):** Data not available
5.5 **Short-Term Inhalation Limits:** Data not available
5.6 **Toxicity by Ingestion:** Grade 2; oral rat LD_{50} = 700 mg/kg
5.7 **Late Toxicity:** Arsenic poisoning
5.8 **Vapor (Gas) Irritant Characteristics:** Data not available
5.9 **Liquid or Solid Irritant Characteristics:** Data not available
5.10 **Odor Threshold:** Not pertinent

6. FIRE HAZARDS

6.1 **Flash Point:** Not flammable
6.2 **Flammable Limits in Air:** Not flammable
6.3 **Fire Extinguishing Agents:** Not pertinent
6.4 **Fire Extinguishing Agents Not to be Used:** Not pertinent
6.5 **Special Hazards of Combustion Products:** Not pertinent
6.6 **Behavior in Fire:** May form toxic oxides of arsenic when heated.
6.7 **Ignition Temperature:** Not pertinent
6.8 **Electrical Hazard:** Not pertinent
6.9 **Burning Rate:** Not pertinent

7. CHEMICAL REACTIVITY

7.1 **Reactivity with Water:** No reaction
7.2 **Reactivity with Common Materials:** No reaction
7.3 **Stability During Transport:** Stable
7.4 **Neutralizing Agents for Acids and Caustics:** Not pertinent
7.5 **Polymerization:** Not pertinent
7.6 **Inhibitor of Polymerization:** Not pertinent

8. WATER POLLUTION

8.1 **Aquatic Toxicity:** 100 ppm/96 hr/scud/not toxic
8.2 **Waterfowl Toxicity:** Data not available
8.3 **Biological Oxygen Demand (BOD):** Data not available
8.4 **Food Chain Concentration Potential:** Data not available

9. SELECTED MANUFACTURERS

The Ansul Company
Chemical Division
1 Stanton Street
Marinette, Wis. 54143

10. SHIPPING INFORMATION

10.1 **Grades or Purity:** Commercial: 50% solution in water, dyed blue
10.2 **Storage Temperature:** Ambient
10.3 **Inert Atmosphere:** No requirement
10.4 **Venting:** Open (flame arrester)

11. HAZARD ASSESSMENT CODE

(See Hazard Assessment Handbook, CG 446-3)

SS

12. HAZARD CLASSIFICATIONS

12.1 **Code of Federal Regulations:** Poisonous, Class B
12.2 **NAS Hazard Rating for Bulk Water Transportation:** Not listed
12.3 **NFPA Hazard Classifications:** Not listed

13. PHYSICAL AND CHEMICAL PROPERTIES

13.1 **Physical State at 15°C and 1 atm:** Solid
13.2 **Molecular Weight:** 138
13.3 **Boiling Point at 1 atm:** >392°F = >200°C = >473°K
13.4 **Freezing Point:** Not pertinent
13.5 **Critical Temperature:** Not pertinent
13.6 **Critical Pressure:** Not pertinent
13.7 **Specific Gravity:** >1.1 (est.) at 20°C (solid)
13.8 **Liquid Surface Tension:** Not pertinent
13.9 **Liquid-Water Interfacial Tension:** Not pertinent
13.10 **Vapor (Gas) Specific Gravity:** Not pertinent
13.11 **Ratio of Specific Heats of Vapor (Gas):** Not pertinent
13.12 **Latent Heat of Vaporization:** Not pertinent
13.13 **Heat of Combustion:** (est.) −6,000 Btu/lb = −3,300 cal/g = -140×10^5 J/kg
13.14 **Heat of Decomposition:** Not pertinent
13.15 **Heat of Solution:** (est.) −54 Btu/lb = −30 cal/g = -1.3×10^5 J/kg
13.16 **Heat of Polymerization:** Not pertinent

NOTES

CAT

CADMIUM ACETATE

Common Synonyms
Cadmium acetate dihydrate

Solid | Colorless | Odorless

Sinks and mixes with water.

AVOID CONTACT WITH SOLID AND DUST. KEEP PEOPLE AWAY.
Wear dust respirator and rubber overclothing (including gloves).
Stop discharge if possible.
Isolate and remove discharged material.
Notify local health and pollution control agencies.

Fire

Not flammable.
POISONOUS GASES MAY BE PRODUCED IN FIRE.

Wear goggles and self-contained breathing apparatus.

Exposure

CALL FOR MEDICAL AID.

DUST
POISONOUS IF INHALED.
If inhaled will cause coughing or difficult breathing.

If in eyes, hold eyelids open and flush with plenty of water.
If breathing has stopped, give artificial respiration.
If breathing is difficult, give oxygen.

SOLID
POISONOUS IF SWALLOWED.

IF SWALLOWED and victim is CONSCIOUS, have victim drink water or milk and have victim induce vomiting.
IF SWALLOWED and victim is UNCONSCIOUS OR HAVING CONVULSIONS, do nothing except keep victim warm.

Water Pollution

Effect of low concentrations on aquatic life is unknown.
May be dangerous if it enters water intakes.

Notify local health and wildlife officials.
Notify operators of nearby water intakes.

1. RESPONSE TO DISCHARGE

(See Response Methods Handbook, CG 446-4)

Issue warning—water contaminant
Disperse and flush

2. LABELS

No hazard label required by Code of Federal Regulations

3. CHEMICAL DESIGNATIONS

3.1 **Synonyms:** Cadmium acetate dihydrate
3.2 **Coast Guard Compatibility Classification:** Not listed
3.3 **Chemical Formula:** $Cd(C_2H_3O_2)_2 \cdot 2H_2O$
3.4 **IMCO/United Nations Numerical Designation:** Not listed

4. OBSERVABLE CHARACTERISTICS

4.1 **Physical State (as shipped):** Solid
4.2 **Color:** White
4.3 **Odor:** None

5. HEALTH HAZARDS

5.1 **Personal Protective Equipment:** Dust mask; goggles or face shield; rubber gloves
5.2 **Symptoms Following Exposure:** Inhalation causes coughing, sneezing, symptoms of lung damage. Ingestion produces severe toxic symptoms; both kidney and liver injuries may occur. Contact with dust causes eye irritation.
5.3 **Treatment for Exposure:** INHALATION: remove victim to fresh air; seek medical attention. INGESTION: induce vomiting; allay gastrointestinal irritation by swallowing milk or egg whites at frequent intervals; perform gastric lavage; seek medical attention. EYES: flush with water for at least 15 min.
5.4 **Toxicity by Inhalation (Threshold Limit Value):** 0.2 mg/m^3 (as cadmium)
5.5 **Short-Term Inhalation Limits:** Data not available
5.6 **Toxicity by Ingestion:** Grade 4; LD_{50} <50 mg/kg
5.7 **Late Toxicity:** Delayed liver, lung, and kidney damage has followed respiratory exposures to cadmium salts in industry.
5.8 **Vapor (Gas) Irritant Characteristics:** Data not available
5.9 **Liquid or Solid Irritant Characteristics:** Data not available
5.10 **Odor Threshold:** Data not available

6. FIRE HAZARDS

6.1 **Flash Point:** Not flammable
6.2 **Flammable Limits in Air:** Not flammable
6.3 **Fire Extinguishing Agents:** Not pertinent
6.4 **Fire Extinguishing Agents Not to be Used:** Not pertinent
6.5 **Special Hazards of Combustion Products:** Toxic cadmium oxide fumes may form in fires.
6.6 **Behavior in Fire:**
6.7 **Ignition Temperature:** Not pertinent
6.8 **Electrical Hazard:** Not pertinent
6.9 **Burning Rate:** Not pertinent

7. CHEMICAL REACTIVITY

7.1 **Reactivity with Water:** No reaction
7.2 **Reactivity with Common Materials:**
7.3 **Stability During Transport:** Stable
7.4 **Neutralizing Agents for Acids and Caustics:** Not pertinent
7.5 **Polymerization:** Not pertinent
7.6 **Inhibitor of Polymerization:** Not pertinent

8. WATER POLLUTION

8.1 **Aquatic Toxicity:** Data not available
8.2 **Waterfowl Toxicity:** Data not available
8.3 **Biological Oxygen Demand (BOD):** Data not available
8.4 **Food Chain Concentration Potential:** Concentrated by shellfish

9. SELECTED MANUFACTURERS

1. American Hoechst Corp.
   Chemicals and Plastics Div.
   Rt. 202—206 North
   Somerville, N. J. 08876
2. Mallinckrodt Chemical Works
   223 Westside Avenue
   P. O. Box 384
   Jersey City, N. J. 07303
3. J. T. Baker Chemical Co.
   Phillipsburg, N. J. 08865

10. SHIPPING INFORMATION

10.1 **Grades or Purity:** Pure, 98%; Reagent
10.2 **Storage Temperature:** Ambient
10.3 **Inert Atmosphere:** No requirement
10.4 **Venting:** Open

11. HAZARD ASSESSMENT CODE

(See Hazard Assessment Handbook, CG 446-3)

SS

12. HAZARD CLASSIFICATIONS

12.1 **Code of Federal Regulations:** Not listed
12.2 **NAS Hazard Rating for Bulk Water Transportation:** Not listed
12.3 **NFPA Hazard Classifications:** Not listed

13. PHYSICAL AND CHEMICAL PROPERTIES

13.1 **Physical State at 15°C and 1 atm:** Solid
13.2 **Molecular Weight:** 266.52
13.3 **Boiling Point at 1 atm:** Not pertinent (decomposes)
13.4 **Freezing Point:** Not pertinent
13.5 **Critical Temperature:** Not pertinent
13.6 **Critical Pressure:** Not pertinent
13.7 **Specific Gravity:** 2.34 at 20°C (solid)
13.8 **Liquid Surface Tension:** Not pertinent
13.9 **Liquid-Water Interfacial Tension:** Not pertinent
13.10 **Vapor (Gas) Specific Gravity:** Not pertinent
13.11 **Ratio of Specific Heats of Vapor (Gas):** Not pertinent
13.12 **Latent Heat of Vaporization:** Not pertinent
13.13 **Heat of Combustion:** Not pertinent
13.14 **Heat of Decomposition:** Not pertinent
13.15 **Heat of Solution:** Data not available
13.16 **Heat of Polymerization:** Not pertinent

NOTES

CMB	# CADMIUM BROMIDE

Common Synonyms			
Cadmium bromide tetrahydrate	Solid	White	Odorless
	Mixes with water.		

AVOID CONTACT WITH SOLID AND DUST. KEEP PEOPLE AWAY.
Wear dust respirator and rubber overclothing (including gloves).
Stop discharge if possible.
Isolate and remove discharged material.
Notify local health and pollution control agencies.

Fire

Not flammable.
POISONOUS GASES MAY BE PRODUCED IN FIRE.
Wear goggles and self-contained breathing apparatus.

Exposure

CALL FOR MEDICAL AID.

DUST
POISONOUS IF INHALED.
If inhaled will cause coughing or difficult breathing.
If in eyes, hold eyelids open and flush with plenty of water.
If breathing has stopped, give artificial respiration.
If breathing is difficult, give oxygen.

SOLID
POISONOUS IF SWALLOWED.
Irritating to skin and eyes.
If swallowed will cause nausea, vomiting and loss of consciousness.
Remove contaminated clothing and shoes.
Flush affected areas with plenty of water.
IF IN EYES, hold eyelids open and flush with plenty of water.
IF SWALLOWED and victim is CONSCIOUS, have victim drink water or milk and have victim induce vomiting.
IF SWALLOWED and victim is UNCONSCIOUS OR HAVING CONVULSIONS, do nothing except keep victim warm.

Water Pollution

Effect of low concentrations on aquatic life is unknown.
May be dangerous if it enters water intakes.
Notify local health and wildlife officials.
Notify operators of nearby water intakes.

1. RESPONSE TO DISCHARGE
(See Response Methods Handbook, CG 446-4)

Issue warning—water contaminant
Disperse and flush

2. LABELS

No hazard label required by Code of Federal Regulations

3. CHEMICAL DESIGNATIONS

3.1 **Synonyms:** Cadmium bromide tetrahydrate
3.2 **Coast Guard Compatibility Classification:** Not listed
3.3 **Chemical Formula:** $CdBr_2 \cdot 4H_2O$
3.4 **IMCO/United Nations Numerical Designation:** Not listed

4. OBSERVABLE CHARACTERISTICS

4.1 **Physical State (as shipped):** Solid
4.2 **Color:** White
4.3 **Odor:** None

5. HEALTH HAZARDS

5.1 **Personal Protective Equipment:** Dust mask; goggles or face shield; rubber gloves
5.2 **Symptoms Following Exposure:** Inhalation causes coughing, sneezing, symptoms of lung damage. Ingestion produces severe toxic symptoms; both kidney and liver injuries may occur. Contact with eyes causes irritation.
5.3 **Treatment for Exposure:** INHALATION: remove patient to fresh air; seek medical attention. INGESTION: induce vomiting; allay GI irritation by giving milk or egg whites at frequent intervals; perform gastric lavage; seek medical attention. EYES: flush with water for at least 15 min.
5.4 **Toxicity by Inhalation (Threshold Limit Value):** 0.2 mg/m^3 (as cadmium)
5.5 **Short-Term Inhalation Limits:** Data not available
5.6 **Toxicity by Ingestion:** Grade 4; LD_{50} < 50 mg/kg
5.7 **Late Toxicity:** Delayed liver, kidney, and lung damage has followed respiratory exposure to cadmium salts in industry.
5.8 **Vapor (Gas) Irritant Characteristics:** Data not available
5.9 **Liquid or Solid Irritant Characteristics:** Data not available
5.10 **Odor Threshold:** Data not available

6. FIRE HAZARDS

6.1 **Flash Point:** Not flammable
6.2 **Flammable Limits in Air:** Not flammable
6.3 **Fire Extinguishing Agents:** Not pertinent
6.4 **Fire Extinguishing Agents Not to be Used:** Not pertinent
6.5 **Special Hazards of Combustion Products:** Toxic cadmium oxide fumes may form in fires.
6.6 **Behavior in Fire:**
6.7 **Ignition Temperature:** Not pertinent
6.8 **Electrical Hazard:** Not pertinent
6.9 **Burning Rate:** Not pertinent

7. CHEMICAL REACTIVITY

7.1 **Reactivity with Water:** No reaction
7.2 **Reactivity with Common Materials:**
7.3 **Stability During Transport:** Stable
7.4 **Neutralizing Agents for Acids and Caustics:** Not pertinent
7.5 **Polymerization:** Not pertinent
7.6 **Inhibitor of Polymerization:** Not pertinent

8. WATER POLLUTION

8.1 **Aquatic Toxicity:** Data not available
8.2 **Waterfowl Toxicity:** Data not available
8.3 **Biological Oxygen Demand (BOD):** None
8.4 **Food Chain Concentration Potential:** Concentrated by shellfish.

9. SELECTED MANUFACTURERS

1. American Hoechst Corp.
   Chemicals and Plastics Div.
   Rt. 202-206 North
   Somerville, N.J. 08876
2. Gallard-Schlesinger Chemical Mfg. Co.
   584 Mineola Ave.
   Carle Place, N.Y. 11514
3. Research Organic/Inorganic Chemical Corp.
   Sun Valley, Calif. 91352

10. SHIPPING INFORMATION

10.1 **Grades or Purity:** Commercial, 99.5%; Anhydrous, 99+%
10.2 **Storage Temperature:** Ambient
10.3 **Inert Atmosphere:** No requirement
10.4 **Venting:** Open

11. HAZARD ASSESSMENT CODE
(See Hazard Assessment Handbook, CG 446-3)

SS

12. HAZARD CLASSIFICATIONS

12.1 **Code of Federal Regulations:** Not listed
12.2 **NAS Hazard Rating for Bulk Water Transportation:** Not listed
12.3 **NFPA Hazard Classifications:** Not listed

13. PHYSICAL AND CHEMICAL PROPERTIES

13.1 **Physical State at 15°C and 1 atm:** Solid
13.2 **Molecular Weight:** 344.27
13.3 **Boiling Point at 1 atm:** Not pertinent (decomposes)
13.4 **Freezing Point:** Not pertinent
13.5 **Critical Temperature:** Not pertinent
13.6 **Critical Pressure:** Not pertinent
13.7 **Specific Gravity:** > 1.1 at 20°C (solid)
13.8 **Liquid Surface Tension:** Not pertinent
13.9 **Liquid-Water Interfacial Tension:** Not pertinent
13.10 **Vapor (Gas) Specific Gravity:** Not pertinent
13.11 **Ratio of Specific Heats of Vapor (Gas):** Not pertinent
13.12 **Latent Heat of Vaporization:** Not pertinent
13.13 **Heat of Combustion:** Not pertinent
13.14 **Heat of Decomposition:** Not pertinent
13.15 **Heat of Solution:** -2.3 Btu/lb = -1.3 cal/g = -0.054×10^5 J/kg
13.16 **Heat of Polymerization:** Not pertinent

NOTES

CDC CADMIUM CHLORIDE

Common Synonyms	Solid crystals	White	Odorless
	Sinks and mixes with water.		

Isolate and remove discharged material.
Notify local health and pollution control agencies.

Fire

Not flammable.

Exposure

CALL FOR MEDICAL AID.

SOLID
Harmful if swallowed.
IF SWALLOWED and victim is CONSCIOUS, have victim drink water or milk, have victim induce vomiting.
IF SWALLOWED and victim is UNCONSCIOUS OR HAVING CONVULSIONS, do nothing except keep victim warm.

Water Pollution

HARMFUL TO AQUATIC LIFE IN VERY LOW CONCENTRATIONS.
May be dangerous if it enters water intakes.
Notify local health and wildlife officials.
Notify operators of nearby water intakes.

1. RESPONSE TO DISCHARGE

(See Response Methods Handbook, CG 446-4)

Issue warning—water contaminant
Should be removed

2. LABEL

3. CHEMICAL DESIGNATIONS

3.1 **Synonyms:** No common synonyms
3.2 **Coast Guard Compatibility Classification:** Not applicable
3.3 **Chemical Formula:** $CdCl_2$
3.4 **IMCO/United Nations Numerical Designation:** Not listed

4. OBSERVABLE CHARACTERISTICS

4.1 **Physical State (as shipped):** Crystalline solid
4.2 **Color:** White
4.3 **Odor:** Odorless

5. HEALTH HAZARDS

5.1 **Personal Protective Equipment:** Safety glasses, rubber gloves, and respirator with proper filter.
5.2 **Symptoms Following Exposure:** Ingestion causes gastroenteric distress, pain, and prostration. Sensory disturbances, liver injury, and convulsions have been observed in severe intoxications.
5.3 **Treatment for Exposure:** INGESTION: induce vomiting and follow with gastric lavage, a saline cathartic, and demulcents. Consider using atropine, opiates, and fluid therapy. $CaNa_2$ EDTA has been effective in acutely poisoned animals and in a few humans. BAL has been found sufficiently effective in animal experiments to justify its use in human intoxication. Since the BAL-cadmium complex has a nephrotoxic action, the physician will have to decide whether or not to use this drug.
5.4 **Toxicity by Inhalation (Threshold Limit Value):** 2 ppm
5.5 **Short-Term Inhalation Limits:** Not pertinent
5.6 **Toxicity by Ingestion:** Grade 4; LD_{50} below 50 mg/kg
5.7 **Late Toxicity:** Data not available
5.8 **Vapor (Gas) Irritant Characteristics:** Not pertinent
5.9 **Liquid or Solid Irritant Characteristics:** Causes smarting of the skin and first degree burns on short exposure; may cause second degree burns on long exposure.
5.10 **Odor Threshold:** Not pertinent

6. FIRE HAZARDS

6.1 **Flash Point:** Not flammable
6.2 **Flammable Limits in Air:** Not flammable
6.3 **Fire Extinguishing Agents:** Not pertinent
6.4 **Fire Extinguishing Agents Not to be Used:** Not pertinent
6.5 **Special Hazards of Combustion Products:** Not pertinent
6.6 **Behavior in Fire:** Not pertinent
6.7 **Ignition Temperature:** Not flammable
6.8 **Electrical Hazard:** Not pertinent
6.9 **Burning Rate:** Not flammable

7. CHEMICAL REACTIVITY

7.1 **Reactivity with Water:** No reaction
7.2 **Reactivity with Common Materials:** No reaction
7.3 **Stability During Transport:** Stable
7.4 **Neutralizing Agents for Acids and Caustics:** Not pertinent
7.5 **Polymerization:** Not pertinent
7.6 **Inhibitor of Polymerization:** Not pertinent

8. WATER POLLUTION

8.1 **Aquatic Toxicity:** 1.94 ppm/96 hr/bluegill/TL_m/fresh water
8.2 **Waterfowl Toxicity:** LC_{50} >5000 ppm.
8.3 **Biological Oxygen Demand (BOD):** Not pertinent
8.4 **Food Chain Concentration Potential:** Not pertinent

9. SELECTED MANUFACTURERS

1. Allied Chemical Corp.
   Specialty Chemicals Division
   Marcus Hook, Pa. 19061
2. Chemetron Corp.
   Inorganic Chemicals Division
   Cleveland, Ohio 44113
3. Mallinckrodt Chemical Works
   Industrial Chemicals Division
   St. Louis, Mo. 63160

10. SHIPPING INFORMATION

10.1 **Grades or Purity:** Data not available
10.2 **Storage Temperature:** Data not available
10.3 **Inert Atmosphere:** Data not available
10.4 **Venting:** Data not available

11. HAZARD ASSESSMENT CODE

(See Hazard Assessment Handbook, CG 446-3)

SS

12. HAZARD CLASSIFICATIONS

12.1 **Code of Federal Regulations:** Poison, Class B
12.2 **NAS Hazard Rating for Bulk Water Transportation:** Not listed
12.3 **NFPA Hazard Classifications:** Not listed

13. PHYSICAL AND CHEMICAL PROPERTIES

13.1 **Physical State at 15°C and 1 atm:** Solid
13.2 **Molecular Weight:** 228.35
13.3 **Boiling Point at 1 atm:** Not pertinent
13.4 **Freezing Point:** Not pertinent
13.5 **Critical Temperature:** Not pertinent
13.6 **Critical Pressure:** Not pertinent
13.7 **Specific Gravity:** 4.05 at 25°C (solid)
13.8 **Liquid Surface Tension:** Not pertinent
13.9 **Liquid-Water Interfacial Tension:** Not pertinent
13.10 **Vapor (Gas) Specific Gravity:** Not pertinent
13.11 **Ratio of Specific Heats of Vapor (Gas):** Not pertinent
13.12 **Latent Heat of Vaporization:** Not pertinent
13.13 **Heat of Combustion:** Not pertinent
13.14 **Heat of Decomposition:** Not pertinent
13.15 **Heat of Solution:** Not pertinent
13.16 **Heat of Polymerization:** Not pertinent

NOTES

REVISED 1978

CFB — CADMIUM FLUOROBORATE

Common Synonyms			
Cadmium fluoborate Cadmium fluoborate solution	Liquid	Colorless	Odorless
	Sinks and mixes with water.		

AVOID CONTACT WITH LIQUID AND VAPOR. KEEP PEOPLE AWAY.
Wear goggles and self-contained breathing apparatus.
Stop discharge if possible.
Isolate and remove discharged material.
Notify local health and pollution control agencies.

Fire

Not flammable.
POISONOUS GASES MAY BE PRODUCED IN FIRE.
Wear goggles and self-contained breathing apparatus.

Exposure

CALL FOR MEDICAL AID.

VAPOR
POISONOUS IF INHALED.
If inhaled will cause coughing or difficult breathing.
If in eyes, hold eyelids open and flush with plenty of water.
If breathing has stopped, give artificial respiration.
If breathing is difficult, give oxygen.

LIQUID
Irritating to skin and eyes.
Harmful if swallowed.
Remove contaminated clothing and shoes.
Flush affected areas with plenty of water.
IF IN EYES, hold eyelids open and flush with plenty of water.
IF SWALLOWED and victim is CONSCIOUS, have victim drink water or milk and have victim induce vomiting.
IF SWALLOWED and victim is UNCONSCIOUS OR HAVING CONVULSIONS, do nothing except keep victim warm.

Water Pollution

Effect of low concentrations on aquatic life is unknown.
May be dangerous if it enters water intakes.
Notify local health and wildlife officials.
Notify operators of nearby water intakes.

1. RESPONSE TO DISCHARGE
(See Response Methods Handbook, CG 446-4)

Issue warning—water contaminant
Disperse and flush

2. LABELS

No hazard label required by Code of Federal Regulations

3. CHEMICAL DESIGNATIONS

3.1 **Synonyms:** Cadmium fluoborate
3.2 **Coast Guard Compatibility Classification:** Not listed
3.3 **Chemical Formula:** $Cd(BF_4)_2-H_2O$
3.4 **IMCO/United Nations Numerical Designation:** Not listed

4. OBSERVABLE CHARACTERISTICS

4.1 **Physical State (as shipped):** Liquid
4.2 **Color:** Colorless
4.3 **Odor:** None

5. HEALTH HAZARDS

5.1 **Personal Protective Equipment:** Rubber gloves and apron; safety glasses and face shield
5.2 **Symptoms Following Exposure:** Ingestion produces severe toxic symptoms; both kidney and liver injuries may occur; may be fatal. Contact with eyes or skin causes irritation.
5.3 **Treatment for Exposure:** INHALATION: remove patient to fresh air; seek medical attention. INGESTION: call a physician at once; if victim is conscious, induce vomiting by giving a tablespoon of salt in a glass of warm water and repeat until vomit is clear; give milk or whites of eggs beaten with water; keep patient warm and quiet. EYES: flush with plenty of water and get medical attention. SKIN: flush with plenty of water.
5.4 **Toxicity by Inhalation (Threshold Limit Value):** 0.2 mg/m^3 (as cadmium)
5.5 **Short-Term Inhalation Limits:** Data not available
5.6 **Toxicity by Ingestion:** Grade 3; LD_{50} 250 mg/kg (rat)
5.7 **Late Toxicity:** Delayed liver, kidney, and lung damage has followed respiratory exposure to cadmium salts in industry.
5.8 **Vapor (Gas) Irritant Characteristics:** Data not available
5.9 **Liquid or Solid Irritant Characteristics:** Data not available
5.10 **Odor Threshold:** Not pertinent

6. FIRE HAZARDS

6.1 **Flash Point:** Not flammable
6.2 **Flammable Limits in Air:** Not flammable
6.3 **Fire Extinguishing Agents:** Not pertinent
6.4 **Fire Extinguishing Agents Not to be Used:** Not pertinent
6.5 **Special Hazards of Combustion Products:** Toxic hydrogen fluoride and cadmium oxide fumes may form in fires.
6.6 **Behavior in Fire:**
6.7 **Ignition Temperature:** Not pertinent
6.8 **Electrical Hazard:** Not pertinent
6.9 **Burning Rate:** Not pertinent

7. CHEMICAL REACTIVITY

7.1 **Reactivity with Water:** No reaction
7.2 **Reactivity with Common Materials:**
7.3 **Stability During Transport:** Stable
7.4 **Neutralizing Agents for Acids and Caustics:** Not pertinent
7.5 **Polymerization:** Not pertinent
7.6 **Inhibitor of Polymerization:** Not pertinent

8. WATER POLLUTION

8.1 **Aquatic Toxicity:** Data not available
8.2 **Waterfowl Toxicity:** Data not available
8.3 **Biological Oxygen Demand (BOD):** Data not available
8.4 **Food Chain Concentration Potential:** Concentrated by shellfish

9. SELECTED MANUFACTURERS

1. Allied Chemical Corp.
P.O. Box 1087R
Morristown, N.Y. 07960
2. The Harshaw Chemical Co.
1945 E. 97 St.
Cleveland, Ohio 44106
3. Gallard-Schlesinger Chemical Mfg. Co.
584 Mineola Avenue
Carle Place, N. Y. 11514

10. SHIPPING INFORMATION

10.1 **Grades or Purity:** Commercial, 50% solution in water
10.2 **Storage Temperature:** Ambient
10.3 **Inert Atmosphere:** No requirement
10.4 **Venting:** Open

11. HAZARD ASSESSMENT CODE
(See Hazard Assessment Handbook, CG 446-3)

A-P

12. HAZARD CLASSIFICATIONS

12.1 **Code of Federal Regulations:** Not listed
12.2 **NAS Hazard Rating for Bulk Water Transportation:** Not listed
12.3 **NFPA Hazard Classifications:**

Category	Classification
Health Hazard (Blue)	1
Flammability (Red)	0
Reactivity (Yellow)	0

13. PHYSICAL AND CHEMICAL PROPERTIES

13.1 **Physical State at 15°C and 1 atm:** Liquid
13.2 **Molecular Weight:** 286 (solute)
13.3 **Boiling Point at 1 atm:** Not pertinent
13.4 **Freezing Point:** Not pertinent
13.5 **Critical Temperature:** Not pertinent
13.6 **Critical Pressure:** Not pertinent
13.7 **Specific Gravity:** 1.60 at 20°C (liquid)
13.8 **Liquid Surface Tension:** Data not available
13.9 **Liquid-Water Interfacial Tension:** Not pertinent
13.10 **Vapor (Gas) Specific Gravity:** Not pertinent
13.11 **Ratio of Specific Heats of Vapor (Gas):** Not pertinent
13.12 **Latent Heat of Vaporization:** Not pertinent
13.13 **Heat of Combustion:** Not pertinent
13.14 **Heat of Decomposition:** Not pertinent
13.15 **Heat of Solution:** Not pertinent
13.16 **Heat of Polymerization:** Not pertinent

NOTES

CMN	CADMIUM NITRATE

Common Synonyms			
Cadmium nitrate tetrahydrate	Solid	White	Odorless
	Sinks in water.		

AVOID CONTACT WITH SOLID AND DUST. KEEP PEOPLE AWAY.
Wear a dust respirator.
Stop discharge if possible.
Isolate and remove discharged material.
Notify local health and pollution control agencies.

Fire

Not flammable.
POISONOUS GASES MAY BE PRODUCED IN FIRE.
Wear goggles and self-contained breathing apparatus.

Exposure

CALL FOR MEDICAL AID.

DUST
POISONOUS IF INHALED.
If inhaled will cause headache, coughing, or difficult breathing.
If in eyes, hold eyelids open and flush with plenty of water.
If breathing has stopped, give artificial respiration.
If breathing is difficult, give oxygen.

SOLID
POISONOUS IF SWALLOWED.
Irritating to skin and eyes.
If swallowed will cause nausea and vomiting.
Remove contaminated clothing and shoes.
Flush affected areas with plenty of water.
IF IN EYES, hold eyelids open and flush with plenty of water.
IF SWALLOWED and victim is CONSCIOUS, have victim drink water or milk and have victim induce vomiting.
IF SWALLOWED and victim is UNCONSCIOUS OR HAVING CONVULSIONS, do nothing except keep victim warm.

Water Pollution

HARMFUL TO AQUATIC LIFE IN VERY LOW CONCENTRATIONS.
May be dangerous if it enters water intakes.
Notify local health and wildlife officials.
Notify operators of nearby water intakes.

1. RESPONSE TO DISCHARGE

(See Response Methods Handbook, CG 446-4)

Issue warning—water contaminant
Disperse and flush

2. LABELS

No hazard label required by Code of Federal Regulations

3. CHEMICAL DESIGNATIONS

3.1 **Synonyms:** Cadmium nitrate tetrahydrate
3.2 **Coast Guard Compatibility Classification:** Not listed
3.3 **Chemical Formula:** $Cd(NO_3)_2 \cdot 4H_2O$
3.4 **IMCO/United Nations Numerical Designation:** Not listed

4. OBSERVABLE CHARACTERISTICS

4.1 **Physical State (as shipped):** Solid
4.2 **Color:** White
4.3 **Odor:** None

5. HEALTH HAZARDS

5.1 **Personal Protective Equipment:** Rubber gloves; safety goggles; dust mask
5.2 **Symptoms Following Exposure:** Inhalation of fumes can produce coughing, chest constriction, headache, nausea, vomiting, pneumonitis. Chronic poisoning is characterized by emphysema and kidney injury. Ingestion causes gastrointestinal disturbance and severe toxic symptoms; both kidney and liver injuries may occur. Contact with eyes causes irritation.
5.3 **Treatment for Exposure:** INHALATION: remove patient to fresh air; seek medical attention. INGESTION: give large amounts of water and induce vomiting; give milk or egg whites; seek medical attention. EYES: flush with copious amounts of water for 15 min.; consult a physician. SKIN: wash with soap and water.
5.4 **Toxicity by Inhalation (Threshold Limit Value):** 0.2 mg/m^3 (as cadmium)
5.5 **Short-Term Inhalation Limits:** Data not available
5.6 **Toxicity by Ingestion:** Grade 3; oral mouse LD_{50} = 100 mg/kg
5.7 **Late Toxicity:** Delayed liver, lung, and kidney damage has followed respiratory exposures to cadmium salts in industry.
5.8 **Vapor (Gas) Irritant Characteristics:** Data not available
5.9 **Liquid or Solid Irritant Characteristics:** Data not available
5.10 **Odor Threshold:** Data not available

6. FIRE HAZARDS

6.1 **Flash Point:** Not flammable
6.2 **Flammable Limits in Air:** Not flammable
6.3 **Fire Extinguishing Agents:** Not pertinent
6.4 **Fire Extinguishing Agents Not to be Used:** Not pertinent
6.5 **Special Hazards of Combustion Products:** Toxic oxides of nitrogen and cadmium oxide fume may form in fires.
6.6 **Behavior in Fire:** Will increase intensity of fire when in contact with combustible material
6.7 **Ignition Temperature:** Not pertinent
6.8 **Electrical Hazard:** Not pertinent
6.9 **Burning Rate:** Not pertinent

7. CHEMICAL REACTIVITY

7.1 **Reactivity with Water:** No reaction
7.2 **Reactivity with Common Materials:** Mixtures with wood or other combustibles may catch fire.
7.3 **Stability During Transport:** Stable
7.4 **Neutralizing Agents for Acids and Caustics:** Not pertinent
7.5 **Polymerization:** Not pertinent
7.6 **Inhibitor of Polymerization:** Not pertinent

8. WATER POLLUTION

8.1 **Aquatic Toxicity:**
0.056 ppm*/**/guppy/LD_{50}/fresh water
0.2 ppm/10 days/stickleback/killed/fresh water
*As cadmium
**Time period not specified
8.2 **Waterfowl Toxicity:** Data not available
8.3 **Biological Oxygen Demand (BOD):** None
8.4 **Food Chain Concentration Potential:** Shellfish concentrate 900–1600 times

9. SELECTED MANUFACTURERS

1. The Harshaw Chemical Co.
   1945 E. 97th Street
   Cleveland, Ohio 44106
2. Mallinckrodt Chemical Works
   223 Westside Avenue
   P.O. Box 384
   Jersey City, N.J. 07303
3. Allied Chemical Corp.
   Specialty Chemicals Div.
   P.O. Box 1087R
   Morristown, N.J. 07960

10. SHIPPING INFORMATION

10.1 **Grades or Purity:** Technical
10.2 **Storage Temperature:** Ambient
10.3 **Inert Atmosphere:** No requirement
10.4 **Venting:** Open

11. HAZARD ASSESSMENT CODE

(See Hazard Assessment Handbook, CG 446-3)

SS

12. HAZARD CLASSIFICATIONS

12.1 **Code of Federal Regulations:** Not listed
12.2 **NAS Hazard Rating for Bulk Water Transportation:** Not listed
12.3 **NFPA Hazard Classifications:**

Category	Classification
Health Hazard (Blue)	2
Flammability (Red)	0
Reactivity (Yellow)	0
	oxy

13. PHYSICAL AND CHEMICAL PROPERTIES

13.1 **Physical State at 15°C and 1 atm:** Solid
13.2 **Molecular Weight:** 308.47
13.3 **Boiling Point at 1 atm:** Not pertinent (decomposes)
13.4 **Freezing Point:** 138°F = 59°C = 332°K
13.5 **Critical Temperature:** Not pertinent
13.6 **Critical Pressure:** Not pertinent
13.7 **Specific Gravity:** 2.45 at 20°C (solid)
13.8 **Liquid Surface Tension:** Not pertinent
13.9 **Liquid-Water Interfacial Tension:** Not pertinent
13.10 **Vapor (Gas) Specific Gravity:** Not pertinent
13.11 **Ratio of Specific Heats of Vapor (Gas):** Not pertinent
13.12 **Latent Heat of Vaporization:** Not pertinent
13.13 **Heat of Combustion:** Not pertinent
13.14 **Heat of Decomposition:** Not pertinent
13.15 **Heat of Solution:** 29.7 Btu/lb = 16.5 cal/g = 0.691×10^5 J/kg
13.16 **Heat of Polymerization:** Not pertinent

NOTES

COX — CADMIUM OXIDE

Common Synonyms
Cadmium fume

Solid — Yellow-brown — Odorless

Sinks in water.

AVOID CONTACT WITH SOLID AND DUST. KEEP PEOPLE AWAY.
Wear a dust respirator.
Stop discharge if possible.
Isolate and remove discharged material.
Notify local health and pollution control agencies.

Fire

Not flammable.
POISONOUS GASES MAY BE PRODUCED IN FIRE.
Wear goggles and self-contained breathing apparatus.

Exposure

CALL FOR MEDICAL AID.

DUST
POISONOUS IF INHALED.
If inhaled will cause coughing.
If in eyes, hold eyelids open and flush with plenty of water.
If breathing has stopped, give artificial respiration.
If breathing is difficult, give oxygen.

SOLID
Irritating to skin and eyes.
If swallowed will cause nausea and vomiting.
Remove contaminated clothing and shoes.
Flush affected areas with plenty of water.
IF IN EYES, hold eyelids open and flush with plenty of water.
IF SWALLOWED and victim is CONSCIOUS, have victim drink water or milk and have victim induce vomiting.
IF SWALLOWED and victim is UNCONSCIOUS OR HAVING CONVULSIONS, do nothing except keep victim warm.

Water Pollution

Effect of low concentrations on aquatic life is unknown.
May be dangerous if it enters water intakes.
Notify local health and wildlife officials.
Notify operators of nearby water intakes.

1. RESPONSE TO DISCHARGE

(See Response Methods Handbook, CG 446-4)

Should be removed
Chemical and physical treatment

2. LABELS

No hazard label required by Code of Federal Regulations

3. CHEMICAL DESIGNATIONS

3.1 **Synonyms:** Cadmium fume
3.2 **Coast Guard Compatibility Classification:** Not listed
3.3 **Chemical Formula:** CdO
3.4 **IMCO/United Nations Numerical Designation:** Not listed

4. OBSERVABLE CHARACTERISTICS

4.1 **Physical State (as shipped):** Solid
4.2 **Color:** Yellow-brown to brown
4.3 **Odor:** None

5. HEALTH HAZARDS

5.1 **Personal Protective Equipment:** Bu. Mines approved respirator; goggles; rubber gloves
5.2 **Symptoms Following Exposure:** A single exposure to cadmium oxide fumes can cause severe or fatal lung irritation; chronic poisoning is characterized by lung injury (emphysema) and kidney disfunction. Ingestion produces severe toxic effects; both kidney and liver injuries may occur. Contact with eyes causes irritation.
5.3 **Treatment for Exposure:** INHALATION: if there has been known exposure to dense cadmium oxide fume or if cough, chest tightness, or respiratory distress occur after possible exposure, place patient at bed rest and call a physician. INGESTION: induce vomiting; stop irritation by giving milk or egg whites at frequent intervals; perform gastric lavage; seek medical attention. EYES: flush with water for at least 15 min.
5.4 **Toxicity by Inhalation (Threshold Limit Value):** 0.1 mg/m^3
5.5 **Short-Term Inhalation Limits:** 0.1 mg/m^3, 30 min.
5.6 **Toxicity by Ingestion:** Grade 3; oral rat LD_{50} = 72 mg/kg
5.7 **Late Toxicity:** Delayed liver, lung, and kidney damage has followed respiratory exposures to cadmium salts in industry.
5.8 **Vapor (Gas) Irritant Characteristics:** Data not available
5.9 **Liquid or Solid Irritant Characteristics:** Data not available
5.10 **Odor Threshold:** Data not available

6. FIRE HAZARDS

6.1 **Flash Point:** Not flammable
6.2 **Flammable Limits in Air:** Not flammable
6.3 **Fire Extinguishing Agents:** Not pertinent
6.4 **Fire Extinguishing Agents Not to be Used:** Not pertinent
6.5 **Special Hazards of Combustion Products:** Toxic cadmium oxide fume may form in fires.
6.6 **Behavior in Fire:**
6.7 **Ignition Temperature:** Not pertinent
6.8 **Electrical Hazard:** Not pertinent
6.9 **Burning Rate:** Not pertinent

7. CHEMICAL REACTIVITY

7.1 **Reactivity with Water:** No reaction
7.2 **Reactivity with Common Materials:**
7.3 **Stability During Transport:** Stable
7.4 **Neutralizing Agents for Acids and Caustics:** Not pertinent
7.5 **Polymerization:** Not pertinent
7.6 **Inhibitor of Polymerization:** Not pertinent

8. WATER POLLUTION

8.1 **Aquatic Toxicity:** Data not available
8.2 **Waterfowl Toxicity:** Data not available
8.3 **Biological Oxygen Demand (BOD):** None
8.4 **Food Chain Concentration Potential:** Concentrated by shellfish

9. SELECTED MANUFACTURERS

1. American Smelting and Refining Co.
   120 Broadway
   New York, N.Y. 10005
2. J. T. Baker Chemical Co.
   Phillipsburg, N.J. 08865
3. Var-lac-oid Chemical Co.
   666 South Front St.
   Elizabeth, N.J. 07202

10. SHIPPING INFORMATION

10.1 **Grades or Purity:** Reagent; technical
10.2 **Storage Temperature:** Ambient
10.3 **Inert Atmosphere:** No requirement
10.4 **Venting:** Open

11. HAZARD ASSESSMENT CODE

(See Hazard Assessment Handbook, CG 446-3)

II

12. HAZARD CLASSIFICATIONS

12.1 **Code of Federal Regulations:** Not listed
12.2 **NAS Hazard Rating for Bulk Water Transportation:** Not listed
12.3 **NFPA Hazard Classifications:** Not listed

13. PHYSICAL AND CHEMICAL PROPERTIES

13.1 **Physical State at 15°C and 1 atm:** Solid
13.2 **Molecular Weight:** 128.4
13.3 **Boiling Point at 1 atm:** Not pertinent (decomposes)
13.4 **Freezing Point:** Not pertinent
13.5 **Critical Temperature:** Not pertinent
13.6 **Critical Pressure:** Not pertinent
13.7 **Specific Gravity:** 6.95 at 20°C (solid)
13.8 **Liquid Surface Tension:** Not pertinent
13.9 **Liquid-Water Interfacial Tension:** Not pertinent
13.10 **Vapor (Gas) Specific Gravity:** Not pertinent
13.11 **Ratio of Specific Heats of Vapor (Gas):** Not pertinent
13.12 **Latent Heat of Vaporization:** Not pertinent
13.13 **Heat of Combustion:** Not pertinent
13.14 **Heat of Decomposition:** Not pertinent
13.15 **Heat of Solution:** Not pertinent
13.16 **Heat of Polymerization:** Not pertinent

NOTES

CMS — CADMIUM SULFATE

Common Synonyms	Solid · White · Odorless
	Sinks and mixes slowly with water.

AVOID CONTACT WITH SOLID AND DUST. KEEP PEOPLE AWAY.
Wear a dust respirator.
Stop discharge if possible.
Isolate and remove discharged material.
Notify local health and pollution control agencies.

Fire

Not flammable.
POISONOUS GASES MAY BE PRODUCED IN FIRE.
Wear goggles and self-contained breathing apparatus.

Exposure

CALL FOR MEDICAL AID.

DUST
POISONOUS IF INHALED.
If inhaled will cause headache, coughing, or difficult breathing.
If in eyes, hold eyelids open and flush with plenty of water.
If breathing has stopped, give artificial respiration.
If breathing is difficult, give oxygen.

SOLID
POISONOUS IF SWALLOWED.
Irritating to skin and eyes.
If swallowed will cause nausea and vomiting.
Remove contaminated clothing and shoes.
Flush affected areas with plenty of water.
IF IN EYES, hold eyelids open and flush with plenty of water.
IF SWALLOWED and victim is CONSCIOUS, have victim drink water or milk and have victim induce vomiting.
IF SWALLOWED and victim is UNCONSCIOUS OR HAVING CONVULSIONS, do nothing except keep victim warm.

Water Pollution

Dangerous to aquatic life in high concentrations.
May be dangerous if it enters water intakes.
Notify local health and wildlife officials.
Notify operators of nearby water intakes.

1. RESPONSE TO DISCHARGE
(See Response Methods Handbook, CG 446-4)

Issue warning—water contaminant
Disperse and flush

2. LABELS

No hazard label required by Code of Federal Regulations

3. CHEMICAL DESIGNATIONS

3.1 **Synonyms:** No common synonyms
3.2 **Coast Guard Compatibility Classification:** Not listed
3.3 **Chemical Formula:** $CdSO_4$
3.4 **IMCO/United Nations Numerical Designation:** Not listed

4. OBSERVABLE CHARACTERISTICS

4.1 **Physical State (as shipped):** Solid
4.2 **Color:** White
4.3 **Odor:** None

5. HEALTH HAZARDS

5.1 **Personal Protective Equipment:** Bu. Mines approved respirator; rubber gloves; safety goggles
5.2 **Symptoms Following Exposure:** Inhalation may cause dryness of throat, coughing, constriction in chest, and headache. Ingestion may cause salivation, vomiting, abdominal pains, or diarrhea. Contact with eyes causes irritation.
5.3 **Treatment for Exposure:** INHALATION: remove victim from exposure and consult a physician. INGESTION: induce vomiting, then allay irritation with milk or egg whites given at frequent intervals; perform gastric lavage; seek medical attention. EYES: flush with water for at least 10 min.; consult a physician. SKIN: wash with soap and water.
5.4 **Toxicity by Inhalation (Threshold Limit Value):** 0.2 mg/m^3 (as cadmium)
5.5 **Short-Term Inhalation Limits:** Data not available
5.6 **Toxicity by Ingestion:** Grade 3; oral mouse LD_{50} = 88 mg/kg
5.7 **Late Toxicity:** Delayed liver, kidney, and lung damage has followed respiratory exposures to cadmium salts in industry.
5.8 **Vapor (Gas) Irritant Characteristics:** Data not available
5.9 **Liquid or Solid Irritant Characteristics:** Data not available
5.10 **Odor Threshold:** Odorless

6. FIRE HAZARDS

6.1 **Flash Point:** Not flammable
6.2 **Flammable Limits in Air:** Not flammable
6.3 **Fire Extinguishing Agents:** Not pertinent
6.4 **Fire Extinguishing Agents Not to be Used:** Not pertinent
6.5 **Special Hazards of Combustion Products:** Toxic cadmium oxide fume may form in fires.
6.6 **Behavior in Fire:**
6.7 **Ignition Temperature:** Not pertinent
6.8 **Electrical Hazard:** Not pertinent
6.9 **Burning Rate:** Not pertinent

7. CHEMICAL REACTIVITY

7.1 **Reactivity with Water:** No reaction
7.2 **Reactivity with Common Materials:**
7.3 **Stability During Transport:** Stable
7.4 **Neutralizing Agents for Acids and Caustics:** Not pertinent
7.5 **Polymerization:** Not pertinent
7.6 **Inhibitor of Polymerization:** Not pertinent

8. WATER POLLUTION

8.1 **Aquatic Toxicity:**
1000 ppm/*/fish/LC_{50}/fresh water
*Time period not specified
8.2 **Waterfowl Toxicity:** Data not available
8.3 **Biological Oxygen Demand (BOD):** None
8.4 **Food Chain Concentration Potential:** Shellfish concentrate cadmium 900—1600 times

9. SELECTED MANUFACTURERS

1. The Harshaw Chemical Co.
   1945 E. 97th Street
   Cleveland, Ohio 44106
2. Mallinckrodt Chemical Works
   223 Westside Avenue
   P.O. Box 384
   Jersey City, N. J. 07303
3. J. T. Baker Chemical Co.
   Phillipsburg, N. J. 08865

10. SHIPPING INFORMATION

10.1 **Grades or Purity:** Technical; 8/3 Hydrate grade; Reagent
10.2 **Storage Temperature:** Ambient
10.3 **Inert Atmosphere:** No requirement
10.4 **Venting:** Open

11. HAZARD ASSESSMENT CODE
(See Hazard Assessment Handbook, CG 446-3)

SS

12. HAZARD CLASSIFICATIONS

12.1 **Code of Federal Regulations:** Not listed
12.2 **NAS Hazard Rating for Bulk Water Transportation:** Not listed
12.3 **NFPA Hazard Classifications:**

Category	Classification
Health Hazard (Blue)	1
Flammability (Red)	0
Reactivity (Yellow)	0

13. PHYSICAL AND CHEMICAL PROPERTIES

13.1 **Physical State at 15°C and 1 atm:** Solid
13.2 **Molecular Weight:** 208.46
13.3 **Boiling Point at 1 atm:** Not pertinent (decomposes)
13.4 **Freezing Point:** Not pertinent
13.5 **Critical Temperature:** Not pertinent
13.6 **Critical Pressure:** Not pertinent
13.7 **Specific Gravity:** 4.7 at 20°C (solid)
13.8 **Liquid Surface Tension:** Not pertinent
13.9 **Liquid-Water Interfacial Tension:** Not pertinent
13.10 **Vapor (Gas) Specific Gravity:** Not pertinent
13.11 **Ratio of Specific Heats of Vapor (Gas):** Not pertinent
13.12 **Latent Heat of Vaporization:** Not pertinent
13.13 **Heat of Combustion:** Not pertinent
13.14 **Heat of Decomposition:** Not pertinent
13.15 **Heat of Solution:** −92.4 Btu/lb = −51.3 cal/g = -2.15×10^5 J/kg
13.16 **Heat of Polymerization:** Not pertinent

NOTES

CCA	# CALCIUM ARSENATE

Common Synonyms Tricalcium arsenate Tricalcium ortho arsenate Cucumber dust	Solid White Odorless Sinks in water.

AVOID CONTACT WITH SOLID AND DUST. KEEP PEOPLE AWAY.
Wear a dust respirator.
Stop discharge if possible.
Isolate and remove discharged material.
Notify local health and pollution control agencies.

Fire	Not flammable. POISONOUS GASES MAY BE PRODUCED IN FIRE. Wear goggles and self-contained breathing apparatus.
Exposure	CALL FOR MEDICAL AID. DUST POISONOUS IF INHALED. If inhaled will cause coughing or difficult breathing. If in eyes, hold eyelids open and flush with plenty of water. If breathing is difficult, give oxygen. SOLID POISONOUS IF SWALLOWED. If swallowed will cause nausea and vomiting. Remove contaminated clothing and shoes. Flush affected areas with plenty of water. IF IN EYES, hold eyelids open and flush with plenty of water. IF SWALLOWED and victim is CONSCIOUS, have victim drink water or milk and have victim induce vomiting. IF SWALLOWED and victim is UNCONSCIOUS OR HAVING CONVULSIONS, do nothing except keep victim warm.
Water Pollution	HARMFUL TO AQUATIC LIFE IN VERY LOW CONCENTRATIONS. May be dangerous if it enters water intakes. Notify local health and wildlife officials. Notify operators of nearby water intakes.

1. RESPONSE TO DISCHARGE

(See Response Methods Handbook, CG 446-4)

Issue warning—poison, water contaminant
Restrict access
Should be removed
Chemical and physical treatment

2. LABEL

3. CHEMICAL DESIGNATIONS

3.1 **Synonyms:** Cucumber dust
Tricalcium arsenate
Tricalcium orthoarsenate

3.2 **Coast Guard Compatibility Classification:** Not listed

3.3 **Chemical Formula:** $Ca_3(AsO_4)_2$

3.4 **IMCO/United Nations Numerical Designation:** 6.1/1573

4. OBSERVABLE CHARACTERISTICS

4.1 **Physical State (as shipped):** Solid

4.2 **Color:** White

4.3 **Odor:** None

5. HEALTH HAZARDS

5.1 **Personal Protective Equipment:** Dust mask; goggles or face shield; protective gloves.

5.2 **Symptoms Following Exposure:** Inhalation causes respiratory irritation. Ingestion causes irritation of mouth and stomach. Contact with eyes causes irritation.

5.3 **Treatment for Exposure:** INHALATION: move to fresh air. INGESTION: give victim one tablespoonful of salt in glass of water; repeat until vomit is clear; then give 2 tablespoonfuls of Epsom salts or milk of magnesia and force fluids; call a physician in all cases of suspected poisoning. EYES: flush with water for at least 15 min. SKIN: flush with water; wash with soap and water.

5.4 **Toxicity by Inhalation (Threshold Limit Value):** $1 mg/m^3$

5.5 **Short-Term Inhalation Limits:** Data not available

5.6 **Toxicity by Ingestion:** Grade 4; oral rat LD_{50} = 20 mg/kg

5.7 **Late Toxicity:** Arsenic compounds may cause skin and lung cancer.

5.8 **Vapor (Gas) Irritant Characteristics:** Data not available

5.9 **Liquid or Solid Irritant Characteristics:** Data not available

5.10 **Odor Threshold:** Data not available

6. FIRE HAZARDS

6.1 **Flash Point:** Not flammable

6.2 **Flammable Limits in Air:** Not flammable

6.3 **Fire Extinguishing Agents:** Not pertinent

6.4 **Fire Extinguishing Agents Not to be Used:** Not pertinent

6.5 **Special Hazards of Combustion Products:** Toxic arsenic fume may be formed in fires.

6.6 **Behavior in Fire:**

6.7 **Ignition Temperature:** Not pertinent

6.8 **Electrical Hazard:** Not pertinent

6.9 **Burning Rate:** Not pertinent

7. CHEMICAL REACTIVITY

7.1 **Reactivity with Water:** No reaction

7.2 **Reactivity with Common Materials:**

7.3 **Stability During Transport:** Stable

7.4 **Neutralizing Agents for Acids and Caustics:** Not pertinent

7.5 **Polymerization:** Not pertinent

7.6 **Inhibitor of Polymerization:** Not pertinent

8. WATER POLLUTION

8.1 **Aquatic Toxicity:**
1.1 ppm/48 hr/perch/toxic/fresh water
4.3 ppm/264 hr/crabs/toxic/fresh water

8.2 **Waterfowl Toxicity:** Data not available

8.3 **Biological Oxygen Demand (BOD):** None

8.4 **Food Chain Concentration Potential:** Possible bioaccumulation problem

9. SELECTED MANUFACTURERS

1. Los Angeles Chemical Co.
4545 Ardine Street
South Gate, Calif. 90280

2. Pfaltz and Bauer, Inc.
126—04 Northern Blvd.
Flushing, N. Y. 11368

10. SHIPPING INFORMATION

10.1 **Grades or Purity:** 70%, containing calcium carbonate and calcium hydroxide (limestone and slaked lime)

10.2 **Storage Temperature:** Ambient

10.3 **Inert Atmosphere:** No requirement

10.4 **Venting:** Open

11. HAZARD ASSESSMENT CODE

(See Hazard Assessment Handbook, CG 446-3)

II

12. HAZARD CLASSIFICATIONS

12.1 **Code of Federal Regulations:** Poisonous solid, Class B

12.2 **NAS Hazard Rating for Bulk Water Transportation:** Not listed

12.3 **NFPA Hazard Classifications:** Not listed

13. PHYSICAL AND CHEMICAL PROPERTIES

13.1 **Physical State at 15°C and 1 atm:** Solid

13.2 **Molecular Weight:** 398

13.3 **Boiling Point at 1 atm:** Not pertinent (decomposes)

13.4 **Freezing Point:** Not pertinent

13.5 **Critical Temperature:** Not pertinent

13.6 **Critical Pressure:** Not pertinent

13.7 **Specific Gravity:** 3.62 at 20°C (solid)

13.8 **Liquid Surface Tension:** Not pertinent

13.9 **Liquid-Water Interfacial Tension:** Not pertinent

13.10 **Vapor (Gas) Specific Gravity:** Not pertinent

13.11 **Ratio of Specific Heats of Vapor (Gas):** Not pertinent

13.12 **Latent Heat of Vaporization:** Not pertinent

13.13 **Heat of Combustion:** Not pertinent

13.14 **Heat of Decomposition:** Not pertinent

13.15 **Heat of Solution:** Not pertinent

13.16 **Heat of Polymerization:** Not pertinent

NOTES

CCB

CALCIUM CARBIDE

Common Synonyms			
Carbide Acetylenogen	Solid granules	Grey to bluish black	Garlic odor
	Sinks in water, and bubbles appear on surface as flammable gas is produced.		

Stop discharge if possible. Stay upwind.
Shut off ignition sources and call fire department.
Avoid contact with solid and gas. Keep people away.
Notify local health and pollution control agencies.

Fire

Not flammable.
FLAMMABLE, EXPLOSIVE GAS IS PRODUCED ON CONTACT WITH WATER.
Let fire burn.
DO NOT USE WATER OR FOAM ON ADJACENT FIRES.

Exposure

CALL FOR MEDICAL AID.

SOLID
Irritating to skin and eyes.
Remove contaminated clothing and shoes.
Flush affected areas with plenty of water.
IF IN EYES, hold eyelids open and flush with plenty of water.

Water Pollution

HARMFUL TO AQUATIC LIFE IN VERY LOW CONCENTRATIONS.
Notify local health and wildlife officials.

1. RESPONSE TO DISCHARGE

(See Response Methods Handbook, CG 446-4)

Issue warning—high flammability
Should be removed
Chemical and physical treatment

2. LABELS

3. CHEMICAL DESIGNATIONS

3.1 **Synonyms:** Acetylenogen; Carbide
3.2 **Coast Guard Compatibility Classification:** Not applicable
3.3 **Chemical Formula:** CaC_2
3.4 **IMCO/United Nations Numerical Designation:** 4.3/1402

4. OBSERVABLE CHARACTERISTICS

4.1 **Physical State (as shipped):** Solid
4.2 **Color:** Gray to bluish black
4.3 **Odor:** Garlic-like

5. HEALTH HAZARDS

5.1 **Personal Protective Equipment:** Chemical safety goggles and (for those exposed to unusually dusty operations) a respirator such as those approved by the U.S. Bureau of Mines for "nuisance dusts."
5.2 **Symptoms Following Exposure:** Eye and skin irritation
5.3 **Treatment for Exposure:** INHALATION OF DUST: remove from further exposure and call a doctor. SKIN: wash with plenty of water. EYES: flush with clean running water at an eye wash fountain for at least 15 min. and get medical attention.
5.4 **Toxicity by Inhalation (Threshold Limit Value):** Not pertinent
5.5 **Short-Term Inhalation Limits:** Not pertinent
5.6 **Toxicity by Ingestion:** Data not available
5.7 **Late Toxicity:** None
5.8 **Vapor (Gas) Irritant Characteristics:** None
5.9 **Liquid or Solid Irritant Characteristics:** Minimum hazard. If spilled on clothing and allowed to remain, may cause smarting and reddening of the skin.
5.10 **Odor Threshold:** Not pertinent

6. FIRE HAZARDS

6.1 **Flash Point:** Not flammable
6.2 **Flammable Limits in Air:** Not flammable
6.3 **Fire Extinguishing Agents:** Dry powder; preferably allow fire to burn
6.4 **Fire Extinguishing Agents Not to be Used:** Water, vaporizing liquid or foam, carbon dioxide
6.5 **Special Hazards of Combustion Products:** Not pertinent
6.6 **Behavior in Fire:** If wet by water, highly flammable acetylene gas is formed.
6.7 **Ignition Temperature:** Not flammable
6.8 **Electrical Hazard:** Not pertinent
6.9 **Burning Rate:** Not flammable

7. CHEMICAL REACTIVITY

7.1 **Reactivity with Water:** Reacts vigorously with water to form the highly flammable acetylene gas, which may ignite spontaneously.
7.2 **Reactivity with Common Materials:** Reacts with copper and brass to form explosive compound.
7.3 **Stability During Transport:** Stable in absence of moisture
7.4 **Neutralizing Agents for Acids and Caustics:** Not pertinent
7.5 **Polymerization:** Not pertinent
7.6 **Inhibitor of Polymerization:** Not pertinent

8. WATER POLLUTION

8.1 **Aquatic Toxicity:** Not pertinent
8.2 **Waterfowl Toxicity:** Not pertinent
8.3 **Biological Oxygen Demand (BOD):** Not pertinent
8.4 **Food Chain Concentration Potential:** None

9. SELECTED MANUFACTURERS

1. Airco Inc.
   Airco Alloys and Carbide Div.
   3801 Highland Ave.
   Niagara Falls, N. Y. 14305
2. Midwest Carbide Corp.
   Keokuk, Iowa 52632
3. Union Carbide Corp.
   Chemicals and Plastics Div.
   270 Park Ave.
   New York, N. Y. 10017

10. SHIPPING INFORMATION

10.1 **Grades or Purity:** Data not available
10.2 **Storage Temperature:** Data not available
10.3 **Inert Atmosphere:** Data not available
10.4 **Venting:** Data not available

11. HAZARD ASSESSMENT CODE

(See Hazard Assessment Handbook, CG 446-3)

RR-C

12. HAZARD CLASSIFICATIONS

12.1 **Code of Federal Regulations:** Flammable Solid
12.2 **NAS Hazard Rating for Bulk Water Transportation:** Not listed
12.3 **NFPA Hazard Classifications:**

Category	Classification
Health Hazard (Blue)	1
Flammability (Red)	4
Reactivity (Yellow)	2 W

13. PHYSICAL AND CHEMICAL PROPERTIES

13.1 **Physical State at 15°C and 1 atm:** Solid
13.2 **Molecular Weight:** 64.10
13.3 **Boiling Point at 1 atm:** Not pertinent
13.4 **Freezing Point:** Not pertinent
13.5 **Critical Temperature:** Not pertinent
13.6 **Critical Pressure:** Not pertinent
13.7 **Specific Gravity:** 2.22 at 18°C (solid)
13.8 **Liquid Surface Tension:** Not pertinent
13.9 **Liquid-Water Interfacial Tension:** Not pertinent
13.10 **Vapor (Gas) Specific Gravity:** Not pertinent
13.11 **Ratio of Specific Heats of Vapor (Gas):** Not pertinent
13.12 **Latent Heat of Vaporization:** Not pertinent
13.13 **Heat of Combustion:** Not pertinent
13.14 **Heat of Decomposition:** Not pertinent
13.15 **Heat of Solution:** Not pertinent
13.16 **Heat of Polymerization:** Not pertinent

NOTES

REVISED 1978

CCC	CALCIUM CHLORATE

Common Synonyms	Solid White Odorless Sinks and mixes with water.

Call fire department.
Keep people away.
Isolate and remove discharged material.
Notify local health and pollution control agencies.

Fire	Not flammable. May explode on contact with combustibles. Irritating gases may be produced when heated. **Combat fires from safe distance or protected location.** **Flood discharge area with water.**
Exposure	**Call for medical aid.** DUST Irritating to eyes, nose and throat. **Move victim to fresh air.** **If in eyes, hold eyelids open and flush with plenty of water.** SOLID Irritating to skin and eyes. Harmful if swallowed. **Remove contaminated clothing and shoes.** **Flush affected areas with plenty of water.** **IF IN EYES, hold eyelids open and flush with plenty of water.** **IF SWALLOWED and victim is CONSCIOUS, have victim drink water or milk and have victim induce vomiting.** **IF SWALLOWED and victim is UNCONSCIOUS OR HAVING CONVULSIONS, do nothing except keep victim warm.**
Water Pollution	Effect of low concentrations on aquatic life is unknown. May be dangerous if it enters water intakes. **Notify local health and wildlife officials.** **Notify operators of nearby water intakes.**

1. RESPONSE TO DISCHARGE

(See Response Methods Handbook, CG 446-4)

Issue warning—oxidizing material, water contaminant
Should be removed
Chemical and physical treatment
Disperse and flush

2. LABEL

Yellow

3. CHEMICAL DESIGNATIONS

3.1 **Synonyms:** No common synonyms
3.2 **Coast Guard Compatibility Classification:** Not applicable
3.3 **Chemical Formula:** $Ca(ClO_3)_2$
3.4 **IMCO/United Nations Numerical Designation:** 5.1/1452

4. OBSERVABLE CHARACTERISTICS

4.1 **Physical State (as shipped):** Solid
4.2 **Color:** White
4.3 **Odor:** None

5. HEALTH HAZARDS

5.1 **Personal Protective Equipment:** Goggles or face shield; dust respirator; coveralls or other protective clothing.
5.2 **Symptoms Following Exposure:** Inhalation of dust causes irritation of upper respiratory system. Dust irritates eyes and skin. Ingestion causes abdominal pain, nausea, vomiting, diarrhea, pallor, shortness of breath, unconsciousness.
5.3 **Treatment for Exposure:** INHALATION: remove to fresh air. EYES: flush with water for 15 min. SKIN: flush with water. INGESTION: induce vomiting and get medical attention.
5.4 **Toxicity by Inhalation (Threshold Limit Value):** Data not available
5.5 **Short-Term Inhalation Limits:** Data not available
5.6 **Toxicity by Ingestion:** Grade 2; oral LD_{50} = 4,500 mg/kg (rat)
5.7 **Late Toxicity:** Data not available
5.8 **Vapor (Gas) Irritant Characteristics:** Not pertinent
5.9 **Liquid or Solid Irritant Characteristics:** Data not available
5.10 **Odor Threshold:** Not pertinent

6. FIRE HAZARDS

6.1 **Flash Point:** Not flammable but may cause fire with other materials.
6.2 **Flammable Limits in Air:** Not pertinent
6.3 **Fire Extinguishing Agents:** Flood with water.
6.4 **Fire Extinguishing Agents Not to be Used:** Not pertinent
6.5 **Special Hazards of Combustion Products:** Not pertinent
6.6 **Behavior in Fire:** When involved in a fire, may cause an explosion. Irritating gases may be generated when heated.
6.7 **Ignition Temperature:** Not pertinent
6.8 **Electrical Hazard:** Not pertinent
6.9 **Burning Rate:** Not pertinent

7. CHEMICAL REACTIVITY

7.1 **Reactivity with Water:** No reaction
7.2 **Reactivity with Common Materials:** May form an explosive mixture with finely divided combustible material. The mixture may ignite when rubbed.
7.3 **Stability During Transport:** Stable
7.4 **Neutralizing Agents for Acids and Caustics:** Not pertinent
7.5 **Polymerization:** Not pertinent
7.6 **Inhibitor of Polymerization:** Not pertinent

8. WATER POLLUTION

8.1 **Aquatic Toxicity:** Data not available
8.2 **Waterfowl Toxicity:** Data not available
8.3 **Biological Oxygen Demand (BOD):** None
8.4 **Food Chain Concentration Potential:** None

9. SELECTED MANUFACTURERS

1. Cerac, Inc.
   Menomenee Falls, Wis. 53051
2. Gallard-Schlesinger Chemical Manufacturing Corp.
   584 Mineola Ave.
   Carle Place, N. Y. 11514
3. Dynamit-Nobel AG
   105 Stonehurst Court
   Northvale, N.J. 07647

10. SHIPPING INFORMATION

10.1 **Grades or Purity:** Commercial. May be shipped as dihydrate.
10.2 **Storage Temperature:** Ambient
10.3 **Inert Atmosphere:** No requirement
10.4 **Venting:** Open

11. HAZARD ASSESSMENT CODE

(See Hazard Assessment Handbook, CG 446-3)

SS

12. HAZARD CLASSIFICATIONS

12.1 **Code of Federal Regulations:** Oxidizing material
12.2 **NAS Hazard Rating for Bulk Water Transportation:** Not listed
12.3 **NFPA Hazard Classifications:** Not listed

13. PHYSICAL AND CHEMICAL PROPERTIES

13.1 **Physical State at 15°C and 1 atm:** Solid
13.2 **Molecular Weight:** 207
13.3 **Boiling Point at 1 atm:** Decomposes
13.4 **Freezing Point:** 644°F = 340°C = 613°K
13.5 **Critical Temperature:** Not pertinent
13.6 **Critical Pressure:** Not pertinent
13.7 **Specific Gravity:** 2.710 at 0°C (solid)
13.8 **Liquid Surface Tension:** Not pertinent
13.9 **Liquid-Water Interfacial Tension:** Not pertinent
13.10 **Vapor (Gas) Specific Gravity:** Not pertinent
13.11 **Ratio of Specific Heats of Vapor (Gas):** Not pertinent
13.12 **Latent Heat of Vaporization:** Not pertinent
13.13 **Heat of Combustion:** Not pertinent
13.14 **Heat of Decomposition:** Not pertinent
13.15 **Heat of Solution:** (est.) −54 Btu/lb = −30 cal/g = -1.3×10^5 J/kg
13.16 **Heat of Polymerization:** Not pertinent

NOTES

CLC	CALCIUM CHLORIDE

Common Synonyms			
Calcium chloride, anhydrous Calcium chloride hydrates	Solid or water solution	White to off-white	Odorless
	Sinks and mixes with water.		

Avoid contact with liquid and solid. Keep people away.
Wear rubber overclothing (including gloves).
Stop discharge if possible.
Isolate and remove discharged material.
Notify local health and pollution control agencies.

Fire

Not flammable.

Exposure

CALL FOR MEDICAL AID.
SOLUTION OR SOLID
Will burn skin and eyes.
If swallowed will cause nausea and vomiting.

Remove contaminated clothing and shoes.
Flush affected areas with plenty of water.
IF IN EYES, hold eyelids open and flush with plenty of water.
IF SWALLOWED and victim is CONSCIOUS, have victim drink water or milk.
IF SWALLOWED and victim is UNCONSCIOUS OR HAVING CONVULSIONS, do nothing except keep victim warm.

Water Pollution

Dangerous to aquatic life in high concentrations.
May be dangerous if it enters water intakes.
Notify local health and wildlife officials.
Notify operators of nearby water intakes.

1. RESPONSE TO DISCHARGE

(See Response Methods Handbook, CG 446-4)

Disperse and flush

2. LABELS

No hazard label required by Code of Federal Regulations

3. CHEMICAL DESIGNATIONS

3.1 **Synonyms:** Calcium chloride, anhydrous; Calcium chloride hydrates
3.2 **Coast Guard Compatibility Classification:** Not listed
3.3 **Chemical Formula:** $CaCl_2 \cdot xH_2O$ where x = 0 to 6
3.4 **IMCO/United Nations Numerical Designation:** Not listed

4. OBSERVABLE CHARACTERISTICS

4.1 **Physical State (as shipped):** Solid, or water solution
4.2 **Color:** White to off-white
4.3 **Odor:** None

5. HEALTH HAZARDS

5.1 **Personal Protective Equipment:** Safety glasses or face shield, dust-type respirator, rubber gloves
5.2 **Symptoms Following Exposure:** Inhalation causes irritation of nose and throat. Ingestion causes irritation of mouth and stomach. Contact with eyes (particularly by dust) causes irritation and possible transient corneal injury. Contact of solid with dry skin causes mild irritation; strong solutions can cause marked irritation, even a superficial burn.
5.3 **Treatment for Exposure:** INHALATION: move to fresh air; if discomfort persists, get medical attention. INGESTION: give large amounts of water. EYES: promptly flood with water and continue washing for at least 15 min.; consult an ophthalmologist. SKIN: flush with water.
5.4 **Toxicity by Inhalation (Threshold Limit Value):** Data not available
5.5 **Short-Term Inhalation Limits:** Data not available
5.6 **Toxicity by Ingestion:** Grade 2; oral LD_{50} = 1,000 mg/kg (rat)
5.7 **Late Toxicity:** Data not available
5.8 **Vapor (Gas) Irritant Characteristics:** Data not available
5.9 **Liquid or Solid Irritant Characteristics:** Data not available
5.10 **Odor Threshold:** Data not available

6. FIRE HAZARDS

6.1 **Flash Point:** Not flammable
6.2 **Flammable Limits in Air:** Not flammable
6.3 **Fire Extinguishing Agents:** Not pertinent
6.4 **Fire Extinguishing Agents Not to be Used:** Not pertinent
6.5 **Special Hazards of Combustion Products:**
6.6 **Behavior in Fire:**
6.7 **Ignition Temperature:** Not pertinent
6.8 **Electrical Hazard:** Not pertinent
6.9 **Burning Rate:** Not pertinent

7. CHEMICAL REACTIVITY

7.1 **Reactivity with Water:** Anhydrous grade dissolves with evolution of some heat.
7.2 **Reactivity with Common Materials:** Metals will slowly corrode in aqueous solutions.
7.3 **Stability During Transport:** Stable
7.4 **Neutralizing Agents for Acids and Caustics:** Not pertinent
7.5 **Polymerization:** Not pertinent
7.6 **Inhibitor of Polymerization:** Not pertinent

8. WATER POLLUTION

8.1 **Aquatic Toxicity:**
555 ppm/168 hr/rock bass/killed/tap water
10,650 ppm/96 hr/sunfish/LD_{50}/fresh water
2,400 ppm/48 hr/marine fish/TL_m/sea water
8.2 **Waterfowl Toxicity:** Data not available
8.3 **Biological Oxygen Demand (BOD):** None
8.4 **Food Chain Concentration Potential:** None

9. SELECTED MANUFACTURERS

1. Dow Chemical Co.
Midland, Mich. 48640
2. Michigan Chemical Corp.
351 E. Ohio St.
Chicago, Ill. 60611
3. Allied Chemical Corp.
Industrial Chemicals Div.
P.O. Box 1139R
Morristown, N.J. 07960

10. SHIPPING INFORMATION

10.1 **Grades or Purity:** Anhydrous 90—97%; water solutions containing 51—86%
10.2 **Storage Temperature:** Ambient
10.3 **Inert Atmosphere:** No requirement
10.4 **Venting:** Open

11. HAZARD ASSESSMENT CODE

(See Hazard Assessment Handbook, CG 446-3)

SS

12. HAZARD CLASSIFICATIONS

12.1 **Code of Federal Regulations:** Not listed
12.2 **NAS Hazard Rating for Bulk Water Transportation:** Not Listed
12.3 **NFPA Hazard Classifications:** Not listed

13. PHYSICAL AND CHEMICAL PROPERTIES

13.1 **Physical State at 15°C and 1 atm:** Solid
13.2 **Molecular Weight:** 110.99 (solute)
13.3 **Boiling Point at 1 atm:** Not pertinent
13.4 **Freezing Point:** Not pertinent
13.5 **Critical Temperature:** Not pertinent
13.6 **Critical Pressure:** Not pertinent
13.7 **Specific Gravity:** 2.15 at 20°C (solid)
13.8 **Liquid Surface Tension:** Not pertinent
13.9 **Liquid-Water Interfacial Tension:** Not pertinent
13.10 **Vapor (Gas) Specific Gravity:** Not pertinent
13.11 **Ratio of Specific Heats of Vapor (Gas):** Not pertinent
13.12 **Latent Heat of Vaporization:** Not pertinent
13.13 **Heat of Combustion:** Not pertinent
13.14 **Heat of Decomposition:** Not pertinent
13.15 **Heat of Solution:** −292 Btu/lb = −162 cal/g = -6.79×10^5 J/kg
13.16 **Heat of Polymerization:** Not pertinent

NOTES

CCR

CALCIUM CHROMATE

Common Synonyms			
Calcium chromate (VI) Gelbin yellow ultramarine Steinbuhl yellow Calcium chromate dihydrate	Solid	Yellow	Odorless
	Sinks and mixes slowly with water.		

Avoid contact with solid and dust. Keep people away.
Wear rubber overclothing (including gloves).
Stop discharge if possible.
Isolate and remove discharged material.
Notify local health and pollution control agencies.

Fire

Not flammable.
Will increase the intensity of a fire.
Flood discharge area with water.

Exposure

CALL FOR MEDICAL AID.

DUST
Irritating to eyes, nose and throat.
If inhaled will cause coughing or difficult breathing.
If in eyes, hold eyelids open and flush with plenty of water.
If breathing has stopped, give artificial respiration.
If breathing is difficult, give oxygen.

SOLID
Will burn skin and eyes.
Harmful if swallowed.
Remove contaminated clothing and shoes.
Flush affected areas with plenty of water.
IF IN EYES, hold eyelids open and flush with plenty of water.
IF SWALLOWED and victim is CONSCIOUS, have victim drink water or milk and have victim induce vomiting.
IF SWALLOWED and victim is UNCONSCIOUS OR HAVING CONVULSIONS, do nothing except keep victim warm.

Water Pollution

Effect of low concentrations on aquatic life is unknown.
May be dangerous if it enters water intakes.
Notify local health and wildlife officials.
Notify operators of nearby water intakes.

1. RESPONSE TO DISCHARGE

(See Response Methods Handbook, CG 446-4)

Issue warning—water contaminant
Disperse and flush

2. LABELS

No hazard label required by Code of Federal Regulations

3. CHEMICAL DESIGNATIONS

3.1 **Synonyms:** Calcium chromate (VI)
Calcium chromate dihydrate
Gelbin Yellow Ultramarine
Steinbuhl Yellow

3.2 **Coast Guard Compatibility Classification:** Not listed

3.3 **Chemical Formula:** $CaCrO_4 \cdot 2H_2O$

3.4 **IMCO/United Nations Numerical Designation:** Not listed

4. OBSERVABLE CHARACTERISTICS

4.1 **Physical State (as shipped):** Solid

4.2 **Color:** Yellow

4.3 **Odor:** None

5. HEALTH HAZARDS

5.1 **Personal Protective Equipment:** Dust mask; goggles or face shield; protective gloves

5.2 **Symptoms Following Exposure:** Inhalation causes irritation of nose and throat. Ingestion causes severe circulatory collapse and chronic nephritis. Contact with eyes causes irritation. Contact with skin may cause dermatitis and ulcers.

5.3 **Treatment for Exposure:** INHALATION: remove to fresh air. INGESTION: give large amounts of water; induce vomiting. EYES: flush with water for at least 15 min. SKIN: treat local injuries like acid burns; scrub with dilute (2%) sodium hyposulfite solution.

5.4 **Toxicity by Inhalation (Threshold Limit Value):** 0.1 mg/m^3

5.5 **Short-Term Inhalation Limits:** Data not available

5.6 **Toxicity by Ingestion:** Grade 3; LD_{50} 50—500 mg/kg

5.7 **Late Toxicity:** Lung cancer may develop.

5.8 **Vapor (Gas) Irritant Characteristics:** Data not available

5.9 **Liquid or Solid Irritant Characteristics:** Data not available

5.10 **Odor Threshold:** Data not available

6. FIRE HAZARDS

6.1 **Flash Point:** Not flammable

6.2 **Flammable Limits in Air:** Not flammable

6.3 **Fire Extinguishing Agents:** Not pertinent

6.4 **Fire Extinguishing Agents Not to be Used:** Not pertinent

6.5 **Special Hazards of Combustion Products:** Toxic chromium fume may be formed in fires.

6.6 **Behavior in Fire:** The hydrated salt loses water when hot and changes color, but no increase in hazard occurs.

6.7 **Ignition Temperature:** Not pertinent

6.8 **Electrical Hazard:** Not pertinent

6.9 **Burning Rate:** Not pertinent

7. CHEMICAL REACTIVITY

7.1 **Reactivity with Water:** No reaction

7.2 **Reactivity with Common Materials:**

7.3 **Stability During Transport:** Stable

7.4 **Neutralizing Agents for Acids and Caustics:** Not pertinent

7.5 **Polymerization:** Not pertinent

7.6 **Inhibitor of Polymerization:** Not pertinent

8. WATER POLLUTION

8.1 **Aquatic Toxicity:** Data not available

8.2 **Waterfowl Toxicity:** Data not available

8.3 **Biological Oxygen Demand (BOD):** None

8.4 **Food Chain Concentration Potential:** Bioconcentration up to 2000-fold possible under constant exposure. Not significant under spill conditions.

9. SELECTED MANUFACTURERS

1. Allied Chemical Corp.
Specialty Chemicals Div.
P.O. Box 1087R
Morristown, N. J. 07960
2. Gallard-Schlesinger Chemical Mfg. Co.
584 Mineola Avenue
Carle Place, N.Y. 11514
3. Pfaltz and Bauer, Inc.
126-04 Northern Blvd.
Flushing, N.Y. 11368

10. SHIPPING INFORMATION

10.1 **Grades or Purity:** Technical

10.2 **Storage Temperature:** Ambient

10.3 **Inert Atmosphere:** No requirement

10.4 **Venting:** Open

11. HAZARD ASSESSMENT CODE

(See Hazard Assessment Handbook, CG 446-3)

SS

12. HAZARD CLASSIFICATIONS

12.1 **Code of Federal Regulations:** Not listed

12.2 **NAS Hazard Rating for Bulk Water Transportation:** Not listed

12.3 **NFPA Hazard Classifications:** Not listed

13. PHYSICAL AND CHEMICAL PROPERTIES

13.1 **Physical State at 15°C and 1 atm:** Solid

13.2 **Molecular Weight:** 192.1

13.3 **Boiling Point at 1 atm:** Not pertinent (decomposes)

13.4 **Freezing Point:** Not pertinent

13.5 **Critical Temperature:** Not pertinent

13.6 **Critical Pressure:** Not pertinent

13.7 **Specific Gravity:** >1 at 20°C (solid)

13.8 **Liquid Surface Tension:** Not pertinent

13.9 **Liquid-Water Interfacial Tension:** Not pertinent

13.10 **Vapor (Gas) Specific Gravity:** Not pertinent

13.11 **Ratio of Specific Heats of Vapor (Gas):** Not pertinent

13.12 **Latent Heat of Vaporization:** Not pertinent

13.13 **Heat of Combustion:** Not pertinent

13.14 **Heat of Decomposition:** Not pertinent

13.15 **Heat of Solution:** -73 Btu/lb $= -41$ cal/g $= -1.7 \times 10^5$ J/kg

13.16 **Heat of Polymerization:** Not pertinent

NOTES

CCN — CALCIUM CYANIDE

Common Synonyms
Cyanide of calcium

Solid — White to gray or black — Almond odor

Sinks and mixes with water.

AVOID CONTACT WITH SOLID AND DUST. KEEP PEOPLE AWAY.
Wear goggles and self-contained breathing apparatus.
Isolate and remove discharged material.
Notify local health and pollution control agencies.

Fire

Not flammable.
POISONOUS GASES ARE PRODUCED WHEN HEATED.
DO NOT USE WATER, FOAM OR CARBON DIOXIDE ON ADJACENT FIRES.

Exposure

CALL FOR MEDICAL AID.

DUST
POISONOUS IF INHALED.
Irritating to eyes, nose and throat.
Move victim to fresh air.
If in eyes, hold eyelids open and flush with plenty of water.
If breathing has stopped, give artificial respiration.
If breathing is difficult, give oxygen.

SOLID
POISONOUS IF SWALLOWED.
Irritating to skin and eyes.
Remove contaminated clothing and shoes.
Flush affected areas with plenty of water.
IF IN EYES, hold eyelids open and flush with plenty of water.
IF SWALLOWED and victim is CONSCIOUS, have victim drink water or milk and have victim induce vomiting.
IF SWALLOWED and victim is UNCONSCIOUS OR HAVING CONVULSIONS, do nothing except keep victim warm.

Water Pollution

HARMFUL TO AQUATIC LIFE IN VERY LOW CONCENTRATIONS.
MAY BE DANGEROUS IF IT ENTERS WATER INTAKES.
Notify local health and wildlife officials.
Notify operators of nearby water intakes.

1. RESPONSE TO DISCHARGE

(See Response Methods Handbook, CG 446-4)

Issue warning—poison, air contaminant, water contaminant
Restrict access
Evacuate area
Should be removed
Chemical and physical treatment

2. LABEL

3. CHEMICAL DESIGNATIONS

3.1 **Synonyms:** Cyanide of calcium; Cyanogas G-Fumigant; Cyanogas A-Dust
3.2 **Coast Guard Compatibility Classification:** Not applicable
3.3 **Chemical Formula:** $Ca(CN)_2$ plus inert ingredients
3.4 **IMCO/United Nations Numerical Designation:** 6.1/1575

4. OBSERVABLE CHARACTERISTICS

4.1 **Physical State (as shipped):** Solid
4.2 **Color:** White to gray to black
4.3 **Odor:** Compound reacts with moisture in air to form hydrogen cyanide gas, which has a characteristic almond-like odor.

5. HEALTH HAZARDS

5.1 **Personal Protective Equipment:** Self-contained breathing apparatus and full protective clothing, including rubber footwear.
5.2 **Symptoms Following Exposure:** Inhalation or ingestion causes headache, nausea, vomiting and weakness; high concentrations are rapidly fatal.
5.3 **Treatment for Exposure:** *Call a doctor immediately.* INHALATION: break amyl nitrite pearl in cloth and hold lightly under nose for 15 sec.; repeat 5 times at 15-sec. intervals; use artificial respiration if breathing stops. EYES: flush with water for 15 min.; do not allow water to enter nose or mouth. SKIN: flush with water; do not allow water to enter nose or mouth. INGESTION: break an amyl nitrite pearl in a cloth and hold lightly under nose for 15 sec.; if patient is conscious, induce vomiting and repeat until vomit is clear; repeat inhalation of amyl nitrite 5 times at 15-sec. intervals; use artificial respiration if breathing has stopped.
5.4 **Toxicity by Inhalation (Threshold Limit Value):** 5 mg/m^3 (as cyanide)
5.5 **Short-Term Inhalation Limits:** 5 mg/m^3 for 30 min.
5.6 **Toxicity by Ingestion:** Grade 4; oral LD_{50} = 39 mg/kg (rat)
5.7 **Late Toxicity:** Data not available
5.8 **Vapor (Gas) Irritant Characteristics:** Data not available
5.9 **Liquid or Solid Irritant Characteristics:** Data not available
5.10 **Odor Threshold:** Not pertinent

6. FIRE HAZARDS

6.1 **Flash Point:** Not flammable
6.2 **Flammable Limits in Air:** Not flammable
6.3 **Fire Extinguishing Agents:** Use dry chemical, sand, or earth on adjacent fires.
6.4 **Fire Extinguishing Agents Not to be Used:** Do not use water or carbon dioxide on adjacent fires.
6.5 **Special Hazards of Combustion Products:** Decomposes in fire to give very toxic gases, including hydrogen cyanide.
6.6 **Behavior in Fire:** Not pertinent
6.7 **Ignition Temperature:** Not pertinent
6.8 **Electrical Hazard:** Not pertinent
6.9 **Burning Rate:** Not pertinent

7. CHEMICAL REACTIVITY

7.1 **Reactivity with Water:** Releases very poisonous hydrogen cyanide gas slowly on contact with water. Release is rapid if acid is also present.
7.2 **Reactivity with Common Materials:** No reaction
7.3 **Stability During Transport:** Stable if kept dry
7.4 **Neutralizing Agents for Acids and Caustics:** Not pertinent
7.5 **Polymerization:** Not pertinent
7.6 **Inhibitor of Polymerization:** Not pertinent

8. WATER POLLUTION

8.1 **Aquatic Toxicity:**
0.12 ppm/96 hr/sunfish/TL_m/fresh water
>25 ppm/48 hr/cockle/LC_{50}/salt water
8.2 **Waterfowl Toxicity:** Data not available
8.3 **Biological Oxygen Demand (BOD):** Data not available
8.4 **Food Chain Concentration Potential:** None

9. SELECTED MANUFACTURERS

1. American Cyanamid Co.
   Agricultural Division
   P. O. Box 400
   Princeton, N. J. 08540
2. Pfaltz and Bauer, Inc.
   126-04 Northern Boulevard
   Flushing, N. Y. 11368

10. SHIPPING INFORMATION

10.1 **Grades or Purity:** 42% with 58% inert ingredients. May contain up to 3% calcium carbide, which releases flammable acetylene gas when wet.
10.2 **Storage Temperature:** Ambient
10.3 **Inert Atmosphere:** No requirement
10.4 **Venting:** Sealed containers in well-ventilated area

11. HAZARD ASSESSMENT CODE

(See Hazard Assessment Handbook, CG 446-3)

SS

12. HAZARD CLASSIFICATIONS

12.1 **Code of Federal Regulations:** Poisonous, class B
12.2 **NAS Hazard Rating for Bulk Water Transportation:** Not listed
12.3 **NFPA Hazard Classifications:**

Category	Classification
Health Hazard (Blue)	3
Flammability (Red)	0
Reactivity (Yellow)	0

13. PHYSICAL AND CHEMICAL PROPERTIES

13.1 **Physical State at 15°C and 1 atm:** Solid
13.2 **Molecular Weight:** 92
13.3 **Boiling Point at 1 atm:** Decomposes
13.4 **Freezing Point:** Not pertinent
13.5 **Critical Temperature:** Not pertinent
13.6 **Critical Pressure:** Not pertinent
13.7 **Specific Gravity:** 1.853 at 20°C (solid)
13.8 **Liquid Surface Tension:** Not pertinent
13.9 **Liquid-Water Interfacial Tension:** Not pertinent
13.10 **Vapor (Gas) Specific Gravity:** Not pertinent
13.11 **Ratio of Specific Heats of Vapor (Gas):** Not pertinent
13.12 **Latent Heat of Vaporization:** Not pertinent
13.13 **Heat of Combustion:** Not pertinent
13.14 **Heat of Decomposition:** Not pertinent
13.15 **Heat of Solution:** –264 Btu/lb = –147 cal/g = -6.14×10^5 J/kg
13.16 **Heat of Polymerization:** Not pertinent

NOTES

CAF — CALCIUM FLUORIDE

Common Synonyms			
Fluospar	Solid powder or granules	Gray	Odorless
	Sinks in water.		

Isolate and remove discharged material.
Notify local health and pollution control agencies.

Fire	Not flammable.
Exposure	**CALL FOR MEDICAL AID.** SOLID Harmful if swallowed. Not irritating to skin or eyes. **IF SWALLOWED and victim is CONSCIOUS, have victim drink water or milk.**
Water Pollution	Dangerous to aquatic life in high concentrations. May be dangerous if it enters water intakes. **Notify local health and pollution control officials.** **Notify operators of nearby water intakes.**

1. RESPONSE TO DISCHARGE
(See Response Methods Handbook, CG 446-4)

Disperse and flush

2. LABELS

No hazard label required by Code of Federal Regulations

3. CHEMICAL DESIGNATIONS

3.1 **Synonyms:** Fluospar; Fluorspar

3.2 **Coast Guard Compatibility Classification:** Not applicable

3.3 **Chemical Formula:** CaF_2

3.4 **IMCO/United Nations Numerical Designation:** Not listed

4. OBSERVABLE CHARACTERISTICS

4.1 **Physical State (as shipped):** Solid

4.2 **Color:** Gray

4.3 **Odor:** Odorless

5. HEALTH HAZARDS

5.1 **Personal Protective Equipment:** For dust only

5.2 **Symptoms Following Exposure:** Little acute toxicity

5.3 **Treatment for Exposure:** Usually no treatment needed

5.4 **Toxicity by Inhalation (Threshold Limit Value):** Not pertinent

5.5 **Short-Term Inhalation Limits:** Not pertinent

5.6 **Toxicity by Ingestion:** Grade 2; LD_{50} 0.5 to 5 g/kg

5.7 **Late Toxicity:** Data not available

5.8 **Vapor (Gas) Irritant Characteristics:** Not pertinent

5.9 **Liquid or Solid Irritant Characteristics:** No appreciable hazard. Practically harmless to the skin.

5.10 **Odor Threshold:** Not pertinent

6. FIRE HAZARDS

6.1 **Flash Point:** Not flammable

6.2 **Flammable Limits in Air:** Not flammable

6.3 **Fire Extinguishing Agents:** Not pertinent

6.4 **Fire Extinguishing Agents Not to be Used:** Not pertinent

6.5 **Special Hazards of Combustion Products:** Not pertinent

6.6 **Behavior in Fire:** Not pertinent

6.7 **Ignition Temperature:** Not flammable

6.8 **Electrical Hazard:** Not pertinent

6.9 **Burning Rate:** Not flammable

7. CHEMICAL REACTIVITY

7.1 **Reactivity with Water:** No reaction

7.2 **Reactivity with Common Materials:** No reaction

7.3 **Stability During Transport:** Stable

7.4 **Neutralizing Agents for Acids and Caustics:** Not pertinent

7.5 **Polymerization:** Not pertinent

7.6 **Inhibitor of Polymerization:** Not pertinent

8. WATER POLLUTION

8.1 **Aquatic Toxicity:**
30000 ppm/*/tinca vulgaris/lethal/fresh water
*Time period not specified

8.2 **Waterfowl Toxicity:** Data not available

8.3 **Biological Oxygen Demand (BOD):** Not pertinent

8.4 **Food Chain Concentration Potential:** Not pertinent

9. SELECTED MANUFACTURERS

1. Allied Chemical Corp.
   Specialty Chemicals Div.
   Marcus Hook, Pa. 19061
2. Combustion Engineering Inc.
   C-E Minerals Div.
   443 South Gulph Rd.
   King of Prussia, Pa. 19406
3. Ozark-Mahoning Co.
   Tulsa, Oklahoma 74119

10. SHIPPING INFORMATION

10.1 **Grades or Purity:** Acid grade: 97.4%; Ceramic grade: 91.5%; Fine powder (dry or damp cake); Gravel fluorspar; Pellet

10.2 **Storage Temperature:** Data not available

10.3 **Inert Atmosphere:** Data not available

10.4 **Venting:** Data not available

11. HAZARD ASSESSMENT CODE
(See Hazard Assessment Handbook, CG 446-3)

II

12. HAZARD CLASSIFICATIONS

12.1 **Code of Federal Regulations:** Not listed

12.2 **NAS Hazard Rating for Bulk Water Transportation:** Not listed

12.3 **NFPA Hazard Classifications:** Not listed

13. PHYSICAL AND CHEMICAL PROPERTIES

13.1 **Physical State at 15°C and 1 atm:** Solid

13.2 **Molecular Weight:** 78.08

13.3 **Boiling Point at 1 atm:** Not pertinent

13.4 **Freezing Point:** Not pertinent

13.5 **Critical Temperature:** Not pertinent

13.6 **Critical Pressure:** Not pertinent

13.7 **Specific Gravity:** 3.18 at 20°C (solid)

13.8 **Liquid Surface Tension:** Not pertinent

13.9 **Liquid-Water Interfacial Tension:** Not pertinent

13.10 **Vapor (Gas) Specific Gravity:** Not pertinent

13.11 **Ratio of Specific Heats of Vapor (Gas):** Not pertinent

13.12 **Latent Heat of Vaporization:** Not pertinent

13.13 **Heat of Combustion:** Not pertinent

13.14 **Heat of Decomposition:** Not pertinent

13.15 **Heat of Solution:** Not pertinent

13.16 **Heat of Polymerization:** Not pertinent

NOTES

CALCIUM HYDROXIDE

Common Synonyms			
Slaked lime	Solid granules	White	Odorless
	Sinks in water.		

Avoid contact with solid and dust. Keep people away.
Wear goggles, self-contained breathing apparatus, and rubber overclothing (including gloves).
Isolate and remove discharged material.
Notify local health and pollution control agencies.

Fire

Not flammable.

Exposure

CALL FOR MEDICAL AID.

DUST
Irritating to nose and throat if inhaled.
Move to fresh air.

SOLID
Will burn skin and eyes.
Harmful if swallowed.
Remove contaminated clothing and shoes.
Flush affected areas with plenty of water.
IF IN EYES, hold eyelids open and flush with plenty of water.
IF SWALLOWED and victim is CONSCIOUS, have victim drink water or milk.
DO NOT INDUCE VOMITING.

Water Pollution

HARMFUL TO AQUATIC LIFE IN VERY LOW CONCENTRATIONS.
May be dangerous if it enters water intakes.
Notify local health and wildlife officials.
Notify operators of nearby water intakes.

1. RESPONSE TO DISCHARGE

(See Response Methods Handbook, CG 446-4)

Disperse and flush

2. LABELS

No hazard label required by Code of Federal Regulations

3. CHEMICAL DESIGNATIONS

3.1 **Synonyms:** Slaked lime
3.2 **Coast Guard Compatibility Classification:** Not applicable
3.3 **Chemical Formula:** $Ca(OH)_2$
3.4 **IMCO/United Nations Numerical Designation:** Not listed

4. OBSERVABLE CHARACTERISTICS

4.1 **Physical State (as shipped):** Solid
4.2 **Color:** White
4.3 **Odor:** None

5. HEALTH HAZARDS

5.1 **Personal Protective Equipment:** Dust-proof goggles and mask.
5.2 **Symptoms Following Exposure:** Dust irritates eyes, nose, and throat.
5.3 **Treatment for Exposure:** INGESTION: have victim drink milk and water. Do NOT induce vomiting. EYES: flush with a gentle stream of water for at least 10 min. and consult an opthalmologist for further treatment without delay. SKIN: wash off the lime and consult a physician.
5.4 **Toxicity by Inhalation (Threshold Limit Value):** Not pertinent
5.5 **Short-Term Inhalation Limits:** Not pertinent
5.6 **Toxicity by Ingestion:** Grade 1; LD_{50} 5 to 15 g/kg (rat)
5.7 **Late Toxicity:** None
5.8 **Vapor (Gas) Irritant Characteristics:** Not pertinent
5.9 **Liquid or Solid Irritant Characteristics:** None
5.10 **Odor Threshold:** Not pertinent

6. FIRE HAZARDS

6.1 **Flash Point:** Not flammable
6.2 **Flammable Limits in Air:** Not flammable
6.3 **Fire Extinguishing Agents:** Not pertinent
6.4 **Fire Extinguishing Agents Not to be Used:** Not pertinent
6.5 **Special Hazards of Combustion Products:** Not pertinent
6.6 **Behavior in Fire:** Not pertinent
6.7 **Ignition Temperature:** Not flammable
6.8 **Electrical Hazard:** Not pertinent
6.9 **Burning Rate:** Not flammable

7. CHEMICAL REACTIVITY

7.1 **Reactivity with Water:** No reaction
7.2 **Reactivity with Common Materials:** No reaction
7.3 **Stability During Transport:** Stable
7.4 **Neutralizing Agents for Acids and Caustics:** Not pertinent
7.5 **Polymerization:** Not pertinent
7.6 **Inhibitor of Polymerization:** Not pertinent

8. WATER POLLUTION

8.1 **Aquatic Toxicity:**
92 ppm/7 hr/trout/toxic/fresh water
240 ppm/24 hr/mosquito fish/TL_m/fresh water
8.2 **Waterfowl Toxicity:** Data not available
8.3 **Biological Oxygen Demand (BOD):** None
8.4 **Food Chain Concentration Potential:** None

9. SELECTED MANUFACTURERS

1. Ash Grove Cement Co.
1000 Ten Main Center
Kansas City, Mo. 64105
2. Engelhard Minerals & Chemical Corp.
Minerals and Chemicals Division
Menlo Park, N. J. 08817
3. United States Gypsum Co.
Chemicals Division
Chicago, Ill. 60606

10. SHIPPING INFORMATION

10.1 **Grades or Purity:** Agricultural: 65-71%; industrial: 70-73%; chemical: 71-73%
10.2 **Storage Temperature:** Data not available
10.3 **Inert Atmosphere:** Data not available
10.4 **Venting:** Data not available

11. HAZARD ASSESSMENT CODE

(See Hazard Assessment Handbook, CG 446-3)

I I

12. HAZARD CLASSIFICATIONS

12.1 **Code of Federal Regulations:** Not listed
12.2 **NAS Hazard Rating for Bulk Water Transportation:** Not listed
12.3 **NFPA Hazard Classifications:** Not listed

13. PHYSICAL AND CHEMICAL PROPERTIES

13.1 **Physical State at 15°C and 1 atm:** Solid
13.2 **Molecular Weight:** 74.09
13.3 **Boiling Point at 1 atm:** Not pertinent
13.4 **Freezing Point:** Not pertinent
13.5 **Critical Temperature:** Not pertinent
13.6 **Critical Pressure:** Not pertinent
13.7 **Specific Gravity:** 2.24 at 20°C (solid)
13.8 **Liquid Surface Tension:** Not pertinent
13.9 **Liquid-Water Interfacial Tension:** Not pertinent
13.10 **Vapor (Gas) Specific Gravity:** Not pertinent
13.11 **Ratio of Specific Heats of Vapor (Gas):** Not pertinent
13.12 **Latent Heat of Vaporization:** Not pertinent
13.13 **Heat of Combustion:** Not pertinent
13.14 **Heat of Decomposition:** Not pertinent
13.15 **Heat of Solution:** Not pertinent
13.16 **Heat of Polymerization:** Not pertinent

NOTES

CALCIUM HYPOCHLORITE

Common Synonyms		
HTH HTH Dry Chlorine Neutral anhydrous calcium hypochlorite Sentry	Solid granules White Household bleaching powder odor Sinks and mixes with water.	

Avoid contact with solid, and dust. Keep people away.
Wear goggles and self-contained breathing apparatus.
Isolate and remove discharged material.
Notify local health and pollution control agencies.

Fire	Not flammable. May cause fire on contact with combustibles. POISONOUS GASES ARE PRODUCED WHEN HEATED. Wear chemical protective suit including self-contained breathing apparatus. Extinguish adjacent fires with water.
Exposure	CALL FOR MEDICAL AID. SOLID Irritating to skin and eyes. If swallowed, will cause nausea, vomiting or loss of consciousness. Remove contaminated clothing and shoes. Flush affected areas with plenty of water. IF IN EYES, hold eyelids open and flush with plenty of water. IF SWALLOWED and victim is CONSCIOUS, have victim drink water or milk. IF SWALLOWED and victim is UNCONSCIOUS OR HAVING CONVULSIONS, do nothing except keep victim warm. DO NOT INDUCE VOMITING.
Water Pollution	HARMFUL TO AQUATIC LIFE IN VERY LOW CONCENTRATIONS. May be dangerous if it enters water intakes. Notify local health and wildlife officials. Notify operators of nearby water intakes.

1. RESPONSE TO DISCHARGE

(See Response Methods Handbook, CG 446-4)

Issue warning—corrosive
Disperse and flush

2. LABEL

Yellow

3. CHEMICAL DESIGNATIONS

3.1 **Synonyms:** HTH; HTH Dry Chlorine; Neutral anhydrous calcium hypochlorite; Sentry
3.2 **Coast Guard Compatibility Classification:** Not applicable
3.3 **Chemical Formula:** $Ca(OCl)_2$
3.4 **IMCO/United Nations Numerical Designation:** 5.1/1748

4. OBSERVABLE CHARACTERISTICS

4.1 **Physical State (as shipped):** Solid
4.2 **Color:** White
4.3 **Odor:** Like bleaching powder

5. HEALTH HAZARDS

5.1 **Personal Protective Equipment:** Protective goggles, dust mask.
5.2 **Symptoms Following Exposure:** INHALATION: hypochlorous acid fumes* cause severe respiratory tract irritation and pulmonary edema. INGESTION: pain and inflammation of mouth, pharynx, esophagus, and stomach; erosion of mucous membranes, chiefly of the stomach; vomiting (hemorrhaging may cause vomitus to resemble coffee grounds); circulatory collapse, with cold and clammy skin, cyanosis, and shallow respirations; confusion, delirium, coma; edema of pharynx, glottis, and larynx, with stridor and obstruction; perforation of esophagus or stomach, with mediastinitis or peritonitis. SKIN CONTACT: may cause vesicular eruptions and eczematoid dermatitis.
5.3 **Treatment for Exposure:** INGESTION: swallow immediately milk, egg white, starch paste, milk of magnesia, aluminum hydroxide gel, or magnesium trisilicate gel. Avoid sodium bicarbonate because of the release of carbon dioxide. Do not use acidic antidotes; cautious gastric lavage with tap water or a 1% solution of sodium thiosulfate; milk of magnesia (1 oz) left in the stomach is useful as a mild antacid, adsorbent, demulcent, and cathartic; demulcents, such as starch, egg white, milk, gruel; opiates for the control of pain. Treat shock vigorously with intravenous fluids. Prompt surgical intervention when indicated, e.g., tracheotomy, gastrectomy. SKIN: wash with liberal quantities of water and apply a paste of baking soda.
5.4 **Toxicity by Inhalation (Threshold Limit Value):** Not pertinent
5.5 **Short-Term Inhalation Limits:** Not pertinent
5.6 **Toxicity by Ingestion:** Grade 0; LD_{50} above 15 g/kg

*Given off only if compound comes in contact with acid.

6. FIRE HAZARDS

6.1 **Flash Point:** Not flammable
6.2 **Flammable Limits in Air:** Not flammable
6.3 **Fire Extinguishing Agents:** Not pertinent
6.4 **Fire Extinguishing Agents Not to be Used:** Not pertinent
6.5 **Special Hazards of Combustion Products:** Not pertinent
6.6 **Behavior in Fire:** Poisonous gases may be produced when heated
6.7 **Ignition Temperature:** Not flammable
6.8 **Electrical Hazard:** Not pertinent
6.9 **Burning Rate:** Not flammable

7. CHEMICAL REACTIVITY

7.1 **Reactivity with Water:** No reaction
7.2 **Reactivity with Common Materials:** May cause fire in contact with wood or straw. Corrosive to most metals.
7.3 **Stability During Transport:** The 70% grade may decompose violently if exposed to heat or direct sunlight. Gives off chlorine and chlorine monoxide above 350°F (poisonous gases).
7.4 **Neutralizing Agents for Acids and Caustics:** Dilute with water.
7.5 **Polymerization:** Not pertinent
7.6 **Inhibitor of Polymerization:** Not pertinent

8. WATER POLLUTION

8.1 **Aquatic Toxicity:** 0.5 ppm/*/trout/killed/fresh water
*Time period not specified
8.2 **Waterfowl Toxicity:** Data not available
8.3 **Biological Oxygen Demand (BOD):** Not pertinent
8.4 **Food Chain Concentration Potential:** Not pertinent

9. SELECTED MANUFACTURERS

1. Olin Corp.
Chemicals Division
Brandenburg, Ky. 40108
2. Pennwalt Corp.
Chemical Division
4655 Biddle Ave.
Wyandotte, Mich. 48193
3. P P G Industries Inc.
Industrial Chemicals Division
Barberton, Ohio 44203

10. SHIPPING INFORMATION

10.1 **Grades or Purity:** 70% (self-propagating); 65% (non propagating)
10.2 **Storage Temperature:** Data not available
10.3 **Inert Atmosphere:** Data not available
10.4 **Venting:** Data not available

11. HAZARD ASSESSMENT CODE

(See Hazard Assessment Handbook, CG 446-3)

SS

12. HAZARD CLASSIFICATIONS

12.1 **Code of Federal Regulations:** Oxidizing material
12.2 **NAS Hazard Rating for Bulk Water Transportation:** Not listed
12.3 **NFPA Hazard Classifications:**

Category	Classification*	
Health Hazard (Blue)	1	2
Flammability (Red)	0	0
Reactivity (Yellow)	2	2
	oxy	oxy

*First column refers to non-fire situation.

13. PHYSICAL AND CHEMICAL PROPERTIES

13.1 **Physical State at 15°C and 1 atm:** Solid
13.2 **Molecular Weight:** 174.98
13.3 **Boiling Point at 1 atm:** Not pertinent
13.4 **Freezing Point:** Not pertinent
13.5 **Critical Temperature:** Not pertinent
13.6 **Critical Pressure:** Not pertinent
13.7 **Specific Gravity:** 2.35 at 20°C (solid)
13.8 **Liquid Surface Tension:** Not pertinent
13.9 **Liquid-Water Interfacial Tension:** Not pertinent
13.10 **Vapor (Gas) Specific Gravity:** Not pertinent
13.11 **Ratio of Specific Heats of Vapor (Gas):** Not pertinent
13.12 **Latent Heat of Vaporization:** Not pertinent
13.13 **Heat of Combustion:** Not pertinent
13.14 **Heat of Decomposition:** Not pertinent
13.15 **Heat of Solution:** Not pertinent
13.16 **Heat of Polymerization:** Not pertinent

5. HEALTH HAZARDS (Cont'd.)

5.7 **Late Toxicity:** None
5.8 **Vapor (Gas) Irritant Characteristics:** Not pertinent
5.9 **Liquid or Solid Irritant Characteristics:** Irritates eyes, skin, and mucous membranes.
5.10 **Odor Threshold:** Not pertinent

REVISED 1978

CAM	CALCIUM, METALLIC

Common Synonyms	Solid Silvery to grayish white Odorless Sinks in water. Reacts slowly with water.

Avoid contact with solid and dust. Keep people away.
Wear rubber overclothing (including gloves).
Call fire department.
Isolate and remove discharged material.
Notify local health and pollution control agencies.

Fire	FLAMMABLE. **Extinguish with dry graphite, soda ash, or other inert powder.** **DO NOT USE WATER, FOAM, CARBON DIOXIDE OR VAPORIZING LIQUIDS ON FIRE.**
Exposure	**Call for medical aid.** SOLID Will burn skin and eyes. **Remove contaminated clothing and shoes.** **Flush affected areas with plenty of water.** **IF IN EYES, hold eyelids open and flush with plenty of water.**
Water Pollution	Dangerous to aquatic life in high concentrations. May be dangerous if it enters water intakes. **Notify local health and wildlife officials.** **Notify operators of nearby water intakes.**

1. RESPONSE TO DISCHARGE
(See Response Methods Handbook, CG 446-4)

Issue warning—high flammability
Disperse and flush

2. LABELS

3. CHEMICAL DESIGNATIONS
3.1 **Synonyms:** No common synonyms
3.2 **Coast Guard Compatibility Classification:** Not applicable
3.3 **Chemical Formula:** Ca
3.4 **IMCO/United Nations Numerical Designation:** 4.3/1401

4. OBSERVABLE CHARACTERISTICS
4.1 **Physical State (as shipped):** Solid
4.2 **Color:** Silvery; turns to grayish-white on exposure to air.
4.3 **Odor:** None

5. HEALTH HAZARDS
5.1 **Personal Protective Equipment:** Goggles and rubber gloves.
5.2 **Symptoms Following Exposure:** Contact with eyes or skin produces caustic burns.
5.3 **Treatment for Exposure:** EYES or SKIN: flush with water.
5.4 **Toxicity by Inhalation (Threshold Limit Value):** Not pertinent
5.5 **Short-Term Inhalation Limits:** Not pertinent
5.6 **Toxicity by Ingestion:** Data not available
5.7 **Late Toxicity:** None
5.8 **Vapor (Gas) Irritant Characteristics:** Not pertinent
5.9 **Liquid or Solid Irritant Characteristics:** Data not available
5.10 **Odor Threshold:** Not pertinent

6. FIRE HAZARDS
6.1 **Flash Point:** Not pertinent (flammable solid)
6.2 **Flammable Limits in Air:** Not pertinent
6.3 **Fire Extinguishing Agents:** Dry graphite, soda ash, powdered sodium chloride, or appropriate metal fire extinguishing dry powder.
6.4 **Fire Extinguishing Agents Not to be Used:** Water, halogenated hydrocarbons, dry chemical, carbon dioxide
6.5 **Special Hazards of Combustion Products:** Not pertinent
6.6 **Behavior in Fire:** Burns violently, especially if finely divided.
6.7 **Ignition Temperature:** 1454 ± 18°F
6.8 **Electrical Hazard:** Not pertinent
6.9 **Burning Rate:** Not pertinent

7. CHEMICAL REACTIVITY
7.1 **Reactivity with Water:** Reacts to form flammable hydrogen gas, which may ignite. The reaction is not violent.
7.2 **Reactivity with Common Materials:** Reacts with moist air to form skin of hydroxide. The reaction is not hazardous.
7.3 **Stability During Transport:** Stable
7.4 **Neutralizing Agents for Acids and Caustics:** Flush with water
7.5 **Polymerization:** Not pertinent
7.6 **Inhibitor of Polymerization:** Not pertinent

8. WATER POLLUTION
8.1 **Aquatic Toxicity:** See Calcium hydroxide (CAH)
8.2 **Waterfowl Toxicity:** See Calcium hydroxide (CAH)
8.3 **Biological Oxygen Demand (BOD):** None
8.4 **Food Chain Concentration Potential:** None

9. SELECTED MANUFACTURERS
1. Pfizer Minerals
   Metals and Composite Products Division
   253 E. 42 Street
   New York, N. Y. 10017
2. Gallard-Schlesinger Chemical Manufacturing Co.
   Atomergic Chemetals Division
   584 Mineola Ave.
   Carle Place, L. I., N. Y. 11514
3. Cerac, Inc.
   13460 West Silver Spring Rd.
   Menomonee Falls, Wis. 53051

10. SHIPPING INFORMATION
10.1 **Grades or Purity:** Commercial, 99.5%; redistilled 99.9%
10.2 **Storage Temperature:** Ambient
10.3 **Inert Atmosphere:** No requirement
10.4 **Venting:** Sealed containers must be in a ventilated area.

11. HAZARD ASSESSMENT CODE
(See Hazard Assessment Handbook, CG 446-3)

II-RR

12. HAZARD CLASSIFICATIONS
12.1 **Code of Federal Regulations:** Flammable solid
12.2 **NAS Hazard Rating for Bulk Water Transportation:** Not listed
12.3 **NFPA Hazard Classifications:**

Category	Classification
Health Hazard (Blue)	1
Flammability (Red)	1
Reactivity (Yellow)	2 W̶

13. PHYSICAL AND CHEMICAL PROPERTIES
13.1 **Physical State at 15°C and 1 atm:** Solid
13.2 **Molecular Weight:** 40.1
13.3 **Boiling Point at 1 atm:** 2,714°F = 1,490°C = 1,763°K
13.4 **Freezing Point:** 1,562°F = 850°C = 1,123°K
13.5 **Critical Temperature:** Not pertinent
13.6 **Critical Pressure:** Not pertinent
13.7 **Specific Gravity:** 1.55 at 20°C (solid)
13.8 **Liquid Surface Tension:** Not pertinent
13.9 **Liquid-Water Interfacial Tension:** Not pertinent
13.10 **Vapor (Gas) Specific Gravity:** Not pertinent
13.11 **Ratio of Specific Heats of Vapor (Gas):** Not pertinent
13.12 **Latent Heat of Vaporization:** Not pertinent
13.13 **Heat of Combustion:** –6790 Btu/lb = –3,770 cal/g = -158×10^5 J/kg
13.14 **Heat of Decomposition:** Not pertinent
13.15 **Heat of Solution:** Not pertinent
13.16 **Heat of Polymerization:** Not pertinent

NOTES

CNT

CALCIUM NITRATE

Common Synonyms

Solid — White — Odorless

Sinks and mixes with water.

Call fire department.
Avoid contact with solid. Keep people away.
Isolate and remove discharged material.
Notify local health and pollution control agencies.

Fire

Not flammable.
May cause fire on contact with combustibles.
POISONOUS GASES MAY BE PRODUCED WHEN HEATED.
Flood discharge area with water.

Exposure

Call for medical aid.

DUST
Irritating to eyes, nose and throat.
Move victim to fresh air.
If in eyes, hold eyelids open and flush with plenty of water.

SOLID
Irritating to skin and eyes.
Harmful if swallowed.
Remove contaminated clothing and shoes.
Flush affected areas with plenty of water.
IF IN EYES, hold eyelids open and flush with plenty of water.
IF SWALLOWED and victim is CONSCIOUS, have victim drink water or milk.

Water Pollution

Dangerous to aquatic life in high concentrations.
May be dangerous if it enters water intakes.
Notify local health and wildlife officials.
Notify operators of nearby water intakes.

1. RESPONSE TO DISCHARGE
(See Response Methods Handbook, CG 446-4)

Issue warning—oxidizing material
Disperse and flush

2. LABEL

Yellow

3. CHEMICAL DESIGNATIONS

3.1 **Synonyms:** Calcium nitrate tetrahydrate
3.2 **Coast Guard Compatibility Classification:** Not applicable
3.3 **Chemical Formula:** $Ca(NO_3)_2 \cdot 4H_2O$
3.4 **IMCO/United Nations Numerical Designation:** 5.1/1454

4. OBSERVABLE CHARACTERISTICS

4.1 **Physical State (as shipped):** Solid
4.2 **Color:** White
4.3 **Odor:** None

5. HEALTH HAZARDS

5.1 **Personal Protective Equipment:** Dust respirator and rubber gloves.
5.2 **Symptoms Following Exposure:** Dust causes mild irritation of eyes.
5.3 **Treatment for Exposure:** EYES or SKIN: flush with water.
5.4 **Toxicity by Inhalation (Threshold Limit Value):** Data not available
5.5 **Short-Term Inhalation Limits:** Data not available
5.6 **Toxicity by Ingestion:** Data not available
5.7 **Late Toxicity:** None
5.8 **Vapor (Gas) Irritant Characteristics:** Not pertinent
5.9 **Liquid or Solid Irritant Characteristics:** Data not available
5.10 **Odor Threshold:** Not pertinent

6. FIRE HAZARDS

6.1 **Flash Point:** Not flammable, but may cause fire on contact with combustibles.
6.2 **Flammable Limits in Air:** Not flammable
6.3 **Fire Extinguishing Agents:** Flood with water.
6.4 **Fire Extinguishing Agents Not to be Used:** Not pertinent
6.5 **Special Hazards of Combustion Products:** May give off toxic oxides of nitrogen when involved in fire.
6.6 **Behavior in Fire:** Greatly intensifies the burning of all combustible materials.
6.7 **Ignition Temperature:** Not pertinent
6.8 **Electrical Hazard:** Not pertinent
6.9 **Burning Rate:** Not pertinent

7. CHEMICAL REACTIVITY

7.1 **Reactivity with Water:** No reaction
7.2 **Reactivity with Common Materials:** Contact with combustible material may cause fire.
7.3 **Stability During Transport:** Stable
7.4 **Neutralizing Agents for Acids and Caustics:** Not pertinent
7.5 **Polymerization:** Not pertinent
7.6 **Inhibitor of Polymerization:** Not pertinent

8. WATER POLLUTION

8.1 **Aquatic Toxicity:** 10,000 ppm/96 hr/sunfish/TL_m/fresh water
8.2 **Waterfowl Toxicity:** Data not available
8.3 **Biological Oxygen Demand (BOD):** None
8.4 **Food Chain Concentration Potential:** None

9. SELECTED MANUFACTURERS

1. J. T. Baker Chemical Co.
Phillipsburg, N. J. 08865
2. Mallinckrodt Chemical Works
223 West Side Avenue
P. O. Box 384
Jersey City, N. J. 07303
3. Barium and Chemicals, Inc.
P. O. Box 218
Steubenville, Ohio 43952

10. SHIPPING INFORMATION

10.1 **Grades or Purity:** Analytical reagent (99.0+%); purified; technical
10.2 **Storage Temperature:** Ambient
10.3 **Inert Atmosphere:** No requirement
10.4 **Venting:** Open

11. HAZARD ASSESSMENT CODE
(See Hazard Assessment Handbook, CG 446-3)

SS

12. HAZARD CLASSIFICATIONS

12.1 **Code of Federal Regulations:** Oxidizing material
12.2 **NAS Hazard Rating for Bulk Water Transportation:** Not listed
12.3 **NFPA Hazard Classifications:** Not listed

13. PHYSICAL AND CHEMICAL PROPERTIES

13.1 **Physical State at 15°C and 1 atm:** Solid
13.2 **Molecular Weight:** 164
13.3 **Boiling Point at 1 atm:** Decomposes
13.4 **Freezing Point:** 1,042°F = 561°C = 834°K
13.5 **Critical Temperature:** Not pertinent
13.6 **Critical Pressure:** Not pertinent
13.7 **Specific Gravity:** 2.50 at 18°C (solid)
13.8 **Liquid Surface Tension:** Not pertinent
13.9 **Liquid-Water Interfacial Tension:** Not pertinent
13.10 **Vapor (Gas) Specific Gravity:** Not pertinent
13.11 **Ratio of Specific Heats of Vapor (Gas):** Not pertinent
13.12 **Latent Heat of Vaporization:** Not pertinent
13.13 **Heat of Combustion:** Not pertinent
13.14 **Heat of Decomposition:** Not pertinent
13.15 **Heat of Solution:** (est.) −90 Btu/lb = −50 cal/g = -2.1×10^5 J/kg
13.16 **Heat of Polymerization:** Not pertinent

NOTES

CAO

CALCIUM OXIDE

Common Synonyms			
Unslaked lime Quicklime	Solid granules	White to grey	Odorless
	Sinks and reacts violently with water. Water appears to boil.		

Avoid contact with s.⁎.d. Keep people away.
Wear rubber overclothing (including gloves).
Notify local health and pollution control agencies.

Fire

Not flammable.
May cause fire on contact with water and combustibles.
Extinguish adjacent fires with dry chemical or carbon dioxide.
DO NOT USE WATER ON ADJACENT FIRES.

Exposure

CALL FOR MEDICAL AID.

DUST
Irritating to nose and throat.
Move to fresh air.

SOLID
Will burn skin and eyes.
Harmful if swallowed.
Remove contaminated clothing and shoes.
Flush affected areas with plenty of water.
IF IN EYES, hold eyelids open and flush with plenty of water.
IF SWALLOWED and victim is CONSCIOUS, have victim drink water or milk.
DO NOT INDUCE VOMITING.

Water Pollution

HARMFUL TO AQUATIC LIFE IN VERY LOW CONCENTRATIONS.
May be dangerous if it enters water intakes.
Notify local health and wildlife officials.
Notify operators of nearby water intakes.

1. RESPONSE TO DISCHARGE

(See Response Methods Handbook, CG 446-4)

Issue warning—corrosive
Restrict access
Chemical and physical treatment

2. LABELS

No hazard label required by Code of Federal Regulations

3. CHEMICAL DESIGNATIONS

3.1 **Synonyms:** Quicklime
Unslaked lime
3.2 **Coast Guard Compatibility Classification:** Not applicable
3.3 **Chemical Formula:** CaO
3.4 **IMCO/United Nations Numerical Designation:** 9.0/1910

4. OBSERVABLE CHARACTERISTICS

4.1 **Physical State (as shipped):** Solid
4.2 **Color:** White to gray
4.3 **Odor:** Odorless

5. HEALTH HAZARDS

5.1 **Personal Protective Equipment:** Protective gloves, goggles, and any type of respirator prescribed for fine dust.
5.2 **Symptoms Following Exposure:** Causes burns on mucous membrane and skin. Inhalation of dust causes sneezing.
5.3 **Treatment for Exposure:** INGESTION: if victim is conscious, have him drink water or milk. Do NOT induce vomiting. SKIN AND EYES: flush with water and seek medical help.
5.4 **Toxicity by Inhalation (Threshold Limit Value):** 5 mg/m³
5.5 **Short-Term Inhalation Limits:** 10 mg/m³ for 30 min.
5.6 **Toxicity by Ingestion:** Data not available
5.7 **Late Toxicity:** None
5.8 **Vapor (Gas) Irritant Characteristics:** Not pertinent
5.9 **Liquid or Solid Irritant Characteristics:** Causes smarting of the skin and first-degree burns on short exposure and may cause secondary burns on long exposure.
5.10 **Odor Threshold:** Not pertinent

6. FIRE HAZARDS

6.1 **Flash Point:** Not flammable
6.2 **Flammable Limits in Air:** Not flammable
6.3 **Fire Extinguishing Agents:** Not pertinent
6.4 **Fire Extinguishing Agents Not to be Used:** Do not use water on adjacent fires.
6.5 **Special Hazards of Combustion Products:** Not pertinent
6.6 **Behavior in Fire:** Not pertinent
6.7 **Ignition Temperature:** Not flammable
6.8 **Electrical Hazard:** Not pertinent
6.9 **Burning Rate:** Not flammable

7. CHEMICAL REACTIVITY

7.1 **Reactivity with Water:** Heat may cause ignition of combustibles. Material swells during reaction.
7.2 **Reactivity with Common Materials:** No reaction unless water present, then chief effect is that of heat liberated.
7.3 **Stability During Transport:** Stable
7.4 **Neutralizing Agents for Acids and Caustics:** Not pertinent
7.5 **Polymerization:** Not pertinent
7.6 **Inhibitor of Polymerization:** Not pertinent

8. WATER POLLUTION

8.1 **Aquatic Toxicity:**
92 ppm/7 hr/trout/toxic/fresh water
240 ppm/24 hr/mosquito fish/TL_m/fresh water
8.2 **Waterfowl Toxicity:** Data not available
8.3 **Biological Oxygen Demand (BOD):** Not pertinent
8.4 **Food Chain Concentration Potential:** Not pertinent

9. SELECTED MANUFACTURERS

1. Allied Chemical Corp.
Specialty Chemicals Division
Marcus Hook, Pa. 19061
2. Ash Grove Cement Co.
1000 Ten Main Center
Kansas City, Mo. 64105
3. United States Gypsum Co.
Chemicals Division
Chicago, Ill. 60606

10. SHIPPING INFORMATION

10.1 **Grades or Purity:** 96—97%
10.2 **Storage Temperature:** Data not available
10.3 **Inert Atmosphere:** Data not available
10.4 **Venting:** Data not available

11. HAZARD ASSESSMENT CODE

(See Hazard Assessment Handbook, CG 446-3)

RR

12. HAZARD CLASSIFICATIONS

12.1 **Code of Federal Regulations:** ORM–B
12.2 **NAS Hazard Rating for Bulk Water Transportation:** Not listed
12.3 **NFPA Hazard Classifications:**

Category	Classification
Health Hazard (Blue)	1
Flammability (Red)	0
Reactivity (Yellow)	1

13. PHYSICAL AND CHEMICAL PROPERTIES

13.1 **Physical State at 15°C and 1 atm:** Solid
13.2 **Molecular Weight:** 56.08
13.3 **Boiling Point at 1 atm:** Not pertinent
13.4 **Freezing Point:** Not pertinent
13.5 **Critical Temperature:** Not pertinent
13.6 **Critical Pressure:** Not pertinent
13.7 **Specific Gravity:** 3.3 at 20°C (solid)
13.8 **Liquid Surface Tension:** Not pertinent
13.9 **Liquid-Water Interfacial Tension:** Not pertinent
13.10 **Vapor (Gas) Specific Gravity:** Not pertinent
13.11 **Ratio of Specific Heats of Vapor (Gas):** Not pertinent
13.12 **Latent Heat of Vaporization:** Not pertinent
13.13 **Heat of Combustion:** Not pertinent
13.14 **Heat of Decomposition:** Not pertinent
13.15 **Heat of Solution:** Not pertinent
13.16 **Heat of Polymerization:** Not pertinent

NOTES

CALCIUM PEROXIDE

Common Synonyms			
Calcium dioxide	Solid powder	Yellow to white	Odorless
	Sinks in water.		

Call fire department. Keep people away.
Isolate and remove discharged material.
Notify local health and pollution control agencies.

Fire

Not flammable.
May cause fire on contact with combustibles.
Containers may explode in fire.
Flood discharge area with water.
Cool exposed containers with water.

Exposure

Call for medical aid.

DUST
Irritating to eyes, nose and throat.
Move victim to fresh air.
If in eyes, hold eyelids open and flush with plenty of water.

SOLID
Irritating to skin and eyes.
Harmful if swallowed.
Remove contaminated clothing and shoes.
Flush affected areas with plenty of water.
IF IN EYES, hold eyelids open and flush with plenty of water.
IF SWALLOWED and victim is CONSCIOUS, have victim drink water or milk.

Water Pollution

Effect of low concentrations on aquatic life is unknown.
May be dangerous if it enters water intakes.
Notify local health and wildlife officials.
Notify operators of nearby water intakes.

1. RESPONSE TO DISCHARGE
(See Response Methods Handbook, CG 446-4)

Issue warning—oxidizing material
Disperse and flush

2. LABEL

Yellow

3. CHEMICAL DESIGNATIONS

3.1 **Synonyms:** Calcium dioxide
3.2 **Coast Guard Compatibility Classification:** Not applicable
3.3 **Chemical Formula:** CaO_2
3.4 **IMCO/United Nations Numerical Designation:** 5.1/1457

4. OBSERVABLE CHARACTERISTICS

4.1 **Physical State (as shipped):** Solid
4.2 **Color:** Yellow-white
4.3 **Odor:** None

5. HEALTH HAZARDS

5.1 **Personal Protective Equipment:** U. S. B. M.-approved toxic dust respirator; general-purpose gloves; chemical safety goggles; full cover clothing.
5.2 **Symptoms Following Exposure:** Inhalation of dust irritates nose and throat. Dust also irritates eyes and skin on contact and irritates mouth and stomach if ingested.
5.3 **Treatment for Exposure:** INHALATION: remove to fresh air. EYES: flush with water for 15 min. and consult physician. SKIN: flush with water. INGESTION: give large amounts of water.
5.4 **Toxicity by Inhalation (Threshold Limit Value):** Data not available
5.5 **Short-Term Inhalation Limits:** Data not available
5.6 **Toxicity by Ingestion:** Data not available
5.7 **Late Toxicity:** Data not available
5.8 **Vapor (Gas) Irritant Characteristics:** Data not available
5.9 **Liquid or Solid Irritant Characteristics:** Data not available
5.10 **Odor Threshold:** Not pertinent

6. FIRE HAZARDS

6.1 **Flash Point:** Not flammable, but may cause fire on contact with combustible material.
6.2 **Flammable Limits in Air:** Not pertinent
6.3 **Fire Extinguishing Agents:** Flood with water, or use dry powder (e.g., graphite or powdered limestone)
6.4 **Fire Extinguishing Agents Not to be Used:** Not pertinent
6.5 **Special Hazards of Combustion Products:** Not pertinent
6.6 **Behavior in Fire:** Can increase severity of fire. Containers may explode.
6.7 **Ignition Temperature:** Not pertinent
6.8 **Electrical Hazard:** Not pertinent
6.9 **Burning Rate:** Not pertinent

7. CHEMICAL REACTIVITY

7.1 **Reactivity with Water:** Reacts very slowly with water at room temperature to form limewater and oxygen gas.
7.2 **Reactivity with Common Materials:** Heavy metals and dirt can accelerate decomposition to lime and oxygen. The reaction is not explosive.
7.3 **Stability During Transport:** Stable
7.4 **Neutralizing Agents for Acids and Caustics:** Flush with water
7.5 **Polymerization:** Not pertinent
7.6 **Inhibitor of Polymerization:** Not pertinent

8. WATER POLLUTION

8.1 **Aquatic Toxicity:** Data not available
8.2 **Waterfowl Toxicity:** Data not available
8.3 **Biological Oxygen Demand (BOD):** None
8.4 **Food Chain Concentration Potential:** None

9. SELECTED MANUFACTURERS

1. FMC Corporation
Industrial Chemicals Division
633 Third Avenue
New York, N. Y. 10017
2. Barium and Chemicals, Inc.
P. O. Box 218
Steubenville, Ohio 43952

10. SHIPPING INFORMATION

10.1 **Grades or Purity:** Commercial: 60+%
10.2 **Storage Temperature:** Stable
10.3 **Inert Atmosphere:** No requirement
10.4 **Venting:** Pressure-vacuum

11. HAZARD ASSESSMENT CODE
(See Hazard Assessment Handbook, CG 446-3)

II-RR

12. HAZARD CLASSIFICATIONS

12.1 **Code of Federal Regulations:** Oxidizing material
12.2 **NAS Hazard Rating for Bulk Water Transportation:** Not listed
12.3 **NFPA Hazard Classifications:** Not listed

13. PHYSICAL AND CHEMICAL PROPERTIES

13.1 **Physical State at 15°C and 1 atm:** Solid
13.2 **Molecular Weight:** 72.1
13.3 **Boiling Point at 1 atm:** Decomposes
13.4 **Freezing Point:** Not pertinent
13.5 **Critical Temperature:** Not pertinent
13.6 **Critical Pressure:** Not pertinent
13.7 **Specific Gravity:** 2.92 at 25°C (solid)
13.8 **Liquid Surface Tension:** Not pertinent
13.9 **Liquid-Water Interfacial Tension:** Not pertinent
13.10 **Vapor (Gas) Specific Gravity:** Not pertinent
13.11 **Ratio of Specific Heats of Vapor (Gas):** Not pertinent
13.12 **Latent Heat of Vaporization:** Not pertinent
13.13 **Heat of Combustion:** Not pertinent
13.14 **Heat of Decomposition:** −135 Btu/lb = −75 cal/g = −3.1 × [illegible]/kg
13.15 **Heat of Solution:** Not pertinent
13.16 **Heat of Polymerization:** Not pertinent

NOTES

CALCIUM PHOSPHATE

Common Synonyms			
MCP Monocalcium phosphate monohydrate Calcium biphosphate DCP Dicalcium phosphate Calcium pyrophosphate	Solid	White	Odorless
	Sinks and mixes with water.		

Stop discharge if possible. Keep people away.
Avoid contact with solid and dust.
Isolate and remove discharged material.
Notify local health and pollution control agencies.

Fire

Not flammable.

Exposure

CALL FOR MEDICAL AID.

DUST
Irritating to eyes, nose and throat.
If inhaled will cause coughing or difficult breathing.
If in eyes, hold eyelids open and flush with plenty of water.
If breathing has stopped, give artificial respiration.
If breathing is difficult, give oxygen.

SOLID
Irritating to skin and eyes.
If swallowed will cause nausea and vomiting.
Remove contaminated clothing and shoes.
Flush affected areas with plenty of water.
IF IN EYES, hold eyelids open and flush with plenty of water.
IF SWALLOWED and victim is CONSCIOUS, have victim drink water or milk and have victim induce vomiting.
IF SWALLOWED and victim is UNCONSCIOUS OR HAVING CONVULSIONS, do nothing except keep victim warm.

Water Pollution

Effect of low concentrations on aquatic life is unknown.
May be dangerous if it enters water intakes.
Notify local health and wildlife officials.
Notify operators of nearby water intakes.

1. RESPONSE TO DISCHARGE

(See Response Methods Handbook, CG 446-4)

Disperse and flush

2. LABELS

No hazard label required by Code of Federal Regulations

3. CHEMICAL DESIGNATIONS

3.1 **Synonyms:** (a) MCP; Monocalcium phosphate, monohydrate; Calcium phosphate, monobasic, monohydrate; Acid calcium phosphate; calcium superphosphate; Calcium biphosphate; Primary calcium phosphate (b) DCP; Dicalcium phosphate (anhydrous or dihydrate); Dibasic calcium phosphate; Calcium monohydrogen phosphate; Secondary calcium phosphate (c) TCP; Calcium

(Continued on page 4)

4. OBSERVABLE CHARACTERISTICS

4.1 **Physical State (as shipped):** Solid
4.2 **Color:** White
4.3 **Odor:** None

5. HEALTH HAZARDS

5.1 **Personal Protective Equipment:** Dust mask, goggles, and gloves
5.2 **Symptoms Following Exposure:** Inhalation of MCP or DCP may cause irritation of upper respiratory tract; prolonged inhalation of concentrated pyrophosphate may cause a deposit of TCP in the lungs. Ingestion of large quantities of any form of calcium phosphate may cause nausea, vomiting, cramps, and diarrhea; MCP may also be corrosive to membranes of mouth, throat, and gastrointestinal tract. Local irritation of the eyes may result from contact with any of these phosphates; prolonged or repeated exposure to MCP may lead to chronic conjunctivitis. Contact with skin by MCP may cause local irritation or chronic dermatitis; prolonged or repeated contact with concentrated aqueous slurries of DCP may cause local irritation.
5.3 **Treatment for Exposure:** INHALATION: move to clean air; see physician in case of persistent coughing, expectoration, chest pain, or shortness of breath. INGESTION: get medical attention quickly; induce vomiting by giving large amounts of water or warm salty water, or by tickling back of patient's throat; continue until vomitus is clear; follow with milk, eggs, or olive oil to soothe stomach. EYES: immediately flush with large quantities of running water for at least 15 min., holding eyelids apart to insure thorough flushing of eyes and lids; do not attempt to neutralize with chemical agents; get medical attention quickly; if physician is not available, continue irrigation for another 15 min. SKIN: immediately flush with water; remove contaminated clothing under shower; do not attempt to neutralize with chemical agent; get medical attention for persistent irritation.
5.4 **Toxicity by Inhalation (Threshold Limit Value):** Data not available
5.5 **Short-Term Inhalation Limits:** Data not available
5.6 **Toxicity by Ingestion:** Grade 0; LD_{50} > 15g/kg

6. FIRE HAZARDS

6.1 **Flash Point:** Not flammable
6.2 **Flammable Limits in Air:** Not flammable
6.3 **Fire Extinguishing Agents:** Not pertinent
6.4 **Fire Extinguishing Agents Not to be Used:** Not pertinent
6.5 **Special Hazards of Combustion Products:**
6.6 **Behavior in Fire:**
6.7 **Ignition Temperature:** Not pertinent
6.8 **Electrical Hazard:** Not pertinent
6.9 **Burning Rate:** Not pertinent

7. CHEMICAL REACTIVITY

7.1 **Reactivity with Water:** No reaction
7.2 **Reactivity with Common Materials:** Some calcium phosphates form acid solutions in water. These may attack metals with formation of flammable hydrogen gas, which may collect in enclosed spaces.
7.3 **Stability During Transport:** Stable
7.4 **Neutralizing Agents for Acids and Caustics:** Flush with water
7.5 **Polymerization:** Not pertinent
7.6 **Inhibitor of Polymerization:** Not pertinent

8. WATER POLLUTION

8.1 **Aquatic Toxicity:** Data not available
8.2 **Waterfowl Toxicity:** Data not available
8.3 **Biological Oxygen Demand (BOD):** None
8.4 **Food Chain Concentration Potential:** None

9. SELECTED MANUFACTURERS

1. Stauffer Chemical Co.
   Industrial Chemical Division
   Westport, Conn. 06880
2. Mallinckrodt Chemical Works
   223 Westside Avenue
   P.O. Box 384
   Jersey City, N.J. 07303
3. Monsanto Company
   800 North Lindbergh Blvd.
   St. Louis, Mo. 63166

10. SHIPPING INFORMATION

10.1 **Grades or Purity:** NF; USP; Dentrifrice; Reagent
10.2 **Storage Temperature:** Ambient
10.3 **Inert Atmosphere:** No requirement
10.4 **Venting:** Open

11. HAZARD ASSESSMENT CODE

(See Hazard Assessment Handbook, CG 446-3)

SS (MCP)
II (DCP, calcium pyrophosphate)

12. HAZARD CLASSIFICATIONS

12.1 **Code of Federal Regulations:** Not listed
12.2 **NAS Hazard Rating for Bulk Water Transportation:** Not listed
12.3 **NFPA Hazard Classifications:** Not listed

3. CHEMICAL DESIGNATIONS (Cont'd.)

phosphate tribasic (d) Calcium pyrophosphate

3.2 **Coast Guard Compatibility Classification:** Not listed
3.3 **Chemical Formula:**
(a) $CaH_4(PO_4)_2 \cdot H_2O$
(b) $CaHPO_4$ or $CaHPO_4 \cdot 2H_2O$
(c) $Ca_3(PO_4)_2$
(d) $Ca_2P_2O_7$
3.4 **IMCO/United Nations Numerical Designation:** Not listed

13. PHYSICAL AND CHEMICAL PROPERTIES

13.1 **Physical State at 15°C and 1 atm:** Solid
13.2 **Molecular Weight:** Monocalcium phosphate: 252.16; Dicalcium phosphate: 136.06; Calcium pyrophosphate: 254
13.3 **Boiling Point at 1 atm:** Not pertinent (decomposes)
13.4 **Freezing Point:** Not pertinent
13.5 **Critical Temperature:** Not pertinent
13.6 **Critical Pressure:** Not pertinent
13.7 **Specific Gravity:** 2—3 at 20°C (solid)
13.8 **Liquid Surface Tension:** Not pertinent
13.9 **Liquid-Water Interfacial Tension:** Not pertinent
13.10 **Vapor (Gas) Specific Gravity:** Not pertinent
13.11 **Ratio of Specific Heats of Vapor (Gas):** Not pertinent
13.12 **Latent Heat of Vaporization:** Not pertinent
13.13 **Heat of Combustion:** Not pertinent
13.14 **Heat of Decomposition:** Not pertinent
13.15 **Heat of Solution:** Not pertinent
13.16 **Heat of Polymerization:** Not pertinent

5. HEALTH HAZARDS (Cont'd.)

5.7 **Late Toxicity:** None
5.8 **Vapor (Gas) Irritant Characteristics:** Data not available
5.9 **Liquid or Solid Irritant Characteristics:** Data not available
5.10 **Odor Threshold:** Tribasic is odorless

CPP

CALCIUM PHOSPHIDE

Common Synonyms			
Photophor	Solid	Gray	Musty odor
	Reacts violently with water. Poisonous, flammable vapor is produced.		

AVOID CONTACT WITH LIQUID AND VAPOR. KEEP PEOPLE AWAY.
Wear goggles and self-contained breathing apparatus.
Call fire department.
Isolate and remove dishcarged material.
Notify local health and pollution control agencies.

Fire

FLAMMABLE.
IGNITES WHEN EXPOSED TO MOISTURE.
POISONOUS GASES ARE PRODUCED IN FIRE.
Wear goggles and self-contained breathing apparatus.
Extinguish with dry chemicals or carbon dioxide.
DO NOT USE WATER OR FOAM ON FIRE.
Cool exposed containers with water.

Exposure

CALL FOR MEDICAL AID.

VAPOR PRODUCED IN REACTION WITH WATER.
POISONOUS IF INHALED.
Irritating to eyes, nose and throat.
Move to fresh air.
If breathing has stopped, give artificial respiration.
If breathing is difficult, give oxygen.

DUST
Irritating to eyes.
Move victim to fresh air.
If in eyes, hold eyelids open and flush with plenty of water.

SOLID
Irritating to skin and eyes.
Harmful if swallowed.
Remove contaminated clothing and shoes.
Flush affected areas with plenty of water.
IF IN EYES, hold eyelids open and flush with plenty of water.
IF SWALLOWED and victim is CONSCIOUS, have victim drink water or milk.

Water Pollution

Effect of low concentrations on aquatic life is unknown.
May be dangerous if it enters water intakes.
Notify local health and wildlife officials.
Notify operators of nearby water intakes.

1. RESPONSE TO DISCHARGE

(See Response Methods Handbook, CG 446-4)

Issue warning—high flammability, poison, air contaminant, water contaminant
Restrict access
Evacuate area (large discharge only)
Should be removed

2. LABELS

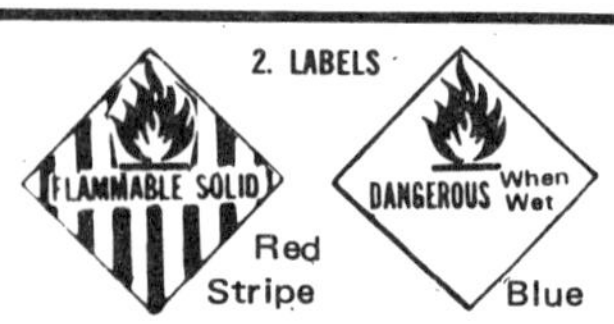

3. CHEMICAL DESIGNATIONS

3.1 **Synonyms:** Photophor
3.2 **Coast Guard Compatibility Classification:** Not applicable
3.3 **Chemical Formula:** Ca_3P_2
3.4 **IMCO/United Nations Numerical Designation:** 4.3/1360

4. OBSERVABLE CHARACTERISTICS

4.1 **Physical State (as shipped):** Solid
4.2 **Color:** Gray
4.3 **Odor:** Musty, like acetylene

5. HEALTH HAZARDS

5.1 **Personal Protective Equipment:** Dust respirator; protective gloves and clothing; goggles.
5.2 **Symptoms Following Exposure:** Inhalation or ingestion causes faintness, weakness, nausea, vomiting. External contact with dust causes irritation of eyes and skin.
5.3 **Treatment for Exposure:** INHALATION: remove to fresh air; call a physician and alert to possibility of phosphine poisoning. EYES or SKIN: flush with water. INGESTION: give large amounts of water; call physician and alert to possibility of phosphine poisoning.
5.4 **Toxicity by Inhalation (Threshold Limit Value):** Data not available
5.5 **Short-Term Inhalation Limits:** Data not available
5.6 **Toxicity by Ingestion:** Data not available
5.7 **Late Toxicity:** Data not available
5.8 **Vapor (Gas) Irritant Characteristics:** Data not available
5.9 **Liquid or Solid Irritant Characteristics:** Data not available
5.10 **Odor Threshold:** 1-100 mg/m³

6. FIRE HAZARDS

6.1 **Flash Point:** Not flammable but may ignite spontaneously if wetted.
6.2 **Flammable Limits in Air:** Not flammable
6.3 **Fire Extinguishing Agents:** Extinguish adjacent fires with dry chemical or carbon dioxide.
6.4 **Fire Extinguishing Agents Not to be Used:** Water, foam
6.5 **Special Hazards of Combustion Products:** Not pertinent
6.6 **Behavior in Fire:** Can cause spontaneous ignition if wet. Contributes dense smoke of phosphoric acid.
6.7 **Ignition Temperature:** Not pertinent
6.8 **Electrical Hazard:** Not pertinent
6.9 **Burning Rate:** Not pertinent

7. CHEMICAL REACTIVITY

7.1 **Reactivity with Water:** Reacts vigorously, generating phosphine (a poisonous, spontaneously flammable gas).
7.2 **Reactivity with Common Materials:** Can react with surface moisture to evolve phosphine, which is toxic and spontaneously flammable.
7.3 **Stability During Transport:** Stable if dry
7.4 **Neutralizing Agents for Acids and Caustics:** Not pertinent
7.5 **Polymerization:** Not pertinent
7.6 **Inhibitor of Polymerization:** Not pertinent

8. WATER POLLUTION

8.1 **Aquatic Toxicity:** Data not available
8.2 **Waterfowl Toxicity:** Data not available
8.3 **Biological Oxygen Demand (BOD):** Data not available
8.4 **Food Chain Concentration Potential:** None

9. SELECTED MANUFACTURERS

1. Ventron Corp.
   Alfa Products
   P. O. Box 159
   Beverly, Mass. 01915
2. Cerac, Inc.
   13460 W. Silver Spring Rd.
   Menomonee Falls, Wis. 53051

10. SHIPPING INFORMATION

10.1 **Grades or Purity:** Commercial
10.2 **Storage Temperature:** Ambient
10.3 **Inert Atmosphere:** No requirement
10.4 **Venting:** Sealed containers must be in well-ventilated area.

11. HAZARD ASSESSMENT CODE

(See Hazard Assessment Handbook, CG 446-3)

RR

12. HAZARD CLASSIFICATIONS

12.1 **Code of Federal Regulations:** Flammable solid
12.2 **NAS Hazard Rating for Bulk Water Transportation:** Not listed
12.3 **NFPA Hazard Classifications:** Not listed

13. PHYSICAL AND CHEMICAL PROPERTIES

13.1 **Physical State at 15°C and 1 atm:** Solid
13.2 **Molecular Weight:** 182.2
13.3 **Boiling Point at 1 atm:** Decomposes
13.4 **Freezing Point:** (approx.) 2,910°F = 1,600 °C = 1,870°K
13.5 **Critical Temperature:** Not pertinent
13.6 **Critical Pressure:** Not pertinent
13.7 **Specific Gravity:** 2.51 at 20°C (solid)
13.8 **Liquid Surface Tension:** Not pertinent
13.9 **Liquid-Water Interfacial Tension:** Not pertinent
13.10 **Vapor (Gas) Specific Gravity:** Not pertinent
13.11 **Ratio of Specific Heats of Vapor (Gas):** Not pertinent
13.12 **Latent Heat of Vaporization:** Not pertinent
13.13 **Heat of Combustion:** Not pertinent
13.14 **Heat of Decomposition:** Not pertinent
13.15 **Heat of Solution:** Data not available
13.16 **Heat of Polymerization:** Not pertinent

NOTES

CRE	CALCIUM RESINATE

Common Synonyms

Metallic resinate
Calcium abietate
Calcium rosin
Calcium resinate, fused

Solid — Yellow to dark brown — Odorless

Sinks in water.

Call fire department.
Stop discharge if possible. Keep people away.
Avoid contact with solid and dust.
Isolate and remove discharged material.
Notify local health and pollution control agencies.

Fire

Combustible.
MAY IGNITE WHEN EXPOSED TO AIR.
Extinguish with dry chemicals, foam or carbon dioxide.

Exposure

CALL FOR MEDICAL AID.

DUST
Irritating to eyes, nose and throat.
If inhaled will cause coughing or difficult breathing.
If in eyes, hold eyelids open and flush with plenty of water.
If breathing has stopped, give artificial respiration.
If breathing is difficult, give oxygen.

SOLID
Harmful if swallowed.
Remove contaminated clothing and shoes.
Flush affected areas with plenty of water.
IF IN EYES, hold eyelids open and flush with plenty of water.
IF SWALLOWED and victim is CONSCIOUS, have victim drink water or milk and have victim induce vomiting.
IF SWALLOWED and victim is UNCONSCIOUS OR HAVING CONVULSIONS, do nothing except keep victim warm.

Water Pollution

Effect of low concentrations on aquatic life is unknown.
May be dangerous if it enters water intakes.
Notify local health and wildlife officials.
Notify operators of nearby water intakes.

1. RESPONSE TO DISCHARGE

(See Response Methods Handbook, CG 446-4)

Should be removed
Chemical and physical treatment

2. LABEL

Red Stripe

3. CHEMICAL DESIGNATIONS

3.1 **Synonyms:** Calcium abietate; Calcium resinate, fused; Calcium rosin; Limed wood rosin; Metallic resinate

3.2 **Coast Guard Compatibility Classification:** Not listed

3.3 **Chemical Formula:** (approx.) $Ca(OOCC_{19}H_{29})_2$

3.4 **IMCO/United Nations Numerical Designation:** Not listed

4. OBSERVABLE CHARACTERISTICS

4.1 **Physical State (as shipped):** Solid

4.2 **Color:** Amber; very dark brown

4.3 **Odor:** None

5. HEALTH HAZARDS

5.1 **Personal Protective Equipment:** Dust mask; goggles or face shield; gloves

5.2 **Symptoms Following Exposure:** Inhalation of fumes from heated chemical may cause irritation of nose and throat. Ingestion causes irritation of nose and throat. Contact with eyes causes irritation. Contact of molten material with skin causes burns.

5.3 **Treatment for Exposure:** INHALATION: move victim to fresh air; get medical help immediately. INGESTION: give large amounts of water; induce vomiting. EYES: flush immediately with cold water; get medical help immediately. SKIN: if molten chemical burns skin, apply cold water immediately; get medical help for burn treatment.

5.4 **Toxicity by Inhalation (Threshold Limit Value):** Data not available

5.5 **Short-Term Inhalation Limits:** Data not available

5.6 **Toxicity by Ingestion:** Data not available

5.7 **Late Toxicity:** Data not available

5.8 **Vapor (Gas) Irritant Characteristics:** Data not available

5.9 **Liquid or Solid Irritant Characteristics:** Data not available

5.10 **Odor Threshold:** Data not available

6. FIRE HAZARDS

6.1 **Flash Point:** Not pertinent (combustible solid)

6.2 **Flammable Limits in Air:** Not pertinent

6.3 **Fire Extinguishing Agents:** Foam, dry chemical, carbon dioxide

6.4 **Fire Extinguishing Agents Not to be Used:** Water

6.5 **Special Hazards of Combustion Products:**

6.6 **Behavior in Fire:**

6.7 **Ignition Temperature:** 480°F (may ignite spontaneously)

6.8 **Electrical Hazard:** Not pertinent

6.9 **Burning Rate:** Not pertinent

7. CHEMICAL REACTIVITY

7.1 **Reactivity with Water:** No reaction

7.2 **Reactivity with Common Materials:**

7.3 **Stability During Transport:** Stable

7.4 **Neutralizing Agents for Acids and Caustics:** Not pertinent

7.5 **Polymerization:** Not pertinent

7.6 **Inhibitor of Polymerization:** Not pertinent

8. WATER POLLUTION

8.1 **Aquatic Toxicity:** Data not available

8.2 **Waterfowl Toxicity:** Data not available

8.3 **Biological Oxygen Demand (BOD):** Data not available

8.4 **Food Chain Concentration Potential:** None

9. SELECTED MANUFACTURERS

1. Crosby Chemicals, Inc.
P.O. Box 460
Picayune, Miss. 39466

2. Reichold Chemicals Inc.
Newport Division
P.O. Box 1433
Pensacola, Fla. 32596

10. SHIPPING INFORMATION

10.1 **Grades or Purity:** Commercial

10.2 **Storage Temperature:** Ambient

10.3 **Inert Atmosphere:** No requirement

10.4 **Venting:** Open

11. HAZARD ASSESSMENT CODE

(See Hazard Assessment Handbook, CG 446-3)

II

12. HAZARD CLASSIFICATIONS

12.1 **Code of Federal Regulations:** Flammable solid

12.2 **NAS Hazard Rating for Bulk Water Transportation:** Not listed

12.3 **NFPA Hazard Classifications:** Not listed

13. PHYSICAL AND CHEMICAL PROPERTIES

13.1 **Physical State at 15°C and 1 atm:** Solid

13.2 **Molecular Weight:** 643 (approx.)

13.3 **Boiling Point at 1 atm:** $>600°F = >316°C = >589°K$

13.4 **Freezing Point:** Not pertinent

13.5 **Critical Temperature:** Not pertinent

13.6 **Critical Pressure:** Not pertinent

13.7 **Specific Gravity:** 1.13 at 25°C (solid)

13.8 **Liquid Surface Tension:** Not pertinent

13.9 **Liquid-Water Interfacial Tension:** Not pertinent

13.10 **Vapor (Gas) Specific Gravity:** Not pertinent

13.11 **Ratio of Specific Heats of Vapor (Gas):** Not pertinent

13.12 **Latent Heat of Vaporization:** Not pertinent

13.13 **Heat of Combustion:** Data not available

13.14 **Heat of Decomposition:** Not pertinent

13.15 **Heat of Solution:** Not pertinent

13.16 **Heat of Polymerization:** Not pertinent

NOTES

CPH — CAMPHENE

Common Synonyms			
3,3-Dimethyl-2-methylene norcamphane 2,2-Dimethyl-3-methylene-norborane	Solid	White	Camphor-like odor
	Floats on water.		

Stop discharge if possible. Keep people away.
Call fire department.
Avoid contact with solid and dust.
Isolate and remove discharged material.
Notify local health and pollution control agencies.

Fire

Combustible.
Extinguish with dry chemicals, foam or carbon dioxide.
Water may be ineffective on fire.

Exposure

CALL FOR MEDICAL AID.

DUST
Irritating to eyes, nose and throat.
If inhaled will cause headache or difficult breathing.
If in eyes, hold eyelids open and flush with plenty of water.
If breathing has stopped, give artificial respiration.
If breathing is difficult, give oxygen.

SOLID
Irritating to skin and eyes.
Harmful if swallowed.
Remove contaminated clothing and shoes.
Flush affected areas with plenty of water.
IF IN EYES, hold eyelids open and flush with plenty of water.
IF SWALLOWED and victim is CONSCIOUS, have victim drink water or milk.
IF SWALLOWED and victim is UNCONSCIOUS OR HAVING CONVULSIONS, do nothing except keep victim warm.

Water Pollution

Effect of low concentrations on aquatic life is unknown.
May be dangerous if it enters water intakes.
Notify local health and wildlife officials.
Notify operators of nearby water intakes.

1. RESPONSE TO DISCHARGE
(See Response Methods Handbook, CG 446-4)

Mechanical containment
Should be removed
Chemical and physical treatment

2. LABELS

No hazard label required by Code of Federal Regulations

3. CHEMICAL DESIGNATIONS

3.1 **Synonyms:** 2,2-Dimethyl-3-methylenenorbornane
3,3-Dimethyl-2-methylenenorcamphane
3.2 **Coast Guard Compatibility Classification:** Not listed
3.3 **Chemical Formula:** $C_{10}H_{16}$
3.4 **IMCO/United Nations Numerical Designation:** Not listed

4. OBSERVABLE CHARACTERISTICS

4.1 **Physical State (as shipped):** Solid
4.2 **Color:** White
4.3 **Odor:** Camphoraceous

5. HEALTH HAZARDS

5.1 **Personal Protective Equipment:** Gloves and face shield
5.2 **Symptoms Following Exposure:** Inhalation causes irritation of nose and throat. Contact with eyes or skin causes irritation.
5.3 **Treatment for Exposure:** INHALATION: move victim to fresh air; call physician immediately. EYES: flush immediately with clean, cool water; call physician immediately. SKIN: wash with alcohol, follow with soap and water wash.
5.4 **Toxicity by Inhalation (Threshold Limit Value):** Data not available
5.5 **Short-Term Inhalation Limits:** Data not available
5.6 **Toxicity by Ingestion:** Data not available
5.7 **Late Toxicity:** Data not available
5.8 **Vapor (Gas) Irritant Characteristics:** Data not available
5.9 **Liquid or Solid Irritant Characteristics:** Data not available
5.10 **Odor Threshold:** Data not available

6. FIRE HAZARDS

6.1 **Flash Point:** 108°F O.C.; 92°F C.C. (typical)
6.2 **Flammable Limits in Air:** Data not available
6.3 **Fire Extinguishing Agents:** Foam, dry chemical, carbon dioxide
6.4 **Fire Extinguishing Agents Not to be Used:** Water
6.5 **Special Hazards of Combustion Products:**
6.6 **Behavior in Fire:**
6.7 **Ignition Temperature:** Data not available
6.8 **Electrical Hazard:** Data not available
6.9 **Burning Rate:** Not pertinent

7. CHEMICAL REACTIVITY

7.1 **Reactivity with Water:** No reaction
7.2 **Reactivity with Common Materials:**
7.3 **Stability During Transport:** Stable
7.4 **Neutralizing Agents for Acids and Caustics:** Not pertinent
7.5 **Polymerization:** Not pertinent
7.6 **Inhibitor of Polymerization:** Not pertinent

8. WATER POLLUTION

8.1 **Aquatic Toxicity:** Data not available
8.2 **Waterfowl Toxicity:** Data not available
8.3 **Biological Oxygen Demand (BOD):** Data not available
8.4 **Food Chain Concentration Potential:** None

9. SELECTED MANUFACTURERS

1. Hercules, Inc.
Organics Department
Wilmington, Del. 19899
2. Reichold Chemicals, Inc.
Newport Division
407 S. Pace Boulevard
Pensacola, Fla. 32596
3. SCM Corporation
Glidden-Durkee Div.
P.O. Box 389
Jacksonville, Fla. 32201

10. SHIPPING INFORMATION

10.1 **Grades or Purity:** Commercial, 75+%
10.2 **Storage Temperature:** Ambient
10.3 **Inert Atmosphere:** No requirement
10.4 **Venting:** Open

11. HAZARD ASSESSMENT CODE
(See Hazard Assessment Handbook, CG 446-3)

II

12. HAZARD CLASSIFICATIONS

12.1 **Code of Federal Regulations:** ORM-A
12.2 **NAS Hazard Rating for Bulk Water Transportation:** Not listed
12.3 **NFPA Hazard Classifications:** Not listed

13. PHYSICAL AND CHEMICAL PROPERTIES

13.1 **Physical State at 15°C and 1 atm:** Solid
13.2 **Molecular Weight:** 136
13.3 **Boiling Point at 1 atm:** 310°F = 154°C = 427°K
13.4 **Freezing Point:** 122°F = 50°C = 323°K
13.5 **Critical Temperature:** Not pertinent
13.6 **Critical Pressure:** Not pertinent
13.7 **Specific Gravity:** 0.87 at 15°C (solid)
13.8 **Liquid Surface Tension:** Not pertinent
13.9 **Liquid-Water Interfacial Tension:** Not pertinent
13.10 **Vapor (Gas) Specific Gravity:** Not pertinent
13.11 **Ratio of Specific Heats of Vapor (Gas):** Not pertinent
13.12 **Latent Heat of Vaporization:** Not pertinent
13.13 **Heat of Combustion:** $-19{,}400$ Btu/lb $= -10{,}800$ cal/g $= -452 \times 10^5$ J/kg
13.14 **Heat of Decomposition:** Not pertinent
13.15 **Heat of Solution:** Not pertinent
13.16 **Heat of Polymerization:** Not pertinent

NOTES

CPO	CAMPHOR OIL

Common Synonyms
Liquid camphor

Oily liquid — Colorless or brown or blue — Penetrating camphor odor

Usually floats on water.

Stop discharge if possible.
Call fire department.
Avoid contact with liquid.
Isolate and remove discharged material.
Notify local health and pollution control agencies.

Fire

Combustible.
Extinguish with dry chemical, foam or carbon dioxide.

Exposure

CALL FOR MEDICAL AID.

VAPOR
Not irritating to eyes, nose or throat.
Move to fresh air.

LIQUID
Irritating to skin and eyes.
If swallowed, will cause nausea, vomiting or loss of consciousness.
Remove contaminated clothing and shoes.
Flush affected areas with plenty of water.
IF IN EYES, hold eyelids open and flush with plenty of water.
IF SWALLOWED and victim is CONSCIOUS, have victim drink water or milk.

Water Pollution

Effect of low concentrations on aquatic life is unknown.
Fouling to shoreline.
May be dangerous if it enters water intakes.
Notify local health and pollution control officials.
Notify operators of nearby water intakes.

1. RESPONSE TO DISCHARGE
(See Response Methods Handbook, CG 446-4)

Issue warning—water contaminant
Mechanical containment
Should be removed

2. LABELS

No hazard label required by Code of Federal Regulations

3. CHEMICAL DESIGNATIONS

3.1 **Synonyms:** Liquid impure camphor; Liquid gum camphor
3.2 **Coast Guard Compatibility Classification:** Ketone
3.3 **Chemical Formula:** $C_{10}H_{16}O$
3.4 **IMCO/United Nations Numerical Designation:** 3.3/1130

4. OBSERVABLE CHARACTERISTICS

4.1 **Physical State (as shipped):** Liquid
4.2 **Color:** Colorless; brown; blue
4.3 **Odor:** Like camphor; fragrant and penetrating

5. HEALTH HAZARDS

5.1 **Personal Protective Equipment:** Eye protection
5.2 **Symptoms Following Exposure:** Within 5 to 90 minutes after swallowing, the following may be noted: nausea and vomiting; feeling of warmth; headache; confusion, vertigo, excitement, restlessness, delirium, and hallucinations; increased muscular excitability, tremors, and jerky movements; epileptiform convulsions, followed by depression (convulsions sometimes occur early in the syndrome and may be severe, but they do not have the grave prognosis of strychnine convulsions); coma; central nervous depression may at times be the primary clinical response; death results from respiratory failure or from status epilepticus; slow convalescence (days or weeks), often with persistent gastric distress.
5.3 **Treatment for Exposure:** For an oral intoxication, administer gastric lavage, cathartics, diuretics, and sedatives. Control convulsions with a short-acting barbiturate, chloral hydrate, or ether. Do NOT use analeptics or opiates.
5.4 **Toxicity by Inhalation (Threshold Limit Value):** 2 ppm
5.5 **Short-Term Inhalation Limits:** 3 ppm for 30 min.
5.6 **Toxicity by Ingestion:** Data not available
5.7 **Late Toxicity:** Data not available
5.8 **Vapor (Gas) Irritant Characteristics:** Vapors are nonirritating to the eyes and throat.
5.9 **Liquid or Solid Irritant Characteristics:** Minimum hazard. If spilled on clothing and allowed to remain, may cause smarting and reddening of the skin.
5.10 **Odor Threshold:** Data not available

6. FIRE HAZARDS

6.1 **Flash Point:** 117°F C.C.
6.2 **Flammable Limits in Air:** Data not available
6.3 **Fire Extinguishing Agents:** Foam, carbon dioxide, or dry chemical
6.4 **Fire Extinguishing Agents Not to be Used:** Not pertinent
6.5 **Special Hazards of Combustion Products:** Not pertinent
6.6 **Behavior in Fire:** The solid often evaporates without first melting.
6.7 **Ignition Temperature:** 466°C
6.8 **Electrical Hazard:** Not pertinent
6.9 **Burning Rate:** Data not available

7. CHEMICAL REACTIVITY

7.1 **Reactivity with Water:** No reaction
7.2 **Reactivity with Common Materials:** No reaction
7.3 **Stability During Transport:** Stable
7.4 **Neutralizing Agents for Acids and Caustics:** Not pertinent
7.5 **Polymerization:** Not pertinent
7.6 **Inhibitor of Polymerization:** Not pertinent

8. WATER POLLUTION

8.1 **Aquatic Toxicity:** Data not available
8.2 **Waterfowl Toxicity:** Data not available
8.3 **Biological Oxygen Demand (BOD):** Data not available
8.4 **Food Chain Concentration Potential:** None

9. SELECTED MANUFACTURERS

1. CPC International, Inc.
S. B. Penick and Co. Division
100 Church St.
New York, N. Y. 10007
2. Glyco Chemicals, Inc.
51 Weaver St.
Greenwich, Conn. 06830
3. R. W. Greef & Co., Inc.
1 Rockefeller Plaza
New York, N. Y. 10020

10. SHIPPING INFORMATION

10.1 **Grades or Purity:** Each lot of camphor oil has a unique composition, which varies with the season of the year and the country of origin. At least a dozen grades are known; all have Chinese names. Most camphor sold in the U. S. is synthetic and is quite pure.
10.2 **Storage Temperature:** Ambient
10.3 **Inert Atmosphere:** No requirement
10.4 **Venting:** Open

11. HAZARD ASSESSMENT CODE
(See Hazard Assessment Handbook, CG 446-3)

A-T-U

12. HAZARD CLASSIFICATIONS

12.1 **Code of Federal Regulations:** Combustible Liquid
12.2 **NAS Hazard Rating for Bulk Water Transportation:**

Category	Rating
Fire	2
Health	
Vapor Irritant	0
Liquid or Solid Irritant	1
Poisons	1
Water Pollution	
Human Toxicity	3
Aquatic Toxicity	1
Aesthetic Effect	2
Reactivity	
Other Chemicals	2
Water	0
Self-Reaction	0

12.3 **NFPA Hazard Classifications:** Not listed

13. PHYSICAL AND CHEMICAL PROPERTIES

13.1 **Physical State at 15°C and 1 atm:** Solid (liquid if even slightly impure)
13.2 **Molecular Weight:** Not pertinent
13.3 **Boiling Point at 1 atm:** >392°F = >200°C = >473°K
13.4 **Freezing Point:** Not pertinent
13.5 **Critical Temperature:** Not pertinent
13.6 **Critical Pressure:** Not pertinent
13.7 **Specific Gravity:** 0.923 at 25°C (liquid)
13.8 **Liquid Surface Tension:** Data not available
13.9 **Liquid-Water Interfacial Tension:** Data not available
13.10 **Vapor (Gas) Specific Gravity:** Not pertinent
13.11 **Ratio of Specific Heats of Vapor (Gas):** Not pertinent
13.12 **Latent Heat of Vaporization:** Not pertinent
13.13 **Heat of Combustion:** Data not available
13.14 **Heat of Decomposition:** Not pertinent
13.15 **Heat of Solution:** Not pertinent
13.16 **Heat of Polymerization:** Not pertinent

NOTES

REVISED 1978

CLS CAPROLACTAM, LIQUID

Common Synonyms			
Aminocaproic lactam 2-Ketohexamethylenimine 2-Oxohexamethylenimine epsilon-Caprolactam Hexahydro-2H-azepine-2-one	Liquid	Colorless	Mild odor
	Sinks and mixes with water.		

Stop discharge if possible. Keep people away.
Call fire department.
Isolate and remove discharged material.
Notify local health and pollution control agencies.

Fire

Combustible.
POISONOUS GASES MAY BE PRODUCED IN FIRE.
Extinguish with water, dry chemicals, foam, or carbon dioxide.
Cool exposed containers with water.

Exposure

Call for medical aid.

LIQUID
Irritating to skin and eyes.
Harmful if swallowed.
Remove contaminated clothing and shoes.
Flush affected areas with plenty of water.
IF IN EYES, hold eyelids open and flush with plenty of water.
IF SWALLOWED and victim is CONSCIOUS, have victim drink water or milk.

Water Pollution

Effect of low concentrations on aquatic life is unknown.
May be dangerous if it enters water intakes.
Notify local health and wildlife officials.
Notify operators of nearby water intakes.

1. RESPONSE TO DISCHARGE

(See Response Methods Handbook, CG 446-4)

Issue warning—water contaminant
Disperse and flush

2. LABELS

No label required by Code of Federal Regulations

3. CHEMICAL DESIGNATIONS

3.1 **Synonyms:** Aminocaproic lactam; epsilon- Caprolactam; Hexahydro-2H-azepin-2-one; 2-Ketohexamethylenimine
3.2 **Coast Guard Compatibility Classification:** Caprolactam solution
3.3 **Chemical Formula:** $HNCH_2(CH_2)_4CO$
3.4 **IMCO/United Nations Numerical Designation:** Not listed

4. OBSERVABLE CHARACTERISTICS

4.1 **Physical State (as shipped):** Liquid
4.2 **Color:** Colorless
4.3 **Odor:** Mild

5. HEALTH HAZARDS

5.1 **Personal Protective Equipment:** Gas mask or self-contained breathing apparatus; rubber gloves and boots; goggles or face shield.
5.2 **Symptoms Following Exposure:** Inhalation causes coughing or mild irritation. Contact with hot liquid will burn eyes and skin.
5.3 **Treatment for Exposure:** INHALATION: remove patient to fresh air. EYES: wash with copious quantities of water for at least 15 min.; call physician. SKIN: wash with water; call physician in case of thermal burn.
5.4 **Toxicity by Inhalation (Threshold Limit Value):** Data not available
5.5 **Short-Term Inhalation Limits:** Not pertinent
5.6 **Toxicity by Ingestion:** Grade 2; oral rat LD_{50} = 2,140 mg/kg
5.7 **Late Toxicity:** Data not available
5.8 **Vapor (Gas) Irritant Characteristics:** Vapors are nonirritating to the eyes and throat.
5.9 **Liquid or Solid Irritant Characteristics:** No appreciable hazard. Practically harmless to the skin.
5.10 **Odor Threshold:** 0.3 mg/m³

6. FIRE HAZARDS

6.1 **Flash Point:** 257°F O.C.; 230°F C.C.
6.2 **Flammable Limits in Air:** 1.84% (LEL)
6.3 **Fire Extinguishing Agents:** Water, dry chemical, foam, carbon dioxide
6.4 **Fire Extinguishing Agents Not to be Used:** Not pertinent
6.5 **Special Hazards of Combustion Products:** Toxic oxides of nitrogen
6.6 **Behavior in Fire:** Not pertinent
6.7 **Ignition Temperature:** Data not available
6.8 **Electrical Hazard:** Not pertinent
6.9 **Burning Rate:** 2.4 mm/min.

7. CHEMICAL REACTIVITY

7.1 **Reactivity with Water:** No reaction
7.2 **Reactivity with Common Materials:** No reaction
7.3 **Stability During Transport:** Stable
7.4 **Neutralizing Agents for Acids and Caustics:** Not pertinent
7.5 **Polymerization:** Not pertinent
7.6 **Inhibitor of Polymerization:** Not pertinent

8. WATER POLLUTION

8.1 **Aquatic Toxicity:** Data not available
8.2 **Waterfowl Toxicity:** Data not available
8.3 **Biological Oxygen Demand (BOD):** Data not available
8.4 **Food Chain Concentration Potential:** None

9. SELECTED MANUFACTURERS

1. Dow Badische Company
Williamsburg, Va. 23185
2. Allied Chemical Corporation
Plastics Division
P. O. Box 2365R
Morristown, N. J. 07960
3. Nipro, Incorporated
P. O. Box 1485
Augusta, Ga. 30903

10. SHIPPING INFORMATION

10.1 **Grades or Purity:** 99%
10.2 **Storage Temperature:** 75°C
10.3 **Inert Atmosphere:** Nitrogen cushion
10.4 **Venting:** Pressure-vacuum

11. HAZARD ASSESSMENT CODE

(See Hazard Assessment Handbook, CG 446-3)

A-P-Q

12. HAZARD CLASSIFICATIONS

12.1 **Code of Federal Regulations:** Not listed
12.2 **NAS Hazard Rating for Bulk Water Transportation:**

Category	Rating
Fire	1
Health	
Vapor Irritant	0
Liquid or Solid Irritant	0
Poisons	4
Water Pollution	
Human Toxicity	2
Aquatic Toxicity	
Aesthetic Effect	0
Reactivity	
Other Chemicals	2
Water	0
Self-Reaction	1

12.3 **NFPA Hazard Classifications:** Not listed

13. PHYSICAL AND CHEMICAL PROPERTIES

13.1 **Physical State at 15°C and 1 atm:** Solid
13.2 **Molecular Weight:** 113
13.3 **Boiling Point at 1 atm:** 515°F = 268°C = 541°K
13.4 **Freezing Point:** 154°F = 68°C = 341°K
13.5 **Critical Temperature:** 944.4°F = 506.9°C = 780.1°K
13.6 **Critical Pressure:** 660 psia = 45 atm = 4.6 MN/m²
13.7 **Specific Gravity:** 1.02 at 77°C (liquid)
13.8 **Liquid Surface Tension:** (est.) 20 dynes/cm = 0.020 N/m at 77°C
13.9 **Liquid-Water Interfacial Tension:** (est.) 40 dynes/cm = 0.040 N/m at 77°C
13.10 **Vapor (Gas) Specific Gravity:** 3.9
13.11 **Ratio of Specific Heats of Vapor (Gas):** Not pertinent
13.12 **Latent Heat of Vaporization:** 209 Btu/lb = 116 cal/g = 4.85×10^5 J/kg
13.13 **Heat of Combustion:** −13,700 Btu/lb = −7,640 cal/g = -320×10^5 J/kg
13.14 **Heat of Decomposition:** Not pertinent
13.15 **Heat of Solution:** Not pertinent
13.16 **Heat of Polymerization:** −324 Btu/lb = −180 cal/g = -7.5×10^5 J/kg

NOTES

CPT — CAPTAN

Common Synonyms			
N-[(Trichloromethyl) thio]-4-cyclohexene-1, 2, -dicarboximide. N-trichloromethylthio-cis-Δ^4-cyclohexene-1, 2, -dicarboximide. Orthocide Vanicide	Solid	White to brown	Slight odor
	Sinks in water.		

AVOID CONTACT WITH SOLID AND DUST. KEEP PEOPLE AWAY.
Wear goggles, self-contained breathing apparatus, and rubber overclothing (including gloves).
Call fire department.
Stay upwind, use water spray to "knock down" dust.
Isolate and remove discharged material.
Notify local health and pollution control agencies.

Fire

FLAMMABLE.
POISONOUS GASES MAY BE PRODUCED IN FIRE.
Wear goggles and self-contained breathing apparatus.
Extinguish with water, dry chemicals, foam, or carbon dioxide.
Cool exposed containers with water.

Exposure

CALL FOR MEDICAL AID.

DUST
POISONOUS IF INHALED OR IF SKIN IS EXPOSED.
Move victim to fresh air.
If in eyes, hold eyelids open and flush with plenty of water.
If breathing has stopped, give artificial respiration.
If breathing is difficult, give oxygen.

SOLID
POISONOUS IF SWALLOWED OR IF SKIN IS EXPOSED.
Remove contaminated clothing and shoes.
Flush affected areas with plenty of water.
IF IN EYES, hold eyelids open and flush with plenty of water.
IF SWALLOWED and victim is CONSCIOUS, have victim drink water or milk and have victim induce vomiting.
IF SWALLOWED and victim is UNCONSCIOUS OR HAVING CONVULSIONS, do nothing except keep victim warm.

Water Pollution

HARMFUL TO AQUATIC LIFE IN VERY LOW CONCENTRATIONS.
May be dangerous if it enters water intakes.
Notify local health and wildlife officials.
Notify operators of nearby water intakes.

1. RESPONSE TO DISCHARGE

(See Response Methods Handbook, CG 446-4)

Issue warning—poison, water contaminant
Restrict access
Should be removed
Chemical and physical treatment

2. LABEL

3. CHEMICAL DESIGNATIONS

3.1 **Synonyms:** N-[(Trichloromethyl)thio]-4-cyclohexene-1, 2-dicarboximide; N-Trichloromethylthio-cis-Δ^4-cyclohexene-1, 2-dicarboximide; Orthocide; Vanicide
3.2 **Coast Guard Compatibility Classification:** Not applicable
3.3 **Chemical Formula:** $C_9H_8Cl_3NO_2S$
3.4 **IMCO/United Nations Numerical Designation:** 6.1/1609

4. OBSERVABLE CHARACTERISTICS

4.1 **Physical State (as shipped):** Solid
4.2 **Color:** White to buff
4.3 **Odor:** Slight pungent

5. HEALTH HAZARDS

5.1 **Personal Protective Equipment:** Dust mask, rubber gloves, and goggles.
5.2 **Symptoms Following Exposure:** Vapor irritates eyes. Ingestion causes depression, lachrymation, labored respiration, diarrhea.
5.3 **Treatment for Exposure:** Remove from exposure; keep airways open; administer artificial respiration if necessary. EYES: flush with water for 15 min. and get medical attention. SKIN: wash with soap and water. INGESTION: maintain respiration; induce vomiting (lavage stomach if patient is unconscious); give symptomatic and supportive treatment; save agent and vomitus for laboratory examination.
5.4 **Toxicity by Inhalation (Threshold Limit Value):** Data not available
5.5 **Short-Term Inhalation Limits:** Data not available
5.6 **Toxicity by Ingestion:** Grade 3; oral rat LD_{50} = 480 mg/kg
5.7 **Late Toxicity:** None observed in several species.
5.8 **Vapor (Gas) Irritant Characteristics:** Data not available
5.9 **Liquid or Solid Irritant Characteristics:** Data not available
5.10 **Odor Threshold:** Data not available

6. FIRE HAZARDS

6.1 **Flash Point:** Flammable solid
6.2 **Flammable Limits in Air:** Not pertinent
6.3 **Fire Extinguishing Agents:** Water, carbon dioxide, foam
6.4 **Fire Extinguishing Agents Not to be Used:** Not pertinent
6.5 **Special Hazards of Combustion Products:** Irritating and toxic gases are produced in a fire; they may include sulfur dioxide, hydrogen chloride, phosgene, and oxides of nitrogen.
6.6 **Behavior in Fire:** Not pertinent
6.7 **Ignition Temperature:** Data not available
6.8 **Electrical Hazard:** Not pertinent
6.9 **Burning Rate:** Not pertinent

7. CHEMICAL REACTIVITY

7.1 **Reactivity with Water:** No reaction
7.2 **Reactivity with Common Materials:** No reaction
7.3 **Stability During Transport:** Stable
7.4 **Neutralizing Agents for Acids and Caustics:** Not pertinent
7.5 **Polymerization:** Not pertinent
7.6 **Inhibitor of Polymerization:** Not pertinent

8. WATER POLLUTION

8.1 **Aquatic Toxicity:** 30 ppm/24 hr/zebrafish/TL_m/fresh water
8.2 **Waterfowl Toxicity:** > 5,000 ppm LC_{50}
8.3 **Biological Oxygen Demand (BOD):** Data not available
8.4 **Food Chain Concentration Potential:** Bioaccumulation not likely

9. SELECTED MANUFACTURERS

1. R. T. Vanderbilt Company, Inc. 230 Park Avenue New York, N. Y. 10017
2. Stauffer Chemical Co. Mountain View, Calif. 94040
3. Chevron Chemical Co. 940 Hensley Street Richmond, Calif. 94804

10. SHIPPING INFORMATION

10.1 **Grades or Purity:** Technical: 90-97%; also available as dusts, wettable powders, and aqueous suspension.
10.2 **Storage Temperature:** Ambient
10.3 **Inert Atmosphere:** No requirement
10.4 **Venting:** Open (flame arrester)

11. HAZARD ASSESSMENT CODE

(See Hazard Assessment Handbook, CG 446-3)

II

12. HAZARD CLASSIFICATIONS

12.1 **Code of Federal Regulations:** Poisonous, class B
12.2 **NAS Hazard Rating for Bulk Water Transportation:** Not listed
12.3 **NFPA Hazard Classifications:** Not listed

13. PHYSICAL AND CHEMICAL PROPERTIES

13.1 **Physical State at 15°C and 1 atm:** Solid
13.2 **Molecular Weight:** 300.6
13.3 **Boiling Point at 1 atm:** Decomposes
13.4 **Freezing Point:** 338°F = 170°C = 443°K
13.5 **Critical Temperature:** Not pertinent
13.6 **Critical Pressure:** Not pertinent
13.7 **Specific Gravity:** 1.74 at 20°C (solid)
13.8 **Liquid Surface Tension:** Not pertinent
13.9 **Liquid-Water Interfacial Tension:** Not pertinent
13.10 **Vapor (Gas) Specific Gravity:** Not pertinent
13.11 **Ratio of Specific Heats of Vapor (Gas):** Not pertinent
13.12 **Latent Heat of Vaporization:** Not pertinent
13.13 **Heat of Combustion:** (est.) −7,100 Btu/lb = −3,940 cal/g = −165 × 10^5 J/kg
13.14 **Heat of Decomposition:** Not pertinent
13.15 **Heat of Solution:** Not pertinent
13.16 **Heat of Polymerization:** Not pertinent

NOTES

CBY	CARBARYL

Common Synonyms		
Sevin 1-Naphthyl N-methyl-carbamate	Solid powder, or in solution White to gray Weak odor Solid sinks in water; solution may float on water.	

Stop discharge if possible.
Call fire department.
Avoid contact with solid and solution.
Isolate and remove discharged material.
Notify local health and pollution control agencies.

Fire	Solid not flammable, but usually dissolved in combustible liquid. Extinguish with water, foam, dry chemical or carbon dioxide.
Exposure	CALL FOR MEDICAL AID. SOLID OR SOLUTION Irritating to skin and eyes. Harmful if swallowed. Flush affected areas with plenty of water. IF IN EYES, hold eyelids open and flush with plenty of water. IF SWALLOWED and victim is CONSCIOUS, have victim drink water or milk and have victim induce vomiting. IF SWALLOWED and victim is UNCONSCIOUS OR HAVING CONVULSIONS, do nothing except keep victim warm.
Water Pollution	HARMFUL TO AQUATIC LIFE IN VERY LOW CONCENTRATIONS. Fouling to shoreline. May be dangerous if it enters water intakes. Notify local health and wildlife officials. Notify operators of nearby water intakes.

1. RESPONSE TO DISCHARGE

(See Response Methods Handbook, CG 446-4)

Issue warning—water contaminant
Should be removed

2. LABELS

No hazard label required by Code of Federal Regulations

3. CHEMICAL DESIGNATIONS

3.1 **Synonyms:** 1-Naphthyl N-methylcarbamate; Sevin
3.2 **Coast Guard Compatibility Classification:** Not applicable
3.3 **Chemical Formula:** $C_{12}H_{11}NO_2$
3.4 **IMCO/United Nations Numerical Designation:** 6.1/6.1/1615

4. OBSERVABLE CHARACTERISTICS

4.1 **Physical State (as shipped):** Solid
4.2 **Color:** Off-white
4.3 **Odor:** Weak

5. HEALTH HAZARDS

5.1 **Personal Protective Equipment:** Data not available
5.2 **Symptoms Following Exposure:** Data not available
5.3 **Treatment for Exposure:** SKIN AND EYE CONTACT: flood affected tissues with water. INGESTION: induce vomiting.
5.4 **Toxicity by Inhalation (Threshold Limit Value):** 5 mg/m³
5.5 **Short-Term Inhalation Limits:** Dust, 30 mg/m³ for 30 min.
5.6 **Toxicity by Ingestion:** Grade 2; LD_{50} 0.5 to 5 g/kg (rat LD_{50} 0.51 g/kg)
5.7 **Late Toxicity:** Liver damage to rats at high dose by mouth.
5.8 **Vapor (Gas) Irritant Characteristics:** None
5.9 **Liquid or Solid Irritant Characteristics:** No hazard
5.10 **Odor Threshold:** Not pertinent

6. FIRE HAZARDS

6.1 **Flash Point:** Not flammable
6.2 **Flammable Limits in Air:** Not flammable
6.3 **Fire Extinguishing Agents:** Solution fires: Water, foam, dry chemical, CO_2
6.4 **Fire Extinguishing Agents Not to be Used:** Not pertinent
6.5 **Special Hazards of Combustion Products:** Not pertinent
6.6 **Behavior in Fire:** Not pertinent
6.7 **Ignition Temperature:** Not flammable
6.8 **Electrical Hazard:** Not pertinent
6.9 **Burning Rate:** Not flammable

7. CHEMICAL REACTIVITY

7.1 **Reactivity with Water:** No reaction
7.2 **Reactivity with Common Materials:** No reaction
7.3 **Stability During Transport:** Stable
7.4 **Neutralizing Agents for Acids and Caustics:** Not pertinent
7.5 **Polymerization:** Not pertinent
7.6 **Inhibitor of Polymerization:** Not pertinent

8. WATER POLLUTION

8.1 **Aquatic Toxicity:**
5.5 ppm/96 hr/bluegill/TL_m/fresh water
0.013 ppm/48 hr/white shrimp/TL_m/salt water. Toxic to shrimp and crab species
8.2 **Waterfowl Toxicity:** LD_{50} = 2179 mg/kg
8.3 **Biological Oxygen Demand (BOD):** Data not available
8.4 **Food Chain Concentration Potential:** None observed

9. SELECTED MANUFACTURERS

Union Carbide Corp.
Chemicals and Plastics Division
270 Park Ave.
New York, N. Y. 10017

10. SHIPPING INFORMATION

10.1 **Grades or Purity:**
Sevin 50 W wettable powder
Sevin sprayable 80% powder
Sevin 4 oil
10.2 **Storage Temperature:** Data not available
10.3 **Inert Atmosphere:** Data not available
10.4 **Venting:** Data not available

11. HAZARD ASSESSMENT CODE

(See Hazard Assessment Handbook, CG 446-3)

I I

12. HAZARD CLASSIFICATIONS

12.1 **Code of Federal Regulations:** ORM–A
12.2 **NAS Hazard Rating for Bulk Water Transportation:** Not listed
12.3 **NFPA Hazard Classifications:** Not listed

13. PHYSICAL AND CHEMICAL PROPERTIES

13.1 **Physical State at 15°C and 1 atm:** Solid
13.2 **Molecular Weight:** Not pertinent
13.3 **Boiling Point at 1 atm:** Not pertinent
13.4 **Freezing Point:** 288°F = 142°C = 415°K
13.5 **Critical Temperature:** Not pertinent
13.6 **Critical Pressure:** Not pertinent
13.7 **Specific Gravity:** 1.23 at 20°C (solid)
13.8 **Liquid Surface Tension:** Not pertinent
13.9 **Liquid-Water Interfacial Tension:** Not pertinent
13.10 **Vapor (Gas) Specific Gravity:** Not pertinent
13.11 **Ratio of Specific Heats of Vapor (Gas):** Not pertinent
13.12 **Latent Heat of Vaporization:** Not pertinent
13.13 **Heat of Combustion:** Not pertinent
13.14 **Heat of Decomposition:** Not pertinent
13.15 **Heat of Solution:** Not pertinent
13.16 **Heat of Polymerization:** Not pertinent

NOTES

REVISED 1978

CBO CARBOLIC OIL

Common Synonyms			
Middle Oil Liquefied phenol	Liquid	Colorless – darkens on exposure to light	Sweet, tar odor
	Sinks and mixes with water.		

AVOID CONTACT WITH LIQUID. Keep people away.
Wear chemical protective suit with self-contained breathing apparatus.
Stop discharge if possible.
Call fire department.
Isolate and remove discharged material.
Notify local health and pollution control agencies.

Fire

Combustible.
POISONOUS GASES ARE PRODUCED WHEN HEATED.
Wear chemical protective suit with self-contained breathing apparatus.
Extinguish with water, dry chemical, foam or carbon dioxide.

Exposure

CALL FOR MEDICAL AID.

LIQUID
POISONOUS IF SWALLOWED OR IF SKIN IS EXPOSED.
Remove contaminated clothing and shoes.
Flush affected area with plenty of water.
IF IN EYES, hold eyelids open and flush with plenty of water.
IF SWALLOWED and victim is CONSCIOUS, have victim drink water or milk.
DO NOT INDUCE VOMITING.

Water Pollution

HARMFUL TO AQUATIC LIFE IN VERY LOW CONCENTRATIONS.
May be dangerous if it enters water intakes.
Notify local health and wildlife officials.
Notify operators of nearby water intakes.

1. RESPONSE TO DISCHARGE

(See Response Methods Handbook, CG 446-4)

Issue warning—poison
Restrict access
Should be removed
Chemical and physical treatment

2. LABEL

3. CHEMICAL DESIGNATIONS

3.1 **Synonyms:** Middle oil; Liquefied phenol
3.2 **Coast Guard Compatibility Classification:** Phenol
3.3 **Chemical Formula:** C_6H_5OH
3.4 **IMCO/United Nations Numerical Designation:** 9.0/1671

4. OBSERVABLE CHARACTERISTICS

4.1 **Physical State (as shipped):** Liquid
4.2 **Color:** Colorless, darkens on exposure to light
4.3 **Odor:** Sweet, tar-like

5. HEALTH HAZARDS

5.1 **Personal Protective Equipment:** Fresh air mask for confined areas; rubber gloves; protective clothing; full face shield
5.2 **Symptoms Following Exposure:** Will burn eyes and skin. The analgesic action may cause loss of pain sensation. Readily absorbed through skin, causing increased heart rate, convulsions, and death.
5.3 **Treatment for Exposure:** INHALATION: remove victim to fresh air, keep quiet and warm. If breathing stops, start artificial respiration. INGESTION: do NOT induce vomiting. Give milk, egg whites, or large amounts of water. Get medical assistance. No known antidote. EYES AND SKIN: remove contaminated clothing. Flush eyes with water for 15 minutes or until physician arrives. Wash skin with soap and water.
5.4 **Toxicity by Inhalation (Threshold Limit Value):** 5 ppm
5.5 **Short-Term Inhalation Limits:** Data not available
5.6 **Toxicity by Ingestion:** Grade 2; LD_{50} 0.5 to 5 g/kg (rat)
5.7 **Late Toxicity:** Causes cancer in experimental animals.
5.8 **Vapor (Gas) Irritant Characteristics:** Vapors cause moderate irritation such that personnel will find high concentrations unpleasant. The effect is temporary.
5.9 **Liquid or Solid Irritant Characteristics:** Fairly severe skin irritant. May cause pain and second-degree burns after a few minutes' contact.
5.10 **Odor Threshold:** 0.05 ppm

6. FIRE HAZARDS

6.1 **Flash Point:** 185°F O.C.; 175°F C.C.
6.2 **Flammable Limits in Air:** 1.7%—8.6%
6.3 **Fire Extinguishing Agents:** Water, dry chemical, foam, or carbon dioxide
6.4 **Fire Extinguishing Agents Not to be Used:** Not pertinent
6.5 **Special Hazards of Combustion Products:** Unburned vapor is toxic
6.6 **Behavior in Fire:** Yields flammable vapors when heated, which will form explosive mixtures with air
6.7 **Ignition Temperature:** 1319°F
6.8 **Electrical Hazard:** Not pertinent
6.9 **Burning Rate:** 3.5 mm/min.

7. CHEMICAL REACTIVITY

7.1 **Reactivity with Water:** No reaction
7.2 **Reactivity with Common Materials:** No reaction
7.3 **Stability During Transport:** Stable
7.4 **Neutralizing Agents for Acids and Caustics:** Not pertinent
7.5 **Polymerization:** Not pertinent
7.6 **Inhibitor of Polymerization:** Not pertinent

8. WATER POLLUTION

8.1 **Aquatic Toxicity:**
11.5-28.5 mg/l/96 hr/bluegill/TL_m/fresh water
1.5 ppm/48 hr/rainbow trout/TL_m/fresh water
8.2 **Waterfowl Toxicity:** Data not available
8.3 **Biological Oxygen Demand (BOD):** 200%, 5 days
8.4 **Food Chain Concentration Potential:** None

9. SELECTED MANUFACTURERS

1. Dow Chemical Co.
Midland, Mich. 48640
2. Monsanto Co.
Monsanto Polymers and Petrochemicals Co.
800 N. Lindbergh Blvd.
St. Louis, Mo. 63166
3. United States Steel Corp.
USS Chemicals Division
Clairton, Pa. 15025

10. SHIPPING INFORMATION

10.1 **Grades or Purity:** 90-92% phenol; 80-82% phenol (Remainder consists of cresols and water)
10.2 **Storage Temperature:** Ambient
10.3 **Inert Atmosphere:** No requirement
10.4 **Venting:** Pressure-vacuum

11. HAZARD ASSESSMENT CODE

(See Hazard Assessment Handbook, CG 446-3)

A-P-Q

12. HAZARD CLASSIFICATIONS

12.1 **Code of Federal Regulations:** Poisonous liquid or solid, Class B
12.2 **NAS Hazard Rating for Bulk Water Transportation:**

Category	Rating
Fire	1
Health	
Vapor Irritant	2
Liquid or Solid Irritant	3
Poisons	3
Water Pollution	
Human Toxicity	2
Aquatic Toxicity	3
Aesthetic Effect	3
Reactivity	
Other Chemicals	2
Water	0
Self-Reaction	0

12.3 **NFPA Hazard Classifications:**

Category	Classification
Health Hazard (Blue)	3
Flammability (Red)	2
Reactivity (Yellow)	0

13. PHYSICAL AND CHEMICAL PROPERTIES*

13.1 **Physical State at 15°C and 1 atm:** Liquid
13.2 **Molecular Weight:** 94.11
13.3 **Boiling Point at 1 atm:** 359.2°F = 181.8°C = 455.0°K
13.4 **Freezing Point:** <105.6°F = <40.9°C = <314.1°K
13.5 **Critical Temperature:** 790.0°F = 421.1°C = 694.3°K
13.6 **Critical Pressure:** 889 psia = 60.5 atm = 6.13 MN/m²
13.7 **Specific Gravity:** 1.04 at 41°C (liquid)
13.8 **Liquid Surface Tension:** 41 dynes/cm = 0.041 N/m at 20°C
13.9 **Liquid-Water Interfacial Tension:** Not pertinent
13.10 **Vapor (Gas) Specific Gravity:** Not pertinent
13.11 **Ratio of Specific Heats of Vapor (Gas):** 1.089
13.12 **Latent Heat of Vaporization:** 129.6 Btu/lb = 72.0 cal/g = 3.014×10^5 J/kg
13.13 **Heat of Combustion:** −13,401 Btu/lb = −7445 cal/g = -311.707×10^5 J/kg
13.14 **Heat of Decomposition:** Not pertinent
13.15 **Heat of Solution:** Not pertinent
13.16 **Heat of Polymerization:** Not pertinent

*Data apply to phenol, the principal constituent.

NOTES

REVISED 1978

CBB	CARBON BISULFIDE

Common Synonyms			
Carbon disulfide	Watery liquid	Colorless to yellow	Rotten egg to sweet odor
	Sinks in water. Flammable, irritating vapor is produced.		

Avoid contact with liquid and vapor. Keep people away.
Wear goggles, self-contained breathing apparatus and rubber overclothing (including gloves).
Shut off ignition sources and call fire department.
Stop discharge if possible.
Stay upwind and use water spray to "knock down" vapor.
Isolate and remove discharged material.
Notify local health and pollution control agencies.

Fire

FLAMMABLE.
Flashback along vapor trail may occur.
Vapor may explode if ignited in an enclosed area.
Wear goggles, self-contained breathing apparatus, and rubber overclothing (including gloves).
Extinguish with dry chemical or carbon dioxide.
Water and foam may be ineffective on fire.
Cool exposed containers with water.

Exposure

CALL FOR MEDICAL AID.

VAPOR
Irritating to eyes, nose and throat.
If inhaled, will cause nausea, vomiting, difficult breathing, or loss of consciousness.
Move to fresh air.
If breathing has stopped, give artificial respiration.
If breathing is difficult, give oxygen.

LIQUID
Will burn skin and eyes.
Harmful if swallowed.
Remove contaminated clothing and shoes.
Flush affected areas with plenty of water.
IF IN EYES, hold eyelids open and flush with plenty of water.
IF SWALLOWED and victim is CONSCIOUS, have victim drink water or milk and have victim induce vomiting.
IF SWALLOWED and victim is UNCONSCIOUS OR HAVING CONVULSIONS, do nothing except keep victim warm.

Water Pollution

HARMFUL TO AQUATIC LIFE IN VERY LOW CONCENTRATIONS.
May be dangerous if it enters water intakes.
Notify local health and wildlife officials.
Notify operators of nearby water intakes.

1. RESPONSE TO DISCHARGE

(See Response Methods Handbook, CG 446-4)

Issue warning—high flammability
Restrict access
Evacuate area

2. LABEL

Red

3. CHEMICAL DESIGNATIONS

3.1 **Synonyms:** Carbon disulfide
3.2 **Coast Guard Compatibility Classification:** Carbon disulfide
3.3 **Chemical Formula:** CS_2
3.4 **IMCO/United Nations Numerical Designation:** 3.1/1131

4. OBSERVABLE CHARACTERISTICS

4.1 **Physical State (as shipped):** Liquid
4.2 **Color:** Colorless
4.3 **Odor:** Faint sweetish; disagreeable; offensive, like that of decaying cabbage

5. HEALTH HAZARDS

5.1 **Personal Protective Equipment:** Only self-contained breathing mask with full face, approved by the United States Bureau of Mines, is recommended. If the vapor concentration exceeds 2% by volume or is unknown, supplied-air respiratory equipment of appropriate design with full face masks should be used by all persons entering contaminated area. Masks should be used only for emergency situations and should be located accordingly. Almost any type of industrial clothing is satisfactory. Splashes of small quantity are not harmful to fabrics, and evaporation from clothing is quite rapid. Clothing should, however, be removed and the skin washed with water. Goggles should be used when there is any danger of CS_2 splashes or spray.

5.2 **Symptoms Following Exposure:** ACUTE EXPOSURE: mild to moderate irritation of skin, eyes, and mucous membranes from liquid or concentrated vapors; headache; garlicky breath, nausea, vomiting, diarrhea (even after vapor exposures), and occasionally abdominal pain; weak pulse, palpitations; fatigue, weakness in the legs, unsteady gait, vertigo; mania, hallucinations of sight, hearing, taste, and smell in acute, massive vapor exposures; central nervous depression with respiratory paralysis; death may occur during coma or after a convulsion.

5.3 **Treatment for Exposure:** INHALATION: remove victim promptly from contaminated area. Administer oxygen and artificial respiration if needed. SKIN CONTACT: wash affected areas with copious quantities of water. INGESTION: induce vomiting and follow with gastric lavage and saline cathartics.

5.4 **Toxicity by Inhalation (Threshold Limit Value):** 20 ppm

5.5 **Short-Term Inhalation Limits:** 100 ppm for 30 minutes, 200 ppm for 10 minutes, 100 ppm for 30 minutes and 50 ppm for 60 minutes.

6. FIRE HAZARDS

6.1 **Flash Point:** −22°F C.C.
6.2 **Flammable Limits in Air:** 1.3%—50%
6.3 **Fire Extinguishing Agents:** Dry chemical, carbon dioxide
6.4 **Fire Extinguishing Agents Not to be Used:** Not pertinent
6.5 **Special Hazards of Combustion Products:** Toxic gases are generated; wear self-contained breathing apparatus.
6.6 **Behavior in Fire:** Not pertinent
6.7 **Ignition Temperature:** 212°F
6.8 **Electrical Hazard:** Contact of the liquid or vapor with the surface of a lighted electric light bulb could result in ignition.
6.9 **Burning Rate:** 2.7 mm/min.

7. CHEMICAL REACTIVITY

7.1 **Reactivity with Water:** No reaction
7.2 **Reactivity with Common Materials:** No reaction
7.3 **Stability During Transport:** Stable
7.4 **Neutralizing Agents for Acids and Caustics:** Not pertinent
7.5 **Polymerization:** Not pertinent
7.6 **Inhibitor of Polymerization:** Not pertinent

8. WATER POLLUTION

8.1 **Aquatic Toxicity:** 35 ppm/48 hr/mosquito fish/TL_m/fresh water
8.2 **Waterfowl Toxicity:** Data not available
8.3 **Biological Oxygen Demand (BOD):** Data not available
8.4 **Food Chain Concentration Potential:** None

9. SELECTED MANUFACTURERS

1. FMC Corp.
Inorganic Chemicals Div.
633 Third Ave.
New York, N. Y. 10017
2. PPG Industries Inc.
Industrial Chemicals Div.
Barberton, Ohio 44203
3. Stauffer Chemical Co.
Industrial Chemicals Div.
Delaware City, Delaware 19706

10. SHIPPING INFORMATION

10.1 **Grades or Purity:** Commercial; technical; USP
10.2 **Storage Temperature:** Ambient
10.3 **Inert Atmosphere:** Inerted
10.4 **Venting:** Pressure-vacuum

11. HAZARD ASSESSMENT CODE

(See Hazard Assessment Handbook, CG 446-3)

A-X-Y

12. HAZARD CLASSIFICATIONS

12.1 **Code of Federal Regulations:** Flammable liquid

12.2 **NAS Hazard Rating for Bulk Water Transportation:**

Category	Rating
Fire	4*
Health	
Vapor Irritant	2
Liquid or Solid Irritant	2
Poisons	3
Water Pollution	
Human Toxicity	1
Aquatic Toxicity	2
Aesthetic Effect	3
Reactivity	
Other Chemicals	2
Water	0
Self-Reaction	0

12.3 **NFPA Hazard Classifications:**

Category	Classification
Health Hazard (Blue)	2
Flammability (Red)	3
Reactivity (Yellow)	0

13. PHYSICAL AND CHEMICAL PROPERTIES

13.1 **Physical State at 15°C and 1 atm:** Liquid
13.2 **Molecular Weight:** 76.14
13.3 **Boiling Point at 1 atm:** 115°F = 46.3°C = 319.5°K
13.4 **Freezing Point:** −168.9°F = −111.6°C = 161.6°K
13.5 **Critical Temperature:** 523°F = 273°C = 546°K
13.6 **Critical Pressure:** 1100 psia = 76 atm = 7.7 MN/m^2
13.7 **Specific Gravity:** 1.26 at 20°C (liquid)
13.8 **Liquid Surface Tension:** 32 dynes/cm = .032 N/m at 20°C
13.9 **Liquid-Water Interfacial Tension:** 48.4 dynes/cm = .0484 N/m at 20°C
13.10 **Vapor (Gas) Specific Gravity:** 2.6
13.11 **Ratio of Specific Heats of Vapor (Gas):** 1.292
13.12 **Latent Heat of Vaporization:** 153 Btu/lb = 85 cal/g = 3.559 × 10^5 J/kg
13.13 **Heat of Combustion:** −5814 Btu/lb = −3230 cal/g = −135.2 × 10^5 J/kg
13.14 **Heat of Decomposition:** Not pertinent
13.15 **Heat of Solution:** Not pertinent
13.16 **Heat of Polymerization:** Not pertinent

5. HEALTH HAZARDS (Cont'd.)

5.6 **Toxicity by Ingestion:** Grade 2; rat LD_{50} = 0.1 - 0.99 g/kg
5.7 **Late Toxicity:** Non-specific liver cell damage in rats; higher incidence of upper respiratory disease in humans.
5.8 **Vapor (Gas) Irritant Characteristics:** Vapors cause moderate irritation such that personnel will find high concentrations unpleasant. The effect is temporary.
5.9 **Liquid or Solid Irritant Characteristics:** Causes smarting of the skin and first-degree burns on short exposure and may cause secondary burns on long exposure.
5.10 **Odor Threshold:** 0.21 ppm

REVISED 1978

CDO	# CARBON DIOXIDE

Common Synonyms			
Carbonic acid gas Carbonic anhydride	Liquefied compressed gas or solid	Colorless gas or white solid	Odorless
	Solid sinks and boils in water. Visible vapor cloud is produced.		

AVOID CONTACT WITH LIQUID AND SOLID. KEEP PEOPLE AWAY.
Wear goggles, self-contained breathing apparatus, and rubber overclothing (including gloves).
Stop discharge if possible.

Fire	Not flammable. Containers may explode in fire. Cool exposed containers with water.
Exposure	Call for medical aid. VAPOR If inhaled will cause dizziness, or difficult breathing. Move victim to fresh air. If breathing is difficult, give oxygen. LIQUID OR SOLID Will cause frostbite. Flush affected areas with plenty of water. DO NOT RUB AFFECTED AREAS.
Water Pollution	Not harmful to aquatic life.

1. RESPONSE TO DISCHARGE

(See Response Methods Handbook, CG 446-4)

Restrict access
Disperse and flush

2. LABEL

Green

3. CHEMICAL DESIGNATIONS

3.1 **Synonyms:** Carbonic acid gas
Carbonic anhydride
3.2 **Coast Guard Compatibility Classification:** Not applicable
3.3 **Chemical Formula:** CO_2
3.4 **IMCO/United Nations Numerical Designation:** 2/1013

4. OBSERVABLE CHARACTERISTICS

4.1 **Physical State (as shipped):** Liquefied compressed gas or solid ("Dry Ice")
4.2 **Color:** Colorless
4.3 **Odor:** None

5. HEALTH HAZARDS

5.1 **Personal Protective Equipment:** Self-contained breathing apparatus in excessively high CO_2 concentration areas. For handling liquid or solid, wear safety goggles or face shield, insulated gloves, long-sleeved shirt, and trousers worn outside boots or over high-top shoes to shed spilled liquid.
5.2 **Symptoms Following Exposure:** Inhalation causes increased respiration rate, headache, subtle physiological changes for up to 5% concentration and prolonged exposure. Higher concentrations can cause unconsciousness and death. Solid can cause cold contact burns. Liquid or cold gas can cause freezing injury to skin or eyes similar to a burn.
5.3 **Treatment for Exposure:** INHALATION: move victim to fresh air. SKIN: treat burns from contact with solid in same way as frostbite.
5.4 **Toxicity by Inhalation (Threshold Limit Value):** 5000 ppm
5.5 **Short-Term Inhalation Limits:** 30,000 ppm for 60 min.
5.6 **Toxicity by Ingestion:** Not pertinent (gas with low boiling point)
5.7 **Late Toxicity:** None
5.8 **Vapor (Gas) Irritant Characteristics:** Data not available
5.9 **Liquid or Solid Irritant Characteristics:** Data not available
5.10 **Odor Threshold:** Not pertinent

6. FIRE HAZARDS

6.1 **Flash Point:** Not flammable
6.2 **Flammable Limits in Air:** Not flammable
6.3 **Fire Extinguishing Agents:** Not pertinent
6.4 **Fire Extinguishing Agents Not to be Used:** Not pertinent
6.5 **Special Hazards of Combustion Products:** Not pertinent
6.6 **Behavior in Fire:** Containers may explode when heated.
6.7 **Ignition Temperature:** Not pertinent
6.8 **Electrical Hazard:** Not pertinent
6.9 **Burning Rate:** Not pertinent

7. CHEMICAL REACTIVITY

7.1 **Reactivity with Water:** No reaction
7.2 **Reactivity with Common Materials:** No reaction
7.3 **Stability During Transport:** Stable
7.4 **Neutralizing Agents for Acids and Caustics:** Not pertinent
7.5 **Polymerization:** Not pertinent
7.6 **Inhibitor of Polymerization:** Not pertinent

8. WATER POLLUTION

8.1 **Aquatic Toxicity:**
100-200 mg/l /*/various organisms/LC/ fresh water
*Time period not specified
8.2 **Waterfowl Toxicity:** Inhalation 5-8%, no effect
8.3 **Biological Oxygen Demand (BOD):** None
8.4 **Food Chain Concentration Potential:** None

9. SELECTED MANUFACTURERS

1. Chemetron Corporation
Cardox Products Division
111 E. Wacker Drive
Chicago, Ill. 60601
2. Union Carbide Corporation
Linde Division
Morristown, N. J. 08057
3. Liquid Carbonic Corporation
135 S. LaSalle St.
Chicago, Ill. 60603

10. SHIPPING INFORMATION

10.1 **Grades or Purity:** Research: 99.995+%; Instrument: 99.99+%; Bone Dry: 99.95+%; Commercial: 99.5+%
10.2 **Storage Temperature:** Ambient
10.3 **Inert Atmosphere:** No requirement
10.4 **Venting:** Liquid—safety relief; solid—open

11. HAZARD ASSESSMENT CODE

(See Hazard Assessment Handbook, CG 446-3)

A-C-II

12. HAZARD CLASSIFICATIONS

12.1 **Code of Federal Regulations:** Nonflammable compressed gas
12.2 **NAS Hazard Rating for Bulk Water Transportation:** Not listed
12.3 **NFPA Hazard Classifications:** Not listed

13. PHYSICAL AND CHEMICAL PROPERTIES

13.1 **Physical State at 15°C and 1 atm:** Gas
13.2 **Molecular Weight:** 44.0
13.3 **Boiling Point at 1 atm:** Not pertinent (sublimes)
13.4 **Freezing Point:** $-109.3°F = -78.5°C = 194.7°K$
13.5 **Critical Temperature:** $88°F = 31°C = 304°K$
13.6 **Critical Pressure:** $1{,}070$ psia $= 72.9$ atm $= 7.40$ MN/m^2
13.7 **Specific Gravity:** 1.56 at −79°C (solid)
13.8 **Liquid Surface Tension:** Not pertinent
13.9 **Liquid-Water Interfacial Tension:** Not pertinent
13.10 **Vapor (Gas) Specific Gravity:** 1.53
13.11 **Ratio of Specific Heats of Vapor (Gas):** 1.0474
13.12 **Latent Heat of Vaporization:** 150 Btu/lb $= 83$ cal/g $= 3.5 \times 10^5$ J/kg
13.13 **Heat of Combustion:** Not pertinent
13.14 **Heat of Decomposition:** Not pertinent
13.15 **Heat of Solution:** Not pertinent
13.16 **Heat of Polymerization:** Not pertinent

NOTES

CMO	# CARBON MONOXIDE

Common Synonyms			
Monoxide	Compressed gas or liquefied compressed gas	Colorless	Odorless
	Liquid floats and boils on water. Poisonous, flammable visible vapor cloud is produced.		

AVOID CONTACT WITH LIQUID AND VAPOR. KEEP PEOPLE AWAY.
Wear goggles, self-contained breathing apparatus, and rubber overclothing (including gloves).
Shut off ignition sources. Call fire department.
Stop discharge if possible.
Evacuate area in case of large discharge.
Notify local health and pollution control agencies.

Fire

FLAMMABLE.
Containers may explode in fire.
Wear goggles and self-contained breathing apparatus.
Let fire burn.
Stop flow of gas if possible.
Cool exposed containers and protect men effecting shutoff with water.

Exposure

CALL FOR MEDICAL AID.

VAPOR
POISONOUS IF INHALED.
Move victim to fresh air.
If breathing has stopped, give artificial respiration.
If breathing is difficult, give oxygen.

LIQUID
Will cause frostbite.
Flush affected areas with plenty of water.
DO NOT RUB AFFECTED AREAS.

Water Pollution

HARMFUL TO AQUATIC LIFE IN VERY LOW CONCENTRATIONS.
May be dangerous if it enters water intakes.
Notify local health and wildlife officials.
Notify operators of nearby water intakes.

1. RESPONSE TO DISCHARGE
(See Response Methods Handbook, CG 446-4)

Issue warning—poison, high flammability
Restrict access
Evacuate area

2. LABEL

Red

3. CHEMICAL DESIGNATIONS

3.1 **Synonyms:** Monoxide
3.2 **Coast Guard Compatibility Classification:** Not applicable
3.3 **Chemical Formula:** CO
3.4 **IMCO/United Nations Numerical Designation:** 2/1016

4. OBSERVABLE CHARACTERISTICS

4.1 **Physical State (as shipped):** Compressed gas or liquefied gas
4.2 **Color:** Colorless
4.3 **Odor:** None

5. HEALTH HAZARDS

5.1 **Personal Protective Equipment:** Self-contained breathing apparatus; safety glasses and safety shoes; Type D or Type N canister mask.
5.2 **Symptoms Following Exposure:** Inhalation causes headache, dizziness, weakness of limbs, confusion, nausea, unconsciousness, and finally death. 0.04% conc., 2-3 hr. or .06% conc., 1 hr.—headache and discomfort; with moderate exercise, 0.1-0.2% will produce throbbing in the head in about ½ hr., a tendency to stagger in about 1½ hr., and confusion of the mind, headache, and nausea in about 2 hrs. 0.20-.25% usually produces unconsciousness in about ½ hr. Inhalation of a 0.4% conc. can prove fatal in less than 1 hr. Inhalation of high concentrations can cause sudden, unexpected collapse. Contact of liquid with skin will cause frostbite.
5.3 **Treatment for Exposure:** INHALATION: remove from exposure; give oxygen if available; support respiration; call a doctor. SKIN: if burned by liquid, treat as frostbite.
5.4 **Toxicity by Inhalation (Threshold Limit Value):** 50 ppm
5.5 **Short-Term Inhalation Limits:** 400 ppm, 15 min.
5.6 **Toxicity by Ingestion:** Not pertinent (gas with low boiling point)
5.7 **Late Toxicity:** Toxicity from overexposure persists for many days.
5.8 **Vapor (Gas) Irritant Characteristics:** Data not available
5.9 **Liquid or Solid Irritant Characteristics:** Data not available
5.10 **Odor Threshold:** Not pertinent

6. FIRE HAZARDS

6.1 **Flash Point:** Not pertinent
6.2 **Flammable Limits in Air:** 12%—75%
6.3 **Fire Extinguishing Agents:** Let fire burn; shut off flow of gas and cool adjacent exposures with water. Extinguish (only if wearing self-contained breathing apparatus) with dry chemicals or carbon dioxide.
6.4 **Fire Extinguishing Agents Not to be Used:** Not pertinent
6.5 **Special Hazards of Combustion Products:** Asphyxiation due to carbon dioxide production may result.
6.6 **Behavior in Fire:** Flame has very little color. Containers may explode in fire.
6.7 **Ignition Temperature:** 1,128°F
6.8 **Electrical Hazard:** Data not available
6.9 **Burning Rate:** Not pertinent

7. CHEMICAL REACTIVITY

7.1 **Reactivity with Water:** No reaction
7.2 **Reactivity with Common Materials:** No reaction
7.3 **Stability During Transport:** Stable
7.4 **Neutralizing Agents for Acids and Caustics:** Not pertinent
7.5 **Polymerization:** Not pertinent
7.6 **Inhibitor of Polymerization:** Not pertinent

8. WATER POLLUTION

8.1 **Aquatic Toxicity:** 1.5 ppm/1-6 hr/minnows and sunfish/killed/fresh water
8.2 **Waterfowl Toxicity:** Data not available
8.3 **Biological Oxygen Demand (BOD):** None
8.4 **Food Chain Concentration Potential:** None

9. SELECTED MANUFACTURERS

1. Air Products and Chemicals, Inc.
   Specialty Gas & Equipment Dept.
   733 W. Broad Street
   Emmaus, Pa. 18049
2. Matheson Gas Products
   East Rutherford, N. J. 07073

10. SHIPPING INFORMATION

10.1 **Grades or Purity:** Liquid: 98.6+%; Gas: Research High Purity; CP (99.5%); Technical (99.0+%); Commercial (97.5+%)
10.2 **Storage Temperature:** Ambient (for gas); −312.7°F (for liquid)
10.3 **Inert Atmosphere:** No requirement
10.4 **Venting:** Safety relief

11. HAZARD ASSESSMENT CODE
(See Hazard Assessment Handbook, CG 446-3)

A-B-C-D-E-F-G

12. HAZARD CLASSIFICATIONS

12.1 **Code of Federal Regulations:** Flammable compressed gas
12.2 **NAS Hazard Rating for Bulk Water Transportation:** Not listed
12.3 **NFPA Hazard Classifications:**

Category	Classification
Health Hazard (Blue)	2
Flammability (Red)	4
Reactivity (Yellow)	0

13. PHYSICAL AND CHEMICAL PROPERTIES

13.1 **Physical State at 15°C and 1 atm:** Gas
13.2 **Molecular Weight:** 28.0
13.3 **Boiling Point at 1 atm:** −312.7°F = −191.5°C = 81.7°K
13.4 **Freezing Point:** −326°F = −199°C = 74°K
13.5 **Critical Temperature:** −220°F = −140°C = 133°K
13.6 **Critical Pressure:** 507.5 psia = 34.51 atm = 3.502 MN/m²
13.7 **Specific Gravity:** 0.791 at −191.5°C (liquid)
13.8 **Liquid Surface Tension:** 9.8 dynes/cm = 0.098 N/m at −193°C
13.9 **Liquid-Water Interfacial Tension:** Not pertinent
13.10 **Vapor (Gas) Specific Gravity:** 0.97
13.11 **Ratio of Specific Heats of Vapor (Gas):** 1.3962
13.12 **Latent Heat of Vaporization:** 92.8 Btu/lb = 51.6 cal/g = 2.16×10^5 J/kg
13.13 **Heat of Combustion:** −4,343 Btu/lb = −2,412 cal/g = -101×10^5 J/kg
13.14 **Heat of Decomposition:** Not pertinent
13.15 **Heat of Solution:** Not pertinent
13.16 **Heat of Polymerization:** Not pertinent

NOTES

CBT	CARBON TETRACHLORIDE

Common Synonyms Carbon Tet Tetrachloromethane	Watery liquid · Colorless · Sweet odor Sinks in water. Poisonous vapor is produced.

Avoid contact with liquid and vapor. Keep people away.
Wear goggles and self-contained breathing apparatus.
Stop discharge if possible.
Stay upwind and use water spray to "knock down" vapor.
Notify local health and pollution control agencies.

Fire	Not flammable. POISONOUS AND IRRITATING GASES ARE PRODUCED WHEN HEATED. Wear goggles and self-contained breathing apparatus.
Exposure	CALL FOR MEDICAL AID. VAPOR POISONOUS IF INHALED. Irritating to eyes Move to fresh air. If breathing has stopped, give artificial respiration. If breathing is difficult, give oxygen. LIQUID POISONOUS IF SWALLOWED. Irritating to skin and eyes. Remove contaminated clothing and shoes. Flush affected areas with plenty of water. IF IN EYES, hold eyelids open and flush with plenty of water. IF SWALLOWED and victim is CONSCIOUS, have victim drink water or milk and have victim induce vomiting. IF SWALLOWED and victim is UNCONSCIOUS OR HAVING CONVULSIONS, do nothing except keep victim warm.
Water Pollution	Effect of low concentrations on aquatic life is unknown. May be dangerous if it enters water intakes. Notify local health and pollution control officials. Notify operators of nearby water intakes.

1. RESPONSE TO DISCHARGE

(See Response Methods Handbook, CG 446-4)

Issue warning—poison
Restrict access
Should be removed

2. LABELS

No hazard label required by Code of Federal Regulations

3. CHEMICAL DESIGNATIONS

3.1 **Synonyms:** Benzinoform; Necatorina; Perchloromethane; Tetrachloromethane

3.2 **Coast Guard Compatibility Classification:** Halogenated hydrocarbon

3.3 **Chemical Formula:** CCl_4

3.4 **IMCO/United Nations Numerical Designation:** 6.1/1846

4. OBSERVABLE CHARACTERISTICS

4.1 **Physical State (as shipped):** Liquid

4.2 **Color:** Colorless

4.3 **Odor:** Sweetish, aromatic; moderately strong ethereal; somewhat resembling that of chloroform.

5. HEALTH HAZARDS

5.1 **Personal Protective Equipment:** Organic vapor canister with full face mask; protective clothing; rubber gloves.

5.2 **Symptoms Following Exposure:** Dizziness, incoordination, anesthesia; may be accompanied by nausea and liver damage. Kidney damage also occurs, often producing decrease or stopping of urinary output.

5.3 **Treatment for Exposure:** EYES AND SKIN: flush with plenty of water; for eyes, get medical attention. Remove contaminated clothing and wash before reuse. INHALATION: immediately remove to fresh air, keep patient warm and quiet and get medical attention promptly. Start artificial respiration if breathing stops. INGESTION: induce vomiting and get medical attention promptly. No specific antidote known.

5.4 **Toxicity by Inhalation (Threshold Limit Value):** 10 ppm

5.5 **Short-Term Inhalation Limits:** 25 ppm for 30 min.

5.6 **Toxicity by Ingestion:** Grade 2; LD_{50} 0.5 to 5 g/kg (rat)

5.7 **Late Toxicity:** Causes severe liver damage and death if ingested.

5.8 **Vapor (Gas) Irritant Characteristics:** Vapors cause moderate irritation such that personnel will find high concentrations unpleasant. The effect is temporary.

5.9 **Liquid or Solid Irritant Characteristics:** Minimum hazard. If spilled on clothing and allowed to remain, may cause smarting and reddening of the skin.

5.10 **Odor Threshold:** Greater than 10 ppm

6. FIRE HAZARDS

6.1 **Flash Point:** Not flammable

6.2 **Flammable Limits in Air:** Not flammable

6.3 **Fire Extinguishing Agents:** Not pertinent

6.4 **Fire Extinguishing Agents Not to be Used:** Not pertinent

6.5 **Special Hazards of Combustion Products:** Forms poisonous phosgene gas when exposed to open flames.

6.6 **Behavior in Fire:** Decomposes to form chlorine and phosgene

6.7 **Ignition Temperature:** Not flammable

6.8 **Electrical Hazard:** Not pertinent

6.9 **Burning Rate:** Not flammable

7. CHEMICAL REACTIVITY

7.1 **Reactivity with Water:** No reaction

7.2 **Reactivity with Common Materials:** No reaction

7.3 **Stability During Transport:** Stable

7.4 **Neutralizing Agents for Acids and Caustics:** Not pertinent

7.5 **Polymerization:** Not pertinent

7.6 **Inhibitor of Polymerization:** Not pertinent

8. WATER POLLUTION

8.1 **Aquatic Toxicity:** Data not available

8.2 **Waterfowl Toxicity:** Data not available

8.3 **Biological Oxygen Demand (BOD):** None

8.4 **Food Chain Concentration Potential:** None

9. SELECTED MANUFACTURERS

1. Dow Chemical Co.
   Midland, Michigan 48640
2. FMC Corp.
   Inorganic Chemicals Div.
   633 Third Ave.
   New York, N. Y. 10017
3. Stauffer Chemical Co.
   Industrial Chemicals Div.
   Le Moyne, Alabama 36505

10. SHIPPING INFORMATION

10.1 **Grades or Purity:** Commercial; technical; USP

10.2 **Storage Temperature:** Ambient

10.3 **Inert Atmosphere:** No requirement

10.4 **Venting:** Pressure-vacuum

11. HAZARD ASSESSMENT CODE

(See Hazard Assessment Handbook, CG 446-3)

A-X

12. HAZARD CLASSIFICATIONS

12.1 **Code of Federal Regulations:** ORM–A

12.2 **NAS Hazard Rating for Bulk Water Transportation:**

Category	Rating
Fire	0
Health	
Vapor Irritant	2
Liquid or Solid Irritant	1
Poisons	4
Water Pollution	
Human Toxicity	2
Aquatic Toxicity	2
Aesthetic Effect	2
Reactivity	
Other Chemicals	1
Water	0
Self-Reaction	0

12.3 **NFPA Hazard Classifications:**

Category	Classification
Health Hazard (Blue)	3
Flammability (Red)	0
Reactivity (Yellow)	0

13. PHYSICAL AND CHEMICAL PROPERTIES

13.1 **Physical State at 15°C and 1 atm:** Liquid

13.2 **Molecular Weight:** 153.83

13.3 **Boiling Point at 1 atm:** 170°F = 76.5°C = 349.7°K

13.4 **Freezing Point:** −9.4°F = −23.0°C = 250.2°K

13.5 **Critical Temperature:** 541°F = 283°C = 556°K

13.6 **Critical Pressure:** 660 psia = 45 atm = 4.6 MN/m²

13.7 **Specific Gravity:** 1.59 at 20°C (liquid)

13.8 **Liquid Surface Tension:** 27.0 dynes/cm = 0.027 N/m at 20°C

13.9 **Liquid-Water Interfacial Tension:** 45.0 dynes/cm = 0.045 N/m at 20°C

13.10 **Vapor (Gas) Specific Gravity:** 5.3

13.11 **Ratio of Specific Heats of Vapor (Gas):** 1.111

13.12 **Latent Heat of Vaporization:** 84.2 Btu/lb = 46.8 cal/g = 1.959×10^5 J/kg

13.13 **Heat of Combustion:** Not pertinent

13.14 **Heat of Decomposition:** Not pertinent

13.15 **Heat of Solution:** Not pertinent

13.16 **Heat of Polymerization:** Not pertinent

NOTES

REVISED 1978

CAR	CARENE

Common Synonyms			
3-Carene Isodiprene 4,7,7-Trimethyl-3-norcarene 3,7,7-Trimethylbicyclo[0,1,4] hept-3-ene	Liquid	Colorless	Sweet, turpentine-like odor
	Floats on water.		

Stop discharge if possible. Keep people away.
Avoid contact with liquid.
Isolate and remove discharged material.
Notify local health and pollution control agencies.

Fire	Combustible. Extinguish with dry chemicals, foam or carbon dioxide. Water may be ineffective on fire. Cool exposed containers with water.
Exposure	CALL FOR MEDICAL AID. VAPOR Irritating to eyes, nose and throat. If inhaled will cause headache, coughing or difficult breathing. If in eyes, hold eyelids open and flush with plenty of water. If breathing has stopped, give artificial respiration. If breathing is difficult, give oxygen. LIQUID Irritating to skin and eyes. If swallowed will cause nausea and vomiting. Remove contaminated clothing and shoes. Flush affected areas with plenty of water. IF IN EYES, hold eyelids open and flush with plenty of water. IF SWALLOWED and victim is CONSCIOUS, have victim drink water or milk and have victim induce vomiting. IF SWALLOWED and victim is UNCONSCIOUS OR HAVING CONVULSIONS, do nothing except keep victim warm.
Water Pollution	Effect of low concentrations on aquatic life is unknown. Fouling to shoreline. May be dangerous if it enters water intakes. Notify local health and wildlife officials. Notify operators of nearby water intakes.

1. RESPONSE TO DISCHARGE

(See Response Methods Handbook, CG 446-4)

Issue warning—high flammability
Mechanical containment
Should be removed
Chemical and physical treatment

2. LABELS

No hazard label required by Code of Federal Regulations

3. CHEMICAL DESIGNATIONS

3.1 **Synonyms:** 3-Carene; Isodiprene; 4, 7, 7-Trimethyl-3-norcarene 3, 7, 7-Trimethylbicyclo[0. 1. 4] hept-3-ene
3.2 **Coast Guard Compatibility Classification:** Not listed
3.3 **Chemical Formula:** $C_{10}H_{16}$
3.4 **IMCO/United Nations Numerical Designation:** Not listed

4. OBSERVABLE CHARACTERISTICS

4.1 **Physical State (as shipped):** Liquid
4.2 **Color:** Colorless
4.3 **Odor:** Sweet, pungent, like turpentine

5. HEALTH HAZARDS

5.1 **Personal Protective Equipment:** Organic canister or air-supplied mask; goggles or face shield; rubber gloves
5.2 **Symptoms Following Exposure:** Inhalation causes headache, confusion, respiratory distress. Ingestion irritates entire digestive system and may injure kidneys; if liquid enters lungs, it causes severe pneumonitis. Contact with eyes or skin causes irritation.
5.3 **Treatment for Exposure:** INHALATION: move victim to fresh air; call a doctor; administer artificial respiration and oxygen if required. INGESTION: give large amounts of water and induce vomiting; call a doctor. EYES: flush with water for at least 15 min. SKIN: wipe off; wash with soap and water.
5.4 **Toxicity by Inhalation (Threshold Limit Value):** Data not available
5.5 **Short-Term Inhalation Limits:** Data not available
5.6 **Toxicity by Ingestion:** Grade 2; oral rat LD_{50} = 4.8 g/kg
5.7 **Late Toxicity:** Data not available
5.8 **Vapor (Gas) Irritant Characteristics:** Data not available
5.9 **Liquid or Solid Irritant Characteristics:** Data not available
5.10 **Odor Threshold:** Data not available

6. FIRE HAZARDS

6.1 **Flash Point:** Data not available
6.2 **Flammable Limits in Air:** Data not available
6.3 **Fire Extinguishing Agents:** Foam, dry chemical, carbon dioxide
6.4 **Fire Extinguishing Agents Not to be Used:** Water may be ineffective on fire
6.5 **Special Hazards of Combustion Products:**
6.6 **Behavior in Fire:**
6.7 **Ignition Temperature:** Data not available
6.8 **Electrical Hazard:** Data not available
6.9 **Burning Rate:** Data not available

7. CHEMICAL REACTIVITY

7.1 **Reactivity with Water:** No reaction
7.2 **Reactivity with Common Materials:** Will attack some forms of plastics
7.3 **Stability During Transport:** Stable
7.4 **Neutralizing Agents for Acids and Caustics:** Not pertinent
7.5 **Polymerization:** Not pertinent
7.6 **Inhibitor of Polymerization:** Not pertinent

8. WATER POLLUTION

8.1 **Aquatic Toxicity:** Data not available
8.2 **Waterfowl Toxicity:** Data not available
8.3 **Biological Oxygen Demand (BOD):** Data not available
8.4 **Food Chain Concentration Potential:** None

9. SELECTED MANUFACTURERS

Northwest Petrochemical Corp.
P.O. Box 99
Anacortes, Wash. 98221

10. SHIPPING INFORMATION

10.1 **Grades or Purity:** Commercial
10.2 **Storage Temperature:** Ambient
10.3 **Inert Atmosphere:** No requirement
10.4 **Venting:** Open

11. HAZARD ASSESSMENT CODE

(See Hazard Assessment Handbook, CG 446-3)

A-T-U

12. HAZARD CLASSIFICATIONS

12.1 **Code of Federal Regulations:** Not listed
12.2 **NAS Hazard Rating for Bulk Water Transportation:** Not listed
12.3 **NFPA Hazard Classifications:** Not listed

13. PHYSICAL AND CHEMICAL PROPERTIES

13.1 **Physical State at 15°C and 1 atm:** Liquid
13.2 **Molecular Weight:** 136
13.3 **Boiling Point at 1 atm:** 338°F = 170°C = 443°K
13.4 **Freezing Point:** Not pertinent
13.5 **Critical Temperature:** Not pertinent
13.6 **Critical Pressure:** Not pertinent
13.7 **Specific Gravity:** 0.860 at 20°C (liquid)
13.8 **Liquid Surface Tension:** Data not available
13.9 **Liquid-Water Interfacial Tension:** Data not available
13.10 **Vapor (Gas) Specific Gravity:** Not pertinent
13.11 **Ratio of Specific Heats of Vapor (Gas):** Not pertinent
13.12 **Latent Heat of Vaporization:** Data not available
13.13 **Heat of Combustion:** (est.) −19,370 Btu/lb = 10,760 cal/g = -450×10^5 J/kg
13.14 **Heat of Decomposition:** Not pertinent
13.15 **Heat of Solution:** Not pertinent
13.16 **Heat of Polymerization:** Not pertinent

NOTES

CTC CATECHOL

Common Synonyms
Oxyphenic acid
Pyrocatechinic acid
Pyrocatechuic acid

Solid White Odorless

Sinks and mixes with water.

Avoid contact with solid and dust. Keep people away.
Wear rubber overclothing (including gloves).
Stop discharge if possible.
Call fire department.
Isolate and remove discharged material.
Notify local health and pollution control agencies.

Fire

Combustible.
POISONOUS GASES MAY BE PRODUCED WHEN HEATED.
Wear goggles and self-contained breathing apparatus.
Extinguish with dry chemicals, alcohol foam, or carbon dioxide.
Water may be ineffective on fire.
Cool exposed containers with water.

Exposure

CALL FOR MEDICAL AID.

DUST
Irritating to eyes, nose and throat.
If inhaled will cause coughing or difficult breathing.
If in eyes, hold eyelids open and flush with plenty of water.
If breathing has stopped, give artificial respiration.
If breathing is difficult, give oxygen.

SOLID
Will burn skin and eyes.
Harmful if swallowed.
Remove contaminated clothing and shoes.
Flush affected areas with plenty of water.
IF IN EYES, hold eyelids open and flush with plenty of water.
IF SWALLOWED and victim is CONSCIOUS, have victim drink water or milk and have victim induce vomiting.
IF SWALLOWED and victim is UNCONSCIOUS OR HAVING CONVULSIONS, do nothing except keep victim warm.

Water Pollution

Effect of low concentrations on aquatic life is unknown.
May be dangerous if it enters water intakes.
Notify local health and wildlife officials.
Notify operators of nearby water intakes.

1. RESPONSE TO DISCHARGE

(See Response Methods Handbook, CG 446-4)

Issue warning—water contaminant
Disperse and flush

2. LABELS

No hazard label required by Code of Federal Regulations

3. CHEMICAL DESIGNATIONS

3.1 **Synonyms:** 1,2-Benzenediol; Catechin; 1,2-Dihydroxybenzene; ortho-Dihydroxybenzene; Oxyphenic acid; Pyrocatechin; Pyrocatechinic acid; Pyrocatechol; Pyrocatechuic acid

3.2 **Coast Guard Compatibility Classification:** Not listed

3.3 **Chemical Formula:** 1,2-HOC_6H_4OH

3.4 **IMCO/United Nations Numerical Designation:** Not listed

4. OBSERVABLE CHARACTERISTICS

4.1 **Physical State (as shipped):** Solid

4.2 **Color:** White

4.3 **Odor:** None

5. HEALTH HAZARDS

5.1 **Personal Protective Equipment:** Dust respirator if required; rubber gloves, apron, and boots; face shield

5.2 **Symptoms Following Exposure:** Inhalation of dusts or mists may cause irritation of eyes, nose, and throat. Ingestion may cause convulsions and respiratory failure. Contact with eyes causes burns and possible permanent impairment of vision. Prolonged or repeated contact with skin may cause burn.

5.3 **Treatment for Exposure:** INHALATION: if ill effects occur, get medical attention. INGESTION: promptly give milk or plenty of water and induce vomiting; get medical attention promptly; no specific antidote known. EYES and SKIN: immediately flush with plenty of water for at least 15 min.; for eyes get medical attention promptly; remove and wash all contaminated clothing before reuse.

5.4 **Toxicity by Inhalation (Threshold Limit Value):** Data not available

5.5 **Short-Term Inhalation Limits:** Data not available

5.6 **Toxicity by Ingestion:** Grade 2; LD_{50} 0.5—5 g/kg (rat)

5.7 **Late Toxicity:** Causes tumors in mice

5.8 **Vapor (Gas) Irritant Characteristics:** Data not available

5.9 **Liquid or Solid Irritant Characteristics:** Data not available

5.10 **Odor Threshold:** Data not available

6. FIRE HAZARDS

6.1 **Flash Point:** (liquid) 278°F O.C. 261°F C.C.

6.2 **Flammable Limits in Air:** Not pertinent (combustible solid)

6.3 **Fire Extinguishing Agents:** Dry chemical, "alcohol" foam, carbon dioxide

6.4 **Fire Extinguishing Agents Not to be Used:** Water and foam may be ineffective.

6.5 **Special Hazards of Combustion Products:** May form toxic fumes at high temperatures

6.6 **Behavior in Fire:**

6.7 **Ignition Temperature:** Data not available

6.8 **Electrical Hazard:** Not pertinent

6.9 **Burning Rate:** Not pertinent

7. CHEMICAL REACTIVITY

7.1 **Reactivity with Water:** No reaction

7.2 **Reactivity with Common Materials:**

7.3 **Stability During Transport:** Stable

7.4 **Neutralizing Agents for Acids and Caustics:** Not pertinent

7.5 **Polymerization:** Not pertinent

7.6 **Inhibitor of Polymerization:** Not pertinent

8. WATER POLLUTION

8.1 **Aquatic Toxicity:** 14 ppm/48 hr/goldfish/TL_m

8.2 **Waterfowl Toxicity:** Data not available

8.3 **Biological Oxygen Demand (BOD):** Data not available

8.4 **Food Chain Concentration Potential:** None

9. SELECTED MANUFACTURERS

1. Crown Zellerbach Corp.
   Chemical Products Div.
   Camas, Wash. 98607
2. Dow Chemical Co.
   Midland, Mich. 48640
3. Eastman Organic Chemicals
   Rochester, N. Y. 14650

10. SHIPPING INFORMATION

10.1 **Grades or Purity:** CP—high purity, 99.3+%; XP—extremely high purity, 99.8+%

10.2 **Storage Temperature:** Ambient

10.3 **Inert Atmosphere:** No requirement

10.4 **Venting:** Open

11. HAZARD ASSESSMENT CODE

(See Hazard Assessment Handbook, CG 446-3)

SS

12. HAZARD CLASSIFICATIONS

12.1 **Code of Federal Regulations:** Not listed

12.2 **NAS Hazard Rating for Bulk Water Transportation:** Not listed

12.3 **NFPA Hazard Classifications:** Not listed

13. PHYSICAL AND CHEMICAL PROPERTIES

13.1 **Physical State at 15°C and 1 atm:** Solid

13.2 **Molecular Weight:** 110.11

13.3 **Boiling Point at 1 atm:** 473.9°F = 245.5°C = 418.7°K

13.4 **Freezing Point:** 219.7°F = 104.3°C = 377.5°K

13.5 **Critical Temperature:** Not pertinent

13.6 **Critical Pressure:** Not pertinent

13.7 **Specific Gravity:** 1.344 at 20°C (solid)

13.8 **Liquid Surface Tension:** Not pertinent

13.9 **Liquid-Water Interfacial Tension:** Not pertinent

13.10 **Vapor (Gas) Specific Gravity:** Not pertinent

13.11 **Ratio of Specific Heats of Vapor (Gas):** Not pertinent

13.12 **Latent Heat of Vaporization:** Not pertinent

13.13 **Heat of Combustion:** −11,200 Btu/lb = −6,220 cal/g = -260×10^5 J/kg

13.14 **Heat of Decomposition:** Not pertinent

13.15 **Heat of Solution:** Not pertinent

13.16 **Heat of Polymerization:** Not pertinent

NOTES

CPS	# CAUSTIC POTASH SOLUTION

Common Synonyms Potassium hydroxide solution Lye	Thick liquid Colorless Odorless Sinks and mixes with water.

AVOID CONTACT WITH LIQUID. Keep people away.
Wear rubber overclothing (including gloves).
Stop discharge if possible.
Isolate and remove discharged material.
Notify local health and pollution control agencies.

Fire	Not flammable.
Exposure	CALL FOR MEDICAL AID. LIQUID Will burn skin and eyes. Harmful if swallowed. Remove contaminated clothing and shoes. Flush affected areas with plenty of water. IF IN EYES, hold eyelids open and flush with plenty of water. IF SWALLOWED and victim is CONSCIOUS, have victim drink water or milk. DO NOT INDUCE VOMITING.
Water Pollution	HARMFUL TO AQUATIC LIFE IN VERY LOW CONCENTRATIONS. May be dangerous if it enters water intakes. Notify local health and wildlife officials. Notify operators of nearby water intakes.

1. RESPONSE TO DISCHARGE
(See Response Methods Handbook, CG 446-4)

Issue warning—corrosive
Restrict access
Disperse and flush

2. LABEL

3. CHEMICAL DESIGNATIONS

3.1 **Synonyms:** Potassium hydroxide solution; Lye
3.2 **Coast Guard Compatibility Classification:** Caustic
3.3 **Chemical Formula:** $KOH-H_2O$
3.4 **IMCO/United Nations Numerical Designation:** 8.0/1814

4. OBSERVABLE CHARACTERISTICS

4.1 **Physical State (as shipped):** Liquid
4.2 **Color:** Colorless
4.3 **Odor:** None

5. HEALTH HAZARDS

5.1 **Personal Protective Equipment:** Wide-brimmed hat and close-fitting safety goggles equipped with rubber side shields; long-sleeved cotton shirt or jacket with buttoned collar and buttoned sleeves; rubber or rubber-coated canvas gloves. (Shirt sleeves should be buttoned over the gloves so that any spilled material will run down the outside.) Rubber safety-toe shoes or boots and cotton coveralls. (Trouser cuffs should be worn outside of boots.) Rubber apron.
5.2 **Symptoms Following Exposure:** Causes severe burns of eyes, skin, and mucous membranes.
5.3 **Treatment for Exposure:** (Act quickly!) EYES: flush with water for at least 15 min. SKIN: flush with water, then rinse with dilute vinegar (acetic acid). INGESTION: give water and milk. Do NOT induce vomiting. Call physician at once, even when injury seems to be slight.
5.4 **Toxicity by Inhalation (Threshold Limit Value):** Not pertinent
5.5 **Short-Term Inhalation Limits:** Not pertinent
5.6 **Toxicity by Ingestion:** Grade 2; oral rat LD_{50} = 365 mg/kg
5.7 **Late Toxicity:** None
5.8 **Vapor (Gas) Irritant Characteristics:** Not pertinent
5.9 **Liquid or Solid Irritant Characteristics:** Severe skin irritant. Causes second- and third-degree burns on short contact and is very injurious to the eyes.
5.10 **Odor Threshold:** Not pertinent

6. FIRE HAZARDS

6.1 **Flash Point:** Not flammable
6.2 **Flammable Limits in Air:** Not flammable
6.3 **Fire Extinguishing Agents:** Not pertinent
6.4 **Fire Extinguishing Agents Not to be Used:** Not pertinent
6.5 **Special Hazards of Combustion Products:** Not pertinent
6.6 **Behavior in Fire:** Not pertinent
6.7 **Ignition Temperature:** Not flammable
6.8 **Electrical Hazard:** Not pertinent
6.9 **Burning Rate:** Not flammable

7. CHEMICAL REACTIVITY

7.1 **Reactivity with Water:** None
7.2 **Reactivity with Common Materials:** Attacks wool, leather and some metals such as aluminum, tin, lead and zinc to produce flammable hydrogen gas. Separate from easily ignitible materials.
7.3 **Stability During Transport:** Stable
7.4 **Neutralizing Agents for Acids and Caustics:** Dilute with water and rinse with dilute acid such as acetic acid.
7.5 **Polymerization:** Not pertinent
7.6 **Inhibitor of Polymerization:** Not pertinent

8. WATER POLLUTION

8.1 **Aquatic Toxicity:** 80 ppm/24 hr/mosquito fish/TL_m/fresh water (these figures are for 100% potassium hydroxide).
8.2 **Waterfowl Toxicity:** Data not available
8.3 **Biological Oxygen Demand (BOD):** None
8.4 **Food Chain Concentration Potential:** None

9. SELECTED MANUFACTURERS

1. Diamond Shamrock Corp.
   Diamond Shamrock Chemical Co.
   Electro Chemicals Div.
   Deer Park, Texas 77536
2. Monsanto Co.
   Monsanto Industrial Chemicals Co.
   800 North Lindbergh Blvd.
   St. Louis, Mo. 63166
3. Occidental Petroleum Corp.
   Hooker Chemical Corp.
   Industrial Chemicals Div.
   Niagara Falls, N. Y. 14302

10. SHIPPING INFORMATION

10.1 **Grades or Purity:** 45—50%
10.2 **Storage Temperature:** Ambient or elevated
10.3 **Inert Atmosphere:** No requirement
10.4 **Venting:** Open

11. HAZARD ASSESSMENT CODE
(See Hazard Assessment Handbook, CG 446-3)

A-P

12. HAZARD CLASSIFICATIONS

12.1 **Code of Federal Regulations:** Corrosive material
12.2 **NAS Hazard Rating for Bulk Water Transportation:**

Category	Rating
Fire	0
Health	
Vapor Irritant	0
Liquid or Solid Irritant	4
Poisons	1
Water Pollution	
Human Toxicity	2
Aquatic Toxicity	3
Aesthetic Effect	2
Reactivity	
Other Chemicals	4
Water	0
Self-Reaction	0

12.3 **NFPA Hazard Classifications:**

Category	Classification
Health Hazard (Blue)	3
Flammability (Red)	0
Reactivity (Yellow)	1

13. PHYSICAL AND CHEMICAL PROPERTIES

13.1 **Physical State at 15°C and 1 atm:** Liquid
13.2 **Molecular Weight:** Not pertinent
13.3 **Boiling Point at 1 atm:** >266°F = >130°C = >403°K
13.4 **Freezing Point:** Not pertinent
13.5 **Critical Temperature:** Not pertinent
13.6 **Critical Pressure:** Not pertinent
13.7 **Specific Gravity:** 1.45—1.50 at 20°C (liquid)
13.8 **Liquid Surface Tension:** Not pertinent
13.9 **Liquid-Water Interfacial Tension:** Not pertinent
13.10 **Vapor (Gas) Specific Gravity:** Not pertinent
13.11 **Ratio of Specific Heats of Vapor (Gas):** Not pertinent
13.12 **Latent Heat of Vaporization:** Not pertinent
13.13 **Heat of Combustion:** Not pertinent
13.14 **Heat of Decomposition:** Not pertinent
13.15 **Heat of Solution:** (est.) −17 Btu/lb = −10 cal/g = -0.4×10^5 J/kg
13.16 **Heat of Polymerization:** Not pertinent

NOTES

REVISED 1978

CAUSTIC SODA SOLUTION

Common Synonyms
Sodium hydroxide solution
Lye

Thick liquid — Colorless — Odorless

Sinks and mixes with water.

AVOID CONTACT WITH LIQUID. Keep people away.
Wear rubber overclothing (including gloves).
Stop discharge if possible.
Isolate and remove discharged material.
Notify local health and pollution control agencies.

Fire

Not flammable.

Exposure

CALL FOR MEDICAL AID.

LIQUID
Will burn skin and eyes.
Harmful if swallowed.

Remove contaminated clothing and shoes.
Flush affected areas with plenty of water.
IF IN EYES, hold eyelids open and flush with plenty of water.
IF SWALLOWED and victim is CONSCIOUS, have victim drink water or milk.
DO NOT INDUCE VOMITING.

Water Pollution

Dangerous to aquatic life in high concentrations.
May be dangerous if it enters water intakes.

Notify local health and pollution control officials.
Notify operators of nearby water intakes.

1. RESPONSE TO DISCHARGE

(See Response Methods Handbook, CG 446-4)

Issue warning—corrosive
Restrict access
Disperse and flush

2. LABEL

3. CHEMICAL DESIGNATIONS

3.1 **Synonyms:** Sodium hydroxide solution; Lye
3.2 **Coast Guard Compatibility Classification:** Caustic
3.3 **Chemical Formula:** $NaOH-H_2O$
3.4 **IMCO/United Nations Numerical Designation:** 8.0/1824

4. OBSERVABLE CHARACTERISTICS

4.1 **Physical State (as shipped):** Liquid
4.2 **Color:** Colorless
4.3 **Odor:** None

5. HEALTH HAZARDS

5.1 **Personal Protective Equipment:** Wide-brimmed hat; safety goggles with rubber side shields; tight-fitting cotton clothing; rubber gloves under shirt cuffs; rubber boots and apron.
5.2 **Symptoms Following Exposure:** Causes severe burns of eyes, skin, and mucous membranes.
5.3 **Treatment for Exposure:** (Act quickly!) EYES: flush with water at once for at least 15 min. SKIN: flush with water, then rinse with dilute vinegar (acetic acid). INGESTION: give water and milk. Do NOT induce vomiting. Call physician at once, even when injury seems to be slight.
5.4 **Toxicity by Inhalation (Threshold Limit Value):** Not pertinent
5.5 **Short-Term Inhalation Limits:** Not pertinent
5.6 **Toxicity by Ingestion:** Grade 2; oral rabbit LD_{50} = 500 mg/kg
5.7 **Late Toxicity:** None
5.8 **Vapor (Gas) Irritant Characteristics:** Not pertinent
5.9 **Liquid or Solid Irritant Characteristics:** Severe skin irritant. Causes second- and third-degree burns on short contact and is very injurious to the eyes.
5.10 **Odor Threshold:** Not pertinent

6. FIRE HAZARDS

6.1 **Flash Point:** Not flammable
6.2 **Flammable Limits in Air:** Not flammable
6.3 **Fire Extinguishing Agents:** Not pertinent
6.4 **Fire Extinguishing Agents Not to be Used:** Not pertinent
6.5 **Special Hazards of Combustion Products:** Not pertinent
6.6 **Behavior in Fire:** Not pertinent
6.7 **Ignition Temperature:** Not flammable
6.8 **Electrical Hazard:** Not pertinent
6.9 **Burning Rate:** Not flammable

7. CHEMICAL REACTIVITY

7.1 **Reactivity with Water:** No reaction
7.2 **Reactivity with Common Materials:** Corrosive to aluminum, zinc, and tin. Contact with some metals may generate hydrogen gas, which is explosive and flammable.
7.3 **Stability During Transport:** Stable
7.4 **Neutralizing Agents for Acids and Caustics:** Dilute with water, rinse with dilute acetic acid.
7.5 **Polymerization:** Not pertinent
7.6 **Inhibitor of Polymerization:** Not pertinent

8. WATER POLLUTION

8.1 **Aquatic Toxicity:**
125 ppm/96 hr/mosquito fish/TL_m/fresh water
180 ppm/23 hr/oysters/lethal/salt water
(These figures are for 100% sodium hydroxide)
8.2 **Waterfowl Toxicity:** Data not available
8.3 **Biological Oxygen Demand (BOD):** None
8.4 **Food Chain Concentration Potential:** None

9. SELECTED MANUFACTURERS

1. Diamond Shamrock Corp.
Electrochemicals Division
Painesville, Ohio 44077
2. Dow Chemical Co.
Midland, Mich. 48640
3. PPG Industries, Inc.
Industrial Chemicals Division
Barberton, Ohio 44203

10. SHIPPING INFORMATION

10.1 **Grades or Purity:** 50-73%
10.2 **Storage Temperature:** Ambient or elevated
10.3 **Inert Atmosphere:** No requirement
10.4 **Venting:** Open

11. HAZARD ASSESSMENT CODE

(See Hazard Assessment Handbook, CG 446-3)

A-P

12. HAZARD CLASSIFICATIONS

12.1 **Code of Federal Regulations:** Corrosive material
12.2 **NAS Hazard Rating for Bulk Water Transportation:**

Category	Rating
Fire	0
Health	
Vapor Irritant	0
Liquid or Solid Irritant	4
Poisons	1
Water Pollution	
Human Toxicity	2
Aquatic Toxicity	3
Aesthetic Effect	2
Reactivity	
Other Chemicals	4
Water	0
Self-Reaction	0

12.3 **NFPA Hazard Classifications:**

Category	Classification
Health Hazard (Blue)	3
Flammability (Red)	0
Reactivity (Yellow)	1

13. PHYSICAL AND CHEMICAL PROPERTIES

13.1 **Physical State at 15°C and 1 atm:** Liquid
13.2 **Molecular Weight:** Not pertinent
13.3 **Boiling Point at 1 atm:** >266°F = >130°C = >403°K
13.4 **Freezing Point:** Not pertinent
13.5 **Critical Temperature:** Not pertinent
13.6 **Critical Pressure:** Not pertinent
13.7 **Specific Gravity:** 1.5 at 20°C
13.8 **Liquid Surface Tension:** Not pertinent
13.9 **Liquid-Water Interfacial Tension:** Not pertinent
13.10 **Vapor (Gas) Specific Gravity:** Not pertinent
13.11 **Ratio of Specific Heats of Vapor (Gas):** Not pertinent
13.12 **Latent Heat of Vaporization:** Not pertinent
13.13 **Heat of Combustion:** Not pertinent
13.14 **Heat of Decomposition:** Not pertinent
13.15 **Heat of Solution:** Not pertinent
13.16 **Heat of Polymerization:** Not pertinent

NOTES

REVISED 1978

CHC	# CHARCOAL

Common Synonyms			
Vegetable carbon Animal carbon Activated charcoal Wood charcoal Shell charcoal	Solid powder, lumps, or grains May float or sink in water.	Black	Odorless

Shut off ignition sources. Call fire department.
Stay upwind. Use water spray to "knock down" dust.
Isolate and remove discharged material.
Notify local health and pollution control agencies.

Fire	Combustible. POISONOUS GASES MAY BE PRODUCED IN FIRE. Flood discharge area with water.
Exposure	DUST Irritating to eyes, nose and throat. Move victim to fresh air. If in eyes, hold eyelids open and flush with plenty of water. SOLID Not harmful.
Water Pollution	Effect of low concentrations on aquatic life is unknown. May be dangerous if it enters water intakes. Notify local health and wildlife officials. Notify operators of nearby water intakes.

1. RESPONSE TO DISCHARGE

(See Response Methods Handbook, CG 446-4)

Issue warning—high flammability
Disperse and flush

2. LABEL

Red Stripe

3. CHEMICAL DESIGNATIONS

3.1 **Synonyms:** Vegetable carbon; Vegetable charcoal; Animal carbon; Animal charcoal; Mineral carbon; Mineral charcoal; Activated charcoal; Wood charcoal; Shell charcoal

3.2 **Coast Guard Compatibility Classification:** Not applicable

3.3 **Chemical Formula:** C

3.4 **IMCO/United Nations Numerical Designation:** 4.2/1361, 1362

4. OBSERVABLE CHARACTERISTICS

4.1 **Physical State (as shipped):** Powder, lumps, grains, or sticks

4.2 **Color:** Black

4.3 **Odor:** None

5. HEALTH HAZARDS

5.1 **Personal Protective Equipment:** Respirator for dust

5.2 **Symptoms Following Exposure:** No significant symptoms

5.3 **Treatment for Exposure:** No treatment required

5.4 **Toxicity by Inhalation (Threshold Limit Value):** Not pertinent

5.5 **Short-Term Inhalation Limits:** Not pertinent

5.6 **Toxicity by Ingestion:** Non-toxic (actually used in therapy of poisoning cases)

5.7 **Late Toxicity:** None

5.8 **Vapor (Gas) Irritant Characteristics:** Not pertinent

5.9 **Liquid or Solid Irritant Characteristics:** Data not available

5.10 **Odor Threshold:** Not pertinent

6. FIRE HAZARDS

6.1 **Flash Point:** Flammable solid; may ignite spontaneously in air.

6.2 **Flammable Limits in Air:** Not pertinent

6.3 **Fire Extinguishing Agents:** Water

6.4 **Fire Extinguishing Agents Not to be Used:** Not pertinent

6.5 **Special Hazards of Combustion Products:** Incomplete combustion forms toxic carbon monoxide.

6.6 **Behavior in Fire:** Not pertinent

6.7 **Ignition Temperature:** 600-750°F

6.8 **Electrical Hazard:** Class I, Group F

6.9 **Burning Rate:** Not pertinent

7. CHEMICAL REACTIVITY

7.1 **Reactivity with Water:** No reaction

7.2 **Reactivity with Common Materials:** No reaction

7.3 **Stability During Transport:** May ignite spontaneously in air.

7.4 **Neutralizing Agents for Acids and Caustics:** Not pertinent

7.5 **Polymerization:** Not pertinent

7.6 **Inhibitor of Polymerization:** Not pertinent

8. WATER POLLUTION

8.1 **Aquatic Toxicity:** None

8.2 **Waterfowl Toxicity:** None

8.3 **Biological Oxygen Demand (BOD):** None

8.4 **Food Chain Concentration Potential:** None

9. SELECTED MANUFACTURERS

1. Barneby-Cheney Co.
   Cassady at Eighth Street
   Columbus, Ohio 43216
2. North American Carbon, Inc.
   P. O. Box 19737
   Columbus, Ohio 43219
3. Union Carbide Corp.
   Carbon Products Division
   270 Park Avenue
   New York, N. Y. 10017

10. SHIPPING INFORMATION

10.1 **Grades or Purity:** Various grades; those containing appreciable volatile material are more likely to catch fire. All shipments must be exposed to air and so certified.

10.2 **Storage Temperature:** Ambient

10.3 **Inert Atmosphere:** No requirement

10.4 **Venting:** Open (flame arrester)

11. HAZARD ASSESSMENT CODE

(See Hazard Assessment Handbook, CG 446-3)

II

12. HAZARD CLASSIFICATIONS

12.1 **Code of Federal Regulations:** Flammable solid

12.2 **NAS Hazard Rating for Bulk Water Transportation:** Not listed

12.3 **NFPA Hazard Classifications:** Not listed

13. PHYSICAL AND CHEMICAL PROPERTIES

13.1 **Physical State at 15°C and 1 atm:** Solid

13.2 **Molecular Weight:** 12

13.3 **Boiling Point at 1 atm:** Very high

13.4 **Freezing Point:** Not pertinent

13.5 **Critical Temperature:** Not pertinent

13.6 **Critical Pressure:** Not pertinent

13.7 **Specific Gravity:** 2 at 20°C (solid)

13.8 **Liquid Surface Tension:** Not pertinent

13.9 **Liquid-Water Interfacial Tension:** Not pertinent

13.10 **Vapor (Gas) Specific Gravity:** Not pertinent

13.11 **Ratio of Specific Heats of Vapor (Gas):** Not pertinent

13.12 **Latent Heat of Vaporization:** Not pertinent

13.13 **Heat of Combustion:** 14,100 Btu/lb = 7,830 cal/g = 328×10^5 J/kg

13.14 **Heat of Decomposition:** Not pertinent

13.15 **Heat of Solution:** Not pertinent

13.16 **Heat of Polymerization:** Not pertinent

NOTES

CDN — CHLORDANE

Common Synonyms			
Chlordan 1, 2, 4, 5, 6, 7, 8, 8-octachloro-2, 3, 3a, 4, 7, 7a-hexahydro-4, 7-methanoindene Toxichlor Octa-Klor Velsicol 1068	Liquid or solution Sinks in water.	Brown	Sharp odor

AVOID CONTACT WITH LIQUID. KEEP PEOPLE AWAY.
Wear goggles, self-contained breathing apparatus, and rubber overclothing (including gloves).
Stop discharge if possible.
Call fire department.
Isolate and remove discharged material.
Notify local health and pollution control agencies.

Fire

Not flammable but solution may be combustible.
POISONOUS GASES MAY BE PRODUCED IN FIRE.
Extinguish with dry chemicals, foam or carbon dioxide.
Water may be ineffective on fire.
Cool exposed containers with water.

Exposure

CALL FOR MEDICAL AID.

LIQUID OR SOLUTION
POISONOUS IF SWALLOWED OR IF SKIN IS EXPOSED.
Irritating to skin and eyes.
Remove contaminated clothing and shoes.
Flush affected areas with plenty of water.
DO NOT RUB AFFECTED AREAS.
IF IN EYES, hold eyelids open and flush with plenty of water.
IF SWALLOWED and victim is CONSCIOUS, have victim drink water or milk and have victim induce vomiting.
IF SWALLOWED and victim is UNCONSCIOUS OR HAVING CONVULSIONS, do nothing except keep victim warm.

Water Pollution

HARMFUL TO AQUATIC LIFE IN VERY LOW CONCENTRATIONS.
May be dangerous if it enters water intakes.
Notify local health and wildlife officials.
Notify operators of nearby water intakes.

1. RESPONSE TO DISCHARGE

(See Response Methods Handbook, CG 446-4)

Issue warning—poison
Restrict access
Should be removed
Chemical and physical treatment

2. LABELS

No hazard label required by Code of Federal Regulations

3. CHEMICAL DESIGNATIONS

3.1 **Synonyms:** Chlordan; 1, 2, 4, 5, 6, 7, 8, 8-Octachloro-2, 3, 3a, 4, 7, 7a-hexahydro-4, 7-methanoindene; Velsicol 1068; CD-68; Toxichlor; Octa-Klor

3.2 **Coast Guard Compatibility Classification:** Not applicable

3.3 **Chemical Formula:** $C_{10}H_6Cl_8$

3.4 **IMCO/United Nations Numerical Designation:** 6.1/1615

4. OBSERVABLE CHARACTERISTICS

4.1 **Physical State (as shipped):** Liquid

4.2 **Color:** Brown

4.3 **Odor:** Penetrating; aromatic; slightly pungent, like chlorine

5. HEALTH HAZARDS

5.1 **Personal Protective Equipment:** Respirator for sprays, fogs, or dust; goggles; rubber gloves.

5.2 **Symptoms Following Exposure:** Moderately irritating to eyes and skin. Ingestion, absorption through skin, or inhalation of mist or dust may cause excitability, convulsions, nausea, vomiting, diarrhea, and some local irritation of the gastrointestinal tract.

5.3 **Treatment for Exposure:** INHALATION: administer oxygen and give fluid therapy; do not give epinephrine, since it may induce ventricular fibrillation; enforce complete rest. EYES: flush with water for at least 15 min. SKIN: wash off skin with adequate quantities of soap and water; do NOT scrub. INGESTION: induce vomiting and follow with gastric lavage and administration of saline cathartics; ether and barbiturates may be used to control convulsions; oxygen and fluid therapy are also recommended; do NOT give epinephrine. Since no specific antidotes are known, symptomatic therapy must be accompanied by complete rest.

5.4 **Toxicity by Inhalation (Threshold Limit Value):** 0.5 mg/m³

5.5 **Short-Term Inhalation Limits:** 2 mg/m³ for 30 min.

5.6 **Toxicity by Ingestion:** Grade 3; oral LD_{50} = 283 mg/kg (rat)

5.7 **Late Toxicity:** Possible liver damage; loss of appetite and weight.

5.8 **Vapor (Gas) Irritant Characteristics:** Data not available

5.9 **Liquid or Solid Irritant Characteristics:** Data not available

5.10 **Odor Threshold:** Data not available

6. FIRE HAZARDS

6.1 **Flash Point:** Solution: 225°F O.C.; 132°F C.C. Solid is not flammable.

6.2 **Flammable Limits in Air:** 0.7%—5% (kerosene solution)

6.3 **Fire Extinguishing Agents:** Dry chemical, foam, carbon dioxide

6.4 **Fire Extinguishing Agents Not to be Used:** Water may be ineffective on solution fire.

6.5 **Special Hazards of Combustion Products:** Irritating and toxic hydrogen chloride and phosgene gases may be formed when kerosene solution of compound burns.

6.6 **Behavior in Fire:** Not pertinent

6.7 **Ignition Temperature:** 410°F (kerosene solvent)

6.8 **Electrical Hazard:** Data not available

6.9 **Burning Rate:** Not pertinent

7. CHEMICAL REACTIVITY

7.1 **Reactivity with Water:** No reaction

7.2 **Reactivity with Common Materials:** No reaction

7.3 **Stability During Transport:** Stable to 160°F

7.4 **Neutralizing Agents for Acids and Caustics:** Not pertinent

7.5 **Polymerization:** Not pertinent

7.6 **Inhibitor of Polymerization:** Not pertinent

8. WATER POLLUTION

8.1 **Aquatic Toxicity:** 0.5 ppm/96 hr/goldfish/TL_m/fresh water

8.2 **Waterfowl Toxicity:** LD_{50} = 1,200 mg/kg

8.3 **Biological Oxygen Demand (BOD):** Data not available

8.4 **Food Chain Concentration Potential:** High

9. SELECTED MANUFACTURERS

1. Velsicol Chemical Corp.
341 East Ohio Street
Chicago, Ill. 60611

2. S. B. Penick & Co.
100 Church Street
New York, N. Y. 10007

3. Chempar Chemical Co.
260 Madison Avenue
New York, N. Y. 10016

10. SHIPPING INFORMATION

10.1 **Grades or Purity:** Technical. A variety of dusts, powders, and solutions in kerosene containing 2-80% chlordane are shipped.

10.2 **Storage Temperature:** Ambient

10.3 **Inert Atmosphere:** No requirement

10.4 **Venting:** Open (flame arrester)

11. HAZARD ASSESSMENT CODE

(See Hazard Assessment Handbook, CG 446-3)

A-X-Y

12. HAZARD CLASSIFICATIONS

12.1 **Code of Federal Regulations:** Combustible Liquid

12.2 **NAS Hazard Rating for Bulk Water Transportation:** Not listed

12.3 **NFPA Hazard Classifications:** Not listed

13. PHYSICAL AND CHEMICAL PROPERTIES*

13.1 **Physical State at 15°C and 1 atm:** Liquid

13.2 **Molecular Weight:** 409.8

13.3 **Boiling Point at 1 atm:** Decomposes

13.4 **Freezing Point:** Not pertinent

13.5 **Critical Temperature:** Not pertinent

13.6 **Critical Pressure:** Not pertinent

13.7 **Specific Gravity:** 1.6 at 25°C (liquid)

13.8 **Liquid Surface Tension:** (est.) 25 dynes/cm = 0.025 N/m at 20°C

13.9 **Liquid-Water Interfacial Tension:** (est.) 50 dynes/cm = 0.05 N/m at 20°C

13.10 **Vapor (Gas) Specific Gravity:** Not pertinent

13.11 **Ratio of Specific Heats of Vapor (Gas):** Not pertinent

13.12 **Latent Heat of Vaporization:** Not pertinent

13.13 **Heat of Combustion:** (est.) −4,000 Btu/lb = −2,200 cal/g = -93×10^5 J/kg

13.14 **Heat of Decomposition:** Not pertinent

13.15 **Heat of Solution:** Not pertinent

13.16 **Heat of Polymerization:** Not pertinent

* Properties refer to undiluted, technical-grade chlordane.

NOTES

CLX	# CHLORINE

Common Synonyms	Liquefied compressed gas Greenish yellow Irritating, bleach-like choking odor Sinks and boils in water. Poisonous, visible vapor cloud is produced.

AVOID CONTACT WITH LIQUID AND VAPOR. Keep people away.
Wear goggles, self-contained breathing apparatus and rubber overclothing (including gloves).
Stop discharge if possible.
Evacuate area in case of large discharge.
Stay upwind and use water spray to "knock down" vapor.
Notify local health and pollution control agencies.

Fire	Not flammable. May cause fire on contact with combustibles. POISONOUS GASES ARE PRODUCED IN FIRES. Stop flow of gas if possible. Cool exposed containers and protect men effecting shutoff with water. Wear goggles, self-contained breathing apparatus, and rubber overclothing (including gloves).
 **Exposure**	CALL FOR MEDICAL AID. VAPOR POISONOUS IF INHALED. Will burn eyes. Move to fresh air. If breathing has stopped, give artificial respiration (but NOT mouth-to-mouth). If breathing is difficult, give oxygen. IF IN EYES, hold eyelids open and flush with plenty of water. LIQUID Will burn skin and eyes. Will cause frostbite. Flush affected areas with plenty of water. IF IN EYES, hold eyelids open and flush with plenty of water. DO NOT RUB AFFECTED AREAS.
Water Pollution	HARMFUL TO AQUATIC LIFE IN VERY LOW CONCENTRATIONS. May be dangerous if it enters water intakes. Notify local health and wildlife officials. Notify operators of nearby water intakes.

1. RESPONSE TO DISCHARGE

(See Response Methods Handbook, CG 446-4)

Issue warning—poison
Restrict access
Evacuate area

2. LABEL

3. CHEMICAL DESIGNATIONS

3.1 **Synonyms:** No common synonyms
3.2 **Coast Guard Compatibility Classification:** Halogen
3.3 **Chemical Formula:** Cl_2
3.4 **IMCO/United Nations Numerical Designation:** 2.0/1017

4. OBSERVABLE CHARACTERISTICS

4.1 **Physical State (as shipped):** Liquefied compressed gas
4.2 **Color:** Greenish yellow
4.3 **Odor:** Characteristic choking

5. HEALTH HAZARDS

5.1 **Personal Protective Equipment:** Quick-opening safety shower and eye fountain; respiratory equipment approved for chlorine service. Wear safety goggles at all times when in vicinity of liquid chlorine.
5.2 **Symptoms Following Exposure:** Eye irritation, sneezing, copious salivation, general excitement and restlessness. Irritation may persist for several days. High concentrations cause respiratory distress and violent coughing, often with retching. Death may result from suffocation.
5.3 **Treatment for Exposure:** INHALATION: remove victim from source of exposure; call a doctor; support respiration; administer oxygen. EYES: flush with copious amounts of water for at least 15 min.
5.4 **Toxicity by Inhalation (Threshold Limit Value):** 1 ppm
5.5 **Short-Term Inhalation Limits:** 3 ppm for 5 min.
5.6 **Toxicity by Ingestion:** Not pertinent; ingestion unlikely (chlorine is a gas above -34.5°C)
5.7 **Late Toxicity:** None
5.8 **Vapor (Gas) Irritant Characteristics:** Vapors cause severe irritation of eyes and throat and can cause eye and lung injury. They cannot be tolerated even at low concentrations.
5.9 **Liquid or Solid Irritant Characteristics:** Causes smarting of the skin and first-degree burns on short exposure; may cause secondary burns on long exposure.
5.10 **Odor Threshold:** 3.5 ppm

6. FIRE HAZARDS

6.1 **Flash Point:** Not flammable
6.2 **Flammable Limits in Air:** Not flammable
6.3 **Fire Extinguishing Agents:** Not pertinent
6.4 **Fire Extinguishing Agents Not to be Used:** Not pertinent
6.5 **Special Hazards of Combustion Products:** Toxic products are generated when combustibles burn in chlorine.
6.6 **Behavior in Fire:** Most combustibles will burn in chlorine, although gas is not flammable.
6.7 **Ignition Temperature:** Not flammable
6.8 **Electrical Hazard:** Data not available
6.9 **Burning Rate:** Not flammable

7. CHEMICAL REACTIVITY

7.1 **Reactivity with Water:** Forms a corrosive solution
7.2 **Reactivity with Common Materials:** Reacts vigorously with most metals at high temperature. Copper may burn spontaneously.
7.3 **Stability During Transport:** Stable
7.4 **Neutralizing Agents for Acids and Caustics:** Not pertinent
7.5 **Polymerization:** Not pertinent
7.6 **Inhibitor of Polymerization:** Not pertinent

8. WATER POLLUTION

8.1 **Aquatic Toxicity:**
0.08 ppm/168 hr/trout/TL_m/fresh water
10 ppm/1 hr/tunicates/killed/salt water
8.2 **Waterfowl Toxicity:** Data not available
8.3 **Biological Oxygen Demand (BOD):** None
8.4 **Food Chain Concentration Potential:** None

9. SELECTED MANUFACTURERS

1. Diamond Shamrock Corp.
Electro Chemicals Division
Deer Park, Texas 77536
2. Dow Chemical Co.
Midland, Mich. 48640
3. PPG Industries Inc.
Industrial Chemicals Division
Barberton, Ohio 44203

10. SHIPPING INFORMATION

10.1 **Grades or Purity:** Research purity; ultra high purity; high purity
10.2 **Storage Temperature:** Ambient
10.3 **Inert Atmosphere:** No requirement
10.4 **Venting:** Safety relief (300 psi)

11. HAZARD ASSESSMENT CODE

(See Hazard Assessment Handbook, CG 446-3)

A-C-I-J

12. HAZARD CLASSIFICATIONS

12.1 **Code of Federal Regulations:** Nonflammable compressed gas
12.2 **NAS Hazard Rating for Bulk Water Transportation:**

Category	Rating
Fire	0
Health	
Vapor Irritant	4
Liquid or Solid Irritant	2
Poisons	4
Water Pollution	
Human Toxicity	2
Aquatic Toxicity	3
Aesthetic Effect	2
Reactivity	
Other Chemicals	4
Water	1
Self-Reaction	0

12.3 **NFPA Hazard Classifications:**

Category	Classification
Health Hazard (Blue)	3
Flammability (Red)	0
Reactivity (Yellow)	0 oxy

13. PHYSICAL AND CHEMICAL PROPERTIES

13.1 **Physical State at 15°C and 1 atm:** Gas
13.2 **Molecular Weight:** 70.91
13.3 **Boiling Point at 1 atm:** -29.4°F = -34.1°C = 239.1°K
13.4 **Freezing Point:** -150°F = -101°C = 172°K
13.5 **Critical Temperature:** 291°F = 144°C = 417°K
13.6 **Critical Pressure:** 1118 psia = 76.05 atm = 7.704 MN/m^2
13.7 **Specific Gravity:** 1.424 at 15°C (liquid)
13.8 **Liquid Surface Tension:** 26.55 dynes/cm at -35.3°C
13.9 **Liquid-Water Interfacial Tension:** Not pertinent
13.10 **Vapor (Gas) Specific Gravity:** 2.4
13.11 **Ratio of Specific Heats of Vapor (Gas):** 1.325
13.12 **Latent Heat of Vaporization:** 124 Btu/lb = 68.7 cal/g = 2.87×10^5 J/kg
13.13 **Heat of Combustion:** Not pertinent
13.14 **Heat of Decomposition:** Not pertinent
13.15 **Heat of Solution:** Not pertinent
13.16 **Heat of Polymerization:** Not pertinent

NOTES

REVISED 1978

CTF — CHLORINE TRIFLUORIDE

Common Synonyms	Liquefied compressed gas	Greenish yellow liquid or colorless gas	Strong sweetish odor
	Sinks and may boil in water. Reacts violently with water to produce poisonous gas. Boiling point is 53° F.		

AVOID CONTACT WITH LIQUID AND VAPOR. KEEP PEOPLE AWAY.
Wear chemical protective suit with self-contained breathing apparatus.
Stop discharge if possible.
Evacuate area in case of large discharge.
Call fire department.
Isolate and remove discharged material.
Notify local health and pollution control agencies.

Fire

Not flammable.
May explode on contact with combustibles.
POISONOUS GASES ARE PRODUCED WHEN HEATED.
Containers may explode in fire.
Wear chemical protective suit with self-contained breathing apparatus.
Combat fires from safe distance or protected location.
Extinguish with dry chemicals or carbon dioxide.
DO NOT USE WATER OR FOAM ON ADJACENT FIRES.
Cool exposed containers with water.

Exposure

CALL FOR MEDICAL AID.

VAPOR
POISONOUS IF INHALED.
Irritating to skin, eyes, nose and throat.
Move victim to fresh air.
If breathing has stopped, give artificial respiration.
If breathing is difficult, give oxygen.

LIQUID
POISONOUS IF SWALLOWED.
Will burn skin and eyes.
Remove contaminated clothing and shoes.
Flush affected areas with plenty of water.
DO NOT RUB AFFECTED AREAS.
IF IN EYES, hold eyelids open and flush with plenty of water.
IF SWALLOWED and victim is CONSCIOUS, have victim drink water or milk.
DO NOT INDUCE VOMITING.

Water Pollution

Effect of low concentrations on aquatic life is unknown.
May be dangerous if it enters water intakes.
Notify local health and wildlife officials.
Notify operators of nearby water intakes.

1. RESPONSE TO DISCHARGE

(See Response Methods Handbook, CG 446-4)

Issue warning—poison, air contaminant, water contaminant, corrosive, oxidizing material
Restrict access
Evacuate area

2. LABELS

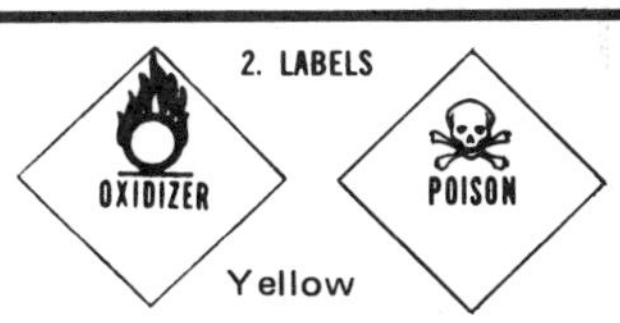

3. CHEMICAL DESIGNATIONS

3.1 **Synonyms:** CTF
3.2 **Coast Guard Compatibility Classification:** Not applicable
3.3 **Chemical Formula:** ClF_3
3.4 **IMCO/United Nations Numerical Designation:** 2/1749

4. OBSERVABLE CHARACTERISTICS

4.1 **Physical State (as shipped):** Liquefied compressed gas
4.2 **Color:** Gas: colorless; Liquid: greenish-yellow
4.3 **Odor:** Acrid; strong, pungent, sweetish; sweet and irritating

5. HEALTH HAZARDS

5.1 **Personal Protective Equipment:** Neoprene gloves and protective clothing made of glass fiber and Teflon, including full hood; self-contained breathing apparatus with full face mask.
5.2 **Symptoms Following Exposure:** Inhalation causes extreme irritation of respiratory tract; pulmonary edema may result. Vapors are very irritating to eyes and skin; liquid causes severe burns.
5.3 **Treatment for Exposure:** Call physician at once after any exposure to this compound. INHALATION: remove victim to fresh air and *keep him quiet;* give artificial respiration if breathing has stopped; give oxygen; enforce rest for 24 hours. EYES: flush with water for at least 15 min.; get medical attention, but do not interrupt flushing for at least 10 min. SKIN: flush with water, then with 2-3% aqueous ammonia, then again with water; apply ice-cold pack of saturated Epsom salt or 70% ethyl alcohol.
5.4 **Toxicity by Inhalation (Threshold Limit Value):** 0.1 ppm (ceiling limit)
5.5 **Short-Term Inhalation Limits:** 0.1 ppm for 5 min.
5.6 **Toxicity by Ingestion:** Grade 4; LD_{50} < 50 mg/kg
5.7 **Late Toxicity:** Data not available
5.8 **Vapor (Gas) Irritant Characteristics:** Vapors cause severe irritation of eyes and throat and can cause eye or lung injury. They cannot be tolerated even at low concentrations.
5.9 **Liquid or Solid Irritant Characteristics:** Severe skin irritant, causes second- and third-degree burns on short contact and is very injurious to the eyes.
5.10 **Odor Threshold:** Data not available

6. FIRE HAZARDS

6.1 **Flash Point:** Not flammable, but may cause fire on contact with some materials.
6.2 **Flammable Limits in Air:** Not pertinent
6.3 **Fire Extinguishing Agents:** Dry chemical
6.4 **Fire Extinguishing Agents Not to be Used:** Do not use water on adjacent fires unless well protected against hydrogen fluoride gas.
6.5 **Special Hazards of Combustion Products:** If released from container, fumes are toxic and irritating.
6.6 **Behavior in Fire:** If released from container, can increase the intensity of fire. Containers may explode.
6.7 **Ignition Temperature:** Not pertinent
6.8 **Electrical Hazard:** Not pertinent
6.9 **Burning Rate:** Not pertinent

7. CHEMICAL REACTIVITY

7.1 **Reactivity with Water:** Reacts explosively with water, evolving hydrogen fluoride (hydrofluoric acid) and chlorine.
7.2 **Reactivity with Common Materials:** Causes ignition of all combustible materials and even sand or concrete. Very similar to fluorine gas.
7.3 **Stability During Transport:** Stable
7.4 **Neutralizing Agents for Acids and Caustics:** Flood with water
7.5 **Polymerization:** Not pertinent
7.6 **Inhibitor of Polymerization:** Not pertinent

8. WATER POLLUTION

8.1 **Aquatic Toxicity:** Data not available
8.2 **Waterfowl Toxicity:** Data not available
8.3 **Biological Oxygen Demand (BOD):** None
8.4 **Food Chain Concentration Potential:** None

9. SELECTED MANUFACTURERS

1. Matheson Gas Products Co.
   East Rutherford, N. J. 07073
2. Air Products and Chemicals, Inc.
   Specialty Gases Department
   P. O. Box 538
   Allentown, Pa. 18105
3. Union Carbide Corp.
   Linde Division
   270 Park Avenue
   New York, N. Y. 10017

10. SHIPPING INFORMATION

10.1 **Grades or Purity:** 99+%
10.2 **Storage Temperature:** Ambient
10.3 **Inert Atmosphere:** No requirement
10.4 **Venting:** Safety relief

11. HAZARD ASSESSMENT CODE

(See Hazard Assessment Handbook, CG 446-3)

A-C-O

12. HAZARD CLASSIFICATIONS

12.1 **Code of Federal Regulations:** Oxidizer
12.2 **NAS Hazard Rating for Bulk Water Transportation:**

Category	Rating
Fire	0
Health	
Vapor Irritant	4
Liquid or Solid Irritant	4
Poisons	4
Water Pollution	
Human Toxicity	4
Aquatic Toxicity	2
Aesthetic Effect	3
Reactivity	
Other Chemicals	4
Water	4
Self-Reaction	0

12.3 **NFPA Hazard Classifications:** Not listed

13. PHYSICAL AND CHEMICAL PROPERTIES

13.1 **Physical State at 15°C and 1 atm:** Gas
13.2 **Molecular Weight:** 92.5
13.3 **Boiling Point at 1 atm:** 53°F = 11.6 °C = 284.8°K
13.4 **Freezing Point:** −105°F = −76.1°C = 197.1°K
13.5 **Critical Temperature:** 307°F = 153°C = 426°K
13.6 **Critical Pressure:** 837 psia = 56.9 atm = 5.77 MN/m^2
13.7 **Specific Gravity:** 1.85 at 11°C (liquid)
13.8 **Liquid Surface Tension:** 26.6 dynes/cm = 0.0266 N/m at 0°C
13.9 **Liquid-Water Interfacial Tension:** Not pertinent
13.10 **Vapor (Gas) Specific Gravity:** 3.2
13.11 **Ratio of Specific Heats of Vapor (Gas):** 1.2832
13.12 **Latent Heat of Vaporization:** 128 Btu/lb = 71.2 cal/g = 2.98×10^5 J/kg
13.13 **Heat of Combustion:** Not pertinent
13.14 **Heat of Decomposition:** Not pertinent
13.15 **Heat of Solution:** Data not available
13.16 **Heat of Polymerization:** Not pertinent

NOTES

CRA CHLOROACETOPHENONE

Common Synonyms			
Phenacyl chloride omega-Chloroacetophenone alpha-Chloroacetophenone Phenyl chloromethyl ketone Chloromethyl phenyl ketone Tear gas	Solid Sinks in water.	White to light yellow	Sharp odor

AVOID CONTACT WITH SOLID AND DUST. KEEP PEOPLE AWAY.
Wear goggles, self-contained breathing apparatus, and rubber overclothing (including gloves).
Call fire department.
Stay upwind. Use water spray to "knock down" dust.
Isolate and remove discharged material.
Notify local health and pollution control agencies.

Fire

Combustible.
POISONOUS GASES MAY BE PRODUCED IN FIRE.
Wear goggles, self-contained breathing apparatus and rubber overclothing (including gloves).
Flood discharge area with water.

Exposure

CALL FOR MEDICAL AID.

DUST
POISONOUS IF INHALED OR IF SKIN IS EXPOSED.
Irritating to eyes, nose and throat.
Move victim to fresh air.
If in eyes, hold eyelids open and flush with plenty of water.
If breathing has stopped, give artificial respiration.
If breathing is difficult, give oxygen.

SOLID
POISONOUS IF SWALLOWED.
Irritating to skin and eyes.
Remove contaminated clothing and shoes.
Flush affected areas with plenty of water.
IF IN EYES, hold eyelids open and flush with plenty of water.
IF SWALLOWED and victim is CONSCIOUS, have victim drink water or milk and have victim induce vomiting.
IF SWALLOWED and victim is UNCONSCIOUS OR HAVING CONVULSIONS, do nothing except keep victim warm.

Water Pollution

Effect of low concentrations on aquatic life is unknown.
May be dangerous if it enters water intakes.
Notify local health and wildlife officials.
Notify operators of nearby water intakes.

1. RESPONSE TO DISCHARGE
(See Response Methods Handbook, CG 446-4)

Issue warning—air contaminant
Restrict access
Disperse and flush

2. LABEL

3. CHEMICAL DESIGNATIONS

3.1 **Synonyms:** Phenacyl chloride; omega-Chloroacetophenone; alpha-Chloroacetophenone; Phenyl chloromethyl ketone; Tear gas; Chloromethyl phenyl ketone
3.2 **Coast Guard Compatibility Classification:** Not applicable
3.3 **Chemical Formula:** $C_6H_5COCH_2Cl$
3.4 **IMCO/United Nations Numerical Designation:** 6.1/1697

4. OBSERVABLE CHARACTERISTICS

4.1 **Physical State (as shipped):** Solid
4.2 **Color:** White to pale yellow
4.3 **Odor:** Pungent

5. HEALTH HAZARDS

5.1 **Personal Protective Equipment:** Full-face organic canister mask; self-contained breathing apparatus; rubber gloves; protective clothing.
5.2 **Symptoms Following Exposure:** Inhalation causes tearing, burning of the eyes and difficulty in breathing; high concentrations may lead to development of acute pulmonary edema after latencies of 8 hrs. to several days; possible systemic manifestations include agitation, coma, contraction of pupils of eyes, loss of reflexes. External contact causes irritation of skin and intense irritation of eyes. Ingestion causes agitation, coma, contraction of pupils of eye, loss of reflexes.
5.3 **Treatment for Exposure:** INHALATION: remove victim from contaminated atmosphere at once; give artificial respiration and oxygen, if necessary; watch for pulmonary edema for several days. EYES: flush with water; do not rub. SKIN: flush with water. INGESTION: get medical attention; watch for development of pulmonary edema for several days.
5.4 **Toxicity by Inhalation (Threshold Limit Value):** 0.05 ppm
5.5 **Short-Term Inhalation Limits:** Data not available
5.6 **Toxicity by Ingestion:** Grade 3; oral LD_{50} = 52 mg/kg (rat)
5.7 **Late Toxicity:** Fatty infiltration of liver
5.8 **Vapor (Gas) Irritant Characteristics:** Data not available
5.9 **Liquid or Solid Irritant Characteristics:** Data not available
5.10 **Odor Threshold:** 0.016 ppm

6. FIRE HAZARDS

6.1 **Flash Point:** Combustible solid 244°F C.C. (solutions only)
6.2 **Flammable Limits in Air:** Not pertinent
6.3 **Fire Extinguishing Agents:** Water
6.4 **Fire Extinguishing Agents Not to be Used:** Not pertinent
6.5 **Special Hazards of Combustion Products:** Irritating hydrogen chloride may form.
6.6 **Behavior in Fire:** Unburned material may become volatile and cause severe eye irritation.
6.7 **Ignition Temperature:** Data not available
6.8 **Electrical Hazard:** Data not available
6.9 **Burning Rate:** Not pertinent

7. CHEMICAL REACTIVITY

7.1 **Reactivity with Water:** Reacts slowly, generating hydrogen chloride. The reaction is not hazardous.
7.2 **Reactivity with Common Materials:** Reacts slowly with metals, causing mild corrosion.
7.3 **Stability During Transport:** Stable
7.4 **Neutralizing Agents for Acids and Caustics:** Not pertinent
7.5 **Polymerization:** Not pertinent
7.6 **Inhibitor of Polymerization:** Not pertinent

8. WATER POLLUTION

8.1 **Aquatic Toxicity:** Data not available
8.2 **Waterfowl Toxicity:** Data not available
8.3 **Biological Oxygen Demand (BOD):** Data not available
8.4 **Food Chain Concentration Potential:** None

9. SELECTED MANUFACTURERS

1. Federal Laboratories, Inc.
   Saltsburg, Pa. 15681
2. Pfaltz and Bauer, Inc.
   126-04 Northern Boulevard
   Flushing, N. Y. 11368
3. MC/B Manufacturing Chemists
   2909 Highland Avenue
   Norwood, Ohio 45212

10. SHIPPING INFORMATION

10.1 **Grades or Purity:** Sometimes shipped as a solution in an organic solvent.
10.2 **Storage Temperature:** Ambient
10.3 **Inert Atmosphere:** No requirement
10.4 **Venting:** Pressure-vacuum

11. HAZARD ASSESSMENT CODE
(See Hazard Assessment Handbook, CG 446-3)

II-RR

12. HAZARD CLASSIFICATIONS

12.1 **Code of Federal Regulations:** Irritant
12.2 **NAS Hazard Rating for Bulk Water Transportation:** Not listed
12.3 **NFPA Hazard Classifications:**

Category	Classification
Health Hazard (Blue)	2
Flammability (Red)	1
Reactivity (Yellow)	0

13. PHYSICAL AND CHEMICAL PROPERTIES

13.1 **Physical State at 15°C and 1 atm:** Solid
13.2 **Molecular Weight:** 154.6
13.3 **Boiling Point at 1 atm:** 477°F = 247°C = 520°K
13.4 **Freezing Point:** 68-138°F = 20-59°C = 293-332°K
13.5 **Critical Temperature:** Not pertinent
13.6 **Critical Pressure:** Not pertinent
13.7 **Specific Gravity:** 1.32 at 15°C (solid)
13.8 **Liquid Surface Tension:** Not pertinent
13.9 **Liquid-Water Interfacial Tension:** Not pertinent
13.10 **Vapor (Gas) Specific Gravity:** Not pertinent
13.11 **Ratio of Specific Heats of Vapor (Gas):** Not pertinent
13.12 **Latent Heat of Vaporization:** Not pertinent
13.13 **Heat of Combustion:** (est.) −9,340 Btu/lb = −5,190 cal/g = -217×10^5 J/kg
13.14 **Heat of Decomposition:** Not pertinent
13.15 **Heat of Solution:** Not pertinent
13.16 **Heat of Polymerization:** Not pertinent

NOTES

CAC

CHLOROACETYL CHLORIDE

Common Synonyms	Liquid	Colorless to light yellow	Sharp, extremely irritating odor
	Reacts violently with water. Irritating vapor is produced.		

Avoid contact with liquid and vapor. Keep people away.
Wear goggles and self-contained breathing apparatus.
Stop discharge if possible.
Isolate and remove discharged material.
Notify local health and pollution control agencies.

Fire

Not flammable.
POISONOUS GASES MAY BE PRODUCED WHEN HEATED.
DO NOT USE WATER ON ADJACENT FIRES.
Cool exposed containers with water.

Exposure

Call for medical aid.

VAPOR
Irritating to eyes, nose and throat.
Move victim to fresh air.
If breathing is difficult, give oxygen.

LIQUID
Will burn skin and eyes.
Harmful if swallowed.
Remove contaminated clothing and shoes.
Flush affected areas with plenty of water.
IF IN EYES, hold eyelids open and flush with plenty of water.
IF SWALLOWED and victim is CONSCIOUS, have victim drink water or milk.
DO NOT INDUCE VOMITING.

Water Pollution

Effect of low concentrations on aquatic life is unknown.
May be dangerous if it enters water intakes.
Notify local health and wildlife officials.
Notify operators of nearby water intakes.

1. RESPONSE TO DISCHARGE
(See Response Methods Handbook, CG 446-4)

Issue warning—air contaminant, corrosive
Restrict access
Disperse and flush

2. LABEL

3. CHEMICAL DESIGNATIONS
3.1 **Synonyms:** Chloracetyl chloride
3.2 **Coast Guard Compatibility Classification:** Not applicable
3.3 **Chemical Formula:** $ClCH_2COCl$
3.4 **IMCO/United Nations Numerical Designation:** 8/1752

4. OBSERVABLE CHARACTERISTICS
4.1 **Physical State (as shipped):** Liquid
4.2 **Color:** Colorless to slightly yellow
4.3 **Odor:** Sharp, pungent; extremely irritating

5. HEALTH HAZARDS
5.1 **Personal Protective Equipment:** Acid-type canister mask; self-contained breathing apparatus (full face); rubber gloves and full protective clothing.
5.2 **Symptoms Following Exposure:** Inhalation causes severe irritation of upper respiratory system. External contact causes severe irritation of eyes and skin. Ingestion causes severe irritation of mouth and stomach.
5.3 **Treatment for Exposure:** INHALATION: remove from exposure; support respiration; call physician. EYES: wash with copious amounts of water for 15 min.; call physician. SKIN: wash with large amounts of water; treat burns as required. INGESTION: do NOT induce vomiting; give large amounts of water; call a physician.
5.4 **Toxicity by Inhalation (Threshold Limit Value):** Data not available
5.5 **Short-Term Inhalation Limits:** Data not available
5.6 **Toxicity by Ingestion:** Grade 2; LD_{50} 0.5 to 5 g/kg
5.7 **Late Toxicity:** Data not available
5.8 **Vapor (Gas) Irritant Characteristics:** Vapors are moderately irritating, such that personnel will not usually tolerate moderate or high vapor concentrations.
5.9 **Liquid or Solid Irritant Characteristics:** Severe skin irritant. Causes second- and third-degree burns on short contact and is very injurious to the eyes.
5.10 **Odor Threshold:** Data not available

6. FIRE HAZARDS
6.1 **Flash Point:** Not flammable
6.2 **Flammable Limits in Air:** Not flammable
6.3 **Fire Extinguishing Agents:** Not pertinent
6.4 **Fire Extinguishing Agents Not to be Used:** Do not use water on adjacent fires.
6.5 **Special Hazards of Combustion Products:** Heat of fire can cause decomposition, with evolution of highly toxic and irritating hydrogen chloride and phosgene vapors.
6.6 **Behavior in Fire:** Highly irritating (tear gas) vapors released when heated. Hydrogen chloride gas is released if in contact with water.
6.7 **Ignition Temperature:** Not pertinent
6.8 **Electrical Hazard:** Not pertinent
6.9 **Burning Rate:** Not pertinent

7. CHEMICAL REACTIVITY
7.1 **Reactivity with Water:** Reacts vigorously to generate hydrogen chloride (hydrochloric acid).
7.2 **Reactivity with Common Materials:** Will react with surface moisture to generate hydrogen chloride, which is corrosive to metals.
7.3 **Stability During Transport:** Stable
7.4 **Neutralizing Agents for Acids and Caustics:** Flush with water, rinse with sodium bicarbonate or lime solution.
7.5 **Polymerization:** Not pertinent
7.6 **Inhibitor of Polymerization:** Not pertinent

8. WATER POLLUTION
8.1 **Aquatic Toxicity:** Data not available
8.2 **Waterfowl Toxicity:** Data not available
8.3 **Biological Oxygen Demand (BOD):** Data not available
8.4 **Food Chain Concentration Potential:** None

9. SELECTED MANUFACTURERS
1. Dow Chemical Co.
   Midland, Mich. 48640
2. White Chemical Corp.
   P. O. Box 278
   Bayonne, N. J. 07002
3. Eastman Kodak Co.
   Eastman Organic Chemicals
   Rochester, N. Y. 14650

10. SHIPPING INFORMATION
10.1 **Grades or Purity:** Commercial
10.2 **Storage Temperature:** Ambient
10.3 **Inert Atmosphere:** No requirement
10.4 **Venting:** Pressure-vacuum

11. HAZARD ASSESSMENT CODE
(See Hazard Assessment Handbook, CG 446-3)

A-O

12. HAZARD CLASSIFICATIONS
12.1 **Code of Federal Regulations:** Corrosive liquid
12.2 **NAS Hazard Rating for Bulk Water Transportation:**

Category	Rating
Fire	0
Health	
Vapor Irritant	3
Liquid or Solid Irritant	4
Poisons	3
Water Pollution	
Human Toxicity	2
Aquatic Toxicity	2
Aesthetic Effect	2
Reactivity	
Other Chemicals	2
Water	3
Self-Reaction	0

12.3 **NFPA Hazard Classifications:**

Category	Classification
Health Hazard (Blue)	3
Flammability (Red)	0
Reactivity (Yellow)	0

13. PHYSICAL AND CHEMICAL PROPERTIES
13.1 **Physical State at 15°C and 1 atm:** Liquid
13.2 **Molecular Weight:** 112.9
13.3 **Boiling Point at 1 atm:** 221°F = 105°C = 378°K
13.4 **Freezing Point:** −8.5°F = −22.5°C = 250.7°K
13.5 **Critical Temperature:** Not pertinent
13.6 **Critical Pressure:** 1.423 at 20°C (liquid)
13.7 **Specific Gravity:** 1.42 at 20°C (liquid)
13.8 **Liquid Surface Tension:** (est.) 25 dynes/cm = 0.025 N/m at 20°C
13.9 **Liquid-Water Interfacial Tension:** Not pertinent
13.10 **Vapor (Gas) Specific Gravity:** 3.9
13.11 **Ratio of Specific Heats of Vapor (Gas):** 1.1191
13.12 **Latent Heat of Vaporization:** 166 Btu/lb = 92.0 cal/g = 3.85×10^5 J/kg
13.13 **Heat of Combustion:** (est.) −4,000 Btu/lb = −2,000 cal/g = -90×10^5 J/kg
13.14 **Heat of Decomposition:** Not pertinent
13.15 **Heat of Solution:** (est.) −54 Btu/lb = −30 cal/g = -1.3×10^5 J/kg
13.16 **Heat of Polymerization:** Not pertinent

NOTES

CAP p-CHLOROANILINE

Common Synonyms			
1-Amino-4-chlorobenzene 4-Chlorophenylamine 4-Chloroaniline	Solid	Yellowish white	Mild sweet odor
	Sinks and mixes slowly with water.		

AVOID CONTACT WITH SOLID AND DUST. KEEP PEOPLE AWAY.
Wear rubber overclothing (including gloves) and dust respirator.
Call fire department.
Stay upwind. Use water spray to "knock down" dust.
Isolate and remove discharged material.
Notify local health and pollution control agencies.

Fire

Combustible.
POISONOUS GASES ARE PRODUCED IN FIRE.
Wear goggles and self-contained breathing apparatus.
Extinguish with water, dry chemicals, foam, or carbon dioxide.
Cool exposed containers with water.

Exposure

SOLID AND DUST
POISONOUS IF SWALLOWED OR IF SKIN IS EXPOSED.
Remove contaminated clothing and shoes.
Flush affected areas with plenty of water.
IF IN EYES, hold eyelids open and flush with plenty of water.
IF SWALLOWED and victim is CONSCIOUS, have victim drink water or milk and have victim induce vomiting.
IF SWALLOWED and victim is UNCONSCIOUS OR HAVING CONVULSIONS, do nothing except keep victim warm.

Water Pollution

Effect of low concentrations on aquatic life is unknown.
May be dangerous if it enters water intakes.
Notify local health and wildlife officials.
Notify operators of nearby water intakes.

1. RESPONSE TO DISCHARGE

(See Response Methods Handbook, CG 446-4)

Issue warning—poison
Restrict access
Should be removed
Chemical and physical treatment

2. LABELS

No label required by Code of Federal Regulations

3. CHEMICAL DESIGNATIONS

3.1 **Synonyms:** 1-Amino-4-chlorobenzene
4-Chlorophenylamine
4-Chloroaniline

3.2 **Coast Guard Compatibility Classification:** Not applicable

3.3 **Chemical Formula:** 4-$ClC_6H_4NH_2$

3.4 **IMCO/United Nations Numerical Designation:** 6.1/1576

4. OBSERVABLE CHARACTERISTICS

4.1 **Physical State (as shipped):** Solid

4.2 **Color:** Yellowish, white; cream to tan color (darkens on storage)

4.3 **Odor:** Slightly sweetish; characteristic amine

5. HEALTH HAZARDS

5.1 **Personal Protective Equipment:** Rubber gloves; chemical goggles; protective clothing; dust respirator.

5.2 **Symptoms Following Exposure:** Inhalation or ingestion causes bluish tint to fingernails, lips, and ears indicative of cyanosis; headache, drowsiness, and nausea, followed by unconsciousness. Liquid can be absorbed through skin and cause similar symptoms. Contact with eyes causes irritation.

5.3 **Treatment for Exposure:** INHALATION: remove victim from exposure immediately; if needed, administer oxygen; refer to physician. EYES: flush with water for at least 15 min. SKIN: remove victim from exposure immediately; remove contaminated clothing; wash contacted area with copious amounts of water and soap; if needed, administer oxygen; refer to physician. INGESTION: induce vomiting; get medical attention.

5.4 **Toxicity by Inhalation (Threshold Limit Value):** Data not available

5.5 **Short-Term Inhalation Limits:** Data not available

5.6 **Toxicity by Ingestion:** Grade 3; oral LD_{50} = 300 mg/kg (rat)

5.7 **Late Toxicity:** Data not available

5.8 **Vapor (Gas) Irritant Characteristics:** Data not available

5.9 **Liquid or Solid Irritant Characteristics:** Data not available

5.10 **Odor Threshold:** Data not available

6. FIRE HAZARDS

6.1 **Flash Point:** (Combustible solid) > 220°F O.C.

6.2 **Flammable Limits in Air:** Not pertinent

6.3 **Fire Extinguishing Agents:** Water, dry chemical, foam or carbon dioxide

6.4 **Fire Extinguishing Agents Not to be Used:** Not pertinent

6.5 **Special Hazards of Combustion Products:** Irritating and toxic hydrogen chloride and oxides of nitrogen may form in fires.

6.6 **Behavior in Fire:** Data not available

6.7 **Ignition Temperature:** Data not available

6.8 **Electrical Hazard:** Not pertinent

6.9 **Burning Rate:** Not pertinent

7. CHEMICAL REACTIVITY

7.1 **Reactivity with Water:** No reaction

7.2 **Reactivity with Common Materials:** No reaction

7.3 **Stability During Transport:** Stable

7.4 **Neutralizing Agents for Acids and Caustics:** Not pertinent

7.5 **Polymerization:** Not pertinent

7.6 **Inhibitor of Polymerization:** Not pertinent

8. WATER POLLUTION

8.1 **Aquatic Toxicity:** Data not available

8.2 **Waterfowl Toxicity:** Data not available

8.3 **Biological Oxygen Demand (BOD):** Data not available

8.4 **Food Chain Concentration Potential:** Data not available

9. SELECTED MANUFACTURERS

1. Monsanto Company
Monsanto Industrial Chemicals
800 North Lindbergh Boulevard
St. Louis, Mo. 63166

2. E. I. duPont de Nemours & Co.
Organic Chemicals Department
Dyes and Chemicals Division
Wilmington, Del. 19898

3. Aldrich Chemical Co.
940 West St. Paul Avenue
Milwaukee, Wis. 53233

10. SHIPPING INFORMATION

10.1 **Grades or Purity:** 99.0%, Technical

10.2 **Storage Temperature:** Ambient

10.3 **Inert Atmosphere:** No requirement

10.4 **Venting:** Open (flame arrester). Store containers in a well-ventilated area.

11. HAZARD ASSESSMENT CODE

(See Hazard Assessment Handbook, CG 446-3)

II

12. HAZARD CLASSIFICATIONS

12.1 **Code of Federal Regulations:** Not listed

12.2 **NAS Hazard Rating for Bulk Water Transportation:** Not listed

12.3 **NFPA Hazard Classifications:** Not listed

13. PHYSICAL AND CHEMICAL PROPERTIES

13.1 **Physical State at 15°C and 1 atm:** Solid

13.2 **Molecular Weight:** 127.6

13.3 **Boiling Point at 1 atm:** 446°F = 230°C = 503°K

13.4 **Freezing Point:** 158°F = 70°C = 343°K

13.5 **Critical Temperature:** Not pertinent

13.6 **Critical Pressure:** Not pertinent

13.7 **Specific Gravity:** 1.43 at 19°C (solid)

13.8 **Liquid Surface Tension:** Not pertinent

13.9 **Liquid-Water Interfacial Tension:** Not pertinent

13.10 **Vapor (Gas) Specific Gravity:** Not pertinent

13.11 **Ratio of Specific Heats of Vapor (Gas):** Not pertinent

13.12 **Latent Heat of Vaporization:** Not pertinent

13.13 **Heat of Combustion:** (est.) −11,000 Btu/lb = −6,000 cal/g = -250×10^5 J/kg

13.14 **Heat of Decomposition:** Not pertinent

13.15 **Heat of Solution:** Not pertinent

13.16 **Heat of Polymerization:** Not pertinent

NOTES

CRB	CHLOROBENZENE

Common Synonyms Monochlorobenzene Phenyl chloride Benzene chloride	Watery liquid Colorless Sweet, almond odor Sinks in water. Flammable vapor is produced.

Avoid contact with liquid and vapor. Keep people away.
Stop discharge if possible.
Call fire department.
Stay upwind and use water spray to "knock down" vapor.
Isolate and remove discharged material.
Notify local health and pollution control agencies.

Fire	FLAMMABLE Flashback along vapor trail may occur. Vapor may explode if ignited in an enclosed area Wear goggles and self-contained breathing apparatus. Extinguish with dry chemical, foam, or carbon dioxide.
Exposure	CALL FOR MEDICAL AID. VAPOR If inhaled, will cause coughing or dizziness. Not irritating to eyes, nose and throat. Move to fresh air. If breathing has stopped, give artificial respiration. If breathing is difficult, give oxygen. LIQUID Irritating to skin and eyes. Harmful if swallowed. Remove contaminated clothing and shoes. Flush affected areas with plenty of water. IF IN EYES, hold eyelids open and flush with plenty of water. IF SWALLOWED and victim is CONSCIOUS, have victim drink water or milk.
Water Pollution	HARMFUL TO AQUATIC LIFE IN VERY LOW CONCENTRATIONS. May be dangerous if it enters water intakes. Notify local health and wildlife officials. Notify operators of nearby water intakes.

1. RESPONSE TO DISCHARGE
(See Response Methods Handbook, CG 446-4)

Should be removed
Chemical and physical treatment

2. LABEL

Red

3. CHEMICAL DESIGNATIONS

3.1 **Synonyms:** Benzene chloride; MCB; Monochlorobenzene; Phenyl chloride
3.2 **Coast Guard Compatibility Classification:** Halogenated hydrocarbon
3.3 **Chemical Formula:** C_6H_5Cl
3.4 **IMCO/United Nations Numerical Designation:** 3.3/1134

4. OBSERVABLE CHARACTERISTICS

4.1 **Physical State (as shipped):** Liquid
4.2 **Color:** Colorless
4.3 **Odor:** Mild amine odor; sweet, almond-like; aromatic

5. HEALTH HAZARDS

5.1 **Personal Protective Equipment:** Organic vapor-acid gas respirator where appropriate; neoprene or vinyl gloves; chemical safety spectacles, plus face shield where appropriate; rubber footwear; apron or impervious clothing for splash protection; hard hat.
5.2 **Symptoms Following Exposure:** Irritating to skin, eyes and mucous membranes. Repeated exposure of skin may cause dermatitis due to defatting action. Chronic inhalation of vapors or mist may result in damage to lungs, liver, and kidneys. Acute vapor exposures can cause symptoms ranging from coughing to transient anesthesia and central nervous system depression.
5.3 **Treatment for Exposure:** Get medical attention for all eye exposures and any serious over-exposures. Treat the symptoms. INHALATION: remove to clean air; administer oxygen as needed. INGESTION: dilute by drinking water; if vomiting occurs, administer more water. Administer saline laxative. EYES: flush thoroughly with water. SKIN: remove contaminated clothing, wash exposed area with soap and water.
5.4 **Toxicity by Inhalation (Threshold Limit Value):** 75 ppm
5.5 **Short-Term Inhalation Limits:** Data not available
5.6 **Toxicity by Ingestion:** Grade 2; LD_{50} 0.5 to 5 g/kg (rat, rabbit)
5.7 **Late Toxicity:** Data not available
5.8 **Vapor (Gas) Irritant Characteristics:** Vapors are nonirritating to the eyes and throat.
5.9 **Liquid or Solid Irritant Characteristics:** Minimum hazard. If spilled on clothing and allowed to remain, may cause smarting and reddening of the skin.
5.10 **Odor Threshold:** 0.21 ppm

6. FIRE HAZARDS

6.1 **Flash Point:** 84°F C.C.; 97°F O.C.
6.2 **Flammable Limits in Air:** 1.3%–7.1%
6.3 **Fire Extinguishing Agents:** Carbon dioxide, dry chemical, foam or water spray
6.4 **Fire Extinguishing Agents Not to be Used:** Not pertinent
6.5 **Special Hazards of Combustion Products:** Burning in open flame can form toxic phosgene and hydrogen chloride gases.
6.6 **Behavior in Fire:** Heavy vapor can travel a considerable distance to a source of ignition and flash back.
6.7 **Ignition Temperature:** 1184°F
6.8 **Electrical Hazard:** Data not available
6.9 **Burning Rate:** (est.) 4.6 mm/min.

7. CHEMICAL REACTIVITY

7.1 **Reactivity with Water:** No reaction
7.2 **Reactivity with Common Materials:** No reaction
7.3 **Stability During Transport:** Stable
7.4 **Neutralizing Agents for Acids and Caustics:** Not pertinent
7.5 **Polymerization:** Not pertinent
7.6 **Inhibitor of Polymerization:** Not pertinent

8. WATER POLLUTION

8.1 **Aquatic Toxicity:** 20 ppm/96 hr/bluegill/TL_m/fresh water
8.2 **Waterfowl Toxicity:** Data not available
8.3 **Biological Oxygen Demand (BOD):** 0.3 lb/lb, 5 days
8.4 **Food Chain Concentration Potential:** Data not available

9. SELECTED MANUFACTURERS

1. Dow Chemical Co.
   Midland, Mich. 48640
2. Monsanto Co.
   Monsanto Industrial Chemicals Co.
   800 North Lindbergh Blvd.
   St. Louis, Mo. 63166
3. Montrose Chemical Corp.
   500 South Virgil Ave.
   Los Angeles, Calif. 90005

10. SHIPPING INFORMATION

10.1 **Grades or Purity:** 99.5%; technical
10.2 **Storage Temperature:** Ambient
10.3 **Inert Atmosphere:** No requirement
10.4 **Venting:** Pressure-vacuum

11. HAZARD ASSESSMENT CODE
(See Hazard Assessment Handbook, CG 446-3)

A-T-X

12. HAZARD CLASSIFICATIONS

12.1 **Code of Federal Regulations:** Flammable Liquid
12.2 **NAS Hazard Rating for Bulk Water Transportation:**

Category	Rating
Fire	3
Health	
Vapor Irritant	0
Liquid or Solid Irritant	1
Poisons	2
Water Pollution	
Human Toxicity	1
Aquatic Toxicity	3
Aesthetic Effect	2
Reactivity	
Other Chemicals	1
Water	0
Self-Reaction	0

12.3 **NFPA Hazard Classifications:**

Category	Classification
Health Hazard (Blue)	2
Flammability (Red)	3
Reactivity (Yellow)	0

13. PHYSICAL AND CHEMICAL PROPERTIES

13.1 **Physical State at 15°C and 1 atm:** Liquid
13.2 **Molecular Weight:** 112.56
13.3 **Boiling Point at 1 atm:** 270°F = 132°C = 405°K
13.4 **Freezing Point:** −50.1°F = −45.6°C = 227.6°K
13.5 **Critical Temperature:** 678°F = 359°C = 632°K
13.6 **Critical Pressure:** 656 psia = 44.6 atm = 4.52 MN/m^2
13.7 **Specific Gravity:** 1.11 at 20°C (liquid)
13.8 **Liquid Surface Tension:** 33 dynes/cm = 0.033 N/m at 25°C
13.9 **Liquid-Water Interfacial Tension:** 37.41 dynes/cm = 0.03741 N/m at 20°C
13.10 **Vapor (Gas) Specific Gravity:** Not pertinent
13.11 **Ratio of Specific Heats of Vapor (Gas):** 1.094
13.12 **Latent Heat of Vaporization:** 135 Btu/lb = 75 cal/g = 3.140×10^5 J/kg
13.13 **Heat of Combustion:** (est.) 12,000 Btu/lb = 6700 cal/g = 280×10^5 J/kg
13.14 **Heat of Decomposition:** Not pertinent
13.15 **Heat of Solution:** Not pertinent
13.16 **Heat of Polymerization:** Not pertinent

NOTES

REVISED 1978

CBN	4-CHLOROBUTYRONITRILE

Common Synonyms		
4-Chlorobutyronitrile	Liquid	White to light yellow
	Sinks in water.	

Stop discharge if possible. Keep people away.
Call fire department.
Avoid contact with liquid.
Isolate and remove discharged material.
Notify local health and pollution control agencies.

Fire

Combustible.
POISONOUS GASES MAY BE PRODUCED IN FIRE.
Irritating gases may be produced when heated.
Wear goggles and self-contained breathing apparatus.
Extinguish with water, dry chemicals, foam, or carbon dioxide.

Exposure

CALL FOR MEDICAL AID.

VAPOR
Irritating to eyes, nose and throat.
If inhaled will cause coughing or difficult breathing.
If in eyes, hold eyelids open and flush with plenty of water.
If breathing has stopped, give artificial respiration.
If breathing is difficult, give oxygen.

LIQUID
Irritating to skin and eyes.
If swallowed will cause nausea and vomiting.
Remove contaminated clothing and shoes.
Flush affected areas with plenty of water.
IF IN EYES, hold eyelids open and flush with plenty of water.
IF SWALLOWED and victim is CONSCIOUS, have victim drink water or milk and have victim induce vomiting.
IF SWALLOWED and victim is UNCONSCIOUS OR HAVING CONVULSIONS, do nothing except keep victim warm.

Water Pollution

Effect of low concentrations on aquatic life is unknown.
May be dangerous if it enters water intakes.
Notify local health and wildlife officials.
Notify operators of nearby water intakes.

1. RESPONSE TO DISCHARGE
(See Response Methods Handbook, CG 446-4)

Issue warning—water contaminant
Should be removed
Chemical and physical treatment

2. LABELS

No hazard label required by Code of Federal Regulations

3. CHEMICAL DESIGNATIONS

3.1 **Synonyms:** 4-Chlorobutyronitrile (Practical, mixture with 4-bromobutyronitrile)
3.2 **Coast Guard Compatibility Classification:** Not listed
3.3 **Chemical Formula:** $CH_2ClCH_2CH_2CN$ + $CH_2BrCH_2CH_2CN$
3.4 **IMCO/United Nations Numerical Designation:** Not listed

4. OBSERVABLE CHARACTERISTICS

4.1 **Physical State (as shipped):** Liquid
4.2 **Color:** White to pale yellow
4.3 **Odor:** Data not available

5. HEALTH HAZARDS

5.1 **Personal Protective Equipment:** Air supply mask or self-contained breathing apparatus for repeated handling large amounts; rubber gloves; safety goggles
5.2 **Symptoms Following Exposure:** Chemical is moderately toxic. Inhalation causes irritation of nose and throat. Ingestion causes irritation of mouth and stomach. Contact with eyes causes irritation. Can penetrate the skin on prolonged contact; only slightly irritating.
5.3 **Treatment for Exposure:** INHALATION: move victim to fresh air; administer artificial respiration if required; call a doctor. INGESTION: give large amount of water; induce vomiting. EYES: flush with water for at least 15 min. SKIN: flush with plenty of water.
5.4 **Toxicity by Inhalation (Threshold Limit Value):** Data not available
5.5 **Short-Term Inhalation Limits:** Data not available
5.6 **Toxicity by Ingestion:** Grade 3; LD_{50} 50—400 mg/kg (rat)
5.7 **Late Toxicity:** Data not available
5.8 **Vapor (Gas) Irritant Characteristics:** Data not available
5.9 **Liquid or Solid Irritant Characteristics:** Data not available
5.10 **Odor Threshold:** Data not available

6. FIRE HAZARDS

6.1 **Flash Point:** Data not available
6.2 **Flammable Limits in Air:** Data not available
6.3 **Fire Extinguishing Agents:** Water, dry chemical, foam, carbon dioxide
6.4 **Fire Extinguishing Agents Not to be Used:**
6.5 **Special Hazards of Combustion Products:** Toxic and irritating hydrogen cyanide, hydrogen bromide, and hydrogen chloride may form in fires.
6.6 **Behavior in Fire:**
6.7 **Ignition Temperature:** Data not available
6.8 **Electrical Hazard:** Data not available
6.9 **Burning Rate:** Data not available

7. CHEMICAL REACTIVITY

7.1 **Reactivity with Water:** No reaction
7.2 **Reactivity with Common Materials:** May attack some forms of plastics.
7.3 **Stability During Transport:** Stable
7.4 **Neutralizing Agents for Acids and Caustics:** Not pertinent
7.5 **Polymerization:** Not pertinent
7.6 **Inhibitor of Polymerization:** Not pertinent

8. WATER POLLUTION

8.1 **Aquatic Toxicity:** Data not available
8.2 **Waterfowl Toxicity:** Data not available
8.3 **Biological Oxygen Demand (BOD):** Data not available
8.4 **Food Chain Concentration Potential:** None

9. SELECTED MANUFACTURERS

Eastman Kodak Co.
343 State St.
Rochester, N. Y. 14650

10. SHIPPING INFORMATION

10.1 **Grades or Purity:** 50%, + 40% 4-bromobutyronitrile + 8% glutaronitrile. Major components have same hazard ratings.
10.2 **Storage Temperature:** Ambient
10.3 **Inert Atmosphere:** No requirement
10.4 **Venting:** Open

11. HAZARD ASSESSMENT CODE
(See Hazard Assessment Handbook, CG 446-3)

A-X-Y

12. HAZARD CLASSIFICATIONS

12.1 **Code of Federal Regulations:** Not listed
12.2 **NAS Hazard Rating for Bulk Water Transportation:** Not listed
12.3 **NFPA Hazard Classifications:** Not listed

13. PHYSICAL AND CHEMICAL PROPERTIES

13.1 **Physical State at 15°C and 1 atm:** Liquid
13.2 **Molecular Weight:** 103.55
13.3 **Boiling Point at 1 atm:** 374°F = 190°C = 463°K
13.4 **Freezing Point:** Not pertinent
13.5 **Critical Temperature:** Not pertinent
13.6 **Critical Pressure:** Not pertinent
13.7 **Specific Gravity:** 1.22 at 20°C (liquid)
13.8 **Liquid Surface Tension:** Data not available
13.9 **Liquid-Water Interfacial Tension:** Data not available
13.10 **Vapor (Gas) Specific Gravity:** 3.57
13.11 **Ratio of Specific Heats of Vapor (Gas):** (est.) 1.080 at 20°C
13.12 **Latent Heat of Vaporization:** (est.) 185 Btu/lb = 103 cal/g = 4.31×10^5 J/kg
13.13 **Heat of Combustion:** Data not available
13.14 **Heat of Decomposition:** Not pertinent
13.15 **Heat of Solution:** Not pertinent
13.16 **Heat of Polymerization:** Not pertinent

NOTES

CRF — CHLOROFORM

Common Synonyms			
Trichloromethane	Watery liquid	Colorless	Sweet odor
	Sinks in water. Irritating vapor is produced.		

Avoid contact with liquid and vapor. Stay upwind.
Wear goggles and self-contained breathing apparatus.
Stop discharge if possible. Keep people away.
Notify local health and pollution control agencies.

Fire

Not flammable.
POISONOUS AND IRRITATING GASES ARE PRODUCED WHEN HEATED.
Wear goggles and self-contained breathing apparatus.

Exposure

CALL FOR MEDICAL AID.

VAPOR
Irritating to eyes, nose and throat.
If inhaled, will cause headache, nausea, dizziness, or loss of consciousness.
Move to fresh air.
If breathing has stopped, give artificial respiration.
If breathing is difficult, give oxygen.

LIQUID
Irritating to skin and eyes.
Harmful if swallowed.
Remove contaminated clothing.
Flush affected areas with plenty of water.
IF IN EYES, hold eyelids open and flush with plenty of water.
IF SWALLOWED and victim is CONSCIOUS, have victim drink water or milk and have victim induce vomiting.
IF SWALLOWED and victim is UNCONSCIOUS AND HAVING CONVULSIONS, do nothing except keep victim warm.

Water Pollution

Effect of low concentrations on aquatic life is unknown.
May be dangerous if it enters water intakes.
Notify local health and pollution control officials.
Notify operators of nearby water intakes.

1. RESPONSE TO DISCHARGE

(See Response Methods Handbook, CG 446-4)

Issue warning—air contaminant
Restrict access
Should be removed

2. LABELS

No hazard label required by Code of Federal Regulations

3. CHEMICAL DESIGNATIONS

3.1 **Synonyms:** Trichloromethane
3.2 **Coast Guard Compatibility Classification:** Halogenated compound
3.3 **Chemical Formula:** $CHCl_3$
3.4 **IMCO/United Nations Numerical Designation:** 9.0/1888

4. OBSERVABLE CHARACTERISTICS

4.1 **Physical State (as shipped):** Liquid
4.2 **Color:** Colorless
4.3 **Odor:** Pleasant, sweet; ethereal

5. HEALTH HAZARDS

5.1 **Personal Protective Equipment:** Chemical goggles. 50 ppm to 2%: suitable full-face gas mask. Above 2%: suitable self-contained system.
5.2 **Symptoms Following Exposure:** Headache, nausea, dizziness, drunkenness, narcosis.
5.3 **Treatment for Exposure:** INHALATION: if ill effects develop, get victim to fresh air, keep him warm and quiet, and get medical attention. If breathing stops, start artificial respiration. INGESTION: induce vomiting and get medical attention. No known antidote; treat symptoms. EYES: flush with plenty of water for at least 15 minutes and get medical attention. SKIN: wash with soap and water, remove contaminated clothing and free of chemical.
5.4 **Toxicity by Inhalation (Threshold Limit Value):** 10 ppm for 10-hour work day
5.5 **Short-Term Inhalation Limits:** 50 ppm for 10 min.
5.6 **Toxicity by Ingestion:** Grade 2; LD_{50} 0.5 to 5 g/kg
5.7 **Late Toxicity:** None
5.8 **Vapor (Gas) Irritant Characteristics:** Vapors cause moderate irritation such that personnel will find high concentrations unpleasant. The effect is temporary.
5.9 **Liquid or Solid Irritant Characteristics:** Minimum hazard. If spilled on clothing and allowed to remain, may cause smarting and reddening of the skin.
5.10 **Odor Threshold:** 205-307 ppm

6. FIRE HAZARDS

6.1 **Flash Point:** Not flammable
6.2 **Flammable Limits in Air:** Not flammable
6.3 **Fire Extinguishing Agents:** Not pertinent
6.4 **Fire Extinguishing Agents Not to be Used:** Not pertinent
6.5 **Special Hazards of Combustion Products:** Poisonous and irritating gases are produced when heated.
6.6 **Behavior in Fire:** Decomposes, producing toxic gases
6.7 **Ignition Temperature:** Not flammable
6.8 **Electrical Hazard:** Not pertinent
6.9 **Burning Rate:** Not flammable

7. CHEMICAL REACTIVITY

7.1 **Reactivity with Water:** No reaction
7.2 **Reactivity with Common Materials:** No reaction
7.3 **Stability During Transport:** Stable
7.4 **Neutralizing Agents for Acids and Caustics:** Not pertinent
7.5 **Polymerization:** Not pertinent
7.6 **Inhibitor of Polymerization:** Not pertinent

8. WATER POLLUTION

8.1 **Aquatic Toxicity:** Data not available
8.2 **Waterfowl Toxicity:** Data not available
8.3 **Biological Oxygen Demand (BOD):** None
8.4 **Food Chain Concentration Potential:** None

9. SELECTED MANUFACTURERS

1. Allied Chemical Corp.
Specialty Chemicals Division
Moundsville, W. Va. 26041
2. Dow Chemical Co.
Midland, Mich. 48640
3. Stauffer Chemical Co.
Industrial Chemicals Division
Louisville, Ky. 40200

10. SHIPPING INFORMATION

10.1 **Grades or Purity:** Technical, USP
10.2 **Storage Temperature:** Ambient
10.3 **Inert Atmosphere:** No requirement
10.4 **Venting:** Open

11. HAZARD ASSESSMENT CODE

(See Hazard Assessment Handbook, CG 446-3)

A-X

12. HAZARD CLASSIFICATIONS

12.1 **Code of Federal Regulations:** ORM-A
12.2 **NAS Hazard Rating for Bulk Water Transportation:**

Category	Rating
Fire	1
Health	
Vapor Irritant	2
Liquid or Solid Irritant	1
Poisons	2
Water Pollution	
Human Toxicity	1
Aquatic Toxicity	2
Aesthetic Effect	2
Reactivity	
Other Chemicals	1
Water	0
Self-Reaction	0

12.3 **NFPA Hazard Classifications:**

Category	Classification
Health Hazard (Blue)	2
Flammability (Red)	0
Reactivity (Yellow)	0

13. PHYSICAL AND CHEMICAL PROPERTIES

13.1 **Physical State at 15°C and 1 atm:** Liquid
13.2 **Molecular Weight:** 119.39
13.3 **Boiling Point at 1 atm:** 142°F = 61.2°C = 334.4°K
13.4 **Freezing Point:** −82.3°F = −63.5°C = 209.7°K
13.5 **Critical Temperature:** 506°F = 263.2°C = 536.4°K
13.6 **Critical Pressure:** 790 psia = 54 atm = 5.5 MN/m^2
13.7 **Specific Gravity:** 1.49 at 20°C (liquid)
13.8 **Liquid Surface Tension:** 27.1 dynes/cm = 0.0271 N/m at 20°C
13.9 **Liquid-Water Interfacial Tension:** 32.8 dynes/cm = 0.0328 N/m at 20°C
13.10 **Vapor (Gas) Specific Gravity:** 4.1
13.11 **Ratio of Specific Heats of Vapor (Gas):** 1.146
13.12 **Latent Heat of Vaporization:** 106.7 Btu/lb = 59.3 cal/g = 2.483×10^5 J/kg
13.13 **Heat of Combustion:** Not pertinent
13.14 **Heat of Decomposition:** Not pertinent
13.15 **Heat of Solution:** Not pertinent
13.16 **Heat of Polymerization:** Not pertinent

NOTES

REVISED 1978

CHD — CHLOROHYDRINS (CRUDE)

Common Synonyms			
Crude epichlorohydrin	Watery liquid	Colorless to yellow	Garlic odor
	Sinks and mixes with water. Poisonous vapor is produced.		

Avoid contact with liquid and vapor. Keep people away.
Wear goggles and self-contained breathing apparatus.
Stop discharge if possible.
Call fire department.
Stay upwind and use water spray to "knock down" vapor.
Isolate and remove discharged material.
Notify local health and pollution control agencies.

Fire

Combustible.
Containers may explode in fire.
Poisonous gases are produced in fire.
Wear goggles and self-contained breathing apparatus
Fight fire from a safe distance or protected location.
Extinguish with water, dry chemical, alcohol foam, or carbon dioxide.
Cool exposed containers with water.

Exposure

CALL FOR MEDICAL AID.

VAPOR
Poisonous if inhaled.
Irritating to eyes, nose and throat.
Move to fresh air.
If breathing has stopped, give artificial respiration.
If breathing is difficult, give oxygen.

LIQUID
Poisonous if swallowed.
Irritating to skin and eyes.
Remove contaminated clothing and shoes.
Flush affected areas with plenty of water.
IF IN EYES, hold eyelids open and flush with plenty of water.
IF SWALLOWED and victim is CONSCIOUS, have victim drink water or milk and have victim induce vomiting.
IF SWALLOWED and victim is UNCONSCIOUS OR HAVING CONVULSIONS, do nothing except keep victim warm.

Water Pollution

Effect of low concentrations on aquatic life is unknown.
May be dangerous if it enters water intakes.
Notify local health and pollution control officials.
Notify operators of nearby water intakes.

1. RESPONSE TO DISCHARGE

(See Response Methods Handbook, CG 446-4)

Issue warning—water contaminant
Restrict access
Disperse and flush

2. LABEL

3. CHEMICAL DESIGNATIONS

3.1 **Synonyms:** Crude epichlorohydrin
3.2 **Coast Guard Compatibility Classification:** Special class
3.3 **Chemical Formula:** $O{\cdot}CH_2{\cdot}CH{\cdot}CH_2Cl$
3.4 **IMCO/United Nations Numerical Designation:** 6.1/2023

4. OBSERVABLE CHARACTERISTICS

4.1 **Physical State (as shipped):** Liquid
4.2 **Color:** Yellow to colorless
4.3 **Odor:** Pungent; like chloroform; garlic

5. HEALTH HAZARDS

5.1 **Personal Protective Equipment:** Organic canister mask or air pack; protective goggles; protective gloves.
5.2 **Symptoms Following Exposure:** May cause central nervous system depression. Vapor is irritating to eyes, nose and throat. Headache, nausea, vomiting, collapse if swallowed. Liquid irritates skin.
5.3 **Treatment for Exposure:** INHALATION: remove to fresh air, keep warm and quiet. Get medical attention at once. If breathing stops, start artificial respiration. INGESTION: Induce vomiting and call physician. Do NOT induce vomiting if unconscious. No specific antidote known. EYES AND SKIN: Flush with water for at least 15 min. and get medical attention. Remove contaminated clothing and wash before reuse.
5.4 **Toxicity by Inhalation (Threshold Limit Value):** 5 ppm
5.5 **Short-Term Inhalation Limits:** 10 ppm for 30 min.
5.6 **Toxicity by Ingestion:** Grade 3; 50-500 mg/kg
5.7 **Late Toxicity:** Data not available
5.8 **Vapor (Gas) Irritant Characteristics:** Vapors cause moderate irritation such that personnel will find high concentrations unpleasant. The effect is temporary.
5.9 **Liquid or Solid Irritant Characteristics:** Minimum hazard. If spilled on clothing and allowed to remain, may cause smarting and reddening of the skin.
5.10 **Odor Threshold:** 10 ppm

6. FIRE HAZARDS

6.1 **Flash Point:** 92°F O.C.; 100°F C.C.
6.2 **Flammable Limits in Air:** 3.8%—21%
6.3 **Fire Extinguishing Agents:** Alcohol foam, dry chemical, carbon dioxide, water spray
6.4 **Fire Extinguishing Agents Not to be Used:** Avoid use of dry chemical if fire occurs in container with confined vent.
6.5 **Special Hazards of Combustion Products:** Toxic irritating vapors are generated when heated.
6.6 **Behavior in Fire:** Containers may explode in fire because of polymerization.
6.7 **Ignition Temperature:** 804°F
6.8 **Electrical Hazard:** Data not available
6.9 **Burning Rate:** 2.6 mm/min.

7. CHEMICAL REACTIVITY

7.1 **Reactivity with Water:** Mild reaction; not likely to be hazardous
7.2 **Reactivity with Common Materials:** No reaction
7.3 **Stability During Transport:** Stable
7.4 **Neutralizing Agents for Acids and Caustics:** Not pertinent
7.5 **Polymerization:** Can polymerize in presence of strong acids and bases, particularly when hot
7.6 **Inhibitor of Polymerization:** None

8. WATER POLLUTION

8.1 **Aquatic Toxicity:** Data not available
8.2 **Waterfowl Toxicity:** Data not available
8.3 **Biological Oxygen Demand (BOD):** Data not available
8.4 **Food Chain Concentration Potential:** None

9. SELECTED MANUFACTURERS

1. Dow Chemical Co.
   Midland, Mich. 48640
2. Shell Chemical Co.
   Industrial Chemicals Division
   Houston, Texas 77001
3. Union Carbide Corp.
   Chemicals and Plastics Division
   270 Park Ave.
   New York, N. Y. 10017

10. SHIPPING INFORMATION

10.1 **Grades or Purity:** 90% epichlorohydrin, the balance being water (2.5%), 1, 2, 3-trichloropropane (5%), glycerol (1.8%), isopropyl chloride (0.5%), n-propyl chloride (0.8%) and others (1.0%).
10.2 **Storage Temperature:** Ambient
10.3 **Inert Atmosphere:** No requirement
10.4 **Venting:** Pressure-vacuum

11. HAZARD ASSESSMENT CODE

(See Hazard Assessment Handbook, CG 446-3)

A-T-U-X-Y

12. HAZARD CLASSIFICATIONS

12.1 **Code of Federal Regulations:** Poison, Class B
12.2 **NAS Hazard Rating for Bulk Water Transportation:**

Category	Rating
Fire	3
Health	
Vapor Irritant	3
Liquid or Solid Irritant	3
Poisons	4
Water Pollution	
Human Toxicity	3
Aquatic Toxicity	3
Aesthetic Effect	2
Reactivity	
Other Chemicals	3
Water	1
Self-Reaction	2

12.3 **NFPA Hazard Classifications:**

Category	Classification
Health Hazard (Blue)	3
Flammability (Red)	3
Reactivity (Yellow)	2

13. PHYSICAL AND CHEMICAL PROPERTIES

13.1 **Physical State at 15°C and 1 atm:** Liquid
13.2 **Molecular Weight:** Not pertinent
13.3 **Boiling Point at 1 atm:** Not pertinent
13.4 **Freezing Point:** Not pertinent
13.5 **Critical Temperature:** Not pertinent
13.6 **Critical Pressure:** Not pertinent
13.7 **Specific Gravity:** 1.18 at 20°C (liquid)
13.8 **Liquid Surface Tension:** Not pertinent
13.9 **Liquid-Water Interfacial Tension:** Not pertinent
13.10 **Vapor (Gas) Specific Gravity:** Not pertinent
13.11 **Ratio of Specific Heats of Vapor (Gas):** Not pertinent
13.12 **Latent Heat of Vaporization:** (est.) 142 Btu/lb = 78.8 cal/g = 3.30×10^5 J/kg
13.13 **Heat of Combustion:** (approx.) −8100 Btu/lb = −4500 cal/g = -190×10^5 J/kg
13.14 **Heat of Decomposition:** Not pertinent
13.15 **Heat of Solution:** Not pertinent
13.16 **Heat of Polymerization:** Not pertinent

NOTES

REVISED 1978

CME	CHLOROMETHYL METHYL ETHER

Common Synonyms			
Monochloromethyl ether Methyl chloromethyl ether, anhydrous	Liquid	Colorless	Irritating odor
	May float or sink in water. Poisonous, flammable vapor is produced.		

Avoid contact with liquid and vapor. Keep people away.
Wear goggles, self-contained breathing apparatus, and rubber overclothing (including gloves).
Shut off ignition sources. Call fire department.
Stop discharge if possible.
Isolate and remove discharged material.
Notify local health and pollution control agencies.

Fire

FLAMMABLE.
POISONOUS GASES MAY BE PRODUCED IN FIRE.
Flashback along vapor trail may occur.
Vapor may explode if ignited in an enclosed area.
Wear goggles and self-contained breathing apparatus.
Extinguish with dry chemicals, foam or carbon dioxide.
Water may be ineffective on fire.
Cool exposed containers with water.

Exposure

Call for medical aid.

VAPOR
Irritating to eyes, nose and throat.
If inhaled will cause difficult breathing.
Move victim to fresh air.
If breathing is difficult, give oxygen.

LIQUID
Will burn skin and eyes.
Harmful if swallowed.
Remove contaminated clothing and shoes.
Flush affected areas with plenty of water.
IF IN EYES, hold eyelids open and flush with plenty of water.
IF SWALLOWED and victim is CONSCIOUS, have victim drink water or milk.
DO NOT INDUCE VOMITING.

Water Pollution

Effect of low concentrations on aquatic life is unknown.
Fouling to shoreline.
May be dangerous if it enters water intakes.
Notify local health and wildlife officials.
Notify operators of nearby water intakes.

1. RESPONSE TO DISCHARGE

(See Response Methods Handbook, CG 446-4)

Issue warning—poison, high flammability, air contaminant, corrosive
Restrict access
Evacuate area

2. LABEL

Red

3. CHEMICAL DESIGNATIONS

3.1 **Synonyms:** Monochloromethyl ether; Methyl chloromethyl ether, anhydrous
3.2 **Coast Guard Compatibility Classification:** Not applicable
3.3 **Chemical Formula:** $ClCH_2OCH_3$
3.4 **IMCO/United Nations Numerical Designation:** 3.1/1239

4. OBSERVABLE CHARACTERISTICS

4.1 **Physical State (as shipped):** Liquid
4.2 **Color:** Colorless
4.3 **Odor:** Irritating

5. HEALTH HAZARDS

5.1 **Personal Protective Equipment:** Self-contained breathing apparatus; goggles; rubber gloves; protective clothing.
5.2 **Symptoms Following Exposure:** Inhalation causes sore throat, fever, chills, difficulty in breathing. Contact of liquid with eyes causes severe burns and necrosis; vapor is a powerful tear gas. Skin contact causes severe burns and necrosis. Ingestion causes severe burns of mouth and stomach.
5.3 **Treatment for Exposure:** INHALATION: remove from exposure; support respiration; call physician. EYES: wash with copious quantities of water for at least 15 min.; call physician. SKIN: wash with large amounts of water. INGESTION: do NOT induce vomiting; give large amounts of water; call a physician.
5.4 **Toxicity by Inhalation (Threshold Limit Value):** Data not available
5.5 **Short-Term Inhalation Limits:** Data not available
5.6 **Toxicity by Ingestion:** Grade 2; oral LD_{50} = 817 mg/kg (rat)
5.7 **Late Toxicity:** Considered to be lung cancer-producing.
5.8 **Vapor (Gas) Irritant Characteristics:** Vapors are moderately irritating such that personnel will not usually tolerate moderate or high concentrations.
5.9 **Liquid or Solid Irritant Characteristics:** Fairly severe skin irritant. May cause pain and second-degree burns after a few minutes' contact.
5.10 **Odor Threshold:** Data not available

6. FIRE HAZARDS

6.1 **Flash Point:** 0°F O.C.
6.2 **Flammable Limits in Air:** Data not available
6.3 **Fire Extinguishing Agents:** Foam, dry chemical, carbon dioxide
6.4 **Fire Extinguishing Agents Not to be Used:** Water may be ineffective.
6.5 **Special Hazards of Combustion Products:** Irritating and toxic hydrogen chloride and phosgene gases may be formed.
6.6 **Behavior in Fire:** Unburned material may form powerful tear gas. When wet, also forms irritating formaldehyde gas.
6.7 **Ignition Temperature:** Data not available
6.8 **Electrical Hazard:** Data not available
6.9 **Burning Rate:** 3.0 mm/min.

7. CHEMICAL REACTIVITY

7.1 **Reactivity with Water:** Reacts to evolve formaldehyde and hydrogen chloride. The reaction is not violent.
7.2 **Reactivity with Common Materials:** Will react with surface moisture to evolve hydrogen chloride, which is corrosive to metal.
7.3 **Stability During Transport:** Stable
7.4 **Neutralizing Agents for Acids and Caustics:** Flood with water. Rinse with sodium bicarbonate or lime solution.
7.5 **Polymerization:** Not pertinent
7.6 **Inhibitor of Polymerization:** Not pertinent

8. WATER POLLUTION

8.1 **Aquatic Toxicity:** Data not available
8.2 **Waterfowl Toxicity:** Data not available
8.3 **Biological Oxygen Demand (BOD):** Data not available
8.4 **Food Chain Concentration Potential:** None

9. SELECTED MANUFACTURERS

1. Dow Chemical Co.
Midland, Mich. 48640
2. Aldrich Chemical Co.
940 West St. Paul Ave.
Milwaukee, Wis. 53233
3. Eastman Kodak Co.
Eastman Organic Chemicals
Rochester, N. Y. 14650

10. SHIPPING INFORMATION

10.1 **Grades or Purity:** Commercial
10.2 **Storage Temperature:** Ambient
10.3 **Inert Atmosphere:** No requirement
10.4 **Venting:** Pressure-vacuum

11. HAZARD ASSESSMENT CODE

(See Hazard Assessment Handbook, CG 446-3)

A-O-T-U-V-W-X-Y

12. HAZARD CLASSIFICATIONS

12.1 **Code of Federal Regulations:** Flammable liquid
12.2 **NAS Hazard Rating for Bulk Water Transportation:**

Category	Rating
Fire	3
Health	
Vapor Irritant	3
Liquid or Solid Irritant	3
Poisons	3
Water Pollution	
Human Toxicity	2
Aquatic Toxicity	2
Aesthetic Effect	2
Reactivity	
Other Chemicals	3
Water	3
Self-Reaction	0

12.3 **NFPA Hazard Classifications:** Not listed

13. PHYSICAL AND CHEMICAL PROPERTIES

13.1 **Physical State at 15°C and 1 atm:** Liquid
13.2 **Molecular Weight:** 80.5
13.3 **Boiling Point at 1 atm:** 140°F = 60°C = 333°K
13.4 **Freezing Point:** −154.3°F = −103.5°C = 169.7°K
13.5 **Critical Temperature:** Not pertinent
13.6 **Critical Pressure:** Not pertinent
13.7 **Specific Gravity:** 1.07 at 25°C (liquid)
13.8 **Liquid Surface Tension:** (est.) 30 dynes/cm = 0.030 N/m at 20°C
13.9 **Liquid-Water Interfacial Tension:** Not pertinent
13.10 **Vapor (Gas) Specific Gravity:** 2.8
13.11 **Ratio of Specific Heats of Vapor (Gas):** 1.1195
13.12 **Latent Heat of Vaporization:** (est.) 154 Btu/lb = 85.6 cal/g = 3.58×10^5 J/kg
13.13 **Heat of Combustion:** (est.) −7,300 Btu/lb = −4,100 cal/g = -170×10^5 J/kg
13.14 **Heat of Decomposition:** Not pertinent
13.15 **Heat of Solution:** Not pertinent
13.16 **Heat of Polymerization:** Not pertinent

NOTES

CPN p-CHLOROPHENOL

Common Synonyms			
4-Chlorophenol	Solid	White to straw	Medicinal odor
	Sinks in water.		

Avoid contact with solid and dust. Keep people away.
Wear rubber overclothing (including gloves).
Stop discharge if possible.
Call fire department.
Isolate and remove discharged material.
Notify local health and pollution control agencies.

Fire

Combustible.
POISONOUS GASES MAY BE PRODUCED IN FIRE.
Irritating gases may be produced when heated.
Wear goggles and self-contained breathing apparatus.
Extinguish with water, dry chemicals, foam, or carbon dioxide.

Exposure

CALL FOR MEDICAL AID.

DUST
Irritating to eyes, nose and throat.
If inhaled will cause headache or dizziness.
If in eyes, hold eyelids open and flush with plenty of water.
If breathing has stopped, give artificial respiration.
If breathing is difficult, give oxygen.

SOLID
Will burn skin and eyes.
If swallowed will cause nausea and vomiting.
Remove contaminated clothing and shoes.
Flush affected areas with plenty of water.
IF IN EYES, hold eyelids open and flush with plenty of water.
IF SWALLOWED and victim is CONSCIOUS, have victim drink water or milk.
IF SWALLOWED and victim is UNCONSCIOUS OR HAVING CONVULSIONS, do nothing except keep victim warm.
DO NOT INDUCE VOMITING.

Water Pollution

HARMFUL TO AQUATIC LIFE IN VERY LOW CONCENTRATIONS.
May be dangerous if it enters water intakes.
Notify local health and wildlife officials.
Notify operators of nearby water intakes.

1. RESPONSE TO DISCHARGE
(See Response Methods Handbook, CG 446-4)

Issue warning—water contaminant
Should be removed
Chemical and physical treatment

2. LABELS

No hazard label required by Code of Federal Regulations

3. CHEMICAL DESIGNATIONS

3.1 **Synonyms:** 4-Chlorophenol
3.2 **Coast Guard Compatibility Classification:** Not listed
3.3 **Chemical Formula:** 1,4-ClC_6H_4OH
3.4 **IMCO/United Nations Numerical Designation:** 6.1/2020

4. OBSERVABLE CHARACTERISTICS

4.1 **Physical State (as shipped):** Solid
4.2 **Color:** Colorless to straw
4.3 **Odor:** Strong medicinal

5. HEALTH HAZARDS

5.1 **Personal Protective Equipment:** Rubber gloves; face shield; boots and apron; respiratory protection
5.2 **Symptoms Following Exposure:** Inhalation causes headache, dizziness, weak pulse. Ingestion causes irritation of mouth and stomach; headache, dizziness, weak pulse. Contact with eyes causes severe irritation and burning. Contact with skin causes irritation and burn; if absorbed, causes same symptoms as inhalation.
5.3 **Treatment for Exposure:** INHALATION: move to fresh air; get medical attention if any symptoms develop. INGESTION: do not induce vomiting unless advised by a physician; give large amounts of milk, egg whites, or water and get medical help immediately; no specific antidote known. EYES: immediately flush with plenty of water for at least 30 min. SKIN: flush in safety shower while removing all contaminated clothing; wash with soap and water.
5.4 **Toxicity by Inhalation (Threshold Limit Value):** Data not available
5.5 **Short-Term Inhalation Limits:** Data not available
5.6 **Toxicity by Ingestion:** Grade 3; oral LD_{50} = 500 mg/kg (rat)
5.7 **Late Toxicity:** Data not available
5.8 **Vapor (Gas) Irritant Characteristics:** Data not available
5.9 **Liquid or Solid Irritant Characteristics:** Data not available
5.10 **Odor Threshold:** 30 ppm

6. FIRE HAZARDS

6.1 **Flash Point:** 250°F C.C.
6.2 **Flammable Limits in Air:** Not pertinent (combustible solid)
6.3 **Fire Extinguishing Agents:** Water, foam, dry chemical, carbon dioxide
6.4 **Fire Extinguishing Agents Not to be Used:**
6.5 **Special Hazards of Combustion Products:** Toxic and irritating hydrogen chloride and chlorine gases may form in fires.
6.6 **Behavior in Fire:**
6.7 **Ignition Temperature:** Data not available
6.8 **Electrical Hazard:** Not pertinent
6.9 **Burning Rate:** Not pertinent

7. CHEMICAL REACTIVITY

7.1 **Reactivity with Water:** No reaction
7.2 **Reactivity with Common Materials:**
7.3 **Stability During Transport:** Stable
7.4 **Neutralizing Agents for Acids and Caustics:** Not pertinent
7.5 **Polymerization:** Not pertinent
7.6 **Inhibitor of Polymerization:** Not pertinent

8. WATER POLLUTION

8.1 **Aquatic Toxicity:**
0.7 ppm/96 hr/crab/lethal range/sea water
0.4 ppm/96 hr/crab/safe range/sea water
14 ppm/24 hr/minnow/TL_m/fresh water
8.2 **Waterfowl Toxicity:** Data not available
8.3 **Biological Oxygen Demand (BOD):** Data not available
8.4 **Food Chain Concentration Potential:** None

9. SELECTED MANUFACTURERS

1. The Dow Chemical Co.
   2030 Dow Center
   Midland, Mich. 48640
2. Progil, Inc.
   509 Madison Avenue
   New York, N. Y. 10022
3. Monsanto Company
   800 North Lindbergh Blvd.
   St. Louis, Mo. 63166

10. SHIPPING INFORMATION

10.1 **Grades or Purity:** Pure, 99%
10.2 **Storage Temperature:** Ambient
10.3 **Inert Atmosphere:** No requirement
10.4 **Venting:** Open

11. HAZARD ASSESSMENT CODE
(See Hazard Assessment Handbook, CG 446-3)

II-SS

12. HAZARD CLASSIFICATIONS

12.1 **Code of Federal Regulations:** Not listed
12.2 **NAS Hazard Rating for Bulk Water Transportation:** Not listed
12.3 **NFPA Hazard Classifications:**

Category	Classification
Health Hazard (Blue)	3
Flammability (Red)	1
Reactivity (Yellow)	0

13. PHYSICAL AND CHEMICAL PROPERTIES

13.1 **Physical State at 15°C and 1 atm:** Solid
13.2 **Molecular Weight:** 128.6
13.3 **Boiling Point at 1 atm:** 428°F = 220°C = 493°K
13.4 **Freezing Point:** 109°F = 43°C = 316°K
13.5 **Critical Temperature:** Not pertinent
13.6 **Critical Pressure:** Not pertinent
13.7 **Specific Gravity:** 1.31 at 20°C (solid)
13.8 **Liquid Surface Tension:** Not pertinent
13.9 **Liquid-Water Interfacial Tension:** Not pertinent
13.10 **Vapor (Gas) Specific Gravity:** Not pertinent
13.11 **Ratio of Specific Heats of Vapor (Gas):** Not pertinent
13.12 **Latent Heat of Vaporization:** (est.) 160 Btu/lb = 89 cal/g = 3.7×10^5 J/kg
13.13 **Heat of Combustion:** –9,330 Btu/lb = –5,180 cal/g = -217×10^5 J/kg
13.14 **Heat of Decomposition:** Not pertinent
13.15 **Heat of Solution:** Not pertinent
13.16 **Heat of Polymerization:** Not pertinent

NOTES

CHLOROPICRIN, LIQUID

Common Synonyms			
Trichloronitromethane Nitrochloroform Picfume Nitrotrichloromethane	Oily liquid	Colorless	Extremely irritating odor
	Sinks in water. Poisonous vapor is produced.		

AVOID CONTACT WITH LIQUID AND VAPOR. KEEP PEOPLE AWAY.
Wear goggles, self-contained breathing apparatus, and rubber overclothing (including gloves).
Stop discharge if possible.
Isolate and remove discharged material.
Notify local health and pollution control agencies.

Fire

Not flammable.
POISONOUS GASES ARE PRODUCED IN FIRE.
Containers may explode in fire.
Wear chemical protective suit with self-contained breathing apparatus.
Cool exposed containers with water.

Exposure

CALL FOR MEDICAL AID.

VAPOR
POISONOUS IF INHALED.
Irritating to eyes, nose and throat.
Move victim to fresh air.
If breathing has stopped, give artificial respiration.
If breathing is difficult, give oxygen.

LIQUID
POISONOUS IF SWALLOWED OR IF SKIN IS EXPOSED.
Will burn skin and eyes.
Remove contaminated clothing and shoes.
Flush affected areas with plenty of water.
IF IN EYES, hold eyelids open and flush with plenty of water.
IF SWALLOWED and victim is CONSCIOUS, have victim drink water or milk.
DO NOT INDUCE VOMITING.

Water Pollution

Effect of low concentrations on aquatic life is unknown.
May be dangerous if it enters water intakes.
Notify local health and wildlife officials.
Notify operators of nearby water intakes.

1. RESPONSE TO DISCHARGE

(See Response Methods Handbook, CG 446-4)

Issue warning—poison, water contaminant, air contaminant
Restrict access
Evacuate area
Should be removed
Chemical and physical treatment

2. LABEL

3. CHEMICAL DESIGNATIONS

3.1 **Synonyms:** Trichloronitromethane
Nitrochloroform
Picfume
Nitrotrichloromethane

3.2 **Coast Guard Compatibility Classification:** Not applicable

3.3 **Chemical Formula:** Cl_3CNO_2

3.4 **IMCO/United Nations Numerical Designation:** 6.1/1580

4. OBSERVABLE CHARACTERISTICS

4.1 **Physical State (as shipped):** Liquid

4.2 **Color:** Colorless

4.3 **Odor:** Intensely irritating, lachrymator; intense and penetrating odor which causes a pronounced secretion of tears.

5. HEALTH HAZARDS

5.1 **Personal Protective Equipment:** Organic-type canister mask; goggles or face shield; rubber gloves; protective clothing.

5.2 **Symptoms Following Exposure:** Inhalation causes nausea, eye watering, vomiting, bronchitis, and pulmonary edema. Vapor is a powerful tear gas. Liquid irritates and burns skin and causes severe burns of eyes. Ingestion causes severe irritation of mouth and stomach.

5.3 **Treatment for Exposure:** Get medical attention following all exposures to this compound. INHALATION: remove from exposure; support respiration. EYES: flush with copious quantities of water for at least 15 min. SKIN: wash with water for 15 min. INGESTION: do NOT induce vomiting; give large amounts of water.

5.4 **Toxicity by Inhalation (Threshold Limit Value):** 0.1 ppm

5.5 **Short-Term Inhalation Limits:** Data not available

5.6 **Toxicity by Ingestion:** Grade 3; oral LD_{50} = 250 mg/kg (rat)

5.7 **Late Toxicity:** Data not available

5.8 **Vapor (Gas) Irritant Characteristics:** Vapors cause severe irritation of eyes and throat and can cause eye or lung injury. They cannot be tolerated even at low concentrations.

5.9 **Liquid or Solid Irritant Characteristics:** Severe skin irritant. Causes second- and third-degree burns on short contact and is very injurious to the eyes.

5.10 **Odor Threshold:** 1.1 ppm

6. FIRE HAZARDS

6.1 **Flash Point:** Not flammable

6.2 **Flammable Limits in Air:** Not flammable

6.3 **Fire Extinguishing Agents:** Cool exposed containers with water.

6.4 **Fire Extinguishing Agents Not to be Used:** Not pertinent

6.5 **Special Hazards of Combustion Products:** Not pertinent

6.6 **Behavior in Fire:** Compound forms a powerful tear gas when heated. Heated material may detonate under fire conditions.

6.7 **Ignition Temperature:** Not pertinent

6.8 **Electrical Hazard:** Not pertinent

6.9 **Burning Rate:** Not pertinent

7. CHEMICAL REACTIVITY

7.1 **Reactivity with Water:** No reaction

7.2 **Reactivity with Common Materials:** No reaction

7.3 **Stability During Transport:** Stable

7.4 **Neutralizing Agents for Acids and Caustics:** Not pertinent

7.5 **Polymerization:** Not pertinent

7.6 **Inhibitor of Polymerization:** Not pertinent

8. WATER POLLUTION

8.1 **Aquatic Toxicity:** Data not available

8.2 **Waterfowl Toxicity:** Data not available

8.3 **Biological Oxygen Demand (BOD):** Data not available

8.4 **Food Chain Concentration Potential:** Data not available

9. SELECTED MANUFACTURERS

1. Great Lakes Chemical Corporation
P. O. Box 2200
West Lafayette, Ind. 47906
2. Dow Chemical Co.
Midland, Mich. 48640
3. Sobin Chemicals, Inc.
Sobin Park
Boston, Mass. 02210

10. SHIPPING INFORMATION

10.1 **Grades or Purity:** 99%

10.2 **Storage Temperature:** Ambient

10.3 **Inert Atmosphere:** No requirement

10.4 **Venting:** Pressure-vacuum

11. HAZARD ASSESSMENT CODE

(See Hazard Assessment Handbook, CG 446-3)

A-X

12. HAZARD CLASSIFICATIONS

12.1 **Code of Federal Regulations:** Poisonous, Class B

12.2 **NAS Hazard Rating for Bulk Water Transportation:**

Category	Rating
Fire	0
Health	
Vapor Irritant	4
Liquid or Solid Irritant	4
Poisons	4
Water Pollution	
Human Toxicity	4
Aquatic Toxicity	3
Aesthetic Effect	4
Reactivity	
Other Chemicals	1
Water	0
Self-Reaction	0

12.3 **NFPA Hazard Classifications:**

Category	Classification
Health Hazard (Blue)	4
Flammability (Red)	0
Reactivity (Yellow)	3

13. PHYSICAL AND CHEMICAL PROPERTIES

13.1 **Physical State at 15°C and 1 atm:** Liquid

13.2 **Molecular Weight:** 164.4

13.3 **Boiling Point at 1 atm:** 234°F = 112°C = 385°K

13.4 **Freezing Point:** −83°F = −64°C = 209°K

13.5 **Critical Temperature:** Not pertinent

13.6 **Critical Pressure:** Not pertinent

13.7 **Specific Gravity:** 1.64 at 25°C (liquid)

13.8 **Liquid Surface Tension:** 32.3 dynes/cm = 0.0323 N/m at 20°C

13.9 **Liquid-Water Interfacial Tension:** (est.) 30 dynes/cm = 0.03 N/m at 20°C

13.10 **Vapor (Gas) Specific Gravity:** 5.7

13.11 **Ratio of Specific Heats of Vapor (Gas):** 1.0991

13.12 **Latent Heat of Vaporization:** 103 Btu/lb = 57.3 cal/g = 2.4×10^5 J/kg

13.13 **Heat of Combustion:** Not pertinent

13.14 **Heat of Decomposition:** Not pertinent

13.15 **Heat of Solution:** Not pertinent

13.16 **Heat of Polymerization:** Not pertinent

NOTES

CSA | CHLOROSULFONIC ACID

Common Synonyms
Chlorsulfonic acid

Liquid — Colorless to light yellow — Sharp, choking odor

Reacts violently with water.
Appears to explode.
Poisonous gas is produced.

AVOID CONTACT WITH LIQUID AND VAPOR. Keep people away.
Wear goggles, self-contained breathing apparatus, and rubber overclothing (including gloves).
Stop discharge if possible. Call fire department.
Notify local health and pollution control agencies.

Fire

May cause fire on contact with combustibles.
Flammable, explosive gases may be formed on contact with metals and moisture.

DO NOT USE WATER.
Use dry chemicals or carbon dioxide.
Wear goggles, self-contained breathing apparatus, and rubber overclothing (including gloves).

Exposure

CALL FOR MEDICAL AID.

VAPOR
Irritating to eyes, nose, and throat.
Harmful if inhaled.

Move to fresh air.
If breathing has stopped, give artificial respiration.
If breathing is difficult, give oxygen.

LIQUID
Will burn skin and eyes.
Harmful if swallowed.

Remove contaminated clothing and shoes.
Flush affected areas with plenty of water.
IF IN EYES, hold eyelids open and flush with plenty of water.
IF SWALLOWED, and victim is CONSCIOUS, have victim drink water or milk.
DO NOT INDUCE VOMITING.

Water Pollution

Dangerous to aquatic life in high concentrations.
May be dangerous if it enters water intakes.

Notify local health and pollution control officials.
Notify operators of nearby water intakes.

1. RESPONSE TO DISCHARGE

(See Response Methods Handbook, CG 446-4)

Issue warning—air contaminant, corrosive
Restrict access
Chemical and physical treatment

2. LABEL

3. CHEMICAL DESIGNATIONS

3.1 **Synonyms:** Chlorsulfonic acid

3.2 **Coast Guard Compatibility Classification:** Inorganic acids

3.3 **Chemical Formula:** $ClSO_3H$

3.4 **IMCO/United Nations Numerical Designation:** 8.0/1754

4. OBSERVABLE CHARACTERISTICS

4.1 **Physical State (as shipped):** Liquid

4.2 **Color:** Colorless to pale yellow

4.3 **Odor:** Sharp, acrid, penetrating, pungent

5. HEALTH HAZARDS

5.1 **Personal Protective Equipment:** Acid-proof goggles or a rubber hood, long rubber gloves, rubber shoes, long rubber apron, shirt and trousers of wool or acrylic fiber, and a hat with a brim. For emergency use involving considerable exposure, a complete rubber suit with hood, gloves and boots of rubber should be used. In case of fire use self-contained breathing apparatus.

5.2 **Symptoms Following Exposure:** INHALATION: vapor extremely irritating to lungs and mucous membranes. Vapor has such a sharp and penetrating odor that inhalation of severely toxic quantities is unlikely unless it is impossible to escape the fumes. CONTACT WITH EYES OR SKIN: liquid acid will severely burn body tissue.

5.3 **Treatment for Exposure:** Call a physician in all cases. INHALATION: remove victim to fresh air; if he is not breathing, apply artificial respiration; give oxygen if breathing is difficult; do NOT induce vomiting. SKIN: flush with plenty of water for at least 15 min. while removing contaminated clothing and shoes.

5.4 **Toxicity by Inhalation (Threshold Limit Value):** 5 ppm

5.5 **Short-Term Inhalation Limits:** 5 ppm for 5 min.; 30 ppm for 10 min.; 10 ppm for 60 min.

5.6 **Toxicity by Ingestion:** Data not available

5.7 **Late Toxicity:** None

5.8 **Vapor (Gas) Irritant Characteristics:** Severe eye and throat irritant. Can cause eye or lung injury and cannot be tolerated even at low concentrations.

5.9 **Liquid or Solid Irritant Characteristics:** Severe skin irritant. Causes second- and third-degree burns on short contact; very injurious to the eyes.

5.10 **Odor Threshold:** 1-5 ppm

6. FIRE HAZARDS

6.1 **Flash Point:** Not flammable

6.2 **Flammable Limits in Air:** Not flammable

6.3 **Fire Extinguishing Agents:** Not pertinent

6.4 **Fire Extinguishing Agents Not to be Used:** Water

6.5 **Special Hazards of Combustion Products:** Decomposes into irritating and toxic gases

6.6 **Behavior in Fire:** Although nonflammable, it may ignite other combustibles. Contact with water AND metal produces explosive hydrogen gas.

6.7 **Ignition Temperature:** Not flammable

6.8 **Electrical Hazard:** Not pertinent

6.9 **Burning Rate:** Not flammable

7. CHEMICAL REACTIVITY

7.1 **Reactivity with Water:** Reacts violently with water, forming hydrochloric acid (vapor) and sulfuric acid.

7.2 **Reactivity with Common Materials:** Hydrogen, a highly flammable and explosive gas, is generated by the action of the acid on most metals. May cause ignition by contact with combustible materials.

7.3 **Stability During Transport:** Stable

7.4 **Neutralizing Agents for Acids and Caustics:** Although the acid reacts violently with water, flooding (from a distance) must be carried out before neutralizing with lime water or sodium bicarbonate solution.

7.5 **Polymerization:** Not pertinent

7.6 **Inhibitor of Polymerization:** Not pertinent

8. WATER POLLUTION

8.1 **Aquatic Toxicity:**
282 ppm/96 hr/mosquito fish/TL_m/fresh water
100-300 ppm/48 hr/shrimp/LC_{50}/salt water

8.2 **Waterfowl Toxicity:** Data not available

8.3 **Biological Oxygen Demand (BOD):** None

8.4 **Food Chain Concentration Potential:** None

9. SELECTED MANUFACTURERS

1. Cities Service Co., Inc.
Industrial Chemicals Division
60 Wall St.
New York, N. Y. 10005
2. E. I. du Pont de Nemours & Co. Inc.
Industrial & Biochemicals Dept.
Wilmington, Del. 19898
3. Monsanto Co.
Monsanto Industrial Chemicals Co.
800 North Lindbergh Blvd.
St. Louis, Mo. 63166

10. SHIPPING INFORMATION

10.1 **Grades or Purity:** Technical

10.2 **Storage Temperature:** Ambient

10.3 **Inert Atmosphere:** No requirement

10.4 **Venting:** Pressure-vacuum

11. HAZARD ASSESSMENT CODE

(See Hazard Assessment Handbook, CG 446-3)

A-O

12. HAZARD CLASSIFICATIONS

12.1 **Code of Federal Regulations:** Corrosive material

12.2 **NAS Hazard Rating for Bulk Water Transportation:**

Category	Rating
Fire	0
Health	
Vapor Irritant	4
Liquid or Solid Irritant	4
Poisons	4
Water Pollution	
Human Toxicity	2
Aquatic Toxicity	3
Aesthetic Effect	2
Reactivity	
Other Chemicals	4
Water	4
Self-Reaction	0

12.3 **NFPA Hazard Classifications:**

Category	Classification
Health Hazard (Blue)	3
Flammability (Red)	0
Reactivity (Yellow)	2
	W, oxy

13. PHYSICAL AND CHEMICAL PROPERTIES

13.1 **Physical State at 15°C and 1 atm:** Liquid

13.2 **Molecular Weight:** 116.53

13.3 **Boiling Point at 1 atm:** 311°F = 155°C = 428°K

13.4 **Freezing Point:** −112°F = −80°C = 193°K

13.5 **Critical Temperature:** Not pertinent

13.6 **Critical Pressure:** Not pertinent

13.7 **Specific Gravity:** 1.75 at 20°C (liquid)

13.8 **Liquid Surface Tension:** Not pertinent

13.9 **Liquid-Water Interfacial Tension:** Not pertinent

13.10 **Vapor (Gas) Specific Gravity:** Not pertinent

13.11 **Ratio of Specific Heats of Vapor (Gas):** Not pertinent

13.12 **Latent Heat of Vaporization:** (est.) 198 Btu/lb = 110 cal/g = 4.6×10^5 J/kg

13.13 **Heat of Combustion:** Not pertinent

13.14 **Heat of Decomposition:** Not pertinent

13.15 **Heat of Solution:** Not pertinent

13.16 **Heat of Polymerization:** Not pertinent

NOTES

REVISED 1978

CTD 4-CHLORO-o-TOLUIDINE

Common Synonyms			
2-Amino-5-chlorotoluene 5-Chloro-2-aminotoluene 4-Chloro-2-methylaniline Fast Red TR base Red TR base	Solid	Gray to white	Weak fishy odor
	Sinks in water. Freezing point is 79° F.		

AVOID CONTACT WITH SOLID AND DUST. KEEP PEOPLE AWAY.
Wear goggles, self-contained breathing apparatus, and rubber overclothing (including gloves).
Call fire department.
Stay upwind. Use water spray to "knock down" dust.
Isolate and remove discharged material.
Notify local health and pollution control agencies.

Fire

Combustible.
POISONOUS GASES MAY BE PRODUCED IN FIRE.
Extinguish with water, dry chemicals, foam, or carbon dioxide.
Cool exposed containers with water.

Exposure

CALL FOR MEDICAL AID.

DUST
POISONOUS IF INHALED.
Move victim to fresh air.
If in eyes, hold eyelids open and flush with plenty of water.
If breathing is difficult, give oxygen.

SOLID
POISONOUS IF SWALLOWED.
Remove contaminated clothing and shoes.
Flush affected areas with plenty of water.
IF IN EYES, hold eyelids open and flush with plenty of water.
IF SWALLOWED and victim is CONSCIOUS, have victim drink water or milk and have victim induce vomiting.
IF SWALLOWED and victim is UNCONSCIOUS OR HAVING CONVULSIONS, do nothing except keep victim warm.

Water Pollution

Effect of low concentrations on aquatic life is unknown.
May be dangerous if it enters water intakes.
Notify local health and wildlife officials.
Notify operators of nearby water intakes.

1. RESPONSE TO DISCHARGE

(See Response Methods Handbook, CG 446-4)

Issue warning—poison, water contaminant
Restrict access
Should be removed
Chemical and physical treatment

2. LABEL

3. CHEMICAL DESIGNATIONS

3.1 **Synonyms:** 2-Amino-5-chlorotoluene
5-Chloro-2-aminotoluene
4-Chloro-2-methylaniline
Fast Red TR base
Red TR base

3.2 **Coast Guard Compatibility Classification:** Not applicable

3.3 **Chemical Formula:** 2-CH_3-4-$ClC_6H_3NH_2$

3.4 **IMCO/United Nations Numerical Designation:** 6.1/1579

4. OBSERVABLE CHARACTERISTICS

4.1 **Physical State (as shipped):** Solid

4.2 **Color:** Grayish-white

4.3 **Odor:** Weak, fishy

5. HEALTH HAZARDS

5.1 **Personal Protective Equipment:** Dust respirator; goggles; rubber gloves; protective clothing.

5.2 **Symptoms Following Exposure:** Inhalation, ingestion, or skin contact causes bluish tint in fingernails, lips, and ears. Headache, drowsiness, and nausea also occur. Contact with eyes causes irritation.

5.3 **Treatment for Exposure:** INHALATION: remove victim to fresh air; give oxygen if needed; get medical attention. EYES: flush with water for at least 15 min. SKIN: wash immediately with soap and water. INGESTION: induce vomiting; get medical attention.

5.4 **Toxicity by Inhalation (Threshold Limit Value):** Data not available

5.5 **Short-Term Inhalation Limits:** Data not available

5.6 **Toxicity by Ingestion:** Grade 3; oral LD_{50} = 464 mg/kg (rat)

5.7 **Late Toxicity:** Data not available

5.8 **Vapor (Gas) Irritant Characteristics:** Data not available

5.9 **Liquid or Solid Irritant Characteristics:** Data not available

5.10 **Odor Threshold:** Data not available

6. FIRE HAZARDS

6.1 **Flash Point:** Combustible solid

6.2 **Flammable Limits in Air:** Not pertinent

6.3 **Fire Extinguishing Agents:** Water, dry chemical

6.4 **Fire Extinguishing Agents Not to be Used:** Not pertinent

6.5 **Special Hazards of Combustion Products:** Toxic oxides of nitrogen and hydrochloric acid fumes may form.

6.6 **Behavior in Fire:** Not pertinent

6.7 **Ignition Temperature:** Data not available

6.8 **Electrical Hazard:** Not pertinent

6.9 **Burning Rate:** Not pertinent

7. CHEMICAL REACTIVITY

7.1 **Reactivity with Water:** No reaction

7.2 **Reactivity with Common Materials:** No reaction

7.3 **Stability During Transport:** Stable

7.4 **Neutralizing Agents for Acids and Caustics:** Not pertinent

7.5 **Polymerization:** Not pertinent

7.6 **Inhibitor of Polymerization:** Not pertinent

8. WATER POLLUTION

8.1 **Aquatic Toxicity:** Data not available

8.2 **Waterfowl Toxicity:** Data not available

8.3 **Biological Oxygen Demand (BOD):** Data not available

8.4 **Food Chain Concentration Potential:** Data not available

9. SELECTED MANUFACTURERS

1. Pfister Chemical Inc.
   Linden Avenue
   Ridgefield, N. J. 07657
2. E. I. duPont de Nemours & Co.
   Dyes and Chemicals Division
   Wilmington, Del. 19898
3. Blackman-Uhler Chemical Co.
   P. O. Box 5627
   Spartanburg, S. C. 29301

10. SHIPPING INFORMATION

10.1 **Grades or Purity:** 99%

10.2 **Storage Temperature:** Ambient

10.3 **Inert Atmosphere:** No requirement

10.4 **Venting:** Pressure-vacuum

11. HAZARD ASSESSMENT CODE

(See Hazard Assessment Handbook, CG 446-3)

II

12. HAZARD CLASSIFICATIONS

12.1 **Code of Federal Regulations:** Poisonous, Class B

12.2 **NAS Hazard Rating for Bulk Water Transportation:** Not listed

12.3 **NFPA Hazard Classifications:** Not listed

13. PHYSICAL AND CHEMICAL PROPERTIES

13.1 **Physical State at 15°C and 1 atm:** Solid

13.2 **Molecular Weight:** 141.6

13.3 **Boiling Point at 1 atm:** 466°F = 241°C = 514°K

13.4 **Freezing Point:** 77°F = 25°C = 298°K

13.5 **Critical Temperature:** Not pertinent

13.6 **Critical Pressure:** Not pertinent

13.7 **Specific Gravity:** (est.) > 1.1 at 20°C (solid)

13.8 **Liquid Surface Tension:** Not pertinent

13.9 **Liquid-Water Interfacial Tension:** Not pertinent

13.10 **Vapor (Gas) Specific Gravity:** Not pertinent

13.11 **Ratio of Specific Heats of Vapor (Gas):** Not pertinent

13.12 **Latent Heat of Vaporization:** Not pertinent

13.13 **Heat of Combustion:** Not pertinent

13.14 **Heat of Decomposition:** Not pertinent

13.15 **Heat of Solution:** Not pertinent

13.16 **Heat of Polymerization:** Not pertinent

NOTES

CMA — CHROMIC ANHYDRIDE

Common Synonyms			
Chromic oxide Chromium trioxide Chromic acid	Solid flakes or powder	Dark red	Odorless
	Sinks and mixes with water.		

Avoid contact with solid and dust. Keep people away.
Wear goggles, self-contained breathing apparatus, and rubber overclothing (including gloves).
Stay upwind and use water spray to "knock down" dust.
Isolate and remove discharged material.
Notify local health and pollution control agencies.

Fire	Not flammable. May cause fire on contact with combustibles. Containers may explode when heated in a fire. Extinguish with water. Cool exposed containers with water.
Exposure	CALL FOR MEDICAL AID. SOLID Will burn skin and eyes. Harmful if swallowed. Remove contaminated clothing and shoes. Flush affected areas with plenty of water. IF IN EYES, hold eyelids open and flush with plenty of water. IF SWALLOWED and victim is CONSCIOUS, have victim drink water or milk. DO NOT INDUCE VOMITING.
Water Pollution	HARMFUL TO AQUATIC LIFE IN VERY LOW CONCENTRATIONS. May be dangerous if it enters water intakes. Notify local health and wildlife officials. Notify operators of nearby water intakes.

1. RESPONSE TO DISCHARGE
(See Response Methods Handbook, CG 446-4)

Issue warning—water contaminant
Disperse and flush

2. LABEL

Yellow

3. CHEMICAL DESIGNATIONS

3.1 **Synonyms:** Chromic acid; Chromic oxide; Chromium trioxide
3.2 **Coast Guard Compatibility Classification:** Not applicable
3.3 **Chemical Formula:** CrO_3
3.4 **IMCO/United Nations Numerical Designation:** 5.1/1463

4. OBSERVABLE CHARACTERISTICS

4.1 **Physical State (as shipped):** Solid
4.2 **Color:** Dark red
4.3 **Odor:** None

5. HEALTH HAZARDS

5.1 **Personal Protective Equipment:** Goggles and respirator. (Special chromic acid filters are available for respirators to prevent inhalation of dust or mist.)
5.2 **Symptoms Following Exposure:** Very irritating to eyes and respiratory tract. Ingestion causes severe gastrointestinal symptoms. Contact with eyes or skin causes burns; prolonged contact produces dermatitis ("chrome sores").
5.3 **Treatment for Exposure:** INGESTION: call a physician; do NOT induce vomiting. SKIN OR EYES: wash eyes thoroughly for at least 15 min.; flush contacted skin areas with water; remove contaminated clothing and wash before reuse.
5.4 **Toxicity by Inhalation (Threshold Limit Value):** Not pertinent
5.5 **Short-Term Inhalation Limits:** Not pertinent
5.6 **Toxicity by Ingestion:** Grade 3; LD_{50} 50 to 500 mg/kg
5.7 **Late Toxicity:** Lung cancer
5.8 **Vapor (Gas) Irritant Characteristics:** Not pertinent
5.9 **Liquid or Solid Irritant Characteristics:** Severe skin irritant. Causes second- and third-degree burns on short contact; very injurious to the eyes.
5.10 **Odor Threshold:** Not pertinent

6. FIRE HAZARDS

6.1 **Flash Point:** Not flammable
6.2 **Flammable Limits in Air:** Not flammable
6.3 **Fire Extinguishing Agents:** Water
6.4 **Fire Extinguishing Agents Not to be Used:** Not pertinent
6.5 **Special Hazards of Combustion Products:** Not pertinent
6.6 **Behavior in Fire:** Containers may explode
6.7 **Ignition Temperature:** May ignite organic materials on contact.
6.8 **Electrical Hazard:** Not pertinent
6.9 **Burning Rate:** Not flammable

7. CHEMICAL REACTIVITY

7.1 **Reactivity with Water:** No reaction
7.2 **Reactivity with Common Materials:** May react with organic materials rapidly enough to generate sufficient heat to cause ignition. Prolonged contact, particularly on wood floors, may produce a fire hazard.
7.3 **Stability During Transport:** Stable
7.4 **Neutralizing Agents for Acids and Caustics:** Flood with water, rinse with sodium bicarbonate solution.
7.5 **Polymerization:** Not pertinent
7.6 **Inhibitor of Polymerization:** Not pertinent

8. WATER POLLUTION

8.1 **Aquatic Toxicity:**
0.01 ppm/48 hr/daphnia/TL_m
52 ppm/96 hr/goldfish/died
8.2 **Waterfowl Toxicity:** Data not available
8.3 **Biological Oxygen Demand (BOD):** None
8.4 **Food Chain Concentration Potential:** None

9. SELECTED MANUFACTURERS

1. Allied Chemical Corp.
   Industrial Chemicals Division
   Morristown, N. J. 07960
2. Diamond Shamrock Corp.
   Soda Products Division
   Castle Haynes, N. C. 28429
3. Essex Chemical Corp.
   Chemicals Division
   Kearny, N. J. 07032

10. SHIPPING INFORMATION

10.1 **Grades or Purity:** Technical; technical flake: 99.75%
10.2 **Storage Temperature:** Data not available
10.3 **Inert Atmosphere:** Data not available
10.4 **Venting:** Data not available

11. HAZARD ASSESSMENT CODE
(See Hazard Assessment Handbook, CG 446-3)

SS

12. HAZARD CLASSIFICATIONS

12.1 **Code of Federal Regulations:** Oxidizing material
12.2 **NAS Hazard Rating for Bulk Water Transportation:** Not listed
12.3 **NFPA Hazard Classifications:**

Category	Classification
Health Hazard (Blue)	3
Flammability (Red)	0
Reactivity (Yellow)	1

13. PHYSICAL AND CHEMICAL PROPERTIES

13.1 **Physical State at 15°C and 1 atm:** Solid
13.2 **Molecular Weight:** 100.01
13.3 **Boiling Point at 1 atm:** Not pertinent
13.4 **Freezing Point:** Not pertinent
13.5 **Critical Temperature:** Not pertinent
13.6 **Critical Pressure:** Not pertinent
13.7 **Specific Gravity:** 2.70 at 20°C (solid)
13.8 **Liquid Surface Tension:** Not pertinent
13.9 **Liquid-Water Interfacial Tension:** Not pertinent
13.10 **Vapor (Gas) Specific Gravity:** Not pertinent
13.11 **Ratio of Specific Heats of Vapor (Gas):** Not pertinent
13.12 **Latent Heat of Vaporization:** Not pertinent
13.13 **Heat of Combustion:** Not pertinent
13.14 **Heat of Decomposition:** Not pertinent
13.15 **Heat of Solution:** Not pertinent
13.16 **Heat of Polymerization:** Not pertinent

NOTES

REVISED 1978

CMC

CHROMYL CHLORIDE

Common Synonyms			
Chromium (VI) dioxychloride Chromium oxychloride	Liquid	Dark red	Unpleasant odor
	Reacts violently with water. Irritating visible vapor cloud is produced.		

Avoid contact with liquid and vapor. Keep people away.
Stop discharge if possible.
Call fire department.
Isolate and remove discharged material.
Notify local health and pollution control agencies.

Fire

Not flammable.
May cause fire on contact with combustibles.
Irritating gases are produced when heated.
Containers may explode in fire.
Extinguish with dry chemicals or carbon dioxide.
DO NOT USE WATER OR FOAM ON FIRE.
Cool exposed containers with water.

Exposure

CALL FOR MEDICAL AID.

VAPOR
Irritating to eyes, nose and throat.
If inhaled will cause difficult breathing.
Move victim to fresh air.
If breathing has stopped, give artificial respiration.
If breathing is difficult, give oxygen.

LIQUID
POISONOUS IF SWALLOWED.
Will burn skin and eyes.
Remove contaminated clothing and shoes.
Flush affected areas with plenty of water.
IF IN EYES, hold eyelids open and flush with plenty of water.
IF SWALLOWED and victim is CONSCIOUS, have victim drink water or milk.
DO NOT INDUCE VOMITING.

Water Pollution

Effect of low concentrations on aquatic life is unknown.
May be dangerous if it enters water intakes.
Notify local health and wildlife officials.
Notify operators of nearby water intakes.

1. RESPONSE TO DISCHARGE

(See Response Methods Handbook, CG 446-4)

Issue warning—poison, corrosive, air contaminant, water contaminant, oxidizing substance
Restrict access
Evacuate area
Disperse and flush with care

2. LABEL

3. CHEMICAL DESIGNATIONS

3.1 **Synonyms:** Chromium (VI) dioxychloride
Chromium oxychloride
3.2 **Coast Guard Compatibility Classification:** Not applicable
3.3 **Chemical Formula:** CrO_2Cl_2
3.4 **IMCO/United Nations Numerical Designation:** 8/1758

4. OBSERVABLE CHARACTERISTICS

4.1 **Physical State (as shipped):** Liquid
4.2 **Color:** Dark red
4.3 **Odor:** Acrid

5. HEALTH HAZARDS

5.1 **Personal Protective Equipment:** Self-contained breathing apparatus (full face); rubber gloves; protective clothing.
5.2 **Symptoms Following Exposure:** Inhalation causes severe irritation of upper respiratory system. Contact with eyes or skin causes irritation and burning. Ingestion causes burning of mouth and stomach.
5.3 **Treatment for Exposure:** Get medical attention following all exposures to this compound. INHALATION: remove from exposure; support respiration. EYES: flush with copious quantities of water for 15 min. SKIN: flush with water for 15 min. INGESTION: do NOT induce vomiting; give large amounts of water.
5.4 **Toxicity by Inhalation (Threshold Limit Value):** Data not available
5.5 **Short-Term Inhalation Limits:** Data not available
5.6 **Toxicity by Ingestion:** Grade 4; $LD_{50} < 50$ mg/kg
5.7 **Late Toxicity:** Data not available
5.8 **Vapor (Gas) Irritant Characteristics:** Vapors cause severe irritation of eyes and throat and can cause eye and lung injury. They cannot be tolerated even at low concentrations.
5.9 **Liquid or Solid Irritant Characteristics:** Severe skin irritant. Causes second- and third-degree burns on short contact and is very injurious to the eyes.
5.10 **Odor Threshold:** Data not available

6. FIRE HAZARDS

6.1 **Flash Point:** Not flammable, but may cause fire on contact with combustible materials.
6.2 **Flammable Limits in Air:** Not flammable
6.3 **Fire Extinguishing Agents:** Dry chemical or carbon dioxide
6.4 **Fire Extinguishing Agents Not to be Used:** Do not use water on adjacent fires unless fully protected against toxic fumes.
6.5 **Special Hazards of Combustion Products:** Not pertinent
6.6 **Behavior in Fire:** Vapors are very irritating to eyes and mucous membranes. May increase severity of fire.
6.7 **Ignition Temperature:** Not pertinent
6.8 **Electrical Hazard:** Not pertinent
6.9 **Burning Rate:** Not pertinent

7. CHEMICAL REACTIVITY

7.1 **Reactivity with Water:** Reacts violently to form hydrogen chloride (hydrochloric acid) and chlorine gases and chromic acid.
7.2 **Reactivity with Common Materials:** Will cause severe corrosion of common metals.
7.3 **Stability During Transport:** Stable
7.4 **Neutralizing Agents for Acids and Caustics:** Flood with water. Rinse with sodium bicarbonate or lime solution.
7.5 **Polymerization:** Not pertinent
7.6 **Inhibitor of Polymerization:** Not pertinent

8. WATER POLLUTION

8.1 **Aquatic Toxicity:** Data not available
8.2 **Waterfowl Toxicity:** Data not available
8.3 **Biological Oxygen Demand (BOD):** None
8.4 **Food Chain Concentration Potential:** None

9. SELECTED MANUFACTURERS

1. Gallard-Schlesinger Chemical Manufacturing Corp.
584 Mineola Avenue
Carle Place, Long Island, N. Y. 11514
2. Ventron Corp., Alfa Products
P. O. Box 159
Beverly, Mass. 01915
3. Research Organic/Inorganic Chemical Corp.
11686 Sheldon Rd.
Sun Valley, Calif. 91352

10. SHIPPING INFORMATION

10.1 **Grades or Purity:** 99.5+%
10.2 **Storage Temperature:** Ambient
10.3 **Inert Atmosphere:** No requirement
10.4 **Venting:** Pressure-vacuum

11. HAZARD ASSESSMENT CODE

(See Hazard Assessment Handbook, CG 446-3)

A-O

12. HAZARD CLASSIFICATIONS

12.1 **Code of Federal Regulations:** Corrosive
12.2 **NAS Hazard Rating for Bulk Water Transportation:**

Category	Rating
Fire	0
Health	
Vapor Irritant	4
Liquid or Solid Irritant	4
Poisons	4
Water Pollution	
Human Toxicity	4
Aquatic Toxicity	4
Aesthetic Effect	2
Reactivity	
Other Chemicals	4
Water	4
Self-Reaction	0

12.3 **NFPA Hazard Classifications:** Not listed

13. PHYSICAL AND CHEMICAL PROPERTIES

13.1 **Physical State at 15°C and 1 atm:** Liquid
13.2 **Molecular Weight:** 154.9
13.3 **Boiling Point at 1 atm:** 241°F = 116°C = 389°K
13.4 **Freezing Point:** −141.7°F = −96.5°C = 176.7°K
13.5 **Critical Temperature:** Not pertinent
13.6 **Critical Pressure:** Not pertinent
13.7 **Specific Gravity:** 1.96 at 20°C (liquid)
13.8 **Liquid Surface Tension:** 36.61 dynes/cm = 0.03661 N/m at 19°C
13.9 **Liquid-Water Interfacial Tension:** Not pertinent
13.10 **Vapor (Gas) Specific Gravity:** 5.3
13.11 **Ratio of Specific Heats of Vapor (Gas):** 1.2832
13.12 **Latent Heat of Vaporization:** 113 Btu/lb = 62.6 cal/g = 2.62×10^5 J/kg
13.13 **Heat of Combustion:** Not pertinent
13.14 **Heat of Decomposition:** Not pertinent
13.15 **Heat of Solution:** −279 Btu/lb = −155 cal/g = -6.48×10^5 J/kg
13.16 **Heat of Polymerization:** Not pertinent

NOTES

CIT CITRIC ACID

Common Synonyms			
beta-Hydroxy-tricarboxylic acid 2-Hydroxy-1,2,3-propane-tricarboxylic acid beta-Hydroxytricarballylic acid	Solid	White	Odorless
	Sinks and mixes with water.		

Stop discharge if possible. Keep people away.
Avoid contact with solid and dust.
Call fire department.
Isolate and remove discharged material.
Notify local health and pollution control agencies.

Fire	Combustible. Extinguish with dry chemicals, water, foam or carbon dioxide.
Exposure	CALL FOR MEDICAL AID. DUST Irritating to eyes, nose and throat. If inhaled will cause coughing or difficult breathing. If in eyes, hold eyelids open and flush with plenty of water. If breathing has stopped, give artificial respiration. If breathing is difficult, give oxygen. SOLID Irritating to skin and eyes. Harmful if swallowed. Remove contaminated clothing and shoes. Flush affected areas with plenty of water. IF IN EYES, hold eyelids open and flush with plenty of water. IF SWALLOWED and victim is CONSCIOUS, have victim drink water or milk. IF SWALLOWED and victim is UNCONSCIOUS OR HAVING CONVULSIONS, do nothing except keep victim warm.
Water Pollution	Dangerous to aquatic life in high concentrations. May be dangerous if it enters water intakes. Notify local health and wildlife officials. Notify operators of nearby water intakes.

1. RESPONSE TO DISCHARGE
(See Response Methods Handbook, CG 446-4)

Disperse and flush

2. LABELS

No hazard label required by Code of Federal Regulations

3. CHEMICAL DESIGNATIONS

3.1 **Synonyms:** 2-Hydroxy-1,2,3-propane-tricarboxylic acid; beta-Hydroxytricarballylic acid; beta-Hydroxy-tricarboxylic acid

3.2 **Coast Guard Compatibility Classification:** Not listed

3.3 **Chemical Formula:** $HOC(CH_2CO_2H)_2CO_2H$

3.4 **IMCO/United Nations Numerical Designation:** Not listed

4. OBSERVABLE CHARACTERISTICS

4.1 **Physical State (as shipped):** Solid

4.2 **Color:** White

4.3 **Odor:** None

5. HEALTH HAZARDS

5.1 **Personal Protective Equipment:** Dust mask; goggles or face shield; protective gloves

5.2 **Symptoms Following Exposure:** Inhalation of dust irritates nose and throat. Contact with eyes causes irritation.

5.3 **Treatment for Exposure:** INHALATION: move to fresh air. EYES: flush immediately with physiological saline or water; get medical care if irritation persists. SKIN: flush with water.

5.4 **Toxicity by Inhalation (Threshold Limit Value):** Data not available

5.5 **Short-Term Inhalation Limits:** Data not available

5.6 **Toxicity by Ingestion:** Grade 1; oral LD_{50} = 11.7 g/kg (rat)

5.7 **Late Toxicity:** Chronic effects in humans are unknown

5.8 **Vapor (Gas) Irritant Characteristics:** Not pertinent

5.9 **Liquid or Solid Irritant Characteristics:** Data not available

5.10 **Odor Threshold:** Odorless

6. FIRE HAZARDS

6.1 **Flash Point:** Not pertinent (combustible solid)

6.2 **Flammable Limits in Air:** 0.28–2.29 kg/m³ (dust)

6.3 **Fire Extinguishing Agents:** Water, foam, dry chemical, carbon dioxide

6.4 **Fire Extinguishing Agents Not to be Used:**

6.5 **Special Hazards of Combustion Products:**

6.6 **Behavior in Fire:** Melts and decomposes. The reaction is not hazardous.

6.7 **Ignition Temperature:** 1,850°F (powder)

6.8 **Electrical Hazard:** Not pertinent

6.9 **Burning Rate:** Not pertinent

7. CHEMICAL REACTIVITY

7.1 **Reactivity with Water:** No reaction

7.2 **Reactivity with Common Materials:** Will corrode copper, zinc, aluminum and their alloys.

7.3 **Stability During Transport:** Stable

7.4 **Neutralizing Agents for Acids and Caustics:** Not pertinent

7.5 **Polymerization:** Not pertinent

7.6 **Inhibitor of Polymerization:** Not pertinent

8. WATER POLLUTION

8.1 **Aquatic Toxicity:**
894 ppm/4 hr/goldfish/killed/fresh water
160 ppm/48 hr/shore crab/TL_m/salt water

8.2 **Waterfowl Toxicity:** Data not available

8.3 **Biological Oxygen Demand (BOD):** 40%, 5 days; 60%, 10–20 days

8.4 **Food Chain Concentration Potential:** None

9. SELECTED MANUFACTURERS

1. Pfizer Chemicals Division
235 E. 42nd St.
New York, N. Y. 10017

2. Allied Chemical Corp.
Specialty Chemicals Div.
P. O. Box 1087R
Morristown, N. J. 07960

3. Mallinckrodt Chemical Works
223 Westside Avenue
P. O. Box 384
Jersey City, N. J. 07303

10. SHIPPING INFORMATION

10.1 **Grades or Purity:** USP; Reagent; Monohydrate grade

10.2 **Storage Temperature:** Ambient

10.3 **Inert Atmosphere:** No requirement

10.4 **Venting:** Open

11. HAZARD ASSESSMENT CODE
(See Hazard Assessment Handbook, CG 446-3)

SS

12. HAZARD CLASSIFICATIONS

12.1 **Code of Federal Regulations:** Not listed

12.2 **NAS Hazard Rating for Bulk Water Transportation:** Not listed

12.3 **NFPA Hazard Classifications:** Not listed

13. PHYSICAL AND CHEMICAL PROPERTIES

13.1 **Physical State at 15°C and 1 atm:** Solid

13.2 **Molecular Weight:** 192.1

13.3 **Boiling Point at 1 atm:** Not pertinent (decomposes)

13.4 **Freezing Point:** 307°F = 153°C = 426°K

13.5 **Critical Temperature:** Not pertinent

13.6 **Critical Pressure:** Not pertinent

13.7 **Specific Gravity:** 1.54 at 20°C (solid)

13.8 **Liquid Surface Tension:** Not pertinent

13.9 **Liquid-Water Interfacial Tension:** Not pertinent

13.10 **Vapor (Gas) Specific Gravity:** Not pertinent

13.11 **Ratio of Specific Heats of Vapor (Gas):** Not pertinent

13.12 **Latent Heat of Vaporization:** Not pertinent

13.13 **Heat of Combustion:** –4,000 Btu/lb = –2,220 cal/g = –93.0 J/kg

13.14 **Heat of Decomposition:** Not pertinent

13.15 **Heat of Solution:** Not pertinent

13.16 **Heat of Polymerization:** Not pertinent

NOTES

CBA

COBALT ACETATE

Common Synonyms			
Cobalt acetate tetrahydrate Cobaltous acetate Cobalt[II] acetate	Solid	Pink	Vinegar-like odor
	Sinks and mixes with water.		

Stop discharge if possible. Keep people away.
Avoid contact with solid and dust.
Isolate and remove discharged material.
Notify local health and pollution control agencies.

Fire	Not flammable. POISONOUS GASES MAY BE PRODUCED IN FIRE. Wear goggles and self-contained breathing apparatus.
Exposure	CALL FOR MEDICAL AID. DUST Irritating to eyes, nose and throat. If inhaled will cause coughing or difficult breathing. If in eyes, hold eyelids open and flush with plenty of water. If breathing has stopped, give artificial respiration. If breathing is difficult, give oxygen. SOLID Irritating to skin and eyes. If swallowed will cause nausea and vomiting. Remove contaminated clothing and shoes. Flush affected areas with plenty of water. IF IN EYES, hold eyelids open and flush with plenty of water. IF SWALLOWED and victim is CONSCIOUS, have victim drink water or milk and have victim induce vomiting. IF SWALLOWED and victim is UNCONSCIOUS OR HAVING CONVULSIONS, do nothing except keep victim warm.
Water Pollution	Effect of low concentrations on aquatic life is unknown. May be dangerous if it enters water intakes. Notify local health and wildlife officials. Notify operators of nearby water intakes.

1. RESPONSE TO DISCHARGE

(See Response Methods Handbook, CG 446-4)

Issue warning—water contaminant
Disperse and flush

2. LABELS

No hazard label required by Code of Federal Regulations

3. CHEMICAL DESIGNATIONS

3.1 **Synonyms:** Cobalt (II) acetate; Cobalt acetate tetrahydrate; Cobaltous acetate
3.2 **Coast Guard Compatibility Classification:** Not listed
3.3 **Chemical Formula:** $Co(C_2H_3O_2)_2 \cdot 4H_2O$
3.4 **IMCO/United Nations Numerical Designation:** Not listed

4. OBSERVABLE CHARACTERISTICS

4.1 **Physical State (as shipped):** Solid
4.2 **Color:** Pink
4.3 **Odor:** Slight acetic acid odor; vinegar-like

5. HEALTH HAZARDS

5.1 **Personal Protective Equipment:** Dust respirator; rubber gloves; goggles or face shield; protective clothing
5.2 **Symptoms Following Exposure:** Inhalation causes shortness of breath and coughing; permanent disability may occur. Ingestion causes pain and vomiting. Contact with eyes causes irritation. Contact with skin may cause dermatitis.
5.3 **Treatment for Exposure:** INHALATION: move to fresh air; if breathing has stopped, begin artificial respiration. INGESTION: give large amounts of water; induce vomiting. EYES: flush with water for at least 15 min. SKIN: wash with soap and water.
5.4 **Toxicity by Inhalation (Threshold Limit Value):** 0.1 mg/m^3 (as cobalt)
5.5 **Short-Term Inhalation Limits:** Data not available
5.6 **Toxicity by Ingestion:** Grade 3; LD_{50} 50—500 mg/kg
5.7 **Late Toxicity:** Data not available
5.8 **Vapor (Gas) Irritant Characteristics:** Data not available
5.9 **Liquid or Solid Irritant Characteristics:** Data not available
5.10 **Odor Threshold:** Data not available

6. FIRE HAZARDS

6.1 **Flash Point:** Not flammable
6.2 **Flammable Limits in Air:** Not flammable
6.3 **Fire Extinguishing Agents:** Not pertinent
6.4 **Fire Extinguishing Agents Not to be Used:** Not pertinent
6.5 **Special Hazards of Combustion Products:** Toxic cobalt oxide fumes may form in fire.
6.6 **Behavior in Fire:**
6.7 **Ignition Temperature:** Not pertinent
6.8 **Electrical Hazard:** Not pertinent
6.9 **Burning Rate:** Not pertinent

7. CHEMICAL REACTIVITY

7.1 **Reactivity with Water:** No reaction
7.2 **Reactivity with Common Materials:**
7.3 **Stability During Transport:** Stable
7.4 **Neutralizing Agents for Acids and Caustics:** Not pertinent
7.5 **Polymerization:** Not pertinent
7.6 **Inhibitor of Polymerization:** Not pertinent

8. WATER POLLUTION

8.1 **Aquatic Toxicity:** Data not available
8.2 **Waterfowl Toxicity:** Data not available
8.3 **Biological Oxygen Demand (BOD):** Data not available
8.4 **Food Chain Concentration Potential:** Bioconcentration of 200—1000 fold only under constant exposure. Not significant in spill conditions.

9. SELECTED MANUFACTURERS

1. Mooney Chemicals, Inc.
   2301 Scranton Road
   Cleveland, Ohio 44113
2. The Harshaw Chemical Co.
   1945 E. 97th St.
   Cleveland, Ohio 44106
3. Mallinckrodt Chemical Works
   223 Westside Avenue
   P. O. Box 384
   Jersey City, N. J. 07303

10. SHIPPING INFORMATION

10.1 **Grades or Purity:** Technical; Reagent
10.2 **Storage Temperature:** Ambient
10.3 **Inert Atmosphere:** No requirement
10.4 **Venting:** Open

11. HAZARD ASSESSMENT CODE

(See Hazard Assessment Handbook, CG 446-3)

SS

12. HAZARD CLASSIFICATIONS

12.1 **Code of Federal Regulations:** Not listed
12.2 **NAS Hazard Rating for Bulk Water Transportation:** Not listed
12.3 **NFPA Hazard Classifications:** Not listed

13. PHYSICAL AND CHEMICAL PROPERTIES

13.1 **Physical State at 15°C and 1 atm:** Solid
13.2 **Molecular Weight:** 249.1
13.3 **Boiling Point at 1 atm:** Not pertinent (decomposes)
13.4 **Freezing Point:** 284°F = 140°C = 413°K
13.5 **Critical Temperature:** Not pertinent
13.6 **Critical Pressure:** Not pertinent
13.7 **Specific Gravity:** 1.71 at 20°C (solid)
13.8 **Liquid Surface Tension:** Not pertinent
13.9 **Liquid-Water Interfacial Tension:** Not pertinent
13.10 **Vapor (Gas) Specific Gravity:** Not pertinent
13.11 **Ratio of Specific Heats of Vapor (Gas):** Not pertinent
13.12 **Latent Heat of Vaporization:** Not pertinent
13.13 **Heat of Combustion:** Not pertinent
13.14 **Heat of Decomposition:** Not pertinent
13.15 **Heat of Solution:** Not pertinent
13.16 **Heat of Polymerization:** Not pertinent

NOTES

CBC COBALT CHLORIDE

Common Synonyms			
Cobaltous chloride hexahydrate Cobaltous chloride Cobaltous chloride dihydrate	Solid	Pink to red	Slight sharp odor
	Sinks and mixes with water.		

Stop discharge if possible. Keep people away.
Avoid contact with solid and dust.
Isolate and remove discharged material.
Notify local health and pollution control agencies.

Fire	Not flammable.
Exposure	**CALL FOR MEDICAL AID.** DUST Irritating to eyes, nose and throat. If inhaled will cause coughing or difficult breathing. **If in eyes, hold eyelids open and flush with plenty of water.** **If breathing has stopped, give artificial respiration.** **If breathing is difficult, give oxygen.** SOLID Irritating to skin and eyes. If swallowed will cause nausea and vomiting. **Remove contaminated clothing and shoes.** **Flush affected areas with plenty of water.** **IF IN EYES, hold eyelids open and flush with plenty of water.** **IF SWALLOWED and victim is CONSCIOUS, have victim drink water or milk and have victim induce vomiting.** **IF SWALLOWED and victim is UNCONSCIOUS OR HAVING CONVULSIONS, do nothing except keep victim warm.**
Water Pollution	Dangerous to aquatic life in high concentrations. May be dangerous if it enters water intakes. **Notify local health and wildlife officials.** **Notify operators of nearby water intakes.**

1. RESPONSE TO DISCHARGE
(See Response Methods Handbook, CG 446-4)

Issue warning—water contaminant
Disperse and flush

2. LABELS

No hazard label required by Code of Federal Regulations

3. CHEMICAL DESIGNATIONS

3.1 **Synonyms:** Cobalt(II) chloride; Cobaltous chloride; Cobaltous chloride dihydrate; Cobaltous chloride hexahydrate
3.2 **Coast Guard Compatibility Classification:** Not listed
3.3 **Chemical Formula:** $CoCl_2$
3.4 **IMCO/United Nations Numerical Designation:** Not listed

4. OBSERVABLE CHARACTERISTICS

4.1 **Physical State (as shipped):** Solid
4.2 **Color:** Pink to red
4.3 **Odor:** Very slight acrid

5. HEALTH HAZARDS

5.1 **Personal Protective Equipment:** Rubber gloves; side-shield goggles; Bu. of Mines respirator; protective clothing
5.2 **Symptoms Following Exposure:** Inhalation causes respiratory disease, shortness of breath, and coughing; permanent disability may occur. Ingestion causes pain, vomiting, and diarrhea. Contact causes irritation of eyes and may cause skin rash.
5.3 **Treatment for Exposure:** INHALATION: move victim to fresh air; if breathing has stopped, begin artificial respiration and call a doctor. INGESTION: give large amount of water; induce vomiting. EYES: flush with water at least 15 min.; consult physician if irritation persists. SKIN: flush with water.
5.4 **Toxicity by Inhalation (Threshold Limit Value):** 0.1 mg/m³ (as cobalt)
5.5 **Short-Term Inhalation Limits:** Data not available
5.6 **Toxicity by Ingestion:** Grade 3; LD_{50} 50—500 mg/kg
5.7 **Late Toxicity:** Data not available
5.8 **Vapor (Gas) Irritant Characteristics:** Data not available
5.9 **Liquid or Solid Irritant Characteristics:** Data not available
5.10 **Odor Threshold:** Data not available

6. FIRE HAZARDS

6.1 **Flash Point:** Not flammable
6.2 **Flammable Limits in Air:** Not flammable
6.3 **Fire Extinguishing Agents:** Not pertinent
6.4 **Fire Extinguishing Agents Not to be Used:** Not pertinent
6.5 **Special Hazards of Combustion Products:** Toxic cobalt oxide fumes may form in fire.
6.6 **Behavior in Fire:**
6.7 **Ignition Temperature:** Not pertinent
6.8 **Electrical Hazard:** Not pertinent
6.9 **Burning Rate:** Not pertinent

7. CHEMICAL REACTIVITY

7.1 **Reactivity with Water:** No reaction
7.2 **Reactivity with Common Materials:**
7.3 **Stability During Transport:** Stable
7.4 **Neutralizing Agents for Acids and Caustics:** Not pertinent
7.5 **Polymerization:** Not pertinent
7.6 **Inhibitor of Polymerization:** Not pertinent

8. WATER POLLUTION

8.1 **Aquatic Toxicity:**
1000 ppm/30–32 hr/goldfish/killed/fresh (hard) water
10 ppm/168 hr/goldfish/killed/fresh (soft) water
200 ppm/*/mummichogs/no effect/sea water
*Time period not specified.
8.2 **Waterfowl Toxicity:** Data not available
8.3 **Biological Oxygen Demand (BOD):** None

9. SELECTED MANUFACTURERS

1. Mooney Chemicals, Inc.
   2301 Scranton Rd.
   Cleveland, Ohio 44113
2. The Harshaw Chemical Co.
   1945 East 97th St.
   Cleveland, Ohio 44106
3. Allied Chemical Corp.
   Specialty Chemicals Div.
   P. O. Box 1087 R
   Morristown, N. J. 07960

10. SHIPPING INFORMATION

10.1 **Grades or Purity:** Anhydrous, 100.1%. May also be shipped as dihydrate.
10.2 **Storage Temperature:** Ambient
10.3 **Inert Atmosphere:** No requirement
10.4 **Venting:** Open

11. HAZARD ASSESSMENT CODE
(See Hazard Assessment Handbook, CG 446-3)

SS

12. HAZARD CLASSIFICATIONS

12.1 **Code of Federal Regulations:** Not listed
12.2 **NAS Hazard Rating for Bulk Water Transportation:** Not listed
12.3 **NFPA Hazard Classifications:** Not listed

13. PHYSICAL AND CHEMICAL PROPERTIES

13.1 **Physical State at 15°C and 1 atm:** Solid
13.2 **Molecular Weight:** 237.9
13.3 **Boiling Point at 1 atm:** Not pertinent (decomposes)
13.4 **Freezing Point:** 187°F = 86°C = 359°K
13.5 **Critical Temperature:** Not pertinent
13.6 **Critical Pressure:** Not pertinent
13.7 **Specific Gravity:** 1.924 at 20°C (solid)
13.8 **Liquid Surface Tension:** Not pertinent
13.9 **Liquid-Water Interfacial Tension:** Not pertinent
13.10 **Vapor (Gas) Specific Gravity:** Not pertinent
13.11 **Ratio of Specific Heats of Vapor (Gas):** Not pertinent
13.12 **Latent Heat of Vaporization:** Not pertinent
13.13 **Heat of Combustion:** Not pertinent
13.14 **Heat of Decomposition:** Not pertinent
13.15 **Heat of Solution:** 22 Btu/lb = 12 cal/g = 0.50×10^5 J/kg
13.16 **Heat of Polymerization:** Not pertinent

8. WATER POLLUTION (Cont'd.)

8.4 **Food Chain Concentration Potential:** Bioconcentration of 200—1000 fold only under constant exposure. Not significant in spill conditions.

CON

COBALT NITRATE

Common Synonyms
Cobaltous nitrate
Cobalt[II] nitrate
Cobaltous nitrate hexahydrate

Solid | Red | Odorless

Sinks and mixes with water.

Stop discharge if possible. Keep people away.
Avoid contact with solid and dust.
Isolate and remove discharged material.
Notify local health and pollution control agencies.

Fire

Not flammable.
Will increase the intensity of a fire.
POISONOUS GASES MAY BE PRODUCED IN FIRE.
Wear goggles and self-contained breathing apparatus.
Flood discharge area with water.

Exposure

CALL FOR MEDICAL AID.

DUST
Irritating to eyes, nose and throat.
If inhaled will cause coughing or difficult breathing.
If in eyes, hold eyelids open and flush with plenty of water.
If breathing has stopped, give artificial respiration.
If breathing is difficult, give oxygen.

SOLID
Irritating to skin and eyes.
If swallowed will cause nausea and vomiting.
Remove contaminated clothing and shoes.
Flush affected areas with plenty of water.
IF IN EYES, hold eyelids open and flush with plenty of water.
IF SWALLOWED and victim is CONSCIOUS, have victim drink water or milk and have victim induce vomiting.
IF SWALLOWED and victim is UNCONSCIOUS OR HAVING CONVULSIONS, do nothing except keep victim warm.

Water Pollution

HARMFUL TO AQUATIC LIFE IN VERY LOW CONCENTRATIONS.
May be dangerous if it enters water intakes.
Notify local health and wildlife officials.
Notify operators of nearby water intakes.

1. RESPONSE TO DISCHARGE

(See Response Methods Handbook, CG 446-4)

Issue warning—water contaminant
Disperse and flush

2. LABELS

No hazard label required by Code of Federal Regulations

3. CHEMICAL DESIGNATIONS

3.1 **Synonyms:** Cobalt (II) nitrate; Cobaltous nitrate; Cobaltous nitrate hexahydrate
3.2 **Coast Guard Compatibility Classification:** Not listed
3.3 **Chemical Formula:** $Co(NO_3)_2 \cdot 6H_2O$
3.4 **IMCO/United Nations Numerical Designation:** Not listed

4. OBSERVABLE CHARACTERISTICS

4.1 **Physical State (as shipped):** Solid
4.2 **Color:** Red
4.3 **Odor:** None

5. HEALTH HAZARDS

5.1 **Personal Protective Equipment:** Bu. Mines approved respirator; rubber gloves; safety goggles; protective clothing
5.2 **Symptoms Following Exposure:** Inhalation causes shortness of breath and coughing; permanent disability may occur. Ingestion causes pain and vomiting. Contact with eyes or skin causes irritation.
5.3 **Treatment for Exposure:** INHALATION: move to fresh air; if breathing has stopped, begin artificial respiration and call a doctor. INGESTION: give large amount of water; induce vomiting; call a doctor. EYES: flush with water for at least 15 min. SKIN: flush with water.
5.4 **Toxicity by Inhalation (Threshold Limit Value):** Data not available
5.5 **Short-Term Inhalation Limits:** Data not available
5.6 **Toxicity by Ingestion:** Grade 3; LD_{50} ~ 400 mg/kg (rabbit)
5.7 **Late Toxicity:** Causes malignant tumors in rabbits
5.8 **Vapor (Gas) Irritant Characteristics:** Data not available
5.9 **Liquid or Solid Irritant Characteristics:** Data not available
5.10 **Odor Threshold:** Not pertinent

6. FIRE HAZARDS

6.1 **Flash Point:** Not flammable
6.2 **Flammable Limits in Air:** Not flammable
6.3 **Fire Extinguishing Agents:** Not pertinent
6.4 **Fire Extinguishing Agents Not to be Used:** Not pertinent
6.5 **Special Hazards of Combustion Products:** Toxic oxides of nitrogen may form in fire.
6.6 **Behavior in Fire:** May increase the intensity of fire
6.7 **Ignition Temperature:** Not pertinent
6.8 **Electrical Hazard:** Not pertinent
6.9 **Burning Rate:** Not pertinent

7. CHEMICAL REACTIVITY

7.1 **Reactivity with Water:** No reaction
7.2 **Reactivity with Common Materials:** Contact with wood and paper may cause fire.
7.3 **Stability During Transport:** Stable
7.4 **Neutralizing Agents for Acids and Caustics:** Not pertinent
7.5 **Polymerization:** Not pertinent
7.6 **Inhibitor of Polymerization:** Not pertinent

8. WATER POLLUTION

8.1 **Aquatic Toxicity:**
10 ppm*/stickleback/lethal conc. limit/fresh water
15 ppm*/68 hr/stickleback/avg. survival time/fresh water
20 ppm*/96 hr/stickleback/avg. survival time/fresh water
*as cobalt
8.2 **Waterfowl Toxicity:** Data not available
8.3 **Biological Oxygen Demand (BOD):** None

9. SELECTED MANUFACTURERS

1. Mooney Chemicals, Inc.
2301 Scranton Road
Cleveland, Ohio 44113
2. The Harshaw Chemical Co.
1945 E. 97th St.
Cleveland, Ohio 44106
3. Allied Chemical Corp.
Specialty Chemicals Div.
P. O. Box 1087R
Morristown, N. J. 07960

10. SHIPPING INFORMATION

10.1 **Grades or Purity:** Technical hexahydrate. May also be shipped as anhydrous solid.
10.2 **Storage Temperature:** Ambient
10.3 **Inert Atmosphere:** No requirement
10.4 **Venting:** Open

11. HAZARD ASSESSMENT CODE

(See Hazard Assessment Handbook, CG 446-3)

SS

12. HAZARD CLASSIFICATIONS

12.1 **Code of Federal Regulations:** Not listed
12.2 **NAS Hazard Rating for Bulk Water Transportation:** Not listed
12.3 **NFPA Hazard Classifications:**

Category	Classification*	
Health Hazard (Blue)	0	1
Flammability (Red)	0	0
Reactivity (Yellow)	0	0
	oxy	oxy

*First column refers to nonfire situation.

13. PHYSICAL AND CHEMICAL PROPERTIES

13.1 **Physical State at 15°C and 1 atm:** Solid
13.2 **Molecular Weight:** 291.04
13.3 **Boiling Point at 1 atm:** Not pertinent (decomposes)
13.4 **Freezing Point:** 131°F = 55°C = 328°K
13.5 **Critical Temperature:** Not pertinent
13.6 **Critical Pressure:** Not pertinent
13.7 **Specific Gravity:** 1.54 at 20°C (solid)
13.8 **Liquid Surface Tension:** Not pertinent
13.9 **Liquid-Water Interfacial Tension:** Not pertinent
13.10 **Vapor (Gas) Specific Gravity:** Not pertinent
13.11 **Ratio of Specific Heats of Vapor (Gas):** Not pertinent
13.12 **Latent Heat of Vaporization:** Not pertinent
13.13 **Heat of Combustion:** Not pertinent
13.14 **Heat of Decomposition:** Not pertinent
13.15 **Heat of Solution:** 31 Btu/lb = 17 cal/g = 0.71×10^5 J/kg
13.16 **Heat of Polymerization:** Not pertinent

8. WATER POLLUTION (Cont'd.)

8.4 **Food Chain Concentration Potential:** Bioconcentration of 200—1000 fold only under constant exposure. Not significant in spill conditions.

CBS COBALT SULFATE

Common Synonyms
Cobaltous sulfate heptahydrate
Cobalt[II] sulfate
Bieberite

Solid | Rose-pink | Odorless

Sinks and mixes with water.

Stop discharge if possible. Keep people away.
Avoid contact with solid and dust.
Isolate and remove discharged material.
Notify local health and pollution control agencies.

Fire

Not flammable.
POISONOUS GASES MAY BE PRODUCED IN FIRE.
Wear goggles and self-contained breathing apparatus.

Exposure

CALL FOR MEDICAL AID.

DUST
Irritating to eyes, nose and throat.
If inhaled will cause coughing or difficult breathing.
If in eyes, hold eyelids open and flush with plenty of water.
If breathing has stopped, give artificial respiration.
If breathing is difficult, give oxygen.

SOLID
Irritating to skin and eyes.
If swallowed will cause nausea and vomiting.
Remove contaminated clothing and shoes.
Flush affected areas with plenty of water.
IF IN EYES, hold eyelids open and flush with plenty of water.
IF SWALLOWED and victim is CONSCIOUS, have victim drink water or milk and have victim induce vomiting.
IF SWALLOWED and victim is UNCONSCIOUS OR HAVING CONVULSIONS, do nothing except keep victim warm.

Water Pollution

HARMFUL TO AQUATIC LIFE IN VERY LOW CONCENTRATIONS.
May be dangerous if it enters water intakes.
Notify local health and wildlife officials.
Notify operators of nearby water intakes.

1. RESPONSE TO DISCHARGE
(See Response Methods Handbook, CG 446-4)

Issue warning – water contaminant
Disperse and flush

2. LABELS

No hazard label required by Code of Federal Regulations

3. CHEMICAL DESIGNATIONS

3.1 **Synonyms:** Bieberite; Cobalt(II) sulfate; Cobaltous sulfate heptahydrate
3.2 **Coast Guard Compatibility Classification:** Not listed
3.3 **Chemical Formula:** $CoSO_4 \cdot 7H_2O$
3.4 **IMCO/United Nations Numerical Designation:** Not listed

4. OBSERVABLE CHARACTERISTICS

4.1 **Physical State (as shipped):** Solid
4.2 **Color:** Rose, pink
4.3 **Odor:** None

5. HEALTH HAZARDS

5.1 **Personal Protective Equipment:** Bu. Mines approved respirator; goggles; protective gloves
5.2 **Symptoms Following Exposure:** Inhalation causes shortness of breath and coughing; permanent disability may occur. Ingestion causes pain and vomiting. Contact with eyes or skin causes irritation.
5.3 **Treatment for Exposure:** INHALATION: move to fresh air; if breathing has stopped, begin artificial respiration and call a doctor. INGESTION: give large amount of water; induce vomiting; call a doctor. EYES: flush with water for at least 15 min.; consult a physician if irritation persists. SKIN: flush with water.
5.4 **Toxicity by Inhalation (Threshold Limit Value):** 0.1 mg/m^3 (as cobalt)
5.5 **Short-Term Inhalation Limits:** Data not available
5.6 **Toxicity by Ingestion:** Grade 3; LD_{50} 50 – 500 mg/kg
5.7 **Late Toxicity:** Data not available
5.8 **Vapor (Gas) Irritant Characteristics:** Data not available
5.9 **Liquid or Solid Irritant Characteristics:** Data not available
5.10 **Odor Threshold:** Data not available

6. FIRE HAZARDS

6.1 **Flash Point:** Not flammable
6.2 **Flammable Limits in Air:** Not flammable
6.3 **Fire Extinguishing Agents:** Not pertinent
6.4 **Fire Extinguishing Agents Not to be Used:** Not pertinent
6.5 **Special Hazards of Combustion Products:** Toxic cobalt oxide fumes may form in fire.
6.6 **Behavior in Fire:**
6.7 **Ignition Temperature:** Not pertinent
6.8 **Electrical Hazard:** Not pertinent
6.9 **Burning Rate:** Not pertinent

7. CHEMICAL REACTIVITY

7.1 **Reactivity with Water:** No reaction
7.2 **Reactivity with Common Materials:**
7.3 **Stability During Transport:** Stable
7.4 **Neutralizing Agents for Acids and Caustics:** Not pertinent
7.5 **Polymerization:** Not pertinent
7.6 **Inhibitor of Polymerization:** Not pertinent

8. WATER POLLUTION

8.1 **Aquatic Toxicity:**
10 ppm*/**/stickleback/lethal/fresh water
*As cobalt
**Time period not specified.
8.2 **Waterfowl Toxicity:** Data not available
8.3 **Biological Oxygen Demand (BOD):** None
8.4 **Food Chain Concentration Potential:** Bioconcentration of 200 – 1000 fold only under constant exposure. Not significant in spill conditions.

9. SELECTED MANUFACTURERS

1. Mooney Chemicals, Inc.
2301 Scranton Road
Cleveland, Ohio 44113
2. The Harshaw Chemical Co.
1945 E. 97th St.
Cleveland, Ohio 44106
3. Mallinckrodt Chemical Works
223 Westside Ave.
P. O. Box 384
Jersey City, N. J. 07303

10. SHIPPING INFORMATION

10.1 **Grades or Purity:** Technical; reagent; may also be shipped as a monohydrate or hexahydrate.
10.2 **Storage Temperature:** Ambient
10.3 **Inert Atmosphere:** No requirement
10.4 **Venting:** Open

11. HAZARD ASSESSMENT CODE
(See Hazard Assessment Handbook, CG 446-3)

SS

12. HAZARD CLASSIFICATIONS

12.1 **Code of Federal Regulations:** Not listed
12.2 **NAS Hazard Rating for Bulk Water Transportation:** Not listed
12.3 **NFPA Hazard Classifications:** Not listed

13. PHYSICAL AND CHEMICAL PROPERTIES

13.1 **Physical State at 15°C and 1 atm:** Solid
13.2 **Molecular Weight:** 281.1
13.3 **Boiling Point at 1 atm:** Not pertinent (decomposes)
13.4 **Freezing Point:** Not pertinent
13.5 **Critical Temperature:** Not pertinent
13.6 **Critical Pressure:** Not pertinent
13.7 **Specific Gravity:** 1.948 at 20°C (solid)
13.8 **Liquid Surface Tension:** Not pertinent
13.9 **Liquid-Water Interfacial Tension:** Not pertinent
13.10 **Vapor (Gas) Specific Gravity:** Not pertinent
13.11 **Ratio of Specific Heats of Vapor (Gas):** Not pertinent
13.12 **Latent Heat of Vaporization:** Not pertinent
13.13 **Heat of Combustion:** Not pertinent
13.14 **Heat of Decomposition:** Not pertinent
13.15 **Heat of Solution:** 23 Btu/lb = 13 cal/g = 0.54×10^5 J/kg
13.16 **Heat of Polymerization:** Not pertinent

NOTES

CLD	# COLLODION

Common Synonyms			
Cellulose nitrate solution Nitrocellulose solution Pyroxylin solution Box toe gum Nitrocellulose gum	Thick liquid	Colorless	Ether-like odor
	Floats on water. Flammable, irritating vapor is produced. Boiling point is around 94° F.		

Shut off ignition sources. Call fire department.
Stop discharge if possible.
Isolate and remove discharged material.
Notify local health and pollution control agencies.

Fire	FLAMMABLE. POISONOUS GASES MAY BE PRODUCED IN FIRE. Containers may explode in fire. Flashback along vapor trail may occur. Vapor may explode if ignited in an enclosed area. **Extinguish with dry chemicals, alcohol foam, or carbon dioxide.** **Water may be ineffective on fire.** **Cool exposed containers with water.**
Exposure	**Call for medical aid.** VAPOR Irritating to eyes, nose and throat. If inhaled will cause dizziness, difficult breathing, or loss of consciousness. **Move victim to fresh air.** **If breathing has stopped, give artificial respiration.** **If breathing is difficult, give oxygen.** LIQUID Not harmful.
Water Pollution	Effect of low concentrations on aquatic life is unknown. Fouling to shoreline. May be dangerous if it enters water intakes. **Notify local health and wildlife officials.** **Notify operators of nearby water intakes.**

1. RESPONSE TO DISCHARGE
(See Response Methods Handbook, CG 446-4)

Issue warning—high flammability
Restrict access
Mechanical containment
Should be removed
Chemical and physical treatment

2. LABEL

Red

3. CHEMICAL DESIGNATIONS

3.1 **Synonyms:** Cellulose nitrate solution; Nitrocellulose solution; Pyroxylin solution; Box toe gum; Nitrocellulose gum
3.2 **Coast Guard Compatibility Classification:** Not applicable
3.3 **Chemical Formula:** Not pertinent
3.4 **IMCO/United Nations Numerical Designation:** 3.2/2059, 2061 3.3/2060, 2062

4. OBSERVABLE CHARACTERISTICS

4.1 **Physical State (as shipped):** Viscous liquid
4.2 **Color:** Colorless
4.3 **Odor:** Depends on solvent used; often that of ether.

5. HEALTH HAZARDS

5.1 **Personal Protective Equipment:** Self-contained breathing apparatus; rubber gloves; goggles or face shield.
5.2 **Symptoms Following Exposure:** High concentration of ether fumes may cause narcosis, loss of consciousness and respiratory paralysis if inhaled. Contact with eyes causes irritation.
5.3 **Treatment for Exposure:** INHALATION: remove victim to fresh air; initiate artificial respiration if breathing has stopped; call physician.
5.4 **Toxicity by Inhalation (Threshold Limit Value):** 400 ppm
5.5 **Short-Term Inhalation Limits:** Data not available
5.6 **Toxicity by Ingestion:** Grade 0; LD_{50} > 15 g/kg
5.7 **Late Toxicity:** Data not available
5.8 **Vapor (Gas) Irritant Characteristics:** Vapors cause a slight smarting of the eyes or respiratory system if present in high concentrations. The effect is temporary.
5.9 **Liquid or Solid Irritant Characteristics:** No appreciable hazard. Practically harmless to the skin.
5.10 **Odor Threshold:** Data not available

6. FIRE HAZARDS

6.1 **Flash Point:** −49°F C.C. (ether)
6.2 **Flammable Limits in Air:** 1.9% 36% (ether solution)
6.3 **Fire Extinguishing Agents:** Dry chemical, alcohol foam, carbon dioxide
6.4 **Fire Extinguishing Agents Not to be Used:** Water may be ineffective.
6.5 **Special Hazards of Combustion Products:** The formation of extremely toxic gases, notably oxides of nitrogen, hydrogen cyanide, and carbon monoxide is possible.
6.6 **Behavior in Fire:** Highly flammable solvent vapors are formed. May travel a long distance to a source of ignition and flash back.
6.7 **Ignition Temperature:** 356°F (ether)
6.8 **Electrical Hazard:** Class I, Group C
6.9 **Burning Rate:** Data not available

7. CHEMICAL REACTIVITY

7.1 **Reactivity with Water:** No reaction
7.2 **Reactivity with Common Materials:** No reaction
7.3 **Stability During Transport:** Stable
7.4 **Neutralizing Agents for Acids and Caustics:** Not pertinent
7.5 **Polymerization:** Not pertinent
7.6 **Inhibitor of Polymerization:** Not pertinent

8. WATER POLLUTION

8.1 **Aquatic Toxicity:** Data not available
8.2 **Waterfowl Toxicity:** Data not available
8.3 **Biological Oxygen Demand (BOD):** Data not available
8.4 **Food Chain Concentration Potential:** None

9. SELECTED MANUFACTURERS

1. Mallinckrodt Chemical Works
Second and Mallinckrodt Streets
St. Louis, Mo. 63160
2. Polysciences, Inc.
Paul Valley Industrial Park
Warrington, Pa. 18976

10. SHIPPING INFORMATION

10.1 **Grades or Purity:** USP. All grades contain less than 60% nitrocellulose by weight.
10.2 **Storage Temperature:** Ambient
10.3 **Inert Atmosphere:** No requirement
10.4 **Venting:** Pressure-vacuum

11. HAZARD ASSESSMENT CODE
(See Hazard Assessment Handbook, CG 446-3)

A-T-U-V-W

12. HAZARD CLASSIFICATIONS

12.1 **Code of Federal Regulations:** Flammable liquid
12.2 **NAS Hazard Rating for Bulk Water Transportation:**

Category	Rating
Fire	4
Health	
Vapor Irritant	1
Liquid or Solid Irritant	0
Poisons	2
Water Pollution	
Human Toxicity	0
Aquatic Toxicity	1
Aesthetic Effect	1
Reactivity	
Other Chemicals	1
Water	0
Self-Reaction	0

12.3 **NFPA Hazard Classifications:**

Category	Classification*	
Health Hazard (Blue)	0	2
Flammability (Red)	3	3
Reactivity (Yellow)	3	3

*First column refers to non-fire situation.

13. PHYSICAL AND CHEMICAL PROPERTIES

13.1 **Physical State at 15°C and 1 atm:** Liquid
13.2 **Molecular Weight:** Not pertinent
13.3 **Boiling Point at 1 atm:** 93°F = 34°C = 307°K (ether solvent)
13.4 **Freezing Point:** Not pertinent
13.5 **Critical Temperature:** Not pertinent
13.6 **Critical Pressure:** Not pertinent
13.7 **Specific Gravity:** 0.772 at 25°C (liquid)
13.8 **Liquid Surface Tension:** Not pertinent
13.9 **Liquid-Water Interfacial Tension:** Not pertinent
13.10 **Vapor (Gas) Specific Gravity:** Not pertinent
13.11 **Ratio of Specific Heats of Vapor (Gas):** Not pertinent
13.12 **Latent Heat of Vaporization:** Not pertinent
13.13 **Heat of Combustion:** Data not available
13.14 **Heat of Decomposition:** Not pertinent
13.15 **Heat of Solution:** Not pertinent
13.16 **Heat of Polymerization:** Not pertinent

NOTES

COPPER ACETATE

Common Synonyms	Solid	Bluish-green	Odorless
Acetic acid, cupric salt Crystallized verdigris Neutral verdigris Cupric acetate monohydrate	Mixes with water.		

Stop discharge if possible. Keep people away.
Avoid contact with solid and dust.
Isolate and remove discharged material.
Notify local health and pollution control agencies.

Fire	Not flammable. Irritating gases may be produced when heated.
Exposure	CALL FOR MEDICAL AID. DUST Irritating to eyes, nose and throat. If inhaled will cause coughing or difficult breathing. If in eyes, hold eyelids open and flush with plenty of water. If breathing has stopped, give artificial respiration. If breathing is difficult, give oxygen. SOLID Will burn eyes. Irritating to eyes. If swallowed will cause nausea, vomiting or loss of consciousness. Remove contaminated clothing and shoes. Flush affected areas with plenty of water. IF IN EYES, hold eyelids open and flush with plenty of water. IF SWALLOWED and victim is CONSCIOUS, have victim drink water or milk and have victim induce vomiting. IF SWALLOWED and victim is UNCONSCIOUS OR HAVING CONVULSIONS, do nothing except keep victim warm.
Water Pollution	Effect of low concentrations on aquatic life is unknown. May be dangerous if it enters water intakes. Notify local health and wildlife officials. Notify operators of nearby water intakes.

1. RESPONSE TO DISCHARGE
(See Response Methods Handbook, CG 446-4)

Issue warning—water contaminant
Disperse and flush

2. LABELS

No hazard label required by Code of Federal Regulations

3. CHEMICAL DESIGNATIONS

3.1 **Synonyms:** Acetic acid, cupric salt; Crystallized verdigris; Cupric acetate monohydrate; Neutral verdigris
3.2 **Coast Guard Compatibility Classification:** Not listed
3.3 **Chemical Formula:** $Cu(C_2H_3O_2)_2 \cdot H_2O$
3.4 **IMCO/United Nations Numerical Designation:** Not listed

4. OBSERVABLE CHARACTERISTICS

4.1 **Physical State (as shipped):** Solid
4.2 **Color:** Bluish green
4.3 **Odor:** None

5. HEALTH HAZARDS

5.1 **Personal Protective Equipment:** Dust mask; goggles or face shield; protective gloves
5.2 **Symptoms Following Exposure:** Inhalation of dust causes irritation of throat and lungs. Ingestion of large amounts causes violent vomiting and purging, intense pain, collapse, coma, convulsions, and paralysis. Contact with solutions irritates eyes; contact with solid causes severe eye surface injury and irritation of skin.
5.3 **Treatment for Exposure:** INHALATION: move to fresh air. INGESTION: give large amount of water; induce vomiting; get medical attention. EYES: flush with water for at least 15 min.; get medical attention if injury was caused by solid. SKIN: flush with water.
5.4 **Toxicity by Inhalation (Threshold Limit Value):** Data not available
5.5 **Short-Term Inhalation Limits:** Data not available
5.6 **Toxicity by Ingestion:** Grade 2; LD_{50} 0.5—5 g/kg (rat)
5.7 **Late Toxicity:** Causes degeneration of liver in dogs
5.8 **Vapor (Gas) Irritant Characteristics:** Data not available
5.9 **Liquid or Solid Irritant Characteristics:** Data not available
5.10 **Odor Threshold:** Data not available

6. FIRE HAZARDS

6.1 **Flash Point:** Not flammable
6.2 **Flammable Limits in Air:** Not flammable
6.3 **Fire Extinguishing Agents:** Not pertinent
6.4 **Fire Extinguishing Agents Not to be Used:** Not pertinent
6.5 **Special Hazards of Combustion Products:** Irritating vapors of acetic acid may form in fires.
6.6 **Behavior in Fire:**
6.7 **Ignition Temperature:** Not pertinent
6.8 **Electrical Hazard:** Not pertinent
6.9 **Burning Rate:** Not pertinent

7. CHEMICAL REACTIVITY

7.1 **Reactivity with Water:** No reaction
7.2 **Reactivity with Common Materials:**
7.3 **Stability During Transport:** Stable
7.4 **Neutralizing Agents for Acids and Caustics:** Not pertinent
7.5 **Polymerization:** Not pertinent
7.6 **Inhibitor of Polymerization:** Not pertinent

8. WATER POLLUTION

8.1 **Aquatic Toxicity:** Data not available
8.2 **Waterfowl Toxicity:** Data not available
8.3 **Biological Oxygen Demand (BOD):** Data not available
8.4 **Food Chain Concentration Potential:** Copper known to be accumulated by shellfish. Hazard to humans unknown.

9. SELECTED MANUFACTURERS

1. Allied Chemical Corp.
   General Chemical Div.
   40 Rector St.
   New York, N. Y. 10006
2. Union Carbide Corp.
   Chemicals and Plastics Div.
   270 Park Ave.
   New York, N. Y. 10017
3. J. T. Baker Chemical Co.
   Phillipsburg, N. J. 08865

10. SHIPPING INFORMATION

10.1 **Grades or Purity:** Technical, 95—99%; Reagent, 99%
10.2 **Storage Temperature:** Ambient
10.3 **Inert Atmosphere:** No requirement
10.4 **Venting:** Open

11. HAZARD ASSESSMENT CODE
(See Hazard Assessment Handbook, CG 446-3)

SS

12. HAZARD CLASSIFICATIONS

12.1 **Code of Federal Regulations:** Not listed
12.2 **NAS Hazard Rating for Bulk Water Transportation:** Not listed
12.3 **NFPA Hazard Classifications:** Not listed

13. PHYSICAL AND CHEMICAL PROPERTIES

13.1 **Physical State at 15°C and 1 atm:** Solid
13.2 **Molecular Weight:** 199.65
13.3 **Boiling Point at 1 atm:** Not pertinent (decomposes)
13.4 **Freezing Point:** 239°F = 115°C = 388°K
13.5 **Critical Temperature:** Not pertinent
13.6 **Critical Pressure:** Not pertinent
13.7 **Specific Gravity:** 1.9 at 20°C (solid)
13.8 **Liquid Surface Tension:** Not pertinent
13.9 **Liquid-Water Interfacial Tension:** Not pertinent
13.10 **Vapor (Gas) Specific Gravity:** Not pertinent
13.11 **Ratio of Specific Heats of Vapor (Gas):** Not pertinent
13.12 **Latent Heat of Vaporization:** Not pertinent
13.13 **Heat of Combustion:** Not pertinent
13.14 **Heat of Decomposition:** Not pertinent
13.15 **Heat of Solution:** Not pertinent
13.16 **Heat of Polymerization:** Not pertinent

NOTES

CAA	COPPER ACETOARSENITE

Common Synonyms	Solid Powder	Green	Odorless
Schweinfurth green Imperial green Paris green Emerald green Moss green	Sinks and mixes slowly with water.		

AVOID CONTACT WITH SOLID AND DUST. KEEP PEOPLE AWAY.
Stay upwind. Use water spray to "knock down" dust.
Wear goggles and dust respirator.
Isolate and remove discharged material.
Notify local health and pollution control agencies.

Fire

Not flammable.
POISONOUS GASES MAY BE PRODUCED WHEN HEATED.
Cool exposed containers with water.

Exposure

CALL FOR MEDICAL AID.

DUST
POISONOUS IF INHALED.
Irritating to eyes, nose and throat.
Move victim to fresh air.
If in eyes, hold eyelids open and flush with plenty of water.
If breathing is difficult, give oxygen.

SOLID
POISONOUS IF SWALLOWED.
Irritating to skin and eyes.
Remove contaminated clothing and shoes.
Flush affected areas with plenty of water.
IF IN EYES, hold eyelids open and flush with plenty of water.
IF SWALLOWED and victim is CONSCIOUS, have victim drink water or milk and have victim induce vomiting.
IF SWALLOWED and victim is UNCONSCIOUS OR HAVING CONVULSIONS, do nothing except keep victim warm.

Water Pollution

Effect of low concentrations on aquatic life is unknown.
May be dangerous if it enters water intakes.
Notify local health and wildlife officials.
Notify operators of nearby water intakes.

1. RESPONSE TO DISCHARGE

(See Response Methods Handbook, CG 446-4)

Issue warning—poison, water contaminant
Restrict access
Should be removed
Chemical and physical treatment

2. LABEL

3. CHEMICAL DESIGNATIONS

3.1 **Synonyms:** Imperial green, Moss green, Vienna green, Meadow green, Paris green, Mitis green, Kings green, Parrot green
3.2 **Coast Guard Compatibility Classification:** Not applicable
3.3 **Chemical Formula:** $3Cu(AsO_2)_2 \cdot Cu(C_2H_3O_2)_2$
3.4 **IMCO/United Nations Numerical Designation:** 6.1/1585

4. OBSERVABLE CHARACTERISTICS

4.1 **Physical State (as shipped):** Solid
4.2 **Color:** Emerald green
4.3 **Odor:** None

5. HEALTH HAZARDS

5.1 **Personal Protective Equipment:** Dust respirator; impervious gloves; goggles or face shield.
5.2 **Symptoms Following Exposure:** Dust causes eye irritation. Ingestion causes gastric disturbance, tremors, muscular cramps, and nervous collapse which may lead to death.
5.3 **Treatment for Exposure:** Following ingestion or unusually severe exposure to dust, get medical attention. Alert doctor to possibility of arsenic poisoning. EYES: flush with water for 15 min. SKIN: wash thoroughly with soap and water. INGESTION: give copious drafts of water and induce repeated vomiting. Give cathartic of 2 oz. Epsom salt in water.
5.4 **Toxicity by Inhalation (Threshold Limit Value):** 0.5 mg/m^3 (as arsenic)
5.5 **Short-Term Inhalation Limits:** Data not available
5.6 **Toxicity by Ingestion:** Grade 4; oral LD_{50} = 22 mg/kg (rat)
5.7 **Late Toxicity:** Arsenic poisoning
5.8 **Vapor (Gas) Irritant Characteristics:** Not pertinent
5.9 **Liquid or Solid Irritant Characteristics:** Data not available
5.10 **Odor Threshold:** Not pertinent

6. FIRE HAZARDS

6.1 **Flash Point:** Not flammable
6.2 **Flammable Limits in Air:** Not flammable
6.3 **Fire Extinguishing Agents:** Not pertinent
6.4 **Fire Extinguishing Agents Not to be Used:** Not pertinent
6.5 **Special Hazards of Combustion Products:** Poisonous, volatile arsenic oxides may be formed in fires.
6.6 **Behavior in Fire:** Not pertinent
6.7 **Ignition Temperature:** Not pertinent
6.8 **Electrical Hazard:** Not pertinent
6.9 **Burning Rate:** Not pertinent

7. CHEMICAL REACTIVITY

7.1 **Reactivity with Water:** No reaction
7.2 **Reactivity with Common Materials:** No reaction
7.3 **Stability During Transport:** Stable
7.4 **Neutralizing Agents for Acids and Caustics:** Not pertinent
7.5 **Polymerization:** Not pertinent
7.6 **Inhibitor of Polymerization:** Not pertinent

8. WATER POLLUTION

8.1 **Aquatic Toxicity:** Data not available
8.2 **Waterfowl Toxicity:** > 5,000 ppm LC_{50}
8.3 **Biological Oxygen Demand (BOD):** Data not available
8.4 **Food Chain Concentration Potential:** Data not available

9. SELECTED MANUFACTURERS

1. Los Angeles Chemical Company
   4545 Ardine Street
   South Gate, Calif. 90280
2. Chempar Chemical Co., Inc.
   260 Madison Avenue
   New York, N. Y. 10016

10. SHIPPING INFORMATION

10.1 **Grades or Purity:** Commercial
10.2 **Storage Temperature:** Ambient
10.3 **Inert Atmosphere:** No requirement
10.4 **Venting:** Open

11. HAZARD ASSESSMENT CODE

(See Hazard Assessment Handbook, CG 446-3)

II-SS

12. HAZARD CLASSIFICATIONS

12.1 **Code of Federal Regulations:** Poisonous, Class B
12.2 **NAS Hazard Rating for Bulk Water Transportation:** Not listed
12.3 **NFPA Hazard Classifications:** Not listed

13. PHYSICAL AND CHEMICAL PROPERTIES

13.1 **Physical State at 15°C and 1 atm:** Solid
13.2 **Molecular Weight:** 1014
13.3 **Boiling Point at 1 atm:** Decomposes
13.4 **Freezing Point:** Not pertinent
13.5 **Critical Temperature:** Not pertinent
13.6 **Critical Pressure:** Not pertinent
13.7 **Specific Gravity:** (est.) > 1.1 at 20°C (solid)
13.8 **Liquid Surface Tension:** Not pertinent
13.9 **Liquid-Water Interfacial Tension:** Not pertinent
13.10 **Vapor (Gas) Specific Gravity:** Not pertinent
13.11 **Ratio of Specific Heats of Vapor (Gas):** Not pertinent
13.12 **Latent Heat of Vaporization:** Not pertinent
13.13 **Heat of Combustion:** Not pertinent
13.14 **Heat of Decomposition:** Not pertinent
13.15 **Heat of Solution:** Not pertinent
13.16 **Heat of Polymerization:** Not pertinent

NOTES

CPA	COPPER ARSENITE

Common Synonyms			
Cupric arsenite Swedish green Cupric green Copper orthoarsenite	Solid Sinks in water.	Green	Odorless

AVOID CONTACT WITH SOLID AND DUST. KEEP PEOPLE AWAY.
Stay upwind. Use water spray to "knock down" dust.
Wear goggles and dust respirator.
Isolate and remove discharged material.
Notify local health and pollution control agencies.

Fire

Not flammable.
POISONOUS GASES MAY BE PRODUCED WHEN HEATED.
Cool exposed containers with water.

Exposure

CALL FOR MEDICAL AID.

DUST
POISONOUS IF INHALED.
Irritating to eyes, nose and throat.
Move victim to fresh air.
If in eyes, hold eyelids open and flush with plenty of water.
If breathing is difficult, give oxygen.

SOLID
POISONOUS IF SWALLOWED.
Irritating to skin and eyes.
Remove contaminated clothing and shoes.
Flush affected areas with plenty of water.
IF IN EYES, hold eyelids open and flush with plenty of water.
IF SWALLOWED and victim is CONSCIOUS, have victim drink water or milk and have victim induce vomiting.
IF SWALLOWED and victim is UNCONSCIOUS OR HAVING CONVULSIONS, do nothing except keep victim warm.

Water Pollution

Effect of low concentrations on aquatic life is unknown.
May be dangerous if it enters water intakes.
Notify local health and wildlife officials.
Notify operators of nearby water intakes.

1. RESPONSE TO DISCHARGE

(See Response Methods Handbook, CG 446-4)

Issue warning—poison, water contaminant
Restrict access
Should be removed
Chemical and physical treatment

2. LABEL

3. CHEMICAL DESIGNATIONS

3.1 **Synonyms:** Cupric arsenite; Swedish green; Scheele's green; Cupric green; Copper orthoarsenite
3.2 **Coast Guard Compatibility Classification:** Not applicable
3.3 **Chemical Formula:** $CuHAsO_3$
3.4 **IMCO/United Nations Numerical Designation:** 6.1/1586

4. OBSERVABLE CHARACTERISTICS

4.1 **Physical State (as shipped):** Solid
4.2 **Color:** Green; yellowish green
4.3 **Odor:** None

5. HEALTH HAZARDS

5.1 **Personal Protective Equipment:** Dust respirator; rubber gloves; goggles or face shield.
5.2 **Symptoms Following Exposure:** Dust irritates eyes. Ingestion causes gastric disturbance, tremors, muscular cramps, and nervous collapse that may cause death.
5.3 **Treatment for Exposure:** Following ingestion or unusually severe exposure to dust, get medical attention. Alert doctor to possibility of arsenic poisoning. EYES: flush with water for 15 min. SKIN: wash with soap and water. INGESTION: give large amounts of water; induce vomiting; give cathartic, such as 2 oz. of Epsom salt in water.
5.4 **Toxicity by Inhalation (Threshold Limit Value):** 0.5 mg/m^3 (as arsenic)
5.5 **Short-Term Inhalation Limits:** Data not available
5.6 **Toxicity by Ingestion:** Grade 3; LD_{50} 50 to 500 mg/kg
5.7 **Late Toxicity:** Arsenic poisoning
5.8 **Vapor (Gas) Irritant Characteristics:** Not pertinent
5.9 **Liquid or Solid Irritant Characteristics:** Data not available
5.10 **Odor Threshold:** Not pertinent

6. FIRE HAZARDS

6.1 **Flash Point:** Not flammable
6.2 **Flammable Limits in Air:** Not flammable
6.3 **Fire Extinguishing Agents:** Not pertinent
6.4 **Fire Extinguishing Agents Not to be Used:** Not pertinent
6.5 **Special Hazards of Combustion Products:** Poisonous, volatile arsenic oxides may be formed in fires.
6.6 **Behavior in Fire:** Not pertinent
6.7 **Ignition Temperature:** Not pertinent
6.8 **Electrical Hazard:** Not pertinent
6.9 **Burning Rate:** Not pertinent

7. CHEMICAL REACTIVITY

7.1 **Reactivity with Water:** No reaction
7.2 **Reactivity with Common Materials:** No reaction
7.3 **Stability During Transport:** Stable
7.4 **Neutralizing Agents for Acids and Caustics:** Not pertinent
7.5 **Polymerization:** Not pertinent
7.6 **Inhibitor of Polymerization:** Not pertinent

8. WATER POLLUTION

8.1 **Aquatic Toxicity:** Data not available
8.2 **Waterfowl Toxicity:** Data not available
8.3 **Biological Oxygen Demand (BOD):** Data not available
8.4 **Food Chain Concentration Potential:** Data not available

9. SELECTED MANUFACTURERS

1. Gallard-Schlesinger Chemical Manufacturing Co.
   584 Mineola Avenue
   Carle Place, N. Y. 11514
2. Cerac, Inc.
   13460 West Silver Spring Rd.
   Menomenee Falls, Wis. 53051

10. SHIPPING INFORMATION

10.1 **Grades or Purity:** Commercial
10.2 **Storage Temperature:** Ambient
10.3 **Inert Atmosphere:** No requirement
10.4 **Venting:** Open

11. HAZARD ASSESSMENT CODE

(See Hazard Assessment Handbook, CG 446-3)

II

12. HAZARD CLASSIFICATIONS

12.1 **Code of Federal Regulations:** Poisonous, Class B
12.2 **NAS Hazard Rating for Bulk Water Transportation:** Not listed
12.3 **NFPA Hazard Classifications:** Not listed

13. PHYSICAL AND CHEMICAL PROPERTIES

13.1 **Physical State at 15°C and 1 atm:** Solid
13.2 **Molecular Weight:** 277.4
13.3 **Boiling Point at 1 atm:** Decomposes
13.4 **Freezing Point:** Not pertinent
13.5 **Critical Temperature:** Not pertinent
13.6 **Critical Pressure:** Not pertinent
13.7 **Specific Gravity:** (est.) > 1.1 at 20°C (solid)
13.8 **Liquid Surface Tension:** Not pertinent
13.9 **Liquid-Water Interfacial Tension:** Not pertinent
13.10 **Vapor (Gas) Specific Gravity:** Not pertinent
13.11 **Ratio of Specific Heats of Vapor (Gas):** Not pertinent
13.12 **Latent Heat of Vaporization:** Not pertinent
13.13 **Heat of Combustion:** Not pertinent
13.14 **Heat of Decomposition:** Not pertinent
13.15 **Heat of Solution:** Not pertinent
13.16 **Heat of Polymerization:** Not pertinent

NOTES

CPB COPPER BROMIDE

Common Synonyms
Cupric bromide, anhydrous

Solid | Black | Odorless

Sinks and mixes with water.

Stop discharge if possible. Keep people away.
Avoid contact with solid and dust.
Isolate and remove discharged material.
Notify local health and pollution control agencies.

Fire

Not flammable.
Irritating gases may be produced when heated.

Exposure

CALL FOR MEDICAL AID.

DUST
Irritating to eyes, nose and throat.
If inhaled will cause coughing or difficult breathing.

If in eyes, hold eyelids open and flush with plenty of water.
If breathing has stopped, give artificial respiration.
If breathing is difficult, give oxygen.

SOLID
Will burn eyes.
Irritating to skin.

Remove contaminated clothing and shoes.
Flush affected areas with plenty of water.
IF IN EYES, hold eyelids open and flush with plenty of water.
IF SWALLOWED and victim is CONSCIOUS, have victim drink water or milk and have victim induce vomiting.
IF SWALLOWED and victim is UNCONSCIOUS OR HAVING CONVULSIONS, do nothing except keep victim warm.

Water Pollution

Effect of low concentrations on aquatic life is unknown.
May be dangerous if it enters water intakes.

Notify local health and wildlife officials.
Notify operators of nearby water intakes.

1. RESPONSE TO DISCHARGE

(See Response Methods Handbook, CG 446-4)

Issue warning—water contaminant
Disperse and flush

2. LABELS

No hazard label required by Code of Federal Regulations

3. CHEMICAL DESIGNATIONS

3.1 **Synonyms:** Cupric bromide, anhydrous
3.2 **Coast Guard Compatibility Classification:** Not listed
3.3 **Chemical Formula:** $CuBr_2$
3.4 **IMCO/United Nations Numerical Designation:** Not listed

4. OBSERVABLE CHARACTERISTICS

4.1 **Physical State (as shipped):** Solid
4.2 **Color:** Black
4.3 **Odor:** None

5. HEALTH HAZARDS

5.1 **Personal Protective Equipment:** Dust mask; goggles or face shield; protective gloves
5.2 **Symptoms Following Exposure:** Inhalation of dust causes irritation of throat and lungs. Ingestion of large amounts causes violent vomiting and purging, intense pain, collapse, coma, convulsions, and paralysis. Contact with solutions causes eye irritation; contact with solid causes severe eye surface injury and skin irritation.
5.3 **Treatment for Exposure:** INHALATION: move to fresh air. INGESTION: give large amount of water; induce vomiting; get medical attention. EYES: flush with water for at least 15 min.; get medical attention if injury was caused by solid. SKIN: flush with water.
5.4 **Toxicity by Inhalation (Threshold Limit Value):** Data not available
5.5 **Short-Term Inhalation Limits:** Data not available
5.6 **Toxicity by Ingestion:** Grade 3; LD_{50} 50—500 mg/kg
5.7 **Late Toxicity:** Data not available
5.8 **Vapor (Gas) Irritant Characteristics:** Data not available
5.9 **Liquid or Solid Irritant Characteristics:** Data not available
5.10 **Odor Threshold:** Data not available

6. FIRE HAZARDS

6.1 **Flash Point:** Not flammable
6.2 **Flammable Limits in Air:** Not flammable
6.3 **Fire Extinguishing Agents:** Not pertinent
6.4 **Fire Extinguishing Agents Not to be Used:** Not pertinent
6.5 **Special Hazards of Combustion Products:** Irritating hydrogen bromide gas may form in fire.
6.6 **Behavior in Fire:**
6.7 **Ignition Temperature:** Not pertinent
6.8 **Electrical Hazard:** Not pertinent
6.9 **Burning Rate:** Not pertinent

7. CHEMICAL REACTIVITY

7.1 **Reactivity with Water:** No reaction
7.2 **Reactivity with Common Materials:**
7.3 **Stability During Transport:** Stable
7.4 **Neutralizing Agents for Acids and Caustics:** Not pertinent
7.5 **Polymerization:** Not pertinent
7.6 **Inhibitor of Polymerization:** Not pertinent

8. WATER POLLUTION

8.1 **Aquatic Toxicity:** Data not available
8.2 **Waterfowl Toxicity:** Data not available
8.3 **Biological Oxygen Demand (BOD):** None
8.4 **Food Chain Concentration Potential:** Copper known to be accumulated by shellfish. Hazard to humans unknown.

9. SELECTED MANUFACTURERS

1. American Hoechst Corp.
   Chemicals and Plastics Div.
   Rt. 202—206 North
   Somerville, N. J. 08876
2. Mallinckrodt Chemical Works
   223 Westside Ave.
   P. O. Box 384
   Jersey City, N. J. 07303
3. J. T. Baker Chemical Co.
   Phillipsburg, N. J. 08865

10. SHIPPING INFORMATION

10.1 **Grades or Purity:** Pure, 99%, Reagent
10.2 **Storage Temperature:** Ambient
10.3 **Inert Atmosphere:** No requirement
10.4 **Venting:** Open

11. HAZARD ASSESSMENT CODE

(See Hazard Assessment Handbook, CG 446-3)

SS

12. HAZARD CLASSIFICATIONS

12.1 **Code of Federal Regulations:** Not listed
12.2 **NAS Hazard Rating for Bulk Water Transportation:** Not listed
12.3 **NFPA Hazard Classifications:** Not listed

13. PHYSICAL AND CHEMICAL PROPERTIES

13.1 **Physical State at 15°C and 1 atm:** Solid
13.2 **Molecular Weight:** 223.35
13.3 **Boiling Point at 1 atm:** Not pertinent (decomposes)
13.4 **Freezing Point:** 928°F = 498°C = 771°K
13.5 **Critical Temperature:** Not pertinent
13.6 **Critical Pressure:** Not pertinent
13.7 **Specific Gravity:** 4.77 at 20°C (solid)
13.8 **Liquid Surface Tension:** Not pertinent
13.9 **Liquid-Water Interfacial Tension:** Not pertinent
13.10 **Vapor (Gas) Specific Gravity:** Not pertinent
13.11 **Ratio of Specific Heats of Vapor (Gas):** Not pertinent
13.12 **Latent Heat of Vaporization:** Not pertinent
13.13 **Heat of Combustion:** Not pertinent
13.14 **Heat of Decomposition:** Not pertinent
13.15 **Heat of Solution:** –70.9 Btu/lb = –39.4 cal/g = -1.65×10^5 J/kg
13.16 **Heat of Polymerization:** Not pertinent

NOTES

CPC	COPPER CHLORIDE

Common Synonyms			
Cupric chloride dihydrate Eriocholcite (anhydrous)	Solid	Blue-green	Odorless
	Sinks and mixes with water.		

Stop discharge if possible. Keep people away.
Avoid contact with solid and dust.
Isolate and remove discharged material.
Notify local health and pollution control agencies.

Fire	Not flammable.
Exposure	CALL FOR MEDICAL AID. DUST Irritating to eyes, nose and throat. If inhaled will cause coughing or difficult breathing. If in eyes, hold eyelids open and flush with plenty of water. If breathing has stopped, give artificial respiration. If breathing is difficult, give oxygen. SOLID Will burn eyes. Irritating to eyes. If swallowed will cause nausea and vomiting. Remove contaminated clothing and shoes. Flush affected areas with plenty of water. IF IN EYES, hold eyelids open and flush with plenty of water. IF SWALLOWED and victim is CONSCIOUS, have victim drink water or milk and have victim induce vomiting. IF SWALLOWED and victim is UNCONSCIOUS OR HAVING CONVULSIONS, do nothing except keep victim warm.
Water Pollution	HARMFUL TO AQUATIC LIFE IN VERY LOW CONCENTRATIONS. May be dangerous if it enters water intakes. Notify local health and wildlife officials. Notify operators of nearby water intakes.

1. RESPONSE TO DISCHARGE
(See Response Methods Handbook, CG 446-4)

Issue warning—water contaminant
Disperse and flush

2. LABELS

No hazard label required by Code of Federal Regulations

3. CHEMICAL DESIGNATIONS

3.1 **Synonyms:** Cupric chloride dihydrate; Eriochalcite (anhydrous)
3.2 **Coast Guard Compatibility Classification:** Not listed
3.3 **Chemical Formula:** $CuCl_2 \cdot 2H_2O$
3.4 **IMCO/United Nations Numerical Designation:** Not listed

4. OBSERVABLE CHARACTERISTICS

4.1 **Physical State (as shipped):** Solid
4.2 **Color:** Green; blue-green; blue
4.3 **Odor:** None

5. HEALTH HAZARDS

5.1 **Personal Protective Equipment:** Bu. Mines approved respirator; rubber gloves; safety goggles
5.2 **Symptoms Following Exposure:** Inhalation causes coughing and sneezing. Ingestion causes pain and vomiting. Contact with solutions irritates eyes; contact with solid causes severe eye surface injury and skin irritation.
5.3 **Treatment for Exposure:** INHALATION: move to fresh air. INGESTION: give large amounts of water; induce vomiting; get medical attention. EYES: flush with water for 15 min.; consult a physician if injury was caused by solid. SKIN: flush with water.
5.4 **Toxicity by Inhalation (Threshold Limit Value):** Data not available
5.5 **Short-Term Inhalation Limits:** Data not available
5.6 **Toxicity by Ingestion:** Grade 3; LD_{50} 50—500 mg/kg
5.7 **Late Toxicity:** Causes liver damage in rabbits
5.8 **Vapor (Gas) Irritant Characteristics:** Data not available
5.9 **Liquid or Solid Irritant Characteristics:** Data not available
5.10 **Odor Threshold:** Data not available

6. FIRE HAZARDS

6.1 **Flash Point:** Not flammable
6.2 **Flammable Limits in Air:** Not flammable
6.3 **Fire Extinguishing Agents:** Not pertinent
6.4 **Fire Extinguishing Agents Not to be Used:** Not pertinent
6.5 **Special Hazards of Combustion Products:** Irritating hydrogen chloride gas may form in fire.
6.6 **Behavior in Fire:**
6.7 **Ignition Temperature:** Not pertinent
6.8 **Electrical Hazard:** Not pertinent
6.9 **Burning Rate:** Not pertinent

7. CHEMICAL REACTIVITY

7.1 **Reactivity with Water:** No reaction
7.2 **Reactivity with Common Materials:** In presence of moisture may corrode metals; the reaction is not hazardous.
7.3 **Stability During Transport:** Stable
7.4 **Neutralizing Agents for Acids and Caustics:** Flush with water, rinse with dilute solution of sodium bicarbonate or soda ash.
7.5 **Polymerization:** Not pertinent
7.6 **Inhibitor of Polymerization:** Not pertinent

8. WATER POLLUTION

8.1 **Aquatic Toxicity:**
0.009 ppm (as Cu)/*/goldfish/rapid death/fresh water
0.1—0.5 ppm/*/oyster/toxic/salt water
0.55 ppm/12 hr/mussel/killed/salt water
*Time period not specified.
8.2 **Waterfowl Toxicity:** Data not available
8.3 **Biological Oxygen Demand (BOD):** None
8.4 **Food Chain Concentration Potential:** Copper known to be accumulated by shellfish. Hazard to humans unknown.

9. SELECTED MANUFACTURERS

1. The Harshaw Chemical Co.
1945 E. 97th St.
Cleveland, Ohio 44106
2. J. T. Baker Chemical Co.
Phillipsburg, N. J. 08865
3. Diamond Shamrock Chemical Co.
Chemetals Division
711 Pittman Road
Baltimore, Md. 21226

10. SHIPPING INFORMATION

10.1 **Grades or Purity:** Reagent; C.P.; Technical
10.2 **Storage Temperature:** Ambient
10.3 **Inert Atmosphere:** No requirement
10.4 **Venting:** Open

11. HAZARD ASSESSMENT CODE
(See Hazard Assessment Handbook, CG 446-3)

SS

12. HAZARD CLASSIFICATIONS

12.1 **Code of Federal Regulations:** ORM–B
12.2 **NAS Hazard Rating for Bulk Water Transportation:** Not listed
12.3 **NFPA Hazard Classifications:**

Category	Classification
Health Hazard (Blue)	0
Flammability (Red)	0
Reactivity (Yellow)	0

13. PHYSICAL AND CHEMICAL PROPERTIES

13.1 **Physical State at 15°C and 1 atm:** Solid
13.2 **Molecular Weight:** 170.48 (dihydrate)
13.3 **Boiling Point at 1 atm:** Not pertinent (decomposes)
13.4 **Freezing Point:** Not pertinent
13.5 **Critical Temperature:** Not pertinent
13.6 **Critical Pressure:** Not pertinent
13.7 **Specific Gravity:** 2.54 at 20°C (solid)
13.8 **Liquid Surface Tension:** Not pertinent
13.9 **Liquid-Water Interfacial Tension:** Not pertinent
13.10 **Vapor (Gas) Specific Gravity:** Not pertinent
13.11 **Ratio of Specific Heats of Vapor (Gas):** Not pertinent
13.12 **Latent Heat of Vaporization:** Not pertinent
13.13 **Heat of Combustion:** Not pertinent
13.14 **Heat of Decomposition:** Not pertinent
13.15 **Heat of Solution:** Not pertinent
13.16 **Heat of Polymerization:** Not pertinent

NOTES

CCY — COPPER CYANIDE

Common Synonyms		
Cuprous cyanide Cupricin	Solid powder Sinks in water.	White

AVOID CONTACT WITH SOLID AND DUST. KEEP PEOPLE AWAY.
Stop discharge if possible.
Isolate and remove discharged material.
Notify local health and pollution control agencies.

Fire

Not flammable.
POISONOUS GASES MAY BE PRODUCED WHEN HEATED.

Exposure

CALL FOR MEDICAL AID.

DUST
Irritating to eyes, nose and throat.
If inhaled will cause dizziness or loss of consciousness.
If in eyes, hold eyelids open and flush with plenty of water.
If breathing has stopped, give artificial respiration.
If breathing is difficult, give oxygen.

SOLID
POISONOUS IF SWALLOWED.
Irritating to skin and eyes.
If swallowed will cause dizziness and loss of consciousness.
Remove contaminated clothing and shoes.
Flush affected areas with plenty of water.
IF IN EYES, hold eyelids open and flush with plenty of water.
IF SWALLOWED and victim is CONSCIOUS, have victim drink water or milk and have victim induce vomiting.
IF SWALLOWED and victim is UNCONSCIOUS OR HAVING CONVULSIONS, do nothing except keep victim warm.

Water Pollution

Effect of low concentrations on aquatic life is unknown.
May be dangerous if it enters water intakes.
Notify local health and wildlife officials.
Notify operators of nearby water intakes.

1. RESPONSE TO DISCHARGE

(See Response Methods Handbook, CG 446-4)

Issue warning—poison, water contaminant
Restrict access
Should be removed
Chemical and physical treatment

2. LABEL

3. CHEMICAL DESIGNATIONS

3.1 **Synonyms:** Cupricin; Cuprous cyanide
3.2 **Coast Guard Compatibility Classification:** Not listed
3.3 **Chemical Formula:** CuCN
3.4 **IMCO/United Nations Numerical Designation:** 6.1/1587

4. OBSERVABLE CHARACTERISTICS

4.1 **Physical State (as shipped):** Powder
4.2 **Color:** White
4.3 **Odor:** Data not available

5. HEALTH HAZARDS

5.1 **Personal Protective Equipment:** Dust respirator; protective goggles or face mask; protective clothing
5.2 **Symptoms Following Exposure:** Following severe exposure to dust, symptoms of cyanide poisoning may develop (see ingestion). Ingestion causes anxiety, confusion, dizziness, sudden loss of consciousness, odor of bitter almonds on breath or in vomitus, rapid weak pulse, convulsions, and paralysis. Contact with eyes causes irritation.
5.3 **Treatment for Exposure:** *Get medical attention after all exposures to this substance.* INHALATION: remove victim to fresh air. INGESTION: if breathing has stopped, begin artificial respiration immediately; administer by inhalation amyl nitrite pearls for 15—30 seconds of every minute, while a sodium nitrite solution is being prepared; discontinue amyl nitrite and immediately inject intravenously 10 ml of a 3% soln. of sodium nitrite (nonsterile if necessary) over a period of 2 to 4 min.; do not remove needle; through same needle infuse 50 ml of a 25% aqueous soln. of sodium thiosulfate; injection should take about 10 min. (Concentrations of 5—50% are permissible if total dose is approx. 12 grams.) Oxygen therapy may be of value in combination with the above. If symptoms recur, repeat injections of nitrite and thiosulfate at half the above doses. EYES: flush with water for at least 15 min. SKIN: flush with water; wash with soap and water.
5.4 **Toxicity by Inhalation (Threshold Limit Value):** 5 mg/m^3 (as cyanide)
5.5 **Short-Term Inhalation Limits:** Data not available
5.6 **Toxicity by Ingestion:** Grade 4; LD_{50} < 50 mg/kg
5.7 **Late Toxicity:** Data not available

6. FIRE HAZARDS

6.1 **Flash Point:** Not flammable
6.2 **Flammable Limits in Air:** Not flammable
6.3 **Fire Extinguishing Agents:** Not pertinent
6.4 **Fire Extinguishing Agents Not to be Used:** Not pertinent
6.5 **Special Hazards of Combustion Products:** Toxic hydrogen cyanide gas may form in fires.
6.6 **Behavior in Fire:**
6.7 **Ignition Temperature:** Not pertinent
6.8 **Electrical Hazard:** Not pertinent
6.9 **Burning Rate:** Not pertinent

7. CHEMICAL REACTIVITY

7.1 **Reactivity with Water:** No reaction
7.2 **Reactivity with Common Materials:**
7.3 **Stability During Transport:** Stable. In presence of moisture, toxic hydrogen cyanide gas may collect in enclosed spaces.
7.4 **Neutralizing Agents for Acids and Caustics:** Not pertinent
7.5 **Polymerization:** Not pertinent
7.6 **Inhibitor of Polymerization:** Not pertinent

8. WATER POLLUTION

8.1 **Aquatic Toxicity:** Data not available
8.2 **Waterfowl Toxicity:** Data not available
8.3 **Biological Oxygen Demand (BOD):** Data not available
8.4 **Food Chain Concentration Potential:** Copper known to be accumulated by shellfish. Hazard to humans unknown.

9. SELECTED MANUFACTURERS

1. J. T. Baker Chemical Co.
   Phillipsburg, N. J. 08865
2. Gallard-Schlesinger Chemical Mfg. Co.
   584 Mineola Ave.
   Carle Place, N. Y. 11514
3. Pfaltz and Bauer, Inc.
   126—04 Northern Blvd.
   Flushing, N. Y. 11368

10. SHIPPING INFORMATION

10.1 **Grades or Purity:** Technical; C.P.
10.2 **Storage Temperature:** Ambient
10.3 **Inert Atmosphere:** No requirement
10.4 **Venting:** Closed container

11. HAZARD ASSESSMENT CODE

(See Hazard Assessment Handbook, CG 446-3)

II

12. HAZARD CLASSIFICATIONS

12.1 **Code of Federal Regulations:** Poisonous solid, Class B
12.2 **NAS Hazard Rating for Bulk Water Transportation:** Not listed
12.3 **NFPA Hazard Classifications:** Not listed

13. PHYSICAL AND CHEMICAL PROPERTIES

13.1 **Physical State at 15°C and 1 atm:** Solid
13.2 **Molecular Weight:** 89.56
13.3 **Boiling Point at 1 atm:** Not pertinent (decomposes)
13.4 **Freezing Point:** Not pertinent
13.5 **Critical Temperature:** Not pertinent
13.6 **Critical Pressure:** Not pertinent
13.7 **Specific Gravity:** 2.92 at 20°C (solid)
13.8 **Liquid Surface Tension:** Not pertinent
13.9 **Liquid-Water Interfacial Tension:** Not pertinent
13.10 **Vapor (Gas) Specific Gravity:** Not pertinent
13.11 **Ratio of Specific Heats of Vapor (Gas):** Not pertinent
13.12 **Latent Heat of Vaporization:** Not pertinent
13.13 **Heat of Combustion:** Not pertinent
13.14 **Heat of Decomposition:** Not pertinent
13.15 **Heat of Solution:** Not pertinent
13.16 **Heat of Polymerization:** Not pertinent

5. HEALTH HAZARDS (Cont'd.)

5.8 **Vapor (Gas) Irritant Characteristics:** Data not available
5.9 **Liquid or Solid Irritant Characteristics:** Data not available
5.10 **Odor Threshold:** Data not available

CPF	# COPPER FLUOROBORATE

Common Synonyms			
Cupric fluoborate solution Copper borofluoride solution Copper[II] fluoborate solution	Liquid	Dark blue	Odorless
	Sinks and mixes with water.		

Stop discharge if possible. Keep people away.
Avoid contact with liquid and vapor.
Isolate and remove discharged material.
Notify local health and pollution control agencies.

Fire	Not flammable. Irritating gases may be produced when heated.
Exposure	CALL FOR MEDICAL AID. VAPOR Irritating to eyes, nose and throat. If inhaled will cause coughing or difficult breathing. If in eyes, hold eyelids open and flush with plenty of water. If breathing has stopped, give artificial respiration. If breathing is difficult, give oxygen. LIQUID Irritating to skin and eyes. If swallowed will cause nausea and vomiting. Remove contaminated clothing and shoes. Flush affected areas with plenty of water. IF IN EYES, hold eyelids open and flush with plenty of water. IF SWALLOWED and victim is CONSCIOUS, have victim drink water or milk and have victim induce vomiting. IF SWALLOWED and victim is UNCONSCIOUS OR HAVING CONVULSIONS, do nothing except keep victim warm.
Water Pollution	Effect of low concentrations on aquatic life is unknown. May be dangerous if it enters water intakes. Notify local health and wildlife officials. Notify operators of nearby water intakes.

1. RESPONSE TO DISCHARGE

(See Response Methods Handbook, CG 446-4)

Issue warning—water contaminant
Disperse and flush

2. LABELS

No hazard label required by Code of Federal Regulations

3. CHEMICAL DESIGNATIONS

3.1 **Synonyms:** Copper borofluoride solution; Copper (II) fluoborate solution; Cupric fluoborate solution
3.2 **Coast Guard Compatibility Classification:** Not listed
3.3 **Chemical Formula:** $Cu(BF_4)_2—H_2O$
3.4 **IMCO/United Nations Numerical Designation:** Not listed

4. OBSERVABLE CHARACTERISTICS

4.1 **Physical State (as shipped):** Liquid
4.2 **Color:** Clear, dark blue
4.3 **Odor:** None

5. HEALTH HAZARDS

5.1 **Personal Protective Equipment:** Goggles or face shield; rubber apron and gloves
5.2 **Symptoms Following Exposure:** Inhalation of mist irritates nose and throat. Ingestion causes pain and vomiting. Contact causes severe irritation of eyes and irritation of skin.
5.3 **Treatment for Exposure:** INHALATION: move to fresh air. INGESTION: give large amounts of water and induce vomiting if required. EYES: flush with water for at least 15 min.; get medical attention if irritation persists. SKIN: flush with water.
5.4 **Toxicity by Inhalation (Threshold Limit Value):** Data not available
5.5 **Short-Term Inhalation Limits:** Data not available
5.6 **Toxicity by Ingestion:** Grade 3; LD_{50} 50—500 mg/kg
5.7 **Late Toxicity:** Data not available
5.8 **Vapor (Gas) Irritant Characteristics:** Data not available
5.9 **Liquid or Solid Irritant Characteristics:** Data not available
5.10 **Odor Threshold:** Data not available

6. FIRE HAZARDS

6.1 **Flash Point:** Not flammable
6.2 **Flammable Limits in Air:** Not flammable
6.3 **Fire Extinguishing Agents:** Not pertinent
6.4 **Fire Extinguishing Agents Not to be Used:** Not pertinent
6.5 **Special Hazards of Combustion Products:** Irritating hydrogen fluoride gas may form in fires.
6.6 **Behavior in Fire:**
6.7 **Ignition Temperature:** Not pertinent
6.8 **Electrical Hazard:** Not pertinent
6.9 **Burning Rate:** Not pertinent

7. CHEMICAL REACTIVITY

7.1 **Reactivity with Water:** No reaction
7.2 **Reactivity with Common Materials:** May corrode some metals
7.3 **Stability During Transport:** Stable
7.4 **Neutralizing Agents for Acids and Caustics:** Flush with water, rinse with dilute solution of sodium bicarbonate or soda ash.
7.5 **Polymerization:** Not pertinent
7.6 **Inhibitor of Polymerization:** Not pertinent

8. WATER POLLUTION

8.1 **Aquatic Toxicity:** Data not available
8.2 **Waterfowl Toxicity:** Data not available
8.3 **Biological Oxygen Demand (BOD):** None
8.4 **Food Chain Concentration Potential:** Copper known to be accumulated by shellfish. Hazard to humans unknown.

9. SELECTED MANUFACTURERS

1. American Hoechst Corp.
Chemicals and Plastics Div.
Rt. 202—206 North
Somerville, N. J. 08876
2. Allied Chemical Corp.
P. O. Box 1087R
Morristown, N. J. 07960
3. The Harshaw Chemical Co.
1945 E. 97 St.
Cleveland, Ohio 44106

10. SHIPPING INFORMATION

10.1 **Grades or Purity:**
Technical } 45—50% solutions in water
C.P. }
10.2 **Storage Temperature:** Ambient
10.3 **Inert Atmosphere:** No requirement
10.4 **Venting:** Open

11. HAZARD ASSESSMENT CODE

(See Hazard Assessment Handbook, CG 446-3)

A-P

12. HAZARD CLASSIFICATIONS

12.1 **Code of Federal Regulations:** Not listed
12.2 **NAS Hazard Rating for Bulk Water Transportation:** Not listed
12.3 **NFPA Hazard Classifications:**

Category	Classification
Health Hazard (Blue)	1
Flammability (Red)	0
Reactivity (Yellow)	0

13. PHYSICAL AND CHEMICAL PROPERTIES

13.1 **Physical State at 15°C and 1 atm:** Liquid
13.2 **Molecular Weight:** 237.16 (solute only)
13.3 **Boiling Point at 1 atm:** (approx.) 212°F = 100°C = 373°K
13.4 **Freezing Point:** Data not available
13.5 **Critical Temperature:** Not pertinent
13.6 **Critical Pressure:** Not pertinent
13.7 **Specific Gravity:** 1.54 at 20°C (liquid)
13.8 **Liquid Surface Tension:** Data not available
13.9 **Liquid-Water Interfacial Tension:** Not pertinent
13.10 **Vapor (Gas) Specific Gravity:** Not pertinent
13.11 **Ratio of Specific Heats of Vapor (Gas):** Not pertinent
13.12 **Latent Heat of Vaporization:** Not pertinent
13.13 **Heat of Combustion:** Not pertinent
13.14 **Heat of Decomposition:** Not pertinent
13.15 **Heat of Solution:** Not pertinent
13.16 **Heat of Polymerization:** Not pertinent

NOTES

CID — COPPER IODIDE

Common Synonyms			
Cuprous iodide	Solid	Tan or off-white	Odorless
	Sinks in water.		

AVOID CONTACT WITH SOLID AND DUST. KEEP PEOPLE AWAY.
Stop discharge if possible.
Isolate and remove discharged material.
Notify local health and pollution control agencies.

Fire

Not flammable.
Irritating gases may be produced when heated.

Exposure

CALL FOR MEDICAL AID.

DUST
Irritating to eyes, nose and throat.
If inhaled will cause coughing or difficult breathing.
If in eyes, hold eyelids open and flush with plenty of water.
If breathing has stopped, give artificial respiration.
If breathing is difficult, give oxygen.

SOLID
POISONOUS IF SWALLOWED.
Irritating to skin and eyes.
If swallowed will cause nausea, vomiting or loss of consciousness.
Remove contaminated clothing and shoes.
Flush affected areas with plenty of water.
IF IN EYES, hold eyelids open and flush with plenty of water.
IF SWALLOWED and victim is CONSCIOUS, have victim drink water or milk and have victim induce vomiting.
IF SWALLOWED and victim is UNCONSCIOUS OR HAVING CONVULSIONS, do nothing except keep victim warm.

Water Pollution

Effect of low concentrations on aquatic life is unknown.
May be dangerous if it enters water intakes.
Notify local health and wildlife officials.
Notify operators of nearby water intakes.

1. RESPONSE TO DISCHARGE
(See Response Methods Handbook, CG 446-4)

Should be removed
Chemical and physical treatment

2. LABELS

No hazard label required by Code of Federal Regulations

3. CHEMICAL DESIGNATIONS

3.1 **Synonyms:** Cuprous iodide; Marshite
3.2 **Coast Guard Compatibility Classification:** Not listed
3.3 **Chemical Formula:** CuI
3.4 **IMCO/United Nations Numerical Designation:** Not listed

4. OBSERVABLE CHARACTERISTICS

4.1 **Physical State (as shipped):** Solid
4.2 **Color:** Beige
4.3 **Odor:** None

5. HEALTH HAZARDS

5.1 **Personal Protective Equipment:** Dust mask; goggles or face shield; protective gloves
5.2 **Symptoms Following Exposure:** Inhalation causes irritation of nose and throat. Ingestion of copper salts produces violent vomiting and purging, intense pain, collapse, coma, convulsions, and paralysis. Contact with eyes or skin causes irritation.
5.3 **Treatment for Exposure:** INHALATION: move to fresh air. INGESTION: give large amounts of water; induce vomiting; get medical attention. EYES: flush with water for at least 15 min. SKIN: flush with water; wash with soap and water.
5.4 **Toxicity by Inhalation (Threshold Limit Value):** Data not available
5.5 **Short-Term Inhalation Limits:** Data not available
5.6 **Toxicity by Ingestion:** Grade 3; LD_{50} 50—500 mg/kg
5.7 **Late Toxicity:** Data not available
5.8 **Vapor (Gas) Irritant Characteristics:** Data not available
5.9 **Liquid or Solid Irritant Characteristics:** Data not available
5.10 **Odor Threshold:** Data not available

6. FIRE HAZARDS

6.1 **Flash Point:** Not flammable
6.2 **Flammable Limits in Air:** Not flammable
6.3 **Fire Extinguishing Agents:** Not pertinent
6.4 **Fire Extinguishing Agents Not to be Used:** Not pertinent
6.5 **Special Hazards of Combustion Products:** Irritating hydrogen iodide or iodine vapors may form in fire.
6.6 **Behavior in Fire:**
6.7 **Ignition Temperature:** Not pertinent
6.8 **Electrical Hazard:** Not pertinent
6.9 **Burning Rate:** Not pertinent

7. CHEMICAL REACTIVITY

7.1 **Reactivity with Water:** No reaction
7.2 **Reactivity with Common Materials:**
7.3 **Stability During Transport:** Stable
7.4 **Neutralizing Agents for Acids and Caustics:** Not pertinent
7.5 **Polymerization:** Not pertinent
7.6 **Inhibitor of Polymerization:** Not pertinent

8. WATER POLLUTION

8.1 **Aquatic Toxicity:** Data not available
8.2 **Waterfowl Toxicity:** Data not available
8.3 **Biological Oxygen Demand (BOD):** Data not available
8.4 **Food Chain Concentration Potential:** Copper known to be accumulated by shellfish. Hazard to humans unknown.

9. SELECTED MANUFACTURERS

1. Gallard-Schlesinger Chemical Mfg. Co.
   584 Mineola Avenue
   Carle Place, N. Y. 11514
2. Pfaltz and Bauer, Inc.
   126—04 Northern Blvd.
   Flushing, N. Y. 11368
3. Ventron Corp.
   P. O. Box 159
   Beverly, Mass. 01915

10. SHIPPING INFORMATION

10.1 **Grades or Purity:** Anhydrous, 99+%
10.2 **Storage Temperature:** Ambient
10.3 **Inert Atmosphere:** No requirement
10.4 **Venting:** Open

11. HAZARD ASSESSMENT CODE
(See Hazard Assessment Handbook, CG 446-3)

II

12. HAZARD CLASSIFICATIONS

12.1 **Code of Federal Regulations:** Not listed
12.2 **NAS Hazard Rating for Bulk Water Transportation:** Not listed
12.3 **NFPA Hazard Classifications:** Not listed

13. PHYSICAL AND CHEMICAL PROPERTIES

13.1 **Physical State at 15°C and 1 atm:** Solid
13.2 **Molecular Weight:** 190.4
13.3 **Boiling Point at 1 atm:** 2,354°F = 1,290°C = 1,563°K
13.4 **Freezing Point:** 1,121°F = 605°C = 878°K
13.5 **Critical Temperature:** Not pertinent
13.6 **Critical Pressure:** Not pertinent
13.7 **Specific Gravity:** 5.62 at 20°C (solid)
13.8 **Liquid Surface Tension:** Not pertinent
13.9 **Liquid-Water Interfacial Tension:** Not pertinent
13.10 **Vapor (Gas) Specific Gravity:** Not pertinent
13.11 **Ratio of Specific Heats of Vapor (Gas):** Not pertinent
13.12 **Latent Heat of Vaporization:** Not pertinent
13.13 **Heat of Combustion:** Not pertinent
13.14 **Heat of Decomposition:** Not pertinent
13.15 **Heat of Solution:** Not pertinent
13.16 **Heat of Polymerization:** Not pertinent

NOTES

CNN | COPPER NAPHTHENATE

Common Synonyms			
Paint drier	Liquid	Dark green	Gasoline-like odor
	May float or sink in water.		

Shut off ignition sources. Call fire department.
Stop discharge if possible.
Isolate and remove discharged material.
Notify local health and pollution control agencies.

Fire	Combustible. **Extinguish with dry chemicals, foam or carbon dioxide.** **Water may be ineffective on fire.** **Cool exposed containers with water.**
Exposure	**Call for medical aid.** LIQUID Irritating to skin and eyes. Harmful if swallowed. **Remove contaminated clothing and shoes.** **Flush affected areas with plenty of water.** **IF IN EYES, hold eyelids open and flush with plenty of water.** **IF SWALLOWED and victim is CONSCIOUS, have victim drink water or milk.** **DO NOT INDUCE VOMITING.**
Water Pollution	HARMFUL TO AQUATIC LIFE IN VERY LOW CONCENTRATIONS. Fouling to shoreline. May be dangerous if it enters water intakes. **Notify local health and wildlife officials.** **Notify operators of nearby water intakes.**

1. RESPONSE TO DISCHARGE
(See Response Methods Handbook, CG 446-4)

Issue warning—water contaminant
Mechanical containment
Should be removed
Chemical and physical treatment

2. LABELS

No label required by Code of Federal Regulations

3. CHEMICAL DESIGNATIONS

3.1 **Synonyms:** Paint drier
3.2 **Coast Guard Compatibility Classification:** Not applicable
3.3 **Chemical Formula:** Mixture
3.4 **IMCO/United Nations Numerical Designation:** 3.3/1168

4. OBSERVABLE CHARACTERISTICS

4.1 **Physical State (as shipped):** Liquid
4.2 **Color:** Dark green
4.3 **Odor:** Like gasoline; slight aromatic

5. HEALTH HAZARDS

5.1 **Personal Protective Equipment:** Goggles or face shield; plastic gloves (as for gasoline).
5.2 **Symptoms Following Exposure:** Vapor causes mild irritation of eyes and mild irritation of respiratory tract if inhaled. Ingestion causes irritation of stomach. Aspiration causes severe lung irritation and rapidly developing pulmonary edema; central nervous system excitement followed by depression.
5.3 **Treatment for Exposure:** INHALATION: remove victim to fresh air. EYES: wash with copious amounts of water for at least 15 min. SKIN: wipe off and wash with soap and water. INGESTION: do NOT induce vomiting; guard against aspiration into lungs. ASPIRATION: enforce bed rest; give oxygen; call a doctor.
5.4 **Toxicity by Inhalation (Threshold Limit Value):** 500 ppm
5.5 **Short-Term Inhalation Limits:** Data not available
5.6 **Toxicity by Ingestion:** Grade 1; oral rat LD_{50} = 4-6 g/kg
5.7 **Late Toxicity:** Data not available
5.8 **Vapor (Gas) Irritant Characteristics:** Vapors are non-irritating to the eyes and throat.
5.9 **Liquid or Solid Irritant Characteristics:** Minimum hazard. If spilled on clothing and allowed to remain, may cause smarting and reddening of skin.
5.10 **Odor Threshold:** Data not available

6. FIRE HAZARDS

6.1 **Flash Point:** 100°F C.C. (typical)
6.2 **Flammable Limits in Air:** 0.8%–5.0% (mineral spirits)
6.3 **Fire Extinguishing Agents:** Dry chemical, foam, carbon dioxide
6.4 **Fire Extinguishing Agents Not to be Used:** Water may be ineffective.
6.5 **Special Hazards of Combustion Products:** Not pertinent
6.6 **Behavior in Fire:** Not pertinent
6.7 **Ignition Temperature:** 540°F (mineral spirits)
6.8 **Electrical Hazard:** Not pertinent
6.9 **Burning Rate:** 4 mm/min.

7. CHEMICAL REACTIVITY

7.1 **Reactivity with Water:** No reaction
7.2 **Reactivity with Common Materials:** No reaction
7.3 **Stability During Transport:** Stable
7.4 **Neutralizing Agents for Acids and Caustics:** Not pertinent
7.5 **Polymerization:** Not pertinent
7.6 **Inhibitor of Polymerization:** Not pertinent

8. WATER POLLUTION

8.1 **Aquatic Toxicity:** 2.0 ppm/72 hr/blue-green algae/100% kill/fresh water
8.2 **Waterfowl Toxicity:** Data not available
8.3 **Biological Oxygen Demand (BOD):** 8%, 5 days
8.4 **Food Chain Concentration Potential:** None

9. SELECTED MANUFACTURERS

1. Witco Chemical Corporation
   Organics Division
   277 Park Avenue
   New York, N. Y. 10017
2. Cincinnati Milacron Chemicals Inc.
   500 Jersey Ave.
   New Brunswick, N. J. 08903
3. Ferro Corporation
   7050 Krick Rd.
   Bedford, Ohio 44146

10. SHIPPING INFORMATION

10.1 **Grades or Purity:** 8% in mineral spirits or mineral oil. 5% in mineral spirits. May float instead of sink in water.
10.2 **Storage Temperature:** Ambient
10.3 **Inert Atmosphere:** No requirement
10.4 **Venting:** Pressure-vacuum or open (flame arrester)

11. HAZARD ASSESSMENT CODE
(See Hazard Assessment Handbook, CG 446-3)

A-T-U-X-Y

12. HAZARD CLASSIFICATIONS

12.1 **Code of Federal Regulations:** Not listed
12.2 **NAS Hazard Rating for Bulk Water Transportation:** Not listed
12.3 **NFPA Hazard Classifications:** Not listed

13. PHYSICAL AND CHEMICAL PROPERTIES

13.1 **Physical State at 15°C and 1 atm:** Liquid
13.2 **Molecular Weight:** Mixture
13.3 **Boiling Point at 1 atm:** 310–395°F = 154–202°C = 427–475°K
13.4 **Freezing Point:** Not pertinent
13.5 **Critical Temperature:** Not pertinent
13.6 **Critical Pressure:** Not pertinent
13.7 **Specific Gravity:** 0.93–1.05 at 25°C (liquid)
13.8 **Liquid Surface Tension:** 20 dynes/cm = 0.020 N/m at 20°C
13.9 **Liquid-Water Interfacial Tension:** 45 dynes/cm = 0.045 N/m at 20°C
13.10 **Vapor (Gas) Specific Gravity:** Not pertinent
13.11 **Ratio of Specific Heats of Vapor (Gas):** Not pertinent
13.12 **Latent Heat of Vaporization:** Not pertinent
13.13 **Heat of Combustion:** (est.) –17,600 Btu/lb = –9,800 cal/g = -410×10^5 J/kg
13.14 **Heat of Decomposition:** Not pertinent
13.15 **Heat of Solution:** Not pertinent
13.16 **Heat of Polymerization:** Not pertinent

NOTES

CNI COPPER NITRATE

Common Synonyms			
Cupric nitrate trihydrate Gerhardite	Solid	Blue	Odorless
	Sinks and mixes with water.		

Avoid contact with solid and dust. Keep people away.
Stop discharge if possible.
Isolate and remove discharged material.
Notify local health and pollution control agencies.

Fire

Not flammable.
Will increase the intensity of a fire.
Irritating gases may be produced when heated.
Flood discharge area with water.

Exposure

CALL FOR MEDICAL AID.

DUST
Irritating to eyes, nose and throat.
If inhaled will cause coughing or difficult breathing.

If in eyes, hold eyelids open and flush with plenty of water.
If breathing has stopped, give artificial respiration.
If breathing is difficult, give oxygen.

SOLID
Will burn eyes.
If swallowed will cause nausea, vomiting or loss of consciousness.

Remove contaminated clothing and shoes.
Flush affected areas with plenty of water.
IF IN EYES, hold eyelids open and flush with plenty of water.
IF SWALLOWED and victim is CONSCIOUS, have victim drink water or milk and have victim induce vomiting.
IF SWALLOWED and victim is UNCONSCIOUS OR HAVING CONVULSIONS, do nothing except keep victim warm.

Water Pollution

HARMFUL TO AQUATIC LIFE IN VERY LOW CONCENTRATIONS.
May be dangerous if it enters water intakes.

Notify local health and wildlife officials.
Notify operators of nearby water intakes.

1. RESPONSE TO DISCHARGE

(See Response Methods Handbook, CG 446-4)

Issue warning—water contaminant
Disperse and flush

2. LABELS

No hazard label required by Code of Federal Regulations

3. CHEMICAL DESIGNATIONS

3.1 **Synonyms:** Cupric nitrate trihydrate; Gerhardite
3.2 **Coast Guard Compatibility Classification:** Not listed
3.3 **Chemical Formula:** $Cu(NO_3)_2 \cdot 3H_2O$
3.4 **IMCO/United Nations Numerical Designation:** Not listed

4. OBSERVABLE CHARACTERISTICS

4.1 **Physical State (as shipped):** Solid
4.2 **Color:** Blue
4.3 **Odor:** None

5. HEALTH HAZARDS

5.1 **Personal Protective Equipment:** Dust mask; goggles or face shield; protective gloves
5.2 **Symptoms Following Exposure:** Inhalation causes irritation of throat and lungs. Ingestion of large amounts causes violent vomiting and purging, intense pain, collapse, coma, convulsions, and paralysis. Solutions irritate eyes; contact with solid causes severe eye surface injury and skin irritation.
5.3 **Treatment for Exposure:** INHALATION: move to fresh air. INGESTION: give large amounts of water; induce vomiting; get medical attention. EYES: flush with water for at least 15 min.; get medical attention if injury was caused by solid. SKIN: flush with water.
5.4 **Toxicity by Inhalation(Threshold Limit Value):** Data not available
5.5 **Short-Term Inhalation Limits:** Data not available
5.6 **Toxicity by Ingestion:** Grade 2; LD_{50} 0.5—5 g/kg
5.7 **Late Toxicity:** Data not available
5.8 **Vapor (Gas) Irritant Characteristics:** Data not available
5.9 **Liquid or Solid Irritant Characteristics:** Data not available
5.10 **Odor Threshold:** Data not available

6. FIRE HAZARDS

6.1 **Flash Point:** Not flammable
6.2 **Flammable Limits in Air:** Not flammable
6.3 **Fire Extinguishing Agents:** Not pertinent
6.4 **Fire Extinguishing Agents Not to be Used:** Not pertinent
6.5 **Special Hazards of Combustion Products:** Toxic and irritating oxides of nitrogen may form in fire.
6.6 **Behavior in Fire:** Can increase intensity of fire if in contact with combustible material.
6.7 **Ignition Temperature:** Not pertinent
6.8 **Electrical Hazard:** Not pertinent
6.9 **Burning Rate:** Not pertinent

7. CHEMICAL REACTIVITY

7.1 **Reactivity with Water:** No reaction
7.2 **Reactivity with Common Materials:** Mixtures with wood, paper, and other combustibles may catch fire.
7.3 **Stability During Transport:** Stable
7.4 **Neutralizing Agents for Acids and Caustics:** Not pertinent
7.5 **Polymerization:** Not pertinent
7.6 **Inhibitor of Polymerization:** Not pertinent

8. WATER POLLUTION

8.1 **Aquatic Toxicity:**
5 ppm/11 hr/rainbow trout/killed/fresh water
0.22—1.0 ppm/48—240 hr/barnacles & related species/killed/salt water
1.9 ppm/96 hr/oyster/TL_m/salt water
8.2 **Waterfowl Toxicity:** Data not available
8.3 **Biological Oxygen Demand (BOD):** None
8.4 **Food Chain Concentration Potential:** Copper known to be accumulated by shellfish. Can be concentrated by food chain. Hazard to humans unknown.

9. SELECTED MANUFACTURERS

1. American Hoechst Corp.
Chemicals and Plastics Div.
Rt. 202—206 North
Somerville, N. J. 08876
2. Allied Chemical Corp.
Specialty Chemicals Div.
P. O. Box 1087R
Morristown, N. J. 07960
3. J. T. Baker Chemical Co.
Phillipsburg, N. J. 08865

10. SHIPPING INFORMATION

10.1 **Grades or Purity:** Pure, 100%; Technical; Reagent. May also be shipped as anhydrous grade.
10.2 **Storage Temperature:** Ambient
10.3 **Inert Atmosphere:** No requirement
10.4 **Venting:** Open

11. HAZARD ASSESSMENT CODE

(See Hazard Assessment Handbook, CG 446-3)

SS

12. HAZARD CLASSIFICATIONS

12.1 **Code of Federal Regulations:** Not listed
12.2 **NAS Hazard Rating for Bulk Water Transportation:** Not listed
12.3 **NFPA Hazard Classifications:**

Category	Classification*	
Health Hazard (Blue)	0	1
Flammability (Red)	0	0
Reactivity (Yellow)	0	0
	oxy	oxy

*First column refers to nonfire situation.

13. PHYSICAL AND CHEMICAL PROPERTIES

13.1 **Physical State at 15°C and 1 atm:** Solid
13.2 **Molecular Weight:** 241.60
13.3 **Boiling Point at 1 atm:** Not pertinent (decomposes)
13.4 **Freezing Point:** 238.1°F = 114.5°C = 387.7°K
13.5 **Critical Temperature:** Not pertinent
13.6 **Critical Pressure:** Not pertinent
13.7 **Specific Gravity:** 2.32 at 20°C (solid)
13.8 **Liquid Surface Tension:** Not pertinent
13.9 **Liquid-Water Interfacial Tension:** Not pertinent
13.10 **Vapor (Gas) Specific Gravity:** Not pertinent
13.11 **Ratio of Specific Heats of Vapor (Gas):** Not pertinent
13.12 **Latent Heat of Vaporization:** Not pertinent
13.13 **Heat of Combustion:** Not pertinent
13.14 **Heat of Decomposition:** Not pertinent
13.15 **Heat of Solution:** Not pertinent
13.16 **Heat of Polymerization:** Not pertinent

NOTES

COPPER OXALATE

COL

Common Synonyms			
Cupric oxalate hemihydrate	Solid	Bluish white	Odorless
	Sinks in water.		

Stop discharge if possible. Keep people away.
Avoid contact with solid and dust.
Isolate and remove discharged material.
Notify local health and pollution control agencies.

Fire

Not flammable.
POISONOUS GASES MAY BE PRODUCED WHEN HEATED.

Exposure

CALL FOR MEDICAL AID.

DUST
Irritating to eyes, nose and throat.
If inhaled will cause coughing or difficult breathing.
If in eyes, hold eyelids open and flush with plenty of water.
If breathing has stopped, give artificial respiration.
If breathing is difficult, give oxygen.

SOLID
Irritating to skin and eyes.
If swallowed will cause nausea, vomiting or loss of consciousness.
Remove contaminated clothing and shoes.
Flush affected areas with plenty of water.
IF IN EYES, hold eyelids open and flush with plenty of water.
IF SWALLOWED and victim is CONSCIOUS, have victim drink water or milk and have victim induce vomiting.
IF SWALLOWED and victim is UNCONSCIOUS OR HAVING CONVULSIONS, do nothing except keep victim warm.

Water Pollution

Effect of low concentrations on aquatic life is unknown.
May be dangerous if it enters water intakes.
Notify local health and wildlife officials.
Notify operators of nearby water intakes.

1. RESPONSE TO DISCHARGE
(See Response Methods Handbook, CG 446-4)

Should be removed
Chemical and physical treatment

2. LABELS

No hazard label required by Code of Federal Regulations

3. CHEMICAL DESIGNATIONS

3.1 **Synonyms:** Cupric oxalate hemihydrate
3.2 **Coast Guard Compatibility Classification:** Not listed
3.3 **Chemical Formula:** $CuC_2O_4 \cdot \frac{1}{2}H_2O$
3.4 **IMCO/United Nations Numerical Designation:** Not listed

4. OBSERVABLE CHARACTERISTICS

4.1 **Physical State (as shipped):** Solid
4.2 **Color:** Bluish white
4.3 **Odor:** None

5. HEALTH HAZARDS

5.1 **Personal Protective Equipment:** Dust mask; goggles or face shield; protective gloves
5.2 **Symptoms Following Exposure:** Inhalation causes irritation of nose and throat. Ingestion of very large amounts may produce symptoms of oxalate poisoning; watch for edema of the glottis and delayed constriction of esophagus. Contact with eyes causes irritation.
5.3 **Treatment for Exposure:** INHALATION: remove to fresh air; if exposure has been prolonged, watch for symptoms of oxalate poisoning (nausea, shock, collapse, and convulsions). INGESTION: give large amount of water; induce vomiting; get medical attention. EYES: flush with water for at least 15 min. SKIN: flush with water.
5.4 **Toxicity by Inhalation (Threshold Limit Value):** Data not available
5.5 **Short-Term Inhalation Limits:** Data not available
5.6 **Toxicity by Ingestion:** Data not available
5.7 **Late Toxicity:** Data not available
5.8 **Vapor (Gas) Irritant Characteristics:** Data not available
5.9 **Liquid or Solid Irritant Characteristics:** Data not available
5.10 **Odor Threshold:** Data not available

6. FIRE HAZARDS

6.1 **Flash Point:** Not flammable
6.2 **Flammable Limits in Air:** Not flammable
6.3 **Fire Extinguishing Agents:** Not pertinent
6.4 **Fire Extinguishing Agents Not to be Used:** Not pertinent
6.5 **Special Hazards of Combustion Products:** Toxic carbon monoxide gas may form in fire.
6.6 **Behavior in Fire:**
6.7 **Ignition Temperature:** Not pertinent
6.8 **Electrical Hazard:** Not pertinent
6.9 **Burning Rate:** Not pertinent

7. CHEMICAL REACTIVITY

7.1 **Reactivity with Water:** No reaction
7.2 **Reactivity with Common Materials:**
7.3 **Stability During Transport:** Stable
7.4 **Neutralizing Agents for Acids and Caustics:** Not pertinent
7.5 **Polymerization:** Not pertinent
7.6 **Inhibitor of Polymerization:** Not pertinent

8. WATER POLLUTION

8.1 **Aquatic Toxicity:** Data not available
8.2 **Waterfowl Toxicity:** Data not available
8.3 **Biological Oxygen Demand (BOD):** Data not available
8.4 **Food Chain Concentration Potential:** Copper known to be accumulated by shellfish. Hazard to humans unknown.

9. SELECTED MANUFACTURERS

1. Gallard-Schlesinger Chemical Mfg. Co.
584 Mineola Ave.
Carle Place, N. Y. 11514

2. Pfaltz and Bauer, Inc.
126—04 Northern Blvd.
Flushing, N. Y. 11368

10. SHIPPING INFORMATION

10.1 **Grades or Purity:** 98+%
10.2 **Storage Temperature:** Ambient
10.3 **Inert Atmosphere:** No requirement
10.4 **Venting:** Open

11. HAZARD ASSESSMENT CODE
(See Hazard Assessment Handbook, CG 446-3)

II

12. HAZARD CLASSIFICATIONS

12.1 **Code of Federal Regulations:** Not listed
12.2 **NAS Hazard Rating for Bulk Water Transportation:** Not listed
12.3 **NFPA Hazard Classifications:** Not listed

13. PHYSICAL AND CHEMICAL PROPERTIES

13.1 **Physical State at 15°C and 1 atm:** Solid
13.2 **Molecular Weight:** 160.6
13.3 **Boiling Point at 1 atm:** Not pertinent (decomposes)
13.4 **Freezing Point:** Not pertinent
13.5 **Critical Temperature:** Not pertinent
13.6 **Critical Pressure:** Not pertinent
13.7 **Specific Gravity:** >1 at 20°C (solid)
13.8 **Liquid Surface Tension:** Not pertinent
13.9 **Liquid-Water Interfacial Tension:** Not pertinent
13.10 **Vapor (Gas) Specific Gravity:** Not pertinent
13.11 **Ratio of Specific Heats of Vapor (Gas):** Not pertinent
13.12 **Latent Heat of Vaporization:** Not pertinent
13.13 **Heat of Combustion:** Not pertinent
13.14 **Heat of Decomposition:** Not pertinent
13.15 **Heat of Solution:** Not pertinent
13.16 **Heat of Polymerization:** Not pertinent

NOTES

CSF	COPPER SULFATE

Common Synonyms			
Blue Vitriol Sulfate of copper Copper sulfate pentahydrate Cupric sulfate	Solid-granules or crystals Sinks and mixes with water.	White to blue	Odorless

Isolate and remove discharged material.
Notify local health and pollution control agencies.

Fire	Not flammable.
Exposure	CALL FOR MEDICAL AID. SOLID If swallowed, will cause nausea, vomiting or loss of consciousness. IF SWALLOWED and victim is CONSCIOUS, have victim drink water or milk and have victim induce vomiting. IF SWALLOWED and victim is UNCONSCIOUS OR HAVING CONVULSIONS, do nothing except keep victim warm.
Water Pollution	HARMFUL TO AQUATIC LIFE IN VERY LOW CONCENTRATIONS. May be dangerous if it enters water intakes. Notify local health and wildlife officials. Notify operators of nearby water intakes.

1. RESPONSE TO DISCHARGE

(See Response Methods Handbook, CG 446-4)

Issue warning – water contaminant
Disperse and flush

2. LABELS

No hazard label required by Code of Federal Regulations

3. CHEMICAL DESIGNATIONS

3.1 **Synonyms:** Blue vitriol; Copper sulfate pentahydrate; Cupric sulfate; Sulfate of copper

3.2 **Coast Guard Compatibility Classification:** Not applicable

3.3 **Chemical Formula:** $CuSO_4 \cdot 5H_2O$

3.4 **IMCO/United Nations Numerical Designation:** Not listed

4. OBSERVABLE CHARACTERISTICS

4.1 **Physical State (as shipped):** Solid

4.2 **Color:** Blue

4.3 **Odor:** None

5. HEALTH HAZARDS

5.1 **Personal Protective Equipment:** Filtering masks to minimize inhalation of dust.

5.2 **Symptoms Following Exposure:** INGESTION: copper sulfate may induce severe gastroenteric distress (vomiting, gastroenteric pain, and local corrosion and hemorrhages), prostration, anuria, hematuria, anemia, increase in white blood cells, icterus, coma, respiratory difficulties, and circulatory failure.

5.3 **Treatment for Exposure:** INGESTION: induce vomiting and administer gastric lavage; give a saline cathartic, fluid therapy, and transfusions if required; calcium disodium EDTA has been found moderately effective. SKIN AND EYES: wash affected tissues with water.

5.4 **Toxicity by Inhalation (Threshold Limit Value):** Not pertinent

5.5 **Short-Term Inhalation Limits:** Not pertinent

5.6 **Toxicity by Ingestion:** Grade 3; LD_{50} 50 to 500 mg/kg (rat)

5.7 **Late Toxicity:** Causes liver, kidney and testicular damage in rats.

5.8 **Vapor (Gas) Irritant Characteristics:** Not pertinent

5.9 **Liquid or Solid Irritant Characteristics:** Causes smarting of the skin and first degree burns on short exposure; may cause second degree burns on long exposure.

5.10 **Odor Threshold:** Not pertinent

6. FIRE HAZARDS

6.1 **Flash Point:** Not flammable

6.2 **Flammable Limits in Air:** Not flammable

6.3 **Fire Extinguishing Agents:** Not pertinent

6.4 **Fire Extinguishing Agents Not to be Used:** Not pertinent

6.5 **Special Hazards of Combustion Products:** Not pertinent

6.6 **Behavior in Fire:** Not pertinent

6.7 **Ignition Temperature:** Not flammable

6.8 **Electrical Hazard:** Not pertinent

6.9 **Burning Rate:** Not flammable

7. CHEMICAL REACTIVITY

7.1 **Reactivity with Water:** No reaction

7.2 **Reactivity with Common Materials:** No reaction

7.3 **Stability During Transport:** Stable

7.4 **Neutralizing Agents for Acids and Caustics:** Not pertinent

7.5 **Polymerization:** Not pertinent

7.6 **Inhibitor of Polymerization:** Not pertinent

8. WATER POLLUTION

8.1 **Aquatic Toxicity:**
3.8 ppm/24 hr/rainbow trout/TL_m/fresh water
0.14 ppm/48 hr/prawn/LC_{50}/salt water

8.2 **Waterfowl Toxicity:** Data not available

8.3 **Biological Oxygen Demand (BOD):** None

8.4 **Food Chain Concentration Potential:** None

9. SELECTED MANUFACTURERS

1. Anaconda Co.
Great Falls, Mont. 59401
2. Cities Service Co., Inc.
Industrial Chemicals Division
60 Wall St.
New York, N. Y. 10005
3. Mallinckrodt Chemical Works
Industrial Chemicals Division
St. Louis, Mo. 63160

10. SHIPPING INFORMATION

10.1 **Grades or Purity:** Data not available

10.2 **Storage Temperature:** Ambient

10.3 **Inert Atmosphere:** No requirement

10.4 **Venting:** Data not available

11. HAZARD ASSESSMENT CODE

(See Hazard Assessment Handbook, CG 446-3)

S S

12. HAZARD CLASSIFICATIONS

12.1 **Code of Federal Regulations:** Not listed

12.2 **NAS Hazard Rating for Bulk Water Transportation:** Not listed

12.3 **NFPA Hazard Classifications:** Not listed

13. PHYSICAL AND CHEMICAL PROPERTIES

13.1 **Physical State at 15°C and 1 atm:** Solid

13.2 **Molecular Weight:** 249.7

13.3 **Boiling Point at 1 atm:** Not pertinent

13.4 **Freezing Point:** Not pertinent

13.5 **Critical Temperature:** Not pertinent

13.6 **Critical Pressure:** Not pertinent

13.7 **Specific Gravity:** 2.29 at 15°C (solid)

13.8 **Liquid Surface Tension:** Not pertinent

13.9 **Liquid-Water Interfacial Tension:** Not pertinent

13.10 **Vapor (Gas) Specific Gravity:** Not pertinent

13.11 **Ratio of Specific Heats of Vapor (Gas):** Not pertinent

13.12 **Latent Heat of Vaporization:** Not pertinent

13.13 **Heat of Combustion:** Not pertinent

13.14 **Heat of Decomposition:** Not pertinent

13.15 **Heat of Solution:** Not pertinent

13.16 **Heat of Polymerization:** Not pertinent

NOTES

REVISED 1978

CSY	CORN SYRUP

Common Synonyms	Thick liquid Colorless Odorless Sinks and mixes with water.

Stop discharge if possible.
Isolate and remove discharged material.
Notify local health and pollution control agencies.

Fire	Not flammable.
Exposure	Not harmful.
Water Pollution	Dangerous to aquatic life in high concentrations. May be dangerous if it enters water intakes. **Notify local health and pollution control officials.** **Notify operators of nearby water intakes.**

1. RESPONSE TO DISCHARGE

(See Response Methods Handbook, CG 446-4)

Disperse and flush

2. LABELS

No hazard label required by Code of Federal Regulations

3. CHEMICAL DESIGNATIONS

3.1 **Synonyms:** No common synonyms
3.2 **Coast Guard Compatibility Classification:** Miscellaneous water solutions
3.3 **Chemical Formula:** Not pertinent
3.4 **IMCO/United Nations Numerical Designation:** Not listed

4. OBSERVABLE CHARACTERISTICS

4.1 **Physical State (as shipped):** Liquid
4.2 **Color:** Colorless
4.3 **Odor:** None

5. HEALTH HAZARDS

5.1 **Personal Protective Equipment:** None
5.2 **Symptoms Following Exposure:** Not toxic
5.3 **Treatment for Exposure:** None
5.4 **Toxicity by Inhalation (Threshold Limit Value):** Not pertinent
5.5 **Short-Term Inhalation Limits:** Not pertinent
5.6 **Toxicity by Ingestion:** None
5.7 **Late Toxicity:** None
5.8 **Vapor (Gas) Irritant Characteristics:** Not pertinent
5.9 **Liquid or Solid Irritant Characteristics:** None
5.10 **Odor Threshold:** Not pertinent

6. FIRE HAZARDS

6.1 **Flash Point:** Not flammable
6.2 **Flammable Limits in Air:** Not flammable
6.3 **Fire Extinguishing Agents:** Not pertinent
6.4 **Fire Extinguishing Agents Not to be Used:** Not pertinent
6.5 **Special Hazards of Combustion Products:** Not pertinent
6.6 **Behavior in Fire:** Not pertinent
6.7 **Ignition Temperature:** Not flammable
6.8 **Electrical Hazard:** Not pertinent
6.9 **Burning Rate:** Not flammable

7. CHEMICAL REACTIVITY

7.1 **Reactivity with Water:** No reaction
7.2 **Reactivity with Common Materials:** No reaction
7.3 **Stability During Transport:** Stable
7.4 **Neutralizing Agents for Acids and Caustics:** Not pertinent
7.5 **Polymerization:** Not pertinent
7.6 **Inhibitor of Polymerization:** Not pertinent

8. WATER POLLUTION

8.1 **Aquatic Toxicity:** None
8.2 **Waterfowl Toxicity:** None
8.3 **Biological Oxygen Demand (BOD):** 53%, 5 days; 99% (theor.), 5 days
8.4 **Food Chain Concentration Potential:** None

9. SELECTED MANUFACTURERS

1. CPC International, Inc.
   Argo, Ill. 06502
2. A. E. Staley Mfg. Co.
   Decatur, Ill. 62525
3. Standard Brands, Inc.
   Clinton Corn Processing Co. Division
   Clinton, Iowa 52732

10. SHIPPING INFORMATION

10.1 **Grades or Purity:** Data not available
10.2 **Storage Temperature:** Ambient
10.3 **Inert Atmosphere:** No requirement
10.4 **Venting:** Data not available

11. HAZARD ASSESSMENT CODE

(See Hazard Assessment Handbook, CG 446-3)

A-P

12. HAZARD CLASSIFICATIONS

12.1 **Code of Federal Regulations:** Not listed
12.2 **NAS Hazard Rating for Bulk Water Transportation:** Not listed
12.3 **NFPA Hazard Classifications:** Not listed

13. PHYSICAL AND CHEMICAL PROPERTIES

13.1 **Physical State at 15°C and 1 atm:** Liquid
13.2 **Molecular Weight:** Not pertinent
13.3 **Boiling Point at 1 atm:** Not pertinent
13.4 **Freezing Point:** Not pertinent
13.5 **Critical Temperature:** Not pertinent
13.6 **Critical Pressure:** Not pertinent
13.7 **Specific Gravity:** 1.4 at 37°C (liquid)
13.8 **Liquid Surface Tension:** Not pertinent
13.9 **Liquid-Water Interfacial Tension:** Not pertinent
13.10 **Vapor (Gas) Specific Gravity:** Not pertinent
13.11 **Ratio of Specific Heats of Vapor (Gas):** Not pertinent
13.12 **Latent Heat of Vaporization:** Not pertinent
13.13 **Heat of Combustion:** Not pertinent
13.14 **Heat of Decomposition:** Not pertinent
13.15 **Heat of Solution:** Not pertinent
13.16 **Heat of Polymerization:** Not pertinent

NOTES

REVISED 1978

COU	COUMAPHOS

Common Synonyms			
CO-RAL O,O-Diethyl-O-(3-chloro-4-methyl-2-oxo-(2H)-1-benzopyran-7-yl) phosphorothioate	Solid	White	Weak sulfurous odor
	Solid sinks in water.		

Stop discharge if possible.
Call fire department.
Avoid contact with solid and dust.
Isolate and remove discharged material.
Notify local health and pollution control agencies.

Fire	Combustible. Extinguish with water, foam, carbon dioxide or dry chemical.
Exposure	CALL FOR MEDICAL AID. SOLID OR DUST Irritating to skin and eyes. Harmful if swallowed or inhaled. Flush affected areas with plenty of water. IF IN EYES, hold eyelids open and flush with plenty of water. IF SWALLOWED and victim is CONSCIOUS, have victim drink water or milk and have victim induce vomiting. IF SWALLOWED and victim is UNCONSCIOUS OR HAVING CONVULSIONS, do nothing except keep victim warm. If inhaled, move to fresh air. If breathing has stopped, give artificial respiration. If breathing is difficult, give oxygen.
Water Pollution	HARMFUL TO AQUATIC LIFE IN VERY LOW CONCENTRATIONS. Fouling to shoreline. May be dangerous if it enters water intakes. Notify local health and wildlife officials. Notify operators of nearby water intakes.

1. RESPONSE TO DISCHARGE
(See Response Methods Handbook, CG 446-4)

Issue warning—water contaminant
Restrict access
Should be removed
Chemical and physical treatment

2. LABELS

No hazard label required by Code of Federal Regulations

3. CHEMICAL DESIGNATIONS

3.1 **Synonyms:** CO—RAL; O,O-Diethyl O-(3-chloro-4-methyl-2-oxo-(2H)-1-benzopyran-7-yl) phosphorothioate
3.2 **Coast Guard Compatibility Classification:** Not listed
3.3 **Chemical Formula:** $C_{14}H_{16}ClO_5PS$
3.4 **IMCO/United Nations Numerical Designation:** 6.1/1893

4. OBSERVABLE CHARACTERISTICS

4.1 **Physical State (as shipped):** Solid
4.2 **Color:** White
4.3 **Odor:** Slight odor of sulfur compounds

5. HEALTH HAZARDS

5.1 **Personal Protective Equipment:** Organic vapor respirator; rubber gloves; goggles
5.2 **Symptoms Following Exposure:** Inhalation or ingestion causes sense of tightness in chest, sweating, contracted pupils of eyes, stomach pains, vomiting, and diarrhea. Pulmonary edema may develop as long as 12 hours after acute exposures. Contact with eyes causes irritation; overexposure may cause same symptoms as inhalation. Irritates skin.
5.3 **Treatment for Exposure:** *Get medical attention after all overexposure to this substance. Atropine sulfate and 2-PAM are antidotes.* INHALATION: remove to fresh air; support respiration; keep patient quiet until medical help arrives; observe for development of pulmonary edema, even after 12 hours. INGESTION: induce vomiting; give water and again induce vomiting; give aqueous slurry of medicinal charcoal. EYES: flush with water for at least 15 min. SKIN: wash with soap and water.
5.4 **Toxicity by Inhalation (Threshold Limit Value):** Data not available
5.5 **Short-Term Inhalation Limits:** Data not available
5.6 **Toxicity by Ingestion:** Grade 4; oral LD_{50} = 16 mg/kg (rat)
5.7 **Late Toxicity:** Data not available
5.8 **Vapor (Gas) Irritant Characteristics:** Data not available
5.9 **Liquid or Solid Irritant Characteristics:** Data not available
5.10 **Odor Threshold:** 0.02 ppm

6. FIRE HAZARDS

6.1 **Flash Point:** Not pertinent (combustible solid)
6.2 **Flammable Limits in Air:** Not pertinent
6.3 **Fire Extinguishing Agents:** Water, foam, carbon dioxide, dry chemical
6.4 **Fire Extinguishing Agents Not to be Used:**
6.5 **Special Hazards of Combustion Products:** Toxic and irritating oxides of sulfur and phosphorus may form in fires.
6.6 **Behavior in Fire:**
6.7 **Ignition Temperature:** Data not available
6.8 **Electrical Hazard:** Not pertinent
6.9 **Burning Rate:** Not pertinent

7. CHEMICAL REACTIVITY

7.1 **Reactivity with Water:** No reaction
7.2 **Reactivity with Common Materials:**
7.3 **Stability During Transport:** Stable
7.4 **Neutralizing Agents for Acids and Caustics:** Not pertinent
7.5 **Polymerization:** Not pertinent
7.6 **Inhibitor of Polymerization:** Not pertinent

8. WATER POLLUTION

8.1 **Aquatic Toxicity:**
0.18 ppm/96 hr/bluegill/TL_m/fresh water
0.11 ppm/48 hr/oyster larvae/TL_m/salt water
8.2 **Waterfowl Toxicity:** 21.5 to 41.5 mg/kg LD_{50}
8.3 **Biological Oxygen Demand (BOD):** Data not available
8.4 **Food Chain Concentration Potential:** None

9. SELECTED MANUFACTURERS

Bayvet Corporation
Box 390
Shawnee Mission, Kan. 66201

10. SHIPPING INFORMATION

10.1 **Grades or Purity:** 5—50% active ingredient, the balance being inert solids
10.2 **Storage Temperature:** Ambient
10.3 **Inert Atmosphere:** No requirement
10.4 **Venting:** Open

11. HAZARD ASSESSMENT CODE
(See Hazard Assessment Handbook, CG 446-3)

II

12. HAZARD CLASSIFICATIONS

12.1 **Code of Federal Regulations:** Not listed
12.2 **NAS Hazard Rating for Bulk Water Transportation:** Not listed
12.3 **NFPA Hazard Classifications:** Not listed

13. PHYSICAL AND CHEMICAL PROPERTIES

13.1 **Physical State at 15°C and 1 atm:** Solid
13.2 **Molecular Weight:** 362.5
13.3 **Boiling Point at 1 atm:** Not pertinent
13.4 **Freezing Point:** 199°F = 93°C = 366°K
13.5 **Critical Temperature:** Not pertinent
13.6 **Critical Pressure:** Not pertinent
13.7 **Specific Gravity:** 1.474 at 20°C (solid)
13.8 **Liquid Surface Tension:** Not pertinent
13.9 **Liquid-Water Interfacial Tension:** Not pertinent
13.10 **Vapor (Gas) Specific Gravity:** Not pertinent
13.11 **Ratio of Specific Heats of Vapor (Gas):** Not pertinent
13.12 **Latent Heat of Vaporization:** Not pertinent
13.13 **Heat of Combustion:** Not pertinent
13.14 **Heat of Decomposition:** Not pertinent
13.15 **Heat of Solution:** Not pertinent
13.16 **Heat of Polymerization:** Not pertinent

NOTES

CCT

CREOSOTE, COAL TAR

Common Synonyms			
Creosote oil Dead oil	Liquid	Yellow to black	Tarry odor
	May float or sink in water.		

Stop discharge if possible.
Call fire department.
Isolate and remove discharged material.
Notify local health and pollution control agencies.

Fire	Combustible. Extinguish with dry chemicals, foam or carbon dioxide. Water may be ineffective on fire.
Exposure	Call for medical aid. LIQUID Irritating to skin and eyes. Harmful if swallowed. Remove contaminated clothing and shoes. Flush affected areas with plenty of water. IF IN EYES, hold eyelids open and flush with plenty of water. IF SWALLOWED and victim is CONSCIOUS, have victim drink water or milk and have victim induce vomiting. IF SWALLOWED and victim is UNCONSCIOUS OR HAVING CONVULSIONS, do nothing except keep victim warm.
Water Pollution	Effect of low concentrations on aquatic life is unknown. Fouling to shoreline. May be dangerous if it enters water intakes. Notify local health and wildlife officials. Notify operators of nearby water intakes.

1. RESPONSE TO DISCHARGE

(See Response Methods Handbook, CG 446-4)

Issue warning—water contaminant
Mechanical containment
Should be removed
Chemical and physical treatment

2. LABELS

No label required by Code of Federal Regulations

3. CHEMICAL DESIGNATIONS

3.1 **Synonyms:** Creosote oil; Dead oil
3.2 **Coast Guard Compatibility Classification:** Cresol
3.3 **Chemical Formula:** Mixture
3.4 **IMCO/United Nations Numerical Designation:** 9/2022

4. OBSERVABLE CHARACTERISTICS

4.1 **Physical State (as shipped):** Liquid
4.2 **Color:** Yellow to brown to black
4.3 **Odor:** Creosote or tarry; aromatic

5. HEALTH HAZARDS

5.1 **Personal Protective Equipment:** All-service canister mask; rubber gloves; chemical safety goggles and/or face shield; overalls or a neoprene apron; barrier creams.
5.2 **Symptoms Following Exposure:** Vapors cause moderate irritation of nose and throat. Liquid causes severe burns of eyes and reddening and itching of skin. Prolonged contact with skin can cause burns. Ingestion causes salivation, vomiting, respiratory difficulties, thready pulse, vertigo, headache, loss of pupillary reflexes, hypothermia, cyanosis, mild convulsions.
5.3 **Treatment for Exposure:** INHALATION: remove victim to fresh air; if he is not breathing, give artificial respiration, preferably mouth-to-mouth; if breathing is difficult, give oxygen; call a physician. EYES: flush immediately with plenty of water for at least 15 min. and call a physician. SKIN: wipe with vegetable oil or margarine, then wash with soap and water. INGESTION: have victim drink water or milk; do NOT induce vomiting.
5.4 **Toxicity by Inhalation (Threshold Limit Value):** Data not available
5.5 **Short-Term Inhalation Limits:** Data not available
5.6 **Toxicity by Ingestion:** Grade 2; LD_{50} 0.5 to 5 g/kg
5.7 **Late Toxicity:** Repeated exposures may cause cancer of skin.
5.8 **Vapor (Gas) Irritant Characteristics:** Vapors cause moderate irritation such that personnel will find high concentrations unpleasant. The effect is temporary.
5.9 **Liquid or Solid Irritant Characteristics:** Fairly severe skin irritant. May cause pain and second-degree burns after a few minutes' contact.
5.10 **Odor Threshold:** Data not available

6. FIRE HAZARDS

6.1 **Flash Point:** > 160°F C.C.
6.2 **Flammable Limits in Air:** Not pertinent
6.3 **Fire Extinguishing Agents:** Dry chemical, carbon dioxide or foam
6.4 **Fire Extinguishing Agents Not to be Used:** Water may be ineffective.
6.5 **Special Hazards of Combustion Products:** Data not available
6.6 **Behavior in Fire:** Heavy, irritating black smoke is formed.
6.7 **Ignition Temperature:** 637°F
6.8 **Electrical Hazard:** Not pertinent
6.9 **Burning Rate:** Data not available

7. CHEMICAL REACTIVITY

7.1 **Reactivity with Water:** No reaction
7.2 **Reactivity with Common Materials:** No reaction
7.3 **Stability During Transport:** Stable
7.4 **Neutralizing Agents for Acids and Caustics:** Not pertinent
7.5 **Polymerization:** Not pertinent
7.6 **Inhibitor of Polymerization:** Not pertinent

8. WATER POLLUTION

8.1 **Aquatic Toxicity:** Data not available
8.2 **Waterfowl Toxicity:** Data not available
8.3 **Biological Oxygen Demand (BOD):** Data not available
8.4 **Food Chain Concentration Potential:** None

9. SELECTED MANUFACTURERS

1. Witco Chemical Corporation
   Pioneer Division
   277 Park Avenue
   New York, N. Y. 10017
2. Koppers Company, Inc.
   Organic Materials Division
   Koppers Building
   Pittsburgh, Pa. 15219
3. Allied Chemical Corporation
   Semet Solvay Division
   Ironton Tar Plant
   3330 South Third Street
   Ironton, Ohio 45638

10. SHIPPING INFORMATION

10.1 **Grades or Purity:** Whole creosote or various fractions, depending on boiling point. All have similar properties.
10.2 **Storage Temperature:** Ambient
10.3 **Inert Atmosphere:** No requirement
10.4 **Venting:** Open (flame arrester)

11. HAZARD ASSESSMENT CODE

(See Hazard Assessment Handbook, CG 446-3)

A-T-U-X-Y

12. HAZARD CLASSIFICATIONS

12.1 **Code of Federal Regulations:** Combustible Liquid
12.2 **NAS Hazard Rating for Bulk Water Transportation:**

Category	Rating
Fire	1
Health	
Vapor Irritant	2
Liquid or Solid Irritant	3
Poisons	2
Water Pollution	
Human Toxicity	2
Aquatic Toxicity	3
Aesthetic Effect	4
Reactivity	
Other Chemicals	1
Water	0
Self-Reaction	0

12.3 **NFPA Hazard Classifications:**

Category	Classification
Health Hazard (Blue)	2
Flammability (Red)	2
Reactivity (Yellow)	0

13. PHYSICAL AND CHEMICAL PROPERTIES

13.1 **Physical State at 15°C and 1 atm:** Liquid
13.2 **Molecular Weight:** Mixture
13.3 **Boiling Point at 1 atm:** > 356°F = > 180°C = > 353°K
13.4 **Freezing Point:** Not pertinent
13.5 **Critical Temperature:** Not pertinent
13.6 **Critical Pressure:** Not pertinent
13.7 **Specific Gravity:** 1.05-1.09 at 15°C (liquid)
13.8 **Liquid Surface Tension:** (est.) 15 dynes/cm = 0.015 N/m at 20°C
13.9 **Liquid-Water Interfacial Tension:** (est.) 20 dynes/cm = 0.020 N/m at 20°C
13.10 **Vapor (Gas) Specific Gravity:** Not pertinent
13.11 **Ratio of Specific Heats of Vapor (Gas):** Not pertinent
13.12 **Latent Heat of Vaporization:** Not pertinent
13.13 **Heat of Combustion:** (est.) −12,500 Btu/lb = −6,900 cal/g = -290×10^5 J/kg
13.14 **Heat of Decomposition:** Not pertinent
13.15 **Heat of Solution:** Not pertinent
13.16 **Heat of Polymerization:** Not pertinent

NOTES

CRS	CRESOLS

Common Synonyms	Watery liquid, or solid crystals	Colorless or yellow	Sweet tarry odor
Cresylic acids Hydroxytoluenes Methylphenols Oxytoluenes	Sinks in water.		

Avoid contact with liquid. Keep people away.
Wear goggles, self-contained breathing apparatus, and rubber overclothing (including gloves).
Stop discharge if possible.
Call fire department.
Isolate and remove discharged material.
Notify local health and pollution control agencies.

Fire

Combustible.
POISONOUS GASES MAY BE PRODUCED IN FIRE.

Wear goggles and self-contained breathing apparatus.
Extinguish with water, dry chemical, foam or carbon dioxide.
Cool exposed containers with water.

Exposure

CALL FOR MEDICAL AID.
LIQUID
Will burn skin and eyes.
Harmful if swallowed.

Remove contaminated clothing and shoes.
Flush affected areas with plenty of water.
IF IN EYES, hold eyelids open and flush with plenty of water.
IF SWALLOWED and victim is CONSCIOUS, have victim drink water or milk.
DO NOT INDUCE VOMITING.

Water Pollution

HARMFUL TO AQUATIC LIFE IN VERY LOW CONCENTRATIONS.
May be dangerous if it enters water intakes.

Notify local health and wildlife officials.
Notify operators of nearby water intakes.

1. RESPONSE TO DISCHARGE

(See Response Methods Handbook, CG 446-4)

Issue warning—water contaminant, poison
Restrict access
Should be removed
Chemical and physical treatment

2. LABEL

3. CHEMICAL DESIGNATIONS

3.1 **Synonyms:**
Cresylic acids
Hydroxytoluenes
Methylphenols
Oxytoluenes
Tar acids

3.2 **Coast Guard Compatibility Classification:** Cresol

3.3 **Chemical Formula:** $CH_3C_6H_4OH$

3.4 **IMCO/United Nations Numerical Designation:** 9.0/2022

4. OBSERVABLE CHARACTERISTICS

4.1 **Physical State (as shipped):** Liquid or solid

4.2 **Color:** Colorless to dark yellow

4.3 **Odor:** Sweet, tarry

5. HEALTH HAZARDS

5.1 **Personal Protective Equipment:** Organic vapor canister unit (USBM Type B) approved by U. S. Bureau of Mines. Rubber gloves; chemical safety goggles; face shield; coveralls and/or rubber apron; rubber shoes or boots.

5.2 **Symptoms Following Exposure:** Vapors cause irritation of eyes, nose, and throat. Contact with skin or eyes causes severe burns. Chemical is rapidly absorbed through skin.

5.3 **Treatment for Exposure:** Call a physician. INHALATION: remove to fresh air. INGESTION: have victim drink water or milk; do NOT induce vomiting. SKIN OR EYES: flush immediately with plenty of water for at least 15 min.; remove contaminated clothing immediately and wash before reuse; discard contaminated shoes.

5.4 **Toxicity by Inhalation (Threshold Limit Value):** 5 ppm

5.5 **Short-Term Inhalation Limits:** Data not available

5.6 **Toxicity by Ingestion:** Grade 2; LD_{50} 0.5 to 5 g/kg (rat, rabbit)

5.7 **Late Toxicity:** Data not available

5.8 **Vapor (Gas) Irritant Characteristics:** Vapors cause moderate irritation such that personnel will find high concentrations unpleasant. The effect is temporary.

5.9 **Liquid or Solid Irritant Characteristics:** Fairly severe skin irritant; may cause pain and second-degree burns after a few minutes' contact.

5.10 **Odor Threshold:** 5 ppm

6. FIRE HAZARDS

6.1 **Flash Point:** 175-185°F O.C.; 178°F C.C.

6.2 **Flammable Limits in Air:** LEL: 1.4% (ortho); 1.1% (meta or para)

6.3 **Fire Extinguishing Agents:** Water, dry chemical, carbon dioxide, and foam

6.4 **Fire Extinguishing Agents Not to be Used:** Not pertinent

6.5 **Special Hazards of Combustion Products:** Flammable toxic vapors given off in a fire.

6.6 **Behavior in Fire:** Sealed closed containers can build up pressure if exposed to heat (fire).

6.7 **Ignition Temperature:** 1110°F (o-cresol) 1038°F (m-or p-cresol)

6.8 **Electrical Hazard:** Data not available

6.9 **Burning Rate:** Data not available

7. CHEMICAL REACTIVITY

7.1 **Reactivity with Water:** No reaction

7.2 **Reactivity with Common Materials:** No reaction

7.3 **Stability During Transport:** Stable

7.4 **Neutralizing Agents for Acids and Caustics:** Not pertinent

7.5 **Polymerization:** Not pertinent

7.6 **Inhibitor of Polymerization:** Not pertinent

8. WATER POLLUTION

8.1 **Aquatic Toxicity:**
24 mg/l /96 hr/bluegill/TL_m/fresh water
10-100 ppm/48 hr/shrimp/LC_{50}/salt water

8.2 **Waterfowl Toxicity:** Data not available

8.3 **Biological Oxygen Demand (BOD):**
m-cresol: 170%, 5 days
o-cresol: 164%, 5 days
p-cresol: 144%, 5 days

8.4 **Food Chain Concentration Potential:** None

9. SELECTED MANUFACTURERS

1. Allied Chemical Corp.
Plastics Division
Morristown, N. J. 07960

2. Continental Oil Co.
Pitt-Consol Chemicals Division
Park 80 Plaza East
Saddle Brook, N. J. 07662

3. Koppers Co., Inc.
Organic Materials Division
Pittsburgh, Pa. 15219

10. SHIPPING INFORMATION

10.1 **Grades or Purity:**
USP Liquid (mixed isomers)
Phenol-cresol mixtures
Ortho-cresol 80 to 98% containing phenol
Meta-cresol 60 to 98% containing other cresols and xylenols
Para-cresol 92 to 98% containing meta-cresol
Meta-para-cresol containing ortho-cresol and xylenols

11. HAZARD ASSESSMENT CODE

(See Hazard Assessment Handbook, CG 446-3)

A-P-Q-T-U-X-Y

12. HAZARD CLASSIFICATIONS

12.1 **Code of Federal Regulations:** Poisonous liquid or solid, Class B

12.2 **NAS Hazard Rating for Bulk Water Transportation:**

Category	Rating
Fire	1
Health	
Vapor Irritant	2
Liquid or Solid Irritant	3
Poisons	2
Water Pollution	
Human Toxicity	1
Aquatic Toxicity	3
Aesthetic Effect	4
Reactivity	
Other Chemicals	2
Water	0
Self-Reaction	0

12.3 **NFPA Hazard Classifications:**

Category	Classification ortho	Classification meta and para
Health Hazard (Blue)	3	3
Flammability (Red)	2	1
Reactivity (Yellow)	0	0

13. PHYSICAL AND CHEMICAL PROPERTIES

13.1 **Physical State at 15°C and 1 atm:** Liquid

13.2 **Molecular Weight:** 108.13

13.3 **Boiling Point at 1 atm:** >350°F = >177°C = >450°K

13.4 **Freezing Point:** Not pertinent

13.5 **Critical Temperature:** Not pertinent

13.6 **Critical Pressure:** Not pertinent

13.7 **Specific Gravity:** 1.03–1.07 at 20°C (liquid)

13.8 **Liquid Surface Tension:** 37 dynes/cm = 0.037 N/m at 20°C

13.9 **Liquid-Water Interfacial Tension:** Data not available

13.10 **Vapor (Gas) Specific Gravity:** Not pertinent

13.11 **Ratio of Specific Heats of Vapor (Gas):** 1.073

13.12 **Latent Heat of Vaporization:** (est.) 200 Btu/lb = 110 cal/g = 4.6×10^5 J/kg

13.13 **Heat of Combustion:**
−14,720 to −14,740 Btu/lb
= −8180 to −8190 cal/g
= −342.5 to −342.9 × 10^5 J/kg

13.14 **Heat of Decomposition:** Not pertinent

13.15 **Heat of Solution:** Not pertinent

13.16 **Heat of Polymerization:** Not pertinent

10. SHIPPING INFORMATION (Cont'd.)

"Resin" cresols containing phenols and xylenols
Cresylic acids containing xylenols, cresols and phenols

10.2 **Storage Temperature:** Ambient

10.3 **Inert Atmosphere:** No requirement

10.4 **Venting:** Open

REVISED 1978

CGE	# CRESYL GLYCIDYL ETHER

Common Synonyms	Solid White Sinks and mixes with water.

Stop discharge if possible. Keep people away.
Shut off ignition sources. Call fire department.
Avoid contact with solid and dust.
Isolate and remove discharged material.
Notify local health and pollution control agencies.

Fire	Combustible. Extinguish with dry chemicals, foam or carbon dioxide. Water may be ineffective on fire. Cool exposed containers with water.
Exposure	CALL FOR MEDICAL AID. DUST Irritating to eyes, nose and throat. Harmful if inhaled. If in eyes, hold eyelids open and flush with plenty of water. If breathing has stopped, give artificial respiration. If breathing is difficult, give oxygen. SOLID Irritating to skin and eyes. Harmful if swallowed. Remove contaminated clothing and shoes. Flush affected areas with plenty of water. IF IN EYES, hold eyelids open and flush with plenty of water. IF SWALLOWED and victim is CONSCIOUS, have victim drink water or milk. IF SWALLOWED and victim is UNCONSCIOUS OR HAVING CONVULSIONS, do nothing except keep victim warm.
Water Pollution	Effect of low concentrations on aquatic life is unknown. May be dangerous if it enters water intakes. Notify local health and wildlife officials. Notify operators of nearby water intakes.

1. RESPONSE TO DISCHARGE
(See Response Methods Handbook, CG 446-4)

Issue warning—water contaminant
Should be removed
Chemical and physical treatment

2. LABELS

No hazard label required by Code of Federal Regulations

3. CHEMICAL DESIGNATIONS

3.1 **Synonyms:** Cresol, epoxypropyl ether; Tolyl epoxypropyl ether; Tolyl glycidyl ether
3.2 **Coast Guard Compatibility Classification:** Not listed
3.3 **Chemical Formula:** $CH_3C_6H_4-O-CH_2-CH-CH_2-O$ (epoxide ring)
3.4 **IMCO/United Nations Numerical Designation:** Not listed

4. OBSERVABLE CHARACTERISTICS

4.1 **Physical State (as shipped):** Liquid
4.2 **Color:** Colorless
4.3 **Odor:** Data not available

5. HEALTH HAZARDS

5.1 **Personal Protective Equipment:** Organic canister mask or air pack; rubber gloves; goggles or face shield; body-covering clothing
5.2 **Symptoms Following Exposure:** Contact with eyes causes irritation. Contact with skin causes primary irritation and allergic sensitization.
5.3 **Treatment for Exposure:** INHALATION: move to fresh air. EYES: flush with water for at least 15 min.; get medical attention. SKIN: immediately wash off with soap and water.
5.4 **Toxicity by Inhalation (Threshold Limit Value):** Data not available
5.5 **Short-Term Inhalation Limits:** Data not available
5.6 **Toxicity by Ingestion:** Data not available
5.7 **Late Toxicity:** Data not available
5.8 **Vapor (Gas) Irritant Characteristics:** Data not available
5.9 **Liquid or Solid Irritant Characteristics:** Data not available
5.10 **Odor Threshold:** Data not available

6. FIRE HAZARDS

6.1 **Flash Point:** 200°F O.C.
6.2 **Flammable Limits in Air:** Data not available
6.3 **Fire Extinguishing Agents:** Dry chemical, foam, carbon dioxide
6.4 **Fire Extinguishing Agents Not to be Used:** Water may be ineffective.
6.5 **Special Hazards of Combustion Products:**
6.6 **Behavior in Fire:**
6.7 **Ignition Temperature:** Data not available
6.8 **Electrical Hazard:** Data not available
6.9 **Burning Rate:** Data not available

7. CHEMICAL REACTIVITY

7.1 **Reactivity with Water:** No reaction
7.2 **Reactivity with Common Materials:** May attack some forms of plastics.
7.3 **Stability During Transport:** Stable
7.4 **Neutralizing Agents for Acids and Caustics:** Not pertinent
7.5 **Polymerization:** Not pertinent
7.6 **Inhibitor of Polymerization:** Not pertinent

8. WATER POLLUTION

8.1 **Aquatic Toxicity:** Data not available
8.2 **Waterfowl Toxicity:** Data not available
8.3 **Biological Oxygen Demand (BOD):** Data not available
8.4 **Food Chain Concentration Potential:** None

9. SELECTED MANUFACTURERS

Shell Chemical Co.
One Shell Plaza
P. O. Box 2463
Houston, Texas 77001

10. SHIPPING INFORMATION

10.1 **Grades or Purity:** Technical, 100%
10.2 **Storage Temperature:** Ambient
10.3 **Inert Atmosphere:** No requirement
10.4 **Venting:** Open

11. HAZARD ASSESSMENT CODE
(See Hazard Assessment Handbook, CG 446-3)

A-T-U-X-Y

12. HAZARD CLASSIFICATIONS

12.1 **Code of Federal Regulations:** Not listed
12.2 **NAS Hazard Rating for Bulk Water Transportation:** Not listed
12.3 **NFPA Hazard Classifications:** Not listed

13. PHYSICAL AND CHEMICAL PROPERTIES

13.1 **Physical State at 15°C and 1 atm:** Liquid
13.2 **Molecular Weight:** 164
13.3 **Boiling Point at 1 atm:** (approx.) 498°F = 259°C = 532°K
13.4 **Freezing Point:** Not pertinent
13.5 **Critical Temperature:** Not pertinent
13.6 **Critical Pressure:** Not pertinent
13.7 **Specific Gravity:** 1.09 at 20°C (liquid)
13.8 **Liquid Surface Tension:** Data not available
13.9 **Liquid-Water Interfacial Tension:** Data not available
13.10 **Vapor (Gas) Specific Gravity:** Not pertinent
13.11 **Ratio of Specific Heats of Vapor (Gas):** Not pertinent
13.12 **Latent Heat of Vaporization:** Data not available
13.13 **Heat of Combustion:** (est.) −16,500 Btu/lb = −9,190 cal/g = -384×10^5 J/kg
13.14 **Heat of Decomposition:** Not pertinent
13.15 **Heat of Solution:** Not pertinent
13.16 **Heat of Polymerization:** Not pertinent

NOTES

CTA — CROTONALDEHYDE

Common Synonyms			
Crotonic aldehyde beta-Methylacrolein trans-2-Butenal	Watery liquid	Yellow	Tar odor
	Floats and mixes slowly with water. Flammable, irritating vapor is produced.		

Avoid contact with liquid and vapor. Keep people away.
Wear goggles, self-contained breathing apparatus, and rubber overclothing (including gloves).
Shut off ignition sources and call fire department.
Stop discharge if possible.
Stay upwind and use water spray to "knock down" vapor.
Isolate and remove discharged material.
Notify local health and pollution control agencies.

Fire

FLAMMABLE.
Flashback along vapor trail may occur.
Vapor may explode if ignited in an enclosed area.
Wear goggles, self-contained breathing apparatus, and rubber overclothing (including gloves).
Combat fires from safe distance or protected location.
Extinguish with dry chemical, foam or carbon dioxide.
Water may be ineffective on fire.
Cool exposed containers with water.

Exposure

CALL FOR MEDICAL AID.

VAPOR
Irritating to eyes, nose and throat.
If inhaled, will cause coughing, nausea, vomiting, or loss of consciousness.
Move to fresh air.
If breathing has stopped, give artificial respiration.
If breathing is difficult, give oxygen.

LIQUID
Will burn skin and eyes.
Harmful if swallowed.
Remove contaminated clothing and shoes.
Flush affected areas with plenty of water.
IF IN EYES, hold eyelids open and flush with plenty of water.
IF SWALLOWED and victim is CONSCIOUS, have victim drink water or milk.
DO NOT INDUCE VOMITING.

Water Pollution

Effect of low concentrations on aquatic life is unknown.
Fouling to shoreline.
May be dangerous if it enters water intakes.
Notify local health and pollution control officials.
Notify operators of nearby water intakes.

1. RESPONSE TO DISCHARGE

(See Response Methods Handbook, CG 446-4)

Issue warning—high flammability, water contaminant
Restrict access
Disperse and flush

2. LABEL

Red

3. CHEMICAL DESIGNATIONS

3.1 **Synonyms:** beta-Methylacrolein; Crotenaldehyde; Crotonic aldehyde; trans-2-Butenal
3.2 **Coast Guard Compatibility Classification:** Aldehyde
3.3 **Chemical Formula:** $CH_3CH{=}CHCHO$
3.4 **IMCO/United Nations Numerical Designation:** 3.2/1143

4. OBSERVABLE CHARACTERISTICS

4.1 **Physical State (as shipped):** Liquid
4.2 **Color:** Yellow
4.3 **Odor:** Tarry

5. HEALTH HAZARDS

5.1 **Personal Protective Equipment:** Air-supplied mask for concentrations above 2% by volume; plastic gloves; monogoggles; eye bath and safety shower
5.2 **Symptoms Following Exposure:** INHALATION: vapor is exceedingly irritating, causing coughing, chest pain, nausea, vomiting, and collapse. CONTACT WITH SKIN OR EYES: may cause burns and systemic illness. Contact of liquid or vapors with eyes causes burns.
5.3 **Treatment for Exposure:** INHALATION: remove victim to fresh air; give oxygen if breathing is difficult; call a physician. INGESTION: have victim drink water or milk; do NOT induce vomiting. SKIN OR EYES: immediately flush with plenty of water for at least 15 min.; physician should see cases of eye irritation from vapor or liquid.
5.4 **Toxicity by Inhalation (Threshold Limit Value):** 2 ppm
5.5 **Short-Term Inhalation Limits:** Data not available
5.6 **Toxicity by Ingestion:** Grade 3; LD_{50} 50 to 500 mg/kg
5.7 **Late Toxicity:** Data not available
5.8 **Vapor (Gas) Irritant Characteristics:** Vapor is moderately irritating such that personnel will not usually tolerate moderate or high vapor concentrations.
5.9 **Liquid or Solid Irritant Characteristics:** Fairly severe skin irritant; may cause pain and second-degree burns after a few minutes' contact.
5.10 **Odor Threshold:** 0.13 ppm

6. FIRE HAZARDS

6.1 **Flash Point:** 59°F O.C.
6.2 **Flammable Limits in Air:** 2.1%—15.5%
6.3 **Fire Extinguishing Agents:** Foam, dry chemical, carbon dioxide
6.4 **Fire Extinguishing Agents Not to be Used:** Not pertinent
6.5 **Special Hazards of Combustion Products:** Vapors are very irritating to nose, eyes, and skin.
6.6 **Behavior in Fire:** Vapor is heavier than air and may travel considerable distance to a source of ignition and flash back.
6.7 **Ignition Temperature:** 450°F
6.8 **Electrical Hazard:** Data not available
6.9 **Burning Rate:** 3.3 mm/min.

7. CHEMICAL REACTIVITY

7.1 **Reactivity with Water:** No reaction
7.2 **Reactivity with Common Materials:** No reaction
7.3 **Stability During Transport:** May polymerize
7.4 **Neutralizing Agents for Acids and Caustics:** Not pertinent
7.5 **Polymerization:** May polymerize or condense with evolution of heat in presence of alkalies, amines, or acids.
7.6 **Inhibitor of Polymerization:** None used

8. WATER POLLUTION

8.1 **Aquatic Toxicity:** Data not available
8.2 **Waterfowl Toxicity:** Data not available
8.3 **Biological Oxygen Demand (BOD):** 1.3 lb/lb, 10 days
8.4 **Food Chain Concentration Potential:** None

9. SELECTED MANUFACTURERS

1. Celanese Corp.
   Celanese Chemical Co. Division
   245 Park Ave.
   New York, N. Y. 10017
2. Union Carbide Corp.
   Chemicals and Plastics Division
   270 Park Ave.
   New York, N. Y. 10017

10. SHIPPING INFORMATION

10.1 **Grades or Purity:** 98.0%
10.2 **Storage Temperature:** Ambient
10.3 **Inert Atmosphere:** No requirement
10.4 **Venting:** Pressure-vacuum

11. HAZARD ASSESSMENT CODE

(See Hazard Assessment Handbook, CG 446-3)

A-P-Q

12. HAZARD CLASSIFICATIONS

12.1 **Code of Federal Regulations:** Flammable liquid
12.2 **NAS Hazard Rating for Bulk Water Transportation:**

Category	Rating
Fire	3
Health	
Vapor Irritant	3
Liquid or Solid Irritant	3
Poisons	3
Water Pollution	
Human Toxicity	3
Aquatic Toxicity	3
Aesthetic Effect	3
Reactivity	
Other Chemicals	2
Water	0
Self-Reaction	1

12.3 **NFPA Hazard Classifications:**

Category	Classification
Health Hazard (Blue)	3
Flammability (Red)	3
Reactivity (Yellow)	2

13. PHYSICAL AND CHEMICAL PROPERTIES

13.1 **Physical State at 15°C and 1 atm:** Liquid
13.2 **Molecular Weight:** 70.09
13.3 **Boiling Point at 1 atm:** 216.0°F = 102.2°C = 375.4°K
13.4 **Freezing Point:** −100°F = −75°C = 198°K
13.5 **Critical Temperature:** 563°F = 295°C = 568°K
13.6 **Critical Pressure:** 630 psia = 43 atm = 4.4 MN/m²
13.7 **Specific Gravity:** 0.852 at 20°C (liquid)
13.8 **Liquid Surface Tension:** Data not available
13.9 **Liquid-Water Interfacial Tension:** Data not available
13.10 **Vapor (Gas) Specific Gravity:** Not pertinent
13.11 **Ratio of Specific Heats of Vapor (Gas):** 1.104
13.12 **Latent Heat of Vaporization:** 200 Btu/lb = 111 cal/g = 4.65×10^5 J/kg
13.13 **Heat of Combustion:** -14,000 Btu/lb = -7760 cal/g = -325×10^5 J/kg
13.14 **Heat of Decomposition:** Not pertinent
13.15 **Heat of Solution:** Not pertinent
13.16 **Heat of Polymerization:** Not pertinent

NOTES

REVISED 1978

CUM — CUMENE

Common Synonyms	Watery liquid	Colorless	Gasoline-like odor
Cumol Isopropylbenzene	Floats on water.		

Stop discharge if possible. Keep people away.
Call fire department.
Avoid contact with liquid.
Isolate and remove discharged material.
Notify local health and pollution control agencies.

Fire

Combustible.
Extinguish with water, dry chemical, foam or carbon dioxide.
Cool exposed containers with water.

Exposure

CALL FOR MEDICAL AID.

LIQUID
Irritating to skin and eyes.
Harmful if swallowed.
Remove contaminated clothing and shoes.
Flush affected areas with plenty of water.
IF IN EYES, hold eyelids open and flush with plenty of water.
IF SWALLOWED and victim is CONSCIOUS, have victim drink water or milk.

Water Pollution

Effect of low concentrations on aquatic life is unknown.
Fouling to shoreline.
May be dangerous if it enters water intakes.
Notify local health and pollution control officials.
Notify operators of nearby water intakes.

1. RESPONSE TO DISCHARGE

(See Response Methods Handbook, CG 446-4)

Mechanical containment
Should be removed
Chemical and physical treatment

2. LABELS

No hazard label required by Code of Federal Regulations

3. CHEMICAL DESIGNATIONS

3.1 **Synonyms:** Cumol; Isopropylbenzene
3.2 **Coast Guard Compatibility Classification:** Aromatic hydrocarbon
3.3 **Chemical Formula:** $C_6H_5CH(CH_3)_2$
3.4 **IMCO/United Nations Numerical Designation:** 3.3/1918

4. OBSERVABLE CHARACTERISTICS

4.1 **Physical State (as shipped):** Liquid
4.2 **Color:** Colorless
4.3 **Odor:** Strong, slightly irritant; fragrant; aromatic

5. HEALTH HAZARDS

5.1 **Personal Protective Equipment:** As necessary to avoid skin exposure. If concentration in air is greater than 1000 ppm, use self-contained breathing apparatus.
5.2 **Symptoms Following Exposure:** Narcotic action with long-lasting effects; depressant to central nervous system.
5.3 **Treatment for Exposure:** INHALATION: move patient immediately to fresh air; administer artificial respiration or oxygen if necessary; seek medical attention.SKIN OR EYES: wash exposed skin surfaces thoroughly; flush eyes thoroughly with water for 15 min.
5.4 **Toxicity by Inhalation (Threshold Limit Value):** 50 ppm
5.5 **Short-Term Inhalation Limits:** Data not available
5.6 **Toxicity by Ingestion:** Grade 3; LD_{50} 50 to 500 mg/kg
5.7 **Late Toxicity:** None reported
5.8 **Vapor (Gas) Irritant Characteristics:** Vapors cause a slight smarting of the eyes or respiratory system if present in high concentrations. The effect is temporary
5.9 **Liquid or Solid Irritant Characteristics:** Minimum hazard. If spilled on clothing and allowed to remain, may cause smarting and reddening of the skin.
5.10 **Odor Threshold:** 1.2 ppm

6. FIRE HAZARDS

6.1 **Flash Point:** 111°F C.C.
6.2 **Flammable Limits in Air:** 0.9%—6.5%
6.3 **Fire Extinguishing Agents:** Foam, carbon dioxide or dry chemical
6.4 **Fire Extinguishing Agents Not to be Used:** Not pertinent
6.5 **Special Hazards of Combustion Products:** Not pertinent
6.6 **Behavior in Fire:** Not pertinent
6.7 **Ignition Temperature:** 797°F
6.8 **Electrical Hazard:** Data not available
6.9 **Burning Rate:** 5.0 mm/min.

7. CHEMICAL REACTIVITY

7.1 **Reactivity with Water:** No reaction
7.2 **Reactivity with Common Materials:** No reaction
7.3 **Stability During Transport:** Stable
7.4 **Neutralizing Agents for Acids and Caustics:** Not pertinent
7.5 **Polymerization:** Not pertinent
7.6 **Inhibitor of Polymerization:** Not pertinent

8. WATER POLLUTION

8.1 **Aquatic Toxicity:** 110 ppm/24 hr/Brine Shrimp/TL_m
8.2 **Waterfowl Toxicity:** Data not available
8.3 **Biological Oxygen Demand (BOD):** 40% of theoretical in 5 days, fresh water
8.4 **Food Chain Concentration Potential:** Data not available

9. SELECTED MANUFACTURERS

1. Gulf Oil Corp.
   Gulf Oil Chemicals Co. Div.
   Cedar Bayou, Texas 77520
2. Monsanto Co.
   Monsanto Polymers & Petrochemicals Co.
   800 North Lindbergh Blvd.
   St. Louis, Mo. 63166
3. Union Carbide Corp.
   Chemjcals & Plastics Div.
   270 Park Ave.
   New York, N. Y. 10017

10. SHIPPING INFORMATION

10.1 **Grades or Purity:** Research grade; pure grade; technical grade
10.2 **Storage Temperature:** Ambient
10.3 **Inert Atmosphere:** No requirement
10.4 **Venting:** Open (flame arrester)

11. HAZARD ASSESSMENT CODE

(See Hazard Assessment Handbook, CG 446-3)

A-T-U

12. HAZARD CLASSIFICATIONS

12.1 **Code of Federal Regulations:** Combustible Liquid

12.2 **NAS Hazard Rating for Bulk Water Transportation:**

Category	Rating
Fire	2
Health	
Vapor Irritant	1
Liquid or Solid Irritant	1
Poisons	1
Water Pollution	
Human Toxicity	1
Aquatic Toxicity	3
Aesthetic Effect	2
Reactivity	
Other Chemicals	1
Water	0
Self-Reaction	0

12.3 **NFPA Hazard Classifications:**

Category	Classification
Health Hazard (Blue)	0
Flammability (Red)	2
Reactivity (Yellow)	0

13. PHYSICAL AND CHEMICAL PROPERTIES

13.1 **Physical State at 15°C and 1 atm:** Liquid
13.2 **Molecular Weight:** 120.19
13.3 **Boiling Point at 1 atm:** 306.3°F = 152.4°C = 425.6°K
13.4 **Freezing Point:** −140.9°F = −96.1°C = 177.1°K
13.5 **Critical Temperature:** 676.2°F = 357.9°C = 631.1°K
13.6 **Critical Pressure:** 465.5 psia = 31.67 atm = 3.208 MN/m^2
13.7 **Specific Gravity:** 0.866 at 15°C (liquid)
13.8 **Liquid Surface Tension:** 28.2 dynes/cm = 0.0282 N/m at 20°C
13.9 **Liquid-Water Interfacial Tension:** 54.6 dynes/cm = 0.0546 N/m at 22.7°
13.10 **Vapor (Gas) Specific Gravity:** Not pertinent
13.11 **Ratio of Specific Heats of Vapor (Gas):** 1.059
13.12 **Latent Heat of Vaporization:** 134 Btu/lb = 74.6 cal/g = 3.12×10^5 J/kg
13.13 **Heat of Combustion:** −17,710 Btu/lb = −9840 cal/g = -412.0×10^5 J/kg
13.14 **Heat of Decomposition:** Not pertinent
13.15 **Heat of Solution:** Not pertinent
13.16 **Heat of Polymerization:** Not pertinent

NOTES

REVISED 1978

CUMENE HYDROPEROXIDE

Common Synonyms			
alpha, alpha-Dimethylbenzene hydroperoxide Dimethylbenzyl hydroperoxide CHP Isopropylbenzene hydroperoxide Cumyl hydroperoxide	Liquid	Colorless to light yellow	Sharp, irritating odor
	Sinks in water.		

Stop discharge if possible. Keep people away.
Call fire department.
Isolate and remove discharged material.
Notify local health and pollution control agencies.

Fire	Combustible. POISONOUS GASES MAY BE PRODUCED IN FIRE. Containers may explode in fire. **Extinguish with dry chemicals, foam or carbon dioxide.** **Water may be ineffective on fire.** **Cool exposed containers with water.**
Exposure	**Call for medical aid.** VAPOR Irritating to eyes, nose and throat. If inhaled will cause headache or coughing. **Move victim to fresh air.** **If breathing is difficult, give oxygen.** LIQUID Irritating to skin and eyes. Harmful if swallowed. **Remove contaminated clothing and shoes.** **Flush affected areas with plenty of water.** **IF IN EYES, hold eyelids open and flush with plenty of water.** **IF SWALLOWED and victim is CONSCIOUS, have victim drink water or milk and have victim induce vomiting.** **IF SWALLOWED and victim is UNCONSCIOUS OR HAVING CONVULSIONS, do nothing except keep victim warm.**
Water Pollution	Effect of low concentrations on aquatic life is unknown. Fouling to shoreline. May be dangerous if it enters water intakes. **Notify local health and wildlife officials.** **Notify operators of nearby water intakes.**

1. RESPONSE TO DISCHARGE

(See Response Methods Handbook, CG 446-4)

Issue warning—oxidizing material, water contaminant
Mechanical containment
Should be removed
Chemical and physical treatment

2. LABEL

Yellow

3. CHEMICAL DESIGNATIONS

3.1 **Synonyms:** alpha, alpha-Dimethylbenzene hydroperoxide; Dimethylbenzyl hydroperoxide; Isopropylbenzene hydroperoxide; Cumyl hydroperoxide; CHP

3.2 **Coast Guard Compatibility Classification:** Not applicable

3.3 **Chemical Formula:** $C_6H_5C(OOH)(CH_3)_2$—$C_6H_5CH(CH_3)_2$ (mixture)

(Continued on page 4)

4. OBSERVABLE CHARACTERISTICS

4.1 **Physical State (as shipped):** Liquid

4.2 **Color:** Colorless to pale yellow

4.3 **Odor:** Sharp, irritant, aromatic

5. HEALTH HAZARDS

5.1 **Personal Protective Equipment:** Self-contained or air-line breathing apparatus; solvent-resistant rubber gloves; chemical splash goggles; rubber apron; rubber or PVC clothing; full face shield.

5.2 **Symptoms Following Exposure:** Inhalation of vapor causes headache and burning throat. Liquid causes severe irritation of eyes; on skin, causes burning, throbbing sensation, irritation, and blisters. Ingestion causes irritation of mouth and stomach.

5.3 **Treatment for Exposure:** Get medical attention after all exposures to this compound. INHALATION: remove victim and administer artificial respiration and oxygen if necessary. EYES: flush with water for 15 min. SKIN: wash several times with soap and water; treat as burn. INGESTION: induce vomiting and follow with gastric lavage.

5.4 **Toxicity by Inhalation (Threshold Limit Value):** Data not available

5.5 **Short-Term Inhalation Limits:** Data not available

5.6 **Toxicity by Ingestion:** Grade 3; oral LD_{50} = 382 mg/kg (rat)

5.7 **Late Toxicity:** Data not available

5.8 **Vapor (Gas) Irritant Characteristics:** Data not available

5.9 **Liquid or Solid Irritant Characteristics:** Data not available

5.10 **Odor Threshold:** Data not available

6. FIRE HAZARDS

6.1 **Flash Point:** 147°F O.C.; 120°F C.C.

6.2 **Flammable Limits in Air:** 0.9%—6.5%

6.3 **Fire Extinguishing Agents:** Foam, dry chemical, or carbon dioxide

6.4 **Fire Extinguishing Agents Not to be Used:** Water may be ineffective.

6.5 **Special Hazards of Combustion Products:** Toxic phenol vapors may form from hot material.

6.6 **Behavior in Fire:** May decompose violently when heated. Burning rate becomes more rapid as fire burns.

6.7 **Ignition Temperature:** Decomposes violently at temperatures above 300°F

6.8 **Electrical Hazard:** Data not available

6.9 **Burning Rate:** Data not available

7. CHEMICAL REACTIVITY

7.1 **Reactivity with Water:** No reaction

7.2 **Reactivity with Common Materials:** Decomposition is catalyzed by metals such as aluminum, copper, brass, zinc, and lead. The reaction is not hazardous unless hot.

7.3 **Stability During Transport:** Stable if kept below 125°F and out of direct sunlight.

7.4 **Neutralizing Agents for Acids and Caustics:** Not pertinent

7.5 **Polymerization:** Not pertinent

7.6 **Inhibitor of Polymerization:** Not pertinent

8. WATER POLLUTION

8.1 **Aquatic Toxicity:** Data not available

8.2 **Waterfowl Toxicity:** Data not available

8.3 **Biological Oxygen Demand (BOD):** Data not available

8.4 **Food Chain Concentration Potential:** None

9. SELECTED MANUFACTURERS

1. Hercules Incorporated
   Organics Department
   Wilmington, Del. 19899
2. Reichhold Chemicals, Inc.
   Specialty Chemicals Division
   P. O. Box 9405
   Austin, Texas 78766
3. Allied Chemical Corporation
   Plastics Division
   P. O. Box 2365R
   Morristown, N. J. 07960

10. SHIPPING INFORMATION

10.1 **Grades or Purity:** 77-85%, the balance being cumene hydrocarbon.

10.2 **Storage Temperature:** Below 125°F

10.3 **Inert Atmosphere:** No requirement

10.4 **Venting:** Containers must be stored in well-ventilated area.

11. HAZARD ASSESSMENT CODE

(See Hazard Assessment Handbook, CG 446-3)

A-T-U-X-Y

12. HAZARD CLASSIFICATIONS

12.1 **Code of Federal Regulations:** Organic Peroxide

12.2 **NAS Hazard Rating for Bulk Water Transportation:** Not listed

12.3 **NFPA Hazard Classifications:**

Category	Classification
Health Hazard (Blue)	1
Flammability (Red)	2
Reactivity (Yellow)	4
	oxy

13. PHYSICAL AND CHEMICAL PROPERTIES

13.1 **Physical State at 15°C and 1 atm:** Liquid

13.2 **Molecular Weight:** Mixture

13.3 **Boiling Point at 1 atm:** Decomposes

13.4 **Freezing Point:** 16°F = −9°C = 264°K

13.5 **Critical Temperature:** Not pertinent

13.6 **Critical Pressure:** Not pertinent

13.7 **Specific Gravity:** 1.03 at 25°C (liquid)

13.8 **Liquid Surface Tension:** (est.) 25 dynes/cm = 0.025 N/m at 20°C

13.9 **Liquid-Water Interfacial Tension:** (est.) 30 dynes/cm = 0.030 N/m at 20°C

13.10 **Vapor (Gas) Specific Gravity:** Not pertinent

13.11 **Ratio of Specific Heats of Vapor (Gas):** Not pertinent

13.12 **Latent Heat of Vaporization:** Not pertinent

13.13 **Heat of Combustion:** (est.) −13,300 Btu/lb = −7,400 cal/g = -310×10^5 J/kg

13.14 **Heat of Decomposition:** −855 Btu/lb = −475 cal/g = -19.9×10^5 J/kg

13.15 **Heat of Solution:** Not pertinent

13.16 **Heat of Polymerization:** Not pertinent

3. CHEMICAL DESIGNATIONS (Cont'd.)

3.4 **IMCO/United Nations Numerical Designation:** 5.2/1524

CES — CUPRIETHYLENEDIAMINE SOLUTION

Common Synonyms	Liquid · Blue to dark purple · Fishy odor Sinks and mixes with water. Irritating vapor is produced.

AVOID CONTACT WITH LIQUID. KEEP PEOPLE AWAY.
Wear rubber overclothing (including gloves).
Stop discharge if possible.
Isolate and remove discharged material.
Notify local health and pollution control agencies.

Fire

Not flammable.
Irritating gases may be produced when heated.

Exposure

CALL FOR MEDICAL AID.

VAPOR
Irritating to eyes, nose and throat.
If inhaled will cause difficult breathing.
Move victim to fresh air.
If breathing is difficult, give oxygen.

LIQUID
POISONOUS IF SWALLOWED OR IF SKIN IS EXPOSED.
Irritating to skin and eyes.
Remove contaminated clothing and shoes.
Flush affected areas with plenty of water.
IF IN EYES, hold eyelids open and flush with plenty of water.
IF SWALLOWED and victim is CONSCIOUS, have victim drink water or milk and have victim induce vomiting.
IF SWALLOWED and victim is UNCONSCIOUS OR HAVING CONVULSIONS, do nothing except keep victim warm.

Water Pollution

Effect of low concentrations on aquatic life is unknown.
May be dangerous if it enters water intakes.
Notify local health and wildlife officials.
Notify operators of nearby water intakes.

1. RESPONSE TO DISCHARGE

(See Response Methods Handbook, CG 446-4)

Issue warning—corrosive, water contaminant
Restrict access
Disperse and flush

2. LABEL

3. CHEMICAL DESIGNATIONS

3.1 **Synonyms:** Cupriethylenediamine hydroxide solution
3.2 **Coast Guard Compatibility Classification:** Not applicable
3.3 **Chemical Formula:** $Cu(OH)_2 \cdot NH_2CH_2CH_2NH_2 \cdot H_2O$
3.4 **IMCO/United Nations Numerical Designation:** 8/1761

4. OBSERVABLE CHARACTERISTICS

4.1 **Physical State (as shipped):** Liquid
4.2 **Color:** Blue or dark purple (May contain red or blue sediment)
4.3 **Odor:** Like ammonia, fishy

5. HEALTH HAZARDS

5.1 **Personal Protective Equipment:** Goggles or face shield; organic canister mask; rubber gloves; protective clothing.
5.2 **Symptoms Following Exposure:** Inhalation of vapor irritates mucous membrane and may cause asthma. Liquid causes severe irritation of eyes and possible corneal injury. Contact with skin causes irritation. Ingestion causes irritation of mouth and stomach.
5.3 **Treatment for Exposure:** INHALATION: remove from exposure; support respiration; call physician. EYES: irrigate with copious quantities of water for at least 15 min.; call physician. SKIN: wash with large amounts of water. INGESTION: give large amounts of water.
5.4 **Toxicity by Inhalation (Threshold Limit Value):** Data not available
5.5 **Short-Term Inhalation Limits:** Data not available
5.6 **Toxicity by Ingestion:** Data not available
5.7 **Late Toxicity:** Data not available
5.8 **Vapor (Gas) Irritant Characteristics:** Data not available
5.9 **Liquid or Solid Irritant Characteristics:** Data not available
5.10 **Odor Threshold:** Data not available

6. FIRE HAZARDS

6.1 **Flash Point:** Non-flammable solution
6.2 **Flammable Limits in Air:** Not flammable
6.3 **Fire Extinguishing Agents:** Not pertinent
6.4 **Fire Extinguishing Agents Not to be Used:** Not pertinent
6.5 **Special Hazards of Combustion Products:** Not pertinent
6.6 **Behavior in Fire:** Irritating vapors of ethylenediamine may be produced when heated.
6.7 **Ignition Temperature:** Not pertinent
6.8 **Electrical Hazard:** Not pertinent
6.9 **Burning Rate:** Not pertinent

7. CHEMICAL REACTIVITY

7.1 **Reactivity with Water:** No reaction
7.2 **Reactivity with Common Materials:** Dissolves cotton, wood, and other cellulosic materials. Corrosive to copper, aluminum, zinc, and tin.
7.3 **Stability During Transport:** Stable
7.4 **Neutralizing Agents for Acids and Caustics:** Flush with water
7.5 **Polymerization:** Not pertinent
7.6 **Inhibitor of Polymerization:** Not pertinent

8. WATER POLLUTION

8.1 **Aquatic Toxicity:** Data not available
8.2 **Waterfowl Toxicity:** Data not available
8.3 **Biological Oxygen Demand (BOD):** Data not available
8.4 **Food Chain Concentration Potential:** None

9. SELECTED MANUFACTURERS

Olin Corporation
Ecusta Paper Division
Pisgah Forest, N. C. 28768

10. SHIPPING INFORMATION

10.1 **Grades or Purity:** Commercial
10.2 **Storage Temperature:** Ambient
10.3 **Inert Atmosphere:** Nitrogen
10.4 **Venting:** Pressure-vacuum

11. HAZARD ASSESSMENT CODE

(See Hazard Assessment Handbook, CG 446-3)

A-P-Q

12. HAZARD CLASSIFICATIONS

12.1 **Code of Federal Regulations:** Corrosive
12.2 **NAS Hazard Rating for Bulk Water Transportation:** Not listed
12.3 **NFPA Hazard Classifications:**

Category	Classification
Health Hazard (Blue)	4
Flammability (Red)	4
Reactivity (Yellow)	2

13. PHYSICAL AND CHEMICAL PROPERTIES

13.1 **Physical State at 15°C and 1 atm:** Liquid
13.2 **Molecular Weight:** Mixture
13.3 **Boiling Point at 1 atm:** (approx.) 212°F = 100°C = 373°K
13.4 **Freezing Point:** Not pertinent
13.5 **Critical Temperature:** Not pertinent
13.6 **Critical Pressure:** Not pertinent
13.7 **Specific Gravity:** (est.) > 1.1 at 20°C (liquid)
13.8 **Liquid Surface Tension:** Data not available
13.9 **Liquid-Water Interfacial Tension:** Not pertinent
13.10 **Vapor (Gas) Specific Gravity:** Not pertinent
13.11 **Ratio of Specific Heats of Vapor (Gas):** Not pertinent
13.12 **Latent Heat of Vaporization:** Not pertinent
13.13 **Heat of Combustion:** Not pertinent
13.14 **Heat of Decomposition:** Not pertinent
13.15 **Heat of Solution:** Not pertinent
13.16 **Heat of Polymerization:** Not pertinent

NOTES

CYA	CYANOACETIC ACID

Common Synonyms			
Cyanacetic acid Malonic mononitrile	Liquid	Yellowish brown	Unpleasant odor
	Sinks and mixes with water.		

Stop discharge if possible. Keep people away.
Shut off ignition sources. Call fire department.
Avoid contact with liquid and vapor.
Isolate and remove discharged material.
Notify local health and pollution control agencies.

Fire

Combustible.
POISONOUS GASES MAY BE PRODUCED IN FIRE.
Extinguish with water, dry chemicals, foam or carbon dioxide.

Exposure

CALL FOR MEDICAL AID.

VAPOR
Irritating to eyes, nose and throat.
Harmful if inhaled.
If in eyes, hold eyelids open and flush with plenty of water.
If breathing has stopped, give artificial respiration.
If breathing is difficult, give oxygen.

LIQUID
Irritating to skin and eyes.
Harmful if swallowed.
Remove contaminated clothing and shoes.
Flush affected areas with plenty of water.
IF IN EYES, hold eyelids open and flush with plenty of water.
IF SWALLOWED and victim is CONSCIOUS, have victim drink water or milk.
IF SWALLOWED and victim is UNCONSCIOUS OR HAVING CONVULSIONS, do nothing except keep victim warm.

Water Pollution

HARMFUL TO AQUATIC LIFE IN VERY LOW CONCENTRATIONS.
May be dangerous if it enters water intakes.
Notify local health and wildlife officials.
Notify operators of nearby water intakes.

1. RESPONSE TO DISCHARGE
(See Response Methods Handbook, CG 446-4)

Issue warning—water contaminant
Disperse and flush

2. LABELS

No hazard label required by Code of Federal Regulations

3. CHEMICAL DESIGNATIONS

3.1 **Synonyms:** Cyanacetic acid; Malonic mononitrile
3.2 **Coast Guard Compatibility Classification:** Not listed
3.3 **Chemical Formula:** $CNCH_2COOH$
3.4 **IMCO/United Nations Numerical Designation:** Not listed

4. OBSERVABLE CHARACTERISTICS

4.1 **Physical State (as shipped):** Solid
4.2 **Color:** White
4.3 **Odor:** Data not available

5. HEALTH HAZARDS

5.1 **Personal Protective Equipment:** Dust mask; goggles or face shield; rubber gloves
5.2 **Symptoms Following Exposure:** Contact irritates eyes and may irritate skin.
5.3 **Treatment for Exposure:** INHALATION: move to fresh air. INGESTION: give large amounts of water; get medical attention. EYES: flush with water for at least 15 min. SKIN: flush with water.
5.4 **Toxicity by Inhalation (Threshold Limit Value):** Data not available
5.5 **Short-Term Inhalation Limits:** Data not available
5.6 **Toxicity by Ingestion:** Data not available
5.7 **Late Toxicity:** Data not available
5.8 **Vapor (Gas) Irritant Characteristics:** Data not available
5.9 **Liquid or Solid Irritant Characteristics:** Data not available
5.10 **Odor Threshold:** Data not available

6. FIRE HAZARDS

6.1 **Flash Point:** Not pertinent (combustible solid)
6.2 **Flammable Limits in Air:** Not pertinent
6.3 **Fire Extinguishing Agents:** Water, foam, dry chemical, carbon dioxide
6.4 **Fire Extinguishing Agents Not to be Used:**
6.5 **Special Hazards of Combustion Products:** Toxic oxides of nitrogen and toxic and flammable acetonitrile vapors may form in fire.
6.6 **Behavior in Fire:**
6.7 **Ignition Temperature:** Not pertinent
6.8 **Electrical Hazard:** Not pertinent
6.9 **Burning Rate:** Not pertinent

7. CHEMICAL REACTIVITY

7.1 **Reactivity with Water:** No reaction
7.2 **Reactivity with Common Materials:**
7.3 **Stability During Transport:** Stable
7.4 **Neutralizing Agents for Acids and Caustics:** Not pertinent
7.5 **Polymerization:** Not pertinent
7.6 **Inhibitor of Polymerization:** Not pertinent

8. WATER POLLUTION

8.1 **Aquatic Toxicity:** Data not available
8.2 **Waterfowl Toxicity:** Data not available
8.3 **Biological Oxygen Demand (BOD):** Data not available
8.4 **Food Chain Concentration Potential:** None

9. SELECTED MANUFACTURERS

1. Kay-Fries Chemicals, Inc.
   360 Lexington Ave.
   New York, N. Y. 10017
2. Eastman Organic Chemicals
   Rochester, N. Y. 14650
3. Aldrich Chemical Co.
   940 W. Saint Paul Ave.
   Milwaukee, Wis. 53233

10. SHIPPING INFORMATION

10.1 **Grades or Purity:** 98+%
10.2 **Storage Temperature:** Ambient
10.3 **Inert Atmosphere:** No requirement
10.4 **Venting:** Open

11. HAZARD ASSESSMENT CODE
(See Hazard Assessment Handbook, CG 446-3)

SS

12. HAZARD CLASSIFICATIONS

12.1 **Code of Federal Regulations:** Not listed
12.2 **NAS Hazard Rating for Bulk Water Transportation:** Not listed
12.3 **NFPA Hazard Classifications:**

Category	Classification*	
Health Hazard (Blue)	0	3
Flammability (Red)	1	1
Reactivity (Yellow)	0	0

*First column refers to nonfire situation.

13. PHYSICAL AND CHEMICAL PROPERTIES

13.1 **Physical State at 15°C and 1 atm:** Solid
13.2 **Molecular Weight:** 85.06
13.3 **Boiling Point at 1 atm:** Not pertinent (decomposes)
13.4 **Freezing Point:** 151°F = 66°C = 339°K
13.5 **Critical Temperature:** Not pertinent
13.6 **Critical Pressure:** Not pertinent
13.7 **Specific Gravity:** >1.1 at 20°C (solid)
13.8 **Liquid Surface Tension:** Not pertinent
13.9 **Liquid-Water Interfacial Tension:** Not pertinent
13.10 **Vapor (Gas) Specific Gravity:** Not pertinent
13.11 **Ratio of Specific Heats of Vapor (Gas):** Not pertinent
13.12 **Latent Heat of Vaporization:** Not pertinent
13.13 **Heat of Combustion:** $-6{,}300$ Btu/lb $= -3{,}500$ cal/g $= -146 \times 10^5$ J/kg
13.14 **Heat of Decomposition:** Not pertinent
13.15 **Heat of Solution:** Data not available
13.16 **Heat of Polymerization:** Not pertinent

NOTES

CYG	# CYANOGEN

Common Synonyms	
Ethane dinitrile Dicyan Oxalic acid dinitrile Oxalonitrile Dicyanogen	Liquefied gas — Colorless — Almond Odor Floats and boils on water. Poisonous, flammable visible vapor cloud is produced.

AVOID CONTACT WITH LIQUID AND VAPOR. KEEP PEOPLE AWAY.
Wear goggles, self-contained breathing apparatus, and rubber overclothing (including gloves).
Shut off ignition sources. Call fire department.
Stop discharge if possible.
Evacuate area in case of large discharge.
Stay upwind. Use water spray to "knock down" vapor.
Isolate and remove discharged material.
Notify local health and pollution control agencies.

Fire

FLAMMABLE.
POISONOUS GASES ARE PRODUCED IN FIRE.
Containers may explode in fire.
Flashback along vapor trail may occur.
Vapor may explode if ignited in an enclosed area.
Wear goggles, self-contained breathing apparatus and rubber overclothing (including gloves).
Let fire burn.
Stop flow of gas if possible.
Cool exposed containers and protect men effecting shutoff with water.

Exposure

CALL FOR MEDICAL AID.

VAPOR
POISONOUS IF INHALED.
Irritating to eyes.
Move victim to fresh air.
If in eyes, hold eyelids open and flush with plenty of water.
If breathing has stopped, give artificial respiration (but not mouth-to-mouth).

LIQUID
POISONOUS IF SWALLOWED.
Will cause frostbite.
Remove contaminated clothing and shoes.
Flush affected areas with plenty of water.
DO NOT RUB AFFECTED AREAS.
IF IN EYES, hold eyelids open and flush with plenty of water.
IF SWALLOWED and victim is CONSCIOUS, have victim drink water or milk and have victim induce vomiting.
IF SWALLOWED and victim is UNCONSCIOUS OR HAVING CONVULSIONS, do nothing except keep victim warm.

Water Pollution

Effect of low concentrations on aquatic life is unknown.
May be dangerous if it enters water intakes.
Notify local health and wildlife officials.
Notify operators of nearby water intakes.

1. RESPONSE TO DISCHARGE

(See Response Methods Handbook, CG 446-4)

Issue warning—poison, high flammability, air contaminant, water contaminant
Restrict access
Evacuate area

2. LABELS

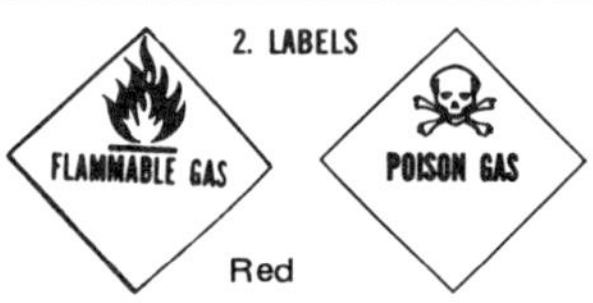

3. CHEMICAL DESIGNATIONS

3.1 **Synonyms:** Ethanedinitrile; Dicyan; Oxalic acid dinitrile; Oxalonitrile; Dicyanogen
3.2 **Coast Guard Compatibility Classification:** Not applicable
3.3 **Chemical Formula:** $(CN)_2$
3.4 **IMCO/United Nations Numerical Designation:** 2/1026

4. OBSERVABLE CHARACTERISTICS

4.1 **Physical State (as shipped):** Liquefied compressed gas
4.2 **Color:** Colorless
4.3 **Odor:** Characteristic almond-like; pungent, penetrating; may not be sufficiently strong to provide an adequate warning.

5. HEALTH HAZARDS

5.1 **Personal Protective Equipment:** Self-contained breathing apparatus; rubber gloves; rubber protective clothing; rubber-soled shoes.
5.2 **Symptoms Following Exposure:** Vapor irritates eyes and causes giddiness, headache, fatigue, and nausea if inhaled.
5.3 **Treatment for Exposure:** In general, treatment is similar to that used following exposure to hydrogen cyanide. INHALATION: move victim to fresh air and have him lie down; do not permit him to exert himself; remove contaminated clothing but keep patient covered and comfortably warm; summon a physician; break an amyl nitrite pearl in a cloth and hold it lightly under the victim's nose for 15 seconds; repeat five times at about 15-sec. intervals; use artificial respiration if breathing has stopped. EYES: flush with water for at least 15 min.
5.4 **Toxicity by Inhalation (Threshold Limit Value):** 10 ppm
5.5 **Short-Term Inhalation Limits:** 5 mg/m³ for 30 min. (as a cyanide)
5.6 **Toxicity by Ingestion:** Data not available
5.7 **Late Toxicity:** Data not available
5.8 **Vapor (Gas) Irritant Characteristics:** Data not available
5.9 **Liquid or Solid Irritant Characteristics:** Data not available
5.10 **Odor Threshold:** Data not available

6. FIRE HAZARDS

6.1 **Flash Point:** Flammable gas
6.2 **Flammable Limits in Air:** 6.6%–43%
6.3 **Fire Extinguishing Agents:** Let fire burn, shut off flow of gas, cool exposed areas with water.
6.4 **Fire Extinguishing Agents Not to be Used:** Not pertinent
6.5 **Special Hazards of Combustion Products:** Unburned vapors are highly toxic.
6.6 **Behavior in Fire:** Vapor is heavier than air and may travel considerable distance to a source of ignition and flash back. Containers may explode in fire, releasing the highly toxic gas.
6.7 **Ignition Temperature:** Data not available
6.8 **Electrical Hazard:** Data not available
6.9 **Burning Rate:** Not pertinent

7. CHEMICAL REACTIVITY

7.1 **Reactivity with Water:** No reaction, but water provides heat to vaporize liquid cyanogen.
7.2 **Reactivity with Common Materials:** No reaction
7.3 **Stability During Transport:** Stable
7.4 **Neutralizing Agents for Acids and Caustics:** Not pertinent
7.5 **Polymerization:** Not pertinent
7.6 **Inhibitor of Polymerization:** Not pertinent

8. WATER POLLUTION

8.1 **Aquatic Toxicity:** Data not available
8.2 **Waterfowl Toxicity:** Data not available
8.3 **Biological Oxygen Demand (BOD):** Data not available
8.4 **Food Chain Concentration Potential:** None

9. SELECTED MANUFACTURERS

1. Matheson Gas Products
   P. O. Box 85
   East Rutherford, N. J. 07073
2. Air Products and Chemicals, Inc.
   Specialty Gas Dept.
   P. O. Box 538
   Allentown, Pa. 18105
3. Union Carbide Corp.
   Linde Division
   270 Park Avenue
   New York, N. Y. 10017

10. SHIPPING INFORMATION

10.1 **Grades or Purity:** 98.5%
10.2 **Storage Temperature:** Cool ambient
10.3 **Inert Atmosphere:** No requirement
10.4 **Venting:** Store containers in well-ventilated area

11. HAZARD ASSESSMENT CODE

(See Hazard Assessment Handbook, CG 446-3)

A-B-C-K-L-M-N

12. HAZARD CLASSIFICATIONS

12.1 **Code of Federal Regulations:** Poisonous, Class A
12.2 **NAS Hazard Rating for Bulk Water Transportation:** Not listed
12.3 **NFPA Hazard Classifications:**

Category	Classification
Health Hazard (Blue)	4
Flammability (Red)	4
Reactivity (Yellow)	2

13. PHYSICAL AND CHEMICAL PROPERTIES

13.1 **Physical State at 15°C and 1 atm:** Gas
13.2 **Molecular Weight:** 52.0
13.3 **Boiling Point at 1 atm:** −6.1°F = −21.1°C = 252.1°K
13.4 **Freezing Point:** −18.2°F = −27.9°C = 245.3°K
13.5 **Critical Temperature:** 259.9°F = 126.6°C = 399.8°K
13.6 **Critical Pressure:** 857 psia = 58.2 atm = 5.91 MN/m²
13.7 **Specific Gravity:** 0.954 at −21°C (liquid)
13.8 **Liquid Surface Tension:** 22 dynes/cm = 0.022 N/m at −21°C
13.9 **Liquid-Water Interfacial Tension:** Not pertinent
13.10 **Vapor (Gas) Specific Gravity:** 1.8
13.11 **Ratio of Specific Heats of Vapor (Gas):** 1.205 at 25°C
13.12 **Latent Heat of Vaporization:** 200 Btu/lb = 111 cal/g = 4.65 × 10⁵ J/kg
13.13 **Heat of Combustion:** −9,059 Btu/lb = −5,033 cal/g = −210.6 × 10⁵ J/kg
13.14 **Heat of Decomposition:** Not pertinent
13.15 **Heat of Solution:** 2,520 Btu/lb = 1,400 cal/g = 58.5 × 10⁵ J/kg
13.16 **Heat of Polymerization:** Not pertinent

NOTES

CBR

CYANOGEN BROMIDE

Common Synonyms	Solid crystals	Colorless	Penetrating odor
	Sinks and mixes with water.		

AVOID CONTACT WITH SOLID. Keep people away.
Wear chemical protective suit with self-contained breathing apparatus.
Isolate and remove discharged material.
Notify local health and pollution control agencies.

Fire

Not flammable.
POISONOUS GASES ARE PRODUCED WHEN HEATED.
Wear chemical protective suit with self-contained breathing apparatus.
Cool exposed containers with water.

Exposure

CALL FOR MEDICAL AID.

DUST
POISONOUS IF INHALED OR IF SKIN IS EXPOSED.
Irritating to eyes.
Move to fresh air.
If breathing has stopped, give artificial respiration.
If breathing is difficult, give oxygen.

SOLIDS
POISONOUS IF SWALLOWED.
Will burn skin and eyes.
Remove contaminated clothing and shoes.
Flush affected areas with plenty of water.
IF IN EYES, hold eyelids open and flush with plenty of water.
IF SWALLOWED and victim is CONSCIOUS, have victim drink water or milk.
DO NOT INDUCE VOMITING.

Water Pollution

HARMFUL TO AQUATIC LIFE IN VERY LOW CONCENTRATIONS.
May be dangerous if it enters water intakes.
Notify local health and wildlife officials.
Notify operators of nearby water intakes.

1. RESPONSE TO DISCHARGE

(See Response Methods Handbook, CG 446-4)

Issue warning—poison
Restrict access
Disperse and flush

2. LABEL

3. CHEMICAL DESIGNATIONS

3.1 **Synonyms:** No common synonyms

3.2 **Coast Guard Compatibility Classification:** Not applicable

3.3 **Chemical Formula:** BrCN

3.4 **IMCO/United Nations Numerical Designation:** 6.1/1889

4. OBSERVABLE CHARACTERISTICS

4.1 **Physical State (as shipped):** Solid

4.2 **Color:** Colorless

4.3 **Odor:** Penetrating

5. HEALTH HAZARDS

5.1 **Personal Protective Equipment:** Chemical cartridge respirator, goggles, protective clothing, rubber gloves

5.2 **Symptoms Following Exposure:** Same symptoms as hydrogen cyanide. Because it irritates the eyes, throat, and lungs severely, it is unlikely that anyone would voluntarily remain in areas with a high enough concentration to exert a cyanide effect.

5.3 **Treatment for Exposure:** Call a physician. INHALATION: remove victim to fresh air; if he is not breathing, give artificial respiration, preferably mouth-to-mouth; if symptoms of cyanide poisoning are observed, administer amyl nitrite as instructed for HCN. INGESTION: have victim drink water or milk; do NOT induce vomiting.

5.4 **Toxicity by Inhalation (Threshold Limit Value):** 0.5 ppm (suggested)

5.5 **Short-Term Inhalation Limits:** Data not available

5.6 **Toxicity by Ingestion:** Data not available

5.7 **Late Toxicity:** Workers exposed to solutions may develop dermatitis.

5.8 **Vapor (Gas) Irritant Characteristics:** Vapors cause severe irritation of eyes and throat and can cause eye and lung injury. They cannot be tolerated even at low concentrations.

5.9 **Liquid or Solid Irritant Characteristics:** Severe skin irritant. Causes second- and third-degree burns on short contact; very injurious to the eyes.

5.10 **Odor Threshold:** Data not available

6. FIRE HAZARDS

6.1 **Flash Point:** Not flammable

6.2 **Flammable Limits in Air:** Not flammable

6.3 **Fire Extinguishing Agents:** Not pertinent

6.4 **Fire Extinguishing Agents Not to be Used:** Not pertinent

6.5 **Special Hazards of Combustion Products:** Poison gases are produced in fire.

6.6 **Behavior in Fire:** Not pertinent

6.7 **Ignition Temperature:** Not flammable

6.8 **Electrical Hazard:** Not pertinent

6.9 **Burning Rate:** Not flammable

7. CHEMICAL REACTIVITY

7.1 **Reactivity with Water:** No reaction

7.2 **Reactivity with Common Materials:** No reaction

7.3 **Stability During Transport:** Stable

7.4 **Neutralizing Agents for Acids and Caustics:** Strong bleaching powder solution; let stand 24 hr.

7.5 **Polymerization:** Does not occur

7.6 **Inhibitor of Polymerization:** Not pertinent

8. WATER POLLUTION

8.1 **Aquatic Toxicity:** Data not available

8.2 **Waterfowl Toxicity:** Data not available

8.3 **Biological Oxygen Demand (BOD):** Data not available

8.4 **Food Chain Concentration Potential:** None

9. SELECTED MANUFACTURERS

1. Apache Chemicals
   Rockford, Ill. 61105
2. Columbia Organic Chemicals, Inc.
   912 Drake St.
   Columbia, S. C. 29205
3. Eastman Kodak Co.
   Eastman Organic Chemicals Division
   Rochester, N. Y. 14650

10. SHIPPING INFORMATION

10.1 **Grades or Purity:** Data not available

10.2 **Storage Temperature:** Data not available

10.3 **Inert Atmosphere:** Data not available

10.4 **Venting:** Data not available

11. HAZARD ASSESSMENT CODE

(See Hazard Assessment Handbook, CG 446-3)

II

12. HAZARD CLASSIFICATIONS

12.1 **Code of Federal Regulations:** Poisonous liquid or solid, Class B

12.2 **NAS Hazard Rating for Bulk Water Transportation:**

Category	Rating
Fire	0
Health	
Vapor Irritant	4
Liquid or Solid Irritant	4
Poisons	4
Water Pollution	
Human Toxicity	4
Aquatic Toxicity	3
Aesthetic Effect	3
Reactivity	
Other Chemicals	3
Water	1
Self-Reaction	1

12.3 **NFPA Hazard Classifications:**

Category	Classification
Health Hazard (Blue)	3
Flammability (Red)	0
Reactivity (Yellow)	2

13. PHYSICAL AND CHEMICAL PROPERTIES

13.1 **Physical State at 15°C and 1 atm:** Solid

13.2 **Molecular Weight:** 105.93

13.3 **Boiling Point at 1 atm:** Not pertinent

13.4 **Freezing Point:** 120 to 124°F = 49 to 51°C = 322 to 324°K

13.5 **Critical Temperature:** Not pertinent

13.6 **Critical Pressure:** Not pertinent

13.7 **Specific Gravity:** 2.015 at 20°C (solid)

13.8 **Liquid Surface Tension:** Not pertinent

13.9 **Liquid-Water Interfacial Tension:** Not pertinent

13.10 **Vapor (Gas) Specific Gravity:** 3.6

13.11 **Ratio of Specific Heats of Vapor (Gas):** Not pertinent

13.12 **Latent Heat of Vaporization:** Not pertinent

13.13 **Heat of Combustion:** Not pertinent

13.14 **Heat of Decomposition:** Not pertinent

13.15 **Heat of Solution:** Not pertinent

13.16 **Heat of Polymerization:** Not pertinent

NOTES

REVISED 1978

CCL	# CYANOGEN CHLORIDE

Common Synonyms	Liquid or compressed gas · Colorless · Sharp, pungent odor Poisonous vapor cloud is produced. Boiling point is 56° F.

AVOID CONTACT WITH LIQUID AND VAPOR.
Keep people away. Evacuate area in case of large discharges.
Wear chemical protective suit with self-contained breathing apparatus.
Stop discharge if possible.
Stay upwind and use water spray to "knock down" vapor.
Isolate and remove discharged material.
Notify local health and pollution control agencies.

Fire	Not flammable. POISONOUS GASES ARE PRODUCED WHEN HEATED IN FIRE. Wear chemical protective suit with self-contained breathing apparatus. Cool exposed containers with water.
Exposure	CALL FOR MEDICAL AID. VAPOR POISONOUS IF INHALED OR IF SKIN IS EXPOSED. Irritating to eyes. Move to fresh air. If breathing has stopped, give artificial respiration. If breathing is difficult, give oxygen. LIQUID POISONOUS IF SWALLOWED. Will burn skin and eyes. Remove contaminated clothing and shoes. Flush affected areas with plenty of water. IF IN EYES, hold eyelids open and flush with plenty of water. IF SWALLOWED and victim is CONSCIOUS, have victim drink water or milk. DO NOT INDUCE VOMITING.
Water Pollution	HARMFUL TO AQUATIC LIFE IN VERY LOW CONCENTRATIONS. May be dangerous if it enters water intakes. Notify local health and wildlife officials. Notify operators of nearby water intakes.

1. RESPONSE TO DISCHARGE
(See Response Methods Handbook, CG 446-4)

Issue warning—poison
Restrict access
Evacuate area

2. LABELS

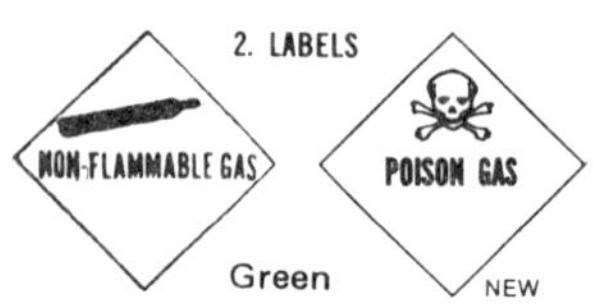

3. CHEMICAL DESIGNATIONS

3.1 **Synonyms:** No common synonyms
3.2 **Coast Guard Compatibility Classification:** Not applicable
3.3 **Chemical Formula:** CNCl
3.4 **IMCO/United Nations Numerical Designation:** 6.1/1589

4. OBSERVABLE CHARACTERISTICS

4.1 **Physical State (as shipped):** Compressed gas
4.2 **Color:** Colorless
4.3 **Odor:** Acrid, choking

5. HEALTH HAZARDS

5.1 **Personal Protective Equipment:** Chemical cartridge respirator, goggles, protective clothing.
5.2 **Symptoms Following Exposure:** Similar in toxicity and mode of action to hydrogen cyanide, but is much more irritating. Can cause a marked irritation of the respiratory tract with a hemorrhagic exudate of the bronchi and trachea and pulmonary edema. It is improbable that anyone would voluntarily remain in areas with a high enough concentration to exert a typical nitrile effect.
5.3 **Treatment for Exposure:** INHALATION: support respiration and administer oxygen; call a doctor; if nitrile effect is seen, administer amyl nitrite. INGESTION: have victim drink water or milk; do NOT induce vomiting.
5.4 **Toxicity by Inhalation (Threshold Limit Value):** >0.5 ppm
5.5 **Short-Term Inhalation Limits:** Data not available
5.6 **Toxicity by Ingestion:** Not pertinent
5.7 **Late Toxicity:** Long term exposure causes dermatitis, loss of appetite, headache, upper respiratory irritation in humans.
5.8 **Vapor (Gas) Irritant Characteristics:** Vapors cause severe irritation of eyes and throat and can cause eye and lung injury. They cannot be tolerated even at low concentrations.
5.9 **Liquid or Solid Irritant Characteristics:** Severe skin irritant. Causes second- and third-degree burns on short contact; very injurious to the eyes.
5.10 **Odor Threshold:** 1 ppm

6. FIRE HAZARDS

6.1 **Flash Point:** Not flammable
6.2 **Flammable Limits in Air:** Not flammable
6.3 **Fire Extinguishing Agents:** Not pertinent
6.4 **Fire Extinguishing Agents Not to be Used:** Not pertinent
6.5 **Special Hazards of Combustion Products:** Not pertinent
6.6 **Behavior in Fire:** Overheated containers can explode
6.7 **Ignition Temperature:** Not flammable
6.8 **Electrical Hazard:** Not pertinent
6.9 **Burning Rate:** Not flammable

7. CHEMICAL REACTIVITY

7.1 **Reactivity with Water:** Very slow reaction
7.2 **Reactivity with Common Materials:** Slow, not immediately hazardous
7.3 **Stability During Transport:** Stable
7.4 **Neutralizing Agents for Acids and Caustics:** Not pertinent
7.5 **Polymerization:** Not pertinent
7.6 **Inhibitor of Polymerization:** Not pertinent

8. WATER POLLUTION

8.1 **Aquatic Toxicity:** 0.08 ppm/*/fish/killed/fresh water
*Duration not specified
8.2 **Waterfowl Toxicity:** Data not available
8.3 **Biological Oxygen Demand (BOD):** Data not available
8.4 **Food Chain Concentration Potential:** None

9. SELECTED MANUFACTURERS

1. Gallard-Schlesinger Chemical Manufacturing Corp.
Atomergic Chemetals Co. Division
584 Mineola Ave.
Carle Place, N. Y. 11514
2. Nilok Chemicals Inc.
5030 Millington Road
Memphis, Tenn. 38127

10. SHIPPING INFORMATION

10.1 **Grades or Purity:** Data not available
10.2 **Storage Temperature:** Data not available
10.3 **Inert Atmosphere:** Data not available
10.4 **Venting:** Data not available

11. HAZARD ASSESSMENT CODE
(See Hazard Assessment Handbook, CG 446-3)

A-C-I-J-X

12. HAZARD CLASSIFICATIONS

12.1 **Code of Federal Regulations:** Poisonous gas or liquid, Class A
12.2 **NAS Hazard Rating for Bulk Water Transportation:** Not listed
12.3 **NFPA Hazard Classifications:** Not listed

13. PHYSICAL AND CHEMICAL PROPERTIES

13.1 **Physical State at 15°C and 1 atm:** Gas
13.2 **Molecular Weight:** 61.48
13.3 **Boiling Point at 1 atm:** 55.6°F = 13.1°C = 286.3°K
13.4 **Freezing Point:** 20°F = −6.9°C = 266.3°K
13.5 **Critical Temperature:** Not pertinent
13.6 **Critical Pressure:** Not pertinent
13.7 **Specific Gravity:** 1.222 at 0°C (liquid)
13.8 **Liquid Surface Tension:** 24.6 dynes/cm = 0.0246 N/m at 10°C
13.9 **Liquid-Water Interfacial Tension:** Data not available
13.10 **Vapor (Gas) Specific Gravity:** 2.1
13.11 **Ratio of Specific Heats of Vapor (Gas):** 1.229
13.12 **Latent Heat of Vaporization:** 191.3 Btu/lb = 106.3 cal/g = 4.451×10^5 J/kg
13.13 **Heat of Combustion:** Not pertinent
13.14 **Heat of Decomposition:** Not pertinent
13.15 **Heat of Solution:** Not pertinent
13.16 **Heat of Polymerization:** Not pertinent

NOTES

REVISED 1978

CHX	CYCLOHEXANE

Common Synonyms			
Hexahydrobenzene	Watery liquid	Colorless	Gasoline-like odor
	Floats on water. Flammable irritating vapor is produced. Freezing point is 44° F.		

Stop discharge if possible. Keep people away.
Shut off ignition sources and call fire department.
Avoid contact with liquid and vapor.
Stay upwind and use water spray to "knock down" vapor.
Isolate and remove discharged material.
Notify local health and pollution control agencies.

Fire

FLAMMABLE.
Flashback along vapor trail may occur.
Vapor may explode if ignited in an enclosed area.
Extinguish with foam, dry chemicals or carbon dioxide.
Water may be ineffective on fire.
Cool exposed containers with water.

Exposure

CALL FOR MEDICAL AID.

VAPOR
Irritating to eyes, nose and throat.
If inhaled, will cause dizziness, nausea, vomiting or loss of consciousness.
Move to fresh air.
If breathing has stopped, give artificial respiration.
If breathing is difficult, give oxygen.

LIQUID
Irritating to skin and eyes.
Harmful if swallowed.
Remove contaminated clothing and shoes.
Flush affected areas with plenty of water.
IF IN EYES, hold eyelids open and flush with plenty of water.
IF SWALLOWED, and victim is CONSCIOUS, have victim drink water or milk.

Water Pollution

Dangerous to aquatic life in high concentrations.
Fouling to shoreline.
May be dangerous if it enters water intakes.
Notify local health and pollution control officials.
Notify operators of nearby water intakes.

1. RESPONSE TO DISCHARGE

(See Response Methods Handbook, CG 446-4)

Issue warning—high flammability
Evacuate area
Disperse and flush

2. LABEL

Red

3. CHEMICAL DESIGNATIONS

3.1 **Synonyms:** Hexahydrobenzene; Hexamethylene; Hexanaphthene
3.2 **Coast Guard Compatibility Classification:** Paraffin
3.3 **Chemical Formula:** $[(CH_2)_6]$
3.4 **IMCO/United Nations Numerical Designation:** 3.1/1145

4. OBSERVABLE CHARACTERISTICS

4.1 **Physical State (as shipped):** Liquid
4.2 **Color:** Colorless
4.3 **Odor:** Resembling benzene; mild, sweet, resembling chloroform

5. HEALTH HAZARDS

5.1 **Personal Protective Equipment:** Hydrocarbon vapor canister, supplied-air or hose mask, hydrocarbon-insoluble rubber or plastic gloves, chemical goggles or face splash shield, hydrocarbon-insoluble rubber or plastic apron.
5.2 **Symptoms Following Exposure:** Dizziness, with nausea and vomiting. Concentrated vapor may cause unconsciousness and collapse.
5.3 **Treatment for Exposure:** INHALATION: remove victim to fresh air; if breathing stops, apply artificial respiration and administer oxygen. SKIN OR EYE CONTACT: remove contaminated clothing and gently flush affected areas with water for 15 min.; call a physician.
5.4 **Toxicity by Inhalation (Threshold Limit Value):** 300 ppm
5.5 **Short-Term Inhalation Limits:** 300 ppm for 60 min.
5.6 **Toxicity by Ingestion:** Grade 2; LD_{50} 0.5 to 5 g/kg
5.7 **Late Toxicity:** None
5.8 **Vapor (Gas) Irritant Characteristics:** Vapors cause a slight smarting of the eyes or respiratory system if present in high concentrations. The effect is temporary.
5.9 **Liquid or Solid Irritant Characteristics:** Minimum hazard. If spilled on clothing and allowed to remain, may cause smarting and reddening of the skin.
5.10 **Odor Threshold:** Data not available

6. FIRE HAZARDS

6.1 **Flash Point:** −4°F C.C.
6.2 **Flammable Limits in Air:** 1.33%—8.35%
6.3 **Fire Extinguishing Agents:** Foam, carbon dioxide, dry chemical.
6.4 **Fire Extinguishing Agents Not to be Used:** Not pertinent
6.5 **Special Hazards of Combustion Products:** Not pertinent
6.6 **Behavior in Fire:** Not pertinent
6.7 **Ignition Temperature:** 518°F
6.8 **Electrical Hazard:** Data not available
6.9 **Burning Rate:** 6.9 mm/min.

7. CHEMICAL REACTIVITY

7.1 **Reactivity with Water:** No reaction
7.2 **Reactivity with Common Materials:** No reaction
7.3 **Stability During Transport:** Stable
7.4 **Neutralizing Agents for Acids and Caustics:** Not pertinent
7.5 **Polymerization:** Not pertinent
7.6 **Inhibitor of Polymerization:** Not pertinent

8. WATER POLLUTION

8.1 **Aquatic Toxicity:** 15,500 ppm/24 hr/mosquito fish/TL_m/fresh water
8.2 **Waterfowl Toxicity:** Data not available
8.3 **Biological Oxygen Demand (BOD):** Data not available
8.4 **Food Chain Concentration Potential:** None

9. SELECTED MANUFACTURERS

1. Exxon Chemical Co.
   Houston, Texas 77001
2. Phillips Petroleum Co.
   Bartlesville, Okla. 74004
3. Texaco, Inc.
   135 East 42nd St.
   New York, N. Y. 10017

10. SHIPPING INFORMATION

10.1 **Grades or Purity:** Research grades: 99.5%, 98.0%; commercial: 85-98%
10.2 **Storage Temperature:** Ambient
10.3 **Inert Atmosphere:** No requirement
10.4 **Venting:** Open (flame arrester) or pressure-vacuum

11. HAZARD ASSESSMENT CODE

(See Hazard Assessment Handbook, CG 446-3)

A-T-U-V-W

12. HAZARD CLASSIFICATIONS

12.1 **Code of Federal Regulations:** Flammable liquid
12.2 **NAS Hazard Rating for Bulk Water Transportation:**

Category	Rating
Fire	3
Health	
Vapor Irritant	1
Liquid or Solid Irritant	1
Poisons	2
Water Pollution	
Human Toxicity	1
Aquatic Toxicity	2
Aesthetic Effect	2
Reactivity	
Other Chemicals	0
Water	0
Self-Reaction	0

12.3 **NFPA Hazard Classifications:**

Category	Classification
Health Hazard (Blue)	1
Flammability (Red)	3
Reactivity (Yellow)	0

13. PHYSICAL AND CHEMICAL PROPERTIES

13.1 **Physical State at 15°C and 1 atm:** Liquid
13.2 **Molecular Weight:** 84.16
13.3 **Boiling Point at 1 atm:** 177.3°F = 80.7°C = 353.9°K
13.4 **Freezing Point:** 43.8°F = 6.6°C = 279.8°K
13.5 **Critical Temperature:** 536.5°F = 280.3°C = 553.5°K
13.6 **Critical Pressure:** 591 psia = 40.2 atm = 4.07 MN/m^2
13.7 **Specific Gravity:** 0.779 at 20°C (liquid)
13.8 **Liquid Surface Tension:** 24.6 dynes/cm = 0.0246 N/m at 20°C
13.9 **Liquid-Water Interfacial Tension:** 50 dynes/cm = 0.050 N/m at 25°C
13.10 **Vapor (Gas) Specific Gravity:** 2.9
13.11 **Ratio of Specific Heats of Vapor (Gas):** 1.087
13.12 **Latent Heat of Vaporization:** 150 Btu/lb = 85 cal/g = 3.6×10^5 J/kg
13.13 **Heat of Combustion:** −18,684 Btu/lb = −10,380 cal/g = -434.59×10^5 J/kg
13.14 **Heat of Decomposition:** Not pertinent
13.15 **Heat of Solution:** Not pertinent
13.16 **Heat of Polymerization:** Not pertinent

NOTES

REVISED 1978

CHN — CYCLOHEXANOL

Common Synonyms			
Hexalin Hexahydrophenol Cyclohexyl alcohol Hydroxycyclohexane	Oily liquid	Colorless to light yellow	Alcohol odor
	Floats and mixes slowly with water. May solidify. Freezing point is 75° F		

Avoid contact with liquid. Keep people away.
Wear rubber overclothing (including gloves).
Stop discharge if possible.
Isolate and remove discharged material.
Notify local health and pollution control agencies.

Fire

Combustible.
Extinguish with water, dry chemical, foam, or carbon dioxide.
Cool exposed containers with water.

Exposure

CALL FOR MEDICAL AID.

LIQUID OR SOLID
Will burn skin and eyes.
Harmful if swallowed.

Remove contaminated clothing and shoes.
Flush affected areas with plenty of water.
IF IN EYES, hold eyelids open and flush with plenty of water.
IF SWALLOWED and victim is CONSCIOUS, have victim drink water or milk.

Water Pollution

Effect of low concentrations on aquatic life is unknown.
May be dangerous if it enters water intakes.
Notify local health and pollution control officials.
Notify operators of nearby water intakes.

1. RESPONSE TO DISCHARGE

(See Response Methods Handbook, CG 446-4)

Disperse and flush

2. LABELS

No hazard label required by Code of Federal Regulations

3. CHEMICAL DESIGNATIONS

3.1 **Synonyms:**

Adronal	Hexalin
Anol	Hexahydrophenol
Cyclohexyl alcohol	Hydroxycyclohexane

3.2 **Coast Guard Compatibility Classification:** Alcohol

3.3 **Chemical Formula:** $\overline{(CH_2)_5CHOH}$ (ring)

3.4 **IMCO/United Nations Numerical Designation:** Not listed

4. OBSERVABLE CHARACTERISTICS

4.1 **Physical State (as shipped):** Solid or liquid

4.2 **Color:** Colorless to faintly yellow

4.3 **Odor:** Like camphor

5. HEALTH HAZARDS

5.1 **Personal Protective Equipment:** Goggles or face shield

5.2 **Symptoms Following Exposure:** Narcosis—depression of the central nervous system tending to produce sleep or unconsciousness.

5.3 **Treatment for Exposure:** Eye contact is more hazardous than inhalation, skin irritation, or ingestion. Flush eyes with water and remove victim to fresh air.

5.4 **Toxicity by Inhalation (Threshold Limit Value):** 50 ppm

5.5 **Short-Term Inhalation Limits:** Data not available

5.6 **Toxicity by Ingestion:** Grade 2; LD_{50} 0.5 to 5 g/kg

5.7 **Late Toxicity:** Data not available

5.8 **Vapor (Gas) Irritant Characteristics:** Vapors cause a slight smarting of the eyes or respiratory system if present in high concentrations. The effect is temporary.

5.9 **Liquid or Solid Irritant Characteristics:** Causes smarting of the skin and first-degree burns on short exposure and may cause secondary burns on long exposure.

5.10 **Odor Threshold:** Data not available

6. FIRE HAZARDS

6.1 **Flash Point:** 160°F O.C.; 154°F C.C.

6.2 **Flammable Limits in Air:** Data not available

6.3 **Fire Extinguishing Agents:** Water, foam, carbon dioxide, or dry chemical.

6.4 **Fire Extinguishing Agents Not to be Used:** Not pertinent

6.5 **Special Hazards of Combustion Products:** Not pertinent

6.6 **Behavior in Fire:** Not pertinent

6.7 **Ignition Temperature:** 572°F

6.8 **Electrical Hazard:** Data not available

6.9 **Burning Rate:** 3.9 mm/min.

7. CHEMICAL REACTIVITY

7.1 **Reactivity with Water:** No reaction

7.2 **Reactivity with Common Materials:** No reaction

7.3 **Stability During Transport:** Stable

7.4 **Neutralizing Agents for Acids and Caustics:** Not pertinent

7.5 **Polymerization:** Not pertinent

7.6 **Inhibitor of Polymerization:** Not pertinent

8. WATER POLLUTION

8.1 **Aquatic Toxicity:** Data not available

8.2 **Waterfowl Toxicity:** Data not available

8.3 **Biological Oxygen Demand (BOD):** 0.08 lb/lb, 5 days

8.4 **Food Chain Concentration Potential:** None

9. SELECTED MANUFACTURERS

1. Allied Chemical Corp.
   Specialty Chemicals Division
   Hopewell, Va. 23860
2. Monsanto Co.
   Monsanto Industrial Chemicals Co.
   St. Louis, Mo. 63166
3. Union Carbide Corp.
   Chemicals and Plastics Division
   270 Park Ave.
   New York, N. Y. 10017

10. SHIPPING INFORMATION

10.1 **Grades or Purity:** Technical; pure

10.2 **Storage Temperature:** Data not available

10.3 **Inert Atmosphere:** Data not available

10.4 **Venting:** Data not available

11. HAZARD ASSESSMENT CODE

(See Hazard Assessment Handbook, CG 446-3)

A-P-Q-T-U

12. HAZARD CLASSIFICATIONS

12.1 **Code of Federal Regulations:** Combustible Liquid

12.2 **NAS Hazard Rating for Bulk Water Transportation:**

Category	Rating
Fire	1
Health	
Vapor Irritant	1
Liquid or Solid Irritant	2
Poisons	1
Water Pollution	
Human Toxicity	1
Aquatic Toxicity	3
Aesthetic Effect	2
Reactivity	
Other Chemicals	2
Water	0
Self-Reaction	0

12.3 **NFPA Hazard Classifications:**

Category	Classification
Health Hazard (Blue)	1
Flammability (Red)	2
Reactivity (Yellow)	0

13. PHYSICAL AND CHEMICAL PROPERTIES

13.1 **Physical State at 15°C and 1 atm:** Solid

13.2 **Molecular Weight:** 100.16

13.3 **Boiling Point at 1 atm:** 322°F = 161°C = 434°K

13.4 **Freezing Point:** 74.5°F = 23.6°C = 296.8°K

13.5 **Critical Temperature:** 666°F = 352°C = 625°K

13.6 **Critical Pressure:** 540 psia = 37 atm = 3.7 MN/m^2

13.7 **Specific Gravity:** 0.947 at 20°C (liquid)

13.8 **Liquid Surface Tension:** 34.2 dynes/cm = 0.0342 N/m at 16.2°C

13.9 **Liquid-Water Interfacial Tension:** 3.9 dynes/cm = 0.0039 N/m at 25°C

13.10 **Vapor (Gas) Specific Gravity:** Not pertinent

13.11 **Ratio of Specific Heats of Vapor (Gas):** 1.071

13.12 **Latent Heat of Vaporization:** 196 Btu/lb = 109 cal/g = 4.56×10^5 J/kg

13.13 **Heat of Combustion:** -16,000 Btu/lb = -8910 cal/g = -373×10^5 J/kg

13.14 **Heat of Decomposition:** Not pertinent

13.15 **Heat of Solution:** Not pertinent

13.16 **Heat of Polymerization:** Not pertinent

NOTES

REVISED 1978

CCH — CYCLOHEXANONE

Common Synonyms			
Cyclohexyl Ketone	Watery liquid	Colorless to light yellow	Sweet, peppermint odor
	Floats and mixes slowly with water.		

Avoid contact with liquid. Keep people away.
Wear goggles, self-contained breathing apparatus, and rubber overclothing (including gloves).
Stop discharge if possible.
Call fire department.
Isolate and remove discharged material.
Notify local health and pollution control agencies.

Fire

Combustible.
Extinguish with dry chemicals, foam, or carbon dioxide.

Exposure

CALL FOR MEDICAL AID.

LIQUID
Will burn skin and eyes.
Harmful if swallowed.
Remove contaminated clothing and shoes.
Flush affected areas with plenty of water.
IF IN EYES, hold eyelids open and flush with plenty of water.
IF SWALLOWED, and victim is CONSCIOUS, have victim drink water or milk.

Water Pollution

Effect of low concentrations on aquatic life is unknown.
May be dangerous if it enters water intakes.
Notify local health and pollution control officials.
Notify operators of nearby water intakes.

1. RESPONSE TO DISCHARGE

(See Response Methods Handbook, CG 446-4)

Disperse and flush

2. LABELS

No hazard label required by Code of Federal Regulations

3. CHEMICAL DESIGNATIONS

3.1 **Synonyms:** Anone
Cyclohexyl ketone
Hytrol O
Nadone
Pimelic ketone
Sextone

3.2 **Coast Guard Compatibility Classification:** Ketone

3.3 **Chemical Formula:** $\overline{(CH_2)_5CO}$

3.4 **IMCO/United Nations Numerical Designation:** 3.3/1915

4. OBSERVABLE CHARACTERISTICS

4.1 **Physical State (as shipped):** Liquid

4.2 **Color:** Colorless to slightly yellow

4.3 **Odor:** Like peppermint and acetone

5. HEALTH HAZARDS

5.1 **Personal Protective Equipment:** Chemical goggles.

5.2 **Symptoms Following Exposure:** Inhalation of vapors from hot material can cause narcosis. The liquid may cause dermatitis.

5.3 **Treatment for Exposure:** Immediately flush eyes with plenty of water; call a physician.

5.4 **Toxicity by Inhalation (Threshold Limit Value):** 50 ppm

5.5 **Short-Term Inhalation Limits:** Data not available

5.6 **Toxicity by Ingestion:** Grade 2; LD_{50} 0.5 to 5 g/kg

5.7 **Late Toxicity:** Data not available

5.8 **Vapor (Gas) Irritant Characteristics:** Vapor is moderately irritating such that personnel will not usually tolerate moderate or high vapor concentrations.

5.9 **Liquid or Solid Characteristics:** Causes smarting of the skin and first-degree burns on short exposure and may cause secondary burns on long exposure.

5.10 **Odor Threshold:** 0.12 ppm

6. FIRE HAZARDS

6.1 **Flash Point:** 129°F O.C.; 111°F C.C.

6.2 **Flammable Limits in Air:** 1.1% LFL

6.3 **Fire Extinguishing Agents:** Water, dry chemical or carbon dioxide

6.4 **Fire Extinguishing Agents Not to be Used:** Not pertinent

6.5 **Special Hazards of Combustion Products:** Not pertinent

6.6 **Behavior in Fire:** Not pertinent

6.7 **Ignition Temperature:** 788°F

6.8 **Electrical Hazard:** Data not available

6.9 **Burning Rate:** 4.2 mm/min.

7. CHEMICAL REACTIVITY

7.1 **Reactivity with Water:** No reaction

7.2 **Reactivity with Common Materials:** No reaction

7.3 **Stability During Transport:** Stable

7.4 **Neutralizing Agents for Acids and Caustics:** Not pertinent

7.5 **Polymerization:** Not pertinent

7.6 **Inhibitor of Polymerization:** Not pertinent

8. WATER POLLUTION

8.1 **Aquatic Toxicity:** Data not available

8.2 **Waterfowl Toxicity:** Data not available

8.3 **Biological Oxygen Demand (BOD):** Data not available

8.4 **Food Chain Concentration Potential:** None

9. SELECTED MANUFACTURERS

1. Allied Chemical Corp.
   Specialty Chemicals Div.
   Hopewell, Virginia 23860
2. E. I. du Pont de Nemours & Co.
   Industrial & Biochemicals Dept.
   Wilmington, Delaware 19898
3. Monsanto Co.
   Monsanto Industrial Chemicals Co.
   800 North Lindbergh Blvd.
   St. Louis, Mo. 63166

10. SHIPPING INFORMATION

10.1 **Grades or Purity:** Technical: 99.87%

10.2 **Storage Temperature:** Data not available

10.3 **Inert Atmosphere:** Data not available

10.4 **Venting:** Data not available

11. HAZARD ASSESSMENT CODE

(See Hazard Assessment Handbook, CG 446-3)

A-P-Q-T-U

12. HAZARD CLASSIFICATIONS

12.1 **Code of Federal Regulations:** Combustible Liquid

12.2 **NAS Hazard Rating for Bulk Water Transportation:**

Category	Rating
Fire	2
Health	
Vapor Irritant	3
Liquid or Solid Irritant	2
Poisons	1
Water Pollution	
Human Toxicity	1
Aquatic Toxicity	3
Aesthetic Effect	2
Reactivity	
Other Chemicals	2
Water	0
Self-Reaction	0

12.3 **NFPA Hazard Classifications:** Not listed

13. PHYSICAL AND CHEMICAL PROPERTIES

13.1 **Physical State at 15°C and 1 atm:** Liquid

13.2 **Molecular Weight:** 98.15

13.3 **Boiling Point at 1 atm:** 312.4°F = 155.8°C = 429.0°K

13.4 **Freezing Point:** −24.2°F = −31.2°C = 242.0°K

13.5 **Critical Temperature:** 673°F = 356°C = 629°K

13.6 **Critical Pressure:** 560 psia = 38 atm = 3.8 MN/m^2

13.7 **Specific Gravity:** 0.945 at 20°C (liquid)

13.8 **Liquid Surface Tension:** 34 dynes/cm = 0.034 N/m at 20°C

13.9 **Liquid-Water Interfacial Tension:** 90 dynes/cm = 0.090 N/m at 22.7°C

13.10 **Vapor (Gas) Specific Gravity:** Not pertinent

13.11 **Ratio of Specific Heats of Vapor (Gas):** 1.084

13.12 **Latent Heat of Vaporization:** 160 Btu/lb = 91 cal/g = 3.8×10^5 J/kg

13.13 **Heat of Combustion:** −15,430 Btu/lb = −8570 cal/g = -358.8×10^5 J/kg

13.14 **Heat of Decomposition:** Not pertinent

13.15 **Heat of Solution:** Not pertinent

13.16 **Heat of Polymerization:** Not pertinent

NOTES

REVISED 1978

CHP | CYCLOHEXANONE PEROXIDE

Common Synonyms			
1-Hydroperoxycyclohexyl 1-hydroxycyclohexyl peroxide Hydroperoxycyclohexyl hydroxycyclohexyl peroxide Dicyclohexanone diperoxide	Thick liquid Sinks in water.	White	Odorless

Stop discharge if possible. Keep people away.
Call fire department.
Isolate and remove discharged material.
Notify local health and pollution control agencies.

Fire	Combustible. Containers may explode in fire. Extinguish with water, dry chemicals, foam, or carbon dioxide. Cool exposed containers with water.
Exposure	Call for medical aid. LIQUID Irritating to skin and eyes. Harmful if swallowed. Remove contaminated clothing and shoes. Flush affected areas with plenty of water. IF IN EYES, hold eyelids open and flush with plenty of water. IF SWALLOWED and victim is CONSCIOUS, have victim drink water or milk.
Water Pollution	Effect of low concentrations on aquatic life is unknown. Fouling to shoreline. May be dangerous if it enters water intakes. Notify local health and wildlife officials. Notify operators of nearby water intakes.

1. RESPONSE TO DISCHARGE

(See Response Methods Handbook, CG 446-4)

Issue warning - oxidizing material, water contaminant
Mechanical containment
Should be removed
Chemical and physical treatment

2. LABEL

Yellow

3. CHEMICAL DESIGNATIONS

3.1 **Synonyms:** 1-Hydroperoxycyclohexyl 1-hydroxycyclohexyl peroxide; Hydroperoxycyclohexyl hydroxycyclohexyl peroxide; Cadox HDP; Luperco JDB-50-T; Dicyclohexanone diperoxide

3.2 **Coast Guard Compatibility Classification:** Not applicable

3.3 **Chemical Formula:** $C_6H_{10}(OOH)\cdot OO\cdot C_6H_{10}OH$ in dibutyl phthalate

(Continued on page 4)

4. OBSERVABLE CHARACTERISTICS

4.1 **Physical State (as shipped):** Liquid or paste

4.2 **Color:** Colorless or white

4.3 **Odor:** None

5. HEALTH HAZARDS

5.1 **Personal Protective Equipment:** Goggles or face shield; rubber gloves; protective clothing.

5.2 **Symptoms Following Exposure:** Irritates eyes and skin on contact. Ingestion causes irritation of mouth and stomach.

5.3 **Treatment for Exposure:** EYES: flush with water for at least 15 min.; get medical attention. SKIN: wipe off and wash with soap and water; get medical attention if irritation occurs. INGESTION: induce vomiting; get medical attention.

5.4 **Toxicity by Inhalation (Threshold Limit Value):** Not pertinent

5.5 **Short-Term Inhalation Limits:** Not pertinent

5.6 **Toxicity by Ingestion:** Grade 2; LD_{50} 0.5 to 5 g/kg

5.7 **Late Toxicity:** Data not available

5.8 **Vapor (Gas) Irritant Characteristics:** Not pertinent

5.9 **Liquid or Solid Irritant Characteristics:** Data not available

5.10 **Odor Threshold:** Not pertinent

6. FIRE HAZARDS

6.1 **Flash Point:** Combustible solution 315°F C.C. (dibutyl phthalate)

6.2 **Flammable Limits in Air:** Not pertinent

6.3 **Fire Extinguishing Agents:** Water, foam, dry chemical, carbon dioxide

6.4 **Fire Extinguishing Agents Not to be Used:** Not pertinent

6.5 **Special Hazards of Combustion Products:** Not pertinent

6.6 **Behavior in Fire:** May explode

6.7 **Ignition Temperature:** 757°F (dibutyl phthalate)

6.8 **Electrical Hazard:** Not pertinent

6.9 **Burning Rate:** Data not available

7. CHEMICAL REACTIVITY

7.1 **Reactivity with Water:** No reaction

7.2 **Reactivity with Common Materials:** No reaction

7.3 **Stability During Transport:** Stable

7.4 **Neutralizing Agents for Acids and Caustics:** Not pertinent

7.5 **Polymerization:** Not pertinent

7.6 **Inhibitor of Polymerization:** Not pertinent

8. WATER POLLUTION

8.1 **Aquatic Toxicity:** Data not available

8.2 **Waterfowl Toxicity:** Data not available

8.3 **Biological Oxygen Demand (BOD):** Data not available

8.4 **Food Chain Concentration Potential:** None

9. SELECTED MANUFACTURERS

1. Pennwalt Corporation
   Lucidol Division
   P. O. Box 1048
   Buffalo, N. Y. 14240
2. Dart Industries, Inc.
   Aztec Chemical Division
   555 Garden Street
   Elyria, Ohio 44035
3. Story Chemical Corp.
   500 Agard Road
   Muskegon, Mich. 49445

10. SHIPPING INFORMATION

10.1 **Grades or Purity:** 45% paste with dibutyl phthalate

10.2 **Storage Temperature:** Cool ambient

10.3 **Inert Atmosphere:** No requirement

10.4 **Venting:** Open (flame arrester)

11. HAZARD ASSESSMENT CODE

(See Hazard Assessment Handbook, CG 446-3)

A-T-U-X-Y

12. HAZARD CLASSIFICATIONS

12.1 **Code of Federal Regulations:** Organic Peroxide

12.2 **NAS Hazard Rating for Bulk Water Transportation:** Not listed

12.3 **NFPA Hazard Classifications:** Not listed

13. PHYSICAL AND CHEMICAL PROPERTIES

13.1 **Physical State at 15°C and 1 atm:** Solid or liquid

13.2 **Molecular Weight:** Mixture

13.3 **Boiling Point at 1 atm:** Decomposes

13.4 **Freezing Point:** Not pertinent

13.5 **Critical Temperature:** Not pertinent

13.6 **Critical Pressure:** Not pertinent

13.7 **Specific Gravity:** 1.05 at 20°C (liquid)

13.8 **Liquid Surface Tension:** (est.) 30 dynes/cm = 0.030 N/m at 20°C

13.9 **Liquid-Water Interfacial Tension:** (est.) 35 dynes/cm = 0.035 N/m at 20°C

13.10 **Vapor (Gas) Specific Gravity:** Not pertinent

13.11 **Ratio of Specific Heats of Vapor (Gas):** Not pertinent

13.12 **Latent Heat of Vaporization:** Not pertinent

13.13 **Heat of Combustion:** (est.) −14,000 Btu/lb = −7,900 cal/g = -330×10^5 J/kg

13.14 **Heat of Decomposition:** Data not available

13.15 **Heat of Solution:** Not pertinent

13.16 **Heat of Polymerization:** Not pertinent

3. CHEMICAL DESIGNATIONS (Cont'd.)

3.4 **IMCO/United Nations Numerical Designation:** 1/0071 (>70%); 5.2/1525 (<70%)

CHT — CYCLOHEXENYLTRICHLOROSILANE

Common Synonyms	Liquid	Colorless	Sharp, irritating odor
	Reacts with water. Poisonous gas is produced on contact with water.		

Avoid contact with liquid. Keep people away.
Stop discharge if possible.
Call fire department.
Isolate and remove discharged material.
Notify local health and pollution control agencies.

Fire

Combustible.
POISONOUS GASES MAY BE PRODUCED IN FIRE.
Wear goggles and self-contained breathing apparatus.
Extinguish with dry chemicals or carbon dioxide.
DO NOT USE WATER OR FOAM ON FIRE.

Exposure

CALL FOR MEDICAL AID.

GAS PRODUCED IN REACTION WITH WATER.
POISONOUS IF INHALED.
Irritating to eyes, nose and throat.
Move victim to fresh air.
If breathing has stopped, give artificial respiration.
If breathing is difficult, give oxygen.

LIQUID
Will burn skin and eyes.
Harmful if swallowed.
Remove contaminated clothing and shoes.
Flush affected areas with plenty of water.
IF IN EYES, hold eyelids open and flush with plenty of water.
IF SWALLOWED and victim is CONSCIOUS, have victim drink water or milk.
DO NOT INDUCE VOMITING.

Water Pollution

Effect of low concentrations on aquatic life is unknown.
May be dangerous if it enters water intakes.
Notify local health and wildlife officials.
Notify operators of nearby water intakes.

1. RESPONSE TO DISCHARGE
(See Response Methods Handbook, CG 446-4)

Issue warning—corrosive, water contaminant
Restrict access
Disperse and flush with care

2. LABEL

3. CHEMICAL DESIGNATIONS

3.1 **Synonyms:** No common synonyms
3.2 **Coast Guard Compatibility Classification:** Not applicable
3.3 **Chemical Formula:** $C_6H_9SiCl_3$
3.4 **IMCO/United Nations Numerical Designation:** 8/1762

4. OBSERVABLE CHARACTERISTICS

4.1 **Physical State (as shipped):** Liquid
4.2 **Color:** Colorless
4.3 **Odor:** Sharp, hydrochloric-acid-like; pungent and irritating

5. HEALTH HAZARDS

5.1 **Personal Protective Equipment:** Acid-vapor type air respirator; rubber gloves; chemical worker's goggles; other protective equipment as necessary to protect skin and eyes.
5.2 **Symptoms Following Exposure:** Inhalation causes irritation of mucous membrane. Contact with eyes or skin causes severe burns. Ingestion causes severe burns of mouth and stomach.
5.3 **Treatment for Exposure:** Get medical attention immediately following all exposures to this compound. INHALATION: remove from exposure; support respiration. EYES: flush with water for 15 min. SKIN: flush with water. INGESTION: give large amounts of water.
5.4 **Toxicity by Inhalation (Threshold Limit Value):** Data not available
5.5 **Short-Term Inhalation Limits:** Data not available
5.6 **Toxicity by Ingestion:** Grade 2; oral LD_{50} = 2,830 mg/kg (rat)
5.7 **Late Toxicity:** Data not available
5.8 **Vapor (Gas) Irritant Characteristics:** Data not available
5.9 **Liquid or Solid Irritant Characteristics:** Data not available
5.10 **Odor Threshold:** Data not available

6. FIRE HAZARDS

6.1 **Flash Point:** >150°F O.C.
6.2 **Flammable Limits in Air:** Not pertinent
6.3 **Fire Extinguishing Agents:** Dry chemical, carbon dioxide
6.4 **Fire Extinguishing Agents Not to be Used:** Water or foam
6.5 **Special Hazards of Combustion Products:** Irritating, toxic hydrogen chloride and phosgene may be generated in a fire.
6.6 **Behavior in Fire:** Difficult to extinguish. Re-ignition may occur. Water applied to adjacent fires will produce hydrogen chloride upon contact with this material.
6.7 **Ignition Temperature:** Data not available
6.8 **Electrical Hazard:** Data not available
6.9 **Burning Rate:** Data not available

7. CHEMICAL REACTIVITY

7.1 **Reactivity with Water:** Reacts to generate hydrogen chloride (hydrochloric acid).
7.2 **Reactivity with Common Materials:** Corrodes metals by reacting with surface moisture and generating hydrogen chloride.
7.3 **Stability During Transport:** Stable
7.4 **Neutralizing Agents for Acids and Caustics:** Flush with water, rinse with sodium bicarbonate or lime solution.
7.5 **Polymerization:** Not pertinent
7.6 **Inhibitor of Polymerization:** Not pertinent

8. WATER POLLUTION

8.1 **Aquatic Toxicity:** Data not available
8.2 **Waterfowl Toxicity:** Data not available
8.3 **Biological Oxygen Demand (BOD):** Data not available
8.4 **Food Chain Concentration Potential:** None

9. SELECTED MANUFACTURERS

Dow Corning Corporation
P. O. Box 592
Midland, Mich. 48640

10. SHIPPING INFORMATION

10.1 **Grades or Purity:** Commercial
10.2 **Storage Temperature:** Ambient
10.3 **Inert Atmosphere:** No requirement
10.4 **Venting:** Pressure-vacuum

11. HAZARD ASSESSMENT CODE
(See Hazard Assessment Handbook, CG 446-3)

A-O

12. HAZARD CLASSIFICATIONS

12.1 **Code of Federal Regulations:** Corrosive
12.2 **NAS Hazard Rating for Bulk Water Transportation:** Not listed
12.3 **NFPA Hazard Classifications:** Not listed

13. PHYSICAL AND CHEMICAL PROPERTIES

13.1 **Physical State at 15°C and 1 atm:** Liquid
13.2 **Molecular Weight:** 215.6
13.3 **Boiling Point at 1 atm:** >300°F = >149°C = >422°K
13.4 **Freezing Point:** (est.) <77°F = <25°C = <248°K
13.5 **Critical Temperature:** Not pertinent
13.6 **Critical Pressure:** Not pertinent
13.7 **Specific Gravity:** 1.23 at 20°C (liquid)
13.8 **Liquid Surface Tension:** (est.) 30 dynes/cm = 0.030 N/m at 20°C
13.9 **Liquid-Water Interfacial Tension:** Not pertinent
13.10 **Vapor (Gas) Specific Gravity:** Not pertinent
13.11 **Ratio of Specific Heats of Vapor (Gas):** Not pertinent
13.12 **Latent Heat of Vaporization:** Not pertinent
13.13 **Heat of Combustion:** (est.) −78 Btu/lb = −43 cal/g = -1.8×10^5 J/kg
13.14 **Heat of Decomposition:** Not pertinent
13.15 **Heat of Solution:** Data not available
13.16 **Heat of Polymerization:** Not pertinent

NOTES

CYCLOHEXYLAMINE

Common Synonyms			
Aminocyclohexane Hexahydroaniline	Liquid	Colorless	Strong, fishy odor
	Floats and mixes with water.		

Avoid contact with liquid. Keep people away.
Wear goggles, self-contained breathing apparatus, and rubber overclothing (including gloves).
Stop discharge if possible.
Call fire department.
Isolate and remove discharged material.
Notify local health and pollution control agencies.

Fire

FLAMMABLE
Flashback along vapor trail may occur.
Vapor may explode if ignited in an enclosed area.
Wear goggles and self-contained breathing apparatus.
Extinguish with dry chemicals, foam, or carbon dioxide.
Cool exposed containers with water.

Exposure

CALL FOR MEDICAL AID.

LIQUID
Will burn skin and eyes.
Harmful if swallowed.
Remove contaminated clothing and shoes.
Flush affected areas with plenty of water.
IF IN EYES, hold eyelids open and flush with plenty of water.
IF SWALLOWED, and victim is CONSCIOUS, have victim drink water or milk.
DO NOT INDUCE VOMITING.

Water Pollution

Effect of low concentrations on aquatic life is unknown.
May be dangerous if it enters water intakes.
Notify local health and pollution control officials.
Notify operators of nearby water intakes.

1. RESPONSE TO DISCHARGE

(See Response Methods Handbook, CG 446-4)

Disperse and flush

2. LABEL

Red

3. CHEMICAL DESIGNATIONS

3.1 **Synonyms:** Aminocyclohexane; Hexahydroaniline
3.2 **Coast Guard Compatibility Classification:** Aliphatic amine
3.3 **Chemical Formula:** $\overline{(CH_2)_5 C}HNH_2$
3.4 **IMCO/United Nations Numerical Designation:** Not listed

4. OBSERVABLE CHARACTERISTICS

4.1 **Physical State (as shipped):** Liquid
4.2 **Color:** Colorless
4.3 **Odor:** Strong fishy

5. HEALTH HAZARDS

5.1 **Personal Protective Equipment:** Rubber gloves, chemical goggles, approved Bureau of Mines respirator for organic vapors.
5.2 **Symptoms Following Exposure:** Cyclohexylamine is strongly caustic. Inhalation of vapors and contact of liquid with skin and eyes causes severe burns.
5.3 **Treatment for Exposure** INGESTION: do NOT induce vomiting. EYES: flush with water for at least 15 min. and obtain immediate medical attention. SKIN: immediately remove contaminated clothing and flush skin with large amounts of water.
5.4 **Toxicity by Inhalation (Threshold Limit Value):** 300 mg/m^3
5.5 **Short-Term Inhalation Limits:** Data not available
5.6 **Toxicity by Ingestion:** Grade 3; LD_{50} 50 to 500 mg/kg
5.7 **Late Toxicity:** Produced cancer of the bladder in the rat.
5.8 **Vapor (Gas) Irritant Characteristics:** Vapor is moderately irritating such that personnel will not usually tolerate moderate or high vapor concentrations.
5.9 **Liquid or Solid Irritant Characteristics:** Severe skin irritant. Causes second- and third-degree burns on short contact; very injurious to the eyes.
5.10 **Odor Threshold:** Data not available

6. FIRE HAZARDS

6.1 **Flash Point:** 90°F O.C.
6.2 **Flammable Limits in Air:** Data not available
6.3 **Fire Extinguishing Agents:** Foam, carbon dioxide, dry chemical
6.4 **Fire Extinguishing Agents Not to be Used:** Not pertinent
6.5 **Special Hazards of Combustion Products:** Not pertinent
6.6 **Behavior in Fire:** Not pertinent
6.7 **Ignition Temperature:** 560°F
6.8 **Electrical Hazard:** Data not available
6.9 **Burning Rate:** 5.0 mm/min

7. CHEMICAL REACTIVITY

7.1 **Reactivity with Water:** No reaction
7.2 **Reactivity with Common Materials:** No reaction
7.3 **Stability During Transport:** Stable
7.4 **Neutralizing Agents for Acids and Caustics:** Flush with water.
7.5 **Polymerization:** Not pertinent
7.6 **Inhibitor of Polymerization:** Not pertinent

8. WATER POLLUTION

8.1 **Aquatic Toxicity:** Data not available
8.2 **Waterfowl Toxicity:** Data not available
8.3 **Biological Oxygen Demand (BOD):** Data not available
8.4 **Food Chain Concentration Potential:** None

9. SELECTED MANUFACTURERS

1. Abbott Laboratories
   Chemical Division
   Wichita, Kansas 67200
2. Monsanto Co.
   Monsanto Industrial Chemicals Co.
   800 North Lindbergh Blvd.
   St. Louis, Mo. 63166
3. Virginia Chemicals Inc.
   3340 West Norfolk Rd.
   Portsmouth, Virginia 23703

10. SHIPPING INFORMATION

10.1 **Grades or Purity:** Data not available
10.2 **Storage Temperature:** Data not available
10.3 **Inert Atmosphere:** Data not available
10.4 **Venting:** Data not available

11. HAZARD ASSESSMENT CODE

(See Hazard Assessment Handbook, CG 446-3)

A-P-Q

12. HAZARD CLASSIFICATIONS

12.1 **Code of Federal Regulations:** Flammable Liquid
12.2 **NAS Hazard Rating for Bulk Water Transportation:** Not listed
12.3 **NFPA Hazard Classifications:**

Category	Classification
Health Hazard (Blue)	2
Flammability (Red)	3
Reactivity (Yellow)	0

13. PHYSICAL AND CHEMICAL PROPERTIES

13.1 **Physical State at 15°C and 1 atm:** Liquid
13.2 **Molecular Weight:** 99.18
13.3 **Boiling Point at 1 atm:** 274.1°F = 134.5°C = 407.7°K
13.4 **Freezing Point:** 0.1°F = −17.7°C = 255.5°K
13.5 **Critical Temperature:** 648°F = 342°C = 615°K
13.6 **Critical Pressure:** Not pertinent
13.7 **Specific Gravity:** 0.865 at 20°C (liquid)
13.8 **Liquid Surface Tension:** Not pertinent
13.9 **Liquid-Water Interfacial Tension:** Not pertinent
13.10 **Vapor (Gas) Specific Gravity:** Not pertinent
13.11 **Ratio of Specific Heats of Vapor (Gas):** Not pertinent
13.12 **Latent Heat of Vaporization:** 158 Btu/lb = 87.6 cal/g = 3.67×10^5 J/kg
13.13 **Heat of Combustion:** (est.) −18,000 Btu/lb = −10,000 cal/g = -420×10^5 J/kg
13.14 **Heat of Decomposition:** Not pertinent
13.15 **Heat of Solution:** (est.) −4 Btu/lb = −2 cal/g = -0.1×10^5 J/kg
13.16 **Heat of Polymerization:** Not pertinent

NOTES

REVISED 1978

CYP	# CYCLOPENTANE

Common Synonyms			
Pentamethylene	Watery liquid	Colorless	Mild, sweet odor
	Floats on water. Flammable, irritating vapor is produced.		

Shut off ignition sources. Call fire department.
Stop discharge if possible. Keep people away.
Evacuate area in case of large discharge.
Isolate and remove discharged material.
Notify local health and pollution control agencies.

Fire

FLAMMABLE.
Containers may explode in fire.
Flashback along vapor trail may occur.
Vapor may explode if ignited in an enclosed area.

Extinguish with dry chemicals, foam or carbon dioxide.
Water may be ineffective on fire.
Cool exposed containers with water.

Exposure

Call for medical aid.

VAPOR
Irritating to eyes, nose and throat.
If inhaled will cause dizziness, nausea, vomiting, difficult breathing, or loss of consciousness.

Move victim to fresh air.
If breathing has stopped, give artificial respiration.
If breathing is difficult, give oxygen.

LIQUID
Irritating to skin and eyes.
Harmful if swallowed.

Remove contaminated clothing and shoes.
Flush affected areas with plenty of water.
IF IN EYES, hold eyelids open and flush with plenty of water.
IF SWALLOWED and victim is CONSCIOUS, have victim drink water or milk.
DO NOT INDUCE VOMITING.

Water Pollution

Effect of low concentrations on aquatic life is unknown.
Fouling to shoreline.
May be dangerous if it enters water intakes.

Notify local health and wildlife officials.
Notify operators of nearby water intakes.

1. RESPONSE TO DISCHARGE

(See Response Methods Handbook, CG 446-4)

Issue warning—high flammability
Evacuate area
Disperse and flush

2. LABEL

Red

3. CHEMICAL DESIGNATIONS

3.1 **Synonyms:** Pentamethylene
3.2 **Coast Guard Compatibility Classification:** Saturated hydrocarbon
3.3 **Chemical Formula:** C_5H_{10}
3.4 **IMCO/United Nations Numerical Designation:** 3.1/1146

4. OBSERVABLE CHARACTERISTICS

4.1 **Physical State (as shipped):** Liquid
4.2 **Color:** Colorless
4.3 **Odor:** Like gasoline; mild, sweet

5. HEALTH HAZARDS

5.1 **Personal Protective Equipment:** Hydrocarbon canister, supplied-air, or hose mask; rubber or plastic gloves; chemical goggles or face shield.
5.2 **Symptoms Following Exposure:** Inhalation causes dizziness, nausea, and vomiting; concentrated vapor may cause unconsciousness and collapse. Vapor causes slight smarting of eyes. Contact with liquid causes irritation of eyes and may irritate skin if allowed to remain. Ingestion causes irritation of stomach. Aspiration produces severe lung irritation and rapidly developing pulmonary edema; central nervous system excitement followed by depression.
5.3 **Treatment for Exposure:** INHALATION: remove to fresh air; if breathing stops, apply artificial respiration and administer oxygen. EYES: flush with water for at least 15 min.; call a physician. SKIN: flush well with water, then wash with soap and water. INGESTION: do NOT induce vomiting; guard against aspiration into lungs. ASPIRATION: enforce bed rest; give oxygen; get medical attention.
5.4 **Toxicity by Inhalation (Threshold Limit Value):** Data not available
5.5 **Short-Term Inhalation Limits:** 300 ppm for 60 min.
5.6 **Toxicity by Ingestion:** Grade 2; LD_{50} 0.5 to 5 g/kg
5.7 **Late Toxicity:** None
5.8 **Vapor (Gas) Irritant Characteristics:** Vapors cause a slight smarting of the eyes or respiratory system if present in high concentrations. The effect is temporary.
5.9 **Liquid or Solid Irritant Characteristics:** Minimum hazard. If spilled on clothing and allowed to remain, may cause smarting and reddening of the skin.
5.10 **Odor Threshold:** Data not available

6. FIRE HAZARDS

6.1 **Flash Point:** <20°F C.C.
6.2 **Flammable Limits in Air:** (approx.) 1.1%—8.7%
6.3 **Fire Extinguishing Agents:** Dry chemical, foam, carbon dioxide
6.4 **Fire Extinguishing Agents Not to be Used:** Water may be ineffective.
6.5 **Special Hazards of Combustion Products:** Not pertinent
6.6 **Behavior in Fire:** Containers may explode.
6.7 **Ignition Temperature:** 716°F
6.8 **Electrical Hazard:** Not pertinent
6.9 **Burning Rate:** 7.9 mm/min.

7. CHEMICAL REACTIVITY

7.1 **Reactivity with Water:** No reaction
7.2 **Reactivity with Common Materials:** No reaction
7.3 **Stability During Transport:** Stable
7.4 **Neutralizing Agents for Acids and Caustics:** Not pertinent
7.5 **Polymerization:** Not pertinent
7.6 **Inhibitor of Polymerization:** Not pertinent

8. WATER POLLUTION

8.1 **Aquatic Toxicity:** Data not available
8.2 **Waterfowl Toxicity:** Data not available
8.3 **Biological Oxygen Demand (BOD):** Data not available
8.4 **Food Chain Concentration Potential:** None

9. SELECTED MANUFACTURERS

1. Phillips Petroleum Company
   Chemical Department
   Special Products Division
   Bartlesville, Okla. 74004
2. Aldrich Chemical Co.
   940 West St. Paul Avenue
   Milwaukee, Wis. 53233
3. Pfaltz & Bauer, Inc.
   126-04 Northern Boulevard
   Flushing, N. Y. 11368

10. SHIPPING INFORMATION

10.1 **Grades or Purity:** Commercial: 60% (remainder consists of hydrocarbons of similar boiling point); Research: 99+%
10.2 **Storage Temperature:** Ambient
10.3 **Inert Atmosphere:** No requirement
10.4 **Venting:** Pressure-vacuum

11. HAZARD ASSESSMENT CODE

(See Hazard Assessment Handbook, CG 446-3)

A-T-U-V-W

12. HAZARD CLASSIFICATIONS

12.1 **Code of Federal Regulations:** Flammable liquid
12.2 **NAS Hazard Rating for Bulk Water Transportation:**

Category	Rating
Fire	3
Health	
Vapor Irritant	1
Liquid or Solid Irritant	1
Poisons	1
Water Pollution	
Human Toxicity	2
Aquatic Toxicity	1
Aesthetic Effect	0
Reactivity	
Other Chemicals	2
Water	0
Self-Reaction	0

12.3 **NFPA Hazard Classifications:**

Category	Classification
Health Hazard (Blue)	1
Flammability (Red)	3
Reactivity (Yellow)	0

13. PHYSICAL AND CHEMICAL PROPERTIES

13.1 **Physical State at 15°C and 1 atm:** Liquid
13.2 **Molecular Weight:** 70.1
13.3 **Boiling Point at 1 atm:** 120.7°F = 49.3°C = 322.5°K
13.4 **Freezing Point:** −137.0°F = −93.9°C = −179.3°K
13.5 **Critical Temperature:** 461.5°F = 238.6°C = 511.8°K
13.6 **Critical Pressure:** 654 psia = 44.4 atm = 4.51 MN/m^2
13.7 **Specific Gravity:** 0.74 at 20°C (liquid)
13.8 **Liquid Surface Tension:** 23 dynes/cm = 0.023 N/m at 20°C
13.9 **Liquid-Water Interfacial Tension:** (est.) 28 dynes/cm = 0.028 N/m at 20°C
13.10 **Vapor (Gas) Specific Gravity:** 2.4
13.11 **Ratio of Specific Heats of Vapor (Gas):** 1.1217
13.12 **Latent Heat of Vaporization:** 170 Btu/lb = 94 cal/g = 3.9×10^5 J/kg
13.13 **Heat of Combustion:** −19,990 Btu/lb = −11,110 cal/g = -465×10^5 J/kg
13.14 **Heat of Decomposition:** Not pertinent
13.15 **Heat of Solution:** Not pertinent
13.16 **Heat of Polymerization:** Not pertinent

NOTES

CYCLOPROPANE

Common Synonyms			
Trimethylene	Liquefied gas	Colorless	Mild sweet odor
	Floats and boils on water. Flammable visible vapor cloud is produced.		

Shut off ignition sources. Call fire department.
Stop discharge if possible. Keep people away.
Stay upwind. Use water spray to "knock down" vapor.
Evacuate area in case of large discharge.
Avoid contact with liquid.
Isolate and remove discharged material.
Notify local health and pollution control agencies.

Fire

FLAMMABLE.
Containers may explode in fire.
Flashback along vapor trail may occur.
Vapor may explode if ignited in an enclosed area.
Let fire burn.
Stop flow of gas if possible.
Cool exposed containers and protect men effecting shutoff with water.

Exposure

Call for medical aid.

VAPOR
If inhaled will cause difficult breathing.
Move victim to fresh air.
If breathing has stopped, give artificial respiration.
If breathing is difficult, give oxygen.

LIQUID
Will cause frostbite.
Flush affected areas with plenty of water.
DO NOT RUB AFFECTED AREAS.

Water Pollution

Not harmful to aquatic life.

1. RESPONSE TO DISCHARGE
(See Response Methods Handbook, CG 446-4)

Issue warning—high flammability
Restrict access
Evacuate area

2. LABEL

Red

3. CHEMICAL DESIGNATIONS
3.1 **Synonyms:** Trimethylene
3.2 **Coast Guard Compatibility Classification:** Not applicable
3.3 **Chemical Formula:** C_3H_6
3.4 **IMCO/United Nations Numerical Designation:** 2/1027

4. OBSERVABLE CHARACTERISTICS
4.1 **Physical State (as shipped):** Liquefied gas
4.2 **Color:** Colorless
4.3 **Odor:** Sweet, slightly pungent, not irritating; characteristic odor, like light petroleum solvent

5. HEALTH HAZARDS
5.1 **Personal Protective Equipment:** Self-contained breathing apparatus for high concentrations of vapor; safety goggles or face shield.
5.2 **Symptoms Following Exposure:** Inhalation causes some analgesia, anesthesia, pupil dilation, shallow depth of respirations, decreasing muscle tone. Contact with liquid may cause frostbite.
5.3 **Treatment for Exposure:** INHALATION: remove promptly to fresh air; if symptoms of asphyxiation persist, administer artificial respiration and oxygen; treat symptomatically thereafter. SKIN: if frostbite has occurred, apply warm water; treat burn.
5.4 **Toxicity by Inhalation (Threshold Limit Value):** 400 ppm (suggested)
5.5 **Short-Term Inhalation Limits:** Data not available
5.6 **Toxicity by Ingestion:** Data not available
5.7 **Late Toxicity:** None
5.8 **Vapor (Gas) Irritant Characteristics:** Data not available
5.9 **Liquid or Solid Irritant Characteristics:** Data not available
5.10 **Odor Threshold:** Data not available

6. FIRE HAZARDS
6.1 **Flash Point:** Flammable gas
6.2 **Flammable Limits in Air:** 2.4%—10.3%
6.3 **Fire Extinguishing Agents:** Shut off flow of gas.
6.4 **Fire Extinguishing Agents Not to be Used:** Not pertinent
6.5 **Special Hazards of Combustion Products:** Not pertinent
6.6 **Behavior in Fire:** Containers may explode.
6.7 **Ignition Temperature:** 932°F
6.8 **Electrical Hazard:** Class I, Group C
6.9 **Burning Rate:** Not pertinent

7. CHEMICAL REACTIVITY
7.1 **Reactivity with Water:** No reaction
7.2 **Reactivity with Common Materials:** No reaction
7.3 **Stability During Transport:** Stable
7.4 **Neutralizing Agents for Acids and Caustics:** Not pertinent
7.5 **Polymerization:** Not pertinent
7.6 **Inhibitor of Polymerization:** Not pertinent

8. WATER POLLUTION
8.1 **Aquatic Toxicity:** None
8.2 **Waterfowl Toxicity:** None
8.3 **Biological Oxygen Demand (BOD):** None
8.4 **Food Chain Concentration Potential:** None

9. SELECTED MANUFACTURERS
1. Airco, Inc.
Ohio Medical Products Division
3030 Airco Drive
Madison, Wis. 53701
2. Matheson Gas Products Co.
East Rutherford, N. J. 07073
3. Air Products and Chemicals, Inc.
Specialty Gas Department
P. O. Box 538
Allentown, Pa. 18105

10. SHIPPING INFORMATION
10.1 **Grades or Purity:** 99.5+%; USP
10.2 **Storage Temperature:** Ambient
10.3 **Inert Atmosphere:** No requirement
10.4 **Venting:** Safety relief

11. HAZARD ASSESSMENT CODE
(See Hazard Assessment Handbook, CG 446-3)

A-B-C-D-E-F-G

12. HAZARD CLASSIFICATIONS
12.1 **Code of Federal Regulations:** Flammable compressed gas
12.2 **NAS Hazard Rating for Bulk Water Transportation:** Not listed
12.3 **NFPA Hazard Classifications:**

Category	Classification
Health Hazard (Blue)	1
Flammability (Red)	4
Reactivity (Yellow)	0

13. PHYSICAL AND CHEMICAL PROPERTIES
13.1 **Physical State at 15°C and 1 atm:** Gas
13.2 **Molecular Weight:** 42.1
13.3 **Boiling Point at 1 atm:** $-27.2°F = -32.9°C = 240.3°K$
13.4 **Freezing Point:** $-197.3°F = -127.4°C = 145.8°K$
13.5 **Critical Temperature:** $256.5°F = 124.7°C = 397.9°K$
13.6 **Critical Pressure:** 798 psia = 54.2 atm = 5.50 MN/m^2
13.7 **Specific Gravity:** 0.676 at −33°C (liquid)
13.8 **Liquid Surface Tension:** 22 dynes/cm = 0.022 N/m at −40°C
13.9 **Liquid-Water Interfacial Tension:** Not pertinent
13.10 **Vapor (Gas) Specific Gravity:** 1.48
13.11 **Ratio of Specific Heats of Vapor (Gas):** 1.1790
13.12 **Latent Heat of Vaporization:** 203 Btu/lb = 113 cal/g = 4.73×10^5 J/kg
13.13 **Heat of Combustion:** −21,247 Btu/lb = −11,804 cal/g = -493.88×10^5 J/kg
13.14 **Heat of Decomposition:** Not pertinent
13.15 **Heat of Solution:** Not pertinent
13.16 **Heat of Polymerization:** Not pertinent

NOTES

CMP	p-CYMENE

Common Synonyms			
p-Isopropyltoluene 1-Methyl-4-isopropylbenzene Cymol Methylpropylbenzene Isopropyltoluol	Liquid	Colorless	Mild pleasant odor
	Floats on water.		

Stop discharge if possible. Keep people away.
Call fire department.
Isolate and remove discharged material.
Notify local health and pollution control agencies.

Fire	Combustible. Extinguish with dry chemicals, foam or carbon dioxide. Water may be ineffective on fire. Cool exposed containers with water.
Exposure	Call for medical aid. LIQUID Irritating to skin and eyes. Remove contaminated clothing and shoes. Flush affected areas with plenty of water. IF IN EYES, hold eyelids open and flush with plenty of water. IF SWALLOWED and victim is CONSCIOUS, have victim drink water or milk.
Water Pollution	Effect of low concentrations on aquatic life is unknown. Fouling to shoreline. May be dangerous if it enters water intakes. Notify local health and wildlife officials. Notify operators of nearby water intakes.

1. RESPONSE TO DISCHARGE

(See Response Methods Handbook, CG 446-4)

Issue warning—water contaminant
Mechanical containment
Should be removed
Chemical and physical treatment

2. LABELS

No label required by Code of Federal Regulations

3. CHEMICAL DESIGNATIONS

3.1 **Synonyms:** Cymol; p-Isopropyltoluene; Isopropyltoluol; 1-Methyl-4-isopropylbenzene; Methyl propyl benzene

3.2 **Coast Guard Compatibility Classification:** Aromatic hydrocarbon

3.3 **Chemical Formula:** $p\text{-}CH_3C_6H_4CH(CH_3)_2$

3.4 **IMCO/United Nations Numerical Designation:** 3.3/2046

4. OBSERVABLE CHARACTERISTICS

4.1 **Physical State (as shipped):** Liquid

4.2 **Color:** Colorless

4.3 **Odor:** Mild, pleasant; aromatic, solvent-type

5. HEALTH HAZARDS

5.1 **Personal Protective Equipment:** Self-contained or air-line breathing apparatus; solvent-resistant rubber gloves; chemical splash goggles.

5.2 **Symptoms Following Exposure:** Inhalation causes impairment of coordination, headache. Contact with liquid causes mild irritation of eyes and skin. Ingestion causes irritation of mouth and stomach.

5.3 **Treatment for Exposure:** INHALATION: remove victim from contaminated area; administer artificial respiration if necessary; call physician. EYES: flush with water for 15 min.; call a physician. SKIN: wipe off liquid; wash well with soap and water. INGESTION: induce vomiting; get medical attention.

5.4 **Toxicity by Inhalation (Threshold Limit Value):** Data not available

5.5 **Short-Term Inhalation Limits:** Data not available

5.6 **Toxicity by Ingestion:** Grade 2; oral rat LD_{50} = 4,750 mg/kg. Oral human TD_{LO} (lowest toxic dose) = 86 mg/kg (affects central nervous system)

5.7 **Late Toxicity:** Data not available

5.8 **Vapor (Gas) Irritant Characteristics:** Vapors are nonirritating to eyes and throat.

5.9 **Liquid or Solid Irritant Characteristics:** Minimum hazard. If spilled on clothing and allowed to remain, may cause smarting and reddening of the skin.

5.10 **Odor Threshold:** Data not available

6. FIRE HAZARDS

6.1 **Flash Point:** 140°F O.C.; 117°F C.C.

6.2 **Flammable Limits in Air:** 0.7%—5.6%

6.3 **Fire Extinguishing Agents:** Foam, dry chemical, carbon dioxide

6.4 **Fire Extinguishing Agents Not to be Used:** Water may be ineffective.

6.5 **Special Hazards of Combustion Products:** Not pertinent

6.6 **Behavior in Fire:** Not pertinent

6.7 **Ignition Temperature:** 817°F

6.8 **Electrical Hazard:** Data not available

6.9 **Burning Rate:** 6.1 mm/min.

7. CHEMICAL REACTIVITY

7.1 **Reactivity with Water:** No reaction

7.2 **Reactivity with Common Materials:** No reaction

7.3 **Stability During Transport:** Stable

7.4 **Neutralizing Agents for Acids and Caustics:** Not pertinent

7.5 **Polymerization:** Not pertinent

7.6 **Inhibitor of Polymerization:** Not pertinent

8. WATER POLLUTION

8.1 **Aquatic Toxicity:** Data not available

8.2 **Waterfowl Toxicity:** Data not available

8.3 **Biological Oxygen Demand (BOD):** Data not available

8.4 **Food Chain Concentration Potential:** None

9. SELECTED MANUFACTURERS

1. Hercules Incorporated
   Organics Department
   Wilmington, Del. 19899
2. Eastman Kodak Co.
   Eastman Organic Chemicals
   Rochester, N. Y. 14650
3. Aldrich Chemical Co.
   940 West St. Paul Avenue
   Milwaukee, Wis. 53233

10. SHIPPING INFORMATION

10.1 **Grades or Purity:** 95+%

10.2 **Storage Temperature:** Ambient

10.3 **Inert Atmosphere:** No requirement

10.4 **Venting:** Open (flame arrester)

11. HAZARD ASSESSMENT CODE

(See Hazard Assessment Handbook, CG 446-3)

A-T-U

12. HAZARD CLASSIFICATIONS

12.1 **Code of Federal Regulations:** Not listed

12.2 **NAS Hazard Rating for Bulk Water Transportation:**

Category	Rating
Fire	2
Health	
Vapor Irritant	0
Liquid or Solid Irritant	1
Poisons	1
Water Pollution	
Human Toxicity	1
Aquatic Toxicity	1
Aesthetic Effect	2
Reactivity	
Other Chemicals	1
Water	0
Self-Reaction	0

12.3 **NFPA Hazard Classifications:**

Category	Classification
Health Hazard (Blue)	2
Flammability (Red)	2
Reactivity (Yellow)	0

13. PHYSICAL AND CHEMICAL PROPERTIES

13.1 **Physical State at 15°C and 1 atm:** Liquid

13.2 **Molecular Weight:** 134.2

13.3 **Boiling Point at 1 atm:** 351°F = 177°C = 450°K

13.4 **Freezing Point:** −90.2°F = −67.9°C = 205.3°K

13.5 **Critical Temperature:** Not pertinent

13.6 **Critical Pressure:** Not pertinent

13.7 **Specific Gravity:** 0.857 at 20°C (liquid)

13.8 **Liquid Surface Tension:** 28.09 dynes/cm = 0.02809 N/m at 20°C

13.9 **Liquid-Water Interfacial Tension:** 36.41 dynes/cm = 0.03641 N/m at 20°C

13.10 **Vapor (Gas) Specific Gravity:** Not pertinent

13.11 **Ratio of Specific Heats of Vapor (Gas):** Not pertinent

13.12 **Latent Heat of Vaporization:** 122 Btu/lb = 67.8 cal/g = 2.84×10^5 J/kg

13.13 **Heat of Combustion:** −18,800 Btu/lb = −10,400 cal/g = -437×10^5 J/kg

13.14 **Heat of Decomposition:** Not pertinent

13.15 **Heat of Solution:** Not pertinent

13.16 **Heat of Polymerization:** Not pertinent

NOTES

DLP	DALAPON

Common Synonyms
2,2-Dichloropropanoic acid
2,2-Dichloropropionic acid

Liquid — Colorless — Acrid odor

Sinks and mixes with water.

AVOID CONTACT WITH LIQUID. Keep people away.
Wear goggles and self-contained breathing apparatus.
Call fire department.
Isolate and remove discharged material.
Notify local health and pollution control agencies.

Fire

Combustible.
Extinguish with dry chemical, alcohol foam, or carbon dioxide.

Exposure

CALL FOR MEDICAL AID.

VAPOR
Irritating to eyes, nose and throat.
Move to fresh air.
IF IN EYES, hold eyelids open and flush with plenty of water.
If breathing has stopped, give artificial respiration.
If breathing is difficult, give oxygen.

LIQUID
Will burn skin and eyes.
Harmful if swallowed.
Remove contaminated clothing and shoes.
Flush affected areas with plenty of water.
IF IN EYES, hold eyelids open and flush with plenty of water.
IF SWALLOWED and victim is CONSCIOUS, have victim drink water or milk.
DO NOT INDUCE VOMITING.

Water Pollution

HARMFUL TO AQUATIC LIFE IN VERY LOW CONCENTRATIONS.
May be dangerous if it enters water intakes.
Notify local health and wildlife officials.
Notify operators of nearby water intakes.

1. RESPONSE TO DISCHARGE
(See Response Methods Handbook, CG 446-4)

Issue warning—corrosive
Restrict access
Disperse and flush

2. LABEL

3. CHEMICAL DESIGNATIONS

3.1 **Synonyms:** 2,2-Dichloropropanoic acid; alpha, alpha-Dichloropropionic acid; 2,2-Dichloropropionic acid
3.2 **Coast Guard Compatibility Classification:** Not listed
3.3 **Chemical Formula:** CH_3CCl_2COOH
3.4 **IMCO/United Nations Numerical Designation:** 8/1760

4. OBSERVABLE CHARACTERISTICS

4.1 **Physical State (as shipped):** Liquid
4.2 **Color:** Colorless
4.3 **Odor:** Acrid

5. HEALTH HAZARDS

5.1 **Personal Protective Equipment:** Protective clothing, including goggles, gloves and boots; self-contained breathing apparatus
5.2 **Symptoms Following Exposure:** Inhalation causes severe irritation of nose, mouth, and lungs. Ingestion causes severe irritation of mouth and stomach. Contact with eyes or skin causes irritation and burns.
5.3 **Treatment for Exposure:** INHALATION: move to fresh air; if patient is not breathing, give artificial respiration; keep patient quiet; get medical attention. INGESTION: give large amounts of water; get medical attention. EYES: flush with water for at least 15 min.; get medical attention. SKIN: flush with water; get medical attention if irritation persists.
5.4 **Toxicity by Inhalation (Threshold Limit Value):** Data not available
5.5 **Short-Term Inhalation Limits:** Data not available
5.6 **Toxicity by Ingestion:** Grade 2; oral LD_{50} = 3.65 g/kg (mouse), 7.57 g/kg (rat)
5.7 **Late Toxicity:** Data not available
5.8 **Vapor (Gas) Irritant Characteristics:** Data not available
5.9 **Liquid or Solid Irritant Characteristics:** Data not available
5.10 **Odor Threshold:** 2500 mg/m³

6. FIRE HAZARDS

6.1 **Flash Point:** Data not available
6.2 **Flammable Limits in Air:** Data not available
6.3 **Fire Extinguishing Agents:** Dry chemical, alcohol foam, carbon dioxide
6.4 **Fire Extinguishing Agents Not to be Used:** Water may be ineffective.
6.5 **Special Hazards of Combustion Products:** Irritating fumes of hydrochloric acid may form in fire.
6.6 **Behavior in Fire:** Volatilizes with steam
6.7 **Ignition Temperature:** Data not available
6.8 **Electrical Hazard:** Data not available
6.9 **Burning Rate:** Data not available

7. CHEMICAL REACTIVITY

7.1 **Reactivity with Water:** Reacts slowly to form hydrochloric and pyruvic acids. The reaction is not hazardous.
7.2 **Reactivity with Common Materials:** Very corrosive to aluminum and copper alloys. Flammable and explosive hydrogen gas may form in enclosed spaces.
7.3 **Stability During Transport:** Stable
7.4 **Neutralizing Agents for Acids and Caustics:** Flush with water; rinse with dilute sodium bicarbonate or soda ash solution.
7.5 **Polymerization:** Not pertinent
7.6 **Inhibitor of Polymerization:** Not pertinent

8. WATER POLLUTION

8.1 **Aquatic Toxicity:**
105 ppm/96 hr/bluegill/TL_m/fresh water
1 ppm/48 hr/brown shrimp/TL_m/salt water
8.2 **Waterfowl Toxicity:** > 5,000 ppm LC_{50}
8.3 **Biological Oxygen Demand (BOD):**
0.04 lb/lb, 5 days, unacclimated seed
0.32 lb/lb, 5 days, acclimated seed
8.4 **Food Chain Concentration Potential:** None

9. SELECTED MANUFACTURERS

Dow Chemical Co.
Midland, Mich. 48640

10. SHIPPING INFORMATION

10.1 **Grades or Purity:** Technical grade, 90%; solid formulations of sodium and magnesium salts are sometimes referred to as Dalapon and are much less corrosive.
10.2 **Storage Temperature:** 70—90°F
10.3 **Inert Atmosphere:** No requirement
10.4 **Venting:** Open

11. HAZARD ASSESSMENT CODE
(See Hazard Assessment Handbook, CG 446-3)

A-P-Q

12. HAZARD CLASSIFICATIONS

12.1 **Code of Federal Regulations:** Corrosive liquid
12.2 **NAS Hazard Rating for Bulk Water Transportation:** Not listed
12.3 **NFPA Hazard Classifications:** Not listed

13. PHYSICAL AND CHEMICAL PROPERTIES

13.1 **Physical State at 15°C and 1 atm:** Liquid
13.2 **Molecular Weight:** 143
13.3 **Boiling Point at 1 atm:** 374°F = 190°C = 463°K
13.4 **Freezing Point:** 46°F = 8°C = 281°K
13.5 **Critical Temperature:** Data not available
13.6 **Critical Pressure:** Data not available
13.7 **Specific Gravity:** 1.39 at 23°C (liquid)
13.8 **Liquid Surface Tension:** Data not available
13.9 **Liquid-Water Interfacial Tension:** Not pertinent
13.10 **Vapor (Gas) Specific Gravity:** 4.9
13.11 **Ratio of Specific Heats of Vapor (Gas):** Not pertinent
13.12 **Latent Heat of Vaporization:** Data not available
13.13 **Heat of Combustion:** Data not available
13.14 **Heat of Decomposition:** Not pertinent
13.15 **Heat of Solution:** Data not available
13.16 **Heat of Polymerization:** Not pertinent

NOTES

DDD	DDD

Common Synonyms		
TDE 1,1-Dichloro-2,2-bis(p-chloro-phenyl) ethane Dichlorodiphenyldichloroethane	Solid Sinks in water.	White

Stop discharge if possible. Keep people away.
Avoid contact with solid and dust.
Isolate and remove discharged material.
Notify local health and pollution control agencies.

Fire	Combustible. Irritating gases may be produced when heated. Wear goggles and self-contained breathing apparatus. Extinguish with water, dry chemicals, foam, or carbon dioxide.
Exposure	CALL FOR MEDICAL AID. DUST Irritating to eyes, nose and throat. Harmful if inhaled. If in eyes, hold eyelids open and flush with plenty of water. If breathing has stopped, give artificial respiration. If breathing is difficult, give oxygen. SOLID Irritating to skin and eyes. Harmful if swallowed. Remove contaminated clothing and shoes. Flush affected areas with plenty of water. IF IN EYES, hold eyelids open and flush with plenty of water. IF SWALLOWED and victim is CONSCIOUS, have victim drink water or milk. IF SWALLOWED and victim is UNCONSCIOUS OR HAVING CONVULSIONS, do nothing except keep victim warm.
Water Pollution	HARMFUL TO AQUATIC LIFE IN VERY LOW CONCENTRATIONS. May be dangerous if it enters water intakes. Notify local health and wildlife officials. Notify operators of nearby water intakes.

1. RESPONSE TO DISCHARGE
(See Response Methods Handbook, CG 446-4)

Issue warning—water contaminant
Should be removed
Chemical and physical treatment

2. LABELS
No hazard label required by Code of Federal Regulations

3. CHEMICAL DESIGNATIONS
3.1 **Synonyms:** 1,1-Dichloro-2,2-bis (p-chlorophenyl) ethane; Dichlorodiphenyldichloroethane; TDE
3.2 **Coast Guard Compatibility Classification:** Not listed
3.3 **Chemical Formula:** $(4\text{-}ClC_6H_4)_2CH{-}CHCl_2$
3.4 **IMCO/United Nations Numerical Designation:** Not listed

4. OBSERVABLE CHARACTERISTICS
4.1 **Physical State (as shipped):** Solid
4.2 **Color:** White
4.3 **Odor:** Data not available

5. HEALTH HAZARDS
5.1 **Personal Protective Equipment:** Dust mask; goggles or face shield; rubber gloves
5.2 **Symptoms Following Exposure:** Ingestion causes vomiting and delayed symptoms similar to those caused by DDT. Contact with eyes causes irritation.
5.3 **Treatment for Exposure:** INGESTION: treatment should be given by a physician and is similar to that given following ingestion of DDT. EYES: flush with water.
5.4 **Toxicity by Inhalation (Threshold Limit Value):** Data not available
5.5 **Short-Term Inhalation Limits:** Data not available
5.6 **Toxicity by Ingestion:** Grade 2; oral LD_{50} = 1.2 g/kg (mouse), 3.4 g/kg (rat)
5.7 **Late Toxicity:** Data not available
5.8 **Vapor (Gas) Irritant Characteristics:** Data not available
5.9 **Liquid or Solid Irritant Characteristics:** Data not available
5.10 **Odor Threshold:** Data not available

6. FIRE HAZARDS
6.1 **Flash Point:** Not pertinent
6.2 **Flammable Limits in Air:** Not pertinent
6.3 **Fire Extinguishing Agents:** Water, foam, dry chemical, carbon dioxide
6.4 **Fire Extinguishing Agents Not to be Used:**
6.5 **Special Hazards of Combustion Products:** Irritating hydrogen chloride fumes may form in fires.
6.6 **Behavior in Fire:**
6.7 **Ignition Temperature:** Data not available
6.8 **Electrical Hazard:** Not pertinent
6.9 **Burning Rate:** Not pertinent

7. CHEMICAL REACTIVITY
7.1 **Reactivity with Water:** No reaction
7.2 **Reactivity with Common Materials:**
7.3 **Stability During Transport:** Stable
7.4 **Neutralizing Agents for Acids and Caustics:** Not pertinent
7.5 **Polymerization:** Not pertinent
7.6 **Inhibitor of Polymerization:** Not pertinent

8. WATER POLLUTION
8.1 **Aquatic Toxicity:**
$<$2.6 ppm/96 hr/catfish/TL_m/fresh water
0.15–0.2 ppm/48 hr/brown shrimp/TL_m/salt water
0.0068 ppm/24 hr/brown shrimp/EC_{50}/salt water
8.2 **Waterfowl Toxicity:** 4,800-5,200 ppm LC_{50}
8.3 **Biological Oxygen Demand (BOD):** Data not available
8.4 **Food Chain Concentration Potential:** High

9. SELECTED MANUFACTURERS
1. Aldrich Chemical Co.
940 W. Saint Paul Ave.
Milwaukee, Wis. 53233
2. Pfaltz and Bauer, Inc.
375 Fairfield Ave.
Stamford, Conn. 06902

10. SHIPPING INFORMATION
10.1 **Grades or Purity:** Technical
10.2 **Storage Temperature:** Ambient
10.3 **Inert Atmosphere:** No requirement
10.4 **Venting:** Open

11. HAZARD ASSESSMENT CODE
(See Hazard Assessment Handbook, CG 446-3)

II

12. HAZARD CLASSIFICATIONS
12.1 **Code of Federal Regulations:** Not listed
12.2 **NAS Hazard Rating for Bulk Water Transportation:** Not listed
12.3 **NFPA Hazard Classifications:** Not listed

13. PHYSICAL AND CHEMICAL PROPERTIES
13.1 **Physical State at 15°C and 1 atm:** Solid
13.2 **Molecular Weight:** 320
13.3 **Boiling Point at 1 atm:** Not pertinent (decomposes)
13.4 **Freezing Point:** 234°F = 112°C = 385°K
13.5 **Critical Temperature:** Not pertinent
13.6 **Critical Pressure:** Not pertinent
13.7 **Specific Gravity:** 1.476 at 20°C (solid)
13.8 **Liquid Surface Tension:** Not pertinent
13.9 **Liquid-Water Interfacial Tension:** Not pertinent
13.10 **Vapor (Gas) Specific Gravity:** Not pertinent
13.11 **Ratio of Specific Heats of Vapor (Gas):** Not pertinent
13.12 **Latent Heat of Vaporization:** Not pertinent
13.13 **Heat of Combustion:** Data not available
13.14 **Heat of Decomposition:** Not pertinent
13.15 **Heat of Solution:** Not pertinent
13.16 **Heat of Polymerization:** Not pertinent

NOTES

DDT	DDT

Common Synonyms			
Dichlorodiphenyltrichloroethane p,p' - DDT	Solid	Colorless	Odorless
	Sinks in water.		

Avoid contact with solid.
Call fire department.
Isolate and remove discharged material.
Notify local health and pollution control agencies.

Fire	Combustible. POISONOUS GASES ARE PRODUCED IN FIRE. Wear goggles and self-contained breathing apparatus. Extinguish with water, dry chemical, foam, or carbon dioxide.
Exposure	CALL FOR MEDICAL AID. SOLIDS Irritating to skin and eyes. If swallowed, will cause nausea, vomiting, headache, or loss of consciousness. Remove contaminated clothing and shoes. Flush affected areas with plenty of water. IF IN EYES, hold eyelids open and flush with plenty of water. IF SWALLOWED and victim is CONSCIOUS, have victim drink water or milk.
Water Pollution	HARMFUL TO AQUATIC LIFE IN VERY LOW CONCENTRATIONS. May be dangerous if it enters water intakes. Notify local health and wildlife officials. Notify operators of nearby water intakes.

1. RESPONSE TO DISCHARGE
(See Response Methods Handbook, CG 446-4)

Issue warning—water contaminant
Should be removed

2. LABELS
No hazard label required by Code of Federal Regulations

3. CHEMICAL DESIGNATIONS
3.1 **Synonyms:** Dichlorodiphenyltrichloroethane
p, p'-DDT
1, 1, 1-Trichloro-2, 2-bis(p-chlorophenyl) ethane
3.2 **Coast Guard Compatibility Classification:** Not applicable
3.3 **Chemical Formula:** $(p\text{-}ClC_6H_4)_2CHCCl_3$
3.4 **IMCO/United Nations Numerical Designation:** 9/

4. OBSERVABLE CHARACTERISTICS
4.1 **Physical State (as shipped):** Solid
4.2 **Color:** White
4.3 **Odor:** None

5. HEALTH HAZARDS
5.1 **Personal Protective Equipment:** Data not available
5.2 **Symptoms Following Exposure:** Very large doses are followed promptly by vomiting, due to local gastric irritation; delayed emesis or diarrhea may occur. With smaller doses, symptoms usually appear 2-3 hours after ingestion. These include tingling of lips, tongue, and face; malaise, headache, sore throat, fatigue, coarse tremors of neck, head, and eyelids; apprehension, ataxia, and confusion. Convulsions may alternate with periods of coma and partial paralysis. Vital signs are essentially normal, but in severe poisoning the pulse may be irregular and abnormally slow; ventricular fibrillation and sudden death may occur at any time during acute phase. Pulmonary edema usually indicates solvent intoxication.
5.3 **Treatment for Exposure:** INGESTION: treatment should be done by a physician. It usually includes gastric lavage and administration of saline cathartic, phenobarbital, and parenteral fluids. Patient should be kept quiet and under observation for at least 24 hours.
5.4 **Toxicity by Inhalation (Threshold Limit Value):** Not pertinent
5.5 **Short-Term Inhalation Limits:** Not pertinent
5.6 **Toxicity by Ingestion:** Grade 3; LD_{50} 50 to 500 mg/kg (rat)
5.7 **Late Toxicity:** Data not available
5.8 **Vapor (Gas) Irritant Characteristics:** Not pertinent
5.9 **Liquid or Solid Irritant Characteristics:** Minimum hazard. If spilled on clothing and allowed to remain, may cause smarting and reddening of the skin.
5.10 **Odor Threshold:** Not pertinent

6. FIRE HAZARDS
6.1 **Flash Point:** 162°F-171°F C.C.
6.2 **Flammable Limits in Air:** Not pertinent
6.3 **Fire Extinguishing Agents:** Water, foam, dry chemical, or carbon dioxide
6.4 **Fire Extinguishing Agents Not to be Used:** Not pertinent
6.5 **Special Hazards of Combustion Products:** Toxic and irritating gases may be generated
6.6 **Behavior in Fire:** Melts and burns
6.7 **Ignition Temperature:** Data not available
6.8 **Electrical Hazard:** Not pertinent
6.9 **Burning Rate:** Data not available

7. CHEMICAL REACTIVITY
7.1 **Reactivity with Water:** No reaction
7.2 **Reactivity with Common Materials:** No reaction
7.3 **Stability During Transport:** Stable
7.4 **Neutralizing Agents for Acids and Caustics:** Not pertinent
7.5 **Polymerization:** Not pertinent
7.6 **Inhibitor of Polymerization:** Not pertinent

8. WATER POLLUTION
8.1 **Aquatic Toxicity:**
0.0039 ppm/24 hr/bass/TL_{50}/fresh water
0.0018 ppm/96 hr/bass/TL_{50}/fresh water
0.0028 ppm/48 hr/killifish/50% kill/salt water
8.2 **Waterfowl Toxicity:** 2240 mg/kg
8.3 **Biological Oxygen Demand (BOD):** Not pertinent
8.4 **Food Chain Concentration Potential:** High

9. SELECTED MANUFACTURERS
1. Lebanon Chemical Corp.
Lebanon, Pa. 17042
2. Montrose Chemical Corp. of Calif.
500 South Virgil Ave.
Los Angeles, Calif. 90005

10. SHIPPING INFORMATION
10.1 **Grades or Purity:** Technical
10.2 **Storage Temperature:** Data not available
10.3 **Inert Atmosphere:** Data not available
10.4 **Venting:** Data not available

11. HAZARD ASSESSMENT CODE
(See Hazard Assessment Handbook, CG 446-3)

II

12. HAZARD CLASSIFICATIONS
12.1 **Code of Federal Regulations:** ORM-A
12.2 **NAS Hazard Rating for Bulk Water Transportation:** Not listed
12.3 **NFPA Hazard Classifications:** Not listed

13. PHYSICAL AND CHEMICAL PROPERTIES
13.1 **Physical State at 15°C and 1 atm:** Solid
13.2 **Molecular Weight:** 354.5
13.3 **Boiling Point at 1 atm:** Not pertinent
13.4 **Freezing Point:** 226°F = 108°C = 381°K
13.5 **Critical Temperature:** Not pertinent
13.6 **Critical Pressure:** Not pertinent
13.7 **Specific Gravity:** 1.56 at 15°C (solid)
13.8 **Liquid Surface Tension:** Not pertinent
13.9 **Liquid-Water Interfacial Tension:** Not pertinent
13.10 **Vapor (Gas) Specific Gravity:** Not pertinent
13.11 **Ratio of Specific Heats of Vapor (Gas):** Not pertinent
13.12 **Latent Heat of Vaporization:** Not pertinent
13.13 **Heat of Combustion:** Not pertinent
13.14 **Heat of Decomposition:** Not pertinent
13.15 **Heat of Solution:** Not pertinent
13.16 **Heat of Polymerization:** Not pertinent

NOTES

DBR — DECABORANE

Common Synonyms			
	Solid	White	Sharp odor
	Floats on water.		

AVOID CONTACT WITH SOLID AND DUST. KEEP PEOPLE AWAY.
Wear goggles, self-contained breathing apparatus, and rubber overclothing (including gloves).
Shut off ignition sources. Call fire department.
Stop discharge if possible.
Stay upwind. Use water spray to "knock down" vapor.
Avoid contact with liquid.
Isolate and remove discharged material.
Notify local health and pollution control agencies.

Fire

FLAMMABLE.
POISONOUS GASES MAY BE PRODUCED IN FIRE.
Containers may explode in fire.
Wear goggles, self-contained breathing apparatus and rubber overclothing (including gloves).
Extinguish with water, dry chemicals, foam, or carbon dioxide.
Do not use vaporizing liquids on fire.
Cool exposed containers with water.

Exposure

CALL FOR MEDICAL AID.

DUST
POISONOUS IF INHALED OR IF SKIN IS EXPOSED.
Move victim to fresh air.
If in eyes, hold eyelids open and flush with plenty of water.
If breathing is difficult, give oxygen.

SOLID
POISONOUS IF SWALLOWED OR IF SKIN IS EXPOSED.
Remove contaminated clothing and shoes.
Flush affected areas with plenty of water.
IF IN EYES, hold eyelids open and flush with plenty of water.
IF SWALLOWED and victim is CONSCIOUS, have victim drink water or milk and have victim induce vomiting.
IF SWALLOWED and victim is UNCONSCIOUS OR HAVING CONVULSIONS, do nothing except keep victim warm.

Water Pollution

Effect of low concentrations on aquatic life is unknown.
Fouling to shoreline.
May be dangerous if it enters water intakes.
Notify local health and wildlife officials.
Notify operators of nearby water intakes.

1. RESPONSE TO DISCHARGE

(See Response Methods Handbook, CG 446-4)

Issue warning—high flammability, water contaminant, poison
Restrict access
Mechanical containment
Should be removed
Chemical and physical treatment

2. LABELS

3. CHEMICAL DESIGNATIONS

3.1 **Synonyms:** No common synonyms
3.2 **Coast Guard Compatibility Classification:** Not applicable
3.3 **Chemical Formula:** $B_{10}H_{14}$
3.4 **IMCO/United Nations Numerical Designation:** 4.1/1868

4. OBSERVABLE CHARACTERISTICS

4.1 **Physical State (as shipped):** Solid
4.2 **Color:** White
4.3 **Odor:** Pungent

5. HEALTH HAZARDS

5.1 **Personal Protective Equipment:** Self-contained breathing apparatus or positive-pressure hose mask; rubber boots or overshoes; clothing made of material resistant to decaborane; rubber gloves; chemical-type goggles or face shield.

5.2 **Symptoms Following Exposure:** (The onset of symptoms is frequently delayed until one or two days after exposure.) Inhalation or ingestion causes headache, nausea, light-headedness, drowsiness, nervousness, lack of coordination, and tremor; muscle spasms and generalized convulsions may occur. Dust irritates eyes and skin and may give same systemic symptoms as for inhalation if left on skin.

5.3 **Treatment for Exposure:** Get medical attention after all exposures to this compound. Symptoms may be delayed for 48 hours. INHALATION: move patient to fresh air; keep him warm and quiet. EYES: flush with water for at least 15 min. SKIN: immediately wash with soap and plenty of water. INGESTION: if victim is conscious, give a tablespoonful of salt in a glass of warm water and repeat until vomit fluid is clear. *Note to physician:* Treat symptomatically; administration of methocarbamol or other muscle relaxant may be helpful immediately following exposure and in the absence of symptoms.

5.4 **Toxicity by Inhalation (Threshold Limit Value):** 0.05 ppm
5.5 **Short-Term Inhalation Limits:** Data not available
5.6 **Toxicity by Ingestion:** Grade 4; oral LD_{50} = 40 mg/kg (mouse)
5.7 **Late Toxicity:** Data not available
5.8 **Vapor (Gas) Irritant Characteristics:** Data not available

6. FIRE HAZARDS

6.1 **Flash Point:** (Flammable solid) 176°F C.C.
6.2 **Flammable Limits in Air:** Not pertinent
6.3 **Fire Extinguishing Agents:** Water, foam, dry chemical, and carbon dioxide.
6.4 **Fire Extinguishing Agents Not to be Used:** Halogenated extinguishing agents
6.5 **Special Hazards of Combustion Products:** May give toxic fumes of unburned material.
6.6 **Behavior in Fire:** May explode when hot. Burns with a green-colored flame.
6.7 **Ignition Temperature:** 300°F
6.8 **Electrical Hazard:** Not pertinent
6.9 **Burning Rate:** Not pertinent

7. CHEMICAL REACTIVITY

7.1 **Reactivity with Water:** Reacts slowly to form flammable hydrogen gas, which can accumulate in closed area.
7.2 **Reactivity with Common Materials:** Corrosive to natural rubber, some synthetic rubbers, some greases, and some lubricants.
7.3 **Stability During Transport:** Stable
7.4 **Neutralizing Agents for Acids and Caustics:** Flush with 3% aqueous ammonia solution, then with water. Methyl alcohol may also be used.
7.5 **Polymerization:** Not pertinent
7.6 **Inhibitor of Polymerization:** Not pertinent

8. WATER POLLUTION

8.1 **Aquatic Toxicity:** Data not available
8.2 **Waterfowl Toxicity:** Data not available
8.3 **Biological Oxygen Demand (BOD):** Data not available
8.4 **Food Chain Concentration Potential:** None

9. SELECTED MANUFACTURERS

1. Callery Cehmical Company
Callery, Pa. 16024
2. Strem Chemicals, Inc.
150 Andover St.
Danvers, Mass. 01923
3. K & K Laboratories, Inc.
121 Express Street
Plainview, N. Y. 11803

10. SHIPPING INFORMATION

10.1 **Grades or Purity:** Technical: 95+%; High purity: 99+%
10.2 **Storage Temperature:** Ambient
10.3 **Inert Atmosphere:** No requirement
10.4 **Venting:** Pressure-vacuum

11. HAZARD ASSESSMENT CODE

(See Hazard Assessment Handbook, CG 446-3)

II-RR

12. HAZARD CLASSIFICATIONS

12.1 **Code of Federal Regulations:** Flammable solid
12.2 **NAS Hazard Rating for Bulk Water Transportation:** Not listed
12.3 **NFPA Hazard Classifications:**

Category	Classification
Health Hazard (Blue)	3
Flammability (Red)	2
Reactivity (Yellow)	1

13. PHYSICAL AND CHEMICAL PROPERTIES

13.1 **Physical State at 15°C and 1 atm:** Solid
13.2 **Molecular Weight:** 122.3
13.3 **Boiling Point at 1 atm:** 415°F = 213°C = 486°K
13.4 **Freezing Point:** 210°F = 99°C = 372°K
13.5 **Critical Temperature:** Not pertinent
13.6 **Critical Pressure:** Not pertinent
13.7 **Specific Gravity:** 0.94 at 25°C (solid)
13.8 **Liquid Surface Tension:** Not pertinent
13.9 **Liquid-Water Interfacial Tension:** Not pertinent
13.10 **Vapor (Gas) Specific Gravity:** Not pertinent
13.11 **Ratio of Specific Heats of Vapor (Gas):** Not pertinent
13.12 **Latent Heat of Vaporization:** Not pertinent
13.13 **Heat of Combustion:** −28,699 Btu/lb = −15,944 cal/g = -667.10×10^5 J/kg
13.14 **Heat of Decomposition:** −279 Btu/lb = −155 cal/g = -6.49×10^5 J/kg
13.15 **Heat of Solution:** Not pertinent
13.16 **Heat of Polymerization:** Not pertinent

5. HEALTH HAZARDS (Cont'd.)

5.9 **Liquid or Solid Irritant Characteristics:** Data not available
5.10 **Odor Threshold:** 0.05 ppm

DHN DECAHYDRONAPHTHALENE

Common Synonyms

Bicyclo [4.4.0]decane
Napthalane
Perhydronapthalene
Dec
Decalin

Liquid Colorless Turpentine-like odor

Floats on water.

Stop discharge if possible. Keep people away.
Call fire department.
Isolate and remove discharged material.
Notify local health and pollution control agencies.

Fire

Combustible.
Extinguish with dry chemicals, foam or carbon dioxide.
Water may be ineffective on fire.

Exposure

Call for medical aid.

LIQUID
Irritating to skin and eyes.
If swallowed will cause headache, nausea, or vomiting.
Remove contaminated clothing and shoes.
Flush affected areas with plenty of water.
IF IN EYES, hold eyelids open and flush with plenty of water.
IF SWALLOWED and victim is CONSCIOUS, have victim drink water or milk.

Water Pollution

Effect of low concentrations on aquatic life is unknown.
Fouling to shoreline.
May be dangerous if it enters water intakes.
Notify local health and wildlife officials.
Notify operators of nearby water intakes.

1. RESPONSE TO DISCHARGE

(See Response Methods Handbook, CG 446-4)

Mechanical containment
Should be removed
Chemical and physical treatment

2. LABELS

No label required by Code of Federal Regulations

3. CHEMICAL DESIGNATIONS

3.1 **Synonyms:** Bicyclo [4.4.0] decane; Naphthalane; Perhydronaphthalene; Dec; Decalin; De Kalin; Naphthane; *cis*- or *trans*-Decahydronaphthalene
3.2 **Coast Guard Compatibility Classification:** Not applicable
3.3 **Chemical Formula:** $C_{10}H_{18}$
3.4 **IMCO/United Nations Numerical Designation:** 3.3/1147

4. OBSERVABLE CHARACTERISTICS

4.1 **Physical State (as shipped):** Liquid
4.2 **Color:** Colorless
4.3 **Odor:** Aromatic; like turpentine; mild, characteristic

5. HEALTH HAZARDS

5.1 **Personal Protective Equipment:** Air mask or self-contained breathing apparatus if in enclosed tank; rubber gloves or protective cream; goggles or face shield.
5.2 **Symptoms Following Exposure:** Inhalation or ingestion irritates nose and throat, causes numbness, headache, vomiting; urine may become blue. Irritates eyes. Liquid de-fats skin and causes cracking and secondary infection; eczema may develop.
5.3 **Treatment for Exposure:** INHALATION: remove to fresh air. EYES: flush with water for at least 15 min. SKIN: wash with water and mild soap. INGESTION: give emetic such as warm salt water, followed by a mild cathartic; direct physician to conserve liver and kidney function.
5.4 **Toxicity by Inhalation (Threshold Limit Value):** 25 ppm (suggested)
5.5 **Short-Term Inhalation Limits:** Data not available
5.6 **Toxicity by Ingestion:** Grade 2; oral LD_{50} = 4,170 mg/kg (rat)
5.7 **Late Toxicity:** Data not available
5.8 **Vapor (Gas) Irritant Characteristics:** Data not available
5.9 **Liquid or Solid Irritant Characteristics:** Data not available
5.10 **Odor Threshold:** Data not available

6. FIRE HAZARDS

6.1 **Flash Point:** 134°F O.C.
6.2 **Flammable Limits in Air:** 0.7%—5.4%
6.3 **Fire Extinguishing Agents:** Water, foam, dry chemical, carbon dioxide
6.4 **Fire Extinguishing Agents Not to be Used:** Water
6.5 **Special Hazards of Combustion Products:** Not pertinent
6.6 **Behavior in Fire:** Not pertinent
6.7 **Ignition Temperature:** 482°F
6.8 **Electrical Hazard:** Data not available
6.9 **Burning Rate:** 5.9 mm/min.

7. CHEMICAL REACTIVITY

7.1 **Reactivity with Water:** No reaction
7.2 **Reactivity with Common Materials:** No reaction
7.3 **Stability During Transport:** Stable
7.4 **Neutralizing Agents for Acids and Caustics:** Not pertinent
7.5 **Polymerization:** Not pertinent
7.6 **Inhibitor of Polymerization:** Not pertinent

8. WATER POLLUTION

8.1 **Aquatic Toxicity:** Data not available
8.2 **Waterfowl Toxicity:** Data not available
8.3 **Biological Oxygen Demand (BOD):** Data not available
8.4 **Food Chain Concentration Potential:** Data not available

9. SELECTED MANUFACTURERS

1. Progil, Inc.
509 Madison Avenue
New York, N. Y. 10022
2. E. I. duPont de Nemours & Co.
Organic Chemicals Department
Dyes and Chemicals Division
Wilmington, Del. 19898
3. Eastman Kodak Co.
Eastman Organic Cehmicals
Rochester, N. Y. 14650

10. SHIPPING INFORMATION

10.1 **Grades or Purity:** Technical: mixture of cis-(35%) and trans-(65%) isomers. Properties of all such mixtures are very similar. Spectro grade.
10.2 **Storage Temperature:** Ambient
10.3 **Inert Atmosphere:** No requirement
10.4 **Venting:** Open (flame arrester)

11. HAZARD ASSESSMENT CODE

(See Hazard Assessment Handbook, CG 446-3)

A-T-U

12. HAZARD CLASSIFICATIONS

12.1 **Code of Federal Regulations:** Combustible Liquid
12.2 **NAS Hazard Rating for Bulk Water Transportation:** Not listed
12.3 **NFPA Hazard Classifications:**

Category	Classification trans-	Classification Decalin
Health Hazard (Blue)	0	2
Flammability (Red)	2	2
Reactivity (Yellow)	0	0

13. PHYSICAL AND CHEMICAL PROPERTIES

13.1 **Physical State at 15°C and 1 atm:** Liquid
13.2 **Molecular Weight:** 138.2
13.3 **Boiling Point at 1 atm:** 383°F = 195°C = 468°K
13.4 **Freezing Point:** −44°F = −42°C = 231°K
13.5 **Critical Temperature:** Not pertinent
13.6 **Critical Pressure:** Not pertinent
13.7 **Specific Gravity:** 0.89 at 20°C (liquid)
13.8 **Liquid Surface Tension:** 30 dynes/cm = 0.030 N/m at 20°C
13.9 **Liquid-Water Interfacial Tension:** 51.5 dynes/cm = 0.0515 N/m at 20°C
13.10 **Vapor (Gas) Specific Gravity:** Not pertinent
13.11 **Ratio of Specific Heats of Vapor (Gas):** Not pertinent
13.12 **Latent Heat of Vaporization:** 130 Btu/lb = 71 cal/g = 3.0×10^5 J/kg
13.13 **Heat of Combustion:** −19,200 Btu/lb = −10,700 cal/g = -447×10^5 J/kg
13.14 **Heat of Decomposition:** Not pertinent
13.15 **Heat of Solution:** Not pertinent
13.16 **Heat of Polymerization:** Not pertinent

NOTES

DAL DECALDEHYDE

Common Synonyms			
Decanal Capric aldehyde Capraldehyde	Liquid	Colorless to light yellow	Pleasant odor
	Floats on water. Freezing point in 64° F.		

Stop discharge if possible.
Avoid contact with liquid.
Call fire department.
Isolate and remove discharged material.
Notify local health and pollution control agencies.

Fire

Combustible.
Extinguish with dry chemical, foam, or carbon dioxide.

Exposure

CALL FOR MEDICAL AID.

LIQUID
Irritating to skin and eyes.
Remove contaminated clothing and shoes.
Flush affected areas with plenty of water.
IF IN EYES, hold eyelids open and flush with plenty of water.

Water Pollution

Effect of low concentrations on aquatic life is unknown.
Fouling to shoreline.
May be dangerous if it enters water intakes.
Notify local health and pollution control officials.
Notify operators of nearby water intakes.

1. RESPONSE TO DISCHARGE

(See Response Methods Handbook, CG 446-4)

Mechanical containment
Should be removed
Chemical and physical treatment

2. LABELS

No hazard label required by Code of Federal Regulations

3. CHEMICAL DESIGNATIONS

3.1 **Synonyms:** Aldehyde C-10
Capraldehyde
Capric aldehyde
Decanal
n-Decyl aldehyde

3.2 **Coast Guard Compatibility Classification:** Aldehyde

3.3 **Chemical Formula:** $CH_3(CH_2)_8CHO$

3.4 **IMCO/United Nations Numerical Designation:** Not listed

4. OBSERVABLE CHARACTERISTICS

4.1 **Physical State (as shipped):** Liquid

4.2 **Color:** Colorless to pale yellow

4.3 **Odor:** Pleasant

5. HEALTH HAZARDS

5.1 **Personal Protective Equipment:** Protective clothing and chemical goggles.

5.2 **Symptoms Following Exposure:** On direct contact can produce eye and skin irritation; low general toxicity.

5.3 **Treatment for Exposure:** CONTACT WITH EYES AND SKIN: wash with water for 15 min.

5.4 **Toxicity by Inhalation (Threshold Limit Value):** Data not available

5.5 **Short-Term Inhalation Limits:** Data not available

5.6 **Toxicity by Ingestion:** Grade 0; LD_{50}>33.3 g/kg (rat)

5.7 **Late Toxicity:** Data not available

5.8 **Vapor (Gas) Irritant Characteristics:** Data not available

5.9 **Liquid or Solid Irritant Characteristics:** Skin effects thought to be minor

5.10 **Odor Threshold:** 0.168 ppm

6. FIRE HAZARDS

6.1 **Flash Point:** 185°F O.C.

6.2 **Flammable Limits in Air:** Data not available

6.3 **Fire Extinguishing Agents:** Foam, dry chemical, carbon dioxide

6.4 **Fire Extinguishing Agents Not to be Used:** Not pertinent

6.5 **Special Hazards of Combustion Products:** Not pertinent

6.6 **Behavior in Fire:** Not pertinent

6.7 **Ignition Temperature:** Data not available

6.8 **Electrical Hazard:** Data not available

6.9 **Burning Rate:** Data not available

7. CHEMICAL REACTIVITY

7.1 **Reactivity with Water:** No reaction

7.2 **Reactivity with Common Materials:** No reaction

7.3 **Stability During Transport:** Stable

7.4 **Neutralizing Agents for Acids and Caustics:** Not pertinent

7.5 **Polymerization:** Not pertinent

7.6 **Inhibitor of Polymerization:** Not pertinent

8. WATER POLLUTION

8.1 **Aquatic Toxicity:** Data not available

8.2 **Waterfowl Toxicity:** Data not available

8.3 **Biological Oxygen Demand (BOD):** Data not available

8.4 **Food Chain Concentration Potential:** None

9. SELECTED MANUFACTURERS

1. Givaudan Corp.
Aroma Chemicals Division
Clifton, N. J. 07014
2. International Flavors and Fragrances, Inc.
Union Beach, N. J. 07735
3. F. Ritter and Co.
4001 Goodwin Ave.
Los Angeles, Calif. 90039

10. SHIPPING INFORMATION

10.1 **Grades or Purity:** Data not available

10.2 **Storage Temperature:** Ambient

10.3 **Inert Atmosphere:** No requirement

10.4 **Venting:** Open (flame arrester)

11. HAZARD ASSESSMENT CODE

(See Hazard Assessment Handbook, CG 446-3)

A-T-U

12. HAZARD CLASSIFICATIONS

12.1 **Code of Federal Regulations:** Combustible Liquid

12.2 **NAS Hazard Rating for Bulk Water Transportation:** Not listed

12.3 **NFPA Hazard Classifications:** Not listed

13. PHYSICAL AND CHEMICAL PROPERTIES

13.1 **Physical State at 15°C and 1 atm:** Liquid

13.2 **Molecular Weight:** 145.3

13.3 **Boiling Point at 1 atm:** 404 to 410°F = 207 to 210°C = 480 to 483°K

13.4 **Freezing Point:** 64°F = 18°C = 291°K

13.5 **Critical Temperature:** Not pertinent

13.6 **Critical Pressure:** Not pertinent

13.7 **Specific Gravity:** 0.830 at 15°C (liquid)

13.8 **Liquid Surface Tension:** 28.0 dynes/cm = 0.0280 N/m at 24°C

13.9 **Liquid-Water Interfacial Tension:** 16.9 dynes/cm = 0.0169 N/m at 22.7°C

13.10 **Vapor (Gas) Specific Gravity:** Not pertinent

13.11 **Ratio of Specific Heats of Vapor (Gas):** 1.036

13.12 **Latent Heat of Vaporization:** Not pertinent

13.13 **Heat of Combustion:** (est.) −18,000 Btu/lb = −10,000 cal/g = -424×10^5 J/kg

13.14 **Heat of Decomposition:** Not pertinent

13.15 **Heat of Solution:** Not pertinent

13.16 **Heat of Polymerization:** Not pertinent

NOTES

REVISED 1978

DCE	1-DECENE

Common Synonyms
alpha-Decene

Watery liquid | Colorless | Pleasant odor

Floats on water.

Stop discharge if possible.
Avoid contact with liquid.
Call fire department.
Isolate and remove discharged material.
Notify local health and pollution control agencies.

Fire

Combustible.
Extinguish with dry chemical, foam, or carbon dioxide.

Exposure

CALL FOR MEDICAL AID.

LIQUID
Irritating to skin and eyes.
Remove contaminated clothing and shoes.
Flush affected areas with plenty of water.
IF IN EYES, hold eyelids open and flush with plenty of water.

Water Pollution

Effect of low concentrations on aquatic life is unknown.
Fouling to shoreline.
May be dangerous if it enters water intakes.
Notify local health and pollution control officials.
Notify operators of nearby water intakes.

1. RESPONSE TO DISCHARGE
(See Response Methods Handbook, CG 446-4)

Mechanical containment
Should be removed
Chemical and physical treatment

2. LABELS

No hazard label required by Code of Federal Regulations

3. CHEMICAL DESIGNATIONS

3.1 **Synonyms:** alpha-Decene
3.2 **Coast Guard Compatibility Classification:** Olefin
3.3 **Chemical Formula:** $CH_2{=}CH(CH_2)_7CH_3$
3.4 **IMCO/United Nations Numerical Designation:** Not listed

4. OBSERVABLE CHARACTERISTICS

4.1 **Physical State (as shipped):** Liquid
4.2 **Color:** Colorless
4.3 **Odor:** Mild, pleasant

5. HEALTH HAZARDS

5.1 **Personal Protective Equipment:** Organic canister or air-supplied mask, goggles or face shield.
5.2 **Symptoms Following Exposure:** Vapors may produce slight irritation of eyes and respiratory tract if present in high concentration. May also act as a slight anesthetic at high concentration.
5.3 **Treatment for Exposure:** CONTACT WITH EYES OR SKIN: splashes in the eye should be removed by thorough flushing with water. Skin areas should be washed with soap and water. Contaminated clothing should be laundered before reuse.
5.4 **Toxicity by Inhalation (Threshold Limit Value):** Data not available
5.5 **Short-Term Inhalation Limits:** Data not available
5.6 **Toxicity by Ingestion:** Data not available
5.7 **Late Toxicity:** Data not available
5.8 **Vapor (Gas) Irritant Characteristics:** Slight smarting of eyes and respiratory system at high concentrations. The effect is temporary.
5.9 **Liquid or Solid Irritant Characteristics:** Minimum hazard. If spilled on clothing and allowed to remain, may cause smarting and reddening of the skin.
5.10 **Odor Threshold:** Data not available

6. FIRE HAZARDS

6.1 **Flash Point:** 128°F O.C.
6.2 **Flammable Limits in Air:** Not pertinent
6.3 **Fire Extinguishing Agents:** Foam, dry chemical, or carbon dioxide
6.4 **Fire Extinguishing Agents Not to be Used:** Not pertinent
6.5 **Special Hazards of Combustion Products:** Not pertinent
6.6 **Behavior in Fire:** Not pertinent
6.7 **Ignition Temperature:** 455°F
6.8 **Electrical Hazard:** Data not available
6.9 **Burning Rate:** 6.0 mm/min.

7. CHEMICAL REACTIVITY

7.1 **Reactivity with Water:** No reaction
7.2 **Reactivity with Common Materials:** No reaction
7.3 **Stability During Transport:** Stable
7.4 **Neutralizing Agents for Acids and Caustics:** Not pertinent
7.5 **Polymerization:** Not pertinent
7.6 **Inhibitor of Polymerization:** Not pertinent

8. WATER POLLUTION

8.1 **Aquatic Toxicity:** Data not available
8.2 **Waterfowl Toxicity:** Data not available
8.3 **Biological Oxygen Demand (BOD):** Data not available
8.4 **Food Chain Concentration Potential:** None

9. SELECTED MANUFACTURERS

1. Gulf Oil Corp.
   Gulf Oil Chemicals Co.
   Petrochemicals Division
   Cedar Bayou, Texas 77520
2. The Humphrey Chemical Co.
   Devine St.
   North Haven, Conn. 06473
3. Phillips Petroleum Co.
   Bartlesville, Okla. 74004

10. SHIPPING INFORMATION

10.1 **Grades or Purity:** Technical: 95-99%
10.2 **Storage Temperature:** Ambient
10.3 **Inert Atmosphere:** No requirement
10.4 **Venting:** Open (flame arrester)

11. HAZARD ASSESSMENT CODE
(See Hazard Assessment Handbook, CG 446-3)

A-T-U

12. HAZARD CLASSIFICATIONS

12.1 **Code of Federal Regulations:** Combustible Liquid
12.2 **NAS Hazard Rating for Bulk Water Transportation:** Not listed
12.3 **NFPA Hazard Classifications:** Not listed

13. PHYSICAL AND CHEMICAL PROPERTIES

13.1 **Physical State at 15°C and 1 atm:** Liquid
13.2 **Molecular Weight:** 140.2
13.3 **Boiling Point at 1 atm:** 339.1°F = 170.6°C = 443.8°K
13.4 **Freezing Point:** −87.3°F = −66.3°C = 206.9°K
13.5 **Critical Temperature:** Not pertinent
13.6 **Critical Pressure:** Not pertinent
13.7 **Specific Gravity:** 0.741 at 20°C (liquid)
13.8 **Liquid Surface Tension:** 24 dynes/cm = 0.024 N/m at 20°C
13.9 **Liquid-Water Interfacial Tension:** 28 dynes/cm = 0.028 N/m at 22.7°C
13.10 **Vapor (Gas) Specific Gravity:** 4.8
13.11 **Ratio of Specific Heats of Vapor (Gas):** 1.039
13.12 **Latent Heat of Vaporization:** 119 Btu/lb = 65.9 cal/g = 2.76×10^5 J/kg
13.13 **Heat of Combustion:** −19,107 Btu/lb = −10,615 cal/g = -444.43×10^5 J/kg
13.14 **Heat of Decomposition:** Not pertinent
13.15 **Heat of Solution:** Not pertinent
13.16 **Heat of Polymerization:** Not pertinent

NOTES

REVISED 1978

DAN	# n-DECYL ALCOHOL

Common Synonyms			
1-Decanol Nonylcarbinol Capric alcohol	Liquid	Colorless to light yellow	Faint alcohol odor
	Floats on water. Freezing point is 44° F.		

Stop discharge if possible. Keep people away.
Call fire department.
Avoid contact with liquid.
Isolate and remove discharged material.
Notify local health and pollution control agencies.

Fire	Combustible. Extinguish with dry chemical.
Exposure	CALL FOR MEDICAL AID. LIQUID Irritating to eyes. If in eyes, hold eyelids open and flush with plenty of water.
Water Pollution	Effect of low concentrations on aquatic life is unknown. Fouling to shoreline. May be dangerous if it enters water intakes. Notify local health and pollution control officials. Notify operators of nearby water intakes.

1. RESPONSE TO DISCHARGE
(See Response Methods Handbook, CG 446-4)

Mechanical containment
Should be removed
Chemical and physical treatment

2. LABELS
No hazard label required by Code of Federal Regulations

3. CHEMICAL DESIGNATIONS
3.1 **Synonyms:** Alcohol C-10; Dytol S-91; Capric alcohol; Lorol-22; 1-Decanol; Nonylcarbinol
3.2 **Coast Guard Compatibility Classification:** Alcohol
3.3 **Chemical Formula:** $CH_3(CH_2)_8CH_2OH$
3.4 **IMCO/United Nations Numerical Designation:** Not listed

4. OBSERVABLE CHARACTERISTICS
4.1 **Physical State (as shipped):** Liquid
4.2 **Color:** Colorless to light yellow
4.3 **Odor:** Faint alcoholic

5. HEALTH HAZARDS
5.1 **Personal Protective Equipment:** Goggles or face shield.
5.2 **Symptoms Following Exposure:** Direct contact can produce eye irritation; low general toxicity.
5.3 **Treatment for Exposure:** CONTACT WITH EYES: flush with water for 15 min.
5.4 **Toxicity by Inhalation (Threshold Limit Value):** Not pertinent
5.5 **Short-Term Inhalation Limits:** Not pertinent
5.6 **Toxicity by Ingestion:** Grade 1; LD_{50} 5 to 15 g/kg (rat)
5.7 **Late Toxicity:** Data not available
5.8 **Vapor (Gas) Irritant Characteristics:** None
5.9 **Liquid or Solid Irritant Characteristics:** No appreciable hazard. Practically harmless to the skin.
5.10 **Odor Threshold:** Data not available

6. FIRE HAZARDS
6.1 **Flash Point:** 180°F O.C.
6.2 **Flammable Limits in Air:** Data not available
6.3 **Fire Extinguishing Agents:** Dry chemical
6.4 **Fire Extinguishing Agents Not to be Used:** Not pertinent
6.5 **Special Hazards of Combustion Products:** Not pertinent
6.6 **Behavior in Fire:** Not pertinent
6.7 **Ignition Temperature:** Data not available
6.8 **Electrical Hazard:** Data not available
6.9 **Burning Rate:** Data not available

7. CHEMICAL REACTIVITY
7.1 **Reactivity with Water:** No reaction
7.2 **Reactivity with Common Materials:** No reaction
7.3 **Stability During Transport:** Stable
7.4 **Neutralizing Agents for Acids and Caustics:** Not pertinent
7.5 **Polymerization:** Not pertinent
7.6 **Inhibitor of Polymerization:** Not pertinent

8. WATER POLLUTION
8.1 **Aquatic Toxicity:** Data not available
8.2 **Waterfowl Toxicity:** Data not available
8.3 **Biological Oxygen Demand (BOD):** 29.3%, 1 day
8.4 **Food Chain Concentration Potential:** None

9. SELECTED MANUFACTURERS
1. Continental Oil Co.
   Conoco Chemicals Division
   Park 80 Plaza East
   Saddle Brook, N. J. 07662
2. The Procter and Gamble Co.
   301 East Sixth St.
   Cincinnati, Ohio 45201

10. SHIPPING INFORMATION
10.1 **Grades or Purity:** 95-99%
10.2 **Storage Temperature:** Ambient
10.3 **Inert Atmosphere:** No requirement
10.4 **Venting:** Open (flame arrester)

11. HAZARD ASSESSMENT CODE
(See Hazard Assessment Handbook, CG 446-3)

A-T-U

12. HAZARD CLASSIFICATIONS
12.1 **Code of Federal Regulations:** Combustible Liquid
12.2 **NAS Hazard Rating for Bulk Water Transportation:**

Category	Rating
Fire	1
Health	
Vapor Irritant	0
Liquid or Solid Irritant	0
Poisons	0
Water Pollution	
Human Toxicity	0
Aquatic Toxicity	0
Aesthetic Effect	2
Reactivity	
Other Chemicals	2
Water	0
Self-Reaction	0

12.3 **NFPA Hazard Classifications:** Not listed

13. PHYSICAL AND CHEMICAL PROPERTIES
13.1 **Physical State at 15°C and 1 atm:** Liquid
13.2 **Molecular Weight:** 158.29
13.3 **Boiling Point at 1 atm:** 446°F = 230°C = 503°K
13.4 **Freezing Point:** 44°F = 6.9°C = 280.1°K
13.5 **Critical Temperature:** 801°F = 427°C = 700°K
13.6 **Critical Pressure:** 320 psia = 22 atm = 2.2 MN/m^2
13.7 **Specific Gravity:** 0.840 at 20°C (liquid)
13.8 **Liquid Surface Tension:** 28.9 dynes/cm = 0.0289 N/m at 20°C
13.9 **Liquid-Water Interfacial Tension:** (est.) 60 dynes/cm = 0.06 N/m at 10°C
13.10 **Vapor (Gas) Specific Gravity:** Not pertinent
13.11 **Ratio of Specific Heats of Vapor (Gas):** 1.035
13.12 **Latent Heat of Vaporization:** (est.) 130 Btu/lb = 74 cal/g = 3.1 × 10^5 J/kg
13.13 **Heat of Combustion:** -18,000 Btu/lb = -9980 cal/g = 418 × 10^5 J/kg
13.14 **Heat of Decomposition:** Not pertinent
13.15 **Heat of Solution:** Not pertinent
13.16 **Heat of Polymerization:** Not pertinent

NOTES

REVISED 1978

DBZ	n-DECYLBENZENE

Common Synonyms		
1-Phenyldecane Decylbenzene	Liquid Floats on water.	White

Stop discharge if possible. Keep people away.
Shut off ignition sources. Call fire department.
Avoid contact with liquid and vapor.
Isolate and remove discharged material.
Notify local health and pollution control agencies.

Fire

Combustible.
Extinguish with dry chemicals or carbon dioxide.
Water and foam may be ineffective on fire.
Cool exposed containers with water.

Exposure

CALL FOR MEDICAL AID.

VAPOR
Irritating to eyes, nose and throat.
If inhaled will cause coughing or difficult breathing.
If in eyes, hold eyelids open and flush with plenty of water.
If breathing has stopped, give artificial respiration.
If breathing is difficult, give oxygen.

LIQUID
Irritating to skin and eyes.
If swallowed will cause nausea and vomiting.
Remove contaminated clothing and shoes.
Flush affected areas with plenty of water.
IF IN EYES, hold eyelids open and flush with plenty of water.
DO NOT INDUCE VOMITING.

Water Pollution

Effect of low concentrations on aquatic life is unknown.
Fouling to shoreline.
May be dangerous if it enters water intakes.
Notify local health and wildlife officials.
Notify operators of nearby water intakes.

1. RESPONSE TO DISCHARGE

(See Response Methods Handbook, CG 446-4)

Mechanical containment
Should be removed
Chemical and physical treatment

2. LABELS

No hazard label required by Code of Federal Regulations

3. CHEMICAL DESIGNATIONS

3.1 **Synonyms:** Decylbenzene; 1-Phenyldecane
3.2 **Coast Guard Compatibility Classification:** Aromatic hydrocarbon
3.3 **Chemical Formula:** $C_6H_5(CH_2)_9CH_3$
3.4 **IMCO/United Nations Numerical Designation:** Not listed

4. OBSERVABLE CHARACTERISTICS

4.1 **Physical State (as shipped):** Liquid
4.2 **Color:** Colorless
4.3 **Odor:** Data not available

5. HEALTH HAZARDS

5.1 **Personal Protective Equipment:** Goggles or face shield; rubber gloves
5.2 **Symptoms Following Exposure:** Inhalation of vapor causes slight irritation of nose and throat. Aspiration of liquid into lungs causes coughing, distress, and pulmonary edema. Ingestion causes irritation of stomach. Contact with vapor or liquid causes irritation of eyes and mild irritation of skin.
5.3 **Treatment for Exposure:** INHALATION: move to fresh air. INGESTION: do NOT induce vomiting; call a doctor. EYES: flush with water. SKIN: wipe off; flush with water; wash with soap and water. ASPIRATION: enforce bed rest; administer oxygen; call a doctor.
5.4 **Toxicity by Inhalation (Threshold Limit Value):** Data not available
5.5 **Short-Term Inhalation Limits:** Data not available
5.6 **Toxicity by Ingestion:** Data not available
5.7 **Late Toxicity:** Data not available
5.8 **Vapor (Gas) Irritant Characteristics:** Data not available
5.9 **Liquid or Solid Irritant Characteristics:** Data not available
5.10 **Odor Threshold:** Data not available

6. FIRE HAZARDS

6.1 **Flash Point:** 225°F C.C.
6.2 **Flammable Limits in Air:** Data not available
6.3 **Fire Extinguishing Agents:** Dry chemical, carbon dioxide
6.4 **Fire Extinguishing Agents Not to be Used:** Water or foam may cause frothing.
6.5 **Special Hazards of Combustion Products:**
6.6 **Behavior in Fire:**
6.7 **Ignition Temperature:** Data not available
6.8 **Electrical Hazard:** Data not available
6.9 **Burning Rate:** 5.04 mm/min.

7. CHEMICAL REACTIVITY

7.1 **Reactivity with Water:** No reaction
7.2 **Reactivity with Common Materials:** May attack some forms of plastics.
7.3 **Stability During Transport:** Stable
7.4 **Neutralizing Agents for Acids and Caustics:** Not pertinent
7.5 **Polymerization:** Not pertinent
7.6 **Inhibitor of Polymerization:** Not pertinent

8. WATER POLLUTION

8.1 **Aquatic Toxicity:** Data not available
8.2 **Waterfowl Toxicity:** Data not available
8.3 **Biological Oxygen Demand (BOD):** Data not available
8.4 **Food Chain Concentration Potential:** None

9. SELECTED MANUFACTURERS

1. The Humphrey Chemical Co.
   Devine Street
   North Haven, Conn. 06473
2. Pfaltz and Bauer, Inc.
   375 Fairfield Ave.
   Stamford, Conn. 06902

10. SHIPPING INFORMATION

10.1 **Grades or Purity:** Commercial
10.2 **Storage Temperature:** Ambient
10.3 **Inert Atmosphere:** No requirement
10.4 **Venting:** Open (flame arrester)

11. HAZARD ASSESSMENT CODE

(See Hazard Assessment Handbook, CG 446-3)

A-T-U

12. HAZARD CLASSIFICATIONS

12.1 **Code of Federal Regulations:** Not listed
12.2 **NAS Hazard Rating for Bulk Water Transportation:** Not listed
12.3 **NFPA Hazard Classifications:**

Category	Classification
Health Hazard (Blue)	2
Flammability (Red)	1
Reactivity (Yellow)	0

13. PHYSICAL AND CHEMICAL PROPERTIES

13.1 **Physical State at 15°C and 1 atm:** Liquid
13.2 **Molecular Weight:** 218
13.3 **Boiling Point at 1 atm:** 572°F = 300°C = 573°K
13.4 **Freezing Point:** Not pertinent
13.5 **Critical Temperature:** Not pertinent
13.6 **Critical Pressure:** Not pertinent
13.7 **Specific Gravity:** 0.855 at 20°C (liquid)
13.8 **Liquid Surface Tension:** 29.95 dynes/cm = 0.02995 N/m at 20°C
13.9 **Liquid-Water Interfacial Tension:** Data not available
13.10 **Vapor (Gas) Specific Gravity:** Not pertinent
13.11 **Ratio of Specific Heats of Vapor (Gas):** Not pertinent
13.12 **Latent Heat of Vaporization:** 103.8 Btu/lb = 57.67 cal/g = 2.413×10^5 J/kg
13.13 **Heat of Combustion:** −18,400 Btu/lb = −10,200 cal/g = -427×10^5 J/kg
13.14 **Heat of Decomposition:** Not pertinent
13.15 **Heat of Solution:** Not pertinent
13.16 **Heat of Polymerization:** Not pertinent

NOTES

DTN — DEMETON

Common Synonyms			
O,O-[Diethyl-O(and S)-[2-(ethylthio)ethyl] phosphorothioates	Liquid	Yellowish-brown	Unpleasant odor
	Sinks in water.		

AVOID CONTACT WITH LIQUID AND VAPOR. KEEP PEOPLE AWAY.
Wear goggles and self-contained breathing apparatus.
Shut off ignition sources. Call fire department.
Stop discharge if possible.
Isolate and remove discharged material.
Notify local health and pollution control agencies.

Fire

Solution in a combustible solvent.
POISONOUS GASES MAY BE PRODUCED IN FIRE.
Flashback along vapor trail may occur.
Vapor may explode if ignited in an enclosed area.
Wear goggles and self-contained breathing apparatus.
Extinguish with dry chemicals, foam or carbon dioxide.

Exposure

CALL FOR MEDICAL AID.

VAPOR
POISONOUS IF INHALED.
If inhaled will cause headache or difficult breathing.
If in eyes, hold eyelids open and flush with plenty of water.
If breathing has stopped, give artificial respiration.
If breathing is difficult, give oxygen.

LIQUID
POISONOUS IF SWALLOWED.
Irritating to skin and eyes.
If swallowed will cause nausea, vomiting or loss of consciousness.
Remove contaminated clothing and shoes.
Flush affected areas with plenty of water.
IF IN EYES, hold eyelids open and flush with plenty of water.
IF SWALLOWED and victim is CONSCIOUS, have victim drink water or milk and have victim induce vomiting.
IF SWALLOWED and victim is UNCONSCIOUS OR HAVING CONVULSIONS, do nothing except keep victim warm.

Water Pollution

Effect of low concentrations on aquatic life is unknown.
May be dangerous if it enters water intakes.
Notify local health and wildlife officials.
Notify operators of nearby water intakes.

1. RESPONSE TO DISCHARGE

(See Response Methods Handbook, CG 446-4)

Issue warning—poison, water contaminant, high flammability
Restrict access
Mechanical containment
Should be removed
Chemical and physical treatment

2. LABEL

3. CHEMICAL DESIGNATIONS

3.1 **Synonyms:** O,O-Diethyl-O (and S)-[2-(ethylthio) ethyl] phosphorothioates; Mixture of O- and S-[2-(Ethylthio) ethyl] phosphorothioic acid, O,O-diethyl ester; Systox and isosystox mixture

3.2 **Coast Guard Compatibility Classification:** Not listed

3.3 **Chemical Formula:** $C_8H_{19}O_3PS_2$—C_8H_{10} mixture

3.4 **IMCO/United Nations Numerical Designation:** Not listed

4. OBSERVABLE CHARACTERISTICS

4.1 **Physical State (as shipped):** Liquid

4.2 **Color:** Amber

4.3 **Odor:** Offensive sulfur-compound odor

5. HEALTH HAZARDS

5.1 **Personal Protective Equipment:** Organic vapor respirator in confined areas; rubber or latex gloves; splash goggles; rubber boots

5.2 **Symptoms Following Exposure:** Inhalation causes headache, vertigo, blurred vision, lachrymation, salivation, sweating, muscular weakness and ataxia, dyspnea, diarrhea, abdominal cramps, vomiting, coma, pulmonary edema, and death. Ingestion causes nausea, vomiting, muscle twitching, coma. Contact with eyes or skin causes irritation.

5.3 **Treatment for Exposure:** *Speed is essential. Call a physician after all overexposure to demeton.* INHALATION: move to fresh air; if needed, begin artificial respiration. INGESTION: administer milk, water, or salt-water and induce vomiting repeatedly. EYES: flush with water for at least 15 min. SKIN: flood and wash exposed skin areas thoroughly with water; remove contaminated clothing under a shower; wash with soap and water.

5.4 **Toxicity by Inhalation (Threshold Limit Value):** 0.1 mg/m^3

5.5 **Short-Term Inhalation Limits:** 0.5 mg/m^3, 30 min.

5.6 **Toxicity by Ingestion:** Grade 4; oral LD_{50} = 1.7 mg/kg (rat)

5.7 **Late Toxicity:** Data not available

5.8 **Vapor (Gas) Irritant Characteristics:** Data not available

5.9 **Liquid or Solid Irritant Characteristics:** Data not available

5.10 **Odor Threshold:** Data not available

6. FIRE HAZARDS

6.1 **Flash Point:** 113°F C.C.

6.2 **Flammable Limits in Air:** 1.0% – 5.3%

6.3 **Fire Extinguishing Agents:** Dry chemical, foam, carbon dioxide

6.4 **Fire Extinguishing Agents Not to be Used:** Water may be ineffective on fire.

6.5 **Special Hazards of Combustion Products:** Irritating fumes of sulfur dioxide and phosphoric acid may form in fire.

6.6 **Behavior in Fire:** Compound may volatilize and form toxic fumes. Vapor of solvent is heavier than air and may travel considerable distance to a source of ignition and flash back.

6.7 **Ignition Temperature:** 867°F (xylene solvent)

6.8 **Electrical Hazard:** (xylene) Class I, Group D

6.9 **Burning Rate:** 5.8 mm/min.

7. CHEMICAL REACTIVITY

7.1 **Reactivity with Water:** No reaction

7.2 **Reactivity with Common Materials:** May attack some forms of plastics.

7.3 **Stability During Transport:** Stable

7.4 **Neutralizing Agents for Acids and Caustics:** Not pertinent

7.5 **Polymerization:** Not pertinent

7.6 **Inhibitor of Polymerization:** Not pertinent

8. WATER POLLUTION

8.1 **Aquatic Toxicity:** 0.1 ppm/2-8 hr/bluegill/reduction of enzyme in brain/fresh water

8.2 **Waterfowl Toxicity:** 7-15 mg/kg

8.3 **Biological Oxygen Demand (BOD):** Data not available

8.4 **Food Chain Concentration Potential:** None

9. SELECTED MANUFACTURERS

Chemagro
Division of Baychem Corp.
Kansas City, Mo. 64120

10. SHIPPING INFORMATION

10.1 **Grades or Purity:** 25%-66% solution in xylenes, which are combustible solvents

10.2 **Storage Temperature:** Ambient

10.3 **Inert Atmosphere:** No requirement

10.4 **Venting:** Open (flame arrester)

11. HAZARD ASSESSMENT CODE

(See Hazard Assessment Handbook, CG 446-3)

A-X-Y

12. HAZARD CLASSIFICATIONS

12.1 **Code of Federal Regulations:** Poison, Class B

12.2 **NAS Hazard Rating for Bulk Water Transportation:** Not listed

12.3 **NFPA Hazard Classifications:** Not listed

13. PHYSICAL AND CHEMICAL PROPERTIES

13.1 **Physical State at 15°C and 1 atm:** Liquid

13.2 **Molecular Weight:** 258

13.3 **Boiling Point at 1 atm:** >284°F = >140°C = >413°K

13.4 **Freezing Point:** Not pertinent

13.5 **Critical Temperature:** Not pertinent

13.6 **Critical Pressure:** Not pertinent

13.7 **Specific Gravity:** 1.1 at 20°C (liquid)

13.8 **Liquid Surface Tension:** Data not available

13.9 **Liquid-Water Interfacial Tension:** Data not available

13.10 **Vapor (Gas) Specific Gravity:** Not pertinent

13.11 **Ratio of Specific Heats of Vapor (Gas):** Not pertinent

13.12 **Latent Heat of Vaporization:** Not pertinent

13.13 **Heat of Combustion:** Data not available

13.14 **Heat of Decomposition:** Not pertinent

13.15 **Heat of Solution:** Not pertinent

13.16 **Heat of Polymerization:** Not pertinent

NOTES

DES — 2,4-D ESTERS

Common Synonyms			
Butyl 2,4-Dichlorophenoxyacetate Isopropyl 2,4-dichlorophenoxy acetate 2,4-Dichlorophenoxyacetic acid, butyl ester 2,4-Dichlorophenoxyacetic acid, butoxyethyl ester	Liquid Sinks in water.	Yellowish brown	Fuel oil-like odor

Stop discharge if possible. Keep people away.
Shut off ignition sources. Call fire department.
Avoid contact with liquid.
Isolate and remove discharged material.
Notify local health and pollution control agencies.

Fire	Combustible. Irritating gases may be produced when heated. Wear goggles and self-contained breathing apparatus. Extinguish with dry chemicals, foam or carbon dioxide. Water may be ineffective on fire. Cool exposed containers with water.
Exposure	CALL FOR MEDICAL AID. LIQUID Irritating to skin and eyes. Harmful if swallowed. Remove contaminated clothing and shoes. Flush affected areas with plenty of water. IF IN EYES, hold eyelids open and flush with plenty of water. IF SWALLOWED and victim is CONSCIOUS, have victim drink water or milk and have victim induce vomiting. IF SWALLOWED and victim is UNCONSCIOUS OR HAVING CONVULSIONS, do nothing except keep victim warm.
Water Pollution	Dangerous to aquatic life in high concentrations. May be dangerous if it enters water intakes. Notify local health and wildlife officials. Notify operators of nearby water intakes.

1. RESPONSE TO DISCHARGE

(See Response Methods Handbook, CG 446-4)

Issue warning—water contaminant
Should be removed
Chemical and physical treatment

2. LABELS

No hazard label required by Code of Federal Regulations

3. CHEMICAL DESIGNATIONS

3.1 **Synonyms:** Butoxyethyl 2,4-dichlorophenoxyacetate; Butyl 2,4-dichlorophenoxyacetate; 2,4-Dichlorophenoxyacetic acid, butoxyethyl ester; 2,4-Dichlorophenoxyacetic acid, butyl ester; 2,4-Dichlorophenoxyacetic acid, isopropyl ester; Isopropyl 2,4-dichlorophenoxy acetate

3.2 **Coast Guard Compatibility Classification:** Not listed

(Continued on page 4)

4. OBSERVABLE CHARACTERISTICS

4.1 **Physical State (as shipped):** Liquid

4.2 **Color:** Brown; amber

4.3 **Odor:** May have odor of fuel oil.

5. HEALTH HAZARDS

5.1 **Personal Protective Equipment:** Face shield or goggles, rubber gloves

5.2 **Symptoms Following Exposure:** Contact with eyes may cause mild irritation.

5.3 **Treatment for Exposure:** INGESTION: if large amounts are swallowed, induce vomiting and get medical help. EYES: flush with plenty of water and see a doctor. SKIN: flush with water, wash with soap and water.

5.4 **Toxicity by Inhalation (Threshold Limit Value):** Data not available

5.5 **Short-Term Inhalation Limits:** Data not available

5.6 **Toxicity by Ingestion:** Grade 2 or 3; LD_{50} 320—617 mg/kg

5.7 **Late Toxicity:** Data not available

5.8 **Vapor (Gas) Irritant Characteristics:** Data not available

5.9 **Liquid or Solid Irritant Characteristics:** Data not available

5.10 **Odor Threshold:** Data not available

6. FIRE HAZARDS

6.1 **Flash Point:** >175°F O.C.

6.2 **Flammable Limits in Air:** Data not available

6.3 **Fire Extinguishing Agents:** Foam, dry chemical, carbon dioxide

6.4 **Fire Extinguishing Agents Not to be Used:** Water may be ineffective.

6.5 **Special Hazards of Combustion Products:** Irritating hydrogen chloride vapor may form in fire.

6.6 **Behavior in Fire:**

6.7 **Ignition Temperature:** Data not available

6.8 **Electrical Hazard:** Data not available

6.9 **Burning Rate:** Data not available

7. CHEMICAL REACTIVITY

7.1 **Reactivity with Water:** No reaction

7.2 **Reactivity with Common Materials:** May attack some forms of plastics

7.3 **Stability During Transport:** Stable

7.4 **Neutralizing Agents for Acids and Caustics:** Not pertinent

7.5 **Polymerization:** Not pertinent

7.6 **Inhibitor of Polymerization:** Not pertinent

8. WATER POLLUTION

8.1 **Aquatic Toxicity:**
350 ppm/24 hr/bass, bluegill/50% kill/fresh water
1.0—5.0 ppm/96 hr/oyster/39% shell growth disease/salt water

8.2 **Waterfowl Toxicity:** LD_{50} 2025.0 mg/kg

8.3 **Biological Oxygen Demand (BOD):** Data not available

8.4 **Food Chain Concentration Potential:** None

9. SELECTED MANUFACTURERS

1. Amchem Products, Inc.
   Brookside Ave.
   Ambler, Pa. 19002
2. Chipman Division of Rhodia, Inc.
   23 Belmont Drive
   Somerset, N. J. 08873
3. The Dow Chemical Co.
   Midland, Mich. 48640

10. SHIPPING INFORMATION

10.1 **Grades or Purity:** Technical, 99%; 64% in petroleum oil

10.2 **Storage Temperature:** Ambient

10.3 **Inert Atmosphere:** No requirement

10.4 **Venting:** Open

11. HAZARD ASSESSMENT CODE

(See Hazard Assessment Handbook, CG 446-3)

A-X-Y

12. HAZARD CLASSIFICATIONS

12.1 **Code of Federal Regulations:** Not listed

12.2 **NAS Hazard Rating for Bulk Water Transportation:** Not listed

12.3 **NFPA Hazard Classifications:** Not listed

13. PHYSICAL AND CHEMICAL PROPERTIES

13.1 **Physical State at 15°C and 1 atm:** Liquid

13.2 **Molecular Weight:** 234—291

13.3 **Boiling Point at 1 atm:** Very high

13.4 **Freezing Point:** Not pertinent

13.5 **Critical Temperature:** Not pertinent

13.6 **Critical Pressure:** Not pertinent

13.7 **Specific Gravity:** 1.088—1.237 at 20°C (liquid)

13.8 **Liquid Surface Tension:** Data not available

13.9 **Liquid-Water Interfacial Tension:** Data not available

13.10 **Vapor (Gas) Specific Gravity:** Not pertinent

13.11 **Ratio of Specific Heats of Vapor (Gas):** Not pertinent

13.12 **Latent Heat of Vaporization:** Data not available

13.13 **Heat of Combustion:** Data not available

13.14 **Heat of Decomposition:** Not pertinent

13.15 **Heat of Solution:** Not pertinent

13.16 **Heat of Polymerization:** Not pertinent

3. CHEMICAL DESIGNATIONS (Cont'd.)

3.3 **Chemical Formula:** 2,4-$Cl_2C_6H_3OCH_2COOR$, where $R=C_4H_9$, C_3H_7, or $CH_2CH_2OC_4H_9$

3.4 **IMCO/United Nations Numerical Designation:** Not listed

DTS

DEXTROSE SOLUTION

Common Synonyms	Watery liquid	Colorless	Odorless
Glucose solution Corn sugar solution Grape sugar solution	Sinks and mixes with water.		

Stop discharge if possible.
Isolate and remove discharged material.
Notify local health and pollution control agencies.

Fire	Not flammable.
Exposure	Not harmful.
Water Pollution	Dangerous to aquatic life in high concentrations. May be dangerous if it enters water intakes. Notify local health and pollution control officials. Notify operators of nearby water intakes.

1. RESPONSE TO DISCHARGE

(See Response Methods Handbook, CG 446-4)

Disperse and flush

2. LABELS

No hazard label required by Code of Federal Regulations

3. CHEMICAL DESIGNATIONS

3.1 **Synonyms:**
Corn sugar solution
Glucose solution
Grape sugar solution

3.2 **Coast Guard Compatibility Classification:** Miscellaneous water solutions

3.3 **Chemical Formula:** $C_6H_{12}O_6-H_2O$

3.4 **IMCO/United Nations Numerical Designation:** Not listed

4. OBSERVABLE CHARACTERISTICS

4.1 **Physical State (as shipped):** Liquid

4.2 **Color:** Clear, colorless

4.3 **Odor:** None

5. HEALTH HAZARDS

5.1 **Personal Protective Equipment:** None needed

5.2 **Symptoms Following Exposure:** No toxicity

5.3 **Treatment for Exposure:** None needed

5.4 **Toxicity by Inhalation (Threshold Limit Value):** Not pertinent

5.5 **Short-Term Inhalation Limits:** Not pertinent

5.6 **Toxicity by Ingestion:** None

5.7 **Late Toxicity:** None

5.8 **Vapor (Gas) Irritant Characteristics:** Not pertinent

5.9 **Liquid or Solid Irritant Characteristics:** None

5.10 **Odor Threshold:** Not pertinent

6. FIRE HAZARDS

6.1 **Flash Point:** Not flammable

6.2 **Flammable Limits in Air:** Not flammable

6.3 **Fire Extinguishing Agents:** Not pertinent

6.4 **Fire Extinguishing Agents Not to be Used:** Not pertinent

6.5 **Special Hazards of Combustion Products:** Not pertinent

6.6 **Behavior in Fire:** Not pertinent

6.7 **Ignition Temperature:** Not flammable

6.8 **Electrical Hazard:** Not pertinent

6.9 **Burning Rate:** Not flammable

7. CHEMICAL REACTIVITY

7.1 **Reactivity with Water:** No reaction

7.2 **Reactivity with Common Materials:** No reaction

7.3 **Stability During Transport:** Stable

7.4 **Neutralizing Agents for Acids and Caustics:** Not pertinent

7.5 **Polymerization:** Not pertinent

7.6 **Inhibitor of Polymerization:** Not pertinent

8. WATER POLLUTION

8.1 **Aquatic Toxicity:** Data not available

8.2 **Waterfowl Toxicity:** Data not available

8.3 **Biological Oxygen Demand (BOD):** 53%, 5 days; 99% (theor), 5 days

8.4 **Food Chain Concentration Potential:** None

9. SELECTED MANUFACTURERS

1. CPC International, Inc.
Argo, Ill. 60502
2. A. E. Staley Mfg. Co.
Decatur, Ill. 62529
3. Standard Brands, Inc.
Clinton Corn Processing Division
Clinton, Iowa 52732

10. SHIPPING INFORMATION

10.1 **Grades or Purity:** 65-69% solids

10.2 **Storage Temperature:** 130° to 145°F

10.3 **Inert Atmosphere:** No requirement

10.4 **Venting:** Data not available

11. HAZARD ASSESSMENT CODE

(See Hazard Assessment Handbook, CG 446-3)

A-P

12. HAZARD CLASSIFICATIONS

12.1 **Code of Federal Regulations:** Not listed

12.2 **NAS Hazard Rating for Bulk Water Transportation:** Not listed

12.3 **NFPA Hazard Classifications:** Not listed

13. PHYSICAL AND CHEMICAL PROPERTIES

13.1 **Physical State at 15°C and 1 atm:** Liquid

13.2 **Molecular Weight:** Not pertinent

13.3 **Boiling Point at 1 atm:** $>212°F = >100°C = >373°K$

13.4 **Freezing Point:** $<32°F = <0°C = <273°K$

13.5 **Critical Temperature:** Not pertinent

13.6 **Critical Pressure:** Not pertinent

13.7 **Specific Gravity:** (est.) 1.20 at 20°C (liquid)

13.8 **Liquid Surface Tension:** (est.) 60 dynes/cm = 0.06 N/m at 20°C

13.9 **Liquid-Water Interfacial Tension:** Not pertinent

13.10 **Vapor (Gas) Specific Gravity:** Not pertinent

13.11 **Ratio of Specific Heats of Vapor (Gas):** Not pertinent

13.12 **Latent Heat of Vaporization:** Not pertinent

13.13 **Heat of Combustion:** Not pertinent

13.14 **Heat of Decomposition:** Not pertinent

13.15 **Heat of Solution:** Not pertinent

13.16 **Heat of Polymerization:** Not pertinent

NOTES

REVISED 1978

DIACETONE ALCOHOL

Common Synonyms			
4-Hydroxy-4-methyl-2-pentanone Diacetone	Watery liquid	Colorless to light yellow	Mild, pleasant odor
	Floats and mixes with water.		

Avoid contact with liquid.
Wear goggles and self-contained breathing apparatus.
Stop discharge if possible.
Call fire department.
Isolate and remove discharged material.
Notify local health and pollution control agencies.

Fire	FLAMMABLE Flashback along vapor trail may occur. Vapor may explode if ignited in an enclosed area. Wear goggles and self-contained breathing apparatus. Extinguish with dry chemical, alcohol foam or carbon dioxide. Water may be ineffective on fire.
Exposure	CALL FOR MEDICAL AID. LIQUID Irritating to skin and eyes. Harmful if swallowed. Remove contaminated clothing and shoes. Flush affected areas with plenty of water. IF IN EYES, hold eyelids open and flush with plenty of water. IF SWALLOWED and victim is CONSCIOUS, have victim drink water or milk.
Water Pollution	Effect of low concentrations on aquatic life is unknown. May be dangerous if it enters water intakes. Notify local health and pollution control officials. Notify operators of nearby water intakes.

1. RESPONSE TO DISCHARGE

(See Response Methods Handbook, CG 446-4)

Disperse and flush

2. LABEL

Red

3. CHEMICAL DESIGNATIONS

3.1 **Synonyms:** Diacetone
4-Hydroxy-4-methyl-2-pentanone
Tyranton

3.2 **Coast Guard Compatibility Classification:** Alcohol

3.3 **Chemical Formula:** $CH_3C(OH)(CH_3)CH_2COCH_3$

3.4 **IMCO/United Nations Numerical Designation:** 3.3/1148

4. OBSERVABLE CHARACTERISTICS

4.1 **Physical State (as shipped):** Liquid

4.2 **Color:** Colorless to yellow

4.3 **Odor:** Mild, pleasant

5. HEALTH HAZARDS

5.1 **Personal Protective Equipment:** Air pack or organic canister, rubber gloves, goggles.

5.2 **Symptoms Following Exposure:** Vapor is irritating to the mucous membrane of the eye and respiratory tract. Inhalation can cause dizziness, nausea, some anesthesia. Very high concentrations have a narcotic effect. The liquid is not highly irritating to the skin but can cause dermatitis.

5.3 **Treatment for Exposure:** INHALATION: remove victim to fresh air. Give artificial respiration if breathing has stopped. CONTACT WITH EYES OR SKIN: wash affected skin areas with water; flush eyes with water and get medical care if discomfort persists.

5.4 **Toxicity by Inhalation (Threshold Limit Value):** 50 ppm

5.5 **Short-Term Inhalation Limits:** 150 ppm for 30 min.

5.6 **Toxicity by Ingestion:** Grade 2; LD_{50} 0.5 to 5 g/kg (rat)

5.7 **Late Toxicity:** Data not available

5.8 **Vapor (Gas) Irritant Characteristics:** Vapors cause moderate irritation such that personnel will find high concentrations unpleasant. The effect is temporary.

5.9 **Liquid or Solid Irritant Characteristics:** Minimum hazard. If spilled on clothing and allowed to remain, may cause smarting and reddening of the skin.

5.10 **Odor Threshold:** Data not available

6. FIRE HAZARDS

6.1 **Flash Point:** 142°F O.C.; 125°F C.C.

6.2 **Flammable Limits in Air:** 1.8%–6.9%

6.3 **Fire Extinguishing Agents:** Dry chemical, alcohol foam, carbon dioxide

6.4 **Fire Extinguishing Agents Not to be Used:** Not pertinent

6.5 **Special Hazards of Combustion Products:** Not pertinent

6.6 **Behavior in Fire:** Not pertinent

6.7 **Ignition Temperature:** 1118°F

6.8 **Electrical Hazard:** Not pertinent

6.9 **Burning Rate:** Data not available

7. CHEMICAL REACTIVITY

7.1 **Reactivity with Water:** No reaction

7.2 **Reactivity with Common Materials:** No reaction

7.3 **Stability During Transport:** Stable

7.4 **Neutralizing Agents for Acids and Caustics:** Not pertinent

7.5 **Polymerization:** Not pertinent

7.6 **Inhibitor of Polymerization:** Not pertinent

8. WATER POLLUTION

8.1 **Aquatic Toxicity:** Data not available

8.2 **Waterfowl Toxicity:** Data not available

8.3 **Biological Oxygen Demand (BOD):** Data not available

8.4 **Food Chain Concentration Potential:** None

9. SELECTED MANUFACTURERS

1. Celanese Corp.
Celanese Chemical Co. Division
245 Park Ave.
New York, N. Y. 10017

2. Shell Chemical Co.
Industrial Chemicals Division
Houston, Texas 77001

3. Union Carbide Corp.
Chemicals and Plastics Division
270 Park Ave.
New York, N. Y. 10017

10. SHIPPING INFORMATION

10.1 **Grades or Purity:** 99.0%

10.2 **Storage Temperature:** Ambient

10.3 **Inert Atmosphere:** No requirement

10.4 **Venting:** Open (flame arrester)

11. HAZARD ASSESSMENT CODE

(See Hazard Assessment Handbook, CG 446-3)

A-P-Q

12. HAZARD CLASSIFICATIONS

12.1 **Code of Federal Regulations:** Flammable Liquid

12.2 **NAS Hazard Rating for Bulk Water Transportation:** Not listed

12.3 **NFPA Hazard Classifications:**

Category	Classification
Health Hazard (Blue)	1
Flammability (Red)	2
Reactivity (Yellow)	0

13. PHYSICAL AND CHEMICAL PROPERTIES

13.1 **Physical State at 15°C and 1 atm:** Liquid

13.2 **Molecular Weight:** 116.16

13.3 **Boiling Point at 1 atm:** 336.6°F = 169.2°C = 442.4°K

13.4 **Freezing Point:** −45.0°F = −42.8°C = 230.4°K

13.5 **Critical Temperature:** 633°F = 334°C = 607°K

13.6 **Critical Pressure:** 380 psia = 36 atm = 3.6 MN/m^2

13.7 **Specific Gravity:** 0.938 at 20°C (liquid)

13.8 **Liquid Surface Tension:** Not pertinent

13.9 **Liquid-Water Interfacial Tension:** Not pertinent

13.10 **Vapor (Gas) Specific Gravity:** Not pertinent

13.11 **Ratio of Specific Heats of Vapor (Gas):** 1.052

13.12 **Latent Heat of Vaporization:** 150 Btu/lb = 85 cal/g = 3.6×10^5 J/kg

13.13 **Heat of Combustion:** (est.) −13,000 Btu/lb = −7,250 cal/g = -303×10^5 J/kg

13.14 **Heat of Decomposition:** Not pertinent

13.15 **Heat of Solution:** Not pertinent

13.16 **Heat of Polymerization:** Not pertinent

(Continued on pages 5 and 6)

NOTES

REVISED 1978

DAP — Di-n-AMYL PHTHALATE

Common Synonyms			
Phthalic acid, dipentyl ester Diamyl phthalate Phthalic acid, diamyl ester Dipentyl phthalate	Liquid	White	Odorless
	Floats on water.		

Stop discharge if possible. Keep people away.
Shut off ignition sources. Call fire department.
Avoid contact with liquid and vapor.
Isolate and remove discharged material.
Notify local health and pollution control agencies.

Fire

Combustible.
Extinguish with dry chemicals or carbon dioxide.
Water and foam may be ineffective on fire.
Cool exposed containers with water.

Exposure

CALL FOR MEDICAL AID.

VAPOR
Irritating to eyes, nose and throat.
If inhaled will cause headache, coughing, or difficult breathing.
If in eyes, hold eyelids open and flush with plenty of water.
If breathing has stopped, give artificial respiration.
If breathing is difficult, give oxygen.

LIQUID
Irritating to skin and eyes.
Harmful if swallowed.
Remove contaminated clothing and shoes.
Flush affected areas with plenty of water.
IF IN EYES, hold eyelids open and flush with plenty of water.
IF SWALLOWED and victim is CONSCIOUS, have victim drink water or milk.
IF SWALLOWED and victim is UNCONSCIOUS OR HAVING CONVULSIONS, do nothing except keep victim warm.

Water Pollution

Effect of low concentrations on aquatic life is unknown.
Fouling to shoreline.
May be dangerous if it enters water intakes.
Notify local health and wildlife officials.
Notify operators of nearby water intakes.

1. RESPONSE TO DISCHARGE
(See Response Methods Handbook, CG 446-4)

Mechanical containment
Should be removed
Chemical and physical treatment

2. LABELS

No hazard label required by Code of Federal Regulations

3. CHEMICAL DESIGNATIONS

3.1 **Synonyms:** Diamyl phthalate; Dipentyl phthalate; Phthalic acid, diamyl ester; Phthalic acid, dipentyl ester
3.2 **Coast Guard Compatibility Classification:** Not listed
3.3 **Chemical Formula:** $C_6H_4(COOC_5H_{11})_2$
3.4 **IMCO/United Nations Numerical Designation:** Not listed

4. OBSERVABLE CHARACTERISTICS

4.1 **Physical State (as shipped):** Liquid
4.2 **Color:** Colorless
4.3 **Odor:** None

5. HEALTH HAZARDS

5.1 **Personal Protective Equipment:** Goggles or face shield; rubber gloves
5.2 **Symptoms Following Exposure:** Inhalation of vapors from very hot material may cause headache, drowsiness, and convulsions. Hot vapors may irritate eyes.
5.3 **Treatment for Exposure:** INHALATION: move to fresh air. EYES: flush with water. SKIN: wipe off; flush with water; wash with soap and water.
5.4 **Toxicity by Inhalation (Threshold Limit Value):** Data not available
5.5 **Short-Term Inhalation Limits:** Data not available
5.6 **Toxicity by Ingestion:** Data not available
5.7 **Late Toxicity:** Causes birth defects in rats (skeletal and gross abnormalities)
5.8 **Vapor (Gas) Irritant Characteristics:** Data not available
5.9 **Liquid or Solid Irritant Characteristics:** Data not available
5.10 **Odor Threshold:** Data not available

6. FIRE HAZARDS

6.1 **Flash Point:** 245°F C.C.
6.2 **Flammable Limits in Air:** Data not available
6.3 **Fire Extinguishing Agents:** Foam, dry chemical, carbon dioxide
6.4 **Fire Extinguishing Agents Not to be Used:** Water or foam may cause frothing.
6.5 **Special Hazards of Combustion Products:**
6.6 **Behavior in Fire:**
6.7 **Ignition Temperature:** Data not available
6.8 **Electrical Hazard:** Data not available
6.9 **Burning Rate:** Data not available

7. CHEMICAL REACTIVITY

7.1 **Reactivity with Water:** No reaction
7.2 **Reactivity with Common Materials:** May attack some forms of plastics
7.3 **Stability During Transport:** Stable
7.4 **Neutralizing Agents for Acids and Caustics:** Not pertinent
7.5 **Polymerization:** Not pertinent
7.6 **Inhibitor of Polymerization:** Not pertinent

8. WATER POLLUTION

8.1 **Aquatic Toxicity:** Data not available
8.2 **Waterfowl Toxicity:** Data not available
8.3 **Biological Oxygen Demand (BOD):** Data not available
8.4 **Food Chain Concentration Potential:** None

9. SELECTED MANUFACTURERS

1. Eastman Organic Chemicals
   Rochester, N. Y. 14650
2. Pfaltz and Bauer, Inc.
   375 Fairfield Ave.
   Stamford, Conn. 06902

10. SHIPPING INFORMATION

10.1 **Grades or Purity:** Technical
10.2 **Storage Temperature:** Ambient
10.3 **Inert Atmosphere:** No requirement
10.4 **Venting:** Open

11. HAZARD ASSESSMENT CODE
(See Hazard Assessment Handbook, CG 446-3)

A-T-U

12. HAZARD CLASSIFICATIONS

12.1 **Code of Federal Regulations:** Not listed
12.2 **NAS Hazard Rating for Bulk Water Transportation:** Not listed
12.3 **NFPA Hazard Classifications:**

Category	Classification
Health Hazard (Blue)	0
Flammability (Red)	1
Reactivity (Yellow)	0

13. PHYSICAL AND CHEMICAL PROPERTIES

13.1 **Physical State at 15°C and 1 atm:** Liquid
13.2 **Molecular Weight:** 306
13.3 **Boiling Point at 1 atm:** Very high
13.4 **Freezing Point:** Not pertinent
13.5 **Critical Temperature:** Not pertinent
13.6 **Critical Pressure:** Not pertinent
13.7 **Specific Gravity:** 0.82 at 20°C (liquid)
13.8 **Liquid Surface Tension:** 31.5 dynes/cm = 0.0315 N/m at 20°C
13.9 **Liquid-Water Interfacial Tension:** Data not available
13.10 **Vapor (Gas) Specific Gravity:** Not pertinent
13.11 **Ratio of Specific Heats of Vapor (Gas):** Not pertinent
13.12 **Latent Heat of Vaporization:** 140 Btu/lb = 76 cal/g = 3.2×10^5 J/kg
13.13 **Heat of Combustion:** −13,900 Btu/lb = −7,720 cal/g = -323×10^5 J/kg
13.14 **Heat of Decomposition:** Not pertinent
13.15 **Heat of Solution:** Not pertinent
13.16 **Heat of Polymerization:** Not pertinent

NOTES

DZN

DIAZINON

Common Synonyms		
0, 0-Diethyl 0-(2-isopropyl-6-methyl-4-pyrimidinyl) phosphorothioate 0, 0-Diethyl 0-(2-isopropyl-4-methyl-6-pyrimidyl) thiophosphate Alfa-tox, Spectracide, Saralex	Liquid Sinks in water.	Light to dark brown

Stop discharge if possible. Keep people away.
Isolate and remove discharged material.
Notify local health and pollution control agencies.

Fire

Not flammable.
POISONOUS GASES ARE PRODUCED WHEN HEATED.

Exposure

CALL FOR MEDICAL AID.

LIQUID
POISONOUS IF SWALLOWED.
Irritating to skin and eyes.
Remove contaminated clothing and shoes.
Flush affected areas with plenty of water.
IF IN EYES, hold eyelids open and flush with plenty of water.
IF SWALLOWED and victim is CONSCIOUS, have victim drink water or milk.

Water Pollution

HARMFUL TO AQUATIC LIFE IN VERY LOW CONCENTRATIONS.
May be dangerous if it enters water intakes.
Notify local health and wildlife officials.
Notify operators of nearby water intakes.

1. RESPONSE TO DISCHARGE

(See Response Methods Handbook, CG 446-4)

Issue warning—poison, water contaminant, high flammability (if solution)
Restrict access
Should be removed
Chemical and physical treatment

2. LABELS

No hazard label required by Code of Federal Regulations

3. CHEMICAL DESIGNATIONS

3.1 **Synonyms:** O, O-Diethyl O-(2-isopropyl-6-methyl-4-pyrimidinyl) phosphorothioate; O, O-Diethyl O-2-isopropyl-4-methyl-6-pyrimidyl thiophosphate; Diethyl 2-isopropyl-4-methyl-6-pyrimidyl thionophosphate; Alfa-tox; Saralex; Spectracide

3.2 **Coast Guard Compatibility Classification:** Not applicable

3.3 **Chemical Formula:** $C_{12}H_{21}N_2O_3PS$

3.4 **IMCO/United Nations Numerical Designation:** 6.1/1615

4. OBSERVABLE CHARACTERISTICS

4.1 **Physical State (as shipped):** Solid, or liquid solution

4.2 **Color:** Amber to dark brown

4.3 **Odor:** Data not available

5. HEALTH HAZARDS

5.1 **Personal Protective Equipment:** Goggles or face shield; rubber gloves; protective clothing.

5.2 **Symptoms Following Exposure:** Ingestion or prolonged inhalation of mist causes headache, giddiness, blurred vision, nervousness, weakness, cramps, diarrhea, discomfort in the chest, sweating, miosis, tearing, salivation and other excessive respiratory tract secretion, vomiting, cyanosis, papilledema, uncontrollable muscle twitches, convulsions, coma, loss of reflexes, and loss of sphincter control. Liquid irritates eyes and skin.

5.3 **Treatment for Exposure:** INHALATION: remove to fresh air; keep warm; get medical attention at once. EYES: flush with plenty of water for at least 15 min. and get medical attention. SKIN: wash contaminated area with soap and water. INGESTION: get medical attention at once; give water slurry of charcoal; do NOT give milk or alcohol.

5.4 **Toxicity by Inhalation (Threshold Limit Value):** Data not available

5.5 **Short-Term Inhalation Limits:** Not pertinent

5.6 **Toxicity by Ingestion:** Grade 3; oral LD_{50} = 76 mg/kg (rat)

5.7 **Late Toxicity:** May be mutagenic

5.8 **Vapor (Gas) Irritant Characteristics:** Data not available

5.9 **Liquid or Solid Irritant Characteristics:** Data not available

5.10 **Odor Threshold:** Data not available

6. FIRE HAZARDS

6.1 **Flash Point:** 82—105°F C.C. (solutions only; pure liquid difficult to burn)

6.2 **Flammable Limits in Air:** Not pertinent

6.3 **Fire Extinguishing Agents:** (for solutions) Foam, dry chemical, or carbon dioxide

6.4 **Fire Extinguishing Agents Not to be Used:** Water may be ineffective.

6.5 **Special Hazards of Combustion Products:** Oxides of sulfur and of phosphorus are generated in fires.

6.6 **Behavior in Fire:** Not pertinent

6.7 **Ignition Temperature:** Not pertinent

6.8 **Electrical Hazard:** Data not available

6.9 **Burning Rate:** (for solutions) 4 mm/min.

7. CHEMICAL REACTIVITY

7.1 **Reactivity with Water:** No reaction

7.2 **Reactivity with Common Materials:** No reaction

7.3 **Stability During Transport:** Stable

7.4 **Neutralizing Agents for Acids and Caustics:** Not pertinent

7.5 **Polymerization:** Not pertinent

7.6 **Inhibitor of Polymerization:** Not pertinent

8. WATER POLLUTION

8.1 **Aquatic Toxicity:**
0.025 ppm/96 hr/stonefly nymph/TL_m/fresh water
30 μ g/l /48 hr/bluegill/TL_m/fresh water (becomes bound to soil when used according to directions)

8.2 **Waterfowl Toxicity:**
LD_{50} = 3.54 mg/kg
LC_{50} = 5 days, 90 ppm mallard duck
LC_{50} = 7 days, 68 ppm quail

8.3 **Biological Oxygen Demand (BOD):** Data not available

8.4 **Food Chain Concentration Potential:** Data not available

9. SELECTED MANUFACTURERS

Ciba-Geigy Corporation
Agricultural Division
P. O. Box 11422
Greensboro, N. C. 27409

10. SHIPPING INFORMATION

10.1 **Grades or Purity:** Technical; wettable powders; a variety of emulsifiable solutions in combustible solvents.

10.2 **Storage Temperature:** Ambient

10.3 **Inert Atmosphere:** No requirement

10.4 **Venting:** Open (flame arrester)

11. HAZARD ASSESSMENT CODE

(See Hazard Assessment Handbook, CG 446-3)

A-X-Y

12. HAZARD CLASSIFICATIONS

12.1 **Code of Federal Regulations:** ORM-A

12.2 **NAS Hazard Rating for Bulk Water Transportation:** Not listed

12.3 **NFPA Hazard Classifications:** Not listed

13. PHYSICAL AND CHEMICAL PROPERTIES

13.1 **Physical State at 15°C and 1 atm:** Liquid

13.2 **Molecular Weight:** 304.4

13.3 **Boiling Point at 1 atm:** Very high; decomposes

13.4 **Freezing Point:** Not pertinent

13.5 **Critical Temperature:** Not pertinent

13.6 **Critical Pressure:** Not pertinent

13.7 **Specific Gravity:** 1.117 at 20°C (liquid)

13.8 **Liquid Surface Tension:** (est.) 35 dynes/cm = 0.035 N/m at 20°C

13.9 **Liquid-Water Interfacial Tension:** (est.) 40 dynes/cm = 0.040 N/m at 20°C

13.10 **Vapor (Gas) Specific Gravity:** Not pertinent

13.11 **Ratio of Specific Heats of Vapor (Gas):** Not pertinent

13.12 **Latent Heat of Vaporization:** Not pertinent

13.13 **Heat of Combustion:** (est.) −12,000 Btu/lb = −6,500 cal/g = -270×10^5 J/kg

13.14 **Heat of Decomposition:** Not pertinent

13.15 **Heat of Solution:** Not pertinent

13.16 **Heat of Polymerization:** Not pertinent

NOTES

DPO — DIBENZOYL PEROXIDE

Common Synonyms			
Benzoyl peroxide Benzoyl superoxide Lucidol Oxylite BPO BP	Solid powder or granules	White	Odorless
	Sinks in water.		

Keep people away. Evacuate area in case of large discharge.
Wear goggles and self-contained breathing apparatus.
Stay upwind and use water spray to "knock down" dust.
Call fire department.
Avoid contact with solid and dust.
Isolate and remove discharged material.
Notify local health and pollution control agencies.

Fire

FLAMMABLE.
MAY EXPLODE IF SUBJECTED TO HEAT, SHOCK OR FRICTION.
May cause fire and explode on contact with combustibles.
Evacuate surrounding area.
Wear goggles and self-contained breathing apparatus.
Combat fires from safe distance.
Cool exposed containers with water.

Exposure

CALL FOR MEDICAL AID.

SOLID
Irritating to skin and eyes.
Harmful if swallowed.
Remove contaminated clothing and shoes.
Flush affected areas with plenty of water.
IF IN EYES, hold eyelids open and flush with plenty of water.
IF SWALLOWED and victim is CONSCIOUS, have victim drink water or milk and have victim induce vomiting.
IF SWALLOWED and victim is UNCONSCIOUS OR HAVING CONVULSIONS, do nothing except keep victim warm.

Water Pollution

Effect of low concentrations on aquatic life is unknown.
May be dangerous if it enters water intakes.
Notify local health and pollution control officials.
Notify operators of nearby water intakes.

1. RESPONSE TO DISCHARGE

(See Response Methods Handbook, CG 446-4)

Issue warning—high flammability
Should be removed
Chemical and physical treatment

2. LABEL

Yellow

3. CHEMICAL DESIGNATIONS

3.1 **Synonyms:**
Benzoyl peroxide — BPO
Benzoyl superoxide — Lucidol-70
BP — Oxylite

3.2 **Coast Guard Compatibility Classification:** Not applicable

3.3 **Chemical Formula:** $C_6H_5CO{\cdot}O{\cdot}O{\cdot}COC_6H_5$

3.4 **IMCO/United Nations Numerical Designation:** 5.2/1521

4. OBSERVABLE CHARACTERISTICS

4.1 **Physical State (as shipped):** Solid

4.2 **Color:** White

4.3 **Odor:** None

5. HEALTH HAZARDS

5.1 **Personal Protective Equipment:** Safety goggles, face shield, rubber gloves.

5.2 **Symptoms Following Exposure:** CONTACT WITH EYES OR SKIN: irritates eyes. Prolonged contact may irritate skin.

5.3 **Treatment for Exposure:** INGESTION: administer an emetic to induce vomiting and call a physician. CONTACT WITH EYES OR SKIN: do not use oils or ointments; flush eyes with plenty of water and get medical attention; wash skin with plenty of soap and water.

5.4 **Toxicity by Inhalation (Threshold Limit Value):** 5 mg/m³

5.5 **Short-Term Inhalation Limits:** Not pertinent

5.6 **Toxicity by Ingestion:** Grade 2; LD_{50} 0.5 to 5 g/kg

5.7 **Late Toxicity:** Data not available

5.8 **Vapor (Gas) Irritant Characteristics:** Not pertinent

5.9 **Liquid or Solid Irritant Characteristics:** Minimum hazard. If spilled on clothing and allowed to remain, may cause smarting and reddening of the skin.

5.10 **Odor Threshold:** Not pertinent

6. FIRE HAZARDS

6.1 **Flash Point:** Highly flammable solid; explosion-sensitive to shock, heat, and friction

6.2 **Flammable Limits in Air:** Not pertinent

6.3 **Fire Extinguishing Agents:** Difficult to extinguish once ignited. Use water spray to cool surrounding area.

6.4 **Fire Extinguishing Agents Not to be Used:** Do not use hand extinguishers

6.5 **Special Hazards of Combustion Products:** Suffocating smoke evolved

6.6 **Behavior in Fire:** May explode

6.7 **Ignition Temperature:** Data not available

6.8 **Electrical Hazard:** Not pertinent

6.9 **Burning Rate:** Not pertinent

7. CHEMICAL REACTIVITY

7.1 **Reactivity with Water:** No reaction

7.2 **Reactivity with Common Materials:** Special care must be taken to avoid contamination with combustible materials (wood, paper, etc.),various inorganic and organic acids, alkalies, alcohols, amines, easily oxidizable materials such as ethers, or materials used as accelerators in polymerization reactions.

7.3 **Stability During Transport:** Extremely explosion-sensitive to shock (impact, blows), heat, and friction. Has been reported to explode for apparently no specific reason. Self-reactive.

7.4 **Neutralizing Agents for Acids and Caustics:** Not pertinent

7.5 **Polymerization:** Not pertinent

7.6 **Inhibitor of Polymerization:** Not pertinent

8. WATER POLLUTION

8.1 **Aquatic Toxicity:** Data not available

8.2 **Waterfowl Toxicity:** Data not available

8.3 **Biological Oxygen Demand (BOD):** Data not available

8.4 **Food Chain Concentration Potential:** None

9. SELECTED MANUFACTURERS

1. Pennwalt Corp.
Lucidol Division
1740 Military Rd.
Buffalo, N. Y. 14240

2. Reichold Chemicals, Inc.
Specialty Chemicals Division
Austin, Texas 78712

3. Witco Chemical Corp.
U. S. Peroxygen Division
850 Morton Ave.
Richmond, Calif. 94804

10. SHIPPING INFORMATION

10.1 **Grades or Purity:** 98% (dry); 70-78% (wet); various pastes of dibenzoyl peroxide and liquid plasticizers such as tricresyl phosphate, silicone oil.

10.2 **Storage Temperature:** 65-85°F

10.3 **Inert Atmosphere:** Data not available

10.4 **Venting:** Data not available

11. HAZARD ASSESSMENT CODE

(See Hazard Assessment Handbook, CG 446-3)

II

12. HAZARD CLASSIFICATIONS

12.1 **Code of Federal Regulations:** Organic Peroxide

12.2 **NAS Hazard Rating for Bulk Water Transportation:** Not listed

12.3 **NFPA Hazard Classifications:**

Category	Classification
Health Hazard (Blue)	1
Flammability (Red)	4
Reactivity (Yellow)	4 oxy

13. PHYSICAL AND CHEMICAL PROPERTIES

13.1 **Physical State at 15°C and 1 atm:** Solid

13.2 **Molecular Weight:** 242.22

13.3 **Boiling Point at 1 atm:** Not pertinent

13.4 **Freezing Point:** 217°F = 103°C = 376°K

13.5 **Critical Temperature:** Not pertinent

13.6 **Critical Pressure:** Not pertinent

13.7 **Specific Gravity:** 1.334 at 15°C (solid)

13.8 **Liquid Surface Tension:** Not pertinent

13.9 **Liquid-Water Interfacial Tension:** Not pertinent

13.10 **Vapor (Gas) Specific Gravity:** Not pertinent

13.11 **Ratio of Specific Heats of Vapor (Gas):** Not pertinent

13.12 **Latent Heat of Vaporization:** Not pertinent

13.13 **Heat of Combustion:** Not pertinent

13.14 **Heat of Decomposition:** Not pertinent

13.15 **Heat of Solution:** Not pertinent

13.16 **Heat of Polymerization:** Not pertinent

NOTES

REVISED 1978

DBA Di-n-BUTYLAMINE

Common Synonyms			
Dibutylamine 1-Butanamine, N-butyl	Liquid	Colorless	Weak ammonia, fishy odor
	Floats and mixes with water.		

Stop discharge if possible. Keep people away.
Shut off ignition sources. Call fire department.
Avoid contact with liquid and vapor.
Isolate and remove discharged material.
Notify local health and pollution control agencies.

Fire

Combustible.
POISONOUS GASES MAY BE PRODUCED IN FIRE.
Wear goggles and self-contained breathing apparatus.
Extinguish with dry chemicals, alcohol foam, or carbon dioxide.
Water may be ineffective on fire.
Cool exposed containers with water.

Exposure

CALL FOR MEDICAL AID.

VAPOR
Irritating to eyes, nose and throat.
If inhaled will cause headache, coughing or difficult breathing.
If in eyes, hold eyelids open and flush with plenty of water.
If breathing has stopped, give artificial respiration.
If breathing is difficult, give oxygen.

LIQUID
Irritating to skin and eyes.
If swallowed will cause nausea and vomiting.
Remove contaminated clothing and shoes.
Flush affected areas with plenty of water.
IF IN EYES, hold eyelids open and flush with plenty of water.
IF SWALLOWED and victim is CONSCIOUS, have victim drink water or milk.
IF SWALLOWED and victim is UNCONSCIOUS OR HAVING CONVULSIONS, do nothing except keep victim warm.

Water Pollution

Effect of low concentrations on aquatic life is unknown.
May be dangerous if it enters water intakes.
Notify local health and wildlife officials.
Notify operators of nearby water intakes.

1. RESPONSE TO DISCHARGE
(See Response Methods Handbook, CG 446-4)

Issue warning—water contaminant
Restrict access
Disperse and flush

2. LABELS

No hazard label required by Code of Federal Regulations

3. CHEMICAL DESIGNATIONS

3.1 **Synonyms:** 1-Butanamine, N-butyl; Dibutylamine
3.2 **Coast Guard Compatibility Classification:** Aliphatic amine
3.3 **Chemical Formula:** $(C_4H_9)_2NH$
3.4 **IMCO/United Nations Numerical Designation:** Not listed

4. OBSERVABLE CHARACTERISTICS

4.1 **Physical State (as shipped):** Liquid
4.2 **Color:** Colorless
4.3 **Odor:** Weak ammonia, fishy

5. HEALTH HAZARDS

5.1 **Personal Protective Equipment:** Goggles or face shield; rubber gloves
5.2 **Symptoms Following Exposure:** Inhalation causes irritation of nose, throat, and lungs; coughing; nausea; headache. Ingestion causes irritation of mouth and stomach. Contact with eyes causes irritation. Contact with skin causes irritation and dermatitis.
5.3 **Treatment for Exposure:** INHALATION: move from exposure; if breathing has stopped, start artificial respiration. INGESTION: give large amount of water. EYES: irrigate with water for 15 min.; get medical attention for possible eye damage. SKIN: wash with large amounts of water for 15 min.
5.4 **Toxicity by Inhalation (Threshold Limit Value):** Data not available
5.5 **Short-Term Inhalation Limits:** Data not available
5.6 **Toxicity by Ingestion:** Grade 3; oral LD_{50} = 360 mg/kg (rat)
5.7 **Late Toxicity:** Data not available
5.8 **Vapor (Gas) Irritant Characteristics:** Vapors cause moderate irritation such that personnel will find high concentrations unpleasant. The effect is temporary.
5.9 **Liquid or Solid Irritant Characteristics:** Severe skin irritant. Causes second- and third-degree burns on short contact and is very injurious to the eyes.
5.10 **Odor Threshold:** Data not available

6. FIRE HAZARDS

6.1 **Flash Point:** 125°F O.C.
6.2 **Flammable Limits in Air:** 1.1% (LFL)
6.3 **Fire Extinguishing Agents:** "Alcohol" foam, dry chemical, carbon dioxide
6.4 **Fire Extinguishing Agents Not to be Used:** Water may be ineffective.
6.5 **Special Hazards of Combustion Products:** Toxic oxides of nitrogen may form in fires.
6.6 **Behavior in Fire:**
6.7 **Ignition Temperature:** Data not available
6.8 **Electrical Hazard:** Data not available
6.9 **Burning Rate:** 5.84 mm/min.

7. CHEMICAL REACTIVITY

7.1 **Reactivity with Water:** No reaction
7.2 **Reactivity with Common Materials:** May corrode some metals and attack some forms of plastics
7.3 **Stability During Transport:** Stable
7.4 **Neutralizing Agents for Acids and Caustics:** Not pertinent
7.5 **Polymerization:** Not pertinent
7.6 **Inhibitor of Polymerization:** Not pertinent

8. WATER POLLUTION

8.1 **Aquatic Toxicity:** Data not available
8.2 **Waterfowl Toxicity:** Data not available
8.3 **Biological Oxygen Demand (BOD):** Data not available
8.4 **Food Chain Concentration Potential:** None

9. SELECTED MANUFACTURERS

1. Union Carbide Corp.
   Chemicals and Plastics Division
   270 Park Ave.
   New York, N. Y. 10017
2. Pennwalt Corporation
   Three Parkway
   Philadelphia, Pa. 19102
3. Virginia Chemicals, Inc.
   340 West Norfolk Road
   Portsmouth, Va. 23703

10. SHIPPING INFORMATION

10.1 **Grades or Purity:** Technical
10.2 **Storage Temperature:** Ambient
10.3 **Inert Atmosphere:** No requirement
10.4 **Venting:** Open (flame arrester)

11. HAZARD ASSESSMENT CODE
(See Hazard Assessment Handbook, CG 446-3)

A-T-U

12. HAZARD CLASSIFICATIONS

12.1 **Code of Federal Regulations:** Not listed
12.2 **NAS Hazard Rating for Bulk Water Transportation:**

Category	Rating
Fire	2
Health	
Vapor Irritant	2
Liquid or Solid Irritant	4
Poisons	2
Water Pollution	
Human Toxicity	2
Aquatic Toxicity	2
Aesthetic Effect	2
Reactivity	
Other Chemicals	3
Water	0
Self-Reaction	0

12.3 **NFPA Hazard Classifications:**

Category	Classification
Health Hazard (Blue)	3
Flammability (Red)	2
Reactivity (Yellow)	0

13. PHYSICAL AND CHEMICAL PROPERTIES

13.1 **Physical State at 15°C and 1 atm:** Liquid
13.2 **Molecular Weight:** 129.25
13.3 **Boiling Point at 1 atm:** 319.3°F = 159.6°C = 432.8°K
13.4 **Freezing Point:** −80°F = −62°C = 211°K
13.5 **Critical Temperature:** Not pertinent
13.6 **Critical Pressure:** Not pertinent
13.7 **Specific Gravity:** 0.759 at 20°C (liquid)
13.8 **Liquid Surface Tension:** 24.76 dynes/cm = 0.02476 N/m at 20°C
13.9 **Liquid-Water Interfacial Tension:** Not pertinent
13.10 **Vapor (Gas) Specific Gravity:** 4.5
13.11 **Ratio of Specific Heats of Vapor (Gas):** Not pertinent
13.12 **Latent Heat of Vaporization:** 130 Btu/lb = 72.3 cal/g = 3.03×10^5 J/kg
13.13 **Heat of Combustion:** −18,800 Btu/lb = −10,440 cal/g = -436.8×10^5 J/kg
13.14 **Heat of Decomposition:** Not pertinent
13.15 **Heat of Solution:** Not pertinent
13.16 **Heat of Polymerization:** Not pertinent

NOTES

DBE	# DI-n-BUTYL ETHER

Common Synonyms			
n-Dibutyl ether n-Butyl ether Butyl ether 1-Butoxy butane Dibutyl ether	Liquid	Colorless	Mild pleasant odor
	Floats on water. Flammable, irritating vapor is produced.		

Stop discharge if possible. Keep people away.
Shut off ignition sources. Call fire department.
Isolate and remove discharged material.
Notify local health and pollution control agencies.

Fire	FLAMMABLE. Containers may explode in fire. Flashback along vapor trail may occur. Vapor may explode if ignited in an enclosed area. Extinguish with dry chemicals, alcohol foam, or carbon dioxide. Water may be ineffective on fire. Cool exposed containers with water.
Exposure	Call for medical aid. VAPOR Irritating to eyes, nose and throat. Move victim to fresh air. If breathing is difficult, give oxygen. LIQUID Irritating to skin and eyes. Remove contaminated clothing and shoes. Flush affected areas with plenty of water. IF IN EYES, hold eyelids open and flush with plenty of water. IF SWALLOWED and victim is CONSCIOUS, have victim drink water or milk.
Water Pollution	Effect of low concentrations on aquatic life is unknown. Fouling to shoreline. May be dangerous if it enters water intakes. Notify local health and wildlife officials. Notify operators of nearby water intakes.

1. RESPONSE TO DISCHARGE

(See Response Methods Handbook, CG 446-4)

Mechanical containment
Should be removed
Chemical and physical treatment

2. LABELS

No label required by Code of Federal Regulations.

3. CHEMICAL DESIGNATIONS

3.1 **Synonyms:** n-Dibutyl ether; n-Butyl ether; Butyl ether; 1-Butoxybutane; Dibutyl ether; Dibutyl oxide

3.2 **Coast Guard Compatibility Classification:** Ethers (21)

3.3 **Chemical Formula:** $C_4H_9OC_4H_9$

3.4 **IMCO/United Nations Numerical Designation:** 3.3/1149

4. OBSERVABLE CHARACTERISTICS

4.1 **Physical State (as shipped):** Liquid

4.2 **Color:** Colorless

4.3 **Odor:** Mild, etherlike

5. HEALTH HAZARDS

5.1 **Personal Protective Equipment:** Goggles or face shield; rubber gloves.

5.2 **Symptoms Following Exposure:** Inhalation causes irritation of nose and throat. Liquid irritates eyes and may irritate skin on prolonged contact. Ingestion causes irritation of mouth and stomach.

5.3 **Treatment for Exposure:** INHALATION: remove to fresh air. EYES: after contact with liquid, flush with water for at least 15 min. SKIN: wipe off, wash well with soap and water. INGESTION: induce vomiting.

5.4 **Toxicity by Inhalation (Threshold Limit Value):** Data not available

5.5 **Short-Term Inhalation Limits:** Data not available

5.6 **Toxicity by Ingestion:** Grade 1; oral LD_{50} = 7,400 mg/kg (rat)

5.7 **Late Toxicity:** Data not available

5.8 **Vapor (Gas) Irritant Characteristics:** Data not available

5.9 **Liquid or Solid Irritant Characteristics:** Data not available

5.10 **Odor Threshold:** Data not available

6. FIRE HAZARDS

6.1 **Flash Point:** 92°F O.C.

6.2 **Flammable Limits in Air:** 1.5% – 7.6%

6.3 **Fire Extinguishing Agents:** Dry chemical, "alcohol" foam, or carbon dioxide.

6.4 **Fire Extinguishing Agents Not to be Used:** Water may be ineffective.

6.5 **Special Hazards of Combustion Products:** Not pertinent

6.6 **Behavior in Fire:** Vapor is heavier than air and may travel a considerable distance to a source of ignition and flash back.

6.7 **Ignition Temperature:** 382°F

6.8 **Electrical Hazard:** Data not available

6.9 **Burning Rate:** 5.7 mm/min.

7. CHEMICAL REACTIVITY

7.1 **Reactivity with Water:** No reaction

7.2 **Reactivity with Common Materials:** No reaction

7.3 **Stability During Transport:** Stable

7.4 **Neutralizing Agents for Acids and Caustics:** Not pertinent

7.5 **Polymerization:** Not pertinent

7.6 **Inhibitor of Polymerization:** Not pertinent

8. WATER POLLUTION

8.1 **Aquatic Toxicity:** Data not available

8.2 **Waterfowl Toxicity:** Data not available

8.3 **Biological Oxygen Demand (BOD):** Data not available

8.4 **Food Chain Concentration Potential:** None

9. SELECTED MANUFACTURERS

1. Publicker Industries Inc.
   1429 Walnut Street
   Philadelphia, Pa. 19102
2. Union Carbide Corporation
   Chemicals and Plastics Division
   270 Park Avenue
   New York, N. Y. 10017
3. MC/B Manufacturing Chemists
   2909 Highland Ave.
   Norwood, Ohio 45212

10. SHIPPING INFORMATION

10.1 **Grades or Purity:** 99+%

10.2 **Storage Temperature:** Ambient

10.3 **Inert Atmosphere:** No requirement

10.4 **Venting:** Open (flame arrester)

11. HAZARD ASSESSMENT CODE

(See Hazard Assessment Handbook, CG 446-3)

A-T-U

12. HAZARD CLASSIFICATIONS

12.1 **Code of Federal Regulations:** Not listed

12.2 **NAS Hazard Rating for Bulk Water Transportation:** Not listed

12.3 **NFPA Hazard Classifications:**

Category	Classification
Health Hazard (Blue)	2
Flammability (Red)	3
Reactivity (Yellow)	0

13. PHYSICAL AND CHEMICAL PROPERTIES

13.1 **Physical State at 15°C and 1 atm:** Liquid

13.2 **Molecular Weight:** 130.2

13.3 **Boiling Point at 1 atm:** 288°F = 142°C = 415°K

13.4 **Freezing Point:** −139.7°F = −95.4°C = 177.8°K

13.5 **Critical Temperature:** Not pertinent

13.6 **Critical Pressure:** Not pertinent

13.7 **Specific Gravity:** 0.767 at 20°C (liquid)

13.8 **Liquid Surface Tension:** 23 dynes/cm = 0.023 N/m at 20°C

13.9 **Liquid-Water Interfacial Tension:** (est.) 30 dynes/cm = 0.030 N/m at 20°C

13.10 **Vapor (Gas) Specific Gravity:** 4.5

13.11 **Ratio of Specific Heats of Vapor (Gas):** 1.0434

13.12 **Latent Heat of Vaporization:** 120 Btu/lb = 68 cal/g = 2.8×10^5 J/kg

13.13 **Heat of Combustion:** −17,670 Btu/lb = −9,820 cal/g = -411×10^5 J/kg

13.14 **Heat of Decomposition:** Not pertinent

13.15 **Heat of Solution:** Not pertinent

13.16 **Heat of Polymerization:** Not pertinent

NOTES

DBK — Di-n-BUTYL KETONE

Common Synonyms
5-Noranone

Liquid — Colorless to light yellow

Floats on water.
Freezing point is 21°F.

Stop discharge if possible. Keep people away.
Shut off ignition sources. Call fire department.
Avoid contact with liquid and vapor.
Isolate and remove discharged material.
Notify local health and pollution control agencies.

Fire

Combustible.
Extinguish with dry chemicals, foam or carbon dioxide.
Water may be ineffective on fire.
Cool exposed containers with water.

Exposure

CALL FOR MEDICAL AID.

VAPOR
Irritating to eyes, nose and throat.
If inhaled will cause coughing or difficult breathing.
If in eyes, hold eyelids open and flush with plenty of water.
If breathing has stopped, give artificial respiration.
If breathing is difficult, give oxygen.

LIQUID
Irritating to skin and eyes.
If swallowed will cause nausea and vomiting.
Remove contaminated clothing and shoes.
Flush affected areas with plenty of water.
IF IN EYES, hold eyelids open and flush with plenty of water.
IF SWALLOWED and victim is CONSCIOUS, have victim drink water or milk.
IF SWALLOWED and victim is UNCONSCIOUS OR HAVING CONVULSIONS, do nothing except keep victim warm.

Water Pollution

Effect of low concentrations on aquatic life is unknown.
Fouling to shoreline.
May be dangerous if it enters water intakes.
Notify local health and wildlife officials.
Notify operators of nearby water intakes.

1. RESPONSE TO DISCHARGE

(See Response Methods Handbook, CG 446-4)

Mechanical containment
Should be removed
Chemical and physical treatment

2. LABELS

No hazard label required by Code of Federal Regulations

3. CHEMICAL DESIGNATIONS

3.1 **Synonyms:** 5-Nonanone
3.2 **Coast Guard Compatibility Classification:** Not listed
3.3 **Chemical Formula:** $CH_3(CH_2)_3CO(CH_2)_3CH_3$
3.4 **IMCO/United Nations Numerical Designation:** Not listed

4. OBSERVABLE CHARACTERISTICS

4.1 **Physical State (as shipped):** Liquid
4.2 **Color:** White to light yellow
4.3 **Odor:** Data not available

5. HEALTH HAZARDS

5.1 **Personal Protective Equipment:** Rubber gloves; goggles or face shield
5.2 **Symptoms Following Exposure:** Inhalation causes irritation of nose and throat. Ingestion causes irritation of mouth and stomach. Contact with eyes or skin causes irritation.
5.3 **Treatment for Exposure:** INHALATION: remove to fresh air; administer artificial respiration if needed. EYES: flush with water for at least 15 min. SKIN: flush with water.
5.4 **Toxicity by Inhalation (Threshold Limit Value):** Data not available
5.5 **Short-Term Inhalation Limits:** Data not available
5.6 **Toxicity by Ingestion:** Data not available
5.7 **Late Toxicity:** Data not available
5.8 **Vapor (Gas) Irritant Characteristics:** Data not available
5.9 **Liquid or Solid Irritant Characteristics:** Data not available
5.10 **Odor Threshold:** Data not available

6. FIRE HAZARDS

6.1 **Flash Point:** Data not available
6.2 **Flammable Limits in Air:** Data not available
6.3 **Fire Extinguishing Agents:** Foam, dry chemical, carbon dioxide
6.4 **Fire Extinguishing Agents Not to be Used:** Water may be ineffective.
6.5 **Special Hazards of Combustion Products:**
6.6 **Behavior in Fire:**
6.7 **Ignition Temperature:** Data not available
6.8 **Electrical Hazard:** Data not available
6.9 **Burning Rate:** Data not available

7. CHEMICAL REACTIVITY

7.1 **Reactivity with Water:** No reaction
7.2 **Reactivity with Common Materials:** May attack some forms of plastics.
7.3 **Stability During Transport:** Stable
7.4 **Neutralizing Agents for Acids and Caustics:** Not pertinent
7.5 **Polymerization:** Not pertinent
7.6 **Inhibitor of Polymerization:** Not pertinent

8. WATER POLLUTION

8.1 **Aquatic Toxicity:** Data not available
8.2 **Waterfowl Toxicity:** Data not available
8.3 **Biological Oxygen Demand (BOD):** Data not available
8.4 **Food Chain Concentration Potential:** None

9. SELECTED MANUFACTURERS

1. Eastman Organic Chemicals
   343 State Street
   Rochester, N. Y. 14650
2. Aldrich Chemical Co.
   940 W. Saint Paul Ave.
   Milwaukee, Wis. 53233
3. Pfaltz and Bauer, Inc.
   375 Fairfield Ave.
   Stamford, Conn. 06902

10. SHIPPING INFORMATION

10.1 **Grades or Purity:** 98+%
10.2 **Storage Temperature:** Ambient
10.3 **Inert Atmosphere:** No requirement
10.4 **Venting:** Open (flame arrester)

11. HAZARD ASSESSMENT CODE

(See Hazard Assessment Handbook, CG 446-3)

A-T-U

12. HAZARD CLASSIFICATIONS

12.1 **Code of Federal Regulations:** Not listed
12.2 **NAS Hazard Rating for Bulk Water Transportation:** Not listed
12.3 **NFPA Hazard Classifications:** Not listed

13. PHYSICAL AND CHEMICAL PROPERTIES

13.1 **Physical State at 15°C and 1 atm:** Liquid
13.2 **Molecular Weight:** 142
13.3 **Boiling Point at 1 atm:** 370°F = 188°C = 461°K
13.4 **Freezing Point:** 21°F = −6°C = 267°K
13.5 **Critical Temperature:** Not pertinent
13.6 **Critical Pressure:** Not pertinent
13.7 **Specific Gravity:** 0.822 at 20°C (liquid)
13.8 **Liquid Surface Tension:** 26.60 dynes/cm = 0.0266 N/m at 21.1°C
13.9 **Liquid-Water Interfacial Tension:** Data not available
13.10 **Vapor (Gas) Specific Gravity:** Not pertinent
13.11 **Ratio of Specific Heats of Vapor (Gas):** Not pertinent
13.12 **Latent Heat of Vaporization:** 161 Btu/lb = 89.6 cal/g = 3.75×10^5 J/kg
13.13 **Heat of Combustion:** −16,080 Btu/lb = −8,930 cal/g = -374×10^5 J/kg
13.14 **Heat of Decomposition:** Not pertinent
13.15 **Heat of Solution:** Not pertinent
13.16 **Heat of Polymerization:** Not pertinent

NOTES

DBT — DIBUTYLPHENOL

Common Synonyms			
2, 6-Di-tert-butylphenol	Solid or liquid	Colorless to light yellow	Odorless
	Floats on water. Freezing point is 97° F.		

Avoid contact with liquid and solid. Keep people away.
Wear rubber overclothing (including gloves).
Stop discharge if possible.
Call fire department.
Isolate and remove discharged material.
Notify local health and pollution control agencies.

Fire

Combustible.
Extinguish with dry chemicals, foam or carbon dioxide.
Water may be ineffective on fire.
Cool exposed containers with water.

Exposure

Call for medical aid.

LIQUID OR SOLID
Will burn skin and eyes.
Harmful if swallowed.
Remove contaminated clothing and shoes.
Flush affected areas with plenty of water.
IF IN EYES, hold eyelids open and flush with plenty of water.
IF SWALLOWED and victim is CONSCIOUS, have victim drink water or milk.
DO NOT INDUCE VOMITING.

Water Pollution

Effect of low concentrations on aquatic life is unknown.
Fouling to shoreline.
May be dangerous if it enters water intakes.
Notify local health and wildlife officials.
Notify operators of nearby water intakes.

1. RESPONSE TO DISCHARGE

(See Response Methods Handbook, CG 446-4)

Issue warning—water contaminant
Mechanical containment
Should be removed
Chemical and physical treatment

2. LABELS

No label required by Code of Federal Regulations

3. CHEMICAL DESIGNATIONS

3.1 **Synonyms:** 2, 6-Di-tert-butylphenol
3.2 **Coast Guard Compatibility Classification:** Phenols (15)
3.3 **Chemical Formula:** 2,6-(t-$C_4H_9)_2C_6H_3OH$
3.4 **IMCO/United Nations Numerical Designation:** Not listed

4. OBSERVABLE CHARACTERISTICS

4.1 **Physical State (as shipped):** Solid or liquid
4.2 **Color:** Colorless; light straw
4.3 **Odor:** None

5. HEALTH HAZARDS

5.1 **Personal Protective Equipment:** Goggles or face shield; rubber gloves.
5.2 **Symptoms Following Exposure:** Irritates eyes and (on prolonged contact) skin. Ingestion causes irritation of mouth and stomach.
5.3 **Treatment for Exposure:** EYES: flush with water for at least 15 min. SKIN: wipe off, wash well with soap and water. INGESTION: induce vomiting; get medical attention.
5.4 **Toxicity by Inhalation (Threshold Limit Value):** Data not available
5.5 **Short-Term Inhalation Limits:** Not pertinent
5.6 **Toxicity by Ingestion:** Grade 2; oral LD_{50} (2,6-Di-sec-butyl phenol) = 1.32 g/kg (rat)
5.7 **Late Toxicity:** Data not available
5.8 **Vapor (Gas) Irritant Characteristics:** Data not available
5.9 **Liquid or Solid Irritant Characteristics:** Data not available
5.10 **Odor Threshold:** Data not available

6. FIRE HAZARDS

6.1 **Flash Point:** >200°F O.C.
6.2 **Flammable Limits in Air:** Not pertinent
6.3 **Fire Extinguishing Agents:** Dry chemical, foam, carbon dioxide
6.4 **Fire Extinguishing Agents Not to be Used:** Water may be ineffective.
6.5 **Special Hazards of Combustion Products:** Not pertinent
6.6 **Behavior in Fire:** Not pertinent
6.7 **Ignition Temperature:** Data not available
6.8 **Electrical Hazard:** Data not available
6.9 **Burning Rate:** Data not available

7. CHEMICAL REACTIVITY

7.1 **Reactivity with Water:** No reaction
7.2 **Reactivity with Common Materials:** No reaction
7.3 **Stability During Transport:** Stable
7.4 **Neutralizing Agents for Acids and Caustics:** Not pertinent
7.5 **Polymerization:** Not pertinent
7.6 **Inhibitor of Polymerization:** Not pertinent

8. WATER POLLUTION

8.1 **Aquatic Toxicity:** Data not available
8.2 **Waterfowl Toxicity:** Data not available
8.3 **Biological Oxygen Demand (BOD):** Data not available
8.4 **Food Chain Concentration Potential:** None

9. SELECTED MANUFACTURERS

1. Ethyl Corporation
Industrial Chemicals Division
Ethyl Tower
451 Florida
Baton Rouge, La. 70801
2. Eastman Kodak Co.
Eastman Organic Chemicals
Rochester, N. Y. 14650
3. Aldrich Chemical Co.
940 West St. Paul Ave.
Milwaukee, Wis. 53233

10. SHIPPING INFORMATION

10.1 **Grades or Purity:** Commercial
10.2 **Storage Temperature:** Ambient (solid); 100°F (liquid)
10.3 **Inert Atmosphere:** No requirement
10.4 **Venting:** Open

11. HAZARD ASSESSMENT CODE

(See Hazard Assessment Handbook, CG 446-3)

A-T-U-II

12. HAZARD CLASSIFICATIONS

12.1 **Code of Federal Regulations:** Not listed
12.2 **NAS Hazard Rating for Bulk Water Transportation:** Not listed
12.3 **NFPA Hazard Classifications:** Not listed

13. PHYSICAL AND CHEMICAL PROPERTIES

13.1 **Physical State at 15°C and 1 atm:** Solid
13.2 **Molecular Weight:** 206.3
13.3 **Boiling Point at 1 atm:** 487°F = 253°C = 526°K
13.4 **Freezing Point:** 97°F = 36°C = 309°K
13.5 **Critical Temperature:** Not pertinent
13.6 **Critical Pressure:** Not pertinent
13.7 **Specific Gravity:** 0.914 at 20°C (solid)
13.8 **Liquid Surface Tension:** Not pertinent
13.9 **Liquid-Water Interfacial Tension:** Not pertinent
13.10 **Vapor (Gas) Specific Gravity:** Not pertinent
13.11 **Ratio of Specific Heats of Vapor (Gas):** Not pertinent
13.12 **Latent Heat of Vaporization:** Not pertinent
13.13 **Heat of Combustion:** (est.) −18,000 Btu/lb −9,800 cal/g = -410×10^5 J/kg
13.14 **Heat of Decomposition:** Not pertinent
13.15 **Heat of Solution:** Not pertinent
13.16 **Heat of Polymerization:** Not pertinent

NOTES

DPA	# DIBUTYL PHTHALATE

Common Synonyms			
DBP Butyl phthalate Phthalic acid, dibutyl ester	Oily liquid	Colorless	Odorless
	Sinks slowly in water.		

Stop discharge if possible.
Call fire department.
Isolate and remove discharged material.
Notify local health and pollution control agencies.

Fire	Combustible. Extinguish with dry chemical, foam, or carbon dioxide.
Exposure	LIQUID No appreciable harm.
Water Pollution	Dangerous to aquatic life in high concentrations. Fouling to shoreline. May be dangerous if it enters water intakes. Notify local health and pollution control officials. Notify operators of nearby water intakes.

1. RESPONSE TO DISCHARGE

(See Response Methods Handbook, CG 446-4)

Mechanical containment
Should be removed
Chemical and physical treatment

2. LABELS

No hazard label required by Code of Federal Regulations

3. CHEMICAL DESIGNATIONS

3.1 **Synonyms:** Butyl phthalate; DBP; Phthalic acid, dibutyl ester; RC Plasticizer DBP; Witcizer 300
3.2 **Coast Guard Compatibility Classification:** Ester
3.3 **Chemical Formula:** $O\text{-}C_6H_4[COO(CH_2)_3CH_3]_2$
3.4 **IMCO/United Nations Numerical Designation:** Not listed

4. OBSERVABLE CHARACTERISTICS

4.1 **Physical State (as shipped):** Liquid
4.2 **Color:** Colorless
4.3 **Odor:** Slight characteristic ester odor; mild; practically none; slightly aromatic

5. HEALTH HAZARDS

5.1 **Personal Protective Equipment:** Eye protection
5.2 **Symptoms Following Exposure:** Vapors from very hot material may irritate eyes and produce headache, drowsiness, and convulsions.
5.3 **Treatment for Exposure:** Remove to fresh air. Wash affected skin areas with water. Flush eyes with water.
5.4 **Toxicity by Inhalation (Threshold Limit Value):** 5 mg/m³
5.5 **Short-Term Inhalation Limits:** Not pertinent
5.6 **Toxicity by Ingestion:** Grade 1; LD_{50} 5 to 15 g/kg (rat)
5.7 **Late Toxicity:** Birth defects in rats; polyneuritis in humans
5.8 **Vapor (Gas) Irritant Characteristics:** Not pertinent
5.9 **Liquid or Solid Irritant Characteristics:** No appreciable hazard. Practically harmless to the skin.
5.10 **Odor Threshold:** Data not available

6. FIRE HAZARDS

6.1 **Flash Point:** 355°F O.C.; 315°F C.C.
6.2 **Flammable Limits in Air:** 0.5%–2.5% (calculated)
6.3 **Fire Extinguishing Agents:** Dry powder, carbon dioxide, foam
6.4 **Fire Extinguishing Agents Not to be Used:** Water or foam may cause frothing.
6.5 **Special Hazards of Combustion Products:** Not pertinent
6.6 **Behavior in Fire:** Not pertinent
6.7 **Ignition Temperature:** 757°F
6.8 **Electrical Hazard:** Not pertinent
6.9 **Burning Rate:** Data not available

7. CHEMICAL REACTIVITY

7.1 **Reactivity with Water:** No reaction
7.2 **Reactivity with Common Materials:** No reaction
7.3 **Stability During Transport:** Stable
7.4 **Neutralizing Agents for Acids and Caustics:** Not pertinent
7.5 **Polymerization:** Not pertinent
7.6 **Inhibitor of Polymerization:** Not pertinent

8. WATER POLLUTION

8.1 **Aquatic Toxicity:** 1230 ppm/24 hr/bluegill/TL_m/fresh water
8.2 **Waterfowl Toxicity:** LC_{50} >5000 ppm
8.3 **Biological Oxygen Demand (BOD):** 0.43 lb/lb, 5 days
8.4 **Food Chain Concentration Potential:** None

9. SELECTED MANUFACTURERS

1. W. R. Grace & Co.
   Hatco Chemical Division
   Fords, N. J. 08863
2. Monsanto Co.
   Monsanto Industrial Chemicals Co.
   800 N. Lindbergh Blvd.
   St. Louis, Mo. 63166
3. United States Steel Corp.
   USS Chemicals Division
   Neville Island, Pa. 15225

10. SHIPPING INFORMATION

10.1 **Grades or Purity:** 99.6%
10.2 **Storage Temperature:** Data not available
10.3 **Inert Atmosphere:** Data not available
10.4 **Venting:** Data not available

11. HAZARD ASSESSMENT CODE

(See Hazard Assessment Handbook, CG 446-3)

A-T-U-X-Y

12. HAZARD CLASSIFICATIONS

12.1 **Code of Federal Regulations:** Not listed
12.2 **NAS Hazard Rating for Bulk Water Transportation:**

Category	Rating
Fire	1
Health	
Vapor Irritant	0
Liquid or Solid Irritant	0
Poisons	0
Water Pollution	
Human Toxicity	0
Aquatic Toxicity	1
Aesthetic Effect	3
Reactivity	
Other Chemicals	1
Water	0
Self-Reaction	0

12.3 **NFPA Hazard Classifications:** Not listed

13. PHYSICAL AND CHEMICAL PROPERTIES

13.1 **Physical State at 15°C and 1 atm:** Liquid
13.2 **Molecular Weight:** 278.35
13.3 **Boiling Point at 1 atm:** 635°F = 335°C = 608°K
13.4 **Freezing Point:** −31°F = −35°C = 238°K
13.5 **Critical Temperature:** 932°F = 500°C = 773°K
13.6 **Critical Pressure:** 250 psia = 17 atm = 1.7 MN/m²
13.7 **Specific Gravity:** 1.049 at 20°C (liquid)
13.8 **Liquid Surface Tension:** 34 dynes/cm = 0.034 N/m at 20°C
13.9 **Liquid-Water Interfacial Tension:** 27 dynes/cm = 0.027 N/m at 22.7°C
13.10 **Vapor (Gas) Specific Gravity:** Not pertinent
13.11 **Ratio of Specific Heats of Vapor (Gas):** Not pertinent
13.12 **Latent Heat of Vaporization:** Not pertinent
13.13 **Heat of Combustion:** -13,300 Btu/lb = -7400 cal/g = -310 × 10⁵ J/kg
13.14 **Heat of Decomposition:** Not pertinent
13.15 **Heat of Solution:** Not pertinent
13.16 **Heat of Polymerization:** Not pertinent

NOTES

REVISED 1978

DBO

o-DICHLOROBENZENE

Common Synonyms
1, 2-Dichlorobenzene
Orthodichlorobenzene

Liquid   Colorless   Pleasant odor

Sinks in water.

Avoid contact with liquid.
Wear goggles and self-contained breathing apparatus.
Stop discharge if possible.
Call fire department.
Isolate and remove discharged material.
Notify local health and pollution control agencies.

Fire

Combustible
POISONOUS GASES ARE PRODUCED IN FIRE.
Wear goggles and self-contained breathing apparatus.
Extinguish with water, dry chemical, foam, or carbon dioxide.
Cool exposed containers with water.

Exposure

CALL FOR MEDICAL AID.

LIQUID
Irritating to skin and eyes.
Harmful if swallowed.
Remove contaminated clothing and shoes.
Flush affected areas with plenty of water.
IF IN EYES, hold eyelids open and flush with plenty of water.
IF SWALLOWED and victim is CONSCIOUS, have victim drink water or milk and have victim induce vomiting.
IF SWALLOWED and victim is UNCONSCIOUS OR HAVING CONVULSIONS, do nothing except keep victim warm.

Water Pollution

Effect of low concentrations on aquatic life is unknown.
May be dangerous if it enters water intakes.
Notify local health and pollution control officials.
Notify operators of nearby water intakes.

1. RESPONSE TO DISCHARGE

(See Response Methods Handbook, CG 446-4)

Issue warning—water contaminant
Should be removed
Chemical and physical treatment

2. LABELS

No hazard label required by Code of Federal Regulations

3. CHEMICAL DESIGNATIONS

3.1 **Synonyms:**
1,2-Dichlorobenzene
Dowtherm E
Orthodichlorobenzene

3.2 **Coast Guard Compatibility Classification:** Halogenated hydrocarbon

3.3 **Chemical Formula:** o-$C_6H_4Cl_2$

3.4 **IMCO/United Nations Numerical Designation:** 6.1/1591

4. OBSERVABLE CHARACTERISTICS

4.1 **Physical State (as shipped):** Liquid

4.2 **Color:** Colorless

4.3 **Odor:** Aromatic; characteristic aromatic

5. HEALTH HAZARDS

5.1 **Personal Protective Equipment:** Organic vapor-acid gas respirator; neoprene or vinyl gloves; chemical safety spectacles, face shield, rubber footwear, apron, protective clothing.

5.2 **Symptoms Following Exposure:** Chronic inhalation of mist or vapors may result in damage to lungs, liver, and kidneys. Acute vapor exposure can cause symptoms ranging from coughing to central nervous system depression and transient anesthesia. Irritating to skin, eyes, and mucous membranes. May cause dermatitis.

5.3 **Treatment for Exposure:** INHALATION: remove victim to fresh air, keep him quiet and warm, and call a physician promptly. INGESTION: no known antidote; treat symptomatically; induce vomiting and get medical attention promptly. EYES AND SKIN: flush with plenty of water; get medical attention for eyes; remove contaminated clothing and wash before reuse.

5.4 **Toxicity by Inhalation (Threshold Limit Value):** 50 ppm

5.5 **Short-Term Inhalation Limits:** 50 ppm for 15 min.

5.6 **Toxicity by Ingestion:** Grade 2; LD_{50} 0.5 to 5 g/kg

5.7 **Late Toxicity:** Causes kidney and liver damage in rats. Effects unknown in humans.

5.8 **Vapor (Gas) Irritant Characteristics:** Vapors cause moderate irritation such that personnel will find high concentrations unpleasant. The effect is temporary.

5.9 **Liquid or Solid Irritant Characteristics:** Minimum hazard. If spilled on clothing and allowed to remain, may cause smarting and reddening of the skin.

5.10 **Odor Threshold:** 4.0 ppm; 50 ppm

6. FIRE HAZARDS

6.1 **Flash Point:** 165°F O.C.; 155°F C.C.

6.2 **Flammable Limits in Air:** 2.2%—9.2%

6.3 **Fire Extinguishing Agents:** Water, foam, dry chemical, or carbon dioxide

6.4 **Fire Extinguishing Agents Not to be Used:** Not pertinent

6.5 **Special Hazards of Combustion Products:** Irritating vapors including hydrogen chloride gas, chlorocarbons, chlorine

6.6 **Behavior in Fire:** Not pertinent

6.7 **Ignition Temperature:** 1198°F

6.8 **Electrical Hazard:** Not pertinent

6.9 **Burning Rate:** 1.3 mm/min.

7. CHEMICAL REACTIVITY

7.1 **Reactivity with Water:** No reaction

7.2 **Reactivity with Common Materials:** No reaction

7.3 **Stability During Transport:** Stable

7.4 **Neutralizing Agents for Acids and Caustics:** Not pertinent

7.5 **Polymerization:** Not pertinent

7.6 **Inhibitor of Polymerization:** Not pertinent

8. WATER POLLUTION

8.1 **Aquatic Toxicity:**
13 ppm/*/marine plankton/no growth/salt water
*Time period not specified.

8.2 **Waterfowl Toxicity:** Data not available

8.3 **Biological Oxygen Demand (BOD):** <0.1% (theor.), 1/8 day

8.4 **Food Chain Concentration Potential:** Data not available

9. SELECTED MANUFACTURERS

1. Dow Chemical Co.
   Midland, Mich. 48640
2. Monsanto Co.
   Monsanto Industrial Chemicals Co.
   800 North Lindbergh Blvd.
   St. Louis, Mo. 63166
3. Standard Chlorine Chemical Co., Inc.
   1015-25 Belleville Turnpike
   Kearney, N. J. 07032

10. SHIPPING INFORMATION

10.1 **Grades or Purity:**
Technical: 99.5% min. dichlorobenzene (ratio-ortho + para/meta: 80 min.)
Technical: 85% orthodichlorobenzene, 14.0% paradichlorobenzene
Technical: 80% ortho, 17% para, 2% meta
Pure: not less than 99.5% ortho, not more than 0.5% para

10.2 **Storage Temperature:** Data not available *(Continued on page 4)*

11. HAZARD ASSESSMENT CODE

(See Hazard Assessment Handbook, CG 446-3)

A-X-Y

12. HAZARD CLASSIFICATIONS

12.1 **Code of Federal Regulations:** ORM-A

12.2 **NAS Hazard Rating for Bulk Water Transportation:**

Category	Rating
Fire	1
Health	
Vapor Irritant	2
Liquid or Solid Irritant	1
Poisons	1
Water Pollution	
Human Toxicity	1
Aquatic Toxicity	3
Aesthetic Effect	2
Reactivity	
Other Chemicals	1
Water	0
Self-Reaction	0

12.3 **NFPA Hazard Classifications:**

Category	Classification
Health Hazard (Blue)	2
Flammability (Red)	2
Reactivity (Yellow)	0

13. PHYSICAL AND CHEMICAL PROPERTIES

13.1 **Physical State at 15°C and 1 atm:** Liquid

13.2 **Molecular Weight:** 147.01

13.3 **Boiling Point at 1 atm:** 356.9°F = 180.5°C = 453.7°K

13.4 **Freezing Point:** 0.3°F = −17.6°C = 255.6°K

13.5 **Critical Temperature:** Not pertinent

13.6 **Critical Pressure:** Not pertinent

13.7 **Specific Gravity:** 1.306 at 20°C (liquid)

13.8 **Liquid Surface Tension:** 37 dynes/cm = 0.037 N/m at 20°C

13.9 **Liquid-Water Interfacial Tension:** (est.) 40 dynes/cm = 0.04 N/m at 20°C

13.10 **Vapor (Gas) Specific Gravity:** Not pertinent

13.11 **Ratio of Specific Heats of Vapor (Gas):** 1.080

13.12 **Latent Heat of Vaporization:** 115 Btu/lb = 63.9 cal/g = 2.68×10^5 J/kg

13.13 **Heat of Combustion:** −7969 Btu/lb = −4427 cal/g = -185.4×10^5 J/kg

13.14 **Heat of Decomposition:** Not pertinent

13.15 **Heat of Solution:** Not pertinent

13.16 **Heat of Polymerization:** Not pertinent

10. SHIPPING INFORMATION (Cont'd.)

10.3 **Inert Atmosphere:** Data not available

10.4 **Venting:** Data not available

REVISED 1978

DBP — p-DICHLOROBENZENE

Common Synonyms			
Paradichlorobenzene Dichloricide Paradi	Solid crystals	White to clear	Mothballs odor
	Sinks in water.		

Avoid contact with solid.
Call fire department.
Isolate and remove discharged material.
Notify local health and pollution control agencies.

Fire	Combustible. POISONOUS GASES ARE PRODUCED IN FIRE. Wear goggles and self-contained breathing apparatus. Extinguish with water, dry chemical, foam, or carbon dioxide. Cool exposed containers with water.
Exposure	CALL FOR MEDICAL AID. SOLID Irritating to skin and eyes. Harmful if swallowed. Remove contaminated clothing and shoes. Flush affected areas with plenty of water. IF IN EYES, hold eyelids open and flush with plenty of water. IF SWALLOWED and victim is CONSCIOUS, have victim drink water or milk.
Water Pollution	HARMFUL TO AQUATIC LIFE IN VERY LOW CONCENTRATIONS. Fouling to shoreline. May be dangerous if it enters water intakes. Notify local health and wildlife officials. Notify operators of nearby water intakes.

1. RESPONSE TO DISCHARGE
(See Response Methods Handbook, CG 446-4)

Issue warning—water contaminant
Should be removed
Chemical and physical treatment

2. LABELS

No hazard label required by Code of Federal Regulations

3. CHEMICAL DESIGNATIONS

3.1 **Synonyms:** Dichloricide, Paradi, Paradichlorobenzene, Paradow, Paramoth, Santochlor
3.2 **Coast Guard Compatibility Classification:** Halogenated hydrocarbon
3.3 **Chemical Formula:** p-$C_6H_4Cl_2$
3.4 **IMCO/United Nations Numerical Designation:** 9.0/1592

4. OBSERVABLE CHARACTERISTICS

4.1 **Physical State (as shipped):** Solid
4.2 **Color:** White
4.3 **Odor:** Aromatic

5. HEALTH HAZARDS

5.1 **Personal Protective Equipment:** Full face mask fitted with organic vapor canister for concentrations over 75 ppm; clean protective clothing; eye protection.
5.2 **Symptoms Following Exposure:** INHALATION: irritation of upper respiratory tract; over-exposure may cause depression and injury to liver and kidney. EYE CONTACT: pain and mild irritation.
5.3 **Treatment for Exposure:** INHALATION: if any ill effects develop, remove patient to fresh air and get medical attention. If breathing stops, give artificial respiration. EYES: flush with plenty of water and get medical attention if ill effects develop. SKIN AND INGESTION: no problem likely.
5.4 **Toxicity by Inhalation (Threshold Limit Value):** 75 ppm
5.5 **Short-Term Inhalation Limits:** 50 ppm for 60 min.
5.6 **Toxicity by Ingestion:** Grade 2; LD_{50} 0.5 to 5 g/kg
5.7 **Late Toxicity:** Data not available
5.8 **Vapor (Gas) Irritant Characteristics:** Vapors cause moderate irritation such that personnel will find high concentrations unpleasant. The effect is temporary.
5.9 **Liquid or Solid Irritant Characteristics:** Minimum hazard. If spilled on clothing and allowed to remain, may cause smarting and reddening of the skin.
5.10 **Odor Threshold:** 15-30 ppm

6. FIRE HAZARDS

6.1 **Flash Point:** 165°F O.C.; 150°F C.C.
6.2 **Flammable Limits in Air:** Data not available
6.3 **Fire Extinguishing Agents:** Water, foam, carbon dioxide or dry chemical.
6.4 **Fire Extinguishing Agents Not to be Used:** Not pertinent
6.5 **Special Hazards of Combustion Products:** Vapors are irritating. Toxic chlorine, hydrogen chloride, and phosgene gases may be generated in fires.
6.6 **Behavior in Fire:** Not pertinent
6.7 **Ignition Temperature:** Data not available
6.8 **Electrical Hazard:** Not pertinent
6.9 **Burning Rate:** 1.3 mm/min. (approx.)

7. CHEMICAL REACTIVITY

7.1 **Reactivity with Water:** No reaction
7.2 **Reactivity with Common Materials:** No reaction
7.3 **Stability During Transport:** Stable
7.4 **Neutralizing Agents for Acids and Caustics:** Not pertinent
7.5 **Polymerization:** Not pertinent
7.6 **Inhibitor of Polymerization:** Not pertinent

8. WATER POLLUTION

8.1 **Aquatic Toxicity:**
50 ppm/*/fish/lethal/fresh water
880 mg/l /48 hr/rainbow trout/TL_m/fresh water
*No time interval specified
8.2 **Waterfowl Toxicity:** Data not available
8.3 **Biological Oxygen Demand (BOD):** Data not available
8.4 **Food Chain Concentration Potential:** Data not available

9. SELECTED MANUFACTURERS

1. Dow Chemical Co.
Midland, Mich. 48640
2. PPG Industries Inc.
Industrial Chemical Division
New Martinsville, W. Va. 26155
3. Standard Chlorine Chemical Co.
1015-25 Belleville Turnpike
Kearny, N. J. 07032

10. SHIPPING INFORMATION

10.1 **Grades or Purity:** Solid: 5 grades, chemical purity close to 100%
Liquid: 1-2% orthodichlorobenzene.
10.2 **Storage Temperature:** Data not available
10.3 **Inert Atmosphere:** Data not available
10.4 **Venting:** Data not available

11. HAZARD ASSESSMENT CODE
(See Hazard Assessment Handbook, CG 446-3)

II

12. HAZARD CLASSIFICATIONS

12.1 **Code of Federal Regulations:** ORM-A
12.2 **NAS Hazard Rating for Bulk Water Transportation:** Not listed
12.3 **NFPA Hazard Classifications:** Not listed

13. PHYSICAL AND CHEMICAL PROPERTIES

13.1 **Physical State at 15°C and 1 atm:** Solid
13.2 **Molecular Weight:** 147.01
13.3 **Boiling Point at 1 atm:** 345.6°F = 174.2°C = 447.4°K
13.4 **Freezing Point:** 130°F = 53°C = 326°K
13.5 **Critical Temperature:** Not pertinent
13.6 **Critical Pressure:** Not pertinent
13.7 **Specific Gravity:** 1.458 at 20°C (solid)
13.8 **Liquid Surface Tension:** Not pertinent
13.9 **Liquid-Water Interfacial Tension:** Not pertinent
13.10 **Vapor (Gas) Specific Gravity:** Not pertinent
13.11 **Ratio of Specific Heats of Vapor (Gas):** Not pertinent
13.12 **Latent Heat of Vaporization:** Not pertinent
13.13 **Heat of Combustion:** Not pertinent
13.14 **Heat of Decomposition:** Not pertinent
13.15 **Heat of Solution:** Not pertinent
13.16 **Heat of Polymerization:** Not pertinent

NOTES

REVISED 1978

DZP — DI-(p-CHLOROBENZOYL) PEROXIDE

Common Synonyms	Solid or paste	White	Odorless
Bis-(p-chlorobenzoyl) peroxide p-Chlorobenzoyl peroxide p, p′-Chlorobenzoyl peroxide Di-(4-chlorobenzoyl) peroxide	Sinks in water.		

Stop discharge if possible. Keep people away.
Call fire department.
Isolate and remove discharged material.
Notify local health and pollution control agencies.

Fire

May explode on contact with combustibles.
POISONOUS GASES ARE PRODUCED IN FIRE.
Combat fires from safe distance or protected location.
Flood discharge area with water.
Cool exposed containers with water.

Exposure

Call for medical aid.

SOLID
Irritating to skin and eyes.
Harmful if swallowed.
Remove contaminated clothing and shoes.
Flush affected areas with plenty of water.
IF IN EYES, hold eyelids open and flush with plenty of water.
IF SWALLOWED and victim is CONSCIOUS, have victim drink water or milk and have victim induce vomiting.
IF SWALLOWED and victim is UNCONSCIOUS OR HAVING CONVULSIONS, do nothing except keep victim warm.

Water Pollution

Effect of low concentrations on aquatic life is unknown.
May be dangerous if it enters water intakes.
Notify local health and wildlife officials.
Notify operators of nearby water intakes.

1. RESPONSE TO DISCHARGE

(See Response Methods Handbook, CG 446-4)

Issue warning—oxidizing material, water contaminant
Should be removed
Chemical and physical treatment

2. LABEL

Yellow

3. CHEMICAL DESIGNATIONS

3.1 **Synonyms:** Bis-(p-chlorobenzoyl) peroxide; p-Chlorobenzoyl peroxide; p,p′-Dichlorobenzoyl peroxide; Di-(4-chlorobenzoyl) peroxide; Cadox PS
3.2 **Coast Guard Compatibility Classification:** Not applicable
3.3 **Chemical Formula:** $(p\text{-}ClC_6H_4COO)_2$
3.4 **IMCO/United Nations Numerical Designation:** 1/0149 (>70%); 5.2/1531 (>10% water or >30% water)

4. OBSERVABLE CHARACTERISTICS

4.1 **Physical State (as shipped):** Solid; or paste in silicone fluid and dibutyl phthalate.
4.2 **Color:** White
4.3 **Odor:** None

5. HEALTH HAZARDS

5.1 **Personal Protective Equipment:** Goggles or face shield; rubber gloves; protective clothing.
5.2 **Symptoms Following Exposure:** Irritates eyes and (on prolonged contact) skin. Ingestion causes irritation of mouth and stomach.
5.3 **Treatment for Exposure:** EYES: wash with water for at least 15 min.; consult a doctor. SKIN: wash with soap and water. INGESTION: induce vomiting and call a doctor.
5.4 **Toxicity by Inhalation (Threshold Limit Value):** Data not available
5.5 **Short-Term Inhalation Limits:** Not pertinent
5.6 **Toxicity by Ingestion:** Data not available
5.7 **Late Toxicity:** Data not available
5.8 **Vapor (Gas) Irritant Characteristics:** Data not available
5.9 **Liquid or Solid Irritant Characteristics:** Data not available
5.10 **Odor Threshold:** Not pertinent

6. FIRE HAZARDS

6.1 **Flash Point:** Not pertinent
6.2 **Flammable Limits in Air:** Not pertinent
6.3 **Fire Extinguishing Agents:** Flood with water, or use dry chemical, foam, carbon dioxide.
6.4 **Fire Extinguishing Agents Not to be Used:** Not pertinent
6.5 **Special Hazards of Combustion Products:** Toxic chlorinated biphenyls are formed in fires.
6.6 **Behavior in Fire:** Solid may explode. Burns very rapidly when ignited. Smoke is unusually heavy when paste form is involved.
6.7 **Ignition Temperature:** Data not available
6.8 **Electrical Hazard:** Data not available
6.9 **Burning Rate:** Not pertinent

7. CHEMICAL REACTIVITY

7.1 **Reactivity with Water:** No reaction
7.2 **Reactivity with Common Materials:** May react vigorously with combustible materials.
7.3 **Stability During Transport:** Stable if below 80°F.
7.4 **Neutralizing Agents for Acids and Caustics:** Not pertinent
7.5 **Polymerization:** Not pertinent
7.6 **Inhibitor of Polymerization:** Not pertinent

8. WATER POLLUTION

8.1 **Aquatic Toxicity:** Data not available
8.2 **Waterfowl Toxicity:** Data not available
8.3 **Biological Oxygen Demand (BOD):** Data not available
8.4 **Food Chain Concentration Potential:** Data not available

9. SELECTED MANUFACTURERS

Noury Chemical Corp.
Burt, New York 14028

10. SHIPPING INFORMATION

10.1 **Grades or Purity:** Dry; wet with more than 30% water; 50% paste with silicone fluid.
10.2 **Storage Temperature:** Below 80°F
10.3 **Inert Atmosphere:** No requirement
10.4 **Venting:** Pressure-vacuum

11. HAZARD ASSESSMENT CODE

(See Hazard Assessment Handbook, CG 446-3)

II

12. HAZARD CLASSIFICATIONS

12.1 **Code of Federal Regulations:** Organic Peroxide
12.2 **NAS Hazard Rating for Bulk Water Transportation:** Not listed
12.3 **NFPA Hazard Classifications:** Not listed

13. PHYSICAL AND CHEMICAL PROPERTIES

13.1 **Physical State at 15°C and 1 atm:** Solid
13.2 **Molecular Weight:** 311.1
13.3 **Boiling Point at 1 atm:** Decomposes
13.4 **Freezing Point:** Not pertinent
13.5 **Critical Temperature:** Not pertinent
13.6 **Critical Pressure:** Not pertinent
13.7 **Specific Gravity:** >1.1 at 20°C (solid)
13.8 **Liquid Surface Tension:** Not pertinent
13.9 **Liquid-Water Interfacial Tension:** Not pertinent
13.10 **Vapor (Gas) Specific Gravity:** Not pertinent
13.11 **Ratio of Specific Heats of Vapor (Gas):** Not pertinent
13.12 **Latent Heat of Vaporization:** Not pertinent
13.13 **Heat of Combustion:** (est.) $-9{,}000$ Btu/lb $= -5{,}000$ cal/g $= -210 \times 10^5$ J/kg
13.14 **Heat of Decomposition:** Data not available
13.15 **Heat of Solution:** Not pertinent
13.16 **Heat of Polymerization:** Not pertinent

NOTES

DCB — DICHLOROBUTENE

Common Synonyms			
1,4-Dichloro-2-butene 2-Butylene dichloride 1,4-Dichloro-2-butylene cis-1,4-dichloro-2-butene trans-1,4-dichloro-2-butene	Liquid Sinks and mixes slowly with water.	Colorless	Sweet odor

Stop discharge if possible. Keep people away.
Call fire department.
Isolate and remove discharged material.
Notify local health and pollution control agencies.

Fire	FLAMMABLE POISONOUS GASES ARE PRODUCED IN FIRE. Flashback along vapor trail may occur. Vapor may explode if ignited in an enclosed area. Wear goggles and self-contained breathing apparatus. Extinguish with water, dry chemicals, foam, or carbon dioxide. Cool exposed containers with water.
Exposure	Call for medical aid. LIQUID Irritating to skin and eyes. Harmful if swallowed. Remove contaminated clothing and shoes. Flush affected areas with plenty of water. IF IN EYES, hold eyelids open and flush with plenty of water. IF SWALLOWED and victim is CONSCIOUS, have victim drink water or milk and have victim induce vomiting. IF SWALLOWED and victim is UNCONSCIOUS OR HAVING CONVULSIONS, do nothing except keep victim warm.
Water Pollution	Effect of low concentrations on aquatic life is unknown. May be dangerous if it enters water intakes. Notify local health and wildlife officials. Notify operators of nearby water intakes.

1. RESPONSE TO DISCHARGE

(See Response Methods Handbook, CG 446-4)

Issue warning—corrosive, water contaminant
Restrict access
Should be removed
Chemical and physical treatment

2. LABEL

Red

3. CHEMICAL DESIGNATIONS

3.1 **Synonyms:** 1,4-Dichloro-2-butene; 2-Butylene dichloride; 1,4-Dichloro-2-butylene; cis-1,4-Dichloro-2-butene; trans-1,4-Dichloro-2-butene
3.2 **Coast Guard Compatibility Classification:** Halogenated compounds (5)
3.3 **Chemical Formula:** $ClCH_2CH{=}CHCH_2Cl$
3.4 **IMCO/United Nations Numerical Designation:** 8/1760

4. OBSERVABLE CHARACTERISTICS

4.1 **Physical State (as shipped):** Liquid
4.2 **Color:** Colorless
4.3 **Odor:** Characteristic; sweet, pungent

5. HEALTH HAZARDS

5.1 **Personal Protective Equipment:** Rubber gloves; chemical splash goggles; rubber boots and apron; barrier cream; organic canister mask.
5.2 **Symptoms Following Exposure:** Inhalation of vapor irritates nose and throat. Contact with eyes causes intense irritation and tears. Contact of liquid with skin causes severe blistering and dermatitis. Ingestion causes severe irritation of mouth and stomach.
5.3 **Treatment for Exposure:** INHALATION: remove from exposure; provide low-pressure oxygen if required; keep under observation until edema is ruled out. EYES: irrigate immediately for 15 min.; call physician. SKIN: wash immediately and thoroughly with soap and water; treat as a chemical burn. INGESTION: induce vomiting; call physician.
5.4 **Toxicity by Inhalation (Threshold Limit Value):** Data not available
5.5 **Short-Term Inhalation Limits:** Data not available
5.6 **Toxicity by Ingestion:** Grade 3; oral LD_{50} (1,4-dichloro-2-butene) = 89 mg/kg (rat)
5.7 **Late Toxicity:** Data not available
5.8 **Vapor (Gas) Irritant Characteristics:** Data not available
5.9 **Liquid or Solid Irritant Characteristics:** Data not available
5.10 **Odor Threshold:** Data not available

6. FIRE HAZARDS

6.1 **Flash Point:** Data not available
6.2 **Flammable Limits in Air:** 1.5%–4%
6.3 **Fire Extinguishing Agents:** Water, foam, dry chemical or carbon dioxide
6.4 **Fire Extinguishing Agents Not to be Used:** Not pertinent
6.5 **Special Hazards of Combustion Products:** Decomposition vapors contain phosgene and hydrogen chloride gases; both are toxic and irritating.
6.6 **Behavior in Fire:** Not pertinent
6.7 **Ignition Temperature:** Data not available
6.8 **Electrical Hazard:** Data not available
6.9 **Burning Rate:** 2.6 mm/min.

7. CHEMICAL REACTIVITY

7.1 **Reactivity with Water:** Reacts slowly to form hydrochloric acid.
7.2 **Reactivity with Common Materials:** Corrodes metal when wet.
7.3 **Stability During Transport:** Stable
7.4 **Neutralizing Agents for Acids and Caustics:** Not pertinent
7.5 **Polymerization:** Not pertinent
7.6 **Inhibitor of Polymerization:** Not pertinent

8. WATER POLLUTION

8.1 **Aquatic Toxicity:** Data not available
8.2 **Waterfowl Toxicity:** Data not available
8.3 **Biological Oxygen Demand (BOD):** Data not available
8.4 **Food Chain Concentration Potential:** None

9. SELECTED MANUFACTURERS

1. Petro-Tex Chemical Corporation
   8600 Park Place Boulevard
   Houston, Texas 77017
2. Pfaltz and Bauer, Inc.
   126-04 Northern Boulevard
   Flushing, N. Y. 11368
3. Eastman Kodak Co.
   Eastman Organic Chemicals
   Rochester, N. Y. 14650

10. SHIPPING INFORMATION

10.1 **Grades or Purity:** *Cis-trans* equilibrium mixture, 98+%
10.2 **Storage Temperature:** Ambient
10.3 **Inert Atmosphere:** No requirement
10.4 **Venting:** Open (flame arrester)

11. HAZARD ASSESSMENT CODE

(See Hazard Assessment Handbook, CG 446-3)

A-X-Y

12. HAZARD CLASSIFICATIONS

12.1 **Code of Federal Regulations:** Flammable Liquid
12.2 **NAS Hazard Rating for Bulk Water Transportation:** Not listed
12.3 **NFPA Hazard Classifications:** Not listed

13. PHYSICAL AND CHEMICAL PROPERTIES

13.1 **Physical State at 15°C and 1 atm:** Liquid
13.2 **Molecular Weight:** 125.0
13.3 **Boiling Point at 1 atm:** 313°F = 156°C = 429°K
13.4 **Freezing Point:** cis: −54°F = −48°C = 225°K; trans: 37°F = 3°C = 276°K
13.5 **Critical Temperature:** Not pertinent
13.6 **Critical Pressure:** Not pertinent
13.7 **Specific Gravity:** 1.112 at 20°C (liquid)
13.8 **Liquid Surface Tension:** (est.) 24 dynes/cm = 0.024 N/m at 20°C
13.9 **Liquid-Water Interfacial Tension:** (est.) 30 dynes/cm = 0.030 N/m at 20°C
13.10 **Vapor (Gas) Specific Gravity:** 4
13.11 **Ratio of Specific Heats of Vapor (Gas):** 1.0874
13.12 **Latent Heat of Vaporization:** (est.) 130 Btu/lb = 73 cal/g = 3.1×10^5 J/kg
13.13 **Heat of Combustion:** −17,500 Btu/lb = −9,720 cal/g = -407×10^5 J/kg
13.14 **Heat of Decomposition:** Not pertinent
13.15 **Heat of Solution:** Not pertinent
13.16 **Heat of Polymerization:** Not pertinent

NOTES

DCF	# DICHLORODIFLUOROMETHANE

Common Synonyms			
Arcton 6 Eskimon 12 F-12 Freon 12 Genetron 12 Isotron 12	Compressed gas Visible vapor cloud is produced.	Colorless	Faint odor

Stop discharge if possible.
Notify local health and pollution control agencies.

Fire	Not flammable. Cool exposed containers with water.
Exposure	CALL FOR MEDICAL AID. VAPOR Not irritating to eyes, nose or throat. If inhaled, will cause dizziness, difficult breathing. Move to fresh air. If breathing has stopped, give artificial respiration. If breathing is difficult, give oxygen.
Water Pollution	Not harmful to aquatic life.

1. RESPONSE TO DISCHARGE

(See Response Methods Handbook, CG 446-4)

Disperse and flush

2. LABEL

Green

3. CHEMICAL DESIGNATIONS

3.1 **Synonyms:** Eskimon 12, Genetron 12, F-12, Halon 122, Freon 12, Isotron 12, Frigen 12, Ucon 12

3.2 **Coast Guard Compatibility Classification:** Halogenated hydrocarbon

3.3 **Chemical Formula:** CCl_2F_2

3.4 **IMCO/United Nations Numerical Designation:** 2.0/1028

4. OBSERVABLE CHARACTERISTICS

4.1 **Physical State (as shipped):** Liquefied compressed gas

4.2 **Color:** Colorless

4.3 **Odor:** Odorless; slight; characteristic

5. HEALTH HAZARDS

5.1 **Personal Protective Equipment:** Rubber gloves, monogoggles.

5.2 **Symptoms Following Exposure:** INHALATION: some narcosis when 10% in air is breathed.

5.3 **Treatment for Exposure:** Remove patient to non-contaminated area and apply artificial respiration if breathing has stopped; call a physician immediately; oxygen may be given.

5.4 **Toxicity by Inhalation (Threshold Limit Value):** 1000 ppm

5.5 **Short-Term Inhalation Limits:** 5000 ppm for 60 min.

5.6 **Toxicity by Ingestion:** Not pertinent

5.7 **Late Toxicity:** None

5.8 **Vapor (Gas) Irritant Characteristics:** None, except at very high concentrations, which may irritate lungs.

5.9 **Liquid or Solid Irritant Characteristics:** No appreciable hazard. Practically harmless to the skin because it is very volatile and evaporates quickly.

5.10 **Odor Threshold:** Data not available

6. FIRE HAZARDS

6.1 **Flash Point:** Not flammable

6.2 **Flammable Limits in Air:** Not flammable

6.3 **Fire Extinguishing Agents:** Not pertinent

6.4 **Fire Extinguishing Agents Not to be Used:** Not pertinent

6.5 **Special Hazards of Combustion Products:** Although nonflammable, dissociation products generated in a fire may be irritating or toxic.

6.6 **Behavior in Fire:** Helps extinguish fire.

6.7 **Ignition Temperature:** Not flammable

6.8 **Electrical Hazard:** Not pertinent

6.9 **Burning Rate:** Not flammable

7. CHEMICAL REACTIVITY

7.1 **Reactivity with Water:** No reaction

7.2 **Reactivity with Common Materials:** No reaction

7.3 **Stability During Transport:** Stable

7.4 **Neutralizing Agents for Acids and Caustics:** Not pertinent

7.5 **Polymerization:** Not pertinent

7.6 **Inhibitor of Polymerization:** Not pertinent

8. WATER POLLUTION

8.1 **Aquatic Toxicity:** Not pertinent

8.2 **Waterfowl Toxicity:** Not pertinent

8.3 **Biological Oxygen Demand (BOD):** Not pertinent

8.4 **Food Chain Concentration Potential:** None

9. SELECTED MANUFACTURERS

1. Allied Chemical Corp.
   Specialty Chemicals Division
   Baton Rouge, La. 70821
2. E. I. du Pont de Nemours & Co., Inc.
   Freon Products Division
   Wilmington, Del. 19898
3. Pennwalt Corp.
   Chemical Division
   Calvert City, Ky. 42029

10. SHIPPING INFORMATION

10.1 **Grades or Purity:** 99.5% (vol.)

10.2 **Storage Temperature:** Ambient

10.3 **Inert Atmosphere:** No requirement

10.4 **Venting:** Safety relief

11. HAZARD ASSESSMENT CODE

(See Hazard Assessment Handbook, CG 446-3)

A-C-I-J

12. HAZARD CLASSIFICATIONS

12.1 **Code of Federal Regulations:** Nonflammable compressed gas

12.2 **NAS Hazard Rating for Bulk Water Transportation:**

Category	Rating
Fire	0
Health	
Vapor Irritant	0
Liquid or Solid Irritant	0
Poisons	1
Water Pollution	
Human Toxicity	0
Aquatic Toxicity	0
Aesthetic Effect	0
Reactivity	
Other Chemicals	1
Water	0
Self-Reaction	0

12.3 **NFPA Hazard Classifications:**

Category	Classification
Health Hazard (Blue)	0
Flammability (Red)	0
Reactivity (Yellow)	0

13. PHYSICAL AND CHEMICAL PROPERTIES

13.1 **Physical State at 15°C and 1 atm:** Gas

13.2 **Molecular Weight:** 120.91

13.3 **Boiling Point at 1 atm:** $-21.6°F = -29.8°C = 243.4°K$

13.4 **Freezing Point:** $-251.9°F = -157.7°C = 115.5°K$

13.5 **Critical Temperature:** $233.2°F = 111.8°C = 385.0°K$

13.6 **Critical Pressure:** $598\ psia = 40.7\ atm = 4.12\ MN/m^2$

13.7 **Specific Gravity:** 1.35 at 15°C (liquid)

13.8 **Liquid Surface Tension:** Not pertinent

13.9 **Liquid-Water Interfacial Tension:** Not pertinent

13.10 **Vapor (Gas) Specific Gravity:** 4.2

13.11 **Ratio of Specific Heats of Vapor (Gas):** 1.129

13.12 **Latent Heat of Vaporization:** $140\ Btu/lb = 77.9\ cal/g = 3.26 \times 10^5\ J/kg$

13.13 **Heat of Combustion:** Not pertinent

13.14 **Heat of Decomposition:** Not pertinent

13.15 **Heat of Solution:** Not pertinent

13.16 **Heat of Polymerization:** Not pertinent

NOTES

REVISED 1978

DEL | 1,2-DICHLOROETHYLENE

Common Synonyms			
Acetylene dichloride sym-dichloroethylene Dioform cis-1, 2-dichloroethylene trans-1, 2-dichloroethylene	Liquid	Colorless	Sweet pleasant odor
	Sinks in water. Flammable, irritating vapor is produced.		

Wear goggles and self-contained breathing apparatus.
Shut off ignition sources. Call fire department.
Stop discharge if possible. Keep people away.
Isolate and remove discharged material.
Notify local health and pollution control agencies.

Fire

FLAMMABLE.
POISONOUS GASES MAY BE PRODUCED IN FIRE.
Containers may explode in fire.
Flashback along vapor trail may occur.
Vapor may explode if ignited in an enclosed area.
Extinguish with dry chemicals, foam or carbon dioxide.
Water may be ineffective on fire.
Cool exposed containers with water.

Exposure

Call for medical aid.

VAPOR
If inhaled will cause dizziness, nausea, vomiting, or difficult breathing.
Move victim to fresh air.
If breathing has stopped, give artificial respiration.
If breathing is difficult, give oxygen.

LIQUID
Harmful if swallowed.
IF SWALLOWED and victim is CONSCIOUS, have victim drink water or milk.

Water Pollution

Effect of low concentrations on aquatic life is unknown.
May be dangerous if it enters water intakes.
Notify local health and wildlife officials.
Notify operators of nearby water intakes.

1. RESPONSE TO DISCHARGE

(See Response Methods Handbook, CG 446-4)

Issue warning—high flammability
Restrict access
Evacuate area
Should be removed
Chemical and physical treatment

2. LABEL

Red

3. CHEMICAL DESIGNATIONS

3.1 **Synonyms:** Acetylene dichloride; sym-Dichloroethylene; Dioform; cis- or trans-1,2-Dichloroethylene
3.2 **Coast Guard Compatibility Classification:** Not applicable
3.3 **Chemical Formula:** CICH=CHCl
3.4 **IMCO/United Nations Numerical Designation:** 3.2/1150

4. OBSERVABLE CHARACTERISTICS

4.1 **Physical State (as shipped):** Liquid
4.2 **Color:** Colorless
4.3 **Odor:** Ethereal, slightly acrid; pleasant, chloroform-like

5. HEALTH HAZARDS

5.1 **Personal Protective Equipment:** Rubber gloves; safety goggles; air supply mask or self-contained breathing apparatus.
5.2 **Symptoms Following Exposure:** Inhalation causes nausea, vomiting, weakness, tremor, epigastric cramps, central nervous depression. Contact with liquid causes irritation of eyes and (on prolonged contact) skin. Ingestion causes slight depression to deep narcosis.
5.3 **Treatment for Exposure:** INHALATION: remove from further exposure; if breathing is difficult, give oxygen; if victim is not breathing, give artificial respiration, preferably mouth-to-mouth; give oxygen when breathing is resumed; call a physician. EYES: flush with water for at least 15 min. SKIN: wash well with soap and water. INGESTION: give gastric lavage and cathartics.
5.4 **Toxicity by Inhalation (Threshold Limit Value):** 200 ppm
5.5 **Short-Term Inhalation Limits:** Data not available
5.6 **Toxicity by Ingestion:** Grade 2; oral LD_{50} = 770 mg/kg (rat)
5.7 **Late Toxicity:** Produces liver and kidney injury in experimental animals
5.8 **Vapor (Gas) Irritant Characteristics:** Data not available
5.9 **Liquid or Solid Irritant Characteristics:** Data not available
5.10 **Odor Threshold:** Data not available

6. FIRE HAZARDS

6.1 **Flash Point:** 37°F C.C.
6.2 **Flammable Limits in Air:** 9.7%–12.8%
6.3 **Fire Extinguishing Agents:** Dry chemical, foam, carbon dioxide
6.4 **Fire Extinguishing Agents Not to be Used:** Water may be ineffective.
6.5 **Special Hazards of Combustion Products:** Phosgene and hydrogen chloride fumes may form in fires.
6.6 **Behavior in Fire:** Vapor is heavier than air and may travel a considerable distance to a source of ignition and flash back.
6.7 **Ignition Temperature:** Data not available
6.8 **Electrical Hazard:** Data not available
6.9 **Burning Rate:** 2.6 mm/min.

7. CHEMICAL REACTIVITY

Reactivity with Water: No reaction
Reactivity with Common Materials: No reaction
7.3 **Stability During Transport:** Stable
7.4 **Neutralizing Agents for Acids and Caustics:** Not pertinent
7.5 **Polymerization:** Will not occur under ordinary conditions of shipment. The reaction is not vigorous.
7.6 **Inhibitor of Polymerization:** None used

8. WATER POLLUTION

8.1 **Aquatic Toxicity:** Data not available
8.2 **Waterfowl Toxicity:** Data not available
8.3 **Biological Oxygen Demand (BOD):** Data not available
8.4 **Food Chain Concentration Potential:** None

9. SELECTED MANUFACTURERS

1. Eastman Kodak Co.
   Eastman Organic Chemicals
   Rochester, N. Y. 14650
2. Dynamit-Nobel A G
   105 Stonehurst Court
   Northvale, N. J. 07647
3. Pfaltz and Bauer, Inc.
   126-04 Northern Boulevard
   Flushing, N. Y. 11368

10. SHIPPING INFORMATION

10.1 **Grades or Purity:** Commercial
10.2 **Storage Temperature:** Ambient
10.3 **Inert Atmosphere:** No requirement
10.4 **Venting:** Pressure-vacuum

11. HAZARD ASSESSMENT CODE

(See Hazard Assessment Handbook, CG 446-3)

A-X-Y

12. HAZARD CLASSIFICATIONS

12.1 **Code of Federal Regulations:** Flammable liquid
12.2 **NAS Hazard Rating for Bulk Water Transportation:** Not listed
12.3 **NFPA Hazard Classifications:**

Category	Classification
Health Hazard (Blue)	2
Flammability (Red)	3
Reactivity (Yellow)	2

13. PHYSICAL AND CHEMICAL PROPERTIES

13.1 **Physical State at 15°C and 1 atm:** Liquid
13.2 **Molecular Weight:** 97.0
13.3 **Boiling Point at 1 atm:**
cis: 140°F = 60°C = 333°K
trans: 118°F = 48°C = 321°K
13.4 **Freezing Point:**
cis: −114°F = −81°C = 192°K
trans: −58°F = −50°C = 223°K
13.5 **Critical Temperature:** Not pertinent
13.6 **Critical Pressure:** Not pertinent
13.7 **Specific Gravity:** 1.27 at 25°C (liquid)
13.8 **Liquid Surface Tension:** 24 dynes/cm = 0.024 N/m at 20°C
13.9 **Liquid-Water Interfacial Tension:** (est.) 30 dynes/cm = 0.030 N/m at 20°C
13.10 **Vapor (Gas) Specific Gravity:** 3.34
13.11 **Ratio of Specific Heats of Vapor (Gas):** 1.1468
13.12 **Latent Heat of Vaporization:** 130 Btu/lb = 72 cal/g = 3.0×10^5 J/kg
13.13 **Heat of Combustion:** −4,847.2 Btu/lb = −2,692.9 cal/g = -112.67×10^5 J/kg
13.14 **Heat of Decomposition:** Not pertinent
13.15 **Heat of Solution:** Not pertinent
13.16 **Heat of Polymerization:** Not pertinent

NOTES

DEE — DICHLOROETHYL ETHER

Common Synonyms
Bis(2-chloroethyl) ether
2, 2'-Dichloroethyl ether
Dichloroether
Di-(2-chloroethyl) ether
Chlorex

Liquid   Colorless   Sweet pleasant odor

Sinks and mixes slowly with water.

AVOID CONTACT WITH LIQUID. KEEP PEOPLE AWAY.
Wear rubber overclothing (including gloves).
Stop discharge if possible.
Call fire department.
Isolate and remove discharged material.
Notify local health and pollution control agencies.

Fire

Combustible.
POISONOUS GASES MAY BE PRODUCED IN FIRE.
Wear goggles and self-contained breathing apparatus.
Extinguish with water, dry chemicals, foam, or carbon dioxide.
Cool exposed containers with water.

Exposure

CALL FOR MEDICAL AID.

LIQUID
POISONOUS IF SWALLOWED OR IF SKIN IS EXPOSED.
Irritating to skin and eyes.
Remove contaminated clothing and shoes.
Flush affected areas with plenty of water.
IF IN EYES, hold eyelids open and flush with plenty of water.
IF SWALLOWED and victim is CONSCIOUS, have victim drink water or milk and have victim induce vomiting.
IF SWALLOWED and victim is UNCONSCIOUS OR HAVING CONVULSIONS, do nothing except keep victim warm.

Water Pollution

Effect of low concentrations on aquatic life is unknown.
May be dangerous if it enters water intakes.
Notify local health and wildlife officials.
Notify operators of nearby water intakes.

1. RESPONSE TO DISCHARGE

(See Response Methods Handbook, CG 446-4)

Issue warning—poison, water contaminant
Restrict access
Should be removed
Chemical and physical treatment

2. LABELS

No label required by Code of Federal Regulations

3. CHEMICAL DESIGNATIONS

3.1 **Synonyms:** Bis (2-chloroethyl) ether; 2,2'-Dichloroethyl ether; Dichlorodiethyl ether; Di-(2-chloroethyl) ether; Chlorex; DCEE; beta, beta'-Dichloroethyl ether
3.2 **Coast Guard Compatibility Classification:** Halogenated hydrocarbon
3.3 **Chemical Formula:** $(ClCH_2CH_2)_2O$
3.4 **IMCO/United Nations Numerical Designation:** 3.3/1916

4. OBSERVABLE CHARACTERISTICS

4.1 **Physical State (as shipped):** Liquid
4.2 **Color:** Colorless
4.3 **Odor:** Sweet, like chloroform

5. HEALTH HAZARDS

5.1 **Personal Protective Equipment:** Goggles or face shield; rubber gloves; protective clothing.
5.2 **Symptoms Following Exposure:** Inhalation of vapor causes irritation of nose, coughing, nausea. Liquid irritates eyes and causes mild irritation of skin. (Can be absorbed in toxic amounts through the skin.) Ingestion causes irritation of mouth and stomach, symptoms of systemic poisoning.
5.3 **Treatment for Exposure:** INHALATION: remove from exposure; support respiration; call physician if needed. EYES: irrigate with copious quantities of water for 15 min.; call physician. SKIN: wipe off, wash well with soap and water. INGESTION: induce vomiting; get medical attention.
5.4 **Toxicity by Inhalation (Threshold Limit Value):** 5 ppm
5.5 **Short-Term Inhalation Limits:** 35 ppm for 30 min.
5.6 **Toxicity by Ingestion:** Grade 3; oral LD_{50} = 75 mg/kg (rat)
5.7 **Late Toxicity:** Said to be carcinogenic
5.8 **Vapor (Gas) Irritant Characteristics:** Vapors are moderately irritating such that personnel will not usually tolerate moderate or high vapor concentrations.
5.9 **Liquid or Solid Irritant Characteristics:** Causes smarting of the skin and first-degree burns on short exposure; may cause second-degree burns on long exposure.
5.10 **Odor Threshold:** Data not available

6. FIRE HAZARDS

6.1 **Flash Point:** 180°F O.C., 131°F C.C.
6.2 **Flammable Limits in Air:** Data not available
6.3 **Fire Extinguishing Agents:** Water, foam, dry chemical, carbon dioxide
6.4 **Fire Extinguishing Agents Not to be Used:** Not pertinent
6.5 **Special Hazards of Combustion Products:** May form phosgene or hydrogen chloride in fires.
6.6 **Behavior in Fire:** Not pertinent
6.7 **Ignition Temperature:** 696°F
6.8 **Electrical Hazard:** Data not available
6.9 **Burning Rate:** 2.4 mm/min.

7. CHEMICAL REACTIVITY

7.1 **Reactivity with Water:** No reaction
7.2 **Reactivity with Common Materials:** No reaction
7.3 **Stability During Transport:** Stable
7.4 **Neutralizing Agents for Acids and Caustics:** Not pertinent
7.5 **Polymerization:** Not pertinent
7.6 **Inhibitor of Polymerization:** Not pertinent

8. WATER POLLUTION

8.1 **Aquatic Toxicity:** Data not available
8.2 **Waterfowl Toxicity:** Data not available
8.3 **Biological Oxygen Demand (BOD):** Data not available
8.4 **Food Chain Concentration Potential:** None

9. SELECTED MANUFACTURERS

1. Nisso Petrochemical Industries Co., Ltd. Tokyo, Japan
2. Aldrich Chemical Co., Inc. 940 West St. Paul Avenue Milwaukee, Wis. 53233
3. Eastman Kodak Co. Eastman Organic Chemicals Rochester, N. Y. 14650

10. SHIPPING INFORMATION

10.1 **Grades or Purity:** Commercial
10.2 **Storage Temperature:** Ambient
10.3 **Inert Atmosphere:** No requirement
10.4 **Venting:** Open (flame arrester)

11. HAZARD ASSESSMENT CODE

(See Hazard Assessment Handbook, CG 446-3)

A-P-Q-X-Y

12. HAZARD CLASSIFICATIONS

12.1 **Code of Federal Regulations:** Not listed
12.2 **NAS Hazard Rating for Bulk Water Transportation:**

Category	Rating
Fire	2
Health	
Vapor Irritant	3
Liquid or Solid Irritant	2
Poisons	3
Water Pollution	
Human Toxicity	3
Aquatic Toxicity	2
Aesthetic Effect	2
Reactivity	
Other Chemicals	1
Water	0
Self-Reaction	0

12.3 **NFPA Hazard Classifications:** Not listed

13. PHYSICAL AND CHEMICAL PROPERTIES

13.1 **Physical State at 15°C and 1 atm:** Liquid
13.2 **Molecular Weight:** 143.0
13.3 **Boiling Point at 1 atm:** 353°F = 178°C = 451°K
13.4 **Freezing Point:** −62°F = −52°C = 221°K
13.5 **Critical Temperature:** Not pertinent
13.6 **Critical Pressure:** Not pertinent
13.7 **Specific Gravity:** 1.22 at 20°C (liquid)
13.8 **Liquid Surface Tension:** 37.9 dynes/cm = 0.0379 N/m at 19°C
13.9 **Liquid-Water Interfacial Tension:** (est.) 40 dynes/cm = 0.040 N/m at 20°C
13.10 **Vapor (Gas) Specific Gravity:** 4.93
13.11 **Ratio of Specific Heats of Vapor (Gas):** 1.0743
13.12 **Latent Heat of Vaporization:** 143 Btu/lb = 79.5 cal/g = 3.33×10^5 J/kg
13.13 **Heat of Combustion:** (est.) −7,530 Btu/lb = −4,180 cal/g = -175×10^5 J/kg
13.14 **Heat of Decomposition:** Not pertinent
13.15 **Heat of Solution:** Not pertinent
13.16 **Heat of Polymerization:** Not pertinent

NOTES

DCM

DICHLOROMETHANE

Common Synonyms
Methylene chloride
Methylene dichloride

Watery liquid | Colorless | Sweet, pleasant odor

Sinks in water. Irritating vapor is produced.

Stop discharge if possible.
Avoid contact with liquid and vapor.
Isolate and remove discharged material.
Notify local health and pollution control agencies.

Fire

Not flammable.
POISONOUS GASES ARE PRODUCED WHEN HEATED.
Wear goggles and self-contained breathing apparatus.
Cool exposed containers with water.

Exposure

CALL FOR MEDICAL AID.

VAPOR
Irritating to eyes, nose and throat.
If inhaled, will cause nausea and dizziness.
Move to fresh air.
If breathing has stopped, give artificial respiration.
If breathing is difficult, give oxygen.

LIQUID
Irritating to skin and eyes.
Harmful if swallowed.
Remove contaminated clothing and shoes.
Flush affected areas with plenty of water.
IF IN EYES, hold eyelids open and flush with plenty of water.
IF SWALLOWED and victim is CONSCIOUS, have victim drink water or milk.

Water Pollution

Effect of low concentrations on aquatic life is unknown.
May be dangerous if it enters water intakes.
Notify local health and pollution control officials.
Notify operators of nearby water intakes.

1. RESPONSE TO DISCHARGE
(See Response Methods Handbook, CG 446-4)

Disperse and flush

2. LABELS

No hazard label required by Code of Federal Regulations

3. CHEMICAL DESIGNATIONS

3.1 **Synonyms:** Methylene chloride; Methylene dichloride
3.2 **Coast Guard Compatibility Classification:** Halogenated hydrocarbon
3.3 **Chemical Formula:** CH_2Cl_2
3.4 **IMCO/United Nations Numerical Designation:** 9.0/1593

4. OBSERVABLE CHARACTERISTICS

4.1 **Physical State (as shipped):** Liquid
4.2 **Color:** Colorless
4.3 **Odor:** Pleasant, aromatic; like chloroform; sweet, ethereal

5. HEALTH HAZARDS

5.1 **Personal Protective Equipment:** Organic vapor canister mask, safety glasses, protective clothing.
5.2 **Symptoms Following Exposure:** INHALATION: anesthetic effects, nausea and drunkenness. CONTACT WITH SKIN AND EYES: skin irritation, irritation of eyes and nose.
5.3 **Treatment for Exposure:** INHALATION: remove from exposure. Give oxygen if needed. INGESTION: no specific antidote. CONTACT WITH SKIN AND EYES: remove contaminated clothing; wash skin or eyes if affected.
5.4 **Toxicity by Inhalation (Threshold Limit Value):** 500 ppm
5.5 **Short-Term Inhalation Limits:** 100 ppm for 60 min.
5.6 **Toxicity by Ingestion:** Grade 2; LD_{50} 0.5 to 5 g/kg
5.7 **Late Toxicity:** None
5.8 **Vapor (Gas) Irritant Characteristics:** Vapors cause moderate irritation such that personnel will find high concentrations unpleasant. The effect is temporary.
5.9 **Liquid or Solid Irritant Characteristics:** Minimum hazard. If spilled on clothing and allowed to remain, may cause smarting and reddening of the skin.
5.10 **Odor Threshold:** 205-307 ppm

6. FIRE HAZARDS

6.1 **Flash Point:** Not flammable under conditions likely to be encountered.
6.2 **Flammable Limits in Air:** 12% – 19%
6.3 **Fire Extinguishing Agents:** Not pertinent
6.4 **Fire Extinguishing Agents Not to be Used:** Not pertinent
6.5 **Special Hazards of Combustion Products:** Dissociation products generated in a fire may be irritating or toxic.
6.6 **Behavior in Fire:** Not pertinent
6.7 **Ignition Temperature:** 1184°F
6.8 **Electrical Hazard:** Not pertinent
6.9 **Burning Rate:** Not pertinent

7. CHEMICAL REACTIVITY

7.1 **Reactivity with Water:** No reaction
7.2 **Reactivity with Common Materials:** No reaction
7.3 **Stability During Transport:** Stable
7.4 **Neutralizing Agents for Acids and Caustics:** Not pertinent
7.5 **Polymerization:** Not pertinent
7.6 **Inhibitor of Polymerization:** Not pertinent

8. WATER POLLUTION

8.1 **Aquatic Toxicity:** Not pertinent
8.2 **Waterfowl Toxicity:** Not pertinent
8.3 **Biological Oxygen Demand (BOD):** Not pertinent
8.4 **Food Chain Concentration Potential:** None

9. SELECTED MANUFACTURERS

1. Ancon Chemical Corp. Westlake, La. 70669
2. Dow Chemical Co. Midland, Mich. 48640
3. Ethyl Corp. Industrial Chemicals Division 451 Florida St. Baton Rouge, La. 70801

10. SHIPPING INFORMATION

10.1 **Grades or Purity:** Aerosol grade; technical grade
10.2 **Storage Temperature:** Data not available
10.3 **Inert Atmosphere:** Inerted
10.4 **Venting:** Data not available

11. HAZARD ASSESSMENT CODE
(See Hazard Assessment Handbook, CG 446-3)

A-P-X

12. HAZARD CLASSIFICATIONS

12.1 **Code of Federal Regulations:** ORM-A
12.2 **NAS Hazard Rating for Bulk Water Transportation:**

Category	Rating
Fire	1
Health	
Vapor Irritant	2
Liquid or Solid Irritant	1
Poisons	2
Water Pollution	
Human Toxicity	1
Aquatic Toxicity	2
Aesthetic Effect	2
Reactivity	
Other Chemicals	1
Water	0
Self-Reaction	0

12.3 **NFPA Hazard Classifications:**

Category	Classification
Health Hazard (Blue)	2
Flammability (Red)	1
Reactivity (Yellow)	0

13. PHYSICAL AND CHEMICAL PROPERTIES

13.1 **Physical State at 15°C and 1 atm:** Liquid
13.2 **Molecular Weight:** 84.93
13.3 **Boiling Point at 1 atm:** 104°F = 39.8°C = 313.0°K
13.4 **Freezing Point:** −142°F = −96.7°C = 176.5°K
13.5 **Critical Temperature:** 473°F = 245°C = 518°K
13.6 **Critical Pressure:** 895 psia = 60.9 atm = 6.17 MN/m²
13.7 **Specific Gravity:** 1.322 at 20°C (liquid)
13.8 **Liquid Surface Tension:** Not pertinent
13.9 **Liquid-Water Interfacial Tension:** Not pertinent
13.10 **Vapor (Gas) Specific Gravity:** 2.9
13.11 **Ratio of Specific Heats of Vapor (Gas):** 1.199
13.12 **Latent Heat of Vaporization:** 142 Btu/lb = 78.7 cal/g = 3.30 × 10^5 J/kg
13.13 **Heat of Combustion:** Not pertinent
13.14 **Heat of Decomposition:** Not pertinent
13.15 **Heat of Solution:** Not pertinent
13.16 **Heat of Polymerization:** Not pertinent

NOTES

REVISED 1978

DCP

2,4-DICHLOROPHENOL

Common Synonyms			
	Solid crystals	Colorless	Medicinal odor
	Sinks in water.		

Avoid contact with solid and dust. Keep people away.
Wear goggles, self-contained breathing apparatus, and rubber overclothing (including gloves).
Call fire department.
Isolate and remove discharged material.
Notify local health and pollution control agencies.

Fire	Combustible. POISONOUS GASES ARE PRODUCED IN FIRE. Wear goggles, self-contained breathing apparatus, and rubber overclothing (including gloves). Extinguish with dry chemical, foam, or carbon dioxide. Cool exposed containers with water.
Exposure	CALL FOR MEDICAL AID. SOLID OR DUST Will burn skin and eyes. Poisonous if swallowed. Remove contaminated clothing and shoes. Flush affected areas with plenty of water. IF IN EYES, hold eyelids open and flush with plenty of water. IF SWALLOWED and victim is CONSCIOUS, have victim drink water or milk.
Water Pollution	Effect of low concentrations on aquatic life is unknown. May be dangerous if it enters water intakes. Notify local health and wildlife officials. Notify operators of nearby water intakes.

1. RESPONSE TO DISCHARGE

(See Response Methods Handbook, CG 446-1)

Issue warning—water contaminant
Should be removed
Chemical and physical treatment

2. LABELS

No hazard label required by Code of Federal Regulations

3. CHEMICAL DESIGNATIONS

3.1 **Synonyms:** No common synonyms
3.2 **Coast Guard Compatibility Classification:** Not applicable
3.3 **Chemical Formula:** $HOC_6H_3Cl_2$-2, 4
3.4 **IMCO/United Nations Numerical Designation:** 6.1/2020

4. OBSERVABLE CHARACTERISTICS

4.1 **Physical State (as shipped):** Solid
4.2 **Color:** White
4.3 **Odor:** Strong medicinal

5. HEALTH HAZARDS

5.1 **Personal Protective Equipment:** Bureau of Mines approved respirator, rubber gloves, chemical goggles.
5.2 **Symptoms Following Exposure:** Tremors, convulsions, shortness of breath, inhibition of respiratory system.
5.3 **Treatment for Exposure:** Inhalation — rest; Ingestion — drink water, epsom salt solution.
5.4 **Toxicity by Inhalation (Threshold Limit Value):** Not pertinent
5.5 **Short-Term Inhalation Limits:** Data not available
5.6 **Toxicity by Ingestion:** Grade 2; LD_{50} 0.5 to 5 g/kg (rat)
5.7 **Late Toxicity:** Data not available
5.8 **Vapor (Gas) Irritant Characteristics:** Not pertinent
5.9 **Liquid or Solid Irritant Characteristics:** Fairly severe skin irritant. May cause pain and second-degree burns after a few minutes' contact.
5.10 **Odor Threshold:** Data not available

6. FIRE HAZARDS

6.1 **Flash Point:** 200°F O.C.; 237°F C.C.
6.2 **Flammable Limits in Air:** Data not available
6.3 **Fire Extinguishing Agents:** Water, foam, carbon dioxide, dry chemical
6.4 **Fire Extinguishing Agents Not to be Used:** Water or foam may cause frothing.
6.5 **Special Hazards of Combustion Products:** Toxic gases can be evolved.
6.6 **Behavior in Fire:** Solid melts and burns.
6.7 **Ignition Temperature:** Data not available
6.8 **Electrical Hazard:** Not pertinent
6.9 **Burning Rate:** Not pertinent

7. CHEMICAL REACTIVITY

7.1 **Reactivity with Water:** No reaction
7.2 **Reactivity with Common Materials:** May react vigorously with oxidizing materials.
7.3 **Stability During Transport:** Stable
7.4 **Neutralizing Agents for Acids and Caustics:** Not pertinent
7.5 **Polymerization:** Not pertinent
7.6 **Inhibitor of Polymerization:** Not pertinent

8. WATER POLLUTION

8.1 **Aquatic Toxicity:**
5 ppm/3 hours/rainbow trout/killed/fresh-water
5 ppm/12 hours/bluegills/killed/fresh-water
8.2 **Waterfowl Toxicity:** Data not available
8.3 **Biological Oxygen Demand (BOD):** 100%, 5 days
8.4 **Food Chain Concentration Potential:** Data not available

9. SELECTED MANUFACTURERS

1. Dow Chemical Co.
Midland, Mich. 48640
2. Monsanto Co.
Monsanto Industrial Chemicals Co.
800 North Lindbergh Blvd.
St. Louis, Mo. 63166

10. SHIPPING INFORMATION

10.1 **Grades or Purity:** Data not available
10.2 **Storage Temperature:** Data not available
10.3 **Inert Atmosphere:** Data not available
10.4 **Venting:** Data not available

11. HAZARD ASSESSMENT CODE

(See Hazard Assessment Handbook, CG 446-3)

I I

12. HAZARD CLASSIFICATIONS

12.1 **Code of Federal Regulations:** Not listed
12.2 **NAS Hazard Rating for Bulk Water Transportation:** Not listed
12.3 **NFPA Hazard Classifications:** Not listed

13. PHYSICAL AND CHEMICAL PROPERTIES

13.1 **Physical State at 15°C and 1 atm:** Solid
13.2 **Molecular Weight:** 163.01
13.3 **Boiling Point at 1 atm:** 421°F = 216°C = 489°K
13.4 **Freezing Point:** 110°F = 45°C = 318°K
13.5 **Critical Temperature:** Not pertinent
13.6 **Critical Pressure:** Not pertinent
13.7 **Specific Gravity:** 1.40 at 15°C (solid)
13.8 **Liquid Surface Tension:** Not pertinent
13.9 **Liquid-Water Interfacial Tension:** Not pertinent
13.10 **Vapor (Gas) Specific Gravity:** Not pertinent
13.11 **Ratio of Specific Heats of Vapor (Gas):** Not pertinent
13.12 **Latent Heat of Vaporization:** Not pertinent
13.13 **Heat of Combustion:** Not pertinent
13.14 **Heat of Decomposition:** Not pertinent
13.15 **Heat of Solution:** Not pertinent
13.16 **Heat of Polymerization:** Not pertinent

NOTES

REVISED 1978

DCA	2,4-DICHLOROPHENOXYACETIC ACID

Common Synonyms	
2, 4-D	Solid · White to tan · Odorless Sinks in water.

Keep people away.
Isolate and remove discharged material.
Notify local health and pollution control agencies.

Fire	Combustible. POISONOUS GASES MAY BE PRODUCED IN FIRE. Wear goggles and self-contained breathing apparatus. Flood discharge area with water.
Exposure	CALL FOR MEDICAL AID. SOLID POISONOUS IF SWALLOWED. IF SWALLOWED and victim is CONSCIOUS, have victim drink water or milk and have victim induce vomiting. IF SWALLOWED and victim is UNCONSCIOUS OR HAVING CONVULSIONS, do nothing except keep victim warm.
Water Pollution	Dangerous to aquatic life in high concentrations. May be dangerous if it enters water intakes. Notify local health and wildlife officials. Notify operators of nearby water intakes.

1. RESPONSE TO DISCHARGE

(See Response Methods Handbook, CG 446-4)

Issue warning—poison, water contaminant
Restrict access
Should be removed
Chemical and physical treatment

2. LABELS

No hazard label required by Code of Federal Regulations

3. CHEMICAL DESIGNATIONS

3.1 **Synonyms:** 2,4-D
3.2 **Coast Guard Compatibility Classification:** Not applicable
3.3 **Chemical Formula:** $2,4\text{-}Cl_2C_6H_3OCH_2COOH$
3.4 **IMCO/United Nations Numerical Designation:** 6.1/1609

4. OBSERVABLE CHARACTERISTICS

4.1 **Physical State (as shipped):** Solid
4.2 **Color:** White to tan
4.3 **Odor:** None

5. HEALTH HAZARDS

5.1 **Personal Protective Equipment:** Protective dust mask; rubber gloves; chemical goggles.
5.2 **Symptoms Following Exposure:** Dust may irritate eyes. Ingestion causes gastroentric distress, diarrhea, mild central nervous system depression, dysphagia, and possible transient liver and kidney injury.
5.3 **Treatment for Exposure:** EYES: flush with water for at least 15 min. SKIN: wash well with soap and water. INGESTION: induce vomiting and follow with gastric lavage and supportive therapy.
5.4 **Toxicity by Inhalation (Threshold Limit Value):** Data not available
5.5 **Short-Term Inhalation Limits:** Data not available
5.6 **Toxicity by Ingestion:** Grade 3; oral rat LD_{50} = 375 mg/kg (rat), 80 mg/kg (human)
5.7 **Late Toxicity:** Causes birth defects in some laboratory animals
5.8 **Vapor (Gas) Irritant Characteristics:** Not pertinent
5.9 **Liquid or Solid Irritant Characteristics:** Data not available
5.10 **Odor Threshold:** Not pertinent

6. FIRE HAZARDS

6.1 **Flash Point:** Not pertinent (combustible solid)
6.2 **Flammable Limits in Air:** Not pertinent
6.3 **Fire Extinguishing Agents:** Water, foam
6.4 **Fire Extinguishing Agents Not to be Used:** Not pertinent
6.5 **Special Hazards of Combustion Products:** Toxic and irritating hydrogen chloride or phosgene gases may form.
6.6 **Behavior in Fire:** Not pertinent
6.7 **Ignition Temperature:** Not pertinent
6.8 **Electrical Hazard:** Data not available
6.9 **Burning Rate:** Not pertinent

7. CHEMICAL REACTIVITY

7.1 **Reactivity with Water:** No reaction
7.2 **Reactivity with Common Materials:** No reaction
7.3 **Stability During Transport:** Stable
7.4 **Neutralizing Agents for Acids and Caustics:** Flush with water, rinse with sodium bicarbonate or lime solution.
7.5 **Polymerization:** Not pertinent
7.6 **Inhibitor of Polymerization:** Not pertinent

8. WATER POLLUTION

8.1 **Aquatic Toxicity:**
5 ppm/48 hr/killifish/50% kill/salt water
375 mg/l/48 hr/bluegill/TL_m/fresh water
8.2 **Waterfowl Toxicity:** LD_{50} = ~2000 mg/kg
8.3 **Biological Oxygen Demand (BOD):** Data not available
8.4 **Food Chain Concentration Potential:** No buildup in food chain

9. SELECTED MANUFACTURERS

1. Dow Chemical Co.
   Midland, Mich. 48640
2. Rhodia, Inc.
   Chipman Division
   120 Jersey Ave.
   New Brunswick, N. J. 08903
3. Monsanto Co.
   800 N. Lindbergh Boulevard
   St. Louis, Mo. 63166

10. SHIPPING INFORMATION

10.1 **Grades or Purity:** 98+%
10.2 **Storage Temperature:** Ambient
10.3 **Inert Atmosphere:** No requirement
10.4 **Venting:** Open

11. HAZARD ASSESSMENT CODE

(See Hazard Assessment Handbook, CG 446-3)

II

12. HAZARD CLASSIFICATIONS

12.1 **Code of Federal Regulations:** ORM-A
12.2 **NAS Hazard Rating for Bulk Water Transportation:** Not listed
12.3 **NFPA Hazard Classifications:** Not listed

13. PHYSICAL AND CHEMICAL PROPERTIES

13.1 **Physical State at 15°C and 1 atm:** Solid
13.2 **Molecular Weight:** 221.0
13.3 **Boiling Point at 1 atm:** Very high
13.4 **Freezing Point:** 286°F = 141°C = 314°K
13.5 **Critical Temperature:** Not pertinent
13.6 **Critical Pressure:** Not pertinent
13.7 **Specific Gravity:** 1.563 at 20°C (solid)
13.8 **Liquid Surface Tension:** Not pertinent
13.9 **Liquid-Water Interfacial Tension:** Not pertinent
13.10 **Vapor (Gas) Specific Gravity:** Not pertinent
13.11 **Ratio of Specific Heats of Vapor (Gas):** Not pertinent
13.12 **Latent Heat of Vaporization:** Not pertinent
13.13 **Heat of Combustion:** (est.) −7,700 Btu/lb = −4,300 cal/g = -180×10^5 J/kg
13.14 **Heat of Decomposition:** Not pertinent
13.15 **Heat of Solution:** Not pertinent
13.16 **Heat of Polymerization:** Not pertinent

NOTES

DPP	# DICHLOROPROPANE

Common Synonyms	Watery liquid	Colorless	Sweet odor
Propylene dichloride 1, 2 - dichloropropane	Sinks in water. Flammable, irritating vapor is produced.		

Stop discharge if possible. Keep people away.
Shut off ignition sources and call fire department.
Stay upwind and use water spray to "knock down" vapor.
Avoid contact with liquid and vapor.
Isolate and remove discharged material.
Notify local health and pollution control agencies.

Fire

FLAMMABLE.
POISONOUS GASES ARE PRODUCED IN FIRE.
Flashback along vapor trail may occur.
Vapor may explode if ignited in an enclosed area.
Wear goggles and self-contained breathing apparatus.
Extinguish with foam, dry chemical, or carbon dioxide.
Cool exposed containers with water.

Exposure

CALL FOR MEDICAL AID.

VAPOR
Irritating to eyes, nose and throat.
Move to fresh air.
If breathing has stopped, give artificial respiration.
If breathing is difficult, give oxygen.

LIQUID
Irritating to skin and eyes.
Harmful if swallowed.
Remove contaminated clothing and shoes.
Flush affected areas with plenty of water.
IF IN EYES, hold eyelids open and flush with plenty of water.
IF SWALLOWED and victim is CONSCIOUS, have victim drink water or milk.

Water Pollution

Effect of low concentrations on aquatic life is unknown.
May be dangerous if it enters water intakes.
Notify local health and wildlife officials.
Notify operators of nearby water intakes.

1. RESPONSE TO DISCHARGE

(See Response Methods Handbook, CG 446-4)

Issue warning—high flammability
Evacuate area

2. LABEL

Red

3. CHEMICAL DESIGNATIONS

3.1 **Synonyms:** 1,2-Dichloropropane; Propylene dichloride
3.2 **Coast Guard Compatibility Classification:** Halogenated hydrocarbon
3.3 **Chemical Formula:** $CH_3CHClCH_2Cl$
3.4 **IMCO/United Nations Numerical Designation:** 3.2/1279

4. OBSERVABLE CHARACTERISTICS

4.1 **Physical State (as shipped):** Liquid
4.2 **Color:** Colorless
4.3 **Odor:** Sweet

5. HEALTH HAZARDS

5.1 **Personal Protective Equipment:** Air supply in confined area, rubber gloves, chemical goggles, protective coveralls and rubber footwear.
5.2 **Symptoms Following Exposure:** Contact with skin or eyes may cause irritation.
5.3 **Treatment for Exposure:** INHALATION: remove to fresh air. CONTACT WITH SKIN OR EYES: wash skin thoroughly with soap and water. Flush eyes with water for 15 min. Call a doctor.
5.4 **Toxicity by Inhalation (Threshold Limit Value):** 75 ppm
5.5 **Short-Term Inhalation Limits:** Data not available
5.6 **Toxicity by Ingestion:** Grade 2; LD_{50} 0.5 to 5 g/kg (guinea pig)
5.7 **Late Toxicity:** Data not available
5.8 **Vapor (Gas) Irritant Characteristics:** Vapors cause a slight smarting of the eyes or respiratory system if present in high concentrations. The effect is temporary.
5.9 **Liquid or Solid Irritant Characteristics:** Minimum hazard. If spilled on clothing and allowed to remain, may cause smarting and reddening of the skin.
5.10 **Odor Threshold:** Data not available

6. FIRE HAZARDS

6.1 **Flash Point:** 70°F O.C.; 60°F C.C.
6.2 **Flammable Limits in Air:** 3.4% - 14.5%
6.3 **Fire Extinguishing Agents:** Foam, carbon dioxide, dry chemical.
6.4 **Fire Extinguishing Agents Not to be Used:** Not pertinent
6.5 **Special Hazards of Combustion Products:** Toxic and irritating gases may be generated.
6.6 **Behavior in Fire:** Not pertinent
6.7 **Ignition Temperature:** 1035°F
6.8 **Electrical Hazard:** Not pertinent
6.9 **Burning Rate:** (est.) 3.2 mm/min.

7. CHEMICAL REACTIVITY

7.1 **Reactivity with Water:** No reaction
7.2 **Reactivity with Common Materials:** No reaction
7.3 **Stability During Transport:** Stable
7.4 **Neutralizing Agents for Acids and Caustics:** Not pertinent
7.5 **Polymerization:** Not pertinent
7.6 **Inhibitor of Polymerization:** Not pertinent

8. WATER POLLUTION

8.1 **Aquatic Toxicity:** >100 ppm/crustacea/TL_m/salt water
8.2 **Waterfowl Toxicity:** Data not available
8.3 **Biological Oxygen Demand (BOD):** Data not available
8.4 **Food Chain Concentration Potential:** None

9. SELECTED MANUFACTURERS

1. Dow Chemical Co.
   Midland, Mich. 48640
2. Jefferson Chemical Co., Inc.
   3336 Richmond Ave.
   Houston, Texas 77052
3. Union Carbide Corp.
   Chemicals and Plastics Division
   270 Park Ave.
   New York, N. Y. 10017

10. SHIPPING INFORMATION

10.1 **Grades or Purity:** Refined
10.2 **Storage Temperature:** Ambient
10.3 **Inert Atmosphere:** No requirement
10.4 **Venting:** Pressure-vacuum

11. HAZARD ASSESSMENT CODE

(See Hazard Assessment Handbook, CG 446-3)

A-X-Y

12. HAZARD CLASSIFICATIONS

12.1 **Code of Federal Regulations:** Flammable liquid
12.2 **NAS Hazard Rating for Bulk Water Transportation:**

Category	Rating
Fire	3
Health	
Vapor Irritant	1
Liquid or Solid Irritant	1
Poisons	3
Water Pollution	
Human Toxicity	1
Aquatic Toxicity	3
Aesthetic Effect	2
Reactivity	
Other Chemicals	1
Water	0
Self-Reaction	0

12.3 **NFPA Hazard Classifications:**

Category	Classification
Health Hazard (Blue)	2
Flammability (Red)	3
Reactivity (Yellow)	0

13. PHYSICAL AND CHEMICAL PROPERTIES

13.1 **Physical State at 15°C and 1 atm:** Liquid
13.2 **Molecular Weight:** 102.9
13.3 **Boiling Point at 1 atm:** 206°F = 96.4°C = 369.6°K
13.4 **Freezing Point:** −148°F = −100°C = 173°K
13.5 **Critical Temperature:** Not pertinent
13.6 **Critical Pressure:** Not pertinent
13.7 **Specific Gravity:** 1.158 at 20°C (liquid)
13.8 **Liquid Surface Tension:** 29 dynes/cm = 0.029 N/m at 20°C
13.9 **Liquid-Water Interfacial Tension:** 37.9 dynes/cm = 0.0379 N/m at 22.7°C
13.10 **Vapor (Gas) Specific Gravity:** 3.5
13.11 **Ratio of Specific Heats of Vapor (Gas):** 1.094
13.12 **Latent Heat of Vaporization:** 122 Btu/lb = 67.7 cal/g = 2.83×10^5 J/kg
13.13 **Heat of Combustion:** (est.) 7300 Btu/lb = 4100 cal/g = 170×10^5 J/kg
13.14 **Heat of Decomposition:** Not pertinent
13.15 **Heat of Solution:** Not pertinent
13.16 **Heat of Polymerization:** Not pertinent

NOTES

REVISED 1978

DPR — DICHLOROPROPENE

Common Synonyms			
1,3-dichloropropene Telone	Liquid	Colorless	Sweet odor
	Sinks in water. Flammable, irritating vapor is produced.		

Avoid contact with liquid and vapor. Keep people away.
Wear rubber overclothing (including gloves).
Stop discharge if possible.
Call fire department.
Isolate and remove discharged material.
Notify local health and pollution control agencies.

Fire

FLAMMABLE
POISONOUS GASES ARE PRODUCED IN FIRE.
Flashback along vapor trail may occur.
Vapor may explode if ignited in an enclosed area.
WEAR GOGGLES AND SELF-CONTAINED BREATHING APPARATUS.
Extinguish with water, dry chemical, foam, or carbon dioxide.
Cool exposed containers with water.

Exposure

CALL FOR MEDICAL AID.

VAPOR
Irritating to eyes, nose and throat.
Move to fresh air.
If breathing has stopped, give artificial respiration.
If breathing is difficult, give oxygen.

LIQUID
Will burn skin and eyes.
Harmful if swallowed.
Remove contaminated clothing and shoes.
Flush affected areas with plenty of water.
IF IN EYES, hold eyelids open and flush with plenty of water.
IF SWALLOWED and victim is CONSCIOUS, have victim drink water or milk and have victim induce vomiting.
IF SWALLOWED and victim is UNCONSCIOUS OR HAVING CONVULSIONS, do nothing, except keep victim warm.

Water Pollution

Effect of low concentrations on aquatic life is unknown.
May be dangerous if it enters water intakes.
Notify local health and wildlife officials.
Notify operators of nearby water intakes.

1. RESPONSE TO DISCHARGE
(See Response Methods Handbook, CG 446-4)

Disperse and flush

2. LABEL

Red

3. CHEMICAL DESIGNATIONS

3.1 **Synonyms:** 1,3-Dichloropropene; Telone

3.2 **Coast Guard Compatibility Classification:** Substituted allyl

3.3 **Chemical Formula:** $ClCH_2CH{=}CHCl$

3.4 **IMCO/United Nations Numerical Designation:** 3.3/2047

4. OBSERVABLE CHARACTERISTICS

4.1 **Physical State (as shipped):** Liquid

4.2 **Color:** Colorless

4.3 **Odor:** Sweet; like chloroform

5. HEALTH HAZARDS

5.1 **Personal Protective Equipment:** An approved full face mask equipped with a fresh black canister meeting specifications of the U. S. Bureau of Mines for organic vapors, a full face self-contained breathing apparatus, or full face air-supplied respirator.

5.2 **Symptoms Following Exposure:** Smarting of skin and eyes. Prolonged contact of liquid with skin may cause second-degree burns.

5.3 **Treatment for Exposure:** INHALATION: remove patient to fresh air, keep warm and quiet; call physician immediately; give artificial respiration if breathing has stopped. INGESTION: call physician immediately. Induce vomiting by giving an emetic, e.g., 2 tablespoons table salt in glass of warm water. CONTACT WITH SKIN OR EYES: immediately remove contaminated clothing and shoes. Wash skin with soap and plenty of water. For eyes, flush immediately with plenty of water for at least 15 min. Call physician.

5.4 **Toxicity by Inhalation (Threshold Limit Value):** Data not available

5.5 **Short-Term Inhalation Limits:** Data not available

5.6 **Toxicity by Ingestion:** Grade 3; LD_{50} 50 to 500 mg/kg

5.7 **Late Toxicity:** Data not available

5.8 **Vapor (Gas) Irritant Characteristics:** Vapors cause moderate irritation such that personnel will find high concentrations unpleasant. The effect is temporary.

5.9 **Liquid or Solid Irritant Characteristics:** Causes smarting of the skin and first-degree burns on short exposure and may cause secondary burns on long exposure.

5.10 **Odor Threshold:** Data not available

6. FIRE HAZARDS

6.1 **Flash Point:** 95°F C.C.

6.2 **Flammable Limits in Air:** Data not available

6.3 **Fire Extinguishing Agents:** Water, dry chemical, foam, carbon dioxide

6.4 **Fire Extinguishing Agents Not to be Used:** Not pertinent

6.5 **Special Hazards of Combustion Products:** Toxic and irritating gases may be generated

6.6 **Behavior in Fire:** Not pertinent

6.7 **Ignition Temperature:** Data not available

6.8 **Electrical Hazard:** Data not available

6.9 **Burning Rate:** (est.) 3.4 mm/min.

7. CHEMICAL REACTIVITY

7.1 **Reactivity with Water:** No reaction

7.2 **Reactivity with Common Materials:** No reaction

7.3 **Stability During Transport:** Stable

7.4 **Neutralizing Agents for Acids and Caustics:** Not pertinent

7.5 **Polymerization:** Not pertinent

7.6 **Inhibitor of Polymerization:** Not pertinent

8. WATER POLLUTION

8.1 **Aquatic Toxicity:** 100 ppm/*/daphnia/toxic/fresh water

*Time period not specified.

8.2 **Waterfowl Toxicity:** Data not available

8.3 **Biological Oxygen Demand (BOD):** Data not available

8.4 **Food Chain Concentration Potential:** None

9. SELECTED MANUFACTURERS

Dow Chemical Co.
Midland, Mich. 48640

10. SHIPPING INFORMATION

10.1 **Grades or Purity:** Telone soil fumigant: 100%; Telone C soil fungicide: 85%, chloropicrin 15%

10.2 **Storage Temperature:** Ambient

10.3 **Inert Atmosphere:** No requirement

10.4 **Venting:** Pressure-vacuum

11. HAZARD ASSESSMENT CODE
(See Hazard Assessment Handbook, CG 446-3)

A-X-Y

12. HAZARD CLASSIFICATIONS

12.1 **Code of Federal Regulations:** Flammable Liquid

12.2 **NAS Hazard Rating for Bulk Water Transportation:**

Category	Rating
Fire	3
Health	
Vapor Irritant	2
Liquid or Solid Irritant	2
Poisons	3
Water Pollution	
Human Toxicity	3
Aquatic Toxicity	3
Aesthetic Effect	2
Reactivity	
Other Chemicals	1
Water	0
Self-Reaction	1

12.3 **NFPA Hazard Classifications:** Not listed

13. PHYSICAL AND CHEMICAL PROPERTIES

13.1 **Physical State at 15°C and 1 atm:** Liquid

13.2 **Molecular Weight:** 110.98

13.3 **Boiling Point at 1 atm:** 170°F = 77°C = 350°K

13.4 **Freezing Point:** Not pertinent

13.5 **Critical Temperature:** Not pertinent

13.6 **Critical Pressure:** Not pertinent

13.7 **Specific Gravity:** 1.2 at 20°C (liquid)

13.8 **Liquid Surface Tension:** 31.2 dynes/cm = 0.0312 N/m at 24°C

13.9 **Liquid-Water Interfacial Tension:** 23.8 dynes/cm = 0.0238 N/m at 24°C

13.10 **Vapor (Gas) Specific Gravity:** Not pertinent

13.11 **Ratio of Specific Heats of Vapor (Gas):** (est.) 1.116

13.12 **Latent Heat of Vaporization:** (est.) 113 Btu/lb = 62.8 cal/g = 2.63×10^5 J/kg

13.13 **Heat of Combustion:** (est.) 6900 Btu/lb = 3900 cal/g = 160×10^5 J/kg

13.14 **Heat of Decomposition:** Not pertinent

13.15 **Heat of Solution:** Not pertinent

13.16 **Heat of Polymerization:** Not pertinent

NOTES

REVISED 1978

DTM	# 4,4'-DICHLORO-alpha-TRICHLOROMETHYLBENZHYDROL

Common Synonyms			
Dicofol Kelthanethanol Kelthane 1,1-Bis(p-chlorophenyl)-2,2,2-trichloroethanol	Solid powder or liquid (solution) May float or sink in water.	Solid is white to grey; liquid is brown	Solid is odorless; liquid has odor of solvent

Stop discharge if possible. Keep people away.
Shut off ignition sources. Call fire department.
Avoid contact with liquid and solid.
Isolate and remove discharged material.
Notify local health and pollution control agencies.

Fire	LIQUID SOLVENT MAY BE FLAMMABLE. SOLID IS COMBUSTIBLE. Irritating gases may be produced when heated. Flashback along vapor trail may occur. Vapor may explode if ignited in an enclosed area. Wear goggles and self-contained breathing apparatus. Extinguish with dry chemicals, foam or carbon dioxide. Water may be ineffective on fire. Cool exposed containers with water.
Exposure	CALL FOR MEDICAL AID. DUST Irritating to eyes, nose and throat. If inhaled will cause headache or dizziness. If in eyes, hold eyelids open and flush with plenty of water. If breathing has stopped, give artificial respiration. If breathing is difficult, give oxygen. LIQUID OR SOLID Irritating to skin and eyes. If swallowed will cause headache, nausea or dizziness. Remove contaminated clothing and shoes. Flush affected areas with plenty of water. IF IN EYES, hold eyelids open and flush with plenty of water. IF SWALLOWED and victim is CONSCIOUS, have victim drink water or milk. IF SWALLOWED and victim is UNCONSCIOUS OR HAVING CONVULSIONS, do nothing except keep victim warm.
Water Pollution	Dangerous to aquatic life in high concentrations. Fouling to shoreline. May be dangerous if it enters water intakes. Notify local health and wildlife officials. Notify operators of nearby water intakes.

1. RESPONSE TO DISCHARGE
(See Response Methods Handbook, CG 446-4)

Issue warning—water contaminant, high flammability (liquid only)
Mechanical containment (liquid only)
Should be removed
Chemical and physical treatment

2. LABELS

No hazard label required by Code of Federal Regulations

3. CHEMICAL DESIGNATIONS

3.1 **Synonyms:** 1,1-Bis(p-chlorophenyl)-2,2,2-trichoroethanol; Di-(p-chlorophenyl) trichloromethylcarbinol; Dicofol; Kelthane; Kelthanethanol
3.2 **Coast Guard Compatibility Classification:** Not listed
3.3 **Chemical Formula:** $(ClC_6H_4)_2C(OH)CCl_3$
3.4 **IMCO/United Nations Numerical Designation:** Not listed

4. OBSERVABLE CHARACTERISTICS

4.1 **Physical State (as shipped):** Powder, semi-solid, or liquid
4.2 **Color:** White to gray solid, red to black semi-solid, brown liquid
4.3 **Odor:** Solids have slight characteristic odor; liquid has odor of the solvent xylene.

5. HEALTH HAZARDS

5.1 **Personal Protective Equipment:** Self-contained breathing apparatus if exposed to vapors; rubber gloves; splash goggles
5.2 **Symptoms Following Exposure:** Inhalation or ingestion causes nausea, headache, weight loss, convulsions, possible kidney and liver damage. Contact with eyes causes irritation.
5.3 **Treatment for Exposure:** INHALATION: move to fresh air; apply artificial respiration and oxygen if indicated. EYES: wash for 15 min. with water; call a physician. SKIN: wash well with soap and water.
5.4 **Toxicity by Inhalation (Threshold Limit Value):** Data not available
5.5 **Short-Term Inhalation Limits:** Data not available
5.6 **Toxicity by Ingestion:** Grade 2; oral LD_{50} = 575 mg/kg (rat), 1,810 mg/kg (rabbit)
5.7 **Late Toxicity:** Suppresses immune reactions in rats
5.8 **Vapor (Gas) Irritant Characteristics:** Data not available
5.9 **Liquid or Solid Irritant Characteristics:** Data not available
5.10 **Odor Threshold:** Data not available

6. FIRE HAZARDS

6.1 **Flash Point:** 75°F O.C. (xylene)
6.2 **Flammable Limits in Air:** 1.1% - 7.0% (for xylene solution)
6.3 **Fire Extinguishing Agents:** Foam, dry chemical, carbon dioxide
6.4 **Fire Extinguishing Agents Not to be Used:** Water may be ineffective.
6.5 **Special Hazards of Combustion Products:** Irritating hydrogen chloride fumes may form in fire.
6.6 **Behavior in Fire:** Xylene solvent vapors may travel to source of ignition and flash back.
6.7 **Ignition Temperature:** 986°F (for xylene solution)
6.8 **Electrical Hazard:** (xylene) Class I, Group D
6.9 **Burning Rate:** (xylene) 5.8 mm/min.

7. CHEMICAL REACTIVITY

7.1 **Reactivity with Water:** No reaction
7.2 **Reactivity with Common Materials:** Contact with steel at elevated temperature causes formation of toxic chlorine and hydrogen chloride gases. Liquid may attack some forms of plastics.
7.3 **Stability During Transport:** Stable
7.4 **Neutralizing Agents for Acids and Caustics:** Not pertinent
7.5 **Polymerization:** Not pertinent
7.6 **Inhibitor of Polymerization:** Not pertinent

8. WATER POLLUTION

8.1 **Aquatic Toxicity:** 110 ppm/24 hr/rainbow trout/TL_m/fresh water
8.2 **Waterfowl Toxicity:** 1,700-1,900 ppm
8.3 **Biological Oxygen Demand (BOD):** Data not available
8.4 **Food Chain Concentration Potential:** Data not available

9. SELECTED MANUFACTURERS

Rohm and Haas Co.
Independence Mall West
Philadelphia, Pa. 19105

10. SHIPPING INFORMATION

10.1 **Grades or Purity:** Technical; 82—88%; 35% plus 65% inert solids; 75% solution in xylene, a combustible solvent.
10.2 **Storage Temperature:** Ambient
10.3 **Inert Atmosphere:** No requirement
10.4 **Venting:** Open (flame arrester) (for liquid form)

11. HAZARD ASSESSMENT CODE
(See Hazard Assessment Handbook, CG 446-3)

Liquid: A-T-U
Solid: II

12. HAZARD CLASSIFICATIONS

12.1 **Code of Federal Regulations:** Not listed
12.2 **NAS Hazard Rating for Bulk Water Transportation:** Not listed
12.3 **NFPA Hazard Classifications:** Not listed

13. PHYSICAL AND CHEMICAL PROPERTIES

13.1 **Physical State at 15°C and 1 atm:** Solid or liquid
13.2 **Molecular Weight:** 470.5
13.3 **Boiling Point at 1 atm:** (Data apply to xylene solution; solid decomposes) 282°F = 139°C = 412°K
13.4 **Freezing Point:** Not pertinent
13.5 **Critical Temperature:** Not pertinent
13.6 **Critical Pressure:** Not pertinent
13.7 **Specific Gravity:** >1.1 at 20°C (solid) <0.9 at 20°C (liquid)
13.8 **Liquid Surface Tension:** Not pertinent
13.9 **Liquid-Water Interfacial Tension:** Not pertinent
13.10 **Vapor (Gas) Specific Gravity:** Not pertinent
13.11 **Ratio of Specific Heats of Vapor (Gas):** Not pertinent
13.12 **Latent Heat of Vaporization:** Not pertinent
13.13 **Heat of Combustion:** Data not available
13.14 **Heat of Decomposition:** Not pertinent
13.15 **Heat of Solution:** Not pertinent
13.16 **Heat of Polymerization:** Not pertinent

NOTES

DPT DICYCLOPENTADIENE

Common Synonyms			
Dicy	Liquid, or solid crystals	Colorless	Camphor odor
	Floats on water. Freezing point is 41° F.		

Stop discharge if possible.
Call fire department.
Avoid contact with liquid.
Stay upwind and use water spray to "knock down" vapor.
Isolate and remove discharged material.
Notify local health and pollution control agencies.

Fire	FLAMMABLE Flashback along vapor trail may occur. Vapor may explode if ignited in an enclosed area. Extinguish with water, dry chemical, foam, or carbon dioxide. Cool exposed containers with water.
Exposure	CALL FOR MEDICAL AID. LIQUID OR SOLID Irritating to skin and eyes. Remove contaminated clothing and shoes. Flush affected areas with plenty of water. IF IN EYES, hold eyelids open and flush with plenty of water.
Water Pollution	Effect of low concentrations on aquatic life is unknown. Fouling to shoreline. May be dangerous if it enters water intakes. Notify local health and wildlife officials. Notify operators of nearby water intakes.

1. RESPONSE TO DISCHARGE
(See Response Methods Handbook, CG 446-4)

Mechanical containment
Should be removed
Chemical and physical treatment

2. LABEL

Red

3. CHEMICAL DESIGNATIONS

3.1 **Synonyms:** Dicy; 3a, 4, 7, 7a-Tetrahydro-4, 7-methanoindene
3.2 **Coast Guard Compatibility Classification:** Olefin
3.3 **Chemical Formula:** $C_{10}H_{12}$
3.4 **IMCO/United Nations Numerical Designation:** 3.3/2048

4. OBSERVABLE CHARACTERISTICS

4.1 **Physical State (as shipped):** Liquid
4.2 **Color:** Colorless
4.3 **Odor:** Camphor-like

5. HEALTH HAZARDS

5.1 **Personal Protective Equipment:** Air-supplied mask in confined areas, rubber gloves, safety glasses.
5.2 **Symptoms Following Exposure:** Vapor irritates mucous membranes and respiratory tract, causes nausea, vomiting, headache, and dizziness. Direct contact irritates skin.
5.3 **Treatment for Exposure:** INHALATION: remove victim from contaminated area and call physician if unconscious; if breathing is irregular or stopped, give oxygen and start resuscitation. EYES OR SKIN: flush with plenty of water for 15 min.
5.4 **Toxicity by Inhalation (Threshold Limit Value):** 75-100 ppm
5.5 **Short-Term Inhalation Limits:** Data not available
5.6 **Toxicity by Ingestion:** Grade 2; oral rat LD_{50} = 0.82 g/kg
5.7 **Late Toxicity:** Data not available
5.8 **Vapor (Gas) Irritant Characteristics:** Vapors cause a slight smarting of the eyes and respiratory system if present in high concentrations. The effect is temporary.
5.9 **Liquid or Solid Irritant Characteristics:** Minimum hazard. If spilled on clothing and allowed to remain, may cause smarting and reddening of the skin.
5.10 **Odor Threshold:** <0.003 ppm

6. FIRE HAZARDS

6.1 **Flash Point:** 90°F O.C.
6.2 **Flammable Limits in Air:** 0.8%–6.3%
6.3 **Fire Extinguishing Agents:** Foam, carbon dioxide, dry chemical, or water spray.
6.4 **Fire Extinguishing Agents Not to be Used:** Not pertinent
6.5 **Special Hazards of Combustion Products:** Not pertinent
6.6 **Behavior in Fire:** Not pertinent
6.7 **Ignition Temperature:** 941°F
6.8 **Electrical Hazard:** Data not available
6.9 **Burning Rate:** Data not available

7. CHEMICAL REACTIVITY

7.1 **Reactivity with Water:** No reaction
7.2 **Reactivity with Common Materials:** No reaction
7.3 **Stability During Transport:** Stable
7.4 **Neutralizing Agents for Acids and Caustics:** Not pertinent
7.5 **Polymerization:** May occur in presence of acids, but not hazardous.
7.6 **Inhibitor of Polymerization:** Not pertinent

8. WATER POLLUTION

8.1 **Aquatic Toxicity:** Data not available
8.2 **Waterfowl Toxicity:** Data not available
8.3 **Biological Oxygen Demand (BOD):** Data not available
8.4 **Food Chain Concentration Potential:** None

9. SELECTED MANUFACTURERS

1. Exxon Chemical Co.
Houston, Texas 77001
2. Northwest Industries Inc.
Velsicol Chemical Corp.
341 East Ohio St.
Chicago, Ill. 60611
3. Union Carbide Corp.
Chemicals and Plastics Division
270 Park Ave.
New York, N. Y. 10017

10. SHIPPING INFORMATION

10.1 **Grades or Purity:** 97%
10.2 **Storage Temperature:** Ambient
10.3 **Inert Atmosphere:** No requirement
10.4 **Venting:** Open (flame arrester)

11. HAZARD ASSESSMENT CODE
(See Hazard Assessment Handbook, CG 446-3)

A-T-U

12. HAZARD CLASSIFICATIONS

12.1 **Code of Federal Regulations:** Flammable Liquid
12.2 **NAS Hazard Rating for Bulk Water Transportation:**

Category	Rating
Fire	3
Health	
Vapor Irritant	1
Liquid or Solid Irritant	1
Poisons	2
Water Pollution	
Human Toxicity	1
Aquatic Toxicity	1
Aesthetic Effect	2
Reactivity	
Other Chemicals	1
Water	0
Self-Reaction	1

12.3 **NFPA Hazard Classifications:** Not listed

13. PHYSICAL AND CHEMICAL PROPERTIES

13.1 **Physical State at 15°C and 1 atm:** Liquid
13.2 **Molecular Weight:** 132.31
13.3 **Boiling Point at 1 atm:** 338°F = 170°C = 443°K
13.4 **Freezing Point:** 41°F = 5°C = 278°K
13.5 **Critical Temperature:** Not pertinent
13.6 **Critical Pressure:** Not pertinent
13.7 **Specific Gravity:** 0.978 at 20°C (liquid)
13.8 **Liquid Surface Tension:** Data not available
13.9 **Liquid-Water Interfacial Tension:** Data not available
13.10 **Vapor (Gas) Specific Gravity:** Not pertinent
13.11 **Ratio of Specific Heats of Vapor (Gas):** Not pertinent
13.12 **Latent Heat of Vaporization:** Not pertinent
13.13 **Heat of Combustion:** -18,800 Btu/lb = -10,400 cal/g = -437 × 10^5 J/kg
13.14 **Heat of Decomposition:** Not pertinent
13.15 **Heat of Solution:** Not pertinent
13.16 **Heat of Polymerization:** Not pertinent

NOTES

REVISED 1978

DED

DIELDRIN

Common Synonyms
HEOD
endo,exo-1,2,3,4,10,10-Hexachloro-6,7-epoxy-1,4,4a,5,6,7,8,8a-octahydro 1,4:5,8-dimethanonaphthalene

Solid | Light brown | Mild chemical odor

Sinks in water.

AVOID CONTACT WITH SOLID AND DUST. KEEP PEOPLE AWAY.
Wear goggles, dust respirator and rubber overclothing (including gloves).
Stop discharge if possible.
Isolate and remove discharged material.
Notify local health and pollution control agencies.

Fire

Not flammable.
POISONOUS GASES MAY BE PRODUCED WHEN HEATED.

Exposure

CALL FOR MEDICAL AID.

DUST
POISONOUS IF INHALED OR IF SKIN IS EXPOSED.
If inhaled will cause headache, dizziness, or loss of consciousness.

If in eyes, hold eyelids open and flush with plenty of water.
If breathing has stopped, give artificial respiration.
If breathing is difficult, give oxygen.

SOLID
POISONOUS IF SWALLOWED OR IF SKIN IS EXPOSED.
If swallowed will cause headache, nausea, dizziness, vomiting, or loss of consciousness.

Remove contaminated clothing and shoes.
Flush affected areas with plenty of water.
IF IN EYES, hold eyelids open and flush with plenty of water.
IF SWALLOWED and victim is CONSCIOUS, have victim drink water or milk and have victim induce vomiting.
IF SWALLOWED and victim is UNCONSCIOUS OR HAVING CONVULSIONS, do nothing except keep victim warm.

Water Pollution

HARMFUL TO AQUATIC LIFE IN VERY LOW CONCENTRATIONS.
May be dangerous if it enters water intakes.

Notify local health and wildlife officials.
Notify operators of nearby water intakes.

1. RESPONSE TO DISCHARGE

(See Response Methods Handbook, CG 446-4)

Issue warning—water contaminant
Restrict access
Should be removed
Chemical and physical treatment

2. LABELS

No hazard label required by Code of Federal Regulations

3. CHEMICAL DESIGNATIONS

3.1 **Synonyms:** HEOD; *endo, exo*-1,2,3,4,10,10-Hexachloro-6,7-expoxy-1,4,4a,5,6,7,8,8a-octahydro-1,4:5,8-dimethanonaphthalene
3.2 **Coast Guard Compatibility Classification:** Not listed
3.3 **Chemical Formula:** $C_{12}H_8Cl_6O$
3.4 **IMCO/United Nations Numerical Designation:** Not listed

4. OBSERVABLE CHARACTERISTICS

4.1 **Physical State (as shipped):** Solid
4.2 **Color:** Buff to light brown.
4.3 **Odor:** Mild chemical

5. HEALTH HAZARDS

5.1 **Personal Protective Equipment:** U. S. Bu. Mines approved respirator; clean rubber gloves; goggles or face shield
5.2 **Symptoms Following Exposure:** Inhalation, ingestion, or skin contact causes irritability, convulsions and/or coma, nausea, vomiting, headache, fainting, tremors. Contact with eyes causes irritation.
5.3 **Treatment for Exposure:** INHALATION: move to fresh air; give oxygen and artificial respiration as required. INGESTION: induce vomiting and get medical attention. EYES: flush with plenty of water; get medical attention. SKIN: flush with plenty of water.
5.4 **Toxicity by Inhalation (Threshold Limit Value):** 0.25 mg/m³
5.5 **Short-Term Inhalation Limits:** 1 mg/m³ for 30 min.
5.6 **Toxicity by Ingestion:** Grade 4; oral LD_{50} = 46 mg/kg (rat), 65 mg/kg (dog)
5.7 **Late Toxicity:** Banned by EPA in October 1974 because of alleged "imminent hazard to human health" as a potential carcinogen in man.
5.8 **Vapor (Gas) Irritant Characteristics:** Data not available
5.9 **Liquid or Solid Irritant Characteristics:** Minimum hazard. If spilled on clothing and allowed to remain, may cause smarting and reddening of the skin.
5.10 **Odor Threshold:** 0.041 ppm

6. FIRE HAZARDS

6.1 **Flash Point:** Not flammable
6.2 **Flammable Limits in Air:** Not flammable
6.3 **Fire Extinguishing Agents:** Not pertinent
6.4 **Fire Extinguishing Agents Not to be Used:**
6.5 **Special Hazards of Combustion Products:** Toxic and irritating hydrogen chloride fumes may form in fire.
6.6 **Behavior in Fire:**
6.7 **Ignition Temperature:** Not pertinent
6.8 **Electrical Hazard:** Not pertinent
6.9 **Burning Rate:** Not pertinent

7. CHEMICAL REACTIVITY

7.1 **Reactivity with Water:** No reaction
7.2 **Reactivity with Common Materials:**
7.3 **Stability During Transport:** Stable
7.4 **Neutralizing Agents for Acids and Caustics:** Not pertinent
7.5 **Polymerization:** Not pertinent
7.6 **Inhibitor of Polymerization:** Not pertinent

8. WATER POLLUTION

8.1 **Aquatic Toxicity:**
0.0079 mg/l /96 hr/bluegill/TL_m/fresh water
.037 ppm/96 hr/goldfish/TL_m/fresh water
0.050 ppm/5 hr/mullet/100% kill/salt water
0.025–.050 ppm/48 hr/brown shrimp/TL_m/salt water
8.2 **Waterfowl Toxicity:** LD_{50} 381.0 mg/kg
8.3 **Biological Oxygen Demand (BOD):** Data not available
8.4 **Food Chain Concentration Potential:** High

9. SELECTED MANUFACTURERS

Shell Chemical Co.
Agricultural Div.
2401 Crow Canyon Rd.
San Ramon, Calif. 94583

10. SHIPPING INFORMATION

10.1 **Grades or Purity:** Technical, 85+% HEOD; 18% emulsifiable concentrates in petroleum hydrocarbons, which are combustible.
10.2 **Storage Temperature:** Ambient
10.3 **Inert Atmosphere:** No requirement
10.4 **Venting:** Open (flame arrester) (for liquid form)

11. HAZARD ASSESSMENT CODE

(See Hazard Assessment Handbook, CG 446-3)

II

12. HAZARD CLASSIFICATIONS

12.1 **Code of Federal Regulations:** ORM–A
12.2 **NAS Hazard Rating for Bulk Water Transportation:** Not listed
12.3 **NFPA Hazard Classifications:** Not listed

13. PHYSICAL AND CHEMICAL PROPERTIES

13.1 **Physical State at 15°C and 1 atm:** Solid
13.2 **Molecular Weight:** 380.93
13.3 **Boiling Point at 1 atm:** Not pertinent (decomposes)
13.4 **Freezing Point:** 349°F = 176°C = 449°K
13.5 **Critical Temperature:** Not pertinent
13.6 **Critical Pressure:** Not pertinent
13.7 **Specific Gravity:** 1.75 at 20°C (solid)
13.8 **Liquid Surface Tension:** Not pertinent
13.9 **Liquid-Water Interfacial Tension:** Not pertinent
13.10 **Vapor (Gas) Specific Gravity:** Not pertinent
13.11 **Ratio of Specific Heats of Vapor (Gas):** Not pertinent
13.12 **Latent Heat of Vaporization:** Not pertinent
13.13 **Heat of Combustion:** Data not available
13.14 **Heat of Decomposition:** Not pertinent
13.15 **Heat of Solution:** Not pertinent
13.16 **Heat of Polymerization:** Not pertinent

NOTES

DEA	DIETHANOLAMINE

Common Synonyms			
2,2'-Iminodiethanol Bis (2-hydroxyethyl) amine	Oily liquid or solid crystals	Liquid is colorless; Solid is white	Slight dead fish or ammonia odor
	Sinks and mixes with water.		

Avoid contact with liquid. Keep people away.
Wear goggles, self-contained breathing apparatus, and rubber overclothing (including gloves).
Stop discharge if possible.
Call fire department.
Isolate and remove discharged material.
Notify local health and pollution control agencies.

Fire

Combustible.
Wear goggles, self-contained breathing apparatus, and rubber overclothing (including gloves).
Extinguish with dry chemical, alcohol foam, or carbon dioxide.
Cool exposed containers with water.

Exposure

CALL FOR MEDICAL AID.

LIQUID
Irritating to skin and eyes.
Remove contaminated clothing and shoes.
Flush affected areas with plenty of water.
IF IN EYES, hold eyelids open and flush with plenty of water.
IF SWALLOWED and victim is CONSCIOUS, have victim drink water or milk and have victim induce vomiting.
IF SWALLOWED and victim is UNCONSCIOUS OR HAVING CONVULSIONS, do nothing except keep victim warm.

Water Pollution

Dangerous to aquatic life in high concentrations.
May be dangerous if it enters water intakes.
Notify local health and wildlife officials.
Notify operators of nearby water intakes.

1. RESPONSE TO DISCHARGE

(See Response Methods Handbook, CG 446-4)

Disperse and flush

2. LABELS

No hazard label required by Code of Federal Regulations

3. CHEMICAL DESIGNATIONS

3.1 **Synonyms:** Bis(2-hydroxyethyl) amine
DEA
2, 2'-Dihydroxydiethyl amine
Di(2-hydroxyethyl) amine
2, 2'-Iminodiethanol

3.2 **Coast Guard Compatibility Classification:** Alkanolamine

3.3 **Chemical Formula:** $(HOCH_2CH_2)_2NH$

3.4 **IMCO/United Nations Numerical Designation:** Not listed

4. OBSERVABLE CHARACTERISTICS

4.1 **Physical State (as shipped):** Liquid

4.2 **Color:** Colorless

4.3 **Odor:** Mild ammoniacal; faint, fishy, characteristic

5. HEALTH HAZARDS

5.1 **Personal Protective Equipment:** Full face mask or amine vapor mask only, if required; clean body-covering clothing, chemical goggles.

5.2 **Symptoms Following Exposure:** Irritation of eyes and skin. Breathing vapors may cause coughing, a smothering sensation, nausea, headache.

5.3 **Treatment for Exposure:** INHALATION: no problem likely. Get medical attention if ill effects develop. INGESTION: induce vomiting if large amounts are swallowed and call a physician. Treat symptomatically. No known antidote. EYES: flush with plenty of water for at least 15 min. and get medical attention promptly. SKIN: flush with plenty of water. Wash contaminated clothing before reuse.

5.4 **Toxicity by Inhalation (Threshold Limit Value):** Not pertinent

5.5 **Short-Term Inhalation Limits:** Not pertinent

5.6 **Toxicity by Ingestion:** Grade 2; LD_{50} 0.5 to 5 g/kg (rat)

5.7 **Late Toxicity:** Data not available

5.8 **Vapor (Gas) Irritant Characteristics:** Vapors cause moderate irritation such that personnel will find high concentrations unpleasant. The effect is temporary.

5.9 **Liquid or Solid Irritant Characteristics:** Causes smarting of the skin and first-degree burns on short exposure and may cause secondary burns on long exposure.

5.10 **Odor Threshold:** Data not available

6. FIRE HAZARDS

6.1 **Flash Point:** 305°F O.C.

6.2 **Flammable Limits in Air:** 1.6% (calc.) 9.8% (est.)

6.3 **Fire Extinguishing Agents:** Water, alcohol foam, carbon dioxide, dry chemical

6.4 **Fire Extinguishing Agents Not to be Used:** Addition of water may cause frothing.

6.5 **Special Hazards of Combustion Products:** Irritating vapors are generated when heated.

6.6 **Behavior in Fire:** Not pertinent

6.7 **Ignition Temperature:** 1224°F

6.8 **Electrical Hazard:** Not pertinent

6.9 **Burning Rate:** 0.74 mm/min.

7. CHEMICAL REACTIVITY

7.1 **Reactivity with Water:** No reaction

7.2 **Reactivity with Common Materials:** No reaction

7.3 **Stability During Transport:** Stable

7.4 **Neutralizing Agents for Acids and Caustics:** Flush with water.

7.5 **Polymerization:** Not pertinent

7.6 **Inhibitor of Polymerization:** Not pertinent

8. WATER POLLUTION

8.1 **Aquatic Toxicity:** 2100 ppm/24 hr/sunfish, bluegill/TL_m/fresh water

8.2 **Waterfowl Toxicity:** Data not available

8.3 **Biological Oxygen Demand (BOD):** 10% (theor.), 5 days

8.4 **Food Chain Concentration Potential:** None

9. SELECTED MANUFACTURERS

1. Jefferson Chemical Co., Inc.
3336 Richmond Ave.
Houston, Texas 77052

2. Olin Corp.
Chemicals Division
120 Long Ridge Rd.
Stamford, Conn. 06904

3. Union Carbide Corp.
Chemicals and Plastics Division
270 Park Ave.
New York, N. Y. 10017

10. SHIPPING INFORMATION

10.1 **Grades or Purity:** 85-99.5%

10.2 **Storage Temperature:** Ambient

10.3 **Inert Atmosphere:** No requirement

10.4 **Venting:** Open

11. HAZARD ASSESSMENT CODE

(See Hazard Assessment Handbook, CG 446-3)

A-P-Q

12. HAZARD CLASSIFICATIONS

12.1 **Code of Federal Regulations:** Not listed

12.2 **NAS Hazard Rating for Bulk Water Transportation:**

Category	Rating
Fire	1
Health	
Vapor Irritant	2
Liquid or Solid Irritant	2
Poisons	2
Water Pollution	
Human Toxicity	1
Aquatic Toxicity	1
Aesthetic Effect	2
Reactivity	
Other Chemicals	3
Water	0
Self-Reaction	0

12.3 **NFPA Hazard Classifications:** Not listed

13. PHYSICAL AND CHEMICAL PROPERTIES

13.1 **Physical State at 15°C and 1 atm:** Solid

13.2 **Molecular Weight:** 105.14

13.3 **Boiling Point at 1 atm:** 515.1°F = 268.4°C = 541.6°K

13.4 **Freezing Point:** 82°F = 28°C = 301°K

13.5 **Critical Temperature:** 828°F = 442°C = 715°K

13.6 **Critical Pressure:** 470 psia = 32 atm = 3.2 MN/m^2

13.7 **Specific Gravity:** 1.095 at 28°C (liquid)

13.8 **Liquid Surface Tension:** Not pertinent

13.9 **Liquid-Water Interfacial Tension:** Not pertinent

13.10 **Vapor (Gas) Specific Gravity:** Not pertinent

13.11 **Ratio of Specific Heats of Vapor (Gas):** 1.053

13.12 **Latent Heat of Vaporization:** 266 Btu/lb = 148 cal/g = 6.20×10^5 J/kg

13.13 **Heat of Combustion:** -10,790 Btu/lb = -6000 cal/g = -251×10^5 J/kg

13.14 **Heat of Decomposition:** Not pertinent

13.15 **Heat of Solution:** (est.) −13 Btu/lb = −7 cal/g = -0.3×10^5 J/kg

13.16 **Heat of Polymerization:** Not pertinent

NOTES

REVISED 1978

DEN	DIETHYLAMINE

Common Synonyms			
DEN	Watery liquid	Colorless	Fishy, ammonia odor
	Floats and mixes with water. Flammable, irritating vapor is produced.		

Avoid contact with liquid and vapor. Keep people away.
Wear goggles, self-contained breathing apparatus, and rubber overclothing (including gloves).
Shut off ignition sources and call fire department.
Stop discharge if possible.
Stay upwind and use water spray to "knock down" vapor.
Isolate and remove discharged material.
Notify local health and pollution control agencies.

Fire

FLAMMABLE.
Flashback along vapor trail may occur.
Irritating vapors are produced when heated.
Vapor may explode if ignited in an enclosed area.
Wear goggles, self-contained breathing apparatus, and rubber overclothing (including gloves).
Extinguish with dry chemical, alcohol foam or carbon dioxide.
Cool exposed containers with water.

Exposure

CALL FOR MEDICAL AID.

VAPOR
Irritating to eyes, nose and throat.
Harmful if inhaled.
Move to fresh air.
If breathing has stopped, give artificial respiration.
If breathing is difficult, give oxygen.

LIQUID
Will burn eyes.
Harmful if swallowed.
Remove contaminated clothing and shoes.
Flush affected areas with plenty of water.
IF IN EYES, hold eyelids open and flush with plenty of water.
IF SWALLOWED and victim is CONSCIOUS, have victim drink water or milk.

Water Pollution

Harmful to aquatic life in very low concentrations.
May be dangerous if it enters water intakes.
Notify local health and wildlife officials.
Notify operators of nearby water intakes.

1. RESPONSE TO DISCHARGE

(See Response Methods Handbook, CG 446-4)

Issue warning—high flammability
Evacuate area
Disperse and flush

2. LABEL

Red

3. CHEMICAL DESIGNATIONS

3.1 **Synonyms:** DEN
3.2 **Coast Guard Compatibility Classification:** Aliphatic amine
3.3 **Chemical Formula:** $(CH_3CH_2)_2NH$
3.4 **IMCO/United Nations Numerical Designation:** 3.1/1154

4. OBSERVABLE CHARACTERISTICS

4.1 **Physical State (as shipped):** Liquid
4.2 **Color:** Colorless
4.3 **Odor:** Ammoniacal; sharp, fishy

5. HEALTH HAZARDS

5.1 **Personal Protective Equipment:** Chemical safety goggles, rubber gloves, and apron.
5.2 **Symptoms Following Exposure:** Irritation and burning of eyes, skin, and respiratory system. High concentration of vapor can cause asphyxiation.
5.3 **Treatment for Exposure:** In case of contact, flush skin or eyes with plenty of water for at least 15 min.; for eyes, get medical attention.
5.4 **Toxicity by Inhalation (Threshold Limit Value):** 25 ppm
5.5 **Short-Term Inhalation Limits:** 100 ppm for 30 min.
5.6 **Toxicity by Ingestion:** Grade 2; 0.5 to 5 g/kg (rat)
5.7 **Late Toxicity:** None
5.8 **Vapor (Gas) Irritant Characteristics:** Vapor is moderately irritating such that personnel will not usually tolerate moderate or high vapor concentrations.
5.9 **Liquid or Solid Irritant Characteristics:** Minimum hazard. If spilled on clothing and allowed to remain, may cause smarting and reddening of the skin.
5.10 **Odor Threshold:** 0.14 ppm

6. FIRE HAZARDS

6.1 **Flash Point:** 5°F O.C.
6.2 **Flammable Limits in Air:** 1.8%–10.1%
6.3 **Fire Extinguishing Agents:** Dry chemical, carbon dioxide, or alcohol foam
6.4 **Fire Extinguishing Agents Not to be Used:** Data not available
6.5 **Special Hazards of Combustion Products:** Vapors are irritating
6.6 **Behavior in Fire:** Vapors are heavier than air and may travel considerable distance to a source of ignition and flash back.
6.7 **Ignition Temperature:** 594°F
6.8 **Electrical Hazard:** Data not available
6.9 **Burning Rate:** 6.7 mm/min.

7. CHEMICAL REACTIVITY

7.1 **Reactivity with Water:** No reaction
7.2 **Reactivity with Common Materials:** No hazardous reaction
7.3 **Stability During Transport:** Stable
7.4 **Neutralizing Agents for Acids and Caustics:** Flush with water
7.5 **Polymerization:** Not pertinent
7.6 **Inhibitor of Polymerization:** Not pertinent

8. WATER POLLUTION

8.1 **Aquatic Toxicity:** 85 mg/l /48 hr/creek chub/TL_m/fresh water
8.2 **Waterfowl Toxicity:** Data not available
8.3 **Biological Oxygen Demand (BOD):** Data not available
8.4 **Food Chain Concentration Potential:** None

9. SELECTED MANUFACTURERS

1. Air Products and Chemicals, Inc.
   Allentown, Pa. 18105
2. E. I. duPont de Nemours & Co., Inc.
   Industrial and Biochemicals Dept.
   Wilmington, Del. 19898
3. Pennwalt Corp.
   Chemical Division
   4655 Biddle Ave.
   Wyandotte, Mich. 48193

10. SHIPPING INFORMATION

10.1 **Grades or Purity:** Technical: 99%
10.2 **Storage Temperature:** Data not available
10.3 **Inert Atmosphere:** Data not available
10.4 **Venting:** Data not available

11. HAZARD ASSESSMENT CODE

(See Hazard Assessment Handbook, CG 446-3)

A-P-Q-R-S

12. HAZARD CLASSIFICATIONS

12.1 **Code of Federal Regulations:** Flammable liquid
12.2 **NAS Hazard Rating for Bulk Water Transportation:**

Category	Rating
Fire	3
Health	
Vapor Irritant	3
Liquid or Solid Irritant	1
Poisons	2
Water Pollution	
Human Toxicity	2
Aquatic Toxicity	3
Aesthetic Effect	1
Reactivity	
Other Chemicals	3
Water	0
Self-Reaction	0

12.3 **NFPA Hazard Classifications:** Not listed

13. PHYSICAL AND CHEMICAL PROPERTIES

13.1 **Physical State at 15°C and 1 atm:** Liquid
13.2 **Molecular Weight:** 73.14
13.3 **Boiling Point at 1 atm:** 132°F = 55.5°C = 328.7°K
13.4 **Freezing Point:** −57.6°F = −49.8°C = 223.4°K
13.5 **Critical Temperature:** 434.3°F = 223.5°C = 496.7°K
13.6 **Critical Pressure:** 538 psia = 36.6 atm = 3.71 MN/m^2
13.7 **Specific Gravity:** 0.708 at 20°C (liquid)
13.8 **Liquid Surface Tension:** 20.05 dynes/cm = 0.02005 N/m at 20°C
13.9 **Liquid-Water Interfacial Tension:** Not pertinent
13.10 **Vapor (Gas) Specific Gravity:** 2.5
13.11 **Ratio of Specific Heats of Vapor (Gas):** 1.079
13.12 **Latent Heat of Vaporization:** 170 Btu/lb = 93 cal/g = 3.9×10^5 J/kg
13.13 **Heat of Combustion:** −17,990 Btu/lb = −9994 cal/g = -418.4×10^5 J/kg
13.14 **Heat of Decomposition:** Not pertinent
13.15 **Heat of Solution:** −202 Btu/lb = −112 cal/g = -4.69×10^5 J/kg
13.16 **Heat of Polymerization:** Not pertinent

NOTES

REVISED 1978

DEB — DIETHYLBENZENE

Common Synonyms	Liquid	Colorless	Sweet, gasoline-like odor
	Floats on water.		

Stop discharge if possible. Keep people away.
Call fire department.
Avoid contact with liquid.
Isolate and remove discharged material.
Notify local health and pollution control agencies.

Fire	Combustible. Extinguish with water, dry chemical, foam, or carbon dioxide. Cool exposed containers with water.
Exposure	CALL FOR MEDICAL AID. LIQUID Irritating to skin and eyes. Remove contaminated clothing and shoes. Flush affected areas with plenty of water. IF IN EYES, hold eyelids open and flush with plenty of water.
Water Pollution	Effect of low concentrations on aquatic life is unknown. Fouling to shoreline. May be dangerous if it enters water intakes. Notify local health and wildlife officials. Notify operators of nearby water intakes.

1. RESPONSE TO DISCHARGE

(See Response Methods Handbook, CG 446-4)

Mechanical containment
Should be removed
Chemical and physical treatment

2. LABELS

No hazard label required by Code of Federal Regulations

3. CHEMICAL DESIGNATIONS

3.1 **Synonyms:** No common synonyms
3.2 **Coast Guard Compatibility Classification:** Aromatic hydrocarbon
3.3 **Chemical Formula:** $C_6H_4(C_2H_5)_2$
3.4 **IMCO/United Nations Numerical Designation:** 3.3/2049

4. OBSERVABLE CHARACTERISTICS

4.1 **Physical State (as shipped):** Liquid
4.2 **Color:** Colorless
4.3 **Odor:** Characteristic aromatic; like benzene; like toluene

5. HEALTH HAZARDS

5.1 **Personal Protective Equipment:** Self-contained breathing apparatus, safety goggles
5.2 **Symptoms Following Exposure:** High vapor concentrations produce eye and respiratory tract irritation, dizziness, depression. Liquid irritates and may blister skin; can cause corneal injury to eye.
5.3 **Treatment for Exposure:** INHALATION: remove to fresh air and start artificial respiration. INGESTION: do NOT induce vomiting; call a doctor. CONTACT WITH EYES AND SKIN: flush with water for 15 min. Wash skin with soap and water.
5.4 **Toxicity by Inhalation (Threshold Limit Value):** Data not available
5.5 **Short-Term Inhalation Limits:** Data not available
5.6 **Toxicity by Ingestion:** Grade 2; oral rat LD_{50} = 1.2 g/kg
5.7 **Late Toxicity:** Data not available
5.8 **Vapor (Gas) Irritant Characteristics:** Vapors cause a slight smarting of the eyes and respiratory system if present in high concentrations. The effect is temporary.
5.9 **Liquid or Solid Irritant Characteristics:** Minimum hazard. If spilled on clothing and allowed to remain, may cause smarting and reddening of the skin.
5.10 **Odor Threshold:** Data not available

6. FIRE HAZARDS

6.1 **Flash Point:** 135°F C.C.
6.2 **Flammable Limits in Air:** Data not available
6.3 **Fire Extinguishing Agents:** Foam, water, carbon dioxide, or dry chemical
6.4 **Fire Extinguishing Agents Not to be Used:** Not pertinent
6.5 **Special Hazards of Combustion Products:** Not pertinent
6.6 **Behavior in Fire:** Not pertinent
6.7 **Ignition Temperature:** 743°F (ortho)
6.8 **Electrical Hazard:** Not pertinent
6.9 **Burning Rate:** Data not available

7. CHEMICAL REACTIVITY

7.1 **Reactivity with Water:** No reaction
7.2 **Reactivity with Common Materials:** No reaction
7.3 **Stability During Transport:** Stable
7.4 **Neutralizing Agents for Acids and Caustics:** Not pertinent
7.5 **Polymerization:** Not pertinent
7.6 **Inhibitor of Polymerization:** Not pertinent

8. WATER POLLUTION

8.1 **Aquatic Toxicity:** Data not available
8.2 **Waterfowl Toxicity:** Data not available
8.3 **Biological Oxygen Demand (BOD):** Data not available
8.4 **Food Chain Concentration Potential:** None

9. SELECTED MANUFACTURERS

1. American Petrofina Inc. Cosden Oil & Chemical Co. Division Big Spring, Texas 79720
2. Dow Chemical Co. Midland, Mich. 48640
3. Sinclair-Koppers Co. Koppers Bldg. Pittsburgh, Pa. 15219

10. SHIPPING INFORMATION

10.1 **Grades or Purity:** Technical (mixture of isomers)
10.2 **Storage Temperature:** Ambient
10.3 **Inert Atmosphere:** No requirement
10.4 **Venting:** Open (flame arrester)

11. HAZARD ASSESSMENT CODE

(See Hazard Assessment Handbook, CG 446-3)

A-T-U

12. HAZARD CLASSIFICATIONS

12.1 **Code of Federal Regulations:** Combustible Liquid
12.2 **NAS Hazard Rating for Bulk Water Transportation:**

Category	Rating
Fire	2
Health	
Vapor Irritant	1
Liquid or Solid Irritant	1
Poisons	1
Water Pollution	
Human Toxicity	1
Aquatic Toxicity	3
Aesthetic Effect	2
Reactivity	
Other Chemicals	1
Water	0
Self-Reaction	0

12.3 **NFPA Hazard Classifications:** Not listed

13. PHYSICAL AND CHEMICAL PROPERTIES

13.1 **Physical State at 15°C and 1 atm:** Liquid
13.2 **Molecular Weight:** 134.21
13.3 **Boiling Point at 1 atm:** 356°F = 180°C = 453°K
13.4 **Freezing Point:** <160°F = <70°C = <343°K
13.5 **Critical Temperature:** Not pertinent
13.6 **Critical Pressure:** Not pertinent
13.7 **Specific Gravity:** 0.86 at 20°C (liquid)
13.8 **Liquid Surface Tension:** 30 dynes/cm = 0.030 N/m at 20°C
13.9 **Liquid-Water Interfacial Tension:** Data not available
13.10 **Vapor (Gas) Specific Gravity:** Not pertinent
13.11 **Ratio of Specific Heats of Vapor (Gas):** Not pertinent
13.12 **Latent Heat of Vaporization:** 140 Btu/lb = 77 cal/g = 3.2×10^5 J/kg
13.13 **Heat of Combustion:** -17,800 Btu/lb = -9890 cal/g = -414×10^5 J/kg
13.14 **Heat of Decomposition:** Not pertinent
13.15 **Heat of Solution:** Not pertinent
13.16 **Heat of Polymerization:** Not pertinent

NOTES

REVISED 1978

DEC	# DIETHYL CARBONATE

Common Synonyms			
Ethyl carbonate Carbonic acid, diethyl ester	Watery liquid	Colorless	Pleasant odor
	Floats on water. Flammable, irritating vapor is produced.		

Shut off ignition sources and call fire department.
Stop discharge if possible. Keep people away.
Avoid contact with liquid and vapor.
Stay upwind and use water spray to "knock down" vapor.
Isolate and remove discharged material.
Notify local health and pollution control agencies.

Fire	FLAMMABLE. Flashback along vapor trail may occur. Vapor may explode if ignited in an enclosed area. **Extinguish with dry chemical, foam, or carbon dioxide.** **Water may be ineffective on fire.** **Cool exposed containers with water.**
Exposure	CALL FOR MEDICAL AID. VAPORS Irritating to eyes, nose and throat. If inhaled will cause headache, dizziness, nausea, or loss of consciousness. **Move to fresh air.** **If breathing has stopped, give artificial respiration.** **If breathing is difficult, give oxygen.** LIQUID Irritating to skin and eyes. Harmful if swallowed. **Remove contaminated clothing and shoes.** **Flush affected areas with plenty of water.** **IF IN EYES, hold eyelids open and flush with plenty of water.** **IF SWALLOWED and victim is CONSCIOUS, have victim drink water or milk.**
Water Pollution	Effect of low concentrations on aquatic life is unknown. Fouling to shoreline. May be dangerous if it enters water intakes. **Notify local health and wildlife officials.** **Notify operators of nearby water intakes.**

1. RESPONSE TO DISCHARGE
(See Response Methods Handbook, CG 446-4)

Disperse and flush

2. LABEL

Red

3. CHEMICAL DESIGNATIONS

3.1 **Synonyms:** Carbonic acid diethyl ester; Ethyl carbonate; Eufin

3.2 **Coast Guard Compatibility Classification:** Ester

3.3 **Chemical Formula:** $(CH_3CH_2)_2CO_3$

3.4 **IMCO/United Nations Numerical Designation:** Not listed

4. OBSERVABLE CHARACTERISTICS

4.1 **Physical State (as shipped):** Liquid

4.2 **Color:** Colorless

4.3 **Odor:** Pleasant, ethereal; mild and nonresidual

5. HEALTH HAZARDS

5.1 **Personal Protective Equipment:** Protective clothing; rubber gloves and goggles; organic vapor canister or air mask.

5.2 **Symptoms Following Exposure:** High vapor concentrations can cause headache, irritation of eyes and respiratory tract, dizziness, nausea, weakness, loss of consciousness.

5.3 **Treatment for Exposure:** INHALATION: remove from exposure; administer artificial respiration and oxygen if needed. EYES: flush with water for at least 15 min.

5.4 **Toxicity by Inhalation (Threshold Limit Value):** Data not available

5.5 **Short-Term Inhalation Limits:** Data not available

5.6 **Toxicity by Ingestion:** Data not available

5.7 **Late Toxicity:** None

5.8 **Vapor (Gas) Irritant Characteristics:** Vapors may cause slight smarting of eyes.

5.9 **Liquid or Solid Irritant Characteristics:** Minimum hazard.

5.10 **Odor Threshold:** Data not available

6. FIRE HAZARDS

6.1 **Flash Point:** 115°F O.C.; 77°F C.C.

6.2 **Flammable Limits in Air:** Data not available

6.3 **Fire Extinguishing Agents:** Foam, carbon dioxide, dry chemical

6.4 **Fire Extinguishing Agents Not to be Used:** Water

6.5 **Special Hazards of Combustion Products:** Not pertinent

6.6 **Behavior in Fire:** Not pertinent

6.7 **Ignition Temperature:** Data not available

6.8 **Electrical Hazard:** Not pertinent

6.9 **Burning Rate:** 3.4 mm/min.

7. CHEMICAL REACTIVITY

7.1 **Reactivity with Water:** Too slow to be hazardous

7.2 **Reactivity with Common Materials:** No reaction

7.3 **Stability During Transport:** Stable

7.4 **Neutralizing Agents for Acids and Caustics:** Not pertinent

7.5 **Polymerization:** Not pertinent

7.6 **Inhibitor of Polymerization:** Not pertinent

8. WATER POLLUTION

8.1 **Aquatic Toxicity:** Data not available

8.2 **Waterfowl Toxicity:** Data not available

8.3 **Biological Oxygen Demand (BOD):** Data not available

8.4 **Food Chain Concentration Potential:** None

9. SELECTED MANUFACTURERS

1. Chemetron Corp.
   Organic Chemicals Division
   La Porte, Texas 77571
2. FMC Corp.
   Organic Chemicals Division
   633 Third Ave.
   New York, N. Y. 10012

10. SHIPPING INFORMATION

10.1 **Grades or Purity:** 99-100%

10.2 **Storage Temperature:** Data not available

10.3 **Inert Atmosphere:** Data not available

10.4 **Venting:** Data not available

11. HAZARD ASSESSMENT CODE
(See Hazard Assessment Handbook, CG 446-3)

A-T-U

12. HAZARD CLASSIFICATIONS

12.1 **Code of Federal Regulations:** Flammable Liquid

12.2 **NAS Hazard Rating for Bulk Water Transportation:** Not listed

12.3 **NFPA Hazard Classifications:** Not listed

13. PHYSICAL AND CHEMICAL PROPERTIES

13.1 **Physical State at 15°C and 1 atm:** Liquid

13.2 **Molecular Weight:** 118.13

13.3 **Boiling Point at 1 atm:** 260.2°F = 126.8°C = 400.0°K

13.4 **Freezing Point:** −45°F = −43°C = 230°K

13.5 **Critical Temperature:** Not pertinent

13.6 **Critical Pressure:** Not pertinent

13.7 **Specific Gravity:** 0.975 at 20°C (liquid)

13.8 **Liquid Surface Tension:** 26.3 dynes/cm = 0.0263 N/m at 20°C

13.9 **Liquid-Water Interfacial Tension:** 12.86 dynes/cm = 0.01286 N/m at 20°C

13.10 **Vapor (Gas) Specific Gravity:** Not pertinent

13.11 **Ratio of Specific Heats of Vapor (Gas):** (est.) 1.110

13.12 **Latent Heat of Vaporization:** 130 Btu/lb = 73 cal/g = 3.1×10^5 J/kg

13.13 **Heat of Combustion:** -9760 Btu/lb = -5420 cal/g = -227×10^5 J/kg

13.14 **Heat of Decomposition:** Not pertinent

13.15 **Heat of Solution:** Not pertinent

13.16 **Heat of Polymerization:** Not pertinent

NOTES

REVISED 1978

DEG — DIETHYLENE GLYCOL

Common Synonyms			
DEG 2,2' Oxybisethanol Diglycol β,β'-Dihydroxydiethyl ether	Oily liquid	Colorless	Odorless
	Sinks and mixes with water.		

Stop discharge if possible.
Call fire department.
Isolate and remove discharged material.
Notify local health and pollution control agencies.

Fire	Combustible. Extinguish with dry chemical, alcohol foam, or carbon dioxide. Water and foam may be ineffective on fire. Cool exposed containers with water.
Exposure	Not harmful.
Water Pollution	Dangerous to aquatic life in high concentrations. May be dangerous if it enters water intakes. Notify local health and wildlife officials. Notify operators of nearby water intakes.

1. RESPONSE TO DISCHARGE

(See Response Methods Handbook, CG 446-4)

Disperse and flush

2. LABELS

No hazard label required by Code of Federal Regulations

3. CHEMICAL DESIGNATIONS

3.1 **Synonyms:** Bis(2-hydroxyethyl) ether
Diglycol
β,β'-Dihydroxydiethyl ether
3-Oxa-1, 5-pentanediol
2, 2'-Oxybisethanol

3.2 **Coast Guard Compatibility Classification:** Glycol

3.3 **Chemical Formula:** $(HOCH_2CH_2)_2O$

3.4 **IMCO/United Nations Numerical Designation:** Not listed

4. OBSERVABLE CHARACTERISTICS

4.1 **Physical State (as shipped):** Liquid

4.2 **Color:** Colorless

4.3 **Odor:** Practically odorless

5. HEALTH HAZARDS

5.1 **Personal Protective Equipment:** Full face mask with canister for short exposures to high vapor levels; rubber gloves; goggles.

5.2 **Symptoms Following Exposure:** Ingestion of large amounts may cause degeneration of kidney and liver and cause death. Liquid may cause slight skin irritation.

5.3 **Treatment for Exposure:** INHALATION: no problem likely. If any ill effects do develop, get medical attention. INGESTION: induce vomiting if ingested. No known antidote; treat symptomatically. EYE AND SKIN: flush with water. If any ill effects occur, get medical attention.

5.4 **Toxicity by Inhalation (Threshold Limit Value):** 100 ppm (suggested)

5.5 **Short-Term Inhalation Limits:** Not pertinent

5.6 **Toxicity by Ingestion:** Grade 0; LD_{50} above 15 g/kg (rat)

5.7 **Late Toxicity:** Kidney and liver damage

5.8 **Vapor (Gas) Irritant Characteristics:** None

5.9 **Liquid or Solid Irritant Characteristics:** None

5.10 **Odor Threshold:** Not pertinent

6. FIRE HAZARDS

6.1 **Flash Point:** 255°F C.C.

6.2 **Flammable Limits in Air:** 1.6% 10.8%

6.3 **Fire Extinguishing Agents:** Alcohol foam, carbon dioxide, dry chemical

6.4 **Fire Extinguishing Agents Not to be Used:** Water or foam may cause frothing

6.5 **Special Hazards of Combustion Products:** Not pertinent

6.6 **Behavior in Fire:** Not pertinent

6.7 **Ignition Temperature:** 444°F

6.8 **Electrical Hazard:** Not pertinent

6.9 **Burning Rate:** 1.5 mm/min.

7. CHEMICAL REACTIVITY

7.1 **Reactivity with Water:** No reaction

7.2 **Reactivity with Common Materials:** No reaction

7.3 **Stability During Transport:** Stable

7.4 **Neutralizing Agents for Acids and Caustics:** Not pertinent

7.5 **Polymerization:** Not pertinent

7.6 **Inhibitor of Polymerization:** Not pertinent

8. WATER POLLUTION

8.1 **Aquatic Toxicity:** > 32,000 ppm/96 hr/mosquito fish/TL_m/fresh water

8.2 **Waterfowl Toxicity:** Data not available

8.3 **Biological Oxygen Demand (BOD):** 6%, 5 days

8.4 **Food Chain Concentration Potential:** None

9. SELECTED MANUFACTURERS

1. Dow Chemical Co.
Midland, Mich. 48640
2. Jefferson Chemical Co., Inc.
3336 Richmond Ave.
Houston, Texas 77052
3. Union Carbide Corp.
Chemicals and Plastics Division
270 Park Ave.
New York, N. Y. 10017

10. SHIPPING INFORMATION

10.1 **Grades or Purity:** Regular grade; polyester grade

10.2 **Storage Temperature:** Ambient

10.3 **Inert Atmosphere:** No requirement

10.4 **Venting:** Open (flame arrester)

11. HAZARD ASSESSMENT CODE

(See Hazard Assessment Handbook, CG 446-3)

A-P-Q

12. HAZARD CLASSIFICATIONS

12.1 **Code of Federal Regulations:** Not listed

12.2 **NAS Hazard Rating for Bulk Water Transportation:**

Category	Rating
Fire	1
Health	
Vapor Irritant	0
Liquid or Solid Irritant	0
Poisons	1
Water Pollution	
Human Toxicity	1
Aquatic Toxicity	1
Aesthetic Effect	0
Reactivity	
Other Chemicals	2
Water	0
Self-Reaction	0

12.3 **NFPA Hazard Classifications:**

Category	Classification
Health Hazard (Blue)	1
Flammability (Red)	1
Reactivity (Yellow)	0

13. PHYSICAL AND CHEMICAL PROPERTIES

13.1 **Physical State at 15°C and 1 atm:** Liquid

13.2 **Molecular Weight:** 106.12

13.3 **Boiling Point at 1 atm:** 473°F = 245°C = 518°K

13.4 **Freezing Point:** 20°F = −8°C = 265°K

13.5 **Critical Temperature:** 766°F = 408°C = 681°K

13.6 **Critical Pressure:** 680 psia = 46 atm = 4.7 MN/m^2

13.7 **Specific Gravity:** 1.118 at 20°C (liquid)

13.8 **Liquid Surface Tension:** Not pertinent

13.9 **Liquid-Water Interfacial Tension:** Not pertinent

13.10 **Vapor (Gas) Specific Gravity:** Not pertinent

13.11 **Ratio of Specific Heats of Vapor (Gas):** Not pertinent

13.12 **Latent Heat of Vaporization:** 270 Btu/lb = 150 cal/g = 6.28×10^5 J/kg

13.13 **Heat of Combustion:** −9617 Btu/lb = −5343 cal/g = -223.7×10^5 J/kg

13.14 **Heat of Decomposition:** Not pertinent

13.15 **Heat of Solution:** Not pertinent

13.16 **Heat of Polymerization:** Not pertinent

(Continued on pages 5 and 6)

NOTES

DGD	DIETHYLENE GLYCOL DIMETHYL ETHER

Common Synonyms
Poly Solv
Bis(2-methoxyethyl)-ether
Diglyme

Watery liquid — Colorless — Pleasant odor

Floats and mixes with water.

Stop discharge if possible.
Call fire department.
Isolate and remove discharged material.
Notify local health and pollution control agencies.

Fire

Combustible.
Extinguish with dry chemical, foam, or carbon dioxide.
Cool exposed containers with water.

Exposure

CALL FOR MEDICAL AID.

LIQUID
If swallowed, will cause nausea, vomiting, or loss of consciousness.
IF SWALLOWED and victim is CONSCIOUS, have victim drink water or milk.

Water Pollution

Effect of low concentrations on aquatic life is unknown.
May be dangerous if it enters water intakes.
Notify local health and wildlife officials.
Notify operators of nearby water intakes.

1. RESPONSE TO DISCHARGE

(See Response Methods Handbook, CG 446-4)

Disperse and flush

2. LABELS

No hazard label required by Code of Federal Regulations

3. CHEMICAL DESIGNATIONS

3.1 **Synonyms:** Bis(2-methoxyethyl) ether; Diglyme; Poly-Solv

3.2 **Coast Guard Compatibility Classification:** Alcohol, glycol, glycol ether

3.3 **Chemical Formula:** $(CH_3OCH_2CH_2)_2O$

3.4 **IMCO/United Nations Numerical Designation:** Not listed

4. OBSERVABLE CHARACTERISTICS

4.1 **Physical State (as shipped):** Liquid

4.2 **Color:** Colorless

4.3 **Odor:** Mild ethereal

5. HEALTH HAZARDS

5.1 **Personal Protective Equipment:** Vinyl (not rubber) gloves; safety goggles.

5.2 **Symptoms Following Exposure:** INGESTION (severe cases): nausea, vomiting, abdominal cramps, weakness progressing to coma.

5.3 **Treatment for Exposure:** INGESTION: give water and induce vomiting; oxygen and artificial respiration as needed.

5.4 **Toxicity by Inhalation (Threshold Limit Value):** Not pertinent

5.5 **Short-Term Inhalation Limits:** Data not available

5.6 **Toxicity by Ingestion:** Data not available

5.7 **Late Toxicity:** Data not available

5.8 **Vapor (Gas) Irritant Characteristics:** None

5.9 **Liquid or Solid Irritant Characteristics:** None

5.10 **Odor Threshold:** Data not available

6. FIRE HAZARDS

6.1 **Flash Point:** 158°F O.C.

6.2 **Flammable Limits in Air:** Data not available

6.3 **Fire Extinguishing Agents:** Dry chemical, foam, carbon dioxide

6.4 **Fire Extinguishing Agents Not to be Used:** Not pertinent

6.5 **Special Hazards of Combustion Products:** Not pertinent

6.6 **Behavior in Fire:** Not pertinent

6.7 **Ignition Temperature:** Data not available

6.8 **Electrical Hazard:** Not pertinent

6.9 **Burning Rate:** Data not available

7. CHEMICAL REACTIVITY

7.1 **Reactivity with Water:** No reaction

7.2 **Reactivity with Common Materials:** No reaction

7.3 **Stability During Transport:** Stable

7.4 **Neutralizing Agents for Acids and Caustics:** Not pertinent

7.5 **Polymerization:** Not pertinent

7.6 **Inhibitor of Polymerization:** Not pertinent

8. WATER POLLUTION

8.1 **Aquatic Toxicity:** Data not available

8.2 **Waterfowl Toxicity:** Data not available

8.3 **Biological Oxygen Demand (BOD):** Data not available

8.4 **Food Chain Concentration Potential:** None

9. SELECTED MANUFACTURERS

1. Olin Corp.
   Chemicals Division
   120 Long Ridge Rd.
   Stamford, Conn. 06904
2. The Ansul Co.
   Chemical Division
   101 Stanton St.
   Marinette, Wis. 54143

10. SHIPPING INFORMATION

10.1 **Grades or Purity:** Technical

10.2 **Storage Temperature:** Data not available

10.3 **Inert Atmosphere:** Data not available

10.4 **Venting:** Data not available

11. HAZARD ASSESSMENT CODE

(See Hazard Assessment Handbook, CG 446-3)

A-P-Q

12. HAZARD CLASSIFICATIONS

12.1 **Code of Federal Regulations:** Combustible Liquid

12.2 **NAS Hazard Rating for Bulk Water Transportation:** Not listed

12.3 **NFPA Hazard Classifications:** Not listed

13. PHYSICAL AND CHEMICAL PROPERTIES

13.1 **Physical State at 15°C and 1 atm:** Liquid

13.2 **Molecular Weight:** 134.12

13.3 **Boiling Point at 1 atm:** 324°F = 162°C = 435°K

13.4 **Freezing Point:** −94°F = −70°C = 203°K

13.5 **Critical Temperature:** Not pertinent

13.6 **Critical Pressure:** Not pertinent

13.7 **Specific Gravity:** 0.945 at 20°C (liquid)

13.8 **Liquid Surface Tension:** Not pertinent

13.9 **Liquid-Water Interfacial Tension:** Not pertinent

13.10 **Vapor (Gas) Specific Gravity:** Not pertinent

13.11 **Ratio of Specific Heats of Vapor (Gas):** Not pertinent

13.12 **Latent Heat of Vaporization:** 130 Btu/lb = 74 cal/g = 3.1×10^5 J/kg

13.13 **Heat of Combustion:** (est.) −11,300 Btu/lb = −6,260 cal/g = -262×10^5 J/kg

13.14 **Heat of Decomposition:** Not pertinent

13.15 **Heat of Solution:** Not pertinent

13.16 **Heat of Polymerization:** Not pertinent

NOTES

REVISED 1978

DME — DIETHYLENE GLYCOL MONOBUTYL ETHER

Common Synonyms	Liquid	Colorless	Mild pleasant odor
2-(2-Butoxyethoxy)ethanol Butoxydiethylene glycol Butyl "Carbitol" Dowanol DB Poly-solv DB	Mixes with water.		

Stop discharge if possible. Keep people away.
Call fire department.
Isolate and remove discharged material.
Notify local health and pollution control agencies.

Fire	Combustible. Extinguish with water, dry chemicals, alcohol foam, or carbon dioxide. Cool exposed containers with water.
Exposure	Call for medical aid. LIQUID Irritating to skin and eyes. Remove contaminated clothing and shoes. Flush affected areas with plenty of water. IF IN EYES, hold eyelids open and flush with plenty of water. IF SWALLOWED and victim is CONSCIOUS, have victim drink water or milk.
Water Pollution	Effect of low concentrations on aquatic life is unknown. May be dangerous if it enters water intakes. Notify local health and wildlife officials. Notify operators of nearby water intakes.

1. RESPONSE TO DISCHARGE

(See Response Methods Handbook, CG 446-4)

Issue warning—water contaminant
Disperse and flush

2. LABELS

No label required by Code of Federal Regulations

3. CHEMICAL DESIGNATIONS

3.1 **Synonyms:** Butoxydiethylene glycol; Butoxydiglycol; Diglycol monobutyl ether; Butyl "Carbitol"; Dowanol DB; Poly-Solv DB

3.2 **Coast Guard Compatibility Classification:** Glycol ether

3.3 **Chemical Formula:** $C_4H_9OCH_2CH_2OCH_2CH_2OH$

3.4 **IMCO/United Nations Numerical Designation:** Not listed

4. OBSERVABLE CHARACTERISTICS

4.1 **Physical State (as shipped):** Liquid

4.2 **Color:** Colorless

4.3 **Odor:** Mild, characteristic; pleasant

5. HEALTH HAZARDS

5.1 **Personal Protective Equipment:** Safety goggles or face shield.

5.2 **Symptoms Following Exposure:** Inhalation for brief periods has no significant effect. Contact with liquid causes moderate irritation of eyes and corneal injury. Prolonged contact with skin causes only minor irritation.

5.3 **Treatment for Exposure:** INHALATION: remove to fresh air; if ill effects are observed, call a doctor. EYES: immediately flush with plenty of water for at least 15 min. SKIN: wash well with soap and water. INGESTION: give large amounts of water.

5.4 **Toxicity by Inhalation (Threshold Limit Value):** Data not available

5.5 **Short-Term Inhalation Limits:** Data not available

5.6 **Toxicity by Ingestion:** Grade 2; oral LD_{50} = 2 g/kg (guinea pig)

5.7 **Late Toxicity:** Data not available

5.8 **Vapor (Gas) Irritant Characteristics:** Vapors cause a slight smarting of the eyes or respiratory system if present in high concentrations. The effect is temporary.

5.9 **Liquid or Solid Irritant Characteristics:** No appreciable hazard. Practically harmless to the skin.

5.10 **Odor Threshold:** Data not available

6. FIRE HAZARDS

6.1 **Flash Point:** 230°F O.C.; 172°F C.C.

6.2 **Flammable Limits in Air:** Not pertinent

6.3 **Fire Extinguishing Agents:** Water, "alcohol" foam, carbon dioxide, dry chemical

6.4 **Fire Extinguishing Agents Not to be Used:** Not pertinent

6.5 **Special Hazards of Combustion Products:** Not pertinent

6.6 **Behavior in Fire:** Not pertinent

6.7 **Ignition Temperature:** 442°F

6.8 **Electrical Hazard:** Not pertinent

6.9 **Burning Rate:** 3.3 mm/min.

7. CHEMICAL REACTIVITY

7.1 **Reactivity with Water:** No reaction

7.2 **Reactivity with Common Materials:** No reaction

7.3 **Stability During Transport:** Stable

7.4 **Neutralizing Agents for Acids and Caustics:** Not pertinent

7.5 **Polymerization:** Not pertinent

7.6 **Inhibitor of Polymerization:** Not pertinent

8. WATER POLLUTION

8.1 **Aquatic Toxicity:** Data not available

8.2 **Waterfowl Toxicity:** Data not available

8.3 **Biological Oxygen Demand (BOD):** 34% of theoretical in 5 days

8.4 **Food Chain Concentration Potential:** None

9. SELECTED MANUFACTURERS

1. Union Carbide Corporation
   Chemicals and Plastics Division
   270 Park Avenue
   New York, N. Y. 10017
2. Shell Chemical Company
   Industrial Chemicals Division
   One Shell Plaza
   Houston, Texas 77001
3. Jefferson Chemical Company, Inc.
   260 Madison Avenue
   New York, N. Y. 10016

10. SHIPPING INFORMATION

10.1 **Grades or Purity:** Commercial

10.2 **Storage Temperature:** Ambient

10.3 **Inert Atmosphere:** No requirement

10.4 **Venting:** Open (flame arrester)

11. HAZARD ASSESSMENT CODE

(See Hazard Assessment Handbook, CG 446-3)

A-P-Q

12. HAZARD CLASSIFICATIONS

12.1 **Code of Federal Regulations:** Not listed

12.2 **NAS Hazard Rating for Bulk Water Transportation:**

Category	Rating
Fire	1
Health	
Vapor Irritant	1
Liquid or Solid Irritant	0
Poisons	1
Water Pollution	
Human Toxicity	1
Aquatic Toxicity	3
Aesthetic Effect	1
Reactivity	
Other Chemicals	2
Water	0
Self-Reaction	0

12.3 **NFPA Hazard Classifications:**

Category	Classification
Health Hazard (Blue)	1
Flammability (Red)	2
Reactivity (Yellow)	0

13. PHYSICAL AND CHEMICAL PROPERTIES

13.1 **Physical State at 15°C and 1 atm:** Liquid

13.2 **Molecular Weight:** 162.2

13.3 **Boiling Point at 1 atm:** 448°F = 231°C = 504°K

13.4 **Freezing Point:** −90°F = −68°C = 205°K

13.5 **Critical Temperature:** Not pertinent

13.6 **Critical Pressure:** Not pertinent

13.7 **Specific Gravity:** 0.954 at 20°C (liquid)

13.8 **Liquid Surface Tension:** 34 dynes/cm = 0.034 N/m at 25°C

13.9 **Liquid-Water Interfacial Tension:** Not pertinent

13.10 **Vapor (Gas) Specific Gravity:** Not pertinent

13.11 **Ratio of Specific Heats of Vapor (Gas):** Not pertinent

13.12 **Latent Heat of Vaporization:** 130 Btu/lb = 74 cal/g = 3.1×10^5 J/kg

13.13 **Heat of Combustion:** (est.) −14,000 Btu/lb = −7,900 cal/g = -330×10^5 J/kg

13.14 **Heat of Decomposition:** Not pertinent

13.15 **Heat of Solution:** (est.) −36 Btu/lb = −20 cal/g = -0.84×10^5 J/kg

13.16 **Heat of Polymerization:** Not pertinent

NOTES

DEM	DIETHYLENE GLYCOL MONOBUTYL ETHER ACETATE

DIETHYLENE GLYCOL MONOBUTYL ETHER ACETATE

Common Synonyms			
2-(2-Butoxyethoxy) ethanol acetate 2-(2-Butoxyethoxy) ethyl acetate Diglycol monobutyl ether acetate Butyl "Carbitol" acetate	Liquid	Colorless	Mild odor
	Floats and mixes slowly with water.		

Stop discharge if possible. Keep people away.
Call fire department.
Isolate and remove discharged material.
Notify local health and pollution control agencies.

Fire	Combustible. Extinguish with water, dry chemicals, alcohol foam, or carbon dioxide. Cool exposed containers with water.
Exposure	Call for medical aid. LIQUID Irritating to skin and eyes. Remove contaminated clothing and shoes. Flush affected areas with plenty of water. IF IN EYES, hold eyelids open and flush with plenty of water. IF SWALLOWED and victim is CONSCIOUS, have victim drink water or milk.
Water Pollution	Effect of low concentrations on aquatic life is unknown. May be dangerous if it enters water intakes. Notify local health and wildlife officials. Notify operators of nearby water intakes.

1. RESPONSE TO DISCHARGE

(See Response Methods Handbook, CG 446-4)

Issue warning—water contaminant
Disperse and flush

2. LABELS

No label required by Code of Federal Regulations

3. CHEMICAL DESIGNATIONS

3.1 **Synonyms:** 2-(2-Butoxyethoxy) ethyl acetate; Diglycol monobutyl ether acetate; Butyl "Carbitol" acetate; Ektasolve DB Acetate

3.2 **Coast Guard Compatibility Classification:** Ester (13)

3.3 **Chemical Formula:** $C_4H_9OCH_2CH_2OCH_2CH_2OCOCH_3$

3.4 **IMCO/United Nations Numerical Designation:** Not listed

4. OBSERVABLE CHARACTERISTICS

4.1 **Physical State (as shipped):** Liquid

4.2 **Color:** Colorless

4.3 **Odor:** Mild, nonresidual

5. HEALTH HAZARDS

5.1 **Personal Protective Equipment:** Face shield or safety glasses; protective gloves; air mask for prolonged exposure to vapor.

5.2 **Symptoms Following Exposure:** Prolonged breathing of vapor may cause irritation and nausea. Contact with liquid may cause mild irritation of eyes and skin. Can be absorbed through skin in toxic amounts.

5.3 **Treatment for Exposure:** INHALATION: move victim to fresh air; if breathing has stopped, administer artificial respiration. EYES: flush with water for at least 15 min. SKIN: wash skin with large amounts of water for 15 min.; call physician if needed. INGESTION: induce vomiting; get medical attention.

5.4 **Toxicity by Inhalation (Threshold Limit Value):** Data not available

5.5 **Short-Term Inhalation Limits:** Because of high boiling point (246°C), hazards from inhalation are minimal

5.6 **Toxicity by Ingestion:** Grade 2; oral LD_{50} = 2.34 g/kg (guinea pig)

5.7 **Late Toxicity:** Kidney damage noted in animals following repeated contact with skin.

5.8 **Vapor (Gas) Irritant Characteristics:** Data not available

5.9 **Liquid or Solid Irritant Characteristics:** Data not available

5.10 **Odor Threshold:** Data not available

6. FIRE HAZARDS

6.1 **Flash Point:** 240°F O.C.

6.2 **Flammable Limits in Air:** 0.8%–5.0%

6.3 **Fire Extinguishing Agents:** Water, alcohol foam, dry chemical, carbon dioxide

6.4 **Fire Extinguishing Agents Not to be Used:** Not pertinent

6.5 **Special Hazards of Combustion Products:** Not pertinent

6.6 **Behavior in Fire:** Not pertinent

6.7 **Ignition Temperature:** 563°F

6.8 **Electrical Hazard:** Data not available

6.9 **Burning Rate:** 3.8 mm/min.

7. CHEMICAL REACTIVITY

7.1 **Reactivity with Water:** No reaction

7.2 **Reactivity with Common Materials:** No reaction

7.3 **Stability During Transport:** Stable

7.4 **Neutralizing Agents for Acids and Caustics:** Not pertinent

7.5 **Polymerization:** Not pertinent

7.6 **Inhibitor of Polymerization:** Not pertinent

8. WATER POLLUTION

8.1 **Aquatic Toxicity:** Data not available

8.2 **Waterfowl Toxicity:** Data not available

8.3 **Biological Oxygen Demand (BOD):** Data not available

8.4 **Food Chain Concentration Potential:** None

9. SELECTED MANUFACTURERS

1. Union Carbide Corporation
   Chemicals and Plastics Division
   270 Park Avenue
   New York, N. Y. 10017
2. Eastman Chemical Products, Inc.
   Kingsport, Tenn. 37662
3. Eastman Kodak Co.
   Eastman Organic Chemicals
   Rochester, N. Y. 14650

10. SHIPPING INFORMATION

10.1 **Grades or Purity:** 98+%

10.2 **Storage Temperature:** Ambient

10.3 **Inert Atmosphere:** No requirement

10.4 **Venting:** Open (flame arrester)

11. HAZARD ASSESSMENT CODE

(See Hazard Assessment Handbook, CG 446-3)

A-P-Q

12. HAZARD CLASSIFICATIONS

12.1 **Code of Federal Regulations:** Not listed

12.2 **NAS Hazard Rating for Bulk Water Transportation:** Not listed

12.3 **NFPA Hazard Classifications:**

Category	Classification
Health Hazard (Blue)	
Flammability (Red)	1
Reactivity (Yellow)	0

13. PHYSICAL AND CHEMICAL PROPERTIES

13.1 **Physical State at 15°C and 1 atm:** Liquid

13.2 **Molecular Weight:** 204.3

13.3 **Boiling Point at 1 atm:** 475°F = 246°C = 519°K

13.4 **Freezing Point:** −27°F = −33°C = 240°K

13.5 **Critical Temperature:** Not pertinent

13.6 **Critical Pressure:** Not pertinent

13.7 **Specific Gravity:** 0.985 at 20°C (liquid)

13.8 **Liquid Surface Tension:** (est.) 22 dynes/cm = 0.022 N/m at 20°C

13.9 **Liquid-Water Interfacial Tension:** Not pertinent

13.10 **Vapor (Gas) Specific Gravity:** Not pertinent

13.11 **Ratio of Specific Heats of Vapor (Gas):** Not pertinent

13.12 **Latent Heat of Vaporization:** 106 Btu/lb = 59 cal/g = 2.5×10^5 J/kg

13.13 **Heat of Combustion:** (est.) −13,000 Btu/lb = −7,400 cal/g = -310×10^5 J/kg

13.14 **Heat of Decomposition:** Not pertinent

13.15 **Heat of Solution:** (est.) −27 Btu/lb = −15 cal/g = -0.63×10^5 J/kg

13.16 **Heat of Polymerization:** Not pertinent

NOTES

DGE — DIETHYLENE GLYCOL MONOETHYL ETHER

Common Synonyms			
Poly-Solv DE Dowanol DE Carbitol 2-(2-ethoxyethoxy) ethanol	Liquid	Colorless	Fruity odor
	Floats and mixes with water.		

Stop discharge if possible.
Call fire department.
Isolate and remove discharged material.
Notify local health and pollution control agencies.

Fire

Combustible.
Extinguish with dry material, alcohol foam, or carbon dioxide.
Cool exposed containers with water.

Exposure

CALL FOR MEDICAL AID.

LIQUID
Harmful if swallowed.
IF SWALLOWED and victim is CONSCIOUS, have victim drink water or milk.

Water Pollution

Effect of low concentrations on aquatic life is unknown.
May be dangerous if it enters water intakes.
Notify local health and wildlife officials.
Notify operators of nearby water intakes.

1. RESPONSE TO DISCHARGE

(See Response Methods Handbook, CG 446-4)

Disperse and flush

2. LABELS

No hazard label required by Code of Federal Regulations

3. CHEMICAL DESIGNATIONS

3.1 **Synonyms:** Carbitol; Diethylene glycol ethyl ether; Dowanol DE; 2-(2-Ethoxyethoxy) ethanol; Ethoxy diglycol; Poly-Solv DE

3.2 **Coast Guard Compatibility Classification:** Glycol ether

3.3 **Chemical Formula:** $CH_3CH_2OCH_2CH_2OCH_2CH_2OH$

3.4 **IMCO/United Nations Numerical Designation:** Not listed

4. OBSERVABLE CHARACTERISTICS

4.1 **Physical State (as shipped):** Liquid

4.2 **Color:** Colorless

4.3 **Odor:** Weakly fruity; mild and characteristic

5. HEALTH HAZARDS

5.1 **Personal Protective Equipment:** Goggles.

5.2 **Symptoms Following Exposure:** None expected.

5.3 **Treatment for Exposure:** SKIN AND EYES: flush with water.

5.4 **Toxicity by Inhalation (Threshold Limit Value):** Not pertinent

5.5 **Short-Term Inhalation Limits:** Data not available

5.6 **Toxicity by Ingestion:** Grade 2; LD_{50} 0.5 to 5 g/kg

5.7 **Late Toxicity:** Data not available

5.8 **Vapor (Gas) Irritant Characteristics:** None

5.9 **Liquid or Solid Irritant Characteristics:** No appreciable hazard. Practically harmless to the skin.

5.10 **Odor Threshold:** Data not available

6. FIRE HAZARDS

6.1 **Flash Point:** 201°F C.C.; 205°F O.C.

6.2 **Flammable Limits in Air:** 1.2% - 8.5% (est.)

6.3 **Fire Extinguishing Agents:** Alcohol foam, dry chemical, or carbon dioxide

6.4 **Fire Extinguishing Agents Not to be Used:** Not pertinent

6.5 **Special Hazards of Combustion Products:** Not pertinent

6.6 **Behavior in Fire:** Not pertinent

6.7 **Ignition Temperature:** 400°F

6.8 **Electrical Hazard:** Not pertinent

6.9 **Burning Rate:** 2.5 mm/min.

7. CHEMICAL REACTIVITY

7.1 **Reactivity with Water:** No reaction

7.2 **Reactivity with Common Materials:** No reaction

7.3 **Stability During Transport:** Stable

7.4 **Neutralizing Agents for Acids and Caustics:** Not pertinent

7.5 **Polymerization:** Not pertinent

7.6 **Inhibitor of Polymerization:** Not pertinent

8. WATER POLLUTION

8.1 **Aquatic Toxicity:** Data not available

8.2 **Waterfowl Toxicity:** Data not available

8.3 **Biological Oxygen Demand (BOD):** 34%, 5 days (theor.)

8.4 **Food Chain Concentration Potential:** None

9. SELECTED MANUFACTURERS

1. Dow Chemical Co.
   Midland, Mich. 48640
2. Jefferson Chemical Co. Inc.
   3336 Richmond Ave.
   Houston, Texas 77052
3. Union Carbide Corp.
   Chemicals and Plastics Division
   270 Park Ave.
   New York, N. Y. 10017

10. SHIPPING INFORMATION

10.1 **Grades or Purity:** Commercial

10.2 **Storage Temperature:** Ambient

10.3 **Inert Atmosphere:** No requirement

10.4 **Venting:** Open (flame arrester)

11. HAZARD ASSESSMENT CODE

(See Hazard Assessment Handbook, CG 446-3)

A-P-Q

12. HAZARD CLASSIFICATIONS

12.1 **Code of Federal Regulations:** Not listed

12.2 **NAS Hazard Rating for Bulk Water Transportation:**

Category	Rating
Fire	1
Health	
Vapor Irritant	0
Liquid or Solid Irritant	0
Poisons	0
Water Pollution	
Human Toxicity	1
Aquatic Toxicity	3
Aesthetic Effect	1
Reactivity	
Other Chemicals	2
Water	0
Self-Reaction	0

12.3 **NFPA Hazard Classifications:** Not listed

13. PHYSICAL AND CHEMICAL PROPERTIES

13.1 **Physical State at 15°C and 1 atm:** Liquid

13.2 **Molecular Weight:** 134.17

13.3 **Boiling Point at 1 atm:** 396°F = 202°C = 475°K

13.4 **Freezing Point:** −105°F = −76°C = 197°K

13.5 **Critical Temperature:** Not pertinent

13.6 **Critical Pressure:** Not pertinent

13.7 **Specific Gravity:** 0.99 at 20°C (liquid)

13.8 **Liquid Surface Tension:** Not pertinent

13.9 **Liquid-Water Interfacial Tension:** Not pertinent

13.10 **Vapor (Gas) Specific Gravity:** Not pertinent

13.11 **Ratio of Specific Heats of Vapor (Gas):** Not pertinent

13.12 **Latent Heat of Vaporization:** 150 Btu/lb = 85 cal/g = 3.6×10^5 J/kg

13.13 **Heat of Combustion:** -11,390 Btu/lb = -6330 cal/g = -265×10^5 J/kg

13.14 **Heat of Decomposition:** Not pertinent

13.15 **Heat of Solution:** Not pertinent

13.16 **Heat of Polymerization:** Not pertinent

NOTES

REVISED 1978

DGM

DIETHYLENE GLYCOL MONOMETHYL ETHER

Common Synonyms
Poly-Solv DM,
Dowanol DM
2-(2-methoxyethoxy)-ethanol
Methyl Carbitol

Liquid | Colorless | Pleasant odor

Floats and mixes with water.

Stop discharge if possible.
Call fire department.
Avoid contact with liquid.
Isolate and remove discharged material.
Notify local health and pollution control agencies.

Fire

Combustible.

Extinguish with dry chemical, water, or carbon dioxide.
Cool exposed containers with water.

Exposure

CALL FOR MEDICAL AID.
LIQUID
Irritating to eyes.
Harmful if swallowed.

IF IN EYES, hold eyelids open and flush with plenty of water.
IF SWALLOWED and victim is CONSCIOUS, have victim drink water or milk.

Water Pollution

Effect of low concentrations on aquatic life is unknown.
May be dangerous if it enters water intakes.

Notify local health and wildlife officials.
Notify operators of nearby water intakes.

1. RESPONSE TO DISCHARGE

(See Response Methods Handbook, CG 446-4)

Disperse and flush

2. LABELS

No hazard label required by Code of Federal Regulations

3. CHEMICAL DESIGNATIONS

3.1 **Synonyms:** Diethylene glycol methyl ether
Dowanol DM
2-(2-Methoxyethoxy) ethanol
Methyl Carbitol
Poly-Solv DM

3.2 **Coast Guard Compatibility Classification:** Glycol ether

3.3 **Chemical Formula:** $CH_3OCH_2CH_2OCH_2CH_2OH$

3.4 **IMCO/United Nations Numerical Designation:** Not listed

4. OBSERVABLE CHARACTERISTICS

4.1 **Physical State (as shipped):** Liquid

4.2 **Color:** Colorless

4.3 **Odor:** Mild, pleasant; faint characteristic

5. HEALTH HAZARDS

5.1 **Personal Protective Equipment:** Safety goggles.

5.2 **Symptoms Following Exposure:** Liquid may irritate eyes.

5.3 **Treatment for Exposure:** SKIN OR EYES: flush with water.

5.4 **Toxicity by Inhalation (Threshold Limit Value):** Not pertinent

5.5 **Short-Term Inhalation Limits:** Not pertinent

5.6 **Toxicity by Ingestion:** Grade 2; LD_{50} 0.5 to 5 g/kg (guinea pig)

5.7 **Late Toxicity:** Data not available

5.8 **Vapor (Gas) Irritant Characteristics:** None

5.9 **Liquid or Solid Irritant Characteristics:** None

5.10 **Odor Threshold:** Data not available

6. FIRE HAZARDS

6.1 **Flash Point:** 200°F O.C.

6.2 **Flammable Limits in Air:** LFL = 1.2%

6.3 **Fire Extinguishing Agents:** Water, carbon dioxide, or dry chemical

6.4 **Fire Extinguishing Agents Not to be Used:** Not pertinent

6.5 **Special Hazards of Combustion Products:** Not pertinent

6.6 **Behavior in Fire:** Not pertinent

6.7 **Ignition Temperature:** Data not available

6.8 **Electrical Hazard:** Not pertinent

6.9 **Burning Rate:** Data not available

7. CHEMICAL REACTIVITY

7.1 **Reactivity with Water:** No reaction

7.2 **Reactivity with Common Materials:** No reaction

7.3 **Stability During Transport:** Stable

7.4 **Neutralizing Agents for Acids and Caustics:** Not pertinent

7.5 **Polymerization:** Not pertinent

7.6 **Inhibitor of Polymerization:** Not pertinent

8. WATER POLLUTION

8.1 **Aquatic Toxicity:** Data not available

8.2 **Waterfowl Toxicity:** Data not available

8.3 **Biological Oxygen Demand (BOD):** 34% of theoretical in 5 days

8.4 **Food Chain Concentration Potential:** None

9. SELECTED MANUFACTURERS

1. Eastman Kodak Co.
Tennessee Eastman Division
Kingsport, Tenn. 37662
2. Jefferson Chemical Co.
3336 Richmond Ave.
Houston, Tex. 77052
3. Union Carbide Corp.
Chemicals & Plastics Division
270 Park Ave.
New York, N. Y. 10017

10. SHIPPING INFORMATION

10.1 **Grades or Purity:** Commercial

10.2 **Storage Temperature:** Ambient

10.3 **Inert Atmosphere:** No requirement

10.4 **Venting:** Open (flame arrester)

11. HAZARD ASSESSMENT CODE

(See Hazard Assessment Handbook, CG 446-3)

A-P-Q

12. HAZARD CLASSIFICATIONS

12.1 **Code of Federal Regulations:** Combustible Liquid

12.2 **NAS Hazard Rating for Bulk Water Transportation:**

Category	Rating
Fire	1
Health	
Vapor Irritant	0
Liquid or Solid Irritant	0
Poisons	0
Water Pollution	
Human Toxicity	1
Aquatic Toxicity	3
Aesthetic Effect	1
Reactivity	
Other Chemicals	2
Water	0
Self-Reaction	0

12.3 **NFPA Hazard Classifications:** Not listed

13. PHYSICAL AND CHEMICAL PROPERTIES

13.1 **Physical State at 15°C and 1 atm:** Liquid

13.2 **Molecular Weight:** 120.15

13.3 **Boiling Point at 1 atm:** 381°F = 194°C = 467°K

13.4 **Freezing Point:** −120°F = −85°C = 188°K

13.5 **Critical Temperature:** Not pertinent

13.6 **Critical Pressure:** Not pertinent

13.7 **Specific Gravity:** 1.025 at 20°C (liquid)

13.8 **Liquid Surface Tension:** Not pertinent

13.9 **Liquid-Water Interfacial Tension:** Not pertinent

13.10 **Vapor (Gas) Specific Gravity:** Not pertinent

13.11 **Ratio of Specific Heats of Vapor (Gas):** Not pertinent

13.12 **Latent Heat of Vaporization:** 160 Btu/lb = 90 cal/g = 3.8×10^5 J/kg

13.13 **Heat of Combustion:** -10,830 Btu/lb = -6020 cal/g = -252×10^5 J/kg

13.14 **Heat of Decomposition:** Not pertinent

13.15 **Heat of Solution:** Not pertinent

13.16 **Heat of Polymerization:** Not pertinent

NOTES

REVISED 1978

DET DIETHYLENETRIAMINE

Common Synonyms			
2,2'-Diaminodiethylamine Bis (2-aminoethyl) amine	Liquid	Colorless to yellow	Ammonia odor
	Floats and mixes with water.		

Avoid contact with liquid. Keep people away.
Wear goggles, self-contained breathing apparatus, and rubber overclothing (including gloves).
Stop discharge if possible.
Call fire department.
Isolate and remove discharged material.
Notify local health and pollution control agencies.

Fire

Combustible.
Wear goggles, self-contained breathing apparatus, and rubber overclothing (including gloves).
Extinguish with water, dry chemical, alcohol foam, or carbon dioxide.
Cool exposed containers with water.

Exposure

CALL FOR MEDICAL AID.

LIQUID
Will burn skin and eyes.
Harmful if swallowed.
Remove contaminated clothing and shoes.
Flush affected areas with plenty of water.
IF IN EYES, hold eyelids open and flush with plenty of water.
IF SWALLOWED and victim is CONSCIOUS, have victim drink water or milk.
DO NOT INDUCE VOMITING.

Water Pollution

Effect of low concentrations on aquatic life is unknown.
May be dangerous if it enters water intakes.
Notify local health and wildlife officials.
Notify operators of nearby water intakes.

1. RESPONSE TO DISCHARGE

(See Response Methods Handbook, CG 446-4)

Disperse and flush

2. LABEL

3. CHEMICAL DESIGNATIONS

3.1 **Synonyms:** Bis(2-aminoethyl)amine; 2, 2'-Diaminodiethylamine

3.2 **Coast Guard Compatibility Classification:** Amine

3.3 **Chemical Formula:** $NH_2(CH_2)_2NH(CH_2)_2NH_2$

3.4 **IMCO/United Nations Numerical Designation:** Not listed

4. OBSERVABLE CHARACTERISTICS

4.1 **Physical State (as shipped):** Liquid

4.2 **Color:** Colorless to light amber; yellow

4.3 **Odor:** Strong ammoniacal; mildly ammoniacal

5. HEALTH HAZARDS

5.1 **Personal Protective Equipment:** Amine respiratory cartridge mask; rubber gloves; splash-proof goggles.

5.2 **Symptoms Following Exposure:** Prolonged breathing of vapors may cause asthma. Liquid burns skin and eyes. A skin rash can form.

5.3 **Treatment for Exposure:** INHALATION: remove victim to fresh air. INGESTION: do NOT induce vomiting; give large quantities of water; give at least one ounce of vinegar in an equal amount of water; get medical attention. SKIN CONTACT: flush with plenty of water. EYE CONTACT: flush with plenty of water for at least 15 min. and get medical attention.

5.4 **Toxicity by Inhalation (Threshold Limit Value):** 1 ppm

5.5 **Short-Term Inhalation Limits:** Data not available

5.6 **Toxicity by Ingestion:** Grade 2; LD_{50} 0.5 to 5 g/kg (rat)

5.7 **Late Toxicity:** Data not available

5.8 **Vapor (Gas) Irritant Characteristics:** Vapors cause moderate irritation such that personnel will find high concentrations unpleasant. The effect is temporary.

5.9 **Liquid or Solid Irritant Characteristics:** Causes smarting of the skin and first-degree burns on short exposure and may cause secondary burns on long exposure.

5.10 **Odor Threshold:** 10 ppm

6. FIRE HAZARDS

6.1 **Flash Point:** 210°F O.C.

6.2 **Flammable Limits in Air:** (calc.) 1%–10%

6.3 **Fire Extinguishing Agents:** Water spray, alcohol foam, carbon dioxide, dry chemical

6.4 **Fire Extinguishing Agents Not to be Used:** Water or foam may cause frothing

6.5 **Special Hazards of Combustion Products:** Irritating vapors are generated when heated.

6.6 **Behavior in Fire:** Not pertinent

6.7 **Ignition Temperature:** 676°F

6.8 **Electrical Hazard:** Not pertinent

6.9 **Burning Rate:** Data not available

7. CHEMICAL REACTIVITY

7.1 **Reactivity with Water:** No reaction

7.2 **Reactivity with Common Materials:** No hazardous reaction

7.3 **Stability During Transport:** Stable

7.4 **Neutralizing Agents for Acids and Caustics:** Flush with water

7.5 **Polymerization:** Not pertinent

7.6 **Inhibitor of Polymerization:** Not pertinent

8. WATER POLLUTION

8.1 **Aquatic Toxicity:** 710 ppm/24 hr/brine shrimp/TL_m

8.2 **Waterfowl Toxicity:** Data not available

8.3 **Biological Oxygen Demand (BOD):** Data not available

8.4 **Food Chain Concentration Potential:** 23% of theoretical in 5 days/freshwater acclimated seed

9. SELECTED MANUFACTURERS

1. Dow Chemical Co.
   Midland, Mich. 48640
2. Jefferson Chemical Co., Inc.
   3336 Richmond Ave.
   Houston, Texas 77052
3. Union Carbide Corp.
   Chemicals and Plastics Division
   270 Park Ave.
   New York, N. Y. 10017

10. SHIPPING INFORMATION

10.1 **Grades or Purity:** 98-99%

10.2 **Storage Temperature:** Ambient

10.3 **Inert Atmosphere:** No requirement

10.4 **Venting:** Open

11. HAZARD ASSESSMENT CODE

(See Hazard Assessment Handbook, CG 446-3)

A-P-Q

12. HAZARD CLASSIFICATIONS

12.1 **Code of Federal Regulations:** Corrosive Material

12.2 **NAS Hazard Rating for Bulk Water Transportation:**

Category	Rating
Fire	1
Health	
Vapor Irritant	2
Liquid or Solid Irritant	2
Poisons	2
Water Pollution	
Human Toxicity	2
Aquatic Toxicity	2
Aesthetic Effect	3
Reactivity	
Other Chemicals	3
Water	0
Self-Reaction	0

12.3 **NFPA Hazard Classifications:**

Category	Classification
Health Hazard (Blue)	3
Flammability (Red)	1
Reactivity (Yellow)	0

13. PHYSICAL AND CHEMICAL PROPERTIES

13.1 **Physical State at 15°C and 1 atm:** Liquid

13.2 **Molecular Weight:** 103.17

13.3 **Boiling Point at 1 atm:** 405°F = 207°C = 480°K

13.4 **Freezing Point:** −38°F = −39°C = 234°K

13.5 **Critical Temperature:** Not pertinent

13.6 **Critical Pressure:** Not pertinent

13.7 **Specific Gravity:** 0.954 at 20°C (liquid)

13.8 **Liquid Surface Tension:** Not pertinent

13.9 **Liquid-Water Interfacial Tension:** Not pertinent

13.10 **Vapor (Gas) Specific Gravity:** Not pertinent

13.11 **Ratio of Specific Heats of Vapor (Gas):** Not pertinent

13.12 **Latent Heat of Vaporization:** Not pertinent

13.13 **Heat of Combustion:** (est.) −13,300 Btu/lb = −7,390 cal/g = -309×10^5 J/kg

13.14 **Heat of Decomposition:** Not pertinent

13.15 **Heat of Solution:** (est.) −13 Btu/lb = −7 cal/g = -0.3×10^5 J/kg

13.16 **Heat of Polymerization:** Not pertinent

NOTES

REVISED 1978

DEP — DI-(2-ETHYLHEXYL) PHOSPHORIC ACID

Common Synonyms			
2-Ethyl-1-hexanol hydrogen phosphate Bis(2-ethylhexyl) hydrogen phosphate Di-(2-ethylhexyl) phosphate Di-(2-ethylhexyl) acid phosphate	Liquid Floats on water.	Light yellow	Odorless

Stop discharge if possible. Keep people away.
Call fire department.
Isolate and remove discharged material.
Notify local health and pollution control agencies.

Fire	Combustible. **Extinguish with dry chemicals, alcohol foam, or carbon dioxide.** **Water and foam may be ineffective on fire.**
Exposure	**Call for medical aid.** LIQUID Irritating to skin and eyes. Harmful if swallowed. **Remove contaminated clothing and shoes.** **Flush affected areas with plenty of water.** **IF IN EYES, hold eyelids open and flush with plenty of water.** **IF SWALLOWED and victim is CONSCIOUS, have victim drink water or milk and have victim induce vomiting.** **IF SWALLOWED and victim is UNCONSCIOUS OR HAVING CONVULSIONS, do nothing except keep victim warm.**
Water Pollution	Effect of low concentrations on aquatic life is unknown. Fouling to shoreline. May be dangerous if it enters water intakes. **Notify local health and wildlife officials.** **Notify operators of nearby water intakes.**

1. RESPONSE TO DISCHARGE

(See Response Methods Handbook, CG 446-4)

Issue warning—corrosive, water contaminant
Restrict access
Mechanical containment
Should be removed
Chemical and physical treatment

2. LABEL

3. CHEMICAL DESIGNATIONS

3.1 **Synonyms:** Bis-(2-ethylhexyl) hydrogen phosphate; Di-(2-ethylhexyl) phosphate; Di-(2-ethylhexyl) acid phosphate; DEHPA
3.2 **Coast Guard Compatibility Classification:** Not applicable
3.3 **Chemical Formula:** $[CH_3CH_2CH_2CH_2CH(C_2H_5)CH_2O]_2POOH$
3.4 **IMCO/United Nations Numerical Designation:** Not listed

4. OBSERVABLE CHARACTERISTICS

4.1 **Physical State (as shipped):** Liquid
4.2 **Color:** Amber
4.3 **Odor:** None

5. HEALTH HAZARDS

5.1 **Personal Protective Equipment:** Goggles or face shield; rubber gloves; protective clothing.
5.2 **Symptoms Following Exposure:** Contact with liquid irritates eyes and may cause serious injury; consult an eye specialist. Also causes skin irritation on contact. Ingestion produces irritation similar to that caused by strong vinegar.
5.3 **Treatment for Exposure:** EYES: immediately flush with plenty of water for at least 15 min.; see a physician. SKIN: immediately flush with plenty of water for at least 15 min. INGESTION: induce vomiting and call a physician.
5.4 **Toxicity by Inhalation (Threshold Limit Value):** Data not available
5.5 **Short-Term Inhalation Limits:** Data not available
5.6 **Toxicity by Ingestion:** Grade 2; LD_{50} 0.5 to 5 g/kg
5.7 **Late Toxicity:** Data not available
5.8 **Vapor (Gas) Irritant Characteristics:** Vapors are nonirritating to the eyes and throat.
5.9 **Liquid or Solid Irritant Characteristics:** Causes smarting of the skin and first-degree burns on short exposure and may cause second-degree burns on long exposure.
5.10 **Odor Threshold:** Data not available

6. FIRE HAZARDS

6.1 **Flash Point:** 385°F O.C.
6.2 **Flammable Limits in Air:** Not pertinent
6.3 **Fire Extinguishing Agents:** Dry chemical, alcohol foam, carbon dioxide
6.4 **Fire Extinguishing Agents Not to be Used:** Water or foam may cause frothing.
6.5 **Special Hazards of Combustion Products:** Irritating phosphorus oxides may be released.
6.6 **Behavior in Fire:** Not pertinent
6.7 **Ignition Temperature:** Data not available
6.8 **Electrical Hazard:** Not pertinent
6.9 **Burning Rate:** Data not available

7. CHEMICAL REACTIVITY

7.1 **Reactivity with Water:** No reaction
7.2 **Reactivity with Common Materials:** Mildly corrosive to most metals; may form flammable hydrogen gas.
7.3 **Stability During Transport:** Stable
7.4 **Neutralizing Agents for Acids and Caustics:** Sodium bicarbonate or lime solution
7.5 **Polymerization:** Not pertinent
7.6 **Inhibitor of Polymerization:** Not pertinent

8. WATER POLLUTION

8.1 **Aquatic Toxicity:** Data not available
8.2 **Waterfowl Toxicity:** Data not available
8.3 **Biological Oxygen Demand (BOD):** Data not available
8.4 **Food Chain Concentration Potential:** None

9. SELECTED MANUFACTURERS

Union Carbide Corporation
Chemicals and Plastics Division
270 Park Avenue
New York, N. Y. 10017

10. SHIPPING INFORMATION

10.1 **Grades or Purity:** 92+%
10.2 **Storage Temperature:** Ambient
10.3 **Inert Atmosphere:** No requirement
10.4 **Venting:** Open

11. HAZARD ASSESSMENT CODE

(See Hazard Assessment Handbook, CG 446-3)

A-T-U

12. HAZARD CLASSIFICATIONS

12.1 **Code of Federal Regulations:** Corrosive liquid
12.2 **NAS Hazard Rating for Bulk Water Transportation:**

Category	Rating
Fire	1
Health	
Vapor Irritant	0
Liquid or Solid Irritant	2
Poisons	3
Water Pollution	
Human Toxicity	—
Aquatic Toxicity	—
Aesthetic Effect	2
Reactivity	
Other Chemicals	1
Water	0
Self-Reaction	0

12.3 **NFPA Hazard Classifications:** Not listed

13. PHYSICAL AND CHEMICAL PROPERTIES

13.1 **Physical State at 15°C and 1 atm:** Liquid
13.2 **Molecular Weight:** 322.4
13.3 **Boiling Point at 1 atm:** Decomposes
13.4 **Freezing Point:** $<-76°F = <-60°C = <213°K$
13.5 **Critical Temperature:** Not pertinent
13.6 **Critical Pressure:** Not pertinent
13.7 **Specific Gravity:** 0.977 at 20°C (liquid)
13.8 **Liquid Surface Tension:** (est.) 20 dynes/cm = 0.020 N/m at 20°C
13.9 **Liquid-Water Interfacial Tension:** (est.) 30 dynes/cm = 0.030 N/m at 20°C
13.10 **Vapor (Gas) Specific Gravity:** Not pertinent
13.11 **Ratio of Specific Heats of Vapor (Gas):** Not pertinent
13.12 **Latent Heat of Vaporization:** Not pertinent
13.13 **Heat of Combustion:** −13,970 Btu/lb = −7,760 cal/g = -325×10^5 J/kg
13.14 **Heat of Decomposition:** Not pertinent
13.15 **Heat of Solution:** Not pertinent
13.16 **Heat of Polymerization:** Not pertinent

NOTES

DPH — DIETHYL PHTHALATE

Common Synonyms			
Phthalic acid, diethyl ester Ethyl phthalate 1,2-Benzenedicarboxylic acid, diethyl ester	Liquid	White	Mild chemical odor
	Sinks in water. Freezing point is 27°F.		

Stop discharge if possible. Keep people away.
Call fire department.
Isolate and remove discharged material.
Notify local health and pollution control agencies.

Fire

Combustible.
Irritating gases may be produced when heated.
Wear goggles and self-contained breathing apparatus.
Extinguish with dry chemicals, foam or carbon dioxide.
Water may be ineffective on fire.

Exposure

LIQUID
Not harmful.

Water Pollution

HARMFUL TO AQUATIC LIFE IN VERY LOW CONCENTRATIONS.
May be dangerous if it enters water intakes.
Notify local health and wildlife officials.
Notify operators of nearby water intakes.

1. RESPONSE TO DISCHARGE

(See Response Methods Handbook, CG 446-4)

Should be removed
Chemical and physical treatment

2. LABELS

No hazard label required by Code of Federal Regulations

3. CHEMICAL DESIGNATIONS

3.1 **Synonyms:** 1,2-Benzenedicarboxylic acid, diethyl ester; Ethyl phthalate; Phthalic acid, diethyl ester

3.2 **Coast Guard Compatibility Classification:** Not listed

3.3 **Chemical Formula:** $C_6H_4(COOC_2H_5)_2$

3.4 **IMCO/United Nations Numerical Designation:** Not listed

4. OBSERVABLE CHARACTERISTICS

4.1 **Physical State (as shipped):** Liquid

4.2 **Color:** Colorless

4.3 **Odor:** Slight characteristic ester odor

5. HEALTH HAZARDS

5.1 **Personal Protective Equipment:** Rubber gloves; goggles or face shield

5.2 **Symptoms Following Exposure:** Symptoms unlikely from any form of exposure.

5.3 **Treatment for Exposure:** INHALATION: remove to fresh air. EYES: flush with water. SKIN: flush with water, wash well with soap and water.

5.4 **Toxicity by Inhalation (Threshold Limit Value):** Data not available

5.5 **Short-Term Inhalation Limits:** Data not available

5.6 **Toxicity by Ingestion:** Grade 2; oral LD_{50} = 1,000 mg/kg (rabbit)

5.7 **Late Toxicity:** Prolonged inhalation of heated vapor produces irritation of upper respiratory tract in humans

5.8 **Vapor (Gas) Irritant Characteristics:** Odorless

5.9 **Liquid or Solid Irritant Characteristics:** Data not available

5.10 **Odor Threshold:** Data not available

6. FIRE HAZARDS

6.1 **Flash Point:** 305°F O.C.

6.2 **Flammable Limits in Air:** LFL 0.75% (at 368°F)

6.3 **Fire Extinguishing Agents:** Dry chemical, foam, carbon dioxide

6.4 **Fire Extinguishing Agents Not to be Used:** Water or foam may cause frothing.

6.5 **Special Hazards of Combustion Products:** Irritating vapors of unburned chemical may form in fire.

6.6 **Behavior in Fire:**

6.7 **Ignition Temperature:** 855°F

6.8 **Electrical Hazard:** Data not available

6.9 **Burning Rate:** Data not available

7. CHEMICAL REACTIVITY

7.1 **Reactivity with Water:** No reaction

7.2 **Reactivity with Common Materials:** May attack some forms of plastics.

7.3 **Stability During Transport:** Stable

7.4 **Neutralizing Agents for Acids and Caustics:** Not pertinent

7.5 **Polymerization:** Not pertinent

7.6 **Inhibitor of Polymerization:** Not pertinent

8. WATER POLLUTION

8.1 **Aquatic Toxicity:** 1.2 ppm/30 min/goldfish/killed/fresh water

8.2 **Waterfowl Toxicity:** Data not available

8.3 **Biological Oxygen Demand (BOD):** Data not available

8.4 **Food Chain Concentration Potential:** None

9. SELECTED MANUFACTURERS

1. Eastman Chemical Products, Inc. Kingsport, Tenn. 37662
2. Monsanto Company 800 North Lindbergh Blvd. St. Louis, Mo. 63166
3. Aldrich Chemical Co. 940 W. Saint Paul Ave. Milwaukee, Wis. 53233

10. SHIPPING INFORMATION

10.1 **Grades or Purity:** Technical; 99+%

10.2 **Storage Temperature:** Ambient

10.3 **Inert Atmosphere:** No requirement

10.4 **Venting:** Open

11. HAZARD ASSESSMENT CODE

(See Hazard Assessment Handbook, CG 446-3)

A-X-Y

12. HAZARD CLASSIFICATIONS

12.1 **Code of Federal Regulations:** Not listed

12.2 **NAS Hazard Rating for Bulk Water Transportation:** Not listed

12.3 **NFPA Hazard Classifications:**

Category	Classification
Health Hazard (Blue)	0
Flammability (Red)	1
Reactivity (Yellow)	0

13. PHYSICAL AND CHEMICAL PROPERTIES

13.1 **Physical State at 15°C and 1 atm:** Liquid

13.2 **Molecular Weight:** 222

13.3 **Boiling Point at 1 atm:** 569.3°F = 298.5°C = 571.7°K

13.4 **Freezing Point:** 27°F = −3°C = 270°K

13.5 **Critical Temperature:** Not pertinent

13.6 **Critical Pressure:** Not pertinent

13.7 **Specific Gravity:** 1.12 at 20°C (liquid)

13.8 **Liquid Surface Tension:** 37.5 dynes/cm = 0.0375 N/m at 20°C

13.9 **Liquid-Water Interfacial Tension:** 16.27 dynes/cm = 0.01627 N/m at 20.5°C

13.10 **Vapor (Gas) Specific Gravity:** Not pertinent

13.11 **Ratio of Specific Heats of Vapor (Gas):** Not pertinent

13.12 **Latent Heat of Vaporization:** 170 Btu/lb = 96 cal/g = 4.0×10^5 J/kg

13.13 **Heat of Combustion:** −10,920 Btu/lb = −6,070 cal/g = -254×10^5 J/kg

13.14 **Heat of Decomposition:** Not pertinent

13.15 **Heat of Solution:** Not pertinent

13.16 **Heat of Polymerization:** Not pertinent

NOTES

DEZ | DIETHYLZINC

Common Synonyms		
Zinc diethyl Ethylzinc Zinc ethyl	Watery liquid	Colorless
	IGNITES WHEN EXPOSED TO AIR. Flammable, irritating vapor is produced.	

Call fire department.
Avoid contact with liquid and vapor. Keep people away.
Stop discharge if possible.
Isolate and remove discharged material.
Notify local health and pollution control agencies.

Fire	IGNITES WHEN EXPOSED TO AIR. Extinguish with dry graphite, soda ash, or other inert powder. DO NOT USE WATER, FOAM, CARBON DIOXIDE, DRY CHEMICALS OR VAPORIZING LIQUIDS ON FIRE. DO NOT USE WATER ON ADJACENT FIRES.
Exposure	Call for medical aid. VAPOR OR DUST Irritating to eyes, nose and throat. If inhaled will cause headache, or difficult breathing. Move victim to fresh air. If in eyes, hold eyelids open and flush with plenty of water. If breathing has stopped, give artificial respiration. If breathing is difficult, give oxygen. LIQUID Will burn skin and eyes. If swallowed will cause nausea, and vomiting. Remove contaminated clothing and shoes. Flush affected areas with plenty of water. IF IN EYES, hold eyelids open and flush with plenty of water. IF SWALLOWED and victim is CONSCIOUS, have victim drink water or milk. DO NOT INDUCE VOMITING.
Water Pollution	Effect of low concentrations on aquatic life is unknown. May be dangerous if it enters water intakes. Notify local health and wildlife officials. Notify operators of nearby water intakes.

1. RESPONSE TO DISCHARGE

(See Response Methods Handbook, CG 446-4)

Issue warning—high flammability
Restrict access
Evacuate area

2. LABELS

3. CHEMICAL DESIGNATIONS

3.1 **Synonyms:** Zinc diethyl; Ethyl zinc; Zinc ethyl
3.2 **Coast Guard Compatibility Classification:** Not applicable
3.3 **Chemical Formula:** $(C_2H_5)_2Zn$
3.4 **IMCO/United Nations Numerical Designation:** 4.2/1366

4. OBSERVABLE CHARACTERISTICS

4.1 **Physical State (as shipped):** Liquid
4.2 **Color:** Colorless
4.3 **Odor:** Not pertinent

5. HEALTH HAZARDS

5.1 **Personal Protective Equipment:** Cartridge-type or fresh air mask for fumes or smoke; PVC fire-retardant or asbestos gloves; full face shield, safety glasses, or goggles; fire-retardant coveralls as standard wear; for special cases, use asbestos coat or rain suit.

5.2 **Symptoms Following Exposure:** Inhalation of mist or vapor causes immediate irritation of nose and throat; excessive or prolonged inhalation of fumes from ignition or decomposition may cause "metal fume fever" (sore throat, headache, fever, chills, nausea, vomiting, muscular aches, perspiration, constricting sensation in lungs, weakness, sometimes prostration); symptoms usually last 12-24 hrs., with complete recovery in 24-48 hrs. Eyes are immediately and severely irritated on contact with liquid, vapor, or dilute solution; without thorough irrigation, cornea may be permanently damaged. Moisture in skin combines with chemical to cause thermal and acid burns; tissue may be scarred without prompt treatment. Ingestion is unlikely but would cause immediate burns at site of contact; pain, nausea, vomiting, cramps, and diarrhea may follow; if untreated, tissue may become ulcerated.

5.3 **Treatment for Exposure:** INHALATION: move victim to fresh air and call doctor immediately; give mouth-to-mouth resuscitation if needed; keep victim warm and comfortable; oxygen should be given only by experienced person, and only on doctor's instructions. EYES: flush with large amounts of running water for at least 15 min., holding eyelids apart to insure thorough washing; get medical attention as soon as possible; do not use chemical neutralizers, and avoid oils or ointments unless prescribed by doctor. SKIN: flush affected area with large amounts of water; do not use chemical neutralizers; get medical attention if irritation persists. INGESTION: do NOT induce vomiting; have victim drink large amounts of water or milk immediately; if vomiting occurs, give more fluids; get medical attention.

6. FIRE HAZARDS

6.1 **Flash Point:** Not pertinent (ignites spontaneously)
6.2 **Flammable Limits in Air:** Not pertinent
6.3 **Fire Extinguishing Agents:** Dry chemical, sand, or powdered limestone
6.4 **Fire Extinguishing Agents Not to be Used:** Water, foam, halogenated agents, carbon dioxide
6.5 **Special Hazards of Combustion Products:** Yields zinc oxide fumes when burning; can cause "metal fume fever" (see 5.2)
6.6 **Behavior in Fire:** Reacts spontaneously with air or oxygen, and violently with water, evolving flammable ethane gas. Contact with water applied to adjacent fires will intensify the fire.
6.7 **Ignition Temperature:** Below 0°F
6.8 **Electrical Hazard:** Not pertinent
6.9 **Burning Rate:** Not pertinent

7. CHEMICAL REACTIVITY

7.1 **Reactivity with Water:** Reacts violently to form flammable ethane gas.
7.2 **Reactivity with Common Materials:** Will react with surface moisture, generating flammable ethane gas.
7.3 **Stability During Transport:** Stable
7.4 **Neutralizing Agents for Acids and Caustics:** Not pertinent
7.5 **Polymerization:** Not pertinent
7.6 **Inhibitor of Polymerization:** Not pertinent

8. WATER POLLUTION

8.1 **Aquatic Toxicity:** Not pertinent
8.2 **Waterfowl Toxicity:** Not pertinent
8.3 **Biological Oxygen Demand (BOD):** None
8.4 **Food Chain Concentration Potential:** None

9. SELECTED MANUFACTURERS

1. Ethyl Corporation
Industrial Chemicals Division
Ethyl Tower
451 Florida
Baton Rouge, La. 70801
2. Texas Alkyls, Incorporated
P. O. Box 600
Deer Park, Texas 77536
3. Ventron Corporation
Alfa Products
P. O. Box 159
Beverly, Mass. 01915

10. SHIPPING INFORMATION

10.1 **Grades or Purity:** 95-98%. Also shipped as 15-25% by weight solutions in hydrocarbon solvents.
10.2 **Storage Temperature:** Ambient
10.3 **Inert Atmosphere:** Inerted with dry nitrogen gas
10.4 **Venting:** Safety relief

11. HAZARD ASSESSMENT CODE

(See Hazard Assessment Handbook, CG 446-3)

A-O-Z

12. HAZARD CLASSIFICATIONS

12.1 **Code of Federal Regulations:** Flammable liquid
12.2 **NAS Hazard Rating for Bulk Water Transportation:** Not listed
12.3 **NFPA Hazard Classifications:**

Category	Classification
Health Hazard (Blue)	0
Flammability (Red)	3
Reactivity (Yellow)	3 W̶

13. PHYSICAL AND CHEMICAL PROPERTIES

13.1 **Physical State at 15°C and 1 atm:** Liquid
13.2 **Molecular Weight:** 123.5
13.3 **Boiling Point at 1 atm:** 255°F = 124°C = 397°K
13.4 **Freezing Point:** −18°F = −28°C = 245°K
13.5 **Critical Temperature:** Not pertinent
13.6 **Critical Pressure:** Not pertinent
13.7 **Specific Gravity:** 1.207 at 20°C (liquid)
13.8 **Liquid Surface Tension:** (est.) 20 dynes/cm = 0.020 N/m at 20°C
13.9 **Liquid-Water Interfacial Tension:** Not pertinent
13.10 **Vapor (Gas) Specific Gravity:** Not pertinent
13.11 **Ratio of Specific Heats of Vapor (Gas):** Not pertinent
13.12 **Latent Heat of Vaporization:** 120 Btu/lb = 68 cal/g = 2.8×10^5 J/kg
13.13 **Heat of Combustion:** −11,700 Btu/lb = −6,495 cal/g = -272×10^5 J/kg
13.14 **Heat of Decomposition:** Not pertinent
13.15 **Heat of Solution:** Not pertinent
13.16 **Heat of Polymerization:** Not pertinent

5. HEALTH HAZARDS (Cont'd.)

5.4 **Toxicity by Inhalation (Threshold Limit Value):** Not pertinent
5.5 **Short-Term Inhalation Limits:** Not pertinent
5.6 **Toxicity by Ingestion:** Not pertinent
5.7 **Late Toxicity:** Not pertinent
5.8 **Vapor (Gas) Irritant Characteristics:** Data not available
5.9 **Liquid or Solid Irritant Characteristics:** Data not available
5.10 **Odor Threshold:** Not pertinent

DFE	1,1-DIFLUOROETHANE

Common Synonyms
Ethylidene fluoride
Ethylidene difluoride
Refrigerant 152a

Liquefied compressed gas — Colorless

Sinks and boils in water. Flammable, irritating vapor is produced.
Visible vapor cloud is produced. Boiling point is 76°F.

Shut off ignition sources. Call fire department.
Avoid contact with liquid and vapor. Keep people away.
Stop discharge if possible.
Stay upwind. Use water spray to "knock down" vapor.
Isolate and remove discharged material.
Notify local health and pollution control agencies.

Fire

FLAMMABLE.
POISONOUS GASES MAY BE PRODUCED IN FIRE.
Containers may explode in fire.
Flashback along vapor trail may occur.
Vapor may explode if ignited in an enclosed area.
Wear goggles and self-contained breathing apparatus.
Stop flow of gas if possible.
Let fire burn.
Cool exposed containers with water.

Exposure

CALL FOR MEDICAL AID.

VAPOR
Irritating to eyes, nose and throat.
Harmful if inhaled.
Move victim to fresh air.
If breathing has stopped, give artificial respiration.
If breathing is difficult, give oxygen.

LIQUID
Will burn eyes.
Will cause frostbite.
Harmful if swallowed.
Remove contaminated clothing and shoes.
Flush affected areas with plenty of water.
DO NOT RUB AFFECTED AREAS.
IF SWALLOWED and victim is CONSCIOUS, have victim drink water or milk.

Water Pollution

Not harmful to aquatic life.

1. RESPONSE TO DISCHARGE

(See Response Methods Handbook, CG 446-4)

Issue warning—high flammability
Restrict access
Evacuate area

2. LABEL

Red

3. CHEMICAL DESIGNATIONS

3.1 **Synonyms:** Ethylidene difluoride; Ethylidene fluoride; Refrigerant 152a
3.2 **Coast Guard Compatibility Classification:** Not listed
3.3 **Chemical Formula:** CH_3CHF_2
3.4 **IMCO/United Nations Numerical Designation:** 2/1030

4. OBSERVABLE CHARACTERISTICS*

4.1 **Physical State (as shipped):** Liquefied compressed gas
4.2 **Color:** Colorless
4.3 **Odor:** Faint

5. HEALTH HAZARDS

5.1 **Personal Protective Equipment:** Individual breathing devices with air supply; neoprene gloves; protective clothing; eye protection
5.2 **Symptoms Following Exposure:** Inhalation of concentrated gas will cause suffocation. Contact with liquid can damage eyes because of low temperature. Frostbite may result from contact with liquid.
5.3 **Treatment for Exposure:** INHALATION: remove to fresh air; use artificial respiration if necessary. EYES: get medical attention if liquid has entered eyes. SKIN: soak in lukewarm water (for frostbite).
5.4 **Toxicity by Inhalation (Threshold Limit Value):** Data not available
5.5 **Short-Term Inhalation Limits:** Data not available
5.6 **Toxicity by Ingestion:** Not pertinent (boils at −24.7°C)
5.7 **Late Toxicity:** Data not available
5.8 **Vapor (Gas) Irritant Characteristics:** Data not available
5.9 **Liquid or Solid Irritant Characteristics:** Data not available
5.10 **Odor Threshold:** Data not available

6. FIRE HAZARDS

6.1 **Flash Point:** Not pertinent
6.2 **Flammable Limits in Air:** 3.7%–18%
6.3 **Fire Extinguishing Agents:** Shut off gas source; use water to cool adjacent combustibles.
6.4 **Fire Extinguishing Agents Not to be Used:**
6.5 **Special Hazards of Combustion Products:** Irritating hydrogen fluoride fumes may form in fire.
6.6 **Behavior in Fire:** Containers may explode. Vapors are heavier than air and may travel a considerable distance to an ignition source and flash back.
6.7 **Ignition Temperature:** Data not available
6.8 **Electrical Hazard:** Data not available
6.9 **Burning Rate:** Not pertinent

7. CHEMICAL REACTIVITY

7.1 **Reactivity with Water:** No reaction
7.2 **Reactivity with Common Materials:**
7.3 **Stability During Transport:** Stable
7.4 **Neutralizing Agents for Acids and Caustics:** Not pertinent
7.5 **Polymerization:** Not pertinent
7.6 **Inhibitor of Polymerization:** Not pertinent

8. WATER POLLUTION

8.1 **Aquatic Toxicity:** Data not available
8.2 **Waterfowl Toxicity:** Data not available
8.3 **Biological Oxygen Demand (BOD):** None
8.4 **Food Chain Concentration Potential:** None

9. SELECTED MANUFACTURERS

E. I. du Pont de Nemours & Co. (Inc.)
Freon Products Div.
Wilmington, Del. 19898

10. SHIPPING INFORMATION

10.1 **Grades or Purity:** Commercial
10.2 **Storage Temperature:** Ambient
10.3 **Inert Atmosphere:** No requirement
10.4 **Venting:** Safety relief

11. HAZARD ASSESSMENT CODE

(See Hazard Assessment Handbook, CG 446-3)

A-B-C-D-E-F-G

12. HAZARD CLASSIFICATIONS

12.1 **Code of Federal Regulations:** Flammable compressed gas
12.2 **NAS Hazard Rating for Bulk Water Transportation:** Not listed
12.3 **NFPA Hazard Classifications:** Not listed

13. PHYSICAL AND CHEMICAL PROPERTIES

13.1 **Physical State at 15°C and 1 atm:** Gas
13.2 **Molecular Weight:** 66.05
13.3 **Boiling Point at 1 atm:** 52.3°F = 11.3°C = 248.5°K
13.4 **Freezing Point:** −179°F = −117°C = 156°K
13.5 **Critical Temperature:** 236.3°F = 113.5°C = 386.6°K
13.6 **Critical Pressure:** 652 psia = 44.37 atm = 4.50 MN/m²
13.7 **Specific Gravity:** 0.95 at 20°C (liquid)
13.8 **Liquid Surface Tension:** 11.25 dynes/cm = 0.01125 N/m at 20°C
13.9 **Liquid-Water Interfacial Tension:** Data not available
13.10 **Vapor (Gas) Specific Gravity:** 2.3
13.11 **Ratio of Specific Heats of Vapor (Gas):** 1.141
13.12 **Latent Heat of Vaporization:** 140.5 Btu/lb = 78.03 cal/g = 3.265×10^5 J/kg
13.13 **Heat of Combustion:** −7,950 Btu/lb = −4,420 cal/g = -185×10^5 J/kg
13.14 **Heat of Decomposition:** Not pertinent
13.15 **Heat of Solution:** Not pertinent
13.16 **Heat of Polymerization:** Not pertinent

NOTES

DFA DIFLUOROPHOSPHORIC ACID, ANHYDROUS

Common Synonyms			
Difluorophosphorus acid	Liquid	Colorless	Sharp, irritating odor
	Reacts violently with water. Irritating gas is produced on contact with water.		

AVOID CONTACT WITH LIQUID AND VAPOR. KEEP PEOPLE AWAY.
Wear goggles, self-contained breathing apparatus, and rubber overclothing (including gloves).
Stop discharge if possible.
Isolate and remove discharged material.
Notify local health and pollution control agencies.

Fire

Not flammable.
POISONOUS GASES MAY BE PRODUCED WHEN HEATED.

Exposure

Call for medical aid.

VAPOR
Irritating to eyes, nose and throat.
Move victim to fresh air.
If in eyes, hold eyelids open and flush with plenty of water.
If breathing is difficult, give oxygen.

LIQUID
Will burn skin and eyes.
Harmful if swallowed.
Remove contaminated clothing and shoes.
Flush affected areas with plenty of water.
IF IN EYES, hold eyelids open and flush with plenty of water.
IF SWALLOWED and victim is CONSCIOUS, have victim drink water or milk.
DO NOT INDUCE VOMITING.

Water Pollution

Effect of low concentrations on aquatic life is unknown.
May be dangerous if it enters water intakes.
Notify local health and wildlife officials.
Notify operators of nearby water intakes.

1. RESPONSE TO DISCHARGE
(See Response Methods Handbook, CG 446-4)

Issue warning—poison, corrosive, water contaminant
Restrict access
Evacuate area
Disperse and flush with care

2. LABEL

3. CHEMICAL DESIGNATIONS

3.1 **Synonyms:** Difluorophosphorus acid
3.2 **Coast Guard Compatibility Classification:** Not applicable
3.3 **Chemical Formula:** $HOPOF_2$
3.4 **IMCO/United Nations Numerical Designation:** 8/1768

4. OBSERVABLE CHARACTERISTICS

4.1 **Physical State (as shipped):** Liquid
4.2 **Color:** Colorless
4.3 **Odor:** Sharp, very irritating

5. HEALTH HAZARDS

5.1 **Personal Protective Equipment:** Air line mask or self-contained breathing apparatus; full protective clothing.
5.2 **Symptoms Following Exposure:** Inhalation causes severe irritation of upper respiratory tract. Contact with liquid causes severe irritation of eyes and skin. Ingestion causes severe burns of mouth and stomach.
5.3 **Treatment for Exposure:** Get medical attention as soon as possible following exposures to this compound. INHALATION: remove from exposure and support respiration. EYES: wash with copious volumes of water for at least 15 min. SKIN: wash with large amounts of water for 15 min. INGESTION: if victim is conscious, have him drink large amounts of water followed by milk or milk of magnesia.
5.4 **Toxicity by Inhalation (Threshold Limit Value):** Data not available
5.5 **Short-Term Inhalation Limits:** Data not available
5.6 **Toxicity by Ingestion:** Data not available
5.7 **Late Toxicity:** Data not available
5.8 **Vapor (Gas) Irritant Characteristics:** Vapors cause moderate irritation such that personnel will find high concentrations unpleasant. The effect is temporary.
5.9 **Liquid or Solid Irritant Characteristics:** Severe skin irritant. Causes second- and third-degree burns on short contact and is very injurious to the eyes.
5.10 **Odor Threshold:** Data not available

6. FIRE HAZARDS

6.1 **Flash Point:** Not flammable
6.2 **Flammable Limits in Air:** Not flammable
6.3 **Fire Extinguishing Agents:** Not pertinent
6.4 **Fire Extinguishing Agents Not to be Used:** Do not use water on adjacent fires.
6.5 **Special Hazards of Combustion Products:** Irritating and toxic fumes of hydrogen fluoride and phosphoric acid may be formed in fires.
6.6 **Behavior in Fire:** Not pertinent
6.7 **Ignition Temperature:** Not pertinent
6.8 **Electrical Hazard:** Not pertinent
6.9 **Burning Rate:** Not pertinent

7. CHEMICAL REACTIVITY

7.1 **Reactivity with Water:** Reacts vigorously to form corrosive and toxic hydrofluoric acid.
7.2 **Reactivity with Common Materials:** In the presence of moisture, is corrosive to glass, other siliceous materials, and most metals.
7.3 **Stability During Transport:** Stable
7.4 **Neutralizing Agents for Acids and Caustics:** Flush with water, rinse with sodium bicarbonate or lime solution.
7.5 **Polymerization:** Not pertinent
7.6 **Inhibitor of Polymerization:** Not pertinent

8. WATER POLLUTION

8.1 **Aquatic Toxicity:** Data not available
8.2 **Waterfowl Toxicity:** Data not available
8.3 **Biological Oxygen Demand (BOD):** None
8.4 **Food Chain Concentration Potential:** None

9. SELECTED MANUFACTURERS

1. Ozark-Mahoning Company
   1870 South Boulder
   Tulsa, Okla. 74119
2. Pfaltz and Bauer, Inc.
   126-04 Northern Boulevard
   Flushing, N. Y. 11368

10. SHIPPING INFORMATION

10.1 **Grades or Purity:** FP Acid No. 2, 90+%; Commercial, 96+% plus 3.5% monofluorophosphoric acid
10.2 **Storage Temperature:** Ambient
10.3 **Inert Atmosphere:** No requirement
10.4 **Venting:** Pressure-vacuum

11. HAZARD ASSESSMENT CODE
(See Hazard Assessment Handbook, CG 446-3)

A-O

12. HAZARD CLASSIFICATIONS

12.1 **Code of Federal Regulations:** Corrosive liquid
12.2 **NAS Hazard Rating for Bulk Water Transportation:**

Category	Rating
Fire	0
Health	
Vapor Irritant	2
Liquid or Solid Irritant	4
Poisons	2
Water Pollution	
Human Toxicity	
Aquatic Toxicity	
Aesthetic Effect	2
Reactivity	
Other Chemicals	3
Water	1
Self-Reaction	0

12.3 **NFPA Hazard Classifications:** Not listed

13. PHYSICAL AND CHEMICAL PROPERTIES

13.1 **Physical State at 15°C and 1 atm:** Liquid
13.2 **Molecular Weight:** 103.0
13.3 **Boiling Point at 1 atm:** 241°F = 116°C = 389°K
13.4 **Freezing Point:** −139°F = −95°C = 178°K
13.5 **Critical Temperature:** Not pertinent
13.6 **Critical Pressure:** Not pertinent
13.7 **Specific Gravity:** 1.583 at 25°C (liquid)
13.8 **Liquid Surface Tension:** Data not available
13.9 **Liquid-Water Interfacial Tension:** Not pertinent
13.10 **Vapor (Gas) Specific Gravity:** Not pertinent
13.11 **Ratio of Specific Heats of Vapor (Gas):** Not pertinent
13.12 **Latent Heat of Vaporization:** 140 Btu/lb = 77 cal/g = 3.2×10^5 J/kg
13.13 **Heat of Combustion:** Not pertinent
13.14 **Heat of Decomposition:** Not pertinent
13.15 **Heat of Solution:** Data not available
13.16 **Heat of Polymerization:** Not pertinent

NOTES

DHP — DIHEPTYL PHTHALATE

Common Synonyms
Phthalic acid, diheptyl ester

Liquid — White — Odorless

May float or sink in water.

Stop discharge if possible. Keep people away.
Avoid contact with liquid.
Call fire department.
Isolate and remove discharged material.
Notify local health and pollution control agencies.

Fire

Combustible.
Extinguish with dry chemicals, foam or carbon dioxide.
Water may be ineffective on fire.

Exposure

CALL FOR MEDICAL AID.

LIQUID
Irritating to skin and eyes.
Harmful if swallowed.

Remove contaminated clothing and shoes.
Flush affected areas with plenty of water.
IF IN EYES, hold eyelids open and flush with plenty of water.
IF SWALLOWED and victim is CONSCIOUS, have victim drink water or milk.
IF SWALLOWED and victim is UNCONSCIOUS OR HAVING CONVULSIONS, do nothing except keep victim warm.

Water Pollution

Effect of low concentrations on aquatic life is unknown.
Fouling to shoreline.
May be dangerous if it enters water intakes.
Notify local health and wildlife officials.
Notify operators of nearby water intakes.

1. RESPONSE TO DISCHARGE

(See Response Methods Handbook, CG 446-4)

Mechanical containment
Should be removed
Chemical and physical treatment

2. LABELS

No hazard label required by Code of Federal Regulations

3. CHEMICAL DESIGNATIONS

3.1 **Synonyms:** Phthalic acid, diheptyl ester
3.2 **Coast Guard Compatibility Classification:** Ester
3.3 **Chemical Formula:** $C_6H_4(COOC_7H_{15})_2$
3.4 **IMCO/United Nations Numerical Designation:** Not listed

4. OBSERVABLE CHARACTERISTICS

4.1 **Physical State (as shipped):** Liquid
4.2 **Color:** Colorless
4.3 **Odor:** None

5. HEALTH HAZARDS

5.1 **Personal Protective Equipment:** Goggles or face shield; rubber gloves
5.2 **Symptoms Following Exposure:** Inhalation of vapors from very hot material may cause headache, drowsiness, and convulsions. Contact with eyes may cause irritation.
5.3 **Treatment for Exposure:** INHALATION: move to fresh air. EYES: flush with water. SKIN: wipe off, flush with water; wash with soap and water.
5.4 **Toxicity by Inhalation (Threshold Limit Value):** Data not available
5.5 **Short-Term Inhalation Limits:** Data not available
5.6 **Toxicity by Ingestion:** Data not available
5.7 **Late Toxicity:** Data not available
5.8 **Vapor (Gas) Irritant Characteristics:** Data not available
5.9 **Liquid or Solid Irritant Characteristics:** Data not available
5.10 **Odor Threshold:** Data not available

6. FIRE HAZARDS

6.1 **Flash Point:** Data not available
6.2 **Flammable Limits in Air:** Data not available
6.3 **Fire Extinguishing Agents:** Foam, dry chemical, carbon dioxide
6.4 **Fire Extinguishing Agents Not to be Used:** Water may be ineffective.
6.5 **Special Hazards of Combustion Products:**
6.6 **Behavior in Fire:**
6.7 **Ignition Temperature:** Data not available
6.8 **Electrical Hazard:** Data not available
6.9 **Burning Rate:** Data not available

7. CHEMICAL REACTIVITY

7.1 **Reactivity with Water:** No reaction
7.2 **Reactivity with Common Materials:** May attack some forms of plastics.
7.3 **Stability During Transport:** Stable
7.4 **Neutralizing Agents for Acids and Caustics:** Not pertinent
7.5 **Polymerization:** Not pertinent
7.6 **Inhibitor of Polymerization:** Not pertinent

8. WATER POLLUTION

8.1 **Aquatic Toxicity:** Data not available
8.2 **Waterfowl Toxicity:** Data not available
8.3 **Biological Oxygen Demand (BOD):** Data not available
8.4 **Food Chain Concentration Potential:** None

9. SELECTED MANUFACTURERS

1. K & K Laboratories, Inc.
121 Express Street
Plainview, N. Y. 11803
2. Research Organic/Inorganic Chemical Corp.
11686 Sheldon
Sun Valley, Calif. 91352

10. SHIPPING INFORMATION

10.1 **Grades or Purity:** Pure
10.2 **Storage Temperature:** Ambient
10.3 **Inert Atmosphere:** No requirement
10.4 **Venting:** Open

11. HAZARD ASSESSMENT CODE

(See Hazard Assessment Handbook, CG 446-3)

A-T-U-X-Y

12. HAZARD CLASSIFICATIONS

12.1 **Code of Federal Regulations:** Not listed
12.2 **NAS Hazard Rating for Bulk Water Transportation:** Not listed
12.3 **NFPA Hazard Classifications:** Not listed

13. PHYSICAL AND CHEMICAL PROPERTIES

13.1 **Physical State at 15°C and 1 atm:** Liquid
13.2 **Molecular Weight:** 362
13.3 **Boiling Point at 1 atm:** Not pertinent (decomposes)
13.4 **Freezing Point:** Not pertinent
13.5 **Critical Temperature:** Not pertinent
13.6 **Critical Pressure:** Not pertinent
13.7 **Specific Gravity:** (est.) 1.0 at 20°C (liquid)
13.8 **Liquid Surface Tension:** Data not available
13.9 **Liquid-Water Interfacial Tension:** Data not available
13.10 **Vapor (Gas) Specific Gravity:** Not pertinent
13.11 **Ratio of Specific Heats of Vapor (Gas):** Not pertinent
13.12 **Latent Heat of Vaporization:** Data not available
13.13 **Heat of Combustion:** (est.) $-16,850$ Btu/lb $= -9,370$ cal/g $= -392 \times 10^5$ J/kg
13.14 **Heat of Decomposition:** Not pertinent
13.15 **Heat of Solution:** Not pertinent
13.16 **Heat of Polymerization:** Not pertinent

NOTES

DBC — DIISOBUTYLCARBINOL

Common Synonyms	Oily liquid	Colorless
2,6-Dimethyl-4-heptanol	Floats on water.	

Stop discharge if possible.
Call fire department.
Isolate and remove discharged material.
Notify local health and pollution control agencies.

Fire	Combustible. Extinguish with dry chemical, alcohol foam, or carbon dioxide. Cool exposed containers with water.
Exposure	CALL FOR MEDICAL AID. LIQUID Harmful if swallowed. IF SWALLOWED and victim is CONSCIOUS, have victim drink water or milk.
Water Pollution	Effect of low concentrations on aquatic life is unknown. Fouling to shoreline. May be dangerous if it enters water intakes. Notify local health and wildlife officials. Notify operators of nearby water intakes.

1. RESPONSE TO DISCHARGE
(See Response Methods Handbook, CG 446-4)

Mechanical containment
Chemical and physical treatment

2. LABELS

No hazard label required by Code of Federal Regulations

3. CHEMICAL DESIGNATIONS

3.1 **Synonyms:** 2,6 Dimethyl-4-heptanol
3.2 **Coast Guard Compatibility Classification:** Alcohol
3.3 **Chemical Formula:** $[(CH_3)_2CHCH_2]_2CHOH$
3.4 **IMCO/United Nations Numerical Designation:** Not listed

4. OBSERVABLE CHARACTERISTICS

4.1 **Physical State (as shipped):** Liquid
4.2 **Color:** Colorless
4.3 **Odor:** Characteristic

5. HEALTH HAZARDS

5.1 **Personal Protective Equipment:** Air-supplied mask for prolonged exposure; plastic gloves; goggles.
5.2 **Symptoms Following Exposure:** None expected
5.3 **Treatment for Exposure:** SKIN AND EYES: flush with water
5.4 **Toxicity by Inhalation (Threshold Limit Value):** Not pertinent
5.5 **Short-Term Inhalation Limits:** Not pertinent
5.6 **Toxicity by Ingestion:** Grade 2; LD_{50} 0.5 to 5 g/kg (rat)
5.7 **Late Toxicity:** Data not available
5.8 **Vapor (Gas) Irritant Characteristics:** None
5.9 **Liquid or Solid Irritant Characteristics:** None
5.10 **Odor Threshold:** Data not available

6. FIRE HAZARDS

6.1 **Flash Point:** 162°F O.C.; 165°F C.C.
6.2 **Flammable Limits in Air:** 0.8%—6.1%
6.3 **Fire Extinguishing Agents:** Carbon dioxide, dry chemical, alcohol foam
6.4 **Fire Extinguishing Agents Not to be Used:** Not pertinent
6.5 **Special Hazards of Combustion Products:** Not pertinent
6.6 **Behavior in Fire:** Not pertinent
6.7 **Ignition Temperature:** 494°F (calc.)
6.8 **Electrical Hazard:** Not pertinent
6.9 **Burning Rate:** Data not available

7. CHEMICAL REACTIVITY

7.1 **Reactivity with Water:** No reaction
7.2 **Reactivity with Common Materials:** No reaction
7.3 **Stability During Transport:** Stable
7.4 **Neutralizing Agents for Acids and Caustics:** Not pertinent
7.5 **Polymerization:** Not pertinent
7.6 **Inhibitor of Polymerization:** Not pertinent

8. WATER POLLUTION

8.1 **Aquatic Toxicity:** Data not available
8.2 **Waterfowl Toxicity:** Data not available
8.3 **Biological Oxygen Demand (BOD):** Data not available
8.4 **Food Chain Concentration Potential:** None

9. SELECTED MANUFACTURERS

Union Carbide Corp.
Chemicals and Plastics Division
270 Park Ave.
New York, N. Y. 10017

10. SHIPPING INFORMATION

10.1 **Grades or Purity:** 98.0%
10.2 **Storage Temperature:** Ambient
10.3 **Inert Atmosphere:** No requirement
10.4 **Venting:** Open (flame arrester)

11. HAZARD ASSESSMENT CODE
(See Hazard Assessment Handbook, CG 446-3)

A-T-U

12. HAZARD CLASSIFICATIONS

12.1 **Code of Federal Regulations:** Combustible Liquid
12.2 **NAS Hazard Rating for Bulk Water Transportation:** Not listed
12.3 **NFPA Hazard Classifications:** Not listed

13. PHYSICAL AND CHEMICAL PROPERTIES

13.1 **Physical State at 15°C and 1 atm:** Liquid
13.2 **Molecular Weight:** 144.26
13.3 **Boiling Point at 1 atm:** 352°F = 178°C = 451°K
13.4 **Freezing Point:** −85°F = −65°C = 208°K
13.5 **Critical Temperature:** Not pertinent
13.6 **Critical Pressure:** Not pertinent
13.7 **Specific Gravity:** 0.812 at 20°C (liquid)
13.8 **Liquid Surface Tension:** Data not available
13.9 **Liquid-Water Interfacial Tension:** Data not available
13.10 **Vapor (Gas) Specific Gravity:** Not pertinent
13.11 **Ratio of Specific Heats of Vapor (Gas):** Not pertinent
13.12 **Latent Heat of Vaporization:** 140 Btu/lb = 76 cal/g = 3.2×10^5 J/kg
13.13 **Heat of Combustion:** (est.) −17,400 Btu/lb = −9,680 cal/g = -405×10^5 J/kg
13.14 **Heat of Decomposition:** Not pertinent
13.15 **Heat of Solution:** Not pertinent
13.16 **Heat of Polymerization:** Not pertinent

NOTES

DBL

DIISOBUTYLENE

Common Synonyms			
2,4,4-Trimethyl-1-pentene	Liquid	Colorless	Gasoline-like odor
	Floats on water. Flammable, irritating vapor is produced.		

Stop discharge if possible. Keep people away.
Shut off ignition sources and call fire department.
Stay upwind and use water spray to "knock down" vapor.
Isolate and remove discharged material.
Notify local health and pollution control agencies.

Fire

FLAMMABLE.
Flashback along vapor trail may occur.
Vapor may explode if ignited in an enclosed area.
Extinguish with dry chemical, foam, or carbon dioxide.
Water may be ineffective on fire.
Cool exposed containers with water.

Exposure

CALL FOR MEDICAL AID.

VAPOR
Irritating to eyes, nose and throat.
If inhaled, will cause dizziness, headache, difficult breathing or loss of consciousness.
Move to fresh air.
If breathing has stopped, give artificial respiration.
If breathing is difficult, give oxygen.

LIQUID
Irritating to skin and eyes.
If swallowed, will cause nausea or vomiting.
Remove contaminated clothing and shoes.
Flush affected areas with plenty of water.
IF IN EYES, hold eyelids open and flush with plenty of water.
IF SWALLOWED and victim is CONSCIOUS, have victim drink water or milk.
DO NOT INDUCE VOMITING.

Water Pollution

Effect of low concentrations on aquatic life is unknown.
Fouling to shoreline.
May be dangerous if it enters water intakes.
Notify local health and wildlife officials.
Notify operators of nearby water intakes.

1. RESPONSE TO DISCHARGE
(See Response Methods Handbook, CG 446-4)

Issue warning—high flammability
Evacuate area

2. LABEL

Red

3. CHEMICAL DESIGNATIONS

3.1 **Synonyms:** 2,4,4-Trimethyl-1-pentene
3.2 **Coast Guard Compatibility Classification:** Olefin
3.3 **Chemical Formula:** $(CH_3)_3CCH_2C(CH_3)=CH_2$
3.4 **IMCO/United Nations Numerical Designation:** 3.2/2050

4. OBSERVABLE CHARACTERISTICS

4.1 **Physical State (as shipped):** Liquid
4.2 **Color:** Colorless
4.3 **Odor:** Like gasoline

5. HEALTH HAZARDS

5.1 **Personal Protective Equipment:** Protective goggles.
5.2 **Symptoms Following Exposure:** Low general toxicity; may act as simple asphyxiant in high vapor concentrations.
5.3 **Treatment for Exposure:** INHALATION: remove from exposure; support respiration.
5.4 **Toxicity by Inhalation (Threshold Limit Value):** Data not available
5.5 **Short-Term Inhalation Limits:** Data not available
5.6 **Toxicity by Ingestion:** Data not available
5.7 **Late Toxicity:** Liver and kidney damage in exp. animals.
5.8 **Vapor (Gas) Irritant Characteristics:** Vapors are nonirritating to the eyes and throat.
5.9 **Liquid or Solid Irritant Characteristics:** Minimum hazard. If spilled on clothing and allowed to remain, may cause smarting and reddening of the skin.
5.10 **Odor Threshold:** Data not available

6. FIRE HAZARDS

6.1 **Flash Point:** 35°F (est.)
6.2 **Flammable Limits in Air:** 0.9% LEL (est.)
6.3 **Fire Extinguishing Agents:** Dry chemicals, foam, or carbon dioxide
6.4 **Fire Extinguishing Agents Not to be Used:** Water may be ineffective
6.5 **Special Hazards of Combustion Products:** Not pertinent
6.6 **Behavior in Fire:** Not pertinent
6.7 **Ignition Temperature:** 788°F
6.8 **Electrical Hazard:** Not pertinent
6.9 **Burning Rate:** 7.9 mm/min.

7. CHEMICAL REACTIVITY

7.1 **Reactivity with Water:** No reaction
7.2 **Reactivity with Common Materials:** No reaction
7.3 **Stability During Transport:** Stable
7.4 **Neutralizing Agents for Acids and Caustics:** Not pertinent
7.5 **Polymerization:** Not pertinent
7.6 **Inhibitor of Polymerization:** Not pertinent

8. WATER POLLUTION

8.1 **Aquatic Toxicity:** Data not available
8.2 **Waterfowl Toxicity:** Data not available
8.3 **Biological Oxygen Demand (BOD):** Data not available
8.4 **Food Chain Concentration Potential:** None

9. SELECTED MANUFACTURERS

1. B. F. Goodrich Co.
6100 Oak Tree Blvd.
Cleveland, Ohio 44131
2. Petro-Tex Chemical Corp.
8600 Park Place
Houston, Texas 77017
3. Texaco, Inc.
135 East 42nd St.
New York, N. Y. 10017

10. SHIPPING INFORMATION

10.1 **Grades or Purity:** Research grade: 99.86%; Pure grade: 99.39%; Technical grade: 98.7%
10.2 **Storage Temperature:** Ambient
10.3 **Inert Atmosphere:** No requirement
10.4 **Venting:** Open (flame arrester) or pressure-vacuum

11. HAZARD ASSESSMENT CODE
(See Hazard Assessment Handbook, CG 446-3)

A-T-U-V-W

12. HAZARD CLASSIFICATIONS

12.1 **Code of Federal Regulations:** Flammable liquid
12.2 **NAS Hazard Rating for Bulk Water Transportation:**

Category	Rating
Fire	3
Health	
Vapor Irritant	0
Liquid or Solid Irritant	1
Poisons	0
Water Pollution	
Human Toxicity	1
Aquatic Toxicity	1
Aesthetic Effect	2
Reactivity	
Other Chemicals	1
Water	0
Self-Reaction	1

12.3 **NFPA Hazard Classifications:** Not listed

13. PHYSICAL AND CHEMICAL PROPERTIES

13.1 **Physical State at 15°C and 1 atm:** Liquid
13.2 **Molecular Weight:** 112.22
13.3 **Boiling Point at 1 atm:** 214.7°F = 101.5°C = 374.7°K
13.4 **Freezing Point:** −136.3°F = −93.5°C = 179.7°K
13.5 **Critical Temperature:** 548°F = 286.7°C = 559.9°K
13.6 **Critical Pressure:** 380 psia = 25.85 atm = 2.619 MN/m²
13.7 **Specific Gravity:** 0.715 at 20°C (liquid)
13.8 **Liquid Surface Tension:** 20.7 dynes/cm = 0.0207 N/m at 20°C
13.9 **Liquid-Water Interfacial Tension:** Data not available
13.10 **Vapor (Gas) Specific Gravity:** Not pertinent
13.11 **Ratio of Specific Heats of Vapor (Gas):** (est.) 1.049
13.12 **Latent Heat of Vaporization:** 110 Btu/lb = 60 cal/g = 2.5×10^5 J/kg
13.13 **Heat of Combustion:** −18,900 Btu/lb = −10,500 cal/g = -440×10^5 J/kg
13.14 **Heat of Decomposition:** Not pertinent
13.15 **Heat of Solution:** Not pertinent
13.16 **Heat of Polymerization:** Not pertinent

NOTES

REVISED 1978

DIK — DIISOBUTYL KETONE

Common Synonyms			
2,6-Dimethyl-4-heptane 5-Diisopropylacetone Isovalerone DIBK	Liquid	Colorless	Mild, sweet odor
	Floats on water.		

Stop discharge if possible. Keep people away.
Avoid contact with liquid and vapor.
Call fire department.
Isolate and remove discharged material.
Notify local health and pollution control agencies.

Fire

Combustible.
Extinguish with dry chemicals, foam or carbon dioxide.
Water may be ineffective on fire.

Exposure

CALL FOR MEDICAL AID.

VAPOR
Irritating to eyes, nose and throat.
If inhaled will cause coughing or difficult breathing.
If in eyes, hold eyelids open and flush with plenty of water.
If breathing has stopped, give artificial respiration.
If breathing is difficult, give oxygen.

LIQUID
Irritating to skin and eyes.
If swallowed will cause nausea and vomiting.
Remove contaminated clothing and shoes.
Flush affected areas with plenty of water.
IF IN EYES, hold eyelids open and flush with plenty of water.
IF SWALLOWED and victim is CONSCIOUS, have victim drink water or milk.
IF SWALLOWED and victim is UNCONSCIOUS OR HAVING CONVULSIONS, do nothing except keep victim warm.

Water Pollution

Effect of low concentrations on aquatic life is unknown.
Fouling to shoreline.
May be dangerous if it enters water intakes.
Notify local health and wildlife officials.
Notify operators of nearby water intakes.

1. RESPONSE TO DISCHARGE

(See Response Methods Handbook, CG 446-4)

Mechanical containment
Should be removed
Chemical and physical treatment

2. LABELS

No hazard label required by Code of Federal Regulations

3. CHEMICAL DESIGNATIONS

3.1 **Synonyms:** DIBK; sym-Diisopropylacetone; 2,6-Dimethyl-4-heptanone; Isovalerone

3.2 **Coast Guard Compatibility Classification:** Ketone

3.3 **Chemical Formula:** $(CH_3)_2CHCH_2COCH_2CH(CH_3)_2$ or $C_9H_{18}O$

3.4 **IMCO/United Nations Numerical Designation:** 3.3/1157

4. OBSERVABLE CHARACTERISTICS

4.1 **Physical State (as shipped):** Liquid

4.2 **Color:** Colorless

4.3 **Odor:** Mild; characteristic ketonic

5. HEALTH HAZARDS

5.1 **Personal Protective Equipment:** Air-supplied mask in confined areas; plastic gloves; face shield and safety glasses

5.2 **Symptoms Following Exposure:** Inhalation of vapor causes irritation of nose and throat. Ingestion causes irritation of mouth and stomach. Vapor irritates eyes. Contact with liquid irritates skin.

5.3 **Treatment for Exposure:** INHALATION: move to fresh air; give oxygen if breathing is difficult; call a physician. EYES: flush with plenty of water. SKIN: wipe off; flush with plenty of water; wash with soap and water.

5.4 **Toxicity by Inhalation (Threshold Limit Value):** 25 ppm

5.5 **Short-Term Inhalation Limits:** 50 ppm, 30 min.

5.6 **Toxicity by Ingestion:** Grade 2; oral LD_{50} = 1.4 g/kg (mouse), 5.75 g/kg (rat)

5.7 **Late Toxicity:** Causes increased liver and kidney weights in rats, decreased liver weights in guinea pigs

5.8 **Vapor (Gas) Irritant Characteristics:** Vapors cause moderate irritation such that personnel will find high concentrations unpleasant. The effect is temporary.

5.9 **Liquid or Solid Irritant Characteristics:** Minimum hazard. If spilled on clothing and allowed to remain, may cause smarting and reddening of skin.

5.10 **Odor Threshold:** Data not available

6. FIRE HAZARDS

6.1 **Flash Point:** 131°F O.C., 120°F C.C.

6.2 **Flammable Limits in Air:** 0.81% – 7.1% at 200°F

6.3 **Fire Extinguishing Agents:** Foam, dry chemical, carbon dioxide

6.4 **Fire Extinguishing Agents Not to be Used:** Water may be ineffective.

6.5 **Special Hazards of Combustion Products:**

6.6 **Behavior in Fire:**

6.7 **Ignition Temperature:** 745

6.8 **Electrical Hazard:** Data not available

6.9 **Burning Rate:** Data not available

7. CHEMICAL REACTIVITY

7.1 **Reactivity with Water:** No reaction

7.2 **Reactivity with Common Materials:** May attack some forms of plastics.

7.3 **Stability During Transport:** Stable

7.4 **Neutralizing Agents for Acids and Caustics:** Not pertinent

7.5 **Polymerization:** Not pertinent

7.6 **Inhibitor of Polymerization:** Not pertinent

8. WATER POLLUTION

8.1 **Aquatic Toxicity:** 65 ppm/24 hr/brine shrimp/TL_m

8.2 **Waterfowl Toxicity:** Data not available

8.3 **Biological Oxygen Demand (BOD):** 4% of theoretical in 5 days, fresh water

8.4 **Food Chain Concentration Potential:** None

9. SELECTED MANUFACTURERS

1. Union Carbide Corp.
   Chemicals and Plastics Div.
   270 Park Avenue
   New York, N. Y. 10017
2. Eastman Chemical Products, Inc.
   Kingsport, Tenn. 37662
3. Aldrich Chemical Co.
   940 W. Saint Paul Ave.
   Milwaukee, Wis. 53233

10. SHIPPING INFORMATION

10.1 **Grades or Purity:** Technical

10.2 **Storage Temperature:** Ambient

10.3 **Inert Atmosphere:** No requirement

10.4 **Venting:** Open (flame arrester)

11. HAZARD ASSESSMENT CODE

(See Hazard Assessment Handbook, CG 446-3)

A-T-U

12. HAZARD CLASSIFICATIONS

12.1 **Code of Federal Regulations:** Combustible Liquid

12.2 **NAS Hazard Rating for Bulk Water Transportation:**

Category	Rating
Fire	2
Health	
Vapor Irritant	2
Liquid or Solid Irritant	1
Poisons	1
Water Pollution	
Human Toxicity	1
Aquatic Toxicity	2
Aesthetic Effect	2
Reactivity	
Other Chemicals	2
Water	0
Self-Reaction	0

12.3 **NFPA Hazard Classifications:**

Category	Classification
Health Hazard (Blue)	1
Flammability (Red)	2
Reactivity (Yellow)	0

13. PHYSICAL AND CHEMICAL PROPERTIES

13.1 **Physical State at 15°C and 1 atm:** Liquid

13.2 **Molecular Weight:** 142.23

13.3 **Boiling Point at 1 atm:** 325°F = 163°C = 436°K

13.4 **Freezing Point:** −43°F = −42°C = 231°K

13.5 **Critical Temperature:** Not pertinent

13.6 **Critical Pressure:** Not pertinent

13.7 **Specific Gravity:** 0.806 at 20°C (liquid)

13.8 **Liquid Surface Tension:** 23.92 dynes/cm = 0.02392 N/m at 22°C

13.9 **Liquid-Water Interfacial Tension:** Data not available

13.10 **Vapor (Gas) Specific Gravity:** 4.9

13.11 **Ratio of Specific Heats of Vapor (Gas):** Not pertinent

13.12 **Latent Heat of Vaporization:** 121 Btu/lb = 67 cal/g = 2.8×10^5 J/kg

13.13 **Heat of Combustion:** −16,040 Btu/lb = −8,910 cal/g = -373×10^5 J/kg

13.14 **Heat of Decomposition:** Not pertinent

13.15 **Heat of Solution:** Not pertinent

13.16 **Heat of Polymerization:** Not pertinent

NOTES

DID — DIISODECYL PHTHALATE

Common Synonyms		
Phthalic acid, bis (8-methylnonyl) ester Phthalic acid, diisodecyl ester Plasticized DDP	Liquid May float or sink in water.	Colorless

Stop discharge if possible.
Isolate and remove discharged material.
Notify local health and pollution control agencies.

Fire	Combustible. Extinguish with dry chemicals, foam or carbon dioxide. Water may be ineffective on fire.
Exposure	LIQUID Not harmful.
Water Pollution	Effect of low concentrations on aquatic life is unknown. Fouling to shoreline. May be dangerous if it enters water intakes. Notify local health and wildlife officials. Notify operators of nearby water intakes.

1. RESPONSE TO DISCHARGE

(See Response Methods Handbook, CG 446-4)

Mechanical containment
Should be removed
Chemical and physical treatment

2. LABELS

No hazard label required by Code of Federal Regulations

3. CHEMICAL DESIGNATIONS

3.1 **Synonyms:** Phthalic acid, bis(8-methylnonyl) ester; Phthalic acid, diisodecyl ester; Plasticizer DDP

3.2 **Coast Guard Compatibility Classification:** Ester

3.3 **Chemical Formula:** $C_{28}H_{46}O_4$

3.4 **IMCO/United Nations Numerical Designation:** Not listed

4. OBSERVABLE CHARACTERISTICS

4.1 **Physical State (as shipped):** Liquid

4.2 **Color:** Colorless

4.3 **Odor:** Faint

5. HEALTH HAZARDS

5.1 **Personal Protective Equipment:** Goggles or face shield; rubber gloves

5.2 **Symptoms Following Exposure:** No symptoms reported for any rate of exposure.

5.3 **Treatment for Exposure:** INGESTION: call physician. EYES: flush with water; call physician. SKIN: wipe off; wash with soap and water.

5.4 **Toxicity by Inhalation (Threshold Limit Value):** Data not available

5.5 **Short-Term Inhalation Limits:** Data not available

5.6 **Toxicity by Ingestion:** Data not available

5.7 **Late Toxicity:** Data not available

5.8 **Vapor (Gas) Irritant Characteristics:** Data not available

5.9 **Liquid or Solid Irritant Characteristics:** Data not available

5.10 **Odor Threshold:** Data not available

6. FIRE HAZARDS

6.1 **Flash Point:** 450°F O.C.

6.2 **Flammable Limits in Air:** LFL 0.27% at 508°F

6.3 **Fire Extinguishing Agents:** Dry chemical, foam, carbon dioxide

6.4 **Fire Extinguishing Agents Not to be Used:** Water may be ineffective.

6.5 **Special Hazards of Combustion Products:**

6.6 **Behavior in Fire:**

6.7 **Ignition Temperature:** 755°F

6.8 **Electrical Hazard:** Data not available

6.9 **Burning Rate:** Data not available

7. CHEMICAL REACTIVITY

7.1 **Reactivity with Water:** No reaction

7.2 **Reactivity with Common Materials:** May attack some forms of plastics

7.3 **Stability During Transport:** Stable

7.4 **Neutralizing Agents for Acids and Caustics:** Not pertinent

7.5 **Polymerization:** Not pertinent

7.6 **Inhibitor of Polymerization:** Not pertinent

8. WATER POLLUTION

8.1 **Aquatic Toxicity:** Data not available

8.2 **Waterfowl Toxicity:** Data not available

8.3 **Biological Oxygen Demand (BOD):** Data not available

8.4 **Food Chain Concentration Potential:** None

9. SELECTED MANUFACTURERS

1. Eastman Chemical Co.
   Kingsport, Tenn. 37662
2. W. R. Grace & Co.
   Hatco Chemical Div.
   King George Post Rd.
   Fords, N. J. 08863
3. Pfaltz and Bauer, Inc.
   375 Fairfield Ave.
   Stamford, Conn. 06902

10. SHIPPING INFORMATION

10.1 **Grades or Purity:** Technical

10.2 **Storage Temperature:** Ambient

10.3 **Inert Atmosphere:** No requirement

10.4 **Venting:** Open

11. HAZARD ASSESSMENT CODE

(See Hazard Assessment Handbook, CG 446-3)

A-T-U

12. HAZARD CLASSIFICATIONS

12.1 **Code of Federal Regulations:** Not listed

12.2 **NAS Hazard Rating for Bulk Water Transportation:** Not listed

12.3 **NFPA Hazard Classifications:** Not listed

13. PHYSICAL AND CHEMICAL PROPERTIES

13.1 **Physical State at 15°C and 1 atm:** Liquid

13.2 **Molecular Weight:** 446.7 (theor.)

13.3 **Boiling Point at 1 atm:** Very high

13.4 **Freezing Point:** −58°F = −50°C = 223°K

13.5 **Critical Temperature:** Not pertinent

13.6 **Critical Pressure:** Not pertinent

13.7 **Specific Gravity:** 0.967 at 20°C (liquid)

13.8 **Liquid Surface Tension:** Data not available

13.9 **Liquid-Water Interfacial Tension:** Data not available

13.10 **Vapor (Gas) Specific Gravity:** Not pertinent

13.11 **Ratio of Specific Heats of Vapor (Gas):** Not pertinent

13.12 **Latent Heat of Vaporization:** Data not available

13.13 **Heat of Combustion:** (est.) −16,600 Btu/lb = −9,220 cal/g = -386×10^5 J/kg

13.14 **Heat of Decomposition:** Not pertinent

13.15 **Heat of Solution:** Not pertinent

13.16 **Heat of Polymerization:** Not pertinent

NOTES

DIP	DIISOPROPANOLAMINE

Common Synonyms			
1,1'-Iminodi-2-Propanol 2,2'-Dihydroxydipropylamine	Liquid or solid crystals	Liquid is colorless Solid is white to light yellow	Dead fish or ammonia odor
	Liquid floats and mixes with water. Solid sinks and mixes in water.		

Avoid contact with liquid. Keep people away.
Wear goggles, self-contained breathing apparatus, and rubber overclothing (including gloves).
Stop discharge if possible.
Call fire department.
Isolate and remove discharged material.
Notify local health and pollution control agencies.

Fire	Combustible. Wear goggles, self-contained breathing apparatus, and rubber overclothing (including gloves). Extinguish with dry chemical, alcohol foam, or carbon dioxide. Water may be ineffective on fire. Cool exposed containers with water.
Exposure	CALL FOR MEDICAL AID. LIQUID OR SOLID Will burn skin and eyes. Harmful if swallowed. Remove contaminated clothing and shoes. Flush affected areas with plenty of water. IF IN EYES, hold eyelids open and flush with plenty of water. IF SWALLOWED and victim is CONSCIOUS, have victim drink water or milk and have victim induce vomiting. IF SWALLOWED and victim is UNCONSCIOUS OR HAVING CONVULSIONS, do nothing except keep victim warm.
Water Pollution	Effect of low concentrations on aquatic life is unknown. May be dangerous if it enters water intakes. Notify local health and wildlife officials. Notify operators of nearby water intakes.

1. RESPONSE TO DISCHARGE

(See Response Methods Handbook, CG 446-4)

Disperse and flush

2. LABELS

No hazard label required by Code of Federal Regulations

3. CHEMICAL DESIGNATIONS

3.1 **Synonyms:** 2,2'-Dihydroxydipropylamine; 1,1'-Iminodi-2-propanol

3.2 **Coast Guard Compatibility Classification:** Alkanolamine

3.3 **Chemical Formula:** $[CH_3CH(OH)CH_2]NH$

3.4 **IMCO/United Nations Numerical Designation:** Not listed

4. OBSERVABLE CHARACTERISTICS

4.1 **Physical State (as shipped):** Liquid or solid

4.2 **Color:** Colorless

4.3 **Odor:** Fishy; ammoniacal

5. HEALTH HAZARDS

5.1 **Personal Protective Equipment:** Full face mask or amine vapor mask only if required; clean, body-covering clothing, rubber gloves, apron, boots and face shield.

5.2 **Symptoms Following Exposure:** Vapor concentrations too low to irritate unless exposure is prolonged. Liquid will burn eyes and skin.

5.3 **Treatment for Exposure:** INHALATION: if ill effects occur, remove person to fresh air and get medical help. INGESTION: if swallowed and patient is conscious and not convulsing, promptly give milk or water, then induce vomiting; get medical help. No specific antidote known. EYE AND SKIN: immediately flush with plenty of water for at least 15 min. For eyes, get medical help promptly. Remove and wash contaminated clothing before reuse.

5.4 **Toxicity by Inhalation (Threshold Limit Value):** Not pertinent

5.5 **Short-Term Inhalation Limits:** Not pertinent

5.6 **Toxicity by Ingestion:** Grade 2; LD_{50} 0.5 to 5 g/kg (rat)

5.7 **Late Toxicity:** Data not available

5.8 **Vapor (Gas) Irritant Characteristics:** Vapors cause moderate irritation such that personnel will find high concentrations unpleasant. The effect is temporary.

5.9 **Liquid or Solid Irritant Characteristics:** Causes smarting of the skin and first-degree burns on short exposure and may cause secondary burns on long exposure.

5.10 **Odor Threshold:** Data not available

6. FIRE HAZARDS

6.1 **Flash Point:** 260°F O.C.

6.2 **Flammable Limits in Air:** 1.1% (calc.)—5.4% (est.)

6.3 **Fire Extinguishing Agents:** Water, alcohol foam, dry chemical, or carbon dioxide

6.4 **Fire Extinguishing Agents Not to be Used:** Water or foam may cause frothing

6.5 **Special Hazards of Combustion Products:** Not pertinent

6.6 **Behavior in Fire:** Not pertinent

6.7 **Ignition Temperature:** 580°F (calc.)

6.8 **Electrical Hazard:** Not pertinent

6.9 **Burning Rate:** Data not available

7. CHEMICAL REACTIVITY

7.1 **Reactivity with Water:** No reaction

7.2 **Reactivity with Common Materials:** No reaction

7.3 **Stability During Transport:** Stable

7.4 **Neutralizing Agents for Acids and Caustics:** Flush with water

7.5 **Polymerization:** Not pertinent

7.6 **Inhibitor of Polymerization:** Not pertinent

8. WATER POLLUTION

8.1 **Aquatic Toxicity:** Data not available

8.2 **Waterfowl Toxicity:** Data not available

8.3 **Biological Oxygen Demand (BOD):** Data not available

8.4 **Food Chain Concentration Potential:** None

9. SELECTED MANUFACTURERS

1. Dow Chemical Co.
   Midland, Mich. 48640
2. Union Carbide Corp.
   Chemicals and Plastics Division
   270 Park Ave.
   New York, N. Y. 10017

10. SHIPPING INFORMATION

10.1 **Grades or Purity:** 97%

10.2 **Storage Temperature:** Ambient

10.3 **Inert Atmosphere:** No requirement

10.4 **Venting:** Open

11. HAZARD ASSESSMENT CODE

(See Hazard Assessment Handbook, CG 446-3)

A-P-Q

12. HAZARD CLASSIFICATIONS

12.1 **Code of Federal Regulations:** Not listed

12.2 **NAS Hazard Rating for Bulk Water Transportation:**

Category	Rating
Fire	1
Health	
Vapor Irritant	2
Liquid or Solid Irritant	2
Poisons	2
Water Pollution	
Human Toxicity	2
Aquatic Toxicity	2
Aesthetic Effect	2
Reactivity	
Other Chemicals	3
Water	0
Self-Reaction	0

12.3 **NFPA Hazard Classifications:**

Category	Classification
Health Hazard (Blue)	3
Flammability (Red)	3
Reactivity (Yellow)	0

13. PHYSICAL AND CHEMICAL PROPERTIES

13.1 **Physical State at 15°C and 1 atm:** Liquid or solid

13.2 **Molecular Weight:** 133.19

13.3 **Boiling Point at 1 atm:** 479.7°F = 248.7°C = 521.9°K

13.4 **Freezing Point:** 108°F = 42°C = 315°K

13.5 **Critical Temperature:** 750°F = 399°C = 672°K

13.6 **Critical Pressure:** 529 psia = 36 atm = 3.6 MN/m^2

13.7 **Specific Gravity:** 0.99 at 42°C (liquid)

13.8 **Liquid Surface Tension:** Not pertinent

13.9 **Liquid-Water Interfacial Tension:** Not pertinent

13.10 **Vapor (Gas) Specific Gravity:** Not pertinent

13.11 **Ratio of Specific Heats of Vapor (Gas):** Not pertinent

13.12 **Latent Heat of Vaporization:** 185 Btu/lb = 103 cal/g = 4.31×10^5 J/kg

13.13 **Heat of Combustion:** (est.) −12,300 Btu/lb = −6860 cal/g = -287×10^5 J/kg

13.14 **Heat of Decomposition:** Not pertinent

13.15 **Heat of Solution:** (est.) −13 Btu/lb = −7 cal/g = -0.3×10^5 J/kg

13.16 **Heat of Polymerization:** Not pertinent

NOTES

REVISED 1978

DIA	DIISOPROPYLAMINE

Common Synonyms	Liquid	Colorless	Fishy odor
	Floats and mixes with water.		

Shut off ignition sources. Call fire department.
AVOID CONTACT WITH SOLID. KEEP PEOPLE AWAY.
Stop discharge if possible.
Stay upwind. Use water spray to "knock down" vapor.
Isolate and remove discharged material.
Notify local health and pollution control agencies.

Fire

FLAMMABLE.
POISONOUS GASES MAY BE PRODUCED IN FIRE.
Containers may explode in fire.
Flashback along vapor trail may occur.
Vapor may explode if ignited in an enclosed area.
Wear goggles and self-contained breathing apparatus.
Extinguish with dry chemicals, alcohol foam, or carbon dioxide.
Water may be ineffective on fire.
Cool exposed containers with water.

Exposure

CALL FOR MEDICAL AID.

VAPOR
Irritating to eyes, nose and throat.
If inhaled will cause coughing or difficult breathing.
If in eyes, hold eyelids open and flush with plenty of water.
If breathing has stopped, give artificial respiration.
If breathing is difficult, give oxygen.

LIQUID
Will burn eyes.
Irritating to eyes.
If swallowed will cause nausea and vomiting.
Remove contaminated clothing and shoes.
Flush affected areas with plenty of water.
IF IN EYES, hold eyelids open and flush with plenty of water.
IF SWALLOWED and victim is CONSCIOUS, have victim drink water or milk and have victim induce vomiting.
IF SWALLOWED and victim is UNCONSCIOUS OR HAVING CONVULSIONS, do nothing except keep victim warm.

Water Pollution

HARMFUL TO AQUATIC LIFE IN VERY LOW CONCENTRATIONS.
May be dangerous if it enters water intakes.
Notify local health and wildlife officials.
Notify operators of nearby water intakes.

1. RESPONSE TO DISCHARGE

(See Response Methods Handbook, CG 446-4)

Issue warning—high flammability, water contaminant, air contaminant
Restrict access
Disperse and flush

2. LABEL

Red

3. CHEMICAL DESIGNATIONS

3.1 **Synonyms:** No common synonyms
3.2 **Coast Guard Compatibility Classification:** Aliphatic amine
3.3 **Chemical Formula:** $[(CH_3)_2CH]_2NH$ or $C_6H_{15}N$
3.4 **IMCO/United Nations Numerical Designation:** 3.2/1158

4. OBSERVABLE CHARACTERISTICS

4.1 **Physical State (as shipped):** Liquid
4.2 **Color:** Colorless
4.3 **Odor:** Amine

5. HEALTH HAZARDS

5.1 **Personal Protective Equipment:** Air-supplied mask; plastic gloves; monogoggles; rubber apron
5.2 **Symptoms Following Exposure:** Inhalation of vapors causes irritation, sometimes with nausea and vomiting; can also cause burns to the respiratory system. Ingestion causes irritation of mouth and stomach. Vapor irritates eyes; liquid causes severe burn, like caustic. Contact with skin causes irritation.
5.3 **Treatment for Exposure:** INHALATION: move victim to fresh air and keep him quiet and comfortably warm; give oxygen if breathing is difficult; call a physician. INGESTION: induce vomiting by giving a large volume of warm salt water; consult a physician. EYES: *immediately* flush eyes with plenty of water for at least 15 min., then get medical care. SKIN: flush with water; remove contaminated clothing and wash skin; if there is any redness or evidence of burning, consult a physician.
5.4 **Toxicity by Inhalation (Threshold Limit Value):** 5 ppm
5.5 **Short-Term Inhalation Limits:** Mouse LCL_O = 5,000 ppm for 20 min.
5.6 **Toxicity by Ingestion:** Grade 2; oral LD_{50} = 0.7 g/kg (rat)
5.7 **Late Toxicity:** Data not available
5.8 **Vapor (Gas) Irritant Characteristics:** Vapors are moderately irritating such that personnel will not usually tolerate moderate or high concentrations.
5.9 **Liquid or Solid Irritant Characteristics:** Causes smarting of the skin and first-degree burns on short exposure; may cause second-degree burns on long exposure.
5.10 **Odor Threshold:** Data not available

6. FIRE HAZARDS

6.1 **Flash Point:** 20°F O.C., 35°F C.C.
6.2 **Flammable Limits in Air:** 0.8%—7.1%
6.3 **Fire Extinguishing Agents:** "Alcohol" foam, dry chemical, carbon dioxide
6.4 **Fire Extinguishing Agents Not to be Used:** Water may be ineffective.
6.5 **Special Hazards of Combustion Products:** Toxic oxides of nitrogen may form in fires.
6.6 **Behavior in Fire:** Vapor is heavier than air and may travel to a source of ignition and flash back.
6.7 **Ignition Temperature:** 600°F
6.8 **Electrical Hazard:** Class I
6.9 **Burning Rate:** Data not available

7. CHEMICAL REACTIVITY

7.1 **Reactivity with Water:** No reaction
7.2 **Reactivity with Common Materials:** May attack some forms of plastics
7.3 **Stability During Transport:** Stable
7.4 **Neutralizing Agents for Acids and Caustics:** Not pertinent
7.5 **Polymerization:** Not pertinent
7.6 **Inhibitor of Polymerization:** Not pertinent

8. WATER POLLUTION

8.1 **Aquatic Toxicity:**
60 ppm/24 hr/creek chub/lethal/fresh water
40—60 ppm/*/creek chub/"critical range"/fresh water
*Time period not specified.
8.2 **Waterfowl Toxicity:** Data not available
8.3 **Biological Oxygen Demand (BOD):** Data not available
8.4 **Food Chain Concentration Potential:** None

9. SELECTED MANUFACTURERS

1. Union Carbide Corp.
Chemicals and Plastics Div.
270 Park Avenue
New York, N. Y. 10017
2. Pennwalt Corporation
Three Parkway
Philadelphia, Pa. 19102
3. Virginia Chemicals, Inc.
3340 W. Norfolk Rd.
Portsmouth, Va. 23703

10. SHIPPING INFORMATION

10.1 **Grades or Purity:** Commercial, 100%
10.2 **Storage Temperature:** Ambient
10.3 **Inert Atmosphere:** No requirement
10.4 **Venting:** Open (flame arrester)

11. HAZARD ASSESSMENT CODE

(See Hazard Assessment Handbook, CG 446-3)

A-P-Q-R-S

12. HAZARD CLASSIFICATIONS

12.1 **Code of Federal Regulations:** Flammable liquid
12.2 **NAS Hazard Rating for Bulk Water Transportation:**

Category	Rating
Fire	3
Health	
Vapor Irritant	3
Liquid or Solid Irritant	2
Poisons	4
Water Pollution	
Human Toxicity	2
Aquatic Toxicity	3
Aesthetic Effect	2
Reactivity	
Other Chemicals	3
Water	0
Self-Reaction	0

12.3 **NFPA Hazard Classifications:**

Category	Classification
Health Hazard (Blue)	3
Flammability (Red)	3
Reactivity (Yellow)	0

13. PHYSICAL AND CHEMICAL PROPERTIES

13.1 **Physical State at 15°C and 1 atm:** Liquid
13.2 **Molecular Weight:** 101.19
13.3 **Boiling Point at 1 atm:** 183.0°F = 83.9°C = 357.1°K
13.4 **Freezing Point:** −141.3°F = −96.3°C = 176.9°K
13.5 **Critical Temperature:** 480.2°F = 249.0°C = 522.2°K
13.6 **Critical Pressure:** (est.) 400 psia = 30 atm = 3 MN/m^2
13.7 **Specific Gravity:** 0.717 at 20°C (liquid)
13.8 **Liquid Surface Tension:** 19.64 dynes/cm = 0.01964 N/m at 20°C
13.9 **Liquid-Water Interfacial Tension:** Not pertinent
13.10 **Vapor (Gas) Specific Gravity:** 3.5
13.11 **Ratio of Specific Heats of Vapor (Gas):** (est.) 1.064
13.12 **Latent Heat of Vaporization:** 121 Btu/lb = 67.5 cal/g = 2.82 × 10^5 J/kg
13.13 **Heat of Combustion:** −19,800 Btu/lb = −11,000 cal/g = −460 × 10^5 J/kg
13.14 **Heat of Decomposition:** Not pertinent
13.15 **Heat of Solution:** −140 Btu/lb = −76 cal/g = −3.2 × 10^5 J/kg
13.16 **Heat of Polymerization:** Not pertinent

NOTES

DIH DIISOPROPYLBENZENE HYDROPEROXIDE

Common Synonyms			
Isopropylcumyl hydroperoxide	Liquid	Colorless to pale yellow	Sharp, unpleasant odor
	May float or sink in water.		

Stop discharge if possible. Keep people away.
Call fire department.
Avoid contact with liquid and vapor.
Isolate and remove discharged material.
Notify local health and pollution control agencies.

Fire

Combustible.
Will increase the intensity of a fire.
May cause fire on contact with combustibles.
Containers may explode in fire.

Extinguish with dry chemicals, foam or carbon dioxide.
Water may be ineffective on fire.
Cool exposed containers with water.

Exposure

CALL FOR MEDICAL AID.

VAPOR
Irritating to eyes, nose and throat.
If inhaled will cause coughing or difficult breathing.

If in eyes, hold eyelids open and flush with plenty of water.
If breathing has stopped, give artificial respiration.
If breathing is difficult, give oxygen.

LIQUID
Irritating to skin and eyes.
Harmful if swallowed.

Remove contaminated clothing and shoes.
Flush affected areas with plenty of water.
IF IN EYES, hold eyelids open and flush with plenty of water.
IF SWALLOWED and victim is CONSCIOUS, have victim drink water or milk.
IF SWALLOWED and victim is UNCONSCIOUS OR HAVING CONVULSIONS, do nothing except keep victim warm.

Water Pollution

Effect of low concentrations on aquatic life is unknown.
Fouling to shoreline.
May be dangerous if it enters water intakes.

Notify local health and wildlife officials.
Notify operators of nearby water intakes.

1. RESPONSE TO DISCHARGE

(See Response Methods Handbook, CG 446-4)

Issue warning—oxidizing material
Restrict access
Mechanical containment
Should be removed
Chemical and physical treatment

2. LABEL

Yellow

3. CHEMICAL DESIGNATIONS

3.1 **Synonyms:** Isopropylcumyl hydroperoxide
3.2 **Coast Guard Compatibility Classification:** Not listed
3.3 **Chemical Formula:** $(CH_3)_2CHC_6H_4C(CH_3)_2OOH$ $+(CH_3)_2CHC_6H_4CH(CH_3)_2$
3.4 **IMCO/United Nations Numerical Designation:** 5.2/1875

4. OBSERVABLE CHARACTERISTICS

4.1 **Physical State (as shipped):** Liquid
4.2 **Color:** Colorless to pale yellow.
4.3 **Odor:** Sharp, disagreeable

5. HEALTH HAZARDS

5.1 **Personal Protective Equipment:** Solvent-resistant gloves; chemical-resistant apron; chemical goggles or face shield; self-contained breathing apparatus
5.2 **Symptoms Following Exposure:** Inhalation causes irritation of nose and throat.Contact with eyes or skin causes throbbing sensation and irritation.
5.3 **Treatment for Exposure:** INHALATION: move to fresh air; call a doctor. EYES: flush with water for 15 min., holding eyelids open; call physician. SKIN: wash several times with soap and water.
5.4 **Toxicity by Inhalation (Threshold Limit Value):** Data not available
5.5 **Short-Term Inhalation Limits:** Data not available
5.6 **Toxicity by Ingestion:** Data not available
5.7 **Late Toxicity:** Data not available
5.8 **Vapor (Gas) Irritant Characteristics:** Data not available
5.9 **Liquid or Solid Irritant Characteristics:** Data not available
5.10 **Odor Threshold:** Data not available

6. FIRE HAZARDS

6.1 **Flash Point:** 175°F C.C.
6.2 **Flammable Limits in Air:** Data not available
6.3 **Fire Extinguishing Agents:** Foam, dry chemical, carbon dioxide
6.4 **Fire Extinguishing Agents Not to be Used:** Water may be ineffective.
6.5 **Special Hazards of Combustion Products:** Flammable alcohol and ketone gases are formed in fire.
6.6 **Behavior in Fire:** Burns with a flare effect. Containers may explode.
6.7 **Ignition Temperature:** Data not available
6.8 **Electrical Hazard:** Data not available
6.9 **Burning Rate:** Data not available

7. CHEMICAL REACTIVITY

7.1 **Reactivity with Water:** No reaction
7.2 **Reactivity with Common Materials:** Aluminum, copper, brass, lead, zinc salts, mineral acids, oxidizing or reducing agents all can cause rapid decomposition.
7.3 **Stability During Transport:** Unstable; slowly evolves oxygen
7.4 **Neutralizing Agents for Acids and Caustics:** Not pertinent
7.5 **Polymerization:** Not pertinent
7.6 **Inhibitor of Polymerization:** Not pertinent

8. WATER POLLUTION

8.1 **Aquatic Toxicity:** Data not available
8.2 **Waterfowl Toxicity:** Data not available
8.3 **Biological Oxygen Demand (BOD):** Data not available
8.4 **Food Chain Concentration Potential:** None

9. SELECTED MANUFACTURERS

Hercules, Inc.
Wilmington, Del. 19899

10. SHIPPING INFORMATION

10.1 **Grades or Purity:** 54.06%, the balance being diisopropylbenzene, a combustible hydrocarbon
10.2 **Storage Temperature:** Ambient
10.3 **Inert Atmosphere:** No requirement
10.4 **Venting:** Open (flame arrester)

11. HAZARD ASSESSMENT CODE

(See Hazard Assessment Handbook, CG 446-3)

A-T-U

12. HAZARD CLASSIFICATIONS

12.1 **Code of Federal Regulations:** Organic Peroxide
12.2 **NAS Hazard Rating for Bulk Water Transportation:** Not listed
12.3 **NFPA Hazard Classifications:** Not listed

13. PHYSICAL AND CHEMICAL PROPERTIES

13.1 **Physical State at 15°C and 1 atm:** Liquid
13.2 **Molecular Weight:** 194.26
13.3 **Boiling Point at 1 atm:** Not pertinent (decomposes)
13.4 **Freezing Point:** $<15°F = <-9°C = <264°K$
13.5 **Critical Temperature:** Not pertinent
13.6 **Critical Pressure:** Not pertinent
13.7 **Specific Gravity:** 0.956 at 15°C (liquid)
13.8 **Liquid Surface Tension:** Data not available
13.9 **Liquid-Water Interfacial Tension:** Data not available
13.10 **Vapor (Gas) Specific Gravity:** Not pertinent
13.11 **Ratio of Specific Heats of Vapor (Gas):** Not pertinent
13.12 **Latent Heat of Vaporization:** Not pertinent
13.13 **Heat of Combustion:** Data not available
13.14 **Heat of Decomposition:** Data not available
13.15 **Heat of Solution:** Not pertinent
13.16 **Heat of Polymerization:** Not pertinent

NOTES

DAC DIMETHYLACETAMIDE

Common Synonyms

N-N-Dimethylacetamide
Acetic acid, dimethylamide

Liquid | Colorless | Weak fishy odor

Mixes with water.

Stop discharge if possible. Keep people away.
Call fire department.
Isolate and remove discharged material.
Notify local health and pollution control agencies.

Fire

Combustible.
Extinguish with water, dry chemicals, alcohol foam, or carbon dioxide.

Exposure

Call for medical aid.

LIQUID
Irritating to skin and eyes.
Harmful if swallowed.
Remove contaminated clothing and shoes.
Flush affected areas with plenty of water.
IF IN EYES, hold eyelids open and flush with plenty of water.
IF SWALLOWED and victim is CONSCIOUS, have victim drink water or milk and have victim induce vomiting.
IF SWALLOWED and victim is UNCONSCIOUS OR HAVING CONVULSIONS, do nothing except keep victim warm.

Water Pollution

Effect of low concentrations on aquatic life is unknown.
May be dangerous if it enters water intakes.
Notify local health and wildlife officials.
Notify operators of nearby water intakes.

1. RESPONSE TO DISCHARGE

(See Response Methods Handbook, CG 446-4)

Issue warning—water contaminant
Disperse and flush

2. LABELS

No label required by Code of Federal Regulations

3. CHEMICAL DESIGNATIONS

3.1 **Synonyms:** N,N-Dimethylacetamide; Acetic acid, dimethylamide
3.2 **Coast Guard Compatibility Classification:** Not applicable
3.3 **Chemical Formula:** $CH_3CON(CH_3)_2$
3.4 **IMCO/United Nations Numerical Designation:** Not listed

4. OBSERVABLE CHARACTERISTICS

4.1 **Physical State (as shipped):** Liquid
4.2 **Color:** Colorless
4.3 **Odor:** Weak, fishy

5. HEALTH HAZARDS

5.1 **Personal Protective Equipment:** Goggles or face shield; rubber gloves.
5.2 **Symptoms Following Exposure:** Liquid causes mild irritation of eyes and skin. Ingestion causes depression, lethargy, confusion and disorientation, visual and auditory hallucinations, perceptual distortions, delusions, emotional detachment, and affective blunting.
5.3 **Treatment for Exposure:** EYES: flush with plenty of water for 15 min.; get medical attention. SKIN: flush with plenty of water for 15 min. INGESTION: induce vomiting and follow with gastric lavage and saline cathartics; treatment for liver and kidney injury is supportive and symptomatic.
5.4 **Toxicity by Inhalation (Threshold Limit Value):** 10 ppm
5.5 **Short-Term Inhalation Limits:** Data not available
5.6 **Toxicity by Ingestion:** Grade 1; oral LD_{50} = 5.63 g/kg (rat)
5.7 **Late Toxicity:** May produce chronic liver and kidney damage.
5.8 **Vapor (Gas) Irritant Characteristics:** Data not available
5.9 **Liquid or Solid Irritant Characteristics:** Data not available
5.10 **Odor Threshold:** 46.8 ppm

6. FIRE HAZARDS

6.1 **Flash Point:** 158°F O.C.
6.2 **Flammable Limits in Air:** 1.5%—11.5%
6.3 **Fire Extinguishing Agents:** Water, dry chemical, alcohol foam, carbon dioxide
6.4 **Fire Extinguishing Agents Not to be Used:** Not pertinent
6.5 **Special Hazards of Combustion Products:** Not pertinent
6.6 **Behavior in Fire:** Not pertinent
6.7 **Ignition Temperature:** 914°F
6.8 **Electrical Hazard:** Not pertinent
6.9 **Burning Rate:** 2.8 mm/min.

7. CHEMICAL REACTIVITY

7.1 **Reactivity with Water:** No reaction
7.2 **Reactivity with Common Materials:** No reaction
7.3 **Stability During Transport:** Stable
7.4 **Neutralizing Agents for Acids and Caustics:** Not pertinent
7.5 **Polymerization:** Not pertinent
7.6 **Inhibitor of Polymerization:** Not pertinent

8. WATER POLLUTION

8.1 **Aquatic Toxicity:** Data not available
8.2 **Waterfowl Toxicity:** Data not available
8.3 **Biological Oxygen Demand (BOD):** Data not available
8.4 **Food Chain Concentration Potential:** None

9. SELECTED MANUFACTURERS

1. E. I. duPont de Nemours & Co., Inc.
   Biochemicals Department
   1007 Market Street
   Wilmington, Del. 19898
2. BASF—Wyandotte Corp.
   1609 Biddle Avenue
   Wyandotte, Mich. 48192
3. Pfaltz and Bauer, Inc.
   126-04 Northern Boulevard
   Flushing, N. Y. 11368

10. SHIPPING INFORMATION

10.1 **Grades or Purity:** Technical
10.2 **Storage Temperature:** Ambient
10.3 **Inert Atmosphere:** No requirement
10.4 **Venting:** Open (flame arrester)

11. HAZARD ASSESSMENT CODE

(See Hazard Assessment Handbook, CG 446-3)

A-P-Q

12. HAZARD CLASSIFICATIONS

12.1 **Code of Federal Regulations:** Not listed
12.2 **NAS Hazard Rating for Bulk Water Transportation:** Not listed
12.3 **NFPA Hazard Classifications:** Not listed

13. PHYSICAL AND CHEMICAL PROPERTIES

13.1 **Physical State at 15°C and 1 atm:** Liquid
13.2 **Molecular Weight:** 87.1
13.3 **Boiling Point at 1 atm:** 331°F = 166°C = 439°K
13.4 **Freezing Point:** −4°F = −20°C = 253°K
13.5 **Critical Temperature:** Not pertinent
13.6 **Critical Pressure:** Not pertinent
13.7 **Specific Gravity:** 0.943 at 20°C (liquid)
13.8 **Liquid Surface Tension:** 34 dynes/cm = 0.034 N/m at 20°C
13.9 **Liquid-Water Interfacial Tension:** Not pertinent
13.10 **Vapor (Gas) Specific Gravity:** Not pertinent
13.11 **Ratio of Specific Heats of Vapor (Gas):** Not pertinent
13.12 **Latent Heat of Vaporization:** 214 Btu/lb = 119 cal/g = 4.98×10^5 J/kg
13.13 **Heat of Combustion:** −12,560 Btu/lb = −6,980 cal/g = -292×10^5 J/kg
13.14 **Heat of Decomposition:** Not pertinent
13.15 **Heat of Solution:** Data not available
13.16 **Heat of Polymerization:** Not pertinent

NOTES

DMA — DIMETHYLAMINE

Common Synonyms	Liquefied compressed gas Colorless Dead fish or ammonia odor Floats and boils on water. Flammable, irritating vapor is produced. Boiling point is 44° F.

Avoid contact with liquid and vapor. Keep people away.
Wear goggles, self-contained breathing apparatus, and rubber overclothing (including gloves).
Shut off ignition sources and call fire department.
Stop discharge if possible.
Stay upwind and use water spray to "knock down" vapor.
Isolate and remove discharged material.
Notify local health and pollution control agencies.

Fire	FLAMMABLE. Flashback along vapor trail may occur. May explode if ignited in an enclosed area. **Wear goggles, self-contained breathing apparatus, and rubber overclothing (including gloves).** **Stop flow of gas if possible.** **Cool exposed containers and protect men effecting shutoff with water.** **Let fire burn.**
Exposure	CALL FOR MEDICAL AID. VAPOR Irritating to eyes, nose and throat. If inhaled, will cause difficult breathing. **Move to fresh air.** **If breathing has stopped, give artificial respiration.** **If breathing is difficult, give oxygen.** LIQUID Will burn skin and eyes. Harmful if swallowed. **Remove contaminated clothing and shoes.** **Flush affected areas with plenty of water.** **IF IN EYES, hold eyelids open and flush with plenty of water.** **IF SWALLOWED, and victim is CONSCIOUS, have victim drink water or milk.**
Water Pollution	HARMFUL TO AQUATIC LIFE IN VERY LOW CONCENTRATIONS. May be dangerous if it enters water intakes. **Notify local health and wildlife officials.** **Notify operators of nearby water intakes.**

1. RESPONSE TO DISCHARGE

(See Response Methods Handbook, CG 446-4)

Issue warning—high flammability
Restrict access
Evacuate area

2. LABEL

Red

3. CHEMICAL DESIGNATIONS

3.1 **Synonyms:** No common synonyms
3.2 **Coast Guard Compatibility Classification:** Amine
3.3 **Chemical Formula:** $(CH_3)_2NH$
3.4 **IMCO/United Nations Numerical Designation:** 2.0/1032

4. OBSERVABLE CHARACTERISTICS

4.1 **Physical State (as shipped):** Compressed gas
4.2 **Color:** Colorless
4.3 **Odor:** Fishy; strongly ammoniacal

5. HEALTH HAZARDS

5.1 **Personal Protective Equipment:** Chemical goggles and full face shield; molded rubber acid gloves; self-contained breathing apparatus.
5.2 **Symptoms Following Exposure:** Inhalation at high concentration (> 100 ppm) causes nose and throat irritation progressing all the way to pulmonary edema. Eye and skin irritation.
5.3 **Treatment for Exposure:** INHALATION: remove victim to fresh air and call a physician; if breathing has stopped, administer artificial respiration and oxygen; keep victim warm and quiet; do not give stimulants. EYES: flush continuously and thoroughly with water for at least 15 min. SKIN: remove contaminated clothing immediately; flush affected area with large amounts of water and then wash with soap and water.
5.4 **Toxicity by Inhalation (Threshold Limit Value):** 10 ppm
5.5 **Short-Term Inhalation Limits:** 20 ppm for 5 min.
5.6 **Toxicity by Ingestion:** Not pertinent
5.7 **Late Toxicity:** None
5.8 **Vapor (Gas) Irritant Characteristics:** Vapors cause moderate irritation such that personnel will find high concentrations unpleasant. The effect is temporary.
5.9 **Liquid or Solid Irritant Characteristics:** Causes smarting of the skin and first-degree burns on short exposure and may cause secondary burns on long exposure.
5.10 **Odor Threshold:** 0.047 ppm

6. FIRE HAZARDS

6.1 **Flash Point:** 20°F C.C.
6.2 **Flammable Limits in Air:** 2.8%—14.4%
6.3 **Fire Extinguishing Agents:** Stop flow of gas. Use water spray, carbon dioxide, or dry chemical for fires in water solutions
6.4 **Fire Extinguishing Agents Not to be Used:** Do not use foam
6.5 **Special Hazards of Combustion Products:** Vapors are eye, skin and respiratory irritants
6.6 **Behavior in Fire:** Not pertinent
6.7 **Ignition Temperature:** 756°F
6.8 **Electrical Hazard:** Data not available
6.9 **Burning Rate:** 4.5 mm/min.

7. CHEMICAL REACTIVITY

7.1 **Reactivity with Water:** No reaction
7.2 **Reactivity with Common Materials:** No hazardous reaction
7.3 **Stability During Transport:** Stable
7.4 **Neutralizing Agents for Acids and Caustics:** Not pertinent
7.5 **Polymerization:** Not pertinent
7.6 **Inhibitor of Polymerization:** Not pertinent

8. WATER POLLUTION

8.1 **Aquatic Toxicity:**
50 ppm/24 hr/chub/died/fresh water
>100 ppm/48 hr/shrimp/LC_{50}/salt water
8.2 **Waterfowl Toxicity:** Data not available
8.3 **Biological Oxygen Demand (BOD):** Data not available
8.4 **Food Chain Concentration Potential:** None

9. SELECTED MANUFACTURERS

1. Commercial Solvents Corp.
245 Park Ave.
New York, N. Y. 10017
2. GAF Corp.
Chemical Division
Calvert City, Ky. 42029
3. Rohm and Haas Co.
Independence Mall West
Philadelphia, Pa. 19105

10. SHIPPING INFORMATION

10.1 **Grades or Purity:** Anhydrous: 99.5%. Aqueous solutions: 25%, 40 %, 50%, 60%.
10.2 **Storage Temperature:** Ambient
10.3 **Inert Atmosphere:** No requirement
10.4 **Venting:** Safety relief

11. HAZARD ASSESSMENT CODE

(See Hazard Assessment Handbook, CG 446-3)

A-B-C-K-L-M-N

12. HAZARD CLASSIFICATIONS

12.1 **Code of Federal Regulations:** Flammable compressed gas
12.2 **NAS Hazard Rating for Bulk Water Transportation:**

Category	Rating
Fire	4
Health	
Vapor Irritant	2
Liquid or Solid Irritant	2
Poisons	2
Water Pollution	
Human Toxicity	2
Aquatic Toxicity	3
Aesthetic Effect	2
Reactivity	
Other Chemicals	3
Water	0
Self-Reaction	0

12.3 **NFPA Hazard Classifications:**

Category	Classification
Health Hazard (Blue)	3
Flammability (Red)	4
Reactivity (Yellow)	0

13. PHYSICAL AND CHEMICAL PROPERTIES

13.1 **Physical State at 15°C and 1 atm:** Gas
13.2 **Molecular Weight:** 45.08
13.3 **Boiling Point at 1 atm:** 44.42°F = 6.9°C = 280.1°K
13.4 **Freezing Point:** −134.0°F = −92.2°C = 181.0°K
13.5 **Critical Temperature:** 328.3°F = 164.6°C = 437.8°K
13.6 **Critical Pressure:** 770 psia = 52.4 atm = 5.31 MN/m²
13.7 **Specific Gravity:** 0.671 at 6.9°C (liquid)
13.8 **Liquid Surface Tension:** Not pertinent
13.9 **Liquid-Water Interfacial Tension:** Not pertinent
13.10 **Vapor (Gas) Specific Gravity:** 1.6
13.11 **Ratio of Specific Heats of Vapor (Gas):** 1.139
13.12 **Latent Heat of Vaporization:** 252.9 Btu/lb = 140.5 cal/g = 5.882×10^5 J/kg
13.13 **Heat of Combustion:** −16,800 Btu/lb = −9340 cal/g = -391.0×10^5 J/kg
13.14 **Heat of Decomposition:** Not pertinent
13.15 **Heat of Solution:** −515 Btu/lb = −286 cal/g = -12.0×10^5 J/kg
13.16 **Heat of Polymerization:** Not pertinent

NOTES

REVISED 1978

DMD — DIMETHYLDICHLOROSILANE

Common Synonyms

Liquid   Colorless   Sharp irritating odor

Reacts violently with water. Irritating gas is produced on contact with water.

Shut off ignition sources. Call fire department.
Avoid contact with liquid and vapor. Keep people away.
Wear goggles and self-contained breathing apparatus.
Stop discharge if possible.
Isolate and remove discharged material.
Notify local health and pollution control agencies.

Fire

FLAMMABLE.
POISONOUS GASES MAY BE PRODUCED IN FIRE.
Flashback along vapor trail may occur.
Vapor may explode if ignited in an enclosed area.
Extinguish with dry chemicals or carbon dioxide.
DO NOT USE WATER OR FOAM ON FIRE.

Exposure

Call for medical aid.

VAPOR
Irritating to eyes, nose and throat.
Move victim to fresh air.
If breathing is difficult, give oxygen.

LIQUID
Will burn skin and eyes.
Harmful if swallowed.
Remove contaminated clothing and shoes.
Flush affected areas with plenty of water.
IF IN EYES, hold eyelids open and flush with plenty of water.
IF SWALLOWED and victim is CONSCIOUS, have victim drink water or milk.
DO NOT INDUCE VOMITING.

Water Pollution

Effect of low concentrations on aquatic life is unknown.
May be dangerous if it enters water intakes.
Notify local health and wildlife officials.
Notify operators of nearby water intakes.

1. RESPONSE TO DISCHARGE

(See Response Methods Handbook, CG 446-4)

Issue warning—high flammability, corrosive
Restrict access
Evacuate area
Disperse and flush with care

2. LABEL

Red

3. CHEMICAL DESIGNATIONS

3.1 **Synonyms:** No common synonyms
3.2 **Coast Guard Compatibility Classification:** Not applicable
3.3 **Chemical Formula:** $(CH_3)_2SiCl_2$
3.4 **IMCO/United Nations Numerical Designation:** 3.2/1162

4. OBSERVABLE CHARACTERISTICS

4.1 **Physical State (as shipped):** Liquid
4.2 **Color:** Colorless
4.3 **Odor:** Sharp, like hydrochloric acid

5. HEALTH HAZARDS

5.1 **Personal Protective Equipment:** Acid-vapor type respiratory protection; rubber gloves; chemical worker's goggles; other protective equipment as necessary to protect skin and eyes.
5.2 **Symptoms Following Exposure:** Inhalation irritates mucous membranes. Contact with liquid causes severe burns of eyes and skin. Ingestion causes severe burns of mouth and stomach.
5.3 **Treatment for Exposure:** INHALATION: remove from exposure and support respiration; call physician if needed. EYES: flush with water for 15 min.; obtain medical attention immediately. SKIN: flush with water; obtain medical attention immediately. INGESTION: if victim is conscious, give large amounts of water followed by milk or milk of magnesia.
5.4 **Toxicity by Inhalation (Threshold Limit Value):** Data not available
5.5 **Short-Term Inhalation Limits:** Data not available
5.6 **Toxicity by Ingestion:** Grade 3; LD_{50} 50 to 500 mg/kg
5.7 **Late Toxicity:** Data not available
5.8 **Vapor (Gas) Irritant Characteristics:** Vapors cause severe irritation of eyes and throat and can cause eye and lung injury. They cannot be tolerated even at low concentrations.
5.9 **Liquid or Solid Irritant Characteristics:** Severe skin irritant. Causes second- and third-degree burns on short contact and is very injurious to the eyes.
5.10 **Odor Threshold:** Data not available

6. FIRE HAZARDS

6.1 **Flash Point:** 15°F O.C.
6.2 **Flammable Limits in Air:** 1.4%–9.5%
6.3 **Fire Extinguishing Agents:** Dry chemical, carbon dioxide
6.4 **Fire Extinguishing Agents Not to be Used:** Water or foam
6.5 **Special Hazards of Combustion Products:** Hydrogen chloride and phosgene gases may form; both are toxic and irritating.
6.6 **Behavior in Fire:** Difficult to extinguish. Re-ignition may occur. Contact with water applied to adjacent fires produces toxic and irritating fumes.
6.7 **Ignition Temperature:** Above 750°F
6.8 **Electrical Hazard:** Data not available
6.9 **Burning Rate:** 3.3 mm/min.

7. CHEMICAL REACTIVITY

7.1 **Reactivity with Water:** Reacts vigorously with water to generate hydrogen chloride (hydrochloric acid).
7.2 **Reactivity with Common Materials:** Will react with surface moisture to generate hydrogen chloride, which is corrosive to common metals.
7.3 **Stability During Transport:** Stable
7.4 **Neutralizing Agents for Acids and Caustics:** Sodium bicarbonate or lime
7.5 **Polymerization:** Not pertinent
7.6 **Inhibitor of Polymerization:** Not pertinent

8. WATER POLLUTION

8.1 **Aquatic Toxicity:** Data not available
8.2 **Waterfowl Toxicity:** Data not available
8.3 **Biological Oxygen Demand (BOD):** Data not available
8.4 **Food Chain Concentration Potential:** None

9. SELECTED MANUFACTURERS

1. Union Carbide Corporation
   Chemicals and Plastics Division
   270 Park Avenue
   New York, N. Y. 10017
2. Dow Corning Corporation
   P. O. Box 592
   Midland, Mich. 48640
3. General Electric Co.
   Silicone Products Department
   Waterford, N. Y. 12188

10. SHIPPING INFORMATION

10.1 **Grades or Purity:** 99+%
10.2 **Storage Temperature:** Ambient
10.3 **Inert Atmosphere:** No requirement
10.4 **Venting:** Pressure-vacuum

11. HAZARD ASSESSMENT CODE

(See Hazard Assessment Handbook, CG 446-3)

A-O

12. HAZARD CLASSIFICATIONS

12.1 **Code of Federal Regulations:** Flammable liquid
12.2 **NAS Hazard Rating for Bulk Water Transportation:**

Category	Rating
Fire	3
Health	
Vapor Irritant	4
Liquid or Solid Irritant	4
Poisons	3
Water Pollution	
Human Toxicity	3
Aquatic Toxicity	3
Aesthetic Effect	2
Reactivity	
Other Chemicals	3
Water	4
Self-Reaction	1

12.3 **NFPA Hazard Classifications:**

Category	Classification
Health Hazard (Blue)	3
Flammability (Red)	3
Reactivity (Yellow)	1

13. PHYSICAL AND CHEMICAL PROPERTIES

13.1 **Physical State at 15°C and 1 atm:** Liquid
13.2 **Molecular Weight:** 129
13.3 **Boiling Point at 1 atm:** 158.8°F = 70.5°C = 343.7°K
13.4 **Freezing Point:** −122°F = −86°C = 187°K
13.5 **Critical Temperature:** Not pertinent
13.6 **Critical Pressure:** Not pertinent
13.7 **Specific Gravity:** 1.07 at 25°C (liquid)
13.8 **Liquid Surface Tension:** 20.1 dynes/cm = 0.0201 N/m at 20°C
13.9 **Liquid-Water Interfacial Tension:** Not pertinent
13.10 **Vapor (Gas) Specific Gravity:** 4.4
13.11 **Ratio of Specific Heats of Vapor (Gas):** Data not available
13.12 **Latent Heat of Vaporization:** 100 Btu/lb = 58 cal/g = 2.4×10^5 J/kg
13.13 **Heat of Combustion:** (est.) −6,000 Btu/lb = −3,300 cal/g = -140×10^5 J/kg
13.14 **Heat of Decomposition:** Not pertinent
13.15 **Heat of Solution:** Data not available
13.16 **Heat of Polymerization:** Not pertinent

NOTES

DIM	DIMETHYL ETHER

Common Synonyms			
Methyl ether	Liquefied gas	Colorless	Pleasant odor
	Floats and boils on water. Flammable, irritating vapor is produced.		

Shut off ignition sources. Call fire department.
Stop discharge if possible. Keep people away.
Stay upwind. Use water spray to "knock down" vapor.
Isolate and remove discharged material.
Notify local health and pollution control agencies.

Fire

FLAMMABLE.
Containers may explode in fire.
Flashback along vapor trail may occur.
Vapor may explode if ignited in an enclosed area.
Let fire burn.
Stop flow of gas if possible.
Cool exposed containers and protect men effecting shutoff with water.

Exposure

Call for medical aid.

VAPOR
Irritating to eyes, nose and throat.
If inhaled will cause headache, dizziness, or loss of consciousness.
Move victim to fresh air.
If breathing has stopped, give artificial respiration.
If breathing is difficult, give oxygen.

LIQUID
Irritating to skin and eyes.
Will cause frostbite.
Remove contaminated clothing and shoes.
Flush affected areas with plenty of water.
DO NOT RUB AFFECTED AREAS.
IF IN EYES, hold eyelids open and flush with plenty of water.
IF SWALLOWED and victim is CONSCIOUS, have victim drink water or milk.

Water Pollution

Effect of low concentrations on aquatic life is unknown.
May be dangerous if it enters water intakes.
Notify local health and wildlife officials.
Notify operators of nearby water intakes.

1. RESPONSE TO DISCHARGE

(See Response Methods Handbook, CG 446-4)

Issue warning—high flammability
Restrict access
Evacuate area

2. LABEL

Red

3. CHEMICAL DESIGNATIONS

3.1 **Synonyms:** Methyl ether; Wood ether
3.2 **Coast Guard Compatibility Classification:** Ethers (21)
3.3 **Chemical Formula:** CH_3OCH_3
3.4 **IMCO/United Nations Numerical Designation:** 2/1033

4. OBSERVABLE CHARACTERISTICS

4.1 **Physical State (as shipped):** Liquid under pressure
4.2 **Color:** Colorless
4.3 **Odor:** Chloroform-like; sweet

5. HEALTH HAZARDS

5.1 **Personal Protective Equipment:** Mask for organic vapors; plastic or rubber gloves; safety glasses.
5.2 **Symptoms Following Exposure:** Inhalation produces some anesthesia (but less than that of ethyl ether), blurring of vision, headache, intoxication, loss of consciousness. Liquid or concentrated vapor irritates eyes. Contact of liquid with skin may cause frostbite.
5.3 **Treatment for Exposure:** INHALATION: remove from exposure and support respiration; call physician. EYES: wash with water for at least 15 min.; consult an eye specialist. SKIN: treat frostbite by use of warm water or by wrapping the affected part in blanket.
5.4 **Toxicity by Inhalation (Threshold Limit Value):** Data not available
5.5 **Short-Term Inhalation Limits:** Data not available
5.6 **Toxicity by Ingestion:** Not pertinent
5.7 **Late Toxicity:** Data not available
5.8 **Vapor (Gas) Irritant Characteristics:** Data not available
5.9 **Liquid or Solid Irritant Characteristics:** Data not available
5.10 **Odor Threshold:** Data not available

6. FIRE HAZARDS

6.1 **Flash Point:** Not pertinent (flammable gas)
6.2 **Flammable Limits in Air:** 2%–50%
6.3 **Fire Extinguishing Agents:** Let fire burn; shut off gas flow; cool exposed surroundings with water.
6.4 **Fire Extinguishing Agents Not to be Used:** Not pertinent
6.5 **Special Hazards of Combustion Products:** Not pertinent
6.6 **Behavior in Fire:** Containers may explode. Vapors are heavier than air and may travel long distance to a source of ignition and flash back.
6.7 **Ignition Temperature:** 662°F
6.8 **Electrical Hazard:** Data not available
6.9 **Burning Rate:** 6.6 mm/min.

7. CHEMICAL REACTIVITY

7.1 **Reactivity with Water:** No reaction
7.2 **Reactivity with Common Materials:** No reaction
7.3 **Stability During Transport:** Stable
7.4 **Neutralizing Agents for Acids and Caustics:** Not pertinent
7.5 **Polymerization:** Not pertinent
7.6 **Inhibitor of Polymerization:** Not pertinent

8. WATER POLLUTION

8.1 **Aquatic Toxicity:** Data not available
8.2 **Waterfowl Toxicity:** Data not available
8.3 **Biological Oxygen Demand (BOD):** Data not available
8.4 **Food Chain Concentration Potential:** None

9. SELECTED MANUFACTURERS

1. Union Carbide Corporation
   Chemicals and Plastics Division
   270 Park Avenue
   New York, N. Y. 10017
2. Matheson Gas Products
   East Rutherford, N. J. 07073
3. Pfaltz and Bauer, Inc.
   126-04 Northern Boulevard
   Flushing, N. Y. 11368

10. SHIPPING INFORMATION

10.1 **Grades or Purity:** 99+%
10.2 **Storage Temperature:** Ambient
10.3 **Inert Atmosphere:** No requirement
10.4 **Venting:** Safety relief

11. HAZARD ASSESSMENT CODE

(See Hazard Assessment Handbook, CG 446-3)

A-B-C-K-L-M-N

12. HAZARD CLASSIFICATIONS

12.1 **Code of Federal Regulations:** Flammable compressed gas
12.2 **NAS Hazard Rating for Bulk Water Transportation:** Not listed
12.3 **NFPA Hazard Classifications:**

Category	Classification
Health Hazard (Blue)	2
Flammability (Red)	4
Reactivity (Yellow)	0

13. PHYSICAL AND CHEMICAL PROPERTIES

13.1 **Physical State at 15°C and 1 atm:** Gas
13.2 **Molecular Weight:** 46.1
13.3 **Boiling Point at 1 atm:** $-12.5°F = -24.7°C = 248.5°K$
13.4 **Freezing Point:** $-222.7°F = -141.5°C = 131.7°K$
13.5 **Critical Temperature:** $260.4°F = 126.9°C = 400.1°K$
13.6 **Critical Pressure:** 780 psia = 53 atm = 5.4 MN/m^2
13.7 **Specific Gravity:** 0.724 at −24.7°C (liquid)
13.8 **Liquid Surface Tension:** 21 dynes/cm = 0.021 N/m at −40°C
13.9 **Liquid-Water Interfacial Tension:** (est.) 15 dynes/cm = 0.015 N/m at −40°C
13.10 **Vapor (Gas) Specific Gravity:** 1.6
13.11 **Ratio of Specific Heats of Vapor (Gas):** 1.1456
13.12 **Latent Heat of Vaporization:** 200 Btu/lb = 111 cal/g = 4.65×10^5 J/kg
13.13 **Heat of Combustion:** −13,450 Btu/lb = −7,480 cal/g = -313×10^5 J/kg
13.14 **Heat of Decomposition:** Not pertinent
13.15 **Heat of Solution:** Not pertinent
13.16 **Heat of Polymerization:** Not pertinent

NOTES

DMF — DIMETHYLFORMAMIDE

Common Synonyms			
N,N-Dimethylformamide DMF	Watery liquid	Colorless	Slight ammonia odor
	Floats and mixes with water.		

Avoid contact with liquid. Keep people away.
Wear goggles, self-contained breathing apparatus, and rubber overclothing (including gloves).
Call fire department.
Stop discharge if possible.
Isolate and remove discharged material.
Notify local health and pollution control agencies.

Fire

Combustible.
Wear goggles, self-contained breathing apparatus, and rubber overclothing (including gloves).
Extinguish with dry chemical, water, alcohol foam or carbon dioxide.

Exposure

CALL FOR MEDICAL AID.

LIQUID
Will burn skin and eyes.
Remove contaminated clothing and shoes.
Flush affected areas with plenty of water.
IF IN EYES, hold eyelids open and flush with plenty of water.

Water Pollution

Effect of low concentrations on aquatic life is unknown.
May be dangerous if it enters water intakes.
Notify local health and wildlife officials.
Notify operators of nearby water intakes.

1. RESPONSE TO DISCHARGE

(See Response Methods Handbook, CG 446-4)

Restrict access
Disperse and flush

2. LABELS

No hazard label required by Code of Federal Regulations

3. CHEMICAL DESIGNATIONS

3.1 **Synonyms:** N, N-Dimethylformamide; DMF
3.2 **Coast Guard Compatibility Classification:** Amide
3.3 **Chemical Formula:** $HCON(CH_3)_2$
3.4 **IMCO/United Nations Numerical Designation:** Not listed

4. OBSERVABLE CHARACTERISTICS

4.1 **Physical State (as shipped):** Liquid
4.2 **Color:** Colorless
4.3 **Odor:** Slight amine odor; essentially odorless

5. HEALTH HAZARDS

5.1 **Personal Protective Equipment:** Safety glasses or face shield; rubber apron and boots.
5.2 **Symptoms Following Exposure:** Irritation of eyes, skin and nose. May cause nausea.
5.3 **Treatment for Exposure:** INHALATION: remove victim to fresh air; if he is not breathing, give artificial respiration; for difficult breathing give oxygen; call a physician. SKIN OR EYES: flush with plenty of water while removing contaminated clothing and shoes.
5.4 **Toxicity by Inhalation (Threshold Limit Value):** 10 ppm
5.5 **Short-Term Inhalation Limits:** Data not available
5.6 **Toxicity by Ingestion:** Grade 1; LD_{50} 5 to 15 g/kg (rat)
5.7 **Late Toxicity:** Causes abortions in pregnant rats, possibly in humans also.
5.8 **Vapor (Gas) Irritant Characteristics:** Vapors cause moderate irritation, such that personnel will find high concentrations unpleasant. The effect is temporary.
5.9 **Liquid or Solid Irritant Characteristics:** Causes smarting of the skin and first-degree burns on short exposure and may cause secondary burns on long exposure.
5.10 **Odor Threshold:** 100 ppm

6. FIRE HAZARDS

6.1 **Flash Point:** 153°F O.C.; 136°F C.C.
6.2 **Flammable Limits in Air:** 2.2% 15.2%
6.3 **Fire Extinguishing Agents:** Water, foam, carbon dioxide, or dry chemical
6.4 **Fire Extinguishing Agents Not to be Used:** Not pertinent
6.5 **Special Hazards of Combustion Products:** Vapors are irritating
6.6 **Behavior in Fire:** Not pertinent
6.7 **Ignition Temperature:** 833°F
6.8 **Electrical Hazard:** Not pertinent
6.9 **Burning Rate:** 2.2 mm/min.

7. CHEMICAL REACTIVITY

7.1 **Reactivity with Water:** No reaction
7.2 **Reactivity with Common Materials:** No reaction
7.3 **Stability During Transport:** Stable
7.4 **Neutralizing Agents for Acids and Caustics:** Not pertinent
7.5 **Polymerization:** Not pertinent
7.6 **Inhibitor of Polymerization:** Not pertinent

8. WATER POLLUTION

8.1 **Aquatic Toxicity:** Data not available
8.2 **Waterfowl Toxicity:** Data not available
8.3 **Biological Oxygen Demand (BOD):** 0.9 lb/lb, 5 days
8.4 **Food Chain Concentration Potential:** None

9. SELECTED MANUFACTURERS

1. Air Products and Chemicals, Inc. Allentown, Pa. 18105
2. E. I. duPont de Nemours & Co., Inc. Organic Chemicals Dept. Wilmington, Del. 19898
3. Houghton Chemical Co. 52 Cambridge St. Boston, Mass. 02134

10. SHIPPING INFORMATION

10.1 **Grades or Purity:** Data not available
10.2 **Storage Temperature:** Ambient
10.3 **Inert Atmosphere:** No requirement
10.4 **Venting:** Open (flame arrester)

11. HAZARD ASSESSMENT CODE

(See Hazard Assessment Handbook, CG 446-3)

A-P-Q

12. HAZARD CLASSIFICATIONS

12.1 **Code of Federal Regulations:** Combustible Liquid
12.2 **NAS Hazard Rating for Bulk Water Transportation:**

Category	Rating
Fire	2
Health	
Vapor Irritant	2
Liquid or Solid Irritant	2
Poisons	3
Water Pollution	
Human Toxicity	2
Aquatic Toxicity	2
Aesthetic Effect	0
Reactivity	
Other Chemicals	1
Water	0
Self-Reaction	0

12.3 **NFPA Hazard Classifications:** Not listed

13. PHYSICAL AND CHEMICAL PROPERTIES

13.1 **Physical State at 15°C and 1 atm:** Liquid
13.2 **Molecular Weight:** 73.09
13.3 **Boiling Point at 1 atm:** 307°F = 153°C = 426°K
13.4 **Freezing Point:** −78°F = −61°C = 212°K
13.5 **Critical Temperature:** Not pertinent
13.6 **Critical Pressure:** Not pertinent
13.7 **Specific Gravity:** 0.950 at 20°C (liquid)
13.8 **Liquid Surface Tension:** Not pertinent
13.9 **Liquid-Water Interfacial Tension:** Not pertinent
13.10 **Vapor (Gas) Specific Gravity:** Not pertinent
13.11 **Ratio of Specific Heats of Vapor (Gas):** 1.101
13.12 **Latent Heat of Vaporization:** 248 Btu/lb = 138 cal/g = 5.78×10^5 J/kg
13.13 **Heat of Combustion:** −11,280 Btu/lb = −6267 cal/g = -262.4×10^5 J/kg
13.14 **Heat of Decomposition:** Not pertinent
13.15 **Heat of Solution:** −63 Btu/lb = −35 cal/g = -1.5×10^5 J/kg
13.16 **Heat of Polymerization:** Not pertinent

NOTES

REVISED 1978

DDW	DIMETHYLHEXANE DIHYDROPEROXIDE, WET

Common Synonyms		
2,5-Dimethylhexane-2,5-dihydroperoxide 2,5-Dihydroperoxy-2,5-dimethylhexane	Wet solid May float or sink in water.	White

Stop discharge if possible. Keep people away.
Call fire department.
Avoid contact with solid.
Isolate and remove discharged material.
Notify local health and pollution control agencies.

Fire

Not flammable when wet.
Will increase the intensity of a fire.
May explode if exposed to heat or flames.
Flood discharge area with water to prevent fire.
Extinguish with dry chemicals, foam or carbon dioxide.
Water may be ineffective on fire.
Cool exposed containers with water.

Exposure

CALL FOR MEDICAL AID.

VAPOR
Irritating to eyes, nose and throat.
If inhaled will cause coughing or difficult breathing.
If in eyes, hold eyelids open and flush with plenty of water.
If breathing has stopped, give artificial respiration.
If breathing is difficult, give oxygen.

SOLID
Irritating to skin and eyes.
Harmful if swallowed.
Remove contaminated clothing and shoes.
Flush affected areas with plenty of water.
IF IN EYES, hold eyelids open and flush with plenty of water.
IF SWALLOWED and victim is CONSCIOUS, have victim drink water or milk and have victim induce vomiting.
IF SWALLOWED and victim is UNCONSCIOUS OR HAVING CONVULSIONS, do nothing except keep victim warm.

Water Pollution

Effect of low concentrations on aquatic life is unknown.
May be dangerous if it enters water intakes.
Notify local health and wildlife officials.
Notify operators of nearby water intakes.

1. RESPONSE TO DISCHARGE

(See Response Methods Handbook, CG 446-4)

Issue warning—oxidizing material
Restrict access
Mechanical containment
Should be removed
Chemical and physical treatment

2. LABEL

Yellow

3. CHEMICAL DESIGNATIONS

3.1 **Synonyms:** 2,5-Dihydroperoxy-2,5-dimethylhexane; 2,5-Dimethylhexane-2,5-dihydroperoxide
3.2 **Coast Guard Compatibility Classification:** Not listed
3.3 **Chemical Formula:** $C_8H_{16}(OOH)_2$—H_2O
3.4 **IMCO/United Nations Numerical Designation:** 5.2/1528

4. OBSERVABLE CHARACTERISTICS

4.1 **Physical State (as shipped):** Wet solid
4.2 **Color:** White
4.3 **Odor:** Data not available

5. HEALTH HAZARDS

5.1 **Personal Protective Equipment:** Goggles or face shield; rubber gloves
5.2 **Symptoms Following Exposure:** Contact with eyes or skin causes irritation.
5.3 **Treatment for Exposure:** INHALATION: remove from exposure; call a doctor. EYES: wash with large amount of water for at least 15 min. SKIN: wash with large amount of water.
5.4 **Toxicity by Inhalation (Threshold Limit Value):** Data not available
5.5 **Short-Term Inhalation Limits:** Data not available
5.6 **Toxicity by Ingestion:** Data not available
5.7 **Late Toxicity:** Data not available
5.8 **Vapor (Gas) Irritant Characteristics:** Data not available
5.9 **Liquid or Solid Irritant Characteristics:** Data not available
5.10 **Odor Threshold:** Data not available

6. FIRE HAZARDS

6.1 **Flash Point:** Not pertinent
6.2 **Flammable Limits in Air:** Not pertinent
6.3 **Fire Extinguishing Agents:** Dry chemical, foam, carbon dioxide
6.4 **Fire Extinguishing Agents Not to be Used:** Water may be ineffective on fire.
6.5 **Special Hazards of Combustion Products:**
6.6 **Behavior in Fire:** Decomposes violently when heated in fire. Can increase intensity of fire when in contact with combustible material. Containers may explode.
6.7 **Ignition Temperature:** Not pertinent
6.8 **Electrical Hazard:** Not pertinent
6.9 **Burning Rate:** Not pertinent

7. CHEMICAL REACTIVITY

7.1 **Reactivity with Water:** No reaction
7.2 **Reactivity with Common Materials:** Decomposes in contact with many metals and acids.
7.3 **Stability During Transport:** Stable below 100°F
7.4 **Neutralizing Agents for Acids and Caustics:** Not pertinent
7.5 **Polymerization:** Not pertinent
7.6 **Inhibitor of Polymerization:** Not pertinent

8. WATER POLLUTION

8.1 **Aquatic Toxicity:** Data not available
8.2 **Waterfowl Toxicity:** Data not available
8.3 **Biological Oxygen Demand (BOD):** Data not available
8.4 **Food Chain Concentration Potential:** None

9. SELECTED MANUFACTURERS

1. Pennwalt Corporation
   Lucidol Division
   P. O. Box 1048
   Buffalo, N. Y. 14240
2. Witco Chemical Corporation
   U. S. Peroxygen Division
   850 Morton Avenue
   Richmond, Calif. 94804
3. Polysciences, Inc.
   Paul Valley Industrial Park
   Warrington, Pa. 18976

10. SHIPPING INFORMATION

10.1 **Grades or Purity:** Approx. 30+% water. The dry chemical is too hazardous to ship.
10.2 **Storage Temperature:** 4-38°C (40-100°F)
10.3 **Inert Atmosphere:** No requirement
10.4 **Venting:** Open

11. HAZARD ASSESSMENT CODE

(See Hazard Assessment Handbook, CG 446-3)

II

12. HAZARD CLASSIFICATIONS

12.1 **Code of Federal Regulations:** Organic Peroxide
12.2 **NAS Hazard Rating for Bulk Water Transportation:** Not listed
12.3 **NFPA Hazard Classifications:** Not listed

13. PHYSICAL AND CHEMICAL PROPERTIES

13.1 **Physical State at 15°C and 1 atm:** Solid
13.2 **Molecular Weight:** 178.2
13.3 **Boiling Point at 1 atm:** Not pertinent (decomposes)
13.4 **Freezing Point:** Not pertinent
13.5 **Critical Temperature:** Not pertinent
13.6 **Critical Pressure:** Not pertinent
13.7 **Specific Gravity:** (est.) 1.0 at 20°C (solid)
13.8 **Liquid Surface Tension:** Not pertinent
13.9 **Liquid-Water Interfacial Tension:** Not pertinent
13.10 **Vapor (Gas) Specific Gravity:** Not pertinent
13.11 **Ratio of Specific Heats of Vapor (Gas):** Not pertinent
13.12 **Latent Heat of Vaporization:** Not pertinent
13.13 **Heat of Combustion:** Data not available
13.14 **Heat of Decomposition:** Data not available
13.15 **Heat of Solution:** Not pertinent
13.16 **Heat of Polymerization:** Not pertinent

NOTES

DMH | 1,1-DIMETHYLHYDRAZINE

Common Synonyms			
UDMH Dimazine unsym-Dimethylhydrazine	Watery liquid	Colorless	Fishy or ammonia-like odor
	Floats and mixes with water.		

AVOID CONTACT WITH LIQUID AND VAPOR. Keep people away.
Wear chemical protective suit with self-contained breathing apparatus.
Shut off ignition sources and call fire department.
Stop discharge if possible.
Stay upwind and use water spray to "knock down" vapor.
Evacuate area in case of large discharge.
Isolate and remove discharged material.
Notify local health and pollution control agencies.

Fire

FLAMMABLE.
POISONOUS GASES ARE PRODUCED WHEN HEATED.
Flashback along vapor trail may occur.
Vapor may explode if ignited in an enclosed area.
Wear chemical protective suit with self-contained breathing apparatus.
Flood discharge area with water.
Extinguish with water.

Exposure

CALL FOR MEDICAL AID.

VAPOR
POISONOUS IF INHALED OR IF SKIN IS EXPOSED.
Irritating to eyes.
Move to fresh air.
If breathing has stopped, give artificial respiration.
If breathing is difficult, give oxygen.

LIQUID
POISONOUS IF SWALLOWED OR IF SKIN IS EXPOSED.
Will burn eyes.
Remove contaminated clothing and shoes.
Flush affected areas with plenty of water.
IF IN EYES, hold eyelids open and flush with plenty of water.
IF SWALLOWED and victim is CONSCIOUS, have victim drink water or milk.
DO NOT INDUCE VOMITING.

Water Pollution

Effect of low concentrations on aquatic life is unknown.
May be dangerous if it enters water intakes.
Notify local health and wildlife officials.
Notify operators of nearby water intakes.

1. RESPONSE TO DISCHARGE

(See Response Methods Handbook, CG 446-4)

Issue warning—high flammability
Restrict access
Evacuate area
Disperse and flush

2. LABELS

Red

3. CHEMICAL DESIGNATIONS

3.1 **Synonyms:** Dimazine
UDMH
unsym-Dimethylhydrazine

3.2 **Coast Guard Compatibility Classification:** Not applicable

3.3 **Chemical Formula:** $(CH_3)_2N{\cdot}NH_2$

3.4 **IMCO/United Nations Numerical Designation:** 3.2/1163

4. OBSERVABLE CHARACTERISTICS

4.1 **Physical State (as shipped):** Liquid

4.2 **Color:** Colorless

4.3 **Odor:** Sharp ammoniacal; fishy

5. HEALTH HAZARDS

5.1 **Personal Protective Equipment:** Rubber gloves, boots, and apron; plastic face shield. Gas mask with ammonia (GMD) canister protects for 30 min. against 1% concentration; for longer periods or higher concentrations, use self-contained breathing apparatus.

5.2 **Symptoms Following Exposure:** Breathing of vapor causes pulmonary irritation, delayed gastro-intestinal irritation, tremors, and convulsions. Contact with skin or mucous membranes causes chemical burns. Can be absorbed through skin to cause systemic intoxication and convulsions.

5.3 **Treatment for Exposure:** INHALATION: remove victim from contaminated area, give artificial respiration and oxygen if needed; watch for signs of pulmonary edema; enforce absolute rest. INGESTION: do NOT induce vomiting; hospitalize. SKIN OR EYES: flood with water and treat as alkaline burn.

5.4 **Toxicity by Inhalation (Threshold Limit Value):** 0.5 ppm

5.5 **Short-Term Inhalation Limits:** 100 ppm for 10 min.; 50 ppm for 30 min.; 30 ppm for 60 min.

5.6 **Toxicity by Ingestion:** Grade 3; LD_{50} 50 to 500 mg/kg (rat, mouse)

5.7 **Late Toxicity:** Data not available
Mild anemia, upper respiratory irritation, and muscle tremors in dogs following chronic exposure.

5.8 **Vapor (Gas) Irritant Characteristics:** Vapor is moderately irritating such that personnel will not usually tolerate moderate or high concentrations.

5.9 **Liquid or Solid Irritant Characteristics:** Severe skin irritant. Causes second- and third-degree burns on short contact and is very injurious to the eyes.

5.10 **Odor Threshold:** 6–14 ppm

6. FIRE HAZARDS

6.1 **Flash Point:** 34°F C.C.

6.2 **Flammable Limits in Air:** 2%–95%

6.3 **Fire Extinguishing Agents:** Flood with water

6.4 **Fire Extinguishing Agents Not to be Used:** In large fires, water fog, carbon dioxide, and bicarbonate types may allow flash-back and explosive re-ignition.

6.5 **Special Hazards of Combustion Products:** None

6.6 **Behavior in Fire:** Tends to re-ignite unless diluted with much water.

6.7 **Ignition Temperature:** 452–482°F

6.8 **Electrical Hazard:** Class I, Group D

6.9 **Burning Rate:** 3.8 mm/min.

7. CHEMICAL REACTIVITY

7.1 **Reactivity with Water:** No reaction

7.2 **Reactivity with Common Materials:** Dissolves, swells, and disintegrates many plastics

7.3 **Stability During Transport:** Stable below 1112°F

7.4 **Neutralizing Agents for Acids and Caustics:** Flush with water

7.5 **Polymerization:** Not pertinent

7.6 **Inhibitor of Polymerization:** Not pertinent

8. WATER POLLUTION

8.1 **Aquatic Toxicity:** Data not available

8.2 **Waterfowl Toxicity:** Data not available

8.3 **Biological Oxygen Demand (BOD):** Data not available

8.4 **Food Chain Concentration Potential:** Data not available

9. SELECTED MANUFACTURERS

FMC Corporation
Organic Chemicals Division
633 Third Ave.
New York, N. Y. 10012

10. SHIPPING INFORMATION

10.1 **Grades or Purity:** Propellant-grade: 98% min.

10.2 **Storage Temperature:** Below 120°F

10.3 **Inert Atmosphere:** Inerted

10.4 **Venting:** Data not available

11. HAZARD ASSESSMENT CODE

(See Hazard Assessment Handbook, CG 446-3)

A-P-Q-R-S

12. HAZARD CLASSIFICATIONS

12.1 **Code of Federal Regulations:** Flammable liquid

12.2 **NAS Hazard Rating for Bulk Water Transportation:**

Category	Rating
Fire	4
Health	
Vapor Irritant	3
Liquid or Solid Irritant	4
Poisons	4
Water Pollution	
Human Toxicity	4
Aquatic Toxicity	2
Aesthetic Effect	2
Reactivity	
Other Chemicals	4
Water	0
Self-Reaction	4

12.3 **NFPA Hazard Classifications:**

Category	Classification
Health Hazard (Blue)	3
Flammability (Red)	3
Reactivity (Yellow)	1

13. PHYSICAL AND CHEMICAL PROPERTIES

13.1 **Physical State at 15°C and 1 atm:** Liquid

13.2 **Molecular Weight:** 60.11

13.3 **Boiling Point at 1 atm:** 146.0°F = 63.3°C = 336.5°K

13.4 **Freezing Point:** −71°F = −57°C = 216°K

13.5 **Critical Temperature:** 480°F = 249°C = 522°K

13.6 **Critical Pressure:** 865 psia = 53.5 atm = 5.40 MN/m^2

13.7 **Specific Gravity:** 0.791 at 20°C (liquid)

13.8 **Liquid Surface Tension:** 28 dynes/cm = 0.028 N/m at 25°C

13.9 **Liquid-Water Interfacial Tension:** Not pertinent

13.10 **Vapor (Gas) Specific Gravity:** 2.1

13.11 **Ratio of Specific Heats of Vapor (Gas):** (est.) 1.152

13.12 **Latent Heat of Vaporization:** 261 Btu/lb = 145 cal/g = 6.07×10^5 J/kg

13.13 **Heat of Combustion:** −14,170 Btu/lb = −7870 cal/g = -329.3×10^5 J/kg

13.14 **Heat of Decomposition:** Not pertinent

13.15 **Heat of Solution:** (est.) −30 Btu/lb = −10 cal/g = -0.6×10^5 J/kg

13.16 **Heat of Polymerization:** Not pertinent

NOTES

REVISED 1978

DMP	DIMETHYLPOLYSILOXANE

Common Synonyms	Liquid	Colorless	Odorless
Poly (dimethylsiloxane) Dimethyl silicone oil Silicone fluids Dimethyl silicone fluids	Floats on water.		

Stop discharge if possible. Keep people away.
Call fire department.
Isolate and remove discharged material.
Notify local health and pollution control agencies.

Fire

Combustible.
Extinguish with dry chemicals, foam or carbon dioxide.
Water may be ineffective on fire.

Exposure

Call for medical aid.

LIQUID
Irritating to eyes.
IF IN EYES, hold eyelids open and flush with plenty of water.
IF SWALLOWED and victim is CONSCIOUS, have victim drink water or milk.

Water Pollution

Effect of low concentrations on aquatic life is unknown.
Fouling to shoreline.
May be dangerous if it enters water intakes.
Notify local health and wildlife officials.
Notify operators of nearby water intakes.

1. RESPONSE TO DISCHARGE

(See Response Methods Handbook, CG 446-4)

Mechanical containment
Should be removed
Chemical and physical treatment

2. LABELS

No label required by Code of Federal Regulations

3. CHEMICAL DESIGNATIONS

3.1 **Synonyms:** Poly (dimethylsiloxane); Dimethylsilicone oil; Silicone fluids; Dimethyl silicone fluids
3.2 **Coast Guard Compatibility Classification:** Not applicable
3.3 **Chemical Formula:** $(CH_3)_3Si\text{-}O\text{-}[Si(CH_3)_2O]_n\text{-}Si(CH_3)_3$
3.4 **IMCO/United Nations Numerical Designation:** Not listed

4. OBSERVABLE CHARACTERISTICS

4.1 **Physical State (as shipped):** Liquid
4.2 **Color:** Colorless
4.3 **Odor:** None

5. HEALTH HAZARDS

5.1 **Personal Protective Equipment:** Safety goggles.
5.2 **Symptoms Following Exposure:** Contact of liquid with eyes may cause temporary discomfort. Does not irritate skin. Harmless when ingested.
5.3 **Treatment for Exposure:** Except for eye contact, exposures generally do not require treatment. EYES: flush with water.
5.4 **Toxicity by Inhalation (Threshold Limit Value):** Data not available
5.5 **Short-Term Inhalation Limits:** Not pertinent
5.6 **Toxicity by Ingestion:** Data not available
5.7 **Late Toxicity:** None
5.8 **Vapor (Gas) Irritant Characteristics:** Data not available
5.9 **Liquid or Solid Irritant Characteristics:** Data not available
5.10 **Odor Threshold:** Not pertinent

6. FIRE HAZARDS

6.1 **Flash Point:** 275-635°F O.C.
6.2 **Flammable Limits in Air:** Not pertinent
6.3 **Fire Extinguishing Agents:** Foam, dry chemical, carbon dioxide.
6.4 **Fire Extinguishing Agents Not to be Used:** Water may be ineffective.
6.5 **Special Hazards of Combustion Products:** Not pertinent
6.6 **Behavior in Fire:** Not pertinent
6.7 **Ignition Temperature:** 820-860°F
6.8 **Electrical Hazard:** Not pertinent
6.9 **Burning Rate:** Not pertinent

7. CHEMICAL REACTIVITY

7.1 **Reactivity with Water:** No reaction
7.2 **Reactivity with Common Materials:** No reaction
7.3 **Stability During Transport:** Stable
7.4 **Neutralizing Agents for Acids and Caustics:** Not pertinent
7.5 **Polymerization:** Not pertinent
7.6 **Inhibitor of Polymerization:** Not pertinent

8. WATER POLLUTION

8.1 **Aquatic Toxicity:** Data not available
8.2 **Waterfowl Toxicity:** Data not available
8.3 **Biological Oxygen Demand (BOD):** Data not available
8.4 **Food Chain Concentration Potential:** Data not available

9. SELECTED MANUFACTURERS

1. Dow Corning Corporation
   South Saginaw Road
   Midland, Mich. 48640
2. Union Carbide Corporation
   Chemicals and Plastics Division
   270 Park Avenue
   New York, N.Y. 10017

10. SHIPPING INFORMATION

10.1 **Grades or Purity:** A series of compounds having viscosities of from 50 to 100,000 cp is available.
10.2 **Storage Temperature:** Ambient
10.3 **Inert Atmosphere:** No requirement
10.4 **Venting:** Open

11. HAZARD ASSESSMENT CODE

(See Hazard Assessment Handbook, CG 446-3)

A-T-U-X-Y

12. HAZARD CLASSIFICATIONS

12.1 **Code of Federal Regulations:** Not listed
12.2 **NAS Hazard Rating for Bulk Water Transportation:** Not listed
12.3 **NFPA Hazard Classifications:** Not listed

13. PHYSICAL AND CHEMICAL PROPERTIES

13.1 **Physical State at 15°C and 1 atm:** Liquid
13.2 **Molecular Weight:** Not pertinent
13.3 **Boiling Point at 1 atm:** $>300°F = >149°C = >422°K$
13.4 **Freezing Point:** Not pertinent
13.5 **Critical Temperature:** Not pertinent
13.6 **Critical Pressure:** Not pertinent
13.7 **Specific Gravity:** 0.98 at 20°C (liquid)
13.8 **Liquid Surface Tension:** 19-21 dynes/cm = 0.019-0.021 N/m at 20°C
13.9 **Liquid-Water Interfacial Tension:** (est.) 30 dynes/cm = 0.030 N/m at 20°C
13.10 **Vapor (Gas) Specific Gravity:** Not pertinent
13.11 **Ratio of Specific Heats of Vapor (Gas):** Not pertinent
13.12 **Latent Heat of Vaporization:** Not pertinent
13.13 **Heat of Combustion:** (est.) $-11{,}000$ Btu/lb $= -6{,}200$ cal/g $= -260 \times 10^5$ J/kg
13.14 **Heat of Decomposition:** Not pertinent
13.15 **Heat of Solution:** Not pertinent
13.16 **Heat of Polymerization:** Not pertinent

NOTES

DSF	DIMETHYL SULFATE

Common Synonyms			
Methyl sulfate	Liquid	Colorless	Mild onion odor
	Sinks and mixes slowly with water.		

AVOID CONTACT WITH LIQUID. Keep people away.
Wear goggles, self-contained breathing apparatus, and rubber overclothing (including gloves).
Stop discharge if possible.
Call fire department.
Isolate and remove discharged material.
Notify local health and pollution control agencies.

Fire

Combustible.
POISONOUS GASES ARE PRODUCED WHEN HEATED.
Wear goggles, self-contained breathing apparatus, and rubber overclothing (including gloves).
Extinguish with water, foam, dry chemical, or carbon dioxide.
Cool exposed containers with water.

Exposure

CALL FOR MEDICAL AID.

LIQUID
POISONOUS IF SWALLOWED OR IF SKIN IS EXPOSED.
Will burn eyes.
Remove contaminated clothing and shoes.
Flush affected areas with plenty of water.
IF IN EYES, hold eyelids open and flush with plenty of water.
IF SWALLOWED and victim is CONSCIOUS, have victim drink water or milk.
DO NOT INDUCE VOMITING.

Water Pollution

Effect of low concentrations on aquatic life is unknown.
May be dangerous if it enters water intakes.
Notify local health and wildlife officials.
Notify operators of nearby water intakes.

1. RESPONSE TO DISCHARGE

(See Response Methods Handbook, CG 446-4)

Issue warning—poison, corrosive
Restrict access
Chemical and physical treatment

2. LABEL

3. CHEMICAL DESIGNATIONS

3.1 **Synonyms:** No common synonyms
3.2 **Coast Guard Compatibility Classification:** Ester
3.3 **Chemical Formula:** $(CH_3)_2SO_4$
3.4 **IMCO/United Nations Numerical Designation:** 6.1/1595

4. OBSERVABLE CHARACTERISTICS

4.1 **Physical State (as shipped):** Liquid
4.2 **Color:** Colorless
4.3 **Odor:** No characteristic odor; slight, not distinctive; weak onion

5. HEALTH HAZARDS

5.1 **Personal Protective Equipment:** Chemical goggles; self-contained breathing apparatus; safety hat; rubber shoes; rubber suit; rubber gloves; safety shower and eye wash fountain.
5.2 **Symptoms Following Exposure:** Severe irritation to eyes, eyelids, respiratory tract and skin. Dry, painful cough; foamy, white sputum; difficulty in breathing; malaise and fever; inflammation and edema of lungs.
5.3 **Treatment for Exposure:** Contact with dimethyl sulfate liquid or vapor (>1 ppm) requires immediate treatment. Call a physician, even if there is no evidence of injury, as symptoms may not appear for several hours. INHALATION: get victim to fresh air immediately; administer 100% oxygen, even if no injury is apparent, and continue for 30 min. each hour for 6 hours; give artificial respiration if breathing is weak or fails, but do not interrupt oxygen therapy; if victim's coughing prevents use of a mask, use oxygen tent under atmospheric pressure. INGESTION: do NOT induce vomiting. SKIN: wash thoroughly. EYE: flush with running water for at least 15 min.
5.4 **Toxicity by Inhalation (Threshold Limit Value):** 1 ppm
5.5 **Short-Term Inhalation Limits:** Data not available
5.6 **Toxicity by Ingestion:** Grade 3; 50 to 500 mg/kg (rat)
5.7 **Late Toxicity:** Causes birth defects in rats (malignant tumors in nervous system).
5.8 **Vapor (Gas) Irritant Characteristics:** Vapors cause severe irritation of eye and throat and can cause eye and lung injury. They cannot be tolerated even at low concentration.
5.9 **Liquid or Solid Irritant Characteristics:** Severe skin irritant. Causes second- and third-degree burns on short contact; very injurious to the eyes.
5.10 **Odor Threshold:** Data not available

6. FIRE HAZARDS

6.1 **Flash Point:** 240°F O.C.; 182°F C.C.
6.2 **Flammable Limits in Air:** Data not available
6.3 **Fire Extinguishing Agents:** Water, foam, carbon dioxide or dry chemical
6.4 **Fire Extinguishing Agents Not to be Used:** Not pertinent
6.5 **Special Hazards of Combustion Products:** Flammable, toxic vapors generated
6.6 **Behavior in Fire:** Not pertinent
6.7 **Ignition Temperature:** 370°F
6.8 **Electrical Hazard:** Not pertinent
6.9 **Burning Rate:** Data not available

7. CHEMICAL REACTIVITY

7.1 **Reactivity with Water:** Slow, non-hazardous reaction
7.2 **Reactivity with Common Materials:** Corrodes metal when wet
7.3 **Stability During Transport:** Stable
7.4 **Neutralizing Agents for Acids and Cuastics:** Sodium bicarbonate or lime
7.5 **Polymerization:** Not pertinent
7.6 **Inhibitor of Polymerization:** Not pertinent

8. WATER POLLUTION

8.1 **Aquatic Toxicity:** Data not available
8.2 **Waterfowl Toxicity:** Data not available
8.3 **Biological Oxygen Demand (BOD):** Data not available
8.4 **Food Chain Concentration Potential:** None

9. SELECTED MANUFACTURERS

E. I. duPont de Nemours & Co., Inc.
Industrial and Biochemicals Dept.
Wilmington, Del. 19898

10. SHIPPING INFORMATION

10.1 **Grades or Purity:** Technical
10.2 **Storage Temperature:** Ambient
10.3 **Inert Atmosphere:** No requirement
10.4 **Venting:** Pressure-Vacuum

11. HAZARD ASSESSMENT CODE

(See Hazard Assessment Handbook, CG 446-3)

A-P-Q-X-Y

12. HAZARD CLASSIFICATIONS

12.1 **Code of Federal Regulations:** Corrosive material
12.2 **NAS Hazard Rating for Bulk Water Transportation:**

Category	Rating
Fire	1
Health	
Vapor Irritant	4
Liquid or Solid Irritant	4
Poisons	4
Water Pollution	
Human Toxicity	4
Aquatic Toxicity	3
Aesthetic Effect	2
Reactivity	
Other Chemicals	3
Water	0
Self-Reaction	0

12.3 **NFPA Hazard Classifications:**

Category	Classification
Health Hazard (Blue)	4
Flammability (Red)	2
Reactivity (Yellow)	0

13. PHYSICAL AND CHEMICAL PROPERTIES

13.1 **Physical State at 15°C and 1 atm:** Liquid
13.2 **Molecular Weight:** 126.13
13.3 **Boiling Point at 1 atm:** 371.8°F = 188.8°C = 462.0°K
13.4 **Freezing Point:** −25.2°F = −31.8°C = 241.4°K
13.5 **Critical Temperature:** Not pertinent
13.6 **Critical Pressure:** Not pertinent
13.7 **Specific Gravity:** 1.33 at 15°C (liquid)
13.8 **Liquid Surface Tension:** 40.1 dynes/cm = 0.0401 N/m at 18°C
13.9 **Liquid-Water Interfacial Tension:** (est.) 20 dynes/cm = 0.02 N/m at 20°C
13.10 **Vapor (Gas) Specific Gravity:** Not pertinent
13.11 **Ratio of Specific Heats of Vapor (Gas):** Not pertinent
13.12 **Latent Heat of Vaporization:** Not pertinent
13.13 **Heat of Combustion:** Not pertinent
13.14 **Heat of Decomposition:** Not pertinent
13.15 **Heat of Solution:** Not pertinent
13.16 **Heat of Polymerization:** Not pertinent

NOTES

REVISED 1978

DSL — DIMETHYL SULFIDE

Common Synonyms			
Methyl sulfide 2-Thiapropane Methanethiomethane	Liquid	Colorless to light yellow	Unpleasant odor
	Floats and mixes slowly with water. Irritating vapor is produced. Boiling point is 99° F.		

Shut off ignition sources. Call fire department.
Stop discharge if possible. Keep people away.
Evacuate area in case of large discharge.
Isolate and remove discharged material.
Notify local health and pollution control agencies.

Fire

FLAMMABLE.
POISONOUS GASES ARE PRODUCED IN FIRE.
Containers may explode in fire.
Flashback along vapor trail may occur.
Vapor may explode if ignited in an enclosed area.
Wear goggles and self-contained breathing apparatus.
Extinguish with dry chemicals, alcohol foam, or carbon dioxide.
Water may be ineffective on fire.
Cool exposed containers with water.

Exposure

Call for medical aid.

VAPOR
Irritating to eyes, nose and throat.
Move victim to fresh air.

LIQUID
Irritating to skin and eyes.
Harmful if swallowed.
Remove contaminated clothing and shoes.
Flush affected areas with plenty of water.
IF IN EYES, hold eyelids open and flush with plenty of water.
IF SWALLOWED and victim is CONSCIOUS, have victim drink water or milk and have victim induce vomiting.
IF SWALLOWED and victim is UNCONSCIOUS OR HAVING CONVULSIONS, do nothing except keep victim warm.

Water Pollution

Effect of low concentrations on aquatic life is unknown.
May be dangerous if it enters water intakes.
Notify local health and wildlife officials.
Notify operators of nearby water intakes.

1. RESPONSE TO DISCHARGE

(See Response Methods Handbook, CG 446-4)

Issue warning—high flammability
Restrict access
Evacuate area
Disperse and flush

2. LABEL

Red

3. CHEMICAL DESIGNATIONS

3.1 **Synonyms:** DMS
Methanethiomethane
Methyl sulfide
2-Thiapropane

3.2 **Coast Guard Compatibility Classification:** Not applicable

3.3 **Chemical Formula:** $(CH_3)_2S$

3.4 **IMCO/United Nations Numerical Designation:** 3.1/1164

4. OBSERVABLE CHARACTERISTICS

4.1 **Physical State (as shipped):** Liquid

4.2 **Color:** Colorless to straw

4.3 **Odor:** Ethereal, permeating; disagreeable; offensive

5. HEALTH HAZARDS

5.1 **Personal Protective Equipment:** Respirator with organic vapor canister; rubber or plastic gloves; goggles or face shield.

5.2 **Symptoms Following Exposure:** Inhalation causes moderate irritation of upper respiratory system. Contact of liquid with eyes causes moderate irritation. Repeated contact with skin may extract oils and result in irritation. Ingestion causes nausea and irritation of mouth and stomach.

5.3 **Treatment for Exposure:** INHALATION: move victim to fresh air at once; enforce rest, and keep warm; get medical attention immediately. EYES: flush with water for at least 15 min.; if irritation persists, get medical attention. SKIN: flush with plenty of water and wash thoroughly; get treatment for any lasting irritation. INGESTION: if large amounts are swallowed, induce vomiting by tickling the back of the throat with the finger or by giving an emetic such as two tablespoons of common salt in a glass of warm water; get medical attention.

5.4 **Toxicity by Inhalation (Threshold Limit Value):** Data not available

5.5 **Short-Term Inhalation Limits:** Data not available

5.6 **Toxicity by Ingestion:** Grade 2; oral LD_{50} = 535 mg/kg (rat)

5.7 **Late Toxicity:** Data not available

5.8 **Vapor (Gas) Irritant Characteristics:** Vapors are moderately irritating such that personnel will not usually tolerate moderate or high vapor concentrations.

5.9 **Liquid or Solid Irritant Characteristics:** Causes smarting of the skin and first-degree burns on short exposure and may cause second-degree burns on long exposure.

5.10 **Odor Threshold:** 0.001 ppm

6. FIRE HAZARDS

6.1 **Flash Point:** −36°F C.C.

6.2 **Flammable Limits in Air:** 2.2%–19.7%

6.3 **Fire Extinguishing Agents:** Dry chemical, foam, alcohol foam, carbon dioxide

6.4 **Fire Extinguishing Agents Not to be Used:** Water may be ineffective.

6.5 **Special Hazards of Combustion Products:** Toxic and irritating sulfur dioxide is formed.

6.6 **Behavior in Fire:** Vapor is heavier than air and may travel considerable distance to source of ignition and flash back.

6.7 **Ignition Temperature:** 403°F

6.8 **Electrical Hazard:** Data not available

6.9 **Burning Rate:** 4.8 mm/min.

7. CHEMICAL REACTIVITY

7.1 **Reactivity with Water:** No reaction

7.2 **Reactivity with Common Materials:** No reaction

7.3 **Stability During Transport:** Stable

7.4 **Neutralizing Agents for Acids and Caustics:** Not pertinent

7.5 **Polymerization:** Not pertinent

7.6 **Inhibitor of Polymerization:** Not pertinent

8. WATER POLLUTION

8.1 **Aquatic Toxicity:** Data not available

8.2 **Waterfowl Toxicity:** Data not available

8.3 **Biological Oxygen Demand (BOD):** Approximately 1 lb/lb

8.4 **Food Chain Concentration Potential:** None

9. SELECTED MANUFACTURERS

1. Crown Zellerbach Corp.
Chemical Products Division
Camus, Wash. 98607
2. Phillips Petroleum Company
Chemical Department
Special Products Division
Bartlesville, Okla. 74004
3. United States Steel Corp.
USS Chemicals Division
P. O. Box 86
Pittsburgh, Pa. 15230

10. SHIPPING INFORMATION

10.1 **Grades or Purity:** 99.8%

10.2 **Storage Temperature:** Ambient

10.3 **Inert Atmosphere:** No requirement

10.4 **Venting:** Pressure-vacuum

11. HAZARD ASSESSMENT CODE

(See Hazard Assessment Handbook, CG 446-3)

A-P-Q-T-U-V-W

12. HAZARD CLASSIFICATIONS

12.1 **Code of Federal Regulations:** Flammable liquid

12.2 **NAS Hazard Rating for Bulk Water Transportation:**

Category	Rating
Fire	4
Health	
Vapor Irritant	3
Liquid or Solid Irritant	2
Poisons	2
Water Pollution	
Human Toxicity	—
Aquatic Toxicity	3
Aesthetic Effect	4
Reactivity	
Other Chemicals	1
Water	0
Self-Reaction	0

12.3 **NFPA Hazard Classifications:**

Category	Classification
Health Hazard (Blue)	4
Flammability (Red)	4
Reactivity (Yellow)	0

13. PHYSICAL AND CHEMICAL PROPERTIES

13.1 **Physical State at 15°C and 1 atm:** Liquid

13.2 **Molecular Weight:** 62.1

13.3 **Boiling Point at 1 atm:** 99°F = 37°C = 310°K

13.4 **Freezing Point:** −144°F = −98°C = 175°K

13.5 **Critical Temperature:** 444°F = 229°C = 502°K

13.6 **Critical Pressure:** 826 psia = 56.1 atm = 5.69 MN/m^2

13.7 **Specific Gravity:** 0.85 at 20°C (liquid)

13.8 **Liquid Surface Tension:** 26.5 dynes/cm = 0.0265 N/m at 11°C

13.9 **Liquid-Water Interfacial Tension:** (est.) 30 dynes/cm = 0.030 N/m at 20°C

13.10 **Vapor (Gas) Specific Gravity:** 2.14

13.11 **Ratio of Specific Heats of Vapor (Gas):** 1.1277 at 16°C

13.12 **Latent Heat of Vaporization:** 194 Btu/lb = 108 cal/g = 4.52 × 10^5 J/kg

13.13 **Heat of Combustion:** −13,200 Btu/lb = −7,340 cal/g = −307 × 10^5 J/kg

13.14 **Heat of Decomposition:** Not pertinent

13.15 **Heat of Solution:** Not pertinent

13.16 **Heat of Polymerization:** Not pertinent

NOTES

DMS	DIMETHYL SULFOXIDE

Common Synonyms			
Methyl Sulfoxide DMSO	Liquid	Colorless	Mild garlic odor
	Sinks and mixes with water.		

Stop discharge if possible.
Call fire department.
Avoid contact with liquid.
Isolate and remove discharged materials.
Notify local health and pollution control agencies.

Fire	Combustible. POISONOUS GASES MAY BE PRODUCED IN FIRE. **Wear goggles, self-contained breathing apparatus, and rubber overclothing (including gloves).** **Extinguish with water, dry chemical, alcohol foam, or carbon dioxide.**
Exposure	**CALL FOR MEDICAL AID.** LIQUID Irritating to skin and eyes. **Flush affected areas with plenty of water.** **IF IN EYES, hold eyelids open and flush with plenty of water.**
Water Pollution	Dangerous to aquatic life in high concentrations. May be dangerous if it enters water intakes. **Notify local health and wildlife officials.** **Notify operators of nearby water intakes.**

1. RESPONSE TO DISCHARGE
(See Response Methods Handbook, CG 446-4)

Disperse and flush

2. LABELS
No hazard label required by Code of Federal Regulations

3. CHEMICAL DESIGNATIONS
3.1 **Synonyms:** DMSO; Methyl sulfoxide
3.2 **Coast Guard Compatibility Classification:** Not applicable
3.3 **Chemical Formula:** CH_3SOCH_3
3.4 **IMCO/United Nations Numerical Designation:** Not listed

4. OBSERVABLE CHARACTERISTICS
4.1 **Physical State (as shipped):** Liquid
4.2 **Color:** Colorless
4.3 **Odor:** Slight; almost odorless

5. HEALTH HAZARDS
5.1 **Personal Protective Equipment:** Butyl rubber gloves, safety goggles. Respiratory filter if airborne sprays or drops are present.
5.2 **Symptoms Following Exposure:** Slight eye irritation.
5.3 **Treatment for Exposure:** Wash eyes and skin with water.
5.4 **Toxicity by Inhalation (Threshold Limit Value):** Data not available
5.5 **Short-Term Inhalation Limits:** Data not available
5.6 **Toxicity by Ingestion:** Grade 0; above 15 g/kg
5.7 **Late Toxicity:** Causes damage to eye in dogs, pigs, rats, and rabbits.
5.8 **Vapor (Gas) Irritant Characteristics:** Vapors cause moderate irritation such that personnel will find high concentrations unpleasant. The effect is temporary.
5.9 **Liquid or Solid Irritant Characteristics:** Minimum hazard. If spilled on clothing and allowed to remain, may cause smarting and reddening of the skin.
5.10 **Odor Threshold:** Data not available

6. FIRE HAZARDS
6.1 **Flash Point:** 203°F O.C.; 190°F C.C.
6.2 **Flammable Limits in Air:** 3%–63%
6.3 **Fire Extinguishing Agents:** Water, foam, dry chemical, or carbon dioxide
6.4 **Fire Extinguishing Agents Not to be Used:** Not pertinent
6.5 **Special Hazards of Combustion Products:** Sulfur dioxide, formaldehyde, and methyl mercaptan can form
6.6 **Behavior in Fire:** Not pertinent
6.7 **Ignition Temperature:** 572°F
6.8 **Electrical Hazard:** Not pertinent
6.9 **Burning Rate:** 2.0 mm/min.

7. CHEMICAL REACTIVITY
7.1 **Reactivity with Water:** No reaction
7.2 **Reactivity with Common Materials:** No reaction
7.3 **Stability During Transport:** Stable
7.4 **Neutralizing Agents for Acids and Caustics:** Not pertinent
7.5 **Polymerization:** Not pertinent
7.6 **Inhibitor of Polymerization:** Not pertinent

8. WATER POLLUTION
8.1 **Aquatic Toxicity:** 33,500 ppm/48 hr/bluegill/TL_m/fresh water
8.2 **Waterfowl Toxicity:** Data not available
8.3 **Biological Oxygen Demand (BOD):** Data not available
8.4 **Food Chain Concentration Potential:** None

9. SELECTED MANUFACTURERS
Crown Zellerbach Corp.
Chemical Products Division
1 Bush St.
San Francisco, Calif. 94104

10. SHIPPING INFORMATION
10.1 **Grades or Purity:** 99%
10.2 **Storage Temperature:** Ambient
10.3 **Inert Atmosphere:** No requirement
10.4 **Venting:** Open (flame arrester) or pressure-vacuum

11. HAZARD ASSESSMENT CODE
(See Hazard Assessment Handbook, CG 446-3)

A-P-Q

12. HAZARD CLASSIFICATIONS
12.1 **Code of Federal Regulations:** Combustible Liquid
12.2 **NAS Hazard Rating for Bulk Water Transportation:** Not listed
12.3 **NFPA Hazard Classifications:** Not listed

13. PHYSICAL AND CHEMICAL PROPERTIES
13.1 **Physical State at 15°C and 1 atm:** Liquid
13.2 **Molecular Weight:** 78.13
13.3 **Boiling Point at 1 atm:** 372°F = 189°C = 462°K
13.4 **Freezing Point:** 65.5°F = 18.6°C = 291.8°K
13.5 **Critical Temperature:** Not pertinent
13.6 **Critical Pressure:** Not pertinent
13.7 **Specific Gravity:** 1.101 at 20°C (liquid)
13.8 **Liquid Surface Tension:** Not pertinent
13.9 **Liquid-Water Interfacial Tension:** Not pertinent
13.10 **Vapor (Gas) Specific Gravity:** Not pertinent
13.11 **Ratio of Specific Heats of Vapor (Gas):** Not pertinent
13.12 **Latent Heat of Vaporization:** 259 Btu/lb = 144 cal/g = 6.03×10^5 J/kg
13.13 **Heat of Combustion:** −10,890 Btu/lb = −6050 cal/g = -253.3×10^5 J/kg
13.14 **Heat of Decomposition:** Not pertinent
13.15 **Heat of Solution:** −97 Btu/lb = −54 cal/g = -2.3×10^5 J/kg
13.16 **Heat of Polymerization:** Not pertinent

NOTES

REVISED 1978

DMT — DIMETHYL TEREPHTHALATE

Common Synonyms			
Terephthalic acid, dimethyl ester	Solid or heated liquid	White solid or colorless liquid	Odorless
	Liquid solidifies. Solid and liquid sink in water.		

Avoid contact with liquid. Keep people away.
Stop discharge if possible.
Isolate and remove discharged material.
Notify local health and pollution control agencies.

Fire	Combustible. **Extinguish with water, dry chemicals, foam, or carbon dioxide.**
Exposure	**Call for medical aid.** DUST Not harmful. **Move victim to fresh air.** **If in eyes, hold eyelids open and flush with plenty of water.** LIQUID OR SOLID Heated liquid will burn skin and eyes. Harmful if swallowed. **Remove contaminated clothing and shoes.** **Flush affected areas with plenty of water.** **IF IN EYES, hold eyelids open and flush with plenty of water.** **IF SWALLOWED and victim is CONSCIOUS, have victim drink water or milk.**
Water Pollution	Effect of low concentrations on aquatic life is unknown. Fouling to shoreline. May be dangerous if it enters water intakes. **Notify local health and wildlife officials.** **Notify operators of nearby water intakes.**

1. RESPONSE TO DISCHARGE

(See Response Methods Handbook, CG 446-4)

Mechanical containment (if floating)
Should be removed
Chemical and physical treatment

2. LABELS

No label required by Code of Federal Regulations

3. CHEMICAL DESIGNATIONS

3.1 **Synonyms:** Terephthalic acid, dimethyl ester
3.2 **Coast Guard Compatibility Classification:** Not applicable
3.3 **Chemical Formula:** 1,4-$CH_3OOCC_6H_4COOCH_3$
3.4 **IMCO/United Nations Numerical Designation:** Not listed

4. OBSERVABLE CHARACTERISTICS

4.1 **Physical State (as shipped):** Solid or liquid
4.2 **Color:** Colorless
4.3 **Odor:** None; weak aromatic

5. HEALTH HAZARDS

5.1 **Personal Protective Equipment:** Molten DMT: goggles, face shield, gauntlets, and protective clothing. Solid: dust mask, goggles.
5.2 **Symptoms Following Exposure:** Molten DMT will cause severe burns of skin on contact.
5.3 **Treatment for Exposure:** EYES: flush dust from eyes with water. SKIN: wash with soap and water. If burned by molten DMT, flush area immediately with cold water for at least 15 min.; apply ice pack for at least 30 min.; do not try to rub DMT off a burn or remove clothing that DMT has penetrated, because this will remove underlying skin; seek prompt medical treatment for significant burns.
5.4 **Toxicity by Inhalation (Threshold Limit Value):** Data not available
5.5 **Short-Term Inhalation Limits:** Not pertinent
5.6 **Toxicity by Ingestion:** Grade 2; oral LD_{50} = 4,390 mg/kg (rat)
5.7 **Late Toxicity:** Data not available
5.8 **Vapor (Gas) Irritant Characteristics:** Not pertinent
5.9 **Liquid or Solid Irritant Characteristics:** Data not available
5.10 **Odor Threshold:** Not pertinent

6. FIRE HAZARDS

6.1 **Flash Point:** 298°F O.C. (molten)
6.2 **Flammable Limits in Air:** Not pertinent
6.3 **Fire Extinguishing Agents:** Water, dry chemical, foam, carbon dioxide
6.4 **Fire Extinguishing Agents Not to be Used:** Not pertinent
6.5 **Special Hazards of Combustion Products:** Not pertinent
6.6 **Behavior in Fire:** Not pertinent
6.7 **Ignition Temperature:** 1,058°F (dust)
6.8 **Electrical Hazard:** Not pertinent
6.9 **Burning Rate:** Not pertinent

7. CHEMICAL REACTIVITY

7.1 **Reactivity with Water:** No reaction
7.2 **Reactivity with Common Materials:** No reaction
7.3 **Stability During Transport:** Stable
7.4 **Neutralizing Agents for Acids and Caustics:** Not pertinent
7.5 **Polymerization:** Not pertinent
7.6 **Inhibitor of Polymerization:** Not pertinent

8. WATER POLLUTION

8.1 **Aquatic Toxicity:** Data not available
8.2 **Waterfowl Toxicity:** Data not available
8.3 **Biological Oxygen Demand (BOD):** Data not available
8.4 **Food Chain Concentration Potential:** None

9. SELECTED MANUFACTURERS

1. Hercules Incorporated
   Synthetics Department
   910 Market Street
   Wilmington, Del. 19899
2. Amoco Chemicals Corporation
   200 East Randolph Drive
   Chicago, Ill. 60601
3. E. I. duPont de Nemours & Company, Inc.
   Polymer Intermediates Department
   Wilmington, Del. 19898

10. SHIPPING INFORMATION

10.1 **Grades or Purity:** 99.9%
10.2 **Storage Temperature:** Ambient
10.3 **Inert Atmosphere:** Inert gas blanket is advisable.
10.4 **Venting:** Pressure-vacuum

11. HAZARD ASSESSMENT CODE

(See Hazard Assessment Handbook, CG 446-3)

A-T-U-X-Y-II

12. HAZARD CLASSIFICATIONS

12.1 **Code of Federal Regulations:** Not listed
12.2 **NAS Hazard Rating for Bulk Water Transportation:** Not listed
12.3 **NFPA Hazard Classifications:** Not listed

13. PHYSICAL AND CHEMICAL PROPERTIES

13.1 **Physical State at 15°C and 1 atm:** Solid
13.2 **Molecular Weight:** 194.2
13.3 **Boiling Point at 1 atm:** 540°F = 282°C = 555°K
13.4 **Freezing Point:** 284°F = 140°C = 413°K
13.5 **Critical Temperature:** Not pertinent
13.6 **Critical Pressure:** Not pertinent
13.7 **Specific Gravity:** 1.2 at 20°C (solid); 1.08 at 145°C (liquid)
13.8 **Liquid Surface Tension:** Not pertinent
13.9 **Liquid-Water Interfacial Tension:** Not pertinent
13.10 **Vapor (Gas) Specific Gravity:** Not pertinent
13.11 **Ratio of Specific Heats of Vapor (Gas):** Not pertinent
13.12 **Latent Heat of Vaporization:** 121 Btu/lb = 67.2 cal/g = 2.81 × 10^5 J/kg
13.13 **Heat of Combustion:** −10,310 Btu/lb = −5,727 cal/g = −239.6 × 10^5 J/kg
13.14 **Heat of Decomposition:** Not pertinent
13.15 **Heat of Solution:** Not pertinent
13.16 **Heat of Polymerization:** Not pertinent

NOTES

DMZ	DIMETHYLZINC

Common Synonyms		
Zinc dimethyl Zinc methyl Methylzinc	Liquid	Colorless
	IGNITES WHEN EXPOSED TO AIR. Reacts violently with water and produces flammable vapor.	

Call fire department.
AVOID CONTACT WITH LIQUID AND VAPOR. KEEP PEOPLE AWAY.
Wear rubber overclothing (including gloves).
Stop discharge if possible.
Isolate and remove discharged material.
Notify local health and pollution control agencies.

Fire

IGNITES WHEN EXPOSED TO AIR.
Irritating gases are produced when heated.
Extinguish with dry graphite, soda ash, or other inert powder.
DO NOT USE WATER, FOAM, CARBON DIOXIDE, DRY CHEMICALS OR VAPORIZING LIQUIDS ON FIRE.
DO NOT USE WATER ON ADJACENT FIRES.

Exposure

Call for medical aid.

VAPOR OR MIST
Irritating to eyes, nose and throat.
If inhaled will cause headache, nausea, vomiting or difficult breathing.
Move victim to fresh air.
If in eyes, hold eyelids open and flush with plenty of water.
If breathing has stopped, give artificial respiration.
If breathing is difficult, give oxygen.

LIQUID
Will burn skin and eyes.
If swallowed will cause nausea, or vomiting.
Remove contaminated clothing and shoes.
Flush affected areas with plenty of water.
IF IN EYES, hold eyelids open and flush with plenty of water.
IF SWALLOWED and victim is CONSCIOUS, have victim drink water or milk.
DO NOT INDUCE VOMITING.

Water Pollution

Effect of low concentrations on aquatic life is unknown.
May be dangerous if it enters water intakes.
Notify local health and wildlife officials.
Notify operators of nearby water intakes.

1. RESPONSE TO DISCHARGE

(See Response Methods Handbook, CG 446-4)

Issue warning—high flammability
Restrict access
Evacuate area

2. LABELS

3. CHEMICAL DESIGNATIONS

3.1 **Synonyms:** Zinc dimethyl; Zinc methyl; Methyl zinc
3.2 **Coast Guard Compatibility Classification:** Not applicable
3.3 **Chemical Formula:** $(CH_3)_2Zn$
3.4 **IMCO/United Nations Numerical Designation:** 4.2/1370

4. OBSERVABLE CHARACTERISTICS

4.1 **Physical State (as shipped):** Liquid
4.2 **Color:** Colorless
4.3 **Odor:** Not pertinent

5. HEALTH HAZARDS

5.1 **Personal Protective Equipment:** Cartridge-type or fresh air mask for fumes or smoke; PVC fire-retardant or asbestos gloves; full face shield, safety glasses, or goggles; fire-retardant coveralls as standard wear; for special cases, use asbestos coat or rain suit.

5.2 **Symptoms Following Exposure:** Inhalation of mist or vapor causes immediate irritation of upper respiratory tract. Excessive or prolonged inhalation of fumes from ignition or decomposition may cause "metal fume fever" (sore throat, headache, fever, chills, nausea, vomiting, muscular aches, perspiration, constricting sensation in lungs, weakness, sometimes prostration). Symptoms usually last 12-24 hrs. Eyes are immediately and severely irritated by liquid, vapor, or dilute solutions. If not removed by thorough flushing with water, chemical may permanently damage cornea. Skin will undergo thermal and acid burns when chemical reacts with moisture in skin. Unless washed quickly, skin may be scarred. Treat dilute solutions with same precautions as concentrated liquid. Ingestion, while unlikely, would cause immediate burns at site of contact. Nausea, vomiting, cramps, and diarrhea may follow. Tissues may ulcerate if not treated.

5.3 **Treatment for Exposure:** INHALATION: highly unlikely, as liquid or vapor either ignites spontaneously or reacts with moisture to form methane and zinc oxide. Move victim to clean air and administer mouth-to-mouth resuscitation if breathing has ceased; give oxygen only when authorized by physician; keep victim warm and comfortable; call physician immediately. EYES: immediately flush with large amounts of water for at least 15 min., holding eyelids apart to insure thorough irrigation; use oils or ointments only when directed by physician, and do not attempt to neutralize with chemicals; get medical attention as soon as possible. SKIN: immediately flush affected area with large volumes of water; do not attempt to neutralize with chemicals; get medical attention if irritation persists. INGESTION: highly unlikely, as liquid or vapor either

6. FIRE HAZARDS

6.1 **Flash Point:** Not pertinent (ignites spontaneously)
6.2 **Flammable Limits in Air:** Not pertinent
6.3 **Fire Extinguishing Agents:** Dry chemical, sand, powdered limestone
6.4 **Fire Extinguishing Agents Not to be Used:** Water, foam, halogenated agents, or carbon dioxide
6.5 **Special Hazards of Combustion Products:** Smoke contains zinc oxide, which can irritate lungs and cause metal fume fever.
6.6 **Behavior in Fire:** Reacts spontaneously with air or oxygen and violently with water, evolving methane. Contact with water applied to adjacent fires will intensify fire.
6.7 **Ignition Temperature:** Below 0°F
6.8 **Electrical Hazard:** Not pertinent
6.9 **Burning Rate:** Not pertinent

7. CHEMICAL REACTIVITY

7.1 **Reactivity with Water:** Reacts vigorously, generating flammable methane gas.
7.2 **Reactivity with Common Materials:** Will react with surface moisture to generate flammable methane.
7.3 **Stability During Transport:** Stable
7.4 **Neutralizing Agents for Acids and Caustics:** Not pertinent
7.5 **Polymerization:** Not pertinent
7.6 **Inhibitor of Polymerization:** Not pertinent

8. WATER POLLUTION

8.1 **Aquatic Toxicity:** Not pertinent
8.2 **Waterfowl Toxicity:** Not pertinent
8.3 **Biological Oxygen Demand (BOD):** None
8.4 **Food Chain Concentration Potential:** None

9. SELECTED MANUFACTURERS

1. Texas Alkyls, Incorporated
   P. O. Box 600
   Deer Park, Texas 77536
2. Ventron Corporation
   Alfa Products
   Beverly, Mass. 01915

10. SHIPPING INFORMATION

10.1 **Grades or Purity:** Technical; Electronic
10.2 **Storage Temperature:** Ambient
10.3 **Inert Atmosphere:** Dry nitrogen gas
10.4 **Venting:** Safety relief

11. HAZARD ASSESSMENT CODE

(See Hazard Assessment Handbook, CG 446-3)

A-O-Z

12. HAZARD CLASSIFICATIONS

12.1 **Code of Federal Regulations:** Flammable liquid
12.2 **NAS Hazard Rating for Bulk Water Transportation:** Not listed
12.3 **NFPA Hazard Classifications:** Not listed

13. PHYSICAL AND CHEMICAL PROPERTIES

13.1 **Physical State at 15°C and 1 atm:** Liquid
13.2 **Molecular Weight:** 95.4
13.3 **Boiling Point at 1 atm:** 113°F = 45°C = 318°K
13.4 **Freezing Point:** −44°F = −42°C = 231°K
13.5 **Critical Temperature:** Not pertinent
13.6 **Critical Pressure:** Not pertinent
13.7 **Specific Gravity:** 1.39 at 10.5°C (liquid)
13.8 **Liquid Surface Tension:** (est.) 18 dynes/cm = 0.018 N/m at 20°C
13.9 **Liquid-Water Interfacial Tension:** Not pertinent
13.10 **Vapor (Gas) Specific Gravity:** Not pertinent
13.11 **Ratio of Specific Heats of Vapor (Gas):** Not pertinent
13.12 **Latent Heat of Vaporization:** 134.9 Btu/lb = 74.95 cal/g = 3.138×10^5 J/kg
13.13 **Heat of Combustion:** Data not available
13.14 **Heat of Decomposition:** Not pertinent
13.15 **Heat of Solution:** Not pertinent
13.16 **Heat of Polymerization:** Not pertinent

5. HEALTH HAZARDS (Cont'd.)

ignites spontaneously or reacts with moisture to form methane and zinc oxide. Do NOT induce vomiting; immediately dilute material by giving large amounts of water or milk; if vomiting occurs, give more fluids; when vomiting ceases, milk or olive oil may be given for their soothing effect; get medical attention.

5.4 **Toxicity by Inhalation (Threshold Limit Value):** Not pertinent
5.5 **Short-Term Inhalation Limits:** Not pertinent
5.6 **Toxicity by Ingestion:** Not pertinent
5.7 **Late Toxicity:** Not pertinent
5.8 **Vapor (Gas) Irritant Characteristics:** Data not available
5.9 **Liquid or Solid Irritant Characteristics:** Data not available
5.10 **Odor Threshold:** Not pertinent

DNT — 2,4-DINITROANILINE

Common Synonyms			
2,4-Dinitraniline	Solid powder or crystals	Yellow	Musty odor
	Sinks in water.		

AVOID CONTACT WITH SOLID AND DUST. Keep people away.
Wear goggles, self-contained breathing apparatus, and rubber overclothing (including gloves).
Evacuate area in case of large discharge.
Call fire department.
Isolate and remove discharged material.
Notify local health and pollution control agencies.

Fire

Combustible.
May explode if subjected to heat or flame.
POISONOUS GAS IS PRODUCED WHEN HEATED.
Evacuate surrounding area.
Wear goggles, self-contained breathing apparatus, and rubber overclothing (including gloves).
Combat fires from safe distance or protected location with unmanned hose holder or monitor nozzle.

Exposure

CALL FOR MEDICAL AID.

DUST
POISONOUS IF INHALED.
Move to fresh air.

SOLIDS
POISONOUS IF SWALLOWED OR IF SKIN IS EXPOSED.
Irritating to eyes.
Remove contaminated clothing and shoes.
Flush affected areas with plenty of water.
IF IN EYES, hold eyelids open and flush with plenty of water.
IF SWALLOWED and victim is CONSCIOUS, have victim drink water or milk and have victim induce vomiting.
IF SWALLOWED and victim is UNCONSCIOUS or HAVING CONVULSIONS, do nothing except keep victim warm.

Water Pollution

Effect of low concentrations on aquatic life is unknown.
May be dangerous if it enters water intakes.
Notify local health and wildlife officials.
Notify operators of nearby water intakes.

1. RESPONSE TO DISCHARGE

(See Response Methods Handbook, CG 446-4)

Should be removed
Chemical and physical treatment

2. LABELS

No hazard label required by Code of Federal Regulations

3. CHEMICAL DESIGNATIONS

3.1 **Synonyms:** 2, 4-Dinitraniline
3.2 **Coast Guard Compatibility Classification:** Not applicable
3.3 **Chemical Formula:** $NH_2C_6H_3(NO_2)_2$-2, 4
3.4 **IMCO/United Nations Numerical Designation:** 6.1/1596

4. OBSERVABLE CHARACTERISTICS

4.1 **Physical State (as shipped):** Solid
4.2 **Color:** Yellow
4.3 **Odor:** Slight musty

5. HEALTH HAZARDS

5.1 **Personal Protective Equipment:** Self-contained breathing apparatus; butyl rubber gloves; eye goggles; plastic lab coat; protective shoes.
5.2 **Symptoms Following Exposure:** May cause headache, nausea, stupor. Irritating to skin and mucous membrane.
5.3 **Treatment for Exposure:** INHALATION: artificial respiration if necessary. INGESTION: induce vomiting; give universal antidote; get prompt medical care. CONTACT WITH SKIN AND EYES: remove victim from exposure; wash exposed skin with warm water and soap; flush eyes with water.
5.4 **Toxicity by Inhalation (Threshold Limit Value):** Not pertinent
5.5 **Short-Term Inhalation Limits:** Not pertinent
5.6 **Toxicity by Ingestion:** Grade 3; oral rate LD_{50} = 418 mg/kg
5.7 **Late Toxicity:** Data not available
5.8 **Vapor (Gas) Irritant Characteristics:** Not pertinent
5.9 **Liquid or Solid Irritant Characteristics:** Causes smarting of the skin and first degree burns on short exposure; may cause second degree burns on long exposure.
5.10 **Odor Threshold:** Not pertinent

6. FIRE HAZARDS

6.1 **Flash Point:** 435°F C.C.
6.2 **Flammable Limits in Air:** Data not available
6.3 **Fire Extinguishing Agents:** For small fires, use water, dry chemical, foam or carbon dioxide
6.4 **Fire Extinguishing Agents Not to be Used:** Water or foam may cause frothing
6.5 **Special Hazards of Combustion Products:** Vapors and combustion gases are irritating
6.6 **Behavior in Fire:** May explode
6.7 **Ignition Temperature:** Data not available
6.8 **Electrical Hazard:** Not pertinent
6.9 **Burning Rate:** Not pertinent

7. CHEMICAL REACTIVITY

7.1 **Reactivity with Water:** No reaction
7.2 **Reactivity with Common Materials:** Reacts with oxidizing materials.
7.3 **Stability During Transport:** May detonate when heated under confinement
7.4 **Neutralizing Agents for Acids and Caustics:** Not pertinent
7.5 **Polymerization:** Not pertinent
7.6 **Inhibitor of Polymerization:** Not pertinent

8. WATER POLLUTION

8.1 **Aquatic Toxicity:** Data not available
8.2 **Waterfowl Toxicity:** Data not available
8.3 **Biological Oxygen Demand (BOD):** Data not available
8.4 **Food Chain Concentration Potential:** None

9. SELECTED MANUFACTURERS

1. American Aniline Products, Inc.
   25 McLean Blvd.
   Paterson, N. J. 07509
2. American Hoechst Corp.
   Chemicals and Plastics Division
   Somerville, N. J. 08876
3. Martin Marietta Corp.
   Southern Dyestuff Co. Division
   Charlotte, N. C. 28201

10. SHIPPING INFORMATION

10.1 **Grades or Purity:** Technical and Pure
10.2 **Storage Temperature:** Ambient
10.3 **Inert Atmosphere:** No requirement
10.4 **Venting:** Data not available

11. HAZARD ASSESSMENT CODE

(See Hazard Assessment Handbook, CG 446-3)

II

12. HAZARD CLASSIFICATIONS

12.1 **Code of Federal Regulations:** Not listed
12.2 **NAS Hazard Rating for Bulk Water Transportation:** Not listed
12.3 **NFPA Hazard Classifications:**

Category	Classification
Health Hazard (Blue)	3
Flammability (Red)	1
Reactivity (Yellow)	3

13. PHYSICAL AND CHEMICAL PROPERTIES

13.1 **Physical State at 15°C and 1 atm:** Solid
13.2 **Molecular Weight:** 183.12
13.3 **Boiling Point at 1 atm:** Not pertinent
13.4 **Freezing Point:** 368°F = 187°C = 460°K
13.5 **Critical Temperature:** Not pertinent
13.6 **Critical Pressure:** Not pertinent
13.7 **Specific Gravity:** 1.615 at 15°C (solid)
13.8 **Liquid Surface Tension:** Not pertinent
13.9 **Liquid-Water Interfacial Tension:** Not pertinent
13.10 **Vapor (Gas) Specific Gravity:** Not pertinent
13.11 **Ratio of Specific Heats of Vapor (Gas):** Not pertinent
13.12 **Latent Heat of Vaporization:** Not pertinent
13.13 **Heat of Combustion:** Not pertinent
13.14 **Heat of Decomposition:** Not pertinent
13.15 **Heat of Solution:** Not pertinent
13.16 **Heat of Polymerization:** Not pertinent

NOTES

REVISED 1978

DNB — m-DINITROBENZENE

Common Synonyms			
1, 3-Dinitrobenzene 1, 3-Dinitrobenzol m-DNB	Solid	Yellow	Slight odor
	Sinks in water.		

AVOID CONTACT WITH SOLID AND DUST. KEEP PEOPLE AWAY.
Wear goggles, self-contained breathing apparatus, and rubber overclothing (including gloves).
Call fire department.
Isolate and remove discharged material.
Notify local health and pollution control agencies.

Fire

Combustible.
May explode if exposed to heat or flames.
Flood discharge area with water.
Combat fires from behind barrier.

Exposure

CALL FOR MEDICAL AID.

VAPOR OR DUST
POISONOUS IF INHALED OR IF SKIN IS EXPOSED.
Move victim to fresh air.
If in eyes, hold eyelids open and flush with plenty of water.
If breathing is difficult, give oxygen.

SOLID
POISONOUS IF SWALLOWED OR IF SKIN IS EXPOSED.
Remove contaminated clothing and shoes.
Flush affected areas with plenty of water.
IF IN EYES, hold eyelids open and flush with plenty of water.
IF SWALLOWED and victim is CONSCIOUS, have victim drink water or milk and have victim induce vomiting.
IF SWALLOWED and victim is UNCONSCIOUS OR HAVING CONVULSIONS, do nothing except keep victim warm.

Water Pollution

HARMFUL TO AQUATIC LIFE IN VERY LOW CONCENTRATIONS.
May be dangerous if it enters water intakes.
Notify local health and wildlife officials.
Notify operators of nearby water intakes.

1. RESPONSE TO DISCHARGE

(See Response Methods Handbook, CG 446-4)

Issue warning—poison, water contaminant
Restrict access
Should be removed
Chemical and physical treatment

2. LABEL

3. CHEMICAL DESIGNATIONS

3.1 **Synonyms:** 1,3-Dinitrobenzene; 1,3-Dinitrobenzol; m-DNB; meta-Dinitrobenzene; Dinitrobenzol
3.2 **Coast Guard Compatibility Classification:** To be developed
3.3 **Chemical Formula:** 1,3-$C_6H_4(NO_2)_2$
3.4 **IMCO/United Nations Numerical Designation:** 6.1/1597

4. OBSERVABLE CHARACTERISTICS

4.1 **Physical State (as shipped):** Solid
4.2 **Color:** Yellow
4.3 **Odor:** Weak

5. HEALTH HAZARDS

5.1 **Personal Protective Equipment:** Dust respirator; rubber gloves; protective clothing.
5.2 **Symptoms Following Exposure:** Inhalation or ingestion causes loss of color, nausea, headache, dizziness, drowsiness, and collapse. Eyes are irritated by liquid. Stains skin yellow; if contact is prolonged, can be absorbed into blood and cause same symptoms as for inhalation.
5.3 **Treatment for Exposure:** INHALATION: remove from exposure; get medical attention for methemoglobinemia. EYES: flush with water for at least 15 min. SKIN: wash well with soap and water. INGESTION: induce vomiting, if conscious; give gastric lavage and saline cathartic; get medical attention.
5.4 **Toxicity by Inhalation (Threshold Limit Value):** 1 mg/m^3
5.5 **Short-Term Inhalation Limits:** Data not available
5.6 **Toxicity by Ingestion:** Grade 4; oral LD_{50} = 42 mg/kg (bird)
5.7 **Late Toxicity:** May cause liver damage, anemia, neuritis.
5.8 **Vapor (Gas) Irritant Characteristics:** Data not available
5.9 **Liquid or Solid Irritant Characteristics:** Data not available
5.10 **Odor Threshold:** Data not available

6. FIRE HAZARDS

6.1 **Flash Point:** Not pertinent (combustible solid)
6.2 **Flammable Limits in Air:** Not pertinent
6.3 **Fire Extinguishing Agents:** Water from protected location
6.4 **Fire Extinguishing Agents Not to be Used:** Not pertinent
6.5 **Special Hazards of Combustion Products:** Not pertinent
6.6 **Behavior in Fire:** May explode
6.7 **Ignition Temperature:** Data not available
6.8 **Electrical Hazard:** Not pertinent
6.9 **Burning Rate:** Not pertinent

7. CHEMICAL REACTIVITY

7.1 **Reactivity with Water:** No reaction
7.2 **Reactivity with Common Materials:** No reaction
7.3 **Stability During Transport:** Stable
7.4 **Neutralizing Agents for Acids and Caustics:** Not pertinent
7.5 **Polymerization:** Not pertinent
7.6 **Inhibitor of Polymerization:** Not pertinent

8. WATER POLLUTION

8.1 **Aquatic Toxicity:** 8-10 mg/l/6 hr/minnows/min. lethal dose/fresh water
8.2 **Waterfowl Toxicity:** Data not available
8.3 **Biological Oxygen Demand (BOD):** Data not available
8.4 **Food Chain Concentration Potential:** Data not available

9. SELECTED MANUFACTURERS

1. Aldrich Chemical Co., Inc.
   940 West St. Paul Avenue
   Milwaukee, Wis. 53233
2. Eastman Kodak Co.
   Eastman Organic Chemicals
   Rochester, N. Y. 14650

10. SHIPPING INFORMATION

10.1 **Grades or Purity:** Commercial
10.2 **Storage Temperature:** Ambient
10.3 **Inert Atmosphere:** No requirement
10.4 **Venting:** Open (flame arrester)

11. HAZARD ASSESSMENT CODE

(See Hazard Assessment Handbook, CG 446-3)

II

12. HAZARD CLASSIFICATIONS

12.1 **Code of Federal Regulations:** Poisonous, Class B
12.2 **NAS Hazard Rating for Bulk Water Transportation:** Not listed
12.3 **NFPA Hazard Classifications:** Not listed

13. PHYSICAL AND CHEMICAL PROPERTIES

13.1 **Physical State at 15°C and 1 atm:** Solid
13.2 **Molecular Weight:** 168.1
13.3 **Boiling Point at 1 atm:** 556°F = 291°C = 564°K
13.4 **Freezing Point:** 194°F = 90°C = 363°K
13.5 **Critical Temperature:** Not pertinent
13.6 **Critical Pressure:** Not pertinent
13.7 **Specific Gravity:** 1.58 at 18°C (solid)
13.8 **Liquid Surface Tension:** Not pertinent
13.9 **Liquid-Water Interfacial Tension:** Not pertinent
13.10 **Vapor (Gas) Specific Gravity:** Not pertinent
13.11 **Ratio of Specific Heats of Vapor (Gas):** Not pertinent
13.12 **Latent Heat of Vaporization:** Not pertinent
13.13 **Heat of Combustion:** −7,378 Btu/lb = −4,099 cal/g = -171.5×10^5 J/kg
13.14 **Heat of Decomposition:** Not pertinent
13.15 **Heat of Solution:** Not pertinent
13.16 **Heat of Polymerization:** Not pertinent

NOTES

DNC	**DINITROCRESOLS**

Common Synonyms
4,6-Dinitro-o-cresol
3,5-Dinitro-o-cresol
2,6-Dinitro-o-cresol

Solid — Yellow

Sinks in water.

AVOID CONTACT WITH SOLID AND DUST. KEEP PEOPLE AWAY.
Wear goggles, dust respirator and rubber overclothing (including gloves).
Stop discharge if possible.
Call fire department.
Isolate and remove discharged material.
Notify local health and pollution control agencies.

Fire

Combustible.
POISONOUS GASES MAY BE PRODUCED IN FIRE.
Containers may explode in fire.
Wear goggles and self-contained breathing apparatus.
Extinguish with water, dry chemicals, foam, or carbon dioxide.

Exposure

CALL FOR MEDICAL AID.

DUST
POISONOUS IF INHALED OR IF SKIN IS EXPOSED.
If inhaled will cause loss of consciousness.
If in eyes, hold eyelids open and flush with plenty of water.
If breathing has stopped, give artificial respiration.
If breathing is difficult, give oxygen.

SOLID
POISONOUS IF SWALLOWED OR IF SKIN IS EXPOSED.
If swallowed will cause nausea, vomiting or loss of consciousness.
Remove contaminated clothing and shoes.
Flush affected areas with plenty of water.
IF IN EYES, hold eyelids open and flush with plenty of water.
IF SWALLOWED and victim is CONSCIOUS, have victim drink water or milk and have victim induce vomiting.
IF SWALLOWED and victim is UNCONSCIOUS OR HAVING CONVULSIONS, do nothing except keep victim warm.

Water Pollution

HARMFUL TO AQUATIC LIFE IN VERY LOW CONCENTRATIONS.
May be dangerous if it enters water intakes.
Notify local health and wildlife officials.
Notify operators of nearby water intakes.

1. RESPONSE TO DISCHARGE
(See Response Methods Handbook, CG 446-4)

Issue warning—water contaminant
Restrict access
Should be removed
Chemical and physical treatment

2. LABELS

No hazard label required by Code of Federal Regulations

3. CHEMICAL DESIGNATIONS

3.1 **Synonyms:** 2,6-Dinitro-p-cresol; 3,5-Dinitro-o-cresol; 4,6-Dinitro-o-cresol
3.2 **Coast Guard Compatibility Classification:** Not listed
3.3 **Chemical Formula:** $CH_3-C_6H_2(NO_2)_2(OH)$
3.4 **IMCO/United Nations Numerical Designation:** 6.1/1598

4. OBSERVABLE CHARACTERISTICS

4.1 **Physical State (as shipped):** Solid
4.2 **Color:** Yellow
4.3 **Odor:** Data not available

5. HEALTH HAZARDS

5.1 **Personal Protective Equipment:** Dust mask; goggles or face shield; protective clothing; rubber gloves
5.2 **Symptoms Following Exposure:** Very high fever is prominent sign of intoxication following absorption of a toxic dose of dinitro-o-cresol. Inhalation of dust may cause same symptoms as ingestion. Ingestion causes a feeling of well-being, profuse sweating, yellow urine, increased basal metabolism, marked thirst, vomiting, convulsions, coma, and death. Contact with eyes causes irritation. Contact with skin causes local necrosis and dangerous systemic effects. *Note:* Some authorities recommend that all exposed workers have blood tests regularly to determine the level of this substance. Further contact should be avoided if the level exceeds 20 micrograms per gram.
5.3 **Treatment for Exposure:** INHALATION: apply ice packs to promote heat loss; replace fluids and electrolytes; allay anxiety. INGESTION: same as for inhalation; also, give large amounts of water and induce vomiting; get medical attention. EYES: flush with water for at least 15 min. SKIN: wash thoroughly with soap and water.
5.4 **Toxicity by Inhalation (Threshold Limit Value):** 0.2 mg/m^3
5.5 **Short-Term Inhalation Limits:** 1 mg/m^3, 30 min.
5.6 **Toxicity by Ingestion:** Grade 4; LD_{50} <50 mg/kg (rat)
5.7 **Late Toxicity:** Data not available
5.8 **Vapor (Gas) Irritant Characteristics:** Data not available
5.9 **Liquid or Solid Irritant Characteristics:** Data not available
5.10 **Odor Threshold:** Data not available

6. FIRE HAZARDS

6.1 **Flash Point:** Not pertinent
6.2 **Flammable Limits in Air:** Not pertinent
6.3 **Fire Extinguishing Agents:** Water, foam, dry chemical, carbon dioxide
6.4 **Fire Extinguishing Agents Not to be Used:**
6.5 **Special Hazards of Combustion Products:** Toxic oxides of nitrogen may form in fire.
6.6 **Behavior in Fire:** Containers may explode.
6.7 **Ignition Temperature:** Not pertinent
6.8 **Electrical Hazard:** Not pertinent
6.9 **Burning Rate:** Not pertinent

7. CHEMICAL REACTIVITY

7.1 **Reactivity with Water:** No reaction
7.2 **Reactivity with Common Materials:**
7.3 **Stability During Transport:** Stable
7.4 **Neutralizing Agents for Acids and Caustics:** Not pertinent
7.5 **Polymerization:** Not pertinent
7.6 **Inhibitor of Polymerization:** Not pertinent

8. WATER POLLUTION

8.1 **Aquatic Toxicity:**
8 ppm/48 hr/daphnia magna/TL_m/fresh water
0.00032 ppm/4 days/naiads/LC_{50}/fresh water
8.2 **Waterfowl Toxicity:** Data not available
8.3 **Biological Oxygen Demand (BOD):** Data not available
8.4 **Food Chain Concentration Potential:** None

9. SELECTED MANUFACTURERS

1. Sherwin Williams Chemicals
P. O. Box 6520
Cleveland, Ohio 44101
2. Pfaltz & Bauer, Inc.
375 Fairfield Ave.
Stamford, Conn. 06902
3. J. T. Baker Chemical Co.
Phillipsburg, N. J. 08865

10. SHIPPING INFORMATION

10.1 **Grades or Purity:** Technical, 90—95%; Paste containing 55—60% water; pure.
10.2 **Storage Temperature:** Ambient
10.3 **Inert Atmosphere:** No requirement
10.4 **Venting:** Open

11. HAZARD ASSESSMENT CODE
(See Hazard Assessment Handbook, CG 446-3)

II

12. HAZARD CLASSIFICATIONS

12.1 **Code of Federal Regulations:** Not listed
12.2 **NAS Hazard Rating for Bulk Water Transportation:** Not listed
12.3 **NFPA Hazard Classifications:** Not listed

13. PHYSICAL AND CHEMICAL PROPERTIES

13.1 **Physical State at 15°C and 1 atm:** Solid
13.2 **Molecular Weight:** 198
13.3 **Boiling Point at 1 atm:** Not pertinent (decomposes)
13.4 **Freezing Point:** 176—187°F = 80—86°C = 353—359°K
13.5 **Critical Temperature:** Not pertinent
13.6 **Critical Pressure:** Not pertinent
13.7 **Specific Gravity:** (est.) > 1.1 at 20°C (solid)
13.8 **Liquid Surface Tension:** Not pertinent
13.9 **Liquid-Water Interfacial Tension:** Not pertinent
13.10 **Vapor (Gas) Specific Gravity:** Not pertinent
13.11 **Ratio of Specific Heats of Vapor (Gas):** Not pertinent
13.12 **Latent Heat of Vaporization:** Not pertinent
13.13 **Heat of Combustion:** –7,050 Btu/lb = –3,920 cal/g = –164 X 10^5 J/kg
13.14 **Heat of Decomposition:** Data not available
13.15 **Heat of Solution:** Not pertinent
13.16 **Heat of Polymerization:** Not pertinent

5. HEALTH HAZARDS (Cont'd.)

DNP	2,4-DINITROPHENOL

Common Synonyms alpha-Dinitrophenol 1-Hydroxy-2,4-dinitro-benzene	Solid crystal	Yellow	Sweet, musty odor
	Sinks in water.		

AVOID CONTACT WITH SOLID AND DUST. Keep people away.
Wear goggles, self-contained breathing apparatus, and rubber overclothing (including gloves).
Evacuate area in case of large discharges.
Call fire department.
Isolate and remove discharged material.
Notify local health and pollution control agencies.

Fire

Combustible
May explode if subjected to heat or flame.
POISONOUS GAS IS PRODUCED WHEN HEATED.
Evacuate surrounding area.
Wear goggles, self-contained breathing apparatus, and rubber overclothing (including gloves).
Combat fires from safe distance or protected location.

Exposure

CALL FOR MEDICAL AID.

DUST
POISONOUS IF INHALED OR IF SKIN IS EXPOSED.
Move to fresh air.
If breathing has stopped, give artificial respiration.
If breathing is difficult, give oxygen.

SOLID
POISONOUS IF SWALLOWED.
Remove contaminated clothing and shoes.
Flush affected areas with plenty of water.
IF IN EYES, hold eyelids open and flush with plenty of water.
IF SWALLOWED, and victim is CONSCIOUS, have victim drink water or milk.

Water Pollution

HARMFUL TO AQUATIC LIFE IN VERY LOW CONCENTRATIONS.
May be dangerous if it enters water intakes.
Notify local health and wildlife officials.
Notify operators of nearby water intakes.

1. RESPONSE TO DISCHARGE

(See Response Methods Handbook, CG 446-4)

Issue warning—poison
Restrict access
Should be removed
Chemical and physical treatment

2. LABEL

3. CHEMICAL DESIGNATIONS

3.1 **Synonyms:** Aldifen; alpha-Dinitrophenol; 1-Hydroxy-2, 4-dinitrobenzene
3.2 **Coast Guard Compatibility Classification:** Not applicable
3.3 **Chemical Formula:** $HOC_6H_3(NO_2)_2$-2, 4
3.4 **IMCO/United Nations Numerical Designation:** 1.0/0076

4. OBSERVABLE CHARACTERISTICS

4.1 **Physical State (as shipped):** Solid
4.2 **Color:** Yellow
4.3 **Odor:** Musty; sweet

5. HEALTH HAZARDS

5.1 **Personal Protective Equipment:** Self-contained breathing apparatus; butyl rubber gloves; goggles; lab coat; protective shoes.
5.2 **Symptoms Following Exposure:** Liver damage, metabolic stimulant, dermatitis, dilation of pupils.
5.3 **Treatment for Exposure:** Remove victim from contaminated area and wash exposed skin with soap and water. Administer oxygen if respiratory problems develop. Refer to a doctor.
5.4 **Toxicity by Inhalation (Threshold Limit Value):** 0.2 mg/m^3
5.5 **Short-Term Inhalation Limits:** Data not available
5.6 **Toxicity by Ingestion:** Grade 4; LD_{50} below 50 mg/kg
5.7 **Late Toxicity:** Produces clouding of lens of eye (cataracts) in animals and humans, birth defects in chick embryos.
5.8 **Vapor (Gas) Irritant Characteristics:** Not pertinent
5.9 **Liquid or Solid Irritant Characteristics:** Causes smarting of the skin and first-degree burns on short exposure; may cause second-degree burns on long exposure.
5.10 **Odor Threshold:** Data not available

6. FIRE HAZARDS

6.1 **Flash Point:** Data not available
6.2 **Flammable Limits in Air:** Not pertinent
6.3 **Fire Extinguishing Agents:** Water, dry chemical, carbon dioxide, foam
6.4 **Fire Extinguishing Agents Not to be Used:** Not pertinent
6.5 **Special Hazards of Combustion Products:** Vapors are toxic
6.6 **Behavior in Fire:** Can detonate or explode when heated under confinement
6.7 **Ignition Temperature:** Data not available
6.8 **Electrical Hazard:** Not pertinent
6.9 **Burning Rate:** Not pertinent

7. CHEMICAL REACTIVITY

7.1 **Reactivity with Water:** No reaction
7.2 **Reactivity with Common Materials:** Reacts with oxidizing materials and combustibles.
7.3 **Stability During Transport:** May detonate when heated under confinement
7.4 **Neutralizing Agents for Acids and Caustics:** Not pertinent
7.5 **Polymerization:** Not pertinent
7.6 **Inhibitor of Polymerization:** Not pertinent

8. WATER POLLUTION

8.1 **Aquatic Toxicity:** 200 ppm/0.3 hr/minnow/died/fresh water
8.2 **Waterfowl Toxicity:** Data not available
8.3 **Biological Oxygen Demand (BOD):** Data not available
8.4 **Food Chain Concentration Potential:** None

9. SELECTED MANUFACTURERS

1. American Aniline Products, Inc.
   25 McLean Blvd.
   Paterson, N. J. 07509
2. Martin Marietta Corp.
   Southern Dyestuff Co., Division
   Charlotte, N. C. 28201

10. SHIPPING INFORMATION

10.1 **Grades or Purity:** Data not available
10.2 **Storage Temperature:** Ambient
10.3 **Inert Atmosphere:** No requirement
10.4 **Venting:** Data not available

11. HAZARD ASSESSMENT CODE

(See Hazard Assessment Handbook, CG 446-3)

II

12. HAZARD CLASSIFICATIONS

12.1 **Code of Federal Regulations:** Poisonous liquid or solid, Class B
12.2 **NAS Hazard Rating for Bulk Water Transportation:** Not listed
12.3 **NFPA Hazard Classifications:** Not listed

13. PHYSICAL AND CHEMICAL PROPERTIES

13.1 **Physical State at 15°C and 1 atm:** Solid
13.2 **Molecular Weight:** 184.1
13.3 **Boiling Point at 1 atm:** Not pertinent
13.4 **Freezing Point:** 235°F = 113°C = 386°K
13.5 **Critical Temperature:** Not pertinent
13.6 **Critical Pressure:** Not pertinent
13.7 **Specific Gravity:** 1.68 at 20°C (solid)
13.8 **Liquid Surface Tension:** Not pertinent
13.9 **Liquid-Water Interfacial Tension:** Not pertinent
13.10 **Vapor (Gas) Specific Gravity:** Not pertinent
13.11 **Ratio of Specific Heats of Vapor (Gas):** Not pertinent
13.12 **Latent Heat of Vaporization:** Not pertinent
13.13 **Heat of Combustion:** Not pertinent
13.14 **Heat of Decomposition:** Not pertinent
13.15 **Heat of Solution:** Not pertinent
13.16 **Heat of Polymerization:** Not pertinent

NOTES

REVISED 1978

DTT	# 2,4-DINITROTOLUENE

Common Synonyms			
DNT 2, 4-Dinitrotoluol 1-Methyl-2, 4-Dinitrobenzene	Solid or heated liquid	Yellow to red solid or yellow liquid	Slight odor
	Liquid solidifies. Solid and liquid sink in water.		

AVOID CONTACT WITH LIQUID AND SOLID. KEEP PEOPLE AWAY.
Wear rubber overclothing (including gloves).
Stop discharge if possible.
Call fire department.
Isolate and remove discharged material.
Notify local health and pollution control agencies.

Fire

Combustible.
POISONOUS GASES MAY BE PRODUCED IN FIRE.
Containers may explode in fire.
Wear goggles and self-contained breathing apparatus.
Extinguish with water, dry chemicals, foam, or carbon dioxide.
Cool exposed containers with water.

Exposure

CALL FOR MEDICAL AID.

LIQUID OR SOLID
POISONOUS IF SWALLOWED OR IF SKIN IS EXPOSED.
Will burn skin and eyes.
If swallowed will cause nausea, vomiting or loss of consciousness.
Remove contaminated clothing and shoes.
Flush affected areas with plenty of water.
IF IN EYES, hold eyelids open and flush with plenty of water.
IF SWALLOWED and victim is CONSCIOUS, have victim drink water or milk and have victim induce vomiting.
IF SWALLOWED and victim is UNCONSCIOUS OR HAVING CONVULSIONS, do nothing except keep victim warm.

Water Pollution

Effect of low concentrations on aquatic life is unknown.
May be dangerous if it enters water intakes.
Notify local health and wildlife officials.
Notify operators of nearby water intakes.

1. RESPONSE TO DISCHARGE

(See Response Methods Handbook, CG 446-4)

Issue warning—poison, water contaminant
Restrict access
Should be removed
Chemical and physical treatment

2. LABELS

No label required by Code of Federal Regulations

3. CHEMICAL DESIGNATIONS

3.1 **Synonyms:** 2,4-Dinitrotoluol; 1-Methyl-2,4-dinitrobenzene; DNT
3.2 **Coast Guard Compatibility Classification:** To be developed
3.3 **Chemical Formula:** $2,4\text{-}(NO_2)_2C_6H_3CH_3$
3.4 **IMCO/United Nations Numerical Designation:** Solid: 6.1/2038; Liquid: 6.1/1600

4. OBSERVABLE CHARACTERISTICS

4.1 **Physical State (as shipped):** Solid or liquid
4.2 **Color:** Yellow (liquid); yellow to red (solid)
4.3 **Odor:** Weak

5. HEALTH HAZARDS

5.1 **Personal Protective Equipment:** Air-line mask or self-contained breathing apparatus; safety goggles and face shield; rubber gloves and boots; protective clothing.
5.2 **Symptoms Following Exposure:** Ingestion or overexposure to vapors from hot liquid can cause loss of color, nausea, headache, dizziness, drowsiness, collapse. Hot liquid can burn eyes and skin. Prolonged skin contact with solid can give same symptoms as after inhalation or ingestion.
5.3 **Treatment for Exposure:** INHALATION: remove victim from exposure; get medical attention for methemoglobinemia. EYES: flush with copious amounts of water and get medical attention. SKIN: wash well with soap and water. INGESTION: induce vomiting, if victim is conscious; give gastric lavage and saline cathartic; get medical attention.
5.4 **Toxicity by Inhalation (Threshold Limit Value):** 1.5 mg/m^3
5.5 **Short-Term Inhalation Limits:** Data not available
5.6 **Toxicity by Ingestion:** Grade 4; oral LD_{50} = 30 mg/kg (rat)
5.7 **Late Toxicity:** May cause liver damage, anemia, neuritis.
5.8 **Vapor (Gas) Irritant Characteristics:** Not pertinent
5.9 **Liquid or Solid Irritant Characteristics:** Data not available
5.10 **Odor Threshold:** Data not available

6. FIRE HAZARDS

6.1 **Flash Point:** 404°F C.C.
6.2 **Flammable Limits in Air:** Not pertinent
6.3 **Fire Extinguishing Agents:** Water, dry chemical, carbon dioxide from protected location
6.4 **Fire Extinguishing Agents Not to be Used:** Data not available
6.5 **Special Hazards of Combustion Products:** Nitrogen oxides and dense black smoke are produced in a fire.
6.6 **Behavior in Fire:** Decomposition is self-sustaining at 280°C. Containers may explode in a fire.
6.7 **Ignition Temperature:** Not pertinent
6.8 **Electrical Hazard:** Not pertinent
6.9 **Burning Rate:** Not pertinent

7. CHEMICAL REACTIVITY

7.1 **Reactivity with Water:** No reaction
7.2 **Reactivity with Common Materials:** No reaction
7.3 **Stability During Transport:** Stable below 482°F (250°C)
7.4 **Neutralizing Agents for Acids and Caustics:** Not pertinent
7.5 **Polymerization:** Not pertinent
7.6 **Inhibitor of Polymerization:** Not pertinent

8. WATER POLLUTION

8.1 **Aquatic Toxicity:** Data not available
8.2 **Waterfowl Toxicity:** Data not available
8.3 **Biological Oxygen Demand (BOD):** Data not available
8.4 **Food Chain Concentration Potential:** Data not available

9. SELECTED MANUFACTURERS

1. Rubicon Chemicals, Inc.
   Geismar, La. 70734
2. Air Products and Chemicals, Inc.
   Chemicals Group
   5 Executive Mall
   Wayne, Pa. 19087
3. E. I. duPont de Nemours & Company, Inc.
   Dyes and Chemicals Division
   Wilmington, Del. 19898

10. SHIPPING INFORMATION

10.1 **Grades or Purity:** Technical. Mixtures such as an 80:20 mixture of 2,4- and 2,6-isomers are also available. The hazard properties are similar.
10.2 **Storage Temperature:** Ambient (solid); >90°C (liquid)
10.3 **Inert Atmosphere:** No requirement
10.4 **Venting:** Open (flame arrester)

11. HAZARD ASSESSMENT CODE

(See Hazard Assessment Handbook, CG 446-3)

A-X-Y-II

12. HAZARD CLASSIFICATIONS

12.1 **Code of Federal Regulations:** Not listed
12.2 **NAS Hazard Rating for Bulk Water Transportation:** Not listed
12.3 **NFPA Hazard Classifications:**

Category	Classification
Health Hazard (Blue)	3
Flammability (Red)	1
Reactivity (Yellow)	3

13. PHYSICAL AND CHEMICAL PROPERTIES

13.1 **Physical State at 15°C and 1 atm:** Solid
13.2 **Molecular Weight:** 182.1
13.3 **Boiling Point at 1 atm:** Decomposes
13.4 **Freezing Point:** 158°F = 70°C = 343°K
13.5 **Critical Temperature:** Not pertinent
13.6 **Critical Pressure:** Not pertinent
13.7 **Specific Gravity:** 1.379 at 20°C (liquid)
13.8 **Liquid Surface Tension:** Data not available
13.9 **Liquid-Water Interfacial Tension:** Data not available
13.10 **Vapor (Gas) Specific Gravity:** Not pertinent
13.11 **Ratio of Specific Heats of Vapor (Gas):** Not pertinent
13.12 **Latent Heat of Vaporization:** 170 Btu/lb = 93 cal/g = 3.9×10^5 J/kg
13.13 **Heat of Combustion:** −8,305 Btu/lb = −4,614 cal/g = -193.0×10^5 J/kg
13.14 **Heat of Decomposition:** Not pertinent
13.15 **Heat of Solution:** Not pertinent
13.16 **Heat of Polymerization:** Not pertinent

NOTES

DOA — DIOCTYL ADIPATE

Common Synonyms			
Adipic acid, bis (2-ethylhexyl) ester Di (2-ethylhexyl) adipate	Oily liquid	Colorless	Odorless
	Floats on water.		

Stop discharge if possible.
Call fire department.
Avoid contact with liquid.
Isolate and remove discharged material.
Notify local health and pollution control agencies.

Fire	Combustible. Extinguish with foam, dry chemical, or carbon dioxide. Water may be ineffective on fire.
Exposure	CALL FOR MEDICAL AID. LIQUID Irritating to eyes. Flush affected areas with plenty of water. IF IN EYES, hold eyelids open and flush with plenty of water.
Water Pollution	Effect of low concentrations on aquatic life is unknown. Fouling to shoreline. May be dangerous if it enters water intakes. Notify local health and wildlife officials. Notify operators of nearby water intakes.

1. RESPONSE TO DISCHARGE

(See Response Methods Handbook, CG 446-4)

Mechanical containment
Chemical and physical treatment

2. LABELS

No hazard label required by Code of Federal Regulations

3. CHEMICAL DESIGNATIONS

3.1 **Synonyms:** Adipic acid, bis(2-ethylhexyl) ester; Adipol 2EH; Di(2-ethylhexyl) adipage; DOA

3.2 **Coast Guard Compatibility Classification:** Ester

3.3 **Chemical Formula:** $C_8H_{17}OOC(CH_2)_4COOC_8H_{17}$

3.4 **IMCO/United Nations Numerical Designation:** Not listed

4. OBSERVABLE CHARACTERISTICS

4.1 **Physical State (as shipped):** Liquid

4.2 **Color:** Colorless

4.3 **Odor:** Mild characteristic

5. HEALTH HAZARDS

5.1 **Personal Protective Equipment:** None required

5.2 **Symptoms Following Exposure:** Low toxicity; no reports of injury in industrial handling.

5.3 **Treatment for Exposure:** CONTACT WITH SKIN AND EYES: wipe off and wash skin with soap and water. Treat like lubricating oil. Flush eyes with water. Remove to fresh air.

5.4 **Toxicity by Inhalation (Threshold Limit Value):** Not pertinent

5.5 **Short-Term Inhalation Limits:** Not pertinent

5.6 **Toxicity by Ingestion:** Grade 1; LD_{50} 5 to 15 g/kg

5.7 **Late Toxicity:** None

5.8 **Vapor (Gas) Irritant Characteristics:** Vapors are nonirritating to the eyes and throat.

5.9 **Liquid or Solid Irritant Characteristics:** Minimum hazard. If spilled on clothing and allowed to remain, may cause smarting and reddening of the skin.

5.10 **Odor Threshold:** Not pertinent

6. FIRE HAZARDS

6.1 **Flash Point:** 390°F O.C.

6.2 **Flammable Limits in Air:** Data not available

6.3 **Fire Extinguishing Agents:** Data not available

6.4 **Fire Extinguishing Agents Not to be Used:** Data not available

6.5 **Special Hazards of Combustion Products:** None

6.6 **Behavior in Fire:** Not pertinent

6.7 **Ignition Temperature:** Data not available

6.8 **Electrical Hazard:** Data not available

6.9 **Burning Rate:** Data not available

7. CHEMICAL REACTIVITY

7.1 **Reactivity with Water:** No reaction

7.2 **Reactivity with Common Materials:** No reaction

7.3 **Stability During Transport:** Stable

7.4 **Neutralizing Agents for Acids and Caustics:** Not pertinent

7.5 **Polymerization:** Not pertinent

7.6 **Inhibitor of Polymerization:** Not pertinent

8. WATER POLLUTION

8.1 **Aquatic Toxicity:** Data not available

8.2 **Waterfowl Toxicity:** Data not available

8.3 **Biological Oxygen Demand (BOD):** Data not available

8.4 **Food Chain Concentration Potential:** None

9. SELECTED MANUFACTURERS

1. Allied Chemical Corp.
   Plastics Division
   Morristown, N. J. 07960
2. Diamond Shamrock Corp.
   Nopco Chemicals Division
   Morristown, N. J. 07960
3. W. R. Grace & Co.
   Hatco Chemical Division
   Fords, N. J. 08863

10. SHIPPING INFORMATION

10.1 **Grades or Purity:** 99.6%

10.2 **Storage Temperature:** Ambient

10.3 **Inert Atmosphere:** No requirement

10.4 **Venting:** Open (flame arrester)

11. HAZARD ASSESSMENT CODE

(See Hazard Assessment Handbook, CG 446-3)

A-T-U

12. HAZARD CLASSIFICATIONS

12.1 **Code of Federal Regulations:** Not listed

12.2 **NAS Hazard Rating for Bulk Water Transportation:** Not listed

12.3 **NFPA Hazard Classifications:** Not listed

13. PHYSICAL AND CHEMICAL PROPERTIES

13.1 **Physical State at 15°C and 1 atm:** Liquid

13.2 **Molecular Weight:** 371

13.3 **Boiling Point at 1 atm:** Very high

13.4 **Freezing Point:** Not pertinent

13.5 **Critical Temperature:** Not pertinent

13.6 **Critical Pressure:** Not pertinent

13.7 **Specific Gravity:** 0.928 at 20°C (liquid)

13.8 **Liquid Surface Tension:** (est.) 15 dynes/cm = 0.015 N/m at 20°C

13.9 **Liquid-Water Interfacial Tension:** (est.) 30 dynes/cm = 0.03 N/m at 20°C

13.10 **Vapor (Gas) Specific Gravity:** Not pertinent

13.11 **Ratio of Specific Heats of Vapor (Gas):** Not pertinent

13.12 **Latent Heat of Vaporization:** Not pertinent

13.13 **Heat of Combustion:** -15,430 Btu/lb = -8580 cal/g = -359×10^5 J/kg

13.14 **Heat of Decomposition:** Not pertinent

13.15 **Heat of Solution:** Not pertinent

13.16 **Heat of Polymerization:** Not pertinent

NOTES

REVISED 1978

DOP — DIOCTYL PHTHALATE

Common Synonyms	Oily liquid	Colorless	Slight odor
Phthalic acid, bis(2-ethylhexyl ester) DOP Bis(2-ethylhexyl) phthalate Di(2-ethylhexyl) phthalate Octoil	Floats on water.		

Stop discharge if possible.
Call fire department.
Isolate discharge if possible.
Notify local health and pollution control agencies.

Fire	Combustible. **Extinguish with dry chemical, foam, or carbon dioxide.**
Exposure	Not harmful.
Water Pollution	Effect of low concentrations on aquatic life is unknown. Fouling to shoreline. May be dangerous if it enters water intakes. **Notify local health and wildlife officials.** **Notify operators of nearby water intakes.**

1. RESPONSE TO DISCHARGE

(See Response Methods Handbook, CG 446-4)

Mechanical containment
Chemical and physical treatment

2. LABELS

No hazard label required by Code of Federal Regulations

3. CHEMICAL DESIGNATIONS

3.1 **Synonyms:**
Bis (2-ethylhexyl) phthalate
Di (2-ethylhexyl) phthalate
DOP
Octoil
Phthalic acid, bis (2-ethylhexyl) ester

3.2 **Coast Guard Compatibility Classification:** Ester

(Continued on page 4)

4. OBSERVABLE CHARACTERISTICS

4.1 **Physical State (as shipped):** Liquid
4.2 **Color:** Colorless
4.3 **Odor:** Very slight, characteristic

5. HEALTH HAZARDS

5.1 **Personal Protective Equipment:** Not required
5.2 **Symptoms Following Exposure:** Produces no ill effects at normal temperatures but may give off irritating vapor at high temperature.
5.3 **Treatment for Exposure:** Leave contaminated area; wash skin with soap and water; flush eyes with water
5.4 **Toxicity by Inhalation (Threshold Limit Value):** Not pertinent
5.5 **Short-Term Inhalation Limits:** Not pertinent
5.6 **Toxicity by Ingestion:** Grade 0; LD_{50} above 15 g/kg (rat)
5.7 **Late Toxicity:** Not established
5.8 **Vapor (Gas) Irritant Characteristics:** Nonirritating to the eyes and throat.
5.9 **Liquid or Solid Irritant Characteristics:** No appreciable hazard. Practically harmless to the skin.
5.10 **Odor Threshold:** Not pertinent

6. FIRE HAZARDS

6.1 **Flash Point:** 425°F O.C.
6.2 **Flammable Limits in Air:** Not pertinent
6.3 **Fire Extinguishing Agents:** Dry powder, carbon dioxide, foam
6.4 **Fire Extinguishing Agents Not to be Used:** Water or foam may cause frothing
6.5 **Special Hazards of Combustion Products:** None
6.6 **Behavior in Fire:** Not pertinent
6.7 **Ignition Temperature:** Data not available
6.8 **Electrical Hazard:** Not pertinent
6.9 **Burning Rate:** Data not available

7. CHEMICAL REACTIVITY

7.1 **Reactivity with Water:** No reaction
7.2 **Reactivity with Common Materials:** No reaction
7.3 **Stability During Transport:** Stable
7.4 **Neutralizing Agents for Acids and Caustics:** Not pertinent
7.5 **Polymerization:** Not pertinent
7.6 **Inhibitor of Polymerization:** Not pertinent

8. WATER POLLUTION

8.1 **Aquatic Toxicity:** Data not available
8.2 **Waterfowl Toxicity:** Data not available
8.3 **Biological Oxygen Demand (BOD):** Data not available
8.4 **Food Chain Concentration Potential:** None

9. SELECTED MANUFACTURERS

1. Allied Chemical Corp.
   Plastics Division
   Morristown, N. J. 07960
2. W. R. Grace & Co.
   Hatco Chemical Division
   Fords, N. J. 08863
3. Monsanto Corp.
   Monsanto Industrial Chemicals Co.
   800 N. Lindbergh Blvd.
   St. Louis, Mo. 63166

10. SHIPPING INFORMATION

10.1 **Grades or Purity:** Data not available
10.2 **Storage Temperature:** Ambient
10.3 **Inert Atmosphere:** No requirement
10.4 **Venting:** Open (flame arrester)

11. HAZARD ASSESSMENT CODE

(See Hazard Assessment Handbook, CG 446-3)

A-T-U-X-Y

12. HAZARD CLASSIFICATIONS

12.1 **Code of Federal Regulations:** Not listed
12.2 **NAS Hazard Rating for Bulk Water Transportation:** Not listed
12.3 **NFPA Hazard Classifications:** Not listed

13. PHYSICAL AND CHEMICAL PROPERTIES

13.1 **Physical State at 15°C and 1 atm:** Liquid
13.2 **Molecular Weight:** 390.6
13.3 **Boiling Point at 1 atm:** 727°F = 386°C = 659°K
13.4 **Freezing Point:** Not pertinent
13.5 **Critical Temperature:** Not pertinent
13.6 **Critical Pressure:** Not pertinent
13.7 **Specific Gravity:** 0.980 at 25°C (liquid)
13.8 **Liquid Surface Tension:** (est.) 15 dynes/cm = 0.015 N/m at 20°C
13.9 **Liquid-Water Interfacial Tension:** (est.) 30 dynes/cm = 0.03 N/m at 20°C
13.10 **Vapor (Gas) Specific Gravity:** Not pertinent
13.11 **Ratio of Specific Heats of Vapor (Gas):** Not pertinent
13.12 **Latent Heat of Vaporization:** Not pertinent
13.13 **Heat of Combustion:** -15,130 Btu/lb = -8410 cal/g = -352×10^5 J/kg
13.14 **Heat of Decomposition:** Not pertinent
13.15 **Heat of Solution:** Not pertinent
13.16 **Heat of Polymerization:** Not pertinent

3. CHEMICAL DESIGNATIONS (Cont'd.)

3.3 **Chemical Formula:** $o\text{-}C_6H_4[COOCH_2CH(C_2H_5)(CH_2)_3CH_3]_2$
3.4 **IMCO/United Nations Numerical Designation:** Not listed

REVISED 1978

DIOCTYL SODIUM SULFOSUCCINATE

Common Synonyms			
Bis(2-ethylhexyl) sodium sulfosuccinate Aerosol surfactant Alrowet D65 Di-(2-ethylhexyl) sulfosuccinate, sodium salt Sodium dioctyl sulfosuccinate	Waxy solid or watery solution Sinks and mixes slowly with water.	Colorless to white	Odorless

Isolate and remove discharged material.
Notify local health and pollution control agencies.

Fire	Not flammable.
Exposure	**Call for medical aid.** LIQUID OR SOLID Irritating to skin and eyes. Harmful if swallowed. **Remove contaminated clothing and shoes.** **Flush affected areas with plenty of water.** **IF IN EYES, hold eyelids open and flush with plenty of water.** **IF SWALLOWED and victim is CONSCIOUS, have victim drink water or milk.**
Water Pollution	Effect of low concentrations on aquatic life is unknown. May be dangerous if it enters water intakes. **Notify local health and wildlife officials.** **Notify operators of nearby water intakes.**

1. RESPONSE TO DISCHARGE

(See Response Methods Handbook, CG 446-4)

Issue warning—water contaminant
Disperse and flush

2. LABELS

No label required by Code of Federal Regulations

3. CHEMICAL DESIGNATIONS

3.1 **Synonyms:** Aerosol Surfactant; Alrowet D65; Bis(2-ethylhexyl)sodium sulfosuccinate; Di(2-ethylhexyl) sulfosuccinate, sodium salt; Sodium dioctyl sulfosuccinate

3.2 **Coast Guard Compatibility Classification:** Not applicable

3.3 **Chemical Formula:** $C_8H_{17}OOCCH_2CH(SO_3Na)COOC_8H_{17}$

3.4 **IMCO/United Nations Numerical Designation:** Not listed

4. OBSERVABLE CHARACTERISTICS

4.1 **Physical State (as shipped):** Waxy solid or water solution

4.2 **Color:** Colorless or off-white

4.3 **Odor:** None

5. HEALTH HAZARDS

5.1 **Personal Protective Equipment:** Chemical goggles; rubber gloves; dust respirator.

5.2 **Symptoms Following Exposure:** Liquid is strong irritant to eye and may irritate skin by removing natural oils. Ingestion causes diarrhea and intestinal bloating.

5.3 **Treatment for Exposure:** EYES: irrigate with copious volumes of water for at least 15 min.; call physician. SKIN: rinse off with water. INGESTION: drink large amounts of water.

5.4 **Toxicity by Inhalation (Threshold Limit Value):** Data not available

5.5 **Short-Term Inhalation Limits:** Not pertinent

5.6 **Toxicity by Ingestion:** Grade 2; oral LD_{50} = 1,900 mg/kg (rat)

5.7 **Late Toxicity:** Data not available

5.8 **Vapor (Gas) Irritant Characteristics:** Not pertinent

5.9 **Liquid or Solid Irritant Characteristics:** Data not available

5.10 **Odor Threshold:** Not pertinent

6. FIRE HAZARDS

6.1 **Flash Point:** Not flammable

6.2 **Flammable Limits in Air:** Not flammable

6.3 **Fire Extinguishing Agents:** Not pertinent

6.4 **Fire Extinguishing Agents Not to be Used:** Not pertinent

6.5 **Special Hazards of Combustion Products:** Not pertinent

6.6 **Behavior in Fire:** Causes foaming and spreading of water. Assists in putting out fires by water.

6.7 **Ignition Temperature:** Not pertinent

6.8 **Electrical Hazard:** Not pertinent

6.9 **Burning Rate:** Not pertinent

7. CHEMICAL REACTIVITY

7.1 **Reactivity with Water:** No reaction

7.2 **Reactivity with Common Materials:** No reaction

7.3 **Stability During Transport:** Stable

7.4 **Neutralizing Agents for Acids and Caustics:** Not pertinent

7.5 **Polymerization:** Not pertinent

7.6 **Inhibitor of Polymerization:** Not pertinent

8. WATER POLLUTION

8.1 **Aquatic Toxicity:** Data not available

8.2 **Waterfowl Toxicity:** Data not available

8.3 **Biological Oxygen Demand (BOD):** Data not available

8.4 **Food Chain Concentration Potential:** None

9. SELECTED MANUFACTURERS

1. Proctor Chemical Company, Inc.
   P. O. Box 399
   Salisbury, N. C. 28144
2. American Cyanamid Co.
   Industrial Chemicals and Plastics Division
   Wayne, N. J. 07470
3. Ciba-Geigy Corporation
   Dyes and Specialty Chemicals Division
   Bayonne, N. J. 07002

10. SHIPPING INFORMATION

10.1 **Grades or Purity:** Since many grades contain inert diluents, concentration of compound may be as low as 50% by weight. Many are water solutions.

10.2 **Storage Temperature:** Ambient

10.3 **Inert Atmosphere:** No requirement

10.4 **Venting:** Open

11. HAZARD ASSESSMENT CODE

(See Hazard Assessment Handbook, CG 446-3)

II-SS

12. HAZARD CLASSIFICATIONS

12.1 **Code of Federal Regulations:** Not listed

12.2 **NAS Hazard Rating for Bulk Water Transportation:** Not listed

12.3 **NFPA Hazard Classifications:** Not listed

13. PHYSICAL AND CHEMICAL PROPERTIES

13.1 **Physical State at 15°C and 1 atm:** Solid or liquid

13.2 **Molecular Weight:** 444

13.3 **Boiling Point at 1 atm:** Not pertinent (decomposes)

13.4 **Freezing Point:** (solid form) 311°F = 155°C = 428°K

13.5 **Critical Temperature:** Not pertinent

13.6 **Critical Pressure:** Not pertinent

13.7 **Specific Gravity:** 1.1 at 20°C (solid or liquid)

13.8 **Liquid Surface Tension:** Not pertinent

13.9 **Liquid-Water Interfacial Tension:** Not pertinent

13.10 **Vapor (Gas) Specific Gravity:** Not pertinent

13.11 **Ratio of Specific Heats of Vapor (Gas):** Not pertinent

13.12 **Latent Heat of Vaporization:** Not pertinent

13.13 **Heat of Combustion:** Not pertinent

13.14 **Heat of Decomposition:** Not pertinent

13.15 **Heat of Solution:** Not pertinent

13.16 **Heat of Polymerization:** Not pertinent

NOTES

DOX	1,4-DIOXANE

Common Synonyms			
Dioxane p-Dioxane	Liquid	Colorless	Slight alcohol odor
	Sinks and mixes with water. Flammable, irritating vapor is produced. Freezing point is 53° F.		

Avoid contact with liquid and vapor. Keep people away.
Wear goggles, self-contained breathing apparatus, and rubber overclothing (including gloves).
Shut off ignition sources and call fire department.
Stop discharge if possible.
Stay upwind and use water spray to "knock down" vapor.
Isolate and remove discharged material.
Notify local health and pollution control agencies.

Fire

FLAMMABLE.
Flashback along vapor trail may occur.
Vapor may explode if ignited in an enclosed area.
Wear goggles, self-contained breathing apparatus, and rubber overclothing (including gloves).
Extinguish with dry chemical, alcohol foam or carbon dioxide.
Water may be ineffective on fire.
Cool exposed containers with water.

Exposure

CALL FOR MEDICAL AID.

VAPOR
Irritating to eyes, nose and throat.
Harmful if inhaled.
Move to fresh air.
If breathing has stopped, give artificial respiration.
If breathing is difficult, give oxygen.

LIQUID
Irritating to skin and eyes.
Harmful if swallowed.
Remove contaminated clothing and shoes.
Flush affected areas with plenty of water.
IF IN EYES, hold eyelids open and flush with plenty of water.
IF SWALLOWED and victim is CONSCIOUS, have victim drink water or milk, have victim induce vomiting.
IF SWALLOWED and victim is UNCONSCIOUS OR HAVING CONVULSIONS, do nothing except keep victim warm.

Water Pollution

Effect of low concentrations on aquatic life is unknown.
May be dangerous if it enters water intakes.
Notify local health and wildlife officials.
Notify operators of nearby water intakes.

1. RESPONSE TO DISCHARGE

(See Response Methods Handbook, CG 446-4)

Issue warning—high flammability
Disperse and flush

2. LABEL

Red

3. CHEMICAL DESIGNATIONS

3.1 **Synonyms:** Di (ethylene oxide)
Dioxan
p-Dioxane

3.2 **Coast Guard Compatibility Classification:** Ether

3.3 **Chemical Formula:** $CH_2CH_2OCH_2CH_2O$

3.4 **IMCO/United Nations Numerical Designation:** 3.2/1165

4. OBSERVABLE CHARACTERISTICS

4.1 **Physical State (as shipped):** Liquid

4.2 **Color:** Colorless

4.3 **Odor:** Mild; somewhat alcoholic; like butyl alcohol; ethereal

5. HEALTH HAZARDS

5.1 **Personal Protective Equipment:** Fresh air mask; rubber gloves; monogoggles; safety shower and eye bath.

5.2 **Symptoms Following Exposure:** No significant irritation from brief exposure of skin; prolonged or repeated exposure may cause a rash or burn and absorption of toxic amounts leading to serious injury of liver and kidney. Chemical has poor warning properties; illness may be delayed. Moderately irritating to eyes; overexposure may cause corneal injury.

5.3 **Treatment for Exposure:** INHALATION: promptly remove victim to fresh air, keep him quiet and warm, and call physician; start artificial respiration if breathing stops. INGESTION: if large amounts are swallowed, quickly induce vomiting and get medical attention; no specific antidote known. SKIN AND EYES: flush with plenty of water for 15 min.; remove contaminated clothing and wash before reuse; get medical attention for eyes and if ill effects occur from skin contact.

5.4 **Toxicity by Inhalation (Threshold Limit Value):** 100 ppm

5.5 **Short-Term Inhalation Limits:** 100 ppm for 60 min.

5.6 **Toxicity by Ingestion:** Grade 2; LD_{50} 0.5 to 5 g/kg (guinea pig: 3.90 g/kg)

5.7 **Late Toxicity:** Causes cancer in rats.

5.8 **Vapor (Gas) Irritant Characteristics:** Vapors cause a slight smarting of the eyes or respiratory system if present in high concentrations. The effect is temporary.

5.9 **Liquid or Solid Irritant Characteristics:** Minimum hazard. If spilled on clothing and allowed to remain, may cause smarting and reddening of the skin.

5.10 **Odor Threshold:** 620 mg/m^3

6. FIRE HAZARDS

6.1 **Flash Point:** 54°F C.C.; 74°F O.C.

6.2 **Flammable Limits in Air:** 1.97%—22.5% by vol.

6.3 **Fire Extinguishing Agents:** Alcohol foam, carbon dioxide, dry chemical

6.4 **Fire Extinguishing Agents Not to be Used:** Not pertinent

6.5 **Special Hazards of Combustion Products:** Toxic vapors are generated when heated.

6.6 **Behavior in Fire:** Vapor is heavier than air and may travel to a source of ignition and flash back.

6.7 **Ignition Temperature:** 356°F

6.8 **Electrical Hazard:** Not pertinent

6.9 **Burning Rate:** Data not available

7. CHEMICAL REACTIVITY

7.1 **Reactivity with Water:** No reaction

7.2 **Reactivity with Common Materials:** No reaction

7.3 **Stability During Transport:** Stable

7.4 **Neutralizing Agents for Acids and Caustics:** Not pertinent

7.5 **Polymerization:** Not pertinent

7.6 **Inhibitor of Polymerization:** Not pertinent

8. WATER POLLUTION

8.1 **Aquatic Toxicity:** Data not available

8.2 **Waterfowl Toxicity:** Data not available

8.3 **Biological Oxygen Demand (BOD):** 0% (theor.), 10 days

8.4 **Food Chain Concentration Potential:** None

9. SELECTED MANUFACTURERS

1. Dow Chemical Co.
Midland, Mich. 48640

2. Ferro Corp.
Grant Chemical Division
12418 Scenic Highway
Baton Rouge, La. 70821

3. Union Carbide Corp.
Chemicals and Plastics Division
270 Park Ave.
New York, N. Y. 10017

10. SHIPPING INFORMATION

10.1 **Grades or Purity:** Data not available

10.2 **Storage Temperature:** Ambient

10.3 **Inert Atmosphere:** No requirement

10.4 **Venting:** Open (flame arrester) or pressure-vacuum

11. HAZARD ASSESSMENT CODE

(See Hazard Assessment Handbook, CG 446-3)

A-P-Q

12. HAZARD CLASSIFICATIONS

12.1 **Code of Federal Regulations:** Flammable liquid

12.2 **NAS Hazard Rating for Bulk Water Transportation:**

Category	Rating
Fire	3
Health	
Vapor Irritant	1
Liquid or Solid Irritant	1
Poisons	3
Water Pollution	
Human Toxicity	1
Aquatic Toxicity	2
Aesthetic Effect	2
Reactivity	
Other Chemicals	1
Water	0
Self-Reaction	0

12.3 **NFPA Hazard Classifications:**

Category	Classification
Health Hazard (Blue)	2
Flammability (Red)	3
Reactivity (Yellow)	1

13. PHYSICAL AND CHEMICAL PROPERTIES

13.1 **Physical State at 15°C and 1 atm:** Liquid

13.2 **Molecular Weight:** 88.11

13.3 **Boiling Point at 1 atm:** 214.3°F = 101.3°C = 374.5°K

13.4 **Freezing Point:** 53.2°F = 11.8°C = 285.2°K

13.5 **Critical Temperature:** 597°F = 314°C = 587°K

13.6 **Critical Pressure:** 755 psia = 51.4 atm = 5.21 MN/m^2

13.7 **Specific Gravity:** 1.036 at 20°C (liquid)

13.8 **Liquid Surface Tension:** Not pertinent

13.9 **Liquid-Water Interfacial Tension:** Not pertinent

13.10 **Vapor (Gas) Specific Gravity:** Not pertinent

13.11 **Ratio of Specific Heats of Vapor (Gas):** 1.1

13.12 **Latent Heat of Vaporization:** 178 Btu/lb = 98.6 cal/g = 4.13×10^5 J/kg

13.13 **Heat of Combustion:** −11,590 Btu/lb = −6440 cal/g = -269.6×10^5 J/kg

13.14 **Heat of Decomposition:** Not pertinent

13.15 **Heat of Solution:** (est.) −9 Btu/lb = −5 cal/g = -0.2×10^5 J/kg

13.16 **Heat of Polymerization:** Not pertinent

NOTES

REVISED 1978

DPN — DIPENTENE

Common Synonyms			
p-Mentha-1, 8-diene Limonene Terpinene delta-1, 8-Terpodiene Phellandrene	Liquid Floats on water.	Colorless to light yellow	Pleasant lemon-like odor

Stop discharge if possible. Keep people away.
Shut off ignition sources. Call fire department.
Isolate and remove discharged material.
Notify local health and pollution control agencies.

Fire	Combustible. Containers may explode in fire. **Extinguish with dry chemicals, foam or carbon dioxide.** **Water may be ineffective on fire.** **Cool exposed containers with water.**
Exposure	**Call for medical aid.** VAPOR Irritating to eyes, nose and throat. **Move victim to fresh air.** **If breathing is difficult, give oxygen.** LIQUID Irritating to skin and eyes. Harmful if swallowed. **Remove contaminated clothing and shoes.** **Flush affected areas with plenty of water.** **IF IN EYES, hold eyelids open and flush with plenty of water.** **IF SWALLOWED and victim is CONSCIOUS, have victim drink water or milk.**
Water Pollution	Effect of low concentrations on aquatic life is unknown. Fouling to shoreline. May be dangerous if it enters water intakes. **Notify local health and wildlife officials.** **Notify operators of nearby water intakes.**

1. RESPONSE TO DISCHARGE

(See Response Methods Handbook, CG 446-4)

Mechanical containment
Should be removed
Chemical and physical treatment

2. LABELS

No label required by Code of Federal Regulations

3. CHEMICAL DESIGNATIONS

3.1 **Synonyms:** Limonene; para-Mentha-1, 8-diene; Phellandrene; Terpinene; delta-1, 8-Terpodiene
3.2 **Coast Guard Compatibility Classification:** Olefins (11)
3.3 **Chemical Formula:** $C_{10}H_{16}$
3.4 **IMCO/United Nations Numerical Designation:** 3.3/2052

4. OBSERVABLE CHARACTERISTICS

4.1 **Physical State (as shipped):** Liquid
4.2 **Color:** Colorless; very pale yellow
4.3 **Odor:** Pleasant, pinelike; lemonlike

5. HEALTH HAZARDS

5.1 **Personal Protective Equipment:** Solvent-resistant gloves; safety glasses or face shield; self-contained breathing apparatus for high vapor concentrations.
5.2 **Symptoms Following Exposure:** Liquid irritates eyes; prolonged contact with skin causes irritation. Ingestion causes irritation of gastrointestinal tract.
5.3 **Treatment for Exposure:** INHALATION: remove victim from contaminated area; administer artificial respiration if necessary; call physician. EYES: flush with water for 15 min.; call physician. SKIN: wash with soap and water. INGESTION: induce vomiting; call physician.
5.4 **Toxicity by Inhalation (Threshold Limit Value):** Data not available
5.5 **Short-Term Inhalation Limits:** Data not available
5.6 **Toxicity by Ingestion:** Grade 2; oral LDL_O = 4,600 mg/kg (rat)
5.7 **Late Toxicity:** Data not available
5.8 **Vapor (Gas) Irritant Characteristics:** Vapors cause a slight smarting of the eyes or respiratory system if present in high concentrations. The effect is temporary.
5.9 **Liquid or Solid Irritant Characteristics:** Minimum hazard. If spilled on clothing and allowed to remain, may cause smarting and reddening of the skin.
5.10 **Odor Threshold:** Data not available

6. FIRE HAZARDS

6.1 **Flash Point:** 115°F C.C.
6.2 **Flammable Limits in Air:** 0.7% 6.1%
6.3 **Fire Extinguishing Agents:** Foam, dry chemical, carbon dioxide
6.4 **Fire Extinguishing Agents Not to be Used:** Water may be ineffective.
6.5 **Special Hazards of Combustion Products:** Not pertinent
6.6 **Behavior in Fire:** Containers may explode.
6.7 **Ignition Temperature:** 458°F
6.8 **Electrical Hazard:** Data not available
6.9 **Burning Rate:** 5.5 mm/min.

7. CHEMICAL REACTIVITY

7.1 **Reactivity with Water:** No reaction
7.2 **Reactivity with Common Materials:** No reaction
7.3 **Stability During Transport:** Stable
7.4 **Neutralizing Agents for Acids and Caustics:** Not pertinent
7.5 **Polymerization:** Not pertinent
7.6 **Inhibitor of Polymerization:** Not pertinent

8. WATER POLLUTION

8.1 **Aquatic Toxicity:** Data not available
8.2 **Waterfowl Toxicity:** Data not available
8.3 **Biological Oxygen Demand (BOD):** Data not available
8.4 **Food Chain Concentration Potential:** None

9. SELECTED MANUFACTURERS

1. Hercules Incorporated
   Organics Department
   Wilmington, Del. 19899
2. Reichhold Chemicals Inc.
   Newport Division
   P. O. Box 1433
   Pensacola, Fla. 32596
3. SCM Corporation
   Glidden-Durkee Division
   P. O. Box 389
   Jacksonville, Fla. 32201

10. SHIPPING INFORMATION

10.1 **Grades or Purity:** Several technical grades, all having same general properties.
10.2 **Storage Temperature:** Ambient
10.3 **Inert Atmosphere:** No requirement
10.4 **Venting:** Open (flame arrester)

11. HAZARD ASSESSMENT CODE

(See Hazard Assessment Handbook, CG 446-3)

A-T-U

12. HAZARD CLASSIFICATIONS

12.1 **Code of Federal Regulations:** Not listed
12.2 **NAS Hazard Rating for Bulk Water Transportation:**

Category	Rating
Fire	2
Health	
Vapor Irritant	1
Liquid or Solid Irritant	1
Poisons	0
Water Pollution	
Human Toxicity	1
Aquatic Toxicity	1
Aesthetic Effect	2
Reactivity	
Other Chemicals	1
Water	0
Self-Reaction	1

12.3 **NFPA Hazard Classifications:**

Category	Classification
Health Hazard (Blue)	0
Flammability (Red)	2
Reactivity (Yellow)	0

13. PHYSICAL AND CHEMICAL PROPERTIES

13.1 **Physical State at 15°C and 1 atm:** Liquid
13.2 **Molecular Weight:** 136.2
13.3 **Boiling Point at 1 atm:** 352°F = 178°C = 451°K
13.4 **Freezing Point:** −40°F = −40°C = 233°K
13.5 **Critical Temperature:** Not pertinent
13.6 **Critical Pressure:** Not pertinent
13.7 **Specific Gravity:** 0.842 at 21°C (liquid)
13.8 **Liquid Surface Tension:** (est.) 26 dynes/cm = 0.026 N/m at 20°C
13.9 **Liquid-Water Interfacial Tension:** 27.45 dynes/cm = 0.02745 N/m at 33.5°C
13.10 **Vapor (Gas) Specific Gravity:** 4.9
13.11 **Ratio of Specific Heats of Vapor (Gas):** Not pertinent
13.12 **Latent Heat of Vaporization:** 140 Btu/lb = 77 cal/g = 3.2×10^5 J/kg
13.13 **Heat of Combustion:** −19,520 Btu/lb = −10,840 cal/g = -454×10^5 J/kg
13.14 **Heat of Decomposition:** Not pertinent
13.15 **Heat of Solution:** Not pertinent
13.16 **Heat of Polymerization:** Not pertinent

NOTES

DAM — DIPHENYLAMINE

Common Synonyms
Anilinobenzene
N-Phenylaniline

Solid or liquid — Light tan to brown — Pleasant odor

Sinks in water.

Stop discharge is possible. Keep people away.
Call fire department.
Avoid contact with liquid and solid.
Isolate and remove discharged material.
Notify local health and pollution control agencies.

Fire

Combustible.
POISONOUS GASES MAY BE PRODUCED IN FIRE.
Wear goggles and self-contained breathing apparatus.
Extinguish with dry checmicals or carbon dioxide.
Water and foam may be ineffective on fire.
Cool exposed containers with water.

Exposure

CALL FOR MEDICAL AID.

DUST
Irritating to eyes, nose and throat.
Harmful if inhaled.
If in eyes, hold eyelids open and flush with plenty of water.
If breathing has stopped, give artificial respiration.
If breathing is difficult, give oxygen.

LIQUID OR SOLID
Irritating to skin and eyes.
Harmful if swallowed.
Remove contaminated clothing and shoes.
Flush affected areas with plenty of water.
IF IN EYES, hold eyelids open and flush with plenty of water.
IF SWALLOWED and victim is CONSCIOUS, have victim drink water or milk.
IF SWALLOWED and victim is UNCONSCIOUS OR HAVING CONVULSIONS, do nothing except keep victim warm.

Water Pollution

Effect of low concentrations on aquatic life is unknown.
May be dangerous if it enters water intakes.
Notify local health and wildlife officials.
Notify operators of nearby water intakes.

1. RESPONSE TO DISCHARGE
(See Response Methods Handbook, CG 446-4)

Issue warning—water contaminant
Should be removed
Chemical and physical treatment

2. LABELS

No hazard label required by Code of Federal Regulations

3. CHEMICAL DESIGNATIONS

3.1 **Synonyms:** Anilinobenzene; N-Phenylaniline
3.2 **Coast Guard Compatibility Classification:** Data not available
3.3 **Chemical Formula:** $(C_6H_5)_2NH$
3.4 **IMCO/United Nations Numerical Designation:** Data not available

4. OBSERVABLE CHARACTERISTICS

4.1 **Physical State (as shipped):** Solid or liquid
4.2 **Color:** Very pale tan; amber to brown
4.3 **Odor:** Characteristic, pleasant

5. HEALTH HAZARDS

5.1 **Personal Protective Equipment:** Respirator; safety goggles or face shield; rubber gloves
5.2 **Symptoms Following Exposure:** Inhalation may irritate mucous membranes. Overexposure, including ingestion of solid or skin contact, may cause fast pulse, hypertension, and bladder trouble. Contact with dust irritates eyes.
5.3 **Treatment for Exposure:** INHALATION: move victim to fresh air. INGESTION: get medical attention; observe for methemoglobinemia. EYES: flush with plenty of water and see physician. SKIN: wash with soap and water.
5.4 **Toxicity by Inhalation (Threshold Limit Value):** 10 mg/m^3
5.5 **Short-Term Inhalation Limits:** Data not available
5.6 **Toxicity by Ingestion:** Grade 2; oral LD_{50} = 2,000 mg/kg (rat)
5.7 **Late Toxicity:** Causes birth defects in rats (polycytic kidneys)
5.8 **Vapor (Gas) Irritant Characteristics:** Data not available
5.9 **Liquid or Solid Irritant Characteristics:** Data not available
5.10 **Odor Threshold:** Data not available

6. FIRE HAZARDS

6.1 **Flash Point:** (liquid) 307°F O.C.
6.2 **Flammable Limits in Air:** Not pertinent
6.3 **Fire Extinguishing Agents:** Foam, dry chemical, carbon dioxide
6.4 **Fire Extinguishing Agents Not to be Used:** Water or foam may cause frothing.
6.5 **Special Hazards of Combustion Products:** Toxic oxides of nitrogen may form in fire.
6.6 **Behavior in Fire:** Dust may be explosive if mixed with air in critical proportions and in the presence of a source of ignition.
6.7 **Ignition Temperature:** 1,175°F
6.8 **Electrical Hazard:** Not pertinent
6.9 **Burning Rate:** Not pertinent

7. CHEMICAL REACTIVITY

7.1 **Reactivity with Water:** No reaction
7.2 **Reactivity with Common Materials:**
7.3 **Stability During Transport:** Stable
7.4 **Neutralizing Agents for Acids and Caustics:** Not pertinent
7.5 **Polymerization:** Not pertinent
7.6 **Inhibitor of Polymerization:** Not pertinent

8. WATER POLLUTION

8.1 **Aquatic Toxicity:** Data not available
8.2 **Waterfowl Toxicity:** Data not available
8.3 **Biological Oxygen Demand (BOD):** Data not available
8.4 **Food Chain Concentration Potential:** None

9. SELECTED MANUFACTURERS

1. Uniroyal Chemical
   Div. of Uniroyal, Inc.
   Elm Street
   Naugatuck, Conn. 06770
2. American Cyanamid Co.
   Intermediates Dept.
   Bound Brook, N. J. 08805
3. Eastman Organic Chemicals
   Rochester, N. Y. 14650

10. SHIPPING INFORMATION

10.1 **Grades or Purity:** Technical; sometimes shipped as liquid
10.2 **Storage Temperature:** Ambient for solid, elevated for liquid
10.3 **Inert Atmosphere:** No requirement
10.4 **Venting:** Open

11. HAZARD ASSESSMENT CODE
(See Hazard Assessment Handbook, CG 446-3)

II

12. HAZARD CLASSIFICATIONS

12.1 **Code of Federal Regulations:** Not listed
12.2 **NAS Hazard Rating for Bulk Water Transportation:** Not listed
12.3 **NFPA Hazard Classifications:**

Category	Classification
Health Hazard (Blue)	3
Flammability (Red)	1
Reactivity (Yellow)	0

13. PHYSICAL AND CHEMICAL PROPERTIES

13.1 **Physical State at 15°C and 1 atm:** Solid or liquid
13.2 **Molecular Weight:** 169.2
13.3 **Boiling Point at 1 atm:** 576°F = 302°C = 575°K
13.4 **Freezing Point:** 127°F = 53°C = 326°K
13.5 **Critical Temperature:** Not pertinent
13.6 **Critical Pressure:** Not pertinent
13.7 **Specific Gravity:** 1.068 at 61°C (liquid)
13.8 **Liquid Surface Tension:** 39.3 dynes/cm = 0.0393 N/m at 60°C
13.9 **Liquid-Water Interfacial Tension:** Not pertinent
13.10 **Vapor (Gas) Specific Gravity:** Not pertinent
13.11 **Ratio of Specific Heats of Vapor (Gas):** Not pertinent
13.12 **Latent Heat of Vaporization:** Not pertinent
13.13 **Heat of Combustion:** $-16{,}300$ Btu/lb = $-9{,}060$ cal/g = -379×10^5 J/kg
13.14 **Heat of Decomposition:** Not pertinent
13.15 **Heat of Solution:** Not pertinent
13.16 **Heat of Polymerization:** Not pertinent

NOTES

DIPHENYLDICHLOROSILANE

Common Synonyms			
Dichlorodiphenylsilane Dichlorodiphenylsilicane Diphenylsilicon dichloride	Liquid	Colorless	Sharp, irritating odor
	Reacts with water. Irritating vapor is produced.		

Wear goggles, self-contained breathing apparatus, and rubber overclothing (including gloves).
Stop discharge if possible. Keep people away.
Call fire department.
Isolate and remove discharged material.
Notify local health and pollution control agencies.

Fire	Combustible. POISONOUS GASES MAY BE PRODUCED IN FIRE. Extinguish with dry chemicals or carbon dioxide. DO NOT USE WATER OR FOAM ON FIRE.
Exposure	Call for medical aid. VAPOR Irritating to eyes, nose and throat. Move victim to fresh air. If breathing is difficult, give oxygen. LIQUID Will burn skin and eyes. Harmful if swallowed. Remove contaminated clothing and shoes. Flush affected areas with plenty of water. IF IN EYES, hold eyelids open and flush with plenty of water. IF SWALLOWED and victim is CONSCIOUS, have victim drink water or milk. DO NOT INDUCE VOMITING.
Water Pollution	Effect of low concentrations on aquatic life is unknown. May be dangerous if it enters water intakes. Notify local health and wildlife officials. Notify operators of nearby water intakes.

1. RESPONSE TO DISCHARGE

(See Response Methods Handbook, CG 446-4)

Issue warning—corrosive, water contaminant
Restrict access
Disperse and flush with care

2. LABEL

3. CHEMICAL DESIGNATIONS

3.1 **Synonyms:** Dichlorodiphenylsilane
Dichlorodiphenylsilicane
Diphenylsilicon dichloride

3.2 **Coast Guard Compatibility Classification:** Not applicable

3.3 **Chemical Formula:** $(C_6H_5)_2SiCl_2$

3.4 **IMCO/United Nations Numerical Designation:** 8/1769

4. OBSERVABLE CHARACTERISTICS

4.1 **Physical State (as shipped):** Liquid

4.2 **Color:** Colorless

4.3 **Odor:** Sharp, hydrochloric acid-like; pungent

5. HEALTH HAZARDS

5.1 **Personal Protective Equipment:** Acid-vapor-type respiratory protection; rubber gloves; chemical worker's goggles; other protective equipment as necessary to protect skin and eyes.

5.2 **Symptoms Following Exposure:** Inhalation irritates mucous membranes. Contact with liquid causes severe burns of eyes and skin. Ingestion causes severe burns of mouth and stomach.

5.3 **Treatment for Exposure:** INHALATION: remove victim from exposure; support respiration; call physician if needed. EYES: flush with water for 15 min.; obtain medical attention immediately. SKIN: flush with water; obtain medical attention for acid burns. INGESTION: give large amounts of water, if victim is conscious; give milk, or milk of magnesia; call physician.

5.4 **Toxicity by Inhalation (Threshold Limit Value):** Data not available

5.5 **Short-Term Inhalation Limits:** Data not available

5.6 **Toxicity by Ingestion:** Grade 3; LD_{50} 50 to 500 mg/kg

5.7 **Late Toxicity:** Data not available

5.8 **Vapor (Gas) Irritant Characteristics:** Vapors cause moderate irritation such that personnel will find high concentrations unpleasant. The effect is temporary.

5.9 **Liquid or Solid Irritant Characteristics:** Severe skin irritant. Causes second- and third-degree burns on short contact and is very injurious to the eyes.

5.10 **Odor Threshold:** Data not available

6. FIRE HAZARDS

6.1 **Flash Point:** 288°F O.C.

6.2 **Flammable Limits in Air:** Data not available

6.3 **Fire Extinguishing Agents:** Dry chemical, carbon dioxide

6.4 **Fire Extinguishing Agents Not to be Used:** Water and foam

6.5 **Special Hazards of Combustion Products:** Hydrochloric acid and phosgene fumes may be formed.

6.6 **Behavior in Fire:** Difficult to extinguish; re-ignition may occur. Contact with water or foam applied to adjacent fires will produce irritating hydrogen chloride fumes.

6.7 **Ignition Temperature:** Data not available

6.8 **Electrical Hazard:** Data not available

6.9 **Burning Rate:** 2.7 mm/min.

7. CHEMICAL REACTIVITY

7.1 **Reactivity with Water:** Reacts with water to generate hydrogen chloride (hydrochloric acid).

7.2 **Reactivity with Common Materials:** Reacts with surface moisture to generate hydrogen chloride, which is corrosive to common metals.

7.3 **Stability During Transport:** Stable

7.4 **Neutralizing Agents for Acids and Caustics:** Flood with water, rinse with sodium bicarbonate or lime solution.

7.5 **Polymerization:** Not pertinent

7.6 **Inhibitor of Polymerization:** Not pertinent

8. WATER POLLUTION

8.1 **Aquatic Toxicity:** Data not available

8.2 **Waterfowl Toxicity:** Data not available

8.3 **Biological Oxygen Demand (BOD):** Data not available

8.4 **Food Chain Concentration Potential:** None

9. SELECTED MANUFACTURERS

1. Dow Corning Corporation
   P. O. Box 592
   Midland, Mich. 48640
2. PCR, Inc.
   P. O. Box 1466
   Gainesville, Fla. 32602
3. Ventron Corp.
   Alfa Products
   P. O. Box 159
   Beverly, Mass. 01915

10. SHIPPING INFORMATION

10.1 **Grades or Purity:** 96+%

10.2 **Storage Temperature:** Ambient

10.3 **Inert Atmosphere:** No requirement

10.4 **Venting:** Pressure-vacuum

11. HAZARD ASSESSMENT CODE

(See Hazard Assessment Handbook, CG 446-3)

A-O

12. HAZARD CLASSIFICATIONS

12.1 **Code of Federal Regulations:** Corrosive

12.2 **NAS Hazard Rating for Bulk Water Transportation:**

Category	Rating
Fire	1
Health	
Vapor Irritant	2
Liquid or Solid Irritant	4
Poisons	3
Water Pollution	
Human Toxicity	3
Aquatic Toxicity	3
Aesthetic Effect	3
Reactivity	
Other Chemicals	3
Water	4
Self-Reaction	1

12.3 **NFPA Hazard Classifications:** Not listed

13. PHYSICAL AND CHEMICAL PROPERTIES

13.1 **Physical State at 15°C and 1 atm:** Liquid

13.2 **Molecular Weight:** 253

13.3 **Boiling Point at 1 atm:** 579°F = 304°C = 577°K

13.4 **Freezing Point:** Not pertinent

13.5 **Critical Temperature:** Not pertinent

13.6 **Critical Pressure:** Not pertinent

13.7 **Specific Gravity:** 1.22 at 25°C (liquid)

13.8 **Liquid Surface Tension:** (est.) 26 dynes/cm = 0.026 N/m at 20°C

13.9 **Liquid-Water Interfacial Tension:** Not pertinent

13.10 **Vapor (Gas) Specific Gravity:** Not pertinent

13.11 **Ratio of Specific Heats of Vapor (Gas):** Not pertinent

13.12 **Latent Heat of Vaporization:** 106 Btu/lb = 59 cal/g = 2.5×10^5 J/kg

13.13 **Heat of Combustion:** (est.) −11,000 Btu/lb = −6,200 cal/g = -260×10^5 J/kg

13.14 **Heat of Decomposition:** Not pertinent

13.15 **Heat of Solution:** Data not available

13.16 **Heat of Polymerization:** Not pertinent

NOTES

DPE — DIPHENYL ETHER

Common Synonyms			
Phenyl ether Diphenyl oxide Phenoxybenzene	Liquid	Colorless	Mild pleasant odor
	May float or sink in water. Freezing point is 81° F.		

Stop discharge if possible. Keep people away.
Call fire department.
Avoid contact with liquid and solid.
Shut off ignition sources. Call fire department.
Stop discharge if possible.

Fire	Combustible. Extinguish with dry chemicals or carbon dioxide. Water and foam may be ineffective on fire.
Exposure	Call for medical aid. LIQUID OR SOLID Irritating to skin and eyes. Harmful if swallowed. Remove contaminated clothing and shoes. Flush affected areas with plenty of water. IF IN EYES, hold eyelids open and flush with plenty of water. IF SWALLOWED and victim is CONSCIOUS, have victim drink water or milk.
Water Pollution	Effect of low concentrations on aquatic life is unknown. Fouling to shoreline. May be dangerous if it enters water intakes. Notify local health and wildlife officials. Notify operators of nearby water intakes.

1. RESPONSE TO DISCHARGE
(See Response Methods Handbook, CG 446-4)

Mechanical containment
Should be removed
Chemical and physical treatment

2. LABELS

No label required by Code of Federal Regulations

3. CHEMICAL DESIGNATIONS

3.1 **Synonyms:** Phenyl ether
Diphenyl oxide
Phenoxybenzene
3.2 **Coast Guard Compatibility Classification:** Not applicable
3.3 **Chemical Formula:** $C_6H_5OC_6H_5$
3.4 **IMCO/United Nations Numerical Designation:** Not listed

4. OBSERVABLE CHARACTERISTICS

4.1 **Physical State (as shipped):** Liquid
4.2 **Color:** Colorless
4.3 **Odor:** Weak geranium

5. HEALTH HAZARDS

5.1 **Personal Protective Equipment:** Goggles or face shield; rubber gloves.
5.2 **Symptoms Following Exposure:** Inhalation may cause nausea because of disagreeable odor. Contact of liquid with eyes causes mild irritation. Prolonged exposure of skin to liquid causes reddening and irritation. Ingestion produces nausea.
5.3 **Treatment for Exposure:** EYES: flush with water for at least 15 min. SKIN: wipe off, wash with soap and water. INGESTION: induce vomiting and get medical attention.
5.4 **Toxicity by Inhalation (Threshold Limit Value):** 1 ppm
5.5 **Short-Term Inhalation Limits:** Data not available
5.6 **Toxicity by Ingestion:** Grade 2; oral LD_{50} = 3,370 mg/kg (rat)
5.7 **Late Toxicity:** Data not available
5.8 **Vapor (Gas) Irritant Characteristics:** Data not available
5.9 **Liquid or Solid Irritant Characteristics:** Data not available
5.10 **Odor Threshold:** 0.1 ppm

6. FIRE HAZARDS

6.1 **Flash Point:** 239°F C.C.
6.2 **Flammable Limits in Air:** 0.8% – 1.5%
6.3 **Fire Extinguishing Agents:** Dry chemical, carbon dioxide
6.4 **Fire Extinguishing Agents Not to be Used:** Water or foam may cause frothing.
6.5 **Special Hazards of Combustion Products:** Not pertinent
6.6 **Behavior in Fire:** Not pertinent
6.7 **Ignition Temperature:** 1,148°F
6.8 **Electrical Hazard:** Data not available
6.9 **Burning Rate:** 3.2 mm/min.

7. CHEMICAL REACTIVITY

7.1 **Reactivity with Water:** No reaction
7.2 **Reactivity with Common Materials:** No reaction
7.3 **Stability During Transport:** Stable
7.4 **Neutralizing Agents for Acids and Caustics:** Not pertinent
7.5 **Polymerization:** Not pertinent
7.6 **Inhibitor of Polymerization:** Not pertinent

8. WATER POLLUTION

8.1 **Aquatic Toxicity:** Data not available
8.2 **Waterfowl Toxicity:** Data not available
8.3 **Biological Oxygen Demand (BOD):** Data not available
8.4 **Food Chain Concentration Potential:** None

9. SELECTED MANUFACTURERS

1. Dow Chemical Co.
Midland, Mich. 48640
2. Givaudan Corp.
100 Delawanna Ave.
Clifton, N. J. 07014
3. Eastman Kodak Co.
Eastman Organic Chemicals
Rochester, N. Y. 14650

10. SHIPPING INFORMATION

10.1 **Grades or Purity:** Pure grade; Technical grade; Perfume grade
10.2 **Storage Temperature:** Ambient
10.3 **Inert Atmosphere:** No requirement
10.4 **Venting:** Open (flame arrester)

11. HAZARD ASSESSMENT CODE
(See Hazard Assessment Handbook, CG 446-3)

A-T-U-X-Y

12. HAZARD CLASSIFICATIONS

12.1 **Code of Federal Regulations:** Not listed
12.2 **NAS Hazard Rating for Bulk Water Transportation:** Not listed
12.3 **NFPA Hazard Classifications:**

Category	Classification
Health Hazard (Blue)	1
Flammability (Red)	1
Reactivity (Yellow)	0

13. PHYSICAL AND CHEMICAL PROPERTIES

13.1 **Physical State at 15°C and 1 atm:** Solid
13.2 **Molecular Weight:** 170.2
13.3 **Boiling Point at 1 atm:** 495°F = 257°C = 530°K
13.4 **Freezing Point:** 81°F = 27°C = 300°K
13.5 **Critical Temperature:** 921°F = 494°C = 767°K
13.6 **Critical Pressure:** 478 psia = 32.5 atm = 3.30 MN/m^2
13.7 **Specific Gravity:** 1.07 at 27°C (liquid)
13.8 **Liquid Surface Tension:** 40.05 dynes/cm = 0.0401 N/m at 20°C
13.9 **Liquid-Water Interfacial Tension:** (est.) 36 dynes/cm = 0.036 N/m at 20°C
13.10 **Vapor (Gas) Specific Gravity:** Not pertinent
13.11 **Ratio of Specific Heats of Vapor (Gas):** Not pertinent
13.12 **Latent Heat of Vaporization:** 130 Btu/lb = 72 cal/g = 3.0×10^5 J/kg
13.13 **Heat of Combustion:** –15,520 Btu/lb = –8,620 cal/g = -361×10^5 J/kg
13.14 **Heat of Decomposition:** Not pertinent
13.15 **Heat of Solution:** Not pertinent
13.16 **Heat of Polymerization:** Not pertinent

NOTES

DPM — DIPHENYLMETHANE DIISOCYANATE

Common Synonyms		
MDI Diphenylmethane-4,4'-diisocyanate	Solid	White to light yellow
	Sinks in water.	

Call fire department.
Isolate and remove discharged material.
Notify local health and pollution control agencies.

Fire

Combustible.
Wear goggles and self-contained breathing apparatus.
Extinguish with foam, dry chemical or carbon dioxide.

Exposure

CALL FOR MEDICAL AID.

SOLID
Irritating to skin and eyes.
Flush affected areas with plenty of water.
IF IN EYES, hold eyelids open and flush with plenty of water.

Water Pollution

Effect of low concentrations on aquatic life is unknown.
May be dangerous if it enters water intakes.
Notify local health and wildlife officials.
Notify operators of nearby water intakes.

1. RESPONSE TO DISCHARGE

(See Response Methods Handbook, CG 446-4)

Issue warning—water contaminant
Should be removed
Chemical and physical treatment

2. LABELS

No hazard label required by Code of Federal Regulations

3. CHEMICAL DESIGNATIONS

3.1 **Synonyms:** Carwinate 125 M; Diphenylmethane 4,4'-diisocyanate; Hylene M50; MDI; Methylenebis (4-phenyl isocyanate); Multrathane M; Nacconate 300; Vilrathane 4300

3.2 **Coast Guard Compatibility Classification:** Isocyanate

3.3 **Chemical Formula:** $(p\text{-}OCNC_6H_4)_2CH_2$

3.4 **IMCO/United Nations Numerical Designation:** Not listed

4. OBSERVABLE CHARACTERISTICS

4.1 **Physical State (as shipped):** Solid

4.2 **Color:** Colorless to light yellow

4.3 **Odor:** Data not available

5. HEALTH HAZARDS

5.1 **Personal Protective Equipment:** Mask or respirator of type approved by U.S. Bureau of Mines (above 135°); clean rubber gloves; chemical goggles; clean waterproof or freshly laundered protective clothing (coveralls, rubber boots, cap, etc.).

5.2 **Symptoms Following Exposure:** Breathlessness, chest discomfort, and reduced pulmonary function.

5.3 **Treatment for Exposure:** INHALATION: treat symptomatically; vaso-dilators; oxygen. Call a physician. SKIN CONTACT: wash with soap and water. Rubbing alcohol helpful. EYE CONTACT: flush with water at least 15 min. Call a physician.

5.4 **Toxicity by Inhalation (Threshold Limit Value):** 0.02 ppm

5.5 **Short-Term Inhalation Limits:** Data not available

5.6 **Toxicity by Ingestion:** Data not available

5.7 **Late Toxicity:** Data not available

5.8 **Vapor (Gas) Irritant Characteristics:** Severe irritation of eyes and throat; can cause eye and lung injury. Cannot be tolerated even at low concentrations.

5.9 **Liquid or Solid Irritant Characteristics:** Minimum hazard. If spilled on clothing and allowed to remain, may cause reddening of the skin.

5.10 **Odor Threshold:** Data not available

6. FIRE HAZARDS

6.1 **Flash Point:** 425°F O.C.

6.2 **Flammable Limits in Air:** Not pertinent

6.3 **Fire Extinguishing Agents:** Carbon dioxide or dry chemical

6.4 **Fire Extinguishing Agents Not to be Used:** Not pertinent

6.5 **Special Hazards of Combustion Products:** Toxic vapors are generated when heated.

6.6 **Behavior in Fire:** Solid melts and burns

6.7 **Ignition Temperature:** Data not available

6.8 **Electrical Hazard:** Not pertinent

6.9 **Burning Rate:** Not pertinent

7. CHEMICAL REACTIVITY

7.1 **Reactivity with Water:** Slow, non-hazardous. Forms carbon dioxide gas.

7.2 **Reactivity with Common Materials:** Data not available

7.3 **Stability During Transport:** Stable

7.4 **Neutralizing Agents for Acids and Caustics:** Not pertinent

7.5 **Polymerization:** May occur slowly. Is not hazardous.

7.6 **Inhibitor of Polymerization:** Not pertinent

8. WATER POLLUTION

8.1 **Aquatic Toxicity:** Data not available

8.2 **Waterfowl Toxicity:** Data not available

8.3 **Biological Oxygen Demand (BOD):** Data not available

8.4 **Food Chain Concentration Potential:** None

9. SELECTED MANUFACTURERS

1. Baychem Corp.
   Mobay Chemical Co. Division
   Penn Lincoln Pkwy. West
   Pittsburgh, Pa. 15205
2. E. I. du Pont de Nemours & Co., Inc.
   Elastomer Chemicals Dept.
   Wilmington, Del. 19898
3. Upjohn Co.
   Polymer Chemicals Division
   La Porte, Tex. 77571

10. SHIPPING INFORMATION

10.1 **Grades or Purity:** Solid grades: 91-99%; liquid grades may contain 50% o-dichlorobenzene

10.2 **Storage Temperature:** 0°–40°F

10.3 **Inert Atmosphere:** No requirement

10.4 **Venting:** Pressure-vacuum

11. HAZARD ASSESSMENT CODE

(See Hazard Assessment Handbook, CG 446-3)

II

12. HAZARD CLASSIFICATIONS

12.1 **Code of Federal Regulations:** Not listed

12.2 **NAS Hazard Rating for Bulk Water Transportation:**

Category	Rating
Fire	1
Health	
Vapor Irritant	3
Liquid or Solid Irritant	2
Poisons	4
Water Pollution	
Human Toxicity	1
Aquatic Toxicity	
Aesthetic Effect	2
Reactivity	
Other Chemicals	1
Water	3
Self-Reaction	1

12.3 **NFPA Hazard Classifications:** Not listed

13. PHYSICAL AND CHEMICAL PROPERTIES

13.1 **Physical State at 15°C and 1 atm:** Solid

13.2 **Molecular Weight:** 250.3

13.3 **Boiling Point at 1 atm:** 738°F = 392°C = 665°K

13.4 **Freezing Point:** 100°F = 37.7°C = 311°K

13.5 **Critical Temperature:** Not pertinent

13.6 **Critical Pressure:** Not pertinent

13.7 **Specific Gravity:** 1.2 at 20°C (solid)

13.8 **Liquid Surface Tension:** Not pertinent

13.9 **Liquid-Water Interfacial Tension:** Not pertinent

13.10 **Vapor (Gas) Specific Gravity:** Not pertinent

13.11 **Ratio of Specific Heats of Vapor (Gas):** Not pertinent

13.12 **Latent Heat of Vaporization:** Not pertinent

13.13 **Heat of Combustion:** Not pertinent

13.14 **Heat of Decomposition:** Not pertinent

13.15 **Heat of Solution:** Not pertinent

13.16 **Heat of Polymerization:** Not pertinent

NOTES

REVISED 1978

DNA — Di-n-PROPYLAMINE

Common Synonyms			
N-propyl-1-propanamine	Liquid	Colorless	Strong ammonia-like odor
	Floats and mixes with water.		

Shut off ignition sources. Call fire department.
Avoid contact with liquid and vapor. Keep people away.
Stop discharge if possible.
Stay upwind. Use water spray to "knock down" vapor.
Isolate and remove discharged material.
Notify local health and pollution control agencies.

Fire

FLAMMABLE.
POISONOUS GASES MAY BE PRODUCED IN FIRE.
Containers may explode in fire.
Flashback along vapor trail may occur.
Vapor may explode if ignited in an enclosed area.
Wear goggles and self-contained breathing apparatus.
Extinguish with dry chemicals, foam or carbon dioxide.
Water may be ineffective on fire.
Cool exposed containers with water.

Exposure

CALL FOR MEDICAL AID.

VAPOR
Irritating to eyes, nose and throat.
If inhaled will cause headache, dizziness, coughing, or difficult breathing.
If in eyes, hold eyelids open and flush with plenty of water.
If breathing has stopped, give artificial respiration.
If breathing is difficult, give oxygen.

LIQUID
Will burn eyes.
If swallowed will cause nausea and vomiting.

Remove contaminated clothing and shoes.
Flush affected areas with plenty of water.
IF IN EYES, hold eyelids open and flush with plenty of water.
IF SWALLOWED and victim is CONSCIOUS, have victim drink water or milk.
IF SWALLOWED and victim is UNCONSCIOUS OR HAVING CONVULSIONS, do nothing except keep victim warm.

Water Pollution

Effect of low concentrations on aquatic life is unknown.
Fouling to shoreline.
May be dangerous if it enters water intakes.
Notify local health and wildlife officials.
Notify operators of nearby water intakes.

1. RESPONSE TO DISCHARGE
(See Response Methods Handbook, CG 446-4)

Issue warning—water contaminant, air contaminant, high flammability
Restrict access
Disperse and flush

2. LABEL

Red

3. CHEMICAL DESIGNATIONS

3.1 **Synonyms:** N-n-Propyl-1-propanamine
3.2 **Coast Guard Compatibility Classification:** Not listed
3.3 **Chemical Formula:** $(CH_3CH_2CH_2)_2NH$
3.4 **IMCO/United Nations Numerical Designation:** Not listed

4. OBSERVABLE CHARACTERISTICS

4.1 **Physical State (as shipped):** Liquid
4.2 **Color:** Colorless
4.3 **Odor:** Strong ammoniacal

5. HEALTH HAZARDS

5.1 **Personal Protective Equipment:** Self-contained breathing apparatus; butyl rubber gloves; butyl rubber apron; face shield
5.2 **Symptoms Following Exposure:** Inhalation causes severe coughing and chest pain due to irritation of air passages; can cause lung edema; may also cause headache, nausea, faintness, and anxiety. Ingestion causes irritation and burning of mouth and stomach. Contact with eyes causes severe irritation and edema of the cornea. Contact with skin causes severe irritation.
5.3 **Treatment for Exposure:** INHALATION: remove victim to fresh air; if he is not breathing, give artificial respiration; if breathing is difficult, give oxygen; call a physician. INGESTION: give large amount of water; get medical attention. EYES: flush with water for 15 min.; get medical attention for burns. SKIN: flush with water for 15 min.
5.4 **Toxicity by Inhalation (Threshold Limit Value):** Data not available
5.5 **Short-Term Inhalation Limits:** Data not available
5.6 **Toxicity by Ingestion:** Grade 3; oral rat LD_{50} = 200 mg/kg (rat), 800 mg/kg (mouse)
5.7 **Late Toxicity:** Causes degenerative changes in liver and kidneys of rats and rabbits
5.8 **Vapor (Gas) Irritant Characteristics:** Data not available
5.9 **Liquid or Solid Irritant Characteristics:** Data not available
5.10 **Odor Threshold:** Data not available

6. FIRE HAZARDS

6.1 **Flash Point:** 45°F O.C.
6.2 **Flammable Limits in Air:** Data not available
6.3 **Fire Extinguishing Agents:** Foam, dry chemical, carbon dioxide
6.4 **Fire Extinguishing Agents Not to be Used:** Water may be ineffective.
6.5 **Special Hazards of Combustion Products:** Toxic oxides of nitrogen may form in fires.
6.6 **Behavior in Fire:**
6.7 **Ignition Temperature:** Data not available
6.8 **Electrical Hazard:** Data not available
6.9 **Burning Rate:** 6.1 mm/min.

7. CHEMICAL REACTIVITY

7.1 **Reactivity with Water:** No reaction
7.2 **Reactivity with Common Materials:** May attack some forms of plastics
7.3 **Stability During Transport:** Stable
7.4 **Neutralizing Agents for Acids and Caustics:** Not pertinent
7.5 **Polymerization:** Not pertinent
7.6 **Inhibitor of Polymerization:** Not pertinent

8. WATER POLLUTION

8.1 **Aquatic Toxicity:** 20–60 ppm/24 hr/creek chub/critical range/fresh water
8.2 **Waterfowl Toxicity:** Data not available
8.3 **Biological Oxygen Demand (BOD):** Data not available
8.4 **Food Chain Concentration Potential:** None

9. SELECTED MANUFACTURERS

1. Air Products and Chemicals, Inc.
U.S. Highway 90
Pace, Fla. 32502
2. Pennwalt Corporation
Three Parkway
Philadelphia, Pa. 19102
3. Eastman Organic Chemicals
Rochester, N.Y. 14650

10. SHIPPING INFORMATION

10.1 **Grades or Purity:** Technical, 98%; Pure, 99+%
10.2 **Storage Temperature:** Ambient
10.3 **Inert Atmosphere:** No requirement
10.4 **Venting:** Open (flame arrester)

11. HAZARD ASSESSMENT CODE
(See Hazard Assessment Handbook, CG 446-3)

A-P-Q-T-U

12. HAZARD CLASSIFICATIONS

12.1 **Code of Federal Regulations:** Flammable liquid
12.2 **NAS Hazard Rating for Bulk Water Transportation:** Not listed
12.3 **NFPA Hazard Classifications:**

Category	Classification
Health Hazard (Blue)	3
Flammability (Red)	3
Reactivity (Yellow)	0

13. PHYSICAL AND CHEMICAL PROPERTIES

13.1 **Physical State at 15°C and 1 atm:** Liquid
13.2 **Molecular Weight:** 101.19
13.3 **Boiling Point at 1 atm:** 228.7°F = 109.3°C = 382.5°K
13.4 **Freezing Point:** −81°F = −63°C = 210°K
13.5 **Critical Temperature:** 531°F = 277°C = 550°K
13.6 **Critical Pressure:** 456 psia = 31.0 atm = 3.14 MN/m^2
13.7 **Specific Gravity:** 0.738 at 20°C (liquid)
13.8 **Liquid Surface Tension:** 6.58 dynes/cm = 0.00658 N/m at 20°C
13.9 **Liquid-Water Interfacial Tension:** Not pertinent
13.10 **Vapor (Gas) Specific Gravity:** 3.5
13.11 **Ratio of Specific Heats of Vapor (Gas):**
13.12 **Latent Heat of Vaporization:** 143 Btu/lb = 79.5 cal/g = 3.33 × 10^5 J/kg
13.13 **Heat of Combustion:** −18,750 Btu/lb = −10,420 cal/g = −436.0 × 10^5 J/kg
13.14 **Heat of Decomposition:** Not pertinent
13.15 **Heat of Solution:** Not pertinent
13.16 **Heat of Polymerization:** Not pertinent

NOTES

DPG — DIPROPYLENE GLYCOL

Common Synonyms			
	Thick liquid	Colorless	Odorless
	Sinks and mixes with water.		

Stop discharge if possible.
Call fire department.
Isolate and remove discharged material.
Notify local health and pollution control agencies.

Fire

Combustible.
Extinguish with water, dry chemical, alcohol foam, or carbon dioxide.

Exposure

CALL FOR MEDICAL AID.

LIQUID
Irritating to eyes.
IF IN EYES, hold eyelids open and flush with plenty of water.

Water Pollution

Effect of low concentrations on aquatic life is unknown.
May be dangerous if it enters water intakes.
Notify local health and wildlife officials.
Notify operators of nearby water intakes.

1. RESPONSE TO DISCHARGE

(See Response Methods Handbook, CG 446-4)

Disperse and flush

2. LABELS

No hazard label required by Code of Federal Regulations

3. CHEMICAL DESIGNATIONS

3.1 **Synonyms:** No common synonyms
3.2 **Coast Guard Compatibility Classification:** Glycol
3.3 **Chemical Formula:** $(CH_3CHOHCH_2)_2O$
3.4 **IMCO/United Nations Numerical Designation:** Not listed

4. OBSERVABLE CHARACTERISTICS

4.1 **Physical State (as shipped):** Liquid
4.2 **Color:** Colorless
4.3 **Odor:** Practically none

5. HEALTH HAZARDS

5.1 **Personal Protective Equipment:** Safety glasses with side shields or goggles, shower and eye bath.
5.2 **Symptoms Following Exposure:** Minor eye irritation
5.3 **Treatment for Exposure:** EYES: irrigate briefly with water; if any ill effects, get medical attention. SKIN & INGESTION: if any ill effects develop, get medical attention.
5.4 **Toxicity by Inhalation (Threshold Limit Value):** Not pertinent
5.5 **Short-Term Inhalation Limits:** Not pertinent
5.6 **Toxicity by Ingestion:** Grade 1; LD_{50} 5 to 15 g/kg (rat)
5.7 **Late Toxicity:** Data not available
5.8 **Vapor (Gas) Irritant Characteristics:** Nonirritating to the eyes and throat.
5.9 **Liquid or Solid Irritant Characteristics:** No appreciable hazard. Practically harmless to the skin.
5.10 **Odor Threshold:** Not pertinent

6. FIRE HAZARDS

6.1 **Flash Point:** 280°F O.C.
6.2 **Flammable Limits in Air:** LFL = 2.2% (approx.)
6.3 **Fire Extinguishing Agents:** Water fog, alcohol foam, carbon dioxide, dry chemical
6.4 **Fire Extinguishing Agents Not to be Used:** Water or foam may cause frothing
6.5 **Special Hazards of Combustion Products:** Not pertinent
6.6 **Behavior in Fire:** Not pertinent
6.7 **Ignition Temperature:** Data not available
6.8 **Electrical Hazard:** Not pertinent
6.9 **Burning Rate:** 2.0 mm/min.

7. CHEMICAL REACTIVITY

7.1 **Reactivity with Water:** No reaction
7.2 **Reactivity with Common Materials:** No reaction
7.3 **Stability During Transport:** Stable
7.4 **Neutralizing Agents for Acids and Caustics:** Not pertinent
7.5 **Polymerization:** Not pertinent
7.6 **Inhibitor of Polymerization:** Not pertinent

8. WATER POLLUTION

8.1 **Aquatic Toxicity:** Data not available
8.2 **Waterfowl Toxicity:** Data not available
8.3 **Biological Oxygen Demand (BOD):** Data not available
8.4 **Food Chain Concentration Potential:** None

9. SELECTED MANUFACTURERS

1. Celanese Corp.
Celanese Chemical Co. Division
245 Park Ave.
New York, N. Y. 10017
2. Dow Chemical Co.
Midland, Mich. 48640
3. Union Carbide Corp.
Chemicals and Plastics Division
270 Park Ave.
New York, N. Y. 10017

10. SHIPPING INFORMATION

10.1 **Grades or Purity:** Commercial: 99%
10.2 **Storage Temperature:** Ambient
10.3 **Inert Atmosphere:** No requirement
10.4 **Venting:** Open (flame arrester)

11. HAZARD ASSESSMENT CODE

(See Hazard Assessment Handbook, CG 446-3)

A-P-O

12. HAZARD CLASSIFICATIONS

12.1 **Code of Federal Regulations:** Not listed
12.2 **NAS Hazard Rating for Bulk Water Transportation:**

Category	Rating
Fire	1
Health	
Vapor Irritant	0
Liquid or Solid Irritant	0
Poisons	1
Water Pollution	
Human Toxicity	0
Aquatic Toxicity	1
Aesthetic Effect	0
Reactivity	
Other Chemicals	2
Water	0
Self-Reaction	0

12.3 **NFPA Hazard Classifications:** Not listed

13. PHYSICAL AND CHEMICAL PROPERTIES

13.1 **Physical State at 15°C and 1 atm:** Liquid
13.2 **Molecular Weight:** 134.17
13.3 **Boiling Point at 1 atm:** 420°F = 232°C = 505°K
13.4 **Freezing Point:** > −40°F = > −40°C = > 233°K
13.5 **Critical Temperature:** 720°F = 382°C = 655°K
13.6 **Critical Pressure:** 529 psia = 36 atm = 3.6 MN/m²
13.7 **Specific Gravity:** 1.023 at 20°C (liquid)
13.8 **Liquid Surface Tension:** Not pertinent
13.9 **Liquid-Water Interfacial Tension:** Not pertinent
13.10 **Vapor (Gas) Specific Gravity:** Not pertinent
13.11 **Ratio of Specific Heats of Vapor (Gas):** 1.0
13.12 **Latent Heat of Vaporization:** 170 Btu/lb = 96 cal/g = 4.0×10^5 J/kg
13.13 **Heat of Combustion:** -11,650 Btu/lb = -6470 cal/g = -271×10^5 J/kg
13.14 **Heat of Decomposition:** Not pertinent
13.15 **Heat of Solution:** (est.) −13 Btu/lb = −7 cal/g = -0.3×10^5 J/kg
13.16 **Heat of Polymerization:** Not pertinent

NOTES

REVISED 1978

DFF	DISTILLATES: FLASHED FEED STOCKS

Common Synonyms	Liquid Colorless Gasoline-like odor Floats on water. Flammable, irritating vapor is produced.

Stop discharge if possible. Keep people away.
Call fire department.
Avoid contact with liquid and vapor.
Stay upwind and use water spray to "knock down" vapor.
Isolate and remove discharged material.
Notify local health and pollution control agencies.

Fire	Flammable. Flashback along vapor trail may occur. Vapor may explode if ignited in an enclosed area. Extinguish with foam or dry chemicals. Water may be ineffective on fire. Cool exposed containers with water.
Exposure	CALL FOR MEDICAL AID. VAPOR Irritating to eyes, nose and throat. Move to fresh air. If breathing has stopped, give artificial respiration. If breathing is difficult, give oxygen. LIQUIDS Irritating to eyes. Harmful if swallowed. Remove contaminated clothing and shoes. Flush affected areas with plenty of water. IF IN EYES, hold eyelids open and flush with plenty of water. IF SWALLOWED and victim is CONSCIOUS, have victim drink water or milk. DO NOT INDUCE VOMITING.
Water Pollution	HARMFUL TO AQUATIC LIFE IN VERY LOW CONCENTRATIONS. Fouling to shoreline. May be dangerous if it enters water intakes. Notify local health and wildlife officials. Notify operators of nearby water intakes.

1. RESPONSE TO DISCHARGE

(See Response Methods Handbook, CG 446-4)

Issue warning—high flammability
Evacuate area
Disperse and flush

2. LABEL

Red

3. CHEMICAL DESIGNATIONS

3.1 **Synonyms:** Petroleum distillate
3.2 **Coast Guard Compatibility Classification:** Miscellaneous hydrocarbon mixtures
3.3 **Chemical Formula:** Not pertinent
3.4 **IMCO/United Nations Numerical Designation:** 3.1, 3.2, 3.3/1268

4. OBSERVABLE CHARACTERISTICS

4.1 **Physical State (as shipped):** Liquid
4.2 **Color:** Colorless
4.3 **Odor:** Gasoline

5. HEALTH HAZARDS

5.1 **Personal Protective Equipment:** Data not available
5.2 **Symptoms Following Exposure:** INHALATION: irritation of upper respiratory tract; dizziness, headache, coma, respiratory arrest; cardiac arrhythmias may occur. ASPIRATION: severe lung irritation, coughing, pulmonary edema, signs of bronchopneumonia; acute central nervous system excitation, followed by depression. INGESTION: irritation of mouth and stomach, other symptoms as above.
5.3 **Treatment for Exposure:** Seek medical attention. INHALATION: maintain respiration, administer oxygen. ASPIRATION: enforce bed rest; administer oxygen. INGESTION: do NOT induce vomiting; lavage carefully if appreciable quantity was swallowed; guard against aspiration into lungs. EYES: wash with copious amounts of water. SKIN: wipe off and wash with soap and water.
5.4 **Toxicity by Inhalation (Threshold Limit Value):** No single TLV applicable.
5.5 **Short-Term Inhalation Limits:** 500 ppm for 30 min.
5.6 **Toxicity by Ingestion:** Grade 2; LD_{50} 0.5–5 g/kg
5.7 **Late Toxicity:** None
5.8 **Vapor (Gas) Irritant Characteristics:** Vapors cause a slight smarting of the eyes or respiratory system if present in high concentrations. The effect is temporary.
5.9 **Liquid or Solid Irritant Characteristics:** Minimum hazard. If spilled on clothing and allowed to remain, may cause smarting and reddening of the skin.
5.10 **Odor Threshold:** 0.25 ppm

6. FIRE HAZARDS

6.1 **Flash Point:**
(a) <0°F C.C.
(b) 0-73°F C.C.
(c) 73-141°F C.C.
6.2 **Flammable Limits in Air:** Data not available
6.3 **Fire Extinguishing Agents:** Foam, carbon dioxide, dry chemical
6.4 **Fire Extinguishing Agents Not to be Used:** Water may be ineffective.
6.5 **Special Hazards of Combustion Products:** Not pertinent
6.6 **Behavior in Fire:** Not pertinent
6.7 **Ignition Temperature:** Data not available
6.8 **Electrical Hazard:** Class I, group D
6.9 **Burning Rate:** Approx. 4 mm/min.

7. CHEMICAL REACTIVITY

7.1 **Reactivity with Water:** No reaction
7.2 **Reactivity with Common Materials:** No reaction
7.3 **Stability During Transport:** Stable
7.4 **Neutralizing Agents for Acids and Caustics:** Not pertinent
7.5 **Polymerization:** Not pertinent
7.6 **Inhibitor of Polymerization:** Not pertinent

8. WATER POLLUTION

8.1 **Aquatic Toxicity:**
90 ppm/24 hr/juvenile American shad/ TL_m/fresh water
91 ppm/24 hr/juvenile American shad/ TL_m/salt water
8.2 **Waterfowl Toxicity:** Data not available
8.3 **Biological Oxygen Demand (BOD):** 8%, 5 days
8.4 **Food Chain Concentration Potential:** None

9. SELECTED MANUFACTURERS

1. Atlantic Richfield Co.
717 Fifth Ave.
New York, N. Y. 10022
2. Mobil Oil Corp.
150 E. 42nd St.
New York, N. Y. 10017
3. Shell Oil Co.
One Shell Plaza
Houston, Tex. 77001

10. SHIPPING INFORMATION

10.1 **Grades or Purity:** Composition varies with range of distillation temperatures used.
10.2 **Storage Temperature:** Ambient
10.3 **Inert Atmosphere:** No requirement
10.4 **Venting:** Data not available

11. HAZARD ASSESSMENT CODE

(See Hazard Assessment Handbook, CG 446-3)

A-T-U-V-W

12. HAZARD CLASSIFICATIONS

12.1 **Code of Federal Regulations:** Flammable liquid
12.2 **NAS Hazard Rating for Bulk Water Transportation:**

Category	Rating
Fire	3
Health	
Vapor Irritant	1
Liquid or Solid Irritant	1
Poisons	2
Water Pollution	
Human Toxicity	1
Aquatic Toxicity	2
Aesthetic Effect	2
Reactivity	
Other Chemicals	0
Water	0
Self-Reaction	0

12.3 **NFPA Hazard Classifications:** Not listed

13. PHYSICAL AND CHEMICAL PROPERTIES

13.1 **Physical State at 15°C and 1 atm:** Liquid
13.2 **Molecular Weight:** Not pertinent
13.3 **Boiling Point at 1 atm:** 58–275°F = 14–135°C = 287–408°K
13.4 **Freezing Point:** Not pertinent
13.5 **Critical Temperature:** Not pertinent
13.6 **Critical Pressure:** Not pertinent
13.7 **Specific Gravity:** 0.71–0.75 at 15°C (liquid)
13.8 **Liquid Surface Tension:** 19–23 dynes/cm = 0.019–0.023 N/m at 20°C
13.9 **Liquid-Water Interfacial Tension:** 49–51 dynes/cm = 0.049–0.051 N/m at 20°C
13.10 **Vapor (Gas) Specific Gravity:** 3.4
13.11 **Ratio of Specific Heats of Vapor (Gas):** (est.) 1.054
13.12 **Latent Heat of Vaporization:** 130–150 Btu/lb = 71–81 cal/g = 3.0–3.4 × 10^5 J/kg
13.13 **Heat of Combustion:** −18,720 Btu/lb = −10,400 cal/g = −435.4 × 10^5 J/kg
13.14 **Heat of Decomposition:** Not pertinent
13.15 **Heat of Solution:** Not pertinent
13.16 **Heat of Polymerization:** Not pertinent

NOTES

REVISED 1978

DISTILLATES: STRAIGHT RUN

Common Synonyms
Straight run gasoline

Watery liquid | Colorless | Gasoline-like odor

Floats on water. Flammable, irritating vapor is produced.

Stop discharge if possible. Keep people away.
Call fire department.
Avoid contact with liquid and vapor.
Stay upwind and use water spray to "knock down" vapor.
Isolate and remove discharged material.
Notify local health and pollution control agencies.

Fire

FLAMMABLE.
Flashback along vapor trail may occur.
Vapor may explode if ignited in an enclosed area.
Extinguish with foam or dry chemicals.
Water may be ineffective on fire.
Cool exposed containers with water.

Exposure

CALL FOR MEDICAL AID.

VAPOR
Irritating to eyes, nose and throat.
Move to fresh air.
If breathing has stopped, give artificial respiration.
If breathing is difficult, give oxygen.

LIQUIDS
Irritating to eyes.
Harmful if swallowed.
Remove contaminated clothing and shoes.
Flush affected areas with plenty of water.
IF IN EYES, hold eyelids open and flush with plenty of water.
IF SWALLOWED and victim is CONSCIOUS, have victim drink water or milk.
DO NOT INDUCE VOMITING.

Water Pollution

HARMFUL TO AQUATIC LIFE IN VERY LOW CONCENTRATIONS.
Fouling to shoreline.
May be dangerous if it enters water intakes.
Notify local health and wildlife officials.
Notify operators of nearby water intakes.

1. RESPONSE TO DISCHARGE

(See Response Methods Handbook, CG 446-4)

Issue warning—high flammability
Evacuate area
Disperse and flush

2. LABEL

Red

3. CHEMICAL DESIGNATIONS

3.1 **Synonyms:** Petroleum distillate
3.2 **Coast Guard Compatibility Classification:** Miscellaneous hydrocarbon mixtures
3.3 **Chemical Formula:** Not pertinent
3.4 **IMCO/United Nations Numerical Designation:** 3.1, 3.2, 3.3/1268

4. OBSERVABLE CHARACTERISTICS

4.1 **Physical State (as shipped):** Liquid
4.2 **Color:** Colorless
4.3 **Odor:** Gasoline

5. HEALTH HAZARDS

5.1 **Personal Protective Equipment:** Data not available
5.2 **Symptoms Following Exposure:** INHALATION: irritation of upper respiratory tract; dizziness, headache, coma, respiratory arrest; cardiac arrhythmias may occur. ASPIRATION: severe lung irritation, coughing, pulmonary edema, signs of bronchopneumonia; acute central nervous system excitation, followed by depression. INGESTION: irritation of mouth and stomach; other symptoms as above.
5.3 **Treatment for Exposure:** Seek medical attention. INHALATION: maintain respiration and administer oxygen. ASPIRATION: enforce bed rest and administer oxygen. INGESTION: do NOT induce vomiting; lavage carefully if appreciable quantity was swallowed; guard against aspiration into lungs. EYES: wash with plenty of water. SKIN: remove by wiping and wash with soap and water.
5.4 **Toxicity by Inhalation (Threshold Limit Value):** No single TLV applicable
5.5 **Short-Term Inhalation Limits:** 500 ppm for 30 min.
5.6 **Toxicity by Ingestion:** Grade 2; LD_{50} 0.5 – 5 g/kg
5.7 **Late Toxicity:** None
5.8 **Vapor (Gas) Irritant Characteristics:** Vapors cause a slight smarting of the eyes or respiratory system if present in high concentrations. The effect is temporary.
5.9 **Liquid or Solid Irritant Characteristics:** Minimum hazard. If spilled on clothing and allowed to remain, may cause smarting and reddening of the skin.
5.10 **Odor Threshold:** 0.25 ppm

6. FIRE HAZARDS

6.1 **Flash Point:**
(a) < 0°F C.C.
(b) 0° – 73°F C.C.
(c) 73° – 141°F C.C.
6.2 **Flammable Limits in Air:** 1.1 – 8.7%
6.3 **Fire Extinguishing Agents:** Foam, carbon dioxide, dry chemical
6.4 **Fire Extinguishing Agents Not to be Used:** Water may be ineffective
6.5 **Special Hazards of Combustion Products:** Not pertinent
6.6 **Behavior in Fire:** Not pertinent
6.7 **Ignition Temperature:** Data not available
6.8 **Electrical Hazard:** Class I, group D
6.9 **Burning Rate:** Approx. 4 mm/min.

7. CHEMICAL REACTIVITY

7.1 **Reactivity with Water:** No reaction
7.2 **Reactivity with Common Materials:** No reaction
7.3 **Stability During Transport:** Stable
7.4 **Neutralizing Agents for Acids and Caustics:** Not pertinent
7.5 **Polymerization:** Not pertinent
7.6 **Inhibitor of Polymerization:** Not pertinent

8. WATER POLLUTION

8.1 **Aquatic Toxicity:**
90 ppm/24 hr/juvenile American shad/TL_m/fresh water
91 ppm/24 hr/juvenile American shad/TL_m/salt water
8.2 **Waterfowl Toxicity:** Data not available
8.3 **Biological Oxygen Demand (BOD):** 8%, 5 days
8.4 **Food Chain Concentration Potential:** None

9. SELECTED MANUFACTURERS

1. Atlantic Richfield Co.
717 Fifth Ave.
New York, N. Y. 10022
2. Mobil Oil Corp.
150 E. 42nd St.
New York, N. Y. 10017
3. Shell Oil Co.
One Shell Plaza
Houston, Tex. 77001

10. SHIPPING INFORMATION

10.1 **Grades or Purity:** Composition varies with range of distillation temperatures used.
10.2 **Storage Temperature:** Ambient
10.3 **Inert Atmosphere:** No requirement
10.4 **Venting:** Data not available

11. HAZARD ASSESSMENT CODE

(See Hazard Assessment Handbook, CG 446-3)

A-T-U-V-W

12. HAZARD CLASSIFICATIONS

12.1 **Code of Federal Regulations:** Flammable liquid
12.2 **NAS Hazard Rating for Bulk Water Transportation:**

Category	Rating
Fire	3
Health	
Vapor Irritant	1
Liquid or Solid Irritant	1
Poisons	2
Water Pollution	
Human Toxicity	1
Aquatic Toxicity	2
Aesthetic Effect	2
Reactivity	
Other Chemicals	0
Water	0
Self-Reaction	0

12.3 **NFPA Hazard Classifications:** Not listed

13. PHYSICAL AND CHEMICAL PROPERTIES

13.1 **Physical State at 15°C and 1 atm:** Liquid
13.2 **Molecular Weight:** Not pertinent
13.3 **Boiling Point at 1 atm:** 58 – 275°F = 14 – 135°C = 287 – 408°K
13.4 **Freezing Point:** Not pertinent
13.5 **Critical Temperature:** Not pertinent
13.6 **Critical Pressure:** Not pertinent
13.7 **Specific Gravity:** 0.731 at 16°C (liquid)
13.8 **Liquid Surface Tension:** 19 – 23 dynes/cm = 0.019 – 0.023 N/m at 20°C
13.9 **Liquid-Water Interfacial Tension:** 49 – 51 dynes/cm = 0.049 – 0.051 N/m at 20°C
13.10 **Vapor (Gas) Specific Gravity:** 3.4
13.11 **Ratio of Specific Heats of Vapor (Gas):** (est.) 1.054
13.12 **Latent Heat of Vaporization:** 130-150 Btu/lb = 71 – 81 cal/g = 3.0 – 3.4 × 10^5 J/kg
13.13 **Heat of Combustion:** −18,720 Btu/lb = −10,400 cal/g = −435.4 × 10^5 J/kg
13.14 **Heat of Decomposition:** Not pertinent
13.15 **Heat of Solution:** Not pertinent
13.16 **Heat of Polymerization:** Not pertinent

NOTES

REVISED 1978

DDN	DODECANOL

Common Synonyms
Lauryl alcohol
Dodecyl alcohol

Thick liquid — Colorless — Sweet odor

Floats on water. Freezing point is 75° F.

Stop discharge if possible.
Call fire department.
Avoid contact with liquid.
Isolate and remove discharged material.
Notify local health and pollution control agencies.

Fire

Combustible.
Extinguish with dry chemical, alcohol foam, or carbon dioxide.
Water may be ineffective on fire.
Cool exposed containers with water.

Exposure

CALL FOR MEDICAL AID.

LIQUID
Irritating to skin.
Will burn eyes.
Flush affected areas with plenty of water.
IF IN EYES, hold eyelids open and flush with plenty of water.

Water Pollution

Effect of low concentrations on aquatic life is unknown.
Fouling to shoreline.
May be dangerous if it enters water intakes.
Notify local health and wildlife officials.
Notify operators of nearby water intakes.

1. RESPONSE TO DISCHARGE

(See Response Methods Handbook, CG 446-4)

Disperse and flush

2. LABELS

No hazard label required by Code of Federal Regulations

3. CHEMICAL DESIGNATIONS

3.1 **Synonyms:** Dodecyl alcohol
Lauryl alcohol
3.2 **Coast Guard Compatibility Classification:** Alcohol
3.3 **Chemical Formula:** $CH_3(CH_2)_{10}CH_2OH$
3.4 **IMCO/United Nations Numerical Designation:** Not listed

4. OBSERVABLE CHARACTERISTICS

4.1 **Physical State (as shipped):** Liquid
4.2 **Color:** Colorless
4.3 **Odor:** Typical fatty alcohol odor; sweet

5. HEALTH HAZARDS

5.1 **Personal Protective Equipment:** Chemical gloves; chemical goggles
5.2 **Symptoms Following Exposure:** Liquid will cause burning of the eyes and may irritate skin.
5.3 **Treatment for Exposure:** SKIN OR EYES: wash exposed areas with water.
5.4 **Toxicity by Inhalation (Threshold Limit Value):** Not pertinent
5.5 **Short-Term Inhalation Limits:** Not pertinent
5.6 **Toxicity by Ingestion:** Grade 1; LD_{50} 5 to 15 g/kg (humans)
5.7 **Late Toxicity:** Data not available
5.8 **Vapor (Gas) Irritant Characteristics:** None
5.9 **Liquid or Solid Irritant Characteristics:** No appreciable hazard. Practically harmless to the skin.
5.10 **Odor Threshold:** Data not available

6. FIRE HAZARDS

6.1 **Flash Point:** 260°F C.C.
6.2 **Flammable Limits in Air:** Data not available
6.3 **Fire Extinguishing Agents:** Alcohol foam, carbon dioxide, dry chemical
6.4 **Fire Extinguishing Agents Not to be Used:** Water or foam may cause frothing
6.5 **Special Hazards of Combustion Products:** Not pertinent
6.6 **Behavior in Fire:** Not pertinent
6.7 **Ignition Temperature:** 527°F
6.8 **Electrical Hazard:** Not pertinent
6.9 **Burning Rate:** Data not available

7. CHEMICAL REACTIVITY

7.1 **Reactivity with Water:** No reaction
7.2 **Reactivity with Common Materials:** No reaction
7.3 **Stability During Transport:** Stable
7.4 **Neutralizing Agents for Acids and Caustics:** Not pertinent
7.5 **Polymerization:** Not pertinent
7.6 **Inhibitor of Polymerization:** Not pertinent

8. WATER POLLUTION

8.1 **Aquatic Toxicity:** Data not available
8.2 **Waterfowl Toxicity:** Data not available
8.3 **Biological Oxygen Demand (BOD):** 30% (theor.), 5 days
8.4 **Food Chain Concentration Potential:** None

9. SELECTED MANUFACTURERS

1. Continental Oil Co.
Conoco Chemicals Division
Park 80 Plaza East
Saddle Brook, N. J. 07662
2. Givaudan Corp.
Aroma Chemicals Division
Clifton, N. J. 07014
3. Procter and Gamble Co.
Ivorydale, Ohio 45217

10. SHIPPING INFORMATION

10.1 **Grades or Purity:** 98.5–99.5%
10.2 **Storage Temperature:** Ambient
10.3 **Inert Atmosphere:** No requirement
10.4 **Venting:** Open (flame arrester)

11. HAZARD ASSESSMENT CODE

(See Hazard Assessment Handbook, CG 446-3)

A-T-U

12. HAZARD CLASSIFICATIONS

12.1 **Code of Federal Regulations:** Not listed
12.2 **NAS Hazard Rating for Bulk Water Transportation:** Not listed
12.3 **NFPA Hazard Classifications:** Not listed

13. PHYSICAL AND CHEMICAL PROPERTIES

13.1 **Physical State at 15°C and 1 atm:** Liquid
13.2 **Molecular Weight:** 186.33
13.3 **Boiling Point at 1 atm:** 498°F = 259°C = 532°K
13.4 **Freezing Point:** 75°F = 24°C = 297°K
13.5 **Critical Temperature:** 763°F = 406°C = 679°K
13.6 **Critical Pressure:** 280 psia = 19 atm = 1.9 MN/m^2
13.7 **Specific Gravity:** 0.831 at 24°C (liquid)
13.8 **Liquid Surface Tension:** 27.4 dynes/cm = 0.0274 N/m at 25°C
13.9 **Liquid-Water Interfacial Tension:** (est.) 30 dynes/cm = 0.03 N/m at 30°C
13.10 **Vapor (Gas) Specific Gravity:** Not pertinent
13.11 **Ratio of Specific Heats of Vapor (Gas):** 1.030
13.12 **Latent Heat of Vaporization:** (est.) 110 Btu/lb = 62 cal/g = 2.6×10^5 J/kg
13.13 **Heat of Combustion:** (est.) −18,000 Btu/lb = −10,000 cal/g = -420×10^5 J/kg
13.14 **Heat of Decomposition:** Not pertinent
13.15 **Heat of Solution:** Not pertinent
13.16 **Heat of Polymerization:** Not pertinent

NOTES

REVISED 1978

DOD — DODECENE

Common Synonyms
Propylene tetramer

Watery liquid — Colorless — Pleasant odor

Floats on water.

Stop discharge if possible.
Call fire department.
Avoid contact with liquid.
Isolate and remove discharged material.
Notify local health and pollution control agencies.

Fire

Combustible.
Extinguish with foam, dry chemical or carbon dioxide.

Exposure

CALL FOR MEDICAL AID.

LIQUID
Irritating to skin and eyes.
Harmful if swallowed.
Remove contaminated clothing and shoes.
Flush affected areas with plenty of water.
IF IN EYES, hold eyelids open and flush with plenty of water.
IF SWALLOWED and victim is CONSCIOUS, have victim drink water or milk.
DO NOT INDUCE VOMITING.

Water Pollution

Effect of low concentrations on aquatic life is unknown.
Fouling to shoreline.
May be dangerous if it enters water intakes.
Notify local health and wildlife officials.
Notify operators of nearby water intakes.

1. RESPONSE TO DISCHARGE

(See Response Methods Handbook, CG 446-4)

Mechanical containment
Chemical and physical treatment

2. LABELS

No hazard label required by Code of Federal Regulations

3. CHEMICAL DESIGNATIONS

3.1 **Synonyms:** Dodecene (non-linear)
Propylene tetramer
Tetrapropylene

3.2 **Coast Guard Compatibility Classification:** Olefin

3.3 **Chemical Formula:** $C_{12}H_{24}$

3.4 **IMCO/United Nations Numerical Designation:** Not listed

4. OBSERVABLE CHARACTERISTICS

4.1 **Physical State (as shipped):** Liquid

4.2 **Color:** Colorless

4.3 **Odor:** Characteristic

5. HEALTH HAZARDS

5.1 **Personal Protective Equipment:** Protective gloves; no respiratory protection needed if ventilation is adequate.

5.2 **Symptoms Following Exposure:** No inhalation hazard expected. Aspiration hazard if ingested. Minor skin and eye irritation.

5.3 **Treatment for Exposure:** INHALATION: Remove victim to fresh air. INGESTION: do NOT induce vomiting! Do NOT lavage! Give vegetable oil and demulcents; call physician. EYE CONTACT: flush with water for 15 min. SKIN CONTACT: wash with soap and water.

5.4 **Toxicity by Inhalation (Threshold Limit Value):** 200 ppm

5.5 **Short-Term Inhalation Limits:** Data not available

5.6 **Toxicity by Ingestion:** Grade 0; LD_{50} above 15 g/kg

5.7 **Late Toxicity:** Data not available

5.8 **Vapor (Gas) Irritant Characteristics:** Slight smarting of eyes and respiratory system at high concentrations. The effect is temporary.

5.9 **Liquid or Solid Irritant Characteristics:** Minimum hazard. If spilled on clothing and allowed to remain, may cause smarting and reddening of skin.

5.10 **Odor Threshold:** Data not available

6. FIRE HAZARDS

6.1 **Flash Point:** 120°F C.C.; 134°F O.C.

6.2 **Flammable Limits in Air:** Data not available

6.3 **Fire Extinguishing Agents:** Water fog, foam, carbon dioxide, dry chemical

6.4 **Fire Extinguishing Agents Not to be Used:** Water may be ineffective.

6.5 **Special Hazards of Combustion Products:** Not pertinent

6.6 **Behavior in Fire:** Not pertinent

6.7 **Ignition Temperature:** 400°F (est.)

6.8 **Electrical Hazard:** Not pertinent

6.9 **Burning Rate:** Data not available

7. CHEMICAL REACTIVITY

7.1 **Reactivity with Water:** No reaction

7.2 **Reactivity with Common Materials:** No reaction

7.3 **Stability During Transport:** Stable

7.4 **Neutralizing Agents for Acids and Caustics:** Not pertinent

7.5 **Polymerization:** Not pertinent

7.6 **Inhibitor of Polymerization:** Not pertinent

8. WATER POLLUTION

8.1 **Aquatic Toxicity:** Data not available

8.2 **Waterfowl Toxicity:** Data not available

8.3 **Biological Oxygen Demand (BOD):** Data not available

8.4 **Food Chain Concentration Potential:** None

9. SELECTED MANUFACTURERS

1. Atlantic Richfield Co.
ARCO Chemical Division
260 S. Broad St.
Philadelphia, Pa. 19101
2. Continental Oil Co.
Conoco Chemicals Division
Park 80 Plaza East
Saddle Brook, N. J. 07662
3. Sun Oil Co.
St. Davids, Pa. 19087

10. SHIPPING INFORMATION

10.1 **Grades or Purity:** 98.5—99+% olefin

10.2 **Storage Temperature:** Ambient

10.3 **Inert Atmosphere:** No requirement

10.4 **Venting:** Open (flame arrester)

11. HAZARD ASSESSMENT CODE

(See Hazard Assessment Handbook, CG 446-3)

A-T-U

12. HAZARD CLASSIFICATIONS

12.1 **Code of Federal Regulations:** Combustible Liquid

12.2 **NAS Hazard Rating for Bulk Water Transportation:**

Category	Rating
Fire	2
Health	
Vapor Irritant	1
Liquid or Solid Irritant	1
Poisons	1
Water Pollution	
Human Toxicity	0
Aquatic Toxicity	1
Aesthetic Effect	2
Reactivity	
Other Chemicals	1
Water	0
Self-Reaction	1

12.3 **NFPA Hazard Classifications:**

Category	Classification
Health Hazard (Blue)	0
Flammability (Red)	2
Reactivity (Yellow)	0

13. PHYSICAL AND CHEMICAL PROPERTIES

13.1 **Physical State at 15°C and 1 atm:** Liquid

13.2 **Molecular Weight:** 168.31

13.3 **Boiling Point at 1 atm:** 365—385°F = 185—196°C = 458—469°K

13.4 **Freezing Point:** Not pertinent

13.5 **Critical Temperature:** Not pertinent

13.6 **Critical Pressure:** Not pertinent

13.7 **Specific Gravity:** 0.77 at 20°C (liquid)

13.8 **Liquid Surface Tension:** 23.0 dynes/cm = 0.0230 N/m at 20°C

13.9 **Liquid-Water Interfacial Tension:** Data not available

13.10 **Vapor (Gas) Specific Gravity:** Not pertinent

13.11 **Ratio of Specific Heats of Vapor (Gas):** Not pertinent

13.12 **Latent Heat of Vaporization:** Not pertinent

13.13 **Heat of Combustion:** -19,100 Btu/lb = -10,600 cal/g = -444 × 10^5 J/kg

13.14 **Heat of Decomposition:** Not pertinent

13.15 **Heat of Solution:** Not pertinent

13.16 **Heat of Polymerization:** Not pertinent

NOTES

REVISED 1978

DDC 1-DODECENE

Common Synonyms
alpha-Dodecylene

Watery liquid | Colorless | Mild, pleasant odor

Floats on water.

Stop discharge if possible.
Call fire department.
Avoid contact with liquid.
Isolate and remove discharged material.
Notify local health and pollution control agencies.

Fire

Combustible.
Extinguish with foam, dry chemical or carbon dioxide.

Exposure

CALL FOR MEDICAL AID.

LIQUID
Irritating to skin and eyes.
Harmful if swallowed.
Remove contaminated clothing and shoes.
Flush affected areas with plenty of water.
IF IN EYES, hold eyelids open and flush with plenty of water.
IF SWALLOWED and victim is CONSCIOUS, have victim drink water or milk.
DO NOT INDUCE VOMITING.

Water Pollution

Effect of low concentrations on aquatic life is unknown.
Fouling to shoreline.
May be dangerous if it enters water intakes.
Notify local health and wildlife officials.
Notify operators of nearby water intakes.

1. RESPONSE TO DISCHARGE

(See Response Methods Handbook, CG 446-4)

Mechanical containment
Chemical and physical treatment

2. LABELS

No hazard label required by Code of Federal Regulations

3. CHEMICAL DESIGNATIONS

3.1 **Synonyms:** Adacene-12
alpha-Dodecylene
3.2 **Coast Guard Compatibility Classification:** Olefin
3.3 **Chemical Formula:** $CH_3(CH_2)_9CH=CH_2$
3.4 **IMCO/United Nations Numerical Designation:** Not listed

4. OBSERVABLE CHARACTERISTICS

4.1 **Physical State (as shipped):** Liquid
4.2 **Color:** Colorless
4.3 **Odor:** Mild; pleasant

5. HEALTH HAZARDS

5.1 **Personal Protective Equipment:** Protective gloves; goggles or face shield
5.2 **Symptoms Following Exposure:** No inhalation hazard expected. Aspiration hazard if ingested. Minor skin and eye irritation.
5.3 **Treatment for Exposure:** INHALATION: remove victim to fresh air. INGESTION: do NOT induce vomiting! Do NOT lavage! Give vegetable oil and demulcents; call physician. EYE CONTACT: flush with water for 15 min. SKIN CONTACT: wash with soap and water.
5.4 **Toxicity by Inhalation (Threshold Limit Value):** Data not available
5.5 **Short-Term Inhalation Limits:** Data not available
5.6 **Toxicity by Ingestion:** Data not available
5.7 **Late Toxicity:** Data not available
5.8 **Vapor (Gas) Irritant Characteristics:** Slight smarting of eyes and respiratory system at high concentrations. The effect is temporary.
5.9 **Liquid or Solid Irritant Characteristics:** Minimum hazard. If spilled on clothing and allowed to remain, may cause smarting and reddening of skin.
5.10 **Odor Threshold:** Data not available

6. FIRE HAZARDS

6.1 **Flash Point:** 174°F (approx.)
6.2 **Flammable Limits in Air:** Data not available
6.3 **Fire Extinguishing Agents:** Foam, dry chemical, carbon dioxide
6.4 **Fire Extinguishing Agents Not to be Used:** Not pertinent
6.5 **Special Hazards of Combustion Products:** Not pertinent
6.6 **Behavior in Fire:** Not pertinent
6.7 **Ignition Temperature:** 491°F
6.8 **Electrical Hazard:** Not pertinent
6.9 **Burning Rate:** 5.8 mm/min.

7. CHEMICAL REACTIVITY

7.1 **Reactivity with Water:** No reaction
7.2 **Reactivity with Common Materials:** No reaction
7.3 **Stability During Transport:** Stable
7.4 **Neutralizing Agents for Acids and Caustics:** Not pertinent
7.5 **Polymerization:** Not pertinent
7.6 **Inhibitor of Polymerization:** Not pertinent

8. WATER POLLUTION

8.1 **Aquatic Toxicity:** Data not available
8.2 **Waterfowl Toxicity:** Data not available
8.3 **Biological Oxygen Demand (BOD):** Data not available
8.4 **Food Chain Concentration Potential:** None

9. SELECTED MANUFACTURERS

1. Gulf Oil Chemicals Co.
Petrochemicals Division
Houston, Tex. 77001
2. Humphrey Chemical Co.
Devine St.
North Haven, Conn. 06473
3. Phillips Petroleum Co.
Bartlesville, Okla. 74004

10. SHIPPING INFORMATION

10.1 **Grades or Purity:** 93—98% normal alpha olefins
10.2 **Storage Temperature:** Ambient
10.3 **Inert Atmosphere:** No requirement
10.4 **Venting:** Open (flame arrester)

11. HAZARD ASSESSMENT CODE

(See Hazard Assessment Handbook, CG 446-3)

A-T-U

12. HAZARD CLASSIFICATIONS

12.1 **Code of Federal Regulations:** Combustible Liquid
12.2 **NAS Hazard Rating for Bulk Water Transportation:** Not listed
12.3 **NFPA Hazard Classifications:** Not listed

13. PHYSICAL AND CHEMICAL PROPERTIES

13.1 **Physical State at 15°C and 1 atm:** Liquid
13.2 **Molecular Weight:** 168.31
13.3 **Boiling Point at 1 atm:** 415°F = 213°C = 486°K
13.4 **Freezing Point:** −31°F = −35°C = 238°K
13.5 **Critical Temperature:** Not pertinent
13.6 **Critical Pressure:** Not pertinent
13.7 **Specific Gravity:** 0.758 at 20°C (liquid)
13.8 **Liquid Surface Tension:** 25.6 dynes/cm = 0.0256 N/m at 20°C
13.9 **Liquid-Water Interfacial Tension:** Data not available
13.10 **Vapor (Gas) Specific Gravity:** Not pertinent
13.11 **Ratio of Specific Heats of Vapor (Gas):** 1.032
13.12 **Latent Heat of Vaporization:** 110 Btu/lb = 61.0 cal/g = 2.55×10^5 J/kg
13.13 **Heat of Combustion:** −18,911 Btu/lb = −10,506 cal/g = -439.87×10^5 J/kg
13.14 **Heat of Decomposition:** Not pertinent
13.15 **Heat of Solution:** Not pertinent
13.16 **Heat of Polymerization:** Not pertinent

NOTES

DDB — DODECYLBENZENE

Common Synonyms			
1-Phenyldodecane n-Dodecylbenzene Laurylbenzene Detergent alkylate #2 Dodecylbenzene (linear)	Liquid	Colorless	Weak oily odor
	Floats on water.		

Stop discharge if possible. Keep people away.
Call fire department.
Avoid contact with liquid.
Isolate and remove discharged material.
Notify local health and pollution control agencies.

Fire

Combustible.
Extinguish with dry chemicals, foam or carbon dioxide.
Water may be ineffective on fire.

Exposure

Call for medical aid.

LIQUID
Irritating to skin and eyes.
Harmful if swallowed.
Remove contaminated clothing and shoes.
Flush affected areas with plenty of water.
IF IN EYES, hold eyelids open and flush with plenty of water.
IF SWALLOWED and victim is CONSCIOUS, have victim drink water or milk.

Water Pollution

Effect of low concentrations on aquatic life is unknown.
Fouling to shoreline.
May be dangerous if it enters water intakes.
Notify local health and wildlife officials.
Notify operators of nearby water intakes.

1. RESPONSE TO DISCHARGE

(See Response Methods Handbook, CG 446-4)

Mechanical containment
Should be removed
Chemical and physical treatment

2. LABELS

No label required by Code of Federal Regulations

3. CHEMICAL DESIGNATIONS

3.1 **Synonyms:** Detergent Alkylate #2; n-Dodecylbenzene; Dodecylbenzene (linear); Lauryl benzene; 1-Phenyldodecane; UCANE Alkylate 12
3.2 **Coast Guard Compatibility Classification:** Aromatic hydrocarbon
3.3 **Chemical Formula:** $C_6H_5(CH_2)_{11}CH_3$
3.4 **IMCO/United Nations Numerical Designation:** Not listed

4. OBSERVABLE CHARACTERISTICS

4.1 **Physical State (as shipped):** Liquid
4.2 **Color:** Colorless
4.3 **Odor:** Weak oily

5. HEALTH HAZARDS

5.1 **Personal Protective Equipment:** Goggles or face shield; rubber gloves.
5.2 **Symptoms Following Exposure:** Liquid causes mild irritation of eyes and may cause allergenic responses on repeated contact with skin. Ingestion causes nausea.
5.3 **Treatment for Exposure:** EYES: flush with water for at least 15 min.; get medical attention for persistent irritation. SKIN: wash with soap and water. INGESTION: do NOT induce vomiting; get medical attention.
5.4 **Toxicity by Inhalation (Threshold Limit Value):** Data not available
5.5 **Short-Term Inhalation Limits:** Not pertinent
5.6 **Toxicity by Ingestion:** Grade 1; LD_{50} 5 to 15 g/kg
5.7 **Late Toxicity:** Data not available
5.8 **Vapor (Gas) Irritant Characteristics:** Vapors are nonirritating to eyes and throat.
5.9 **Liquid or Solid Irritant Characteristics:** No appreciable hazard. Practically harmless to the skin.
5.10 **Odor Threshold:** Data not available

6. FIRE HAZARDS

6.1 **Flash Point:** 275°F O.C.
6.2 **Flammable Limits in Air:** Data not available
6.3 **Fire Extinguishing Agents:** Dry chemical, foam
6.4 **Fire Extinguishing Agents Not to be Used:** Water may be ineffective.
6.5 **Special Hazards of Combustion Products:** Not pertinent
6.6 **Behavior in Fire:** Not pertinent
6.7 **Ignition Temperature:** Data not available
6.8 **Electrical Hazard:** Data not available
6.9 **Burning Rate:** 3.7 mm/min.

7. CHEMICAL REACTIVITY

7.1 **Reactivity with Water:** No reaction
7.2 **Reactivity with Common Materials:** No reaction
7.3 **Stability During Transport:** Stable
7.4 **Neutralizing Agents for Acids and Caustics:** Not pertinent
7.5 **Polymerization:** Not pertinent
7.6 **Inhibitor of Polymerization:** Not pertinent

8. WATER POLLUTION

8.1 **Aquatic Toxicity:** Data not available
8.2 **Waterfowl Toxicity:** Data not available
8.3 **Biological Oxygen Demand (BOD):** Data not available
8.4 **Food Chain Concentration Potential:** None

9. SELECTED MANUFACTURERS

1. Witco Chemical Corporation
   Witfield Division
   2100 E. 223rd Street
   Carson, Calif. 90745
2. Union Carbide Corporation
   Chemicals and Plastics Division
   270 Park Avenue
   New York, N. Y. 10017
3. Continental Oil Company
   Conoco Chemicals
   Park-Eighty Plaza West-One
   Saddle Brook, N. J. 07662

10. SHIPPING INFORMATION

10.1 **Grades or Purity:** Various mixtures containing 70—80% of undecyl- plus dodecyl-benzene, along with 10% decylbenzene and other analogous hydrocarbons.
10.2 **Storage Temperature:** Ambient
10.3 **Inert Atmosphere:** No requirement
10.4 **Venting:** Open (flame arrester)

11. HAZARD ASSESSMENT CODE

(See Hazard Assessment Handbook, CG 446-3)

A-T-U

12. HAZARD CLASSIFICATIONS

12.1 **Code of Federal Regulations:** Not listed
12.2 **NAS Hazard Rating for Bulk Water Transportation:**

Category	Rating
Fire	1
Health	
Vapor Irritant	0
Liquid or Solid Irritant	0
Poisons	0
Water Pollution	
Human Toxicity	1
Aquatic Toxicity	3
Aesthetic Effect	3
Reactivity	
Other Chemicals	1
Water	0
Self-Reaction	0

12.3 **NFPA Hazard Classifications:** Not listed

13. PHYSICAL AND CHEMICAL PROPERTIES

13.1 **Physical State at 15°C and 1 atm:** Liquid
13.2 **Molecular Weight:** 240
13.3 **Boiling Point at 1 atm:** 550°F = 288°C = 561°K
13.4 **Freezing Point:** Not pertinent
13.5 **Critical Temperature:** Not pertinent
13.6 **Critical Pressure:** Not pertinent
13.7 **Specific Gravity:** 0.860 at 20°C (liquid)
13.8 **Liquid Surface Tension:** 30.12 dynes/cm = 0.0301 N/m at 20°C
13.9 **Liquid-Water Interfacial Tension:** (est.) 30 dynes/cm = 0.030 N/m at 20°C
13.10 **Vapor (Gas) Specific Gravity:** Not pertinent
13.11 **Ratio of Specific Heats of Vapor (Gas):** Not pertinent
13.12 **Latent Heat of Vaporization:** 150 Btu/lb = 82 cal/g = 3.4×10^5 J/kg
13.13 **Heat of Combustion:** −18,100 Btu/lb = −10,000 cal/g = -418×10^5 J/kg
13.14 **Heat of Decomposition:** Not pertinent
13.15 **Heat of Solution:** Not pertinent
13.16 **Heat of Polymerization:** Not pertinent

NOTES

DCS — DODECYLBENZENESULFONIC ACID, CALCIUM SALT

Common Synonyms			
Calcium alkylaromatic sulfonate Calcium alkylbenzenesulfonate	Liquid	Yellowish-brown	Solvent odor
	May float or sink in water.		

Shut off ignition sources. Call fire department.
Avoid contact with liquid. Keep people away.
Stop discharge if possible.
Isolate and remove discharged material.
Notify local health and pollution control agencies.

Fire

FLAMMABLE.
Containers may explode in fire.
Flashback along vapor trail may occur.
Vapor may explode if ignited in an enclosed area.
Extinguish with dry chemicals, foam or carbon dioxide.
Water may be ineffective on fire.
Cool exposed containers with water.

Exposure

CALL FOR MEDICAL AID.

LIQUID
Irritating to skin and eyes.
If swallowed will cause nausea.
Remove contaminated clothing and shoes.
Flush affected areas with plenty of water.
IF IN EYES, hold eyelids open and flush with plenty of water.
IF SWALLOWED and victim is CONSCIOUS, have victim drink water or milk.
IF SWALLOWED and victim is UNCONSCIOUS OR HAVING CONVULSIONS, do nothing except keep victim warm.
DO NOT INDUCE VOMITING.

Water Pollution

Effect of low concentrations on aquatic life is unknown.
Fouling to shoreline.
May be dangerous if it enters water intakes.
Notify local health and wildlife officials.
Notify operators of nearby water intakes.

1. RESPONSE TO DISCHARGE
(See Response Methods Handbook, CG 446-4)

Issue warning—high flammability
Mechanical containment
Should be removed
Chemical and physical treatment

2. LABEL

Red

3. CHEMICAL DESIGNATIONS

3.1 **Synonyms:** Calcium alkylaromatic sulfonate; Calcium alkylbenzenesulfonate
3.2 **Coast Guard Compatibility Classification:** Not listed
3.3 **Chemical Formula:** $(C_{12}H_{25}C_6H_4SO_3)_2Ca$—solvent
3.4 **IMCO/United Nations Numerical Designation:** Not listed

4. OBSERVABLE CHARACTERISTICS

4.1 **Physical State (as shipped):** Liquid
4.2 **Color:** Amber
4.3 **Odor:** Hydrocarbon solvent

5. HEALTH HAZARDS

5.1 **Personal Protective Equipment:** Goggles or face shield; rubber gloves
5.2 **Symptoms Following Exposure:** Ingestion causes irritation of mouth and stomach. Contact with eyes causes irritation. Has a drying effect on skin; prolonged contact may cause irritation.
5.3 **Treatment for Exposure:** INGESTION: give large amount of water; do NOT induce vomiting. EYES: flush with water for at least 15 min. SKIN: wipe off, flush with water.
5.4 **Toxicity by Inhalation (Threshold Limit Value):** Data not available
5.5 **Short-Term Inhalation Limits:** Data not available
5.6 **Toxicity by Ingestion:** Data not available
5.7 **Late Toxicity:** Data not available
5.8 **Vapor (Gas) Irritant Characteristics:** Data not available
5.9 **Liquid or Solid Irritant Characteristics:** Data not available
5.10 **Odor Threshold:** Data not available

6. FIRE HAZARDS

6.1 **Flash Point:** $<100°F$ C.C.
6.2 **Flammable Limits in Air:** Data not available
6.3 **Fire Extinguishing Agents:** Dry chemical, foam, carbon dioxide
6.4 **Fire Extinguishing Agents Not to be Used:** Water may be ineffective.
6.5 **Special Hazards of Combustion Products:**
6.6 **Behavior in Fire:**
6.7 **Ignition Temperature:** Data not available
6.8 **Electrical Hazard:** Data not available
6.9 **Burning Rate:** $\simeq 4$ mm/min.

7. CHEMICAL REACTIVITY

7.1 **Reactivity with Water:** No reaction
7.2 **Reactivity with Common Materials:**
7.3 **Stability During Transport:** Stable
7.4 **Neutralizing Agents for Acids and Caustics:** Not pertinent
7.5 **Polymerization:** Not pertinent
7.6 **Inhibitor of Polymerization:** Not pertinent

8. WATER POLLUTION

8.1 **Aquatic Toxicity:** Data not available
8.2 **Waterfowl Toxicity:** Data not available
8.3 **Biological Oxygen Demand (BOD):** Data not available
8.4 **Food Chain Concentration Potential:** None

9. SELECTED MANUFACTURERS

1. The Richardson Company
   Organic Chemicals Division
   Melrose Park, Ill. 60160
2. Witco Chemical Corp.
   Organics Division
   277 Park Avenue
   New York, N. Y. 10017
3. ICI United States, Inc.
   Concord Pike and New Murphy Road
   Wilmington, Del. 19899

10. SHIPPING INFORMATION

10.1 **Grades or Purity:** 59—67%; remainder is a hydrocarbon solvent that is combustible or flammable.
10.2 **Storage Temperature:** Ambient
10.3 **Inert Atmosphere:** No requirement
10.4 **Venting:** Open (flame arrester)

11. HAZARD ASSESSMENT CODE
(See Hazard Assessment Handbook, CG 446-3)

A-T-U

12. HAZARD CLASSIFICATIONS

12.1 **Code of Federal Regulations:** Flammable liquid
12.2 **NAS Hazard Rating for Bulk Water Transportation:** Not listed
12.3 **NFPA Hazard Classifications:** Not listed

13. PHYSICAL AND CHEMICAL PROPERTIES

13.1 **Physical State at 15°C and 1 atm:** Liquid
13.2 **Molecular Weight:** Not pertinent (mixture)
13.3 **Boiling Point at 1 atm:** Not pertinent (decomposes)
13.4 **Freezing Point:** Not pertinent
13.5 **Critical Temperature:** Not pertinent
13.6 **Critical Pressure:** Not pertinent
13.7 **Specific Gravity:** 1.04 at 25°C (liquid solution) 0.9 at 25°C (solvent)
13.8 **Liquid Surface Tension:** Data not available
13.9 **Liquid-Water Interfacial Tension:** Data not available
13.10 **Vapor (Gas) Specific Gravity:** Not pertinent
13.11 **Ratio of Specific Heats of Vapor (Gas):** Not pertinent
13.12 **Latent Heat of Vaporization:** Not pertinent
13.13 **Heat of Combustion:** Not pertinent
13.14 **Heat of Decomposition:** Not pertinent
13.15 **Heat of Solution:** Not pertinent
13.16 **Heat of Polymerization:** Not pertinent

NOTES

DAI	DODECYLBENZENESULFONIC ACID, ISOPROPYLAMINE SALT

Common Synonyms		
Isopropylamine dodecylbenzenesulfonate	Solid	Off-white
	May float or sink and mix with water.	

Avoid contact with solid. Keep people away.
Stop discharge if possible.
Isolate and remove discharged material.
Notify local health and pollution control agencies.

Fire	Combustible. Extinguish with dry chemicals, foam or carbon dioxide. Water may be ineffective on fire.
Exposure	CALL FOR MEDICAL AID. SOLID Irritating to skin and eyes. If swallowed will cause nausea and vomiting. Remove contaminated clothing and shoes. Flush affected areas with plenty of water. IF IN EYES, hold eyelids open and flush with plenty of water. IF SWALLOWED and victim is CONSCIOUS, have victim drink water or milk. IF SWALLOWED and victim is UNCONSCIOUS OR HAVING CONVULSIONS, do nothing except keep victim warm.
Water Pollution	Effect of low concentrations on aquatic life is unknown. May be dangerous if it enters water intakes. Notify local health and wildlife officials. Notify operators of nearby water intakes.

1. RESPONSE TO DISCHARGE
(See Response Methods Handbook, CG 446-4)

Disperse and flush

2. LABELS

No hazard label required by Code of Federal Regulations

3. CHEMICAL DESIGNATIONS

3.1 **Synonyms:** Isopropylamine dodecylbenzenesulfonate

3.2 **Coast Guard Compatibility Classification:** Not listed

3.3 **Chemical Formula:** $C_{12}H_{25}C_6H_4SO_3H \cdot NH_2CH(CH_3)_2$

3.4 **IMCO/United Nations Numerical Designation:** Not listed

4. OBSERVABLE CHARACTERISTICS

4.1 **Physical State (as shipped):** Solid, or viscous liquid

4.2 **Color:** Off-white

4.3 **Odor:** Liquid form has sweet, petroleum-type odor.

5. HEALTH HAZARDS

5.1 **Personal Protective Equipment:** Rubber gloves; goggles; mask (for liquid form)

5.2 **Symptoms Following Exposure:** Ingestion causes mild irritation of mouth and stomach. Contact with liquid causes irritation of eyes and (on prolonged contact) mild irritation of skin.

5.3 **Treatment for Exposure:** INGESTION: give large amount of water; consult a doctor if large amount was ingested. EYES: flush with water for at least 15 min.; consult a doctor if irritation persists. SKIN: flush with water.

5.4 **Toxicity by Inhalation (Threshold Limit Value):** Data not available

5.5 **Short-Term Inhalation Limits:** Data not available

5.6 **Toxicity by Ingestion:** Grade 2; LD_{50} 3.54 g/kg (rat)

5.7 **Late Toxicity:** Data not available

5.8 **Vapor (Gas) Irritant Characteristics:** Data not available

5.9 **Liquid or Solid Irritant Characteristics:** Data not available

5.10 **Odor Threshold:** Data not available

6. FIRE HAZARDS

6.1 **Flash Point:** (liquid) > 300°F C.C.

6.2 **Flammable Limits in Air:** Not pertinent

6.3 **Fire Extinguishing Agents:** Water, foam, carbon dioxide, dry chemical

6.4 **Fire Extinguishing Agents Not to be Used:**

6.5 **Special Hazards of Combustion Products:** Toxic oxides of nitrogen may form in fire.

6.6 **Behavior in Fire:**

6.7 **Ignition Temperature:** Data not available

6.8 **Electrical Hazard:** Not pertinent

6.9 **Burning Rate:** Not pertinent

7. CHEMICAL REACTIVITY

7.1 **Reactivity with Water:** No reaction

7.2 **Reactivity with Common Materials:**

7.3 **Stability During Transport:** Stable

7.4 **Neutralizing Agents for Acids and Caustics:** Not pertinent

7.5 **Polymerization:** Not pertinent

7.6 **Inhibitor of Polymerization:** Not pertinent

8. WATER POLLUTION

8.1 **Aquatic Toxicity:** Data not available

8.2 **Waterfowl Toxicity:** Data not available

8.3 **Biological Oxygen Demand (BOD):** Data not available

8.4 **Food Chain Concentration Potential:** None

9. SELECTED MANUFACTURERS

1. The Richardson Company
   100 New Street
   Paterson, N. J. 07501
2. ICI United States, Inc.
   Concord Pike and New Murphy Rd.
   Wilmington, Del. 19899

10. SHIPPING INFORMATION

10.1 **Grades or Purity:** Technical, 96+%; may also be shipped as a concentrated solution in a combustible petroleum solvent.

10.2 **Storage Temperature:** Ambient

10.3 **Inert Atmosphere:** No requirement

10.4 **Venting:** Open

11. HAZARD ASSESSMENT CODE
(See Hazard Assessment Handbook, CG 446-3)

SS

12. HAZARD CLASSIFICATIONS

12.1 **Code of Federal Regulations:** Not listed

12.2 **NAS Hazard Rating for Bulk Water Transportation:** Not listed

12.3 **NFPA Hazard Classifications:** Not listed

13. PHYSICAL AND CHEMICAL PROPERTIES

13.1 **Physical State at 15°C and 1 atm:** Solid

13.2 **Molecular Weight:** 385.5

13.3 **Boiling Point at 1 atm:** Not pertinent (decomposes)

13.4 **Freezing Point:** Not pertinent

13.5 **Critical Temperature:** Not pertinent

13.6 **Critical Pressure:** Not pertinent

13.7 **Specific Gravity:** 1.03 at 20°C (solid) 1.03 at 25°C (liquid)

13.8 **Liquid Surface Tension:** Not pertinent

13.9 **Liquid-Water Interfacial Tension:** Not pertinent

13.10 **Vapor (Gas) Specific Gravity:** Not pertinent

13.11 **Ratio of Specific Heats of Vapor (Gas):** Not pertinent

13.12 **Latent Heat of Vaporization:** Not pertinent

13.13 **Heat of Combustion:** Not pertinent

13.14 **Heat of Decomposition:** Not pertinent

13.15 **Heat of Solution:** Data not available

13.16 **Heat of Polymerization:** Not pertinent

NOTES

DBS	DODECYLBENZENESULFONIC ACID, TRIETHANOLAMINE SALT

Common Synonyms		
Triethanolamine dodecylbenzenesulfonate	Liquid	Yellowish brown
	Sinks and mixes with water.	

Avoid contact with liquid. Keep people away.
Stop discharge if possible.
Isolate and remove discharged material.
Notify local health and pollution control agencies.

Fire	Combustible. Extinguish with water, dry chemicals, foam or carbon dioxide.
Exposure	CALL FOR MEDICAL AID. LIQUID Irritating to skin and eyes. If swallowed will cause nausea and vomiting. Remove contaminated clothing and shoes. Flush affected areas with plenty of water. IF IN EYES, hold eyelids open and flush with plenty of water. IF SWALLOWED and victim is CONSCIOUS, have victim drink water or milk. IF SWALLOWED and victim is UNCONSCIOUS OR HAVING CONVULSIONS, do nothing except keep victim warm.
Water Pollution	Effect of low concentrations on aquatic life is unknown. May be dangerous if it enters water intakes. Notify local health and wildlife officials. Notify operators of nearby water intakes.

1. RESPONSE TO DISCHARGE
(See Response Methods Handbook, CG 446-4)

Disperse and flush

2. LABELS
No hazard label required by Code of Federal Regulations

3. CHEMICAL DESIGNATIONS
3.1 **Synonyms:** Triethanolamine dodecylbenzenesulfonate
3.2 **Coast Guard Compatibility Classification:** Not listed
3.3 **Chemical Formula:** $C_{12}H_{25}C_6H_4SO_3H \cdot N(CH_2CH_2OH)_3 - H_2O$
3.4 **IMCO/United Nations Numerical Designation:** Not listed

4. OBSERVABLE CHARACTERISTICS
4.1 **Physical State (as shipped):** Liquid
4.2 **Color:** Amber
4.3 **Odor:** Data not available

5. HEALTH HAZARDS
5.1 **Personal Protective Equipment:** Rubber gloves; face mask or goggles
5.2 **Symptoms Following Exposure:** Ingestion may cause irritation of mouth and stomach. Contact with eyes or prolonged contact with skin may cause irritation.
5.3 **Treatment for Exposure:** INGESTION: give large amount of water; consult a doctor if large amount was ingested. EYES: flush with water; consult a doctor if irritation persists. SKIN: flush with water.
5.4 **Toxicity by Inhalation (Threshold Limit Value):** Data not available
5.5 **Short-Term Inhalation Limits:** Data not available
5.6 **Toxicity by Ingestion:** Data not available
5.7 **Late Toxicity:** Data not available
5.8 **Vapor (Gas) Irritant Characteristics:** Data not available
5.9 **Liquid or Solid Irritant Characteristics:** Data not available
5.10 **Odor Threshold:** Data not available

6. FIRE HAZARDS
6.1 **Flash Point:** Not pertinent
6.2 **Flammable Limits in Air:** Not pertinent
6.3 **Fire Extinguishing Agents:** Not pertinent
6.4 **Fire Extinguishing Agents Not to be Used:** Not pertinent
6.5 **Special Hazards of Combustion Products:** Toxic oxides of nitrogen and irritating oxides of sulfur may form in fire.
6.6 **Behavior in Fire:**
6.7 **Ignition Temperature:** Not pertinent
6.8 **Electrical Hazard:** Not pertinent
6.9 **Burning Rate:** Not pertinent

7. CHEMICAL REACTIVITY
7.1 **Reactivity with Water:** No reaction
7.2 **Reactivity with Common Materials:**
7.3 **Stability During Transport:** Stable
7.4 **Neutralizing Agents for Acids and Caustics:** Not pertinent
7.5 **Polymerization:** Not pertinent
7.6 **Inhibitor of Polymerization:** Not pertinent

8. WATER POLLUTION
8.1 **Aquatic Toxicity:** Data not available
8.2 **Waterfowl Toxicity:** Data not available
8.3 **Biological Oxygen Demand (BOD):** Data not available
8.4 **Food Chain Concentration Potential:** None

9. SELECTED MANUFACTURERS
1. The Richardson Company
   Organic Chemicals Division
   Melrose Park, Ill. 60160
2. Ashland Chemical Co.
   P. O. Box 2219
   Columbus, Ohio 43216
3. Stepan Chemical Co.
   Edens and Winnetka
   Northfield, Ill. 60093

10. SHIPPING INFORMATION
10.1 **Grades or Purity:** 60% solution in water
10.2 **Storage Temperature:** Ambient
10.3 **Inert Atmosphere:** No requirement
10.4 **Venting:** Open

11. HAZARD ASSESSMENT CODE
(See Hazard Assessment Handbook, CG 446-3)

A-P

12. HAZARD CLASSIFICATIONS
12.1 **Code of Federal Regulations:** Not listed
12.2 **NAS Hazard Rating for Bulk Water Transportation:** Not listed
12.3 **NFPA Hazard Classifications:** Not listed

13. PHYSICAL AND CHEMICAL PROPERTIES
13.1 **Physical State at 15°C and 1 atm:** Liquid
13.2 **Molecular Weight:** 475.6 (solute)
13.3 **Boiling Point at 1 atm:** Not pertinent
13.4 **Freezing Point:** Not pertinent
13.5 **Critical Temperature:** Not pertinent
13.6 **Critical Pressure:** Not pertinent
13.7 **Specific Gravity:** (est.) 1.2 at 20°C (liquid)
13.8 **Liquid Surface Tension:** Data not available
13.9 **Liquid-Water Interfacial Tension:** Not pertinent
13.10 **Vapor (Gas) Specific Gravity:** Not pertinent
13.11 **Ratio of Specific Heats of Vapor (Gas):** Not pertinent
13.12 **Latent Heat of Vaporization:** Not pertinent
13.13 **Heat of Combustion:** Not pertinent
13.14 **Heat of Decomposition:** Not pertinent
13.15 **Heat of Solution:** Not pertinent
13.16 **Heat of Polymerization:** Not pertinent

NOTES

DODECYL SULFATE, DIETHANOLAMINE SALT

Common Synonyms			
Diethanolamine lauryl sulfate solution Lauryl sulfate, diethanolamine salt solution	Liquid	Clear to pale yellow	Mild odor
	May float or sink and mix with water.		

Avoid contact with liquid. Keep people away.
Stop discharge if possible.
Isolate and remove discharged material.
Notify local health and pollution control agencies.

Fire	Not flammable.
Exposure	CALL FOR MEDICAL AID. LIQUID Irritating to skin and eyes. If swallowed will cause nausea and vomiting. Remove contaminated clothing and shoes. Flush affected areas with plenty of water. IF IN EYES, hold eyelids open and flush with plenty of water. IF SWALLOWED and victim is CONSCIOUS, have victim drink water or milk. IF SWALLOWED and victim is UNCONSCIOUS OR HAVING CONVULSIONS, do nothing except keep victim warm.
Water Pollution	Effect of low concentrations on aquatic life is unknown. May be dangerous if it enters water intakes. Notify local health and wildlife officials. Notify operators of nearby water intakes.

1. RESPONSE TO DISCHARGE

(See Response Methods Handbook, CG 446-4)

Disperse and flush

2. LABELS

No hazard label required by Code of Federal Regulations

3. CHEMICAL DESIGNATIONS

3.1 **Synonyms:** Diethanolamine lauryl sulfate solution
Lauryl sulfate, diethanolamine salt solution

3.2 **Coast Guard Compatibility Classification:** Not listed

3.3 **Chemical Formula:** $C_{12}H_{25}OSO_3H \cdot (HOCH_2CH_2)_2NH-H_2O$

3.4 **IMCO/United Nations Numerical Designation:** Not listed

4. OBSERVABLE CHARACTERISTICS

4.1 **Physical State (as shipped):** Liquid

4.2 **Color:** Clear, pale yellow

4.3 **Odor:** Mild fatty

5. HEALTH HAZARDS

5.1 **Personal Protective Equipment:** Chemical goggles or face shield; protective gloves

5.2 **Symptoms Following Exposure:** Ingestion causes mild irritation of stomach. Liquid irritates eyes and causes some corneal damage after prolonged contact. Prolonged contact with skin causes mild irritation.

5.3 **Treatment for Exposure:** INGESTION: consult a doctor if large amount was ingested. EYES: wash well with water; consult a doctor if irritation persists. SKIN: flush with water.

5.4 **Toxicity by Inhalation (Threshold Limit Value):** Data not available

5.5 **Short-Term Inhalation Limits:** Data not available

5.6 **Toxicity by Ingestion:** Data not available

5.7 **Late Toxicity:** Data not available

5.8 **Vapor (Gas) Irritant Characteristics:** Data not available

5.9 **Liquid or Solid Irritant Characteristics:** Data not available

5.10 **Odor Threshold:** Data not available

6. FIRE HAZARDS

6.1 **Flash Point:** Not pertinent

6.2 **Flammable Limits in Air:** Not pertinent

6.3 **Fire Extinguishing Agents:** Not pertinent

6.4 **Fire Extinguishing Agents Not to be Used:** Not pertinent

6.5 **Special Hazards of Combustion Products:** Toxic vapors of diethanolamine and oxides nitrogen may form in fires.

6.6 **Behavior in Fire:**

6.7 **Ignition Temperature:** Not pertinent

6.8 **Electrical Hazard:** Not pertinent

6.9 **Burning Rate:** Not pertinent

7. CHEMICAL REACTIVITY

7.1 **Reactivity with Water:** No reaction

7.2 **Reactivity with Common Materials:**

7.3 **Stability During Transport:** Stable

7.4 **Neutralizing Agents for Acids and Caustics:** Not pertinent

7.5 **Polymerization:** Not pertinent

7.6 **Inhibitor of Polymerization:** Not pertinent

8. WATER POLLUTION

8.1 **Aquatic Toxicity:** Data not available

8.2 **Waterfowl Toxicity:** Data not available

8.3 **Biological Oxygen Demand (BOD):** Data not available

8.4 **Food Chain Concentration Potential:** None

9. SELECTED MANUFACTURERS

1. Henkel Inc.
   Chemical Specialties Div.
   480 Alfred Avenue
   Teaneck, N.J. 07666
2. E. I. duPont de Nemours & Co., Inc.
   Dyes and Chemicals Division
   1007 Market Street
   Wilmington, Del. 19898
3. Ashland Chemical Co.
   P.O. Box 2219
   Columbus, Ohio 43216

10. SHIPPING INFORMATION

10.1 **Grades or Purity:** 35—40% solution in water

10.2 **Storage Temperature:** Below 38°C (100°F)

10.3 **Inert Atmosphere:** No requirement

10.4 **Venting:** Open

11. HAZARD ASSESSMENT CODE

(See Hazard Assessment Handbook, CG 446-3)

A-P

12. HAZARD CLASSIFICATIONS

12.1 **Code of Federal Regulations:** Not listed

12.2 **NAS Hazard Rating for Bulk Water Transportation:** Not listed

12.3 **NFPA Hazard Classifications:** Not listed

13. PHYSICAL AND CHEMICAL PROPERTIES

13.1 **Physical State at 15°C and 1 atm:** Liquid

13.2 **Molecular Weight:** Not pertinent (solution)

13.3 **Boiling Point at 1 atm:** Not pertinent (decomposes)

13.4 **Freezing Point:** Not pertinent

13.5 **Critical Temperature:** Not pertinent

13.6 **Critical Pressure:** Not pertinent

13.7 **Specific Gravity:** 1.01 at 20°C (liquid)

13.8 **Liquid Surface Tension:** Data not available

13.9 **Liquid-Water Interfacial Tension:** Not pertinent

13.10 **Vapor (Gas) Specific Gravity:** Not pertinent

13.11 **Ratio of Specific Heats of Vapor (Gas):** Not pertinent

13.12 **Latent Heat of Vaporization:** Not pertinent

13.13 **Heat of Combustion:** Not pertinent

13.14 **Heat of Decomposition:** Not pertinent

13.15 **Heat of Solution:** Not pertinent

13.16 **Heat of Polymerization:** Not pertinent

NOTES

DSM — DODECYL SULFATE, MAGNESIUM SALT

Common Synonyms			
Magnesium dodecyl sulfate Magnesium lauryl sulfate Lauryl sulfate, magnesium salt	Liquid	Light yellow	Mild odor
	May float or sink and mix with water.		

Avoid contact with liquid. Keep people away.
Stop discharge if possible.
Isolate and remove discharged material.
Notify local health and pollution control agencies.

Fire	Not flammable.
Exposure	CALL FOR MEDICAL AID. LIQUID Irritating to skin and eyes. If swallowed will cause nausea and vomiting. Remove contaminated clothing and shoes. Flush affected areas with plenty of water. IF IN EYES, hold eyelids open and flush with plenty of water. IF SWALLOWED and victim is CONSCIOUS, have victim drink water or milk. IF SWALLOWED and victim is UNCONSCIOUS OR HAVING CONVULSIONS, do nothing except keep victim warm.
Water Pollution	Effect of low concentrations on aquatic life is unknown. May be dangerous if it enters water intakes. Notify local health and wildlife officials. Notify operators of nearby water intakes.

1. RESPONSE TO DISCHARGE

(See Response Methods Handbook, CG 446-4)

Disperse and flush

2. LABELS

No hazard label required by Code of Federal Regulations

3. CHEMICAL DESIGNATIONS

3.1 **Synonyms:**
Lauryl magnesium sulfate
Lauryl sulfate, magnesium salt
Magnesium dodecyl sulfate
Magnesium lauryl sulfate

3.2 **Coast Guard Compatibility Classification:** Not listed

3.3 **Chemical Formula:** $(C_{12}H_{25}OSO_3)_2Mg-H_2O$

3.4 **IMCO/United Nations Numerical Designation:** Not listed

4. OBSERVABLE CHARACTERISTICS

4.1 **Physical State (as shipped):** Liquid

4.2 **Color:** Light yellow

4.3 **Odor:** Bland

5. HEALTH HAZARDS

5.1 **Personal Protective Equipment:** Goggles or face shield; rubber gloves

5.2 **Symptoms Following Exposure:** Ingestion causes mild irritation of stomach. Contact with liquid causes mild irritation of eyes. Skin is mildly irritated with prolonged contact.

5.3 **Treatment for Exposure:** INGESTION: consult a doctor if large amount was ingested. EYES: flush with water; consult a doctor if irritation persists. SKIN: flush with water.

5.4 **Toxicity by Inhalation (Threshold Limit Value):** Data not available

5.5 **Short-Term Inhalation Limits:** Data not available

5.6 **Toxicity by Ingestion:** Data not available

5.7 **Late Toxicity:** Data not available

5.8 **Vapor (Gas) Irritant Characteristics:** Data not available

5.9 **Liquid or Solid Irritant Characteristics:** Data not available

5.10 **Odor Threshold:** Data not available

6. FIRE HAZARDS

6.1 **Flash Point:** Not flammable

6.2 **Flammable Limits in Air:** Not flammable

6.3 **Fire Extinguishing Agents:** Not pertinent

6.4 **Fire Extinguishing Agents Not to be Used:** Not pertinent

6.5 **Special Hazards of Combustion Products:**

6.6 **Behavior in Fire:**

6.7 **Ignition Temperature:** Not pertinent

6.8 **Electrical Hazard:** Not pertinent

6.9 **Burning Rate:** Not pertinent

7. CHEMICAL REACTIVITY

7.1 **Reactivity with Water:** No reaction

7.2 **Reactivity with Common Materials:**

7.3 **Stability During Transport:** Stable

7.4 **Neutralizing Agents for Acids and Caustics:** Not pertinent

7.5 **Polymerization:** Not pertinent

7.6 **Inhibitor of Polymerization:** Not pertinent

8. WATER POLLUTION

8.1 **Aquatic Toxicity:** Data not available

8.2 **Waterfowl Toxicity:** Data not available

8.3 **Biological Oxygen Demand (BOD):** Data not available

8.4 **Food Chain Concentration Potential:** None

9. SELECTED MANUFACTURERS

1. Alcolac
3440 Fairfield Rd.
Baltimore, Md. 21226
2. Surfact-Co., Inc.
14000 South Seeley Avenue
Blue Island, Ill. 60406
3. Stepan Chemical Co.
Edens and Winnetka
Northfield, Ill. 60093

10. SHIPPING INFORMATION

10.1 **Grades or Purity:** 27—30% solution in water

10.2 **Storage Temperature:** Ambient

10.3 **Inert Atmosphere:** No requirement

10.4 **Venting:** Open

11. HAZARD ASSESSMENT CODE

(See Hazard Assessment Handbook, CG 446-3)

A-P

12. HAZARD CLASSIFICATIONS

12.1 **Code of Federal Regulations:** Not listed

12.2 **NAS Hazard Rating for Bulk Water Transportation:** Not listed

12.3 **NFPA Hazard Classifications:** Not listed

13. PHYSICAL AND CHEMICAL PROPERTIES

13.1 **Physical State at 15°C and 1 atm:** Liquid

13.2 **Molecular Weight:** 555 (solute)

13.3 **Boiling Point at 1 atm:** Not pertinent

13.4 **Freezing Point:** Not pertinent

13.5 **Critical Temperature:** Not pertinent

13.6 **Critical Pressure:** Not pertinent

13.7 **Specific Gravity:** 1.04 at 20°C (liquid)

13.8 **Liquid Surface Tension:** Data not available

13.9 **Liquid-Water Interfacial Tension:** Not pertinent

13.10 **Vapor (Gas) Specific Gravity:** Not pertinent

13.11 **Ratio of Specific Heats of Vapor (Gas):** Not pertinent

13.12 **Latent Heat of Vaporization:** Not pertinent

13.13 **Heat of Combustion:** Not pertinent

13.14 **Heat of Decomposition:** Not pertinent

13.15 **Heat of Solution:** Not pertinent

13.16 **Heat of Polymerization:** Not pertinent

NOTES

DDS — DODECYL SULFATE, SODIUM SALT

Common Synonyms			
Sodium lauryl sulfate Lauryl sodium sulfate Sodium dodecyl sulfate	Solid paste or liquid	White to pale yellow	Mild odor
	Sinks and mixes with water.		

Avoid contact with solid and dust. Keep people away.
Stop discharge if possible.
Isolate and remove discharged material.
Notify local health and pollution control agencies.

Fire

Not flammable.

Exposure

CALL FOR MEDICAL AID.

DUST
Irritating to eyes, nose and throat.
If inhaled will cause coughing.

If in eyes, hold eyelids open and flush with plenty of water.
If breathing has stopped, give artificial respiration.
If breathing is difficult, give oxygen.

SOLID OR LIQUID
Will burn eyes.
Irritating to eyes.
If swallowed will cause nausea and vomiting.

Remove contaminated clothing and shoes.
Flush affected areas with plenty of water.
IF IN EYES, hold eyelids open and flush with plenty of water.
IF SWALLOWED and victim is UNCONSCIOUS OR HAVING CONVULSIONS, do nothing except keep victim warm.

Water Pollution

Effect of low concentrations on aquatic life is unknown.
May be dangerous if it enters water intakes.

Notify local health and wildlife officials.
Notify operators of nearby water intakes.

1. RESPONSE TO DISCHARGE
(See Response Methods Handbook, CG 446-4)

Disperse and flush

2. LABELS

No hazard label required by Code of Federal Regulations

3. CHEMICAL DESIGNATIONS

3.1 **Synonyms:** Lauryl sodium sulfate; Lauryl sulfate, sodium salt; Sodium dodecyl sulfate; Sodium lauryl sulfate

3.2 **Coast Guard Compatibility Classification:** Not listed

3.3 **Chemical Formula:** $C_{12}H_{25}OSO_3Na$

3.4 **IMCO/United Nations Numerical Designation:** Not listed

4. OBSERVABLE CHARACTERISTICS

4.1 **Physical State (as shipped):** Solid

4.2 **Color:** White to pale yellow

4.3 **Odor:** Mild fatty

5. HEALTH HAZARDS

5.1 **Personal Protective Equipment:** Protective gloves; dust mask or face shield

5.2 **Symptoms Following Exposure:** Inhalation of dust causes sneezing and coughing. Ingestion of large amounts causes irritation of stomach. Dust irritates eyes and may cause burns on prolonged contact. Contact with skin causes some irritation; continued exposure to water solution causes drying out and cracking.

5.3 **Treatment for Exposure:** INHALATION: remove from exposure. INGESTION: consult a doctor if large amounts have been ingested. EYES: flush well with water; consult doctor if irritation persists. SKIN: flush with water.

5.4 **Toxicity by Inhalation (Threshold Limit Value):** Data not available

5.5 **Short-Term Inhalation Limits:** Data not available

5.6 **Toxicity by Ingestion:** Grade 2; oral LD_{50} = 1 g/kg (rat)

5.7 **Late Toxicity:** Data not available

5.8 **Vapor (Gas) Irritant Characteristics:** Data not available

5.9 **Liquid or Solid Irritant Characteristics:** Data not available

5.10 **Odor Threshold:** Data not available

6. FIRE HAZARDS

6.1 **Flash Point:** Not flammable

6.2 **Flammable Limits in Air:** Not flammable

6.3 **Fire Extinguishing Agents:** Not pertinent

6.4 **Fire Extinguishing Agents Not to be Used:** Not pertinent

6.5 **Special Hazards of Combustion Products:**

6.6 **Behavior in Fire:**

6.7 **Ignition Temperature:** Not pertinent

6.8 **Electrical Hazard:** Not pertinent

6.9 **Burning Rate:** Not pertinent

7. CHEMICAL REACTIVITY

7.1 **Reactivity with Water:** No reaction

7.2 **Reactivity with Common Materials:**

7.3 **Stability During Transport:** Stable

7.4 **Neutralizing Agents for Acids and Caustics:** Not pertinent

7.5 **Polymerization:** Not pertinent

7.6 **Inhibitor of Polymerization:** Not pertinent

8. WATER POLLUTION

8.1 **Aquatic Toxicity:** 5.8 ppm/24 hr/plaice/LC_{50}

8.2 **Waterfowl Toxicity:** Data not available

8.3 **Biological Oxygen Demand (BOD):** 99% of theoretical in 5 days

8.4 **Food Chain Concentration Potential:** None

9. SELECTED MANUFACTURERS

1. Henkel, Inc.
   Chemical Specialties Division
   480 Alfred Avenue
   Teaneck, N. J. 07666
2. E. I. du Pont de Nemours & Co., Inc.
   Dyes and Chemicals Division
   1007 Market Street
   Wilmington, Del. 19899
3. Ashland Chemical Co.
   P. O. Box 2219
   Columbus, Ohio 43216

10. SHIPPING INFORMATION

10.1 **Grades or Purity:** Technical, 89—96%; pharmaceutical grade; also shipped as 28—50% solutions in water.

10.2 **Storage Temperature:** Below 38°C (100°F)

10.3 **Inert Atmosphere:** No requirement

10.4 **Venting:** Open

11. HAZARD ASSESSMENT CODE
(See Hazard Assessment Handbook, CG 446-3)

SS

12. HAZARD CLASSIFICATIONS

12.1 **Code of Federal Regulations:** Not listed

12.2 **NAS Hazard Rating for Bulk Water Transportation:** Not listed

12.3 **NFPA Hazard Classifications:** Not listed

13. PHYSICAL AND CHEMICAL PROPERTIES

13.1 **Physical State at 15°C and 1 atm:** Solid

13.2 **Molecular Weight:** 288

13.3 **Boiling Point at 1 atm:** Not pertinent (decomposes)

13.4 **Freezing Point:** Not pertinent

13.5 **Critical Temperature:** Not pertinent

13.6 **Critical Pressure:** Not pertinent

13.7 **Specific Gravity:** >1.1 at 20°C (solid)

13.8 **Liquid Surface Tension:** Not pertinent

13.9 **Liquid-Water Interfacial Tension:** Not pertinent

13.10 **Vapor (Gas) Specific Gravity:** Not pertinent

13.11 **Ratio of Specific Heats of Vapor (Gas):** Not pertinent

13.12 **Latent Heat of Vaporization:** Not pertinent

13.13 **Heat of Combustion:** Not pertinent

13.14 **Heat of Decomposition:** Not pertinent

13.15 **Heat of Solution:** Not pertinent

13.16 **Heat of Polymerization:** Not pertinent

NOTES

DST DODECYL SULFATE, TRIETHANOLAMINE SALT

Common Synonyms			
Triethanolamine lauryl sulfate Lauryl sulfate, triethanolamine salt	Liquid	Colorless	Mild odor
	Sinks and mixes with water.		

Avoid contact with liquid. Keep people away.
Stop discharge if possible.
Isolate and remove discharged material.
Notify local health and pollution control agencies.

Fire	Not flammable.
Exposure	CALL FOR MEDICAL AID. LIQUID Irritating to skin and eyes. If swallowed will cause nausea and vomiting. Remove contaminated clothing and shoes. Flush affected areas with plenty of water. IF IN EYES, hold eyelids open and flush with plenty of water. IF SWALLOWED and victim is CONSCIOUS, have victim drink water or milk. IF SWALLOWED and victim is UNCONSCIOUS OR HAVING CONVULSIONS, do nothing except keep victim warm.
Water Pollution	Effect of low concentrations on aquatic life is unknown. May be dangerous if it enters water intakes. Notify local health and wildlife officials. Notify operators of nearby water intakes.

1. RESPONSE TO DISCHARGE

(See Response Methods Handbook, CG 446-4)

Disperse and flush

2. LABELS

No hazard label required by Code of Federal Regulations

3. CHEMICAL DESIGNATIONS

3.1 **Synonyms:** Lauryl sulfate, triethanolamine salt; Triethanolamine lauryl sulfate

3.2 **Coast Guard Compatibility Classification:** Not listed

3.3 **Chemical Formula:** $C_{12}H_{25}OSO_3H \cdot (HOCH_2CH_2)_3N-H_2O$

3.4 **IMCO/United Nations Numerical Designation:** Not listed

4. OBSERVABLE CHARACTERISTICS

4.1 **Physical State (as shipped):** Liquid

4.2 **Color:** Water-white

4.3 **Odor:** Mild fatty

5. HEALTH HAZARDS

5.1 **Personal Protective Equipment:** Rubber gloves; goggles or face shield

5.2 **Symptoms Following Exposure:** Ingestion causes mild irritation of stomach. Contact with liquid irritates eyes and causes some corneal damage if prolonged. Skin is mildly irritated on prolonged contact.

5.3 **Treatment for Exposure:** INGESTION: consult a doctor if large amount was ingested. EYES: flush with water; consult a doctor if irritation persists. SKIN: flush with water.

5.4 **Toxicity by Inhalation (Threshold Limit Value):** Data not available

5.5 **Short-Term Inhalation Limits:** Data not available

5.6 **Toxicity by Ingestion:** Data not available

5.7 **Late Toxicity:** Data not available

5.8 **Vapor (Gas) Irritant Characteristics:** Data not available

5.9 **Liquid or Solid Irritant Characteristics:** Data not available

5.10 **Odor Threshold:** Data not available

6. FIRE HAZARDS

6.1 **Flash Point:** Not flammable

6.2 **Flammable Limits in Air:** Not flammable

6.3 **Fire Extinguishing Agents:** Not pertinent

6.4 **Fire Extinguishing Agents Not to be Used:** Not pertinent

6.5 **Special Hazards of Combustion Products:** Toxic vapors of triethanolamine and oxides of nitrogen may form in fire.

6.6 **Behavior in Fire:**

6.7 **Ignition Temperature:** Not pertinent

6.8 **Electrical Hazard:** Not pertinent

6.9 **Burning Rate:** Not pertinent

7. CHEMICAL REACTIVITY

7.1 **Reactivity with Water:** No reaction

7.2 **Reactivity with Common Materials:**

7.3 **Stability During Transport:** Stable

7.4 **Neutralizing Agents for Acids and Caustics:** Not pertinent

7.5 **Polymerization:** Not pertinent

7.6 **Inhibitor of Polymerization:** Not pertinent

8. WATER POLLUTION

8.1 **Aquatic Toxicity:** Data not available

8.2 **Waterfowl Toxicity:** Data not available

8.3 **Biological Oxygen Demand (BOD):** Data not available

8.4 **Food Chain Concentration Potential:** None

9. SELECTED MANUFACTURERS

1. Henkel Inc.
   Chemical Specialties Div.
   480 Alfred Avenue
   Teaneck, N.J. 07666
2. Ashland Chemical Co.
   P.O. Box 2219
   Columbus, Ohio 43216
3. Stepan Chemical Co.
   Edens and Winnetka
   Northfield, Ill. 60093

10. SHIPPING INFORMATION

10.1 **Grades or Purity:** 40-45% solution in water

10.2 **Storage Temperature:** Ambient

10.3 **Inert Atmosphere:** No requirement

10.4 **Venting:** Open

11. HAZARD ASSESSMENT CODE

(See Hazard Assessment Handbook, CG 446-3)

A-P

12. HAZARD CLASSIFICATIONS

12.1 **Code of Federal Regulations:** Not listed

12.2 **NAS Hazard Rating for Bulk Water Transportation:** Not listed

12.3 **NFPA Hazard Classifications:** Not listed

13. PHYSICAL AND CHEMICAL PROPERTIES

13.1 **Physical State at 15°C and 1 atm:** Liquid

13.2 **Molecular Weight:** 415 (solute)

13.3 **Boiling Point at 1 atm:** Not pertinent

13.4 **Freezing Point:** Not pertinent

13.5 **Critical Temperature:** Not pertinent

13.6 **Critical Pressure:** Not pertinent

13.7 **Specific Gravity:** (est.) > 1.1 at 20°C (liquid)

13.8 **Liquid Surface Tension:** Data not available

13.9 **Liquid-Water Interfacial Tension:** Not pertinent

13.10 **Vapor (Gas) Specific Gravity:** Not pertinent

13.11 **Ratio of Specific Heats of Vapor (Gas):** Not pertinent

13.12 **Latent Heat of Vaporization:** Not pertinent

13.13 **Heat of Combustion:** Not pertinent

13.14 **Heat of Decomposition:** Not pertinent

13.15 **Heat of Solution:** Not pertinent

13.16 **Heat of Polymerization:** Not pertinent

NOTES

DTC	# DODECYLTRICHLOROSILANE

Common Synonyms	Liquid	Colorless	Sharp, irritating odor
	Reacts with water. Irritating gas is produced on contact with water.		

Avoid contact with liquid and vapor. Keep people away.
Wear goggles and self-contained breathing apparatus.
Stop discharge if possible.
Call fire department.
Isolate and remove discharged material.
Notify local health and pollution control agencies.

Fire

Combustible.
POISONOUS GASES MAY BE PRODUCED IN FIRE.
Extinguish with dry chemicals or carbon dioxide.
DO NOT USE WATER OR FOAM ON FIRE.

Exposure

Call for medical aid.

VAPOR
Irritating to eyes, nose and throat.
Move victim to fresh air.
If breathing is difficult, give oxygen.

LIQUID
Will burn skin and eyes.
Harmful if swallowed.
Remove contaminated clothing and shoes.
Flush affected areas with plenty of water.
IF IN EYES, hold eyelids open and flush with plenty of water.
IF SWALLOWED and victim is CONSCIOUS, have victim drink water or milk.
DO NOT INDUCE VOMITING.

Water Pollution

Effect of low concentrations on aquatic life is unknown.
May be dangerous if it enters water intakes.
Notify local health and wildlife officials.
Notify operators of nearby water intakes.

1. RESPONSE TO DISCHARGE

(See Response Methods Handbook, CG 446-4)

Issue warning—corrosive, water contaminant
Restrict access
Disperse and flush with care

2. LABEL

3. CHEMICAL DESIGNATIONS

3.1 **Synonyms:** No common synonyms
3.2 **Coast Guard Compatibility Classification:** Not applicable
3.3 **Chemical Formula:** $CH_3(CH_2)_{11}SiCl_3$
3.4 **IMCO/United Nations Numerical Designation:** 8/1771

4. OBSERVABLE CHARACTERISTICS

4.1 **Physical State (as shipped):** Liquid
4.2 **Color:** Colorless
4.3 **Odor:** Sharp, like hydrochloric acid; pungent and irritating

5. HEALTH HAZARDS

5.1 **Personal Protective Equipment:** Acid-vapor type respiratory protection; rubber gloves; chemical worker's goggles; other protective equipment as necessary to protect eyes and skin.
5.2 **Symptoms Following Exposure:** Inhalation irritates mucous membrane. Contact with liquid causes severe burns of eyes and skin. Ingestion causes severe burns of mouth and stomach.
5.3 **Treatment for Exposure:** INHALATION: remove from exposure; support respiration; call physician if needed. EYES: flush with water for 15 min.; obtain medical attention immediately. SKIN: flush with water; obtain medical attention if skin is burned. INGESTION: if victim is conscious, give large amounts of water, then milk or milk of magnesia.
5.4 **Toxicity by Inhalation (Threshold Limit Value):** Data not available
5.5 **Short-Term Inhalation Limits:** Data not available
5.6 **Toxicity by Ingestion:** Grade 3; LD_{50} 50 to 500 mg/kg
5.7 **Late Toxicity:** Data not available
5.8 **Vapor (Gas) Irritant Characteristics:** Vapors cause moderate irritation such that personnel will find high concentrations unpleasant. The effect is temporary.
5.9 **Liquid or Solid Irritant Characteristics:** Severe skin irritant. Causes second- and third-degree burns on short contact and is very injurious to the eyes.
5.10 **Odor Threshold:** Data not available

6. FIRE HAZARDS

6.1 **Flash Point:** > 150°F O.C.
6.2 **Flammable Limits in Air:** Data not available
6.3 **Fire Extinguishing Agents:** Dry chemical, carbon dioxide
6.4 **Fire Extinguishing Agents Not to be Used:** Water, foam
6.5 **Special Hazards of Combustion Products:** Hydrochloric acid and phosgene fumes may form in fires.
6.6 **Behavior in Fire:** Difficult to extinguish; re-ignition may occur. Contact with water applied to adjacent fires produces irritating hydrogen chloride fumes.
6.7 **Ignition Temperature:** Data not available
6.8 **Electrical Hazard:** Data not available
6.9 **Burning Rate:** Data not available

7. CHEMICAL REACTIVITY

7.1 **Reactivity with Water:** Generates hydrogen chloride (hydrochloric acid).
7.2 **Reactivity with Common Materials:** Reacts with surface moisture to generate hydrogen chloride, which is corrosive to common metals.
7.3 **Stability During Transport:** Stable
7.4 **Neutralizing Agents for Acids and Caustics:** Flush with water, rinse with sodium bicarbonate or lime solution.
7.5 **Polymerization:** Not pertinent
7.6 **Inhibitor of Polymerization:** Not pertinent

8. WATER POLLUTION

8.1 **Aquatic Toxicity:** Data not available
8.2 **Waterfowl Toxicity:** Data not available
8.3 **Biological Oxygen Demand (BOD):** Data not available
8.4 **Food Chain Concentration Potential:** None

9. SELECTED MANUFACTURERS

1. Dow Corning Corporation
   P. O. Box 592
   Midland, Mich. 48640
2. PCR, Inc.
   P. O. Box 1466
   Gainesville, Fla. 32602

10. SHIPPING INFORMATION

10.1 **Grades or Purity:** Commercial
10.2 **Storage Temperature:** Ambient
10.3 **Inert Atmosphere:** No requirement
10.4 **Venting:** Pressure-vacuum

11. HAZARD ASSESSMENT CODE

(See Hazard Assessment Handbook, CG 446-3)

A-O

12. HAZARD CLASSIFICATIONS

12.1 **Code of Federal Regulations:** Corrosive liquid
12.2 **NAS Hazard Rating for Bulk Water Transportation:**

Category	Rating
Fire	1
Health	
Vapor Irritant	2
Liquid or Solid Irritant	4
Poisons	3
Water Pollution	
Human Toxicity	3
Aquatic Toxicity	3
Aesthetic Effect	3
Reactivity	
Other Chemicals	3
Water	4
Self-Reaction	1

12.3 **NFPA Hazard Classifications:** Not listed

13. PHYSICAL AND CHEMICAL PROPERTIES

13.1 **Physical State at 15°C and 1 atm:** Liquid
13.2 **Molecular Weight:** 303.7
13.3 **Boiling Point at 1 atm:** > 300°F = > 149°C = > 422°K
13.4 **Freezing Point:** Not pertinent
13.5 **Critical Temperature:** Not pertinent
13.6 **Critical Pressure:** Not pertinent
13.7 **Specific Gravity:** 1.03 at 20°C (liquid)
13.8 **Liquid Surface Tension:** Data not available
13.9 **Liquid-Water Interfacial Tension:** Not pertinent
13.10 **Vapor (Gas) Specific Gravity:** Not pertinent
13.11 **Ratio of Specific Heats of Vapor (Gas):** Not pertinent
13.12 **Latent Heat of Vaporization:** Not pertinent
13.13 **Heat of Combustion:** (est.) −11,000 Btu/lb = −6,200 cal/g = -260×10^5 J/kg
13.14 **Heat of Decomposition:** Not pertinent
13.15 **Heat of Solution:** Data not available
13.16 **Heat of Polymerization:** Not pertinent

NOTES

DTH — DOWTHERM

Common Synonyms	Liquid Light to dark brown Fragrant odor May float or sink in water. Freezing point is 54° F.

Stop discharge if possible.
Call fire department.
Avoid contact with liquid.
Isolate and remove discharged material.
Notify local health and pollution control agencies.

Fire	Combustible. **Wear self-contained breathing apparatus.** **Extinguish with water, dry chemical, foam or carbon dioxide.**
Exposure	**CALL FOR MEDICAL AID.** LIQUID Irritating to skin and eyes. Harmful if swallowed. **Remove contaminated clothing and shoes.** **Flush affected areas with plenty of water.** **IF IN EYES, hold eyelids open and flush with plenty of water.** **IF SWALLOWED and victim is CONSCIOUS, have victim drink water or milk, have victim induce vomiting.** **IF SWALLOWED and victim is UNCONSCIOUS OR HAVING CONVULSIONS, do nothing except keep victim warm.**
Water Pollution	Effect of low concentrations on aquatic life is unknown. May be dangerous if it enters water intakes. **Notify local health and wildlife officials.** **Notify operators of nearby water intakes.**

1. RESPONSE TO DISCHARGE

(See Response Methods Handbook, CG 446-4)

Issue warning—water contaminant
Should be removed
Chemical and physical treatment

2. LABELS

No hazard label required by Code of Federal Regulations

3. CHEMICAL DESIGNATIONS

3.1 **Synonyms:** Biphenyl-diphenyl ether

3.2 **Coast Guard Compatibility Classification:** Aromatic hydrocarbon

3.3 **Chemical Formula:** $(C_6H_5\text{-}O\text{-}C_6H_5)(C_6H_5\text{-}C_6H_5)$

3.4 **IMCO/United Nations Numerical Designation:** Not listed

4. OBSERVABLE CHARACTERISTICS

4.1 **Physical State (as shipped):** Liquid

4.2 **Color:** Straw

4.3 **Odor:** Aromatic; distinctive, objectionable

5. HEALTH HAZARDS

5.1 **Personal Protective Equipment:** Safety glasses

5.2 **Symptoms Following Exposure:** Odor of hot material may cause discomfort. Mildly irritating to eyes and skin.

5.3 **Treatment for Exposure:** INHALATION: if ill effects are experienced, remove to fresh air and get medical attention. INGESTION: no known antidote; treat the symptoms; induce vomiting if large amounts are swallowed and get medical attention. EYES OR SKIN: flush with plenty of water; get medical attention if ill effects develop.

5.4 **Toxicity by Inhalation (Threshold Limit Value):** Data not available

5.5 **Short-Term Inhalation Limits:** Data not available

5.6 **Toxicity by Ingestion:** Grade 2; 0.5 to 5 g/kg

5.7 **Late Toxicity:** Data not available

5.8 **Vapor (Gas) Irritant Characteristics:** Vapors cause a slight smarting of the eyes or respiratory system if present in high concentrations. The effect is temporary.

5.9 **Liquid or Solid Irritant Characteristics:** Minimum hazard. If spilled on clothing and allowed to remain, may cause smarting and reddening of the skin.

5.10 **Odor Threshold:** 0.1-1.0 ppm

6. FIRE HAZARDS

6.1 **Flash Point:** 255°F O.C.

6.2 **Flammable Limits in Air:**
At 500°F: 0.5% – 6.2%
At 300°F: 0.8% – 3.3%

6.3 **Fire Extinguishing Agents:** Water fog, foam, carbon dioxide, dry chemical

6.4 **Fire Extinguishing Agents Not to be Used:** Data not available

6.5 **Special Hazards of Combustion Products:** Irritating gases generated when heated

6.6 **Behavior in Fire:** Not pertinent

6.7 **Ignition Temperature:** 1150°F

6.8 **Electrical Hazard:** Not pertinent

6.9 **Burning Rate:** Data not available

7. CHEMICAL REACTIVITY

7.1 **Reactivity with Water:** No reaction

7.2 **Reactivity with Common Materials:** No reaction

7.3 **Stability During Transport:** Stable

7.4 **Neutralizing Agents for Acids and Caustics:** Not pertinent

7.5 **Polymerization:** Not pertinent

7.6 **Inhibitor of Polymerization:** Not pertinent

8. WATER POLLUTION

8.1 **Aquatic Toxicity:** Data not available

8.2 **Waterfowl Toxicity:** Data not available

8.3 **Biological Oxygen Demand (BOD):** Data not available

8.4 **Food Chain Concentration Potential:** Data not available

9. SELECTED MANUFACTURERS

Dow Chemical Co.
Midland, Mich. 48640

10. SHIPPING INFORMATION

10.1 **Grades or Purity:** 73.5% Diphenyl oxide, 26.5% Diphenyl (eutectic)

10.2 **Storage Temperature:** Ambient

10.3 **Inert Atmosphere:** No requirement

10.4 **Venting:** Open (flame arrester)

11. HAZARD ASSESSMENT CODE

(See Hazard Assessment Handbook, CG 446-3)

A-T-U-X-Y

12. HAZARD CLASSIFICATIONS

12.1 **Code of Federal Regulations:** Not listed

12.2 **NAS Hazard Rating for Bulk Water Transportation:**

Category	Rating
Fire	1
Health	
Vapor Irritant	1
Liquid or Solid Irritant	1
Poisons	1
Water Pollution	
Human Toxicity	0
Aquatic Toxicity	0
Aesthetic Effect	3
Reactivity	
Other Chemicals	0
Water	0
Self-Reaction	0

12.3 **NFPA Hazard Classifications:** Not listed

13. PHYSICAL AND CHEMICAL PROPERTIES

13.1 **Physical State at 15°C and 1 atm:** Liquid

13.2 **Molecular Weight:** 166

13.3 **Boiling Point at 1 atm:** 494°F = 257°C = 530°K

13.4 **Freezing Point:** 54°F = 12°C = 285°K

13.5 **Critical Temperature:** 932°F = 500°C = 773°K

13.6 **Critical Pressure:** 456 psia = 31 atm = 3.1 MN/m^2

13.7 **Specific Gravity:** 1.06 at 21°C (liquid)

13.8 **Liquid Surface Tension:** 40.1 dynes/cm = 0.0401 N/m at 20°C

13.9 **Liquid-Water Interfacial Tension:** (est.) 30 dynes/cm = 0.03 N/m at 20°C

13.10 **Vapor (Gas) Specific Gravity:** Not pertinent

13.11 **Ratio of Specific Heats of Vapor (Gas):** 1.046

13.12 **Latent Heat of Vaporization:** Not pertinent

13.13 **Heat of Combustion:** $-14{,}000$ Btu/lb = -7778 cal/g = -325.6×10^5 J/kg

13.14 **Heat of Decomposition:** Not pertinent

13.15 **Heat of Solution:** Not pertinent

13.16 **Heat of Polymerization:** Not pertinent

NOTES

ENDRIN

Common Synonyms			
1, 2, 3, 4, 10, 10-Hexachloro-6, 7-epoxy-1, 4, 4a, 5, 6, 7, 8, 8a-octahydroendo, endo-1, 4:5, 8-dimethanonapthalene Hexadrin Mendrin	Solid or solution	Colorless to tan	Odorless
	Sinks in water.		

AVOID CONTACT WITH LIQUID, SOLID AND DUST. KEEP PEOPLE AWAY.
Wear goggles, self-contained breathing apparatus, and rubber overclothing (including gloves).
Stop discharge if possible.
Call fire department.
Isolate and remove discharged material.
Notify local health and pollution control agencies.

Fire

Combustible solution or non-flammable solid.
POISONOUS GASES ARE PRODUCED IN FIRE.
Extinguish with dry chemicals, foam or carbon dioxide.
Water may be ineffective on fire.

Exposure

CALL FOR MEDICAL AID.

DUST
POISONOUS IF INHALED OR IF SKIN IS EXPOSED.
Irritating to eyes, nose and throat.
Move victim to fresh air.
If in eyes, hold eyelids open and flush with plenty of water.
If breathing is difficult, give oxygen.

LIQUID OR SOLID
POISONOUS IF SWALLOWED OR IF SKIN IS EXPOSED.
Irritating to skin and eyes.
Remove contaminated clothing and shoes.
Flush affected areas with plenty of water.
IF IN EYES, hold eyelids open and flush with plenty of water.
IF SWALLOWED and victim is CONSCIOUS, have victim drink water or milk and have victim induce vomiting.
IF SWALLOWED and victim is UNCONSCIOUS OR HAVING CONVULSIONS, do nothing except keep victim warm.

Water Pollution

HARMFUL TO AQUATIC LIFE IN VERY LOW CONCENTRATIONS.
May be dangerous if it enters water intakes.
Notify local health and wildlife officials.
Notify operators of nearby water intakes.

1. RESPONSE TO DISCHARGE

(See Response Methods Handbook, CG 446-4)

Issue warning—poison, water contaminant
Restrict access
Should be removed
Chemical and physical treatment

2. LABEL

3. CHEMICAL DESIGNATIONS

3.1 **Synonyms:** 1,2,3,4,10,10-Hexachloro-6,7-epoxy-1,4,4a,5,6,7,8,8a-octahydro-endo, endo-1,4:5,8-dimethanonaphthalene; Hexadrin; Mendrin
3.2 **Coast Guard Compatibility Classification:** Not applicable
3.3 **Chemical Formula:** $C_{12}H_8Cl_6O$
3.4 **IMCO/United Nations Numerical Designation:** 6.1/1615

4. OBSERVABLE CHARACTERISTICS

4.1 **Physical State (as shipped):** Solid (Sometimes shipped as an emulsifiable concentrate in xylene solution)
4.2 **Color:** Colorless to tan
4.3 **Odor:** None

5. HEALTH HAZARDS

5.1 **Personal Protective Equipment:** Respirator for spray, fog, or dust; rubber gloves and boots.
5.2 **Symptoms Following Exposure:** Inhalation causes moderate irritation of nose and throat; prolonged breathing may cause same toxic symptoms as for ingestion. Contact with liquid causes moderate irritation of eyes and skin. Prolonged contact with skin may cause same toxic symptoms as for ingestion. Ingestion causes frothing of the mouth, facial congestion, convulsions, violent muscular contractions, dizziness, weakness, nausea.
5.3 **Treatment for Exposure:** Get medical attention after all exposures to this compound. INHALATION: remove from exposure. EYES: flush with water for at least 15 min. SKIN: wash with plenty of soap and water, but do not scrub. INGESTION: remove from the gastrointestinal tract, either by inducing vomiting (unless hydrocarbon solvents are involved and the amount of insecticide is well below the toxic amount) or by gastric lavage with saline solution; saline cathartics may also be beneficial; fats and oils should be avoided; sedation with barbituates is indicated if signs of CNS irritation are present; patient should have absolute quiet, expert nursing care, and a minimum of external stimuli to reduce danger of convulsions; epinephrine is contraindicated in view of the danger of precipitating ventricular fibrillation; if material ingested was dissolved in a hydrocarbon solvent, observe patient for possible development of hydrocarbon pneumonitis.
5.4 **Toxicity by Inhalation (Threshold Limit Value):** 0.1 mg/m³
5.5 **Short-Term Inhalation Limits:** 0.5 mg/m³ for 30 min.
5.6 **Toxicity by Ingestion:** Grade 4; oral LD_{50} = 3 mg/kg (rat)
5.7 **Late Toxicity:** None known

6. FIRE HAZARDS

6.1 **Flash Point:** Non flammable solid or combustible solution >80°F O.C. (xylene)
6.2 **Flammable Limits in Air:** 1.1%–7% (xylene)
6.3 **Fire Extinguishing Agents:** (Solution) Dry chemical, foam, carbon dioxide
6.4 **Fire Extinguishing Agents Not to be Used:** Water may be ineffective on solution fire.
6.5 **Special Hazards of Combustion Products:** Toxic hydrogen chloride and phosgene may be generated when solution burns.
6.6 **Behavior in Fire:** Not pertinent
6.7 **Ignition Temperature:** Not pertinent
6.8 **Electrical Hazard:** Not pertinent
6.9 **Burning Rate:** 4 mm/min. (xylene)

7. CHEMICAL REACTIVITY

7.1 **Reactivity with Water:** No reaction
7.2 **Reactivity with Common Materials:** No reaction
7.3 **Stability During Transport:** Stable
7.4 **Neutralizing Agents for Acids and Caustics:** Not pertinent
7.5 **Polymerization:** Not pertinent
7.6 **Inhibitor of Polymerization:** Not pertinent

8. WATER POLLUTION

8.1 **Aquatic Toxicity:**
0.005 ppm/48 hr/carp/TL_m/fresh water
0.0025 ppm/48 hr/brown shrimp/TL_m/salt water
8.2 **Waterfowl Toxicity:** LD_{50} = 5.64 mg/kg
8.3 **Biological Oxygen Demand (BOD):** Data not available
8.4 **Food Chain Concentration Potential:** Probable

9. SELECTED MANUFACTURERS

1. Velsicol Chemical Corp.
341 East Ohio St.
Chicago, Ill. 60611
2. Shell Chemical Co.
1 Shell Plaza
Houston, Texas 77002

10. SHIPPING INFORMATION

10.1 **Grades or Purity:** Technical, 95–98%; Dry formulations, up to 75% endrin; liquid formulations, up to 25% in flammable xylene
10.2 **Storage Temperature:** Ambient
10.3 **Inert Atmosphere:** No requirement
10.4 **Venting:** Open

11. HAZARD ASSESSMENT CODE

(See Hazard Assessment Handbook, CG 446-3)

II

12. HAZARD CLASSIFICATIONS

12.1 **Code of Federal Regulations:** Poisonous, Class B
12.2 **NAS Hazard Rating for Bulk Water Transportation:** Not listed
12.3 **NFPA Hazard Classifications:**

Category	Classification Solution	Dry
Health Hazard (Blue)	3	2
Flammability (Red)	1	0
Reactivity (Yellow)	0	0

13. PHYSICAL AND CHEMICAL PROPERTIES

13.1 **Physical State at 15°C and 1 atm:** Solid
13.2 **Molecular Weight:** 380.92
13.3 **Boiling Point at 1 atm:** Not pertinent (decomposes)
13.4 **Freezing Point:** 392°F = 200°C = 573°K
13.5 **Critical Temperature:** Not pertinent
13.6 **Critical Pressure:** Not pertinent
13.7 **Specific Gravity:** 1.65 at 25°C (solid)
13.8 **Liquid Surface Tension:** Not pertinent
13.9 **Liquid-Water Interfacial Tension:** Not pertinent
13.10 **Vapor (Gas) Specific Gravity:** Not pertinent
13.11 **Ratio of Specific Heats of Vapor (Gas):** Not pertinent
13.12 **Latent Heat of Vaporization:** Not pertinent
13.13 **Heat of Combustion:** Not pertinent
13.14 **Heat of Decomposition:** Not pertinent
13.15 **Heat of Solution:** Not pertinent
13.16 **Heat of Polymerization:** Not pertinent

5. HEALTH HAZARDS (Cont'd.)

5.8 **Vapor (Gas) Irritant Characteristics:** Data not available
5.9 **Liquid or Solid Irritant Characteristics:** Data not available
5.10 **Odor Threshold:** Not pertinent (solid)

EPC	# EPICHLOROHYDRIN

Common Synonyms	Watery liquid	Colorless	Sweet garlic odor
1-Chloro-2,3-epoxypropane 3-Chloro-1,2-propylene oxide	Sinks and mixes with water. Poisonous, flammable vapor is produced.		

AVOID CONTACT WITH LIQUID AND VAPOR. Keep people away.
Wear goggles, self-contained breathing apparatus, and rubber overclothing (including gloves).
Stop discharge if possible.
Call fire department.
Stay upwind and use water spray to "knock down" vapor.
Isolate and remove discharged material.
Notify local health and pollution control agencies.

Fire

Combustible.
POISONOUS GASES ARE PRODUCED IN FIRE.
Vapor may explode if ignited in an enclosed area.
Wear goggles, self-contained breathing apparatus, and rubber overclothing (including gloves).
Combat fires from safe distance or protected location.
Extinguish with water, dry chemical, alcohol foam or carbon dioxide.
Cool exposed containers with water.

Exposure

CALL FOR MEDICAL AID.

VAPOR
POISONOUS IF INHALED.
Irritating to eyes.
Move to fresh air.
If breathing has stopped, give artificial respiration.
If breathing is difficult, give oxygen.

LIQUID
POISONOUS IF SWALLOWED.
Will burn skin and eyes.
Remove contaminated clothing and shoes.
Flush affected areas with plenty of water.
IF IN EYES, hold eyelids open and flush with plenty of water.
IF SWALLOWED and victim is CONSCIOUS, have victim drink water or milk, and have victim induce vomiting.
IF SWALLOWED and victim is UNCONSCIOUS OR HAVING CONVULSIONS, do nothing except keep victim warm.

Water Pollution

Effect of low concentrations on aquatic life is unknown.
May be dangerous if it enters water intakes.
Notify local health and wildlife officials.
Notify operators of nearby water intakes.

1. RESPONSE TO DISCHARGE

(See Response Methods Handbook, CG 446-4)

Issue warning—water contaminant
Restrict access
Disperse and flush

2. LABEL

3. CHEMICAL DESIGNATIONS

3.1 **Synonyms:** 1-Chloro-2, 3-epoxypropane; Chloromethyloxirane; gamma-Chloropropylene oxide; 3-Chloro-1, 2-propylene oxide
3.2 **Coast Guard Compatibility Classification:** Epichlorohydrin
3.3 **Chemical Formula:** OCH_2CHCH_2Cl
3.4 **IMCO/United Nations Numerical Designation:** 6.1/2023

4. OBSERVABLE CHARACTERISTICS

4.1 **Physical State (as shipped):** Liquid
4.2 **Color:** Colorless
4.3 **Odor:** Pungent, garlic; sweet, pungent; like chloroform

5. HEALTH HAZARDS

5.1 **Personal Protective Equipment:** Air pack or organic canister mask; protective gloves and goggles.
5.2 **Symptoms Following Exposure:** Vapor is irritating to eyes, nose, and throat; may cause headache, nausea, vomiting, and central nervous system depression. Rapidly fatal if swallowed, i.e., nausea, vomiting, and collapse. Skin contact is irritating.
5.3 **Treatment for Exposure:** INHALATION: remove victim to fresh air, keep him warm and quiet, and get medical attention immediately; if breathing stops, start artificial respiration. INGESTION: induce vomiting (but only if victim is conscious and without convulsions) and call a physician promptly; no specific antidote known. EYES OR SKIN: immediately flush with water for at least 15 min. and get medical attention; remove contaminated clothing and wash before reuse.
5.4 **Toxicity by Inhalation (Threshold Limit Value):** 5 ppm
5.5 **Short-Term Inhalation Limits:** 10 ppm for 30 min.
5.6 **Toxicity by Ingestion:** Grade 3; LD_{50} 50 to 500 mg/kg
5.7 **Late Toxicity:** Causes cancer in experimental animals.
5.8 **Vapor (Gas) Irritant Characteristics:** Vapor is moderately irritating such that personnel will not usually tolerate moderate or high vapor concentrations.
5.9 **Liquid or Solid Irritant Characteristics:** Fairly severe skin irritant. May cause pain and second-degree burns after a few minutes' contact.
5.10 **Odor Threshold:** 10 ppm

6. FIRE HAZARDS

6.1 **Flash Point:** 92°F O.C.; 100°F C.C.
6.2 **Flammable Limits in Air:** 3.8%—21.0%
6.3 **Fire Extinguishing Agents:** Alcohol foam, dry chemical, carbon dioxide, water spray
6.4 **Fire Extinguishing Agents Not to be Used:** Avoid use of dry chemical if fire occurs in container with confined vent.
6.5 **Special Hazards of Combustion Products:** Toxic, irritating vapors are generated when heated.
6.6 **Behavior in Fire:** Containers may explode in fire because of polymerization.
6.7 **Ignition Temperature:** 804°F
6.8 **Electrical Hazard:** Not pertinent
6.9 **Burning Rate:** 2.6 mm/min.

7. CHEMICAL REACTIVITY

7.1 **Reactivity with Water:** Mild reaction; not likely to be hazardous.
7.2 **Reactivity with Common Materials:** No reaction
7.3 **Stability During Transport:** Stable
7.4 **Neutralizing Agents for Acids and Caustics:** Not pertinent
7.5 **Polymerization:** Can polymerize in presence of strong acids and bases, particularly when hot.
7.6 **Inhibitor of Polymerization:** None used

8. WATER POLLUTION

8.1 **Aquatic Toxicity:** 36 ppm/48 hrs/Rasbora (fish)/TL_m/freshwater
8.2 **Waterfowl Toxicity:** Data not available
8.3 **Biological Oxygen Demand (BOD):** Data not available
8.4 **Food Chain Concentration Potential:** None

9. SELECTED MANUFACTURERS

1. Dow Chemical Co.
Midland, Mich. 48640
2. Shell Chemical Co.
Industrial Chemicals Division
Houston, Texas 77001
3. Union Carbide Corp.
Chemicals and Plastics Division
270 Park Ave.
New York, N. Y. 10017

10. SHIPPING INFORMATION

10.1 **Grades or Purity:** 99.0%
10.2 **Storage Temperature:** Ambient
10.3 **Inert Atmosphere:** No requirement
10.4 **Venting:** Pressure-vacuum

11. HAZARD ASSESSMENT CODE

(See Hazard Assessment Handbook, CG 446-3)

A-P-Q

12. HAZARD CLASSIFICATIONS

12.1 **Code of Federal Regulations:** Poison, Class B
12.2 **NAS Hazard Rating for Bulk Water Transportation:**

Category	Rating
Fire	3
Health	
Vapor Irritant	3
Liquid or Solid Irritant	3
Poisons	4
Water Pollution	
Human Toxicity	3
Aquatic Toxicity	3
Aesthetic Effect	2
Reactivity	
Other Chemicals	3
Water	1
Self-Reaction	2

12.3 **NFPA Hazard Classifications:**

Category	Classification
Health Hazard (Blue)	3
Flammability (Red)	3
Reactivity (Yellow)	2

13. PHYSICAL AND CHEMICAL PROPERTIES

13.1 **Physical State at 15°C and 1 atm:** Liquid
13.2 **Molecular Weight:** 92.53
13.3 **Boiling Point at 1 atm:** 239.4°F = 115.2°C = 388.4°K
13.4 **Freezing Point:** −72.6°F = −58.1°C = 215.1°K
13.5 **Critical Temperature:** Not pertinent
13.6 **Critical Pressure:** Not pertinent
13.7 **Specific Gravity:** 1.18 at 20°C (liquid)
13.8 **Liquid Surface Tension:** 37.0 dynes/cm = 0.037 N/m at 20°C
13.9 **Liquid-Water Interfacial Tension:** Not pertinent
13.10 **Vapor (Gas) Specific Gravity:** Not pertinent
13.11 **Ratio of Specific Heats of Vapor (Gas):** 1.155
13.12 **Latent Heat of Vaporization:** 176 Btu/lb = 97.9 cal/g = 4.10×10^5 J/kg
13.13 **Heat of Combustion:** −8143 Btu/lb = −4524 cal/g = -189.4×10^5 J/kg
13.14 **Heat of Decomposition:** Not pertinent
13.15 **Heat of Solution:** Not pertinent
13.16 **Heat of Polymerization:** Not pertinent

NOTES

REVISED 1978

EVO	EPOXIDIZED VEGETABLE OILS

Common Synonyms			
Epoxidized drying oils Epoxidized oils	Oily liquid	Pale yellow	Odorless
	Floats on water.		

Stop discharge if possible.
Call fire department.
Isolate and remove discharged material.
Notify local health and pollution control agencies.

Fire	Combustible. Extinguish with foam, dry chemical or carbon dioxide.
Exposure	Not harmful.
Water Pollution	Effect of low concentrations on aquatic life is unknown. Fouling to shoreline. May be dangerous if it enters water intakes. Notify local health and wildlife officials. Notify operators of nearby water intakes.

1. RESPONSE TO DISCHARGE
(See Response Methods Handbook, CG 446-4)

Mechanical containment
Chemical and physical treatment

2. LABELS

No hazard label required by Code of Federal Regulations

3. CHEMICAL DESIGNATIONS

3.1 **Synonyms:** Drying oil epoxides; Epoxidized drying oils; Epoxidized oils
3.2 **Coast Guard Compatibility Classification:** Glycol ethers
3.3 **Chemical Formula:** $(O{\cdot}CRH{\cdot}CHR'COO)_3C_3H_5$
3.4 **IMCO/United Nations Numerical Designation:** Not listed

4. OBSERVABLE CHARACTERISTICS

4.1 **Physical State (as shipped):** Liquid
4.2 **Color:** Colorless to yellow
4.3 **Odor:** Very weak

5. HEALTH HAZARDS

5.1 **Personal Protective Equipment:** Data not available
5.2 **Symptoms Following Exposure:** Data not available
5.3 **Treatment for Exposure:** Data not available
5.4 **Toxicity by Inhalation (Threshold Limit Value):** Not pertinent
5.5 **Short-Term Inhalation Limits:** Not pertinent
5.6 **Toxicity by Ingestion:** Grade 0; LD_{50} above 15 g/kg (rat)
5.7 **Late Toxicity:** Data not available
5.8 **Vapor (Gas) Irritant Characteristics:** None expected
5.9 **Liquid or Solid Irritant Characteristics:** Data not available
5.10 **Odor Threshold:** Not pertinent

6. FIRE HAZARDS

6.1 **Flash Point:** 585°F O.C.
6.2 **Flammable Limits in Air:** Not pertinent
6.3 **Fire Extinguishing Agents:** Foam, dry chemical, carbon dioxide
6.4 **Fire Extinguishing Agents Not to be Used:** Not pertinent
6.5 **Special Hazards of Combustion Products:** Not pertinent
6.6 **Behavior in Fire:** Not pertinent
6.7 **Ignition Temperature:** Data not available
6.8 **Electrical Hazard:** Not pertinent
6.9 **Burning Rate:** Data not available

7. CHEMICAL REACTIVITY

7.1 **Reactivity with Water:** No reaction
7.2 **Reactivity with Common Materials:** No reaction
7.3 **Stability During Transport:** Stable
7.4 **Neutralizing Agents for Acids and Caustics:** Not pertinent
7.5 **Polymerization:** Not pertinent
7.6 **Inhibitor of Polymerization:** Not pertinent

8. WATER POLLUTION

8.1 **Aquatic Toxicity:** 240 ppm/24 hr/brine shrimp/TL_m
8.2 **Waterfowl Toxicity:**
8.3 **Biological Oxygen Demand (BOD):** 4% of theoretical in 5 days, fresh water, acclimated seed
8.4 **Food Chain Concentration Potential:**

9. SELECTED MANUFACTURERS

1. Ashland Oil, Inc.
   Chemical Products Division
   Columbus, Ohio 43216
2. Swift & Co.
   Swift Agricultural Chemicals Corp.
   Wilmington, N. C. 28401
3. Witco Chemical Corp.
   Argus Chemical Corp. Division
   633 Court St.
   Brooklyn, N. Y. 11231

10. SHIPPING INFORMATION

10.1 **Grades or Purity:** Epoxidized vegetable oil; Epoxidized soybean oil
10.2 **Storage Temperature:** Ambient
10.3 **Inert Atmosphere:** No requirement
10.4 **Venting:** Open

11. HAZARD ASSESSMENT CODE
(See Hazard Assessment Handbook, CG 446-3)

A-T-U

12. HAZARD CLASSIFICATIONS

12.1 **Code of Federal Regulations:** Not listed
12.2 **NAS Hazard Rating for Bulk Water Transportation:** Not listed
12.3 **NFPA Hazard Classifications:** Not listed

13. PHYSICAL AND CHEMICAL PROPERTIES

13.1 **Physical State at 15°C and 1 atm:** Liquid
13.2 **Molecular Weight:** Not pertinent
13.3 **Boiling Point at 1 atm:** Very high
13.4 **Freezing Point:** Not pertinent
13.5 **Critical Temperature:** Not pertinent
13.6 **Critical Pressure:** Not pertinent
13.7 **Specific Gravity:** 1.0 at 20°C (liquid)
13.8 **Liquid Surface Tension:** 36.2 dynes/cm = 0.0362 N/m at 24°C
13.9 **Liquid-Water Interfacial Tension:** 50 dynes/cm = 0.05 N/m at 22.7°C
13.10 **Vapor (Gas) Specific Gravity:** Not pertinent
13.11 **Ratio of Specific Heats of Vapor (Gas):** Not pertinent
13.12 **Latent Heat of Vaporization:** Not pertinent
13.13 **Heat of Combustion:** (est.) $-13{,}000$ Btu/lb $= -7{,}000$ cal/g $= -300 \times 10^5$ J/kg
13.14 **Heat of Decomposition:** Not pertinent
13.15 **Heat of Solution:** Not pertinent
13.16 **Heat of Polymerization:** Not pertinent

NOTES

REVISED 1978

ETHANE

ETH

Common Synonyms

Liquefied compressed gas — Colorless — Mild gasoline-like odor

Floats and boils on water. Flammable visible vapor cloud is produced.

Stop discharge if possible. Keep people away.
Shut off ignition sources and call fire department.
Stay upwind and use water spray to "knock down" vapor.
Avoid contact with liquid.
Notify local health and pollution control agencies.

Fire

FLAMMABLE.
Flashback along vapor trail may occur.
Vapor may explode if ignited in an enclosed area.

Stop flow of gas if possible.
Cool exposed containers and protect men effecting shutoff with water.
Let fire burn.

Exposure

CALL FOR MEDICAL AID.

VAPOR
If inhaled will cause difficult breathing.
Not irritating to eyes, nose or throat.

Move to fresh air.
If breathing has stopped, give artificial respiration.
If breathing is difficult, give oxygen.

LIQUID
Will cause frostbite.

Flush affected areas with plenty of water.
DO NOT RUB AFFECTED AREAS.

Water Pollution

Not harmful.

1. RESPONSE TO DISCHARGE

(See Response Methods Handbook, CG 446-4)

Issue warning—high flammability
Evacuate area

2. LABEL

Red

3. CHEMICAL DESIGNATIONS

3.1 **Synonyms:** Methylmethane
3.2 **Coast Guard Compatibility Classification:** Paraffin
3.3 **Chemical Formula:** C_2H_6
3.4 **IMCO/United Nations Numerical Designation:** 2.0/1035

4. OBSERVABLE CHARACTERISTICS

4.1 **Physical State (as shipped):** Liquid or compressed gas
4.2 **Color:** Colorless
4.3 **Odor:** Weak, sweetish

5. HEALTH HAZARDS

5.1 **Personal Protective Equipment:** Self-contained breathing apparatus for high vapor concentrations.
5.2 **Symptoms Following Exposure:** In high vapor concentrations, can act as simple asphyxiant. Liquid causes severe frostbite.
5.3 **Treatment for Exposure:** Remove from exposure; support respiration.
5.4 **Toxicity by Inhalation (Threshold Limit Value):** Not pertinent
5.5 **Short-Term Inhalation Limits:** Not pertinent
5.6 **Toxicity by Ingestion:** Not pertinent
5.7 **Late Toxicity:** None
5.8 **Vapor (Gas) Irritant Characteristics:** Vapors are nonirritating to the eyes and throat.
5.9 **Liquid or Solid Irritant Characteristics:** No appreciable hazard. Practically harmless to the skin because it is very volatile and evaporates quickly.
5.10 **Odor Threshold:** 899 ppm

6. FIRE HAZARDS

6.1 **Flash Point:** −211°F
6.2 **Flammable Limits in Air:** 2.9%—13.0%
6.3 **Fire Extinguishing Agents:** Stop flow of gas
6.4 **Fire Extinguishing Agents Not to be Used:** Data not available
6.5 **Special Hazards of Combustion Products:** Not pertinent
6.6 **Behavior in Fire:** Not pertinent
6.7 **Ignition Temperature:** 940°F
6.8 **Electrical Hazard:** Class I, group D
6.9 **Burning Rate:** 7.3 mm/min.

7. CHEMICAL REACTIVITY

7.1 **Reactivity with Water:** No reaction
7.2 **Reactivity with Common Materials:** No reaction
7.3 **Stability During Transport:** Stable
7.4 **Neutralizing Agents for Acids and Caustics:** Not pertinent
7.5 **Polymerization:** Not pertinent
7.6 **Inhibitor of Polymerization:** Not pertinent

8. WATER POLLUTION

8.1 **Aquatic Toxicity:** None
8.2 **Waterfowl Toxicity:** None
8.3 **Biological Oxygen Demand (BOD):** None
8.4 **Food Chain Concentration Potential:** None

9. SELECTED MANUFACTURERS

1. Atlantic Richfield Co.
ARCO Chemical Co. Division
260 South Broad St.
Philadelphia, Pa. 19101
2. Cities Service Co., Inc.
Petrochemicals Division
60 Wall St.
New York, N. Y. 10005
3. Phillips Petroleum Co.
Bartlesville, Okla. 74004

10. SHIPPING INFORMATION

10.1 **Grades or Purity:** Research; pure
10.2 **Storage Temperature:** −128°F
10.3 **Inert Atmosphere:** No requirement
10.4 **Venting:** Safety relief

11. HAZARD ASSESSMENT CODE

(See Hazard Assessment Handbook, CG 446-3)

A-B-C-D-E-F-G

12. HAZARD CLASSIFICATIONS

12.1 **Code of Federal Regulations:** Flammable compressed gas
12.2 **NAS Hazard Rating for Bulk Water Transportation:** Not listed
12.3 **NFPA Hazard Classifications:** Not listed

13. PHYSICAL AND CHEMICAL PROPERTIES

13.1 **Physical State at 15°C and 1 atm:** Gas
13.2 **Molecular Weight:** 30.07
13.3 **Boiling Point at 1 atm:** −127.5°F = −88.6°C = 264.6°K
13.4 **Freezing Point:** −279.9°F = −183.3°C = 89.9°K
13.5 **Critical Temperature:** 90.1°F = 32.3°C = 305.5°K
13.6 **Critical Pressure:** 708.0 psia = 48.16 atm = 4.879 MN/m^2
13.7 **Specific Gravity:** 0.546 at −88.6°C (liquid)
13.8 **Liquid Surface Tension:** 16 dynes/cm = 0.016 N/m at −88°C
13.9 **Liquid-Water Interfacial Tension:** (est.) 45 dynes/cm = 0.045 N/m at −88°C
13.10 **Vapor (Gas) Specific Gravity:** 1.1
13.11 **Ratio of Specific Heats of Vapor (Gas):** 1.191
13.12 **Latent Heat of Vaporization:** 211 Btu/lb = 117 cal/g = 4.90 × 10^5 J/kg
13.13 **Heat of Combustion:** −20,293 Btu/lb = −11,274 cal/g = −472.02 × 10^5 J/kg
13.14 **Heat of Decomposition:** Not pertinent
13.15 **Heat of Solution:** Not pertinent
13.16 **Heat of Polymerization:** Not pertinent

NOTES

REVISED 1978

ETHOXYDIHYDROPYRAN

Common Synonyms		
2-Ethoxy-3,4-dihydro-2H-pyran	Liquid	Colorless
	Floats on water.	

Stop discharge if possible. Keep people away.
Shut off ignition sources. Call fire department.
Avoid contact with liquid.
Isolate and remove discharged material.
Notify local health and pollution control agencies.

Fire	Combustible. Extinguish with dry chemicals, foam or carbon dioxide. Water may be ineffective on fire.
Exposure	Call for medical aid. Exposure data not available.
Water Pollution	Effect of low concentrations on aquatic life is unknown. Fouling to shoreline. May be dangerous if it enters water intakes. Notify local health and wildlife officials. Notify operators of nearby water intakes.

1. RESPONSE TO DISCHARGE
(See Response Methods Handbook, CG 446-4)

Mechanical containment
Should be removed
Chemical and physical treatment

2. LABELS

No label required by Code of Federal Regulations

3. CHEMICAL DESIGNATIONS

3.1 **Synonyms:** 2-Ethoxy-3,4-dihydro-2H-pyran
3.2 **Coast Guard Compatibility Classification:** Not applicable
3.3 **Chemical Formula:** $OCH{=}CHCH_2CH_2CHOC_2H_5$ (ring: O bonded to the CH carrying OC_2H_5)
3.4 **IMCO/United Nations Numerical Designation:** Not listed

4. OBSERVABLE CHARACTERISTICS

4.1 **Physical State (as shipped):** Liquid
4.2 **Color:** Colorless
4.3 **Odor:** Data not available

5. HEALTH HAZARDS

5.1 **Personal Protective Equipment:** Goggles or face shield; rubber gloves.
5.2 **Symptoms Following Exposure:** Contact with liquid irritates eyes. Also irritates skin on prolonged contact.
5.3 **Treatment for Exposure:** EYES: flush with water for at least 15 min. SKIN: wipe off; wash well with soap and water. INGESTION: induce vomiting; get medical attention.
5.4 **Toxicity by Inhalation (Threshold Limit Value):** Data not available
5.5 **Short-Term Inhalation Limits:** Not pertinent
5.6 **Toxicity by Ingestion:** Data not available
5.7 **Late Toxicity:** Data not available
5.8 **Vapor (Gas) Irritant Characteristics:** Data not available
5.9 **Liquid or Solid Irritant Characteristics:** Data not available
5.10 **Odor Threshold:** Data not available

6. FIRE HAZARDS

6.1 **Flash Point:** 98°F O.C.
6.2 **Flammable Limits in Air:** Data not available
6.3 **Fire Extinguishing Agents:** Foam, dry chemical, carbon dioxide
6.4 **Fire Extinguishing Agents Not to be Used:** Water may be ineffective.
6.5 **Special Hazards of Combustion Products:** Not pertinent
6.6 **Behavior in Fire:** Not pertinent
6.7 **Ignition Temperature:** Data not available
6.8 **Electrical Hazard:** Data not available
6.9 **Burning Rate:** 4.8 mm/min.

7. CHEMICAL REACTIVITY

7.1 **Reactivity with Water:** No reaction
7.2 **Reactivity with Common Materials:** No reaction
7.3 **Stability During Transport:** Stable
7.4 **Neutralizing Agents for Acids and Caustics:** Not pertinent
7.5 **Polymerization:** Not pertinent
7.6 **Inhibitor of Polymerization:** Not pertinent

8. WATER POLLUTION

8.1 **Aquatic Toxicity:** Data not available
8.2 **Waterfowl Toxicity:** Data not available
8.3 **Biological Oxygen Demand (BOD):** Data not available
8.4 **Food Chain Concentration Potential:** None

9. SELECTED MANUFACTURERS

1. Union Carbide Corporation
   Chemicals and Plastics Division
   270 Park Avenue
   New York, N. Y. 10017
2. Guardian Chemical Corporation
   Eastern Chemical Division
   P. O. Box 2500-B
   Hauppauge, N. Y. 11787

10. SHIPPING INFORMATION

10.1 **Grades or Purity:** Commercial
10.2 **Storage Temperature:** Ambient
10.3 **Inert Atmosphere:** No requirement
10.4 **Venting:** Open (flame arrester)

11. HAZARD ASSESSMENT CODE
(See Hazard Assessment Handbook, CG 446-3)

A-T-U

12. HAZARD CLASSIFICATIONS

12.1 **Code of Federal Regulations:** Not listed
12.2 **NAS Hazard Rating for Bulk Water Transportation:** Not listed
12.3 **NFPA Hazard Classifications:**

Category	Classification
Health Hazard (Blue)	2
Flammability (Red)	2
Reactivity (Yellow)	1

13. PHYSICAL AND CHEMICAL PROPERTIES

13.1 **Physical State at 15°C and 1 atm:** Liquid
13.2 **Molecular Weight:** 128.17
13.3 **Boiling Point at 1 atm:** 289°F = 143°C = 416°K
13.4 **Freezing Point:** −148°F = −100°C = 173°K
13.5 **Critical Temperature:** Not pertinent
13.6 **Critical Pressure:** Not pertinent
13.7 **Specific Gravity:** 0.875 at 20°C (liquid)
13.8 **Liquid Surface Tension:** (est.) 25 dynes/cm = 0.025 N/m at 20°C
13.9 **Liquid-Water Interfacial Tension:** (est.) 30 dynes/cm = 0.030 N/m at 20°C
13.10 **Vapor (Gas) Specific Gravity:** Not pertinent
13.11 **Ratio of Specific Heats of Vapor (Gas):** Not pertinent
13.12 **Latent Heat of Vaporization:** (est.) 120 Btu/lb = 69 cal/g = 2.9×10^5 J/kg
13.13 **Heat of Combustion:** (est.) −14,000 Btu/lb = −7,900 cal/g = -330×10^5 J/kg
13.14 **Heat of Decomposition:** Not pertinent
13.15 **Heat of Solution:** Not pertinent
13.16 **Heat of Polymerization:** Not pertinent

NOTES

EOD ETHOXYLATED DODECANOL

Common Synonyms
Ethoxylated dodecyl alcohol
Ethoxylated lauryl alcohol
Tergitol nonionic TMN

Oily liquid | Colorless to yellow | Pleasant odor

Mixes slowly with water. Freezing point is 61° F.

Avoid contact with liquid.
Stop discharge if possible.
Call fire department.
Isolate and remove discharged material.
Notify local health and pollution control agencies.

Fire

Combustible.
Extinguish with water, foam, dry chemical, or carbon dioxide.

Exposure

CALL FOR MEDICAL AID.

LIQUID
Irritating to skin.
Will burn eyes.
Harmful if swallowed.
Flush affected areas with plenty of water.
IF IN EYES, hold eyelids open and flush with plenty of water.
IF SWALLOWED and victim is CONSCIOUS, have victim drink water or milk.

Water Pollution

Effect of low concentrations on aquatic life is unknown.
May be dangerous if it enters water intakes.
Notify local health and wildlife officials.
Notify operators of nearby water intakes.

1. RESPONSE TO DISCHARGE

(See Response Methods Handbook, CG 446-4)

Disperse and flush

2. LABELS

No hazard label required by Code of Federal Regulations

3. CHEMICAL DESIGNATIONS

3.1 **Synonyms:**
Ethoxylated dodecyl alcohol
Ethoxylated lauryl alcohol
Poly(oxyethyl) dodecyl ether
Poly(oxyethyl) lauryl ether
Tergitol Nonionic TMN

3.2 **Coast Guard Compatibility Classification:** Alcohol

(Continued on page 4)

4. OBSERVABLE CHARACTERISTICS

4.1 **Physical State (as shipped):** Liquid
4.2 **Color:** Colorless
4.3 **Odor:** Mild and pleasant

5. HEALTH HAZARDS

5.1 **Personal Protective Equipment:** Plastic gloves, goggles.
5.2 **Symptoms Following Exposure:** Liquid causes eye injury and de-fats the skin, causing irritation.
5.3 **Treatment for Exposure:** Flush eyes with water for at least 15 min. Wash skin well with water. Get medical attention.
5.4 **Toxicity by Inhalation (Threshold Limit Value):** Not pertinent
5.5 **Short-Term Inhalation Limits:** Not pertinent
5.6 **Toxicity by Ingestion:** Grade 1; 5-15 g/kg (rat)
5.7 **Late Toxicity:** Data not available
5.8 **Vapor (Gas) Irritant Characteristics:** None
5.9 **Liquid or Solid Irritant Characteristics:** Liquid causes eye injury. Contact with skin may cause irritation.
5.10 **Odor Threshold:** Not pertinent

6. FIRE HAZARDS

6.1 **Flash Point:** 470°F O.C.
6.2 **Flammable Limits in Air:** Not pertinent
6.3 **Fire Extinguishing Agents:** Dry chemical, carbon dioxide, or alcohol foam.
6.4 **Fire Extinguishing Agents Not to be Used:** Not pertinent
6.5 **Special Hazards of Combustion Products:** Not pertinent
6.6 **Behavior in Fire:** Not pertinent
6.7 **Ignition Temperature:** Data not available
6.8 **Electrical Hazard:** Not pertinent
6.9 **Burning Rate:** Data not available

7. CHEMICAL REACTIVITY

7.1 **Reactivity with Water:** No reaction
7.2 **Reactivity with Common Materials:** No reaction
7.3 **Stability During Transport:** Stable
7.4 **Neutralizing Agents for Acids and Caustics:** Not pertinent
7.5 **Polymerization:** Not pertinent
7.6 **Inhibitor of Polymerization:** Not pertinent

8. WATER POLLUTION

8.1 **Aquatic Toxicity:** Data not available
8.2 **Waterfowl Toxicity:** Data not available
8.3 **Biological Oxygen Demand (BOD):** Data not available
8.4 **Food Chain Concentration Potential:** None

9. SELECTED MANUFACTURERS

1. BASF Wyandotte Corp.
Industrial Chemicals Group
Wyandotte, Mich. 48192
2. Jefferson Chemical Co.
3336 Richmond Ave.
Houston, Tex. 77052
3. Union Carbide Corp.
Chemicals and Plastics Division
270 Park Ave.
New York, N. Y. 10017

10. SHIPPING INFORMATION

10.1 **Grades or Purity:** Usually 100%
10.2 **Storage Temperature:** Ambient
10.3 **Inert Atmosphere:** No requirement
10.4 **Venting:** Open (flame arrester)

11. HAZARD ASSESSMENT CODE

(See Hazard Assessment Handbook, CG 446-3)

A-P-Q

12. HAZARD CLASSIFICATIONS

12.1 **Code of Federal Regulations:** Not listed
12.2 **NAS Hazard Rating for Bulk Water Transportation:** Not listed
12.3 **NFPA Hazard Classifications:** Not listed

13. PHYSICAL AND CHEMICAL PROPERTIES

13.1 **Physical State at 15°C and 1 atm:** Liquid
13.2 **Molecular Weight:** 450—626
13.3 **Boiling Point at 1 atm:** Very high
13.4 **Freezing Point:** 61°F = 16°C = 289°K
13.5 **Critical Temperature:** Not pertinent
13.6 **Critical Pressure:** Not pertinent
13.7 **Specific Gravity:** 1.02 at 20°C (liquid)
13.8 **Liquid Surface Tension:** Not pertinent
13.9 **Liquid-Water Interfacial Tension:** Not pertinent
13.10 **Vapor (Gas) Specific Gravity:** Not pertinent
13.11 **Ratio of Specific Heats of Vapor (Gas):** Not pertinent
13.12 **Latent Heat of Vaporization:** Not pertinent
13.13 **Heat of Combustion:** (est.) −11,200 Btu/lb = −6,200 cal/g = -260×10^5 J/kg
13.14 **Heat of Decomposition:** Not pertinent
13.15 **Heat of Solution:** Not pertinent
13.16 **Heat of Polymerization:** Not pertinent

3. CHEMICAL DESIGNATIONS (Cont'd.)

3.3 **Chemical Formula:**
$C_{12}H_{25}O(CH_2CH_2O)_nCH_2CH_2OH$
n=6-10 (average)

3.4 **IMCO/United Nations Numerical Designation:** Not listed

ENP — ETHOXYLATED NONYLPHENOL

Common Synonyms

Liquid or solid — White — Mild odor

May float or sink in water.

Avoid contact with liquid and solid. Keep people away.
Stop discharge if possible.
Call fire department.
Isolate and remove discharged material.
Notify local health and pollution control agencies.

Fire

Combustible.
Extinguish with dry chemicals, foam, or carbon dioxide.
Water may be ineffective on fire.
Cool exposed containers with water.

Exposure

CALL FOR MEDICAL AID.

LIQUID OR SOLID
Irritating to skin and eyes.
Harmful if swallowed.
Remove contaminated clothing and shoes.
Flush affected areas with plenty of water.
IF IN EYES, hold eyelids open and flush with plenty of water.
IF SWALLOWED and victim is CONSCIOUS, have victim drink water or milk.
IF SWALLOWED and victim is UNCONSCIOUS OR HAVING CONVULSIONS, do nothing except keep victim warm.

Water Pollution

Effect of low concentrations on aquatic life is unknown.
May be dangerous if it enters water intakes.
Notify local health and wildlife officials.
Notify operators of nearby water intakes.

1. RESPONSE TO DISCHARGE

(See Response Methods Handbook, CG 446-4)

Disperse and flush

2. LABELS

No hazard label required by Code of Federal Regulations

3. CHEMICAL DESIGNATIONS

3.1 **Synonyms:** No common synonyms

3.2 **Coast Guard Compatibility Classification:** Glycol ether

3.3 **Chemical Formula:** $C_9H_{19}C_6H_4O(C_2H_4O)_nH$ where n = 4 – 100

3.4 **IMCO/United Nations Numerical Designation:** Not listed

4. OBSERVABLE CHARACTERISTICS

4.1 **Physical State (as shipped):** Liquid or solid

4.2 **Color:** White

4.3 **Odor:** Mild aromatic

5. HEALTH HAZARDS

5.1 **Personal Protective Equipment:** Gloves and safety glasses

5.2 **Symptoms Following Exposure:** Contact with eyes causes irritation. Prolonged contact with skin causes irritation.

5.3 **Treatment for Exposure:** No treatment required for inhalation or ingestion. EYES: flush with copious quantities of tap water for 15 min. and seek appropriate medical attention. SKIN: wash affected areas with soap and water.

5.4 **Toxicity by Inhalation (Threshold Limit Value):** Data not available

5.5 **Short-Term Inhalation Limits:** Data not available

5.6 **Toxicity by Ingestion:** Grade 2; oral LD_{50} = 1,310 mg/kg (rat)

5.7 **Late Toxicity:** Data not available

5.8 **Vapor (Gas) Irritant Characteristics:** Data not available

5.9 **Liquid or Solid Irritant Characteristics:** Data not available

5.10 **Odor Threshold:** Data not available

6. FIRE HAZARDS

6.1 **Flash Point:** (burns with difficulty) 338—600°F O.C.; > 140°F C.C.

6.2 **Flammable Limits in Air:** Not pertinent

6.3 **Fire Extinguishing Agents:** Water, foam, carbon dioxide

6.4 **Fire Extinguishing Agents Not to be Used:**

6.5 **Special Hazards of Combustion Products:**

6.6 **Behavior in Fire:**

6.7 **Ignition Temperature:** Data not available

6.8 **Electrical Hazard:** Not pertinent

6.9 **Burning Rate:** Not pertinent

7. CHEMICAL REACTIVITY

7.1 **Reactivity with Water:** No reaction

7.2 **Reactivity with Common Materials:**

7.3 **Stability During Transport:** Stable

7.4 **Neutralizing Agents for Acids and Caustics:** Not pertinent

7.5 **Polymerization:** Not pertinent

7.6 **Inhibitor of Polymerization:** Not pertinent

8. WATER POLLUTION

8.1 **Aquatic Toxicity:** 1.3 to > 1,000 ppm/96 hr/bluegills/LC_{50} (increased ethyleneoxide chain length gives decreased toxicity)

8.2 **Waterfowl Toxicity:** Data not available

8.3 **Biological Oxygen Demand (BOD):** 0.5-2.0% of theoretical in 5 days

8.4 **Food Chain Concentration Potential:** None

9. SELECTED MANUFACTURERS

1. Diamond Shamrock Chemical Co.
1100 Superior Avenue
Cleveland, Ohio 44114

2. GAF Corp.
140 West 51 St.
New York, N.Y. 10020

3. Monsanto Company
800 North Lindbergh Blvd.
St. Louis, Mo. 63166

10. SHIPPING INFORMATION

10.1 **Grades or Purity:** Commercial

10.2 **Storage Temperature:** Ambient

10.3 **Inert Atmosphere:** No requirement

10.4 **Venting:** Open

11. HAZARD ASSESSMENT CODE

(See Hazard Assessment Handbook, CG 446-3)

(liquid) A-P
(solid) SS

12. HAZARD CLASSIFICATIONS

12.1 **Code of Federal Regulations:** Not listed

12.2 **NAS Hazard Rating for Bulk Water Transportation:**

Category	Rating
Fire	1
Health	
Vapor Irritant	0
Liquid or Solid Irritant	2
Poisons	1
Water Pollution	
Human Toxicity	2
Aquatic Toxicity	1
Aesthetic Effect	0
Reactivity	
Other Chemicals	1
Water	0
Self-Reaction	0

12.3 **NFPA Hazard Classifications:** Not listed

13. PHYSICAL AND CHEMICAL PROPERTIES

13.1 **Physical State at 15°C and 1 atm:** Solid or liquid

13.2 **Molecular Weight:** > 500

13.3 **Boiling Point at 1 atm:** Not pertinent (decomposes)

13.4 **Freezing Point:** Not pertinent

13.5 **Critical Temperature:** Not pertinent

13.6 **Critical Pressure:** Not pertinent

13.7 **Specific Gravity:** 0.99—1.07 at 25°C (liquid)

13.8 **Liquid Surface Tension:** Data not available

13.9 **Liquid-Water Interfacial Tension:** Not pertinent

13.10 **Vapor (Gas) Specific Gravity:** Not pertinent

13.11 **Ratio of Specific Heats of Vapor (Gas):** Not pertinent

13.12 **Latent Heat of Vaporization:** Not pertinent

13.13 **Heat of Combustion:** Data not available

13.14 **Heat of Decomposition:** Not pertinent

13.15 **Heat of Solution:** Data not available

13.16 **Heat of Polymerization:** Not pertinent

NOTES

EOP ETHOXYLATED PENTADECANOL

Common Synonyms			
Ethoxylated pentadecylalcohol Tergitol nonionic 45-S-10	Liquid	Colorless to yellow	Pleasant odor
	Mixes slowly with water. Freezing point is 59° F.		

Avoid contact with liquid.
Stop discharge if possible.
Call fire department.
Isolate and remove discharged material.
Notify local health and pollution control agencies.

Fire	Combustible. **Extinguish with water, dry chemical, alcohol foam, or carbon dioxide.** **Cool exposed containers with water.**
Exposure	**CALL FOR MEDICAL AID.** **LIQUID** Irritating to skin. Will burn eyes. **Flush affected areas with plenty of water.** **IF IN EYES, hold eyelids open and flush with plenty of water.**
Water Pollution	Effect of low concentrations on aquatic life is unknown. May be dangerous if it enters water intakes. **Notify local health and wildlife officials.** **Notify operators of nearby water intakes.**

1. RESPONSE TO DISCHARGE
(See Response Methods Handbook, CG 446-4)

Disperse and flush

2. LABELS

No hazard label required by Code of Federal Regulations

3. CHEMICAL DESIGNATIONS

3.1 **Synonyms:** Ethoxylated pentadecyl alcohol
Poly(oxyethyl) pentadecyl ether
Tergitol Nonionic 45-S-10

3.2 **Coast Guard Compatibility Classification:** Alcohol

3.3 **Chemical Formula:** $C_{15}H_{31}O(CH_2CH_2O)_nCH_2CH_2OH$
n = 10 (average)

3.4 **IMCO/United Nations Numerical Designation:** Not listed

4. OBSERVABLE CHARACTERISTICS

4.1 **Physical State (as shipped):** Liquid

4.2 **Color:** Colorless

4.3 **Odor:** Mild, pleasant

5. HEALTH HAZARDS

5.1 **Personal Protective Equipment:** Plastic gloves, goggles

5.2 **Symptoms Following Exposure:** Liquid causes eye injury and de-fats the skin, causing irritation.

5.3 **Treatment for Exposure:** Flush eyes with water for at least 15 min. Wash skin well with water. Get medical attention.

5.4 **Toxicity by Inhalation (Threshold Limit Value):** Not pertinent

5.5 **Short-Term Inhalation Limits:** Not pertinent

5.6 **Toxicity by Ingestion:** Data not available

5.7 **Late Toxicity:** Data not available

5.8 **Vapor (Gas) Irritant Characteristics:** None

5.9 **Liquid or Solid Irritant Characteristics:** Liquid causes eye injury. Contact with skin may cause irritation.

5.10 **Odor Threshold:** Not pertinent

6. FIRE HAZARDS

6.1 **Flash Point:** 470°F O.C.

6.2 **Flammable Limits in Air:** Not pertinent

6.3 **Fire Extinguishing Agents:** Carbon dioxide or dry chemical for small fires; alcohol foam and water for large fires

6.4 **Fire Extinguishing Agents Not to be Used:** Not pertinent

6.5 **Special Hazards of Combustion Products:** Not pertinent

6.6 **Behavior in Fire:** Not pertinent

6.7 **Ignition Temperature:** Data not available

6.8 **Electrical Hazard:** Not pertinent

6.9 **Burning Rate:** Data not available

7. CHEMICAL REACTIVITY

7.1 **Reactivity with Water:** No reaction

7.2 **Reactivity with Common Materials:** No reaction

7.3 **Stability During Transport:** Stable

7.4 **Neutralizing Agents for Acids and Caustics:** Not pertinent

7.5 **Polymerization:** Not pertinent

7.6 **Inhibitor of Polymerization:** Not pertinent

8. WATER POLLUTION

8.1 **Aquatic Toxicity:** Data not available

8.2 **Waterfowl Toxicity:** Data not available

8.3 **Biological Oxygen Demand (BOD):** Data not available

8.4 **Food Chain Concentration Potential:** None

9. SELECTED MANUFACTURERS

1. BASF Wyandotte Corp.
   Industrial Chemicals Group
   Wyandotte, Mich. 48192
2. Jefferson Chemical Co.
   3336 Richmond Ave.
   Houston, Texas 77052
3. Union Carbide Corp.
   Chemicals and Plastics Division
   270 Park Ave.
   New York, N. Y. 10017

10. SHIPPING INFORMATION

10.1 **Grades or Purity:** Usually 100%

10.2 **Storage Temperature:** Ambient

10.3 **Inert Atmosphere:** No requirement

10.4 **Venting:** Open (flame arrester)

11. HAZARD ASSESSMENT CODE
(See Hazard Assessment Handbook, CG 446-3)

A-P-Q

12. HAZARD CLASSIFICATIONS

12.1 **Code of Federal Regulations:** Not listed

12.2 **NAS Hazard Rating for Bulk Water Transportation:** Not listed

12.3 **NFPA Hazard Classifications:** Not listed

13. PHYSICAL AND CHEMICAL PROPERTIES

13.1 **Physical State at 15°C and 1 atm:** Liquid

13.2 **Molecular Weight:** 660

13.3 **Boiling Point at 1 atm:** Very high

13.4 **Freezing Point:** 59°F = 15°C = 288°K

13.5 **Critical Temperature:** Not pertinent

13.6 **Critical Pressure:** Not pertinent

13.7 **Specific Gravity:** 1.007 at 15°C (liquid)

13.8 **Liquid Surface Tension:** Not pertinent

13.9 **Liquid-Water Interfacial Tension:** Not pertinent

13.10 **Vapor (Gas) Specific Gravity:** Not pertinent

13.11 **Ratio of Specific Heats of Vapor (Gas):** Not pertinent

13.12 **Latent Heat of Vaporization:** Not pertinent

13.13 **Heat of Combustion:** (est.) $-11{,}000$ Btu/lb $= -6{,}200$ cal/g $= -260 \times 10^5$ J/kg

13.14 **Heat of Decomposition:** Not pertinent

13.15 **Heat of Solution:** (est.) -9 Btu/lb $= -5$ cal/g $= -0.2 \times 10^5$ J/kg

13.16 **Heat of Polymerization:** Not pertinent

NOTES

ETHOXYLATED TETRADECANOL

Common Synonyms
Ethoxylated tetradecyl alcohol
Ethoxylated myristyl alcohol
Tergitol nonionic 45-S-10

Liquid | Colorless to yellow | Pleasant odor

Mixes slowly with water. Freezing point is 59° F.

Avoid contact with liquid.
Stop discharge if possible.
Call fire department.
Isolate and remove discharged material.
Notify local health and pollution control agencies.

Fire

Combustible.
Extinguish with water, dry chemical, alcohol foam, or carbon dioxide.
Cool exposed containers with water.

Exposure

CALL FOR MEDICAL AID.

LIQUID
Irritating to skin.
Will burn eyes.
Flush affected areas with plenty of water.
IF IN EYES, hold eyelids open and flush with plenty of water.

Water Pollution

Effect of low concentrations on aquatic life is unknown.
May be dangerous if it enters water intakes.
Notify local health and wildlife officials.
Notify operators of nearby water intakes.

1. RESPONSE TO DISCHARGE

(See Response Methods Handbook, CG 446-4)

Disperse and flush

2. LABELS

No hazard label required by Code of Federal Regulations

3. CHEMICAL DESIGNATIONS

3.1 **Synonyms:** Ethoxylated myristyl alcohol
Ethoxylated tetradecyl alcohol
Poly(oxyethyl) myristyl ether
Poly(oxyethyl) tetradecyl ether
Tergitol Nonionic 45-S-10

3.2 **Coast Guard Compatibility Classification:** Alcohol

3.3 **Chemical Formula:** $C_{14}H_{29}O(CH_2CH_2O)_nCH_2CH_2OH$
n = 10 (average)

(Continued on page 4)

4. OBSERVABLE CHARACTERISTICS

4.1 **Physical State (as shipped):** Liquid
4.2 **Color:** Colorless
4.3 **Odor:** Mild, pleasant

5. HEALTH HAZARDS

5.1 **Personal Protective Equipment:** Plastic gloves; goggles.
5.2 **Symptoms Following Exposure:** Liquid causes eye injury and de-fats the skin, causing irritation.
5.3 **Treatment for Exposure:** Flush eyes with water for at least 15 min. Wash skin well with water. Get medical attention.
5.4 **Toxicity by Inhalation (Threshold Limit Value):** Not pertinent
5.5 **Short-Term Inhalation Limits:** Not pertinent
5.6 **Toxicity by Ingestion:** Data not available
5.7 **Late Toxicity:** Data not available
5.8 **Vapor (Gas) Irritant Characteristics:** None
5.9 **Liquid or Solid Irritant Characteristics:** Liquid causes eye injury. Contact with skin may cause irritation.
5.10 **Odor Threshold:** Not pertinent

6. FIRE HAZARDS

6.1 **Flash Point:** 470°F O.C.
6.2 **Flammable Limits in Air:** Not pertinent
6.3 **Fire Extinguishing Agents:** Carbon dioxide or dry chemical for small fires; alcohol foam and water for large fires.
6.4 **Fire Extinguishing Agents Not to be Used:** Not pertinent
6.5 **Special Hazards of Combustion Products:** Not pertinent
6.6 **Behavior in Fire:** Not pertinent
6.7 **Ignition Temperature:** Data not available
6.8 **Electrical Hazard:** Not pertinent
6.9 **Burning Rate:** Data not available

7. CHEMICAL REACTIVITY

7.1 **Reactivity with Water:** No reaction
7.2 **Reactivity with Common Materials:** No reaction
7.3 **Stability During Transport:** Stable
7.4 **Neutralizing Agents for Acids and Caustics:** Not pertinent
7.5 **Polymerization:** Not pertinent
7.6 **Inhibitor of Polymerization:** Not pertinent

8. WATER POLLUTION

8.1 **Aquatic Toxicity:** Data not available
8.2 **Waterfowl Toxicity:** Data not available
8.3 **Biological Oxygen Demand (BOD):** Data not available
8.4 **Food Chain Concentration Potential:** None

9. SELECTED MANUFACTURERS

1. BASF Wyandotte Corp.
Industrial Chemicals Group
Wyandotte, Mich. 48192
2. Jefferson Chemical Co.
3336 Richmond Ave.
Houston, Texas 77052
3. Union Carbide Corp.
Chemicals and Plastics Div.
270 Park Ave.
New York, N. Y. 10017

10. SHIPPING INFORMATION

10.1 **Grades or Purity:** Usually 100%
10.2 **Storage Temperature:** Ambient
10.3 **Inert Atmosphere:** No requirement
10.4 **Venting:** Open (flame arrester)

11. HAZARD ASSESSMENT CODE

(See Hazard Assessment Handbook, CG 446-3)

A-P-Q

12. HAZARD CLASSIFICATIONS

12.1 **Code of Federal Regulations:** Not listed
12.2 **NAS Hazard Rating for Bulk Water Transportation:** Not listed
12.3 **NFPA Hazard Classifications:** Not listed

13. PHYSICAL AND CHEMICAL PROPERTIES

13.1 **Physical State at 15°C and 1 atm:** Liquid
13.2 **Molecular Weight:** 660
13.3 **Boiling Point at 1 atm:** Very high
13.4 **Freezing Point:** 59°F = 15°C = 288°K
13.5 **Critical Temperature:** Not pertinent
13.6 **Critical Pressure:** Not pertinent
13.7 **Specific Gravity:** 1.007 at 15°C (liquid)
13.8 **Liquid Surface Tension:** Not pertinent
13.9 **Liquid-Water Interfacial Tension:** Not pertinent
13.10 **Vapor (Gas) Specific Gravity:** Not pertinent
13.11 **Ratio of Specific Heats of Vapor (Gas):** Not pertinent
13.12 **Latent Heat of Vaporization:** Not pertinent
13.13 **Heat of Combustion:** (est.) −11,000 Btu/lb = −6,200 cal/g = -260×10^5 J/kg
13.14 **Heat of Decomposition:** Not pertinent
13.15 **Heat of Solution:** (est.) −9 Btu/lb = −5 cal/g = -0.2×10^5 J/kg
13.16 **Heat of Polymerization:** Not pertinent

3. CHEMICAL DESIGNATIONS (Cont'd.)

3.4 **IMCO/United Nations Numerical Designation:** Not listed

ETD — ETHOXYLATED TRIDECANOL

Common Synonyms			
Ethoxylated tridecyl alcohol Tergitol nonionic 3-A-6	Liquid	Colorless to yellow	Pleasant odor

Avoid contact with liquid.
Stop discharge if possible.
Call fire department.
Isolate and remove discharged material.
Notify local health and pollution agencies.

Fire

Combustible.
Extinguish with water, dry chemical, alcohol foam, or carbon dioxide.
Cool exposed containers with water.

Exposure

CALL FOR MEDICAL AID.

LIQUID
Irritating to skin.
Will burn eyes.
Harmful if swallowed.
Flush affected areas with plenty of water.
IF IN EYES, hold eyelids open and flush with plenty of water.
IF SWALLOWED and victim is CONSCIOUS, have victim drink water or milk.

Water Pollution

Effect of low concentrations on aquatic life is unknown.
May be dangerous if it enters water intakes.
Notify local health and wildlife officials.
Notify operators of nearby water intakes.

1. RESPONSE TO DISCHARGE

(See Response Methods Handbook, CG 446-4)

Disperse and flush

2. LABELS

No hazard label required by Code of Federal Regulations

3. CHEMICAL DESIGNATIONS

3.1 **Synonyms:** Ethoxylated tridecyl alcohol
Poly(oxyethyl) tridecyl ether
Tergitol 3-A-6 Nonionic

3.2 **Coast Guard Compatibility Classification:** Alcohol

3.3 **Chemical Formula:** $C_{13}H_{27}O(CH_2CH_2O)_nCH_2CH_2OH$
n = 5 (average)

3.4 **IMCO/United Nations Numerical Designation:** Not listed

4. OBSERVABLE CHARACTERISTICS

4.1 **Physical State (as shipped):** Liquid

4.2 **Color:** Colorless

4.3 **Odor:** Mild, pleasant

5. HEALTH HAZARDS

5.1 **Personal Protective Equipment:** Plastic gloves; goggles.

5.2 **Symptoms Following Exposure:** Liquid causes eye injury and de-fats the skin, causing irritation.

5.3 **Treatment for Exposure:** In case of contact with eyes, immediately flush with plenty of water for at least 15 min.; get medical attention.

5.4 **Toxicity by Inhalation (Threshold Limit Value):** Not pertinent

5.5 **Short-Term Inhalation Limits:** Not pertinent

5.6 **Toxicity by Ingestion:** Grade 2; LD_{50} 0.5 to 5 g/kg (rat)

5.7 **Late Toxicity:** Data not available

5.8 **Vapor (Gas) Irritant Characteristics:** None

5.9 **Liquid or Solid Irritant Characteristics:** Liquid causes eye injury. Contact with skin may cause irritation.

5.10 **Odor Threshold:** Not pertinent

6. FIRE HAZARDS

6.1 **Flash Point:** 385°F O.C.

6.2 **Flammable Limits in Air:** Not pertinent

6.3 **Fire Extinguishing Agents:** Carbon dioxide or dry chemical for small fires; alcohol foam and water for large fires.

6.4 **Fire Extinguishing Agents Not to be Used:** Data not available

6.5 **Special Hazards of Combustion Products:** Not pertinent

6.6 **Behavior in Fire:** Not pertinent

6.7 **Ignition Temperature:** Data not available

6.8 **Electrical Hazard:** Not pertinent

6.9 **Burning Rate:** Data not available

7. CHEMICAL REACTIVITY

7.1 **Reactivity with Water:** No reaction

7.2 **Reactivity with Common Materials:** No reaction

7.3 **Stability During Transport:** Stable

7.4 **Neutralizing Agents for Acids and Caustics:** Not pertinent

7.5 **Polymerization:** Not pertinent

7.6 **Inhibitor of Polymerization:** Not pertinent

8. WATER POLLUTION

8.1 **Aquatic Toxicity:** Data not available

8.2 **Waterfowl Toxicity:** Data not available

8.3 **Biological Oxygen Demand (BOD):** Data not available

8.4 **Food Chain Concentration Potential:** None

9. SELECTED MANUFACTURERS

1. BASF Wyandotte Corp.
Industrial Chemicals Group
Wyandotte, Mich. 48192

2. Jefferson Chemical Co.
3336 Richmond Ave.
Houston, Texas 77052

3. Union Carbide Corp.
Chemicals and Plastics Division
270 Park Ave.
New York, N. Y. 10017

10. SHIPPING INFORMATION

10.1 **Grades or Purity:** Usually 100%

10.2 **Storage Temperature:** Ambient

10.3 **Inert Atmosphere:** No requirement

10.4 **Venting:** Open (flame arrester)

11. HAZARD ASSESSMENT CODE

(See Hazard Assessment Handbook, CG 446-3)

A-P-Q

12. HAZARD CLASSIFICATIONS

12.1 **Code of Federal Regulations:** Not listed

12.2 **NAS Hazard Rating for Bulk Water Transportation:** Not listed

12.3 **NFPA Hazard Classifications:** Not listed

13. PHYSICAL AND CHEMICAL PROPERTIES

13.1 **Physical State at 15°C and 1 atm:** Liquid

13.2 **Molecular Weight:** 464

13.3 **Boiling Point at 1 atm:** Very high

13.4 **Freezing Point:** Data not available

13.5 **Critical Temperature:** Not pertinent

13.6 **Critical Pressure:** Not pertinent

13.7 **Specific Gravity:** 1.00 at 15°C (liquid)

13.8 **Liquid Surface Tension:** Not pertinent

13.9 **Liquid-Water Interfacial Tension:** Not pertinent

13.10 **Vapor (Gas) Specific Gravity:** Not pertinent

13.11 **Ratio of Specific Heats of Vapor (Gas):** Not pertinent

13.12 **Latent Heat of Vaporization:** Not pertinent

13.13 **Heat of Combustion:** (est.) $-11{,}000$ Btu/lb $= -6{,}200$ cal/g $= -260 \times 10^5$ J/kg

13.14 **Heat of Decomposition:** Not pertinent

13.15 **Heat of Solution:** (est.) $-9 = -5$ cal/g $= -0.2 \times 10^5$ J/kg

13.16 **Heat of Polymerization:** Not pertinent

NOTES

ETG

ETHOXY TRIGLYCOL

Common Synonyms
Ethoxytriethylene glycol
Triethylene glycol monoethyl ether

Liquid | Colorless | Odorless

Sinks and mixes with water.

Stop discharge if possible.
Call fire department.
Isolate and remove discharged material.
Notify local health and pollution control agencies.

Fire

Combustible.
Extinguish with dry chemical, alcohol foam, or carbon dioxide.
Water and foam may be ineffective on fire.
Cool exposed containers with water.

Exposure

Not harmful.

Water Pollution

Effect of low concentrations on aquatic life is unknown.
May be dangerous if it enters water intakes.
Notify local health and wildlife officials.
Notify operators of nearby water intakes.

1. RESPONSE TO DISCHARGE
(See Response Methods Handbook, CG 446-4)

Disperse and flush

2. LABELS

No hazard label required by Code of Federal Regulations

3. CHEMICAL DESIGNATIONS

3.1 **Synonyms:** Dowanol TE
Ethoxytriethylene glycol
Triethylene glycol monoethyl ether
Triglycol monoethyl ether

3.2 **Coast Guard Compatibility Classification:** Glycol

3.3 **Chemical Formula:** $C_2H_5O(CH_2)_2O(CH_2)_2OCH_2CH_2OH$

3.4 **IMCO/United Nations Numerical Designation:** Not listed

4. OBSERVABLE CHARACTERISTICS

4.1 **Physical State (as shipped):** Liquid
4.2 **Color:** Colorless
4.3 **Odor:** Practically odorless; characteristic

5. HEALTH HAZARDS

5.1 **Personal Protective Equipment:** Chemical safety goggles and adequate protective clothing.
5.2 **Symptoms Following Exposure:** No appreciable hazard in ordinary handling or use.
5.3 **Treatment for Exposure:** Wash affected parts with water.
5.4 **Toxicity by Inhalation (Threshold Limit Value):** Not pertinent
5.5 **Short-Term Inhalation Limits:** Not pertinent
5.6 **Toxicity by Ingestion:** Grade 1; LD_{50} 5 to 15 g/kg (rat)
5.7 **Late Toxicity:** Data not available
5.8 **Vapor (Gas) Irritant Characteristics:** Vapors are nonirritating to the eyes and throat.
5.9 **Liquid or Solid Irritant Characteristics:** No appreciable hazard. Practically harmless to the skin.
5.10 **Odor Threshold:** Not pertinent

6. FIRE HAZARDS

6.1 **Flash Point:** 275°F O.C.
6.2 **Flammable Limits in Air:** Not pertinent
6.3 **Fire Extinguishing Agents:** Dry chemical, carbon dioxide, or alcohol foam
6.4 **Fire Extinguishing Agents Not to be Used:** Water or foam may cause frothing.
6.5 **Special Hazards of Combustion Products:** Not pertinent
6.6 **Behavior in Fire:** Not pertinent
6.7 **Ignition Temperature:** Data not available
6.8 **Electrical Hazard:** Not pertinent
6.9 **Burning Rate:** Data not available

7. CHEMICAL REACTIVITY

7.1 **Reactivity with Water:** No reaction
7.2 **Reactivity with Common Materials:** No reaction
7.3 **Stability During Transport:** Stable
7.4 **Neutralizing Agents for Acids and Caustics:** Not pertinent
7.5 **Polymerization:** Not pertinent
7.6 **Inhibitor of Polymerization:** Not pertinent

8. WATER POLLUTION

8.1 **Aquatic Toxicity:** Data not available
8.2 **Waterfowl Toxicity:** Data not available
8.3 **Biological Oxygen Demand (BOD):** Data not available
8.4 **Food Chain Concentration Potential:** None

9. SELECTED MANUFACTURERS

1. Dow Chemical Co.
   Midland, Mich. 48640
2. Olin Corp.
   Chemicals Division
   120 Long Ridge Rd.
   Stamford, Conn. 06904
3. Union Carbide Corp.
   Chemicals and Plastics Division
   270 Park Ave.
   New York, N. Y. 10017

10. SHIPPING INFORMATION

10.1 **Grades or Purity:** Data not available
10.2 **Storage Temperature:** Ambient
10.3 **Inert Atmosphere:** No requirement
10.4 **Venting:** Open (flame arrester)

11. HAZARD ASSESSMENT CODE
(See Hazard Assessment Handbook, CG 446-3)

A-P-Q

12. HAZARD CLASSIFICATIONS

12.1 **Code of Federal Regulations:** Not listed

12.2 **NAS Hazard Rating for Bulk Water Transportation:**

Category	Rating
Fire	1
Health	
Vapor Irritant	0
Liquid or Solid Irritant	0
Poisons	0
Water Pollution	
Human Toxicity	0
Aquatic Toxicity	1
Aesthetic Effect	1
Reactivity	
Other Chemicals	2
Water	0
Self-Reaction	0

12.3 **NFPA Hazard Classifications:** Not listed

13. PHYSICAL AND CHEMICAL PROPERTIES

13.1 **Physical State at 15°C and 1 atm:** Liquid
13.2 **Molecular Weight:** 178
13.3 **Boiling Point at 1 atm:** 493°F = 256°C = 529°K
13.4 **Freezing Point:** −1.7°F = −18.7°C = 254.5°K
13.5 **Critical Temperature:** Not pertinent
13.6 **Critical Pressure:** Not pertinent
13.7 **Specific Gravity:** 1.020 at 20°C (liquid)
13.8 **Liquid Surface Tension:** Not pertinent
13.9 **Liquid-Water Interfacial Tension:** Not pertinent
13.10 **Vapor (Gas) Specific Gravity:** Not pertinent
13.11 **Ratio of Specific Heats of Vapor (Gas):** 1.033
13.12 **Latent Heat of Vaporization:** (est.) 125 Btu/lb = 69 cal/g = 2.9×10^5 J/kg
13.13 **Heat of Combustion:** (est.) −11,100 Btu/lb = −6,170 cal/g = -258×10^5 J/kg
13.14 **Heat of Decomposition:** Not pertinent
13.15 **Heat of Solution:** Not pertinent
13.16 **Heat of Polymerization:** Not pertinent

NOTES

ETA	# ETHYL ACETATE

Common Synonyms
Acetic acid, ethyl ester
Acetic ester
Vinegar naphtha

Watery liquid — Colorless — Pleasant fruity odor

Floats on water. Flammable, irritating vapor is produced.

Stop discharge if possible. Keep people away.
Shut off ignition sources and call fire department.
Stay upwind and use water spray to "knock down" vapor.
Avoid contact with liquid and vapor.
Isolate and remove discharged material.
Notify local health and pollution control agencies.

Fire

FLAMMABLE.
Flashback along vapor trail may occur.
Vapor may explode if ignited in an enclosed area.
Extinguish with dry chemical, alcohol foam, or carbon dioxide.
Water may be ineffective on fire.
Cool exposed containers with water.

Exposure

CALL FOR MEDICAL AID.

VAPOR
Irritating to eyes, nose and throat.
If inhaled will cause headache, dizziness, nausea, or loss of consciousness.
Move to fresh air.
If breathing has stopped, give artificial respiration.
If breathing is difficult, give oxygen.

LIQUID
Irritating to skin and eyes.
Harmful if swallowed.
Remove contaminated clothing and shoes.
Flush affected areas with plenty of water.
IF IN EYES, hold eyelids open and flush with plenty of water.
IF SWALLOWED and victim is CONSCIOUS, have victim drink water or milk.

Water Pollution

Effect of low concentrations on aquatic life is unknown.
May be dangerous if it enters water intakes.
Notify local health and wildlife officials.
Notify operators of nearby water intakes.

1. RESPONSE TO DISCHARGE

(See Response Methods Handbook, CG 446-4)

Issue warning—high flammability
Evacuate area
Disperse and flush

2. LABEL

Red

3. CHEMICAL DESIGNATIONS

3.1 **Synonyms:** Acetic acid, ethyl ester
Acetic ester
Acetic ether
Ethyl ethanoate

3.2 **Coast Guard Compatibility Classification:** Ester

3.3 **Chemical Formula:** $CH_3COOCH_2CH_3$

3.4 **IMCO/United Nations Numerical Designation:** 3.2/1173

4. OBSERVABLE CHARACTERISTICS

4.1 **Physical State (as shipped):** Liquid

4.2 **Color:** Colorless

4.3 **Odor:** Pleasant, fruity

5. HEALTH HAZARDS

5.1 **Personal Protective Equipment:** Organic vapor canister or air mask; goggles or face shield.

5.2 **Symptoms Following Exposure:** Headache, irritation of respiratory passages and eyes, dizziness and nausea, weakness, loss of consciousness.

5.3 **Treatment for Exposure:** INHALATION: if victim is overcome, move him to fresh air immediately and call a physician; if breathing is irregular or stopped, start resuscitation and administer oxygen. EYES: flush with water for at least 15 min.

5.4 **Toxicity by Inhalation (Threshold Limit Value):** 400 ppm

5.5 **Short-Term Inhalation Limits:** 1000 ppm for 15 min.

5.6 **Toxicity by Ingestion:** Grade 2; LD_{50} 0.5 to 5 g/kg

5.7 **Late Toxicity:** Data not available

5.8 **Vapor (Gas) Irritant Characteristics:** Vapors cause a slight smarting of the eyes or respiratory system if present in high concentrations. The effect is temporary.

5.9 **Liquid or Solid Irritant Characteristics:** Minimum hazard. If spilled on clothing and allowed to remain, may cause smarting and reddening of the skin.

5.10 **Odor Threshold:** 1 ppm

6. FIRE HAZARDS

6.1 **Flash Point:** 24°F C.C.; 55°F O.C.

6.2 **Flammable Limits in Air:** 2.2%–9.0%

6.3 **Fire Extinguishing Agents:** Alcohol foam, carbon dioxide or dry chemicals

6.4 **Fire Extinguishing Agents Not to be Used:** Not pertinent

6.5 **Special Hazards of Combustion Products:** Not pertinent

6.6 **Behavior in Fire:** Not pertinent

6.7 **Ignition Temperature:** 800°F

6.8 **Electrical Hazard:** Class I, group D

6.9 **Burning Rate:** 3.7 mm/min.

7. CHEMICAL REACTIVITY

7.1 **Reactivity with Water:** No reaction

7.2 **Reactivity with Common Materials:** No reaction

7.3 **Stability During Transport:** Stable

7.4 **Neutralizing Agents for Acids and Caustics:** Not pertinent

7.5 **Polymerization:** Not pertinent

7.6 **Inhibitor of Polymerization:** Not pertinent

8. WATER POLLUTION

8.1 **Aquatic Toxicity:** Data not available

8.2 **Waterfowl Toxicity:** Data not available

8.3 **Biological Oxygen Demand (BOD):** (Theor.) 66%, 5 days

8.4 **Food Chain Concentration Potential:** None

9. SELECTED MANUFACTURERS

1. Eastman Kodak Co.
Tennessee Eastman Co., Div.
Kingsport, Tenn. 37662
2. Monsanto Co.
Monsanto Industrial Chemicals Co.
800 North Lindbergh Blvd.
St. Louis, Mo. 63166
3. Union Carbide Corp.
Chemicals and Plastics Div.
270 Park Ave.
New York, N. Y. 10017

10. SHIPPING INFORMATION

10.1 **Grades or Purity:** 85–100%

10.2 **Storage Temperature:** Ambient

10.3 **Inert Atmosphere:** No requirement

10.4 **Venting:** Open (flame arrester) or pressure-vacuum

11. HAZARD ASSESSMENT CODE

(See Hazard Assessment Handbook, CG 446-3)

A-P-Q

12. HAZARD CLASSIFICATIONS

12.1 **Code of Federal Regulations:** Flammable liquid

12.2 **NAS Hazard Rating for Bulk Water Transportation:**

Category	Rating
Fire	3
Health	
Vapor Irritant	1
Liquid or Solid Irritant	1
Poisons	2
Water Pollution	
Human Toxicity	1
Aquatic Toxicity	2
Aesthetic Effect	2
Reactivity	
Other Chemicals	1
Water	0
Self-Reaction	0

12.3 **NFPA Hazard Classifications:**

Category	Classification
Health Hazard (Blue)	1
Flammability (Red)	3
Reactivity (Yellow)	0

13. PHYSICAL AND CHEMICAL PROPERTIES

13.1 **Physical State at 15°C and 1 atm:** Liquid

13.2 **Molecular Weight:** 88.11

13.3 **Boiling Point at 1 atm:** 171°F = 77°C = 350°K

13.4 **Freezing Point:** −117°F = −83°C = 190°K

13.5 **Critical Temperature:** 482°F = 250°C = 523°K

13.6 **Critical Pressure:** 558 psia = 38 atm = 3.8 MN/m²

13.7 **Specific Gravity:** 0.902 at 20°C (liquid)

13.8 **Liquid Surface Tension:** 24 dynes/cm = 0.024 N/m at 20°C

13.9 **Liquid-Water Interfacial Tension:** 6.79 dynes/cm = 0.00679 N/m at 30°C

13.10 **Vapor (Gas) Specific Gravity:** 3.0

13.11 **Ratio of Specific Heats of Vapor (Gas):** 1.080

13.12 **Latent Heat of Vaporization:** 158 Btu/lb = 87.6 cal/g = 3.67×10^5 J/kg

13.13 **Heat of Combustion:** −10,110 Btu/lb = −5616 cal/g = -235.1×10^5 J/kg

13.14 **Heat of Decomposition:** Not pertinent

13.15 **Heat of Solution:** Not pertinent

13.16 **Heat of Polymerization:** Not pertinent

NOTES

REVISED 1978

EAA	ETHYL ACETOACETATE

Common Synonyms			
Acetoacetic acid, ethyl ester Acetoacetic ester Diacetic ether Ethyl 3-oxobutanoate EAA	Liquid Mixes with water.	Colorless	Pleasant fruity odor

Stop discharge if possible. Keep people away.
Call fire department.
Isolate and remove discharged material.
Notify local health and pollution control agencies.

Fire	Combustible. Extinguish with dry chemicals, alcohol foam, or carbon dioxide.
Exposure	Call for medical aid. LIQUID Irritating to eyes. Harmful if swallowed. Remove contaminated clothing and shoes. Flush affected areas with plenty of water. IF IN EYES, hold eyelids open and flush with plenty of water. IF SWALLOWED and victim is CONSCIOUS, have victim drink water or milk.
Water Pollution	Effect of low concentrations on aquatic life is unknown. May be dangerous if it enters water intakes. Notify local health and wildlife officials. Notify operators of nearby water intakes.

1. RESPONSE TO DISCHARGE

(See Response Methods Handbook, CG 446-4)

Issue warning—water contaminant
Disperse and flush

2. LABELS

No label required by Code of Federal Regulations

3. CHEMICAL DESIGNATIONS

3.1 **Synonyms:** Acetoacetic acid, ethyl ester; Acetoacetic ester; Diacetic ether; Ethyl 3-oxobutanoate; EAA

3.2 **Coast Guard Compatibility Classification:** Not applicable

3.3 **Chemical Formula:** $CH_3COCH_2COOC_2H_5$

3.4 **IMCO/United Nations Numerical Designation:** Not listed

4. OBSERVABLE CHARACTERISTICS

4.1 **Physical State (as shipped):** Liquid

4.2 **Color:** Colorless

4.3 **Odor:** Agreeable, fruity

5. HEALTH HAZARDS

5.1 **Personal Protective Equipment:** Goggles or face shield; rubber gloves.

5.2 **Symptoms Following Exposure:** Liquid may cause mild irritation of eyes.

5.3 **Treatment for Exposure:** EYES: flush with water for 15 min.

5.4 **Toxicity by Inhalation (Threshold Limit Value):** Data not available

5.5 **Short-Term Inhalation Limits:** Data not available

5.6 **Toxicity by Ingestion:** Grade 2; oral LD_{50} = 3,980 mg/kg (rat)

5.7 **Late Toxicity:** Data not available

5.8 **Vapor (Gas) Irritant Characteristics:** Data not available

5.9 **Liquid or Solid Irritant Characteristics:** Data not available

5.10 **Odor Threshold:** Data not available

6. FIRE HAZARDS

6.1 **Flash Point:** 176°F O.C.; 135°F C.C.

6.2 **Flammable Limits in Air:** 1.4%–9.5%

6.3 **Fire Extinguishing Agents:** Dry chemical, alcohol foam, carbon dioxide

6.4 **Fire Extinguishing Agents Not to be Used:** Water may be ineffective.

6.5 **Special Hazards of Combustion Products:** Not pertinent

6.6 **Behavior in Fire:** Not pertinent

6.7 **Ignition Temperature:** 563°F

6.8 **Electrical Hazard:** Data not available

6.9 **Burning Rate:** 2.4 mm/min.

7. CHEMICAL REACTIVITY

7.1 **Reactivity with Water:** No reaction

7.2 **Reactivity with Common Materials:** No reaction

7.3 **Stability During Transport:** Stable

7.4 **Neutralizing Agents for Acids and Caustics:** Not pertinent

7.5 **Polymerization:** Not pertinent

7.6 **Inhibitor of Polymerization:** Not pertinent

8. WATER POLLUTION

8.1 **Aquatic Toxicity:** Data not available

8.2 **Waterfowl Toxicity:** Data not available

8.3 **Biological Oxygen Demand (BOD):** Data not available

8.4 **Food Chain Concentration Potential:** None

9. SELECTED MANUFACTURERS

1. Eastman Chemical Products, Inc.
Kingsport, Tenn. 37662
2. Lonza, Inc.
22-10 Route 208
Fair Lawn, N. J. 07410
3. Aldrich Chemical Co.
940 West St. Paul Ave.
Milwaukee, Wis. 53233

10. SHIPPING INFORMATION

10.1 **Grades or Purity:** 98+%

10.2 **Storage Temperature:** Ambient

10.3 **Inert Atmosphere:** No requirement

10.4 **Venting:** Open (flame arrester)

11. HAZARD ASSESSMENT CODE

(See Hazard Assessment Handbook, CG 446-3)

A-P-Q

12. HAZARD CLASSIFICATIONS

12.1 **Code of Federal Regulations:** Not listed

12.2 **NAS Hazard Rating for Bulk Water Transportation:** Not listed

12.3 **NFPA Hazard Classifications:**

Category	Classification
Health Hazard (Blue)	2
Flammability (Red)	2
Reactivity (Yellow)	0

13. PHYSICAL AND CHEMICAL PROPERTIES

13.1 **Physical State at 15°C and 1 atm:** Liquid

13.2 **Molecular Weight:** 130.1

13.3 **Boiling Point at 1 atm:** 363°F = 184°C = 457°K

13.4 **Freezing Point:** <−112°F = <−80°C = <193°K

13.5 **Critical Temperature:** Not pertinent

13.6 **Critical Pressure:** Not pertinent

13.7 **Specific Gravity:** 1.028 at 20°C (liquid)

13.8 **Liquid Surface Tension:** 32.5 dynes/cm = 0.0325 N/m at 20°C

13.9 **Liquid-Water Interfacial Tension:** (est.) 35 dynes/cm = 0.035 N/m at 20°C

13.10 **Vapor (Gas) Specific Gravity:** 4.48

13.11 **Ratio of Specific Heats of Vapor (Gas):** Not pertinent

13.12 **Latent Heat of Vaporization:** 160 Btu/lb = 91 cal/g = 3.8×10^5 J/kg

13.13 **Heat of Combustion:** −9,349 Btu/lb = −5,194 cal/g = -217.3×10^5 J/kg

13.14 **Heat of Decomposition:** Not pertinent

13.15 **Heat of Solution:** Not pertinent

13.16 **Heat of Polymerization:** Not pertinent

NOTES

EAC

ETHYL ACRYLATE

Common Synonyms			
Acrylic acid, ethyl ester Ethyl 2-propenoate	Liquid	Colorless	Fruity odor
	Floats on water. Flammable, irritating vapor is produced.		

Avoid contact with liquid and vapor. Keep people away.
Wear goggles, self-contained breathing apparatus, and rubber overclothing (including gloves).
Shut off ignition sources and call fire department.
Stop discharge if possible.
Stay upwind and use water spray to "knock down" vapor.
Isolate and remove discharged material.
Notify local health and pollution control agencies.

Fire

FLAMMABLE.
Containers may explode in fire.
Flashback along vapor trail may occur.
Vapor may explode if ignited in an enclosed area.
Wear goggles, self-contained breathing apparatus, and rubber overclothing (including gloves).
Combat fires from safe distance or protected location.
Extinguish with dry chemical, foam, or carbon dioxide.
Water may be ineffective on fire.
Cool exposed containers with water.

Exposure

CALL FOR MEDICAL AID.

VAPOR
Irritating to eyes, nose and throat.
If inhaled, will cause headache or nausea.
Move to fresh air.
If breathing has stopped, give artificial respiration.
If breathing is difficult, give oxygen.

LIQUID
Will burn skin and eyes.
Harmful if swallowed.
Remove contaminated clothing and shoes.
Flush affected areas with plenty of water.
IF IN EYES, hold eyelids open and flush with plenty of water.
IF SWALLOWED and victim is CONSCIOUS, have victim drink water or milk.

Water Pollution

Effect of low concentrations on aquatic life is unknown.
Fouling to shoreline.
May be dangerous if it enters water intakes.
Notify local health and wildlife officials.
Notify operators of nearby water intakes.

1. RESPONSE TO DISCHARGE

(See Response Methods Handbook, CG 446-4)

Issue warning—high flammability
Restrict access
Evacuate area
Disperse and flush

2. LABEL

Red

3. CHEMICAL DESIGNATIONS

3.1 **Synonyms:** Acrylic acid, ethyl ester
Ethyl 2-propenoate
3.2 **Coast Guard Compatibility Classification:** Acrylate
3.3 **Chemical Formula:** $CH_2=CHCOOCH_2CH_3$
3.4 **IMCO/United Nations Numerical Designation:** 3.2/1917

4. OBSERVABLE CHARACTERISTICS

4.1 **Physical State (as shipped):** Liquid
4.2 **Color:** Colorless
4.3 **Odor:** Characteristic acrylic odor; sharp, fragrant; acrid; slightly nauseating; sharp, ester-type

5. HEALTH HAZARDS

5.1 **Personal Protective Equipment:** Organic canister or air-supplied mask; acid goggles; impervious gloves.
5.2 **Symptoms Following Exposure:** May cause irritation and burns of eyes and skin. Exposure to excessive vapor concentrations can also cause drowsiness accompanied by nausea, headache, or extreme irritation of the respiratory tract.
5.3 **Treatment for Exposure:** INHALATION: remove victim to fresh air and administer artificial respiration if necessary. SKIN AND EYES: wash for 15 min. with copious quantities of water. Call a physician.
5.4 **Toxicity by Inhalation (Threshold Limit Value):** 25 ppm
5.5 **Short-Term Inhalation Limits:** 50 ppm for 15 min.
5.6 **Toxicity by Ingestion:** Grade 2; LD_{50} 0.5 to 5 g/kg (rat)
5.7 **Late Toxicity:** Repeated exposure may develop sensitivity.
5.8 **Vapor (Gas) Irritant Characteristics:** Vapor is moderately irritating such that personnel will not usually tolerate moderate or high vapor concentrations.
5.9 **Liquid or Solid Irritant Characteristics:** Causes smarting of the skin and first-degree burns on short exposure and may cause secondary burns on long exposure.
5.10 **Odor Threshold:** 0.00024 ppm

6. FIRE HAZARDS

6.1 **Flash Point:** 44°F O.C.
6.2 **Flammable Limits in Air:** 1.8%—9.5% (calc.)
6.3 **Fire Extinguishing Agents:** Dry chemical, foam or carbon dioxide
6.4 **Fire Extinguishing Agents Not to be Used:** Not pertinent
6.5 **Special Hazards of Combustion Products:** Toxic and irritating vapors generated when heated.
6.6 **Behavior in Fire:** Vapor is heavier than air and may travel considerable distance to a source of ignition and flash back. May polymerize and cause container to explode.
6.7 **Ignition Temperature:** 721°F
6.8 **Electrical Hazard:** Data not available
6.9 **Burning Rate:** 4.3 mm/min.

7. CHEMICAL REACTIVITY

7.1 **Reactivity with Water:** No reaction
7.2 **Reactivity with Common Materials:** No reaction
7.3 **Stability During Transport:** Stable
7.4 **Neutralizing Agents for Acids and Caustics:** Not pertinent
7.5 **Polymerization:** May occur; exclude moisture, light; avoid exposure to high temperatures; store in presence of air.
7.6 **Inhibitor of Polymerization:** 13-17 ppm monomethyl ether of hydroquinone

8. WATER POLLUTION

8.1 **Aquatic Toxicity:** 12 ppm/24 hrs/brine shrimp/TL_m
8.2 **Waterfowl Toxicity:** Data not available
8.3 **Biological Oxygen Demand (BOD):** 66% of theoretical in 5 days, freshwater, acclimated seed
8.4 **Food Chain Concentration Potential:** None

9. SELECTED MANUFACTURERS

1. Celanese Corp.
Celanese Chemical Co. Division
245 Park Ave.
New York, N. Y. 10017
2. Dow Badische Co.
Williamsburg, Va. 23185
3. Rohm and Haas Co.
Independence Mall West
Philadelphia, Pa. 19105

10. SHIPPING INFORMATION

10.1 **Grades or Purity:** 98.5—99.5%
10.2 **Storage Temperature:** Ambient
10.3 **Inert Atmosphere:** No requirement
10.4 **Venting:** Pressure-vacuum

11. HAZARD ASSESSMENT CODE

(See Hazard Assessment Handbook, CG 446-3)

A-P-Q-T-U-Z

12. HAZARD CLASSIFICATIONS

12.1 **Code of Federal Regulations:** Flammable liquid
12.2 **NAS Hazard Rating for Bulk Water Transportation:**

Category	Rating
Fire	3
Health	
Vapor Irritant	3
Liquid or Solid Irritant	2
Poisons	3
Water Pollution	
Human Toxicity	2
Aquatic Toxicity	2
Aesthetic Effect	2
Reactivity	
Other Chemicals	2
Water	0
Self-Reaction	3

12.3 **NFPA Hazard Classifications:**

Category	Classification
Health Hazard (Blue)	2
Flammability (Red)	3
Reactivity (Yellow)	2

13. PHYSICAL AND CHEMICAL PROPERTIES

13.1 **Physical State at 15°C and 1 atm:** Liquid
13.2 **Molecular Weight:** 100.12
13.3 **Boiling Point at 1 atm:** 211.3°F = 99.6°C = 372.8°K
13.4 **Freezing Point:** −98°F = −72°C = 201°K
13.5 **Critical Temperature:** 534°F = 279°C = 552°K
13.6 **Critical Pressure:** 544 psia = 37 atm = 3.7 MN/m²
13.7 **Specific Gravity:** 0.923 at 20°C (liquid)
13.8 **Liquid Surface Tension:** 25 dynes/cm = 0.025 N/m at 20°C
13.9 **Liquid-Water Interfacial Tension:** (est.) 40 dynes/cm = 0.04 N/m at 20°C
13.10 **Vapor (Gas) Specific Gravity:** Not pertinent
13.11 **Ratio of Specific Heats of Vapor (Gas):** 1.080
13.12 **Latent Heat of Vaporization:** 149 Btu/lb = 82.9 cal/g = 3.47×10^5 J/kg
13.13 **Heat of Combustion:** −11,880 Btu/lb = −6600 cal/g = -276.3×10^5 J/kg
13.14 **Heat of Decomposition:** Not pertinent
13.15 **Heat of Solution:** Not pertinent
13.16 **Heat of Polymerization:** −335 Btu/lb = −186 cal/g = -7.79×10^5 J/kg

NOTES

REVISED 1978

ETHYL ALCOHOL

Common Synonyms			
Ethanol Alcohol Grain alcohol Denatured alcohol	Watery liquid	Colorless	Alcohol odor
	Floats and mixes with water. Flammable, irritating vapor is produced.		

Stop discharge if possible. Keep people away.
Shut off ignition sources and call fire department.
Stay upwind and use water spray to "knock down" vapor.
Isolate and remove discharged material.
Notify local health and pollution control agencies.

Fire	FLAMMABLE. Flashback along vapor trail may occur. Vapor may explode if ignited in an enclosed area. Extinguish with dry chemical, alcohol foam, or carbon dioxide. Water may be ineffective on fire. Cool exposed containers with water.
Exposure	CALL FOR MEDICAL AID. VAPOR Irritating to eyes, nose and throat. Move to fresh air. LIQUID Not harmful.
Water Pollution	Dangerous to aquatic life in high concentrations. May be dangerous if it enters water intakes. Notify local health and wildlife officials. Notify operators of nearby water intakes.

1. RESPONSE TO DISCHARGE

(See Response Methods Handbook, CG 446-4)

Issue warning—high flammability
Disperse and flush

2. LABEL

Red

3. CHEMICAL DESIGNATIONS

3.1 **Synonyms:**

Alcohol	Fermentation alcohol
Cologne spirit	Grain alcohol
Denatured alcohol	Spirit
Ethanol	Spirits of wine

3.2 **Coast Guard Compatibility Classification:** Alcohol

3.3 **Chemical Formula:** C_2H_5OH

3.4 **IMCO/United Nations Numerical Designation:** 3.2/1170

4. OBSERVABLE CHARACTERISTICS

4.1 **Physical State (as shipped):** Liquid

4.2 **Color:** Colorless

4.3 **Odor:** Mild, rather pleasant; like wine or whiskey. (Denatured alcohol may be unpleasant.)

5. HEALTH HAZARDS

5.1 **Personal Protective Equipment:** All-purpose canister; safety goggles. Avoid contact with liquid and inhalation of vapors.

5.2 **Symptoms Following Exposure:** Irritation of eyes, nose and throat. Headache and drowsiness may occur. Liquid causes intoxication.

5.3 **Treatment for Exposure:** INHALATION: if breathing is affected, remove victim to fresh air; call physician; administer oxygen. Speed is of primary importance. EYES OR SKIN: flush with water.

5.4 **Toxicity by Inhalation (Threshold Limit Value):** 1000 ppm

5.5 **Short-Term Inhalation Limits:** 5000 ppm for 15 min.

5.6 **Toxicity by Ingestion:** Grade 1; LD_{50} 5 to 15 g/kg

5.7 **Late Toxicity:** None

5.8 **Vapor (Gas) Irritant Characteristics:** Vapors cause a slight smarting of the eyes or respiratory system if present in high concentrations. The effect is temporary.

5.9 **Liquid or Solid Irritant Characteristics:** No appreciable hazard. Practically harmless to the skin.

5.10 **Odor Threshold:** 10 ppm

6. FIRE HAZARDS

6.1 **Flash Point:** 55°F C.C.; 64°F O.C.

6.2 **Flammable Limits in Air:** 3.3%–19%

6.3 **Fire Extinguishing Agents:** Carbon dioxide, dry chemical, water spray, alcohol foam

6.4 **Fire Extinguishing Agents Not to be Used:** None

6.5 **Special Hazards of Combustion Products:** None

6.6 **Behavior in Fire:** Not pertinent

6.7 **Ignition Temperature:** 689°F

6.8 **Electrical Hazard:** Class I, Group D

6.9 **Burning Rate:** 3.9 mm/min.

7. CHEMICAL REACTIVITY

7.1 **Reactivity with Water:** No reaction

7.2 **Reactivity with Common Materials:** No reaction

7.3 **Stability During Transport:** Stable

7.4 **Neutralizing Agents for Acids and Caustics:** Not pertinent

7.5 **Polymerization:** Not pertinent

7.6 **Inhibitor of Polymerization:** Not pertinent

8. WATER POLLUTION

8.1 **Aquatic Toxicity:** 250 ppm/6 hr/goldfish/lethal/fresh water

8.2 **Waterfowl Toxicity:** Data not available

8.3 **Biological Oxygen Demand (BOD):** 125%, 5 days; 44.2% (theor.), 5 days; 71.2% (theor.), 20 days

8.4 **Food Chain Concentration Potential:** None

9. SELECTED MANUFACTURERS

1. Commercial Solvents Corp.
   245 Park Ave.
   New York, N. Y. 10017
2. National Distillers and Chemical Corp.
   U. S. Industrial Chemicals Co. Division
   Tuscola, Ill. 61953
3. Publicker Industries, Inc.
   1429 Walnut St.
   Philadelphia, Pa. 19102

10. SHIPPING INFORMATION

10.1 **Grades or Purity:** Anhydrous (200 proof); 190 proof; specially denatured; completely denatured

10.2 **Storage Temperature:** Ambient

10.3 **Inert Atmosphere:** No requirement

10.4 **Venting:** Open (flame arrester) or pressure-vacuum

11. HAZARD ASSESSMENT CODE

(See Hazard Assessment Handbook, CG 446-3)

A-P-Q-R-S

12. HAZARD CLASSIFICATIONS

12.1 **Code of Federal Regulations:** Flammable liquid

12.2 **NAS Hazard Rating for Bulk Water Transportation:**

Category	Rating
Fire	3
Health	
Vapor Irritant	1
Liquid or Solid Irritant	0
Poisons	1
Water Pollution	
Human Toxicity	1
Aquatic Toxicity	1
Aesthetic Effect	1
Reactivity	
Other Chemicals	2
Water	0
Self-Reaction	0

12.3 **NFPA Hazard Classifications:**

Category	Classification
Health Hazard (Blue)	0
Flammability (Red)	3
Reactivity (Yellow)	0

13. PHYSICAL AND CHEMICAL PROPERTIES

13.1 **Physical State at 15°C and 1 atm:** Liquid

13.2 **Molecular Weight:** 46.07

13.3 **Boiling Point at 1 atm:** 172.9°F = 78.3°C = 351.5°K

13.4 **Freezing Point:** −173°F = −114°C = 159°K

13.5 **Critical Temperature:** 469.6°F = 243.1°C = 516.3°K

13.6 **Critical Pressure:** 926 psia = 63.0 atm = 6.38 MN/m^2

13.7 **Specific Gravity:** 0.790 at 20°C (liquid)

13.8 **Liquid Surface Tension:** Not pertinent

13.9 **Liquid-Water Interfacial Tension:** Not pertinent

13.10 **Vapor (Gas) Specific Gravity:** 1.6

13.11 **Ratio of Specific Heats of Vapor (Gas):** 1.128

13.12 **Latent Heat of Vaporization:** 360 Btu/lb = 200 cal/g = 8.37×10^5 J/kg

13.13 **Heat of Combustion:** −11,570 Btu/lb = 6425 cal/g = -268.8×10^5 J/kg

13.14 **Heat of Decomposition:** Not pertinent

13.15 **Heat of Solution:** −99 Btu/lb = −55 cal/g = -2.3×10^5 J/kg

13.16 **Heat of Polymerization:** Not pertinent

NOTES

REVISED 1978

EAD	ETHYLALUMINUM DICHLORIDE

Common Synonyms	
EADC Aluminum ethyl dichloride	Liquid heated Colorless to light yellow IGNITES WHEN EXPOSED TO AIR. Reacts violently with water. Poisonous gas is produced on contact with water. Freezing point is 90° F.

Call fire department.
Wear goggles and self-contained breathing apparatus.
Stop discharge if possible. Keep people away.
Isolate and remove discharged material.
Notify local health and pollution control agencies.

Fire

IGNITES WHEN EXPOSED TO AIR.
Irritating gases are produced when heated.
Extinguish with dry graphite, soda ash, or other inert powder.
DO NOT USE WATER, FOAM, CARBON DIOXIDE, DRY CHEMICAL OR VAPORIZING LIQUID ON FIRE.
DO NOT USE WATER ON ADJACENT FIRES.

Exposure

CALL FOR MEDICAL AID.

GAS PRODUCED IN REACTION WITH WATER.
POISONOUS IF INHALED.
Irritating to eyes, nose and throat.
Move to fresh air.
If breathing has stopped, give artificial respiration.
If breathing is difficult, give oxygen.

LIQUID (HEATED)
Will burn skin and eyes.
Harmful if swallowed.
Remove contaminated clothing and shoes.
Flush affected areas with plenty of water.
IF IN EYES, hold eyelids open and flush with plenty of water.
IF SWALLOWED and victim is CONSCIOUS, have victim drink water or milk.
DO NOT INDUCE VOMITING.

Water Pollution

Effect of low concentrations on aquatic life is unknown.
May be dangerous if it enters water intakes.
Notify local health and wildlife officials.
Notify operators of nearby water intakes.

1. RESPONSE TO DISCHARGE

(See Response Methods Handbook, CG 446-4)

Issue warning—high flammability, corrosive
Restrict access
Evacuate area

2. LABELS

Blue

3. CHEMICAL DESIGNATIONS

3.1 **Synonyms:** Aluminum ethyl dichloride EADC
3.2 **Coast Guard Compatibility Classification:** Not applicable
3.3 **Chemical Formula:** $C_2H_5AlCl_2$
3.4 **IMCO/United Nations Numerical Designation:** 4.2/1924

4. OBSERVABLE CHARACTERISTICS

4.1 **Physical State (as shipped):** Liquid
4.2 **Color:** Colorless to light amber; yellow
4.3 **Odor:** Not pertinent

5. HEALTH HAZARDS

5.1 **Personal Protective Equipment:** Full protective clothing, preferably of aluminized glass cloth; goggles, face shield, gloves; in case of fire, all-purpose canister or self-contained breathing apparatus.
5.2 **Symptoms Following Exposure:** Inhalation of smoke from fire causes metal-fume fever (flu-like symptoms); acid fumes irritate nose and throat. Contact with liquid (which is spontaneously flammable) causes severe burns of eyes and skin.
5.3 **Treatment for Exposure:** INHALATION: only fumes from fire need be considered; metal-fume fever is not critical and lasts less than 36 hrs.; irritation of nose and throat by acid vapors may require treatment by a physician. EYES: flush gently with water for 15 min.; treat burns if fire occurred; get medical attention. SKIN: wash with water; treat burns caused by fire; get medical attention.
5.4 **Toxicity by Inhalation (Threshold Limit Value):** Data not available
5.5 **Short-Term Inhalation Limits:** Not pertinent
5.6 **Toxicity by Ingestion:** Data not available
5.7 **Late Toxicity:** Metal-fume fever may develop after breathing smoke from fire.
5.8 **Vapor (Gas) Irritant Characteristics:** Data not available
5.9 **Liquid or Solid Irritant Characteristics:** Severe skin irritant. Causes second- and third-degree burns on short contact and is very injurious to the eyes.
5.10 **Odor Threshold:** Not pertinent

6. FIRE HAZARDS

6.1 **Flash Point:** Not pertinent (ignites spontaneously)
6.2 **Flammable Limits in Air:** Not pertinent
6.3 **Fire Extinguishing Agents:** Dry chemical, inert dry powders such as sand, limestone.
6.4 **Fire Extinguishing Agents Not to be Used:** Water, foam, halogenated agents, or carbon dioxide
6.5 **Special Hazards of Combustion Products:** Intense smoke may cause metal-fume fever. Irritating hydrogen chloride also formed.
6.6 **Behavior in Fire:** Contact with water applied to adjacent fires will cause formation of irritating smoke containing aluminum oxide and hydrogen chloride.
6.7 **Ignition Temperature:** Ignites spontaneously in air at ambient temperature.
6.8 **Electrical Hazard:** Not pertinent
6.9 **Burning Rate:** Not pertinent

7. CHEMICAL REACTIVITY

7.1 **Reactivity with Water:** Reacts violently to form hydrogen chloride fumes and flammable ethane gas.
7.2 **Reactivity with Common Materials:** Reacts with surface moisture to generate hydrogen chloride, which is corrosive to common metals.
7.3 **Stability During Transport:** Stable
7.4 **Neutralizing Agents for Acids and Caustics:** Rinse with sodium bicarbonate or lime solution.
7.5 **Polymerization:** Not pertinent
7.6 **Inhibitor of Polymerization:** Not pertinent

8. WATER POLLUTION

8.1 **Aquatic Toxicity:** Data not available
8.2 **Waterfowl Toxicity:** Data not available
8.3 **Biological Oxygen Demand (BOD):** None
8.4 **Food Chain Concentration Potential:** None

9. SELECTED MANUFACTURERS

1. Ethyl Corporation
Industrial Chemicals Division
Ethyl Tower
451 Florida
Baton Rouge, La. 70801
2. Texas Alkyls, Incorporated
P. O. Box 600
Deer Park, Texas 77536
3. Ventron Corp.
Alfa Products
P. O. Box 159
Beverly, Mass. 01915

10. SHIPPING INFORMATION

10.1 **Grades or Purity:** Pure (neat). 25% or less by weight in benzene, hexane, or heptane. Solutions are not pyrophoric.
10.2 **Storage Temperature:** 35—40°C
10.3 **Inert Atmosphere:** Inerted; dry nitrogen at 5 psig.
10.4 **Venting:** Safety relief with rupture disc.

11. HAZARD ASSESSMENT CODE

(See Hazard Assessment Handbook, CG 446-3)

A-O-Z

12. HAZARD CLASSIFICATIONS

12.1 **Code of Federal Regulations:** Flammable liquid
12.2 **NAS Hazard Rating for Bulk Water Transportation:** Not listed
12.3 **NFPA Hazard Classifications:**

Category	Classification
Health Hazard (Blue)	3
Flammability (Red)	3
Reactivity (Yellow)	3 ₩

13. PHYSICAL AND CHEMICAL PROPERTIES

13.1 **Physical State at 15°C and 1 atm:** Solid
13.2 **Molecular Weight:** 130.0
13.3 **Boiling Point at 1 atm:** 381°F = 194°C = 467°K
13.4 **Freezing Point:** 90°F = 32°C = 305°K
13.5 **Critical Temperature:** Not pertinent
13.6 **Critical Pressure:** Not pertinent
13.7 **Specific Gravity:** 1.227 at 35°C (liquid)
13.8 **Liquid Surface Tension:** (est.) 30 dynes/cm = 0.030 N/m at 35°C
13.9 **Liquid-Water Interfacial Tension:** Not pertinent
13.10 **Vapor (Gas) Specific Gravity:** Not pertinent
13.11 **Ratio of Specific Heats of Vapor (Gas):** Not pertinent
13.12 **Latent Heat of Vaporization:** Not pertinent
13.13 **Heat of Combustion:** (est.) −5,600 Btu/lb = −3,100 cal/g = −130 × 10^5 J/kg
13.14 **Heat of Decomposition:** Not pertinent
13.15 **Heat of Solution:** Data not available
13.16 **Heat of Polymerization:** Not pertinent

NOTES

ETHYLALUMINUM SESQUICHLORIDE

Common Synonyms		
EASC	Liquid	Colorless to yellow
	IGNITES WHEN EXPOSED TO AIR. Reacts violently with water. Poisonous and flammable gases are produced on contact with water.	

Call fire department.
Wear goggles, self-contained breathing apparatus, and rubber overclothing (including gloves).
Stop discharge if possible. Keep people away.
Isolate and remove discharged material.
Notify local health and pollution control agencies.

Fire	IGNITES WHEN EXPOSED TO AIR. If in water, let fire burn. If not in water, extinguish with dry graphite, soda ash, or other inert powder. DO NOT USE WATER, FOAM, CARBON DIOXIDE, DRY CHEMICALS OR VAPORIZING LIQUIDS ON FIRE. DO NOT USE WATER ON ADJACENT FIRES.
Exposure	Call for medical aid. VAPOR Irritating to eyes, nose and throat. Harmful if inhaled. Move victim to fresh air. If breathing is difficult, give oxygen. LIQUID Will burn skin and eyes. Harmful if swallowed. Remove contaminated clothing and shoes. Flush affected areas with plenty of water. IF IN EYES, hold eyelids open and flush with plenty of water. IF SWALLOWED and victim is CONSCIOUS, have victim drink water or milk. DO NOT INDUCE VOMITING.
Water Pollution	Effect of low concentrations on aquatic life is unknown. May be dangerous if it enters water intakes. Notify local health and wildlife officials. Notify operators of nearby water intakes.

1. RESPONSE TO DISCHARGE

(See Response Methods Handbook, CG 446-4)

Issue warning—high flammability, corrosive
Restrict access
Evacuate area

2. LABELS

3. CHEMICAL DESIGNATIONS

3.1 **Synonyms:** EASC
3.2 **Coast Guard Compatibility Classification:** Not applicable
3.3 **Chemical Formula:** $(C_2H_5)_3Al_2Cl_3$
3.4 **IMCO/United Nations Numerical Designation:** 4.2/1925

4. OBSERVABLE CHARACTERISTICS

4.1 **Physical State (as shipped):** Liquid
4.2 **Color:** Yellow; colorless
4.3 **Odor:** Not pertinent

5. HEALTH HAZARDS

5.1 **Personal Protective Equipment:** Full protective clothing, preferably of aluminized glass cloth; goggles, face shield, gloves; in case of fire, all-purpose canister or self-contained breathing apparatus.
5.2 **Symptoms Following Exposure:** Inhalation of smoke from fire causes metal-fume fever (flu-like symptoms); acid fumes irritate nose and throat. Contact with liquid, which is spontaneously flammable, causes severe burns of eyes and skin.
5.3 **Treatment for Exposure:** INHALATION: only fumes from fire need be considered; metal-fume fever is not critical and lasts less than 36 hrs.; irritation of nose and throat by acid vapors may require treatment by a physician. EYES: flush gently with water for 15 min.; treat burns if fire occurred; get medical attention. SKIN: wash with water; treat burns caused by fire; get medical attention.
5.4 **Toxicity by Inhalation (Threshold Limit Value):** Data not available
5.5 **Short-Term Inhalation Limits:** Not pertinent
5.6 **Toxicity by Ingestion:** Data not available
5.7 **Late Toxicity:** Metal-fume fever may develop after breathing smoke from fire.
5.8 **Vapor (Gas) Irritant Characteristics:** Data not available
5.9 **Liquid or Solid Irritant Characteristics:** Severe skin irritant. Causes second- and third-degree burns on short contact and is very injurious to the eyes.
5.10 **Odor Threshold:** Not pertinent

6. FIRE HAZARDS

6.1 **Flash Point:** Not pertinent (ignites spontaneously)
6.2 **Flammable Limits in Air:** Not pertinent
6.3 **Fire Extinguishing Agents:** Dry chemical, inert dry powders such as sand or limestone
6.4 **Fire Extinguishing Agents Not to be Used:** Water, foam, halogenated agents, or carbon dioxide
6.5 **Special Hazards of Combustion Products:** Intense smoke may cause metal-fume fever. Irritating hydrogen chloride also formed.
6.6 **Behavior in Fire:** Contact with water from adjacent fires will cause formation of irritating smoke containing aluminum oxide and hydrogen chloride.
6.7 **Ignition Temperature:** Not pertinent
6.8 **Electrical Hazard:** Not pertinent
6.9 **Burning Rate:** Not pertinent

7. CHEMICAL REACTIVITY

7.1 **Reactivity with Water:** Reacts violently to form hydrogen chloride fumes and flammable ethane gas.
7.2 **Reactivity with Common Materials:** Reacts with surface moisture to generate hydrogen chloride, which is corrosive to common metals.
7.3 **Stability During Transport:** Stable
7.4 **Neutralizing Agents for Acids and Caustics:** Rinse with sodium bicarbonate or lime solution.
7.5 **Polymerization:** Not pertinent
7.6 **Inhibitor of Polymerization:** Not pertinent

8. WATER POLLUTION

8.1 **Aquatic Toxicity:** Data not available
8.2 **Waterfowl Toxicity:** Data not available
8.3 **Biological Oxygen Demand (BOD):** None
8.4 **Food Chain Concentration Potential:** None

9. SELECTED MANUFACTURERS

1. Ethyl Corporation
   Industrial Chemicals Division
   Ethyl Tower
   451 Florida
   Baton Rouge, La. 70801
2. Texas Alkyls, Incorporated
   P. O. Box 600
   Deer Park, Texas 77536
3. Ventron Corp.
   Alfa Products
   P. O. Box 159
   Beverly, Mass. 01915

10. SHIPPING INFORMATION

10.1 **Grades or Purity:** Pure (neat); 25% or less by weight in benzene, hexane, or heptane. Solutions are not pyrophoric.
10.2 **Storage Temperature:** Ambient
10.3 **Inert Atmosphere:** Inerted; dry nitrogen at 5 psig.
10.4 **Venting:** Safety relief with rupture disc.

11. HAZARD ASSESSMENT CODE

(See Hazard Assessment Handbook, CG 446-3)

A-O-Z

12. HAZARD CLASSIFICATIONS

12.1 **Code of Federal Regulations:** Flammable liquid
12.2 **NAS Hazard Rating for Bulk Water Transportation:** Not listed
12.3 **NFPA Hazard Classifications:**

Category	Classification
Health Hazard (Blue)	
Flammability (Red)	3
Reactivity (Yellow)	3
	~~W~~

13. PHYSICAL AND CHEMICAL PROPERTIES

13.1 **Physical State at 15°C and 1 atm:** Liquid
13.2 **Molecular Weight:** 247.5
13.3 **Boiling Point at 1 atm:** 399°F = 204°C = 477°K
13.4 **Freezing Point:** −4°F = −20°C = 253°K
13.5 **Critical Temperature:** Not pertinent
13.6 **Critical Pressure:** Not pertinent
13.7 **Specific Gravity:** 1.092 at 25°C (liquid)
13.8 **Liquid Surface Tension:** (est.) 32 dynes/cm = 0.032 N/m at 20°C
13.9 **Liquid-Water Interfacial Tension:** Not pertinent
13.10 **Vapor (Gas) Specific Gravity:** Not pertinent
13.11 **Ratio of Specific Heats of Vapor (Gas):** Not pertinent
13.12 **Latent Heat of Vaporization:** Not pertinent
13.13 **Heat of Combustion:** (est.) −8,600 Btu/lb = −4,800 cal/g = -200×10^5 J/kg
13.14 **Heat of Decomposition:** Not pertinent
13.15 **Heat of Solution:** Data not available
13.16 **Heat of Polymerization:** Not pertinent

NOTES

EAM — ETHYLAMINE

Common Synonyms			
Aminoethane Monoethylamine	Liquid	Colorless	Strong ammonia-like odor
	Mixes with water. Boiling point is 62° F. Flammable, irritating vapor is produced.		

Shut off ignition sources. Call fire department.
Stop discharge if possible.
Evacuate area in case of large discharge.
Stay upwind. Use water spray to "knock down" vapor.
Isolate and remove discharged material.
Notify local health and pollution control agencies.

Fire

FLAMMABLE.
Irritating gases are produced when heated.
Containers may explode in fire.
Flashback along vapor trail may occur.
Vapor may explode if ignited in an enclosed area.
Extinguish with dry chemicals, alcohol foam, or carbon dioxide.
Water may be ineffective on fire.
Cool exposed containers with water.

Exposure

Call for medical aid.

VAPOR
Irritating to eyes, nose and throat.
Harmful if inhaled.
Move victim to fresh air.
If breathing is difficult, give oxygen.

LIQUID
Irritating to skin and eyes.
Harmful if swallowed.
Remove contaminated clothing and shoes.
Flush affected areas with plenty of water.
IF IN EYES, hold eyelids open and flush with plenty of water.
IF SWALLOWED and victim is CONSCIOUS, have victim drink water or milk.

Water Pollution

HARMFUL TO AQUATIC LIFE IN VERY LOW CONCENTRATIONS.
May be dangerous if it enters water intakes.
Notify local health and wildlife officials.
Notify operators of nearby water intakes.

1. RESPONSE TO DISCHARGE

(See Response Methods Handbook, CG 446-4)

Issue warning—high flammability, air contaminant, water contaminant
Restrict access
Evacuate area
Disperse and flush

2. LABEL

Red

3. CHEMICAL DESIGNATIONS

3.1 **Synonyms:** Aminoethane
Monoethylamine
3.2 **Coast Guard Compatibility Classification:** Amines (4)
3.3 **Chemical Formula:** $C_2H_5NH_2$
3.4 **IMCO/United Nations Numerical Designation:** 2/1036

4. OBSERVABLE CHARACTERISTICS

4.1 **Physical State (as shipped):** Liquid
4.2 **Color:** Colorless
4.3 **Odor:** Pungent; strong ammoniacal

5. HEALTH HAZARDS

5.1 **Personal Protective Equipment:** Amine-type or ammonia-type mask; plastic gloves; face shield and goggles.
5.2 **Symptoms Following Exposure:** Inhalation causes irritation of respiratory tract and lungs; pulmonary edema may result. Liquid causes severe irritation and burns of eyes and skin, and can permanently injure eyes after 15 seconds' contact. Ingestion causes severe burns of mouth and stomach; can be fatal.
5.3 **Treatment for Exposure:** Get prompt medical attention for anyone overcome or injured by exposure to this compound. INHALATION: remove victim to fresh air, keep him warm, and administer oxygen until medical help arrives. EYES: wash for 15 min. with water; avoid pressure on eyelids. SKIN: wash with soap and water; do not use ointments for at least 24 hrs.; do not cover burned area with dry clothing; keep moist with physiological saline solution. INGESTION: if victim is conscious, give large amount of water, then induce vomiting.
5.4 **Toxicity by Inhalation (Threshold Limit Value):** 10 ppm
5.5 **Short-Term Inhalation Limits:** 25 ppm for 30 min.
5.6 **Toxicity by Ingestion:** Grade 3; oral LD_{50} = 400 mg/kg (rat)
5.7 **Late Toxicity:** None
5.8 **Vapor (Gas) Irritant Characteristics:** Vapors are moderately irritating such that personnel will not usually tolerate moderate or high concentrations.
5.9 **Liquid or Solid Irritant Characteristics:** Causes smarting of the skin and first-degree burns on short exposure and may cause second-degree burns on long exposure.
5.10 **Odor Threshold:** Data not available

6. FIRE HAZARDS

6.1 **Flash Point:** 0°F O.C.
6.2 **Flammable Limits in Air:** 3.5%—14%
6.3 **Fire Extinguishing Agents:** Dry chemical, alcohol foam, carbon dioxide
6.4 **Fire Extinguishing Agents Not to be Used:** Water may be ineffective.
6.5 **Special Hazards of Combustion Products:** Irritating and toxic oxides of nitrogen may be formed.
6.6 **Behavior in Fire:** Vapor is heavier than air and may travel a considerable distance to a source of ignition and flash back. Containers may explode when heated.
6.7 **Ignition Temperature:** 724°F
6.8 **Electrical Hazard:** Data not available
6.9 **Burning Rate:** 5.0 mm/min.

7. CHEMICAL REACTIVITY

7.1 **Reactivity with Water:** No reaction
7.2 **Reactivity with Common Materials:** Will strip and dissolve paint; dissolves most plastic materials; can cause swelling of rubber by absorption. The reactions are not hazardous.
7.3 **Stability During Transport:** Stable
7.4 **Neutralizing Agents for Acids and Caustics:** Flush with water
7.5 **Polymerization:** Not pertinent
7.6 **Inhibitor of Polymerization:** Not pertinent

8. WATER POLLUTION

8.1 **Aquatic Toxicity:** 40 ppm/24 hr/chub/TL_m/fresh water
8.2 **Waterfowl Toxicity:** Data not available
8.3 **Biological Oxygen Demand (BOD):** Data not available
8.4 **Food Chain Concentration Potential:** None

9. SELECTED MANUFACTURERS

1. Virginia Chemicals Inc.
3340 West Norfolk Road
Portsmouth, Va. 23703
2. Pennwalt Corporation
Three Parkway
Philadelphia, Pa. 19102
3. Union Carbide Corporation
Chemicals and Plastics Division
270 Park Avenue
New York, N. Y. 10017

10. SHIPPING INFORMATION

10.1 **Grades or Purity:** Anhydrous (98.5+%); 70—72% in water
10.2 **Storage Temperature:** Ambient
10.3 **Inert Atmosphere:** No requirement
10.4 **Venting:** Pressure-vacuum

11. HAZARD ASSESSMENT CODE

(See Hazard Assessment Handbook, CG 446-3)

A-B-C-K-L-M-N

12. HAZARD CLASSIFICATIONS

12.1 **Code of Federal Regulations:** Flammable liquid
12.2 **NAS Hazard Rating for Bulk Water Transportation:**

Category	Rating
Fire	4
Health	
Vapor Irritant	3
Liquid or Solid Irritant	2
Poisons	3
Water Pollution	
Human Toxicity	3
Aquatic Toxicity	3
Aesthetic Effect	3
Reactivity	
Other Chemicals	2
Water	3
Self-Reaction	0

12.3 **NFPA Hazard Classifications:**

Category	Classification
Health Hazard (Blue)	3
Flammability (Red)	4
Reactivity (Yellow)	0

13. PHYSICAL AND CHEMICAL PROPERTIES

13.1 **Physical State at 15°C and 1 atm:** Liquid
13.2 **Molecular Weight:** 45.1
13.3 **Boiling Point at 1 atm:** 61.7°F = 16.5°C = 289.7°K
13.4 **Freezing Point:** −114°F = −81°C = 192°K
13.5 **Critical Temperature:** 361°F = 183°C = 456°K
13.6 **Critical Pressure:** 827 psia = 56.2 atm = 5.70 MN/m^2
13.7 **Specific Gravity:** 0.687 at 15°C (liquid)
13.8 **Liquid Surface Tension:** 20.5 dynes/cm = 0.0205 N/m at 15°C
13.9 **Liquid-Water Interfacial Tension:** Not pertinent
13.10 **Vapor (Gas) Specific Gravity:** 1.5
13.11 **Ratio of Specific Heats of Vapor (Gas):** 1.1181
13.12 **Latent Heat of Vaporization:** 253 Btu/lb = 146 cal/g = 6.11×10^5 J/kg
13.13 **Heat of Combustion:** −16,180 Btu/lb = −8,990 cal/g = -376×10^5 J/kg
13.14 **Heat of Decomposition:** Not pertinent
13.15 **Heat of Solution:** Data not available
13.16 **Heat of Polymerization:** Not pertinent

NOTES

ETB | ETHYLBENZENE

Common Synonyms			
Phenylethane EB	Liquid	Colorless	Sweet, gasoline-like odor
	Floats on water. Flammable, irritating vapor is produced.		

Avoid contact with liquid and vapor. Keep people away.
Wear goggles, self-contained breathing apparatus, and rubber overclothing (including gloves).
Shut off ignition sources and call fire department.
Stop discharge if possible.
Stay upwind and use water spray to "knock down" vapor.
Isolate and remove discharged material.
Notify local health and pollution control agencies.

Fire

FLAMMABLE.
Flashback along vapor trail may occur.
Vapor may explode if ignited in an enclosed area.
Wear goggles, self-contained breathing apparatus, and rubber overclothing (including gloves).
Extinguish with dry chemical, foam, or carbon dioxide.
Water may be ineffective on fire.
Cool exposed containers with water.

Exposure

CALL FOR MEDICAL AID.

VAPOR
Irritating to eyes, nose and throat.
If inhaled, will cause dizziness or difficult breathing.
Move to fresh air.
If breathing has stopped, give artificial respiration.
If breathing is difficult, give oxygen.

LIQUID
Will burn skin and eyes.
Harmful if swallowed.
Remove contaminated clothing and shoes.
Flush affected areas with plenty of water.
IF IN EYES, hold eyelids open and flush with plenty of water.
IF SWALLOWED and victim is CONSCIOUS, have victim drink water or milk.
DO NOT INDUCE VOMITING.

Water Pollution

HARMFUL TO AQUATIC LIFE IN VERY LOW CONCENTRATIONS.
Fouling to shoreline.
May be dangerous if it enters water intakes.
Notify local health and wildlife officials.
Notify operators of nearby water intakes.

1. RESPONSE TO DISCHARGE

(See Response Methods Handbook, CG 446-4)

Mechanical containment
Should be removed
Chemical and physical treatment

2. LABEL

Red

3. CHEMICAL DESIGNATIONS

3.1 **Synonyms:** EB
Phenylethane

3.2 **Coast Guard Compatibility Classification:** Aromatic hydrocarbon

3.3 **Chemical Formula:** $C_6H_5CH_2CH_3$

3.4 **IMCO/United Nations Numerical Designation:** 3.3/1175

4. OBSERVABLE CHARACTERISTICS

4.1 **Physical State (as shipped):** Liquid

4.2 **Color:** Colorless

4.3 **Odor:** Aromatic

5. HEALTH HAZARDS

5.1 **Personal Protective Equipment:** Self-contained breathing apparatus; safety goggles.

5.2 **Symptoms Following Exposure:** Inhalation may cause irritation of nose, dizziness, depression. Moderate irritation of eye with corneal injury possible. Irritates skin and may cause blisters.

5.3 **Treatment for Exposure:** INHALATION: if ill effects occur, remove victim to fresh air, keep him warm and quiet, and get medical help promptly; if breathing stops, give artificial respiration. INGESTION: induce vomiting only upon physician's approval; material in lung may cause chemical pneumonitis. SKIN AND EYES: promptly flush with plenty of water (15 min. for eyes) and get medical attention; remove and wash contaminated clothing before reuse.

5.4 **Toxicity by Inhalation (Threshold Limit Value):** 100 ppm

5.5 **Short-Term Inhalation Limits:** 200 ppm for 30 min.

5.6 **Toxicity by Ingestion:** Grade 2; LD_{50} 0.5 to 5 g/kg (rat)

5.7 **Late Toxicity:** Data not available

5.8 **Vapor (Gas) Irritant Characteristics:** Vapors cause moderate irritation such that personnel will find high concentrations unpleasant. The effect is temporary.

5.9 **Liquid or Solid Irritant Characteristics:** Causes smarting of the skin and first-degree burns on short exposure; may cause secondary burns on long exposure.

5.10 **Odor Threshold:** 140 ppm

6. FIRE HAZARDS

6.1 **Flash Point:** 80°F O.C.; 59°F C.C.

6.2 **Flammable Limits in Air:** 1.0%–6.7%

6.3 **Fire Extinguishing Agents:** Foam (most effective), water fog, carbon dioxide or dry chemical.

6.4 **Fire Extinguishing Agents Not to be Used:** Not pertinent

6.5 **Special Hazards of Combustion Products:** Irritating vapors are generated when heated.

6.6 **Behavior in Fire:** Vapor is heavier than air and may travel considerable distance to the source of ignition and flash back.

6.7 **Ignition Temperature:** 860°F

6.8 **Electrical Hazard:** Not pertinent

6.9 **Burning Rate:** 5.8 mm/min.

7. CHEMICAL REACTIVITY

7.1 **Reactivity with Water:** No reaction

7.2 **Reactivity with Common Materials:** No reaction

7.3 **Stability During Transport:** Stable

7.4 **Neutralizing Agents for Acids and Caustics:** Not pertinent

7.5 **Polymerization:** Not pertinent

7.6 **Inhibitor of Polymerization:** Not pertinent

8. WATER POLLUTION

8.1 **Aquatic Toxicity:** 29 ppm/96 hr/bluegill/TL_m/fresh water

8.2 **Waterfowl Toxicity:** Data not available

8.3 **Biological Oxygen Demand (BOD):** 2.8% (theor.), 5 days

8.4 **Food Chain Concentration Potential:** None

9. SELECTED MANUFACTURERS

1. Amoco Chemicals Corp.
130 East Randolph Drive
Chicago, Ill. 60601
2. Dow Chemical Co.
Midland, Mich. 48640
3. Monsanto Co.
800 North Lindbergh Blvd.
St. Louis, Mo. 63166

10. SHIPPING INFORMATION

10.1 **Grades or Purity:** Research grade: 99.98%; pure grade: 99.5%; technical grade: 99.0%

10.2 **Storage Temperature:** Ambient

10.3 **Inert Atmosphere:** No requirement

10.4 **Venting:** Open (flame arrester) or pressure-vacuum

11. HAZARD ASSESSMENT CODE

(See Hazard Assessment Handbook, CG 446-3)

A-T-U

12. HAZARD CLASSIFICATIONS

12.1 **Code of Federal Regulations:** Flammable Liquid

12.2 **NAS Hazard Rating for Bulk Water Transportation:**

Category	Rating
Fire	3
Health	
Vapor Irritant	2
Liquid or Solid Irritant	2
Poisons	2
Water Pollution	
Human Toxicity	1
Aquatic Toxicity	3
Aesthetic Effect	2
Reactivity	
Other Chemicals	1
Water	0
Self-Reaction	0

12.3 **NFPA Hazard Classifications:**

Category	Classification
Health Hazard (Blue)	2
Flammability (Red)	3
Reactivity (Yellow)	0

13. PHYSICAL AND CHEMICAL PROPERTIES

13.1 **Physical State at 15°C and 1 atm:** Liquid

13.2 **Molecular Weight:** 106.17

13.3 **Boiling Point at 1 atm:** 277.2°F = 136.2°C = 409.4°K

13.4 **Freezing Point:** −139°F = −95°C = 178°K

13.5 **Critical Temperature:** 651.0°F = 343.9°C = 617.1°K

13.6 **Critical Pressure:** 523 psia = 35.6 atm = 3.61 MN/m^2

13.7 **Specific Gravity:** 0.867 at 20°C (liquid)

13.8 **Liquid Surface Tension:** 29.2 dynes/cm = 0.0292 N/m at 20°C

13.9 **Liquid-Water Interfacial Tension:** 35.48 dynes/cm = 0.03548 N/m at 20°C

13.10 **Vapor (Gas) Specific Gravity:** Not pertinent

13.11 **Ratio of Specific Heats of Vapor (Gas):** 1.071

13.12 **Latent Heat of Vaporization:** 144 Btu/lb = 80.1 cal/g = 3.35×10^5 J/kg

13.13 **Heat of Combustion:** −17,780 Btu/lb = −9877 cal/g = -413.5×10^5 J/kg

13.14 **Heat of Decomposition:** Not pertinent

13.15 **Heat of Solution:** Not pertinent

13.16 **Heat of Polymerization:** Not pertinent

NOTES

REVISED 1978

EBT — ETHYL BUTANOL

Common Synonyms			
Pseudohexyl alcohol sec-Pentylcarbinol	Liquid	Colorless	Mild alcohol odor
	Floats on water.		

Stop discharge if possible.
Call fire department.
Avoid contact with liquid.
Isolate and remove discharged material.
Notify local health and pollution control agencies.

Fire

Combustible.
Extinguish with dry chemical, foam, or carbon dioxide.
Cool exposed containers with water.

Exposure

CALL FOR MEDICAL AID.

LIQUID
Will burn eyes.
Harmful if swallowed.
IF IN EYES, hold eyelids open and flush with plenty of water.
IF SWALLOWED and victim is CONSCIOUS, have victim drink water or milk.

Water Pollution

Effect of low concentrations on aquatic life is unknown.
Fouling to shoreline.
May be dangerous if it enters water intakes.
Notify local health and wildlife officials.
Notify operators of nearby water intakes.

1. RESPONSE TO DISCHARGE

(See Response Methods Handbook, CG 446-4)

Dispense and flush

2. LABELS

No hazard label required by Code of Federal Regulations

3. CHEMICAL DESIGNATIONS

3.1 **Synonyms:** 2-Ethyl-1-butanol
2-Ethylbutyl alcohol
sec-Hexyl alcohol
sec-Pentyl carbinol
Pseudohexyl alcohol

3.2 **Coast Guard Compatibility Classification:** Alcohol

3.3 **Chemical Formula:** $(C_2H_5)_2CHCH_2OH$

3.4 **IMCO/United Nations Numerical Designation:** Not listed

4. OBSERVABLE CHARACTERISTICS

4.1 **Physical State (as shipped):** Liquid

4.2 **Color:** Colorless

4.3 **Odor:** Mild and nonresidual

5. HEALTH HAZARDS

5.1 **Personal Protective Equipment:** Fresh-air mask; plastic gloves; coverall goggles; safety shower and eye bath.

5.2 **Symptoms Following Exposure:** Liquid causes eye burns. Vapors may be mildly irritating to nose and throat.

5.3 **Treatment for Exposure:** Remove to fresh air. Remove contaminated clothing. Wash affected skin areas with water. Flush eyes with water for at least 15 min. and get medical care.

5.4 **Toxicity by Inhalation (Threshold Limit Value):** Data not available

5.5 **Short-Term Inhalation Limits:** Data not available

5.6 **Toxicity by Ingestion:** Grade 2; LD_{50} 0.5 to 5 g/kg (rat)

5.7 **Late Toxicity:** Data not available

5.8 **Vapor (Gas) Irritant Characteristics:** Vapors cause a slight smarting of the eyes or respiratory system if present in high concentrations. The effect is temporary.

5.9 **Liquid or Solid Irritant Characteristics:** Irritates eyes; moderate irritation of skin.

5.10 **Odor Threshold:** Data not available

6. FIRE HAZARDS

6.1 **Flash Point:** 128°F O.C.

6.2 **Flammable Limits in Air:** 1.9%–8.8%

6.3 **Fire Extinguishing Agents:** Carbon dioxide or dry chemical for small fires; alcohol foam for large fires.

6.4 **Fire Extinguishing Agents Not to be Used:** Not pertinent

6.5 **Special Hazards of Combustion Products:** Not pertinent

6.6 **Behavior in Fire:** Not pertinent

6.7 **Ignition Temperature:** 580°F (calc.)

6.8 **Electrical Hazard:** Data not available

6.9 **Burning Rate:** Data not available

7. CHEMICAL REACTIVITY

7.1 **Reactivity with Water:** No reaction

7.2 **Reactivity with Common Materials:** No reaction

7.3 **Stability During Transport:** Stable

7.4 **Neutralizing Agents for Acids and Caustics:** Not pertinent

7.5 **Polymerization:** Not pertinent

7.6 **Inhibitor of Polymerization:** Not pertinent

8. WATER POLLUTION

8.1 **Aquatic Toxicity:** Data not available

8.2 **Waterfowl Toxicity:** Data not available

8.3 **Biological Oxygen Demand (BOD):** Data not available

8.4 **Food Chain Concentration Potential:** None

9. SELECTED MANUFACTURERS

Union Carbide Corp.
Chemicals and Plastics Division
270 Park Ave.
New York, N. Y. 10017

10. SHIPPING INFORMATION

10.1 **Grades or Purity:** Data not available

10.2 **Storage Temperature:** Ambient

10.3 **Inert Atmosphere:** No requirement

10.4 **Venting:** Open (flame arrester) or pressure-vacuum

11. HAZARD ASSESSMENT CODE

(See Hazard Assessment Handbook, CG 446-3)

A-T-U

12. HAZARD CLASSIFICATIONS

12.1 **Code of Federal Regulations:** Combustible Liquid

12.2 **NAS Hazard Rating for Bulk Water Transportation:** Not listed

12.3 **NFPA Hazard Classifications:**

Category	Classification
Health Hazard (Blue)	1
Flammability (Red)	2
Reactivity (Yellow)	0

13. PHYSICAL AND CHEMICAL PROPERTIES

13.1 **Physical State at 15°C and 1 atm:** Liquid

13.2 **Molecular Weight:** 102.17

13.3 **Boiling Point at 1 atm:** 293°F = 146°C = 419°K

13.4 **Freezing Point:** −173°F = −114°C = 159°K

13.5 **Critical Temperature:** Not pertinent

13.6 **Critical Pressure:** Not pertinent

13.7 **Specific Gravity:** 0.834 at 20°C (liquid)

13.8 **Liquid Surface Tension:** 24.3 dynes/cm = 0.0243 N/m at 25°C

13.9 **Liquid-Water Interfacial Tension:** (est.) 40 dynes/cm = 0.04 N/m at 20°C

13.10 **Vapor (Gas) Specific Gravity:** Not pertinent

13.11 **Ratio of Specific Heats of Vapor (Gas):** Not pertinent

13.12 **Latent Heat of Vaporization:** 196.0 Btu/lb = 108.9 cal/g = 4.559×10^5 J/kg

13.13 **Heat of Combustion:** (est.) −16,600 Btu/lb = −9,250 cal/g = -387×10^5 J/kg

13.14 **Heat of Decomposition:** Not pertinent

13.15 **Heat of Solution:** Not pertinent

13.16 **Heat of Polymerization:** Not pertinent

NOTES

ETHYL BUTYRATE

Common Synonyms			
Ethyl butanoate Butyric acid, ethyl ester Butyric ether	Liquid	Colorless	Fruity odor
	Floats on water.		

Shut off ignition sources. Call fire department.
Avoid contact with liquid and vapor. Keep people away.
Stop discharge if possible.
Stay upwind. Use water spray to "knock down" vapor.
Isolate and remove discharged material.
Notify local health and pollution control agencies.

Fire

FLAMMABLE.
Containers may explode in fire.
Flashback along vapor trail may occur.
Vapor may explode if ignited in an enclosed area.
Extinguish with dry chemicals, alcohol foam, or carbon dioxide.
Water may be ineffective on fire.
Cool exposed containers with water.

Exposure

CALL FOR MEDICAL AID.

VAPOR
Irritating to eyes, nose and throat.
If inhaled will cause headache or dizziness.
If in eyes, hold eyelids open and flush with plenty of water.
If breathing has stopped, give artificial respiration.
If breathing is difficult, give oxygen.

LIQUID
Irritating to skin and eyes.
If swallowed will cause nausea, vomiting, dizziness or headache.
Remove contaminated clothing and shoes.
Flush affected areas with plenty of water.
IF IN EYES, hold eyelids open and flush with plenty of water.
IF SWALLOWED and victim is CONSCIOUS, have victim drink water or milk and have victim induce vomiting.
IF SWALLOWED and victim is UNCONSCIOUS OR HAVING CONVULSIONS, do nothing except keep victim warm.

Water Pollution

Effect of low concentrations on aquatic life is unknown.
Fouling to shoreline.
May be dangerous if it enters water intakes.
Notify local health and wildlike officials.
Notify operators of nearby water intakes.

1. RESPONSE TO DISCHARGE

(See Response Methods Handbook, CG 446-4)

Issue warning—high flammability
Restrict access
Mechanical containment
Should be removed
Chemical and physical treatment

2. LABEL

Red

3. CHEMICAL DESIGNATIONS

3.1 **Synonyms:** Butyric acid, ethyl ester; Butyric ether; Ethyl butanoate
3.2 **Coast Guard Compatibility Classification:** Not listed
3.3 **Chemical Formula:** $CH_3CH_2CH_2COOC_2H_5$
3.4 **IMCO/United Nations Numerical Designation:** 3.3/1180

4. OBSERVABLE CHARACTERISTICS

4.1 **Physical State (as shipped):** Liquid
4.2 **Color:** Colorless
4.3 **Odor:** Like apple or pineapple

5. HEALTH HAZARDS

5.1 **Personal Protective Equipment:** All-purpose canister mask or chemical cartridge respirator; glass or face shield; rubber gloves
5.2 **Symptoms Following Exposure:** Inhalation or ingestion causes headache, dizziness, nausea, vomiting, and narcosis. Contact with liquid irritates eyes.
5.3 **Treatment for Exposure:** INHALATION: move victim to fresh air and call a physician; give artificial respiration if necessary. INGESTION: induce vomiting and call a physician. EYES: flush with water for at least 15 min. SKIN: flush with water; wash with soap and water.
5.4 **Toxicity by Inhalation (Threshold Limit Value):** Data not available
5.5 **Short-Term Inhalation Limits:** Data not available
5.6 **Toxicity by Ingestion:** Grade 1; oral LD_{50} = 13 g/kg (rat)
5.7 **Late Toxicity:** Data not available
5.8 **Vapor (Gas) Irritant Characteristics:** Data not available
5.9 **Liquid or Solid Irritant Characteristics:** Data not available
5.10 **Odor Threshold:** 0.015 ppm

6. FIRE HAZARDS

6.1 **Flash Point:** 85°F O.C.; 75°F C.C.
6.2 **Flammable Limits in Air:** Data not available
6.3 **Fire Extinguishing Agents:** Dry chemical, alcohol foam, carbon dioxide
6.4 **Fire Extinguishing Agents Not to be Used:** Water may be ineffective
6.5 **Special Hazards of Combustion Products:**
6.6 **Behavior in Fire:** Vapor is heavier than air and may travel to a source of ignition and flash back. Containers may explode in fire.
6.7 **Ignition Temperature:** 865°F
6.8 **Electrical Hazard:** Data not available
6.9 **Burning Rate:** 4.72 mm/min.

7. CHEMICAL REACTIVITY

7.1 **Reactivity with Water:** No reaction
7.2 **Reactivity with Common Materials:** May attack some forms of plastics
7.3 **Stability During Transport:** Stable
7.4 **Neutralizing Agents for Acids and Caustics:** Not pertinent
7.5 **Polymerization:** Not pertinent
7.6 **Inhibitor of Polymerization:** Not pertinent

8. WATER POLLUTION

8.1 **Aquatic Toxicity:** Data not available
8.2 **Waterfowl Toxicity:** Data not available
8.3 **Biological Oxygen Demand (BOD):** Data not available
8.4 **Food Chain Concentration Potential:** None

9. SELECTED MANUFACTURERS

1. Fritzsche Dodge and Olcott Inc.
76 Ninth Avenue
New York, N.Y. 10011
2. Northwestern Chemical Co.
120 N. Aurora St.
West Chicago, Ill. 60185
3. Aldrich Chemical Co.
940 W. Saint Paul Ave.
Milwaukee, Wis. 53233

10. SHIPPING INFORMATION

10.1 **Grades or Purity:** Commercial, 98+%
10.2 **Storage Temperature:** Ambient
10.3 **Inert Atmosphere:** No requirement
10.4 **Venting:** Open (flame arrester)

11. HAZARD ASSESSMENT CODE

(See Hazard Assessment Handbook, CG 446-3)

A-T-U

12. HAZARD CLASSIFICATIONS

12.1 **Code of Federal Regulations:** Flammable liquid
12.2 **NAS Hazard Rating for Bulk Water Transportation:** Not listed
12.3 **NFPA Hazard Classifications:**

Category	Classification
Health Hazard (Blue)	0
Flammability (Red)	3
Reactivity (Yellow)	0

13. PHYSICAL AND CHEMICAL PROPERTIES

13.1 **Physical State at 15°C and 1 atm:** Liquid
13.2 **Molecular Weight:** 116.16
13.3 **Boiling Point at 1 atm:** 250°F = 121°C = 394°K
13.4 **Freezing Point:** −135°F = −93°C = 180°K
13.5 **Critical Temperature:** 559°F = 293°C = 566°K
13.6 **Critical Pressure:** 460 psia = 31 atm = 3.2 MN/m^2
13.7 **Specific Gravity:** 0.879 at 20°C (liquid)
13.8 **Liquid Surface Tension:** 24.5 dynes/cm = 0.0245 N/m at 20°C
13.9 **Liquid-Water Interfacial Tension:** Data not available
13.10 **Vapor (Gas) Specific Gravity:** 4.0
13.11 **Ratio of Specific Heats of Vapor (Gas):** Not pertinent
13.12 **Latent Heat of Vaporization:** 128 Btu/lb = 71 cal/g = 3.0×10^5 J/kg
13.13 **Heat of Combustion:** −13,200 Btu/lb = −7330 cal/g = -306×10^5 J/kg
13.14 **Heat of Decomposition:** Not pertinent
13.15 **Heat of Solution:** Not pertinent
13.16 **Heat of Polymerization:** Not pertinent

NOTES

ECL	# ETHYL CHLORIDE

Common Synonyms			
Chloroethane Ether, hydrochloric	Liquid	Colorless	Pleasant odor
	Floats and may boil on water. Flammable, irritating vapor is produced. Boiling point is 54° F.		

Stop discharge if possible. Keep people away.
Shut off ignition sources and call fire department.
Avoid contact with liquid.
Stay upwind and use water spray to "knock down" vapor.
Isolate and remove discharged material.
Notify local health and pollution control agencies.

Fire	FLAMMABLE. Flashback along vapor trail may occur. Vapor may explode if ignited in an enclosed area. Wear goggles and self-contained breathing apparatus. Stop flow of gas or liquid if possible. Cool exposed containers and men effecting shutoff with water. Let fire burn.
Exposure	CALL FOR MEDICAL AID. VAPOR Irritating to eyes, nose and throat. If inhaled, will cause dizziness or loss of consciousness. Move to fresh air. If breathing has stopped, give artificial respiration. If breathing is difficult, give oxygen. LIQUID Irritating to skin and eyes. Will cause frostbite. Flush affected areas with plenty of water. IF IN EYES, hold eyelids open and flush with plenty of water. DO NOT RUB AFFECTED AREAS.
Water Pollution	Not harmful to aquatic life. May be dangerous if it enters water intakes. Notify operators of nearby water intakes.

1. RESPONSE TO DISCHARGE

(See Response Methods Handbook, CG 446-4)

Issue warning—high flammability
Restrict access
Evacuate area

2. LABEL

Red

3. CHEMICAL DESIGNATIONS

3.1 **Synonyms:** Chloroethane
Monochlorethane
3.2 **Coast Guard Compatibility Classification:** Halogenated hydrocarbon
3.3 **Chemical Formula:** C_2H_5Cl
3.4 **IMCO/United Nations Numerical Designation:** 2.0/1037

4. OBSERVABLE CHARACTERISTICS

4.1 **Physical State (as shipped):** Liquid
4.2 **Color:** Colorless
4.3 **Odor:** Ethereal; pungent, ethereal; ether-like

5. HEALTH HAZARDS

5.1 **Personal Protective Equipment:** Neoprene rubber clothing where liquid contact is likely; chemical worker's goggles. RESPIRATORY PROTECTION: for 1000 ppm to 2% for 1/2 hr or less, full face mask and organic vapor canister; for greater levels, self-contained breathing apparatus or equivalent.
5.2 **Symptoms Following Exposure:** Vapor causes drunkenness, anesthesia, possible lung injury. Liquid may cause frostbite on eyes and skin.
5.3 **Treatment for Exposure:** INHALATION: get person to fresh air, keep warm and quiet. Get medical attention. SKIN: treat frostbite.
5.4 **Toxicity by Inhalation (Threshold Limit Value):** 1000 ppm
5.5 **Short-Term Inhalation Limits:** Data not available
5.6 **Toxicity by Ingestion:** Not pertinent
5.7 **Late Toxicity:** Data not available
5.8 **Vapor (Gas) Irritant Characteristics:** Vapors cause a slight smarting of the eyes or respiratory system if present in high concentrations. The effect is temporary.
5.9 **Liquid or Solid Irritant Characteristics:** Minimum hazard. If spilled on clothing and allowed to remain, may cause smarting and reddening of the skin.
5.10 **Odor Threshold:** Data not available

6. FIRE HAZARDS

6.1 **Flash Point:** −58°F C.C.
−45°F O.C.
6.2 **Flammable Limits in Air:** 3.6%—12%
6.3 **Fire Extinguishing Agents:** Water fog, carbon dioxide, dry chemical. For large fires it is best to allow material to burn while cooling surrounding equipment. Stop flow of ethyl chloride.
6.4 **Fire Extinguishing Agents Not to be Used:** Not pertinent
6.5 **Special Hazards of Combustion Products:** Toxic and irritating gases are generated in fires.
6.6 **Behavior in Fire:** Containers may explode.
6.7 **Ignition Temperature:** 966°F
6.8 **Electrical Hazard:** Not pertinent
6.9 **Burning Rate:** 3.8 mm/min.

7. CHEMICAL REACTIVITY

7.1 **Reactivity with Water:** No reaction
7.2 **Reactivity with Common Materials:** No reaction
7.3 **Stability During Transport:** Stable
7.4 **Neutralizing Agents for Acids and Caustics:** Not pertinent
7.5 **Polymerization:** Not pertinent
7.6 **Inhibitor of Polymerization:** Not pertinent

8. WATER POLLUTION

8.1 **Aquatic Toxicity:** Data not available
8.2 **Waterfowl Toxicity:** Data not available
8.3 **Biological Oxygen Demand (BOD):** Data not available
8.4 **Food Chain Concentration Potential:** None

9. SELECTED MANUFACTURERS

1. Ethyl Corp.
Industrial Chemicals Division
451 Florida
Baton Rouge, La. 70801
2. PPG Industries Inc.
Industrial Chemical Division
Lake Charles, La. 70601

10. SHIPPING INFORMATION

10.1 **Grades or Purity:** Technical: 98—100%;
USP: 100%
10.2 **Storage Temperature:** Ambient
10.3 **Inert Atmosphere:** No requirement
10.4 **Venting:** Safety relief

11. HAZARD ASSESSMENT CODE

(See Hazard Assessment Handbook, CG 446-3)

A-B-C-D-E-F-G

12. HAZARD CLASSIFICATIONS

12.1 **Code of Federal Regulations:** Flammable liquid
12.2 **NAS Hazard Rating for Bulk Water Transportation:**

Category	Rating
Fire	4
Health	
Vapor Irritant	1
Liquid or Solid Irritant	1
Poisons	1
Water Pollution	
Human Toxicity	0
Aquatic Toxicity	1
Aesthetic Effect	1
Reactivity	
Other Chemicals	1
Water	0
Self-Reaction	0

12.3 **NFPA Hazard Classifications:**

Category	Classification
Health Hazard (Blue)	2
Flammability (Red)	4
Reactivity (Yellow)	0

13. PHYSICAL AND CHEMICAL PROPERTIES

13.1 **Physical State at 15°C and 1 atm:** Gas
13.2 **Molecular Weight:** 64.52
13.3 **Boiling Point at 1 atm:** 54.0°F = 12.2°C = 285.4°K
13.4 **Freezing Point:** −213°F = −136°C = 137°K
13.5 **Critical Temperature:** 369°F = 187.2°C = 460.4°K
13.6 **Critical Pressure:** 758 psia = 51.6 atm = 5.23 MN/m2
13.7 **Specific Gravity:** 0.906 at 12.2°C (liquid)
13.8 **Liquid Surface Tension:** 19.5 dynes/cm = 0.0195 N/m at 20°C
13.9 **Liquid-Water Interfacial Tension:** (est.) 40 dynes/cm = 0.04 N/m at 0°C
13.10 **Vapor (Gas) Specific Gravity:** 2.2
13.11 **Ratio of Specific Heats of Vapor (Gas):** 1.155
13.12 **Latent Heat of Vaporization:** 163 Btu/lb = 90.6 cal/g = 3.79×10^5 J/kg
13.13 **Heat of Combustion:** −8100 Btu/lb = −4500 cal/g = -188.4×10^5 J/kg
13.14 **Heat of Decomposition:** Not pertinent
13.15 **Heat of Solution:** Not pertinent
13.16 **Heat of Polymerization:** Not pertinent

NOTES

REVISED 1978

ECA — ETHYL CHLOROACETATE

Common Synonyms			
Chloroacetic acid, ethyl ester Ethyl chloracetate Ethyl chloroethanoate Monochloroacetic acid, ethyl ester Monochlorethanoic acid, ethyl ester	Liquid	Colorless to light brown	Irritating or fruity odor
	Sinks in water. Irritating vapor is produced.		

AVOID CONTACT WITH LIQUID AND VAPOR. KEEP PEOPLE AWAY.
Wear goggles, self-contained breathing apparatus, and rubber overclothing (including gloves).
Shut off ignition sources. Call fire department.
Stop discharge if possible.
Isolate and remove discharged material.
Notify local health and pollution control agencies.

Fire

Combustible.
POISONOUS GASES MAY BE PRODUCED IN FIRE.
Extinguish with water, dry chemicals, foam, or carbon dioxide.

Exposure

CALL FOR MEDICAL AID.

VAPOR
Irritating to eyes, nose and throat.
If inhaled will cause headache, or nausea.
Move victim to fresh air.
If breathing is difficult, give oxygen.

LIQUID
POISONOUS IF SWALLOWED.
Irritating to skin and eyes.
Remove contaminated clothing and shoes.
Flush affected areas with plenty of water.
IF IN EYES, hold eyelids open and flush with plenty of water.
IF SWALLOWED and victim is CONSCIOUS, have victim drink water or milk and have victim induce vomiting.
IF SWALLOWED and victim is UNCONSCIOUS OR HAVING CONVULSIONS, do nothing except keep victim warm.

Water Pollution

Effect of low concentrations on aquatic life is unknown.
May be dangerous if it enters water intakes.
Notify local health and wildlife officials.
Notify operators of nearby water intakes.

1. RESPONSE TO DISCHARGE

(See Response Methods Handbook, CG 446-4)

Issue warning—water contaminant
Should be removed
Chemical and physical treatment

2. LABELS

No label required by Code of Federal Regulations

3. CHEMICAL DESIGNATIONS

3.1 **Synonyms:** Chloroacetic acid, ethyl ester; Ethyl chloracetate; Ethyl chloroethanoate; Monochloracetic acid, ethyl ester; Monochlorethanoic acid, ethyl ester
3.2 **Coast Guard Compatibility Classification:** Not applicable
3.3 **Chemical Formula:** $ClCH_2COOC_2H_5$
3.4 **IMCO/United Nations Numerical Designation:** 3.3/1181

4. OBSERVABLE CHARACTERISTICS

4.1 **Physical State (as shipped):** Liquid
4.2 **Color:** Colorless; light straw to tan
4.3 **Odor:** Extremely irritating; fruity; pungent

5. HEALTH HAZARDS

5.1 **Personal Protective Equipment:** Organic canister mask; rubber gloves; chemical goggles.
5.2 **Symptoms Following Exposure:** Inhalation causes irritation of mucous membrane, headache, and nausea. Contact with liquid causes extreme eye irritation and conjunctivitis; irritates skin if not removed at once. Ingestion causes irritation of mouth and stomach.
5.3 **Treatment for Exposure:** INHALATION: remove patient to fresh air; get medical attention. EYES: flush with copious quantities of water for at least 15 min.; get medical attention if irritation persists. SKIN: wash with soap and water. INGESTION: give large amount of water and induce vomiting; get medical attention.
5.4 **Toxicity by Inhalation (Threshold Limit Value):** Data not available
5.5 **Short-Term Inhalation Limits:** Data not available
5.6 **Toxicity by Ingestion:** Grade 4; LD_{50} < 50 mg/kg
5.7 **Late Toxicity:** Data not available
5.8 **Vapor (Gas) Irritant Characteristics:** Data not available
5.9 **Liquid or Solid Irritant Characteristics:** Data not available
5.10 **Odor Threshold:** Data not available

6. FIRE HAZARDS

6.1 **Flash Point:** 129°F O.C., 100°F C.C.
6.2 **Flammable Limits in Air:** Data not available
6.3 **Fire Extinguishing Agents:** Water fog, foam, dry chemical, carbon dioxide
6.4 **Fire Extinguishing Agents Not to be Used:** Not pertinent
6.5 **Special Hazards of Combustion Products:** Irritating, toxic hydrogen chloride and phosgene may be generated in fires.
6.6 **Behavior in Fire:** Not pertinent
6.7 **Ignition Temperature:** Data not available
6.8 **Electrical Hazard:** Data not available
6.9 **Burning Rate:** 2.3 mm/min.

7. CHEMICAL REACTIVITY

7.1 **Reactivity with Water:** Very slow, not hazardous.
7.2 **Reactivity with Common Materials:** Slow hydrolysis to acidic products will cause slow corrosion of common metals. No hazard involved.
7.3 **Stability During Transport:** Stable
7.4 **Neutralizing Agents for Acids and Caustics:** Not pertinent
7.5 **Polymerization:** Not pertinent
7.6 **Inhibitor of Polymerization:** Not pertinent

8. WATER POLLUTION

8.1 **Aquatic Toxicity:** Data not available
8.2 **Waterfowl Toxicity:** Data not available
8.3 **Biological Oxygen Demand (BOD):** Data not available
8.4 **Food Chain Concentration Potential:** None

9. SELECTED MANUFACTURERS

1. Kay-Fries Chemicals, Inc.
   360 Lexington Avenue
   New York, N. Y. 10017
2. Monsanto Company
   Monsanto Industrial Chemicals
   800 North Lindbergh Boulevard
   St. Louis, Mo. 63166
3. Dow Chemical Co.
   Midland, Mich. 48640

10. SHIPPING INFORMATION

10.1 **Grades or Purity:** 99+%
10.2 **Storage Temperature:** Ambient
10.3 **Inert Atmosphere:** No requirement
10.4 **Venting:** Open (flame arrester)

11. HAZARD ASSESSMENT CODE

(See Hazard Assessment Handbook, CG 446-3)

A-X-Y

12. HAZARD CLASSIFICATIONS

12.1 **Code of Federal Regulations:** Combustible Liquid
12.2 **NAS Hazard Rating for Bulk Water Transportation:** Not listed
12.3 **NFPA Hazard Classifications:**

Category	Classification
Health Hazard (Blue)	—
Flammability (Red)	3
Reactivity (Yellow)	0

13. PHYSICAL AND CHEMICAL PROPERTIES

13.1 **Physical State at 15°C and 1 atm:** Liquid
13.2 **Molecular Weight:** 122.6
13.3 **Boiling Point at 1 atm:** 289°F = 143°C = 416°K
13.4 **Freezing Point:** −15°F = −26°C = 247°K
13.5 **Critical Temperature:** Not pertinent
13.6 **Critical Pressure:** Not pertinent
13.7 **Specific Gravity:** 1.15 at 20°C (liquid)
13.8 **Liquid Surface Tension:** (est.) 26 dynes/cm = 0.026 N/m at 20°C
13.9 **Liquid-Water Interfacial Tension:** (est.) 24 dynes/cm = 0.024 N/m at 20°C
13.10 **Vapor (Gas) Specific Gravity:** 4.3
13.11 **Ratio of Specific Heats of Vapor (Gas):** Data not available
13.12 **Latent Heat of Vaporization:** 155 Btu/lb = 86 cal/g = 3.6×10^5 J/kg
13.13 **Heat of Combustion:** −7,250 Btu/lb = −4,028 cal/g = -168×10^5 J/kg
13.14 **Heat of Decomposition:** Not pertinent
13.15 **Heat of Solution:** Not pertinent
13.16 **Heat of Polymerization:** Not pertinent

NOTES

ECF	ETHYL CHLOROFORMATE

Common Synonyms			
Chloroformic acid, ethyl ester Ethyl chlorocarbonate	Liquid	Colorless to light yellow	Irritating odor
	Reacts slowly with water. Irritating vapor is produced.		

AVOID CONTACT WITH LIQUID AND VAPOR. KEEP PEOPLE AWAY.
Wear goggles, self-contained breathing apparatus, and rubber overclothing (including gloves).
Shut off ignition sources. Call fire department.
Stop discharge if possible.
Evacuate area in case of large discharge.
Stay upwind. Use water spray to "knock down" vapor.
Isolate and remove discharged material.
Notify local health and pollution control agencies.

Fire

FLAMMABLE.
POISONOUS GASES MAY BE PRODUCED IN FIRE.
Flashback along vapor trail may occur.
Vapor may explode if ignited in an enclosed area.
Wear goggles, self-contained breathing apparatus and rubber overclothing (including gloves).
Extinguish with dry chemicals or carbon dioxide.
Cool exposed containers with water.

Exposure

CALL FOR MEDICAL AID.

VAPOR
POISONOUS IF INHALED OR IF SKIN IS EXPOSED.
Irritating to eyes, nose and throat.
Move victim to fresh air.
If in eyes, hold eyelids open and flush with plenty of water.
If breathing has stopped, give artificial respiration.
If breathing is difficult, give oxygen.

LIQUID
POISONOUS IF SWALLOWED OR IF SKIN IS EXPOSED.
Remove contaminated clothing and shoes.
Flush affected areas with plenty of water.
IF IN EYES, hold eyelids open and flush with plenty of water.
IF SWALLOWED and victim is CONSCIOUS, have victim drink water or milk.
DO NOT INDUCE VOMITING.

Water Pollution

Effect of low concentrations on aquatic life is unknown.
May be dangerous if it enters water intakes.
Notify local health and wildlife officials.
Notify operators of nearby water intakes.

1. RESPONSE TO DISCHARGE

(See Response Methods Handbook, CG 446-4)

Issue warning—corrosive, high flammability, poison
Restrict access
Disperse and flush

2. LABELS

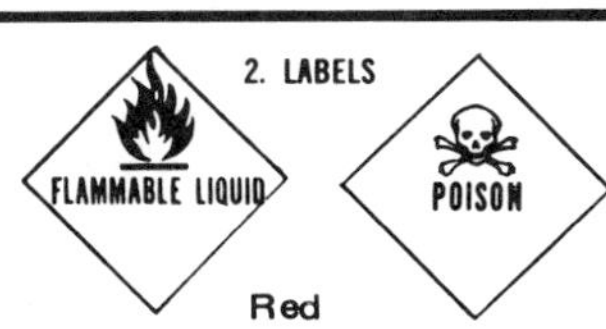

3. CHEMICAL DESIGNATIONS

3.1 **Synonyms:** Chloroformic acid, ethyl ester; Ethyl chlorocarbonate
3.2 **Coast Guard Compatibility Classification:** Not applicable
3.3 **Chemical Formula:** $ClCOOC_2H_5$
3.4 **IMCO/United Nations Numerical Designation:** 3.2/1182

4. OBSERVABLE CHARACTERISTICS

4.1 **Physical State (as shipped):** Liquid
4.2 **Color:** Colorless to pale yellow
4.3 **Odor:** Irritating; sharp, like hydrochloric acid

5. HEALTH HAZARDS

5.1 **Personal Protective Equipment:** Air-line mask, self-contained breathing apparatus, or organic and acid canister mask; full protective clothing.
5.2 **Symptoms Following Exposure:** Inhalation causes mucous membrane irritation, coughing, and sneezing. Vapor causes severe lachrymation; liquid causes acid-type burns of eyes and skin, like those of hydrochloric acid. Ingestion causes severe burns of mouth and stomach.
5.3 **Treatment for Exposure:** INHALATION: remove to fresh air; use artificial respiration if breathing has stopped; call a doctor; keep victim quiet and administer oxygen if needed. EYES: flush with water for at least 15 min.; see a doctor. SKIN: wash liberally with water for at least 15 min., then apply dilute solution of sodium bicarbonate or commercially prepared neutralizer. INGESTION: do NOT induce vomiting; give large amount of water; get medical attention.
5.4 **Toxicity by Inhalation (Threshold Limit Value):** Data not available
5.5 **Short-Term Inhalation Limits:** Data not available
5.6 **Toxicity by Ingestion:** Grade 4; oral $LD_{50} < 50$ mg/kg (rat)
5.7 **Late Toxicity:** Data not available
5.8 **Vapor (Gas) Irritant Characteristics:** Vapors are moderately irritating such that personnel will not usually tolerate moderate or high concentrations.
5.9 **Liquid or Solid Irritant Characteristics:** Causes smarting of the skin and first-degree burns on short exposure and may cause second-degree burns on long exposure.
5.10 **Odor Threshold:** Data not available

6. FIRE HAZARDS

6.1 **Flash Point:** 82°F O.C.; 61°F C.C.
6.2 **Flammable Limits in Air:** Data not available
6.3 **Fire Extinguishing Agents:** Water, dry chemical, carbon dioxide
6.4 **Fire Extinguishing Agents Not to be Used:** Not pertinent
6.5 **Special Hazards of Combustion Products:** Toxic chlorine and phosgene gases may be formed in fires.
6.6 **Behavior in Fire:** Not pertinent
6.7 **Ignition Temperature:** 932°F
6.8 **Electrical Hazard:** Data not available
6.9 **Burning Rate:** 2.6 mm/min.

7. CHEMICAL REACTIVITY

7.1 **Reactivity with Water:** Slow reaction with water, evolving hydrogen chloride (hydrochloric acid)
7.2 **Reactivity with Common Materials:** Slow evolution of hydrogen chloride from surface moisture reaction can cause slow corrosion.
7.3 **Stability During Transport:** Stable
7.4 **Neutralizing Agents for Acids and Caustics:** Flush with water, rinse with sodium bicarbonate or lime solution.
7.5 **Polymerization:** Not pertinent
7.6 **Inhibitor of Polymerization:** Not pertinent

8. WATER POLLUTION

8.1 **Aquatic Toxicity:** Data not available
8.2 **Waterfowl Toxicity:** Data not available
8.3 **Biological Oxygen Demand (BOD):** Data not available
8.4 **Food Chain Concentration Potential:** None

9. SELECTED MANUFACTURERS

1. Story Chemical Corporation
   Ott Division
   500 Agard Road
   Muskegon, Mich. 49445
2. Chemetron Corporation
   Chemicals Group
   393 Seventh Avenue
   New York, N. Y. 10001
3. FMC Corporation
   Industrial Chemical Division
   633 Third Avenue
   New York, N. Y. 10017

10. SHIPPING INFORMATION

10.1 **Grades or Purity:** Technical: 94+%
10.2 **Storage Temperature:** Ambient
10.3 **Inert Atmosphere:** No requirement
10.4 **Venting:** Pressure-vacuum

11. HAZARD ASSESSMENT CODE

(See Hazard Assessment Handbook, CG 446-3)

A-O-X-Y

12. HAZARD CLASSIFICATIONS

12.1 **Code of Federal Regulations:** Flammable Liquid
12.2 **NAS Hazard Rating for Bulk Water Transportation:**

Category	Rating
Fire	3
Health	
Vapor Irritant	3
Liquid or Solid Irritant	3
Poisons	3
Water Pollution	
Human Toxicity	3
Aquatic Toxicity	3
Aesthetic Effect	3
Reactivity	
Other Chemicals	2
Water	2
Self-Reaction	0

12.3 **NFPA Hazard Classifications:**

Category	Classification
Health Hazard (Blue)	—
Flammability (Red)	3
Reactivity (Yellow)	1

13. PHYSICAL AND CHEMICAL PROPERTIES

13.1 **Physical State at 15°C and 1 atm:** Liquid
13.2 **Molecular Weight:** 108.5
13.3 **Boiling Point at 1 atm:** 201°F = 94°C = 367°K
13.4 **Freezing Point:** −114°F = −81°C = 192°K
13.5 **Critical Temperature:** Not pertinent
13.6 **Critical Pressure:** Not pertinent
13.7 **Specific Gravity:** 1.135 at 20°C (liquid)
13.8 **Liquid Surface Tension:** 27.5 dynes/cm = 0.0275 N/m at 15°C
13.9 **Liquid-Water Interfacial Tension:** Not pertinent
13.10 **Vapor (Gas) Specific Gravity:** 3.7
13.11 **Ratio of Specific Heats of Vapor (Gas):** 1.1044
13.12 **Latent Heat of Vaporization:** (est.) 140 Btu/lb = 79 cal/g = 3.3×10^5 J/kg
13.13 **Heat of Combustion:** (est.) −6,900 Btu/lb = −3,800 cal/g = -160×10^5 J/kg
13.14 **Heat of Decomposition:** Not pertinent
13.15 **Heat of Solution:** Data not available
13.16 **Heat of Polymerization:** Not pertinent

NOTES

ECS ETHYLDICHLOROSILANE

Common Synonyms	Liquid Colorless Sharp, irritating odor Reacts violently with water. Irritating gas is produced on contact with water.

Shut off ignition sources. Call fire department.
Avoid contact with liquid. Keep people away.
Stop discharge if possible.
Isolate and remove discharged material.
Notify local health and pollution control agencies.

Fire	FLAMMABLE. POISONOUS GASES MAY BE PRODUCED IN FIRE. Flashback along vapor trail may occur. Vapor may explode if ignited in an enclosed area. Wear chemical protective suit with self-contained breathing apparatus. DO NOT USE WATER OR FOAM ON FIRE.
Exposure	Call for medical aid. VAPOR Irritating to eyes, nose and throat. Move victim to fresh air. If breathing is difficult, give oxygen. LIQUID Will burn skin and eyes. Harmful if swallowed. Remove contaminated clothing and shoes. Flush affected areas with plenty of water. IF IN EYES, hold eyelids open and flush with plenty of water. IF SWALLOWED and victim is CONSCIOUS, have victim drink water or milk. DO NOT INDUCE VOMITING.
Water Pollution	Effect of low concentrations on aquatic life is unknown. May be dangerous if it enters water intakes. Notify local health and wildlife officials. Notify operators of nearby water intakes.

1. RESPONSE TO DISCHARGE
(See Response Methods Handbook, CG 446-4)

Issue warning—high flammability, corrosive
Restrict access
Evacuate area
Disperse and flush with care

2. LABEL

Red

3. CHEMICAL DESIGNATIONS

3.1 **Synonyms:** No common synonyms
3.2 **Coast Guard Compatibility Classification:** Not applicable
3.3 **Chemical Formula:** $C_2H_5SiHCl_2$
3.4 **IMCO/United Nations Numerical Designation:** 3.2/1183

4. OBSERVABLE CHARACTERISTICS

4.1 **Physical State (as shipped):** Liquid
4.2 **Color:** Colorless
4.3 **Odor:** Sharp, hydrochloric acid-like; acrid

5. HEALTH HAZARDS

5.1 **Personal Protective Equipment:** Acid-vapor-type respiratory protection; rubber gloves; chemical worker's goggles; other equipment as necessary to protect skin and eyes.
5.2 **Symptoms Following Exposure:** Inhalation irritates mucous membranes. Contact with liquid causes severe burns of eyes and skin. Ingestion causes severe burns of mouth and stomach.
5.3 **Treatment for Exposure:** Get medical attention following all exposures to this compound. INHALATION: remove to fresh air; give artificial respiration if required. EYES: flush with water for 15 min. SKIN: flush with water. INGESTION: do NOT induce vomiting; give large amounts of water, followed by milk or milk of magnesia.
5.4 **Toxicity by Inhalation (Threshold Limit Value):** Data not available
5.5 **Short-Term Inhalation Limits:** Data not available
5.6 **Toxicity by Ingestion:** Grade 3; LD_{50} 50 to 500 mg/kg
5.7 **Late Toxicity:** Data not available
5.8 **Vapor (Gas) Irritant Characteristics:** Vapors cause severe irritation of eyes and throat and can cause eye and lung injury. They cannot be tolerated even at low concentrations.
5.9 **Liquid or Solid Irritant Characteristics:** Severe skin irritant. Causes second- and third-degree burns on short contact and is very injurious to the eyes.
5.10 **Odor Threshold:** Data not available

6. FIRE HAZARDS

6.1 **Flash Point:** 30°F O.C.
6.2 **Flammable Limits in Air:** 2.9% (LFL)
6.3 **Fire Extinguishing Agents:** Dry chemical
6.4 **Fire Extinguishing Agents Not to be Used:** Water, foam
6.5 **Special Hazards of Combustion Products:** Toxic hydrogen chloride and phosgene gases may be formed.
6.6 **Behavior in Fire:** Difficult to extinguish; re-ignition may occur. Contact with water applied to adjacent fires produces irritating hydrogen chloride fumes and flammable hydrogen gas.
6.7 **Ignition Temperature:** Data not available
6.8 **Electrical Hazard:** Data not available
6.9 **Burning Rate:** 3.2 mm/min.

7. CHEMICAL REACTIVITY

7.1 **Reactivity with Water:** Reacts vigorously, evolving hydrogen chloride (hydrochloric acid)
7.2 **Reactivity with Common Materials:** Reaction with surface moisture will generate hydrogen chloride, which corrodes common metals.
7.3 **Stability During Transport:** Stable
7.4 **Neutralizing Agents for Acids and Caustics:** Flood with water, rinse with sodium bicarbonate or lime solution.
7.5 **Polymerization:** Not pertinent
7.6 **Inhibitor of Polymerization:** Not pertinent

8. WATER POLLUTION

8.1 **Aquatic Toxicity:** Data not available
8.2 **Waterfowl Toxicity:** Data not available
8.3 **Biological Oxygen Demand (BOD):** Data not available
8.4 **Food Chain Concentration Potential:** None

9. SELECTED MANUFACTURERS

Dow Corning Corporation
P. O. Box 592
Midland, Mich. 48640

10. SHIPPING INFORMATION

10.1 **Grades or Purity:** Commercial
10.2 **Storage Temperature:** Ambient
10.3 **Inert Atmosphere:** No requirement
10.4 **Venting:** Pressure vacuum

11. HAZARD ASSESSMENT CODE
(See Hazard Assessment Handbook, CG 446-3)

A-O

12. HAZARD CLASSIFICATIONS

12.1 **Code of Federal Regulations:** Flammable liquid
12.2 **NAS Hazard Rating for Bulk Water Transportation:**

Category	Rating
Fire	3
Health	
Vapor Irritant	4
Liquid or Solid Irritant	4
Poisons	3
Water Pollution	
Human Toxicity	3
Aquatic Toxicity	3
Aesthetic Effect	2
Reactivity	
Other Chemicals	3
Water	4
Self-Reaction	1

12.3 **NFPA Hazard Classifications:** Not listed

13. PHYSICAL AND CHEMICAL PROPERTIES

13.1 **Physical State at 15°C and 1 atm:** Liquid
13.2 **Molecular Weight:** 129.1
13.3 **Boiling Point at 1 atm:** 165°F = 74°C = 347°K
13.4 **Freezing Point:** Not pertinent
13.5 **Critical Temperature:** Not pertinent
13.6 **Critical Pressure:** Not pertinent
13.7 **Specific Gravity:** 1.092 at 20°C (liquid)
13.8 **Liquid Surface Tension:** 21.7 dynes/cm = 0.0217 N/m at 20°C
13.9 **Liquid-Water Interfacial Tension:** Not pertinent
13.10 **Vapor (Gas) Specific Gravity:** 4.5
13.11 **Ratio of Specific Heats of Vapor (Gas):** Data not available
13.12 **Latent Heat of Vaporization:** (est.) 104 Btu/lb = 57.8 cal/g = 2.42×10^5 J/kg
13.13 **Heat of Combustion:** (est.) −6,500 Btu/lb = −3,600 cal/g = -150×10^5 J/kg
13.14 **Heat of Decomposition:** Not pertinent
13.15 **Heat of Solution:** Data not available
13.16 **Heat of Polymerization:** Not pertinent

NOTES

ETL	ETHYLENE

Common Synonyms Ethene Olefiant gas	Liquefied compressed gas Colorless Sweet odor Floats and boils on water. Flammable visible vapor cloud is produced.

Wear self-contained breathing apparatus.
Stop discharge if possible. Keep people away.
Shut off ignition sources and call fire department.
Avoid contact with liquid.
Stay upwind and use water spray to "knock down" vapor.
Notify local health and pollution control agencies.

Fire	FLAMMABLE. Flashback along vapor trail may occur. May explode if ignited in an enclosed area. Wear self-contained breathing apparatus. Stop flow of gas if possible. Cool exposed containers and men effecting shutoff with water. Let fire burn.
Exposure	CALL FOR MEDICAL AID. VAPOR Not irritating to eyes, nose or throat. If inhaled, will cause headache, dizziness, or loss of consciousness. Move to fresh air. If breathing has stopped, give artificial respiration. If breathing is difficult, give oxygen. LIQUID Will cause frostbite. Flush affected areas with plenty of water. DO NOT RUB AFFECTED AREAS.
Water Pollution	Not harmful to aquatic life.

1. RESPONSE TO DISCHARGE

(See Response Methods Handbook, CG 446-4)

Issue warning—high flammability
Restrict access
Evacuate area

2. LABEL

Red

3. CHEMICAL DESIGNATIONS

3.1 **Synonyms:** Ethene
Olefiant gas
3.2 **Coast Guard Compatibility Classification:** Olefin
3.3 **Chemical Formula:** C_2H_4
3.4 **IMCO/United Nations Numerical Designation:**
2.0/1962 (compressed)
2.0/1038 (liquefied)

4. OBSERVABLE CHARACTERISTICS

4.1 **Physical State (as shipped):** Compressed gas or liquefied gas
4.2 **Color:** Colorless
4.3 **Odor:** Slightly sweet; faint sweet; slight ethereal

5. HEALTH HAZARDS

5.1 **Personal Protective Equipment:** Organic vapor canister or air-supplied mask.
5.2 **Symptoms Following Exposure:** Moderate concentration in air causes drowsiness, dizziness, and unconsciousness. Overexposure causes headache, drowsiness, muscular weakness.
5.3 **Treatment for Exposure:** Remove victim to fresh air, give artificial respiration and oxygen if breathing has stopped, and call a physician.
5.4 **Toxicity by Inhalation (Threshold Limit Value):** Simple asphyxiant.
5.5 **Short-Term Inhalation Limits:** Not pertinent
5.6 **Toxicity by Ingestion:** Not pertinent
5.7 **Late Toxicity:** None
5.8 **Vapor (Gas) Irritant Characteristics:** Vapors are nonirritating to the eyes and throat.
5.9 **Liquid or Solid Irritant Characteristics:** No appreciable hazard. Practically harmless to the skin, but may cause frostbite.
5.10 **Odor Threshold:** Data not available.

6. FIRE HAZARDS

6.1 **Flash Point:** −213°F (approx.) C.C.
6.2 **Flammable Limits in Air:** 2.75%—28.6%
6.3 **Fire Extinguishing Agents:** Stop flow of gas if possible. Use carbon dioxide, dry chemical, water fog.
6.4 **Fire Extinguishing Agents Not to be Used:** Not pertinent
6.5 **Special Hazards of Combustion Products:** Vapors are anesthetic.
6.6 **Behavior in Fire:** Container may explode.
6.7 **Ignition Temperature:** 842°F
6.8 **Electrical Hazard:** Class I, group C
6.9 **Burning Rate:** 7.4 mm/min.

7. CHEMICAL REACTIVITY

7.1 **Reactivity with Water:** No reaction
7.2 **Reactivity with Common Materials:** No reaction
7.3 **Stability During Transport:** Stable
7.4 **Neutralizing Agents for Acids and Caustics:** Not pertinent
7.5 **Polymerization:** Not pertinent
7.6 **Inhibitor of Polymerization:** Not pertinent

8. WATER POLLUTION

8.1 **Aquatic Toxicity:** 22 ppm/1 hr/sunfish/killed/fresh water
8.2 **Waterfowl Toxicity:** Data not available
8.3 **Biological Oxygen Demand (BOD):** Data not available
8.4 **Food Chain Concentration Potential:** None

9. SELECTED MANUFACTURERS

1. Exxon Chemical Co.
Houston, Tex. 77001
2. Phillips Petroleum Co.
Bartlesville, Okla. 74004
3. Union Carbide Corp.
Chemicals and Plastics Division
270 Park Ave.
New York, N. Y. 10017

10. SHIPPING INFORMATION

10.1 **Grades or Purity:** 99—100%
10.2 **Storage Temperature:** Ambient (gas); −155°F (liquid)
10.3 **Inert Atmosphere:** No requirement
10.4 **Venting:** Safety relief

11. HAZARD ASSESSMENT CODE

(See Hazard Assessment Handbook, CG 446-3)

A-B-C-D-E-F-G

12. HAZARD CLASSIFICATIONS

12.1 **Code of Federal Regulations:** Flammable compressed gas
12.2 **NAS Hazard Rating for Bulk Water Transportation:**

Category	Rating
Fire	4
Health	
Vapor Irritant	0
Liquid or Solid Irritant	0
Poisons	1
Water Pollution	
Human Toxicity	0
Aquatic Toxicity	1
Aesthetic Effect	0
Reactivity	
Other Chemicals	1
Water	0
Self-Reaction	2

12.3 **NFPA Hazard Classifications:**

Category	Classification
Health Hazard (Blue)	1
Flammability (Red)	4
Reactivity (Yellow)	2

13. PHYSICAL AND CHEMICAL PROPERTIES

13.1 **Physical State at 15°C and 1 atm:** Gas
13.2 **Molecular Weight:** 28.05
13.3 **Boiling Point at 1 atm:** −154.7°F = −103.7°C = 169.5°K
13.4 **Freezing Point:** −272.4°F = −169.1°C = 104.1°K
13.5 **Critical Temperature:** 49.8°F = 9.9°C = 283.1°K
13.6 **Critical Pressure:** 742 psia = 50.5 atm = 5.11 MN/m^2
13.7 **Specific Gravity:** 0.569 at −103.8°C (liquid)
13.8 **Liquid Surface Tension:** 16 dynes/cm = 0.016 N/m at −104°C
13.9 **Liquid-Water Interfacial Tension:** (est.) 50 dynes/cm = 0.05 N/m at −104°C
13.10 **Vapor (Gas) Specific Gravity:** 1.0
13.11 **Ratio of Specific Heats of Vapor (Gas):** 1.240
13.12 **Latent Heat of Vaporization:** 207.7 Btu/lb = 115.4 cal/g = 4.832×10^5 J/kg
13.13 **Heat of Combustion:** −20,290 Btu/lb = −11,272 cal/g = -471.94×10^5 J/kg
13.14 **Heat of Decomposition:** Not pertinent
13.15 **Heat of Solution:** Not pertinent
13.16 **Heat of Polymerization:** Not pertinent

NOTES

REVISED 1978

ECH | ETHYLENE CHLOROHYDRIN

Common Synonyms
2-Chloroethanol
2-Chlorethanol
2-Chloroethyl alcohol
Ethylene chlorhydrin
Glycol chlorohydrin

Liquid   Colorless   Faint sweet pleasant odor

Mixes with water. Irritating vapor is produced.

Stop discharge if possible. Keep people away.
Stay upwind. Use water spray to "knock down" vapor.
Call fire department.
Isolate and remove discharged material.
Notify local health and pollution control agencies.

Fire

Combustible.
POISONOUS GASES MAY BE PRODUCED IN FIRE.
Extinguish with water, dry chemicals, alcohol foam, or carbon dioxide.

Exposure

Call for medical aid.

VAPOR
Irritating to eyes, nose and throat.
If inhaled will cause coughing or difficult breathing.
Move victim to fresh air.
If breathing has stopped, give artificial respiration.
If breathing is difficult, give oxygen.

LIQUID
Irritating to skin and eyes.
Harmful if swallowed.
Remove contaminated clothing and shoes.
Flush affected areas with plenty of water.
IF IN EYES, hold eyelids open and flush with plenty of water.
IF SWALLOWED, and victim is CONSCIOUS, have victim drink water or milk.
DO NOT INDUCE VOMITING.

Water Pollution

Effect of low concentrations on aquatic life is unknown.
May be dangerous if it enters water intakes.
Notify local health and wildlife officials.
Notify operators of nearby water intakes.

1. RESPONSE TO DISCHARGE

(See Response Methods Handbook, CG 446-4)

Issue warning—water contaminant
Restrict access
Disperse and flush

2. LABEL

3. CHEMICAL DESIGNATIONS

3.1 **Synonyms:** 2-Chloroethanol; 2-Chlorethanol; 2-Chloroethyl alcohol; Ethylene chlorhydrin; Glycol chlorohydrin
3.2 **Coast Guard Compatibility Classification:** To be developed
3.3 **Chemical Formula:** $ClCH_2CH_2OH$
3.4 **IMCO/United Nations Numerical Designation:** 3.3/1135

4. OBSERVABLE CHARACTERISTICS

4.1 **Physical State (as shipped):** Liquid
4.2 **Color:** Colorless
4.3 **Odor:** Faint, ethereal

5. HEALTH HAZARDS

5.1 **Personal Protective Equipment:** Organic canister mask or self-contained breathing apparatus; goggles or face shield; rubber gloves.
5.2 **Symptoms Following Exposure:** Inhalation causes irritation of upper respiratory system, nausea, headache, delirium, coma, collapse. Liquid causes irritation of eyes and skin; prolonged contact with skin may allow penetration into body and cause same symptoms as following ingestion or inhalation. Ingestion causes nausea, headache, delirium, coma, and collapse.
5.3 **Treatment for Exposure:** INHALATION: remove from exposure; give artificial respiration if breathing has stopped; call physician. EYES: flush with water for at least 15 min.; get medical attention if irritation persists. SKIN: wash off with copious amounts of water; call physician if contact has been prolonged. INGESTION: give large amounts of water; get medical attention.
5.4 **Toxicity by Inhalation (Threshold Limit Value):** 5 ppm
5.5 **Short-Term Inhalation Limits:** Data not available
5.6 **Toxicity by Ingestion:** Grade 3; oral LD_{50} = 71 mg/kg (rat)
5.7 **Late Toxicity:** Damage to central nervous system and liver in humans
5.8 **Vapor (Gas) Irritant Characteristics:** Data not available
5.9 **Liquid or Solid Irritant Characteristics:** Data not available
5.10 **Odor Threshold:** Odorless

6. FIRE HAZARDS

6.1 **Flash Point:** 139°F O.C.
6.2 **Flammable Limits in Air:** 4.9%–15.9%
6.3 **Fire Extinguishing Agents:** Water, alcohol foam, dry chemical, or carbon dioxide
6.4 **Fire Extinguishing Agents Not to be Used:** Not pertinent
6.5 **Special Hazards of Combustion Products:** Toxic hydrogen chloride and phosgene fumes may be formed.
6.6 **Behavior in Fire:** Vapors are heavier than air and may flash back to a source of ignition.
6.7 **Ignition Temperature:** 797°F
6.8 **Electrical Hazard:** Data not available
6.9 **Burning Rate:** 1.7 mm/min.

7. CHEMICAL REACTIVITY

7.1 **Reactivity with Water:** No reaction
7.2 **Reactivity with Common Materials:** No reaction
7.3 **Stability During Transport:** Stable
7.4 **Neutralizing Agents for Acids and Caustics:** Not pertinent
7.5 **Polymerization:** Not pertinent
7.6 **Inhibitor of Polymerization:** Not pertinent

8. WATER POLLUTION

8.1 **Aquatic Toxicity:** Data not available
8.2 **Waterfowl Toxicity:** Data not available
8.3 **Biological Oxygen Demand (BOD):** 0.50 lb/lb, 10 days
8.4 **Food Chain Concentration Potential:** None

9. SELECTED MANUFACTURERS

1. Union Carbide Corporation
Chemicals and Plastics Division
270 Park Avenue
New York, N. Y. 10017
2. Pfaltz and Bauer, Inc.
126-04 Northern Boulevard
Flushing, N. Y. 11368
3. Eastman Kodak Co.
Eastman Organic Chemicals
Rochester, N. Y. 14650

10. SHIPPING INFORMATION

10.1 **Grades or Purity:** 99+%
10.2 **Storage Temperature:** Ambient
10.3 **Inert Atmosphere:** No requirement
10.4 **Venting:** Open (flame arrester)

11. HAZARD ASSESSMENT CODE

(See Hazard Assessment Handbook, CG 446-3)

A-P-Q

12. HAZARD CLASSIFICATIONS

12.1 **Code of Federal Regulations:** Poison, Class B
12.2 **NAS Hazard Rating for Bulk Water Transportation:** Not listed
12.3 **NFPA Hazard Classifications:**

Category	Classification
Health Hazard (Blue)	3
Flammability (Red)	2
Reactivity (Yellow)	0

13. PHYSICAL AND CHEMICAL PROPERTIES

13.1 **Physical State at 15°C and 1 atm:** Liquid
13.2 **Molecular Weight:** 80.51
13.3 **Boiling Point at 1 atm:** 263.7°F = 128.7°C = 401.9°K
13.4 **Freezing Point:** −80.7°F = −62.6°C = 210.6°K
13.5 **Critical Temperature:** Not pertinent
13.6 **Critical Pressure:** Not pertinent
13.7 **Specific Gravity:** 1.197 at 20°C (liquid)
13.8 **Liquid Surface Tension:** Data not available
13.9 **Liquid-Water Interfacial Tension:** Not pertinent
13.10 **Vapor (Gas) Specific Gravity:** 2.8
13.11 **Ratio of Specific Heats of Vapor (Gas):** Data not available
13.12 **Latent Heat of Vaporization:** 221 Btu/lb = 123 cal/g = 5.15×10^5 J/kg
13.13 **Heat of Combustion:** −6,487 Btu/lb = −3,604 cal/g = -150.8×10^5 J/kg
13.14 **Heat of Decomposition:** Not pertinent
13.15 **Heat of Solution:** Not pertinent
13.16 **Heat of Polymerization:** Not pertinent

NOTES

ETC — ETHYLENE CYANOHYDRIN

Common Synonyms			
Hydracrylonitrile 2-Cyanoethanol Glycol cyanohydrin 3-Hydroxypropanenitrile	Liquid	Colorless to yellow-brown	Weak odor to odorless
	Sinks and mixes with water.		

Stop discharge if possible.
Call fire department.
Avoid contact with liquid.
Isolate and remove discharged material.
Notify local health and pollution control agencies.

Fire

Combustible.
POISONOUS GASES MAY BE PRODUCED WHEN HEATED.
Wear self-contained breathing apparatus.
Extinguish with dry chemical, alcohol foam, or carbon dioxide.
Water may be ineffective on fire.
Cool exposed containers with water.

Exposure

CALL FOR MEDICAL AID.

LIQUID
Irritating to skin and eyes.
Harmful if swallowed.
Remove contaminated clothing and shoes.
Flush affected areas with plenty of water.
IF IN EYES, hold eyelids open and flush with plenty of water.
IF SWALLOWED and victim is CONSCIOUS, have victim drink water or milk. Have victim induce vomiting.
IF SWALLOWED and victim is UNCONSCIOUS OR HAVING CONVULSIONS, do nothing except keep victim warm.

Water Pollution

Effect of low concentrations on aquatic life is unknown.
May be dangerous if it enters water intakes.
Notify local health and wildlife officials.
Notify operators of nearby water intakes.

1. RESPONSE TO DISCHARGE

(See Response Methods Handbook, CG 446-4)

Disperse and flush

2. LABELS

No hazard label required by Code of Federal Regulations

3. CHEMICAL DESIGNATIONS

3.1 **Synonyms:** 2-Cyanoethanol
Glycol cyanohydrin
1-Hydroxy-2-cyanoethane
Hydracrylonitrile
3-Hydroxypropanenitrile

3.2 **Coast Guard Compatibility Classification:** Alcohol

3.3 **Chemical Formula:** $HOCH_2CH_2CN$

3.4 **IMCO/United Nations Numerical Designation:** Not listed

4. OBSERVABLE CHARACTERISTICS

4.1 **Physical State (as shipped):** Liquid

4.2 **Color:** Colorless or straw-colored

4.3 **Odor:** Practically odorless; characteristic

5. HEALTH HAZARDS

5.1 **Personal Protective Equipment:** Air-supplied mask; plastic gloves; rubber clothing; vapor-proof goggles.

5.2 **Symptoms Following Exposure:** Liquid causes eye irritation. If swallowed, may cause severe kidney injury.

5.3 **Treatment for Exposure:** INGESTION: induce vomiting at once and call a physician. EYES: wash with flowing water for at least 15 min. SKIN: flush exposed areas with plenty of water.

5.4 **Toxicity by Inhalation (Threshold Limit Value):** Not pertinent

5.5 **Short-Term Inhalation Limits:** Not pertinent

5.6 **Toxicity by Ingestion:** Grade 2; LD_{50} 0.5 to 5 g/kg

5.7 **Late Toxicity:** Ingestion of liquid may cause severe kidney damage.

5.8 **Vapor (Gas) Irritant Characteristics:** Vapors are nonirritating to the eyes and throat.

5.9 **Liquid or Solid Irritant Characteristics:** No appreciable hazard. Practically harmless to the skin.

5.10 **Odor Threshold:** Not pertinent

6. FIRE HAZARDS

6.1 **Flash Point:** 265°F O.C.

6.2 **Flammable Limits in Air:** 2.3% (calc.)—12.1% (est.)

6.3 **Fire Extinguishing Agents:** Carbon dioxide or dry chemical for small fires; alcohol-type foam for large fires.

6.4 **Fire Extinguishing Agents Not to be Used:** Water or foam may cause frothing.

6.5 **Special Hazards of Combustion Products:** Toxic gases are generated when heated.

6.6 **Behavior in Fire:** Decomposes, generating toxic gases

6.7 **Ignition Temperature:** 922°F

6.8 **Electrical Hazard:** Not pertinent

6.9 **Burning Rate:** Data not available

7. CHEMICAL REACTIVITY

7.1 **Reactivity with Water:** No reaction

7.2 **Reactivity with Common Materials:** No reaction

7.3 **Stability During Transport:** Stable

7.4 **Neutralizing Agents for Acids and Caustics:** Not pertinent

7.5 **Polymerization:** Not pertinent

7.6 **Inhibitor of Polymerization:** Not pertinent

8. WATER POLLUTION

8.1 **Aquatic Toxicity:** Data not available

8.2 **Waterfowl Toxicity:** Data not available

8.3 **Biological Oxygen Demand (BOD):** Data not available

8.4 **Food Chain Concentration Potential:** None

9. SELECTED MANUFACTURERS

1. American Aniline & Extract Co., Inc.
Venango and F Streets
Philadelphia, Pa. 19134
2. Union Carbide Corp.
Chemicals and Plastics Division
270 Park Avenue
New York, N. Y. 10017

10. SHIPPING INFORMATION

10.1 **Grades or Purity:** Data not available

10.2 **Storage Temperature:** Ambient

10.3 **Inert Atmosphere:** No requirement

10.4 **Venting:** Open (flame arrester)

11. HAZARD ASSESSMENT CODE

(See Hazard Assessment Handbook, CG 446-3)

A-P-Q

12. HAZARD CLASSIFICATIONS

12.1 **Code of Federal Regulations:** Not listed

12.2 **NAS Hazard Rating for Bulk Water Transportation:**

Category	Rating
Fire	1
Health	
Vapor Irritant	0
Liquid or Solid Irritant	0
Poisons	2
Water Pollution	
Human Toxicity	1
Aquatic Toxicity	2
Aesthetic Effect	2
Reactivity	
Other Chemicals	2
Water	0
Self-Reaction	0

12.3 **NFPA Hazard Classifications:**

Category	Classification
Health Hazard (Blue)	2
Flammability (Red)	1
Reactivity (Yellow)	1

13. PHYSICAL AND CHEMICAL PROPERTIES

13.1 **Physical State at 15°C and 1 atm:** Liquid

13.2 **Molecular Weight:** 71.08

13.3 **Boiling Point at 1 atm:** 445.5°F = 229.7°C = 502.9°K

13.4 **Freezing Point:** −51.2°F = −46.2°C = 227.0°K

13.5 **Critical Temperature:** 804°F = 429°C = 702°K

13.6 **Critical Pressure:** 720 psia = 49 atm = 4.9 MN/m^2

13.7 **Specific Gravity:** 1.047 at 20°C (liquid)

13.8 **Liquid Surface Tension:** Not pertinent

13.9 **Liquid-Water Interfacial Tension:** Not pertinent

13.10 **Vapor (Gas) Specific Gravity:** Not pertinent

13.11 **Ratio of Specific Heats of Vapor (Gas):** Not pertinent

13.12 **Latent Heat of Vaporization:** Not pertinent

13.13 **Heat of Combustion:** Data not available

13.14 **Heat of Decomposition:** Not pertinent

13.15 **Heat of Solution:** Not pertinent

13.16 **Heat of Polymerization:** Not pertinent

NOTES

REVISED 1978

EDA	# ETHYLENEDIAMINE

Common Synonyms
1,2-Diaminoethane
1,2-Ethanediamine

Liquid | Colorless | Mild ammonia odor

Floats and mixes with water. Irritating vapor is produced. Freezing point is 52° F.

Avoid contact with liquid and vapor. Keep people away.
Wear goggles, self-contained breathing apparatus, and rubber overclothing (including gloves).
Stop discharge if possible.
Call fire department.
Isolate and remove discharged material.
Notify local health and pollution control agencies.

Fire

Combustible.
Wear goggles, self-contained breathing apparatus, and rubber overclothing (including gloves).
Extinguish with water, dry chemical, alcohol foam, or carbon dioxide.
Cool exposed containers with water.

Exposure

CALL FOR MEDICAL AID.

VAPOR
Irritating to eyes, nose and throat.
Harmful if inhaled.
Move to fresh air.
If breathing has stopped, give artificial respiration.
If breathing is difficult, give oxygen.

LIQUID
Will burn skin and eyes.
Harmful if swallowed.
Remove contaminated clothing and shoes.
Flush affected areas with plenty of water.
IF IN EYES, hold eyelids open and flush with plenty of water.
IF SWALLOWED and victim is CONSCIOUS, have victim drink water or milk.
DO NOT INDUCE VOMITING.

Water Pollution

HARMFUL TO AQUATIC LIFE IN VERY LOW CONCENTRATIONS.
May be dangerous if it enters water intakes.
Notify local health and wildlife officials.
Notify operators of nearby water intakes.

1. RESPONSE TO DISCHARGE
(See Response Methods Handbook, CG 446-4)

Disperse and flush

2. LABELS

No hazard label required by Code of Federal Regulations

3. CHEMICAL DESIGNATIONS

3.1 **Synonyms:** 1,2-Diaminoethane
1,2-Ethanediamine
3.2 **Coast Guard Compatibility Classification:** Aliphatic amine
3.3 **Chemical Formula:** $NH_2CH_2CH_2NH_2$
3.4 **IMCO/United Nations Numerical Designation:** 8.0/1604

4. OBSERVABLE CHARACTERISTICS

4.1 **Physical State (as shipped):** Liquid
4.2 **Color:** Colorless
4.3 **Odor:** Strong ammoniacal odor; ammonia-like; mild and ammoniacal odor

5. HEALTH HAZARDS

5.1 **Personal Protective Equipment:** Full rubber protective clothing, incl. gloves and boots; chemical worker's goggles; face shield where contact with face is likely. If necessary to enter closed area for 1/2 hr or less with mist, wear full-faced gas mask with canister approved by Bureau of Standards for use with ammonia.
5.2 **Symptoms Following Exposure:** High concentration of vapor burns eyes and irritates nose and throat. Liquid burns eyes and skin.
5.3 **Treatment for Exposure:** Get medical help immediately! INGESTION: drink large amounts of water or milk quickly, induce vomiting only if instructed by physician. EYES: flush immediately and thoroughly with flowing water for at least 15 min. SKIN: remove clothing and flush affected area with copious amounts of flowing water, then wash with soap and water; severe exposure may require showering.
5.4 **Toxicity by Inhalation (Threshold Limit Value):** 10 ppm
5.5 **Short-Term Inhalation Limits:** 20 ppm for 5 min.
5.6 **Toxicity by Ingestion:** Grade 2; LD_{50} 0.5 to 5 g/kg (female rat)
5.7 **Late Toxicity:** Data not available
5.8 **Vapor (Gas) Irritant Characteristics:** Vapor is moderately irritating such that personnel will not usually tolerate moderate or high vapor concentrations.
5.9 **Liquid or Solid Irritant Characteristics:** Fairly severe skin irritant; may cause pain and second-degree burns after a few minutes' contact.
5.10 **Odor Threshold:** 10 ppm

6. FIRE HAZARDS

6.1 **Flash Point:** 99°F O.C.; 150°F C.C.
6.2 **Flammable Limits in Air:** 5.8%–11.1%
6.3 **Fire Extinguishing Agents:** Carbon dioxide, dry chemicals, foam or water
6.4 **Fire Extinguishing Agents Not to be Used:** Do not use water in case of drum or tank fires.
6.5 **Special Hazards of Combustion Products:** Irritating vapors are generated when heated.
6.6 **Behavior in Fire:** Not pertinent
6.7 **Ignition Temperature:** 715°F
6.8 **Electrical Hazard:** Not pertinent
6.9 **Burning Rate:** 2.2 mm/min.

7. CHEMICAL REACTIVITY

7.1 **Reactivity with Water:** Gives off heat, but reaction is not hazardous.
7.2 **Reactivity with Common Materials:** No reaction
7.3 **Stability During Transport:** Stable
7.4 **Neutralizing Agents for Acids and Caustics:** Flush with water
7.5 **Polymerization:** Not pertinent
7.6 **Inhibitor of Polymerization:** Not pertinent

8. WATER POLLUTION

8.1 **Aquatic Toxicity:** 60 ppm/24 hr/chub/killed/fresh water
8.2 **Waterfowl Toxicity:** Data not available
8.3 **Biological Oxygen Demand (BOD):** 75% (theor.), 5 days
8.4 **Food Chain Concentration Potential:** None

9. SELECTED MANUFACTURERS

1. Dow Chemical Co.
Midland, Michigan 48640
2. Jefferson Chemical Co., Inc.
3336 Richmond Ave.
Houston, Texas 77052
3. Union Carbide Corp.
Chemicals and Plastics Div.
270 Park Ave.
New York, N. Y. 10017

10. SHIPPING INFORMATION

10.1 **Grades or Purity:** 99+%
10.2 **Storage Temperature:** Ambient
10.3 **Inert Atmosphere:** No requirement
10.4 **Venting:** Pressure-vacuum

11. HAZARD ASSESSMENT CODE
(See Hazard Assessment Handbook, CG 446-3)

A-P-Q

12. HAZARD CLASSIFICATIONS

12.1 **Code of Federal Regulations:** Not listed
12.2 **NAS Hazard Rating for Bulk Water Transportation:**

Category	Rating
Fire	3
Health	
Vapor Irritant	3
Liquid or Solid Irritant	3
Poisons	3
Water Pollution	
Human Toxicity	2
Aquatic Toxicity	3
Aesthetic Effect	2
Reactivity	
Other Chemicals	3
Water	0
Self-Reaction	0

12.3 **NFPA Hazard Classifications:**

Category	Classification
Health Hazard (Blue)	3
Flammability (Red)	2
Reactivity (Yellow)	0

13. PHYSICAL AND CHEMICAL PROPERTIES

13.1 **Physical State at 15°C and 1 atm:** Liquid
13.2 **Molecular Weight:** 60.10
13.3 **Boiling Point at 1 atm:** 243°F = 117°C = 390°K
13.4 **Freezing Point:** 51.8°F = 11.0°C = 284.2°K
13.5 **Critical Temperature:** 608°F = 320°C = 593°K
13.6 **Critical Pressure:** 941 psia = 64 atm = 6.4 MN/m^2
13.7 **Specific Gravity:** 0.909 at 20°C (liquid)
13.8 **Liquid Surface Tension:** Not pertinent
13.9 **Liquid-Water Interfacial Tension:** Not pertinent
13.10 **Vapor (Gas) Specific Gravity:** Not pertinent
13.11 **Ratio of Specific Heats of Vapor (Gas):** 1.087
13.12 **Latent Heat of Vaporization:** 288 Btu/lb = 160 cal/g = 6.70×10^5 J/kg
13.13 **Heat of Combustion:** −12,290 Btu/lb = −6830 cal/g = -286.0×10^5 J/kg
13.14 **Heat of Decomposition:** Not pertinent
13.15 **Heat of Solution:** (est.) −9 Btu/lb = −5 cal/g = -0.2×10^5 J/kg
13.16 **Heat of Polymerization:** Not pertinent

NOTES

REVISED 1978

EDT	ETHYLENEDIAMINE TETRACETIC ACID

Common Synonyms			
Versene acid Tetrine acid Ethylene bis (iminodiacetic acid) EDTA	Solid powder Floats on water.	White	Odorless

Isolate and remove discharged material.
Notify local health and pollution control agencies.

Fire

Not flammable.

Exposure

CALL FOR MEDICAL AID.

SOLID
Irritating to skin and eyes.
Remove contaminated clothing and shoes.
Flush affected areas with plenty of water.
IF IN EYES, hold eyelids open and flush with plenty of water.

Water Pollution

Effect of low concentrations on aquatic life is unknown.
Fouling to shoreline.
May be dangerous if it enters water intakes.
Notify local health and wildlife officials.
Notify operators of nearby water intakes.

1. RESPONSE TO DISCHARGE

(See Response Methods Handbook, CG 446-4)

Issue warning—water contaminant
Should be removed

2. LABELS

No hazard label required by Code of Federal Regulations

3. CHEMICAL DESIGNATIONS

3.1 **Synonyms:** ETDA; Endrate; Ethylenebis [iminodiacetic acid]; (Ethylenedinitrilo)tetraacetic acid; Tetrine acid; Versene acid

3.2 **Coast Guard Compatibility Classification:** Not applicable

3.3 **Chemical Formula:** $(HOOCCH_2)_2NCH_2CH_2N(CH_2COOH)_2$

3.4 **IMCO/United Nations Numerical Designation:** Not listed

4. OBSERVABLE CHARACTERISTICS

4.1 **Physical State (as shipped):** Solid

4.2 **Color:** White

4.3 **Odor:** None

5. HEALTH HAZARDS

5.1 **Personal Protective Equipment:** Data not available

5.2 **Symptoms Following Exposure:** Data not available

5.3 **Treatment for Exposure:** Data not available

5.4 **Toxicity by Inhalation (Threshold Limit Value):** Not pertinent

5.5 **Short-Term Inhalation Limits:** Not pertinent

5.6 **Toxicity by Ingestion:** Grade 1; LD_{50} 5 to 15 g/kg (as sodium or calcium salt)

5.7 **Late Toxicity:** Data not available

5.8 **Vapor (Gas) Irritant Characteristics:** Not pertinent

5.9 **Liquid or Solid Irritant Characteristics:** Minimum hazard. If spilled on clothing and allowed to remain, may cause smarting and reddening of the skin.

5.10 **Odor Threshold:** Not pertinent

6. FIRE HAZARDS

6.1 **Flash Point:** Not flammable

6.2 **Flammable Limits in Air:** Not pertinent

6.3 **Fire Extinguishing Agents:** Not pertinent

6.4 **Fire Extinguishing Agents Not to be Used:** Not pertinent

6.5 **Special Hazards of Combustion Products:** Not pertinent

6.6 **Behavior in Fire:** Not pertinent

6.7 **Ignition Temperature:** Not pertinent

6.8 **Electrical Hazard:** Not pertinent

6.9 **Burning Rate:** Not pertinent

7. CHEMICAL REACTIVITY

7.1 **Reactivity with Water:** No reaction

7.2 **Reactivity with Common Materials:** No reaction

7.3 **Stability During Transport:** Stable

7.4 **Neutralizing Agents for Acids and Caustics:** Not pertinent

7.5 **Polymerization:** Not pertinent

7.6 **Inhibitor of Polymerization:** Not pertinent

8. WATER POLLUTION

8.1 **Aquatic Toxicity:** 129 ppm/96 hr/catfish/tapwater

8.2 **Waterfowl Toxicity:** Data not available

8.3 **Biological Oxygen Demand (BOD):** 1%, 5 days

8.4 **Food Chain Concentration Potential:** None

9. SELECTED MANUFACTURERS

1. Ciba-Geigy Corp.
   Geigy Industrial Chemicals Division
   McIntosh, Ala. 36553
2. Dow Chemical Co.
   Midland, Mich. 48640
3. W. R. Grace & Co.
   Dewey & Almy Chemical Division
   Cambridge, Mass. 02140

10. SHIPPING INFORMATION

10.1 **Grades or Purity:** Data not available

10.2 **Storage Temperature:** Ambient

10.3 **Inert Atmosphere:** No requirement

10.4 **Venting:** Open

11. HAZARD ASSESSMENT CODE

(See Hazard Assessment Handbook, CG 446-3)

II

12. HAZARD CLASSIFICATIONS

12.1 **Code of Federal Regulations:** Not listed

12.2 **NAS Hazard Rating for Bulk Water Transportation:**

Category	Rating
Fire	0
Health	
Vapor Irritant	0
Liquid or Solid Irritant	1
Poisons	0
Water Pollution	
Human Toxicity	0
Aquatic Toxicity	2
Aesthetic Effect	1
Reactivity	
Other Chemicals	0
Water	0
Self-Reaction	0

12.3 **NFPA Hazard Classifications:** Not listed

13. PHYSICAL AND CHEMICAL PROPERTIES

13.1 **Physical State at 15°C and 1 atm:** Powder

13.2 **Molecular Weight:** 164

13.3 **Boiling Point at 1 atm:** Not pertinent

13.4 **Freezing Point:** Data not available

13.5 **Critical Temperature:** Not pertinent

13.6 **Critical Pressure:** Not pertinent

13.7 **Specific Gravity:** 0.86 at 20°C (solid)

13.8 **Liquid Surface Tension:** Not pertinent

13.9 **Liquid-Water Interfacial Tension:** Not pertinent

13.10 **Vapor (Gas) Specific Gravity:** Not pertinent

13.11 **Ratio of Specific Heats of Vapor (Gas):** Not pertinent

13.12 **Latent Heat of Vaporization:** Not pertinent

13.13 **Heat of Combustion:** Not pertinent

13.14 **Heat of Decomposition:** Not pertinent

13.15 **Heat of Solution:** Not pertinent

13.16 **Heat of Polymerization:** Not pertinent

NOTES

REVISED 1978

EDB	ETHYLENE DIBROMIDE

Common Synonyms			
1,2-Dibromoethane Ethylene bromide	Liquid	Colorless	Sweet odor
	Sinks in water. Poisonous vapor is produced. Freezing point is 50° F.		

Stop discharge if possible. Keep people away.
Avoid contact with liquid and vapor.
Isolate and remove discharged material.
Notify local health and pollution control agencies.

Fire

Not flammable.
POISONOUS GASES ARE PRODUCED WHEN HEATED.
Wear goggles, self-contained breathing apparatus, and rubber overclothing (including gloves).
Cool exposed containers with water.

Exposure

CALL FOR MEDICAL AID.

VAPOR
POISONOUS IF INHALED.
Irritating to eyes, nose and throat.
Move to fresh air.
If breathing has stopped, give artificial respiration.
If breathing is difficult, give oxygen.

LIQUID
POISONOUS IF SWALLOWED OR IF SKIN IS EXPOSED.
Irritating to skin and eyes.
Remove contaminated clothing and shoes.
Flush affected areas with plenty of water.
IF IN EYES, hold eyelids open and flush with plenty of water.
IF SWALLOWED and victim is CONSCIOUS, have victim drink water or milk.

Water Pollution

HARMFUL TO AQUATIC LIFE IN VERY LOW CONCENTRATIONS.
May be dangerous if it enters water intakes.
Notify local health and wildlife officials.
Notify operators of nearby water intakes.

1. RESPONSE TO DISCHARGE
(See Response Methods Handbook, CG 446-4)

Should be removed
Chemical and physical treatment

2. LABELS

No hazard label required by Code of Federal Regulations

3. CHEMICAL DESIGNATIONS

3.1 **Synonyms:** Bromofume; 1, 2-Dibromoethane; sym-Dibromoethane; Dowfume 40, W-10, W-15, W-40; Ethylene bromide; Glycol dibromide
3.2 **Coast Guard Compatibility Classification:** Halogenated hydrocarbon
3.3 **Chemical Formula:** $BrCH_2CH_2Br$
3.4 **IMCO/United Nations Numerical Designation:** 6.1/1605

4. OBSERVABLE CHARACTERISTICS

4.1 **Physical State (as shipped):** Liquid
4.2 **Color:** Colorless
4.3 **Odor:** Mildly sweet; like chloroform

5. HEALTH HAZARDS

5.1 **Personal Protective Equipment:** Canister type mask or self-contained air mask; neoprene gloves; chemical safety goggles.
5.2 **Symptoms Following Exposure:** Local inflammation, blisters and ulcers on skin; irritation in lungs and organic injury to liver and kidneys; may be absorbed through skin.
5.3 **Treatment for Exposure:** Remove from exposure. Remove contaminated clothing. Wash skin with soap and water. Flush eyes with plenty of water. Consult physician.
5.4 **Toxicity by Inhalation (Threshold Limit Value):** 3 mg/m^3
5.5 **Short-Term Inhalation Limits:** 50 ppm for 5 min.
5.6 **Toxicity by Ingestion:** Grade 3; LD_{50} 50 to 500 mg/kg
5.7 **Late Toxicity:** Data not available
5.8 **Vapor (Gas) Irritant Characteristics:** Vapors cause a slight smarting of the eyes or respiratory system if present in high concentrations. The effect is temporary.
5.9 **Liquid or Solid Irritant Characteristics:** Minimum hazard. If spilled on clothing and allowed to remain, may cause smarting and reddening of the skin.
5.10 **Odor Threshold:** Data not available

6. FIRE HAZARDS

6.1 **Flash Point:** Not flammable
6.2 **Flammable Limits in Air:** Not flammable
6.3 **Fire Extinguishing Agents:** Not pertinent
6.4 **Fire Extinguishing Agents Not to be Used:** Not pertinent
6.5 **Special Hazards of Combustion Products:** Decomposition gases are toxic and irritating.
6.6 **Behavior in Fire:** Decomposes into toxic irritating gases. Reacts with hot metals such as aluminum and magnesium.
6.7 **Ignition Temperature:** Not flammable
6.8 **Electrical Hazard:** Not pertinent
6.9 **Burning Rate:** Not flammable

7. CHEMICAL REACTIVITY

7.1 **Reactivity with Water:** No reaction
7.2 **Reactivity with Common Materials:** No reaction
7.3 **Stability During Transport:** Stable
7.4 **Neutralizing Agents for Acids and Caustics:** Not pertinent
7.5 **Polymerization:** Not pertinent
7.6 **Inhibitor of Polymerization:** Not pertinent

8. WATER POLLUTION

8.1 **Aquatic Toxicity:** 18 mg/l /48 hr/bluegill/fresh water
8.2 **Waterfowl Toxicity:** Data not available
8.3 **Biological Oxygen Demand (BOD):** Data not available
8.4 **Food Chain Concentration Potential:** None

9. SELECTED MANUFACTURERS

1. Dow Chemical Co.
   Midland, Mich. 48640
2. Great Lakes Chemical Corp.
   West Lafayette, Ind. 47906
3. PPG Industries, Inc.
   Houston Chemical Co. Division
   1 Gateway Center
   Pittsburgh, Pa. 15222

10. SHIPPING INFORMATION

10.1 **Grades or Purity:** Commercial
10.2 **Storage Temperature:** Ambient
10.3 **Inert Atmosphere:** No requirement
10.4 **Venting:** Pressure-vacuum

11. HAZARD ASSESSMENT CODE
(See Hazard Assessment Handbook, CG 446-3)

A-X

12. HAZARD CLASSIFICATIONS

12.1 **Code of Federal Regulations:** ORM-A
12.2 **NAS Hazard Rating for Bulk Water Transportation:**

Category	Rating
Fire	0
Health	
Vapor Irritant	1
Liquid or Solid Irritant	1
Poisons	3
Water Pollution	
Human Toxicity	3
Aquatic Toxicity	3
Aesthetic Effect	2
Reactivity	
Other Chemicals	1
Water	0
Self-Reaction	0

12.3 **NFPA Hazard Classifications:**

Category	Classification
Health Hazard (Blue)	3
Flammability (Red)	0
Reactivity (Yellow)	0

13. PHYSICAL AND CHEMICAL PROPERTIES

13.1 **Physical State at 15°C and 1 atm:** Liquid
13.2 **Molecular Weight:** 187.86
13.3 **Boiling Point at 1 atm:** 268°F = 131°C = 404°K
13.4 **Freezing Point:** 49.6°F = 9.8°C = 283.0°K
13.5 **Critical Temperature:** Not pertinent
13.6 **Critical Pressure:** Not pertinent
13.7 **Specific Gravity:** 2.180 at 20°C (liquid)
13.8 **Liquid Surface Tension:** 38.75 dynes/cm = 0.03875 N/m at 20°C
13.9 **Liquid-Water Interfacial Tension:** 36.54 dynes/cm = 0.03654 N/m at 20°C
13.10 **Vapor (Gas) Specific Gravity:** Not pertinent
13.11 **Ratio of Specific Heats of Vapor (Gas):** 1.109
13.12 **Latent Heat of Vaporization:** 82.1 Btu/lb = 45.6 cal/g = 1.91×10^5 J/kg
13.13 **Heat of Combustion:** Not pertinent
13.14 **Heat of Decomposition:** Not pertinent
13.15 **Heat of Solution:** Not pertinent
13.16 **Heat of Polymerization:** Not pertinent

NOTES

REVISED 1978

EDC — ETHYLENE DICHLORIDE

Common Synonyms			
1,2-Dichloroethane Ethylene chloride EDC	Liquid	Colorless	Sweet odor
	Sinks in water. Flammable, irritating vapor is produced.		

Avoid contact with liquid and vapor. Keep people away.
Wear goggles, self-contained breathing apparatus, and rubber overclothing (including gloves).
Shut off ignition sources and call fire department.
Stop discharge if possible.
Stay upwind and use water spray to "knock down" vapor.
Isolate and remove discharged material.
Notify local health and pollution control agencies.

Fire

FLAMMABLE.
POISONOUS GASES ARE PRODUCED IN FIRE.
Flashback along vapor trail may occur.
Vapor may explode if ignited in an enclosed area.
Wear goggles, self-contained breathing apparatus, and rubber overclothing (including gloves).
Extinguish with dry chemical, foam, or carbon dioxide.
Water may be ineffective on fire.
Cool exposed containers with water.

Exposure

CALL FOR MEDICAL AID.

VAPOR
Irritating to eyes, nose and throat.
If inhaled, will cause nausea, dizziness or difficult breathing.
Move to fresh air.
If breathing has stopped, give artificial respiration.
If breathing is difficult, give oxygen.

LIQUID
Will burn skin and eyes.
Harmful if swallowed.
Remove contaminated clothing and shoes.
Flush affected areas with plenty of water.
IF IN EYES, hold eyelids open and flush with plenty of water.
IF SWALLOWED and victim is CONSCIOUS, have victim drink water or milk and have victim induce vomiting.
IF SWALLOWED and victim is UNCONSCIOUS OR HAVING CONVULSIONS, do nothing except keep victim warm.

Water Pollution

Dangerous to aquatic life in high concentrations.
May be dangerous if it enters water intakes.
Notify local health and wildlife officials.
Notify operators of nearby water intakes.

1. RESPONSE TO DISCHARGE

(See Response Methods Handbook, CG 446-4)

Issue warning—high flammability
Disperse and flush

2. LABEL

Red

3. CHEMICAL DESIGNATIONS

3.1 **Synonyms:**
Brocide; EDC; 1, 2-Dichloroethane; Ethylene chloride; Dutch liquid; Glycol dichloride

3.2 **Coast Guard Compatibility Classification:** Halogenated hydrocarbon

3.3 **Chemical Formula:** $ClCH_2CH_2Cl$

3.4 **IMCO/United Nations Numerical Designation:** 3.2/1184

4. OBSERVABLE CHARACTERISTICS

4.1 **Physical State (as shipped):** Liquid

4.2 **Color:** Colorless

4.3 **Odor:** Ethereal; chloroform-like; ether-like

5. HEALTH HAZARDS

5.1 **Personal Protective Equipment:** Clean, body-covering clothing and safety glasses with side shields. Respiratory protection: up to 50 ppm, none; 50 ppm to 2%, 1/2 hr or less, full face mask and canister; greater than 2%, self-contained breathing apparatus.

5.2 **Symptoms Following Exposure:** Inhalation of vapors causes nausea, drunkenness, depression. Contact of liquid with eyes may produce corneal injury. Prolonged contact with skin may cause a burn.

5.3 **Treatment for Exposure:** INHALATION: if victim is overcome, remove him to fresh air, keep him quiet and warm, and get medical attention immediately; if breathing stops, give artificial respiration. INGESTION: induce vomiting; call a physician; treat the symptoms. EYES: flush immediately with copious amounts of flowing water for at least 15 min. SKIN: remove clothing and wash skin thoroughly with soap and water; wash contaminated clothing before reuse.

5.4 **Toxicity by Inhalation (Threshold Limit Value):** 50 ppm

5.5 **Short-Term Inhalation Limits:** 200 ppm for 5 min. during any 3-hour period.

5.6 **Toxicity by Ingestion:** Grade 2; LD_{50} 0.5 to 5 g/kg (rat)

5.7 **Late Toxicity:** Data not available

5.8 **Vapor (Gas) Irritant Characteristics:** Vapors cause moderate irritation such that personnel will find high concentrations unpleasant. The effect is temporary.

5.9 **Liquid or Solid Irritant Characteristics:** Causes smarting of the skin and first-degree burns on short exposure; may cause secondary burns on long exposure.

5.10 **Odor Threshold:** 100 ppm

6. FIRE HAZARDS

6.1 **Flash Point:** 60°F O.C.; 55°F C.C.

6.2 **Flammable Limits in Air:** 6.2%–15.6%

6.3 **Fire Extinguishing Agents:** Foam, carbon dioxide, dry chemical

6.4 **Fire Extinguishing Agents Not to be Used:** Water may be ineffective.

6.5 **Special Hazards of Combustion Products:** Toxic and irritating gases (hydrogen chloride, phosgene) are generated.

6.6 **Behavior in Fire:** Vapor is heavier than air and may travel considerable distance to a source of ignition and flash back.

6.7 **Ignition Temperature:** 775°F

6.8 **Electrical Hazard:** Class I, group D

6.9 **Burning Rate:** 1.6 mm/min

7. CHEMICAL REACTIVITY

7.1 **Reactivity with Water:** No reaction

7.2 **Reactivity with Common Materials:** No reaction

7.3 **Stability During Transport:** Stable

7.4 **Neutralizing Agents for Acids and Caustics:** Not pertinent

7.5 **Polymerization:** Not pertinent

7.6 **Inhibitor of Polymerization:** Not pertinent

8. WATER POLLUTION

8.1 **Aquatic Toxicity:**
150 ppm/*/pin perch/TL_m/salt water
*Time period not specified.

8.2 **Water Fowl Toxicity:** Data not available

8.3 **Biological Oxygen Demand (BOD):** 0.002 lb/lb, 5 days

8.4 **Food Chain Concentration Potential:** None

9. SELECTED MANUFACTURERS

1. Continental Oil Co.
   Conoco Chemicals Division
   Park 80 Plaza East
   Saddle Brook, N. J. 07662
2. Dow Chemical Co.
   Midland, Mich. 48640
3. Shell Chemical Co.
   Industrial Chemicals Division
   Houston, Texas 77001

10. SHIPPING INFORMATION

10.1 **Grades or Purity:** Commercial

10.2 **Storage Temperature:** Ambient

10.3 **Inert Atmosphere:** No requirement

10.4 **Venting:** Pressure-vacuum

11. HAZARD ASSESSMENT CODE

(See Hazard Assessment Handbook, CG 446-3)

A-X

12. HAZARD CLASSIFICATIONS

12.1 **Code of Federal Regulations:** Flammable liquid

12.2 **NAS Hazard Rating for Bulk Water Transportation:**

Category	Rating
Fire	3
Health	
Vapor Irritant	2
Liquid or Solid Irritant	2
Poisons	3
Water Pollution	
Human Toxicity	3
Aquatic Toxicity	2
Aesthetic Effect	2
Reactivity	
Other Chemicals	1
Water	0
Self-Reaction	0

12.3 **NFPA Hazard Classifications:**

Category	Classification
Health Hazard (Blue)	2
Flammability (Red)	3
Reactivity (Yellow)	0

13. PHYSICAL AND CHEMICAL PROPERTIES

13.1 **Physical State at 15°C and 1 atm:** Liquid

13.2 **Molecular Weight:** 98.96

13.3 **Boiling Point at 1 atm:** 182.3°F = 83.5°C = 356.7°K

13.4 **Freezing Point:** −32.3°F = −35.7°C = 237.5°K

13.5 **Critical Temperature:** 550°F = 288°C = 561°K

13.6 **Critical Pressure:** 735 psia = 50 atm = 5.1 MN/m^2

13.7 **Specific Gravity:** 1.253 at 20°C (liquid)

13.8 **Liquid Surface Tension:** 32.2 dynes/cm = 0.0322 N/m at 20°C

13.9 **Liquid-Water Interfacial Tension:** (est.) 30 dynes/cm = 0.03 N/m at 25°C

13.10 **Vapor (Gas) Specific Gravity:** 3.4

13.11 **Ratio of Specific Heats of Vapor (Gas):** 1.118

13.12 **Latent Heat of Vaporization:** 138 Btu/lb = 76.4 cal/g = 3.20×10^5 J/kg

13.13 **Heat of Combustion:** (est.) 3400 Btu/lb = 1900 cal/g = 80×10^5 J/kg

13.14 **Heat of Decomposition:** Not pertinent

13.15 **Heat of Solution:** Not pertinent

13.16 **Heat of Polymerization:** Not pertinent

NOTES

REVISED 1978

EGL

ETHYLENE GLYCOL

Common Synonyms			
Glycol Monoethylene glycol 1,2-Ethanediol	Thick liquid	Colorless	Odorless
	Sinks and mixes with water.		

Stop discharge if possible.
Call fire department.
Isolate and remove discharged material.
Notify local health and pollution control agencies.

Fire

Combustible.
Extinguish with dry chemical, alcohol foam, or carbon dioxide.
Water may be ineffective on fire.
Cool exposed containers with water.

Exposure

CALL FOR MEDICAL AID.

LIQUID
Irritating to skin and eyes.
If swallowed, will cause loss of consciousness.
Remove contaminated clothing and shoes.
Flush affected areas with plenty of water.
IF IN EYES, hold eyelids open and flush with plenty of water.
IF SWALLOWED and victim is CONSCIOUS, have victim drink water or milk and have victim induce vomiting.
IF SWALLOWED and victim is UNCONSCIOUS OR HAVING CONVULSIONS, do nothing except keep victim warm.

Water Pollution

Effect of low concentrations on aquatic life is unknown.
May be dangerous if it enters water intakes.
Notify local health and wildlife officials.
Notify operators of nearby water intakes.

1. RESPONSE TO DISCHARGE

(See Response Methods Handbook, CG 446-4)

Disperse and flush

2. LABELS

No hazard label required by Code of Federal Regulations

3. CHEMICAL DESIGNATIONS

3.1 **Synonyms:** 1, 2-Dihydroxyethane
1, 2-Ethanediol
Ethylene dihydrate
Glycol
Monoethylene glycol

3.2 **Coast Guard Compatibility Classification:** Glycol

3.3 **Chemical Formula:** $HOCH_2CH_2OH$

3.4 **IMCO/United Nations Numerical Designation:** Not listed

4. OBSERVABLE CHARACTERISTICS

4.1 **Physical State (as shipped):** Liquid

4.2 **Color:** Colorless

4.3 **Odor:** Slight odor

5. HEALTH HAZARDS

5.1 **Personal Protective Equipment:** Goggles; shower and eye bath.

5.2 **Symptoms Following Exposure:** Inhalation of vapor is not hazardous. Ingestion causes stupor or coma, sometimes leading to fatal kidney injury.

5.3 **Treatment for Exposure:** INGESTION: induce vomiting and call a physician. SKIN AND EYES: flush with water.

5.4 **Toxicity by Inhalation (Threshold Limit Value):** 100 ppm

5.5 **Short-Term Inhalation Limits:** Not pertinent

5.6 **Toxicity by Ingestion:** Grade 1; LD_{50} 5 to 15 g/kg (rat, guinea pig, mouse)

5.7 **Late Toxicity:** Fatal kidney injury may result if ingested.

5.8 **Vapor (Gas) Irritant Characteristics:** Vapors are nonirritating to the eyes and throat.

5.9 **Liquid or Solid Irritant Characteristics:** No appreciable hazard. Practically harmless to the skin.

5.10 **Odor Threshold:** Not pertinent

6. FIRE HAZARDS

6.1 **Flash Point:** 240°F O.C.; 232°F C.C.

6.2 **Flammable Limits in Air:** LEL = 3.2%; UEL not listed

6.3 **Fire Extinguishing Agents:** Water fog, alcohol foam, carbon dioxide, or dry chemical

6.4 **Fire Extinguishing Agents Not to be Used:** Water or foam may cause frothing.

6.5 **Special Hazards of Combustion Products:** Not pertinent

6.6 **Behavior in Fire:** Not pertinent

6.7 **Ignition Temperature:** 775°F

6.8 **Electrical Hazard:** Not pertinent

6.9 **Burning Rate:** 1.0 mm/min.

7. CHEMICAL REACTIVITY

7.1 **Reactivity with Water:** No reaction

7.2 **Reactivity with Common Materials:** No reaction

7.3 **Stability During Transport:** Stable

7.4 **Neutralizing Agents for Acids and Caustics:** Not pertinent

7.5 **Polymerization:** Not pertinent

7.6 **Inhibitor of Polymerization:** Not pertinent

8. WATER POLLUTION

8.1 **Aquatic Toxicity:** >100 ppm/48 hr/shrimp/LC_{50}/salt water

8.2 **Waterfowl Toxicity:** Data not available

8.3 **Biological Oxygen Demand (BOD):** 12.5% (theor.), 5 days; 78% (theor.), 20 days

8.4 **Food Chain Concentration Potential:** None

9. SELECTED MANUFACTURERS

1. Dow Chemical Co.
Midland, Mich. 48640

2. PPG Industries, Inc.
Chemical Division
Guayanilla, Puerto Rico 00656

3. Union Carbide Corp.
Chemicals and Plastics Division
270 Park Ave.
New York, N. Y. 10017

10. SHIPPING INFORMATION

10.1 **Grades or Purity:** Industrial grade; low-conductivity grade

10.2 **Storage Temperature:** Ambient

10.3 **Inert Atmosphere:** No requirement

10.4 **Venting:** Open (flame arrester)

11. HAZARD ASSESSMENT CODE

(See Hazard Assessment Handbook, CG 446-3)

A-P-Q

12. HAZARD CLASSIFICATIONS

12.1 **Code of Federal Regulations:** Not listed

12.2 **NAS Hazard Rating for Bulk Water Transportation:**

Category	Rating
Fire	1
Health	
Vapor Irritant	0
Liquid or Solid Irritant	0
Poisons	1
Water Pollution	
Human Toxicity	2
Aquatic Toxicity	1
Aesthetic Effect	1
Reactivity	
Other Chemicals	2
Water	0
Self-Reaction	0

12.3 **NFPA Hazard Classifications:**

Category	Classification
Health Hazard (Blue)	1
Flammability (Red)	1
Reactivity (Yellow)	0

13. PHYSICAL AND CHEMICAL PROPERTIES

13.1 **Physical State at 15°C and 1 atm:** Liquid

13.2 **Molecular Weight:** 62.07

13.3 **Boiling Point at 1 atm:** 387.7°F = 197.6°C = 470.8°K

13.4 **Freezing Point:** 8.6°F = −13°C = 260°K

13.5 **Critical Temperature:** Not pertinent

13.6 **Critical Pressure:** Not pertinent

13.7 **Specific Gravity:** 1.115 at 20°C (liquid)

13.8 **Liquid Surface Tension:** Not pertinent

13.9 **Liquid-Water Interfacial Tension:** Not pertinent

13.10 **Vapor (Gas) Specific Gravity:** Not pertinent

13.11 **Ratio of Specific Heats of Vapor (Gas):** 1.095

13.12 **Latent Heat of Vaporization:** 344 Btu/lb = 191 cal/g = 8.00×10^5 J/kg

13.13 **Heat of Combustion:** −7259 Btu/lb = −4033 cal/g = -168.9×10^5 J/kg

13.14 **Heat of Decomposition:** Not pertinent

13.15 **Heat of Solution:** (est.) −20 Btu/lb = −12 cal/g = -0.5×10^5 J/kg

13.16 **Heat of Polymerization:** Not pertinent

NOTES

EGY ETHYLENE GLYCOL DIACETATE

Common Synonyms			
Ethylene acetate Glycol diacetate Ethylene diacetate	Liquid	Colorless	Weak fruity odor
	Sinks and mixes with water.		

Stop discharge if possible. Keep people away.
Call fire department.
Isolate and remove discharged material.
Notify local health and pollution control agencies.

Fire

Combustible.
Extinguish with water, dry chemicals, alcohol foam, or carbon dioxide.

Exposure

Call for medical aid.

LIQUID
Irritating to skin and eyes.
Remove contaminated clothing and shoes.
Flush affected areas with plenty of water.
IF IN EYES, hold eyelids open and flush with plenty of water.
IF SWALLOWED and victim is CONSCIOUS, have victim drink water or milk.

Water Pollution

Effect of low concentrations on aquatic life is unknown.
May be dangerous if it enters water intakes.
Notify local health and wildlife officials.
Notify operators of nearby water intakes.

1. RESPONSE TO DISCHARGE

(See Response Methods Handbook, CG 446-4)

Issue warning—water contaminant
Disperse and flush

2. LABELS

No label required by Code of Federal Regulations

3. CHEMICAL DESIGNATIONS

3.1 **Synonyms:** Ethylene acetate; Ethylene diacetate; Glycol diacetate
3.2 **Coast Guard Compatibility Classification:** Esters (13)
3.3 **Chemical Formula:** $CH_3COOCH_2CH_2OCOCH_3$
3.4 **IMCO/United Nations Numerical Designation:** Not listed

4. OBSERVABLE CHARACTERISTICS

4.1 **Physical State (as shipped):** Liquid
4.2 **Color:** Colorless
4.3 **Odor:** Weak; fruity

5. HEALTH HAZARDS

5.1 **Personal Protective Equipment:** Goggles or face shield; rubber gloves.
5.2 **Symptoms Following Exposure:** Inhalation is not hazardous. Liquid causes mild irritation of eyes. Ingestion causes stupor or coma.
5.3 **Treatment for Exposure:** INHALATION: remove to fresh air. EYES and SKIN: flush well with water. INGESTION: induce vomiting; call a physician.
5.4 **Toxicity by Inhalation (Threshold Limit Value):** Data not available
5.5 **Short-Term Inhalation Limits:** Not pertinent
5.6 **Toxicity by Ingestion:** Grade 1; oral LD_{50} = 6,860 mg/kg (rat)
5.7 **Late Toxicity:** Ingestion may cause severe injury to kidneys.
5.8 **Vapor (Gas) Irritant Characteristics:** Vapors are nonirritating to eyes and throat.
5.9 **Liquid or Solid Irritant Characteristics:** No appreciable hazard. Practically harmless to the skin.
5.10 **Odor Threshold:** Data not available

6. FIRE HAZARDS

6.1 **Flash Point:** 205°F O.C.; 191°F C.C.
6.2 **Flammable Limits in Air:** 1.6%–8.4%
6.3 **Fire Extinguishing Agents:** Water, alcohol foam, dry chemical, or carbon dioxide
6.4 **Fire Extinguishing Agents Not to be Used:** Not pertinent
6.5 **Special Hazards of Combustion Products:** Not pertinent
6.6 **Behavior in Fire:** Not pertinent
6.7 **Ignition Temperature:** 900°F
6.8 **Electrical Hazard:** Data not available
6.9 **Burning Rate:** 2.9 mm/min.

7. CHEMICAL REACTIVITY

7.1 **Reactivity with Water:** No reaction
7.2 **Reactivity with Common Materials:** No reaction
7.3 **Stability During Transport:** Stable
7.4 **Neutralizing Agents for Acids and Caustics:** Not pertinent
7.5 **Polymerization:** Not pertinent
7.6 **Inhibitor of Polymerization:** Not pertinent

8. WATER POLLUTION

8.1 **Aquatic Toxicity:** Data not available
8.2 **Waterfowl Toxicity:** Data not available
8.3 **Biological Oxygen Demand (BOD):** Data not available
8.4 **Food Chain Concentration Potential:** None

9. SELECTED MANUFACTURERS

1. Union Carbide Corporation
   Chemicals and Plastics Division
   270 Park Avenue
   New York, N. Y. 10017
2. Eastman Chemical Products, Inc.
   Kingsport, Tenn. 37662.
3. Pfaltz and Bauer, Inc.
   126-04 Northern Boulevard
   Flushing, N. Y. 11368

10. SHIPPING INFORMATION

10.1 **Grades or Purity:** 98+%
10.2 **Storage Temperature:** Ambient
10.3 **Inert Atmosphere:** No requirement
10.4 **Venting:** Open (flame arrester)

11. HAZARD ASSESSMENT CODE

(See Hazard Assessment Handbook, CG 446-3)

A-P-Q

12. HAZARD CLASSIFICATIONS

12.1 **Code of Federal Regulations:** Not listed
12.2 **NAS Hazard Rating for Bulk Water Transportation:**

Category	Rating
Fire	1
Health	
Vapor Irritant	0
Liquid or Solid Irritant	0
Poisons	0
Water Pollution	
Human Toxicity	1
Aquatic Toxicity	2
Aesthetic Effect	2
Reactivity	
Other Chemicals	1
Water	0
Self-Reaction	0

12.3 **NFPA Hazard Classifications:**

Category	Classification
Health Hazard (Blue)	1
Flammability (Red)	1
Reactivity (Yellow)	0

13. PHYSICAL AND CHEMICAL PROPERTIES

13.1 **Physical State at 15°C and 1 atm:** Liquid
13.2 **Molecular Weight:** 146.1
13.3 **Boiling Point at 1 atm:** 375.6°F = 190.9°C = 464.1°K
13.4 **Freezing Point:** −42.7°F = −41.5°C = 231.7°K
13.5 **Critical Temperature:** Not pertinent
13.6 **Critical Pressure:** Not pertinent
13.7 **Specific Gravity:** 1.104 at 20°C (liquid)
13.8 **Liquid Surface Tension:** (est.) 20 dynes/cm = 0.020 N/m at 20°C
13.9 **Liquid-Water Interfacial Tension:** Not pertinent
13.10 **Vapor (Gas) Specific Gravity:** Not pertinent
13.11 **Ratio of Specific Heats of Vapor (Gas):** Not pertinent
13.12 **Latent Heat of Vaporization:** 133 Btu/lb = 74 cal/g = 3.1×10^5 J/kg
13.13 **Heat of Combustion:** (est.) −11,000 Btu/lb = −6,000 cal/g = -250×10^5 J/kg
13.14 **Heat of Decomposition:** Not pertinent
13.15 **Heat of Solution:** Not pertinent
13.16 **Heat of Polymerization:** Not pertinent

NOTES

EEE

ETHYLENE GLYCOL DIETHYL ETHER

Common Synonyms

Diethyl "Cellosolve"
1,2-Diethoxyethane

Liquid Colorless Mild pleasant odor

Floats and mixes slowly with water. Irritating vapor is produced.

Stop discharge if possible. Keep people away.
Call fire department.
Isolate and remove discharged material.
Notify local health and pollution control agencies.

Fire

Combustible.
Extinguish with dry chemicals, alcohol foam, or carbon dioxide.
Water may be ineffective on fire.

Exposure

Call for medical aid.

VAPOR
Irritating to eyes, nose and throat.
Move victim to fresh air.

LIQUID
Irritating to skin and eyes.
Harmful if swallowed.
Remove contaminated clothing and shoes.
Flush affected areas with plenty of water.
IF IN EYES, hold eyelids open and flush with plenty of water.
IF SWALLOWED and victim is CONSCIOUS, have victim drink water or milk.

Water Pollution

Effect of low concentrations on aquatic life is unknown.
Fouling to shoreline.
May be dangerous if it enters water intakes.
Notify local health and wildlife officials.
Notify operators of nearby water intakes.

1. RESPONSE TO DISCHARGE

(See Response Methods Handbook, CG 446-4)

Issue warning—water contaminant
Disperse and flush

2. LABELS

No label required by Code of Federal Regulations

3. CHEMICAL DESIGNATIONS

3.1 **Synonyms:** Diethyl "Cellosolve"; 1,2-Diethoxyethane
3.2 **Coast Guard Compatibility Classification:** Alcohols (6)
3.3 **Chemical Formula:** $C_2H_5OCH_2CH_2OC_2H_5$
3.4 **IMCO/United Nations Numerical Designation:** 3.3/1153

4. OBSERVABLE CHARACTERISTICS

4.1 **Physical State (as shipped):** Liquid
4.2 **Color:** Colorless
4.3 **Odor:** Faint, ethereal

5. HEALTH HAZARDS

5.1 **Personal Protective Equipment:** Protective goggles or face shield; rubber gloves.
5.2 **Symptoms Following Exposure:** Inhalation causes irritation of nose and throat. Contact with liquid irritates eyes but has little or no effect on skin. Ingestion causes irritation of mouth and stomach.
5.3 **Treatment for Exposure:** INHALATION: remove from exposure. EYES: flush with water for at least 15 min. SKIN: wash with copious amounts of water. INGESTION: drink water and get medical attention.
5.4 **Toxicity by Inhalation (Threshold Limit Value):** Data not available
5.5 **Short-Term Inhalation Limits:** Data not available
5.6 **Toxicity by Ingestion:** Grade 2; oral LD_{50} = 4,390 mg/kg (rat)
5.7 **Late Toxicity:** Data not available
5.8 **Vapor (Gas) Irritant Characteristics:** Data not available
5.9 **Liquid or Solid Irritant Characteristics:** Data not available
5.10 **Odor Threshold:** Data not available

6. FIRE HAZARDS

6.1 **Flash Point:** 90°F O.C.
6.2 **Flammable Limits in Air:** Data not available
6.3 **Fire Extinguishing Agents:** Dry chemical, alcohol foam, carbon dioxide
6.4 **Fire Extinguishing Agents Not to be Used:** Water may be ineffective.
6.5 **Special Hazards of Combustion Products:** Not pertinent
6.6 **Behavior in Fire:** Not pertinent
6.7 **Ignition Temperature:** 406°F
6.8 **Electrical Hazard:** Data not available
6.9 **Burning Rate:** 4.1 mm/min.

7. CHEMICAL REACTIVITY

7.1 **Reactivity with Water:** No reaction
7.2 **Reactivity with Common Materials:** No reaction
7.3 **Stability During Transport:** Stable
7.4 **Neutralizing Agents for Acids and Caustics:** Not pertinent
7.5 **Polymerization:** Not pertinent
7.6 **Inhibitor of Polymerization:** Not pertinent

8. WATER POLLUTION

8.1 **Aquatic Toxicity:** Data not available
8.2 **Waterfowl Toxicity:** Data not available
8.3 **Biological Oxygen Demand (BOD):** 0.10 lb/lb, 10 days
8.4 **Food Chain Concentration Potential:** None

9. SELECTED MANUFACTURERS

1. Eastern Chemical Division
   230 Marcus Boulevard
   Hauppauge, N. Y. 11787
2. Pfaltz and Bauer, Inc.
   126-04 Northern Boulevard
   Flushing, N. Y. 11368

10. SHIPPING INFORMATION

10.1 **Grades or Purity:** Commercial
10.2 **Storage Temperature:** Ambient
10.3 **Inert Atmosphere:** No requirement
10.4 **Venting:** Open (flame arrester)

11. HAZARD ASSESSMENT CODE

(See Hazard Assessment Handbook, CG 446-3)

A-P-Q-T-U

12. HAZARD CLASSIFICATIONS

12.1 **Code of Federal Regulations:** Combustible Liquid
12.2 **NAS Hazard Rating for Bulk Water Transportation:** Not listed
12.3 **NFPA Hazard Classifications:** Not listed

13. PHYSICAL AND CHEMICAL PROPERTIES

13.1 **Physical State at 15°C and 1 atm:** Liquid
13.2 **Molecular Weight:** 118.2
13.3 **Boiling Point at 1 atm:** 252°F = 122°C = 395°K
13.4 **Freezing Point:** −101°F = −74°C = 199°K
13.5 **Critical Temperature:** Not pertinent
13.6 **Critical Pressure:** Not pertinent
13.7 **Specific Gravity:** 0.8484 at 20°C (liquid)
13.8 **Liquid Surface Tension:** (est.) 26 dynes/cm = 0.026 N/m at 20°C
13.9 **Liquid-Water Interfacial Tension:** Not pertinent
13.10 **Vapor (Gas) Specific Gravity:** 4.1
13.11 **Ratio of Specific Heats of Vapor (Gas):** 1.0504
13.12 **Latent Heat of Vaporization:** 192 Btu/lb = 107 cal/g = 4.48×10^5 J/kg
13.13 **Heat of Combustion:** (est.) −15,000 Btu/lb = −8,100 cal/g = -340×10^5 J/kg
13.14 **Heat of Decomposition:** Not pertinent
13.15 **Heat of Solution:** Not pertinent
13.16 **Heat of Polymerization:** Not pertinent

NOTES

EGD	# ETHYLENE GLYCOL DIMETHYL ETHER

Common Synonyms
Monoglyme
1,2-Dimethoxyethane
Dimethyl Cellosolve
Ansul Ether 12'

Liquid | Colorless | Fragrant odor

Floats and mixes with water. Irritating vapor is produced.

Stop discharge if possible.
Call fire department.
Isolate and remove discharged material.
Notify local health and pollution control agencies.

Fire

Combustible.
Extinguish with dry chemical, alcohol foam, or carbon dioxide.
Cool exposed containers with water.

Exposure

CALL FOR MEDICAL AID.

VAPOR
If inhaled, will cause dizziness or difficult breathing.
Move to fresh air.
If breathing has stopped, give artificial respiration.
If breathing is difficult, give oxygen.

LIQUID
Not irritating to skin.
If swallowed, will cause nausea, vomiting or loss of consciousness.
IF SWALLOWED and victim is CONSCIOUS, have victim drink water or milk.
IF SWALLOWED and victim is UNCONSCIOUS OR HAVING CONVULSIONS, do nothing except keep victim warm.

Water Pollution

Effect of low concentrations on aquatic life is unknown.
May be dangerous if it enters water intakes.
Notify local health and wildlife officials.
Notify operators of nearby water intakes.

1. RESPONSE TO DISCHARGE

(See Response Methods Handbook, CG 446-4)

Issue warning—high flammability
Disperse and flush

2. LABELS

No hazard label required by Code of Federal Regulations

3. CHEMICAL DESIGNATIONS

3.1 **Synonyms:** Ansul Ether 121
1, 2-Dimethoxyethane
Dimethyl Cellosolve
Monoglyme

3.2 **Coast Guard Compatibility Classification:** Glycol ether

3.3 **Chemical Formula:** $CH_3OCH_2CH_2OCH_3$

3.4 **IMCO/United Nations Numerical Designation:** Not listed

4. OBSERVABLE CHARACTERISTICS

4.1 **Physical State (as shipped):** Liquid

4.2 **Color:** Colorless

4.3 **Odor:** Mild etheral

5. HEALTH HAZARDS

5.1 **Personal Protective Equipment:** Vinyl (not rubber) protective gloves; safety glasses or goggles.

5.2 **Symptoms Following Exposure:** If ingested causes nausea, vomiting, cramps, weakness, coma.

5.3 **Treatment for Exposure:** INHALATION: oxygen and artificial respiration as needed. INGESTION gastric lavage with water or water-mineral oil.

5.4 **Toxicity by Inhalation (Threshold Limit Value):** Data not available

5.5 **Short-Term Inhalation Limits:** Data not available

5.6 **Toxicity by Ingestion:** Grade 1; LD_{50} 5 to 15 g/kg (adult albino rat)

5.7 **Late Toxicity:** Data not available

5.8 **Vapor (Gas) Irritant Characteristics:** None

5.9 **Liquid or Solid Irritant Characteristics:** None

5.10 **Odor Threshold:** Data not available

6. FIRE HAZARDS

6.1 **Flash Point:** 104°F C.C.

6.2 **Flammable Limits in Air:** Data not available

6.3 **Fire Extinguishing Agents:** Dry chemical, foam, carbon dioxide

6.4 **Fire Extinguishing Agents Not to be Used:** Not pertinent

6.5 **Special Hazards of Combustion Products:** Not pertinent

6.6 **Behavior in Fire:** Containers may explode in fires.

6.7 **Ignition Temperature:** 1373°F

6.8 **Electrical Hazard:** Not pertinent

6.9 **Burning Rate:** 4.9 mm/min.

7. CHEMICAL REACTIVITY

7.1 **Reactivity with Water:** No reaction

7.2 **Reactivity with Common Materials:** No reaction

7.3 **Stability During Transport:** Stable

7.4 **Neutralizing Agents for Acids and Caustics:** Not pertinent

7.5 **Polymerization:** Not pertinent

7.6 **Inhibitor of Polymerization:** Not pertinent

8. WATER POLLUTION

8.1 **Aquatic Toxicity:** Data not available

8.2 **Waterfowl Toxicity:** Data not available

8.3 **Biological Oxygen Demand (BOD):** Data not available

8.4 **Food Chain Concentration Potential:** None

9. SELECTED MANUFACTURERS

The Ansul Company
Chemical Division
One Stanton St.
Marinette, Wis. 54143

10. SHIPPING INFORMATION

10.1 **Grades or Purity:** Commercial

10.2 **Storage Temperature:** Ambient

10.3 **Inert Atmosphere:** No requirement

10.4 **Venting:** Pressure-vacuum

11. HAZARD ASSESSMENT CODE

(See Hazard Assessment Handbook, CG 446-3)

A-P-Q-R-S

12. HAZARD CLASSIFICATIONS

12.1 **Code of Federal Regulations:** Combustible Liquid

12.2 **NAS Hazard Rating for Bulk Water Transportation:** Not listed

12.3 **NFPA Hazard Classifications:** Not listed

13. PHYSICAL AND CHEMICAL PROPERTIES

13.1 **Physical State at 15°C and 1 atm:** Liquid

13.2 **Molecular Weight:** 90.12

13.3 **Boiling Point at 1 atm:** 185.4°F = 85.2°C = 358.4°K

13.4 **Freezing Point:** −92°F = −69°C = 204°K

13.5 **Critical Temperature:** 505°F = 263°C = 536°K

13.6 **Critical Pressure:** 561 psia = 38.2 atm = 3.87 MN/m^2

13.7 **Specific Gravity:** 0.868 at 20°C (liquid)

13.8 **Liquid Surface Tension:** Not pertinent

13.9 **Liquid-Water Interfacial Tension:** Not pertinent

13.10 **Vapor (Gas) Specific Gravity:** 3.1

13.11 **Ratio of Specific Heats of Vapor (Gas):** 1.071

13.12 **Latent Heat of Vaporization:** 134 Btu/lb = 74.6 cal/g = 3.12×10^5 J/kg

13.13 **Heat of Combustion:** −12,020 Btu/lb = −6680 cal/g = -279.7×10^5 J/kg

13.14 **Heat of Decomposition:** Not pertinent

13.15 **Heat of Solution:** (est.) −9 Btu/lb = −5 cal/g = -0.2×10^5 J/kg

13.16 **Heat of Polymerization:** Not pertinent

NOTES

REVISED 1978

EGM ETHYLENE GLYCOL MONOBUTYL ETHER

Common Synonyms			
2-Butoxyethanol Butyl Cellosolve Dowanol EB Poly-Solv EB	Oily liquid	Colorless	Mild rancid odor
	Floats and mixes with water.		

Stop discharge if possible.
Call fire department.
Avoid contact with liquid.
Isolate and remove discharged material.
Notify local health and pollution control agencies.

Fire	Combustible. **Extinguish with dry chemical, alcohol foam, or carbon dioxide.** **Cool exposed containers with water.**
Exposure	CALL FOR MEDICAL AID. LIQUID Irritating to skin and eyes. Harmful if swallowed. **Remove contaminated clothing and shoes.** **Flush affected areas with plenty of water.** **IF IN EYES, hold eyelids open and flush with plenty of water.** **IF SWALLOWED, and victim is CONSCIOUS, have victim drink water or milk.**
Water Pollution	Effect of low concentrations on aquatic life is unknown. May be dangerous if it enters water intakes. **Notify local health and wildlife officials.** **Notify operators of nearby water intakes.**

1. RESPONSE TO DISCHARGE

(See Response Methods Handbook, CG 446-4)

Disperse and flush

2. LABELS

No hazard label required by Code of Federal Regulations

3. CHEMICAL DESIGNATIONS

3.1 **Synonyms:** 2-Butoxyethanol; Butyl Cellosolve; Dowanol EB; Glycol butyl ether; Poly-Solv EB

3.2 **Coast Guard Compatibility Classification:** Glycol ether

3.3 **Chemical Formula:** $CH_3(CH_2)_3OCH_2CH_2OH$

3.4 **IMCO/United Nations Numerical Designation:** Not listed

4. OBSERVABLE CHARACTERISTICS

4.1 **Physical State (as shipped):** Liquid

4.2 **Color:** Colorless

4.3 **Odor:** Mild, characteristic; slightly rancid; mild ethereal

5. HEALTH HAZARDS

5.1 **Personal Protective Equipment:** Air pack or organic canister respirator; rubber gloves; goggles; clothing to prevent body contact with liquid.

5.2 **Symptoms Following Exposure:** Vapors irritate eyes and nose. Ingestion or skin contact causes headache, nausea, vomiting, dizziness.

5.3 **Treatment for Exposure:** INHALATION: remove to fresh air and call a physician. SKIN OR EYES: immediately flush with plenty of water; get medical care for eyes.

5.4 **Toxicity by Inhalation (Threshold Limit Value):** 50 ppm

5.5 **Short-Term Inhalation Limits:** Data not available

5.6 **Toxicity by Ingestion:** Grade 2; LD_{50} 0.5 to 5 g/kg (rat)

5.7 **Late Toxicity:** Data not available

5.8 **Vapor (Gas) Irritant Characteristics:** Vapors cause a slight smarting of the eyes or respiratory system if present in high concentrations. The effect is temporary.

5.9 **Liquid or Solid Irritant Characteristics:** Minimum hazard. If spilled on clothing and allowed to remain, may cause smarting and reddening of the skin.

5.10 **Odor Threshold:** Data not available

6. FIRE HAZARDS

6.1 **Flash Point:** 165°F O.C.; 155°F C.C.

6.2 **Flammable Limits in Air:** 1.1%–10.6%

6.3 **Fire Extinguishing Agents:** Carbon dioxide or dry chemical for small fires; alcohol-type foam for large fires.

6.4 **Fire Extinguishing Agents Not to be Used:** Data not available

6.5 **Special Hazards of Combustion Products:** Not pertinent

6.6 **Behavior in Fire:** Not pertinent

6.7 **Ignition Temperature:** 472°F

6.8 **Electrical Hazard:** Not pertinent

6.9 **Burning Rate:** 6.7 mm/min.

7. CHEMICAL REACTIVITY

7.1 **Reactivity with Water:** No reaction

7.2 **Reactivity with Common Materials:** No reaction

7.3 **Stability During Transport:** Stable

7.4 **Neutralizing Agents for Acids and Caustics:** Not pertinent

7.5 **Polymerization:** Not pertinent

7.6 **Inhibitor of Polymerization:** Not pertinent

8. WATER POLLUTION

8.1 **Aquatic Toxicity:** 1000 ppm/24 hr/brine shrimp/TL_m

8.2 **Waterfowl Toxicity:** Data not available

8.3 **Biological Oxygen Demand (BOD):** 26% of theoretical in 5 days, fresh water

8.4 **Food Chain Concentration Potential:** None

9. SELECTED MANUFACTURERS

1. Jefferson Chemical Co.
   3336 Richmond Ave.
   Houston, Texas 77052
2. Olin Corp.
   Chemicals Division
   Brandenburg, Ky. 40108
3. Union Carbide Corp.
   Chemicals and Plastics Division
   270 Park Ave.
   New York, N. Y. 10017

10. SHIPPING INFORMATION

10.1 **Grades or Purity:** Commercial

10.2 **Storage Temperature:** Ambient

10.3 **Inert Atmosphere:** No requirement

10.4 **Venting:** Open (flame arrester)

11. HAZARD ASSESSMENT CODE

(See Hazard Assessment Handbook, CG 446-3)

A-P-Q

12. HAZARD CLASSIFICATIONS

12.1 **Code of Federal Regulations:** Combustible Liquid

12.2 **NAS Hazard Rating for Bulk Water Transportation:**

Category	Rating
Fire	1
Health	
Vapor Irritant	1
Liquid or Solid Irritant	1
Poisons	2
Water Pollution	
Human Toxicity	2
Aquatic Toxicity	2
Aesthetic Effect	1
Reactivity	
Other Chemicals	2
Water	0
Self-Reaction	0

12.3 **NFPA Hazard Classifications:**

Category	Classification
Health Hazard (Blue)	2
Flammability (Red)	2
Reactivity (Yellow)	0

13. PHYSICAL AND CHEMICAL PROPERTIES

13.1 **Physical State at 15°C and 1 atm:** Liquid

13.2 **Molecular Weight:** 118.18

13.3 **Boiling Point at 1 atm:** 340.2°F = 171.2°C = 444.4°K

13.4 **Freezing Point:** −103°F = −75°C = 198°K

13.5 **Critical Temperature:** 694°F = 368°C = 641°K

13.6 **Critical Pressure:** 470 psia = 32 atm = 3.2 MN/m^2

13.7 **Specific Gravity:** 0.902 at 20°C (liquid)

13.8 **Liquid Surface Tension:** Not pertinent

13.9 **Liquid-Water Interfacial Tension:** Not pertinent

13.10 **Vapor (Gas) Specific Gravity:** Not pertinent

13.11 **Ratio of Specific Heats of Vapor (Gas):** 1.047

13.12 **Latent Heat of Vaporization:** 157 Btu/lb = 87.1 cal/g = 3.65×10^5 J/kg

13.13 **Heat of Combustion:** -13,890 Btu/lb = -7720 cal/g = -323×10^5 J/kg

13.14 **Heat of Decomposition:** Not pertinent

13.15 **Heat of Solution:** (est.) −9 Btu/lb = −5 cal/g = -0.2×10^5 J/kg

13.16 **Heat of Polymerization:** Not pertinent

NOTES

REVISED 1978

EMA — ETHYLENE GLYCOL MONOBUTYL ETHER ACETATE

Common Synonyms	Liquid	Colorless	Weak fruity odor
2-Butoxyethanol acetate 2-Butoxyethyl acetate Glycol monobutyl ether acetate Butyl "Cellosolve" acetate	Floats and mixes slowly with water.		

Stop discharge if possible. Keep people away.
Call fire department.
Isolate and remove discharged material.
Notify local health and pollution control agencies.

Fire

Combustible.
Extinguish with dry chemicals, alcohol foam, or carbon dioxide.
Water may be ineffective on fire.

Exposure

Call for medical aid.

LIQUID
Irritating to skin and eyes.
Harmful if swallowed.
Remove contaminated clothing and shoes.
Flush affected areas with plenty of water.
IF IN EYES, hold eyelids open and flush with plenty of water.
IF SWALLOWED and victim is CONSCIOUS, have victim drink water or milk.

Water Pollution

Effect of low concentrations on aquatic life is unknown.
Fouling to shoreline.
May be dangerous if it enters water intakes.
Notify local health and wildlife officials.
Notify operators of nearby water intakes.

1. RESPONSE TO DISCHARGE

(See Response Methods Handbook, CG 446-4)

Issue warning—water contaminant
Disperse and flush

2. LABELS

No label required by Code of Federal Regulations

3. CHEMICAL DESIGNATIONS

3.1 **Synonyms:** 2-Butoxyethanol, acetate; 2-Butoxyethyl acetate; Butyl "Cellosolve" acetate; Glycol monobutyl ether acetate

3.2 **Coast Guard Compatibility Classification:** Esters (13)

3.3 **Chemical Formula:** n-$C_4H_9OCH_2CH_2OCOCH_3$

3.4 **IMCO/United Nations Numerical Designation:** Not listed

4. OBSERVABLE CHARACTERISTICS

4.1 **Physical State (as shipped):** Liquid

4.2 **Color:** Colorless

4.3 **Odor:** Weak; fruity

5. HEALTH HAZARDS

5.1 **Personal Protective Equipment:** Goggles or face shield; rubber gloves.

5.2 **Symptoms Following Exposure:** Inhalation of concentrated vapor may cause headache, nausea, dizziness. Liquid causes irritation of eyes and mild irritation of skin. Ingestion produces same symptoms as inhalation.

5.3 **Treatment for Exposure:** INHALATION: remove to fresh air. EYES: flush thoroughly with water for at least 15 min. SKIN: flush thoroughly with water. INGESTION: induce vomiting; get medical attention.

5.4 **Toxicity by Inhalation (Threshold Limit Value):** Data not available

5.5 **Short-Term Inhalation Limits:** Data not available

5.6 **Toxicity by Ingestion:** Grade 2; oral LD_{50} = 3,200 mg/kg (mouse)

5.7 **Late Toxicity:** Data not available

5.8 **Vapor (Gas) Irritant Characteristics:** Data not available

5.9 **Liquid or Solid Irritant Characteristics:** Data not available

5.10 **Odor Threshold:** Data not available

6. FIRE HAZARDS

6.1 **Flash Point:** 190°F O.C.; 160°F C.C.

6.2 **Flammable Limits in Air:** 0.9%–8.5%

6.3 **Fire Extinguishing Agents:** Dry chemical, alcohol foam, carbon dioxide

6.4 **Fire Extinguishing Agents Not to be Used:** Water may be ineffective.

6.5 **Special Hazards of Combustion Products:** Not pertinent

6.6 **Behavior in Fire:** Not pertinent

6.7 **Ignition Temperature:** 645°F

6.8 **Electrical Hazard:** Data not available

6.9 **Burning Rate:** 4.1 mm/min.

7. CHEMICAL REACTIVITY

7.1 **Reactivity with Water:** No reaction

7.2 **Reactivity with Common Materials:** No reaction

7.3 **Stability During Transport:** Stable

7.4 **Neutralizing Agents for Acids and Caustics:** Not pertinent

7.5 **Polymerization:** Not pertinent

7.6 **Inhibitor of Polymerization:** Not pertinent

8. WATER POLLUTION

8.1 **Aquatic Toxicity:** Data not available

8.2 **Waterfowl Toxicity:** Data not available

8.3 **Biological Oxygen Demand (BOD):** Data not available

8.4 **Food Chain Concentration Potential:** None

9. SELECTED MANUFACTURERS

1. Union Carbide Corporation
   Chemicals and Plastics Division
   270 Park Avenue
   New York, N. Y. 10017
2. Eastman Chemical Products, Inc.
   Kingsport, Tenn. 37662

10. SHIPPING INFORMATION

10.1 **Grades or Purity:** 98+%

10.2 **Storage Temperature:** Ambient

10.3 **Inert Atmosphere:** No requirement

10.4 **Venting:** Open (flame arrester)

11. HAZARD ASSESSMENT CODE

(See Hazard Assessment Handbook, CG 446-3)

A-P-Q-T-U

12. HAZARD CLASSIFICATIONS

12.1 **Code of Federal Regulations:** Combustible Liquid

12.2 **NAS Hazard Rating for Bulk Water Transportation:** Not listed

12.3 **NFPA Hazard Classifications:** Not listed

13. PHYSICAL AND CHEMICAL PROPERTIES

13.1 **Physical State at 15°C and 1 atm:** Liquid

13.2 **Molecular Weight:** 160.21

13.3 **Boiling Point at 1 atm:** 378.0°F = 192.2°C = 465.4°K

13.4 **Freezing Point:** −82.3°F = −63.5°C = 209.7°K

13.5 **Critical Temperature:** Not pertinent

13.6 **Critical Pressure:** Not pertinent

13.7 **Specific Gravity:** 0.942 at 20°C (liquid)

13.8 **Liquid Surface Tension:** (est.) 26 dynes/cm = 0.026 N/m at 20°C

13.9 **Liquid-Water Interfacial Tension:** Not pertinent

13.10 **Vapor (Gas) Specific Gravity:** Not pertinent

13.11 **Ratio of Specific Heats of Vapor (Gas):** Not pertinent

13.12 **Latent Heat of Vaporization:** 120 Btu/lb = 65 cal/g = 2.7×10^5 J/kg

13.13 **Heat of Combustion:** (est.) −14,000 Btu/lb = −7,700 cal/g = -320×10^5 J/kg

13.14 **Heat of Decomposition:** Not pertinent

13.15 **Heat of Solution:** Not pertinent

13.16 **Heat of Polymerization:** Not pertinent

NOTES

EGE — ETHYLENE GLYCOL MONOETHYL ETHER

Common Synonyms			
Poly-Solv EE Dowanol EE Cellosolve Glycol monoethyl ether 2-Ethoxyethanol	Oily liquid	Colorless	Sweet odor
	Floats and mixes with water.		

Stop discharge if possible.
Call fire department.
Avoid contact with liquid.
Isolate and remove discharged material.
Notify local health and pollution control agencies.

Fire	Combustible. Extinguish with dry chemical, alcohol foam, or carbon dioxide. Cool exposed containers with water.
Exposure	CALL FOR MEDICAL AID. LIQUID Irritating to skin and eyes. Harmful if swallowed. Remove contaminated clothing and shoes. Flush affected areas with plenty of water. IF IN EYES, hold eyelids open and flush with plenty of water. IF SWALLOWED and victim is CONSCIOUS, have victim drink water or milk.
Water Pollution	Effect of low concentrations on aquatic life is unknown. May be dangerous if it enters water intakes. Notify local health and wildlife officials. Notify operators of nearby water intakes.

1. RESPONSE TO DISCHARGE
(See Response Methods Handbook, CG 446-4)

Disperse and flush

2. LABELS

No hazard label required by Code of Federal Regulations

3. CHEMICAL DESIGNATIONS

3.1 **Synonyms:** Cellosolve; Dowanol EE; 2-Ethoxyethanol; Ethylene glycol ethyl ether; Glycol monoethyl ether; Poly-Solv EE

3.2 **Coast Guard Compatibility Classification:** Glycol ether

3.3 **Chemical Formula:** $HOCH_2CH_2OCH_2CH_3$

3.4 **IMCO/United Nations Numerical Designation:** 3.3/1171

4. OBSERVABLE CHARACTERISTICS

4.1 **Physical State (as shipped):** Liquid

4.2 **Color:** Colorless

4.3 **Odor:** Sweetish; mild, pleasant, ethereal

5. HEALTH HAZARDS

5.1 **Personal Protective Equipment:** Organic gas mask; goggles or face shield; rubber gloves.

5.2 **Symptoms Following Exposure:** Some eye irritation. Inhalation of vapors causes irritation of nose.

5.3 **Treatment for Exposure:** Flush eyes with water for 15 min. Flush skin with large volumes of water. Call a physician.

5.4 **Toxicity by Inhalation (Threshold Limit Value):** 200 ppm

5.5 **Short-Term Inhalation Limits:** Not pertinent

5.6 **Toxicity by Ingestion:** Grade 2; LD_{50} 0.5 to 5 g/kg (rat, rabbit, guinea pig)

5.7 **Late Toxicity:** Data not available

5.8 **Vapor (Gas) Irritant Characteristics:** Vapors cause a slight smarting of the eyes or respiratory system if present in high concentrations. The effect is temporary.

5.9 **Liquid or Solid Irritant Characteristics:** Minimum hazard. If spilled on clothing and allowed to remain, may cause smarting and reddening of the skin.

5.10 **Odor Threshold:** Data not available

6. FIRE HAZARDS

6.1 **Flash Point:** 120°F O.C.; 202°F C.C.

6.2 **Flammable Limits in Air:** 1.8%–14.0%

6.3 **Fire Extinguishing Agents:** Alcohol foam, carbon dioxide or dry chemical

6.4 **Fire Extinguishing Agents Not to be Used:** Not pertinent

6.5 **Special Hazards of Combustion Products:** Not pertinent

6.6 **Behavior in Fire:** Not pertinent

6.7 **Ignition Temperature:** 455°F

6.8 **Electrical Hazard:** Not pertinent

6.9 **Burning Rate:** 2.4 mm/min

7. CHEMICAL REACTIVITY

7.1 **Reactivity with Water:** No reaction

7.2 **Reactivity with Common Materials:** No reaction

7.3 **Stability During Transport:** Stable

7.4 **Neutralizing Agents for Acids and Caustics:** Not pertinent

7.5 **Polymerization:** Not pertinent

7.6 **Inhibitor of Polymerization:** Not pertinent

8. WATER POLLUTION

8.1 **Aquatic Toxicity:** Data not available

8.2 **Waterfowl Toxicity:** Data not available

8.3 **Biological Oxygen Demand (BOD):** 1.58 lb/lb, 5 days; 54% (theor.), 5 days

8.4 **Food Chain Concentration Potential:** None

9. SELECTED MANUFACTURERS

1. Jefferson Chemical Co.
   3336 Richmond Ave.
   Houston, Texas 77052
2. Olin Corp.
   Chemicals Division
   Brandenburg, Ky. 40108
3. Union Carbide Corp.
   Chemicals and Plastics Division
   270 Park Ave.
   New York, N. Y. 10017

10. SHIPPING INFORMATION

10.1 **Grades or Purity:** Commercial

10.2 **Storage Temperature:** Ambient

10.3 **Inert Atmosphere:** No requirement

10.4 **Venting:** Open (flame arrester)

11. HAZARD ASSESSMENT CODE
(See Hazard Assessment Handbook, CG 446-3)

A-P-Q

12. HAZARD CLASSIFICATIONS

12.1 **Code of Federal Regulations:** Combustible Liquid

12.2 **NAS Hazard Rating for Bulk Water Transportation:**

Category	Rating
Fire	2
Health	
Vapor Irritant	1
Liquid or Solid Irritant	1
Poisons	2
Water Pollution	
Human Toxicity	1
Aquatic Toxicity	2
Aesthetic Effect	1
Reactivity	
Other Chemicals	2
Water	0
Self-Reaction	0

12.3 **NFPA Hazard Classifications:** Not listed

13. PHYSICAL AND CHEMICAL PROPERTIES

13.1 **Physical State at 15°C and 1 atm:** Liquid

13.2 **Molecular Weight:** 90.12

13.3 **Boiling Point at 1 atm:** 275.2°F = 135.1°C = 408.3°K

13.4 **Freezing Point:** Data not available

13.5 **Critical Temperature:** Not pertinent

13.6 **Critical Pressure:** Not pertinent

13.7 **Specific Gravity:** 0.931 at 20°C (liquid)

13.8 **Liquid Surface Tension:** Not pertinent

13.9 **Liquid-Water Interfacial Tension:** Not pertinent

13.10 **Vapor (Gas) Specific Gravity:** Not pertinent

13.11 **Ratio of Specific Heats of Vapor (Gas):** 1.064

13.12 **Latent Heat of Vaporization:** 191 Btu/lb = 106 cal/g = 4.44×10^5 J/kg

13.13 **Heat of Combustion:** (est.) −13,000 Btu/lb = −7,400 cal/g = -310×10^5 J/kg

13.14 **Heat of Decomposition:** Not pertinent

13.15 **Heat of Solution:** (est.) −9 Btu/lb = −5 cal/g = -0.2×10^5 J/kg

13.16 **Heat of Polymerization:** Not pertinent

NOTES

REVISED 1978

EGA — ETHYLENE GLYCOL MONOETHYL ETHER ACETATE

Common Synonyms			
Poly-Solv EE acetate Cellosolve acetate 2-Ethoxyethyl acetate	Liquid	Colorless	Pleasant odor
	Floats and mixes slowly with water.		

Stop discharge if possible.
Call fire department.
Avoid contact with liquid.
Isolate and remove discharged material.
Notify local health and pollution control agencies.

Fire

Combustible.
Extinguish with dry chemical, alcohol foam, or carbon dioxide.
Cool exposed containers with water.

Exposure

CALL FOR MEDICAL AID.

LIQUID
Irritating to skin and eyes.
Harmful if swallowed.
Remove contaminated clothing and shoes.
Flush affected areas with plenty of water.
IF IN EYES, hold eyelids open and flush with plenty of water.
IF SWALLOWED and victim is CONSCIOUS, have victim drink water or milk.

Water Pollution

Effect of low concentrations on aquatic life is unknown.
May be dangerous if it enters water intakes.
Notify local health and wildlife officials.
Notify operators of nearby water intakes.

1. RESPONSE TO DISCHARGE

(See Response Methods Handbook, CG 446-4)

Disperse and flush

2. LABELS

No hazard label required by Code of Federal Regulations

3. CHEMICAL DESIGNATIONS

3.1 **Synonyms:** Cellosolve acetate
2-Ethoxyethanol, acetate
2-Ethoxyethyl acetate
Glycol monoethyl ether acetate
Poly-Solv EE acetate

3.2 **Coast Guard Compatibility Classification:** Ester

3.3 **Chemical Formula:** $CH_3COOCH_2CH_2OC_2H_5$

3.4 **IMCO/United Nations Numerical Designation:** 3.3/1172

4. OBSERVABLE CHARACTERISTICS

4.1 **Physical State (as shipped):** Liquid

4.2 **Color:** Colorless

4.3 **Odor:** Mild, pleasant; ester-like

5. HEALTH HAZARDS

5.1 **Personal Protective Equipment:** Chemical safety goggles

5.2 **Symptoms Following Exposure:** Vapors irritate nose and eyes in high concentrations. Liquid irritates skin in prolonged or repeated contact.

5.3 **Treatment for Exposure:** INHALATION: if victim is overcome, remove him to fresh air and call physician. EYES: flush with large amounts of water. SKIN: wash exposed areas.

5.4 **Toxicity by Inhalation (Threshold Limit Value):** 100 ppm

5.5 **Short-Term Inhalation Limits:** Data not available

5.6 **Toxicity by Ingestion:** Grade 2; LD_{50} 0.5 to 5 g/kg (rabbit)

5.7 **Late Toxicity:** Causes kidney damage in laboratory animals. Effects unknown in humans.

5.8 **Vapor (Gas) Irritant Characteristics:** Vapors cause a slight smarting of the eyes and respiratory system if present in high concentrations. The effect is temporary.

5.9 **Liquid or Solid Irritant Characteristics:** Minimum hazard. If spilled on clothing and allowed to remain, may cause smarting and reddening of the skin.

5.10 **Odor Threshold:** 0.056 ppm

6. FIRE HAZARDS

6.1 **Flash Point:** 135°F O.C.

6.2 **Flammable Limits in Air:** 1.7%–6.7%

6.3 **Fire Extinguishing Agents:** Alcohol foam, dry chemical or carbon dioxide

6.4 **Fire Extinguishing Agents Not to be Used:** Not pertinent

6.5 **Special Hazards of Combustion Products:** Not pertinent

6.6 **Behavior in Fire:** Not pertinent

6.7 **Ignition Temperature:** 715°F

6.8 **Electrical Hazard:** Not pertinent

6.9 **Burning Rate:** Data not available

7. CHEMICAL REACTIVITY

7.1 **Reactivity with Water:** No reaction

7.2 **Reactivity with Common Materials:** No reaction

7.3 **Stability During Transport:** Stable

7.4 **Neutralizing Agents for Acids and Caustics:** Not pertinent

7.5 **Polymerization:** Not pertinent

7.6 **Inhibitor of Polymerization:** Not pertinent

8. WATER POLLUTION

8.1 **Aquatic Toxicity:** 4000 ppm/24 hr/brine shrimp/TL_m

8.2 **Waterfowl Toxicity:** Data not available

8.3 **Biological Oxygen Demand (BOD):** 36% of theoretical in 5 days, freshwater

8.4 **Food Chain Concentration Potential:** None

9. SELECTED MANUFACTURERS

1. Eastman Kodak Co.
Texas Eastman Co. Division
Longview, Texas 75601
2. Olin Corp.
Chemicals Division
Brandenburg, Ky. 40108
3. Union Carbide Corp.
Chemicals and Plastics Div.
270 Park Ave.
New York, N. Y. 10017

10. SHIPPING INFORMATION

10.1 **Grades or Purity:** Commercial

10.2 **Storage Temperature:** Ambient

10.3 **Inert Atmosphere:** No requirement

10.4 **Venting:** Open (flame arrester)

11. HAZARD ASSESSMENT CODE

(See Hazard Assessment Handbook, CG 446-3)

A-P-Q

12. HAZARD CLASSIFICATIONS

12.1 **Code of Federal Regulations:** Combustible Liquid

12.2 **NAS Hazard Rating for Bulk Water Transportation:**

Category	Rating
Fire	2
Health	
Vapor Irritant	1
Liquid or Solid Irritant	1
Poisons	1
Water Pollution	
Human Toxicity	1
Aquatic Toxicity	2
Aesthetic Effect	2
Reactivity	
Other Chemicals	1
Water	0
Self-Reaction	0

12.3 **NFPA Hazard Classifications:** Not listed

13. PHYSICAL AND CHEMICAL PROPERTIES

13.1 **Physical State at 15°C and 1 atm:** Liquid

13.2 **Molecular Weight:** 132.16

13.3 **Boiling Point at 1 atm:** 313°F = 156°C = 429°K

13.4 **Freezing Point:** −79.1°F = −61.7°C = 211.5°K

13.5 **Critical Temperature:** 633°F = 334°C = 607°K

13.6 **Critical Pressure:** 440 psia = 30 atm = 3.0 MN/m^2

13.7 **Specific Gravity:** 0.974 at 20°C (liquid)

13.8 **Liquid Surface Tension:** Not pertinent

13.9 **Liquid-Water Interfacial Tension:** Not pertinent

13.10 **Vapor (Gas) Specific Gravity:** Not pertinent

13.11 **Ratio of Specific Heats of Vapor (Gas):** 1.054

13.12 **Latent Heat of Vaporization:** 130 Btu/lb = 74 cal/g = 3.1×10^5 J/kg

13.13 **Heat of Combustion:** (est.) −10,700 Btu/lb = −6,000 cal/g = -250×10^5 J/kg

13.14 **Heat of Decomposition:** Not pertinent

13.15 **Heat of Solution:** Not pertinent

13.16 **Heat of Polymerization:** Not pertinent

NOTES

REVISED 1978

EME — ETHYLENE GLYCOL MONOMETHYL ETHER

Common Synonyms
Poly-Solv EM
Dowanol EM
Methyl Cellosolve
Glycol monomethylether
2-Methoxyethanol

Liquid | Colorless | Odorless

Floats and mixes with water.

Stop discharge if possible.
Call fire department.
Avoid contact with liquid.
Isolate and remove discharged material.
Notify local health and pollution control agencies.

Fire

Combustible.
Extinguish with dry chemical, alcohol foam, or carbon dioxide.
Cool exposed containers with water.

Exposure

CALL FOR MEDICAL AID.

LIQUID
Irritating to skin and eyes.
Harmful if swallowed.
Remove contaminated clothing and shoes.
Flush affected areas with plenty of water.
IF IN EYES, hold eyelids open and flush with plenty of water.
IF SWALLOWED, and victim is CONSCIOUS, have victim drink water or milk.

Water Pollution

Effect of low concentrations on aquatic life is unknown.
May be dangerous if it enters water intakes.
Notify local health and wildlife officials.
Notify operators of nearby water intakes.

1. RESPONSE TO DISCHARGE
(See Response Methods Handbook, CG 446-4)

Disperse and flush

2. LABELS

No hazard label required by Code of Federal Regulations

3. CHEMICAL DESIGNATIONS

3.1 **Synonyms:** Dowanol EM
Glycol monomethyl ether
2-Methoxyethanol
Methyl Cellosolve
Poly-Solv EM

3.2 **Coast Guard Compatibility Classification:** Glycol ether

3.3 **Chemical Formula:** $CH_3OCH_2CH_2OH$

3.4 **IMCO/United Nations Numerical Designation:** 3.3/1188

4. OBSERVABLE CHARACTERISTICS

4.1 **Physical State (as shipped):** Liquid
4.2 **Color:** Colorless
4.3 **Odor:** Mild ethereal

5. HEALTH HAZARDS

5.1 **Personal Protective Equipment:** Chemical safety goggles; protective clothing; supplied-air respirator for high concentrations; safety shower and eye bath.

5.2 **Symptoms Following Exposure:** Irritation of skin and eyes. Chronic exposure may also cause weakness, sleepiness, headache, gastrointestinal upset, weight loss, change of personality.

5.3 **Treatment for Exposure:** SKIN OR EYES: wash affected area with water for 15 min.

5.4 **Toxicity by Inhalation (Threshold Limit Value):** 25 ppm

5.5 **Short-Term Inhalation Limits:** Not pertinent

5.6 **Toxicity by Ingestion:** Grade 2; LD_{50} 0.5 to 5 g/kg (rat, rabbit, guinea pig)

5.7 **Late Toxicity:** Causes blood disorders and damage to central nervous system in humans.

5.8 **Vapor (Gas) Irritant Characteristics:** Vapors cause a slight smarting of the eyes or respiratory system if present in high concentrations. The effect is temporary.

5.9 **Liquid or Solid Irritant Characteristics:** Minimum hazard. If spilled on clothing and allowed to remain, may cause smarting and reddening of the skin.

5.10 **Odor Threshold:** 0.9 ppm

6. FIRE HAZARDS

6.1 **Flash Point:** 120°F O.C.; 107°F C.C.

6.2 **Flammable Limits in Air:** 2.5%–19.8%

6.3 **Fire Extinguishing Agents:** Dry chemical, carbon dioxide or alcohol foam

6.4 **Fire Extinguishing Agents Not to be Used:** Not pertinent

6.5 **Special Hazards of Combustion Products:** Not pertinent

6.6 **Behavior in Fire:** Not pertinent

6.7 **Ignition Temperature:** 551°F

6.8 **Electrical Hazard:** Not pertinent

6.9 **Burning Rate:** 1.8 mm/min.

7. CHEMICAL REACTIVITY

7.1 **Reactivity with Water:** No reaction

7.2 **Reactivity with Common Materials:** No reaction

7.3 **Stability During Transport:** Stable

7.4 **Neutralizing Agents for Acids and Caustics:** Not pertinent

7.5 **Polymerization:** Not pertinent

7.6 **Inhibitor of Polymerization:** Not pertinent

8. WATER POLLUTION

8.1 **Aquatic Toxicity:** Data not available

8.2 **Waterfowl Toxicity:** Data not available

8.3 **Biological Oxygen Demand (BOD):** 15% (theor.), 1-10 days

8.4 **Food Chain Concentration Potential:** None

9. SELECTED MANUFACTURERS

1. Jefferson Chemical Co., Inc.
3336 Richmond Ave.
Houston, Texas 77052
2. Olin Corp.
Chemicals Division
Brandenburg, Ky. 40108
3. Union Carbide Corp.
Chemicals and Plastics Division
270 Park Ave.
New York, N. Y. 10017

10. SHIPPING INFORMATION

10.1 **Grades or Purity:** Commercial

10.2 **Storage Temperature:** Ambient

10.3 **Inert Atmosphere:** No requirement

10.4 **Venting:** Open (flame arrester)

11. HAZARD ASSESSMENT CODE
(See Hazard Assessment Handbook, CG 446-3)

A-P-Q

12. HAZARD CLASSIFICATIONS

12.1 **Code of Federal Regulations:** Combustible Liquid

12.2 **NAS Hazard Rating for Bulk Water Transportation:**

Category	Rating
Fire	2
Health	
Vapor Irritant	1
Liquid or Solid Irritant	1
Poisons	2
Water Pollution	
Human Toxicity	2
Aquatic Toxicity	2
Aesthetic Effect	1
Reactivity	
Other Chemicals	2
Water	0
Self-Reaction	0

12.3 **NFPA Hazard Classifications:** Not listed

13. PHYSICAL AND CHEMICAL PROPERTIES

13.1 **Physical State at 15°C and 1 atm:** Liquid

13.2 **Molecular Weight:** 76.10

13.3 **Boiling Point at 1 atm:** 256.1°F = 124.5°C = 397.7°K

13.4 **Freezing Point:** −121.2°F = −85.1°C = 188.1°K

13.5 **Critical Temperature:** 558°F = 292°C = 565°K

13.6 **Critical Pressure:** 735 psia = 50 atm = 5.1 MN/m²

13.7 **Specific Gravity:** 0.966 at 20°C (liquid)

13.8 **Liquid Surface Tension:** 33 dynes/cm = 0.033 N/m at 20°C

13.9 **Liquid-Water Interfacial Tension:** Not pertinent

13.10 **Vapor (Gas) Specific Gravity:** Not pertinent

13.11 **Ratio of Specific Heats of Vapor (Gas):** 1.079

13.12 **Latent Heat of Vaporization:** 223 Btu/lb = 124 cal/g = 5.19×10^5 J/kg

13.13 **Heat of Combustion:** -9460 Btu/lb = -5250 cal/g = -220×10^5 J/kg

13.14 **Heat of Decomposition:** Not pertinent

13.15 **Heat of Solution:** Data not available

13.16 **Heat of Polymerization:** Not pertinent

NOTES

REVISED 1978

ETI	# ETHYLENEIMINE

Common Synonyms			
Aziridine	Oily liquid	Colorless	Ammonia odor
	Floats and mixes with water. Poisonous, flammable vapor is produced.		

AVOID CONTACT WITH LIQUID AND VAPOR. Keep people away.
Wear goggles, self-contained breathing apparatus, and rubber overclothing (including gloves).
Shut off ignition sources and call fire department.
Stop discharge if possible.
Stay upwind and use water spray to "knock down" vapor.
Isolate and remove discharged material.
Notify local health and pollution control agencies.

Fire

FLAMMABLE.
Containers may explode when heated.
Flashback along vapor trail may occur.
Vapor may explode if ignited in an enclosed area.
Wear goggles, self-contained breathing apparatus, and rubber overclothing (including gloves).
Combat fires from behind barrier.
Extinguish with dry chemical, alcohol foam, or carbon dioxide.
Water may be ineffective on fire.
Cool exposed containers with water.

Exposure

CALL FOR MEDICAL AID.

VAPOR
POISONOUS IF INHALED OR IF SKIN IS EXPOSED.
Irritating to eyes.
Move to fresh air.
If breathing has stopped, give artificial respiration.
If breathing is difficult, give oxygen.

LIQUID
POISONOUS IF SWALLOWED OR IF SKIN IS EXPOSED.
Will burn eyes.
Remove contaminated clothing and shoes.
Flush affected areas with plenty of water.
IF IN EYES, hold eyelids open and flush with plenty of water.
IF SWALLOWED, and victim is CONSCIOUS, have victim drink water or milk.

Water Pollution

Effect of low concentrations on aquatic life is unknown.
May be dangerous if it enters water intakes.
Notify local health and wildlife officials.
Notify operators of nearby water intakes.

1. RESPONSE TO DISCHARGE
(See Response Methods Handbook, CG 446-4)

Issue warning—high flammability, poison
Restrict access
Evacuate area
Disperse and flush

2. LABELS

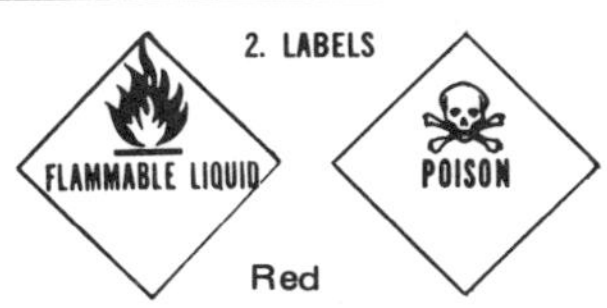

Red

3. CHEMICAL DESIGNATIONS

3.1 **Synonyms:** Azirane
Aziridine
3.2 **Coast Guard Compatibility Classification:** To be developed
3.3 **Chemical Formula:** $\overline{CH_2CH_2}NH$
3.4 **IMCO/United Nations Numerical Designation:** 3.2/1185

4. OBSERVABLE CHARACTERISTICS

4.1 **Physical State (as shipped):** Liquid
4.2 **Color:** Colorless
4.3 **Odor:** Fishy; ammoniacal

5. HEALTH HAZARDS

5.1 **Personal Protective Equipment:** If exposure is possible, wear full protective clothing (neoprene slicker suit, rubber boots, rubber gloves, chemical goggles). If vapors may be present, wear all-purpose canister or gas mask; if vapors are known to be present, use self-contained breathing apparatus.
5.2 **Symptoms Following Exposure:** Material gives inadequate warning of overexposure by respiration or skin contact. May cause nausea, vomiting, and possibly death when inhaled, ingested, or absorbed through skin. Severe blistering agent; can produce third-degree chemical burns of skin. Has corrosive effect on mucous menbranes and may cause scarring of esophagus if swallowed. Corrosive to eye tissue; may cause permanent corneal opacity and conjunctival scarring. Effects on eye tissue, mucous membrane, and skin may be delayed.
5.3 **Treatment for Exposure:** INHALATION: remove victim from exposure and administer oxygen; steriod therapy (by physician) is recommended. SKIN OR EYES: prompt and adequate irrigation with water (within 60 seconds of exposure) can prevent serious injury.
5.4 **Toxicity by Inhalation (Threshold Limit Value):** Data not available
5.5 **Short-Term Inhalation Limits:** 5 ppm for 30 min.
5.6 **Toxicity by Ingestion:** Grade 4; LD_{50} below 50 mg/kg (rat)
5.7 **Late Toxicity:** Causes cancer in mice. Effects on man unknown.
5.8 **Vapor (Gas) Irritant Characteristics:** Vapor is moderately irritating such that personnel will not usually tolerate moderate or high concentrations.

6. FIRE HAZARDS

6.1 **Flash Point:** 1°F O.C.; 12°F C.C.
6.2 **Flammable Limits in Air:** 3.3%—54.8%
6.3 **Fire Extinguishing Agents:** Dry chemical, alcohol foam, carbon dioxide
6.4 **Fire Extinguishing Agents Not to be Used:** Not pertinent
6.5 **Special Hazards of Combustion Products:** Irritating vapors generated when heated.
6.6 **Behavior in Fire:** Vapor is heavier than air and may travel a considerable distance to a source of ignition and flash back. May polymerize in fires with evolution of heat and container rupture.
6.7 **Ignition Temperature:** 608°F
6.8 **Electrical Hazard:** Not pertinent
6.9 **Burning Rate:** Data not available

7. CHEMICAL REACTIVITY

7.1 **Reactivity with Water:** Mild reaction, non-hazardous
7.2 **Reactivity with Common Materials:** Contact with silver or aluminum may cause polymerization.
7.3 **Stability During Transport:** Stable unless heated under pressure.
7.4 **Neutralizing Agents for Acids and Caustics:** Flush with water
7.5 **Polymerization:** Explosive polymerization can occur when in contact with acids.
7.6 **Inhibitor of Polymerization:** None used

8. WATER POLLUTION

8.1 **Aquatic Toxicity:** Data not available
8.2 **Waterfowl Toxicity:** Data not available
8.3 **Biological Oxygen Demand (BOD):** Data not available
8.4 **Food Chain Concentration Potential:** None

9. SELECTED MANUFACTURERS

Dow Chemical Co.
Midland, Michigan 48640

10. SHIPPING INFORMATION

10.1 **Grades or Purity:** 99.0%
10.2 **Storage Temperature:** Ambient
10.3 **Inert Atmosphere:** Inerted
10.4 **Venting:** Safety relief

11. HAZARD ASSESSMENT CODE
(See Hazard Assessment Handbook, CG 446-3)

A-P-Q-R-S-Z

12. HAZARD CLASSIFICATIONS

12.1 **Code of Federal Regulations:** Flammable liquid
12.2 **NAS Hazard Rating for Bulk Water Transportation:**

Category	Rating
Fire	3
Health	
Vapor Irritant	3
Liquid or Solid Irritant	2
Poisons	4
Water Pollution	
Human Toxicity	4
Aquatic Toxicity	3
Aesthetic Effect	3
Reactivity	
Other Chemicals	3
Water	1
Self-Reaction	3

12.3 **NFPA Hazard Classifications:**

Category	Classification
Health Hazard (Blue)	3
Flammability (Red)	3
Reactivity (Yellow)	2

13. PHYSICAL AND CHEMICAL PROPERTIES

13.1 **Physical State at 15°C and 1 atm:** Liquid
13.2 **Molecular Weight:** 43.07
13.3 **Boiling Point at 1 atm:** 133°F = 56°C = 329°K
13.4 **Freezing Point:** −108°F = −78°C = 195°K
13.5 **Critical Temperature:** Not pertinent
13.6 **Critical Pressure:** Not pertinent
13.7 **Specific Gravity:** 0.832 at 20°C (liquid)
13.8 **Liquid Surface Tension:** 34.5 dynes/cm = 0.0345 N/m at 20°C
13.9 **Liquid-Water Interfacial Tension:** Not pertinent
13.10 **Vapor (Gas) Specific Gravity:** 1.5
13.11 **Ratio of Specific Heats of Vapor (Gas):** 1.192
13.12 **Latent Heat of Vaporization:** 333 Btu/lb = 185 cal/g = 7.75×10^5 J/kg
13.13 **Heat of Combustion:** −15,930 Btu/lb = −8850 cal/g = -370.5×10^5 J/kg
13.14 **Heat of Decomposition:** Not pertinent
13.15 **Heat of Solution:** (est.) −26 Btu/lb = −14 cal/g = -0.6×10^5 J/kg
13.16 **Heat of Polymerization:** (est.) −900 Btu/lb = −500 cal/g = -20×10^5 J/kg

5. HEALTH HAZARDS (Cont'd.)

5.9 **Liquid or Solid Irritant Characteristics:** Causes smarting of the skin and first-degree burns on short exposure and may cause secondary burns on long exposure.
5.10 **Odor Threshold:** Data not available

REVISED 1978

EOX — ETHYLENE OXIDE

Common Synonyms
Oxirane
1,2-Epoxyethane

Liquefied gas — Colorless — Sweet odor

Floats and mixes with water. Flammable, irritating vapor is produced. Boiling point is 51° F.

Avoid contact with liquid. Keep people away.
Wear goggles, self-contained breathing apparatus, and rubber overclothing (including gloves).
Shut off ignition sources and call fire department.
Stop discharge if possible.
Stay upwind and use water spray to "knock down" vapor.
Isolate and remove discharged material.
Notify local health and pollution control agencies.

Fire

FLAMMABLE.
Containers may explode when heated.
Flashback along vapor trail may occur.
Vapor may explode if ignited in an enclosed area.
Wear goggles, self-contained breathing apparatus, and rubber overclothing (including gloves).
Stop flow of gas if possible.
Combat fires from behind barrier, with unmanned hose holder or monitor noz
Flood discharge area with water.
Cool exposed containers and protect men effecting shutoff with water.
Extinguish with alcohol foam, dry chemical, or carbon dioxide.

Exposure

CALL FOR MEDICAL AID.

VAPOR
Irritating to eyes, nose and throat.
If inhaled, will cause nausea, vomiting and difficult breathing.
Move to fresh air.
If breathing has stopped, give artificial respiration.
If breathing is difficult, give oxygen.

LIQUID
Will burn skin and eyes.
Harmful if swallowed.
Remove contaminated clothing and shoes.
Flush affected areas with plenty of water.
IF IN EYES, hold eyelids open, and flush with plenty of water.
IF SWALLOWED, and victim is CONSCIOUS, have victim drink water or milk.

Water Pollution

Effect of low concentrations on aquatic life is unknown.
May be dangerous if it enters water intakes.
Notify local health and wildlife officials.
Notify operators of nearby water intakes.

1. RESPONSE TO DISCHARGE

(See Response Methods Handbook, CG 446-4)

Issue warning high flammability
Evacuate area

2. LABEL

Red

3. CHEMICAL DESIGNATIONS

3.1 **Synonyms:** 1,2-Epoxyethane
Oxirane
3.2 **Coast Guard Compatibility Classification:** Alkylene oxide
3.3 **Chemical Formula:** CH_2CH_2O
3.4 **IMCO/United Nations Numerical Designation:** 2.0/1040

4. OBSERVABLE CHARACTERISTICS

4.1 **Physical State (as shipped):** Liquid
4.2 **Color:** Colorless
4.3 **Odor:** Sweet; characteristic ether-like

5. HEALTH HAZARDS

5.1 **Personal Protective Equipment:** Air-supplied mask; goggles or face shield; rubber shoes and coveralls.
5.2 **Symptoms Following Exposure:** Exposure to low vapor concentrations often results in delayed nausea and vomiting. Higher concentrations produce irritation of eyes, nose, and throat; high concentrations may cause edema of lungs. Contact with skin causes blistering and burns.
5.3 **Treatment for Exposure:** INHALATION: leave contaminated area immediately; if nausea and vomiting start, call a physician. SKIN OR EYES: flush immediately with plenty of water for at least 15 min. and seek medical attention.
5.4 **Toxicity by Inhalation (Threshold Limit Value):** 50 ppm
5.5 **Short-Term Inhalation Limits:** 200 ppm for 30 min.
5.6 **Toxicity by Ingestion:** Grade 3; oral rat LD_{50} = .33 g/kg
5.7 **Late Toxicity:** Causes cancer in mice. Effects on humans unknown.
5.8 **Vapor (Gas) Irritant Characteristics:** Vapor is moderately irritating such that personnel will not usually tolerate moderate or high vapor concentrations.
5.9 **Liquid or Solid Irritant Characteristics:** Fairly severe skin irritant; may cause pain and second-degree burns after a few minutes' contact.
5.10 **Odor Threshold:** 50 ppm

6. FIRE HAZARDS

6.1 **Flash Point:** < 0°F O.C.
6.2 **Flammable Limits in Air:** 3%—100%
6.3 **Fire Extinguishing Agents:** Stop flow of gas. Use water, carbon dioxide, dry chemical or alcohol foam.
6.4 **Fire Extinguishing Agents Not to be Used:** Not pertinent
6.5 **Special Hazards of Combustion Products:** Irritating vapors generated when heated.
6.6 **Behavior in Fire:** Vapor is heavier than air and may travel considerable distance to a source of ignition and flash back. Containers may explode when heated.
6.7 **Ignition Temperature:** 804°F
6.8 **Electrical Hazard:** Class I, group B
6.9 **Burning Rate:** 3.5 mm/min.

7. CHEMICAL REACTIVITY

7.1 **Reactivity with Water:** Slow reaction, not hazardous
7.2 **Reactivity with Common Materials:** No reaction
7.3 **Stability During Transport:** Stable
7.4 **Neutralizing Agents for Acids and Caustics:** Not pertinent
7.5 **Polymerization:** May polymerize violently if contaminated with alkaline or acidic materials and metal oxides or chlorides.
7.6 **Inhibitor of Polymerization:** None used

8. WATER POLLUTION

8.1 **Aquatic Toxicity:** Data not available
8.2 **Waterfowl Toxicity:** Data not available
8.3 **Biological Oxygen Demand (BOD):** Data not available
8.4 **Food Chain Concentration Potential:** None

9. SELECTED MANUFACTURERS

1. Dow Chemical Co.
Midland, Mich. 48640
2. Jefferson Chemical Co., Inc.
3336 Richmond Ave.
Houston, Texas 77052
3. Union Carbide Corp.
Chemicals and Plastics Div.
270 Park Ave.
New York, N. Y. 10017

10. SHIPPING INFORMATION

10.1 **Grades or Purity:** Commercial: 100% Must contain no acetylene
10.2 **Storage Temperature:** Ambient
10.3 **Inert Atmosphere:** Inerted
10.4 **Venting:** Safety relief

11. HAZARD ASSESSMENT CODE

(See Hazard Assessment Handbook, CG 446-3)

A-B-C-K-L-M-N

12. HAZARD CLASSIFICATIONS

12.1 **Code of Federal Regulations:** Flammable liquid
12.2 **NAS Hazard Rating for Bulk Water Transportation:**

Category	Rating
Fire	4
Health	
Vapor Irritant	3
Liquid or Solid Irritant	3
Poisons	2
Water Pollution	
Human Toxicity	3
Aquatic Toxicity	2
Aesthetic Effect	1
Reactivity	
Other Chemicals	3
Water	1
Self-Reaction	4

12.3 **NFPA Hazard Classifications:**

Category	Classification
Health Hazard (Blue)	2
Flammability (Red)	4
Reactivity (Yellow)	3

13. PHYSICAL AND CHEMICAL PROPERTIES

13.1 **Physical State at 15°C and 1 atm:** Gas
13.2 **Molecular Weight:** 44.05
13.3 **Boiling Point at 1 atm:** 51.1°F = 10.6°C = 283.8°K
13.4 **Freezing Point:** −170.7°F = −112.6°C = 160.6°K
13.5 **Critical Temperature:** 385°F = 196°C = 469°K
13.6 **Critical Pressure:** 1040 psia = 71.0 atm = 7.2 MN/m²
13.7 **Specific Gravity:** 0.869 at 20°C (liquid)
13.8 **Liquid Surface Tension:** 24.3 dynes/cm = 0.0243 N/m at 20°C
13.9 **Liquid-Water Interfacial Tension:** Not pertinent
13.10 **Vapor (Gas) Specific Gravity:** 1.5
13.11 **Ratio of Specific Heats of Vapor (Gas):** 1.212
13.12 **Latent Heat of Vaporization:** 249.3 Btu/lb = 138.5 cal/g = 5.799 × 10^5 J/kg
13.13 **Heat of Combustion:** −11,480 Btu/lb = −6380 cal/g = −267.1 × 10^5 J/kg
13.14 **Heat of Decomposition:** Not pertinent
13.15 **Heat of Solution:** −61 Btu/lb = −34 cal/g = −1.4 × 10^5 J/kg
13.16 **Heat of Polymerization:** Not pertinent

NOTES

REVISED 1978

EET	ETHYL ETHER

Common Synonyms			
Anesthetic ether Sulfuric ether Ether Diethyl ether Diethyl oxide	Watery liquid	Colorless	Sweet odor
	Floats on water. Flammable, irritating vapor is produced. Boiling point is 94° F.		

Avoid contact with liquid and vapor. Keep people away.
Wear goggles and self-contained breathing apparatus.
Shut off ignition sources and call fire department.
Stop discharge if possible.
Stay upwind and use water spray to "knock down" vapor.
Isolate and remove discharged material.
Notify local health and pollution control agencies.

Fire

FLAMMABLE.
Flashback along vapor trail may occur.
Vapor may explode if ignited in an enclosed area.
Wear goggles and self-contained breathing apparatus.
Extinguish with dry chemical, foam, or carbon dioxide.
Water may be ineffective on fire.
Cool exposed containers with water.

Exposure

CALL FOR MEDICAL AID.

VAPOR
Irritating to eyes, nose and throat.
If inhaled, will cause nausea, vomiting, headache, or loss of consciousness.
Move to fresh air.
If breathing has stopped, give artificial respiration.
If breathing is difficult, give oxygen.

LIQUID
Irritating to skin.
Harmful if swallowed.
Remove contaminated clothing and shoes.
Flush affected areas with plenty of water.
IF IN EYES, hold eyelids open and flush with plenty of water.
IF SWALLOWED, and victim is CONSCIOUS, have victim drink water or milk.

Water Pollution

Effect of low concentrations on aquatic life is unknown.
May be dangerous if it enters water intakes.
Notify local health and wildlife officials.
Notify operators of nearby water intakes.

1. RESPONSE TO DISCHARGE

(See Response Methods Handbook, CG 446-4)

Issue warning—high flammability
Restrict access
Evacuate area

2. LABEL

Red

3. CHEMICAL DESIGNATIONS

3.1 **Synonyms:** Anesthesia ether; Diethyl ether; Diethyl oxide; Ether; Ethoxyethane; Sulfuric ether

3.2 **Coast Guard Compatibility Classification:** Ether

3.3 **Chemical Formula:** $C_2H_5OC_2H_5$

3.4 **IMCO/United Nations Numerical Designation:** 3.1/1155

4. OBSERVABLE CHARACTERISTICS

4.1 **Physical State (as shipped):** Liquid

4.2 **Color:** Colorless

4.3 **Odor:** Sweet, pungent

5. HEALTH HAZARDS

5.1 **Personal Protective Equipment:** Approved organic vapor canister mask; chemical goggles; synthetic rubber or plastic gloves.

5.2 **Symptoms Following Exposure:** Vapor inhalation may cause headache, nausea, vomiting, and loss of consciousness. Contact with eyes will be irritating. Skin contact from clothing wet with the chemical may cause burns.

5.3 **Treatment for Exposure:** INHALATION: remove victim to fresh air; if breathing has stopped, apply artificial respiration; if breathing is irregular, give oxygen; call a physician. EYES: flush immediately with water for 15 min.

5.4 **Toxicity by Inhalation (Threshold Limit Value):** 400 ppm

5.5 **Short-Term Inhalation Limits:** 1000 ppm for 30 min.

5.6 **Toxicity by Ingestion:** Grade 2; LD_{50} 0.5 to 5 g/kg

5.7 **Late Toxicity:** None

5.8 **Vapor (Gas) Irritant Characteristics:** Vapors cause a slight smarting of the eyes or respiratory system if present in high concentrations. The effect is temporary.

5.9 **Liquid or Solid Irritant Characteristics:** No appreciable hazard. Practically harmless to the skin because it is very volatile and evaporates quickly.

5.10 **Odor Threshold:** 0.83 ppm

6. FIRE HAZARDS

6.1 **Flash Point:** −40°F O.C.; −49°F C.C.

6.2 **Flammable Limits in Air:** 1.85%–36.5%

6.3 **Fire Extinguishing Agents:** Dry chemical, carbon dioxide or foam

6.4 **Fire Extinguishing Agents Not to be Used:** Not pertinent

6.5 **Special Hazards of Combustion Products:** Not pertinent

6.6 **Behavior in Fire:** Vapor is heavier than air and may travel considerable distance to a source of ignition and flash back. Decomposes violently when heated.

6.7 **Ignition Temperature:** 356°F

6.8 **Electrical Hazard:** Class I, group C

6.9 **Burning Rate:** 6.7 mm/min.

7. CHEMICAL REACTIVITY

7.1 **Reactivity with Water:** No reaction

7.2 **Reactivity with Common Materials:** No reaction

7.3 **Stability During Transport:** Stable

7.4 **Neutralizing Agents for Acids and Caustics:** Not pertinent

7.5 **Polymerization:** Not pertinent

7.6 **Inhibitor of Polymerization:** Not pertinent

8. WATER POLLUTION

8.1 **Aquatic Toxicity:** Data not available

8.2 **Waterfowl Toxicity:** Data not available

8.3 **Biological Oxygen Demand (BOD):** 3%, 5 days

8.4 **Food Chain Concentration Potential:** None

9. SELECTED MANUFACTURERS

1. Exxon Chemical Co.
   Houston, Texas 77001
2. Publicker Industries Inc.
   1429 Walnut Street
   Philadelphia, Pa. 19102
3. Union Carbide Corp.
   Chemicals and Plastics Division
   270 Park Ave.
   New York, N. Y. 10017

10. SHIPPING INFORMATION

10.1 **Grades or Purity:** Reagent; absolute; purified; anesthesia; USP; concentrated

10.2 **Storage Temperature:** Ambient

10.3 **Inert Atmosphere:** Inerted

10.4 **Venting:** Pressure-vacuum

11. HAZARD ASSESSMENT CODE

(See Hazard Assessment Handbook, CG 446-3)

A-P-Q-T-U-V-W

12. HAZARD CLASSIFICATIONS

12.1 **Code of Federal Regulations:** Flammable liquid

12.2 **NAS Hazard Rating for Bulk Water Transportation:**

Category	Rating
Fire	4
Health	
Vapor Irritant	1
Liquid or Solid Irritant	0
Poisons	2
Water Pollution	
Human Toxicity	0
Aquatic Toxicity	1
Aesthetic Effect	1
Reactivity	
Other Chemicals	1
Water	0
Self-Reaction	0

12.3 **NFPA Hazard Classifications:**

Category	Classification
Health Hazard (Blue)	2
Flammability (Red)	4
Reactivity (Yellow)	1

13. PHYSICAL AND CHEMICAL PROPERTIES

13.1 **Physical State at 15°C and 1 atm:** Liquid

13.2 **Molecular Weight:** 74.12

13.3 **Boiling Point at 1 atm:** 94.3°F = 34.6°C = 307.8°K

13.4 **Freezing Point:** −177.3°F = −116.3°C = 156.9°K

13.5 **Critical Temperature:** 380.3°F = 193.5°C = 466.7°K

13.6 **Critical Pressure:** 527 psia = 35.9 atm = 3.64 MN/m^2

13.7 **Specific Gravity:** 0.714 at 20°C (liquid)

13.8 **Liquid Surface Tension:** 17.0 dynes/cm = 0.0170 N/m at 20°C

13.9 **Liquid-Water Interfacial Tension:** Not pertinent

13.10 **Vapor (Gas) Specific Gravity:** 2.6

13.11 **Ratio of Specific Heats of Vapor (Gas):** 1.081

13.12 **Latent Heat of Vaporization:** 153 Btu/lb = 84.9 cal/g = 3.56×10^5 J/kg

13.13 **Heat of Combustion:** −14,550 Btu/lb = −8082 cal/g = -338.4×10^5 J/kg

13.14 **Heat of Decomposition:** Not pertinent

13.15 **Heat of Solution:** Not pertinent

13.16 **Heat of Polymerization:** Not pertinent

NOTES

REVISED 1978

EFM ETHYL FORMATE

Common Synonyms			
Formic acid, ethyl ester Ethyl formic ester Ethyl methanoate Formic ether	Liquid	Colorless	Pleasant odor
	Floats and mixes with water. Flammable, irritating vapor is produced.		

Shut off ignition sources. Call fire department.
Stop discharge if possible. Keep people away.
Stay upwind. Use water spray to "knock down" vapor.
Isolate and remove discharged material.
Notify local health and pollution control agencies.

Fire	FLAMMABLE. Containers may explode in fire. Flashback along vapor trail may occur. Vapor may explode if ignited in an enclosed area. **Extinguish with dry chemicals, alcohol foam, or carbon dioxide.** **Water may be ineffective on fire.** **Cool exposed containers with water.**
Exposure	**Call for medical aid.** VAPOR Irritating to eyes, nose and throat. If inhaled will cause difficult breathing. **Move victim to fresh air.** **If breathing has stopped, give artificial respiration.** **If breathing is difficult, give oxygen.** LIQUID Irritating to skin and eyes. Harmful if swallowed. **Remove contaminated clothing and shoes.** **Flush affected areas with plenty of water.** **IF IN EYES, hold eyelids open and flush with plenty of water.** **IF SWALLOWED and victim is CONSCIOUS, have victim drink water or milk.**
Water Pollution	Effect of low concentrations on aquatic life is unknown. May be dangerous if it enters water intakes. **Notify local health and wildlife officials.** **Notify operators of nearby water intakes.**

1. RESPONSE TO DISCHARGE

(See Response Methods Handbook, CG 446-4)

Issue warning—high flammability, water contaminant
Restrict access
Evacuate area
Disperse and flush

2. LABEL

Red

3. CHEMICAL DESIGNATIONS

3.1 **Synonyms:** Ethyl formic ester; Ethyl methanoate; Formic acid, ethyl ester; Formic ether
3.2 **Coast Guard Compatibility Classification:** Esters (13)
3.3 **Chemical Formula:** $HCOOC_2H_5$
3.4 **IMCO/United Nations Numerical Designation:** 3.1/1190

4. OBSERVABLE CHARACTERISTICS

4.1 **Physical State (as shipped):** Liquid
4.2 **Color:** Colorless
4.3 **Odor:** Characteristic; pleasant aromatic

5. HEALTH HAZARDS

5.1 **Personal Protective Equipment:** Organic canister gas mask; goggles or face shield; rubber gloves.
5.2 **Symptoms Following Exposure:** Inhalation of vapor causes slight irritation of the eyes and rapidly increasing irritation of the nose. High concentrations cause deep narcosis within a few minutes followed by death within a few hours. Contact with liquid causes moderate irritation of eyes and mild irritation of skin. Ingestion causes irritation of mouth and stomach; may cause deep narcosis and death if not treated.
5.3 **Treatment for Exposure:** INHALATION: remove from exposure; begin artificial respiration if breathing has stopped; call physician. EYES: wash with water for 15 min.; call physician if needed. SKIN: wash with water for 15 min.; call physician if irritation persists. INGESTION: do NOT induce vomiting; get medical attention at once.
5.4 **Toxicity by Inhalation (Threshold Limit Value):** 100 ppm
5.5 **Short-Term Inhalation Limits:** Data not available
5.6 **Toxicity by Ingestion:** Grade 2; oral LD_{50} = 1,850 mg/kg (rat)
5.7 **Late Toxicity:** Data not available
5.8 **Vapor (Gas) Irritant Characteristics:** Vapors cause moderate irritation such that personnel will find high concentrations unpleasant. The effect is temporary.
5.9 **Liquid or Solid Irritant Characteristics:** Fairly severe skin irritant. May cause pain and second-degree burns after a few minutes' contact.
5.10 **Odor Threshold:** Data not available

6. FIRE HAZARDS

6.1 **Flash Point:** 10°F O.C.; −4°F C.C.
6.2 **Flammable Limits in Air:** 2.8%–16.0%
6.3 **Fire Extinguishing Agents:** Dry chemical, alcohol foam, carbon dioxide
6.4 **Fire Extinguishing Agents Not to be Used:** Water may be ineffective.
6.5 **Special Hazards of Combustion Products:** Not pertinent
6.6 **Behavior in Fire:** Vapor is heavier than air and may travel long distance to a source of ignition and flash back.
6.7 **Ignition Temperature:** 851°F
6.8 **Electrical Hazard:** Data not available
6.9 **Burning Rate:** 3.6 mm/min.

7. CHEMICAL REACTIVITY

7.1 **Reactivity with Water:** No reaction
7.2 **Reactivity with Common Materials:** No reaction
7.3 **Stability During Transport:** Stable
7.4 **Neutralizing Agents for Acids and Caustics:** Not pertinent
7.5 **Polymerization:** Not pertinent
7.6 **Inhibitor of Polymerization:** Not pertinent

8. WATER POLLUTION

8.1 **Aquatic Toxicity:** Data not available
8.2 **Waterfowl Toxicity:** Data not available
8.3 **Biological Oxygen Demand (BOD):** 30%, 10 days; 0.5 lb/lb, 10 days
8.4 **Food Chain Concentration Potential:** None

9. SELECTED MANUFACTURERS

1. Fritzsche, Dodge & Olcott, Inc.
   76 Ninth Avenue
   New York, N. Y. 10001
2. Eastman Kodak Co.
   Eastman Organic Chemicals
   Rochester, N. Y. 14650
3. Aldrich Chemical Co.
   940 West St. Paul Avenue
   Milwaukee, Wis. 53233

10. SHIPPING INFORMATION

10.1 **Grades or Purity:** 95+%
10.2 **Storage Temperature:** Ambient
10.3 **Inert Atmosphere:** No requirement
10.4 **Venting:** Pressure-vacuum

11. HAZARD ASSESSMENT CODE

(See Hazard Assessment Handbook, CG 446-3)

A-P-Q-R-S

12. HAZARD CLASSIFICATIONS

12.1 **Code of Federal Regulations:** Flammable liquid
12.2 **NAS Hazard Rating for Bulk Water Transportation:**

Category	Rating
Fire	3
Health	
Vapor Irritant	2
Liquid or Solid Irritant	3
Poisons	3
Water Pollution	
Human Toxicity	2
Aquatic Toxicity	2
Aesthetic Effect	2
Reactivity	
Other Chemicals	1
Water	1
Self-Reaction	0

12.3 **NFPA Hazard Classifications:**

Category	Classification
Health Hazard (Blue)	2
Flammability (Red)	3
Reactivity (Yellow)	0

13. PHYSICAL AND CHEMICAL PROPERTIES

13.1 **Physical State at 15°C and 1 atm:** Liquid
13.2 **Molecular Weight:** 74.1
13.3 **Boiling Point at 1 atm:** 129.6°F = 54.2°C = 327.4°K
13.4 **Freezing Point:** −110°F = −79°C = 194°K
13.5 **Critical Temperature:** 455°F = 235°C = 508°K
13.6 **Critical Pressure:** 686 psia = 46.6 atm = 4.73 MN/m^2
13.7 **Specific Gravity:** 0.922 at 20°C (liquid)
13.8 **Liquid Surface Tension:** 24 dynes/cm = 0.024 N/m at 20°C
13.9 **Liquid-Water Interfacial Tension:** (est.) 28 dynes/cm = 0.028 N/m at 20°C
13.10 **Vapor (Gas) Specific Gravity:** 2.6
13.11 **Ratio of Specific Heats of Vapor (Gas):** 1.1014
13.12 **Latent Heat of Vaporization:** 176 Btu/lb = 98 cal/g = 4.1 × 10^5 J/kg
13.13 **Heat of Combustion:** −9,500 Btu/lb = −5,300 cal/g = −220 × 10^5 J/kg
13.14 **Heat of Decomposition:** Not pertinent
13.15 **Heat of Solution:** −50 Btu/lb = −28 cal/g = −1.2 × 10^5 J/kg
13.16 **Heat of Polymerization:** Not pertinent

NOTES

EHA — ETHYLHEXALDEHYDE

Common Synonyms			
2-Ethylhexanal Butylethylacetaldehyde alpha-Ethylcaproaldehyde Octyl aldehyde 2-Ethyl hexaldehyde	Liquid	White	Mild odor
	Floats on water.		

Avoid contact with liquid. Keep people away.
Wear goggles, self-contained breathing apparatus, and rubber overclothing.
Stop discharge if possible.
Isolate and remove discharged material.
Notify local health and pollution control agencies.

Fire	Combustible. Extinguish with dry chemicals, alcohol foam, or carbon dioxide. Water may be ineffective on fire. Cool exposed containers with water.
Exposure	CALL FOR MEDICAL AID. VAPOR Irritating to eyes, nose and throat. If inhaled will cause coughing or difficult breathing. If in eyes, hold eyelids open and flush with plenty of water. If breathing has stopped, give artificial respiration. If breathing is difficult, give oxygen. LIQUID Irritating to skin and eyes. If swallowed will cause nausea and vomiting. Remove contaminated clothing and shoes. Flush affected areas with plenty of water. IF IN EYES, hold eyelids open and flush with plenty of water. IF SWALLOWED and victim is CONSCIOUS, have victim drink water or milk. IF SWALLOWED and victim is UNCONSCIOUS OR HAVING CONVULSIONS, do nothing except keep victim warm.
Water Pollution	Effect of low concentrations on aquatic life is unknown. Fouling to shoreline. May be dangerous if it enters water intakes. Notify local health and wildlife officials. Notify operators of nearby water intakes.

1. RESPONSE TO DISCHARGE

(See Response Methods Handbook, CG 446-4)

Mechanical containment
Should be removed
Chemical and physical treatment

2. LABELS

No hazard label required by Code of Federal Regulations

3. CHEMICAL DESIGNATIONS

3.1 **Synonyms:** Butylethylacetaldehyde; 2-Ethylcaproaldehyde; 2-Ethylhexaldehyde; 2-Ethylhexanal; Octyl aldehyde

3.2 **Coast Guard Compatibility Classification:** Aldehyde

3.3 **Chemical Formula:** $C_4H_9CH(C_2H_5)CHO$

3.4 **IMCO/United Nations Numerical Designation:** 3.3/1191

4. OBSERVABLE CHARACTERISTICS

4.1 **Physical State (as shipped):** Liquid

4.2 **Color:** Colorless

4.3 **Odor:** Mild characteristic

5. HEALTH HAZARDS

5.1 **Personal Protective Equipment:** Rubber gloves; safety goggles or face shield

5.2 **Symptoms Following Exposure:** Inhalation may be irritating to mucous membrane; overexposure may cause dizziness and collapse. Ingestion causes irritation of mouth and stomach. Contact with eyes or skin causes irritation.

5.3 **Treatment for Exposure:** INHALATION: remove victim to fresh air; give oxygen if breathing is difficult; call a doctor. EYES: irrigate immediately for 15 min., then get medical attention. SKIN: flush with water; wash with soap and water.

5.4 **Toxicity by Inhalation (Threshold Limit Value):** Data not available

5.5 **Short-Term Inhalation Limits:** Data not available

5.6 **Toxicity by Ingestion:** Grade 2; oral rat LD_{50} = 3,730 mg/kg

5.7 **Late Toxicity:** Data not available

5.8 **Vapor (Gas) Irritant Characteristics:** Vapors cause a slight smarting of the eyes or respiratory system if present in high concentrations. The effect is temporary.

5.9 **Liquid or Solid Irritant Characteristics:** Causes smarting of the skin and first-degree burns on short exposure; may cause second-degree burns on long exposure.

5.10 **Odor Threshold:** Data not available

6. FIRE HAZARDS

6.1 **Flash Point:** 127°F O.C.; 112°F C.C.

6.2 **Flammable Limits in Air:** Data not available

6.3 **Fire Extinguishing Agents:** Dry chemical, alcohol foam, carbon dioxide

6.4 **Fire Extinguishing Agents Not to be Used:** Water may be ineffective

6.5 **Special Hazards of Combustion Products:**

6.6 **Behavior in Fire:**

6.7 **Ignition Temperature:** 387°F

6.8 **Electrical Hazard:** Data not available

6.9 **Burning Rate:** Data not available

7. CHEMICAL REACTIVITY

7.1 **Reactivity with Water:** No reaction

7.2 **Reactivity with Common Materials:** May ignite spontaneously when spilled on clothing, paper, or other absorbent materials

7.3 **Stability During Transport:** Stable

7.4 **Neutralizing Agents for Acids and Caustics:** Not pertinent

7.5 **Polymerization:** Not pertinent

7.6 **Inhibitor of Polymerization:** Not pertinent

8. WATER POLLUTION

8.1 **Aquatic Toxicity:** Data not available

8.2 **Waterfowl Toxicity:** Data not available

8.3 **Biological Oxygen Demand (BOD):** Data not available

8.4 **Food Chain Concentration Potential:** None

9. SELECTED MANUFACTURERS

1. Eastman Chemical Products, Inc.
   Kingsport, Tenn. 37662
2. Union Carbide Corp.
   Chemicals and Plastics
   270 Park Avenue
   New York, N.Y. 10017
3. Pfaltz and Bauer, Inc.
   375 Fairfield Ave.
   Stamford, Conn. 06902

10. SHIPPING INFORMATION

10.1 **Grades or Purity:** Commercial, 95.0+%

10.2 **Storage Temperature:** Ambient

10.3 **Inert Atmosphere:** No requirement

10.4 **Venting:** Open (flame arrester)

11. HAZARD ASSESSMENT CODE

(See Hazard Assessment Handbook, CG 446-3)

A-T-U

12. HAZARD CLASSIFICATIONS

12.1 **Code of Federal Regulations:** Combustible Liquid

12.2 **NAS Hazard Rating for Bulk Water Transportation:**

Category	Rating
Fire	2
Health	
Vapor Irritant	1
Liquid or Solid Irritant	2
Poisons	1
Water Pollution	
Human Toxicity	2
Aquatic Toxicity	2
Aesthetic Effect	2
Reactivity	
Other Chemicals	2
Water	0
Self-Reaction	1

12.3 **NFPA Hazard Classifications:**

Category	Classification
Health Hazard (Blue)	2
Flammability (Red)	2
Reactivity (Yellow)	1

13. PHYSICAL AND CHEMICAL PROPERTIES

13.1 **Physical State at 15°C and 1 atm:** Liquid

13.2 **Molecular Weight:** 128.22

13.3 **Boiling Point at 1 atm:** 327°F = 164°C = 437°K

13.4 **Freezing Point:** Not pertinent

13.5 **Critical Temperature:** Not pertinent

13.6 **Critical Pressure:** Not pertinent

13.7 **Specific Gravity:** 0.820 at 20°C (liquid)

13.8 **Liquid Surface Tension:** Data not available

13.9 **Liquid-Water Interfacial Tension:** Data not available

13.10 **Vapor (Gas) Specific Gravity:** Not pertinent

13.11 **Ratio of Specific Heats of Vapor (Gas):** Not pertinent

13.12 **Latent Heat of Vaporization:** 164 Btu/lb = 91.2 cal/g = 3.82×10^5 J/kg

13.13 **Heat of Combustion:** –15,860 Btu/lb = –8,810 cal/g = -369×10^5 J/kg

13.14 **Heat of Decomposition:** Not pertinent

13.15 **Heat of Solution:** Not pertinent

13.16 **Heat of Polymerization:** Not pertinent

NOTES

EAI 2-ETHYLHEXYL ACRYLATE, INHIBITED

Common Synonyms			
Acrylic acid, 2-ethylhexylester 2-Ethylhexyl, 2-propenoate	Liquid	Colorless	Sharp odor
	Floats on water.		

Stop discharge if possible. Keep people away.
Call fire department.
Isolate and remove discharged material.
Notify local health and pollution control agencies.

Fire	Combustible. Containers may explode in fire. **Combat fires from safe distance or protected location.** **Extinguish with dry chemicals or carbon dioxide.** **Water may be ineffective on fire.** **Cool exposed containers with water.**
Exposure	**Call for medical aid.** LIQUID Irritating to skin and eyes. Harmful if swallowed. **Remove contaminated clothing and shoes.** **Flush affected areas with plenty of water.** **IF IN EYES, hold eyelids open and flush with plenty of water.** **IF SWALLOWED and victim is CONSCIOUS, have victim drink water or milk and have victim induce vomiting.** **IF SWALLOWED and victim is UNCONSCIOUS OR HAVING CONVULSIONS, do nothing except keep victim warm.**
Water Pollution	Effect of low concentrations on aquatic life is unknown. Fouling to shoreline. May be dangerous if it enters water intakes. **Notify local health and wildlife officials.** **Notify operators of nearby water intakes.**

1. RESPONSE TO DISCHARGE

(See Response Methods Handbook, CG 446-4)

Issue warning—water contaminant
Restrict access
Mechanical containment
Should be removed
Chemical and physical treatment

2. LABELS

No label required by Code of Federal Regulations

3. CHEMICAL DESIGNATIONS

3.1 **Synonyms:** Acrylic acid, 2-ethylhexyl ester; 2-Ethylhexyl 2-propenoate
3.2 **Coast Guard Compatibility Classification:** Acrylate
3.3 **Chemical Formula:** $CH_2{=}CHCOOCH_2CH(C_2H_5)(CH_2)_3CH_3$
3.4 **IMCO/United Nations Numerical Designation:** Not listed

4. OBSERVABLE CHARACTERISTICS

4.1 **Physical State (as shipped):** Liquid
4.2 **Color:** Colorless
4.3 **Odor:** Sharp; mild ester-type; inoffensive

5. HEALTH HAZARDS

5.1 **Personal Protective Equipment:** Self-contained breathing apparatus; rubber gloves; vapor-proof chemical safety goggles; impervious apron and boots.
5.2 **Symptoms Following Exposure:** Inhalation of concentrated vapor causes drowsiness and convulsions. Liquid causes irritation of eyes and may irritate skin on prolonged exposure. Ingestion produces same symptoms as inhalation.
5.3 **Treatment for Exposure:** INHALATION: give artificial respiration and oxygen if necessary; call a physician. EYES: immediately flush with plenty of water for at least 15 min.; get medical attention. SKIN: immediately flush with plenty of water for at least 15 min. INGESTION: induce vomiting and consult a physician.
5.4 **Toxicity by Inhalation (Threshold Limit Value):** Data not available
5.5 **Short-Term Inhalation Limits:** Data not available
5.6 **Toxicity by Ingestion:** Grade 2; oral rat LD_{50} = 1,540 mg/kg
5.7 **Late Toxicity:** Data not available
5.8 **Vapor (Gas) Irritant Characteristics:** Vapors are nonirritating to eyes and throat.
5.9 **Liquid or Solid Irritant Characteristics:** Minimum hazard. If spilled on clothing and allowed to remain, may cause smarting and reddening of the skin.
5.10 **Odor Threshold:** Data not available

6. FIRE HAZARDS

6.1 **Flash Point:** 195°F O.C.
6.2 **Flammable Limits in Air:** 0.8% 6.4%
6.3 **Fire Extinguishing Agents:** Dry chemical or carbon dioxide
6.4 **Fire Extinguishing Agents Not to be Used:** Water or foam may cause frothing.
6.5 **Special Hazards of Combustion Products:** Not pertinent
6.6 **Behavior in Fire:** Heat can result in a severe polymerization with rapid release of energy. Sealed containers may rupture explosively if hot.
6.7 **Ignition Temperature:** 496°F
6.8 **Electrical Hazard:** Data not available
6.9 **Burning Rate:** 4.6 mm/min.

7. CHEMICAL REACTIVITY

7.1 **Reactivity with Water:** No reaction
7.2 **Reactivity with Common Materials:** No reaction
7.3 **Stability During Transport:** Stable
7.4 **Neutralizing Agents for Acids and Caustics:** Not pertinent
7.5 **Polymerization:** Will polymerize in the absence of inhibitor and when heated.
7.6 **Inhibitor of Polymerization:** Monomethyl ether of hydroquinone, 13–120 ppm. Hydroquinone, 90–120 ppm.

8. WATER POLLUTION

8.1 **Aquatic Toxicity:** 72 ppm/24 hr/brine shrimp/TL_m
8.2 **Waterfowl Toxicity:** Data not available
8.3 **Biological Oxygen Demand (BOD):** 9% of theoretical in 5 days, fresh water, acclimated seed
8.4 **Food Chain Concentration Potential:** None

9. SELECTED MANUFACTURERS

1. Celanese Chemical Company
   245 Park Avenue
   New York, N. Y. 10017
2. Rohm and Haas Co., Inc.
   Independence Mall West
   Philadelphia, Pa. 19105
3. Union Carbide Corporation
   Chemicals and Plastics Division
   270 Park Avenue
   New York, N. Y. 10017

10. SHIPPING INFORMATION

10.1 **Grades or Purity:** 99+%
10.2 **Storage Temperature:** <100°F (38°C)
10.3 **Inert Atmosphere:** No requirement
10.4 **Venting:** Open (flame arrester)

11. HAZARD ASSESSMENT CODE

(See Hazard Assessment Handbook, CG 446-3)

A-T-U-Z

12. HAZARD CLASSIFICATIONS

12.1 **Code of Federal Regulations:** Not listed
12.2 **NAS Hazard Rating for Bulk Water Transportation:**

Category	Rating
Fire	1
Health	
Vapor Irritant	0
Liquid or Solid Irritant	1
Poisons	1
Water Pollution	
Human Toxicity	1
Aquatic Toxicity	2
Aesthetic Effect	2
Reactivity	
Other Chemicals	2
Water	0
Self-Reaction	3

12.3 **NFPA Hazard Classifications:**

Category	Classification
Health Hazard (Blue)	2
Flammability (Red)	2
Reactivity (Yellow)	1

13. PHYSICAL AND CHEMICAL PROPERTIES

13.1 **Physical State at 15°C and 1 atm:** Liquid
13.2 **Molecular Weight:** 184.2
13.3 **Boiling Point at 1 atm:** (polymerizes) 417°F = 214°C = 487°K
13.4 **Freezing Point:** −130°F = −90°C = 183°K
13.5 **Critical Temperature:** Not pertinent
13.6 **Critical Pressure:** Not pertinent
13.7 **Specific Gravity:** 0.885 at 20°C (liquid)
13.8 **Liquid Surface Tension:** (est.) 26 dynes/cm = 0.026 N/m at 20°C
13.9 **Liquid-Water Interfacial Tension:** (est.) 30 dynes/cm = 0.030 N/m at 20°C
13.10 **Vapor (Gas) Specific Gravity:** Not pertinent
13.11 **Ratio of Specific Heats of Vapor (Gas):** Not pertinent
13.12 **Latent Heat of Vaporization:** 110 Btu/lb = 61 cal/g = 2.6×10^5 J/kg
13.13 **Heat of Combustion:** −15,500 Btu/lb = −8,600 cal/g = -360×10^5 J/kg
13.14 **Heat of Decomposition:** Not pertinent
13.15 **Heat of Solution:** Not pertinent
13.16 **Heat of Polymerization:** −142 Btu/lb = −79 cal/g = -3.3×10^5 J/kg

NOTES

2-ETHYL HEXANOL

EHX

Common Synonyms			
2-Ethyl-1-hexanol 2-Ethylhexyl alcohol	Oily liquid	Colorless	Faint odor
	Floats on water.		

Stop discharge if possible.
Call fire department.
Avoid contact with liquid.
Isolate and remove discharged material.
Notify local health and pollution control agencies.

Fire

Combustible.
Extinguish with dry chemical, alcohol foam, or carbon dioxide.
Cool exposed containers with water.

Exposure

CALL FOR MEDICAL AID.

LIQUID
Irritating to skin and eyes.
Harmful if swallowed.
Remove contaminated clothing and shoes.
Flush affected areas with plenty of water.
IF IN EYES, hold eyelids open and flush with plenty of water.
IF SWALLOWED and victim is CONSCIOUS, have victim drink water or milk.

Water Pollution

Effect of low concentrations on aquatic life is unknown.
Fouling to shoreline.
May be dangerous if it enters water intakes.
Notify local health and wildlife officials.
Notify operators of nearby water intakes.

1. RESPONSE TO DISCHARGE

(See Response Methods Handbook, CG 446-4)

Mechanical containment
Chemical and physical treatment

2. LABELS

No hazard label required by Code of Federal Regulations

3. CHEMICAL DESIGNATIONS

3.1 **Synonyms:** 2-Ethyl-1-hexanol
2-Ethylhexyl alcohol

3.2 **Coast Guard Compatibility Classification:** Alcohol

3.3 **Chemical Formula:** $CH_3(CH_2)_3CH(C_2H_5)CH_2OH$

3.4 **IMCO/United Nations Numerical Designation:** Not listed

4. OBSERVABLE CHARACTERISTICS

4.1 **Physical State (as shipped):** Liquid

4.2 **Color:** Colorless

4.3 **Odor:** Strong

5. HEALTH HAZARDS

5.1 **Personal Protective Equipment:** Air pack or organic canister; goggles; rubber gloves.

5.2 **Symptoms Following Exposure:** Anesthesia, nausea, headache, dizziness; mildly irritating to skin and eyes.

5.3 **Treatment for Exposure:** INHALATION: move victim to fresh air. SKIN: wash affected areas with water. EYES: flush with water for 15 min. Get medical care.

5.4 **Toxicity by Inhalation (Threshold Limit Value):** Not pertinent

5.5 **Short-Term Inhalation Limits:** Not pertinent

5.6 **Toxicity by Ingestion:** Grade 2; LD_{50} 0.5 to 5 g/kg (lab animals)

5.7 **Late Toxicity:** Increased excitability of central nervous system in rats and rabbits.

5.8 **Vapor (Gas) Irritant Characteristics:** Vapors cause a slight smarting of the eyes or respiratory system if present in high concentrations. The effect is temporary.

5.9 **Liquid or Solid Irritant Characteristics:** Minimum hazard. If spilled on clothing and allowed to remain, may cause smarting and reddening of the skin.

5.10 **Odor Threshold:** Data not available.

6. FIRE HAZARDS

6.1 **Flash Point:** 185°F O.C.; 175°F C.C.

6.2 **Flammable Limits in Air:** Data not available

6.3 **Fire Extinguishing Agents:** Foam, carbon dioxide, dry chemical

6.4 **Fire Extinguishing Agents Not to be Used:** Not pertinent

6.5 **Special Hazards of Combustion Products:** Not pertinent

6.6 **Behavior in Fire:** Not pertinent

6.7 **Ignition Temperature:** 581°F

6.8 **Electrical Hazard:** Not pertinent

6.9 **Burning Rate:** 4.0 mm/min.

7. CHEMICAL REACTIVITY

7.1 **Reactivity with Water:** No reaction

7.2 **Reactivity with Common Materials:** No reaction

7.3 **Stability During Transport:** Stable

7.4 **Neutralizing Agents for Acids and Caustics:** Not pertinent

7.5 **Polymerization:** Not pertinent

7.6 **Inhibitor of Polymerization:** Not pertinent

8. WATER POLLUTION

8.1 **Aquatic Toxicity:** 19 ppm/24 hr/brine shrimp/TLM

8.2 **Waterfowl Toxicity:** Data not available

8.3 **Biological Oxygen Demand (BOD):** 88% of theoretical in 5 days

8.4 **Food Chain Concentration Potential:** None

9. SELECTED MANUFACTURERS

1. Eastman Chemical Products, Inc. Kingsport, Tenn. 37662
2. Oxochem Enterprise Penuelas, Puerto Rico 00724
3. Shell Chemical Co. Industrial Chemicals Division Houston, Tex. 77001
4. Union Carbide Corp. Chemical and Plastics Division 270 Park Avenue New York, N. Y. 10017

10. SHIPPING INFORMATION

10.1 **Grades or Purity:** 99-99.7%

10.2 **Storage Temperature:** Ambient

10.3 **Inert Atmosphere:** No requirement

10.4 **Venting:** Open (flame arrester)

11. HAZARD ASSESSMENT CODE

(See Hazard Assessment Handbook, CG 446-3)

A-T-U

12. HAZARD CLASSIFICATIONS

12.1 **Code of Federal Regulations:** Combustible Liquid

12.2 **NAS Hazard Rating for Bulk Water Transportation:**

Category	Rating
Fire	1
Health	
Vapor Irritant	1
Liquid or Solid Irritant	1
Poisons	1
Water Pollution	
Human Toxicity	1
Aquatic Toxicity	1
Aesthetic Effect	2
Reactivity	
Other Chemicals	2
Water	0
Self-Reaction	0

12.3 **NFPA Hazard Classifications:**

Category	Classification
Health Hazard (Blue)	2
Flammability (Red)	2
Reactivity (Yellow)	0

13. PHYSICAL AND CHEMICAL PROPERTIES

13.1 **Physical State at 15°C and 1 atm:** Liquid

13.2 **Molecular Weight:** 130.23

13.3 **Boiling Point at 1 atm:** 364.5°F = 184.7°C = 457.9°K

13.4 **Freezing Point:** $<$158°F = $<$70°C = $<$343°K

13.5 **Critical Temperature:** 711°F = 377°C = 650°K

13.6 **Critical Pressure:** 512 psia = 34.8 atm = 3.53 MN/m^2

13.7 **Specific Gravity:** 0.834 at 20°C (liquid)

13.8 **Liquid Surface Tension:** 27.6 dynes/cm = 0.0276 N/m at 20°C

13.9 **Liquid-Water Interfacial Tension:** 22 dynes/cm = 0.022 N/m at 22.7°C

13.10 **Vapor (Gas) Specific Gravity:** Not pertinent

13.11 **Ratio of Specific Heats of Vapor (Gas):** Not pertinent

13.12 **Latent Heat of Vaporization:** 167 Btu/lb = 92.8 cal/g = 3.89×10^5 J/kg

13.13 **Heat of Combustion:** −17,480 Btu/lb = −9710 cal/g = -406.5×10^5 J/kg

13.14 **Heat of Decomposition:** Not pertinent

13.15 **Heat of Solution:** Not pertinent

13.16 **Heat of Polymerization:** Not pertinent

NOTES

REVISED 1978

EHT — ETHYL HEXYL TALLATE

Common Synonyms	Liquid	Light yellow	Mild odor
	Floats on water.		

Stop discharge if possible.
Call fire department.
Isolate and remove discharged material.
Notify local health and pollution control agencies.

Fire	Combustible. **Extinguish with dry chemical, foam, or carbon dioxide.**
Exposure	Not harmful.
Water Pollution	Effect of low concentrations on aquatic life is unknown. Fouling to shoreline. May be dangerous if it enters water intakes. **Notify local health and wildlife officials. Notify operators of nearby water intakes.**

1. RESPONSE TO DISCHARGE
(See Response Methods Handbook, CG 446-4)

Mechanical containment
Chemical and physical treatment

2. LABELS
No hazard label required by Code of Federal Regulations

3. CHEMICAL DESIGNATIONS
3.1 **Synonyms:** Croplas EH
3.2 **Coast Guard Compatibility Classification:** Ester
3.3 **Chemical Formula:** (Mixture)
3.4 **IMCO/United Nations Numerical Designation:** Not listed

4. OBSERVABLE CHARACTERISTICS
4.1 **Physical State (as shipped):** Liquid
4.2 **Color:** Pale yellow
4.3 **Odor:** Mild characteristic

5. HEALTH HAZARDS
5.1 **Personal Protective Equipment:** Data not available
5.2 **Symptoms Following Exposure:** Data not available
5.3 **Treatment for Exposure:** Data not available
5.4 **Toxicity by Inhalation (Threshold Limit Value):** Not pertinent
5.5 **Short-Term Inhalation Limits:** Not pertinent
5.6 **Toxicity by Ingestion:** Data not available
5.7 **Late Toxicity:** Data not available
5.8 **Vapor (Gas) Irritant Characteristics:** None
5.9 **Liquid or Solid Irritant Characteristics:** None
5.10 **Odor Threshold:** Not pertinent

6. FIRE HAZARDS
6.1 **Flash Point:** 395°F O.C.
6.2 **Flammable Limits in Air:** Data not available
6.3 **Fire Extinguishing Agents:** Data not available
6.4 **Fire Extinguishing Agents Not to be Used:** Data not available
6.5 **Special Hazards of Combustion Products:** Not pertinent
6.6 **Behavior in Fire:** Not pertinent
6.7 **Ignition Temperature:** Data not available
6.8 **Electrical Hazard:** Not pertinent
6.9 **Burning Rate:** Data not available

7. CHEMICAL REACTIVITY
7.1 **Reactivity with Water:** No reaction
7.2 **Reactivity with Common Materials:** No reaction
7.3 **Stability During Transport:** Stable
7.4 **Neutralizing Agents for Acids and Caustics:** Not pertinent
7.5 **Polymerization:** Not pertinent
7.6 **Inhibitor of Polymerization:** Not pertinent

8. WATER POLLUTION
8.1 **Aquatic Toxicity:** Data not available
8.2 **Waterfowl Toxicity:** Data not available
8.3 **Biological Oxygen Demand (BOD):** Data not available
8.4 **Food Chain Concentration Potential:** None

9. SELECTED MANUFACTURERS
1. Arizona Chemical Co.
   111 W. 50th St.
   New York, N. Y. 10020
2. Crosby Chemicals, Inc.
   Picayune, Miss. 39466
3. Union Carbide Corp.
   Chemicals and Plastics Division
   270 Park Ave.
   New York, N. Y. 10017

10. SHIPPING INFORMATION
10.1 **Grades or Purity:** Data not available
10.2 **Storage Temperature:** Ambient
10.3 **Inert Atmosphere:** No requirement
10.4 **Venting:** Open (flame arrester)

11. HAZARD ASSESSMENT CODE
(See Hazard Assessment Handbook, CG 446-3)

A-T-U-X-Y

12. HAZARD CLASSIFICATIONS
12.1 **Code of Federal Regulations:** Not listed
12.2 **NAS Hazard Rating for Bulk Water Transportation:** Not listed
12.3 **NFPA Hazard Classifications:** Not listed

13. PHYSICAL AND CHEMICAL PROPERTIES
13.1 **Physical State at 15°C and 1 atm:** Liquid
13.2 **Molecular Weight:** Not pertinent
13.3 **Boiling Point at 1 atm:** Very high
13.4 **Freezing Point:** Not pertinent
13.5 **Critical Temperature:** Not pertinent
13.6 **Critical Pressure:** Not pertinent
13.7 **Specific Gravity:** (est.) 0.95 at 20°C (liquid)
13.8 **Liquid Surface Tension:** 30.4 dynes/cm = 0.0304 N/m at 24°C
13.9 **Liquid-Water Interfacial Tension:** 28 dynes/cm = 0.028 N/m at 21.3°C
13.10 **Vapor (Gas) Specific Gravity:** Not pertinent
13.11 **Ratio of Specific Heats of Vapor (Gas):** Not pertinent
13.12 **Latent Heat of Vaporization:** Not pertinent
13.13 **Heat of Combustion:** (est.) $-17{,}000$ Btu/lb $= -10{,}000$ cal/g $= -400 \times 10^5$ J/kg
13.14 **Heat of Decomposition:** Not pertinent
13.15 **Heat of Solution:** Not pertinent
13.16 **Heat of Polymerization:** Not pertinent

NOTES

REVISED 1978

ENB	# ETHYLIDENENORBORNENE

Common Synonyms			
5-Ethylidenebicyclo (2,2,1) hept-2-ene Ethylidenenorcamphene Ethylidenenorbornylene	Liquid	White	Turpentine-like odor
	Floats on water.		

AVOID CONTACT WITH LIQUID AND VAPOR. KEEP PEOPLE AWAY.
Shut off ignition sources. Call fire department.
Stop discharge if possible.
Stay upwind. Use water spray to "knock down" vapor.
Isolate and remove discharged material.
Notify local health and pollution control agencies.

Fire

Combustible.
Extinguish with dry chemicals, alcohol foam, or carbon dioxide.
Water may be ineffective on fire.
Cool exposed containers with water.

Exposure

CALL FOR MEDICAL AID.

VAPOR
Irritating to eyes, nose and throat.
If inhaled will cause headache, coughing or difficult breathing.
If in eyes, hold eyelids open and flush with plenty of water.
If breathing has stopped, give artificial respiration.
If breathing is difficult, give oxygen.

LIQUID
POISONOUS IF SWALLOWED.
Irritating to skin and eyes.
If swallowed will cause nausea and vomiting.
Remove contaminated clothing and shoes.
Flush affected areas with plenty of water.
IF IN EYES, hold eyelids open and flush with plenty of water.
IF SWALLOWED and victim is CONSCIOUS, have victim drink water or milk and have victim induce vomiting.
IF SWALLOWED and victim is UNCONSCIOUS OR HAVING CONVULSIONS, do nothing except keep victim warm.

Water Pollution

Effect of low concentrations on aquatic life is unknown.
Fouling to shoreline.
May be dangerous if it enters water intakes.
Notify local health and wildlife officials.
Notify operators of nearby water intakes.

1. RESPONSE TO DISCHARGE

(See Response Methods Handbook, CG 446-4)

Mechanical containment
Should be removed
Chemical and physical treatment

2. LABELS

No hazard label required by Code of Federal Regulations

3. CHEMICAL DESIGNATIONS

3.1 **Synonyms:** 5-Ethylidenebicyclo(2,2.1)hept-2-ene
Ethylidenenorbornylene
Ethylidenenorcamphene

3.2 **Coast Guard Compatibility Classification:** Not listed

3.3 **Chemical Formula:** C_9H_{12}

3.4 **IMCO/United Nations Numerical Designation:** Not listed

4. OBSERVABLE CHARACTERISTICS

4.1 **Physical State (as shipped):** Liquid

4.2 **Color:** Colorless

4.3 **Odor:** Like turpentine

5. HEALTH HAZARDS

5.1 **Personal Protective Equipment:** Organic canister or air-supplied mask; goggles or face shield; rubber gloves

5.2 **Symptoms Following Exposure:** Inhalation of vapors causes headache, confusion, and respiratory distress. Ingestion causes irritation of entire digestive system. Aspiration causes severe pneumonia. Contact with liquid causes irritation of eyes and skin.

5.3 **Treatment for Exposure:** INHALATION: remove victim to fresh air; administer artificial respiration and oxygen if required; call a doctor. INGESTION: give large amount of water and induce vomiting; get medical attention at once. EYES: flush with water for at least 15 min. SKIN: wipe off, wash with soap and water.

5.4 **Toxicity by Inhalation (Threshold Limit Value):** Data not available

5.5 **Short-Term Inhalation Limits:** Data not available

5.6 **Toxicity by Ingestion:** Grade 2; oral LD_{50} = 2.83 g/kg (rat)

5.7 **Late Toxicity:** Causes kidney lesions and gain in kidney and liver weights in rats

5.8 **Vapor (Gas) Irritant Characteristics:** Vapors are moderately irritating such that personnel will not usually tolerate moderate or high concentrations.

5.9 **Liquid or Solid Irritant Characteristics:** Minimum hazard. If spilled on clothing and allowed to remain, may cause smarting and reddening of skin.

5.10 **Odor Threshold:** 0.007 ppm

6. FIRE HAZARDS

6.1 **Flash Point:** 98°F O.C.

6.2 **Flammable Limits in Air:** Data not available

6.3 **Fire Extinguishing Agents:** Dry chemical, foam, carbon dioxide

6.4 **Fire Extinguishing Agents Not to be Used:** Water may be ineffective

6.5 **Special Hazards of Combustion Products:**

6.6 **Behavior in Fire:**

6.7 **Ignition Temperature:** Data not available

6.8 **Electrical Hazard:** Data not available

6.9 **Burning Rate:** Data not available

7. CHEMICAL REACTIVITY

7.1 **Reactivity with Water:** No reaction

7.2 **Reactivity with Common Materials:**

7.3 **Stability During Transport:** Stable

7.4 **Neutralizing Agents for Acids and Caustics:** Not pertinent

7.5 **Polymerization:** Data not available

7.6 **Inhibitor of Polymerization:** Data not available

8. WATER POLLUTION

8.1 **Aquatic Toxicity:** Data not available

8.2 **Waterfowl Toxicity:** Data not available

8.3 **Biological Oxygen Demand (BOD):** Data not available

8.4 **Food Chain Concentration Potential:** None

9. SELECTED MANUFACTURERS

1. Union Carbide Corp.
   Chemicals and Plastics Div.
   270 Park Avenue
   New York, N.Y. 10017
2. Aldrich Chemical Co.
   940 W. Saint Paul Ave.
   Milwaukee, Wis. 53233
3. Pfaltz and Bauer, Inc.
   375 Fairfield Ave.
   Stamford, Conn. 06902

10. SHIPPING INFORMATION

10.1 **Grades or Purity:** Commercial, 99+%

10.2 **Storage Temperature:** Ambient

10.3 **Inert Atmosphere:** No requirement

10.4 **Venting:** Open (flame arrester)

11. HAZARD ASSESSMENT CODE

(See Hazard Assessment Handbook, CG 446-3)

A-T-U

12. HAZARD CLASSIFICATIONS

12.1 **Code of Federal Regulations:** Not listed

12.2 **NAS Hazard Rating for Bulk Water Transportation:**

Category	Rating
Fire	3
Health	
Vapor Irritant	3
Liquid or Solid Irritant	1
Poisons	4
Water Pollution	
Human Toxicity	4
Aquatic Toxicity	
Aesthetic Effect	3
Reactivity	
Other Chemicals	2
Water	0
Self-Reaction	3

12.3 **NFPA Hazard Classifications:** Not listed

13. PHYSICAL AND CHEMICAL PROPERTIES

13.1 **Physical State at 15°C and 1 atm:** Liquid

13.2 **Molecular Weight:** 120.2

13.3 **Boiling Point at 1 atm:** 297.7°F = 147.6°C = 420.8°K

13.4 **Freezing Point:** −112°F = −80°C = 193°K

13.5 **Critical Temperature:** Not pertinent

13.6 **Critical Pressure:** Not pertinent

13.7 **Specific Gravity:** 0.896 at 20°C (liquid)

13.8 **Liquid Surface Tension:** Data not available

13.9 **Liquid-Water Interfacial Tension:** Data not available

13.10 **Vapor (Gas) Specific Gravity:** 4.1

13.11 **Ratio of Specific Heats of Vapor (Gas):** Not pertinent

13.12 **Latent Heat of Vaporization:** Data not available

13.13 **Heat of Combustion:** (est.) −18,800 Btu/lb = −10,450 cal/g = -437×10^5 J/kg

13.14 **Heat of Decomposition:** Not pertinent

13.15 **Heat of Solution:** Not pertinent

13.16 **Heat of Polymerization:** Data not available

NOTES

ELT — ETHYL LACTATE

Common Synonyms			
Lactic acid, ethyl ester Ethyl 2-hydroxypropionate Ethyl 2-hydroxypropanoate	Liquid	Colorless	Mild odor
	Mixes with water.		

Stop discharge if possible.
Call fire department.
Isolate and remove discharged material.
Notify local health and pollution control agencies.

Fire	Combustible. Extinguish with water, dry chemicals, foam, or carbon dioxide. Cool exposed containers with water.
Exposure	Call for medical aid. LIQUID Harmful if swallowed. IF SWALLOWED and victim is CONSCIOUS, have victim drink water or milk.
Water Pollution	Effect of low concentrations on aquatic life is unknown. May be dangerous if it enters water intakes. Notify local health and wildlife officials. Notify operators of nearby water intakes.

1. RESPONSE TO DISCHARGE

(See Response Methods Handbook, CG 446-4)

Issue warning—water contaminant
Disperse and flush

2. LABELS

No label required by Code of Federal Regulations

3. CHEMICAL DESIGNATIONS

3.1 **Synonyms:** Ethyl alpha-hydroxy-propionate; Ethyl 2-hydroxypropanoate; Ethyl DL-Lactate; Lactic acid, ethyl ester
3.2 **Coast Guard Compatibility Classification:** Not applicable
3.3 **Chemical Formula:** $CH_3CHOHCOOC_2H_5$
3.4 **IMCO/United Nations Numerical Designation:** 3.3/1192

4. OBSERVABLE CHARACTERISTICS

4.1 **Physical State (as shipped):** Liquid
4.2 **Color:** Colorless
4.3 **Odor:** Mild characteristic

5. HEALTH HAZARDS

5.1 **Personal Protective Equipment:** Goggles or face shield; rubber gloves.
5.2 **Symptoms Following Exposure:** Inhalation of concentrated vapor may cause drowsiness. Contact with liquid causes mild irritation of eyes and (on prolonged contact) skin. Ingestion may cause narcosis.
5.3 **Treatment for Exposure:** INHALATION: remove victim to fresh air. EYES and SKIN: flush well with water. INGESTION: induce vomiting; get medical attention.
5.4 **Toxicity by Inhalation (Threshold Limit Value):** Data not available
5.5 **Short-Term Inhalation Limits:** Data not available
5.6 **Toxicity by Ingestion:** Grade 2; oral LD_{50} = 2,580 mg/kg (mouse)
5.7 **Late Toxicity:** Data not available
5.8 **Vapor (Gas) Irritant Characteristics:** Data not available
5.9 **Liquid or Solid Irritant Characteristics:** Data not available
5.10 **Odor Threshold:** Data not available

6. FIRE HAZARDS

6.1 **Flash Point:** 158°F O.C.; 115°F C.C.
6.2 **Flammable Limits in Air:** 1.5%–11.4%
6.3 **Fire Extinguishing Agents:** Water, dry chemical, alcohol foam, carbon dioxide
6.4 **Fire Extinguishing Agents Not to be Used:** Not pertinent
6.5 **Special Hazards of Combustion Products:** Not pertinent
6.6 **Behavior in Fire:** Not pertinent
6.7 **Ignition Temperature:** 752°F
6.8 **Electrical Hazard:** Data not available
6.9 **Burning Rate:** Data not available

7. CHEMICAL REACTIVITY

7.1 **Reactivity with Water:** No reaction
7.2 **Reactivity with Common Materials:** No reaction
7.3 **Stability During Transport:** Stable
7.4 **Neutralizing Agents for Acids and Caustics:** Not pertinent
7.5 **Polymerization:** Not pertinent
7.6 **Inhibitor of Polymerization:** Not pertinent

8. WATER POLLUTION

8.1 **Aquatic Toxicity:** Data not available
8.2 **Waterfowl Toxicity:** Data not available
8.3 **Biological Oxygen Demand (BOD):** Data not available
8.4 **Food Chain Concentration Potential:** None

9. SELECTED MANUFACTURERS

1. Kay-Fries Chemicals, Inc.
   360 Lexington Avenue
   New York, N. Y. 10017
2. Aceto Chemical Co.
   126-02 Northern Boulevard
   Flushing, N. Y. 11368
3. Eastman Kodak Co.
   Eastman Organic Chemicals
   Rochester, N. Y. 14650

10. SHIPPING INFORMATION

10.1 **Grades or Purity:** Commercial
10.2 **Storage Temperature:** Ambient
10.3 **Inert Atmosphere:** No requirement
10.4 **Venting:** Open (flame arrester)

11. HAZARD ASSESSMENT CODE

(See Hazard Assessment Handbook, CG 446-3)

A-P-Q

12. HAZARD CLASSIFICATIONS

12.1 **Code of Federal Regulations:** Combustible Liquid
12.2 **NAS Hazard Rating for Bulk Water Transportation:** Not listed
12.3 **NFPA Hazard Classifications:**

Category	Classification
Health Hazard (Blue)	2
Flammability (Red)	2
Reactivity (Yellow)	0

13. PHYSICAL AND CHEMICAL PROPERTIES

13.1 **Physical State at 15°C and 1 atm:** Liquid
13.2 **Molecular Weight:** 118.1
13.3 **Boiling Point at 1 atm:** 309°F = 154°C = 427°K
13.4 **Freezing Point:** Not pertinent
13.5 **Critical Temperature:** Not pertinent
13.6 **Critical Pressure:** Not pertinent
13.7 **Specific Gravity:** 1.03 at 20°C (liquid)
13.8 **Liquid Surface Tension:** 29.20 dynes/cm = 0.0292 N/m at 20°C
13.9 **Liquid-Water Interfacial Tension:** Not pertinent
13.10 **Vapor (Gas) Specific Gravity:** Not pertinent
13.11 **Ratio of Specific Heats of Vapor (Gas):** Not pertinent
13.12 **Latent Heat of Vaporization:** Not pertinent
13.13 **Heat of Combustion:** (est.) −11,600 Btu/lb = −6,500 cal/g = -270×10^5 J/kg
13.14 **Heat of Decomposition:** Not pertinent
13.15 **Heat of Solution:** Data not available
13.16 **Heat of Polymerization:** Not pertinent

NOTES

EMC	ETHYL MERCAPTAN

Common Synonyms			
Ethanethiol Mercaptoethane Ethyl sulfhydrate Thioethyl alcohol	Liquid	Colorless to yellow	Strong skunk-like odor
	Floats and mixes slowly with water. Poisonous, flammable vapor is produced. Boiling point is 95° F.		

AVOID CONTACT WITH LIQUID AND VAPOR. KEEP PEOPLE AWAY.
Wear goggles and self-contained breathing apparatus.
Shut off ignition sources. Call fire department.
Stop discharge if possible.
Stay upwind. Use water spray to "knock down" vapor.
Isolate and remove discharged material.
Notify local health and pollution control agencies.

Fire

FLAMMABLE.
POISONOUS GASES ARE PRODUCED IN FIRE.
Containers may explode in fire.
Flashback along vapor trail may occur.
Vapor may explode if ignited in an enclosed area.
Extinguish with dry chemicals, foam or carbon dioxide.
Water may be ineffective on fire.
Cool exposed containers with water.

Exposure

CALL FOR MEDICAL AID.

VAPOR
POISONOUS IF INHALED.
Move victim to fresh air.
If breathing has stopped, give artificial respiration.
If breathing is difficult, give oxygen.

LIQUID
POISONOUS IF SWALLOWED.
IF SWALLOWED and victim is CONSCIOUS, have victim drink water or milk and have victim induce vomiting.
IF SWALLOWED and victim is UNCONSCIOUS OR HAVING CONVULSIONS, do nothing except keep victim warm.

Water Pollution

Effect of low concentrations on aquatic life is unknown.
May be dangerous if it enters water intakes.
Notify local health and wildlife officials.
Notify operators of nearby water intakes.

1. RESPONSE TO DISCHARGE

(See Response Methods Handbook, CG 446-4)

Issue warning—high flammability, water contaminant, air contaminant
Restrict access
Evacuate area
Disperse and flush

2. LABEL

Red

3. CHEMICAL DESIGNATIONS

3.1 **Synonyms:** Ethanethiol; Ethyl sulfhydrate; Mercaptoethane; Thioethyl alcohol
3.2 **Coast Guard Compatibility Classification:** Not applicable
3.3 **Chemical Formula:** C_2H_5SH
3.4 **IMCO/United Nations Numerical Designation:** 3.1/1228

4. OBSERVABLE CHARACTERISTICS

4.1 **Physical State (as shipped):** Liquid
4.2 **Color:** Colorless to yellow
4.3 **Odor:** Strong skunk; offensive garlic

5. HEALTH HAZARDS

5.1 **Personal Protective Equipment:** Plastic gloves; goggles or face shield.
5.2 **Symptoms Following Exposure:** Inhalation of vapor causes muscular weakness, convulsions, respiratory paralysis. High concentrations may cause pulmonary irritation. Liquid irritates eyes and skin. Ingestion causes nausea and irritation of mouth and stomach.
5.3 **Treatment for Exposure:** INHALATION: move victim to fresh air; if he is unconscious, give artificial respiration and oxygen; get medical attention. EYES: flush with water for at least 15 min. following contact with liquid; get medical attention if irritation persists. SKIN: wash well with water. INGESTION: induce vomiting and follow with gastric lavage; get medical attention.
5.4 **Toxicity by Inhalation (Threshold Limit Value):** 0.5 ppm
5.5 **Short-Term Inhalation Limits:** Data not available
5.6 **Toxicity by Ingestion:** Grade 2; oral LD_{50} = 682 mg/kg (rat)
5.7 **Late Toxicity:** May impair respiratory muscle function in warm-blooded experimental animals
5.8 **Vapor (Gas) Irritant Characteristics:** Data not available
5.9 **Liquid or Solid Irritant Characteristics:** Data not available
5.10 **Odor Threshold:** 0.001 ppm

6. FIRE HAZARDS

6.1 **Flash Point:** <0°F O.C.
6.2 **Flammable Limits in Air:** 2.8%–18%
6.3 **Fire Extinguishing Agents:** Dry chemical, foam, carbon dioxide
6.4 **Fire Extinguishing Agents Not to be Used:** Water may be ineffective.
6.5 **Special Hazards of Combustion Products:** Irritating fumes of sulfur dioxide are generated.
6.6 **Behavior in Fire:** Vapor is heavier than air and may travel long distance to a source of ignition and flash back; containers may explode in a fire; offensive fumes are released when heated.
6.7 **Ignition Temperature:** 572°F
6.8 **Electrical Hazard:** Data not available
6.9 **Burning Rate:** 5.7 mm/min.

7. CHEMICAL REACTIVITY

7.1 **Reactivity with Water:** No reaction
7.2 **Reactivity with Common Materials:** No reaction
7.3 **Stability During Transport:** Stable
7.4 **Neutralizing Agents for Acids and Caustics:** Not pertinent
7.5 **Polymerization:** Not pertinent
7.6 **Inhibitor of Polymerization:** Not pertinent

8. WATER POLLUTION

8.1 **Aquatic Toxicity:** Data not available
8.2 **Waterfowl Toxicity:** Data not available
8.3 **Biological Oxygen Demand (BOD):** Data not available
8.4 **Food Chain Concentration Potential:** None

9. SELECTED MANUFACTURERS

1. Pennwalt Corporation
Chemicals Division
Three Parkway
Philadelphia, Pa. 19102
2. Phillips Petroleum Company
Chemical Department
Special Products Division
Bartlesville, Okla. 74004
3. Eastman Kodak Co.
Eastman Organic Chemicals
Rochester, N. Y. 14650

10. SHIPPING INFORMATION

10.1 **Grades or Purity:** 98.5+%
10.2 **Storage Temperature:** Below 30°C
10.3 **Inert Atmosphere:** No requirement
10.4 **Venting:** Pressure-vacuum

11. HAZARD ASSESSMENT CODE

(See Hazard Assessment Handbook, CG 446-3)

A-P-Q-T-U-V-W

12. HAZARD CLASSIFICATIONS

12.1 **Code of Federal Regulations:** Flammable liquid
12.2 **NAS Hazard Rating for Bulk Water Transportation:** Not listed
12.3 **NFPA Hazard Classifications:**

Category	Classification
Health Hazard (Blue)	2
Flammability (Red)	4
Reactivity (Yellow)	0

13. PHYSICAL AND CHEMICAL PROPERTIES

13.1 **Physical State at 15°C and 1 atm:** Liquid
13.2 **Molecular Weight:** 62.1
13.3 **Boiling Point at 1 atm:** 93.9°F = 34.4°C = 307.6°K
13.4 **Freezing Point:** −234°F = −147°C = 126°K
13.5 **Critical Temperature:** 439°F = 226°C = 499°K
13.6 **Critical Pressure:** 798 psia = 54.2 atm = 5.50 MN/m^2
13.7 **Specific Gravity:** 0.826 at 20°C (liquid)
13.8 **Liquid Surface Tension:** 23.5 dynes/cm = 0.0235 N/m at 20°C
13.9 **Liquid-Water Interfacial Tension:** (est.) 25 dynes/cm = 0.025 N/m at 20°C
13.10 **Vapor (Gas) Specific Gravity:** 2.1
13.11 **Ratio of Specific Heats of Vapor (Gas):** 1.1308 at 16°C
13.12 **Latent Heat of Vaporization:** 189 Btu/lb = 105 cal/g = 4.39 × 10^5 J/kg
13.13 **Heat of Combustion:** −15,000 Btu/lb = −8,300 cal/g = −350 × 10^5 J/kg
13.14 **Heat of Decomposition:** Not pertinent
13.15 **Heat of Solution:** Not pertinent
13.16 **Heat of Polymerization:** Not pertinent

NOTES

ETM	ETHYL METHACRYLATE

Common Synonyms			
Ethyl methacrylate-inhibited Methacrylic acid, ethyl ester Ethyl 2-methacrylate Ethyl alpha-methylmethacrylate Ethyl 2-methyl – 2-propenoate	Liquid	Colorless	Sharp unpleasant odor
	Floats on water.		

Shut off ignition sources. Call fire department.
Avoid contact with liquid and vapor. Keep people away.
Stop discharge if possible.
Stay upwind. Use water spray to "knock down" vapor.
Isolate and remove discharged material.
Notify local health and pollution control agencies.

Fire

FLAMMABLE.
Containers may explode in fire.
Flashback along vapor trail may occur.
Vapor may explode if ignited in an enclosed area.
Extinguish with dry chemicals, foam or carbon dioxide.
Water may be ineffective on fire.
Cool exposed containers with water.

Exposure

CALL FOR MEDICAL AID.

VAPOR
Irritating to eyes, nose and throat.
If inhaled will cause coughing or difficult breathing.
If in eyes, hold eyelids open and flush with plenty of water.
If breathing has stopped, give artificial respiration.
If breathing is difficult, give oxygen.

LIQUID
Irritating to skin and eyes.
If swallowed will cause nausea and vomiting.
Remove contaminated clothing and shoes.
Flush affected areas with plenty of water.
IF IN EYES, hold eyelids open and flush with plenty of water.
IF SWALLOWED and victim is CONSCIOUS, have victim drink water or milk and have victim induce vomiting.
IF SWALLOWED and victim is UNCONSCIOUS OR HAVING CONVULSIONS, do nothing except keep victim warm.

Water Pollution

Effect of low concentrations on aquatic life is unknown.
Fouling to shoreline.
May be dangerous if it enters water intakes.
Notify local health and wildlife officials.
Notify operators of nearby water intakes.

1. RESPONSE TO DISCHARGE

(See Response Methods Handbook, CG 446-4)

Issue warning—high flammability
Restrict access
Mechanical containment
Should be removed
Chemical and physical treatment

2. LABEL

Red

3. CHEMICAL DESIGNATIONS

3.1 **Synonyms:** Ethyl 2-methacrylate; Ethyl methacrylate—inhibited; Ethyl *alpha*-methylmethacrylate; Ethyl 2-methyl-2-propenoate; Methacrylic acid, ethyl ester
3.2 **Coast Guard Compatibility Classification:** Acrylate
3.3 **Chemical Formula:** $CH_2=C(CH_3)COOC_2H_5$
3.4 **IMCO/United Nations Numerical Designation:** Not listed

4. OBSERVABLE CHARACTERISTICS

4.1 **Physical State (as shipped):** Liquid
4.2 **Color:** Colorless
4.3 **Odor:** Acrid acrylic

5. HEALTH HAZARDS

5.1 **Personal Protective Equipment:** Impervious gloves; splash goggles; self-contained breathing apparatus if exposed to vapors; coveralls
5.2 **Symptoms Following Exposure:** Inhalation may cause irritation of the mucous membrane. Ingestion causes irritation of mouth and stomach. Contact with liquid irritates eyes and skin.
5.3 **Treatment for Exposure:** INHALATION: remove victim to fresh air; apply artificial respiration and oxygen if indicated. INGESTION: induce vomiting; call a physician. EYES: wash with copious quantities of water for 15 min.; call a physician. SKIN: flush with water; wash with soap and water.
5.4 **Toxicity by Inhalation (Threshold Limit Value):** Data not available
5.5 **Short-Term Inhalation Limits:** Data not available
5.6 **Toxicity by Ingestion:** Grade 2; oral LD_{50} = 4 g/kg (rabbit)
5.7 **Late Toxicity:** Causes birth defects in experimental animals
5.8 **Vapor (Gas) Irritant Characteristics:** Data not available
5.9 **Liquid or Solid Irritant Characteristics:** Data not available
5.10 **Odor Threshold:** Data not available

6. FIRE HAZARDS

6.1 **Flash Point:** 85°F O.C.; 80°F C.C.
6.2 **Flammable Limits in Air:** 1.8%(LFL)
6.3 **Fire Extinguishing Agents:** Foam, dry chemical, carbon dioxide
6.4 **Fire Extinguishing Agents Not to be Used:** Water may be ineffective
6.5 **Special Hazards of Combustion Products:**
6.6 **Behavior in Fire:** Sealed containers may rupture explosively if hot. Heat can cause a violent polymerization reaction with rapid release of energy. Vapors are heavier than air and can travel to a source of ignition and flash back.
6.7 **Ignition Temperature:** 740°F
6.8 **Electrical Hazard:** Data not available
6.9 **Burning Rate:** 4.56 mm/min.

7. CHEMICAL REACTIVITY

7.1 **Reactivity with Water:** No reaction
7.2 **Reactivity with Common Materials:**
7.3 **Stability During Transport:** Stable
7.4 **Neutralizing Agents for Acids and Caustics:** Not pertinent
7.5 **Polymerization:** If proper concentration of inhibitor is not present or when material is hot, a violent polymerization reaction may occur.
7.6 **Inhibitor of Polymerization:** Oxygen in the air inhibits polymerization.

8. WATER POLLUTION

8.1 **Aquatic Toxicity:** Data not available
8.2 **Waterfowl Toxicity:** Data not available
8.3 **Biological Oxygen Demand (BOD):** Data not available
8.4 **Food Chain Concentration Potential:** None

9. SELECTED MANUFACTURERS

1. Polysciences, Inc.
   Paul Valley Industrial Park
   Warrington, Pa. 18976
2. Rohm and Haas
   Independence Mall West
   Philadelphia, Pa. 19105
3. E. I. duPont de Nemours & Co., Inc.
   Plastics Dept.
   Polymer Products Division
   Wilmington, Del. 19898

10. SHIPPING INFORMATION

10.1 **Grades or Purity:** Technical
10.2 **Storage Temperature:** Below 38°C (100°F)
10.3 **Inert Atmosphere:** Ventilated (natural)
10.4 **Venting:** Open (flame arrester)

11. HAZARD ASSESSMENT CODE

(See Hazard Assessment Handbook, CG 446-3)

A-T-U-Z

12. HAZARD CLASSIFICATIONS

12.1 **Code of Federal Regulations:** Flammable liquid
12.2 **NAS Hazard Rating for Bulk Water Transportation:** Not listed
12.3 **NFPA Hazard Classifications:**

Category	Classification
Health Hazard (Blue)	2
Flammability (Red)	3
Reactivity (Yellow)	0

13. PHYSICAL AND CHEMICAL PROPERTIES

13.1 **Physical State at 15°C and 1 atm:** Liquid
13.2 **Molecular Weight:** 114
13.3 **Boiling Point at 1 atm:** 243°F = 117°C = 390°K
13.4 **Freezing Point:** <-58°F = <-50°C = <223°K
13.5 **Critical Temperature:** Not pertinent
13.6 **Critical Pressure:** Not pertinent
13.7 **Specific Gravity:** 0.9151 at 20°C (liquid)
13.8 **Liquid Surface Tension:** Data not available
13.9 **Liquid-Water Interfacial Tension:** Data not available
13.10 **Vapor (Gas) Specific Gravity:** 3.9
13.11 **Ratio of Specific Heats of Vapor (Gas):** 1.064
13.12 **Latent Heat of Vaporization:** 170 Btu/lb = 96 cal/g = 4.0×10^5 J/kg
13.13 **Heat of Combustion:** −12,670 Btu/lb = −7,040 cal/g = -294×10^5 J/kg
13.14 **Heat of Decomposition:** Not pertinent
13.15 **Heat of Solution:** Not pertinent
13.16 **Heat of Polymerization:** −218 Btu/lb = −121 cal/g = -5.06×10^5 J/kg

NOTES

ETN	ETHYL NITRITE

Common Synonyms			
Nitrous ether Sweet spirit of nitre Spirit of ether nitrite	Liquid	Colorless to light yellow	Pleasant odor
	Floats on water. May boil on water. Boiling point is 63°F.		

Shut off ignition sources. Call fire department.
Stop discharge if possible. Keep people away.
Isolate and remove discharged material.
Notify local health and pollution control agencies.

Fire	FLAMMABLE. POISONOUS GASES ARE PRODUCED IN FIRE. Containers may explode in fire. Flashback along vapor trail may occur. Vapor may explode if ignited in an enclosed area. **Wear goggles and self-contained breathing apparatus.** **Combat fires from safe distance or protected location.** **Extinguish with water, dry chemicals, foam, or carbon dioxide.** **Cool exposed containers with water.**
Exposure	**Call for medical aid.** VAPOR If inhaled will cause headache, dizziness, or loss of consciousness. **Move victim to fresh air.** **If breathing has stopped, give artificial respiration.** **If breathing is difficult, give oxygen.** LIQUID If swallowed will cause headache, or loss of consciousness. **IF SWALLOWED and victim is CONSCIOUS, have victim drink water or milk.**
Water Pollution	Effect of low concentrations on aquatic life is unknown. May be dangerous if it enters water intakes. **Notify local health and wildlife officials.** **Notify operators of nearby water intakes.**

1. RESPONSE TO DISCHARGE

(See Response Methods Handbook, CG 446-4)

Issue warning—high flammability
Restrict access
Evacuate area
Disperse and flush

2. LABEL

Red

3. CHEMICAL DESIGNATIONS

3.1 **Synonyms:** Nitrous ether; Spirit of ether nitrite; Sweet spirit of nitre
3.2 **Coast Guard Compatibility Classification:** Not applicable
3.3 **Chemical Formula:** C_2H_5ONO
3.4 **IMCO/United Nations Numerical Designation:** 3.1/1194

4. OBSERVABLE CHARACTERISTICS

4.1 **Physical State (as shipped):** Liquid
4.2 **Color:** Colorless to pale yellow
4.3 **Odor:** Aromatic; ethereal; characteristic

5. HEALTH HAZARDS

5.1 **Personal Protective Equipment:** Self-contained breathing apparatus; goggles or face shield; rubber gloves.
5.2 **Symptoms Following Exposure:** Inhalation or ingestion causes headache, increased pulse rate, decreased blood pressure, and unconsciousness. Contact with liquid irritates eyes and skin.
5.3 **Treatment for Exposure:** INHALATION: remove victim from exposure; if breathing has stopped, give artificial respiration; call physician. EYES: flush with water for at least 15 min.; get medical attention if irritation persists. SKIN: flush with water, wash with soap and water. INGESTION: do NOT induce vomiting; call physician.
5.4 **Toxicity by Inhalation (Threshold Limit Value):** Data not available
5.5 **Short-Term Inhalation Limits:** Data not available
5.6 **Toxicity by Ingestion:** Data not available
5.7 **Late Toxicity:** Data not available
5.8 **Vapor (Gas) Irritant Characteristics:** Data not available
5.9 **Liquid or Solid Irritant Characteristics:** Data not available
5.10 **Odor Threshold:** Data not available

6. FIRE HAZARDS

6.1 **Flash Point:** −31°F C.C.
6.2 **Flammable Limits in Air:** 3%—>50%
6.3 **Fire Extinguishing Agents:** Water, dry chemical, foam, carbon dioxide
6.4 **Fire Extinguishing Agents Not to be Used:** Not pertinent
6.5 **Special Hazards of Combustion Products:** Toxic oxides of nitrogen are generated.
6.6 **Behavior in Fire:** Vapors are heavier than air and may travel a considerable distance to a source of ignition and flash back; can decompose violently above 194°F; containers may explode in a fire.
6.7 **Ignition Temperature:** 194°F
6.8 **Electrical Hazard:** Data not available
6.9 **Burning Rate:** 2.6 mm/min.

7. CHEMICAL REACTIVITY

7.1 **Reactivity with Water:** No reaction
7.2 **Reactivity with Common Materials:** No reaction
7.3 **Stability During Transport:** Stable if stored in a cool place and not exposed to strong light.
7.4 **Neutralizing Agents for Acids and Caustics:** Not pertinent
7.5 **Polymerization:** Not pertinent
7.6 **Inhibitor of Polymerization:** Not pertinent

8. WATER POLLUTION

8.1 **Aquatic Toxicity:** Data not available
8.2 **Waterfowl Toxicity:** Data not available
8.3 **Biological Oxygen Demand (BOD):** Data not available
8.4 **Food Chain Concentration Potential:** None

9. SELECTED MANUFACTURERS

1. Mallinckrodt Chemical Works
   223 West Side Avenue
   P. O. Box 384
   Jersey City, N. J. 07303
2. Pfaltz and Bauer, Inc.
   126-04 Northern Boulevard
   Flushing, N. Y. 11368

10. SHIPPING INFORMATION

10.1 **Grades or Purity:** Often shipped as a 85—92% (by volume) solution in ethyl alcohol. Properties very similar.
10.2 **Storage Temperature:** Cool ambient
10.3 **Inert Atmosphere:** No requirement
10.4 **Venting:** Safety relief

11. HAZARD ASSESSMENT CODE

(See Hazard Assessment Handbook, CG 446-3)

A-B-C-D-E-F-G

12. HAZARD CLASSIFICATIONS

12.1 **Code of Federal Regulations:** Flammable liquid
12.2 **NAS Hazard Rating for Bulk Water Transportation**

Category	Rating
Fire	4
Health	
Vapor Irritant	
Liquid or Solid Irritant	
Poisons	3
Water Pollution	
Human Toxicity	
Aquatic Toxicity	1
Aesthetic Effect	
Reactivity	
Other Chemicals	
Water	0
Self-Reaction	4

12.3 **NFPA Hazard Classifications:**

Category	Classification
Health Hazard (Blue)	2
Flammability (Red)	4
Reactivity (Yellow)	4

13. PHYSICAL AND CHEMICAL PROPERTIES

13.1 **Physical State at 15°C and 1 atm:** Liquid
13.2 **Molecular Weight:** 75.1
13.3 **Boiling Point at 1 atm:** 63°F = 17°C = 290°K
13.4 **Freezing Point:** −58°F = −50°C = 223°K
13.5 **Critical Temperature:** Not pertinent
13.6 **Critical Pressure:** Not pertinent
13.7 **Specific Gravity:** 0.900 at 15°C (liquid)
13.8 **Liquid Surface Tension:** (est.) 30 dynes/cm = 0.030 N/m at 20°C
13.9 **Liquid-Water Interfacial Tension:** (est.) 35 dynes/cm = 0.035 N/m at 20°C
13.10 **Vapor (Gas) Specific Gravity:** 2.6
13.11 **Ratio of Specific Heats of Vapor (Gas):** Data not available
13.12 **Latent Heat of Vaporization:** 229 Btu/lb = 127 cal/g = 5.32×10^5 J/kg
13.13 **Heat of Combustion:** (est.) −7,800 Btu/lb = −4,300 cal/g = -180×10^5 J/kg
13.14 **Heat of Decomposition:** Not pertinent
13.15 **Heat of Solution:** Not pertinent
13.16 **Heat of Polymerization:** Not pertinent

NOTES

EPS | ETHYLPHENYLDICHLOROSILANE

Common Synonyms	Liquid Colorless Sharp irritating odor Reacts with water. Poisonous gas is produced on contact with water.

AVOID CONTACT WITH LIQUID AND VAPOR. KEEP PEOPLE AWAY.
Wear rubber overclothing (including gloves).
Stop discharge if possible.
Call fire department.
Isolate and remove discharged material.
Notify local health and pollution control agencies.

Fire	Combustible. POISONOUS GASES MAY BE PRODUCED IN FIRE. **Extinguish with dry chemicals or carbon dioxide.** **DO NOT USE WATER OR FOAM ON FIRE.**
Exposure	CALL FOR MEDICAL AID. GAS PRODUCED IN REACTION WITH WATER. POISONOUS IF INHALED. Irritating to eyes, nose and throat. **Move to fresh air.** **If breathing has stopped, give artificial respiration.** **If breathing is difficult, give oxygen.** LIQUID Will burn skin and eyes. Harmful if swallowed. **Remove contaminated clothing and shoes.** **Flush affected areas with plenty of water.** **IF IN EYES, hold eyelids open and flush with plenty of water.** **IF SWALLOWED and victim is CONSCIOUS, have victim drink water or milk.**
Water Pollution	Effect of low concentrations on aquatic life is unknown. May be dangerous if it enters water intakes. **Notify local health and wildlife officials.** **Notify operators of nearby water intakes.**

1. RESPONSE TO DISCHARGE
(See Response Methods Handbook, CG 446-4)

Issue warning—corrosive, water contaminant
Restrict access
Disperse and flush with care

2. LABEL

3. CHEMICAL DESIGNATIONS

3.1 **Synonyms:** No common synonyms
3.2 **Coast Guard Compatibility Classification:** Not applicable
3.3 **Chemical Formula:** $(C_2H_5)(C_6H_5)SiCl_2$
3.4 **IMCO/United Nations Numerical Designation:** 8/1760

4. OBSERVABLE CHARACTERISTICS

4.1 **Physical State (as shipped):** Liquid
4.2 **Color:** Colorless
4.3 **Odor:** Sharp, like hydrochloric acid

5. HEALTH HAZARDS

5.1 **Personal Protective Equipment:** Acid-vapor-type respiratory protection; rubber gloves; chemical worker's goggles; other equipment as necessary to protect skin and eyes.
5.2 **Symptoms Following Exposure:** Inhalation irritates nose and throat. Contact with liquid causes severe burns of eyes and skin. Ingestion causes severe burns of mouth and stomach.
5.3 **Treatment for Exposure:** INHALATION: remove victim to fresh air; give artificial respiration if needed; call physician. EYES: flush with water for 15 min.; obtain medical attention immediately. SKIN: flush with water; obtain medical attention if burning has occurred. INGESTION: if victim is conscious, give large amounts of water, then induce vomiting; get medical attention.
5.4 **Toxicity by Inhalation (Threshold Limit Value):** Data not available
5.5 **Short-Term Inhalation Limits:** Data not available
5.6 **Toxicity by Ingestion:** Grade 3; LD_{50} 50 to 500 mg/kg
5.7 **Late Toxicity:** Data not available
5.8 **Vapor (Gas) Irritant Characteristics:** Vapors cause moderate irritation such that personnel will find high concentrations unpleasant. The effect is temporary.
5.9 **Liquid or Solid Irritant Characteristics:** Severe skin irritant. Causes second- and third-degree burns on short contact and is very injurious to the eyes.
5.10 **Odor Threshold:** Data not available

6. FIRE HAZARDS

6.1 **Flash Point:** >150°F O.C.
6.2 **Flammable Limits in Air:** Data not available
6.3 **Fire Extinguishing Agents:** Dry chemical
6.4 **Fire Extinguishing Agents Not to be Used:** Water, foam
6.5 **Special Hazards of Combustion Products:** Toxic hydrogen chloride and phosgene fumes may be formed.
6.6 **Behavior in Fire:** Difficult to extinguish; re-ignition may occur. Contact with water applied to adjacent fires will generate irritating hydrogen chloride gas.
6.7 **Ignition Temperature:** Data not available
6.8 **Electrical Hazard:** Data not available
6.9 **Burning Rate:** 3.7 mm/min.

7. CHEMICAL REACTIVITY

7.1 **Reactivity with Water:** Reacts with water to generate hydrogen chloride (hydrochloric acid)
7.2 **Reactivity with Common Materials:** Will react with surface moisture to evolve hydrogen chloride, which is corrosive to common metals.
7.3 **Stability During Transport:** Stable
7.4 **Neutralizing Agents for Acids and Caustics:** Flush with water, rinse with sodium bicarbonate or lime solution.
7.5 **Polymerization:** Not pertinent
7.6 **Inhibitor of Polymerization:** Not pertinent

8. WATER POLLUTION

8.1 **Aquatic Toxicity:** Data not available
8.2 **Waterfowl Toxicity:** Data not available
8.3 **Biological Oxygen Demand (BOD):** Data not available
8.4 **Food Chain Concentration Potential:** None

9. SELECTED MANUFACTURERS

Dow Corning Corporation
P. O. Box 592
Midland, Mich. 48640

10. SHIPPING INFORMATION

10.1 **Grades or Purity:** Commercial
10.2 **Storage Temperature:** Ambient
10.3 **Inert Atmosphere:** No requirement
10.4 **Venting:** Pressure-vacuum

11. HAZARD ASSESSMENT CODE
(See Hazard Assessment Handbook, CG 446-3)

A-O

12. HAZARD CLASSIFICATIONS

12.1 **Code of Federal Regulations:** Corrosive
12.2 **NAS Hazard Rating for Bulk Water Transportation:**

Category	Rating
Fire	2
Health	
Vapor Irritant	2
Liquid or Solid Irritant	4
Poisons	3
Water Pollution	
Human Toxicity	3
Aquatic Toxicity	3
Aesthetic Effect	3
Reactivity	
Other Chemicals	3
Water	4
Self-Reaction	1

12.3 **NFPA Hazard Classifications:** Not listed

13. PHYSICAL AND CHEMICAL PROPERTIES

13.1 **Physical State at 15°C and 1 atm:** Liquid
13.2 **Molecular Weight:** 205.1
13.3 **Boiling Point at 1 atm:** >300°F = >149°C = >422°K
13.4 **Freezing Point:** Not pertinent
13.5 **Critical Temperature:** Not pertinent
13.6 **Critical Pressure:** Not pertinent
13.7 **Specific Gravity:** 1.159 at 15°C (liquid)
13.8 **Liquid Surface Tension:** (est.) 25 dynes/cm = 0.025 N/m at 20°C
13.9 **Liquid-Water Interfacial Tension:** Not pertinent
13.10 **Vapor (Gas) Specific Gravity:** Not pertinent
13.11 **Ratio of Specific Heats of Vapor (Gas):** Not pertinent
13.12 **Latent Heat of Vaporization:** 103 Btu/lb = 57 cal/g = 2.4×10^5 J/kg
13.13 **Heat of Combustion:** (est.) −9,900 Btu/lb = −5,500 cal/g = -230×10^5 J/kg
13.14 **Heat of Decomposition:** Not pertinent
13.15 **Heat of Solution:** Data not available
13.16 **Heat of Polymerization:** Not pertinent

NOTES

EPD	ETHYL PHOSPHONOTHIOIC DICHLORIDE, ANHYDROUS

Common Synonyms			
Ethyl thionophosphoryl dichloride	Liquid	Colorless	Choking odor
	Reacts with water. Poisonous gas is produced on contact with water.		

Stop discharge if possible. Keep people away.
Call fire department.
Isolate and remove discharged material.
Notify local health and pollution control agencies.

Fire

Combustible.
POISONOUS GASES ARE PRODUCED IN FIRE.
Wear goggles and self-contained breathing apparatus.
Extinguish with dry chemicals or carbon dioxide.
DO NOT USE WATER OR FOAM ON FIRE.

Exposure

CALL FOR MEDICAL AID.

GAS PRODUCED IN REACTION WITH WATER.
POISONOUS IF INHALED.
Irritating to eyes, nose and throat.
Move to fresh air.
If breathing has stopped, give artificial respiration.
If breathing is difficult, give oxygen.

LIQUID
Irritating to skin and eyes.
Harmful if swallowed.
Remove contaminated clothing and shoes.
Flush affected areas with plenty of water.
IF IN EYES, hold eyelids open and flush with plenty of water.
IF SWALLOWED and victim is CONSCIOUS, have victim drink water or milk.

Water Pollution

Effect of low concentrations on aquatic life is unknown.
May be dangerous if it enters water intakes.
Notify local health and wildlife officials.
Notify operators of nearby water intakes.

1. RESPONSE TO DISCHARGE

(See Response Methods Handbook, CG 446-4)

Issue warning—corrosive, water contaminant
Restrict access
Disperse and flush with care

2. LABEL

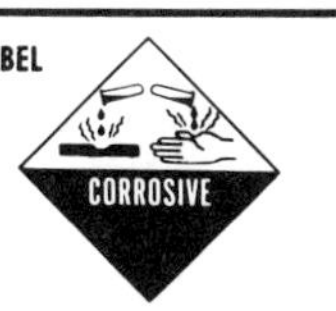

3. CHEMICAL DESIGNATIONS

3.1 **Synonyms:** Ethyl phosphorodichlorido-thionate; Ethyl thionophosphonyl dichloride
3.2 **Coast Guard Compatibility Classification:** Not applicable
3.3 **Chemical Formula:** $CH_3CH_2PSCl_2$
3.4 **IMCO/United Nations Numerical Designation:** 8/1760

4. OBSERVABLE CHARACTERISTICS

4.1 **Physical State (as shipped):** Liquid
4.2 **Color:** Colorless
4.3 **Odor:** Acrid and choking

5. HEALTH HAZARDS

5.1 **Personal Protective Equipment:** Air mask; rubber or neoprene gloves; vapor-tight goggles.
5.2 **Symptoms Following Exposure:** Inhalation of vapor causes pulmonary and eye irritation; effects on lungs may be delayed 24 hours; very similar to phosgene poisoning. Contact with liquid causes painful irritation of eyes and lachrymation; also causes severe irritation and possible damage to skin. Ingestion causes severe irritation of mouth and stomach.
5.3 **Treatment for Exposure:** INHALATION: remove victim from exposure; oxygen can be used for pulmonary symptoms with decongestants; enforce complete rest, because effects may be delayed 24 hours; similar to phosgene poisoning. EYES: flush thoroughly with water and seek medical attention; apply Pontocaine drops (1/2%) and cortisone ointment (1%). SKIN: wash thoroughly with soap and water. INGESTION: give large amounts of water; induce vomiting; get medical attention; enforce rest for 24-36 hours.
5.4 **Toxicity by Inhalation (Threshold Limit Value):** Data not available
5.5 **Short-Term Inhalation Limits:** Data not available
5.6 **Toxicity by Ingestion:** Grade 2; LD_{50} 0.5 to 5 g/kg
5.7 **Late Toxicity:** Data not available
5.8 **Vapor (Gas) Irritant Characteristics:** Data not available
5.9 **Liquid or Solid Irritant Characteristics:** Data not available
5.10 **Odor Threshold:** Data not available

6. FIRE HAZARDS

6.1 **Flash Point:** 203°F O.C.
6.2 **Flammable Limits in Air:** Data not available
6.3 **Fire Extinguishing Agents:** Dry chemical or carbon dioxide
6.4 **Fire Extinguishing Agents Not to be Used:** Water or foam
6.5 **Special Hazards of Combustion Products:** Oxides of sulfur, phosphorus; hydrogen chloride and phosgene.
6.6 **Behavior in Fire:** Contact with water applied to adjacent fires will produce irritating fumes of hydrogen chloride.
6.7 **Ignition Temperature:** Data not available
6.8 **Electrical Hazard:** Data not available
6.9 **Burning Rate:** Data not available

7. CHEMICAL REACTIVITY

7.1 **Reactivity with Water:** Reacts with water to evolve hydrogen chloride (hydrochloric acid)
7.2 **Reactivity with Common Materials:** Will react with surface moisture to evolve hydrogen chloride, which is corrosive to common metals.
7.3 **Stability During Transport:** Stable
7.4 **Neutralizing Agents for Acids and Caustics:** Flood with water, rinse with sodium bicarbonate or lime solution.
7.5 **Polymerization:** Not pertinent
7.6 **Inhibitor of Polymerization:** Not pertinent

8. WATER POLLUTION

8.1 **Aquatic Toxicity:** Data not available
8.2 **Waterfowl Toxicity:** Data not available
8.3 **Biological Oxygen Demand (BOD):** Data not available
8.4 **Food Chain Concentration Potential:** None

9. SELECTED MANUFACTURERS

1. Ethyl Corporation
   Industrial Chemicals Division
   Ethyl Tower
   451 Florida
   Baton Rouge, La. 70801
2. Pfaltz and Bauer, Inc.
   126-04 Northern Boulevard
   Flushing, N. Y. 11368

10. SHIPPING INFORMATION

10.1 **Grades or Purity:** Commercial
10.2 **Storage Temperature:** Ambient
10.3 **Inert Atmosphere:** Inerted with dry nitrogen.
10.4 **Venting:** Pressure-vacuum

11. HAZARD ASSESSMENT CODE

(See Hazard Assessment Handbook, CG 446-3)

A-O

12. HAZARD CLASSIFICATIONS

12.1 **Code of Federal Regulations:** Corrosive liquid
12.2 **NAS Hazard Rating for Bulk Water Transportation:** Not listed
12.3 **NFPA Hazard Classifications:** Not listed

13. PHYSICAL AND CHEMICAL PROPERTIES

13.1 **Physical State at 15°C and 1 atm:** Liquid
13.2 **Molecular Weight:** 163
13.3 **Boiling Point at 1 atm:** 342°F = 172°C = 445°K
13.4 **Freezing Point:** <−58°F = <−50°C = <223°K
13.5 **Critical Temperature:** Not pertinent
13.6 **Critical Pressure:** Not pertinent
13.7 **Specific Gravity:** 1.35 at 20°C (liquid)
13.8 **Liquid Surface Tension:** (est.) 28 dynes/cm = 0.028 N/m at 20°C
13.9 **Liquid-Water Interfacial Tension:** Not pertinent
13.10 **Vapor (Gas) Specific Gravity:** Not pertinent
13.11 **Ratio of Specific Heats of Vapor (Gas):** Not pertinent
13.12 **Latent Heat of Vaporization:** Not pertinent
13.13 **Heat of Combustion:** −7,700 Btu/lb = −4,280 cal/g = -179×10^5 J/kg
13.14 **Heat of Decomposition:** Not pertinent
13.15 **Heat of Solution:** Data not available
13.16 **Heat of Polymerization:** Not pertinent

NOTES

EPP

ETHYL PHOSPHORODICHLORIDATE

Common Synonyms

Phosphorodichloridic acid, ethyl ester
Ethyl dichlorophosphate

Liquid | Colorless | Choking odor

Reacts with water. Irritating gas is produced on contact with water.

AVOID CONTACT WITH LIQUID AND VAPOR. KEEP PEOPLE AWAY.
Wear goggles, self-contained breathing apparatus and rubber overclothing (including gloves).
Stop discharge if possible.
Isolate and remove discharged material.
Notify local health and pollution control agencies.

Fire

Fire data not available.

Exposure

Call for medical aid.

VAPOR
Irritating to eyes, nose and throat.
Harmful if inhaled.
Move victim to fresh air.
If breathing is difficult, give oxygen.

LIQUID
Will burn skin and eyes.
Harmful if swallowed.
Remove contaminated clothing and shoes.
Flush affected areas with plenty of water.
IF IN EYES, hold eyelids open and flush with plenty of water.
IF SWALLOWED and victim is CONSCIOUS, have victim drink water or milk.
DO NOT INDUCE VOMITING.

Water Pollution

Effect of low concentrations on aquatic life is unknown.
May be dangerous if it enters water intakes.
Notify local health and wildlife officials.
Notify operators of nearby water intakes.

1. RESPONSE TO DISCHARGE

(See Response Methods Handbook, CG 446-4)

Issue warning—corrosive, water contaminant
Restrict access
Disperse and flush with care

2. LABEL

3. CHEMICAL DESIGNATIONS

3.1 **Synonyms:** Ethyl dichlorophosphate; Phosphorodichloridic acid, ethyl ester
3.2 **Coast Guard Compatibility Classification:** Not applicable
3.3 **Chemical Formula:** $Cl_2(OC_2H_5)PO$
3.4 **IMCO/United Nations Numerical Designation:** 8/1760

4. OBSERVABLE CHARACTERISTICS

4.1 **Physical State (as shipped):** Liquid
4.2 **Color:** Colorless
4.3 **Odor:** Acrid, choking

5. HEALTH HAZARDS

5.1 **Personal Protective Equipment:** Goggles and face shield; self-contained or air-line respirator; rubber gloves, boots, and clothing.
5.2 **Symptoms Following Exposure:** Inhalation of vapor irritates nose and throat. Contact with liquid causes severe burns of eyes and skin. Ingestion causes severe burns of mouth and stomach.
5.3 **Treatment for Exposure:** INHALATION: remove victim from exposure; if his breathing has stopped, start artificial respiration, call a doctor. EYES: flush with water for at least 15 min.; get medical attention for burns. SKIN: flush with water; get medical attention for burns. INGESTION: do NOT induce vomiting; give large amounts of water, followed by milk or milk of magnesia.
5.4 **Toxicity by Inhalation (Threshold Limit Value):** Data not available
5.5 **Short-Term Inhalation Limits:** Data not available
5.6 **Toxicity by Ingestion:** Data not available
5.7 **Late Toxicity:** Data not available
5.8 **Vapor (Gas) Irritant Characteristics:** Data not available
5.9 **Liquid or Solid Irritant Characteristics:** Data not available
5.10 **Odor Threshold:** Data not available

6. FIRE HAZARDS

6.1 **Flash Point:** Data not available
6.2 **Flammable Limits in Air:** Data not available
6.3 **Fire Extinguishing Agents:** Dry chemical or carbon dioxide
6.4 **Fire Extinguishing Agents Not to be Used:** Water or foam
6.5 **Special Hazards of Combustion Products:** Irritating fumes of hydrogen chloride and phosphoric acid may be formed.
6.6 **Behavior in Fire:** Contact with water applied to adjacent fires produces irritating fumes of hydrogen chloride.
6.7 **Ignition Temperature:** Data not available
6.8 **Electrical Hazard:** Data not available
6.9 **Burning Rate:** Data not available

7. CHEMICAL REACTIVITY

7.1 **Reactivity with Water:** Reacts with water to evolve hydrogen chloride (hydrochloric acid).
7.2 **Reactivity with Common Materials:** Will react with surface moisture to evolve hydrogen chloride, which is corrosive to common metals.
7.3 **Stability During Transport:** Stable
7.4 **Neutralizing Agents for Acids and Caustics:** Flood with water, rinse with sodium bicarbonate or lime solution.
7.5 **Polymerization:** Not pertinent
7.6 **Inhibitor of Polymerization:** Not pertinent

8. WATER POLLUTION

8.1 **Aquatic Toxicity:** Data not available
8.2 **Waterfowl Toxicity:** Data not available
8.3 **Biological Oxygen Demand (BOD):** Data not available
8.4 **Food Chain Concentration Potential:** None

9. SELECTED MANUFACTURERS

1. Aldrich Chemical Co.
   940 West Saint Paul Ave.
   Milwaukee, Wis. 53233
2. Pfaltz and Bauer, Inc.
   126-04 Northern Boulevard
   Flushing, N. Y. 11368

10. SHIPPING INFORMATION

10.1 **Grades or Purity:** 97%
10.2 **Storage Temperature:** Ambient
10.3 **Inert Atmosphere:** No requirement
10.4 **Venting:** Pressure-vacuum

11. HAZARD ASSESSMENT CODE

(See Hazard Assessment Handbook, CG 446-3)

A-O

12. HAZARD CLASSIFICATIONS

12.1 **Code of Federal Regulations:** Corrosive
12.2 **NAS Hazard Rating for Bulk Water Transportation:** Not listed
12.3 **NFPA Hazard Classifications:** Not listed

13. PHYSICAL AND CHEMICAL PROPERTIES

13.1 **Physical State at 15°C and 1 atm:** Liquid
13.2 **Molecular Weight:** 162.9
13.3 **Boiling Point at 1 atm:** 333°F = 167°C = 440°K
13.4 **Freezing Point:** Not pertinent
13.5 **Critical Temperature:** Not pertinent
13.6 **Critical Pressure:** Not pertinent
13.7 **Specific Gravity:** 1.35 at 19°C (liquid)
13.8 **Liquid Surface Tension:** (est.) 32.8 dynes/cm = 0.0328 N/m at 20°C
13.9 **Liquid-Water Interfacial Tension:** Not pertinent
13.10 **Vapor (Gas) Specific Gravity:** Not pertinent
13.11 **Ratio of Specific Heats of Vapor (Gas):** Not pertinent
13.12 **Latent Heat of Vaporization:** Not pertinent
13.13 **Heat of Combustion:** (est.) −4,700 Btu/lb = −2,600 cal/g = -110×10^5 J/kg
13.14 **Heat of Decomposition:** Not pertinent
13.15 **Heat of Solution:** Data not available
13.16 **Heat of Polymerization:** Not pertinent

NOTES

EPA	# 2-ETHYL-3-PROPYLACROLEIN

Common Synonyms		
2-Ethyl-2-hexenal 2-Ethyl-3-propyl acrylaldehyde	Liquid	Yellow
	Floats on water.	

Avoid contact with liquid.
Wear goggles, self-contained breathing apparatus, and rubber overclothing (including gloves).
Stop discharge if possible.
Call fire department.
Isolate and remove discharged material.
Notify local health and pollution control agencies.

Fire	**Combustible.** **Wear goggles, self-contained breathing apparatus, and rubber overclothing (including gloves).** **Extinguish with dry chemical, alcohol foam or carbon dioxide.** **Cool exposed containers with water.**
Exposure	**CALL FOR MEDICAL AID.** **LIQUID** **Will burn skin and eyes.** **Harmful if swallowed.** **Remove contaminated clothing and shoes.** **Flush affected areas with plenty of water.** **IF IN EYES, hold eyelids open and flush with plenty of water.** **IF SWALLOWED and victim is CONSCIOUS, have victim drink water or milk.**
Water Pollution	**Effect of low concentrations on aquatic life is unknown.** **Fouling to shoreline.** **May be dangerous if it enters water intakes.** **Notify local health and wildlife officials.** **Notify operators of nearby water intakes.**

1. RESPONSE TO DISCHARGE
(See Response Methods Handbook, CG 446-4)

Mechanical containment
Should be removed
Chemical and physical treatment

2. LABELS

No hazard label required by Code of Federal Regulations

3. CHEMICAL DESIGNATIONS

3.1 **Synonyms:** 2-Ethyl-2-hexenal
2-Ethyl-3-propylacrylaldehyde
3.2 **Coast Guard Compatibility Classification:** Aldehyde
3.3 **Chemical Formula:** $CH_3(CH_2)_2CH{=}C(C_2H_5)CHO$
3.4 **IMCO/United Nations Numerical Designation:** 3.3/1191

4. OBSERVABLE CHARACTERISTICS

4.1 **Physical State (as shipped):** Liquid
4.2 **Color:** Yellow
4.3 **Odor:** Sharp, powerful, irritating

5. HEALTH HAZARDS

5.1 **Personal Protective Equipment:** Protective clothing; eye protection; approved respirator for high vapor concentrations.
5.2 **Symptoms Following Exposure:** Vapor is irritating. Contact produces skin and eye irritation.
5.3 **Treatment for Exposure:** Remove from exposure. Wash affected areas of body with water for 15 min.
5.4 **Toxicity by Inhalation (Threshold Limit Value):** Data not available
5.5 **Short-Term Inhalation Limits:** Data not available
5.6 **Toxicity by Ingestion:** Grade 2; LD_{50} 0.5 to 5 g/kg (rat)
5.7 **Late Toxicity:** Data not available
5.8 **Vapor (Gas) Irritant Characteristics:** Vapor is moderately irritating such that personnel will not usually tolerate moderate or high vapor concentrations.
5.9 **Liquid or Solid Irritant Characteristics:** Causes smarting of the skin and first-degree burns on short exposure; may cause secondary burns on long exposure.
5.10 **Odor Threshold:** Data not available

6. FIRE HAZARDS

6.1 **Flash Point:** 155°F O.C.
6.2 **Flammable Limits in Air:** Data not available
6.3 **Fire Extinguishing Agents:** Alcohol foam, dry chemical or carbon dioxide
6.4 **Fire Extinguishing Agents Not to be Used:** Not pertinent
6.5 **Special Hazards of Combustion Products:** Not pertinent
6.6 **Behavior in Fire:** Not pertinent
6.7 **Ignition Temperature:** 200°C
6.8 **Electrical Hazard:** Not pertinent
6.9 **Burning Rate:** Data not available

7. CHEMICAL REACTIVITY

7.1 **Reactivity with Water:** No reaction
7.2 **Reactivity with Common Materials:** No reaction
7.3 **Stability During Transport:** Stable
7.4 **Neutralizing Agents for Acids and Caustics:** Not pertinent
7.5 **Polymerization:** Not pertinent
7.6 **Inhibitor of Polymerization:** Not pertinent

8. WATER POLLUTION

8.1 **Aquatic Toxicity:** Data not available
8.2 **Waterfowl Toxicity:** Data not available
8.3 **Biological Oxygen Demand (BOD):** 52% (theor.), 10 days
8.4 **Food Chain Concentration Potential:** None

9. SELECTED MANUFACTURERS

Union Carbide Corp.
Chemicals and Plastics Division
270 Park Ave.
New York, N. Y. 10017

10. SHIPPING INFORMATION

10.1 **Grades or Purity:** Technical
10.2 **Storage Temperature:** Ambient
10.3 **Inert Atmosphere:** No requirement
10.4 **Venting:** Pressure-vacuum

11. HAZARD ASSESSMENT CODE
(See Hazard Assessment Handbook, CG 446-3)

A-T-U

12. HAZARD CLASSIFICATIONS

12.1 **Code of Federal Regulations:** Not listed
12.2 **NAS Hazard Rating for Bulk Water Transportation:**

Category	Rating
Fire	2
Health	
Vapor Irritant	3
Liquid or Solid Irritant	2
Poisons	3
Water Pollution	
Human Toxicity	3
Aquatic Toxicity	3
Aesthetic Effect	2
Reactivity	
Other Chemicals	2
Water	0
Self-Reaction	1

12.3 **NFPA Hazard Classifications:** Not listed

13. PHYSICAL AND CHEMICAL PROPERTIES

13.1 **Physical State at 15°C and 1 atm:** Liquid
13.2 **Molecular Weight:** 126.2
13.3 **Boiling Point at 1 atm:** 283°F = 175°C = 448°K
13.4 **Freezing Point:** Not pertinent
13.5 **Critical Temperature:** Not pertinent
13.6 **Critical Pressure:** Not pertinent
13.7 **Specific Gravity:** 0.857 at 15°C (liquid)
13.8 **Liquid Surface Tension:** 28.2 dynes/cm = 0.0282 N/m at 20°C
13.9 **Liquid-Water Interfacial Tension:** (est.) 40 dynes/cm = 0.04 N/m at 20°C
13.10 **Vapor (Gas) Specific Gravity:** Not pertinent
13.11 **Ratio of Specific Heats of Vapor (Gas):** Not pertinent
13.12 **Latent Heat of Vaporization:** Not pertinent
13.13 **Heat of Combustion:** -15,610 Btu/lb = -8670 cal/g = -363×10^5 J/kg
13.14 **Heat of Decomposition:** Not pertinent
13.15 **Heat of Solution:** Not pertinent
13.16 **Heat of Polymerization:** Not pertinent

NOTES

REVISED 1978

ESC — ETHYL SILICATE

Common Synonyms			
Tetraethyl silicate Silibond Tetraethyl orthosilicate Ethyl orthosilicate Ethyl silicate 40 Ethyl silicate condensed	Watery liquid	Colorless	Mild odor
	May float or sink in water. Reacts slowly with water.		

Stop discharge if possible. Keep people away.
Call fire department.
Isolate and remove discharged material.
Notify local health and pollution control agencies.

Fire

Combustible.
Extinguish with water, dry chemicals, foam, or carbon dioxide.

Exposure

Call for medical aid.

LIQUID
Irritating to eyes.
If swallowed will cause nausea and vomiting.
Remove contaminated clothing and shoes.
Flush affected areas with plenty of water.
IF IN EYES, hold eyelids open and flush with plenty of water.
IF SWALLOWED and victim is CONSCIOUS, have victim drink water or milk and have victim induce vomiting.
IF SWALLOWED and victim is UNCONSCIOUS OR HAVING CONVULSIONS, do nothing except keep victim warm.

Water Pollution

Effect of low concentrations on aquatic life is unknown.
Fouling to shoreline.
May be dangerous if it enters water intakes.
Notify local health and wildlife officials.
Notify operators of nearby water intakes.

1. RESPONSE TO DISCHARGE
(See Response Methods Handbook, CG 446-4)

Disperse and flush

2. LABELS

No label required by Code of Federal Regulations

3. CHEMICAL DESIGNATIONS

3.1 **Synonyms:** Ethyl orthosilicate; Ethyl silicate 40; Ethyl silicate condensed; Silibond; Tetraethyl orthosilicate; Tetraethyl silicate
3.2 **Coast Guard Compatibility Classification:** Not applicable
3.3 **Chemical Formula:** $(C_2H_5O)_4Si$
3.4 **IMCO/United Nations Numerical Designation:** 3.3/1292

4. OBSERVABLE CHARACTERISTICS

4.1 **Physical State (as shipped):** Liquid
4.2 **Color:** Colorless
4.3 **Odor:** Mild and nonresidual; faint

5. HEALTH HAZARDS

5.1 **Personal Protective Equipment:** Rubber or polyethylene gloves; safety glasses or other form of eye protection; self-contained breathing apparatus or one that absorbs organic vapors.
5.2 **Symptoms Following Exposure:** Inhalation of vapor causes eye and nose irritation, unsteadiness, tremors, salivation, respiratory difficulty, and unconsciousness. Contact with liquid irritates eyes and may cause dryness, cracking, and inflammation of skin. Ingestion may produce nausea, vomiting, and cramps.
5.3 **Treatment for Exposure:** INHALATION: move patient from contaminated atmosphere; if his breathing has ceased, start mouth-to-mouth artificial respiration; oxygen, if available, should be administered only by an experienced person when authorized by a physician; keep patient warm and comfortable; call physician immediately. EYES: flush immediately with large quantities of running water for at least 15 min.; obtain medical attention if irritation persists. SKIN: immediately flush affected areas with large volumes of water; obtain medical attention if irritation persists. INGESTION: give large amounts of water or warm salty water and induce vomiting; milk, eggs or olive oil may then be given; obtain medical attention if abdominal discomfort persists.
5.4 **Toxicity by Inhalation (Threshold Limit Value):** 100 ppm
5.5 **Short-Term Inhalation Limits:** 200 ppm for 30 min.
5.6 **Toxicity by Ingestion:** Data not available
5.7 **Late Toxicity:** Liver, kidney, and lung damage may result from overexposures by inhalation or ingestion.
5.8 **Vapor (Gas) Irritant Characteristics:** Data not available

6. FIRE HAZARDS

6.1 **Flash Point:** 125°F O.C.; 99°F C.C.
6.2 **Flammable Limits in Air:** 1.3% – 23%
6.3 **Fire Extinguishing Agents:** Water, foam, dry chemical, carbon dioxide
6.4 **Fire Extinguishing Agents Not to be Used:** Data not available
6.5 **Special Hazards of Combustion Products:** Not pertinent
6.6 **Behavior in Fire:** Not pertinent
6.7 **Ignition Temperature:** Data not available
6.8 **Electrical Hazard:** Data not available
6.9 **Burning Rate:** 4.4 mm/min.

7. CHEMICAL REACTIVITY

7.1 **Reactivity with Water:** Reacts slowly, forming non-toxic silica and ethyl alcohol.
7.2 **Reactivity with Common Materials:** Causes swelling and hardening of some plastics.
7.3 **Stability During Transport:** Stable
7.4 **Neutralizing Agents for Acids and Caustics:** Not pertinent
7.5 **Polymerization:** Not pertinent
7.6 **Inhibitor of Polymerization:** Not pertinent

8. WATER POLLUTION

8.1 **Aquatic Toxicity:** Data not available
8.2 **Waterfowl Toxicity:** Data not available
8.3 **Biological Oxygen Demand (BOD):** Data not available
8.4 **Food Chain Concentration Potential:** None

9. SELECTED MANUFACTURERS

1. Stauffer Chemical Company
Specialty Chemical Division
Westport, Conn. 06880
2. Union Carbide Corporation
Chemicals and Plastics Div.
270 Park Avenue
New York, N. Y. 10017
3. Dynamit-Nobel of America, Inc.
105 Stonehurst Court
Northvale, N. J. 07647

10. SHIPPING INFORMATION

10.1 **Grades or Purity:** 90 – 99%
10.2 **Storage Temperature:** Ambient
10.3 **Inert Atmosphere:** No requirement
10.4 **Venting:** Open (flame arrester)

11. HAZARD ASSESSMENT CODE
(See Hazard Assessment Handbook, CG 446-3)

A-O-T-U-X-Y

12. HAZARD CLASSIFICATIONS

12.1 **Code of Federal Regulations:** Combustible Liquid
12.2 **NAS Hazard Rating for Bulk Water Transportation:** Not listed
12.3 **NFPA Hazard Classifications:**

Category	Classification
Health Hazard (Blue)	2
Flammability (Red)	2
Reactivity (Yellow)	0

13. PHYSICAL AND CHEMICAL PROPERTIES

13.1 **Physical State at 15°C and 1 atm:** Liquid
13.2 **Molecular Weight:** 208.3
13.3 **Boiling Point at 1 atm:** 336°F = 169°C = 442°K
13.4 **Freezing Point:** −121.9°F = −85.5°C = 187.7°K
13.5 **Critical Temperature:** Not pertinent
13.6 **Critical Pressure:** Not pertinent
13.7 **Specific Gravity:** 0.933 at 20°C (liquid)
13.8 **Liquid Surface Tension:** 22.8 dynes/cm = 0.0228 N/m at 20°C
13.9 **Liquid-Water Interfacial Tension:** Not pertinent
13.10 **Vapor (Gas) Specific Gravity:** Not pertinent
13.11 **Ratio of Specific Heats of Vapor (Gas):** Not pertinent
13.12 **Latent Heat of Vaporization:** 95 Btu/lb = 53 cal/g = 2.2×10^5 J/kg
13.13 **Heat of Combustion:** (est.) −12,000 Btu/lb = −6,700 cal/g = -280×10^5 J/kg
13.14 **Heat of Decomposition:** Not pertinent
13.15 **Heat of Solution:** Not pertinent
13.16 **Heat of Polymerization:** Not pertinent

5. HEALTH HAZARDS (Cont'd.)

5.9 **Liquid or Solid Irritant Characteristics:** Data not available
5.10 **Odor Threshold:** 85 ppm

ETS	ETHYLTRICHLOROSILANE

Common Synonyms			
Trichloroethyl silane Ethylsilicon trichloride Ethyl silicone trichloride Trichloroethyl silicone	Liquid	Colorless	Sharp irritating odor
	Reacts violently with water. Irritating gas is produced on contact with water.		

AVOID CONTACT WITH LIQUID AND VAPOR. KEEP PEOPLE AWAY.
Wear goggles, self-contained breathing apparatus, and rubber overclothing (including gloves).
Shut off ignition sources. Call fire department.
Stop discharge if possible.
Isolate and remove discharged material.
Notify local health and pollution control agencies.

Fire	FLAMMABLE. POISONOUS GASES ARE PRODUCED IN FIRE. Containers may explode in fire. Flashback along vapor trail may occur. Vapor may explode if ignited in an enclosed area. **Wear goggles and self-contained breathing apparatus.** **Extinguish with dry chemicals, alcohol foam, or carbon dioxide.** **Water may be ineffective on fire.** **Cool exposed containers with water.**
Exposure	**Call for medical aid.** VAPOR Irritating to eyes, nose and throat. If inhaled will cause difficult breathing. **Move victim to fresh air.** **If breathing has stopped, give artificial respiration.** **If breathing is difficult, give oxygen.** LIQUID Will burn skin and eyes. Harmful if swallowed. **Remove contaminated clothing and shoes.** **Flush affected areas with plenty of water.** **IF IN EYES, hold eyelids open and flush with plenty of water.** **IF SWALLOWED and victim is CONSCIOUS, have victim drink water or milk.** **DO NOT INDUCE VOMITING.**
Water Pollution	Effect of low concentrations on aquatic life is unknown. May be dangerous if it enters water intakes. **Notify local health and wildlife officials.** **Notify operators of nearby water intakes.**

1. RESPONSE TO DISCHARGE

(See Response Methods Handbook, CG 446-4)

Issue warning—high flammability, corrosive
Restrict access
Disperse and flush with care

2. LABEL

Red

3. CHEMICAL DESIGNATIONS

3.1 **Synonyms:** Ethyl silicone trichloride; Ethyl silicon trichloride; Trichloroethylsilane; Trichloroethylsilicane
3.2 **Coast Guard Compatibility Classification:** Not applicable
3.3 **Chemical Formula:** $C_2H_5SiCl_3$
3.4 **IMCO/United Nations Numerical Designation:** 3.2/1196

4. OBSERVABLE CHARACTERISTICS

4.1 **Physical State (as shipped):** Liquid
4.2 **Color:** Colorless
4.3 **Odor:** Sharp, hydrochloric acid-like; pungent; acrid

5. HEALTH HAZARDS

5.1 **Personal Protective Equipment:** Full protective clothing; acid-vapor-type respiratory protection; rubber gloves; chemical worker's goggles; other equipment as necessary to protect skin and eyes.
5.2 **Symptoms Following Exposure:** Inhalation causes irritation of nose and throat. Contact with liquid causes severe burns of eyes and skin. Ingestion causes burns of mouth and stomach.
5.3 **Treatment for Exposure:** INHALATION: remove victim from exposure; administer artificial respiration if breathing has stopped; call physician. EYES: flush with water for 15 min.; obtain medical attention immediately. SKIN: flush with water; obtain medical attention immediately if irritation persists. INGESTION: give large amounts of water; get medical attention.
5.4 **Toxicity by Inhalation (Threshold Limit Value):** Data not available
5.5 **Short-Term Inhalation Limits:** Data not available
5.6 **Toxicity by Ingestion:** Grade 2; oral LD_{50} = 1,330 mg/kg (rat)
5.7 **Late Toxicity:** Data not available
5.8 **Vapor (Gas) Irritant Characteristics:** Vapors cause severe irritation of the eyes and throat and can cause eye and lung injury. They cannot be tolerated even at low concentrations.
5.9 **Liquid or Solid Irritant Characteristics:** Severe skin irritant. Causes second- and third-degree burns on short contact and is very injurious to the eyes.
5.10 **Odor Threshold:** Data not available

6. FIRE HAZARDS

6.1 **Flash Point:** 57°F C.C.
6.2 **Flammable Limits in Air:** Data not available
6.3 **Fire Extinguishing Agents:** Dry chemical, alcohol foam, carbon dioxide
6.4 **Fire Extinguishing Agents Not to be Used:** Water may be ineffective.
6.5 **Special Hazards of Combustion Products:** Toxic hydrogen chloride and phosgene gases may form.
6.6 **Behavior in Fire:** Vapors are heavier than air and may travel a considerable distance to a source of ignition and flash back. Difficult to extinguish; re-ignition may occur. Contact with water applied to adjacent fires will produce irritating hydrogen chloride fumes.
6.7 **Ignition Temperature:** Data not available
6.8 **Electrical Hazard:** Data not available
6.9 **Burning Rate:** 2.0 mm/min.

7. CHEMICAL REACTIVITY

7.1 **Reactivity with Water:** Reacts vigorously, evolving hydrogen chloride (hydrochloric acid).
7.2 **Reactivity with Common Materials:** Reacts with surface moisture to form hydrogen chloride, which is corrosive to common metals.
7.3 **Stability During Transport:** Stable
7.4 **Neutralizing Agents for Acids and Caustics:** Flood with water, rinse with sodium bicarbonate or lime solution.
7.5 **Polymerization:** Not pertinent
7.6 **Inhibitor of Polymerization:** Not pertinent

8. WATER POLLUTION

8.1 **Aquatic Toxicity:** Data not available
8.2 **Waterfowl Toxicity:** Data not available
8.3 **Biological Oxygen Demand (BOD):** Data not available
8.4 **Food Chain Concentration Potential:** None

9. SELECTED MANUFACTURERS

1. Union Carbide Corporation
   Chemicals and Plastics Div.
   270 Park Avenue
   New York, N. Y. 10017
2. Dow Corning Corporation
   P. O. Box 592
   Midland, Mich. 48640
3. Petrarch Systems
   P. O. Box 141
   Levittown, Pa. 15059

10. SHIPPING INFORMATION

10.1 **Grades or Purity:** 98+%
10.2 **Storage Temperature:** Ambient
10.3 **Inert Atmosphere:** No requirement
10.4 **Venting:** Pressure-vacuum

11. HAZARD ASSESSMENT CODE

(See Hazard Assessment Handbook, CG 446-3)

A-O

12. HAZARD CLASSIFICATIONS

12.1 **Code of Federal Regulations:** Flammable liquid
12.2 **NAS Hazard Rating for Bulk Water Transportation:**

Category	Rating
Fire	3
Health	
Vapor Irritant	4
Liquid or Solid Irritant	4
Poisons	4
Water Pollution	
Human Toxicity	
Aquatic Toxicity	
Aesthetic Effect	2
Reactivity	
Other Chemicals	2
Water	3
Self-Reaction	1

12.3 **NFPA Hazard Classifications:**

Category	Classification
Health Hazard (Blue)	3
Flammability (Red)	3
Reactivity (Yellow)	0

13. PHYSICAL AND CHEMICAL PROPERTIES

13.1 **Physical State at 15°C and 1 atm:** Liquid
13.2 **Molecular Weight:** 163.5
13.3 **Boiling Point at 1 atm:** 210°F = 99°C = 372°K
13.4 **Freezing Point:** Not pertinent
13.5 **Critical Temperature:** Not pertinent
13.6 **Critical Pressure:** Not pertinent
13.7 **Specific Gravity:** 1.24 at 25°C (liquid)
13.8 **Liquid Surface Tension:** (est.) 25 dynes/cm = 0.025 N/m at 20°C
13.9 **Liquid-Water Interfacial Tension:** Not pertinent
13.10 **Vapor (Gas) Specific Gravity:** 5.6
13.11 **Ratio of Specific Heats of Vapor (Gas):** Not pertinent
13.12 **Latent Heat of Vaporization:** 104 Btu/lb = 58 cal/g = 2.4×10^5 J/kg
13.13 **Heat of Combustion:** (est.) −4,300 Btu/lb = −2,400 cal/g = -100×10^5 J/kg
13.14 **Heat of Decomposition:** Not pertinent
13.15 **Heat of Solution:** Data not available
13.16 **Heat of Polymerization:** Not pertinent

NOTES

FAC FERRIC AMMONIUM CITRATE

Common Synonyms			
Ammonium ferric citrate Ferric ammonium citrate, brown Ferric ammonium citrate, green	Solid	Red, green or brown	Odorless
	Sinks and mixes slowly with water.		

Avoid contact with solid and dust. Keep people away.
Stop discharge if possible.
Isolate and remove discharged material.
Notify local health and pollution control agencies.

Fire	Not flammable.
Exposure	CALL FOR MEDICAL AID. DUST Irritating to eyes, nose and throat. If inhaled will cause coughing or difficult breathing. **If in eyes, hold eyelids open and flush with plenty of water.** **If breathing has stopped, give artificial respiration.** **If breathing is difficult, give oxygen.** SOLID Irritating to skin and eyes. If swallowed will cause nausea and vomiting. **Remove contaminated clothing and shoes.** **Flush affected areas with plenty of water.** **IF IN EYES, hold eyelids open and flush with plenty of water.** **IF SWALLOWED and victim is CONSCIOUS, have victim drink water or milk.** **IF SWALLOWED and victim is UNCONSCIOUS OR HAVING CONVULSIONS, do nothing except keep victim warm.**
Water Pollution	Effect of low concentrations on aquatic life is unknown. May be dangerous if it enters water intakes. **Notify local health and wildlife officials.** **Notify operators of nearby water intakes.**

1. RESPONSE TO DISCHARGE
(See Response Methods Handbook, CG 446-4)

Disperse and flush

2. LABELS

No hazard label required by Code of Federal Regulations

3. CHEMICAL DESIGNATIONS

3.1 **Synonyms:** Ammonium ferric citrate; Ferric ammonium citrate, brown; Ferric ammonium citrate, green
3.2 **Coast Guard Compatibility Classification:** Not listed
3.3 **Chemical Formula:** Mixture of $FeC_6H_5O_7$, $(NH_4)_2HC_6H_5O_7$, and water of hydration
3.4 **IMCO/United Nations Numerical Designation:** Not listed

4. OBSERVABLE CHARACTERISTICS

4.1 **Physical State (as shipped):** Solid
4.2 **Color:** Red, green, or brown
4.3 **Odor:** None

5. HEALTH HAZARDS

5.1 **Personal Protective Equipment:** Approved respirator for nuisance dust; chemical goggles or face shield
5.2 **Symptoms Following Exposure:** Inhalation of dust irritates nose and throat. Ingestion causes irritation of mouth and stomach. Dust irritates eyes and causes mild irritation of skin on prolonged contact.
5.3 **Treatment for Exposure:** INGESTION: give large amount of water. EYES or SKIN: flush with water.
5.4 **Toxicity by Inhalation (Threshold Limit Value):** 1 mg/m^3 (as iron)
5.5 **Short-Term Inhalation Limits:** Data not available
5.6 **Toxicity by Ingestion:** Data not available
5.7 **Late Toxicity:** Data not available
5.8 **Vapor (Gas) Irritant Characteristics:** Data not available
5.9 **Liquid or Solid Irritant Characteristics:** Data not available
5.10 **Odor Threshold:** Data not available

6. FIRE HAZARDS

6.1 **Flash Point:** Not flammable
6.2 **Flammable Limits in Air:** Not flammable
6.3 **Fire Extinguishing Agents:** Not pertinent
6.4 **Fire Extinguishing Agents Not to be Used:**
6.5 **Special Hazards of Combustion Products:** Toxic oxides of nitrogen or ammonia gas may be formed in fires.
6.6 **Behavior in Fire:**
6.7 **Ignition Temperature:** Not pertinent
6.8 **Electrical Hazard:** Not pertinent
6.9 **Burning Rate:** Not pertinent

7. CHEMICAL REACTIVITY

7.1 **Reactivity with Water:** No reaction
7.2 **Reactivity with Common Materials:**
7.3 **Stability During Transport:** Stable
7.4 **Neutralizing Agents for Acids and Caustics:** Not pertinent
7.5 **Polymerization:** Not pertinent
7.6 **Inhibitor of Polymerization:** Not pertinent

8. WATER POLLUTION

8.1 **Aquatic Toxicity:** Data not available
8.2 **Waterfowl Toxicity:** Data not available
8.3 **Biological Oxygen Demand (BOD):** Data not available
8.4 **Food Chain Concentration Potential:** None

9. SELECTED MANUFACTURERS

1. Pfizer Chemicals Div., 235 E. 42nd St., New York, N.Y. 10017
2. Mallinckrodt Chemical Works, 223 Westside Ave., P.O. Box 384, Jersey City, N.J. 07303
3. Gallard-Schlesinger Chemical Mfg. Co., 584 Mineola Ave., Carle Place, N.Y. 11514

10. SHIPPING INFORMATION

10.1 **Grades or Purity:** Commercial
10.2 **Storage Temperature:** Ambient
10.3 **Inert Atmosphere:** No requirement
10.4 **Venting:** Open

11. HAZARD ASSESSMENT CODE
(See Hazard Assessment Handbook, CG 446-3)

SS

12. HAZARD CLASSIFICATIONS

12.1 **Code of Federal Regulations:** Not listed
12.2 **NAS Hazard Rating for Bulk Water Transportation:** Not listed
12.3 **NFPA Hazard Classifications:** Not listed

13. PHYSICAL AND CHEMICAL PROPERTIES

13.1 **Physical State at 15°C and 1 atm:** Solid
13.2 **Molecular Weight:** Not pertinent (mixture)
13.3 **Boiling Point at 1 atm:** Not pertinent (decomposes)
13.4 **Freezing Point:** Not pertinent
13.5 **Critical Temperature:** Not pertinent
13.6 **Critical Pressure:** Not pertinent
13.7 **Specific Gravity:** 1.8 at 20°C (solid)
13.8 **Liquid Surface Tension:** Not pertinent
13.9 **Liquid-Water Interfacial Tension:** Not pertinent
13.10 **Vapor (Gas) Specific Gravity:** Not pertinent
13.11 **Ratio of Specific Heats of Vapor (Gas):** Not pertinent
13.12 **Latent Heat of Vaporization:** Not pertinent
13.13 **Heat of Combustion:** Not pertinent
13.14 **Heat of Decomposition:** Not pertinent
13.15 **Heat of Solution:** Not pertinent
13.16 **Heat of Polymerization:** Not pertinent

NOTES

FAO — FERRIC AMMONIUM OXALATE

Common Synonyms			
Ammonium ferric oxalate trihydrate Ammonium trioxalatoferrate III trihydrate	Solid powder	Yellowish-green	Light burnt-sugar odor
	Sinks and mixes with water.		

Avoid contact with solid and dust. Keep people away.
Stop discharge if possible.
Isolate and remove discharged material.
Notify local health and pollution control agencies.

Fire

Not flammable.
POISONOUS GASES MAY BE PRODUCED WHEN HEATED.

Exposure

CALL FOR MEDICAL AID.

DUST
Irritating to eyes, nose and throat.
If inhaled will cause coughing or difficult breathing.
If in eyes, hold eyelids open and flush with plenty of water.
If breathing has stopped, give artificial respiration.
If breathing is difficult, give oxygen.

SOLID
Will burn skin and eyes.
If swallowed will cause nausea and vomiting.
Remove contaminated clothing and shoes.
Flush affected areas with plenty of water.
IF IN EYES, hold eyelids open and flush with plenty of water.
IF SWALLOWED and victim is CONSCIOUS, have victim drink water or milk and have victim induce vomiting.
IF SWALLOWED and victim is UNCONSCIOUS OR HAVING CONVULSIONS, do nothing except keep victim warm.

Water Pollution

Effect of low concentrations on aquatic life is unknown.
May be dangerous if it enters water intakes.
Notify local health and wildlife officials.
Notify operators of nearby water intakes.

1. RESPONSE TO DISCHARGE
(See Response Methods Handbook, CG 446-4)

Issue warning—water contaminant
Disperse and flush

2. LABELS

No hazard label required by Code of Federal Regulations

3. CHEMICAL DESIGNATIONS

3.1 **Synonyms:** Ammonium ferric oxalate trihydrate; Ammonium trioxalatoferrate (III) trihydrate
3.2 **Coast Guard Compatibility Classification:** Not listed
3.3 **Chemical Formula:** $Fe(NH_4)_3(C_2O_4)_3 \cdot 3H_2O$
3.4 **IMCO/United Nations Numerical Designation:** Not listed

4. OBSERVABLE CHARACTERISTICS

4.1 **Physical State (as shipped):** Solid
4.2 **Color:** Yellowish-green
4.3 **Odor:** Slight burnt sugar odor

5. HEALTH HAZARDS

5.1 **Personal Protective Equipment:** Approved dust respirator; rubber or plastic-coated gloves; chemical goggles or face shield
5.2 **Symptoms Following Exposure:** Inhalation of dust may cause irritation of nose and throat. Ingestion causes burning pain in throat and stomach; mucous membranes become white; may also cause vomiting, weak pulse, cardiovascular collapse, and death. Contact with dust irritates eyes and skin; may cause severe skin burns.
5.3 **Treatment for Exposure:** *(treat victim promptly)* INHALATION: move to fresh air; get medical attention if any symptoms persist. INGESTION: give immediately a dilute solution of any soluble calcium salt such as calcium lactate, limewater, chalk, or milk; induce vomiting; get medical attention. (Watch for edema of the glottis and delayed constriction of esophagus.) EYES: flush with water and get medical attention. SKIN: flush with water.
5.4 **Toxicity by Inhalation (Threshold Limit Value):** 1 mg/m³ (as iron)
5.5 **Short-Term Inhalation Limits:** Data not available
5.6 **Toxicity by Ingestion:** Data not available
5.7 **Late Toxicity:** Data not available
5.8 **Vapor (Gas) Irritant Characteristics:** Data not available
5.9 **Liquid or Solid Irritant Characteristics:** Data not available
5.10 **Odor Threshold:** Data not available

6. FIRE HAZARDS

6.1 **Flash Point:** Not flammable
6.2 **Flammable Limits in Air:** Not flammable
6.3 **Fire Extinguishing Agents:** Not pertinent
6.4 **Fire Extinguishing Agents Not to be Used:** Not pertinent
6.5 **Special Hazards of Combustion Products:** Toxic oxides of nitrogen, ammonia, and carbon monoxide may form in fires.
6.6 **Behavior in Fire:**
6.7 **Ignition Temperature:** Not pertinent
6.8 **Electrical Hazard:** Not pertinent
6.9 **Burning Rate:** Not pertinent

7. CHEMICAL REACTIVITY

7.1 **Reactivity with Water:** No reaction
7.2 **Reactivity with Common Materials:**
7.3 **Stability During Transport:** Stable
7.4 **Neutralizing Agents for Acids and Caustics:** Not pertinent
7.5 **Polymerization:** Not pertinent
7.6 **Inhibitor of Polymerization:** Not pertinent

8. WATER POLLUTION

8.1 **Aquatic Toxicity:** Data not available
8.2 **Waterfowl Toxicity:** Data not available
8.3 **Biological Oxygen Demand (BOD):** Data not available
8.4 **Food Chain Concentration Potential:** None

9. SELECTED MANUFACTURERS

1. Pfizer Chemicals Div.
   235 E. 42nd St.
   New York, N. Y. 10017
2. Pfaltz and Bauer, Inc.
   375 Fairfield Ave.
   Stamford, Conn. 06902

10. SHIPPING INFORMATION

10.1 **Grades or Purity:** Commercial
10.2 **Storage Temperature:** Ambient
10.3 **Inert Atmosphere:** No requirement
10.4 **Venting:** Open

11. HAZARD ASSESSMENT CODE
(See Hazard Assessment Handbook, CG 446-3)

SS

12. HAZARD CLASSIFICATIONS

12.1 **Code of Federal Regulations:** Not listed
12.2 **NAS Hazard Rating for Bulk Water Transportation:** Not listed
12.3 **NFPA Hazard Classifications:** Not listed

13. PHYSICAL AND CHEMICAL PROPERTIES

13.1 **Physical State at 15°C and 1 atm:** Solid
13.2 **Molecular Weight:** 428
13.3 **Boiling Point at 1 atm:** Not pertinent (decomposes)
13.4 **Freezing Point:** Not pertinent
13.5 **Critical Temperature:** Not pertinent
13.6 **Critical Pressure:** Not pertinent
13.7 **Specific Gravity:** 1.78 at 20°C (solid)
13.8 **Liquid Surface Tension:** Not pertinent
13.9 **Liquid-Water Interfacial Tension:** Not pertinent
13.10 **Vapor (Gas) Specific Gravity:** Not pertinent
13.11 **Ratio of Specific Heats of Vapor (Gas):** Not pertinent
13.12 **Latent Heat of Vaporization:** Not pertinent
13.13 **Heat of Combustion:** Not pertinent
13.14 **Heat of Decomposition:** Not pertinent
13.15 **Heat of Solution:** Not pertinent
13.16 **Heat of Polymerization:** Not pertinent

NOTES

FCL — FERRIC CHLORIDE

Common Synonyms			
Iron III chloride Iron perchloride Iron trichloride Ferric chloride, anhydrous Ferric chloride, hexahydrate	Solid	Greenish black	Odorless
	Sinks and mixes with water.		

Avoid contact with solid and dust. Keep people away.
Stop discharge if possible.
Isolate and remove discharged material.
Notify local health and pollution control agencies.

Fire	Not flammable.
Exposure	CALL FOR MEDICAL AID. DUST Irritating to eyes, nose and throat. If inhaled will cause coughing or difficult breathing. If in eyes, hold eyelids open and flush with plenty of water. If breathing has stopped, give artificial respiration. If breathing is difficult, give oxygen. SOLID Will burn skin and eyes. If swallowed will cause nausea and vomiting. Remove contaminated clothing and shoes. Flush affected areas with plenty of water. IF IN EYES, hold eyelids open and flush with plenty of water. IF SWALLOWED and victim is CONSCIOUS, have victim drink water or milk and have victim induce vomiting. IF SWALLOWED and victim is UNCONSCIOUS OR HAVING CONVULSIONS, do nothing except keep victim warm.
Water Pollution	HARMFUL TO AQUATIC LIFE IN VERY LOW CONCENTRATIONS. May be dangerous if it enters water intakes. Notify local health and wildlife officials. Notify operators of nearby water intakes.

1. RESPONSE TO DISCHARGE
(See Response Methods Handbook, CG 446-4)

Disperse and flush

2. LABELS

No hazard label required by Code of Federal Regulations

3. CHEMICAL DESIGNATIONS

3.1 **Synonyms:** Ferric chloride, anhydrous; Ferric chloride, hexahydrate; Iron (III) chloride; Iron perchloride; Iron trichloride

3.2 **Coast Guard Compatibility Classification:** Not listed

3.3 **Chemical Formula:** $FeCl_3$ or $FeCl_3 \cdot 6H_2O$

3.4 **IMCO/United Nations Numerical Designation:** 8/1773

4. OBSERVABLE CHARACTERISTICS

4.1 **Physical State (as shipped):** Solid

4.2 **Color:** Anhydrous: greenish black; Hydrate: brown

4.3 **Odor:** None

5. HEALTH HAZARDS

5.1 **Personal Protective Equipment:** Dust respirator if required; rubber apron and boots; chemical worker's goggles or face shield

5.2 **Symptoms Following Exposure:** Inhalation of dust may irritate nose and throat. Ingestion causes irritation of mouth and stomach. Dust irritates eyes. Prolonged contact with skin causes irritation and burns.

5.3 **Treatment for Exposure:** INGESTION: give large amounts of water; induce vomiting if large amounts have been swallowed. EYES: immediately flush with plenty of water for at least 15 min.; get medical attention promptly. SKIN: flush with water.

5.4 **Toxicity by Inhalation (Threshold Limit Value):** 1 mg/m^3 (as iron)

5.5 **Short-Term Inhalation Limits:** Data not available

5.6 **Toxicity by Ingestion:** Grade 2; LD_{50} 0.5 – 5 g/kg (rat)

5.7 **Late Toxicity:** Data not available

5.8 **Vapor (Gas) Irritant Characteristics:** Not pertinent

5.9 **Liquid or Solid Irritant Characteristics:** Data not available

5.10 **Odor Threshold:** Data not available

6. FIRE HAZARDS

6.1 **Flash Point:** Not flammable

6.2 **Flammable Limits in Air:** Not flammable

6.3 **Fire Extinguishing Agents:** Not pertinent

6.4 **Fire Extinguishing Agents Not to be Used:**

6.5 **Special Hazards of Combustion Products:** Irritating hydrogen chloride fumes may form in fire.

6.6 **Behavior in Fire:**

6.7 **Ignition Temperature:** Not pertinent

6.8 **Electrical Hazard:** Not pertinent

6.9 **Burning Rate:** Not pertinent

7. CHEMICAL REACTIVITY

7.1 **Reactivity with Water:** No reaction

7.2 **Reactivity with Common Materials:** Water solutions are acidic and corrosive to most metals.

7.3 **Stability During Transport:** Stable

7.4 **Neutralizing Agents for Acids and Caustics:** Flush with water, rinse with dilute sodium bicarbonate or soda ash solutions.

7.5 **Polymerization:** Not pertinent

7.6 **Inhibitor of Polymerization:** Not pertinent

8. WATER POLLUTION

8.1 **Aquatic Toxicity:**
1.2 ppm/144 hr/stickleback/harmful/fresh water
1 ppm/240 hr/stickleback/safe/fresh water
15 ppm/96 hr/daphnia/TL_m/fresh water

8.2 **Waterfowl Toxicity:** Data not available

8.3 **Biological Oxygen Demand (BOD):** None

8.4 **Food Chain Concentration Potential:** None

9. SELECTED MANUFACTURERS

1. Pennwalt Corp.
   Three Parkway
   Philadelphia, Pa. 19102
2. Dow Chemical Co.
   Midland, Mich. 48640
3. J. T. Baker Chemical Co.
   Phillipsburg, N. J. 08865

10. SHIPPING INFORMATION

10.1 **Grades or Purity:** Anhydrous; Hydrate; Reagent; 46% solution in water

10.2 **Storage Temperature:** Ambient

10.3 **Inert Atmosphere:** No requirement

10.4 **Venting:** Open

11. HAZARD ASSESSMENT CODE
(See Hazard Assessment Handbook, CG 446-3)

SS

12. HAZARD CLASSIFICATIONS

12.1 **Code of Federal Regulations:** ORM-B

12.2 **NAS Hazard Rating for Bulk Water Transportation:** Not listed

12.3 **NFPA Hazard Classifications:** Not listed

13. PHYSICAL AND CHEMICAL PROPERTIES

13.1 **Physical State at 15°C and 1 atm:** Solid

13.2 **Molecular Weight:** 162.22 (anhydrous)

13.3 **Boiling Point at 1 atm:** Not pertinent (decomposes)

13.4 **Freezing Point:** Not pertinent

13.5 **Critical Temperature:** Not pertinent

13.6 **Critical Pressure:** Not pertinent

13.7 **Specific Gravity:** 2.8 at 20°C (anhydrous solid)

13.8 **Liquid Surface Tension:** Not pertinent

13.9 **Liquid-Water Interfacial Tension:** Not pertinent

13.10 **Vapor (Gas) Specific Gravity:** Not pertinent

13.11 **Ratio of Specific Heats of Vapor (Gas):** Not pertinent

13.12 **Latent Heat of Vaporization:** Not pertinent

13.13 **Heat of Combustion:** Not pertinent

13.14 **Heat of Decomposition:** Not pertinent

13.15 **Heat of Solution:** (anhydrous) -360 Btu/lb $= -200$ cal/g $= -8.4 \times 10^5$ J/kg

13.16 **Heat of Polymerization:** Not pertinent

NOTES

FCP — FERRIC GLYCEROPHOSPHATE

Common Synonyms	Solid	Greenish-brown to greenish yellow
	Sinks and mixes with water.	

Avoid contact with solid and dust. Keep people away.
Stop discharge if possible.
Isolate and remove discharged material.
Notify local health and pollution control agencies.

Fire

Not flammable.

Exposure

CALL FOR MEDICAL AID.

DUST
Irritating to eyes, nose and throat.
If inhaled will cause coughing or difficult breathing.

If in eyes, hold eyelids open and flush with plenty of water.
If breathing has stopped, give artificial respiration.
If breathing is difficult, give oxygen.

SOLID
Irritating to skin and eyes.
Harmful if swallowed.

Remove contaminated clothing and shoes.
Flush affected areas with plenty of water.
IF IN EYES, hold eyelids open and flush with plenty of water.
IF SWALLOWED and victim is CONSCIOUS, have victim drink water or milk and have victim induce vomiting.
IF SWALLOWED and victim is UNCONSCIOUS OR HAVING CONVULSIONS, do nothing except keep victim warm.

Water Pollution

Effect of low concentrations on aquatic life is unknown.
May be dangerous if it enters water intakes.

Notify local health and wildlife officials.
Notify operators of nearby water intakes.

1. RESPONSE TO DISCHARGE

(See Response Methods Handbook, CG 446-4)

Disperse and flush

2. LABELS

No hazard label required by Code of Federal Regulations

3. CHEMICAL DESIGNATIONS

3.1 **Synonyms:** No common synonyms
3.2 **Coast Guard Compatibility Classification:** Not listed
3.3 **Chemical Formula:** (approx.) $Fe_2[C_3H_5(OH)_2PO_4]_3 \cdot xH_2O$
3.4 **IMCO/United Nations Numerical Designation:** Not listed

4. OBSERVABLE CHARACTERISTICS

4.1 **Physical State (as shipped):** Solid
4.2 **Color:** Greenish-brown, greenish-yellow
4.3 **Odor:** None

5. HEALTH HAZARDS

5.1 **Personal Protective Equipment:** Goggles or face shield; dust mask; rubber gloves
5.2 **Symptoms Following Exposure:** Inhalation of dust may irritate nose and throat. Contact with dust irritates eyes and (on prolonged contact) skin.
5.3 **Treatment for Exposure:** INHALATION: move to fresh air. INGESTION: give large amount of water; induce vomiting if large amounts have been swallowed. EYES: flush with water for at least 15 min. SKIN: flush with water.
5.4 **Toxicity by Inhalation (Threshold Limit Value):** 1 mg/m³ (as iron)
5.5 **Short-Term Inhalation Limits:** Data not available
5.6 **Toxicity by Ingestion:** Data not available
5.7 **Late Toxicity:** Data not available
5.8 **Vapor (Gas) Irritant Characteristics:** Not pertinent
5.9 **Liquid or Solid Irritant Characteristics:** Data not available
5.10 **Odor Threshold:** Odorless

6. FIRE HAZARDS

6.1 **Flash Point:** Not flammable
6.2 **Flammable Limits in Air:** Not flammable
6.3 **Fire Extinguishing Agents:** Not pertinent
6.4 **Fire Extinguishing Agents Not to be Used:** Not pertinent
6.5 **Special Hazards of Combustion Products:**
6.6 **Behavior in Fire:**
6.7 **Ignition Temperature:** Not pertinent
6.8 **Electrical Hazard:** Not pertinent
6.9 **Burning Rate:** Not pertinent

7. CHEMICAL REACTIVITY

7.1 **Reactivity with Water:** No reaction
7.2 **Reactivity with Common Materials:**
7.3 **Stability During Transport:** Stable
7.4 **Neutralizing Agents for Acids and Caustics:** Not pertinent
7.5 **Polymerization:** Not pertinent
7.6 **Inhibitor of Polymerization:** Not pertinent

8. WATER POLLUTION

8.1 **Aquatic Toxicity:** Data not available
8.2 **Waterfowl Toxicity:** Data not available
8.3 **Biological Oxygen Demand (BOD):** Data not available
8.4 **Food Chain Concentration Potential:** None

9. SELECTED MANUFACTURERS

1. Pfaltz and Bauer, Inc.
375 Fairfield Ave.
Stamford, Conn. 06902
2. Shepherd Chemical Co.
4900 Beech St.
Cincinnati, Ohio 45212

10. SHIPPING INFORMATION

10.1 **Grades or Purity:** Commercial
10.2 **Storage Temperature:** Ambient
10.3 **Inert Atmosphere:** No requirement
10.4 **Venting:** Open

11. HAZARD ASSESSMENT CODE

(See Hazard Assessment Handbook, CG 446-3)

SS

12. HAZARD CLASSIFICATIONS

12.1 **Code of Federal Regulations:** Not listed
12.2 **NAS Hazard Rating for Bulk Water Transportation:** Not listed
12.3 **NFPA Hazard Classifications:** Not listed

13. PHYSICAL AND CHEMICAL PROPERTIES

13.1 **Physical State at 15°C and 1 atm:** Solid
13.2 **Molecular Weight:** 470 (approx.)
13.3 **Boiling Point at 1 atm:** Not pertinent (decomposes)
13.4 **Freezing Point:** Not pertinent
13.5 **Critical Temperature:** Not pertinent
13.6 **Critical Pressure:** Not pertinent
13.7 **Specific Gravity:** 1.5 at 20°C (solid)
13.8 **Liquid Surface Tension:** Not pertinent
13.9 **Liquid-Water Interfacial Tension:** Not pertinent
13.10 **Vapor (Gas) Specific Gravity:** Not pertinent
13.11 **Ratio of Specific Heats of Vapor (Gas):** Not pertinent
13.12 **Latent Heat of Vaporization:** Not pertinent
13.13 **Heat of Combustion:** Not pertinent
13.14 **Heat of Decomposition:** Not pertinent
13.15 **Heat of Solution:** Not pertinent
13.16 **Heat of Polymerization:** Not pertinent

NOTES

FNT	# FERRIC NITRATE

Common Synonyms			
Ferric nitrate nonahydrate Nitric acid, iron III salt	Solid	Green, colorless to pale violet	Odorless
	Sinks and mixes with water.		

Avoid contact with solid and dust. Keep people away.
Stop discharge if possible.
Isolate and remove discharged material.
Notify local health and pollution control agencies.

Fire	Not flammable. Will increase the intensity of a fire. **Flood discharge area with water.**
Exposure	**CALL FOR MEDICAL AID.** DUST Irritating to eyes, nose and throat. If inhaled will cause coughing or difficult breathing. **If in eyes, hold eyelids open and flush with plenty of water.** **If breathing has stopped, give artificial respiration.** **If breathing is difficult, give oxygen.** SOLID Irritating to skin and eyes. Harmful if swallowed. **Remove contaminated clothing and shoes.** **Flush affected areas with plenty of water.** **IF IN EYES, hold eyelids open and flush with plenty of water.** **IF SWALLOWED and victim is CONSCIOUS, have victim drink water or milk and have victim induce vomiting.** **IF SWALLOWED and victim is UNCONSCIOUS OR HAVING CONVULSIONS, do nothing except keep victim warm.**
Water Pollution	Effect of low concentrations on aquatic life is unknown. May be dangerous if it enters water intakes. **Notify local health and wildlife officials.** **Notify operators of nearby water intakes.**

1. RESPONSE TO DISCHARGE

(See Response Methods Handbook, CG 446-4)

Disperse and flush

2. LABELS

No hazard label required by Code of Federal Regulations

3. CHEMICAL DESIGNATIONS

3.1 **Synonyms:** Ferric nitrate nonahydrate; Nitric acid, iron (+3) salt
3.2 **Coast Guard Compatibility Classification:** Not listed
3.3 **Chemical Formula:** $Fe(NO_3)_3 \cdot 9H_2O$
3.4 **IMCO/United Nations Numerical Designation:** 5.1/1466

4. OBSERVABLE CHARACTERISTICS

4.1 **Physical State (as shipped):** Solid
4.2 **Color:** Green; colorless to pale violet
4.3 **Odor:** None

5. HEALTH HAZARDS

5.1 **Personal Protective Equipment:** Dust mask; goggles or face shield; protective gloves
5.2 **Symptoms Following Exposure:** Inhalation of dust irritates nose and throat. Ingestion causes irritation of mouth and stomach. Dust irritates eyes and can irritate skin on prolonged contact.
5.3 **Treatment for Exposure:** INHALATION: move to fresh air. INGESTION: give large amounts of water; induce vomiting if large amounts have been swallowed. EYES: flush with water; get medical attention if irritation persists. SKIN: flush with water.
5.4 **Toxicity by Inhalation (Threshold Limit Value):** 1 mg/m^3 (as iron)
5.5 **Short-Term Inhalation Limits:** Data not available
5.6 **Toxicity by Ingestion:** Grade 2; LD_{50} 0.5 – 5 g/kg
5.7 **Late Toxicity:** Data not available
5.8 **Vapor (Gas) Irritant Characteristics:** Not pertinent
5.9 **Liquid or Solid Irritant Characteristics:** Data not available
5.10 **Odor Threshold:** Data not available

6. FIRE HAZARDS

6.1 **Flash Point:** Not flammable
6.2 **Flammable Limits in Air:** Not flammable
6.3 **Fire Extinguishing Agents:** Not pertinent
6.4 **Fire Extinguishing Agents Not to be Used:**
6.5 **Special Hazards of Combustion Products:** Toxic oxides of nitrogen and nitric acid vapor may form in fires.
6.6 **Behavior in Fire:** In contact with combustible materials, will increase the intensity of a fire. Apply water to cool containers or spilled material.
6.7 **Ignition Temperature:** Not pertinent
6.8 **Electrical Hazard:** Not pertinent
6.9 **Burning Rate:** Not pertinent

7. CHEMICAL REACTIVITY

7.1 **Reactivity with Water:** No reaction
7.2 **Reactivity with Common Materials:** Solutions are corrosive to most metals. Contact of solid with wood or paper may cause fire.
7.3 **Stability During Transport:** Stable
7.4 **Neutralizing Agents for Acids and Caustics:** Not pertinent
7.5 **Polymerization:** Not pertinent
7.6 **Inhibitor of Polymerization:** Not pertinent

8. WATER POLLUTION

8.1 **Aquatic Toxicity:** Data not available
8.2 **Waterfowl Toxicity:** Data not available
8.3 **Biological Oxygen Demand (BOD):** None
8.4 **Food Chain Concentration Potential:** None

9. SELECTED MANUFACTURERS

1. Allied Chemical Corp.
   Industrial Chemicals Div.
   P. O. Box 70
   Morristown, N. J. 07960
2. J. T. Baker Chemical Co.
   Phillipsburg, N. J. 08865
3. Gallard-Schlesinger Chemical Mfg. Co.
   584 Mineola Ave.
   Carle Place, N. Y. 11514

10. SHIPPING INFORMATION

10.1 **Grades or Purity:** Technical, 99.0%; Analytical
10.2 **Storage Temperature:** Ambient
10.3 **Inert Atmosphere:** No requirement
10.4 **Venting:** Open

11. HAZARD ASSESSMENT CODE

(See Hazard Assessment Handbook, CG 446-3)

SS

12. HAZARD CLASSIFICATIONS

12.1 **Code of Federal Regulations:** Not listed
12.2 **NAS Hazard Rating for Bulk Water Transportation:** Not listed
12.3 **NFPA Hazard Classifications:** Not listed

13. PHYSICAL AND CHEMICAL PROPERTIES

13.1 **Physical State at 15°C and 1 atm:** Solid
13.2 **Molecular Weight:** 404.02
13.3 **Boiling Point at 1 atm:** Not pertinent (decomposes)
13.4 **Freezing Point:** 117°F = 47°C = 320°K
13.5 **Critical Temperature:** Not pertinent
13.6 **Critical Pressure:** Not pertinent
13.7 **Specific Gravity:** 1.7 at 20°C (solid)
13.8 **Liquid Surface Tension:** Not pertinent
13.9 **Liquid-Water Interfacial Tension:** Not pertinent
13.10 **Vapor (Gas) Specific Gravity:** Not pertinent
13.11 **Ratio of Specific Heats of Vapor (Gas):** Not pertinent
13.12 **Latent Heat of Vaporization:** Not pertinent
13.13 **Heat of Combustion:** Not pertinent
13.14 **Heat of Decomposition:** Not pertinent
13.15 **Heat of Solution:** 40 Btu/lb = 22 cal/g = 0.92×10^5 J/kg
13.16 **Heat of Polymerization:** Not pertinent

NOTES

FSF — FERRIC SULFATE

Common Synonyms	Solid	Gray to white	Odorless
Iron III sulfate Iron tersulfate Iron sesquisulfate	Sinks and mixes slowly with water.		

Avoid contact with solid and dust. Keep people away.
Stop discharge if possible.
Isolate and remove discharged material.
Notify local health and pollution control agencies.

Fire	Not flammable.
Exposure	CALL FOR MEDICAL AID. DUST Irritating to eyes, nose and throat. If inhaled will cause coughing or difficult breathing. If in eyes, hold eyelids open and flush with plenty of water. If breathing has stopped, give artificial respiration. If breathing is difficult, give oxygen. SOLID Irritating to skin and eyes. If swallowed will cause nausea and vomiting. Remove contaminated clothing and shoes. Flush affected areas with plenty of water. IF IN EYES, hold eyelids open and flush with plenty of water. IF SWALLOWED and victim is CONSCIOUS, have victim drink water or milk and have victim induce vomiting. IF SWALLOWED and victim is UNCONSCIOUS OR HAVING CONVULSIONS, do nothing except keep victim warm.
Water Pollution	Dangerous to aquatic life in high concentrations. May be dangerous if it enters water intakes. Notify local health and wildlife officials. Notify operators of nearby water intakes.

1. RESPONSE TO DISCHARGE
(See Response Methods Handbook, CG 446-4)

Disperse and flush

2. LABELS

No hazard label required by Code of Federal Regulations

3. CHEMICAL DESIGNATIONS

3.1 **Synonyms:** Iron sesquisulfate; Iron (III) sulfate; Iron tersulfate

3.2 **Coast Guard Compatibility Classification:** Not listed

3.3 **Chemical Formula:** $Fe_2(SO_4)_3$

3.4 **IMCO/United Nations Numerical Designation:** Not listed

4. OBSERVABLE CHARACTERISTICS

4.1 **Physical State (as shipped):** Solid

4.2 **Color:** Gray white

4.3 **Odor:** None

5. HEALTH HAZARDS

5.1 **Personal Protective Equipment:** Dust mask; goggles or face shield; protective gloves

5.2 **Symptoms Following Exposure:** Inhalation of dust irritates nose and throat. Ingestion causes irritation of mouth and stomach. Dust irritates eyes and can irritate skin on prolonged contact.

5.3 **Treatment for Exposure:** INHALATION: move to fresh air. INGESTION: give large amount of water; induce vomiting if large amounts have been swallowed. EYES: flush with water; get medical attention if irritation persists. SKIN: flush with water.

5.4 **Toxicity by Inhalation (Threshold Limit Value):** 1 mg/m^3 (as iron)

5.5 **Short-Term Inhalation Limits:** Data not available

5.6 **Toxicity by Ingestion:** Data not available

5.7 **Late Toxicity:** Data not available

5.8 **Vapor (Gas) Irritant Characteristics:** Data not available

5.9 **Liquid or Solid Irritant Characteristics:** Data not available

5.10 **Odor Threshold:** Data not available

6. FIRE HAZARDS

6.1 **Flash Point:** Not flammable

6.2 **Flammable Limits in Air:** Not flammable

6.3 **Fire Extinguishing Agents:** Not pertinent

6.4 **Fire Extinguishing Agents Not to be Used:** Not pertinent

6.5 **Special Hazards of Combustion Products:**

6.6 **Behavior in Fire:**

6.7 **Ignition Temperature:** Not pertinent

6.8 **Electrical Hazard:** Not pertinent

6.9 **Burning Rate:** Not pertinent

7. CHEMICAL REACTIVITY

7.1 **Reactivity with Water:** No reaction

7.2 **Reactivity with Common Materials:** Corrosive to copper, copper alloys, mild steel, and galvanized steel.

7.3 **Stability During Transport:** Stable

7.4 **Neutralizing Agents for Acids and Caustics:** Flush with water.

7.5 **Polymerization:** Not pertinent

7.6 **Inhibitor of Polymerization:** Not pertinent

8. WATER POLLUTION

8.1 **Aquatic Toxicity:**
0.716 ppm/12—24 hr/shiners, carp, suckers/killed/fresh water
133 ppm/48 hr/mosquito fish/TL_m/fresh water

8.2 **Waterfowl Toxicity:** Data not available

8.3 **Biological Oxygen Demand (BOD):** None

8.4 **Food Chain Concentration Potential:** None

9. SELECTED MANUFACTURERS

1. J. T. Baker Chemical Co.
Phillipsburg, N. J. 08865
2. Cities Service Co.
Industrial Chemicals Div.
3445 Peachtree Road, N. E.
Atlanta, Ga. 30326
3. Mallinckrodt Chemical Works
223 Westside Ave.
P. O. Box 384
Jersey City, N. J. 07303

10. SHIPPING INFORMATION

10.1 **Grades or Purity:** Anhydrous; Hydrate, 73%; sometimes shipped as water solutions, which are acidic.

10.2 **Storage Temperature:** Ambient

10.3 **Inert Atmosphere:** No requirement

10.4 **Venting:** Open

11. HAZARD ASSESSMENT CODE
(See Hazard Assessment Handbook, CG 446-3)

SS

12. HAZARD CLASSIFICATIONS

12.1 **Code of Federal Regulations:** Not listed

12.2 **NAS Hazard Rating for Bulk Water Transportation:** Not listed

12.3 **NFPA Hazard Classifications:** Not listed

13. PHYSICAL AND CHEMICAL PROPERTIES

13.1 **Physical State at 15°C and 1 atm:** Solid

13.2 **Molecular Weight:** 399.88

13.3 **Boiling Point at 1 atm:** Not pertinent (decomposes)

13.4 **Freezing Point:** Not pertinent

13.5 **Critical Temperature:** Not pertinent

13.6 **Critical Pressure:** Not pertinent

13.7 **Specific Gravity:** 3.1 at 20°C (solid)

13.8 **Liquid Surface Tension:** Not pertinent

13.9 **Liquid-Water Interfacial Tension:** Not pertinent

13.10 **Vapor (Gas) Specific Gravity:** Not pertinent

13.11 **Ratio of Specific Heats of Vapor (Gas):** Not pertinent

13.12 **Latent Heat of Vaporization:** Not pertinent

13.13 **Heat of Combustion:** Not pertinent

13.14 **Heat of Decomposition:** Not pertinent

13.15 **Heat of Solution:** –53.3 Btu/lb = –29.6 cal/g = -1.24×10^5 J/kg

13.16 **Heat of Polymerization:** Not pertinent

NOTES

FAS

FERROUS AMMONIUM SULFATE

Common Synonyms			
Mohr's salt Iron ammonium sulfate Ferrous ammonium sulfate hexahydrate	Solid	Pale blue-green	Odorless
	Sinks and mixes slowly with water.		

Avoid contact with solid and dust. Keep people away.
Stop discharge if possible.
Isolate and remove discharged material.
Notify local health and pollution control agencies.

Fire

Not flammable.

Exposure

CALL FOR MEDICAL AID.

DUST
Irritating to eyes, nose and throat.
If inhaled will cause coughing or difficult breathing.
If in eyes, hold eyelids open and flush with plenty of water.
If breathing has stopped, give artificial respiration.
If breathing is difficult, give oxygen.

SOLID
Irritating to skin and eyes.
If swallowed will cause nausea and vomiting.
Remove contaminated clothing and shoes.
Flush affected areas with plenty of water.
IF IN EYES, hold eyelids open and flush with plenty of water.
IF SWALLOWED and victim is CONSCIOUS, have victim drink water or milk and have victim induce vomiting.
IF SWALLOWED and victim is UNCONSCIOUS OR HAVING CONVULSIONS, do nothing except keep victim warm.

Water Pollution

Effect of low concentrations on aquatic life is unknown.
May be dangerous if it enters water intakes.
Notify local health and wildlife officials.
Notify operators of nearby water intakes.

1. RESPONSE TO DISCHARGE

(See Response Methods Handbook, CG 446-4)

Disperse and flush

2. LABELS

No hazard label required by Code of Federal Regulations

3. CHEMICAL DESIGNATIONS

3.1 **Synonyms:** Ammonium ferrous sulfate; Ammonium iron sulfate; Ferrous ammonium sulfate hexahydrate; Iron ammonium sulfate; Mohr's salt

3.2 **Coast Guard Compatibility Classification:** Not listed

3.3 **Chemical Formula:** $Fe(NH_4)_2(SO_4)_2 \cdot 6H_2O$

3.4 **IMCO/United Nations Numerical Designation:** Not listed

4. OBSERVABLE CHARACTERISTICS

4.1 **Physical State (as shipped):** Solid

4.2 **Color:** Pale bluish-green

4.3 **Odor:** None

5. HEALTH HAZARDS

5.1 **Personal Protective Equipment:** Dust mask; goggles or face shield; protective gloves

5.2 **Symptoms Following Exposure:** Inhalation of dust irritates nose and throat. Ingestion causes irritation of mouth and stomach. Dust irritates eyes and can irritate skin on prolonged contact.

5.3 **Treatment for Exposure:** INGESTION: give large amount of water; induce vomiting if large amounts have been swallowed. EYES or SKIN: flush with water.

5.4 **Toxicity by Inhalation (Threshold Limit Value):** 1 mg/m^3 (as iron)

5.5 **Short-Term Inhalation Limits:** Data not available

5.6 **Toxicity by Ingestion:** Grade 2; LD_{50} 0.5 – 5 g/kg (rat)

5.7 **Late Toxicity:** May cause eye degeneration in rabbits

5.8 **Vapor (Gas) Irritant Characteristics:** Data not available

5.9 **Liquid or Solid Irritant Characteristics:** Data not available

5.10 **Odor Threshold:** Data not available

6. FIRE HAZARDS

6.1 **Flash Point:** Not flammable

6.2 **Flammable Limits in Air:** Not flammable

6.3 **Fire Extinguishing Agents:** Not pertinent

6.4 **Fire Extinguishing Agents Not to be Used:** Not pertinent

6.5 **Special Hazards of Combustion Products:** Irritating and toxic ammonia and oxides of nitrogen may form in fires.

6.6 **Behavior in Fire:** Not pertinent

6.7 **Ignition Temperature:** Not pertinent

6.8 **Electrical Hazard:** Not pertinent

6.9 **Burning Rate:** Not pertinent

7. CHEMICAL REACTIVITY

7.1 **Reactivity with Water:** No reaction

7.2 **Reactivity with Common Materials:**

7.3 **Stability During Transport:** Stable

7.4 **Neutralizing Agents for Acids and Caustics:** Not pertinent

7.5 **Polymerization:** Not pertinent

7.6 **Inhibitor of Polymerization:** Not pertinent

8. WATER POLLUTION

8.1 **Aquatic Toxicity:** Data not available

8.2 **Waterfowl Toxicity:** Data not available

8.3 **Biological Oxygen Demand (BOD):** Data not available

8.4 **Food Chain Concentration Potential:** None

9. SELECTED MANUFACTURERS

1. Allied Chemical Corp.
   General Chemical Div.
   40 Rector St.
   New York, N. Y. 10006
2. J. T. Baker Chemical Co.
   Phillipsburg, N. J. 08865
3. Mallinckrodt Chemical Works
   223 Westside Ave.
   P. O. Box 384
   Jersey City, N. J. 07303

10. SHIPPING INFORMATION

10.1 **Grades or Purity:** Technical, 99 – 102%; Reagent

10.2 **Storage Temperature:** Ambient

10.3 **Inert Atmosphere:** No requirement

10.4 **Venting:** Open

11. HAZARD ASSESSMENT CODE

(See Hazard Assessment Handbook, CG 446-3)

SS

12. HAZARD CLASSIFICATIONS

12.1 **Code of Federal Regulations:** Not listed

12.2 **NAS Hazard Rating for Bulk Water Transportation:** Not listed

12.3 **NFPA Hazard Classifications:** Not listed

13. PHYSICAL AND CHEMICAL PROPERTIES

13.1 **Physical State at 15°C and 1 atm:** Solid

13.2 **Molecular Weight:** 392.16

13.3 **Boiling Point at 1 atm:** Not pertinent (decomposes)

13.4 **Freezing Point:** Not pertinent

13.5 **Critical Temperature:** Not pertinent

13.6 **Critical Pressure:** Not pertinent

13.7 **Specific Gravity:** 1.86 at 20°C (solid)

13.8 **Liquid Surface Tension:** Not pertinent

13.9 **Liquid-Water Interfacial Tension:** Not pertinent

13.10 **Vapor (Gas) Specific Gravity:** Not pertinent

13.11 **Ratio of Specific Heats of Vapor (Gas):** Not pertinent

13.12 **Latent Heat of Vaporization:** Not pertinent

13.13 **Heat of Combustion:** Not pertinent

13.14 **Heat of Decomposition:** Not pertinent

13.15 **Heat of Solution:** Data not available

13.16 **Heat of Polymerization:** Not pertinent

NOTES

FEC — FERROUS CHLORIDE

Common Synonyms
Iron dichloride
Iron protochloride
Ferrous chloride tetrahydrate

Solid — Pale green — Odorless

Sinks and mixes slowly with water.

Stop discharge if possible.
Avoid contact with solid and dust.
Isolate and remove discharged material.
Notify local health and pollution control agencies.

Fire

Not flammable.

Exposure

CALL FOR MEDICAL AID.

DUST
Irritating to eyes, nose and throat.
If inhaled will cause difficult breathing.
If in eyes, hold eyelids open and flush with plenty of water.
If breathing has stopped, give artificial respiration.
If breathing is difficult, give oxygen.

SOLID
Irritating to skin and eyes.
If swallowed will cause nausea and vomiting.
Remove contaminated clothing and shoes.
Flush affected areas with plenty of water.
IF IN EYES, hold eyelids open and flush with plenty of water.
IF SWALLOWED and victim is CONSCIOUS, have victim drink water or milk and have victim induce vomiting.
IF SWALLOWED and victim is UNCONSCIOUS OR HAVING CONVULSIONS, do nothing except keep victim warm.

Water Pollution

HARMFUL TO AQUATIC LIFE IN VERY LOW CONCENTRATIONS.
May be dangerous if it enters water intakes.
Notify local health and wildlife officials.
Notify operators of nearby water intakes.

1. RESPONSE TO DISCHARGE

(See Response Methods Handbook, CG 446-4)

Disperse and flush

2. LABELS

No hazard label required by Code of Federal Regulations

3. CHEMICAL DESIGNATIONS

3.1 **Synonyms:** Ferrous chloride tetrahydrate; Iron dichloride; Iron protochloride
3.2 **Coast Guard Compatibility Classification:** Not listed
3.3 **Chemical Formula:** $FeCl_2 \cdot 4H_2O$
3.4 **IMCO/United Nations Numerical Designation:** Not listed

4. OBSERVABLE CHARACTERISTICS

4.1 **Physical State (as shipped):** Solid
4.2 **Color:** Pale green
4.3 **Odor:** None

5. HEALTH HAZARDS

5.1 **Personal Protective Equipment:** Dust mask; goggles or face shield; rubber gloves
5.2 **Symptoms Following Exposure:** Inhalation of dust irritates nose and throat. Ingestion causes irritation of mouth and stomach. Dust irritates eyes and may cause skin irritation on prolonged contact.
5.3 **Treatment for Exposure:** INHALATION: move to fresh air. INGESTION: if large amounts are swallowed, promptly induce vomiting and get medical help. EYES: flush with plenty of water for at least 15 min.; get medical help *promptly* if ill effects develop. SKIN: wash with soap and water.
5.4 **Toxicity by Inhalation (Threshold Limit Value):** 1 mg/m^3 (as iron)
5.5 **Short-Term Inhalation Limits:** Data not available
5.6 **Toxicity by Ingestion:** Grade 2; LD_{50} 0.5 – 5 g/kg (rat)
5.7 **Late Toxicity:** Data not available
5.8 **Vapor (Gas) Irritant Characteristics:** Data not available
5.9 **Liquid or Solid Irritant Characteristics:** Data not available
5.10 **Odor Threshold:** Data not available

6. FIRE HAZARDS

6.1 **Flash Point:** Not flammable
6.2 **Flammable Limits in Air:** Not flammable
6.3 **Fire Extinguishing Agents:** Not pertinent
6.4 **Fire Extinguishing Agents Not to be Used:** Not pertinent
6.5 **Special Hazards of Combustion Products:** Irritating hydrogen chloride fumes may form in fire.
6.6 **Behavior in Fire:**
6.7 **Ignition Temperature:** Not pertinent
6.8 **Electrical Hazard:** Not pertinent
6.9 **Burning Rate:** Not pertinent

7. CHEMICAL REACTIVITY

7.1 **Reactivity with Water:** No reaction
7.2 **Reactivity with Common Materials:** Solutions may corrode metals.
7.3 **Stability During Transport:** Stable
7.4 **Neutralizing Agents for Acids and Caustics:** Flush with water, rinse with dilute solution of sodium bicarbonate or soda ash.
7.5 **Polymerization:** Not pertinent
7.6 **Inhibitor of Polymerization:** Not pertinent

8. WATER POLLUTION

8.1 **Aquatic Toxicity:** <38 ppm/64 hr/daphnia/toxic/fresh water
8.2 **Waterfowl Toxicity:** Data not available
8.3 **Biological Oxygen Demand (BOD):** Data not available
8.4 **Food Chain Concentration Potential:** None

9. SELECTED MANUFACTURERS

1. Dow Chemical Co.
Midland, Mich. 48640
2. Mallinckrodt Chemical Works
223 Westside Ave.
P. O. Box 384
Jersey City, N. J. 07303
3. K. A. Chemicals, Inc.
2720 Des Plaines Ave.
Des Plaines, Ill. 60018

10. SHIPPING INFORMATION

10.1 **Grades or Purity:** Technical; 35% solution in water
10.2 **Storage Temperature:** Ambient
10.3 **Inert Atmosphere:** No requirement
10.4 **Venting:** Open

11. HAZARD ASSESSMENT CODE

(See Hazard Assessment Handbook, CG 446-3)

SS

12. HAZARD CLASSIFICATIONS

12.1 **Code of Federal Regulations:** Not listed
12.2 **NAS Hazard Rating for Bulk Water Transportation:** Not listed
12.3 **NFPA Hazard Classifications:** Not listed

13. PHYSICAL AND CHEMICAL PROPERTIES

13.1 **Physical State at 15°C and 1 atm:** Solid
13.2 **Molecular Weight:** 198
13.3 **Boiling Point at 1 atm:** Not pertinent (decomposes)
13.4 **Freezing Point:** Not pertinent
13.5 **Critical Temperature:** Not pertinent
13.6 **Critical Pressure:** Not pertinent
13.7 **Specific Gravity:** 1.93 at 20°C (solid)
13.8 **Liquid Surface Tension:** Not pertinent
13.9 **Liquid-Water Interfacial Tension:** Not pertinent
13.10 **Vapor (Gas) Specific Gravity:** Not pertinent
13.11 **Ratio of Specific Heats of Vapor (Gas):** Not pertinent
13.12 **Latent Heat of Vaporization:** Not pertinent
13.13 **Heat of Combustion:** Not pertinent
13.14 **Heat of Decomposition:** Not pertinent
13.15 **Heat of Solution:** −18 Btu/lb = −10 cal/g = -0.42×10^5 J/kg
13.16 **Heat of Polymerization:** Not pertinent

NOTES

FFB	FERROUS FLUOROBORATE

Common Synonyms		
Ferrous borofluoride	Liquid	Yellow-green
	Sinks and mixes with water.	

AVOID CONTACT WITH LIQUID AND VAPOR. KEEP PEOPLE AWAY.
Wear goggles and self-contained breathing apparatus.
Stop discharge if possible.
Isolate and remove discharged material.
Notify local health and pollution control agencies.

Fire	Not flammable. Irritating gases may be produced when heated.
Exposure	**CALL FOR MEDICAL AID.** VAPOR POISONOUS IF INHALED. **If in eyes, hold eyelids open and flush with plenty of water.** **If breathing has stopped, give artificial respiration.** **If breathing is difficult, give oxygen.** LIQUID Irritating to skin and eyes. If swallowed will cause nausea and vomiting. **Remove contaminated clothing and shoes.** **Flush affected areas with plenty of water.** **IF IN EYES, hold eyelids open and flush with plenty of water.** **IF SWALLOWED and victim is CONSCIOUS, have victim drink water or milk and have victim induce vomiting.** **IF SWALLOWED and victim is UNCONSCIOUS OR HAVING CONVULSIONS, do nothing except keep victim warm.**
Water Pollution	Effect of low concentrations on aquatic life is unknown. May be dangerous if it enters water intakes. **Notify local health and wildlife officials.** **Notify operators of nearby water intakes.**

1. RESPONSE TO DISCHARGE
(See Response Methods Handbook, CG 446-4)

Issue warning—water contaminant
Disperse and flush

2. LABELS

No hazard label required by Code of Federal Regulations

3. CHEMICAL DESIGNATIONS

3.1 **Synonyms:** Ferrous borofluoride
3.2 **Coast Guard Compatibility Classification:** Not listed
3.3 **Chemical Formula:** $Fe(BF_4)_2-H_2O$
3.4 **IMCO/United Nations Numerical Designation:** Not listed

4. OBSERVABLE CHARACTERISTICS

4.1 **Physical State (as shipped):** Liquid
4.2 **Color:** Yellow-green
4.3 **Odor:** Data not available

5. HEALTH HAZARDS

5.1 **Personal Protective Equipment:** Goggles or face shield; rubber gloves
5.2 **Symptoms Following Exposure:** Ingestion causes irritation of mouth and stomach. Contact with eyes or skin causes irritation.
5.3 **Treatment for Exposure:** INGESTION: give large amount of water; induce vomiting; get medical attention. EYES: flush with water for at least 15 min. SKIN: flush with water.
5.4 **Toxicity by Inhalation (Threshold Limit Value):** 1 mg/m^3 (as iron)
5.5 **Short-Term Inhalation Limits:** Data not available
5.6 **Toxicity by Ingestion:** Data not available
5.7 **Late Toxicity:** Data not available
5.8 **Vapor (Gas) Irritant Characteristics:** Data not available
5.9 **Liquid or Solid Irritant Characteristics:** Data not available
5.10 **Odor Threshold:** Data not available

6. FIRE HAZARDS

6.1 **Flash Point:** Not flammable
6.2 **Flammable Limits in Air:** Not flammable
6.3 **Fire Extinguishing Agents:** Not pertinent
6.4 **Fire Extinguishing Agents Not to be Used:** Not pertinent
6.5 **Special Hazards of Combustion Products:**
6.6 **Behavior in Fire:** Not pertinent
6.7 **Ignition Temperature:** Not pertinent
6.8 **Electrical Hazard:** Not pertinent
6.9 **Burning Rate:** Not pertinent

7. CHEMICAL REACTIVITY

7.1 **Reactivity with Water:** No reaction
7.2 **Reactivity with Common Materials:**
7.3 **Stability During Transport:** Stable
7.4 **Neutralizing Agents for Acids and Caustics:** Not pertinent
7.5 **Polymerization:** Not pertinent
7.6 **Inhibitor of Polymerization:** Not pertinent

8. WATER POLLUTION

8.1 **Aquatic Toxicity:** Data not available
8.2 **Waterfowl Toxicity:** Data not available
8.3 **Biological Oxygen Demand (BOD):** Data not available
8.4 **Food Chain Concentration Potential:** None

9. SELECTED MANUFACTURERS

1. American Hoechst Corp.
   Chemicals and Plastics Division
   Somerville, N. J. 08876
2. Harstan Chemical Corp.
   1247 38th Street
   Brooklyn, N. Y. 11218
3. Pfaltz and Bauer, Inc.
   375 Fairfield Ave.
   Stamford, Conn. 06902

10. SHIPPING INFORMATION

10.1 **Grades or Purity:** 50% solution in water
10.2 **Storage Temperature:** Ambient
10.3 **Inert Atmosphere:** No requirement
10.4 **Venting:** Open

11. HAZARD ASSESSMENT CODE
(See Hazard Assessment Handbook, CG 446-3)

A-P

12. HAZARD CLASSIFICATIONS

12.1 **Code of Federal Regulations:** Not listed
12.2 **NAS Hazard Rating for Bulk Water Transportation:** Not listed
12.3 **NFPA Hazard Classifications:** Not listed

13. PHYSICAL AND CHEMICAL PROPERTIES

13.1 **Physical State at 15°C and 1 atm:** Liquid
13.2 **Molecular Weight:** 229.5 (solute only)
13.3 **Boiling Point at 1 atm:** Not pertinent (decomposes)
13.4 **Freezing Point:** Not pertinent
13.5 **Critical Temperature:** Not pertinent
13.6 **Critical Pressure:** Not pertinent
13.7 **Specific Gravity:** (est.) > 1.1 at 20°C (liquid)
13.8 **Liquid Surface Tension:** Not pertinent
13.9 **Liquid-Water Interfacial Tension:** Not pertinent
13.10 **Vapor (Gas) Specific Gravity:** Not pertinent
13.11 **Ratio of Specific Heats of Vapor (Gas):** Not pertinent
13.12 **Latent Heat of Vaporization:** Not pertinent
13.13 **Heat of Combustion:** Not pertinent
13.14 **Heat of Decomposition:** Not pertinent
13.15 **Heat of Solution:** Not pertinent
13.16 **Heat of Polymerization:** Not pertinent

NOTES

FOX	FERROUS OXALATE

Common Synonyms			
Oxalic acid, ferrous salt Ferrox Iron protoxalate Ferrous oxalate dihydrate	Solid	Yellow	Odorless
	Sinks in water.		

Stop discharge if possible. Keep people away.
Avoid contact with solid and dust.
Isolate and remove discharged material.
Notify local health and pollution control agencies.

Fire	Not flammable.
Exposure	CALL FOR MEDICAL AID. DUST Irritating to eyes, nose and throat. If inhaled will cause coughing or difficult breathing. If in eyes, hold eyelids open and flush with plenty of water. If breathing has stopped, give artificial respiration. If breathing is difficult, give oxygen. SOLID Irritating to skin and eyes. If swallowed will cause nausea, vomiting or loss of consciousness. Remove contaminated clothing and shoes. Flush affected areas with plenty of water. IF IN EYES, hold eyelids open and flush with plenty of water. IF SWALLOWED and victim is CONSCIOUS, have victim drink water or milk. IF SWALLOWED and victim is UNCONSCIOUS OR HAVING CONVULSIONS, do nothing except keep victim warm.
Water Pollution	Effect of low concentrations on aquatic life is unknown. May be dangerous if it enters water intakes. Notify local health and wildlife officials. Notify operators of nearby water intakes.

1. RESPONSE TO DISCHARGE
(See Response Methods Handbook, CG 446-4)

Should be removed
Chemical and physical treatment

2. LABELS

No hazard label required by Code of Federal Regulations

3. CHEMICAL DESIGNATIONS

3.1 **Synonyms:** Ferrous oxalate dihydrate; Ferrox; Iron protoxalate; Oxalic acid, ferrous salt
3.2 **Coast Guard Compatibility Classification:** Not listed
3.3 **Chemical Formula:** $FeC_2O_4 \cdot 2H_2O$
3.4 **IMCO/United Nations Numerical Designation:** Not listed

4. OBSERVABLE CHARACTERISTICS

4.1 **Physical State (as shipped):** Solid
4.2 **Color:** Pale yellow
4.3 **Odor:** None

5. HEALTH HAZARDS

5.1 **Personal Protective Equipment:** Dust mask; goggles or face shield; protective gloves
5.2 **Symptoms Following Exposure:** Inhalation of dust may cause irritation of nose and throat. Ingestion causes burning pain in throat and stomach; mucous membranes turn white; can also cause vomiting, weak pulse, collapse, and death. Dust irritates eyes and may irritate skin on prolonged contact.
5.3 **Treatment for Exposure:** *(must be prompt)* INHALATION: move to fresh air; get medical attention if any symptoms persist. INGESTION: give immediately by mouth a dilute solution of any soluble calcium salt (calcium lactate, lime water, chalk solution, or even milk); large amounts of calcium are required; give gastric lavage with dilute lime water; consult physician. Watch for edema of the glottis and delayed constriction of esophagus. EYES: flush with water for at least 15 min. SKIN: flush with water.
5.4 **Toxicity by Inhalation (Threshold Limit Value):** Data not available
5.5 **Short-Term Inhalation Limits:** Data not available
5.6 **Toxicity by Ingestion:** Data not available
5.7 **Late Toxicity:** Data not available
5.8 **Vapor (Gas) Irritant Characteristics:** Data not available
5.9 **Liquid or Solid Irritant Characteristics:** Data not available
5.10 **Odor Threshold:** Data not available

6. FIRE HAZARDS

6.1 **Flash Point:** Not flammable
6.2 **Flammable Limits in Air:** Not flammable
6.3 **Fire Extinguishing Agents:** Not pertinent
6.4 **Fire Extinguishing Agents Not to be Used:** Not pertinent
6.5 **Special Hazards of Combustion Products:** Iron fume or iron oxide fume may form in fire.
6.6 **Behavior in Fire:** Not pertinent
6.7 **Ignition Temperature:** Not pertinent
6.8 **Electrical Hazard:** Not pertinent
6.9 **Burning Rate:** Not pertinent

7. CHEMICAL REACTIVITY

7.1 **Reactivity with Water:** No reaction
7.2 **Reactivity with Common Materials:**
7.3 **Stability During Transport:** Stable
7.4 **Neutralizing Agents for Acids and Caustics:** Not pertinent
7.5 **Polymerization:** Not pertinent
7.6 **Inhibitor of Polymerization:** Not pertinent

8. WATER POLLUTION

8.1 **Aquatic Toxicity:** Data not available
8.2 **Waterfowl Toxicity:** Data not available
8.3 **Biological Oxygen Demand (BOD):** Data not available
8.4 **Food Chain Concentration Potential:** None

9. SELECTED MANUFACTURERS

1. Cities Service Co.
3445 Peachtree Road, N.E.
Atlanta, Ga. 30302
2. Pfaltz and Bauer, Inc.
375 Fairfield Ave.
Stamford, Conn. 06902

10. SHIPPING INFORMATION

10.1 **Grades or Purity:** Commercial
10.2 **Storage Temperature:** Ambient
10.3 **Inert Atmosphere:** No requirement
10.4 **Venting:** Open

11. HAZARD ASSESSMENT CODE
(See Hazard Assessment Handbook, CG 446-3)

II

12. HAZARD CLASSIFICATIONS

12.1 **Code of Federal Regulations:** Not listed
12.2 **NAS Hazard Rating for Bulk Water Transportation:** Not listed
12.3 **NFPA Hazard Classifications:** Not listed

13. PHYSICAL AND CHEMICAL PROPERTIES

13.1 **Physical State at 15°C and 1 atm:** Solid
13.2 **Molecular Weight:** 179.9
13.3 **Boiling Point at 1 atm:** Not pertinent (decomposes)
13.4 **Freezing Point:** Not pertinent
13.5 **Critical Temperature:** Not pertinent
13.6 **Critical Pressure:** Not pertinent
13.7 **Specific Gravity:** 2.3 at 20°C (solid)
13.8 **Liquid Surface Tension:** Not pertinent
13.9 **Liquid-Water Interfacial Tension:** Not pertinent
13.10 **Vapor (Gas) Specific Gravity:** Not pertinent
13.11 **Ratio of Specific Heats of Vapor (Gas):** Not pertinent
13.12 **Latent Heat of Vaporization:** Not pertinent
13.13 **Heat of Combustion:** Not pertinent
13.14 **Heat of Decomposition:** Not pertinent
13.15 **Heat of Solution:** Not pertinent
13.16 **Heat of Polymerization:** Not pertinent

NOTES

FRS	FERROUS SULFATE

Common Synonyms			
Iron vitriol Green vitriol Copperas Iron (ous) sulfate	Solid	Green	Odorless
	Sinks and mixes with water.		

Isolate and remove discharged material.
Notify local health and pollution control agencies.

Fire	Not flammable.
Exposure	CALL FOR MEDICAL AID. SOLID If swallowed will cause nausea, vomiting or loss of consciousness. IF SWALLOWED and victim is CONSCIOUS, have victim drink water or milk, and have victim induce vomiting. IF SWALLOWED and victim is UNCONSCIOUS OR HAVING CONVULSIONS, do nothing except keep victim warm.
Water Pollution	HARMFUL TO AQUATIC LIFE IN VERY LOW CONCENTRATIONS. May be dangerous if it enters water intakes. Notify local health and wildlife officials. Notify operators of nearby water intakes.

1. RESPONSE TO DISCHARGE

(See Response Methods Handbook, CG 446-4)

Disperse and flush

2. LABELS

No hazard label required by Code of Federal Regulations

3. CHEMICAL DESIGNATIONS

3.1 **Synonyms:** Copperas
Green vitriol
Iron(ous) sulfate
Iron vitriol

3.2 **Coast Guard Compatibility Classification:** Not applicable

3.3 **Chemical Formula:** $FeSO_4 \cdot 7H_2O$

3.4 **IMCO/United Nations Numerical Designation:** Not listed

4. OBSERVABLE CHARACTERISTICS

4.1 **Physical State (as shipped):** Solid

4.2 **Color:** Green

4.3 **Odor:** Odorless

5. HEALTH HAZARDS

5.1 **Personal Protective Equipment:** Mask if dust is present.

5.2 **Symptoms Following Exposure:** INGESTION: abdominal pain, retching, diarrhea, dehydration, shock, pallor, cyanosis, rapid or weak pulse, shallow respiration, low blood pressure.

5.3 **Treatment for Exposure:** INGESTION: give milk immediately and then induce vomiting by stroking the pharynx with a blunt object such as a spoon handle. Gastric lavage with 1 pint of 5% aqueous solution of mono- or disodium phosphate if promptly available; otherwise use water. Get medical attention.

5.4 **Toxicity by Inhalation (Threshold Limit Value):** Not pertinent

5.5 **Short-Term Inhalation Limits:** Not pertinent

5.6 **Toxicity by Ingestion:** Grade 2; LD_{50} 0.5 to 5 g/kg (rat)

5.7 **Late Toxicity:** Data not available

5.8 **Vapor (Gas) Irritant Characteristics:** Not pertinent

5.9 **Liquid or Solid Irritant Characteristics:** None

5.10 **Odor Threshold:** Not pertinent

6. FIRE HAZARDS

6.1 **Flash Point:** Not flammable

6.2 **Flammable Limits in Air:** Not flammable

6.3 **Fire Extinguishing Agents:** Not pertinent

6.4 **Fire Extinguishing Agents Not to be Used:** Not pertinent

6.5 **Special Hazards of Combustion Products:** Not pertinent

6.6 **Behavior in Fire:** Not pertinent

6.7 **Ignition Temperature:** Not flammable

6.8 **Electrical Hazard:** Not pertinent

6.9 **Burning Rate:** Not flammable

7. CHEMICAL REACTIVITY

7.1 **Reactivity with Water:** No reaction

7.2 **Reactivity with Common Materials:** No reaction

7.3 **Stability During Transport:** Stable

7.4 **Neutralizing Agents for Acids and Caustics:** Not pertinent

7.5 **Polymerization:** Not pertinent

7.6 **Inhibitor of Polymerization:** Not pertinent

8. WATER POLLUTION

8.1 **Aquatic Toxicity:**
100 ppm/4.2 hr/bass/lethal/fresh water
56 ppm/48 hr/shrimp/LC_{50}/salt water

8.2 **Waterfowl Toxicity:** Data not available

8.3 **Biological Oxygen Demand (BOD):** Data not available

8.4 **Food Chain Concentration Potential:** None

9. SELECTED MANUFACTURERS

1. Cosmin Corp.
830 North East Loop
San Antonio, Tex. 78209
2. N L Industries, Inc.
Titanium Pigment Division
St. Louis, Mo. 63111
3. Pfizer, Inc.
C. K. Williams & Co. Division
640 N. 13th St.
Easton, Pa. 18042

10. SHIPPING INFORMATION

10.1 **Grades or Purity:** USP; Commercial

10.2 **Storage Temperature:** Ambient

10.3 **Inert Atmosphere:** No requirement

10.4 **Venting:** Open

11. HAZARD ASSESSMENT CODE

(See Hazard Assessment Handbook, CG 446-3)

SS

12. HAZARD CLASSIFICATIONS

12.1 **Code of Federal Regulations:** Not listed

12.2 **NAS Hazard Rating for Bulk Water Transportation:** Not listed

12.3 **NFPA Hazard Classifications:** Not listed

13. PHYSICAL AND CHEMICAL PROPERTIES

13.1 **Physical State at 15°C and 1 atm:** Solid

13.2 **Molecular Weight:** 169.96

13.3 **Boiling Point at 1 atm:** Not pertinent

13.4 **Freezing Point:** Not pertinent

13.5 **Critical Temperature:** Not pertinent

13.6 **Critical Pressure:** Not pertinent

13.7 **Specific Gravity:** 1.90 at 15°C (solid)

13.8 **Liquid Surface Tension:** Not pertinent

13.9 **Liquid-Water Interfacial Tension:** Not pertinent

13.10 **Vapor (Gas) Specific Gravity:** Not pertinent

13.11 **Ratio of Specific Heats of Vapor (Gas):** Not pertinent

13.12 **Latent Heat of Vaporization:** Not pertinent

13.13 **Heat of Combustion:** Not pertinent

13.14 **Heat of Decomposition:** Not pertinent

13.15 **Heat of Solution:** Not pertinent

13.16 **Heat of Polymerization:** Not pertinent

NOTES

FLUORINE

Common Synonyms	Liquefied gas	Clear to yellow	Very irritating odor
	Liquid sinks and boils in water. Poisonous visible vapor cloud is produced.		

AVOID CONTACT WITH LIQUID AND VAPOR. KEEP PEOPLE AWAY.
Wear chemical protective suit.
Stop discharge if possible.
Evacuate area in case of large discharge.
Notify local health and pollution control agencies.

Fire

Not flammable.
WILL CAUSE FIRE AND REACT VIOLENTLY WITH COMBUSTIBLES.
POISONOUS GAS IS PRODUCED IN FIRE.
Wear chemical protective suit with self-contained breathing apparatus.
Combat fires from behind barrier, with unmanned hose holder or monitor nozzle.
Cool exposed containers with water.

Exposure

CALL FOR MEDICAL AID.

VAPOR
POISONOUS IF INHALED.
Irritating to eyes.
Move to fresh air.
If breathing has stopped, give artificial respiration.
If breathing is difficult, give oxygen.

LIQUID
Will burn skin and eyes.
Will cause frostbite.
Flush affected areas with plenty of water.
IF IN EYES, hold eyelids open and flush with plenty of water.
DO NOT RUB AFFECTED AREAS.

Water Pollution

HARMFUL TO AQUATIC LIFE IN VERY LOW CONCENTRATIONS.
May be dangerous if it enters water intakes.
Notify local health and wildlife officials.
Notify operators of nearby water intakes.

1. RESPONSE TO DISCHARGE

(See Response Methods Handbook, CG 446-4)

Issue warning poison
Restrict access
Evacuate area

2. LABELS

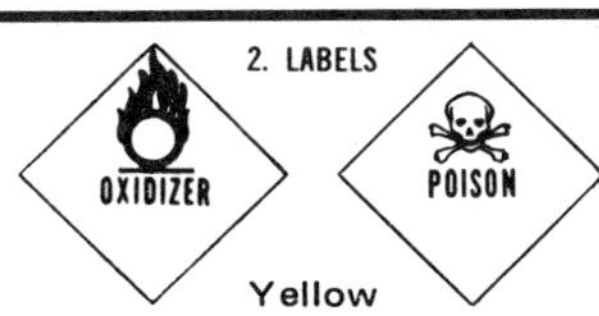

Yellow

3. CHEMICAL DESIGNATIONS

3.1 **Synonyms:** No common synonyms
3.2 **Coast Guard Compatibility Classification:** Not applicable
3.3 **Chemical Formula:** F_2
3.4 **IMCO/United Nations Numerical Designation:** 2.0/1045

4. OBSERVABLE CHARACTERISTICS

4.1 **Physical State (as shipped):** Compressed gas
4.2 **Color:** Pale yellow
4.3 **Odor:** Strong, choking, intense

5. HEALTH HAZARDS

5.1 **Personal Protective Equipment:** Tight-fitting chemical goggles; special clothing, not easily ignited by fluorine gas.
5.2 **Symptoms Following Exposure:** Severe burning of eyes, skin and respiratory system. The burns may develop slowly after exposure.
5.3 **Treatment for Exposure:** Flush all affected parts with water for at least 15 min. Do NOT use ointments. Administer artificial respiration and oxygen if required.
5.4 **Toxicity by Inhalation (Threshold Limit Value):** 1 ppm
5.5 **Short-Term Inhalation Limits:** 0.5 ppm for 5 min.
5.6 **Toxicity by Ingestion:** Not pertinent
5.7 **Late Toxicity:** Severe burns may develop slowly after exposure.
5.8 **Vapor (Gas) Irritant Characteristics:** Vapors cause severe irritation of eye and throat and can cause eye and lung injury. They cannot be tolerated even at low concentrations.
5.9 **Liquid or Solid Irritant Characteristics:** Severe skin irritant. Causes second- and third-degree burns on short contact and is very injurious to the eyes.
5.10 **Odor Threshold:** 0.035 ppm

6. FIRE HAZARDS

6.1 **Flash Point:** Not flammable
6.2 **Flammable Limits in Air:** Not flammable
6.3 **Fire Extinguishing Agents:** Not pertinent
6.4 **Fire Extinguishing Agents Not to be Used:** Do not direct water onto fluorine leaks.
6.5 **Special Hazards of Combustion Products:** Toxic gases generated in fires involving fluorine.
6.6 **Behavior in Fire:** Dangerously reactive gas. Ignites most combustibles.
6.7 **Ignition Temperature:** Not flammable
6.8 **Electrical Hazard:** Not pertinent
6.9 **Burning Rate:** Not flammable

7. CHEMICAL REACTIVITY

7.1 **Reactivity with Water:** Reacts with water to form hydrogen fluoride, oxygen, and oxygen difluoride.
7.2 **Reactivity with Common Materials:** Reacts violently with all combustible materials, except the metal cylinders in which it is shipped.
7.3 **Stability During Transport:** Stable
7.4 **Neutralizing Agents for Acids and Caustics:** Not pertinent
7.5 **Polymerization:** Not pertinent
7.6 **Inhibitor of Polymerization:** Not pertinent

8. WATER POLLUTION

8.1 **Aquatic Toxicity:**
2.3 ppm/*/trout/TL_m/fresh water
*Time period not specified.
8.2 **Waterfowl Toxicity:** Data not available
8.3 **Biological Oxygen Demand (BOD):** None
8.4 **Food Chain Concentration Potential:** None

9. SELECTED MANUFACTURERS

1. Air Products and Chemicals, Inc.
   Allentown, Pa. 18105
2. Allied Chemical Corp.
   Specialty Chemicals Division
   Baton Rouge, La. 70821
3. Union Carbide Corp.
   Nuclear Division
   270 Park Ave.
   New York, N. Y. 10017

10. SHIPPING INFORMATION

10.1 **Grades or Purity:** 98%
10.2 **Storage Temperature:** Ambient
10.3 **Inert Atmosphere:** No requirement
10.4 **Venting:** Safety relief

11. HAZARD ASSESSMENT CODE

(See Hazard Assessment Handbook, CG 446-3)

A-C

12. HAZARD CLASSIFICATIONS

12.1 **Code of Federal Regulations:** Non-Flammable Gas
12.2 **NAS Hazard Rating for Bulk Water Transportation:** Not listed
12.3 **NFPA Hazard Classifications:**

Category	Classification
Health Hazard (Blue)	4
Flammability (Red)	0
Reactivity (Yellow)	3 W, oxy

13. PHYSICAL AND CHEMICAL PROPERTIES

13.1 **Physical State at 15°C and 1 atm:** Gas
13.2 **Molecular Weight:** 37.99
13.3 **Boiling Point at 1 atm:** $-306°F = -188°C = 85°K$
13.4 **Freezing Point:** $-362°F = -219°C = 54°K$
13.5 **Critical Temperature:** $-199.5°F = -128.6°C = 144.6°K$
13.6 **Critical Pressure:** 809.7 psia = 55.08 atm = 5.58 MN/m^2
13.7 **Specific Gravity:** 1.5 at −188°C (liquid)
13.8 **Liquid Surface Tension:** Not pertinent
13.9 **Liquid-Water Interfacial Tension:** Not pertinent
13.10 **Vapor (Gas) Specific Gravity:** Not pertinent
13.11 **Ratio of Specific Heats of Vapor (Gas):** 1.362
13.12 **Latent Heat of Vaporization:** 71.6 Btu/lb = 39.8 cal/g = 1.67×10^5 J/kg
13.13 **Heat of Combustion:** Not pertinent
13.14 **Heat of Decomposition:** Not pertinent
13.15 **Heat of Solution:** Not pertinent
13.16 **Heat of Polymerization:** Not pertinent

NOTES

REVISED 1978

FSL	FLUOSILICIC ACID

Common Synonyms			
Hydrofluosilicic acid Hydrogen hexafluorosilicate Hexafluosilicic acid Silicofluoric acid Fluorosilicic acid Sand acid	Liquid	Colorless	Sharp unpleasant odor
	Sinks and mixes with water.		

Avoid contact with liquid and vapor. Keep people away.
Stop discharge if possible.
Isolate and remove discharged material.
Notify local health and pollution control agencies.

Fire	Not Flammable. Irritating gases may be produced when heated.
Exposure	CALL FOR MEDICAL AID. VAPOR Irritating to eyes, nose and throat. If inhaled will cause coughing or difficult breathing. **If in eyes, hold eyelids open and flush with plenty of water.** **If breathing has stopped, give artificial respiration.** **If breathing is difficult, give oxygen.** LIQUID Will burn skin and eyes. If swallowed will cause nausea. **Remove contaminated clothing and shoes.** **Flush affected areas with plenty of water.** **IF IN EYES, hold eyelids open and flush with plenty of water.** **IF SWALLOWED and victim is CONSCIOUS, have victim drink water or milk.** **IF SWALLOWED and victim is UNCONSCIOUS OR HAVING CONVULSIONS, do nothing except keep victim warm.**
Water Pollution	Effect of low concentrations on aquatic life is unknown. May be dangerous if it enters water intakes. **Notify local health and wildlife officials.** **Notify operators of nearby water intakes.**

1. RESPONSE TO DISCHARGE

(See Response Methods Handbook, CG 446-4)

Issue warning—corrosive
Restrict access
Disperse and flush

2. LABEL

3. CHEMICAL DESIGNATIONS

3.1 **Synonyms:** Fluorosilicic acid; Hexafluosilicic acid; Hydrofluosilicic acid; Hydrogen hexafluorosilicate; Sand acid; Silicofluoric acid

3.2 **Coast Guard Compatibility Classification:** Not listed

3.3 **Chemical Formula:** $H_2SiF_6-H_2O$

3.4 **IMCO/United Nations Numerical Designation:** 8/1778

4. OBSERVABLE CHARACTERISTICS

4.1 **Physical State (as shipped):** Liquid

4.2 **Color:** Transparent, straw colored; colorless

4.3 **Odor:** Acrid, sharp

5. HEALTH HAZARDS

5.1 **Personal Protective Equipment:** Rubber gloves; safety glasses; protective clothing

5.2 **Symptoms Following Exposure:** Inhalation of vapor produces severe corrosive effect on mucous membrane. Ingestion causes severe burns of mouth and stomach. Contact with liquid or vapor causes severe burns of eyes and skin.

5.3 **Treatment for Exposure:** INHALATION: remove victim to fresh air; get medical attention. INGESTION: give large amounts of water; do NOT induce vomiting. EYES: immediately wash with water for 15 min.; call physician. SKIN: wash affected parts with water; treat as for hydrogen fluoride burn with iced benzalkonium chloride soaks.

5.4 **Toxicity by Inhalation (Threshold Limit Value):** Data not available

5.5 **Short-Term Inhalation Limits:** Data not available

5.6 **Toxicity by Ingestion:** Data not available

5.7 **Late Toxicity:** Data not available

5.8 **Vapor (Gas) Irritant Characteristics:** Data not available

5.9 **Liquid or Solid Irritant Characteristics:** Data not available

5.10 **Odor Threshold:** Data not available

6. FIRE HAZARDS

6.1 **Flash Point:** Not flammable

6.2 **Flammable Limits in Air:** Not flammable

6.3 **Fire Extinguishing Agents:** Not pertinent

6.4 **Fire Extinguishing Agents Not to be Used:** Not pertinent

6.5 **Special Hazards of Combustion Products:** Irritating fumes of hydrogen fluoride may form in fire.

6.6 **Behavior in Fire:** Not pertinent

6.7 **Ignition Temperature:** Not pertinent

6.8 **Electrical Hazard:** Not pertinent

6.9 **Burning Rate:** Not pertinent

7. CHEMICAL REACTIVITY

7.1 **Reactivity with Water:** No reaction

7.2 **Reactivity with Common Materials:** Will corrode most metals, producing flammable hydrogen gas, which may collect in enclosed spaces.

7.3 **Stability During Transport:** Stable

7.4 **Neutralizing Agents for Acids and Caustics:** Flush with water, rinse with dilute solution of sodium carbonate or soda ash.

7.5 **Polymerization:** Not pertinent

7.6 **Inhibitor of Polymerization:** Not pertinent

8. WATER POLLUTION

8.1 **Aquatic Toxicity:** Data not available

8.2 **Waterfowl Toxicity:** Data not available

8.3 **Biological Oxygen Demand (BOD):** None

8.4 **Food Chain Concentration Potential:** None

9. SELECTED MANUFACTURERS

1. W. R. Grace & Co.
   Agricultural Chemicals Group
   Baltimore, Md. 21203
2. The Harshaw Chemical Co.
   1945 E. 97 St.
   Cleveland, Ohio 44106
3. Pfaltz and Bauer, Inc.
   375 Fairfield Ave.
   Stamford, Conn. 06902

10. SHIPPING INFORMATION

10.1 **Grades or Purity:** 22–30% solutions in water

10.2 **Storage Temperature:** Ambient

10.3 **Inert Atmosphere:** No requirement

10.4 **Venting:** Open

11. HAZARD ASSESSMENT CODE

(See Hazard Assessment Handbook, CG 446-3)

A-P

12. HAZARD CLASSIFICATIONS

12.1 **Code of Federal Regulations:** Corrosive liquid

12.2 **NAS Hazard Rating for Bulk Water Transportation:**

Category	Rating
Fire	0
Health	
Vapor Irritant	0
Liquid or Solid Irritant	3
Poisons	3
Water Pollution	
Human Toxicity	
Aquatic Toxicity	3
Aesthetic Effect	0
Reactivity	
Other Chemicals	1
Water	0
Self-Reaction	0

12.3 **NFPA Hazard Classifications:**

Category	Classification
Health Hazard (Blue)	3
Flammability (Red)	0
Reactivity (Yellow)	0

13. PHYSICAL AND CHEMICAL PROPERTIES

13.1 **Physical State at 15°C and 1 atm:** Liquid

13.2 **Molecular Weight:** 144.09 (solute only)

13.3 **Boiling Point at 1 atm:** (water) $\sim 212°F = \sim 100°C = \sim 373°K$

13.4 **Freezing Point:** (typical) -24 to $-4°F = -31$ to $-20°C = 242$ to $253°K$

13.5 **Critical Temperature:** Not pertinent

13.6 **Critical Pressure:** Not pertinent

13.7 **Specific Gravity:** (approx.) 1.3 at 25°C (liquid)

13.8 **Liquid Surface Tension:** Not pertinent

13.9 **Liquid-Water Interfacial Tension:** Not pertinent

13.10 **Vapor (Gas) Specific Gravity:** Not pertinent

13.11 **Ratio of Specific Heats of Vapor (Gas):** Not pertinent

13.12 **Latent Heat of Vaporization:** Not pertinent

13.13 **Heat of Combustion:** Not pertinent

13.14 **Heat of Decomposition:** Not pertinent

13.15 **Heat of Solution:** Not pertinent

13.16 **Heat of Polymerization:** Not pertinent

NOTES

FSA	# FLUOSULFONIC ACID

Common Synonyms			
Fluorosulfuric acid Fluorosulfonic acid	Liquid	Colorless to light yellow	Choking odor
	Reacts violently with water. Irritating mist and gas are produced on contact with water.		

AVOID CONTACT WITH LIQUID, GAS AND MIST. KEEP PEOPLE AWAY.
Wear goggles, self-contained breathing apparatus, and rubber overclothing (including gloves).
Stop discharge if possible.
Isolate and remove discharged material.
Notify local health and pollution control agencies.

Fire	Not flammable. **DO NOT USE WATER OR FOAM ON ADJACENT FIRES.**
Exposure	**Call for medical aid.** VAPOR OR MIST Irritating to eyes, nose and throat. Harmful if inhaled. **Move victim to fresh air.** **If breathing is difficult, give oxygen.** LIQUID Will burn skin and eyes. Harmful if swallowed. **Remove contaminated clothing and shoes.** **Flush affected areas with plenty of water.** **IF IN EYES, hold eyelids open and flush with plenty of water.** **IF SWALLOWED and victim is CONSCIOUS, have victim drink water or milk.** **DO NOT INDUCE VOMITING.**
Water Pollution	Effect of low concentrations on aquatic life is unknown. May be dangerous if it enters water intakes. **Notify local health and wildlife officials.** **Notify operators of nearby water intakes.**

1. RESPONSE TO DISCHARGE

(See Response Methods Handbook, CG 446-4)

Issue warning—corrosive, air contaminant, water contaminant
Restrict access
Evacuate area
Disperse and flush with care

2. LABEL

3. CHEMICAL DESIGNATIONS

3.1 **Synonyms:** Fluorosulfuric acid; Fluorosulfonic acid
3.2 **Coast Guard Compatibility Classification:** Not applicable
3.3 **Chemical Formula:** FSO_3H
3.4 **IMCO/United Nations Numerical Designation:** 8/1777

4. OBSERVABLE CHARACTERISTICS

4.1 **Physical State (as shipped):** Liquid
4.2 **Color:** Somewhat cloudy, colorless to slightly yellow
4.3 **Odor:** Choking, irritating

5. HEALTH HAZARDS

5.1 **Personal Protective Equipment:** Rubber gloves, shoes, and clothing; goggles and face shield; acid-type canister mask or air-line mask.
5.2 **Symptoms Following Exposure:** Inhalation of fumes causes severe irritation of nose and throat. Contact of liquid with eyes or skin causes very severe burns. Ingestion causes very severe burns of mouth and stomach.
5.3 **Treatment for Exposure:** Get medical attention quickly following all exposures to this compound. INHALATION: remove victim to fresh air; if he is unconscious, give artificial respiration. EYES: flush with water until medical help arrives. SKIN: flush with water until medical help arrives; soak burned area in strong Epsom salt solution; pay particular attention to area around fingernails. INGESTION: give large amounts of water.
5.4 **Toxicity by Inhalation (Threshold Limit Value):** Data not available
5.5 **Short-Term Inhalation Limits:** Data not available
5.6 **Toxicity by Ingestion:** Data not available
5.7 **Late Toxicity:** Data not available
5.8 **Vapor (Gas) Irritant Characteristics:** Vapors cause severe irritation of eyes and throat and can cause eye and lung injury. They cannot be tolerated even at low concentrations.
5.9 **Liquid or Solid Irritant Characteristics:** Severe skin irritant. Causes second- and third-degree burns on short contact and is very injurious to the eyes.
5.10 **Odor Threshold:** Data not available

6. FIRE HAZARDS

6.1 **Flash Point:** Not flammable
6.2 **Flammable Limits in Air:** Not flammable
6.3 **Fire Extinguishing Agents:** Not pertinent
6.4 **Fire Extinguishing Agents Not to be Used:** Do not use water or foam on adjacent fires.
6.5 **Special Hazards of Combustion Products:** Toxic and irritating fumes of hydrogen fluoride and sulfuric acid may form in fires.
6.6 **Behavior in Fire:** Contact with water applied to adjacent fires produces toxic, irritating fumes of hydrogen fluoride.
6.7 **Ignition Temperature:** Not pertinent
6.8 **Electrical Hazard:** Not pertinent
6.9 **Burning Rate:** Not pertinent

7. CHEMICAL REACTIVITY

7.1 **Reactivity with Water:** Reacts violently with water to generate hydrogen fluoride and sulfuric acid mists.
7.2 **Reactivity with Common Materials:** Reacts with metals, generating flammable hydrogen.
7.3 **Stability During Transport:** Stable
7.4 **Neutralizing Agents for Acids and Caustics:** Flood with water, rinse with sodium bicarbonate or lime solution.
7.5 **Polymerization:** Not pertinent
7.6 **Inhibitor of Polymerization:** Not pertinent

8. WATER POLLUTION

8.1 **Aquatic Toxicity:** Data not available
8.2 **Waterfowl Toxicity:** Data not available
8.3 **Biological Oxygen Demand (BOD):** Data not available
8.4 **Food Chain Concentration Potential:** None

9. SELECTED MANUFACTURERS

1. Allied Chemical Corporation
   Specialty Chemicals Division
   P. O. Box 1087R
   Morristown, N. J. 07960
2. PCR, Inc.
   P. O. Box 1466
   Gainesville, Fla. 32602

10. SHIPPING INFORMATION

10.1 **Grades or Purity:** 98.5%
10.2 **Storage Temperature:** Ambient
10.3 **Inert Atmosphere:** No requirement
10.4 **Venting:** Pressure-vacuum, with protection from moisture in air.

11. HAZARD ASSESSMENT CODE

(See Hazard Assessment Handbook, CG 446-3)

A-O

12. HAZARD CLASSIFICATIONS

12.1 **Code of Federal Regulations:** Corrosive
12.2 **NAS Hazard Rating for Bulk Water Transportation:**

Category	Rating
Fire	0
Health	
Vapor Irritant	4
Liquid or Solid Irritant	4
Poisons	3
Water Pollution	
Human Toxicity	
Aquatic Toxicity	2
Aesthetic Effect	2
Reactivity	
Other Chemicals	4
Water	4
Self-Reaction	0

12.3 **NFPA Hazard Classifications:** Not listed

13. PHYSICAL AND CHEMICAL PROPERTIES

13.1 **Physical State at 15°C and 1 atm:** Liquid
13.2 **Molecular Weight:** 100.07
13.3 **Boiling Point at 1 atm:** 324.9°F = 162.7°C = 435.9°K
13.4 **Freezing Point:** −125.1°F = −87.3°C = 185.9°K
13.5 **Critical Temperature:** Not pertinent
13.6 **Critical Pressure:** Not pertinent
13.7 **Specific Gravity:** 1.73 at 25°C (liquid)
13.8 **Liquid Surface Tension:** Data not available
13.9 **Liquid-Water Interfacial Tension:** Not pertinent
13.10 **Vapor (Gas) Specific Gravity:** Not pertinent
13.11 **Ratio of Specific Heats of Vapor (Gas):** Not pertinent
13.12 **Latent Heat of Vaporization:** 170 Btu/lb = 94 cal/g = 3.9×10^5 J/kg
13.13 **Heat of Combustion:** Not pertinent
13.14 **Heat of Decomposition:** Not pertinent
13.15 **Heat of Solution:** Data not available
13.16 **Heat of Polymerization:** Not pertinent

NOTES

FMS — FORMALDEHYDE SOLUTION

Common Synonyms	Watery liquid	Colorless	Irritating odor
Formalith Fyde Formalin Formic aldehyde solution Methanal solution	Sinks and mixes with water.		

Avoid contact with liquid.
Wear goggles, self-contained breathing apparatus, and rubber overclothing (including gloves).
Stop discharge if possible.
Call fire department.
Isolate and remove discharged material.
Notify local health and pollution control agencies.

Fire

Combustible.
Wear goggles, self-contained breathing apparatus, and rubber overclothing (including gloves).
Extinguish with water, dry chemical, alcohol foam, or carbon dioxide.
Cool exposed containers with water.

Exposure

CALL FOR MEDICAL AID.

LIQUID
Will burn skin and eyes.
If swallowed, will cause nausea, vomiting or loss of consciousness.
Remove contaminated clothing and shoes.
Flush affected areas with plenty of water.
IF IN EYES, hold eyelids open and flush with plenty of water.
IF SWALLOWED and victim is CONSCIOUS, have victim drink water or milk, and have victim induce vomiting.
IF SWALLOWED and victim is UNCONSCIOUS OR HAVING CONVULSIONS, do nothing except keep victim warm.

Water Pollution

HARMFUL TO AQUATIC LIFE IN VERY LOW CONCENTRATIONS.
May be dangerous if it enters water intakes.
Notify local health and wildlife officials.
Notify operators of nearby water intakes.

1. RESPONSE TO DISCHARGE

(See Response Methods Handbook, CG 446-4)

Issue warning—water contaminant
Disperse and flush

2. LABELS

No hazard label required by Code of Federal Regulations

3. CHEMICAL DESIGNATIONS

3.1 **Synonyms:** Formalin; Fyde; Formalith; Methanal; Formic aldehyde
3.2 **Coast Guard Compatibility Classification:** Aldehyde
3.3 **Chemical Formula:** $HCHO/H_2O/CH_3OH$
3.4 **IMCO/United Nations Numerical Designation:** 3.3/1198

4. OBSERVABLE CHARACTERISTICS

4.1 **Physical State (as shipped):** Liquid
4.2 **Color:** Colorless
4.3 **Odor:** Pungent, irritating; characteristic, pungent

5. HEALTH HAZARDS

5.1 **Personal Protective Equipment:** Self-contained breathing apparatus; chemical goggles; protective clothing; synthetic rubber or plastic gloves.
5.2 **Symptoms Following Exposure:** INHALATION: vapors are irritating and will cause coughing, chest pain, nausea, and vomiting. INGESTION: causes nausea, vomiting, abdominal pain, and collapse. Contact with skin and eyes causes severe irritation.
5.3 **Treatment for Exposure:** INHALATION: remove victim to fresh air; give oxygen if breathing is difficult; call a physician. INGESTION: induce vomiting at once and repeat until vomit is clear; then give milk or raw egg and call a physician. SKIN OR EYES: flush immediately with plenty of water for at least 15 min.; remove contaminated clothing; call a physician for eyes.
5.4 **Toxicity by Inhalation (Threshold Limit Value):** 2 ppm
5.5 **Short-Term Inhalation Limits:** 5 ppm for 5 min., 3 ppm for 60 min. (tentative).
5.6 **Toxicity by Ingestion:** (Formaldehyde solution) Grade 2; LD_{50} 0.5 to 5 g/kg
5.7 **Late Toxicity:** None
5.8 **Vapor (Gas) Irritant Characteristics:** Vapor is moderately irritating such that personnel will not usually tolerate moderate or high concentrations.
5.9 **Liquid or Solid Irritant Characteristics:** Causes smarting of the skin and first-degree burns on short exposure. May cause secondary burns on long exposure.
5.10 **Odor Threshold:** 0.8 ppm

6. FIRE HAZARDS

6.1 **Flash Point:** (37% formaldehyde) Methanol-free: 182°F C.C. 15% methanol: 122°F C.C.
6.2 **Flammable Limits in Air:** 7.0%–73%
6.3 **Fire Extinguishing Agents:** Water, dry chemical, carbon dioxide, or alcohol foam
6.4 **Fire Extinguishing Agents Not to be Used:** Data not available
6.5 **Special Hazards of Combustion Products:** Toxic vapors are generated.
6.6 **Behavior in Fire:** Not pertinent
6.7 **Ignition Temperature:** 806°F
6.8 **Electrical Hazard:** Not pertinent
6.9 **Burning Rate:** Not pertinent

7. CHEMICAL REACTIVITY

7.1 **Reactivity with Water:** No reaction
7.2 **Reactivity with Common Materials:** No reaction
7.3 **Stability During Transport:** Stable
7.4 **Neutralizing Agents for Acids and Caustics:** Not pertinent
7.5 **Polymerization:** Not pertinent
7.6 **Inhibitor of Polymerization:** Not pertinent

8. WATER POLLUTION

8.1 **Aquatic Toxicity:** (formaldehyde)
25 mg/l/96 hr/channel cat/TL_m/fresh water
32 ppm/24 hr/catfish/TL_m/fresh water
100-330 ppm/48 hr/flounder/TL_m/salt water
8.2 **Waterfowl Toxicity:** Data not available
8.3 **Biological Oxygen Demand (BOD):** 37%, 5 days; 47% (theor.), 5 days
8.4 **Food Chain Concentration Potential:** None

9. SELECTED MANUFACTURERS

1. Borden, Inc.
Borden Chemical Division
50 W. Broad Street
Columbus, Ohio 43215
2. Celanese Corp.
Celanese Chemical Co. Division
245 Park Ave.
New York, N. Y. 10017
3. E. I. duPont de Nemours & Co., Inc.
Industrial & Biochemicals Dept.
Wilmington, Del. 19898

10. SHIPPING INFORMATION

10.1 **Grades or Purity:** 30-50% formaldehyde in water containing 0-15% methyl alcohol
10.2 **Storage Temperature:** Ambient
10.3 **Inert Atmosphere:** No requirement
10.4 **Venting:** Pressure-vacuum

11. HAZARD ASSESSMENT CODE

(See Hazard Assessment Handbook, CG 446-3)

A-P-Q

12. HAZARD CLASSIFICATIONS

12.1 **Code of Federal Regulations:** Combustible Liquid
12.2 **NAS Hazard Rating for Bulk Water Transportation:**

Category	Rating
Fire	2
Health	
Vapor Irritant	3
Liquid or Solid Irritant	2
Poisons	3
Water Pollution	
Human Toxicity	3
Aquatic Toxicity	3
Aesthetic Effect	2
Reactivity	
Other Chemicals	2
Water	0
Self-Reaction	1

12.3 **NFPA Hazard Classifications:**

Category	Classification
Health Hazard (Blue)	2
Flammability (Red)	2
Reactivity (Yellow)	0

13. PHYSICAL AND CHEMICAL PROPERTIES

13.1 **Physical State at 15°C and 1 atm:** Liquid
13.2 **Molecular Weight:** 18 30
13.3 **Boiling Point at 1 atm:** Not pertinent
13.4 **Freezing Point:** Not pertinent
13.5 **Critical Temperature:** Not pertinent
13.6 **Critical Pressure:** Not pertinent
13.7 **Specific Gravity:** 1.1 at 25°C (liquid)
13.8 **Liquid Surface Tension:** Not pertinent
13.9 **Liquid-Water Interfacial Tension:** Not pertinent
13.10 **Vapor (Gas) Specific Gravity:** Not pertinent
13.11 **Ratio of Specific Heats of Vapor (Gas):** Not pertinent
13.12 **Latent Heat of Vaporization:** Not pertinent
13.13 **Heat of Combustion:** Not pertinent
13.14 **Heat of Decomposition:** Not pertinent
13.15 **Heat of Solution:** (est.) −9 Btu/lb = −5 cal/g = -0.2×10^5 J/kg
13.16 **Heat of Polymerization:** Not pertinent

NOTES

FMA	FORMIC ACID

Common Synonyms			
Methanoic acid Formylic acid	Liquid	Colorless	Penetrating odor
	Sinks and mixes with water. Freezing point is 47° F.		

AVOID CONTACT WITH LIQUID. Keep people away.
Wear goggles, self-contained breathing apparatus, and rubber overclothing (including gloves).
Stop discharge if possible.
Call fire department.
Isolate and remove discharged material.
Notify local health and pollution control agencies.

Fire	Combustible. **Wear goggles, self-contained breathing apparatus, and rubber overclothing (including gloves).** **Extinguish with water, dry chemical, alcohol foam, or carbon dioxide.** **Cool exposed containers with water.**
Exposure	**CALL FOR MEDICAL AID.** LIQUID Will burn skin and eyes. Harmful if swallowed. **Remove contaminated clothing and shoes.** **Flush affected areas with plenty of water.** **IF IN EYES, hold eyelids open and flush with plenty of water.** **IF SWALLOWED and victim is CONSCIOUS, have victim drink water or milk.** **IF SWALLOWED and victim is UNCONSCIOUS OR HAVING CONVULSIONS, do nothing except keep victim warm.** **DO NOT INDUCE VOMITING.**
Water Pollution	Dangerous to aquatic life in high concentrations. May be dangerous if it enters water intakes. **Notify local health and wildlife officials.** **Notify operators of nearby water intakes.**

1. RESPONSE TO DISCHARGE

(See Response Methods Handbook, CG 446-4)

Issue warning—corrosive
Restrict access
Disperse and flush

2. LABEL

3. CHEMICAL DESIGNATIONS

3.1 **Synonyms:** Methanoic acid; Formylic acid
3.2 **Coast Guard Compatibility Classification:** Organic acid
3.3 **Chemical Formula:** HCOOH
3.4 **IMCO/United Nations Numerical Designation:** 8.0/1779

4. OBSERVABLE CHARACTERISTICS

4.1 **Physical State (as shipped):** Liquid
4.2 **Color:** Colorless
4.3 **Odor:** Pungent, penetrating

5. HEALTH HAZARDS

5.1 **Personal Protective Equipment:** Self-contained breathing apparatus; chemical goggles or face shield; rubber suit, gloves, and shoes.
5.2 **Symptoms Following Exposure:** Liquid causes skin and eye burns. Vapors are irritating and painful to breathe. Vapor exposure may cause nausea and vomiting.
5.3 **Treatment for Exposure:** INHALATION: move victim to fresh air; give oxygen if breathing is difficult; call a physician. INGESTION: do NOT induce vomiting; give water or milk. SKIN OR EYES: immediately flush affected area with plenty of water for at least 15 min.; get medical care for eyes.
5.4 **Toxicity by Inhalation (Threshold Limit Value):** 5 ppm
5.5 **Short-Term Inhalation Limits:** Data not available
5.6 **Toxicity by Ingestion:** Grade 2; oral rat LD_{50} = 1.21 g/kg
5.7 **Late Toxicity:** None
5.8 **Vapor (Gas) Irritant Characteristics:** Vapor is moderately irritating such that personnel will not usually tolerate moderate or high vapor concentrations.
5.9 **Liquid or Solid Irritant Characteristics:** Fairly severe skin irritant; may cause pain and second-degree burns after a few minutes' contact.
5.10 **Odor Threshold:** Data not available

6. FIRE HAZARDS

6.1 **Flash Point:** 138°F O.C.
6.2 **Flammable Limits in Air:** 18%–57%
6.3 **Fire Extinguishing Agents:** Water, carbon dioxide, dry chemical, or alcohol foam
6.4 **Fire Extinguishing Agents Not to be Used:** Not pertinent
6.5 **Special Hazards of Combustion Products:** Toxic vapor generated in fires
6.6 **Behavior in Fire:** Not pertinent
6.7 **Ignition Temperature:** 1114°F
6.8 **Electrical Hazard:** Not pertinent
6.9 **Burning Rate:** 0.5 mm/min.

7. CHEMICAL REACTIVITY

7.1 **Reactivity with Water:** No reaction
7.2 **Reactivity with Common Materials:** No reaction
7.3 **Stability During Transport:** Stable
7.4 **Neutralizing Agents for Acids and Caustics:** Flush with water, then neutralize with lime.
7.5 **Polymerization:** Not pertinent
7.6 **Inhibitor of Polymerization:** Not pertinent

8. WATER POLLUTION

8.1 **Aquatic Toxicity:**
175 mg/l /24 hr/bluegill/TL_m/fresh water
120 ppm/48 hr/daphnia/TL_m/fresh water
8.2 **Waterfowl Toxicity:** Data not available
8.3 **Biological Oxygen Demand (BOD):** 2%, 5 days; 40% (theor.), 5 days
8.4 **Food Chain Concentration Potential:** None

9. SELECTED MANUFACTURERS

1. E. I. duPont de Nemours & Co., Inc.
Industrial & Biochemicals Dept.
Wilmington, Del. 19898
2. Middleboro Industries
Orono, Maine 04473
3. Union Carbide Corp.
Chemicals and Plastics Division
270 Park Ave.
New York, N. Y. 10017

10. SHIPPING INFORMATION

10.1 **Grades or Purity:** Technical, pharmaceutical: 85–95%
10.2 **Storage Temperature:** Ambient
10.3 **Inert Atmosphere:** No requirement
10.4 **Venting:** Pressure-vacuum

11. HAZARD ASSESSMENT CODE

(See Hazard Assessment Handbook, CG 446-3)

A-P-Q

12. HAZARD CLASSIFICATIONS

12.1 **Code of Federal Regulations:** Corrosive material
12.2 **NAS Hazard Rating for Bulk Water Transportation:**

Category	Rating
Fire	2
Health	
Vapor Irritant	3
Liquid or Solid Irritant	3
Poisons	3
Water Pollution	
Human Toxicity	3
Aquatic Toxicity	2
Aesthetic Effect	2
Reactivity	
Other Chemicals	2
Water	0
Self-Reaction	0

12.3 **NFPA Hazard Classifications:**

Category	Classification
Health Hazard (Blue)	3
Flammability (Red)	2
Reactivity (Yellow)	0

13. PHYSICAL AND CHEMICAL PROPERTIES

13.1 **Physical State at 15°C and 1 atm:** Liquid
13.2 **Molecular Weight:** 46.03
13.3 **Boiling Point at 1 atm:** 214°F = 101°C = 374°K
13.4 **Freezing Point:** 47.1°F = 8.4°C = 281.6°K
13.5 **Critical Temperature:** Not pertinent
13.6 **Critical Pressure:** Not pertinent
13.7 **Specific Gravity:** 1.22 at 20°C (liquid)
13.8 **Liquid Surface Tension:** 38 dynes/cm = 0.038 N/m at 15°C
13.9 **Liquid-Water Interfacial Tension:** Not pertinent
13.10 **Vapor (Gas) Specific Gravity:** Not pertinent
13.11 **Ratio of Specific Heats of Vapor (Gas):** 1.228
13.12 **Latent Heat of Vaporization:** 216 Btu/lb = 120 cal/g = 5.02×10^5 J/kg
13.13 **Heat of Combustion:** −2045 Btu/lb = −1136 cal/g = -47.56×10^5 J/kg
13.14 **Heat of Decomposition:** Not pertinent
13.15 **Heat of Solution:** (est.) −26 Btu/lb = −14 cal/g = -0.6×10^5 J/kg
13.16 **Heat of Polymerization:** Not pertinent

NOTES

REVISED 1978

FUM — FUMARIC ACID

Common Synonyms			
Trans-Butenedioic acid Allomaleic acid Trans-1,2,-ethylenedicarboxylic acid Boletic acid Lichenic acid	Solid	White	Odorless
	Sinks and mixes with water.		

Stop discharge if possible. Keep people away.
Shut off ignition sources. Call fire department.
Avoid contact with solid and dust.
Isolate and remove discharged material.
Notify local health and pollution control agencies.

Fire

Combustible.
Dust cloud may explode if ignited in an enclosed area.
Irritating gases may be produced when heated.
Extinguish with water, dry chemicals, foam, or carbon dioxide.

Exposure

CALL FOR MEDICAL AID

DUST
Irritating to eyes, nose and throat.
If inhaled will cause coughing or difficult breathing.
If in eyes, hold eyelids open and flush with plenty of water.
If breathing has stopped, give artificial respiration.
If breathing is difficult, give oxygen.

SOLID
Irritating to skin and eyes.
Remove contaminated clothing and shoes.
Flush affected areas with plenty of water.
IF IN EYES, hold eyelids open and flush with plenty of water.

Water Pollution

Effect of low concentrations on aquatic life is unknown.
May be dangerous if it enters water intakes.
Notify local health and wildlife officials.
Notify operators of nearby water intakes.

1. RESPONSE TO DISCHARGE
(See Response Methods Handbook, CG 446-4)

Disperse and flush

2. LABELS

No hazard label required by Code of Federal Regulations

3. CHEMICAL DESIGNATIONS

3.1 **Synonyms:** Allomaleic acid; Boletic acid; *trans*-Butenedioic acid; *trans*-1,2-Ethylenedicarboxylic acid; Lichenic acid

3.2 **Coast Guard Compatibility Classification:** Not listed

3.3 **Chemical Formula:** $HO_2CCH{=}CHCO_2H$

3.4 **IMCO/United Nations Numerical Designation:** Not listed

4. OBSERVABLE CHARACTERISTICS

4.1 **Physical State (as shipped):** Solid

4.2 **Color:** White

4.3 **Odor:** None

5. HEALTH HAZARDS

5.1 **Personal Protective Equipment:** Dust mask; gloves; safety glasses; dust cap

5.2 **Symptoms Following Exposure:** Inhalation of dust may cause respiratory irritation. Compound is non-toxic when ingested. Prolonged contact with eyes or skin may cause irritation.

5.3 **Treatment for Exposure:** INHALATION: move to fresh air. EYES: flush with water; get medical attention if irritation persists. SKIN: flush with water.

5.4 **Toxicity by Inhalation (Threshold Limit Value):** Data not available

5.5 **Short-Term Inhalation Limits:** Data not available

5.6 **Toxicity by Ingestion:** Data not available

5.7 **Late Toxicity:** Data not available

5.8 **Vapor (Gas) Irritant Characteristics:** Data not available

5.9 **Liquid or Solid Irritant Characteristics:** Data not available

5.10 **Odor Threshold:** Odorless

6. FIRE HAZARDS

6.1 **Flash Point:** Not pertinent

6.2 **Flammable Limits in Air:** Not pertinent

6.3 **Fire Extinguishing Agents:** Water spray, dry chemical, foam, carbon dioxide

6.4 **Fire Extinguishing Agents Not to be Used:**

6.5 **Special Hazards of Combustion Products:** Irritating fumes of maleic anhydride may form in fires.

6.6 **Behavior in Fire:** Dust presents explosion hazard; knock down dust with water fog.

6.7 **Ignition Temperature:** 1364°F (powder)

6.8 **Electrical Hazard:** Not pertinent

6.9 **Burning Rate:** Not pertinent

7. CHEMICAL REACTIVITY

7.1 **Reactivity with Water:** No reaction

7.2 **Reactivity with Common Materials:**

7.3 **Stability During Transport:** Stable

7.4 **Neutralizing Agents for Acids and Caustics:** Not pertinent

7.5 **Polymerization:** Not pertinent

7.6 **Inhibitor of Polymerization:** Not pertinent

8. WATER POLLUTION

8.1 **Aquatic Toxicity:** Data not available

8.2 **Waterfowl Toxicity:** Data not available

8.3 **Biological Oxygen Demand (BOD):** 61%, 5 days

8.4 **Food Chain Concentration Potential:** None

9. SELECTED MANUFACTURERS

1. Tenneco Chemicals, Inc.
   Intermediates Div.
   Turner Place
   P. O. Box 2
   Piscataway, N. J. 08854
2. Pfizer, Inc.
   Chemicals Div.
   235 East 42 St.
   New York, N. Y. 10017
3. Monsanto Company
   800 North Lindbergh Blvd.
   St. Louis, Mo. 63166

10. SHIPPING INFORMATION

10.1 **Grades or Purity:** Technical; Purified food grade

10.2 **Storage Temperature:** Ambient

10.3 **Inert Atmosphere:** No requirement

10.4 **Venting:** Open

11. HAZARD ASSESSMENT CODE
(See Hazard Assessment Handbook, CG 446-3)

II

12. HAZARD CLASSIFICATIONS

12.1 **Code of Federal Regulations:** Not listed

12.2 **NAS Hazard Rating for Bulk Water Transportation:** Not listed

12.3 **NFPA Hazard Classifications:** Not listed

13. PHYSICAL AND CHEMICAL PROPERTIES

13.1 **Physical State at 15°C and 1 atm:** Solid

13.2 **Molecular Weight:** 116.07

13.3 **Boiling Point at 1 atm:** Very high

13.4 **Freezing Point:** Not pertinent

13.5 **Critical Temperature:** Not pertinent

13.6 **Critical Pressure:** Not pertinent

13.7 **Specific Gravity:** 1.635 at 20°C (solid)

13.8 **Liquid Surface Tension:** Not pertinent

13.9 **Liquid-Water Interfacial Tension:** Not pertinent

13.10 **Vapor (Gas) Specific Gravity:** Not pertinent

13.11 **Ratio of Specific Heats of Vapor (Gas):** Not pertinent

13.12 **Latent Heat of Vaporization:** Not pertinent

13.13 **Heat of Combustion:** $-4{,}970$ Btu/lb $= -2{,}760$ cal/g $= -116 \times 10^5$ J/kg

13.14 **Heat of Decomposition:** Not pertinent

13.15 **Heat of Solution:** Not pertinent

13.16 **Heat of Polymerization:** Not pertinent

NOTES

FFA FURFURAL

Common Synonyms			
Pyromucic aldehyde 2-Furaldehyde Furfuraldehyde Fural	Oily liquid Sinks in water.	Colorless to reddish-brown	Almond odor

Avoid contact with liquid. Keep people away.
Wear goggles, self-contained breathing apparatus, and rubber overclothing (including gloves).
Call fire department.
Stop discharge if possible.
Isolate and remove discharged material.
Notify local health and pollution control agencies.

Fire

Combustible.
Wear goggles, self-contained breathing apparatus, and rubber overclothing (including gloves).
Extinguish with water, dry chemical, alcohol foam, or carbon dioxide.
Cool exposed containers with water.

Exposure

CALL FOR MEDICAL AID.

LIQUID
Will burn skin and eyes.
Harmful if swallowed.
Remove contaminated clothing and shoes.
Flush affected areas with plenty of water.
IF IN EYES, hold eyelids open and flush with plenty of water.
IF SWALLOWED and victim is CONSCIOUS, have victim drink water or milk, have victim induce vomiting.
IF SWALLOWED and victim is UNCONSCIOUS OR HAVING CONVULSIONS, do nothing except keep victim warm.

Water Pollution

HARMFUL TO AQUATIC LIFE IN VERY LOW CONCENTRATIONS.
May be dangerous if it enters water intakes.
Notify local health and wildlife officials.
Notify operators of nearby water intakes.

1. RESPONSE TO DISCHARGE

(See Response Methods Handbook, CG 446-4)

Issue warning—water contaminant
Should be removed
Chemical and physical treatment

2. LABELS

No hazard label required by Code of Federal Regulations

3. CHEMICAL DESIGNATIONS

3.1 **Synonyms:** 2-Furaldehyde; Fural; Furfuraldehyde; Furfurole; Pyromucic aldehyde; Quakeral
3.2 **Coast Guard Compatibility Classification:** Aldehyde
3.3 **Chemical Formula:** $\underline{O\text{-}CH_2CH_2CH_2C}HCHO$
3.4 **IMCO/United Nations Numerical Designation:** 3.3/1199

4. OBSERVABLE CHARACTERISTICS

4.1 **Physical State (as shipped):** Liquid
4.2 **Color:** Colorless to reddish brown
4.3 **Odor:** Almond-like

5. HEALTH HAZARDS

5.1 **Personal Protective Equipment:** Skin and eye protection.
5.2 **Symptoms Following Exposure:** Vapor may irritate eyes and respiratory system. Liquid irritates skin and may cause dermatitis.
5.3 **Treatment for Exposure:** INHALATION: general treatment for overexposure to vapors of toxic chemicals; keep airway open, give respiration and oxygen if necessary; observe for premonitory signs and symptoms of pulmonary edema. INGESTION: induce vomiting, then give gastric lavage and saline cathartics. SKIN AND MUCOUS MEMBRANES: flood affected tissues with water.
5.4 **Toxicity by Inhalation (Threshold Limit Value):** 5 ppm
5.5 **Short-Term Inhalation Limits:** 15 ppm for 15 min.
5.6 **Toxicity by Ingestion:** Grade 3; LD_{50} 50 to 500 mg/kg
5.7 **Late Toxicity:** Causes liver damage in rats.
5.8 **Vapor (Gas) Irritant Characteristics:** Vapors cause moderate irritation such that personnel will find high concentrations unpleasant. The effect is temporary.
5.9 **Liquid or Solid Irritant Characteristics:** Causes smarting of the skin and first-degree burns on short exposure; may cause secondary burns on long exposure.
5.10 **Odor Threshold:** Data not available

6. FIRE HAZARDS

6.1 **Flash Point:** 153°F O.C.; 140°F C.C.
6.2 **Flammable Limits in Air:** 2.1%—19.3%
6.3 **Fire Extinguishing Agents:** Water, foam, carbon dioxide, dry chemical or alcohol foam
6.4 **Fire Extinguishing Agents Not to be Used:** Not pertinent
6.5 **Special Hazards of Combustion Products:** Irritating vapors are generated when heated
6.6 **Behavior in Fire:** Not pertinent
6.7 **Ignition Temperature:** 739°F
6.8 **Electrical Hazard:** Not pertinent
6.9 **Burning Rate:** 2.6 mm/min.

7. CHEMICAL REACTIVITY

7.1 **Reactivity with Water:** No reaction
7.2 **Reactivity with Common Materials:** No reaction
7.3 **Stability During Transport:** Stable
7.4 **Neutralizing Agents for Acids and Caustics:** Not pertinent
7.5 **Polymerization:** Not pertinent
7.6 **Inhibitor of Polymerization:** Not pertinent

8. WATER POLLUTION

8.1 **Aquatic Toxicity:**
24 ppm/96 hr/bluegill/TL_m/fresh water
32 ppm/24 hr/sunfish/TL_m/fresh water
8.2 **Waterfowl Toxicity:** Data not available
8.3 **Biological Oxygen Demand (BOD):** 0.28—0.77 lb/lb, 5 days
8.4 **Food Chain Concentration Potential:** None

9. SELECTED MANUFACTURERS

The Quaker Oats Co.
Chemicals Division
Merchandise Mart Plaza
Chicago, Ill. 60654

10. SHIPPING INFORMATION

10.1 **Grades or Purity:** Commercial
10.2 **Storage Temperature:** Ambient
10.3 **Inert Atmosphere:** No requirement
10.4 **Venting:** Pressure-vacuum

11. HAZARD ASSESSMENT CODE

(See Hazard Assessment Handbook, CG 446-3)

A-P-Q

12. HAZARD CLASSIFICATIONS

12.1 **Code of Federal Regulations:** Combustible Liquid
12.2 **NAS Hazard Rating for Bulk Water Transportation:**

Category	Rating
Fire	2
Health	
Vapor Irritant	2
Liquid or Solid Irritant	2
Poisons	3
Water Pollution	
Human Toxicity	3
Aquatic Toxicity	3
Aesthetic Effect	2
Reactivity	
Other Chemicals	2
Water	0
Self-Reaction	1

12.3 **NFPA Hazard Classifications:**

Category	Classification
Health Hazard (Blue)	2
Flammability (Red)	2
Reactivity (Yellow)	0

13. PHYSICAL AND CHEMICAL PROPERTIES

13.1 **Physical State at 15°C and 1 atm:** Liquid
13.2 **Molecular Weight:** 96.08
13.3 **Boiling Point at 1 atm:** 323.1°F = 161.7°C = 434.9°K
13.4 **Freezing Point:** −33.7°F = −36.5°C = 236.7°K
13.5 **Critical Temperature:** 745°F = 397°C = 670°K
13.6 **Critical Pressure:** 798 psia = 54.3 atm = 5.50 MN/m²
13.7 **Specific Gravity:** 1.159 at 20°C (liquid)
13.8 **Liquid Surface Tension:** 43.5 dynes/cm = 0.0435 N/m at 20°C
13.9 **Liquid-Water Interfacial Tension:** (est.) 30 dynes/cm = 0.03 N/m at 20°C
13.10 **Vapor (Gas) Specific Gravity:** Not pertinent
13.11 **Ratio of Specific Heats of Vapor (Gas):** Not pertinent
13.12 **Latent Heat of Vaporization:** 191 Btu/lb = 106 cal/g = 4.44 × 10^5 J/kg
13.13 **Heat of Combustion:** −10,490 Btu/lb = −5830 cal/g = −244.1 × 10^5 J/kg
13.14 **Heat of Decomposition:** Not pertinent
13.15 **Heat of Solution:** Not pertinent
13.16 **Heat of Polymerization:** Not pertinent

NOTES

REVISED 1978

FAL — FURFURYL ALCOHOL

Common Synonyms	Liquid	Colorless to light yellow	Mild irritating
2-Furancarbinol Furfuralcohol 2-Furylcarbinol 2-Hydroxymethylfuran	Mixes with water.		

Stop discharge if possible. Keep people away.
Call fire department.
Isolate and remove discharged material.
Notify local health and pollution control agencies.

Fire

Combustible.
Extinguish with water, dry chemicals, foam, or carbon dioxide.

Exposure

Call for medical aid.

LIQUID
Irritating to skin and eyes.
Harmful if swallowed.
Remove contaminated clothing and shoes.
Flush affected areas with plenty of water.
IF IN EYES, hold eyelids open and flush with plenty of water.
IF SWALLOWED and victim is CONSCIOUS, have victim drink water or milk and have victim induce vomiting.
IF SWALLOWED and victim is UNCONSCIOUS OR HAVING CONVULSIONS, do nothing except keep victim warm.

Water Pollution

Effect of low concentrations on aquatic life is unknown.
May be dangerous if it enters water intakes.
Notify local health and wildlife officials.
Notify operators of nearby water intakes.

1. RESPONSE TO DISCHARGE
(See Response Methods Handbook, CG 446-4)

Issue warning—water contaminant
Disperse and flush

2. LABELS

No label required by Code of Federal Regulations

3. CHEMICAL DESIGNATIONS

3.1 **Synonyms:** 2-Furancarbinol; Furfuralcohol; alpha-Furylcarbinol; 2-Furylcarbinol; 2-Hydroxymethylfuran
3.2 **Coast Guard Compatibility Classification:** Alcohol
3.3 **Chemical Formula:** $C_5H_6O_2$
3.4 **IMCO/United Nations Numerical Designation:** Not listed

4. OBSERVABLE CHARACTERISTICS

4.1 **Physical State (as shipped):** Liquid
4.2 **Color:** Colorless or amber
4.3 **Odor:** Mildly irritating

5. HEALTH HAZARDS

5.1 **Personal Protective Equipment:** Goggles or face shield; rubber gloves.
5.2 **Symptoms Following Exposure:** Inhalation causes headache, nausea, and irritation of nose and throat. Vapor irritates eyes; liquid causes inflammation and corneal opacity. Contact of skin with liquid causes dryness and irritation. Ingestion causes headache, nausea, and irritation of mouth and stomach.
5.3 **Treatment for Exposure:** INHALATION: remove victim to fresh air; if breathing is difficult, call a physician. EYES: immediately flush with water for 15 min.; get medical attention. SKIN: wash promptly with soap and water. INGESTION: give large amount of water, and induce vomiting; follow with gastric lavage and saline cathartics; get medical attention.
5.4 **Toxicity by Inhalation (Threshold Limit Value):** 5 ppm
5.5 **Short-Term Inhalation Limits:** 50 ppm/30 min.
5.6 **Toxicity by Ingestion:** Grade 3; oral LD_{50} = 132 mg/kg (rat)
5.7 **Late Toxicity:** Data not available
5.8 **Vapor (Gas) Irritant Characteristics:** Vapors cause moderate irritation such that personnel will find high concentrations unpleasant. The effect is temporary.
5.9 **Liquid or Solid Irritant Characteristics:** Minimum hazard. If spilled on clothing and allowed to remain, may cause smarting and reddening of the skin.
5.10 **Odor Threshold:** 8 ppm

6. FIRE HAZARDS

6.1 **Flash Point:** 167°F O.C., 149°F C.C.
6.2 **Flammable Limits in Air:** 1.8%–16.3%
6.3 **Fire Extinguishing Agents:** Water, dry chemical, foam, carbon dioxide
6.4 **Fire Extinguishing Agents Not to be Used:** Not pertinent
6.5 **Special Hazards of Combustion Products:** Not pertinent
6.6 **Behavior in Fire:** Not pertinent
6.7 **Ignition Temperature:** 736°F
6.8 **Electrical Hazard:** Data not available
6.9 **Burning Rate:** 2.3 mm/min.

7. CHEMICAL REACTIVITY

7.1 **Reactivity with Water:** No reaction
7.2 **Reactivity with Common Materials:** No reaction
7.3 **Stability During Transport:** Darkens and forms water-insoluble material on exposure to air or acids, particularly when hot.
7.4 **Neutralizing Agents for Acids and Caustics:** Not pertinent
7.5 **Polymerization:** Not pertinent
7.6 **Inhibitor of Polymerization:** Not pertinent

8. WATER POLLUTION

8.1 **Aquatic Toxicity:** Data not available
8.2 **Waterfowl Toxicity:** Data not available
8.3 **Biological Oxygen Demand (BOD):** Data not available
8.4 **Food Chain Concentration Potential:** None

9. SELECTED MANUFACTURERS

1. The Quaker Oats Company
   Chemicals Division
   345 Merchandise Mart Plaza
   Chicago, Ill. 60654
2. Eastman Kodak Co.
   Eastman Organic Chemicals
   Rochester, N. Y. 14650
3. Aldrich Chemical Co.
   940 West St. Paul Ave.
   Milwaukee, Wis. 53233

10. SHIPPING INFORMATION

10.1 **Grades or Purity:** Technical
10.2 **Storage Temperature:** Ambient
10.3 **Inert Atmosphere:** No requirement
10.4 **Venting:** Open (flame arrester)

11. HAZARD ASSESSMENT CODE
(See Hazard Assessment Handbook, CG 446-3)

A-P-Q

12. HAZARD CLASSIFICATIONS

12.1 **Code of Federal Regulations:** Not listed
12.2 **NAS Hazard Rating for Bulk Water Transportation:**

Category	Rating
Fire	1
Health	
Vapor Irritant	2
Liquid or Solid Irritant	1
Poisons	2
Water Pollution	
Human Toxicity	2
Aquatic Toxicity	3
Aesthetic Effect	1
Reactivity	
Other Chemicals	2
Water	0
Self-Reaction	0

12.3 **NFPA Hazard Classifications:**

Category	Classification
Health Hazard (Blue)	1
Flammability (Red)	2
Reactivity (Yellow)	1

13. PHYSICAL AND CHEMICAL PROPERTIES

13.1 **Physical State at 15°C and 1 atm:** Liquid
13.2 **Molecular Weight:** 98.1
13.3 **Boiling Point at 1 atm:** 338°F = 170°C = 443°K
13.4 **Freezing Point:** 5°F = −15°C = 258°K
13.5 **Critical Temperature:** Not pertinent
13.6 **Critical Pressure:** Not pertinent
13.7 **Specific Gravity:** 1.13 at 20°C (liquid)
13.8 **Liquid Surface Tension:** 38 dynes/cm = 0.038 N/m at 20°C
13.9 **Liquid-Water Interfacial Tension:** Not pertinent
13.10 **Vapor (Gas) Specific Gravity:** 3.4
13.11 **Ratio of Specific Heats of Vapor (Gas):** Not pertinent
13.12 **Latent Heat of Vaporization:** 230 Btu/lb = 130 cal/g = 5.4×10^5 J/kg
13.13 **Heat of Combustion:** −11,200 Btu/lb = −6,200 cal/g = -260×10^5 J/kg
13.14 **Heat of Decomposition:** Not pertinent
13.15 **Heat of Solution:** Not pertinent
13.16 **Heat of Polymerization:** Not pertinent

NOTES

GLA — GALLIC ACID

Common Synonyms
3,4,5-Trihydroxybenzoic acid
Gallic acid monohydrate

Solid — White — Odorless

Sinks in water.

Stop discharge if possible. Keep people away.
Shut off ignition sources. Call fire department.
Avoid contact with solid and dust.
Isolate and remove discharged material.
Notify local health and pollution control agencies.

Fire

Combustible.
Extinguish with water, dry chemicals, foam, or carbon dioxide.

Exposure

CALL FOR MEDICAL AID.

DUST
Irritating to eyes, nose and throat.
If inhaled will cause coughing or difficult breathing.
If in eyes, hold eyelids open and flush with plenty of water.
If breathing has stopped, give artificial respiration.
If breathing is difficult, give oxygen.

SOLID
Irritating to skin and eyes.
Harmful if swallowed.
Remove contaminated clothing and shoes.
Flush affected areas with plenty of water.
IF IN EYES, hold eyelids open and flush with plenty of water.
IF SWALLOWED and victim is CONSCIOUS, have victim drink water or milk and have victim induce vomiting.
IF SWALLOWED and victim is UNCONSCIOUS OR HAVING CONVULSIONS, do nothing except keep victim warm.

Water Pollution

HARMFUL TO AQUATIC LIFE IN VERY LOW CONCENTRATIONS.
May be dangerous if it enters water intakes.
Notify local health and wildlife officials.
Notify operators of nearby water intakes.

1. RESPONSE TO DISCHARGE

(See Response Methods Handbook, CG 446-4)

Disperse and flush

2. LABELS

No hazard label required by Code of Federal Regulations

3. CHEMICAL DESIGNATIONS

3.1 **Synonyms:** Gallic acid monohydrate; 3,4,5-Trihydroxybenzoic acid
3.2 **Coast Guard Compatibility Classification:** Not listed
3.3 **Chemical Formula:** $3,4,5\text{-}(HO)_3C_6H_2COOH \cdot H_2O$
3.4 **IMCO/United Nations Numerical Designation:** Not listed

4. OBSERVABLE CHARACTERISTICS

4.1 **Physical State (as shipped):** Solid
4.2 **Color:** White
4.3 **Odor:** None

5. HEALTH HAZARDS

5.1 **Personal Protective Equipment:** Bu. Mines approved respirator; rubber gloves; safety goggles
5.2 **Symptoms Following Exposure:** Inhalation of dust may irritate nose and throat. Contact with eyes or skin causes irritation.
5.3 **Treatment for Exposure:** INGESTION: give large amount of water; induce vomiting. EYES: flush with water for at least 10 min.; consult a physician if irritation persists. SKIN: wash with soap and water.
5.4 **Toxicity by Inhalation (Threshold Limit Value):** Data not available
5.5 **Short-Term Inhalation Limits:** Data not available
5.6 **Toxicity by Ingestion:** Grade 2; LD_{50} 0.5–5 g/kg (rat)
5.7 **Late Toxicity:** Data not available
5.8 **Vapor (Gas) Irritant Characteristics:** Data not available
5.9 **Liquid or Solid Irritant Characteristics:** Data not available
5.10 **Odor Threshold:** Data not available

6. FIRE HAZARDS

6.1 **Flash Point:** Not pertinent
6.2 **Flammable Limits in Air:** Not pertinent
6.3 **Fire Extinguishing Agents:** Water, foam, dry chemical, carbon dioxide
6.4 **Fire Extinguishing Agents Not to be Used:**
6.5 **Special Hazards of Combustion Products:**
6.6 **Behavior in Fire:**
6.7 **Ignition Temperature:** Not pertinent
6.8 **Electrical Hazard:** Not pertinent
6.9 **Burning Rate:** Not pertinent

7. CHEMICAL REACTIVITY

7.1 **Reactivity with Water:** No reaction
7.2 **Reactivity with Common Materials:**
7.3 **Stability During Transport:** Stable
7.4 **Neutralizing Agents for Acids and Caustics:** Not pertinent
7.5 **Polymerization:** Not pertinent
7.6 **Inhibitor of Polymerization:** Not pertinent

8. WATER POLLUTION

8.1 **Aquatic Toxicity:**
30–35 ppm/6 hr/minnow/MLD/fresh water
15–20 ppm/6 hr/minnow/MLD/fresh water
8.2 **Waterfowl Toxicity:** Data not available
8.3 **Biological Oxygen Demand (BOD):** 8%, 5 days
8.4 **Food Chain Concentration Potential:** None

9. SELECTED MANUFACTURERS

1. The Harshaw Chemical Co.
1945 E. 97 St.
Cleveland, Ohio 44106
2. Aldrich Chemical Co.
940 W. Saint Paul Ave.
Milwaukee, Wis. 53233
3. Eastman Organic Chemicals
Rochester, N. Y. 14650

10. SHIPPING INFORMATION

10.1 **Grades or Purity:** N.F.; Practical
10.2 **Storage Temperature:** Ambient
10.3 **Inert Atmosphere:** No requirement
10.4 **Venting:** Open

11. HAZARD ASSESSMENT CODE

(See Hazard Assessment Handbook, CG 446-3)

II-SS

12. HAZARD CLASSIFICATIONS

12.1 **Code of Federal Regulations:** Not listed
12.2 **NAS Hazard Rating for Bulk Water Transportation:** Not listed
12.3 **NFPA Hazard Classifications:**

Category	Classification
Health Hazard (Blue)	0
Flammability (Red)	1
Reactivity (Yellow)	0

13. PHYSICAL AND CHEMICAL PROPERTIES

13.1 **Physical State at 15°C and 1 atm:** Solid
13.2 **Molecular Weight:** 188
13.3 **Boiling Point at 1 atm:** Not pertinent (decomposes)
13.4 **Freezing Point:** Not pertinent
13.5 **Critical Temperature:** Not pertinent
13.6 **Critical Pressure:** Not pertinent
13.7 **Specific Gravity:** 1.7 at 20°C (solid)
13.8 **Liquid Surface Tension:** Not pertinent
13.9 **Liquid-Water Interfacial Tension:** Not pertinent
13.10 **Vapor (Gas) Specific Gravity:** Not pertinent
13.11 **Ratio of Specific Heats of Vapor (Gas):** Not pertinent
13.12 **Latent Heat of Vaporization:** Not pertinent
13.13 **Heat of Combustion:** −6,060 Btu/lb = −3,370 cal/g = -141×10^5 J/kg
13.14 **Heat of Decomposition:** Not pertinent
13.15 **Heat of Solution:** Not pertinent
13.16 **Heat of Polymerization:** Not pertinent

NOTES

GOC

GAS OIL: CRACKED

Common Synonyms			
	Liquid	Colorless	Gasoline-like odor
	Floats on water.		

Stop discharge if possible. Keep people away.
Call fire department.
Isolate and remove discharged material.
Notify local health and pollution control agencies.

Fire

Combustible.
Extinguish with dry chemical, foam, or carbon dioxide.

Exposure

CALL FOR MEDICAL AID.

LIQUID
Harmful if swallowed.
IF SWALLOWED and victim is CONSCIOUS, have victim drink water or milk.
DO NOT INDUCE VOMITING.

Water Pollution

HARMFUL TO AQUATIC LIFE IN VERY LOW CONCENTRATIONS.
Fouling to shoreline.
May be dangerous if it enters water intakes.
Notify local health and wildlife officials.
Notify operators of nearby water intakes.

1. RESPONSE TO DISCHARGE

(See Response Methods Handbook, CG 446-4)

Mechanical containment
Should be removed
Physical and chemical treatment

2. LABELS

No hazard label required by Code of Federal Regulations

3. CHEMICAL DESIGNATIONS

3.1 **Synonyms:** No common synonyms

3.2 **Coast Guard Compatibility Classification:** Miscellaneous hydrocarbon mixtures

3.3 **Chemical Formula:** Not pertinent (mixture)

3.4 **IMCO/United Nations Numerical Designation:** 3.3/1202 (gas oil)

4. OBSERVABLE CHARACTERISTICS

4.1 **Physical State (as shipped):** Liquid

4.2 **Color:** Yellow to brown

4.3 **Odor:** Like gasoline and petroleum

5. HEALTH HAZARDS

5.1 **Personal Protective Equipment:** Protective goggles, gloves.

5.2 **Symptoms Following Exposure:** INHALATION: causes irritation of upper respiratory tract; stimulation, then depression; dizziness, headache, incoordination, anesthesia, coma, respiratory arrest; irregular heartbeat is a complication. ASPIRATION: causes severe coughing, gagging, distress, rapid development of pulmonary edema. INGESTION: causes irritation of throat and stomach; stimulation, then depression.

5.3 **Treatment for Exposure:** Get medical attention. INHALATION: maintain respiration; administer oxygen if needed. ASPIRATION: enforce bed rest and administer oxygen. INGESTION: give victim water or milk; do NOT induce vomiting; guard against aspiration into lungs. EYES: wash with copious quantity of water. SKIN: remove by wiping, then wash with soap and water.

5.4 **Toxicity by Inhalation (Threshold Limit Value):** No single value applicable.

5.5 **Short-Term Inhalation Limits:** Not pertinent

5.6 **Toxicity by Ingestion:** Grade 2; LD_{50} 0.5 to 5 g/kg

5.7 **Late Toxicity:** None

5.8 **Vapor (Gas) Irritant Characteristics:** Vapors cause a slight smarting of the eyes or respiratory system if present in high concentrations. The effect is temporary.

5.9 **Liquid or Solid Irritant Characteristics:** Minimum hazard. If spilled on clothing and allowed to remain, may cause smarting and reddening of skin.

5.10 **Odor Threshold:** 0.25 ppm

6. FIRE HAZARDS

6.1 **Flash Point:** 150°F C.C.

6.2 **Flammable Limits in Air:** 6.0%—13.5%

6.3 **Fire Extinguishing Agents:** Water, foam, carbon dioxide, or dry chemical

6.4 **Fire Extinguishing Agents Not to be Used:** Not pertinent

6.5 **Special Hazards of Combustion Products:** Not pertinent

6.6 **Behavior in Fire:** Not pertinent

6.7 **Ignition Temperature:** 640°F

6.8 **Electrical Hazard:** Not pertinent

6.9 **Burning Rate:** 4 mm/min.

7. CHEMICAL REACTIVITY

7.1 **Reactivity with Water:** No reaction

7.2 **Reactivity with Common Materials:** No reaction

7.3 **Stability During Transport:** Stable

7.4 **Neutralizing Agents for Acids and Caustics:** Not pertinent

7.5 **Polymerization:** Not pertinent

7.6 **Inhibitor of Polymerization:** Not pertinent

8. WATER POLLUTION

8.1 **Aquatic Toxicity:**
90 ppm/24 hr/juvenile shad/TL_m/fresh water
91 ppm/24 hr/juvenile shad/TL_m/salt water

8.2 **Waterfowl Toxicity:** Data not available

8.3 **Biological Oxygen Demand (BOD):** 8%, 5 days

8.4 **Food Chain Concentration Potential:** None

9. SELECTED MANUFACTURERS

1. Exxon Co.
   30 Rockefeller Plaza
   New York, N. Y. 10020
2. Shell Oil Co.
   1 Shell Plaza
   Houston, Texas 77001
3. Sun Oil Co.
   St. Davids, Pa. 19087

10. SHIPPING INFORMATION

10.1 **Grades or Purity:** Composition varies widely with the refinery operation involved.

10.2 **Storage Temperature:** Ambient

10.3 **Inert Atmosphere:** No requirement

10.4 **Venting:** Open (flame arrester)

11. HAZARD ASSESSMENT CODE

(See Hazard Assessment Handbook, CG 446-3)

A-T-U

12. HAZARD CLASSIFICATIONS

12.1 **Code of Federal Regulations:** Combustible Liquid

12.2 **NAS Hazard Rating for Bulk Water Transportation:**

Category	Rating
Fire	3
Health	
Vapor Irritant	1
Liquid or Solid Irritant	1
Poisons	2
Water Pollution	
Human Toxicity	1
Aquatic Toxicity	2
Aesthetic Effect	2
Reactivity	
Other Chemicals	0
Water	0
Self-Reaction	0

12.3 **NFPA Hazard Classifications:**

Category	Classification
Health Hazard (Blue)	1
Flammability (Red)	3
Reactivity (Yellow)	0

13. PHYSICAL AND CHEMICAL PROPERTIES

13.1 **Physical State at 15°C and 1 atm:** Liquid

13.2 **Molecular Weight:** Not pertinent

13.3 **Boiling Point at 1 atm:** 375 – 750°F = 190 – 399°C = 463 – 672°K

13.4 **Freezing Point:** Not pertinent

13.5 **Critical Temperature:** Not pertinent

13.6 **Critical Pressure:** Not pertinent

13.7 **Specific Gravity:** 0.848 16°C (liquid)

13.8 **Liquid Surface Tension:** (est.) 25 dynes/cm = 0.025 N/m at 20°C

13.9 **Liquid-Water Interfacial Tension:** (est.) 50 dynes/cm = 0.05 N/m at 20°C

13.10 **Vapor (Gas) Specific Gravity:** 3.4

13.11 **Ratio of Specific Heats of Vapor (Gas):** Not pertinent

13.12 **Latent Heat of Vaporization:** Not pertinent

13.13 **Heat of Combustion:** -18,400 Btu/lb = -10,200 cal/g = 428 × 10^5 J/kg

13.14 **Heat of Decomposition:** Not pertinent

13.15 **Heat of Solution:** Not pertinent

13.16 **Heat of Polymerization:** Not pertinent

NOTES

REVISED 1978

GASOLINE BLENDING STOCKS: ALKYLATES

Common Synonyms	Watery liquid	Colorless	Gasoline odor
	Floats on water. Flammable, irritating vapor is produced.		

Stop discharge if possible. Keep people away.
Shut off ignition sources and call fire department.
Stay upwind and use water spray to "knock down" vapor.
Isolate and remove discharged material.
Notify local health and pollution control agencies.

Fire

FLAMMABLE.
Flashback along vapor trail may occur.
Vapor may explode if ignited in an enclosed area.
Extinguish with dry chemical, foam, or carbon dioxide.
Water may be ineffective on fire.
Cool exposed containers with water.

Exposure

CALL FOR MEDICAL AID.

VAPOR
Irritating to eyes, nose and throat.
If inhaled, will cause dizziness, headache, difficult breathing or loss of consciousness.
Move to fresh air.
If breathing has stopped, give artificial respiration.
If breathing is difficult, give oxygen.

LIQUID
Irritating to skin and eyes.
If swallowed, will cause nausea and vomiting.
Remove contaminated clothing and shoes.
Flush affected areas with plenty of water.
IF IN EYES, hold eyelids open and flush with plenty of water.
IF SWALLOWED and victim is CONSCIOUS, have victim drink water or milk.
DO NOT INDUCE VOMITING.

Water Pollution

HARMFUL TO AQUATIC LIFE IN VERY LOW CONCENTRATIONS.
Fouling to shoreline.
May be dangerous if it enters water intakes.
Notify local health and wildlife officials.
Notify operators of nearby water intakes.

1. RESPONSE TO DISCHARGE

(See Response Methods Handbook, CG 446-4)

Issue warning—high flammability
Evacuate area
Disperse and flush

2. LABEL

Red

3. CHEMICAL DESIGNATIONS

3.1 **Synonyms:** No common synonyms
3.2 **Coast Guard Compatibility Classification:** Miscellaneous hydrocarbon mixtures
3.3 **Chemical Formula:** Not pertinent
3.4 **IMCO/United Nations Numerical Designation:** 3.1, 3.2/1203

4. OBSERVABLE CHARACTERISTICS

4.1 **Physical State (as shipped):** Liquid
4.2 **Color:** Colorless
4.3 **Odor:** Gasoline

5. HEALTH HAZARDS

5.1 **Personal Protective Equipment:** Protective goggles, gloves.
5.2 **Symptoms Following Exposure:** INHALATION causes irritation of upper respiratory tract; central nervous system stimulation followed by depression of varying degrees ranging from dizziness, headache, and incoordination to anesthesia, coma, and respiratory arrest; irregular heartbeat is dangerous complication. ASPIRATION causes severe lung irritation with coughing, gagging, dyspnea, substernal distress, and rapidly developing pulmonary edema; later, signs of bronchopneumonia and pneumonitis; acute onset of central nervous system excitement followed by depression. INGESTION causes irritation of mucous membranes of throat, esophagus, and stomach; stimulation followed by depression of central nervous system; irregular heartbeat.
5.3 **Treatment for Exposure:** Seek medical attention. INHALATION: maintain respiration; give oxygen if needed. ASPIRATION: enforce bed rest; administer oxygen. INGESTION: do NOT induce vomiting; lavage carefully if appreciable quantity was ingested; guard against aspiration into lungs. EYES: wash with copious quantity of water. SKIN: wipe off and wash with soap and water.
5.4 **Toxicity by Inhalation (Threshold Limit Value):** No single TLV applicable.
5.5 **Short-Term Inhalation Limits:** 500 ppm for 30 min.
5.6 **Toxicity by Ingestion:** Grade 2; LD_{50} 0.5 to 5 g/kg
5.7 **Late Toxicity:** None

6. FIRE HAZARDS

6.1 **Flash Point:**
(a) <0°F C.C.
(b) 0-73°F C.C.
6.2 **Flammable Limits in Air:**
(a) 1.1%—8.7%
6.3 **Fire Extinguishing Agents:** Dry chemical, foam, carbon dioxide
6.4 **Fire Extinguishing Agents Not to be Used:** Water may be ineffective
6.5 **Special Hazards of Combustion Products:** None
6.6 **Behavior in Fire:** Vapor is heavier than air and may travel a considerable distance to a source of ignition and flash back.
6.7 **Ignition Temperature:** Data not available
6.8 **Electrical Hazard:** Class I, group D
6.9 **Burning Rate:** 4 mm/min.

7. CHEMICAL REACTIVITY

7.1 **Reactivity with Water:** No reaction
7.2 **Reactivity with Common Materials:** No reaction
7.3 **Stability During Transport:** Stable
7.4 **Neutralizing Agents for Acids and Caustics:** Not pertinent
7.5 **Polymerization:** Not pertinent
7.6 **Inhibitor of Polymerization:** Not pertinent

8. WATER POLLUTION

8.1 **Aquatic Toxicity:**
90 ppm/24 hr/juvenile American shad/ TL_m/fresh water
91 ppm/24 hr/juvenile American shad/ TL_m/salt water
8.2 **Waterfowl Toxicity:** Data not available
8.3 **Biological Oxygen Demand (BOD):** 8%, 5 days
8.4 **Food Chain Concentration Potential:** None

9. SELECTED MANUFACTURERS

1. Exxon Co.
30 Rockefeller Plaza
New York, N. Y. 10020
2. Shell Oil Co.
1 Shell Plaza
Houston, Tex. 77001
3. Sun Oil Co.
St. Davids, Pa. 19087

10. SHIPPING INFORMATION

10.1 **Grades or Purity:** Composition varies with range of distillation temperatures used.
10.2 **Storage Temperature:** Ambient
10.3 **Inert Atmosphere:** No requirement
10.4 **Venting:** Open (flame arrester) or pressure-vacuum

11. HAZARD ASSESSMENT CODE

(See Hazard Assessment Handbook, CG 446-3)

A-T-U-V-W

12. HAZARD CLASSIFICATIONS

12.1 **Code of Federal Regulations:** Flammable liquid
12.2 **NAS Hazard Rating for Bulk Water Transportation:**

Category	Rating
Fire	3
Health	
Vapor Irritant	1
Liquid or Solid Irritant	1
Poisons	2
Water Pollution	
Human Toxicity	1
Aquatic Toxicity	2
Aesthetic Effect	2
Reactivity	
Other Chemicals	0
Water	0
Self-Reaction	0

12.3 **NFPA Hazard Classifications:**

Category	Classification
Health Hazard (Blue)	1
Flammability (Red)	3
Reactivity (Yellow)	0

13. PHYSICAL AND CHEMICAL PROPERTIES

13.1 **Physical State at 15°C and 1 atm:** Liquid
13.2 **Molecular Weight:** Not pertinent
13.3 **Boiling Point at 1 atm:** 58–275°F = 14–135°C = 287–408°K
13.4 **Freezing Point:** Not pertinent
13.5 **Critical Temperature:** Not pertinent
13.6 **Critical Pressure:** Not pertinent
13.7 **Specific Gravity:** 0.71–0.75 at 15°C (liquid)
13.8 **Liquid Surface Tension:** 19–23 dynes/cm = 0.019–0.023 N/m at 20°C
13.9 **Liquid-Water Interfacial Tension:** 49–51 dynes/cm = 0.049–0.051 N/m at 20°C
13.10 **Vapor (Gas) Specific Gravity:** 3.4
13.11 **Ratio of Specific Heats of Vapor (Gas):** Not pertinent
13.12 **Latent Heat of Vaporization:** 130–150 Btu/lb = 71–81 cal/g = $3.0-3.4 \times 10^5$ J/kg
13.13 **Heat of Combustion:** −18,720 Btu/lb = −10,400 cal/g = -435.4×10^5 J/kg
13.14 **Heat of Decomposition:** Not pertinent
13.15 **Heat of Solution:** Not pertinent
13.16 **Heat of Polymerization:** Not pertinent

5. HEALTH HAZARD (Cont'd.)

5.8 **Vapor (Gas) Irritant Characteristics:** Vapors cause a slight smarting of the eyes or respiratory system if present in high concentrations. The effect is temporary.
5.9 **Liquid or Solid Irritant Characteristics:** Minimum hazard. If spilled on clothing and allowed to remain, may cause smarting and reddening of the skin.
5.10 **Odor Threshold:** 0.25 ppm

REVISED 1978

GRF	GASOLINE BLENDING STOCKS: REFORMATES

Common Synonyms	Watery liquid Colorless Gasoline odor Floats on water. Flammable, irritating vapor is produced.

Stop discharge if possible. Keep people away.
Shut off ignition sources and call fire department.
Stay upwind and use water spray to "knock down" vapor.
Isolate and remove discharged material.
Notify local health and pollution control agencies.

Fire

FLAMMABLE.
Flashback along vapor trail may occur.
Vapor may explode if ignited in an enclosed area.
Extinguish with dry chemical, foam or carbon dioxide.
Water may be ineffective on fire.
Cool exposed containers with water.

Exposure

CALL FOR MEDICAL AID.

VAPOR
Irritating to eyes, nose and throat.
If inhaled will cause dizziness, headache, difficult breathing or loss of consciousness.
Move to fresh air.
If breathing has stopped, give artificial respiration.
If breathing is difficult, give oxygen.

LIQUID
Irritating to skin and eyes.
If swallowed, will cause nausea or vomiting.
Remove contaminated clothing and shoes.
Flush affected areas with plenty of water.
IF IN EYES, hold eyelids open and flush with plenty of water.
IF SWALLOWED and victim is CONSCIOUS, have victim drink water or milk.
DO NOT INDUCE VOMITING.

Water Pollution

HARMFUL TO AQUATIC LIFE IN VERY LOW CONCENTRATIONS.
Fouling to shoreline.
May be dangerous if it enters water intakes.
Notify local health and wildlife officials.
Notify operators of nearby water intakes.

1. RESPONSE TO DISCHARGE

(See Response Methods Handbook, CG 446-4)

Issue warning – high flammability
Evacuate area
Disperse and flush

2. LABEL

Red

3. CHEMICAL DESIGNATIONS

3.1 **Synonyms:** No common synonyms
3.2 **Coast Guard Compatibility Classification:** Miscellaneous hydrocarbon mixtures
3.3 **Chemical Formula:** Not pertinent
3.4 **IMCO/United Nations Numerical Designation:** 3.1, 3.2/1203

4. OBSERVABLE CHARACTERISTICS

4.1 **Physical State (as shipped):** Liquid
4.2 **Color:** Colorless
4.3 **Odor:** Gasoline

5. HEALTH HAZARDS

5.1 **Personal Protective Equipment:** Protective goggles, gloves.
5.2 **Symptoms Following Exposure:** INHALATION causes irritation of upper respiratory tract; central nervous system stimulation followed by depression of varying degrees ranging from dizziness, headache, and incoordination to anesthesia, coma, and respiratory arrest; irregular heartbeat is dangerous complication. ASPIRATION causes severe lung irritation with coughing, gagging, dyspnea, substernal distress, and rapidly developing pulmonary edema; later, signs of bronchopneumonia and pneumonitis; acute onset of central nervous system followed by depression. INGESTION causes irritation of mucous membranes of throat, esophagus, and stomach; stimulation followed by depression of central nervous system; irregular heartbeat.
5.3 **Treatment for Exposure:** Seek medical attention. INHALATION: maintain respiration; give oxygen if needed. ASPIRATION: enforce bed rest; administer oxygen. INGESTION: do NOT induce vomiting; lavage carefully if appreciable quantity was ingested; guard against aspiration into lungs. EYES: wash with copious quantity of water. SKIN: wipe off and wash with soap and water.
5.4 **Toxicity by Inhalation (Threshold Limit Value):** No single TLV applicable.
5.5 **Short-Term Inhalation Limits:** 500 ppm for 30 min.
5.6 **Toxicity by Ingestion:** Grade 2; LD_{50} 0.5 to 5 g/kg
5.7 **Late Toxicity:** None
5.8 **Vapor (Gas) Irritant Characteristics:** Vapors cause a slight smarting of the eyes or respiratory system if present in high concentrations. The effect is temporary.

6. FIRE HAZARDS

6.1 **Flash Point:**
(a) <0°F C.C.
(b) 0-73°F C.C.
6.2 **Flammable Limits in Air:**
(a) 1.1%–8.7%
6.3 **Fire Extinguishing Agents:** Dry chemical, foam, carbon dioxide
6.4 **Fire Extinguishing Agents Not to be Used:** Water may be ineffective
6.5 **Special Hazards of Combustion Products:** None
6.6 **Behavior in Fire:** Vapor is heavier than air and may travel a considerable distance to a source of ignition and flash back.
6.7 **Ignition Temperature:** Data not available
6.8 **Electrical Hazard:** Class I, group D
6.9 **Burning Rate:** 4 mm/min.

7. CHEMICAL REACTIVITY

7.1 **Reactivity with Water:** No reaction
7.2 **Reactivity with Common Materials:** No reaction
7.3 **Stability During Transport:** Stable
7.4 **Neutralizing Agents for Acids and Caustics:** Not pertinent
7.5 **Polymerization:** Not pertinent
7.6 **Inhibitor of Polymerization:** Not pertinent

8. WATER POLLUTION

8.1 **Aquatic Toxicity:**
90 ppm/24 hr/juvenile American shad/TL_m/fresh water
91 ppm/24 hr/juvenile American shad/TL_m/salt water
8.2 **Waterfowl Toxicity:** Data not available
8.3 **Biological Oxygen Demand (BOD):** 8%, 5 days
8.4 **Food Chain Concentration Potential:** None

9. SELECTED MANUFACTURERS

1. Exxon Co.
30 Rockefeller Plaza
New York, N. Y. 10020
2. Shell Oil Co.
1 Shell Plaza
Houston, Texas 77001
3. Sun Oil Co.
St. Davids, Pa. 19087

10. SHIPPING INFORMATION

10.1 **Grades or Purity:** Composition varies with range of distillation temperatures used
10.2 **Storage Temperature:** Ambient
10.3 **Inert Atmosphere:** No requirement
10.4 **Venting:** Open (flame arrester) or pressure-vacuum

11. HAZARD ASSESSMENT CODE

(See Hazard Assessment Handbook, CG 446-3)

A-T-U-V-W

12. HAZARD CLASSIFICATIONS

12.1 **Code of Federal Regulations:** Flammable liquid
12.2 **NAS Hazard Rating for Bulk Water Transportation:**

Category	Rating
Fire	3
Health	
Vapor Irritant	1
Liquid or Solid Irritant	1
Poisons	2
Water Pollution	
Human Toxicity	1
Aquatic Toxicity	2
Aesthetic Effect	2
Reactivity	
Other Chemicals	0
Water	0
Self-Reaction	0

12.3 **NFPA Hazard Classifications:**

Category	Classification
Health Hazard (Blue)	1
Flammability (Red)	3
Reactivity (Yellow)	0

13. PHYSICAL AND CHEMICAL PROPERTIES

13.1 **Physical State at 15°C and 1 atm:** Liquid
13.2 **Molecular Weight:** Not pertinent
13.3 **Boiling Point at 1 atm:** 58–275°F = 14–135°C = 287–408°K
13.4 **Freezing Point:** Not pertinent
13.5 **Critical Temperature:** Not pertinent
13.6 **Critical Pressure:** Not pertinent
13.7 **Specific Gravity:** 0.7934 at 20°C (liquid)
13.8 **Liquid Surface Tension:** 19–23 dynes/cm = 0.019–0.023 N/m at 20°C
13.9 **Liquid-Water Interfacial Tension:** 49–51 dynes/cm = 0.049–0.051 N/m at 20°C
13.10 **Vapor (Gas) Specific Gravity:** 3.4
13.11 **Ratio of Specific Heats of Vapor (Gas):** Not pertinent
13.12 **Latent Heat of Vaporization:** 130–150 Btu/lb = 71–81 cal/g = 3.0–3.4 × 10^5 J/kg
13.13 **Heat of Combustion:** −18,720 Btu/lb = −10,400 cal/g = −435.4 × 10^5 J/kg
13.14 **Heat of Decomposition:** Not pertinent
13.15 **Heat of Solution:** Not pertinent
13.16 **Heat of Polymerization:** Not pertinent

5. HEALTH HAZARDS (Cont'd.)

5.9 **Liquid or Solid Irritant Characteristics:** Minimum hazard. If spilled on clothing and allowed to remain, may cause smarting and reddening of the skin.
5.10 **Odor Threshold:** 0.25 ppm

GAT	# GASOLINES: AUTOMOTIVE (< 4.23g lead/gal)

Common Synonyms	Watery liquid Colorless to pale brown or pink Gasoline odor Floats on water. Flammable, irritating vapor is produced.

Stop discharge if possible. Keep people away.
Shut off ignition sources and call fire department.
Stay upwind and use water spray to "knock down" vapor.
Isolate and remove discharged material.
Notify local health and pollution control agencies.

Fire

FLAMMABLE.
Flashback along vapor trail may occur.
Vapor may explode if ignited in an enclosed area.
Extinguish with dry chemical, foam, or carbon dioxide.
Water may be ineffective on fire.
Cool exposed containers with water.

Exposure

CALL FOR MEDICAL AID.

VAPOR
Irritating to eyes, nose and throat.
If inhaled, will cause dizziness, headache, difficult breathing or loss of consciousness.
Move to fresh air.
If breathing has stopped, give artificial respiration.
If breathing is difficult, give oxygen.

LIQUID
Irritating to skin and eyes.
If swallowed, will cause nausea or vomiting.
Remove contaminated clothing and shoes.
Flush affected areas with plenty of water.
IF IN EYES, hold eyelids open and flush with plenty of water.
IF SWALLOWED and victim is CONSCIOUS, have victim drink water or milk.
DO NOT INDUCE VOMITING.

Water Pollution

HARMFUL TO AQUATIC LIFE IN VERY LOW CONCENTRATIONS.
Fouling to shoreline.
May be dangerous if it enters water intakes.
Notify local health and wildlife officials.
Notify operators of nearby water intakes.

1. RESPONSE TO DISCHARGE

(See Response Methods Handbook, CG 446-4)

Issue warning — high flammability
Evacuate area
Disperse and flush

2. LABEL

Red

3. CHEMICAL DESIGNATIONS

3.1 **Synonyms:** Motor spirit
Petrol

3.2 **Coast Guard Compatibility Classification:** Miscellaneous hydrocarbon mixtures

3.3 **Chemical Formula:** (Mixture of hydrocarbons)

3.4 **IMCO/United Nations Numerical Designation:** 3.1/1203

4. OBSERVABLE CHARACTERISTICS

4.1 **Physical State (as shipped):** Liquid

4.2 **Color:** Colorless to brown

4.3 **Odor:** Gasoline

5. HEALTH HAZARDS

5.1 **Personal Protective Equipment:** Protective goggles, gloves.

5.2 **Symptoms Following Exposure:** Irritation of mucous membranes and stimulation followed by depression of central nervous system. Breathing of vapor may also cause dizziness, headache, and incoordination or, in more severe cases, anesthesia, coma, and respiratory arrest. If liquid enters lungs, it will cause severe irritation, coughing, gagging, pulmonary edema, and, later, signs of bronchopneumonia and pneumonitis. Swallowing may cause irregular heartbeat.

5.3 **Treatment for Exposure:** INHALATION: maintain respiration and administer oxygen; enforce bed rest if liquid is in lungs. INGESTION: do NOT induce vomiting; stomach should be lavaged (by doctor) if appreciable quantity is swallowed. EYES: wash with copious quantity of water. SKIN: wipe off and wash with soap and water.

5.4 **Toxicity by Inhalation (Threshold Limit Value):** No single TLV applies.

5.5 **Short-Term Inhalation Limits:** 500 ppm for 30 min.

5.6 **Toxicity by Ingestion:** Grade 2; LD_{50} 0.5 to 5 g/kg.

5.7 **Late Toxicity:** None

5.8 **Vapor (Gas) Irritant Characteristics:** Vapors cause a slight smarting of the eyes or respiratory system if present in high concentrations. The effect is temporary.

5.9 **Liquid or Solid Irritant Characteristics:** Minimum hazard. If spilled on clothing and allowed to remain, may cause smarting and reddening of the skin.

5.10 **Odor Threshold:** 0.25 ppm

6. FIRE HAZARDS

6.1 **Flash Point:** −36°F C.C.

6.2 **Flammable Limits in Air:** 1.4% 7.4%

6.3 **Fire Extinguishing Agents:** Foam, carbon dioxide, dry chemical

6.4 **Fire Extinguishing Agents Not to be Used:** Water may be ineffective

6.5 **Special Hazards of Combustion Products:** None

6.6 **Behavior in Fire:** Vapor is heavier than air and may travel considerable distance to a source of ignition and flash back.

6.7 **Ignition Temperature:** 853°F

6.8 **Electrical Hazard:** Class I, Group D

6.9 **Burning Rate:** 4 mm/min.

7. CHEMICAL REACTIVITY

7.1 **Reactivity with Water:** No reaction

7.2 **Reactivity with Common Materials:** No reaction

7.3 **Stability During Transport:** Stable

7.4 **Neutralizing Agents for Acids and Caustics:** Not pertinent

7.5 **Polymerization:** Not pertinent

7.6 **Inhibitor of Polymerization:** Not pertinent

8. WATER POLLUTION

8.1 **Aquatic Toxicity:**
90 ppm/24 hr/juvenile American shad/ TL_m/fresh water
91 mg/l /24 hr/juvenile American shad/ TL_m/salt water

8.2 **Waterfowl Toxicity:** Data not available

8.3 **Biological Oxygen Demand (BOD):** 8%, 5 days

8.4 **Food Chain Concentration Potential:** None

9. SELECTED MANUFACTURERS

1. Exxon Co.
30 Rockefeller Plaza
New York, N. Y. 10020
2. Shell Oil Co.
1 Shell Plaza
Houston, Texas 77001
3. Sun Oil Co.
St. Davids, Pa. 19087

10. SHIPPING INFORMATION

10.1 **Grades or Purity:** Various octane ratings; military specifications

10.2 **Storage Temperature:** Ambient

10.3 **Inert Atmosphere:** No requirement

10.4 **Venting:** Open (flame arrester) or pressure-vacuum

11. HAZARD ASSESSMENT CODE

(See Hazard Assessment Handbook, CG 446-3)

A-T-U-V-W

12. HAZARD CLASSIFICATIONS

12.1 **Code of Federal Regulations:** Flammable liquid

12.2 **NAS Hazard Rating for Bulk Water Transportation:**

Category	Rating
Fire	3
Health	
Vapor Irritant	1
Liquid or Solid Irritant	1
Poisons	2
Water Pollution	
Human Toxicity	1
Aquatic Toxicity	2
Aesthetic Effect	2
Reactivity	
Other Chemicals	0
Water	0
Self-Reaction	0

12.3 **NFPA Hazard Classifications:**

Category	Classification
Health Hazard (Blue)	1
Flammability (Red)	3
Reactivity (Yellow)	0

13. PHYSICAL AND CHEMICAL PROPERTIES

13.1 **Physical State at 15°C and 1 atm:** Liquid

13.2 **Molecular Weight:** Not pertinent

13.3 **Boiling Point at 1 atm:** 140 390°F = 60 199°C = 333 472°K

13.4 **Freezing Point:** Not pertinent

13.5 **Critical Temperature:** Not pertinent

13.6 **Critical Pressure:** Not pertinent

13.7 **Specific Gravity:** 0.7321 at 20°C (liquid)

13.8 **Liquid Surface Tension:** 19 23 dynes/cm = 0.019 0.023 N/m at 20°C

13.9 **Liquid-Water Interfacial Tension:** 49 51 dynes/cm = 0.049 0.051 N/m at 20°C

13.10 **Vapor (Gas) Specific Gravity:** 3.4

13.11 **Ratio of Specific Heats of Vapor (Gas):** (est.) 1.054

13.12 **Latent Heat of Vaporization:** 130 150 Btu/lb = 71 81 cal/g = 3.0 3.4×10^5 J/kg

13.13 **Heat of Combustion:** −18,720 Btu/lb = −10,400 cal/g = -435.1×10^5 J/kg

13.14 **Heat of Decomposition:** Not pertinent

13.15 **Heat of Solution:** Not pertinent

13.16 **Heat of Polymerization:** Not pertinent

NOTES

REVISED 1978

GAV

GASOLINES: AVIATION

(< 4.86g lead/gal)

Common Synonyms	Watery liquid	Red, blue, green, brown or purple	Gasoline odor
	Floats on water. Flammable, irritating vapor is produced.		

Stop discharge if possible. Keep people away.
Shut off ignition sources and call fire department.
Stay upwind and use water spray to "knock down" vapor.
Isolate and remove discharged material.
Notify local health and pollution control agencies.

Fire

FLAMMABLE.
Flashback along vapor trail may occur.
Vapor may explode if ignited in an enclosed area.
Extinguish with dry chemical, foam or carbon dioxide.
Water may be ineffective on fire.
Cool exposed containers with water.

Exposure

CALL FOR MEDICAL AID.

VAPOR
Irritating to eyes, nose and throat.
If inhaled will cause dizziness, headache, difficult breathing or loss of consciousness.
Move to fresh air.
If breathing has stopped, give artificial respiration.
If breathing is difficult, give oxygen.

LIQUID
Irritating to skin and eyes.
If swallowed, will cause nausea or vomiting.
Remove contaminated clothing and shoes.
Flush affected areas with plenty of water.
IF IN EYES, hold eyelids open and flush with plenty of water.
IF SWALLOWED and victim is CONSCIOUS, have victim drink water or milk.
DO NOT INDUCE VOMITING.

Water Pollution

HARMFUL TO AQUATIC LIFE IN VERY LOW CONCENTRATIONS.
Fouling to shoreline.
May be dangerous if it enters water intakes.
Notify local health and wildlife officials.
Notify operators of nearby water intakes.

1. RESPONSE TO DISCHARGE

(See Response Methods Handbook, CG 446-4)

Issue warning—high flammability
Evacuate area
Disperse and flush

2. LABEL

Red

3. CHEMICAL DESIGNATIONS

3.1 **Synonyms:** No common synonyms
3.2 **Coast Guard Compatibility Classification:** Miscellaneous hydrocarbon mixtures
3.3 **Chemical Formula:** Not pertinent
3.4 **IMCO/United Nations Numerical Designation:** 3.1/1203

4. OBSERVABLE CHARACTERISTICS

4.1 **Physical State (as shipped):** Liquid
4.2 **Color:** Red, blue, green, brown, purple
4.3 **Odor:** Gasoline

5. HEALTH HAZARDS

5.1 **Personal Protective Equipment:** Protective goggles, gloves.
5.2 **Symptoms Following Exposure:** INHALATION causes irritation of upper respiratory tract; central nervous system stimulation followed by depression of varying degrees ranging from dizziness, headache, and incoordination to anesthesia, coma, and respiratory arrest; irregular heartbeat is dangerous complication. ASPIRATION causes severe lung irritation with coughing, gagging, dyspnea, substernal distress, and rapidly developing pulmonary edema; later, signs of bronchopneumonia and pneumonitis; acute onset of central nervous system excitement followed by depression. INGESTION causes irritation of mucous membranes of throat, esophagus, and stomach; stimulation followed by depression of central nervous system; irregular heartbeat.
5.3 **Treatment for Exposure:** Seek medical attention. INHALATION: maintain respiration; give oxygen if needed. ASPIRATION: enforce bed rest; administer oxygen. INGESTION: do NOT induce vomiting; lavage carefully if appreciable quantity was ingested; guard against aspiration into lungs. EYES: wash with copious quantity of water. SKIN: wipe off and wash with soap and water.
5.4 **Toxicity by Inhalation (Threshold Limit Value):** No single TLV applicable.
5.5 **Short-Term Inhalation Limits:** 500 ppm for 30 min.
5.6 **Toxicity by Ingestion:** Grade 2; LD_{50} 0.5 to 5 g/kg
5.7 **Late Toxicity:** None

6. FIRE HAZARDS

6.1 **Flash Point:** −50°F C.C.
6.2 **Flammable Limits in Air:** 1.2%—7.1%
6.3 **Fire Extinguishing Agents:** Foam, carbon dioxide, dry chemical
6.4 **Fire Extinguishing Agents Not to be Used:** Water may be ineffective
6.5 **Special Hazards of Combustion Products:** None
6.6 **Behavior in Fire:** Vapor is heavier than air and may travel a considerable distance to a source of ignition and flash back.
6.7 **Ignition Temperature:** 824°F
6.8 **Electrical Hazard:** Class I, group D
6.9 **Burning Rate:** 4 mm/min.

7. CHEMICAL REACTIVITY

7.1 **Reactivity with Water:** No reaction
7.2 **Reactivity with Common Materials:** No reaction
7.3 **Stability During Transport:** Stable
7.4 **Neutralizing Agents for Acids and Caustics:** Not pertinent
7.5 **Polymerization:** Not pertinent
7.6 **Inhibitor of Polymerization:** Not pertinent

8. WATER POLLUTION

8.1 **Aquatic Toxicity:**
90 ppm/24 hr/juvenile American shad/TL_m/fresh water
91 ppm/24 hr/juvenile American shad/TL_m/salt water
8.2 **Waterfowl Toxicity:** Data not available
8.3 **Biological Oxygen Demand (BOD):** 8%, 5 days
8.4 **Food Chain Concentration Potential:** None

9. SELECTED MANUFACTURERS

1. Exxon Co.
30 Rockefeller Plaza
New York, N. Y. 10020
2. Shell Oil Co.
1 Shell Plaza
Houston, Texas 77001
3. Sun Oil Co.
St. Davids, Pa. 19087

10. SHIPPING INFORMATION

10.1 **Grades or Purity:** Grades 80/87, 100/130, and 115/145: Specification MIL-G-5572e
10.2 **Storage Temperature:** Ambient
10.3 **Inert Atmosphere:** No requirement
10.4 **Venting:** Open (flame arrester) or pressure-vacuum

11. HAZARD ASSESSMENT CODE

(See Hazard Assessment Handbook, CG 446-3)

A-T-U-V-W

12. HAZARD CLASSIFICATIONS

12.1 **Code of Federal Regulations:** Flammable liquid
12.2 **NAS Hazard Rating for Bulk Water Transportation:**

Category	Rating
Fire	3
Health	
Vapor Irritant	1
Liquid or Solid Irritant	1
Poisons	2
Water Pollution	
Human Toxicity	1
Aquatic Toxicity	2
Aesthetic Effect	2
Reactivity	
Other Chemicals	0
Water	0
Self-Reaction	0

12.3 **NFPA Hazard Classifications:**

Category	Classification
Health Hazard (Blue)	1
Flammability (Red)	3
Reactivity (Yellow)	0

13. PHYSICAL AND CHEMICAL PROPERTIES

13.1 **Physical State at 15°C and 1 atm:** Liquid
13.2 **Molecular Weight:** Not pertinent
13.3 **Boiling Point at 1 atm:** 160–340°F = 71–171°C = 344–444°K
13.4 **Freezing Point:** <76°F = <24.4°C = <297.6°K
13.5 **Critical Temperature:** Not pertinent
13.6 **Critical Pressure:** Not pertinent
13.7 **Specific Gravity:** 0.711 at 15°C (liquid)
13.8 **Liquid Surface Tension:** 19–23 dynes/cm = 0.019–0.023 N/m at 20°C
13.9 **Liquid-Water Interfacial Tension:** 49–51 dynes/cm = 0.049–0.051 N/m at 20°C
13.10 **Vapor (Gas) Specific Gravity:** 3.4
13.11 **Ratio of Specific Heats of Vapor (Gas):** (est.) 1.054
13.12 **Latent Heat of Vaporization:** 130–150 Btu/lb = 71–81 cal/g = 3.0–3.4 × 10^5 J/kg
13.13 **Heat of Combustion:** −18,720 Btu/lb = −10,400 cal/g = −435.4 × 10^5 J/kg
13.14 **Heat of Decomposition:** Not pertinent
13.15 **Heat of Solution:** Not pertinent
13.16 **Heat of Polymerization:** Not pertinent

5. HEALTH HAZARDS (Cont'd.)

5.8 **Vapor (Gas) Irritant Characteristics:** Vapors cause a slight smarting of the eyes or respiratory system if present in high concentrations. The effect is temporary.
5.9 **Liquid or Solid Irritant Characteristics:** Minimum hazard. If spilled on clothing and allowed to remain, may cause smarting and reddening of the skin.
5.10 **Odor Threshold:** 0.25 ppm

REVISED 1978

GCS — GASOLINES: CASINGHEAD

Common Synonyms	Watery liquid — Colorless — Gasoline odor Floats on water. Flammable, irritating vapor is produced.

Stop discharge if possible. Keep people away.
Shut off ignition sources and call fire department.
Stay upwind and use water spray to "knock down" vapor.
Isolate and remove discharged material.
Notify local health and pollution control agencies.

Fire

FLAMMABLE.
Flashback along vapor trail may occur.
Vapor may explode if ignited in an enclosed area.
Extinguish with dry chemical, foam or carbon dioxide.
Water may be ineffective on fire.
Cool exposed containers with water.

Exposure

CALL FOR MEDICAL AID.

VAPOR
Irritating to eyes, nose and throat.
If inhaled will cause dizziness, headache, difficult breathing or loss of consciousness.
Move to fresh air.
If breathing has stopped, give artificial respiration.
If breathing is difficult, give oxygen.

LIQUID
Irritating to skin and eyes.
If swallowed, will cause nausea or vomiting.
Flush affected areas with plenty of water.
IF IN EYES, hold eyelids open and flush with plenty of water.
IF SWALLOWED and victim is CONSCIOUS, have victim drink water or milk.
DO NOT INDUCE VOMITING.

Water Pollution

HARMFUL TO AQUATIC LIFE IN VERY LOW CONCENTRATIONS.
Fouling to shoreline.
May be dangerous if it enters water intakes.
Notify local health and wildlife officials.
Notify operators of nearby water intakes.

1. RESPONSE TO DISCHARGE

(See Response Methods Handbook, CG 446-4)

Issue warning—high flammability
Evacuate area
Disperse and flush

2. LABEL

Red

3. CHEMICAL DESIGNATIONS

3.1 **Synonyms:** Natural gasoline
3.2 **Coast Guard Compatibility Classification:** Miscellaneous hydrocarbon mixtures
3.3 **Chemical Formula:** Not pertinent
3.4 **IMCO/United Nations Numerical Designation:** 3.1/1257

4. OBSERVABLE CHARACTERISTICS

4.1 **Physical State (as shipped):** Liquid
4.2 **Color:** Colorless
4.3 **Odor:** Gasoline

5. HEALTH HAZARDS

5.1 **Personal Protective Equipment:** Protective goggles, gloves.
5.2 **Symptoms Following Exposure:** INHALATION causes irritation of upper respiratory tract; central nervous system stimulation followed by depression of varying degrees ranging from dizziness, headache, and incoordination to anesthesia, coma, and respiratory arrest; irregular heartbeat is dangerous complication. ASPIRATION causes severe lung irritation with coughing, gagging, dyspnea, substernal distress, and rapidly developing pulmonary edema; later, signs of bronchopneumonia and pneumonitis, acute onset of central nervous system excitement followed by depression. INGESTION causes irritation of mucous membranes of throat, esophagus, and stomach; stimulation followed by depression of central nervous system; irregular heartbeat.
5.3 **Treatment for Exposure:** Seek medical attention. INHALATION: maintain respiration; give oxygen if needed. ASPIRATION: enforce bed rest; administer oxygen. INGESTION: do NOT induce vomiting; lavage carefully if appreciable quantity was ingested; guard against aspiration into lungs. EYES: wash with copious quantity of water. SKIN: wipe off and wash with soap and water.
5.4 **Toxicity by Inhalation (Threshold Limit Value):** No single TLV applicable
5.5 **Short-Term Inhalation Limits:** 500 ppm for 30 min.
5.6 **Toxicity by Ingestion:** Grade 2; LD_{50} 0.5 to 5 g/kg
5.7 **Late Toxicity:** None

6. FIRE HAZARDS

6.1 **Flash Point:** <0°F O.C.
6.2 **Flammable Limits in Air:** 1.3%–7.1%
6.3 **Fire Extinguishing Agents:** Dry chemical, foam, or carbon dioxide
6.4 **Fire Extinguishing Agents Not to be Used:** Water may be ineffective
6.5 **Special Hazards of Combustion Products:** None
6.6 **Behavior in Fire:** Vapor is heavier than air and may travel a considerable distance to a source of ignition and flash back.
6.7 **Ignition Temperature:** Data not available
6.8 **Electrical Hazard:** Class I, group D
6.9 **Burning Rate:** 4 mm/min.

7. CHEMICAL REACTIVITY

7.1 **Reactivity with Water:** No reaction
7.2 **Reactivity with Common Materials:** No reaction
7.3 **Stability During Transport:** Stable
7.4 **Neutralizing Agents for Acids and Caustics:** Not pertinent
7.5 **Polymerization:** Not pertinent
7.6 **Inhibitor of Polymerization:** Not pertinent

8. WATER POLLUTION

8.1 **Aquatic Toxicity:**
90 ppm/24 hr/juvenile American shad/ TL_m/fresh water
91 ppm/24 hr/juvenile American shad/ TL_m/salt water
8.2 **Waterfowl Toxicity:** Data not available
8.3 **Biological Oxygen Demand (BOD):** 8%, 5 days
8.4 **Food Chain Concentration Potential:** None

9. SELECTED MANUFACTURERS

1. Exxon Co.
   30 Rockefeller Plaza
   New York, N. Y. 10020
2. Shell Oil Co.
   1 Shell Plaza
   Houston, Texas 77001
3. Sun Oil Co.
   St. Davids, Pa. 19087

10. SHIPPING INFORMATION

10.1 **Grades or Purity:** Composition depends on location of oil well
10.2 **Storage Temperature:** Ambient
10.3 **Inert Atmosphere:** No requirement
10.4 **Venting:** Open (flame arrester) or pressure-vacuum

11. HAZARD ASSESSMENT CODE

(See Hazard Assessment Handbook, CG 446-3)

A-T-U-V-W

12. HAZARD CLASSIFICATIONS

12.1 **Code of Federal Regulations:** Flammable liquid
12.2 **NAS Hazard Rating for Bulk Water Transportation:**

Category	Rating
Fire	4
Health	
Vapor Irritant	1
Liquid or Solid Irritant	0
Poisons	1
Water Pollution	
Human Toxicity	1
Aquatic Toxicity	2
Aesthetic Effect	1
Reactivity	
Other Chemicals	0
Water	0
Self-Reaction	0

12.3 **NFPA Hazard Classifications:**

Category	Classification
Health Hazard (Blue)	1
Flammability (Red)	3
Reactivity (Yellow)	0

13. PHYSICAL AND CHEMICAL PROPERTIES

13.1 **Physical State at 15°C and 1 atm:** Liquid
13.2 **Molecular Weight:** Not pertinent
13.3 **Boiling Point at 1 atm:** 58–275°F = 14–135°C = 287–408°K
13.4 **Freezing Point:** Not pertinent
13.5 **Critical Temperature:** Not pertinent
13.6 **Critical Pressure:** Not pertinent
13.7 **Specific Gravity:** 0.671 at 15°C (liquid)
13.8 **Liquid Surface Tension:** 19–23 dynes/cm = 0.019–0.023 N/m at 20°C
13.9 **Liquid-Water Interfacial Tension:** 49–51 dynes/cm = 0.049–0.051 N/m at 20°C
13.10 **Vapor (Gas) Specific Gravity:** 3.4
13.11 **Ratio of Specific Heats of Vapor (Gas):** Not pertinent
13.12 **Latent Heat of Vaporization:** 130–150 Btu/lb = 71–81 cal/g = $3.0–3.4 \times 10^5$ J/kg
13.13 **Heat of Combustion:** −18,720 Btu/lb = −10,400 cal/g = -435.4×10^5 J/kg
13.14 **Heat of Decomposition:** Not pertinent
13.15 **Heat of Solution:** Not pertinent
13.16 **Heat of Polymerization:** Not pertinent

5. HEALTH HAZARDS (Cont'd.)

5.8 **Vapor (Gas) Irritant Characteristics:** Vapors cause a slight smarting of the eyes or respiratory system if present in high concentrations. The effect is temporary.
5.9 **Liquid or Solid Irritant Characteristics:** Minumum hazard. If spilled on clothing and allowed to remain, may cause smarting and reddening of the skin.
5.10 **Odor Threshold:** 0.25 ppm

REVISED 1978

GPL	# GASOLINES: POLYMER

Common Synonyms	Watery liquid — Colorless — Gasoline odor Floats on water. Flammable, irritating vapor is produced.

Stop discharge if possible. Keep people away.
Shut off ignition sources and call fire department.
Stay upwind and use water spray to "knock down" vapor.
Isolate and remove discharged material.
Notify local health and pollution control agencies.

Fire

FLAMMABLE.
Flashback along vapor trail may occur.
Vapor may explode if ignited in an enclosed area.
Extinguish with dry chemical, foam, or carbon dioxide.
Water may be ineffective on fire.
Cool exposed containers with water.

Exposure

CALL FOR MEDICAL AID.

VAPOR
Irritating to eyes, nose and throat.
If inhaled, will cause dizziness, headaches, difficult breathing or loss of consciousness.
Move to fresh air.
If breathing has stopped, give artificial respiration.
If breathing is difficult, give oxygen.

LIQUID
Irritating to skin and eyes.
If swallowed, will cause nausea or vomiting.
Remove contaminated clothing and shoes.
Flush affected areas with plenty of water.
IF IN EYES, hold eyelids open and flush with plenty of water.
IF SWALLOWED and victim is CONSCIOUS, have victim drink water or milk.
DO NOT INDUCE VOMITING.

Water Pollution

Notify local health and wildlife officials.
Notify operators of nearby water intakes.

1. RESPONSE TO DISCHARGE

(See Response Methods Handbook, CG 446-4)

Issue warning—high flammability
Evacuate area
Disperse and flush

2. LABEL

Red

3. CHEMICAL DESIGNATIONS

3.1 **Synonyms:** No common synonyms
3.2 **Coast Guard Compatibility Classification:** Miscellaneous hydrocarbon mixtures
3.3 **Chemical Formula:** Not pertinent
3.4 **IMCO/United Nations Numerical Designation:** 3.2/1215

4. OBSERVABLE CHARACTERISTICS

4.1 **Physical State (as shipped):** Liquid
4.2 **Color:** Colorless
4.3 **Odor:** Gasoline

5. HEALTH HAZARDS

5.1 **Personal Protective Equipment:** Protective goggles, gloves.
5.2 **Symptoms Following Exposure:** INHALATION causes irritation of upper respiratory tract; central nervous system stimulation followed by depression of varying degrees ranging from dizziness, headache, and incoordination to anesthesia, coma, and respiratory arrest; irregular heartbeat is dangerous complication. ASPIRATION causes severe lung irritation with coughing, gagging, dyspnea, substernal distress, and rapidly developing pulmonary edema; later, signs of bronchopneumonia and pneumonitis; acute onset of central nervous system excitement followed by depression. INGESTION causes irritation of mucous membranes of throat, esophagus, and stomach; stimulation followed by depression of central nervous system; irregular heartbeat.
5.3 **Treatment for Exposure:** Seek medical attention. INHALATION: maintain respiration; give oxygen if needed. ASPIRATION: enforce bed rest; administer oxygen. INGESTION: do NOT induce vomiting; lavage carefully if appreciable quantity was ingested; guard against aspiration into lungs. EYES: wash with copious quantity of water. SKIN: wipe off and wash with soap and water.
5.4 **Toxicity by Inhalation (Threshold Limit Value):** No single TLV applicable
5.5 **Short-Term Inhalation Limits:** 500 ppm for 30 min.
5.6 **Toxicity by Ingestion:** Grade 2; LD_{50} 0.5 to 5 g/kg
5.7 **Late Toxicity:** None
5.8 **Vapor (Gas) Irritant Characteristics:** Vapors cause a slight smarting of the eyes or respiratory system if present in high concentrations. The effect is temporary.

6. FIRE HAZARDS

6.1 **Flash Point:** 0-73°F C.C.
6.2 **Flammable Limits in Air:** 1.3%—7.1%
6.3 **Fire Extinguishing Agents:** Dry chemical, foam, carbon dioxide
6.4 **Fire Extinguishing Agents Not to be Used:** Water may be ineffective
6.5 **Special Hazards of Combustion Products:** None
6.6 **Behavior in Fire:** Vapor is heavier than air and may travel a considerable distance to a source of ignition and flash back.
6.7 **Ignition Temperature:** Data not available
6.8 **Electrical Hazard:** Class I, group D
6.9 **Burning Rate:** 4 mm/min.

7. CHEMICAL REACTIVITY

7.1 **Reactivity with Water:** No reaction
7.2 **Reactivity with Common Materials:** No reaction
7.3 **Stability During Transport:** Stable
7.4 **Neutralizing Agents for Acids and Caustics:** Not pertinent
7.5 **Polymerization:** Not pertinent
7.6 **Inhibitor of Polymerization:** Not pertinent

8. WATER POLLUTION

8.1 **Aquatic Toxicity:**
90 ppm/24 hr/juvenile American shad/ TL_m/fresh water
91 ppm/24 hr/juvenile American shad/ TL_m/salt water
8.2 **Waterfowl Toxicity:** Data not available
8.3 **Biological Oxygen Demand (BOD):** 8%, 5 days
8.4 **Food Chain Concentration Potential:** None

9. SELECTED MANUFACTURERS

1. Exxon Co.
30 Rockefeller Plaza
New York, N. Y. 10020
2. Shell Oil Co.
1 Shell Plaza
Houston, Texas 77001
3. Sun Oil Co.
St. Davids, Pa. 19087

10. SHIPPING INFORMATION

10.1 **Grades or Purity:** Composition varies with range of distillation temperatures used. Contains mostly isohexane-isooctane
10.2 **Storage Temperature:** Ambient
10.3 **Inert Atmosphere:** No requirement
10.4 **Venting:** Open (flame arrester) or pressure-vacuum

11. HAZARD ASSESSMENT CODE

(See Hazard Assessment Handbook, CG 446-3)

A-T-U-V-W

12. HAZARD CLASSIFICATIONS

12.1 **Code of Federal Regulations:** Flammable liquid
12.2 **NAS Hazard Rating for Bulk Water Transportation:**

Category	Rating
Fire	3
Health	
Vapor Irritant	1
Liquid or Solid Irritant	1
Poisons	2
Water Pollution	
Human Toxicity	1
Aquatic Toxicity	2
Aesthetic Effect	2
Reactivity	
Other Chemicals	0
Water	0
Self-Reaction	0

12.3 **NFPA Hazard Classifications:**

Category	Classification
Health Hazard (Blue)	1
Flammability (Red)	3
Reactivity (Yellow)	0

13. PHYSICAL AND CHEMICAL PROPERTIES

13.1 **Physical State at 15°C and 1 atm:** Liquid
13.2 **Molecular Weight:** Not pertinent
13.3 **Boiling Point at 1 atm:** 58–275°F = 14–135°C = 287–408°K
13.4 **Freezing Point:** Not pertinent
13.5 **Critical Temperature:** Not pertinent
13.6 **Critical Pressure:** Not pertinent
13.7 **Specific Gravity:** 0.71–0.75 at 15°C (liquid)
13.8 **Liquid Surface Tension:** 19–23 dynes/cm = 0.019–0.023 N/m at 20°C
13.9 **Liquid-Water Interfacial Tension:** 49–51 dynes/cm = 0.049–0.051 N/m at 20°C
13.10 **Vapor (Gas) Specific Gravity:** 3.4
13.11 **Ratio of Specific Heats of Vapor (Gas):** Not pertinent
13.12 **Latent Heat of Vaporization:** 130–150 Btu/lb = 71–81 cal/g = $3.0–3.4 \times 10^5$ J/kg
13.13 **Heat of Combustion:** −18,720 Btu/lb = −10,400 cal/g = -435.4×10^5 J/kg
13.14 **Heat of Decomposition:** Not pertinent
13.15 **Heat of Solution:** Not pertinent
13.16 **Heat of Polymerization:** Not pertinent

5. HEALTH HAZARDS (Cont'd.)

5.9 **Liquid or Solid Irritant Characteristics:** Minimum hazard. If spilled on clothing and allowed to remain, may cause smarting and reddening of the skin.
5.10 **Odor Threshold:** 0.25 ppm

REVISED 1978

GSR

GASOLINES: STRAIGHT RUN

Common Synonyms	Watery liquid — Colorless — Gasoline odor Floats on water. Flammable, irritating vapor is produced.

Stop discharge if possible. Keep people away.
Shut off ignition sources and call fire department.
Stay upwind and use water spray to "knock down" vapor.
Isolate and remove discharged material.
Notify local health and pollution control agencies.

Fire	FLAMMABLE. Flashback along vapor trail may occur. Vapor may explode if ignited in an enclosed area. **Extinguish with dry chemical, foam, or carbon dioxide.** **Water may be ineffective on fire.** **Cool exposed containers with water.**
Exposure	**CALL FOR MEDICAL AID.** VAPOR Irritating to eyes, nose and throat. If inhaled, will cause dizziness, headache, difficult breathing or loss of consciousness. **Move to fresh air.** **If breathing has stopped, give artificial respiration.** **If breathing is difficult, give oxygen.** LIQUID Irritating to skin and eyes. If swallowed, will cause nausea or vomiting. **Remove contaminated clothing and shoes.** **Flush affected areas with plenty of water.** **IF IN EYES, hold eyelids open and flush with plenty of water.** **IF SWALLOWED and victim is CONSCIOUS, have victim drink water or milk.** **DO NOT INDUCE VOMITING.** **Notify local health and wildlife officials.** **Notify operators of nearby water intakes.**
Water Pollution	HARMFUL TO AQUATIC LIFE IN VERY LOW CONCENTRATIONS. Fouling to shoreline. May be dangerous if it enters water intakes.

1. RESPONSE TO DISCHARGE

(See Response Methods Handbook, CG 446-4)

Issue warning — high flammability
Evacuate area
Disperse and flush

2. LABEL

Red

3. CHEMICAL DESIGNATIONS

3.1 **Synonyms:** No common synonyms
3.2 **Coast Guard Compatibility Classification:** Miscellaneous hydrocarbon mixtures
3.3 **Chemical Formula:** Not pertinent
3.4 **IMCO/United Nations Numerical Designation:** 3.1, 3.2/1203

4. OBSERVABLE CHARACTERISTICS

4.1 **Physical State (as shipped):** Liquid
4.2 **Color:** Colorless
4.3 **Odor:** Gasoline

5. HEALTH HAZARDS

5.1 **Personal Protective Equipment:** Protective goggles, gloves.
5.2 **Symptoms Following Exposure:** INHALATION causes irritation of upper respiratory tract; central nervous system stimulation followed by depression of varying degrees ranging from dizziness, headache, and incoordination to anesthesia, coma, and respiratory arrest; irregular heartbeat is dangerous complication. ASPIRATION causes severe lung irritation with coughing, gagging, dyspnea, substernal distress, and rapidly developing pulmonary edema; later, signs of bronchopneumonia and pneumonitis; acute onset of central nervous system excitement followed by depression. INGESTION causes irritation of mucous membranes of throat, esophagus, and stomach; stimulation followed by depression of central nervous system; irregular heartbeat.
5.3 **Treatment for Exposure:** Seek medical attention. INHALATION: maintain respiration; give oxygen if needed. ASPIRATION: enforce bed rest; administer oxygen. INGESTION: do NOT induce vomiting; lavage carefully if appreciable quantity was ingested; guard against aspiration into lungs. EYES: wash with copious quantity of water. SKIN: wipe off and wash with soap and water.
5.4 **Toxicity by Inhalation (Threshold Limit Value):** No single TLV applicable.
5.5 **Short-Term Inhalation Limits:** 500 ppm for 30 min.
5.6 **Toxicity by Ingestion:** Grade 2; LD_{50} 0.5 to 5 g/kg
5.7 **Late Toxicity:** None

6. FIRE HAZARDS

6.1 **Flash Point:**
(a) <0°F C.C.
(b) 0-73°F C.C.
6.2 **Flammable Limits in Air:**
(a) 1.3% 7.1%
6.3 **Fire Extinguishing Agents:** Dry chemical, foam, carbon dioxide
6.4 **Fire Extinguishing Agents Not to be Used:** Water may be ineffective
6.5 **Special Hazards of Combustion Products:** None
6.6 **Behavior in Fire:** Vapor is heavier than air and may travel a considerable distance to a source of ignition and flash back.
6.7 **Ignition Temperature:** Data not available
6.8 **Electrical Hazard:** Class I, group D
6.9 **Burning Rate:** 4 mm/min.

7. CHEMICAL REACTIVITY

7.1 **Reactivity with Water:** No reaction
7.2 **Reactivity with Common Materials:** No reaction
7.3 **Stability During Transport:** Stable
7.4 **Neutralizing Agents for Acids and Caustics:** Not pertinent
7.5 **Polymerization:** Not pertinent
7.6 **Inhibitor of Polymerization:** Not pertinent

8. WATER POLLUTION

8.1 **Aquatic Toxicity:**
90 ppm/24 hr/juvenile American shad/ TL_m/fresh water
91 ppm/24 hr/juvenile American shad/ TL_m/salt water
8.2 **Waterfowl Toxicity:** Data not available
8.3 **Biological Oxygen Demand (BOD):** 8%, 5 days
8.4 **Food Chain Concentration Potential:** None

9. SELECTED MANUFACTURERS

1. Exxon Co.
30 Rockefeller Plaza
New York, N. Y. 10020
2. Shell Oil Co.
1 Shell Plaza
Houston, Texas 77001
3. Sun Oil Co.
St. Davids, Pa. 19087

10. SHIPPING INFORMATION

10.1 **Grades or Purity:** Composition varies with range of distillation temperatures used.
10.2 **Storage Temperature:** Ambient
10.3 **Inert Atmosphere:** No requirement
10.4 **Venting:** Open (flame arrester) or pressure-vacuum

11. HAZARD ASSESSMENT CODE

(See Hazard Assessment Handbook, CG 446-3)

A-T-U-V-W

12. HAZARD CLASSIFICATIONS

12.1 **Code of Federal Regulations:** Flammable liquid
12.2 **NAS Hazard Rating for Bulk Water Transportation:**

Category	Rating
Fire	3
Health	
Vapor Irritant	1
Liquid or Solid Irritant	1
Poisons	2
Water Pollution	
Human Toxicity	1
Aquatic Toxicity	2
Aesthetic Effect	2
Reactivity	
Other Chemicals	0
Water	0
Self-Reaction	0

12.3 **NFPA Hazard Classifications:**

Category	Classification
Health Hazard (Blue)	1
Flammability (Red)	3
Reactivity (Yellow)	0

13. PHYSICAL AND CHEMICAL PROPERTIES

13.1 **Physical State at 15°C and 1 atm:** Liquid
13.2 **Molecular Weight:** Not pertinent
13.3 **Boiling Point at 1 atm:** 58 275°F = 14 135°C = 287 408°K
13.4 **Freezing Point:** Not pertinent
13.5 **Critical Temperature:** Not pertinent
13.6 **Critical Pressure:** Not pertinent
13.7 **Specific Gravity:** 0.71 0.747 at 15°C (liquid)
13.8 **Liquid Surface Tension:** 19 23 dynes/cm = 0.019 0.023 N/m at 20°C
13.9 **Liquid-Water Interfacial Tension:** 49 51 dynes/cm = 0.049 0.051 N/m at 20°C
13.10 **Vapor (Gas) Specific Gravity:** 3.4
13.11 **Ratio of Specific Heats of Vapor (Gas):** Not pertinent
13.12 **Latent Heat of Vaporization:** 130 150 Btu/lb = 71 81 cal/g = 3.0 3.4 × 10^5 J/kg
13.13 **Heat of Combustion:** −18,720 Btu/lb = −10,400 cal/g = −435.4 × 10^5 J/kg
13.14 **Heat of Decomposition:** Not pertinent
13.15 **Heat of Solution:** Not pertinent
13.16 **Heat of Polymerization:** Not pertinent

5. HEALTH HAZARDS (Cont'd.)

5.8 **Vapor (Gas) Irritant Characteristics:** Vapors cause a slight smarting of the eyes or respiratory system if present in high concentrations. The effect is temporary.
5.9 **Liquid or Solid Irritant Characteristics:** Minimum hazard. If spilled on clothing and allowed to remain, may cause smarting and reddening of the skin.
5.10 **Odor Threshold:** 0.25 ppm

REVISED 1978

GTA — GLUTARALDEHYDE SOLUTION

Common Synonyms		
1,5-Pentanedial	Liquid	Light yellow
	Mixes with water.	

Stop discharge if possible. Keep people away.
Isolate and remove discharged material.
Notify local health and pollution control agencies.

Fire

Not flammable.

Exposure

Call for medical aid.

LIQUID
Irritating to skin and eyes.
Harmful if swallowed.
Remove contaminated clothing and shoes.
Flush affected areas with plenty of water.
IF IN EYES, hold eyelids open and flush with plenty of water.
IF SWALLOWED and victim is CONSCIOUS, have victim drink water or milk.

Water Pollution

Effect of low concentrations on aquatic life is unknown.
May be dangerous if it enters water intakes.
Notify local health and wildlife officials.
Notify operators of nearby water intakes.

1. RESPONSE TO DISCHARGE

(See Response Methods Handbook, CG 446-4)

Issue warning water contaminant
Restrict access
Disperse and flush

2. LABELS

No label required by Code of Federal Regulations

3. CHEMICAL DESIGNATIONS

3.1 **Synonyms:** 1,5-Pentanedial solution
3.2 **Coast Guard Compatibility Classification:** Aldehyde
3.3 **Chemical Formula:** OHC–$(CH_2)_3$–CHO (in water)
3.4 **IMCO/United Nations Numerical Designation:** Not listed

4. OBSERVABLE CHARACTERISTICS

4.1 **Physical State (as shipped):** Liquid
4.2 **Color:** Pale yellow
4.3 **Odor:** Like rotten apples

5. HEALTH HAZARDS

5.1 **Personal Protective Equipment:** Goggles or face shield; rubber gloves.
5.2 **Symptoms Following Exposure:** Contact with liquid causes severe irritation of eyes and irritation of skin. Chemical readily penetrates skin in harmful amounts. Ingestion causes irritation of mouth and stomach.
5.3 **Treatment for Exposure:** EYES: immediately flush with plenty of water for at least 15 min.; get medical attention. SKIN: immediately flush with plenty of water for at least 15 min. INGESTION: give large amounts of water and induce vomiting; get medical attention.
5.4 **Toxicity by Inhalation (Threshold Limit Value):** Data not available
5.5 **Short-Term Inhalation Limits:** Data not available
5.6 **Toxicity by Ingestion:** Grade 2; oral rat LD_{50} = 2,380 mg/kg
5.7 **Late Toxicity:** Induces contact dermatitis in some people
5.8 **Vapor (Gas) Irritant Characteristics:** Data not available
5.9 **Liquid or Solid Irritant Characteristics:** Data not available
5.10 **Odor Threshold:** Data not available

6. FIRE HAZARDS

6.1 **Flash Point:** Non-flammable solution
6.2 **Flammable Limits in Air:** Not pertinent
6.3 **Fire Extinguishing Agents:** Not pertinent
6.4 **Fire Extinguishing Agents Not to be Used:** Not pertinent
6.5 **Special Hazards of Combustion Products:** Not pertinent
6.6 **Behavior in Fire:** Not pertinent
6.7 **Ignition Temperature:** Not pertinent
6.8 **Electrical Hazard:** Not pertinent
6.9 **Burning Rate:** Not pertinent

7. CHEMICAL REACTIVITY

7.1 **Reactivity with Water:** No reaction
7.2 **Reactivity with Common Materials:** No reaction
7.3 **Stability During Transport:** Stable
7.4 **Neutralizing Agents for Acids and Caustics:** Not pertinent
7.5 **Polymerization:** Not pertinent
7.6 **Inhibitor of Polymerization:** Not pertinent

8. WATER POLLUTION

8.1 **Aquatic Toxicity:** Data not available
8.2 **Waterfowl Toxicity:** Data not available
8.3 **Biological Oxygen Demand (BOD):** Data not available
8.4 **Food Chain Concentration Potential:** None

9. SELECTED MANUFACTURERS

1. Union Carbide Corporation
Chemicals and Plastics Division
270 Park Avenue
New York, N. Y. 10017
2. Polysciences, Inc.
Paul Valley Industrial Park
Warrington, Pa. 18976

10. SHIPPING INFORMATION

10.1 **Grades or Purity:** 25% aqueous solution; 50% aqueous solution
10.2 **Storage Temperature:** Ambient
10.3 **Inert Atmosphere:** No requirement
10.4 **Venting:** Open

11. HAZARD ASSESSMENT CODE

(See Hazard Assessment Handbook, CG 446-3)

A-P

12. HAZARD CLASSIFICATIONS

12.1 **Code of Federal Regulations:** Not listed
12.2 **NAS Hazard Rating for Bulk Water Transportation:** Not listed
12.3 **NFPA Hazard Classifications:** Not listed

13. PHYSICAL AND CHEMICAL PROPERTIES

13.1 **Physical State at 15°C and 1 atm:** Liquid
13.2 **Molecular Weight:** Mixture
13.3 **Boiling Point at 1 atm:** >212°F = >100°C = >373°K
13.4 **Freezing Point:** <20°F = <−7°C = <266°K
13.5 **Critical Temperature:** Not pertinent
13.6 **Critical Pressure:** Not pertinent
13.7 **Specific Gravity:** 1.062–1.124 at 20°C (liquid)
13.8 **Liquid Surface Tension:** (est.) <80 dynes/cm = <0.080 N/m at 20°C
13.9 **Liquid-Water Interfacial Tension:** Not pertinent
13.10 **Vapor (Gas) Specific Gravity:** Not pertinent
13.11 **Ratio of Specific Heats of Vapor (Gas):** Not pertinent
13.12 **Latent Heat of Vaporization:** Not pertinent
13.13 **Heat of Combustion:** Not pertinent
13.14 **Heat of Decomposition:** Not pertinent
13.15 **Heat of Solution:** Not pertinent
13.16 **Heat of Polymerization:** Not pertinent

NOTES

GCR

GLYCERINE

Common Synonyms			
1,2,3-Propanetriol 1,2,3-Trihydroxypropane Glycerol	Oily liquid	Colorless	Odorless
	Sinks and mixes with water. Freezing point is 64° F.		

Stop discharge if possible.
Call fire department.
Isolate and remove discharged material.
Notify local health and pollution control agencies.

Fire	Combustible. Extinguish with dry chemical, alcohol foam, or carbon dioxide. Water may be ineffective on fire. Cool exposed containers with water.
Exposure	Not harmful.
Water Pollution	Effect of low concentrations on aquatic life is unknown. May be dangerous if it enters water intakes. Notify local health and wildlife officials. Notify operators of nearby water intakes.

1. RESPONSE TO DISCHARGE

(See Response Methods Handbook, CG 446-4)

Disperse and flush

2. LABELS

No hazard label required by Code of Federal Regulations

3. CHEMICAL DESIGNATIONS

3.1 **Synonyms:** Glycerol; 1, 2, 3-Propanetriol; 1, 2, 3-Trihydroxypropane

3.2 **Coast Guard Compatibility Classification:** Alcohol

3.3 **Chemical Formula:** $HOCH_2CH(OH)CH_2OH$

3.4 **IMCO/United Nations Numerical Designation:** Not listed

4. OBSERVABLE CHARACTERISTICS

4.1 **Physical State (as shipped):** Liquid

4.2 **Color:** Colorless

4.3 **Odor:** Odorless

5. HEALTH HAZARDS

5.1 **Personal Protective Equipment:** Rubber gloves, goggles.

5.2 **Symptoms Following Exposure:** No hazard

5.3 **Treatment for Exposure:** No hazard

5.4 **Toxicity by Inhalation (Threshold Limit Value):** Not pertinent

5.5 **Short-Term Inhalation Limits:** Not pertinent

5.6 **Toxicity by Ingestion:** Grade 0; LD_{50} above 15 g/kg

5.7 **Late Toxicity:** None

5.8 **Vapor (Gas) Irritant Characteristics:** Vapors are nonirritating to the eyes and throat.

5.9 **Liquid or Solid Irritant Characteristics:** No appreciable hazard. Practically harmless to the skin.

5.10 **Odor Threshold:** Not pertinent

6. FIRE HAZARDS

6.1 **Flash Point:** 350°F O.C.; 320°F C.C.

6.2 **Flammable Limits in Air:** Not pertinent

6.3 **Fire Extinguishing Agents:** Alcohol foam, dry chemical, carbon dioxide, water fog

6.4 **Fire Extinguishing Agents Not to be Used:** Water or foam may cause frothing.

6.5 **Special Hazards of Combustion Products:** Not pertinent

6.6 **Behavior in Fire:** Not pertinent

6.7 **Ignition Temperature:** 698°F

6.8 **Electrical Hazard:** Not pertinent

6.9 **Burning Rate:** 0.9 mm/min.

7. CHEMICAL REACTIVITY

7.1 **Reactivity with Water:** No reaction

7.2 **Reactivity with Common Materials:** No reaction

7.3 **Stability During Transport:** Stable

7.4 **Neutralizing Agents for Acids and Caustics:** Not pertinent

7.5 **Polymerization:** Not pertinent

7.6 **Inhibitor of Polymerization:** Not pertinent

8. WATER POLLUTION

8.1 **Aquatic Toxicity:** Data not available

8.2 **Waterfowl Toxicity:** Data not available

8.3 **Biological Oxygen Demand (BOD):** 20% (theor.), 5 days

8.4 **Food Chain Concentration Potential:** None

9. SELECTED MANUFACTURERS

1. Lever Brothers Co.
   390 Park Ave.
   New York, N. Y. 10022
2. Procter & Gamble Co.
   301 East Sixth St.
   Cincinnati, Ohio 45201
3. Shell Chemical Co.
   Industrial Chemicals Division
   Houston, Tex. 77001

10. SHIPPING INFORMATION

10.1 **Grades or Purity:** CP: 99.5%; USP: 96%

10.2 **Storage Temperature:** Ambient

10.3 **Inert Atmosphere:** No requirement

10.4 **Venting:** Open (flame arrester) or pressure-vacuum

11. HAZARD ASSESSMENT CODE

(See Hazard Assessment Handbook, CG 446-3)

A-P-Q

12. HAZARD CLASSIFICATIONS

12.1 **Code of Federal Regulations:** Not listed

12.2 **NAS Hazard Rating for Bulk Water Transportation:**

Category	Rating
Fire	1
Health	
Vapor Irritant	0
Liquid or Solid Irritant	0
Poisons	0
Water Pollution	
Human Toxicity	0
Aquatic Toxicity	0
Aesthetic Effect	0
Reactivity	
Other Chemicals	2
Water	0
Self-Reaction	0

12.3 **NFPA Hazard Classifications:**

Category	Classification
Health Hazard (Blue)	1
Flammability (Red)	1
Reactivity (Yellow)	0

13. PHYSICAL AND CHEMICAL PROPERTIES

13.1 **Physical State at 15°C and 1 atm:** Liquid

13.2 **Molecular Weight:** 92.10

13.3 **Boiling Point at 1 atm:** 554°F = 290°C = 563°K

13.4 **Freezing Point:** 64.2°F = 17.9°C = 291.1°K

13.5 **Critical Temperature:** Not pertinent

13.6 **Critical Pressure:** Not pertinent

13.7 **Specific Gravity:** 1.261 at 20°C (liquid)

13.8 **Liquid Surface Tension:** Not pertinent

13.9 **Liquid-Water Interfacial Tension:** Not pertinent

13.10 **Vapor (Gas) Specific Gravity:** Not pertinent

13.11 **Ratio of Specific Heats of Vapor (Gas):** Not pertinent

13.12 **Latent Heat of Vaporization:** 288 Btu/lb = 160 cal/g = 6.70×10^5 J/kg

13.13 **Heat of Combustion:** −7758 Btu/lb = −4310 cal/g = -180.5×10^5 J/kg

13.14 **Heat of Decomposition:** Not pertinent

13.15 **Heat of Solution:** (est.) −9 Btu/lb = −5 cal/g = -0.2×10^5 J/kg

13.16 **Heat of Polymerization:** Not pertinent

NOTES

GCM	GLYCIDYL METHACRYLATE

Common Synonyms			
Glycidyl alpha-methyl acrylate Methacrylic acid, 2,3-epoxy-propyl ester	Liquid	Colorless	Fruity odor
	Floats on water.		

Avoid contact with liquid. Keep people away.
Wear rubber overclothing (including gloves).
Call fire department.
Stop discharge if possible.
Isolate and remove discharged material.
Notify local health and pollution control agencies.

Fire	Combustible. **Wear goggles, self-contained breathing apparatus, and rubber overclothing (including gloves).** **Extinguish with foam, dry chemical, or carbon dioxide.** **Water may be ineffective on fire.**
Exposure	**CALL FOR MEDICAL AID.** **LIQUID** Will burn skin and eyes. Harmful if swallowed. **Remove contaminated clothing and shoes.** **Flush affected areas with plenty of water.** **IF IN EYES, hold eyelids open and flush with plenty of water.** **IF SWALLOWED and victim is CONSCIOUS, have victim drink water or milk.**
Water Pollution	Effect of low concentrations on aquatic life is unknown. Fouling to shoreline. May be dangerous if it enters water intakes. **Notify local health and wildlife officials.** **Notify operators of nearby water intakes.**

1. RESPONSE TO DISCHARGE
(See Response Methods Handbook, CG 446-4)

Mechanical containment
Should be removed
Chemical and physical treatment

2. LABELS

No hazard label required by Code of Federal Regulations

3. CHEMICAL DESIGNATIONS

3.1 **Synonyms:**
Glycidyl alpha-methyl acrylate
Methacrylic acid, 2, 3-epoxypropyl ester

3.2 **Coast Guard Compatibility Classification:** Not applicable

3.3 **Chemical Formula:** $CH_2{=}CH(CH_3)COOCH_2\underline{CHCH_2O}$

3.4 **IMCO/United Nations Numerical Designation:** Not listed

4. OBSERVABLE CHARACTERISTICS

4.1 **Physical State (as shipped):** Liquid

4.2 **Color:** Colorless

4.3 **Odor:** Data not available

5. HEALTH HAZARDS

5.1 **Personal Protective Equipment:** Polyethylene-coated apron and gloves and close-fitting goggles.

5.2 **Symptoms Following Exposure:** The liquid irritates eyes about as much as soap. Prolonged contact with skin produces irritation and dermatitis.

5.3 **Treatment for Exposure:** SKIN: wash thoroughly with soap and water and treat as a chemical burn. EYES: irrigate with clear water for 15 min. and get médical attention.

5.4 **Toxicity by Inhalation (Threshold Limit Value):** 25 ppm

5.5 **Short-Term Inhalation Limits:** Data not available

5.6 **Toxicity by Ingestion:** Grade 2; LD_{50} 0.5 to 5 g/kg (rat)

5.7 **Late Toxicity:** Data not available

5.8 **Vapor (Gas) Irritant Characteristics:** Vapor is moderately irritating such that personnel will not usually tolerate moderate or high vapor concentrations.

5.9 **Liquid or Solid Irritant Characteristics:** Causes smarting of the skin and first-degree burns on short exposure; may cause secondary burns on long exposure. In eyes the irritation is similar to that caused by ordinary soap.

5.10 **Odor Threshold:** Data not available

6. FIRE HAZARDS

6.1 **Flash Point:** 183°F O.C.

6.2 **Flammable Limits in Air:** Data not available

6.3 **Fire Extinguishing Agents:** Data not available

6.4 **Fire Extinguishing Agents Not to be Used:** Data not available

6.5 **Special Hazards of Combustion Products:** Irritating vapors are generated when heated

6.6 **Behavior in Fire:** Not pertinent

6.7 **Ignition Temperature:** Data not available

6.8 **Electrical Hazard:** Not pertinent

6.9 **Burning Rate:** Data not available

7. CHEMICAL REACTIVITY

7.1 **Reactivity with Water:** No reaction

7.2 **Reactivity with Common Materials:** No reaction

7.3 **Stability During Transport:** Stable

7.4 **Neutralizing Agents for Acids and Caustics:** Not pertinent

7.5 **Polymerization:** Heat, peroxides, and caustics all cause polymerization; the reaction is not considered hazardous.

7.6 **Inhibitor of Polymerization:** Hydroquinone monomethyl ether: 50 ppm

8. WATER POLLUTION

8.1 **Aquatic Toxicity:** Data not available

8.2 **Waterfowl Toxicity:** Data not available

8.3 **Biological Oxygen Demand (BOD):** Data not available

8.4 **Food Chain Concentration Potential:** None

9. SELECTED MANUFACTURERS

1. Alcolac Inc.
   3440 Fairfield Rd.
   Baltimore Md. 21226
2. American Aniline & Extract Co., Inc.
   Venango and F Sts.
   Philadelphia, Pa. 19134
3. Polysciences, Inc.
   Warrington, Pa. 18976

10. SHIPPING INFORMATION

10.1 **Grades or Purity:** Technical: 92%

10.2 **Storage Temperature:** Ambient

10.3 **Inert Atmosphere:** No requirement

10.4 **Venting:** Open (flame arrester)

11. HAZARD ASSESSMENT CODE
(See Hazard Assessment Handbook, CG 446-3)

A-T-U-X-Y-Z

12. HAZARD CLASSIFICATIONS

12.1 **Code of Federal Regulations:** Combustible Liquid

12.2 **NAS Hazard Rating for Bulk Water Transportation:** Not listed

12.3 **NFPA Hazard Classifications:** Not listed

13. PHYSICAL AND CHEMICAL PROPERTIES

13.1 **Physical State at 15°C and 1 atm:** Liquid

13.2 **Molecular Weight:** 142.2

13.3 **Boiling Point at 1 atm:** Very high

13.4 **Freezing Point:** Data not available

13.5 **Critical Temperature:** Not pertinent

13.6 **Critical Pressure:** Not pertinent

13.7 **Specific Gravity:** 1.073 at 20°C (liquid)

13.8 **Liquid Surface Tension:** (est.) 25 dynes/cm = 0.025 N/m at 20°C

13.9 **Liquid-Water Interfacial Tension:** (est.) 40 dynes/cm = 0.04 N/m at 20°C

13.10 **Vapor (Gas) Specific Gravity:** Not pertinent

13.11 **Ratio of Specific Heats of Vapor (Gas):** (est.) 1.043

13.12 **Latent Heat of Vaporization:** Not pertinent

13.13 **Heat of Combustion:** (est.) $-10{,}800$ Btu/lb $= -5{,}980$ cal/g $= -250 \times 10^5$ J/kg

13.14 **Heat of Decompcsition:** Not pertinent

13.15 **Heat of Solution:** Not pertinent

13.16 **Heat of Polymerization:** (est.) -900 Btu/lb $= -500$ cal/g $= -20 \times 10^5$ J/kg

NOTES

GOS | GLYOXAL, 40% SOLUTION

Common Synonyms

Biformyl
Diformyl
Ethanedial
Oxal
Oxaldehyde

Liquid — Light yellow — Weak sour odor

Mixes with water.

Stop discharge if possible. Keep people away.
Isolate and remove discharged material.
Notify local health and pollution control agencies.

Fire

Not flammable.

Exposure

Call for medical aid.

LIQUID
Irritating to skin and eyes.
Harmful if swallowed.
Remove contaminated clothing and shoes.
Flush affected areas with plenty of water.
IF IN EYES, hold eyelids open and flush with plenty of water.
IF SWALLOWED and victim is CONSCIOUS, have victim drink water or milk.

Water Pollution

Effect of low concentrations on aquatic life is unknown.
May be dangerous if it enters water intakes.
Notify local health and wildlife officials.
Notify operators of nearby water intakes.

1. RESPONSE TO DISCHARGE

(See Response Methods Handbook, CG 446-4)

Issue warning—water contaminant
Disperse and flush

2. LABELS

No label required by Code of Federal Regulations

3. CHEMICAL DESIGNATIONS

3.1 **Synonyms:** Biformal; Biformyl; Diformyl; Ethanedial; Oxal; Oxaldehyde
3.2 **Coast Guard Compatibility Classification:** To be developed
3.3 **Chemical Formula:** CHO–CHO (in water)
3.4 **IMCO/United Nations Numerical Designation:** Not listed

4. OBSERVABLE CHARACTERISTICS

4.1 **Physical State (as shipped):** Liquid
4.2 **Color:** Pale yellow
4.3 **Odor:** Weak sour

5. HEALTH HAZARDS

5.1 **Personal Protective Equipment:** Goggles or face shield; rubber gloves.
5.2 **Symptoms Following Exposure:** Inhalation causes some irritation of nose and throat. Contact with liquid irritates eyes and causes mild irritation of skin; stains skin yellow. (No information available on symptoms of ingestion.)
5.3 **Treatment for Exposure:** INHALATION: remove from exposure. EYES or SKIN: flood with water for 15 min. INGESTION: no information on treatment.
5.4 **Toxicity by Inhalation (Threshold Limit Value):** Data not available
5.5 **Short-Term Inhalation Limits:** Data not available
5.6 **Toxicity by Ingestion:** Grade 2; oral rat LD_{50} = 2,020 mg/kg
5.7 **Late Toxicity:** Data not available
5.8 **Vapor (Gas) Irritant Characteristics:** Vapors cause a slight smarting of the eyes or respiratory system if present in high concentrations. The effect is temporary.
5.9 **Liquid or Solid Irritant Characteristics:** Minimum hazard. If spilled on clothing and allowed to remain, may cause smarting and reddening of the skin.
5.10 **Odor Threshold:** Data not available

6. FIRE HAZARDS

6.1 **Flash Point:** Non-flammable solution
6.2 **Flammable Limits in Air:** Not pertinent
6.3 **Fire Extinguishing Agents:** Not pertinent
6.4 **Fire Extinguishing Agents Not to be Used:** Not pertinent
6.5 **Special Hazards of Combustion Products:** Not pertinent
6.6 **Behavior in Fire:** Heat may cause polymerization to a combustible, viscous material.
6.7 **Ignition Temperature:** Not pertinent
6.8 **Electrical Hazard:** Not pertinent
6.9 **Burning Rate:** Not pertinent

7. CHEMICAL REACTIVITY

7.1 **Reactivity with Water:** No reaction
7.2 **Reactivity with Common Materials:** Corrosive to most metals. The reaction is slow.
7.3 **Stability During Transport:** Stable
7.4 **Neutralizing Agents for Acids and Caustics:** Not pertinent
7.5 **Polymerization:** Not pertinent
7.6 **Inhibitor of Polymerization:** Not pertinent

8. WATER POLLUTION

8.1 **Aquatic Toxicity:** Data not available
8.2 **Waterfowl Toxicity:** Data not available
8.3 **Biological Oxygen Demand (BOD):** Data not available
8.4 **Food Chain Concentration Potential:** None

9. SELECTED MANUFACTURERS

1. Union Carbide Corporation
   Chemicals and Plastics Division
   270 Park Avenue
   New York, N. Y. 10017
2. Eastman Kodak Co.
   Eastman Organic Products
   Rochester, N. Y. 14650
3. Aldrich Chemical Co.
   940 West St. Paul Avenue
   Milwaukee, Wis. 53233

10. SHIPPING INFORMATION

10.1 **Grades or Purity:** 40% in water
10.2 **Storage Temperature:** 10—120°F
10.3 **Inert Atmosphere:** No requirement
10.4 **Venting:** Open

11. HAZARD ASSESSMENT CODE

(See Hazard Assessment Handbook, CG 446-3)

A-P

12. HAZARD CLASSIFICATIONS

12.1 **Code of Federal Regulations:** Not listed
12.2 **NAS Hazard Rating for Bulk Water Transportation:**

Category	Rating
Fire	1
Health	
Vapor Irritant	1
Liquid or Solid Irritant	1
Poisons	1
Water Pollution	
Human Toxicity	1
Aquatic Toxicity	1
Aesthetic Effect	1
Reactivity	
Other Chemicals	2
Water	0
Self-Reaction	1

12.3 **NFPA Hazard Classifications:** Not listed

13. PHYSICAL AND CHEMICAL PROPERTIES

13.1 **Physical State at 15°C and 1 atm:** Liquid
13.2 **Molecular Weight:** Mixture
13.3 **Boiling Point at 1 atm:** Data not available
13.4 **Freezing Point:** 5°F = −15°C = 258°K
13.5 **Critical Temperature:** Not pertinent
13.6 **Critical Pressure:** Not pertinent
13.7 **Specific Gravity:** 1.29 at 20°C (liquid)
13.8 **Liquid Surface Tension:** Data not available
13.9 **Liquid-Water Interfacial Tension:** Not pertinent
13.10 **Vapor (Gas) Specific Gravity:** Not pertinent
13.11 **Ratio of Specific Heats of Vapor (Gas):** Not pertinent
13.12 **Latent Heat of Vaporization:** Not pertinent
13.13 **Heat of Combustion:** Not pertinent
13.14 **Heat of Decomposition:** Not pertinent
13.15 **Heat of Solution:** Not pertinent
13.16 **Heat of Polymerization:** Not pertinent

NOTES

HTC — HEPTACHLOR

Common Synonyms			
1,4,5,6,7,8,8-Heptachloro-3a,4,7,7a-tetrahydro-4,7,-methanoindene 1,4,5,6,7,8,8a-Heptachlorodicyclopentadiene E3314	Solid	White to light tan	Camphor-like odor
	Sinks in water.		

AVOID CONTACT WITH SOLID AND DUST. KEEP PEOPLE AWAY.
Wear goggles and self-contained breathing apparatus.
Stop discharge if possible.
Isolate and remove discharged material.
Notify local health and pollution control agencies.

Fire	Not Flammable. Irritating gases may be produced when heated.
Exposure	DUST POISONOUS IF INHALED. If inhaled will cause headache or loss of consciousness. **If in eyes, hold eyelids open and flush with plenty of water.** **If breathing has stopped, give artificial respiration.** **If breathing is difficult, give oxygen.** SOLID POISONOUS IF SWALLOWED. Irritating to skin and eyes. If swallowed will cause nausea and vomiting. **Remove contaminated clothing and shoes.** **Flush affected areas with plenty of water.** **IF IN EYES, hold eyelids open and flush with plenty of water.** **IF SWALLOWED and victim is CONSCIOUS, have victim drink water or milk.** **IF SWALLOWED and victim is UNCONSCIOUS OR HAVING CONVULSIONS, do nothing except keep victim warm.**
Water Pollution	HARMFUL TO AQUATIC LIFE IN VERY LOW CONCENTRATIONS. May be dangerous if it enters water intakes. **Notify local health and wildlife officials.** **Notify operators of nearby water intakes.**

1. RESPONSE TO DISCHARGE

(See Response Methods Handbook, CG 446-4)

Issue warning—water contaminant
Should be removed
Chemical and physical treatment

2. LABELS

No hazard label required by Code of Federal Regulations

3. CHEMICAL DESIGNATIONS

3.1 **Synonyms:** E 3314; 1,4,5,6,7,8,8a-Heptachlorodicyclopentadiene; 1,4,5,6,7,8,8-Heptachloro-3a,4,7,7a-tetrahydro-4,7-methanoindene; Velsicol

3.2 **Coast Guard Compatibility Classification:** Not listed

3.3 **Chemical Formula:** $C_{10}H_5Cl_7$

3.4 **IMCO/United Nations Numerical Designation:** Not listed

4. OBSERVABLE CHARACTERISTICS

4.1 **Physical State (as shipped):** Solid

4.2 **Color:** White; light tan

4.3 **Odor:** Camphor-like

5. HEALTH HAZARDS

5.1 **Personal Protective Equipment:** Protective respirator; rubber gloves; clean clothes

5.2 **Symptoms Following Exposure:** Inhalation of dust causes irritability, tremors, and collapse. Ingestion causes nausea, vomiting, diarrhea, and irritation of the gastrointestinal tract. Contact with dust causes irritation of eyes and moderate irritation of skin.

5.3 **Treatment for Exposure:** *Get medical attention following all overexposures to heptachlor.* INHALATION: move to fresh air; if exposure to dust was severe, get medical attention. INGESTION: lavage stomach with warm tap water (unless convulsions are imminent); fats and oils should be avoided, as they increase the rate of absorption of all chlorinated hydrocarbons. EYES: wash repeatedly with water. SKIN: flush with water, then wash with soap and water.

5.4 **Toxicity by Inhalation (Threshold Limit Value):** 0.5 mg/m^3

5.5 **Short-Term Inhalation Limits:** 2 mg/m^3 for 30 min.

5.6 **Toxicity by Ingestion:** Grade 4; oral LD_{50} = 40 mg/kg (rat)

5.7 **Late Toxicity:** Liver damage may develop.

5.8 **Vapor (Gas) Irritant Characteristics:** Data not available

5.9 **Liquid or Solid Irritant Characteristics:** Data not available

5.10 **Odor Threshold:** 0.02 ppm

6. FIRE HAZARDS

6.1 **Flash Point:** Not flammable

6.2 **Flammable Limits in Air:** Not flammable

6.3 **Fire Extinguishing Agents:** Not pertinent

6.4 **Fire Extinguishing Agents Not to be Used:** Not pertinent

6.5 **Special Hazards of Combustion Products:** Irritating hydrogen chloride fumes may form in fire.

6.6 **Behavior in Fire:**

6.7 **Ignition Temperature:** Not pertinent

6.8 **Electrical Hazard:** Not pertinent

6.9 **Burning Rate:** Not pertinent

7. CHEMICAL REACTIVITY

7.1 **Reactivity with Water:** No reaction

7.2 **Reactivity with Common Materials:**

7.3 **Stability During Transport:** Stable

7.4 **Neutralizing Agents for Acids and Caustics:** Not pertinent

7.5 **Polymerization:** Not pertinent

7.6 **Inhibitor of Polymerization:** Not pertinent

8. WATER POLLUTION

8.1 **Aquatic Toxicity:**
0.1 ppm/24 hr/trout/80% kill/fresh water
0.230 ppm/96 hr/goldfish/TL_m/fresh water
0.25 ppm/48 hr/white shrimp/TL_m/salt water

8.2 **Waterfowl Toxicity:** 2000 mg/kg

8.3 **Biological Oxygen Demand (BOD):** Data not available

8.4 **Food Chain Concentration Potential:** Bioconcentration of up to 17,600 in oysters and 300 in bluegills. A spill could cause potential problem with shellfish.

9. SELECTED MANUFACTURERS

Velsicol Chemical Corp.
341 E. Ohio St.
Chicago, Ill. 60611

10. SHIPPING INFORMATION

10.1 **Grades or Purity:** Commercial, 72+%

10.2 **Storage Temperature:** Ambient

10.3 **Inert Atmosphere:** No requirement

10.4 **Venting:** Open

11. HAZARD ASSESSMENT CODE

(See Hazard Assessment Handbook, CG 446-3)

II

12. HAZARD CLASSIFICATIONS

12.1 **Code of Federal Regulations:** Not listed

12.2 **NAS Hazard Rating for Bulk Water Transportation:** Not listed

12.3 **NFPA Hazard Classifications:** Not listed

13. PHYSICAL AND CHEMICAL PROPERTIES

13.1 **Physical State at 15°C and 1 atm:** Solid

13.2 **Molecular Weight:** 373.5

13.3 **Boiling Point at 1 atm:** Not pertinent (decomposes)

13.4 **Freezing Point:** 115—165°F = 46—74°C = 319—347°K

13.5 **Critical Temperature:** Not pertinent

13.6 **Critical Pressure:** Not pertinent

13.7 **Specific Gravity:** 1.66 at 20°C (solid)

13.8 **Liquid Surface Tension:** Not pertinent

13.9 **Liquid-Water Interfacial Tension:** Not pertinent

13.10 **Vapor (Gas) Specific Gravity:** Not pertinent

13.11 **Ratio of Specific Heats of Vapor (Gas):** Not pertinent

13.12 **Latent Heat of Vaporization:** Not pertinent

13.13 **Heat of Combustion:** Not pertinent

13.14 **Heat of Decomposition:** Not pertinent

13.15 **Heat of Solution:** Not pertinent

13.16 **Heat of Polymerization:** Not pertinent

NOTES

HPT	HEPTANE

Common Synonyms			
n-Heptane	Watery liquid	Colorless	Gasoline-like odor
	Floats on water. Flammable vapor is produced.		

Stop discharge if possible. Keep people away.
Shut off ignition sources and call fire department.
Stay upwind and use water spray to "knock down" vapor.
Isolate and remove discharged material.
Notify local health and pollution control agencies.

Fire

FLAMMABLE.
Flashback along vapor trail may occur.
Vapor may explode if ignited in an enclosed area.
Extinguish with dry chemical, foam or carbon dioxide.
Water may be ineffective on fire.
Cool exposed containers with water.

Exposure

CALL FOR MEDICAL AID.

VAPOR
Not irritating to eyes, nose or throat.
If inhaled, will cause coughing or difficult breathing.
Move to fresh air.
If breathing has stopped, give artificial respiration.
If breathing is difficult, give oxygen.

LIQUID
Irritating to skin and eyes.
If swallowed, will cause nausea or vomiting.
Remove contaminated clothing and shoes.
Flush affected areas with plenty of water.
IF IN EYES, hold eyelids open and flush with plenty of water.
IF SWALLOWED and victim is CONSCIOUS, have victim drink water or milk.
DO NOT INDUCE VOMITING.

Water Pollution

Dangerous to aquatic life in high concentrations.
Fouling to shoreline.
May be dangerous if it enters water intakes.
Notify local health and wildlife officials.
Notify operators of nearby water intakes.

1. RESPONSE TO DISCHARGE

(See Response Methods Handbook, CG 446-4)

Issue warning—high flammability
Evacuate area
Disperse and flush

2. LABEL

Red

3. CHEMICAL DESIGNATIONS

3.1 **Synonyms:** n-Heptane
3.2 **Coast Guard Compatibility Classification:** Paraffin
3.3 **Chemical Formula:** $CH_3(CH_2)_5CH_3$
3.4 **IMCO/United Nations Numerical Designation:** 3.2/1206

4. OBSERVABLE CHARACTERISTICS

4.1 **Physical State (as shipped):** Liquid
4.2 **Color:** Colorless
4.3 **Odor:** Gasoline

5. HEALTH HAZARDS

5.1 **Personal Protective Equipment:** Safety glasses; gloves; similar to gasoline.
5.2 **Symptoms Following Exposure:** INHALATION: irritation of respiratory tract, coughing, depression, cardiac arrhythmias. ASPIRATION: severe lung irritation, pulmonary edema, mild excitement followed by depression. INGESTION: nausea, vomiting, swelling of abdomen, depression, headache.
5.3 **Treatment for Exposure:** INHALATION: maintain respiration; give oxygen if needed. ASPIRATION: enforce bed rest; administer oxygen. INGESTION: do NOT induce vomiting. SKIN OR EYES: remove contaminated clothing, wipe and wash skin area with soap and water; wash eyes with plenty of water.
5.4 **Toxicity by Inhalation (Threshold Limit Value):** 500 ppm
5.5 **Short-Term Inhalation Limits:** 500 ppm for 30 min.
5.6 **Toxicity by Ingestion:** Grade 0; LD_{50} above 15 g/kg
5.7 **Late Toxicity:** None
5.8 **Vapor (Gas) Irritant Characteristics:** Vapors are nonirritating to the eyes and throat.
5.9 **Liquid or Solid Irritant Characteristics:** Minimum hazard. If spilled on clothing and allowed to remain, may cause smarting and reddening of the skin.
5.10 **Odor Threshold:** 220 ppm

6. FIRE HAZARDS

6.1 **Flash Point:** 25° F C.C.
6.2 **Flammable Limits in Air:** 1.0%—7.0%
6.3 **Fire Extinguishing Agents:** Foam, dry chemical, carbon dioxide
6.4 **Fire Extinguishing Agents Not to be Used:** Not pertinent
6.5 **Special Hazards of Combustion Products:** Not pertinent
6.6 **Behavior in Fire:** Not pertinent
6.7 **Ignition Temperature:** 433°F
6.8 **Electrical Hazard:** Class I, group D
6.9 **Burning Rate:** 6.8 mm/min.

7. CHEMICAL REACTIVITY

7.1 **Reactivity with Water:** No reaction
7.2 **Reactivity with Common Materials:** No reaction
7.3 **Stability During Transport:** Stable
7.4 **Neutralizing Agents for Acids and Caustics:** Not pertinent
7.5 **Polymerization:** Not pertinent
7.6 **Inhibitor of Polymerization:** Not pertinent

8. WATER POLLUTION

8.1 **Aquatic Toxicity:** 4924 ppm/24 hr/mosquito fish/TL_m/fresh water
8.2 **Waterfowl Toxicity:** Data not available
8.3 **Biological Oxygen Demand (BOD):** 0% (theor.), 7 days
8.4 **Food Chain Concentration Potential:** None

9. SELECTED MANUFACTURERS

1. Humphrey Chemical Co.
   Devine St.
   North Haven, Conn. 06473
2. Phillips Petroleum Co.
   Bartlesville, Okla. 74004
3. Skelly Oil Co.
   Kansas City, Mo. 64141

10. SHIPPING INFORMATION

10.1 **Grades or Purity:** Various grades, all greater than 99%
10.2 **Storage Temperature:** Ambient
10.3 **Inert Atmosphere:** No requirement
10.4 **Venting:** Open (flame arrester) or pressure-vacuum

11. HAZARD ASSESSMENT CODE

(See Hazard Assessment Handbook, CG 446-3)

A-T-U-V-W

12. HAZARD CLASSIFICATIONS

12.1 **Code of Federal Regulations:** Flammable liquid
12.2 **NAS Hazard Rating for Bulk Water Transportation:**

Category	Rating
Fire	3
Health	
Vapor Irritant	0
Liquid or Solid Irritant	1
Poisons	1
Water Pollution	
Human Toxicity	1
Aquatic Toxicity	1
Aesthetic Effect	2
Reactivity	
Other Chemicals	0
Water	0
Self-Reaction	0

12.3 **NFPA Hazard Classifications:** Not listed

13. PHYSICAL AND CHEMICAL PROPERTIES

13.1 **Physical State at 15°C and 1 atm:** Liquid
13.2 **Molecular Weight:** 100.21
13.3 **Boiling Point at 1 atm:** 209.1°F = 98.4°C = 371.6°K
13.4 **Freezing Point:** −131°F = −90.6°C = 182.6°K
13.5 **Critical Temperature:** 513°F = 267°C = 540°K
13.6 **Critical Pressure:** 400 psia = 27 atm = 2.7 MN/m^2
13.7 **Specific Gravity:** 0.6838 at 20°C (liquid)
13.8 **Liquid Surface Tension:** 19.3 dynes/cm = 0.0193 N/m at 20°C
13.9 **Liquid-Water Interfacial Tension:** 51 dynes/cm = 0.051 N/m at 20°C
13.10 **Vapor (Gas) Specific Gravity:** Data not available
13.11 **Ratio of Specific Heats of Vapor (Gas):** 1.054
13.12 **Latent Heat of Vaporization:** 136.1 Btu/lb = 75.61 cal/g = 3.166×10^5 J/kg
13.13 **Heat of Combustion:** −19,170 Btu/lb = −10,650 cal/g = -445.9×10^5 J/kg
13.14 **Heat of Decomposition:** Not pertinent
13.15 **Heat of Solution:** Not pertinent
13.16 **Heat of Polymerization:** Not pertinent

NOTES

REVISED 1978

HTN HEPTANOL

Common Synonyms			
Enanthic alcohol 1-Hydroxyheptane 1-Heptanol Heptyl alcohol	Watery liquid Floats on water.	Colorless	Weak alcohol odor

Stop discharge if possible.
Call fire department.
Isolate and remove discharged material.
Notify local health and pollution control agencies.

Fire

Combustible.

Extinguish with dry chemical, foam, or carbon dioxide.
Cool exposed containers with water.

Exposure

Not harmful.

Water Pollution

Effect of low concentrations on aquatic life is unknown.
Fouling to shoreline.
May be dangerous if it enters water intakes.
Notify local health and wildlife officials.
Notify operators of nearby water intakes.

1. RESPONSE TO DISCHARGE

(See Response Methods Handbook, CG 446-4)

Mechanical containment
Should be removed
Chemical and physical treatment

2. LABELS

No hazard label required by Code of Federal Regulations

3. CHEMICAL DESIGNATIONS

3.1 **Synonyms:** Enanthic alcohol
1-Heptanol
Heptyl alcohol
1-Hydroxyheptane

3.2 **Coast Guard Compatibility Classification:** Alcohol

3.3 **Chemical Formula:** $CH_3(CH_2)_5CH_2OH$

3.4 **IMCO/United Nations Numerical Designation:** Not listed

4. OBSERVABLE CHARACTERISTICS

4.1 **Physical State (as shipped):** Liquid

4.2 **Color:** Colorless

4.3 **Odor:** Weak alcoholic

5. HEALTH HAZARDS

5.1 **Personal Protective Equipment:** Chemical goggles or face shield.

5.2 **Symptoms Following Exposure:** Low toxicity; liquid may irritate eyes.

5.3 **Treatment for Exposure:** Flush all affected parts with plenty of water.

5.4 **Toxicity by Inhalation (Threshold Limit Value):** Data not available

5.5 **Short-Term Inhalation Limits:** Data not available

5.6 **Toxicity by Ingestion:** Grade 2; oral rat LD_{50} = 1.87 g/kg

5.7 **Late Toxicity:** Data not available

5.8 **Vapor (Gas) Irritant Characteristics:** Nonirritating

5.9 **Liquid or Solid Irritant Characteristics:** Liquid may irritate eyes; it is not irritating to skin.

5.10 **Odor Threshold:** 0.49 ppm

6. FIRE HAZARDS

6.1 **Flash Point:** 170°F O.C.

6.2 **Flammable Limits in Air:** Data not available

6.3 **Fire Extinguishing Agents:** Foam, carbon dioxide, or dry chemical

6.4 **Fire Extinguishing Agents Not to be Used:** Not pertinent

6.5 **Special Hazards of Combustion Products:** Not pertinent

6.6 **Behavior in Fire:** Not pertinent

6.7 **Ignition Temperature:** Data not available

6.8 **Electrical Hazard:** Not pertinent

6.9 **Burning Rate:** 3.2 mm/min.

7. CHEMICAL REACTIVITY

7.1 **Reactivity with Water:** No reaction

7.2 **Reactivity with Common Materials:** No reaction

7.3 **Stability During Transport:** Stable

7.4 **Neutralizing Agents for Acids and Caustics:** Not pertinent

7.5 **Polymerization:** Not pertinent

7.6 **Inhibitor of Polymerization:** Not pertinent

8. WATER POLLUTION

8.1 **Aquatic Toxicity:** Data not available

8.2 **Waterfowl Toxicity:** Data not available

8.3 **Biological Oxygen Demand (BOD):** 20.4% (theor.), 5 days

8.4 **Food Chain Concentration Potential:** None

9. SELECTED MANUFACTURERS

1. NL Industries, Inc.
Baker Castor Oil Co. Division
Bayonne, N. J. 07002

2. Union Carbide Corp.
Chemicals and Plastics Division
270 Park Ave.
New York, N. Y. 10017

10. SHIPPING INFORMATION

10.1 **Grades or Purity:** Data not available

10.2 **Storage Temperature:** Ambient

10.3 **Inert Atmosphere:** No requirement

10.4 **Venting:** Open (flame arrester)

11. HAZARD ASSESSMENT CODE

(See Hazard Assessment Handbook, CG 446-3)

A-T-U

12. HAZARD CLASSIFICATIONS

12.1 **Code of Federal Regulations:** Not listed

12.2 **NAS Hazard Rating for Bulk Water Transportation:** Not listed

12.3 **NFPA Hazard Classifications:** Not listed

13. PHYSICAL AND CHEMICAL PROPERTIES

13.1 **Physical State at 15°C and 1 atm:** Liquid

13.2 **Molecular Weight:** 116.20

13.3 **Boiling Point at 1 atm:** 349°F = 176°C = 449°K

13.4 **Freezing Point:** −29°F = −34°C = 239°K

13.5 **Critical Temperature:** 680°F = 360°C = 633°K

13.6 **Critical Pressure:** 440 psia = 30 atm = 3.0 MN/m^2

13.7 **Specific Gravity:** 0.822 at 20°C (liquid)

13.8 **Liquid Surface Tension:** 26.2 dynes/cm = 0.0262 N/mn at 15°C

13.9 **Liquid-Water Interfacial Tension:** 7.7 dynes/cm = 0.0077 N/m at 25°C

13.10 **Vapor (Gas) Specific Gravity:** Not pertinent

13.11 **Ratio of Specific Heats of Vapor (Gas):** 1.049

13.12 **Latent Heat of Vaporization:** 189 Btu/lb = 105 cal/g = 4.40×10^5 J/kg

13.13 **Heat of Combustion:** −15,810 Btu/lb = −8784 cal/g = -367.8×10^5 J/kg

13.14 **Heat of Decomposition:** Not pertinent

13.15 **Heat of Solution:** Not pertinent

13.16 **Heat of Polymerization:** Not pertinent

NOTES

REVISED 1978

HTE — 1-HEPTENE

Common Synonyms			
Heptylene	Watery liquid	Colorless	Gasoline-like odor
	Floats on water. Flammable, irritating vapor is produced.		

Stop discharge if possible. Keep people away.
Shut off ignition sources and call fire department.
Stay upwind and use water spray to "knock down" vapor.
Avoid contact with liquid and vapor.
Isolate and remove discharged material.
Notify local health and pollution control agencies.

Fire

FLAMMABLE.
Flashback along vapor trail may occur.
Vapor may explode if ignited in an enclosed area.
Extinguish with dry chemical, foam, or carbon dioxide.
Water may be ineffective on fire.
Cool exposed containers with water.

Exposure

CALL FOR MEDICAL AID.

VAPOR
Irritating to eyes, nose and throat.
If inhaled, will cause dizziness or difficult breathing.
Move to fresh air.
If breathing has stopped, give artificial respiration.
If breathing is difficult, give oxygen.

LIQUID
Irritating to skin and eyes.
Harmful if swallowed.
Remove contaminated clothing and shoes.
Flush affected areas with plenty of water.
IF IN EYES, hold eyelids open and flush with plenty of water.
IF SWALLOWED and victim is CONSCIOUS, have victim drink water or milk.
DO NOT INDUCE VOMITING.

Water Pollution

Effect of low concentrations on aquatic life is unknown.
Fouling to shoreline.
May be dangerous if it enters water intakes.
Notify local health and wildlife officials.
Notify operators of nearby water intakes.

1. RESPONSE TO DISCHARGE

(See Response Methods Handbook, CG 446-4)

Issue warning—high flammability
Restrict access
Evacuate area

2. LABEL

Red

3. CHEMICAL DESIGNATIONS

3.1 **Synonyms:** Heptylene
3.2 **Coast Guard Compatibility Classification:** Olefin
3.3 **Chemical Formula:** $CH_3(CH_2)_4CH=CH_2$
3.4 **IMCO/United Nations Numerical Designation:** Not listed

4. OBSERVABLE CHARACTERISTICS

4.1 **Physical State (as shipped):** Liquid
4.2 **Color:** Colorless
4.3 **Odor:** Like gasoline

5. HEALTH HAZARDS

5.1 **Personal Protective Equipment:** Safety goggles or face shield; similar to gasoline.
5.2 **Symptoms Following Exposure:** High concentrations may produce slight irritation of eyes and respiratory tract; may also act as simple asphyxiant and slight anesthetic.
5.3 **Treatment for Exposure:** Remove from exposure. Administer artificial respiration if needed.
5.4 **Toxicity by Inhalation (Threshold Limit Value):** Data not available
5.5 **Short-Term Inhalation Limits:** Data not available
5.6 **Toxicity by Ingestion:** Data not available
5.7 **Late Toxicity:** Data not available
5.8 **Vapor (Gas) Irritant Characteristics:** Vapors cause a slight smarting of the eyes or respiratory system if present in high concentrations. The effect is temporary.
5.9 **Liquid or Solid Irritant Characteristics:** Minimum hazard. If spilled on clothing and allowed to remain, may cause smarting and reddening of the skin.
5.10 **Odor Threshold:** Data not available

6. FIRE HAZARDS

6.1 **Flash Point:** 25°F C.C. (est.)
6.2 **Flammable Limits in Air:** LEL = 1.0%; UEL not listed
6.3 **Fire Extinguishing Agents:** Foam, dry chemical, or carbon dioxide
6.4 **Fire Extinguishing Agents Not to be Used:** Not pertinent
6.5 **Special Hazards of Combustion Products:** Not pertinent
6.6 **Behavior in Fire:** Not pertinent
6.7 **Ignition Temperature:** 500°F
6.8 **Electrical Hazard:** Not pertinent
6.9 **Burning Rate:** 6.4 mm/min.

7. CHEMICAL REACTIVITY

7.1 **Reactivity with Water:** No reaction
7.2 **Reactivity with Common Materials:** No reaction
7.3 **Stability During Transport:** Stable
7.4 **Neutralizing Agents for Acids and Caustics:** Not pertinent
7.5 **Polymerization:** Not pertinent
7.6 **Inhibitor of Polymerization:** Not pertinent

8. WATER POLLUTION

8.1 **Aquatic Toxicity:** Data not available
8.2 **Waterfowl Toxicity:** Data not available
8.3 **Biological Oxygen Demand (BOD):** Data not available
8.4 **Food Chain Concentration Potential:** None

9. SELECTED MANUFACTURERS

1. American Petrofina, Inc.
   Cosden Oil and Chemical Co.
   Big Spring, Texas 79720
2. Exxon Chemical Co.
   Houston, Texas 77001
3. Phillips Petroleum Co.
   Bartlesville, Okla. 74004

10. SHIPPING INFORMATION

10.1 **Grades or Purity:** Technical
10.2 **Storage Temperature:** Ambient
10.3 **Inert Atmosphere:** No requirement
10.4 **Venting:** Open (flame arrester)

11. HAZARD ASSESSMENT CODE

(See Hazard Assessment Handbook, CG 446-3)

A-T-U-V-W

12. HAZARD CLASSIFICATIONS

12.1 **Code of Federal Regulations:** Flammable liquid
12.2 **NAS Hazard Rating for Bulk Water Transportation:**

Category	Rating
Fire	3
Health	
Vapor Irritant	1
Liquid or Solid Irritant	1
Poisons	0
Water Pollution	
Human Toxicity	0
Aquatic Toxicity	1
Aesthetic Effect	2
Reactivity	
Other Chemicals	1
Water	0
Self-Reaction	1

12.3 **NFPA Hazard Classifications:** Not listed

13. PHYSICAL AND CHEMICAL PROPERTIES

13.1 **Physical State at 15°C and 1 atm:** Liquid
13.2 **Molecular Weight:** 98.18
13.3 **Boiling Point at 1 atm:** 200.5°F = 93.6°C = 366.8°K
13.4 **Freezing Point:** −182°F = −119°C = 154°K
13.5 **Critical Temperature:** 507.4°F = 264.1°C = 537.3°K
13.6 **Critical Pressure:** 420 psia = 28.57 atm = 2.89 MN/m^2
13.7 **Specific Gravity:** 0.697 at 20°C (liquid)
13.8 **Liquid Surface Tension:** 20.5 dynes/cm = 0.0205 N/m at 20°C
13.9 **Liquid-Water Interfacial Tension:** (est.) 50 dynes/cm = 0.05 N/m at 20°C
13.10 **Vapor (Gas) Specific Gravity:** 3.4
13.11 **Ratio of Specific Heats of Vapor (Gas):** 1.057
13.12 **Latent Heat of Vaporization:** 137 Btu/lb = 76.3 cal/g = 3.20×10^5 J/kg
13.13 **Heat of Combustion:** −19,377 Btu/lb = −10,765 cal/g = -450.71×10^5 J/kg
13.14 **Heat of Decomposition:** Not pertinent
13.15 **Heat of Solution:** Not pertinent
13.16 **Heat of Polymerization:** Not pertinent

NOTES

REVISED 1978

HCC	# HEXACHLOROCYCLOPENTADIENE

Common Synonyms Perchlorocyclopentadiene	Liquid Greenish yellow Harsh, unpleasant odor Sinks in water.

AVOID CONTACT WITH LIQUID AND VAPOR. KEEP PEOPLE AWAY.
Wear goggles, self-contained breathing apparatus, and rubber overclothing (including gloves).
Stop discharge if possible.
Isolate and remove discharged material.
Notify local health and pollution control agencies.

Fire	Not flammable. Poisonous gases may be produced when heated.
 **Exposure**	CALL FOR MEDICAL AID. VAPOR POISONOUS IF INHALED. If inhaled will cause coughing or difficult breathing. **If in eyes, hold eyelids open and flush with plenty of water.** **If breathing has stopped, give artificial respiration.** **If breathing is difficult, give oxygen.** LIQUID Will burn skin and eyes. If swallowed will cause nausea and vomiting.. **Remove contaminated clothing and shoes.** **Flush affected areas with plenty of water.** **IF IN EYES, hold eyelids open and flush with plenty of water.** **IF SWALLOWED and victim is CONSCIOUS, have victim drink water or milk and have victim induce vomiting.** **IF SWALLOWED and victim is UNCONSCIOUS OR HAVING CONVULSIONS, do nothing except keep victim warm.**
Water Pollution	Dangerous to aquatic life in high concentrations. May be dangerous if it enters water intakes. **Notify local health and wildlife officials.** **Notify operators of nearby water intakes.**

1. RESPONSE TO DISCHARGE

(See Response Methods Handbook, CG 446-4)

Issue warning—poison, water contaminant, air contaminant
Restrict access
Should be removed
Chemical and physical treatment

2. LABEL

3. CHEMICAL DESIGNATIONS

3.1 **Synonyms:** Perchlorocyclopentadiene
3.2 **Coast Guard Compatibility Classification:** Not listed
3.3 **Chemical Formula:** C_5Cl_6
3.4 **IMCO/United Nations Numerical Designation:** Not listed

4. OBSERVABLE CHARACTERISTICS

4.1 **Physical State (as shipped):** Liquid
4.2 **Color:** Yellow or greenish yellow
4.3 **Odor:** Harsh, pungent

5. HEALTH HAZARDS

5.1 **Personal Protective Equipment:** Protective clothing, including rubber gloves and rubber shoes or boots; self-contained breathing apparatus; face shield
5.2 **Symptoms Following Exposure:** Inhalation of mist is highly irritating to mucous membranes, causing lachrymation, sneezing, and salivation; pulmonary edema may occur. Ingestion causes nausea, vomiting, diarrhea, depression. Contact with eyes causes severe irritation. Liquid is extremely irritating to the skin, causing blistering and burning.
5.3 **Treatment for Exposure:** INHALATION: remove victim to fresh air; give artificial respiration and/or oxygen as needed. INGESTION: give large amounts of water and induce vomiting; give saline laxative. EYES: flush with water for at least 15 min.; if irritation remains, get medical attention. SKIN: wash with soap and water until no odor remains.
5.4 **Toxicity by Inhalation (Threshold Limit Value):** 0.1 ppm (proposed)
5.5 **Short-Term Inhalation Limits:** Data not available
5.6 **Toxicity by Ingestion:** Grade 4; oral LD_{50} = 0.505 mg/kg (mouse), 113 mg/kg (rat)
5.7 **Late Toxicity:** Data not available
5.8 **Vapor (Gas) Irritant Characteristics:** Data not available
5.9 **Liquid or Solid Irritant Characteristics:** Data not available
5.10 **Odor Threshold:** 0.15—0.33 ppm

6. FIRE HAZARDS

6.1 **Flash Point:** Not flammable
6.2 **Flammable Limits in Air:** Not flammable
6.3 **Fire Extinguishing Agents:** Not pertinent
6.4 **Fire Extinguishing Agents Not to be Used:** If water is used on adjacent fires, do not allow water to enter drums or storage tanks.
6.5 **Special Hazards of Combustion Products:** Toxic hydrogen chloride, chlorine, and phosgene gases may form in fires.
6.6 **Behavior in Fire:**
6.7 **Ignition Temperature:** Not pertinent
6.8 **Electrical Hazard:** Not pertinent
6.9 **Burning Rate:** Not pertinent

7. CHEMICAL REACTIVITY

7.1 **Reactivity with Water:** Reacts slowly to form hydrochloric acid. The reaction is not hazardous.
7.2 **Reactivity with Common Materials:** In presence of moisture, will corrode iron and other metals. Flammable and explosive hydrogen gas may collect in enclosed space.
7.3 **Stability During Transport:** Stable
7.4 **Neutralizing Agents for Acids and Caustics:** Rinse with dilute solution of sodium bicarbonate or soda ash.
7.5 **Polymerization:** Not pertinent
7.6 **Inhibitor of Polymerization:** Not pertinent

8. WATER POLLUTION

8.1 **Aquatic Toxicity:** Highly toxic
8.2 **Waterfowl Toxicity:** Highly toxic
8.3 **Biological Oxygen Demand (BOD):** Data not available
8.4 **Food Chain Concentration Potential:** Possible accumulation of breakdown products

9. SELECTED MANUFACTURERS

1. Hooker Chemical Corporation
   Industrial Chemicals Division
   Niagara Falls, N. Y. 14302
2. Velsicol Chemical Corporation
   341 East Ohio Street
   Chicago, Ill. 60611

10. SHIPPING INFORMATION

10.1 **Grades or Purity:** Commercial, 97+%; Synthesis grade
10.2 **Storage Temperature:** Ambient
10.3 **Inert Atmosphere:** No requirement
10.4 **Venting:** Open

11. HAZARD ASSESSMENT CODE

(See Hazard Assessment Handbook, CG 446-3)

A-X

12. HAZARD CLASSIFICATIONS

12.1 **Code of Federal Regulations:** Poison, Class B
12.2 **NAS Hazard Rating for Bulk Water Transportation:** Not listed
12.3 **NFPA Hazard Classifications:** Not listed

13. PHYSICAL AND CHEMICAL PROPERTIES

13.1 **Physical State at 15°C and 1 atm:** Liquid
13.2 **Molecular Weight:** 272.7
13.3 **Boiling Point at 1 atm:** 462°F = 239°C = 512°K
13.4 **Freezing Point:** 50°F = 10°C = 283°K
13.5 **Critical Temperature:** Not pertinent
13.6 **Critical Pressure:** Not pertinent
13.7 **Specific Gravity:** 1.71 at 20°C (liquid)
13.8 **Liquid Surface Tension:** 37.5 dynes/cm = 0.0375 N/m at 20°C
13.9 **Liquid-Water Interfacial Tension:** Data not available
13.10 **Vapor (Gas) Specific Gravity:** 9.42
13.11 **Ratio of Specific Heats of Vapor (Gas):** Data not available
13.12 **Latent Heat of Vaporization:** (est.) 76 Btu/lb = 42 cal/g = 1.8×10^5 J/kg
13.13 **Heat of Combustion:** Data not available
13.14 **Heat of Decomposition:** Not pertinent
13.15 **Heat of Solution:** Not pertinent
13.16 **Heat of Polymerization:** Not pertinent

NOTES

HSS — HEXADECYL SULFATE, SODIUM SALT

Common Synonyms			
Sodium cetyl sulfate solution Cetyl sodium sulfate	Solid paste or liquid	White	Mild odor
	May float or sink in water.		

Stop discharge if possible. Keep people away.
Isolate and remove discharged material.
Notify local health and pollution control agencies.

Fire	Not Flammable.
Exposure	CALL FOR MEDICAL AID. LIQUID OR SOLID Irritating to skin and eyes. Harmful if swallowed. Remove contaminated clothing and shoes. Flush affected areas with plenty of water. IF IN EYES, hold eyelids open and flush with plenty of water. IF SWALLOWED and victim is CONSCIOUS, have victim drink water or milk. IF SWALLOWED and victim is UNCONSCIOUS OR HAVING CONVULSIONS, do nothing except keep victim warm.
Water Pollution	Effect of low concentrations on aquatic life is unknown. May be dangerous if it enters water intakes. Notify local health and wildlife officials. Notify operators of nearby water intakes.

1. RESPONSE TO DISCHARGE
(See Response Methods Handbook, CG 446-4)

Disperse and flush

2. LABELS

No hazard label required by Code of Federal Regulations

3. CHEMICAL DESIGNATIONS

3.1 **Synonyms:** Cetyl sodium sulfate; Sodium cetyl sulfate solution
3.2 **Coast Guard Compatibility Classification:** Not listed
3.3 **Chemical Formula:** $CH_3(CH_2)_{14}CH_2OSO_3Na-H_2O$
3.4 **IMCO/United Nations Numerical Designation:** Not listed

4. OBSERVABLE CHARACTERISTICS

4.1 **Physical State (as shipped):** Pasty solid or liquid
4.2 **Color:** White
4.3 **Odor:** Mild

5. HEALTH HAZARDS

5.1 **Personal Protective Equipment:** Plastic or rubber gloves; goggles or face shield
5.2 **Symptoms Following Exposure:** Contact with eyes causes mild irritation. May cause skin to dry out and become irritated.
5.3 **Treatment for Exposure:** EYES or SKIN: flush with water.
5.4 **Toxicity by Inhalation (Threshold Limit Value):** Data not available
5.5 **Short-Term Inhalation Limits:** Data not available
5.6 **Toxicity by Ingestion:** Data not available
5.7 **Late Toxicity:** Data not available
5.8 **Vapor (Gas) Irritant Characteristics:** Data not available
5.9 **Liquid or Solid Irritant Characteristics:** Data not available
5.10 **Odor Threshold:** Data not available

6. FIRE HAZARDS

6.1 **Flash Point:** Not flammable
6.2 **Flammable Limits in Air:** Not flammable
6.3 **Fire Extinguishing Agents:** Not pertinent
6.4 **Fire Extinguishing Agents Not to be Used:** Not pertinent
6.5 **Special Hazards of Combustion Products:**
6.6 **Behavior in Fire:**
6.7 **Ignition Temperature:** Not pertinent
6.8 **Electrical Hazard:** Not pertinent
6.9 **Burning Rate:** Not pertinent

7. CHEMICAL REACTIVITY

7.1 **Reactivity with Water:** No reaction
7.2 **Reactivity with Common Materials:**
7.3 **Stability During Transport:** Stable
7.4 **Neutralizing Agents for Acids and Caustics:** Not pertinent
7.5 **Polymerization:** Not pertinent
7.6 **Inhibitor of Polymerization:** Not pertinent

8. WATER POLLUTION

8.1 **Aquatic Toxicity:** 0.78 ppm/24 hr/oryzias/TL_m
8.2 **Waterfowl Toxicity:** Data not available
8.3 **Biological Oxygen Demand (BOD):** Data not available
8.4 **Food Chain Concentration Potential:** None

9. SELECTED MANUFACTURERS

Alcolac Inc.
3440 Fairfield Rd.
Baltimore, Md. 21226

10. SHIPPING INFORMATION

10.1 **Grades or Purity:** Commercial
10.2 **Storage Temperature:** Ambient
10.3 **Inert Atmosphere:** No requirement
10.4 **Venting:** Open

11. HAZARD ASSESSMENT CODE
(See Hazard Assessment Handbook, CG 446-3)

(Liquid) A-P
(Solid) SS

12. HAZARD CLASSIFICATIONS

12.1 **Code of Federal Regulations:** Not listed
12.2 **NAS Hazard Rating for Bulk Water Transportation:** Not listed
12.3 **NFPA Hazard Classifications:** Not listed

13. PHYSICAL AND CHEMICAL PROPERTIES

13.1 **Physical State at 15°C and 1 atm:** Solid or liquid
13.2 **Molecular Weight:** Not pertinent (mixture)
13.3 **Boiling Point at 1 atm:** Not pertinent
13.4 **Freezing Point:** Not pertinent
13.5 **Critical Temperature:** Not pertinent
13.6 **Critical Pressure:** Not pertinent
13.7 **Specific Gravity:** 1 at 20°C (liquid)
13.8 **Liquid Surface Tension:** Not pertinent
13.9 **Liquid-Water Interfacial Tension:** Not pertinent
13.10 **Vapor (Gas) Specific Gravity:** Not pertinent
13.11 **Ratio of Specific Heats of Vapor (Gas):** Not pertinent
13.12 **Latent Heat of Vaporization:** Not pertinent
13.13 **Heat of Combustion:** Not pertinent
13.14 **Heat of Decomposition:** Not pertinent
13.15 **Heat of Solution:** Not pertinent
13.16 **Heat of Polymerization:** Not pertinent

NOTES

HAC — HEXADECYLTRIMETHYLAMMONIUM CHLORIDE

Common Synonyms			
Cetyltrimethylammonium chloride solution	Liquid	Clear to pale yellow	Rubbing alcohol odor
	Floats or sinks in water.		

Stop discharge if possible. Keep people away.
Shut off ignition sources. Call fire department.
Avoid contact with liquid and vapor.
Stay upwind and use water spray to knock down vapor.
Isolate and remove discharged material.
Notify local health and pollution control agencies.

Fire

FLAMMABLE.
Irritating gases may be produced when heated.
Flashback vapor along vapor trail may occur.
Vapor may explode if ignited in an enclosed area.

Extinguish with dry chemicals, foam, or carbon dioxide.
Water may be ineffective on fire.
Cool exposed containers with water.

Exposure

CALL FOR MEDICAL AID.

VAPOR
Irritating to eyes, nose and throat.
Harmful if inhaled.

If in eyes, hold eyelids open and flush with plenty of water.
If breathing has stopped, give artificial respiration.
If breathing is difficult, give oxygen.

LIQUID
Irritating to skin and eyes.
Harmful if swallowed.

Remove contaminated clothing and shoes.
Flush affected areas with plenty of water.
IF IN EYES, hold eyelids open and flush with plenty of water.
IF SWALLOWED and victim is CONSCIOUS, have victim drink water or milk.
IF SWALLOWED and victim is UNCONSCIOUS OR HAVING CONVULSIONS, do nothing except keep victim warm.
DO NOT INDUCE VOMITING.

Water Pollution

Effect of low concentrations on aquatic life is unknown.
May be dangerous if it enters water intakes.

Notify local health and wildlife officials.
Notify operators of nearby water intakes.

1. RESPONSE TO DISCHARGE
(See Response Methods Handbook, CG 446-4)

Issue warning—water contaminant
Disperse and flush

2. LABELS

No hazard label required by Code of Federal Regulations

3. CHEMICAL DESIGNATIONS

3.1 **Synonyms:** Cetyltrimethylammonium chloride solution

3.2 **Coast Guard Compatibility Classification:** Not listed

3.3 **Chemical Formula:** $C_{16}H_{33}(CH_3)_3NCl - H_2O - (CH_3)_2CHOH$

3.4 **IMCO/United Nations Numerical Designation:** Not listed

4. OBSERVABLE CHARACTERISTICS

4.1 **Physical State (as shipped):** Liquid

4.2 **Color:** Almost clear to pale yellow

4.3 **Odor:** Like rubbing alcohol

5. HEALTH HAZARDS

5.1 **Personal Protective Equipment:** Goggles or face shield; rubber gloves

5.2 **Symptoms Following Exposure:** Ingestion may produce toxic effects. Contact with eyes or skin may cause severe damage.

5.3 **Treatment for Exposure:** INGESTION: do NOT induce vomiting; drink large quantities of fluid and call a physician immediately. EYES: flush with water for at least 15 min. and call a physician. SKIN: flush with water.

5.4 **Toxicity by Inhalation (Threshold Limit Value):** Data not available

5.5 **Short-Term Inhalation Limits:** Data not available

5.6 **Toxicity by Ingestion:** Grade 3; LD_{50} 250 mg/kg(rat)

5.7 **Late Toxicity:** Data not available

5.8 **Vapor (Gas) Irritant Characteristics:** Data not available

5.9 **Liquid or Solid Irritant Characteristics:** Data not available

5.10 **Odor Threshold:** Data not available

6. FIRE HAZARDS

6.1 **Flash Point:** 69°F C.C.; (for isopropyl alcohol solutions only)

6.2 **Flammable Limits in Air:** 2%—12% (isopropyl alcohol)

6.3 **Fire Extinguishing Agents:** Dry chemical, alcohol foam, carbon dioxide

6.4 **Fire Extinguishing Agents Not to be Used:** Water may be ineffective

6.5 **Special Hazards of Combustion Products:** Irritating fumes of hydrogen chloride may form in fires.

6.6 **Behavior in Fire:** Solvent vapors are heavier than air and may travel to a source of ignition and flash back.

6.7 **Ignition Temperature:** 750°F (isopropyl alcohol)

6.8 **Electrical Hazard:** (for isopropyl alcohol) Class I, Group D

6.9 **Burning Rate:** (isopropyl alcohol solutions) 2.3 mm/min.

7. CHEMICAL REACTIVITY

7.1 **Reactivity with Water:** No reaction

7.2 **Reactivity with Common Materials:**

7.3 **Stability During Transport:** Stable

7.4 **Neutralizing Agents for Acids and Caustics:** Not pertinent

7.5 **Polymerization:** Not pertinent

7.6 **Inhibitor of Polymerization:** Not pertinent

8. WATER POLLUTION

8.1 **Aquatic Toxicity:** Data not available

8.2 **Waterfowl Toxicity:** Data not available

8.3 **Biological Oxygen Demand (BOD):** Data not available

8.4 **Food Chain Concentration Potential:** None

9. SELECTED MANUFACTURERS

1. Armak Chemicals Div.
   Box 1805
   Chicago, Ill. 60690
2. Lonza, Inc.
   Fair Lawn, N.J. 07410
3. Ashland Chemical Co.
   P.O. Box 2219
   Columbus, Ohio 43216

10. SHIPPING INFORMATION

10.1 **Grades or Purity:** 25% solution in water; 50% solution in a mixture of isopropyl alcohol and water, which is flammable. If the chemical (or a solution with concentration greater than 50%) is shipped, contact with skin and eyes should be avoided.

10.2 **Storage Temperature:** Ambient

10.3 **Inert Atmosphere:** No requirement

10.4 **Venting:** Open

11. HAZARD ASSESSMENT CODE
(See Hazard Assessment Handbook, CG 446-3)

A-P

12. HAZARD CLASSIFICATIONS

12.1 **Code of Federal Regulations:** Not listed

12.2 **NAS Hazard Rating for Bulk Water Transportation:** Not listed

12.3 **NFPA Hazard Classifications:** Not listed

13. PHYSICAL AND CHEMICAL PROPERTIES

13.1 **Physical State at 15°C and 1 atm:** Liquid

13.2 **Molecular Weight:** 319 (solute only)

13.3 **Boiling Point at 1 atm:** (isopropyl alcohol) 180°F = 82.3°C = 355.5°K

13.4 **Freezing Point:** Not pertinent

13.5 **Critical Temperature:** Not pertinent

13.6 **Critical Pressure:** Not pertinent

13.7 **Specific Gravity:** (approx.) 0.9 at 25°C (liquid)

13.8 **Liquid Surface Tension:** Not pertinent

13.9 **Liquid-Water Interfacial Tension:** Not pertinent

13.10 **Vapor (Gas) Specific Gravity:** Not pertinent

13.11 **Ratio of Specific Heats of Vapor (Gas):** Not pertinent

13.12 **Latent Heat of Vaporization:** Not pertinent

13.13 **Heat of Combustion:** Not pertinent

13.14 **Heat of Decomposition:** Not pertinent

13.15 **Heat of Solution:** Not pertinent

13.16 **Heat of Polymerization:** Not pertinent

NOTES

HAL	n-HEXALDEHYDE

Common Synonyms			
Capronic aldehyde Caproaldehyde Hexanal Capronaldehyde n-Caproylaldehyde	Liquid	Colorless	Sharp unpleasant odor
	Floats on water. Flammable, irritating vapor is produced.		

Stop discharge if possible. Keep people away.
Shut off ignition sources. Call fire department.
Avoid contact with liquid and vapor.
Isolate and remove discharged material.
Notify local health and pollution control agencies.

Fire	Combustible. Flashback along vapor trail may occur. Vapor may explode if ignited in an enclosed area. **Extinguish with dry chemicals, foam or carbon dioxide.** **Water may be ineffective on fire.** **Cool exposed containers with water.**
Exposure	VAPOR Irritating to eyes, nose and throat. Harmful if inhaled. **If in eyes, hold eyelids open and flush with plenty of water.** **If breathing has stopped, give artificial respiration.** **If breathing is difficult, give oxygen.** LIQUID Irritating to skin and eyes. If swallowed will cause nausea and vomiting. **Remove contaminated clothing and shoes.** **Flush affected areas with plenty of water.** **IF IN EYES, hold eyelids open and flush with plenty of water.** **IF SWALLOWED and victim is CONSCIOUS, have victim drink water or milk and have victim induce vomiting.** **IF SWALLOWED and victim is UNCONSCIOUS OR HAVING CONVULSIONS, do nothing except keep victim warm.**
Water Pollution	Effect of low concentrations on aquatic life is unknown. Fouling to shoreline. May be dangerous if it enters water intakes. **Notify local health and wildlife officials.** **Notify operators of nearby water intakes.**

1. RESPONSE TO DISCHARGE
(See Response Methods Handbook, CG 446-4)

Issue warning—high flammability
Restrict access
Mechanical containment
Should be removed
Chemical and physical treatment

2. LABEL

Red

3. CHEMICAL DESIGNATIONS

3.1 **Synonyms:** Caproaldehyde; Caproic aldehyde; Capronaldehyde; n-Caproylaldehyde; Hexanal
3.2 **Coast Guard Compatibility Classification:** Not listed
3.3 **Chemical Formula:** $CH_3(CH_2)_4CHO$
3.4 **IMCO/United Nations Numerical Designation:** 3.3/1207

4. OBSERVABLE CHARACTERISTICS

4.1 **Physical State (as shipped):** Liquid
4.2 **Color:** Colorless
4.3 **Odor:** Pungent

5. HEALTH HAZARDS

5.1 **Personal Protective Equipment:** Goggles or face shield; rubber gloves
5.2 **Symptoms Following Exposure:** Ingestion causes irritation of mouth and stomach. Contact with vapor or liquid irritates eyes. Liquid irritates skin.
5.3 **Treatment for Exposure:** INGESTION: give large amount of water and induce vomiting. EYES: flush with water for at least 15 min. SKIN: wipe off; wash with soap and water.
5.4 **Toxicity by Inhalation (Threshold Limit Value):** Data not available
5.5 **Short-Term Inhalation Limits:** Data not available
5.6 **Toxicity by Ingestion:** Grade 2; oral LD_{50} = 4,890 mg/kg (rat)
5.7 **Late Toxicity:** Data not available
5.8 **Vapor (Gas) Irritant Characteristics:** Data not available
5.9 **Liquid or Solid Irritant Characteristics:** Data not available
5.10 **Odor Threshold:** Data not available

6. FIRE HAZARDS

6.1 **Flash Point:** 90°F O.C.
6.2 **Flammable Limits in Air:** Data not available
6.3 **Fire Extinguishing Agents:** Dry chemical, foam, carbon dioxide
6.4 **Fire Extinguishing Agents Not to be Used:** Water may be ineffective.
6.5 **Special Hazards of Combustion Products:**
6.6 **Behavior in Fire:** Vapor is heavier than air and may travel to a source of ignition and flash back.
6.7 **Ignition Temperature:** Data not available
6.8 **Electrical Hazard:** Data not available
6.9 **Burning Rate:** 5.21 mm/min.

7. CHEMICAL REACTIVITY

7.1 **Reactivity with Water:** No reaction
7.2 **Reactivity with Common Materials:** May attack some forms of plastics.
7.3 **Stability During Transport:** Stable
7.4 **Neutralizing Agents for Acids and Caustics:** Not pertinent
7.5 **Polymerization:** Not pertinent
7.6 **Inhibitor of Polymerization:** Not pertinent

8. WATER POLLUTION

8.1 **Aquatic Toxicity:** Data not available
8.2 **Waterfowl Toxicity:** Data not available
8.3 **Biological Oxygen Demand (BOD):** Data not available
8.4 **Food Chain Concentration Potential:** None

9. SELECTED MANUFACTURERS

1. Aldrich Chemical Co.
   940 W. Saint Paul Ave.
   Milwaukee, Wis. 53233
2. Pfaltz & Bauer, Inc.
   375 Fairfield Ave.
   Stamford, Conn. 06902

10. SHIPPING INFORMATION

10.1 **Grades or Purity:** 99+%; Commercial
10.2 **Storage Temperature:** Ambient
10.3 **Inert Atmosphere:** No requirement
10.4 **Venting:** Open (flame arrester)

11. HAZARD ASSESSMENT CODE
(See Hazard Assessment Handbook, CG 446-3)

A-T-U

12. HAZARD CLASSIFICATIONS

12.1 **Code of Federal Regulations:** Flammable liquid
12.2 **NAS Hazard Rating for Bulk Water Transportation:** Not listed
12.3 **NFPA Hazard Classifications:**

Category	Classification
Health Hazard (Blue)	2
Flammability (Red)	3
Reactivity (Yellow)	1

13. PHYSICAL AND CHEMICAL PROPERTIES

13.1 **Physical State at 15°C and 1 atm:** Liquid
13.2 **Molecular Weight:** 100
13.3 **Boiling Point at 1 atm:** 262°F = 128°C = 401°K
13.4 **Freezing Point:** Not pertinent
13.5 **Critical Temperature:** Not pertinent
13.6 **Critical Pressure:** Not pertinent
13.7 **Specific Gravity:** 0.83 at 20°C (liquid)
13.8 **Liquid Surface Tension:** Data not available
13.9 **Liquid-Water Interfacial Tension:** Data not available
13.10 **Vapor (Gas) Specific Gravity:** 3.5
13.11 **Ratio of Specific Heats of Vapor (Gas):** (est.) 1.061 at 20°
13.12 **Latent Heat of Vaporization:** (est.) 153 Btu/lb = 85 cal/g = 3.6×10^5 J/kg
13.13 **Heat of Combustion:** (est.) −17,000 Btu/lb = −9,430 cal/g = -394×10^5 J/kg
13.14 **Heat of Decomposition:** Not pertinent
13.15 **Heat of Solution:** Not pertinent
13.16 **Heat of Polymerization:** Not pertinent

NOTES

HMD HEXAMETHYLENEDIAMINE

Common Synonyms
1,6-Hexanediamine
1,6-Diaminohexane

Solid or watery liquid   Colorless   Weak ammonia odor

Floats and mixes with water.

Avoid contact with liquid. Keep people away.
Wear goggles, self-contained breathing apparatus, and rubber overclothing (including gloves).
Stop discharge if possible.
Call fire department.
Isolate and remove discharged material.
Notify local health and pollution control agencies.

Fire

Combustible.
Wear goggles, self-contained breathing apparatus, and rubber overclothing (including gloves).
Extinguish with water, foam, dry chemical, or carbon dioxide.

Exposure

CALL FOR MEDICAL AID.

LIQUID OR SOLID
POISONOUS IF SWALLOWED OR IF SKIN IS EXPOSED.
Will burn eyes.
Remove contaminated clothing and shoes.
Flush affected areas with plenty of water.
IF IN EYES, hold eyelids open and flush with plenty of water.
IF SWALLOWED and victim is CONSCIOUS, have victim drink water or milk.

Water Pollution

Effect of low concentrations on aquatic life is unknown.
May be dangerous if it enters water intakes.
Notify local health and wildlife officials.
Notify operators of nearby water intakes.

1. RESPONSE TO DISCHARGE
(See Response Methods Handbook, CG 446-4)

Issue warning—corrosive, water contaminant
Disperse and flush

2. LABEL

3. CHEMICAL DESIGNATIONS

3.1 **Synonyms:** 1,6-Diaminohexane; 1,6-Hexanediamine; HMDA
3.2 **Coast Guard Compatibility Classification:** Amine
3.3 **Chemical Formula:** $NH_2(CH_2)_6NH_2$
3.4 **IMCO/United Nations Numerical Designation:** 8.0/1783

4. OBSERVABLE CHARACTERISTICS

4.1 **Physical State (as shipped):** Solid (anhydrous) or liquid (70% solution)
4.2 **Color:** Glassy solid; clear liquid
4.3 **Odor:** Weak, fishy

5. HEALTH HAZARDS

5.1 **Personal Protective Equipment:** Protective clothing; eye protection.
5.2 **Symptoms Following Exposure:** Vapors cause irritation of eyes and respiratory tract. Liquid irritates eyes and skin, may cause dermatitis.
5.3 **Treatment for Exposure:** SKIN OR EYES: flush immediately with water for 15 min.; call a physician.
5.4 **Toxicity by Inhalation (Threshold Limit Value):** Data not available
5.5 **Short-Term Inhalation Limits:** Data not available
5.6 **Toxicity by Ingestion:** Data not available
5.7 **Late Toxicity:** Repeated exposure can cause anemia and damage kidney and liver.
5.8 **Vapor (Gas) Irritant Characteristics:** Vapors cause a slight smarting of the eyes or respiratory system if present in high concentrations. The effect is temporary.
5.9 **Liquid or Solid Irritant Characteristics:** Causes smarting of the skin and first-degree burns on short exposure; may cause secondary burns on long exposure.
5.10 **Odor Threshold:** 0.0041 mg/m^3

6. FIRE HAZARDS

6.1 **Flash Point:** 160°F O.C.
6.2 **Flammable Limits in Air:** 0.7%—6.3%
6.3 **Fire Extinguishing Agents:** Data not available
6.4 **Fire Extinguishing Agents Not to be Used:** Data not available
6.5 **Special Hazards of Combustion Products:** Data not available
6.6 **Behavior in Fire:** Data not available
6.7 **Ignition Temperature:** Data not available
6.8 **Electrical Hazard:** Data not available
6.9 **Burning Rate:** Data not available

7. CHEMICAL REACTIVITY

7.1 **Reactivity with Water:** No reaction
7.2 **Reactivity with Common Materials:** No reaction
7.3 **Stability During Transport:** Stable
7.4 **Neutralizing Agents for Acids and Caustics:** Flush with water.
7.5 **Polymerization:** Not pertinent
7.6 **Inhibitor of Polymerization:** Not pertinent

8. WATER POLLUTION

8.1 **Aquatic Toxicity:** Data not available
8.2 **Waterfowl Toxicity:** Data not available
8.3 **Biological Oxygen Demand (BOD):** Data not available
8.4 **Food Chain Concentration Potential:** None

9. SELECTED MANUFACTURERS

1. Celanese Corp.
   Celanese Chemical Co. Division
   245 Park Ave.
   New York, N. Y. 10017
2. E. I. duPont de Nemours & Co., Inc.
   Plastics Department
   Wilmington, Del. 19898
3. Monsanto Co.
   Monsanto Textiles Co.
   800 N. Lindbergh Blvd.
   St. Louis, Mo. 63166

10. SHIPPING INFORMATION

10.1 **Grades or Purity:** Anhydrous: 99.8%; 70% solution
10.2 **Storage Temperature:** Ambient
10.3 **Inert Atmosphere:** Nitrogen
10.4 **Venting:** Pressure-vacuum

11. HAZARD ASSESSMENT CODE
(See Hazard Assessment Handbook, CG 446-3)

A-P-Q

12. HAZARD CLASSIFICATIONS

12.1 **Code of Federal Regulations:** Corrosive material
12.2 **NAS Hazard Rating for Bulk Water Transportation:**

Category	Rating
Fire	1
Health	
Vapor Irritant	1
Liquid or Solid Irritant	2
Poisons	3
Water Pollution	
Human Toxicity	2
Aquatic Toxicity	3
Aesthetic Effect	1
Reactivity	
Other Chemicals	3
Water	0
Self-Reaction	0

12.3 **NFPA Hazard Classifications:** Not listed

13. PHYSICAL AND CHEMICAL PROPERTIES

13.1 **Physical State at 15°C and 1 atm:** Solid (anhydrous); Liquid (70% solution)
13.2 **Molecular Weight:** 116.21
13.3 **Boiling Point at 1 atm:** 478°K = 205°C = 401°F
13.4 **Freezing Point:** (anhyd.) 104.9°F = 40.5°C = 313.7°K; (70% soln.) 28°F = −2°C = 269°K
13.5 **Critical Temperature:** Not pertinent
13.6 **Critical Pressure:** Not pertinent
13.7 **Specific Gravity:** (anhyd.) 0.799 at 60°C (liquid); (70% soln.) 0.933 at 20°C (liquid)
13.8 **Liquid Surface Tension:** (anhyd.) 34.6 dynes/cm = 0.0346 N/m at 60°C
13.9 **Liquid-Water Interfacial Tension:** Not pertinent
13.10 **Vapor (Gas) Specific Gravity:** Not pertinent
13.11 **Ratio of Specific Heats of Vapor (Gas):** Not pertinent
13.12 **Latent Heat of Vaporization:** 203 Btu/lb = 113 cal/g = 4.73 × 10^5 J/kg
13.13 **Heat of Combustion:** (est.) −12,200 Btu/lb = −6,790 cal/g = −284 × 10^5 J/kg
13.14 **Heat of Decomposition:** Not pertinent
13.15 **Heat of Solution:** (est.) −9 Btu/lb = −5 cal/g = −0.2 × 10^5 J/kg
13.16 **Heat of Polymerization:** Not pertinent

NOTES

REVISED 1978

HMI — HEXAMETHYLENEIMINE

Common Synonyms			
Hexahydroazepine Azacycloheptane	Liquid	Colorless to light yellow	Ammonia-like odor
	Floats and mixes slowly with water. Irritating vapor is produced.		

AVOID CONTACT WITH LIQUID AND VAPOR. KEEP PEOPLE AWAY.
Shut off ignition sources. Call fire depfrtment.
Stop discharge if possible.
Isolate and remove discharged material.
Notify local health and pollution control agencies.

Fire

Combustible.
POISONOUS GASES MAY BE PRODUCED IN FIRE.
Flashback along vapor trail may occur.
Vapor may explode if ignited in an enclosed area.
Wear goggles and self-contained breathing apparatus.
Extinguish with dry chemicals, alcohol foam, or carbon dioxide.
Water may be ineffective on fire.
Cool exposed containers with water.

Exposure

CALL FOR MEDICAL AID.

VAPOR
Irritating to eyes, nose and throat.
If inhaled will cause coughing, difficult breathing, or loss of consciousness.
If in eyes, hold eyelids open and flush with plenty of water.
If breathing has stopped, give artificial respiration.
If breathing is difficult, give oxygen.

LIQUID
POISONOUS IF SWALLOWED.
Will burn skin and eyes.
If swallowed will cause nausea.
Remove contaminated clothing and shoes.
Flush affected areas with plenty of water.
IF IN EYES, hold eyelids open and flush with plenty of water.
IF SWALLOWED and victim is CONSCIOUS, have victim drink water or milk.
IF SWALLOWED and victim is UNCONSCIOUS OR HAVING CONVULSIONS, do nothing except keep victim warm.
DO NOT INDUCE VOMITING.

Water Pollution

Effect of low concentrations on aquatic life is unknown.
May be dangerous if it enters water intakes.
Notify local health and wildlife officials.
Notify operators of nearby water intakes.

1. RESPONSE TO DISCHARGE
(See Response Methods Handbook, CG 446-4)

Issue warning—corrosive, air contaminant, water contaminant
Restrict access
Disperse and flush

2. LABEL

3. CHEMICAL DESIGNATIONS
3.1 **Synonyms:** Azacycloheptane; Hexahydroazepine; Homopiperidine
3.2 **Coast Guard Compatibility Classification:** Aliphatic amine
3.3 **Chemical Formula:** $CH_2CH_2CH_2CH_2CH_2CH_2NH$ (cyclic)
3.4 **IMCO/United Nations Numerical Designation:** Not listed

4. OBSERVABLE CHARACTERISTICS
4.1 **Physical State (as shipped):** Liquid
4.2 **Color:** Colorless to light yellow
4.3 **Odor:** Ammonia-like

5. HEALTH HAZARDS
5.1 **Personal Protective Equipment:** Self-contained breathing apparatus; impervious gloves; chemical safety goggles; impervious apron and boots
5.2 **Symptoms Following Exposure:** Inhalation of vapor irritates respiratory tract; high concentrations may cause disturbance of the central nervous system. Ingestion causes burns of mouth and stomach. Contact with concentrated vapor may cause severe eye injury. Contact with liquid causes burns of eyes and skin.
5.3 **Treatment for Exposure:** INHALATION: remove victim to uncontaminated atmosphere; get medical attention. INGESTION: give large amount of water; do NOT induce vomiting; get medical attention if large amount was swallowed. EYES: flush with water for 15 min. and get medical attention. SKIN: flush with water; wash with soap and water.
5.4 **Toxicity by Inhalation (Threshold Limit Value):** Data not available
5.5 **Short-Term Inhalation Limits:** Data not available
5.6 **Toxicity by Ingestion:** Grade 4; oral LD_{50} = 32 mg/kg (rat)
5.7 **Late Toxicity:** Data not available
5.8 **Vapor (Gas) Irritant Characteristics:** Data not available
5.9 **Liquid or Solid Irritant Characteristics:** Data not available
5.10 **Odor Threshold:** Data not available

6. FIRE HAZARDS
6.1 **Flash Point:** 99°F O.C.
6.2 **Flammable Limits in Air:** 1.6%—2.3%
6.3 **Fire Extinguishing Agents:** Dry chemical, alcohol foam, carbon dioxide
6.4 **Fire Extinguishing Agents Not to be Used:** Water may be ineffective.
6.5 **Special Hazards of Combustion Products:** Toxic oxides of nitrogen may form in fire.
6.6 **Behavior in Fire:** Vapor is heavier than air and may travel to a source of ignition and flash back.
6.7 **Ignition Temperature:** Data not available
6.8 **Electrical Hazard:** Data not available
6.9 **Burning Rate:** Data not available

7. CHEMICAL REACTIVITY
7.1 **Reactivity with Water:** No reaction
7.2 **Reactivity with Common Materials:**
7.3 **Stability During Transport:** Stable
7.4 **Neutralizing Agents for Acids and Caustics:** Not pertinent
7.5 **Polymerization:** Not pertinent
7.6 **Inhibitor of Polymerization:** Not pertinent

8. WATER POLLUTION
8.1 **Aquatic Toxicity:** Data not available
8.2 **Waterfowl Toxicity:** Data not available
8.3 **Biological Oxygen Demand (BOD):** Data not available
8.4 **Food Chain Concentration Potential:** None

9. SELECTED MANUFACTURERS
1. Celanese Chemical Company
245 Park Avenue
New York, N. Y. 10017
2. Aldrich Chemical Co.
940 W. Saint Paul Ave.
Milwaukee, Wis. 53233
3. Pfaltz and Bauer, Inc.
375 Fairfield Ave.
Stamford, Conn. 06902

10. SHIPPING INFORMATION
10.1 **Grades or Purity:** Commercial; Pure
10.2 **Storage Temperature:** Ambient
10.3 **Inert Atmosphere:** No requirement
10.4 **Venting:** Open (flame arrester)

11. HAZARD ASSESSMENT CODE
(See Hazard Assessment Handbook, CG 446-3)

A-P-Q

12. HAZARD CLASSIFICATIONS
12.1 **Code of Federal Regulations:** Corrosive liquid
12.2 **NAS Hazard Rating for Bulk Water Transportation:** Not listed
12.3 **NFPA Hazard Classifications:** Not listed

13. PHYSICAL AND CHEMICAL PROPERTIES
13.1 **Physical State at 15°C and 1 atm:** Liquid
13.2 **Molecular Weight:** 99
13.3 **Boiling Point at 1 atm:** 270°F = 132°C = 405°K
13.4 **Freezing Point:** Not pertinent
13.5 **Critical Temperature:** Not pertinent
13.6 **Critical Pressure:** Not pertinent
13.7 **Specific Gravity:** 0.880 at 20°C (liquid)
13.8 **Liquid Surface Tension:** Data not available
13.9 **Liquid-Water Interfacial Tension:** Not pertinent
13.10 **Vapor (Gas) Specific Gravity:** Not pertinent
13.11 **Ratio of Specific Heats of Vapor (Gas):** Not pertinent
13.12 **Latent Heat of Vaporization:** Data not available
13.13 **Heat of Combustion:** Data not available
13.14 **Heat of Decomposition:** Not pertinent
13.15 **Heat of Solution:** Data not available
13.16 **Heat of Polymerization:** Not pertinent

NOTES

HMT — HEXAMETHYLENETETRAMINE

Common Synonyms			
HEXA Urotropin Hexamine Metheneamine Ammonioformaldehyde	Solid crystals or powder	White	Mild ammonia odor
	Sinks and mixes with water.		

Avoid contact with solid.
Call fire department.
Isolate and remove discharged material.
Notify local health and pollution control agencies.

Fire

Combustible.
Extinguish with water, foam, dry chemical, or carbon dioxide.
Cool exposed containers with water.

Exposure

CALL FOR MEDICAL AID.

SOLID
Irritating to skin and eyes.
Remove contaminated clothing and shoes.
Flush affected areas with plenty of water.
IF IN EYES, hold eyelids open and flush with plenty of water.

Water Pollution

Effect of low concentrations on aquatic life is unknown.
May be dangerous if it enters water intakes.
Notify local health and wildlife officials.
Notify operators of nearby water intakes.

1. RESPONSE TO DISCHARGE

(See Response Methods Handbook, CG 446-4)

Disperse and flush

2. LABELS

No hazard label required by Code of Federal Regulations

3. CHEMICAL DESIGNATIONS

3.1 **Synonyms:** Aminoform
Ammoform
Hexamine
Metheneamine
Urotropin

3.2 **Coast Guard Compatibility Classification:** Aldehyde or amine

3.3 **Chemical Formula:** $C_6H_{12}N_4$

3.4 **IMCO/United Nations Numerical Designation:** 4.1/1328

4. OBSERVABLE CHARACTERISTICS

4.1 **Physical State (as shipped):** Solid

4.2 **Color:** White

4.3 **Odor:** Mild ammonia-like

5. HEALTH HAZARDS

5.1 **Personal Protective Equipment:** Gloves; for dusty or spatter conditions, use dust filter respirator and goggles.

5.2 **Symptoms Following Exposure:** Prolonged and repeated contact may cause skin irritation.

5.3 **Treatment for Exposure:** Wash skin or eyes thoroughly with water. Call a physician.

5.4 **Toxicity by Inhalation (Threshold Limit Value):** Not pertinent

5.5 **Short-Term Inhalation Limits:** Not pertinent

5.6 **Toxicity by Ingestion:** Grade 2; LD_{50} 0.5 to 5 g/kg (human)

5.7 **Late Toxicity:** None

5.8 **Vapor (Gas) Irritant Characteristics:** Vapors are nonirritating to the eyes and throat.

5.9 **Liquid or Solid Irritant Characteristics:** Minimum hazard. If spilled on clothing and allowed to remain, may cause smarting and reddening of the skin.

5.10 **Odor Threshold:** Not pertinent

6. FIRE HAZARDS

6.1 **Flash Point:** 482°F C.C.

6.2 **Flammable Limits in Air:** Not pertinent

6.3 **Fire Extinguishing Agents:** Water, foam, carbon dioxide, dry chemical

6.4 **Fire Extinguishing Agents Not to be Used:** Not pertinent

6.5 **Special Hazards of Combustion Products:** Formaldehyde gas and ammonia may be given off when hot.

6.6 **Behavior in Fire:** Not pertinent

6.7 **Ignition Temperature:** >700°F

6.8 **Electrical Hazard:** Not pertinent

6.9 **Burning Rate:** Not pertinent

7. CHEMICAL REACTIVITY

7.1 **Reactivity with Water:** No reaction

7.2 **Reactivity with Common Materials:** No reaction

7.3 **Stability During Transport:** Stable

7.4 **Neutralizing Agents for Acids and Caustics:** Not pertinent

7.5 **Polymerization:** Not pertinent

7.6 **Inhibitor of Polymerization:** Not pertinent

8. WATER POLLUTION

8.1 **Aquatic Toxicity:** Data not available

8.2 **Waterfowl Toxicity:** Data not available

8.3 **Biological Oxygen Demand (BOD):** Data not available

8.4 **Food Chain Concentration Potential:** None

9. SELECTED MANUFACTURERS

1. Borden, Inc.
Borden Chemical Division
50 W. Broad St.
Columbus, Ohio 43215
2. Hooker Chemical Corp.
Walck Rd.
N. Tonawanda, N. Y. 14120
3. Wright Chemical Corp.
Riegelwood, N. C. 28456

10. SHIPPING INFORMATION

10.1 **Grades or Purity:** Technical; USP

10.2 **Storage Temperature:** Ambient

10.3 **Inert Atmosphere:** No requirement

10.4 **Venting:** Open (flame arrester)

11. HAZARD ASSESSMENT CODE

(See Hazard Assessment Handbook, CG 446-3)

SS

12. HAZARD CLASSIFICATIONS

12.1 **Code of Federal Regulations:** Not listed

12.2 **NAS Hazard Rating for Bulk Water Transportation:** Not listed

12.3 **NFPA Hazard Classifications:** Not listed

13. PHYSICAL AND CHEMICAL PROPERTIES

13.1 **Physical State at 15°C and 1 atm:** Solid

13.2 **Molecular Weight:** 140.19

13.3 **Boiling Point at 1 atm:** Not pertinent

13.4 **Freezing Point:** Not pertinent

13.5 **Critical Temperature:** Not pertinent

13.6 **Critical Pressure:** Not pertinent

13.7 **Specific Gravity:** 1.35 at 20°C (solid)

13.8 **Liquid Surface Tension:** Not pertinent

13.9 **Liquid-Water Interfacial Tension:** Not pertinent

13.10 **Vapor (Gas) Specific Gravity:** Not pertinent

13.11 **Ratio of Specific Heats of Vapor (Gas):** Not pertinent

13.12 **Latent Heat of Vaporization:** Not pertinent

13.13 **Heat of Combustion:** $-13{,}300$ Btu/lb $= -7400$ cal/g $= -310 \times 10^5$ J/kg

13.14 **Heat of Decomposition:** Not pertinent

13.15 **Heat of Solution:** Not pertinent

13.16 **Heat of Polymerization:** Not pertinent

NOTES

HEXANE

Common Synonyms
n-Hexane

Watery liquid | Colorless | Gasoline-like odor

Floats on water. Flammable, irritating vapor is produced.

Stop discharge if possible. Keep people away.
Shut off ignition sources and call fire department.
Stay upwind and use water spray to "knock down" vapor.
Isolate and remove discharged material.
Notify local health and pollution control agencies.

Fire

FLAMMABLE.
Flashback along vapor trail may occur.
Vapor may explode if ignited in an enclosed area.
Extinguish with dry chemical, foam, or carbon dioxide.
Water may be ineffective on fire.
Cool exposed containers with water.

Exposure

CALL FOR MEDICAL AID.

VAPOR
Irritating to nose and throat.
If inhaled, will cause coughing or dizziness.
Move to fresh air.
If breathing has stopped, give artificial respiration.
If breathing is difficult, give oxygen.

LIQUID
Irritating to skin and eyes.
If swallowed, will cause nausea or vomiting.
Remove contaminated clothing and shoes.
Flush affected areas with plenty of water.
IF IN EYES, hold eyelids open and flush with plenty of water.
IF SWALLOWED and victim is CONSCIOUS, have victim drink water or milk.
DO NOT INDUCE VOMITING.

Water Pollution

Effect of low concentrations on aquatic life is unknown.
Fouling to shoreline.
May be dangerous if it enters water intakes.
Notify local health and wildlife officials.
Notify operators of nearby water intakes.

1. RESPONSE TO DISCHARGE

(See Response Methods Handbook, CG 446-4)

Issue warning high flammability
Evacuate area
Disperse and flush

2. LABEL

Red

3. CHEMICAL DESIGNATIONS

3.1 **Synonyms:** No common synonyms
3.2 **Coast Guard Compatibility Classification:** Paraffin
3.3 **Chemical Formula:** $CH_3(CH_2)_4CH_3$
3.4 **IMCO/United Nations Numerical Designation:** 3.1/1208

4. OBSERVABLE CHARACTERISTICS

4.1 **Physical State (as shipped):** Liquid
4.2 **Color:** Colorless
4.3 **Odor:** Like gasoline

5. HEALTH HAZARDS

5.1 **Personal Protective Equipment:** Eye protection (like gasoline).
5.2 **Symptoms Following Exposure:** INHALATION causes irritation of respiratory tract, cough, mild depression, cardiac arrhythmias. ASPIRATION causes severe lung irritation, coughing, pulmonary edema; excitement followed by depression. INGESTION causes nausea, vomiting, swelling of abdomen, headache, depression.
5.3 **Treatment for Exposure:** Call a doctor. INHALATION: maintain respiration, give oxygen if needed. ASPIRATION: enforce bed rest; give oxygen if needed. INGESTION: do NOT induce vomiting. SKIN OR EYES: wipe off; wash skin with soap and water; wash eyes with copious amounts of water.
5.4 **Toxicity by Inhalation (Threshold Limit Value):** 500 ppm
5.5 **Short-Term Inhalation Limits:** 500 ppm for 30 min.
5.6 **Toxicity by Ingestion:** Very slight
5.7 **Late Toxicity:** None
5.8 **Vapor (Gas) Irritant Characteristics:** Vapors are nonirritating to the eyes and throat.
5.9 **Liquid or Solid Irritant Characteristics:** No appreciable hazard. Practically harmless to the skin.
5.10 **Odor Threshold:** Data not available

6. FIRE HAZARDS

6.1 **Flash Point:** −7°F C.C.
6.2 **Flammable Limits in Air:** 1.2%—7.7%
6.3 **Fire Extinguishing Agents:** Foam, dry chemical, carbon dioxide
6.4 **Fire Extinguishing Agents Not to be Used:** Not pertinent
6.5 **Special Hazards of Combustion Products:** Not pertinent
6.6 **Behavior in Fire:** Vapors may explode
6.7 **Ignition Temperature:** 437°F
6.8 **Electrical Hazard:** Class I, group D
6.9 **Burning Rate:** 7.3 mm/min.

7. CHEMICAL REACTIVITY

7.1 **Reactivity with Water:** No reaction
7.2 **Reactivity with Common Materials:** No reaction
7.3 **Stability During Transport:** Stable
7.4 **Neutralizing Agents for Acids and Caustics:** Not pertinent
7.5 **Polymerization:** Not pertinent
7.6 **Inhibitor of Polymerization:** Not pertinent

8. WATER POLLUTION

8.1 **Aquatic Toxicity:** Data not available
8.2 **Waterfowl Toxicity:** Data not available
8.3 **Biological Oxygen Demand (BOD):** 0% (theor.), 7 days
8.4 **Food Chain Concentration Potential:** None

9. SELECTED MANUFACTURERS

1. Humphrey Chemical Co.
   Devine St.
   North Haven, Conn. 06473
2. Phillips Petroleum Co.
   Bartlesville, Okla. 74004
3. Skelly Oil Co.
   Kansas City, Mo. 64141

10. SHIPPING INFORMATION

10.1 **Grades or Purity:** Research grade; technical grade
10.2 **Storage Temperature:** Ambient
10.3 **Inert Atmosphere:** No requirement
10.4 **Venting:** Open (flame arrester) or pressure-vacuum

11. HAZARD ASSESSMENT CODE

(See Hazard Assessment Handbook, CG 446-3)

A-T-U-V-W

12. HAZARD CLASSIFICATIONS

12.1 **Code of Federal Regulations:** Flammable liquid
12.2 **NAS Hazard Rating for Bulk Water Transportation:**

Category	Rating
Fire	3
Health	
Vapor Irritant	0
Liquid or Solid Irritant	0
Poisons	1
Water Pollution	
Human Toxicity	1
Aquatic Toxicity	1
Aesthetic Effect	1
Reactivity	
Other Chemicals	0
Water	0
Self-Reaction	0

12.3 **NFPA Hazard Classifications:**

Category	Classification
Health Hazard (Blue)	1
Flammability (Red)	3
Reactivity (Yellow)	0

13. PHYSICAL AND CHEMICAL PROPERTIES

13.1 **Physical State at 15°C and 1 atm:** Liquid
13.2 **Molecular Weight:** 86.17
13.3 **Boiling Point at 1 atm:** 155.7°F = 68.7°C = 341.9°K
13.4 **Freezing Point:** −219.3°F = −139.6°C = 133.6°K
13.5 **Critical Temperature:** 453.6°F = 234.2°C = 507.4°K
13.6 **Critical Pressure:** 436.6 psia = 29.7 atm = 3.01 MN/m²
13.7 **Specific Gravity:** 0.659 at 20 °C (liquid)
13.8 **Liquid Surface Tension:** 18.4 dynes/cm = 0.0184 N/m at 20°C
13.9 **Liquid-Water Interfacial Tension:** 51.1 dynes/cm = 0.0511 N/m at 20°C
13.10 **Vapor (Gas) Specific Gravity:** 3.0
13.11 **Ratio of Specific Heats of Vapor (Gas):** 1.063
13.12 **Latent Heat of Vaporization:** 144 Btu/lb = 80.0 cal/g = 3.35×10^5 J/kg
13.13 **Heat of Combustion:** −19,246 Btu/lb = −10,692 cal/g = -447.65×10^5 J/kg
13.14 **Heat of Decomposition:** Not pertinent
13.15 **Heat of Solution:** Not pertinent
13.16 **Heat of Polymerization:** Not pertinent

NOTES

REVISED 1978

HXN	HEXANOL

Common Synonyms			
1-Hexanol n-Hexyl alcohol n-Amylcarbinol	Clear liquid Floats on water.	Colorless	Sweet odor

Stop discharge if possible. Keep people away.
Call fire department.
Avoid contact with liquid.
Isolate and remove discharged material.
Notify local health and pollution control agencies.

Fire	Combustible. Extinguish with dry chemical, alcohol foam, or carbon dioxide. Water may be ineffective on fire.
Exposure	CALL FOR MEDICAL AID. LIQUID Will burn eyes. Harmful if swallowed. Remove contaminated clothing and shoes. Flush affected areas with plenty of water. IF IN EYES, hold eyelids open and flush with plenty of water. IF SWALLOWED and victim is CONSCIOUS, have victim drink water or milk.
Water Pollution	Effect of low concentrations on aquatic life is unknown. Fouling to shoreline. May be dangerous if it enters water intakes. Notify local health and wildlife officials. Notify operators of nearby water intakes.

1. RESPONSE TO DISCHARGE

(See Response Methods Handbook, CG 446-4)

Mechanical containment
Chemical and physical treatment

2. LABELS

No hazard label required by Code of Federal Regulations

3. CHEMICAL DESIGNATIONS

3.1 **Synonyms:** Amylcarbinol
1-Hexanol
Hexyl alcohol
1-Hydroxyhexane

3.2 **Coast Guard Compatibility Classification:** Alcohol

3.3 **Chemical Formula:** $CH_3(CH_2)_4CH_2OH$

3.4 **IMCO/United Nations Numerical Designation:** Not listed

4. OBSERVABLE CHARACTERISTICS

4.1 **Physical State (as shipped):** Liquid

4.2 **Color:** Colorless

4.3 **Odor:** Sweet; mild

5. HEALTH HAZARDS

5.1 **Personal Protective Equipment:** Chemical gloves, chemical goggles.

5.2 **Symptoms Following Exposure:** Liquid causes eye burns and skin irritation. Breathing vapors is not expected to cause systemic illness.

5.3 **Treatment for Exposure:** In case of contact, immediately flush skin and eyes with plenty of water. Wash eyes at least 15 min. and get medical care.

5.4 **Toxicity by Inhalation (Threshold Limit Value):** Data not available

5.5 **Short-Term Inhalation Limits:** Data not available

5.6 **Toxicity by Ingestion:** Grade 2; LD_{50} 0.5 to 5 g/kg (rat)

5.7 **Late Toxicity:** Data not available

5.8 **Vapor (Gas) Irritant Characteristics:** Data not available

5.9 **Liquid or Solid Irritant Characteristics:** Causes smarting of the skin and first-degree burns on short exposure; may cause second-degree burns on long exposure.

5.10 **Odor Threshold:** Data not available

6. FIRE HAZARDS

6.1 **Flash Point:** 149°F O.C.; 145°F C.C.

6.2 **Flammable Limits in Air:** 1.2%—7.7% (calc.)

6.3 **Fire Extinguishing Agents:** Alcohol foam, dry chemical, carbon dioxide

6.4 **Fire Extinguishing Agents Not to be Used:** Not pertinent

6.5 **Special Hazards of Combustion Products:** Not pertinent

6.6 **Behavior in Fire:** Not pertinent

6.7 **Ignition Temperature:** 580°F (est.)

6.8 **Electrical Hazard:** Not pertinent

6.9 **Burning Rate:** Data not available

7. CHEMICAL REACTIVITY

7.1 **Reactivity with Water:** No reaction

7.2 **Reactivity with Common Materials:** No reaction

7.3 **Stability During Transport:** Stable

7.4 **Neutralizing Agents for Acids and Caustics:** Not pertinent

7.5 **Polymerization:** Not pertinent

7.6 **Inhibitor of Polymerization:** Not pertinent

8. WATER POLLUTION

8.1 **Aquatic Toxicity:** Data not available

8.2 **Waterfowl Toxicity:** Data not available

8.3 **Biological Oxygen Demand (BOD):** 57%, 1-10 days

8.4 **Food Chain Concentration Potential:** None

9. SELECTED MANUFACTURERS

1. Continental Oil Co.
   Conoco Chemicals Division
   Park 80 Plaza East
   Saddle Brook, N. J. 07662
2. Ethyl Corp.
   Industrial Chemicals Division
   451 Florida St.
   Baton Rouge, La. 70801
3. Union Carbide Corp.
   Chemicals and Plastics Division
   270 Park Ave.
   New York, N. Y. 10017

10. SHIPPING INFORMATION

10.1 **Grades or Purity:** 99+%

10.2 **Storage Temperature:** Ambient

10.3 **Inert Atmosphere:** No requirement

10.4 **Venting:** Open (flame arrester)

11. HAZARD ASSESSMENT CODE

(See Hazard Assessment Handbook, CG 446-3)

A-T-U

12. HAZARD CLASSIFICATIONS

12.1 **Code of Federal Regulations:** Combustible liquid

12.2 **NAS Hazard Rating for Bulk Water Transportation:** Not listed

12.3 **NFPA Hazard Classifications:** Not listed

13. PHYSICAL AND CHEMICAL PROPERTIES

13.1 **Physical State at 15°C and 1 atm:** Liquid

13.2 **Molecular Weight:** 102.18

13.3 **Boiling Point at 1 atm:** 314.8°F = 157.1°C = 430.3°K

13.4 **Freezing Point:** −48.3°F = −44.6°C = 228.6°K

13.5 **Critical Temperature:** 638.6°F = 337°C = 610.2°K

13.6 **Critical Pressure:** 485 psia = 33 atm = 3.34 MN/m^2

13.7 **Specific Gravity:** 0.850 at 20°C (liquid)

13.8 **Liquid Surface Tension:** 24.5 dynes/cm = 0.0245 N/m at 20°C

13.9 **Liquid-Water Interfacial Tension:** 6.8 dynes/cm = 0.0068 N/m at 25°C

13.10 **Vapor (Gas) Specific Gravity:** Not pertinent

13.11 **Ratio of Specific Heats of Vapor (Gas):** 1.057

13.12 **Latent Heat of Vaporization:** 209 Btu/lb = 116 cal/g = 4.86×10^5 J/kg

13.13 **Heat of Combustion:** −16,810 Btu/lb = −9340 cal/g = -391.0×10^5 J/kg

13.14 **Heat of Decomposition:** Not pertinent

13.15 **Heat of Solution:** Not pertinent

13.16 **Heat of Polymerization:** Not pertinent

NOTES

REVISED 1978

HXE — 1-HEXENE

Common Synonyms	Watery liquid	Colorless	Mild pleasant odor
alpha-Hexene Hexylene	Floats on water. Flammable, irritating vapor is produced.		

Stop discharge if possible. Keep people away.
Shut off ignition sources and call fire department.
Avoid contact with liquid and vapor.
Stay upwind and use water spray to "knock down" vapor.
Isolate and remove discharged material.
Notify local health and pollution control agencies

Fire

FLAMMABLE.
Flashback along vapor trail may occur.
Vapor may explode if ignited in an enclosed area.
Extinguish with dry chemical, foam, or carbon dioxide.
Water may be ineffective on fire.
Cool exposed containers with water.

Exposure

CALL FOR MEDICAL AID.

VAPOR
If inhaled, will cause dizziness, difficult breathing, or loss of consciousness.
Move to fresh air.
If breathing has stopped, give artificial respiration.
If breathing is difficult, give oxygen.

LIQUID
Irritating to skin and eyes.
Harmful if swallowed.
Remove contaminated clothing and shoes.
Flush affected areas with plenty of water.
IF IN EYES, hold eyelids open and flush with plenty of water.
IF SWALLOWED and victim is CONSCIOUS, have victim drink water or milk.
DO NOT INDUCE VOMITING.

Water Pollution

Effect of low concentrations on aquatic life is unknown.
Fouling to shoreline.
May be dangerous if it enters water intakes.
Notify local health and wildlife officials.
Notify operators of nearby water intakes.

1. RESPONSE TO DISCHARGE

(See Response Methods Handbook, CG 446-4)

Issue warning—high flammability
Evacuate area
Disperse and flush

2. LABEL

Red

3. CHEMICAL DESIGNATIONS

3.1 **Synonyms:** alpha-Hexene
3.2 **Coast Guard Compatibility Classification:** Olefin
3.3 **Chemical Formula:** $CH_3(CH_2)_3CH{=}CH_2$
3.4 **IMCO/United Nations Numerical Designation:** Not listed

4. OBSERVABLE CHARACTERISTICS

4.1 **Physical State (as shipped):** Liquid
4.2 **Color:** Colorless
4.3 **Odor:** Mild, pleasant

5. HEALTH HAZARDS

5.1 **Personal Protective Equipment:** Approved organic vapor respirator or air-line mask; protective goggles or face shield.
5.2 **Symptoms Following Exposure:** Inhalation may cause giddiness or incoordination similar to that from gasoline vapor. Prolonged exposure to high concentrations may induce loss of consciousness or death.
5.3 **Treatment for Exposure:** SKIN OR EYES: wash exposed skin areas with soap and water; thoroughly flush eyes with water to remove any splashes; launder contaminated clothing before reuse.
5.4 **Toxicity by Inhalation (Threshold Limit Value):** 500 ppm (suggested)
5.5 **Short-Term Inhalation Limits:** Data not available
5.6 **Toxicity by Ingestion:** Data not available
5.7 **Late Toxicity:** Data not available
5.8 **Vapor (Gas) Irritant Characteristics:** Slight smarting of the eyes or respiratory system if present in high concentrations. Effect is temporary.
5.9 **Liquid or Solid Irritant Characteristics:** Data not available
5.10 **Odor Threshold:** Data not available

6. FIRE HAZARDS

6.1 **Flash Point:** −15°F C.C.
6.2 **Flammable Limits in Air:** LEL (est.) = 1.2%; UEL not listed
6.3 **Fire Extinguishing Agents:** Foam, dry chemical, or carbon dioxide
6.4 **Fire Extinguishing Agents Not to be Used:** Water may be ineffective
6.5 **Special Hazards of Combustion Products:** Not pertinent
6.6 **Behavior in Fire:** Not pertinent
6.7 **Ignition Temperature:** 521°F
6.8 **Electrical Hazard:** Not pertinent
6.9 **Burning Rate:** 8.1 mm/min.

7. CHEMICAL REACTIVITY

7.1 **Reactivity with Water:** No reaction
7.2 **Reactivity with Common Materials:** No reaction
7.3 **Stability During Transport:** Stable
7.4 **Neutralizing Agents for Acids and Caustics:** Not pertinent
7.5 **Polymerization:** Not pertinent
7.6 **Inhibitor of Polymerization:** Not pertinent

8. WATER POLLUTION

8.1 **Aquatic Toxicity:** Data not available
8.2 **Waterfowl Toxicity:** Data not available
8.3 **Biological Oxygen Demand (BOD):** 1.5%, 7 days
8.4 **Food Chain Concentration Potential:** None

9. SELECTED MANUFACTURERS

1. Gulf Oil Corp. Petrochemicals Division Cedar Bayou, Texas 77520
2. The Humphrey Chemical Co. Devine St. North Haven, Conn. 06473
3. Phillips Petroleum Co. Bartlesville, Okla. 74004

10. SHIPPING INFORMATION

10.1 **Grades or Purity:** Technical, 95-98%; Pure, 99+%
10.2 **Storage Temperature:** Ambient
10.3 **Inert Atmosphere:** No requirement
10.4 **Venting:** Open (flame arrester) or pressure-vacuum

11. HAZARD ASSESSMENT CODE

(See Hazard Assessment Handbook, CG 446-3)

A-T-U-V-W

12. HAZARD CLASSIFICATIONS

12.1 **Code of Federal Regulations:** Flammable liquid
12.2 **NAS Hazard Rating for Bulk Water Transportation:** Not listed
12.3 **NFPA Hazard Classifications:** Not listed

13. PHYSICAL AND CHEMICAL PROPERTIES

13.1 **Physical State at 15°C and 1 atm:** Liquid
13.2 **Molecular Weight:** 84.16
13.3 **Boiling Point at 1 atm:** 146.3°F = 63.5°C = 336.7°K
13.4 **Freezing Point:** −219.6°F = −139.8°C = 133.4°K
13.5 **Critical Temperature:** 447.4°F = 230.8°C = 504.0°K
13.6 **Critical Pressure:** 460 psia = 31.3 atm = 3.17 MN/m²
13.7 **Specific Gravity:** 0.673 at 20°C (liquid)
13.8 **Liquid Surface Tension:** 18.8 dynes/cm = 0.0188 N/m at 20°C
13.9 **Liquid-Water Interfacial Tension:** 31.6 dynes/cm = 0.0316 N/m at 22.7°C
13.10 **Vapor (Gas) Specific Gravity:** 2.9
13.11 **Ratio of Specific Heats of Vapor (Gas):** 1.068
13.12 **Latent Heat of Vaporization:** 140 Btu/lb = 80 cal/g = 3.3×10^5 J/kg
13.13 **Heat of Combustion:** −19,134 Btu/lb = −10,630 cal/g = -445.06×10^5 J/kg
13.14 **Heat of Decomposition:** Not pertinent
13.15 **Heat of Solution:** Not pertinent
13.16 **Heat of Polymerization:** Not pertinent

NOTES

REVISED 1978

HXG	HEXYLENE GLYCOL

Common Synonyms			
2-Methyl-2,4-pentanediol	Oily liquid	Colorless	Mild sweet odor
	Floats and mixes slowly with water.		

Stop discharge if possible.
Call fire department.
Isolate and remove discharged material.
Notify local health and pollution control agencies.

Fire	Combustible. Extinguish with dry chemical, alcohol foam, or carbon dioxide. Water may be ineffective on fire. Cool exposed containers with water.
Exposure	CALL FOR MEDICAL AID. LIQUID Irritating to skin and eyes. Remove contaminated clothing and shoes. Flush affected areas with plenty of water. IF IN EYES, hold eyelids open and flush with plenty of water.
Water Pollution	Effect of low concentrations on aquatic life is unknown. May be dangerous if it enters water intakes. Notify local health and wildlife officials. Notify operators of nearby water intakes.

1. RESPONSE TO DISCHARGE

(See Response Methods Handbook, CG 446-4)

Issue warning—high flammability, corrosive
Restrict access
Chemical and physical treatment

2. LABEL

3. CHEMICAL DESIGNATIONS

3.1 **Synonyms:** No common synonyms
3.2 **Coast Guard Compatibility Classification:** Glycol
3.3 **Chemical Formula:** N_2H_4
3.4 **IMCO/United Nations Numerical Designation:** 8.0/2030

4. OBSERVABLE CHARACTERISTICS

4.1 **Physical State (as shipped):** Liquid
4.2 **Color:** Colorless
4.3 **Odor:** Ammonia-like

5. HEALTH HAZARDS

5.1 **Personal Protective Equipment:** Ammonia-type gas mask; self-contained breathing apparatus; plastic-coated or rubber gloves, clothes, and apron; safety shower must be available.
5.2 **Symptoms Following Exposure:** Vapors cause itching, swelling, and blistering of eyelids, skin, nose and throat; symptoms may be delayed for several hours. Temporary blindness may occur. Liquid causes a caustic-like burn if not washed off at once. Ingestion or absorption through skin causes nausea, dizziness, headache. Severe exposure may cause death.
5.3 **Treatment for Exposure:** Call a doctor at once. INHALATION: remove to fresh air; observe for development of delayed symptoms. Keep quiet. INGESTION: do NOT induce vomiting; give egg whites or other emollient. SKIN OR EYES: wash with large amounts of water for at least 15 min.
5.4 **Toxicity by Inhalation (Threshold Limit Value):** 1 ppm
5.5 **Short-Term Inhalation Limits:** 1 ppm for 30 min.
5.6 **Toxicity by Ingestion:** Grade 3; LD_{50} 50 to 500 mg/kg (rat)
5.7 **Late Toxicity:** Data not available
5.8 **Vapor (Gas) Irritant Characteristics:** Vapor is moderately irritating such that personnel will not usually tolerate moderate or high vapor concentrations.
5.9 **Liquid or Solid Irritant Characteristics:** Severe skin irritant. Causes second- and third-degree burns on short contact; very injurious to the eyes.
5.10 **Odor Threshold:** Data not available

6. FIRE HAZARDS

6.1 **Flash Point:** 126°F O.C.
6.2 **Flammable Limits in Air:** 4.7%—100%
6.3 **Fire Extinguishing Agents:** Water, alcohol foam, carbon dioxide, or dry chemical
6.4 **Fire Extinguishing Agents Not to be Used:** Not pertinent
6.5 **Special Hazards of Combustion Products:** Toxic vapor is generated when heated.
6.6 **Behavior in Fire:** May explode if confined.
6.7 **Ignition Temperature:** May ignite spontaneously 518°F (glass)
6.8 **Electrical Hazard:** Not pertinent
6.9 **Burning Rate:** 1 mm/min. (est.)

7. CHEMICAL REACTIVITY

7.1 **Reactivity with Water:** No reaction
7.2 **Reactivity with Common Materials:** Can catch fire when in contact with porous materials such as wood, asbestos, cloth, earth and rusty metals.
7.3 **Stability During Transport:** Stable at ordinary temperatures. When heated, can decompose to nitrogen and ammonia gases, but decomposition is not hazardous unless material is confined.
7.4 **Neutralizing Agents for Acids and Caustics:** Flush with water. Neutralize the resulting solution with calcium hypochlorite (HTH) (7 lbs per lb of hydrazine).
7.5 **Polymerization:** Not pertinent
7.6 **Inhibitor of Polymerization:** Not pertinent

8. WATER POLLUTION

8.1 **Aquatic Toxicity:** 146 ppm/0.5 hr/rainbow trout/died/fresh water
8.2 **Waterfowl Toxicity:** Data not available
8.3 **Biological Oxygen Demand (BOD):** 100%
8.4 **Food Chain Concentration Potential:** None

9. SELECTED MANUFACTURERS

1. Fairmount Chemical Co., Inc.
   117 Blanchard St.
   Newark, N. J. 07105
2. Olin Corp.
   Chemicals Division
   120 Long Ridge Rd.
   Stamford, Conn. 06904
3. Uniroyal, Inc.
   Uniroyal Chemical Division
   Spencer St.
   Naugatuck, Conn. 06770

10. SHIPPING INFORMATION

10.1 **Grades or Purity:** Anhydrous; 35—64% water solutions
10.2 **Storage Temperature:** Ambient
10.3 **Inert Atmosphere:** Padded
10.4 **Venting:** Pressure-vacuum

11. HAZARD ASSESSMENT CODE

(See Hazard Assessment Handbook, CG 446-3)

A-P-Q

12. HAZARD CLASSIFICATIONS

12.1 **Code of Federal Regulations:** Corrosive material
12.2 **NAS Hazard Rating for Bulk Water Transportation:**

Category	Rating
Fire	4
Health	
Vapor Irritant	3
Liquid or Solid Irritant	4
Poisons	4
Water Pollution	
Human Toxicity	4
Aquatic Toxicity	3
Aesthetic Effect	2
Reactivity	
Other Chemicals	4
Water	0
Self-Reaction	4

12.3 **NFPA Hazard Classifications:**

Category	Classification
Health Hazard (Blue)	3
Flammability (Red)	3
Reactivity (Yellow)	2

13. PHYSICAL AND CHEMICAL PROPERTIES

13.1 **Physical State at 15°C and 1 atm:** Liquid
13.2 **Molecular Weight:** 32.05
13.3 **Boiling Point at 1 atm:** 236.3°F = 113.5°C = 386.7°K
13.4 **Freezing Point:** 34.7°F = 1.5°C = 274.7°K
13.5 **Critical Temperature:** 716°F = 380°C = 653°K
13.6 **Critical Pressure:** 2130 psia = 145 atm = 14.7 MN/m^2
13.7 **Specific Gravity:** 1.008 at 20°C (liquid)
13.8 **Liquid Surface Tension:** Not pertinent
13.9 **Liquid-Water Interfacial Tension:** Not pertinent
13.10 **Vapor (Gas) Specific Gravity:** Not pertinent
13.11 **Ratio of Specific Heats of Vapor (Gas):** 1.191
13.12 **Latent Heat of Vaporization:** 538 Btu/lb = 299 cal/g = 12.5×10^5 J/kg
13.13 **Heat of Combustion:** −8345 Btu/lb = −4636 cal/g = -194.1×10^5 J/kg
13.14 **Heat of Decomposition:** Not pertinent
13.15 **Heat of Solution:** −218 Btu/lb = −121 cal/g = -5.07×10^5 J/kg
13.16 **Heat of Polymerization:** Not pertinent

NOTES

REVISED 1978

HDZ	HYDRAZINE

Common Synonyms

Watery liquid — Colorless — Ammonia odor

Mixes with water. Poisonous, flammable vapor is produced. Freezing point is 35° F.

AVOID CONTACT WITH LIQUID AND VAPOR. Keep people away.
Wear chemical protective suit with self-contained breathing apparatus.
Stop discharge if possible. Call fire department.
Stay upwind and use water spray to "knock down" vapor.
Isolate and remove discharged material.
Notify local health and pollution control agencies.

Fire

FLAMMABLE
Flashback along vapor trail may occur.
Vapor may explode if ignited in an enclosed area.
Wear chemical protective suit with self-contained breathing apparatus.
Combat fires from safe distance or protected location.
Flood discharge area with water.
Extinguish with dry chemical, alcohol foam, or carbon dioxide.
Cool exposed containers with water.
Continue cooling after fire has been extinguished.

Exposure

CALL FOR MEDICAL AID.

VAPOR
POISONOUS IF INHALED OR IF SKIN IS EXPOSED.
Irritating to eyes.
Move to fresh air.
If breathing has stopped, give artificial respiration.
If breathing is difficult, give oxygen.

LIQUID
POISONOUS IF SWALLOWED OR IF SKIN IS EXPOSED.
Will burn eyes.
Remove contaminated clothing and shoes.
Flush affected areas with plenty of water.
IF IN EYES, hold eyelids open and flush with plenty of water.
IF SWALLOWED and victim is CONSCIOUS, have victim drink water or milk.
DO NOT INDUCE VOMITING.

Water Pollution

HARMFUL TO AQUATIC LIFE IN VERY LOW CONCENTRATIONS.
May be dangerous if it enters water intakes.
Notify local health and wildlife officials.
Notify operators of nearby water intakes.

1. RESPONSE TO DISCHARGE

(See Response Methods Handbook, CG 446-4)

Issue warning—high flammability, corrosive
Restrict access
Chemical and physical treatment

2. LABELS

Red

3. CHEMICAL DESIGNATIONS

3.1 **Synonyms:** No common synonyms
3.2 **Coast Guard Compatibility Classification:** Not applicable
3.3 **Chemical Formula:** N_2H_4
3.4 **IMCO/United Nations Numerical Designation:** 8.0/2030

4. OBSERVABLE CHARACTERISTICS

4.1 **Physical State (as shipped):** Liquid
4.2 **Color:** Colorless
4.3 **Odor:** Ammonia-like

5. HEALTH HAZARDS

5.1 **Personal Protective Equipment:** Ammonia-type gas mask; self-contained breathing apparatus; plastic-coated or rubber gloves, clothes, and apron; safety shower must be available.
5.2 **Symptoms Following Exposure:** Vapors cause itching, swelling, and blistering of eyelids, skin, nose and throat; symptoms may be delayed for several hours. Temporary blindness may occur. Liquid causes a caustic-like burn if not washed off at once. Ingestion or absorption through skin causes nausea, dizziness, headache. Severe exposure may cause death.
5.3 **Treatment for Exposure:** Call a doctor at once. INHALATION: remove to fresh air; observe for development of delayed symptoms. Keep quiet. INGESTION: do NOT induce vomiting; give egg whites or other emollient. SKIN OR EYES: wash with large amounts of water for at least 15 min.
5.4 **Toxicity by Inhalation (Threshold Limit Value):** 1 ppm
5.5 **Short-Term Inhalation Limits:** 1 ppm for 30 min.
5.6 **Toxicity by Ingestion:** Grade 3; LD_{50} 50 to 500 mg/kg (rat)
5.7 **Late Toxicity:** Causes lung cancer in mice.
5.8 **Vapor (Gas) Irritant Characteristics:** Vapor is moderately irritating such that personnel will not usually tolerate moderate or high vapor concentrations.
5.9 **Liquid or Solid Irritant Characteristics:** Severe skin irritant. Causes second- and third-degree burns on short contact; very injurious to the eyes.
5.10 **Odor Threshold:** 3 - 4 ppm

6. FIRE HAZARDS

6.1 **Flash Point:** 100°F O.C.
6.2 **Flammable Limits in Air:** 4.7% 100%
6.3 **Fire Extinguishing Agents:** Water, alcohol foam, carbon dioxide, or dry chemical
6.4 **Fire Extinguishing Agents Not to be Used:** Not pertinent
6.5 **Special Hazards of Combustion Products:** Toxic vapor is generated when heated.
6.6 **Behavior in Fire:** May explode if confined.
6.7 **Ignition Temperature:** May ignite spontaneously 518°F (glass)
6.8 **Electrical Hazard:** Not pertinent
6.9 **Burning Rate:** 1 mm/min. (est.)

7. CHEMICAL REACTIVITY

7.1 **Reactivity with Water:** No reaction
7.2 **Reactivity with Common Materials:** Can catch fire when in contact with porous materials such as wood, asbestos, cloth, earth and rusty metals.
7.3 **Stability During Transport:** Stable at ordinary temperatures. When heated, can decompose to nitrogen and ammonia gases, but decomposition is not hazardous unless material is confined.
7.4 **Neutralizing Agents for Acids and Caustics:** Flush with water. Neutralize the resulting solution with calcium hypochlorite (HTH) (7 lbs per lb of hydrazine).
7.5 **Polymerization:** Not pertinent
7.6 **Inhibitor of Polymerization:** Not pertinent

8. WATER POLLUTION

8.1 **Aquatic Toxicity:** 146 ppm/0.5 hr/rainbow trout/died/fresh water
8.2 **Waterfowl Toxicity:** Data not available
8.3 **Biological Oxygen Demand (BOD):** 100%
8.4 **Food Chain Concentration Potential:** None

9. SELECTED MANUFACTURERS

1. Fairmount Chemical Co., Inc.
 117 Blanchard St.
 Newark, N. J. 07105
2. Olin Corp.
 Chemicals Division
 120 Long Ridge Rd.
 Stamford, Conn. 06904
3. Uniroyal, Inc.
 Uniroyal Chemical Division
 Spencer St.
 Naugatuck, Conn. 06770

10. SHIPPING INFORMATION

10.1 **Grades or Purity:** Anhydrous; 35–64% water solutions
10.2 **Storage Temperature:** Ambient
10.3 **Inert Atmosphere:** Padded
10.4 **Venting:** Pressure-vacuum

11. HAZARD ASSESSMENT CODE

(See Hazard Assessment Handbook, CG 446-3)

A-P-Q

12. HAZARD CLASSIFICATIONS

12.1 **Code of Federal Regulations:** Flammable Liquid
12.2 **NAS Hazard Rating for Bulk Water Transportation:**

Category	Rating
Fire	4
Health	
Vapor Irritant	3
Liquid or Solid Irritant	4
Poisons	4
Water Pollution	
Human Toxicity	4
Aquatic Toxicity	3
Aesthetic Effect	2
Reactivity	
Other Chemicals	4
Water	0
Self-Reaction	4

12.3 **NFPA Hazard Classifications:**

Category	Classification
Health Hazard (Blue)	3
Flammability (Red)	3
Reactivity (Yellow)	2

13. PHYSICAL AND CHEMICAL PROPERTIES

13.1 **Physical State at 15°C and 1 atm:** Liquid
13.2 **Molecular Weight:** 32.05
13.3 **Boiling Point at 1 atm:** 236.3°F = 113.5°C = 386.7°K
13.4 **Freezing Point:** 34.7°F = 1.5°C = 274.7°K
13.5 **Critical Temperature:** 716°F = 380°C = 653°K
13.6 **Critical Pressure:** 2130 psia = 145 atm = 14.7 MN/m^2
13.7 **Specific Gravity:** 1.008 at 20°C (liquid)
13.8 **Liquid Surface Tension:** Not pertinent
13.9 **Liquid-Water Interfacial Tension:** Not pertinent
13.10 **Vapor (Gas) Specific Gravity:** Not pertinent
13.11 **Ratio of Specific Heats of Vapor (Gas):** 1.191
13.12 **Latent Heat of Vaporization:** 538 Btu/lb = 299 cal/g = 12.5×10^5 J/kg
13.13 **Heat of Combustion:** −8345 Btu/lb = −4636 cal/g = -194.1×10^5 J/kg
13.14 **Heat of Decomposition:** Not pertinent
13.15 **Heat of Solution:** −218 Btu/lb = −121 cal/g = -5.07×10^5 J/kg
13.16 **Heat of Polymerization:** Not pertinent

NOTES

REVISED 1978

HCL	HYDROCHLORIC ACID

Common Synonyms Muriatic Acid	Watery liquid Colorless Sharp, irritating odor Sinks and mixes with water. Irritating vapor is produced.

AVOID CONTACT WITH LIQUID AND VAPOR. Keep people away.
Wear chemical protective suit with self-contained breathing apparatus.
Stop discharge if possible.
Stay upwind and use water spray to "knock down" vapor.
Isolate and remove discharged material.
Notify local health and pollution control agencies.

Fire	Not flammable. Flammable gas may be produced on contact with metals. Wear chemical protective suit with self-contained breathing apparatus.
Exposure	CALL FOR MEDICAL AID. VAPOR Irritating to eyes, nose and throat. If inhaled, will cause coughing or difficult breathing. Move to fresh air. If breathing has stopped, give artificial respiration. If breathing is difficult, give oxygen. LIQUID Will burn skin and eyes. Harmful if swallowed. Remove contaminated clothing and shoes. Flush affected areas with plenty of water. IF IN EYES, hold eyelids open and flush with plenty of water. IF SWALLOWED and victim is CONSCIOUS, have victim drink water or milk. DO NOT INDUCE VOMITING.
Water Pollution	Dangerous to aquatic life in high concentrations. May be dangerous if it enters water intakes. Notify local health and wildlife officials. Notify operators of nearby water intakes.

1. RESPONSE TO DISCHARGE

(See Response Methods Handbook, CG 446-4)

Issue warning—corrosive
Restrict access
Disperse and flush

2. LABEL

3. CHEMICAL DESIGNATIONS

3.1 **Synonyms:** Muriatic acid

3.2 **Coast Guard Compatibility Classification:** Non-oxidizing mineral acid

3.3 **Chemical Formula:** $HCl\text{-}H_2O$

3.4 **IMCO/United Nations Numerical Designation:** 8.0/1789

4. OBSERVABLE CHARACTERISTICS

4.1 **Physical State (as shipped):** Liquid

4.2 **Color:** Colorless to light yellow

4.3 **Odor:** Pungent; sharp, pungent, irritating

5. HEALTH HAZARDS

5.1 **Personal Protective Equipment:** Self-contained breathing equipment, air-line mask, or industrial canister-type gas mask; rubber or rubber-coated gloves, apron, coat, overalls, shoes.

5.2 **Symptoms Following Exposure:** Inhalation of fumes results in coughing and choking sensation, and irritation of nose and lungs. Liquid causes burns.

5.3 **Treatment for Exposure:** INHALATION: remove person to fresh air; keep him warm and quiet and get medical attention immediately; start artificial respiration if breathing stops. INGESTION: have person drink water or milk; do NOT induce vomiting. EYES: immediately flush with plenty of water for at least 15 min. and get medical attention; continue flushing for another 15 min. if physician does not arrive promptly. SKIN: immediately flush skin while removing contaminated clothing; get medical attention promptly; use soap and wash area for at least 15 min.

5.4 **Toxicity by Inhalation (Threshold Limit Value):** 5 ppm

5.5 **Short-Term Inhalation Limits:** 5 ppm for 5 min.

5.6 **Toxicity by Ingestion:** Data not available

5.7 **Late Toxicity:** None

5.8 **Vapor (Gas) Irritant Characteristics:** Vapor is moderately irritating such that personnel will not usually tolerate moderate or high vapor concentrations.

5.9 **Liquid or Solid Irritant Characteristics:** Fairly severe skin irritant; may cause pain and second-degree burns after a few minutes' contact.

5.10 **Odor Threshold:** 1-5 ppm

6. FIRE HAZARDS

6.1 **Flash Point:** Not flammable

6.2 **Flammable Limits in Air:** Not flammable

6.3 **Fire Extinguishing Agents:** Not pertinent

6.4 **Fire Extinguishing Agents Not to be Used:** Not pertinent

6.5 **Special Hazards of Combustion Products:** Toxic and irritating vapors are generated when heated.

6.6 **Behavior in Fire:** Not pertinent

6.7 **Ignition Temperature:** Not flammable

6.8 **Electrical Hazard:** Not pertinent

6.9 **Burning Rate:** Not flammable

7. CHEMICAL REACTIVITY

7.1 **Reactivity with Water:** No reaction

7.2 **Reactivity with Common Materials:** Corrosive to most metals with evolution of hydrogen gas, which may form explosive mixtures with air.

7.3 **Stability During Transport:** Stable

7.4 **Neutralizing Agents for Acids and Caustics:** Flush with water; apply powdered limestone, slaked lime, soda ash, or sodium bicarbonate.

7.5 **Polymerization:** Not pertinent

7.6 **Inhibitor of Polymerization:** Not pertinent

8. WATER POLLUTION

8.1 **Aquatic Toxicity:**
282 ppm/96 hr/mosquito fish/TL_m/fresh water
100-330 ppm/48 hr/shrimp/LC_{50}/salt water

8.2 **Waterfowl Toxicity:** Data not available

8.3 **Biological Oxygen Demand (BOD):** None

8.4 **Food Chain Concentration Potential:** None

9. SELECTED MANUFACTURERS

1. Diamond Shamrock Corp.
   Electro Chemicals Division
   300 Union Commerce Bldg.
   Cleveland, Ohio 44115
2. Stauffer Chemical Co.
   Industrial Chemicals Division
   299 Park Ave.
   New York, N. Y. 10017
3. Vulcan Materials Co.
   Chemicals Division
   Wichita, Kan. 67201

10. SHIPPING INFORMATION

10.1 **Grades or Purity:**
Food processing or technical:
18° Be—27.9%, 20 Be—31.5%,
22° Be—35.2%;
Reagent, ACS, and USP: 23° Be—37.1%

10.2 **Storage Temperature:** Ambient

10.3 **Inert Atmosphere:** No requirement

10.4 **Venting:** Open

11. HAZARD ASSESSMENT CODE

(See Hazard Assessment Handbook, CG 446-3)

A-P

12. HAZARD CLASSIFICATIONS

12.1 **Code of Federal Regulations:** Corrosive material

12.2 **NAS Hazard Rating for Bulk Water Transportation:**

Category	Rating
Fire	0
Health	
Vapor Irritant	3
Liquid or Solid Irritant	3
Poisons	2
Water Pollution	
Human Toxicity	2
Aquatic Toxicity	2
Aesthetic Effect	2
Reactivity	
Other Chemicals	3
Water	0
Self-Reaction	0

12.3 **NFPA Hazard Classifications:**

Category	Classification
Health Hazard (Blue)	3
Flammability (Red)	0
Reactivity (Yellow)	0

13. PHYSICAL AND CHEMICAL PROPERTIES*

13.1 **Physical State at 15°C and 1 atm:** Liquid

13.2 **Molecular Weight:** Not pertinent

13.3 **Boiling Point at 1 atm:** 123°F = 50.5°C = 323.8°K

13.4 **Freezing Point:** Not pertinent

13.5 **Critical Temperature:** Not pertinent

13.6 **Critical Pressure:** Not pertinent

13.7 **Specific Gravity:** 1.19 at 20°C (liquid)

13.8 **Liquid Surface Tension:** Not pertinent

13.9 **Liquid-Water Interfacial Tension:** Not pertinent

13.10 **Vapor (Gas) Specific Gravity:** Not pertinent

13.11 **Ratio of Specific Heats of Vapor (Gas):** Not pertinent

13.12 **Latent Heat of Vaporization:** 178 Btu/lb = 98.6 cal/g = 4.13×10^5 J/kg

13.13 **Heat of Combustion:** Not pertinent

13.14 **Heat of Decomposition:** Not pertinent

13.15 **Heat of Solution:** -860 Btu/lb = -480 cal/g = -20×10^5 J/kg

13.16 **Heat of Polymerization:** Not pertinent

*Physical properties apply to 37% solution.

NOTES

REVISED 1978

HFA — HYDROFLUORIC ACID

Common Synonyms	Watery liquid / Colorless to green / Irritating odor
	Sinks and mixes with water. Harmful vapor is produced.

AVOID CONTACT WITH LIQUID AND VAPOR. Keep people away.
Wear chemical protective suit with self-contained breathing apparatus.
Stop discharge if possible.
Stay upwind and use water spray to "knock down" vapor.
Isolate and remove discharged material.
Notify local health and pollution control agencies.

Fire

Not flammable.
Flammable gas may be produced on contact with metals.
Wear chemical protective suit with self-contained breathing apparatus.

Exposure

CALL FOR MEDICAL AID.

VAPOR
Will burn eyes, nose and throat.
Harmful if inhaled.
Move to fresh air.
If breathing has stopped, give artificial respiration.
If breathing is difficult, give oxygen.
IF IN EYES, hold eyelids open and flush with plenty of water.

LIQUID
Will burn skin and eyes.
Harmful if swallowed.
Remove contaminated clothing and shoes.
Flush affected areas with plenty of water.
IF IN EYES, hold eyelids open and flush with plenty of water.
IF SWALLOWED and victim is CONSCIOUS, have victim drink water or milk.
DO NOT INDUCE VOMITING.

Water Pollution

HARMFUL TO AQUATIC LIFE IN VERY LOW CONCENTRATIONS.
May be dangerous if it enters water intakes.
Notify local health and wildlife officials.
Notify operators of nearby water intakes.

1. RESPONSE TO DISCHARGE

(See Response Methods Handbook, CG 446-4)

Issue warning—corrosive
Restrict access
Disperse and flush

2. LABEL

3. CHEMICAL DESIGNATIONS

3.1 **Synonyms:** No common synonyms
3.2 **Coast Guard Compatibility Classification:** Non-oxidizing mineral acid
3.3 **Chemical Formula:** $HF\text{-}H_2O$
3.4 **IMCO/United Nations Numerical Designation:** 8.0/1790

4. OBSERVABLE CHARACTERISTICS

4.1 **Physical State (as shipped):** Liquid
4.2 **Color:** Colorless to slightly yellow
4.3 **Odor:** Pungent, irritating

5. HEALTH HAZARDS

5.1 **Personal Protective Equipment:** Proper protective clothing must be worn that encapsulates the body including the face. All persons handling this product must be familiar with and must observe all the precautions outlined in the Manufacturing Chemists' Association Chemical Safety Data Sheet SD-25. A shower and an eye wash must be available.
5.2 **Symptoms Following Exposure:** Serious and painful burns of eyes and skin.
5.3 **Treatment for Exposure:** INGESTION: have victim drink water or milk; do NOT induce vomiting. SKIN: if victim has come in contact with liquid or vapor, put him in a shower and call a physician. EYES: flush with water for at least 15 min. and consult physician.
5.4 **Toxicity by Inhalation (Threshold Limit Value):** 3 ppm
5.5 **Short-Term Inhalation Limits:** 50 ppm for 60 min.
5.6 **Toxicity by Ingestion:** Data not available
5.7 **Late Toxicity:** Data not available
5.8 **Vapor (Gas) Irritant Characteristics:** Vapors cause severe irritation of eye and throat and can cause eye and lung injury. They cannot be tolerated even at low concentrations.
5.9 **Liquid or Solid Irritant Characteristics:** Severe skin irritant. Causes second- and third-degree burns on short contact; very injurious to the eyes.
5.10 **Odor Threshold:** Data not available

6. FIRE HAZARDS

6.1 **Flash Point:** Not flammable
6.2 **Flammable Limits in Air:** Not flammable
6.3 **Fire Extinguishing Agents:** Not pertinent
6.4 **Fire Extinguishing Agents Not to be Used:** Not pertinent
6.5 **Special Hazards of Combustion Products:** Toxic and irritating vapors are generated when heated.
6.6 **Behavior in Fire:** Not pertinent
6.7 **Ignition Temperature:** Not flammable
6.8 **Electrical Hazard:** Not pertinent
6.9 **Burning Rate:** Not flammable

7. CHEMICAL REACTIVITY

7.1 **Reactivity with Water:** No reaction
7.2 **Reactivity with Common Materials:** Will attack glass, concrete and certain metals containing silica, such as cast iron. Will attack natural rubber, leather, and many organic materials. May generate flammable hydrogen in contact with some metals.
7.3 **Stability During Transport:** Stable
7.4 **Neutralizing Agents for Acids and Caustics:** Flush with water; apply powdered limestone, slaked lime, soda ash, or sodium bicarbonate.
7.5 **Polymerization:** Not pertinent
7.6 **Inhibitor of Polymerization:** Not pertinent

8. WATER POLLUTION

8.1 **Aquatic Toxicity:** 60 ppm/*/fish/lethal/fresh water
*Time period not specified
8.2 **Waterfowl Toxicity:** Data not available
8.3 **Biological Oxygen Demand (BOD):** None
8.4 **Food Chain Concentration Potential:** None

9. SELECTED MANUFACTURERS

1. Allied Chemical Corp.
Industrial Chemicals Division
P. O. Box 1139 R
Morristown, N. J. 07960
2. E. I. duPont de Nemours & Co., Inc.
Industrial & Biochemicals Dept.
DuPont Bldg.
Wilmington, Del. 19898
3. Kaiser Aluminum & Chemical Corp.
Kaiser Chemicals Division
P. O. Box 337
Gramercy, La. 70052

10. SHIPPING INFORMATION

10.1 **Grades or Purity:** Reagent: 48-51%; technical: 52-55%; 70% grade
10.2 **Storage Temperature:** Ambient
10.3 **Inert Atmosphere:** No requirement
10.4 **Venting:** Pressure-vacuum

11. HAZARD ASSESSMENT CODE

(See Hazard Assessment Handbook, CG 446-3)

A-P

12. HAZARD CLASSIFICATIONS

12.1 **Code of Federal Regulations:** Corrosive material
12.2 **NAS Hazard Rating for Bulk Water Transportation:**

Category	Rating
Fire	0
Health	
Vapor Irritant	4
Liquid or Solid Irritant	4
Poisons	4
Water Pollution	
Human Toxicity	4
Aquatic Toxicity	3
Aesthetic Effect	2
Reactivity	
Other Chemicals	3
Water	0
Self-Reaction	0

12.3 **NFPA Hazard Classifications:**

Category	Classification
Health Hazard (Blue)	4
Flammability (Red)	0
Reactivity (Yellow)	0

13. PHYSICAL AND CHEMICAL PROPERTIES*

13.1 **Physical State at 15°C and 1 atm:** Liquid
13.2 **Molecular Weight:** Not pertinent
13.3 **Boiling Point at 1 atm:** 152°F = 67°C = 340°K
13.4 **Freezing Point:** Not pertinent
13.5 **Critical Temperature:** Not pertinent
13.6 **Critical Pressure:** Not pertinent
13.7 **Specific Gravity:** 1.258 at 25°C (liquid)
13.8 **Liquid Surface Tension:** Not pertinent
13.9 **Liquid-Water Interfacial Tension:** Not pertinent
13.10 **Vapor (Gas) Specific Gravity:** Not pertinent
13.11 **Ratio of Specific Heats of Vapor (Gas):** Not pertinent
13.12 **Latent Heat of Vaporization:** 649 Btu/lb = 361 cal/g = 15.1×10^5 J/kg
13.13 **Heat of Combustion:** Not pertinent
13.14 **Heat of Decomposition:** Not pertinent
13.15 **Heat of Solution:** -66.6 Btu/lb = -37.0 cal/g = -1.55×10^5 J/kg
13.16 **Heat of Polymerization:** Not pertinent

*Physical properties apply to 70% solution.

NOTES

REVISED 1978

HBR	# HYDROGEN BROMIDE

Common Synonyms			
Hydrobromic Acid, anhydrous	Liquefied compressed gas	Colorless	Irritating odor
	Sinks and mixes with water. Poisonous visible vapor cloud is produced.		

AVOID CONTACT WITH LIQUID AND VAPOR. Keep people away.
Wear chemical protective suit with self-contained breathing apparatus.
Stop discharge if possible.
Stay upwind and use water spray to "knock down" vapor.
Isolate and remove discharged material.
Notify local health and pollution control agencies.

Fire

Not flammable.
Flammable gas may be produced on contact with metals.
Wear chemical protective suit with self-contained breathing apparatus.

Exposure

CALL FOR MEDICAL AID.
VAPOR
POISONOUS IF INHALED.
Irritating to eyes, nose and throat.
Move to fresh air.
If breathing has stopped, give artificial respiration.
If breathing is difficult, give oxygen.
LIQUID
POISONOUS IF SWALLOWED.
Will burn skin and eyes.
Will cause frostbite.
Remove contaminated clothing and shoes.
Flush affected areas with plenty of water.
IF IN EYES, hold eyelids open and flush with plenty of water.
IF SWALLOWED and victim is CONSCIOUS, have victim drink water or milk.
DO NOT INDUCE VOMITING.
DO NOT RUB AFFECTED AREAS.

Water Pollution

Dangerous to aquatic life in high concentrations.
May be dangerous if it enters water intakes.
Notify local health and wildlife officials.
Notify operators of nearby water intakes.

1. RESPONSE TO DISCHARGE

(See Response Methods Handbook, CG 446-4)

Issue warning—poison, corrosive, air contaminant
Restrict access
Evacuate area
Disperse and flush

2. LABEL

Green

3. CHEMICAL DESIGNATIONS

3.1 **Synonyms:** Hydrobromic acid, anhydrous; Hydrogen bromide, anhydrous
3.2 **Coast Guard Compatibility Classification:** Not listed
3.3 **Chemical Formula:** HBr
3.4 **IMCO/United Nations Numerical Designation:** 2/1048

4. OBSERVABLE CHARACTERISTICS

4.1 **Physical State (as shipped):** Compressed liquefied gas
4.2 **Color:** Colorless
4.3 **Odor:** Sharp, pungent, irritating

5. HEALTH HAZARDS

5.1 **Personal Protective Equipment:** Full face mask and acid gas canister; self-contained breathing apparatus; chemical goggles; rubber apron and gloves; acid-proof clothing; safety shower
5.2 **Symptoms Following Exposure:** Inhalation causes severe irritation of nose and upper respiratory tract, lung injury. Ingestion causes burns of mouth and stomach. Contact with eyes causes severe irritation and burns. Contact with skin causes irritation and burns.
5.3 **Treatment for Exposure:** *Get medical attention after all overexposures to this chemical.* INHALATION: move victim to fresh air and keep him warm and quiet; if a qualified person is available to give oxygen, such treatment may be helpful. INGESTION: give large amounts of water or milk; do NOT induce vomiting. EYES: flush with water for at least 15 min. SKIN: flush with water; treat acid burns.
5.4 **Toxicity by Inhalation (Threshold Limit Value):** 3 ppm
5.5 **Short-Term Inhalation Limits:** 5 ppm for 5 min.
5.6 **Toxicity by Ingestion:** Data not available
5.7 **Late Toxicity:** Data not available
5.8 **Vapor (Gas) Irritant Characteristics:** Data not available
5.9 **Liquid or Solid Irritant Characteristics:** Data not available
5.10 **Odor Threshold:** Data not available

6. FIRE HAZARDS

6.1 **Flash Point:** Not flammable
6.2 **Flammable Limits in Air:** Not flammable
6.3 **Fire Extinguishing Agents:** Not pertinent
6.4 **Fire Extinguishing Agents Not to be Used:** Not pertinent
6.5 **Special Hazards of Combustion Products:** Not pertinent
6.6 **Behavior in Fire:** Pressurized container may explode and release toxic, irritating vapor.
6.7 **Ignition Temperature:** Not pertinent
6.8 **Electrical Hazard:** Not pertinent
6.9 **Burning Rate:** Not pertinent

7. CHEMICAL REACTIVITY

7.1 **Reactivity with Water:** Moderate reaction with evolution of heat.
7.2 **Reactivity with Common Materials:** Rapidly absorbs moisture, forming hydrobromic acid. Highly corrosive to most metals, with evolution of flammable hydrogen gas.
7.3 **Stability During Transport:** Stable
7.4 **Neutralizing Agents for Acids and Caustics:** Flush with water; apply powdered limestone, slaked lime, soda ash, or sodium bicarbonate.
7.5 **Polymerization:** Not pertinent
7.6 **Inhibitor of Polymerization:** Not pertinent

8. WATER POLLUTION

8.1 **Aquatic Toxicity:** 10—100 ppm/96 hr/*/TL_m/*
*Species and water type not specified.
8.2 **Waterfowl Toxicity:** Data not available
8.3 **Biological Oxygen Demand (BOD):** None
8.4 **Food Chain Concentration Potential:** None

9. SELECTED MANUFACTURERS

1. Union Carbide Corp.
Linde Division
270 Park Avenue
New York, N. Y. 10017
2. Matheson Gas Products
932 Paterson Plank Road
East Rutherford, N. J. 07073
3. Air Products and Chemicals, Inc.
Industrial Gas Division
Box 538
Allentown, Pa. 18105

10. SHIPPING INFORMATION

10.1 **Grades or Purity:** 99.8+%
10.2 **Storage Temperature:** Ambient or lower
10.3 **Inert Atmosphere:** No requirement
10.4 **Venting:** Safety relief

11. HAZARD ASSESSMENT CODE

(See Hazard Assessment Handbook, CG 446-3)

A-C-K-M-N-O

12. HAZARD CLASSIFICATIONS

12.1 **Code of Federal Regulations:** Nonflammable compressed gas
12.2 **NAS Hazard Rating for Bulk Water Transportation:** Not listed
12.3 **NFPA Hazard Classifications:**

Category	Classification
Health Hazard (Blue)	3
Flammability (Red)	0
Reactivity (Yellow)	0

13. PHYSICAL AND CHEMICAL PROPERTIES

13.1 **Physical State at 15°C and 1 atm:** Gas
13.2 **Molecular Weight:** 80.92
13.3 **Boiling Point at 1 atm:** −88.2°F = −66.8°C = 206.4°K
13.4 **Freezing Point:** Not pertinent
13.5 **Critical Temperature:** 193.6°F = 89.8°C = 363.0°K
13.6 **Critical Pressure:** 1,235 psia = 84 atm = 8.52 MN/m^2
13.7 **Specific Gravity:** 2.14 at −67°C (liquid)
13.8 **Liquid Surface Tension:** 27.1 dynes/cm = 0.0271 N/m at −67.1°C
13.9 **Liquid-Water Interfacial Tension:** Not pertinent
13.10 **Vapor (Gas) Specific Gravity:** 2.71
13.11 **Ratio of Specific Heats of Vapor (Gas):** 1.38
13.12 **Latent Heat of Vaporization:** 92.3 Btu/lb = 51.3 cal/g = 2.15 × 10^5 J/kg
13.13 **Heat of Combustion:** Not pertinent
13.14 **Heat of Decomposition:** Not pertinent
13.15 **Heat of Solution:** 445 Btu/lb = 247 cal/g = 10.3 × 10^5 J/kg
13.16 **Heat of Polymerization:** Not pertinent

NOTES

HDC — HYDROGEN CHLORIDE

Common Synonyms			
Hydrochloric Acid, anhydrous	Liquefied compressed gas	Colorless to slightly yellow	Irritating odor
	Sinks and mixes with water. Poisonous visible vapor cloud is produced.		

AVOID CONTACT WITH LIQUID AND VAPOR. Keep people away.
Wear chemical protective suit with self-contained breathing apparatus.
Stop discharge if possible.
Stay upwind and use water spray to "knock down" vapor.
Isolate and remove discharged material.
Notify local health and pollution control agencies.

Fire

Not flammable.
Flammable gas may be produced on contact with metals.
Wear chemical protective suit with self-contained breathing apparatus.

Exposure

CALL FOR MEDICAL AID.

VAPOR
POISONOUS IF INHALED.
Irritating to eyes, nose and throat.
Move to fresh air.
If breathing has stopped, give artificial respiration.
If breathing is difficult, give oxygen.

LIQUID
POISONOUS IF SWALLOWED.
Will burn skin and eyes.
Will cause frostbite.
Remove contaminated clothing and shoes.
Flush affected areas with plenty of water.
IF IN EYES, hold eyelids open and flush with plenty of water.
IF SWALLOWED and victim is CONSCIOUS, have victim drink water or milk.
DO NOT INDUCE VOMITING.
DO NOT RUB AFFECTED AREAS.

Water Pollution

Dangerous to aquatic life in high concentrations.
May be dangerous if it enters water intakes.
Notify local health and wildlife officials.
Notify operators of nearby water intakes.

1. RESPONSE TO DISCHARGE

(See Response Methods Handbook, CG 446-4)

Issue warning—poison, corrosive, air contaminant
Restrict access
Evacuate area
Disperse and flush

2. LABEL

Green

3. CHEMICAL DESIGNATIONS

3.1 **Synonyms:** Hydrochloric acid, anhydrous
3.2 **Coast Guard Compatibility Classification:** Inorganic acid
3.3 **Chemical Formula:** HCl
3.4 **IMCO/United Nations Numerical Designation:** 2.0/1050

4. OBSERVABLE CHARACTERISTICS

4.1 **Physical State (as shipped):** Compressed, liquefied gas
4.2 **Color:** Colorless to slightly yellow
4.3 **Odor:** Sharp, pungent, irritating

5. HEALTH HAZARDS

5.1 **Personal Protective Equipment:** Full face mask and acid gas canister; self-contained breathing apparatus; chemical goggles; rubber apron and gloves; acid-proof clothing; safety shower.
5.2 **Symptoms Following Exposure:** Severely irritating to nose and upper respiratory tract; lung injury.
5.3 **Treatment for Exposure:** INHALATION: immediately remove patient to fresh air, keep him warm and quiet, and call a physician immediately; if a qualified person is available to give oxygen, such treatment may be helpful. INGESTION: have victim drink water or milk; do NOT induce vomiting. EYES OR SKIN: immediately flush with plenty of water for at least 15 min.; for eyes get medical attention promptly; air contaminated clothing and wash before reuse.
5.4 **Toxicity by Inhalation (Threshold Limit Value):** 5 ppm
5.5 **Short-Term Inhalation Limits:** 5 ppm for 5 min.
5.6 **Toxicity by Ingestion:** Data not available
5.7 **Late Toxicity:** None
5.8 **Vapor (Gas) Irritant Characteristics:** Vapors cause severe irritation of eye and throat and can cause eye and lung injury. They cannot be tolerated even at low concentrations.
5.9 **Liquid or Solid Irritant Characteristics:** Fairly severe skin irritant; may cause pain and second-degree burns after a few minutes' contact.
5.10 **Odor Threshold:** 1-5 ppm

6. FIRE HAZARDS

6.1 **Flash Point:** Not flammable
6.2 **Flammable Limits in Air:** Not flammable
6.3 **Fire Extinguishing Agents:** Not pertinent
6.4 **Fire Extinguishing Agents Not to be Used:** Not pertinent
6.5 **Special Hazards of Combustion Products:** Not pertinent
6.6 **Behavior in Fire:** Pressurized container may explode and release toxic, irritating vapors.
6.7 **Ignition Temperature:** Not flammable
6.8 **Electrical Hazard:** Not pertinent
6.9 **Burning Rate:** Not flammable

7. CHEMICAL REACTIVITY

7.1 **Reactivity with Water:** Moderate reaction with evolution of heat.
7.2 **Reactivity with Common Materials:** Rapidly absorbs moisture, forming hydrochloric acid. Highly corrosive to most metals with evolution of flammable hydrogen gas.
7.3 **Stability During Transport:** Stable
7.4 **Neutralizing Agents for Acids and Caustics:** Flush with water; apply powdered limestone, slaked lime, soda ash, or sodium bicarbonate
7.5 **Polymerization:** Not pertinent
7.6 **Inhibitor of Polymerization:** Not pertinent

8. WATER POLLUTION

8.1 **Aquatic Toxicity:**
282 ppm/96 hr/mosquito fish/TL_m/fresh water
100-330 ppm/48 hr/shrimp/LC_{50}/salt water
8.2 **Waterfowl Toxicity:** Data not available
8.3 **Biological Oxygen Demand (BOD):** None
8.4 **Food Chain Concentration Potential:** None

9. SELECTED MANUFACTURERS

1. Diamond Shamrock Corp.
Electro Chemicals Division
300 Union Commerce Bldg.
Cleveland, Ohio 44115
2. Stauffer Chemical Co.
Industrial Chemicals Division
299 Park Ave.
New York, N. Y. 10017
3. Vulcan Materials Co.
Chemical Division
Wichita, Kan. 67201

10. SHIPPING INFORMATION

10.1 **Grades or Purity:** Technical: 97.5-99%
10.2 **Storage Temperature:** Ambient or lower
10.3 **Inert Atmosphere:** No requirement
10.4 **Venting:** Safety relief

11. HAZARD ASSESSMENT CODE

(See Hazard Assessment Handbook, CG 446-3)

A-C-K-M-N-O

12. HAZARD CLASSIFICATIONS

12.1 **Code of Federal Regulations:** Nonflammable compressed gas
12.2 **NAS Hazard Rating for Bulk Water Transportation:**

Category	Rating
Fire	0
Health	
Vapor Irritant	4
Liquid or Solid Irritant	3
Poisons	3
Water Pollution	
Human Toxicity	2
Aquatic Toxicity	2
Aesthetic Effect	2
Reactivity	
Other Chemicals	4
Water	2
Self-Reaction	0

12.3 **NFPA Hazard Classifications:**

Category	Classification
Health Hazard (Blue)	3
Flammability (Red)	0
Reactivity (Yellow)	0

13. PHYSICAL AND CHEMICAL PROPERTIES

13.1 **Physical State at 15°C and 1 atm:** Gas
13.2 **Molecular Weight:** 36.46
13.3 **Boiling Point at 1 atm:** −121°F = −85.0°C = 188.2°K
13.4 **Freezing Point:** −175°F = −115°C = 158°K
13.5 **Critical Temperature:** 124.5°F = 51.4°C = 324.6°K
13.6 **Critical Pressure:** 1200 psia = 81.6 atm = 8.27 MN/m^2
13.7 **Specific Gravity:** 1.191 at −85°C (liquid)
13.8 **Liquid Surface Tension:** Not pertinent
13.9 **Liquid-Water Interfacial Tension:** Not pertinent
13.10 **Vapor (Gas) Specific Gravity:** 1.3
13.11 **Ratio of Specific Heats of Vapor (Gas):** 1.398
13.12 **Latent Heat of Vaporization:** 185 Btu/lb = 103 cal/g = 4.31×10^5 J/kg
13.13 **Heat of Combustion:** Not pertinent
13.14 **Heat of Decomposition:** Not pertinent
13.15 **Heat of Solution:** −884 Btu/lb = −491 cal/g = -20.6×10^5 J/kg
13.16 **Heat of Polymerization:** Not pertinent

NOTES

REVISED 1978

HCN | HYDROGEN CYANIDE

Common Synonyms
Hydrocyanic acid
Prussic acid

Watery liquid, or gas   Colorless   Bitter almond odor

Sinks and mixes with water. Poisonous, flammable vapor is produced and rises. Boiling point is 78° F.

AVOID CONTACT WITH LIQUID AND VAPOR. EVACUATE AREA.
WEAR CHEMICAL PROTECTIVE SUIT WITH SELF-CONTAINED BREATHING APPARATUS.
Stop discharge if possible. Call fire department.
Shut off ignition sources.
Stay upwind and use water spray to "knock down" vapor.
Isolate and remove discharged material.
Notify local health and pollution control agencies.

Fire

FLAMMABLE.
Flashback along vapor trail may occur.
Vapor may explode if ignited in an enclosed area.
WEAR CHEMICAL PROTECTIVE SUIT WITH SELF-CONTAINED BREATHING APPARATUS.
Stop discharge if possible.
Cool exposed containers and protect men effecting shutoff with water.
Let fire burn.

Exposure

CALL FOR MEDICAL AID.

VAPOR
POISONOUS IF INHALED OR IF SKIN IS EXPOSED.
Irritating to eyes.
Move to fresh air.
If breathing has stopped, give artificial respiration (but NOT mouth to mouth).
If breathing is difficult, give oxygen.

LIQUID
POISONOUS IF SWALLOWED OR IF SKIN IS EXPOSED.
Irritating to eyes.
Remove contaminated clothing and shoes.
Flush affected areas with plenty of water.
IF IN EYES, hold eyelids open and flush with plenty of water.
IF SWALLOWED and victim is CONSCIOUS, have victim drink water or milk and have victim induce vomiting.
IF SWALLOWED and victim is UNCONSCIOUS OR HAVING CONVULSIONS, do nothing except keep victim warm.

Water Pollution

HARMFUL TO AQUATIC LIFE IN VERY LOW CONCENTRATIONS.
May be dangerous if it enters water intakes.
Notify local health and wildlife officials.
Notify operators of nearby water intakes.

1. RESPONSE TO DISCHARGE

(See Response Methods Handbook, CG 446-4)

Issue warning—high flammability, water contaminant
Restrict access
Evacuate area

2. LABELS

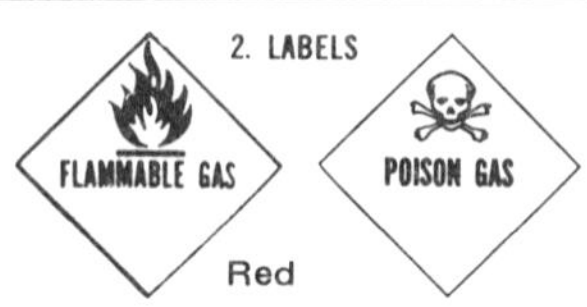

3. CHEMICAL DESIGNATIONS

3.1 **Synonyms:** Hydrocyanic acid
Prussic acid
3.2 **Coast Guard Compatibility Classification:** Not applicable
3.3 **Chemical Formula:** HCN
3.4 **IMCO/United Nations Numerical Designation:** 2.0/1051

4. OBSERVABLE CHARACTERISTICS

4.1 **Physical State (as shipped):** Liquid
4.2 **Color:** Colorless to bluish white
4.3 **Odor:** Characteristic sweetish, like almond

5. HEALTH HAZARDS

CAUTION—Class A poison; asphyxiation can be caused by ingestion, inhalation, or absorption of liquid or vapor through skin (particularly eyes, mucous membranes, and feet).

5.1 **Personal Protective Equipment:** Escape purposes only—air escape mask with 5-minute air cylinder. Work purposes—vapor-proof emergency suit or vinyl-coated coverall, plus air mask with clear-view facepiece, speaking diaphragm, demand regulator, and 30-minute air cylinder. Rubber gloves; chemical safety goggles; quick-opening safety shower.
5.2 **Symptoms Following Exposure:** Irritation of throat, palpitation, difficult breathing, reddening of eyes, salivation, nausea, headache, weakness of arms and legs, giddiness—followed by collapse and convulsions.
5.3 **Treatment for Exposure:** Call a doctor. If breathing has stopped, give artificial respiration until doctor arrives. INHALATION: remove patient to fresh air. SKIN CONTACT: remove contaminated clothing and wash skin thoroughly with copious quantities of water and soap. EYE CONTACT: hold eyelids apart and wash eye with continuous gentle stream of water for at least 15 min.
If patient is unconscious, administer amyl nitrite by crushing a pearl (ampule) in a cloth and hoding this under patient's nose for 15 seconds in every minute. Do not interrupt artificial respiration. Replace amyl nitrite pearl when its strength is spent. Continue treatment until patient's condition improves or doctor arrives.
5.4 **Toxicity by Inhalation (Threshold Limit Value):** 10 ppm
5.5 **Short-Term Inhalation Limits:** 20 ppm for 30 min.
5.6 **Toxicity by Ingestion:** Grade 4; LD_{50} less than 50 mg/kg

6. FIRE HAZARDS

6.1 **Flash Point:** 0°F C.C.
6.2 **Flammable Limits in Air:** 5.6%–40.0%
6.3 **Fire Extinguishing Agents:** Stop flow of gas
6.4 **Fire Extinguishing Agents Not to be Used:** None
6.5 **Special Hazards of Combustion Products:** Extremely toxic vapors are generated even at ordinary temperatures.
6.6 **Behavior in Fire:** Containers may explode with ignition of contents.
6.7 **Ignition Temperature:** 1004°F
6.8 **Electrical Hazard:** Data not available
6.9 **Burning Rate:** 1.8 mm/min.

7. CHEMICAL REACTIVITY

7.1 **Reactivity with Water:** Dissolves with a moderate reaction.
7.2 **Reactivity with Common Materials:** None
7.3 **Stability During Transport:** May become unstable and subject to explosion if stored for extended time or exposed to high temp. and pressure.
7.4 **Neutralizing Agents for Acids and Caustics:** The weak acidity can be neutralized by slaked lime, but this does not destroy the poisonous property.
7.5 **Polymerization:** Not pertinent
7.6 **Inhibitor of Polymerization:** Not pertinent

8. WATER POLLUTION

8.1 **Aquatic Toxicity:**
0.16 ppm/72 hr/young bass/TL_m/fresh water
0.069 ppm/24 hr/pin perch/TL_m/salt water
8.2 **Waterfowl Toxicity:** Data not available
8.3 **Biological Oxygen Demand (BOD):** Data not available
8.4 **Food Chain Concentration Potential:** None

9. SELECTED MANUFACTURERS

1. E. I. duPont de Nemours & Co., Inc.
Electrochemicals Dept.
Wilmington, Del. 19898
2. Monsanto Polymers & Petrochemicals Co.
800 North Lindbergh Blvd.
St. Louis, Mo. 63166
3. Rohm and Haas Co.
Independence Mall West
Philadelphia, Pa. 19105

10. SHIPPING INFORMATION

10.1 **Grades or Purity:** 96%; sometimes shipped as a water solution, or absorbed on an inert solid. All forms are extremely toxic.
10.2 **Storage Temperature:** Data not available
10.3 **Inert Atmosphere:** May be padded
10.4 **Venting:** Data not available

11. HAZARD ASSESSMENT CODE

(See Hazard Assessment Handbook, CG 446-3)

A-B-C-K-L-M-N

12. HAZARD CLASSIFICATIONS

12.1 **Code of Federal Regulations:** Poisonous gas or liquid, Class A
12.2 **NAS Hazard Rating for Bulk Water Transportation:**

Category	Rating
Fire	4
Health	
Vapor Irritant	2
Liquid or Solid Irritant	1
Poisons	4
Water Pollution	
Human Toxicity	4
Aquatic Toxicity	4
Aesthetic Effect	1
Reactivity	
Other Chemicals	3
Water	0
Self-Reaction	3

12.3 **NFPA Hazard Classifications:**

Category	Classification
Health Hazard (Blue)	4
Flammability (Red)	4
Reactivity (Yellow)	2

13. PHYSICAL AND CHEMICAL PROPERTIES

13.1 **Physical State at 15°C and 1 atm:** Liquid
13.2 **Molecular Weight:** 27.03
13.3 **Boiling Point at 1 atm:** 78.3°F = 25.7°C = 298.9°K
13.4 **Freezing Point:** 8.1°F = −13.3°C = 259.9°K
13.5 **Critical Temperature:** 362.3°F = 183.5°C = 456.7°K
13.6 **Critical Pressure:** 735 psia = 50 atm = 5.07 MN/m^2
13.7 **Specific Gravity:** 0.689 at 20°C (liquid)
13.8 **Liquid Surface Tension:** Not pertinent
13.9 **Liquid-Water Interfacial Tension:** Not pertinent
13.10 **Vapor (Gas) Specific Gravity:** 0.9
13.11 **Ratio of Specific Heats of Vapor (Gas):** 1.303
13.12 **Latent Heat of Vaporization:** 444 Btu/lb = 247 cal/g = 10.3×10^5 J/kg
13.13 **Heat of Combustion:** −10,560 Btu/lb = −5864 cal/g = -245.3×10^5 J/kg
13.14 **Heat of Decomposition:** Not pertinent
13.15 **Heat of Solution:** Not pertinent
13.16 **Heat of Polymerization:** Not pertinent

5. HEALTH HAZARDS (Cont'd.)

5.7 **Late Toxicity:** Data not available
5.8 **Vapor (Gas) Irritant Characteristics:** Vapor is not very irritating but is extremely poisonous.
5.9 **Liquid or Solid Irritant Characteristics:** Liquid is not irritating but is extremely poisonous if absorbed through skin or eyes.
5.10 **Odor Threshold:** 1 mg/m^3

REVISED 1978

HYDROGEN FLUORIDE

Common Synonyms			
Hydrofluoric acid, anhydrous	Liquid	Colorless	Sharp, irritating odor
	Sinks and mixes with water. Poisonous vapor is produced and slowly rises. Boiling point is 67° F.		

AVOID CONTACT WITH LIQUID AND VAPOR. Keep people away.
Wear chemical protective suit including self-contained breathing apparatus.
Stop discharge if possible.
Stay upwind and use water spray to "knock down" vapor.
Isolate and remove discharged material.
Notify local health and pollution control agencies.

Fire

Not flammable.
Flammable gas may be produced on contact with metals.

Wear chemical protective suit including self-contained breathing apparatus.

Exposure

CALL FOR MEDICAL AID.

VAPOR
POISONOUS IF INHALED.
Irritating to eyes, nose and throat.

Move to fresh air.
If breathing has stopped, give artificial respiration.
If breathing is difficult, give oxygen.

LIQUID
POISONOUS IF SWALLOWED.
Will burn skin and eyes.

Remove contaminated clothing and shoes.
Flush affected areas with plenty of water.
IF IN EYES, hold eyelids open and flush with plenty of water.
IF SWALLOWED and victim is CONSCIOUS, have victim drink water or milk.
DO NOT INDUCE VOMITING.

Water Pollution

HARMFUL TO AQUATIC LIFE IN VERY LOW CONCENTRATIONS.
May be dangerous if it enters water intakes.

Notify local health and wildlife officials.
Notify operators of nearby water intakes.

1. RESPONSE TO DISCHARGE

(See Response Methods Handbook, CG 446-4)

Issue warning—poison, corrosive
Restrict access
Evacuate area
Disperse and flush

2. LABEL

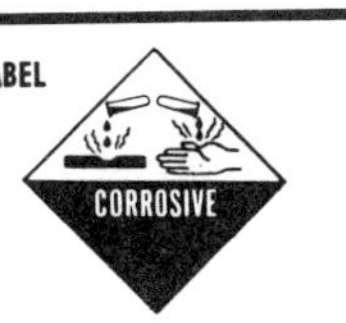

3. CHEMICAL DESIGNATIONS

3.1 **Synonyms:** Hydrofluoric acid, anhydrous
3.2 **Coast Guard Compatibility Classification:** Inorganic acid
3.3 **Chemical Formula:** H_2F_2
3.4 **IMCO/United Nations Numerical Designation:** 2.0/1052

4. OBSERVABLE CHARACTERISTICS

4.1 **Physical State (as shipped):** Liquid
4.2 **Color:** Colorless
4.3 **Odor:** Pungent, irritating

5. HEALTH HAZARDS

5.1 **Personal Protective Equipment:** Acid-resistant hat, safety goggles, face shield, jacket, overalls, gauntlet-type gloves, and boots. The goggles and face shield must have plastic lenses. There must be a shower and an eye wash. Observe all precautions contained in the Manufacturing Chemists' Association Chemical Safety Data Sheet SD-25.
5.2 **Symptoms Following Exposure:** Serious and painful burns of eyes, skin and respiratory tract; pulmonary edema.
5.3 **Treatment for Exposure:** INGESTION: have victim drink water or milk; do NOT induce vomiting. SKIN: flush with water; consult physician. EYES: flush with water for at least 15 min.; consult physician.
5.4 **Toxicity by Inhalation (Threshold Limit Value):** 3 ppm
5.5 **Short-Term Inhalation Limits:** 3 ppm for 15 min.
5.6 **Toxicity by Ingestion:** Oral LDL_O = 80 mg/kg (guinea pig)
5.7 **Late Toxicity:** Data not available
5.8 **Vapor (Gas) Irritant Characteristics:** Vapors cause severe irritation of eye and throat and can cause eye and lung injury. They cannot be tolerated even at low concentrations.
5.9 **Liquid or Solid Irritant Characteristics:** Severe skin irritant. Causes second- and third-degree burns on short contact; very injurious to the eyes.
5.10 **Odor Threshold:** 0.03 mg/m^3

6. FIRE HAZARDS

6.1 **Flash Point:** Not flammable
6.2 **Flammable Limits in Air:** Not flammable
6.3 **Fire Extinguishing Agents:** Not pertinent
6.4 **Fire Extinguishing Agents Not to be Used:** Not pertinent
6.5 **Special Hazards of Combustion Products:** Toxic and irritating vapors are generated when heated.
6.6 **Behavior in Fire:** Not pertinent
6.7 **Ignition Temperature:** Not flammable
6.8 **Electrical Hazard:** Not pertinent
6.9 **Burning Rate:** Not flammable

7. CHEMICAL REACTIVITY

7.1 **Reactivity with Water:** Dissolves with liberation of heat.
7.2 **Reactivity with Common Materials:** Will attack glass, concrete and certain metals, especially those containing silica, such as cast iron. Will attack natural rubber, leather, and many organic materials. May generate flammable hydrogen gas in contact with some metals.
7.3 **Stability During Transport:** Stable
7.4 **Neutralizing Agents for Acids and Caustics:** Flush with water; apply powdered limestone, slaked lime, soda ash, or sodium bicarbonate.
7.5 **Polymerization:** Not pertinent
7.6 **Inhibitor of Polymerization:** Not pertinent

8. WATER POLLUTION

8.1 **Aquatic Toxicity:** 60 ppm/*/fish/lethal/fresh water
*Time period not specified
8.2 **Waterfowl Toxicity:** Data not available
8.3 **Biological Oxygen Demand (BOD):** None
8.4 **Food Chain Concentration Potential:** None

9. SELECTED MANUFACTURERS

1. Allied Chemical Corp.
Industrial Chemicals Division
Morristown, N. J. 07960
2. E. I. duPont de Nemours & Co. Inc.
Industrial & Biochemicals Dept.
Wilmington, Del. 19898
3. Kaiser Aluminum & Chemical Corp.
Kaiser Chemicals Division
Gramercy, La. 70052

10. SHIPPING INFORMATION

10.1 **Grades or Purity:** 99-99.97%
10.2 **Storage Temperature:** Ambient
10.3 **Inert Atmosphere:** No requirement
10.4 **Venting:** Safety relief

11. HAZARD ASSESSMENT CODE

(See Hazard Assessment Handbook, CG 446-3)

A-C-K-M-N-O

12. HAZARD CLASSIFICATIONS

12.1 **Code of Federal Regulations:** Corrosive material
12.2 **NAS Hazard Rating for Bulk Water Transportation:**

Category	Rating
Fire	0
Health	
Vapor Irritant	4
Liquid or Solid Irritant	4
Poisons	4
Water Pollution	
Human Toxicity	4
Aquatic Toxicity	3
Aesthetic Effect	2
Reactivity	
Other Chemicals	4
Water	2
Self-Reaction	0

12.3 **NFPA Hazard Classifications:**

Category	Classification
Health Hazard (Blue)	4
Flammability (Red)	0
Reactivity (Yellow)	0

13. PHYSICAL AND CHEMICAL PROPERTIES

13.1 **Physical State at 15°C and 1 atm:** Liquid
13.2 **Molecular Weight:** 20.01
13.3 **Boiling Point at 1 atm:** 67.1°F = 19.5°C = 292.7°K
13.4 **Freezing Point:** −134°F = −92.2°C = 181.0°K
13.5 **Critical Temperature:** 447°F = 230.6°C = 503.8°K
13.6 **Critical Pressure:** 1100 psia = 74.8 atm = 7.58 MN/m^2
13.7 **Specific Gravity:** 0.992 at 19°C (liquid)
13.8 **Liquid Surface Tension:** Not pertinent
13.9 **Liquid-Water Interfacial Tension:** Not pertinent
13.10 **Vapor (Gas) Specific Gravity:** 0.7
13.11 **Ratio of Specific Heats of Vapor (Gas):** 1.399
13.12 **Latent Heat of Vaporization:** 145 Btu/lb = 80.5 cal/g = 3.37×10^5 J/kg
13.13 **Heat of Combustion:** Not pertinent
13.14 **Heat of Decomposition:** Not pertinent
13.15 **Heat of Solution:** −1322 Btu/lb = −734.6 cal/g = -30.76×10^5 J/kg
13.16 **Heat of Polymerization:** Not pertinent

NOTES

REVISED 1978

HXX HYDROGEN, LIQUEFIED

Common Synonyms

Liquid hydrogen

Liquefied compressed gas Colorless Odorless

Floats and boils on water. Flammable visible vapor cloud is produced.

Shut off ignition sources. Call fire department.
Avoid contact with liquid. Keep people away.
Stop discharge if possible.
Stay upwind. Use water spray to "knock down" vapor.

Fire

FLAMMABLE.
Flame is almost invisible.
Flashback along vapor trail may occur.
Vapor may explode if ignited in an enclosed area.
Evacuate surrounding area.
Stop flow of gas if possible.
Cool exposed containers and protect men effecting shutoff with water.

Exposure

VAPOR
If inhaled in high concentrations will cause difficult breathing, or loss of consciousness.
Move victim to fresh air.
If breathing has stopped, give artificial respiration.

LIQUID
Will cause frostbite.
Flush affected areas with plenty of water.
DO NOT RUB AFFECTED AREAS.

Water Pollution

Not harmful to aquatic life.

1. RESPONSE TO DISCHARGE

(See Response Methods Handbook, CG 446-4)

Issue warning—high flammability
Restrict access
Evacuate area

2. LABEL

Red

3. CHEMICAL DESIGNATIONS

3.1 **Synonyms:** Liquid hydrogen; para Hydrogen
3.2 **Coast Guard Compatibility Classification:** Not applicable
3.3 **Chemical Formula:** H_2
3.4 **IMCO/United Nations Numerical Designation:** 2/1966

4. OBSERVABLE CHARACTERISTICS

4.1 **Physical State (as shipped):** Liquid
4.2 **Color:** Colorless
4.3 **Odor:** None

5. HEALTH HAZARDS

5.1 **Personal Protective Equipment:** Safety goggles or face shield; insulated gloves and long sleeves; cuffless trousers worn outside boots or over high-top shoes to shed spilled liquid; self-contained breathing apparatus containing air (*never* use oxygen).
5.2 **Symptoms Following Exposure:** If atmosphere does not contain enough oxygen, inhalation can cause dizziness, unconsciousness, or even death. Contact of liquid with eyes or skin causes freezing similar to a burn.
5.3 **Treatment for Exposure:** The only effect of exposure to liquid hydrogen is that caused by its unusually low temperature and its action as a simple asphyxiant. INHALATION: if victim is unconscious (due to oxygen deficiency), move him to fresh air and apply resuscitation methods; call physician. EYES: treat for frostbite. SKIN: treat for frostbite; soak in lukewarm water; get medical attention if burn is severe.
5.4 **Toxicity by Inhalation (Threshold Limit Value):** Gas is non-poisonous but can act as a simple asphyxiant.
5.5 **Short-Term Inhalation Limits:** Not pertinent
5.6 **Toxicity by Ingestion:** Not pertinent (gas with low boiling point)
5.7 **Late Toxicity:** None
5.8 **Vapor (Gas) Irritant Characteristics:** Data not available
5.9 **Liquid or Solid Irritant Characteristics:** Data not available
5.10 **Odor Threshold:** Not pertinent

6. FIRE HAZARDS

6.1 **Flash Point:** Not pertinent
6.2 **Flammable Limits in Air:** 4.0%–75.0%
6.3 **Fire Extinguishing Agents:** Let fire burn; shut off gas supply.
6.4 **Fire Extinguishing Agents Not to be Used:** Carbon dioxide
6.5 **Special Hazards of Combustion Products:** Not pertinent
6.6 **Behavior in Fire:** Burns with an almost invisible flame.
6.7 **Ignition Temperature:** 1,065°F
6.8 **Electrical Hazard:** Class I, Group B
6.9 **Burning Rate:** 9.9 mm/min.

7. CHEMICAL REACTIVITY

7.1 **Reactivity with Water:** Ambient temperature of water will cause vigorous vaporization of hydrogen.
7.2 **Reactivity with Common Materials:** No chemical reaction, but low temperature causes most materials to become very brittle.
7.3 **Stability During Transport:** Stable
7.4 **Neutralizing Agents for Acids and Caustics:** Not pertinent
7.5 **Polymerization:** Not pertinent
7.6 **Inhibitor of Polymerization:** Not pertinent

8. WATER POLLUTION

8.1 **Aquatic Toxicity:** None
8.2 **Waterfowl Toxicity:** None
8.3 **Biological Oxygen Demand (BOD):** None
8.4 **Food Chain Concentration Potential:** None

9. SELECTED MANUFACTURERS

1. Union Carbide Corporation
Linde Division
Morristown, N. J. 03057
2. Air Products and Chemicals, Inc.
Specialty Gases Division
P. O. Box 538
Allentown, Pa. 18105
3. Chemetron Corp.
Industrial Gases Division
111 E. Wacker Drive
Chicago, Ill. 60601

10. SHIPPING INFORMATION

10.1 **Grades or Purity:** Commercial
10.2 **Storage Temperature:** −434°F
10.3 **Inert Atmosphere:** No requirement
10.4 **Venting:** Safety relief

11. HAZARD ASSESSMENT CODE

(See Hazard Assessment Handbook, CG 446-3)

A-B-C-D-E-F-G

12. HAZARD CLASSIFICATIONS

12.1 **Code of Federal Regulations:** Flammable compressed gas
12.2 **NAS Hazard Rating for Bulk Water Transportation:** Not listed
12.3 **NFPA Hazard Classifications:**

Category	Classification
Health Hazard (Blue)	0
Flammability (Red)	4
Reactivity (Yellow)	0

13. PHYSICAL AND CHEMICAL PROPERTIES

13.1 **Physical State at 15°C and 1 atm:** Gas
13.2 **Molecular Weight:** 2.0
13.3 **Boiling Point at 1 atm:** −423°F = −253°C = 20°K
13.4 **Freezing Point:** −434°F = −259°C = 14°K
13.5 **Critical Temperature:** −400°F = −240°C = 33°K
13.6 **Critical Pressure:** 188 psia = 12.8 atm = 1.30 MN/m^2
13.7 **Specific Gravity:** 0.071 at −253°C (liquid)
13.8 **Liquid Surface Tension:** 2.3 dynes/cm = 0.023 N/m at −255°C
13.9 **Liquid-Water Interfacial Tension:** Not pertinent
13.10 **Vapor (Gas) Specific Gravity:** 0.067
13.11 **Ratio of Specific Heats of Vapor (Gas):** 1.3962
13.12 **Latent Heat of Vaporization:** 190.5 Btu/lb = 105.8 cal/g = 4.427 × 10^5 J/kg
13.13 **Heat of Combustion:** −50,080 Btu/lb = −27,823 cal/g = −1164.1 × 10^5 J/kg
13.14 **Heat of Decomposition:** Not pertinent
13.15 **Heat of Solution:** Not pertinent
13.16 **Heat of Polymerization:** Not pertinent

NOTES

HPO — HYDROGEN PEROXIDE

Common Synonyms			
Peroxide Albone Superoxol	Watery liquid	Colorless	Slightly sharp odor
	Sinks and mixes with water. Irritating vapor is produced.		

AVOID CONTACT WITH LIQUID AND VAPOR. Keep people away.
Wear chemical protective suit including self-contained breathing apparatus.
Stop discharge if possible. Call fire department.
Isolate and remove discharged material.
Notify local health and pollution control agencies.

Fire

Not flammable.
May cause fire and explode on contact with combustibles and metals.
Containers may explode when heated.
Wear chemical protective suit including self-contained breathing apparatus.
Flood discharge area with water.

Exposure

CALL FOR MEDICAL AID.

VAPOR
Irritating to eyes, nose and throat.
Harmful if inhaled.
Move to fresh air.
If breathing has stopped, give artificial respiration.
If breathing is difficult, give oxygen.

LIQUID
Will burn skin and eyes.
Harmful if swallowed.
Remove contaminated clothing and shoes.
Flush affected areas with plenty of water.
IF IN EYES, hold eyelids open and flush with plenty of water.
IF SWALLOWED and victim is CONSCIOUS, have victim drink water or milk.

Water Pollution

Effect of low concentrations on aquatic life is unknown.
May be dangerous if it enters water intakes.
Notify local health and wildlife officials.
Notify operators of nearby water intakes.

1. RESPONSE TO DISCHARGE

(See Response Methods Handbook, CG 446-4)

Issue warning—corrosive
Restrict access
Disperse and flush

2. LABEL

Yellow

3. CHEMICAL DESIGNATIONS

3.1 **Synonyms:** Peroxide, Albone, Superoxol
3.2 **Coast Guard Compatibility Classification:** Not applicable
3.3 **Chemical Formula:** H_2O_2–H_2O
3.4 **IMCO/United Nations Numerical Designation:** 5.1/2015

4. OBSERVABLE CHARACTERISTICS

4.1 **Physical State (as shipped):** Liquid
4.2 **Color:** Colorless
4.3 **Odor:** Slightly sharp

5. HEALTH HAZARDS

5.1 **Personal Protective Equipment:** Protective garments, both outer and inner, made of a woven polyester fabric or of modacrylic or polyvinylidene fabrics; impermeable apron made of polyvinyl chloride or polyethylene film; neoprene gloves and boots; goggles.
5.2 **Symptoms Following Exposure:** Although solutions and vapors are nontoxic, they are irritating. Vapor causes discomfort of eyes and nose. Moderately concentrated liquid causes whitening of the skin and severe stinging sensation. In most cases the stinging subsides quickly and the skin gradually returns to normal without any damage. Highly concentrated liquid can cause blistering of skin if left on for any length of time; can also cause eye damage.
5.3 **Treatment for Exposure:** Contact should be avoided, but immediate flushing with water will prevent any reaction in case of accidental contact.
5.4 **Toxicity by Inhalation (Threshold Limit Value):** 1 ppm
5.5 **Short-Term Inhalation Limits:** Data not available
5.6 **Toxicity by Ingestion:** Data not available
5.7 **Late Toxicity:** None
5.8 **Vapor (Gas) Irritant Characteristics:** Vapors cause moderate irritation, such that personnel will find high concentrations unpleasant. The effect is temporary.
5.9 **Liquid or Solid Irritant Characteristics:** Fairly severe skin irritant. May cause pain and second-degree burns after a few minutes' contact.
5.10 **Odor Threshold:** Not pertinent

6. FIRE HAZARDS

6.1 **Flash Point:** Not flammable but may cause fire and react violently on contact with combustibles and metals.
6.2 **Flammable Limits in Air:** Not flammable
6.3 **Fire Extinguishing Agents:** Water for fires resulting from spillage
6.4 **Fire Extinguishing Agents Not to be Used:** Not pertinent
6.5 **Special Hazards of Combustion Products:** Not pertinent
6.6 **Behavior in Fire:** May explode in fire
6.7 **Ignition Temperature:** Not flammable
6.8 **Electrical Hazard:** Not pertinent
6.9 **Burning Rate:** Not flammable

7. CHEMICAL REACTIVITY

7.1 **Reactivity with Water:** No reaction
7.2 **Reactivity with Common Materials:** Dirt and many metals cause a rapid decomposition with liberation of oxygen gas; occurs particularly if concentration is above 40%.
7.3 **Stability During Transport:** Pure grades are quite stable, but contamination with metals or dirt can cause rapid or violent decomposition.
7.4 **Neutralizing Agents for Acids and Caustics:** Not pertinent
7.5 **Polymerization:** Not pertinent
7.6 **Inhibitor of Polymerization:** Not pertinent

8. WATER POLLUTION

8.1 **Aquatic Toxicity:**
> 40 ppm/*/fingerling trout/toxic/salt water
*Time period not specified
8.2 **Waterfowl Toxicity:** Data not available
8.3 **Biological Oxygen Demand (BOD):** None
8.4 **Food Chain Concentration Potential:** None

9. SELECTED MANUFACTURERS

1. E. I. du Pont de Nemours & Co., Inc.
   Electrochemicals Dept.
   Wilmington, Del. 19898
2. FMC Corp.
   Inorganic Chemicals Division
   633 Third Ave.
   New York, N. Y. 10017
3. Shell Chemical Co.
   Industrial Chemicals Division
   Houston, Texas 77001

10. SHIPPING INFORMATION

10.1 **Grades or Purity:** Common commercial strengths are 27.5%, 35%, 50%, 70%, 90% and 98%. "High Strength" means greater than 52%. Purity: Technical; Mil. Spec.; ACS. The hazard increases with the strength.
10.2 **Storage Temperature:** Ambient
10.3 **Inert Atmosphere:** No requirement
10.4 **Venting:** Safety relief or pressure-vacuum

11. HAZARD ASSESSMENT CODE

(See Hazard Assessment Handbook, CG 446-3)

A-P-Z

12. HAZARD CLASSIFICATIONS

12.1 **Code of Federal Regulations:** Oxidizer
12.2 **NAS Hazard Rating for Bulk Water Transportation:**

Category	Rating
Fire	0
Health	
Vapor Irritant	2
Liquid or Solid Irritant	3
Poisons	1
Water Pollution	
Human Toxicity	1
Aquatic Toxicity	3
Aesthetic Effect	1
Reactivity	
Other Chemicals	4
Water	1
Self-Reaction	3

12.3 **NFPA Hazard Classifications:**

Category	Classification	
% by wt =	32-52	≥ 52
Health Hazard (Blue)	2	2
Flammability (Red)	0	0
Reactivity (Yellow)	1 oxy	3 oxy

13. PHYSICAL AND CHEMICAL PROPERTIES*

13.1 **Physical State at 15°C and 1 atm:** Liquid
13.2 **Molecular Weight:** 34.01
13.3 **Boiling Point at 1 atm:** 257°F = 125°C = 398°K
13.4 **Freezing Point:** −40.5°F = −40.3°C = 232.9°K
13.5 **Critical Temperature:** Not pertinent
13.6 **Critical Pressure:** Not pertinent
13.7 **Specific Gravity:** 1.29 at 20°C (liquid)
13.8 **Liquid Surface Tension:** Not pertinent
13.9 **Liquid-Water Interfacial Tension:** Not pertinent
13.10 **Vapor (Gas) Specific Gravity:** Not pertinent
13.11 **Ratio of Specific Heats of Vapor (Gas):** 1.241
13.12 **Latent Heat of Vaporization:** 542 Btu/lb = 301 cal/g = 12.6×10^5 J/kg
13.13 **Heat of Combustion:** Not pertinent
13.14 **Heat of Decomposition:** −1220 Btu/lb = −676 cal/g = -28.3×10^5 J/kg
13.15 **Heat of Solution:** −20.2 Btu/lb = −11.2 cal/g = -0.469×10^5 J/kg
13.16 **Heat of Polymerization:** Not pertinent

*Physical properties apply to 70% solution.

NOTES

REVISED 1978

HDS | HYDROGEN SULFIDE

Common Synonyms			
Sulphuretted hydrogen	Liquefied compressed gas	Colorless	Rotten egg odor, but odorless at poisonous concentrations

Sinks and boils in water. Poisonous, flammable, visible vapor cloud is produced.

Avoid contact with gas. Keep people away.
Wear goggles and self-contained breathing apparatus.
Shut off ignition sources and call fire department.
Stop discharge if possible.
Evacuate area in case of large discharges.
Stay upwind and use water spray to "knock down" vapor.
Isolate and remove discharged material.
Notify local health and pollution control agencies.

Fire

FLAMMABLE.
Flashback along vapor trail may occur.
May explode if ignited in an enclosed area.
Wear goggles and self-contained breathing apparatus.
Stop flow of gas if possible.
Cool exposed containers and men effecting shutoff with water.
Let fire burn.

Exposure

CALL FOR MEDICAL AID.

VAPOR
POISONOUS IF INHALED.
Irritating to eyes.
Move to fresh air.
If breathing has stopped, give artificial respiration.
If breathing is difficult, give oxygen.
IF IN EYES, hold eyelids open and flush with plenty of water.

Water Pollution

HARMFUL TO AQUATIC LIFE IN VERY LOW CONCENTRATIONS.
May be dangerous if it enters water intakes.
Notify local health and wildlife officials.
Notify operators of nearby water intakes.

1. RESPONSE TO DISCHARGE

(See Response Methods Handbook, CG 446-4)

Issue warning—high flammability, poison
Restrict access
Evacuate area

2. LABEL

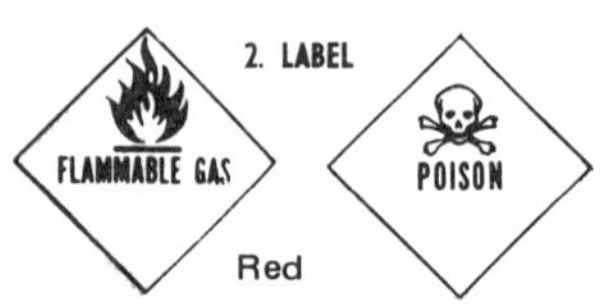

Red

3. CHEMICAL DESIGNATIONS

3.1 **Synonyms:** Sulfuretted hydrogen
3.2 **Coast Guard Compatibility Classification:** Not applicable
3.3 **Chemical Formula:** H_2S
3.4 **IMCO/United Nations Numerical Designation:** 2.0/1053

4. OBSERVABLE CHARACTERISTICS

4.1 **Physical State (as shipped):** Liquid under pressure
4.2 **Color:** Colorless
4.3 **Odor:** Offensive odor, like rotten eggs

5. HEALTH HAZARDS

5.1 **Personal Protective Equipment:** Rubber-framed goggles; approved respiratory protection.
5.2 **Symptoms Following Exposure:** Irritation of eyes, nose and throat. If high concentrations are inhaled, hyperpnea and respiratory paralysis may occur. Very high concentrations may produce pulmonary edema.
5.3 **Treatment for Exposure:** INHALATION: remove victim from exposure; if breathing has stopped, give artificial respiration; administer oxygen if needed; consult physician. EYES: wash with plenty of water.
5.4 **Toxicity by Inhalation (Threshold Limit Value):** 10 ppm
5.5 **Short-Term Inhalation Limits:** 200 ppm for 10 min.; 100 ppm for 30 min. and 50 ppm for 60 min.
5.6 **Toxicity by Ingestion:** Hydrogen sulfide is present as a gas at room temperature, so ingestion not likely.
5.7 **Late Toxicity:** Data not available
5.8 **Vapor (Gas) Irritant Characteristics:** Vapor is moderately irritating such that personnel will not usually tolerate moderate or high vapor concentrations.
5.9 **Liquid or Solid Irritant Characteristics:** Minimum hazard. If spilled on clothing and allowed to remain, may cause smarting and reddening of the skin.
5.10 **Odor Threshold:** 0.0047 ppm

6. FIRE HAZARDS

6.1 **Flash Point:** Flammable gas
6.2 **Flammable Limits in Air:** 4.3%—45%
6.3 **Fire Extinguishing Agents:** Stop flow of gas
6.4 **Fire Extinguishing Agents Not to be Used:** Not pertinent
6.5 **Special Hazards of Combustion Products:** Toxic gases are generated in fires.
6.6 **Behavior in Fire:** Vapor is heavier than air and may travel considerable distance to a source of ignition and flash back.
6.7 **Ignition Temperature:** 500°F
6.8 **Electrical Hazard:** Not pertinent
6.9 **Burning Rate:** 2.3 mm/min. (liquid)

7. CHEMICAL REACTIVITY

7.1 **Reactivity with Water:** No reaction
7.2 **Reactivity with Common Materials:** No reaction
7.3 **Stability During Transport:** Stable
7.4 **Neutralizing Agents for Acids and Caustics:** Not pertinent
7.5 **Polymerization:** Not pertinent
7.6 **Inhibitor of Polymerization:** Not pertinent

8. WATER POLLUTION

8.1 **Aquatic Toxicity:**
1.38 ppm/48 hr/fathead minnows/TL_m/fresh water
sat./0.5 hr/bullia/lethal/salt water
8.2 **Waterfowl Toxicity:** Data not available
8.3 **Biological Oxygen Demand (BOD):** Data not available
8.4 **Food Chain Concentration Potential:** None

9. SELECTED MANUFACTURERS

1. P P G Industries, Inc.
Industrial Chemicals Division
One Gateway Center
Pittsburgh, Pa. 90670
2. Union Oil Co. of California
Union Oil Center
Los Angeles, Cal. 90017
3. Will Ross, Inc.
Matheson Gas Products Division
East Rutherford, N. J. 07073

10. SHIPPING INFORMATION

10.1 **Grades or Purity:** Purified; technical
10.2 **Storage Temperature:** Ambient
10.3 **Inert Atmosphere:** No requirement
10.4 **Venting:** Safety relief

11. HAZARD ASSESSMENT CODE

(See Hazard Assessment Handbook, CG 446-3)

A-B-C-D-E-F-G

12. HAZARD CLASSIFICATIONS

12.1 **Code of Federal Regulations:** Flammable compressed gas
12.2 **NAS Hazard Rating for Bulk Water Transportation:** Not listed
12.3 **NFPA Hazard Classifications:**

Category	Classification
Health Hazard (Blue)	3
Flammability (Red)	4
Reactivity (Yellow)	0

13. PHYSICAL AND CHEMICAL PROPERTIES

13.1 **Physical State at 15°C and 1 atm:** Gas
13.2 **Molecular Weight:** 34.08
13.3 **Boiling Point at 1 atm:** −76.7°F = −60.4°C = 212.8°K
13.4 **Freezing Point:** −117°F = −82.8°C = 190.4°K
13.5 **Critical Temperature:** 212.7°F = 100.4°C = 373.6°K
13.6 **Critical Pressure:** 1300 psia = 88.9 atm = 9.01 MN/m^2
13.7 **Specific Gravity:** 0.916 at −60°C (liquid)
13.8 **Liquid Surface Tension:** (est.) 30 dynes/cm = 0.03 N/m at −61°C
13.9 **Liquid-Water Interfacial Tension:** Data not available
13.10 **Vapor (Gas) Specific Gravity:** 1.2
13.11 **Ratio of Specific Heats of Vapor (Gas):** 1.322
13.12 **Latent Heat of Vaporization:** 234 Btu/lb = 130 cal/g = 5.44×10^5 J/kg
13.13 **Heat of Combustion:** −6552 Btu/lb = −3640 cal/g = -152.4×10^5 J/kg
13.14 **Heat of Decomposition:** Not pertinent
13.15 **Heat of Solution:** Not pertinent
13.16 **Heat of Polymerization:** Not pertinent

NOTES

REVISED 1978

HDQ — HYDROQUINONE

Common Synonyms		
1,4-Benzenediol p-Dihydroxybenzene Hydroquinol Pyrogentisic acid Quinol	Solid	White, light tan to gray
	Sinks and mixes with water.	

Avoid contact with solid and dust. Keep people away.
Shut off ignition sources. Call fire department.
Stop discharge if possible.
Isolate and remove discharged material.
Notify local health and pollution control agencies.

Fire

Combustible.
Dust cloud may explode if ignited in an enclosed area.
Extinguish with water, dry chemicals, foam, or carbon dioxide.

Exposure

CALL FOR MEDICAL AID.

DUST
Irritating to eyes, nose and throat.
Harmful if inhaled.
If in eyes, hold eyelids open and flush with plenty of water.
If breathing has stopped, give artificial respiration.
If breathing is difficult, give oxygen.

SOLID
Will burn eyes.
Irritating to eyes.
If swallowed will cause headache, dizziness, nausea, vomiting, or loss of consciousness.
Remove contaminated clothing and shoes.
Flush affected areas with plenty of water.
IF IN EYES, hold eyelids open and flush with plenty of water.
IF SWALLOWED and victim is CONSCIOUS, have victim drink water or milk and have victim induce vomiting.
IF SWALLOWED and victim is UNCONSCIOUS OR HAVING CONVULSIONS, do nothing except keep victim warm.

Water Pollution

Effect of low concentrations on aquatic life is unknown.
May be dangerous if it enters water intakes.
Notify local health and wildlife officials.
Notify operators of nearby water intakes.

1. RESPONSE TO DISCHARGE

(See Response Methods Handbook, CG 446-4)

Issue warning—water contaminant
Disperse and flush

2. LABELS

No hazard label required by Code of Federal Regulations

3. CHEMICAL DESIGNATIONS

3.1 **Synonyms:** 1,4-Benzenediol; p-Dihydroxybenzene; Hydroquinol; Pyrogentisic acid; Quinol
3.2 **Coast Guard Compatibility Classification:** Not listed
3.3 **Chemical Formula:** 1,4-$C_6H_4(OH)_2$
3.4 **IMCO/United Nations Numerical Designation:** Not listed

4. OBSERVABLE CHARACTERISTICS

4.1 **Physical State (as shipped):** Solid
4.2 **Color:** Light tan to light gray; white
4.3 **Odor:** None

5. HEALTH HAZARDS

5.1 **Personal Protective Equipment:** Goggles; respiratory protection if dust is present
5.2 **Symptoms Following Exposure:** Ingestion can cause ringing in the ears, nausea, dizziness, a sense of suffocation, increased respiration rate, vomiting, pallor, muscular twitchings, headache, dyspnea, cyanosis, delirium, and collapse; the urine is green or brownish-green. Lethal adult dose is 2 grams. Direct contamination of the eye with particles of hydroquinone can cause immediate irritation and may result in ulceration of the cornea. Contact with skin may cause dermatitis.
5.3 **Treatment for Exposure:** INGESTION: induce vomiting; perform gastric lavage, and follow with a saline cathartic and demulcents; get medical attention. EYES: flush immediately with plenty of water for 15 min. and get medical attention. SKIN: wash with soap and water.
5.4 **Toxicity by Inhalation (Threshold Limit Value):** 2 mg/m^3
5.5 **Short-Term Inhalation Limits:** Data not available
5.6 **Toxicity by Ingestion:** Grade 3; LD_{50} 370 mg/kg (rat)
5.7 **Late Toxicity:** Causes bladder cancer in mice, discoloration of eyelids and eye changes in men
5.8 **Vapor (Gas) Irritant Characteristics:** Data not available
5.9 **Liquid or Solid Irritant Characteristics:** Data not available
5.10 **Odor Threshold:** Data not available

6. FIRE HAZARDS

6.1 **Flash Point:** (molten) 350°F O.C.
6.2 **Flammable Limits in Air:** Not pertinent
6.3 **Fire Extinguishing Agents:** Water, foam, dry chemical, carbon dioxide
6.4 **Fire Extinguishing Agents Not to be Used:**
6.5 **Special Hazards of Combustion Products:**
6.6 **Behavior in Fire:** Dust explosion is possible.
6.7 **Ignition Temperature:** Data not available
6.8 **Electrical Hazard:** Data not available
6.9 **Burning Rate:** Not pertinent

7. CHEMICAL REACTIVITY

7.1 **Reactivity with Water:** No reaction
7.2 **Reactivity with Common Materials:**
7.3 **Stability During Transport:** Stable
7.4 **Neutralizing Agents for Acids and Caustics:** Not pertinent
7.5 **Polymerization:** Not pertinent
7.6 **Inhibitor of Polymerization:** Not pertinent

8. WATER POLLUTION

8.1 **Aquatic Toxicity:** 0.287 ppm/48 hr/goldfish/TL_m/fresh water
8.2 **Waterfowl Toxicity:** Data not available
8.3 **Biological Oxygen Demand (BOD):** 53%, 5 days; 25% (theo.), 0.5 days, as catechol
8.4 **Food Chain Concentration Potential:** None

9. SELECTED MANUFACTURERS

1. Eastman Chemical Products, Inc.
   Kingsport, Tenn. 37662
2. Allied Chemical Corp.
   Specialty Chemicals Div.
   P. O. Box 1087R
   Morristown, N. J. 07960
3. Aldrich Chemical Co.
   940 W. Saint Paul Ave.
   Milwaukee, Wis. 53233

10. SHIPPING INFORMATION

10.1 **Grades or Purity:** Pure; Technical
10.2 **Storage Temperature:** Ambient
10.3 **Inert Atmosphere:** No requirement
10.4 **Venting:** Open

11. HAZARD ASSESSMENT CODE

(See Hazard Assessment Handbook, CG 446-3)

SS

12. HAZARD CLASSIFICATIONS

12.1 **Code of Federal Regulations:** Not listed
12.2 **NAS Hazard Rating for Bulk Water Transportation:** Not listed
12.3 **NFPA Hazard Classifications:** Not listed

13. PHYSICAL AND CHEMICAL PROPERTIES

13.1 **Physical State at 15°C and 1 atm:** Solid
13.2 **Molecular Weight:** 110.11
13.3 **Boiling Point at 1 atm:** 545°F = 285°C = 558°K
13.4 **Freezing Point:** 338°F = 170°C = 443°K
13.5 **Critical Temperature:** Not pertinent
13.6 **Critical Pressure:** Not pertinent
13.7 **Specific Gravity:** 1.33 at 20°C (solid)
13.8 **Liquid Surface Tension:** Not pertinent
13.9 **Liquid-Water Interfacial Tension:** Not pertinent
13.10 **Vapor (Gas) Specific Gravity:** Not pertinent
13.11 **Ratio of Specific Heats of Vapor (Gas):** Not pertinent
13.12 **Latent Heat of Vaporization:** Not pertinent
13.13 **Heat of Combustion:** −11,200 Btu/lb = −6,220 cal/g = -260×10^5 J/kg
13.14 **Heat of Decomposition:** Not pertinent
13.15 **Heat of Solution:** Not pertinent
13.16 **Heat of Polymerization:** Not pertinent

NOTES

HAI

2-HYDROXYETHYL ACRYLATE, INHIBITED

Common Synonyms	Liquid	Colorless	Sweet pleasant odor
beta-hydroxyethyl acrylate 2-Hydroxyethyl 2-propenoate	Mixes with water.		

Avoid contact with liquid. Keep people away.
Wear rubber overclothing (including gloves).
Stop discharge if possible.
Call fire department.
Isolate and remove discharged material.
Notify local health and pollution control agencies.

Fire

Combustible.
Containers may explode in fire.
Extinguish with water, dry chemicals, alcohol foam, or carbon dioxide.
Cool exposed containers with water.

Exposure

Call for medical aid.

LIQUID
Will burn skin and eyes.
Harmful if swallowed.
Remove contaminated clothing and shoes.
Flush affected areas with plenty of water.
IF IN EYES, hold eyelids open and flush with plenty of water.
IF SWALLOWED and victim is CONSCIOUS, have victim drink water or milk.
DO NOT INDUCE VOMITING.

Water Pollution

Effect of low concentrations on aquatic life is unknown.
May be dangerous if it enters water intakes.
Notify local health and wildlife officials.
Notify operators of nearby water intakes.

1. RESPONSE TO DISCHARGE

(See Response Methods Handbook, CG 446-4)

Issue warning—water contaminant
Restrict access
Disperse and flush

2. LABELS

No label required by Code of Federal Regulations

3. CHEMICAL DESIGNATIONS

3.1 **Synonyms:** beta-Hydroxyethyl acrylate; 2-Hydroxyethyl 2-propenoate
3.2 **Coast Guard Compatibility Classification:** Monomers (14)
3.3 **Chemical Formula:** $CH_2=CHCOOCH_2CH_2OH$
3.4 **IMCO/United Nations Numerical Designation:** Not listed

4. OBSERVABLE CHARACTERISTICS

4.1 **Physical State (as shipped):** Liquid
4.2 **Color:** Colorless
4.3 **Odor:** Sweet, pleasant

5. HEALTH HAZARDS

5.1 **Personal Protective Equipment:** Goggles or face shield; rubber gloves.
5.2 **Symptoms Following Exposure:** Inhalation causes irritation of nose and throat. Contact with liquid irritates eyes and skin.
5.3 **Treatment for Exposure:** INHALATION: remove victim from exposure; support respiration; call physician if needed. EYES: wash with large amounts of water for 15 min.; call physician. SKIN: flush with water.
5.4 **Toxicity by Inhalation (Threshold Limit Value):** Data not available
5.5 **Short-Term Inhalation Limits:** Data not available
5.6 **Toxicity by Ingestion:** Grade 2; oral LD_{50} = 1,070 mg/kg (rat)
5.7 **Late Toxicity:** Data not available
5.8 **Vapor (Gas) Irritant Characteristics:** Vapors cause severe irritation of eyes and throat and can cause eye and lung injury. They cannot be tolerated even at low concentrations.
5.9 **Liquid or Solid Irritant Characteristics:** Severe skin irritant. Causes second- and third-degree burns on short contact and is very injurious to the eyes.
5.10 **Odor Threshold:** Data not available

6. FIRE HAZARDS

6.1 **Flash Point:** 220°F O.C.
6.2 **Flammable Limits in Air:** Data not available
6.3 **Fire Extinguishing Agents:** Water, dry chemical, alcohol foam, carbon dioxide
6.4 **Fire Extinguishing Agents Not to be Used:** Not pertinent
6.5 **Special Hazards of Combustion Products:** Not pertinent
6.6 **Behavior in Fire:** Containers may explode
6.7 **Ignition Temperature:** Data not available
6.8 **Electrical Hazard:** Data not available
6.9 **Burning Rate:** 2.0 mm/min.

7. CHEMICAL REACTIVITY

7.1 **Reactivity with Water:** No reaction
7.2 **Reactivity with Common Materials:** No reaction
7.3 **Stability During Transport:** Stable
7.4 **Neutralizing Agents for Acids and Caustics:** Not pertinent
7.5 **Polymerization:** In the absence of inhibitor, polymerization will occur, especially when heated.
7.6 **Inhibitor of Polymerization:** Monomethyl ether of hydroquinone, 400 ppm

8. WATER POLLUTION

8.1 **Aquatic Toxicity:** Data not available
8.2 **Waterfowl Toxicity:** Data not available
8.3 **Biological Oxygen Demand (BOD):** Data not available
8.4 **Food Chain Concentration Potential:** None

9. SELECTED MANUFACTURERS

1. Dow Chemical Co.
   Midland, Mich. 48640
2. Pfaltz and Bauer, Inc.
   126-04 Northern Boulevard
   Flushing, N. Y. 11368

10. SHIPPING INFORMATION

10.1 **Grades or Purity:** Commercial
10.2 **Storage Temperature:** Ambient
10.3 **Inert Atmosphere:** No requirement
10.4 **Venting:** Open (flame arrester)

11. HAZARD ASSESSMENT CODE

(See Hazard Assessment Handbook, CG 446-3)

A-P-Q-Z

12. HAZARD CLASSIFICATIONS

12.1 **Code of Federal Regulations:** Not listed
12.2 **NAS Hazard Rating for Bulk Water Transportation:**

Category	Rating
Fire	1
Health	
Vapor Irritant	4
Liquid or Solid Irritant	4
Poisons	4
Water Pollution	
Human Toxicity	3
Aquatic Toxicity	3
Aesthetic Effect	1
Reactivity	
Other Chemicals	1
Water	0
Self-Reaction	3

12.3 **NFPA Hazard Classifications:** Not listed

13. PHYSICAL AND CHEMICAL PROPERTIES

13.1 **Physical State at 15°C and 1 atm:** Liquid
13.2 **Molecular Weight:** 116.1
13.3 **Boiling Point at 1 atm:** >346°F = >210°C = >583°K
13.4 **Freezing Point:** −76°F = −60°C = 213°K
13.5 **Critical Temperature:** Not pertinent
13.6 **Critical Pressure:** Not pertinent
13.7 **Specific Gravity:** 1.10 at 25°C (liquid)
13.8 **Liquid Surface Tension:** (est.) 28 dynes/cm = 0.028 N/m at 20°C
13.9 **Liquid-Water Interfacial Tension:** Not pertinent
13.10 **Vapor (Gas) Specific Gravity:** Not pertinent
13.11 **Ratio of Specific Heats of Vapor (Gas):** Not pertinent
13.12 **Latent Heat of Vaporization:** Not pertinent
13.13 **Heat of Combustion:** (est.) −10,800 Btu/lb = −6,000 cal/g = -250×10^5 J/kg
13.14 **Heat of Decomposition:** Not pertinent
13.15 **Heat of Solution:** Not pertinent
13.16 **Heat of Polymerization:** (est.) −218 Btu/lb = −121 cal/g = -5.06×10^5 J/kg

NOTES

HAS	HYDROXYLAMINE SULFATE

Common Synonyms	Solid	White	Odorless
Oxammonium sulfate	Sinks and mixes with water.		

AVOID CONTACT WITH SOLID AND DUST. KEEP PEOPLE AWAY.
Stop discharge if possible.
Isolate and remove discharged material.
Notify local health and pollution control agencies.

Fire

Not Flammable.
POISONOUS GASES MAY BE PRODUCED IN FIRE.

Exposure

CALL FOR MEDICAL AID.

DUST
Irritating to eyes, nose and throat.
If inhaled will cause difficult breathing or loss of consciousness.
If in eyes, hold eyelids open and flush with plenty of water.
If breathing has stopped, give artificial respiration.
If breathing is difficult, give oxygen.

SOLID
POISONOUS IF SWALLOWED.
Irritating to skin and eyes.
If swallowed will cause nausea or loss of consciousness.
Remove contaminated clothing and shoes.
Flush affected areas with plenty of water.
IF IN EYES, hold eyelids open and flush with plenty of water.
IF SWALLOWED and victim is CONSCIOUS, have victim drink water or milk and have victim induce vomiting.
IF SWALLOWED and victim is UNCONSCIOUS OR HAVING CONVULSIONS, do nothing except keep victim warm.

Water Pollution

Effect of low concentrations on aquatic life is unknown.
May be dangerous if it enters water intakes.
Notify local health and wildlife officials.
Notify operators of nearby water intakes.

1. RESPONSE TO DISCHARGE
(See Response Methods Handbook, CG 446-4)

Issue warning—water contaminant
Restrict access
Disperse and flush

2. LABELS

No hazard label required by Code of Federal Regulations

3. CHEMICAL DESIGNATIONS

3.1 **Synonyms:** Oxammonium sulfate
3.2 **Coast Guard Compatibility Classification:** Not listed
3.3 **Chemical Formula:** $(NH_2OH)_2 \cdot H_2SO_4$
3.4 **IMCO/United Nations Numerical Designation:** Not listed

4. OBSERVABLE CHARACTERISTICS

4.1 **Physical State (as shipped):** Crystals
4.2 **Color:** White
4.3 **Odor:** None

5. HEALTH HAZARDS

5.1 **Personal Protective Equipment:** Acid-resistant protective clothing, including coveralls, wrist-length gloves, cap, goggles, and dust mask
5.2 **Symptoms Following Exposure:** Inhalation of dust or ingestion may cause systemic poisoning characterized by cyanosis, methemoglobinemia, convulsions, and coma. Contact with eyes or skin causes irritation.
5.3 **Treatment for Exposure:** INHALATION: remove victim to fresh air; get medical attention if symptoms occur. INGESTION: give large amount of water; induce vomiting; get medical attention. EYES: flush with water for at least 15 min., and get medical attention. SKIN: flush immediately with plenty of water, then wash with soap and water.
5.4 **Toxicity by Inhalation (Threshold Limit Value):** Data not available
5.5 **Short-Term Inhalation Limits:** Data not available
5.6 **Toxicity by Ingestion:** Grade 3; LD_{50} 50–500 mg/kg
5.7 **Late Toxicity:** Data not available
5.8 **Vapor (Gas) Irritant Characteristics:** Data not available
5.9 **Liquid or Solid Irritant Characteristics:** Data not available
5.10 **Odor Threshold:** Data not available

6. FIRE HAZARDS

6.1 **Flash Point:** Not flammable
6.2 **Flammable Limits in Air:** Not flammable
6.3 **Fire Extinguishing Agents:** Not pertinent
6.4 **Fire Extinguishing Agents Not to be Used:** Not pertinent
6.5 **Special Hazards of Combustion Products:** Sulfuric acid fumes may form in fires.
6.6 **Behavior in Fire:** Not pertinent
6.7 **Ignition Temperature:** Not pertinent
6.8 **Electrical Hazard:** Not pertinent
6.9 **Burning Rate:** Not pertinent

7. CHEMICAL REACTIVITY

7.1 **Reactivity with Water:** No reaction
7.2 **Reactivity with Common Materials:** May be corrosive to metals in presence of moisture
7.3 **Stability During Transport:** Stable
7.4 **Neutralizing Agents for Acids and Caustics:** Flush with water.
7.5 **Polymerization:** Not pertinent
7.6 **Inhibitor of Polymerization:** Not pertinent

8. WATER POLLUTION

8.1 **Aquatic Toxicity:** Data not available
8.2 **Waterfowl Toxicity:** Data not available
8.3 **Biological Oxygen Demand (BOD):** Data not available
8.4 **Food Chain Concentration Potential:** None

9. SELECTED MANUFACTURERS

1. Commercial Solvents Corp.
   245 Park Avenue
   New York, N. Y. 10017
2. Virginia Chemicals, Inc.
   3340 W. Norfolk Road
   Portsmouth, Va. 23703
3. Dow-Badische Co.
   Williamsburg, Va. 23185

10. SHIPPING INFORMATION

10.1 **Grades or Purity:** Commercial, 97—99%
10.2 **Storage Temperature:** Ambient
10.3 **Inert Atmosphere:** No requirement
10.4 **Venting:** Open

11. HAZARD ASSESSMENT CODE
(See Hazard Assessment Handbook, CG 446-3)

SS

12. HAZARD CLASSIFICATIONS

12.1 **Code of Federal Regulations:** Not listed
12.2 **NAS Hazard Rating for Bulk Water Transportation:** Not listed
12.3 **NFPA Hazard Classifications:** Not listed

13. PHYSICAL AND CHEMICAL PROPERTIES

13.1 **Physical State at 15°C and 1 atm:** Solid
13.2 **Molecular Weight:** 164.14
13.3 **Boiling Point at 1 atm:** Not pertinent (decomposes)
13.4 **Freezing Point:** Not pertinent
13.5 **Critical Temperature:** Not pertinent
13.6 **Critical Pressure:** Not pertinent
13.7 **Specific Gravity:** >1 at 20°C (solid)
13.8 **Liquid Surface Tension:** Not pertinent
13.9 **Liquid-Water Interfacial Tension:** Not pertinent
13.10 **Vapor (Gas) Specific Gravity:** Not pertinent
13.11 **Ratio of Specific Heats of Vapor (Gas):** Not pertinent
13.12 **Latent Heat of Vaporization:** Not pertinent
13.13 **Heat of Combustion:** Not pertinent
13.14 **Heat of Decomposition:** Not pertinent
13.15 **Heat of Solution:** Not pertinent
13.16 **Heat of Polymerization:** Not pertinent

NOTES

HPA	HYDROXYPROPYL ACRYLATE

Common Synonyms
Propylene glycol mono-acrylate
1,2-Propanediol-1-acrylate

Liquid — Colorless — Faint unpleasant odor

May float or sink in water.

Avoid contact with liquid and vapor. Keep people away.
Shut off ignition sources. Call fire department.
Stop discharge if possible.
Isolate and remove discharged material.
Notify local health and pollution control agencies.

Fire

Combustible.
CONTAINERS MAY EXPLODE IN FIRE.
Extinguish with dry chemicals, foam, or carbon dioxide.
Water may be ineffective on fire.
Cool exposed containers with water.
COMBAT FIRES FROM SAFE DISTANCE OR PROTECTED LOCATION.

Exposure

CALL FOR MEDICAL AID.

VAPOR
Irritating to eyes, nose and throat.
If inhaled will cause coughing or difficult breathing.
If in eyes, hold eyelids open and flush with plenty of water.
If breathing has stopped, give artificial respiration.
If breathing is difficult, give oxygen.

LIQUID
Will burn skin and eyes.
If swallowed will cause nausea.
Remove contaminated clothing and shoes.
Flush affected areas with plenty of water.
IF IN EYES, hold eyelids open and flush with plenty of water.
IF SWALLOWED and victim is CONSCIOUS, have victim drink water or milk.
IF SWALLOWED and victim is UNCONSCIOUS OR HAVING CONVULSIONS, do nothing except keep victim warm.

Water Pollution

Effect of low concentrations on aquatic life is unknown.
May be dangerous if it enters water intakes.
Notify local health and wildlife officials.
Notify operators of nearby water intakes.

1. RESPONSE TO DISCHARGE
(See Response Methods Handbook, CG 446-4)

Issue warning—water contaminant
Restrict access
Disperse and flush

2. LABELS

No hazard label required by Code of Federal Regulations

3. CHEMICAL DESIGNATIONS

3.1 **Synonyms:** 1,2-Propanediol 1-acrylate; Propylene glycol mono-acrylate
3.2 **Coast Guard Compatibility Classification:** Not listed
3.3 **Chemical Formula:** $CH_3CHOHCH_2OCOCH{=}CH_2$
3.4 **IMCO/United Nations Numerical Designation:** Not listed

4. OBSERVABLE CHARACTERISTICS

4.1 **Physical State (as shipped):** Liquid
4.2 **Color:** Colorless
4.3 **Odor:** Slightly acrylic

5. HEALTH HAZARDS

5.1 **Personal Protective Equipment:** Rubber gloves, apron, and boots; worker's goggles or face shield
5.2 **Symptoms Following Exposure:** Inhalation irritates nose and throat and causes coughing; lung injury may occur. Ingestion causes irritation and burning of mouth and stomach. Vapor irritates eyes. Contact with liquid causes severe burns of eyes and burns of skin.
5.3 **Treatment for Exposure:** INHALATION: if ill effects occur, get patient to fresh air, keep him quiet and warm, and get medical attention; if breathing stops, start artificial respiration. INGESTION: force milk or water immediately; induce vomiting only at physician's recommendation. EYES: promptly flush with plenty of water and get medical attention. SKIN: promptly flush with plenty of water; get medical attention if burning occurs.
5.4 **Toxicity by Inhalation (Threshold Limit Value):** Data not available
5.5 **Short-Term Inhalation Limits:** Data not available
5.6 **Toxicity by Ingestion:** Grade 2; oral LD_{50} = 1,230 mg/kg (rat)
5.7 **Late Toxicity:** Data not available
5.8 **Vapor (Gas) Irritant Characteristics:** Vapors are moderately irritating such that personnel will not usually tolerate moderate or high concentrations.
5.9 **Liquid or Solid Irritant Characteristics:** Severe skin irritant. Causes second- and third-degree burns on short contact and is very injurious to the eyes.
5.10 **Odor Threshold:** Data not available

6. FIRE HAZARDS

6.1 **Flash Point:** 212°F O.C.
6.2 **Flammable Limits in Air:** 1.8% (LFL)
6.3 **Fire Extinguishing Agents:** Dry chemical, "alcohol" foam, carbon dioxide
6.4 **Fire Extinguishing Agents Not to be Used:** Water may be ineffective.
6.5 **Special Hazards of Combustion Products:**
6.6 **Behavior in Fire:**
6.7 **Ignition Temperature:** Data not available
6.8 **Electrical Hazard:** Data not available
6.9 **Burning Rate:** Data not available

7. CHEMICAL REACTIVITY

7.1 **Reactivity with Water:** No reaction
7.2 **Reactivity with Common Materials:**
7.3 **Stability During Transport:** Stable
7.4 **Neutralizing Agents for Acids and Caustics:** Not pertinent
7.5 **Polymerization:** May occur; avoid exposure to high temperatures, ultraviolet light, free-radical initiators.
7.6 **Inhibitor of Polymerization:** 200 ppm hydroquinone

8. WATER POLLUTION

8.1 **Aquatic Toxicity:** Data not available
8.2 **Waterfowl Toxicity:** Data not available
8.3 **Biological Oxygen Demand (BOD):** Data not available
8.4 **Food Chain Concentration Potential:** None

9. SELECTED MANUFACTURERS

1. Dow Chemical Co.
Design Products Dept.
Midland, Mich. 48640
2. Polysciences, Inc.
Paul Valley Industrial Park
Warrington, Pa. 18976

10. SHIPPING INFORMATION

10.1 **Grades or Purity:** Commercial, 97%
10.2 **Storage Temperature:** Ambient
10.3 **Inert Atmosphere:** No requirement
10.4 **Venting:** Open

11. HAZARD ASSESSMENT CODE
(See Hazard Assessment Handbook, CG 446-3)

A-P-Q-Z

12. HAZARD CLASSIFICATIONS

12.1 **Code of Federal Regulations:** Not listed
12.2 **NAS Hazard Rating for Bulk Water Transportation:**

Category	Rating
Fire	1
Health	
Vapor Irritant	3
Liquid or Solid Irritant	4
Poisons	3
Water Pollution	
Human Toxicity	2
Aquatic Toxicity	2
Aesthetic Effect	1
Reactivity	
Other Chemicals	2
Water	1
Self-Reaction	3

12.3 **NFPA Hazard Classifications:**

Category	Classification
Health Hazard (Blue)	1
Flammability (Red)	1
Reactivity (Yellow)	0

13. PHYSICAL AND CHEMICAL PROPERTIES

13.1 **Physical State at 15°C and 1 atm:** Liquid
13.2 **Molecular Weight:** 130
13.3 **Boiling Point at 1 atm:** Not pertinent (decomposes)
13.4 **Freezing Point:** Not pertinent
13.5 **Critical Temperature:** Not pertinent
13.6 **Critical Pressure:** Not pertinent
13.7 **Specific Gravity:** 1.06 at 25°C (liquid)
13.8 **Liquid Surface Tension:** Data not available
13.9 **Liquid-Water Interfacial Tension:** Data not available
13.10 **Vapor (Gas) Specific Gravity:** 4.5
13.11 **Ratio of Specific Heats of Vapor (Gas):** Not pertinent
13.12 **Latent Heat of Vaporization:** Data not available
13.13 **Heat of Combustion:** (est.) −12,300 Btu/lb = −6,850 cal/g = -287×10^5 J/kg
13.14 **Heat of Decomposition:** Not pertinent
13.15 **Heat of Solution:** Not pertinent
13.16 **Heat of Polymerization:** Data not available

NOTES

HPM	HYDROXYPROPYL METHACRYLATE

Common Synonyms			
Propylene glycol monomethacrylate 1,2-Propanediol 1-methacrylate	Liquid	White	Slight unpleasant odor
	May float or sink in water.		

Avoid contact with liquid and vapor. Keep people away.
Shut off ignition sources. Call fire department.
Stop discharge if possible.
Isolate and remove discharged material.
Notify local health and pollution control agencies.

Fire

Combustible.
CONTAINERS MAY EXPLODE IN FIRE.
Extinguish with dry chemicals, foam or carbon dioxide.
Water may be ineffective on fire.
Cool exposed containers with water.
COMBAT FIRES FROM SAFE DISTANCE OR PROTECTED LOCATION.

Exposure

CALL FOR MEDICAL AID.

VAPOR
Irritating to eyes, nose and throat.
If inhaled will cause coughing or difficult breathing.
If in eyes, hold eyelids open and flush with plenty of water.
If breathing has stopped, give artificial respiration.
If breathing is difficult, give oxygen.

LIQUID
Will burn eyes.
Irritating to eyes.
If swallowed will cause nausea and vomiting.
Remove contaminated clothing and shoes.
Flush affected areas with plenty of water.
IF IN EYES, hold eyelids open and flush with plenty of water.
IF SWALLOWED and victim is CONSCIOUS, have victim drink water or milk.
IF SWALLOWED and victim is UNCONSCIOUS OR HAVING CONVULSIONS, do nothing except keep victim warm.

Water Pollution

Effect of low concentrations on aquatic life is unknown.
May be dangerous if it enters water intakes.
Notify local health and wildlife officials.
Notify operators of nearby water intakes.

1. RESPONSE TO DISCHARGE

(See Response Methods Handbook, CG 446-4)

Issue warning—water contaminant
Restrict access
Disperse and flush

2. LABELS

No hazard label required by Code of Federal Regulations

3. CHEMICAL DESIGNATIONS

3.1 **Synonyms:** 1,2-propanediol 1-methacrylate; Propylene glycol monomethacrylate

3.2 **Coast Guard Compatibility Classification:** Not listed

3.3 **Chemical Formula:** $CH_3CHOHCH_2OCOC(CH_3)=CH_2$

3.4 **IMCO/United Nations Numerical Designation:** Not listed

4. OBSERVABLE CHARACTERISTICS

4.1 **Physical State (as shipped):** Liquid

4.2 **Color:** Colorless

4.3 **Odor:** Slight acrylic

5. HEALTH HAZARDS

5.1 **Personal Protective Equipment:** Goggles or face shield; rubber gloves

5.2 **Symptoms Following Exposure:** Inhalation causes coughing and irritation of nose and throat; lung injury may occur. Ingestion causes irritation and burning of mouth and stomach. Contact of vapor with eyes causes irritation. Liquid may cause severe eye burns and irritation of skin.

5.3 **Treatment for Exposure:** INHALATION: remove victim to fresh air; if required, start artificial respiration and call a doctor. INGESTION: force milk or water at once; get medical attention. EYES: flush with water for at least 15 min.; get medical attention if irritation persists. SKIN: flush with water; get medical attention for burns.

5.4 **Toxicity by Inhalation (Threshold Limit Value):** Data not available

5.5 **Short-Term Inhalation Limits:** Data not available

5.6 **Toxicity by Ingestion:** Grade 1; LD_{50} 5—15 g/kg (mouse)

5.7 **Late Toxicity:** Data not available

5.8 **Vapor (Gas) Irritant Characteristics:** Data not available

5.9 **Liquid or Solid Irritant Characteristics:** Data not available

5.10 **Odor Threshold:** Data not available

6. FIRE HAZARDS

6.1 **Flash Point:** 250°F O.C.

6.2 **Flammable Limits in Air:** Data not available

6.3 **Fire Extinguishing Agents:** Foam, dry chemical, carbon dioxide

6.4 **Fire Extinguishing Agents Not to be Used:** Water may be ineffective.

6.5 **Special Hazards of Combustion Products:**

6.6 **Behavior in Fire:** Compound may polymerize when hot and burst container.

6.7 **Ignition Temperature:** Data not available

6.8 **Electrical Hazard:** Data not available

6.9 **Burning Rate:** Data not available

7. CHEMICAL REACTIVITY

7.1 **Reactivity with Water:** No reaction

7.2 **Reactivity with Common Materials:**

7.3 **Stability During Transport:** Stable

7.4 **Neutralizing Agents for Acids and Caustics:** Not pertinent

7.5 **Polymerization:** May polymerize when hot or when exposed to ultraviolet light and free-radical catalysts.

7.6 **Inhibitor of Polymerization:** 200 ppm hydroquinone

8. WATER POLLUTION

8.1 **Aquatic Toxicity:** Data not available

8.2 **Waterfowl Toxicity:** Data not available

8.3 **Biological Oxygen Demand (BOD):** Data not available

8.4 **Food Chain Concentration Potential:** None

9. SELECTED MANUFACTURERS

1. Rohm and Haas
   Independence Mall
   W. Philadelphia, Pa. 19105
2. Polysciences, Inc.
   Paul Valley Industrial Park
   Warrington, Pa. 18976

10. SHIPPING INFORMATION

10.1 **Grades or Purity:** Commercial, 95+%

10.2 **Storage Temperature:** Ambient

10.3 **Inert Atmosphere:** No requirement

10.4 **Venting:** Open

11. HAZARD ASSESSMENT CODE

(See Hazard Assessment Handbook, CG 446-3)

A-P-Q-Z

12. HAZARD CLASSIFICATIONS

12.1 **Code of Federal Regulations:** Not listed

12.2 **NAS Hazard Rating for Bulk Water Transportation:** Not listed

12.3 **NFPA Hazard Classifications:** Not listed

13. PHYSICAL AND CHEMICAL PROPERTIES

13.1 **Physical State at 15°C and 1 atm:** Liquid

13.2 **Molecular Weight:** 144

13.3 **Boiling Point at 1 atm:** Not pertinent (decomposes)

13.4 **Freezing Point:** Not pertinent

13.5 **Critical Temperature:** Not pertinent

13.6 **Critical Pressure:** Not pertinent

13.7 **Specific Gravity:** 1.06 at 20°C (liquid)

13.8 **Liquid Surface Tension:** Data not available

13.9 **Liquid-Water Interfacial Tension:** Data not available

13.10 **Vapor (Gas) Specific Gravity:** Not pertinent

13.11 **Ratio of Specific Heats of Vapor (Gas):** Not pertinent

13.12 **Latent Heat of Vaporization:** Data not available

13.13 **Heat of Combustion:** Data not available

13.14 **Heat of Decomposition:** Not pertinent

13.15 **Heat of Solution:** Not pertinent

13.16 **Heat of Polymerization:** Data not available

NOTES

IAA — ISOAMYL ALCOHOL

Common Synonyms			
Fusel oil Fermentation amyl alcohol Isopentyl alcohol 3-Methyl-1-butanol	Liquid	Colorless	Mild, choking alcohol odor
	Floats and mixes with water. Irritating vapor is produced.		

Stop discharge if possible. Keep people away.
Call fire department.
Avoid contact with liquid.
Isolate and remove discharged material.
Notify local health and pollution control agencies.

Fire

Combustible.
Extinguish with water, dry chemical, alcohol foam, or carbon dioxide.
Cool exposed containers with water.

Exposure

CALL FOR MEDICAL AID.

VAPOR
Irritating to eyes, nose, and throat.
Harmful if inhaled.
Move to fresh air.
If breathing has stopped, give artificial respiration.
If breathing is difficult, give oxygen.

LIQUID
Irritating to eyes.
IF IN EYES, hold eyelids open and flush with plenty of water.

Water Pollution

Dangerous to aquatic life in high concentrations.
May be dangerous if it enters water intakes.
Notify local health and wildlife officials.
Notify operators of nearby water intakes.

1. RESPONSE TO DISCHARGE
(See Response Methods Handbook, CG 446-4)

Disperse and flush

2. LABELS

No hazard label required by Code of Federal Regulations

3. CHEMICAL DESIGNATIONS

3.1 **Synonyms:** Fermentation amyl alcohol; Fusel oil; Isobutylcarbinol; Isopentyl alcohol; 3-Methyl-1-butanol; Potato-spirit oil
3.2 **Coast Guard Compatibility Classification:** Alcohol
3.3 **Chemical Formula:** $(CH_3)_2CHCH_2CH_2OH$
3.4 **IMCO/United Nations Numerical Designation:** 3.3/1105, 3.2/1201

4. OBSERVABLE CHARACTERISTICS

4.1 **Physical State (as shipped):** Liquid
4.2 **Color:** Colorless
4.3 **Odor:** Mild odor; alcoholic, non-residual

5. HEALTH HAZARDS

5.1 **Personal Protective Equipment:** Face shield to avoid splash.
5.2 **Symptoms Following Exposure:** Very high vapor concentrations irritate eyes and upper respiratory tract. Continued contact with skin may cause irritation.
5.3 **Treatment for Exposure:** EYES: immediately flush with plenty of water for at least 15 min.; get medical attention. SKIN: flush with water; wash with soap and water.
5.4 **Toxicity by Inhalation (Threshold Limit Value):** 100 ppm
5.5 **Short-Term Inhalation Limits:** Data not available
5.6 **Toxicity by Ingestion:** Grade 2; LD_{50} 0.5 to 5 g/kg
5.7 **Late Toxicity:** None
5.8 **Vapor (Gas) Irritant Characteristics:** Vapor is moderately irritating such that personnel will not usually tolerate moderate or high vapor concentrations.
5.9 **Liquid or Solid Irritant Characteristics:** Liquid may irritate skin.
5.10 **Odor Threshold:** Data not available

6. FIRE HAZARDS

6.1 **Flash Point:** 114°F O.C.
6.2 **Flammable Limits in Air:** 1.2%–9.0% (212°F)
6.3 **Fire Extinguishing Agents:** Water spray, dry chemical, alcohol foam, or carbon dioxide.
6.4 **Fire Extinguishing Agents Not to be Used:** Not pertinent
6.5 **Special Hazards of Combustion Products:** Not pertinent
6.6 **Behavior in Fire:** Not pertinent
6.7 **Ignition Temperature:** 662°F
6.8 **Electrical Hazard:** Class I, Group C
6.9 **Burning Rate:** 3.6 mm/min.

7. CHEMICAL REACTIVITY

7.1 **Reactivity with Water:** No reaction
7.2 **Reactivity with Common Materials:** No reaction
7.3 **Stability During Transport:** Stable
7.4 **Neutralizing Agents for Acids and Caustics:** Not pertinent
7.5 **Polymerization:** Not pertinent
7.6 **Inhibitor of Polymerization:** Not pertinent

8. WATER POLLUTION

8.1 **Aquatic Toxicity:**
100 ppm/82 hr/goldfish/lethal/fresh water
400-600 ppm/24 hr/creek chub/critical range/river water
8.2 **Waterfowl Toxicity:** Data not available
8.3 **Biological Oxygen Demand (BOD):** 156%, 5 days
8.4 **Food Chain Concentration Potential:** None

9. SELECTED MANUFACTURERS

1. Publicker Industries, Inc.
1429 Walnut St.
Philadelphia, Pa. 19102
2. Union Carbide Corp.
Chemicals and Plastics Division
270 Park Ave.
New York, N.Y. 10017

10. SHIPPING INFORMATION

10.1 **Grades or Purity:** Pure; fusel oil
10.2 **Storage Temperature:** Ambient
10.3 **Inert Atmosphere:** No requirement
10.4 **Venting:** Open (flame arrester)

11. HAZARD ASSESSMENT CODE
(See Hazard Assessment Handbook, CG 446-3)

A-P-Q-T-U

12. HAZARD CLASSIFICATIONS

12.1 **Code of Federal Regulations:** Combustible liquid
12.2 **NAS Hazard Rating for Bulk Water Transportation:** Not listed
12.3 **NFPA Hazard Classifications:**

Category	Classification
Health Hazard (Blue)	1
Flammability (Red)	2
Reactivity (Yellow)	0

13. PHYSICAL AND CHEMICAL PROPERTIES

13.1 **Physical State at 15°C and 1 atm:** Liquid
13.2 **Molecular Weight:** 88.15
13.3 **Boiling Point at 1 atm:** 270°F = 132°C = 405°K
13.4 **Freezing Point:** Not pertinent
13.5 **Critical Temperature:** 585°F = 307°C = 580°K
13.6 **Critical Pressure:** Not pertinent
13.7 **Specific Gravity:** 0.81 at 20°C (liquid)
13.8 **Liquid Surface Tension:** 23.8 dynes/cm = 0.0238 N/m at 20°C
13.9 **Liquid-Water Interfacial Tension:** 5 dynes/cm = 0.005 N/m at 18°C
13.10 **Vapor (Gas) Specific Gravity:** Not pertinent
13.11 **Ratio of Specific Heats of Vapor (Gas):** (est.) 1.062
13.12 **Latent Heat of Vaporization:** 215.6 Btu/lb = 119.8 cal/g = 5.016×10^5 J/kg
13.13 **Heat of Combustion:** −16,200 Btu/lb = −9,000 cal/g = -376.8×10^5 J/kg
13.14 **Heat of Decomposition:** Not pertinent
13.15 **Heat of Solution:** −57.1 Btu/lb = −31.7 cal/g = -1.33×10^5 J/kg
13.16 **Heat of Polymerization:** Not pertinent

NOTES

IBT — ISOBUTANE

Common Synonyms
2-Methylpropane

Liquefied compressed gas — Colorless — Odorless

Floats and boils on water. Flammable visible vapor cloud is produced.

Stop discharge if possible. Keep people away.
Shut off ignition sources and call fire department.
Stay upwind and use water spray to "knock down" vapor.
Isolate and remove discharged material.
Notify local health and pollution control agencies.

Fire

FLAMMABLE.
Flashback along vapor trail may occur.
Vapor may explode if ignited in an enclosed area.
Stop flow of gas if possible.
Cool exposed containers and men effecting shutoff with water.
Let fire burn.

Exposure

CALL FOR MEDICAL AID.

VAPOR
Irritating to eyes.
If inhaled, will cause dizziness, difficult breathing or loss of consciousness.
Move to fresh air.
If breathing has stopped, give artificial respiration.
If breathing is difficult, give oxygen.
IF IN EYES, hold eyelids open and flush with plenty of water.

Water Pollution

Not harmful to aquatic life.

1. RESPONSE TO DISCHARGE

(See Response Methods Handbook, CG 446-4)

Issue warning—high flammability
Restrict access
Evacuate area

2. LABEL

Red

3. CHEMICAL DESIGNATIONS

3.1 **Synonyms:** 2-Methylpropane
3.2 **Coast Guard Compatibility Classification:** Paraffin
3.3 **Chemical Formula:** $CH_3CH(CH_3)_2$
3.4 **IMCO/United Nations Numerical Designation:** 2.0/1969

4. OBSERVABLE CHARACTERISTICS

4.1 **Physical State (as shipped):** Liquid under pressure
4.2 **Color:** Colorless
4.3 **Odor:** Like gasoline

5. HEALTH HAZARDS

5.1 **Personal Protective Equipment:** Self-contained breathing apparatus; safety goggles.
5.2 **Symptoms Following Exposure:** Central nervous system depression ranging from dizziness and incoordination to anesthesia and respiratory arrest, depending on concentration and extent of inhalation. Irregular heartbeat is rare but is a dangerous complication at anesthetic levels.
5.3 **Treatment for Exposure:** INHALATION: protect victim against self-injury if he is stuporous, confused, or anesthetized; apply artificial respiration if breathing has stopped; avoid administration of epinephrine or other sympathomimetic amines; prevent aspiration of vomitus by proper positioning of head; give symptomatic and supportive treatment. INGESTION OR ASPIRATION: no treatment required.
5.4 **Toxicity by Inhalation (Threshold Limit Value):** Data not available
5.5 **Short-Term Inhalation Limits:** Data not available
5.6 **Toxicity by Ingestion:** Not pertinent
5.7 **Late Toxicity:** None
5.8 **Vapor (Gas) Irritant Characteristics:** None
5.9 **Liquid or Solid Irritant Characteristics:** No appreciable hazard. Pratically harmless to skin because it is very volatile and evaporates quickly. Some frostbite possible.
5.10 **Odor Threshold:** Data not available

6. FIRE HAZARDS

6.1 **Flash Point:** −117°F C.C.
6.2 **Flammable Limits in Air:** 1.8%–8.4%
6.3 **Fire Extinguishing Agents:** Stop flow of gas
6.4 **Fire Extinguishing Agents Not to be Used:** Not pertinent
6.5 **Special Hazards of Combustion Products:** Not pertinent
6.6 **Behavior in Fire:** Not pertinent
6.7 **Ignition Temperature:** 890°F
6.8 **Electrical Hazard:** Not pertinent
6.9 **Burning Rate:** 9.3 mm/min.

7. CHEMICAL REACTIVITY

7.1 **Reactivity with Water:** No reaction
7.2 **Reactivity with Common Materials:** No reaction
7.3 **Stability During Transport:** Stable
7.4 **Neutralizing Agents for Acids and Caustics:** Not pertinent
7.5 **Polymerization:** Not pertinent
7.6 **Inhibitor of Polymerization:** Not pertinent

8. WATER POLLUTION

8.1 **Aquatic Toxicity:** None
8.2 **Waterfowl Toxicity:** None
8.3 **Biological Oxygen Demand (BOD):** None
8.4 **Food Chain Concentration Potential:** None

9. SELECTED MANUFACTURERS

1. Atlantic Richfield Co.
   ARCO Chemical Co. Division
   260 South Broad St.
   Philadelphia, Pa. 19101
2. Cities Service Co., Inc.
   Petrochemicals Division
   60 Wall St.
   New York, N. Y. 10005
3. Phillips Petroleum Co.
   Bartlesville, Okla. 74004

10. SHIPPING INFORMATION

10.1 **Grades or Purity:** Pure; technical
10.2 **Storage Temperature:** Ambient
10.3 **Inert Atmosphere:** No requirement
10.4 **Venting:** Safety relief

11. HAZARD ASSESSMENT CODE

(See Hazard Assessment Handbook, CG 446-3)

A-B-C-D-E-F-G

12. HAZARD CLASSIFICATIONS

12.1 **Code of Federal Regulations:** Flammable compressed gas
12.2 **NAS Hazard Rating for Bulk Water Transportation:** Not listed
12.3 **NFPA Hazard Classifications:** Not listed

13. PHYSICAL AND CHEMICAL PROPERTIES

13.1 **Physical State at 15°C and 1 atm:** Gas
13.2 **Molecular Weight:** 58.12
13.3 **Boiling Point at 1 atm:** 10.8°F = −11.8°C = 261.4°K
13.4 **Freezing Point:** −427.5°F = −255.3°C = 17.9°K
13.5 **Critical Temperature:** 275°F = 135°C = 408°K
13.6 **Critical Pressure:** 529 psia = 36.0 atm = 3.65 MN/m²
13.7 **Specific Gravity:** 0.557 at 20°C (liquid)
13.8 **Liquid Surface Tension:** 14 dynes/cm = 0.014 N/m at −10°C
13.9 **Liquid-Water Interfacial Tension:** (est.) 50 dynes/cm = 0.05 N/m at −10°C
13.10 **Vapor (Gas) Specific Gravity:** 2.0
13.11 **Ratio of Specific Heats of Vapor (Gas):** 1.095
13.12 **Latent Heat of Vaporization:** 158 Btu/lb = 87.5 cal/g = 3.66 × 10⁵ J/kg
13.13 **Heat of Combustion:** −19,458 Btu/lb = −10,810 cal/g = −452.59 × 10⁵ J/kg
13.14 **Heat of Decomposition:** Not pertinent
13.15 **Heat of Solution:** Not pertinent
13.16 **Heat of Polymerization:** Not pertinent

NOTES

REVISED 1978

IBA	ISOBUTYL ACETATE

ISOBUTYL ACETATE

Common Synonyms
Acetic acid, isobutyl ester

Watery liquid — Colorless — Pleasant fruity odor

Floats on water. Flammable, irritating vapor is produced.

Stop discharge if possible. Keep people away.
Shut off ignition sources and call fire department.
Avoid contact with liquid and vapor.
Stay upwind and use water spray to "knock down" vapor.
Isolate and remove discharged material.
Notify local health and pollution control agencies.

Fire

FLAMMABLE.
Flashback along vapor trail may occur.
May explode if ignited in an enclosed area.
Extinguish with dry chemical, alcohol foam, or carbon dioxide.
Water may be ineffective on fire.
Cool exposed containers with water.

Exposure

CALL FOR MEDICAL AID.

VAPOR
Irritating to eyes, nose and throat.
If inhaled, will cause nausea, vomiting, dizziness, or loss of consciousness.
Move to fresh air.
If breathing has stopped, give artificial respiration.
If breathing is difficult, give oxygen.

LIQUID
Irritating to skin and eyes.
Remove contaminated clothing and shoes.
Flush affected areas with plenty of water.
IF IN EYES, hold eyelids open and flush with plenty of water.

Water Pollution

Effect of low concentrations on aquatic life is unknown.
Fouling to shoreline.
May be dangerous if it enters water intakes.
Notify local health and wildlife officials.
Notify operators of nearby water intakes.

1. RESPONSE TO DISCHARGE

(See Response Methods Handbook, CG 446-4)

Issue warning—high flammability
Evacuate area
Disperse and flush

2. LABEL

Red

3. CHEMICAL DESIGNATIONS

3.1 **Synonyms:** Acetic acid, isobutyl ester
2-Methyl-1-propyl acetate
beta-Methylpropyl ethanoate

3.2 **Coast Guard Compatibility Classification:** Ester

3.3 **Chemical Formula:** $CH_3COOCH_2CH(CH_3)_2$

3.4 **IMCO/United Nations Numerical Designation:** 3.2/1213

4. OBSERVABLE CHARACTERISTICS

4.1 **Physical State (as shipped):** Liquid

4.2 **Color:** Colorless

4.3 **Odor:** Agreeable fruity odor in low concentrations, disagreeable in higher concentrations; mild, characteristic ester; nonresidual

5. HEALTH HAZARDS

5.1 **Personal Protective Equipment:** Air pack or organic canister mask; chemical goggles.

5.2 **Symptoms Following Exposure:** Vapors may irritate upper respiratory tract and cause nausea, vomiting, dizziness and loss of consciousness. Liquid irritates eyes and may irritate skin.

5.3 **Treatment for Exposure:** INHALATION: remove from exposure; if breathing is irregular or has stopped, start resuscitation and give oxygen; call a doctor. EYES: flush with water for at least 15 minutes.

5.4 **Toxicity by Inhalation (Threshold Limit Value):** 150 ppm

5.5 **Short-Term Inhalation Limits:** Data not available

5.6 **Toxicity by Ingestion:** Data not available

5.7 **Late Toxicity:** Data not available

5.8 **Vapor (Gas) Irritant Characteristics:** Vapors cause a slight smarting of the eyes or respiratory system if present in high concentrations. The effect is temporary.

5.9 **Liquid or Solid Irritant Characteristics:** Minimum hazard. If spilled on clothing and allowed to remain, may cause smarting and reddening of the skin.

5.10 **Odor Threshold:** Data not available

6. FIRE HAZARDS

6.1 **Flash Point:** 62°F C.C.; 85°F O.C.

6.2 **Flammable Limits in Air:** 2.4%—10.5%

6.3 **Fire Extinguishing Agents:** Foam, carbon dioxide and dry chemical

6.4 **Fire Extinguishing Agents Not to be Used:** Water may be ineffective

6.5 **Special Hazards of Combustion Products:** Not pertinent

6.6 **Behavior in Fire:** Not pertinent

6.7 **Ignition Temperature:** 793°F

6.8 **Electrical Hazard:** Class I, group D

6.9 **Burning Rate:** Data not available

7. CHEMICAL REACTIVITY

7.1 **Reactivity with Water:** No reaction

7.2 **Reactivity with Common Materials:** Softens and dissolves many plastics

7.3 **Stability During Transport:** Stable

7.4 **Neutralizing Agents for Acids and Caustics:** Not pertinent

7.5 **Polymerization:** Not pertinent

7.6 **Inhibitor of Polymerization:** Not pertinent

8. WATER POLLUTION

8.1 **Aquatic Toxicity:** 1200 ppm/24 hr/brine shrimp/TLm

8.2 **Waterfowl Toxicity:** Data not available

8.3 **Biological Oxygen Demand (BOD):** 47% of theoretical in 5 days, freshwater, acclimated seed

8.4 **Food Chain Concentration Potential:** None

9. SELECTED MANUFACTURERS

1. Eastman Kodak Co.
Tennessee Eastman Division
Kingsport, Tenn. 37662

2. Fritzsche Dodge and Olcott, Inc.
76 Ninth Ave.
New York, N. Y. 10011

3. Union Carbide Corp.
Chemicals and Plastics Division
270 Park Ave.
New York, N. Y. 10017

10. SHIPPING INFORMATION

10.1 **Grades or Purity:** 95-99+%

10.2 **Storage Temperature:** Ambient

10.3 **Inert Atmosphere:** No requirement

10.4 **Venting:** Open (flame arrester)

11. HAZARD ASSESSMENT CODE

(See Hazard Assessment Handbook, CG 446-3)

A-T-U

12. HAZARD CLASSIFICATIONS

12.1 **Code of Federal Regulations:** Flammable liquid

12.2 **NAS Hazard Rating for Bulk Water Transportation:**

Category	Rating
Fire	3
Health	
Vapor Irritant	1
Liquid or Solid Irritant	1
Poisons	2
Water Pollution	
Human Toxicity	1
Aquatic Toxicity	1
Aesthetic Effect	2
Reactivity	
Other Chemicals	1
Water	0
Self-Reaction	0

12.3 **NFPA Hazard Classifications:**

Category	Classification
Health Hazard (Blue)	1
Flammability (Red)	3
Reactivity (Yellow)	0

13. PHYSICAL AND CHEMICAL PROPERTIES

13.1 **Physical State at 15°C and 1 atm:** Liquid

13.2 **Molecular Weight:** 116.16

13.3 **Boiling Point at 1 atm:** 243.1°F = 117.3°C = 390.5°K

13.4 **Freezing Point:** −142.8°F = −97.1°C = 176.1°K

13.5 **Critical Temperature:** 565°F = 296°C = 569°K

13.6 **Critical Pressure:** 470 psia = 32 atm = 3.2 MN/m^2

13.7 **Specific Gravity:** 0.871 at 20°C (liquid)

13.8 **Liquid Surface Tension:** 23.7 dynes/cm = 0.0237 N/m at 20°C

13.9 **Liquid-Water Interfacial Tension:** (est.) 40 dynes/cm = 0.04 N/m at 20°C

13.10 **Vapor (Gas) Specific Gravity:** Not pertinent

13.11 **Ratio of Specific Heats of Vapor (Gas):** Not pertinent

13.12 **Latent Heat of Vaporization:** 133 Btu/lb = 73.7 cal/g = 3.09 × 10^5 J/kg

13.13 **Heat of Combustion:** (est.) −13,000 Btu/lb = −7220 cal/g = −302 × 10^5 J/kg

13.14 **Heat of Decomposition:** Not pertinent

13.15 **Heat of Solution:** Not pertinent

13.16 **Heat of Polymerization:** Not pertinent

NOTES

REVISED 1978

IAL — ISOBUTYL ALCOHOL

Common Synonyms			
Isobutanol Isopropylcarbinol 2-Methyl-1-propanol	Oily liquid	Colorless	Mild alcohol, choking odor
	Floats and mixes slowly with water. Irritating vapor is produced.		

Stop discharge if possible.
Call fire department.
Avoid contact with liquid and vapor.
Stay upwind and use water spray to "knock down" vapor.
Isolate and remove discharged material.
Notify local health and pollution control agencies.

Fire

FLAMMABLE
Flashback along vapor trail may occur.
Vapor may explode if ignited in an enclosed area.
Extinguish with dry chemical, alcohol foam, or carbon dioxide.
Water may be ineffective on fire.
Cool exposed containers with water.

Exposure

CALL FOR MEDICAL AID.

VAPOR
Irritating to eyes, nose and throat.
If inhaled, will cause nausea, dizziness, or headache.
Move to fresh air.
If breathing has stopped, give artificial respiration.
If breathing is difficult, give oxygen.

LIQUID
Irritating to eyes.
Harmful if swallowed.
IF IN EYES, hold eyelids open and flush with plenty of water.
IF SWALLOWED and victim is CONSCIOUS, have victim drink water or milk.

Water Pollution

Effect of low concentrations on aquatic life is unknown.
May be dangerous if it enters water intakes.
Notify local health and wildlife officials.
Notify operators of nearby water intakes.

1. RESPONSE TO DISCHARGE

(See Response Methods Handbook, CG 446-4)

Disperse and flush

2. LABEL

Red

3. CHEMICAL DESIGNATIONS

3.1 **Synonyms:** Isobutanol
Isopropylcarbinol
2-methyl-1-propanol
Fermentation butyl alcohol

3.2 **Coast Guard Compatibility Classification:** Alcohol

3.3 **Chemical Formula:** $(CH_3)_2CHCH_2OH$

3.4 **IMCO/United Nations Numerical Designation:** 3.3/1212

4. OBSERVABLE CHARACTERISTICS

4.1 **Physical State (as shipped):** Liquid

4.2 **Color:** Colorless

4.3 **Odor:** Slightly suffocating; nonresidual alcoholic

5. HEALTH HAZARDS

5.1 **Personal Protective Equipment:** Air pack or organic canister; chemical goggles.

5.2 **Symptoms Following Exposure:** Contact with eyes is extremely irritating and may cause burns. Breathing vapors will be irritating to the nose and throat. In high concentrations, may cause nausea, dizziness, headache, and stupor.

5.3 **Treatment for Exposure:** INHALATION: if victim is overcome by vapors, remove him from exposure immediately; call a physician; if breathing is irregular or has stopped, start resuscitation; administer oxygen. EYES: flush with water for at least 15 min.

5.4 **Toxicity by Inhalation (Threshold Limit Value):** 100 ppm

5.5 **Short-Term Inhalation Limits:** 200 ppm for 60 min.

5.6 **Toxicity by Ingestion:** Grade 2; LD_{50} 0.5 to 5 g/kg (rat)

5.7 **Late Toxicity:** Data not available

5.8 **Vapor (Gas) Irritant Characteristics:** Vapors cause a slight smarting of the eyes or respiratory system if present in high concentrations. The effect is temporary.

5.9 **Liquid or Solid Irritant Characteristics:** No appreciable hazard. Practically harmless to the skin.

5.10 **Odor Threshold:** Data not available.

6. FIRE HAZARDS

6.1 **Flash Point:** 82°F C.C.; 90°F O.C.

6.2 **Flammable Limits in Air:** 1.6%–10.9%

6.3 **Fire Extinguishing Agents:** Alcohol foam, dry chemical, carbon dioxide

6.4 **Fire Extinguishing Agents Not to be Used:** Water may be ineffective

6.5 **Special Hazards of Combustion Products:** Not pertinent

6.6 **Behavior in Fire:** Not pertinent

6.7 **Ignition Temperature:** 800°F

6.8 **Electrical Hazard:** Class I, group D

6.9 **Burning Rate:** 3.5 mm/min.

7. CHEMICAL REACTIVITY

7.1 **Reactivity with Water:** No reaction

7.2 **Reactivity with Common Materials:** No reaction

7.3 **Stability During Transport:** Stable

7.4 **Neutralizing Agents for Acids and Caustics:** Not pertinent

7.5 **Polymerization:** Not pertinent

7.6 **Inhibitor of Polymerization:** Not pertinent

8. WATER POLLUTION

8.1 **Aquatic Toxicity:** 4680 ppm/1 hr/fish/lethal/fresh water

8.2 **Waterfowl Toxicity:** Data not available

8.3 **Biological Oxygen Demand (BOD):** 162%, 5 days

8.4 **Food Chain Concentration Potential:** None

9. SELECTED MANUFACTURERS

1. Eastman Kodak Co.
Tennessee Eastman Division
Kingsport, Tenn. 37662
2. Oxochem Enterprises
Penuelas, Puerto Rico 00724
3. Union Carbide Corp.
Chemicals and Plastics Division
270 Park Ave.
New York, N. Y. 10017

10. SHIPPING INFORMATION

10.1 **Grades or Purity:** 99+%

10.2 **Storage Temperature:** Ambient

10.3 **Inert Atmosphere:** No requirement

10.4 **Venting:** Open (flame arrester)

11. HAZARD ASSESSMENT CODE

(See Hazard Assessment Handbook, CG 446-3)

A-P-Q

12. HAZARD CLASSIFICATIONS

12.1 **Code of Federal Regulations:** Flammable Liquid

12.2 **NAS Hazard Rating for Bulk Water Transportation:**

Category	Rating
Fire	3
Health	
Vapor Irritant	1
Liquid or Solid Irritant	0
Poisons	1
Water Pollution	
Human Toxicity	2
Aquatic Toxicity	1
Aesthetic Effect	2
Reactivity	
Other Chemicals	2
Water	0
Self-Reaction	0

12.3 **NFPA Hazard Classifications:**

Category	Classification
Health Hazard (Blue)	1
Flammability (Red)	3
Reactivity (Yellow)	0

13. PHYSICAL AND CHEMICAL PROPERTIES

13.1 **Physical State at 15°C and 1 atm:** Liquid

13.2 **Molecular Weight:** 74.12

13.3 **Boiling Point at 1 atm:** 226.2°F = 107.9°C = 381.1°K

13.4 **Freezing Point:** −162°F = −108°C = 165°K

13.5 **Critical Temperature:** 526.3°F = 274.6°C = 547.8°K

13.6 **Critical Pressure:** 623 psia = 42.4 atm = 4.30 MN/m^2

13.7 **Specific Gravity:** 0.802 at 20°C (liquid)

13.8 **Liquid Surface Tension:** Not pertinent

13.9 **Liquid-Water Interfacial Tension:** Not pertinent

13.10 **Vapor (Gas) Specific Gravity:** Not pertinent

13.11 **Ratio of Specific Heats of Vapor (Gas):** Not pertinent

13.12 **Latent Heat of Vaporization:** 248 Btu/lb = 138 cal/g = 5.78 × 10^5 J/kg

13.13 **Heat of Combustion:** −14,220 Btu/lb = −7,900 cal/g = −330.8 × 10^5 J/kg

13.14 **Heat of Decomposition:** Not pertinent

13.15 **Heat of Solution:** (est.) −9 Btu/lb = −5 cal/g = −0.2 × 10^5 J/kg

13.16 **Heat of Polymerization:** Not pertinent

NOTES

REVISED 1978

ISOBUTYLAMINE

Common Synonyms			
Monoisobutylamine 1-Amino-2-methylpropane	Liquid	Colorless	Strong ammonia odor
	Floats and mixes with water.		

Stop discharge if possible. Keep people away.
Shut off ignition sources. Call fire department.
Avoid contact wih liquid and vapor.
Isolate and remove discharged material.
Notify local health and pollution control agencies.

Fire

FLAMMABLE
POISONOUS GASES MAY BE PRODUCED IN FIRE.
Flashback along vapor trail may occur.
Vapor may explode if ignited in an enclosed area.
Wear goggles and self-contained breathing apparatus.
Extinguish with dry chemicals, alcohol foam, or carbon dioxide.
Water may be ineffective on fire.
Cool exposed containers with water.

Exposure

CALL FOR MEDICAL AID.

VAPOR
Irritating to eyes, nose and throat.
If inhaled will cause coughing, difficult breathing or loss of consciousness.
If in eyes, hold eyelids open and flush with plenty of water.
If breathing has stopped, give artificial respiration.
If breathing is difficult, give oxygen.

LIQUID
Will burn skin and eyes.
If swallowed will cause nausea or loss of consciousness.
Remove contaminated clothing and shoes.
Flush affected areas with plenty of water.
IF IN EYES, hold eyelids open and flush with plenty of water.
IF SWALLOWED and victim is CONSCIOUS, have victim drink water or milk.
IF SWALLOWED and victim is CONSCIOUS, have victim drink water or milk and have victim induce vomiting.

Water Pollution

Effect of low concentrations on aquatic life is unknown.
May be dangerous if it enters water intakes.
Notify local health and wildlife officials.
Notify operators of nearby water intakes.

1. RESPONSE TO DISCHARGE

(See Response Methods Handbook, CG 446-4)

Issue warning—high flammability, air contaminant, water contaminant
Restrict access
Evacuate area
Disperse and flush

2. LABEL

Red

3. CHEMICAL DESIGNATIONS

3.1 **Synonyms:** 1-Amino-2-methylpropane; iso-Butylamine; Monoisobutylamine
3.2 **Coast Guard Compatibility Classification:** Not listed
3.3 **Chemical Formula:** $(CH_3)_2CHCH_2NH_2$
3.4 **IMCO/United Nations Numerical Designation:** 3.2/1214

4. OBSERVABLE CHARACTERISTICS

4.1 **Physical State (as shipped):** Liquid
4.2 **Color:** Colorless
4.3 **Odor:** Strong ammoniacal

5. HEALTH HAZARDS

5.1 **Personal Protective Equipment:** Self-contained breathing apparatus; butyl rubber gloves; chemical face shield; butyl rubber apron
5.2 **Symptoms Following Exposure:** Inhalation causes severe coughing and chest pain due to irritation of air passages; can cause lung edema. Compound is sympathomimetic and is also a cardiac depressant and convulsant; ingestion causes nausea and profuse salivation. Contact with eyes causes severe irritation and edema of the cornea. Contact with skin causes severe irritation.
5.3 **Treatment for Exposure:** INHALATION: remove victim to fresh air; if he is not breathing, give artificial respiration; if breathing is difficult, give oxygen; call a physician. INGESTION: give large amount of water followed by dilute vinegar or lemon juice; keep patient warm. EYES: flush with water for 15 min. SKIN: flush with water.
5.4 **Toxicity by Inhalation (Threshold Limit Value):** 5 ppm
5.5 **Short-Term Inhalation Limits:** Data not available
5.6 **Toxicity by Ingestion:** Grade 3; oral LD_{50} = 120 mg/kg (rabbit), 250 mg/kg (rat)
5.7 **Late Toxicity:** Data not available
5.8 **Vapor (Gas) Irritant Characteristics:** Vapors are moderately irritating such that personnel will not usually tolerate moderate or high concentrations.
5.9 **Liquid or Solid Irritant Characteristics:** Severe skin irritant. Causes second- and third-degree burns on short contact and is very injurious to the eyes.
5.10 **Odor Threshold:** Data not available

6. FIRE HAZARDS

6.1 **Flash Point:** 15°F C.C.
6.2 **Flammable Limits in Air:** 3.4%–9%
6.3 **Fire Extinguishing Agents:** Dry chemical, "alcohol" foam, carbon dioxide
6.4 **Fire Extinguishing Agents Not to be Used:** Water may be ineffective.
6.5 **Special Hazards of Combustion Products:** Toxic oxides of nitrogen may be formed in fire.
6.6 **Behavior in Fire:** Vapor is heavier than air and may travel to a source of ignition and flash back. Containers may explode.
6.7 **Ignition Temperature:** 712°F
6.8 **Electrical Hazard:** Data not available
6.9 **Burning Rate:** 6.03 mm/min.

7. CHEMICAL REACTIVITY

7.1 **Reactivity with Water:** No reaction
7.2 **Reactivity with Common Materials:**
7.3 **Stability During Transport:** Stable
7.4 **Neutralizing Agents for Acids and Caustics:** Not pertinent
7.5 **Polymerization:** Not pertinent
7.6 **Inhibitor of Polymerization:** Not pertinent

8. WATER POLLUTION

8.1 **Aquatic Toxicity:** Data not available
8.2 **Waterfowl Toxicity:** Data not available
8.3 **Biological Oxygen Demand (BOD):** Data not available
8.4 **Food Chain Concentration Potential:** None

9. SELECTED MANUFACTURERS

1. Air Products and Chemicals, Inc.
U. S. Highway 90
Pace, Fla. 32502
2. Pennwalt Corp.
Three Parkway
Philadelphia, Pa. 19102
3. Aldrich Chemical Co.
940 W. Saint Paul Ave.
Milwaukee, Wis. 53233

10. SHIPPING INFORMATION

10.1 **Grades or Purity:** Technical; 99+%
10.2 **Storage Temperature:** Ambient
10.3 **Inert Atmosphere:** No requirement
10.4 **Venting:** Open (flame arrester)

11. HAZARD ASSESSMENT CODE

(See Hazard Assessment Handbook, CG 446-3)

A-P-Q-R-S

12. HAZARD CLASSIFICATIONS

12.1 **Code of Federal Regulations:** Flammable liquid
12.2 **NAS Hazard Rating for Bulk Water Transportation:**

Category	Rating
Fire	3
Health	
Vapor Irritant	3
Liquid or Solid Irritant	4
Poisons	4
Water Pollution	
Human Toxicity	2
Aquatic Toxicity	2
Aesthetic Effect	1
Reactivity	
Other Chemicals	3
Water	0
Self-Reaction	0

12.3 **NFPA Hazard Classifications:**

Category	Classification
Health Hazard (Blue)	2
Flammability (Red)	3
Reactivity (Yellow)	0

13. PHYSICAL AND CHEMICAL PROPERTIES

13.1 **Physical State at 15°C and 1 atm:** Liquid
13.2 **Molecular Weight:** 73.1
13.3 **Boiling Point at 1 atm:** 153.3°F = 67.4°C = 340.6°K
13.4 **Freezing Point:** −121.9°F = −85.5°C = 187.7°K
13.5 **Critical Temperature:** 469.4°F = 243.0°C = 516.2°K
13.6 **Critical Pressure:** 620 psia = 42 atm = 4.3 MN/m^2
13.7 **Specific Gravity:** 0.739 at 20°C (liquid)
13.8 **Liquid Surface Tension:** 23.70 dynes/cm = 0.0237 N/m at 20°C
13.9 **Liquid-Water Interfacial Tension:** Not pertinent
13.10 **Vapor (Gas) Specific Gravity:** 2.5
13.11 **Ratio of Specific Heats of Vapor (Gas):** 1.073 at 20°C
13.12 **Latent Heat of Vaporization:** 182 Btu/lb = 101 cal/g = 4.23×10^5 J/kg
13.13 **Heat of Combustion:** −17,550 Btu/lb = −9,760 cal/g = -408×10^5 J/kg
13.14 **Heat of Decomposition:** Not pertinent
13.15 **Heat of Solution:** −148 Btu/lb = −82 cal/g = -3.4×10^5 J/kg
13.16 **Heat of Polymerization:** Not pertinent

NOTES

IBL	# ISOBUTYLENE

Common Synonyms	
Isobutene 2-Methylpropene	Liquefied compressed gas Colorless Sweet gasoline-like odor Floats and boils on water. Flammable visible vapor cloud is produced.

Stop discharge if possible. Keep people away.
Shut off ignition sources and call fire department.
Stay upwind and use water spray to "knock down" vapor.
Avoid contact with liquid.
Isolate and remove discharged material.
Notify local health and pollution control agencies.

Fire	FLAMMABLE. Flashback along vapor trail may occur. Vapor may explode if ignited in an enclosed area. Stop flow of gas if possible. Cool exposed containers and protect men effecting shutoff with water. Let fire burn. Extinguish small fires with water, dry chemical, or carbon dioxide.
Exposure	CALL FOR MEDICAL AID. VAPOR Irritating to eyes, nose, and throat. If inhaled, will cause dizziness, or loss of consciousness. Move to fresh air. If breathing has stopped, give artificial respiration. If breathing is difficult, give oxygen. LIQUID Will cause frostbite. Flush affected areas with plenty of water. DO NOT RUB AFFECTED AREAS.
Water Pollution	Not harmful to aquatic life.

1. RESPONSE TO DISCHARGE

(See Response Methods Handbook, CG 446-4)

Issue warning—high flammability
Restrict access
Evacuate area

2. LABEL

Red

3. CHEMICAL DESIGNATIONS

3.1 **Synonyms:** Isobutene
2-Methylpropene
3.2 **Coast Guard Compatibility Classification:** Olefin
3.3 **Chemical Formula:** $(CH_3)_2C{=}CH_2$
3.4 **IMCO/United Nations Numerical Designation:** 2./1055

4. OBSERVABLE CHARACTERISTICS

4.1 **Physical State (as shipped):** Liquid under pressure
4.2 **Color:** Colorless
4.3 **Odor:** Mild sweetish

5. HEALTH HAZARDS

5.1 **Personal Protective Equipment:** Chemical gloves and eye protection; organic vapor canister or self-contained breathing apparatus.
5.2 **Symptoms Following Exposure:** Inhalation of moderate concentrations causes dizziness, drowsiness, and unconsciousness. Contact with eyes or skin may cause irritation; the liquid may cause frostbite.
5.3 **Treatment for Exposure:** INHALATION: remove victim to fresh air and apply resuscitation; call a physician promptly if victim is unconscious. EYES: if irritated, wash with water. SKIN: if irritated, wash with soap and water.
5.4 **Toxicity by Inhalation (Threshold Limit Value):** 1000 ppm (8 hr)
5.5 **Short-Term Inhalation Limits:** Data not available
5.6 **Toxicity by Ingestion:** Not pertinent
5.7 **Late Toxicity:** None
5.8 **Vapor (Gas) Irritant Characteristics:** Vapors are non-irritating to eyes and throat.
5.9 **Liquid or Solid Irritant Characteristics:** No appreciable hazard. Practically harmless to skin because it is very volatile and evaporates quickly. May cause frostbite.
5.10 **Odor Threshold:** Data not available

6. FIRE HAZARDS

6.1 **Flash Point:** −105°F C.C.
6.2 **Flammable Limits in Air:** 1.8%–9.6%
6.3 **Fire Extinguishing Agents:** Let fire burn, stop flow of gas. Water fog, dry chemical, or carbon dioxide may be used for small fires.
6.4 **Fire Extinguishing Agents Not to be Used:** Not pertinent
6.5 **Special Hazards of Combustion Products:** Not pertinent
6.6 **Behavior in Fire:** Containers may explode in fire. Vapor is heavier than air and may travel a long distance to a source of ignition and flash back.
6.7 **Ignition Temperature:** 869°F
6.8 **Electrical Hazard:** Not pertinent
6.9 **Burning Rate:** Data not available

7. CHEMICAL REACTIVITY

7.1 **Reactivity with Water:** No reaction
7.2 **Reactivity with Common Materials:** No reaction
7.3 **Stability During Transport:** Stable
7.4 **Neutralizing Agents for Acids and Caustics:** Not pertinent
7.5 **Polymerization:** Not pertinent
7.6 **Inhibitor of Polymerization:** Not pertinent

8. WATER POLLUTION

8.1 **Aquatic Toxicity:** None
8.2 **Waterfowl Toxicity:** None
8.3 **Biological Oxygen Demand (BOD):** None
8.4 **Food Chain Concentration Potential:** None

9. SELECTED MANUFACTURERS

1. Petro-Tex Chemical Corp.
8600 Park Place
Houston, Texas 77017
2. Exxon Chemical Co.
Houston, Texas 77001

10. SHIPPING INFORMATION

10.1 **Grades or Purity:** Commercial; polymerization
10.2 **Storage Temperature:** Ambient
10.3 **Inert Atmosphere:** No requirement
10.4 **Venting:** Safety relief

11. HAZARD ASSESSMENT CODE

(See Hazard Assessment Handbook, CG 446-3)

A-B-C-D-E-F-G

12. HAZARD CLASSIFICATIONS

12.1 **Code of Federal Regulations:** Flammable compressed gas
12.2 **NAS Hazard Rating for Bulk Water Transportation:** Not listed
12.3 **NFPA Hazard Classifications:**

Category	Classification
Health Hazard (Blue)	1
Flammability (Red)	4
Reactivity (Yellow)	0

13. PHYSICAL AND CHEMICAL PROPERTIES

13.1 **Physical State at 15°C and 1 atm:** Gas
13.2 **Molecular Weight:** 56.10
13.3 **Boiling Point at 1 atm:** 19.6°F = −6.9°C = 266.3°K
13.4 **Freezing Point:** −220°F = −140.3°C = 132.9°K
13.5 **Critical Temperature:** 292.5°F = −144.7°C = 417.9°K
13.6 **Critical Pressure:** 580 psia = 39.48 atm = 3.99 MN/m^2
13.7 **Specific Gravity:** 0.59 at 20°C (liquid)
13.8 **Liquid Surface Tension:** 15.8 dynes/cm = 0.0158 N/m at 20°C
13.9 **Liquid-Water Interfacial Tension:** (est.) 40 dynes/cm = 0.04 N/m at −10°C
13.10 **Vapor (Gas) Specific Gravity:** 1.9
13.11 **Ratio of Specific Heats of Vapor (Gas):** 1.061
13.12 **Latent Heat of Vaporization:** 170 Btu/lb = 94.3 cal/g = 3.95 × 10^5 J/kg
13.13 **Heat of Combustion:** −19,359 Btu/lb = −10,755 cal/g = −450.29 × 10^5 J/kg
13.14 **Heat of Decomposition:** Not pertinent
13.15 **Heat of Solution:** Not pertinent
13.16 **Heat of Polymerization:** Not pertinent

NOTES

REVISED 1978

IBR	ISOBUTYRIC ACID

Common Synonyms			
Dimethylacetic acid Isopropylformic acid 2-Methylpropanoic acid Propane-2-carboxylic acid	Liquid	Colorless	Unpleasant, acrid odor
	Floats and mixes with water.		

Avoid contact with liquid and vapor. Keep people away.
Wear goggles, self-contained breathing apparatus, and rubber overclothing (including gloves).
Stop discharge if possible.
Call fire department.
Isolate and remove discharged material.
Notify local health and pollution control agencies.

Fire

Combustible.
Extinguish with dry chemicals, alcohol foam, or carbon dioxide.
Cool exposed containers with water.

Exposure

CALL FOR MEDICAL AID.
VAPOR
Irritating to eyes, nose and throat.
Move to fresh air.
If breathing has stopped, give artificial respiration.
If breathing is difficult, give oxygen.
LIQUID
Will burn skin and eyes.
Harmful if swallowed.
Remove contaminated clothing and shoes.
Flush affected areas with plenty of water.
IF IN EYES, hold eyelids open and flush with plenty of water.
IF SWALLOWED and victim is CONSCIOUS, have victim drink water or milk.
DO NOT INDUCE VOMITING.

Water Pollution

Effect of low concentrations on aquatic life is unknown.
May be dangerous if it enters water intakes.
Notify local health and pollution control officials.
Notify operators of nearby water intakes.

1. RESPONSE TO DISCHARGE

(See Response Methods Handbook, CG 446-4)

Issue warning—corrosive
Restrict access
Disperse and flush

2. LABEL

3. CHEMICAL DESIGNATIONS

3.1 **Synonyms:** Dimethylacetic acid; Isopropylformic acid; 2-Methylpropanoic acid; alpha-Methylpropionic acid; Propane-2-carboxylic acid
3.2 **Coast Guard Compatibility Classification:** Not listed
3.3 **Chemical Formula:** $(CH_3)_2CHCOOH$
3.4 **IMCO/United Nations Numerical Designation:** Not listed

4. OBSERVABLE CHARACTERISTICS

4.1 **Physical State (as shipped):** Liquid
4.2 **Color:** Colorless
4.3 **Odor:** Unpleasant, acrid

5. HEALTH HAZARDS

5.1 **Personal Protective Equipment:** Organic chemical respirator; goggles or face shield; rubber gloves
5.2 **Symptoms Following Exposure:** Inhalation causes irritation of nose and throat. Ingestion causes irritation of mouth and stomach. Contact with eyes or skin causes irritation.
5.3 **Treatment for Exposure:** INHALATION: move to fresh air. INGESTION: give large amounts of water. EYES: flush with water for at least 15 min.; get medical attention if irritation persists. SKIN: flush with water.
5.4 **Toxicity by Inhalation (Threshold Limit Value):** Data not available
5.5 **Short-Term Inhalation Limits:** Data not available
5.6 **Toxicity by Ingestion:** Grade 3; oral LD_{50} = 280 mg/kg (rat)
5.7 **Late Toxicity:** Data not available
5.8 **Vapor (Gas) Irritant Characteristics:** Data not available
5.9 **Liquid or Solid Irritant Characteristics:** Data not available
5.10 **Odor Threshold:** Data not available

6. FIRE HAZARDS

6.1 **Flash Point:** 170°F O.C.
6.2 **Flammable Limits in Air:** Data not available
6.3 **Fire Extinguishing Agents:** Dry chemical, alcohol foam, carbon dioxide
6.4 **Fire Extinguishing Agents Not to be Used:** Water may be ineffective.
6.5 **Special Hazards of Combustion Products:**
6.6 **Behavior in Fire:**
6.7 **Ignition Temperature:** 935°F
6.8 **Electrical Hazard:** Data not available
6.9 **Burning Rate:** 2.6 mm/min.

7. CHEMICAL REACTIVITY

7.1 **Reactivity with Water:** No reaction
7.2 **Reactivity with Common Materials:** Corrosive to aluminum and other metals. Flammable hydrogen gas may accumulate in enclosed spaces.
7.3 **Stability During Transport:** Stable
7.4 **Neutralizing Agents for Acids and Caustics:** Flush with water.
7.5 **Polymerization:** Not pertinent
7.6 **Inhibitor of Polymerization:** Not pertinent

8. WATER POLLUTION

8.1 **Aquatic Toxicity:** Data not available
8.2 **Waterfowl Toxicity:** Data not available
8.3 **Biological Oxygen Demand (BOD):** Data not available
8.4 **Food Chain Concentration Potential:** None

9. SELECTED MANUFACTURERS

1. American Hoechst Corp.
   Route 202—206 North
   Somerville, N. J. 08876
2. BASF Wyandotte
   1609 Biddle Avenue
   Wyandotte, Mich. 48192
3. Eastman Organic Chemicals
   343 State Street
   Rochester, N. Y. 14650

10. SHIPPING INFORMATION

10.1 **Grades or Purity:** 99+%
10.2 **Storage Temperature:** Ambient
10.3 **Inert Atmosphere:** No requirement
10.4 **Venting:** Open

11. HAZARD ASSESSMENT CODE

(See Hazard Assessment Handbook, CG 446-3)

A-P-Q

12. HAZARD CLASSIFICATIONS

12.1 **Code of Federal Regulations:** Corrosive liquid
12.2 **NAS Hazard Rating for Bulk Water Transportation:** Not listed
12.3 **NFPA Hazard Classifications:** Not listed

13. PHYSICAL AND CHEMICAL PROPERTIES

13.1 **Physical State at 15°C and 1 atm:** Liquid
13.2 **Molecular Weight:** 88
13.3 **Boiling Point at 1 atm:** 309°F = 154°C = 427°K
13.4 **Freezing Point:** −51°F = −46°C = 227°K
13.5 **Critical Temperature:** 637°F = 336°C = 609°K
13.6 **Critical Pressure:** 588 psia = 40 atm = 4.06 MN/m^2
13.7 **Specific Gravity:** 0.949 at 20°C (liquid)
13.8 **Liquid Surface Tension:** 25.1 dynes/cm = 0.0251 N/m at 20°C
13.9 **Liquid-Water Interfacial Tension:** Not pertinent
13.10 **Vapor (Gas) Specific Gravity:** 3.0
13.11 **Ratio of Specific Heats of Vapor (Gas):** Not pertinent
13.12 **Latent Heat of Vaporization:** 202 Btu/lb = 112 cal/g = 4.68×10^5 J/kg
13.13 **Heat of Combustion:** −10,600 Btu/lb = −5,880 cal/g = -246×10^5 J/kg
13.14 **Heat of Decomposition:** Not pertinent
13.15 **Heat of Solution:** −20.5 Btu/lb = −11.4 cal/g = -0.477×10^5 J/kg
13.16 **Heat of Polymerization:** Not pertinent

NOTES

IBN — ISOBUTYRONITRILE

Common Synonyms			
Isopropyl Cyanide IBN 2-Methylpropanenitrile 2-Methylpropionitrile	Liquid	Colorless	Almond-like
	Floats on water. Flammable vapor is produced.		

AVOID CONTACT WITH LIQUID AND VAPOR. KEEP PEOPLE AWAY.
Wear goggles and self-contained breathing apparatus.
Shut off ignition sources and call fire department.
Stay upwind and use water spray to "knock down" vapor.
Stop discharge if possible.
Isolate and remove discharged material.
Notify local health and pollution control agencies.

Fire

FLAMMABLE.
Flashback along vapor trail may occur.
Vapor may explode if ignited in an enclosed area.
Wear goggles and self-contained breathing apparatus.
Combat fires from safe distance or protected location.
Extinguish with dry chemical, alcohol foam, or carbon dioxide.
Water may be ineffective on fire.
Cool exposed containers with water.

Exposure

CALL FOR MEDICAL AID.

VAPOR
POISONOUS IF INHALED.
Irritating to eyes, nose and throat.
Move to fresh air.
If breathing has stopped, give artificial respiration.
If breathing is difficult, give oxygen.

LIQUID
POISONOUS IF SWALLOWED OR IF SKIN IS EXPOSED.
Irritating to skin and eyes.
Remove contaminated clothing and shoes.
Flush affected areas with plenty of water.
IF IN EYES, hold eyelids open and flush with plenty of water.
IF SWALLOWED and victim is CONSCIOUS, have victim drink water or milk.

Water Pollution

Effect of low concentrations on aquatic life is unknown.
May be dangerous if it enters water intakes.
Notify local health and wildlife officials.
Notify operators of nearby water intakes.

1. RESPONSE TO DISCHARGE

Issue warning—high flammability, poison
Restrict access
Evacuate area
Mechanical containment
Should be removed
Chemical and physical treatment

2. LABEL

Red

3. CHEMICAL DESIGNATIONS

3.1 **Synonyms:** IBN; Isopropyl cyanide; 2-Methylpropanenitrile; 2-Methylpropionitrile
3.2 **Coast Guard Compatibility Classification:** Not listed
3.3 **Chemical Formula:** $(CH_3)_2CHCN$
3.4 **IMCO/United Nations Numerical Designation:** Not listed

4. OBSERVABLE CHARACTERISTICS

4.1 **Physical State (as shipped):** Liquid
4.2 **Color:** Colorless
4.3 **Odor:** Like almonds or benzaldehyde

5. HEALTH HAZARDS

5.1 **Personal Protective Equipment:** Self-contained breathing apparatus; goggles; rubber gloves
5.2 **Symptoms Following Exposure:** Inhalation, ingestion, or skin contact causes weakness, headache, confusion, nausea, vomiting; acute cyanide poisoning may result. Contact with eyes causes irritation.
5.3 **Treatment for Exposure:** *Get medical attention following all overexposures to this chemical. Watch for symptoms of cyanide poisoning.* INHALATION: move patient to fresh air; apply artificial respiration if breathing stops. INGESTION: break an amyl nitrite pearl in a cloth and hold lightly under patient's nose for 15 sec.; if he is conscious, induce vomiting and repeat until vomit is clear; repeat inhalation of amyl nitrite 5 times at 15-sec. intervals. EYES: flush with water for at least 15 min. SKIN: flush with water; remove contaminated clothing; destroy contaminated shoes.
5.4 **Toxicity by Inhalation (Threshold Limit Value):** Data not available
5.5 **Short-Term Inhalation Limits:** Data not available
5.6 **Toxicity by Ingestion:** Grade 3; oral LD_{50} = 100 mg/kg (rat)
5.7 **Late Toxicity:** Data not available
5.8 **Vapor (Gas) Irritant Characteristics:** Data not available
5.9 **Liquid or Solid Irritant Characteristics:** Data not available
5.10 **Odor Threshold:** Data not available

6. FIRE HAZARDS

6.1 **Flash Point:** 47°F C.C.
6.2 **Flammable Limits in Air:** Data not available
6.3 **Fire Extinguishing Agents:** Dry chemical, foam, carbon dioxide
6.4 **Fire Extinguishing Agents Not to be Used:** Water may be ineffective.
6.5 **Special Hazards of Combustion Products:** Toxic oxides of nitrogen may form in fire.
6.6 **Behavior in Fire:**
6.7 **Ignition Temperature:** Data not available
6.8 **Electrical Hazard:** Data not available
6.9 **Burning Rate:** Data not available

7. CHEMICAL REACTIVITY

7.1 **Reactivity with Water:** No reaction
7.2 **Reactivity with Common Materials:**
7.3 **Stability During Transport:** Stable
7.4 **Neutralizing Agents for Acids and Caustics:** Not pertinent
7.5 **Polymerization:** Not pertinent
7.6 **Inhibitor of Polymerization:** Not pertinent

8. WATER POLLUTION

8.1 **Aquatic Toxicity:** Data not available
8.2 **Waterfowl Toxicity:** Data not available
8.3 **Biological Oxygen Demand (BOD):** Data not available
8.4 **Food Chain Concentration Potential:** None

9. SELECTED MANUFACTURERS

1. Air Products and Chemicals, Inc.
   P. O. Box 538
   Allentown, Pa. 18105
2. Kemstar Corp.
   P. O. Box 68
   Scarborough, N. Y. 10510
3. Eastman Chemical Products, Inc.
   Kingsport, Tenn. 37662

10. SHIPPING INFORMATION

10.1 **Grades or Purity:** Technical; Pure
10.2 **Storage Temperature:** Ambient
10.3 **Inert Atmosphere:** No requirement
10.4 **Venting:** Open

11. HAZARD ASSESSMENT CODE

(See Hazard Assessment Handbook, CG 446-3)

A-T-U-V-W

12. HAZARD CLASSIFICATIONS

12.1 **Code of Federal Regulations:** Flammable liquid
12.2 **NAS Hazard Rating for Bulk Water Transportation:** Not listed
12.3 **NFPA Hazard Classifications:** Not listed

13. PHYSICAL AND CHEMICAL PROPERTIES

13.1 **Physical State at 15°C and 1 atm:** Liquid
13.2 **Molecular Weight:** 69.1
13.3 **Boiling Point at 1 atm:** 219°F = 104°C = 377°K
13.4 **Freezing Point:** Not pertinent
13.5 **Critical Temperature:** Data not available
13.6 **Critical Pressure:** Data not available
13.7 **Specific Gravity:** 0.774 at 20°C (liquid)
13.8 **Liquid Surface Tension:** 24.9 dynes/cm = 0.0249 N/m at 20°C
13.9 **Liquid-Water Interfacial Tension:** Data not available
13.10 **Vapor (Gas) Specific Gravity:** 2.4
13.11 **Ratio of Specific Heats of Vapor (Gas):** Data not available
13.12 **Latent Heat of Vaporization:** 200 Btu/lb = 110 cal/g = 4.7×10^5 J/kg
13.13 **Heat of Combustion:** −14,960 Btu/lb = −8,310 cal/g = -348×10^5 J/kg
13.14 **Heat of Decomposition:** Not pertinent
13.15 **Heat of Solution:** Not pertinent
13.16 **Heat of Polymerization:** Not pertinent

NOTES

IDA	ISODECALDEHYDE

Common Synonyms			
Isodecaldehyde, mixed isomers	Liquid	Colorless	Fruity odor
	Floats on water.		

Stop discharge if possible. Keep people away.
Call fire department.
Avoid contact with liquid.
Isolate and remove discharged material.
Notify local health and pollution control agencies.

Fire

Combustible.
Extinguish with dry chemical, foam, or carbon dioxide.
Water may be ineffective on fire.
Cool exposed containers with water.

Exposure

CALL FOR MEDICAL AID.

LIQUID
Irritating to skin and eyes.
Remove contaminated clothing and shoes.
Flush affected areas with plenty of water.
IF IN EYES, hold eyelids open and flush with plenty of water.

Water Pollution

Effect of low concentrations on aquatic life is unknown.
Fouling to shoreline.
May be dangerous if it enters water intakes.
Notify local health and wildlife officials.
Notify operators of nearby water intakes.

1. RESPONSE TO DISCHARGE

(See Response Methods Handbook, CG 446-4)

Mechanical containment
Should be removed
Chemical and physical treatment

2. LABELS

No hazard label required by Code of Federal Regulations

3. CHEMICAL DESIGNATIONS

3.1 **Synonyms:** Isodecaldehyde, mixed isomers; Trimethylheptanals
3.2 **Coast Guard Compatibility Classification:** Aldehyde
3.3 **Chemical Formula:** $C_9H_{19}CHO$
3.4 **IMCO/United Nations Numerical Designation:** Not listed

4. OBSERVABLE CHARACTERISTICS

4.1 **Physical State (as shipped):** Liquid
4.2 **Color:** Colorless
4.3 **Odor:** Somewhat fruity

5. HEALTH HAZARDS

5.1 **Personal Protective Equipment:** Protective clothing; chemical goggles.
5.2 **Symptoms Following Exposure:** Low general toxicity. Liquid may irritate eyes and skin.
5.3 **Treatment for Exposure:** Wash eyes and skin with plenty of water for at least 15 min.
5.4 **Toxicity by Inhalation (Threshold Limit Value):** Data not available
5.5 **Short-Term Inhalation Limits:** Data not available
5.6 **Toxicity by Ingestion:** Data not available
5.7 **Late Toxicity:** Data not available
5.8 **Vapor (Gas) Irritant Characteristics:** Vapors cause a slight smarting of the eyes or respiratory system if present in high concentrations. The effect is temporary.
5.9 **Liquid or Solid Irritant Characteristics:** Minimum hazard. If spilled on clothing and allowed to remain, may cause smarting and reddening of the skin.
5.10 **Odor Threshold:** Data not available

6. FIRE HAZARDS

6.1 **Flash Point:** 185°F O.C.
6.2 **Flammable Limits in Air:** Data not available
6.3 **Fire Extinguishing Agents:** Foam, dry chemical, carbon dioxide
6.4 **Fire Extinguishing Agents Not to be Used:** Not pertinent
6.5 **Special Hazards of Combustion Products:** Not pertinent
6.6 **Behavior in Fire:** Not pertinent
6.7 **Ignition Temperature:** Data not available
6.8 **Electrical Hazard:** Not pertinent
6.9 **Burning Rate:** Data not available

7. CHEMICAL REACTIVITY

7.1 **Reactivity with Water:** No reaction
7.2 **Reactivity with Common Materials:** No reaction
7.3 **Stability During Transport:** Stable
7.4 **Neutralizing Agents for Acids and Caustics:** Not pertinent
7.5 **Polymerization:** Not pertinent
7.6 **Inhibitor of Polymerization:** Not pertinent

8. WATER POLLUTION

8.1 **Aquatic Toxicity:** Data not available
8.2 **Waterfowl Toxicity:** Data not available
8.3 **Biological Oxygen Demand (BOD):** Data not available
8.4 **Food Chain Concentration Potential:** None

9. SELECTED MANUFACTURERS

Union Carbide Corp.
Chemicals and Plastics Division
270 Park Ave.
New York, N. Y. 10017

10. SHIPPING INFORMATION

10.1 **Grades or Purity:** Commercial
10.2 **Storage Temperature:** Ambient
10.3 **Inert Atmosphere:** No requirement
10.4 **Venting:** Open (flame arrester)

11. HAZARD ASSESSMENT CODE

(See Hazard Assessment Handbook, CG 446-3)

A-T-U

12. HAZARD CLASSIFICATIONS

12.1 **Code of Federal Regulations:** Combustible Liquid
12.2 **NAS Hazard Rating for Bulk Water Transportation:**

Category	Rating
Fire	1
Health	
Vapor Irritant	1
Liquid or Solid Irritant	1
Poisons	1
Water Pollution	
Human Toxicity	1
Aquatic Toxicity	1
Aesthetic Effect	2
Reactivity	
Other Chemicals	2
Water	0
Self-Reaction	1

12.3 **NFPA Hazard Classifications:** Not listed

13. PHYSICAL AND CHEMICAL PROPERTIES

13.1 **Physical State at 15°C and 1 atm:** Liquid
13.2 **Molecular Weight:** 156.28
13.3 **Boiling Point at 1 atm:** Data not available
13.4 **Freezing Point:** Data not available
13.5 **Critical Temperature:** Not pertinent
13.6 **Critical Pressure:** Not pertinent
13.7 **Specific Gravity:** (est.) 0.84 at 15° (liquid)
13.8 **Liquid Surface Tension:** (est.) 20 dynes/cm = 0.02 N/m at 20°C
13.9 **Liquid-Water Interfacial Tension:** (est.) 40 dynes/cm = 0.04 N/m at 20°C
13.10 **Vapor (Gas) Specific Gravity:** Not pertinent
13.11 **Ratio of Specific Heats of Vapor (Gas):** Not pertinent
13.12 **Latent Heat of Vaporization:** Data not available
13.13 **Heat of Combustion:** Data not available
13.14 **Heat of Decomposition:** Not pertinent
13.15 **Heat of Solution:** Not pertinent
13.16 **Heat of Polymerization:** Not pertinent

NOTES

REVISED 1978

IAI	# ISODECYL ACRYLATE, INHIBITED

Common Synonyms			
Isodecyl acrylate	Liquid	Colorless	Weak odor
	Floats on water.		

Stop discharge if possible. Keep people away.
Call fire department.
Isolate and remove discharged material.
Notify local health and pollution control agencies.

Fire	Combustible. **Extinguish with dry chemicals, foam or carbon dioxide.** **Water may be ineffective on fire.** **Cool exposed containers with water.**
Exposure	**Call for medical aid.** LIQUID Irritating to skin and eyes. **Remove contaminated clothing and shoes.** **Flush affected areas with plenty of water.** **IF IN EYES, hold eyelids open and flush with plenty of water.** **IF SWALLOWED and victim is CONSCIOUS, have victim drink water or milk.**
Water Pollution	Effect of low concentrations on aquatic life is unknown. Fouling to shoreline. May be dangerous if it enters water intakes. **Notify local health and wildlife officials.** **Notify operators of nearby water intakes.**

1. RESPONSE TO DISCHARGE

(See Response Methods Handbook, CG 446-4)

Issue warning—water contaminant
Mechanical containment
Should be removed
Chemical and physical treatment

2. LABELS

No label required by Code of Federal Regulations

3. CHEMICAL DESIGNATIONS

3.1 **Synonyms:** iso-Decyl acrylate
3.2 **Coast Guard Compatibility Classification:** Monomers (14)
3.3 **Chemical Formula:** $CH_2{=}CHCOOC_{10}H_{21}$
3.4 **IMCO/United Nations Numerical Designation:** Not listed

4. OBSERVABLE CHARACTERISTICS

4.1 **Physical State (as shipped):** Liquid
4.2 **Color:** Colorless
4.3 **Odor:** Weak acrylate

5. HEALTH HAZARDS

5.1 **Personal Protective Equipment:** Goggles or face shield; rubber gloves.
5.2 **Symptoms Following Exposure:** Inhalation causes mild irritation of nose and throat. Eyes are mildly irritated by vapor, more severely by liquid. Prolonged contact of liquid with skin may cause irritation.
5.3 **Treatment for Exposure:** INHALATION: move to fresh air. EYES: flush with water for at least 15 min. after contact with liquid. SKIN: wipe off, wash well with soap and water.
5.4 **Toxicity by Inhalation (Threshold Limit Value):** Data not available
5.5 **Short-Term Inhalation Limits:** Data not available
5.6 **Toxicity by Ingestion:** Grade 1; LD_{50} 5 to 15 g/kg
5.7 **Late Toxicity:** Data not available
5.8 **Vapor (Gas) Irritant Characteristics:** Vapors cause a slight smarting of the eyes or respiratory system if present in high concentrations. The effect is temporary.
5.9 **Liquid or Solid Irritant Characteristics:** Minimum hazard. If spilled on clothing and allowed to remain, may cause smarting and reddening of the skin.
5.10 **Odor Threshold:** Data not available

6. FIRE HAZARDS

6.1 **Flash Point:** 240°F O.C.
6.2 **Flammable Limits in Air:** Data not available
6.3 **Fire Extinguishing Agents:** Foam, dry chemical, carbon dioxide
6.4 **Fire Extinguishing Agents Not to be Used:** Water may be ineffective.
6.5 **Special Hazards of Combustion Products:** Not pertinent
6.6 **Behavior in Fire:** May polymerize to gummy solid. Reaction is not violent.
6.7 **Ignition Temperature:** Data not available
6.8 **Electrical Hazard:** Data not available
6.9 **Burning Rate:** Not pertinent

7. CHEMICAL REACTIVITY

7.1 **Reactivity with Water:** No reaction
7.2 **Reactivity with Common Materials:** No reaction
7.3 **Stability During Transport:** Stable if inhibited.
7.4 **Neutralizing Agents for Acids and Caustics:** Not pertinent
7.5 **Polymerization:** In the absence of inhibitor, polymerization will occur, especially when heated.
7.6 **Inhibitor of Polymerization:** Monomethyl ether of hydroquinone, 25 ppm

8. WATER POLLUTION

8.1 **Aquatic Toxicity:** Data not available
8.2 **Waterfowl Toxicity:** Data not available
8.3 **Biological Oxygen Demand (BOD):** Data not available
8.4 **Food Chain Concentration Potential:** None

9. SELECTED MANUFACTURERS

Union Carbide Corporation
Chemicals and Plastics Division
270 Park Avenue
New York, N. Y. 10017

10. SHIPPING INFORMATION

10.1 **Grades or Purity:** 97.5+%
10.2 **Storage Temperature:** Ambient
10.3 **Inert Atmosphere:** No requirement
10.4 **Venting:** Open (flame arrester)

11. HAZARD ASSESSMENT CODE

(See Hazard Assessment Handbook, CG 446-3)

A-T-U-Z

12. HAZARD CLASSIFICATIONS

12.1 **Code of Federal Regulations:** Not listed
12.2 **NAS Hazard Rating for Bulk Water Transportation:**

Category	Rating
Fire	1
Health	
Vapor Irritant	1
Liquid or Solid Irritant	1
Poisons	1
Water Pollution	
Human Toxicity	1
Aquatic Toxicity	2
Aesthetic Effect	2
Reactivity	
Other Chemicals	2
Water	0
Self-Reaction	3

12.3 **NFPA Hazard Classifications:** Not listed

13. PHYSICAL AND CHEMICAL PROPERTIES

13.1 **Physical State at 15°C and 1 atm:** Liquid
13.2 **Molecular Weight:** 212.4
13.3 **Boiling Point at 1 atm:** Not pertinent (polymerizes)
13.4 **Freezing Point:** −148°F = −100°C = 173°K
13.5 **Critical Temperature:** Not pertinent
13.6 **Critical Pressure:** Not pertinent
13.7 **Specific Gravity:** 0.885 at 20°C (liquid)
13.8 **Liquid Surface Tension:** (est.) 30 dynes/cm = 0.030 N/m at 20°C
13.9 **Liquid-Water Interfacial Tension:** (est.) 30 dynes/cm = 0.030 N/m at 20°C
13.10 **Vapor (Gas) Specific Gravity:** Not pertinent
13.11 **Ratio of Specific Heats of Vapor (Gas):** Not pertinent
13.12 **Latent Heat of Vaporization:** 110 Btu/lb = 61 cal/g = 2.6×10^5 J/kg
13.13 **Heat of Combustion:** (est.) −16,300 Btu/lb = −9,100 cal/g = -380×10^5 J/kg
13.14 **Heat of Decomposition:** Not pertinent
13.15 **Heat of Solution:** Not pertinent
13.16 **Heat of Polymerization:** (est.) −119 Btu/lb = −66 cal/g = -2.8×10^5 J/kg

NOTES

ISA — ISODECYL ALCOHOL

Common Synonyms	Liquid	Colorless	Mild alcohol odor
	Floats on water.		

Avoid contact with liquid. Keep people away.
Wear goggles and rubber overclothing (including gloves).
Stop discharge if possible.
Call fire department.
Isolate and remove discharged material.
Notify local health and pollution control agencies.

Fire

Combustible.
Extinguish with dry chemical, alcohol foam, or carbon dioxide.
Water may be ineffective on fire.

Exposure

CALL FOR MEDICAL AID.

LIQUID
Will burn skin and eyes.
Remove contaminated clothing and shoes.
Flush affected areas with plenty of water.
IF SWALLOWED and victim is CONSCIOUS, have victim drink water or milk.

Water Pollution

Effect of low concentrations on aquatic life is unknown.
Fouling to shoreline.
May be dangerous if it enters water intakes.
Notify local health and wildlife officials.
Notify operators of nearby water intakes.

1. RESPONSE TO DISCHARGE

(See Response Methods Handbook, CG 446-4)

Mechanical containment
Should be removed
Chemical and physical treatment

2. LABELS

No hazard label required by Code of Federal Regulations

3. CHEMICAL DESIGNATIONS

3.1 **Synonyms:** No common synonyms
3.2 **Coast Guard Compatibility Classification:** Alcohol
3.3 **Chemical Formula:** $C_{10}H_{21}OH$
3.4 **IMCO/United Nations Numerical Designation:** Not listed

4. OBSERVABLE CHARACTERISTICS

4.1 **Physical State (as shipped):** Liquid
4.2 **Color:** Colorless
4.3 **Odor:** Weak alcoholic

5. HEALTH HAZARDS

5.1 **Personal Protective Equipment:** Chemical goggles.
5.2 **Symptoms Following Exposure:** Direct contact with skin can produce irritation.
5.3 **Treatment for Exposure:** Wash affected area with water for 15 min.
5.4 **Toxicity by Inhalation (Threshold Limit Value):** Data not available
5.5 **Short-Term Inhalation Limits:** Data not available
5.6 **Toxicity by Ingestion:** Data not available
5.7 **Late Toxicity:** Data not available
5.8 **Vapor (Gas) Irritant Characteristics:** Vapors are non-irritating to the eyes and throat.
5.9 **Liquid or Solid Irritant Characteristics:** Causes smarting of the skin and first-degree burns on short exposure; may cause secondary burns on long exposure.
5.10 **Odor Threshold:** Data not available

6. FIRE HAZARDS

6.1 **Flash Point:** 220°F O.C.
6.2 **Flammable Limits in Air:** Data not available
6.3 **Fire Extinguishing Agents:** Alcohol foam, dry chemicals or carbon dioxide
6.4 **Fire Extinguishing Agents Not to be Used:** Water or foam may cause frothing
6.5 **Special Hazards of Combustion Products:** Not pertinent
6.6 **Behavior in Fire:** Not pertinent
6.7 **Ignition Temperature:** Data not available
6.8 **Electrical Hazard:** Not pertinent
6.9 **Burning Rate:** Data not available

7. CHEMICAL REACTIVITY

7.1 **Reactivity with Water:** No reaction
7.2 **Reactivity with Common Materials:** No reaction
7.3 **Stability During Transport:** Stable
7.4 **Neutralizing Agents for Acids and Caustics:** Not pertinent
7.5 **Polymerization:** Not pertinent
7.6 **Inhibitor of Polymerization:** Not pertinent

8. WATER POLLUTION

8.1 **Aquatic Toxicity:** Data not available
8.2 **Waterfowl Toxicity:** Data not available
8.3 **Biological Oxygen Demand (BOD):** Data not available
8.4 **Food Chain Concentration Potential:** None

9. SELECTED MANUFACTURERS

1. Exxon Chemical Co.
   Houston, Texas 77001
2. Getty Oil Co.
   Los Angeles, Calif. 90054
3. Union Carbide Corp.
   Chemicals and Plastics Division
   270 Park Ave.
   New York, N. Y. 10017

10. SHIPPING INFORMATION

10.1 **Grades or Purity:** Technical: mixed isomers
10.2 **Storage Temperature:** Ambient
10.3 **Inert Atmosphere:** No requirement
10.4 **Venting:** Open (flame arrester)

11. HAZARD ASSESSMENT CODE

(See Hazard Assessment Handbook, CG 446-3)

A-T-U

12. HAZARD CLASSIFICATIONS

12.1 **Code of Federal Regulations:** Not listed
12.2 **NAS Hazard Rating for Bulk Water Transportation:**

Category	Rating
Fire	1
Health	
Vapor Irritant	0
Liquid or Solid Irritant	2
Poisons	0
Water Pollution	
Human Toxicity	0
Aquatic Toxicity	1
Aesthetic Effect	2
Reactivity	
Other Chemicals	2
Water	0
Self-Reaction	0

12.3 **NFPA Hazard Classifications:** Not listed

13. PHYSICAL AND CHEMICAL PROPERTIES

13.1 **Physical State at 15°C and 1 atm:** Liquid
13.2 **Molecular Weight:** 158.29
13.3 **Boiling Point at 1 atm:** 428°F = 220°C = 493°K
13.4 **Freezing Point:** <140°F = <60°C = <333°K
13.5 **Critical Temperature:** Not pertinent
13.6 **Critical Pressure:** Not pertinent
13.7 **Specific Gravity:** 0.841 at 20°C (liquid)
13.8 **Liquid Surface Tension:** Data not available
13.9 **Liquid-Water Interfacial Tension:** Data not available
13.10 **Vapor (Gas) Specific Gravity:** Not pertinent
13.11 **Ratio of Specific Heats of Vapor (Gas):** (est.) 1.032
13.12 **Latent Heat of Vaporization:** (est.) 120 Btu/lb = 67 cal/g = 2.8×10^5 J/kg
13.13 **Heat of Combustion:** Data not available
13.14 **Heat of Decomposition:** Not pertinent
13.15 **Heat of Solution:** Not pertinent
13.16 **Heat of Polymerization:** Not pertinent

NOTES

REVISED 1978

IHA	# ISOHEXANE

Common Synonyms		
2-Methylpentane	Watery liquid	Gasoline-like odor
	Floats on water. Flammable, irritating vapor is produced.	

Stop discharge if possible. Keep people away.
Shut off ignition sources and call fire department.
Stay upwind and use water spray to "knock down" vapor.
Isolate and remove discharged material.
Notify local health and pollution control agencies.

Fire

FLAMMABLE.
Flashback along vapor trail may occur.
Vapor may explode if ignited in an enclosed area.
Extinguish with dry chemical, foam, or carbon dioxide.
Water may be ineffective on fire.
Cool exposed containers with water.

Exposure

CALL FOR MEDICAL AID.

VAPOR
Irritating to eyes, nose and throat.
If inhaled, will cause dizziness, headache, difficult breathing or loss of consciousness.
Move to fresh air.
If breathing has stopped, give artificial respiration.
If breathing is difficult, give oxygen.

LIQUID
Irritating to skin and eyes.
If swallowed, will cause nausea or vomiting.
Remove contaminated clothing and shoes.
Flush affected areas with plenty of water.
IF IN EYES, hold eyelids open and flush with plenty of water.
IF SWALLOWED and victim is CONSCIOUS, have victim drink water or milk.
DO NOT INDUCE VOMITING.

Water Pollution

Effect of low concentrations on aquatic life is unknown.
May be dangerous if it enters water intakes.
Notify local health and wildlife officials.
Notify operators of nearby water intakes.

1. RESPONSE TO DISCHARGE
(See Response Methods Handbook, CG 446-4)

Issue warning—high flammability
Evacuate area
Disperse and flush

2. LABEL

Red

3. CHEMICAL DESIGNATIONS

3.1 **Synonyms:** 2-Methylpentane
3.2 **Coast Guard Compatibility Classification:** Paraffin
3.3 **Chemical Formula:** $CH_3CH(CH_3)CH_2CH_2CH_3$
3.4 **IMCO/United Nations Numerical Designation:** 3.1/1208

4. OBSERVABLE CHARACTERISTICS

4.1 **Physical State (as shipped):** Liquid
4.2 **Color:** Colorless
4.3 **Odor:** Gasoline

5. HEALTH HAZARDS

5.1 **Personal Protective Equipment:** Eye protection (as for gasoline).
5.2 **Symptoms Following Exposure:** Inhalation causes irritation of respiratory tract, cough, mild depression, cardiac arrhythmias. Aspiration causes severe lung irritation, coughing, pulmonary edema; excitement followed by depression. Ingestion causes nausea, vomiting, swelling of abdomen, headache, depression.
5.3 **Treatment for Exposure:** INHALATION: maintain respiration, give oxygen if needed. ASPIRATION: enforce bed rest; give oxygen. INGESTION: do NOT induce vomiting; call a doctor. EYES: wash with copious amount of water. SKIN: wipe off, wash with soap and water.
5.4 **Toxicity by Inhalation (Threshold Limit Value):** Data not available
5.5 **Short-Term Inhalation Limits:** Data not available
5.6 **Toxicity by Ingestion:** Data not available
5.7 **Late Toxicity:** None
5.8 **Vapor (Gas) Irritant Characteristics:** Vapors are nonirritating to the eyes and throat.
5.9 **Liquid or Solid Irritant Characteristics:** No appreciable hazard. Practically harmless to the skin.
5.10 **Odor Threshold:** Data not available

6. FIRE HAZARDS

6.1 **Flash Point:** −20°F C.C.
6.2 **Flammable Limits in Air:** 1.2%—7.7%
6.3 **Fire Extinguishing Agents:** Foam, carbon dioxide, or dry chemical
6.4 **Fire Extinguishing Agents Not to be Used:** Water may be ineffective
6.5 **Special Hazards of Combustion Products:** Not pertinent
6.6 **Behavior in Fire:** Not pertinent
6.7 **Ignition Temperature:** 585°F
6.8 **Electrical Hazard:** Not pertinent
6.9 **Burning Rate:** 8.2 mm/min.

7. CHEMICAL REACTIVITY

7.1 **Reactivity with Water:** No reaction
7.2 **Reactivity with Common Materials:** No reaction
7.3 **Stability During Transport:** Stable
7.4 **Neutralizing Agents for Acids and Caustics:** Not pertinent
7.5 **Polymerization:** Not pertinent
7.6 **Inhibitor of Polymerization:** Not pertinent

8. WATER POLLUTION

8.1 **Aquatic Toxicity:** Data not available
8.2 **Waterfowl Toxicity:** Data not available
8.3 **Biological Oxygen Demand (BOD):** Data not available
8.4 **Food Chain Concentration Potential:** None

9. SELECTED MANUFACTURERS

1. The Humphrey Chemical Co.
   Devine Street
   North Haven, Conn. 06473
2. Pennzoil United, Inc.
   Atlas Processing Co.
   3546 Midway St.
   Shreveport, La. 71109
3. Phillips Petroleum Co.
   Bartlesville, Okla. 74004

10. SHIPPING INFORMATION

10.1 **Grades or Purity:** Research: 99.95%; pure: 99.0%; technical: 95.0%
10.2 **Storage Temperature:** Ambient
10.3 **Inert Atmosphere:** No requirement
10.4 **Venting:** Open (flame arrester) or pressure-vacuum

11. HAZARD ASSESSMENT CODE
(See Hazard Assessment Handbook, CG 446-3)

A-T-U-V-W

12. HAZARD CLASSIFICATIONS

12.1 **Code of Federal Regulations:** Flammable liquid
12.2 **NAS Hazard Rating for Bulk Water Transportation:**

Category	Rating
Fire	3
Health	
Vapor Irritant	0
Liquid or Solid Irritant	0
Poisons	1
Water Pollution	
Human Toxicity	1
Aquatic Toxicity	1
Aesthetic Effect	1
Reactivity	
Other Chemicals	0
Water	0
Self-Reaction	0

12.3 **NFPA Hazard Classifications:** Not listed

13. PHYSICAL AND CHEMICAL PROPERTIES

13.1 **Physical State at 15°C and 1 atm:** Liquid
13.2 **Molecular Weight:** 86.18
13.3 **Boiling Point at 1 atm:** 140.5°F = 60.3°C = 333.5°K
13.4 **Freezing Point:** −244.6°F = −153.7°C = 119.5°K
13.5 **Critical Temperature:** 435.7°F = 224.3°C = 497.5°K
13.6 **Critical Pressure:** 437 psia = 29.7 atm = 3.01 MN/m^2
13.7 **Specific Gravity:** 0.653 at 20°C (liquid)
13.8 **Liquid Surface Tension:** 17.38 dynes/cm = 0.01738 N/m at 20°C
13.9 **Liquid-Water Interfacial Tension:** (est.) 40 dynes/cm = 0.04 N/m at 20°C
13.10 **Vapor (Gas) Specific Gravity:** 2.9
13.11 **Ratio of Specific Heats of Vapor (Gas):** 1.062
13.12 **Latent Heat of Vaporization:** 139 Btu/lb = 77.1 cal/g = 3.23×10^5 J/kg
13.13 **Heat of Combustion:** −19,147 Btu/lb = −10,637 cal/g = -445.35×10^5 J/kg
13.14 **Heat of Decomposition:** Not pertinent
13.15 **Heat of Solution:** Not pertinent
13.16 **Heat of Polymerization:** Not pertinent

NOTES

REVISED 1978

IOC ISOOCTALDEHYDE

Common Synonyms			
6-Methyl-1-heptanal Oxo octaldehyde Isooctylaldehyde	Liquid	Colorless	Mild fruity odor
	Floats on water.		

Stop discharge if possible.
Call fire department.
Isolate and remove discharged material.
Notify local health and pollution control agencies.

Fire	Combustible. Extinguish with dry chemical, foam, or carbon dioxide.
Exposure	CALL FOR MEDICAL AID. LIQUID Irritating to eyes. IF IN EYES, hold eyelids open and flush with plenty of water.
Water Pollution	Effect of low concentrations on aquatic life is unknown. Fouling to shoreline. May be dangerous if it enters water intakes. Notify local health and wildlife officials. Notify operators of nearby water intakes.

1. RESPONSE TO DISCHARGE

(See Response Methods Handbook, CG 446-1)

Mechanical containment
Should be removed
Chemical and physical treatment

2. LABELS

No hazard label required by Code of Federal Regulations

3. CHEMICAL DESIGNATIONS

3.1 **Synonyms:** Dimethylhexanals
Isooctyl aldehyde
6-Methyl-1-heptanal

3.2 **Coast Guard Compatibility Classification:** Aldehyde

3.3 **Chemical Formula:** $(CH_3)_2CH(CH_2)_4CHO$

3.4 **IMCO/United Nations Numerical Designation:** Not listed

4. OBSERVABLE CHARACTERISTICS

4.1 **Physical State (as shipped):** Liquid

4.2 **Color:** Colorless

4.3 **Odor:** Mild fruity

5. HEALTH HAZARDS

5.1 **Personal Protective Equipment:** Chemical goggles.

5.2 **Symptoms Following Exposure:** High vapor concentrations produce eye irritation. Liquid may irritate eyes.

5.3 **Treatment for Exposure:** Remove from exposure. Wash eyes with water for 15 min.

5.4 **Toxicity by Inhalation (Threshold Limit Value):** Data not available

5.5 **Short-Term Inhalation Limits:** Data not available

5.6 **Toxicity by Ingestion:** Data not available

5.7 **Late Toxicity:** Data not available

5.8 **Vapor (Gas) Irritant Characteristics:** Vapors cause a slight smarting of the eyes or respiratory system if present in high concentrations. The effect is temporary.

5.9 **Liquid or Solid Irritant Characteristics:** No appreciable hazard. Practically harmless to the skin.

5.10 **Odor Threshold:** Data not available

6. FIRE HAZARDS

6.1 **Flash Point:** 104°F C.C.

6.2 **Flammable Limits in Air:** Data not available

6.3 **Fire Extinguishing Agents:** Data not available

6.4 **Fire Extinguishing Agents Not to be Used:** Data not available

6.5 **Special Hazards of Combustion Products:** Not pertinent

6.6 **Behavior in Fire:** Not pertinent

6.7 **Ignition Temperature:** 320°F

6.8 **Electrical Hazard:** Not pertinent

6.9 **Burning Rate:** Data not available

7. CHEMICAL REACTIVITY

7.1 **Reactivity with Water:** No reaction

7.2 **Reactivity with Common Materials:** No reaction

7.3 **Stability During Transport:** Stable

7.4 **Neutralizing Agents for Acids and Caustics:** Not pertinent

7.5 **Polymerization:** Not pertinent

7.6 **Inhibitor of Polymerization:** Not pertinent

8. WATER POLLUTION

8.1 **Aquatic Toxicity:** Data not available

8.2 **Waterfowl Toxicity:** Data not available

8.3 **Biological Oxygen Demand (BOD):** Data not available

8.4 **Food Chain Concentration Potential:** None

9. SELECTED MANUFACTURERS

1. Exxon Chemical Co.
Houston, Texas 77001
2. Getty Oil Co.
Los Angeles, Calif. 90054
3. Union Carbide Corp.
Chemicals and Plastics Division
270 Park Ave.
New York, N. Y. 10017

10. SHIPPING INFORMATION

10.1 **Grades or Purity:** Data not available

10.2 **Storage Temperature:** Ambient

10.3 **Inert Atmosphere:** No requirement

10.4 **Venting:** Open (flame arrester)

11. HAZARD ASSESSMENT CODE

(See Hazard Assessment Handbook, CG 446-3)

A-T-U

12. HAZARD CLASSIFICATIONS

12.1 **Code of Federal Regulations:** Combustible Liquid

12.2 **NAS Hazard Rating for Bulk Water Transportation:**

Category	Rating
Fire	2
Health	
Vapor Irritant	1
Liquid or Solid Irritant	0
Poisons	1
Water Pollution	
Human Toxicity	1
Aquatic Toxicity	1
Aesthetic Effect	2
Reactivity	
Other Chemicals	2
Water	0
Self-Reaction	1

12.3 **NFPA Hazard Classifications:** Not listed

13. PHYSICAL AND CHEMICAL PROPERTIES

13.1 **Physical State at 15°C and 1 atm:** Liquid

13.2 **Molecular Weight:** 128.22

13.3 **Boiling Point at 1 atm:** 307—352°F = 153—178°C = 426—451°K

13.4 **Freezing Point:** −180°F = −118°C = 155°K

13.5 **Critical Temperature:** Not pertinent

13.6 **Critical Pressure:** Not pertinent

13.7 **Specific Gravity:** 0.825 at 20°C (liquid)

13.8 **Liquid Surface Tension:** 26.9 dynes/cm = 0.0269 N/m at 20°C

13.9 **Liquid-Water Interfacial Tension:** (est.) 40 dynes/cm = 0.04 N/m at 20°C

13.10 **Vapor (Gas) Specific Gravity:** Not pertinent

13.11 **Ratio of Specific Heats of Vapor (Gas):** (est.) 1.040

13.12 **Latent Heat of Vaporization:** 140 Btu/lb = 77 cal/g = 3.2×10^5 J/kg

13.13 **Heat of Combustion:** (est.) −17,000 Btu/lb = −9600 cal/g = -400×10^5 J/kg

13.14 **Heat of Decomposition:** Not pertinent

13.15 **Heat of Solution:** Not pertinent

13.16 **Heat of Polymerization:** Not pertinent

NOTES

10A ISOOCTYL ALCOHOL

Common Synonyms			
6-Methyl-1-heptanol Oxo octyl alcohol	Liquid	Colorless	Mild odor
	Floats on water.		

Stop discharge if possible.
Call fire department.
Avoid contact with liquid.
Isolate and remove discharged material.
Notify local health and pollution control agencies.

Fire

Combustible.
Extinguish with dry chemical, foam, water, or carbon dioxide.
Cool exposed containers with water.

Exposure

CALL FOR MEDICAL AID.

LIQUID
Irritating to skin and eyes.
Harmful if swallowed.
Remove contaminated clothing.
Flush affected areas with plenty of water.
IF IN EYES, hold eyelids open and flush with plenty of water.
IF SWALLOWED and victim is CONSCIOUS, have victim drink water or milk.

Water Pollution

Effect of low concentrations on aquatic life is unknown.
Fouling to shoreline.
May be dangerous if it enters water intakes.
Notify local health and wildlife officials.
Notify operators of nearby water intakes.

1. RESPONSE TO DISCHARGE

(See Response Methods Handbook, CG 446-4)

Mechanical containment
Should be removed
Chemical and physical treatment

2. LABELS

No hazard label required by Code of Federal Regulations

3. CHEMICAL DESIGNATIONS

3.1 **Synonyms:** Dimethyl-1-hexanols
6-Methyl-1-heptanol

3.2 **Coast Guard Compatibility Classification:** Alcohol

3.3 **Chemical Formula:** $(CH_3)_2CH(CH_2)_4CH_2OH$

3.4 **IMCO/United Nations Numerical Designation:** Not listed

4. OBSERVABLE CHARACTERISTICS

4.1 **Physical State (as shipped):** Liquid

4.2 **Color:** Colorless

4.3 **Odor:** Mild; characteristic

5. HEALTH HAZARDS

5.1 **Personal Protective Equipment:** Air-supplied mask in confined areas; plastic gloves; goggles; eye bath and safety shower.

5.2 **Symptoms Following Exposure:** Inhalation hazard slight. Skin contact results in moderate irritation. Liquid contact with eyes causes severe irritation and possible eye damage.

5.3 **Treatment for Exposure:** Remove to fresh air. Flush skin and eye contact area at once for at least 15 min. Get medical care for eyes.

5.4 **Toxicity by Inhalation (Threshold Limit Value):** Data not available

5.5 **Short-Term Inhalation Limits:** Data not available

5.6 **Toxicity by Ingestion:** Grade 2; LD_{50} 0.5 to 5 g/kg (lab animals)

5.7 **Late Toxicity:** Data not available

5.8 **Vapor (Gas) Irritant Characteristics:** Vapors are nonirritating to the eyes and throat.

5.9 **Liquid or Solid Irritant Characteristics:** No appreciable hazard. Practically harmless to the skin.

5.10 **Odor Threshold:** Data not available

6. FIRE HAZARDS

6.1 **Flash Point:** 180°F O.C.

6.2 **Flammable Limits in Air:** 0.9% (calc.)—5.7% (est.)

6.3 **Fire Extinguishing Agents:** Water, foam, dry chemical, or carbon dioxide

6.4 **Fire Extinguishing Agents Not to be Used:** Not pertinent

6.5 **Special Hazards of Combustion Products:** Not pertinent

6.6 **Behavior in Fire:** Not pertinent

6.7 **Ignition Temperature:** 530°F (est.)

6.8 **Electrical Hazard:** Not pertinent

6.9 **Burning Rate:** Data not available

7. CHEMICAL REACTIVITY

7.1 **Reactivity with Water:** No reaction

7.2 **Reactivity with Common Materials:** No reaction

7.3 **Stability During Transport:** Stable

7.4 **Neutralizing Agents for Acids and Caustics:** Not pertinent

7.5 **Polymerization:** Not pertinent

7.6 **Inhibitor of Polymerization:** Not pertinent

8. WATER POLLUTION

8.1 **Aquatic Toxicity:** Data not available

8.2 **Waterfowl Toxicity:** Data not available

8.3 **Biological Oxygen Demand (BOD):** Data not available

8.4 **Food Chain Concentration Potential:** None

9. SELECTED MANUFACTURERS

1. Exxon Chemical Co.
Houston, Texas 77001
2. Getty Oil Co.
Los Angeles, Calif. 90054
3. Union Carbide Corp.
Chemicals and Plastics Division
270 Park Ave.
New York, N. Y. 10017

10. SHIPPING INFORMATION

10.1 **Grades or Purity:** 99+% (mixed isomers)

10.2 **Storage Temperature:** Ambient

10.3 **Inert Atmosphere:** No requirement

10.4 **Venting:** Open (flame arrester)

11. HAZARD ASSESSMENT CODE

(See Hazard Assessment Handbook, CG 446-3)

A-T-U

12. HAZARD CLASSIFICATIONS

12.1 **Code of Federal Regulations:** Combustible Liquid

12.2 **NAS Hazard Rating for Bulk Water Transportation:**

Category	Rating
Fire	1
Health	
Vapor Irritant	0
Liquid or Solid Irritant	0
Poisons	1
Water Pollution	
Human Toxicity	1
Aquatic Toxicity	3
Aesthetic Effect	2
Reactivity	
Other Chemicals	2
Water	0
Self-Reaction	0

12.3 **NFPA Hazard Classifications:** Not listed

13. PHYSICAL AND CHEMICAL PROPERTIES

13.1 **Physical State at 15°C and 1 atm:** Liquid

13.2 **Molecular Weight:** 130.22

13.3 **Boiling Point at 1 atm:** 367°F = 186°C = 459°K

13.4 **Freezing Point:** <212°F = <100°C = <373°K

13.5 **Critical Temperature:** Not pertinent

13.6 **Critical Pressure:** Not pertinent

13.7 **Specific Gravity:** 0.832 at 20°C (liquid)

13.8 **Liquid Surface Tension:** 29.5 dynes/cm = 0.0295 N/m at 20°C

13.9 **Liquid-Water Interfacial Tension:** (est.) 40 dynes/cm = 0.04 N/m at 20°C

13.10 **Vapor (Gas) Specific Gravity:** Not pertinent

13.11 **Ratio of Specific Heats of Vapor (Gas):** (est.) 1.040

13.12 **Latent Heat of Vaporization:** (est.) 140 Btu/lb = 77 cal/g = 3.2×10^5 J/kg

13.13 **Heat of Combustion:** (est.) −17,400 Btu/lb = −9650 cal/g = -404×10^5 J/kg

13.14 **Heat of Decomposition:** Not pertinent

13.15 **Heat of Solution:** Not pertinent

13.16 **Heat of Polymerization:** Not pertinent

NOTES

IPT	ISOPENTANE

Common Synonyms 2-Methylbutane	Watery liquid Colorless Gasoline-like odor Floats on water. Flammable, irritating vapor is produced. Boiling point is 82° F.

Wear goggles and self-contained breathing apparatus.
Stop discharge if possible. Keep people away.
Shut off ignition sources and call fire department.
Avoid contact with liquid and vapor.
Stay upwind and use water spray to "knock down" vapor.
Isolate and remove discharged material.
Notify local health and pollution control agencies.

Fire	FLAMMABLE. Flashback along vapor trail may occur. Vapor may explode if ignited in an enclosed area. Extinguish with dry chemical, foam, or carbon dioxide. Water may be ineffective on fire. Cool exposed containers and protect men effecting shutoff with water.
Exposure	CALL FOR MEDICAL AID. VAPOR Irritating to nose and throat. If inhaled, will cause coughing, difficult breathing, or loss of consciousness. Move to fresh air. If breathing has stopped, give artificial respiration. If breathing is difficult, give oxygen. LIQUID Irritating to skin and eyes. If swallowed, will cause nausea or vomiting. Remove contaminated clothing and shoes. Flush affected areas with plenty of water. IF IN EYES, hold eyelids open and flush with plenty of water. IF SWALLOWED and victim is CONSCIOUS, have victim drink water or milk. DO NOT INDUCE VOMITING.
Water Pollution	Effect of low concentrations on aquatic life is unknown. May be dangerous if it enters water intakes. Notify local health and wildlife officials. Notify operators of nearby water intakes.

1. RESPONSE TO DISCHARGE
(See Response Methods Handbook, CG 446-4)

Issue warning—high flammability
Evacuate area
Disperse and flush

2. LABEL

Red

3. CHEMICAL DESIGNATIONS

3.1 **Synonyms:** 2-Methylbutane
3.2 **Coast Guard Compatibility Classification:** Paraffin
3.3 **Chemical Formula:** $(CH_3)_2CHCH_2CH_3$
3.4 **IMCO/United Nations Numerical Designation:** 3.1/1217

4. OBSERVABLE CHARACTERISTICS

4.1 **Physical State (as shipped):** Liquid
4.2 **Color:** Colorless
4.3 **Odor:** Like gasoline

5. HEALTH HAZARDS

5.1 **Personal Protective Equipment:** Eye protection (as for gasoline)
5.2 **Symptoms Following Exposure:** Inhalation causes irritation of respiratory tract, cough, mild depression, irregular heartbeat. Aspiration causes severe lung irritation, coughing, pulmonary edema; excitement followed by depression. Ingestion causes nausea, vomiting, swelling of abdomen, headache, depression.
5.3 **Treatment for Exposure:** INHALATION: maintain respiration, give oxygen if needed. ASPIRATION: enforce bed rest; give oxygen. INGESTION: do NOT induce vomiting; call a doctor. EYES: wash with copious amounts of water. SKIN: wipe off, wash with soap and water.
5.4 **Toxicity by Inhalation (Threshold Limit Value):** Data not available
5.5 **Short-Term Inhalation Limits:** Data not available
5.6 **Toxicity by Ingestion:** Grade 1; LD_{50} 5 to 15 g/kg
5.7 **Late Toxicity:** None
5.8 **Vapor (Gas) Irritant Characteristics:** Vapors are nonirritating to the eyes and throat.
5.9 **Liquid or Solid Irritant Characteristics:** No appreciable hazard. Practically harmless to the skin.
5.10 **Odor Threshold:** Data not available

6. FIRE HAZARDS

6.1 **Flash Point:** −70°F C.C. (approx.)
6.2 **Flammable Limits in Air:** 1.4%—8.3%
6.3 **Fire Extinguishing Agents:** Dry chemical, foam, or carbon dioxide
6.4 **Fire Extinguishing Agents Not to be Used:** Water may be ineffective
6.5 **Special Hazards of Combustion Products:** Not pertinent
6.6 **Behavior in Fire:** Highly volatile liquid. Vapors may explode when mixed with air.
6.7 **Ignition Temperature:** 800°F
6.8 **Electrical Hazard:** Not pertinent
6.9 **Burning Rate:** 7.4 mm/min.

7. CHEMICAL REACTIVITY

7.1 **Reactivity with Water:** No reaction
7.2 **Reactivity with Common Materials:** No reaction
7.3 **Stability During Transport:** Stable
7.4 **Neutralizing Agents for Acids and Caustics:** Not pertinent
7.5 **Polymerization:** Not pertinent
7.6 **Inhibitor of Polymerization:** Not pertinent

8. WATER POLLUTION

8.1 **Aquatic Toxicity:** Data not available
8.2 **Waterfowl Toxicity:** Data not available
8.3 **Biological Oxygen Demand (BOD):** Data not available
8.4 **Food Chain Concentration Potential:** None

9. SELECTED MANUFACTURERS

1. Pennzoil United, Inc.
   Atlas Processing Co.
   3546 Midway St.
   Shreveport, La. 71109
2. Phillips Petroleum Co.
   Bartlesville, Okla. 74004

10. SHIPPING INFORMATION

10.1 **Grades or Purity:** Research: 99.99%; pure: 99.4%; technical: 97%
10.2 **Storage Temperature:** Ambient
10.3 **Inert Atmosphere:** No requirement
10.4 **Venting:** Open (flame arrester) or pressure-vacuum

11. HAZARD ASSESSMENT CODE
(See Hazard Assessment Handbook, CG 446-3)

A-T-U-V-W

12. HAZARD CLASSIFICATIONS

12.1 **Code of Federal Regulations:** Flammable liquid
12.2 **NAS Hazard Rating for Bulk Water Transportation:**

Category	Rating
Fire	4
Health	
Vapor Irritant	0
Liquid or Solid Irritant	0
Poisons	1
Water Pollution	
Human Toxicity	1
Aquatic Toxicity	2
Aesthetic Effect	1
Reactivity	
Other Chemicals	0
Water	0
Self-Reaction	0

12.3 **NFPA Hazard Classifications:** Not listed

13. PHYSICAL AND CHEMICAL PROPERTIES

13.1 **Physical State at 15°C and 1 atm:** Liquid
13.2 **Molecular Weight:** 72.15
13.3 **Boiling Point at 1 atm:** 82.2°F = 27.9°C = 301.1°K
13.4 **Freezing Point:** −255.8°F = −159.9°C = 113.3°K
13.5 **Critical Temperature:** 369.0°F = 187.2°C = 460.4°K
13.6 **Critical Pressure:** 491.0 psia = 33.4 atm = 3.38 MN/m^2
13.7 **Specific Gravity:** 0.620 at 20°C (liquid)
13.8 **Liquid Surface Tension:** 16.05 dynes/cm = 0.01605 N/m at 20°C
13.9 **Liquid-Water Interfacial Tension:** 31 dynes/cm = 0.031 N/m at 22.7°C
13.10 **Vapor (Gas) Specific Gravity:** 2.5
13.11 **Ratio of Specific Heats of Vapor (Gas):** 1.076
13.12 **Latent Heat of Vaporization:** 146 Btu/lb = 81.0 cal/g = 3.39×10^5 J/kg
13.13 **Heat of Combustion:** −19,314 Btu/lb = −10,730 cal/g = -449.24×10^5 J/kg
13.14 **Heat of Decomposition:** Not pertinent
13.15 **Heat of Solution:** Not pertinent
13.16 **Heat of Polymerization:** Not pertinent

NOTES

REVISED 1978

IPH ISOPHORONE

Common Synonyms
3,5,5-Trimethyl-2-cyclohexane-1-one

Liquid | Colorless | Camphor-like odor

Floats and mixes slowly with water.

Stop discharge if possible. Keep people away.
Call fire department.
Isolate and remove discharged material.
Notify local health and pollution control agencies.

Fire

Combustible.
Extinguish with dry chemicals, foam or carbon dioxide.
Water may be ineffective on fire.

Exposure

Call for medical aid.

LIQUID
Irritating to skin and eyes.
Harmful if swallowed.
Remove contaminated clothing and shoes.
Flush affected areas with plenty of water.
IF IN EYES, hold eyelids open and flush with plenty of water.
IF SWALLOWED and victim is CONSCIOUS, have victim drink water or milk.
DO NOT INDUCE VOMITING.

Water Pollution

Effect of low concentrations on aquatic life is unknown.
Fouling to shoreline.
May be dangerous if it enters water intakes.
Notify local health and wildlife officials.
Notify operators of nearby water intakes.

1. RESPONSE TO DISCHARGE

(See Response Methods Handbook, CG 446-4)

Issue warning—water contaminant
Restrict access
Mechanical containment
Should be removed
Chemical and physical treatment

2. LABELS

No label required by Code of Federal Regulations

3. CHEMICAL DESIGNATIONS

3.1 **Synonyms:** 3,5,5-Trimethyl-2-cyclohexene-1-one
3.2 **Coast Guard Compatibility Classification:** Ketone
3.3 **Chemical Formula:** $\underline{C}OCH{=}C(CH_3)CH_2C(CH_3)_2\underline{C}H_2$
3.4 **IMCO/United Nations Numerical Designation:** Not listed

4. OBSERVABLE CHARACTERISTICS

4.1 **Physical State (as shipped):** Liquid
4.2 **Color:** Colorless
4.3 **Odor:** Like camphor

5. HEALTH HAZARDS

5.1 **Personal Protective Equipment:** Self-contained breathing apparatus with full face mask; rubber gloves
5.2 **Symptoms Following Exposure:** Inhalation irritates eye, nose and throat; causes central depression and has some anesthetic effect. Contact of liquid with eyes causes severe irritation and possible tissue damage. Skin is irritated by liquid and may crack on prolonged contact. Ingestion causes irritation of mouth and stomach.
5.3 **Treatment for Exposure:** INHALATION: remove victim promptly from contaminated atmosphere; if breathing has stopped, give artificial respiration and oxygen. EYES: flood with water for at least 15 min.; consult an eye specialist as soon as possible. SKIN: flood with water. INGESTION: do NOT induce vomiting; call a doctor.
5.4 **Toxicity by Inhalation (Threshold Limit Value):** 10 ppm
5.5 **Short-Term Inhalation Limits:** Data not available
5.6 **Toxicity by Ingestion:** Grade 2; oral LD_{50} = 2,330 mg/kg (rat)
5.7 **Late Toxicity:** Data not available
5.8 **Vapor (Gas) Irritant Characteristics:** Vapors cause moderate irritation such that personnel will find high concentrations unpleasant. The effect is temporary.
5.9 **Liquid or Solid Irritant Characteristics:** Causes smarting of the skin and first-degree burns on short exposure; may cause second-degree burns on long exposure.
5.10 **Odor Threshold:** Data not available

6. FIRE HAZARDS

6.1 **Flash Point:** 205°F O.C., 184°F C.C.
6.2 **Flammable Limits in Air:** 0.84%–3.8%
6.3 **Fire Extinguishing Agents:** Dry chemical, foam, carbon dioxide
6.4 **Fire Extinguishing Agents Not to be Used:** Water may be ineffective
6.5 **Special Hazards of Combustion Products:** Not pertinent
6.6 **Behavior in Fire:** Not pertinent
6.7 **Ignition Temperature:** 864°F
6.8 **Electrical Hazard:** Data not available
6.9 **Burning Rate:** 4.0 mm/min.

7. CHEMICAL REACTIVITY

7.1 **Reactivity with Water:** No reaction
7.2 **Reactivity with Common Materials:** No reaction
7.3 **Stability During Transport:** Stable
7.4 **Neutralizing Agents for Acids and Caustics:** Not pertinent
7.5 **Polymerization:** Not pertinent
7.6 **Inhibitor of Polymerization:** Not pertinent

8. WATER POLLUTION

8.1 **Aquatic Toxicity:** 430 ppm/24 hr/brine shrimp/TL_m
8.2 **Waterfowl Toxicity:** Data not available
8.3 **Biological Oxygen Demand (BOD):** Data not available
8.4 **Food Chain Concentration Potential:** None

9. SELECTED MANUFACTURERS

1. Union Carbide Corporation
   Chemicals and Plastics Division
   270 Park Avenue
   New York, N. Y. 10017
2. Exxon Chemical Co.
   P. O. Box 3272
   Houston, Texas 77001

10. SHIPPING INFORMATION

10.1 **Grades or Purity:** 99+%
10.2 **Storage Temperature:** Ambient
10.3 **Inert Atmosphere:** No requirement
10.4 **Venting:** Open (flame arrester)

11. HAZARD ASSESSMENT CODE

(See Hazard Assessment Handbook, CG 446-3)

A-P-Q-T-U

12. HAZARD CLASSIFICATIONS

12.1 **Code of Federal Regulations:** Not listed
12.2 **NAS Hazard Rating for Bulk Water Transportation:**

Category	Rating
Fire	1
Health	
Vapor Irritant	2
Liquid or Solid Irritant	2
Poisons	2
Water Pollution	
Human Toxicity	1
Aquatic Toxicity	3
Aesthetic Effect	2
Reactivity	
Other Chemicals	2
Water	0
Self-Reaction	0

12.3 **NFPA Hazard Classifications:**

Category	Classification
Health Hazard (Blue)	2
Flammability (Red)	2
Reactivity (Yellow)	0

13. PHYSICAL AND CHEMICAL PROPERTIES

13.1 **Physical State at 15°C and 1 atm:** Liquid
13.2 **Molecular Weight:** 138.2
13.3 **Boiling Point at 1 atm:** 419.5°F=215.3°C=488.5°K
13.4 **Freezing Point:** 17.4°F = −8.1°C = 265.1°K
13.5 **Critical Temperature:** Not pertinent
13.6 **Critical Pressure:** Not pertinent
13.7 **Specific Gravity:** 0.921 at 25°C (liquid)
13.8 **Liquid Surface Tension:** 32.3 dynes/cm=0.0323 N/m at 20°C
13.9 **Liquid-Water Interfacial Tension:** Not pertinent
13.10 **Vapor (Gas) Specific Gravity:** 4.75
13.11 **Ratio of Specific Heats of Vapor (Gas):** Not pertinent
13.12 **Latent Heat of Vaporization:** 135 Btu/lb = 75 cal/g = 3.14×10^5 J/kg
13.13 **Heat of Combustion:** −16,170 Btu/lb = −8,980 cal/g = -376×10^5 J/kg
13.14 **Heat of Decomposition:** Not pertinent
13.15 **Heat of Solution:** Not pertinent
13.16 **Heat of Polymerization:** Not pertinent

NOTES

IPL	ISOPHTHALIC ACID

Common Synonyms Benzene-1,3-dicarboxylic acid m-Phthalic acid	Solid White Slight unpleasant odor Sinks in water.

Avoid contact with solid and dust. Keep people away.
Stop discharge if possible.
Shut off ignition sources. Call fire department.
Isolate and remove discharged material.
Notify local health and pollution control agencies.

Fire	Combustible. Dust cloud may explode if ignited in an enclosed area. Exgintuish with water, dry chemicals, foam, or carbon dioxide.
Exposure	CALL FOR MEDICAL AID. DUST Irritating to eyes, nose and throat. If inhaled will cause coughing or difficult breathing. If in eyes, hold eyelids open and flush with plenty of water. If breathing has stopped, give artificial respiration. If breathing is difficult, give oxygen. SOLID Irritating to skin and eyes. If swallowed will cause nausea. Remove contaminated clothing and shoes. Flush affected areas with plenty of water. IF IN EYES, hold eyelids open and flush with plenty of water. IF SWALLOWED and victim is CONSCIOUS, have victim drink water or milk. IF SWALLOWED and victim is UNCONSCIOUS OR HAVING CONVULSIONS, do nothing except keep victim warm.
Water Pollution	Effect of low concentrations on aquatic life is unknown. May be dangerous if it enters water intakes. Notify local health and wildlife officials. Notify operators of nearby water intakes.

1. RESPONSE TO DISCHARGE
(See Response Methods Handbook, CG 446-4)

Should be removed
Chemical and physical treatment

2. LABELS

No hazard label required by Code of Federal Regulations

3. CHEMICAL DESIGNATIONS

3.1 **Synonyms:** Benzene-1,3-dicarboxylic acid; m-Phthalic acid
3.2 **Coast Guard Compatibility Classification:** Not listed
3.3 **Chemical Formula:** 1,3-$C_6H_4(COOH)_2$
3.4 **IMCO/United Nations Numerical Designation:** Not listed

4. OBSERVABLE CHARACTERISTICS

4.1 **Physical State (as shipped):** Solid
4.2 **Color:** White
4.3 **Odor:** Slightly acrid

5. HEALTH HAZARDS

5.1 **Personal Protective Equipment:** If without adequate ventilation, use respirator with dust filter, goggles, and gloves
5.2 **Symptoms Following Exposure:** May cause slight to moderate irritation of eyes, skin, and mucous membranes on prolonged contact. Ingestion may cause gastrointestinal irritation.
5.3 **Treatment for Exposure:** INHALATION: remove victim to uncontaminated area; get medical attention if complications arise. INGESTION: get medical attention if complications arise. EYES: flush with large amounts of water for 15 min.; get prompt medical attention. SKIN: wash with water.
5.4 **Toxicity by Inhalation (Threshold Limit Value):** Data not available
5.5 **Short-Term Inhalation Limits:** Data not available
5.6 **Toxicity by Ingestion:** Grade 1; LD_{50} 12.2 g/kg (rat)
5.7 **Late Toxicity:** Data not available
5.8 **Vapor (Gas) Irritant Characteristics:** Data not available
5.9 **Liquid or Solid Irritant Characteristics:** Data not available
5.10 **Odor Threshold:** Data not available

6. FIRE HAZARDS

6.1 **Flash Point:** Not pertinent (combustible solid)
6.2 **Flammable Limits in Air:** Not pertinent
6.3 **Fire Extinguishing Agents:** Water, dry powder, foam, carbon dioxide
6.4 **Fire Extinguishing Agents Not to be Used:**
6.5 **Special Hazards of Combustion Products:**
6.6 **Behavior in Fire:** Dust forms explosive mixture in air.
6.7 **Ignition Temperature:** Data not available
6.8 **Electrical Hazard:** Not pertinent
6.9 **Burning Rate:** Not pertinent

7. CHEMICAL REACTIVITY

7.1 **Reactivity with Water:** No reaction
7.2 **Reactivity with Common Materials:**
7.3 **Stability During Transport:** Stable
7.4 **Neutralizing Agents for Acids and Caustics:** Not pertinent
7.5 **Polymerization:** Not pertinent
7.6 **Inhibitor of Polymerization:** Not pertinent

8. WATER POLLUTION

8.1 **Aquatic Toxicity:** Data not available
8.2 **Waterfowl Toxicity:** Data not available
8.3 **Biological Oxygen Demand (BOD):** Data not available
8.4 **Food Chain Concentration Potential:** None

9. SELECTED MANUFACTURERS

1. Amoco Chemicals Corp.
   130 East Randolph Drive
   Chicago, Ill. 60601
2. Arco Chemical Co.
   260 South Broad St.
   Philadelphia, Pa. 19101
3. Eastman Organic Chemicals
   Rochester, N. Y. 14650

10. SHIPPING INFORMATION

10.1 **Grades or Purity:** Technical, 82–95%
10.2 **Storage Temperature:** Ambient
10.3 **Inert Atmosphere:** Inerted
10.4 **Venting:** Safety relief

11. HAZARD ASSESSMENT CODE
(See Hazard Assessment Handbook, CG 446-3)

II

12. HAZARD CLASSIFICATIONS

12.1 **Code of Federal Regulations:** Not listed
12.2 **NAS Hazard Rating for Bulk Water Transportation:** Not listed
12.3 **NFPA Hazard Classifications:** Not listed

13. PHYSICAL AND CHEMICAL PROPERTIES

13.1 **Physical State at 15°C and 1 atm:** Solid
13.2 **Molecular Weight:** 166
13.3 **Boiling Point at 1 atm:** Not pertinent (sublimes)
13.4 **Freezing Point:** 653°F = 345°C = 618°K
13.5 **Critical Temperature:** Not pertinent
13.6 **Critical Pressure:** Not pertinent
13.7 **Specific Gravity:** 1.54 at 25°C (solid)
13.8 **Liquid Surface Tension:** Not pertinent
13.9 **Liquid-Water Interfacial Tension:** Not pertinent
13.10 **Vapor (Gas) Specific Gravity:** Not pertinent
13.11 **Ratio of Specific Heats of Vapor (Gas):** Not pertinent
13.12 **Latent Heat of Vaporization:** Not pertinent
13.13 **Heat of Combustion:** −8,340 Btu/lb = −4,630 cal/g = -194×10^5 J/kg
13.14 **Heat of Decomposition:** Not pertinent
13.15 **Heat of Solution:** Not pertinent
13.16 **Heat of Polymerization:** Not pertinent

NOTES

IPR	ISOPRENE

Common Synonyms 2-Methyl-1, 3-butadiene beta-Methylbivinyl	Watery liquid	Colorless	Mild odor
	Floats on water. Flammable, irritating vapor is formed. Boiling point is 93° F.		

Stop discharge if possible. Keep people away.
Shut off ignition sources and call fire department.
Stay upwind and use water spray to "knock down" vapor.
Avoid contact with liquid and vapor.
Isolate and remove discharged material.
Notify local health and pollution control agencies.

Fire	FLAMMABLE. Flashback along vapor trail may occur. Containers may explode in fire. Vapor may explode if ignited in an enclosed area. **Wear self-contained breathing apparatus.** **Combat fires from behind barrier or protected location.** **Extinguish with dry chemical, foam, or carbon dioxide.** **Water may be ineffective on fire.** **Cool exposed containers with water.**
Exposure	CALL FOR MEDICAL AID. VAPOR Irritating to eyes, nose and throat. **Move to fresh air.** **If breathing has stopped, give artificial respiration.** **If breathing is difficult, give oxygen.** LIQUID Irritating to skin and eyes. **Remove contaminated clothing and shoes.** **Flush affected areas with plenty of water.** **IF IN EYES, hold eyelids open and flush with plenty of water.**
Water Pollution	HARMFUL TO AQUATIC LIFE IN VERY LOW CONCENTRATIONS. May be dangerous if it enters water intakes. **Notify local health and wildlife officials.** **Notify operators of nearby water intakes.**

1. RESPONSE TO DISCHARGE

(See Response Methods Handbook, CG 446-4)

Issue warning—high flammability
Restrict access
Evacuate area

2. LABEL

Red

3. CHEMICAL DESIGNATIONS

3.1 **Synonyms:** beta-Methylbivinyl
2-Methyl-1, 3-butadiene

3.2 **Coast Guard Compatibility Classification:** Olefin

3.3 **Chemical Formula:** $CH_2=C(CH_3)CH=CH_2$

3.4 **IMCO/United Nations Numerical Designation:** 3.1/1218

4. OBSERVABLE CHARACTERISTICS

4.1 **Physical State (as shipped):** Liquid

4.2 **Color:** Colorless

4.3 **Odor:** Mild, aromatic

5. HEALTH HAZARDS

5.1 **Personal Protective Equipment:** Vapor-proof goggles; self-contained breathing apparatus; leather or rubber safety shoes; rubber gloves.

5.2 **Symptoms Following Exposure:** Vapor produces no effects other than slight irritation of the eyes and upper respiratory tract. Liquid may irritate eyes; like gasoline.

5.3 **Treatment for Exposure:** INHALATION: remove victim promptly from irritating or asphyxiating atmosphere; if symptoms of asphyxiation persist, administer artificial respiration and oxygen; treat symptomatically thereafter; call a physician. EYES: flush with water for at least 15 min.

5.4 **Toxicity by Inhalation (Threshold Limit Value):** Data not available

5.5 **Short-Term Inhalation Limits:** Data not available

5.6 **Toxicity by Ingestion:** Data not available

5.7 **Late Toxicity:** None

5.8 **Vapor (Gas) Irritant Characteristics:** Vapors cause a slight smarting of the eyes or respiratory system if present in high concentrations. The effect is temporary.

5.9 **Liquid or Solid Irritant Characteristics:** Minumum hazard. If spilled on clothing and allowed to remain, may cause smarting and reddening of the skin.

5.10 **Odor Threshold:** 0.005 ppm

6. FIRE HAZARDS

6.1 **Flash Point:** −65°F C.C.

6.2 **Flammable Limits in Air:** 2%—9%

6.3 **Fire Extinguishing Agents:** Dry chemical, foam, or carbon dioxide

6.4 **Fire Extinguishing Agents Not to be Used:** Water may be ineffective

6.5 **Special Hazards of Combustion Products:** Toxic vapors are generated when heated

6.6 **Behavior in Fire:** May polymerize in containers and explode

6.7 **Ignition Temperature:** 428°F

6.8 **Electrical Hazard:** Class I, Group C

6.9 **Burning Rate:** 8.6 mm/min.

7. CHEMICAL REACTIVITY

7.1 **Reactivity with Water:** No reaction

7.2 **Reactivity with Common Materials:** No reaction

7.3 **Stability During Transport:** Stable

7.4 **Neutralizing Agents for Acids and Caustics:** Not pertinent

7.5 **Polymerization:** Polymerization is accelerated by heat and by oxygen, even by the presence of rusty iron. Iron surfaces should be treated with a suitable reducing agent, such as sodium nitrite, before they are placed into isoprene service.

7.6 **Inhibitor of Polymerization:** Tertiary butyl catechol (0.06%). Di-n-butylamine, phenyl-beta-naphthylamine and phenyl-alpha-naphthylamine are also used. IPR must be inhibited when transported interstate

8. WATER POLLUTION

8.1 **Aquatic Toxicity:** 75 ppm/96 hr/fathead minnow/TL_m/fresh water

8.2 **Waterfowl Toxicity:** Data not available

8.3 **Biological Oxygen Demand (BOD):** Data not available

8.4 **Food Chain Concentration Potential:** None

9. SELECTED MANUFACTURERS

1. B. F. Goodrich Co.
   B. F. Goodrich Chemical Co. Division
   6100 Oak Tree Blvd.
   Cleveland, Ohio 44131
2. Goodyear Tire & Rubber Co.
   Chemical Division
   Beaumont, Texas 77704
3. Shell Chemical Co.
   Petrochemical Division
   Houston, Texas 77001

10. SHIPPING INFORMATION

10.1 **Grades or Purity:** Research grade: 99.99%; polymerization grade: 99.8%

10.2 **Storage Temperature:** Ambient

10.3 **Inert Atmosphere:** No requirement

10.4 **Venting:** Pressure-vacuum

11. HAZARD ASSESSMENT CODE

(See Hazard Assessment Handbook, CG 446-3)

A-T-U-V-W-Z

12. HAZARD CLASSIFICATIONS

12.1 **Code of Federal Regulations:** Flammable liquid

12.2 **NAS Hazard Rating for Bulk Water Transportation:**

Category	Rating
Fire	4
Health	
Vapor Irritant	1
Liquid or Solid Irritant	1
Poisons	1
Water Pollution	
Human Toxicity	0
Aquatic Toxicity	1
Aesthetic Effect	1
Reactivity	
Other Chemicals	2
Water	0
Self-Reaction	3

12.3 **NFPA Hazard Classifications:**

Category	Classification
Health Hazard (Blue)	2
Flammability (Red)	4
Reactivity (Yellow)	2

13. PHYSICAL AND CHEMICAL PROPERTIES

13.1 **Physical State at 15°C and 1 atm:** Liquid

13.2 **Molecular Weight:** 68.12

13.3 **Boiling Point at 1 atm:** 93.4°F = 34.1°C = 307.3°K

13.4 **Freezing Point:** −230.7°F = −145.9°C = 127.3°K

13.5 **Critical Temperature:** 412°F = 211.1°C = 484.3°K

13.6 **Critical Pressure:** 550 psia = 37.4 atm = 3.79 MN/m^2

13.7 **Specific Gravity:** 0.681 at 20°C (liquid)

13.8 **Liquid Surface Tension:** 16.9 dynes/cm = 0.0169 N/m at 20°C

13.9 **Liquid-Water Interfacial Tension:** (est.) 40 dynes/cm = 0.04 N/m at 20°C

13.10 **Vapor (Gas) Specific Gravity:** 2.3

13.11 **Ratio of Specific Heats of Vapor (Gas):** 1.091

13.12 **Latent Heat of Vaporization:** 150 Btu/lb = 85 cal/g = 3.6×10^5 J/kg

13.13 **Heat of Combustion:** −18,848 Btu/lb = −10,471 cal/g = -438.40×10^5 J/kg

13.14 **Heat of Decomposition:** Not pertinent

13.15 **Heat of Solution:** Not pertinent

13.16 **Heat of Polymerization:** −499 Btu/lb = −277 cal/g = -11.6×10^5 J/kg

NOTES

REVISED 1978

IAC	ISOPROPYL ACETATE

Common Synonyms			
Acetic acid, isopropyl ester 2-Propyl acetate	Watery liquid	Colorless	Pleasant fruity odor
	Floats and mixes slowly with water. Flammable, irritating vapor is produced.		

Stop discharge if possible. Keep people away.
Shut off ignition sources and call fire department.
Stay upwind and use water spray to "knock down" vapor.
Avoid contact with liquid and vapor.
Isolate and remove discharged material.
Notify local health and pollution control agencies.

Fire

FLAMMABLE.
Flashback along vapor trail may occur.
Vapor may explode if ignited in an enclosed area.

Extinguish with dry chemical, foam, or carbon dioxide.
Water may be ineffective on fire.
Cool exposed containers with water.

Exposure

CALL FOR MEDICAL AID.
VAPOR
Irritating to eyes, nose and throat.

Move to fresh air.
If breathing has stopped, give artificial respiration.
If breathing is difficult, give oxygen.

LIQUID
Irritating to skin and eyes.
Harmful if swallowed.

Remove contaminated clothing.
Flush affected areas with plenty of water.
IF IN EYES, hold eyelids open and flush with plenty of water.
IF SWALLOWED and victim is CONSCIOUS, have victim drink water or milk.

Water Pollution

Effect of low concentrations on aquatic life is unknown.
May be dangerous if it enters water intakes.

Notify local health and wildlife officials.
Notify operators of nearby water intakes.

1. RESPONSE TO DISCHARGE

(See Response Methods Handbook, CG 446-4)

Issue warning—high flammability
Evacuate area
Disperse and flush

2. LABEL

Red

3. CHEMICAL DESIGNATIONS

3.1 **Synonyms:** Acetic acid, isopropyl ester
2-Propyl acetate
3.2 **Coast Guard Compatibility Classification:** Ester
3.3 **Chemical Formula:** $CH_3COOCH(CH_3)_2$
3.4 **IMCO/United Nations Numerical Designation:** 3.2/1220

4. OBSERVABLE CHARACTERISTICS

4.1 **Physical State (as shipped):** Liquid
4.2 **Color:** Colorless
4.3 **Odor:** Pleasant, fruity; nonresidual

5. HEALTH HAZARDS

5.1 **Personal Protective Equipment:** Organic vapor canister or air-supplied mask; chemical goggles or face splash shield.
5.2 **Symptoms Following Exposure:** Vapors irritate eyes and respiratory tract; high concentrations can be anesthetic. Liquid irritates eyes but causes no serious injury; may cause dermatitis; no serious effects if swallowed.
5.3 **Treatment for Exposure:** INHALATION: if victim is overcome by vapors, remove from exposure immediately; call a physician; if breathing is irregular or stopped, start resuscitation and administer oxygen. EYES: flush with water for at least 15 min.
5.4 **Toxicity by Inhalation (Threshold Limit Value):** 250 ppm
5.5 **Short-Term Inhalation Limits:** Data not available
5.6 **Toxicity by Ingestion:** Grade 2; LD_{50} 0.5 to 5 g/kg (rat)
5.7 **Late Toxicity:** Data not available
5.8 **Vapor (Gas) Irritant Characteristics:** Vapors cause a slight smarting of the eyes or respiratory system if present in high concentrations. The effect is temporary.
5.9 **Liquid or Solid Irritant Characteristics:** Minimum hazard. If spilled on clothing and allowed to remain, may cause smarting and reddening of the skin.
5.10 **Odor Threshold:** Data not available

6. FIRE HAZARDS

6.1 **Flash Point:** 37°F C.C.; 60°F O.C.
6.2 **Flammable Limits in Air:** 1.8%–8.0%
6.3 **Fire Extinguishing Agents:** Alcohol foam, dry chemical, carbon dioxide
6.4 **Fire Extinguishing Agents Not to be Used:** Not pertinent
6.5 **Special Hazards of Combustion Products:** Not pertinent
6.6 **Behavior in Fire:** Not pertinent
6.7 **Ignition Temperature:** 860°F
6.8 **Electrical Hazard:** Not pertinent
6.9 **Burning Rate:** Data not available

7. CHEMICAL REACTIVITY

7.1 **Reactivity with Water:** No reaction
7.2 **Reactivity with Common Materials:** No reaction
7.3 **Stability During Transport:** Stable
7.4 **Neutralizing Agents for Acids and Caustics:** Not pertinent
7.5 **Polymerization:** Not pertinent
7.6 **Inhibitor of Polymerization:** Not pertinent

8. WATER POLLUTION

8.1 **Aquatic Toxicity:** Data not available
8.2 **Waterfowl Toxicity:** Data not available
8.3 **Biological Oxygen Demand (BOD):** 26%, 5 days (theor.)
8.4 **Food Chain Concentration Potential:** None

9. SELECTED MANUFACTURERS

1. Eastman Kodak Co.
Tennessee Eastman Co. Division
Kingsport, Tenn. 37662
2. Exxon Chemical Co.
Houston, Texas 77001
3. Union Carbide Corp.
Chemicals and Plastics Division
270 Park Ave.
New York, N. Y. 10017

10. SHIPPING INFORMATION

10.1 **Grades or Purity:** 95-99+%
10.2 **Storage Temperature:** Ambient
10.3 **Inert Atmosphere:** No requirement
10.4 **Venting:** Open (flame arrester) or pressure-vacuum

11. HAZARD ASSESSMENT CODE

(See Hazard Assessment Handbook, CG 446-3)

A-P-Q-T-U-V-W

12. HAZARD CLASSIFICATIONS

12.1 **Code of Federal Regulations:** Flammable liquid
12.2 **NAS Hazard Rating for Bulk Water Transportation:**

Category	Rating
Fire	3
Health	
Vapor Irritant	1
Liquid or Solid Irritant	1
Poisons	2
Water Pollution	
Human Toxicity	1
Aquatic Toxicity	1
Aesthetic Effect	2
Reactivity	
Other Chemicals	1
Water	0
Self-Reaction	0

12.3 **NFPA Hazard Classifications:**

Category	Classification
Health Hazard (Blue)	1
Flammability (Red)	3
Reactivity (Yellow)	0

13. PHYSICAL AND CHEMICAL PROPERTIES

13.1 **Physical State at 15°C and 1 atm:** Liquid
13.2 **Molecular Weight:** 102.13
13.3 **Boiling Point at 1 atm:** 191.3°F = 88.5°C = 361.7°K
13.4 **Freezing Point:** −92.7°F = −69.3°C = 203.9°K
13.5 **Critical Temperature:** 509°F = 265°C = 538°K
13.6 **Critical Pressure:** 529 psia = 36 atm = 3.65 MN/m²
13.7 **Specific Gravity:** 0.874 at 20°C (liquid)
13.8 **Liquid Surface Tension:** 26 dynes/cm = 0.026 N/m at 20°C
13.9 **Liquid-Water Interfacial Tension:** Data not available
13.10 **Vapor (Gas) Specific Gravity:** 3.5
13.11 **Ratio of Specific Heats of Vapor (Gas):** (est.) 1.074
13.12 **Latent Heat of Vaporization:** 150 Btu/lb = 81 cal/g = 3.4×10^5 J/kg
13.13 **Heat of Combustion:** -9420 Btu/lb = -5230 cal/g = -219×10^5 J/kg
13.14 **Heat of Decomposition:** Not pertinent
13.15 **Heat of Solution:** Not pertinent
13.16 **Heat of Polymerization:** Not pertinent

NOTES

REVISED 1978

IPA	ISOPROPYL ALCOHOL

Common Synonyms			
Isopropanol 2-Propanol Dimethylcarbinol sec-Propyl alcohol Rubbing alcohol	Watery liquid	Colorless	Unpleasant alcohol odor like rubbing alcohol
	Floats and mixes with water. Flammable, irritating vapor is produced.		

Stop discharge if possible. Keep people away.
Shut off ignition sources and call fire department.
Stay upwind and use water spray to "knock down" vapor.
Avoid contact with liquid and vapor.
Isolate and remove discharged material.
Notify local health and pollution control agencies.

Fire	FLAMMABLE. Flashback along vapor trail may occur. Vapor may explode if ignited in an enclosed area. Extinguish with dry chemical, alcohol foam, or carbon dioxide. Water may be ineffective on fire. Cool exposed containers with water.
Exposure	CALL FOR MEDICAL AID. VAPOR Irritating to eyes, nose and throat. Move to fresh air. If breathing has stopped, give artificial respiration. If breathing is difficult, give oxygen. LIQUID Irritating to eyes. Harmful if swallowed. IF IN EYES, hold eyelids open and flush with plenty of water. IF SWALLOWED and victim is CONSCIOUS, have victim drink water or milk.
Water Pollution	Dangerous to aquatic life in high concentrations. May be dangerous if it enters water intakes. Notify local health and wildlife officials. Notify operators of nearby water intakes.

1. RESPONSE TO DISCHARGE

(See Response Methods Handbook, CG 446-4)

Issue warning—high flammability
Disperse and flush

2. LABEL

Red

3. CHEMICAL DESIGNATIONS

3.1 **Synonyms:** Dimethylcarbinol; Isopropanol; Petrohol; 2-Propanol; sec-Propyl alcohol; Rubbing alcohol

3.2 **Coast Guard Compatibility Classification:** Alcohol

3.3 **Chemical Formula:** $CH_3CH(OH)CH_3$

3.4 **IMCO/United Nations Numerical Designation:** 3.2/1219

4. OBSERVABLE CHARACTERISTICS

4.1 **Physical State (as shipped):** Liquid

4.2 **Color:** Colorless

4.3 **Odor:** Like ethyl alcohol; sharp, somewhat unpleasant; characteristic mild alcoholic; nonresidual

5. HEALTH HAZARDS

5.1 **Personal Protective Equipment:** Organic vapor canister or air-supplied mask; chemical goggles or face splash shield.

5.2 **Symptoms Following Exposure:** Vapors cause mild irritation of eyes and upper respiratory tract; high concentrations may be anesthetic. Liquid irritates eyes and may cause injury; harmless to skin; if ingested causes drunkenness and vomiting.

5.3 **Treatment for Exposure:** INHALATION: if victim is overcome by vapors, remove from exposure immediately; call a physician; if breathing is irregular or has stopped, start resuscitation and administer oxygen. EYES: flush with water for at least 15 min.

5.4 **Toxicity by Inhalation (Threshold Limit Value):** 400 ppm

5.5 **Short-Term Inhalation Limits:** 400 ppm for 10 min.

5.6 **Toxicity by Ingestion:** Grade 1; LD_{50} 5 to 15 g/kg (rat: LD_{50}: 5.84 g/kg)

5.7 **Late Toxicity:** Data not available

5.8 **Vapor (Gas) Irritant Characteristics:** Vapors cause a slight smarting of the eyes or respiratory system if present in high concentrations. The effect is temporary.

5.9 **Liquid or Solid Irritant Characteristics:** No appreciable hazard. Practically harmless to the skin.

5.10 **Odor Threshold:** 90 mg/m³

6. FIRE HAZARDS

6.1 **Flash Point:** 65°F O.C.; 53°F C.C.

6.2 **Flammable Limits in Air:** 2.3%—12.7%

6.3 **Fire Extinguishing Agents:** Alcohol foam, dry chemical, or carbon dioxide

6.4 **Fire Extinguishing Agents Not to be Used:** Water may be ineffective

6.5 **Special Hazards of Combustion Products:** Not pertinent

6.6 **Behavior in Fire:** Not pertinent

6.7 **Ignition Temperature:** 750°F

6.8 **Electrical Hazard:** Class I, Group D

6.9 **Burning Rate:** 2.3 mm/min.

7. CHEMICAL REACTIVITY

7.1 **Reactivity with Water:** No reaction

7.2 **Reactivity with Common Materials:** No reaction

7.3 **Stability During Transport:** Stable

7.4 **Neutralizing Agents for Acids and Caustics:** Not pertinent

7.5 **Polymerization:** Not pertinent

7.6 **Inhibitor of Polymerization:** Not pertinent

8. WATER POLLUTION

8.1 **Aquatic Toxicity:** 900-1100 ppm/24 hr/chub/critical range/fresh water

8.2 **Waterfowl Toxicity:** Data not available

8.3 **Biological Oxygen Demand (BOD):** 133%, 5 days

8.4 **Food Chain Concentration Potential:** None

9. SELECTED MANUFACTURERS

1. ARCO Chemical Co.
   260 South Broad St.
   Philadelphia, Pa. 19101
2. Shell Chemical Co.
   Industrial Chemical Division
   Houston, Tex. 77001
3. Union Carbide Corp.
   Chemicals and Plastics Division
   270 Park Ave.
   New York, N. Y. 10017

10. SHIPPING INFORMATION

10.1 **Grades or Purity:** 91%, 95%, Anhydrous

10.2 **Storage Temperature:** Ambient

10.3 **Inert Atmosphere:** No requirement

10.4 **Venting:** Open (flame arrester) or pressure-vacuum

11. HAZARD ASSESSMENT CODE

(See Hazard Assessment Handbook, CG 446-3)

A-P-Q-R-S

12. HAZARD CLASSIFICATIONS

12.1 **Code of Federal Regulations:** Flammable liquid

12.2 **NAS Hazard Rating for Bulk Water Transportation:**

Category	Rating
Fire	3
Health	
Vapor Irritant	1
Liquid or Solid Irritant	0
Poisons	2
Water Pollution	
Human Toxicity	2
Aquatic Toxicity	2
Aesthetic Effect	1
Reactivity	
Other Chemicals	2
Water	0
Self-Reaction	0

12.3 **NFPA Hazard Classifications:**

Category	Classification
Health Hazard (Blue)	1
Flammability (Red)	3
Reactivity (Yellow)	0

13. PHYSICAL AND CHEMICAL PROPERTIES

13.1 **Physical State at 15°C and 1 atm:** Liquid

13.2 **Molecular Weight:** 60.10

13.3 **Boiling Point at 1 atm:** 180.1°F = 82.3°C = 355.5°K

13.4 **Freezing Point:** −127.3°F = −88.5°C = 184.7°K

13.5 **Critical Temperature:** 455.4°F = 235.2°C = 508.4°K

13.6 **Critical Pressure:** 691 psia = 47.0 atm = 4.76 MN/m²

13.7 **Specific Gravity:** 0.785 at 20°C (liquid)

13.8 **Liquid Surface Tension:** Not pertinent

13.9 **Liquid-Water Interfacial Tension:** Not pertinent

13.10 **Vapor (Gas) Specific Gravity:** 2.1

13.11 **Ratio of Specific Heats of Vapor (Gas):** 1.105

13.12 **Latent Heat of Vaporization:** 286 Btu/lb = 159 cal/g = 6.66×10^5 J/kg

13.13 **Heat of Combustion:** −12,960 Btu/lb = −7,201 cal/g = -301.5×10^5 J/kg

13.14 **Heat of Decomposition:** Not pertinent

13.15 **Heat of Solution:** (est.) −9 Btu/lb = −5 cal/g = -0.2×10^5 J/kg

13.16 **Heat of Polymerization:** Not pertinent

NOTES

REVISED 1978

IPP	# ISOPROPYLAMINE

Common Synonyms			
2-Aminopropane Monoisopropylamine	Liquid	Colorless	Strong ammonia odor
	Floats and mixes with water. Flammable, irritating vapor is produced. Boiling point is 91°F.		

Stop discharge if possible. Keep people away.
Shut off ignition sources. Call fire department.
Stay upwind. Use water spray to "knock down" vapor.
Avoid contact with liquid and vapor.
Isolate and remove discharged material.
Notify local health and pollution control agencies.

Fire

FLAMMABLE.
POISONOUS GASES MAY BE PRODUCED IN FIRE.
Flashback along vapor trail may occur.
Vapor may explode if ignited in an enclosed area.

Wear goggles and self-contained breathing apparatus.
Extinguish with dry chemicals, alcohol foam, or carbon dioxide.
Water may be ineffective on fire.
Cool exposed containers with water.

Exposure

CALL FOR MEDICAL AID.

VAPOR
Irritating to eyes, nose and throat.
If inhaled will cause coughing, difficult breathing or loss of consciousness.

If in eyes, hold eyelids open and flush with plenty of water.
If breathing has stopped, give artificial respiration.
If breathing is difficult, give oxygen.

LIQUID
Will burn skin and eyes.
If swallowed will cause nausea.

Remove contaminated clothing and shoes.
Flush affected areas with plenty of water.
IF IN EYES, hold eyelids open and flush with plenty of water.
IF SWALLOWED and victim is CONSCIOUS, have victim drink water or milk.
IF SWALLOWED and victim is UNCONSCIOUS OR HAVING CONVULSIONS, do nothing except keep victim warm.

Water Pollution

HARMFUL TO AQUATIC LIFE IN VERY LOW CONCENTRATIONS.
May be dangerous if it enters water intakes.

Notify local health and wildlife officials.
Notify operators of nearby water intakes.

1. RESPONSE TO DISCHARGE

(See Response Methods Handbook, CG 446-4)

Issue warning—high flammability, air contaminant, water contaminant
Restrict access
Evacuate area
Disperse and flush

2. LABEL

Red

3. CHEMICAL DESIGNATIONS

3.1 **Synonyms:** 2-Aminopropane; Monoisopropylamine; iso-Propylamine
3.2 **Coast Guard Compatibility Classification:** Data not available
3.3 **Chemical Formula:** $(CH_3)_2CHNH_2$
3.4 **IMCO/United Nations Numerical Designation:** 3.1/1221

4. OBSERVABLE CHARACTERISTICS

4.1 **Physical State (as shipped):** Liquid
4.2 **Color:** Colorless
4.3 **Odor:** Strong ammoniacal; pungent, irritating, typical amine

5. HEALTH HAZARDS

5.1 **Personal Protective Equipment:** Self-contained breathing apparatus; butyl rubber gloves and apron; chemical face shield or safety goggles
5.2 **Symptoms Following Exposure:** Inhalation causes nose and throat irritation, severe coughing, and chest pain due to irritation of air passages; can cause lung edema and loss of consciousness. Ingestion causes nausea, salivation and severe irritation of mouth and stomach. Contact with eyes causes severe irritation and possible edema of the cornea. Contact with skin causes severe irritation.
5.3 **Treatment for Exposure:** INHALATION: remove victim to fresh air; if he is not breathing, give artificial respiration; if breathing is difficult, give oxygen; call a physician. INGESTION: call a physician immediately; encourage the drinking of large quantities of water followed by dilute vinegar, lemon juice, cider, or other weak acids; keep patient warm. EYES: flush with water for 15 min., holding eyelids apart; call physician as soon as possible, preferably an eye specialist. SKIN: flush with water.
5.4 **Toxicity by Inhalation (Threshold Limit Value):** 5 ppm
5.5 **Short-Term Inhalation Limits:** Data not available
5.6 **Toxicity by Ingestion:** Grade 2; oral LD_{50} = 820 mg/kg (rat), 600 mg/kg (mouse)
5.7 **Late Toxicity:** Data not available
5.8 **Vapor (Gas) Irritant Characteristics:** Vapors are moderately irritating, such that personnel will not usually tolerate moderate or high concentrations.
5.9 **Liquid or Solid Irritant Characteristics:** Causes smarting of the skin and first-degree burns on short exposure; may cause second-degree burns on long exposure.
5.10 **Odor Threshold:** 5 ppm

6. FIRE HAZARDS

6.1 **Flash Point:** −15°F O.C.
6.2 **Flammable Limits in Air:** 2.3%–12%
6.3 **Fire Extinguishing Agents:** Dry chemical, "alcohol" foam, carbon dioxide
6.4 **Fire Extinguishing Agents Not to be Used:** Water may be ineffective.
6.5 **Special Hazards of Combustion Products:** Toxic oxides of nitrogen may form in fire.
6.6 **Behavior in Fire:** Burning isopropylamine is difficult to control because of the ease of reignition of the vapor. Vapors are heavier than air and may travel to a source of ignition and flash back. Containers may explode.
6.7 **Ignition Temperature:** 756°F
6.8 **Electrical Hazard:** Data not available
6.9 **Burning Rate:** 6.33 mm/min.

7. CHEMICAL REACTIVITY

7.1 **Reactivity with Water:** No reaction
7.2 **Reactivity with Common Materials:**
7.3 **Stability During Transport:** Stable
7.4 **Neutralizing Agents for Acids and Caustics:** Not pertinent
7.5 **Polymerization:** Not pertinent
7.6 **Inhibitor of Polymerization:** Not pertinent

8. WATER POLLUTION

8.1 **Aquatic Toxicity:** 40–80 ppm/24 hr/creek chub/critical range/fresh water
8.2 **Waterfowl Toxicity:** Data not available
8.3 **Biological Oxygen Demand (BOD):** Data not available
8.4 **Food Chain Concentration Potential:** None

9. SELECTED MANUFACTURERS

1. Air Products and Chemicals, Inc.
U. S. Highway 90
Pace, Fla. 32502
2. Pennwalt Corporation
Three Parkway
Philadelphia, Pa. 19102
3. Virginia Chemicals, Inc.
3340 W. Norfolk Rd.
Portsmouth, Va. 23703

10. SHIPPING INFORMATION

10.1 **Grades or Purity:** Technical, 99.0%
10.2 **Storage Temperature:** Ambient
10.3 **Inert Atmosphere:** No requirement
10.4 **Venting:** Open (flame arrester)

11. HAZARD ASSESSMENT CODE

(See Hazard Assessment Handbook, CG 446-3)

A-P-Q-R-S

12. HAZARD CLASSIFICATIONS

12.1 **Code of Federal Regulations:** Flammable liquid
12.2 **NAS Hazard Rating for Bulk Water Transportation:**

Category	Rating
Fire	4
Health	
Vapor Irritant	3
Liquid or Solid Irritant	2
Poisons	4
Water Pollution	
Human Toxicity	2
Aquatic Toxicity	3
Aesthetic Effect	2
Reactivity	
Other Chemicals	3
Water	0
Self-Reaction	0

12.3 **NFPA Hazard Classifications:**

Category	Classification
Health Hazard (Blue)	3
Flammability (Red)	4
Reactivity (Yellow)	0

13. PHYSICAL AND CHEMICAL PROPERTIES

13.1 **Physical State at 15°C and 1 atm:** Liquid
13.2 **Molecular Weight:** 59.11
13.3 **Boiling Point at 1 atm:** 90.3°F = 32.4°C = 305.6°K
13.4 **Freezing Point:** −139°F = −95°C = 178°K
13.5 **Critical Temperature:** 396°F = 202°C = 475°K
13.6 **Critical Pressure:** 740 psia = 50 atm = 5.1 MN/m²
13.7 **Specific Gravity:** (est.) 0.691 at 20°C (liquid)
13.8 **Liquid Surface Tension:** 16.8 dynes/cm = 0.0168 N/m at 20°C
13.9 **Liquid-Water Interfacial Tension:** Not pertinent
13.10 **Vapor (Gas) Specific Gravity:** 2.04
13.11 **Ratio of Specific Heats of Vapor (Gas):**
13.12 **Latent Heat of Vaporization:** 193 Btu/lb = 107 cal/g = 4.48 × 10⁵ J/kg
13.13 **Heat of Combustion:** −16,940 Btu/lb = −9,420 cal/g = −394 × 10⁵ J/kg
13.14 **Heat of Decomposition:** Not pertinent
13.15 **Heat of Solution:** −210 Btu/lb = −110 cal/g = −4.8 × 10⁵ J/kg
13.16 **Heat of Polymerization:** Not pertinent

NOTES

IPE — ISOPROPYL ETHER

Common Synonyms			
Diisopropyl ether Diisopropyl oxide 2-Isopropoxypropane	Liquid	Colorless	Sweet odor
	Floats and mixes slowly with water. Flammable, irritating vapor is produced		

Shut off ignition sources. Call fire department.
Stop discharge if possible. Keep people away.
Stay upwind. Use water spray to "knock down" vapor.
Isolate and remove discharged material.
Notify local health and pollution control agencies.

Fire

FLAMMABLE.
Containers may explode in fire.
Flashback along vapor trail may occur.
Vapor may explode if ignited in an enclosed area.
Extinguish with dry chemicals, alcohol foam, or carbon dioxide
Water may be ineffective on fire.
Cool exposed containers with water.

Exposure

Call for medical aid.

VAPOR
Irritating to eyes, nose and throat.
If inhaled will cause headache, dizziness, or nausea.
Move victim to fresh air.
If breathing has stopped, give artificial respiration.
If breathing is difficult, give oxygen.

LIQUID
Irritating to skin and eyes.
Remove contaminated clothing and shoes.
Flush affected areas with plenty of water.
IF IN EYES, hold eyelids open and flush with plenty of water.
IF SWALLOWED and victim is CONSCIOUS, have victim drink water or milk.

Water Pollution

Effect of low concentrations on aquatic life is unknown.
Fouling to shoreline.
May be dangerous if it enters water intakes.
Notify local health and wildlife officials.
Notify operators of nearby water intakes.

1. RESPONSE TO DISCHARGE

(See Response Methods Handbook, CG 446-4)

Issue warning—high flammability
Restrict access
Evacuate area
Disperse and flush

2. LABEL

Red

3. CHEMICAL DESIGNATIONS

3.1 **Synonyms:** Diisopropyl ether; Diisopropyl oxide; 2-Isopropoxypropane
3.2 **Coast Guard Compatibility Classification:** Ethers (21)
3.3 **Chemical Formula:** $(CH_3)_2CHOCH(CH_3)_2$
3.4 **IMCO/United Nations Numerical Designation:** 3.1/1159

4. OBSERVABLE CHARACTERISTICS

4.1 **Physical State (as shipped):** Liquid
4.2 **Color:** Colorless
4.3 **Odor:** Sweet, slightly sharp; characteristic pungent; ethereal; like camphor and ethyl ether

5. HEALTH HAZARDS

5.1 **Personal Protective Equipment:** Air pack or organic canister mask; rubber gloves; goggles.
5.2 **Symptoms Following Exposure:** Inhalation causes anesthesia, nausea, headache, dizziness, and irritation of the eyes and nose. Contact of liquid with eyes causes only minor injury; repeated contact with skin will remove natural oils and may cause dermatitis.
5.3 **Treatment for Exposure:** INHALATION: remove victim to fresh air and obtain medical attention immediately; keep him warm and at rest, and give artificial respiration if breathing stops; maintain an open airway. EYES: flush with water for 15 min. SKIN: flush with water. INGESTION: do NOT induce vomiting; get medical attention.
5.4 **Toxicity by Inhalation (Threshold Limit Value):** 250 ppm (tentative)
5.5 **Short-Term Inhalation Limits:** Data not available
5.6 **Toxicity by Ingestion:** Grade 1; oral LD_{50} = 8,470 mg/kg (rat)
5.7 **Late Toxicity:** Data not available
5.8 **Vapor (Gas) Irritant Characteristics:** Vapors cause a slight smarting of the eyes or respiratory system if present in high concentrations. The effect is temporary.
5.9 **Liquid or Solid Irritant Characteristics:** Minimum hazard. If spilled on clothing and allowed to remain, may cause smarting and reddening of the skin.
5.10 **Odor Threshold:** Data not available

6. FIRE HAZARDS

6.1 **Flash Point:** −15°F O.C., −18°F C.C.
6.2 **Flammable Limits in Air:** 1.4%–7.9%
6.3 **Fire Extinguishing Agents:** Dry chemical, alcohol foam, carbon dioxide
6.4 **Fire Extinguishing Agents Not to be Used:** Water may be ineffective.
6.5 **Special Hazards of Combustion Products:** Not pertinent
6.6 **Behavior in Fire:** Vapor is heavier than air and may travel a considerable distance to a source of ignition and flash back. Containers may explode when heated.
6.7 **Ignition Temperature:** 830°F
6.8 **Electrical Hazard:** Data not available
6.9 **Burning Rate:** 5.0 mm/min.

7. CHEMICAL REACTIVITY

7.1 **Reactivity with Water:** No reaction
7.2 **Reactivity with Common Materials:** No reaction
7.3 **Stability During Transport:** Unstable peroxides may form on long standing in contact with air; these may explode spontaneously or when heated.
7.4 **Neutralizing Agents for Acids and Caustics:** Not pertinent
7.5 **Polymerization:** Not pertinent
7.6 **Inhibitor of Polymerization:** Not pertinent

8. WATER POLLUTION

8.1 **Aquatic Toxicity:** Data not available
8.2 **Waterfowl Toxicity:** Data not available
8.3 **Biological Oxygen Demand (BOD):** Data not available
8.4 **Food Chain Concentration Potential:** None

9. SELECTED MANUFACTURERS

1. Shell Chemical Company
   Industrial Chemicals Division
   P. O. Box 2463
   Houston, Texas 77001
2. Exxon Chemical Company
   P. O. Box 2180
   Houston, Texas 77001
3. Union Carbide Corporation
   Chemicals and Plastics Division
   270 Park Avenue
   New York, N. Y. 10017

10. SHIPPING INFORMATION

10.1 **Grades or Purity:** 94+% May contain 0.01% hydroquinone or other inhibitor to prevent peroxide formation.
10.2 **Storage Temperature:** Ambient
10.3 **Inert Atmosphere:** No requirement
10.4 **Venting:** Pressure-vacuum

11. HAZARD ASSESSMENT CODE

(See Hazard Assessment Handbook, CG 446-3)

A-P-Q-T-U-V-W

12. HAZARD CLASSIFICATIONS

12.1 **Code of Federal Regulations:** Flammable liquid
12.2 **NAS Hazard Rating for Bulk Water Transportation:**

Category	Rating
Fire	3
Health	
Vapor Irritant	1
Liquid or Solid Irritant	1
Poisons	1
Water Pollution	
Human Toxicity	1
Aquatic Toxicity	2
Aesthetic Effect	2
Reactivity	
Other Chemicals	1
Water	0
Self-Reaction	0

12.3 **NFPA Hazard Classifications:**

Category	Classification
Health Hazard (Blue)	2
Flammability (Red)	3
Reactivity (Yellow)	1

13. PHYSICAL AND CHEMICAL PROPERTIES

13.1 **Physical State at 15°C and 1 atm:** Liquid
13.2 **Molecular Weight:** 102.2
13.3 **Boiling Point at 1 atm:** 156°F = 69°C = 342°K
13.4 **Freezing Point:** −123°F = −86°C = 187°K
13.5 **Critical Temperature:** 440.4°F = 226.9°C = 500.1°K
13.6 **Critical Pressure:** 418 psia = 28.4 atm = 2.88 MN/m^2
13.7 **Specific Gravity:** 0.724 at 20°C (liquid)
13.8 **Liquid Surface Tension:** 17.1 dynes/cm = 0.0171 N/m at 25°C
13.9 **Liquid-Water Interfacial Tension:** 17.1 dynes/cm = 0.0171 N/m at 25°C
13.10 **Vapor (Gas) Specific Gravity:** 3.5
13.11 **Ratio of Specific Heats of Vapor (Gas):** 1.0590
13.12 **Latent Heat of Vaporization:** 131 Btu/lb = 73 cal/g = 3.1×10^5 J/kg
13.13 **Heat of Combustion:** −16,900 Btu/lb = −9,390 cal/g = -393×10^5 J/kg
13.14 **Heat of Decomposition:** Not pertinent
13.15 **Heat of Solution:** Not pertinent
13.16 **Heat of Polymerization:** Not pertinent

NOTES

IPM	ISOPROPYL MERCAPTAN

Common Synonyms			
2-Propanethiol Propane-2-thiol	Liquid	White	Strong Skunk Odor
	Floats and mixes with water. Flammable, irritating vapor is produced.		

Shut off ignition sources. Call fire department.
Stop discharge if possible. Keep people away.
Avoid contact with liquid and vapor.
Stay upwind. Use water spray to "knock down" vapor.
Isolate and remove discharged material.
Notify local health and pollution control agencies.

Fire	FLAMMABLE. Irritating gases may be produced when heated. Flashback along vapor trail may occur. Vapor may explode if ignited in an enclosed area. Extinguish with dry chemicals, alcohol foam, or carbon dioxide. Water may be ineffective on fire. Cool exposed containers with water.
Exposure	CALL FOR MEDICAL AID. VAPOR Irritating to eyes, nose and throat. If inhaled will cause difficult breathing, or loss of consciousness. If in eyes, hold eyelids open and flush with plenty of water. If breathing has stopped, give artificial respiration. If breathing is difficult, give oxygen. LIQUID Irritating to skin and eyes. If swallowed will cause nausea and vomiting. Remove contaminated clothing and shoes. Flush affected areas with plenty of water. IF IN EYES, hold eyelids open and flush with plenty of water. IF SWALLOWED and victim is CONSCIOUS, have victim drink water or milk and have victim induce vomiting. IF SWALLOWED and victim is UNCONSCIOUS OR HAVING CONVULSIONS, do nothing except keep victim warm.
Water Pollution	Effect of low concentrations on aquatic life is unknown. May be dangerous if it enters water intakes. Notify local health and wildlife officials. Notify operators of nearby water intakes.

1. RESPONSE TO DISCHARGE

(See Response Methods Handbook, CG 446-4)

Issue warning—high flammability, air contaminant, water contaminant
Restrict access
Evacuate area
Disperse and flush

2. LABEL

Red

3. CHEMICAL DESIGNATIONS

3.1 **Synonyms:** 2-Propanethiol; Propane-2-thiol
3.2 **Coast Guard Compatibility Classification:** Not listed
3.3 **Chemical Formula:** $(CH_3)_2CHSH$
3.4 **IMCO/United Nations Numerical Designation:** Not listed

4. OBSERVABLE CHARACTERISTICS

4.1 **Physical State (as shipped):** Liquid
4.2 **Color:** Colorless
4.3 **Odor:** Powerful skunk

5. HEALTH HAZARDS

5.1 **Personal Protective Equipment:** Self-contained breathing apparatus; goggles or face shield; rubber gloves
5.2 **Symptoms Following Exposure:** Inhalation causes loss of sense of smell, muscular weakness, convulsions, respiratory paralysis. Ingestion causes nausea and vomiting. Contact with eyes or skin causes irritation.
5.3 **Treatment for Exposure:** INHALATION: remove victim to fresh air; start artificial respiration and give oxygen if required; observe for signs of pulmonary edema; get medical attention. INGESTION: give large amount of water and induce vomiting. EYES or SKIN: flush with water.
5.4 **Toxicity by Inhalation (Threshold Limit Value):** Data not available
5.5 **Short-Term Inhalation Limits:** Data not available
5.6 **Toxicity by Ingestion:** Grade 2; oral LD_{50} = 1,790 mg/kg (rat)
5.7 **Late Toxicity:** Data not available
5.8 **Vapor (Gas) Irritant Characteristics:** Data not available
5.9 **Liquid or Solid Irritant Characteristics:** Data not available
5.10 **Odor Threshold:** 0.25 ppb

6. FIRE HAZARDS

6.1 **Flash Point:** −30°F O.C.
6.2 **Flammable Limits in Air:** Data not available
6.3 **Fire Extinguishing Agents:** Dry chemical, alcohol foam, carbon dioxide
6.4 **Fire Extinguishing Agents Not to be Used:** Water may be ineffective.
6.5 **Special Hazards of Combustion Products:** Irritating sulfur dioxide gas is formed in fire.
6.6 **Behavior in Fire:** Vapor is heavier than air and may travel to a source of ignition and flash back. Containers may explode.
6.7 **Ignition Temperature:** Data not available
6.8 **Electrical Hazard:** Data not available
6.9 **Burning Rate:** Data not available

7. CHEMICAL REACTIVITY

7.1 **Reactivity with Water:** No reaction
7.2 **Reactivity with Common Materials:**
7.3 **Stability During Transport:** Stable
7.4 **Neutralizing Agents for Acids and Caustics:** Not pertinent
7.5 **Polymerization:** Not pertinent
7.6 **Inhibitor of Polymerization:** Not pertinent

8. WATER POLLUTION

8.1 **Aquatic Toxicity:** Data not available
8.2 **Waterfowl Toxicity:** Data not available
8.3 **Biological Oxygen Demand (BOD):** Data not available
8.4 **Food Chain Concentration Potential:** None

9. SELECTED MANUFACTURERS

1. Pennwalt Corporation
Three Parkway
Philadelphia, Pa. 19102
2. Phillips Petroleum Co.
Special Products Div.
Chemical Dept.
Bartlesville, Okla. 74004
3. Aldrich Chemical Co.
940 W. Saint Paul Ave.
Milwaukee, Wis. 53233

10. SHIPPING INFORMATION

10.1 **Grades or Purity:** Technical, 98.0+%
10.2 **Storage Temperature:** Ambient
10.3 **Inert Atmosphere:** No requirement
10.4 **Venting:** Pressure-vacuum

11. HAZARD ASSESSMENT CODE

(See Hazard Assessment Handbook, CG 446-3)

A-P-Q-R-S

12. HAZARD CLASSIFICATIONS

12.1 **Code of Federal Regulations:** Flammable liquid
12.2 **NAS Hazard Rating for Bulk Water Transportation:** Not listed
12.3 **NFPA Hazard Classifications:** Not listed

13. PHYSICAL AND CHEMICAL PROPERTIES

13.1 **Physical State at 15°C and 1 atm:** Liquid
13.2 **Molecular Weight:** 76.2
13.3 **Boiling Point at 1 atm:** 126.6°F = 52.5°C = 325.8°K
13.4 **Freezing Point:** −202.8°F = −130.5°C = 142.7°K
13.5 **Critical Temperature:** Data not available
13.6 **Critical Pressure:** Data not available
13.7 **Specific Gravity:** 0.814 at 20°C (liquid)
13.8 **Liquid Surface Tension:** 22.0 dynes/cm = 0.022 N/m at 20°C
13.9 **Liquid-Water Interfacial Tension:** Not pertinent
13.10 **Vapor (Gas) Specific Gravity:** 2.6
13.11 **Ratio of Specific Heats of Vapor (Gas):** 1.0964 at 15.6°C
13.12 **Latent Heat of Vaporization:** 165.7 Btu/lb = 92.1 cal/g = 3.83×10^5 J/kg
13.13 **Heat of Combustion:** −14,920 Btu/lb = −8,290 cal/g = -347×10^5 J/kg
13.14 **Heat of Decomposition:** Not pertinent
13.15 **Heat of Solution:** Not pertinent
13.16 **Heat of Polymerization:** Not pertinent

NOTES

IPC — ISOPROPYL PERCARBONATE

Common Synonyms			
Peroxydicarbonic acid, diisopropyl ester Diisopropyl percarbonate Diisopropyl peroxydicarbonate Isopropyl peroxydicarbonate Peroxydicarbonic acid, bis (1-methylethyl) ester	Solid (Packed in Dry Ice)	White	Sharp Unpleasant Odor
	Sinks in water. Freezing point is 48°F.		

Stop discharge if possible. Keep people away.
Shut off ignition sources. Call fire department.
Avoid contact with solid and dust.
Isolate and remove discharged material.
Notify local health and pollution control agencies.

Fire

COMBUSTIBLE.
May cause fire on contact with combustibles.
Will increase the intensity of a fire.
Containers may explode in fire.

DO NOT USE DRY CHEMICALS, CARBON DIOXIDE, OR FOAM ON FIRE.
Flood discharge area with water from safe distance or protected location.
Cool exposed containers with water.

Exposure

CALL FOR MEDICAL AID.

DUST
Irritating to eyes, nose and throat.
If inhaled will cause coughing or difficult breathing.
If in eyes, hold eyelids open and flush with plenty of water.
If breathing has stopped, give artificial respiration.
If breathing is difficult, give oxygen.

SOLID
Irritating to skin and eyes.
Harmful if swallowed.
Remove contaminated clothing and shoes.
Flush affected areas with plenty of water.
IF IN EYES, hold eyelids open and flush with plenty of water.
IF SWALLOWED and victim is CONSCIOUS, have victim drink water or milk.
IF SWALLOWED and victim is UNCONSCIOUS OR HAVING CONVULSIONS, do nothing except keep victim warm.

Water Pollution

Effect of low concentrations on aquatic life is unknown.
May be dangerous if it enters water intakes.
Notify local health and wildlife officials.
Notify operators of nearby water intakes.

1. RESPONSE TO DISCHARGE

(See Response Methods Handbook, CG 446-4)

Issue warning—high flammability, oxidizing material
Restrict access
Should be removed
Chemical and physical treatment

2. LABEL

Yellow

3. CHEMICAL DESIGNATIONS

3.1 **Synonyms:** Diisopropyl percarbonate; Diisopropyl peroxydicarbonate; Isopropyl peroxydicarbonate; Peroxydicarbonic acid, bis (1-methylethyl) ester; Peroxydicarbonic acid, diisopropyl ester
3.2 **Coast Guard Compatibility Classification:** Not listed
3.3 **Chemical Formula:** $C_3H_7OOCOOCOOC_3H_7$
3.4 **IMCO/United Nations Numerical Designation:** Not listed

4. OBSERVABLE CHARACTERISTICS

4.1 **Physical State (as shipped):** Solid (containers packed in "Dry Ice")
4.2 **Color:** White
4.3 **Odor:** Disagreeable; pungent

5. HEALTH HAZARDS

5.1 **Personal Protective Equipment:** Rubber gloves and shoes; hard hat; chemical splash goggles; plastic apron; respirator (depending on solvent used)
5.2 **Symptoms Following Exposure:** Inhalation overexposure unlikely, but prolonged exposure may cause lung edema. Contact with eyes may cause irritation. Solutions are severe primary skin irritants.
5.3 **Treatment for Exposure:** INHALATION: move to uncontaminated atmosphere; if breathing is difficult, give oxygen. EYES: flush with copious amounts of water. SKIN: wash off with isopropyl alcohol and water; call a physician.
5.4 **Toxicity by Inhalation (Threshold Limit Value):** Data not available
5.5 **Short-Term Inhalation Limits:** Data not available
5.6 **Toxicity by Ingestion:** Grade 2; LD_{50} 0.5–5 g/kg
5.7 **Late Toxicity:** Data not available
5.8 **Vapor (Gas) Irritant Characteristics:** Data not available
5.9 **Liquid or Solid Irritant Characteristics:** Data not available
5.10 **Odor Threshold:** Data not available

6. FIRE HAZARDS

6.1 **Flash Point:** Not pertinent (combustible solid)
6.2 **Flammable Limits in Air:** Not pertinent
6.3 **Fire Extinguishing Agents:** Water
6.4 **Fire Extinguishing Agents Not to be Used:** All extinguishing agents may be ineffective.
6.5 **Special Hazards of Combustion Products:** Flammable and/or toxic gases formed in fires include acetone, isopropyl alcohol, acetaldehyde, and ethane.
6.6 **Behavior in Fire:** Undergoes auto-accelerative decomposition and may self-ignite. Confinement may lead to detonation. Fires very difficult to extinguish because air not needed.
6.7 **Ignition Temperature:** Data not available
6.8 **Electrical Hazard:** Not pertinent
6.9 **Burning Rate:** Not pertinent

7. CHEMICAL REACTIVITY

7.1 **Reactivity with Water:** No reaction
7.2 **Reactivity with Common Materials:** May decompose with formation of oxygen when in contact with metals
7.3 **Stability During Transport:** Unstable above 0°F with formation of oxygen gas
7.4 **Neutralizing Agents for Acids and Caustics:** Not pertinent
7.5 **Polymerization:** Not pertinent
7.6 **Inhibitor of Polymerization:** Not pertinent

8. WATER POLLUTION

8.1 **Aquatic Toxicity:** Data not available
8.2 **Waterfowl Toxicity:** Data not available
8.3 **Biological Oxygen Demand (BOD):** Data not available
8.4 **Food Chain Concentration Potential:** None

9. SELECTED MANUFACTURERS

1. PPG Industries, Inc.
Chemical Division
One Gateway Center
Pittsburgh, Pa. 15222
2. Pennwalt Corp.—Lucidol Division
1740 Military Road
Buffalo, N. Y. 14240

10. SHIPPING INFORMATION

10.1 **Grades or Purity:** Technical, 98.5–99+%
10.2 **Storage Temperature:** Below −18°C (0°F)
10.3 **Inert Atmosphere:** No requirement
10.4 **Venting:** Pressure-vacuum

11. HAZARD ASSESSMENT CODE

(See Hazard Assessment Handbook, CG 446-3)

II

12. HAZARD CLASSIFICATIONS

12.1 **Code of Federal Regulations:** Organic Peroxide
12.2 **NAS Hazard Rating for Bulk Water Transportation:** Not listed
12.3 **NFPA Hazard Classifications:** Not listed

13. PHYSICAL AND CHEMICAL PROPERTIES

13.1 **Physical State at 15°C and 1 atm:** Liquid
13.2 **Molecular Weight:** 206.2
13.3 **Boiling Point at 1 atm:** Not pertinent (decomposes)
13.4 **Freezing Point:** 46–50°F = 8–10°C = 281–283°K
13.5 **Critical Temperature:** Not pertinent
13.6 **Critical Pressure:** Not pertinent
13.7 **Specific Gravity:** 1.08 at 15°C (solid)
13.8 **Liquid Surface Tension:** Not pertinent
13.9 **Liquid-Water Interfacial Tension:** Not pertinent
13.10 **Vapor (Gas) Specific Gravity:** Not pertinent
13.11 **Ratio of Specific Heats of Vapor (Gas):** Not pertinent
13.12 **Latent Heat of Vaporization:** Not pertinent
13.13 **Heat of Combustion:** −8,500 Btu/lb = −4,720 cal/g = -198×10^5 J/kg
13.14 **Heat of Decomposition:** −670 Btu/lb = −370 cal/g = -15.5×10^5 J/kg
13.15 **Heat of Solution:** Not pertinent
13.16 **Heat of Polymerization:** Not pertinent

NOTES

IVA ISOVALERALDEHYDE

Common Synonyms			
3-Methylbutyraldehyde Isovaleric aldehyde Isovaleral 3-Methylbutanal	Liquid	Colorless	Weak suffocating odor
	Floats on water. Flammable, irritating vapor is produced.		

Shut off ignition sources. Call fire department.
Stop discharge if possible. Keep people away.
Stay upwind. Use water spray to "knock down" vapor.
Isolate and remove discharged material.
Notify local health and pollution control agencies.

Fire

FLAMMABLE.
Flashback along vapor trail may occur.
Vapor may explode if ignited in an enclosed area.
Extinguish with dry chemicals, foam or carbon dioxide.
Water may be ineffective on fire.
Cool exposed containers with water.

Exposure

Call for medical aid.

VAPOR
Irritating to eyes, nose and throat.
If inhaled will cause headache, nausea, vomiting or difficult breathing.
Move victim to fresh air.
If breathing has stopped, give artificial respiration.
If breathing is difficult, give oxygen.

LIQUID
Irritating to skin and eyes.
Harmful if swallowed.
IF SWALLOWED and victim is CONSCIOUS, have victim drink water or milk.

Water Pollution

Effect of low concentrations on aquatic life is unknown.
Fouling to shoreline.
May be dangerous if it enters water intakes.
Notify local health and wildlife officials.
Notify operators of nearby water intakes.

1. RESPONSE TO DISCHARGE

(See Response Methods Handbook, CG 446-4)

Issue warning—high flammability, water contaminant
Restrict access
Mechanical containment
Should be removed
Chemical and physical treatment

2. LABEL

Red

3. CHEMICAL DESIGNATIONS

3.1 **Synonyms:** Isovaleral; Isovaleric aldehyde; 3-Methylbutanal; 3-Methylbutyraldehyde
3.2 **Coast Guard Compatibility Classification:** Aldehyde (7)
3.3 **Chemical Formula:** $(CH_3)_2CHCH_2CHO$
3.4 **IMCO/United Nations Numerical Designation:** 3.2/1989

4. OBSERVABLE CHARACTERISTICS

4.1 **Physical State (as shipped):** Liquid
4.2 **Color:** Colorless
4.3 **Odor:** Weakly suffocating

5. HEALTH HAZARDS

5.1 **Personal Protective Equipment:** Goggles or face shield; rubber gloves; air mask or self-contained breathing apparatus for high vapor concentrations.
5.2 **Symptoms Following Exposure:** Inhalation causes chest discomfort, nausea, vomiting, and headache. Contact of liquid with eyes or skin causes irritation. Ingestion causes irritation of mouth and stomach.
5.3 **Treatment for Exposure:** INHALATION: remove victim to fresh air; apply artificial respiration if required; get medical attention. EYES: flush with water for at least 15 min. SKIN: wipe off, wash well with soap and water. INGESTION: induce vomiting; get medical attention.
5.4 **Toxicity by Inhalation (Threshold Limit Value):** Data not available
5.5 **Short-Term Inhalation Limits:** Data not available
5.6 **Toxicity by Ingestion:** Grade 2; oral $LD_{50} \geq 3{,}200$ mg/kg (rat)
5.7 **Late Toxicity:** Data not available
5.8 **Vapor (Gas) Irritant Characteristics:** Data not available
5.9 **Liquid or Solid Irritant Characteristics:** Data not available
5.10 **Odor Threshold:** Data not available

6. FIRE HAZARDS

6.1 **Flash Point:** (est.) 55°F O.C.
6.2 **Flammable Limits in Air:** Data not available
6.3 **Fire Extinguishing Agents:** Dry chemical, foam, carbon dioxide
6.4 **Fire Extinguishing Agents Not to be Used:** Water may be ineffective.
6.5 **Special Hazards of Combustion Products:** Not pertinent
6.6 **Behavior in Fire:** Not pertinent
6.7 **Ignition Temperature:** Data not available
6.8 **Electrical Hazard:** Data not available
6.9 **Burning Rate:** 5.3 mm/min.

7. CHEMICAL REACTIVITY

7.1 **Reactivity with Water:** No reaction
7.2 **Reactivity with Common Materials:** No reaction
7.3 **Stability During Transport:** Stable
7.4 **Neutralizing Agents for Acids and Caustics:** Not pertinent
7.5 **Polymerization:** Not pertinent
7.6 **Inhibitor of Polymerization:** Not pertinent

8. WATER POLLUTION

8.1 **Aquatic Toxicity:** Data not available
8.2 **Waterfowl Toxicity:** Data not available
8.3 **Biological Oxygen Demand (BOD):** Data not available
8.4 **Food Chain Concentration Potential:** None

9. SELECTED MANUFACTURERS

1. Aldrich Chemical Co., Inc.
940 West St. Paul Avenue
Milwaukee, Wis. 53233
2. Eastern Chemical Division
230 Marcus Boulevard
Hauppauge, L. I., N. Y. 11787
3. Pfaltz and Bauer, Inc.
126-04 Northern Boulevard
Flushing, N. Y. 11368

10. SHIPPING INFORMATION

10.1 **Grades or Purity:** Commercial
10.2 **Storage Temperature:** Ambient
10.3 **Inert Atmosphere:** No requirement
10.4 **Venting:** Open (flame arrester)

11. HAZARD ASSESSMENT CODE

(See Hazard Assessment Handbook, CG 446-3)

A-T-U-V-W

12. HAZARD CLASSIFICATIONS

12.1 **Code of Federal Regulations:** Flammable liquid
12.2 **NAS Hazard Rating for Bulk Water Transportation:** Not listed
12.3 **NFPA Hazard Classifications:** Not listed

13. PHYSICAL AND CHEMICAL PROPERTIES

13.1 **Physical State at 15°C and 1 atm:** Liquid
13.2 **Molecular Weight:** 86.1
13.3 **Boiling Point at 1 atm:** 198.5°F = 92.5°C = 365.7°K
13.4 **Freezing Point:** −60°F = −51°C = 222°K
13.5 **Critical Temperature:** Not pertinent
13.6 **Critical Pressure:** Not pertinent
13.7 **Specific Gravity:** 0.785 at 20°C (liquid)
13.8 **Liquid Surface Tension:** 23.7 dynes/cm = 0.0237 N/m at 20°C
13.9 **Liquid-Water Interfacial Tension:** (est.) 30 dynes/cm = 0.030 N/m at 20°C
13.10 **Vapor (Gas) Specific Gravity:** 3
13.11 **Ratio of Specific Heats of Vapor (Gas):** (est.) 1.0736
13.12 **Latent Heat of Vaporization:** (est.) 167 Btu/lb = 93 cal/g = 3.9×10^5 J/kg
13.13 **Heat of Combustion:** −15,500 Btu/lb = −8,620 cal/g = -360×10^5 J/kg
13.14 **Heat of Decomposition:** Not pertinent
13.15 **Heat of Solution:** Not pertinent
13.16 **Heat of Polymerization:** Not pertinent

NOTES

JET FUELS: JP-1

JPO

Common Synonyms
Kerosine
Range oil
Fuel Oil No. 1

Watery liquid | Colorless | Fuel oil odor

Floats on water.

Stop discharge if possible.
Call fire department.
Avoid contact with liquid.
Isolate and remove discharged material.
Notify local health and pollution control agencies.

Fire

Combustible.
Extinguish with dry chemical, foam, or carbon dioxide.
Water may be ineffective on fire.
Cool exposed containers with water.

Exposure

CALL FOR MEDICAL AID.

LIQUID
Irritating to skin and eyes.
Harmful if swallowed.
Remove contaminated clothing and shoes.
Flush affected areas with plenty of water.
IF IN EYES, hold eyelids open and flush with plenty of water.
IF SWALLOWED and victim is CONSCIOUS, have victim drink water or milk.
DO NOT INDUCE VOMITING.

Water Pollution

Dangerous to aquatic life in high concentrations.
Fouling to shoreline.
May be dangerous if it enters water intakes.
Notify local health and wildlife officials.
Notify operators of nearby water intakes.

1. RESPONSE TO DISCHARGE

(See Response Methods Handbook, CG 446-4)

Mechanical containment
Should be removed
Chemical and physical treatment

2. LABELS

No hazard label required by Code of Federal Regulations

3. CHEMICAL DESIGNATIONS

3.1 **Synonyms:** Kerosene
Kerosine
Range Oil
Fuel Oil No. 1

3.2 **Coast Guard Compatibility Classification:** Miscellaneous hydrocarbon mixtures

3.3 **Chemical Formula:** C_nH_{2n+2}

3.4 **IMCO/United Nations Numerical Designation:** 3.3/1223

4. OBSERVABLE CHARACTERISTICS

4.1 **Physical State (as shipped):** Liquid

4.2 **Color:** Colorless to light brown

4.3 **Odor:** Like fuel oil

5. HEALTH HAZARDS

5.1 **Personal Protective Equipment:** Protective gloves; goggles or face shield.

5.2 **Symptoms Following Exposure:** Vapor causes slight irritation of eyes and nose. Liquid irritates stomach; if taken into lungs causes coughing, distress, and rapidly developing pulmonary edema.

5.3 **Treatment for Exposure:** ASPIRATION: enforce bed rest; administer oxygen; call a doctor. INGESTION: do NOT induce vomiting; call a doctor. EYES: wash with plenty of water. SKIN: wipe off and wash with soap and water.

5.4 **Toxicity by Inhalation (Threshold Limit Value):** 200 ppm

5.5 **Short-Term Inhalation Limits:** 2500 mg/m^3 for 60 min.

5.6 **Toxicity by Ingestion:** Grade 1; LD_{50} 5 to 15 g/kg

5.7 **Late Toxicity:** Data not available

5.8 **Vapor (Gas) Irritant Characteristics:** Vapors cause a slight smarting of the eyes or respiratory system if present in high concentrations. The effect is temporary.

5.9 **Liquid or Solid Irritant Characteristics:** Minimum hazard. If spilled on clothing and allowed to remain, may cause smarting and reddening of the skin.

5.10 **Odor Threshold:** 1 ppm

6. FIRE HAZARDS

6.1 **Flash Point:** 100°F C.C.

6.2 **Flammable Limits in Air:** 0.7%–5%

6.3 **Fire Extinguishing Agents:** Foam, dry chemical, or carbon dioxide

6.4 **Fire Extinguishing Agents Not to be Used:** Not pertinent

6.5 **Special Hazards of Combustion Products:** Not pertinent

6.6 **Behavior in Fire:** Not pertinent

6.7 **Ignition Temperature:** 444°F

6.8 **Electrical Hazard:** Not pertinent

6.9 **Burning Rate:** 4 mm/min.

7. CHEMICAL REACTIVITY

7.1 **Reactivity with Water:** No reaction

7.2 **Reactivity with Common Materials:** No reaction

7.3 **Stability During Transport:** Stable

7.4 **Neutralizing Agents for Acids and Caustics:** Not pertinent

7.5 **Polymerization:** Not pertinent

7.6 **Inhibitor of Polymerization:** Not pertinent

8. WATER POLLUTION

8.1 **Aquatic Toxicity:** 2990 ppm/24 hr/bluegill/TL_m/fresh water

8.2 **Waterfowl Toxicity:** Data not available

8.3 **Biological Oxygen Demand (BOD):** 53%, 5 days

8.4 **Food Chain Concentration Potential:** None

9. SELECTED MANUFACTURERS

1. Exxon Co.
30 Rockefeller Plaza
New York, N. Y. 10020

2. Shell Oil Co.
1 Shell Plaza
Houston, Tex. 77001

3. Sun Oil Co.
240 Radnor Rd.
St. Davids, Pa. 19087

10. SHIPPING INFORMATION

10.1 **Grades or Purity:** 100%

10.2 **Storage Temperature:** Ambient

10.3 **Inert Atmosphere:** No requirement

10.4 **Venting:** Open (flame arrester)

11. HAZARD ASSESSMENT CODE

(See Hazard Assessment Handbook, CG 446-3)

A-T-U

12. HAZARD CLASSIFICATIONS

12.1 **Code of Federal Regulations:** Combustible liquid

12.2 **NAS Hazard Rating for Bulk Water Transportation:**

Category	Rating
Fire	2
Health	
Vapor Irritant	1
Liquid or Solid Irritant	1
Poisons	1
Water Pollution	
Human Toxicity	1
Aquatic Toxicity	1
Aesthetic Effect	3
Reactivity	
Other Chemicals	0
Water	0
Self-Reaction	0

12.3 **NFPA Hazard Classifications:**

Category	Classification
Health Hazard (Blue)	0
Flammability (Red)	2
Reactivity (Yellow)	0

13. PHYSICAL AND CHEMICAL PROPERTIES

13.1 **Physical State at 15°C and 1 atm:** Liquid

13.2 **Molecular Weight:** Not pertinent

13.3 **Boiling Point at 1 atm:** 392–500°F = 200–260°C = 473–533°K

13.4 **Freezing Point:** −45 to −55°F = −43 to −49°C = 316 to 321°K

13.5 **Critical Temperature:** Not pertinent

13.6 **Critical Pressure:** Not pertinent

13.7 **Specific Gravity:** 0.80 at 15°C (liquid)

13.8 **Liquid Surface Tension:** 23–32 dynes/cm = 0.023–0.032 N/m at 20°C

13.9 **Liquid-Water Interfacial Tension:** 47–49 dynes/cm = 0.047–0.049 N/m at 20°C

13.10 **Vapor (Gas) Specific Gravity:** Not pertinent

13.11 **Ratio of Specific Heats of Vapor (Gas):** (est.) 1.030

13.12 **Latent Heat of Vaporization:** 110 Btu/lb = 60 cal/g = 2.5×10^5 J/kg

13.13 **Heat of Combustion:** −18,540 Btu/lb = −10,300 cal/g = -431.24×10^5 J/kg

13.14 **Heat of Decomposition:** Not pertinent

13.15 **Heat of Solution:** Not pertinent

13.16 **Heat of Polymerization:** Not pertinent

NOTES

REVISED 1978

JET FUELS: JP-3

Common Synonyms

Watery liquid — Colorless — Fuel oil odor

Floats on water. Irritating, combustible vapor may be produced.

Stop discharge if possible.
Call fire department.
Stay upwind and use water spray to "knock down" vapor.
Avoid contact with liquid and vapor.
Isolate and remove discharged material.
Notify local health and pollution control agencies.

Fire

Combustible.

Extinguish with dry chemical, foam, or carbon dioxide.
Water may be ineffective on fire.
Cool exposed containers with water.

Exposure

CALL FOR MEDICAL AID.

VAPOR
Irritating to eyes, nose and throat.
If inhaled, will cause coughing, dizziness, headache, or difficult breathing.

Move to fresh air.
If breathing has stopped, give artificial respiration.
If breathing is difficult, give oxygen.

LIQUID
Irritating to skin and eyes.
Harmful if swallowed.

Remove contaminated clothing and shoes.
Flush affected areas with plenty of water.
IF IN EYES, hold eyelids open and flush with plenty of water.
IF SWALLOWED and victim is CONSCIOUS, have victim drink water or milk.
DO NOT INDUCE VOMITING.

Water Pollution

Dangerous to aquatic life in high concentrations.
Fouling to shoreline.
May be dangerous if it enters water intakes.

Notify local health and wildlife officials.
Notify operators of nearby water intakes.

1. RESPONSE TO DISCHARGE

(See Response Methods Handbook, CG 446-4)

Mechanical containment
Should be removed
Chemical and physical treatment

2. LABELS

No hazard label required by Code of Federal Regulations

3. CHEMICAL DESIGNATIONS

3.1 **Synonyms:** No common synonyms
3.2 **Coast Guard Compatibility Classification:** Miscellaneous hydrocarbon mixtures
3.3 **Chemical Formula:** C_nH_{2n+2}
3.4 **IMCO/United Nations Numerical Designation:** 3.3/1223

4. OBSERVABLE CHARACTERISTICS

4.1 **Physical State (as shipped):** Liquid
4.2 **Color:** Colorless to light brown
4.3 **Odor:** Like fuel oil

5. HEALTH HAZARDS

5.1 **Personal Protective Equipment:** Protective gloves; goggles or face shield.
5.2 **Symptoms Following Exposure:** Vapor causes slight irritation of eyes and nose. Liquid irritates stomach; if taken into lungs, causes coughing, distress, and rapidly developing pulmonary edema.
5.3 **Treatment for Exposure:** ASPIRATION: enforce bed rest; administer oxygen; call a doctor. INGESTION: do NOT induce vomiting; call a doctor. EYES: wash with plenty of water. SKIN: wipe off and wash with soap and water.
5.4 **Toxicity by Inhalation (Threshold Limit Value):** 200 ppm
5.5 **Short-Term Inhalation Limits:** 2500 mg/m^3 for 60 min.
5.6 **Toxicity by Ingestion:** Grade 2; LD_{50} 0.5 to 5 g/kg
5.7 **Late Toxicity:** Data not available
5.8 **Vapor (Gas) Irritant Characteristics:** Vapors cause a slight smarting of the eyes or respiratory system if present in high concentrations. The effect is temporary.
5.9 **Liquid or Solid Irritant Characteristics:** Minimum hazard. If spilled on clothing and allowed to remain, may cause smarting and reddening of the skin.
5.10 **Odor Threshold:** 1 ppm

6. FIRE HAZARDS

6.1 **Flash Point:** 110-150°F
6.2 **Flammable Limits in Air:** Data not available
6.3 **Fire Extinguishing Agents:** Foam, dry chemical, or carbon dioxide
6.4 **Fire Extinguishing Agents Not to be Used:** Not pertinent
6.5 **Special Hazards of Combustion Products:** Not pertinent
6.6 **Behavior in Fire:** Not pertinent
6.7 **Ignition Temperature:** Data not available
6.8 **Electrical Hazard:** Not pertinent
6.9 **Burning Rate:** 4 mm/min.

7. CHEMICAL REACTIVITY

7.1 **Reactivity with Water:** No reaction
7.2 **Reactivity with Common Materials:** No reaction
7.3 **Stability During Transport:** Stable
7.4 **Neutralizing Agents for Acids and Caustics:** Not pertinent
7.5 **Polymerization:** Not pertinent
7.6 **Inhibitor of Polymerization:** Not pertinent

8. WATER POLLUTION

8.1 **Aquatic Toxicity:** 2990 ppm/24 hr/bluegill/TL_m/fresh water
8.2 **Waterfowl Toxicity:** Data not available
8.3 **Biological Oxygen Demand (BOD):** 53%, 5 days
8.4 **Food Chain Concentration Potential:** None

9. SELECTED MANUFACTURERS

1. Exxon Co.
   30 Rockefeller Plaza
   New York, N. Y. 10020
2. Shell Oil Co.
   1 Shell Plaza
   Houston, Tex. 77001
3. Sun Oil Co.
   240 Radnor Rd.
   St. Davids, Pa. 19087

10. SHIPPING INFORMATION

10.1 **Grades or Purity:** 100%
10.2 **Storage Temperature:** Ambient
10.3 **Inert Atmosphere:** No requirement
10.4 **Venting:** Open (flame arrester)

11. HAZARD ASSESSMENT CODE

(See Hazard Assessment Handbook, CG 446-3)

A-T-U

12. HAZARD CLASSIFICATIONS

12.1 **Code of Federal Regulations:** Combustible Liquid
12.2 **NAS Hazard Rating for Bulk Water Transportation:**

Category	Rating
Fire	3
Health	
Vapor Irritant	1
Liquid or Solid Irritant	1
Poisons	1
Water Pollution	
Human Toxicity	1
Aquatic Toxicity	1
Aesthetic Effect	2
Reactivity	
Other Chemicals	0
Water	0
Self-Reaction	0

12.3 **NFPA Hazard Classifications:**

Category	Classification
Health Hazard (Blue)	0
Flammability (Red)	2
Reactivity (Yellow)	0

13. PHYSICAL AND CHEMICAL PROPERTIES

13.1 **Physical State at 15°C and 1 atm:** Liquid
13.2 **Molecular Weight:** Not pertinent
13.3 **Boiling Point at 1 atm:** 86 – 500°F = 30 – 260°C = 303 – 533°K
13.4 **Freezing Point:** Not pertinent
13.5 **Critical Temperature:** Not pertinent
13.6 **Critical Pressure:** Not pertinent
13.7 **Specific Gravity:** 0.80 at 20°C (liquid)
13.8 **Liquid Surface Tension:** Data not available
13.9 **Liquid-Water Interfacial Tension:** Data not available
13.10 **Vapor (Gas) Specific Gravity:** Not pertinent
13.11 **Ratio of Specific Heats of Vapor (Gas):** (est.) 1.030
13.12 **Latent Heat of Vaporization:** 140 Btu/lb = 78 cal/g = 3.3×10^5 J/kg
13.13 **Heat of Combustion:** −18,540 Btu/lb = −10,300 cal/g = -431.24×10^5 J/kg
13.14 **Heat of Decomposition:** Not pertinent
13.15 **Heat of Solution:** Not pertinent
13.16 **Heat of Polymerization:** Not pertinent

NOTES

REVISED 1978

JET FUELS: JP-4

Common Synonyms			
	Watery liquid	Colorless	Fuel oil odor
	Floats on water.		

Stop discharge if possible. Keep people away.
Shut off ignition sources and call fire department.
Avoid contact with liquid.
Isolate and remove discharged material.
Notify local health and pollution control agencies.

Fire

FLAMMABLE.
Extinguish with dry chemical, foam, or carbon dioxide.
Water may be ineffective on fire.
Cool exposed containers with water.

Exposure

CALL FOR MEDICAL AID.

LIQUID
Irritating to skin and eyes.
Harmful if swallowed.
Remove contaminated clothing and shoes.
Flush affected areas with plenty of water.
IF IN EYES, hold eyelids open and flush with plenty of water.
IF SWALLOWED and victim is CONSCIOUS, have victim drink water or milk.
DO NOT INDUCE VOMITING.

Water Pollution

Dangerous to aquatic life in high concentrations.
Fouling to shoreline.
May be dangerous if it enters water intakes.
Notify local health and wildlife officials.
Notify operators of nearby water intakes.

1. RESPONSE TO DISCHARGE

(See Response Methods Handbook, CG 446-4)

Issue warning—high flammability
Mechanical containment
Should be removed
Chemical and physical treatment

2. LABEL

Red

3. CHEMICAL DESIGNATIONS

3.1 **Synonyms:** No common synonyms
3.2 **Coast Guard Compatibility Classification:** Miscellaneous hydrocarbon mixtures
3.3 **Chemical Formula:** C_nH_{2n+2}
3.4 **IMCO/United Nations Numerical Designation:** 3.2/1863

4. OBSERVABLE CHARACTERISTICS

4.1 **Physical State (as shipped):** Liquid
4.2 **Color:** Colorless to light brown
4.3 **Odor:** Like fuel oil

5. HEALTH HAZARDS

5.1 **Personal Protective Equipment:** Protective gloves; goggles or face shield.
5.2 **Symptoms Following Exposure:** Vapor causes slight irritation of eyes and nose. Liquid irritates stomach; if taken into lungs, causes coughing, distress, and rapidly developing pulmonary edema.
5.3 **Treatment for Exposure:** ASPIRATION: enforce bed rest; administer oxygen; call a doctor. INGESTION: do NOT induce vomiting; call a doctor. EYES: wash with plenty of water. SKIN: wipe off and wash with soap and water.
5.4 **Toxicity by Inhalation (Threshold Limit Value):** 200 ppm
5.5 **Short-Term Inhalation Limits:** 2500 mg/m^3 for 60 min.
5.6 **Toxicity by Ingestion:** Grade 2; LD_{50} 0.5 to 5 g/kg
5.7 **Late Toxicity:** Data not available
5.8 **Vapor (Gas) Irritant Characteristics:** Vapors cause a slight smarting of the eyes or respiratory system if present in high concentrations. The effect is temporary.
5.9 **Liquid or Solid Irritant Characteristics:** Minimum hazard. If spilled on clothing and allowed to remain, may cause smarting and reddening of the skin.
5.10 **Odor Threshold:** 1 ppm

6. FIRE HAZARDS

6.1 **Flash Point:** −10°F to +30°F C.C.
6.2 **Flammable Limits in Air:** 1.3%–8.0%
6.3 **Fire Extinguishing Agents:** Foam, dry chemical, or carbon dioxide
6.4 **Fire Extinguishing Agents Not to be Used:** Not pertinent
6.5 **Special Hazards of Combustion Products:** Not pertinent
6.6 **Behavior in Fire:** Not pertinent
6.7 **Ignition Temperature:** 464°F
6.8 **Electrical Hazard:** Not pertinent
6.9 **Burning Rate:** 4 mm/min.

7. CHEMICAL REACTIVITY

7.1 **Reactivity with Water:** No reaction
7.2 **Reactivity with Common Materials:** No reaction
7.3 **Stability During Transport:** Stable
7.4 **Neutralizing Agents for Acids and Caustics:** Not pertinent
7.5 **Polymerization:** Not pertinent
7.6 **Inhibitor of Polymerization:** Not pertinent

8. WATER POLLUTION

8.1 **Aquatic Toxicity:** 500 ppm/*/salmon fingerling/lethal/fresh water
*Time period not specified
8.2 **Waterfowl Toxicity:** Data not available
8.3 **Biological Oxygen Demand (BOD):** 53%, 5 days
8.4 **Food Chain Concentration Potential:** None

9. SELECTED MANUFACTURERS

1. Exxon Co.
30 Rockefeller Plaza
New York, N. Y. 10020
2. Shell Oil Co.
1 Shell Plaza
Houston, Tex. 77001
3. Sun Oil Co.
240 Radnor Rd.
St. Davids, Pa. 19087

10. SHIPPING INFORMATION

10.1 **Grades or Purity:** 100%
10.2 **Storage Temperature:** Ambient
10.3 **Inert Atmosphere:** No requirement
10.4 **Venting:** Open (flame arrester) or pressure-vacuum

11. HAZARD ASSESSMENT CODE

(See Hazard Assessment Handbook, CG 446-3)

A-T-U

12. HAZARD CLASSIFICATIONS

12.1 **Code of Federal Regulations:** Flammable liquid
12.2 **NAS Hazard Rating for Bulk Water Transportation:**

Category	Rating
Fire	3
Health	
Vapor Irritant	1
Liquid or Solid Irritant	1
Poisons	1
Water Pollution	
Human Toxicity	1
Aquatic Toxicity	1
Aesthetic Effect	3
Reactivity	
Other Chemicals	0
Water	0
Self-Reaction	0

12.3 **NFPA Hazard Classifications:**

Category	Classification
Health Hazard (Blue)	0
Flammability (Red)	2
Reactivity (Yellow)	0

13. PHYSICAL AND CHEMICAL PROPERTIES

13.1 **Physical State at 15°C and 1 atm:** Liquid
13.2 **Molecular Weight:** Not pertinent
13.3 **Boiling Point at 1 atm:** 349 – 549°F = 176 – 287°C = 449 – 560°K
13.4 **Freezing Point:** < −54°F = < −48°C = < 225°K
13.5 **Critical Temperature:** Not pertinent
13.6 **Critical Pressure:** Not pertinent
13.7 **Specific Gravity:** 0.81 at 20°C (liquid)
13.8 **Liquid Surface Tension:** (est.) 25 dynes/cm = 0.025 N/m at 20°C
13.9 **Liquid-Water Interfacial Tension:** (est.) 50 dynes/cm = 0.05 N/m at 20°C
13.10 **Vapor (Gas) Specific Gravity:** Not pertinent
13.11 **Ratio of Specific Heats of Vapor (Gas):** (est.) 1.030
13.12 **Latent Heat of Vaporization:** 140 Btu/lb = 78 cal/g = 3.3×10^5 J/kg
13.13 **Heat of Combustion:** −18,540 Btu/lb = −10,300 cal/g = -431.24×10^5 J/kg
13.14 **Heat of Decomposition:** Not pertinent
13.15 **Heat of Solution:** Not pertinent
13.16 **Heat of Polymerization:** Not pertinent

NOTES

REVISED 1978

JPV

JET FUELS: JP-5

Common Synonyms

Liquid | Colorless | Fuel oil odor

Floats on water.

Stop discharge if possible.
Call fire department.
Avoid contact with liquid.
Isolate and remove discharged material.
Notify local health and pollution control agencies.

Fire

Combustible.
Extinguish with dry chemical, foam, or carbon dioxide.
Water may be ineffective on fire.
Cool exposed containers with water.

Exposure

CALL FOR MEDICAL AID.

LIQUID
Irritating to skin and eyes.
Harmful if swallowed.
Remove contaminated clothing and shoes.
Flush affected areas with plenty of water.
IF IN EYES, hold eyelids open and flush with plenty of water.
IF SWALLOWED and victim is CONSCIOUS, have victim drink water or milk.
DO NOT INDUCE VOMITING.

Water Pollution

Dangerous to aquatic life in high concentrations.
Fouling to shoreline.
May be dangerous if it enters water intakes.
Notify local health and wildlife officials.
Notify operators of nearby water intakes.

1. RESPONSE TO DISCHARGE

(See Response Methods Handbook, CG 446-4)

Mechanical containment
Should be removed
Chemical and physical treatment

2. LABELS

No hazard label required by Code of Federal Regulations

3. CHEMICAL DESIGNATIONS

3.1 **Synonyms:** Kerosene, heavy

3.2 **Coast Guard Compatibility Classification:** Miscellaneous hydrocarbon mixtures

3.3 **Chemical Formula:** Not pertinent

3.4 **IMCO/United Nations Numerical Designation:** 3.3/1223

4. OBSERVABLE CHARACTERISTICS

4.1 **Physical State (as shipped):** Liquid

4.2 **Color:** Colorless to light brown

4.3 **Odor:** Like fuel oil

5. HEALTH HAZARDS

5.1 **Personal Protective Equipment:** Protective gloves; goggles or face shield.

5.2 **Symptoms Following Exposure:** Vapor causes slight irritation of eyes and nose. Liquid irritates stomach; if taken into lungs, causes coughing, distress, and rapidly developing pulmonary edema.

5.3 **Treatment for Exposure:** ASPIRATION: enforce bed rest; administer oxygen; call a doctor. INGESTION: do NOT induce vomiting; call a doctor. EYES: wash with plenty of water. SKIN: wipe off and wash with soap and water.

5.4 **Toxicity by Inhalation (Threshold Limit Value):** 200 ppm

5.5 **Short-Term Inhalation Limits:** 2500 mg/m^3 for 60 min.

5.6 **Toxicity by Ingestion:** Grade 2; LD_{50} 0.5 to 5 g/kg

5.7 **Late Toxicity:** Data not available

5.8 **Vapor (Gas) Irritant Characteristics:** Vapors cause a slight smarting of the eyes or respiratory system if present in high concentrations. The effect is temporary.

5.9 **Liquid or Solid Irritant Characteristics:** Minimum hazard. If spilled on clothing and allowed to remain, may cause smarting and reddening of the skin.

5.10 **Odor Threshold:** 1 ppm

6. FIRE HAZARDS

6.1 **Flash Point:** 140°F (min.) C.C.

6.2 **Flammable Limits in Air:** 0.6% 4.6%

6.3 **Fire Extinguishing Agents:** Foam, dry chemical, or carbon dioxide

6.4 **Fire Extinguishing Agents Not to be Used:** Water may be ineffective

6.5 **Special Hazards of Combustion Products:** Not pertinent

6.6 **Behavior in Fire:** Not pertinent

6.7 **Ignition Temperature:** 475°F

6.8 **Electrical Hazard:** Not pertinent

6.9 **Burning Rate:** 4 mm/min.

7. CHEMICAL REACTIVITY

7.1 **Reactivity with Water:** No reaction

7.2 **Reactivity with Common Materials:** No reaction

7.3 **Stability During Transport:** Stable

7.4 **Neutralizing Agents for Acids and Caustics:** Not pertinent

7.5 **Polymerization:** Not pertinent

7.6 **Inhibitor of Polymerization:** Not pertinent

8. WATER POLLUTION

8.1 **Aquatic Toxicity:** 500 ppm/*/salmon fingerling/lethal/fresh water
*Time period not specified

8.2 **Waterfowl Toxicity:** Data not available

8.3 **Biological Oxygen Demand (BOD):** 53%, 5 days

8.4 **Food Chain Concentration Potential:** None

9. SELECTED MANUFACTURERS

1. Shell Oil Co.
1 Shell Plaza
Houston, Tex. 77001
2. Exxon Co.
30 Rockefeller Plaza
New York, N. Y. 10020
3. Sun Oil Co.
240 Radnor Rd.
St. Davids, Pa. 19087

10. SHIPPING INFORMATION

10.1 **Grades or Purity:** 100%

10.2 **Storage Temperature:** Ambient

10.3 **Inert Atmosphere:** No requirement

10.4 **Venting:** Open (flame arrester)

11. HAZARD ASSESSMENT CODE

(See Hazard Assessment Handbook, CG 446-3)

A-T-U

12. HAZARD CLASSIFICATIONS

12.1 **Code of Federal Regulations:** Combustible Liquid

12.2 **NAS Hazard Rating for Bulk Water Transportation:**

Category	Rating
Fire	1-2
Health	
Vapor Irritant	1
Liquid or Solid Irritant	1
Poisons	1
Water Pollution	
Human Toxicity	1
Aquatic Toxicity	1
Aesthetic Effect	3
Reactivity	
Other Chemicals	0
Water	0
Self-Reaction	0

12.3 **NFPA Hazard Classifications:**

Category	Classification
Health Hazard (Blue)	0
Flammability (Red)	2
Reactivity (Yellow)	0

13. PHYSICAL AND CHEMICAL PROPERTIES

13.1 **Physical State at 15°C and 1 atm:** Liquid

13.2 **Molecular Weight:** Not pertinent

13.3 **Boiling Point at 1 atm:** 349-549°F = 176-287°C = 449-560°K

13.4 **Freezing Point:** < −54°F = < −48°C = < −225°K

13.5 **Critical Temperature:** Not pertinent

13.6 **Critical Pressure:** Not pertinent

13.7 **Specific Gravity:** 0.82 at 15°C (liquid)

13.8 **Liquid Surface Tension:** (est.) 25 dynes/cm = 0.025 N/m at 20°C

13.9 **Liquid-Water Interfacial Tension:** (est.) 50 dynes/cm = 0.05 N/m at 20°C

13.10 **Vapor (Gas) Specific Gravity:** Not pertinent

13.11 **Ratio of Specific Heats of Vapor (Gas):** Not pertinent

13.12 **Latent Heat of Vaporization:** 140 Btu/lb = 78 cal/g = 3.3×10^5 J/kg

13.13 **Heat of Combustion:** −18,540 Btu/lb = −10,300 cal/g = -431.24×10^5 J/kg

13.14 **Heat of Decomposition:** Not pertinent

13.15 **Heat of Solution:** Not pertinent

13.16 **Heat of Polymerization:** Not pertinent

NOTES

REVISED 1978

KRS	KEROSENE

Common Synonyms			
Illuminating oil Kerosine Range oil Fuel oil No. 1 Jet Fuel: JP-1	Watery liquid	Colorless	Fuel oil odor
	Floats on water.		

Stop discharge if possible.
Call fire department.
Avoid contact with liquid.
Isolate and remove discharged material.
Notify local health and pollution control agencies.

Fire	Combustible. Extinguish with dry chemical, foam, or carbon dioxide. Water may be ineffective on fire. Cool exposed containers with water.
Exposure	CALL FOR MEDICAL AID. LIQUID Irritating to skin and eyes. Harmful if swallowed. Remove contaminated clothing and shoes. Flush affected areas with plenty of water. IF IN EYES, hold eyelids open and flush with plenty of water. IF SWALLOWED and victim is CONSCIOUS, have victim drink water or milk. DO NOT INDUCE VOMITING.
Water Pollution	Dangerous to aquatic life in high concentrations. Fouling to shoreline. May be dangerous if it enters water intakes. Notify local health and wildlife officials. Notify operators of nearby water intakes.

1. RESPONSE TO DISCHARGE

(See Response Methods Handbook, CG 446-4)

Mechanical containment
Should be removed
Chemical and physical treatment

2. LABELS

No hazard label required by Code of Federal Regulations

3. CHEMICAL DESIGNATIONS

3.1 **Synonyms:** No. 1 Fuel oil; Kerosine; Illuminating oil; Range oil; JP-1

3.2 **Coast Guard Compatibility Classification:** Miscellaneous hydrocarbon mixtures

3.3 **Chemical Formula:** C_nH_{2n+2}

3.4 **IMCO/United Nations Numerical Designation:** 3.3/1223

4. OBSERVABLE CHARACTERISTICS

4.1 **Physical State (as shipped):** Liquid

4.2 **Color:** Colorless to light brown

4.3 **Odor:** Characteristic; like fuel oil

5. HEALTH HAZARDS

5.1 **Personal Protective Equipment:** Protective gloves; goggles or face shield.

5.2 **Symptoms Following Exposure:** Vapor causes slight irritation of eyes and nose. Liquid irritates stomach; if taken into lungs, causes coughing, distress, and rapidly developing pulmonary edema.

5.3 **Treatment for Exposure:** ASPIRATION: enforce bed rest; administer oxygen; call a doctor. INGESTION: do NOT induce vomiting; call a doctor. EYES: wash with plenty of water. SKIN: wipe off and wash with soap and water.

5.4 **Toxicity by Inhalation (Threshold Limit Value):** 200 ppm

5.5 **Short-Term Inhalation Limits:** 2500 mg/m^3 for 60 min.

5.6 **Toxicity by Ingestion:** Grade 1; LD_{50} 5 to 15 g/kg

5.7 **Late Toxicity:** Data not available

5.8 **Vapor (Gas) Irritant Characteristics:** Vapors cause a slight smarting of the eyes or respiratory system if present in high concentrations. The effect is temporary.

5.9 **Liquid or Solid Irritant Characteristics:** Minimum hazard. If spilled on clothing and allowed to remain, may cause smarting and reddening of the skin.

5.10 **Odor Threshold:** 1 ppm

6. FIRE HAZARDS

6.1 **Flash Point:** 100°F (min.)C.C.

6.2 **Flammable Limits in Air:** 0.7% – 5%

6.3 **Fire Extinguishing Agents:** Foam, dry chemical, or carbon dioxide

6.4 **Fire Extinguishing Agents Not to be Used:** Water may be ineffective

6.5 **Special Hazards of Combustion Products:** Not pertinent

6.6 **Behavior in Fire:** Not pertinent

6.7 **Ignition Temperature:** 444°F

6.8 **Electrical Hazard:** Not pertinent

6.9 **Burning Rate:** 4 mm/min.

7. CHEMICAL REACTIVITY

7.1 **Reactivity with Water:** No reaction

7.2 **Reactivity with Common Materials:** No reaction

7.3 **Stability During Transport:** Stable

7.4 **Neutralizing Agents for Acids and Caustics:** Not pertinent

7.5 **Polymerization:** Not pertinent

7.6 **Inhibitor of Polymerization:** Not pertinent

8. WATER POLLUTION

8.1 **Aquatic Toxicity:** 2990 ppm/24 hr/bluegill/TL_m/fresh water

8.2 **Waterfowl Toxicity:** Data not available

8.3 **Biological Oxygen Demand (BOD):** 53%, 5 days

8.4 **Food Chain Concentration Potential:** None

9. SELECTED MANUFACTURERS

1. Atlantic Richfield Co.
717 Fifth Ave.
New York, N. Y. 10022
2. Shell Oil Co.
1 Shell Plaza
Houston, Tex. 77001
3. Sun Oil Co.
240 Radnor Rd.
St. Davids, Pa. 19087

10. SHIPPING INFORMATION

10.1 **Grades or Purity:** Light hydrocarbon distillate: 100%

10.2 **Storage Temperature:** Ambient

10.3 **Inert Atmosphere:** No requirement

10.4 **Venting:** Open (flame arrester)

11. HAZARD ASSESSMENT CODE

(See Hazard Assessment Handbook, CG 446-3)

A-T-U

12. HAZARD CLASSIFICATIONS

12.1 **Code of Federal Regulations:** Combustible Liquid

12.2 **NAS Hazard Rating for Bulk Water Transportation:**

Category	Rating
Fire	2
Health	
Vapor Irritant	1
Liquid or Solid Irritant	1
Poisons	1
Water Pollution	
Human Toxicity	1
Aquatic Toxicity	1
Aesthetic Effect	3
Reactivity	
Other Chemicals	0
Water	0
Self-Reaction	0

12.3 **NFPA Hazard Classifications:**

Category	Classification
Health Hazard (Blue)	0
Flammability (Red)	2
Reactivity (Yellow)	0

13. PHYSICAL AND CHEMICAL PROPERTIES

13.1 **Physical State at 15°C and 1 atm:** Liquid

13.2 **Molecular Weight:** Not pertinent

13.3 **Boiling Point at 1 atm:** 392-500°F = 200-260°C = 473-533°K

13.4 **Freezing Point:** −50°F = −45.6°C = 227.6°K

13.5 **Critical Temperature:** Not pertinent

13.6 **Critical Pressure:** Not pertinent

13.7 **Specific Gravity:** 0.80 at 15°C (liquid)

13.8 **Liquid Surface Tension:** 23-32 dynes/cm = 0.023-0.032 N/m at 20°C

13.9 **Liquid-Water Interfacial Tension:** 47-49 dynes/cm = 0.047-0.049 N/m at 20°C

13.10 **Vapor (Gas) Specific Gravity:** Not pertinent

13.11 **Ratio of Specific Heats of Vapor (Gas):** Not pertinent

13.12 **Latent Heat of Vaporization:** 110 Btu/lb = 60 cal/g = 2.5 × 10^5 J/kg

13.13 **Heat of Combustion:** −18,540 Btu/lb = −10,300 cal/g = −431.24 × 10^5 J/kg

13.14 **Heat of Decomposition:** Not pertinent

13.15 **Heat of Solution:** Not pertinent

13.16 **Heat of Polymerization:** Not pertinent

NOTES

REVISED 1978

LTA — LACTIC ACID

Common Synonyms
2-Hydroxypropanoic acid
Racemic lactic acid
Milk acid
Alpha-Hydroxypropionic acid

Thick Liquid — Colorless to Yellow — Weak Unpleasant Odor

Sinks and Mixes with water.

Avoid contact with liquid and vapor, keep people away.
Stop discharge if possible.
Isolate and remove discharged material.
Notify local health and pollution control agencies.

Fire

Combustible.
Extinguish with water, dry chemicals, or carbon dioxide.

Exposure

CALL FOR MEDICAL AID.

VAPOR
Irritating to eyes, nose and throat.
If inhaled will cause coughing or difficult breathing.

If in eyes, hold eyelids open and flush with plenty of water.
If breathing has stopped, give artificial respiration.
If breathing is difficult, give oxygen.

LIQUID
Will burn skin and eyes.
If swallowed will cause nausea.

Remove contaminated clothing and shoes.
Flush affected areas with plenty of water.
IF IN EYES, hold eyelids open and flush with plenty of water.
IF SWALLOWED and victim is CONSCIOUS, have victim drink water or milk.
IF SWALLOWED and victim is UNCONSCIOUS OR HAVING CONVULSIONS, do nothing except keep victim warm.

Water Pollution

Dangerous to aquatic life in high concentrations.
May be dangerous if it enters water intakes.

Notify local health and wildlife officials.
Notify operators of nearby water intakes.

1. RESPONSE TO DISCHARGE

(See Response Methods Handbook, CG 446-4)

Disperse and flush

2. LABELS

No hazard label required by Code of Federal Regulations

3. CHEMICAL DESIGNATIONS

3.1 **Synonyms:** 2-Hydroxypropanoic acid; alpha-Hydroxypropionic acid; Milk acid; Racemic lactic acid
3.2 **Coast Guard Compatibility Classification:** Not listed
3.3 **Chemical Formula:** $CH_3CHOHCOOH \cdot H_2O$
3.4 **IMCO/United Nations Numerical Designation:** Not listed

4. OBSERVABLE CHARACTERISTICS

4.1 **Physical State (as shipped):** Syrupy liquid
4.2 **Color:** Yellow to colorless
4.3 **Odor:** None or weak acidic

5. HEALTH HAZARDS

5.1 **Personal Protective Equipment:** Rubber gloves; goggles; self-contained breathing apparatus where high concentrations of mist are present
5.2 **Symptoms Following Exposure:** Inhalation of mist causes coughing and irritation of mucous membranes. Ingestion, even of diluted preparations, has a corrosive effect on the esophagus and stomach. Contact with more concentrated solutions can cause severe burns of skin or eye.
5.3 **Treatment for Exposure:** INHALATION: move to fresh air. INGESTION: give large amount of water. EYES: flush with water for at least 15 min. SKIN: flush with water; wash well with soap and water.
5.4 **Toxicity by Inhalation (Threshold Limit Value):** Data not available
5.5 **Short-Term Inhalation Limits:** Data not available
5.6 **Toxicity by Ingestion:** Grade 2; oral LD_{50} = 1,810 mg/kg (guinea pig)
5.7 **Late Toxicity:** Data not available
5.8 **Vapor (Gas) Irritant Characteristics:** Data not available
5.9 **Liquid or Solid Irritant Characteristics:** Data not available
5.10 **Odor Threshold:** 4.0×10^{-7} ppm

6. FIRE HAZARDS

6.1 **Flash Point:** Not pertinent (not flammable, or burns with difficulty)
6.2 **Flammable Limits in Air:** Not pertinent
6.3 **Fire Extinguishing Agents:** Water, foam, dry chemical, carbon dioxide
6.4 **Fire Extinguishing Agents Not to be Used:**
6.5 **Special Hazards of Combustion Products:**
6.6 **Behavior in Fire:**
6.7 **Ignition Temperature:** Not pertinent
6.8 **Electrical Hazard:** Not pertinent
6.9 **Burning Rate:** Not pertinent

7. CHEMICAL REACTIVITY

7.1 **Reactivity with Water:** No reaction
7.2 **Reactivity with Common Materials:** Slowly corrodes most metals
7.3 **Stability During Transport:** Stable
7.4 **Neutralizing Agents for Acids and Caustics:** Dilute with water; rinse with sodium bicarbonate or lime solution.
7.5 **Polymerization:** Not pertinent
7.6 **Inhibitor of Polymerization:** Not pertinent

8. WATER POLLUTION

8.1 **Aquatic Toxicity:**
430 ppm/> 100 hr/goldfish/no effect/fresh water
654 ppm/6 hr/goldfish/killed/fresh water
8.2 **Waterfowl Toxicity:** Data not available
8.3 **Biological Oxygen Demand (BOD):** 72%, 5 days
8.4 **Food Chain Concentration Potential:** None

9. SELECTED MANUFACTURERS

1. Clinton Corn Processing Co.
   P. O. Box 340
   Clinton, Iowa 52732
2. Mallinckrodt Chemical Works
   223 Westside Avenue
   P. O. Box 384
   Jersey City, N. J. 07303
3. Monsanto Company
   800 North Lindbergh Blvd.
   St. Louis, Mo. 63166

10. SHIPPING INFORMATION

10.1 **Grades or Purity:** USP; Reagent; Technical, 88%; Food processing, 50%, 80%. The balance is water in all cases.
10.2 **Storage Temperature:** Ambient
10.3 **Inert Atmosphere:** No requirement
10.4 **Venting:** Open

11. HAZARD ASSESSMENT CODE

(See Hazard Assessment Handbook, CG 446-3)

A-P

12. HAZARD CLASSIFICATIONS

12.1 **Code of Federal Regulations:** Not listed
12.2 **NAS Hazard Rating for Bulk Water Transportation:** Not listed
12.3 **NFPA Hazard Classifications:** Not listed

13. PHYSICAL AND CHEMICAL PROPERTIES

13.1 **Physical State at 15°C and 1 atm:** Liquid
13.2 **Molecular Weight:** 90
13.3 **Boiling Point at 1 atm:** Not pertinent (decomposes)
13.4 **Freezing Point:** Not pertinent
13.5 **Critical Temperature:** Not pertinent
13.6 **Critical Pressure:** Not pertinent
13.7 **Specific Gravity:** 1.20 at 20° (liquid)
13.8 **Liquid Surface Tension:** Data not available
13.9 **Liquid-Water Interfacial Tension:** Not pertinent
13.10 **Vapor (Gas) Specific Gravity:** Not pertinent
13.11 **Ratio of Specific Heats of Vapor (Gas):** Not pertinent
13.12 **Latent Heat of Vaporization:** Not pertinent
13.13 **Heat of Combustion:** −6,520 Btu/lb = −3,620 cal/g = -152×10^5 J/kg
13.14 **Heat of Decomposition:** Not pertinent
13.15 **Heat of Solution:** Not pertinent
13.16 **Heat of Polymerization:** Not pertinent

NOTES

LLS	# LATEX, LIQUID SYNTHETIC

Common Synonyms Plastic latex Synthetic rubber latex	Liquid White Mixes with water.

Stop discharge if possible.
Isolate and remove discharged material.
Notify local health and pollution control agencies.

Fire	Not flammable. Combustible solid is produced when heated.
Exposure	CALL FOR MEDICAL AID. LIQUID Irritating to eyes. IF IN EYES, hold eyelids open and flush with plenty of water.
Water Pollution	Effect of low concentrations on aquatic life is unknown. Fouling to shoreline. May be dangerous if it enters water intakes. Notify local health and wildlife officials. Notify operators of nearby water intakes.

1. RESPONSE TO DISCHARGE

(See Response Methods Handbook, CG 446-4)

Disperse and flush (if not coagulated)
Should be removed (if coagulated)
Chemical and physical treatment

2. LABELS

No hazard label required by Code of Federal Regulations

3. CHEMICAL DESIGNATIONS

3.1 **Synonyms:** Plastic latex
Synthetic rubber latex
3.2 **Coast Guard Compatibility Classification:** Miscellaneous water solutions
3.3 **Chemical Formula:** Not pertinent
3.4 **IMCO/United Nations Numerical Designation:** Not listed

4. OBSERVABLE CHARACTERISTICS

4.1 **Physical State (as shipped):** Liquid
4.2 **Color:** Milky
4.3 **Odor:** Each type has a characteristic odor

5. HEALTH HAZARDS

5.1 **Personal Protective Equipment:** Chemical goggles or face shield.
5.2 **Symptoms Following Exposure:** Irritation of eyes.
5.3 **Treatment for Exposure:** Flush eyes with water for at least 15 min.
5.4 **Toxicity by Inhalation (Threshold Limit Value):** Not pertinent
5.5 **Short-Term Inhalation Limits:** Not pertinent
5.6 **Toxicity by Ingestion:** Data not available
5.7 **Late Toxicity:** Data not available
5.8 **Vapor (Gas) Irritant Characteristics:** None
5.9 **Liquid or Solid Irritant Characteristics:** Contact with eyes can cause irritation.
5.10 **Odor Threshold:** Not pertinent

6. FIRE HAZARDS

6.1 **Flash Point:** Not flammable unless coagulated.
6.2 **Flammable Limits in Air:** Not flammable
6.3 **Fire Extinguishing Agents:** Not pertinent
6.4 **Fire Extinguishing Agents Not to be Used:** Not pertinent
6.5 **Special Hazards of Combustion Products:** If the latex dries out and then burns, hydrochloric acid, hydrogen cyanide and styrene gases may be evolved. All are irritating and poisonous.
6.6 **Behavior in Fire:** Heat may coagulate the latex and form sticky plastic lumps which may burn.
6.7 **Ignition Temperature:** Not flammable
6.8 **Electrical Hazard:** Data not available
6.9 **Burning Rate:** Not flammable

7. CHEMICAL REACTIVITY

7.1 **Reactivity with Water:** No reaction
7.2 **Reactivity with Common Materials:** No reaction
7.3 **Stability During Transport:** Coagulated by heat and acids to gummy, flammable material.
7.4 **Neutralizing Agents for Acids and Caustics:** Not pertinent
7.5 **Polymerization:** Not pertinent
7.6 **Inhibitor of Polymerization:** Not pertinent

8. WATER POLLUTION

8.1 **Aquatic Toxicity:** Data not available
8.2 **Waterfowl Toxicity:** Data not available
8.3 **Biological Oxygen Demand (BOD):** 1%, 5 days
8.4 **Food Chain Concentration Potential:** None

9. SELECTED MANUFACTURERS

1. Borg-Warner Corp.
Marbon Division
Washington, W. Va. 26181
2. W. R. Grace & Co.
Dewey and Almy Chemical Division
Cambridge, Mass. 02140
3. Rohm and Haas Co.
Independence Mall West
Philadelphia, Pa. 19105

10. SHIPPING INFORMATION

10.1 **Grades or Purity:** All commercial latices are shipped in a variety of concentrations in water, depending on the particular polymer involved and the intended use of the latex. None are particularly hazardous except in fires, where all coagulate to gummy, flammable material.
10.2 **Storage Temperature:** Ambient
10.3 **Inert Atmosphere:** No requirement
10.4 **Venting:** Open

HAZARD ASSESSMENT CODE

(See Hazard Assessment Handbook, CG 446-3)

A-P

12. HAZARD CLASSIFICATIONS

12.1 **Code of Federal Regulations:** Not listed
12.2 **NAS Hazard Rating for Bulk Water Transportation:** Not listed
12.3 **NFPA Hazard Classifications:** Not listed

13. PHYSICAL AND CHEMICAL PROPERTIES

13.1 **Physical State at 15°C and 1 atm:** Liquid
13.2 **Molecular Weight:** Not pertinent
13.3 **Boiling Point at 1 atm:** Very high
13.4 **Freezing Point:** Not pertinent
13.5 **Critical Temperature:** Not pertinent
13.6 **Critical Pressure:** Not pertinent
13.7 **Specific Gravity:** 1.057 at 25°C (liquid)
13.8 **Liquid Surface Tension:** Not pertinent
13.9 **Liquid-Water Interfacial Tension:** Not pertinent
13.10 **Vapor (Gas) Specific Gravity:** Not pertinent
13.11 **Ratio of Specific Heats of Vapor (Gas):** Not pertinent
13.12 **Latent Heat of Vaporization:** Not pertinent
13.13 **Heat of Combustion:** Not pertinent
13.14 **Heat of Decomposition:** Not pertinent
13.15 **Heat of Solution:** Not pertinent
13.16 **Heat of Polymerization:** Not pertinent

NOTES

REVISED 1978

LPO	LAUROYL PEROXIDE

Common Synonyms			
Dilauroyl peroxide Dodecanoyl peroxide	Solid	White	Faint soapy odor
	Floats on water.		

Call fire department.
Keep people away.
Isolate and remove discharged material.
Notify local health and pollution control agencies.

Fire	Combustible. May cause fire on contact with combustibles. Containers may explode in fire. May explode if exposed to heat or flames. Combat fires from behind barrier. Flood discharge area with water. Cool exposed containers with water.
Exposure	Call for medical aid. SOLID Irritating to skin and eyes. Harmful if swallowed. Remove contaminated clothing and shoes. Flush affected areas with plenty of water. IF IN EYES, hold eyelids open and flush with plenty of water. IF SWALLOWED and victim is CONSCIOUS, have victim drink water or milk and have victim induce vomiting. IF SWALLOWED and victim is UNCONSCIOUS OR HAVING CONVULSIONS, do nothing except keep victim warm.
Water Pollution	Effect of low concentrations on aquatic life is unknown. May be dangerous if it enters water intakes. Notify local health and wildlife officials. Notify operators of nearby water intakes.

1. RESPONSE TO DISCHARGE

(See Response Methods Handbook, CG 446-4)

Issue warning—oxidizing material, water contaminant
Mechanical containment
Should be removed
Chemical and physical treatment

2. LABEL

Yellow

3. CHEMICAL DESIGNATIONS

3.1 **Synonyms:** Dilauroyl peroxide
Dodecanoyl peroxide

3.2 **Coast Guard Compatibility Classification:** Not applicable

3.3 **Chemical Formula:** $[CH_3(CH_2)_{10}COO]_2$

3.4 **IMCO/United Nations Numerical Designation:** 5.2/1530

4. OBSERVABLE CHARACTERISTICS

4.1 **Physical State (as shipped):** Solid

4.2 **Color:** White

4.3 **Odor:** Faint pungent; bland, soapy

5. HEALTH HAZARDS

5.1 **Personal Protective Equipment:** Protective gloves, goggles

5.2 **Symptoms Following Exposure:** Contact with liquid irritates eyes and skin. Ingestion causes irritation of mouth and stomach.

5.3 **Treatment for Exposure:** EYES: wash with plenty of water for 15 min. and get medical attention. SKIN: wash with plenty of soap and water. INGESTION: administer an emetic to induce vomiting and call a physician

5.4 **Toxicity by Inhalation (Threshold Limit Value):** Data not available

5.5 **Short-Term Inhalation Limits:** Not pertinent

5.6 **Toxicity by Ingestion:** Data not available

5.7 **Late Toxicity:** Weak carcinogen in mice

5.8 **Vapor (Gas) Irritant Characteristics:** Data not available

5.9 **Liquid or Solid Irritant Characteristics:** Data not available

5.10 **Odor Threshold:** Data not available

6. FIRE HAZARDS

6.1 **Flash Point:** Not pertinent (oxidizing combustible solid)

6.2 **Flammable Limits in Air:** Not pertinent

6.3 **Fire Extinguishing Agents:** Water, dry chemical, foam, or carbon dioxide

6.4 **Fire Extinguishing Agents Not to be Used:** Not pertinent

6.5 **Special Hazards of Combustion Products:** Not pertinent

6.6 **Behavior in Fire:** Can increase the severity of a fire. Becomes sensitive to shock when hot. Containers may explode in a fire. May ignite or explode spontaneously if mixed with flammable materials.

6.7 **Ignition Temperature:** Not pertinent

6.8 **Electrical Hazard:** Not pertinent

6.9 **Burning Rate:** Not pertinent

7. CHEMICAL REACTIVITY

7.1 **Reactivity with Water:** No reaction

7.2 **Reactivity with Common Materials:** May ignite or explode spontaneously when mixed with combustible materials.

7.3 **Stability During Transport:** Stable if not overheated

7.4 **Neutralizing Agents for Acids and Caustics:** Not pertinent

7.5 **Polymerization:** Not pertinent

7.6 **Inhibitor of Polymerization:** Not pertinent

8. WATER POLLUTION

8.1 **Aquatic Toxicity:** Data not available

8.2 **Waterfowl Toxicity:** Data not available

8.3 **Biological Oxygen Demand (BOD):** Data not available

8.4 **Food Chain Concentration Potential:** None

9. SELECTED MANUFACTURERS

1. Pennwalt Corporation
   Lucidol Division
   1740 Military Road
   Buffalo, N. Y. 14240
2. Witco Chemical Corporation
   U. S. Peroxygen Division
   850 Morton Avenue
   Richmond, Calif. 94804
3. Aztec Chemicals Division
   Dart Industries, Inc.
   555 Garden Street
   Elyria, Ohio 44035

10. SHIPPING INFORMATION

10.1 **Grades or Purity:** 97–98%; dry or wetted with water

10.2 **Storage Temperature:** <80°F

10.3 **Inert Atmosphere:** No requirement

10.4 **Venting:** Open

11. HAZARD ASSESSMENT CODE

(See Hazard Assessment Handbook, CG 446-3)

II-Z

12. HAZARD CLASSIFICATIONS

12.1 **Code of Federal Regulations:** Organic Peroxide

12.2 **NAS Hazard Rating for Bulk Water Transportation:** Not listed

12.3 **NFPA Hazard Classifications:**

	Classification
Health Hazard (Blue)	0
Flammability (Red)	2
Reactivity (Yellow)	3
	oxy

13. PHYSICAL AND CHEMICAL PROPERTIES

13.1 **Physical State at 15°C and 1 atm:** Solid

13.2 **Molecular Weight:** 399

13.3 **Boiling Point at 1 atm:** Decomposes

13.4 **Freezing Point:** 129°F = 54°C = 327°K

13.5 **Critical Temperature:** Not pertinent

13.6 **Critical Pressure:** Not pertinent

13.7 **Specific Gravity:** 0.91 at 25°C (solid)

13.8 **Liquid Surface Tension:** Not pertinent

13.9 **Liquid-Water Interfacial Tension:** Not pertinent

13.10 **Vapor (Gas) Specific Gravity:** Not pertinent

13.11 **Ratio of Specific Heats of Vapor (Gas):** Not pertinent

13.12 **Latent Heat of Vaporization:** Not pertinent

13.13 **Heat of Combustion:** (est.) −16,300 Btu/lb = −9,100 cal/g = -380×10^5 J/kg

13.14 **Heat of Decomposition:** Data not available

13.15 **Heat of Solution:** Not pertinent

13.16 **Heat of Polymerization:** Not pertinent

NOTES

LRM	# LAURYL MERCAPTAN

Common Synonyms			
1-Dodecanethiol Dodecyl mercaptan	Oily liquid	Colorless	Mild skunk odor
	Floats on water. Freezing point is 19° F.		

Wear goggles and self-contained breathing apparatus.
Stop discharge if possible. Keep people away.
Avoid contact with liquid.
Shut off ignition sources and call fire department.
Isolate and remove discharged material.
Notify local health and pollution control agencies.

Fire

Combustible
POISONOUS GASES ARE PRODUCED IN FIRE.
Wear goggles and self-contained breathing apparatus.
Extinguish with dry chemical, foam, or carbon dioxide.
Water may be ineffective on fire.
Cool exposed containers with water.

Exposure

CALL FOR MEDICAL AID.

LIQUID OR SOLID
Irritating to skin and eyes.
If swallowed, will cause nausea.

Remove contaminated clothing and shoes.
Flush affected areas with plenty of water.
IF IN EYES, hold eyelids open and flush with plenty of water.
IF SWALLOWED and victim is CONSCIOUS, have victim drink water or milk.

Water Pollution

Effect of low concentrations on aquatic life is unknown.
Fouling to shoreline.
May be dangerous if it enters water intakes.
Notify local health and wildlife officials.
Notify operators of nearby water intakes.

1. RESPONSE TO DISCHARGE
(See Response Methods Handbook, CG 446-4)

Mechanical containment
Should be removed
Chemical and physical treatment

2. LABELS

No hazard label required by Code of Federal Regulations

3. CHEMICAL DESIGNATIONS

3.1 **Synonyms:** 1-Dodecanethiol
Dodecyl mercaptan
3.2 **Coast Guard Compatibility Classification:** Not applicable
3.3 **Chemical Formula:** $CH_3(CH_2)_{10}CH_2SH$
3.4 **IMCO/United Nations Numerical Designation:** Not listed

4. OBSERVABLE CHARACTERISTICS

4.1 **Physical State (as shipped):** Liquid
4.2 **Color:** Colorless
4.3 **Odor:** Mild skunk

5. HEALTH HAZARDS

5.1 **Personal Protective Equipment:** Respirator when mist is present; rubber or vinyl gloves; chemical goggles; rubber shoes and apron.
5.2 **Symptoms Following Exposure:** Liquid is irritating to skin, eyes, and mucous membranes. Ingestion may cause nausea. Repeated skin exposure can cause dermatitis and may produce a sensitizing effect.
5.3 **Treatment for Exposure:** Get medical attention for all eye exposures and any other serious over-exposures. INHALATION (mist): rinse mouth repeatedly with cold water; treatment is symptomatic. INGESTION: dilute by drinking water; if vomiting occurs, drink more water; administer saline laxative. EYES: flush thoroughly with water; ventilation by electric fan is helpful in removing last traces, especially around eyes and eyelids. SKIN: remove contaminated clothing; flush skin with water; wash exposed area with soap and water.
5.4 **Toxicity by Inhalation (Threshold Limit Value):** Data not available
5.5 **Short-Term Inhalation Limits:** Data not available
5.6 **Toxicity by Ingestion:** Data not available
5.7 **Late Toxicity:** Causes decline in kidney and liver function in rats.
5.8 **Vapor (Gas) Irritant Characteristics:** Irritating concentrations of vapor unlikely, but mist can cause irritation of eyes and upper respiratory tract.
5.9 **Liquid or Solid Irritant Characteristics:** Minimum hazard. If spilled on clothing and allowed to remain, may cause smarting and reddening of the skin.
5.10 **Odor Threshold:** 4 mg/m^3

6. FIRE HAZARDS

6.1 **Flash Point:** 262°F O.C.
6.2 **Flammable Limits in Air:** Data not available
6.3 **Fire Extinguishing Agents:** Dry chemical, foam, or carbon dioxide
6.4 **Fire Extinguishing Agents Not to be Used:** Water or foam may cause frothing.
6.5 **Special Hazards of Combustion Products:** Poisonous and irritating gases (e.g., sulfur dioxide) are generated in fires.
6.6 **Behavior in Fire:** Not pertinent
6.7 **Ignition Temperature:** Data not available
6.8 **Electrical Hazard:** Not pertinent
6.9 **Burning Rate:** Data not available

7. CHEMICAL REACTIVITY

7.1 **Reactivity with Water:** No reaction
7.2 **Reactivity with Common Materials:** No reaction
7.3 **Stability During Transport:** Stable
7.4 **Neutralizing Agents for Acids and Caustics:** Not pertinent
7.5 **Polymerization:** Not pertinent
7.6 **Inhibitor of Polymerization:** Not pertinent

8. WATER POLLUTION

8.1 **Aquatic Toxicity:** Data not available
8.2 **Waterfowl Toxicity:** Data not available
8.3 **Biological Oxygen Demand (BOD):** Data not available
8.4 **Food Chain Concentration Potential:** None

9. SELECTED MANUFACTURERS

1. Occidental Petroleum Corp.
Hooker Chemical Corp.
Stamford, Conn. 06905
2. Phillips Petroleum Co.
Bartlesville, Okla. 74004
3. Uniroyal, Inc.
Uniroyal Chemical Co. Division
Naugatuck, Conn. 06770

10. SHIPPING INFORMATION

10.1 **Grades or Purity:** 95% minimum
10.2 **Storage Temperature:** Ambient
10.3 **Inert Atmosphere:** No requirement
10.4 **Venting:** Open (flame arrester)

11. HAZARD ASSESSMENT CODE
(See Hazard Assessment Handbook, CG 446-3)

A-T-U

12. HAZARD CLASSIFICATIONS

12.1 **Code of Federal Regulations:** Not listed
12.2 **NAS Hazard Rating for Bulk Water Transportation:** Not listed
12.3 **NFPA Hazard Classifications:**

Category	Classification
Health Hazard (Blue)	2
Flammability (Red)	1
Reactivity (Yellow)	0

13. PHYSICAL AND CHEMICAL PROPERTIES

13.1 **Physical State at 15°C and 1 atm:** Liquid
13.2 **Molecular Weight:** 202
13.3 **Boiling Point at 1 atm:** Very high
13.4 **Freezing Point:** 19.4°F = −7.0°C = 266.2°K
13.5 **Critical Temperature:** Not pertinent
13.6 **Critical Pressure:** Not pertinent
13.7 **Specific Gravity:** 0.85 at 15°C (liquid)
13.8 **Liquid Surface Tension:** (est.) 30 dynes/cm = 0.03 N/m at 20°C
13.9 **Liquid-Water Interfacial Tension:** (est.) 30 dynes/cm = 0.03 N/m at 20°C
13.10 **Vapor (Gas) Specific Gravity:** Not pertinent
13.11 **Ratio of Specific Heats of Vapor (Gas):** Not pertinent
13.12 **Latent Heat of Vaporization:** (est.) 110 Btu/lb = 60 cal/g = 2.5×10^5 J/kg
13.13 **Heat of Combustion:** (est.) −18,200 Btu/lb = −10,100 cal/g = -422×10^5 J/kg
13.14 **Heat of Decomposition:** Not pertinent
13.15 **Heat of Solution:** Not pertinent
13.16 **Heat of Polymerization:** Not pertinent

NOTES

REVISED 1978

LAC — LEAD ACETATE

Common Synonyms			
Lead acetate trihydrate Neutral Lead acetate Normal lead acetate	Solid	White	Odorless
	Sinks and mixes with water.		

AVOID CONTACT WITH SOLID AND DUST. KEEP PEOPLE AWAY.
Wear dust respirator.
Stop discharge if possible.
Isolate and remove discharged material.
Notify local health and pollution control agencies.

Fire	NOT FLAMMABLE.
Exposure	**CALL FOR MEDICAL AID.** DUST POISONOUS IF INHALED. If inhaled will cause dizziness or loss of consciousness. **If in eyes, hold eyelids open and flush with plenty of water..** **If breathing has stopped, give artificial respiration.** **If breathing is difficult, give oxygen.** SOLID Irritating to skin and eyes If swallowed will cause nausea, vomiting or loss of consciousness. **Remove contaminated clothing and shoes.** **Flush affected areas with plenty of water.** **IF IN EYES, hold eyelids open and flush with plenty of water.** **IF SWALLOWED and victim is CONSCIOUS, have victim drink water or milk and have victim induce vomiting.** **IF SWALLOWED and victim is UNCONSCIOUS OR HAVING CONVULSIONS, do nothing except keep victim warm.**
Water Pollution	Dangerous to aquatic life in high concentrations. May be dangerous if it enters water intakes. **Notify local health and wildlife officials.** **Notify operators of nearby water intakes.**

1. RESPONSE TO DISCHARGE

(See Response Methods Handbook, CG 446-4)

Issue warning—water contaminant
Restrict access
Disperse and flush

2. LABELS

No hazard label required by Code of Federal Regulations

3. CHEMICAL DESIGNATIONS

3.1 **Synonyms:** Lead acetate trihydrate; Neutral lead acetate; Normal lead acetate; Salt of Saturn; Sugar of lead

3.2 **Coast Guard Compatibility Classification:** Not listed

3.3 **Chemical Formula:** $Pb(C_2H_3O_2)_2 \cdot 3H_2O$

3.4 **IMCO/United Nations Numerical Designation:** 9/1616

4. OBSERVABLE CHARACTERISTICS

4.1 **Physical State (as shipped):** Solid

4.2 **Color:** White (commercial grades are frequently brown or grey lumps)

4.3 **Odor:** None

5. HEALTH HAZARDS

5.1 **Personal Protective Equipment:** Dust mask and protective gloves

5.2 **Symptoms Following Exposure:** Early symptoms of lead intoxication via inhalation or ingestion are most commonly gastrointestinal disorders, colic, constipation, etc.; weakness, which may go on to paralysis, chiefly of the extensor muscles of the wrists and less often of the ankles, is noticeable in the most serious cases. Ingestion of a large amount causes local irritation of the alimentary tract; pain, leg cramps, muscle weakness, paresthesias, depression, coma, and death may follow in 1 or 2 days. Contact with eyes causes irritation.

5.3 **Treatment for Exposure:** Remove at once all cases of lead intoxication from further exposure until the blood level is reduced to a safe value; immediately place the individual under medical care. INGESTION: give gastric lavage using 1% solution of sodium or magnesium sulfate; leave 15–30 gm magnesium sulfate in 6–8 oz. of water in the stomach as antidote and cathartic; egg white, milk, and tannin are useful demulcents; atropine sulfate and other antispasmodics may relieve abdominal pain, but morphine may be necessary. EYES or SKIN: flush with water.

5.4 **Toxicity by Inhalation (Threshold Limit Value):** 0.2 mg/m^3 (as lead)

5.5 **Short-Term Inhalation Limits:** Data not available

5.6 **Toxicity by Ingestion:** Grade 2; LD_{50} 0.5–5 g/kg

5.7 **Late Toxicity:** Data not available

5.8 **Vapor (Gas) Irritant Characteristics:** Data not available

5.9 **Liquid or Solid Irritant Characteristics:** Data not available

5.10 **Odor Threshold:** Data not available

6. FIRE HAZARDS

6.1 **Flash Point:** Not flammable

6.2 **Flammable Limits in Air:** Not flammable

6.3 **Fire Extinguishing Agents:** Not pertinent

6.4 **Fire Extinguishing Agents Not to be Used:** Not pertinent

6.5 **Special Hazards of Combustion Products:** Irritating acid fumes may be formed in fires.

6.6 **Behavior in Fire:**

6.7 **Ignition Temperature:** Not pertinent

6.8 **Electrical Hazard:** Not pertinent

6.9 **Burning Rate:** Not pertinent

7. CHEMICAL REACTIVITY

7.1 **Reactivity with Water:** No reaction

7.2 **Reactivity with Common Materials:**

7.3 **Stability During Transport:** Stable

7.4 **Neutralizing Agents for Acids and Caustics:** Not pertinent

7.5 **Polymerization:** Not pertinent

7.6 **Inhibitor of Polymerization:** Not pertinent

8. WATER POLLUTION

8.1 **Aquatic Toxicity:**
50 ppm/16–183 days/catfish/bloodcell injury/tap water
7.48 mg/l (as Pb)/4 days/minnow/TL_m/soft water

8.2 **Waterfowl Toxicity:** Data not available

8.3 **Biological Oxygen Demand (BOD):** Data not available

8.4 **Food Chain Concentration Potential:** Fish and terrestrial animals are capable of concentrating lead.

9. SELECTED MANUFACTURERS

1. Filo Color and Chemical Corp.
   347 Madison Ave.
   New York, N. Y. 10017
2. Mallinckrodt Chemical Works
   223 Westside Ave.
   P. O. Box 384
   Jersey City, N. J. 07303
3. Allied Chemical Corp.
   Specialty Chemicals Div.
   P. O. Box 1087R
   Morristown, N. J. 07960

10. SHIPPING INFORMATION

10.1 **Grades or Purity:** NF; Reagent; Technical, 97%

10.2 **Storage Temperature:** Ambient

10.3 **Inert Atmosphere:** No requirement

10.4 **Venting:** Open

11. HAZARD ASSESSMENT CODE

(See Hazard Assessment Handbook, CG 446-3)

SS

12. HAZARD CLASSIFICATIONS

12.1 **Code of Federal Regulations:** Not listed

12.2 **NAS Hazard Rating for Bulk Water Transportation:** Not listed

12.3 **NFPA Hazard Classifications:** Not listed

13. PHYSICAL AND CHEMICAL PROPERTIES

13.1 **Physical State at 15°C and 1 atm:** Solid

13.2 **Molecular Weight:** 379.3

13.3 **Boiling Point at 1 atm:** Not pertinent (decomposes)

13.4 **Freezing Point:** Not pertinent

13.5 **Critical Temperature:** Not pertinent

13.6 **Critical Pressure:** Not pertinent

13.7 **Specific Gravity:** 2.55 at 20°C (solid)

13.8 **Liquid Surface Tension:** Not pertinent

13.9 **Liquid-Water Interfacial Tension:** Not pertinent

13.10 **Vapor (Gas) Specific Gravity:** Not pertinent

13.11 **Ratio of Specific Heats of Vapor (Gas):** Not pertinent

13.12 **Latent Heat of Vaporization:** Not pertinent

13.13 **Heat of Combustion:** Not pertinent

13.14 **Heat of Decomposition:** Not pertinent

13.15 **Heat of Solution:** Not pertinent

13.16 **Heat of Polymerization:** Not pertinent

NOTES

LAR — LEAD ARSENATE

Common Synonyms: Lead arsenate, acid; Plumbous arsenate

Solid — White — Odorless

Sinks in water.

AVOID CONTACT WITH SOLID. Keep people away.
Stop discharge if possible.
Isolate and remove discharged material.
Notify local health and pollution control agencies.

Fire

Not flammable

Exposure

CALL FOR MEDICAL AID.
SOLID
POISONOUS IF SWALLOWED.
IF SWALLOWED and victim is CONSCIOUS, have victim drink water or milk and have victim induce vomiting.
IF SWALLOWED and victim is UNCONSCIOUS OR HAVING CONVULSIONS, do nothing except keep victim warm.

Water Pollution

HARMFUL TO AQUATIC LIFE IN VERY LOW CONCENTRATIONS.
May be dangerous if it enters water intakes.
Notify local health and wildlife officials.
Notify operators of nearby water intakes.

1. RESPONSE TO DISCHARGE

(See Response Methods Handbook, CG 446-4)

Issue warning—poison, water contaminant
Should be removed
Chemical and physical treatment

2. LABEL

3. CHEMICAL DESIGNATIONS

3.1 **Synonyms:** Lead arsenate, acid; Plumbous arsenate
3.2 **Coast Guard Compatibility Classification:** Not applicable
3.3 **Chemical Formula:** $PbHAsO_4$
3.4 **IMCO/United Nations Numerical Designation:** 6.1/1617

4. OBSERVABLE CHARACTERISTICS

4.1 **Physical State (as shipped):** Solid
4.2 **Color:** White
4.3 **Odor:** None

5. HEALTH HAZARDS

5.1 **Personal Protective Equipment:** Dust respirator; protective clothing to prevent accidental inhalation or ingestion of dust.
5.2 **Symptoms Following Exposure:** Inhalation or ingestion causes dizziness, headache, paralysis, cramps, constipation, collapse, coma. Subacute doses cause irritability, loss of weight, anemia, constipation. Blood and urine concentrations of lead increase.
5.3 **Treatment for Exposure:** A specific medical treatment is used for exposure to this chemical; call a physician immediately! Give victim a tablespoon of salt in glass of warm water and repeat until vomit is clear. Then give two tablespoons of epsom salt or milk of magnesia in water, and plenty of milk and water. Have victim lie down and keep quiet.
5.4 **Toxicity by Inhalation (Threshold Limit Value):** (dust) 0.15 mg/m^3
5.5 **Short-Term Inhalation Limits:** Not pertinent
5.6 **Toxicity by Ingestion:** Grade 4; LD_{50} below 50 mg/kg (rabbit, rat)
5.7 **Late Toxicity:** Lead poisoning
5.8 **Vapor (Gas) Irritant Characteristics:** Not pertinent
5.9 **Liquid or Solid Irritant Characteristics:** None
5.10 **Odor Threshold:** Not pertinent

6. FIRE HAZARDS

6.1 **Flash Point:** Not flammable
6.2 **Flammable Limits in Air:** Not flammable
6.3 **Fire Extinguishing Agents:** Not pertinent
6.4 **Fire Extinguishing Agents Not to be Used:** Not pertinent
6.5 **Special Hazards of Combustion Products:** Not pertinent
6.6 **Behavior in Fire:** Not pertinent
6.7 **Ignition Temperature:** Not flammable
6.8 **Electrical Hazard:** Not pertinent
6.9 **Burning Rate:** Not flammable

7. CHEMICAL REACTIVITY

7.1 **Reactivity with Water:** No reaction
7.2 **Reactivity with Common Materials:** No reaction
7.3 **Stability During Transport:** Stable
7.4 **Neutralizing Agents for Acids and Caustics:** Not pertinent
7.5 **Polymerization:** Not pertinent
7.6 **Inhibitor of Polymerization:** Not pertinent

8. WATER POLLUTION

8.1 **Aquatic Toxicity:** 1.4 ppm/48 hr/bluegill/TL_m/fresh water
8.2 **Waterfowl Toxicity:** Data not available
8.3 **Biological Oxygen Demand (BOD):** None
8.4 **Food Chain Concentration Potential:** Data not available

9. SELECTED MANUFACTURERS

1. Chempar Chemical Co., Inc.
   260 Madison Ave.
   New York, N. Y. 10016
2. Chevron Chemical Co.
   Ortho Division
   940 Hensley St.
   Richmond, Calif. 94804
3. FMC Corp.
   Niagara Chemical Division
   Middleport, N. Y. 14105

10. SHIPPING INFORMATION

10.1 **Grades or Purity:** 94%
10.2 **Storage Temperature:** Ambient
10.3 **Inert Atmosphere:** No requirement
10.4 **Venting:** Open

11. HAZARD ASSESSMENT CODE

(See Hazard Assessment Handbook, CG 446-3)

II

12. HAZARD CLASSIFICATIONS

12.1 **Code of Federal Regulations:** Poisonous liquid or solid, Class B
12.2 **NAS Hazard Rating for Bulk Water Transportation:** Not listed
12.3 **NFPA Hazard Classifications:**

Category	Classification
Health Hazard (Blue)	2
Flammability (Red)	0
Reactivity (Yellow)	0

13. PHYSICAL AND CHEMICAL PROPERTIES

13.1 **Physical State at 15°C and 1 atm:** Solid
13.2 **Molecular Weight:** 347.12
13.3 **Boiling Point at 1 atm:** Decomposes
13.4 **Freezing Point:** Not pertinent
13.5 **Critical Temperature:** Not pertinent
13.6 **Critical Pressure:** Not pertinent
13.7 **Specific Gravity:** 5.79 at 15°C (solid)
13.8 **Liquid Surface Tension:** Not pertinent
13.9 **Liquid-Water Interfacial Tension:** Not pertinent
13.10 **Vapor (Gas) Specific Gravity:** Not pertinent
13.11 **Ratio of Specific Heats of Vapor (Gas):** Not pertinent
13.12 **Latent Heat of Vaporization:** Not pertinent
13.13 **Heat of Combustion:** Not pertinent
13.14 **Heat of Decomposition:** Not pertinent
13.15 **Heat of Solution:** Not pertinent
13.16 **Heat of Polymerization:** Not pertinent

NOTES

LFR	# LEAD FLUORIDE

Common Synonyms			
Lead difluoride Plumbous fluoride	Solid Sinks in water.	White	Odorless

AVOID CONTACT WITH SOLID AND DUST. KEEP PEOPLE AWAY.
Wear dust respirator.
Stop discharge if possible.
Isolate and remove discharged material.
Notify local health and pollution control agencies.

Fire

Not flammable.

Exposure

CALL FOR MEDICAL AID.

DUST
POISONOUS IF INHALED.
If inhaled will cause dizziness or loss of consciousness.

If in eyes, hold eyelids open and flush with plenty of water.
If breathing has stopped, give artificial respiration.
If breathing is difficult, give oxygen.

SOLID
Irritating to skin and eyes.
If swallowed will cause nausea, vomiting or loss of consciousness.

Remove contaminated clothing and shoes.
Flush affected areas with plenty of water.
IF IN EYES, hold eyelids open and flush with plenty of water.
IF SWALLOWED and victim is CONSCIOUS have victim drink water or milk and have victim induce vomiting.
IF SWALLOWED and victim is UNCONSCIOUS OR HAVING CONVULSIONS, do nothing except keep victim warm.

Water Pollution

Dangerous to aquatic life in high concentrations.
May be dangerous if it enters water intakes.

Notify local health and wildlife officials.
Notify operators of nearby water intakes.

1. RESPONSE TO DISCHARGE

(See Response Methods Handbook, CG 446-4)

Issue warning–water contaminant
Restrict access
Should be removed
Chemical and physical treatment

2. LABELS

No hazard label required by Code of Federal Regulations

3. CHEMICAL DESIGNATIONS

3.1 **Synonyms:** Lead difluoride; Plumbous fluoride
3.2 **Coast Guard Compatibility Classification:** Not listed
3.3 **Chemical Formula:** PbF_2
3.4 **IMCO/United Nations Numerical Designation:** Not listed

4. OBSERVABLE CHARACTERISTICS

4.1 **Physical State (as shipped):** Solid
4.2 **Color:** White
4.3 **Odor:** None

5. HEALTH HAZARDS

Dust and fumes of all but the most insoluble lead compounds are readily absorbed on inhalation and, to a lesser degree, after ingestion.

5.1 **Personal Protective Equipment:** Respirator for heavy dust exposure; safety goggles
5.2 **Symptoms Following Exposure:** Not irritating to skin or mucuous membranes; protect against chronic poisoning. Early symptoms of lead intoxication via inhalation or ingestion are most commonly gastrointestinal disorders, colic, constipation, etc.; weakness, which may go on to paralysis chiefly of the extensor muscles of the wrists and less often the ankles, is noticeable in the most serious cases. Ingestion of a large amount causes local irritation of the alimentary tract; pain, leg cramps, muscle weakness, paresthesias, depression, coma, and death may follow in 1 or 2 days. Contact with eyes causes irritation.
5.3 **Treatment for Exposure:** Remove at once all cases of lead intoxication from further exposure until the blood level is reduced to a safe value; immediately place the individual under medical care. INGESTION: give gastric lavage using 1% solution of sodium or magnesium sulfate; leave 15–30 gm magnesium sulfate in 6–8 oz. of water in the stomach as antidote and cathartic; egg white, milk, and tannin are useful demulcents; atropine sulfate and other antispasmodics may relieve abdominal pain, but morphine may be necessary. EYES or SKIN: flush with water.
5.4 **Toxicity by Inhalation (Threshold Limit Value):** 0.15 mg/m^3 (as lead)
5.5 **Short-Term Inhalation Limits:** Data not available
5.6 **Toxicity by Ingestion:** Grade 2; LD_{50} 0.5–5 g/kg
5.7 **Late Toxicity:** Data not available

6. FIRE HAZARDS

6.1 **Flash Point:** Not flammable
6.2 **Flammable Limits in Air:** Not flammable
6.3 **Fire Extinguishing Agents:** Not pertinent
6.4 **Fire Extinguishing Agents Not to be Used:** Not pertinent
6.5 **Special Hazards of Combustion Products:**
6.6 **Behavior in Fire:**
6.7 **Ignition Temperature:** Not pertinent
6.8 **Electrical Hazard:** Not pertinent
6.9 **Burning Rate:** Not pertinent

7. CHEMICAL REACTIVITY

7.1 **Reactivity with Water:** No reaction
7.2 **Reactivity with Common Materials:**
7.3 **Stability During Transport:** Stable
7.4 **Neutralizing Agents for Acids and Caustics:** Not pertinent
7.5 **Polymerization:** Not pertinent
7.6 **Inhibitor of Polymerization:** Not pertinent

8. WATER POLLUTION

8.1 **Aquatic Toxicity:** 7.48 mg/l (as Pb)/4 days/minnow/TL_m/soft water
8.2 **Waterfowl Toxicity:** May be toxic
8.3 **Biological Oxygen Demand (BOD):** None
8.4 **Food Chain Concentration Potential:** Bioconcentrative

9. SELECTED MANUFACTURERS

1. Ozark-Mahoning Co.
   1870 South Boulder
   Tulsa, Okla. 74119
2. American Hoechst Corp.
   Chemicals and Plastics Div.
   Rt. 202–206 North
   Somerville, N. J. 08876
3. Gallard-Schlesinger Chemical Mfg. Co.
   584 Mineola Ave.
   Carle Place, N. Y. 11514

10. SHIPPING INFORMATION

10.1 **Grades or Purity:** Technical; C.P.; Optical grade
10.2 **Storage Temperature:** Ambient
10.3 **Inert Atmosphere:** No requirement
10.4 **Venting:** Open

11. HAZARD ASSESSMENT CODE

(See Hazard Assessment Handbook, CG 446-3)

I I

12. HAZARD CLASSIFICATIONS

12.1 **Code of Federal Regulations:** Not listed
12.2 **NAS Hazard Rating for Bulk Water Transportation:** Not listed
12.3 **NFPA Hazard Classifications:** Not listed

13. PHYSICAL AND CHEMICAL PROPERTIES

13.1 **Physical State at 15°C and 1 atm:** Solid
13.2 **Molecular Weight:** 245.19
13.3 **Boiling Point at 1 atm:** Not pertinent
13.4 **Freezing Point:** Not pertinent
13.5 **Critical Temperature:** Not pertinent
13.6 **Critical Pressure:** Not pertinent
13.7 **Specific Gravity:** 8.24 at 20°C (solid)
13.8 **Liquid Surface Tension:** Not pertinent
13.9 **Liquid-Water Interfacial Tension:** Not pertinent
13.10 **Vapor (Gas) Specific Gravity:** Not pertinent
13.11 **Ratio of Specific Heats of Vapor (Gas):** Not pertinent
13.12 **Latent Heat of Vaporization:** Not pertinent
13.13 **Heat of Combustion:** Not pertinent
13.14 **Heat of Decomposition:** Not pertinent
13.15 **Heat of Solution:** Not pertinent
13.16 **Heat of Polymerization:** Not pertinent

5. HEALTH HAZARDS (Cont'd.)

5.8 **Vapor (Gas) Irritant Characteristics:** Data not available
5.9 **Liquid or Solid Irritant Characteristics:** Data not available
5.10 **Odor Threshold:** Data not available

LFB — LEAD FLUOROBORATE

Common Synonyms			
Lead fluoborate Lead fluoroborate solution	Liquid	Colorless	Odorless
	Sinks and mixes with water.		

AVOID CONTACT WITH LIQUID AND VAPOR. KEEP PEOPLE AWAY.
Wear goggles and self-contained breathing apparatus.
Stop discharge if possible.
Isolate and remove discharged material.
Notify local health and pollution control agencies.

Fire	Not flammable. Irritating gases may be produced when heated. **Wear goggles and self-contained breathing apparatus.**
 **Exposure**	**CALL FOR MEDICAL AID.** VAPOR POISONOUS IF INHALED. If inhaled will cause dizziness or loss of consciousness. **If in eyes, hold eyelids open and flush with plenty of water.** **If breathing has stopped, give artificial respiration.** **If breathing is difficult, give oxygen.** LIQUID Will burn skin and eyes. If swallowed will cause nausea, vomiting or loss of consciousness. **Remove contaminated clothing and shoes.** **Flush affected areas with plenty of water.** **IF IN EYES, hold eyelids open and flush with plenty of water.** **IF SWALLOWED and victim is CONSCIOUS, have victim drink water or milk and have victim induce vomiting.** **IF SWALLOWED and victim is UNCONSCIOUS OR HAVING CONVULSIONS, do nothing except keep victim warm.**
Water Pollution	Effect of low concentrations on aquatic life is unknown. May be dangerous if it enters water intakes. **Notify local health and wildlife officials.** **Notify operators of nearby water intakes.**

1. RESPONSE TO DISCHARGE
(See Response Methods Handbook, CG 446-4)

Issue warning — water contaminant
Restrict access
Disperse and flush

2. LABELS
No hazard label required by Code of Federal Regulations

3. CHEMICAL DESIGNATIONS
3.1 **Synonyms:** Lead fluoborate; Lead fluoroborate solution
3.2 **Coast Guard Compatibility Classification:** Not listed
3.3 **Chemical Formula:** $Pb(BF_4)_2 - H_2O$
3.4 **IMCO/United Nations Numerical Designation:** Not listed

4. OBSERVABLE CHARACTERISTICS
4.1 **Physical State (as shipped):** Liquid
4.2 **Color:** Colorless
4.3 **Odor:** Faint

5. HEALTH HAZARDS
5.1 **Personal Protective Equipment:** Rubber gloves; face shield; rubber apron
5.2 **Symptoms Following Exposure:** Early symptoms of lead intoxication via inhalation or ingestion are most commonly gastrointestinal disorders, colic, constipation, etc.; weakness, which may go on to paralysis, chiefly of the extensor muscles of the wrists and less often of the ankles, is noticeable in the most serious cases. Ingestion of a large amount causes local irritation of the alimentary tract; pain, leg cramps, muscle weakness, paresthesias, depression, coma, and death may follow in 1 or 2 days. Contact with eyes or skin may cause burns and/or irritation.
5.3 **Treatment for Exposure:** Remove at once all cases of lead intoxication from further exposure until the blood level is reduced to a safe value; immediately place the individual under medical care. INGESTION: give gastric lavage using 1% solution of sodium or magnesium sulfate; leave 15–30 gm magnesium sulfate in 6–8 oz. of water in the stomach as antidote and cathartic; egg white, milk, and tannin are useful demulcents; atropine sulfate and other antispasmodics may relieve abdominal pain, but morphine may be necessary. EYES: flush with copious quantities of water for 15 min. SKIN: wash area with soap and water; treat as an acid burn.
5.4 **Toxicity by Inhalation (Threshold Limit Value):** 0.2 mg/m^3 (as lead)
5.5 **Short-Term Inhalation Limits:** Data not available
5.6 **Toxicity by Ingestion:** Grade 2; LD_{50} 0.5–5 g/kg
5.7 **Late Toxicity:** Data not available
5.8 **Vapor (Gas) Irritant Characteristics:** Data not available
5.9 **Liquid or Solid Irritant Characteristics:** Data not available
5.10 **Odor Threshold:** Data not available

6. FIRE HAZARDS
6.1 **Flash Point:** Not flammable
6.2 **Flammable Limits in Air:** Not flammable
6.3 **Fire Extinguishing Agents:** Not pertinent
6.4 **Fire Extinguishing Agents Not to be Used:** Not pertinent
6.5 **Special Hazards of Combustion Products:** Toxic and irritating hydrogen fluoride gas may form in fire.
6.6 **Behavior in Fire:** Not pertinent
6.7 **Ignition Temperature:** Not pertinent
6.8 **Electrical Hazard:** Not pertinent
6.9 **Burning Rate:** Not pertinent

7. CHEMICAL REACTIVITY
7.1 **Reactivity with Water:** No reaction
7.2 **Reactivity with Common Materials:** Solution is acidic and will corrode most metals.
7.3 **Stability During Transport:** Stable
7.4 **Neutralizing Agents for Acids and Caustics:** Flush with water, rinse with dilute solution of sodium bicarbonate or soda ash.
7.5 **Polymerization:** Not pertinent
7.6 **Inhibitor of Polymerization:** Not pertinent

8. WATER POLLUTION
8.1 **Aquatic Toxicity:** Data not available
8.2 **Waterfowl Toxicity:** May be toxic.
8.3 **Biological Oxygen Demand (BOD):** None
8.4 **Food Chain Concentration Potential:** Bioconcentrative.

9. SELECTED MANUFACTURERS
1. Allied Chemical Corp.
   P. O. Box 1087R
   Morristown, N. J. 07960
2. The Harshaw Chemical Co.
   1945 E. 97th St.
   Cleveland, Ohio 44106
3. Pfaltz and Bauer, Inc.
   375 Fairfield Ave.
   Stamford, Conn. 06902

10. SHIPPING INFORMATION
10.1 **Grades or Purity:** 50–62% solutions in water
10.2 **Storage Temperature:** Ambient
10.3 **Inert Atmosphere:** No requirement
10.4 **Venting:** Open

11. HAZARD ASSESSMENT CODE
(See Hazard Assessment Handbook, CG 446-3)

A-P

12. HAZARD CLASSIFICATIONS
12.1 **Code of Federal Regulations:** Not listed
12.2 **NAS Hazard Rating for Bulk Water Transportation:** Not listed
12.3 **NFPA Hazard Classifications:**

Category	Classification
Health Hazard (Blue)	1
Flammability (Red)	0
Reactivity (Yellow)	0

13. PHYSICAL AND CHEMICAL PROPERTIES
13.1 **Physical State at 15°C and 1 atm:** Liquid
13.2 **Molecular Weight:** Mixture
13.3 **Boiling Point at 1 atm:** Not pertinent
13.4 **Freezing Point:** Not pertinent
13.5 **Critical Temperature:** Not pertinent
13.6 **Critical Pressure:** Not pertinent
13.7 **Specific Gravity:** 1.75 at 20°C (liquid)
13.8 **Liquid Surface Tension:** Data not available
13.9 **Liquid-Water Interfacial Tension:** Not pertinent
13.10 **Vapor (Gas) Specific Gravity:** Not pertinent
13.11 **Ratio of Specific Heats of Vapor (Gas):** Not pertinent
13.12 **Latent Heat of Vaporization:** Not pertinent
13.13 **Heat of Combustion:** Not pertinent
13.14 **Heat of Decomposition:** Not pertinent
13.15 **Heat of Solution:** Not pertinent
13.16 **Heat of Polymerization:** Not pertinent

NOTES

LID

LEAD IODIDE

Common Synonyms

Solid — Bright Yellow — Odorless

Sinks in water.

AVOID CONTACT WITH SOLID AND DUST. KEEP PEOPLE AWAY.
Wear dust respirator.
Stop discharge if possible.
Isolate and remove discharged material.
Notify local health and pollution control agencies.

Fire

Not flammable.

Exposure

CALL FOR MEDICAL AID.

DUST
POISONOUS IF INHALED.
If inhaled will cause dizziness or loss of consciousness.
If in eyes, hold eyelids open and flush with plenty of water.
If breathing has stopped, give artificial respiration.
If breathing is difficult, give oxygen.

SOLID
Irritating to skin and eyes.
If swallowed will cause nausea, vomiting or loss of consciousness.

Remove contaminated clothing and shoes.
Flush affected areas with plenty of water.
IF IN EYES, hold eyelids open and flush with plenty of water.
IF SWALLOWED and victim is CONSCIOUS, have victim drink water or milk and have victim induce vomiting.
IF SWALLOWED and victim is UNCONSCIOUS OR HAVING CONVULSIONS, do nothing except keep victim warm.

Water Pollution

Dangerous to aquatic life in high concentrations.
May be dangerous if it enters water intakes.

Notify local health and wildlife officials.
Notify operators of nearby water intakes.

1. RESPONSE TO DISCHARGE

(See Response Methods Handbook, CG 446-4)

Issue warning—water contaminant
Restrict access
Should be removed
Chemical and physical treatment

2. LABELS

No hazard label required by Code of Federal Regulations

3. CHEMICAL DESIGNATIONS

3.1 **Synonyms:** No common synonyms
3.2 **Coast Guard Compatibility Classification:** Not listed
3.3 **Chemical Formula:** PbI_2
3.4 **IMCO/United Nations Numerical Designation:** Not listed

4. OBSERVABLE CHARACTERISTICS

4.1 **Physical State (as shipped):** Solid
4.2 **Color:** Bright yellow
4.3 **Odor:** None

5. HEALTH HAZARDS

5.1 **Personal Protective Equipment:** Dust mask and protective gloves.

5.2 **Symptoms Following Exposure:** Early symptoms of lead intoxication via inhalation or ingestion are most commonly gastrointestinal disorders, colic, constipation, etc.; weakness, which may go on to paralysis, chiefly of the extensor muscles of the wrists and less often the ankles, is noticeable in the most serious cases. Ingestion of a large amount causes local irritation of the alimentary tract. Pain, leg cramps, muscle weakness, paresthesias, depression, coma, and death may follow in 1 or 2 days. Contact with eyes causes irritation.

5.3 **Treatment for Exposure:** Remove at once all cases of lead intoxication from further exposure until the blood level is reduced to a safe value; immediately place the individual under medical care. INGESTION: give gastric lavage using 1% solution of sodium or magnesium sulfate; leave 15–30 mg magnesium sulfate in 6–8 oz. of water in the stomach as antidote and cathartic; egg white, milk, and tannin are useful demulcents; atropine sulfate and other antispasmodics may relieve abdominal pain, but morphine may be necessary. EYES or SKIN: flush with water.

5.4 **Toxicity by Inhalation (Threshold Limit Value):** 0.2 mg/m^3 (as lead)
5.5 **Short-Term Inhalation Limits:** Data not available
5.6 **Toxicity by Ingestion:** Grade 2; LD_{50} 0.5–5 g/kg
5.7 **Late Toxicity:** Data not available
5.8 **Vapor (Gas) Irritant Characteristics:** Data not available
5.9 **Liquid or Solid Irritant Characteristics:** Data not available
5.10 **Odor Threshold:** Data not available

6. FIRE HAZARDS

6.1 **Flash Point:** Not flammable
6.2 **Flammable Limits in Air:** Not pertinent
6.3 **Fire Extinguishing Agents:** Not pertinent
6.4 **Fire Extinguishing Agents Not to be Used:** Not pertinent
6.5 **Special Hazards of Combustion Products:**
6.6 **Behavior in Fire:**
6.7 **Ignition Temperature:** Not pertinent
6.8 **Electrical Hazard:** Not pertinent
6.9 **Burning Rate:** Not pertinent

7. CHEMICAL REACTIVITY

7.1 **Reactivity with Water:** No reaction
7.2 **Reactivity with Common Materials:**
7.3 **Stability During Transport:** Stable
7.4 **Neutralizing Agents for Acids and Caustics:** Not pertinent
7.5 **Polymerization:** Not pertinent
7.6 **Inhibitor of Polymerization:** Not pertinent

8. WATER POLLUTION

8.1 **Aquatic Toxicity:** Data not available
8.2 **Waterfowl Toxicity:** May be toxic
8.3 **Biological Oxygen Demand (BOD):** Data not available
8.4 **Food Chain Concentration Potential:** Bioconcentrative

9. SELECTED MANUFACTURERS

1. RSA Corporation
   690 Sawmill River Road
   Ardsley, N. Y. 10502
2. Ventron Corp.
   P. O. Box 159
   Beverly, Mass. 01915
3. Mallinckrodt Chemical Works
   St. Louis, Mo. 63160

10. SHIPPING INFORMATION

10.1 **Grades or Purity:** 98.5+%; NF
10.2 **Storage Temperature:** Ambient
10.3 **Inert Atmosphere:** No requirement
10.4 **Venting:** Open

11. HAZARD ASSESSMENT CODE

(See Hazard Assessment Handbook, CG 446-3)

II

12. HAZARD CLASSIFICATIONS

12.1 **Code of Federal Regulations:** Not listed
12.2 **NAS Hazard Rating for Bulk Water Transportation:** Not listed
12.3 **NFPA Hazard Classifications:** Not listed

13. PHYSICAL AND CHEMICAL PROPERTIES

13.1 **Physical State at 15°C and 1 atm:** Solid
13.2 **Molecular Weight:** 461.03
13.3 **Boiling Point at 1 atm:** Not pertinent
13.4 **Freezing Point:** Not pertinent
13.5 **Critical Temperature:** Not pertinent
13.6 **Critical Pressure:** Not pertinent
13.7 **Specific Gravity:** 6.16 at 20°C (solid)
13.8 **Liquid Surface Tension:** Not pertinent
13.9 **Liquid-Water Interfacial Tension:** Not pertinent
13.10 **Vapor (Gas) Specific Gravity:** Not pertinent
13.11 **Ratio of Specific Heats of Vapor (Gas):** Not pertinent
13.12 **Latent Heat of Vaporization:** Not pertinent
13.13 **Heat of Combustion:** Not pertinent
13.14 **Heat of Decomposition:** Not pertinent
13.15 **Heat of Solution:** Not pertinent
13.16 **Heat of Polymerization:** Not pertinent

NOTES

LEAD NITRATE

Common Synonyms			
Nitric acid, Lead II salt	Solid	White	Odorless
	Sinks and mixes with water.		

AVOID CONTACT WITH SOLID AND DUST. KEEP PEOPLE AWAY.
Wear dust respirator.
Stop discharge if possible.
Isolate and remove discharged material.
Notify local health and pollution control agencies.

Fire

Not flammable.
Will increase the intensity of a fire.
POISONOUS GASES MAY BE PRODUCED WHEN HEATED.

Flood discharge area with water.
Cool exposed containers with water.

Exposure

CALL FOR MEDICAL AID.

DUST
POISONOUS IF INHALED.
If inhaled will cause dizziness or loss of consciousness.

If in eyes, hold eyelids open and flush with plenty of water.
If breathing has stopped, give artificial respiration.
If breathing is difficult, give oxygen.

SOLID
Irritating to skin and eyes.
If swallowed will cause nausea, vomiting or loss of consciousness.

Remove contaminated clothing and shoes.
Flush affected areas with plenty of water.
IF IN EYES, hold eyelids open and flush with plenty of water.
IF SWALLOWED and victim is CONSCIOUS, have victim drink water or milk and have victim induce vomiting.
IF SWALLOWED and victim is UNCONSCIOUS OR HAVING CONVULSIONS, do nothing except keep victim warm.

Water Pollution

Dangerous to aquatic life in high concentrations.
May be dangerous if it enters water intakes.

Notify local health and wildlife officials.
Notify operators of nearby water intakes.

1. RESPONSE TO DISCHARGE

(See Response Methods Handbook, CG 446-4)

Issue warning — water contaminant, oxidizing material
Restrict access
Disperse and flush

2. LABEL

Yellow

3. CHEMICAL DESIGNATIONS

3.1 **Synonyms:** Nitric acid, lead (2+) salt
3.2 **Coast Guard Compatibility Classification:** Not listed
3.3 **Chemical Formula:** $Pb(NO_3)_2$
3.4 **IMCO/United Nations Numerical Designation:** 5.1/1469

4. OBSERVABLE CHARACTERISTICS

4.1 **Physical State (as shipped):** Solid
4.2 **Color:** White
4.3 **Odor:** None

5. HEALTH HAZARDS

5.1 **Personal Protective Equipment:** Dust mask and protective gloves
5.2 **Symptoms Following Exposure:** Early symptoms of lead intoxication via inhalation or ingestion are most commonly gastrointestinal disorders, colic, constipation, etc.; weakness, which may go on to paralysis, chiefly of the extensor muscles of the wrists and less often the ankles, is noticeable in the most serious cases. Ingestion of a large amount causes local irritation of the alimentary tract; pain, leg cramps, muscle weakness, paresthesias, depression, coma, and death may follow in 1 or 2 days. Contact with eyes causes irritation.
5.3 **Treatment for Exposure:** Remove at once all cases of lead intoxication from further exposure until the blood level is reduced to a safe value; immediately place the individual under medical care. INGESTION: give gastric lavage using 1% solution of sodium or magnesium sulfate; leave 15 – 30 gm magnesium sulfate in 6 – 8 oz. of water in the stomach as antidote and cathartic; egg white, milk, and tannin are useful demulcents; atropine sulfate and other antispasmodics may relieve abdominal pain, but morphine may be necessary. EYES or SKIN: flush with water.
5.4 **Toxicity by Inhalation (Threshold Limit Value):** 0.2 mg/m^3 (as lead)
5.5 **Short-Term Inhalation Limits:** Data not available
5.6 **Toxicity by Ingestion:** Grade 2; LD_{50} 0.5 – 5 g/kg
5.7 **Late Toxicity:** Data not available
5.8 **Vapor (Gas) Irritant Characteristics:** Data not available
5.9 **Liquid or Solid Irritant Characteristics:** Data not available
5.10 **Odor Threshold:** Data not available

6. FIRE HAZARDS

6.1 **Flash Point:** Not flammable
6.2 **Flammable Limits in Air:** Not flammable
6.3 **Fire Extinguishing Agents:** Not pertinent
6.4 **Fire Extinguishing Agents Not to be Used:** Not pertinent
6.5 **Special Hazards of Combustion Products:** Toxic oxides of nitrogen may form in fire.
6.6 **Behavior in Fire:** Increases the intensity of a fire when in contact with burning material. Use plenty of water to cool containers or spilled material.
6.7 **Ignition Temperature:** Not pertinent
6.8 **Electrical Hazard:** Not pertinent
6.9 **Burning Rate:** Not pertinent

7. CHEMICAL REACTIVITY

7.1 **Reactivity with Water:** No reaction
7.2 **Reactivity with Common Materials:** Contact with wood and paper may cause fire.
7.3 **Stability During Transport:** Stable
7.4 **Neutralizing Agents for Acids and Caustics:** Not pertinent
7.5 **Polymerization:** Not pertinent
7.6 **Inhibitor of Polymerization:** Not pertinent

8. WATER POLLUTION

8.1 **Aquatic Toxicity:**
1.6 ppm/*/tadpole/harms growth/tap water
200 mg/l/*/eggs of urchins/abnormalities/salt water
240 ppm/48 hr/mosquito fish/TL_m/fresh water
*Time period not specified.
8.2 **Waterfowl Toxicity:** May be toxic
8.3 **Biological Oxygen Demand (BOD):** None
8.4 **Food Chain Concentration Potential:** Both fish and animals can concentrate lead.

9. SELECTED MANUFACTURERS

1. Filo Color and Chemical Corp.
   347 Madison Ave.
   New York, N. Y. 10017
2. Mallinckrodt Chemical Works
   223 Westside Ave.
   P. O. Box 384
   Jersey City, N. J. 07303
3. J. T. Baker Chemical Co.
   Phillipsburg, N. J. 08865

10. SHIPPING INFORMATION

10.1 **Grades or Purity:** Reagent; Technical, 98+%
10.2 **Storage Temperature:** Ambient
10.3 **Inert Atmosphere:** No requirement
10.4 **Venting:** Open

11. HAZARD ASSESSMENT CODE

(See Hazard Assessment Handbook, CG 446-3)

SS

12. HAZARD CLASSIFICATIONS

12.1 **Code of Federal Regulations:** Oxidizing material
12.2 **NAS Hazard Rating for Bulk Water Transportation:** Not listed
12.3 **NFPA Hazard Classifications:**

Category	Classification*	
Health Hazard (Blue)	0	1
Flammability (Red)	0	0
Reactivity (Yellow)	0	0
	oxy	oxy

*First column refers to non-fire situation.

13. PHYSICAL AND CHEMICAL PROPERTIES

13.1 **Physical State at 15°C and 1 atm:** Solid
13.2 **Molecular Weight:** 331.2
13.3 **Boiling Point at 1 atm:** Not pertinent (decomposes)
13.4 **Freezing Point:** Not pertinent
13.5 **Critical Temperature:** Not pertinent
13.6 **Critical Pressure:** Not pertinent
13.7 **Specific Gravity:** 4.53 at 20°C (solid)
13.8 **Liquid Surface Tension:** Not pertinent
13.9 **Liquid-Water Interfacial Tension:** Not pertinent
13.10 **Vapor (Gas) Specific Gravity:** Not pertinent
13.11 **Ratio of Specific Heats of Vapor (Gas):** Not pertinent
13.12 **Latent Heat of Vaporization:** Not pertinent
13.13 **Heat of Combustion:** Not pertinent
13.14 **Heat of Decomposition:** Not pertinent
13.15 **Heat of Solution:** 41 Btu/lb = 23 cal/g = 0.96 × 10^5 J/kg
13.16 **Heat of Polymerization:** Not pertinent

NOTES

LTT	# LEAD TETRAACETATE

Common Synonyms			
Lead IV acetate	Wet Crystals	Faintly Pink	Vinegar-like Odor
	Reacts with water.		

AVOID CONTACT WITH SOLID. KEEP PEOPLE AWAY.
Wear dust respirator.
Stop discharge if possible.
Isolate and remove discharged material.
Notify local health and pollution control agencies.

Fire

Not flammable.
Will increase the intensity of a fire.

Exposure

CALL FOR MEDICAL AID.
LIQUID/SOLID
Irritating to skin and eyes.
If swallowed will cause nausea, vomiting or loss of consciousness.
Remove contaminated clothing and shoes.
Flush affected areas with plenty of water.
IF IN EYES, hold eyelids open and flush with plenty of water.
IF SWALLOWED and victim is CONSCIOUS, have victim drink water or milk and have victim induce vomiting.
IF SWALLOWED and victim is UNCONSCIOUS OR HAVING CONVULSIONS, do nothing except keep victim warm.

Water Pollution

Effect of low concentrations on aquatic life is unknown.
May be dangerous if it enters water intakes.
Notify local health and wildlife officials.
Notify operators of nearby water intakes.

1. RESPONSE TO DISCHARGE

(See Response Methods Handbook, CG 446-4)

Issue warning – oxidizing material, water contaminant
Restrict access
Disperse and flush

2. LABEL

Yellow

3. CHEMICAL DESIGNATIONS

3.1 **Synonyms:** Lead (IV) acetate
3.2 **Coast Guard Compatibility Classification:** Not listed
3.3 **Chemical Formula:** $Pb(C_2H_3O_2)_4 - CH_3COOH$
3.4 **IMCO/United Nations Numerical Designation:** Not listed

4. OBSERVABLE CHARACTERISTICS

4.1 **Physical State (as shipped):** Crystals wet with glacial acetic acid.
4.2 **Color:** Faintly pink
4.3 **Odor:** Like acetic acid or vinegar

5. HEALTH HAZARDS

5.1 **Personal Protective Equipment:** Goggles or face shield; rubber gloves
5.2 **Symptoms Following Exposure:** Early symptoms of lead intoxication by ingestion are most commonly gastrointestinal disorders, colic, constipation, etc.; weakness, which may go on to paralysis chiefly of the extensor muscles of the wrists and less often of the ankles, is noticeable in the most serious cases. Ingestion of a large amount causes local irritation of the alimentary tract; pain, leg cramps, muscle weakness, paresthesias, depression, coma, and death may follow in 1 or 2 days. Contact causes severe irritation of eyes and can burn skin.
5.3 **Treatment for Exposure:** Remove at once all cases of lead intoxication from further exposure until the blood level is reduced to a safe value; immediately place the individual under medical care. INGESTION: give gastric lavage using 1% solution of sodium or magnesium sulfate; leave 15 – 30 gm magnesium sulfate in 6 – 8 oz. of water in the stomach as antidote and cathartic; egg white, milk, and tannin are useful demulcents; atropine sulfate and other antispasmodics may relieve abdominal pain, but morphine may be necessary. EYES: flush with water for at least 15 min. SKIN: wash contaminated skin with large amounts of water for 15 min.
5.4 **Toxicity by Inhalation (Threshold Limit Value):** 0.2 mg/m^3 (as lead)
5.5 **Short-Term Inhalation Limits:** Data not available
5.6 **Toxicity by Ingestion:** Grade 2; LD_{50} 0.5 – 5 g/kg
5.7 **Late Toxicity:** Data not available
5.8 **Vapor (Gas) Irritant Characteristics:** Data not available
5.9 **Liquid or Solid Irritant Characteristics:** Data not available
5.10 **Odor Threshold:** Data not available

6. FIRE HAZARDS

6.1 **Flash Point:** Not flammable
6.2 **Flammable Limits in Air:** Not flammable
6.3 **Fire Extinguishing Agents:** Not pertinent
6.4 **Fire Extinguishing Agents Not to be Used:** Not pertinent
6.5 **Special Hazards of Combustion Products:**
6.6 **Behavior in Fire:** Can increase the intensity of a fire when in contact with combustible material. Cool containers with plenty of water.
6.7 **Ignition Temperature:** Not pertinent
6.8 **Electrical Hazard:** Not pertinent
6.9 **Burning Rate:** Not pertinent

7. CHEMICAL REACTIVITY

7.1 **Reactivity with Water:** Forms lead dioxide and acetic acid in a reaction that is not violent
7.2 **Reactivity with Common Materials:** May corrode metals when moist
7.3 **Stability During Transport:** Stable
7.4 **Neutralizing Agents for Acids and Caustics:** Dilute with water, rinse with dilute sodium bicarbonate or lime solution.
7.5 **Polymerization:** Not pertinent
7.6 **Inhibitor of Polymerization:** Not pertinent

8. WATER POLLUTION

8.1 **Aquatic Toxicity:** Data not available
8.2 **Waterfowl Toxicity:** May be toxic
8.3 **Biological Oxygen Demand (BOD):** Data not available
8.4 **Food Chain Concentration Potential:** Bioconcentrative

9. SELECTED MANUFACTURERS

1. Arapahoe Chemical Co.
2855 Walnut St.
Boulder, Colo. 80302
2. Ventron Corp., Alfa Products
8 Congress Street
Beverly, Mass. 01915
3. Gallard-Schlesinger Chemical Mfg. Co.
584 Mineola Ave.
Carle Place, N. Y. 11514

10. SHIPPING INFORMATION

10.1 **Grades or Purity:** Commercial, 80 – 90%
10.2 **Storage Temperature:** Ambient
10.3 **Inert Atmosphere:** No requirement
10.4 **Venting:** Open

11. HAZARD ASSESSMENT CODE

(See Hazard Assessment Handbook, CG 446-3)

RR

12. HAZARD CLASSIFICATIONS

12.1 **Code of Federal Regulations:** Oxidizing material
12.2 **NAS Hazard Rating for Bulk Water Transportation:** Not listed
12.3 **NFPA Hazard Classifications:** Not listed

13. PHYSICAL AND CHEMICAL PROPERTIES

13.1 **Physical State at 15°C and 1 atm:** Solid
13.2 **Molecular Weight:** 443.39
13.3 **Boiling Point at 1 atm:** Not pertinent (decomposes)
13.4 **Freezing Point:** 347°F = 175°C = 448°K
13.5 **Critical Temperature:** Not pertinent
13.6 **Critical Pressure:** Not pertinent
13.7 **Specific Gravity:** 2.2 at 20°C (solid)
13.8 **Liquid Surface Tension:** Not pertinent
13.9 **Liquid-Water Interfacial Tension:** Not pertinent
13.10 **Vapor (Gas) Specific Gravity:** Not pertinent
13.11 **Ratio of Specific Heats of Vapor (Gas):** Not pertinent
13.12 **Latent Heat of Vaporization:** Not pertinent
13.13 **Heat of Combustion:** Not pertinent
13.14 **Heat of Decomposition:** Not pertinent
13.15 **Heat of Solution:** Data not available
13.16 **Heat of Polymerization:** Not pertinent

NOTES

LTC	# LEAD THIOCYANATE

Common Synonyms			
Lead sulfocyanate	Solid	White	Odorless
	Sinks and mixes with water.		

AVOID CONTACT WITH SOLID AND DUST. KEEP PEOPLE AWAY.
Wear dust respirator.
Stop discharge if possible.
Isolate and remove discharged material.
Notify local health and pollution control agencies.

Fire

Not flammable.
POISONOUS GASES MAY BE PRODUCED WHEN HEATED.

Exposure

CALL FOR MEDICAL AID.

DUST
POISONOUS IF INHALED.
If inhaled will cause dizziness or loss of consciousness.
If in eyes, hold eyelids open and flush with plenty of water.
If breathing has stopped, give artificial respiration.
If breathing is difficult, give oxygen.

SOLID
Irritating to skin and eyes.
If swallowed will cause nausea, vomiting or loss of consciousness.
Remove contaminated clothing and shoes.
Flush affected areas with plenty of water.
IF IN EYES, hold eyelids open and flush with plenty of water.
IF SWALLOWED victim is CONSCIOUS, have victim drink water or milk and have victim induce vomiting.
IF SWALLOWED and victim is UNCONSCIOUS OR HAVING CONVULSIONS, do nothing except keep victim warm.

Water Pollution

Effect of low concentrations on aquatic life is unknown.
May be dangerous if it enters water intakes.
Notify local health and wildlife officials.
Notify operators of nearby water intakes.

1. RESPONSE TO DISCHARGE

(See Response Methods Handbook, CG 446-4)

Issue warning - water contaminant
Restrict access
Should be removed
Chemical and physical treatment

2. LABEL

3. CHEMICAL DESIGNATIONS

3.1 **Synonyms:** Lead sulfocyanate
3.2 **Coast Guard Compatibility Classification:** Not listed
3.3 **Chemical Formula:** $Pb(SCN)_2$
3.4 **IMCO/United Nations Numerical Designation:** Not listed

4. OBSERVABLE CHARACTERISTICS

4.1 **Physical State (as shipped):** Solid
4.2 **Color:** White
4.3 **Odor:** None

5. HEALTH HAZARDS

5.1 **Personal Protective Equipment:** Dust mask; goggles or face shield; rubber gloves
5.2 **Symptoms Following Exposure:** Early symptoms of lead intoxication via inhalation or ingestion are most commonly gastrointestinal disorders, colic, constipation, etc.; weakness, which may go on to paralysis, chiefly of the extensor muscles of the wrists and less often of the ankles, is noticeable in the most serious cases. Ingestion of a large amount causes local irritation of the alimentary tract; pain, leg cramps, muscle weakness, paresthesias, depression, coma, and death may follow in 1 or 2 days. Contact causes irritation of eyes and mild irritation of skin.
5.3 **Treatment for Exposure:** Remove at once all cases of lead intoxication from further exposure until the blood level is reduced to a safe value; immediately place the individual under medical care. INGESTION: give gastric lavage using 1% solution of sodium or magnesium sulfate; leave 15 - 30 gm magnesium sulfate in 6 - 8 oz. of water in the stomach as antidote and cathartic; egg white, milk, and tannin are useful demulcents; atropine sulfate and other antispasmodics may relieve abdominal pain, but morphine may be necessary. EYES: flush with water for at least 15 min. SKIN: wash well with soap and water.
5.4 **Toxicity by Inhalation (Threshold Limit Value):** 0.2 mg/m^3 (as lead)
5.5 **Short-Term Inhalation Limits:** Data not available
5.6 **Toxicity by Ingestion:** Grade 2; LD_{50} 0.5 - 5 g/kg
5.7 **Late Toxicity:** Data not available
5.8 **Vapor (Gas) Irritant Characteristics:** Data not available
5.9 **Liquid or Solid Irritant Characteristics:** Data not available
5.10 **Odor Threshold:** Data not available

6. FIRE HAZARDS

6.1 **Flash Point:** Not flammable
6.2 **Flammable Limits in Air:** Not flammable
6.3 **Fire Extinguishing Agents:** Not pertinent
6.4 **Fire Extinguishing Agents Not to be Used:** Not pertinent
6.5 **Special Hazards of Combustion Products:** Irritating sulfur dioxide gas may form in fire.
6.6 **Behavior in Fire:**
6.7 **Ignition Temperature:** Not pertinent
6.8 **Electrical Hazard:** Not pertinent
6.9 **Burning Rate:** Not pertinent

7. CHEMICAL REACTIVITY

7.1 **Reactivity with Water:** No reaction
7.2 **Reactivity with Common Materials:**
7.3 **Stability During Transport:** Stable
7.4 **Neutralizing Agents for Acids and Caustics:** Not pertinent
7.5 **Polymerization:** Not pertinent
7.6 **Inhibitor of Polymerization:** Not pertinent

8. WATER POLLUTION

8.1 **Aquatic Toxicity:** Data not available
8.2 **Waterfowl Toxicity:** May be toxic
8.3 **Biological Oxygen Demand (BOD):** Data not available
8.4 **Food Chain Concentration Potential:** Bioconcentrative

9. SELECTED MANUFACTURERS

1. Fielding Chemical Co.
   P. O. Box 21
   Jersey City, N. J. 07303
2. Gallard-Schlesinger Chemical Mfg. Co.
   584 Mineola Ave.
   Carle Place, N. Y. 11514
3. Eastman Organic Chemicals
   Rochester, N. Y. 14650

10. SHIPPING INFORMATION

10.1 **Grades or Purity:** Practical; Commercial
10.2 **Storage Temperature:** Ambient
10.3 **Inert Atmosphere:** No requirement
10.4 **Venting:** Open

11. HAZARD ASSESSMENT CODE

(See Hazard Assessment Handbook, CG 446-3)

II

12. HAZARD CLASSIFICATIONS

12.1 **Code of Federal Regulations:** Poison, Class B
12.2 **NAS Hazard Rating for Bulk Water Transportation:** Not listed
12.3 **NFPA Hazard Classifications:**

Category	Classification
Health Hazard (Blue)	I
Flammability (Red)	I
Reactivity (Yellow)	I

13. PHYSICAL AND CHEMICAL PROPERTIES

13.1 **Physical State at 15°C and 1 atm:** Solid
13.2 **Molecular Weight:** 323.4
13.3 **Boiling Point at 1 atm:** Not pertinent (decomposes)
13.4 **Freezing Point:** Not pertinent
13.5 **Critical Temperature:** Not pertinent
13.6 **Critical Pressure:** Not pertinent
13.7 **Specific Gravity:** 3.82 at 20°C (solid)
13.8 **Liquid Surface Tension:** Not pertinent
13.9 **Liquid-Water Interfacial Tension:** Not pertinent
13.10 **Vapor (Gas) Specific Gravity:** Not pertinent
13.11 **Ratio of Specific Heats of Vapor (Gas):** Not pertinent
13.12 **Latent Heat of Vaporization:** Not pertinent
13.13 **Heat of Combustion:** Not pertinent
13.14 **Heat of Decomposition:** Not pertinent
13.15 **Heat of Solution:** Not pertinent
13.16 **Heat of Polymerization:** Not pertinent

NOTES

LAL — LINEAR ALCOHOLS

Common Synonyms
Dodecanol
Tridecanol
Tetradecanol
Pentadecanol

Liquid — Colorless — Mild alcohol odor

Floats on water.

Stop discharge if possible.
Call fire department.
Avoid contact with liquid.
Isolate and remove discharged material.
Notify local health and pollution control agencies.

Fire

Combustible.
Extinguish with alcohol foam, dry chemical, or carbon dioxide.
Cool exposed containers with water.

Exposure

CALL FOR MEDICAL AID.

LIQUID
Irritating to skin.
Will burn eyes.
Remove contaminated clothing and shoes.
Flush affected areas with plenty of water.
IF IN EYES, hold eyelids open and flush with plenty of water.

Water Pollution

Effect of low concentrations on aquatic life is unknown.
Fouling to shoreline.
May be dangerous if it enters water intakes.
Notify local health and wildlife officials.
Notify operators of nearby water intakes.

1. RESPONSE TO DISCHARGE

(See Response Methods Handbook, CG 446-4)

Mechanical containment
Should be removed
Chemical and physical treatment

2. LABELS

No hazard label required by Code of Federal Regulations

3. CHEMICAL DESIGNATIONS

3.1 **Synonyms:** Dodecanol, Tetradecanol, Tridecanol, Pentadecanol (Could be any of the above or mixtures thereof)
3.2 **Coast Guard Compatibility Classification:** Alcohol
3.3 **Chemical Formula:** $CH_3(CH_2)_{10-13}CH_2OH$
3.4 **IMCO/United Nations Numerical Designation:** Not listed

4. OBSERVABLE CHARACTERISTICS

4.1 **Physical State (as shipped):** Liquid
4.2 **Color:** Colorless
4.3 **Odor:** Mild

5. HEALTH HAZARDS

5.1 **Personal Protective Equipment:** Eye protection.
5.2 **Symptoms Following Exposure:** Direct contact can produce eye irritation. Low general toxicity.
5.3 **Treatment for Exposure:** Wash eyes with water for at least 15 min.
5.4 **Toxicity by Inhalation (Threshold Limit Value):** Not pertinent
5.5 **Short-Term Inhalation Limits:** Not pertinent
5.6 **Toxicity by Ingestion:** Grade 1; LD_{50} 5 to 15 g/kg (rat)
5.7 **Late Toxicity:** Data not available
5.8 **Vapor (Gas) Irritant Characteristics:** None
5.9 **Liquid or Solid Irritant Characteristics:** No appreciable hazard. Practically harmless to skin.
5.10 **Odor Threshold:** Data not available

6. FIRE HAZARDS

6.1 **Flash Point:** 180°F—285°F O.C.
6.2 **Flammable Limits in Air:** Data not available
6.3 **Fire Extinguishing Agents:** Alcohol foam, dry chemical, or carbon dioxide
6.4 **Fire Extinguishing Agents Not to be Used:** Water or foam may cause frothing
6.5 **Special Hazards of Combustion Products:** Not pertinent
6.6 **Behavior in Fire:** Not pertinent
6.7 **Ignition Temperature:** Data not available
6.8 **Electrical Hazard:** Not pertinent
6.9 **Burning Rate:** Data not available

7. CHEMICAL REACTIVITY

7.1 **Reactivity with Water:** No reaction
7.2 **Reactivity with Common Materials:** No reaction
7.3 **Stability During Transport:** Stable
7.4 **Neutralizing Agents for Acids and Caustics:** Not pertinent
7.5 **Polymerization:** Not pertinent
7.6 **Inhibitor of Polymerization:** Not pertinent

8. WATER POLLUTION

8.1 **Aquatic Toxicity:** Data not available
8.2 **Waterfowl Toxicity:** Data not available
8.3 **Biological Oxygen Demand (BOD):** 29%, 1 day
8.4 **Food Chain Concentration Potential:** None

9. SELECTED MANUFACTURERS

1. Ethyl Corp.
Industrial Chemicals Division
451 Florida St.
Baton Rouge, La. 70801
2. Shell Chemical Co.
Industrial Chemicals Division
Houston, Tex. 77001
3. Union Carbide Corp.
Chemicals and Plastics Division
270 Park Ave.
New York, N. Y. 10017

10. SHIPPING INFORMATION

10.1 **Grades or Purity:** Data not available
10.2 **Storage Temperature:** Ambient
10.3 **Inert Atmosphere:** No requirement
10.4 **Venting:** Open (flame arrester)

11. HAZARD ASSESSMENT CODE

(See Hazard Assessment Handbook, CG 446-3)

A-T-U

12. HAZARD CLASSIFICATIONS

12.1 **Code of Federal Regulations:** Not listed
12.2 **NAS Hazard Rating for Bulk Water Transportation:** Not listed
12.3 **NFPA Hazard Classifications:** Not listed

13. PHYSICAL AND CHEMICAL PROPERTIES

13.1 **Physical State at 15°C and 1 atm:** Liquid or solid
13.2 **Molecular Weight:** >186
13.3 **Boiling Point at 1 atm:** >486°F = >252°C = >525°K
13.4 **Freezing Point:** >66°F = >19°C = >292°K
13.5 **Critical Temperature:** Not pertinent
13.6 **Critical Pressure:** Not pertinent
13.7 **Specific Gravity:** 0.84 at 20°C (liquid)
13.8 **Liquid Surface Tension:** (est.) 30 dynes/cm = 0.03 N/m at 30°C
13.9 **Liquid-Water Interfacial Tension:** (est.) 30 dynes/cm = 0.03 N/m at 30°C
13.10 **Vapor (Gas) Specific Gravity:** Not pertinent
13.11 **Ratio of Specific Heats of Vapor (Gas):** Not pertinent
13.12 **Latent Heat of Vaporization:** Not pertinent
13.13 **Heat of Combustion:** (est.) −18,500 Btu/lb = −10,300 cal/g = -429×10^5 J/kg
13.14 **Heat of Decomposition:** Not pertinent
13.15 **Heat of Solution:** Not pertinent
13.16 **Heat of Polymerization:** Not pertinent

NOTES

LNG	# LIQUEFIED NATURAL GAS

Common Synonyms			
LNG	Liquefied gas	Colorless	Odorless or weak skunk odor
	Floats and boils on water. Flammable visible vapor cloud is produced.		

Stop discharge if possible. Keep people away.
Shut off ignition sources and call fire department.
Stay upwind and use water spray to "knock down" vapor.
Evacuate area in case of large discharge.
Avoid contact with liquid.
Notify local health and pollution control agencies.

Fire	FLAMMABLE. Flashback along vapor trail may occur. May explode if ignited in an enclosed area. Stop discharge if possible. Cool exposed area and men effecting shutoff with water. Let fire burn.
Exposure	CALL FOR MEDICAL AID. VAPOR Not irritating to eyes, nose or throat. If inhaled, will cause dizziness, difficult breathing, or loss of consciousness. Move to fresh air. If breathing has stopped, give artificial respiration. If breathing is difficult, give oxygen. LIQUID Will cause frostbite. Flush affected areas with plenty of water. DO NOT RUB AFFECTED AREAS.
Water Pollution	Not harmful to aquatic life.

1. RESPONSE TO DISCHARGE

(See Response Methods Handbook, CG 446-4)

Issue warning — high flammability
Restrict access
Evacuate area

2. LABEL

Red

3. CHEMICAL DESIGNATIONS

3.1 **Synonyms:** LNG
3.2 **Coast Guard Compatibility Classification:** Paraffin
3.3 **Chemical Formula:** $CH_4+C_2H_6$
3.4 **IMCO/United Nations Numerical Designation:** 2.0/2043

4. OBSERVABLE CHARACTERISTICS

4.1 **Physical State (as shipped):** Liquefied gas
4.2 **Color:** Colorless
4.3 **Odor:** Mild, sweet

5. HEALTH HAZARDS

5.1 **Personal Protective Equipment:** Self-contained breathing apparatus; protective clothing if exposed to liquid.
5.2 **Symptoms Following Exposure:** If concentration of gas is high enough, may cause asphyxiation. No detectable systemic effects, even at 5% concentration in air.
5.3 **Treatment for Exposure:** Remove victim to open air. If he is overcome by gas, apply artificial resuscitation.
5.4 **Toxicity by Inhalation (Threshold Limit Value):** Data not available
5.5 **Short-Term Inhalation Limits:** Data not available
5.6 **Toxicity by Ingestion:** Not pertinent
5.7 **Late Toxicity:** None
5.8 **Vapor (Gas) Irritant Characteristics:** Vapors are nonirritating to the eyes and throat.
5.9 **Liquid or Solid Irritant Characteristics:** No appreciable hazard. Practically harmless to the skin because it is very volatile and evaporates quickly. May cause some frostbite.
5.10 **Odor Threshold:** Data not available

6. FIRE HAZARDS

6.1 **Flash Point:** Flammable gas
6.2 **Flammable Limits in Air:** 5.3%—14.0%
6.3 **Fire Extinguishing Agents:** Do not extinguish large spill fires. Allow to burn while cooling adjacent equipment with water spray. Shut off leak if possible. Extinguish small fires with dry chemicals.
6.4 **Fire Extinguishing Agents Not to be Used:** Water
6.5 **Special Hazards of Combustion Products:** Not pertinent
6.6 **Behavior in Fire:** Not pertinent
6.7 **Ignition Temperature:** 999°F
6.8 **Electrical Hazard:** Class I, Group D
6.9 **Burning Rate:** 12.5 mm/min.

7. CHEMICAL REACTIVITY

7.1 **Reactivity with Water:** No reaction
7.2 **Reactivity with Common Materials:** No reaction
7.3 **Stability During Transport:** Stable
7.4 **Neutralizing Agents for Acids and Caustics:** Not pertinent
7.5 **Polymerization:** Not pertinent
7.6 **Inhibitor of Polymerization:** Not pertinent

8. WATER POLLUTION

8.1 **Aquatic Toxicity:** None
8.2 **Waterfowl Toxicity:** None
8.3 **Biological Oxygen Demand (BOD):** None
8.4 **Food Chain Concentration Potential:** None

9. SELECTED MANUFACTURERS

1. Exxon Chemical Co.
   Houston, Tex. 77001
2. Phillips Petroleum Co.
   Bartlesville, Okla. 74004
3. Shell Oil Co.
   1 Shell Plaza
   Houston, Tex. 77001

10. SHIPPING INFORMATION

10.1 **Grades or Purity:** Varies with the point of origin. Usually contains at least 90% methane, with smaller quantities of ethane, propane, butanes and pentanes, carbon dioxide and nitrogen.
10.2 **Storage Temperature:** −260°F (−162°C)
10.3 **Inert Atmosphere:** No requirement
10.4 **Venting:** Safety relief

11. HAZARD ASSESSMENT CODE

(See Hazard Assessment Handbook, CG 446-3)

A-B-C-D-E-F-G

12. HAZARD CLASSIFICATIONS

12.1 **Code of Federal Regulations:** Flammable compressed gas
12.2 **NAS Hazard Rating for Bulk Water Transportation:** Not listed

Category	Rating
Fire	4
Health	
Vapor Irritant	0
Liquid or Solid Irritant	0
Poisons	0
Water Pollution	
Human Toxicity	0
Aquatic Toxicity	0
Aesthetic Effect	0
Reactivity	
Other Chemicals	0
Water	0
Self-Reaction	0

12.3 **NFPA Hazard Classifications:** Not listed

13. PHYSICAL AND CHEMICAL PROPERTIES*

13.1 **Physical State at 15°C and 1 atm:** Gas
13.2 **Molecular Weight:** >16
13.3 **Boiling Point at 1 atm:** −258°F = −161°C = 112°K
13.4 **Freezing Point:** −296°F = −182.2°C = 91.0°K
13.5 **Critical Temperature:** −116°F = −82.2°C = 191.0°K
13.6 **Critical Pressure:** 673 psia = 45.78 atm = 4.64 MN/m^2
13.7 **Specific Gravity:** 0.415 0.45 at −162°C (liquid)
13.8 **Liquid Surface Tension:** 14 dynes/cm = 0.014 N/m at −161°C
13.9 **Liquid-Water Interfacial Tension:** Data not available
13.10 **Vapor (Gas) Specific Gravity:** 0.55 1.0
13.11 **Ratio of Specific Heats of Vapor (Gas):** 1.306
13.12 **Latent Heat of Vaporization:** (est.) 220 Btu/lb = 120 cal/g = 5.1×10^5 J/kg
13.13 **Heat of Combustion:** −21,600 to −23,400 Btu/lb = −12,000 to −13,000 cal/g = −502.4 to −544.3 × 10^5 J/kg
13.14 **Heat of Decomposition:** Not pertinent
13.15 **Heat of Solution:** Not pertinent
13.16 **Heat of Polymerization:** Not pertinent

*Physical properties apply to methane; no "standard" LNG exists.

NOTES

REVISED 1978

LPG — LIQUEFIED PETROLEUM GAS

Common Synonyms			
Propane-butane-(propylene) Pyrofax LPG	Liquefied compressed gas	Colorless	Weak odor; may have skunk odor added
	Floats and boils on water. Flammable vapor cloud is produced.		

Stop discharge if possible. Keep people away.
Shut off ignition sources and call fire department.
Stay upwind and use water spray to "knock down" vapor.
Evacuate area in case of large discharge.
Avoid contact with liquid.
Notify local health and pollution control agencies.

Fire

FLAMMABLE.
Flashback along vapor trail may occur.
Containers may explode in fire.
May explode if ignited in an enclosed area.

Stop discharge if possible.
Cool exposed containers and protect men effecting shutoff with water.
Let fire burn.

Exposure

VAPOR
Not irritating to eyes, nose and throat.
If inhaled, will cause dizziness, difficult breathing, or loss of consciousness.

CALL FOR MEDICAL AID.

LIQUID
Will cause frostbite.

Move to fresh air.
If breathing has stopped, give artificial respiration.
If breathing is difficult, give oxygen.

Flush affected areas with plenty of water.
DO NOT RUB AFFECTED AREAS.

Water Pollution

Not harmful to aquatic life.

1. RESPONSE TO DISCHARGE

(See Response Methods Handbook, CG 446-4)

Issue warning—high flammability
Restrict access
Evacuate area

2. LABEL

Red

3. CHEMICAL DESIGNATIONS

3.1 **Synonyms:** Bottled gas; Propane-butane-(propylene) Pyrofax; LPG
3.2 **Coast Guard Compatibility Classification:** Paraffin or olefin
3.3 **Chemical Formula:** C_3H_8-C_3H_6-C_4H_{10} (mixture)
3.4 **IMCO/United Nations Numerical Designation:** 2.0/1075

4. OBSERVABLE CHARACTERISTICS

4.1 **Physical State (as shipped):** Liquefied compressed gas
4.2 **Color:** Colorless
4.3 **Odor:** Mild, but commercial LPG has a skunk-like odorant added as a warning.

5. HEALTH HAZARDS

5.1 **Personal Protective Equipment:** Self-contained breathing apparatus for high concentrations of gas.
5.2 **Symptoms Following Exposure:** Concentrations in air greater than 10% cause dizziness in a few minutes. 1% concentrations give the same symptom in 10 min. High concentrations cause asphyxiation.
5.3 **Treatment for Exposure:** Remove victim to open air. If he is overcome by gas, apply artificial resuscitation. Guard against self-injury if confused.
5.4 **Toxicity by Inhalation (Threshold Limit Value):** 1000 ppm
5.5 **Short-Term Inhalation Limits:** Data not available
5.6 **Toxicity by Ingestion:** Not pertinent
5.7 **Late Toxicity:** None
5.8 **Vapor (Gas) Irritant Characteristics:** Vapors are nonirritating to the eyes and throat.
5.9 **Liquid or Solid Irritant Characteristics:** No appreciable hazard. Practically harmless to the skin because it is very volatile and evaporates quickly. May cause frostbite.
5.10 **Odor Threshold:** 5000-20,000 ppm

6. FIRE HAZARDS

6.1 **Flash Point:** Propane: −156°F C.C.; butane: −76°F C.C.
6.2 **Flammable Limits in Air:** Propane:2.2%–9.5%; butane: 1.8%–8.4%
6.3 **Fire Extinguishing Agents:** Allow to burn while cooling adjacent equipment with water spray. Extinguish small fires with dry chemicals. Shut off leak if possible.
6.4 **Fire Extinguishing Agents Not to be Used:** Water (let fire burn)
6.5 **Special Hazards of Combustion Products:** Not pertinent
6.6 **Behavior in Fire:** Containers may explode in fire. Vapor heavier than air and may travel a long distance to a source of ignition and flash back.

(Continued on page 4)

7. CHEMICAL REACTIVITY

7.1 **Reactivity with Water:** No reaction
7.2 **Reactivity with Common Materials:** No reaction
7.3 **Stability During Transport:** Stable
7.4 **Neutralizing Agents for Acids and Caustics:** Not pertinent
7.5 **Polymerization:** Not pertinent
7.6 **Inhibitor of Polymerization:** Not pertinent

8. WATER POLLUTION

8.1 **Aquatic Toxicity:** None
8.2 **Waterfowl Toxicity:** None
8.3 **Biological Oxygen Demand (BOD):** None
8.4 **Food Chain Concentration Potential:** None

9. SELECTED MANUFACTURERS

1. Exxon Co.
   Houston, Tex. 77001
2. Phillips Petroleum Co.
   Bartlesville, Okla. 74004
3. Sun Oil Co.
   St. Davids, Pa. 19087

10. SHIPPING INFORMATION

10.1 **Grades or Purity:** Various grades, mostly propane. In some areas propylene may be included. The proportion may be varied with the season.
10.2 **Storage Temperature:** Ambient
10.3 **Inert Atmosphere:** No requirement
10.4 **Venting:** Safety relief

11. HAZARD ASSESSMENT CODE

(See Hazard Assessment Handbook, CG 446-3)

A-B-C-D-E-F-G

12. HAZARD CLASSIFICATIONS

12.1 **Code of Federal Regulations:** Flammable compressed gas
12.2 **NAS Hazard Rating for Bulk Water Transportation:**

Category	Rating
Fire	4
Health	
Vapor Irritant	0
Liquid or Solid Irritant	0
Poisons	0
Water Pollution	
Human Toxicity	0
Aquatic Toxicity	0
Aesthetic Effect	0
Reactivity	
Other Chemicals	0
Water	0
Self-Reaction	0

12.3 **NFPA Hazard Classifications:**

Category	Classification
Health Hazard (Blue)	1
Flammability (Red)	4
Reactivity (Yellow)	0

13. PHYSICAL AND CHEMICAL PROPERTIES*

13.1 **Physical State at 15°C and 1 atm:** Gas
13.2 **Molecular Weight:** >44
13.3 **Boiling Point at 1 atm:** > −40°F = > −40°C = > 233°K
13.4 **Freezing Point:** Not pertinent
13.5 **Critical Temperature:** −142.01°F = −96.67°C = 176.53°K
13.6 **Critical Pressure:** 616.5 psia = 41.94 atm = 4.249 MN/m²
13.7 **Specific Gravity:** 0.51–0.58 at −50°C (liquid)
13.8 **Liquid Surface Tension:** 16 dynes/cm = 0.016 N/m at −47°C
13.9 **Liquid-Water Interfacial Tension:** (est.) 50 dynes/cm = 0.05 N/m at −38°C
13.10 **Vapor (Gas) Specific Gravity:** 1.5
13.11 **Ratio of Specific Heats of Vapor (Gas):** 1.130
13.12 **Latent Heat of Vaporization:** 183.2 Btu/lb = 101.8 cal/g = 4.262×10^5 J/kg
13.13 **Heat of Combustion:** −19,782 Btu/lb = −10,990 cal/g = -460.13×10^5 J/kg
13.14 **Heat of Decomposition:** Not pertinent
13.15 **Heat of Solution:** Not pertinent
13.16 **Heat of Polymerization:** Not pertinent

*Physical properties apply to propane; no "standard" LPG exists.

6. FIRE HAZARDS (Cont'd.)

6.7 **Ignition Temperature:** Propane: 871°F; butane: 761°F
6.8 **Electrical Hazard:** Class I, Group D
6.9 **Burning Rate:** 8.2 mm/min.

REVISED 1978

LTH — LITHARGE

Common Synonyms	Solid	Gray or yellow green or red-brown	Odorless
Lead monoxide Lead oxide yellow Plumbous oxide Lead Protoxide	Sinks in water.		

Keep people away.
Isolate and remove discharged material.
Notify local health and pollution control agencies.

Fire

Not flammable.

Exposure

Call for medical aid.

SOLID OR DUST
Irritating to eyes.
Harmful if inhaled.
Move victim to fresh air.
If in eyes, hold eyelids open and flush with plenty of water.
IF SWALLOWED and victim is CONSCIOUS, have victim drink water or milk and have victim induce vomiting.
IF SWALLOWED and victim is UNCONSCIOUS OR HAVING CONVULSIONS, do nothing except keep victim warm.

Water Pollution

Effect of low concentrations on aquatic life is unknown.
May be dangerous if it enters water intakes.
Notify local health and wildlife officials.
Notify operators of nearby water intakes.

1. RESPONSE TO DISCHARGE

(See Response Methods Handbook, CG 446-4)

Should be removed
Chemical and physical treatment

2. LABELS

No label required by Code of Federal Regulations

3. CHEMICAL DESIGNATIONS

3.1 **Synonyms:** Lead monoxide; Lead oxide, yellow; Lead protoxide; Massicot; Plumbous oxide
3.2 **Coast Guard Compatibility Classification:** Not applicable
3.3 **Chemical Formula:** PbO
3.4 **IMCO/United Nations Numerical Designation:** Not listed

4. OBSERVABLE CHARACTERISTICS

4.1 **Physical State (as shipped):** Solid
4.2 **Color:** Yellow to red; low-metal-content oxides: yellow to green to brown; high-metal-content oxides: gray to brown.
4.3 **Odor:** None

5. HEALTH HAZARDS

5.1 **Personal Protective Equipment:** Dust or metal fume respirator; gloves; goggles.
5.2 **Symptoms Following Exposure:** General symptoms of lead poisoning (delayed). Inhalation or ingestion causes abdominal pain (lead colic), metallic taste in mouth, loss of weight, pain in muscles, and muscular weakness. Dust may irritate eyes.
5.3 **Treatment for Exposure:** Consult physician after ingestion or exposure to high concentrations of dust. INGESTION: call physician at once; as first aid, induce vomiting and give milk and magnesium sulfate (Epsom salt).
5.4 **Toxicity by Inhalation (Threshold Limit Value):** 0.2 mg/m^3
5.5 **Short-Term Inhalation Limits:** Data not available
5.6 **Toxicity by Ingestion:** Data not available
5.7 **Late Toxicity:** Impairs development of human fetal connective tissue cells
5.8 **Vapor (Gas) Irritant Characteristics:** Not pertinent
5.9 **Liquid or Solid Irritant Characteristics:** Data not available
5.10 **Odor Threshold:** Not pertinent

6. FIRE HAZARDS

6.1 **Flash Point:** Not flammable
6.2 **Flammable Limits in Air:** Not flammable
6.3 **Fire Extinguishing Agents:** Not pertinent
6.4 **Fire Extinguishing Agents Not to be Used:** Not pertinent
6.5 **Special Hazards of Combustion Products:** Not pertinent
6.6 **Behavior in Fire:** Not pertinent
6.7 **Ignition Temperature:** Not pertinent
6.8 **Electrical Hazard:** Not pertinent
6.9 **Burning Rate:** Not pertinent

7. CHEMICAL REACTIVITY

7.1 **Reactivity with Water:** No reaction
7.2 **Reactivity with Common Materials:** No reaction
7.3 **Stability During Transport:** Stable
7.4 **Neutralizing Agents for Acids and Caustics:** Not pertinent
7.5 **Polymerization:** Not pertinent
7.6 **Inhibitor of Polymerization:** Not pertinent

8. WATER POLLUTION

8.1 **Aquatic Toxicity:**
> 56,000 ppm/96 hr/mosquito fish/ TL_m/turbid water
8.2 **Waterfowl Toxicity:** Data not available
8.3 **Biological Oxygen Demand (BOD):** Data not available
8.4 **Food Chain Concentration Potential:** Data not available

9. SELECTED MANUFACTURERS

1. NL Industries, Inc.
   Pigments and Chemicals Division
   425 University Avenue
   Westwood, Mass. 02090
2. J. T. Baker Chemical Co.
   Phillipsburg, N. J. 08865
3. Eagle-Picher Industries, Inc.
   Chemicals and Fibers Division
   American Building
   Cincinnati, Ohio 45202

10. SHIPPING INFORMATION

10.1 **Grades or Purity:** Low-metal-content oxides contain 98 to 99.8%. High metal or battery grades contain 50 to 95%. Reagent; purified. Most grades available in several particle sizes.
10.2 **Storage Temperature:** Ambient
10.3 **Inert Atmosphere:** No requirement
10.4 **Venting:** Open

11. HAZARD ASSESSMENT CODE

(See Hazard Assessment Handbook, CG 446-3)

II

12. HAZARD CLASSIFICATIONS

12.1 **Code of Federal Regulations:** Not listed
12.2 **NAS Hazard Rating for Bulk Water Transportation:** Not listed
12.3 **NFPA Hazard Classifications:** Not listed

13. PHYSICAL AND CHEMICAL PROPERTIES

13.1 **Physical State at 15°C and 1 atm:** Solid
13.2 **Molecular Weight:** 223.2
13.3 **Boiling Point at 1 atm:** Not pertinent (Decomposes)
13.4 **Freezing Point:** Not pertinent
13.5 **Critical Temperature:** Not pertinent
13.6 **Critical Pressure:** Not pertinent
13.7 **Specific Gravity:** 9.5 at 20°C (solid)
13.8 **Liquid Surface Tension:** Not pertinent
13.9 **Liquid-Water Interfacial Tension:** Not pertinent
13.10 **Vapor (Gas) Specific Gravity:** Not pertinent
13.11 **Ratio of Specific Heats of Vapor (Gas):** Not pertinent
13.12 **Latent Heat of Vaporization:** Not pertinent
13.13 **Heat of Combustion:** Not pertinent
13.14 **Heat of Decomposition:** Not pertinent
13.15 **Heat of Solution:** Not pertinent
13.16 **Heat of Polymerization:** Not pertinent

NOTES

LAH	# LITHIUM ALUMINUM HYDRIDE

Common Synonyms LAH	Solid powder	White to gray	Odorless
	Reacts violently with water. Flammable gas is produced.		

Avoid contact with solid and dust.
Wear rubber overclothing (including gloves).
Stop discharge if possible. Keep people away.
Shut off ignition sources and call fire department.
Isolate and remove discharged material.
Notify local health and pollution control agencies.

Fire	FLAMMABLE. Flammable gas is released on contact with water, metals, or acids. Wear rubber overclothing (including gloves). DO NOT USE WATER, DRY CHEMICALS, CARBON DIOXIDE, OR FOAM. Extinguish with powdered limestone or powdered graphite.
Exposure	CALL FOR MEDICAL AID. SOLID Will burn skin and eyes. Harmful if swallowed. Remove contaminated clothing and shoes. Flush affected areas with plenty of water. IF IN EYES, hold eyelids open and flush with plenty of water. IF SWALLOWED and victim is CONSCIOUS, have victim drink plenty of water.
Water Pollution	Effect of low concentrations on aquatic life is unknown. May be dangerous if it enters water intakes. Notify local health and wildlife officials. Notify operators of nearby water intakes.

1. RESPONSE TO DISCHARGE

(See Response Methods Handbook, CG 446-4)

Issue warning—high flammability, corrosive
Restrict access
Disperse and flush (with care)

2. LABELS

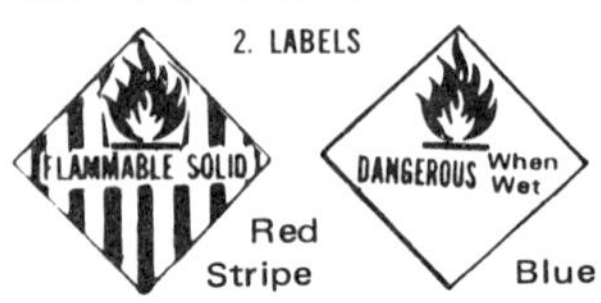

3. CHEMICAL DESIGNATIONS

3.1 **Synonyms:** LAH
3.2 **Coast Guard Compatibility Classification:** Not applicable
3.3 **Chemical Formula:** $LiAlH_4$
3.4 **IMCO/United Nations Numerical Designation:** 4.3/1410

4. OBSERVABLE CHARACTERISTICS

4.1 **Physical State (as shipped):** Solid
4.2 **Color:** White to gray
4.3 **Odor:** None

5. HEALTH HAZARDS

5.1 **Personal Protective Equipment:** Rubberized gloves; full face shield.
5.2 **Symptoms Following Exposure:** Contact of solid with eyes and skin causes severe burns similar to those caused by caustic soda.
5.3 **Treatment for Exposure:** In case of accidental contact with the skin, wipe off excess with a dry paper towel. Wash the affected area with a large volume of water to prevent localized heating of the skin.
5.4 **Toxicity by Inhalation (Threshold Limit Value):** Not pertinent
5.5 **Short-Term Inhalation Limits:** Not pertinent
5.6 **Toxicity by Ingestion:** Data not available
5.7 **Late Toxicity:** Data not available
5.8 **Vapor (Gas) Irritant Characteristics:** Not pertinent
5.9 **Liquid or Solid Irritant Characteristics:** Moisture of skin causes caustic burns.
5.10 **Odor Threshold:** Not pertinent

6. FIRE HAZARDS

6.1 **Flash Point:** Solid
6.2 **Flammable Limits in Air:** Not pertinent
6.3 **Fire Extinguishing Agents:** Powdered graphite, powdered salt, or powdered limestone
6.4 **Fire Extinguishing Agents Not to be Used:** Do NOT use water, soda acid, carbon dioxide or dry chemical.
6.5 **Special Hazards of Combustion Products:** Data not available
6.6 **Behavior in Fire:** Decomposes at 257°F to form hydrogen gas. The heat generated may cause ignition and/or explosion.
6.7 **Ignition Temperature:** Data not available
6.8 **Electrical Hazard:** Class I, Group B
6.9 **Burning Rate:** Not pertinent

7. CHEMICAL REACTIVITY

7.1 **Reactivity with Water:** Reacts violently with water as a dry solid or when dissolved in ether. The hydrogen produced by the reaction with water is a major hazard and necessitates adequate ventilation.
7.2 **Reactivity with Common Materials:** Can burn in heated or moist air.
7.3 **Stability During Transport:** Normally stable; unstable at high temperatures.
7.4 **Neutralizing Agents for Acids and Caustics:** Not pertinent
7.5 **Polymerization:** Not pertinent
7.6 **Inhibitor of Polymerization:** Not pertinent

8. WATER POLLUTION

8.1 **Aquatic Toxicity:** Data not available
8.2 **Waterfowl Toxicity:** Data not available
8.3 **Biological Oxygen Demand (BOD):** Data not available
8.4 **Food Chain Concentration Potential:** None

9. SELECTED MANUFACTURERS

1. Belmont Smelting & Refining Works, Inc.
   330 Belmont Ave.
   Brooklyn, N. Y. 11207
2. Foote Mineral Co.
   Route 100
   Exton, Pa. 19341
3. Ventron Corp.
   Chemical Division
   Beverly, Mass. 01915

10. SHIPPING INFORMATION

10.1 **Grades or Purity:** 95-98%
10.2 **Storage Temperature:** Ambient
10.3 **Inert Atmosphere:** Dry air
10.4 **Venting:** Store container in well-ventilated area.

11. HAZARD ASSESSMENT CODE

(See Hazard Assessment Handbook, CG 446-3)

R-R-C

12. HAZARD CLASSIFICATIONS

12.1 **Code of Federal Regulations:** Flammable solid
12.2 **NAS Hazard Rating for Bulk Water Transportation:** Not listed
12.3 **NFPA Hazard Classifications:**

Category	Classification
Health Hazard (Blue)	3
Flammability (Red)	1
Reactivity (Yellow)	2 W

13. PHYSICAL AND CHEMICAL PROPERTIES

13.1 **Physical State at 15°C and 1 atm:** Solid
13.2 **Molecular Weight:** 37.94
13.3 **Boiling Point at 1 atm:** Decomposes
13.4 **Freezing Point:** Not pertinent
13.5 **Critical Temperature:** Not pertinent
13.6 **Critical Pressure:** Not pertinent
13.7 **Specific Gravity:** 0.917 at 15°C (solid)
13.8 **Liquid Surface Tension:** Not pertinent
13.9 **Liquid-Water Interfacial Tension:** Not pertinent
13.10 **Vapor (Gas) Specific Gravity:** Not pertinent
13.11 **Ratio of Specific Heats of Vapor (Gas):** Not pertinent
13.12 **Latent Heat of Vaporization:** Not pertinent
13.13 **Heat of Combustion:** Not pertinent
13.14 **Heat of Decomposition:** Not pertinent
13.15 **Heat of Solution:** Not pertinent
13.16 **Heat of Polymerization:** Not pertinent

NOTES

REVISED 1978

LHD — LITHIUM HYDRIDE

Common Synonyms	Solid — Crystals are gray or blue; powder is white — Odorless
	Reacts violently with water. Flammable gas is produced.

AVOID CONTACT WITH SOLID AND DUST. KEEP PEOPLE AWAY.
Wear dust respirator and rubber overclothing (including gloves).
Stop discharge if possible.
Isolate and remove discharged material.
Notify local health and pollution control agencies.

Fire

Combustible.
Irritating flammable gas may be produced when heated.
Extinguish with dry graphite, soda ash, or other inert powder.
DO NOT USE WATER, FOAM, CARBON DIOXIDE OR DRY CHEMICALS ON FIRE OR ADJACENT FIRES.

Exposure

CALL FOR MEDICAL AID.

DUST
POISONOUS IF INHALED.
If inhaled will cause coughing or difficult breathing.
If in eyes, hold eyelids open and flush with plenty of water.
If breathing has stopped, give artificial respiration.
If breathing is difficult, give oxygen.

SOLID
Will burn skin and eyes.
If swallowed will cause nausea or loss of consciousness.
Remove contaminated clothing and shoes.
Flush affected areas with plenty of water.
IF IN EYES, hold eyelids open and flush with plenty of water.
IF SWALLOWED and victim is CONSCIOUS, have victim drink water or milk.
IF SWALLOWED and victim is UNCONSCIOUS OR HAVING CONVULSIONS, do nothing except keep victim warm.

Water Pollution

Effect of low concentrations on aquatic life is unknown.
May be dangerous if it enters water intakes.
Notify local health and wildlife officials.
Notify operators of nearby water intakes.

1. RESPONSE TO DISCHARGE
(See Response Methods Handbook, CG 446-4)

Issue warning - high flammability, corrosive
Restrict access
Disperse and flush with care

2. LABELS

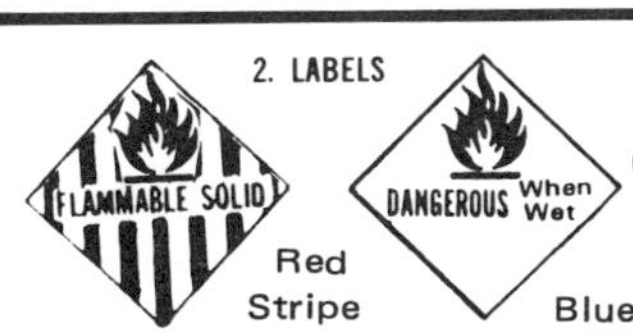

3. CHEMICAL DESIGNATIONS

3.1 **Synonyms:** No common synonyms
3.2 **Coast Guard Compatibility Classification:** Not listed
3.3 **Chemical Formula:** LiH
3.4 **IMCO/United Nations Numerical Designation:** 4.3/1414

4. OBSERVABLE CHARACTERISTICS

4.1 **Physical State (as shipped):** Solid
4.2 **Color:** Gray-blue crystalline mass; finely ground material ranges in color from white to gray.
4.3 **Odor:** None

5. HEALTH HAZARDS

5.1 **Personal Protective Equipment:** Goggles or face shield; rubberized gloves; flame proof outer clothing; respirator; high boots or shoes
5.2 **Symptoms Following Exposure:** Inhalation of dust causes coughing, sneezing, and burning of nose and throat. Ingestion causes severe burns of mouth and stomach; symptoms of central nervous system damage may occur. Contact with eyes or skin causes severe caustic burns.
5.3 **Treatment for Exposure:** *Lithium hydride burns of the eyes, skin, or respiratory tract appear to be worse than those caused by an equivalent amount of sodium hydroxide.* INHALATION: remove victim to fresh air; if irritation persists get medical attention at once. INGESTION: give large volumes of water and milk; gastric lavage may be indicated. EYES: flush with copious quantities of running water for at least 15 min.; get medical attention. SKIN: flush with water; treat as a caustic burn.
5.4 **Toxicity by Inhalation (Threshold Limit Value):** 0.025 mg/m³
5.5 **Short-Term Inhalation Limits:** Data not available
5.6 **Toxicity by Ingestion:** Data not available
5.7 **Late Toxicity:** Data not available
5.8 **Vapor (Gas) Irritant Characteristics:** Data not available
5.9 **Liquid or Solid Irritant Characteristics:** Data not available
5.10 **Odor Threshold:** Data not available

6. FIRE HAZARDS

6.1 **Flash Point:** Not pertinent (combustible solid)
6.2 **Flammable Limits in Air:** Not pertinent
6.3 **Fire Extinguishing Agents:** Dry nitrogen, graphite, or lithium chloride
6.4 **Fire Extinguishing Agents Not to be Used:** Never use water, foam, halogenated hydrocarbons, soda acid, dry chemical, or carbon dioxide.
6.5 **Special Hazards of Combustion Products:** Irritating alkali fumes may form in fire.
6.6 **Behavior in Fire:** May decompose when hot to form flammable hydrogen gas. Reacts violently with water to produce hydrogen, which may explode in air.
6.7 **Ignition Temperature:** 392°F
6.8 **Electrical Hazard:** Not pertinent
6.9 **Burning Rate:** Not pertinent

7. CHEMICAL REACTIVITY

7.1 **Reactivity with Water:** Reacts violently with water to form flammable hydrogen gas and a strong caustic solution; ignition may occur, especially with powder.
7.2 **Reactivity with Common Materials:** May ignite combustible materials if they are damp
7.3 **Stability During Transport:** Stable, if air and moisture are excluded
7.4 **Neutralizing Agents for Acids and Caustics:** Residues should be washed well with water, then rinsed with dilute acetic acid.
7.5 **Polymerization:** Not pertinent
7.6 **Inhibitor of Polymerization:** Not pertinent

8. WATER POLLUTION

8.1 **Aquatic Toxicity:** Data not available
8.2 **Waterfowl Toxicity:** Data not available
8.3 **Biological Oxygen Demand (BOD):** None
8.4 **Food Chain Concentration Potential:** None

9. SELECTED MANUFACTURERS

1. Lithium Corporation of America, P. O. Box 8, Bessemer City, N. C. 28016
2. Ventron Corp., P. O. Box 159, Beverly, Mass. 01915
3. Pfaltz and Bauer, Inc., 375 Fairfield Ave., Stamford, Conn. 06902

10. SHIPPING INFORMATION

10.1 **Grades or Purity:** Commercial, 96.5%
10.2 **Storage Temperature:** Ambient
10.3 **Inert Atmosphere:** Inerted
10.4 **Venting:** Safety-relief

11. HAZARD ASSESSMENT CODE
(See Hazard Assessment Handbook, CG 446-3)

RR

12. HAZARD CLASSIFICATIONS

12.1 **Code of Federal Regulations:** Flammable solid
12.2 **NAS Hazard Rating for Bulk Water Transportation:** Not listed
12.3 **NFPA Hazard Classifications:**

Category	Classification
Health Hazard (Blue)	1
Flammability (Red)	4
Reactivity (Yellow)	2 W̶

13. PHYSICAL AND CHEMICAL PROPERTIES

13.1 **Physical State at 15°C and 1 atm:** Solid
13.2 **Molecular Weight:** 7.95
13.3 **Boiling Point at 1 atm:** Not pertinent (decomposes)
13.4 **Freezing Point:** Not pertinent
13.5 **Critical Temperature:** Not pertinent
13.6 **Critical Pressure:** Not pertinent
13.7 **Specific Gravity:** 0.78 at 20°C (solid)
13.8 **Liquid Surface Tension:** Not pertinent
13.9 **Liquid-Water Interfacial Tension:** Not pertinent
13.10 **Vapor (Gas) Specific Gravity:** Not pertinent
13.11 **Ratio of Specific Heats of Vapor (Gas):** Not pertinent
13.12 **Latent Heat of Vaporization:** Not pertinent
13.13 **Heat of Combustion:** Data not available
13.14 **Heat of Decomposition:** Data not available
13.15 **Heat of Solution:** $-7{,}200$ Btu/lb $= -4{,}000$ cal/g $= -170 \times 10^5$ J/kg
13.16 **Heat of Polymerization:** Not pertinent

NOTES

LITHIUM, METALLIC

Common Synonyms	Soft solid	White to light silver	Odorless
	Reacts violently with water. Flammable gas is produced.		

AVOID CONTACT WITH SOLID. KEEP PEOPLE AWAY.
Wear rubber overclothing (including gloves).
Call fire department.
Isolate and remove discharged material.
Notify local health and pollution control agencies.

Fire	Combustible. Irritating gases may be produced when heated. **Wear goggles and self-contained breathing apparatus.** **Extinguish with dry graphite, soda ash, or other inert powder.** **DO NOT USE WATER, FOAM, DRY CHEMICALS OR CARBON DIOXIDE ON FIRE.** **DO NOT USE WATER OR FOAM ON ADJACENT FIRES.**
Exposure	CALL FOR MEDICAL AID. SOLID Will burn skin and eyes. Harmful if swallowed. **Remove contaminated clothing and shoes.** **Flush affected areas with plenty of water.** **IF IN EYES, hold eyelids open and flush with plenty of water.** **IF SWALLOWED and victim is CONSCIOUS, have victim drink water or milk.** **IF SWALLOWED and victim is UNCONSCIOUS OR HAVING CONVULSIONS, do nothing except keep victim warm.**
Water Pollution	Effect of low concentrations on aquatic life is unknown. May be dangerous if it enters water intakes. **Notify local health and wildlife officials.** **Notify operators of nearby water intakes.**

1. RESPONSE TO DISCHARGE

(See Response Methods Handbook, CG 446-4)

Issue warning—high flammability, corrosive
Restrict access
Disperse and flush with care

2. LABELS

3. CHEMICAL DESIGNATIONS

3.1 **Synonyms:** No common synonyms
3.2 **Coast Guard Compatibility Classification:** Not listed
3.3 **Chemical Formula:** Li
3.4 **IMCO/United Nations Numerical Designation:** 4.3/1415

4. OBSERVABLE CHARACTERISTICS

4.1 **Physical State (as shipped):** Solid
4.2 **Color:** White; light silvery
4.3 **Odor:** None

5. HEALTH HAZARDS

5.1 **Personal Protective Equipment:** Rubber or plastic gloves; face shield; respirator; fire-retardant clothing
5.2 **Symptoms Following Exposure:** Contact with eyes causes caustic irritation or burn. In contact with skin lithium reacts with body moisture to cause chemical burns; foil, ribbon, and wire react relatively slowly.
5.3 **Treatment for Exposure:** EYES or SKIN: flush with water and treat with boric acid.
5.4 **Toxicity by Inhalation (Threshold Limit Value):** Data not available
5.5 **Short-Term Inhalation Limits:** Data not available
5.6 **Toxicity by Ingestion:** Data not available
5.7 **Late Toxicity:** Data not available
5.8 **Vapor (Gas) Irritant Characteristics:** Data not available
5.9 **Liquid or Solid Irritant Characteristics:** Data not available
5.10 **Odor Threshold:** Data not available

6. FIRE HAZARDS

6.1 **Flash Point:** Not pertinent (combustible solid)
6.2 **Flammable Limits in Air:** Not pertinent
6.3 **Fire Extinguishing Agents:** Graphite, lithium chloride
6.4 **Fire Extinguishing Agents Not to be Used:** Water, sand, halogenated hydrocarbons, carbon dioxide, soda-acid, or dry chemical
6.5 **Special Hazards of Combustion Products:** Strong alkali fumes are formed in fire.
6.6 **Behavior in Fire:** Molten lithium is quite easily ignited and is then difficult to extinguish. Hot or burning lithium will react with all gases except those of the helium-argon group. It also reacts violently with concrete, wood, asphalt, sand, asbestos, and in fact, nearly everything except metal. Do not apply water to adjacent fires. Hydrogen explosion may result. *(Continued on page 4)*

7. CHEMICAL REACTIVITY

7.1 **Reactivity with Water:** Reacts violently to form flammable hydrogen gas and strong caustic solution. Ignition usually occurs.
7.2 **Reactivity with Common Materials:** May ignite combustible materials if they are damp.
7.3 **Stability During Transport:** Stable, if air and moisture are excluded.
7.4 **Neutralizing Agents for Acids and Caustics:** Residues should be flushed with water, then rinsed with dilute acetic acid.
7.5 **Polymerization:** Not pertinent
7.6 **Inhibitor of Polymerization:** Not pertinent

8. WATER POLLUTION

8.1 **Aquatic Toxicity:** Data not available
8.2 **Waterfowl Toxicity:** Data not available
8.3 **Biological Oxygen Demand (BOD):** None
8.4 **Food Chain Concentration Potential:** None

9. SELECTED MANUFACTURERS

1. Foote Mineral Co.
   Route 100
   Exton, Pa. 19341
2. Lithium Corporation of America
   Box 795
   Bessemer City, N. C. 28016
3. Ventron Corp.
   P. O. Box 159
   Beverly, Mass. 01915

10. SHIPPING INFORMATION

10.1 **Grades or Purity:** Pure, 99.9%; Powder, shot, wire, ribbon, rod.
10.2 **Storage Temperature:** Ambient
10.3 **Inert Atmosphere:** Inerted
10.4 **Venting:** Safety-relief

11. HAZARD ASSESSMENT CODE

(See Hazard Assessment Handbook, CG 446-3)

RR

12. HAZARD CLASSIFICATIONS

12.1 **Code of Federal Regulations:** Flammable solid
12.2 **NAS Hazard Rating for Bulk Water Transportation:** Not listed
12.3 **NFPA Hazard Classifications:**

Category	Classification
Health Hazard (Blue)	1
Flammability (Red)	1
Reactivity (Yellow)	2
	W̶

13. PHYSICAL AND CHEMICAL PROPERTIES

13.1 **Physical State at 15°C and 1 atm:** Solid
13.2 **Molecular Weight:** 6.939
13.3 **Boiling Point at 1 atm:** Not pertinent
13.4 **Freezing Point:** Not pertinent
13.5 **Critical Temperature:** Not pertinent
13.6 **Critical Pressure:** Not pertinent
13.7 **Specific Gravity:** 0.53 at 20°C (solid)
13.8 **Liquid Surface Tension:** Not pertinent
13.9 **Liquid-Water Interfacial Tension:** Not pertinent
13.10 **Vapor (Gas) Specific Gravity:** Not pertinent
13.11 **Ratio of Specific Heats of Vapor (Gas):** Not pertinent
13.12 **Latent Heat of Vaporization:** Not pertinent
13.13 **Heat of Combustion:** −18,470 Btu/lb = −10,260 cal/g = -429.3×10^5 J/kg
13.14 **Heat of Decomposition:** Not pertinent
13.15 **Heat of Solution:** −31,500 Btu/lb = −17,500 cal/g = -733×10^5 J/kg
13.16 **Heat of Polymerization:** Not pertinent

6. FIRE HAZARDS (Cont'd.)

6.7 **Ignition Temperature:** 354°F
6.8 **Electrical Hazard:** Not pertinent
6.9 **Burning Rate:** Not pertinent

MGX	MAGNESIUM

Common Synonyms	Solid Silvery Odorless Sinks in water.

Call fire department.
Stop discharge if possible. Keep people away.
Isolate and remove discharged material.
Notify local health and pollution control agencies.

Fire	FLAMMABLE. **Extinguish with dry graphite, soda ash, or other inert powder.** **DO NOT USE WATER, FOAM, CARBON DIOXIDE, DRY CHEMICALS, OR VAPORIZING LIQUID ON FIRE.**
Exposure	**Call for medical aid.** SOLID Irritating to eyes. Harmful if swallowed. **IF IN EYES, hold eyelids open and flush with plenty of water.** **IF SWALLOWED and victim is CONSCIOUS, have victim drink water or milk.**
Water Pollution	Effect of low concentrations on aquatic life is unknown. May be dangerous if it enters water intakes. **Notify local health and wildlife officials.** **Notify operators of nearby water intakes.**

1. RESPONSE TO DISCHARGE

(See Response Methods Handbook, CG 446-4)

Issue warning—high flammability
Should be removed
Chemical and physical treatment

2. LABELS

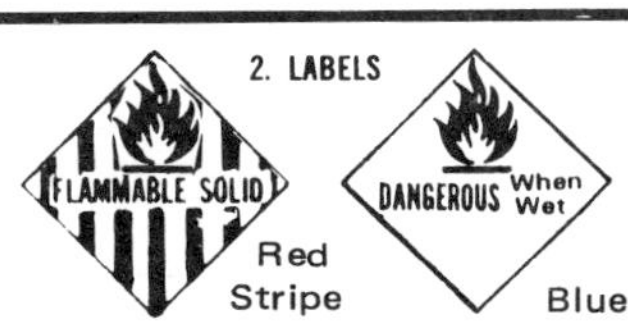

3. CHEMICAL DESIGNATIONS

3.1 **Synonyms:** No common synonyms
3.2 **Coast Guard Compatibility Classification:** Not applicable
3.3 **Chemical Formula:** Mg
3.4 **IMCO/United Nations Numerical Designation:**
Pellets, turnings, or ribbon: 4.1/1869
Powder, non-pyrophoric: 4.3/1418

4. OBSERVABLE CHARACTERISTICS

4.1 **Physical State (as shipped):** Solid
4.2 **Color:** Silvery; looks like aluminum
4.3 **Odor:** None

5. HEALTH HAZARDS

5.1 **Personal Protective Equipment:** Eye protection
5.2 **Symptoms Following Exposure:** Dust irritates eyes in same way as any foreign material. Penetration of skin by fragments of metal is likely to produce local irritation, blisters, and ulcers which may become infected.
5.3 **Treatment for Exposure:** EYES: flush with water to remove dust. SKIN: treat as any puncture.
5.4 **Toxicity by Inhalation (Threshold Limit Value):** Data not available
5.5 **Short-Term Inhalation Limits:** Not pertinent
5.6 **Toxicity by Ingestion:** Oral LDL_0 (lowest lethal dose) = 230 mg/kg (dog)
5.7 **Late Toxicity:** Data not available
5.8 **Vapor (Gas) Irritant Characteristics:** Not pertinent
5.9 **Liquid or Solid Irritant Characteristics:** Data not available
5.10 **Odor Threshold:** Not pertinent

6. FIRE HAZARDS

6.1 **Flash Point:** Not pertinent (solid). Flammable when in the form of turnings or powder.
6.2 **Flammable Limits in Air:** Not pertinent
6.3 **Fire Extinguishing Agents:** Inert dry powders (e.g., graphite, limestone, salt)
6.4 **Fire Extinguishing Agents Not to be Used:** Water, foam, halogenated agents, carbon dioxide.
6.5 **Special Hazards of Combustion Products:** Not pertinent
6.6 **Behavior in Fire:** Forms dense white smoke. Flame is very bright.
6.7 **Ignition Temperature:** 883°F
6.8 **Electrical Hazard:** Class I, Group E
6.9 **Burning Rate:** Not pertinent

7. CHEMICAL REACTIVITY

7.1 **Reactivity with Water:** In finely divided form, reacts with water and acids to release flammable hydrogen gas.
7.2 **Reactivity with Common Materials:** No reaction
7.3 **Stability During Transport:** Stable
7.4 **Neutralizing Agents for Acids and Caustics:** Not pertinent
7.5 **Polymerization:** Not pertinent
7.6 **Inhibitor of Polymerization:** Not pertinent

8. WATER POLLUTION

8.1 **Aquatic Toxicity:** None
8.2 **Waterfowl Toxicity:** None
8.3 **Biological Oxygen Demand (BOD):** None
8.4 **Food Chain Concentration Potential:** None

9. SELECTED MANUFACTURERS

1. NL Industries, Inc.
Magnesium Division
238 North 2200 West
Salt Lake City, Utah 84116
2. Dow Chemical Co.
Midland, Mich. 48640
3. Belmont Smelting and Refining Works, Inc.
330 Belmont Ave.
Brooklyn, N. Y. 11207

10. SHIPPING INFORMATION

10.1 **Grades or Purity:** Pigs, ingots, turnings, sticks; all high purity.
10.2 **Storage Temperature:** Ambient
10.3 **Inert Atmosphere:** No requirement
10.4 **Venting:** Open (flame arrester)

11. HAZARD ASSESSMENT CODE

(See Hazard Assessment Handbook, CG 446-3)

II

12. HAZARD CLASSIFICATIONS

12.1 **Code of Federal Regulations:** Flammable solid
12.2 **NAS Hazard Rating for Bulk Water Transportation:** Not listed
12.3 **NFPA Hazard Classifications:**

Category	Classification
Health Hazard (Blue)	0
Flammability (Red)	1
Reactivity (Yellow)	2 W

13. PHYSICAL AND CHEMICAL PROPERTIES

13.1 **Physical State at 15°C and 1 atm:** Solid
13.2 **Molecular Weight:** 24.3
13.3 **Boiling Point at 1 atm:** 2,012°F = 1,100°C = 1,373°K
13.4 **Freezing Point:** 1,202°F = 650°C = 923°K
13.5 **Critical Temperature:** Not pertinent
13.6 **Critical Pressure:** Not pertinent
13.7 **Specific Gravity:** 1.74 at 20°C (solid)
13.8 **Liquid Surface Tension:** Not pertinent
13.9 **Liquid-Water Interfacial Tension:** Not pertinent
13.10 **Vapor (Gas) Specific Gravity:** Not pertinent
13.11 **Ratio of Specific Heats of Vapor (Gas):** Not pertinent
13.12 **Latent Heat of Vaporization:** Not pertinent
13.13 **Heat of Combustion:** −11,950 Btu/lb = −6,650 cal/g = -278×10^5 J/kg
13.14 **Heat of Decomposition:** Not pertinent
13.15 **Heat of Solution:** Not pertinent
13.16 **Heat of Polymerization:** Not pertinent

NOTES

MPC — MAGNESIUM PERCHLORATE

Common Synonyms			
Anhydrone Dehydrite Magnesium perchlorate, anhydrous Magnesium perchlorate hexahydrate	Solid Sinks and mixes with water.	White	Odorless

Avoid contact with solid and dust. Keep people away.
Stop discharge if possible.
Isolate and remove discharged material.
Notify local health and pollution control agencies.

Fire

Not flammable.
May cause fire on contact with combustibles.
Will increase the intensity of a fire.
Flood discharge area with water.

Exposure

CALL FOR MEDICAL AID.

DUST
Irritating to eyes, nose and throat.
If inhaled will cause difficult breathing.
If in eyes, hold eyelids open and flush with plenty of water.
If breathing has stopped, give artificial respiration.
If breathing is difficult, give oxygen.

SOLID
Irritating to skin and eyes.
If swallowed will cause nausea, vomiting or loss of consciousness.
Remove contaminated clothing and shoes.
Flush affected areas with plenty of water.
IF IN EYES, hold eyelids open and flush with plenty of water.
IF SWALLOWED and victim is CONSCIOUS, have victim drink water or milk and have victim induce vomiting.
IF SWALLOWED and victim is UNCONSCIOUS OR HAVING CONVULSIONS, do nothing except keep victim warm.

Water Pollution

Effect of low concentrations on aquatic life is unknown.
May be dangerous if it enters water intakes.
Notify local health and wildlife officials.
Notify operators of nearby water intakes.

1. RESPONSE TO DISCHARGE

(See Response Methods Handbook, CG 446-4)

Issue warning—oxidizing material
Restrict access
Disperse and flush

2. LABEL

Yellow

3. CHEMICAL DESIGNATIONS

3.1 **Synonyms:** Anhydrone; Dehydrite; Magnesium perchlorate, anhydrous; Magnesium perchlorate hexahydrate
3.2 **Coast Guard Compatibility Classification:** Not listed
3.3 **Chemical Formula:** $Mg(ClO_4)_2$
3.4 **IMCO/United Nations Numerical Designation:** 5.1/1475

4. OBSERVABLE CHARACTERISTICS

4.1 **Physical State (as shipped):** Solid
4.2 **Color:** White
4.3 **Odor:** None

5. HEALTH HAZARDS

5.1 **Personal Protective Equipment:** U. S. Bu. Mines approved respirator; chemical safety goggles; face shield
5.2 **Symptoms Following Exposure:** Inhalation of dust irritates mucous membranes. Ingestion of large amounts may be fatal; immediate symptoms include abdominal pains, nausea and vomiting, diarrhea, pallor, blueness, shortness of breath, unconsciousness. Contact with eyes or skin causes irritation.
5.3 **Treatment for Exposure:** INHALATION: remove victim to fresh air; get medical attention if irritation persists. INGESTION: give large amount of water; induce vomiting; call a physician. EYES: flush with copious quantities of water for at least 15 min.; call physician. SKIN: flush with water.
5.4 **Toxicity by Inhalation (Threshold Limit Value):** Data not available
5.5 **Short-Term Inhalation Limits:** Data not available
5.6 **Toxicity by Ingestion:** Data not available
5.7 **Late Toxicity:** Data not available
5.8 **Vapor (Gas) Irritant Characteristics:** Data not available
5.9 **Liquid or Solid Irritant Characteristics:** Data not available
5.10 **Odor Threshold:** Data not available

6. FIRE HAZARDS

6.1 **Flash Point:** Not flammable, but may cause or increase the intensity of a fire
6.2 **Flammable Limits in Air:** Not flammable
6.3 **Fire Extinguishing Agents:** Not pertinent
6.4 **Fire Extinguishing Agents Not to be Used:** Not pertinent
6.5 **Special Hazards of Combustion Products:**
6.6 **Behavior in Fire:** Can form explosive mixture with combustible material or finely powdered metals. Increases the intensity of fires.
6.7 **Ignition Temperature:** Not pertinent
6.8 **Electrical Hazard:** Not pertinent
6.9 **Burning Rate:** Not pertinent

7. CHEMICAL REACTIVITY

7.1 **Reactivity with Water:** Dissolves with liberation of heat. May cause spattering.
7.2 **Reactivity with Common Materials:** Contact with wood, paper, oils, grease, or finely divided metals may cause fires and explosions.
7.3 **Stability During Transport:** Stable
7.4 **Neutralizing Agents for Acids and Caustics:** Not pertinent
7.5 **Polymerization:** Not pertinent
7.6 **Inhibitor of Polymerization:** Not pertinent

8. WATER POLLUTION

8.1 **Aquatic Toxicity:** Data not available
8.2 **Waterfowl Toxicity:** Data not available
8.3 **Biological Oxygen Demand (BOD):** None
8.4 **Food Chain Concentration Potential:** None

9. SELECTED MANUFACTURERS

1. Hooker Chemical Corporation
   Industrial Chemicals Division
   Niagara Falls, N. Y. 14302
2. Kerr-McGee Chemical Corp.
   Kerr-McGee Center
   Oklahoma City, Okla. 73102
3. D. F. Goldsmith Chemical & Metal Corp.
   909 Pitner Ave.
   Evanston, Ill. 60202

10. SHIPPING INFORMATION

10.1 **Grades or Purity:** Pure anhydrous; 65–68% solution of hexahydrate in water.
10.2 **Storage Temperature:** Ambient
10.3 **Inert Atmosphere:** No requirement
10.4 **Venting:** Safety relief

11. HAZARD ASSESSMENT CODE

(See Hazard Assessment Handbook, CG 446-3)

SS

12. HAZARD CLASSIFICATIONS

12.1 **Code of Federal Regulations:** Oxidizing material
12.2 **NAS Hazard Rating for Bulk Water Transportation:** Not listed
12.3 **NFPA Hazard Classifications:**

Category	Classification
Health Hazard (Blue)	1
Flammability (Red)	0
Reactivity (Yellow)	0
	oxy

13. PHYSICAL AND CHEMICAL PROPERTIES

13.1 **Physical State at 15°C and 1 atm:** Solid
13.2 **Molecular Weight:** 223.2
13.3 **Boiling Point at 1 atm:** Decomposes above 250°C
13.4 **Freezing Point:** Not pertinent
13.5 **Critical Temperature:** Not pertinent
13.6 **Critical Pressure:** Not pertinent
13.7 **Specific Gravity:** 2.21 at 20°C (solid)
13.8 **Liquid Surface Tension:** Not pertinent
13.9 **Liquid-Water Interfacial Tension:** Not pertinent
13.10 **Vapor (Gas) Specific Gravity:** Not pertinent
13.11 **Ratio of Specific Heats of Vapor (Gas):** Not pertinent
13.12 **Latent Heat of Vaporization:** Not pertinent
13.13 **Heat of Combustion:** Not pertinent
13.14 **Heat of Decomposition:** Data not available
13.15 **Heat of Solution:** –260 Btu/lb = –140 cal/g = -6.0×10^5 J/kg
13.16 **Heat of Polymerization:** Not pertinent

NOTES

MLT	**MALATHION**

Common Synonyms			
Cythion Insecticide	Liquid	Yellow to dark brown	Skunk-like odor
	Sinks in water. Freezing point is 37° F.		

AVOID CONTACT WITH LIQUID. Keep people away.
Wear chemical protective suit with self-contained breathing apparatus.
Stop discharge if possible.
Call fire department.
Isolate and remove discharged material.
Notify local health and pollution control agencies.

Fire

Combustible.
POISONOUS GASES ARE PRODUCED IN FIRE AND WHEN HEATED.
Containers may explode in fire.
Wear chemical protective suit with self-contained breathing apparatus.
Extinguish with dry chemical, carbon dioxide, water, or foam.
Cool exposed containers with water.

Exposure

CALL FOR MEDICAL AID.

LIQUID
POISONOUS IF SWALLOWED OR IF SKIN IS EXPOSED.
Irritating to eyes.
Remove contaminated clothing and shoes.
Flush affected areas with plenty of water.
IF IN EYES, hold eyelids open and flush with plenty of water.
IF SWALLOWED and victim is CONSCIOUS, have victim drink water or milk and have victim induce vomiting.
IF SWALLOWED and victim is UNCONSCIOUS OR HAVING CONVULSIONS, do nothing except keep victim warm.

Water Pollution

HARMFUL TO AQUATIC LIFE IN VERY LOW CONCENTRATIONS.
May be dangerous if it enters water intakes.
Notify local health and wildlife officials.
Notify operators of nearby water intakes.

1. RESPONSE TO DISCHARGE

(See Response Methods Handbook, CG 446-4)

Issue warning—poison, water contaminant
Restrict access
Should be removed
Chemical and physical treatment

2. LABEL

3. CHEMICAL DESIGNATIONS

3.1 **Synonyms:** CYTHION Insecticide S-[1,2-Bis (ethoxycarbonyl) ethyl] 0,0-dimethyl phosphorodithioate
3.2 **Coast Guard Compatibility Classification:** Not applicable
3.3 **Chemical Formula:** $C_{10}H_{19}O_6PS_2$
3.4 **IMCO/United Nations Numerical Designation:** 6.1/1893

4. OBSERVABLE CHARACTERISTICS

4.1 **Physical State (as shipped):** Liquid
4.2 **Color:** Yellow to dark brown.
4.3 **Odor:** Characteristic skunk-like mercaptan

5. HEALTH HAZARDS

5.1 **Personal Protective Equipment:** Wear self-contained breathing apparatus (or respirator for organophosphate pesticides) and rubber clothing while fighting fires of malathion with chlorine bleach solution. All clothing contaminated by fumes and vapors must be decontaminated.
5.2 **Symptoms Following Exposure:** Exposure to fumes from a fire or to liquid causes headache, blurred vision, constricted pupils of the eyes, weakness, nausea, cramps, diarrhea, and tightness in the chest. Muscles twitch and convulsions may follow. The symptoms may develop over a period of 8 hours.
5.3 **Treatment for Exposure:** *Speed is essential.* INHALATION: in the nonbreathing victim *immediately* institute artificial respiration, using the mouth-to-mouth, the mouth-to-nose, or the mouth-to-oropharyngeal method. *Call physician!* INGESTION: administer milk, water or salt-water and induce vomiting repeatedly. SKIN OR EYE CONTACT: flood and wash exposed skin areas *thoroughly* with water. Remove contaminated clothing under a shower. Administer atropine, 2 mg (1/30 gr) intramuscularly or intravenously as soon as any local or systemic signs or symptoms of an intoxication are noted; repeat the administration of atropine every 3-8 min. until signs of atropinization (mydriasis, dry mouth, rapid pulse, hot and dry skin) occur; initiate treatment in children with 1 mg of atropine. Watch respiration, and remove bronchial secretions if they appear to be obstructing the airway; intubate if necessary. Give 2-PAM (Pralidoxime; *Protopam*), 2.5 gm in 100 ml of sterile water or in 5% dextrose and water, intravenously, slowly, in 15-30 min.; if sufficient fluid is not available, give 1 gm of 2-PAM in 3 ml of distilled water by deep intramuscular injection; repeat this every half hour if respiration weakens or if muscle fasciculation or convulsions recur.

6. FIRE HAZARDS

6.1 **Flash Point:** >325°F
6.2 **Flammable Limits in Air:** Data not available
6.3 **Fire Extinguishing Agents:** Dry chemical, carbon dioxide, water spray, foam
6.4 **Fire Extinguishing Agents Not to be Used:** Not pertinent
6.5 **Special Hazards of Combustion Products:** Vapors and fumes from fires are hazardous. They include sulfur dioxide and phosphoric acid.
6.6 **Behavior in Fire:** Gives off hazardous fumes. Area surrounding fire should be diked to prevent water runoff.
6.7 **Ignition Temperature:** Data not available
6.8 **Electrical Hazard:** Not pertinent
6.9 **Burning Rate:** Data not available

7. CHEMICAL REACTIVITY

7.1 **Reactivity with Water:** None
7.2 **Reactivity with Common Materials:** No hazardous reaction
7.3 **Stability During Transport:** Not pertinent
7.4 **Neutralizing Agents for Acids and Caustics:** Liquid bleach solution for decontamination.
7.5 **Polymerization:** Not pertinent
7.6 **Inhibitor of Polymerization:** Not pertinent

8. WATER POLLUTION

8.1 **Aquatic Toxicity:** 0.09 ppm/96 hr/bluegill/TL_m/fresh water 0.033-0.083 ppm/96 hr/marine crustaceae/LC_{50}
8.2 **Waterfowl Toxicity:** LD_{50}=1485 mg/kg
8.3 **Biological Oxygen Demand (BOD):** Data not available
8.4 **Food Chain Concentration Potential:** None

9. SELECTED MANUFACTURERS

American Cyanamid Co.
Agricultural Division
P. O. Box 400
Princeton, N. J. 08540

10. SHIPPING INFORMATION

10.1 **Grades or Purity:** CYTHION Insecticide; Malathion ULV Concentrate Insecticide. Many powders, dusts, and spray solutions are sold under a variety of trade names.
10.2 **Storage Temperature:** Below 120°F. Decomposition (non-hazardous) occurs at higher temperatures.
10.3 **Inert Atmosphere:** Data not available
10.4 **Venting:** Data not available

11. HAZARD ASSESSMENT CODE

(See Hazard Assessment Handbook, CG 446-3)

A-X-Y

12. HAZARD CLASSIFICATIONS

12.1 **Code of Federal Regulations:** Poisonous liquid or solid, Class B
12.2 **NAS Hazard Rating for Bulk Water Transportation:** Not listed
12.3 **NFPA Hazard Classifications:** Not listed

13. PHYSICAL AND CHEMICAL PROPERTIES

13.1 **Physical State at 15°C and 1 atm:** Liquid
13.2 **Molecular Weight:** 330.36
13.3 **Boiling Point at 1 atm:** Very high
13.4 **Freezing Point:** 37°F = 2.9°C = 276°K
13.5 **Critical Temperature:** Not pertinent
13.6 **Critical Pressure:** Not pertinent
13.7 **Specific Gravity:** 1.234 at 25°C (liquid)
13.8 **Liquid Surface Tension:** 37.1 dynes/cm = 0.0371 N/m at 24°C
13.9 **Liquid-Water Interfacial Tension:** 19 dynes/cm = 0.019 N/m at 24°C
13.10 **Vapor (Gas) Specific Gravity:** Not pertinent
13.11 **Ratio of Specific Heats of Vapor (Gas):** Not pertinent
13.12 **Latent Heat of Vaporization:** Not pertinent
13.13 **Heat of Combustion:** Data not available
13.14 **Heat of Decomposition:** Not pertinent
13.15 **Heat of Solution:** Not pertinent
13.16 **Heat of Polymerization:** Not pertinent

5. HEALTH HAZARDS (Cont'd.)

5.4 **Toxicity by Inhalation (Threshold Limit Value):** 10 mg/m³
5.5 **Short-Term Inhalation Limits:** Data not available
5.6 **Toxicity by Ingestion:** Grade 2; LD_{50} 0.5 to 5 g/kg (rat)
5.7 **Late Toxicity:** Data not available
5.8 **Vapor (Gas) Irritant Characteristics:** None likely
5.9 **Liquid or Solid Irritant Characteristics:** Minimum hazard. If spilled on clothing and allowed to remain, may cause smarting and reddening of the skin.
5.10 **Odor Threshold:** Data not available

REVISED 1978

MLI	# MALEIC ACID

Common Synonyms cis-Butenedioic acid cis-1,2-Ethylenedicarboxylic acid Maleinic acid Malenic acid Toxilic acid	Solid White Odorless Sinks and mixes with water.

Avoid contact with solid and dust. Keep people away.
Stop discharge if possible.
Call fire department.
Isolate and remove discharged material.
Notify local health and pollution control agencies.

Fire	Combustible. **Extinguish with water, dry chemicals, foam, or carbon dioxide.**
Exposure	**CALL FOR MEDICAL AID.** DUST Irritating to eyes, nose and throat. If inhaled will cause coughing or difficult breathing. **If in eyes, hold eyelids open and flush with plenty of water.** **If breathing has stopped, give artificial respiration.** **If breathing is difficult, give oxygen.** SOLID Irritating to skin and eyes. Harmful if swallowed. **Remove contaminated clothing and shoes.** **Flush affected areas with plenty of water.** **IF IN EYES, hold eyelids open and flush with plenty of water.** **IF SWALLOWED and victim is CONSCIOUS, have victim drink water or milk.** **IF SWALLOWED and victim is UNCONSCIOUS OR HAVING CONVULSIONS, do nothing except keep victim warm.**
Water Pollution	Effect of low concentrations on aquatic life is unknown. May be dangerous if it enters water intakes. **Notify local health and wildlife officials.** **Notify operators of nearby water intakes.**

1. RESPONSE TO DISCHARGE
(See Response Methods Handbook, CG 446-4)

Disperse and flush

2. LABELS

No hazard label required by Code of Federal Regulations

3. CHEMICAL DESIGNATIONS

3.1 **Synonyms:** *cis*-Butenedioic acid; *cis*-1,2-Ethylenedicarboxylic acid; Maleinic acid; Malenic acid; Toxilic acid

3.2 **Coast Guard Compatibility Classification:** Not listed

3.3 **Chemical Formula:** HOOC CH=CH COOH

3.4 **IMCO/United Nations Numerical Designation:** Not listed

4. OBSERVABLE CHARACTERISTICS

4.1 **Physical State (as shipped):** Solid

4.2 **Color:** White

4.3 **Odor:** None

5. HEALTH HAZARDS

5.1 **Personal Protective Equipment:** Dust mask; goggles or face shield; protective gloves

5.2 **Symptoms Following Exposure:** Inhalation causes irritation of nose and throat. Contact with eyes or skin causes irritation.

5.3 **Treatment for Exposure:** INHALATION: remove to fresh air. EYES: immediately flush with plenty of water for 15 min.; get medical attention if irritation persists. SKIN: wash with soap and water.

5.4 **Toxicity by Inhalation (Threshold Limit Value):** Data not available

5.5 **Short-Term Inhalation Limits:** Data not available

5.6 **Toxicity by Ingestion:** Grade 2; oral LD_{50} = 708 mg/kg (rat)

5.7 **Late Toxicity:** Data not available

5.8 **Vapor (Gas) Irritant Characteristics:** Data not available

5.9 **Liquid or Solid Irritant Characteristics:** Data not available

5.10 **Odor Threshold:** Data not available

6. FIRE HAZARDS

6.1 **Flash Point:** Not pertinent (combustible solid)

6.2 **Flammable Limits in Air:** Not pertinent

6.3 **Fire Extinguishing Agents:** Water, foam, dry chemical, carbon dioxide

6.4 **Fire Extinguishing Agents Not to be Used:**

6.5 **Special Hazards of Combustion Products:** Irritating smoke containing maleic anhydride may form in fire.

6.6 **Behavior in Fire:**

6.7 **Ignition Temperature:** Data not available

6.8 **Electrical Hazard:** Not pertinent

6.9 **Burning Rate:** Not pertinent

7. CHEMICAL REACTIVITY

7.1 **Reactivity with Water:** No reaction

7.2 **Reactivity with Common Materials:** May corrode metals when wet.

7.3 **Stability During Transport:** Stable

7.4 **Neutralizing Agents for Acids and Caustics:** Flush with water, rinse with dilute solution of sodium bicarbonate or soda ash.

7.5 **Polymerization:** Not pertinent

7.6 **Inhibitor of Polymerization:** Not pertinent

8. WATER POLLUTION

8.1 **Aquatic Toxicity:**
240 ppm/24 48 hr/mosquito fish/TL_m/fresh water
5 ppm/96 hr/fathead minnow/TL_m/fresh water

8.2 **Waterfowl Toxicity:** Data not available

8.3 **Biological Oxygen Demand (BOD):** 38%, 5 days

8.4 **Food Chain Concentration Potential:** None

9. SELECTED MANUFACTURERS

1. Allied Chemical Corp.
Specialty Chemicals Div.
P. O. Box 1087R
Morristown, N. J. 07960

2. Mallinckrodt Chemical Works
223 Westside Ave.
P. O. Box 384
Jersey City, N. J. 07303

3. Eastman Organic Chemicals
Rochester, N. Y. 14650

10. SHIPPING INFORMATION

10.1 **Grades or Purity:** Reagent; Technical

10.2 **Storage Temperature:** Ambient

10.3 **Inert Atmosphere:** No requirement

10.4 **Venting:** Open

11. HAZARD ASSESSMENT CODE
(See Hazard Assessment Handbook, CG 446-3)

SS

12. HAZARD CLASSIFICATIONS

12.1 **Code of Federal Regulations:** Not listed

12.2 **NAS Hazard Rating for Bulk Water Transportation:** Not listed

12.3 **NFPA Hazard Classifications:** Not listed

13. PHYSICAL AND CHEMICAL PROPERTIES

13.1 **Physical State at 15°C and 1 atm:** Solid

13.2 **Molecular Weight:** 116.1

13.3 **Boiling Point at 1 atm:** Not pertinent (decomposes)

13.4 **Freezing Point:** 266°F = 130°C = 403°K

13.5 **Critical Temperature:** Not pertinent

13.6 **Critical Pressure:** Not pertinent

13.7 **Specific Gravity:** 1.59 at 20°C (solid)

13.8 **Liquid Surface Tension:** Not pertinent

13.9 **Liquid-Water Interfacial Tension:** Not pertinent

13.10 **Vapor (Gas) Specific Gravity:** Not pertinent

13.11 **Ratio of Specific Heats of Vapor (Gas):** Not pertinent

13.12 **Latent Heat of Vaporization:** Not pertinent

13.13 **Heat of Combustion:** −5,000 Btu/lb = −2,800 cal/g = -117×10^5 J/kg

13.14 **Heat of Decomposition:** Not pertinent

13.15 **Heat of Solution:** Not pertinent

13.16 **Heat of Polymerization:** Not pertinent

NOTES

MLA

MALEIC ANHYDRIDE

Common Synonyms			
Toxilic anhydride cis-Butenedioic anhydride	Molten; or solid crystals or tablets	Colorless	Choking odor
	Sinks and mixes slowly with water.		

Avoid contact with solid and liquid. Keep people away.
Wear goggles, self-contained breathing apparatus, and rubber overclothing (including gloves).
Stop discharge if possible.
Call fire department.
Isolate and remove discharged material.
Notify local health and pollution control agencies.

Fire

Combustible.
Dust cloud may be ignited by spark or flame.
Wear goggles, self-contained breathing apparatus, and rubber overclothing (including gloves).
Extinguish with dry chemical, alcohol foam, or carbon dioxide.
Water may be ineffective on fire.

Exposure

CALL FOR MEDICAL AID.

LIQUID OR SOLID
Will burn skin and eyes.
Harmful if swallowed.
Remove contaminated clothing and shoes.
Flush affected areas with plenty of water.
IF IN EYES, hold eyelids open and flush with plenty of water.
IF SWALLOWED and victim is CONSCIOUS, have victim drink water or milk.

Water Pollution

Dangerous to aquatic life in high concentrations.
May be dangerous if it enters water intakes.
Notify local health and wildlife officials.
Notify operators of nearby water intakes.

1. RESPONSE TO DISCHARGE

(See Response Methods Handbook, CG 446-4)

Disperse and flush

2. LABELS

No hazard label required by Code of Federal Regulations

3. CHEMICAL DESIGNATIONS

3.1 **Synonyms:** cis-Butenedioic anhydride; 2,5-Furanedione; Toxilic anhydride

3.2 **Coast Guard Compatibility Classification:** Not applicable

3.3 **Chemical Formula:** OCOCH=CHCO

3.4 **IMCO/United Nations Numerical Designation:** 9.0/

4. OBSERVABLE CHARACTERISTICS

4.1 **Physical State (as shipped):** Solid or liquid

4.2 **Color:** Colorless

4.3 **Odor:** Acrid; choking

5. HEALTH HAZARDS

5.1 **Personal Protective Equipment:** Approved organic vapor-acid gas canister; chemical goggles and face shield; rubber gloves and boots; coveralls or rubber apron.

5.2 **Symptoms Following Exposure:** Inhalation causes coughing, sneezing, throat irritation. Skin contact causes irritation and redness. Vapors cause severe eye irritation; photophobia and double vision may occur.

5.3 **Treatment for Exposure:** INHALATION: give oxygen. EYE OR SKIN CONTACT: flush with lots of water for at least 15 min.; for eyes, call a physician. For molten maleic burns, remove crust and treat as chemical and thermal burn.

5.4 **Toxicity by Inhalation (Threshold Limit Value):** 0.25 ppm

5.5 **Short-Term Inhalation Limits:** Data not available

5.6 **Toxicity by Ingestion:** Grade 2; LD_{50} 0.5 to 5 g/kg

5.7 **Late Toxicity:** None

5.8 **Vapor (Gas) Irritant Characteristics:** Vapors cause moderate irritation, such that personnel will find high concentrations unpleasant. The effect is temporary.

5.9 **Liquid or Solid Irritant Characteristics:** Causes smarting of the skin and first-degree burns on short exposure; may cause secondary burns on long exposure.

5.10 **Odor Threshold:** 1.3 - 2.0 mg/m^3

6. FIRE HAZARDS

6.1 **Flash Point:** (Liquid) 215°F C.C.; 230°F O.C.

6.2 **Flammable Limits in Air:** 1.4%–7.1%

6.3 **Fire Extinguishing Agents:** Alcohol foam, dry chemical or carbon dioxide

6.4 **Fire Extinguishing Agents Not to be Used:** Water or foam may cause frothing

6.5 **Special Hazards of Combustion Products:** Not pertinent

6.6 **Behavior in Fire:** When heated above 300°F in the presence of various materials may generate heat and carbon dioxide. Will explode if confined.

6.7 **Ignition Temperature:** 878°F

6.8 **Electrical Hazard:** Class I, Group D

6.9 **Burning Rate:** 1.4 mm/min.

7. CHEMICAL REACTIVITY

7.1 **Reactivity with Water:** Hot water may cause frothing. Reaction with cold water is slow and non-hazardous.

7.2 **Reactivity with Common Materials:** No reaction

7.3 **Stability During Transport:** Stable

7.4 **Neutralizing Agents for Acids and Caustics:** Solid spills can usually be recovered before any significant reaction with water occurs. Flush area of spill with water.

7.5 **Polymerization:** Very unlikely at ordinary temperatures, even in the molten state.

7.6 **Inhibitor of Polymerization:** None

8. WATER POLLUTION

8.1 **Aquatic Toxicity:** 150 ppm/24 hr/sunfish/TL_m/fresh water

8.2 **Waterfowl Toxicity:** Data not available

8.3 **Biological Oxygen Demand (BOD):** 50%, 5 days

8.4 **Food Chain Concentration Potential:** None

9. SELECTED MANUFACTURERS

1. Monsanto Co.
   Monsanto Industrial Chemicals Co.
   800 North Lindbergh Blvd.
   St. Louis, Mo. 63166
2. Petro-Tex Chemical Corp.
   8600 Park Place
   Houston, Tex.77017
3. United States Steel Corp.
   USS Chemicals Division
   Neville Island, Pa. 15225

10. SHIPPING INFORMATION

10.1 **Grades or Purity:** Commercial: 99.5%

10.2 **Storage Temperature:** Ambient

10.3 **Inert Atmosphere:** No requirement

10.4 **Venting:** Open

11. HAZARD ASSESSMENT CODE

(See Hazard Assessment Handbook, CG 446-3)

RR

12. HAZARD CLASSIFICATIONS

12.1 **Code of Federal Regulations:** Not listed

12.2 **NAS Hazard Rating for Bulk Water Transportation:**

Category	Rating
Fire	1
Health	
Vapor Irritant	2
Liquid or Solid Irritant	2
Poisons	1
Water Pollution	
Human Toxicity	2
Aquatic Toxicity	2
Aesthetic Effect	1
Reactivity	
Other Chemicals	3
Water	2
Self-Reaction	0

12.3 **NFPA Hazard Classifications:**

Category	Classification
Health Hazard (Blue)	3
Flammability (Red)	1
Reactivity (Yellow)	1

13. PHYSICAL AND CHEMICAL PROPERTIES

13.1 **Physical State at 15°C and 1 atm:** Solid

13.2 **Molecular Weight:** 98.06

13.3 **Boiling Point at 1 atm:** 392°F = 200°C = 473°K

13.4 **Freezing Point:** 127°F = 53°C = 326°K

13.5 **Critical Temperature:** Not pertinent

13.6 **Critical Pressure:** Not pertinent

13.7 **Specific Gravity:** 1.43 at 15°C (solid)

13.8 **Liquid Surface Tension:** Not pertinent

13.9 **Liquid-Water Interfacial Tension:** Not pertinent

13.10 **Vapor (Gas) Specific Gravity:** Not pertinent

13.11 **Ratio of Specific Heats of Vapor (Gas):** Not pertinent

13.12 **Latent Heat of Vaporization:** Not pertinent

13.13 **Heat of Combustion:** −5936 Btu/lb = −3298 cal/g = -138.1×10^5 J/kg

13.14 **Heat of Decomposition:** Not pertinent

13.15 **Heat of Solution:** −153 Btu/lb = −85.0 cal/g = -3.56×10^5 J/kg

13.16 **Heat of Polymerization:** Not pertinent

NOTES

REVISED 1978

MLH	# MALEIC HYDRAZIDE

Common Synonyms			
1, 2-Dihydro-3, 6-pyridazinedione Maleic acid hydrazide MH Malazide Regulox	Solid	White	Odorless
	Sinks in water.		

Call fire department.
Keep people away.
Isolate and remove discharged material.
Notify local health and pollution control agencies.

Fire	Combustible. POISONOUS GASES MAY BE PRODUCED IN FIRE. Extinguish with water, dry chemicals, foam, or carbon dioxide.
Exposure	Call for medical aid. DUST Irritating to eyes, nose and throat. Move victim to fresh air. If in eyes, hold eyelids open and flush with plenty of water. If breathing is difficult, give oxygen. SOLID Irritating to skin and eyes. Harmful if swallowed. Remove contaminated clothing and shoes. Flush affected areas with plenty of water. IF IN EYES, hold eyelids open and flush with plenty of water. IF SWALLOWED and victim is CONSCIOUS, have victim drink water or milk.
Water Pollution	Effect of low concentrations on aquatic life is unknown. May be dangerous if it enters water intakes. Notify local health and wildlife officials. Notify operators of nearby water intakes.

1. RESPONSE TO DISCHARGE

(See Response Methods Handbook, CG 446-4)

Issue warning—water contaminant
Should be removed
Chemical and physical treatment

2. LABELS

No label required by Code of Federal Regulations

3. CHEMICAL DESIGNATIONS

3.1 **Synonyms:** 1,2-Dihydro-3,6-pyridazinedione; 6-Hydroxy-3-(2H)-pyridazinone; Maleic acid hydrazide; Malazide; MH; Regulox

3.2 **Coast Guard Compatibility Classification:** Not applicable

3.3 **Chemical Formula:** $C_4H_4N_2O_2$

3.4 **IMCO/United Nations Numerical Designation:** Not listed

4. OBSERVABLE CHARACTERISTICS

4.1 **Physical State (as shipped):** Solid

4.2 **Color:** White

4.3 **Odor:** None

5. HEALTH HAZARDS

5.1 **Personal Protective Equipment:** Goggles or face shield; dust mask.

5.2 **Symptoms Following Exposure:** Inhalation of dust causes irritation of nose and throat. Contact with eyes or skin causes irritation. Ingestion has been observed to cause tremors and muscle spasms in test animals.

5.3 **Treatment for Exposure:** INHALATION: move to fresh air. EYES: flush with water for at least 15 min. SKIN: flush with water, wash with soap and water. INGESTION: get medical attention.

5.4 **Toxicity by Inhalation (Threshold Limit Value):** Data not available

5.5 **Short-Term Inhalation Limits:** Data not available

5.6 **Toxicity by Ingestion:** Grade 2; oral LD_{50} = 3,800 mg/kg (rat)

5.7 **Late Toxicity:** Causes cancer in rats

5.8 **Vapor (Gas) Irritant Characteristics:** Data not available

5.9 **Liquid or Solid Irritant Characteristics:** Data not available

5.10 **Odor Threshold:** Not pertinent

6. FIRE HAZARDS

6.1 **Flash Point:** Not pertinent (combustible solid)

6.2 **Flammable Limits in Air:** Not pertinent

6.3 **Fire Extinguishing Agents:** Water, dry chemical, foam, carbon dioxide

6.4 **Fire Extinguishing Agents Not to be Used:** Not pertinent

6.5 **Special Hazards of Combustion Products:** Toxic nitrogen oxides are produced.

6.6 **Behavior in Fire:** Not pertinent

6.7 **Ignition Temperature:** Not pertinent

6.8 **Electrical Hazard:** Not pertinent

6.9 **Burning Rate:** Not pertinent

7. CHEMICAL REACTIVITY

7.1 **Reactivity with Water:** No reaction

7.2 **Reactivity with Common Materials:** No reaction

7.3 **Stability During Transport:** Stable

7.4 **Neutralizing Agents for Acids and Caustics:** Not pertinent

7.5 **Polymerization:** Not pertinent

7.6 **Inhibitor of Polymerization:** Not pertinent

8. WATER POLLUTION

8.1 **Aquatic Toxicity:** Data not available

8.2 **Waterfowl Toxicity:** Data not available

8.3 **Biological Oxygen Demand (BOD):** Data not available

8.4 **Food Chain Concentration Potential:** None

9. SELECTED MANUFACTURERS

1. The Ansul Company
   Chemical Division
   1 Stanton Street
   Marinette, Wis. 54143
2. Uniroyal Chemical Co.
   Spencer Street
   Naugatuck, Conn. 06770
3. Chemical Formulators, Inc.
   Box 26
   Nitro, W. Va. 25143

10. SHIPPING INFORMATION

10.1 **Grades or Purity:** Technical: 97+%

10.2 **Storage Temperature:** Ambient

10.3 **Inert Atmosphere:** No requirement

10.4 **Venting:** Open

11. HAZARD ASSESSMENT CODE

(See Hazard Assessment Handbook, CG 446-3)

II

12. HAZARD CLASSIFICATIONS

12.1 **Code of Federal Regulations:** Not listed

12.2 **NAS Hazard Rating for Bulk Water Transportation:** Not listed

12.3 **NFPA Hazard Classifications:** Not listed

13. PHYSICAL AND CHEMICAL PROPERTIES

13.1 **Physical State at 15°C and 1 atm:** Solid

13.2 **Molecular Weight:** 112.1

13.3 **Boiling Point at 1 atm:** Not pertinent (decomposes)

13.4 **Freezing Point:** 558°F = 292°C = 565°K

13.5 **Critical Temperature:** Not pertinent

13.6 **Critical Pressure:** Not pertinent

13.7 **Specific Gravity:** 1.60 at 25°C (solid)

13.8 **Liquid Surface Tension:** Not pertinent

13.9 **Liquid-Water Interfacial Tension:** Not pertinent

13.10 **Vapor (Gas) Specific Gravity:** Not pertinent

13.11 **Ratio of Specific Heats of Vapor (Gas):** Not pertinent

13.12 **Latent Heat of Vaporization:** Not pertinent

13.13 **Heat of Combustion:** (est.) −8,200 Btu/lb = −4,500 cal/g = -190×10^5 J/kg

13.14 **Heat of Decomposition:** Not pertinent

13.15 **Heat of Solution:** Not pertinent

13.16 **Heat of Polymerization:** Not pertinent

NOTES

MAT | MERCURIC ACETATE

Common Synonyms	Solid	White	Mild vinegar-like odor
	Sinks and mixes with water.		

AVOID CONTACT WITH SOLID AND DUST. KEEP PEOPLE AWAY.
Wear goggles and dust respirator.
Stay upwind. Use water spray to "knock down" dust.
Isolate and remove discharged material.
Notify local health and pollution control agencies.

Fire	Not flammable. POISONOUS GASES MAY BE PRODUCED WHEN HEATED.
Exposure	CALL FOR MEDICAL AID. DUST POISONOUS IF INHALED. Move victim to fresh air. If in eyes, hold eyelids open and flush with plenty of water. If breathing is difficult, give oxygen. SOLID POISONOUS IF SWALLOWED. IF SWALLOWED and victim is CONSCIOUS, have victim drink water or milk and have victim induce vomiting. IF SWALLOWED and victim is UNCONSCIOUS OR HAVING CONVULSIONS, do nothing except keep victim warm.
Water Pollution	HARMFUL TO AQUATIC LIFE IN VERY LOW CONCENTRATIONS. May be dangerous if it enters water intakes. Notify local health and wildlife officials. Notify operators of nearby water intakes.

1. RESPONSE TO DISCHARGE
(See Response Methods Handbook, CG 446-4)

Issue warning—poison, water contaminant
Restrict access
Should be removed
Chemical and physical treatment

2. LABEL

3. CHEMICAL DESIGNATIONS

3.1 **Synonyms:** No common synonyms
3.2 **Coast Guard Compatibility Classification:** Not applicable
3.3 **Chemical Formula:** $(CH_3COO)_2Hg$
3.4 **IMCO/United Nations Numerical Designation:** 6.1/1629

4. OBSERVABLE CHARACTERISTICS

4.1 **Physical State (as shipped):** Solid
4.2 **Color:** White
4.3 **Odor:** Slight acetic

5. HEALTH HAZARDS

5.1 **Personal Protective Equipment:** Rubber gloves, dust mask, goggles
5.2 **Symptoms Following Exposure:** The general symptoms are those of mercury poisoning, developing rapidly after ingestion but more slowly after low repeated exposures. Contact with eyes causes irritation and ulceration. Skin contact may cause dermatitis. Ingestion causes pain, vomiting, ulceration of mouth and stomach, kidney failure, metallic taste, pallor, and rapid, weak pulse.
5.3 **Treatment for Exposure:** Have physician treat for mercury poisoning. EYES or SKIN: flush with water. INGESTION: call physician; poison should be removed from stomach as soon as possible; give milk or white of eggs beaten with water, then tablespoon of salt in a glass of warm water and repeat until vomit fluid is clear; repeat milk or white of eggs beaten with water.
5.4 **Toxicity by Inhalation (Threshold Limit Value):** 0.05 mg/m^3 (as mercury)
5.5 **Short-Term Inhalation Limits:** Data not available
5.6 **Toxicity by Ingestion:** Grade 3; oral LD_{50} = 76 mg/kg (rat)
5.7 **Late Toxicity:** Intestinal bleeding and kidney damage may develop.
5.8 **Vapor (Gas) Irritant Characteristics:** Not pertinent
5.9 **Liquid or Solid Irritant Characteristics:** Data not available
5.10 **Odor Threshold:** Not pertinent

6. FIRE HAZARDS

6.1 **Flash Point:** Not flammable
6.2 **Flammable Limits in Air:** Not flammable
6.3 **Fire Extinguishing Agents:** Not pertinent
6.4 **Fire Extinguishing Agents Not to be Used:** Not pertinent
6.5 **Special Hazards of Combustion Products:** Smoke may contain toxic mercury or mercury oxide fumes.
6.6 **Behavior in Fire:** Not pertinent
6.7 **Ignition Temperature:** Not pertinent
6.8 **Electrical Hazard:** Not pertinent
6.9 **Burning Rate:** Not pertinent

7. CHEMICAL REACTIVITY

7.1 **Reactivity with Water:** No reaction
7.2 **Reactivity with Common Materials:** No reaction
7.3 **Stability During Transport:** Stable
7.4 **Neutralizing Agents for Acids and Caustics:** Not pertinent
7.5 **Polymerization:** Not pertinent
7.6 **Inhibitor of Polymerization:** Not pertinent

8. WATER POLLUTION

8.1 **Aquatic Toxicity:**
>0.05/*/young salmon/lethal/fresh water
1.7/48 hr/stickleback/TL_m/fresh water
*Time interval not specified.
8.2 **Waterfowl Toxicity:** Data not available
8.3 **Biological Oxygen Demand (BOD):** Data not available
8.4 **Food Chain Concentration Potential:** Fish can accumulate mercury and transfer it to higher levels in the food chain

9. SELECTED MANUFACTURERS

1. Ventron Corporation
Chemicals Division
Congress Street
Beverly, Mass. 01915
2. Mallinckrodt Chemical Works
3600 N. Second Street
St. Louis, Mo. 63160
3. Gallard-Schlesinger Chemical Manufacturing Co.
584 Mineola Ave.
Carle Place, N. Y. 11514

10. SHIPPING INFORMATION

10.1 **Grades or Purity:** C.P.: 99+%
10.2 **Storage Temperature:** Ambient
10.3 **Inert Atmosphere:** No requirement
10.4 **Venting:** Open

11. HAZARD ASSESSMENT CODE
(See Hazard Assessment Handbook, CG 446-3)

SS

12. HAZARD CLASSIFICATIONS

12.1 **Code of Federal Regulations:** Poisonous, Class B
12.2 **NAS Hazard Rating for Bulk Water Transportation:** Not listed
12.3 **NFPA Hazard Classifications:** Not listed

13. PHYSICAL AND CHEMICAL PROPERTIES

13.1 **Physical State at 15°C and 1 atm:** Solid
13.2 **Molecular Weight:** 318.7
13.3 **Boiling Point at 1 atm:** Not pertinent (decomposes)
13.4 **Freezing Point:** Not pertinent
13.5 **Critical Temperature:** Not pertinent
13.6 **Critical Pressure:** Not pertinent
13.7 **Specific Gravity:** 3.27 at 20°C (solid)
13.8 **Liquid Surface Tension:** Not pertinent
13.9 **Liquid-Water Interfacial Tension:** Not pertinent
13.10 **Vapor (Gas) Specific Gravity:** Not pertinent
13.11 **Ratio of Specific Heats of Vapor (Gas):** Not pertinent
13.12 **Latent Heat of Vaporization:** Not pertinent
13.13 **Heat of Combustion:** Not pertinent
13.14 **Heat of Decomposition:** Not pertinent
13.15 **Heat of Solution:** Not pertinent
13.16 **Heat of Polymerization:** Not pertinent

NOTES

MCC

MERCURIC AMMONIUM CHLORIDE

Common Synonyms			
Mercuric chloride, ammoniated Mercury ammonium chloride Mercury amide chloride Ammoniated mercury	Solid Sinks in water.	White	Odorless

AVOID CONTACT WITH SOLID AND DUST. KEEP PEOPLE AWAY.
Wear goggles and dust respirator.
Isolate and remove discharged material.
Notify local health and pollution control agencies.

Fire

Not flammable.
POISONOUS GASES MAY BE PRODUCED WHEN HEATED.

Exposure

CALL FOR MEDICAL AID.

DUST
POISONOUS IF INHALED.
Move victim to fresh air.
If in eyes, hold eyelids open and flush with plenty of water.
If breathing is difficult, give oxygen.

SOLID
POISONOUS IF SWALLOWED.
IF SWALLOWED and victim is CONSCIOUS, have victim drink water or milk and have victim induce vomiting.
IF SWALLOWED and victim is UNCONSCIOUS OR HAVING CONVULSIONS, do nothing except keep victim warm.

Water Pollution

HARMFUL TO AQUATIC LIFE IN VERY LOW CONCENTRATIONS.
May be dangerous if it enters water intakes.
Notify local health and wildlife officials.
Notify operators of nearby water intakes.

1. RESPONSE TO DISCHARGE

(See Response Methods Handbook, CG 446-4)

Issue warning—poison, water contaminant
Restrict access
Should be removed
Chemical and physical treatment

2. LABEL

3. CHEMICAL DESIGNATIONS

3.1 **Synonyms:** Albus; Aminomercuric chloride; Ammoniated mercury; Mercuric chloride, ammoniated; Mercury ammonium chloride; Mercury (II) chloride ammonobasic
3.2 **Coast Guard Compatibility Classification:** Not applicable
3.3 **Chemical Formula:** $HgNH_2Cl$
3.4 **IMCO/United Nations Numerical Designation:** 6.1/1630

4. OBSERVABLE CHARACTERISTICS

4.1 **Physical State (as shipped):** Solid
4.2 **Color:** White
4.3 **Odor:** None

5. HEALTH HAZARDS

5.1 **Personal Protective Equipment:** Gloves, goggles, respirator
5.2 **Symptoms Following Exposure:** The general symptoms are those of mercury poisoning, developing rapidly after ingestion but more slowly after low repeated exposures. Contact with eyes causes irritation and ulceration. Skin contact may cause dermatitis. Ingestion causes pain, vomiting, metallic taste, ulceration of mouth and stomach, pallor, and rapid, weak pulse.
5.3 **Treatment for Exposure:** Have physician treat for mercury poisoning. EYES or SKIN: flush with water. INGESTION: call a physician; give milk or white of egg beaten with water, then a tablespoon of salt in glass of water; repeat until vomit fluid is clear; repeat milk or white of egg.
5.4 **Toxicity by Inhalation (Threshold Limit Value):** 0.05 mg/m^3 (for mercury)
5.5 **Short-Term Inhalation Limits:** Data not available
5.6 **Toxicity by Ingestion:** Data not available
5.7 **Late Toxicity:** Intestinal bleeding and kidney damage may develop.
5.8 **Vapor (Gas) Irritant Characteristics:** Data not available
5.9 **Liquid or Solid Irritant Characteristics:** Data not available
5.10 **Odor Threshold:** Not pertinent

6. FIRE HAZARDS

6.1 **Flash Point:** Not flammable
6.2 **Flammable Limits in Air:** Not flammable
6.3 **Fire Extinguishing Agents:** Not pertinent
6.4 **Fire Extinguishing Agents Not to be Used:** Not pertinent
6.5 **Special Hazards of Combustion Products:** Smoke may contain toxic mercury compounds.
6.6 **Behavior in Fire:** Not pertinent
6.7 **Ignition Temperature:** Not pertinent
6.8 **Electrical Hazard:** Not pertinent
6.9 **Burning Rate:** Not pertinent

7. CHEMICAL REACTIVITY

7.1 **Reactivity with Water:** No reaction
7.2 **Reactivity with Common Materials:** No reaction
7.3 **Stability During Transport:** Stable
7.4 **Neutralizing Agents for Acids and Caustics:** Not pertinent
7.5 **Polymerization:** Not pertinent
7.6 **Inhibitor of Polymerization:** Not pertinent

8. WATER POLLUTION

8.1 **Aquatic Toxicity:**
0.12 ppm/8 hr/minnow/lethal/fresh water
0.023 ppm/*/bivalve larvae/lethal/salt water
*Time period not specified.
8.2 **Waterfowl Toxicity:** Data not available
8.3 **Biological Oxygen Demand (BOD):** Not pertinent
8.4 **Food Chain Concentration Potential:** Data not available

9. SELECTED MANUFACTURERS

Ventron Corporation
Chemicals Division
Congress Street
Beverly, Mass. 01915

10. SHIPPING INFORMATION

10.1 **Grades or Purity:** U.S.P.: 98+%
10.2 **Storage Temperature:** Ambient
10.3 **Inert Atmosphere:** No requirement
10.4 **Venting:** Open

11. HAZARD ASSESSMENT CODE

(See Hazard Assessment Handbook, CG 446-3)

II

12. HAZARD CLASSIFICATIONS

12.1 **Code of Federal Regulations:** Poisonous, Class B
12.2 **NAS Hazard Rating for Bulk Water Transportation:** Not listed
12.3 **NFPA Hazard Classifications:** Not listed

13. PHYSICAL AND CHEMICAL PROPERTIES

13.1 **Physical State at 15°C and 1 atm:** Solid
13.2 **Molecular Weight:** 252.1
13.3 **Boiling Point at 1 atm:** Not pertinent (sublimes at red heat)
13.4 **Freezing Point:** Not pertinent (infusible)
13.5 **Critical Temperature:** Not pertinent
13.6 **Critical Pressure:** Not pertinent
13.7 **Specific Gravity:** 5.7 at 20°C (solid)
13.8 **Liquid Surface Tension:** Not pertinent
13.9 **Liquid-Water Interfacial Tension:** Not pertinent
13.10 **Vapor (Gas) Specific Gravity:** Not pertinent
13.11 **Ratio of Specific Heats of Vapor (Gas):** Not pertinent
13.12 **Latent Heat of Vaporization:** Not pertinent
13.13 **Heat of Combustion:** Not pertinent
13.14 **Heat of Decomposition:** Not pertinent
13.15 **Heat of Solution:** Not pertinent
13.16 **Heat of Polymerization:** Not pertinent

NOTES

MRC — MERCURIC CHLORIDE

Common Synonyms	
Mercury bichloride Corrosive mercury chloride Mercury perchloride Mercury (II) chloride Calochlor	Solid White Sinks and mixes slowly with water.

AVOID CONTACT WITH SOLID AND DUST. KEEP PEOPLE AWAY.
Wear a dust respirator and rubber overclothing (including gloves).
Stop discharge if possible.
Isolate and remove discharged material.
Notify local health and pollution control agencies.

Fire

Not flammable.
POISONOUS GASES MAY BE PRODUCED WHEN HEATED.

Exposure

CALL FOR MEDICAL AID.

DUST
POISONOUS IF INHALED OR IF SKIN IS EXPOSED.
If inhaled will cause coughing or difficult breathing.

If in eyes, hold eyelids open and flush with plenty of water.
If breathing has stopped, give artificial respiration.
If breathing is difficult, give oxygen.

SOLID
POISONOUS IF SWALLOWED OR IF SKIN IS EXPOSED.
Irritating to skin and eyes.
If swallowed will cause nausea and vomiting.

Remove contaminated clothing and shoes.
Flush affected areas with plenty of water.
IF IN EYES, hold eyelids open and flush with plenty of water.
IF SWALLOWED and victim is CONSCIOUS, have victim drink water or milk and have victim induce vomiting.
IF SWALLOWED and victim is UNCONSCIOUS OR HAVING CONVULSIONS, do nothing except keep victim warm.

Water Pollution

HARMFUL TO AQUATIC LIFE IN VERY LOW CONCENTRATIONS.
May be dangerous if it enters water intakes.

Notify local health and wildlife officials.
Notify operators of nearby water intakes.

1. RESPONSE TO DISCHARGE

(See Response Methods Handbook, CG 446-4)

Issue warning—poison, water contaminant
Restrict access
Disperse and flush

2. LABEL

3. CHEMICAL DESIGNATIONS

3.1 **Synonyms:** Calochlor; Corrosive mercury chloride; Corrosive sublimate; Mercury bichloride; Mercury (II) chloride; Mercury perchloride
3.2 **Coast Guard Compatibility Classification:** Not listed
3.3 **Chemical Formula:** $HgCl_2$
3.4 **IMCO/United Nations Numerical Designation:** 6.1/1624

4. OBSERVABLE CHARACTERISTICS

4.1 **Physical State (as shipped):** Solid
4.2 **Color:** White; colorless
4.3 **Odor:** None

5. HEALTH HAZARDS

5.1 **Personal Protective Equipment:** Bu. Mines approved airline respirator; impervious suit; appropriate eye protection
5.2 **Symptoms Following Exposure:** *All forms of exposure to this compound are hazardous; acute systemic mercurialism may be fatal within a few minutes, but death by uremic poisoning is usually delayed 5—12 days.* Acute poisoning has resulted from inhaling dust contentrations of 1.2—8.5 mg/m^3 of air; symptoms include tightness and pain in chest, coughing, and difficulty in breathing. Ingestion causes necrosis, pain, vomiting, and severe purging; as little as 0.5 gm can be fatal. Contact with eyes causes ulceration of conjunctiva and cornea. Contact with skin causes irritation and possible dermatitis; systemic poisoning can occur by absorption through skin.
5.3 **Treatment for Exposure:** *Act promptly! Alimentary absorption is very rapid, and the first 10—15 minutes determine the prognosis.* INHALATION: remove victim to fresh air; get medical attention. INGESTION: give egg whites, milk, or activated charcoal; induce vomiting; consult physician. EYES or SKIN: wash with water for 15 min.
5.4 **Toxicity by Inhalation (Threshold Limit Value):** 0.05 mg/m^3
5.5 **Short-Term Inhalation Limits:** Data not available
5.6 **Toxicity by Ingestion:** Grade 4; oral LD_{50} = 1 mg/kg (rat)
5.7 **Late Toxicity:** Data not available
5.8 **Vapor (Gas) Irritant Characteristics:** Data not available
5.9 **Liquid or Solid Irritant Characteristics:** Data not available
5.10 **Odor Threshold:** Data not available

6. FIRE HAZARDS

6.1 **Flash Point:** Not flammable
6.2 **Flammable Limits in Air:** Not flammable
6.3 **Fire Extinguishing Agents:** Not pertinent
6.4 **Fire Extinguishing Agents Not to be Used:** Not pertinent
6.5 **Special Hazards of Combustion Products:** Heat of fire may cause material to form fumes of mercuric chloride, which are toxic.
6.6 **Behavior in Fire:**
6.7 **Ignition Temperature:** Not pertinent
6.8 **Electrical Hazard:** Not pertinent
6.9 **Burning Rate:** Not pertinent

7. CHEMICAL REACTIVITY

7.1 **Reactivity with Water:** No reaction
7.2 **Reactivity with Common Materials:**
7.3 **Stability During Transport:** Stable
7.4 **Neutralizing Agents for Acids and Caustics:** Not pertinent
7.5 **Polymerization:** Not pertinent
7.6 **Inhibitor of Polymerization:** Not pertinent

8. WATER POLLUTION

8.1 **Aquatic Toxicity:**
0.82 ppm/7 days/goldfish/TL_m/fresh water
0.075 ppm/48 hr/pink shrimp/TL_m/salt water
4.2 ppm/48 hr/oyster/TL_m/sea water
8.2 **Waterfowl Toxicity:** Data not available
8.3 **Biological Oxygen Demand (BOD):** None
8.4 **Food Chain Concentration Potential:** Many organisms are capable of accumulating mercury from water. Bioconcentrative up to 10,000-fold.

9. SELECTED MANUFACTURERS

1. Mallinckrodt Chemicals Works
223 Westside Ave.
P. O. Box 384
Jersey City, N. J. 07303
2. J. T. Baker Chemical Co.
Phillipsburg, N. J. 08865
3. Gallard-Schlesinger Chemical Mfg. Co.
584 Mineola Ave.
Carle Place, N. Y. 11514

10. SHIPPING INFORMATION

10.1 **Grades or Purity:** Reagent; Analytical
10.2 **Storage Temperature:** Ambient
10.3 **Inert Atmosphere:** No requirement
10.4 **Venting:** Open

11. HAZARD ASSESSMENT CODE

(See Hazard Assessment Handbook, CG 446-3)

SS

12. HAZARD CLASSIFICATIONS

12.1 **Code of Federal Regulations:** Poisonous solid, Class B
12.2 **NAS Hazard Rating for Bulk Water Transportation:** Not listed
12.3 **NFPA Hazard Classifications:** Not listed

13. PHYSICAL AND CHEMICAL PROPERTIES

13.1 **Physical State at 15°C and 1 atm:** Solid
13.2 **Molecular Weight:** 271.50
13.3 **Boiling Point at 1 atm:** 576°F = 302°C = 575°K
13.4 **Freezing Point:** 531°F = 277°C = 550°K
13.5 **Critical Temperature:** Not pertinent
13.6 **Critical Pressure:** Not pertinent
13.7 **Specific Gravity:** 5.4 at 20°C (solid)
13.8 **Liquid Surface Tension:** Not pertinent
13.9 **Liquid-Water Interfacial Tension:** Not pertinent
13.10 **Vapor (Gas) Specific Gravity:** Not pertinent
13.11 **Ratio of Specific Heats of Vapor (Gas):** Not pertinent
13.12 **Latent Heat of Vaporization:** Not pertinent
13.13 **Heat of Combustion:** Not pertinent.
13.14 **Heat of Decomposition:** Not pertinent
13.15 **Heat of Solution:** Not pertinent
13.16 **Heat of Polymerization:** Not pertinent

NOTES

MCN	MERCURIC CYANIDE

Common Synonyms	
Cianurina Mercury cyanide Mercury (II) cyanide	Solid (crystals or powder) White Sinks and mixes slowly with water.

AVOID CONTACT WITH SOLID AND DUST. KEEP PEOPLE AWAY.
Wear a dust respirator and rubber overclothing (including gloves).
Stop discharge if possible.
Isolate and remove discharged material.
Notify local health and pollution control agencies.

Fire	Not flammable. POISONOUS GASES MAY BE PRODUCED WHEN HEATED.
Exposure	CALL FOR MEDICAL AID. DUST POISONOUS IF INHALED OR IF SKIN IS EXPOSED. If inhaled will cause coughing or difficult breathing. **If in eyes, hold eyelids open and flush with plenty of water.** **If breathing has stopped, give artificial respiration.** **If breathing is difficult, give oxygen.** SOLID POISONOUS IF SWALLOWED OR IF SKIN IS EXPOSED. Irritating to skin and eyes. If swallowed will cause nausea and vomiting. **Remove contaminated clothing and shoes.** **Flush affected areas with plenty of water.** **IF IN EYES, hold eyelids open and flush with plenty of water.** **IF SWALLOWED and victim is CONSCIOUS, have victim drink water or milk and have victim induce vomiting.** **IF SWALLOWED and victim is UNCONSCIOUS OR HAVING CONVULSIONS, do nothing except keep victim warm.**
Water Pollution	HARMFUL TO AQUATIC LIFE IN VERY LOW CONCENTRATIONS. May be dangerous if it enters water intakes. **Notify local health and wildlife officials.** **Notify operators of nearby water intakes.**

1. RESPONSE TO DISCHARGE
(See Response Methods Handbook, CG 446-4)

Issue warning—poison, water contaminant
Restrict access
Disperse and flush

2. LABEL

POISON

3. CHEMICAL DESIGNATIONS

3.1 **Synonyms:** Cianurina; Mercury cyanide; Mercury (II) cyanide
3.2 **Coast Guard Compatibility Classification:** Not listed
3.3 **Chemical Formula:** $Hg(CN)_2$
3.4 **IMCO/United Nations Numerical Designation:** 6.1/1636

4. OBSERVABLE CHARACTERISTICS

4.1 **Physical State (as shipped):** Solid
4.2 **Color:** Colorless or white
4.3 **Odor:** None

5. HEALTH HAZARDS

5.1 **Personal Protective Equipment:** Dust mask; goggles or face shield; rubber gloves
5.2 **Symptoms Following Exposure:** Symptoms of both cyanide and mercury intoxication can occur. Acute poisoning has resulted from inhaling dust concentrations of 1.2–8.5 mg/m³ of air; symptoms include tightness and pain in chest, coughing, and difficulty in breathing; cyanide poisoning can cause anxiety, confusion, dizziness, and shortness of breath, with possible unconsciousness, convulsions, and paralysis; breath may smell like bitter almonds. Ingestion causes necrosis, pain, vomiting, and severe purging, plus the above symptoms. Contact with eyes causes ulceration of conjunctiva and cornea. Contact with skin causes irritation and possible dermatitis; systemic poisoning can occur by absorption through skin.
5.3 **Treatment for Exposure:** *Act quickly; call physician.* INHALATION: if victim has stopped breathing, start artificial respiration immediately; using amyl nitrite pearls, administer amyl nitrite by inhalation for 15–30 seconds of every minute while sodium nitrite solution is being prepared; discontinue amyl nitrite and immediately inject intravenously 10 ml of a 3% solution of sodium nitrite (nonsterile if necessary) over a period of 2–4 min.; without removing needle, infuse intravenously 50 ml of a 25% aqueous solution of sodium thiosulfate; injection should take about 10 min. (concentrations of 5–50% may be used, but keep total dose approx. 12 gm). Oxygen therapy may be helpful in combination with the above. INGESTION: *alimentary absorption is very rapid, action during first 10–15 min. determines prognosis.* Give egg whites, milk, or activated charcoal and induce vomiting; treat for cyanide poisoning as above. EYES or SKIN: wash with water for 15 min.
5.4 **Toxicity by Inhalation (Threshold Limit Value):** 0.05 mg/m³ (as mercury)
5.5 **Short-Term Inhalation Limits:** Data not available

6. FIRE HAZARDS

6.1 **Flash Point:** Not flammable
6.2 **Flammable Limits in Air:** Not flammable
6.3 **Fire Extinguishing Agents:** Not pertinent
6.4 **Fire Extinguishing Agents Not to be Used:** Not pertinent
6.5 **Special Hazards of Combustion Products:** Fumes from fire may contain toxic mercury and hydrogen cyanide.
6.6 **Behavior in Fire:** Not pertinent
6.7 **Ignition Temperature:** Not pertinent
6.8 **Electrical Hazard:** Not pertinent
6.9 **Burning Rate:** Not pertinent

7. CHEMICAL REACTIVITY

7.1 **Reactivity with Water:** No reaction
7.2 **Reactivity with Common Materials:** Contact with any acidic material will form poisonous hydrogen cyanide gas, which may collect in enclosed spaces.
7.3 **Stability During Transport:** Stable
7.4 **Neutralizing Agents for Acids and Caustics:** Not pertinent
7.5 **Polymerization:** Not pertinent
7.6 **Inhibitor of Polymerization:** Not pertinent

8. WATER POLLUTION

8.1 **Aquatic Toxicity:** 0.02 ppm/48 hr/daphnia magna/TL_m/fresh water
8.2 **Waterfowl Toxicity:** Data not available
8.3 **Biological Oxygen Demand (BOD):** Data not available
8.4 **Food Chain Concentration Potential:** Possible bioaccumulation problem. Many organisms can accumulate mercury from water. Bioconcentrative up to 10,000 fold.

9. SELECTED MANUFACTURERS

1. J. T. Baker Chemical Co.
   Phillipsburg, N. J. 08865
2. Pfaltz and Bauer, Inc.
   375 Fairfield Ave.
   Stamford, Conn. 06902
3. Gallard-Schlesinger Chemical Mfg. Co.
   584 Mineola Ave.
   Carle Place, N. Y. 11514

10. SHIPPING INFORMATION

10.1 **Grades or Purity:** Reagent
10.2 **Storage Temperature:** Ambient
10.3 **Inert Atmosphere:** No requirement
10.4 **Venting:** Open

11. HAZARD ASSESSMENT CODE
(See Hazard Assessment Handbook, CG 446-3)

SS

12. HAZARD CLASSIFICATIONS

12.1 **Code of Federal Regulations:** Poisonous solid, Class B
12.2 **NAS Hazard Rating for Bulk Water Transportation:** Not listed
12.3 **NFPA Hazard Classifications:** Not listed

13. PHYSICAL AND CHEMICAL PROPERTIES

13.1 **Physical State at 15°C and 1 atm:** Solid
13.2 **Molecular Weight:** 252.63
13.3 **Boiling Point at 1 atm:** Not pertinent (decomposes)
13.4 **Freezing Point:** Not pertinent
13.5 **Critical Temperature:** Not pertinent
13.6 **Critical Pressure:** Not pertinent
13.7 **Specific Gravity:** 4.0 at 20°C (solid)
13.8 **Liquid Surface Tension:** Not pertinent
13.9 **Liquid-Water Interfacial Tension:** Not pertinent
13.10 **Vapor (Gas) Specific Gravity:** Not pertinent
13.11 **Ratio of Specific Heats of Vapor (Gas):** Not pertinent
13.12 **Latent Heat of Vaporization:** Not pertinent
13.13 **Heat of Combustion:** Not pertinent
13.14 **Heat of Decomposition:** Not pertinent
13.15 **Heat of Solution:** Not pertinent
13.16 **Heat of Polymerization:** Not pertinent

5. HEALTH HAZARDS (Cont'd.)

5.6 **Toxicity by Ingestion:** Grade 4; oral LD_{50} = 25 mg/kg (rat)
5.7 **Late Toxicity:** Data not available
5.8 **Vapor (Gas) Irritant Characteristics:** Data not available
5.9 **Liquid or Solid Irritant Characteristics:** Data not available
5.10 **Odor Threshold:** Odorless

MID — MERCURIC IODIDE

Common Synonyms
Mercuric iodide, red
Mercury biniodide

Solid — Red — Odorless

Sinks in water.

AVOID CONTACT WITH SOLID AND DUST. KEEP PEOPLE AWAY.
Wear dust respirator and rubber overclothing (including gloves).
Stop discharge if possible.
Isolate and remove discharged material.
Notify local health and pollution control agencies.

Fire

Not flammable.
POISONOUS GASES MAY BE PRODUCED WHEN HEATED.

Exposure

CALL FOR MEDICAL AID.

DUST
POISONOUS IF INHALED OR IF SKIN IS EXPOSED.
If inhaled will cause coughing or difficult breathing.
If in eyes, hold eyelids open and flush with plenty of water.
If breathing has stopped, give artificial respiration.
If breathing is difficult, give oxygen.

SOLID
POISONOUS IF SWALLOWED OR IF SKIN IS EXPOSED.
Irritating to skin and eyes.
If swallowed will cause nausea and vomiting.
Remove contaminated clothing and shoes.
Flush affected areas with plenty of water.
IF IN EYES, hold eyelids open and flush with plenty of water.
IF SWALLOWED and victim is CONSCIOUS, have victim drink water or milk and have victim induce vomiting.
IF SWALLOWED and victim is UNCONSCIOUS OR HAVING CONVULSIONS, do nothing except keep victim warm.

Water Pollution

Effect of low concentrations on aquatic life is unknown.
May be dangerous if it enters water intakes.
Notify local health and wildlife officials.
Notify operators of nearby water intakes.

1. RESPONSE TO DISCHARGE

(See Response Methods Handbook, CG 446-4)

Issue warning—poison, water contaminant
Restrict access
Should be removed
Chemical and physical treatment

2. LABEL

3. CHEMICAL DESIGNATIONS

3.1 **Synonyms:** Mercuric iodide, red; Mercury biniodide
3.2 **Coast Guard Compatibility Classification:** Not listed
3.3 **Chemical Formula:** HgI_2
3.4 **IMCO/United Nations Numerical Designation:** 6.1/6115

4. OBSERVABLE CHARACTERISTICS

4.1 **Physical State (as shipped):** Solid
4.2 **Color:** Red
4.3 **Odor:** None

5. HEALTH HAZARDS

5.1 **Personal Protective Equipment:** Dust mask; goggles or face shield; protective gloves
5.2 **Symptoms Following Exposure:** *All forms of exposure to this compound are hazardous. Acute systemic mercurialism may be fatal within a few minutes; death by uremic poisoning is usually delayed 5—12 days.* Acute poisoning has resulted from inhaling dust concentrations of 1.2—8.5 mg/m^3 of air; symptoms include tightness and pain in chest, coughing, and difficulty in breathing. Ingestion causes necrosis, pain, vomiting, and severe purging. Contact with eyes causes ulceration of conjunctiva and cornea. Contact with skin causes irritation and possible dermatitis; systemic poisoning can occur by absorption through skin.
5.3 **Treatment for Exposure:** INHALATION: remove to fresh air; get medical attention. INGESTION: *alimentary absorption is very rapid; action during first 10—15 min. determines prognosis.* Give egg whites, milk, or activated charcoal and induce vomiting; consult physician. EYES: flush with water for at least 15 min. SKIN: flush with water; wash with soap and water.
5.4 **Toxicity by Inhalation (Threshold Limit Value):** 0.05 mg/m^3 (as mercury)
5.5 **Short-Term Inhalation Limits:** Data not available
5.6 **Toxicity by Ingestion:** Grade 4; oral LD_{50} = 40 mg/kg (rat)
5.7 **Late Toxicity:** Data not available
5.8 **Vapor (Gas) Irritant Characteristics:** Data not available
5.9 **Liquid or Solid Irritant Characteristics:** Data not available
5.10 **Odor Threshold:** Odorless

6. FIRE HAZARDS

6.1 **Flash Point:** Not flammable
6.2 **Flammable Limits in Air:** Not flammable
6.3 **Fire Extinguishing Agents:** Not pertinent
6.4 **Fire Extinguishing Agents Not to be Used:** Not pertinent
6.5 **Special Hazards of Combustion Products:** Fumes from fire may contain toxic mercury vapor.
6.6 **Behavior in Fire:**
6.7 **Ignition Temperature:** Not pertinent
6.8 **Electrical Hazard:** Not pertinent
6.9 **Burning Rate:** Not pertinent

7. CHEMICAL REACTIVITY

7.1 **Reactivity with Water:** No reaction
7.2 **Reactivity with Common Materials:**
7.3 **Stability During Transport:** Stable
7.4 **Neutralizing Agents for Acids and Caustics:** Not pertinent
7.5 **Polymerization:** Not pertinent
7.6 **Inhibitor of Polymerization:** Not pertinent

8. WATER POLLUTION

8.1 **Aquatic Toxicity:** Data not available
8.2 **Waterfowl Toxicity:** Data not available
8.3 **Biological Oxygen Demand (BOD):** Data not available
8.4 **Food Chain Concentration Potential:** Many organisms can accumulate mercury from water. Bioconcentrative up to 10,000 fold.

9. SELECTED MANUFACTURERS

1. Mallinckrodt Chemical Works
   223 Westside Ave.
   P. O. Box 384
   Jersey City, N. J. 07303
2. J. T. Baker Chemical Co.
   Phillipsburg, N. J. 08865
3. Gallard-Schlesinger Chemical Mfg. Co.
   584 Mineola Ave.
   Carle Place, N. Y. 11514

10. SHIPPING INFORMATION

10.1 **Grades or Purity:** Reagent, 99+%
10.2 **Storage Temperature:** Ambient
10.3 **Inert Atmosphere:** No requirement
10.4 **Venting:** Open

11. HAZARD ASSESSMENT CODE

(See Hazard Assessment Handbook, CG 446-3)

II

12. HAZARD CLASSIFICATIONS

12.1 **Code of Federal Regulations:** Poisonous solid, Class B
12.2 **NAS Hazard Rating for Bulk Water Transportation:** Not listed
12.3 **NFPA Hazard Classifications:** Not listed

13. PHYSICAL AND CHEMICAL PROPERTIES

13.1 **Physical State at 15°C and 1 atm:** Solid
13.2 **Molecular Weight:** 454.90
13.3 **Boiling Point at 1 atm:** 669°F = 354°C = 627°K
13.4 **Freezing Point:** 495°F = 257°C = 530°K
13.5 **Critical Temperature:** Not pertinent
13.6 **Critical Pressure:** Not pertinent
13.7 **Specific Gravity:** 6.3 at 20°C (solid)
13.8 **Liquid Surface Tension:** Not pertinent
13.9 **Liquid-Water Interfacial Tension:** Not pertinent
13.10 **Vapor (Gas) Specific Gravity:** Not pertinent
13.11 **Ratio of Specific Hats of Vapor (Gas):** Not pertinent
13.12 **Latent Heat of Vaporization:** Not pertinent
13.13 **Heat of Combustion:** Not pertinent
13.14 **Heat of Decomposition:** Not pertinent
13.15 **Heat of Solution:** Not pertinent
13.16 **Heat of Polymerization:** Not pertinent

NOTES

MNT — MERCURIC NITRATE

Common Synonyms			
Mercury pernitrate Mercury (II) nitrate Mercury nitrate monohydrate	Solid Sinks in water.	White	Sharp odor

AVOID CONTACT WITH SOLID AND DUST. KEEP PEOPLE AWAY.
Wear dust respirator and rubber overclothing (including gloves).
Stop discharge if possible.
Isolate and remove discharged material.
Notify local health and pollution control agencies.

Fire

Not flammable.
Will increase the intensity of a fire.
POISONOUS GASES MAY BE PRODUCED WHEN HEATED.

Exposure

CALL FOR MEDICAL AID.

DUST
POISONOUS IF INHALED OR IF SKIN IS EXPOSED.
If inhaled will cause coughing or difficult breathing.
If in eyes, hold eyelids open and flush with plenty of water.
If breathing has stopped, give artificial respiration.
If breathing is difficult, give oxygen.

SOLID
POISONOUS IF SWALLOWED OF IF SKIN IS EXPOSED.
Irritating to skin and eyes.
If swallowed will cause nausea and vomiting.
Remove contaminated clothing and shoes.
Flush affected areas with plenty of water.
IF IN EYES, hold eyelids open and flush with plenty of water.
IF SWALLOWED and victim is CONSCIOUS, have victim drink water or milk and have victim induce vomiting.
IF SWALLOWED and victim is UNCONSCIOUS OR HAVING CONVULSIONS, do nothing except keep victim warm.

Water Pollution

Effect of low concentrations on aquatic life is unknown.
May be dangerous if it enters water intakes.
Notify local health and wildlife officials.
Notify operators of nearby water intakes.

1. RESPONSE TO DISCHARGE

(See Response Methods Handbook, CG 446-4)

Issue warning—poison, water contaminant, oxidizing material
Restrict access
Disperse and flush

2. LABEL

3. CHEMICAL DESIGNATIONS

3.1 **Synonyms:** Mercuric nitrate monohydrate; Mercury (II) nitrate; Mercury pernitrate
3.2 **Coast Guard Compatibility Classification:** Not listed
3.3 **Chemical Formula:** $Hg(NO_3)_2 \cdot H_2O$
3.4 **IMCO/United Nations Numerical Designation:** 6.1/1625

4. OBSERVABLE CHARACTERISTICS

4.1 **Physical State (as shipped):** Solid
4.2 **Color:** White; colorless
4.3 **Odor:** Sharp odor of nitric acid

5. HEALTH HAZARDS

5.1 **Personal Protective Equipment:** Dust mask; goggles or face shield; protective gloves
5.2 **Symptoms Following Exposure:** *Acute systemic poisoning may be fatal within a few minutes; death by uremic poisoning is usually delayed 5—12 days.* Acute poisoning has resulted from inhaling dust concentrations of 1.2—8.5 mg/m³ of air; symptoms include tightness and pain in chest, coughing, and difficulty in breathing. Ingestion causes necrosis, pain, vomiting, and severe purging. Contact with eyes causes ulceration of conjunctiva and cornea. Contact with skin causes irritation and possible dermatitis; systemic poisoning can occur by absorption through skin.
5.3 **Treatment for Exposure:** INHALATION: remove to fresh air; get medical attention. INGESTION: *alimentary absorption is very rapid; action during first 10—15 min. determines prognosis.* Give egg whites, milk, or activated charcoal and induce vomiting; consult physician. EYES or SKIN: flush with water.
5.4 **Toxicity by Inhalation (Threshold Limit Value):** 0.05 mg/m³ (as mercury)
5.5 **Short-Term Inhalation Limits:** Data not available
5.6 **Toxicity by Ingestion:** Data not available
5.7 **Late Toxicity:** Data not available
5.8 **Vapor (Gas) Irritant Characteristics:** Data not available
5.9 **Liquid or Solid Irritant Characteristics:** Data not available
5.10 **Odor Threshold:** Data not available

6. FIRE HAZARDS

6.1 **Flash Point:** Not flammable, but may intensify fire
6.2 **Flammable Limits in Air:** Not flammable
6.3 **Fire Extinguishing Agents:** Not pertinent
6.4 **Fire Extinguishing Agents Not to be Used:** Not pertinent
6.5 **Special Hazards of Combustion Products:** Vapors from fire may contain toxic mercury and oxides of nitrogen.
6.6 **Behavior in Fire:** May increase intensity of fire if in contact with burning material
6.7 **Ignition Temperature:** Not pertinent
6.8 **Electrical Hazard:** Not pertinent
6.9 **Burning Rate:** Not pertinent

7. CHEMICAL REACTIVITY

7.1 **Reactivity with Water:** Dissolves, then forms cloudy acid solution. The reaction is not hazardous.
7.2 **Reactivity with Common Materials:** Solution will corrode most metals. Solid in contact with wood or paper may cause fire.
7.3 **Stability During Transport:** Stable
7.4 **Neutralizing Agents for Acids and Caustics:** Flush well with water, rinse with dilute solution of sodium bicarbonate or soda ash.
7.5 **Polymerization:** Not pertinent
7.6 **Inhibitor of Polymerization:** Not pertinent

8. WATER POLLUTION

8.1 **Aquatic Toxicity:** Data not available
8.2 **Waterfowl Toxicity:** Data not available
8.3 **Biological Oxygen Demand (BOD):** None
8.4 **Food Chain Concentration Potential:** Possible bioaccumulation problem. Many organisms can accumulate mercury from water. Bioconcentrative up to 10,000 fold.

9. SELECTED MANUFACTURERS

1. Mallinckrodt Chemical Works
   223 Westside Ave.
   P.O. Box 384
   Jersey City, N. J. 07303
2. J. T. Baker Chemical Co.
   Phillipsburg, N. J. 08865
3. Pfaltz and Bauer, Inc.
   375 Fairfield Ave.
   Stamford, Conn. 06902

10. SHIPPING INFORMATION

10.1 **Grades or Purity:** Reagent, 99.0%
10.2 **Storage Temperature:** Ambient
10.3 **Inert Atmosphere:** No requirement
10.4 **Venting:** Open

11. HAZARD ASSESSMENT CODE

(See Hazard Assessment Handbook, CG 446-3)

SS

12. HAZARD CLASSIFICATIONS

12.1 **Code of Federal Regulations:** Poisonous solid, Class B
12.2 **NAS Hazard Rating for Bulk Water Transportation:** Not listed
12.3 **NFPA Hazard Classifications:** Not listed

13. PHYSICAL AND CHEMICAL PROPERTIES

13.1 **Physical State at 15°C and 1 atm:** Solid
13.2 **Molecular Weight:** 342.6
13.3 **Boiling Point at 1 atm:** Not pertinent (decomposes)
13.4 **Freezing Point:** Not pertinent
13.5 **Critical Temperature:** Not pertinent
13.6 **Critical Pressure:** Not pertinent
13.7 **Specific Gravity:** 4.3 at 20°C (solid)
13.8 **Liquid Surface Tension:** Not pertinent
13.9 **Liquid-Water Interfacial Tension:** Not pertinent
13.10 **Vapor (Gas) Specific Gravity:** Not pertinent
13.11 **Ratio of Specific Heats of Vapor (Gas):** Not pertinent
13.12 **Latent Heat of Vaporization:** Not pertinent
13.13 **Heat of Combustion:** Not pertinent
13.14 **Heat of Decomposition:** Not pertinent
13.15 **Heat of Solution:** Not pertinent
13.16 **Heat of Polymerization:** Not pertinent

NOTES

MOX — MERCURIC OXIDE

Common Synonyms			
Mercuric oxide, red Mercuric oxide, yellow Mercury oxide	Solid Sinks in water.	Red, orange or yellow	Odorless

AVOID CONTACT WITH SOLID AND DUST. KEEP PEOPLE AWAY.
Wear dust respirator and rubber overclothing (including gloves).
Stop discharge if possible.
Isolate and remove discharged material.
Notify local health and pollution control agencies.

Fire

Not flammable.
Will increase the intensity of a fire.
POISONOUS GASES MAY BE PRODUCED WHEN HEATED.

Exposure

CALL FOR MEDICAL AID.

DUST
POISONOUS IF INHALED OR IF SKIN IS EXPOSED.
If inhaled will cause coughing or difficult breathing.
If in eyes, hold eyelids open and flush with plenty of water.
If breathing has stopped, give artificial respiration.
If breathing is difficult, give oxygen.

SOLID
POISONOUS IF SWALLOWED OR IF SKIN IS EXPOSED.
Irritating to skin and eyes.
If swallowed will cause nausea or vomiting.
Remove contaminated clothing and shoes.
Flush affected areas with plenty of water.
IF IN EYES, hold eyelids open and flush with plenty of water.
IF SWALLOWED and victim is CONSCIOUS, have victim drink water or milk and have victim induce vomiting.
IF SWALLOWED and victim is UNCONSCIOUS OR HAVING CONVULSIONS, do nothing except keep victim warm.

Water Pollution

Effect of low concentrations on aquatic life is unknown.
May be dangerous if it enters water intakes.
Notify local health and wildlife officials.
Notify operators of nearby water intakes.

1. RESPONSE TO DISCHARGE

(See Response Methods Handbook, CG 446-4)

Issue warning—poison, water contaminant
Restrict access
Should be removed
Chemical and physical treatment

2. LABEL

3. CHEMICAL DESIGNATIONS

3.1 **Synonyms:** Mercuric oxide, red ("Red precipitate"); Mercuric oxide, yellow ("Yellow precipitate"); Mercury oxide

3.2 **Coast Guard Compatibility Classification:** Not listed

3.3 **Chemical Formula:** HgO

3.4 **IMCO/United Nations Numerical Designation:** 6.1/1641

4. OBSERVABLE CHARACTERISTICS

4.1 **Physical State (as shipped):** Solid

4.2 **Color:** Red, orange, or yellow

4.3 **Odor:** None

5. HEALTH HAZARDS

5.1 **Personal Protective Equipment:** Dust mask; goggles or face shield; protective gloves

5.2 **Symptoms Following Exposure:** *Acute systemic poisoning may be fatal within a few minutes; death by uremic poisoning is usually delayed 5—12 days.* Acute poisoning has resulted from inhaling dust concentrations of 1.2—8.5 mg/m³ of air; symptoms include tightness and pain in chest, coughing, and difficulty in breathing. Ingestion causes necrosis, pain, vomiting, and severe purging. Contact with eyes causes ulceration of conjunctiva and cornea. Contact with skin causes irritation and possible dermatitis; systemic poisoning can occur by absorption through skin.

5.3 **Treatment for Exposure:** INHALATION: remove victim to fresh air; get medical attention. INGESTION: alimentary absorption is very rapid; action during first 10—15 min. determines prognosis. Give egg whites, milk, or activated charcoal and induce vomiting; consult physician. EYES: flush with water for at least 15 min. SKIN: wash with soap and water.

5.4 **Toxicity by Inhalation (Threshold Limit Value):** 0.05 mg/m³ (as mercury)

5.5 **Short-Term Inhalation Limits:** Data not available

5.6 **Toxicity by Ingestion:** Grade 4; oral LD_{50} = 18 mg/kg (rat)

5.7 **Late Toxicity:** Causes birth defects in rats

5.8 **Vapor (Gas) Irritant Characteristics:** Data not available

5.9 **Liquid or Solid Irritant Characteristics:** Data not available

5.10 **Odor Threshold:** Odorless

6. FIRE HAZARDS

6.1 **Flash Point:** Not flammable, but may intensify fire

6.2 **Flammable Limits in Air:** Not flammable

6.3 **Fire Extinguishing Agents:** Not pertinent

6.4 **Fire Extinguishing Agents Not to be Used:** Not pertinent

6.5 **Special Hazards of Combustion Products:** Fumes from fire may contain poisonous mercury vapor.

6.6 **Behavior in Fire:** Decomposes at 500°C into mercury and oxygen, which can increase intensity of fire. Solid changes color when hot.

6.7 **Ignition Temperature:** Not pertinent

6.8 **Electrical Hazard:** Not pertinent

6.9 **Burning Rate:** Not pertinent

7. CHEMICAL REACTIVITY

7.1 **Reactivity with Water:** No reaction

7.2 **Reactivity with Common Materials:**

7.3 **Stability During Transport:** Stable

7.4 **Neutralizing Agents for Acids and Caustics:** Not pertinent

7.5 **Polymerization:** Not pertinent

7.6 **Inhibitor of Polymerization:** Not pertinent

8. WATER POLLUTION

8.1 **Aquatic Toxicity:** 0.29 ppm/48 hr/marine fish/TL_m

8.2 **Waterfowl Toxicity:** Data not available

8.3 **Biological Oxygen Demand (BOD):** None

8.4 **Food Chain Concentration Potential:** Many organisms can accumulate mercury from water. Bioconcentrative up to 10,000 fold.

9. SELECTED MANUFACTURERS

1. Mallinckrodt Chemical Works
   223 Westside Ave.
   P. O. Box 384
   Jersey City, N. J. 07303
2. J. T. Baker Chemical Co.
   Phillipsburg, N. J. 08865
3. Gallard-Schlesinger Chemical Mfg. Co.
   584 Mineola Ave.
   Carle Place, N. Y. 11514

10. SHIPPING INFORMATION

10.1 **Grades or Purity:** Red—technical; reagent; purified; Yellow—technical; NF; reagent

10.2 **Storage Temperature:** Ambient

10.3 **Inert Atmosphere:** No requirement

10.4 **Venting:** Open

11. HAZARD ASSESSMENT CODE

(See Hazard Assessment Handbook, CG 446-3)

II

12. HAZARD CLASSIFICATIONS

12.1 **Code of Federal Regulations:** Poisonous solid, Class B

12.2 **NAS Hazard Rating for Bulk Water Transportation:** Not listed

12.3 **NFPA Hazard Classifications:** Not listed

13. PHYSICAL AND CHEMICAL PROPERTIES

13.1 **Physical State at 15°C and 1 atm:** Solid

13.2 **Molecular Weight:** 216.61

13.3 **Boiling Point at 1 atm:** Not pertinent (decomposes)

13.4 **Freezing Point:** Not pertinent

13.5 **Critical Temperature:** Not pertinent

13.6 **Critical Pressure:** Not pertinent

13.7 **Specific Gravity:** 11.1 at 20°C (solid)

13.8 **Liquid Surface Tension:** Not pertinent

13.9 **Liquid-Water Interfacial Tension:** Not pertinent

13.10 **Vapor (Gas) Specific Gravity:** Not pertinent

13.11 **Ratio of Specific Heats of Vapor (Gas):** Not pertinent

13.12 **Latent Heat of Vaporization:** Not pertinent

13.13 **Heat of Combustion:** Not pertinent

13.14 **Heat of Decomposition:** Not pertinent

13.15 **Heat of Solution:** Not pertinent

13.16 **Heat of Polymerization:** Not pertinent

NOTES

MSF — MERCURIC SULFIDE

Common Synonyms			
Mercuric sulfide, black Ethiops mineral Mercuric sulfide, red Vermilion Artificial cinnabar Chinese red	Solid	Red or black	Odorless
	Sinks in water.		

AVOID CONTACT WITH SOLID AND DUST. KEEP PEOPLE AWAY.
Wear dust respirator and rubber overclothing (including gloves).
Stop discharge if possible.
Call fire department.
Isolate and remove discharged material.
Notify local health and pollution control agencies.

Fire

Combustible.
POISONOUS GASES MAY BE PRODUCED IN FIRE.
Wear goggles and self-contained breathing apparatus.
Extinguish with water, foam, or sand.
Other extinguishing agents may be ineffective on fire.

Exposure

CALL FOR MEDICAL AID.

DUST
POISONOUS IF INHALED OR IF SKIN IS EXPOSED.
If inhaled will cause coughing or difficult breathing.
If in eyes, hold eyelids open and flush with plenty of water.
If breathing has stopped, give artificial respiration.
If breathing is difficult, give oxygen.

SOLID
POISONOUS IF SWALLOWED OR IF SKIN IS EXPOSED.
Irritating to skin and eyes.
If swallowed will cause coughing, nausea and vomiting.
Remove contaminated clothing and shoes.
Flush affected areas with plenty of water.
IF IN EYES, hold eyelids open and flush with plenty of water.
IF SWALLOWED and victim is CONSCIOUS, have victim drink water or milk and have victim induce vomiting.
IF SWALLOWED and victim is UNCONSCIOUS OR HAVING CONVULSIONS, do nothing except keep victim warm.

Water Pollution

Effect of low concentrations on aquatic life is unknown.
May be dangerous if it enters water intakes.
Notify local health and wildlife officials.
Notify operators of nearby water intakes.

1. RESPONSE TO DISCHARGE

(See Response Methods Handbook, CG 446-4)

Issue warning – water contaminant
Should be removed
Chemical and physical treatment

2. LABELS

No hazard label required by Code of Federal Regulations

3. CHEMICAL DESIGNATIONS

3.1 **Synonyms:** Mercuric sulfide, black (Ethiops mineral); Mercuric sulfide, red (Vermilion; Artificial cinnabar; Chinese red)
3.2 **Coast Guard Compatibility Classification:** Not listed
3.3 **Chemical Formula:** HgS
3.4 **IMCO/United Nations Numerical Designation:** 6.1/2024

4. OBSERVABLE CHARACTERISTICS

4.1 **Physical State (as shipped):** Solid
4.2 **Color:** Red or black
4.3 **Odor:** None

5. HEALTH HAZARDS

5.1 **Personal Protective Equipment:** Dust mask; goggles or face shield; protective gloves
5.2 **Symptoms Following Exposure:** Acute poisoning can result from inhaling dust concentrations of 1.2 – 8.5 mg/m^3 in air; symptoms include pain and tightness in chest, coughing, and difficulty in breathing. If ingested, toxicity depends on release of the Hg^{++} ion; chronic mercury poisoning can cause kidney, mental, and nervous disturbances. Dust irritates eyes and frequently causes allergic dermatitis; absorption through skin can cause systemic poisoning.
5.3 **Treatment for Exposure:** INHALATION: remove to fresh air; give artificial respiration if breathing has stopped. INGESTION: give egg whites, milk, or activated charcoal; induce vomiting; consult physician. EYES: flush with water. SKIN: wash with soap and water.
5.4 **Toxicity by Inhalation (Threshold Limit Value):** 0.05 mg/m^3 (as mercury)
5.5 **Short-Term Inhalation Limits:** Data not available
5.6 **Toxicity by Ingestion:** Data not available
5.7 **Late Toxicity:** Central nervous system affects, tremors, psychological disturbances in humans
5.8 **Vapor (Gas) Irritant Characteristics:** Data not available
5.9 **Liquid or Solid Irritant Characteristics:** Data not available
5.10 **Odor Threshold:** Odorless

6. FIRE HAZARDS

6.1 **Flash Point:** Not pertinent (combustible solid)
6.2 **Flammable Limits in Air:** Not pertinent
6.3 **Fire Extinguishing Agents:** Water, foam, sand
6.4 **Fire Extinguishing Agents Not to be Used:** Other agents may be ineffective.
6.5 **Special Hazards of Combustion Products:** Smoke from fire contains poisonous mercury vapor and irritating sulfur dioxide gas.
6.6 **Behavior in Fire:** Changes color when hot. Decomposes at burning temperature. The black form may soften, and molten sulfur may flow out and burn.
6.7 **Ignition Temperature:** Data not available
6.8 **Electrical Hazard:** Not pertinent
6.9 **Burning Rate:** Not pertinent

7. CHEMICAL REACTIVITY

7.1 **Reactivity with Water:** No reaction
7.2 **Reactivity with Common Materials:**
7.3 **Stability During Transport:** Stable
7.4 **Neutralizing Agents for Acids and Caustics:** Not pertinent
7.5 **Polymerization:** Not pertinent
7.6 **Inhibitor of Polymerization:** Not pertinent

8. WATER POLLUTION

8.1 **Aquatic Toxicity:** Data not available
8.2 **Waterfowl Toxicity:** Data not available
8.3 **Biological Oxygen Demand (BOD):** Data not available
8.4 **Food Chain Concentration Potential:** Many organisms can accumulate mercury from water. Bioconcentrative up to 10,000 fold.

9. SELECTED MANUFACTURERS

1. Ventron Corp.
   P. O. Box 159
   Beverly, Mass. 01915
2. Gallard-Schlesinger Chemical Mfg. Co.
   584 Mineola Ave.
   Carle Place, N. Y. 11514
3. Merck & Co.
   Merck Chemical Division
   126 East Lincoln Ave.
   Rahway, N. J. 07065

10. SHIPPING INFORMATION

10.1 **Grades or Purity:** The black form may contain up to 40% free sulfur.
10.2 **Storage Temperature:** Ambient
10.3 **Inert Atmosphere:** No requirement
10.4 **Venting:** Open

11. HAZARD ASSESSMENT CODE

(See Hazard Assessment Handbook, CG 446-3)

II

12. HAZARD CLASSIFICATIONS

12.1 **Code of Federal Regulations:** Not listed
12.2 **NAS Hazard Rating for Bulk Water Transportation:** Not listed
12.3 **NFPA Hazard Classifications:** Not listed

13. PHYSICAL AND CHEMICAL PROPERTIES

13.1 **Physical State at 15°C and 1 atm:** Solid
13.2 **Molecular Weight:** 232.7
13.3 **Boiling Point at 1 atm:** Not pertinent
13.4 **Freezing Point:** Not pertinent
13.5 **Critical Temperature:** Not pertinent
13.6 **Critical Pressure:** Not pertinent
13.7 **Specific Gravity:** 8 at 20°C (solid)
13.8 **Liquid Surface Tension:** Not pertinent
13.9 **Liquid-Water Interfacial Tension:** Not pertinent
13.10 **Vapor (Gas) Specific Gravity:** Not pertinent
13.11 **Ratio of Specific Heats of Vapor (Gas):** Not pertinent
13.12 **Latent Heat of Vaporization:** Not pertinent
13.13 **Heat of Combustion:** –1,200 Btu/lb = –670 cal/g = –28 × 10^5 J/kg
13.14 **Heat of Decomposition:** Not pertinent
13.15 **Heat of Solution:** Not pertinent
13.16 **Heat of Polymerization:** Not pertinent

NOTES

MRR — MERCUROUS CHLORIDE

Common Synonyms			
Calomel Mild mercury chloride Mercury monochloride Mercury protochloride Mercury subchloride	Solid	White	Odorless
	Sinks in water.		

AVOID CONTACT WITH SOLID AND DUST. KEEP PEOPLE AWAY.
Wear dust respirator.
Stop discharge if possible.
Isolate and remove discharged material.
Notify local health and pollution control agencies.

Fire	Not flammable. POISONOUS GASES MAY BE PRODUCED WHEN HEATED.
Exposure	CALL FOR MEDICAL AID. DUST POISONOUS IF INHALED. If inhaled will cause coughing or difficult breathing. **If in eyes, hold eyelids open and flush with plenty of water.** **If breathing has stopped, give artificial respiration.** **If breathing is difficult, give oxygen.** SOLID POISONOUS IF SWALLOWED. Irritating to skin and eyes. If swallowed will cause nausea and vomiting. **Remove contaminated clothing and shoes.** **Flush affected areas with plenty of water.** **IF IN EYES, hold eyelids open and flush with plenty of water.** **IF SWALLOWED and victim is CONSCIOUS, have victim drink water or milk and have victim induce vomiting.** **IF SWALLOWED and victim is UNCONSCIOUS OR HAVING CONVULSIONS, do nothing except keep victim warm.**
Water Pollution	Effect of low concentrations on aquatic life is unknown. May be dangerous if it enters water intakes. **Notify local health and wildlife officials.** **Notify operators of nearby water intakes.**

1. RESPONSE TO DISCHARGE

(See Response Methods Handbook, CG 446-4)

Issue warning—water contaminant
Restrict access
Should be removed
Chemical and physical treatment

2. LABELS

No hazard label required by Code of Federal Regulations

3. CHEMICAL DESIGNATIONS

3.1 **Synonyms:** Calomel; Mercury monochloride; Mercury protochloride; Mercury subchloride; Mild mercury chloride

3.2 **Coast Guard Compatibility Classification:** Not listed

3.3 **Chemical Formula:** HgCl or Hg_2Cl_2

3.4 **IMCO/United Nations Numerical Designation:** 6.1/2024

4. OBSERVABLE CHARACTERISTICS

4.1 **Physical State (as shipped):** Solid

4.2 **Color:** White

4.3 **Odor:** None

5. HEALTH HAZARDS

5.1 **Personal Protective Equipment:** Dust mask; goggles or face shield; protective gloves

5.2 **Symptoms Following Exposure:** Acute poisoning can result from inhaling dust concentrations of 1.2—8.5 mg/m^3 in air; symptoms include pain and tightness in chest, coughing, and difficulty in breathing. Compound is an irritant, cathartic, or purgative; rarely, "calomel sickness," a benign reaction with fever and rash, appears after about 1 week; seldom causes systemic poisoning but may be fatal if retained to 30—40 mg/kg. Contact with eyes causes mild irritation.

5.3 **Treatment for Exposure:** INHALATION: remove to fresh air; give artificial respiration if breathing has stopped. INGESTION: give egg whites, milk, or activated charcoal; induce vomiting; consult physician. EYES: flush with water. SKIN: wash with soap and water.

5.4 **Toxicity by Inhalation (Threshold Limit Value):** 0.05 mg/m^3 (as mercury)

5.5 **Short-Term Inhalation Limits:** Data not available

5.6 **Toxicity by Ingestion:** Grade 3; oral LD_{50} = 210 mg/kg (rat)

5.7 **Late Toxicity:** Central nervous system effects, tremors, psychological disturbances in humans

5.8 **Vapor (Gas) Irritant Characteristics:** Data not available

5.9 **Liquid or Solid Irritant Characteristics:** Data not available

5.10 **Odor Threshold:** Odorless

6. FIRE HAZARDS

6.1 **Flash Point:** Not flammable

6.2 **Flammable Limits in Air:** Not flammable

6.3 **Fire Extinguishing Agents:** Not pertinent

6.4 **Fire Extinguishing Agents Not to be Used:** Not pertinent

6.5 **Special Hazards of Combustion Products:** Fumes from fire may contain toxic vapors of substance.

6.6 **Behavior in Fire:** Vaporizes and escapes as a sublimate

6.7 **Ignition Temperature:** Not pertinent

6.8 **Electrical Hazard:** Not pertinent

6.9 **Burning Rate:** Not pertinent

7. CHEMICAL REACTIVITY

7.1 **Reactivity with Water:** No reaction

7.2 **Reactivity with Common Materials:**

7.3 **Stability During Transport:** Stable

7.4 **Neutralizing Agents for Acids and Caustics:** Not pertinent

7.5 **Polymerization:** Not pertinent

7.6 **Inhibitor of Polymerization:** Not pertinent

8. WATER POLLUTION

8.1 **Aquatic Toxicity:** Data not available

8.2 **Waterfowl Toxicity:** Data not available

8.3 **Biological Oxygen Demand (BOD):** Data not available

8.4 **Food Chain Concentration Potential:** Many organisms can accumulate mercury from water. Bioconcentrative up to 10,000 fold.

9. SELECTED MANUFACTURERS

1. Mallinckrodt Chemical Works
   223 Westside Ave.
   P. O. Box 384
   Jersey City, N. J. 07303
2. Gallard-Schlesinger Chemical Mfg. Co.
   584 Mineola Ave.
   Carle Place, N. Y. 11514
3. Pfaltz and Bauer, Inc.
   375 Fairfield Ave.
   Stamford, Conn. 06902

10. SHIPPING INFORMATION

10.1 **Grades or Purity:** NF; Technical, 99.6%; Reagent

10.2 **Storage Temperature:** Ambient

10.3 **Inert Atmosphere:** No requirement

10.4 **Venting:** Open

11. HAZARD ASSESSMENT CODE

(See Hazard Assessment Handbook, CG 446-3)

II

12. HAZARD CLASSIFICATIONS

12.1 **Code of Federal Regulations:** Not listed

12.2 **NAS Hazard Rating for Bulk Water Transportation:** Not listed

12.3 **NFPA Hazard Classifications:** Not listed

13. PHYSICAL AND CHEMICAL PROPERTIES

13.1 **Physical State at 15°C and 1 atm:** Solid

13.2 **Molecular Weight:** 236.1

13.3 **Boiling Point at 1 atm:** Not pertinent

13.4 **Freezing Point:** Not pertinent

13.5 **Critical Temperature:** Not pertinent

13.6 **Critical Pressure:** Not pertinent

13.7 **Specific Gravity:** 7.15 at 20°C (solid)

13.8 **Liquid Surface Tension:** Not pertinent

13.9 **Liquid-Water Interfacial Tension:** Not pertinent

13.10 **Vapor (Gas) Specific Gravity:** Not pertinent

13.11 **Ratio of Specific Heats of Vapor (Gas):** Not pertinent

13.12 **Latent Heat of Vaporization:** Not pertinent

13.13 **Heat of Combustion:** Not pertinent

13.14 **Heat of Decomposition:** Not pertinent

13.15 **Heat of Solution:** Not pertinent

13.16 **Heat of Polymerization:** Not pertinent

NOTES

MRN — MERCUROUS NITRATE

Common Synonyms			
Mercury protonitrate Mercurous nitrate monohydrate	Solid	White	Slight odor
	Sinks in water.		

AVOID CONTACT WITH SOLID AND DUST. KEEP PEOPLE AWAY.
Wear dust respirator and rubber overclothing (including gloves).
Stop discharge if possible.
Isolate and remove discharged material.
Notify local health and pollution control agencies.

Fire

Not flammable.
Will increase the intensity of a fire.
POISONOUS GASES MAY BE PRODUCED WHEN HEATED.

Exposure

CALL FOR MEDICAL AID.

DUST
POISONOUS IF INHALED OR IF SKIN IS EXPOSED.
If inhaled will cause coughing or difficult breathing.
If in eyes, hold eyelids open and flush with plenty of water.
If breathing has stopped, give artificial respiration.
If breathing is difficult, give oxygen.

SOLID
POISONOUS IF SWALLOWED OR IF SKIN IS EXPOSED.
Irritating to skin and eyes.
If swallowed will cause nausea and vomiting.
Remove contaminated clothing and shoes.
Flush affected areas with plenty of water.
IF IN EYES, hold eyelids open and flush with plenty of water.
IF SWALLOWED and victim is CONSCIOUS, have victim drink water or milk and have victim induce vomiting.
IF SWALLOWED and victim is UNCONSCIOUS OR HAVING CONVULSIONS, do nothing except keep victim warm.

Water Pollution

Effect of low concentrations on aquatic life is unknown.
May be dangerous if it enters water intakes.
Notify local health and wildlife officials.
Notify operators of nearby water intakes.

1. RESPONSE TO DISCHARGE

(See Response Methods Handbook, CG 446-4)

Issue warning — poison, water contaminant, oxidizing material
Restrict access
Disperse and flush

2. LABEL

Yellow

3. CHEMICAL DESIGNATIONS

3.1 **Synonyms:** Mercurous nitrate monohydrate; Mercury protonitrate
3.2 **Coast Guard Compatibility Classification:** Not listed
3.3 **Chemical Formula:** $HgNO_3 \cdot H_2O$
3.4 **IMCO/United Nations Numerical Designation:** 6.1/1627

4. OBSERVABLE CHARACTERISTICS

4.1 **Physical State (as shipped):** Solid
4.2 **Color:** White
4.3 **Odor:** Slight odor of nitric acid

5. HEALTH HAZARDS

5.1 **Personal Protective Equipment:** Dust mask; goggles or face shield; protective gloves
5.2 **Symptoms Following Exposure:** *Acute systemic poisoning may be fatal within a few minutes; death by uremic poisoning is usually delayed 5—12 days.* Acute poisoning has resulted from inhaling dust concentrations of 1.2—8.5 mg/m^3 of air; symptoms include tightness and pain in chest, coughing, and difficulty in breathing. Ingestion causes necrosis, pain, vomiting, and severe purging. Contact with eyes causes ulceration of conjunctiva and cornea. Contact with skin causes irritation and possible dermatitis; systemic poisoning can occur by absorption through skin.
5.3 **Treatment for Exposure:** INHALATION: remove victim to fresh air; get medical attention. INGESTION: give egg whites, milk, or activated charcoal; induce vomiting; consult physician. EYES: flush with water for at least 15 min. SKIN: flush with water.
5.4 **Toxicity by Inhalation (Threshold Limit Value):** 0.05 mg/m^3 (as mercury)
5.5 **Short-Term Inhalation Limits:** Data not available
5.6 **Toxicity by Ingestion:** Grade 3; oral LD_{50} = 297 mg/kg (rat)
5.7 **Late Toxicity:** Data not available
5.8 **Vapor (Gas) Irritant Characteristics:** Data not available
5.9 **Liquid or Solid Irritant Characteristics:** Data not available
5.10 **Odor Threshold:** Data not available

6. FIRE HAZARDS

6.1 **Flash Point:** Not flammable, but may intensify fire.
6.2 **Flammable Limits in Air:** Not flammable
6.3 **Fire Extinguishing Agents:** Not pertinent
6.4 **Fire Extinguishing Agents Not to be Used:** Not pertinent
6.5 **Special Hazards of Combustion Products:** Smoke from fire may contain toxic mercury vapor and oxides of nitrogen.
6.6 **Behavior in Fire:** May increase intensity of fire
6.7 **Ignition Temperature:** Not pertinent
6.8 **Electrical Hazard:** Not pertinent
6.9 **Burning Rate:** Not pertinent

7. CHEMICAL REACTIVITY

7.1 **Reactivity with Water:** Dissolves, then forms cloudy acid solution. The reaction is not hazardous.
7.2 **Reactivity with Common Materials:** Solution may corrode most metals. Solid in contact with wood or paper may cause fire.
7.3 **Stability During Transport:** Stable
7.4 **Neutralizing Agents for Acids and Caustics:** Flush with water, rinse with dilute solution of sodium bicarbonate or soda ash.
7.5 **Polymerization:** Not pertinent
7.6 **Inhibitor of Polymerization:** Not pertinent

8. WATER POLLUTION

8.1 **Aquatic Toxicity:** Data not available
8.2 **Waterfowl Toxicity:** Data not available
8.3 **Biological Oxygen Demand (BOD):** Data not available
8.4 **Food Chain Concentration Potential:** Possible bioaccumulation problem. Many organisms can accumulate mercury from water. Bioconcentrative up to 10,000 fold.

9. SELECTED MANUFACTURERS

1. J. T. Baker Chemical Co.
   Phillipsburg, N. J. 08865
2. Gallard-Schlesinger Chemical Mfg. Co.
   584 Mineola Ave.
   Carle Place, N. Y. 11514
3. Pfaltz and Bauer, Inc.
   375 Fairfield Ave.
   Stamford, Conn. 06902

10. SHIPPING INFORMATION

10.1 **Grades or Purity:** Reagent; Purified
10.2 **Storage Temperature:** Ambient
10.3 **Inert Atmosphere:** No requirement
10.4 **Venting:** Open

11. HAZARD ASSESSMENT CODE

(See Hazard Assessment Handbook, CG 446-3)

II

12. HAZARD CLASSIFICATIONS

12.1 **Code of Federal Regulations:** Oxidizer
12.2 **NAS Hazard Rating for Bulk Water Transportation:** Not listed
12.3 **NFPA Hazard Classifications:** Not listed

13. PHYSICAL AND CHEMICAL PROPERTIES

13.1 **Physical State at 15°C and 1 atm:** Solid
13.2 **Molecular Weight:** 280.6
13.3 **Boiling Point at 1 atm:** Not pertinent (decomposes)
13.4 **Freezing Point:** Not pertinent
13.5 **Critical Temperature:** Not pertinent
13.6 **Critical Pressure:** Not pertinent
13.7 **Specific Gravity:** 4.78 at 20°C (solid)
13.8 **Liquid Surface Tension:** Not pertinent
13.9 **Liquid-Water Interfacial Tension:** Not pertinent
13.10 **Vapor (Gas) Specific Gravity:** Not pertinent
13.11 **Ratio of Specific Heats of Vapor (Gas):** Not pertinent
13.12 **Latent Heat of Vaporization:** Not pertinent
13.13 **Heat of Combustion:** Not pertinent
13.14 **Heat of Decomposition:** Not pertinent
13.15 **Heat of Solution:** Not pertinent
13.16 **Heat of Polymerization:** Not pertinent

NOTES

MCR	MERCURY

Common Synonyms
Quicksilver

Liquid | Silver | Odorless

Sinks in water.

AVOID CONTACT WITH LIQUID. Keep people away.
Stop discharge if possible.
Isolate and remove discharged material.
Notify local health and pollution control agencies.

Fire

Not flammable.

Exposure

CALL FOR MEDICAL AID.
LIQUID
Effects of exposure may be delayed.

Water Pollution

HARMFUL TO AQUATIC LIFE IN VERY LOW CONCENTRATIONS.
May be dangerous if it enters water intakes.

Notify local health and wildlife officials.
Notify operators of nearby water intakes.

1. RESPONSE TO DISCHARGE

(See Response Methods Handbook, CG 446-4)

Should be removed
Chemical and physical treatment

2. LABELS

No hazard label required by Code of Federal Regulations

3. CHEMICAL DESIGNATIONS

3.1 **Synonyms:** Quicksilver
3.2 **Coast Guard Compatibility Classification:** Not applicable
3.3 **Chemical Formula:** Hg
3.4 **IMCO/United Nations Numerical Designation:** Not listed

4. OBSERVABLE CHARACTERISTICS

4.1 **Physical State (as shipped):** Liquid
4.2 **Color:** Silvery
4.3 **Odor:** None

5. HEALTH HAZARDS

5.1 **Personal Protective Equipment:** Avoid contact of liquid with skin. For vapor use chemical cartridge (Hopcalite) respirator.
5.2 **Symptoms Following Exposure:** No immediate symptoms. As poisoning becomes established, slight muscular tremor, loss of appetite, nausea, and diarrhea are observed. Psychic, kidney, and cardiovascular disturbances may occur.
5.3 **Treatment for Exposure:** Consult a doctor.
5.4 **Toxicity by Inhalation (Threshold Limit Value):** 0.05 ng/m^3
5.5 **Short-Term Inhalation Limits:** Data not available
5.6 **Toxicity by Ingestion:** No immediate toxicity
5.7 **Late Toxicity:** Development of mercury poisoning
5.8 **Vapor (Gas) Irritant Characteristics:** None
5.9 **Liquid or Solid Irritant Characteristics:** None
5.10 **Odor Threshold:** Not pertinent

6. FIRE HAZARDS

6.1 **Flash Point:** Not flammable
6.2 **Flammable Limits in Air:** Not flammable
6.3 **Fire Extinguishing Agents:** Not pertinent
6.4 **Fire Extinguishing Agents Not to be Used:** Not pertinent
6.5 **Special Hazards of Combustion Products:** Not pertinent
6.6 **Behavior in Fire:** Not pertinent
6.7 **Ignition Temperature:** Not flammable
6.8 **Electrical Hazard:** Not pertinent
6.9 **Burning Rate:** Not flammable

7. CHEMICAL REACTIVITY

7.1 **Reactivity with Water:** No reaction
7.2 **Reactivity with Common Materials:** No reaction
7.3 **Stability During Transport:** Stable
7.4 **Neutralizing Agents for Acids and Caustics:** Not pertinent
7.5 **Polymerization:** Not pertinent
7.6 **Inhibitor of Polymerization:** Not pertinent

8. WATER POLLUTION

8.1 **Aquatic Toxicity:**
0.5—1 ppm/48 hr/caragius ardium/TL_m/fresh water
0.29 ppm/48 hr/marine fish/TL_m/salt water
8.2 **Waterfowl Toxicity:** Data not available
8.3 **Biological Oxygen Demand (BOD):** None
8.4 **Food Chain Concentration Potential:** Mercury concentrates in liver and kidneys of ducks and geese to levels above FDA limit of 0.5 ppm. Muscle tissue usually well below the limit.

9. SELECTED MANUFACTURERS

1. Belmont Smelting and Refining Works, Inc.
330 Belmont Ave.
Brooklyn, N. Y. 11207
2. Engelhard Minerals and Chemical Corp.
Philipp Bros. Division
299 Park Ave.
New York, N. Y. 10017
3. NL Industries
Goldsmith Division
900 W. 18th St.
Chicago, Ill. 60608

10. SHIPPING INFORMATION

10.1 **Grades or Purity:** Pure
10.2 **Storage Temperature:** Ambient
10.3 **Inert Atmosphere:** No requirement
10.4 **Venting:** Open

11. HAZARD ASSESSMENT CODE

(See Hazard Assessment Handbook, CG 446-3)

A-X

12. HAZARD CLASSIFICATIONS

12.1 **Code of Federal Regulations:** ORM-B
12.2 **NAS Hazard Rating for Bulk Water Transportation:** Not listed
12.3 **NFPA Hazard Classifications:** Not listed

13. PHYSICAL AND CHEMICAL PROPERTIES

13.1 **Physical State at 15°C and 1 atm:** Liquid
13.2 **Molecular Weight:** 200.59
13.3 **Boiling Point at 1 atm:** 675°F = 357°C = 630°K
13.4 **Freezing Point:** −38.0°F = −38.9°C = 234.3°K
13.5 **Critical Temperature:** 2664°F = 1462°C = 1735°K
13.6 **Critical Pressure:** 23,300 psia = 1587 atm = 160.8 MN/m^2
13.7 **Specific Gravity:** 13.55 at 20°C (liquid)
13.8 **Liquid Surface Tension:** 470 dynes/cm = 0.470 N/m at 20°C
13.9 **Liquid-Water Interfacial Tension:** 375 dynes/cm = 0.375 N/m at 20°C
13.10 **Vapor (Gas) Specific Gravity:** Not pertinent
13.11 **Ratio of Specific Heats of Vapor (Gas):** Not pertinent
13.12 **Latent Heat of Vaporization:** Not pertinent
13.13 **Heat of Combustion:** Not pertinent
13.14 **Heat of Decomposition:** Not pertinent
13.15 **Heat of Solution:** Not pertinent
13.16 **Heat of Polymerization:** Not pertinent

NOTES

REVISED 1978

MSO

MESITYL OXIDE

Common Synonyms			
4-methyl-3-pentene-2-one Isopropylideneacetone Methyl isobutenyl ketone	Liquid	Colorless to light yellow	Strong peppermint or honey odor
	Floats and mixes with water. Flammable, irritating vapor is produced.		

Shut off ignition sources, call fire department.
Stop discharge if possible. Keep people away.
Stay upwind, use water spray to "knock down" vapor.
Isolate and remove discharged material.
Notify local health and pollution control agencies.

Fire

FLAMMABLE.
Containers may explode in fire.
Flashback along vapor trail may occur.
Vapor may explode if ignited in an enclosed area.

Extinguish with dry chemicals, alcohol foam, or carbon dioxide.
Water may be ineffective on fire.
Cool exposed containers with water.

Exposure

Call for medical aid.

VAPOR
Irritating to eyes, nose and throat.
If inhaled will cause headache, dizziness or difficult breathing.

Move victim to fresh air.
If breathing has stopped, give artificial respiration.
If breathing is difficult, give oxygen.

LIQUID
Irritating to skin and eyes.
Harmful if swallowed.

Remove contaminated clothing and shoes.
Flush affected areas with plenty of water.
IF IN EYES, hold eyelids open and flush with plenty of water.
IF SWALLOWED and victim is CONSCIOUS, have victim drink water or milk.

Water Pollution

Effect of low concentrations on aquatic life is unknown.
Fouling to shoreline.
May be dangerous if it enters water intakes.

Notify local health and wildlife officials.
Notify operators of nearby water intakes.

1. RESPONSE TO DISCHARGE

(See Response Methods Handbook, CG 446-4)

Issue warning—water contaminant
Restrict access
Disperse and flush

2. LABEL

Red

3. CHEMICAL DESIGNATIONS

3.1 **Synonyms:** Isobutenyl methyl ketone; Isopropylideneacetone; Methyl isobutenyl ketone; 4-Methyl-3-pentene-2-one

3.2 **Coast Guard Compatibility Classification:** Ketone

3.3 **Chemical Formula:** $CH_3COCH=C(CH_3)_2$

3.4 **IMCO/United Nations Numerical Designation:** 3.3/1229

4. OBSERVABLE CHARACTERISTICS

4.1 **Physical State (as shipped):** Liquid

4.2 **Color:** Colorless to slightly yellow

4.3 **Odor:** Strong; peppermint; honeylike

5. HEALTH HAZARDS

5.1 **Personal Protective Equipment:** Air pack or organic canister mask; rubber gloves; goggles.

5.2 **Symptoms Following Exposure:** Inhalation causes irritation of nose and throat, headache, dizziness, difficult breathing. Contact with liquid or concentrated vapor causes severe eye irritation. Liquid irritates skin. Ingestion causes irritation of mouth and stomach.

5.3 **Treatment for Exposure:** INHALATION: move victim to fresh air and restore breathing; call physician. EYES: immediately flush with plenty of water for at least 15 min. SKIN: wash with water. INGESTION: give large amount of water; call physician.

5.4 **Toxicity by Inhalation (Threshold Limit Value):** 25 ppm

5.5 **Short-Term Inhalation Limits:** 1,000 ppm for 60 min.

5.6 **Toxicity by Ingestion:** Grade 2; oral LD_{50} = 1,120 mg/kg (rat)

5.7 **Late Toxicity:** Data not available

5.8 **Vapor (Gas) Irritant Characteristics:** Vapors cause moderate irritation such that personnel will find high concentrations unpleasant. The effect is temporary.

5.9 **Liquid or Solid Irritant Characteristics:** Causes smarting of the skin and first-degree burns on short exposure; may cause second-degree burns on long exposure.

5.10 **Odor Threshold:** 12 ppm

6. FIRE HAZARDS

6.1 **Flash Point:** 84°F O.C., 73°F C.C.

6.2 **Flammable Limits in Air:** Data not available

6.3 **Fire Extinguishing Agents:** Alcohol foam, dry chemical, carbon dioxide

6.4 **Fire Extinguishing Agents Not to be Used:** Water may be ineffective.

6.5 **Special Hazards of Combustion Products:** Not pertinent

6.6 **Behavior in Fire:** Vapor is heavier than air and may travel a considerable distance to a source of ignition and flash back.

6.7 **Ignition Temperature:** 652°F

6.8 **Electrical Hazard:** Data not available

6.9 **Burning Rate:** 4.2 mm/min.

7. CHEMICAL REACTIVITY

7.1 **Reactivity with Water:** No reaction

7.2 **Reactivity with Common Materials:** No reaction

7.3 **Stability During Transport:** Stable

7.4 **Neutralizing Agents for Acids and Caustics:** Not pertinent

7.5 **Polymerization:** Not pertinent

7.6 **Inhibitor of Polymerization:** Not pertinent

8. WATER POLLUTION

8.1 **Aquatic Toxicity:** Data not available

8.2 **Waterfowl Toxicity:** Data not available

8.3 **Biological Oxygen Demand (BOD):** Data not available

8.4 **Food Chain Concentration Potential:** None

9. SELECTED MANUFACTURERS

1. Union Carbide Corporation
Chemicals and Plastics Division
270 Park Avenue
New York, N. Y. 10017

2. Shell Chemical Company
Industrial Chemicals Division
P. O. Box 2463
Houston, Texas 77001

3. Eastman Kodak Co.
Eastman Organic Chemicals
Rochester, N. Y. 14650

10. SHIPPING INFORMATION

10.1 **Grades or Purity:** 97+%

10.2 **Storage Temperature:** Ambient

10.3 **Inert Atmosphere:** No requirement

10.4 **Venting:** Open (flame arrester)

11. HAZARD ASSESSMENT CODE

(See Hazard Assessment Handbook, CG 446-3)

A-P-Q-T-U

12. HAZARD CLASSIFICATIONS

12.1 **Code of Federal Regulations:** Flammable Liquid

12.2 **NAS Hazard Rating for Bulk Water Transportation:**

Category	Rating
Fire	3
Health	
Vapor Irritant	2
Liquid or Solid Irritant	2
Poisons	2
Water Pollution	
Human Toxicity	2
Aquatic Toxicity	3
Aesthetic Effect	2
Reactivity	
Other Chemicals	2
Water	0
Self-Reaction	1

12.3 **NFPA Hazard Classifications:**

Category	Classification
Health Hazard (Blue)	3
Flammability (Red)	3
Reactivity (Yellow)	0

13. PHYSICAL AND CHEMICAL PROPERTIES

13.1 **Physical State at 15°C and 1 atm:** Liquid

13.2 **Molecular Weight:** 98.2

13.3 **Boiling Point at 1 atm:** 266°F = 130°C = 403°K

13.4 **Freezing Point:** −51°F = −46°C = 227°K

13.5 **Critical Temperature:** Not pertinent

13.6 **Critical Pressure:** Not pertinent

13.7 **Specific Gravity:** 0.853 at 20°C (liquid)

13.8 **Liquid Surface Tension:** 22.9 dynes/cm = 0.0229 N/m at 20°C

13.9 **Liquid-Water Interfacial Tension:** Not pertinent

13.10 **Vapor (Gas) Specific Gravity:** 3.4

13.11 **Ratio of Specific Heats of Vapor (Gas):** Not pertinent

13.12 **Latent Heat of Vaporization:** 157 Btu/lb = 87 cal/g = 3.7×10^5 J/kg

13.13 **Heat of Combustion:** −14,400 Btu/lb = −8,000 cal/g = -330×10^5 J/kg

13.14 **Heat of Decomposition:** Not pertinent

13.15 **Heat of Solution:** Not pertinent

13.16 **Heat of Polymerization:** Not pertinent

NOTES

MCL — METHALLYL CHLORIDE

Common Synonyms			
3-Chloro-2-methylpropene gamma-Chloroisobutylene beta-Methallyl chloride beta-Methylallyl chloride	Liquid	Colorless to yellow	Sharp penetrating odor
	Floats on water. Flammable, irritating vapor is produced.		

Shut off ignition sources, call fire department.
Stop discharge if possible. Keep people away.
Stay upwind, use water spray to "knock down" vapor.
Isolate and remove discharged material.
Notify local health and pollution control agencies.

Fire

FLAMMABLE.
Irritating gases may be produced when heated.
Containers may explode in fire.
Flashback along vapor trail may occur.
Vapor may explode if ignited in an enclosed area.
Extinguish with dry chemicals, foam or carbon dioxide.
Water may be ineffective on fire.
Cool exposed containers with water.

Exposure

Call for medical aid.

VAPOR
Harmful if inhaled.
Move victim to fresh air.
If breathing has stopped, give artificial respiration.
If breathing is difficult, give oxygen.

LIQUID
Irritating to skin and eyes.
Harmful if swallowed.
Remove Contaminated clothing and shoes.
Flush affected areas with plenty of water.
IF IN EYES, hold eyelids open and flush with plenty of water.
IF SWALLOWED and victim is CONSCIOUS, have victim drink water or milk and have victim induce vomiting.
IF SWALLOWED and victim is UNCONSCIOUS OR HAVING CONVULSIONS, do nothing except keep victim warm.

Water Pollution

Effect of low concentrations on aquatic life is unknown.
Fouling to shoreline.
May be dangerous if it enters water intakes.
Notify local health and wildlife officials.
Notify operators of nearby water intakes.

1. RESPONSE TO DISCHARGE

(See Response Methods Handbook, CG 446-4)

Issue warning—high flammability
Restrict access
Mechanical containment
Should be removed
Chemical and physical treatment

2. LABEL

Red

3. CHEMICAL DESIGNATIONS

3.1 **Synonyms:** gamma-Chloroisobutylene; 3-Chloro-2-methylpropene; beta-Methallyl chloride; beta-Methylallyl chloride
3.2 **Coast Guard Compatibility Classification:** Not applicable
3.3 **Chemical Formula:** $CH_2=C(CH_3)CH_2Cl$
3.4 **IMCO/United Nations Numerical Designation:** 3.2/1993

4. OBSERVABLE CHARACTERISTICS

4.1 **Physical State (as shipped):** Liquid
4.2 **Color:** Colorless to straw
4.3 **Odor:** Pungent, penetrating

5. HEALTH HAZARDS

5.1 **Personal Protective Equipment:** Organic canister mask; goggles; rubber gloves.
5.2 **Symptoms Following Exposure:** Inhalation causes irritation of nose and throat. Contact of vapor or liquid with eyes causes irritation. Liquid irritates skin. Ingestion causes irritation of mouth and stomach.
5.3 **Treatment for Exposure:** INHALATION: remove victim to fresh air; if breathing stops, give artificial respiration and oxygen; subsequent treatment is symptomatic and supportive. EYES: flush with water for at least 15 min.; get medical attention if exposure has been to liquid. SKIN: flush with water; get medical attention if skin is burned. INGESTION: induce vomiting and follow with gastric lavage, demulcents, and saline cathartics; get medical attention.
5.4 **Toxicity by Inhalation (Threshold Limit Value):** Data not available
5.5 **Short-Term Inhalation Limits:** Data not available
5.6 **Toxicity by Ingestion:** Data not available
5.7 **Late Toxicity:** Data not available
5.8 **Vapor (Gas) Irritant Characteristics:** Data not available
5.9 **Liquid or Solid Irritant Characteristics:** Data not available
5.10 **Odor Threshold:** Data not available

6. FIRE HAZARDS

6.1 **Flash Point:** 14°F O.C.
6.2 **Flammable Limits in Air:** 2.3%–9.3%
6.3 **Fire Extinguishing Agents:** Dry chemical, foam, carbon dioxide
6.4 **Fire Extinguishing Agents Not to be Used:** Water may be ineffective.
6.5 **Special Hazards of Combustion Products:** Irritating and toxic hydrogen chloride and phosgene vapors may be formed.
6.6 **Behavior in Fire:** Vapor is heavier than air and may travel considerable distance to a source of ignition and flash back.
6.7 **Ignition Temperature:** Data not available
6.8 **Electrical Hazard:** Data not available
6.9 **Burning Rate:** 4.4 mm/min.

7. CHEMICAL REACTIVITY

7.1 **Reactivity with Water:** No reaction
7.2 **Reactivity with Common Materials:** No reaction
7.3 **Stability During Transport:** Stable
7.4 **Neutralizing Agents for Acids and Caustics:** Not pertinent
7.5 **Polymerization:** Not pertinent
7.6 **Inhibitor of Polymerization:** Not pertinent

8. WATER POLLUTION

8.1 **Aquatic Toxicity:** Data not available
8.2 **Waterfowl Toxicity:** Data not available
8.3 **Biological Oxygen Demand (BOD):** Data not available
8.4 **Food Chain Concentration Potential:** None

9. SELECTED MANUFACTURERS

1. FMC Corporation
   Industrial Chemical Division
   633 Third Avenue
   New York, N. Y. 10017
2. Progil, Inc.
   509 Madison Ave.
   New York, N. Y. 10022
3. Aldrich Chemical Co.
   940 West St. Paul Ave.
   Milwaukee, Wis. 53233

10. SHIPPING INFORMATION

10.1 **Grades or Purity:** 95+%
10.2 **Storage Temperature:** Ambient
10.3 **Inert Atmosphere:** No requirement
10.4 **Venting:** Pressure-vacuum

11. HAZARD ASSESSMENT CODE

(See Hazard Assessment Handbook, CG 446-3)

A-T-U-V-W

12. HAZARD CLASSIFICATIONS

12.1 **Code of Federal Regulations:** Flammable liquid
12.2 **NAS Hazard Rating for Bulk Water Transportation:** Not listed
12.3 **NFPA Hazard Classifications:** Not listed

13. PHYSICAL AND CHEMICAL PROPERTIES

13.1 **Physical State at 15°C and 1 atm:** Liquid
13.2 **Molecular Weight:** 90.55
13.3 **Boiling Point at 1 atm:** 162.0°F = 72.2°C = 345.4°K
13.4 **Freezing Point:** $<-112°F = <-80°C = <193°K$
13.5 **Critical Temperature:** Not pertinent
13.6 **Critical Pressure:** Not pertinent
13.7 **Specific Gravity:** 0.928 at 20°C (liquid)
13.8 **Liquid Surface Tension:** (est.) 25 dynes/cm = 0.025 N/m at 20°C
13.9 **Liquid-Water Interfacial Tension:** (est.) 32 dynes/cm = 0.032 N/m at 20°C
13.10 **Vapor (Gas) Specific Gravity:** 3.12
13.11 **Ratio of Specific Heats of Vapor (Gas):** 1.0893
13.12 **Latent Heat of Vaporization:** 160 Btu/lb = 89 cal/g = 3.7×10^5 J/kg
13.13 **Heat of Combustion:** (est.) −11,600 Btu/lb = −6,500 cal/g = -270×10^5 J/kg
13.14 **Heat of Decomposition:** Not pertinent
13.15 **Heat of Solution:** Not pertinent
13.16 **Heat of Polymerization:** Not pertinent

NOTES

MTH	METHANE

Common Synonyms			
Marsh gas Natural gas	Liquid or gas	Colorless	Weak odor
	Liquid floats and boils on water. Flammable visible vapor cloud is produced.		

Stop discharge if possible. Keep people away.
Shut off ignition sources and call fire department.
Stay upwind and use water spray to "knock down" vapor.
Evacuate area in case of large discharge.
Avoid contact with liquid and vapor.
Notify local health and pollution control agencies.

Fire	FLAMMABLE. Flashback along vapor trail may occur. May explode if ignited in an enclosed area. Stop discharge if possible. Cool exposed containers and protect men effecting shutoff with water. Let fire burn.
Exposure	CALL FOR MEDICAL AID. VAPOR Not irritating to eyes, nose or throat. If inhaled, will cause dizziness, difficult breathing, and loss of consciousness. Move to fresh air. If breathing has stopped, give artificial respiration. If breathing is difficult, give oxygen. LIQUID Will cause frostbite. Flush affected areas with plenty of water. DO NOT RUB AFFECTED AREAS.
Water Pollution	Not harmful to aquatic life.

1. RESPONSE TO DISCHARGE

(See Response Methods Handbook, CG 446-4)

Issue warning—high flammability
Restrict access
Evacuate area

2. LABEL

Red

3. CHEMICAL DESIGNATIONS

3.1 **Synonyms:** Marsh gas
3.2 **Coast Guard Compatibility Classification:** Paraffin
3.3 **Chemical Formula:** CH_4
3.4 **IMCO/United Nations Numerical Designation:** 2.0/1971

4. OBSERVABLE CHARACTERISTICS

4.1 **Physical State (as shipped):** Liquefied gas
4.2 **Color:** Colorless
4.3 **Odor:** Mild, sweet

5. HEALTH HAZARDS

5.1 **Personal Protective Equipment:** Self-contained breathing apparatus for high concentrations; protective clothing if exposed to liquid.
5.2 **Symptoms Following Exposure:** High concentrations may cause asphyxiation. No systemic effects, even at 5% concentration in air.
5.3 **Treatment for Exposure:** Remove to fresh air. Support respiration.
5.4 **Toxicity by Inhalation (Threshold Limit Value):** Not pertinent (methane is an asphyxiant, and limiting factor is available oxygen)
5.5 **Short-Term Inhalation Limits:** Data not available
5.6 **Toxicity by Ingestion:** Not pertinent
5.7 **Late Toxicity:** None
5.8 **Vapor (Gas) Irritant Characteristics:** Vapors are nonirritating to the eyes and throat.
5.9 **Liquid or Solid Irritant Characteristics:** No appreciable hazard. Practically harmless to the skin, because it evaporates quickly, but may cause some frostbite.
5.10 **Odor Threshold:** 200 ppm

6. FIRE HAZARDS

6.1 **Flash Point:** Flammable gas
6.2 **Flammable Limits in Air:** 5.0%–15.0%
6.3 **Fire Extinguishing Agents:** Stop flow of gas
6.4 **Fire Extinguishing Agents Not to be Used:** Water
6.5 **Special Hazards of Combustion Products:** None
6.6 **Behavior in Fire:** Not pertinent
6.7 **Ignition Temperature:** 1004°F
6.8 **Electrical Hazard:** Class I, Group D
6.9 **Burning Rate:** 12.5 mm/min.

7. CHEMICAL REACTIVITY

7.1 **Reactivity with Water:** No reaction
7.2 **Reactivity with Common Materials:** No reaction
7.3 **Stability During Transport:** Stable
7.4 **Neutralizing Agents for Acids and Caustics:** Not pertinent
7.5 **Polymerization:** Not pertinent
7.6 **Inhibitor of Polymerization:** Not pertinent

8. WATER POLLUTION

8.1 **Aquatic Toxicity:** None
8.2 **Waterfowl Toxicity:** None
8.3 **Biological Oxygen Demand (BOD):** None
8.4 **Food Chain Concentration Potential:** None

9. SELECTED MANUFACTURERS

1. Air Products and Chemicals, Inc.
Allentown, Pa. 18105
2. Phillips Petroleum Co.
Bartlesville, Okla. 74004
3. Union Carbide Corp.
Linde Division
270 Park Ave.
New York, N. Y. 10017

10. SHIPPING INFORMATION

10.1 **Grades or Purity:** Research grade; pure grade
10.2 **Storage Temperature:** −260°F
10.3 **Inert Atmosphere:** No requirement
10.4 **Venting:** Safety relief

11. HAZARD ASSESSMENT CODE

(See Hazard Assessment Handbook, CG 446-3)

A-B-C-D-E-F-G

12. HAZARD CLASSIFICATIONS

12.1 **Code of Federal Regulations:** Flammable compressed gas
12.2 **NAS Hazard Rating for Bulk Water Transportation:**

Category	Rating
Fire	4
Health	
Vapor Irritant	0
Liquid or Solid Irritant	0
Poisons	0
Water Pollution	
Human Toxicity	0
Aquatic Toxicity	0
Aesthetic Effect	0
Reactivity	
Other Chemicals	0
Water	0
Self-Reaction	0

12.3 **NFPA Hazard Classifications:** Not listed

13. PHYSICAL AND CHEMICAL PROPERTIES

13.1 **Physical State at 15°C and 1 atm:** Gas
13.2 **Molecular Weight:** 16.04
13.3 **Boiling Point at 1 atm:** −258.7°F = −161.5°C = 111.7°K
13.4 **Freezing Point:** −296.5°F = −182.5°C = 90.7°K
13.5 **Critical Temperature:** −116.5°F = −82.5°C = 190.7°K
13.6 **Critical Pressure:** 668 psia = 45.44 atm = 4.60 MN/m²
13.7 **Specific Gravity:** 0.422 at −160°C (liquid)
13.8 **Liquid Surface Tension:** 14 dynes/cm = 0.014 N/m at −161°C
13.9 **Liquid-Water Interfacial Tension:** (est.) 50 dynes/cm = 0.050 N/m at −161°C
13.10 **Vapor (Gas) Specific Gravity:** 0.55–1.0
13.11 **Ratio of Specific Heats of Vapor (Gas):** 1.306
13.12 **Latent Heat of Vaporization:** 219.4 Btu/lb = 121.9 cal/g = 5.100 × 10^5 J/kg
13.13 **Heat of Combustion:** −21,517 Btu/lb = −11,954 cal/g = −500.2 × 10^5 J/kg
13.14 **Heat of Decomposition:** Not pertinent
13.15 **Heat of Solution:** Not pertinent
13.16 **Heat of Polymerization:** Not pertinent

NOTES

REVISED 1978

MSA	METHANEARSONIC ACID, SODIUM SALTS

Common Synonyms Disodium methane arsonate Monosodium methane arsonate MSMA DSMA	Solid, or water solution	Colorless solid; solution may be red or green	Odorless
	Solid may float or sink in water; solid and solution mix with water.		

Stop discharge if possible.
Avoid contact with solid and solution.
Isolate and remove discharged material.
Notify local health and pollution control agencies.

Fire	Not flammable. POISONOUS GASES ARE PRODUCED WHEN HEATED. Wear goggles and self-contained breathing apparatus.
Exposure	CALL FOR MEDICAL AID. SOLID OR SOLUTION Irritating to skin and eyes. If swallowed, will cause nausea, vomiting, or loss of consciousness. Remove contaminated clothing and shoes. Flush affected areas with plenty of water. IF IN EYES, hold eyelids open and flush with plenty of water. IF SWALLOWED and victim is CONSCIOUS, have victim drink water or milk and have victim induce vomiting. IF SWALLOWED and victim is UNCONSCIOUS OR HAVING CONVULSIONS, do nothing except keep victim warm.
Water Pollution	Dangerous to aquatic life in high concentrations. May be dangerous if it enters water intakes. Notify local health and wildlife officials. Notify operators of nearby water intakes.

1. RESPONSE TO DISCHARGE

(See Response Methods Handbook, CG 446-4)

Issue warning—poison, water contaminant
Should be removed
Chemical and physical treatment

2. LABEL

3. CHEMICAL DESIGNATIONS

3.1 **Synonyms:** Disodium methanearsonate
Disodium methyl arsonate
DSMA
Monosodium methanearsonate
Monosodium methyl arsonate
MSMA

3.2 **Coast Guard Compatibility Classification:** Not applicable

(Continued on page 4)

4. OBSERVABLE CHARACTERISTICS

4.1 **Physical State (as shipped):** Solid or water solution
4.2 **Color:** Colorless; solutions may contain red or green dyes
4.3 **Odor:** None

5. HEALTH HAZARDS

5.1 **Personal Protective Equipment:** Protective clothing to prevent contact with skin; chemical goggles.
5.2 **Symptoms Following Exposure:** Subacute poisoning by arsenicals causes salty taste, burning in throat and stomach, and intestinal pain. Acute toxicity indicated by headache, vomiting, stupor, convulsions, paralysis. About 1 ounce to 1 pound must be taken to cause these symptoms.
5.3 **Treatment for Exposure:** INGESTION: cause vomiting; give water and repeat; give a saline cathartic such as sodium sulfate. SKIN: wash with soap and water. EYES: wash with water; consult physician if irritation remains.
5.4 **Toxicity by Inhalation (Threshold Limit Value):** Not pertinent
5.5 **Short-Term Inhalation Limits:** Not pertinent
5.6 **Toxicity by Ingestion:** Grade 2; LD_{50} 0.5-5 g/kg (rat)
5.7 **Late Toxicity:** Data not available
5.8 **Vapor (Gas) Irritant Characteristics:** None
5.9 **Liquid or Solid Irritant Characteristics:** Repeated contact may cause skin sensitivity.
5.10 **Odor Threshold:** Not pertinent

6. FIRE HAZARDS

6.1 **Flash Point:** Not flammable
6.2 **Flammable Limits in Air:** Not flammable
6.3 **Fire Extinguishing Agents:** Not pertinent
6.4 **Fire Extinguishing Agents Not to be Used:** Not pertinent
6.5 **Special Hazards of Combustion Products:** Toxic gases may be generated in fires.
6.6 **Behavior in Fire:** Not pertinent
6.7 **Ignition Temperature:** Not flammable
6.8 **Electrical Hazard:** Not pertinent
6.9 **Burning Rate:** Not flammable

7. CHEMICAL REACTIVITY

7.1 **Reactivity with Water:** None
7.2 **Reactivity with Common Materials:** None
7.3 **Stability During Transport:** Stable
7.4 **Neutralizing Agents for Acids and Caustics:** Not pertinent
7.5 **Polymerization:** Not pertinent
7.6 **Inhibitor of Polymerization:** Not pertinent

8. WATER POLLUTION

8.1 **Aquatic Toxicity:** >1000 ppm/48 hr/bluegill sunfish/LC_{50}/fresh water
8.2 **Waterfowl Toxicity:** Data not available
8.3 **Biological Oxygen Demand (BOD):** Data not available
8.4 **Food Chain Concentration Potential:** Data not available

9. SELECTED MANUFACTURERS

1. The Ansul Co.
Chemical Division
1 Stanton St.
Marinette, Wis. 54143
2. Diamond Shamrock Corp.
Biochemicals Division
300 Union Commerce Bldg.
Cleveland, Ohio 44115
3. Vineland Chemical Co.
P. O. Box 745
Vineland, N. J. 08360

10. SHIPPING INFORMATION

10.1 **Grades or Purity:** The solid disodium salt (DSMA) contains water crystallization. Salts are often shipped as solutions in water with concentrations up to about 50% solids.
10.2 **Storage Temperature:** Ambient
10.3 **Inert Atmosphere:** No requirement
10.4 **Venting:** Open

11. HAZARD ASSESSMENT CODE

(See Hazard Assessment Handbook, CG 446-3)

SS

12. HAZARD CLASSIFICATIONS

12.1 **Code of Federal Regulations:** Poisonous liquid or solid, Class B
12.2 **NAS Hazard Rating for Bulk Water Transportation:** Not listed
12.3 **NFPA Hazard Classifications:** Not listed

13. PHYSICAL AND CHEMICAL PROPERTIES

13.1 **Physical State at 15°C and 1 atm:** Solid or water solution
13.2 **Molecular Weight:** 162 (MSMA); 292 (DSMA hexahydrate)
13.3 **Boiling Point at 1 atm:** Decomposes
13.4 **Freezing Point:** (MSMA) 243°F = 117°C = 390°K
(DSMA) 137°F = 58°C = 332°K
13.5 **Critical Temperature:** Not pertinent
13.6 **Critical Pressure:** Not pertinent
13.7 **Specific Gravity:** (DSMA) 1.0 at 20°C (solid)
(MSMA solutions) 1.4—1.6 at 20°C (liquid)
13.8 **Liquid Surface Tension:** Not pertinent
13.9 **Liquid-Water Interfacial Tension:** Not pertinent
13.10 **Vapor (Gas) Specific Gravity:** Not pertinent
13.11 **Ratio of Specific Heats of Vapor (Gas):** Not pertinent
13.12 **Latent Heat of Vaporization:** Not pertinent
13.13 **Heat of Combustion:** Not pertinent
13.14 **Heat of Decomposition:** Not pertinent
13.15 **Heat of Solution:** Not pertinent
13.16 **Heat of Polymerization:** Not pertinent

3. CHEMICAL DESIGNATIONS (Cont'd.)

3.3 **Chemical Formula:**
$CH_3AsO(OH)(ONa)$
$CH_3AsO(ONa)_2 \cdot 6H_2O$
3.4 **IMCO/United Nations Numerical Designation:** 6.1/1557

MOC — METHOXYCHLOR

Common Synonyms			
2,2-Di (p-anisyl)-1,1,1-trichloroethane DMDT Methoxy DDT 2,2-Bis (p-methoxyphenyl)-1,1,1-trichloroethane	Solid Sinks in water.	White to light yellow	Mild fruity odor

AVOID CONTACT WITH SOLID AND DUST. KEEP PEOPLE AWAY.
Wear goggles and dust respirator.
Call fire department.
Stay upwind, use water spray to "knock down" dust.
Isolate and remove discharged material.
Notify local health and pollution control agencies.

Fire

Combustible.
POISONOUS GASES MAY BE PRODUCED IN FIRE.
Extinguish with water, dry chemicals, foam, or carbon dioxide.

Exposure

CALL FOR MEDICAL AID.

DUST
POISONOUS IF INHALED.
Move victim to fresh air.
If in eyes, hold eyelids open and flush with plenty of water.
If breathing is difficult, give oxygen.

SOLID
POISONOUS IF SWALLOWED.
Irritating to skin and eyes.
Remove contaminated clothing and shoes.
Flush affected areas with plenty of water.
IF IN EYES, hold eyelids open and flush with plenty of water.
IF SWALLOWED and victim is CONSCIOUS, have victim drink water or milk.

Water Pollution

HARMFUL TO AQUATIC LIFE IN VERY LOW CONCENTRATIONS.
May be dangerous if it enters water intakes.
Notify local health and wildlife officials.
Notify operators of nearby water intakes.

1. RESPONSE TO DISCHARGE

(See Response Methods Handbook, CG 446-4)

Issue warning—poison, water contaminant
Restrict access
Should be removed
Chemical and physical treatment

2. LABELS

No hazard label required by Code of Federal Regulations

3. CHEMICAL DESIGNATIONS

3.1 **Synonyms:** 2,2-Bis (p-methoxyphenyl)-1,1,1-trichloroethane; 2,2-Di-(p-anisyl)-1,1,1-trichloroethane; DMDT; Marlate 50; Methoxy-DDT

3.2 **Coast Guard Compatibility Classification:** Not applicable

3.3 **Chemical Formula:** $C_{16}H_{15}Cl_3O_2$

3.4 **IMCO/United Nations Numerical Designation:** 6.1/1615

4. OBSERVABLE CHARACTERISTICS

4.1 **Physical State (as shipped):** Solid

4.2 **Color:** Light cream; white to light yellow

4.3 **Odor:** Slightly fruity

5. HEALTH HAZARDS

5.1 **Personal Protective Equipment:** Dust respirator if needed; gloves and goggles.

5.2 **Symptoms Following Exposure:** Toxicity is relatively low. Inhalation or ingestion causes generalized depression.

5.3 **Treatment for Exposure:** EYES: flush with water if irritated. SKIN: wash well with soap and water. INGESTION: consult physician.

5.4 **Toxicity by Inhalation (Threshold Limit Value):** 10 mg/m^3

5.5 **Short-Term Inhalation Limits:** Data not available

5.6 **Toxicity by Ingestion:** Grade 1; LD_{50} 5 to 15 g/kg

5.7 **Late Toxicity:** Data not available

5.8 **Vapor (Gas) Irritant Characteristics:** Data not available

5.9 **Liquid or Solid Irritant Characteristics:** Data not available

5.10 **Odor Threshold:** Not pertinent

6. FIRE HAZARDS

6.1 **Flash Point:** Burns only at high temperatures. For liquid forms, see Kerosene.

6.2 **Flammable Limits in Air:** Not pertinent

6.3 **Fire Extinguishing Agents:** Water, foam, dry chemical, carbon dioxide.

6.4 **Fire Extinguishing Agents Not to be Used:** Not pertinent

6.5 **Special Hazards of Combustion Products:** Irritating and toxic hydrogen chloride gas may be formed in fire.

6.6 **Behavior in Fire:** Not pertinent

6.7 **Ignition Temperature:** Not pertinent

6.8 **Electrical Hazard:** Not pertinent

6.9 **Burning Rate:** Not pertinent

7. CHEMICAL REACTIVITY

7.1 **Reactivity with Water:** No reaction

7.2 **Reactivity with Common Materials:** No reaction

7.3 **Stability During Transport:** Stable

7.4 **Neutralizing Agents for Acids and Caustics:** Not pertinent

7.5 **Polymerization:** Not pertinent

7.6 **Inhibitor of Polymerization:** Not pertinent

8. WATER POLLUTION

8.1 **Aquatic Toxicity:**
0.035 ppm/96 hr/fathead/TL_m/fresh water
0.004-.012 ppm/96 hr/marine crustacea/TL_m/salt water

8.2 **Waterfowl Toxicity:** LD_{50} = 2,000 mg/kg

8.3 **Biological Oxygen Demand (BOD):** Data not available

8.4 **Food Chain Concentration Potential:** Data not available

9. SELECTED MANUFACTURERS

1. Ansul Co.
   1 Stanton Street
   Marinette, Wis. 54143
2. Chemical Formulators, Inc.
   P. O. Box 26
   Nitro, W. Va. 25143
3. E. I. duPont de Nemours & Co.
   Biochemicals Department
   308 East Lancaster Ave.
   Wynnewood, Pa. 19096

10. SHIPPING INFORMATION

10.1 **Grades or Purity:** Technical flake or chip: 88% plus 12% isomers; Wettable powders: 50—75%; Dust concentrate: 40%; Emulsifiable concentrate (liquid): 25% solution in petroleum distillate.

10.2 **Storage Temperature:** Ambient

10.3 **Inert Atmosphere:** No requirement

10.4 **Venting:** Open

11. HAZARD ASSESSMENT CODE

(See Hazard Assessment Handbook, CG 446-3)

II

12. HAZARD CLASSIFICATIONS

12.1 **Code of Federal Regulations:** Not Listed

12.2 **NAS Hazard Rating for Bulk Water Transportation:** Not listed

12.3 **NFPA Hazard Classifications:** Not listed

13. PHYSICAL AND CHEMICAL PROPERTIES

13.1 **Physical State at 15°C and 1 atm:** Solid

13.2 **Molecular Weight:** 345.7

13.3 **Boiling Point at 1 atm:** Not pertinent (decomposes)

13.4 **Freezing Point:** 171–192°F = 77–89°C = 350–362°K

13.5 **Critical Temperature:** Not pertinent

13.6 **Critical Pressure:** Not pertinent

13.7 **Specific Gravity:** 1.41 at 25°C (solid)

13.8 **Liquid Surface Tension:** Not pertinent

13.9 **Liquid-Water Interfacial Tension:** Not pertinent

13.10 **Vapor (Gas) Specific Gravity:** Not pertinent

13.11 **Ratio of Specific Heats of Vapor (Gas):** Not pertinent

13.12 **Latent Heat of Vaporization:** Not pertinent

13.13 **Heat of Combustion:** Not pertinent

13.14 **Heat of Decomposition:** Not pertinent

13.15 **Heat of Solution:** Not pertinent

13.16 **Heat of Polymerization:** Not pertinent

NOTES

MTT

METHYL ACETATE

Common Synonyms			
Acetic acid, methyl ester	Liquid	Colorless	Mild sweet odor
	Mixes with water. Flammable, irritating vapor is produced.		

Shut off ignition sources, call fire department.
Stop discharge if possible. Keep people away.
Stay upwind, use water spray to "knock down" vapor.
Isolate and remove discharged material.
Notify local health and pollution control agencies.

Fire

FLAMMABLE.
Containers may explode in fire.
Flashback along vapor trail may occur.
Vapor may explode if ignited in an enclosed area.
Extinguish with dry chemicals, alcohol foam, or carbon dioxide.
Water may be ineffective on fire.
Cool exposed containers with water.

Exposure

Call for medical aid.

VAPOR
Irritating to eyes, nose and throat.
If inhaled will cause headache, or dizziness.
Move victim to fresh air.
If breathing has stopped, give artificial respiration.
If breathing is difficult, give oxygen.

LIQUID
Irritating to skin and eyes.
Remove contaminated clothing and shoes.
Flush affected areas with plenty of water.
IF IN EYES, hold eyelids open and flush with plenty of water.
IF SWALLOWED and victim is CONSCIOUS, have victim drink water or milk.

Water Pollution

Effect of low concentrations on aquatic life is unknown.
May be dangerous if it enters water intakes.
Notify local health and wildlife officials.
Notify operators of nearby water intakes.

1. RESPONSE TO DISCHARGE

(See Response Methods Handbook, CG 446-4)

Issue warning—high flammability
Restrict access
Disperse and flush

2. LABEL

Red

3. CHEMICAL DESIGNATIONS

3.1 **Synonyms:** Acetic acid, methyl ester
3.2 **Coast Guard Compatibility Classification:** Esters (13)
3.3 **Chemical Formula:** CH_3COOCH_3
3.4 **IMCO/United Nations Numerical Designation:** 3.2/1231

4. OBSERVABLE CHARACTERISTICS

4.1 **Physical State (as shipped):** Liquid
4.2 **Color:** Colorless
4.3 **Odor:** Slightly acrid, sweet; fragrant

5. HEALTH HAZARDS

5.1 **Personal Protective Equipment:** Air mask or organic canister mask; goggles or face shield.
5.2 **Symptoms Following Exposure:** (Very similar to those of methyl alcohol, which constitutes 20% of commercial grade.) Inhalation causes headache, fatigue, and drowsiness; high concentrations can produce central nervous system depression and optic nerve damage. Liquid irritates eyes and may cause defatting and cracking of skin. Ingestion causes headache, dizziness, drowsiness, fatigue; may cause severe eye damage.
5.3 **Treatment for Exposure:** INHALATION: remove victim from affected area; if breathing has ceased, apply artificial respiration; call doctor. EYES: irrigate thoroughly with water for 15 min. and call doctor. SKIN: wash affected area with water. INGESTION: get medical attention for methyl alcohol poisoning.
5.4 **Toxicity by Inhalation (Threshold Limit Value):** 200 ppm
5.5 **Short-Term Inhalation Limits:** 400 ppm for 5 min.
5.6 **Toxicity by Ingestion:** Grade 2; oral LD_{50} = 3,700 mg/kg (rabbit)
5.7 **Late Toxicity:** Optic nerve may be damaged following overexposure to vapor or liquid.
5.8 **Vapor (Gas) Irritant Characteristics:** Vapors cause moderate irritation such that personnel will find high concentrations unpleasant. The effect is temporary.
5.9 **Liquid or Solid Irritant Characteristics:** No appreciable hazard. Practically harmless to the skin.
5.10 **Odor Threshold:** Data not available

6. FIRE HAZARDS

6.1 **Flash Point:** 22°F O.C., 14°F C.C.
6.2 **Flammable Limits in Air:** 3.1% 16%
6.3 **Fire Extinguishing Agents:** Dry chemical, alcohol foam, carbon dioxide
6.4 **Fire Extinguishing Agents Not to be Used:** Water may be ineffective.
6.5 **Special Hazards of Combustion Products:** Not pertinent
6.6 **Behavior in Fire:** Vapor is heavier than air and may travel a considerable distance to a source of ignition and flash back.
6.7 **Ignition Temperature:** 935°F
6.8 **Electrical Hazard:** Data not available
6.9 **Burning Rate:** 3.7 mm/min.

7. CHEMICAL REACTIVITY

7.1 **Reactivity with Water:** Reacts slowly to form acetic acid and methyl alcohol; the reaction is not violent.
7.2 **Reactivity with Common Materials:** No reaction
7.3 **Stability During Transport:** Stable
7.4 **Neutralizing Agents for Acids and Caustics:** Not pertinent
7.5 **Polymerization:** Not pertinent
7.6 **Inhibitor of Polymerization:** Not pertinent

8. WATER POLLUTION

8.1 **Aquatic Toxicity:** Data not available
8.2 **Waterfowl Toxicity:** Data not available
8.3 **Biological Oxygen Demand (BOD):** Data not available
8.4 **Food Chain Concentration Potential:** None

9. SELECTED MANUFACTURERS

1. Union Carbide Corporation
Chemicals and Plastics Division
270 Park Avenue
New York, N. Y. 10017
2. Monsanto Polymers & Petrochemicals Co.
800 North Lindbergh Boulevard
St. Louis, Mo. 63166
3. Eastman Kodak Co.
Eastman Organic Chemicals
Rochester, N. Y. 14650

10. SHIPPING INFORMATION

10.1 **Grades or Purity:** 78—82%; remainder is methyl alcohol.
10.2 **Storage Temperature:** Ambient
10.3 **Inert Atmosphere:** No requirement
10.4 **Venting:** Pressure-vacuum

11. HAZARD ASSESSMENT CODE

(See Hazard Assessment Handbook, CG 446-3)

A-P-Q-R-S

12. HAZARD CLASSIFICATIONS

12.1 **Code of Federal Regulations:** Flammable liquid
12.2 **NAS Hazard Rating for Bulk Water Transportation:**

Category	Rating
Fire	3
Health	
Vapor Irritant	2
Liquid or Solid Irritant	0
Poisons	1
Water Pollution	
Human Toxicity	1
Aquatic Toxicity	2
Aesthetic Effect	1
Reactivity	
Other Chemicals	1
Water	0
Self-Reaction	0

12.3 **NFPA Hazard Classifications:**

Category	Classification
Health Hazard (Blue)	1
Flammability (Red)	3
Reactivity (Yellow)	0

13. PHYSICAL AND CHEMICAL PROPERTIES

13.1 **Physical State at 15°C and 1 atm:** Liquid
13.2 **Molecular Weight:** 74.1
13.3 **Boiling Point at 1 atm:** 134.6°F = 57.0°C = 330.2°K
13.4 **Freezing Point:** −145.3°F = −98.5°C = 174.7°K
13.5 **Critical Temperature:** 452.7°F = 233.7°C = 506.9°K
13.6 **Critical Pressure:** 666 psia = 45.3 atm = 4.60 MN/m^2
13.7 **Specific Gravity:** 0.927 at 20°C (liquid)
13.8 **Liquid Surface Tension:** 24 dynes/cm = 0.024 N/m at 20°C
13.9 **Liquid-Water Interfacial Tension:** (est.) 30 dynes/cm = 0.030 N/m at 20°C
13.10 **Vapor (Gas) Specific Gravity:** 2.8
13.11 **Ratio of Specific Heats of Vapor (Gas):** 1.1192
13.12 **Latent Heat of Vaporization:** 174 Btu/lb = 97 cal/g = 4.1 × 10^5 J/kg
13.13 **Heat of Combustion:** 9,260 Btu/lb = 5,150 cal/g = 215 × 10^5 J/kg
13.14 **Heat of Decomposition:** Not pertinent
13.15 **Heat of Solution:** Not pertinent
13.16 **Heat of Polymerization:** Not pertinent

NOTES

MAP

METHYL ACETYLENE - PROPADIENE MIXTURE

Common Synonyms

Liquified compressed gas Colorless Garlic-like odor

Floats and boils on water. Flammable visible vapor cloud is produced.

Shut off ignition sources. Call fire department.
Avoid contact with liquid. Keep people away.
Stop discharge if possible.
Evacuate area in case of large discharge.
Notify local health and pollution control agencies.

Fire

FLAMMABLE.
Containers may explode in fire.
Flashback along vapor trail may occur.
Vapor may explode if ignited in an enclosed area.
Let fire burn.
Stop flow of gas if possible.
Cool exposed containers and protect men effecting shutoff with water.

Exposure

Call for medical aid.

VAPOR
If inhaled will cause difficult breathing.
Move victim to fresh air.
If breathing is difficult, give oxygen.

LIQUID
Will cause frostbite.
Flush affected areas with plenty of water.
DO NOT RUB AFFECTED AREAS.

Water Pollution

Not harmful to aquatic life.

1. RESPONSE TO DISCHARGE

(See Response Methods Handbook, CG 446-4)

Issue warning – high flammability
Restrict access
Evacuate area

2. LABEL

Red

3. CHEMICAL DESIGNATIONS

3.1 **Synonyms:** Allene-methylacetylene mixture; MAPP gas; Methylacetylene-allene mixture; Propadiene-methylacetylene mixture

3.2 **Coast Guard Compatibility Classification:** Olefin

3.3 **Chemical Formula:** $CH_3C{\equiv}CH + CH_2{=}C{=}CH_2$

3.4 **IMCO/United Nations Numerical Designation:** 2/1060

4. OBSERVABLE CHARACTERISTICS

4.1 **Physical State (as shipped):** Liquefied compressed gas

4.2 **Color:** Colorless

4.3 **Odor:** Offensive, like acetylene

5. HEALTH HAZARDS

5.1 **Personal Protective Equipment:** Self-contained breathing apparatus for high concentrations; safety goggles; protective gloves.

5.2 **Symptoms Following Exposure:** Simple asphyxiant. Toxicology of propadiene component not fully established. Contact with liquid may burn eyes and cause frostbite of skin.

5.3 **Treatment for Exposure:** INHALATION: remove victim to fresh air; give artificial respiration if necessary. EYES or SKIN: treat burns caused by cold liquid.

5.4 **Toxicity by Inhalation (Threshold Limit Value):** 1000 ppm

5.5 **Short-Term Inhalation Limits:** Data not available

5.6 **Toxicity by Ingestion:** Not pertinent

5.7 **Late Toxicity:** Lung irritation in rats and dogs

5.8 **Vapor (Gas) Irritant Characteristics:** Data not available

5.9 **Liquid or Solid Irritant Characteristics:** Data not available

5.10 **Odor Threshold:** 100 ppm

6. FIRE HAZARDS

6.1 **Flash Point:** Not pertinent (flammable, liquefied compressed gas)

6.2 **Flammable Limits in Air:** 3% – 11%

6.3 **Fire Extinguishing Agents:** Let fire burn; shut off gas supply; cool adjacent exposures.

6.4 **Fire Extinguishing Agents Not to be Used:** Not pertinent

6.5 **Special Hazards of Combustion Products:** Not pertinent

6.6 **Behavior in Fire:** Containers may explode.

6.7 **Ignition Temperature:** 850°F

6.8 **Electrical Hazard:** Data not available

6.9 **Burning Rate:** Not pertinent

7. CHEMICAL REACTIVITY

7.1 **Reactivity with Water:** No reaction

7.2 **Reactivity with Common Materials:** No reaction, except forms explosive compounds in contact with alloys containing more than 67% copper at high pressures.

7.3 **Stability During Transport:** Stable

7.4 **Neutralizing Agents for Acids and Caustics:** Not pertinent

7.5 **Polymerization:** Not pertinent

7.6 **Inhibitor of Polymerization:** Not pertinent

8. WATER POLLUTION

8.1 **Aquatic Toxicity:** Data not available

8.2 **Waterfowl Toxicity:** Data not available

8.3 **Biological Oxygen Demand (BOD):** None

8.4 **Food Chain Concentration Potential:** None

9. SELECTED MANUFACTURERS

1. Dow Chemical Co.
Midland, Mich. 48640

2. MAPP Products
100 Mountain Avenue
New Providence, N. J. 07974

10. SHIPPING INFORMATION

10.1 **Grades or Purity:** Commercial: 65% of a mixture of methylacetylene (85%) and propadiene (15%) plus 35% of a mixture of C_3 and C_4 saturated and unsaturated hydrocarbons.

10.2 **Storage Temperature:** Ambient but less than 125°F

10.3 **Inert Atmosphere:** No requirement

10.4 **Venting:** Safety relief

11. HAZARD ASSESSMENT CODE

(See Hazard Assessment Handbook, CG 446-3)

A-B-C-D-E-F-G

12. HAZARD CLASSIFICATIONS

12.1 **Code of Federal Regulations:** Flammable compressed gas

12.2 **NAS Hazard Rating for Bulk Water Transportation:** Not listed

12.3 **NFPA Hazard Classifications:** Not listed

13. PHYSICAL AND CHEMICAL PROPERTIES

13.1 **Physical State at 15°C and 1 atm:** Gas

13.2 **Molecular Weight:** 40.1

13.3 **Boiling Point at 1 atm:** −36 to −4°F = −38 to −20°C = 235 to 253°K

13.4 **Freezing Point:** Not pertinent

13.5 **Critical Temperature:** Not pertinent

13.6 **Critical Pressure:** Not pertinent

13.7 **Specific Gravity:** 0.576 at 15°C (liquid)

13.8 **Liquid Surface Tension:** 18 dynes/cm = 0.018 N/m at −24°C

13.9 **Liquid-Water Interfacial Tension:** Not pertinent

13.10 **Vapor (Gas) Specific Gravity:** 1.48

13.11 **Ratio of Specific Heats of Vapor (Gas):** 1.1686

13.12 **Latent Heat of Vaporization:** 227 Btu/lb = 126 cal/g = 5.28×10^5 J/kg

13.13 **Heat of Combustion:** −19,800 Btu/lb = −11,000 cal/g = -460×10^5 J/kg

13.14 **Heat of Decomposition:** Not pertinent

13.15 **Heat of Solution:** Not pertinent

13.16 **Heat of Polymerization:** Not pertinent

NOTES

MAM	METHYL ACRYLATE

Common Synonyms			
Acrylic acid, methyl ester Methyl 2-propenoate	Watery liquid	Colorless	Sweet sharp odor
	Floats and mixes slowly with water. Flammable, irritating vapor is produced.		

Avoid contact with liquid and vapor. Keep people away.
Wear chemical protective suit with self-contained breathing apparatus.
Stop discharge if possible.
Shut off ignition sources and call fire department.
Stay upwind and use water spray to "knock down" vapor.
Isolate and remove discharged material.
Notify local health and pollution control agencies.

Fire

FLAMMABLE.
Flashback along vapor trail may occur.
Containers may explode when heated.
Vapor may explode if ignited in an enclosed area.
Wear chemical protective suit with self-contained breathing apparatus.
Combat fires from safe distance or protected location.
Extinguish with dry chemical, foam, or carbon dioxide.
Water may be ineffective on fire.
Cool exposed containers with water.

Exposure

CALL FOR MEDICAL AID.

VAPOR
Irritating to eyes, nose and throat.
If inhaled, will cause dizziness or difficult breathing.
Move to fresh air.
If breathing has stopped, give artificial respiration.
If breathing is difficult, give oxygen.

LIQUID
Will burn skin and eyes.
Harmful if swallowed.
Remove contaminated clothing.
Flush affected areas with plenty of water.
IF IN EYES, hold eyelids open and flush with plenty of water.
IF SWALLOWED and victim is CONSCIOUS, have victim drink water or milk.

Water Pollution

Effect of low concentrations on aquatic life is unknown.
Fouling to shoreline.
May be dangerous if it enters water intakes.
Notify local health and wildlife officials.
Notify operators of nearby water intakes.

1. RESPONSE TO DISCHARGE

(See Response Methods Handbook, CG 446-4)

Issue warning—high flammability
Evacuate area

2. LABEL

Red

3. CHEMICAL DESIGNATIONS

3.1 **Synonyms:** Acrylic acid, methyl ester; Methyl 2-propenoate
3.2 **Coast Guard Compatibility Classification:** Acrylate
3.3 **Chemical Formula:** $CH_2{=}CHCOOCH_3$
3.4 **IMCO/United Nations Numerical Designation:** 3.2/1919

4. OBSERVABLE CHARACTERISTICS

4.1 **Physical State (as shipped):** Liquid
4.2 **Color:** Colorless
4.3 **Odor:** Characteristic acrylic; sweet, sharp; sharp, fragrant

5. HEALTH HAZARDS

5.1 **Personal Protective Equipment:** Organic canister for high vapor concentrations; rubber gloves; chemical goggles or face shield.
5.2 **Symptoms Following Exposure:** May irritate skin, eyes, respiratory system, and gastro-intestinal tract. Fumes cause tears.
5.3 **Treatment for Exposure:** INHALATION: remove to fresh air; lay patient down; keep him warm; administer artificial respiration if breathing has stopped; administer oxygen. SKIN OR EYES: flush with plenty of water for 15 min.; consult physician for eye exposure.
5.4 **Toxicity by Inhalation (Threshold Limit Value):** 10 ppm
5.5 **Short-Term Inhalation Limits:** 25 ppm for 30 min.
5.6 **Toxicity by Ingestion:** Grade 3; LD_{50} 50 to 500 mg/kg (rabbit)
5.7 **Late Toxicity:** Data not available
5.8 **Vapor (Gas) Irritant Characteristics:** Vapor is moderately irritating such that personnel will not usually tolerate moderate or high vapor concentrations.
5.9 **Liquid or Solid Irritant Characteristics:** Causes smarting of the skin and first-degree burns on short exposure; may cause secondary burns on long exposure.
5.10 **Odor Threshold:** Data not available

6. FIRE HAZARDS

6.1 **Flash Point:** 27°F C.C.; 44°F O.C.
6.2 **Flammable Limits in Air:** 2.8%–25%
6.3 **Fire Extinguishing Agents:** Foam, dry chemical, or carbon dioxide
6.4 **Fire Extinguishing Agents Not to be Used:** Water may be ineffective
6.5 **Special Hazards of Combustion Products:** Irritating vapors are generated in fires.
6.6 **Behavior in Fire:** May polymerize. Vapor is heavier than air and may travel considerable distance to a source of ignition and flash back.
6.7 **Ignition Temperature:** Data not available
6.8 **Electrical Hazard:** Not pertinent
6.9 **Burning Rate:** Data not available

7. CHEMICAL REACTIVITY

7.1 **Reactivity with Water:** No reaction
7.2 **Reactivity with Common Materials:** No reaction
7.3 **Stability During Transport:** Stable
7.4 **Neutralizing Agents for Acids and Caustics:** Not pertinent
7.5 **Polymerization:** Heat may cause an explosive polymerization. Strong ultraviolet light can also initiate polymerization.
7.6 **Inhibitor of Polymerization:** Hydroquinone and its methyl ether, in presence of air.

8. WATER POLLUTION

8.1 **Aquatic Toxicity:** Data not available
8.2 **Waterfowl Toxicity:** Data not available
8.3 **Biological Oxygen Demand (BOD):** Data not available
8.4 **Food Chain Concentration Potential:** None

9. SELECTED MANUFACTURERS

1. Celanese Corp.
   Celanese Chemical Co. Division
   245 Park Ave.
   New York, N. Y. 10017
2. Dow Badische Co.
   Freeport, Tex. 77541
3. Rohm and Haas Co.
   Independence Mall West
   Philadelphia, Pa. 19105

10. SHIPPING INFORMATION

10.1 **Grades or Purity:** 99.9%
10.2 **Storage Temperature:** Ambient if material is inhibited; under 40°F if no inhibitor.
10.3 **Inert Atmosphere:** Air MUST be present.
10.4 **Venting:** Open (flame arrester)

11. HAZARD ASSESSMENT CODE

(See Hazard Assessment Handbook, CG 446-3)

A-P-Q-T-U-V-W-Z

12. HAZARD CLASSIFICATIONS

12.1 **Code of Federal Regulations:** Flammable liquid
12.2 **NAS Hazard Rating for Bulk Water Transportation:**

Category	Rating
Fire	3
Health	
Vapor Irritant	3
Liquid or Solid Irritant	2
Poisons	3
Water Pollution	
Human Toxicity	2
Aquatic Toxicity	2
Aesthetic Effect	2
Reactivity	
Other Chemicals	2
Water	0
Self-Reaction	3

12.3 **NFPA Hazard Classifications:**

Category	Classification
Health Hazard (Blue)	2
Flammability (Red)	3
Reactivity (Yellow)	2

13. PHYSICAL AND CHEMICAL PROPERTIES

13.1 **Physical State at 15°C and 1 atm:** Liquid
13.2 **Molecular Weight:** 86.09
13.3 **Boiling Point at 1 atm:** 177°F = 80.6°C = 353.8°K
13.4 **Freezing Point:** −105.7°F = −76.5°C = 196.7°K
13.5 **Critical Temperature:** 505°F = 263°C = 536°K
13.6 **Critical Pressure:** 630 psia = 43 atm = 4.3 MN/m²
13.7 **Specific Gravity:** 0.956 at 20°C (liquid)
13.8 **Liquid Surface Tension:** 24.2 dynes/cm = 0.0242 N/m at 20°C
13.9 **Liquid-Water Interfacial Tension:** (est.) 30 dynes/cm = 0.03 N/m at 20°C
13.10 **Vapor (Gas) Specific Gravity:** 3.0
13.11 **Ratio of Specific Heats of Vapor (Gas):** 1.102
13.12 **Latent Heat of Vaporization:** 160 Btu/lb = 90 cal/g = 3.8×10^5 J/kg
13.13 **Heat of Combustion:** (est.) −9900 Btu/lb = −5500 cal/g = -230×10^5 J/kg
13.14 **Heat of Decomposition:** Not pertinent
13.15 **Heat of Solution:** Not pertinent
13.16 **Heat of Polymerization:** −392 Btu/lb = −218 cal/g = -9.13×10^5 J/kg

NOTES

REVISED 1978

MAL — METHYL ALCOHOL

Common Synonyms			
Methanol Wood alcohol Wood naphtha Wood spirit Pyroxylic spirit	Watery liquid	Colorless	Alcohol odor
	Floats and mixes with water. Flammable, irritating vapor is produced.		

Stop discharge if possible. Keep people away.
Shut off ignition sources and call fire department.
Stay upwind and use water spray to "knock down" vapor.
Avoid contact with liquid and vapor.
Isolate and remove discharged material.
Notify local health and pollution control agencies.

Fire

FLAMMABLE.
Vapor may explode if ignited in an enclosed area.
Flashback along vapor trail may occur.
Extinguish with dry chemical, alcohol foam, or carbon dioxide.
Water may be ineffective on fire.
Cool exposed containers with water.

Exposure

CALL FOR MEDICAL AID.

VAPOR
Irritating to eyes, nose and throat.
If inhaled, will cause dizziness, headache, difficult breathing, or loss of consciousness.
Move to fresh air.
If breathing has stopped, give artificial respiration.
If breathing is difficult, give oxygen.

LIQUID
POISONOUS IF SWALLOWED.
Irritating to skin and eyes.
Remove contaminated clothing and shoes.
Flush affected areas with plenty of water.
IF IN EYES, hold eyelids open and flush with plenty of water.
IF SWALLOWED and victim is CONSCIOUS, have victim drink water or milk and have victim induce vomiting.
IF SWALLOWED and victim is UNCONSCIOUS OR HAVING CONVULSIONS, do nothing except keep victim warm.

Water Pollution

Dangerous to aquatic life in high concentrations.
May be dangerous if it enters water intakes.
Notify local health and wildlife officials.
Notify operators of nearby water intakes.

1. RESPONSE TO DISCHARGE

(See Response Methods Handbook, CG 446-4)

Issue warning – high flammability
Restrict access
Evacuate area
Disperse and flush

2. LABEL

Red

3. CHEMICAL DESIGNATIONS

3.1 **Synonyms:** Colonial spirit, Columbian spirit, Methanol, Wood alcohol, Wood naphtha, Wood spirit

3.2 **Coast Guard Compatibility Classification:** Alcohol

3.3 **Chemical Formula:** CH_3OH

3.4 **IMCO/United Nations Numerical Designation:** 3.2/1230

4. OBSERVABLE CHARACTERISTICS

4.1 **Physical State (as shipped):** Liquid

4.2 **Color:** Colorless

4.3 **Odor:** Faint alcohol; like ethyl alcohol; faintly sweet; characteristic pungent

5. HEALTH HAZARDS

5.1 **Personal Protective Equipment:** Approved canister mask for high vapor concentrations; safety goggles; rubber gloves.

5.2 **Symptoms Following Exposure:** Exposure to excessive vapor causes eye irritation, headache, fatigue and drowsiness. High concentrations can produce central nervous system depression and optic nerve damage. 50,000 ppm will probably cause death in 1 to 2 hrs. Can be absorbed through skin. Swallowing may cause death or eye damage.

5.3 **Treatment for Exposure:** Remove victim from exposure and apply artifical respiration if breathing has ceased. INGESTION: induce vomiting, then give 2 teaspoons of baking soda in glass of water; call a physician. SKIN OR EYES: flush with water for 15 min.

5.4 **Toxicity by Inhalation (Threshold Limit Value):** 200 ppm

5.5 **Short-Term Inhalation Limits:** 260 mg/m^3 for 60 min.

5.6 **Toxicity by Ingestion:** Grade 1; 5 to 15 g/kg (rat)

5.7 **Late Toxicity:** None

5.8 **Vapor (Gas) Irritant Characteristics:** Vapors cause a slight smarting of the eyes or respiratory system if present in high concentrations. The effect is temporary.

5.9 **Liquid or Solid Irritant Characteristics:** Minimum hazard. If spilled on clothing and allowed to remain, may cause smarting and reddening of the skin.

5.10 **Odor Threshold:** 100 ppm

6. FIRE HAZARDS

6.1 **Flash Point:** 54°F C.C.; 61°F O.C.

6.2 **Flammable Limits in Air:** 6.0%–36.5%

6.3 **Fire Extinguishing Agents:** Alcohol foam, dry chemical, or carbon dioxide

6.4 **Fire Extinguishing Agents Not to be Used:** Water may be ineffective.

6.5 **Special Hazards of Combustion Products:** Not pertinent

6.6 **Behavior in Fire:** Containers may explode.

6.7 **Ignition Temperature:** 867°F

6.8 **Electrical Hazard:** Class I, Group D

6.9 **Burning Rate:** 1.7 mm/min.

7. CHEMICAL REACTIVITY

7.1 **Reactivity with Water:** No reaction

7.2 **Reactivity with Common Materials:** No reaction

7.3 **Stability During Transport:** Stable

7.4 **Neutralizing Agents for Acids and Caustics:** Not pertinent

7.5 **Polymerization:** Not pertinent

7.6 **Inhibitor of Polymerization:** Not pertinent

8. WATER POLLUTION

8.1 **Aquatic Toxicity:** 250 ppm/11 hr/goldfish/died/fresh water

8.2 **Waterfowl Toxicity:** Data not available

8.3 **Biological Oxygen Demand (BOD):** 0.6 to 1.12 lb/lb in 5 days

8.4 **Food Chain Concentration Potential:** None

9. SELECTED MANUFACTURERS

1. Borden Inc.
   Borden Chemical Division
   Geismar, La. 70734
2. Celanese Corp.
   Celanese Chemical Co. Division
   245 Park Ave.
   New York, N. Y. 10017
3. E. I. du Pont de Nemours & Co., Inc.
   Industrial and Biochemical Dept.
   Wilmington, Del. 19898

10. SHIPPING INFORMATION

10.1 **Grades or Purity:** CP, Crude, ACS: all 99.9%

10.2 **Storage Temperature:** Ambient

10.3 **Inert Atmosphere:** No requirement

10.4 **Venting:** Open (flame arrester) or pressure-vacuum

11. HAZARD ASSESSMENT CODE

(See Hazard Assessment Handbook, CG 446-3)

A-P-Q-R-S

12. HAZARD CLASSIFICATIONS

12.1 **Code of Federal Regulations:** Flammable liquid

12.2 **NAS Hazard Rating for Bulk Water Transportation:**

Category	Rating
Fire	3
Health	
Vapor Irritant	1
Liquid or Solid Irritant	1
Poisons	2
Water Pollution	
Human Toxicity	1
Aquatic Toxicity	1
Aesthetic Effect	1
Reactivity	
Other Chemicals	2
Water	0
Self-Reaction	0

12.3 **NFPA Hazard Classifications:**

Category	Classification
Health Hazard (Blue)	1
Flammability (Red)	3
Reactivity (Yellow)	0

13. PHYSICAL AND CHEMICAL PROPERTIES

13.1 **Physical State at 15°C and 1 atm:** Liquid

13.2 **Molecular Weight:** 32.04

13.3 **Boiling Point at 1 atm:** 148.1°F = 64.5°C = 337.7°K

13.4 **Freezing Point:** −144.0°F = −97.8°C = 175.4°K

13.5 **Critical Temperature:** 464°F = 240°C = 513°K

13.6 **Critical Pressure:** 1142.0 psia = 77.7 atm = 7.87 MN/m^2

13.7 **Specific Gravity:** 0.792 at 20°C (liquid)

13.8 **Liquid Surface Tension:** Not pertinent

13.9 **Liquid-Water Interfacial Tension:** Not pertinent

13.10 **Vapor (Gas) Specific Gravity:** 1.1

13.11 **Ratio of Specific Heats of Vapor (Gas):** 1.254

13.12 **Latent Heat of Vaporization:** 473.0 Btu/lb = 262.8 cal/g = 11.00×10^5 J/kg

13.13 **Heat of Combustion:** −8419 Btu/lb = −4677 cal/g = -195.8×10^5 J/kg

13.14 **Heat of Decomposition:** Not pertinent

13.15 **Heat of Solution:** (est.) −9 Btu/lb = −5 cal/g = -0.2×10^5 J/kg

13.16 **Heat of Polymerization:** Not pertinent

NOTES

REVISED 1978

MTA — METHYLAMINE

Common Synonyms	
Aminomethane Monomethylamine Mercurialin	Liquefied compressed gas — Colorless — Ammonia-like odor Mixes with water and boils.

Shut off ignition sources. Call fire department.
Avoid contact with liquid. Keep people away.
Stop discharge if possible.
Evacuate area in case of large discharge.
Notify local health and pollution control agencies.

Fire	FLAMMABLE. POISONOUS GASES MAY BE PRODUCED IN FIRE. Containers may explode in fire. Flashback along vapor trail may occur. Vapor may explode if ignited in an enclosed area. Wear goggles and self-contained breathing apparatus. Let fire burn. Stop flow of gas if possible. Cool exposed containers and protect men effecting shutoff with water.
Exposure	Call for medical aid. VAPOR Irritating to eyes, nose and throat. If inhaled will cause coughing or difficult breathing. Move victim to fresh air. If breathing has stopped, give artificial respiration. If breathing is difficult, give oxygen. LIQUID Will burn skin and eyes. Remove contaminated clothing and shoes. Flush affected areas with plenty of water. IF IN EYES, hold eyelids open and flush with plenty of water. IF SWALLOWED and victim is CONSCIOUS, have victim drink water or milk. DO NOT INDUCE VOMITING.
Water Pollution	HARMFUL TO AQUATIC LIFE IN VERY LOW CONCENTRATIONS. May be dangerous if it enters water intakes. Notify local health and wildlife officials. Notify operators of nearby water intakes.

1. RESPONSE TO DISCHARGE

(See Response Methods Handbook, CG 446-4)

Issue warning—high flammability, air contaminant, water contaminant
Restrict access
Evacuate area
Disperse and flush

2. LABEL

Red

3. CHEMICAL DESIGNATIONS

3.1 **Synonyms:** Aminomethane; Mercurialin; Monomethylamine, anhydrous
3.2 **Coast Guard Compatibility Classification:** Amines (4)
3.3 **Chemical Formula:** CH_3NH_2
3.4 **IMCO/United Nations Numerical Designation:** Anhydrous: 2/1061; Aqueous soln.: 3.1/1235

4. OBSERVABLE CHARACTERISTICS

4.1 **Physical State (as shipped):** Liquefied compressed gas
4.2 **Color:** Colorless
4.3 **Odor:** Like ammonia, pungent, fishy, suffocating

5. HEALTH HAZARDS

5.1 **Personal Protective Equipment:** Goggles or face mask; rubber suit, apron, sleeves, and/or gloves; rubber or leather safety shoes; air-line mask, positive-pressure hose mask, self-contained breathing apparatus, or industrial canister-type gas mask.
5.2 **Symptoms Following Exposure:** Inhalation causes irritation of nose and throat, followed by violent sneezing, burning sensation in throat, coughing, constriction of larynx and difficulty in breathing, pulmonary congestion, edema of the lungs, and conjunctivitis. Contact with liquid burns skin and eyes. (Severe exposure may cause blindness.) Vapors may cause dermatitis. Ingestion causes burns of the mouth, throat, and esophagus.
5.3 **Treatment for Exposure:** Get medical attention for anyone overcome or injured by exposure to this compound. INHALATION: remove victim to fresh air at once; apply artificial respiration if breathing has stopped; administer oxygen. EYES: flush with water for at least 15 min. SKIN: flush with water; if skin is burned do not use ointments or cover for 24 hours. INGESTION: do NOT induce vomiting; give large amount of water.
5.4 **Toxicity by Inhalation (Threshold Limit Value):** 10 ppm
5.5 **Short-Term Inhalation Limits:** Data not available
5.6 **Toxicity by Ingestion:** Grade 2; LD_{50} 0.5 to 5 g/kg
5.7 **Late Toxicity:** Data not available
5.8 **Vapor (Gas) Irritant Characteristics:** Vapors are moderately irritating such that personnel will not usually tolerate moderate or high concentrations.

6. FIRE HAZARDS

6.1 **Flash Point:** Not pertinent (flammable, liquefied compressed gas)
6.2 **Flammable Limits in Air:** 4.3%–21%
6.3 **Fire Extinguishing Agents:** Let gas fire burn; stop flow of gas. Extinguish solution fires with dry chemical, alcohol foam, or carbon dioxide.
6.4 **Fire Extinguishing Agents Not to be Used:** Not pertinent
6.5 **Special Hazards of Combustion Products:** Toxic nitrogen oxides may be formed.
6.6 **Behavior in Fire:** Vapors are heavier than air and may travel considerable distance to a source of ignition and flash back.
6.7 **Ignition Temperature:** 806°F
6.8 **Electrical Hazard:** Data not available
6.9 **Burning Rate:** Not pertinent

7. CHEMICAL REACTIVITY

7.1 **Reactivity with Water:** Dissolves completely
7.2 **Reactivity with Common Materials:** Corrosive to copper, copper alloys, zinc alloys, aluminum, and galvanized surfaces
7.3 **Stability During Transport:** Stable
7.4 **Neutralizing Agents for Acids and Caustics:** Not pertinent
7.5 **Polymerization:** Not pertinent
7.6 **Inhibitor of Polymerization:** Not pertinent

8. WATER POLLUTION

8.1 **Aquatic Toxicity:** > 10 and < 30 ppm/24 hr/creek chub/TL_m
8.2 **Waterfowl Toxicity:** Data not available
8.3 **Biological Oxygen Demand (BOD):** 67.8% of theoretical in 13 days
8.4 **Food Chain Concentration Potential:** None

9. SELECTED MANUFACTURERS

1. Commercial Solvents Corporation
Terre Haute, Ind. 47808
2. E. I. duPont de Nemours & Co., Inc.
Biochemicals Department
1007 Market Street
Wilmington, Del. 19898
3. Matheson Gas Products Co.
East Rutherford, N. J. 07073

10. SHIPPING INFORMATION

10.1 **Grades or Purity:** Anhydrous, 99.3+%. Water solutions, 30—50% by weight.
10.2 **Storage Temperature:** Ambient
10.3 **Inert Atmosphere:** No requirement
10.4 **Venting:** Safety relief

11. HAZARD ASSESSMENT CODE

(See Hazard Assessment Handbook, CG 446-3)

A-B-C-K-L-M-N

12. HAZARD CLASSIFICATIONS

12.1 **Code of Federal Regulations:** Flammable compressed gas
12.2 **NAS Hazard Rating for Bulk Water Transportation:**

Category	Rating
Fire	4
Health	
Vapor Irritant	3
Liquid or Solid Irritant	2
Poisons	3
Water Pollution	
Human Toxicity	2
Aquatic Toxicity	
Aesthetic Effect	2
Reactivity	
Other Chemicals	3
Water	0
Self-Reaction	0

12.3 **NFPA Hazard Classifications:**

Category	Classification
Health Hazard (Blue)	3
Flammability (Red)	4
Reactivity (Yellow)	0

13. PHYSICAL AND CHEMICAL PROPERTIES*

13.1 **Physical State at 15°C and 1 atm:** Gas
13.2 **Molecular Weight:** 31.1
13.3 **Boiling Point at 1 atm:** 20.3°F = −6.5°C = 266.7°K
13.4 **Freezing Point:** −134.5°F = −92.5°C = 180.7°K
13.5 **Critical Temperature:** 318°F = 159°C = 432°K
13.6 **Critical Pressure:** 1,080 psia = 73.6 atm = 7.47 MN/m^2
13.7 **Specific Gravity:** 0.693 at −6.5°C (liquid)
13.8 **Liquid Surface Tension:** 100.59 dynes/cm = 0.1006 N/m at 20°C
13.9 **Liquid-Water Interfacial Tension:** Not pertinent
13.10 **Vapor (Gas) Specific Gravity:** 1.1
13.11 **Ratio of Specific Heats of Vapor (Gas):** 1.1946
13.12 **Latent Heat of Vaporization:** 358 Btu/lb = 199 cal/g = 8.33 × 10^5 J/kg
13.13 **Heat of Combustion:** −15,000 Btu/lb = −8,340 cal/g = −34.9 × 10^6 J/kg
13.14 **Heat of Decomposition:** Not pertinent
13.15 **Heat of Solution:** Data not available
13.16 **Heat of Polymerization:** Not pertinent

*Properties apply to anhydrous material.

5. HEALTH HAZARDS (Cont'd.)

5.9 **Liquid or Solid Irritant Characteristics:** Causes smarting of the skin and first-degree burns on short exposure and may cause second-degree burns on long exposure.
5.10 **Odor Threshold:** 0.021 ppm

METHYL AMYL ACETATE

Common Synonyms			
4-Methyl-2-pentanol acetate MAAc 4-Methyl-2-pentyl acetate Hexyl acetate	Watery liquid Floats on water.	Colorless	Pleasant fruity odor

Stop discharge if possible. Keep people away.
Call fire department.
Avoid contact with liquid.
Isolate and remove discharged material.
Notify local health and pollution control agencies.

Fire	FLAMMABLE Flashback along vapor trail may occur. Vapor may explode if ignited in an enclosed area. Extinguish with dry chemical, alcohol foam, or carbon dioxide. Water may be ineffective on fire. Cool exposed containers with water.
Exposure	CALL FOR MEDICAL AID. LIQUID Irritating to skin and eyes. Remove contaminated clothing and shoes. Flush affected areas with plenty of water. IF IN EYES, hold eyelids open and flush with plenty of water.
Water Pollution	Effect of low concentrations on aquatic life is unknown. Fouling to shoreline. May be dangerous if it enters water intakes. Notify local health and wildlife officials. Notify operators of nearby water intakes.

1. RESPONSE TO DISCHARGE

(See Response Methods Handbook, CG 446-4)

Mechanical containment
Chemical and physical treatment

2. LABEL

Red

3. CHEMICAL DESIGNATIONS

3.1 **Synonyms:** Hexyl acetate
MAAc
Methylisobutylcarbinyl acetate
4-Methyl-2-pentanol, acetate
4-Methyl-2-pentyl acetate

3.2 **Coast Guard Compatibility Classification:** Ester

3.3 **Chemical Formula:** $CH_3COOCH(CH_3)CH_2CH(CH_3)_2$

(Continued on page 4)

4. OBSERVABLE CHARACTERISTICS

4.1 **Physical State (as shipped):** Liquid

4.2 **Color:** Colorless

4.3 **Odor:** Fruity; mild, pleasant; mild and nonresidual

5. HEALTH HAZARDS

5.1 **Personal Protective Equipment:** Organic canister or air pack; rubber gloves; goggles.

5.2 **Symptoms Following Exposure:** Headache, dizziness, nausea, irritation to respiratory passages. Irritates eyes.

5.3 **Treatment for Exposure:** INHALATION: remove from exposure immediately; call a physician; if breathing is irregular or has stopped, start resuscitation and administer oxygen. EYE CONTACT: flush with water for at least 15 min.

5.4 **Toxicity by Inhalation (Threshold Limit Value):** Data not available

5.5 **Short-Term Inhalation Limits:** Data not available

5.6 **Toxicity by Ingestion:** Grade 1; LD_{50} 5 to 15 g/kg

5.7 **Late Toxicity:** None

5.8 **Vapor (Gas) Irritant Characteristics:** Vapors cause moderate irritation, such that personnel will find high concentrations unpleasant. The effect is temporary.

5.9 **Liquid or Solid Irritant Characteristics:** Minimum hazard. If spilled on clothing and allowed to remain, may cause smarting and reddening of the skin.

5.10 **Odor Threshold:** Data not available

6. FIRE HAZARDS

6.1 **Flash Point:** 113°F C.C.; 110°F O.C.

6.2 **Flammable Limits in Air:** 0.9%–5.7% (calc.)

6.3 **Fire Extinguishing Agents:** Alcohol foam, carbon dioxide, or dry chemical

6.4 **Fire Extinguishing Agents Not to be Used:** Not pertinent

6.5 **Special Hazards of Combustion Products:** Not pertinent

6.6 **Behavior in Fire:** Not pertinent

6.7 **Ignition Temperature:** 510°F (calc.)

6.8 **Electrical Hazard:** Not pertinent

6.9 **Burning Rate:** Data not available

7. CHEMICAL REACTIVITY

7.1 **Reactivity with Water:** No reaction

7.2 **Reactivity with Common Materials:** No reaction

7.3 **Stability During Transport:** Stable

7.4 **Neutralizing Agents for Acids and Caustics:** Not pertinent

7.5 **Polymerization:** Not pertinent

7.6 **Inhibitor of Polymerization:** Not pertinent

8. WATER POLLUTION

8.1 **Aquatic Toxicity:** 230 ppm/24 hr/brine shrimp/TL_m

8.2 **Waterfowl Toxicity:** Data not available

8.3 **Biological Oxygen Demand (BOD):** 20% of theoretical in 5 days/freshwater

8.4 **Food Chain Concentration Potential:** None

9. SELECTED MANUFACTURERS

1. Exxon Chemical Co.
Houston, Tex. 77001
2. Shell Chemical Co.
Polymers Division
Houston, Tex. 77001
3. Union Carbide Corp.
Chemicals and Plastics Division
270 Park Ave.
New York, N. Y. 10017

10. SHIPPING INFORMATION

10.1 **Grades or Purity:** 95–99+%

10.2 **Storage Temperature:** Ambient

10.3 **Inert Atmosphere:** No requirement

10.4 **Venting:** Open (flame arrester)

11. HAZARD ASSESSMENT CODE

(See Hazard Assessment Handbook, CG 446-3)

A-T-U

12. HAZARD CLASSIFICATIONS

12.1 **Code of Federal Regulations:** Flammable Liquid

12.2 **NAS Hazard Rating for Bulk Water Transportation:**

Category	Rating
Fire	2
Health	
Vapor Irritant	2
Liquid or Solid Irritant	1
Poisons	1
Water Pollution	
Human Toxicity	1
Aquatic Toxicity	1
Aesthetic Effect	2
Reactivity	
Other Chemicals	1
Water	0
Self-Reaction	0

12.3 **NFPA Hazard Classifications:**

Category	Classification
Health Hazard (Blue)	1
Flammability (Red)	2
Reactivity (Yellow)	0

13. PHYSICAL AND CHEMICAL PROPERTIES

13.1 **Physical State at 15°C and 1 atm:** Liquid

13.2 **Molecular Weight:** 144.22

13.3 **Boiling Point at 1 atm:** 295.2°F = 146.2°C = 419.4°K

13.4 **Freezing Point:** −82.8°F = −63.8°C = 209.4°K

13.5 **Critical Temperature:** 606°F = 319°C = 592°K

13.6 **Critical Pressure:** 382 psia = 26 atm = 2.6 MN/m^2

13.7 **Specific Gravity:** 0.860 at 20°C (liquid)

13.8 **Liquid Surface Tension:** (est.) 25 dynes/cm = 0.025 N/m at 25°C

13.9 **Liquid-Water Interfacial Tension:** (est.) 40 dynes/cm = 0.04 N/m at 25°C

13.10 **Vapor (Gas) Specific Gravity:** Not pertinent

13.11 **Ratio of Specific Heats of Vapor (Gas):** 1.046

13.12 **Latent Heat of Vaporization:** 225 Btu/lb = 125 cal/g = 5.23×10^5 J/kg

13.13 **Heat of Combustion:** (est.) −14,400 Btu/lb = −8,000 cal/g = -335×10^5 J/kg

13.14 **Heat of Decomposition:** Not pertinent

13.15 **Heat of Solution:** Not pertinent

13.16 **Heat of Polymerization:** Not pertinent

3. CHEMICAL DESIGNATIONS (Cont'd.)

3.4 **IMCO/United Nations Numerical Designation:** 3.3/1233

REVISED 1978

MAA — METHYL AMYL ALCOHOL

Common Synonyms			
Isobutylmethylcarbinol 4-Methyl-2-pentanol Methylisobutylcarbinol	Oily liquid	Colorless	Mild alcohol odor
	Floats on water. Irritating vapor is produced.		

Stop discharge if possible. Keep people away.
Call fire department.
Avoid contact with liquid and vapor.
Isolate and remove discharged material.
Notify local health and pollution control agencies.

Fire	Combustible. **Extinguish with dry chemical, alcohol foam, or carbon dioxide.** **Cool exposed containers with water.**
Exposure	**CALL FOR MEDICAL AID.** VAPOR Irritating to eyes, nose and throat. If inhaled, will cause dizziness or difficult breathing. Harmful if skin is exposed. **Move to fresh air.** **If breathing has stopped, give artificial respiration.** **If breathing is difficult, give oxygen.** LIQUID Irritating to skin and eyes. Harmful if swallowed. **Remove contaminated clothing and shoes.** **Flush affected areas with plenty of water.** **IF IN EYES, hold eyelids open and flush with plenty of water.** **IF SWALLOWED and victim is CONSCIOUS, have victim drink water or milk.**
Water Pollution	Effect of low concentrations on aquatic life is unknown. Fouling to shoreline. May be dangerous if it enters water intakes. **Notify local health and wildlife officials.** **Notify operators of nearby water intakes.**

1. RESPONSE TO DISCHARGE

(See Response Methods Handbook, CG 446-4)

Mechanical containment
Chemical and physical treatment

2. LABELS

No hazard label required by Code of Federal Regulations

3. CHEMICAL DESIGNATIONS

3.1 **Synonyms:** Isobutylmethyl carbinol; Isobutyl methylmethanol; MAOH; Methylisobutyl carbinol; 4-Methyl-2-pentanol; MIC

3.2 **Coast Guard Compatibility Classification:** Alcohol

3.3 **Chemical Formula:** $(CH_3)_2CHCH_2CH(OH)CH_3$

3.4 **IMCO/United Nations Numerical Designation:** 3.3/2053

4. OBSERVABLE CHARACTERISTICS

4.1 **Physical State (as shipped):** Liquid

4.2 **Color:** Colorless

4.3 **Odor:** Sharp; mild and nonresidual

5. HEALTH HAZARDS

5.1 **Personal Protective Equipment:** Air pack or organic canister mask; rubber gloves; goggles or face shield.

5.2 **Symptoms Following Exposure:** Vapor irritates eyes and nose. May cause anesthesia. Prolonged contact with liquid causes irritation and cracking of skin, and irritates eyes.

5.3 **Treatment for Exposure:** INHALATION: remove victim to fresh air; give artificial respiration if needed; call a doctor. SKIN: flush with water. EYES: flush with water for at least 15 min.; consult a doctor.

5.4 **Toxicity by Inhalation (Threshold Limit Value):** 25 ppm

5.5 **Short-Term Inhalation Limits:** Data not available

5.6 **Toxicity by Ingestion:** Grade 2; 0.5 to 5 g/kg (rat)

5.7 **Late Toxicity:** None

5.8 **Vapor (Gas) Irritant Characteristics:** Vapors cause moderate irritation such that personnel will find high concentrations unpleasant. The effect is temporary.

5.9 **Liquid or Solid Irritant Characteristics:** Minimum hazard. If spilled on clothing and allowed to remain, may cause smarting and reddening of the skin.

5.10 **Odor Threshold:** Data not available

6. FIRE HAZARDS

6.1 **Flash Point:** 120 – 130°F O.C.; 106°F C.C.

6.2 **Flammable Limits in Air:** 1.0% – 5.5%

6.3 **Fire Extinguishing Agents:** Alcohol foam, dry chemical, or carbon dioxide

6.4 **Fire Extinguishing Agents Not to be Used:** Not pertinent

6.5 **Special Hazards of Combustion Products:** Not pertinent

6.6 **Behavior in Fire:** Not pertinent

6.7 **Ignition Temperature:** 583°F (calc.)

6.8 **Electrical Hazard:** Not pertinent

6.9 **Burning Rate:** 4.7 mm/min.

7. CHEMICAL REACTIVITY

7.1 **Reactivity with Water:** No reaction

7.2 **Reactivity with Common Materials:** No reaction

7.3 **Stability During Transport:** Stable

7.4 **Neutralizing Agents for Acids and Caustics:** Not pertinent

7.5 **Polymerization:** Not pertinent

7.6 **Inhibitor of Polymerization:** Not pertinent

8. WATER POLLUTION

8.1 **Aquatic Toxicity:** 370 ppm/24 hr/brine shrimp/TL_m

8.2 **Waterfowl Toxicity:**

8.3 **Biological Oxygen Demand (BOD):** 50% of theoretical in 5 days, freshwater

8.4 **Food Chain Concentration Potential:**

9. SELECTED MANUFACTURERS

1. Exxon Chemical Co.
   Houston, Tex. 77001
2. Shell Chemical Co.
   Industrial Chemical Division
   Houston, Tex. 77001
3. Union Carbide Corp.
   Chemicals and Plastics Division
   270 Park Ave.
   New York, N. Y. 10017

10. SHIPPING INFORMATION

10.1 **Grades or Purity:** Data not available

10.2 **Storage Temperature:** Ambient

10.3 **Inert Atmosphere:** No requirement

10.4 **Venting:** Open (flame arrester)

11. HAZARD ASSESSMENT CODE

(See Hazard Assessment Handbook, CG 446-3)

A-P-Q-T-U

12. HAZARD CLASSIFICATIONS

12.1 **Code of Federal Regulations:** Combustible liquid

12.2 **NAS Hazard Rating for Bulk Water Transportation:**

Category	Rating
Fire	2
Health	
Vapor Irritant	2
Liquid or Solid Irritant	1
Poisons	2
Water Pollution	
Human Toxicity	2
Aquatic Toxicity	2
Aesthetic Effect	2
Reactivity	
Other Chemicals	2
Water	0
Self-Reaction	0

12.3 **NFPA Hazard Classifications:**

Category	Classification
Health Hazard (Blue)	2
Flammability (Red)	2
Reactivity (Yellow)	0

13. PHYSICAL AND CHEMICAL PROPERTIES

13.1 **Physical State at 15°C and 1 atm:** Liquid

13.2 **Molecular Weight:** 102.18

13.3 **Boiling Point at 1 atm:** 269.2°F = 131.8°C = 405°K

13.4 **Freezing Point:** <-130°F = <-90°C = <183°K

13.5 **Critical Temperature:** 556°F = 291°C = 564°K

13.6 **Critical Pressure:** Not pertinent

13.7 **Specific Gravity:** 0.807 at 20°C (liquid)

13.8 **Liquid Surface Tension:** 22.8 dynes/cm = 0.0228 N/m at 20°C

13.9 **Liquid-Water Interfacial Tension:** (est.) 40 dynes/cm = 0.04 N/m at 20°C

13.10 **Vapor (Gas) Specific Gravity:** Not pertinent

13.11 **Ratio of Specific Heats of Vapor (Gas):** 1.053

13.12 **Latent Heat of Vaporization:** 162 Btu/lb = 90.1 cal/g = 3.77×10^5 J/kg

13.13 **Heat of Combustion:** -16,640 Btu/lb = -9240 cal/g = -387×10^5 J/kg

13.14 **Heat of Decomposition:** Not pertinent

13.15 **Heat of Solution:** Not pertinent

13.16 **Heat of Polymerization:** Not pertinent

NOTES

REVISED 1978

MAN — N-METHYLANILINE

Common Synonyms			
Anilinomethane N-Methylaminobenzene Methylaniline (mono) Methylphenylamine	Liquid	Yellow to light brown	Chemical Odor
	May float or sink in water.		

Avoid contact with liquid. Keep people away.
Stop discharge if possible.
Call fire department.
Isolate and remove discharged material.
Notify local health and pollution control agencies.

Fire	Combustible. POISONOUS GASES MAY BE PRODUCED IN FIRE. Wear goggles and self-contained breathing apparatus. Extinguish with dry chemicals, foam or carbon dioxide. Water may be ineffective on fire. Cool exposed containers with water.
Exposure	CALL FOR MEDICAL AID. LIQUID Irritating to skin and eyes. Harmful if swallowed. Remove contaminated clothing and shoes. Flush affected areas with plenty of water. IF IN EYES, hold eyelids open and flush with plenty of water. IF SWALLOWED and victim is CONSCIOUS, have victim drink water or milk. IF SWALLOWED and victim is UNCONSCIOUS OR HAVING CONVULSIONS, do nothing except keep victim warm.
Water Pollution	Effect of low concentrations on aquatic life is unknown. Fouling to shoreline. May be dangerous if it enters water intakes. Notify local health and wildlife officials. Notify operators of nearby water intakes.

1. RESPONSE TO DISCHARGE
(See Response Methods Handbook, CG 446-4)

Issue warning—poison, water contaminant
Restrict access
Mechanical containment
Should be removed
Chemical and physical treatment

2. LABELS

No hazard label required by Code of Federal Regulations

3. CHEMICAL DESIGNATIONS

3.1 **Synonyms:** Anilinomethane; N-Methylaminobenzene; Methylaniline (mono); Methylphenylamine
3.2 **Coast Guard Compatibility Classification:** Not listed
3.3 **Chemical Formula:** $C_6H_5NHCH_3$
3.4 **IMCO/United Nations Numerical Designation:** Not listed

4. OBSERVABLE CHARACTERISTICS

4.1 **Physical State (as shipped):** Liquid
4.2 **Color:** Yellow to light brown
4.3 **Odor:** Moderate aniline-type

5. HEALTH HAZARDS

5.1 **Personal Protective Equipment:** Approved respirator; rubber gloves; splash proof goggles
5.2 **Symptoms Following Exposure:** Inhalation causes dizziness and headache. Ingestion causes bluish discoloration (cyanosis) of lips, ear lobes, and fingernail beds. Liquid irritates eyes. Absorption through skin produces same symptoms as for ingestion.
5.3 **Treatment for Exposure:** INHALATION: remove victim to fresh air and call a physician at once; administer oxygen until physician arrives. INGESTION: give large amount of water; get medical attention at once. EYES or SKIN: flush with plenty of water for at least 15 min.; if cyanosis is present, shower with soap and warm water, with special attention to scalp and finger nails; remove any contaminated clothing.
5.4 **Toxicity by Inhalation (Threshold Limit Value):** Data not available
5.5 **Short-Term Inhalation Limits:** Data not available
5.6 **Toxicity by Ingestion:** Data not available
5.7 **Late Toxicity:** Data not available
5.8 **Vapor (Gas) Irritant Characteristics:** Data not available
5.9 **Liquid or Solid Irritant Characteristics:** Data not available
5.10 **Odor Threshold:** Data not available

6. FIRE HAZARDS

6.1 **Flash Point:** 175°F C.C.
6.2 **Flammable Limits in Air:** Data not available
6.3 **Fire Extinguishing Agents:** Dry chemical, foam, carbon dioxide
6.4 **Fire Extinguishing Agents Not to be Used:** Water may be ineffective.
6.5 **Special Hazards of Combustion Products:** Toxic vapors are generated when heated.
6.6 **Behavior in Fire:**
6.7 **Ignition Temperature:** Data not available
6.8 **Electrical Hazard:** Data not available
6.9 **Burning Rate:** 3.65 mm/min.

7. CHEMICAL REACTIVITY

7.1 **Reactivity with Water:** No reaction
7.2 **Reactivity with Common Materials:** May attack some forms of plastics
7.3 **Stability During Transport:** Stable
7.4 **Neutralizing Agents for Acids and Caustics:** Not pertinent
7.5 **Polymerization:** Not pertinent
7.6 **Inhibitor of Polymerization:** Not pertinent

8. WATER POLLUTION

8.1 **Aquatic Toxicity:** Data not available
8.2 **Waterfowl Toxicity:** Data not available
8.3 **Biological Oxygen Demand (BOD):** Data not available
8.4 **Food Chain Concentration Potential:** None

9. SELECTED MANUFACTURERS

1. American Cyanamid Co.
   Bound Brook, N. J. 08805
2. Eastman Organic Chemicals
   Rochester, N. Y. 14650
3. Aldrich Chemical Co.
   940 W. Saint Paul Ave.
   Milwaukee, Wis. 53233

10. SHIPPING INFORMATION

10.1 **Grades or Purity:** Technical; Pure, 99+%
10.2 **Storage Temperature:** Ambient
10.3 **Inert Atmosphere:** No requirement
10.4 **Venting:** Open

11. HAZARD ASSESSMENT CODE
(See Hazard Assessment Handbook, CG 446-3)

A-T-U-X-Y

12. HAZARD CLASSIFICATIONS

12.1 **Code of Federal Regulations:** Not Listed
12.2 **NAS Hazard Rating for Bulk Water Transportation:** Not listed
12.3 **NFPA Hazard Classifications:** Not listed

13. PHYSICAL AND CHEMICAL PROPERTIES

13.1 **Physical State at 15°C and 1 atm:** Liquid
13.2 **Molecular Weight:** 107.2
13.3 **Boiling Point at 1 atm:** 384.6°F = 195.9°C = 469.1°K
13.4 **Freezing Point:** −71°F = −57°C = 216°K
13.5 **Critical Temperature:** 802°F = 428°C = 701°K
13.6 **Critical Pressure:** 754 psia = 51.3 atm = 5.20 MN/m^2
13.7 **Specific Gravity:** 0.989 at 20°C (liquid)
13.8 **Liquid Surface Tension:** 39.6 dynes/cm = 0.0396 N/m at 20°C
13.9 **Liquid-Water Interfacial Tension:** Data not available
13.10 **Vapor (Gas) Specific Gravity:** 3.70
13.11 **Ratio of Specific Heats of Vapor (Gas):** Not pertinent
13.12 **Latent Heat of Vaporization:** 180 Btu/lb = 100 cal/g = 4.20×10^5 J/kg
13.13 **Heat of Combustion:** −16,350 Btu/lb = −9,085 cal/g = -380.1×10^5 J/kg
13.14 **Heat of Decomposition:** Not pertinent
13.15 **Heat of Solution:** Not pertinent
13.16 **Heat of Polymerization:** Not pertinent

NOTES

MTB	METHYL BROMIDE

Common Synonyms			
Bromomethane Embafume	Liquefied gas	Colorless	Odorless to sweet odor
	Sinks and boils in water. Poisonous vapor cloud is formed. Boiling point is 39° F.		

AVOID CONTACT WITH LIQUID AND VAPOR. Keep people away.
Wear goggles, self-contained breathing apparatus, and protective impervious overclothing (including gloves).
Stop discharge if possible. Call fire department.
Evacuate area in case of large leaks.
Stay upwind and use water spray to "knock down" vapor.
Isolate and remove discharged material.
Notify local health and pollution control agencies.

Fire	Combustible. POISONOUS AND IRRITATING GASES ARE PRODUCED IN FIRE. Wear goggles, self-contained breathing apparatus, and protective overclothing (including gloves). EXTINGUISH WITH WATER, FOAM, OR CARBON DIOXIDE. Cool exposed containers with water.
Exposure	CALL FOR MEDICAL AID. VAPOR POISONOUS IF INHALED. Irritating to eyes. Move to fresh air. If breathing has stopped, give artificial respiration (but NOT mouth-to-mouth). If breathing is difficult, give oxygen. LIQUID Will burn skin and eyes. Harmful if swallowed. Remove contaminated clothing and shoes. Flush affected areas with plenty of water. DO NOT RUB AFFECTED AREAS. IF IN EYES, hold eyelids open and flush with plenty of water. IF SWALLOWED and victim is CONSCIOUS, have victim drink water or milk. DO NOT INDUCE VOMITING.
Water Pollution	Not harmful to aquatic life. May be dangerous if it enters water intakes. Notify local health and wildlife officials. Notify operators of nearby water intakes.

1. RESPONSE TO DISCHARGE

(See Response Methods Handbook, CG 446-4)

Issue warning—poison
Restrict access

2. LABEL

POISON

3. CHEMICAL DESIGNATIONS

3.1 **Synonyms:** Bromomethane
Embafume
M-B-C Fumigant
Monobromomethane

3.2 **Coast Guard Compatibility Classification:** Halogenated hydrocarbon

3.3 **Chemical Formula:** CH_3Br

3.4 **IMCO/United Nations Numerical Designation:** 2.0/1062

4. OBSERVABLE CHARACTERISTICS

4.1 **Physical State (as shipped):** Liquefied gas

4.2 **Color:** Colorless

4.3 **Odor:** Relatively odorless; sweet, chloroform-like

5. HEALTH HAZARDS

5.1 **Personal Protective Equipment:** Self-contained breathing apparatus; goggles.

5.2 **Symptoms Following Exposure:** Inhalation of vapor causes lung congestion and pulmonary edema. Higher concentrations cause rapid narcosis and death. Contact with liquid irritates eyes and burns skin.

5.3 **Treatment for Exposure:** INHALATION: remove victim to fresh air; give artificial respiration if needed. SKIN OR EYES: flush with water for at least 15 min.

5.4 **Toxicity by Inhalation (Threshold Limit Value):** 15 ppm

5.5 **Short-Term Inhalation Limits:** 20 ppm for 5 min.

5.6 **Toxicity by Ingestion:** Data not available

5.7 **Late Toxicity:** Data not available

5.8 **Vapor (Gas) Irritant Characteristics:** Vapor is moderately irritating such that personnel will not usually tolerate moderate or high vapor concentrations.

5.9 **Liquid or Solid Irritant Characteristics:** Fairly severe skin irritant; may cause pain and second-degree burns after a few minutes' contact.

5.10 **Odor Threshold:** Odorless

6. FIRE HAZARDS

6.1 **Flash Point:** Practically not flammable

6.2 **Flammable Limits in Air:** 10%—15%

6.3 **Fire Extinguishing Agents:** Not pertinent

6.4 **Fire Extinguishing Agents Not to be Used:** Not pertinent

6.5 **Special Hazards of Combustion Products:** Toxic and irritating gases are generated when exposed to fire or heat.

6.6 **Behavior in Fire:** Containers may explode

6.7 **Ignition Temperature:** 999°F

6.8 **Electrical Hazard:** Not pertinent

6.9 **Burning Rate:** Not pertinent

7. CHEMICAL REACTIVITY

7.1 **Reactivity with Water:** No reaction

7.2 **Reactivity with Common Materials:** No reaction

7.3 **Stability During Transport:** Stable

7.4 **Neutralizing Agents for Acids and Caustics:** Not pertinent

7.5 **Polymerization:** Not pertinent

7.6 **Inhibitor of Polymerization:** Not pertinent

8. WATER POLLUTION

8.1 **Aquatic Toxicity:** None

8.2 **Waterfowl Toxicity:** None

8.3 **Biological Oxygen Demand (BOD):** None

8.4 **Food Chain Concentration Potential:** None

9. SELECTED MANUFACTURERS

1. Dow Chemical Co.
Midland, Mich. 48640
2. Great Lakes Chemical Corp.
West Lafayette, Ind. 47906
3. Northwest Industries, Inc.
Michigan Chemical Corp.
351 E. Ohio St.
Chicago, Ill. 60611

10. SHIPPING INFORMATION

10.1 **Grades or Purity:** Commercial: not less than 99.5%

10.2 **Storage Temperature:** Ambient

10.3 **Inert Atmosphere:** No requirement

10.4 **Venting:** Safety relief

11. HAZARD ASSESSMENT CODE

(See Hazard Assessment Handbook, CG 446-3)

A-B-C-I-J

12. HAZARD CLASSIFICATIONS

12.1 **Code of Federal Regulations:** Poisonous liquid or solid, Class B

12.2 **NAS Hazard Rating for Bulk Water Transportation:**

Category	Rating
Fire	1
Health	
Vapor Irritant	3
Liquid or Solid Irritant	3
Poisons	4
Water Pollution	
Human Toxicity	0
Aquatic Toxicity	1
Aesthetic Effect	2
Reactivity	
Other Chemicals	1
Water	0
Self-Reaction	0

12.3 **NFPA Hazard Classifications:**

Category	Classification
Health Hazard (Blue)	3
Flammability (Red)	1
Reactivity (Yellow)	0

13. PHYSICAL AND CHEMICAL PROPERTIES

13.1 **Physical State at 15°C and 1 atm:** Gas

13.2 **Molecular Weight:** 94.95

13.3 **Boiling Point at 1 atm:** 38.5°F = 3.6°C = 276.8°K

13.4 **Freezing Point:** −135°F = −93°C = 180°K

13.5 **Critical Temperature:** 376°F = 191°C = 464°K

13.6 **Critical Pressure:** Not pertinent

13.7 **Specific Gravity:** 1.68 at 20°C (liquid)

13.8 **Liquid Surface Tension:** 24.5 dynes/cm = 0.0245 N/m at 15°C

13.9 **Liquid-Water Interfacial Tension:** Not pertinent

13.10 **Vapor (Gas) Specific Gravity:** 3.3

13.11 **Ratio of Specific Heats of Vapor (Gas):** 1.247

13.12 **Latent Heat of Vaporization:** 108 Btu/lb = 59.7 cal/g = 2.50×10^5 J/kg

13.13 **Heat of Combustion:** −3188 Btu/lb = −1771 cal/g = -74.15×10^5 J/kg

13.14 **Heat of Decomposition:** Not pertinent

13.15 **Heat of Solution:** Not pertinent

13.16 **Heat of Polymerization:** Not pertinent

NOTES

REVISED 1978

MBK | METHYL n-BUTYL KETONE

Common Synonyms 2-Hexanone n-Butyl methyl ketone	Liquid / Clear / Disagreeable Odor Floats and mixes with water.

Shut off ignition sources. Call fire department.
Avoid contact with liquid and vapor. Keep people away.
Stop discharge if possible.
Stay upwind. Use water spray to "knock down" vapor.
Isolate and remove discharged material.
Notify local health and pollution control agencies.

Fire

FLAMMABLE.
Containers may explode in fire.
Flashback along vapor trail may occur.
Vapor may explode if ignited in an enclosed area.
Extinguish with dry chemicals, alcohol foam, or carbon dioxide.
Water may be ineffective on fire.
Cool exposed containers with water.

Exposure

CALL FOR MEDICAL AID.

VAPOR
Irritating to eyes, nose and throat.
Harmful if inhaled.
If in eyes, hold eyelids open and flush with plenty of water.
If breathing has stopped, give artificial respiration.
If breathing is difficult, give oxygen.

LIQUID
Irritating to skin and eyes.
Harmful if swallowed.
Remove contaminated clothing and shoes.
Flsuh affected areas with plenty of water.
IF IN EYES, hold eyelids open and flush with plenty of water.
IF SWALLOWED and victim is CONSCIOUS, have victim drink water of milk and have victim induce vomiting.
IF SWALLOWED and victim is UNCONSCIOUS OR HAVING CONVULSIONS, do nothing except keep victim warm.

Water Pollution

Effect of low concentrations on aquatic life is unknown.
May be dangerous if it enters water intakes.
Notify local health and wildlife officials.
Notify operators of nearby water intakes.

1. RESPONSE TO DISCHARGE

(See Response Methods Handbook, CG 446-4)

Issue warning—air contaminant, water contaminant, high flammability
Restrict access
Mechanical containment
Should be removed
Chemical and physical treatment

2. LABEL

Red

3. CHEMICAL DESIGNATIONS

3.1 **Synonyms:** n-Butyl methyl ketone; 2-Hexanone
3.2 **Coast Guard Compatibility Classification:** Not listed
3.3 **Chemical Formula:** $CH_3(CH_2)_3COCH_3$
3.4 **IMCO/United Nations Numerical Designation:** Not listed

4. OBSERVABLE CHARACTERISTICS

4.1 **Physical State (as shipped):** Liquid
4.2 **Color:** Clear
4.3 **Odor:** Characteristic; strong, disagreeable odor resembling acetone

5. HEALTH HAZARDS

5.1 **Personal Protective Equipment:** Protective gloves; goggles or face shield; approved respirator (for major spills)
5.2 **Symptoms Following Exposure:** Inhalation of high concentrations of vapor may result in narcosis; peripheral neuropathy may develop. Ingestion of large amounts may cause some systemic injury. Contact with eyes causes mild to moderate irritation. Liquid irritates skin; prolonged or repeated contact may cause defatting of the skin with resultant dermatitis.
5.3 **Treatment for Exposure:** INHALATION: move to uncontaminated atmosphere and treat symptomatically; alert physician to possible development of peripheral neuropathy. INGESTION: give large amount of water and induce vomiting. EYES: irrigate immediately and thoroughly with water for 15 min. and get medical attention. SKIN: flush exposed areas thoroughly with water.
5.4 **Toxicity by Inhalation (Threshold Limit Value):** 100 ppm
5.5 **Short-Term Inhalation Limits:** Data not available
5.6 **Toxicity by Ingestion:** Grade 2; oral LD_{50} = 2,590 mg/kg (rat)
5.7 **Late Toxicity:** Peripheral neuropathy in experimental animals and man (disease of motor and/or sensory nerves)
5.8 **Vapor (Gas) Irritant Characteristics:** Data not available
5.9 **Liquid or Solid Irritant Characteristics:** Data not available
5.10 **Odor Threshold:** Data not available

6. FIRE HAZARDS

6.1 **Flash Point:** 83°F O.C.; 77°F C.C.
6.2 **Flammable Limits in Air:** 1.3% 8.0%
6.3 **Fire Extinguishing Agents:** Dry chemical, "alcohol" foam, carbon dioxide
6.4 **Fire Extinguishing Agents Not to be Used:** Water may be ineffective.
6.5 **Special Hazards of Combustion Products:**
6.6 **Behavior in Fire:**
6.7 **Ignition Temperature:** 795°F
6.8 **Electrical Hazard:** Data not available
6.9 **Burning Rate:** 4.8 mm/min.

7. CHEMICAL REACTIVITY

7.1 **Reactivity with Water:** No reaction
7.2 **Reactivity with Common Materials:**
7.3 **Stability During Transport:** Stable
7.4 **Neutralizing Agents for Acids and Caustics:** Not pertinent
7.5 **Polymerization:** Not pertinent
7.6 **Inhibitor of Polymerization:** Not pertinent

8. WATER POLLUTION

8.1 **Aquatic Toxicity:** Data not available
8.2 **Waterfowl Toxicity:** Data not available
8.3 **Biological Oxygen Demand (BOD):** Data not available
8.4 **Food Chain Concentration Potential:** None

9. SELECTED MANUFACTURERS

1. Eastman Chemical Products, Inc.
Kingsport, Tenn. 37662
2. Aldrich Chemical Co.
940 W. Saint Paul Ave.
Milwaukee, Wis. 53233
3. Pfaltz and Bauer, Inc.
375 Fairfield Ave.
Stamford, Conn. 06902

10. SHIPPING INFORMATION

10.1 **Grades or Purity:** Commercial, 95%; Pure, 99%
10.2 **Storage Temperature:** Ambient
10.3 **Inert Atmosphere:** No requirement
10.4 **Venting:** Open (flame arrester)

11. HAZARD ASSESSMENT CODE

(See Hazard Assessment Handbook, CG 446-3)

A-P-Q-T-U

12. HAZARD CLASSIFICATIONS

12.1 **Code of Federal Regulations:** Not listed
12.2 **NAS Hazard Rating for Bulk Water Transportation:** Not listed
12.3 **NFPA Hazard Classifications:**

Category	Classification
Health Hazard (Blue)	2
Flammability (Red)	3
Reactivity (Yellow)	0

13. PHYSICAL AND CHEMICAL PROPERTIES

13.1 **Physical State at 15°C and 1 atm:** Liquid
13.2 **Molecular Weight:** 100.16
13.3 **Boiling Point at 1 atm:** 261°F = 127°C = 400°K
13.4 **Freezing Point:** −70.4°F = −56.9°C = 216.3°K
13.5 **Critical Temperature:** Not pertinent
13.6 **Critical Pressure:** Not pertinent
13.7 **Specific Gravity:** 0.812 at 20°C (liquid)
13.8 **Liquid Surface Tension:** 25.49 dynes/cm = 0.02549 N/m at 20°C
13.9 **Liquid-Water Interfacial Tension:** 9.73 dynes/cm = 0.00973 N/m at 20°C
13.10 **Vapor (Gas) Specific Gravity:** 3.5
13.11 **Ratio of Specific Heats of Vapor (Gas):** Not pertinent
13.12 **Latent Heat of Vaporization:** 148 Btu/lb = 82 cal/g = 3.4×10^5 J/kg
13.13 **Heat of Combustion:** −16,100 Btu/lb = −8,940 cal/g = -374×10^5 J/kg
13.14 **Heat of Decomposition:** Not pertinent
13.15 **Heat of Solution:** Not pertinent
13.16 **Heat of Polymerization:** Not pertinent

NOTES

MTC	# METHYL CHLORIDE

Common Synonyms	
Chloromethane	Liquefied compressed gas Colorless Odorless or sweet odor Floats and boils on water. Flammable, visible vapor cloud is formed.

Avoid contact with liquid and vapor. Keep people away.
Wear goggles and self-contained breathing apparatus.
Stop discharge if possible.
Shut off ignition sources and call fire department.
Stay upwind and use water spray to "knock down" vapor.
Isolate and remove discharged material.
Notify local health and pollution control agencies.

Fire	FLAMMABLE. POISONOUS GASES ARE PRODUCED IN FIRE. Flashback along vapor trail may occur. Vapor may explode if ignited in an enclosed area. Wear goggles and self-contained breathing apparatus. Stop discharge if possible. Cool exposed containers and protect men effecting shutoff with water. Let fire burn.
Exposure	CALL FOR MEDICAL AID. VAPOR Not irritating to eyes, nose or throat. If inhaled, will cause nausea, vomiting, headache, difficult breathing, or loss of consciousness. Move to fresh air. If breathing has stopped, give artificial respiration. If breathing is difficult, give oxygen. LIQUID Will cause frostbite. Flush affected areas with plenty of water. DO NOT RUB AFFECTED AREAS.
Water Pollution	Not harmful to aquatic life.

1. RESPONSE TO DISCHARGE

(See Response Methods Handbook, CG 446-4)

Issue warning—high flammability, air contaminant
Restrict access
Evacuate area

2. LABEL

Red

3. CHEMICAL DESIGNATIONS

3.1 **Synonyms:** Artic
Chloromethane
3.2 **Coast Guard Compatibility Classification:** Halogenated hydrocarbon
3.3 **Chemical Formula:** CH_3Cl
3.4 **IMCO/United Nations Numerical Designation:** 2.0/1063

4. OBSERVABLE CHARACTERISTICS

4.1 **Physical State (as shipped):** Liquefied gas
4.2 **Color:** Colorless
4.3 **Odor:** Faint, sweet, non-irritating; faint ether-like

5. HEALTH HAZARDS

5.1 **Personal Protective Equipment:** Approved canister mask; leather or vinyl gloves; goggles or face shield.
5.2 **Symptoms Following Exposure:** Inhalation causes nausea, vomiting, weakness, headache, emotional disturbances; high concentrations cause mental confusion, eye disturbances, muscular tremors, cyanosis, convulsions. Contact of liquid with skin may cause frostbite.
5.3 **Treatment for Exposure:** Remove to fresh air. Call a doctor and have patient hospitalized for observation of slowly developing symptoms.
5.4 **Toxicity by Inhalation (Threshold Limit Value):** 100 ppm
5.5 **Short-Term Inhalation Limits:** 100 ppm for 5 min.
5.6 **Toxicity by Ingestion:** Not pertinent
5.7 **Late Toxicity:** None
5.8 **Vapor (Gas) Irritant Characteristics:** Vapors are nonirritating to the eyes and throat.
5.9 **Liquid or Solid Irritant Characteristics:** No appreciable hazard. Practically harmless to the skin because it evaporates quickly. May cause frostbite.
5.10 **Odor Threshold:** Data not available

6. FIRE HAZARDS

6.1 **Flash Point:** $<$ 32°F C.C.
6.2 **Flammable Limits in Air:** 8.1%–17.2%
6.3 **Fire Extinguishing Agents:** Dry chemical or carbon dioxide. Stop flow of gas.
6.4 **Fire Extinguishing Agents Not to be Used:** Not pertinent
6.5 **Special Hazards of Combustion Products:** Toxic and irritating gases are generated in fires.
6.6 **Behavior in Fire:** Containers may explode
6.7 **Ignition Temperature:** 1170°F
6.8 **Electrical Hazard:** Not pertinent
6.9 **Burning Rate:** 2.2 mm/min.

7. CHEMICAL REACTIVITY

7.1 **Reactivity with Water:** No reaction
7.2 **Reactivity with Common Materials:** Reacts with zinc, aluminum, magnesium, and their alloys; reaction is not violent.
7.3 **Stability During Transport:** Stable
7.4 **Neutralizing Agents for Acids and Caustics:** Not pertinent
7.5 **Polymerization:** Not pertinent
7.6 **Inhibitor of Polymerization:** Not pertinent

8. WATER POLLUTION

8.1 **Aquatic Toxicity:** None
8.2 **Waterfowl Toxicity:** None
8.3 **Biological Oxygen Demand (BOD):** None
8.4 **Food Chain Concentration Potential:** None

9. SELECTED MANUFACTURERS

1. Ancon Chemical Corp.
Westlake, La. 70669
2. Dow Chemical Co.
Midland, Mich. 48640
3. Ethyl Corp.
Industrial Chemicals Division
451 Florida St.
Baton Rouge, La. 70801

10. SHIPPING INFORMATION

10.1 **Grades or Purity:** Technical grade; "Artic" refrigerant grade
10.2 **Storage Temperature:** Ambient
10.3 **Inert Atmosphere:** No requirement
10.4 **Venting:** Safety relief

11. HAZARD ASSESSMENT CODE

(See Hazard Assessment Handbook, CG 446-3)

A-B-C-D-E-F-G

12. HAZARD CLASSIFICATIONS

12.1 **Code of Federal Regulations:** Flammable compressed gas
12.2 **NAS Hazard Rating for Bulk Water Transportation:**

Category	Rating
Fire	4
Health	
Vapor Irritant	0
Liquid or Solid Irritant	0
Poisons	2
Water Pollution	
Human Toxicity	0
Aquatic Toxicity	1
Aesthetic Effect	0
Reactivity	
Other Chemicals	1
Water	0
Self-Reaction	0

12.3 **NFPA Hazard Classifications:**

Category	Classification
Health Hazard (Blue)	2
Flammability (Red)	4
Reactivity (Yellow)	0

13. PHYSICAL AND CHEMICAL PROPERTIES

13.1 **Physical State at 15°C and 1 atm:** Gas
13.2 **Molecular Weight:** 50.49
13.3 **Boiling Point at 1 atm:** -11.6°F = -24.2°C = 249°K
13.4 **Freezing Point:** -143.9°F = -97.7°C = 175.5°K
13.5 **Critical Temperature:** 290.5°F = 143.6°C = 416.8°K
13.6 **Critical Pressure:** 969 psia = 65.9 atm = 6.68 MN/m^2
13.7 **Specific Gravity:** 0.997 at -24°C (liquid)
13.8 **Liquid Surface Tension:** 16.2 dynes/cm = 0.0162 N/m at 20°C
13.9 **Liquid-Water Interfacial Tension:** (est.) 50 dynes/cm = 0.05 N/m at -24°C
13.10 **Vapor (Gas) Specific Gravity:** 1.7
13.11 **Ratio of Specific Heats of Vapor (Gas):** 1.259
13.12 **Latent Heat of Vaporization:** 182.3 Btu/lb = 101.3 cal/g = 4.241×10^5 J/kg
13.13 **Heat of Combustion:** -5290 Btu/lb = -2939 cal/g = -123.1×10^5 J/kg
13.14 **Heat of Decomposition:** Not pertinent
13.15 **Heat of Solution:** Not pertinent
13.16 **Heat of Polymerization:** Not pertinent

NOTES

REVISED 1978

METHYL CHLOROFORMATE

Common Synonyms			
Chloroformic acid, methyl ester Methyl chlorocarbonate	Liquid	Colorless to light yellow	Unpleasant odor
	Sinks and reacts in water. Flammable, irritating vapor is produced.		

Shut off ignition sources. Call fire department.
Avoid contact with liquid and vapor. Keep people away.
Stop discharge if possible.
Stay upwind. Use water spray to "knock down" vapor.
Isolate and remove discharged material.
Notify local health and pollution control agencies.

Fire

FLAMMABLE.
POISONOUS GASES MAY BE PRODUCED IN FIRE.
Containers may explode in fire.
Flashback along vapor trail may occur.
Vapor may explode if ignited in an enclosed area.
Extinguish with water, dry chemicals, foam, or carbon dioxide.
Cool exposed containers with water.

Exposure

CALL FOR MEDICAL AID.

VAPOR
Irritating to eyes, nose and throat.
If inhaled will cause difficult breathing.
Move victim to fresh air.
If breathing has stopped, give artificial respiration.
If breathing is difficult, give oxygen.

LIQUID
POISONOUS IF SWALLOWED.
Will burn skin and eyes.
Remove contaminated clothing and shoes.
Flush affected areas with plenty of water.
IF IN EYES, hold eyelids open and flush with plenty of water.
IF SWALLOWED and victim is CONSCIOUS, have victim drink water or milk.
DO NOT INDUCE VOMITING.

Water Pollution

Effect of low concentrations on aquatic life is unknown.
May be dangerous if it enters water intakes.
Notify local health and wildlife officials.
Notify operators of nearby water intakes.

1. RESPONSE TO DISCHARGE

(See Response Methods Handbook, CG 446-4)

Issue warning—corrosive, poison, high flammability
Restrict access
Disperse and flush

2. LABELS

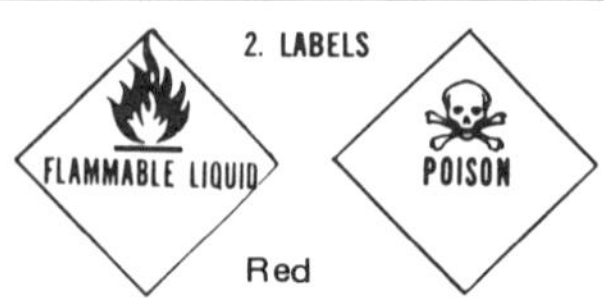

3. CHEMICAL DESIGNATIONS

3.1 **Synonyms:** Chlorocarbonic acid, methyl ester; Chloroformic acid, methyl ester; Methyl chlorocarbonate

3.2 **Coast Guard Compatibility Classification:** Not applicable

3.3 **Chemical Formula:** $ClCOOCH_3$

3.4 **IMCO/United Nations Numerical Designation:** 3.2/1238

4. OBSERVABLE CHARACTERISTICS

4.1 **Physical State (as shipped):** Liquid

4.2 **Color:** Colorless to light yellow

4.3 **Odor:** Acrid

5. HEALTH HAZARDS

5.1 **Personal Protective Equipment:** Acid- or organic-canister mask or self-contained breathing apparatus; goggles or face shield; plastic gloves

5.2 **Symptoms Following Exposure:** Inhalation of vapor irritates nose and throat and can cause delayed pulmonary edema. Liquid irritates eyes and causes severe skin burns if allowed to remain. Ingestion causes burns of mouth and stomach.

5.3 **Treatment for Exposure:** INHALATION: remove victim from exposure; if breathing stops, administer artificial respiration; call physician. EYES: irrigate with copious amounts of water for at least 15 min.; call physician if needed. SKIN: flush with water for 15 min.; get medical attention for burns. INGESTION: give large amounts of water; do NOT induce vomiting; get medical attention.

5.4 **Toxicity by Inhalation (Threshold Limit Value):** Data not available

5.5 **Short-Term Inhalation Limits:** Data not available

5.6 **Toxicity by Ingestion:** Grade 4; oral $LD_{50} < 50$ mg/kg (rat)

5.7 **Late Toxicity:** Data not available

5.8 **Vapor (Gas) Irritant Characteristics:** Data not available

5.9 **Liquid or Solid Irritant Characteristics:** Data not available

5.10 **Odor Threshold:** Data not available

6. FIRE HAZARDS

6.1 **Flash Point:** 76°F O.C., 73°F C.C.

6.2 **Flammable Limits in Air:** LFL = 6.7%

6.3 **Fire Extinguishing Agents:** Water, dry chemical, foam, carbon dioxide

6.4 **Fire Extinguishing Agents Not to be Used:** Not pertinent

6.5 **Special Hazards of Combustion Products:** Irritating and toxic hydrogen chloride and phosgene may be formed.

6.6 **Behavior in Fire:** Vapor is heavier than air and may travel a considerable distance to a source of ignition and flash back.

6.7 **Ignition Temperature:** Data not available

6.8 **Electrical Hazard:** Data not available

6.9 **Burning Rate:** 2.0 mm/min.

7. CHEMICAL REACTIVITY

7.1 **Reactivity with Water:** Reacts slowly, evolving hydrogen chloride (hydrochloric acid). Reaction can be hazardous if water is hot.

7.2 **Reactivity with Common Materials:** Corrodes rubber

7.3 **Stability During Transport:** Stable

7.4 **Neutralizing Agents for Acids and Caustics:** Flood with water, rinse with sodium bicarbonate or lime solution.

7.5 **Polymerization:** Not pertinent

7.6 **Inhibitor of Polymerization:** Not pertinent

8. WATER POLLUTION

8.1 **Aquatic Toxicity:** Data not available

8.2 **Waterfowl Toxicity:** Data not available

8.3 **Biological Oxygen Demand (BOD):** Data not available

8.4 **Food Chain Concentration Potential:** None

9. SELECTED MANUFACTURERS

1. Chemetron Corporation
   Chemicals Group
   393 Seventh Avenue
   New York, N. Y. 10001
2. Air Products and Chemicals, Inc.
   Specialty Chemicals Department
   P. O. Box 538
   Allentown, Pa. 18105
3. Aldrich Chemical Co.
   940 West St. Paul Avenue
   Milwaukee, Wis. 53233

10. SHIPPING INFORMATION

10.1 **Grades or Purity:** 97+%

10.2 **Storage Temperature:** Ambient

10.3 **Inert Atmosphere:** No requirement

10.4 **Venting:** Pressure-vacuum

11. HAZARD ASSESSMENT CODE

(See Hazard Assessment Handbook, CG 446-3)

A-O-X-Y

12. HAZARD CLASSIFICATIONS

12.1 **Code of Federal Regulations:** Flammable Liquid

12.2 **NAS Hazard Rating for Bulk Water Transportation:** Not listed

12.3 **NFPA Hazard Classifications:** Not listed

13. PHYSICAL AND CHEMICAL PROPERTIES

13.1 **Physical State at 15°C and 1 atm:** Liquid

13.2 **Molecular Weight:** 94.5

13.3 **Boiling Point at 1 atm:** 160°F = 71°C = 344°K

13.4 **Freezing Point:** <-114°F = <-81°C = <192°K

13.5 **Critical Temperature:** Not pertinent

13.6 **Critical Pressure:** Not pertinent

13.7 **Specific Gravity:** 1.22 at 20°C (liquid)

13.8 **Liquid Surface Tension:** (est.) 26 dynes/cm = 0.026 N/m at 20°C

13.9 **Liquid-Water Interfacial Tension:** Not pertinent

13.10 **Vapor (Gas) Specific Gravity:** 3.25

13.11 **Ratio of Specific Heats of Vapor (Gas):** 1.1544

13.12 **Latent Heat of Vaporization:** (est.) 153 Btu/lb = 85 cal/g = 3.6×10^5 J/kg

13.13 **Heat of Combustion:** –4,690 Btu/lb = –2,600 cal/g = -109×10^5 J/kg

13.14 **Heat of Decomposition:** Not pertinent

13.15 **Heat of Solution:** Data not available

13.16 **Heat of Polymerization:** Not pertinent

NOTES

MCT	METHYLCYCLOPENTADIENYLMANGANESE TRICARBONYL

Common Synonyms			
Combustion Improver C-12	Liquid	Yellow to dark orange	Faint pleasant odor
	Sinks in water.		

AVOID CONTACT WITH LIQUID. KEEP PEOPLE AWAY.
Wear rubber overclothing (including gloves).
Stop discharge if possible.
Call fire department.
Isolate and remove discharged material.
Notify local health and pollution control agencies.

Fire

Combustible.
POISONOUS GASES ARE PRODUCED IN FIRE.
Wear goggles and self-contained breathing apparatus.
Extinguish with water, dry chemicals, foam, or carbon dioxide.

Exposure

CALL FOR MEDICAL AID.

LIQUID
POISONOUS IF SWALLOWED OR IF SKIN IS EXPOSED.
If swallowed will cause loss of consciousness.
Remove contaminated clothing and shoes.
Flush affected areas with plenty of water.
IF IN EYES, hold eyelids open and flush with plenty of water.
IF SWALLOWED and victim is CONSCIOUS, have victim drink water or milk and have victim induce vomiting.
IF SWALLOWED and victim is UNCONSCIOUS OR HAVING CONVULSIONS, do nothing except keep victim warm.

Water Pollution

Effect of low concentrations on aquatic life is unknown.
May be dangerous if it enters water intakes.
Notify local health and wildlife officials.
Notify operators of nearby water intakes.

1. RESPONSE TO DISCHARGE
(See Response Methods Handbook, CG 446-4)

Issue warning—poison, water contaminant
Restrict access
Should be removed
Chemical and physical treatment

2. LABEL

3. CHEMICAL DESIGNATIONS

3.1 **Synonyms:** Combustion Improver C12
3.2 **Coast Guard Compatibility Classification:** To be developed
3.3 **Chemical Formula:** $C_9H_7O_3Mn$
3.4 **IMCO/United Nations Numerical Designation:** 6.1/—

4. OBSERVABLE CHARACTERISTICS

4.1 **Physical State (as shipped):** Liquid
4.2 **Color:** Straw to dark orange
4.3 **Odor:** Faint, pleasant, herbaceous

5. HEALTH HAZARDS

5.1 **Personal Protective Equipment:** Organic vapor canister mask; rubber gloves and apron; protective goggles or face shield.
5.2 **Symptoms Following Exposure:** Inhalation, ingestion, or skin contact affect central nervous system, causing convulsions, respiratory depression, cyanosis, and coma. Liquid irritates eyes.
5.3 **Treatment for Exposure:** Get medical attention following all exposures to this compound. INHALATION: remove victim from exposure; administer artificial respiration if necessary. EYES: flush with plenty of water for at least 15 min. SKIN: wash well with soap and water. INGESTION: induce vomiting.
5.4 **Toxicity by Inhalation (Threshold Limit Value):** 0.1 ppm (as manganese)
5.5 **Short-Term Inhalation Limits:** Data not available
5.6 **Toxicity by Ingestion:** Grade 4; oral LD_{50} = 23 mg/kg (rat)
5.7 **Late Toxicity:** Data not available
5.8 **Vapor (Gas) Irritant Characteristics:** Data not available
5.9 **Liquid or Solid Irritant Characteristics:** Data not available
5.10 **Odor Threshold:** Data not available

6. FIRE HAZARDS

6.1 **Flash Point:** > 200°F C.C.
6.2 **Flammable Limits in Air:** Data not available
6.3 **Fire Extinguishing Agents:** Dry chemical, foam, water spray, carbon dioxide
6.4 **Fire Extinguishing Agents Not to be Used:** Not pertinent
6.5 **Special Hazards of Combustion Products:** Toxic vapors are formed in a fire.
6.6 **Behavior in Fire:** Not pertinent
6.7 **Ignition Temperature:** Data not available
6.8 **Electrical Hazard:** Data not available
6.9 **Burning Rate:** Data not available

7. CHEMICAL REACTIVITY

7.1 **Reactivity with Water:** No reaction
7.2 **Reactivity with Common Materials:** Data not available
7.3 **Stability During Transport:** Stable
7.4 **Neutralizing Agents for Acids and Caustics:** Not pertinent
7.5 **Polymerization:** Not pertinent
7.6 **Inhibitor of Polymerization:** Not pertinent

8. WATER POLLUTION

8.1 **Aquatic Toxicity:** Data not available
8.2 **Waterfowl Toxicity:** Data not available
8.3 **Biological Oxygen Demand (BOD):** Data not available
8.4 **Food Chain Concentration Potential:** Data not available

9. SELECTED MANUFACTURERS

1. Ethyl Corporation
   Industrial Chemicals Division
   Ethyl Tower, 451 Florida
   Baton Rouge, La. 70801
2. Ventron Corporation
   Alfa Products
   Box 159
   Beverly, Mass. 01915
3. PCR, Inc.
   P. O. Box 1466
   Gainesville, Fla. 32602

10. SHIPPING INFORMATION

10.1 **Grades or Purity:** 99.8%
10.2 **Storage Temperature:** Ambient
10.3 **Inert Atmosphere:** No requirement
10.4 **Venting:** Pressure-vacuum

11. HAZARD ASSESSMENT CODE
(See Hazard Assessment Handbook, CG 446-3)

A-X-Y

12. HAZARD CLASSIFICATIONS

12.1 **Code of Federal Regulations:** Poisonous, Class B
12.2 **NAS Hazard Rating for Bulk Water Transportation:** Not listed
12.3 **NFPA Hazard Classifications:** Not listed

13. PHYSICAL AND CHEMICAL PROPERTIES

13.1 **Physical State at 15°C and 1 atm:** Liquid
13.2 **Molecular Weight:** 218.1
13.3 **Boiling Point at 1 atm:** 451°F = 233°C = 506°K
13.4 **Freezing Point:** 34°F = 1°C = 274°K
13.5 **Critical Temperature:** Not pertinent
13.6 **Critical Pressure:** Not pertinent
13.7 **Specific Gravity:** 1.39 at 20°C (liquid)
13.8 **Liquid Surface Tension:** Data not available
13.9 **Liquid-Water Interfacial Tension:** Data not available
13.10 **Vapor (Gas) Specific Gravity:** Not pertinent
13.11 **Ratio of Specific Heats of Vapor (Gas):** Not pertinent
13.12 **Latent Heat of Vaporization:** Not pertinent
13.13 **Heat of Combustion:** (est.) −9,900 Btu/lb = −5,500 cal/g = -230×10^5 J/kg
13.14 **Heat of Decomposition:** Not pertinent
13.15 **Heat of Solution:** Not pertinent
13.16 **Heat of Polymerization:** Not pertinent

NOTES

MCP — METHYL CYCLOPENTANE

Common Synonyms

Cyclopentane, methyl

Liquid — Colorless — Gasoline-like odor

Floats on water. Flammable, irritating vapor is produced.

Shut off ignition sources. Call fire department.
Stop discharge if possible. Keep people away.
Isolate and remove discharged material.
Notify local health and pollution control agencies.

Fire

FLAMMABLE.
Containers may explode in fire.
Flashback along vapor trail may occur.
Vapor may explode if ignited in an enclosed area.
Extinguish with dry chemicals, foam or carbon dioxide.
Water may be ineffective on fire.
Cool exposed containers with water.

Exposure

Call for medical aid.

VAPOR
If inhaled will cause dizziness or difficult breathing.
Move victim to fresh air.
If breathing has stopped, give artificial respiration.
If breathing is difficult, give oxygen.

LIQUID
Irritating to skin and eyes.
Remove contaminated clothing and shoes.
Flush affected areas with plenty of water.
IF IN EYES, hold eyelids open and flush with plenty of water.
IF SWALLOWED and victim is CONSCIOUS, have victim drink water or milk.

Water Pollution

Effect of low concentrations on aquatic life is unknown.
Fouling to shoreline.
May be dangerous if it enters water intakes.
Notify local health and wildlife officials.
Notify operators of nearby water intakes.

1. RESPONSE TO DISCHARGE

(See Response Methods Handbook, CG 446-4)

Issue warning—high flammability
Evacuate area
Mechanical containment
Chemical and physical treatment

2. LABEL

Red

3. CHEMICAL DESIGNATIONS

3.1 **Synonyms:** Cyclopentane, methyl
3.2 **Coast Guard Compatibility Classification:** Not applicable
3.3 **Chemical Formula:** C_6H_{12}
3.4 **IMCO/United Nations Numerical Designation:** 3.1/1241

4. OBSERVABLE CHARACTERISTICS

4.1 **Physical State (as shipped):** Liquid
4.2 **Color:** Colorless
4.3 **Odor:** Like gasoline

5. HEALTH HAZARDS

5.1 **Personal Protective Equipment:** Self-contained breathing apparatus; goggles or face shield; rubber gloves.
5.2 **Symptoms Following Exposure:** Inhalation causes dizziness, nausea, and vomiting; concentrated vapor may cause unconsciousness and collapse. Liquid causes irritation of eyes and mild irritation of skin if allowed to remain. Ingestion causes irritation of stomach. Aspiration causes severe lung irritation, rapidly developing pulmonary edema, and central nervous system excitement followed by depression.
5.3 **Treatment for Exposure:** INHALATION: remove victim from exposure; if breathing has stopped, begin artificial respiration; call physician. EYES: flush with water for 15 min.; call physician. SKIN: flush well with water, then wash with soap and water. INGESTION: do NOT induce vomiting; guard against aspiration into lungs. ASPIRATION: enforce bed rest; give oxygen; get medical attention.
5.4 **Toxicity by Inhalation (Threshold Limit Value):** Data not available
5.5 **Short-Term Inhalation Limits:** 300 ppm for 60 min.
5.6 **Toxicity by Ingestion:** Grade 1; LD_{50} 5 to 15 g/kg
5.7 **Late Toxicity:** Data not available
5.8 **Vapor (Gas) Irritant Characteristics:** Vapors are nonirritating to eyes and throat.
5.9 **Liquid or Solid Irritant Characteristics:** Minimum hazard. If spilled on clothing and allowed to remain, may cause smarting and reddening of skin.
5.10 **Odor Threshold:** Data not available

6. FIRE HAZARDS

6.1 **Flash Point:** <0°F C.C.
6.2 **Flammable Limits in Air:** 1.1%–8.7% (approx.)
6.3 **Fire Extinguishing Agents:** Dry chemical, foam, carbon dioxide.
6.4 **Fire Extinguishing Agents Not to be Used:** Water may be ineffective.
6.5 **Special Hazards of Combustion Products:** Not pertinent
6.6 **Behavior in Fire:** Vapor is heavier than air and may travel a considerable distance to a source of ignition and flash back.
6.7 **Ignition Temperature:** 624°F
6.8 **Electrical Hazard:** Data not available
6.9 **Burning Rate:** 7.1 mm/min.

7. CHEMICAL REACTIVITY

7.1 **Reactivity with Water:** No reaction
7.2 **Reactivity with Common Materials:** No reaction
7.3 **Stability During Transport:** Stable
7.4 **Neutralizing Agents for Acids and Caustics:** Not pertinent
7.5 **Polymerization:** Not pertinent
7.6 **Inhibitor of Polymerization:** Not pertinent

8. WATER POLLUTION

8.1 **Aquatic Toxicity:** Data not available
8.2 **Waterfowl Toxicity:** Data not available
8.3 **Biological Oxygen Demand (BOD):** Data not available
8.4 **Food Chain Concentration Potential:** None

9. SELECTED MANUFACTURERS

1. Phillips Petroleum Company
   Chemical Department
   Special Products Division
   Bartlesville, Okla. 74004
2. Aldrich Chemical Co.
   940 West St. Paul Ave.
   Milwaukee, Wis. 53233
3. Pfaltz and Bauer, Inc.
   126-04 Northern Boulevard
   Flushing, N. Y. 11368

10. SHIPPING INFORMATION

10.1 **Grades or Purity:** Research: 99.94%; Pure: 99.5%; Technical: 96.5%
10.2 **Storage Temperature:** Ambient
10.3 **Inert Atmosphere:** No requirement
10.4 **Venting:** Pressure-vacuum

11. HAZARD ASSESSMENT CODE

(See Hazard Assessment Handbook, CG 446-3)

A-T-U-V-W

12. HAZARD CLASSIFICATIONS

12.1 **Code of Federal Regulations:** Flammable liquid
12.2 **NAS Hazard Rating for Bulk Water Transportation:**

Category	Rating
Fire	3
Health	
Vapor Irritant	0
Liquid or Solid Irritant	1
Poisons	2
Water Pollution	
Human Toxicity	1
Aquatic Toxicity	2
Aesthetic Effect	2
Reactivity	
Other Chemicals	0
Water	0
Self-Reaction	0

12.3 **NFPA Hazard Classifications:**

Category	Classification
Health Hazard (Blue)	2
Flammability (Red)	3
Reactivity (Yellow)	0

13. PHYSICAL AND CHEMICAL PROPERTIES

13.1 **Physical State at 15°C and 1 atm:** Liquid
13.2 **Molecular Weight:** 84.2
13.3 **Boiling Point at 1 atm:** 161.3°F = 71.8°C = 345.0°K
13.4 **Freezing Point:** −224°F = −142°C = 131°K
13.5 **Critical Temperature:** 499.3°F = 259.6°C = 532.8°K
13.6 **Critical Pressure:** 550 psia = 37.4 atm = 3.79 MN/m^2
13.7 **Specific Gravity:** 0.749 at 20°C (liquid)
13.8 **Liquid Surface Tension:** 21.60 dynes/cm = 0.0216 N/m at 20°C
13.9 **Liquid-Water Interfacial Tension:** (est.) 35 dynes/cm = 0.035 N/m at 20°C
13.10 **Vapor (Gas) Specific Gravity:** 2.9
13.11 **Ratio of Specific Heats of Vapor (Gas):** 1.0834
13.12 **Latent Heat of Vaporization:** 162 Btu/lb = 90 cal/g = 3.8×10^5 J/kg
13.13 **Heat of Combustion:** (liquid) −18,900 Btu/lb = −10,500 cal/g = -440×10^5 J/kg
13.14 **Heat of Decomposition:** Not pertinent
13.15 **Heat of Solution:** Not pertinent
13.16 **Heat of Polymerization:** Not pertinent

NOTES

MCS METHYLDICHLOROSILANE

Common Synonyms	Liquid · Colorless · Sharp irritating odor Reacts violently with water. Irritating gas is produced on contact with water.

AVOID CONTACT WITH LIQUID AND VAPOR. KEEP PEOPLE AWAY.
Wear goggles and self-contained breathing apparatus.
Shut off ignition sources. Call fire department.
Stop discharge if possible.
Isolate and remove discharged material.
Notify local health and pollution control agencies.

Fire	FLAMMABLE. POISONOUS GASES MAY BE PRODUCED IN FIRE. Containers may explode in fire. Flashback along vapor trail may occur. Vapor may explode if ignited in an enclosed area. Extinguish with dry chemicals or carbon dioxide. DO NOT USE WATER OR FOAM ON FIRE. DO NOT USE WATER OR FOAM ON ADJACENT FIRES.
Exposure	Call for medical aid. VAPOR Irritating to eyes, nose and throat. If inhaled will cause difficult breathing. Move victim to fresh air. If breathing has stopped, give artificial respiration. If breathing is difficult, give oxygen. LIQUID Will burn skin and eyes. Harmful if swallowed. Remove contaminated clothing and shoes. Flush affected areas with plenty of water. IF IN EYES, hold eyelids open and flush with plenty of water. IF SWALLOWED and victim is CONSCIOUS, have victim drink water or milk. DO NOT INDUCE VOMITING.
Water Pollution	Effect of low concentrations on aquatic life is unknown. May be dangerous if it enters water intakes. Notify local health and wildlife officials. Notify operators of nearby water intakes.

1. RESPONSE TO DISCHARGE

(See Response Methods Handbook, CG 446-4)

Issue warning—high flammability
Restrict access
Evacuate area
Disperse and flush with care

2. LABEL

Red

3. CHEMICAL DESIGNATIONS

3.1 **Synonyms:** No common synonyms
3.2 **Coast Guard Compatibility Classification:** Not applicable
3.3 **Chemical Formula:** CH_3SiHCl_2
3.4 **IMCO/United Nations Numerical Designation:** 3.2/1242

4. OBSERVABLE CHARACTERISTICS

4.1 **Physical State (as shipped):** Liquid
4.2 **Color:** Colorless
4.3 **Odor:** Acrid; sharp, hydrochloric acid-like

5. HEALTH HAZARDS

5.1 **Personal Protective Equipment:** Full protective clothing; acid-vapor-type respiratory protection; rubber gloves; chemical worker's goggles; other protective equipment as necessary to protect skin and eyes.
5.2 **Symptoms Following Exposure:** Inhalation causes irritation of respiratory tract; heavy exposure can cause pulmonary edema. Contact of liquid with skin or eyes causes severe burns. Ingestion causes burns of mouth and stomach.
5.3 **Treatment for Exposure:** Get medical attention following all exposures to this compound. INHALATION: remove victim from exposure; if breathing has stopped, begin artificial respiration. EYES: flush with water for 15 min. SKIN: flush with water. INGESTION: do NOT induce vomiting; give large amounts of water.
5.4 **Toxicity by Inhalation (Threshold Limit Value):** Data not available
5.5 **Short-Term Inhalation Limits:** Data not available
5.6 **Toxicity by Ingestion:** Grade 3; LD_{50} 50 to 500 mg/kg
5.7 **Late Toxicity:** Data not available
5.8 **Vapor (Gas) Irritant Characteristics:** Vapors cause severe irritation of eyes and throat and can cause eye and lung injury. They cannot be tolerated even at low concentrations.
5.9 **Liquid or Solid Irritant Characteristics:** Severe skin irritant. Causes second- and third-degree burns on short contact and is very injurious to the eyes.
5.10 **Odor Threshold:** Data not available

6. FIRE HAZARDS

6.1 **Flash Point:** −14°F O.C.
6.2 **Flammable Limits in Air:** 6%—55%
6.3 **Fire Extinguishing Agents:** Dry chemical or carbon dioxide
6.4 **Fire Extinguishing Agents Not to be Used:** Water, foam
6.5 **Special Hazards of Combustion Products:** Toxic hydrogen chloride and phosgene gases may be formed.
6.6 **Behavior in Fire:** Difficult to extinguish; re-ignition may occur. Contact with water applied to adjacent fires will generate irritating hydrogen chloride gas.
6.7 **Ignition Temperature:** >600°F
6.8 **Electrical Hazard:** Data not available
6.9 **Burning Rate:** 3.0 mm/min.

7. CHEMICAL REACTIVITY

7.1 **Reactivity with Water:** Reacts violently to form hydrogen chloride (hydrochloric acid)
7.2 **Reactivity with Common Materials:** Reacts with surface moisture to evolve hydrogen chloride, which is corrosive to common metals.
7.3 **Stability During Transport:** Stable
7.4 **Neutralizing Agents for Acids and Caustics:** Flood with water and rinse with sodium bicarbonate or lime solution.
7.5 **Polymerization:** Not pertinent
7.6 **Inhibitor of Polymerization:** Not pertinent

8. WATER POLLUTION

8.1 **Aquatic Toxicity:** Data not available
8.2 **Waterfowl Toxicity:** Data not available
8.3 **Biological Oxygen Demand (BOD):** Data not available
8.4 **Food Chain Concentration Potential:** None

9. SELECTED MANUFACTURERS

1. Union Carbide Corporation
   Chemicals and Plastics Division
   270 Park Avenue
   New York, N. Y. 10017
2. Dow Corning Corporation
   P. O. Box 592
   Midland, Mich. 48640
3. PCR, Inc.
   P. O. Box 1466
   Gainesville, Fla. 32602

10. SHIPPING INFORMATION

10.1 **Grades or Purity:** 97%
10.2 **Storage Temperature:** Ambient
10.3 **Inert Atmosphere:** No requirement
10.4 **Venting:** Pressure-vacuum

11. HAZARD ASSESSMENT CODE

(See Hazard Assessment Handbook, CG 446-3)

A-O

12. HAZARD CLASSIFICATIONS

12.1 **Code of Federal Regulations:** Flammable liquid
12.2 **NAS Hazard Rating for Bulk Water Transportation:**

Category	Rating
Fire	3
Health	
Vapor Irritant	4
Liquid or Solid Irritant	4
Poisons	3
Water Pollution	
Human Toxicity	3
Aquatic Toxicity	3
Aesthetic Effect	2
Reactivity	
Other Chemicals	3
Water	4
Self-Reaction	1

12.3 **NFPA Hazard Classifications:**

Category	Classification
Health Hazard (Blue)	3
Flammability (Red)	3
Reactivity (Yellow)	2 W̶

13. PHYSICAL AND CHEMICAL PROPERTIES

13.1 **Physical State at 15°C and 1 atm:** Liquid
13.2 **Molecular Weight:** 115
13.3 **Boiling Point at 1 atm:** 106.7°F = 41.5°C = 314.7°K
13.4 **Freezing Point:** −135°F = −93°C = 180°K
13.5 **Critical Temperature:** Not pertinent
13.6 **Critical Pressure:** Not pertinent
13.7 **Specific Gravity:** 1.11 at 25°C (liquid)
13.8 **Liquid Surface Tension:** (est.) 35 dynes/cm = 0.035 N/m at 20°C
13.9 **Liquid-Water Interfacial Tension:** Not pertinent
13.10 **Vapor (Gas) Specific Gravity:** 4
13.11 **Ratio of Specific Heats of Vapor (Gas):** Data not available
13.12 **Latent Heat of Vaporization:** 106 Btu/lb = 59 cal/g = 2.5×10^5 J/kg
13.13 **Heat of Combustion:** (est.) −4,700 Btu/lb = −2,600 cal/g = -110×10^5 J/kg
13.14 **Heat of Decomposition:** Not pertinent
13.15 **Heat of Solution:** Data not available
13.16 **Heat of Polymerization:** Not pertinent

NOTES

MEK	METHYL ETHYL KETONE

Common Synonyms			
MEK 2-Butanone Ethyl methyl ketone	Liquid	Colorless	Sweet odor
	Floats and mixes with water. Flammable, irritating vapor is produced.		

Stop discharge if possible. Keep people away.
Shut off ignition sources and call fire department.
Stay upwind and use water spray to "knock down" vapor.
Avoid contact with liquid and vapor.
Isolate and remove discharged material.
Notify local health and pollution control agencies.

Fire

FLAMMABLE.
Flashback along vapor trail may occur.
Vapor may explode if ignited in an enclosed area.
Extinguish with dry chemical, alcohol foam, or carbon dioxide.
Water may be ineffective on fire.
Cool exposed containers with water.

Exposure

CALL FOR MEDICAL AID.

VAPOR
Irritating to eyes, nose and throat.
If inhaled, will cause nausea, vomiting, headache, dizziness, difficult breathing, or loss of consciousness.
Move to fresh air.
If breathing has stopped, give artificial respiration.
If breathing is difficult, give oxygen.

LIQUID
Will burn eyes.
Harmful if swallowed.
Remove contaminated clothing and shoes.
Flush affected areas with plenty of water.
IF IN EYES, hold eyelids open and flush with plenty of water.
IF SWALLOWED and victim is CONSCIOUS, have victim drink water or milk.

Water Pollution

Dangerous to aquatic life in high concentrations.
May be dangerous if it enters water intakes.
Notify local health and wildlife officials.
Notify operators of nearby water intakes.

1. RESPONSE TO DISCHARGE
(See Response Methods Handbook, CG 446-4)

Issue warning—high flammability
Disperse and flush

2. LABEL

Red

3. CHEMICAL DESIGNATIONS

3.1 **Synonyms:** 2-Butanone
Ethyl methyl ketone
MEK
3.2 **Coast Guard Compatibility Classification:** Ketone
3.3 **Chemical Formula:** $CH_3COCH_2CH_3$
3.4 **IMCO/United Nations Numerical Designation:** 3.2/1193

4. OBSERVABLE CHARACTERISTICS

4.1 **Physical State (as shipped):** Liquid
4.2 **Color:** Colorless
4.3 **Odor:** Like acetone; pleasant; pungent

5. HEALTH HAZARDS

5.1 **Personal Protective Equipment:** Organic canister or air pack; plastic gloves; goggles or face shield.
5.2 **Symptoms Following Exposure:** Liquid causes eye burn. Vapor irritates eyes, nose, and throat; can cause headache, dizziness, nausea, weakness, and loss of consciousness.
5.3 **Treatment for Exposure:** INHALATION: remove victim to fresh air; if breathing is irregular or has stopped, start resuscitation and administer oxygen. EYES: wash with plenty of water for at least 15 min. and call physician.
5.4 **Toxicity by Inhalation (Threshold Limit Value):** 200 ppm
5.5 **Short-Term Inhalation Limits:** 290 mg/m^3 for 60 min.
5.6 **Toxicity by Ingestion:** Grade 2; 0.5 to 5 g/kg (rat)
5.7 **Late Toxicity:** None
5.8 **Vapor (Gas) Irritant Characteristics:** Vapors cause a slight smarting of the eyes or respiratory system if present in high concentrations. The effect is temporary.
5.9 **Liquid or Solid Irritant Characteristics:** Minimum hazard. If spilled on clothing and allowed to remain, may cause smarting and reddening of the skin.
5.10 **Odor Threshold:** 10 ppm

6. FIRE HAZARDS

6.1 **Flash Point:** 20°F C.C.; 22°F O.C.
6.2 **Flammable Limits in Air:** 1.8%—11.5%
6.3 **Fire Extinguishing Agents:** Alcohol foam, dry chemical, or carbon dioxide
6.4 **Fire Extinguishing Agents Not to be Used:** Water may be ineffective
6.5 **Special Hazards of Combustion Products:** Not pertinent
6.6 **Behavior in Fire:** Not pertinent
6.7 **Ignition Temperature:** 961°F
6.8 **Electrical Hazard:** Class I, Group D
6.9 **Burning Rate:** 4.1 mm/min.

7. CHEMICAL REACTIVITY

7.1 **Reactivity with Water:** No reaction
7.2 **Reactivity with Common Materials:** No reaction
7.3 **Stability During Transport:** Stable
7.4 **Neutralizing Agents for Acids and Caustics:** Not pertinent
7.5 **Polymerization:** Not pertinent
7.6 **Inhibitor of Polymerization:** Not pertinent

8. WATER POLLUTION

8.1 **Aquatic Toxicity:** 5640 mg/l /48 hr/bluegill/TL$_m$/fresh water
8.2 **Waterfowl Toxicity:** Data not available
8.3 **Biological Oxygen Demand (BOD):** 214%, 5 days
8.4 **Food Chain Concentration Potential:** None

9. SELECTED MANUFACTURERS

1. Celanese Corp.
Celanese Chemical Co. Division
245 Park Ave.
New York, N. Y. 10017
2. Exxon Chemical Co.
Houston, Tex. 77001
3. Shell Chemical Co.
Industrial Chemicals Division
Houston, Tex. 77001

10. SHIPPING INFORMATION

10.1 **Grades or Purity:** 99.5+%
10.2 **Storage Temperature:** Ambient
10.3 **Inert Atmosphere:** No requirement
10.4 **Venting:** Open (flame arrester) or pressure-vacuum

11. HAZARD ASSESSMENT CODE
(See Hazard Assessment Handbook, CG 446-3)

A-P-Q-R-S

12. HAZARD CLASSIFICATIONS

12.1 **Code of Federal Regulations:** Flammable liquid
12.2 **NAS Hazard Rating for Bulk Water Transportation:**

Category	Rating
Fire	3
Health	
Vapor Irritant	1
Liquid or Solid Irritant	1
Poisons	2
Water Pollution	
Human Toxicity	2
Aquatic Toxicity	1
Aesthetic Effect	1
Reactivity	
Other Chemicals	2
Water	0
Self-Reaction	0

12.3 **NFPA Hazard Classifications:**

Category	Classification
Health Hazard (Blue)	1
Flammability (Red)	3
Reactivity (Yellow)	0

13. PHYSICAL AND CHEMICAL PROPERTIES

13.1 **Physical State at 15°C and 1 atm:** Liquid
13.2 **Molecular Weight:** 72.11
13.3 **Boiling Point at 1 atm:** 175.3°F = 79.6°C = 352.8°K
13.4 **Freezing Point:** −123.3°F = −86.3°C = 186.9°K
13.5 **Critical Temperature:** 504.5°F = 262.5°C = 535.7°K
13.6 **Critical Pressure:** 603 psia = 41.0 atm = 4.15 MN/m^2
13.7 **Specific Gravity:** 0.806 at 20°C (liquid)
13.8 **Liquid Surface Tension:** Not pertinent
13.9 **Liquid-Water Interfacial Tension:** Not pertinent
13.10 **Vapor (Gas) Specific Gravity:** 2.5
13.11 **Ratio of Specific Heats of Vapor (Gas):** 1.075
13.12 **Latent Heat of Vaporization:** 191 Btu/lb = 106 cal/g = 4.44×10^5 J/kg
13.13 **Heat of Combustion:** −13,480 Btu/lb = −7491 cal/g = -313.6×10^5 J/kg
13.14 **Heat of Decomposition:** Not pertinent
13.15 **Heat of Solution:** (est.) −9 Btu/lb = −5 cal/g = -0.2×10^5 J/kg
13.16 **Heat of Polymerization:** Not pertinent

NOTES

REVISED 1978

MEP	METHYLETHYLPYRIDINE

Common Synonyms 5-Ethyl-2-methyl pyridine 5-Ethyl-2-picoline MEP	Liquid	Colorless	Sharp odor
	Floats on water.		

Avoid contact with liquid. Keep people away.
Wear goggles, self-contained breathing apparatus, and rubber overclothing (including gloves).
Stop discharge if possible.
Call fire department.
Isolate and remove discharged material.
Notify local health and pollution control agencies.

Fire	Combustible. Wear goggles, self-contained breathing apparatus, and rubber overclothing (including gloves). Extinguish with water, dry chemical, alcohol foam, or carbon dioxide. Cool exposed containers with water.
Exposure	CALL FOR MEDICAL AID. LIQUID Will burn skin and eyes. Harmful if swallowed. Remove contaminated clothing and shoes. Flush affected areas with plenty of water. IF IN EYES, hold eyelids open and flush with plenty of water. IF SWALLOWED and victim is CONSCIOUS, have victim drink water or milk.
Water Pollution	Effect of low concentrations on aquatic life is unknown. Fouling to shoreline. May be dangerous if it enters water intakes. Notify local health and wildlife officials. Notify operators of nearby water intakes.

1. RESPONSE TO DISCHARGE
(See Response Methods Handbook, CG 446-4)

Mechanical containment
Should be removed
Chemical and physical treatment

2. LABEL

3. CHEMICAL DESIGNATIONS

3.1 **Synonyms:** Aldehyde-collidine
Aldehydine
5-Ethyl-2-methylpyridine
5-Ethyl-2-picoline
MEP

3.2 **Coast Guard Compatibility Classification:** Aromatic amine

3.3 **Chemical Formula:** $C_8H_{11}N$

3.4 **IMCO/United Nations Numerical Designation:** Not listed

4. OBSERVABLE CHARACTERISTICS

4.1 **Physical State (as shipped):** Liquid

4.2 **Color:** Colorless

4.3 **Odor:** Sharp, penetrating

5. HEALTH HAZARDS

5.1 **Personal Protective Equipment:** Air-supplied mask for high vapor concentrations; plastic gloves; goggles or face shield.

5.2 **Symptoms Following Exposure:** Breathing of vapors will cause vomiting and chest discomfort. Contact with liquid causes skin and eye burns.

5.3 **Treatment for Exposure:** INHALATION: remove victim to fresh air; give oxygen if breathing is difficult; call a physician. SKIN OR EYES: immediately flush with plenty of water for at least 15 min.; get medical care for eyes.

5.4 **Toxicity by Inhalation (Threshold Limit Value):** 1700 ppm/3.7 hr/lethal (rat)

5.5 **Short-Term Inhalation Limits:** Data not available

5.6 **Toxicity by Ingestion:** Grade 2; LD_{50} 0.5 to 5 g/kg (rat)

5.7 **Late Toxicity:** Data not available

5.8 **Vapor (Gas) Irritant Characteristics:** Vapors cause moderate irritation such that personnel will find high concentrations unpleasant. The effect is temporary.

5.9 **Liquid or Solid Irritant Characteristics:** Causes smarting of the skin and first-degree burns on short exposure; may cause secondary burns on long exposure.

5.10 **Odor Threshold:** Data not available

6. FIRE HAZARDS

6.1 **Flash Point:** 155°F O.C.

6.2 **Flammable Limits in Air:** 1.1%–6.6%

6.3 **Fire Extinguishing Agents:** Foam, carbon dioxide, dry chemical

6.4 **Fire Extinguishing Agents Not to be Used:** Not pertinent

6.5 **Special Hazards of Combustion Products:** Irritating vapors are generated when heated

6.6 **Behavior in Fire:** Not pertinent

6.7 **Ignition Temperature:** 939°F

6.8 **Electrical Hazard:** Not pertinent

6.9 **Burning Rate:** Data not available

7. CHEMICAL REACTIVITY

7.1 **Reactivity with Water:** No reaction

7.2 **Reactivity with Common Materials:** No reaction

7.3 **Stability During Transport:** Stable

7.4 **Neutralizing Agents for Acids and Caustics:** Flush with water, neutralize with dilute acetic acid

7.5 **Polymerization:** Not pertinent

7.6 **Inhibitor of Polymerization:** Not pertinent

8. WATER POLLUTION

8.1 **Aquatic Toxicity:** Data not available

8.2 **Waterfowl Toxicity:** Data not available

8.3 **Biological Oxygen Demand (BOD):** (theor.): 4.4 %, 5 days; 56.6%, 20 days; 0.12-2.14 lb/lb, 5 days

8.4 **Food Chain Concentration Potential:** None

9. SELECTED MANUFACTURERS

1. Lonza, Inc.
22-10 Route 208
Fairlawn, N. J. 07410
2. Union Carbide Corp.
Chemicals and Plastics Division
270 Park Ave.
New York, N. Y. 10017

10. SHIPPING INFORMATION

10.1 **Grades or Purity:** 99.9%

10.2 **Storage Temperature:** Ambient

10.3 **Inert Atmosphere:** No requirement

10.4 **Venting:** Open (flame arrester)

11. HAZARD ASSESSMENT CODE
(See Hazard Assessment Handbook, CG 446-3)

A-P-Q-T-U

12. HAZARD CLASSIFICATIONS

12.1 **Code of Federal Regulations:** Corrosive Material

12.2 **NAS Hazard Rating for Bulk Water Transportation:**

Category	Rating
Fire	1
Health	
Vapor Irritant	2
Liquid or Solid Irritant	2
Poisons	2
Water Pollution	
Human Toxicity	2
Aquatic Toxicity	2
Aesthetic Effect	3
Reactivity	
Other Chemicals	3
Water	0
Self-Reaction	0

12.3 **NFPA Hazard Classifications:**

Category	Classification
Health Hazard (Blue)	2
Flammability (Red)	2
Reactivity (Yellow)	0

13. PHYSICAL AND CHEMICAL PROPERTIES

13.1 **Physical State at 15°C and 1 atm:** Liquid

13.2 **Molecular Weight:** 121.18

13.3 **Boiling Point at 1 atm:** 352°F = 178°C = 451°K

13.4 **Freezing Point:** −94.5°F = −70.3°C = 202.9°K

13.5 **Critical Temperature:** Not pertinent

13.6 **Critical Pressure:** Not pertinent

13.7 **Specific Gravity:** 0.922 at 20°C (liquid)

13.8 **Liquid Surface Tension:** 36 dynes/cm = 0.036 N/m at 20°C

13.9 **Liquid-Water Interfacial Tension:** Data not available

13.10 **Vapor (Gas) Specific Gravity:** Not pertinent

13.11 **Ratio of Specific Heats of Vapor (Gas):** Not pertinent

13.12 **Latent Heat of Vaporization:** Data not available

13.13 **Heat of Combustion:** Data not available

13.14 **Heat of Decomposition:** Not pertinent

13.15 **Heat of Solution:** Data not available

13.16 **Heat of Polymerization:** Not pertinent

NOTES

REVISED 1978

MTF	METHYL FORMAL

Common Synonyms			
Methylal Methylene dimethyl ether Dimethoxymethane Dimethyl formal Formaldehyde dimethylacetol	Liquid	Colorless	Mild sweet odor
	Mixes with water. Flammable, irritating vapor is produced.		

Shut off ignition sources. Call fire department.
Stop discharge if possible. Keep people away.
Isolate and remove discharged material.
Notify local health and pollution control agencies.

Fire

FLAMMABLE.
Irritating gases may be produced when heated.
Containers may explode in fire.
Flashback along vapor trail may occur.
Vapor may explode if ignited in an enclosed area.

Extinguish with dry chemicals, foam or carbon dioxide.
Water may be ineffective on fire.
Cool exposed containers with water.

Exposure

Call for medical aid.

VAPOR
Irritating to eyes, nose and throat.
Harmful if inhaled.

Move victim to fresh air.
If breathing has stopped, give artificial respiration.
If breathing is difficult, give oxygen.

LIQUID
Irritating to skin and eyes.
Harmful if swallowed.

Remove contaminated clothing and shoes.
Flush affected areas with plenty of water.
IF IN EYES, hold eyelids open and flush with plenty of water.
IF SWALLOWED and victim is CONSCIOUS, having victim drink water or milk and have victim induce vomiting.
IF SWALLOWED and victim is UNCONSCIOUS OR HAVING CONVULSIONS, do nothing except keep victim warm.

Water Pollution

Effect of low concentrations on aquatic life is unknown.
May be dangerous if it enters water intakes.

Notify local health and wildlife officials.
Notify operators of nearby water intakes.

1. RESPONSE TO DISCHARGE

(See Response Methods Handbook, CG 446-4)

Issue warning—high flammability, water contaminant
Restrict access
Disperse and flush

2. LABEL

Red

3. CHEMICAL DESIGNATIONS

3.1 **Synonyms:** Dimethoxymethane; Dimethylformal; Formaldehyde dimethylacetal; Methylal; Methylene dimethyl ether

3.2 **Coast Guard Compatibility Classification:** Ether

3.3 **Chemical Formula:** $CH_2(OCH_3)_2$

3.4 **IMCO/United Nations Numerical Designation:** 3.1/1234

4. OBSERVABLE CHARACTERISTICS

4.1 **Physical State (as shipped):** Liquid

4.2 **Color:** Colorless

4.3 **Odor:** Mild, ethereal; chloroform-like

5. HEALTH HAZARDS

5.1 **Personal Protective Equipment:** Self-contained breathing apparatus or all-purpose canister mask; rubber gloves; chemical safety goggles; impervious apron and boots.

5.2 **Symptoms Following Exposure:** Inhalation causes irritation of respiratory system and depression of central nervous system. Liquid causes irritation of eyes and will irritate skin if allowed to remain. Ingestion causes depression of central nervous system.

5.3 **Treatment for Exposure:** INHALATION: remove victim from contaminated area and administer artificial respiration and oxygen if necessary. EYES: flush with plenty of water; get medical attention. SKIN: flush with plenty of water. INGESTION: induce vomiting; administer gastric lavage and saline cathartics; subsequent treatment is symptomatic and supportive.

5.4 **Toxicity by Inhalation (Threshold Limit Value):** 1,000 ppm

5.5 **Short-Term Inhalation Limits:** Data not available

5.6 **Toxicity by Ingestion:** Grade 1; LD_{50} 5 to 15 g/kg

5.7 **Late Toxicity:** Liver and kidney injury may follow high exposures.

5.8 **Vapor (Gas) Irritant Characteristics:** Vapors cause a slight smarting of the eyes or respiratory system if present in high concentrations. The effect is temporary.

5.9 **Liquid or Solid Irritant Characteristics:** Minimum hazard. If spilled on clothing and allowed to remain, may cause smarting and reddening of the skin.

5.10 **Odor Threshold:** Data not available

6. FIRE HAZARDS

6.1 **Flash Point:** 0°F O.C.

6.2 **Flammable Limits in Air:** 1.6%—17.6%

6.3 **Fire Extinguishing Agents:** Dry chemical, foam, carbon dioxide

6.4 **Fire Extinguishing Agents Not to be Used:** Water may be ineffective.

6.5 **Special Hazards of Combustion Products:** Irritating formaldehyde gas may be present in smoke.

6.6 **Behavior in Fire:** Not pertinent

6.7 **Ignition Temperature:** 459°F

6.8 **Electrical Hazard:** Data not available

6.9 **Burning Rate:** 5.5 mm/min.

7. CHEMICAL REACTIVITY

7.1 **Reactivity with Water:** No reaction

7.2 **Reactivity with Common Materials:** No reaction

7.3 **Stability During Transport:** Stable

7.4 **Neutralizing Agents for Acids and Caustics:** Not pertinent

7.5 **Polymerization:** Not pertinent

7.6 **Inhibitor of Polymerization:** Not pertinent

8. WATER POLLUTION

8.1 **Aquatic Toxicity:** Data not available

8.2 **Waterfowl Toxicity:** Data not available

8.3 **Biological Oxygen Demand (BOD):** Data not available

8.4 **Food Chain Concentration Potential:** None

9. SELECTED MANUFACTURERS

1. Celanese Chemical Company
245 Park Avenue
New York, N. Y. 10017

2. Aldrich Chemical Co.
940 West Saint Paul Ave.
Milwaukee, Wis. 53233

3. Eastman Kodak Co.
Eastman Organic Chemicals
Rochester, N. Y. 14650

10. SHIPPING INFORMATION

10.1 **Grades or Purity:** 97+%

10.2 **Storage Temperature:** Ambient

10.3 **Inert Atmosphere:** No requirement

10.4 **Venting:** Pressure-vacuum

11. HAZARD ASSESSMENT CODE

(See Hazard Assessment Handbook, CG 446-3)

A-P-Q-R-S

12. HAZARD CLASSIFICATIONS

12.1 **Code of Federal Regulations:** Flammable liquid

12.2 **NAS Hazard Rating for Bulk Water Transportation:**

Category	Rating
Fire	3
Health	
Vapor Irritant	1
Liquid or Solid Irritant	1
Poisons	1
Water Pollution	
Human Toxicity	1
Aquatic Toxicity	2
Aesthetic Effect	1
Reactivity	
Other Chemicals	2
Water	0
Self-Reaction	0

12.3 **NFPA Hazard Classifications:**

Category	Classification
Health Hazard (Blue)	2
Flammability (Red)	3
Reactivity (Yellow)	2

13. PHYSICAL AND CHEMICAL PROPERTIES

13.1 **Physical State at 15°C and 1 atm:** Liquid

13.2 **Molecular Weight:** 76.1

13.3 **Boiling Point at 1 atm:** 108°F = 42°C = 315°K

13.4 **Freezing Point:** −157°F = −105°C = 168°K

13.5 **Critical Temperature:** 419°F = 215°C = 488°K

13.6 **Critical Pressure:** Not pertinent

13.7 **Specific Gravity:** 0.861 at 20°C (liquid)

13.8 **Liquid Surface Tension:** 21.1 dynes/cm = 0.0211 N/m at 20°C

13.9 **Liquid-Water Interfacial Tension:** Not pertinent

13.10 **Vapor (Gas) Specific Gravity:** 2.6

13.11 **Ratio of Specific Heats of Vapor (Gas):** 1.0888

13.12 **Latent Heat of Vaporization:** 161.5 Btu/lb = 89.8 cal/g = 3.76×10^5 J/kg

13.13 **Heat of Combustion:** −10,970 Btu/lb = −6,100 cal/g = -255×10^5 J/kg

13.14 **Heat of Decomposition:** Not pertinent

13.15 **Heat of Solution:** Not pertinent

13.16 **Heat of Polymerization:** Not pertinent

NOTES

MFM	# METHYL FORMATE

Common Synonyms	Liquid	Colorless	Pleasant odor
Formic acid, methyl ester	Mixes with water. Flammable, irritating vapor is produced. Boiling point is 88° F.		

Shut off ignition sources. Call fire department.
Stop discharge if possible. Keep people away.
Isolate and remove discharged material.
Notify local health and pollution control agencies.

Fire

FLAMMABLE.
Containers may explode in fire.
Flashback along vapor trail may occur.
Vapor may explode if ignited in an enclosed area.
Extinguish with dry chemicals, alcohol foam, or carbon dioxide.
Water may be ineffective on fire.
Cool exposed containers with water.

Exposure

Call for medical aid.

VAPOR
Irritating to eyes, nose and throat.
Harmful if inhaled.
Move victim to fresh air.
If breathing has stopped, give artificial respiration.
If breathing is difficult, give oxygen.

LIQUID
Irritating to skin and eyes.
Harmful if swallowed.
Remove contaminated clothing and shoes.
Flush affected areas with plenty of water.
IF IN EYES, hold eyelids open and flush with plenty of water.
IF SWALLOWED and victim is CONSCIOUS, have victim drink water or milk and have victim induce vomiting.
IF SWALLOWED and victim is UNCONSCIOUS OR HAVING CONVULSIONS, do nothing except keep victim warm.

Water Pollution

Effect of low concentrations on aquatic life is unknown.
May be dangerous if it enters water intakes.
Notify local health and wildlife officials.
Notify operators of nearby water intakes.

1. RESPONSE TO DISCHARGE

(See Response Methods Handbook, CG 446-4)

Issue warning—high flammability, water contaminant
Restrict access
Disperse and flush

2. LABEL

Red

3. CHEMICAL DESIGNATIONS

3.1 **Synonyms:** Formic acid, methyl ester
3.2 **Coast Guard Compatibility Classification:** Esters (13)
3.3 **Chemical Formula:** $HCOOCH_3$
3.4 **IMCO/United Nations Numerical Designation:** 3.1/1243

4. OBSERVABLE CHARACTERISTICS

4.1 **Physical State (as shipped):** Liquid
4.2 **Color:** Colorless
4.3 **Odor:** Pleasant; agreeable

5. HEALTH HAZARDS

5.1 **Personal Protective Equipment:** Goggles or safety glasses; self-contained breathing apparatus; rubber gloves.
5.2 **Symptoms Following Exposure:** Inhalation causes irritation of mucous membranes. Prolonged inhalation can produce narcosis and central nervous symptoms, including some temporary visual disturbance. Contact with liquid irritates eyes and may irritate skin if allowed to remain. Ingestion causes irritation of mouth and stomach and central nervous system depression, including visual disturbances.
5.3 **Treatment for Exposure:** INHALATION: move to fresh air and rest; if pulmonary edema develops, administer oxygen; call physician. EYES: irrigate with water for 15 min. SKIN: wash thoroughly with soap and water. INGESTION: do NOT induce vomiting; get medical attention.
5.4 **Toxicity by Inhalation (Threshold Limit Value):** 100 ppm
5.5 **Short-Term Inhalation Limits:** Data not available
5.6 **Toxicity by Ingestion:** Grade 1; LD_{50} 5 to 15 g/kg
5.7 **Late Toxicity:** Data not available
5.8 **Vapor (Gas) Irritant Characteristics:** Vapors are moderately irritating such that personnel will not usually tolerate moderate or high concentrations.
5.9 **Liquid or Solid Irritant Characteristics:** Minimum hazard. If spilled on clothing and allowed to remain, may cause smarting and reddening of the skin.
5.10 **Odor Threshold:** Data not available

6. FIRE HAZARDS

6.1 **Flash Point:** −26°F C.C.
6.2 **Flammable Limits in Air:** 5%—22.7%
6.3 **Fire Extinguishing Agents:** Dry chemical, alcohol foam, carbon dioxide
6.4 **Fire Extinguishing Agents Not to be Used:** Water may be ineffective.
6.5 **Special Hazards of Combustion Products:** Not pertinent
6.6 **Behavior in Fire:** Vapor is heavier than air and may travel considerable distance to a source of ignition and flash back.
6.7 **Ignition Temperature:** 853°F
6.8 **Electrical Hazard:** Data not available
6.9 **Burning Rate:** 2.5 mm/min.

7. CHEMICAL REACTIVITY

7.1 **Reactivity with Water:** Slow reaction to form formic acid and methyl alcohol; reaction is not hazardous.
7.2 **Reactivity with Common Materials:** No reaction
7.3 **Stability During Transport:** Stable
7.4 **Neutralizing Agents for Acids and Caustics:** Not pertinent
7.5 **Polymerization:** Not pertinent
7.6 **Inhibitor of Polymerization:** Not pertinent

8. WATER POLLUTION

8.1 **Aquatic Toxicity:** Data not available
8.2 **Waterfowl Toxicity:** Data not available
8.3 **Biological Oxygen Demand (BOD):** Data not available
8.4 **Food Chain Concentration Potential:** None

9. SELECTED MANUFACTURERS

1. E. I. duPont de Nemours & Co., Inc.
   Biochemicals Department
   1007 Market Street
   Wilmington, Del. 19898
2. Aldrich Chemical Co.
   940 West Saint Paul Ave.
   Milwaukee, Wis. 53233
3. Eastman Kodak Co.
   Eastman Organic Chemicals
   Rochester, N. Y. 14650

10. SHIPPING INFORMATION

10.1 **Grades or Purity:** Technical, practical, and spectro grades: all 97.5+%
10.2 **Storage Temperature:** <85°F
10.3 **Inert Atmosphere:** No requirement
10.4 **Venting:** Pressure-vacuum

11. HAZARD ASSESSMENT CODE

(See Hazard Assessment Handbook, CG 446-3)

A-P-Q-R-S

12. HAZARD CLASSIFICATIONS

12.1 **Code of Federal Regulations:** Flammable liquid
12.2 **NAS Hazard Rating for Bulk Water Transportation:**

Category	Rating
Fire	4
Health	
Vapor Irritant	3
Liquid or Solid Irritant	1
Poisons	1
Water Pollution	
Human Toxicity	1
Aquatic Toxicity	3
Aesthetic Effect	1
Reactivity	
Other Chemicals	1
Water	0
Self-Reaction	0

12.3 **NFPA Hazard Classifications:**

Category	Classification
Health Hazard (Blue)	2
Flammability (Red)	4
Reactivity (Yellow)	0

13. PHYSICAL AND CHEMICAL PROPERTIES

13.1 **Physical State at 15°C and 1 atm:** Liquid
13.2 **Molecular Weight:** 60.1
13.3 **Boiling Point at 1 atm:** 89.2°F = 31.8°C = 305°K
13.4 **Freezing Point:** −147.6°F = −99.8°C = 173.4°K
13.5 **Critical Temperature:** 417°F = 214°C = 487°K
13.6 **Critical Pressure:** 870 psia = 59.2 atm = 6.00 MN/m^2
13.7 **Specific Gravity:** 0.977 at 20°C (liquid)
13.8 **Liquid Surface Tension:** 25 dynes/cm = 0.025 N/m at 20°C
13.9 **Liquid-Water Interfacial Tension:** Not pertinent
13.10 **Vapor (Gas) Specific Gravity:** 2.07
13.11 **Ratio of Specific Heats of Vapor (Gas):** 1.1446
13.12 **Latent Heat of Vaporization:** 202 Btu/lb = 112 cal/g = 4.69×10^5 J/kg
13.13 **Heat of Combustion:** −6,980 Btu/lb = −3,880 cal/g = -162×10^5 J/kg
13.14 **Heat of Decomposition:** Not pertinent
13.15 **Heat of Solution:** Not pertinent
13.16 **Heat of Polymerization:** Not pertinent

NOTES

MHZ METHYLHYDRAZINE

Common Synonyms			
Monomethylhydrazine MMH	Liquid	Colorless	Ammonia-like odor
	Mixes with water. Poisonous, flammable vapor is produced.		

AVOID CONTACT WITH LIQUID AND VAPOR. KEEP PEOPLE AWAY.
Wear goggles and self-contained breathing apparatus.
Shut off ignition sources. Call fire department.
Stop discharge if possible.
Stay upwind. Use water spray to "knock down" vapor.
Isolate and remove discharged material.
Notify local health and pollution control agencies.

Fire

FLAMMABLE.
Containers may explode in fire.
Flashback along vapor trail may occur.
Vapor may explode if ignited in an enclosed area.
May explode if exposed to heat or flames.

Extinguish with dry chemicals, alcohol foam, or carbon dioxide.
Water may be ineffective on fire.
Cool exposed containers with water.

Exposure

Call for medical aid.

VAPOR
POISONOUS IF INHALED OR IF SKIN IS EXPOSED.
Irritating to eyes, nose and throat.
Move victim to fresh air.
If breathing has stopped, give artificial respiration.
If breathing is difficult, give oxygen.

LIQUID
POISONOUS IF SWALLOWED OR IF SKIN IS EXPOSED.
Will burn skin and eyes.
Remove contaminated clothing and shoes.
Flush affected areas with plenty of water.
IF IN EYES, hold eyelids open and flush with plenty of water.
IF SWALLOWED and victim is CONSCIOUS, have victim drink water or milk.
DO NOT INDUCE VOMITING.

Water Pollution

Effect of low concentrations on aquatic life is unknown.
May be dangerous if it enters water intakes.
Notify local health and wildlife officials.
Notify operators of nearby water intakes.

1. RESPONSE TO DISCHARGE

(See Response Methods Handbook, CG 446-4)

Issue warning—poison, high flammability, water contaminant, air contaminant
Restrict access
Evacuate area
Disperse and flush

2. LABELS

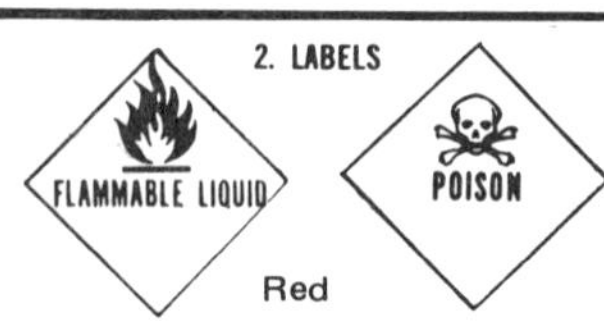

Red

3. CHEMICAL DESIGNATIONS

3.1 **Synonyms:** Monomethylhydrazine; MMH
3.2 **Coast Guard Compatibility Classification:** Not applicable
3.3 **Chemical Formula:** CH_3NHNH_2
3.4 **IMCO/United Nations Numerical Designation:** 3.2/1244

4. OBSERVABLE CHARACTERISTICS

4.1 **Physical State (as shipped):** Liquid
4.2 **Color:** Colorless
4.3 **Odor:** Like ammonia

5. HEALTH HAZARDS

5.1 **Personal Protective Equipment:** Organic canister mask or self-contained breathing apparatus; goggles or face shield; rubber gloves; protective clothing
5.2 **Symptoms Following Exposure:** Tremors and convulsions follow absorption by any route. Inhalation causes local irritation of respiratory tract, respiratory distress, and systemic effects. Contact of liquid with eyes or skin causes irritation and burns. Ingestion causes irritation of mouth and stomach.
5.3 **Treatment for Exposure:** Get medical attention at once following all exposures to this compound. INHALATION: move victim to fresh air and keep him quiet; give artificial respiration if breathing stops. EYES: flush for at least 15 min. with large quantities of water. SKIN: immediately wash with large quantities of water and treat as for alkali burn. INGESTION: give egg whites or other emollient, followed by a 5% salt solution or other mild emetic. Keep patient as quiet as possible. To control convulsions, short-acting barbituates may be administered parenterally by a physician with due regard for depression of respiration.
5.4 **Toxicity by Inhalation (Threshold Limit Value):** 0.2 ppm
5.5 **Short-Term Inhalation Limits:** 90 ppm for 10 min.; 30 ppm for 30 min.; 15 ppm for 60 min.
5.6 **Toxicity by Ingestion:** Grade 4; oral LD_{50} = 33 mg/kg (rat)
5.7 **Late Toxicity:** Hemolytic anemia may result from large doses by any route.
5.8 **Vapor (Gas) Irritant Characteristics:** Vapors are moderately irritating such that personnel will not usually tolerate moderate or high concentrations.

6. FIRE HAZARDS

6.1 **Flash Point:** 62°F O.C.
6.2 **Flammable Limits in Air:** 2.5%–98%
6.3 **Fire Extinguishing Agents:** Water or dry chemical
6.4 **Fire Extinguishing Agents Not to be Used:** Not pertinent
6.5 **Special Hazards of Combustion Products:** Irritating nitrogen oxides are produced.
6.6 **Behavior in Fire:** May explode
6.7 **Ignition Temperature:** 382°F
6.8 **Electrical Hazard:** Data not available
6.9 **Burning Rate:** 2.0 mm/min.

7. CHEMICAL REACTIVITY

7.1 **Reactivity with Water:** No reaction
7.2 **Reactivity with Common Materials:** Reacts slowly with air, but heat may cause ignition of rags, rust, or other combustibles.
7.3 **Stability During Transport:** Stable if not in contact with iron, copper, or their alloys.
7.4 **Neutralizing Agents for Acids and Caustics:** Flush with water.
7.5 **Polymerization:** Not pertinent
7.6 **Inhibitor of Polymerization:** Not pertinent

8. WATER POLLUTION

8.1 **Aquatic Toxicity:** Data not available
8.2 **Waterfowl Toxicity:** Data not available
8.3 **Biological Oxygen Demand (BOD):** Data not available
8.4 **Food Chain Concentration Potential:** None

9. SELECTED MANUFACTURERS

1. Olin Corporation
120 Long Ridge Road
Stamford, Conn. 06904
2. Eastman Kodak Co.
Eastman Organic Chemicals
Rochester, N. Y. 14650
3. Aldrich Chemical Co.
940 West St. Paul Ave.
Milwaukee, Wis. 53233

10. SHIPPING INFORMATION

10.1 **Grades or Purity:** Propellant grade, 99+%; Laboratory grade, 98+%
10.2 **Storage Temperature:** Ambient
10.3 **Inert Atmosphere:** Padded with nitrogen
10.4 **Venting:** Safety relief

11. HAZARD ASSESSMENT CODE

(See Hazard Assessment Handbook, CG 446-3)

A-P-Q-R-S

12. HAZARD CLASSIFICATIONS

12.1 **Code of Federal Regulations:** Flammable liquid
12.2 **NAS Hazard Rating for Bulk Water Transportation:**

Category	Rating
Fire	4
Health	
Vapor Irritant	3
Liquid or Solid Irritant	4
Poisons	4
Water Pollution	
Human Toxicity	4
Aquatic Toxicity	—
Aesthetic Effect	2
Reactivity	
Other Chemicals	4
Water	0
Self-Reaction	4

12.3 **NFPA Hazard Classifications:**

Category	Classification
Health Hazard (Blue)	3
Flammability (Red)	3
Reactivity (Yellow)	1

13. PHYSICAL AND CHEMICAL PROPERTIES

13.1 **Physical State at 15°C and 1 atm:** Liquid
13.2 **Molecular Weight:** 46.1
13.3 **Boiling Point at 1 atm:** 189.5°F = 87.5°C = 360.7°K
13.4 **Freezing Point:** −62.3°F = −52.4°C = 220.8°K
13.5 **Critical Temperature:** 594°F = 312°C = 585°K
13.6 **Critical Pressure:** 1,195 psia = 81.3 atm = 8.25 MN/m²
13.7 **Specific Gravity:** 0.878 at 20°C (liquid)
13.8 **Liquid Surface Tension:** 34.3 dynes/cm = 0.0343 N/m at 20°C
13.9 **Liquid-Water Interfacial Tension:** Not pertinent
13.10 **Vapor (Gas) Specific Gravity:** 1.59
13.11 **Ratio of Specific Heats of Vapor (Gas):** 1.1326
13.12 **Latent Heat of Vaporization:** 376 Btu/lb = 209 cal/g = 8.75×10^5 J/kg
13.13 **Heat of Combustion:** −12,178 Btu/lb = −6,766 cal/g = -283.1×10^5 J/kg
13.14 **Heat of Decomposition:** Not pertinent
13.15 **Heat of Solution:** Not pertinent
13.16 **Heat of Polymerization:** Not pertinent

5. HEALTH HAZARDS (Cont'd.)

5.9 **Liquid or Solid Irritant Characteristics:** Severe skin irritant. Causes second- and third-degree burns on short contact and is very injurious to the eyes.
5.10 **Odor Threshold:** 1–3 ppm

METHYL ISOBUTYL CARBINOL

MIC

Common Synonyms	Oily liquid	Colorless	Mild alcohol odor
Methylamyl alcohol 4-Methyl-2-pentanol Isobutylmethylcarbinol	Floats on water. Irritating vapor is produced.		

Stop discharge if possible. Keep people away.
Avoid contact with liquid and vapor.
Call fire department.
Isolate and remove discharged material.
Notify local health and pollution control agencies.

Fire	Combustible. Extinguish with dry chemical, alcohol foam, or carbon dioxide. Cool exposed containers with water.
Exposure	CALL FOR MEDICAL AID. VAPOR Irritating to eyes, nose and throat. Harmful if skin is exposed. If inhaled, will cause dizziness or difficult breathing. Move to fresh air. If breathing has stopped, give artificial respiration. If breathing is difficult, give oxygen. LIQUID Irritating to skin and eyes. Harmful if swallowed. Remove contaminated clothing and shoes. Flush affected areas with plenty of water. IF IN EYES, hold eyelids open and flush with plenty of water. IF SWALLOWED and victim is CONSCIOUS, have victim drink water or milk.
Water Pollution	Effect of low concentrations on aquatic life is unknown. Fouling to shoreline. May be dangerous if it enters water intakes. Notify local health and wildlife officials. Notify operators of nearby water intakes.

1. RESPONSE TO DISCHARGE

(See Response Methods Handbook, CG 446-4)

Mechanical containment
Chemical and physical treatment

2. LABELS

No hazard label required by Code of Federal Regulations

3. CHEMICAL DESIGNATIONS

3.1 **Synonyms:** Isobutyl methyl carbinol; Methylamyl Alcohol; MAOH; 4-Methyl-2-pentanol; MIBC; MIC

3.2 **Coast Guard Compatibility Classification:** Alcohol

3.3 **Chemical Formula:** $(CH_3)_2CHCH_2CH(OH)CH_3$

3.4 **IMCO/United Nations Numerical Designation:** 3.3/2053

4. OBSERVABLE CHARACTERISTICS

4.1 **Physical State (as shipped):** Liquid

4.2 **Color:** Colorless

4.3 **Odor:** Mild, sharp, non-residual

5. HEALTH HAZARDS

5.1 **Personal Protective Equipment:** Air pack or organic canister mask; rubber gloves; goggles or face shield

5.2 **Symptoms Following Exposure:** Vapor irritates eyes and nose; may cause anesthesia. Prolonged contact with liquid causes irritation and cracking of skin; also irritates eyes.

5.3 **Treatment for Exposure:** INHALATION: remove to fresh air; give artificial respiration if needed; call a doctor. SKIN: flush with water. EYES: flush with water for at least 15 min; consult a doctor.

5.4 **Toxicity by Inhalation (Threshold Limit Value):** 25 ppm

5.5 **Short-Term Inhalation Limits:** Data not available

5.6 **Toxicity by Ingestion:** Grade 2; LD_{50} 0.5 to 5 g/kg (rat)

5.7 **Late Toxicity:** None

5.8 **Vapor (Gas) Irritant Characteristics:** Vapors cause moderate irritation such that personnel will find high concentrations unpleasant. The effect is temporary.

5.9 **Liquid or Solid Irritant Characteristics:** Minimum hazard. If spilled on clothing and allowed to remain, may cause smarting and reddening of the skin.

5.10 **Odor Threshold:** Data not available

6. FIRE HAZARDS

6.1 **Flash Point:** 120-130°F O.C.; 106°F C.C.

6.2 **Flammable Limits in Air:** 1.0%—5.5%

6.3 **Fire Extinguishing Agents:** Alcohol foam, dry chemical, or carbon dioxide

6.4 **Fire Extinguishing Agents Not to be Used:** Not pertinent

6.5 **Special Hazards of Combustion Products:** Not pertinent

6.6 **Behavior in Fire:** Not pertinent

6.7 **Ignition Temperature:** Data not available

6.8 **Electrical Hazard:** Not pertinent

6.9 **Burning Rate:** Data not available

7. CHEMICAL REACTIVITY

7.1 **Reactivity with Water:** No reaction

7.2 **Reactivity with Common Materials:** No reaction

7.3 **Stability During Transport:** Stable

7.4 **Neutralizing Agents for Acids and Caustics:** Not pertinent

7.5 **Polymerization:** Not pertinent

7.6 **Inhibitor of Polymerization:** Not pertinent

8. WATER POLLUTION

8.1 **Aquatic Toxicity:** 370 ppm/24 hr/brine shrimp/TL_m

8.2 **Waterfowl Toxicity:** Data not available

8.3 **Biological Oxygen Demand (BOD):** 50% of theoretical in 5 days, freshwater

8.4 **Food Chain Concentration Potential:** None

9. SELECTED MANUFACTURERS

1. Exxon Chemical Co.
   Houston, Tex. 77001
2. Shell Chemical Co.
   Industrial Chemicals Division
   Houston, Tex. 77001
3. Union Carbide Corp.
   Chemicals & Plastics Division
   270 Park Ave.
   New York, N. Y. 10017

10. SHIPPING INFORMATION

10.1 **Grades or Purity:** Data not available

10.2 **Storage Temperature:** Ambient

10.3 **Inert Atmosphere:** No requirement

10.4 **Venting:** Open (flame arrester)

11. HAZARD ASSESSMENT CODE

(See Hazard Assessment Handbook, CG 446-3)

A-P-Q-T-U

12. HAZARD CLASSIFICATIONS

12.1 **Code of Federal Regulations:** Combustible Liquid

12.2 **NAS Hazard Rating for Bulk Water Transportation:**

Category	Rating
Fire	2
Health	
Vapor Irritant	2
Liquid or Solid Irritant	1
Poisons	2
Water Pollution	
Human Toxicity	2
Aquatic Toxicity	2
Aesthetic Effect	2
Reactivity	
Other Chemicals	2
Water	0
Self-Reaction	0

12.3 **NFPA Hazard Classifications:**

Category	Classification
Health Hazard (Blue)	2
Flammability (Red)	2
Reactivity (Yellow)	0

13. PHYSICAL AND CHEMICAL PROPERTIES

13.1 **Physical State at 15°C and 1 atm:** Liquid

13.2 **Molecular Weight:** 102.18

13.3 **Boiling Point at 1 atm:** 269.2°F = 131.8°C = 405°K

13.4 **Freezing Point:** <-130°F = <-90°C = <183°K

13.5 **Critical Temperature:** 556°F = 291°C = 564°K

13.6 **Critical Pressure:** Not pertinent

13.7 **Specific Gravity:** 0.807 at 20°C (liquid)

13.8 **Liquid Surface Tension:** 22.8 dynes/cm = 0.0228 N/m at 20°C

13.9 **Liquid-Water Interfacial Tension:** (est.) 25 dynes/cm = 0.025 N/m at 20°C

13.10 **Vapor (Gas) Specific Gravity:** Not pertinent

13.11 **Ratio of Specific Heats of Vapor (Gas):** 1.053

13.12 **Latent Heat of Vaporization:** 162 Btu/lb = 90.1 cal/g = 3.77×10^5 J/kg

13.13 **Heat of Combustion:** (est.) −16,600 Btu/lb = −9,300 cal/g = -387×10^5 J/kg

13.14 **Heat of Decomposition:** Not pertinent

13.15 **Heat of Solution:** Not pertinent

13.16 **Heat of Polymerization:** Not pertinent

NOTES

REVISED 1978

MIK	METHYL ISOBUTYL KETONE

Common Synonyms		
4-Methyl-2-pentanone Isobutyl methyl ketone MIBK	Watery liquid	Colorless	Mild pleasant odor

Floats and mixes slowly with water. Flammable, irritating vapor is produced.

Stop discharge if possible. Keep people away.
Shut off ignition sources and call fire department.
Stay upwind and use water spray to "knock down" vapor.
Avoid contact with liquid and vapor.
Isolate and remove discharged material.
Notify local health and pollution control agencies.

Fire

FLAMMABLE.
Flashback along vapor trail may occur.
Vapor may explode if ignited in an enclosed area.
Wear goggles and self-contained breathing apparatus.
Extinguish with dry chemical, alcohol foam, or carbon dioxide.
Water may be ineffective on fire.
Cool exposed containers with water.

Exposure

CALL FOR MEDICAL AID.

VAPOR
Irritating to eyes, nose and throat.
If inhaled, will cause dizziness or loss of consciousness.
Move to fresh air.
If breathing has stopped, give artificial respiration.
If breathing is difficult, give oxygen.

LIQUID
Irritating to skin and eyes.
Harmful if swallowed.

Remove contaminated clothing and shoes.
Flush affected areas with plenty of water.
IF IN EYES, hold eyelids open and flush with plenty of water.
IF SWALLOWED and victim is CONSCIOUS, have victim drink water or milk.

Water Pollution

Effect of low concentrations on aquatic life is unknown.
Fouling to shoreline.
May be dangerous if it enters water intakes.
Notify local health and wildlife officials.
Notify operators of nearby water intakes.

1. RESPONSE TO DISCHARGE

(See Response Methods Handbook, CG 446-4)

Issue warning—high flammability
Evacuate area
Disperse and flush

2. LABEL

Red

3. CHEMICAL DESIGNATIONS

3.1 **Synonyms:** Hexone; Isobutyl methyl ketone; Isopropylacetone; 4-Methyl-2-pentanone; MIBK; MIK
3.2 **Coast Guard Compatibility Classification:** Ketone
3.3 **Chemical Formula:** $(CH_3)_2CHCH_2COCH_3$
3.4 **IMCO/United Nations Numerical Designation:** 3.2/1245

4. OBSERVABLE CHARACTERISTICS

4.1 **Physical State (as shipped):** Liquid
4.2 **Color:** Colorless
4.3 **Odor:** Pleasant; mild, characteristic; sharp; non-residual; ketonic

5. HEALTH HAZARDS

5.1 **Personal Protective Equipment:** Organic canister or air pack; rubber gloves; goggles or face shield.
5.2 **Symptoms Following Exposure:** Vapor causes irritation of eyes and nose; high concentrations cause anesthesia and depression. Liquid dries out skin and may cause dermatitis; irritates eyes but does not injure them.
5.3 **Treatment for Exposure:** INHALATION: remove to fresh air, give artificial respiration if needed; call a doctor. SKIN OR EYES: wash eyes thoroughly with water; wash skin with water until irritation stops.
5.4 **Toxicity by Inhalation (Threshold Limit Value):** 100 ppm
5.5 **Short-Term Inhalation Limits:** 100 ppm for 60 min.
5.6 **Toxicity by Ingestion:** Grade 2; LD_{50} 0.5 to 5 g/kg (rat)
5.7 **Late Toxicity:** None
5.8 **Vapor (Gas) Irritant Characteristics:** Vapors cause a slight smarting of the eyes or respiratory system if present in high concentrations. The effect is temporary.
5.9 **Liquid or Solid Irritant Characteristics:** Minimum hazard. If spilled on clothing and allowed to remain, may cause smarting and reddening of the skin.
5.10 **Odor Threshold:** 0.47 ppm

6. FIRE HAZARDS

6.1 **Flash Point:** 73°F C.C.; 75°F O.C.
6.2 **Flammable Limits in Air:** 1.4%–7.5%
6.3 **Fire Extinguishing Agents:** Alcohol foam, dry chemical, or carbon dioxide
6.4 **Fire Extinguishing Agents Not to be Used:** Water may be ineffective
6.5 **Special Hazards of Combustion Products:** Irritating vapors are generated when heated.
6.6 **Behavior in Fire:** Vapors may travel a considerable distance and ignite.
6.7 **Ignition Temperature:** 854°F
6.8 **Electrical Hazard:** Class I, Group D
6.9 **Burning Rate:** Data not available

7. CHEMICAL REACTIVITY

7.1 **Reactivity with Water:** No reaction
7.2 **Reactivity with Common Materials:** No reaction
7.3 **Stability During Transport:** Stable
7.4 **Neutralizing Agents for Acids and Caustics:** Not pertinent
7.5 **Polymerization:** Not pertinent
7.6 **Inhibitor of Polymerization:** Not pertinent

8. WATER POLLUTION

8.1 **Aquatic Toxicity:** Data not available
8.2 **Waterfowl Toxicity:** Data not available
8.3 **Biological Oxygen Demand (BOD):** (theor.) 1.8%, 0.5 day; (theor.) 12%, 5 days
8.4 **Food Chain Concentration Potential:** None

9. SELECTED MANUFACTURERS

1. Exxon Chemical Co.
Houston, Tex. 77001
2. Shell Chemical Co.
Industrial Chemicals Division
Houston, Tex. 77001
3. Union Carbide Corp.
Chemicals & Plastics Division
270 Park Ave.
New York, N. Y. 10017

10. SHIPPING INFORMATION

10.1 **Grades or Purity:** 99+%
10.2 **Storage Temperature:** Ambient
10.3 **Inert Atmosphere:** No requirement
10.4 **Venting:** Open (flame arrester) or pressure-vacuum

11. HAZARD ASSESSMENT CODE

(See Hazard Assessment Handbook, CG 446-3)

A-P-Q-T-U

12. HAZARD CLASSIFICATIONS

12.1 **Code of Federal Regulations:** Flammable liquid
12.2 **NAS Hazard Rating for Bulk Water Transportation:**

Category	Rating
Fire	3
Health	
Vapor Irritant	1
Liquid or Solid Irritant	1
Poisons	1
Water Pollution	
Human Toxicity	2
Aquatic Toxicity	1
Aesthetic Effect	2
Reactivity	
Other Chemicals	2
Water	0
Self-Reaction	0

12.3 **NFPA Hazard Classifications:**

Category	Classification
Health Hazard (Blue)	2
Flammability (Red)	3
Reactivity (Yellow)	0

13. PHYSICAL AND CHEMICAL PROPERTIES

13.1 **Physical State at 15°C and 1 atm:** Liquid
13.2 **Molecular Weight:** 100.16
13.3 **Boiling Point at 1 atm:** 241.2°F = 116.2 = 389.4°K
13.4 **Freezing Point:** −119°F = −84°C = 189°K
13.5 **Critical Temperature:** 568.9°F = 298.3°C = 571.5°K
13.6 **Critical Pressure:** 475 psia = 32.3 atm = 3.27 MN/m²
13.7 **Specific Gravity:** 0.802 at 20°C (liquid)
13.8 **Liquid Surface Tension:** 23.6 dynes/cm = 0.0236 N/m at 20°C
13.9 **Liquid-Water Interfacial Tension:** 15.7 dynes/cm = 0.0157 N/m at 22.7°C
13.10 **Vapor (Gas) Specific Gravity:** Not pertinent
13.11 **Ratio of Specific Heats of Vapor (Gas):** 1.061
13.12 **Latent Heat of Vaporization:** 149 Btu/lb = 82.5 cal/g = 3.45×10^5 J/kg
13.13 **Heat of Combustion:** (est.) −10,400 Btu/lb = −5,800 cal/g = -242×10^5 J/kg
13.14 **Heat of Decomposition:** Not pertinent
13.15 **Heat of Solution:** (est.) −9 Btu/lb = −5 cal/g = -0.2×10^5 J/kg
13.16 **Heat of Polymerization:** Not pertinent

NOTES

REVISED 1978

MPK — METHYL ISOPROPENYL KETONE, INHIBITED

Common Synonyms	Liquid	Colorless	Sweet pleasant odor
2-Methyl-1-butene-3-one	Floats on water. Flammable, irritating vapor is produced.		

Shut off ignition sources. Call fire department.
Stop discharge if possible. Keep people away.
Isolate and remove discharged material.
Notify local health and pollution control agencies.

Fire

FLAMMABLE.
Containers may explode in fire.
Flashback along vapor trail may occur.
Vapor may explode if ignited in an enclosed area.
Combat fires from safe distance or protected location.
Extinguish with dry chemicals, foam or carbon dioxide.
Water may be ineffective on fire.
Cool exposed containers with water.

Exposure

Call for medical aid.

VAPOR
Irritating to eyes, nose and throat.
Move victim to fresh air.
If breathing has stopped, give artificial respiration.
If breathing is difficult, give oxygen.

LIQUID
Irritating to skin and eyes.
Harmful if swallowed.
Remove contaminated clothing and shoes.
Flush affected areas with plenty of water.
IF IN EYES, hold eyelids open and flush with plenty of water.
IF SWALLOWED and victim is CONSCIOUS, have victim drink water or milk and have victim induce vomiting.
IF SWALLOWED and victim is UNCONSCIOUS OR HAVING CONVULSIONS, do nothing except keep victim warm.

Water Pollution

Effect of low concentrations on aquatic life is unknown.
Fouling to shoreline.
May be dangerous if it enters water intakes.
Notify local health and wildlife officials.
Notify operators of nearby water intakes.

1. RESPONSE TO DISCHARGE

(See Response Methods Handbook, CG 446-4)
Issue warning—high flammability, water contaminant
Restrict access
Mechanical containment
Should be removed
Chemical and physical treatment

2. LABEL

Red

3. CHEMICAL DESIGNATIONS

3.1 **Synonyms:** Isopropenyl methyl ketone; 2-Methyl-1-butene-3-one
3.2 **Coast Guard Compatibility Classification:** Not applicable
3.3 **Chemical Formula:** $CH_3COC(CH_3){=}CH_2$
3.4 **IMCO/United Nations Numerical Designation:** 3.2/1246

4. OBSERVABLE CHARACTERISTICS

4.1 **Physical State (as shipped):** Liquid
4.2 **Color:** Colorless
4.3 **Odor:** Very pungent; pleasant; sweet

5. HEALTH HAZARDS

5.1 **Personal Protective Equipment:** Goggles or face shield; rubber gloves.
5.2 **Symptoms Following Exposure:** Inhalation causes irritation of nose and throat. Liquid may cause severe damage to eyes, resulting possibly in some permanent impairment of vision; vapor produces tears. If not removed promptly from skin, liquid may cause delayed pain and blistering. Ingestion causes irritation of mouth and stomach.
5.3 **Treatment for Exposure:** INHALATION: remove victim from exposure; give artificial respiration if needed; call physician. EYES: immediately irrigate with copious amounts of water for 15 min.; call physician. SKIN: wash off skin with large volumes of water for 15 min.; call physician if burn has occurred. INGESTION: induce vomiting; call physician.
5.4 **Toxicity by Inhalation (Threshold Limit Value):** Data not available
5.5 **Short-Term Inhalation Limits:** Data not available
5.6 **Toxicity by Ingestion:** Grade 3; oral LD_{50} = 180 mg/kg (rat)
5.7 **Late Toxicity:** Data not available
5.8 **Vapor (Gas) Irritant Characteristics:** Vapors are moderately irritating such that personnel will not usually tolerate moderate or high concentrations.
5.9 **Liquid or Solid Irritant Characteristics:** Causes smarting and first-degree burns on short exposure; may cause second-degree burns on long exposure.
5.10 **Odor Threshold:** Data not available

6. FIRE HAZARDS

6.1 **Flash Point:** <73°F C.C.
6.2 **Flammable Limits in Air:** 1.8% 9.0%
6.3 **Fire Extinguishing Agents:** Dry chemical, foam, carbon dioxide
6.4 **Fire Extinguishing Agents Not to be Used:** Water may be ineffective.
6.5 **Special Hazards of Combustion Products:** Not pertinent
6.6 **Behavior in Fire:** May polymerize and explode.
6.7 **Ignition Temperature:** Data not available
6.8 **Electrical Hazard:** Data not available
6.9 **Burning Rate:** 4.7 mm/min.

7. CHEMICAL REACTIVITY

7.1 **Reactivity with Water:** No reaction
7.2 **Reactivity with Common Materials:** No reaction
7.3 **Stability During Transport:** Stable
7.4 **Neutralizing Agents for Acids and Caustics:** Not pertinent
7.5 **Polymerization:** Will polymerize in the absence of inhibitor, especially when heated.
7.6 **Inhibitor of Polymerization:** Up to 1% hydroquinone

8. WATER POLLUTION

8.1 **Aquatic Toxicity:** Data not available
8.2 **Waterfowl Toxicity:** Data not available
8.3 **Biological Oxygen Demand (BOD):** Data not available
8.4 **Food Chain Concentration Potential:** None

9. SELECTED MANUFACTURERS

1. Eastern Chemical Division
   230 Marcus Boulevard
   Hauppauge, L. I., N. Y. 11787
2. Pfaltz and Bauer, Inc.
   126-04 Northern Boulevard
   Flushing, N. Y. 11368

10. SHIPPING INFORMATION

10.1 **Grades or Purity:** Commercial
10.2 **Storage Temperature:** Ambient
10.3 **Inert Atmosphere:** No requirement
10.4 **Venting:** Pressure-vacuum

11. HAZARD ASSESSMENT CODE

(See Hazard Assessment Handbook, CG 446-3)
A-T-U-V-W

12. HAZARD CLASSIFICATIONS

12.1 **Code of Federal Regulations:** Flammable liquid
12.2 **NAS Hazard Rating for Bulk Water Transportation:**

Category	Rating
Fire	3
Health	
Vapor Irritant	3
Liquid or Solid Irritant	2
Poisons	3
Water Pollution	
Human Toxicity	
Aquatic Toxicity	
Aesthetic Effect	2
Reactivity	
Other Chemicals	3
Water	0
Self-Reaction	3

12.3 **NFPA Hazard Classifications:**

Category	Classification
Health Hazard (Blue)	2
Flammability (Red)	
Reactivity (Yellow)	0

13. PHYSICAL AND CHEMICAL PROPERTIES

13.1 **Physical State at 15°C and 1 atm:** Liquid
13.2 **Molecular Weight:** 84.1
13.3 **Boiling Point at 1 atm:** 208°F = 98°C = 371°K
13.4 **Freezing Point:** −65°F = −54°C = 219°K
13.5 **Critical Temperature:** Not pertinent
13.6 **Critical Pressure:** Not pertinent
13.7 **Specific Gravity:** 0.85 at 20°C (liquid)
13.8 **Liquid Surface Tension:** (est.) 26 dynes/cm = 0.026 N/m at 20°C
13.9 **Liquid-Water Interfacial Tension:** (est.) 30 dynes/cm = 0.030 N/m at 20°C
13.10 **Vapor (Gas) Specific Gravity:** 2.9
13.11 **Ratio of Specific Heats of Vapor (Gas):** 1.0796 at 20°C (68°F)
13.12 **Latent Heat of Vaporization:** (est.) 182 Btu/lb = 101 cal/g = 4.23×10^5 J/kg
13.13 **Heat of Combustion:** (est.) −15,500 Btu/lb = −8,600 cal/g = -360×10^5 J/kg
13.14 **Heat of Decomposition:** Not pertinent
13.15 **Heat of Solution:** Not pertinent
13.16 **Heat of Polymerization:** (est.) −380 Btu/lb = −210 cal/g = -8.8×10^5 J/kg

NOTES

MMC

METHYL MERCAPTAN

Common Synonyms			
Methaneethiol Mercaptomethane Thiomethyl alcohol Methyl sulfhydrate	Liquefied compressed gas	Colorless	Strong garlic odor
	Floats and boils on water. Poisonous, flammable vapor is produced.		

AVOID CONTACT WITH LIQUID AND VAPOR. KEEP PEOPLE AWAY.
Wear goggles, self-contained breathing apparatus, and rubber overclothing (including gloves).
Shut off ignition sources. Call fire department.
Stop discharge if possible.
Evacuate area in case of large discharge.
Notify local health and pollution control agencies.

Fire

FLAMMABLE.
POISONOUS GASES ARE PRODUCED IN FIRE.
Containers may explode in fire.
Flashback along vapor trail may occur.
Vapor may explode if ignited in an enclosed area.
Wear goggles and self-contained breathing apparatus.
Let fire burn.
Stop flow of gas if possible.
Cool exposed containers and protect men effecting shutoff with water.

Exposure

CALL FOR MEDICAL AID.

VAPOR
POISONOUS IF INHALED.
Irritating to eyes, nose and throat.
Move victim to fresh air.
If in eyes, hold eyelids open and flush with plenty of water.
If breathing has stopped, give artificial respiration.
If breathing is difficult, give oxygen.

LIQUID
POISONOUS IF SWALLOWED.
Irritating to skin and eyes.
Remove contaminated clothing and shoes.
Flush affected areas with plenty of water.
IF IN EYES, hold eyelids open and flush with plenty of water.
IF SWALLOWED and victim is CONSCIOUS, have victim drink water or milk and have victim induce vomiting.
IF SWALLOWED and victim is UNCONSCIOUS OR HAVING CONVULSIONS, do nothing except keep victim warm.

Water Pollution

HARMFUL TO AQUATIC LIFE IN VERY LOW CONCENTRATIONS.
May be dangerous if it enters water intakes.
Notify local health and wildlife officials.
Notify operators of nearby water intakes.

1. RESPONSE TO DISCHARGE

(See Response Methods Handbook, CG 446-4)

Issue warning—high flammability, water contaminant, air contaminant
Restrict access
Evacuate area
Disperse and flush

2. LABEL

Red

3. CHEMICAL DESIGNATIONS

3.1 **Synonyms:** Mercaptomethane; Methanethiol; Methyl sulfhydrate; Thiomethyl alcohol
3.2 **Coast Guard Compatibility Classification:** Not applicable
3.3 **Chemical Formula:** CH_3SH
3.4 **IMCO/United Nations Numerical Designation:** 2/1064

4. OBSERVABLE CHARACTERISTICS

4.1 **Physical State (as shipped):** Liquefied compressed gas
4.2 **Color:** Colorless
4.3 **Odor:** Garlic; foul; strong offensive

5. HEALTH HAZARDS

5.1 **Personal Protective Equipment:** Rubber gloves; goggles or face shield; air-line or self-contained breathing apparatus.
5.2 **Symptoms Following Exposure:** Inhalation causes irritation of respiratory system, tremors, paralysis, unconsciousness; death may follow respiratory paralysis. Contact with liquid irritates eyes and skin. Ingestion causes irritation of mouth and stomach plus symptoms described for inhalation.
5.3 **Treatment for Exposure:** INHALATION: remove patient immediately from the contaminated area; keep him warm and at complete rest; if necessary give artificial respiration until medical assistance can be obtained; oxygen or oxygen-CO_2 inhalation is recommended, continuing after spontaneous breathing has returned. EYES: for exposure to vapor, apply hot and cold compresses to reduce pain of conjunctivitis; for exposure to liquid, wash with water and obtain medical assistance. SKIN: wash with water. INGESTION: induce vomiting and follow with gastric lavage.
5.4 **Toxicity by Inhalation (Threshold Limit Value):** 0.5 ppm
5.5 **Short-Term Inhalation Limits:** 20 ppm for 5 min.
5.6 **Toxicity by Ingestion:** Data not available
5.7 **Late Toxicity:** Data not available
5.8 **Vapor (Gas) Irritant Characteristics:** Data not available
5.9 **Liquid or Solid Irritant Characteristics:** Data not available
5.10 **Odor Threshold:** 0.0021 ppm

6. FIRE HAZARDS

6.1 **Flash Point:** Not pertinent (flammable, liquefied compressed gas)
6.2 **Flammable Limits in Air:** 3.9%–21.8%
6.3 **Fire Extinguishing Agents:** Preferably let fire burn, stop gas flow. Fires may be extinguished with dry chemical, foam, or carbon dioxide.
6.4 **Fire Extinguishing Agents Not to be Used:** Water may be ineffective.
6.5 **Special Hazards of Combustion Products:** Irritating sulfur dioxide is produced.
6.6 **Behavior in Fire:** Containers may explode.
6.7 **Ignition Temperature:** Data not available
6.8 **Electrical Hazard:** Data not available
6.9 **Burning Rate:** 3.8 mm/min.

7. CHEMICAL REACTIVITY

7.1 **Reactivity with Water:** No reaction
7.2 **Reactivity with Common Materials:** No reaction
7.3 **Stability During Transport:** Stable
7.4 **Neutralizing Agents for Acids and Caustics:** Not pertinent
7.5 **Polymerization:** Not pertinent
7.6 **Inhibitor of Polymerization:** Not pertinent

8. WATER POLLUTION

8.1 **Aquatic Toxicity:** 1.0 ppm/105 min/white bass/death/fresh water
8.2 **Waterfowl Toxicity:** Data not available
8.3 **Biological Oxygen Demand (BOD):** Data not available
8.4 **Food Chain Concentration Potential:** None

9. SELECTED MANUFACTURERS

1. Amoco Chemicals Corporation
130 East Randolph Drive
Chicago, Ill. 60601
2. Pennwalt Corporation
Chemicals Division
Three Parkway
Philadelphia, Pa. 19102
3. Matheson Gas Products Co.
East Rutherford, N. J. 07073

10. SHIPPING INFORMATION

10.1 **Grades or Purity:** 99.5+%
10.2 **Storage Temperature:** Ambient
10.3 **Inert Atmosphere:** No requirement
10.4 **Venting:** Safety relief

11. HAZARD ASSESSMENT CODE

(See Hazard Assessment Handbook, CG 446-3)

A-B-C-K-L-M-N

12. HAZARD CLASSIFICATIONS

12.1 **Code of Federal Regulations:** Flammable compressed gas
12.2 **NAS Hazard Rating for Bulk Water Transportation:** Not listed
12.3 **NFPA Hazard Classifications:**

Category	Classification
Health Hazard (Blue)	2
Flammability (Red)	4
Reactivity (Yellow)	0

13. PHYSICAL AND CHEMICAL PROPERTIES

13.1 **Physical State at 15°C and 1 atm:** Gas
13.2 **Molecular Weight:** 48.1
13.3 **Boiling Point at 1 atm:** 43.2°F = 6.2°C = 279.4°K
13.4 **Freezing Point:** −189°F = −123°C = 150°K
13.5 **Critical Temperature:** 386.2°F = 196.8°C = 470.0°K
13.6 **Critical Pressure:** 1,050 psia = 71.4 atm = 7.25 MN/m^2
13.7 **Specific Gravity:** 0.892 at 6°C (liquid)
13.8 **Liquid Surface Tension:** 31 dynes/cm = 0.031 N/m at 5°C
13.9 **Liquid-Water Interfacial Tension:** Not pertinent
13.10 **Vapor (Gas) Specific Gravity:** 1.66
13.11 **Ratio of Specific Heats of Vapor (Gas):** 1.1988
13.12 **Latent Heat of Vaporization:** 220 Btu/lb = 122 cal/g = 5.10×10^5 J/kg
13.13 **Heat of Combustion:** −11,054 Btu/lb = −6,141 cal/g = -257.0×10^5 J/kg
13.14 **Heat of Decomposition:** Not pertinent
13.15 **Heat of Solution:** Not pertinent
13.16 **Heat of Polymerization:** Not pertinent

NOTES

MMM	# METHYL METHACRYLATE

Common Synonyms Methacrylic acid, methyl ester	Liquid Colorless Pleasant sharp odor Floats on water. Flammable, irritating vapor is produced.

Stop discharge if possible. Keep people away.
Wear goggles, self-contained breathing apparatus, and rubber overclothing (including gloves).
Shut off ignition sources and call fire department.
Stay upwind and use water spray to "knock down" vapor.
Avoid contact with liquid and vapor.
Isolate and remove discharged material.
Notify local health and pollution control agencies.

Fire	FLAMMABLE. Flashback along vapor trail may occur. Container may explode when heated. Vapor may explode if ignited in an enclosed area. Wear goggles, self-contained breathing apparatus, and rubber overclothing (including gloves). Combat fires from safe distance or protected location. Extinguish with dry chemical, foam, or carbon dioxide. Water may be ineffective on fire. Cool exposed containers with water.
Exposure	CALL FOR MEDICAL AID. VAPOR Irritating to eyes, nose and throat. If inhaled, will cause dizziness, headache, difficult breathing or loss of consciousness. Move to fresh air. If breathing has stopped, give artificial respiration. If breathing is difficult, give oxygen. LIQUID Will burn skin and eyes. Harmful if swallowed. Remove contaminated clothing and shoes. Flush affected areas with plenty of water. IF IN EYES, hold eyelids open and flush with plenty of water. IF SWALLOWED and victim is CONSCIOUS, have victim drink water or milk.
Water Pollution	Dangerous to aquatic life in high concentrations. Fouling to shoreline. May be dangerous if it enters water intakes. Notify local health and wildlife officials. Notify operators of nearby water intakes.

1. RESPONSE TO DISCHARGE

(See Response Methods Handbook, CG 446-4)

Issue warning—high flammability
Evacuate area
Disperse and flush

2. LABEL

Red

3. CHEMICAL DESIGNATIONS

3.1 **Synonyms:** Methacrylate monomer; Methacrylic acid, methyl ester; Methyl α-methylacrylate; Methyl 2-methyl-2-propenoate
3.2 **Coast Guard Compatibility Classification:** Acrylate
3.3 **Chemical Formula:** $CH_2=C(CH_3)COOCH_3$
3.4 **IMCO/United Nations Numerical Designation:** 3.2/1247

4. OBSERVABLE CHARACTERISTICS

4.1 **Physical State (as shipped):** Liquid
4.2 **Color:** Colorless
4.3 **Odor:** Sharp, fragrant; pleasant smelling; pungent ester

5. HEALTH HAZARDS

5.1 **Personal Protective Equipment:** Air mask; plastic gloves; goggles.
5.2 **Symptoms Following Exposure:** Irritation of eyes, nose, and throat. Nausea and vomiting. Liquid may cause skin irritation.
5.3 **Treatment for Exposure:** INHALATION: remove to fresh air; apply artificial respiration and oxygen if needed; refer to physician. SKIN OR EYES: flush with plenty of water for 15 min.; refer to physician for eye exposure.
5.4 **Toxicity by Inhalation (Threshold Limit Value):** 100 ppm
5.5 **Short-Term Inhalation Limits:** Data not available
5.6 **Toxicity by Ingestion:** Grade 1; LD_{50} 5 to 15 g/kg (rat)
5.7 **Late Toxicity:** Data not available
5.8 **Vapor (Gas) Irritant Characteristics:** Vapor is moderately irritating such that personnel will not usually tolerate moderate or high vapor concentrations.
5.9 **Liquid or Solid Irritant Characteristics:** Causes smarting of the skin and first-degree burns on short exposure; may cause secondary burns on long exposure.
5.10 **Odor Threshold:** 0.05 ppm

6. FIRE HAZARDS

6.1 **Flash Point:** 50°F O.C.
6.2 **Flammable Limits in Air:** 2.1%—12.5%
6.3 **Fire Extinguishing Agents:** Foam, carbon dioxide, dry chemical
6.4 **Fire Extinguishing Agents Not to be Used:** Water may be ineffective
6.5 **Special Hazards of Combustion Products:** Not pertinent
6.6 **Behavior in Fire:** Vapor is heavier than air and may travel a considerable distance to a source of ignition and flash back. Containers may explode in fire or when heated because of polymerization.
6.7 **Ignition Temperature:** 790°F
6.8 **Electrical Hazard:** Not pertinent
6.9 **Burning Rate:** 2.5 mm/min.

7. CHEMICAL REACTIVITY

7.1 **Reactivity with Water:** No reaction
7.2 **Reactivity with Common Materials:** No reaction
7.3 **Stability During Transport:** Stable
7.4 **Neutralizing Agents for Acids and Caustics:** Not pertinent
7.5 **Polymerization:** Heat, oxidizing agents, and ultraviolet light may cause polymerization.
7.6 **Inhibitor of Polymerization:** Hydroquinone, 22-65 ppm; hydroquinone methyl ether, 22-120 ppm; dimethyl tert-butylphenol, 45–65 ppm

8. WATER POLLUTION

8.1 **Aquatic Toxicity:** 250 ppm/96 hr/bluegill/TL_m/fresh water
8.2 **Waterfowl Toxicity:** Data not available
8.3 **Biological Oxygen Demand (BOD):** (theor.) 47%, 10 days
8.4 **Food Chain Concentration Potential:** None

9. SELECTED MANUFACTURERS

1. American Cyanamid Co.
   Industrial Chemicals & Plastics Division
   Wayne, N. J. 07470
2. E. I. duPont de Nemours & Co. Inc.
   Industrial and Biochemicals Dept.
   Wilmington, Del. 19898
3. Rohm and Haas Co.
   Independence Mall West
   Philadelphia, Pa. 19105

10. SHIPPING INFORMATION

10.1 **Grades or Purity:** 99.8%
10.2 **Storage Temperature:** Ambient
10.3 **Inert Atmosphere:** No requirement
10.4 **Venting:** Pressure-vacuum

11. HAZARD ASSESSMENT CODE

(See Hazard Assessment Handbook, CG 446-3)

A-P-Q-T-U-V-W-Z

12. HAZARD CLASSIFICATIONS

12.1 **Code of Federal Regulations:** Flammable liquid
12.2 **NAS Hazard Rating for Bulk Water Transportation:**

Category	Rating
Fire	3
Health	
Vapor Irritant	3
Liquid or Solid Irritant	2
Poisons	3
Water Pollution	
Human Toxicity	2
Aquatic Toxicity	2
Aesthetic Effect	2
Reactivity	
Other Chemicals	2
Water	0
Self-Reaction	3

12.3 **NFPA Hazard Classifications:**

Category	Classification
Health Hazard (Blue)	2
Flammability (Red)	3
Reactivity (Yellow)	2

13. PHYSICAL AND CHEMICAL PROPERTIES

13.1 **Physical State at 15°C and 1 atm:** Liquid
13.2 **Molecular Weight:** 100.12
13.3 **Boiling Point at 1 atm:** 214°F = 101°C = 374°K
13.4 **Freezing Point:** −54°F = −48°C = 225°K
13.5 **Critical Temperature:** 561°F = 294°C = 567°K
13.6 **Critical Pressure:** 485 psia = 33 atm = 3.3 MN/m²
13.7 **Specific Gravity:** 0.945 at 20°C (liquid)
13.8 **Liquid Surface Tension:** 28 dynes/cm = 0.028 N/m at 20°C
13.9 **Liquid-Water Interfacial Tension:** 14.3 dynes/cm = 0.0143 N/m at 22.7°C
13.10 **Vapor (Gas) Specific Gravity:** Not pertinent
13.11 **Ratio of Specific Heats of Vapor (Gas):** 1.059
13.12 **Latent Heat of Vaporization:** 140 Btu/lb = 77 cal/g = 3.2×10^5 J/kg
13.13 **Heat of Combustion:** (est.) −11,400 Btu/lb = −6,310 cal/g = -264×10^5 J/kg
13.14 **Heat of Decomposition:** Not pertinent
13.15 **Heat of Solution:** Not pertinent
13.16 **Heat of Polymerization:** −248 Btu/lb = −138 cal/g = -5.78×10^5 J/kg

NOTES

REVISED 1978

MPT	# METHYL PARATHION

Common Synonyms			
Metron MPT Paridol Alkron Nitran Wolfatox	Solid crystals or liquid or solution	White solid or brown liquid	Rotten eggs or garlic odor
	Solid and liquid sink in water, solution floats on water. Melting (freezing) point is 65° F.		

AVOID CONTACT WITH LIQUID. Keep people away.
Wear chemical protective suit with self-contained breathing apparatus.
Stop discharge if possible.
Call fire department.
Isolate and remove discharged material.
Notify local health and pollution control agencies.

Fire	Combustible. POISONOUS GASES ARE PRODUCED IN FIRE AND WHEN HEATED. Containers may explode in fire. Wear chemical protective suit with self-contained breathing apparatus. Combat fires from safe distance or protected location. Extinguish with water. Cool exposed containers with water.
Exposure	CALL FOR MEDICAL AID. LIQUID OR SOLID POISONOUS IF SWALLOWED OR IF SKIN IS EXPOSED. Irritating to eyes. Remove contaminated clothing and shoes. Flush affected areas with plenty of water. IF IN EYES, hold eyelids open and flush with plenty of water. IF SWALLOWED and victim is CONSCIOUS, have victim drink water or milk and have victim induce vomiting. IF SWALLOWED and victim is UNCONSCIOUS OR HAVING CONVULSIONS, do nothing except keep victim warm.
Water Pollution	HARMFUL TO AQUATIC LIFE IN VERY LOW CONCENTRATIONS. Solution is fouling to shoreline. May be dangerous if it enters water intakes. Notify local health and wildlife officials. Notify operators of nearby water intakes.

1. RESPONSE TO DISCHARGE
(See Response Methods Handbook, CG 446-4)

Issue warning—poison, water contaminant
Restrict access
Should be removed
Chemical and physical treatment

2. LABEL

3. CHEMICAL DESIGNATIONS

3.1 **Synonyms:** MPT; O,O-Dimethyl O-(p-Nitrophenyl) Phosphorothioate; O,O-Dimethyl O-p-Nitrophenyl thiophosphate; Parathion-methyl
3.2 **Coast Guard Compatibility Classification:** Not applicable
3.3 **Chemical Formula:** $(CH_3O)_2PSOC_6H_4NO_2$-p
3.4 **IMCO/United Nations Numerical Designation:** 6.1/1668

4. OBSERVABLE CHARACTERISTICS

4.1 **Physical State (as shipped):** Solid or liquid
4.2 **Color:** Colorless solid or brownish liquid
4.3 **Odor:** Characteristic; like rotten eggs or garlic

5. HEALTH HAZARDS

5.1 **Personal Protective Equipment:** Approved mask or respirator; natural rubber gloves, overshoes; protective clothing; goggles.
5.2 **Symptoms Following Exposure:** Exposure to fumes from a fire, or to the liquid, causes headache, blurred vision, constricted pupils of the eyes, weakness, nausea, cramps, diarrhea, and tightness in the chest. Muscle twitch and convulsions may follow. Symptoms may develop over a period of 8 hrs.
5.3 **Treatment for Exposure:** *Speed is essential.* INGESTION: call a doctor! If victim is not breathing, immediately institute artificial respiration by mouth-to-mouth, mouth-to-nose, or mouth-to-oropharyngeal method; when victim is conscious, give milk, water, or salt-water and induce vomiting repeatedly. SKIN OR EYES: flood and wash exposed areas thoroughly with water; remove contaminated clothing under a shower.
5.4 **Toxicity by Inhalation (Threshold Limit Value):** 0.2 mg/m³ (solid); 100 ppm (solution)
5.5 **Short-Term Inhalation Limits:** Data not available
5.6 **Toxicity by Ingestion:** Grade 4; LD_{50} below 50 mg/kg (rat)
5.7 **Late Toxicity:** Data not available
5.8 **Vapor (Gas) Irritant Characteristics:** Not pertinent
5.9 **Liquid or Solid Irritant Characteristics:** Poisonous when absorbed through skin.
5.10 **Odor Threshold:** Data not available

6. FIRE HAZARDS

6.1 **Flash Point:** 115°F O.C.
6.2 **Flammable Limits in Air:** Data not available
6.3 **Fire Extinguishing Agents:** Water
6.4 **Fire Extinguishing Agents Not to be Used:** Not pertinent
6.5 **Special Hazards of Combustion Products:** Toxic gases are produced in fires.
6.6 **Behavior in Fire:** Drums may rupture violently.
6.7 **Ignition Temperature:** Data not available
6.8 **Electrical Hazard:** Not pertinent
6.9 **Burning Rate:** Data not available

7. CHEMICAL REACTIVITY

7.1 **Reactivity with Water:** Half decomposed in 8 days at 40°C
7.2 **Reactivity with Common Materials:** Is absorbed in wood, etc., which must be replaced to eliminate poison hazard.
7.3 **Stability During Transport:** Decomposes above 50°C with possible explosive force.
7.4 **Neutralizing Agents for Acids and Caustics:** Apply caustic or soda ash slurry until yellow stains disappear.
7.5 **Polymerization:** Not pertinent
7.6 **Inhibitor of Polymerization:** Not pertinent

8. WATER POLLUTION

8.1 **Aquatic Toxicity:**
1.9 ppm/96 hr/bluegill/TL_m/fresh water
8.3 ppm/96 hr/fathead/TL_m/fresh water
8.2 **Waterfowl Toxicity:** LD_{50} = 10 mg/kg
8.3 **Biological Oxygen Demand (BOD):** Data not available
8.4 **Food Chain Concentration Potential:** Data not available

9. SELECTED MANUFACTURERS

1. Hercules, Inc.
Synthetics Dept.
Plaquemine, La. 70764
2. Monsanto Co.
Agricultural Division
800 N. Lindbergh Blvd.
St. Louis, Mo. 63166
3. Stauffer Chemical Co.
Agricultural Chemicals Division
Green Farms Rd.
Westport, Conn. 06880

10. SHIPPING INFORMATION

10.1 **Grades or Purity:** Pure (solid); technical (liquid); 80% in xylene
10.2 **Storage Temperature:** Below 50°
10.3 **Inert Atmosphere:** No requirement
10.4 **Venting:** Pressure-vacuum

11. HAZARD ASSESSMENT CODE
(See Hazard Assessment Handbook, CG 446-3)

A-X-Y

12. HAZARD CLASSIFICATIONS

12.1 **Code of Federal Regulations:** Poisonous liquid or solid, Class B
12.2 **NAS Hazard Rating for Bulk Water Transportation:** Not listed
12.3 **NFPA Hazard Classifications:**

Category	Classification Xylene sol'n	Solid
Health Hazard (Blue)	4	4
Flammability (Red)	3	1
Reactivity (Yellow)	2	2

13. PHYSICAL AND CHEMICAL PROPERTIES

13.1 **Physical State at 15°C and 1 atm:** Liquid or solid
13.2 **Molecular Weight:** 263.2
13.3 **Boiling Point at 1 atm:** Very high
13.4 **Freezing Point:** 65°F = 18°C = 291°K
13.5 **Critical Temperature:** Not pertinent
13.6 **Critical Pressure:** Not pertinent
13.7 **Specific Gravity:** 1.360 at 20°C (liquid)
13.8 **Liquid Surface Tension:** Data not available
13.9 **Liquid-Water Interfacial Tension:** Data not available
13.10 **Vapor (Gas) Specific Gravity:** Not pertinent
13.11 **Ratio of Specific Heats of Vapor (Gas):** Not pertinent
13.12 **Latent Heat of Vaporization:** Not pertinent
13.13 **Heat of Combustion:** Data not available
13.14 **Heat of Decomposition:** Not pertinent
13.15 **Heat of Solution:** Not pertinent
13.16 **Heat of Polymerization:** Not pertinent

(Continued on pages 5 and 6)

NOTES

REVISED 1978

MPD — METHYL PHOSPHONOTHIOIC DICHLORIDE (ANHYD)

Common Synonyms			
MPTD	Liquid	Colorless	Sharp Unpleasant Odor
	Sinks and mixes violently with water.		

Avoid contact with liquid and vapor. Keep people away.
Stop discharge if possible.
Isolate and remove discharged material.
Notify local health and pollution control agencies.
Stop discharge if possible. Keep people away.

Fire	Combustible. Irritating gases may be produced when heated. Extinguish with dry chemicals or carbon dioxide. DO NOT USE WATER OF FOAM ON FIRE. Cool exposed containers with water.
Exposure	CALL FOR MEDICAL AID. VAPOR Irritating to eyes, nose and thorat. If inhaled will cause coughing or difficult breathing. If in eyes, hold eyelids open and flush with plenty of water. If breathing has stopped, give artificial respiration. If breathing is difficult, give oxygen. LIQUID Will burn skin and eyes. If swallowed will cause nausea and vomiting. Remove contaminated clothing and shoes. Flush affected areas with plenty of water. IF IN EYES, hold eyelids open and flush with plenty of water. IF SWALLOWED and victim is CONSCIOUS, have victim drink water or milk and have victim induce vomiting. IF SWALLOWED and victim is UNCONSCIOUS OR HAVING CONVULSIONS, do nothing except keep victim warm.
Water Pollution	Effect of low concentrations on aquatic life is unknown. May be dangerous if it enters water intakes. Notify local health and wildlife officials. Notify operators of nearby water intakes.

1. RESPONSE TO DISCHARGE

(See Response Methods Handbook, CG 446-4)

Issue warning—corrosive, water contaminant
Restrict access
Disperse and flush

2. LABEL

3. CHEMICAL DESIGNATIONS

3.1 **Synonyms:** MPTD
3.2 **Coast Guard Compatibility Classification:** Not listed
3.3 **Chemical Formula:** CH_3PSCl_2
3.4 **IMCO/United Nations Numerical Designation:** Not listed

4. OBSERVABLE CHARACTERISTICS

4.1 **Physical State (as shipped):** Liquid
4.2 **Color:** Clear
4.3 **Odor:** Acrid

5. HEALTH HAZARDS

Use extreme care when handling this compound. Avoid any contact with liquid or vapor.

5.1 **Personal Protective Equipment:** Rubber or neoprene gloves; respiratory protection; goggles
5.2 **Symptoms Following Exposure:** Inhalation causes irritation of nose and throat; effects are quite similar to those of phosgene. Ingestion causes irritation of mouth and stomach. Delayed, painful eye irritation may occur from exposure to vapor; liquid causes severe irritation. Contact with skin causes irritation and burns.
5.3 **Treatment for Exposure:** *Get medical attention after all exposures to this compound.* INHALATION: remove victim to fresh air; alert physician to delayed effects similar to those of phosgene. INGESTION: give large amount of water and induce vomiting. EYES: flush with water for at least 15 min. SKIN: flush with water.
5.4 **Toxicity by Inhalation (Threshold Limit Value):** Data not available
5.5 **Short-Term Inhalation Limits:** Data not available
5.6 **Toxicity by Ingestion:** Data not available
5.7 **Late Toxicity:** Data not available
5.8 **Vapor (Gas) Irritant Characteristics:** Data not available
5.9 **Liquid or Solid Irritant Characteristics:** Data not available
5.10 **Odor Threshold:** Data not available

6. FIRE HAZARDS

6.1 **Flash Point:** > 122°F O.C.
6.2 **Flammable Limits in Air:** Data not available
6.3 **Fire Extinguishing Agents:** Dry chemical or carbon dioxide
6.4 **Fire Extinguishing Agents Not to be Used:** Water or foam
6.5 **Special Hazards of Combustion Products:** Irritating hydrogen chloride, sulfur dioxide and other fumes may be formed in fire.
6.6 **Behavior in Fire:**
6.7 **Ignition Temperature:** Data not available
6.8 **Electrical Hazard:** Data not available
6.9 **Burning Rate:** Data not available

7. CHEMICAL REACTIVITY

7.1 **Reactivity with Water:** Reacts with water to form hydrochloric acid and/or hydrogen chloride vapor. The reaction may be violent.
7.2 **Reactivity with Common Materials:** Corrosive to metals because of its high acidity
7.3 **Stability During Transport:** Stable
7.4 **Neutralizing Agents for Acids and Caustics:** Flush with water, rinse with dilute sodium bicarbonate or soda ash solution.
7.5 **Polymerization:** Not pertinent
7.6 **Inhibitor of Polymerization:** Not pertinent

8. WATER POLLUTION

8.1 **Aquatic Toxicity:** Data not available
8.2 **Waterfowl Toxicity:** Data not available
8.3 **Biological Oxygen Demand (BOD):** Data not available
8.4 **Food Chain Concentration Potential:** None

9. SELECTED MANUFACTURERS

Ethyl Corporation
Industrial Chemicals Div.
451 Florida
Baton Rouge, La. 70801

10. SHIPPING INFORMATION

10.1 **Grades or Purity:** Technical
10.2 **Storage Temperature:** Ambient
10.3 **Inert Atmosphere:** No requirement
10.4 **Venting:** Open

11. HAZARD ASSESSMENT CODE

(See Hazard Assessment Handbook, CG 446-3)

A-O

12. HAZARD CLASSIFICATIONS

12.1 **Code of Federal Regulations:** Corrosive liquid
12.2 **NAS Hazard Rating for Bulk Water Transportation:** Not listed
12.3 **NFPA Hazard Classifications:** Not listed

13. PHYSICAL AND CHEMICAL PROPERTIES

13.1 **Physical State at 15°C and 1 atm:** Liquid
13.2 **Molecular Weight:** 149
13.3 **Boiling Point at 1 atm:** Data not available
13.4 **Freezing Point:** −14.1°F = −25.6°C = 247.6°K
13.5 **Critical Temperature:** Not pertinent
13.6 **Critical Pressure:** Not pertinent
13.7 **Specific Gravity:** 1.42 at 20°C (liquid)
13.8 **Liquid Surface Tension:** Data not available
13.9 **Liquid-Water Interfacial Tension:** Not pertinent
13.10 **Vapor (Gas) Specific Gravity:** Not pertinent
13.11 **Ratio of Specific Heats of Vapor (Gas):** Not pertinent
13.12 **Latent Heat of Vaporization:** (est.) 110 Btu/lb = 60 cal/g = 2.5×10^5 J/kg
13.13 **Heat of Combustion:** Data not available
13.14 **Heat of Decomposition:** Not pertinent
13.15 **Heat of Solution:** Data not available
13.16 **Heat of Polymerization:** Not pertinent

(Continued on pages 5 and 6)

NOTES

MPY — 1-METHYLPYRROLIDONE

Common Synonyms	Liquid	White	Mild Fishy odor
N-Methylpyrrolidone 1-Methyl-2-pyrrolidinone N-Methylpyrrolidinone N-Methyl-alpha-pyrrolidone	May float or sink in water.		

Avoid contact with liquid. Keep people away.
Stop discharge if possible.
Call fire department.
Isolate and remove discharged material.
Notify local health and pollution control agencies.

Fire

Combustible.
POISONOUS GASES MAY BE PRODUCED WHEN HEATED.
Wear goggles and self-contained breathing apparatus.
Extinguish with dry chemicals, foam, or carbon dioxide.
Water may be ineffective on fire.
Cool exposed containers with water.

Exposure

CALL FOR MEDICAL AID.

LIQUID
Irritating to skin and eyes.
If swallowed will cause nausea and vomiting.
Remove contaminated clothing and shoes.
Flush affected areas with plenty of water.
IF IN EYES, hold eyelids open and flush with plenty of water.
IF SWALLOWED and victim is CONSCIOUS, have victim drink water or milk and have victim induce vomiting.
IF SWALLOWED and victim is UNCONSCIOUS OR HAVING CONVULSIONS, do nothing except keep victim warm.

Water Pollution

Effect of low concentrations on aquatic life is unknown.
May be dangerous if it enters water intakes.
Notify local health and wildlife officials.
Notify operators of nearby water intakes.

1. RESPONSE TO DISCHARGE

(See Response Methods Handbook, CG 446-4)

Issue warning—water contaminant
Disperse and flush

2. LABELS

No hazard label required by Code of Federal Regulations

3. CHEMICAL DESIGNATIONS

3.1 **Synonyms:** 1-Methyl-2-pyrrolidinone; N-Methylpyrrolidinone; N-Methyl-pyrrolidone; N-Methyl-*alpha*-pyrrolidone
3.2 **Coast Guard Compatibility Classification:** Not listed
3.3 **Chemical Formula:** C_5H_9NO
3.4 **IMCO/United Nations Numerical Designation:** Not listed

4. OBSERVABLE CHARACTERISTICS

4.1 **Physical State (as shipped):** Liquid
4.2 **Color:** Colorless
4.3 **Odor:** Mild amine-like

5. HEALTH HAZARDS

5.1 **Personal Protective Equipment:** Goggles or face shield; rubber gloves
5.2 **Symptoms Following Exposure:** Inhalation of hot vapors can irritate nose and throat. Ingestion causes irritation of mouth and stomach. Contact with eyes causes irritation. Repeated and prolonged skin contact produces a mild, transient irritation.
5.3 **Treatment for Exposure:** INHALATION: move to fresh air. INGESTION: give large amount of water and induce vomiting. EYES: flush with water for at least 15 min. SKIN: remove from skin and eyes by flooding the affected tissues with water; wash with soap and water.
5.4 **Toxicity by Inhalation (Threshold Limit Value):** Data not available
5.5 **Short-Term Inhalation Limits:** Data not available
5.6 **Toxicity by Ingestion:** Grade 2; oral LD_{50} = 3.5 mg/kg (rabbit)
5.7 **Late Toxicity:** Causes blood abnormalities in rats
5.8 **Vapor (Gas) Irritant Characteristics:** Data not available
5.9 **Liquid or Solid Irritant Characteristics:** Data not available
5.10 **Odor Threshold:** Data not available

6. FIRE HAZARDS

6.1 **Flash Point:** 204°F O.C.
6.2 **Flammable Limits in Air:** Data not available
6.3 **Fire Extinguishing Agents:** Dry chemical, "alcohol" foam, carbon dioxide
6.4 **Fire Extinguishing Agents Not to be Used:** Water may be ineffective.
6.5 **Special Hazards of Combustion Products:** Toxic oxides of nitrogen may be formed in fire.
6.6 **Behavior in Fire:**
6.7 **Ignition Temperature:** Data not available
6.8 **Electrical Hazard:** Data not available
6.9 **Burning Rate:** Data not available

7. CHEMICAL REACTIVITY

7.1 **Reactivity with Water:** No reaction
7.2 **Reactivity with Common Materials:**
7.3 **Stability During Transport:** Stable
7.4 **Neutralizing Agents for Acids and Caustics:** Not pertinent
7.5 **Polymerization:** Not pertinent
7.6 **Inhibitor of Polymerization:** Not pertinent

8. WATER POLLUTION

8.1 **Aquatic Toxicity:** Data not available
8.2 **Waterfowl Toxicity:** Data not available
8.3 **Biological Oxygen Demand (BOD):** Data not available
8.4 **Food Chain Concentration Potential:** None

9. SELECTED MANUFACTURERS

1. Aldrich Chemical Co.
940 W. Saint Paul Ave.
Milwaukee, Wis. 53233
2. Eastman Organic Chemicals
Rochester, N. Y. 14650
3. Pfaltz and Bauer, Inc.
375 Fairfield Ave.
Stamford, Conn. 06902

10. SHIPPING INFORMATION

10.1 **Grades or Purity:** Technical
10.2 **Storage Temperature:** Ambient
10.3 **Inert Atmosphere:** No requirement
10.4 **Venting:** Open (flame arrester)

11. HAZARD ASSESSMENT CODE

(See Hazard Assessment Handbook, CG 446-3)

A-P-Q

12. HAZARD CLASSIFICATIONS

12.1 **Code of Federal Regulations:** Not listed
12.2 **NAS Hazard Rating for Bulk Water Transportation:** Not listed
12.3 **NFPA Hazard Classifications:**

Category	Classification
Health Hazard (Blue)	2
Flammability (Red)	1
Reactivity (Yellow)	0

13. PHYSICAL AND CHEMICAL PROPERTIES

13.1 **Physical State at 15°C and 1 atm:** Liquid
13.2 **Molecular Weight:** 99
13.3 **Boiling Point at 1 atm:** 396°F = 202°C = 475°K
13.4 **Freezing Point:** 1°F = −17°C = 256°K
13.5 **Critical Temperature:** Not pertinent
13.6 **Critical Pressure:** Not pertinent
13.7 **Specific Gravity:** 1.03 at 25°C (liquid)
13.8 **Liquid Surface Tension:** Data not available
13.9 **Liquid-Water Interfacial Tension:** Not pertinent
13.10 **Vapor (Gas) Specific Gravity:** 3.4
13.11 **Ratio of Specific Heats of Vapor (Gas):** Not pertinent
13.12 **Latent Heat of Vaporization:** Data not available
13.13 **Heat of Combustion:** −13,000 Btu/lb = −7,220 cal/g = -302×10^5 J/kg
13.14 **Heat of Decomposition:** Not pertinent
13.15 **Heat of Solution:** Data not available
13.16 **Heat of Polymerization:** Not pertinent

NOTES

MSR

alpha-METHYLSTYRENE

Common Synonyms
Isopropenylbenzene
Phenylpropylene
1-Methyl-1-phenylethylene

Liquid — Colorless

Floats on water.

Avoid contact with liquid. Keep people away.
Stop discharge if possible.
Call fire department.
Isolate and remove discharged material.
Notify local health and pollution control agencies.

Fire

Combustible.
Extinguish with dry chemicals, foam or carbon dioxide.
Water may be ineffective on fire.
Cool exposed containers with water.

Exposure

CALL FOR MEDICAL AID.

LIQUID
Will burn skin and eyes.
If swallowed will cause nausea and vomiting.
Remove contaminated clothing and shoes.
Flush affected areas with plenty of water.
IF IN EYES, hold eyelids open and flush with plenty of water.
IF SWALLOWED and victim is CONSCIOUS, have victim drink water or milk.
IF SWALLOWED and victim is UNCONSCIOUS OR HAVING CONVULSIONS, do nothing except keep victim warm.
DO NOT INDUCE VOMITING.

Water Pollution

HARMFUL TO AQUATIC LIFE IN VERY LOW CONCENTRATIONS.
Fouling to shoreline.
May be dangerous if it enters water intakes.
Notify local health and wildlife officials.
Notify operators of nearby water intakes.

1. RESPONSE TO DISCHARGE

(See Response Methods Handbook, CG 446-4)

Mechanical containment
Should be removed
Chemical and physical treatment

2. LABELS

No hazard label required by Code of Federal Regulations

3. CHEMICAL DESIGNATIONS

3.1 **Synonyms:** Isopropenylbenzene; 1-Methyl-1-phenylethylene; Phenylpropylene
3.2 **Coast Guard Compatibility Classification:** Olefin
3.3 **Chemical Formula:** $C_6H_5C(CH_3)=CH_2$
3.4 **IMCO/United Nations Numerical Designation:** Not listed

4. OBSERVABLE CHARACTERISTICS

4.1 **Physical State (as shipped):** Liquid
4.2 **Color:** Colorless
4.3 **Odor:** Characteristic

5. HEALTH HAZARDS

5.1 **Personal Protective Equipment:** Neoprene gloves; splashproof goggles or face shield
5.2 **Symptoms Following Exposure:** Inhalation causes irritation of respiratory tract, headache, dizziness, light-headedness, and breathlessness. Ingestion causes irritation of mouth and stomach. Contact with liquid irritates eyes. Prolonged skin contact can cause severe rashes, swelling, and blistering.
5.3 **Treatment for Exposure:** INHALATION: remove victim to fresh air; if he is not breathing, give artificial respiration; contact a physician; keep victim quiet and warm. INGESTION: do NOT induce vomiting; call physician. EYES: flush with water for at least 15 min.; get medical attention. SKIN: wash area with soap and water.
5.4 **Toxicity by Inhalation (Threshold Limit Value):** 100 ppm
5.5 **Short-Term Inhalation Limits:** 100 ppm, 30 min.
5.6 **Toxicity by Ingestion:** Grade 2; LD_{50} 0.5—5 g/kg
5.7 **Late Toxicity:** Data not available
5.8 **Vapor (Gas) Irritant Characteristics:** Data not available
5.9 **Liquid or Solid Irritant Characteristics:** Data not available
5.10 **Odor Threshold:** <10 ppm

6. FIRE HAZARDS

6.1 **Flash Point:** 137°F C.C.
6.2 **Flammable Limits in Air:** 1.9%—6.1%
6.3 **Fire Extinguishing Agents:** Dry chemical, foam, carbon dioxide
6.4 **Fire Extinguishing Agents Not to be Used:** Water may be ineffective.
6.5 **Special Hazards of Combustion Products:**
6.6 **Behavior in Fire:**
6.7 **Ignition Temperature:** 1,066°F
6.8 **Electrical Hazard:** Data not available
6.9 **Burning Rate:** Data not available

7. CHEMICAL REACTIVITY

7.1 **Reactivity with Water:** No reaction
7.2 **Reactivity with Common Materials:** May attack some forms of plastics.
7.3 **Stability During Transport:** Stable
7.4 **Neutralizing Agents for Acids and Caustics:** Not pertinent
7.5 **Polymerization:** Hazardous polymerization unlikely to occur except when in contact with alkali metals or metallo-organic compounds.
7.6 **Inhibitor of Polymerization:** 10—20 ppm tert-butylcatechol

8. WATER POLLUTION

8.1 **Aquatic Toxicity:** 10 ppm/96 hr/fathead minnow/LC_{50}
8.2 **Waterfowl Toxicity:** Data not available
8.3 **Biological Oxygen Demand (BOD):** Data not available
8.4 **Food Chain Concentration Potential:** None

9. SELECTED MANUFACTURERS

1. Allied Chemical Corp.
   Plastics Division
   P. O. Box 2365R
   Morristown, N. J. 07960
2. United States Steel Co.
   Chemicals Division
   600 Grant St.
   Pittsburgh, Pa. 15230
3. Dow Chemical Co.
   2030 Dow Center
   Midland, Mich. 48640

10. SHIPPING INFORMATION

10.1 **Grades or Purity:** Commercial
10.2 **Storage Temperature:** Ambient
10.3 **Inert Atmosphere:** No requirement
10.4 **Venting:** Open (flame arrester)

11. HAZARD ASSESSMENT CODE

(See Hazard Assessment Handbook, CG 446-3)

A-T-U

12. HAZARD CLASSIFICATIONS

12.1 **Code of Federal Regulations:** Not listed
12.2 **NAS Hazard Rating for Bulk Water Transportation:** Not listed
12.3 **NFPA Hazard Classifications:**

Category	Classification
Health Hazard (Blue)	1
Flammability (Red)	2
Reactivity (Yellow)	1

13. PHYSICAL AND CHEMICAL PROPERTIES

13.1 **Physical State at 15°C and 1 atm:** Liquid
13.2 **Molecular Weight:** 118.17
13.3 **Boiling Point at 1 atm:** 329°F = 165°C = 438°K
13.4 **Freezing Point:** −9.8°F = −23.2°C = 250.0°K
13.5 **Critical Temperature:** 719.1°F = 381.7°C = 654.9°K
13.6 **Critical Pressure:** 494 psia = 33.6 atm = 3.41 MN/m^2
13.7 **Specific Gravity:** 0.91 at 20°C (liquid)
13.8 **Liquid Surface Tension:** 33.88 dynes/cm = 0.03388 N/m at 20°C
13.9 **Liquid-Water Interfacial Tension:** Data not available
13.10 **Vapor (Gas) Specific Gravity:** 4.08
13.11 **Ratio of Specific Heats of Vapor (Gas):** 1.060 at 27°C
13.12 **Latent Heat of Vaporization:** 140.4 Btu/lb = 78.0 cal/g = 3.26×10^5 J/kg
13.13 **Heat of Combustion:** −17,690 Btu/lb = −9,830 cal/g = -411×10^5 J/kg
13.14 **Heat of Decomposition:** Not pertinent
13.15 **Heat of Solution:** Not pertinent
13.16 **Heat of Polymerization:** Data not available

NOTES

MTS	METHYLTRICHLOROSILANE

Common Synonyms	Liquid Colorless Sharp irritating odor Reacts violently with water. Irritating gas is produced on contact with water.

AVOID CONTACT WITH LIQUID AND VAPOR. KEEP PEOPLE AWAY.
Wear goggles, self-contained breathing apparatus, and rubber overclothing (including gloves).
Shut off ignition sources. Call fire department.
Stop discharge if possible.
Isolate and remove discharged material.
Notify local health and pollution control agencies.

Fire	FLAMMABLE. POISONOUS GASES MAY BE PRODUCED IN FIRE. Containers may explode in fire. Flashback along vapor trail may occur. Vapor may explode if ignited in an enclosed area. Wear goggles and self-contained breathing apparatus. Extinguish with dry chemicals or carbon dioxide. DO NOT USE WATER OR FOAM ON FIRE. DO NOT USE WATER OR FOAM ON ADJACENT FIRES.
Exposure	Call for medical aid. VAPOR Irritating to eyes, nose and throat. If inhaled will cause difficult breathing. Move victim to fresh air. If breathing has stopped, give artificial respiration. If breathing is difficult, give oxygen. LIQUID Will burn skin and eyes. Harmful if swallowed. Remove contaminated clothing and shoes. Flush affected areas with plenty of water. IF IN EYES, hold eyelids open and flush with plenty of water. IF SWALLOWED and victim is CONSCIOUS, have victim drink water or milk. DO NOT INDUCE VOMITING.
Water Pollution	Effect of low concentrations on aquatic life is unknown. May be dangerous if it enters water intakes. Notify local health and wildlife officials. Notify operators of nearby water intakes.

1. RESPONSE TO DISCHARGE
(See Response Methods Handbook, CG 446-4)

Issue warning—high flammability, air contaminant
Restrict access
Evacuate area
Disperse and flush with care

2. LABEL

Red

3. CHEMICAL DESIGNATIONS
3.1 **Synonyms:** Trichloromethylsilane
3.2 **Coast Guard Compatibility Classification:** Not applicable
3.3 **Chemical Formula:** CH_3SiCl_3
3.4 **IMCO/United Nations Numerical Designation:** 3.2/1250

4. OBSERVABLE CHARACTERISTICS
4.1 **Physical State (as shipped):** Liquid
4.2 **Color:** Colorless
4.3 **Odor:** Acrid; sharp, like hydrochloric acid

5. HEALTH HAZARDS
5.1 **Personal Protective Equipment:** Full protective clothing; acid-vapor-type respiratory protection; rubber gloves; chemical worker's goggles; other protective equipment as necessary to protect skin and eyes.
5.2 **Symptoms Following Exposure:** Inhalation causes irritation of mucous membrane. Contact with liquid causes severe burns of eyes and skin. Ingestion causes severe burns of mouth and stomach.
5.3 **Treatment for Exposure:** Get medical attention at once following all exposures to this compound. INHALATION: remove victim from exposure; give artificial respiration if breathing has ceased. EYES: flush with water for 15 min. SKIN: flush with water. INGESTION: do NOT induce vomiting; give large amounts of water.
5.4 **Toxicity by Inhalation (Threshold Limit Value):** Data not available
5.5 **Short-Term Inhalation Limits:** Data not available
5.6 **Toxicity by Ingestion:** Grade 3; LD_{50} 50 to 500 mg/kg
5.7 **Late Toxicity:** Data not available
5.8 **Vapor (Gas) Irritant Characteristics:** Vapors cause severe irritation of eyes and throat and can cause eye and lung injury. They cannot be tolerated even at low concentrations.
5.9 **Liquid or Solid Irritant Characteristics:** Severe skin irritant. Causes second- and third-degree burns on short contact and is very injurious to the eyes.
5.10 **Odor Threshold:** Decomposes in moist air, creating HCl with odor threshold of 1 ppm

6. FIRE HAZARDS
6.1 **Flash Point:** 45°F O.C.; 15°F C.C.
6.2 **Flammable Limits in Air:** 5.1%—>20%
6.3 **Fire Extinguishing Agents:** Dry chemical, carbon dioxide
6.4 **Fire Extinguishing Agents Not to be Used:** Water, foam
6.5 **Special Hazards of Combustion Products:** Toxic hydrogen chloride and phosgene gases may form in fires.
6.6 **Behavior in Fire:** Difficult to extinguish; re-ignition may occur. Contact with water applied to adjacent fires produces irritating hydrogen chloride.
6.7 **Ignition Temperature:** >760°F
6.8 **Electrical Hazard:** Data not available
6.9 **Burning Rate:** 1.9 mm/min.

7. CHEMICAL REACTIVITY
7.1 **Reactivity with Water:** Reacts violently to form hydrogen chloride (hydrochloric acid).
7.2 **Reactivity with Common Materials:** Reacts with surface moisture to evolve hydrogen chloride, which is corrosive to metals.
7.3 **Stability During Transport:** Stable
7.4 **Neutralizing Agents for Acids and Caustics:** Flood with water, rinse with sodium bicarbonate or lime solution.
7.5 **Polymerization:** Not pertinent
7.6 **Inhibitor of Polymerization:** Not pertinent

8. WATER POLLUTION
8.1 **Aquatic Toxicity:** Data not available
8.2 **Waterfowl Toxicity:** Data not available
8.3 **Biological Oxygen Demand (BOD):** Data not available
8.4 **Food Chain Concentration Potential:** None

9. SELECTED MANUFACTURERS
1. Union Carbide Corporation
Chemicals and Plastics Division
270 Park Avenue
New York, N. Y. 10017
2. Dow Corning Corporation
P. O. Box 592
Midland, Mich. 48640
3. General Electric Company
Silicone Products Department
Waterford, N. Y. 12188

10. SHIPPING INFORMATION
10.1 **Grades or Purity:** 98+%
10.2 **Storage Temperature:** Ambient
10.3 **Inert Atmosphere:** No requirement
10.4 **Venting:** Safety relief

11. HAZARD ASSESSMENT CODE
(See Hazard Assessment Handbook, CG 446-3)

A-O

12. HAZARD CLASSIFICATIONS
12.1 **Code of Federal Regulations:** Flammable liquid
12.2 **NAS Hazard Rating for Bulk Water Transportation:**

Category	Rating
Fire	3
Health	
Vapor Irritant	4
Liquid or Solid Irritant	4
Poisons	3
Water Pollution	
Human Toxicity	3
Aquatic Toxicity	3
Aesthetic Effect	2
Reactivity	
Other Chemicals	3
Water	4
Self-Reaction	0

12.3 **NFPA Hazard Classifications:**

Category	Classification
Health Hazard (Blue)	3
Flammability (Red)	3
Reactivity (Yellow)	2
	W

13. PHYSICAL AND CHEMICAL PROPERTIES
13.1 **Physical State at 15°C and 1 atm:** Liquid
13.2 **Molecular Weight:** 149.5
13.3 **Boiling Point at 1 atm:** 151.5°F = 66.4°C = 339.6°K
13.4 **Freezing Point:** −130°F = −90°C = 183°K
13.5 **Critical Temperature:** Not pertinent
13.6 **Critical Pressure:** Not pertinent
13.7 **Specific Gravity:** 1.27 at 25°C (liquid)
13.8 **Liquid Surface Tension:** 20.3 dynes/cm = 0.0203 N/m at 20°C
13.9 **Liquid-Water Interfacial Tension:** Not pertinent
13.10 **Vapor (Gas) Specific Gravity:** 5.16
13.11 **Ratio of Specific Heats of Vapor (Gas):** Data not available
13.12 **Latent Heat of Vaporization:** 89.3 Btu/lb = 49.6 cal/g = 2.08×10^5 J/kg
13.13 **Heat of Combustion:** (est.) −3,000 Btu/lb = −1,700 cal/g = -70×10^5 J/kg
13.14 **Heat of Decomposition:** Not pertinent
13.15 **Heat of Solution:** Data not available
13.16 **Heat of Polymerization:** Not pertinent

NOTES

MVK | METHYL VINYL KETONE

Common Synonyms			
3-Buten-2-one	Liquid	Colorless to light yellow	Strong irritating odor
	Mixes with water. Irritating vapor is produced.		

AVOID CONTACT WITH LIQUID. KEEP PEOPLE AWAY.
Wear rubber overclothing (including gloves).
Shut off ignition sources. Call fire department.
Stop discharge if possible.
Stay upwind. Use water spray to "knock down" vapor.
Isolate and remove discharged material.
Notify local health and pollution control agencies.

Fire

FLAMMABLE.
Containers may explode in fire.
Flashback along vapor trail may occur.
Vapor may explode if ignited in an enclosed area.

Extinguish with dry chemicals, alcohol foam, or carbon dioxide.
Water may be ineffective on fire.
Cool exposed containers with water.

Exposure

CALL FOR MEDICAL AID.

VAPOR
Irritating to eyes, nose and throat.
If inhaled will cause coughing or difficult breathing.

Move victim to fresh air.
If breathing has stopped, give artificial respiration.
If breathing is difficult, give oxygen.

LIQUID
POISONOUS IF SWALLOWED.
Will burn skin and eyes.

Remove contaminated clothing and shoes.
Flush affected areas with plenty of water.
IF IN EYES, hold eyelids open and flush with plenty of water.
IF SWALLOWED and victim is CONSCIOUS, have victim drink water or milk.
DO NOT INDUCE VOMITING.

Water Pollution

Effect of low concentrations on aquatic life is unknown.
May be dangerous if it enters water intakes.

Notify local health and wildlife officials.
Notify operators of nearby water intakes.

1. RESPONSE TO DISCHARGE

(See Response Methods Handbook, CG 446-4)

Issue warning—high flammability, water contaminant
Restrict access
Evacuate area
Disperse and flush

2. LABEL

Red

3. CHEMICAL DESIGNATIONS

3.1 **Synonyms:** 3-Buten-2-one
3.2 **Coast Guard Compatibility Classification:** Not applicable
3.3 **Chemical Formula:** $CH_3COCH=CH_2$
3.4 **IMCO/United Nations Numerical Designation:** 3.2/1251

4. OBSERVABLE CHARACTERISTICS

4.1 **Physical State (as shipped):** Liquid
4.2 **Color:** Colorless to light yellow
4.3 **Odor:** Powerfully irritating

5. HEALTH HAZARDS

5.1 **Personal Protective Equipment:** Self-contained breathing apparatus with full face piece; rubber gloves; chemical goggles or face piece of breathing aparatus.
5.2 **Symptoms Following Exposure:** Inhalation causes irritation of nose and throat. Vapor causes tears; contact with liquid can burn eyes. Liquid irritates skin and will cause burn if not removed at once. Ingestion causes irritation of mouth and stomach.
5.3 **Treatment for Exposure:** Get medical attention for all exposures to this compound. INHALATION: move victim to fresh air; administer artificial respiration if necessary. EYES or SKIN: flush with copious quantity of water for 15 min. INGESTION: do NOT induce vomiting.
5.4 **Toxicity by Inhalation (Threshold Limit Value):** Data not available
5.5 **Short-Term Inhalation Limits:** Data not available
5.6 **Toxicity by Ingestion:** Grade 4; LD_{50} <50 mg/kg
5.7 **Late Toxicity:** Data not available
5.8 **Vapor (Gas) Irritant Characteristics:** Vapors cause severe irritation of eyes and throat and can cause eye and lung injury. They cannot be tolerated even at low concentrations.
5.9 **Liquid or Solid Irritant Characteristics:** Severe skin irritant. Causes second- and third-degree burns on short contact and is very injurious to the eyes.
5.10 **Odor Threshold:** 0.5 mg/m³

6. FIRE HAZARDS

6.1 **Flash Point:** 30°F O.C.; 20°F C.C.
6.2 **Flammable Limits in Air:** 2.1% 15.6%
6.3 **Fire Extinguishing Agents:** Dry chemical, alcohol foam, cabon dioxide
6.4 **Fire Extinguishing Agents Not to be Used:** Water may be ineffective.
6.5 **Special Hazards of Combustion Products:** Not pertinent
6.6 **Behavior in Fire:** Vapor is heavier than air and may travel a considerable distance to a source of ignition and flash back. At elevated temperatures (fire conditions) polymerization may take place in containers, causing violent rupture. Unburned vapors are very irritating.
6.7 **Ignition Temperature:** 915°F
6.8 **Electrical Hazard:** Data not available
6.9 **Burning Rate:** 4.5 mm/min.

7. CHEMICAL REACTIVITY

7.1 **Reactivity with Water:** No reaction
7.2 **Reactivity with Common Materials:** No reaction
7.3 **Stability During Transport:** Stable
7.4 **Neutralizing Agents for Acids and Caustics:** Not pertinent
7.5 **Polymerization:** Polymerizes spontaneously upon exposure to heat or sunlight
7.6 **Inhibitor of Polymerization:** Up to 1% hydroquinone

8. WATER POLLUTION

8.1 **Aquatic Toxicity:** Data not available
8.2 **Waterfowl Toxicity:** Data not available
8.3 **Biological Oxygen Demand (BOD):** Data not available
8.4 **Food Chain Concentration Potential:** None

9. SELECTED MANUFACTURERS

1. Pfizer, Incorporated
   Pfizer Chemical Division
   235 East 42nd Street
   New York, N. Y. 10017
2. Aldrich Chemical Co.
   940 West Saint Paul Ave.
   Milwaukee, Wis. 53233
3. Pfaltz and Bauer, Inc.
   126-04 Northern Boulevard
   Flushing, N. Y. 11368

10. SHIPPING INFORMATION

10.1 **Grades or Purity:** 98.5+%
10.2 **Storage Temperature:** Cool ambient
10.3 **Inert Atmosphere:** No requirement
10.4 **Venting:** Pressure-vacuum

11. HAZARD ASSESSMENT CODE

(See Hazard Assessment Handbook, CG 446-3)

A-P-Q-R-S

12. HAZARD CLASSIFICATIONS

12.1 **Code of Federal Regulations:** Flammable liquid
12.2 **NAS Hazard Rating for Bulk Water Transportation:**

Category	Rating
Fire	3
Health	
Vapor Irritant	4
Liquid or Solid Irritant	4
Poisons	4
Water Pollution	
Human Toxicity	4
Aquatic Toxicity	2
Aesthetic Effect	3
Reactivity	
Other Chemicals	3
Water	0
Self-Reaction	3

12.3 **NFPA Hazard Classifications:**

Category	Classification
Health Hazard (Blue)	2
Flammability (Red)	3
Reactivity (Yellow)	2

13. PHYSICAL AND CHEMICAL PROPERTIES

13.1 **Physical State at 15°C and 1 atm:** Liquid
13.2 **Molecular Weight:** 70.1
13.3 **Boiling Point at 1 atm:** 178.5°F = 81.4°C = 354.6°K
13.4 **Freezing Point:** 20°F = −7°C = 266°K
13.5 **Critical Temperature:** Not pertinent
13.6 **Critical Pressure:** Not pertinent
13.7 **Specific Gravity:** 0.864 at 20°C (liquid)
13.8 **Liquid Surface Tension:** (est.) 24 dynes/cm = 0.024 N/m at 20°C
13.9 **Liquid-Water Interfacial Tension:** Not pertinent
13.10 **Vapor (Gas) Specific Gravity:** 2.4
13.11 **Ratio of Specific Heats of Vapor (Gas):** 1.1053
13.12 **Latent Heat of Vaporization:** (est.) 203 Btu/lb = 113 cal/g = 4.73×10^5 J/kg
13.13 **Heat of Combustion:** (est.) −14,600 Btu/lb = −8,100 cal/g = -340×10^5 J/kg
13.14 **Heat of Decomposition:** Not pertinent
13.15 **Heat of Solution:** Not pertinent
13.16 **Heat of Polymerization:** −455 Btu/lb = −253 cal/g = -10.6×10^5 J/kg

NOTES

MNS	MINERAL SPIRITS

Common Synonyms			
Petroleum spirits Naphtha	Watery liquid	Colorless	Gasoline-like odor
	Floats on water.		

Stop discharge if possible.
Call fire department.
Avoid contact with liquid.
Isolate and remove discharged material.
Notify local health and pollution control agencies.

Fire

Combustible.
Extinguish with water, dry chemical, foam, or carbon dioxide.
Cool exposed containers with water.

Exposure

CALL FOR MEDICAL AID.

LIQUID
Irritating to skin and eyes.
Harmful if swallowed.
Remove contaminated clothing and shoes.
Flush affected areas with plenty of water.
IF IN EYES, hold eyelids open and flush with plenty of water.
IF SWALLOWED and victim is CONSCIOUS, have victim drink water or milk.
DO NOT INDUCE VOMITING.

Water Pollution

Effect of low concentrations on aquatic life is unknown.
Fouling to shoreline.
May be dangerous if it enters water intakes.
Notify local health and wildlife officials.
Notify operators of nearby water intakes.

1. RESPONSE TO DISCHARGE

(See Response Methods Handbook, CG 446-4)

Mechanical containment
Should be removed
Chemical and physical treatment

2. LABELS

No hazard label required by Code of Federal Regulations

3. CHEMICAL DESIGNATIONS

3.1 **Synonyms:** Naphtha
Petroleum spirits
3.2 **Coast Guard Compatibility Classification:** Miscellaneous hydrocarbon mixtures
3.3 **Chemical Formula:** Not applicable
3.4 **IMCO/United Nations Numerical Designation:** 3.3/1300

4. OBSERVABLE CHARACTERISTICS

4.1 **Physical State (as shipped):** Liquid
4.2 **Color:** Colorless
4.3 **Odor:** Like gasoline

5. HEALTH HAZARDS

5.1 **Personal Protective Equipment:** Plastic gloves; goggles or face shield (as for gasoline).
5.2 **Symptoms Following Exposure:** INHALATION: mild irritation of respiratory tract. ASPIRATION: severe lung irritation and rapidly developing pulmonary edema; central nervous system excitement followed by depression. INGESTION: irritation of stomach.
5.3 **Treatment for Exposure:** INHALATION: remove victim to fresh air. ASPIRATION: enforce bed rest; give oxygen; call a doctor. INGESTION: do NOT induce vomiting; guard against aspiration into lungs. EYES: wash with copious amounts of water. SKIN: wipe off and wash with soap and water.
5.4 **Toxicity by Inhalation (Threshold Limit Value):** 200 ppm
5.5 **Short-Term Inhalation Limits:** 4000 - 7000 ppm for 60 min.
5.6 **Toxicity by Ingestion:** Grade 2; LD_{50} 0.5 to 5 g/kg
5.7 **Late Toxicity:** Data not available
5.8 **Vapor (Gas) Irritant Characteristics:** Vapors are nonirritating to the eyes and throat.
5.9 **Liquid or Solid Irritant Characteristics:** Minimum hazard. If spilled on clothing and allowed to remain, may cause smarting and reddening of the skin.
5.10 **Odor Threshold:** Data not available

6. FIRE HAZARDS

6.1 **Flash Point:** 105-140°F C.C., depending on grade
6.2 **Flammable Limits in Air:** 0.8%—5.0%
6.3 **Fire Extinguishing Agents:** Foam, carbon dioxide, dry chemical
6.4 **Fire Extinguishing Agents Not to be Used:** Do not use straight hose water stream.
6.5 **Special Hazards of Combustion Products:** Not pertinent
6.6 **Behavior in Fire:** Not pertinent
6.7 **Ignition Temperature:** 540°F
6.8 **Electrical Hazard:** Not pertinent
6.9 **Burning Rate:** 4 mm/min.

7. CHEMICAL REACTIVITY

7.1 **Reactivity with Water:** No reaction
7.2 **Reactivity with Common Materials:** No reaction
7.3 **Stability During Transport:** Stable
7.4 **Neutralizing Agents for Acids and Caustics:** Not pertinent
7.5 **Polymerization:** Not pertinent
7.6 **Inhibitor of Polymerization:** Not pertinent

8. WATER POLLUTION

8.1 **Aquatic Toxicity:** Data not available
8.2 **Waterfowl Toxicity:** Data not available
8.3 **Biological Oxygen Demand (BOD):** 8%, 5 days
8.4 **Food Chain Concentration Potential:** None

9. SELECTED MANUFACTURERS

1. Atlantic Richfield Co.
717 Fifth Ave.
New York, N. Y. 10022
2. Phillips Petroleum
Bartlesville, Okla. 74004
3. Shell Oil Co.
1 Shell Plaza
Houston, Tex. 77001

10. SHIPPING INFORMATION

10.1 **Grades or Purity:** Various grades available. 70-100% of the materials are derived from petroleum, and 0-30% are aromatic hydrocarbons like benzene and toluene. Flash points vary with the exact composition but are usually above 100°F.
10.2 **Storage Temperature:** Ambient
10.3 **Inert Atmosphere:** No requirement
10.4 **Venting:** Open (flame arrester)

11. HAZARD ASSESSMENT CODE

(See Hazard Assessment Handbook, CG 446-3)

A-T-U

12. HAZARD CLASSIFICATIONS

12.1 **Code of Federal Regulations:** Combustible liquid
12.2 **NAS Hazard Rating for Bulk Water Transportation:** Not listed
12.3 **NFPA Hazard Classifications:**

Category	Classification
Health Hazard (Blue)	0
Flammability (Red)	2
Reactivity (Yellow)	0

13. PHYSICAL AND CHEMICAL PROPERTIES

13.1 **Physical State at 15°C and 1 atm:** Liquid
13.2 **Molecular Weight:** Not pertinent
13.3 **Boiling Point at 1 atm:** 310—395°F = 154—202°C = 428—475°K
13.4 **Freezing Point:** Not pertinent
13.5 **Critical Temperature:** Not pertinent
13.6 **Critical Pressure:** Not pertinent
13.7 **Specific Gravity:** 0.78 at 20°C (liquid)
13.8 **Liquid Surface Tension:** Data not available
13.9 **Liquid-Water Interfacial Tension:** Data not available
13.10 **Vapor (Gas) Specific Gravity:** Not pertinent
13.11 **Ratio of Specific Heats of Vapor (Gas):** (est.) 1.030
13.12 **Latent Heat of Vaporization:** Data not available
13.13 **Heat of Combustion:** Data not available
13.14 **Heat of Decomposition:** Not pertinent
13.15 **Heat of Solution:** Not pertinent
13.16 **Heat of Polymerization:** Not pertinent

NOTES

REVISED 1978

MTO

MOLYBDIC TRIOXIDE

Common Synonyms
Molybdenum trioxide
Molybdic anhydride

Solid — Colorless to white or yellow — Odorless

Sinks in water.

Avoid contact with solid. Keep people away.
Stop discharge if possible.
Isolate and remove discharged material.
Notify local health and pollution control agencies.

Fire

Not flammable.

Exposure

CALL FOR MEDICAL AID.

SOLID
Irritating to skin and eyes.
Harmful if swallowed.
Remove contaminated clothing and shoes.
Flush affected areas with plenty of water.
IF IN EYES, hold eyelids open and flush with plenty of water.
IF SWALLOWED and victim is CONSCIOUS, have victim drink water or milk.
IF SWALLOWED and victim is UNCONSCIOUS OR HAVING CONVULSIONS, do nothing except keep victim warm.

Water Pollution

HARMFUL TO AQUATIC LIFE IN VERY LOW CONCENTRATIONS.
May be dangerous if it enters water intakes.
Notify local health and wildlife officials.
Notify operators of nearby water intakes.

1. RESPONSE TO DISCHARGE

(See Response Methods Handbook, CG 446-4)

Issue warning—water contaminant
Should be removed
Chemical and physical treatment

2. LABELS

No hazard label required by Code of Federal Regulations

3. CHEMICAL DESIGNATIONS

3.1 **Synonyms:** Molybdenum trioxide; Molybdic anhydride
3.2 **Coast Guard Compatibility Classification:** Not listed
3.3 **Chemical Formula:** MoO_3
3.4 **IMCO/United Nations Numerical Designation:** Not listed

4. OBSERVABLE CHARACTERISTICS

4.1 **Physical State (as shipped):** Solid
4.2 **Color:** Colorless or white-yellow to yellow.
4.3 **Odor:** None

5. HEALTH HAZARDS

5.1 **Personal Protective Equipment:** U. S. Bu. Mines approved respirator; safety glasses or face shield; protective gloves
5.2 **Symptoms Following Exposure:** Compound is relatively nontoxic. Dust irritates eyes.
5.3 **Treatment for Exposure:** No treatment necessary except those applicable to any nontoxic dust. EYES: flush with water.
5.4 **Toxicity by Inhalation (Threshold Limit Value):** 5 mg/m^3
5.5 **Short-Term Inhalation Limits:** Data not available
5.6 **Toxicity by Ingestion:** Grade 3; LD_{50} 50—500 mg/kg
5.7 **Late Toxicity:** Data not available
5.8 **Vapor (Gas) Irritant Characteristics:** Data not available
5.9 **Liquid or Solid Irritant Characteristics:** Data not available
5.10 **Odor Threshold:** Data not available

6. FIRE HAZARDS

6.1 **Flash Point:** Not flammable
6.2 **Flammable Limits in Air:** Not flammable
6.3 **Fire Extinguishing Agents:** Not pertinent
6.4 **Fire Extinguishing Agents Not to be Used:** Not pertinent
6.5 **Special Hazards of Combustion Products:**
6.6 **Behavior in Fire:**
6.7 **Ignition Temperature:** Not pertinent
6.8 **Electrical Hazard:** Not pertinent
6.9 **Burning Rate:** Not pertinent

7. CHEMICAL REACTIVITY

7.1 **Reactivity with Water:** No reaction
7.2 **Reactivity with Common Materials:**
7.3 **Stability During Transport:** Stable
7.4 **Neutralizing Agents for Acids and Caustics:** Not pertinent
7.5 **Polymerization:** Not pertinent
7.6 **Inhibitor of Polymerization:** Not pertinent

8. WATER POLLUTION

8.1 **Aquatic Toxicity:** 70 ppm/96 hr/fathead minnow/TL_m/soft water
8.2 **Waterfowl Toxicity:** Data not available
8.3 **Biological Oxygen Demand (BOD):** None
8.4 **Food Chain Concentration Potential:** None listed

9. SELECTED MANUFACTURERS

1. Kennecott Copper Corp.
   Utah Copper Div.
   P. O. Box 11299
   Salt Lake City, Utah 84111
2. Allied Chemical Corp.
   Specialty Chemicals Div.
   P. O. Box 1087R
   Morristown, N. J. 07960
3. Climax Molybdenum Company
   One Greenwich Plaza
   Greenwich, Conn. 06830

10. SHIPPING INFORMATION

10.1 **Grades or Purity:** Technical, 59.8—61.6%; Reagent
10.2 **Storage Temperature:** Ambient
10.3 **Inert Atmosphere:** No requirement
10.4 **Venting:** Open

11. HAZARD ASSESSMENT CODE

(See Hazard Assessment Handbook, CG 446-3)

II

12. HAZARD CLASSIFICATIONS

12.1 **Code of Federal Regulations:** Not listed
12.2 **NAS Hazard Rating for Bulk Water Transportation:** Not listed
12.3 **NFPA Hazard Classifications:** Not listed

13. PHYSICAL AND CHEMICAL PROPERTIES

13.1 **Physical State at 15°C and 1 atm:** Solid
13.2 **Molecular Weight:** 143.94
13.3 **Boiling Point at 1 atm:** Not pertinent
13.4 **Freezing Point:** Not pertinent
13.5 **Critical Temperature:** Not pertinent
13.6 **Critical Pressure:** Not pertinent
13.7 **Specific Gravity:** 4.69 at 20°C (solid)
13.8 **Liquid Surface Tension:** Not pertinent
13.9 **Liquid-Water Interfacial Tension:** Not pertinent
13.10 **Vapor (Gas) Specific Gravity:** Not pertinent
13.11 **Ratio of Specific Heats of Vapor (Gas):** Not pertinent
13.12 **Latent Heat of Vaporization:** Not pertinent
13.13 **Heat of Combustion:** Not pertinent
13.14 **Heat of Decomposition:** Not pertinent
13.15 **Heat of Solution:** Not pertinent
13.16 **Heat of Polymerization:** Not pertinent

NOTES

MCA MONOCHLOROACETIC ACID

Common Synonyms			
Chloroacetic acid Chloracetic acid	Solid or heated liquid	Cloudy white solid; liquid is colorless to light yellow	Strong vinegar-like odor
	Mixes with water.		

Avoid contact with liquid. Keep people away.
Stop discharge if possible.
Call fire department.
Isolate and remove discharged material.
Notify local health and pollution control agencies.

Fire	Combustible. POISONOUS GASES MAY BE PRODUCED IN FIRE. Flood discharge area with water.
Exposure	Call for medical aid. LIQUID OR SOLID Will burn skin and eyes. Harmful if swallowed. Remove contaminated clothing and shoes. Flush affected areas with plenty of water. IF IN EYES, hold eyelids open and flush with plenty of water. IF SWALLOWED and victim is CONSCIOUS, have victim drink water or milk. DO NOT INDUCE VOMITING.
Water Pollution	Effect of low concentrations on aquatic life is unknown. May be dangerous if it enters water intakes. Notify local health and wildlife officials. Notify operators of nearby water intakes.

1. RESPONSE TO DISCHARGE

(See Response Methods Handbook, CG 446-4)

Issue warning—corrosive
Restrict access
Disperse and flush

2. LABEL

3. CHEMICAL DESIGNATIONS

3.1 **Synonyms:** Chloracetic acid; Chloroacetic acid
3.2 **Coast Guard Compatibility Classification:** Not listed
3.3 **Chemical Formula:** $ClCH_2COOH$
3.4 **IMCO/United Nations Numerical Designation:** Liquid: 8/1750; Solid: 8/1751

4. OBSERVABLE CHARACTERISTICS

4.1 **Physical State (as shipped):** Liquid or solid
4.2 **Color:** Solid: translucent white; Liquid: clear to light amber
4.3 **Odor:** Characteristic penetrating odor similar to vinegar

5. HEALTH HAZARDS

5.1 **Personal Protective Equipment:** Self-contained breathing apparatus; vinyl or neoprene rubber gloves; goggles and protective face shield; rubberized or acid-resistant clothing
5.2 **Symptoms Following Exposure:** Inhalation causes mucous membrane irritation. Contact with liquid causes severe irritation and burns of the eyes and irritation and burns of skin. Ingestion causes burns of mouth and stomach.
5.3 **Treatment for Exposure:** Get medical attention for all exposures to this compound. INHALATION: remove victim to fresh air and enforce rest until medical attention is obtained. EYES: flush with running water for 15 min. SKIN: flush with water; get treatment for burns. INGESTION: give large amount of water.
5.4 **Toxicity by Inhalation (Threshold Limit Value):** Data not available
5.5 **Short-Term Inhalation Limits:** Data not available
5.6 **Toxicity by Ingestion:** Grade 3; oral LD_{50} = 76.2 mg/kg (rat)
5.7 **Late Toxicity:** Data not available
5.8 **Vapor (Gas) Irritant Characteristics:** Vapors cause severe irritation of eyes and throat and can cause eye and lung injury. They cannot be tolerated even at low concentrations.
5.9 **Liquid or Solid Irritant Characteristics:** Severe skin irritant. Causes second- and third-degree burns on short contact and is very injurious to the eyes.
5.10 **Odor Threshold:** 0.15 mg/m^3

6. FIRE HAZARDS

6.1 **Flash Point:** (almost nonflammable) 259°F C.C.
6.2 **Flammable Limits in Air:** 8% (LFL)
6.3 **Fire Extinguishing Agents:** Dry chemical, carbon dioxide, water
6.4 **Fire Extinguishing Agents Not to be Used:** Not pertinent
6.5 **Special Hazards of Combustion Products:** Hydrogen chloride and phosgene may be generated.
6.6 **Behavior in Fire:** Not pertinent
6.7 **Ignition Temperature:** Data not available; difficult to ignite
6.8 **Electrical Hazard:** Not pertinent
6.9 **Burning Rate:** Not pertinent

7. CHEMICAL REACTIVITY

7.1 **Reactivity with Water:** No reaction
7.2 **Reactivity with Common Materials:** Causes mild corrosion of common metals
7.3 **Stability During Transport:** Stable
7.4 **Neutralizing Agents for Acids and Caustics:** Flush with water, rinse with sodium bicarbonate or lime solution
7.5 **Polymerization:** Not pertinent
7.6 **Inhibitor of Polymerization:** Not pertinent

8. WATER POLLUTION

8.1 **Aquatic Toxicity:** Data not available
8.2 **Waterfowl Toxicity:** Data not available
8.3 **Biological Oxygen Demand (BOD):** Data not available
8.4 **Food Chain Concentration Potential:** None

9. SELECTED MANUFACTURERS

1. Hercules Incorporated
   Coatings & Specialty Products Dept.
   910 Market Street
   Wilmington, Del. 19899
2. Dow Chemical Co.
   Midland, Mich. 48640
3. McKesson Chemical Co.
   Crocker Plaza
   One Post St.
   San Francisco, Calif. 94104

10. SHIPPING INFORMATION

10.1 **Grades or Purity:** Commercial: 97.5+%
10.2 **Storage Temperature:** Solid: ambient; Liquid: 70°C (158°F)
10.3 **Inert Atmosphere:** No requirement
10.4 **Venting:** Open (flame arrester)

11. HAZARD ASSESSMENT CODE

(See Hazard Assessment Handbook, CG 446-3)

A-P

12. HAZARD CLASSIFICATIONS

12.1 **Code of Federal Regulations:** Corrosive liquid
12.2 **NAS Hazard Rating for Bulk Water Transportation:**

Category	Rating
Fire	1
Health	
Vapor Irritant	4
Liquid or Solid Irritant	4
Poisons	4
Water Pollution	
Human Toxicity	3
Aquatic Toxicity	2
Aesthetic Effect	3
Reactivity	
Other Chemicals	4
Water	2
Self-Reaction	0

12.3 **NFPA Hazard Classifications:**

Category	Classification
Health Hazard (Blue)	3
Flammability (Red)	1
Reactivity (Yellow)	0

13. PHYSICAL AND CHEMICAL PROPERTIES

13.1 **Physical State at 15°C and 1 atm:** Solid
13.2 **Molecular Weight:** 94.5
13.3 **Boiling Point at 1 atm:** 372°F = 189°C = 462°K
13.4 **Freezing Point:** 140°F = 60°C = 333°K
13.5 **Critical Temperature:** Not pertinent
13.6 **Critical Pressure:** Not pertinent
13.7 **Specific Gravity:** 1.58 at 20°C (solid)
13.8 **Liquid Surface Tension:** 33 dynes/cm = 0.033 N/m at 80°C
13.9 **Liquid-Water Interfacial Tension:** Not pertinent
13.10 **Vapor (Gas) Specific Gravity:** Not pertinent
13.11 **Ratio of Specific Heats of Vapor (Gas):** Not pertinent
13.12 **Latent Heat of Vaporization:** 250 Btu/lb = 139 cal/g = 5.82×10^5 J/kg
13.13 **Heat of Combustion:** (solid) −1,814 Btu/lb = −1,008 cal/g = -42.17×10^5 J/kg
13.14 **Heat of Decomposition:** Not pertinent
13.15 **Heat of Solution:** −63 Btu/lb = −35 cal/g = -1.5×10^5 J/kg
13.16 **Heat of Polymerization:** Not pertinent

NOTES

MCF	MONOCHLORODIFLUOROMETHANE

Common Synonyms		
Freon-22 Genetron-22 Isotron-22 Uconn-22	Liquefied compressed gas Colorless Faint odor Liquid sinks and boils in water. Visible vapor cloud is formed.	

Stop discharge if possible.
Avoid contact with liquid.
Isolate and remove discharged material.
Notify local health and pollution control agencies.

Fire	Not flammable. POISONOUS GASES ARE PRODUCED WHEN HEATED. Wear goggles and self-contained breathing apparatus. Cool exposed containers with water.
Exposure	CALL FOR MEDICAL AID. VAPOR Not irritating to eyes, nose or throat. If inhaled, will cause dizziness or loss of consciousness. Move to fresh air. If breathing has stopped, give artificial respiration. If breathing is difficult, give oxygen. LIQUID Will cause frostbite. Flush affected areas with plenty of water. DO NOT RUB AFFECTED AREAS.
Water Pollution	Not harmful to aquatic life.

1. RESPONSE TO DISCHARGE

(See Response Methods Handbook, CG 446-4)

Disperse and flush

2. LABEL

Green

3. CHEMICAL DESIGNATIONS

3.1 **Synonyms:** Chlorodifluoromethane
Eskimon-22 Genetron 22
F-22 Isotron 22
Freon-22 Uconn 22

3.2 **Coast Guard Compatibility Classification:** Halogenated hydrocarbon

3.3 **Chemical Formula:** $CHClF_2$

3.4 **IMCO/United Nations Numerical Designation:** 2.0/1018

4. OBSERVABLE CHARACTERISTICS

4.1 **Physical State (as shipped):** Liquefied gas

4.2 **Color:** Colorless

4.3 **Odor:** Faint ethereal; like carbon tetrachloride

5. HEALTH HAZARDS

5.1 **Personal Protective Equipment:** Rubber gloves; goggles.

5.2 **Symptoms Following Exposure:** Inhalation at greater than 10% concentration in air may cause narcosis. Liquid may cause frostbite.

5.3 **Treatment for Exposure:** Remove victim to non-contaminated area and apply artificial respiration if breathing has stopped. Call a physician immediately. Oxygen inhalation may be utilized.

5.4 **Toxicity by Inhalation (Threshold Limit Value):** 1000 ppm

5.5 **Short-Term Inhalation Limits:** Data not available

5.6 **Toxicity by Ingestion:** Not pertinent

5.7 **Late Toxicity:** None

5.8 **Vapor (Gas) Irritant Characteristics:** Vapors are nonirritating to the eyes and throat.

5.9 **Liquid or Solid Irritant Characteristics:** No appreciable hazard. Practically harmless to the skin because it evaporates quickly. Liquid may cause frostbite.

5.10 **Odor Threshold:** Data not available.

6. FIRE HAZARDS

6.1 **Flash Point:** Not flammable

6.2 **Flammable Limits in Air:** Not flammable

6.3 **Fire Extinguishing Agents:** Not pertinent

6.4 **Fire Extinguishing Agents Not to be Used:** Not pertinent

6.5 **Special Hazards of Combustion Products:** Decomposition gases are toxic and irritating.

6.6 **Behavior in Fire:** Not pertinent

6.7 **Ignition Temperature:** Not flammable

6.8 **Electrical Hazard:** Not pertinent

6.9 **Burning Rate:** Not flammable

7. CHEMICAL REACTIVITY

7.1 **Reactivity with Water:** No reaction

7.2 **Reactivity with Common Materials:** No reaction

7.3 **Stability During Transport:** Stable

7.4 **Neutralizing Agents for Acids and Caustics:** Not pertinent

7.5 **Polymerization:** Not pertinent

7.6 **Inhibitor of Polymerization:** Not pertinent

8. WATER POLLUTION

8.1 **Aquatic Toxicity:** None

8.2 **Waterfowl Toxicity:** None

8.3 **Biological Oxygen Demand (BOD):** None

8.4 **Food Chain Concentration Potential:** None

9. SELECTED MANUFACTURERS

1. E. I. duPont de Nemours & Co., Inc.
Freon Products Division
Wilmington, Del. 19898

2. Pennwalt Corp.
Chemicals Division
3 Penn Center
Philadelphia, Pa. 19102

3. Union Carbide Corp.
Chemicals and Plastics Division
270 Park Ave.
New York, N.Y. 10017

10. SHIPPING INFORMATION

10.1 **Grades or Purity:** Propellant grade

10.2 **Storage Temperature:** Ambient

10.3 **Inert Atmosphere:** No requirement

10.4 **Venting:** Safety relief

11. HAZARD ASSESSMENT CODE

(See Hazard Assessment Handbook, CG 446-3)

A-C-I-J

12. HAZARD CLASSIFICATIONS

12.1 **Code of Federal Regulations:** Nonflammable compressed gas

12.2 **NAS Hazard Rating for Bulk Water Transportation:**

Category	Rating
Fire	0
Health	
Vapor Irritant	0
Liquid or Solid Irritant	0
Poisons	1
Water Pollution	
Human Toxicity	0
Aquatic Toxicity	0
Aesthetic Effect	0
Reactivity	
Other Chemicals	1
Water	0
Self-Reaction	0

12.3 **NFPA Hazard Classifications:** Not listed

13. PHYSICAL AND CHEMICAL PROPERTIES

13.1 **Physical State at 15°C and 1 atm:** Gas

13.2 **Molecular Weight:** 86.48

13.3 **Boiling Point at 1 atm:** $-40.9°F = -40.5°C = 232.7°K$

13.4 **Freezing Point:** Not pertinent

13.5 **Critical Temperature:** $204.8°F = 96.0°C = 369.2°K$

13.6 **Critical Pressure:** $716 \text{ psia} = 48.7 \text{ atm} = 4.93 \text{ MN/m}^2$

13.7 **Specific Gravity:** 1.41 at −40°C (liquid)

13.8 **Liquid Surface Tension:** (est.) 15 dynes/cm = 0.015 N/m at −41°C

13.9 **Liquid-Water Interfacial Tension:** Data not available

13.10 **Vapor (Gas) Specific Gravity:** 3.0

13.11 **Ratio of Specific Heats of Vapor (Gas):** (est.) 1.13

13.12 **Latent Heat of Vaporization:** $101 \text{ Btu/lb} = 55.9 \text{ cal/g} = 2.34 \times 10^5 \text{ J/kg}$

13.13 **Heat of Combustion:** Not pertinent

13.14 **Heat of Decomposition:** Not pertinent

13.15 **Heat of Solution:** Not pertinent

13.16 **Heat of Polymerization:** Not pertinent

NOTES

REVISED 1978

MEA — MONOETHANOLAMINE

Common Synonyms
beta-Aminoethyl alcohol
2-Hydroxyethylamine
Ethanolamine
2-Aminoethanol

Oily liquid — Colorless — Slight ammonia odor

Sinks and mixes with water. Freezing point is 51° F.

Avoid contact with liquid.
Wear goggles and self-contained breathing apparatus.
Stop discharge if possible.
Call fire department.
Isolate and remove discharged material.
Notify local health and pollution control agencies.

Fire

Combustible.
Wear goggles and self-contained breathing apparatus.
Extinguish with dry chemical, alcohol foam, or carbon dioxide.
Cool exposed containers with water.

Exposure

CALL FOR MEDICAL AID.

LIQUID OR SOLID
Irritating to skin and eyes.
Harmful if swallowed.
Remove contaminated clothing and shoes.
Flush affected areas with plenty of water.
IF IN EYES, hold eyelids open and flush with plenty of water.
IF SWALLOWED and victim is CONSCIOUS, have victim drink water or milk.

Water Pollution

Dangerous to aquatic life in high concentrations.
May be dangerous if it enters water intakes.
Notify local health and wildlife officials.
Notify operators of nearby water intakes.

1. RESPONSE TO DISCHARGE

(See Response Methods Handbook, CG 446-4)

Disperse and flush

2. LABEL

3. CHEMICAL DESIGNATIONS

3.1 **Synonyms:** 2-Aminoethanol
beta-Aminoethyl alcohol
Ethanolamine
2-Hydroxyethylamine

3.2 **Coast Guard Compatibility Classification:** Alkanolamines

3.3 **Chemical Formula:** $HOCH_2CH_2NH_2$

3.4 **IMCO/United Nations Numerical Designation:** Not listed

4. OBSERVABLE CHARACTERISTICS

4.1 **Physical State (as shipped):** Liquid

4.2 **Color:** Colorless

4.3 **Odor:** Mildly ammoniacal

5. HEALTH HAZARDS

5.1 **Personal Protective Equipment:** Full face shield; goggles; eye wash facility.

5.2 **Symptoms Following Exposure:** Vapor irritates eyes and nose. Liquid causes local injury to mouth, throat, digestive tract, skin, and eyes.

5.3 **Treatment for Exposure:** INGESTION: induce vomiting by giving large volumes of warm salt water (2 tablespoons per glass); call a doctor. EYES: flush with water for at least 15 min.; call a doctor. SKIN: flush with water.

5.4 **Toxicity by Inhalation (Threshold Limit Value):** 3 ppm

5.5 **Short-Term Inhalation Limits:** Data not available

5.6 **Toxicity by Ingestion:** Grade 2; LD_{50} 0.5 to 5 g/kg (rat)

5.7 **Late Toxicity:** Data not available

5.8 **Vapor (Gas) Irritant Characteristics:** Vapors cause moderate irritation such that personnel will find high concentrations unpleasant. The effect is temporary.

5.9 **Liquid or Solid Irritant Characteristics:** Causes smarting of the skin and first-degree burns on short exposure; may cause secondary burns on long exposure.

5.10 **Odor Threshold:** Data not available

6. FIRE HAZARDS

6.1 **Flash Point:** 185°F C.C.; 200°F O.C.

6.2 **Flammable Limits in Air:** Data not available

6.3 **Fire Extinguishing Agents:** Water spray, alcohol foam, dry chemical or carbon dioxide

6.4 **Fire Extinguishing Agents Not to be Used:** Not pertinent

6.5 **Special Hazards of Combustion Products:** Irritating vapors generated when heated.

6.6 **Behavior in Fire:** Not pertinent

6.7 **Ignition Temperature:** Data not available

6.8 **Electrical Hazard:** Not pertinent

6.9 **Burning Rate:** Data not available

7. CHEMICAL REACTIVITY

7.1 **Reactivity with Water:** No reaction

7.2 **Reactivity with Common Materials:** No reaction

7.3 **Stability During Transport:** Stable

7.4 **Neutralizing Agents for Acids and Caustics:** Flush with water.

7.5 **Polymerization:** Not pertinent

7.6 **Inhibitor of Polymerization:** Not pertinent

8. WATER POLLUTION

8.1 **Aquatic Toxicity:** 7100 ppm/48 hr/shrimp/LC_{50}/salt water

8.2 **Waterfowl Toxicity:** Data not available

8.3 **Biological Oxygen Demand (BOD):** 78%, 5 days; (theor.) 0%, 5 days; 64%, 20 days

8.4 **Food Chain Concentration Potential:** None

9. SELECTED MANUFACTURERS

1. Dow Chemical Co.
Midland, Mich. 48640
2. Jefferson Chemical Co.
3336 Richmond Ave.
Houston, Tex. 77052
3. Union Carbide Corp.
Chemicals and Plastics Division
270 Park Ave.
New York, N. Y. 10017

10. SHIPPING INFORMATION

10.1 **Grades or Purity:** NF: 85% (15% water); commercial: 99+%

10.2 **Storage Temperature:** Ambient

10.3 **Inert Atmosphere:** No requirement

10.4 **Venting:** Open

11. HAZARD ASSESSMENT CODE

(See Hazard Assessment Handbook, CG 446-3)

A-P-Q

12. HAZARD CLASSIFICATIONS

12.1 **Code of Federal Regulations:** Corrosive Material

12.2 **NAS Hazard Rating for Bulk Water Transportation:**

Category	Rating
Fire	1
Health	
Vapor Irritant	2
Liquid or Solid Irritant	2
Poisons	2
Water Pollution	
Human Toxicity	2
Aquatic Toxicity	1
Aesthetic Effect	2
Reactivity	
Other Chemicals	3
Water	0
Self-Reaction	0

12.3 **NFPA Hazard Classifications:**

Category	Classification
Health Hazard (Blue)	2
Flammability (Red)	2
Reactivity (Yellow)	0

13. PHYSICAL AND CHEMICAL PROPERTIES

13.1 **Physical State at 15°C and 1 atm:** Liquid

13.2 **Molecular Weight:** 61.08

13.3 **Boiling Point at 1 atm:** 338°F = 170°C = 443°K

13.4 **Freezing Point:** 50.5°F = 10.3°C = 283.5°K

13.5 **Critical Temperature:** 646°F = 341°C = 614°K

13.6 **Critical Pressure:** 647 psia = 44 atm = 4.45 MN/m^2

13.7 **Specific Gravity:** 1.016 at 20°C (liquid)

13.8 **Liquid Surface Tension:** Not pertinent

13.9 **Liquid-Water Interfacial Tension:** Not pertinent

13.10 **Vapor (Gas) Specific Gravity:** Not pertinent

13.11 **Ratio of Specific Heats of Vapor (Gas):** Not pertinent

13.12 **Latent Heat of Vaporization:** 360 Btu/lb = 200 cal/g = 8.37×10^5 J/kg

13.13 **Heat of Combustion:** -10,710 Btu/lb = -5950 cal/g = -249×10^5 J/kg

13.14 **Heat of Decomposition:** Not pertinent

13.15 **Heat of Solution:** (est.) −17 Btu/lb = −10 cal/g = -0.4×10^5 J/kg

13.16 **Heat of Polymerization:** Not pertinent

NOTES

REVISED 1978

MPA MONOISOPROPANOLAMINE

Common Synonyms
Isopropanolamine
1-Amino-2-propanol
2-Hydroxypropylamine

Thick liquid Colorless Slight ammonia odor

Floats and mixes with water. Freezing point is 35° F.

Avoid contact with liquid.
Wear goggles, self-contained breathing apparatus, and rubber overclothing (including gloves).
Stop discharge if possible.
Call fire department.
Isolate and remove discharged material.
Notify local health and pollution control agencies.

Fire

Combustible.
Wear goggles, self-contained breathing apparatus, and rubber overclothing (including gloves).
Extinguish with dry chemical, water, alcohol foam, or carbon dioxide.

Exposure

CALL FOR MEDICAL AID.

LIQUID OR SOLID
Irritating to skin and eyes.
Harmful if swallowed.
Remove contaminated clothing and shoes.
Flush affected areas with plenty of water.
IF IN EYES, hold eyelids open and flush with plenty of water.
IF SWALLOWED and victim is CONSCIOUS, have victim drink water or milk and have victim induce vomiting.
IF SWALLOWED and victim is UNCONSCIOUS OR HAVING CONVULSIONS, do nothing except keep victim warm.

Water Pollution

Effect of low concentrations on aquatic life is unknown.
May be dangerous if it enters water intakes.
Notify local health and wildlife officials.
Notify operators of nearby water intakes.

1. RESPONSE TO DISCHARGE
(See Response Methods Handbook, CG 446-4)

Disperse and flush

2. LABELS

No hazard label required by Code of Federal Regulations

3. CHEMICAL DESIGNATIONS

3.1 **Synonyms:** 1-Amino-2-propanol
2-Hydroxypropylamine
Isopropanolamine
3.2 **Coast Guard Compatibility Classification:** Alkanolamines
3.3 **Chemical Formula:** $CH_3CH(OH)CH_2NH_2$
3.4 **IMCO/United Nations Numerical Designation:** Not listed

4. OBSERVABLE CHARACTERISTICS

4.1 **Physical State (as shipped):** Liquid
4.2 **Color:** Colorless
4.3 **Odor:** Mildly ammoniacal

5. HEALTH HAZARDS

5.1 **Personal Protective Equipment:** Full face shield; goggles; eye wash facility.
5.2 **Symptoms Following Exposure:** Vapor irritates eyes and nose. Liquid causes local injury to mouth, throat, digestive tract, skin, and eyes.
5.3 **Treatment for Exposure:** INGESTION: induce vomiting by giving large volumes of warm salt water (2 tablespoons per glass); call a doctor. EYES: flush with water for at least 15 min. and call a doctor. SKIN: flush with water.
5.4 **Toxicity by Inhalation (Threshold Limit Value):** 5 ppm
5.5 **Short-Term Inhalation Limits:** Data not available
5.6 **Toxicity by Ingestion:** Grade 2; LD_{50} 0.5 to 5 g/kg (rat)
5.7 **Late Toxicity:** Data not available
5.8 **Vapor (Gas) Irritant Characteristics:** Vapors cause a slight smarting of the eyes or respiratory system if present in high concentrations. The effect is temporary.
5.9 **Liquid or Solid Irritant Characteristics:** Causes smarting of the skin and first-degree burns on short exposure and may cause secondary burns on long exposure.
5.10 **Odor Threshold:** Data not available

6. FIRE HAZARDS

6.1 **Flash Point:** 165°F O.C.; 171°F C.C.
6.2 **Flammable Limits in Air:** 2.2%(calc.)-12%(est.)
6.3 **Fire Extinguishing Agents:** Dry chemical, water spray, alcohol foam, or carbon dioxide.
6.4 **Fire Extinguishing Agents Not to be Used:** Not pertinent
6.5 **Special Hazards of Combustion Products:** Irritating vapors generated when heated.
6.6 **Behavior in Fire:** Not pertinent
6.7 **Ignition Temperature:** 706°F (est.)
6.8 **Electrical Hazard:** Not pertinent
6.9 **Burning Rate:** 1.1 mm/min

7. CHEMICAL REACTIVITY

7.1 **Reactivity with Water:** No reaction
7.2 **Reactivity with Common Materials:** No reaction
7.3 **Stability During Transport:** Stable
7.4 **Neutralizing Agents for Acids and Caustics:** Flush with water
7.5 **Polymerization:** Not pertinent
7.6 **Inhibitor of Polymerization:** Not pertinent

8. WATER POLLUTION

8.1 **Aquatic Toxicity:** Data not available
8.2 **Waterfowl Toxicity:** Data not available
8.3 **Biological Oxygen Demand (BOD):** (theor.) 5.1%, 5 days; 46%, 20 days
8.4 **Food Chain Concentration Potential:** None

9. SELECTED MANUFACTURERS

1. Dow Chemical Co.
Midland, Mich. 48640
2. Union Carbide Corp.
Chemicals and Plastics Division
270 Park Ave.
New York, N. Y. 10017

10. SHIPPING INFORMATION

10.1 **Grades or Purity:** 98.5+%
10.2 **Storage Temperature:** Ambient
10.3 **Inert Atmosphere:** No requirement
10.4 **Venting:** Open

11. HAZARD ASSESSMENT CODE
(See Hazard Assessment Handbook, CG 446-3)

A-P-Q

12. HAZARD CLASSIFICATIONS

12.1 **Code of Federal Regulations:** Combustible Liquid
12.2 **NAS Hazard Rating for Bulk Water Transportation:**

Category	Rating
Fire	1
Health	
Vapor Irritant	1
Liquid or Solid Irritant	2
Poisons	1
Water Pollution	
Human Toxicity	2
Aquatic Toxicity	1
Aesthetic Effect	2
Reactivity	
Other Chemicals	3
Water	0
Self-Reaction	0

12.3 **NFPA Hazard Classifications:** Not listed

13. PHYSICAL AND CHEMICAL PROPERTIES

13.1 **Physical State at 15°C and 1 atm:** Liquid
13.2 **Molecular Weight:** 75.11
13.3 **Boiling Point at 1 atm:** 320°F = 160°C = 433°K
13.4 **Freezing Point:** 35.4°F = 1.9°C = 275.1°K
13.5 **Critical Temperature:** 622°F = 328°C = 601°K
13.6 **Critical Pressure:** 850 psia = 58 atm = 5.9 MN/m^2
13.7 **Specific Gravity:** 0.961 at 20°C (liquid)
13.8 **Liquid Surface Tension:** Not pertinent
13.9 **Liquid-Water Interfacial Tension:** Not pertinent
13.10 **Vapor (Gas) Specific Gravity:** Not pertinent
13.11 **Ratio of Specific Heats of Vapor (Gas):** Not pertinent
13.12 **Latent Heat of Vaporization:** 272 Btu/lb = 151 cal/g = 6.32 × 10^5 J/kg
13.13 **Heat of Combustion:** (est.) −13,900 Btu/lb = −7,700 cal/g = −322 × 10^5 J/kg
13.14 **Heat of Decomposition:** Not pertinent
13.15 **Heat of Solution:** (est.) −17 Btu/lb = −10 cal/g = −0.4 × 10^5 J/kg
13.16 **Heat of Polymerization:** Not pertinent

NOTES

REVISED 1978

MPL — MORPHOLINE

Common Synonyms			
Tetrahydro-2H-1,4-oxazine Tetrahydro-p-oxazine	Oily liquid	Colorless	Fishy, ammonia odor
	Floats and mixes with water. Irritating vapor is produced.		

Stop discharge if possible.
Avoid contact with liquid and vapor.
Call fire department.
Isolate and remove discharged material.
Notify local health and pollution control agencies.

Fire

FLAMMABLE
Flashback along vapor trail may occur.
Vapor may explode if ignited in an enclosed area.
Extinguish with water, dry chemical, or carbon dioxide.
Cool exposed containers with water.

Exposure

CALL FOR MEDICAL AID.

VAPOR
Irritating to eyes, nose and throat.
If inhaled, will cause nausea, headache, or difficult breathing.
Move to fresh air.
If breathing has stopped, give artificial respiration.
If breathing is difficult, give oxygen.

LIQUID
Irritating to skin and eyes.
Remove contaminated clothing and shoes.
Flush affected areas with plenty of water.
IF IN EYES, hold eyelids open and flush with plenty of water.
IF SWALLOWED and victim is CONSCIOUS, have victim drink water or milk and have victim induce vomiting.
IF SWALLOWED and victim is UNCONSCIOUS OR HAVING CONVULSIONS, do nothing except keep victim warm.

Water Pollution

Effect of low concentrations on aquatic life is unknown.
May be dangerous if it enters water intakes.
Notify local health and wildlife officials.
Notify operators of nearby water intakes.

1. RESPONSE TO DISCHARGE
(See Response Methods Handbook, CG 446-4)

Disperse and flush

2. LABEL

Red

3. CHEMICAL DESIGNATIONS

3.1 **Synonyms:** Diethyleneimide oxide
Diethylene imidoxide
Diethylene oximide
Tetrahydro-2H-1,4-oxazine
Tetrahydro-p-oxazine

3.2 **Coast Guard Compatibility Classification:** Aliphatic amine

3.3 **Chemical Formula:** $OCH_2CH_2NHCH_2CH_2$

3.4 **IMCO/United Nations Numerical Designation:** 3.3/2054

4. OBSERVABLE CHARACTERISTICS

4.1 **Physical State (as shipped):** Liquid

4.2 **Color:** Colorless

4.3 **Odor:** Fishy; ammoniacal

5. HEALTH HAZARDS

5.1 **Personal Protective Equipment:** Organic vapor canister or self-contained breathing apparatus; rubber boots and gloves; goggles or face shield.

5.2 **Symptoms Following Exposure:** Liquid causes skin and eye burns. Breathing vapors or absorption through skin may cause nausea and headache.

5.3 **Treatment for Exposure:** INHALATION: if ill effects occur, move patient to fresh air, keep him quiet and warm, and call a physician; if breathing stops, start artificial respiration. INGESTION: force milk or water, then immediately induce vomiting; treat symptomatically; no known antidote. SKIN OR EYES: immediately flush with plenty of water for at least 15 min.; for eyes get medical attention promptly.

5.4 **Toxicity by Inhalation (Threshold Limit Value):** 20 ppm

5.5 **Short-Term Inhalation Limits:** 20 ppm for 15 min.

5.6 **Toxicity by Ingestion:** Grade 2; LD_{50} 0.5 to 5 g/kg (guinea pig, rat)

5.7 **Late Toxicity:** Data not available

5.8 **Vapor (Gas) Irritant Characteristics:** Vapors cause a slight smarting of the eyes or respiratory system if present in high concentrations. The effect is temporary.

5.9 **Liquid or Solid Irritant Characteristics:** Minimum hazard. If spilled on clothing and allowed to remain, may cause smarting and reddening of the skin.

5.10 **Odor Threshold:** 0.01 ppm

6. FIRE HAZARDS

6.1 **Flash Point:** 100°F O.C.

6.2 **Flammable Limits in Air:** 1.8%—10.8%

6.3 **Fire Extinguishing Agents:** Water fog, alcohol foam, dry chemical, or carbon dioxide

6.4 **Fire Extinguishing Agents Not to be Used:** Not pertinent

6.5 **Special Hazards of Combustion Products:** Irritating vapors are generated when heated.

6.6 **Behavior in Fire:** Vapor is heavier than air and may travel some distance to source of ignition and flash back.

6.7 **Ignition Temperature:** 590°F

6.8 **Electrical Hazard:** Not pertinent

6.9 **Burning Rate:** 1.9 mm/min

7. CHEMICAL REACTIVITY

7.1 **Reactivity with Water:** No reaction

7.2 **Reactivity with Common Materials:** No reaction

7.3 **Stability During Transport:** Stable

7.4 **Neutralizing Agents for Acids and Caustics:** Flush with water

7.5 **Polymerization:** Not pertinent

7.6 **Inhibitor of Polymerization:** Not pertinent

8. WATER POLLUTION

8.1 **Aquatic Toxicity:** Data not available

8.2 **Waterfowl Toxicity:** Data not available

8.3 **Biological Oxygen Demand (BOD):** (theor.) 0.9%, 5 days; 5.1%, 20 days

8.4 **Food Chain Concentration Potential:** None

9. SELECTED MANUFACTURERS

1. Dow Chemical Co.
Midland, Mich. 48640
2. Jefferson Chemical Co., Inc.
3336 Richmond Ave.
P. O. Box 53300
Houston, Tex. 77052
3. Union Carbide Corp.
Chemicals and Plastics Division
270 Park Ave.
New York, N. Y. 10017

10. SHIPPING INFORMATION

10.1 **Grades or Purity:** Several grades available, most above 99%.

10.2 **Storage Temperature:** Ambient

10.3 **Inert Atmosphere:** No requirement

10.4 **Venting:** Open

11. HAZARD ASSESSMENT CODE
(See Hazard Assessment Handbook, CG 446-3)

A-P-Q

12. HAZARD CLASSIFICATIONS

12.1 **Code of Federal Regulations:** Flammable Liquid

12.2 **NAS Hazard Rating for Bulk Water Transportation:**

Category	Rating
Fire	3
Health	
Vapor Irritant	1
Liquid or Solid Irritant	1
Poisons	1
Water Pollution	
Human Toxicity	2
Aquatic Toxicity	2
Aesthetic Effect	2
Reactivity	
Other Chemicals	3
Water	0
Self-Reaction	0

12.3 **NFPA Hazard Classifications:**

Category	Classification
Health Hazard (Blue)	2
Flammability (Red)	3
Reactivity (Yellow)	0

13. PHYSICAL AND CHEMICAL PROPERTIES

13.1 **Physical State at 15°C and 1 atm:** Liquid

13.2 **Molecular Weight:** 87.12

13.3 **Boiling Point at 1 atm:** 262.8°F = 128.2°C = 401.4°K

13.4 **Freezing Point:** 23.4°F = −4.8°C = 268.4°K

13.5 **Critical Temperature:** 653°F = 345°C = 618°K

13.6 **Critical Pressure:** 794 psia = 54 atm = 5.47 MN/m^2

13.7 **Specific Gravity:** 1.00 at 20°C (liquid)

13.8 **Liquid Surface Tension:** Not pertinent

13.9 **Liquid-Water Interfacial Tension:** Not pertinent

13.10 **Vapor (Gas) Specific Gravity:** Not pertinent

13.11 **Ratio of Specific Heats of Vapor (Gas):** (est.) 1.091

13.12 **Latent Heat of Vaporization:** 182.9 Btu/lb = 101.6 cal/g = 4.254 × 10^5 J/kg

13.13 **Heat of Combustion:** Data not available

13.14 **Heat of Decomposition:** Not pertinent

13.15 **Heat of Solution:** Data not availab

13.16 **Heat of Polymerization:** Not pertinent

NOTES

REVISED 1978

MFA — MOTOR FUEL ANTI-KNOCK COMPOUNDS CONTAINING LEAD ALKYLS

Common Synonyms	Oily liquid Dyed red, orange or blue Sweet fruity odor Sinks in water.

AVOID CONTACT WITH LIQUID AND VAPOR. Keep people away.
Wear goggles, self-contained breathing apparatus, and rubber overclothing (including gloves).
Stop discharge if possible.
Call fire department.
Stay upwind and use water spray to "knock down" vapor.
Isolate and remove discharged material.
Notify local health and pollution control agencies.

Fire	Combustible. POISONOUS GASES ARE PRODUCED IN FIRE. Container may explode when heated. Wear goggles, self-contained breathing apparatus, and rubber overclothing (including gloves). Combat fires from behind barrier or protected location. Flood discharge area with water. Extinguish with water, dry chemical, foam or carbon dioxide. Cool exposed containers with water.
Exposure	CALL FOR MEDICAL AID. LIQUID POISONOUS IF SWALLOWED OR IF SKIN IS EXPOSED. Will burn eyes. Remove contaminated clothes and shoes. Flush affected areas with plenty of water. IF IN EYES, hold eyelids open and flush with plenty of water. IF SWALLOWED and victim is CONSCIOUS, have victim drink water or milk. IF SWALLOWED and victim is UNCONSCIOUS OR HAVING CONVULSIONS, do nothing except keep victim warm.
Water Pollution	Effect of low concentrations on aquatic life is unknown. May be dangerous if it enters water intakes. Notify local health and wildlife officials. Notify operators of nearby water intakes.

1. RESPONSE TO DISCHARGE
(See Response Methods Handbook, CG 446-4)

Issue warning—poison
Restrict access
Evacuate area
Should be removed
Chemical and physical treatment

2. LABEL

3. CHEMICAL DESIGNATIONS

3.1 **Synonyms:** No common synonyms
3.2 **Coast Guard Compatibility Classification:** Motor fuel antiknock compound
3.3 **Chemical Formula:** Not applicable
3.4 **IMCO/United Nations Numerical Designation:** 6.1/1649

4. OBSERVABLE CHARACTERISTICS

4.1 **Physical State (as shipped):** Liquid
4.2 **Color:** Red, orange, or blue
4.3 **Odor:** Sweet, musty; fruity odor

5. HEALTH HAZARDS

5.1 **Personal Protective Equipment:** Organic vapor cartridge-type face mask for emergency or short duration; fresh air mask for longer duration; impervious protective gloves; goggles as required; boots and light-colored clothing.
5.2 **Symptoms Following Exposure:** Increased urinary output of lead. Large degree of absorption from inhalation or skin contact may cause insomnia, excitability, delirium, coma and death.
5.3 **Treatment for Exposure:** Call a physician for any exposure. INHALATION: remove from exposure. INGESTION: no specific antidote. EYES: flush with plenty of water for about 15 min. SKIN: flush with kerosine, wash with soap and water.
5.4 **Toxicity by Inhalation (Threshold Limit Value):** 0.1—0.15 mg/m^3 (as lead).
5.5 **Short-Term Inhalation Limits:** Data not available
5.6 **Toxicity by Ingestion:** Data not available
5.7 **Late Toxicity:** Lead poisoning
5.8 **Vapor (Gas) Irritant Characteristics:** Vapors cause a slight smarting of the eyes or respiratory system if present in high concentrations. The effect is temporary.
5.9 **Liquid or Solid Irritant Characteristics:** Minimum hazard. If spilled on clothing and allowed to remain, may cause smarting and reddening of the skin. Toxic absorption through skin may occur.
5.10 **Odor Threshold:** Data not available

6. FIRE HAZARDS

6.1 **Flash Point:** 89°F—265°F O.C.
6.2 **Flammable Limits in Air:** None established
6.3 **Fire Extinguishing Agents:** Water, foam, dry chemical, carbon dioxide
6.4 **Fire Extinguishing Agents Not to be Used:** Not pertinent
6.5 **Special Hazards of Combustion Products:** Toxic lead-containing gases are generated in fires.
6.6 **Behavior in Fire:** Containers may explode
6.7 **Ignition Temperature:** Begins to decompose above 212°F
6.8 **Electrical Hazard:** Not pertinent
6.9 **Burning Rate:** Data not available

7. CHEMICAL REACTIVITY

7.1 **Reactivity with Water:** No reaction
7.2 **Reactivity with Common Materials:** Reacts with oxidizing materials, active metals and rust, but not considered hazardous.
7.3 **Stability During Transport:** A self-sustaining decomposition occurs if the temperature of the bulk liquid is above 212°F and a flame or hot metal surface serves to ignite the mass. The presence of ethylene dibromide makes the compound stable at 300°F for 15 hrs.
7.4 **Neutralizing Agents for Acids and Caustics:** Not pertinent
7.5 **Polymerization:** Not pertinent
7.6 **Inhibitor of Polymerization:** Not pertinent

8. WATER POLLUTION

8.1 **Aquatic Toxicity:** See Tetraethyl Lead
8.2 **Waterfowl Toxicity:** Data not available
8.3 **Biological Oxygen Demand (BOD):** Data not available
8.4 **Food Chain Concentration Potential:** Data not available

9. SELECTED MANUFACTURERS

1. E.I. duPont de Nemours & Co., Inc.
   Petroleum Chemicals Division
   Wilmington, Del. 19898
2. Ethyl Corp.
   Industrial Chemicals Division
   451 Florida St.
   Baton Rouge, La. 70801
3. PPG Industries, Inc.
   Houston Chemical Co. Division
   One Gateway Center
   Pittsburgh, Pa. 15222

10. SHIPPING INFORMATION

10.1 **Grades or Purity:**
50-60% mixed lead alkyls
18-36% ethylene dibromide
0-19% ethylene dichloride
2-12 % toluene, other solvents, dyes.
10.2 **Storage Temperature:** Ambient
10.3 **Inert Atmosphere:** No requirement
10.4 **Venting:** Pressure-vacuum

11. HAZARD ASSESSMENT CODE
(See Hazard Assessment Handbook, CG 446-3)

A-X-Y

12. HAZARD CLASSIFICATIONS

12.1 **Code of Federal Regulations:** Poisonous liquid or solid, Class B
12.2 **NAS Hazard Rating for Bulk Water Transportation:** Not listed
12.3 **NFPA Hazard Classifications:** Not listed

13. PHYSICAL AND CHEMICAL PROPERTIES

13.1 **Physical State at 15°C and 1 atm:** Liquid
13.2 **Molecular Weight:** Not pertinent
13.3 **Boiling Point at 1 atm:** >200°F = >93°C = >367°K
13.4 **Freezing Point:** Not pertinent
13.5 **Critical Temperature:** Not pertinent
13.6 **Critical Pressure:** Not pertinent
13.7 **Specific Gravity:** 1.5—1.7 at 15°C (liquid)
13.8 **Liquid Surface Tension:** (est.) 20 dynes/cm = 0.020 N/m at 20°C
13.9 **Liquid-Water Interfacial Tension:** (est.) 45 dynes/cm = 0.045 N/m at 20°C
13.10 **Vapor (Gas) Specific Gravity:** Not pertinent
13.11 **Ratio of Specific Heats of Vapor (Gas):** (est.) 1.030
13.12 **Latent Heat of Vaporization:** (est.) 101 Btu/lb = 56.2 cal/g = 2.35×10^5 J/kg
13.13 **Heat of Combustion:** (est.) −18,200 Btu/lb = −10,100 cal/g = -424×10^5 J/kg
13.14 **Heat of Decomposition:** Not pertinent
13.15 **Heat of Solution:** Not pertinent
13.16 **Heat of Polymerization:** Not pertinent

NOTES

REVISED 1978

NAB NABAM

Common Synonyms			
EBDC, sodium salt Ethylenebis [dithiocarbamic acid], disodium salt Disodium ethylenebis [dithiocarbamate] Chem Bam Dithane	Solid or solution Mixes with water.	Colorless to light amber	Slight odor

AVOID CONTACT WITH SOLUTION AND SOLID. KEEP PEOPLE AWAY.
Wear rubber overclothing (including gloves).
Stop discharge if possible.
Isolate and remove discharged material.
Notify local health and pollution control agencies.

Fire

Not flammable.
Poisonous and flammable gases are produced if the solution boils.
Cool exposed containers with water.

Exposure

CALL FOR MEDICAL AID.

DUST
POISONOUS IF INHALED.
Irritating to eyes, nose and throat.
Move victim to fresh air.
If in eyes, hold eyelids open and flush with plenty of water.
If breathing has stopped, give artificial respiration.
If breathing is difficult, give oxygen.

LIQUID OR SOLID
POISONOUS IF SWALLOWED.
Irritating to skin and eyes.
Remove contaminated clothing and shoes.
Flush affected areas with plenty of water.
IF IN EYES, hold eyelids open and flush with plenty of water.
IF SWALLOWED and victim is CONSCIOUS, have victim drink water or milk and have victim induce vomiting.
IF SWALLOWED and victim is UNCONSCIOUS OR HAVING CONVULSIONS, do nothing except keep victim warm.

Water Pollution

HARMFUL TO AQUATIC LIFE IN VERY LOW CONCENTRATIONS.
May be dangerous if it enters water intakes.
Notify local health and wildlife officials.
Notify operators of nearby water intakes.

1. RESPONSE TO DISCHARGE

(See Response Methods Handbook, CG 446-4)

Issue warning—poison, water contaminant
Restrict access
Should be removed
Chemical and physical treatment

2. LABEL

3. CHEMICAL DESIGNATIONS

3.1 **Synonyms:** Chem Bam; Disodium ethylenebis [dithiocarbamate]; Dithane; EBDC, sodium salt; Ethylenebis [dithiocarbamic acid], disodium salt

3.2 **Coast Guard Compatibility Classification:** Not applicable

3.3 **Chemical Formula:** $C_4H_6N_2S_4Na_2$

3.4 **IMCO/United Nations Numerical Designation:** 6.1/1609

4. OBSERVABLE CHARACTERISTICS

4.1 **Physical State (as shipped):** Solid, or 22% solution in water

4.2 **Color:** Colorless to yellow-amber

4.3 **Odor:** Slight sulfide

5. HEALTH HAZARDS

5.1 **Personal Protective Equipment:** Dust mask; self-contained breathing apparatus if compound is hot; goggles; rubber gloves.

5.2 **Symptoms Following Exposure:** Contact with liquid irritates eyes and may cause mild to severe erythema of skin as well as sensitization reactions.

5.3 **Treatment for Exposure:** INHALATION: remove to fresh air; administer artificial respiration and oxygen, if indicated. EYES: flush with copious quantities of water for 15 min.; call a physician. SKIN: wash thoroughly with soap and water. INGESTION: induce vomiting and follow with gastric lavage; get medical attention.

5.4 **Toxicity by Inhalation (Threshold Limit Value):** Data not available

5.5 **Short-Term Inhalation Limits:** Data not available

5.6 **Toxicity by Ingestion:** Grade 3; oral LD_{50} = 395 mg/kg (rat)

5.7 **Late Toxicity:** Degrades to ethylenethiourea, which may affect thyroid gland of animals.

5.8 **Vapor (Gas) Irritant Characteristics:** Data not available

5.9 **Liquid or Solid Irritant Characteristics:** Data not available

5.10 **Odor Threshold:** Data not available

6. FIRE HAZARDS

6.1 **Flash Point:** Not flammable

6.2 **Flammable Limits in Air:** Not flammable

6.3 **Fire Extinguishing Agents:** Not pertinent

6.4 **Fire Extinguishing Agents Not to be Used:** Not pertinent

6.5 **Special Hazards of Combustion Products:** Not pertinent

6.6 **Behavior in Fire:** If water solution boils, poisonous hydrogen sulfide and highly flammable carbon disulfide vapors form.

6.7 **Ignition Temperature:** Not pertinent

6.8 **Electrical Hazard:** Not pertinent

6.9 **Burning Rate:** Not pertinent

7. CHEMICAL REACTIVITY

7.1 **Reactivity with Water:** No reaction unless water is boiling hot, when poisonous hydrogen sulfide and flammable carbon disulfide vapors form.

7.2 **Reactivity with Common Materials:** No reaction

7.3 **Stability During Transport:** Stable

7.4 **Neutralizing Agents for Acids and Caustics:** Not pertinent

7.5 **Polymerization:** Not pertinent

7.6 **Inhibitor of Polymerization:** Not pertinent

8. WATER POLLUTION

8.1 **Aquatic Toxicity:**
2.1 ppm/24 hr/fingerling channel catfish/lethal/fresh water
1 10 ppm /*/marine plankton/killed or no growth/salt water
*Time period not specified.

8.2 **Waterfowl Toxicity:** LD_{50}=2560 ppm (acute exposure)

8.3 **Biological Oxygen Demand (BOD):** Data not available

8.4 **Food Chain Concentration Potential:** None

9. SELECTED MANUFACTURERS

Rohm & Haas Company
Independence Mall West
Philadelphia, Pa. 19105

10. SHIPPING INFORMATION

10.1 **Grades or Purity:** Technical; 22% solution in water

10.2 **Storage Temperature:** Ambient

10.3 **Inert Atmosphere:** No requirement

10.4 **Venting:** Open

11. HAZARD ASSESSMENT CODE

(See Hazard Assessment Handbook, CG 446-3)

SS

12. HAZARD CLASSIFICATIONS

12.1 **Code of Federal Regulations:** Poisonous, Class B

12.2 **NAS Hazard Rating for Bulk Water Transportation:** Not listed

12.3 **NFPA Hazard Classifications:** Not listed

13. PHYSICAL AND CHEMICAL PROPERTIES*

13.1 **Physical State at 15°C and 1 atm:** Solid

13.2 **Molecular Weight:** 256.3

13.3 **Boiling Point at 1 atm:** Not pertinent (decomposes)

13.4 **Freezing Point:** Not pertinent

13.5 **Critical Temperature:** Not pertinent

13.6 **Critical Pressure:** Not pertinent

13.7 **Specific Gravity:** 1.14 at 20°C (solid)

13.8 **Liquid Surface Tension:** Not pertinent

13.9 **Liquid-Water Interfacial Tension:** Not pertinent

13.10 **Vapor (Gas) Specific Gravity:** Not pertinent

13.11 **Ratio of Specific Heats of Vapor (Gas):** Not pertinent

13.12 **Latent Heat of Vaporization:** Not pertinent

13.13 **Heat of Combustion:** Not pertinent

13.14 **Heat of Decomposition:** Not pertinent

13.15 **Heat of Solution:** (est.) 0

13.16 **Heat of Polymerization:** Not pertinent

*Data refer to the solid material.

NOTES

NCT

NAPHTHA: COAL TAR

Common Synonyms			
Mixture of benzene, toluene, xylenes	Watery liquid	Colorless to pale yellow	Gasoline-like odor
	Floats on water. Irritating vapor is produced.		

Stop discharge if possible. Keep people away.
Call fire department.
Avoid contact with liquid and vapor.
Stay upwind and use water spray to "knock down" vapor.
Isolate and remove discharged material.
Notify local health and pollution control agencies.

Fire

Combustible.
Extinguish with foam, dry chemical or carbon dioxide.
Cool exposed containers with water.

Exposure

CALL FOR MEDICAL AID.

VAPOR
Irritating to eyes, nose and throat.
If inhaled, will cause dizziness, headache, difficult breathing or loss of consciousness.
Move to fresh air.
If breathing has stopped, give artificial respiration.
If breathing is difficult, give oxygen.

LIQUID
Irritating to skin and eyes.
If swallowed, will cause nausea or vomiting.
Remove contaminated clothing and shoes.
Flush affected areas with plenty of water.
IF IN EYES, hold eyelids open and flush with plenty of water.
IF SWALLOWED and victim is CONSCIOUS, have victim drink water or milk.
DO NOT INDUCE VOMITING.

Water Pollution

Effect of low concentrations on aquatic life is unknown.
Fouling to shoreline.
May be dangerous if it enters water intakes.
Notify local health and wildlife officials.
Notify operators of nearby water intakes.

1. RESPONSE TO DISCHARGE

(See Response Methods Handbook, CG 446-4)

Mechanical containment
Should be removed
Chemical and physical treatment

2. LABELS

No hazard label required by Code of Federal Regulations

3. CHEMICAL DESIGNATIONS

3.1 **Synonyms:** Mixture of benzene, toluene, and xylenes
3.2 **Coast Guard Compatibility Classification:** Miscellaneous hydrocarbon mixtures
3.3 **Chemical Formula:** Not applicable
3.4 **IMCO/United Nations Numerical Designation:** 3.2/1138

4. OBSERVABLE CHARACTERISTICS

4.1 **Physical State (as shipped):** Liquid
4.2 **Color:** Colorless to pale yellow
4.3 **Odor:** Like benzene, toluene, and xylene

5. HEALTH HAZARDS

5.1 **Personal Protective Equipment:** Hydrocarbon vapor canister or air pack; plastic gloves; goggles or face shield.
5.2 **Symptoms Following Exposure:** Primarily a narcotic, causing unconsciousness in high concentrations. The symptoms of acute benzene poisoning are not likely, since the compound has components other than benzene.
5.3 **Treatment for Exposure:** Remove from exposure. Support respiration. Call physician.
5.4 **Toxicity by Inhalation (Threshold Limit Value):** 100 ppm
5.5 **Short-Term Inhalation Limits:** 75 ppm for 30 min.
5.6 **Toxicity by Ingestion:** Grade 3; LD_{50} 50 to 500 mg/kg
5.7 **Late Toxicity:** Leukemia
5.8 **Vapor (Gas) Irritant Characteristics:** Vapors cause a slight smarting of the eyes or respiratory system if present in high concentrations. The effect is temporary.
5.9 **Liquid or Solid Irritant Characteristics:** Minimum hazard. If spilled on clothing and allowed to remain, may cause a smarting and reddening of the skin.
5.10 **Odor Threshold:** 4.68 ppm

6. FIRE HAZARDS

6.1 **Flash Point:** 107°F C.C.
6.2 **Flammable Limits in Air:** Data not available
6.3 **Fire Extinguishing Agents:** Foam, carbon dioxide, or dry chemical
6.4 **Fire Extinguishing Agents Not to be Used:** Not pertinent
6.5 **Special Hazards of Combustion Products:** Not pertinent
6.6 **Behavior in Fire:** Not pertinent
6.7 **Ignition Temperature:** 900-950°F
6.8 **Electrical Hazard:** Class I, Group D
6.9 **Burning Rate:** 4 mm/min.

7. CHEMICAL REACTIVITY

7.1 **Reactivity with Water:** No reaction
7.2 **Reactivity with Common Materials:** No reaction
7.3 **Stability During Transport:** Stable
7.4 **Neutralizing Agents for Acids and Caustics:** Not pertinent
7.5 **Polymerization:** Not pertinent
7.6 **Inhibitor of Polymerization:** Not pertinent

8. WATER POLLUTION

8.1 **Aquatic Toxicity:** Data not available
8.2 **Waterfowl Toxicity:** Data not available
8.3 **Biological Oxygen Demand (BOD):** Data not available
8.4 **Food Chain Concentration Potential:** None

9. SELECTED MANUFACTURERS

1. Crowley Tar Products, Inc.
   271 Madison Ave.
   New York, N. Y. 10016
2. Koppers Co.
   Organic Materials Division
   Koppers Bldg.
   Pittsburgh, Pa. 15219
3. Neville Chemical Co.
   Neville Island
   Pittsburgh, Pa. 15225

10. SHIPPING INFORMATION

10.1 **Grades or Purity:** Purity varies with coal used and distillation range taken.
10.2 **Storage Temperature:** Ambient
10.3 **Inert Atmosphere:** No requirement
10.4 **Venting:** Open (flame arrester)

11. HAZARD ASSESSMENT CODE

(See Hazard Assessment Handbook, CG 446-3)

A-T-U-V-W

12. HAZARD CLASSIFICATIONS

12.1 **Code of Federal Regulations:** Combustible Liquid
12.2 **NAS Hazard Rating for Bulk Water Transportation:** Not listed
12.3 **NFPA Hazard Classifications:** Not listed

13. PHYSICAL AND CHEMICAL PROPERTIES

13.1 **Physical State at 15°C and 1 atm:** Liquid
13.2 **Molecular Weight:** Not pertinent
13.3 **Boiling Point at 1 atm:** 200-500°F = 93-260°C = 366-533°K
13.4 **Freezing Point:** Not pertinent
13.5 **Critical Temperature:** Not pertinent
13.6 **Critical Pressure:** Not pertinent
13.7 **Specific Gravity:** 0.86-0.88 at 20°C (liquid)
13.8 **Liquid Surface Tension:** (est.) 20 dynes/cm = 0.020 N/m at 20°C
13.9 **Liquid-Water Interfacial Tension:** (est.) 45 dynes/cm = 0.045 N/m at 20°C
13.10 **Vapor (Gas) Specific Gravity:** Data not available
13.11 **Ratio of Specific Heats of Vapor (Gas):** (est.) 1.030
13.12 **Latent Heat of Vaporization:** (est.) 101 Btu/lb = 56.2 cal/g = 2.35×10^5 J/kg
13.13 **Heat of Combustion:** (est.) −18,200 Btu/lb = −10,100 cal/g = -424×10^5 J/kg
13.14 **Heat of Decomposition:** Not pertinent
13.15 **Heat of Solution:** Not pertinent
13.16 **Heat of Polymerization:** Not pertinent

NOTES

REVISED 1978

NTM — NAPHTHALENE, MOLTEN

Common Synonyms			
Naphthalin Tar camphor	Molten solid	Colorless	Mothballs odor
	Solidifies and floats or sinks in water.		

Stop discharge if possible. Keep people away.
Call fire department.
Avoid contact with liquid and solid.
Isolate and remove discharged material.
Notify local health and pollution control agencies.

Fire

Combustible.
Wear goggles and self-contained breathing apparatus.
Extinguish with water, foam, dry chemical or carbon dioxide.
Cool exposed containers with water.

Exposure

CALL FOR MEDICAL AID.
SOLID OR LIQUID
Irritating to skin and eyes.
Remove contaminated clothing and shoes.
Flush affected areas with plenty of water.
IF IN EYES, hold eyelids open and flush with plenty of water.

Water Pollution

HARMFUL TO AQUATIC LIFE IN VERY LOW CONCENTRATIONS.
Fouling to shoreline.
May be dangerous if it enters water intakes.
Notify local health and wildlife officials.
Notify operators of nearby water intakes.

1. RESPONSE TO DISCHARGE

(See Response Methods Handbook, CG 446-4)

Should be removed
Chemical and physical treatment

2. LABELS

No hazard label required by Code of Federal Regulations

3. CHEMICAL DESIGNATIONS

3.1 **Synonyms:** Naphthalin
Tar Camphor
3.2 **Coast Guard Compatibility Classification:** Aromatic hydrocarbon
3.3 **Chemical Formula:** $C_{10}H_8$
3.4 **IMCO/United Nations Numerical Designation:** 4.1/1334

4. OBSERVABLE CHARACTERISTICS

4.1 **Physical State (as shipped):** Liquid
4.2 **Color:** Colorless
4.3 **Odor:** Coal tar; moth balls

5. HEALTH HAZARDS

5.1 **Personal Protective Equipment:** U.S. Bureau of Mines approved organic vapor canister unit (USBM Type B); rubber gloves; chemical safety goggles; face shield; coveralls and/or rubber apron; rubber shoes or boots.
5.2 **Symptoms Following Exposure:** Vapors or fumes are irritating to eyes, nose, and throat and may cause headaches, dizziness, nausea, etc. Solid may be irritating to skin.
5.3 **Treatment for Exposure:** INHALATION: remove to fresh air. SKIN OR EYES: flush immediately with plenty of water for at least 15 min.; remove contaminated clothing immediately; call a physician.
5.4 **Toxicity by Inhalation (Threshold Limit Value):** 10 ppm
5.5 **Short-Term Inhalation Limits:** 15 ppm for 5 min.
5.6 **Toxicity by Ingestion:** Grade 2; oral rat LD_{50} = 1780 mg/kg
5.7 **Late Toxicity:** Data not available
5.8 **Vapors (Gas) Irritant Characteristics:** Vapors cause moderate irritation such that personnel will find high concentrations unpleasant. The effect is temporary.
5.9 **Liquid or Solid Irritant Characteristics:** Hot liquid can cause severe burn. The solid may irritate the skin.
5.10 **Odor Threshold:** Data not available.

6. FIRE HAZARDS

6.1 **Flash Point:** 174°F C.C.; 190°F O.C.
6.2 **Flammable Limits in Air:** 0.9%–5.9%
6.3 **Fire Extinguishing Agents:** Water fog, carbon dioxide, dry chemical, or foam
6.4 **Fire Extinguishing Agents Not to be Used:** Not pertinent
6.5 **Special Hazards of Combustion Products:** Toxic vapors given off in a fire.
6.6 **Behavior in Fire:** Not pertinent
6.7 **Ignition Temperature:** 979°F
6.8 **Electrical Hazard:** Not pertinent
6.9 **Burning Rate:** 4.3 mm/min.

7. CHEMICAL REACTIVITY

7.1 **Reactivity with Water:** Molten naphthalene spatters and foams in contact with water. No chemical reaction is involved.
7.2 **Reactivity with Common Materials:** None
7.3 **Stability During Transport:** Stable
7.4 **Neutralizing Agents for Acids and Caustics:** Not pertinent
7.5 **Polymerization:** Not pertinent
7.6 **Inhibitor of Polymerization:** Not pertinent

8. WATER POLLUTION

8.1 **Aquatic Toxicity:**
150 mg/l /96 hr/sunfish/TL_m/fresh water
1.8 ppm/72 hr/fingerling salmon/critical/salt water
8.2 **Waterfowl Toxicity:** Data not available
8.3 **Biological Oxygen Demand (BOD):** (theor.) 59.5%, 6 days
8.4 **Food Chain Concentration Potential:** None

9. SELECTED MANUFACTURERS

1. Allied Chemical Corp.
Industrial Chemicals Division
Morristown, N. J. 07960
2. Ashland Oil, Inc.
Ashland Chemical Co. Division
Columbus, Ohio 43216
3. United States Steel Corp.
USS Chemicals Division
525 William Penn Place
Pittsburgh, Pa. 15230

10. SHIPPING INFORMATION

10.1 **Grades or Purity:** Pure; crude: 95%
Pure: mp=176°F
Crude: mp=165-176°F
10.2 **Storage Temperature:** Elevated
10.3 **Inert Atmosphere:** No requirement
10.4 **Venting:** Open (flame arrester) or pressure-vacuum

11. HAZARD ASSESSMENT CODE

(See Hazard Assessment Handbook, CG 446-3)

A-T-U-X

12. HAZARD CLASSIFICATIONS

12.1 **Code of Federal Regulations:** ORM-A
12.2 **NAS Hazard Rating for Bulk Water Transportation:**

Category	Rating
Fire	1
Health	
Vapor Irritant	2
Liquid or Solid Irritant	1
Poisons	2
Water Pollution	
Human Toxicity	1
Aquatic Toxicity	3
Aesthetic Effect	3
Reactivity	
Other Chemicals	1
Water	0
Self-Reaction	0

12.3 **NFPA Hazard Classifications:**

Category	Classification
Health Hazard (Blue)	2
Flammability (Red)	2
Reactivity (Yellow)	0

13. PHYSICAL AND CHEMICAL PROPERTIES

13.1 **Physical State at 15°C and 1 atm:** Solid
13.2 **Molecular Weight:** 128.18
13.3 **Boiling Point at 1 atm:** 424°F = 218°C = 491°K
13.4 **Freezing Point:** 176.4°F = 80.2°C = 353.4°K
13.5 **Critical Temperature:** 887.4°F = 475.2°C = 748.4°K
13.6 **Critical Pressure:** 588 psia = 40.0 atm = 4.05 MN/m^2
13.7 **Specific Gravity:** 1.145 at 20°C (solid)
13.8 **Liquid Surface Tension:** 31.8 dynes/cm = 0.0318 N/m at 100°C
13.9 **Liquid-Water Interfacial Tension:** Data not available
13.10 **Vapor (Gas) Specific Gravity:** Not pertinent
13.11 **Ratio of Specific Heats of Vapor (Gas):** 1.068
13.12 **Latent Heat of Vaporization:** 145 Btu/lb = 80.7 cal/g = 3.38×10^5 J/kg
13.13 **Heat of Combustion:** −16,720 Btu/lb = −9287 cal/g = -388.8×10^5 J/kg
13.14 **Heat of Decomposition:** Not pertinent
13.15 **Heat of Solution:** Not pertinent
13.16 **Heat of Polymerization:** Not pertinent

NOTES

REVISED 1978

NSV — NAPHTHA: SOLVENT

Common Synonyms			
Petroleum solvent	Watery liquid	Colorless	Gasoline-like odor
	Floats on water. Vapor is produced.		

Stop discharge if possible. Keep people away.
Call fire department.
Avoid contact with liquid and vapor.
Isolate and remove discharged material.
Notify local health and pollution control agencies.

Fire

Combustible.
Extinguish with foam, dry chemical or carbon dioxide.
Cool exposed containers with water.

Exposure

CALL FOR MEDICAL AID.

VAPOR
Not irritating to eyes, nose or throat.
If inhaled, will cause dizziness or loss of consciousness.
Move to fresh air.
If breathing has stopped, give artificial respiration.
If breathing is difficult, give oxygen.

LIQUID
Irritating to skin and eyes.
Harmful if swallowed.
Remove contaminated clothing and shoes.
Flush affected areas with plenty of water.
IF IN EYES, hold eyelids open and flush with plenty of water.
IF SWALLOWED and victim is CONSCIOUS, have victim drink water or milk.
DO NOT INDUCE VOMITING.

Water Pollution

Effect of low concentrations on aquatic life is unknown.
Fouling to shoreline.
May be dangerous if it enters water intakes.
Notify local health and wildlife officials.
Notify operators of nearby water intakes.

1. RESPONSE TO DISCHARGE

(See Response Methods Handbook, CG 446-4)

Mechanical containment
Should be removed
Chemical and physical treatment

2. LABELS

No hazard label required by Code of Federal Regulations

3. CHEMICAL DESIGNATIONS

3.1 **Synonyms:** Light naphtha
Petroleum solvent

3.2 **Coast Guard Compatibility Classification:** Miscellaneous hydrocarbon mixtures

3.3 **Chemical Formula:** Not applicable

3.4 **IMCO/United Nations Numerical Designation:** 3.2/1256

4. OBSERVABLE CHARACTERISTICS

4.1 **Physical State (as shipped):** Liquid

4.2 **Color:** Colorless

4.3 **Odor:** Like kerosine and gasoline

5. HEALTH HAZARDS

5.1 **Personal Protective Equipment:** Goggles or face shield (as for gasoline).

5.2 **Symptoms Following Exposure:** High concentration of vapors may cause intoxication. If liquid is swallowed, it may get into lungs by aspiration; not very irritating to skin or eyes.

5.3 **Treatment for Exposure:** INHALATION: remove to fresh air, treat symptoms. INGESTION: do NOT induce vomiting; call a doctor. EYES: flush with water for 15 min. SKIN: wipe off, wash with soap and water.

5.4 **Toxicity by Inhalation (Threshold Limit Value):** 200 ppm

5.5 **Short-Term Inhalation Limits:** 500 ppm for 30 min.

5.6 **Toxicity by Ingestion:** Grade 2; LD_{50} 0.5 to 5 g/kg

5.7 **Late Toxicity:** None

5.8 **Vapor (Gas) Irritant Characteristics:** Vapors are nonirritating to eyes and throat.

5.9 **Liquid or Solid Irritant Characteristics:** Minimum hazard. If spilled on clothing and allowed to remain, may cause smarting and reddening of skin.

5.10 **Odor Threshold:** Data not available

6. FIRE HAZARDS

6.1 **Flash Point:** >100°F C.C.

6.2 **Flammable Limits in Air:** 0.8%—5.0%

6.3 **Fire Extinguishing Agents:** Foam, carbon dioxide, or dry chemical

6.4 **Fire Extinguishing Agents Not to be Used:** Not pertinent

6.5 **Special Hazards of Combustion Products:** Not pertinent

6.6 **Behavior in Fire:** Not pertinent

6.7 **Ignition Temperature:** 444°F

6.8 **Electrical Hazard:** Class I, Group D

6.9 **Burning Rate:** 4 mm/min.

7. CHEMICAL REACTIVITY

7.1 **Reactivity with Water:** No reaction

7.2 **Reactivity with Common Materials:** No reaction

7.3 **Stability During Transport:** Stable

7.4 **Neutralizing Agents for Acids and Caustics:** Not pertinent

7.5 **Polymerization:** Not pertinent

7.6 **Inhibitor of Polymerization:** Not pertinent

8. WATER POLLUTION

8.1 **Aquatic Toxicity:** Data not available

8.2 **Waterfowl Toxicity:** Data not available

8.3 **Biological Oxygen Demand (BOD):** Data not available

8.4 **Food Chain Concentration Potential:** None

9. SELECTED MANUFACTURERS

1. Pennsylvania Refining Co.
Butler, Pa. 16001
2. Sun Oil Co.
St. Davids, Pa. 19087
3. Union Oil Co.
Amsco Division
3100 S. Meachem Rd.
Palatine, Ill. 60067

10. SHIPPING INFORMATION

10.1 **Grades or Purity:** Refined solvent; crude light solvent; crude heavy solvent

10.2 **Storage Temperature:** Ambient

10.3 **Inert Atmosphere:** No requirement

10.4 **Venting:** Open (flame arrester) or pressure-vacuum

11. HAZARD ASSESSMENT CODE

(See Hazard Assessment Handbook, CG 446-3)

A-T-U

12. HAZARD CLASSIFICATIONS

12.1 **Code of Federal Regulations:** Combustible Liquid

12.2 **NAS Hazard Rating for Bulk Water Transportation:** Not listed

12.3 **NFPA Hazard Classifications:** Not listed

13. PHYSICAL AND CHEMICAL PROPERTIES

13.1 **Physical State at 15°C and 1 atm:** Liquid

13.2 **Molecular Weight:** Not pertinent

13.3 **Boiling Point at 1 atm:** 266—311°F = 130—155°C = 403—428°K

13.4 **Freezing Point:** Not pertinent

13.5 **Critical Temperature:** Not pertinent

13.6 **Critical Pressure:** Not pertinent

13.7 **Specific Gravity:** 0.85—0.87 at 20°C (liquid)

13.8 **Liquid Surface Tension:** 19—23 dynes/cm = 0.019—0.023 N/m at 20°C

13.9 **Liquid-Water Interfacial Tension:** 39—51 dynes/cm = 0.039—0.051 N/m at 20°C

13.10 **Vapor (Gas) Specific Gravity:** Not pertinent

13.11 **Ratio of Specific Heats of Vapor (Gas):** (est.) 1.030

13.12 **Latent Heat of Vaporization:** 130—150 Btu/lb = 71—81 cal/g = $3.0—3.4 \times 10^5$ J/kg

13.13 **Heat of Combustion:** (est.) −18,200 Btu/lb = −10,100 cal/g = -424×10^5 J/kg

13.14 **Heat of Decomposition:** Not pertinent

13.15 **Heat of Solution:** Not pertinent

13.16 **Heat of Polymerization:** Not pertinent

NOTES

REVISED 1978

NSS — NAPHTHA: STODDARD SOLVENT

Common Synonyms
Petroleum solvent
Drycleaner naphtha
Spotting naphtha

Watery liquid — Colorless — Gasoline-like odor

Floats on water.

Stop discharge if possible. Keep people away.
Call fire department.
Avoid contact with liquid.
Isolate and remove discharged material.
Notify local health and pollution control agencies.

Fire

Combustible.
Extinguish with foam, dry chemical or carbon dioxide.
Cool exposed containers with water.

Exposure

CALL FOR MEDICAL AID.

LIQUID
Irritating to skin and eyes.
Harmful if swallowed.
Remove contaminated clothing and shoes.
Flush affected areas with plenty of water.
IF IN EYES, hold eyelids open and flush with plenty of water.
IF SWALLOWED and victim is CONSCIOUS, have victim drink water or milk.
DO NOT INDUCE VOMITING.

Water Pollution

Effect of low concentrations on aquatic life is unknown.
Fouling to shoreline.
May be dangerous if it enters water intakes.
Notify local health and wildlife officials.
Notify operators of nearby water intakes.

1. RESPONSE TO DISCHARGE

(See Response Methods Handbook, CG 446-4)

Mechanical containment
Should be removed
Chemical and physical treatment

2. LABELS

No hazard label is required by Code of Federal Regulations

3. CHEMICAL DESIGNATIONS

3.1 **Synonyms:** Drycleaners' naphtha
Petroleum solvent
Spotting naphtha

3.2 **Coast Guard Compatibility Classification:** Miscellaneous hydrocarbon mixtures

3.3 **Chemical Formula:** Not applicable

3.4 **IMCO/United Nations Numerical Designation:** 3.3/1268

4. OBSERVABLE CHARACTERISTICS

4.1 **Physical State (as shipped):** Liquid

4.2 **Color:** Colorless

4.3 **Odor:** Like gasoline

5. HEALTH HAZARDS

5.1 **Personal Protective Equipment:** Goggles or face shield (as for gasoline).

5.2 **Symptoms Following Exposure:** High concentration of vapors may cause intoxication. If liquid is swallowed, it may get into lungs by aspiration; not very irritating to skin or eyes.

5.3 **Treatment for Exposure:** INHALATION: remove patient from exposure; treat symptoms. INGESTION: do NOT induce vomiting! Call a doctor. EYES: flush with water for 15 min. SKIN: wipe off and wash with soap and water.

5.4 **Toxicity by Inhalation (Threshold Limit Value):** 200 ppm

5.5 **Short-Term Inhalation Limits:** 500 ppm for 30 min.

5.6 **Toxicity by Ingestion:** Grade 2; LD_{50} 0.5 to 5 g/kg

5.7 **Late Toxicity:** None

5.8 **Vapor (Gas) Irritant Characteristics:** Vapors are nonirritating to the eyes and throat.

5.9 **Liquid or Solid Irritant Characteristics:** Minimum hazard. If spilled on clothing and allowed to remain, may cause smarting and reddening of the skin.

5.10 **Odor Threshold:** Data not available

6. FIRE HAZARDS

6.1 **Flash Point:** 110°F C.C.

6.2 **Flammable Limits in Air:** 0.8%–5.0%

6.3 **Fire Extinguishing Agents:** Foam, dry chemical, or carbon dioxide.

6.4 **Fire Extinguishing Agents Not to be Used:** Not pertinent

6.5 **Special Hazards of Combustion Products:** Not pertinent

6.6 **Behavior in Fire:** Not pertinent

6.7 **Ignition Temperature:** 540°F (est.)

6.8 **Electrical Hazard:** Class I, Group D

6.9 **Burning Rate:** 4 mm/min.

7. CHEMICAL REACTIVITY

7.1 **Reactivity with Water:** No reaction

7.2 **Reactivity with Common Materials:** No reaction

7.3 **Stability During Transport:** Stable

7.4 **Neutralizing Agents for Acids and Caustics:** Not pertinent

7.5 **Polymerization:** Not pertinent

7.6 **Inhibitor of Polymerization:** Not pertinent

8. WATER POLLUTION

8.1 **Aquatic Toxicity:** Data not available

8.2 **Waterfowl Toxicity:** Data not available

8.3 **Biological Oxygen Demand (BOD):** Data not available

8.4 **Food Chain Concentration Potential:** None

9. SELECTED MANUFACTURERS

1. Pennsylvania Refining Co.
Butler, Pa. 16001
2. Sun Oil Co.
St. Davids, Pa. 19087
3. Union Oil Co.
Amsco Division
3100 S. Meachem Rd.
Palatine, Ill. 60067

10. SHIPPING INFORMATION

10.1 **Grades or Purity:** Data not available

10.2 **Storage Temperature:** Ambient

10.3 **Inert Atmosphere:** No requirement

10.4 **Venting:** Open (flame arrester)

11. HAZARD ASSESSMENT CODE

(See Hazard Assessment Handbook, CG 446-3)

A-T-U

12. HAZARD CLASSIFICATIONS

12.1 **Code of Federal Regulations:** Combustible Liquid

12.2 **NAS Hazard Rating for Bulk Water Transportation:** Not listed

12.3 **NFPA Hazard Classifications:**

Category	Classification
Health Hazard (Blue)	0
Flammability (Red)	2
Reactivity (Yellow)	0

13. PHYSICAL AND CHEMICAL PROPERTIES

13.1 **Physical State at 15°C and 1 atm:** Liquid

13.2 **Molecular Weight:** Not pertinent

13.3 **Boiling Point at 1 atm:** 320-390°F = 160-199°C = 433-472°K

13.4 **Freezing Point:** Not pertinent

13.5 **Critical Temperature:** Not pertinent

13.6 **Critical Pressure:** Not pertinent

13.7 **Specific Gravity:** 0.78 at 20°C (liquid)

13.8 **Liquid Surface Tension:** 19-23 dynes/cm = 0.019-0.023 N/m at 20°C

13.9 **Liquid-Water Interfacial Tension:** 39-51 dynes/cm = 0.039-0.051 N/m at 20°C

13.10 **Vapor (Gas) Specific Gravity:** Data not available

13.11 **Ratio of Specific Heats of Vapor (Gas):** (est.) 1.030

13.12 **Latent Heat of Vaporization:** 130-150 Btu/lb = 71-81 cal/g = $3.0\text{-}3.4 \times 10^5$ J/kg

13.13 **Heat of Combustion:** (est.) −18,200 Btu/lb = −10,100 cal/g = -424×10^5 J/kg

13.14 **Heat of Decomposition:** Not pertinent

13.15 **Heat of Solution:** Not pertinent

13.16 **Heat of Polymerization:** Not pertinent

NOTES

REVISED 1978

NVM	**NAPHTHA: VM & P** (75% NAPHTHA)

Common Synonyms			
Petroleum solvent Light naphtha	Watery liquid	Colorless	Gasoline-like odor
	Floats on water. Flammable, irritating vapor is produced.		

Stop discharge if possible. Keep people away.
Shut off ignition sources and call fire department.
Stay upwind and use water spray to "knock down" vapor.
Avoid contact with liquid and vapor.
Isolate and remove discharged material.
Notify local health and pollution control agencies.

Fire

FLAMMABLE.
Flashback along vapor trail may occur.
Vapor may explode if ignited in an enclosed area.
Extinguish with foam, dry chemical or carbon dioxide.
Cool exposed containers with water.

Exposure

CALL FOR MEDICAL AID.

VAPOR
Irritating to eyes, nose and throat.
If inhaled, will cause dizziness or loss of consciousness.
Move to fresh air.
If breathing has stopped, give artificial respiration.
If breathing is difficult, give oxygen.

LIQUID
Irritating to skin and eyes.
If swallowed, will cause nausea or vomiting.
Remove contaminated clothing and shoes.
Flush affected areas with plenty of water.
IF IN EYES, hold eyelids open and flush with plenty of water.
IF SWALLOWED and victim is CONSCIOUS, have victim drink water or milk.
DO NOT INDUCE VOMITING.

Water Pollution

Effect of low concentrations on aquatic life is unknown.
Fouling to shoreline.
May be dangerous if it enters water intakes.
Notify local health and wildlife officials.
Notify operators of nearby water intakes.

1. RESPONSE TO DISCHARGE

(See Response Methods Handbook, CG 446-4)

Issue warning—high flammability
Evacuate area
Disperse and flush

2. LABEL

Red

3. CHEMICAL DESIGNATIONS

3.1 **Synonyms:** Light naphtha
Painter's naphtha
Petroleum solvent

3.2 **Coast Guard Compatibility Classification:** Miscellaneous hydrocarbon mixtures

3.3 **Chemical Formula:** Not applicable

3.4 **IMCO/United Nations Numerical Designation:** 3.2/1255

4. OBSERVABLE CHARACTERISTICS

4.1 **Physical State (as shipped):** Liquid

4.2 **Color:** Colorless

4.3 **Odor:** Like gasoline

5. HEALTH HAZARDS

5.1 **Personal Protective Equipment:** Goggles or face shield (as for gasoline).

5.2 **Symptoms Following Exposure:** Vapor irritates respiratory tract, causes coughing and mild depression. Aspiration causes severe lung irritation with coughing, gagging, and rapidly developing pulmonary edema. Ingestion irritates mouth and stomach, causes nausea, vomiting, swelling of abdomen, cardiac arrhythmias.

5.3 **Treatment for Exposure:** INHALATION: maintain respiration if required. INGESTION: do NOT induce vomiting; observe for pneumonia; support if central nervous system depression occurs. ASPIRATION: enforce rest; administer oxygen; call a doctor. EYES: flush with water for at least 15 min. SKIN: wipe off, wash with soap and water.

5.4 **Toxicity by Inhalation (Threshold Limit Value):** Data not available

5.5 **Short-Term Inhalation Limits:** 500 ppm for 30 min.

5.6 **Toxicity by Ingestion:** Grade 1; LD_{50} 5 to 15 g/kg

5.7 **Late Toxicity:** None

5.8 **Vapor (Gas) Irritant Characteristics:** Vapors cause a slight smarting of the eyes or respiratory system if present in high concentrations. The effect is temporary.

5.9 **Liquid or Solid Irritant Characteristics:** Minimum hazard. If spilled on clothing and allowed to remain, may cause smarting and reddening of the skin.

5.10 **Odor Threshold:** Data not available

6. FIRE HAZARDS

6.1 **Flash Point:** 20-55°F C.C.

6.2 **Flammable Limits in Air:** 0.9%—6.7%

6.3 **Fire Extinguishing Agents:** Foam, carbon dioxide, or dry chemical

6.4 **Fire Extinguishing Agents Not to be Used:** Water may be ineffective

6.5 **Special Hazards of Combustion Products:** Not pertinent

6.6 **Behavior in Fire:** Vapor is heavier than air and may travel long distances to a source of ignition and flash back.

6.7 **Ignition Temperature:** 450°F

6.8 **Electrical Hazard:** Class I, Group D

6.9 **Burning Rate:** 4 mm/min.

7. CHEMICAL REACTIVITY

7.1 **Reactivity with Water:** No reaction

7.2 **Reactivity with Common Materials:** No reaction

7.3 **Stability During Transport:** Stable

7.4 **Neutralizing Agents for Acids and Caustics:** Not pertinent

7.5 **Polymerization:** Not pertinent

7.6 **Inhibitor of Polymerization:** Not pertinent

8. WATER POLLUTION

8.1 **Aquatic Toxicity:** Data not available

8.2 **Waterfowl Toxicity:** Data not available

8.3 **Biological Oxygen Demand (BOD):** Data not available

8.4 **Food Chain Concentration Potential:** None

9. SELECTED MANUFACTURERS

1. Pennsylvania Refining Co.
Butler, Pa. 16001
2. Sun Oil Co.
St. Davids, Pa. 19087
3. Union Oil Co.
Amsco Division
3100 S. Meachem Rd.
Palatine, Ill. 60067

10. SHIPPING INFORMATION

10.1 **Grades or Purity:** Petroleum hydrocarbons (90%) plus aromatic hydrocarbons such as benzene and toluene (10%)

10.2 **Storage Temperature:** Ambient

10.3 **Inert Atmosphere:** No requirement

10.4 **Venting:** Open (flame arrester) or pressure-vacuum

11. HAZARD ASSESSMENT CODE

(See Hazard Assessment Handbook, CG 446-3)

A-T-U-V-W

12. HAZARD CLASSIFICATIONS

12.1 **Code of Federal Regulations:** Flammable liquid

12.2 **NAS Hazard Rating for Bulk Water Transportation:** Not listed

12.3 **NFPA Hazard Classifications:** Not listed

13. PHYSICAL AND CHEMICAL PROPERTIES

13.1 **Physical State at 15°C and 1 atm:** Liquid

13.2 **Molecular Weight:** Not pertinent

13.3 **Boiling Point at 1 atm:** 200-300°F = 93-149°C = 366-422°K

13.4 **Freezing Point:** Not pertinent

13.5 **Critical Temperature:** Not pertinent

13.6 **Critical Pressure:** Not pertinent

13.7 **Specific Gravity:** 0.75 at 20°C (liquid)

13.8 **Liquid Surface Tension:** 19-23 dynes/cm = 0.019-0.023 N/m at 20°C

13.9 **Liquid-Water Interfacial Tension:** 39-51 dynes/cm = 0.039-0.051 N/m at 20°C

13.10 **Vapor (Gas) Specific Gravity:** Data not available

13.11 **Ratio of Specific Heats of Vapor (Gas):** (est.) 1.030

13.12 **Latent Heat of Vaporization:** 130-150 Btu/lb = 71-81 cal/g = $3.0\text{-}3.4 \times 10^5$ J/kg

13.13 **Heat of Combustion:** (est.) −18,200 Btu/lb = −10,100 cal/g = -424×10^5 J/kg

13.14 **Heat of Decomposition:** Not pertinent

13.15 **Heat of Solution:** Not pertinent

13.16 **Heat of Polymerization:** Not pertinent

NOTES

REVISED 1978

NTI — NAPHTHENIC ACIDS

Common Synonyms

Liquid — Gold to black

May float or sink in water.

Avoid contact with liquid and vapor. Keep people away.
Stop discharge if possible.
Call fire department.
Isolate and remove discharged material.
Notify local health and pollution control agencies.

Combustible.
Extinguish with dry chemicals, foam or carbon dioxide.
Water may be ineffective on fire.
Cool exposed containers with water.

Exposure

CALL FOR MEDICAL AID.

VAPOR
Irritating to eyes, nose and throat.
If inhaled will cause coughing or difficult breathing.

If in eyes, hold eyelids open and flush with plenty of water.
If breathing has stopped, give artificial respiration.
If breathing is difficult, give oxygen.

LIQUID
Irritating to skin and eyes.
If swallowed will cause nausea.

Remove contaminated clothing and shoes.
Flush affected areas with plenty of water.
IF IN EYES, hold eyelids open and flush with plenty of water.
IF SWALLOWED and victim is CONSCIOUS, have victim drink water or milk.
IF SWALLOWED and victim is UNCONSCIOUS OR HAVING CONVULSIONS, do nothing except keep victim warm.

Water Pollution

HARMFUL TO AQUATIC LIFE IN VERY LOW CONCENTRATIONS.
Fouling to shoreline.
May be dangerous if it enters water intakes.

Notify local health and wildlife officials.
Notify operators of nearby water intakes.

1. RESPONSE TO DISCHARGE

(See Response Methods Handbook, CG 446-4)

Mechanical containment
Should be removed
Chemical and physical treatment

2. LABELS

No hazard label required by Code of Federal Regulations

3. CHEMICAL DESIGNATIONS

3.1 **Synonyms:** No common synonyms
3.2 **Coast Guard Compatibility Classification:** Not listed
3.3 **Chemical Formula:** $R_2\underline{C}\text{-}CR_2\text{-}CR_2\text{-}CR_2\text{-}\underline{C}R\text{-}(CH_2)_n\text{-}COOH$ where n = 2-6
3.4 **IMCO/United Nations Numerical Designation:** Not listed

4. OBSERVABLE CHARACTERISTICS

4.1 **Physical State (as shipped):** Liquid
4.2 **Color:** Golden to black
4.3 **Odor:** Data not available

5. HEALTH HAZARDS

5.1 **Personal Protective Equipment:** Safety glasses or face mask
5.2 **Symptoms Following Exposure:** Principal effect is that of mild primary irritation when encountered in high concentrations. Inhalation of vapor causes coughing. Liquid is moderately irritating to eyes and slightly to moderately irritating to skin; excessive exposure could result in dermatitis.
5.3 **Treatment for Exposure:** INHALATION: remove to fresh air. INGESTION: give large amounts of water. EYES: flush with water until irritation subsides. SKIN: wash with soap and water; remove contaminated clothing and launder before reuse.
5.4 **Toxicity by Inhalation (Threshold Limit Value):** 10 mg/m^3
5.5 **Short-Term Inhalation Limits:** Data not available
5.6 **Toxicity by Ingestion:** Grade 2; oral LD_{50} = 3,000 mg/kg (rat)
5.7 **Late Toxicity:** Data not available
5.8 **Vapor (Gas) Irritant Characteristics:** Vapors cause a slight smarting of the eyes or respiratory system if present in high concentrations. The effect is temporary.
5.9 **Liquid or Solid Irritant Characteristics:** Minimum hazard. If spilled on clothing and allowed to remain, may cause smarting and reddening of skin.
5.10 **Odor Threshold:** Odorless

6. FIRE HAZARDS

6.1 **Flash Point:** 300°F O.C.
6.2 **Flammable Limits in Air:** 1.0% (LFL)
6.3 **Fire Extinguishing Agents:** Foam, dry chemical, or carbon dioxide
6.4 **Fire Extinguishing Agents Not to be Used:** Water may be ineffective
6.5 **Special Hazards of Combustion Products:**
6.6 **Behavior in Fire:**
6.7 **Ignition Temperature:** Data not available
6.8 **Electrical Hazard:** Data not available
6.9 **Burning Rate:** Data not available

7. CHEMICAL REACTIVITY

7.1 **Reactivity with Water:** No reaction
7.2 **Reactivity with Common Materials:**
7.3 **Stability During Transport:** Stable
7.4 **Neutralizing Agents for Acids and Caustics:** Not pertinent
7.5 **Polymerization:** Not pertinent
7.6 **Inhibitor of Polymerization:** Not pertinent

8. WATER POLLUTION

8.1 **Aquatic Toxicity:**
5-50 ppm/18-60 hr/crayfish/lethal/fresh water
5.6 ppm/96 hr/bluegill/TL_m/fresh water
8.2 **Waterfowl Toxicity:** Data not available
8.3 **Biological Oxygen Demand (BOD):** Data not available
8.4 **Food Chain Concentration Potential:** None

9. SELECTED MANUFACTURERS

1. Exxon Chemical Co.
1333 West Loop South
Houston, Texas 77027
2. Eastman Organic Chemicals
Rochester, N.Y. 14650
3. Sun Oil Co.
1608 Walnut St.
Philadelphia, Pa. 19103

10. SHIPPING INFORMATION

10.1 **Grades or Purity:** Commercial, 100%
10.2 **Storage Temperature:** Ambient
10.3 **Inert Atmosphere:** No requirement
10.4 **Venting:** Open

11. HAZARD ASSESSMENT CODE

(See Hazard Assessment Handbook, CG 446-3)

A-T-U

12. HAZARD CLASSIFICATIONS

12.1 **Code of Federal Regulations:** Not listed
12.2 **NAS Hazard Rating for Bulk Water Transportation:**

Category	Rating
Fire	1
Health	
Vapor Irritant	1
Liquid or Solid Irritant	1
Poisons	3
Water Pollution	
Human Toxicity	1
Aquatic Toxicity	3
Aesthetic Effect	4
Reactivity	
Other Chemicals	3
Water	0
Self-Reaction	0

12.3 **NFPA Hazard Classifications:** Not listed

13. PHYSICAL AND CHEMICAL PROPERTIES

13.1 **Physical State at 15°C and 1 atm:** Liquid
13.2 **Molecular Weight:** 200—250 (mixture)
13.3 **Boiling Point at 1 atm:** 270—470°F = 132—243°C = 405—516°K
13.4 **Freezing Point:** Not pertinent
13.5 **Critical Temperature:** Not pertinent
13.6 **Critical Pressure:** Not pertinent
13.7 **Specific Gravity:** 0.982 at 20°C (liquid)
13.8 **Liquid Surface Tension:** Data not available
13.9 **Liquid-Water Interfacial Tension:** Data not available
13.10 **Vapor (Gas) Specific Gravity:** Not pertinent
13.11 **Ratio of Specific Heats of Vapor (Gas):** Not pertinent
13.12 **Latent Heat of Vaporization:** Data not available
13.13 **Heat of Combustion:** Data not available
13.14 **Heat of Decomposition:** Not pertinent
13.15 **Heat of Solution:** Not pertinent
13.16 **Heat of Polymerization:** Not pertinent

NOTES

NAO — 1-NAPHTHYLAMINE

Common Synonyms			
1-Aminonaphthalene alpha-Naphthylamine	Solid	Light to dark brown	Weak ammonia-like odor
	Sinks in water.		

AVOID CONTACT WITH SOLID AND DUST. KEEP PEOPLE AWAY.
Wear dust respirator and rubber overclothing (including gloves).
Isolate and remove discharged material.
Notify local health and pollution control agencies.

Fire

Combustible.
POISONOUS GASES ARE PRODUCED IN FIRE.
Irritating gases are produced when heated.
Extinguish with water, dry chemicals, foam, or carbon dioxide.

Exposure

CALL FOR MEDICAL AID.

DUST
POISONOUS IF INHALED OR IF SKIN IS EXPOSED.
Irritating to eyes.
Move victim to fresh air.
If in eyes, hold eyelids open and flush with plenty of water.
If breathing is difficult, give oxygen.

SOLID
POISONOUS IF SWALLOWED OR IF SKIN IS EXPOSED.
Remove contaminated clothing and shoes.
Flush affected areas with plenty of water.
IF IN EYES, hold eyelids open and flush with plenty of water.
IF SWALLOWED and victim is CONSCIOUS, have victim drink water or milk and have victim induce vomiting.
IF SWALLOWED and victim is UNCONSCIOUS OR HAVING CONVULSIONS, do nothing except keep victim warm.

Water Pollution

Effect of low concentrations on aquatic life is unknown.
May be dangerous if it enters water intakes.
Notify local health and wildlife officials.
Notify operators of nearby water intakes.

1. RESPONSE TO DISCHARGE
(See Response Methods Handbook, CG 446-4)

Issue warning—poison, water contaminant
Restrict access
Should be removed
Chemical and physical treatment

2. LABELS

None required by Code of Federal Regulations

3. CHEMICAL DESIGNATIONS

3.1 **Synonyms:** 1-Aminonaphthalene alpha-Naphthylamine
3.2 **Coast Guard Compatibility Classification:** Not applicable
3.3 **Chemical Formula:** $1-C_{10}H_7NH_2$
3.4 **IMCO/United Nations Numerical Designation:** 6.1/1650

4. OBSERVABLE CHARACTERISTICS

4.1 **Physical State (as shipped):** Solid
4.2 **Color:** Tan to brown; darkens on storage
4.3 **Odor:** Characteristic amine

5. HEALTH HAZARDS

5.1 **Personal Protective Equipment:** Complete protection for respiratory system, eyes, and skin
5.2 **Symptoms Following Exposure:** Inhalation may cause cyanosis (blue color in lips and under finger nails). Contact with liquid causes local irritation of eyes. Neither ingestion nor contact with skin produces any recognized immediate effects.
5.3 **Treatment for Exposure:** *Persons undergoing severe exposure to this compound should have continuing medical attention for possible development of cancer.* INHALATION: obtain medical attention for cyanosis. EYES: flush with water for at least 15 min. SKIN: wash carefully with soap and water. INGESTION: get medical attention.
5.4 **Toxicity by Inhalation (Threshold Limit Value):** Data not available
5.5 **Short-Term Inhalation Limits:** Data not available
5.6 **Toxicity by Ingestion:** Grade 2; oral LD_{50} = 779 mg/kg (rat), 4,000 mg/kg (mammal)
5.7 **Late Toxicity:** Considered cancer-producing, particularly since it may contain up to 0.5% of 2-naphthylamine.
5.8 **Vapor (Gas) Irritant Characteristics:** Data not available
5.9 **Liquid or Solid Irritant Characteristics:** Data not available
5.10 **Odor Threshold:** Data not available

6. FIRE HAZARDS

6.1 **Flash Point:** (combustible solid) 315°F C.C. (molten solid)
6.2 **Flammable Limits in Air:** Not pertinent
6.3 **Fire Extinguishing Agents:** Water, dry chemical, carbon dioxide, foam
6.4 **Fire Extinguishing Agents Not to be Used:** Water or foam may cause frothing.
6.5 **Special Hazards of Combustion Products:** Toxic nitrogen oxides are produced in a fire.
6.6 **Behavior in Fire:** Not pertinent
6.7 **Ignition Temperature:** Data not available
6.8 **Electrical Hazard:** Not pertinent
6.9 **Burning Rate:** Not pertinent

7. CHEMICAL REACTIVITY

7.1 **Reactivity with Water:** No reaction
7.2 **Reactivity with Common Materials:** No reaction
7.3 **Stability During Transport:** Stable
7.4 **Neutralizing Agents for Acids and Caustics:** Not pertinent
7.5 **Polymerization:** Not pertinent
7.6 **Inhibitor of Polymerization:** Not pertinent

8. WATER POLLUTION

8.1 **Aquatic Toxicity:** Data not available
8.2 **Waterfowl Toxicity:** Data not available
8.3 **Biological Oxygen Demand (BOD):** Data not available
8.4 **Food Chain Concentration Potential:** None

9. SELECTED MANUFACTURERS

1. Aldrich Chemical Co.
   940 West Saint Paul Ave.
   Milwaukee, Wis. 53233
2. Eastman Kodak Co.
   Eastman Organic Chemicals
   Rochester, N. Y. 14650
3. E. I. duPont de Nemours & Co.
   Organic Chemicals Dept.
   Dyes and Chemicals Division
   Wilmington, Del. 19898

10. SHIPPING INFORMATION

10.1 **Grades or Purity:** Pure; Technical
10.2 **Storage Temperature:** Cool ambient
10.3 **Inert Atmosphere:** No requirement
10.4 **Venting:** Open. Store containers in well-ventilated area.

11. HAZARD ASSESSMENT CODE
(See Hazard Assessment Handbook, CG 446-3)

II

12. HAZARD CLASSIFICATIONS

12.1 **Code of Federal Regulations:** Not listed
12.2 **NAS Hazard Rating for Bulk Water Transportation:** Not listed
12.3 **NFPA Hazard Classifications:**

Category	Classification
Health Hazard (Blue)	2
Flammability (Red)	1
Reactivity (Yellow)	0

13. PHYSICAL AND CHEMICAL PROPERTIES

13.1 **Physical State at 15°C and 1 atm:** Solid
13.2 **Molecular Weight:** 143.2
13.3 **Boiling Point at 1 atm:** 572°F = 300°C = 573°K
13.4 **Freezing Point:** 118—122°F = 48—50°C = 321—323°K
13.5 **Critical Temperature:** Not pertinent
13.6 **Critical Pressure:** Not pertinent
13.7 **Specific Gravity:** 1.12 at 25°C (solid)
13.8 **Liquid Surface Tension:** Not pertinent
13.9 **Liquid-Water Interfacial Tension:** Not pertinent
13.10 **Vapor (Gas) Specific Gravity:** Not pertinent
13.11 **Ratio of Specific Heats of Vapor (Gas):** Not pertinent
13.12 **Latent Heat of Vaporization:** Not pertinent
13.13 **Heat of Combustion:** −15,290 Btu/lb = −8,495 cal/g = -355.4×10^5 J/kg
13.14 **Heat of Decomposition:** Not pertinent
13.15 **Heat of Solution:** Not pertinent
13.16 **Heat of Polymerization:** Not pertinent

NOTES

NHX	NEOHEXANE

Common Synonyms			
2,2-Dimethylbutane	Liquid	Colorless	Gasoline-like odor
	Floats on water. Flammable, irritating vapor is produced.		

Shut off ignition sources. Call fire department.
Stop discharge if possible. Keep people away.
Isolate and remove discharged material.
Notify local health and pollution control agencies.

Fire

FLAMMABLE.
Containers may explode in fire.
Flashback along vapor trail may occur.
Vapor may explode if ignited in an enclosed area.
Extinguish with dry chemicals, foam or carbon dioxide.
Water may be ineffective on fire.
Cool exposed containers with water.

Exposure

Call for medical aid.

VAPOR
Irritating to eyes, nose and throat.
If inhaled will cause dizziness, coughing, or difficult breathing.
Move victim to fresh air.
If breathing has stopped, give artificial respiration.
If breathing is difficult, give oxygen.

LIQUID
Irritating to skin and eyes.
If swallowed will cause nausea, or vomiting.
Remove contaminated clothing and shoes.
Flush affected areas with plenty of water.
IF IN EYES, hold eyelids open and flush with plenty of water.
IF SWALLOWED and victim is CONSCIOUS, have victim drink water or milk.
DO NOT INDUCE VOMITING.

Water Pollution

Effect of low concentrations on aquatic life is unknown.
Fouling to shoreline.
May be dangerous if it enters water intakes.
Notify local health and wildlife officials.
Notify operators or nearby water intakes.

1. RESPONSE TO DISCHARGE

(See Response Methods Handbook, CG 446-4)

Issue warning—high flammability
Restrict access
Evacuate area
Disperse and flush

2. LABEL

Red

3. CHEMICAL DESIGNATIONS

3.1 **Synonyms:** 2,2-Dimethylbutane
3.2 **Coast Guard Compatibility Classification:** Not applicable
3.3 **Chemical Formula:** $CH_3C(CH_3)_2CH_2CH_3$
3.4 **IMCO/United Nations Numerical Designation:** 3.1/1258

4. OBSERVABLE CHARACTERISTICS

4.1 **Physical State (as shipped):** Liquid
4.2 **Color:** Colorless
4.3 **Odor:** Like gasoline

5. HEALTH HAZARDS

5.1 **Personal Protective Equipment:** Air-supplied apparatus or organic vapor cartridge; goggles or face shield; rubber gloves.
5.2 **Symptoms Following Exposure:** Inhalation causes dizziness, nausea, and vomiting; concentrated vapor may cause unconsciousness and collapse. Contact with liquid causes irritation of eyes; repeated contact may produce irritation of skin. Ingestion causes irritation of stomach. Aspiration causes severe lung irritation, rapidly developing pulmonary edema, and central nervous system excitement followed by depression.
5.3 **Treatment for Exposure:** INHALATION: remove from exposure; if breathing has stopped, begin artificial respiration; call a physician. EYES: flush with water for 15 min.; call physician if needed. SKIN: flush well with water, then wash with soap and water. INGESTION: do NOT induce vomiting; guard against aspiration into lungs; call a doctor. ASPIRATION: enforce bed rest; give oxygen; get medical attention.
5.4 **Toxicity by Inhalation (Threshold Limit Value):** Data not available
5.5 **Short-Term Inhalation Limits:** Data not available
5.6 **Toxicity by Ingestion:** Data not available
5.7 **Late Toxicity:** None
5.8 **Vapor (Gas) Irritant Characteristics:** Data not available
5.9 **Liquid or Solid Irritant Characteristics:** Data not available
5.10 **Odor Threshold:** Data not available

6. FIRE HAZARDS

6.1 **Flash Point:** −54°F C.C.
6.2 **Flammable Limits in Air:** 1.2% – 7.7%
6.3 **Fire Extinguishing Agents:** Dry chemical, foam, carbon dioxide
6.4 **Fire Extinguishing Agents Not to be Used:** Water may be ineffective.
6.5 **Special Hazards of Combustion Products:** Not pertinent
6.6 **Behavior in Fire:** Not pertinent
6.7 **Ignition Temperature:** 797°F
6.8 **Electrical Hazard:** Data not available
6.9 **Burning Rate:** 9.2 mm/min.

7. CHEMICAL REACTIVITY

7.1 **Reactivity with Water:** No reaction
7.2 **Reactivity with Common Materials:** No reaction
7.3 **Stability During Transport:** Stable
7.4 **Neutralizing Agents for Acids and Caustics:** Not pertinent
7.5 **Polymerization:** Not pertinent
7.6 **Inhibitor of Polymerization:** Not pertinent

8. WATER POLLUTION

8.1 **Aquatic Toxicity:** Data not available
8.2 **Waterfowl Toxicity:** Data not available
8.3 **Biological Oxygen Demand (BOD):** Data not available
8.4 **Food Chain Concentration Potential:** None

9. SELECTED MANUFACTURERS

1. Phillips Petroleum Company
Chemical Department
Special Products Division
Bartlesville, Okla. 74004
2. Aldrich Chemical Co.
940 West Saint Paul Ave.
Milwaukee, Wis. 53233
3. Pfaltz and Bauer, Inc.
126-04 Northern Boulevard
Flushing, N. Y. 11368

10. SHIPPING INFORMATION

10.1 **Grades or Purity:** Research: 99.98%; Pure: 99.5%; Technical: 96.4%
10.2 **Storage Temperature:** Ambient
10.3 **Inert Atmosphere:** No requirement
10.4 **Venting:** Open (flame arrester)

11. HAZARD ASSESSMENT CODE

(See Hazard Assessment Handbook, CG 446-3)

A-T-U-V-W

12. HAZARD CLASSIFICATIONS

12.1 **Code of Federal Regulations:** Flammable liquid
12.2 **NAS Hazard Rating for Bulk Water Transportation:** Not listed
12.3 **NFPA Hazard Classifications:**

Category	Classification
Health Hazard (Blue)	1
Flammability (Red)	3
Reactivity (Yellow)	0

13. PHYSICAL AND CHEMICAL PROPERTIES

13.1 **Physical State at 15°C and 1 atm:** Liquid
13.2 **Molecular Weight:** 86.2
13.3 **Boiling Point at 1 atm:** 121.5°F = 49.7°C = 322.9°K
13.4 **Freezing Point:** −147.8°F = −99.9°C = 173.3°K
13.5 **Critical Temperature:** 420.1°F = 215.6°C = 488.8°K
13.6 **Critical Pressure:** 447 psia = 30.4 atm = 3.08 MN/m^2
13.7 **Specific Gravity:** 0.649 at 20°C (liquid)
13.8 **Liquid Surface Tension:** 16.3 dynes/cm = 0.0163 N/m at 20°C
13.9 **Liquid-Water Interfacial Tension:** (est.) 35 dynes/cm = 0.035 N/m at 20°C
13.10 **Vapor (Gas) Specific Gravity:** 3.0
13.11 **Ratio of Specific Heats of Vapor (Gas):** 1.064 at 16°C
13.12 **Latent Heat of Vaporization:** 131 Btu/lb = 72.9 cal/g = 3.05×10^5 J/kg
13.13 **Heat of Combustion:** −19,310 Btu/lb = −10,730 cal/g = -448.9×10^5 J/kg
13.14 **Heat of Decomposition:** Not pertinent
13.15 **Heat of Solution:** Not pertinent
13.16 **Heat of Polymerization:** Not pertinent

NOTES

NKA — NICKEL ACETATE

Common Synonyms			
Nickel acetate tetrahydrate Acetic acid, Nickel (II) salt Nickelous acetate	Solid	Dull green	Odorless
	Sinks and mixes slowly with water.		

Avoid contact with solid and dust. Keep people away.
Stop discharge if possible.
Isolate and remove discharged material.
Notify local health and pollution control agencies.

Fire	Not flammable.
Exposure	CALL FOR MEDICAL AID. DUST Irritating to eyes, nose and throat. If inhaled will cause coughing or difficult breathing. If in eyes, hold eyelids open and flush with plenty of water. If breathing has stopped, give artificial respiration. If breathing is difficult, give oxygen. SOLID Irritating to skin and eyes. If swallowed will cause nausea and vomiting. Remove contaminated clothing and shoes. Flush affected areas with plenty of water. IF IN EYES, hold eyelids open and flush with plenty of water. IF SWALLOWED and victim is CONSCIOUS, have victim drink water of milk. IF SWALLOWED and victim is UNCONSCIOUS OR HAVING CONVULSIONS, do nothing except keep victim warm.
Water Pollution	Effect of low concentrations on aquatic life is unknown. May be dangerous if it enters water intakes. Notify local health and wildlife officials. Notify operators of nearby water intakes.

1. RESPONSE TO DISCHARGE

(See Response Methods Handbook, CG 446-4)

Disperse and flush

2. LABELS

No hazard label required by Code of Federal Regulations

3. CHEMICAL DESIGNATIONS

3.1 **Synonyms:**
Acetic acid, nickel (2+) salt
Nickel acetate tetrahydrate
Nickelous acetate

3.2 **Coast Guard Compatibility Classification:** Not listed

3.3 **Chemical Formula:** $Ni(C_2H_3O_2)_2 \cdot 4H_2O$

3.4 **IMCO/United Nations Numerical Designation:** Not listed

4. OBSERVABLE CHARACTERISTICS

4.1 **Physical State (as shipped):** Solid

4.2 **Color:** Dull green

4.3 **Odor:** None

5. HEALTH HAZARDS

5.1 **Personal Protective Equipment:** Bu. Mines approved respirator; rubber gloves; safety goggles; protective clothing

5.2 **Symptoms Following Exposure:** Inhalation causes irritation of nose and throat. Ingestion causes vomiting. Contact with eyes causes irritation. May cause dermatitis in contact with skin.

5.3 **Treatment for Exposure:** INHALATION: remove to fresh air; get medical attention if exposure has been severe. INGESTION: give large amount of water. EYES: flush with water 15 min.; consult physician if irritation persists. SKIN: wash with soap and water.

5.4 **Toxicity by Inhalation (Threshold Limit Value):** 1 mg/m^3 (as nickel)

5.5 **Short-Term Inhalation Limits:** Data not available

5.6 **Toxicity by Ingestion:** Grade 2; LD_{50} 0.5—5 g/kg

5.7 **Late Toxicity:** Possible lung cancer

5.8 **Vapor (Gas) Irritant Characteristics:** Data not available

5.9 **Liquid or Solid Irritant Characteristics:** Data not available

5.10 **Odor Threshold:** Data not available

6. FIRE HAZARDS

6.1 **Flash Point:** Not flammable

6.2 **Flammable Limits in Air:** Not flammable

6.3 **Fire Extinguishing Agents:** Not pertinent

6.4 **Fire Extinguishing Agents Not to be Used:** Not pertinent

6.5 **Special Hazards of Combustion Products:**

6.6 **Behavior in Fire:**

6.7 **Ignition Temperature:** Not pertinent

6.8 **Electrical Hazard:** Not pertinent

6.9 **Burning Rate:** Not pertinent

7. CHEMICAL REACTIVITY

7.1 **Reactivity with Water:** No reaction

7.2 **Reactivity with Common Materials:**

7.3 **Stability During Transport:** Stable

7.4 **Neutralizing Agents for Acids and Caustics:** Not pertinent

7.5 **Polymerization:** Not pertinent

7.6 **Inhibitor of Polymerization:** Not pertinent

8. WATER POLLUTION

8.1 **Aquatic Toxicity:** Data not available

8.2 **Waterfowl Toxicity:** Data not available

8.3 **Biological Oxygen Demand (BOD):** Data not available

8.4 **Food Chain Concentration Potential:** None

9. SELECTED MANUFACTURERS

1. American Hoechst Corp.
Chemicals and Plastics Div.
Somerville, N.J. 08876
2. The Harshaw Chemical Co.
1945 E. 97th St.
Cleveland, Ohio
3. J. T. Baker Chemical Co.
Phillipsburg, N.J. 08865

10. SHIPPING INFORMATION

10.1 **Grades or Purity:** Commercial, 99%; Reagent

10.2 **Storage Temperature:** Ambient

10.3 **Inert Atmosphere:** No requirement

10.4 **Venting:** Open

11. HAZARD ASSESSMENT CODE

(See Hazard Assessment Handbook, CG 446-3)

SS

12. HAZARD CLASSIFICATIONS

12.1 **Code of Federal Regulations:** Not listed

12.2 **NAS Hazard Rating for Bulk Water Transportation:** Not listed

12.3 **NFPA Hazard Classifications:**

Category	Classification
Health Hazard (Blue)	1
Flammability (Red)	1
Reactivity (Yellow)	0

13. PHYSICAL AND CHEMICAL PROPERTIES

13.1 **Physical State at 15°C and 1 atm:** Solid

13.2 **Molecular Weight:** 248.86

13.3 **Boiling Point at 1 atm:** Not pertinent (decomposes)

13.4 **Freezing Point:** Not pertinent

13.5 **Critical Temperature:** Not pertinent

13.6 **Critical Pressure:** Not pertinent

13.7 **Specific Gravity:** 1.74 at 20°C (solid)

13.8 **Liquid Surface Tension:** Not pertinent

13.9 **Liquid-Water Interfacial Tension:** Not pertinent

13.10 **Vapor (Gas) Specific Gravity:** Not pertinent

13.11 **Ratio of Specific Heats of Vapor (Gas):** Not pertinent

13.12 **Latent Heat of Vaporization:** Not pertinent

13.13 **Heat of Combustion:** Not pertinent

13.14 **Heat of Decomposition:** Not pertinent

13.15 **Heat of Solution:** Not pertinent

13.16 **Heat of Polymerization:** Not pertinent

NOTES

NAS — NICKEL AMMONIUM SULFATE

Common Synonyms			
Ammonium nickel sulfate Ammonium disulfatonickelate (II) Nickel ammonium sulfate hexahydrate	Solid	Dark green-blue	Odorless
	Sinks and mixes slowly with water.		

Avoid contact with solid and dust. Keep people away.
Stop discharge if possible.
Isolate and remove discharged material.
Notify local health and pollution control agencies.

Fire	Not flammable. POISONOUS GASES MAY BE PRODUCED WHEN HEATED.
Exposure	CALL FOR MEDICAL AID. DUST Irritating to eyes, nose and throat. If inhaled will cause coughing or difficult breathing. If in eyes, hold eyelids open and flush with plenty of water. If breathing has stopped, give artificial respiration. If breathing is difficult, give oxygen. SOLID Irritating to skin and eyes. If swallowed will cause nausea and vomiting. Remove contaminated clothing and shoes. Flush affected areas with plenty of water. IF IN EYES, hold eyelids open and flush with plenty of water. IF SWALLOWED and victim is CONSCIOUS, have victim drink water or milk. IF SWALLOWED and victim is UNCONSCIOUS OR HAVING CONVULSIONS, do nothing except keep victim warm.
Water Pollution	Effect of low concentrations on aquatic life is unknown. May be dangerous if it enters water intakes. Notify local health and wildlife officials. Notify operators of nearby water intakes.

1. RESPONSE TO DISCHARGE
(See Response Methods Handbook, CG 446-4)

Disperse and flush

2. LABELS

No hazard label required by Code of Federal Regulations

3. CHEMICAL DESIGNATIONS

3.1 **Synonyms:** Ammonium disulfato-nickelate (II); Ammonium nickel sulfate; Nickel ammonium sulfate hexahydrate

3.2 **Coast Guard Compatibility Classification:** Not listed

3.3 **Chemical Formula:** $NiSO_4 \cdot (NH_4)_2SO_4 \cdot 6H_2O$

3.4 **IMCO/United Nations Numerical Designation:** Not listed

4. OBSERVABLE CHARACTERISTICS

4.1 **Physical State (as shipped):** Solid

4.2 **Color:** Dark blue-green

4.3 **Odor:** None

5. HEALTH HAZARDS

5.1 **Personal Protective Equipment:** Bu. Mines approved respirator; rubber gloves; face shield or safety goggles; protective clothing

5.2 **Symptoms Following Exposure:** Inhalation causes irritation of nose and throat. Ingestion causes vomiting. Contact with eyes causes irritation. Contact with skin may cause dermatitis.

5.3 **Treatment for Exposure:** INHALATION: move to fresh air; get medical attention if exposure has been severe. INGESTION: give large amount of water. EYES: flush with water for at least 15 min.; get medical attention if irritation persists. SKIN: wash with soap and water.

5.4 **Toxicity by Inhalation (Threshold Limit Value):** 1 mg/m^3 (as nickel)

5.5 **Short-Term Inhalation Limits:** Data not available

5.6 **Toxicity by Ingestion:** Grade 2; LD_{50} 0.5—5 g/kg

5.7 **Late Toxicity:** Possible lung cancer

5.8 **Vapor (Gas) Irritant Characteristics:** Data not available

5.9 **Liquid or Solid Irritant Characteristics:** Data not available

5.10 **Odor Threshold:** Data not available

6. FIRE HAZARDS

6.1 **Flash Point:** Not flammable

6.2 **Flammable Limits in Air:** Not flammable

6.3 **Fire Extinguishing Agents:** Not pertinent

6.4 **Fire Extinguishing Agents Not to be Used:** Not pertinent

6.5 **Special Hazards of Combustion Products:** Toxic oxides of nitrogen may be formed in fire.

6.6 **Behavior in Fire:**

6.7 **Ignition Temperature:** Not pertinent

6.8 **Electrical Hazard:** Not pertinent

6.9 **Burning Rate:** Not pertinent

7. CHEMICAL REACTIVITY

7.1 **Reactivity with Water:** No reaction

7.2 **Reactivity with Common Materials:**

7.3 **Stability During Transport:** Stable

7.4 **Neutralizing Agents for Acids and Caustics:** Not pertinent

7.5 **Polymerization:** Not pertinent

7.6 **Inhibitor of Polymerization:** Not pertinent

8. WATER POLLUTION

8.1 **Aquatic Toxicity:** 6 ppm as Ni/ /daphnia magna/deleterious effect/fresh water

8.2 **Waterfowl Toxicity:** Data not available

8.3 **Biological Oxygen Demand (BOD):** Data not available

8.4 **Food Chain Concentration Potential:** None

9. SELECTED MANUFACTURERS

1. J. T. Baker Chemical Co. Phillipsburg, N. J. 08865
2. McGean Chemical Co. 1250 Terminal Tower Cleveland, Ohio 44113
3. Gallard-Schlesinger Chemical Mfg. Co. 584 Mineola Ave. Carle Place, N. Y. 11514

10. SHIPPING INFORMATION

10.1 **Grades or Purity:** Reagent; Technical

10.2 **Storage Temperature:** Ambient

10.3 **Inert Atmosphere:** No requirement

10.4 **Venting:** Open

11. HAZARD ASSESSMENT CODE
(See Hazard Assessment Handbook, CG 446-3)

SS

12. HAZARD CLASSIFICATIONS

12.1 **Code of Federal Regulations:** Not listed

12.2 **NAS Hazard Rating for Bulk Water Transportation:** Not listed

12.3 **NFPA Hazard Classifications:** Not listed

13. PHYSICAL AND CHEMICAL PROPERTIES

13.1 **Physical State at 15°C and 1 atm:** Solid

13.2 **Molecular Weight:** 395.00

13.3 **Boiling Point at 1 atm:** Not pertinent (decomposes)

13.4 **Freezing Point:** Not pertinent

13.5 **Critical Temperature:** Not pertinent

13.6 **Critical Pressure:** Not pertinent

13.7 **Specific Gravity:** 1.92 at 20°C (solid)

13.8 **Liquid Surface Tension:** Not pertinent

13.9 **Liquid-Water Interfacial Tension:** Not pertinent

13.10 **Vapor (Gas) Specific Gravity:** Not pertinent

13.11 **Ratio of Specific Heats of Vapor (Gas):** Not pertinent

13.12 **Latent Heat of Vaporization:** Not pertinent

13.13 **Heat of Combustion:** Not pertinent

13.14 **Heat of Decomposition:** Not pertinent

13.15 **Heat of Solution:** Not pertinent

13.16 **Heat of Polymerization:** Not pertinent

NOTES

NBR NICKEL BROMIDE

Common Synonyms			
Nickel bromide trihydrate	Solid	Yellowish-green	Odorless
	Sinks and mixes with water.		

Avoid contact with solid and dust. Keep people away.
Stop discharge if possible.
Isolate and remove discharged material.
Notify local health and pollution control agencies.

Fire	Not flammable. Irritating gases may be produced when heated.
Exposure	CALL FOR MEDICAL AID. DUST Irritating to eyes, nose and throat. If inhaled will cause coughing and difficult breathing. If in eyes, hold eyelids open and flush with plenty of water. If breathing has stopped, give artificial respiration. If breathing is difficult, give oxygen. SOLID Irritating to skin and eyes. If swallowed will cause nausea and vomiting. Remove contaminated clothing and shoes. Flush affected areas with plenty of water. IF IN EYES, hold eyelids open and flush with plenty of water. IF SWALLOWED and victim is CONSCIOUS, have victim drink water or milk. IF SWALLOWED and victim is UNCONSCIOUS OR HAVING CONVULSIONS, do nothing except keep victim warm.
Water Pollution	Effect of low concentrations on aquatic life is unknown. May be dangerous if it enters water intakes. Notify local health and wildlife officials. Notify operators of nearby water intakes.

1. RESPONSE TO DISCHARGE

(See Response Methods Handbook, CG 446-4)

Issue warning—water contaminant
Disperse and flush

2. LABELS

No hazard label required by Code of Federal Regulations

3. CHEMICAL DESIGNATIONS

3.1 **Synonyms:** Nickel bromide trihydrate
3.2 **Coast Guard Compatibility Classification:** Not listed
3.3 **Chemical Formula:** $NiBr_2 \cdot 3H_2O$
3.4 **IMCO/United Nations Numerical Designation:** Not listed

4. OBSERVABLE CHARACTERISTICS

4.1 **Physical State (as shipped):** Solid
4.2 **Color:** Yellowish-green
4.3 **Odor:** None

5. HEALTH HAZARDS

5.1 **Personal Protective Equipment:** Bu. Mines approved respirator; rubber gloves; face shield or safety goggles; protective clothing
5.2 **Symptoms Following Exposure:** Inhalation of dust causes irritation of nose and throat. Ingestion causes vomiting; if large amount is swallowed and not thrown up, it can cause drowsiness and other symptoms of bromide poisoning. Dust irritates eyes and may cause dermatitis in contact with skin.
5.3 **Treatment for Exposure:** INHALATION: move to fresh air; get medical attention if exposure has been severe. INGESTION: give large amounts of water. EYES: flush with water for at least 15 min. SKIN: wash with soap and water.
5.4 **Toxicity by Inhalation (Threshold Limit Value):** 1 mg/m^3 (as nickel)
5.5 **Short-Term Inhalation Limits:** Data not available
5.6 **Toxicity by Ingestion:** Grade 2; LD_{50} 0.5—5 g/kg
5.7 **Late Toxicity:** Possible lung cancer
5.8 **Vapor (Gas) Irritant Characteristics:** Data not available
5.9 **Liquid or Solid Irritant Characteristics:** Data not available
5.10 **Odor Threshold:** Data not available

6. FIRE HAZARDS

6.1 **Flash Point:** Not flammable
6.2 **Flammable Limits in Air:** Not flammable
6.3 **Fire Extinguishing Agents:** Not pertinent
6.4 **Fire Extinguishing Agents Not to be Used:** Not pertinent
6.5 **Special Hazards of Combustion Products:** Irritating hydrogen bromide vapors may form in fire.
6.6 **Behavior in Fire:**
6.7 **Ignition Temperature:** Not pertinent
6.8 **Electrical Hazard:** Not pertinent
6.9 **Burning Rate:** Not pertinent

7. CHEMICAL REACTIVITY

7.1 **Reactivity with Water:** No reaction
7.2 **Reactivity with Common Materials:**
7.3 **Stability During Transport:** Stable
7.4 **Neutralizing Agents for Acids and Caustics:** Not pertinent
7.5 **Polymerization:** Not pertinent
7.6 **Inhibitor of Polymerization:** Not pertinent

8. WATER POLLUTION

8.1 **Aquatic Toxicity:** Data not available
8.2 **Waterfowl Toxicity:** Data not available
8.3 **Biological Oxygen Demand (BOD):** Data not available
8.4 **Food Chain Concentration Potential:** None

9. SELECTED MANUFACTURERS

1. D. F. Goldsmith Chemical and Metal Co.
   909 Pitner Ave.
   Evanston, Ill. 60202
2. Gallard-Schlesinger Chemical Mfg. Co.
   584 Mineola Ave.
   Carle Place, N. Y. 11514
3. Ventron Corp.
   P. O. Box 159
   Beverly, Mass. 01915

10. SHIPPING INFORMATION

10.1 **Grades or Purity:** Reagent; Technical
10.2 **Storage Temperature:** Ambient
10.3 **Inert Atmosphere:** No requirement
10.4 **Venting:** Open

11. HAZARD ASSESSMENT CODE

(See Hazard Assessment Handbook, CG 446-3)

SS

12. HAZARD CLASSIFICATIONS

12.1 **Code of Federal Regulations:** Not listed
12.2 **NAS Hazard Rating for Bulk Water Transportation:** Not listed
12.3 **NFPA Hazard Classifications:** Not listed

13. PHYSICAL AND CHEMICAL PROPERTIES

13.1 **Physical State at 15°C and 1 atm:** Solid
13.2 **Molecular Weight:** 272.6
13.3 **Boiling Point at 1 atm:** Not pertinent (decomposes)
13.4 **Freezing Point:** Not pertinent
13.5 **Critical Temperature:** Not pertinent
13.6 **Critical Pressure:** Not pertinent
13.7 **Specific Gravity:** (est.) 4 at 20°C (solid)
13.8 **Liquid Surface Tension:** Not pertinent
13.9 **Liquid-Water Interfacial Tension:** Not pertinent
13.10 **Vapor (Gas) Specific Gravity:** Not pertinent
13.11 **Ratio of Specific Heats of Vapor (Gas):** Not pertinent
13.12 **Latent Heat of Vaporization:** Not pertinent
13.13 **Heat of Combustion:** Not pertinent
13.14 **Heat of Decomposition:** Not pertinent
13.15 **Heat of Solution:** Not pertinent
13.16 **Heat of Polymerization:** Not pertinent

NOTES

NKC | NICKEL CARBONYL

Common Synonyms			
Nickel tetracarbonyl	Liquid	Colorless to yellow	Musty, stale odor
	Sinks in water. Poisonous, flammable vapor is produced.		

AVOID CONTACT WITH LIQUID AND VAPOR. KEEP PEOPLE AWAY.
Wear goggles, self-contained breathing apparatus, and rubber overclothing (including gloves).
Shut off ignition sources. Call fire department.
Stop discharge if possible.
Evacuate area in case of large discharge.
Stay upwind. Use water spray to "knock down" vapor.
Isolate and remove discharged material.
Notify local health and pollution control agencies.

Fire

FLAMMABLE.
POISONOUS GASES MAY BE PRODUCED IN FIRE.
Containers may explode in fire.
Flashback along vapor trail may occur.
Vapor may explode if ignited in an enclosed area.
Wear goggles and self-contained breathing apparatus.
Cool exposed containers with water.
Flood discharge area with water.

Exposure

CALL FOR MEDICAL AID.

VAPOR
POISONOUS IF INHALED.
Irritating to eyes, nose and throat.
Move victim to fresh air.
If breathing is difficult, give oxygen.

LIQUID
POISONOUS IF SWALLOWED OR IF SKIN IS EXPOSED.
Will burn skin and eyes.
Remove contaminated clothing and shoes.
Flush affected areas with plenty of water.
IF IN EYES, hold eyelids open and flush with plenty of water.
IF SWALLOWED and victim is CONSCIOUS, have victim drink water or milk.
DO NOT INDUCE VOMITING.

Water Pollution

Effect of low concentrations on aquatic life is unknown.
May be dangerous if it enters water intakes.
Notify local health and wildlife officials.
Notify operators of nearby water intakes.

1. RESPONSE TO DISCHARGE

(See Response Methods Handbook, CG 446-4)

Issue warning—poison, high flammability, air contaminant, water contaminant
Restrict access
Evacuate area
Should be removed
Chemical and physical treatment

2. LABELS

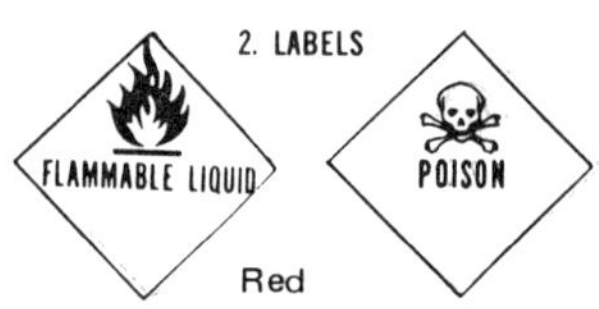

Red

3. CHEMICAL DESIGNATIONS

3.1 **Synonyms:** Nickel tetracarbonyl
3.2 **Coast Guard Compatibility Classification:** Not applicable
3.3 **Chemical Formula:** $Ni(CO)_4$
3.4 **IMCO/United Nations Numerical Designation:** 3.1/1259

4. OBSERVABLE CHARACTERISTICS

4.1 **Physical State (as shipped):** Liquid
4.2 **Color:** Colorless or yellow
4.3 **Odor:** Musty; characteristic

5. HEALTH HAZARDS

5.1 **Personal Protective Equipment:** Self-contained breathing apparatus; complete protective clothing
5.2 **Symptoms Following Exposure:** Inhalation causes giddiness, headache, shortness of breath, vomiting; if victim is removed from exposure, symptoms may disappear but recur 12-36 hours later, along with blue pallor of skin, fever, and cough; death may occur. Ingestion or contact with skin may also produce these symptoms. Abnormal nickel content of urine and blood is a measure of the severity of exposure. Contact of liquid with eyes causes severe irritation.
5.3 **Treatment for Exposure:** *Medical help must be obtained following all exposures to vapor or liquid.* INHALATION: oral administration of sodium diethyldithiocarbamate trihydrate (Dithiocarb); complete bed rest and positive-pressure oxygen are indicated for pulmonary edema; treatment otherwise is symptomatic. EYES: flush with water for at least 15 min. SKIN: wipe off, wash with soap and water. INGESTION: do NOT induce vomiting.
5.4 **Toxicity by Inhalation (Threshold Limit Value):** 0.001 ppm
5.5 **Short-Term Inhalation Limits:** 0.04 ppm for 5 min.
5.6 **Toxicity by Ingestion:** Data not available
5.7 **Late Toxicity:** May produce cancer
5.8 **Vapor (Gas) Irritant Characteristics:** Vapors cause severe irritation of eyes and throat and can cause eye and lung injury. They cannot be tolerated even at low concentrations.
5.9 **Liquid or Solid Irritant Characteristics:** Severe skin irritant. Causes second- and third-degree burns on short contact and is very injurious to the eyes.
5.10 **Odor Threshold:** 1–3 ppm

6. FIRE HAZARDS

6.1 **Flash Point:** <-4°F C.C.
6.2 **Flammable Limits in Air:** 2% (LFL)
6.3 **Fire Extinguishing Agents:** Water
6.4 **Fire Extinguishing Agents Not to be Used:** Data not available
6.5 **Special Hazards of Combustion Products:** Unusually toxic gases formed by incomplete combustion.
6.6 **Behavior in Fire:** Containers may explode when heated.
6.7 **Ignition Temperature:** <200°F (vapor)
6.8 **Electrical Hazard:** Data not available
6.9 **Burning Rate:** 2.7 mm/min.

7. CHEMICAL REACTIVITY

7.1 **Reactivity with Water:** No reaction
7.2 **Reactivity with Common Materials:** No reaction
7.3 **Stability During Transport:** Stable below 100°C
7.4 **Neutralizing Agents for Acids and Caustics:** Not pertinent
7.5 **Polymerization:** Not pertinent
7.6 **Inhibitor of Polymerization:** Not pertinent

8. WATER POLLUTION

8.1 **Aquatic Toxicity:** Data not available
8.2 **Waterfowl Toxicity:** Data not available
8.3 **Biological Oxygen Demand (BOD):** Data not available
8.4 **Food Chain Concentration Potential:** None

9. SELECTED MANUFACTURERS

1. Pressure Chemical Company
3419 Smallman Street
Pittsburgh, Pa. 15201
2. Gallard-Schlessinger Chemical Mfg. Co.
Atomergic Chemetals Division
584 Mineola Ave.
Carle Place, L. I., N. Y. 11514
3. Matheson Gas Products Co.
East Rutherford, N. J. 07073

10. SHIPPING INFORMATION

10.1 **Grades or Purity:** 99.9+%
10.2 **Storage Temperature:** Cool ambient
10.3 **Inert Atmosphere:** Carbon monoxide at 15 psi; carbon dioxide
10.4 **Venting:** Cylinders must be stored in a well-ventilated area.

11. HAZARD ASSESSMENT CODE

(See Hazard Assessment Handbook, CG 446-3)

A-X-Y

12. HAZARD CLASSIFICATIONS

12.1 **Code of Federal Regulations:** Flammable liquid
12.2 **NAS Hazard Rating for Bulk Water Transportation:**

Category	Rating
Fire	3
Health	
Vapor Irritant	4
Liquid or Solid Irritant	4
Poisons	4
Water Pollution	
Human Toxicity	
Aquatic Toxicity	3
Aesthetic Effect	1
Reactivity	
Other Chemicals	4
Water	0
Self-Reaction	4

12.3 **NFPA Hazard Classifications:**

Category	Classification
Health Hazard (Blue)	
Flammability (Red)	3
Reactivity (Yellow)	1
	W

13. PHYSICAL AND CHEMICAL PROPERTIES

13.1 **Physical State at 15°C and 1 atm:** Liquid
13.2 **Molecular Weight:** 170.7
13.3 **Boiling Point at 1 atm:** 109°F = 43°C = 316°K
13.4 **Freezing Point:** -13°F = -25°C = 248°K
13.5 **Critical Temperature:** Not pertinent
13.6 **Critical Pressure:** Not pertinent
13.7 **Specific Gravity:** 1.322 at 17°C (liquid)
13.8 **Liquid Surface Tension:** 15.9 dynes/cm = 0.0159 N/m at 20°C
13.9 **Liquid-Water Interfacial Tension:** Data not available
13.10 **Vapor (Gas) Specific Gravity:** 5.9
13.11 **Ratio of Specific Heats of Vapor (Gas):** Data not available
13.12 **Latent Heat of Vaporization:** 72 Btu/lb = 40 cal/g = 1.7×10^5 J/kg
13.13 **Heat of Combustion:** $-2{,}970$ Btu/lb = $-1{,}650$ cal/g = -69.0×10^5 J/kg
13.14 **Heat of Decomposition:** Not pertinent
13.15 **Heat of Solution:** Not pertinent
13.16 **Heat of Polymerization:** Not pertinent

NOTES

NCL	# NICKEL CHLORIDE

Common Synonyms			
Nickel chloride hexahydrate	Solid	Green	Odorless
	Sinks and mixes with water.		

Avoid contact with solid and dust. Keep people away.
Stop discharge if possible.
Isolate and remove discharged material.
Notify local health and pollution control agencies.

Fire

Not flammable.

Exposure

CALL FOR MEDICAL AID.

DUST
Irritating to eyes, nose and throat.
If inhaled will cause coughing or difficult breathing.
If in eyes, hold eyelids open and flush with plenty of water.
If breathing has stopped, give artificial respiration.
If breathing is difficult, give oxygen.

SOLID
Irritating to skin and eyes.
If swallowed will cause nausea and vomiting.
Remove contaminated clothing and shoes.
Flush affected areas with plenty of water.
IF IN EYES, hold eyelids open and flush with plenty of water.
IF SWALLOWED and victim is CONSCIOUS, have victim drink water or milk.
IF SWALLOWED and victim is UNCONSCIOUS OR HAVING CONVULSIONS, do nothing except keep victim warm.

Water Pollution

HARMFUL TO AQUATIC LIFE IN VERY LOW CONCENTRATIONS.
May be dangerous if it enters water intakes.
Notify local health and wildlife officials.
Notify operators of nearby water intakes.

1. RESPONSE TO DISCHARGE

(See Response Methods Handbook, CG 446-4)

Disperse and flush

2. LABELS

No hazard label required by Code of Federal Regulations

3. CHEMICAL DESIGNATIONS

3.1 **Synonyms:** Nickel chloride hexahydrate
3.2 **Coast Guard Compatibility Classification:** Not listed
3.3 **Chemical Formula:** $NiCl_2 \cdot 6H_2O$
3.4 **IMCO/United Nations Numerical Designation:** Not listed

4. OBSERVABLE CHARACTERISTICS

4.1 **Physical State (as shipped):** Solid
4.2 **Color:** Green
4.3 **Odor:** None

5. HEALTH HAZARDS

5.1 **Personal Protective Equipment:** Goggles or face shield; protective gloves; Bu. Mines approved respirator; protective clothing
5.2 **Symptoms Following Exposure:** Inhalation of dust causes irritation of nose and throat. Ingestion causes vomiting. Dust irritates eyes and may cause dermatitis in contact with skin.
5.3 **Treatment for Exposure:** INHALATION: move to fresh air; get medical attention if exposure has been severe. INGESTION: give large amount of water. EYES: flush with plenty of water for at least 15 min. SKIN: flush with water.
5.4 **Toxicity by Inhalation (Threshold Limit Value):** 1 mg/m^3 (as nickel)
5.5 **Short-Term Inhalation Limits:** Data not available
5.6 **Toxicity by Ingestion:** Grade 2; LD_{50} 0.5—5 g/kg
5.7 **Late Toxicity:** Possible lung cancer
5.8 **Vapor (Gas) Irritant Characteristics:** Data not available
5.9 **Liquid or Solid Irritant Characteristics:** Data not available
5.10 **Odor Threshold:** Data not available

6. FIRE HAZARDS

6.1 **Flash Point:** Not flammable
6.2 **Flammable Limits in Air:** Not flammable
6.3 **Fire Extinguishing Agents:** Not pertinent
6.4 **Fire Extinguishing Agents Not to be Used:** Not pertinent
6.5 **Special Hazards of Combustion Products:** Not pertinent
6.6 **Behavior in Fire:**
6.7 **Ignition Temperature:** Not pertinent
6.8 **Electrical Hazard:** Not pertinent
6.9 **Burning Rate:** Not pertinent

7. CHEMICAL REACTIVITY

7.1 **Reactivity with Water:** No reaction
7.2 **Reactivity with Common Materials:**
7.3 **Stability During Transport:** Stable
7.4 **Neutralizing Agents for Acids and Caustics:** Not pertinent
7.5 **Polymerization:** Not pertinent
7.6 **Inhibitor of Polymerization:** Not pertinent

8. WATER POLLUTION

8.1 **Aquatic Toxicity:**
259 mg/l/*/minnow fundulus/survived/salt water
5.18 ppm/96 hr/fathead minnow/TL_m/soft water
42.4 ppm/96 hr/fathead minnow/TL_m/hard water
*Time period not specified
8.2 **Waterfowl Toxicity:** Data not available
8.3 **Biological Oxygen Demand (BOD):** None
8.4 **Food Chain Concentration Potential:** None

9. SELECTED MANUFACTURERS

1. Tenneco Chemicals, Inc.
Intermediates Div.
Turner Place, P. O. Box 2
Piscataway, N. J. 08854
2. The Harshaw Chemical Co.
1945 East 97 Street
Cleveland, Ohio 44106
3. Allied Chemical Corp.
Specialty Chemicals Div.
P. O. Box 1087R
Morristown, N. J. 07960

10. SHIPPING INFORMATION

10.1 **Grades or Purity:** Technical, 99+%
10.2 **Storage Temperature:** Ambient
10.3 **Inert Atmosphere:** No requirement
10.4 **Venting:** Open

11. HAZARD ASSESSMENT CODE

(See Hazard Assessment Handbook, CG 446-3)

SS

12. HAZARD CLASSIFICATIONS

12.1 **Code of Federal Regulations:** Not listed
12.2 **NAS Hazard Rating for Bulk Water Transportation:** Not listed
12.3 **NFPA Hazard Classifications:**

Category	Classification
Health Hazard (Blue)	1
Flammability (Red)	0
Reactivity (Yellow)	0

13. PHYSICAL AND CHEMICAL PROPERTIES

13.1 **Physical State at 15°C and 1 atm:** Solid
13.2 **Molecular Weight:** 237.7
13.3 **Boiling Point at 1 atm:** Not pertinent (decomposes)
13.4 **Freezing Point:** Not pertinent
13.5 **Critical Temperature:** Not pertinent
13.6 **Critical Pressure:** Not pertinent
13.7 **Specific Gravity:** 3.55 at 15°C (solid)
13.8 **Liquid Surface Tension:** Not pertinent
13.9 **Liquid-Water Interfacial Tension:** Not pertinent
13.10 **Vapor (Gas) Specific Gravity:** Not pertinent
13.11 **Ratio of Specific Heats of Vapor (Gas):** Not pertinent
13.12 **Latent Heat of Vaporization:** Not pertinent
13.13 **Heat of Combustion:** Not pertinent
13.14 **Heat of Decomposition:** Not pertinent
13.15 **Heat of Solution:** 8.8 Btu/lb = 4.9 cal/g = 0.21×10^5 J/kg
13.16 **Heat of Polymerization:** Not pertinent

NOTES

NICKEL CYANIDE

Common Synonyms	
	Solid — Light green or yellow-brown — Weak almond odor Sinks in water.

AVOID CONTACT WITH SOLID AND DUST. KEEP PEOPLE AWAY.
Wear dust respirator and rubber overclothing (including gloves).
Stay upwind. Use water spray to "knock down" dust.
Isolate and remove discharged material.
Notify local health and pollution control agencies.

Fire

Not flammable.

Exposure

CALL FOR MEDICAL AID.

DUST
POISONOUS IF INHALED.
Irritating to eyes, nose and throat.
Move victim to fresh air.
If in eyes, hold eyelids open and flush with plenty of water.
If breathing has stopped, give artificial respiration.
If breathing is difficult, give oxygen.

SOLID
POISONOUS IF SWALLOWED.
Irritating to skin and eyes.
Remove contaminated clothing and shoes.
Flush affected areas with plenty of water.
IF IN EYES, hold eyelids open and flush with plenty of water.
IF SWALLOWED and victim is CONSCIOUS, have victim drink water or milk and have victim induce vomiting.
IF SWALLOWED and victim is UNCONSCIOUS OR HAVING CONVULSIONS, do nothing except keep victim warm.

Water Pollution

Dangerous to aquatic life in high concentrations.
May be dangerous if it enters water intakes.
Notify local health and wildlife officials.
Notify operators of nearby water intakes.

1. RESPONSE TO DISCHARGE

(See Response Methods Handbook, CG 446-4)

Issue warning—poison, water contaminant
Restrict access
Should be removed
Chemical and physical treatment

2. LABEL

3. CHEMICAL DESIGNATIONS

3.1 **Synonyms:** No common synonyms
3.2 **Coast Guard Compatibility Classification:** Not applicable
3.3 **Chemical Formula:** $Ni(CN)_2 \cdot nH_2O$ n=0,2,3,4
3.4 **IMCO/United Nations Numerical Designation:** 6.1/1653

4. OBSERVABLE CHARACTERISTICS

4.1 **Physical State (as shipped):** Solid
4.2 **Color:** Anhydrous: yellow-brown; hydrates: pale green
4.3 **Odor:** Weak characteristic cyanide

5. HEALTH HAZARDS

5.1 **Personal Protective Equipment:** Dust mask; rubber gloves; goggles.
5.2 **Symptoms Following Exposure:** Inhalation of dust or ingestion can produce bitter taste, salivation, numbness of throat, anxiety, irregular respiration, rapid pulse, convulsions, loss of consciousness, paralysis, death.
5.3 **Treatment for Exposure:** Get medical help; first-aid treatment must be prompt; the compound is much less soluble than sodium or potassium cyanides, but cyanide poisoning is a possibility. EYES or SKIN: flush with water; avoid getting solution in mouth. INGESTION: start artificial respiration if breathing has stopped; inhalation of amyl nitrite and intravenous doses of sodium nitrite and sodium thiosulfate should be administered by physician.
5.4 **Toxicity by Inhalation (Threshold Limit Value):** Data not available
5.5 **Short-Term Inhalation Limits:** 5 mg/m³ for 30 min. (as cyanide)
5.6 **Toxicity by Ingestion:** Data not available
5.7 **Late Toxicity:** Data not available
5.8 **Vapor (Gas) Irritant Characteristics:** Data not available
5.9 **Liquid or Solid Irritant Characteristics:** Data not available
5.10 **Odor Threshold:** Not pertinent

6. FIRE HAZARDS

6.1 **Flash Point:** Not flammable
6.2 **Flammable Limits in Air:** Not flammable
6.3 **Fire Extinguishing Agents:** Not pertinent
6.4 **Fire Extinguishing Agents Not to be Used:** Not pertinent
6.5 **Special Hazards of Combustion Products:** Not pertinent
6.6 **Behavior in Fire:** Not pertinent
6.7 **Ignition Temperature:** Not pertinent
6.8 **Electrical Hazard:** Not pertinent
6.9 **Burning Rate:** Not pertinent

7. CHEMICAL REACTIVITY

7.1 **Reactivity with Water:** No reaction
7.2 **Reactivity with Common Materials:** No reaction
7.3 **Stability During Transport:** Stable
7.4 **Neutralizing Agents for Acids and Caustics:** Not pertinent
7.5 **Polymerization:** Not pertinent
7.6 **Inhibitor of Polymerization:** Not pertinent

8. WATER POLLUTION

8.1 **Aquatic Toxicity:**
> 100 mg/l /*/fish/harmless/moderately alkaline water
In acid water, releases hydrogen cyanide, which is fatal to all species.
*Time period not specified.
8.2 **Waterfowl Toxicity:** Data not available
8.3 **Biological Oxygen Demand (BOD):** Data not available
8.4 **Food Chain Concentration Potential:** None

9. SELECTED MANUFACTURERS

1. Pfaltz and Bauer, Inc.
   126—04 Northern Boulevard
   Flushing, N. Y. 11368
2. Var-lac-oid Chemical Co.
   666 S. Front Street
   Elizabeth, N. J. 07202
3. Fielding Chemical Co.
   Forrest and Halladay Streets
   Jersey City, N. J. 07305

10. SHIPPING INFORMATION

10.1 **Grades or Purity:** Commercial
10.2 **Storage Temperature:** Ambient
10.3 **Inert Atmosphere:** No requirement
10.4 **Venting:** Sealed containers in well-ventilated area.

11. HAZARD ASSESSMENT CODE

(See Hazard Assessment Handbook, CG 446-3)

II

12. HAZARD CLASSIFICATIONS

12.1 **Code of Federal Regulations:** Poisonous, Class B
12.2 **NAS Hazard Rating for Bulk Water Transportation:** Not listed
12.3 **NFPA Hazard Classifications:** Not listed

13. PHYSICAL AND CHEMICAL PROPERTIES

13.1 **Physical State at 15°C and 1 atm:** Solid
13.2 **Molecular Weight:** Mixture
13.3 **Boiling Point at 1 atm:** Not pertinent (decomposes)
13.4 **Freezing Point:** Not pertinent
13.5 **Critical Temperature:** Not pertinent
13.6 **Critical Pressure:** Not pertinent
13.7 **Specific Gravity:** 2.4 at 25°C (solid)
13.8 **Liquid Surface Tension:** Not pertinent
13.9 **Liquid-Water Interfacial Tension:** Not pertinent
13.10 **Vapor (Gas) Specific Gravity:** Not pertinent
13.11 **Ratio of Specific Heats of Vapor (Gas):** Not pertinent
13.12 **Latent Heat of Vaporization:** Not pertinent
13.13 **Heat of Combustion:** Not pertinent
13.14 **Heat of Decomposition:** Not pertinent
13.15 **Heat of Solution:** Not pertinent
13.16 **Heat of Polymerization:** Not pertinent

NOTES

NFB	NICKEL FLUOROBORATE

Common Synonyms		
Nickel (II) fluoborate Nickel fluoroborate solution	Liquid Sink and mixes with water.	Green

Avoid contact with liquid and vapor. Keep people away.
Stop discharge if possible.
Isolate and remove discharged material.
Notify local health and pollution control agencies.

Fire	Not flammable.
Exposure	CALL FOR MEDICAL AID. VAPOR Irritating to eyes, nose and throat. Harmful if inhaled. If in eyes, hold eyelids open and flush with plenty of water. If breathing has stopped, give artificial respiration. If breathing is difficult, give oxygen. LIQUID Irritating to skin and eyes. If swallowed will cause nausea and vomiting. Remove contaminated clothing and shoes. Flush affected areas with plenty of water. IF IN EYES, hold eyelids open and flush with plenty of water. IF SWALLOWED and victim is CONSCIOUS, have victim drink water or milk and have victim induce vomiting. IF SWALLOWED and victim is UNCONSCIOUS OR HAVING CONVULSIONS, do nothing except keep victim warm.
Water Pollution	Effect of low concentrations on aquatic life is unknown. May be dangerous if it enters water intakes. Notify local health and wildlife officials. Notify operators of nearby water intakes.

1. RESPONSE TO DISCHARGE

(See Response Methods Handbook, CG 446-4)

Issue warning—water contaminant
Disperse and flush

2. LABELS

No hazard label required by Code of Federal Regulations

3. CHEMICAL DESIGNATIONS

3.1 **Synonyms:** Nickel (II) fluoborate; Nickel fluoroborate solution
3.2 **Coast Guard Compatibility Classification:** Not listed
3.3 **Chemical Formula:** $Ni(BF_4)_2 \cdot H_2O$
3.4 **IMCO/United Nations Numerical Designation:** Not listed

4. OBSERVABLE CHARACTERISTICS

4.1 **Physical State (as shipped):** Liquid
4.2 **Color:** Clear, green
4.3 **Odor:** Data not available

5. HEALTH HAZARDS

5.1 **Personal Protective Equipment:** Safety glasses and face shield; rubber gloves; rubber apron
5.2 **Symptoms Following Exposure:** Ingestion causes irritation of mouth and stomach. Liquid irritates eyes and skin and may cause dermatitis.
5.3 **Treatment for Exposure:** INGESTION: give large amount of water; induce vomiting; get medical attention. EYES: flush with water for at least 15 min. SKIN: flush with water.
5.4 **Toxicity by Inhalation (Threshold Limit Value):** 1 mg/m^3 (as nickel)
5.5 **Short-Term Inhalation Limits:** Data not available
5.6 **Toxicity by Ingestion:** Grade 2; LD_{50} 0.5—5 g/kg
5.7 **Late Toxicity:** Possible lung cancer
5.8 **Vapor (Gas) Irritant Characteristics:** Data not available
5.9 **Liquid or Solid Irritant Characteristics:** Data not available
5.10 **Odor Threshold:** Data not available

6. FIRE HAZARDS

6.1 **Flash Point:** Not flammable
6.2 **Flammable Limits in Air:** Not flammable
6.3 **Fire Extinguishing Agents:** Not pertinent
6.4 **Fire Extinguishing Agents Not to be Used:** Not pertinent
6.5 **Special Hazards of Combustion Products:** Not pertinent
6.6 **Behavior in Fire:**
6.7 **Ignition Temperature:** Not pertinent
6.8 **Electrical Hazard:** Not pertinent
6.9 **Burning Rate:** Not pertinent

7. CHEMICAL REACTIVITY

7.1 **Reactivity with Water:** No reaction
7.2 **Reactivity with Common Materials:**
7.3 **Stability During Transport:** Stable
7.4 **Neutralizing Agents for Acids and Caustics:** Not pertinent
7.5 **Polymerization:** Not pertinent
7.6 **Inhibitor of Polymerization:** Not pertinent

8. WATER POLLUTION

8.1 **Aquatic Toxicity:** Data not available
8.2 **Waterfowl Toxicity:** Data not available
8.3 **Biological Oxygen Demand (BOD):** Data not available
8.4 **Food Chain Concentration Potential:** None

9. SELECTED MANUFACTURERS

1. Allied Chemical Corp.
   P. O. Box 1087R
   Morristown, N. J. 07960
2. The Harshaw Chemical Co.
   1945 East 97 St.
   Cleveland, Ohio 44106
3. Harstan Chemical Corp.
   1247 38 Street
   Brooklyn, N. Y. 11218

10. SHIPPING INFORMATION

10.1 **Grades or Purity:** 44.2—45.2% in water
10.2 **Storage Temperature:** Ambient
10.3 **Inert Atmosphere:** No requirement
10.4 **Venting:** Open

11. HAZARD ASSESSMENT CODE

(See Hazard Assessment Handbook, CG 446-3)

A-P

12. HAZARD CLASSIFICATIONS

12.1 **Code of Federal Regulations:** Not listed
12.2 **NAS Hazard Rating for Bulk Water Transportation:** Not listed
12.3 **NFPA Hazard Classifications:**

Category	Classification
Health Hazard (Blue)	1
Flammability (Red)	0
Reactivity (Yellow)	0

13. PHYSICAL AND CHEMICAL PROPERTIES

13.1 **Physical State at 15°C and 1 atm:** Liquid
13.2 **Molecular Weight:** Mixture
13.3 **Boiling Point at 1 atm:** Not pertinent
13.4 **Freezing Point:** Not pertinent
13.5 **Critical Temperature:** Not pertinent
13.6 **Critical Pressure:** Not pertinent
13.7 **Specific Gravity:** 1.5 at 20°C (liquid)
13.8 **Liquid Surface Tension:** Data not available
13.9 **Liquid-Water Interfacial Tension:** Not pertinent
13.10 **Vapor (Gas) Specific Gravity:** Not pertinent
13.11 **Ratio of Specific Heats of Vapor (Gas):** Not pertinent
13.12 **Latent Heat of Vaporization:** Not pertinent
13.13 **Heat of Combustion:** Not pertinent
13.14 **Heat of Decomposition:** Not pertinent
13.15 **Heat of Solution:** Not pertinent
13.16 **Heat of Polymerization:** Not pertinent

NOTES

NFM

NICKEL FORMATE

Common Synonyms			
Nickel formate dihydrate	Solid	Green	Odorless
	Sinks and mixes with water.		

Avoid contact with solid and dust. Keep people away.
Stop discharge if possible.
Isolate and remove discharged material.
Notify local health and pollution control agencies.

Fire	Not flammable.
Exposure	CALL FOR MEDICAL AID. DUST Irritating to eyes, nose and throat. If inhaled will cause coughing or difficult breathing. If in eyes, hold eyelids open and flush with plenty of water. If breathing has stopped, give artificial respiration. If breathing is difficult, give oxygen. SOLID Irritating to skin and eyes. If swallowed will cause nausea and vomiting. Remove contaminated clothing and shoes. Flush affected areas with plenty of water. IF IN EYES, hold eyelids open and flush with plenty of water. IF SWALLOWED and victim is CONSCIOUS, have victim drink water or milk. IF SWALLOWED and victim is UNCONSCIOUS OR HAVING CONVULSIONS, do nothing except keep victim warm.
Water Pollution	Effect of low concentrations on aquatic life is unknown. May be dangerous if it enters water intakes. Notify local health and wildlife officials. Notify operators of nearby water intakes.

1. RESPONSE TO DISCHARGE

(See Response Methods Handbook, CG 446-4)

Disperse and flush

2. LABELS

No hazard label required by Code of Federal Regulations

3. CHEMICAL DESIGNATIONS

3.1 **Synonyms:** Nickel formate dihydrate
3.2 **Coast Guard Compatibility Classification:** Not listed
3.3 **Chemical Formula:** $Ni(HCO_2)_2 \cdot 2H_2O$
3.4 **IMCO/United Nations Numerical Designation:** Not listed

4. OBSERVABLE CHARACTERISTICS

4.1 **Physical State (as shipped):** Solid
4.2 **Color:** Greenish
4.3 **Odor:** None

5. HEALTH HAZARDS

5.1 **Personal Protective Equipment:** Bu. Mines approved respirator; rubber gloves; face shield or safety goggles; protective clothing
5.2 **Symptoms Following Exposure:** Inhalation of dust causes irritation of nose and throat. Ingestion causes vomiting. Dust irritates eyes and may cause dermatitis in contact with skin.
5.3 **Treatment for Exposure:** INHALATION: move to fresh air; get medical attention if exposure has been severe. INGESTION: give large amounts of water. EYES: flush with water for at least 15 min. SKIN: flush with water.
5.4 **Toxicity by Inhalation (Threshold Limit Value):** 1 mg/m^3 (as nickel)
5.5 **Short-Term Inhalation Limits:** Data not available
5.6 **Toxicity by Ingestion:** Grade 2; LD_{50} 0.5—5 g/kg
5.7 **Late Toxicity:** Possible lung cancer
5.8 **Vapor (Gas) Irritant Characteristics:** Data not available
5.9 **Liquid or Solid Irritant Characteristics:** Data not available
5.10 **Odor Threshold:** Data not available

6. FIRE HAZARDS

6.1 **Flash Point:** Not flammable
6.2 **Flammable Limits in Air:** Not flammable
6.3 **Fire Extinguishing Agents:** Not pertinent
6.4 **Fire Extinguishing Agents Not to be Used:** Not pertinent
6.5 **Special Hazards of Combustion Products:** Irritating formic acid vapors may form in fire.
6.6 **Behavior in Fire:**
6.7 **Ignition Temperature:** Not pertinent
6.8 **Electrical Hazard:** Not pertinent
6.9 **Burning Rate:** Not pertinent

7. CHEMICAL REACTIVITY

7.1 **Reactivity with Water:** No reaction
7.2 **Reactivity with Common Materials:**
7.3 **Stability During Transport:** Stable
7.4 **Neutralizing Agents for Acids and Caustics:** Not pertinent
7.5 **Polymerization:** Not pertinent
7.6 **Inhibitor of Polymerization:** Not pertinent

8. WATER POLLUTION

8.1 **Aquatic Toxicity:** Data not available
8.2 **Waterfowl Toxicity:** Data not available
8.3 **Biological Oxygen Demand (BOD):** Data not available
8.4 **Food Chain Concentration Potential:** None

9. SELECTED MANUFACTURERS

McGean Chemical Co.
1250 Terminal Tower
Cleveland, Ohio 44113

10. SHIPPING INFORMATION

10.1 **Grades or Purity:** Commercial
10.2 **Storage Temperature:** Ambient
10.3 **Inert Atmosphere:** No requirement
10.4 **Venting:** Open

11. HAZARD ASSESSMENT CODE

(See Hazard Assessment Handbook, CG 446-3)

SS

12. HAZARD CLASSIFICATIONS

12.1 **Code of Federal Regulations:** Not listed
12.2 **NAS Hazard Rating for Bulk Water Transportation:** Not listed
12.3 **NFPA Hazard Classifications:** Not listed

13. PHYSICAL AND CHEMICAL PROPERTIES

13.1 **Physical State at 15°C and 1 atm:** Solid
13.2 **Molecular Weight:** 184.8
13.3 **Boiling Point at 1 atm:** Not pertinent (decomposes)
13.4 **Freezing Point:** Not pertinent
13.5 **Critical Temperature:** Not pertinent
13.6 **Critical Pressure:** Not pertinent
13.7 **Specific Gravity:** 2.15 at 20°C (solid)
13.8 **Liquid Surface Tension:** Not pertinent
13.9 **Liquid-Water Interfacial Tension:** Not pertinent
13.10 **Vapor (Gas) Specific Gravity:** Not pertinent
13.11 **Ratio of Specific Heats of Vapor (Gas):** Not pertinent
13.12 **Latent Heat of Vaporization:** Not pertinent
13.13 **Heat of Combustion:** Not pertinent
13.14 **Heat of Decomposition:** Not pertinent
13.15 **Heat of Solution:** Not pertinent
13.16 **Heat of Polymerization:** Not pertinent

NOTES

NNT — NICKEL NITRATE

Common Synonyms			
Nickel nitrate hexahydrate	Solid	Green	Odorless
	Sinks and mixes with water.		

Avoid contact with solid and dust. Keep people away.
Stop discharge if possible.
Isolate and remove discharged material.
Notify local health and pollution control agencies.

Fire	Not flammable. Will increase the intensity of a fire. POISONOUS GASES MAY BE PRODUCED WHEN HEATED. Flood discharge area with water.
Exposure	CALL FOR MEDICAL AID. DUST Irritating to eyes, nose and throat. If inhaled will cause coughing or difficult breathing. If in eyes, hold eyelids open and flush with plenty of water. If breathing has stopped, give artificial respiration. If breathing is difficult, give oxygen. SOLID Irritating to skin and eyes. If swallowed will cause nausea and vomiting. Remove contaminated clothing and shoes. Flush affected areas with plenty of water. IF IN EYES, hold eyelids open and flush with plenty of water. IF SWALLOWED and victim is CONSCIOUS, have victim drink water or milk. IF SWALLOWED and victim is UNCONSCIOUS OR HAVING CONVULSIONS, do nothing except keep victim warm.
Water Pollution	Effect of low concentrations on aquatic life is unknown. May be dangerous if it enters water intakes. Notify local health and wildlife officials. Notify operators of nearby water intakes.

1. RESPONSE TO DISCHARGE
(See Response Methods Handbook, CG 446-4)

Disperse and flush

2. LABELS

No hazard label required by Code of Federal Regulations

3. CHEMICAL DESIGNATIONS

3.1 **Synonyms:** Nickel nitrate hexahydrate
3.2 **Coast Guard Compatibility Classification:** Not listed
3.3 **Chemical Formula:** $Ni(NO_3)_2 \cdot 6H_2O$
3.4 **IMCO/United Nations Numerical Designation:** Not listed

4. OBSERVABLE CHARACTERISTICS

4.1 **Physical State (as shipped):** Solid
4.2 **Color:** Green
4.3 **Odor:** None

5. HEALTH HAZARDS

5.1 **Personal Protective Equipment:** Bu. Mines approved respirator; rubber gloves; face shield or safety goggles; protective clothing
5.2 **Symptoms Following Exposure:** Inhalation of dust causes irritation of nose and throat. Ingestion causes vomiting. Dust irritates eyes and may cause dermatitis in contact with skin.
5.3 **Treatment for Exposure:** INHALATION: move to fresh air; get medical attention if exposure has been severe. INGESTION: give large amount of water. EYES: flush with water for at least 15 min. SKIN: wash with soap and water.
5.4 **Toxicity by Inhalation (Threshold Limit Value):** 1 mg/m^3 (as nickel)
5.5 **Short-Term Inhalation Limits:** Data not available
5.6 **Toxicity by Ingestion:** Grade 2; LD_{50} 0.5—5 g/kg
5.7 **Late Toxicity:** Possible lung cancer
5.8 **Vapor (Gas) Irritant Characteristics:** Data not available
5.9 **Liquid or Solid Irritant Characteristics:** Data not available
5.10 **Odor Threshold:** Data not available

6. FIRE HAZARDS

6.1 **Flash Point:** Not flammable, but may intensify fire
6.2 **Flammable Limits in Air:** Not flammable
6.3 **Fire Extinguishing Agents:** Not pertinent
6.4 **Fire Extinguishing Agents Not to be Used:** Not pertinent
6.5 **Special Hazards of Combustion Products:** Toxic oxides of nitrogen may form in fire.
6.6 **Behavior in Fire:** May increase intensity of fire if in contact with combustible material
6.7 **Ignition Temperature:** Not pertinent
6.8 **Electrical Hazard:** Not pertinent
6.9 **Burning Rate:** Not pertinent

7. CHEMICAL REACTIVITY

7.1 **Reactivity with Water:** No reaction
7.2 **Reactivity with Common Materials:** Contact of solid with wood or paper may cause fires.
7.3 **Stability During Transport:** Stable
7.4 **Neutralizing Agents for Acids and Caustics:** Not pertinent
7.5 **Polymerization:** Not pertinent
7.6 **Inhibitor of Polymerization:** Not pertinent

8. WATER POLLUTION

8.1 **Aquatic Toxicity:**
2.44 ppm/*/stickleback/threshold conc. tap water
0.8 ppm/10 days/stickleback/TL_m/fresh water
*Time period not specified.
8.2 **Waterfowl Toxicity:** Data not available
8.3 **Biological Oxygen Demand (BOD):** None
8.4 **Food Chain Concentration Potential:** None

9. SELECTED MANUFACTURERS

1. The Harshaw Chemical Co.
1945 East 97 St.
Cleveland, Ohio 44106
2. Allied Chemical Corp.
Specialty Chemicals Div.
P. O. Box 1087R
Morristown, N. J. 07960
3. McGean Chemical Co.
1250 Terminal Tower
Cleveland, Ohio 44113

10. SHIPPING INFORMATION

10.1 **Grades or Purity:** Purified, 99.1%
10.2 **Storage Temperature:** Ambient
10.3 **Inert Atmosphere:** No requirement
10.4 **Venting:** Open

11. HAZARD ASSESSMENT CODE
(See Hazard Assessment Handbook, CG 446-3)

SS

12. HAZARD CLASSIFICATIONS

12.1 **Code of Federal Regulations:** Not listed
12.2 **NAS Hazard Rating for Bulk Water Transportation:** Not listed
12.3 **NFPA Hazard Classifications:**

Category	Classification*	
Health Hazard (Blue)	0	1
Flammability (Red)	0	0
Reactivity (Yellow)	0	0
	oxy	oxy

*First column refers to nonfire situation.

13. PHYSICAL AND CHEMICAL PROPERTIES

13.1 **Physical State at 15°C and 1 atm:** Solid
13.2 **Molecular Weight:** 290.8
13.3 **Boiling Point at 1 atm:** Not pertinent (decomposes)
13.4 **Freezing Point:** Not pertinent
13.5 **Critical Temperature:** Not pertinent
13.6 **Critical Pressure:** Not pertinent
13.7 **Specific Gravity:** 2.05 at 20°C (solid)
13.8 **Liquid Surface Tension:** Not pertinent
13.9 **Liquid-Water Interfacial Tension:** Not pertinent
13.10 **Vapor (Gas) Specific Gravity:** Not pertinent
13.11 **Ratio of Specific Heats of Vapor (Gas):** Not pertinent
13.12 **Latent Heat of Vaporization:** Not pertinent
13.13 **Heat of Combustion:** Not pertinent
13.14 **Heat of Decomposition:** Not pertinent
13.15 **Heat of Solution:** 47 Btu/lb = 26 cal/g = 1.1×10^5 J/kg
13.16 **Heat of Polymerization:** Not pertinent

NOTES

NKS — NICKEL SULFATE

Common Synonyms
Nickelous sulfate

Solid crystals — Pale green — Odorless

Sinks and mixes slowly with water.

Isolate and remove discharged material.
Notify local health and pollution control agencies.

Fire

Not flammable.

Exposure

CALL FOR MEDICAL AID.

SOLID
Irritating to skin and eyes.
Remove contaminated clothing and shoes.
Flush affected areas with plenty of water.
IF IN EYES, hold eyelids open and flush with plenty of water.

Water Pollution

Dangerous to aquatic life in high concentrations.
May be dangerous if it enters water intakes.
Notify local health and wildlife officials.
Notify operators of nearby water intakes.

1. RESPONSE TO DISCHARGE

(See Response Methods Handbook, CG 446-4)

Disperse and flush

2. LABELS

No hazard label required by Code of Federal Regulations

3. CHEMICAL DESIGNATIONS

3.1 **Synonyms:** Nickelous sulfate
3.2 **Coast Guard Compatibility Classification:** Not applicable
3.3 **Chemical Formula:** $NiSO_4$
3.4 **IMCO/United Nations Numerical Designation:** Not listed

4. OBSERVABLE CHARACTERISTICS

4.1 **Physical State (as shipped):** Solid
4.2 **Color:** Pale green
4.3 **Odor:** Odorless

5. HEALTH HAZARDS

5.1 **Personal Protective Equipment:** Goggles; gloves.
5.2 **Symptoms Following Exposure:** Dermatitis.
5.3 **Treatment for Exposure:** Wash all affected parts with plenty of water.
5.4 **Toxicity by Inhalation (Threshold Limit Value):** 1 mg/m^3
5.5 **Short-Term Inhalation Limits:** Data not available
5.6 **Toxicity by Ingestion:** Data not available
5.7 **Late Toxicity:** Nose and lung cancer.
5.8 **Vapor (Gas) Irritant Characteristics:** None
5.9 **Liquid or Solid Irritant Characteristics:** Repeated contact can cause dermatitis.
5.10 **Odor Threshold:** Not pertinent

6. FIRE HAZARDS

6.1 **Flash Point:** Not flammable
6.2 **Flammable Limits in Air:** Not flammable
6.3 **Fire Extinguishing Agents:** Not pertinent
6.4 **Fire Extinguishing Agents Not to be Used:** Not pertinent
6.5 **Special Hazards of Combustion Products:** Not pertinent
6.6 **Behavior in Fire:** Not pertinent
6.7 **Ignition Temperature:** Not flammable
6.8 **Electrical Hazard:** Not pertinent
6.9 **Burning Rate:** Not flammable

7. CHEMICAL REACTIVITY

7.1 **Reactivity with Water:** No reaction
7.2 **Reactivity with Common Materials:** No reaction
7.3 **Stability During Transport:** Stable
7.4 **Neutralizing Agents for Acids and Caustics:** Not pertinent
7.5 **Polymerization:** Not pertinent
7.6 **Inhibitor of Polymerization:** Not pertinent

8. WATER POLLUTION

8.1 **Aquatic Toxicity:**
160 ppm/48 hr/rainbow trout/TL_m/fresh water
13.9 ppm/48 hr/prawn/LC_{50}/salt water
8.2 **Waterfowl Toxicity:** Data not available
8.3 **Biological Oxygen Demand (BOD):** None
8.4 **Food Chain Concentration Potential:** None

9. SELECTED MANUFACTURERS

1. Chemetron Corp.
Inorganic Chemicals Division
1250 Terminal Tower
Cleveland, Ohio 44113
2. Mallinckrodt Chemical Works
Industrial Chemicals Division
Second and Mallinckrodt Sts.
St. Louis, Mo. 63160
3. Shepherd Chemical Co.
4900 Beech St.
Cincinnati, Ohio 45212

10. SHIPPING INFORMATION

10.1 **Grades or Purity:** Technical; reagent
10.2 **Storage Temperature:** Ambient
10.3 **Inert Atmosphere:** No requirement
10.4 **Venting:** Open

11. HAZARD ASSESSMENT CODE

(See Hazard Assessment Handbook, CG 446-3)

SS

12. HAZARD CLASSIFICATIONS

12.1 **Code of Federal Regulations:** Not listed
12.2 **NAS Hazard Rating for Bulk Water Transportation:** Not listed
12.3 **NFPA Hazard Classifications:** Not listed

13. PHYSICAL AND CHEMICAL PROPERTIES

13.1 **Physical State at 15°C and 1 atm:** Solid
13.2 **Molecular Weight:** 154.78
13.3 **Boiling Point at 1 atm:** Decomposes
13.4 **Freezing Point:** Not pertinent
13.5 **Critical Temperature:** Not pertinent
13.6 **Critical Pressure:** Not pertinent
13.7 **Specific Gravity:** 3.68 at 20°C (solid)
13.8 **Liquid Surface Tension:** Not pertinent
13.9 **Liquid-Water Interfacial Tension:** Not pertinent
13.10 **Vapor (Gas) Specific Gravity:** Not pertinent
13.11 **Ratio of Specific Heats of Vapor (Gas):** Not pertinent
13.12 **Latent Heat of Vaporization:** Not pertinent
13.13 **Heat of Combustion:** Not pertinent
13.14 **Heat of Decomposition:** Not pertinent
13.15 **Heat of Solution:** Not pertinent
13.16 **Heat of Polymerization:** Not pertinent

NOTES

NIC — NICOTINE

Common Synonyms			
1-Methyl-2-(3-pyridyl) pyrrolidine 3-(1-methyl-2-pyrrolidyl) pyridine	Liquid	Colorless to brown	Fishy odor
	Mixes with water.		

AVOID CONTACT WITH LIQUID. KEEP PEOPLE AWAY.
Wear rubber overclothing (including gloves).
Stop discharge if possible.
Call fire department.
Isolate and remove discharged material.
Notify local health and pollution control agencies.

Fire

Combustible.
POISONOUS GASES ARE PRODUCED WHEN HEATED.
Wear goggles and self-contained breathing apparatus.
Extinguish with dry chemicals, alcohol foam, or carbon dioxide.
Water may be ineffective on fire.
Cool exposed containers with water.

Exposure

CALL FOR MEDICAL AID.

LIQUID
POISONOUS IF SWALLOWED OR IF SKIN IS EXPOSED.
Irritating to eyes.
Remove contaminated clothing and shoes.
Flush affected areas with plenty of water.
IF IN EYES, hold eyelids open and flush with plenty of water.
IF SWALLOWED and victim is CONSCIOUS, have victim drink water or milk and have victim induce vomiting.
IF SWALLOWED and victim is UNCONSCIOUS OR HAVING CONVULSIONS, do nothing except keep victim warm.

Water Pollution

HARMFUL TO AQUATIC LIFE IN VERY LOW CONCENTRATIONS.
May be dangerous if it enters water intakes.
Notify local health and wildlife officials.
Notify operators of nearby water intakes.

1. RESPONSE TO DISCHARGE

(See Response Methods Handbook, CG 446-4)

Issue warning—poison, water contaminant
Restrict access
Disperse and flush

2. LABEL

3. CHEMICAL DESIGNATIONS

3.1 **Synonyms:** 1-Methyl-2-(3-pyridyl) pyrrolidine; 3-(1-Methyl-2-pyrrolidyl) pyridine
3.2 **Coast Guard Compatibility Classification:** Not applicable
3.3 **Chemical Formula:** $C_{10}H_{14}N_2$
3.4 **IMCO/United Nations Numerical Designation:** 6.1/1654

4. OBSERVABLE CHARACTERISTICS

4.1 **Physical State (as shipped):** Liquid
4.2 **Color:** Colorless to brown
4.3 **Odor:** Fishy; develops apyridine or tobacco-like odor on exposure to air

5. HEALTH HAZARDS

5.1 **Personal Protective Equipment:** Goggles or face shield; rubber gloves; protective clothing
5.2 **Symptoms Following Exposure:** Inhalation causes burning sensation in mouth and throat, nausea, headache, confusion, visual disturbances. Contact with liquid irritates eyes and causes local irritation of skin. Can be absorbed through skin in toxic amounts. Ingestion causes burning of mouth and stomach, vomiting, excitement, faintness, paralysis of lungs.
5.3 **Treatment for Exposure:** *Speed of treatment is important following exposure to this compound. Ingestion of as little as 40 mg can be fatal.* EYES: flush with water for at least 15 min. SKIN: wash thoroughly and immediately with cold water. INGESTION: call for physician at once; give 6—8 tablespoons of activated charcoal as a slurry in water; give artificial respiration if breathing has stopped.
5.4 **Toxicity by Inhalation (Threshold Limit Value):** 0.5 mg/m^3
5.5 **Short-Term Inhalation Limits:** Data not available
5.6 **Toxicity by Ingestion:** Grade 4; oral LD_{50} = 53 mg/kg (rat), 1 mg/kg (human)
5.7 **Late Toxicity:** Birth defects (skeletal) in rats
5.8 **Vapor (Gas) Irritant Characteristics:** Vapors are nonirritating to eyes and throat.
5.9 **Liquid or Solid Irritant Characteristics:** No appreciable hazard. Practically harmless to the skin.
5.10 **Odor Threshold:** Data not available

6. FIRE HAZARDS

6.1 **Flash Point:** Data not available
6.2 **Flammable Limits in Air:** 0.7% 4.0%
6.3 **Fire Extinguishing Agents:** Dry chemical, alcohol foam, carbon dioxide
6.4 **Fire Extinguishing Agents Not to be Used:** Water or foam may cause frothing.
6.5 **Special Hazards of Combustion Products:** Smoke may contain toxic vapors of unburned compound.
6.6 **Behavior in Fire:** Not pertinent
6.7 **Ignition Temperature:** 471°F
6.8 **Electrical Hazard:** Data not available
6.9 **Burning Rate:** Data not available

7. CHEMICAL REACTIVITY

7.1 **Reactivity with Water:** No reaction
7.2 **Reactivity with Common Materials:** No reaction
7.3 **Stability During Transport:** Stable
7.4 **Neutralizing Agents for Acids and Caustics:** Not pertinent
7.5 **Polymerization:** Not pertinent
7.6 **Inhibitor of Polymerization:** Not pertinent

8. WATER POLLUTION

8.1 **Aquatic Toxicity:**
3-29 ppm/*/fish/toxic/fresh water
*Time period not specified.
8.2 **Waterfowl Toxicity:** LD_{50} = 587 mg/kg
8.3 **Biological Oxygen Demand (BOD):** Data not available
8.4 **Food Chain Concentration Potential:** None

9. SELECTED MANUFACTURERS

1. Chemical Formulators, Inc.
   Box 26
   Nitro, W. Va. 25143
2. Eastman Kodak Co.
   Eastman Organic Chemicals
   Rochester, N. Y. 14650
3. Pfaltz and Bauer, Inc.
   126—04 Northern Boulevard
   Flushing, N. Y. 11368

10. SHIPPING INFORMATION

10.1 **Grades or Purity:** 93—98%
10.2 **Storage Temperature:** Ambient
10.3 **Inert Atmosphere:** No requirement
10.4 **Venting:** Pressure-vacuum

11. HAZARD ASSESSMENT CODE

(See Hazard Assessment Handbook, CG 446-3)

A-P-Q

12. HAZARD CLASSIFICATIONS

12.1 **Code of Federal Regulations:** Poisonous, Class B
12.2 **NAS Hazard Rating for Bulk Water Transportation:**

Category	Rating
Fire	1
Health	
Vapor Irritant	1
Liquid or Solid Irritant	1
Poisons	4
Water Pollution	
Human Toxicity	
Aquatic Toxicity	3
Aesthetic Effect	4
Reactivity	
Other Chemicals	3
Water	0
Self-Reaction	0

12.3 **NFPA Hazard Classifications:**

Category	Classification
Health Hazard (Blue)	4
Flammability (Red)	1
Reactivity (Yellow)	0

13. PHYSICAL AND CHEMICAL PROPERTIES

13.1 **Physical State at 15°C and 1 atm:** Liquid
13.2 **Molecular Weight:** 162.2
13.3 **Boiling Point at 1 atm:** (decomposes) 482°F = 250°C = 523°K
13.4 **Freezing Point:** Not pertinent
13.5 **Critical Temperature:** Not pertinent
13.6 **Critical Pressure:** Not pertinent
13.7 **Specific Gravity:** 1.016 at 20°C (liquid)
13.8 **Liquid Surface Tension:** 38.61 dynes/cm = 0.03861 N/m at 20°C
13.9 **Liquid-Water Interfacial Tension:** (est.) 20 dynes/cm = 0.020 N/m at 20°C
13.10 **Vapor (Gas) Specific Gravity:** Not pertinent
13.11 **Ratio of Specific Heats of Vapor (Gas):** Not pertinent
13.12 **Latent Heat of Vaporization:** Not pertinent
13.13 **Heat of Combustion:** −15,836 Btu/lb = −8,798 cal/g = -368.1×10^5 J/kg
13.14 **Heat of Decomposition:** Not pertinent
13.15 **Heat of Solution:** Not pertinent
13.16 **Heat of Polymerization:** Not pertinent

NOTES

NCS — NICOTINE SULFATE

Common Synonyms			
Neutral nicotine sulfate Black Leaf 40 (40% water solution)	Solid or solution	White to light brown solid or colorless solution	Odorless solid or tobacco odored solution
	Mixes with water.		

AVOID CONTACT WITH LIQUID AND SOLID. KEEP PEOPLE AWAY.
Wear goggles and dust respirator.
Stop discharge if possible.
Isolate and remove discharged material.
Notify local health and pollution control agencies.

Fire

Not flammable.
POISONOUS GASES MAY BE PRODUCED WHEN HEATED.

Exposure

CALL FOR MEDICAL AID.

VAPOR OR DUST
POISONOUS IF INHALED.
Move victim to fresh air.
If in eyes, hold eyelids open and flush with plenty of water.
If breathing has stopped, give artificial respiration.
If breathing is difficult, give oxygen.

LIQUID OR SOLID
POISONOUS IF SWALLOWED OR IF SKIN IS EXPOSED.
Irritating to eyes.
Remove contaminated clothing and shoes.
Flush affected areas with plenty of water.
IF IN EYES, hold eyelids open and flush with plenty of water.
IF SWALLOWED and victim is CONSCIOUS, have victim drink water or milk and have victim induce vomiting.
IF SWALLOWED and victim is UNCONSCIOUS OR HAVING CONVULSIONS, do nothing except keep victim warm.

Water Pollution

HARMFUL TO AQUATIC LIFE IN VERY LOW CONCENTRATIONS.
May be dangerous if it enters water intakes.
Notify local health and wildlife officials.
Notify operators of nearby water intakes.

1. RESPONSE TO DISCHARGE

(See Response Methods Handbook, CG 446-4)

Issue warning-poison, water contaminant
Restrict access
Disperse and flush

2. LABEL

3. CHEMICAL DESIGNATIONS

3.1 **Synonyms:** Black Leaf 40 (40% water solution); Neutral nicotine sulfate
3.2 **Coast Guard Compatibility Classification:** Not applicable
3.3 **Chemical Formula:** $(C_{10}H_{14}N_2)_2 \cdot H_2SO_4$
3.4 **IMCO/United Nations Numerical Designation:** 6.1/1658

4. OBSERVABLE CHARACTERISTICS

4.1 **Physical State (as shipped):** Solid or water solution.
4.2 **Color:** Solid is off-white to yellow tan; darkens on exposure to light. Water solution may be colorless.
4.3 **Odor:** Solid is odorless; water solution has odor of tobacco.

5. HEALTH HAZARDS

5.1 **Personal Protective Equipment:** Dust mask; rubber gloves.
5.2 **Symptoms Following Exposure:** Inhalation causes irritation of nose and throat. Liquid irritates eyes and (on prolonged contact) skin. Ingestion causes burning of mouth and stomach, vomiting, excitement, faintness, paralysis of lungs.
5.3 **Treatment for Exposure:** EYES: flush with water. SKIN: wash thoroughly and immediately with water. INGESTION: call for medical aid; induce immediate and repeated vomiting; perform gastric lavage with dilute (1:10,000) solution of potassium permanganate or activated charcoal in water or milk; apply artificial respiration if breathing has stopped.
5.4 **Toxicity by Inhalation (Threshold Limit Value):** Data not available
5.5 **Short-Term Inhalation Limits:** Data not available
5.6 **Toxicity by Ingestion:** Grade 3; oral LD_{50} = 55 mg/kg (rat)
5.7 **Late Toxicity:** Data not available
5.8 **Vapor (Gas) Irritant Characteristics:** Data not available
5.9 **Liquid or Solid Irritant Characteristics:** Data not available
5.10 **Odor Threshold:** Not pertinent

6. FIRE HAZARDS

6.1 **Flash Point:** Nonflammable as solid or water solution.
6.2 **Flammable Limits in Air:** Not pertinent
6.3 **Fire Extinguishing Agents:** Not pertinent
6.4 **Fire Extinguishing Agents Not to be Used:** Not pertinent
6.5 **Special Hazards of Combustion Products:** Toxic decomposition products are released in a fire.
6.6 **Behavior in Fire:** Not pertinent
6.7 **Ignition Temperature:** Not pertinent
6.8 **Electrical Hazard:** Not pertinent
6.9 **Burning Rate:** Not pertinent

7. CHEMICAL REACTIVITY

7.1 **Reactivity with Water:** No reaction
7.2 **Reactivity with Common Materials:** No reaction
7.3 **Stability During Transport:** Stable
7.4 **Neutralizing Agents for Acids and Caustics:** Not pertinent
7.5 **Polymerization:** Not pertinent
7.6 **Inhibitor of Polymerization:** Not pertinent

8. WATER POLLUTION

8.1 **Aquatic Toxicity:**
3-29 ppm/*/fish/toxic/fresh water (for nicotine)
*Time period not specified.
8.2 **Waterfowl Toxicity:** LD_{50} = 587 mg/kg (for nicotine)
8.3 **Biological Oxygen Demand (BOD):** Data not available
8.4 **Food Chain Concentration Potential:** None

9. SELECTED MANUFACTURERS

1. Chemical Formulators, Inc.
   Box 26
   Nitro, W. Va. 25143
2. Prentiss Drug and Chemical Co.
   363 Seventh Ave.
   New York, N. Y. 10001
3. Pfaltz and Bauer, Inc.
   126-04 Northern Boulevard
   Flushing, N. Y. 11368

10. SHIPPING INFORMATION

10.1 **Grades or Purity:** Solid: Commercial; Solution: 40%
10.2 **Storage Temperature:** Ambient
10.3 **Inert Atmosphere:** No requirement
10.4 **Venting:** Open

11. HAZARD ASSESSMENT CODE

(See Hazard Assessment Handbook, CG 446-3)

SS

12. HAZARD CLASSIFICATIONS

12.1 **Code of Federal Regulations:** Poisonous, Class B
12.2 **NAS Hazard Rating for Bulk Water Transportation:** Not listed
12.3 **NFPA Hazard Classifications:** Not listed

13. PHYSICAL AND CHEMICAL PROPERTIES

13.1 **Physical State at 15°C and 1 atm:** Solid
13.2 **Molecular Weight:** 422.5
13.3 **Boiling Point at 1 atm:** Not pertinent (decomposes)
13.4 **Freezing Point:** Not pertinent
13.5 **Critical Temperature:** Not pertinent
13.6 **Critical Pressure:** Not pertinent
13.7 **Specific Gravity:** 1.15 at 20°C (solid)
13.8 **Liquid Surface Tension:** Not pertinent
13.9 **Liquid-Water Interfacial Tension:** Not pertinent
13.10 **Vapor (Gas) Specific Gravity:** Not pertinent
13.11 **Ratio of Specific Heats of Vapor (Gas):** Not pertinent
13.12 **Latent Heat of Vaporization:** Not pertinent
13.13 **Heat of Combustion:** Not pertinent
13.14 **Heat of Decomposition:** Not pertinent
13.15 **Heat of Solution:** Not pertinent
13.16 **Heat of Polymerization:** Not pertinent

NOTES

NTL	# NITRALIN

Common Synonyms			
4-(Methylsulfonyl)-2,6-dinitro N,N-dipropylaniline Planavin	Solid	Light yellow to orange	Mild odor
	Sinks in water.		

AVOID CONTACT WITH SOLID AND DUST. KEEP PEOPLE AWAY.
Wear goggles, self-contained breathing apparatus, and rubber overclothing (including gloves).
Stop discharge if possible.
Call fire department.
Isolate and remove discharged material.
Notify local health and pollution control agencies.

Fire

Combustible.
Irritating gases are produced when heated.
Containers may explode in fire.
Combat fires from safe distance or protected location.
Flood discharge area with water.

Exposure

CALL FOR MEDICAL AID.

DUST
POISONOUS IF INHALED.
Move victim to fresh air.
If breathing is difficult, give oxygen.

SOLID
POISONOUS IF SWALLOWED.
Irritating to skin and eyes.
Remove contaminated clothing and shoes.
Flush affected areas with plenty of water.
IF IN EYES, hold eyelids open and flush with plenty of water.
IF SWALLOWED and victim is CONSCIOUS, have victim drink water or milk and have victim induce vomiting.
IF SWALLOWED and victim is UNCONSCIOUS OR HAVING CONVULSIONS, do nothing except keep victim warm.

Water Pollution

Effect of low concentrations on aquatic life is unknown.
May be dangerous if it enters water intakes.
Notify local health and wildlife officials.
Notify operators of nearby water intakes.

1. RESPONSE TO DISCHARGE

(See Response Methods Handbook, CG 446-4)

Issue warning—poison, water contaminant
Restrict access
Should be removed
Chemical and physical treatment

2. LABELS

No label required by Code of Federal Regulations

3. CHEMICAL DESIGNATIONS

3.1 **Synonyms:** 4-(methylsulfonyl)-2,6-dinitro-N,N—dipropylaniline; Planavin
3.2 **Coast Guard Compatibility Classification:** Not applicable
3.3 **Chemical Formula:** $C_{13}H_{19}N_3O_6S$
3.4 **IMCO/United Nations Numerical Designation:** 6.1/1609

4. OBSERVABLE CHARACTERISTICS

4.1 **Physical State (as shipped):** Solid
4.2 **Color:** Light yellow to orange
4.3 **Odor:** Mild chemical

5. HEALTH HAZARDS

5.1 **Personal Protective Equipment:** Dust mask; rubber gloves.
5.2 **Symptoms Following Exposure:** Dust irritates eyes. Other forms of exposure produce no observable symptoms.
5.3 **Treatment for Exposure:** No cases of clinical toxicity on record. Supportive and symptomatic medical treatment recommended if massive overexposure occurs. EYES: flush with water if irritation occurs. SKIN: wash with soap and water. INGESTION: induce vomiting.
5.4 **Toxicity by Inhalation (Threshold Limit Value):** Data not available
5.5 **Short-Term Inhalation Limits:** Not pertinent
5.6 **Toxicity by Ingestion:** Grade 2; oral $LD_{50} \geq$ 2,000 mg/kg (rat)
5.7 **Late Toxicity:** None observed
5.8 **Vapor (Gas) Irritant Characteristics:** Data not available
5.9 **Liquid or Solid Irritant Characteristics:** Data not available
5.10 **Odor Threshold:** Data not available

6. FIRE HAZARDS

6.1 **Flash Point:** Not pertinent (combustible solid)
6.2 **Flammable Limits in Air:** Not pertinent
6.3 **Fire Extinguishing Agents:** Water
6.4 **Fire Extinguishing Agents Not to be Used:** Data not available
6.5 **Special Hazards of Combustion Products:** Irritating oxides of sulfur and nitrogen are formed in fire.
6.6 **Behavior in Fire:** Decomposes vigorously in a self-sustaining reaction at or above 225°C
6.7 **Ignition Temperature:** 435°F
6.8 **Electrical Hazard:** Not pertinent
6.9 **Burning Rate:** Not pertinent

7. CHEMICAL REACTIVITY

7.1 **Reactivity with Water:** No reaction
7.2 **Reactivity with Common Materials:** No reaction
7.3 **Stability During Transport:** Stable
7.4 **Neutralizing Agents for Acids and Caustics:** Not pertinent
7.5 **Polymerization:** Not pertinent
7.6 **Inhibitor of Polymerization:** Not pertinent

8. WATER POLLUTION

8.1 **Aquatic Toxicity:**
27 ppm/96 hr/rainbow trout/LC_{50}/fresh water
1 ppm/96 hr/oysters/no effect/salt water
8.2 **Waterfowl Toxicity:** No toxicity after 2,000 mg/kg (ducks)
8.3 **Biological Oxygen Demand (BOD):** Data not available
8.4 **Food Chain Concentration Potential:** None

9. SELECTED MANUFACTURERS

Shell Chemical Company
Agricultural Division
2401 Crow Canyon Road
San Ramon, Calif. 94583

10. SHIPPING INFORMATION

10.1 **Grades or Purity:** Technical: 94+%; Wettable powder: 75%; Emulsifiable concentrate
10.2 **Storage Temperature:** Ambient
10.3 **Inert Atmosphere:** No requirement
10.4 **Venting:** Open

11. HAZARD ASSESSMENT CODE

(See Hazard Assessment Handbook, CG 446-3)

II

12. HAZARD CLASSIFICATIONS

12.1 **Code of Federal Regulations:** Not Listed
12.2 **NAS Hazard Rating for Bulk Water Transportation:** Not listed
12.3 **NFPA Hazard Classifications:** Not listed

13. PHYSICAL AND CHEMICAL PROPERTIES

13.1 **Physical State at 15°C and 1 atm:** Solid
13.2 **Molecular Weight:** 345.2
13.3 **Boiling Point at 1 atm:** (decomposes) >437°F = >225°C = >498°K
13.4 **Freezing Point:** 304°F = 151°C = 424°K
13.5 **Critical Temperature:** Not pertinent
13.6 **Critical Pressure:** Not pertinent
13.7 **Specific Gravity:** (est.) >1 at 20°C (solid)
13.8 **Liquid Surface Tension:** Not pertinent
13.9 **Liquid-Water Interfacial Tension:** Not pertinent
13.10 **Vapor (Gas) Specific Gravity:** Not pertinent
13.11 **Ratio of Specific Heats of Vapor (Gas):** Not pertinent
13.12 **Latent Heat of Vaporization:** Not pertinent
13.13 **Heat of Combustion:** Data not available
13.14 **Heat of Decomposition:** –450 Btu/lb = –250 cal/g = -10.5×10^5 J/kg
13.15 **Heat of Solution:** Not pertinent
13.16 **Heat of Polymerization:** Not pertinent

NOTES

NITRIC ACID

Common Synonyms	Watery liquid Colorless to light brown Choking odor Sinks and mixes with water. Harmful vapor is produced.

AVOID CONTACT WITH LIQUID AND VAPOR. Keep people away.
Wear chemical protective suit with self-contained breathing apparatus.
Stop discharge if possible.
Isolate and remove discharged material.
Notify local health and pollution control agencies.

Fire	Not flammable. May cause fire on contact with combustibles. Flammable gas may be formed on contact with metals. Poisonous gases are produced when heated. Wear chemical protective suit with self-contained breathing apparatus. Cool exposed containers with water.
Exposure	CALL FOR MEDICAL AID. VAPOR Will burn eyes, nose and throat. If inhaled, will cause difficult breathing or loss of consciousness. Move to fresh air. If breathing has stopped, give artificial respiration. If breathing is difficult, give oxygen. LIQUID Will burn skin and eyes. Harmful if swallowed. Remove contaminated clothing and shoes. Flush affected areas with plenty of water. IF IN EYES, hold eyelids open and flush with plenty of water. IF SWALLOWED and victim is CONSCIOUS, have victim drink water or milk. DO NOT INDUCE VOMITING.
Water Pollution	HARMFUL TO AQUATIC LIFE IN VERY LOW CONCENTRATIONS. May be dangerous if it enters water intakes. Notify local health and wildlife officials. Notify operators of nearby water intakes.

1. RESPONSE TO DISCHARGE

(See Response Methods Handbook, CG 446-4)

Issue warning—corrosive
Restrict access
Evacuate area
Disperse and flush

2. LABELS

3. CHEMICAL DESIGNATIONS

3.1 **Synonyms:** No common synonyms
3.2 **Coast Guard Compatibility Classification:** Nitric acid
3.3 **Chemical Formula:** HNO_3-H_2O
3.4 **IMCO/United Nations Numerical Designation:** 8.0/2031

4. OBSERVABLE CHARACTERISTICS

4.1 **Physical State (as shipped):** Liquid
4.2 **Color:** Colorless
4.3 **Odor:** Acrid; sweet to acrid

5. HEALTH HAZARDS

5.1 **Personal Protective Equipment:** Air mask; rubber acid suit, hood, boots and gloves; chemical goggles; safety shower and eye bath.
5.2 **Symptoms Following Exposure:** Vapors irritate eyes and respiratory tract; lung injury may not become apparent for several hours following exposure. Liquid may cause severe burns to eyes and skin.
5.3 **Treatment for Exposure:** INHALATION: remove to fresh air, administer artificial respiration if required. INGESTION: drink large volumes of water; do NOT induce vomiting. SKIN OR EYES: flush with water for at least 15 min.
5.4 **Toxicity by Inhalation (Threshold Limit Value):** 2 ppm
5.5 **Short-Term Inhalation Limits:** 15 ppm for 5 min.
5.6 **Toxicity by Ingestion:** Grade 3; LD_{50} 50 to 500 mg/kg
5.7 **Late Toxicity:** None
5.8 **Vapor (Gas) Irritant Characteristics:** 58-68%: Vapor is moderately irritating such that personnel will not usually tolerate moderate or high vapor concentrations. 95%: Vapors cause severe irritation of eye and throat and can cause eye and lung injury. They cannot be tolerated even at low concentrations.
5.9 **Liquid or Solid Irritant Characteristics:** Severe skin irritant. Causes second- and third-degree burns on short contact and is very injurious to the eyes.
5.10 **Odor Threshold:** Data not available

6. FIRE HAZARDS

6.1 **Flash Point:** Not flammable
6.2 **Flammable Limits in Air:** Not flammable
6.3 **Fire Extinguishing Agents:** Use water on adjacent fires.
6.4 **Fire Extinguishing Agents Not to be Used:** Not pertinent
6.5 **Special Hazards of Combustion Products:** May give off poisonous oxides of nitrogen and acid fumes when heated in fires.
6.6 **Behavior in Fire:** Decomposes and gives off poisonous oxides of nitrogen.
6.7 **Ignition Temperature:** Not flammable
6.8 **Electrical Hazard:** Not pertinent
6.9 **Burning Rate:** Not flammable

7. CHEMICAL REACTIVITY

7.1 **Reactivity with Water:** May heat up on mixing, but explosion or formation of steam unlikely.
7.2 **Reactivity with Common Materials:** Very corrosive to wood, paper, cloth and most metals. Toxic red oxides of nitrogen are formed.
7.3 **Stability During Transport:** When heated may give off toxic red oxides of nitrogen.
7.4 **Neutralizing Agents for Acids and Caustics:** Flush with water.
7.5 **Polymerization:** Not pertinent
7.6 **Inhibitor of Polymerization:** Not pertinent

8. WATER POLLUTION

8.1 **Aquatic Toxicity:**
72 ppm/96 hr/mosquito fish/TL_m/fresh water
330-1000 ppm/48 hr/cockle/LC_{50}/salt water
8.2 **Waterfowl Toxicity:** Data not available
8.3 **Biological Oxygen Demand (BOD):** None
8.4 **Food Chain Concentration Potential:** None

9. SELECTED MANUFACTURERS

1. Allied Chemical Corp.
Agricultural Division
Morristown, N. J. 07960
2. E. I. duPont de Nemours & Co., Inc.
Explosives Dept.
Wilmington, Del. 19898
3. Hercules, Inc.
Explosives & Chemical Propulsion Dept.
Bessemer, Ala. 35020

10. SHIPPING INFORMATION

10.1 **Grades or Purity:** Various grades: 52-98%
10.2 **Storage Temperature:** Ambient
10.3 **Inert Atmosphere:** No requirement
10.4 **Venting:** Open or pressure-vacuum

11. HAZARD ASSESSMENT CODE

(See Hazard Assessment Handbook, CG 446-3)

A-P

12. HAZARD CLASSIFICATIONS

12.1 **Code of Federal Regulations:** Oxidizer
12.2 **NAS Hazard Rating for Bulk Water Transportation:**

Category	Rating
Fire	0
Health	
Vapor Irritant	3
Liquid or Solid Irritant	4
Poisons	3
Water Pollution	
Human Toxicity	3
Aquatic Toxicity	3
Aesthetic Effect	2
Reactivity	
Other Chemicals	4
Water	0
Self-Reaction	0

12.3 **NFPA Hazard Classifications:**

Category	Classification
Health Hazard (Blue)	2
Flammability (Red)	0
Reactivity (Yellow)	0 oxy

13. PHYSICAL AND CHEMICAL PROPERTIES

13.1 **Physical State at 15°C and 1 atm:** Liquid
13.2 **Molecular Weight:** Not pertinent
13.3 **Boiling Point at 1 atm:** 192.0°F = 88.9°C = 362.1°K
13.4 **Freezing Point:** −50°F = −45.6°C = 227.6°K
13.5 **Critical Temperature:** Not pertinent
13.6 **Critical Pressure:** Not pertinent
13.7 **Specific Gravity:** 1.49 at 20°C (liquid)
13.8 **Liquid Surface Tension:** Not pertinent
13.9 **Liquid-Water Interfacial Tension:** Not pertinent
13.10 **Vapor (Gas) Specific Gravity:** Not pertinent
13.11 **Ratio of Specific Heats of Vapor (Gas):** (est.) 1.248
13.12 **Latent Heat of Vaporization:** 214 Btu/lb = 119 cal/g = 4.98×10^5 J/kg
13.13 **Heat of Combustion:** Not pertinent
13.14 **Heat of Decomposition:** Not pertinent
13.15 **Heat of Solution:** -205 Btu/lb = -114 cal/g = -4.76×10^5 J/kg
13.16 **Heat of Polymerization:** Not pertinent

NOTES

REVISED 1978

NTX — NITRIC OXIDE

Common Synonyms			
Mononitrogen monoxide Nitrogen monoxide	Compressed gas	Colorless	Sharp unpleasant odor
	Reacts with water. Poisonous red-brown vapor cloud is produced.		

AVOID CONTACT WITH GAS. KEEP PEOPLE AWAY.
Wear goggles and self-contained breathing apparatus.
Stop discharge if possible.
Notify local health and pollution control agencies.

Fire

Not flammable.

Exposure

CALL FOR MEDICAL AID
VAPOR
POISONOUS IF INHALED.
Irritating to eyes, nose and throat.
Move victim to fresh air.
If breathing has stopped, give artificial respiration.
If breathing is difficult, give oxygen.

Water Pollution

Effect of low concentrations on aquatic life is unknown.
May be dangerous if it enters water intakes.
Notify local health and wildlife officials.
Notify operators of nearby water intakes.

1. RESPONSE TO DISCHARGE

(See Response Methods Handbook, CG 446-4)

Issue warning—poison, air contaminant
Restrict access
Evacuate area

2. LABEL

3. CHEMICAL DESIGNATIONS

3.1 **Synonyms:** Mononitrogen monoxide; Nitrogen monoxide
3.2 **Coast Guard Compatibility Classification:** Not applicable
3.3 **Chemical Formula:** NO
3.4 **IMCO/United Nations Numerical Designation:** 2/1660

4. OBSERVABLE CHARACTERISTICS

4.1 **Physical State (as shipped):** Compressed gas
4.2 **Color:** Colorless, but becomes red-brown on contact with air because of formation of nitrogen tetroxide
4.3 **Odor:** (nitrogen tetroxide) Pungent, acrid, mildly irritating

5. HEALTH HAZARDS

5.1 **Personal Protective Equipment:** Self-contained breathing apparatus or gas mask with universal canister
5.2 **Symptoms Following Exposure:** Continued inhalation of low concentrations causes chronic irritation, cough, headache, corrosion of teeth, loss of strength; symptoms from overexposure to higher concentrations (which may be delayed for 6—24 hours) include irritation of nose and throat, tightness in chest, difficult breathing, pallor, loss of consciousness, and death; pulmonary edema occurs; if patient recovers, pneumonia may develop.
5.3 **Treatment for Exposure:** *Get medical attention at once following inhalation of this gas.* INHALATION: if breathing has stopped, give artificial respiration with 100% oxygen; keep victim quiet and warm; keep head and chest lower than hips, to aid in drainage from lungs; alert physician to possibility of delayed pulmonary edema during 6—24 hours.
5.4 **Toxicity by Inhalation (Threshold Limit Value):** 25 ppm
5.5 **Short-Term Inhalation Limits:** $<$100 ppm: 60 min.; $<$150 ppm:30 min.; $<$200 ppm: 5 min.
5.6 **Toxicity by Ingestion:** Not pertinent (gas at normal temperatures)
5.7 **Late Toxicity:** Data not available
5.8 **Vapor (Gas) Irritant Characteristics:** Data not available
5.9 **Liquid or Solid Irritant Characteristics:** Data not available
5.10 **Odor Threshold:** Data not available

6. FIRE HAZARDS

6.1 **Flash Point:** Not pertinent (nonflammable compressed gas)
6.2 **Flammable Limits in Air:** Not pertinent
6.3 **Fire Extinguishing Agents:** Not pertinent
6.4 **Fire Extinguishing Agents Not to be Used:** Not pertinent
6.5 **Special Hazards of Combustion Products:** Not pertinent
6.6 **Behavior in Fire:** Supports combustion, so all fires burn more vigorously.
6.7 **Ignition Temperature:** Not pertinent
6.8 **Electrical Hazard:** Not pertinent
6.9 **Burning Rate:** Not pertinent

7. CHEMICAL REACTIVITY

7.1 **Reactivity with Water:** Reacts with water to form nitric acid. The reaction is not violent.
7.2 **Reactivity with Common Materials:** Reacts rapidly with air to form nitrogen tetroxide; see this compound.
7.3 **Stability During Transport:** Stable
7.4 **Neutralizing Agents for Acids and Caustics:** Flood with water, rinse with sodium bicarbonate or lime solution.
7.5 **Polymerization:** Not pertinent
7.6 **Inhibitor of Polymerization:** Not pertinent

8. WATER POLLUTION

8.1 **Aquatic Toxicity:** Data not available
8.2 **Waterfowl Toxicity:** Data not available
8.3 **Biological Oxygen Demand (BOD):** Not pertinent
8.4 **Food Chain Concentration Potential:** None

9. SELECTED MANUFACTURERS

1. Matheson Gas Products
P. O. Box 85
East Rutherford, N. J. 07073
2. Union Carbide Corporation
Linde Division
Morristown, N. J. 08057
3. Air Products and Chemicals, Inc.
Specialty Gas Department
P. O. Box 538
Allentown, Pa. 18105

10. SHIPPING INFORMATION

10.1 **Grades or Purity:** C. P.: 99+%
10.2 **Storage Temperature:** Cool ambient
10.3 **Inert Atmosphere:** No requirement
10.4 **Venting:** Safety relief. Containers must be in well-ventilated area.

11. HAZARD ASSESSMENT CODE

(See Hazard Assessment Handbook, CG 446-3)

A-C

12. HAZARD CLASSIFICATIONS

12.1 **Code of Federal Regulations:** Poisonous, Class A
12.2 **NAS Hazard Rating for Bulk Water Transportation:** Not listed
12.3 **NFPA Hazard Classifications:** Not listed

13. PHYSICAL AND CHEMICAL PROPERTIES

13.1 **Physical State at 15°C and 1 atm:** Gas
13.2 **Molecular Weight:** 30.0
13.3 **Boiling Point at 1 atm:** $-241.1°F = -151.7°C = 121.5°K$
13.4 **Freezing Point:** $-262.5°F = -163.6°C = 109.6°K$
13.5 **Critical Temperature:** $847°F = 453°C = 180°K$
13.6 **Critical Pressure:** 940 psia = 64 atm = 6.5 MN/m^2
13.7 **Specific Gravity:** Not pertinent
13.8 **Liquid Surface Tension:** Not pertinent
13.9 **Liquid-Water Interfacial Tension:** Not pertinent
13.10 **Vapor (Gas) Specific Gravity:** 1.6 (nitrogen dioxide)
13.11 **Ratio of Specific Heats of Vapor (Gas):** 1.400 at 15°C
13.12 **Latent Heat of Vaporization:** Not pertinent
13.13 **Heat of Combustion:** Not pertinent
13.14 **Heat of Decomposition:** Not pertinent
13.15 **Heat of Solution:** -257 Btu/lb $= -143$ cal/g $= -5.98 \times 10^5$ J/kg
13.16 **Heat of Polymerization:** Not pertinent

NOTES

NAA — NITRILOTRIACETIC ACID AND SALTS

Common Synonyms
Triglycine
NTA
Disodium nitrilotriacetate
Trisodium nitrilotriacetate

Solid — White — Odorless

Sinks and mixes with water

Avoid contact with solid and dust. Keep people away.
Stop discharge if possible.
Isolate and remove discharged material.
Notify local health and pollution control agencies.

Fire

Not flammable.

Exposure

CALL FOR MEDICAL AID.

DUST
Irritating to eyes, nose and throat.
If in eyes, hold eyelids open and flush with plenty of water.
If breathing has stopped, give artificial respiration.
If breathing is difficult, give oxygen.

SOLID
Irritating to skin and eyes.
Remove contaminated clothing and shoes.
Flush affected areas with plenty of water.
IF IN EYES, hold eyelids open and flush with plenty of water.
IF SWALLOWED and victim is CONSCIOUS, have victim drink water or milk.
IF SWALLOWED and victim is UNCONSCIOUS OR HAVING CONVULSIONS, do nothing except keep victim warm.

Water Pollution

Dangerous to aquatic life in high concentrations.
May be dangerous if it enters water intakes.
Notify local health and wildlife officials.
Notify operators of nearby water intakes.

1. RESPONSE TO DISCHARGE

(See Response Methods Handbook, CG 446-4)

Disperse and flush

2. LABELS

No hazard label required by Code of Federal Regulations

3. CHEMICAL DESIGNATIONS

3.1 **Synonyms:** Disodium nitrilotriacetate; NTA; Triglycine; Trisodium nitrilotriacetate
3.2 **Coast Guard Compatibility Classification:** Not listed
3.3 **Chemical Formula:** $C_6H_6O_6NNa_3$, $C_6H_7O_6NNa_2$, $C_6H_9O_6N$
3.4 **IMCO/United Nations Numerical Designation:** Not listed

4. OBSERVABLE CHARACTERISTICS

4.1 **Physical State (as shipped):** Solid
4.2 **Color:** White
4.3 **Odor:** None

5. HEALTH HAZARDS

5.1 **Personal Protective Equipment:** Dust mask; rubber gloves; chemical safety goggles
5.2 **Symptoms Following Exposure:** Toxicity and health hazard of these compounds are low. Contact with eyes causes irritation.
5.3 **Treatment for Exposure:** EYES: flush with plenty of water; treat as mild alkaline irritation with boric acid solution; if eyes are irritated, get medical attention. SKIN: flush with plenty of water.
5.4 **Toxicity by Inhalation (Threshold Limit Value):** Data not available
5.5 **Short-Term Inhalation Limits:** Data not available
5.6 **Toxicity by Ingestion:** Grade 2; LD_{50} 0.5—5 g/kg. Disodium Grade 2; LD_{50} 1.2 g/kg (rat). Trisodium Grade 2; LD_{50} 4 g/kg (rat)
5.7 **Late Toxicity:** Data not available
5.8 **Vapor (Gas) Irritant Characteristics:** Data not available
5.9 **Liquid or Solid Irritant Characteristics:** Data not available
5.10 **Odor Threshold:** Data not available

6. FIRE HAZARDS

6.1 **Flash Point:** Not flammable
6.2 **Flammable Limits in Air:** Not flammable
6.3 **Fire Extinguishing Agents:** Not pertinent
6.4 **Fire Extinguishing Agents Not to be Used:** Not pertinent
6.5 **Special Hazards of Combustion Products:** Not pertinent
6.6 **Behavior in Fire:**
6.7 **Ignition Temperature:** Not pertinent
6.8 **Electrical Hazard:** Not pertinent
6.9 **Burning Rate:** Not pertinent

7. CHEMICAL REACTIVITY

7.1 **Reactivity with Water:** No reaction
7.2 **Reactivity with Common Materials:**
7.3 **Stability During Transport:** Stable
7.4 **Neutralizing Agents for Acids and Caustics:** Not pertinent
7.5 **Polymerization:** Not pertinent
7.6 **Inhibitor of Polymerization:** Not pertinent

8. WATER POLLUTION

8.1 **Aquatic Toxicity:** 340 ppm/24 hr/guppy/lethal conc./fresh water
8.2 **Waterfowl Toxicity:** Data not available
8.3 **Biological Oxygen Demand (BOD):** 0%, 5 days
8.4 **Food Chain Concentration Potential:** None

9. SELECTED MANUFACTURERS

W. R. Grace and Co.
Hampshire Chemical Div.
Poisson Avenue
Nashua, N. H. 03060

10. SHIPPING INFORMATION

10.1 **Grades or Purity:** Commercial, 99.5+%; a water solution of trisodium salt containing 43% solids is also shipped.
10.2 **Storage Temperature:** Ambient
10.3 **Inert Atmosphere:** No requirement
10.4 **Venting:** Open

11. HAZARD ASSESSMENT CODE

(See Hazard Assessment Handbook, CG 446-3)

(Acid) II
(Salts) SS

12. HAZARD CLASSIFICATIONS

12.1 **Code of Federal Regulations:** Not listed
12.2 **NAS Hazard Rating for Bulk Water Transportation:** Not listed
12.3 **NFPA Hazard Classifications:** Not listed

13. PHYSICAL AND CHEMICAL PROPERTIES

13.1 **Physical State at 15°C and 1 atm:** Solid
13.2 **Molecular Weight:** Acid 191; Disodium 253; Trisodium 275
13.3 **Boiling Point at 1 atm:** Not pertinent (decomposes)
13.4 **Freezing Point:** Not pertinent
13.5 **Critical Temperature:** Not pertinent
13.6 **Critical Pressure:** Not pertinent
13.7 **Specific Gravity:** > 1 at 20°C (solid)
13.8 **Liquid Surface Tension:** Not pertinent
13.9 **Liquid-Water Interfacial Tension:** Not pertinent
13.10 **Vapor (Gas) Specific Gravity:** Not pertinent
13.11 **Ratio of Specific Heats of Vapor (Gas):** Not pertinent
13.12 **Latent Heat of Vaporization:** Not pertinent
13.13 **Heat of Combustion:** Not pertinent
13.14 **Heat of Decomposition:** Not pertinent
13.15 **Heat of Solution:** Not pertinent
13.16 **Heat of Polymerization:** Not pertinent

NOTES

NTA | 2-NITROANILINE

Common Synonyms			
o-Nitroaniline 1-Amino-2-nitrobenzene o-Nitraniline Azoic Diazo Component 6	Solid Sinks and mixes slowly with water.	Orange	Musty odor

Avoid contact with solid and dust. Keep people away.
Stop discharge if possible.
Call fire department.
Isolate and remove discharged material.
Notify local health and pollution control agencies.

Fire

Combustible.
POISONOUS GASES MAY BE PRODUCED IN FIRE.
Wear goggles and self-contained breathing apparatus.
Extinguish with water, dry chemicals, foam, or carbon dioxide.
Cool exposed containers with water.

Exposure

CALL FOR MEDICAL AID.

DUST
Irritating to eyes, nose and throat.
If inhaled will cause headache, dizziness, or loss of consciousness.
If in eyes, hold eyelids open and flush with plenty of water.
If breathing has stopped, give artificial respiration.
If breathing is difficult, give oxygen.

SOLID
Irritating to skin and eyes.
If swallowed will cause headache, dizziness, nausea, vomiting or loss of consciousness.
Remove contaminated clothing and shoes.
Flush affected areas with plenty of water.
IF IN EYES, hold eyelids open and flush with plenty of water.
IF SWALLOWED and victim is CONSCIOUS, have victim drink water or milk and have victim induce vomiting.
IF SWALLOWED and victim is UNCONSCIOUS OR HAVING CONVULSIONS, do nothing except keep victim warm.

Water Pollution

HARMFUL TO AQUATIC LIFE IN VERY LOW CONCENTRATIONS.
May be dangerous if it enters water intakes.
Notify local health and wildlife officials.
Notify operators of nearby water intakes.

1. RESPONSE TO DISCHARGE

(See Response Methods Handbook, CG 446-4)

Issue warning—poison, water contaminant
Restrict access
Should be removed
Chemical and physical treatment

2. LABEL

3. CHEMICAL DESIGNATIONS

3.1 **Synonyms:** 1-Amino-2-nitrobenzene; Azoic Diazo Component 6; o-Nitraniline; o-Nitroaniline; ONA

3.2 **Coast Guard Compatibility Classification:** Not listed

3.3 **Chemical Formula:** 1,2-$C_6H_4NO_2NH_2$

3.4 **IMCO/United Nations Numerical Designation:** 6.1/1661

4. OBSERVABLE CHARACTERISTICS

4.1 **Physical State (as shipped):** Solid

4.2 **Color:** Orange

4.3 **Odor:** Musty

5. HEALTH HAZARDS

5.1 **Personal Protective Equipment:** Self-contained breathing apparatus; neoprene gauntlets; safety goggles; rubber or neoprene sealed-tongue work shoes and apron; close-weave cotton coveralls capable of closing at wrist and ankle

5.2 **Symptoms Following Exposure:** Inhalation or ingestion causes headache, nausea, methemoglobinemia, vomiting, weakness, and stupor; cyanosis caused by contact usually develops in 4—6 hrs.; prolonged and excessive exposures may also cause liver damage. Contact with eyes or skin causes irritation; continued exposure may cause same symptoms as inhalation or ingestion.

5.3 **Treatment for Exposure:** INHALATION: remove victim to fresh air; administer oxygen if required; get medical attention. INGESTION: induce vomiting; get medical attention. EYES: flush with water for at least 15 min. SKIN: flush with water, wash with soap and water; be sure that no solid remains under fingernails or in hair.

5.4 **Toxicity by Inhalation (Threshold Limit Value):** Data not available

5.5 **Short-Term Inhalation Limits:** Data not available

5.6 **Toxicity by Ingestion:** Grade 2; LD_{50} 0.5—5 g/kg

5.7 **Late Toxicity:** Data not available

5.8 **Vapor (Gas) Irritant Characteristics:** Data not available

5.9 **Liquid or Solid Irritant Characteristics:** Data not available

5.10 **Odor Threshold:** Data not available

6. FIRE HAZARDS

6.1 **Flash Point:** Not pertinent (combustible solid)

6.2 **Flammable Limits in Air:** Not pertinent

6.3 **Fire Extinguishing Agents:** Water, foam, dry chemical, carbon dioxide

6.4 **Fire Extinguishing Agents Not to be Used:**

6.5 **Special Hazards of Combustion Products:** Toxic oxides of nitrogen may form in fire.

6.6 **Behavior in Fire:**

6.7 **Ignition Temperature:** 970°F

6.8 **Electrical Hazard:** Not pertinent

6.9 **Burning Rate:** Not pertinent

7. CHEMICAL REACTIVITY

7.1 **Reactivity with Water:** No reaction

7.2 **Reactivity with Common Materials:**

7.3 **Stability During Transport:** Stable

7.4 **Neutralizing Agents for Acids and Caustics:** Not pertinent

7.5 **Polymerization:** Not pertinent

7.6 **Inhibitor of Polymerization:** Not pertinent

8. WATER POLLUTION

8.1 **Aquatic Toxicity:**
24 ppm/*/daphnia/threshold toxicity/fresh water
*Time period not specified.

8.2 **Waterfowl Toxicity:** Data not available

8.3 **Biological Oxygen Demand (BOD):** Data not available

8.4 **Food Chain Concentration Potential:** None

9. SELECTED MANUFACTURERS

1. American Aniline Products, Inc.
   Mount Vernon St.
   Lock Haven, Pa. 17745
2. Monsanto Company
   800 North Lindbergh Blvd.
   St. Louis, Mo. 63166
3. Eastman Organic Chemicals
   Rochester, N. Y. 14650

10. SHIPPING INFORMATION

10.1 **Grades or Purity:** Commercial, 100%

10.2 **Storage Temperature:** Ambient

10.3 **Inert Atmosphere:** No requirement

10.4 **Venting:** Open

11. HAZARD ASSESSMENT CODE

(See Hazard Assessment Handbook, CG 446-3)

II

12. HAZARD CLASSIFICATIONS

12.1 **Code of Federal Regulations:** Poisonous solid, Class B

12.2 **NAS Hazard Rating for Bulk Water Transportation:** Not listed

12.3 **NFPA Hazard Classifications:** Not listed

13. PHYSICAL AND CHEMICAL PROPERTIES

13.1 **Physical State at 15°C and 1 atm:** Solid

13.2 **Molecular Weight:** 138.1

13.3 **Boiling Point at 1 atm:** 543°F = 284°C = 557°K

13.4 **Freezing Point:** 160°F = 71°C = 344°K

13.5 **Critical Temperature:** Not pertinent

13.6 **Critical Pressure:** Not pertinent

13.7 **Specific Gravity:** 1.44 at 20°C (solid)

13.8 **Liquid Surface Tension:** Not pertinent

13.9 **Liquid-Water Interfacial Tension:** Not pertinent

13.10 **Vapor (Gas) Specific Gravity:** Not pertinent

13.11 **Ratio of Specific Heats of Vapor (Gas):** Not pertinent

13.12 **Latent Heat of Vaporization:** Not pertinent

13.13 **Heat of Combustion:** −10,000 Btu/lb = −5,550 cal/g = -232×10^5 J/kg

13.14 **Heat of Decomposition:** Not pertinent

13.15 **Heat of Solution:** Not pertinent

13.16 **Heat of Polymerization:** Not pertinent

NOTES

NAL	4-NITROANILINE

Common Synonyms			
p-Nitroaniline 1-Amino-4-nitrobenzene Azoic Diazo Component 37 Fast Red IG base Fast Red GG base PNA	Solid Sinks in water.	Yellow	Mild Odor

AVOID CONTACT WITH SOLID AND DUST. KEEP PEOPLE AWAY.
Wear dust respirator.
Stop discharge if possible.
Call fire department.
Isolate and remove discharged material.
Notify local health and pollution control agencies.

Fire

Combustible.
POISONOUS GASES MAY BE PRODUCED IN FIRE.
Wear goggles and self-contained breathing apparatus.
Extinguish with water, dry chemicals, foam, or carbon dioxide.
Cool exposed containers with water.

Exposure

CALL FOR MEDICAL AID.

DUST
POISONOUS IF INHALED.
If inhaled will cause headache, coughing, difficult breathing, or loss of consciousness.
If in eyes, hold eyelids open and flush with plenty of water.
If breathing has stopped, give artificial respiration.
If breathing is difficult, give oxygen.

SOLID
Irritating to skin and eyes.
If swallowed will cause headache, coughing, or loss of consciousness.
Remove contaminated clothing and shoes.
Flush affected areas with plenty of water.
IF IN EYES, hold eyelids open and flush with plenty of water.
IF SWALLOWED and victim is CONSCIOUS, have victim drink water or milk and have victim induce vomiting.
IF SWALLOWED and victim is UNCONSCIOUS OR HAVING CONVULSIONS, do nothing except keep victim warm.

Water Pollution

HARMFUL TO AQUATIC LIFE IN VERY LOW CONCENTRATIONS.
May be dangerous if it enters water intakes.
Notify local health and wildlife officials.
Notify operators of nearby water intakes.

1. RESPONSE TO DISCHARGE

(See Response Methods Handbook, CG 446-4)

Issue warning—poison, water contaminant
Restrict access
Should be removed
Chemical and physical treatment

2. LABEL

POISON

3. CHEMICAL DESIGNATIONS

3.1 **Synonyms:** 1-Amino-4-nitrobenzene; Azoic Diazo Component 37; Fast Red 2G base; Fast Red GG base; p-Nitraniline; p-Nitroaniline; PNA

3.2 **Coast Guard Compatibility Classification:** Not listed

3.3 **Chemical Formula:** $1,4\text{-}C_6H_4NO_2NH_2$

3.4 **IMCO/United Nations Numerical Designation:** 6.1/1661

4. OBSERVABLE CHARACTERISTICS

4.1 **Physical State (as shipped):** Solid

4.2 **Color:** Yellow

4.3 **Odor:** Faint ammonia; slightly pungent

5. HEALTH HAZARDS

5.1 **Personal Protective Equipment:** Bu. Mines dust canister; rubber gloves; chemical safety goggles; face shield; rubber safety shoes

5.2 **Symptoms Following Exposure:** Inhalation or ingestion causes headache, drowsiness, shortness of breath, nausea, methemoglobinemia, and unconsciousness; fingernails, lips, and ears become bluish; prolonged and excessive exposures may also cause liver damage. Contact with eyes causes irritation and possible corneal damage. Contact with skin causes irritation; continued exposure may cause same symptoms as inhalation or ingestion.

5.3 **Treatment for Exposure:** INHALATION: remove victim to fresh air; administer oxygen if required; get medical attention. INGESTION: induce vomiting; get medical attention. EYES: flush with water for at least 15 min. SKIN: flush with water, wash with soap and water; be sure that no compound remains in the hair or under the fingernails.

5.4 **Toxicity by Inhalation (Threshold Limit Value):** 1 ppm

5.5 **Short-Term Inhalation Limits:** Data not available

5.6 **Toxicity by Ingestion:** Grade 2; LD_{50} 0.5—5 g/kg

5.7 **Late Toxicity:** Data not available

5.8 **Vapor (Gas) Irritant Characteristics:** Data not available

5.9 **Liquid or Solid Irritant Characteristics:** Data not available

5.10 **Odor Threshold:** Data not available

6. FIRE HAZARDS

6.1 **Flash Point:** 329°F O.C. (molten solid)

6.2 **Flammable Limits in Air:** Not pertinent

6.3 **Fire Extinguishing Agents:** Water, foam, dry chemical or carbon dioxide

6.4 **Fire Extinguishing Agents Not to be Used:**

6.5 **Special Hazards of Combustion Products:** Toxic oxides of nitrogen may form in fire.

6.6 **Behavior in Fire:** Melts and burns

6.7 **Ignition Temperature:** Data not available

6.8 **Electrical Hazard:** Not pertinent

6.9 **Burning Rate:** Not pertinent

7. CHEMICAL REACTIVITY

7.1 **Reactivity with Water:** No reaction

7.2 **Reactivity with Common Materials:**

7.3 **Stability During Transport:** Stable

7.4 **Neutralizing Agents for Acids and Caustics:** Not pertinent

7.5 **Polymerization:** Not pertinent

7.6 **Inhibitor of Polymerization:** Not pertinent

8. WATER POLLUTION

8.1 **Aquatic Toxicity:**
24 ppm/*/daphnia/threshold toxicity/fresh water
*Time period not specified.

8.2 **Waterfowl Toxicity:** Data not available

8.3 **Biological Oxygen Demand (BOD):** Data not available

8.4 **Food Chain Concentration Potential:** None

9. SELECTED MANUFACTURERS

1. Monsanto Company
800 North Lindbergh Blvd.
St. Louis, Mo. 63166

2. American Aniline Products, Inc.
Mount Vernon St.
Lock Haven, Pa. 17745

3. Eastman Organic Chemicals
Rochester, N. Y. 14650

10. SHIPPING INFORMATION

10.1 **Grades or Purity:** Technical, 100%

10.2 **Storage Temperature:** Ambient

10.3 **Inert Atmosphere:** No requirement

10.4 **Venting:** Open

11. HAZARD ASSESSMENT CODE

(See Hazard Assessment Handbook, CG 446-3)

II

12. HAZARD CLASSIFICATIONS

12.1 **Code of Federal Regulations:** Poisonous solid, Class B

12.2 **NAS Hazard Rating for Bulk Water Transportation:** Not listed

12.3 **NFPA Hazard Classifications:**

Category	Classification
Health Hazard (Blue)	3
Flammability (Red)	1
Reactivity (Yellow)	0

13. PHYSICAL AND CHEMICAL PROPERTIES

13.1 **Physical State at 15°C and 1 atm:** Solid

13.2 **Molecular Weight:** 138.1

13.3 **Boiling Point at 1 atm:** 636°F = 336°C = 609°K

13.4 **Freezing Point:** 295°F = 146°C = 419°K

13.5 **Critical Temperature:** Not pertinent

13.6 **Critical Pressure:** Not pertinent

13.7 **Specific Gravity:** 1.44 at 20°C (solid)

13.8 **Liquid Surface Tension:** Not pertinent

13.9 **Liquid-Water Interfacial Tension:** Not pertinent

13.10 **Vapor (Gas) Specific Gravity:** Not pertinent

13.11 **Ratio of Specific Heats of Vapor (Gas):** Not pertinent

13.12 **Latent Heat of Vaporization:** Not pertinent

13.13 **Heat of Combustion:** −9,920 Btu/lb = −5,510 cal/g = -231×10^5 J/kg

13.14 **Heat of Decomposition:** Not pertinent

13.15 **Heat of Solution:** Not pertinent

13.16 **Heat of Polymerization:** Not pertinent

NOTES

NTB	NITROBENZENE

Common Synonyms	
Nitrobenzol	Oily liquid Light yellow to brown Almond or shoe polish odor Sinks in water. Freezing point is 41° F.

AVOID CONTACT WITH LIQUID. Keep people away.
Wear chemical protective suit with self-contained breathing apparatus.
Stop discharge if possible.
Call fire department.
Isolate and remove discharged material.
Notify local health and pollution control agencies.

Fire

Combustible.
POISONOUS VAPOR IS PRODUCED WHEN HEATED.
Wear chemical protective suit with self-contained breathing apparatus.
Extinguish with water, dry chemical, foam, or carbon dioxide.
Cool exposed containers in water.

Exposure

CALL FOR MEDICAL AID.

LIQUID
POISONOUS IF SWALLOWED OR IF SKIN IS EXPOSED.
Will burn eyes.
Remove contaminated clothing and shoes.
Flush affected areas with plenty of water.
IF IN EYES, hold eyelids open and flush with plenty of water.
IF SWALLOWED and victim is CONSCIOUS, have victim drink water or milk.

Water Pollution

HARMFUL TO AQUATIC LIFE IN VERY LOW CONCENTRATIONS.
May be dangerous if it enters water intakes.
Notify local health and wildlife officials.
Notify operators of nearby water intakes.

1. RESPONSE TO DISCHARGE

(See Response Methods Handbook, CG 446-4)

Issue warning—poison
Restrict access
Should be removed
Chemical and physical treatment

2. LABEL

3. CHEMICAL DESIGNATIONS

3.1 **Synonyms:** Essence of Mirbane
Nitrobenzol
Oil of Mirbane

3.2 **Coast Guard Compatibility Classification:** Nitrocompounds

3.3 **Chemical Formula:** $C_6H_5NO_2$

3.4 **IMCO/United Nations Numerical Designation:** 6.1/1662

4. OBSERVABLE CHARACTERISTICS

4.1 **Physical State (as shipped):** Liquid

4.2 **Color:** Light greenish-yellow

4.3 **Odor:** Like paste shoe polish

5. HEALTH HAZARDS

5.1 **Personal Protective Equipment:** Respirator approved by U.S. Bureau of Mines for organic vapors; rubber gloves; splashproof goggles; eyewash fountain, safety shower and medical oxygen supply.

5.2 **Symptoms Following Exposure:** Highly toxic when absorbed through the skin, inhaled as vapor, or swallowed. First symptoms are a blue discoloration of the lips, nails, and skin. Acute poisoning produces headache, giddiness, weakness, nausea, vomiting, and coma.

5.3 **Treatment for Exposure:** Remove to fresh air and call a physician at once. In case of contact, immediately flush skin or eyes with plenty of water for at least 15 min. If cyanosis (blue discoloration) is present, shower with soap and warm water, with special attention to scalp and fingernails. Administer oxygen until physician arrives.

5.4 **Toxicity by Inhalation (Threshold Limit Value):** 1 ppm

5.5 **Short-Term Inhalation Limits:** 10 ppm for 30 min.

5.6 **Toxicity by Ingestion:** Grade 3; 50-500 mg/kg (dog)

5.7 **Late Toxicity:** Data not available

5.8 **Vapor (Gas) Irritant Characteristics:** Vapor is moderately irritating such that personnel will not usually tolerate moderate or high vapor concentrations.

5.9 **Liquid or Solid Irritant Characteristics:** Causes smarting of the skin and first-degree burns on short exposure; may cause secondary burns on long exposure.

5.10 **Odor Threshold:** 5.94 ppm

6. FIRE HAZARDS

6.1 **Flash Point:** 171°F O.C.; 190°F C.C.

6.2 **Flammable Limits in Air:** 1.8% LEL (UEL not available)

6.3 **Fire Extinguishing Agents:** Water, foam, carbon dioxide, or dry chemical

6.4 **Fire Extinguishing Agents Not to be Used:** Not pertinent

6.5 **Special Hazards of Combustion Products:** Not pertinent

6.6 **Behavior in Fire:** Not pertinent

6.7 **Ignition Temperature:** 924°F

6.8 **Electrical Hazard:** Not pertinent

6.9 **Burning Rate:** 2.9 mm/min.

7. CHEMICAL REACTIVITY

7.1 **Reactivity with Water:** No reaction

7.2 **Reactivity with Common Materials:** No reaction

7.3 **Stability During Transport:** Stable

7.4 **Neutralizing Agents for Acids and Caustics:** Not pertinent

7.5 **Polymerization:** Not pertinent

7.6 **Inhibitor of Polymerization:** Not pertinent

8. WATER POLLUTION

8.1 **Aquatic Toxicity:** 20 ppm/6 hr/minnow/lethal/fresh water

8.2 **Waterfowl Toxicity:** Data not available

8.3 **Biological Oxygen Demand (BOD):** 0%, 5 days

8.4 **Food Chain Concentration Potential:** None

9. SELECTED MANUFACTURERS

1. American Cyanamid Co.
Organic Chemicals Division
Bound Brook, N. J. 08805
2. E. I. duPont de Nemours & Co., Inc.
Explosives Dept.
Wilmington, Del. 19898
3. First Chemical Corp.
656 North State St.
Jackson, Miss. 39205

10. SHIPPING INFORMATION

10.1 **Grades or Purity:** Technical: 99.5-100%

10.2 **Storage Temperature:** Ambient

10.3 **Inert Atmosphere:** No requirement

10.4 **Venting:** Open (flame arrester)

11. HAZARD ASSESSMENT CODE

(See Hazard Assessment Handbook, CG 446-3)

A-X-Y

12. HAZARD CLASSIFICATIONS

12.1 **Code of Federal Regulations:** Poisonous liquid, Class B

12.2 **NAS Hazard Rating for Bulk Water Transportation:**

Category	Rating
Fire	1
Health	
Vapor Irritant	3
Liquid or Solid Irritant	2
Poisons	4
Water Pollution	
Human Toxicity	3
Aquatic Toxicity	3
Aesthetic Effect	3
Reactivity	
Other Chemicals	2
Water	0
Self-Reaction	1

12.3 **NFPA Hazard Classifications:**

Category	Classification
Health Hazard (Blue)	3
Flammability (Red)	2
Reactivity (Yellow)	0

13. PHYSICAL AND CHEMICAL PROPERTIES

13.1 **Physical State at 15°C and 1 atm:** Liquid

13.2 **Molecular Weight:** 123.11

13.3 **Boiling Point at 1 atm:** 411.6°F = 210.9°C = 484.1°K

13.4 **Freezing Point:** 41.2°F = 5.1°C = 278.3°K

13.5 **Critical Temperature:** 836°F = 447°C = 720°K

13.6 **Critical Pressure:** 700 psia = 47.62 atm = 4.824 MN/m²

13.7 **Specific Gravity:** 1.204 at 20°C (liquid)

13.8 **Liquid Surface Tension:** 43.9 dynes/cm = 0.0439 N/m at 20°C

13.9 **Liquid-Water Interfacial Tension:** 25.66 dynes/cm = 0.02566 N/m at 20°C

13.10 **Vapor (Gas) Specific Gravity:** Not pertinent

13.11 **Ratio of Specific Heats of Vapor (Gas):** Not pertinent

13.12 **Latent Heat of Vaporization:** 150 Btu/lb = 85 cal/g = 3.6×10^5 J/kg

13.13 **Heat of Combustion:** −10,420 Btu/lb = −5,791 cal/g = -242.5×10^5 J/kg

13.14 **Heat of Decomposition:** Not pertinent

13.15 **Heat of Solution:** Not pertinent

13.16 **Heat of Polymerization:** Not pertinent

NOTES

REVISED 1978

NTE NITROETHANE

Common Synonyms	Liquid	Colorless	Fruity Odor
	May float or sink in water.		

Shut off ignition sources. Call fire department.
Avoid contact with liquid and vapor. Keep people away.
Stop discharge if possible.
Stay upwind. Use water spray to "knock down" vapor.
Isolate and remove discharged material.
Notify local health and pbllution control agencies.

Fire

Combustible.
POISONOUS GASES MAY BE PRODUCED IN FIRE.
Containers may explode in fire.
Flashback along vapor trail may occur.
Vapor may explode if ignited in an enclosed area.
Wear goggles and self-contained breathing apparatus.
Extinguish with dry chemicals, foam or carbon dioxide.
Water may be ineffective on fire.
Cool exposed containers with water.

Exposure

CALL FOR MEDICAL AID.

VAPOR
Irritating to eyes, nose and throat.
If inhaled will cause coughing or difficult breathing.
If in eyes, hold eyelids open and flush with plenty of water.
If breathing has stopped, give artificial respiration.
If breathing is difficult, give oxygen.

LIQUID
Irritating to skin and eyes.
If swallowed will cause nausea and vomiting.
Remove contaminated clothing and shoes.
Flush affected areas with plenty of water.
IF IN EYES, hold eyelids open and flush with plenty of water.
IF SWALLOWED and victim is CONSCIOUS, have victim drink water or milk.
IF SWALLOWED and victim is UNCONSCIOUS OR HAVING CONVULSIONS, do nothing except keep victim warm.

Water Pollution

Effect of low concentrations on aquatic life is unknown.
Fouling to shoreline.
May be dangerous if it enters water intakes.
Notify local health and wildlife officials.
Notify operators of nearby water intakes.

1. RESPONSE TO DISCHARGE

(See Response Methods Handbook, CG 446-4)

Issue warning—high flammability
Restrict access
Disperse and flush

2. LABEL

Red

3. CHEMICAL DESIGNATIONS

3.1 **Synonyms:** No common synonyms
3.2 **Coast Guard Compatibility Classification:** Not listed
3.3 **Chemical Formula:** $CH_3CH_2NO_2$
3.4 **IMCO/United Nations Numerical Designation:** Not listed

4. OBSERVABLE CHARACTERISTICS

4.1 **Physical State (as shipped):** Liquid
4.2 **Color:** Colorless
4.3 **Odor:** Mild, fruity; somewhat disagreeable, chloroform-like

5. HEALTH HAZARDS

5.1 **Personal Protective Equipment:** Supplied air or self-contained respirator; goggles or face shield; rubber gloves
5.2 **Symptoms Following Exposure:** Inhalation causes moderate irritation of respiratory tract. Ingestion causes irritation of mouth and stomach. Contact with liquid causes irritation of eyes and mild irritation of skin.
5.3 **Treatment for Exposure:** INHALATION: in case of pulmonary symptoms, give bed rest and oxygen; obtain medical attention at once. INGESTION: give large amount of water. EYES or SKIN: flush with water.
5.4 **Toxicity by Inhalation (Threshold Limit Value):** 100 ppm
5.5 **Short-Term Inhalation Limits:** Data not available
5.6 **Toxicity by Ingestion:** Grade 2; oral LD_{50} = 860 mg/kg (mouse)
5.7 **Late Toxicity:** Data not available
5.8 **Vapor (Gas) Irritant Characteristics:** Vapors cause moderate irritation such that personnel will find high concentration unpleasant. The effect is temporary.
5.9 **Liquid or Solid Irritant Characteristics:** Minimum hazard. If spilled on clothing and allowed to remain, may cause smarting and reddening of skin.
5.10 **Odor Threshold:** 163 ppm

6. FIRE HAZARDS

6.1 **Flash Point:** 105°F O.C.; 87°F C.C.
6.2 **Flammable Limits in Air:** 3.4% (LFL)
6.3 **Fire Extinguishing Agents:** Foam, carbon dioxide, or dry chemical
6.4 **Fire Extinguishing Agents Not to be Used:** Water may be ineffective; "alcohol" foam is not effective.
6.5 **Special Hazards of Combustion Products:** Toxic oxides of nitrogen may form in fire.
6.6 **Behavior in Fire:**
6.7 **Ignition Temperature:** 778°F
6.8 **Electrical Hazard:** Data not available
6.9 **Burning Rate:** Data not available

7. CHEMICAL REACTIVITY

7.1 **Reactivity with Water:** No reaction
7.2 **Reactivity with Common Materials:** May attack some forms of plastics
7.3 **Stability During Transport:** Stable
7.4 **Neutralizing Agents for Acids and Caustics:** Not pertinent
7.5 **Polymerization:** Not pertinent
7.6 **Inhibitor of Polymerization:** Not pertinent

8. WATER POLLUTION

8.1 **Aquatic Toxicity:** Data not available
8.2 **Waterfowl Toxicity:** Data not available
8.3 **Biological Oxygen Demand (BOD):** Data not available
8.4 **Food Chain Concentration Potential:** None

9. SELECTED MANUFACTURERS

1. Commercial Solvents Corp.
   245 Park Avenue
   New York, N. Y. 10017
2. Aldrich Chemical Co.
   940 W. Saint Paul Ave.
   Milwaukee, Wis. 53233
3. Pfaltz and Bauer, Inc.
   375 Fairfield Ave.
   Stamford, Conn. 06902

10. SHIPPING INFORMATION

10.1 **Grades or Purity:** Commercial, 92.5+%
10.2 **Storage Temperature:** Ambient
10.3 **Inert Atmosphere:** No requirement
10.4 **Venting:** Open (flame arrester)

11. HAZARD ASSESSMENT CODE

(See Hazard Assessment Handbook, CG 446-3)

A-P-Q-T-U-X-Y

12. HAZARD CLASSIFICATIONS

12.1 **Code of Federal Regulations:** Flammable liquid
12.2 **NAS Hazard Rating for Bulk Water Transportation:**

Category	Rating
Fire	3
Health	
Vapor Irritant	2
Liquid or Solid Irritant	1
Poisons	1
Water Pollution	
Human Toxicity	2
Aquatic Toxicity	
Aesthetic Effect	2
Reactivity	
Other Chemicals	1
Water	0
Self-Reaction	4

12.3 **NFPA Hazard Classifications:**

Category	Classification
Health Hazard (Blue)	1
Flammability (Red)	3
Reactivity (Yellow)	3

13. PHYSICAL AND CHEMICAL PROPERTIES

13.1 **Physical State at 15°C and 1 atm:** Liquid
13.2 **Molecular Weight:** 75.07
13.3 **Boiling Point at 1 atm:** 237°F = 114°C = 387°K
13.4 **Freezing Point:** −130°F = −90°C = 183°K
13.5 **Critical Temperature:** Data not available
13.6 **Critical Pressure:** Data not available
13.7 **Specific Gravity:** 1.05 at 20°C (liquid)
13.8 **Liquid Surface Tension:** 31.3 dynes/cm = 0.0313 N/m at 20°C
13.9 **Liquid-Water Interfacial Tension:** Not pertinent
13.10 **Vapor (Gas) Specific Gravity:** 2.6
13.11 **Ratio of Specific Heats of Vapor (Gas):** (est.) 1.115 at 20°C
13.12 **Latent Heat of Vaporization:** 211 Btu/lb = 117 cal/g = 4.90×10^5 J/kg
13.13 **Heat of Combustion:** −7,720 Btu/lb = −4,290 cal/g = -179×10^5 J/kg
13.14 **Heat of Decomposition:** Not pertinent
13.15 **Heat of Solution:** Not pertinent
13.16 **Heat of Polymerization:** Not pertinent

NOTES

NXX — NITROGEN, LIQUEFIED

Common Synonyms	Liquefied gas Colorless Odorless Floats and boils on water.

Avoid contact with liquid.

Fire	Not flammable.
Exposure	Call for medical aid. VAPOR Not harmful. In high concentrations may cause dizziness, difficult breathing, or loss of consciousness. LIQUID Will cause frostbite. Flush affected areas with plenty of water. DO NOT RUB AFFECTED AREAS.
Water Pollution	Not harmful to aquatic life.

1. RESPONSE TO DISCHARGE

(See Response Methods Handbook, CG 446-4)

Restrict access

2. LABEL

Green

3. CHEMICAL DESIGNATIONS

3.1 **Synonyms:** Liquid nitrogen
3.2 **Coast Guard Compatibility Classification:** Not applicable
3.3 **Chemical Formula:** N_2
3.4 **IMCO/United Nations Numerical Designation:** 2/2040

4. OBSERVABLE CHARACTERISTICS

4.1 **Physical State (as shipped):** Liquefied gas
4.2 **Color:** Colorless to faint yellow
4.3 **Odor:** None

5. HEALTH HAZARDS

5.1 **Personal Protective Equipment:** Safety glasses or face shield; insulated gloves; long sleeves; trousers worn outside boots or over high-top shoes to shed spilled liquid; self-contained breathing apparatus where insufficient air is present.
5.2 **Symptoms Following Exposure:** Inhalation can cause asphyxiation, if atmosphere does not contain oxygen; dizziness, unconsciousness, or even death can result. Contact of liquid with skin or eyes causes frostbite burns.
5.3 **Treatment for Exposure:** INHALATION: remove to fresh air; apply artificial respiration if breathing has stopped; call physician. EYES: treat for frostbite burns caused by liquid. SKIN: treat for frostbite; soak in lukewarm water.
5.4 **Toxicity by Inhalation (Threshold Limit Value):** Non-toxic
5.5 **Short-Term Inhalation Limits:** Not pertinent
5.6 **Toxicity by Ingestion:** Not pertinent
5.7 **Late Toxicity:** None
5.8 **Vapor (Gas) Irritant Characteristics:** None
5.9 **Liquid or Solid Irritant Characteristics:** Frostbite
5.10 **Odor Threshold:** Not pertinent

6. FIRE HAZARDS

6.1 **Flash Point:** Not pertinent (nonflammable compressed gas)
6.2 **Flammable Limits in Air:** Not pertinent
6.3 **Fire Extinguishing Agents:** Not pertinent
6.4 **Fire Extinguishing Agents Not to be Used:** Not pertinent
6.5 **Special Hazards of Combustion Products:** Not pertinent
6.6 **Behavior in Fire:** Containers may explode when heated.
6.7 **Ignition Temperature:** Not pertinent
6.8 **Electrical Hazard:** Not pertinent
6.9 **Burning Rate:** Not pertinent

7. CHEMICAL REACTIVITY

7.1 **Reactivity with Water:** Heat of water will vigorously vaporize liquid nitrogen.
7.2 **Reactivity with Common Materials:** No chemical reaction. Low temperature may cause brittleness in rubber and plastics.
7.3 **Stability During Transport:** Stable
7.4 **Neutralizing Agents for Acids and Caustics:** Not pertinent
7.5 **Polymerization:** Not pertinent
7.6 **Inhibitor of Polymerization:** Not pertinent

8. WATER POLLUTION

8.1 **Aquatic Toxicity:** None
8.2 **Waterfowl Toxicity:** None
8.3 **Biological Oxygen Demand (BOD):** None
8.4 **Food Chain Concentration Potential:** None

9. SELECTED MANUFACTURERS

1. Union Carbide Corporation
Linde Division
Morristown, N. J. 08057
2. Airco, Inc.
575 Mountain Ave.
Murray Hill, N. J. 07974
3. Liquid Carbonic Corporation
135 S. Lasalle Street
Chicago, Ill. 60603

10. SHIPPING INFORMATION

10.1 **Grades or Purity:** 99.5+%
10.2 **Storage Temperature:** −320°F
10.3 **Inert Atmosphere:** No requirement
10.4 **Venting:** Open

11. HAZARD ASSESSMENT CODE

(See Hazard Assessment Handbook, CG 446-3)

A-C-D-F-G

12. HAZARD CLASSIFICATIONS

12.1 **Code of Federal Regulations:** Nonflammable compressed gas
12.2 **NAS Hazard Rating for Bulk Water Transportation:** Not listed
12.3 **NFPA Hazard Classifications:** Not listed

13. PHYSICAL AND CHEMICAL PROPERTIES

13.1 **Physical State at 15°C and 1 atm:** Gas
13.2 **Molecular Weight:** 28.0
13.3 **Boiling Point at 1 atm:** −320.1°F = −195.6°C = 77.6°K
13.4 **Freezing Point:** −354°F = −215°C = 58°K
13.5 **Critical Temperature:** −232.6°F = −147.0°C = 126.2°K
13.6 **Critical Pressure:** 493 psia = 33.5 atm = 3.40 MN/m^2
13.7 **Specific Gravity:** 0.807 at −195.5°C (liquid)
13.8 **Liquid Surface Tension:** 8.3 dynes/cm = 0.083 N/m at −193°C
13.9 **Liquid-Water Interfacial Tension:** Not pertinent
13.10 **Vapor (Gas) Specific Gravity:** 0.965
13.11 **Ratio of Specific Heats of Vapor (Gas):** 1.3962
13.12 **Latent Heat of Vaporization:** 95 Btu/lb = 53 cal/g = 2.2×10^5 J/kg
13.13 **Heat of Combustion:** Not pertinent
13.14 **Heat of Decomposition:** Not pertinent
13.15 **Heat of Solution:** Not pertinent
13.16 **Heat of Polymerization:** Not pertinent

NOTES

NITROGEN TETROXIDE

Common Synonyms			
Nitrogen peroxide Nitrogen dioxide Red oxide of nitrogen Oxides of nitrogen Dinitrogen tetroxide	Liquefied compressed gas	Red-brown	Sharp, unpleasant chemical odor
	Sinks and reacts with water. Poisonous brown vapor is produced.		

AVOID CONTACT WITH LIQUID AND VAPOR. Keep people away.
Wear goggles, self-contained breathing apparatus, and rubber overclothing (including gloves).
Stop discharge if possible.
Evacuate area in case of large discharge.
Notify local health and pollution control agencies.

Fire

Not flammable.
May cause fire and explode on contact with combustibles.
POISONOUS GASES ARE PRODUCED IN FIRE AND WHEN HEATED.
Wear goggles, self-contained breathing apparatus, and rubber overclothing (including gloves).
Flood discharge area with water.
Stop flow of gas or liquid.
Cool exposed containers with water.

Exposure

CALL FOR MEDICAL AID.

VAPOR
POISONOUS IF INHALED.
Irritating to eyes.
Move to fresh air.
If breathing has stopped, give artificial respiration.
If breathing is difficult, give oxygen.

LIQUID
Will burn skin and eyes.
POISONOUS IF SWALLOWED.
Remove contaminated clothing and shoes.
Flush affected areas with plenty of water.
IF IN EYES, hold eyelids open and flush with plenty of water.
IF SWALLOWED and victim is CONSCIOUS, have victim drink water or milk.
DO NOT INDUCE VOMITING.

Water Pollution

HARMFUL TO AQUATIC LIFE IN VERY LOW CONCENTRATIONS.
May be dangerous if it enters water intakes.
Notify local health and wildlife officials.
Notify operators of nearby water intakes.

1. RESPONSE TO DISCHARGE

(See Response Methods Handbook, CG 446-4)

Issue warning—poison, air contaminant, water contaminant, corrosive
Restrict access
Evacuate area

2. LABELS

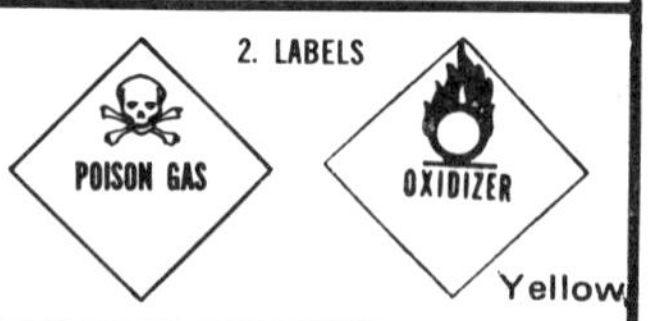

3. CHEMICAL DESIGNATIONS

3.1 **Synonyms:** Dinitrogen tetroxide; Nitrogen peroxide; Nitrogen dioxide; Oxides of nitrogen; Red oxide of nitrogen
3.2 **Coast Guard Compatibility Classification:** Not applicable
3.3 **Chemical Formula:** N_2O_4
3.4 **IMCO/United Nations Numerical Designation:** 2/1067

4. OBSERVABLE CHARACTERISTICS

4.1 **Physical State (as shipped):** Liquid
4.2 **Color:** Red-brown at ambient temperatures; colorless below about 14°F
4.3 **Odor:** Pungent, acrid, mildly irritating

5. HEALTH HAZARDS

5.1 **Personal Protective Equipment:** Rubber gloves; safety goggles and face shield; protective clothing; acid gas canister respirator or self-contained breathing apparatus.
5.2 **Symptoms Following Exposure:** Very concentrated fumes produce coughing, choking, headache, nausea, pain in chest and abdomen; otherwise, few symtoms appear at time of exposure. After symptom-free period of 5—72 hours, pulmonary edema gradually develops, causing fatigue, restlessness, coughing, difficulty in breathing, frothy expectoration, mental confusion, lethargy, bluish skin, and weak, rapid pulse. Since NOX interferes with gas exchange in lungs, unconsciousness and death by asphyxiation can result, usually within a few hours after onset of pulmonary edema.
5.3 **Treatment for Exposure:** INHALATION: remove patient to fresh air and have him breathe as deeply as possible; call a doctor; enforce complete rest for 24—48 hours; keep warm; give oxygen if coughing starts; physician may administer morphine (10 mg.) EYES AND SKIN: flush with water for at least 15 min.
5.4 **Toxicity by Inhalation (Threshold Limit Value):** 5 ppm
5.5 **Short-Term Inhalation Limits:** 25 ppm for 5 min.
5.6 **Toxicity by Ingestion:** Data not available
5.7 **Late Toxicity:** Data not available
5.8 **Vapor (Gas) Irritant Characteristics:** Vapors cause severe irritation of eyes and throat and can cause eye and lung injury. They cannot be tolerated even at low concentrations.
5.9 **Liquid or Solid Irritant Characteristics:** Severe skin irritant. Causes second- and third-degree burns on short contact and is very injurious to the eyes.
5.10 **Odor Threshold:** 5 ppm

6. FIRE HAZARDS

6.1 **Flash Point:** Not flammable
6.2 **Flammable Limits in Air:** Not flammable
6.3 **Fire Extinguishing Agents:** Stop flow of gas
6.4 **Fire Extinguishing Agents Not to be Used:** Not pertinent
6.5 **Special Hazards of Combustion Products:** Produces toxic gas when heated.
6.6 **Behavior in Fire:** Does not burn, but supports combustion of combustible materials such as wood. May cause fire or explode on contact with other materials.
6.7 **Ignition Temperature:** Not flammable
6.8 **Electrical Hazard:** Data not available
6.9 **Burning Rate:** Not flammable

7. CHEMICAL REACTIVITY

7.1 **Reactivity with Water:** Dissolves to form nitric acid and nitric oxide. Nitric oxide reacts with air to form more nitrogen tetroxide.
7.2 **Reactivity with Common Materials:** Very corrosive to metals when wet. Reacts vigorously with combustible materials such as wood.
7.3 **Stability During Transport:** Stable
7.4 **Neutralizing Agents for Acids and Caustics:** Flush with water, then use soda ash or lime.
7.5 **Polymerization:** Not pertinent
7.6 **Inhibitor of Polymerization:** Not pertinent

8. WATER POLLUTION

8.1 **Aquatic Toxicity:**
72 ppm/96 hr/mosquito fish/TL_m/fresh water
330—1000 ppm/48 hr/cockle/LC_{50}/salt water
8.2 **Waterfowl Toxicity:** Data not available
8.3 **Biological Oxygen Demand (BOD):** None
8.4 **Food Chain Concentration Potential:** None

9. SELECTED MANUFACTURERS

Hercules, Inc.
Explosives and Chemical Propulsion Dept.
Hercules, Calif. 94547

10. SHIPPING INFORMATION

10.1 **Grades or Purity:**
10.2 **Storage Temperature:** Ambient. Storage and transfer structures shall be equipped with mechanical ventilation systems.
10.3 **Inert Atmosphere:** No requirement
10.4 **Venting:** Pressure relief valves on containers

11. HAZARD ASSESSMENT CODE

(See Hazard Assessment Handbook, CG 446-3)

A-I-J-O

12. HAZARD CLASSIFICATIONS

12.1 **Code of Federal Regulations:** Poisonous gas or liquid, Class A
12.2 **NAS Hazard Rating for Bulk Water Transportation:**

Category	Rating
Fire	0
Health	
Vapor Irritant	4
Liquid or Solid Irritant	4
Poisons	4
Water Pollution	
Human Toxicity	3
Aquatic Toxicity	
Aesthetic Effect	3
Reactivity	
Other Chemicals	4
Water	2
Self-Reaction	0

12.3 **NFPA Hazard Classifications:**

Category	Classification
Health Hazard (Blue)	3
Flammability (Red)	0
Reactivity (Yellow)	0 oxy

13. PHYSICAL AND CHEMICAL PROPERTIES

13.1 **Physical State at 15°C and 1 atm:** Liquid or gas
13.2 **Molecular Weight:** 92.02
13.3 **Boiling Point at 1 atm:** 70.1°F = 21.2°C = 294°K
13.4 **Freezing Point:** 11.8°F = -11.2°C = 262°K
13.5 **Critical Temperature:** 317.0°F = 158.2°C = 431.4°K
13.6 **Critical Pressure:** 1470 psia = 100 atm = 10.1 MN/m^2
13.7 **Specific Gravity:** 1.45 at 20°C (liquid)
13.8 **Liquid Surface Tension:** Not pertinent
13.9 **Liquid-Water Interfacial Tension:** Not pertinent
13.10 **Vapor (Gas) Specific Gravity:** 3.2
13.11 **Ratio of Specific Heats of Vapor (Gas):** (est.) 1.262
13.12 **Latent Heat of Vaporization:** 178 Btu/lb = 99.1 cal/g = 4.15×10^5 J/kg
13.13 **Heat of Combustion:** Not pertinent
13.14 **Heat of Decomposition:** Not pertinent
13.15 **Heat of Solution:** Data not available
13.16 **Heat of Polymerization:** Not pertinent

NOTES

REVISED 1978

NITROMETHANE

Common Synonyms			
Nitrocarbol	Watery liquid	Colorless	Strong odor
	Sinks and mixes slowly with water. Irritating vapor is produced.		

Stop discharge if possible. Keep people away.
Call fire department.
Avoid contact with vapor.
Isolate and remove discharged material.
Notify local health and pollution control agencies.

Fire

FLAMMABLE
Flashback along vapor trail may occur.
Vapor may explode if ignited in an enclosed area.
Wear self-contained breathing apparatus.
Evacuate surrounding area in large fire.
Combat fires from behind barrier.
Extinguish with water, foam, dry chemicals, or carbon dioxide.
Cool exposed containers with water.
Continue cooling after fire has been extinguished.

Exposure

CALL FOR MEDICAL AID.

VAPOR
Irritating to eyes, nose and throat.
Harmful if inhaled.
Move to fresh air.
If breathing has stopped, give artificial respiration.
If breathing is difficult, give oxygen.

LIQUID
No appreciable harm to skin or eyes.
Harmful if swallowed.
Flush affected areas with plenty of water.
IF SWALLOWED and victim is CONSCIOUS, have victim drink water or milk.

Water Pollution

Effect of low concentrations on aquatic life is unknown.
May be dangerous if it enters water intakes.
Notify local health and wildlife officials.
Notify operators of nearby water intakes.

1. RESPONSE TO DISCHARGE

(See Response Methods Handbook, CG 446-4)

Disperse and flush

2. LABEL

Red

3. CHEMICAL DESIGNATIONS

3.1 **Synonyms:** Nitrocarbol
3.2 **Coast Guard Compatibility Classification:** Not applicable
3.3 **Chemical Formula:** CH_3NO_2
3.4 **IMCO/United Nations Numerical Designation:** 3.3/1261

4. OBSERVABLE CHARACTERISTICS

4.1 **Physical State (as shipped):** Liquid
4.2 **Color:** Colorless
4.3 **Odor:** Fairly strong characteristic

5. HEALTH HAZARDS

5.1 **Personal Protective Equipment:** Air mask (do NOT use organic canister); goggles.
5.2 **Symptoms Following Exposure:** Liquid may dry out skin and cause irritation.
5.3 **Treatment for Exposure:** EYES: flush with water for at least 15 min. SKIN: flush with water.
5.4 **Toxicity by Inhalation (Threshold Limit Value):** 100 ppm
5.5 **Short-Term Inhalation Limits:** Data not available
5.6 **Toxicity by Ingestion:** Grade 2; LD_{50} 0.5 to 5 g/kg (rat)
5.7 **Late Toxicity:** Data not available
5.8 **Vapor (Gas) Irritant Characteristics:** Vapors cause a slight smarting of the eyes or respiratory system if present in high concentrations. The effect is temporary.
5.9 **Liquid or Solid Irritant Characteristics:** No appreciable hazard. Practically harmless to the skin.
5.10 **Odor Threshold:** Less than 200 ppm

6. FIRE HAZARDS

6.1 **Flash Point:** 110°F O.C.; 95°F C.C.
6.2 **Flammable Limits in Air:** 7.3% (LEL)
6.3 **Fire Extinguishing Agents:** Water, foam, dry chemical, carbon dioxide
6.4 **Fire Extinguishing Agents Not to be Used:** Not pertinent
6.5 **Special Hazards of Combustion Products:** Not pertinent
6.6 **Behavior in Fire:** Containers may explode
6.7 **Ignition Temperature:** 785°F
6.8 **Electrical Hazard:** Not pertinent
6.9 **Burning Rate:** 1.1 mm/min.

7. CHEMICAL REACTIVITY

7.1 **Reactivity with Water:** No reaction
7.2 **Reactivity with Common Materials:** Wet material corrodes steel and copper, but the reaction is slow.
7.3 **Stability During Transport:** Considered stable, but may become sensitized by organic bases (amines) and some metal oxides, such as lead pigments.
7.4 **Neutralizing Agents for Acids and Caustics:** Not pertinent
7.5 **Polymerization:** Not pertinent
7.6 **Inhibitor of Polymerization:** Not pertinent

8. WATER POLLUTION

8.1 **Aquatic Toxicity:** Data not available
8.2 **Waterfowl Toxicity:** Data not available
8.3 **Biological Oxygen Demand (BOD):** Data not available
8.4 **Food Chain Concentration Potential:** None

9. SELECTED MANUFACTURERS

Commercial Solvents Corp.
245 Park Ave.
New York, N. Y. 10017

10. SHIPPING INFORMATION

10.1 **Grades or Purity:** 95-99%
10.2 **Storage Temperature:** Ambient
10.3 **Inert Atmosphere:** No requirement
10.4 **Venting:** Open or pressure-vacuum

11. HAZARD ASSESSMENT CODE

(See Hazard Assessment Handbook, CG 446-3)

A-P-Q-R-S

12. HAZARD CLASSIFICATIONS

12.1 **Code of Federal Regulations:** Flammable Liquid
12.2 **NAS Hazard Rating for Bulk Water Transportation:** Not listed
12.3 **NFPA Hazard Classifications:**

Category	Classification
Health Hazard (Blue)	1
Flammability (Red)	3
Reactivity (Yellow)	3

13. PHYSICAL AND CHEMICAL PROPERTIES

13.1 **Physical State at 15°C and 1 atm:** Liquid
13.2 **Molecular Weight:** 61.04
13.3 **Boiling Point at 1 atm:** 214.2°F = 101.2°C = 374.4°K
13.4 **Freezing Point:** −20°F = −29°C = 244°K
13.5 **Critical Temperature:** 599 °F = 315°C = 588°K
13.6 **Critical Pressure:** 915.8 psia = 62.3 atm = 6.311 MN/m^2
13.7 **Specific Gravity:** 1.139 at 20°C (liquid)
13.8 **Liquid Surface Tension:** 37.0 dynes/cm = 0.0370 N/m at 20°C
13.9 **Liquid-Water Interfacial Tension:** Not pertinent
13.10 **Vapor (Gas) Specific Gravity:** Not pertinent
13.11 **Ratio of Specific Heats of Vapor (Gas):** 1.172
13.12 **Latent Heat of Vaporization:** 241 Btu/lb = 134 cal/g = 5.61 × 10^5 J/kg
13.13 **Heat of Combustion:** −4531 Btu/lb = −2517 cal/g = −105.4 × 10^5 J/kg
13.14 **Heat of Decomposition:** Not pertinent
13.15 **Heat of Solution:** (est.) −9 Btu/lb = −5 cal/g = −0.2 × 10^5 J/kg
13.16 **Heat of Polymerization:** Not pertinent

NOTES

REVISED 1978

NTP	2-NITROPHENOL

Common Synonyms		
o-Nitrophenol 2-Hydroxynitrobenzene ONP	Solid	Yellow
	Sinks and mixes slowly with water.	

Stop discharge if possible. Keep people away.
Avoid contact with solid and dust.
Isolate and remove discharged material.
Stop discharge if possible.

Fire

Combustible.
POISONOUS GASES MAY BE PRODUCED IN FIRE.
Wear goggles and self-contained breathing apparatus.
Extinguish with water, dry chemicals, foam, or carbon dioxide.

Exposure

CALL FOR MEDICAL AID.

DUST
Irritating to eyes, nose and throat.
If inhaled will cause headache or loss of consciousness.
If in eyes, hold eyelids open and flush with plenty of water.
If breathing has stopped, give artificial respiration.
If breathing is difficult, give oxygen.

SOLID
Irritating to skin and eyes.
If swallowed will cause headache, nausea, or loss of consciousness.
Remove contaminated clothing and shoes.
Flush affected areas with plenty of water.
IF IN EYES, hold eyelids open and flush with plenty of water.
IF SWALLOWED and victim is CONSCIOUS, have victim drink water or milk.
IF SWALLOWED and victim is UNCONSCIOUS OR HAVING CONVULSIONS, do nothing except keep victim warm.

Water Pollution

HARMFUL TO AQUATIC LIFE IN VERY LOW CONCENTRATIONS.
May be dangerous if it enters water intakes.
Notify local health and wildlife officials.
Notify operators of nearby water intakes.

1. RESPONSE TO DISCHARGE

(See Response Methods Handbook, CG 446-4)

Issue warning—water contaminant
Should be removed
Chemical and physical treatment

2. LABELS

No hazard label required by Code of Federal Regulations

3. CHEMICAL DESIGNATIONS

3.1 **Synonyms:** 2-Hydroxynitrobenzene; o-Nitrophenol; ONP
3.2 **Coast Guard Compatibility Classification:** Not listed
3.3 **Chemical Formula:** 1,2-$HOC_6H_4NO_2$
3.4 **IMCO/United Nations Numerical Designation:** 6.1/1663

4. OBSERVABLE CHARACTERISTICS

4.1 **Physical State (as shipped):** Solid
4.2 **Color:** Yellow
4.3 **Odor:** Peculiar aromatic

5. HEALTH HAZARDS

5.1 **Personal Protective Equipment:** Self-contained breathing apparatus for fumes; rubber gloves; goggles
5.2 **Symptoms Following Exposure:** Inhalation or ingestion causes headache, drowsiness, nausea, and blue color in lips, ears, and fingernails (cyanosis). Contact with eyes causes irritation. Can be absorbed through the intact skin to give same symptoms as for inhalation.
5.3 **Treatment for Exposure:** INHALATION or INGESTION: remove victim to fresh air; give artificial respiration; call a doctor if symptoms persist. EYES: flush with water for at least 15 min.; get medical attention. SKIN: cleanse thoroughly with soap and water.
5.4 **Toxicity by Inhalation (Threshold Limit Value):** Data not available
5.5 **Short-Term Inhalation Limits:** Data not available
5.6 **Toxicity by Ingestion:** Grade 2; oral LD_{50} = 1,297 mg/kg (rat)
5.7 **Late Toxicity:** Data not available
5.8 **Vapor (Gas) Irritant Characteristics:** Data not available
5.9 **Liquid or Solid Irritant Characteristics:** Data not available
5.10 **Odor Threshold:** Data not available

6. FIRE HAZARDS

6.1 **Flash Point:** Not pertinent (combustible solid)
6.2 **Flammable Limits in Air:** Not pertinent
6.3 **Fire Extinguishing Agents:** Water, foam, dry chemical, carbon dioxide
6.4 **Fire Extinguishing Agents Not to be Used:**
6.5 **Special Hazards of Combustion Products:** Toxic and irritating fumes of unburned material and oxides of nitrogen can form in fire.
6.6 **Behavior in Fire:**
6.7 **Ignition Temperature:** Data not available
6.8 **Electrical Hazard:** Not pertinent
6.9 **Burning Rate:** Not pertinent

7. CHEMICAL REACTIVITY

7.1 **Reactivity with Water:** No reaction
7.2 **Reactivity with Common Materials:**
7.3 **Stability During Transport:** Stable
7.4 **Neutralizing Agents for Acids and Caustics:** Not pertinent
7.5 **Polymerization:** Not pertinent
7.6 **Inhibitor of Polymerization:** Not pertinent

8. WATER POLLUTION

8.1 **Aquatic Toxicity:** 46.3–51.6 ppm/48 hr/bluegill/TL_m/fresh water
8.2 **Waterfowl Toxicity:** Data not available
8.3 **Biological Oxygen Demand (BOD):** 4.2% in 0.94 days, acclimated culture
8.4 **Food Chain Concentration Potential:** None

9. SELECTED MANUFACTURERS

1. Monsanto Company
800 North Lindbergh Blvd.
St. Louis, Mo. 63166
2. J. T. Baker Chemical Co.
Phillipsburg, N. J. 08865
3. Eastman Organic Chemicals
Rochester, N. Y. 14650

10. SHIPPING INFORMATION

10.1 **Grades or Purity:** Commercial; Pure
10.2 **Storage Temperature:** Ambient
10.3 **Inert Atmosphere:** No requirement
10.4 **Venting:** Open

11. HAZARD ASSESSMENT CODE

(See Hazard Assessment Handbook, CG 446-3)

II

12. HAZARD CLASSIFICATIONS

12.1 **Code of Federal Regulations:** Not listed
12.2 **NAS Hazard Rating for Bulk Water Transportation:** Not listed
12.3 **NFPA Hazard Classifications:** Not listed

13. PHYSICAL AND CHEMICAL PROPERTIES

13.1 **Physical State at 15°C and 1 atm:** Solid
13.2 **Molecular Weight:** 139.1
13.3 **Boiling Point at 1 atm:** 417°F = 214°C = 487°K
13.4 **Freezing Point:** 111°F = 44°C = 313°K
13.5 **Critical Temperature:** Not pertinent
13.6 **Critical Pressure:** Not pertinent
13.7 **Specific Gravity:** 1.49 at 20°C (solid)
13.8 **Liquid Surface Tension:** Not pertinent
13.9 **Liquid-Water Interfacial Tension:** Not pertinent
13.10 **Vapor (Gas) Specific Gravity:** Not pertinent
13.11 **Ratio of Specific Heats of Vapor (Gas):** Not pertinent
13.12 **Latent Heat of Vaporization:** Not pertinent
13.13 **Heat of Combustion:** −8,910 Btu/lb = −4,950 cal/g = -207×10^5 J/kg
13.14 **Heat of Decomposition:** Not pertinent
13.15 **Heat of Solution:** Not pertinent
13.16 **Heat of Polymerization:** Not pertinent

NOTES

NPH	# 4-NITROPHENOL

Common Synonyms p-Nitrophenol 4-Hydroxynitrobenzene PNP	Solid	Yellow to brown	Sweet odor
	Sinks and mixes with water.		

Stop discharge if possible. Keep people away.
Avoid contact with solid and dust.
Isolate and remove discharged material.
Notify local health and pollution control agencies.

Fire	Combustible. POISONOUS GASES MAY BE PRODUCED IN FIRE. Wear goggles and self-contained breathing apparatus. Extinguish with water, dry chemicals, foam, or carbon dioxide.
Exposure	CALL FOR MEDICAL AID. DUST Irritating to eyes, nost and throat. If inhaled will cause headache or loss of consciousness. If in eyes, hold eyelids open and flush with plenty of water. If breathing has stopped, give artificial respiration. If breathing is difficult, give oxygen. SOLID Irritating to skin and eyes. If swallowed will cause headache, nausea, or loss of consciousness. Remove contaminated clothing and shoes. Flush affected areas with plenty of water. IF IN EYES, hold eyelids open and flush with plenty of water. IF SWALLOWED and victim is CONSCIOUS, have victim drink water or milk. IF SWALLOWED and victim is UNCONSCIOUS OR HAVING CONVULSIONS, do nothing except keep victim warm.
Water Pollution	Effect of low concentrations on aquatic life is unknown. May be dangerous if it enters water intakes. Notify local health and wildlife officials. Notify operators of nearby water intakes.

1. RESPONSE TO DISCHARGE

(See Response Methods Handbook, CG 446-4)

Issue warning—water contaminant
Should be removed
Chemical and physical treatment

2. LABELS

No hazard label required by Code of Federal Regulations

3. CHEMICAL DESIGNATIONS

3.1 **Synonyms:** 4-Hydroxynitrobenzene; p-Nitrophenol; PNP
3.2 **Coast Guard Compatibility Classification:** Not listed
3.3 **Chemical Formula:** 1,4-$HOC_6H_4NO_2$
3.4 **IMCO/United Nations Numerical Designation:** 6.1/1663

4. OBSERVABLE CHARACTERISTICS

4.1 **Physical State (as shipped):** Solid
4.2 **Color:** Yellow to brown
4.3 **Odor:** Slight characteristic, sweet

5. HEALTH HAZARDS

5.1 **Personal Protective Equipment:** Butyl rubber gloves; side-shield safety glasses; dust mask or self-contained breathing apparatus
5.2 **Symptoms Following Exposure:** Inhalation or ingestion causes headache, drowsiness, nausea, and blue color in lips, ears, and fingernails (cyanosis). Contact with eyes or skin causes irritation; can be absorbed through skin to give same symptoms as for inhalation.
5.3 **Treatment for Exposure:** INHALATION or INGESTION: remove victim to fresh air; give artificial respiration; call a doctor if symptoms persist. EYES: flush with water for at least 15 min.; get medical attention. SKIN: flush with water and wash well with soap and water.
5.4 **Toxicity by Inhalation (Threshold Limit Value):** Data not available
5.5 **Short-Term Inhalation Limits:** Data not available
5.6 **Toxicity by Ingestion:** Grade 3; LD_{50} 50—500 mg/kg
5.7 **Late Toxicity:** Data not available
5.8 **Vapor (Gas) Irritant Characteristics:** Data not available
5.9 **Liquid or Solid Irritant Characteristics:** Data not available
5.10 **Odor Threshold:** Odorless

6. FIRE HAZARDS

6.1 **Flash Point:** Not pertinent (combustible solid)
6.2 **Flammable Limits in Air:** Not pertinent
6.3 **Fire Extinguishing Agents:** Water, foam, dry chemical, carbon dioxide
6.4 **Fire Extinguishing Agents Not to be Used:**
6.5 **Special Hazards of Combustion Products:** Toxic oxides of nitrogen and fumes of unburned material may form in fires.
6.6 **Behavior in Fire:** Decomposes violently at 279°C and will burn even in absence of air
6.7 **Ignition Temperature:** Data not available
6.8 **Electrical Hazard:** Not pertinent
6.9 **Burning Rate:** Not pertinent

7. CHEMICAL REACTIVITY

7.1 **Reactivity with Water:** No reaction
7.2 **Reactivity with Common Materials:**
7.3 **Stability During Transport:** Stable
7.4 **Neutralizing Agents for Acids and Caustics:** Not pertinent
7.5 **Polymerization:** Not pertinent
7.6 **Inhibitor of Polymerization:** Not pertinent

8. WATER POLLUTION

8.1 **Aquatic Toxicity:** Data not available
8.2 **Waterfowl Toxicity:** Data not available
8.3 **Biological Oxygen Demand (BOD):** Data not available
8.4 **Food Chain Concentration Potential:** None

9. SELECTED MANUFACTURERS

1. E. I. du Pont de Nemours & Co., Inc.
   Wilmington, Del. 19898
2. Monsanto Company
   800 North Lindbergh Blvd.
   St. Louis, Mo. 63166
3. Aldrich Chemical Co.
   940 W. Saint Paul Ave.
   Milwaukee, Wis. 53233

10. SHIPPING INFORMATION

10.1 **Grades or Purity:** Technical; Pure
10.2 **Storage Temperature:** Ambient
10.3 **Inert Atmosphere:** No requirement
10.4 **Venting:** Open

11. HAZARD ASSESSMENT CODE

(See Hazard Assessment Handbook, CG 446-3)

II-SS

12. HAZARD CLASSIFICATIONS

12.1 **Code of Federal Regulations:** Not listed
12.2 **NAS Hazard Rating for Bulk Water Transportation:** Not listed
12.3 **NFPA Hazard Classifications:**

Category	Classification
Health Hazard (Blue)	3
Flammability (Red)	1
Reactivity (Yellow)	0

13. PHYSICAL AND CHEMICAL PROPERTIES

13.1 **Physical State at 15°C and 1 atm:** Solid
13.2 **Molecular Weight:** 139.1
13.3 **Boiling Point at 1 atm:** Not pertinent (decomposes)
13.4 **Freezing Point:** 235°F = 113°C = 386°K
13.5 **Critical Temperature:** Not pertinent
13.6 **Critical Pressure:** Not pertinent
13.7 **Specific Gravity:** 1.48 at 20°C (solid)
13.8 **Liquid Surface Tension:** Not pertinent
13.9 **Liquid-Water Interfacial Tension:** Not pertinent
13.10 **Vapor (Gas) Specific Gravity:** Not pertinent
13.11 **Ratio of Specific Heats of Vapor (Gas):** Not pertinent
13.12 **Latent Heat of Vaporization:** Not pertinent
13.13 **Heat of Combustion:** −8,870 Btu/lb = −4,930 cal/g = -206×10^5 J/kg
13.14 **Heat of Decomposition:** Not pertinent
13.15 **Heat of Solution:** Not pertinent
13.16 **Heat of Polymerization:** Not pertinent

NOTES

NPP	# 2-NITROPROPANE

Common Synonyms	Liquid	Colorless	Mild, fruity odor
Isonitropropane 2-NP sec-Nitropropane	May float or sink in water.		

Shut off ignition sources. Call fire department.
Avoid contact with liquid and vapor. Keep people away.
Stop discharge if possible.
Stay upwind. Use water spray to "knock down" vapor.
Isolate and remove discharged material.
Notify local health and pollution control agencies.

Fire

Combustible.
Containers may explode in fire.
Flashback along vapor trail may occur.
Vapor may explode if ignited in an enclosed area.
Extinguish with dry chemicals, foam or carbon dioxide.
Water may be ineffective on fire.
Cool exposed containers with water.

Exposure

CALL FOR MEDICAL AID.

VAPOR
Irritating to eyes, nose and throat.
If inhaled will cause headache, dizziness, coughing, or difficult breathing.
If in eyes, hold eyelids open and flush with plenty of water.
If breathing has stopped, give artificial respiration.
If breathing is difficult, give oxygen.

LIQUID
Irritating to skin and eyes.
If swallowed will cause nausea,and vomiting.
Remove contaminated clothing and shoes.
Flush affected areas with plenty of water.
IF IN EYES, hold eyelids open and flush with plenty of water.
IF SWALLOWED and victim is CONSCIOUS, have victim drink water or milk and have victim induce vomiting.
IF SWALLOWED and victim is UNCONSCIOUS OR HAVING CONVULSIONS, do nothing except keep victim warm.

Water Pollution

Effect of low concentrations on aquatic life is unknown.
Fouling to shoreline.
May be dangerous if it enters water intakes.
Notify local health and wildlife officials.
Notify operators of nearby water intakes.

1. RESPONSE TO DISCHARGE

(See Response Methods Handbook, CG 446-4)

Issue warning—high flammability
Restrict access
Disperse and flush

2. LABEL

Red

3. CHEMICAL DESIGNATIONS

3.1 **Synonyms:** Isonitropropane; sec-Nitropropane; 2-NP
3.2 **Coast Guard Compatibility Classification:** Nitrocompound
3.3 **Chemical Formula:** $CH_3CH(NO_2)CH_3$
3.4 **IMCO/United Nations Numerical Designation:** Not listed

4. OBSERVABLE CHARACTERISTICS

4.1 **Physical State (as shipped):** Liquid
4.2 **Color:** Colorless
4.3 **Odor:** Mild, fruity

5. HEALTH HAZARDS

5.1 **Personal Protective Equipment:** Self-contained breathing apparatus; goggles or face shield; rubber gloves
5.2 **Symptoms Following Exposure:** Inhalation causes respiratory tract irritation, headache, dizziness, nausea, and diarrhea. Ingestion causes irritation of mouth and stomach. Contact with liquid irritates eyes and causes mild irritation of skin.
5.3 **Treatment for Exposure:** INHALATION: in case of pulmonary symptoms, enforce bed rest and give oxygen; get medical attention at once. INGESTION: give large amount of water and induce vomiting. EYES or SKIN: flush with water.
5.4 **Toxicity by Inhalation (Threshold Limit Value):** 25 ppm
5.5 **Short-Term Inhalation Limits:** Data not available
5.6 **Toxicity by Ingestion:** Grade 2; oral rat LD_{50} = 720 mg/kg
5.7 **Late Toxicity:** Causes liver cancer in rats
5.8 **Vapor (Gas) Irritant Characteristics:** Vapors cause a slight smarting of the eyes or respiratory system if present in high concentrations. The effect is temporary.
5.9 **Liquid or Solid Irritant Characteristics:** Minimum hazard. If spilled on clothing and allowed to remain, may cause smarting and reddening of skin.
5.10 **Odor Threshold:** 300 ppm

6. FIRE HAZARDS

6.1 **Flash Point:** 100°F O.C.; 82°F C.C.
6.2 **Flammable Limits in Air:** 2.6% (LFL)
6.3 **Fire Extinguishing Agents:** Foam, dry chemical, carbon dioxide
6.4 **Fire Extinguishing Agents Not to be Used:** "Alcohol" foam; water may be ineffective.
6.5 **Special Hazards of Combustion Products:** Toxic oxides of nitrogen may form in fire.
6.6 **Behavior in Fire:**
6.7 **Ignition Temperature:** 802°F
6.8 **Electrical Hazard:** Data not available
6.9 **Burning Rate:** Data not available

7. CHEMICAL REACTIVITY

7.1 **Reactivity with Water:** No reaction
7.2 **Reactivity with Common Materials:** May attack some forms of plastics
7.3 **Stability During Transport:** Stable
7.4 **Neutralizing Agents for Acids and Caustics:** Not pertinent
7.5 **Polymerization:** Not pertinent
7.6 **Inhibitor of Polymerization:** Not pertinent

8. WATER POLLUTION

8.1 **Aquatic Toxicity:** Data not available
8.2 **Waterfowl Toxicity:** Data not available
8.3 **Biological Oxygen Demand (BOD):** Data not available
8.4 **Food Chain Concentration Potential:** None

9. SELECTED MANUFACTURERS

1. Commercial Solvents Corp.
   245 Park Avenue
   New York, N. Y. 10017
2. Aldrich Chemical Co.
   940 W. Saint Paul Ave.
   Milwaukee, Wis. 53233
3. Eastman Organic Chemicals
   Rochester, N. Y. 14650

10. SHIPPING INFORMATION

10.1 **Grades or Purity:** Technical, 94+%
10.2 **Storage Temperature:** Ambient
10.3 **Inert Atmosphere:** No requirement
10.4 **Venting:** Open (flame arrester)

11. HAZARD ASSESSMENT CODE

(See Hazard Assessment Handbook, CG 446-3)

A-P-Q-T-U-X-Y

12. HAZARD CLASSIFICATIONS

12.1 **Code of Federal Regulations:** Flammable liquid
12.2 **NAS Hazard Rating for Bulk Water Transportation:**

Category	Rating
Fire	3
Health	
Vapor Irritant	1
Liquid or Solid Irritant	1
Poisons	1
Water Pollution	
Human Toxicity	2
Aquatic Toxicity	3
Aesthetic Effect	2
Reactivity	
Other Chemicals	3
Water	0
Self-Reaction	4

12.3 **NFPA Hazard Classifications:**

Category	Classification
Health Hazard (Blue)	1
Flammability (Red)	3
Reactivity (Yellow)	3

13. PHYSICAL AND CHEMICAL PROPERTIES

13.1 **Physical State at 15°C and 1 atm:** Liquid
13.2 **Molecular Weight:** 89.09
13.3 **Boiling Point at 1 atm:** 248.5°F = 120.3°C = 393.5°K
13.4 **Freezing Point:** −132°F = −91°C = 182°K
13.5 **Critical Temperature:** Data not available
13.6 **Critical Pressure:** Data not available
13.7 **Specific Gravity:** 0.99 at 20°C (liquid)
13.8 **Liquid Surface Tension:** 30 dynes/cm = 0.030 N/m at 20°C
13.9 **Liquid-Water Interfacial Tension:** Data not available
13.10 **Vapor (Gas) Specific Gravity:** 3.06 at 16°C
13.11 **Ratio of Specific Heats of Vapor (Gas):** 1.090 at 20°C
13.12 **Latent Heat of Vaporization:** 178 Btu/lb = 99 cal/g = 4.1×10^5 J/kg
13.13 **Heat of Combustion:** −9,650 Btu/lb = −5,360 cal/g = -224×10^5 J/kg
13.14 **Heat of Decomposition:** Not pertinent
13.15 **Heat of Solution:** Not pertinent
13.16 **Heat of Polymerization:** Not pertinent

NOTES

NTC	NITROSYL CHLORIDE

Common Synonyms	Liquefied compressed gas — Orange red — Choking odor Liquid sinks and reacts with water. Poisonous visible vapor cloud is produced.

AVOID CONTACT WITH LIQUID AND VAPOR. Keep people away.
Wear goggles, self-contained breathing apparatus, and rubber overclothing (including gloves).
Stop discharge if possible.
Evacuate area in case of large discharge.
Stay upwind and use water spray to "knock down" vapor.
Isolate and remove discharged material.
Notify local health and pollution control agencies.

Fire

Not flammable.
POISONOUS GASES ARE PRODUCED WHEN HEATED.
Wear goggles and self-contained breathing apparatus.

Exposure

CALL FOR MEDICAL AID.

VAPOR
POISONOUS IF INHALED OR IF SKIN IS EXPOSED.
Irritating to eyes.
Move to fresh air.
If breathing has stopped, give artificial respiration.
If breathing is difficult, give oxygen.

LIQUID
POISONOUS IF SWALLOWED OR IF SKIN IS EXPOSED.
Irritating to eyes.
Remove contaminated clothing and shoes.
Flush affected areas with plenty of water.
IF IN EYES, hold eyelids open and flush with plenty of water.

Water Pollution

HARMFUL TO AQUATIC LIFE IN VERY LOW CONCENTRATIONS.
May be dangerous if it enters water intakes.
Notify local health and wildlife officials.
Notify operators of nearby water intakes.

1. RESPONSE TO DISCHARGE

(See Response Methods Handbook, CG 446-4)

Issue warning—poison
Restrict access
Disperse and flush

2. LABEL

Green

3. CHEMICAL DESIGNATIONS

3.1 **Synonyms:** No common synonyms
3.2 **Coast Guard Compatibility Classification:** Not applicable
3.3 **Chemical Formula:** NOCl
3.4 **IMCO/United Nations Numerical Designation:** 2.0/1069

4. OBSERVABLE CHARACTERISTICS

4.1 **Physical State (as shipped):** Liquefied gas
4.2 **Color:** Yellow to red
4.3 **Odor:** Irritating, choking

5. HEALTH HAZARDS

5.1 **Personal Protective Equipment:** Self-contained breathing apparatus (approved mask may be used for short exposures only); rubberized clothing; gloves; shoes; chemical goggles.
5.2 **Symptoms Following Exposure:** Gas is highly toxic. Inhalation causes severe irritation of respiratory tract and damage to mucous membranes. Delayed effects, which include severe pulmonary edema, may not be apparent for several hours.
5.3 **Treatment for Exposure:** INHALATION: remove victim to fresh air; call a doctor; enforce complete rest until doctor arrives; observe at least 24 hours for delayed effects. SKIN OR EYES: flush with water for at least 15 min.; consult physician.
5.4 **Toxicity by Inhalation (Threshold Limit Value):** 1 ppm (recommended)
5.5 **Short-Term Inhalation Limits:** Data not available
5.6 **Toxicity by Ingestion:** Data not available
5.7 **Late Toxicity:** None
5.8 **Vapor (Gas) Irritant Characteristics:** Vapors cause severe irritation of eye and throat and can cause eye and lung injury. They cannot be tolerated even at low concentrations.
5.9 **Liquid or Solid Irritant Characteristics:** Severe burns to eyes and skin.
5.10 **Odor Threshold:** Data not available

6. FIRE HAZARDS

6.1 **Flash Point:** Not flammable
6.2 **Flammable Limits in Air:** Not flammable
6.3 **Fire Extinguishing Agents:** Not pertinent
6.4 **Fire Extinguishing Agents Not to be Used:** Not pertinent
6.5 **Special Hazards of Combustion Products:** Very toxic gases are generated when heated
6.6 **Behavior in Fire:** Not pertinent
6.7 **Ignition Temperature:** Not flammable
6.8 **Electrical Hazard:** Not pertinent
6.9 **Burning Rate:** Not flammable

7. CHEMICAL REACTIVITY

7.1 **Reactivity with Water:** Dissolves and reacts to form acid solution and toxic red oxides of nitrogen.
7.2 **Reactivity with Common Materials:** Corrosive to most metals, but reaction is not hazardous.
7.3 **Stability During Transport:** Stable
7.4 **Neutralizing Agents for Acids and Caustics:** Flush with water. Residual acid may be neutralized with soda ash.
7.5 **Polymerization:** Not pertinent
7.6 **Inhibitor of Polymerization:** Not pertinent

8. WATER POLLUTION

8.1 **Aquatic Toxicity:** Data not available
8.2 **Waterfowl Toxicity:** Data not available
8.3 **Biological Oxygen Demand (BOD):** Data not available
8.4 **Food Chain Concentration Potential:** None

9. SELECTED MANUFACTURERS

1. Air Products and Chemicals Inc.
   Specialty Gas Dept.
   733 West Broad St.
   Emmaus, Pa. 18049
2. Hercules Inc.
   Explosives & Chemical Propulsion Dept.
   Hercules, Calif. 94547

10. SHIPPING INFORMATION

10.1 **Grades or Purity:** 97%
10.2 **Storage Temperature:** Ambient
10.3 **Inert Atmosphere:** No requirement
10.4 **Venting:** Safety relief

11. HAZARD ASSESSMENT CODE

(See Hazard Assessment Handbook, CG 446-3)

A-C-O

12. HAZARD CLASSIFICATIONS

12.1 **Code of Federal Regulations:** Nonflammable compressed gas
12.2 **NAS Hazard Rating for Bulk Water Transportation:** Not listed
12.3 **NFPA Hazard Classifications:** Not listed

13. PHYSICAL AND CHEMICAL PROPERTIES

13.1 **Physical State at 15°C and 1 atm:** Gas
13.2 **Molecular Weight:** 65.46
13.3 **Boiling Point at 1 atm:** 21.6°F = −5.8°C = 267.4°K
13.4 **Freezing Point:** −74°F = −59°C = 214°K
13.5 **Critical Temperature:** 334°F = 168°C = 441°K
13.6 **Critical Pressure:** 1300 psia = 90 atm = 9.1 MN/m^2
13.7 **Specific Gravity:** 1.36 at −5.7°C (liquid)
13.8 **Liquid Surface Tension:** Not pertinent
13.9 **Liquid-Water Interfacial Tension:** Not pertinent
13.10 **Vapor (Gas) Specific Gravity:** 2.3
13.11 **Ratio of Specific Heats of Vapor (Gas):** 1.229
13.12 **Latent Heat of Vaporization:** 164 Btu/lb = 91.0 cal/g = 3.81×10^5 J/kg
13.13 **Heat of Combustion:** Not pertinent
13.14 **Heat of Decomposition:** Not pertinent
13.15 **Heat of Solution:** Not pertinent
13.16 **Heat of Polymerization:** Not pertinent

NOTES

REVISED 1978

NTO	NITROUS OXIDE

Common Synonyms	
Laughing gas Dinitrogen monoxide	Liquefied compressed gas · Colorless · Odorless or mild sweet odor Sinks and boils in water. Visible vapor cloud is produced.

Avoid contact with liquid. Keep people away.
Stop discharge if possible.
Notify local health and pollution control agencies.

Fire	Not flammable but will intensify fires.
Exposure	Call for medical aid. VAPOR If inhaled will cause dizziness, difficult breathing, or loss of consciousness. Move victim to fresh air. If breathing is difficult, give oxygen. LIQUID Will cause frostbite. Flush affected areas with plenty of water. DO NOT RUB AFFECTED AREAS.
Water Pollution	Not harmful to aquatic life.

1. RESPONSE TO DISCHARGE

(See Response Methods Handbook, CG 446-4)

Restrict access

2. LABEL

Green

3. CHEMICAL DESIGNATIONS

3.1 **Synonyms:** Dinitrogen monoxide; Laughing gas

3.2 **Coast Guard Compatibility Classification:** Not applicable

3.3 **Chemical Formula:** N_2O

3.4 **IMCO/United Nations Numerical Designation:** 2/1070

4. OBSERVABLE CHARACTERISTICS

4.1 **Physical State (as shipped):** Liquefied compressed gas

4.2 **Color:** Colorless

4.3 **Odor:** None; slightly sweetish

5. HEALTH HAZARDS

5.1 **Personal Protective Equipment:** Self-contained breathing apparatus for high vapor concentrations

5.2 **Symptoms Following Exposure:** Inhalation causes intense analgesia; concentrations of over 40—60% cause loss of consciousness preceded by hysteria. Contact of liquid with eyes or skin causes frostbite burn.

5.3 **Treatment for Exposure:** INHALATION: remove to fresh air. EYES: get medical attention for frostbite burn. SKIN: treat frostbite burn; soak in lukewarm water.

5.4 **Toxicity by Inhalation (Threshold Limit Value):** Data not available

5.5 **Short-Term Inhalation Limits:** 56,000 ppm for 60 min.

5.6 **Toxicity by Ingestion:** Grade 0; LD_{50} > 15 g/kg

5.7 **Late Toxicity:** Causes birth defects in rats; can cause lethal effects in chick eggs

5.8 **Vapor (Gas) Irritant Characteristics:** Vapors are nonirritating to eyes and throat.

5.9 **Liquid or Solid Irritant Characteristics:** No appreciable hazard; practically harmless to the skin.

5.10 **Odor Threshold:** Data not available

6. FIRE HAZARDS

6.1 **Flash Point:** Not pertinent (nonflammable compressed gas)

6.2 **Flammable Limits in Air:** Not pertinent

6.3 **Fire Extinguishing Agents:** Not pertinent

6.4 **Fire Extinguishing Agents Not to be Used:** Not pertinent

6.5 **Special Hazards of Combustion Products:** Not pertinent

6.6 **Behavior in Fire:** Will support combustion, and may increase intensity of fire. Containers may explode when heated.

6.7 **Ignition Temperature:** Not pertinent

6.8 **Electrical Hazard:** Not pertinent

6.9 **Burning Rate:** Not pertinent

7. CHEMICAL REACTIVITY

7.1 **Reactivity with Water:** No reaction

7.2 **Reactivity with Common Materials:** Supports combustion but does not cause spontaneous ignition.

7.3 **Stability During Transport:** Stable

7.4 **Neutralizing Agents for Acids and Caustics:** Not pertinent

7.5 **Polymerization:** Not pertinent

7.6 **Inhibitor of Polymerization:** Not pertinent

8. WATER POLLUTION

8.1 **Aquatic Toxicity:** None

8.2 **Waterfowl Toxicity:** None

8.3 **Biological Oxygen Demand (BOD):** None

8.4 **Food Chain Concentration Potential:** None

9. SELECTED MANUFACTURERS

1. Matheson Gas Products
   P. O. Box 85
   East Rutherford, N. J. 07073
2. Airco, Incorporated
   Ohio Medical Products Division
   3030 Airco Drive
   Madison, Wis. 53701
3. Air Products and Chemicals, Inc.
   Specialty Gas Department
   P. O. Box 538
   Allentown, Pa. 18105

10. SHIPPING INFORMATION

10.1 **Grades or Purity:** 98.0+%

10.2 **Storage Temperature:** Ambient

10.3 **Inert Atmosphere:** No requirement

10.4 **Venting:** Safety relief

11. HAZARD ASSESSMENT CODE

(See Hazard Assessment Handbook, CG 446-3)

A-C-I-J

12. HAZARD CLASSIFICATIONS

12.1 **Code of Federal Regulations:** Nonflammable compressed gas

12.2 **NAS Hazard Rating for Bulk Water Transportation:**

Category	Rating
Fire	0
Health	
Vapor Irritant	0
Liquid or Solid Irritant	0
Poisons	1
Water Pollution	
Human Toxicity	0
Aquatic Toxicity	0
Aesthetic Effect	0
Reactivity	
Other Chemicals	3
Water	0
Self-Reaction	0

12.3 **NFPA Hazard Classifications:** Not listed

13. PHYSICAL AND CHEMICAL PROPERTIES

13.1 **Physical State at 15°C and 1 atm:** Gas

13.2 **Molecular Weight:** 44.0

13.3 **Boiling Point at 1 atm:** $-129.1°F = -89.5°C = 183.7°K$

13.4 **Freezing Point:** $-131.5°F = -90.8°C = 182.4°K$

13.5 **Critical Temperature:** $97.7°F = 36.5°C = 309.7°K$

13.6 **Critical Pressure:** 1,054 psia = 71.7 atm = 7.28 MN/m^2

13.7 **Specific Gravity:** 1.266 at −89°C (liquid)

13.8 **Liquid Surface Tension:** 10.1 dynes/cm = 0.0101 N/m at −25°C

13.9 **Liquid-Water Interfacial Tension:** Data not available

13.10 **Vapor (Gas) Specific Gravity:** 1.53

13.11 **Ratio of Specific Heats of Vapor (Gas):** 1.303 at 25°C

13.12 **Latent Heat of Vaporization:** 161.7 Btu/lb = 89.9 cal/g = 3.76×10^5 J/kg

13.13 **Heat of Combustion:** Not pertinent

13.14 **Heat of Decomposition:** Not pertinent

13.15 **Heat of Solution:** Not pertinent

13.16 **Heat of Polymerization:** Not pertinent

NOTES

NAN	NONANE

Common Synonyms			
n-Nonane	Liquid	Colorless	Gasoline-like odor
	Floats on water.		

Shut off ignition sources. Call fire department.
Avoid contact with liquid. Keep people away.
Stop discharge if possible.
Isolate and remove discharged material.
Notify local health and pollution control agencies.

Fire

Combustible.
Extinguish with dry chemicals, foam or carbon dioxide.
Water may be ineffective on fire.
Cool exposed containers with water.

Exposure

Call for medical aid.

LIQUID
Irritating to skin and eyes.
If swallowed will cause nausea, and vomiting.
Remove contaminated clothing and shoes.
Flush affected areas with plenty of water.
IF IN EYES, hold eyelids open and flush with plenty of water.
IF SWALLOWED and victim is CONSCIOUS, have victim drink water or milk.
DO NOT INDUCE VOMITING.

Water Pollution

Effect of low concentrations on aquatic life is unknown.
Fouling to shoreline.
May be dangerous if it enters water intakes.
Notify local health and wildlife officials.
Notify operators of nearby water intakes.

1. RESPONSE TO DISCHARGE

(See Response Methods Handbook, CG 446-4)

Mechanical containment
Should be removed
Chemical and physical treatment

2. LABELS

No label required by Code of Federal Regulations

3. CHEMICAL DESIGNATIONS

3.1 **Synonyms:** n-Nonane
3.2 **Coast Guard Compatibility Classification:** Paraffin
3.3 **Chemical Formula:** C_9H_{20}
3.4 **IMCO/United Nations Numerical Designation:** 3.3/1920

4. OBSERVABLE CHARACTERISTICS

4.1 **Physical State (as shipped):** Liquid
4.2 **Color:** Colorless
4.3 **Odor:** Like gasoline

5. HEALTH HAZARDS

5.1 **Personal Protective Equipment:** Self-contained breathing apparatus for high vapor concentrations; goggles or face shield; rubber gloves.
5.2 **Symptoms Following Exposure:** Inhalation of concentrated vapor causes depression, irritation of respiratory tract, and pulmonary edema. Liquid can irritate eyes and (on prolonged contact) skin. Ingestion causes irritation of mouth and stomach. Aspiration causes severe lung irritation, rapidly developing pulmonary edema, and central nervous system excitement followed by depression.
5.3 **Treatment for Exposure:** INHALATION: remove victim from exposure; give artificial respiration if needed. EYES: irrigate with large amounts of water for 15 min. SKIN: flush with water; wash with soap and water. INGESTION: do NOT induce vomiting; call physician. ASPIRATION: enforce bed rest; give oxygen; get medical attention.
5.4 **Toxicity by Inhalation (Threshold Limit Value):** Data not available
5.5 **Short-Term Inhalation Limits:** Data not available
5.6 **Toxicity by Ingestion:** Grade 0; $LD_{50} > 15$ g/kg
5.7 **Late Toxicity:** Data not available
5.8 **Vapor (Gas) Irritant Characteristics:** Vapors are nonirritating to eyes and throat.
5.9 **Liquid or Solid Irritant Characteristics:** No appreciable hazard. Practically harmless to the skin.
5.10 **Odor Threshold:** Data not available

6. FIRE HAZARDS

6.1 **Flash Point:** 88°F C.C.
6.2 **Flammable Limits in Air:** 0.87% – 2.9%
6.3 **Fire Extinguishing Agents:** Dry chemical, foam, carbon dioxide
6.4 **Fire Extinguishing Agents Not to be Used:** Water may be ineffective
6.5 **Special Hazards of Combustion Products:** Not pertinent
6.6 **Behavior in Fire:** Not pertinent
6.7 **Ignition Temperature:** 401°F
6.8 **Electrical Hazard:** Class I, Group D
6.9 **Burning Rate:** 5.8 mm/min.

7. CHEMICAL REACTIVITY

7.1 **Reactivity with Water:** No reaction
7.2 **Reactivity with Common Materials:** No reaction
7.3 **Stability During Transport:** Stable
7.4 **Neutralizing Agents for Acids and Caustics:** Not pertinent
7.5 **Polymerization:** Not pertinent
7.6 **Inhibitor of Polymerization:** Not pertinent

8. WATER POLLUTION

8.1 **Aquatic Toxicity:** Data not available
8.2 **Waterfowl Toxicity:** Data not available
8.3 **Biological Oxygen Demand (BOD):** 1.1%, 1 day
8.4 **Food Chain Concentration Potential:** None

9. SELECTED MANUFACTURERS

1. Phillips Petroleum Company
   Chemical Department
   Special Products Division
   Bartlesville, Okla. 74004
2. The Humphrey Chemical Company
   Devine Street
   North Haven, Conn. 06473
3. Eastman Kodak Co.
   Eastman Organic Chemicals
   Rochester, N. Y. 14650

10. SHIPPING INFORMATION

10.1 **Grades or Purity:** Research, Pure, Technical: all 99.5+%
10.2 **Storage Temperature:** Ambient
10.3 **Inert Atmosphere:** No requirement
10.4 **Venting:** Open (flame arrester)

11. HAZARD ASSESSMENT CODE

(See Hazard Assessment Handbook, CG 446-3)

A-T-U

12. HAZARD CLASSIFICATIONS

12.1 **Code of Federal Regulations:** Not listed
12.2 **NAS Hazard Rating for Bulk Water Transportation:**

Category	Rating
Fire	3
Health	
Vapor Irritant	0
Liquid or Solid Irritant	0
Poisons	0
Water Pollution	
Human Toxicity	0
Aquatic Toxicity	2
Aesthetic Effect	2
Reactivity	
Other Chemicals	0
Water	0
Self-Reaction	0

12.3 **NFPA Hazard Classifications:**

Category	Classification
Health Hazard (Blue)	0
Flammability (Red)	3
Reactivity (Yellow)	0

13. PHYSICAL AND CHEMICAL PROPERTIES

13.1 **Physical State at 15°C and 1 atm:** Liquid
13.2 **Molecular Weight:** 128.3
13.3 **Boiling Point at 1 atm:** 304°F = 151°C = 424°K
13.4 **Freezing Point:** −64.3°F = −53.5°C = 219.7°K
13.5 **Critical Temperature:** 610.5°F = 321.4°C = 594.6°K
13.6 **Critical Pressure:** 335 psia = 22.8 atm = 2.31 MN/m^2
13.7 **Specific Gravity:** 0.718 at 20°C (liquid)
13.8 **Liquid Surface Tension:** 22.9 dynes/cm = 0.0229 N/m at 20°C
13.9 **Liquid-Water Interfacial Tension:** (est.) 35 dynes/cm = 0.035 N/m at 20°C
13.10 **Vapor (Gas) Specific Gravity:** 4.4
13.11 **Ratio of Specific Heats of Vapor (Gas):** 1.042 at 16°C
13.12 **Latent Heat of Vaporization:** 127 Btu/lb = 70.6 cal/g = 2.95×10^5 J/kg
13.13 **Heat of Combustion:** −19,067 Btu/lb = −10,593 cal/g = -443.21×10^5 J/kg
13.14 **Heat of Decomposition:** Not pertinent
13.15 **Heat of Solution:** Not pertinent
13.16 **Heat of Polymerization:** Not pertinent

NOTES

NNN — NONANOL

Common Synonyms			
1-Nonanol Nonyl alcohol Octycarbinol	Liquid	Colorless	Rose or fruity odor
	Floats on water. Freezing point is 23° F.		

Stop discharge if possible.
Call fire department.
Isolate and remove discharged material.
Notify local health and pollution control agencies.

Fire

Combustible.
Extinguish with dry chemical, alcohol foam, or carbon dioxide.
Water may be ineffective on fire.
Cool exposed containers with water.

Exposure

CALL FOR MEDICAL AID.

LIQUID
Harmful if swallowed.
IF SWALLOWED and victim is CONSCIOUS, have victim drink water or milk.

Water Pollution

Effect of low concentrations on aquatic life is unknown.
Fouling to shoreline.
May be dangerous if it enters water intakes.
Notify local health and wildlife officials.
Notify operators of nearby water intakes.

1. RESPONSE TO DISCHARGE

(See Response Methods Handbook, CG 446-4)

Mechanical containment
Should be removed
Chemical and physical treatment

2. LABELS

No hazard label required by Code of Federal Regulations

3. CHEMICAL DESIGNATIONS

3.1 **Synonyms:** 1-Nonanol; Nonyl alcohol; Octylcarbinol; Pelargonic alcohol
3.2 **Coast Guard Compatibility Classification:** Alcohol
3.3 **Chemical Formula:** $CH_3(CH_2)_7CH_2OH$
3.4 **IMCO/United Nations Numerical Designation:** Not listed

4. OBSERVABLE CHARACTERISTICS

4.1 **Physical State (as shipped):** Liquid
4.2 **Color:** Colorless
4.3 **Odor:** Rose-citrus

5. HEALTH HAZARDS

5.1 **Personal Protective Equipment:** Goggles or face shield; rubber gloves.
5.2 **Symptoms Following Exposure:** Liquid irritates eyes.
5.3 **Treatment for Exposure:** Flush eyes and skin with water for at least 15 min.
5.4 **Toxicity by Inhalation (Threshold Limit Value):** Data not available
5.5 **Short-Term Inhalation Limits:** Data not available
5.6 **Toxicity by Ingestion:** Grade 2; 0.5 to 5 g/kg (rat)
5.7 **Late Toxicity:** None
5.8 **Vapor (Gas) Irritant Characteristics:** Vapors are nonirritating to the eyes and throat.
5.9 **Liquid or Solid Irritant Characteristics:** No appreciable hazard. Practically harmless to the skin.
5.10 **Odor Threshold:** Data not available

6. FIRE HAZARDS

6.1 **Flash Point:** 210°F O.C.; 165°F C.C.
6.2 **Flammable Limits in Air:** 0.8%–6.1%
6.3 **Fire Extinguishing Agents:** Alcohol foam, dry chemical, or carbon dioxide
6.4 **Fire Extinguishing Agents Not to be Used:** Water may be ineffective
6.5 **Special Hazards of Combustion Products:** Not pertinent
6.6 **Behavior in Fire:** Not pertinent
6.7 **Ignition Temperature:** Data not available
6.8 **Electrical Hazard:** Not pertinent
6.9 **Burning Rate:** Data not available

7. CHEMICAL REACTIVITY

7.1 **Reactivity with Water:** No reaction
7.2 **Reactivity with Common Materials:** No reaction
7.3 **Stability During Transport:** Stable
7.4 **Neutralizing Agents for Acids and Caustics:** Not pertinent
7.5 **Polymerization:** Not pertinent
7.6 **Inhibitor of Polymerization:** Not pertinent

8. WATER POLLUTION

8.1 **Aquatic Toxicity:** Data not available
8.2 **Waterfowl Toxicity:** Data not available
8.3 **Biological Oxygen Demand (BOD):** (theor.) 9.9%, 1 day
8.4 **Food Chain Concentration Potential:** None

9. SELECTED MANUFACTURERS

Givaudan Corp.
Aroma Chemicals Division
125 Delawanna Ave.
Clifton, N. J. 07014

10. SHIPPING INFORMATION

10.1 **Grades or Purity:** 97%
10.2 **Storage Temperature:** Ambient
10.3 **Inert Atmosphere:** No requirement
10.4 **Venting:** Open (flame arrester)

11. HAZARD ASSESSMENT CODE

(See Hazard Assessment Handbook, CG 446-3)

A-T-U

12. HAZARD CLASSIFICATIONS

12.1 **Code of Federal Regulations:** Combustible Liquid
12.2 **NAS Hazard Rating for Bulk Water Transportation:**

Category	Rating
Fire	1
Health	
Vapor Irritant	0
Liquid or Solid Irritant	0
Poisons	0
Water Pollution	
Human Toxicity	1
Aquatic Toxicity	3
Aesthetic Effect	2
Reactivity	
Other Chemicals	2
Water	0
Self-Reaction	0

12.3 **NFPA Hazard Classifications:** Not listed

13. PHYSICAL AND CHEMICAL PROPERTIES

13.1 **Physical State at 15°C and 1 atm:** Liquid
13.2 **Molecular Weight:** 144.26
13.3 **Boiling Point at 1 atm:** 415°F = 213°C = 486°K
13.4 **Freezing Point:** 23°F = −5°C = 268°K
13.5 **Critical Temperature:** 759°F = 404°C = 677°K
13.6 **Critical Pressure:** 350 psia = 24 atm = 2.4 MN/m^2
13.7 **Specific Gravity:** 0.827 at 20°C (liquid)
13.8 **Liquid Surface Tension:** 28 dynes/cm = 0.028 N/m at 24°C
13.9 **Liquid-Water Interfacial Tension:** 9.0 dynes/cm = 0.0090 N/m at 21.3°C
13.10 **Vapor (Gas) Specific Gravity:** Not pertinent
13.11 **Ratio of Specific Heats of Vapor (Gas):** 1.039
13.12 **Latent Heat of Vaporization:** 131 Btu/lb = 72.5 cal/g = 3.04×10^5 J/kg
13.13 **Heat of Combustion:** -17,800 Btu/lb = -9860 cal/g = -413×10^5 J/kg
13.14 **Heat of Decomposition:** Not pertinent
13.15 **Heat of Solution:** Not pertinent
13.16 **Heat of Polymerization:** Not pertinent

NOTES

REVISED 1978

NON	# NONENE

Common Synonyms			
Propylene trimer Nonene (non-linear)	Liquid	Colorless	Gasoline-like odor
	Floats on water. Flammable, irritating vapor is produced.		

Stop discharge if possible. Keep people away.
Shut off ignition sources and call fire department.
Stay upwind and use water spray to "knock down" vapor.
Avoid contact with liquid and vapor.
Isolate and remove discharged material.
Notify local health and pollution control agencies.

Fire

FLAMMABLE.
Flashback along vapor trail may occur.
Vapor may explode if ignited in an enclosed area.
Extinguish with dry chemical, foam, or carbon dioxide.
Water may be ineffective on fire.
Cool exposed containers with water.

Exposure

CALL FOR MEDICAL AID.

VAPOR
Irritating to eyes, nose and throat.
If inhaled, will cause dizziness or loss of consciousness.
Move to fresh air.
If breathing has stopped, give artificial respiration.
If breathing is difficult, give oxygen.

LIQUID
Irritating to skin and eyes.
Remove contaminated clothing and shoes.
Flush affected areas with plenty of water.
IF IN EYES, hold eyelids open and flush with plenty of water.
IF SWALLOWED and victim is CONSCIOUS, have victim drink water or milk.
DO NOT INDUCE VOMITING.

Water Pollution

Effect of low concentrations on aquatic life is unknown.
Fouling to shoreline.
May be dangerous if it enters water intakes.
Notify local health and wildlife officials.
Notify operators of nearby water intakes.

1. RESPONSE TO DISCHARGE

(See Response Methods Handbook, CG 446-4)

Mechanical containment
Should be removed
Chemical and physical treatment

2. LABEL

Red

3. CHEMICAL DESIGNATIONS

3.1 **Synonyms:** Nonene (non-linear)
Propylene trimer
Tripropylene

3.2 **Coast Guard Compatibility Classification:** Olefin

3.3 **Chemical Formula:** C_9H_{18}

3.4 **IMCO/United Nations Numerical Designation:** Not listed

4. OBSERVABLE CHARACTERISTICS

4.1 **Physical State (as shipped):** Liquid

4.2 **Color:** Colorless

4.3 **Odor:** Characteristic; like gasoline

5. HEALTH HAZARDS

5.1 **Personal Protective Equipment:** Respiratory organic vapor canister or air-supplied mask; face splash shield.

5.2 **Symptoms Following Exposure:** High vapor concentrations irritate eyes and respiratory tract and act as an anesthetic.

5.3 **Treatment for Exposure:** INHALATION: remove patient to fresh air; if breathing stops, apply artificial respiration and administer oxygen; call a physician. INGESTION: do NOT induce vomiting because of aspiration hazard.

5.4 **Toxicity by Inhalation (Threshold Limit Value):** Data not available

5.5 **Short-Term Inhalation Limits:** Data not available

5.6 **Toxicity by Ingestion:** Data not available

5.7 **Late Toxicity:** Data not available

5.8 **Vapor (Gas) Irritant Characteristics:** Vapors cause a slight smarting of the eyes or respiratory system if present at high concentrations. The effect is temporary.

5.9 **Liquid or Solid Irritant Characteristics:** Minimum hazard. If spilled on clothing and allowed to remain, may cause smarting and reddening of the skin.

5.10 **Odor Threshold:** Data not available

6. FIRE HAZARDS

6.1 **Flash Point:** 78°F O.C.

6.2 **Flammable Limits in Air:** 0.7%–3.9% (est.)

6.3 **Fire Extinguishing Agents:** Foam, carbon dioxide, or dry chemical

6.4 **Fire Extinguishing Agents Not to be Used:** Water may be ineffective

6.5 **Special Hazards of Combustion Products:** Not pertinent

6.6 **Behavior in Fire:** Not pertinent

6.7 **Ignition Temperature:** Data not available

6.8 **Electrical Hazard:** Not pertinent

6.9 **Burning Rate:** 6.0 mm/min.

7. CHEMICAL REACTIVITY

7.1 **Reactivity with Water:** No reaction

7.2 **Reactivity with Common Materials:** No reaction

7.3 **Stability During Transport:** Stable

7.4 **Neutralizing Agents for Acids and Caustics:** Not pertinent

7.5 **Polymerization:** Not pertinent

7.6 **Inhibitor of Polymerization:** Not pertinent

8. WATER POLLUTION

8.1 **Aquatic Toxicity:** Data not available

8.2 **Waterfowl Toxicity:** Data not available

8.3 **Biological Oxygen Demand (BOD):** Data not available

8.4 **Food Chain Concentration Potential:** None

9. SELECTED MANUFACTURERS

1. Atlantic Richfield Co.
ARCO Chemical Co. Division
260 S. Broad St.
Philadelphia, Pa. 19101
2. Exxon Chemical Co.
Houston, Tex. 77001
3. Sun Oil Co.
St. Davids, Pa. 19087

10. SHIPPING INFORMATION

10.1 **Grades or Purity:** Technical

10.2 **Storage Temperature:** Ambient

10.3 **Inert Atmosphere:** No requirement

10.4 **Venting:** Open (flame arrester) or pressure-vacuum

11. HAZARD ASSESSMENT CODE

(See Hazard Assessment Handbook, CG 446-3)

A-T-U

12. HAZARD CLASSIFICATIONS

12.1 **Code of Federal Regulations:** Flammable Liquid

12.2 **NAS Hazard Rating for Bulk Water Transportation:**

Category	Rating
Fire	3
Health	
Vapor Irritant	1
Liquid or Solid Irritant	1
Poisons	0
Water Pollution	
Human Toxicity	0
Aquatic Toxicity	1
Aesthetic Effect	2
Reactivity	
Other Chemicals	1
Water	0
Self-Reaction	1

12.3 **NFPA Hazard Classifications:** Not listed

13. PHYSICAL AND CHEMICAL PROPERTIES

13.1 **Physical State at 15°C and 1 atm:** Liquid

13.2 **Molecular Weight:** 126.2

13.3 **Boiling Point at 1 atm:** 275–284°F = 135–140°C = 408–413°K

13.4 **Freezing Point:** Not pertinent

13.5 **Critical Temperature:** Data not available

13.6 **Critical Pressure:** Data not available

13.7 **Specific Gravity:** 0.739 at 20°C (liquid)

13.8 **Liquid Surface Tension:** 22 dynes/cm = 0.022 N/m at 24°C

13.9 **Liquid-Water Interfacial Tension:** 35.5 dynes/cm = 0.0355 N/m at 21.3°C

13.10 **Vapor (Gas) Specific Gravity:** Not pertinent

13.11 **Ratio of Specific Heats of Vapor (Gas):** (est.) 1.044

13.12 **Latent Heat of Vaporization:** (est.) 124 Btu/lb = 68.9 cal/g = 2.88×10^5 J/kg

13.13 **Heat of Combustion:** -19,100 Btu/lb = -10,600 cal/g = -445×10^5 J/kg

13.14 **Heat of Decomposition:** Not pertinent

13.15 **Heat of Solution:** Not pertinent

13.16 **Heat of Polymerization:** Not pertinent

NOTES

REVISED 1978

NNE	# 1-NONENE

Common Synonyms 1-Nonylene n-Heptylethylene	Liquid Colorless Gasoline-like odor Floats on water. Flammable, irritating vapor is produced.

Stop discharge if possible. Keep people away.
Shut off ignition sources and call fire department.
Stay upwind and use water spray to "knock down" vapor.
Avoid contact with liquid and vapor.
Isolate and remove discharged material.
Notify local health and pollution control agencies.

Fire

FLAMMABLE.
Flashback along vapor trail may occur.
Vapor may explode if ignited in an enclosed area.
Extinguish with dry chemical, foam, or carbon dioxide.
Water may be ineffective on fire.
Cool exposed containers with water.

Exposure

CALL FOR MEDICAL AID.

VAPOR
Irritating to eyes, nose and throat.
If inhaled, will cause dizziness, headache, difficult breathing or loss of consciousness.
Move to fresh air.
If breathing has stopped, give artificial respiration.
If breathing is difficult, give oxygen.

LIQUID
Irritating to skin and eyes.
Remove contaminated clothing and shoes.
Flush affected areas with plenty of water.
IF IN EYES, hold eyelids open and flush with plenty of water.
IF SWALLOWED and victim is CONSCIOUS, have victim drink water or milk.
DO NOT INDUCE VOMITING.

Water Pollution

Effect of low concentrations on aquatic life is unknown.
Fouling to shoreline.
May be dangerous if it enters water intakes.
Notify local health and wildlife officials.
Notify operators of nearby water intakes.

1. RESPONSE TO DISCHARGE

(See Response Methods Handbook, CG 446-4)

Mechanical containment
Should be removed
Chemical and physical treatment

2. LABEL

Red

3. CHEMICAL DESIGNATIONS

3.1 **Synonyms:** n-Heptylethylene 1-Nonylene
3.2 **Coast Guard Compatibility Classification:** Olefin
3.3 **Chemical Formula:** $CH_3(CH_2)_6CH{=}CH_2$
3.4 **IMCO/United Nations Numerical Designation:** Not listed

4. OBSERVABLE CHARACTERISTICS

4.1 **Physical State (as shipped):** Liquid
4.2 **Color:** Colorless
4.3 **Odor:** Hydrocarbon odor; like gasoline

5. HEALTH HAZARDS

5.1 **Personal Protective Equipment:** Respiratory organic vapor canister or air-supplied mask; face splash shield.
5.2 **Symptoms Following Exposure:** High vapor concentrations irritate eyes and respiratory tract and act as an anesthetic.
5.3 **Treatment for Exposure:** INHALATION: remove to fresh air; if breathing stops, apply artificial respiration; administer oxygen; call a physician. INGESTION: if swallowed, do NOT induce vomiting because of aspiration hazard.
5.4 **Toxicity by Inhalation (Threshold Limit Value):** Data not available
5.5 **Short-Term Inhalation Limits:** Data not available
5.6 **Toxicity by Ingestion:** Data not available
5.7 **Late Toxicity:** Data not available
5.8 **Vapor (Gas) Irritant Characteristics:** Vapors cause a slight smarting of the eyes or respiratory system if present at high concentrations. The effect is temporary.
5.9 **Liquid or Solid Irritant Characteristics:** Minimum hazard. If spilled on clothing and allowed to remain, may cause smarting and reddening of the skin.
5.10 **Odor Threshold:** Data not available

6. FIRE HAZARDS

6.1 **Flash Point:** Data not available
6.2 **Flammable Limits in Air:** 0.8% (LEL)
6.3 **Fire Extinguishing Agents:** Foam, carbon dioxide, or dry chemical
6.4 **Fire Extinguishing Agents Not to be Used:** Water may be ineffective
6.5 **Special Hazards of Combustion Products:** Not pertinent
6.6 **Behavior in Fire:** Not pertinent
6.7 **Ignition Temperature:** Data not available
6.8 **Electrical Hazard:** Not pertinent
6.9 **Burning Rate:** 6.0 mm/min.

7. CHEMICAL REACTIVITY

7.1 **Reactivity with Water:** No reaction
7.2 **Reactivity with Common Materials:** No reaction
7.3 **Stability During Transport:** Stable
7.4 **Neutralizing Agents for Acids and Caustics:** Not pertinent
7.5 **Polymerization:** Not pertinent
7.6 **Inhibitor of Polymerization:** Not pertinent

8. WATER POLLUTION

8.1 **Aquatic Toxicity:** Data not available
8.2 **Waterfowl Toxicity:** Data not available
8.3 **Biological Oxygen Demand (BOD):** Data not available
8.4 **Food Chain Concentration Potential:** None

9. SELECTED MANUFACTURERS

1. The Humphrey Chemical Co.
   Devine St.
   North Haven, Conn. 06473
2. Phillips Petroleum Co.
   Bartlesville, Okla. 74004
3. Exxon Chemical Co.
   Houston, Tex. 77001

10. SHIPPING INFORMATION

10.1 **Grades or Purity:** Technical
10.2 **Storage Temperature:** Ambient
10.3 **Inert Atmosphere:** No requirement
10.4 **Venting:** Open (flame arrester)

11. HAZARD ASSESSMENT CODE

(See Hazard Assessment Handbook, CG 446-3)

A-T-U

12. HAZARD CLASSIFICATIONS

12.1 **Code of Federal Regulations:** Flammable Liquid
12.2 **NAS Hazard Rating for Bulk Water Transportation:**

Category	Rating
Fire	3
Health	
Vapor Irritant	1
Liquid or Solid Irritant	1
Poisons	0
Water Pollution	
Human Toxicity	0
Aquatic Toxicity	1
Aesthetic Effect	2
Reactivity	
Other Chemicals	1
Water	0
Self-Reaction	1

12.3 **NFPA Hazard Classifications:** Not listed

13. PHYSICAL AND CHEMICAL PROPERTIES

13.1 **Physical State at 15°C and 1 atm:** Liquid
13.2 **Molecular Weight:** 126.2
13.3 **Boiling Point at 1 atm:** 297°F = 147°C = 420°K
13.4 **Freezing Point:** −115°F = −81.7°C = 191.5°K
13.5 **Critical Temperature:** 622°F = 327.8°C = 601.0°K
13.6 **Critical Pressure:** 360 psia = 24.5 atm = 2.98 MN/m^2
13.7 **Specific Gravity:** 0.733 at 20°C (liquid)
13.8 **Liquid Surface Tension:** 23.0 dynes/cm = 0.0230 N/m at 20°C
13.9 **Liquid-Water Interfacial Tension:** Data not available
13.10 **Vapor (Gas) Specific Gravity:** Not pertinent
13.11 **Ratio of Specific Heats of Vapor (Gas):** 1.044
13.12 **Latent Heat of Vaporization:** 124 Btu/lb = 68.8 cal/g = 2.88×10^5 J/kg
13.13 **Heat of Combustion:** −18,979 Btu/lb = −10,544 cal/g = -441.46×10^5 J/kg
13.14 **Heat of Decomposition:** Not pertinent
13.15 **Heat of Solution:** Not pertinent
13.16 **Heat of Polymerization:** Not pertinent

NOTES

REVISED 1978

NNP — NONYLPHENOL

Common Synonyms	Thick liquid	Light yellow, straw color	Medicinal odor
	Floats on water.		

Avoid contact with liquid. Keep people away.
Wear rubber overclothing (including gloves).
Stop discharge if possible.
Call fire department.
Isolate and remove discharged material.
Notify local health and pollution control agencies.

Fire

Combustible.
Extinguish with dry chemical, alcohol foam, or carbon dioxide.
Water may be ineffective on fire.
Cool exposed containers with water.

Exposure

CALL FOR MEDICAL AID.

LIQUID
Will burn skin and eyes.
Harmful if swallowed.
Remove contaminated clothing and shoes.
Flush affected areas with plenty of water.
IF IN EYES, hold eyelids open and flush with plenty of water.
IF SWALLOWED and victim is CONSCIOUS, have victim drink water or milk.

Water Pollution

Effect of low concentrations on aquatic life is unknown.
Fouling to shoreline.
May be dangerous if it enters water intakes.
Notify local health and wildlife officials.
Notify operators of nearby water intakes.

1. RESPONSE TO DISCHARGE

(See Response Methods Handbook, CG 446-4)

Mechanical containment
Should be removed
Chemical and physical treatment

2. LABELS

No hazard label required by Code of Federal Regulations

3. CHEMICAL DESIGNATIONS

3.1 **Synonyms:** No common synonyms
3.2 **Coast Guard Compatibility Classification:** Phenol
3.3 **Chemical Formula:** p-$HOC_6H_4(CH_2)_8CH_3$
3.4 **IMCO/United Nations Numerical Designation:** Not listed

4. OBSERVABLE CHARACTERISTICS

4.1 **Physical State (as shipped):** Liquid
4.2 **Color:** Pale yellow
4.3 **Odor:** Phenolic; like disinfectant

5. HEALTH HAZARDS

5.1 **Personal Protective Equipment:** Rubber gloves and splash-proof goggles.
5.2 **Symptoms Following Exposure:** Moderately toxic if swallowed. Severely irritating to skin and eyes.
5.3 **Treatment for Exposure:** EYES: wash with water for 15 min. and get medical attention. SKIN: wash with soap and water.
5.4 **Toxicity by Inhalation (Threshold Limit Value):** Data not available
5.5 **Short-Term Inhalation Limits:** Data not available
5.6 **Toxicity by Ingestion:** Grade 2; LD_{50} 0.5 to 5 g/kg
5.7 **Late Toxicity:** Data not available
5.8 **Vapor (Gas) Irritant Characteristics:** Vapors cause a slight smarting of the eyes or respiratory system if present in high concentrations. The effect is temporary.
5.9 **Liquid or Solid Irritant Characteristics:** Causes smarting of the skin and first-degree burns on short exposure; may cause secondary burns on long exposure.
5.10 **Odor Threshold:** Data not available

6. FIRE HAZARDS

6.1 **Flash Point:** 300°F O.C., 285°F C.C.
6.2 **Flammable Limits in Air:** Approx. 1% (calc. LEL)
6.3 **Fire Extinguishing Agents:** Alcohol foam, dry chemical, or carbon dioxide
6.4 **Fire Extinguishing Agents Not to be Used:** Water or foam may cause frothing.
6.5 **Special Hazards of Combustion Products:** Not pertinent
6.6 **Behavior in Fire:** Not pertinent
6.7 **Ignition Temperature:** Data not available
6.8 **Electrical Hazard:** Not pertinent
6.9 **Burning Rate:** Data not available

7. CHEMICAL REACTIVITY

7.1 **Reactivity with Water:** No reaction
7.2 **Reactivity with Common Materials:** No reaction
7.3 **Stability During Transport:** Stable
7.4 **Neutralizing Agents for Acids and Caustics:** Not pertinent
7.5 **Polymerization:** Not pertinent
7.6 **Inhibitor of Polymerization:** Not pertinent

8. WATER POLLUTION

8.1 **Aquatic Toxicity:** Data not available
8.2 **Waterfowl Toxicity:** Data not available
8.3 **Biological Oxygen Demand (BOD):** Data not available
8.4 **Food Chain Concentration Potential:** None

9. SELECTED MANUFACTURERS

1. GAF Corp.
   Chemical Division
   Calvert City, Ky. 42029
2. Jefferson Chemical Co., Inc.
   3336 Richmond Ave.
   Houston, Tex. 77052
3. Rohm & Haas Co.
   Independence Mall West
   Philadelphia, Pa. 19105

10. SHIPPING INFORMATION

10.1 **Grades or Purity:** 90% para-isomer, plus 4% ortho-isomer and 5% 2, 4-dinonylphenol
10.2 **Storage Temperature:** Ambient
10.3 **Inert Atmosphere:** No requirement
10.4 **Venting:** Open (flame arrester)

11. HAZARD ASSESSMENT CODE

(See Hazard Assessment Handbook, CG 446-3)

A-T-U

12. HAZARD CLASSIFICATIONS

12.1 **Code of Federal Regulations:** Not listed

12.2 **NAS Hazard Rating for Bulk Water Transportation:**

Category	Rating
Fire	1
Health	
Vapor Irritant	1
Liquid or Solid Irritant	2
Poisons	1
Water Pollution	
Human Toxicity	2
Aquatic Toxicity	3
Aesthetic Effect	3
Reactivity	
Other Chemicals	2
Water	0
Self-Reaction	0

12.3 **NFPA Hazard Classifications:**

Category	Classification
Health Hazard (Blue)	1
Flammability (Red)	1
Reactivity (Yellow)	0

13. PHYSICAL AND CHEMICAL PROPERTIES

13.1 **Physical State at 15°C and 1 atm:** Liquid
13.2 **Molecular Weight:** 220.36
13.3 **Boiling Point at 1 atm:** 579°F = 304°C = 577°K
13.4 **Freezing Point:** Not pertinent
13.5 **Critical Temperature:** 878°F = 470°C = 743°K
13.6 **Critical Pressure:** Not pertinent
13.7 **Specific Gravity:** 0.9494 at 25°C (liquid)
13.8 **Liquid Surface Tension:** (est.) 30 dynes/cm = 0.03 N/m at 20°C
13.9 **Liquid-Water Interfacial Tension:** (est.) 30 dynes/cm = 0.03 N/m at 20°C
13.10 **Vapor (Gas) Specific Gravity:** Not pertinent
13.11 **Ratio of Specific Heats of Vapor (Gas):** Not pertinent
13.12 **Latent Heat of Vaporization:** Not pertinent
13.13 **Heat of Combustion:** (est.) −17,500 Btu/lb = −9730 cal/g = -407×10^5 J/kg
13.14 **Heat of Decomposition:** Not pertinent
13.15 **Heat of Solution:** Not pertinent
13.16 **Heat of Polymerization:** Not pertinent

NOTES

REVISED 1978

OAN OCTANE

Common Synonyms	
n-Octane	Liquid Colorless Gasoline-like odor Floats on water. Flammable, irritating vapor is produced.

Shut off ignition sources. Call fire department.
Stop discharge if possible. Keep people away.
Isolate and remove discharged material.
Notify local health and pollution control agencies.

Fire	FLAMMABLE. Containers may explode in fire. Flashback along vapor trail may occur. Vapor may explode if ignited in an enclosed area. Extinguish with dry chemicals, foam or carbon dioxide. Water may be ineffective on fire. Cool exposed containers with water.
Exposure	Call for medical aid. VAPOR Irritating to eyes, nose and throat. If inhaled will cause headache, dizziness, difficult breathing, or loss of consciousness. Move victim to fresh air. If breathing has stopped, give artificial respiration. If breathing is difficult, give oxygen. LIQUID Irritating to skin and eyes. If swallowed will cause nausea, and vomiting. Remove contaminated clothing and shoes. Flush affected areas with plenty of water. IF IN EYES, hold eyelids open and flush with plenty of water. IF SWALLOWED and victim is CONSCIOUS, have victim drink water or milk. DO NOT INDUCE VOMITING.
Water Pollution	Effect of low concentrations on aquatic life is unknown. Fouling to shoreline. May be dangerous if it enters water intakes. Notify local health and wildlife officials. Notify operators of nearby water intakes.

1. RESPONSE TO DISCHARGE

(See Response Methods Handbook, CG 446-4)

Issue warning—high flammability
Restrict access
Mechanical containment
Should be removed
Chemical and physical treatment

2. LABEL

Red

3. CHEMICAL DESIGNATIONS

3.1 **Synonyms:** n-Octane
3.2 **Coast Guard Compatibility Classification:** Paraffin
3.3 **Chemical Formula:** C_8H_{18}
3.4 **IMCO/United Nations Numerical Designation:** 3.2/1262

4. OBSERVABLE CHARACTERISTICS

4.1 **Physical State (as shipped):** Liquid
4.2 **Color:** Colorless
4.3 **Odor:** Like gasoline

5. HEALTH HAZARDS

5.1 **Personal Protective Equipment:** Self-contained breathing apparatus for high vapor concentrations; goggles or face shield; rubber gloves.
5.2 **Symptoms Following Exposure:** Inhalation of concentrated vapor may cause irritation of respiratory tract, depression, and pulmonary edema. Liquid can cause irritation of eyes and (on prolonged contact) irritation and cracking of skin. Ingestion causes irritation of mouth and stomach. Aspiration causes severe lung irritation, rapidly developing pulmonary edema, and central nervous system excitement, followed by depression.
5.3 **Treatment for Exposure:** INHALATION: remove victim from exposure; apply artificial respiration if breathing has stopped; call physician if needed. EYES: irrigate with copious quantities of water for 15 min. SKIN: flush with water; wash with soap and water. INGESTION: do NOT induce vomiting; call physician. ASPIRATION: enforce bed rest; give oxygen; get medical attention.
5.4 **Toxicity by Inhalation (Threshold Limit Value):** 400 ppm
5.5 **Short-Term Inhalation Limits:** 500 ppm for 30 min.
5.6 **Toxicity by Ingestion:** Data not available
5.7 **Late Toxicity:** Data not available
5.8 **Vapor (Gas) Irritant Characteristics:** Data not available
5.9 **Liquid or Solid Irritant Characteristics:** Data not available
5.10 **Odor Threshold:** 4 ppm

6. FIRE HAZARDS

6.1 **Flash Point:** 56°F C.C.
6.2 **Flammable Limits in Air:** 1.0% 6.5%
6.3 **Fire Extinguishing Agents:** Dry chemical, foam, carbon dioxide
6.4 **Fire Extinguishing Agents Not to be Used:** Water may be ineffective.
6.5 **Special Hazards of Combustion Products:** Not pertinent
6.6 **Behavior in Fire:** Vapor is heavier than air and may travel a considerable distance to a source of ignition and flash back.
6.7 **Ignition Temperature:** 428°F
6.8 **Electrical Hazard:** Class I, Group D
6.9 **Burning Rate:** 6.3 mm/min.

7. CHEMICAL REACTIVITY

7.1 **Reactivity with Water:** No reaction
7.2 **Reactivity with Common Materials:** No reaction
7.3 **Stability During Transport:** Stable
7.4 **Neutralizing Agents for Acids and Caustics:** Not pertinent
7.5 **Polymerization:** Not pertinent
7.6 **Inhibitor of Polymerization:** Not pertinent

8. WATER POLLUTION

8.1 **Aquatic Toxicity:** Data not available
8.2 **Waterfowl Toxicity:** Data not available
8.3 **Biological Oxygen Demand (BOD):** Data not available
8.4 **Food Chain Concentration Potential:** None

9. SELECTED MANUFACTURERS

1. Phillips Petroleum Company
   Chemical Department
   Special Products Division
   Bartlesville, Okla. 74004
2. The Humphrey Chemical Company
   Devine Street
   North Haven, Conn. 06473
3. Charter International Oil Co.
   Charter Chemicals
   P. O. Box 5008
   Houston, Texas 77012

10. SHIPPING INFORMATION

10.1 **Grades or Purity:** Research: 99.92%; Pure: 99.6%; Tehcnical: 98.7%
10.2 **Storage Temperature:** Ambient
10.3 **Inert Atmosphere:** No requirement
10.4 **Venting:** Open (flame arrester)

11. HAZARD ASSESSMENT CODE

(See Hazard Assessment Handbook, CG 446-3)

A-T-U

12. HAZARD CLASSIFICATIONS

12.1 **Code of Federal Regulations:** Flammable liquid
12.2 **NAS Hazard Rating for Bulk Water Transportation:** Not listed
12.3 **NFPA Hazard Classifications:**

Category	Classification
Health Hazard (Blue)	0
Flammability (Red)	3
Reactivity (Yellow)	0

13. PHYSICAL AND CHEMICAL PROPERTIES

13.1 **Physical State at 15°C and 1 atm:** Liquid
13.2 **Molecular Weight:** 114.2
13.3 **Boiling Point at 1 atm:** 258.1°F = 125.6°C = 398.9°K
13.4 **Freezing Point:** −70.2°F = −56.8°C = 216.4°K
13.5 **Critical Temperature:** 563.7°F = 295.4°C = 568.6°K
13.6 **Critical Pressure:** 361 psia = 24.5 atm = 2.49 MN/m^2
13.7 **Specific Gravity:** 0.703 at 20°C (liquid)
13.8 **Liquid Surface Tension:** 21.7 dynes/cm = 0.0217 N/m at 20°C
13.9 **Liquid-Water Interfacial Tension:** (est.) 35 dynes/cm = 0.035 N/m at 20°C
13.10 **Vapor (Gas) Specific Gravity:** 3.9
13.11 **Ratio of Specific Heats of Vapor (Gas):** 1.047 at 16°C
13.12 **Latent Heat of Vaporization:** 130.4 Btu/lb = 72.5 cal/g = 3.03×10^5 J/kg
13.13 **Heat of Combustion:** −19,112 Btu/lb = −10,618 cal/g = -444.26×10^5 J/kg
13.14 **Heat of Decomposition:** Not pertinent
13.15 **Heat of Solution:** Not pertinent
13.16 **Heat of Polymerization:** Not pertinent

NOTES

OTA OCTANOL

Common Synonyms
1-Octanol
Octyl alcohol
Heptylcarbinol

Thick liquid | Colorless | Sweet odor

Floats on water.

Stop discharge if possible. Keep people away.
Call fire department.
Avoid contact with liquid.
Isolate and remove discharged material.
Notify local health and pollution control agencies.

Fire

Combustible.
Extinguish with dry chemical, foam, or carbon dioxide.
Cool exposed containers with water.

Exposure

CALL FOR MEDICAL AID.

LIQUID
Irritating to skin.
Will burn eyes.
Remove contaminated clothing and shoes.
Flush affected areas with plenty of water.
IF IN EYES, hold eyelids open and flush with plenty of water.

Water Pollution

Effect of low concentrations on aquatic life is unknown.
Fouling to shoreline.
May be dangerous if it enters water intakes.
Notify local health and wildlife officials.
Notify operators of nearby water intakes.

1. RESPONSE TO DISCHARGE

(See Response Methods Handbook, CG 446-4)

Mechanical containment
Should be removed
Chemical and physical treatment

2. LABELS

No hazard label required by Code of Federal Regulations

3. CHEMICAL DESIGNATIONS

3.1 **Synonyms:** Alcohol C-8
Heptylcarbinol
1-Octanol
Octyl alcohol

3.2 **Coast Guard Compatibility Classification:** Alcohol

3.3 **Chemical Formula:** $CH_3(CH_2)_6CH_2OH$

3.4 **IMCO/United Nations Numerical Designation:** Not listed

4. OBSERVABLE CHARACTERISTICS

4.1 **Physical State (as shipped):** Liquid

4.2 **Color:** Colorless

4.3 **Odor:** Sweet

5. HEALTH HAZARDS

5.1 **Personal Protective Equipment:** Chemical gloves and chemical goggles.

5.2 **Symptoms Following Exposure:** Irritates skin and eyes.

5.3 **Treatment for Exposure:** Flush with copious amounts of water.

5.4 **Toxicity by Inhalation (Threshold Limit Value):** Data not available

5.5 **Short-Term Inhalation Limits:** Data not available

5.6 **Toxicity by Ingestion:** Grade 1; oral rat LD_{50} >3.2 g/kg

5.7 **Late Toxicity:** None

5.8 **Vapor (Gas) Irritant Characteristics:** Vapors are nonirritating to the eyes and throat.

5.9 **Liquid or Solid Irritant Characteristics:** No appreciable hazard. Practically harmless to the skin.

5.10 **Odor Threshold:** 0.49 ppm

6. FIRE HAZARDS

6.1 **Flash Point:** 178°F C.C.

6.2 **Flammable Limits in Air:** Data not available

6.3 **Fire Extinguishing Agents:** Foam, carbon dioxide, or dry chemical

6.4 **Fire Extinguishing Agents Not to be Used:** Not pertinent

6.5 **Special Hazards of Combustion Products:** Not pertinent

6.6 **Behavior in Fire:** Not pertinent

6.7 **Ignition Temperature:** Data not available

6.8 **Electrical Hazard:** Not pertinent

6.9 **Burning Rate:** 3.7 mm/min. (approx.)

7. CHEMICAL REACTIVITY

7.1 **Reactivity with Water:** No reaction

7.2 **Reactivity with Common Materials:** No reaction

7.3 **Stability During Transport:** Stable

7.4 **Neutralizing Agents for Acids and Caustics:** Not pertinent

7.5 **Polymerization:** Not pertinent

7.6 **Inhibitor of Polymerization:** Not pertinent

8. WATER POLLUTION

8.1 **Aquatic Toxicity:** Data not available

8.2 **Waterfowl Toxicity:** Data not available

8.3 **Biological Oxygen Demand (BOD):** 109%, 5 days

8.4 **Food Chain Concentration Potential:** None

9. SELECTED MANUFACTURERS

1. Continental Oil Co.
Conoco Chemicals Division
Park 80 Plaza East
Saddle Brook, N. J. 07662
2. Givaudan Corp.
Aroma Chemicals Division
125 Delawanna Ave.
Clifton, N. J. 07014
3. Procter & Gamble Co.
301 East 6th St.
Cincinnati, Ohio 45201

10. SHIPPING INFORMATION

10.1 **Grades or Purity:** Data not available

10.2 **Storage Temperature:** Ambient

10.3 **Inert Atmosphere:** No requirement

10.4 **Venting:** Open (flame arrester)

11. HAZARD ASSESSMENT CODE

(See Hazard Assessment Handbook, CG 446-3)

A-T-U

12. HAZARD CLASSIFICATIONS

12.1 **Code of Federal Regulations:** Combustible Liquid

12.2 **NAS Hazard Rating for Bulk Water Transportation:** Not listed

12.3 **NFPA Hazard Classifications:** Not listed

13. PHYSICAL AND CHEMICAL PROPERTIES

13.1 **Physical State at 15°C and 1 atm:** Liquid

13.2 **Molecular Weight:** 130.23

13.3 **Boiling Point at 1 atm:** 383°F = 195°C = 468°K

13.4 **Freezing Point:** 5°F = −15°C = 258°K

13.5 **Critical Temperature:** 725°F = 385°C = 658°K

13.6 **Critical Pressure:** 400 psia = 27 atm = 2.7 MN/m^2

13.7 **Specific Gravity:** 0.829 at 20°C (liquid)

13.8 **Liquid Surface Tension:** 27.5 dynes/cm = 0.0275 N/m at 20°C

13.9 **Liquid-Water Interfacial Tension:** 8.52 dynes/cm = 0.00852 N/m at 20°C

13.10 **Vapor (Gas) Specific Gravity:** Not pertinent

13.11 **Ratio of Specific Heats of Vapor (Gas):** 1.044

13.12 **Latent Heat of Vaporization:** 176 Btu/lb = 97.5 cal/g = 4.08×10^5 J/kg

13.13 **Heat of Combustion:** −16,130 Btu/lb = −8,963 cal/g = -375.3×10^5 J/kg

13.14 **Heat of Decomposition:** Not pertinent

13.15 **Heat of Solution:** Not pertinent

13.16 **Heat of Polymerization:** Not pertinent

NOTES

REVISED 1978

OTE	# 1-OCTENE

Common Synonyms Caprylene alpha-Octylene	Liquid — Colorless — Gasoline-like odor Floats on water. Flammable, harmful vapor is produced.

Stop discharge if possible. Keep people away.
Shut off ignition sources and call fire department.
Stay upwind and use water spray to "knock down" vapor.
Avoid contact with liquid and vapor.
Isolate and remove discharged material.
Notify local health and pollution control agencies.

Fire

FLAMMABLE.
Flashback along vapor trail may occur.
Vapor may explode if ignited in an enclosed area.
Extinguish with dry chemical, foam or carbon dioxide.
Water may be ineffective on fire.
Cool exposed containers with water.

Exposure

CALL FOR MEDICAL AID.

VAPOR
If inhaled, will cause dizziness.
Move to fresh air.
If breathing has stopped, give artificial respiration.
If breathing is difficult, give oxygen.

LIQUID
Irritating to skin and eyes.
If swallowed, will cause nausea or vomiting.
Remove contaminated clothing and shoes.
Flush affected areas with plenty of water.
IF IN EYES, hold eyelids open and flush with plenty of water.
IF SWALLOWED and victim is CONSCIOUS, have victim drink water or milk.
DO NOT INDUCE VOMITING.

Water Pollution

Effect of low concentrations on aquatic life is unknown.
Fouling to shoreline.
May be dangerous if it enters water intakes.
Notify local health and wildlife officials.
Notify operators of nearby water intakes.

1. RESPONSE TO DISCHARGE

(See Response Methods Handbook, CG 446-4)

Issue warning—high flammability
Evacuate area
Disperse and flush

2. LABEL

Red

3. CHEMICAL DESIGNATIONS

3.1 **Synonyms:** Caprylene
alpha-Octylene
3.2 **Coast Guard Compatibility Classification:** Olefin
3.3 **Chemical Formula:** $CH_3(CH_2)_5CH{=}CH_2$
3.4 **IMCO/United Nations Numerical Designation:** Not listed

4. OBSERVABLE CHARACTERISTICS

4.1 **Physical State (as shipped):** Liquid
4.2 **Color:** Colorless
4.3 **Odor:** Like gasoline

5. HEALTH HAZARDS

5.1 **Personal Protective Equipment:** Organic vapor canister; goggles or face shield.
5.2 **Symptoms Following Exposure:** Generally low toxicity. Mildly anesthetic at high vapor concentrations. May irritate eyes.
5.3 **Treatment for Exposure:** INHALATION: remove from exposure; support respiration. INGESTION: do NOT induce vomiting.
5.4 **Toxicity by Inhalation (Threshold Limit Value):** Data not available
5.5 **Short-Term Inhalation Limits:** Data not available
5.6 **Toxicity by Ingestion:** Data not available
5.7 **Late Toxicity:** Data not available
5.8 **Vapor (Gas) Irritant Characteristics:** Vapors cause a slight smarting of the eyes or respiratory system if present at high concentrations. The effect is temporary.
5.9 **Liquid or Solid Irritant Characteristics:** Minimum hazard. If spilled on clothing and allowed to remain, may cause smarting and reddening of the skin.
5.10 **Odor Threshold:** Data not available

6. FIRE HAZARDS

6.1 **Flash Point:** 70°F O.C.
6.2 **Flammable Limits in Air:** 0.9% (LEL)
6.3 **Fire Extinguishing Agents:** Dry chemical, foam, or carbon dioxide
6.4 **Fire Extinguishing Agents Not to be Used:** Water may be ineffective
6.5 **Special Hazards of Combustion Products:** Not pertinent
6.6 **Behavior in Fire:** Not pertinent
6.7 **Ignition Temperature:** 493°F
6.8 **Electrical Hazard:** Not pertinent
6.9 **Burning Rate:** 6.5 mm/min.

7. CHEMICAL REACTIVITY

7.1 **Reactivity with Water:** No reaction
7.2 **Reactivity with Common Materials:** No reaction
7.3 **Stability During Transport:** Stable
7.4 **Neutralizing Agents for Acids and Caustics:** Not pertinent
7.5 **Polymerization:** Not pertinent
7.6 **Inhibitor of Polymerization:** Not pertinent

8. WATER POLLUTION

8.1 **Aquatic Toxicity:** Data not available
8.2 **Waterfowl Toxicity:** Data not available
8.3 **Biological Oxygen Demand (BOD):** 0.9% (theor.), 1 day
8.4 **Food Chain Concentration Potential:** None

9. SELECTED MANUFACTURERS

1. Air Products and Chemicals, Inc.
Houdry Division
Paulsboro, N. J. 08066
2. The Humphrey Chemical Co.
Devine St.
North Haven, Conn. 06473
3. Phillips Petroleum Co.
Bartlesville, Okla. 74004

10. SHIPPING INFORMATION

10.1 **Grades or Purity:** Research: 99.7%; pure: 99.3%; technical: 95%
10.2 **Storage Temperature:** Ambient
10.3 **Inert Atmosphere:** No requirement
10.4 **Venting:** Open (flame arrester) or pressure-vacuum

11. HAZARD ASSESSMENT CODE

(See Hazard Assessment Handbook, CG 446-3)

A-T-U

12. HAZARD CLASSIFICATIONS

12.1 **Code of Federal Regulations:** Flammable liquid
12.2 **NAS Hazard Rating for Bulk Water Transportation:** Not listed
12.3 **NFPA Hazard Classifications:** Not listed

13. PHYSICAL AND CHEMICAL PROPERTIES

13.1 **Physical State at 15°C and 1 atm:** Liquid
13.2 **Molecular Weight:** 112.22
13.3 **Boiling Point at 1 atm:** 250.3°F = 121.3°C = 394.5°K
13.4 **Freezing Point:** −151°F = −102°C = 172°K
13.5 **Critical Temperature:** 560.1°F = 293.4°C = 566.6°K
13.6 **Critical Pressure:** 400 psia = 27.2 atm = 2.76 MN/m²
13.7 **Specific Gravity:** 0.715 at 20°C (liquid)
13.8 **Liquid Surface Tension:** 21.76 dynes/cm = 0.02176 N/m at 20°C
13.9 **Liquid-Water Interfacial Tension:** (est.) 50 dynes/cm = 0.05 N/m at 20°C
13.10 **Vapor (Gas) Specific Gravity:** Not pertinent
13.11 **Ratio of Specific Heats of Vapor (Gas):** 1.050
13.12 **Latent Heat of Vaporization:** 129 Btu/lb = 71.9 cal/g = 3.01×10^5 J/kg
13.13 **Heat of Combustion:** −19,170 Btu/lb = −10,650 cal/g = -445.89×10^5 J/kg
13.14 **Heat of Decomposition:** Not pertinent
13.15 **Heat of Solution:** Not pertinent
13.16 **Heat of Polymerization:** Not pertinent

NOTES

REVISED 1978

OET — OCTYL EPOXY TALLATE

Common Synonyms			
Epoxidized tall oil, octyl ester	Liquid	Pale yellow	Mild odor
	Floats on water.		

Avoid contact with liquid. Keep people away.
Stop discharge if possible.
Isolate and remove discharged material.
Notify local health and pollution control agencies.

Fire	Combustible. Extinguish with dry chemicals, foam or carbon dioxide. Water may be ineffective on fire. Cool exposed containers with water.
Exposure	CALL FOR MEDICAL AID. LIQUID Irritating to skin and eyes. Harmful if swallowed. Remove contaminated clothing and shoes. Flush affected areas with plenty of water. IF IN EYES, hold eyelids open and flush with plenty of water. IF SWALLOWED and victim is CONSCIOUS, have victim drink water or milk. IF SWALLOWED and victim is UNCONSCIOUS OR HAVING CONVULSIONS, do nothing except keep victim warm.
Water Pollution	Effect of low concentrations on aquatic life is unknown. Fouling to shoreline. May be dangerous if it enters water intakes. Notify local health and wildlife officials. Notify operators of nearby water intakes.

1. RESPONSE TO DISCHARGE
(See Response Methods Handbook, CG 446-4)

Mechanical containment
Should be removed
Chemical and physical treatment

2. LABELS

No hazard label required by Code of Federal Regulations

3. CHEMICAL DESIGNATIONS

3.1 **Synonyms:** Epoxidized tall oil, octyl ester
3.2 **Coast Guard Compatibility Classification:** Ester
3.3 **Chemical Formula:** Mixture
3.4 **IMCO/United Nations Numerical Designation:** Not listed

4. OBSERVABLE CHARACTERISTICS

4.1 **Physical State (as shipped):** Liquid
4.2 **Color:** Pale yellow
4.3 **Odor:** Mild

5. HEALTH HAZARDS

5.1 **Personal Protective Equipment:** Chemical goggles; face shield; oil-resistant gloves
5.2 **Symptoms Following Exposure:** Contact with eyes causes mild inflammation. Contact with skin may produce allergic response.
5.3 **Treatment for Exposure:** EYES or SKIN: remove excess oil with cloth or absorbent paper; then wash with soapy water and flush with clear water; consult a physician.
5.4 **Toxicity by Inhalation (Threshold Limit Value):** Data not available
5.5 **Short-Term Inhalation Limits:** Data not available
5.6 **Toxicity by Ingestion:** Grade 0; LD_{50} > 15 g/kg
5.7 **Late Toxicity:** Data not available
5.8 **Vapor (Gas) Irritant Characteristics:** Data not available
5.9 **Liquid or Solid Irritant Characteristics:** Data not available
5.10 **Odor Threshold:** Data not available

6. FIRE HAZARDS

6.1 **Flash Point:** 450°F O.C.
6.2 **Flammable Limits in Air:** Not pertinent
6.3 **Fire Extinguishing Agents:** Foam, dry chemical, carbon dioxide
6.4 **Fire Extinguishing Agents Not to be Used:** Water may be ineffective.
6.5 **Special Hazards of Combustion Products:**
6.6 **Behavior in Fire:**
6.7 **Ignition Temperature:** Data not available
6.8 **Electrical Hazard:** Data not available
6.9 **Burning Rate:** Data not available

7. CHEMICAL REACTIVITY

7.1 **Reactivity with Water:** No reaction
7.2 **Reactivity with Common Materials:** May attack some forms of plastics.
7.3 **Stability During Transport:** Stable
7.4 **Neutralizing Agents for Acids and Caustics:** Not pertinent
7.5 **Polymerization:** Not pertinent
7.6 **Inhibitor of Polymerization:** Not pertinent

8. WATER POLLUTION

8.1 **Aquatic Toxicity:** Data not available
8.2 **Waterfowl Toxicity:** Data not available
8.3 **Biological Oxygen Demand (BOD):** Data not available
8.4 **Food Chain Concentration Potential:** None

9. SELECTED MANUFACTURERS

USS Chemicals
600 Grant St.
Pittsburgh, Pa. 15230

10. SHIPPING INFORMATION

10.1 **Grades or Purity:** Commercial
10.2 **Storage Temperature:** Ambient
10.3 **Inert Atmosphere:** No requirement
10.4 **Venting:** Open (flame arrester)

11. HAZARD ASSESSMENT CODE
(See Hazard Assessment Handbook, CG 446-3)

A-T-U

12. HAZARD CLASSIFICATIONS

12.1 **Code of Federal Regulations:** Not listed
12.2 **NAS Hazard Rating for Bulk Water Transportation:** Not listed
12.3 **NFPA Hazard Classifications:** Not listed

13. PHYSICAL AND CHEMICAL PROPERTIES

13.1 **Physical State at 15°C and 1 atm:** Liquid
13.2 **Molecular Weight:** 420 (approx.)
13.3 **Boiling Point at 1 atm:** Not pertinent (decomposes)
13.4 **Freezing Point:** Not pertinent
13.5 **Critical Temperature:** Not pertinent
13.6 **Critical Pressure:** Not pertinent
13.7 **Specific Gravity:** (est.) 1.002 at 20°C (liquid)
13.8 **Liquid Surface Tension:** 40.1 dynes/cm = 0.0401 N/m at 20°C
13.9 **Liquid-Water Interfacial Tension:** Data not available
13.10 **Vapor (Gas) Specific Gravity:** Not pertinent
13.11 **Ratio of Specific Heats of Vapor (Gas):** Not pertinent
13.12 **Latent Heat of Vaporization:** Data not available
13.13 **Heat of Combustion:** Data not available
13.14 **Heat of Decomposition:** Not pertinent
13.15 **Heat of Solution:** Not pertinent
13.16 **Heat of Polymerization:** Not pertinent

NOTES

OCF	OILS: CLARIFIED

Common Synonyms	Oily liquid Colorless Floats on water.

Stop discharge if possible.
Call fire department.
Isolate and remove discharged material.
Notify local health and pollution control agencies.

Fire	Combustible. Extinguish with dry chemical, foam, or carbon dioxide. Water may be ineffective on fire. Cool exposed containers with water.
Exposure	Not harmful.
Water Pollution	Effect of low concentrations on aquatic life is unknown. Fouling to shoreline. May be dangerous if it enters water intakes. Notify local health and wildlife officials. Notify operators of nearby water intakes.

1. RESPONSE TO DISCHARGE

(See Response Methods Handbook, CG 446-4)

Mechanical containment
Should be removed
Chemical and physical treatment

2. LABELS

No hazard label required by Code of Federal Regulations

3. CHEMICAL DESIGNATIONS

3.1 **Synonyms:** No common synonyms
3.2 **Coast Guard Compatibility Classification:** Miscellaneous hydrocarbon mixtures
3.3 **Chemical Formula:** Not applicable
3.4 **IMCO/United Nations Numerical Designation:** Not listed

4. OBSERVABLE CHARACTERISTICS

4.1 **Physical State (as shipped):** Liquid
4.2 **Color:** Data not available
4.3 **Odor:** Data not available

5. HEALTH HAZARDS

5.1 **Personal Protective Equipment:** Goggles or face shield
5.2 **Symptoms Following Exposure:** If liquid is ingested, an increased frequency of bowel movements will occur.
5.3 **Treatment for Exposure:** INGESTION: Do not induce vomiting; SKIN: Wipe off, wash with soap and water; EYES: Wash with water for at least 15 min.
5.4 **Toxicity by Inhalation (Threshold Limit Value):** No single TLV applicable
5.5 **Short-Term Inhalation Limits:** Data not available
5.6 **Toxicity by Ingestion:** Grade 1; LD_{50} 5 to 15g/kg
5.7 **Late Toxicity:** Data not available
5.8 **Vapor (Gas) Irritant Characteristics:** Data not available
5.9 **Liquid or Solid Irritant Characteristics:** Data not available
5.10 **Odor Threshold:** Data not available

6. FIRE HAZARDS

6.1 **Flash Point:** Data not available
6.2 **Flammable Limits in Air:** Data not available
6.3 **Fire Extinguishing Agents:** Dry chemical, foam, or carbon dioxide
6.4 **Fire Extinguishing Agents Not to be Used:** Water may be ineffective
6.5 **Special Hazards of Combustion Products:** Not pertinent
6.6 **Behavior in Fire:** Not pertinent
6.7 **Ignition Temperature:** Data not available
6.8 **Electrical Hazard:** Not pertinent
6.9 **Burning Rate:** 4 mm/min.

7. CHEMICAL REACTIVITY

7.1 **Reactivity with Water:** No reaction
7.2 **Reactivity with Common Materials:** No reaction
7.3 **Stability During Transport:** Stable
7.4 **Neutralizing Agents for Acids and Caustics:** Not pertinent
7.5 **Polymerization:** Not pertinent
7.6 **Inhibitor of Polymerization:** Not pertinent

8. WATER POLLUTION

8.1 **Aquatic Toxicity:** Data not available
8.2 **Waterfowl Toxicity:** Data not available
8.3 **Biological Oxygen Demand (BOD):** Data not available
8.4 **Food Chain Concentration Potential:** None

9. SELECTED MANUFACTURERS

1. Shell Oil Co.
   1 Shell Plaza
   Houston, Tex. 77001
2. Exxon Co.
   Houston, Tex. 77001
3. Sun Oil Co.
   St. Davids, Pa. 19087

10. SHIPPING INFORMATION

10.1 **Grades or Purity:** Data not available
10.2 **Storage Temperature:** Ambient
10.3 **Inert Atmosphere:** No requirement
10.4 **Venting:** Open (flame arrester)

11. HAZARD ASSESSMENT CODE

(See Hazard Assessment Handbook, CG 446-3)

A-T-U

12. HAZARD CLASSIFICATIONS

12.1 **Code of Federal Regulations:** Not listed
12.2 **NAS Hazard Rating for Bulk Water Transportation:** Not listed
12.3 **NFPA Hazard Classifications:** Not listed

13. PHYSICAL AND CHEMICAL PROPERTIES

13.1 **Physical State at 15°C and 1 atm:** Liquid
13.2 **Molecular Weight:** Not pertinent
13.3 **Boiling Point at 1 atm:** Data not available
13.4 **Freezing Point:** Not pertinent
13.5 **Critical Temperature:** Not pertinent
13.6 **Critical Pressure:** Not pertinent
13.7 **Specific Gravity:** (est.) 0.85 at 20°C (liquid)
13.8 **Liquid Surface Tension:** (est.) 25 dynes/cm = 0.025 N/m at 20°C
13.9 **Liquid-Water Interfacial Tension:** (est.) 50 dynes/cm = 0.05 N/m at 20°C
13.10 **Vapor (Gas) Specific Gravity:** Not pertinent
13.11 **Ratio of Specific Heats of Vapor (Gas):** Not pertinent
13.12 **Latent Heat of Vaporization:** Not pertinent
13.13 **Heat of Combustion:** (est.) −18,000 Btu/lb = −10,000 cal/g = -420×10^5 J/kg
13.14 **Heat of Decomposition:** Not pertinent
13.15 **Heat of Solution:** Not pertinent
13.16 **Heat of Polymerization:** Not pertinent

NOTES

REVISED 1978

OILS: CRUDE

Common Synonyms			
Petroleum	Oily liquid	Dark	Acrid odor
	Floats on water. Flammable vapor may be produced.		

Stop discharge if possible. Keep people away.
Call fire department.
Avoid contact with liquid.
Isolate and remove discharged material.
Notify local health and pollution control agencies.

Fire

Combustible.
Extinguish with dry chemical, foam, or carbon dioxide.
Water may be ineffective on fire.
Cool exposed containers with water.

Exposure

CALL FOR MEDICAL AID.

VAPOR
Not irritating to eyes, nose, or throat.

LIQUID
Irritating to skin and eyes.
Remove contaminated clothing and shoes.
Flush affected areas with plenty of water.
IF IN EYES, hold eyelids open and flush with plenty of water.

Water Pollution

HARMFUL TO AQUATIC LIFE IN VERY LOW CONCENTRATIONS.
Fouling to shoreline.
May be dangerous if it enters water intakes.
Notify local health and wildlife officials.
Notify operators of nearby water intakes.

1. RESPONSE TO DISCHARGE

(See Response Methods Handbook, CG 446-4)

Mechanical containment
Should be removed
Chemical and physical treatment

2. LABELS

No hazard label required by Code of Federal Regulations

3. CHEMICAL DESIGNATIONS

3.1 **Synonyms:** Petroleum
3.2 **Coast Guard Compatibility Classification:** Miscellaneous hydrocarbon mixtures
3.3 **Chemical Formula:** Not applicable
3.4 **IMCO/United Nations Numerical Designation:** 3.1/1267

4. OBSERVABLE CHARACTERISTICS

4.1 **Physical State (as shipped):** Liquid
4.2 **Color:** Dark
4.3 **Odor:** Offensive; tarry

5. HEALTH HAZARDS

5.1 **Personal Protective Equipment:** Goggles or face shield; rubber gloves and boots.
5.2 **Symptoms Following Exposure:** May irritate eyes and skin.
5.3 **Treatment for Exposure:** EYES: flush with water for at least 15 min. SKIN: wipe off and wash with soap and water.
5.4 **Toxicity by Inhalation (Threshold Limit Value):** Data not available
5.5 **Short-Term Inhalation Limits:** Data not available
5.6 **Toxicity by Ingestion:** Data not available
5.7 **Late Toxicity:** Data not available
5.8 **Vapor (Gas) Irritant Characteristics:** Vapors are nonirritating to the eyes and throat.
5.9 **Liquid or Solid Irritant Characteristics:** Minimum hazard. If spilled on clothing and allowed to remain, may cause smarting and reddening of the skin.
5.10 **Odor Threshold:** Data not available

6. FIRE HAZARDS

6.1 **Flash Point:** Data not available
6.2 **Flammable Limits in Air:** Data not available
6.3 **Fire Extinguishing Agents:** Dry chemical, foam, or carbon dioxide
6.4 **Fire Extinguishing Agents Not to be Used:** Water may be ineffective
6.5 **Special Hazards of Combustion Products:** Not pertinent
6.6 **Behavior in Fire:** Not pertinent
6.7 **Ignition Temperature:** Data not available
6.8 **Electrical Hazard:** Not pertinent
6.9 **Burning Rate:** 4 mm/min.

7. CHEMICAL REACTIVITY

7.1 **Reactivity with Water:** No reaction
7.2 **Reactivity with Common Materials:** No reaction
7.3 **Stability During Transport:** Stable
7.4 **Neutralizing Agents for Acids and Caustics:** Not pertinent
7.5 **Polymerization:** Not pertinent
7.6 **Inhibitor of Polymerization:** Not pertinent

8. WATER POLLUTION

8.1 **Aquatic Toxicity:**
3 ppm/*/fresh water fish/toxic/fresh water
200 ppm/24 hr/corals: porites/20-90% normal response/salt water
*Time period not specified.
8.2 **Waterfowl Toxicity:** Data not available
8.3 **Biological Oxygen Demand (BOD):** Data not available
8.4 **Food Chain Concentration Potential:** None

9. SELECTED MANUFACTURERS

1. Exxon Co.
Houston, Tex. 77001
2. Shell Oil Co.
1 Shell Plaza
Houston, Tex. 77001
3. Sun Oil Co.
St. Davids, Pa. 19087

10. SHIPPING INFORMATION

10.1 **Grades or Purity:** Wide variety, depending on oil field where produced.
10.2 **Storage Temperature:** Ambient
10.3 **Inert Atmosphere:** No requirement
10.4 **Venting:** Open (flame arrester)

11. HAZARD ASSESSMENT CODE

(See Hazard Assessment Handbook, CG 446-3)

A-T-U

12. HAZARD CLASSIFICATIONS

12.1 **Code of Federal Regulations:** Combustible liquid
12.2 **NAS Hazard Rating for Bulk Water Transportation:**

Category	Rating
Fire	1-3
Health	
Vapor Irritant	0
Liquid or Solid Irritant	1
Poisons	1
Water Pollution	
Human Toxicity	1
Aquatic Toxicity	2
Aesthetic Effect	4
Reactivity	
Other Chemicals	0
Water	0
Self-Reaction	0

12.3 **NFPA Hazard Classifications:**

Category	Classification
Health Hazard (Blue)	1
Flammability (Red)	3
Reactivity (Yellow)	0

13. PHYSICAL AND CHEMICAL PROPERTIES

13.1 **Physical State at 15°C and 1 atm:** Liquid
13.2 **Molecular Weight:** Not pertinent
13.3 **Boiling Point at 1 atm:** 90 – >750°F = 32 – >400°C = 305 – >673°K
13.4 **Freezing Point:** Not pertinent
13.5 **Critical Temperature:** Not pertinent
13.6 **Critical Pressure:** Not pertinent
13.7 **Specific Gravity:** 0.70 – 0.98 at 15°C (liquid)
13.8 **Liquid Surface Tension:** 24 – 38 dynes/cm = 0.024 – 0.038 N/m at 20°C
13.9 **Liquid-Water Interfacial Tension:** Data not available
13.10 **Vapor (Gas) Specific Gravity:** Not pertinent
13.11 **Ratio of Specific Heats of Vapor (Gas):** Not pertinent
13.12 **Latent Heat of Vaporization:** 140 – 150 Btu/lb = 76 – 86 cal/g = $3.2 - 3.6 \times 10^5$ J/kg
13.13 **Heat of Combustion:** −18,252 Btu/lb = −10,140 cal/g = -424.54×10^5 J/kg
13.14 **Heat of Decomposition:** Not pertinent
13.15 **Heat of Solution:** Not pertinent
13.16 **Heat of Polymerization:** Not pertinent

NOTES

REVISED 1978

ODS

OILS: DIESEL

Common Synonyms	Oily liquid	Yellow-brown	Lube or fuel oil odor
Fuel oil 1-D Fuel oil 2-D	Floats on water.		

Stop discharge if possible.
Call fire department.
Avoid contact with liquid.
Isolate and remove discharged material.
Notify local health and pollution control agencies.

Fire

Combustible.
Extinguish with dry chemical, foam, or carbon dioxide.
Water may be ineffective on fire.
Cool exposed containers with water.

Exposure

CALL FOR MEDICAL AID.
LIQUID
Irritating to skin and eyes.
Harmful if swallowed.
Remove contaminated clothing and shoes.
Flush affected areas with plenty of water.
IF IN EYES, hold eyelids open and flush with plenty of water.
IF SWALLOWED and victim is CONSCIOUS, have victim drink water or milk.
DO NOT INDUCE VOMITING.

Water Pollution

Dangerous to aquatic life in high concentrations.
Fouling to shoreline.
May be dangerous if it enters water intakes.
Notify local health and wildlife officials.
Notify operators of nearby water intakes

1. RESPONSE TO DISCHARGE

(See Response Methods Handbook, CG 446-4)

Mechanical containment
Should be removed
Chemical and physical treatment

2. LABELS

No hazard label required by Code of Federal Regulations

3. CHEMICAL DESIGNATIONS

3.1 **Synonyms:** Fuel Oil 1-D
Fuel Oil 2-D
3.2 **Coast Guard Compatibility Classification:** Miscellaneous hydrocarbon mixtures
3.3 **Chemical Formula:** Not applicable
3.4 **IMCO/United Nations Numerical Designation:** 3.1/1270

4. OBSERVABLE CHARACTERISTICS

4.1 **Physical State (as shipped):** Liquid
4.2 **Color:** Light brown
4.3 **Odor:** Like fuel oil

5. HEALTH HAZARDS

5.1 **Personal Protective Equipment:** Goggles or face shield.
5.2 **Symptoms Following Exposure:** If liquid is ingested, an increased frequency of bowel movements will occur.
5.3 **Treatment for Exposure:** INGESTION: do NOT induce vomiting. SKIN: wipe off, wash with soap and water. EYES: wash with copious amounts of water for at least 15 min.
5.4 **Toxicity by Inhalation (Threshold Limit Value):** No single TLV applicable.
5.5 **Short-Term Inhalation Limits:** Data not available
5.6 **Toxicity by Ingestion:** Grade 1; LD_{50} 5 to 15 g/kg
5.7 **Late Toxicity:** Data not available
5.8 **Vapor (Gas) Irritant Characteristics:** Vapors cause a slight smarting of the eyes or respiratory system if present in high concentrations. The effect is temporary.
5.9 **Liquid or Solid Irritant Characteristics:** Minimum hazard. If spilled on clothing and allowed to remain, may cause smarting and reddening of the skin.
5.10 **Odor Threshold:** Data not available

6. FIRE HAZARDS

6.1 **Flash Point:** (1-D) 100°F C.C.; (2-D) 125°F C.C.
6.2 **Flammable Limits in Air:** 1.3—6.0 vol. %
6.3 **Fire Extinguishing Agents:** Dry chemical, foam, or carbon dioxide
6.4 **Fire Extinguishing Agents Not to be Used:** Water may be ineffective
6.5 **Special Hazards of Combustion Products:** Not pertinent
6.6 **Behavior in Fire:** Not pertinent
6.7 **Ignition Temperature:** (1-D) 350-625°F (2-D) 490-545°F
6.8 **Electrical Hazard:** Not pertinent
6.9 **Burning Rate:** 4 mm/min.

7. CHEMICAL REACTIVITY

7.1 **Reactivity with Water:** No reaction
7.2 **Reactivity with Common Materials:** No reaction
7.3 **Stability During Transport:** Stable
7.4 **Neutralizing Agents for Acids and Caustics:** Not pertinent
7.5 **Polymerization:** Not pertinent
7.6 **Inhibitor of Polymerization:** Not pertinent

8. WATER POLLUTION

8.1 **Aquatic Toxicity:** 204 mg/l/24 hr/juvenile American shad/TL_m/salt water
8.2 **Waterfowl Toxicity:** >20 ml/kg /LD_{50}/mallards
8.3 **Biological Oxygen Demand (BOD):** Data not available
8.4 **Food Chain Concentration Potential:** None

9. SELECTED MANUFACTURERS

1. Exxon Co.
Houston, Tex. 77001
2. Shell Oil Co.
1 Shell Plaza
Houston, Tex. 77001
3. Sun Oil Co.
St. Davids, Pa. 19087

10. SHIPPING INFORMATION

10.1 **Grades or Purity:** Diesel Fuel 1-D (ASTM); Diesel Fuel 2-D (ASTM)
10.2 **Storage Temperature:** Ambient
10.3 **Inert Atmosphere:** No requirement
10.4 **Venting:** Open (flame arrester)

11. HAZARD ASSESSMENT CODE

(See Hazard Assessment Handbook, CG 446-3)

A-T-U

12. HAZARD CLASSIFICATIONS

12.1 **Code of Federal Regulations:** Combustible liquid
12.2 **NAS Hazard Rating for Bulk Water Transportation:** Not listed
12.3 **NFPA Hazard Classifications:**

Category	Classification
Health Hazard (Blue)	0
Flammability (Red)	2
Reactivity (Yellow)	0

13. PHYSICAL AND CHEMICAL PROPERTIES

13.1 **Physical State at 15°C and 1 atm:** Liquid
13.2 **Molecular Weight:** Not pertinent
13.3 **Boiling Point at 1 atm:** 550—640°F = 288—338°C = 561—612°K
13.4 **Freezing Point:** 0 to −30°F = −18 to −34°C = 255 to 239°K
13.5 **Critical Temperature:** Not pertinent
13.6 **Critical Pressure:** Not pertinent
13.7 **Specific Gravity:** 0.841 at 16°C (liquid)
13.8 **Liquid Surface Tension:** (est.) 25 dynes/cm = 0.025 N/m at 20°C
13.9 **Liquid-Water Interfacial Tension:** (est.) 50 dynes/cm = 0.05 N/m at 20°C
13.10 **Vapor (Gas) Specific Gravity:** Not pertinent
13.11 **Ratio of Specific Heats of Vapor (Gas):** Not pertinent
13.12 **Latent Heat of Vaporization:** Not pertinent
13.13 **Heat of Combustion:** -18,400 Btu/lb = -10,200 cal/g = 429×10^5 J/kg
13.14 **Heat of Decomposition:** Not pertinent
13.15 **Heat of Solution:** Not pertinent
13.16 **Heat of Polymerization:** Not pertinent

NOTES

REVISED 1978

OCA	OILS, EDIBLE: CASTOR

Common Synonyms	Oily liquid	Light yellow to green	Weak odor
	Floats on water.		

Stop discharge if possible.
Call fire department.
Isolate and remove discharged material.
Notify local health and pollution control agencies.

Fire	Combustible. Extinguish with dry chemical, foam or carbon dioxide. Water may be ineffective on fire. Cool exposed containers with water.
Exposure	Not harmful.
Water Pollution	Effect of low concentrations on aquatic life is unknown. Fouling to shoreline. May be dangerous if it enters water intakes. Notify local health and wildlife officials. Notify operators of nearby water intakes.

1. RESPONSE TO DISCHARGE

(See Response Methods Handbook, CG 446-4)

Mechanical containment
Should be removed
Chemical and physical treatment

2. LABELS

No hazard label required by Code of Federal Regulations

3. CHEMICAL DESIGNATIONS

3.1 **Synonyms:** No common synonyms
3.2 **Coast Guard Compatibility Classification:** Ester
3.3 **Chemical Formula:** Not applicable
3.4 **IMCO/United Nations Numerical Designation:** Not listed

4. OBSERVABLE CHARACTERISTICS

4.1 **Physical State (as shipped):** Liquid
4.2 **Color:** Light yellow to green
4.3 **Odor:** Characteristic; odorless

5. HEALTH HAZARDS

5.1 **Personal Protective Equipment:** Goggles or face shield.
5.2 **Symptoms Following Exposure:** If ingested causes severe diarrhea.
5.3 **Treatment for Exposure:** INGESTION: if more than 2 tablespoons, consult physician. EYES: flush with water for at least 15 min. SKIN: wipe off, wash with soap and water.
5.4 **Toxicity by Inhalation (Threshold Limit Value):** None
5.5 **Short-Term Inhalation Limits:** Not pertinent
5.6 **Toxicity by Ingestion:** Grade 1; LD_{50} 5 to 15 g/kg (Fatal dose unknown but presumably large).
5.7 **Late Toxicity:** None
5.8 **Vapor (Gas) Irritant Characteristics:** None
5.9 **Liquid or Solid Irritant Characteristics:** None
5.10 **Odor Threshold:** Not pertinent

6. FIRE HAZARDS

6.1 **Flash Point:** 445°F C.C.
6.2 **Flammable Limits in Air:** Data not available
6.3 **Fire Extinguishing Agents:** Dry chemical, foam, or carbon dioxide
6.4 **Fire Extinguishing Agents Not to be Used:** Water or foam may cause frothing.
6.5 **Special Hazards of Combustion Products:** Not pertinent
6.6 **Behavior in Fire:** Not pertinent
6.7 **Ignition Temperature:** 840°F
6.8 **Electrical Hazard:** Not pertinent
6.9 **Burning Rate:** Data not available

7. CHEMICAL REACTIVITY

7.1 **Reactivity with Water:** No reaction
7.2 **Reactivity with Common Materials:** No reaction
7.3 **Stability During Transport:** Stable
7.4 **Neutralizing Agents for Acids and Caustics:** Not pertinent
7.5 **Polymerization:** Not pertinent
7.6 **Inhibitor of Polymerization:** Not pertinent

8. WATER POLLUTION

8.1 **Aquatic Toxicity:** Data not available
8.2 **Waterfowl Toxicity:** Data not available
8.3 **Biological Oxygen Demand (BOD):** Data not available
8.4 **Food Chain Concentration Potential:** None

9. SELECTED MANUFACTURERS

1. N. L. Industries, Inc.
   Baker Castor Oil Co. Division
   20 Avenue A
   Bayonne, N. J. 07002
2. Textron, Inc.
   Spencer Kellogg Division
   Buffalo, N. Y. 14240
3. Arthur C. Trask Corp.
   7666 W. 63rd St.
   Argo, Ill. 60501

10. SHIPPING INFORMATION

10.1 **Grades or Purity:** Commercial: meets Mil. Specs. and ASTM; USP; USP Odorless; Refined: meets NAVORD Specs.; Technical. All grades differ only in color and acid values.
10.2 **Storage Temperature:** Ambient
10.3 **Inert Atmosphere:** No requirement
10.4 **Venting:** Open (flame arrester)

11. HAZARD ASSESSMENT CODE

(See Hazard Assessment Handbook, CG 446-3)

A-T-U

12. HAZARD CLASSIFICATIONS

12.1 **Code of Federal Regulations:** Not listed
12.2 **NAS Hazard Rating for Bulk Water Transportation:** Not listed
12.3 **NFPA Hazard Classifications:** Not listed

13. PHYSICAL AND CHEMICAL PROPERTIES

13.1 **Physical State at 15°C and 1 atm:** Liquid
13.2 **Molecular Weight:** Not pertinent
13.3 **Boiling Point at 1 atm:** Varies, depending on composition
13.4 **Freezing Point:** 10°F = −12°C = 261°K
13.5 **Critical Temperature:** Not pertinent
13.6 **Critical Pressure:** Not pertinent
13.7 **Specific Gravity:** 0.96 at 25°C (liquid)
13.8 **Liquid Surface Tension:** 39 dynes/cm = 0.039 N/m at 20°C
13.9 **Liquid-Water Interfacial Tension:** 19.2 dynes/cm = 0.0192 N/m at 22°C
13.10 **Vapor (Gas) Specific Gravity:** Not pertinent
13.11 **Ratio of Specific Heats of Vapor (Gas):** Not pertinent
13.12 **Latent Heat of Vaporization:** Not pertinent
13.13 **Heat of Combustion:** −15,950 Btu/lb = −8,860 cal/g = -371.0×10^5 J/kg
13.14 **Heat of Decomposition:** Not pertinent
13.15 **Heat of Solution:** Not pertinent
13.16 **Heat of Polymerization:** Not pertinent

NOTES

REVISED 1978

OCC

OILS, EDIBLE: COCONUT

Common Synonyms			
Coconut butter Copra oil Coconut oil	Solid or liquid	Light yellow to orange	Weak odor
	Floats on water.		

Stop discharge if possible.
Call fire department.
Isolate and remove discharged material.
Notify local health and pollution control agencies.

Fire	Combustible. Extinguish with dry chemicals, foam or carbon dioxide. Water may be ineffective on fire.
Exposure	LIQUID OR SOLID Not harmful. DO NOT INDUCE VOMITING.
Water Pollution	Effect of low concentrations on aquatic life is unknown. Fouling to shoreline. May be dangerous if it enters water intakes. Notify local health and wildlife officials. Notify operators of nearby water intakes.

1. RESPONSE TO DISCHARGE

(See Response Methods Handbook, CG 446-4)

Mechanical containment
Should be removed
Chemical and physical treatment

2. LABELS

No label required by Code of Federal Regulations

3. CHEMICAL DESIGNATIONS

3.1 **Synonyms:** Coconut butter; Coconut oil; Copra oil
3.2 **Coast Guard Compatibility Classification:** Esters
3.3 **Chemical Formula:** Not applicable
3.4 **IMCO/United Nations Numerical Designation:** Not listed

4. OBSERVABLE CHARACTERISTICS

4.1 **Physical State (as shipped):** Liquid or solid
4.2 **Color:** Light yellow-orange
4.3 **Odor:** Weak acid

5. HEALTH HAZARDS

5.1 **Personal Protective Equipment:** Goggles or face shield; rubber gloves.
5.2 **Symptoms Following Exposure:** Oil is essentially nontoxic, but can cause mild irritation of eyes on contact.
5.3 **Treatment for Exposure:** EYES: flush with water for at least 15 min. INGESTION: do NOT induce vomiting.
5.4 **Toxicity by Inhalation (Threshold Limit Value):** Data not available
5.5 **Short-Term Inhalation Limits:** Not pertinent
5.6 **Toxicity by Ingestion:** Data not available
5.7 **Late Toxicity:** None
5.8 **Vapor (Gas) Irritant Characteristics:** Data not available
5.9 **Liquid or Solid Irritant Characteristics:** Data not available
5.10 **Odor Threshold:** Not pertinent

6. FIRE HAZARDS

6.1 **Flash Point:** 420°F C.C. (crude) 580°F C.C. (refined)
6.2 **Flammable Limits in Air:** Not pertinent
6.3 **Fire Extinguishing Agents:** Foam, carbon dioxide, dry chemical
6.4 **Fire Extinguishing Agents Not to be Used:** Water or foam may cause frothing; water may be ineffective.
6.5 **Special Hazards of Combustion Products:** Not pertinent
6.6 **Behavior in Fire:** Not pertinent
6.7 **Ignition Temperature:** Data not available
6.8 **Electrical Hazard:** Not pertinent
6.9 **Burning Rate:** 4 mm/min.

7. CHEMICAL REACTIVITY

7.1 **Reactivity with Water:** No reaction
7.2 **Reactivity with Common Materials:** No reaction
7.3 **Stability During Transport:** Stable
7.4 **Neutralizing Agents for Acids and Caustics:** Not pertinent
7.5 **Polymerization:** Not pertinent
7.6 **Inhibitor of Polymerization:** Not pertinent

8. WATER POLLUTION

8.1 **Aquatic Toxicity:** Data not available
8.2 **Waterfowl Toxicity:** Data not available
8.3 **Biological Oxygen Demand (BOD):** Data not available
8.4 **Food Chain Concentration Potential:** None

9. SELECTED MANUFACTURERS

1. PVO International, Inc.
   1145 South Tenth Street
   Richmond, Calif. 94804
2. Cargill, Inc.
   Cargill Building
   Minneapolis, Minn. 55402
3. Capital City Products Co.
   P. O. Box 569
   Columbus, Ohio 43216

10. SHIPPING INFORMATION

10.1 **Grades or Purity:** Crude; Cochin. All grades contain 3—5% free fatty acids.
10.2 **Storage Temperature:** Ambient
10.3 **Inert Atmosphere:** No requirement
10.4 **Venting:** Open (flame arrester)

11. HAZARD ASSESSMENT CODE

(See Hazard Assessment Handbook, CG 446-3)

A-T-U

12. HAZARD CLASSIFICATIONS

12.1 **Code of Federal Regulations:** Not listed
12.2 **NAS Hazard Rating for Bulk Water Transportation:** Not listed
12.3 **NFPA Hazard Classifications:**

Category	Classification
Health Hazard (Blue)	0
Flammability (Red)	1
Reactivity (Yellow)	0

13. PHYSICAL AND CHEMICAL PROPERTIES

13.1 **Physical State at 15°C and 1 atm:** Solid or liquid
13.2 **Molecular Weight:** Not pertinent
13.3 **Boiling Point at 1 atm:** Not pertinent (very high)
13.4 **Freezing Point:** (approx.) 76°F = 24°C = 297°K
13.5 **Critical Temperature:** Not pertinent
13.6 **Critical Pressure:** Not pertinent
13.7 **Specific Gravity:** 0.922 at 25°C (liquid)
13.8 **Liquid Surface Tension:** 33.4 dynes/cm = 0.0334 N/m at 20°C
13.9 **Liquid-Water Interfacial Tension:** (est.) 50 dynes/cm = 0.050 N/m at 25°C
13.10 **Vapor (Gas) Specific Gravity:** Not pertinent
13.11 **Ratio of Specific Heats of Vapor (Gas):** Not pertinent
13.12 **Latent Heat of Vaporization:** Not pertinent
13.13 **Heat of Combustion:** (est.) −15,500 Btu/lb = −8,600 cal/g = -360×10^5 J/kg
13.14 **Heat of Decomposition:** Not pertinent
13.15 **Heat of Solution:** Not pertinent
13.16 **Heat of Polymerization:** Not pertinent

NOTES

OCS

OILS, EDIBLE: COTTONSEED

Common Synonyms

Oily liquid — Pale yellow — Odorless

Floats on water. Freezing point is 32° F.

Stop discharge if possible.
Call fire department.
Isolate and remove discharged material.
Notify local health and pollution control agencies.

Fire

Combustible.
Extinguish with dry chemical, foam or carbon dioxide.
Water may be ineffective on fire.
Cool exposed containers with water.

Exposure

Not harmful.

Water Pollution

Effect of low concentrations on aquatic life is unknown.
Fouling to shoreline.
May be dangerous if it enters water intakes.
Notify local health and wildlife officials.
Notify operators of nearby water intakes.

1. RESPONSE TO DISCHARGE

(See Response Methods Handbook, CG 446-4)

Mechanical containment
Should be removed
Chemical and physical treatment

2. LABELS

No hazard label required by Code of Federal Regulations

3. CHEMICAL DESIGNATIONS

3.1 **Synonyms:** No common synonyms
3.2 **Coast Guard Compatibility Classification:** Ester
3.3 **Chemical Formula:** Not applicable
3.4 **IMCO/United Nations Numerical Designation:** Not listed

4. OBSERVABLE CHARACTERISTICS

4.1 **Physical State (as shipped):** Liquid
4.2 **Color:** Pale yellow
4.3 **Odor:** Odorless

5. HEALTH HAZARDS

5.1 **Personal Protective Equipment:** Goggles or face shield.
5.2 **Symptoms Following Exposure:** None; is used as a food.
5.3 **Treatment for Exposure:** EYES: wash with water for at least 15 min.
5.4 **Toxicity by Inhalation (Threshold Limit Value):** None
5.5 **Short-Term Inhalation Limits:** Not pertinent
5.6 **Toxicity by Ingestion:** None
5.7 **Late Toxicity:** None
5.8 **Vapor (Gas) Irritant Characteristics:** None
5.9 **Liquid or Solid Irritant Characteristics:** None
5.10 **Odor Threshold:** Not pertinent

6. FIRE HAZARDS

6.1 **Flash Point:** 486°F C.C. (refined oil); 610°F C.C. (cooking oil)
6.2 **Flammable Limits in Air:** Data not available
6.3 **Fire Extinguishing Agents:** Dry chemical, foam, or carbon dioxide
6.4 **Fire Extinguishing Agents Not to be Used:** Water or foam may cause frothing.
6.5 **Special Hazards of Combustion Products:** Not pertinent
6.6 **Behavior in Fire:** Not pertinent
6.7 **Ignition Temperature:** 650°F (refined oil)
6.8 **Electrical Hazard:** Not pertinent
6.9 **Burning Rate:** Data not available

7. CHEMICAL REACTIVITY

7.1 **Reactivity with Water:** No reaction
7.2 **Reactivity with Common Materials:** No reaction
7.3 **Stability During Transport:** Stable
7.4 **Neutralizing Agents for Acids and Caustics:** Not pertinent
7.5 **Polymerization:** Not pertinent
7.6 **Inhibitor of Polymerization:** Not pertinent

8. WATER POLLUTION

8.1 **Aquatic Toxicity:** Data not available
8.2 **Waterfowl Toxicity:** Data not available
8.3 **Biological Oxygen Demand (BOD):** Data not available
8.4 **Food Chain Concentration Potential:** None

9. SELECTED MANUFACTURERS

1. Diamond Shamrock Corp.
   Nopco Chemicals Division
   Morristown, N. J. 07960
2. Pacific Vegetable Oil Corp.
   World Trade Center
   San Francisco, Calif. 94111
3. Stokely-Van Camp, Inc.
   Capital City Products Co.
   Columbus, Ohio 43216

10. SHIPPING INFORMATION

10.1 **Grades or Purity:** Refined; cooking
10.2 **Storage Temperature:** Ambient
10.3 **Inert Atmosphere:** No requirement
10.4 **Venting:** Open (flame arrester)

11. HAZARD ASSESSMENT CODE

(See Hazard Assessment Handbook, CG 446-3)

A-T-U

12. HAZARD CLASSIFICATIONS

12.1 **Code of Federal Regulations:** Not listed
12.2 **NAS Hazard Rating for Bulk Water Transportation:** Not listed
12.3 **NFPA Hazard Classifications:**

Category	Classification
Health Hazard (Blue)	0
Flammability (Red)	1
Reactivity (Yellow)	0

13. PHYSICAL AND CHEMICAL PROPERTIES

13.1 **Physical State at 15°C and 1 atm:** Liquid
13.2 **Molecular Weight:** Not pertinent
13.3 **Boiling Point at 1 atm:** Very high
13.4 **Freezing Point:** 32°F = 0°C = 273°K
13.5 **Critical Temperature:** Not pertinent
13.6 **Critical Pressure:** Not pertinent
13.7 **Specific Gravity:** 0.922 at 20°C (liquid)
13.8 **Liquid Surface Tension:** 35 dynes/cm = 0.035 N/m at 20°C
13.9 **Liquid-Water Interfacial Tension:** (est.) 50 dynes/cm = 0.05 N/m at 20°C
13.10 **Vapor (Gas) Specific Gravity:** Not pertinent
13.11 **Ratio of Specific Heats of Vapor (Gas):** Not pertinent
13.12 **Latent Heat of Vaporization:** Not pertinent
13.13 **Heat of Combustion:** (est.) −16,000 Btu/lb = −8,870 cal/g = -371×10^5 J/kg
13.14 **Heat of Decomposition:** Not pertinent
13.15 **Heat of Solution:** Not pertinent
13.16 **Heat of Polymerization:** Not pertinent

NOTES

REVISED 1978

OFS — OILS, EDIBLE: FISH

Common Synonyms	Oily liquid	Pale yellow	Fishy odor
	Floats on water.		

Stop discharge if possible.
Call fire department.
Isolate and remove discharged material.
Notify local health and pollution control agencies.

Fire	Combustible. Extinguish with dry chemical, foam or carbon dioxide. Water may be ineffective on fire. Cool exposed containers with water.
Exposure	Not harmful.
Water Pollution	Effect of low concentrations on aquatic life is unknown. Fouling to shoreline. May be dangerous if it enters water intakes. Notify local health and wildlife officials. Notify operators of nearby water intakes.

1. RESPONSE TO DISCHARGE
(See Response Methods Handbook, CG 446-4)

Mechanical containment
Should be removed
Chemical and physical treatment

2. LABELS

No hazard label required by Code of Federal Regulations

3. CHEMICAL DESIGNATIONS

3.1 **Synonyms:** No common synonyms
3.2 **Coast Guard Compatibility Classification:** Ester
3.3 **Chemical Formula:** Not applicable
3.4 **IMCO/United Nations Numerical Designation:** Not listed

4. OBSERVABLE CHARACTERISTICS

4.1 **Physical State (as shipped):** Liquid
4.2 **Color:** Pale yellow
4.3 **Odor:** Fishy

5. HEALTH HAZARDS

5.1 **Personal Protective Equipment:** Goggles or face shield.
5.2 **Symptoms Following Exposure:** None—is a food.
5.3 **Treatment for Exposure:** EYES: flush with water for at least 15 min.
5.4 **Toxicity by Inhalation (Threshold Limit Value):** Not pertinent
5.5 **Short-Term Inhalation Limits:** Not pertinent
5.6 **Toxicity by Ingestion:** None
5.7 **Late Toxicity:** None
5.8 **Vapor (Gas) Irritant Characteristics:** None
5.9 **Liquid or Solid Irritant Characteristics:** None
5.10 **Odor Threshold:** Data not available

6. FIRE HAZARDS

6.1 **Flash Point:** 420°F C.C.
6.2 **Flammable Limits in Air:** Data not available
6.3 **Fire Extinguishing Agents:** Dry chemical, foam, or carbon dioxide
6.4 **Fire Extinguishing Agents Not to be Used:** Water or foam may cause frothing.
6.5 **Special Hazards of Combustion Products:** Not pertinent
6.6 **Behavior in Fire:** Not pertinent
6.7 **Ignition Temperature:** Data not available
6.8 **Electrical Hazard:** Not pertinent
6.9 **Burning Rate:** Data not available

7. CHEMICAL REACTIVITY

7.1 **Reactivity with Water:** No reaction
7.2 **Reactivity with Common Materials:** No reaction
7.3 **Stability During Transport:** Stable
7.4 **Neutralizing Agents for Acids and Caustics:** Not pertinent
7.5 **Polymerization:** Not pertinent
7.6 **Inhibitor of Polymerization:** Not pertinent

8. WATER POLLUTION

8.1 **Aquatic Toxicity:** Data not available
8.2 **Waterfowl Toxicity:** Data not available
8.3 **Biological Oxygen Demand (BOD):** Data not available
8.4 **Food Chain Concentration Potential:** None

9. SELECTED MANUFACTURERS

1. Ashland Oil, Inc.
   Chemical Products Division
   Columbus, Ohio 43216
2. Diamond Shamrock Corp.
   Nopco Chemicals Division
   Morristown, N. J. 07960
3. Marine Products Co.
   333 W. 1st St.
   Boston, Mass. 02127

10. SHIPPING INFORMATION

10.1 **Grades or Purity:** Data not available
10.2 **Storage Temperature:** Ambient
10.3 **Inert Atmosphere:** No requirement
10.4 **Venting:** Open (flame arrester)

11. HAZARD ASSESSMENT CODE
(See Hazard Assessment Handbook, CG 446-3)

A-T-U

12. HAZARD CLASSIFICATIONS

12.1 **Code of Federal Regulations:** Not listed
12.2 **NAS Hazard Rating for Bulk Water Transportation:** Not listed
12.3 **NFPA Hazard Classifications:** Not listed

13. PHYSICAL AND CHEMICAL PROPERTIES

13.1 **Physical State at 15°C and 1 atm:** Liquid
13.2 **Molecular Weight:** Not pertinent
13.3 **Boiling Point at 1 atm:** Very high
13.4 **Freezing Point:** Not pertinent
13.5 **Critical Temperature:** Not pertinent
13.6 **Critical Pressure:** Not pertinent
13.7 **Specific Gravity:** 0.93 at 20°C (liquid)
13.8 **Liquid Surface Tension:** 38 dynes/cm = 0.038 N/m at 20°C
13.9 **Liquid-Water Interfacial Tension:** (est.) 50 dynes/cm = 0.05 N/m at 20°C
13.10 **Vapor (Gas) Specific Gravity:** Not pertinent
13.11 **Ratio of Specific Heats of Vapor (Gas):** Not pertinent
13.12 **Latent Heat of Vaporization:** Not pertinent
13.13 **Heat of Combustion:** (est.) −16,000 Btu/lb = −8,870 cal/g = -371×10^5 J/kg
13.14 **Heat of Decomposition:** Not pertinent
13.15 **Heat of Solution:** Not pertinent
13.16 **Heat of Polymerization:** Not pertinent

NOTES

REVISED 1978

OLD	OILS, EDIBLE: LARD

Common Synonyms	Liquid or solid	Colorless to Light yellow	Fatty odor
Prime steam lard Kettle rendered lard Leaf lard Lard	Floats on water.		

Stop discharge if possible.
Call fire department.
Isolate and remove discharged material.
Notify local health and pollution control agencies.

Fire	Combustible. Extinguish with dry chemicals, foam or carbon dioxide. Water may be ineffective on fire.
Exposure	LIQUID OR SOLID Not harmful. DO NOT INDUCE VOMITING.
Water Pollution	Effect of low concentrations on aquatic life is unknown. Fouling to shoreline. May be dangerous if it enters water intakes. Notify local health and wildlife officials. Notify operators of nearby water intakes.

1. RESPONSE TO DISCHARGE

(See Response Methods Handbook, CG 446-4)

Mechanical containment
Should be removed
Chemical and physical treatment

2. LABELS

No label required by Code of Federal Regulations

3. CHEMICAL DESIGNATIONS

3.1 **Synonyms:** Kettle-rendered lard; Leaf lard; Prime steam lard
3.2 **Coast Guard Compatibility Classification:** Esters (13)
3.3 **Chemical Formula:** Not applicable
3.4 **IMCO/United Nations Numerical Designation:** Not listed

4. OBSERVABLE CHARACTERISTICS

4.1 **Physical State (as shipped):** Solid or liquid
4.2 **Color:** Colorless to pale yellow
4.3 **Odor:** Fatty

5. HEALTH HAZARDS

5.1 **Personal Protective Equipment:** Goggles or face shield; rubber gloves.
5.2 **Symptoms Following Exposure:** Substance is essentially nontoxic. Prolonged contact with skin may cause dermatitis (oil acne). Hot liquid can burn eyes or skin.
5.3 **Treatment for Exposure:** EYES: flush with water for at least 15 min.; get medical attention for burn. SKIN: wipe off; get medical attention for burn. INGESTION: do NOT induce vomiting.
5.4 **Toxicity by Inhalation (Threshold Limit Value):** Not pertinent
5.5 **Short-Term Inhalation Limits:** Not pertinent
5.6 **Toxicity by Ingestion:** Data not available
5.7 **Late Toxicity:** None
5.8 **Vapor (Gas) Irritant Characteristics:** Data not available
5.9 **Liquid or Solid Irritant Characteristics:** Data not available
5.10 **Odor Threshold:** Not pertinent

6. FIRE HAZARDS

6.1 **Flash Point:** 395°F C.C.
6.2 **Flammable Limits in Air:** Not pertinent
6.3 **Fire Extinguishing Agents:** Dry chemical, foam, carbon dioxide
6.4 **Fire Extinguishing Agents Not to be Used:** Water or foam may cause frothing; water may be ineffective.
6.5 **Special Hazards of Combustion Products:** Not pertinent
6.6 **Behavior in Fire:** Not pertinent
6.7 **Ignition Temperature:** 833°F
6.8 **Electrical Hazard:** Not pertinent
6.9 **Burning Rate:** 4 mm/min.

7. CHEMICAL REACTIVITY

7.1 **Reactivity with Water:** No reaction
7.2 **Reactivity with Common Materials:** No reaction
7.3 **Stability During Transport:** Stable
7.4 **Neutralizing Agents for Acids and Caustics:** Not pertinent
7.5 **Polymerization:** Not pertinent
7.6 **Inhibitor of Polymerization:** Not pertinent

8. WATER POLLUTION

8.1 **Aquatic Toxicity:** Data not available
8.2 **Waterfowl Toxicity:** Data not available
8.3 **Biological Oxygen Demand (BOD):** Data not available
8.4 **Food Chain Concentration Potential:** None

9. SELECTED MANUFACTURERS

1. Swift Edible Oil Company
   115 West Jackson Boulevard
   Chicago, Ill. 60604
2. Mayco Oil and Chemical Co., Inc.
   Beaver and Canal Streets
   Bristol, Pa. 19007
3. Colfax, Inc.
   34 Colfax St.
   Pawtucket, R. I. 02860

10. SHIPPING INFORMATION

10.1 **Grades or Purity:** Various grades, depending on source of animal fat used and method of rendering
10.2 **Storage Temperature:** Ambient, or elevated (for liquid)
10.3 **Inert Atmosphere:** No requirement
10.4 **Venting:** Open (flame arrester)

11. HAZARD ASSESSMENT CODE

(See Hazard Assessment Handbook, CG 446-3)

A-T-U

12. HAZARD CLASSIFICATIONS

12.1 **Code of Federal Regulations:** Not listed
12.2 **NAS Hazard Rating for Bulk Water Transportation:** Not listed
12.3 **NFPA Hazard Classifications:**

Category	Classification
Health Hazard (Blue)	0
Flammability (Red)	1
Reactivity (Yellow)	0

13. PHYSICAL AND CHEMICAL PROPERTIES

13.1 **Physical State at 15°C and 1 atm:** Solid
13.2 **Molecular Weight:** Not pertinent
13.3 **Boiling Point at 1 atm:** Not pertinent
13.4 **Freezing Point:** 66—99°F = 19—37°C = 292—310°K
13.5 **Critical Temperature:** Not pertinent
13.6 **Critical Pressure:** Not pertinent
13.7 **Specific Gravity:** 0.861 at 15°C (liquid)
13.8 **Liquid Surface Tension:** (est.) 25 dynes/cm = 0.025 N/m at 30°C
13.9 **Liquid-Water Interfacial Tension:** (est.) 50 dynes/cm = 0.050 N/m at 30°C
13.10 **Vapor (Gas) Specific Gravity:** Not pertinent
13.11 **Ratio of Specific Heats of Vapor (Gas):** Not pertinent
13.12 **Latent Heat of Vaporization:** Not pertinent
13.13 **Heat of Combustion:** −16,750 Btu/lb = −9,320 cal/g = -390×10^5 J/kg
13.14 **Heat of Decomposition:** Not pertinent
13.15 **Heat of Solution:** Not pertinent
13.16 **Heat of Polymerization:** Not pertinent

NOTES

OOL — OILS, EDIBLE: OLIVE

Common Synonyms

Oily liquid — Pale yellow

Floats on water.

Stop discharge if possible.
Call fire department.
Isolate and remove discharged material.
Notify local health and pollution control agencies.

Fire

Combustible.
Extinguish with dry chemical, foam or carbon dioxide.
Water may be ineffective on fire.
Cool exposed containers with water.

Exposure

Not harmful.

Water Pollution

Effect of low concentrations on aquatic life is unknown.
Fouling to shoreline.
May be dangerous if it enters water intakes.
Notify local health and wildlife officials.
Notify operators of nearby water intakes.

1. RESPONSE TO DISCHARGE

(See Response Methods Handbook, CG 446-4)

Mechanical containment
Should be removed
Chemical and physical treatment

2. LABELS

No hazard label required by Code of Federal Regulations

3. CHEMICAL DESIGNATIONS

3.1 **Synonyms:** No common synonyms
3.2 **Coast Guard Compatibility Classification:** Ester
3.3 **Chemical Formula:** Not applicable
3.4 **IMCO/United Nations Numerical Designation:** Not listed

4. OBSERVABLE CHARACTERISTICS

4.1 **Physical State (as shipped):** Liquid
4.2 **Color:** Pale yellow-green
4.3 **Odor:** Weak, characteristic

5. HEALTH HAZARDS

5.1 **Personal Protective Equipment:** Goggles or face shield
5.2 **Symptoms Following Exposure:** None—is a food.
5.3 **Treatment for Exposure:** No treatment necessary.
5.4 **Toxicity by Inhalation (Threshold Limit Value):** Not pertinent
5.5 **Short-Term Inhalation Limits:** Not pertinent
5.6 **Toxicity by Ingestion:** None
5.7 **Late Toxicity:** None
5.8 **Vapor (Gas) Irritant Characteristics:** None
5.9 **Liquid or Solid Irritant Characteristics:** None
5.10 **Odor Threshold:** Data not available

6. FIRE HAZARDS

6.1 **Flash Point:** 437°F C.C.
6.2 **Flammable Limits in Air:** Data not available
6.3 **Fire Extinguishing Agents:** Dry chemical, foam, carbon dioxide
6.4 **Fire Extinguishing Agents Not to be Used:** Water or foam may cause frothing.
6.5 **Special Hazards of Combustion Products:** Not pertinent
6.6 **Behavior in Fire:** Not pertinent
6.7 **Ignition Temperature:** 650°F
6.8 **Electrical Hazard:** Not pertinent
6.9 **Burning Rate:** Data not available

7. CHEMICAL REACTIVITY

7.1 **Reactivity with Water:** No reaction
7.2 **Reactivity with Common Materials:** No reaction
7.3 **Stability During Transport:** Stable
7.4 **Neutralizing Agents for Acids and Caustics:** Not pertinent
7.5 **Polymerization:** Not pertinent
7.6 **Inhibitor of Polymerization:** Not pertinent

8. WATER POLLUTION

8.1 **Aquatic Toxicity:** Data not available
8.2 **Waterfowl Toxicity:** Data not available
8.3 **Biological Oxygen Demand (BOD):** Data not available
8.4 **Food Chain Concentration Potential:** None

9. SELECTED MANUFACTURERS

1. Finetex, Inc.
   418 Falmouth Ave.
   East Patterson, N. J. 07407
2. Reilly-Whiteman, Inc.
   Conshohocken, Pa. 19428
3. Welch, Holme and Clark Co., Inc.
   1000 S. 4th St.
   Harrison, N. J. 07029

10. SHIPPING INFORMATION

10.1 **Grades or Purity:** Data not available
10.2 **Storage Temperature:** Ambient
10.3 **Inert Atmosphere:** No requirement
10.4 **Venting:** Open (flame arrester)

11. HAZARD ASSESSMENT CODE

(See Hazard Assessment Handbook, CG 446-3)

A-T-U

12. HAZARD CLASSIFICATIONS

12.1 **Code of Federal Regulations:** Not listed
12.2 **NAS Hazard Rating for Bulk Water Transportation:** Not listed
12.3 **NFPA Hazard Classifications:** Not listed

13. PHYSICAL AND CHEMICAL PROPERTIES

13.1 **Physical State at 15°C and 1 atm:** Liquid
13.2 **Molecular Weight:** Not pertinent
13.3 **Boiling Point at 1 atm:** Very high
13.4 **Freezing Point:** Not pertinent
13.5 **Critical Temperature:** Not pertinent
13.6 **Critical Pressure:** Not pertinent
13.7 **Specific Gravity:** 0.915 at 20°C (liquid)
13.8 **Liquid Surface Tension:** 36 dynes/cm = 0.036 N/m at 20°C
13.9 **Liquid-Water Interfacial Tension:** (est.) 50 dynes/cm = 0.05 N/m at 20°C
13.10 **Vapor (Gas) Specific Gravity:** Not pertinent
13.11 **Ratio of Specific Heats of Vapor (Gas):** Not pertinent
13.12 **Latent Heat of Vaporization:** Not pertinent
13.13 **Heat of Combustion:** (est.) = 16,000 Btu/lb = −8,870 cal/g = -371×10^5 J/kg
13.14 **Heat of Decomposition:** Not pertinent
13.15 **Heat of Solution:** Not pertinent
13.16 **Heat of Polymerization:** Not pertinent

NOTES

REVISED 1978

OPM — OILS, EDIBLE: PALM

Common Synonyms
Palm fruit oil
Palm butter
Palm oil

Liquid or solid — Orange-red — Pleasant odor

Floats on water.

Stop discharge if possible.
Call fire department.
Isolate and remove discharged material.
Notify local health and pollution control agencies.

Fire

Combustible.
Extinguish with dry chemicals, foam or carbon dioxide.
Water may be ineffective on fire.

Exposure

LIQUID OR SOLID
Not harmful.
DO NOT INDUCE VOMITING.

Water Pollution

Effect of low concentrations on aquatic life is unknown.
Fouling to shoreline.
May be dangerous if it enters water intakes.
Notify local health and wildlife officials.
Notify operators of nearby water intakes.

1. RESPONSE TO DISCHARGE

(See Response Methods Handbook, CG 446-4)

Mechanical containment
Should be removed
Chemical and physical treatment

2. LABELS

No label required by Code of Federal Regulations

3. CHEMICAL DESIGNATIONS

3.1 **Synonyms:** Palm butter; Palm fruit oil
3.2 **Coast Guard Compatibility Classification:** Ester
3.3 **Chemical Formula:** Not applicable
3.4 **IMCO/United Nations Numerical Designation:** Not listed

4. OBSERVABLE CHARACTERISTICS

4.1 **Physical State (as shipped):** Semi-solid to liquid
4.2 **Color:** Orange-red
4.3 **Odor:** Pleasant, characteristic

5. HEALTH HAZARDS

5.1 **Personal Protective Equipment:** Goggles or face shield; rubber gloves
5.2 **Symptoms Following Exposure:** Oil is essentially nontoxic; may cause mild irritation of eyes.
5.3 **Treatment for Exposure:** EYES: flush with water for at least 15 min. INGESTION: do NOT induce vomiting.
5.4 **Toxicity by Inhalation (Threshold Limit Value):** Data not available
5.5 **Short-Term Inhalation Limits:** Not pertinent
5.6 **Toxicity by Ingestion:** Data not available
5.7 **Late Toxicity:** None
5.8 **Vapor (Gas) Irritant Characteristics:** Data not available
5.9 **Liquid or Solid Irritant Characteristics:** Data not available
5.10 **Odor Threshold:** Not pertinent

6. FIRE HAZARDS

6.1 **Flash Point:** 373°F C.C.
6.2 **Flammable Limits in Air:** Not pertinent
6.3 **Fire Extinguishing Agents:** Foam, dry chemical, carbon dioxide
6.4 **Fire Extinguishing Agents Not to be Used:** Water or foam may cause frothing; water may be ineffective.
6.5 **Special Hazards of Combustion Products:** Not pertinent
6.6 **Behavior in Fire:** Not pertinent
6.7 **Ignition Temperature:** 600°F
6.8 **Electrical Hazard:** Not pertinent
6.9 **Burning Rate:** 4 mm/min.

7. CHEMICAL REACTIVITY

7.1 **Reactivity with Water:** No reaction
7.2 **Reactivity with Common Materials:** No reaction
7.3 **Stability During Transport:** Stable
7.4 **Neutralizing Agents for Acids and Caustics:** Not pertinent
7.5 **Polymerization:** Not pertinent
7.6 **Inhibitor of Polymerization:** Not pertinent

8. WATER POLLUTION

8.1 **Aquatic Toxicity:** Data not available
8.2 **Waterfowl Toxicity:** Data not available
8.3 **Biological Oxygen Demand (BOD):** Data not available
8.4 **Food Chain Concentration Potential:** None

9. SELECTED MANUFACTURERS

1. Capital City Products Co.
   P. O. Box 569
   Columbus, Ohio 43216
2. Welch, Holme & Clark Co., Inc.
   1000 South Fourth Street
   Harrison, N. J. 07029
3. PVO International, Inc.
   1145 South Tenth Street
   Richmond, Calif. 94804

10. SHIPPING INFORMATION

10.1 **Grades or Purity:** Various grades, depending on source. Contains 3—45% fatty acids.
10.2 **Storage Temperature:** Ambient
10.3 **Inert Atmosphere:** No requirement
10.4 **Venting:** Open (flame arrester)

11. HAZARD ASSESSMENT CODE

(See Hazard Assessment Handbook, CG 446-3)

A-T-U

12. HAZARD CLASSIFICATIONS

12.1 **Code of Federal Regulations:** Not listed
12.2 **NAS Hazard Rating for Bulk Water Transportation:** Not listed
12.3 **NFPA Hazard Classifications:**

Category	Classification
Health Hazard (Blue)	0
Flammability (Red)	1
Reactivity (Yellow)	0

13. PHYSICAL AND CHEMICAL PROPERTIES

13.1 **Physical State at 15°C and 1 atm:** Solid to liquid
13.2 **Molecular Weight:** Not pertinent
13.3 **Boiling Point at 1 atm:** Not pertinent (very high)
13.4 **Freezing Point:** 70—80°F = 21—27°C = 294 300°K
13.5 **Critical Temperature:** Not pertinent
13.6 **Critical Pressure:** Not pertinent
13.7 **Specific Gravity:** 0.906 at 38°C (liquid)
13.8 **Liquid Surface Tension:** (est.) 25 dynes/cm = 0.025 N/m at 37°C
13.9 **Liquid-Water Interfacial Tension:** (est.) 50 dynes/cm = 0.050 N/m at 37°C
13.10 **Vapor (Gas) Specific Gravity:** Not pertinent
13.11 **Ratio of Specific Heats of Vapor (Gas):** Not pertinent
13.12 **Latent Heat of Vaporization:** Not pertinent
13.13 **Heat of Combustion:** (est.) −15,500 Btu/lb = −8,600 cal/g = -360×10^5 J/kg
13.14 **Heat of Decomposition:** Not pertinent
13.15 **Heat of Solution:** Not pertinent
13.16 **Heat of Polymerization:** Not pertinent

NOTES

OPN	OILS, EDIBLE: PEANUT

Common Synonyms	Oily liquid	Pale yellow	Weak peanut odor
	Floats on water.		

Stop discharge if possible.
Call fire department.
Isolate and remove discharged material.
Notify local health and pollution control agencies.

Fire	Combustible. Extinguish with dry chemical, foam or carbon dioxide. Water may be ineffective on fire. Cool exposed containers with water.
Exposure	Not harmful.
Water Pollution	Effect of low concentrations on aquatic life is unknown. Fouling to shoreline. May be dangerous if it enters water intakes. Notify local health and wildlife officials. Notify operators of nearby water intakes.

1. RESPONSE TO DISCHARGE

(See Response Methods Handbook, CG 446-4)

Mechanical containment
Should be removed
Chemical and physical treatment

2. LABELS

No hazard label required by Code of Federal Regulations

3. CHEMICAL DESIGNATIONS

3.1 **Synonyms:** No common synonyms
3.2 **Coast Guard Compatibility Classification:** Ester
3.3 **Chemical Formula:** Not applicable
3.4 **IMCO/United Nations Numerical Designation:** Not listed

4. OBSERVABLE CHARACTERISTICS

4.1 **Physical State (as shipped):** Liquid
4.2 **Color:** Pale yellow
4.3 **Odor:** Characteristic slight nutty odor

5. HEALTH HAZARDS

5.1 **Personal Protective Equipment:** Goggles or face shield.
5.2 **Symptoms Following Exposure:** None—is a food.
5.3 **Treatment for Exposure:** EYES: flush with water for at least 15 min.
5.4 **Toxicity by Inhalation (Threshold Limit Value):** Not pertinent
5.5 **Short-Term Inhalation Limits:** Not pertinent
5.6 **Toxicity by Ingestion:** None
5.7 **Late Toxicity:** None
5.8 **Vapor (Gas) Irritant Characteristics:** None
5.9 **Liquid or Solid Irritant Characteristics:** None
5.10 **Odor Threshold:** Data not available

6. FIRE HAZARDS

6.1 **Flash Point:** 640°F O.C.; 540°F C.C.
6.2 **Flammable Limits in Air:** Data not available
6.3 **Fire Extinguishing Agents:** Dry chemical, foam, or carbon dioxide
6.4 **Fire Extinguishing Agents Not to be Used:** Water or foam may cause frothing.
6.5 **Special Hazards of Combustion Products:** Not pertinent
6.6 **Behavior in Fire:** Not pertinent
6.7 **Ignition Temperature:** 833°F
6.8 **Electrical Hazard:** Not pertinent
6.9 **Burning Rate:** Data not available

7. CHEMICAL REACTIVITY

7.1 **Reactivity with Water:** No reaction
7.2 **Reactivity with Common Materials:** No reaction
7.3 **Stability During Transport:** Stable
7.4 **Neutralizing Agents for Acids and Caustics:** Not pertinent
7.5 **Polymerization:** Not pertinent
7.6 **Inhibitor of Polymerization:** Not pertinent

8. WATER POLLUTION

8.1 **Aquatic Toxicity:** Data not available
8.2 **Waterfowl Toxicity:** Data not available
8.3 **Biological Oxygen Demand (BOD):** Data not available
8.4 **Food Chain Concentration Potential:** None

9. SELECTED MANUFACTURERS

1. Reilly-Whiteman, Inc.
   Conshohocken, Pa. 19428
2. Stokely-Van Camp, Inc.
   Capital City Products Co.
   Columbus, Ohio 43216
3. Welch, Holme and Clark Co., Inc.
   1000 S. 4th St.
   Harrison, N. J. 07029

10. SHIPPING INFORMATION

10.1 **Grades or Purity:** Data not available
10.2 **Storage Temperature:** Ambient
10.3 **Inert Atmosphere:** No requirement
10.4 **Venting:** Open (flame arrester)

11. HAZARD ASSESSMENT CODE

(See Hazard Assessment Handbook, CG 446-3)

A-T-U

12. HAZARD CLASSIFICATIONS

12.1 **Code of Federal Regulations:** Not listed
12.2 **NAS Hazard Rating for Bulk Water Transportation:** Not listed
12.3 **NFPA Hazard Classifications:**

Category	Classification
Health Hazard (Blue)	0
Flammability (Red)	1
Reactivity (Yellow)	0

13. PHYSICAL AND CHEMICAL PROPERTIES

13.1 **Physical State at 15°C and 1 atm:** Liquid
13.2 **Molecular Weight:** Not pertinent
13.3 **Boiling Point at 1 atm:** Very high
13.4 **Freezing Point:** 28°F = −2°C = 271°K
13.5 **Critical Temperature:** Not pertinent
13.6 **Critical Pressure:** Not pertinent
13.7 **Specific Gravity:** 0.919 at 20°C (liquid)
13.8 **Liquid Surface Tension:** 35.5 dynes/cm = 0.0355 N/m at 20°C
13.9 **Liquid-Water Interfacial Tension:** 30 dynes/cm = 0.030 N/m at 70°C
13.10 **Vapor (Gas) Specific Gravity:** Not pertinent
13.11 **Ratio of Specific Heats of Vapor (Gas):** Not pertinent
13.12 **Latent Heat of Vaporization:** Not pertinent
13.13 **Heat of Combustion:** (est.) −16,000 Btu/lb = −8,870 cal/g = -371×10^5 J/kg
13.14 **Heat of Decomposition:** Not pertinent
13.15 **Heat of Solution:** Not pertinent
13.16 **Heat of Polymerization:** Not pertinent

NOTES

REVISED 1978

OSF — OILS, EDIBLE: SAFFLOWER

Common Synonyms			
Safflower oil Safflower seed oil Carthamus tinctorius oil	Liquid	Light yellow	Bland fatty odor
	Floats on water.		

Stop discharge if possible.
Call fire department.
Isolate and remove discharged material.
Notify local health and pollution control agencies.

Fire	Combustible. Extinguish with dry chemicals, foam or carbon dioxide. Water may be ineffective on fire.
Exposure	LIQUID Not harmful. DO NOT INDUCE VOMITING.
Water Pollution	Effect of low concentrations on aquatic life is unknown. Fouling to shoreline. May be dangerous if it enters water intakes. Notify local health and wildlife officials. Notify operators of nearby water intakes.

1. RESPONSE TO DISCHARGE
(See Response Methods Handbook, CG 446-4)

Mechanical containment
Should be removed
Chemical and physical treatment

2. LABELS

No label required by Code of Federal Regulations

3. CHEMICAL DESIGNATIONS

3.1 **Synonyms:** *Carthamus tinctorius* oil; Safflower seed oil
3.2 **Coast Guard Compatibility Classification:** Ester
3.3 **Chemical Formula:** Not applicable
3.4 **IMCO/United Nations Numerical Designation:** Not listed

4. OBSERVABLE CHARACTERISTICS

4.1 **Physical State (as shipped):** Liquid
4.2 **Color:** Pale yellow
4.3 **Odor:** Bland, fatty

5. HEALTH HAZARDS

5.1 **Personal Protective Equipment:** Goggles or face shield; rubber gloves
5.2 **Symptoms Following Exposure:** Oil is essentially nontoxic. Contact with eyes can cause mild irritation.
5.3 **Treatment for Exposure:** EYES: flush with water for at least 15 min. INGESTION: do NOT induce vomiting.
5.4 **Toxicity by Inhalation (Threshold Limit Value):** Data not available
5.5 **Short-Term Inhalation Limits:** Not pertinent
5.6 **Toxicity by Ingestion:** Data not available
5.7 **Late Toxicity:** None
5.8 **Vapor (Gas) Irritant Characteristics:** Data not available
5.9 **Liquid or Solid Irritant Characteristics:** Data not available
5.10 **Odor Threshold:** Not pertinent

6. FIRE HAZARDS

6.1 **Flash Point:** Data not available
6.2 **Flammable Limits in Air:** Not pertinent
6.3 **Fire Extinguishing Agents:** Dry chemical, foam, or carbon dioxide
6.4 **Fire Extinguishing Agents Not to be Used:** Water may be ineffective; water or foam may cause frothing.
6.5 **Special Hazards of Combustion Products:** Not pertinent
6.6 **Behavior in Fire:** Not pertinent
6.7 **Ignition Temperature:** Data not available
6.8 **Electrical Hazard:** Not pertinent
6.9 **Burning Rate:** 4 mm/min.

7. CHEMICAL REACTIVITY

7.1 **Reactivity with Water:** No reaction
7.2 **Reactivity with Common Materials:** No reaction
7.3 **Stability During Transport:** Stable
7.4 **Neutralizing Agents for Acids and Caustics:** Not pertinent
7.5 **Polymerization:** Not pertinent
7.6 **Inhibitor of Polymerization:** Not pertinent

8. WATER POLLUTION

8.1 **Aquatic Toxicity:** Data not available
8.2 **Waterfowl Toxicity:** Data not available
8.3 **Biological Oxygen Demand (BOD):** Data not available
8.4 **Food Chain Concentration Potential:** None

9. SELECTED MANUFACTURERS

1. PVO International Inc.
   1145 South Tenth Street
   Richmond, Calif. 94804
2. Capital City Products Co.
   P. O. Box 569
   Columbus, Ohio 43216

10. SHIPPING INFORMATION

10.1 **Grades or Purity:** Food grade: contains 0.02% propyl gallate, 0.01% citric acid, or may contain no additives. Technical: non-break and alkali-refined
10.2 **Storage Temperature:** Ambient
10.3 **Inert Atmosphere:** No requirement
10.4 **Venting:** Open (flame arrester)

11. HAZARD ASSESSMENT CODE
(See Hazard Assessment Handbook, CG 446-3)

A-T-U

12. HAZARD CLASSIFICATIONS

12.1 **Code of Federal Regulations:** Not listed
12.2 **NAS Hazard Rating for Bulk Water Transportation:** Not listed
12.3 **NFPA Hazard Classifications:** Not listed

13. PHYSICAL AND CHEMICAL PROPERTIES

13.1 **Physical State at 15°C and 1 atm:** Liquid
13.2 **Molecular Weight:** Not pertinent
13.3 **Boiling Point at 1 atm:** Not pertinent (very high)
13.4 **Freezing Point:** Not pertinent
13.5 **Critical Temperature:** Not pertinent
13.6 **Critical Pressure:** Not pertinent
13.7 **Specific Gravity:** 0.923 at 25°C (liquid)
13.8 **Liquid Surface Tension:** (est.) 25 dynes/cm = 0.025 N/m at 20°C
13.9 **Liquid-Water Interfacial Tension:** (est.) 50 dynes/cm = 0.050 N/m at 20°C
13.10 **Vapor (Gas) Specific Gravity:** Not pertinent
13.11 **Ratio of Specific Heats of Vapor (Gas):** Not pertinent
13.12 **Latent Heat of Vaporization:** Not pertinent
13.13 **Heat of Combustion:** (est.) −15,500 Btu/lb = −8,600 cal/g = -360×10^5 J/kg
13.14 **Heat of Decomposition:** Not pertinent
13.15 **Heat of Solution:** Not pertinent
13.16 **Heat of Polymerization:** Not pertinent

NOTES

OSB	OILS, EDIBLE: SOYA BEAN

Common Synonyms	Oily liquid	Pale yellow	Weak odor
	Floats on water.		

Stop discharge if possible.
Call fire department.
Isolate and remove discharged material.
Notify local health and pollution control agencies.

Fire	Combustible. Extinguish with dry chemical, foam or carbon dioxide. Water may be ineffective on fire. Cool exposed containers with water.
Exposure	Not harmful.
Water Pollution	Effect of low concentrations on aquatic life is unknown. Fouling to shoreline. May be dangerous if it enters water intakes. Notify local health and w'.dlife o:ficials. Notify operators of near'iv water intakes.

1. RESPONSE TO DISCHARGE
(See Response Methods Handbook, CG 446-4)

Mechanical containment
Should be removed
Chemical and physical treatment

2. LABELS

No hazard label required by Code of Federal Regulations

3. CHEMICAL DESIGNATIONS

3.1 **Synonyms:** Soybean oil
3.2 **Coast Guard Compatibility Classification:** Ester
3.3 **Chemical Formula:** Not applicable
3.4 **IMCO/United Nations Numerical Designation:** Not listed

4. OBSERVABLE CHARACTERISTICS

4.1 **Physical State (as shipped):** Liquid
4.2 **Color:** Pale yellow
4.3 **Odor:** Weak

5. HEALTH HAZARDS

5.1 **Personal Protective Equipment:** Goggles or face shield.
5.2 **Symptoms Following Exposure:** None – is a food.
5.3 **Treatment for Exposure:** EYES: flush with water for at least 15 min.
5.4 **Toxicity by Inhalation (Threshold Limit Value):** Not pertinent
5.5 **Short-Term Inhalation Limits:** Not pertinent
5.6 **Toxicity by Ingestion:** None
5.7 **Late Toxicity:** None
5.8 **Vapor (Gas) Irritant Characteristics:** None
5.9 **Liquid or Solid Irritant Characteristics:** None
5.10 **Odor Threshold:** Data not available

6. FIRE HAZARDS

6.1 **Flash Point:** 540°F C.C.
6.2 **Flammable Limits in Air:** Data not available
6.3 **Fire Extinguishing Agents:** Dry chemical, foam, or carbon dioxide
6.4 **Fire Extinguishing Agents Not to be Used:** Water or foam may cause frothing.
6.5 **Special Hazards of Combustion Products:** Not pertinent
6.6 **Behavior in Fire:** Not pertinent
6.7 **Ignition Temperature:** 833°F
6.8 **Electrical Hazard:** Not pertinent
6.9 **Burning Rate:** Data not available

7. CHEMICAL REACTIVITY

7.1 **Reactivity with Water:** No reaction
7.2 **Reactivity with Common Materials:** No reaction
7.3 **Stability During Transport:** Stable
7.4 **Neutralizing Agents for Acids and Caustics:** Not pertinent
7.5 **Polymerization:** Not pertinent
7.6 **Inhibitor of Polymerization:** Not pertinent

8. WATER POLLUTION

8.1 **Aquatic Toxicity:** Data not available
8.2 **Waterfowl Toxicity:** Data not available
8.3 **Biological Oxygen Demand (BOD):** 39%, 5 days
8.4 **Food Chain Concentration Potential:** None

9. SELECTED MANUFACTURERS

1. Ashland Oil Co.
   Chemical Products Division
   Peoria, Ill. 61601
2. A. E. Staley Manufacturing Co.
   Decatur, Ill. 62525
3. Arthur C. Trask Corp.
   7666 W. 63rd St.
   Argo, Ill. 60501

10. SHIPPING INFORMATION

10.1 **Grades or Purity:** Refined, crude
10.2 **Storage Temperature:** Ambient
10.3 **Inert Atmosphere:** No requirement
10.4 **Venting:** Open (flame arrester)

11. HAZARD ASSESSMENT CODE
(See Hazard Assessment Handbook, CG 446-3)

A-T-U

12. HAZARD CLASSIFICATIONS

12.1 **Code of Federal Regulations:** Not listed
12.2 **NAS Hazard Rating for Bulk Water Transportation:** Not listed
12.3 **NFPA Hazard Classifications:**

Category	Classification
Health Hazard (Blue)	0
Flammability (Red)	1
Reactivity (Yellow)	0

13. PHYSICAL AND CHEMICAL PROPERTIES

13.1 **Physical State at 15°C and 1 atm:** Liquid
13.2 **Molecular Weight:** Not pertinent
13.3 **Boiling Point at 1 atm:** Very high
13.4 **Freezing Point:** −4°F = −20°C = 253°K
13.5 **Critical Temperature:** Not pertinent
13.6 **Critical Pressure:** Not pertinent
13.7 **Specific Gravity:** 0.922 at 20°C (liquid)
13.8 **Liquid Surface Tension:** (est.) 25 dynes/cm = 0.025 N/m at 20°C
13.9 **Liquid-Water Interfacial Tension:** (est.) 50 dynes/cm = 0.05 N/m at 20°C
13.10 **Vapor (Gas) Specific Gravity:** Not pertinent
13.11 **Ratio of Specific Heats of Vapor (Gas):** Not pertinent
13.12 **Latent Heat of Vaporization:** Not pertinent
13.13 **Heat of Combustion:** (est.) −16,000 Btu/lb = −8,870 cal/g = -371×10^5 J/kg
13.14 **Heat of Decomposition:** Not pertinent
13.15 **Heat of Solution:** Not pertinent
13.16 **Heat of Polymerization:** Not pertinent

NOTES

REVISED 1978

OTC — OILS, EDIBLE: TUCUM

Common Synonyms			
Aouara oil American palm kernel oil Palm seed oil Tucum oil	Liquid	Light yellow	Weak odor
	Floats on water.		

Stop discharge if possible
Call fire department.
Isolate and remove discharged material.
Notify local health and pollution control agencies.

Fire	Combustible. Extinguish with dry chemicals, foam or carbon dioxide. Water may be ineffective on fire.
Exposure	LIQUID Not harmful. DO NOT INDUCE VOMITING.
Water Pollution	Effect of low concentrations on aquatic life is unknown. Fouling to shoreline. May be dangerous if it enters water intakes. Notify local health and wildlife officials. Notify operators of nearby water intakes.

1. RESPONSE TO DISCHARGE

(See Response Methods Handbook, CG 446-4)

Mechanical containment
Should be removed
Chemical and physical treatment

2. LABELS

No label required by Code of Federal Regulations

3. CHEMICAL DESIGNATIONS

3.1 **Synonyms:** American palm kernel oil; Aouara oil; Palm seed oil
3.2 **Coast Guard Compatibility Classification:** Esters (13)
3.3 **Chemical Formula:** Not applicable
3.4 **IMCO/United Nations Numerical Designation:** Not listed

4. OBSERVABLE CHARACTERISTICS

4.1 **Physical State (as shipped):** Liquid
4.2 **Color:** Light yellow
4.3 **Odor:** Weak acid

5. HEALTH HAZARDS

5.1 **Personal Protective Equipment:** Goggles or face shield; rubber gloves.
5.2 **Symptoms Following Exposure:** Oil is essentially nontoxic. Contact with eyes causes mild irritation, and prolonged contact with skin may cause dermatitis.
5.3 **Treatment for Exposure:** EYES: flush with water for at least 15 min. INGESTION: do NOT induce vomiting.
5.4 **Toxicity by Inhalation (Threshold Limit Value):** Data not available
5.5 **Short-Term Inhalation Limits:** Not pertinent
5.6 **Toxicity by Ingestion:** Data not available
5.7 **Late Toxicity:** None known
5.8 **Vapor (Gas) Irritant Characteristics:** Data not available
5.9 **Liquid or Solid Irritant Characteristics:** Data not available
5.10 **Odor Threshold:** Not pertinent

6. FIRE HAZARDS

6.1 **Flash Point:** 398°F C.C.
6.2 **Flammable Limits in Air:** Not pertinent
6.3 **Fire Extinguishing Agents:** Dry chemical, foam, carbon dioxide
6.4 **Fire Extinguishing Agents Not to be Used:** Water or foam may cause frothing; water may be ineffective.
6.5 **Special Hazards of Combustion Products:** Not pertinent
6.6 **Behavior in Fire:** Not pertinent
6.7 **Ignition Temperature:** Data not available
6.8 **Electrical Hazard:** Not pertinent
6.9 **Burning Rate:** 4 mm/min.

7. CHEMICAL REACTIVITY

7.1 **Reactivity with Water:** No reaction
7.2 **Reactivity with Common Materials:** No reaction
7.3 **Stability During Transport:** Stable
7.4 **Neutralizing Agents for Acids and Caustics:** Not pertinent
7.5 **Polymerization:** Not pertinent
7.6 **Inhibitor of Polymerization:** Not pertinent

8. WATER POLLUTION

8.1 **Aquatic Toxicity:** Data not available
8.2 **Waterfowl Toxicity:** Data not available
8.3 **Biological Oxygen Demand (BOD):** Data not available
8.4 **Food Chain Concentration Potential:** None

9. SELECTED MANUFACTURERS

Capital City Products, Inc.
P. O. Box 569
Columbus, Ohio 43216

10. SHIPPING INFORMATION

10.1 **Grades or Purity:** Commercial
10.2 **Storage Temperature:** Ambient
10.3 **Inert Atmosphere:** No requirement
10.4 **Venting:** Open (flame arrester)

11. HAZARD ASSESSMENT CODE

(See Hazard Assessment Handbook, CG 446-3)

A-T-U

12. HAZARD CLASSIFICATIONS

12.1 **Code of Federal Regulations:** Not listed
12.2 **NAS Hazard Rating for Bulk Water Transportation:** Not listed
12.3 **NFPA Hazard Classifications:**

Category	Classification
Health Hazard (Blue)	0
Flammability (Red)	1
Reactivity (Yellow)	0

13. PHYSICAL AND CHEMICAL PROPERTIES

13.1 **Physical State at 15°C and 1 atm:** Liquid
13.2 **Molecular Weight:** Not applicable
13.3 **Boiling Point at 1 atm:** Not pertinent (very high)
13.4 **Freezing Point:** 86°F = 30°C = 303°K
13.5 **Critical Temperature:** Not pertinent
13.6 **Critical Pressure:** Not pertinent
13.7 **Specific Gravity:** 0.908 at 60°C (liquid)
13.8 **Liquid Surface Tension:** (est.) 25 dynes/cm = 0.025 N/m at 30°C
13.9 **Liquid-Water Interfacial Tension:** (est.) 50 dynes/cm = 0.050 N/m at 30°C
13.10 **Vapor (Gas) Specific Gravity:** Not pertinent
13.11 **Ratio of Specific Heats of Vapor (Gas):** Not pertinent
13.12 **Latent Heat of Vaporization:** Not pertinent
13.13 **Heat of Combustion:** (est.) −15,500 Btu/lb = −8,600 cal/g = -360×10^5 J/kg
13.14 **Heat of Decomposition:** Not pertinent
13.15 **Heat of Solution:** Not pertinent
13.16 **Heat of Polymerization:** Not pertinent

NOTES

OVG	OILS, EDIBLE: VEGETABLE

Common Synonyms	Oily liquid	Pale yellow	Weak, fatty odor
	Floats on water.		

Stop discharge if possible.
Call fire department.
Isolate and remove discharged material.
Notify local health and pollution control agencies.

Fire	Combustible. Extinguish with dry chemical, foam or carbon dioxide. Water may be ineffective on fire. Cool exposed containers with water.
Exposure	Not harmful.
Water Pollution	Effect of low concentrations on aquatic life is unknown. Fouling to shoreline. May be dangerous if it enters water intakes. Notify local health and wildlife officials. Notify operators of nearby water intakes.

1. RESPONSE TO DISCHARGE

(See Response Methods Handbook, CG 446-4)

Mechanical containment
Should be removed
Chemical and physical treatment

2. LABELS

No hazard label required by Code of Federal Regulations

3. CHEMICAL DESIGNATIONS

3.1 **Synonyms:** No common synonyms
3.2 **Coast Guard Compatibility Classification:** Ester
3.3 **Chemical Formula:** Not applicable
3.4 **IMCO/United Nations Numerical Designation:** Not listed

4. OBSERVABLE CHARACTERISTICS

4.1 **Physical State (as shipped):** Liquid
4.2 **Color:** Colorless to pale yellow
4.3 **Odor:** Weak fatty

5. HEALTH HAZARDS

5.1 **Personal Protective Equipment:** Goggles or face shield.
5.2 **Symptoms Following Exposure:** None—is a food.
5.3 **Treatment for Exposure:** EYES: flush with water for at least 15 min.
5.4 **Toxicity by Inhalation (Threshold Limit Value):** Not pertinent
5.5 **Short-Term Inhalation Limits:** Not pertinent
5.6 **Toxicity by Ingestion:** None
5.7 **Late Toxicity:** None
5.8 **Vapor (Gas) Irritant Characteristics:** None
5.9 **Liquid or Solid Irritant Characteristics:** None
5.10 **Odor Threshold:** Data not available

6. FIRE HAZARDS

6.1 **Flash Point:** 610°F O.C.
6.2 **Flammable Limits in Air:** Data not available
6.3 **Fire Extinguishing Agents:** Dry chemical, foam, or carbon dioxide
6.4 **Fire Extinguishing Agents Not to be Used:** Water may be ineffective.
6.5 **Special Hazards of Combustion Products:** Not pertinent
6.6 **Behavior in Fire:** Not pertinent
6.7 **Ignition Temperature:** Data not available
6.8 **Electrical Hazard:** Not pertinent
6.9 **Burning Rate:** Data not available

7. CHEMICAL REACTIVITY

7.1 **Reactivity with Water:** No reaction
7.2 **Reactivity with Common Materials:** No reaction
7.3 **Stability During Transport:** Stable
7.4 **Neutralizing Agents for Acids and Caustics:** Not pertinent
7.5 **Polymerization:** Not pertinent
7.6 **Inhibitor of Polymerization:** Not pertinent

8. WATER POLLUTION

8.1 **Aquatic Toxicity:** Data not available
8.2 **Waterfowl Toxicity:** Data not available
8.3 **Biological Oxygen Demand (BOD):** Data not available
8.4 **Food Chain Concentration Potential:** None

9. SELECTED MANUFACTURERS

1. Ashland Oil Co.
   Chemical Products Division
   Peoria, Ill. 61601
2. Sherwin-Williams Co.
   101 Prospect Ave.
   Cleveland, Ohio 44101
3. A. E. Staley Manufacturing Co.
   Decatur, Ill. 62525

10. SHIPPING INFORMATION

10.1 **Grades or Purity:** Data not available
10.2 **Storage Temperature:** Ambient
10.3 **Inert Atmosphere:** No requirement
10.4 **Venting:** Open (flame arrester)

11. HAZARD ASSESSMENT CODE

(See Hazard Assessment Handbook, CG 446-3)

A-T-U

12. HAZARD CLASSIFICATIONS

12.1 **Code of Federal Regulations:** Not listed
12.2 **NAS Hazard Rating for Bulk Water Transportation:** Not listed
12.3 **NFPA Hazard Classifications:**

Category	Classification
Health Hazard (Blue)	0
Flammability (Red)	1
Reactivity (Yellow)	0

13. PHYSICAL AND CHEMICAL PROPERTIES

13.1 **Physical State at 15°C and 1 atm:** Liquid
13.2 **Molecular Weight:** Not pertinent
13.3 **Boiling Point at 1 atm:** Very high
13.4 **Freezing Point:** Not pertinent
13.5 **Critical Temperature:** Not pertinent
13.6 **Critical Pressure:** Not pertinent
13.7 **Specific Gravity:** 0.923 at 25°C (liquid)
13.8 **Liquid Surface Tension:** (est.) 25 dynes/cm = 0.025 N/m at 20°C
13.9 **Liquid-Water Interfacial Tension:** (est.) 50 dynes/cm = 0.05 N/m at 20°C
13.10 **Vapor (Gas) Specific Gravity:** Not pertinent
13.11 **Ratio of Specific Heats of Vapor (Gas):** Not pertinent
13.12 **Latent Heat of Vaporization:** Not pertinent
13.13 **Heat of Combustion:** (est.) −16,000 Btu/lb = −8,870 cal/g = -371×10^5 J/kg
13.14 **Heat of Decomposition:** Not pertinent
13.15 **Heat of Solution:** Not pertinent
13.16 **Heat of Polymerization:** Not pertinent

NOTES

REVISED 1978

OON	OILS, FUEL: 1

Common Synonyms Kerosene Kerosine Range oil JP-1	Watery liquid Colorless Kerosene odor Floats on water.

Stop discharge if possible.
Call fire department.
Avoid contact with liquid.
Isolate and remove discharged material.
Notify local health and pollution control agencies.

Fire	Combustible. Extinguish with dry chemical, foam or carbon dioxide. Water may be ineffective on fire. Cool exposed containers with water.
Exposure	CALL FOR MEDICAL AID. LIQUID Irritating to skin and eyes. Harmful if swallowed. Remove contaminated clothing and shoes. Flush affected areas with plenty of water. IF IN EYES, hold eyelids open and flush with plenty of water. IF SWALLOWED and victim is CONSCIOUS, have victim drink water or milk. DO NOT INDUCE VOMITING.
Water Pollution	Dangerous to aquatic life in high concentrations. Fouling to shoreline. May be dangerous if it enters water intakes. Notify local health and wildlife officials. Notify operators of nearby water intakes.

1. RESPONSE TO DISCHARGE

(See Response Methods Handbook, CG 446-4)

Mechanical containment
Should be removed
Chemical and physical treatment

2. LABELS

No hazard label required by Code of Federal Regulations

3. CHEMICAL DESIGNATIONS

3.1 **Synonyms:** JP-1
Kerosene
Kerosine
Range oil

3.2 **Coast Guard Compatibility Classification:** Miscellaneous hydrocarbon mixtures

3.3 **Chemical Formula:** Not applicable

3.4 **IMCO/United Nations Numerical Designation:** 3.3/1223

4. OBSERVABLE CHARACTERISTICS

4.1 **Physical State (as shipped):** Liquid

4.2 **Color:** Colorless to light brown

4.3 **Odor:** Characteristic

5. HEALTH HAZARDS

5.1 **Personal Protective Equipment:** Protective gloves; goggles or face shield.

5.2 **Symptoms Following Exposure:** INGESTION causes irritation of gastrointestinal tract; pulmonary tract irritation secondary to exhalation of vapors. ASPIRATION causes severe lung irritation with coughing, gagging, dyspnea, substernal distress, and rapidly developing pulmonary edema; signs of bronchopneumonia and pneumonitis appear later; minimal central nervous system depression.

5.3 **Treatment for Exposure:** INGESTION: do NOT lavage or induce vomiting; call physician. ASPIRATION: enforce bed rest; administer oxygen; call physician. EYES: wash with plenty of water. SKIN: wipe off and wash with soap and water.

5.4 **Toxicity by Inhalation (Threshold Limit Value):** 200 ppm (suggested)

5.5 **Short-Term Inhalation Limits:** Data not available

5.6 **Toxicity by Ingestion:** Grade 1; LD_{50} 5-15 g/kg

5.7 **Late Toxicity:** Data not available

5.8 **Vapor (Gas) Irritant Characteristics:** Vapors cause a slight smarting of the eyes or respiratory system if present in high concentrations. The effect is temporary.

5.9 **Liquid or Solid Irritant Characteristics:** Minimum hazard. If spilled on clothing and allowed to remain, may cause smarting and reddening of the skin.

5.10 **Odor Threshold:** 1 ppm

6. FIRE HAZARDS

6.1 **Flash Point:** 100°F C.C.

6.2 **Flammable Limits in Air:** 0.7%–5%

6.3 **Fire Extinguishing Agents:** Dry chemical, foam, or carbon dioxide

6.4 **Fire Extinguishing Agents Not to be Used:** Water may be ineffective

6.5 **Special Hazards of Combustion Products:** Not pertinent

6.6 **Behavior in Fire:** Not pertinent

6.7 **Ignition Temperature:** 444°F

6.8 **Electrical Hazard:** Not pertinent

6.9 **Burning Rate:** 4 mm/min.

7. CHEMICAL REACTIVITY

7.1 **Reactivity with Water:** No reaction

7.2 **Reactivity with Common Materials:** No reaction

7.3 **Stability During Transport:** Stable

7.4 **Neutralizing Agents for Acids and Caustics:** Not pertinent

7.5 **Polymerization:** Not pertinent

7.6 **Inhibitor of Polymerization:** Not pertinent

8. WATER POLLUTION

8.1 **Aquatic Toxicity:** 2990 ppm/24 hr/bluegill/TL_m/fresh water

8.2 **Waterfowl Toxicity:** Data not available

8.3 **Biological Oxygen Demand (BOD):** 53%, 5 days

8.4 **Food Chain Concentration Potential:** None

9. SELECTED MANUFACTURERS

1. Atlantic Richfield
   717 Fifth Ave.
   New York, N. Y. 10022
2. Shell Oil Co.
   1 Shell Plaza
   Houston, Tex. 77001
3. Sun Oil Co.
   St. Davids, Pa. 19087

10. SHIPPING INFORMATION

10.1 **Grades or Purity:** Light hydrocarbon distillate: 100%

10.2 **Storage Temperature:** Ambient

10.3 **Inert Atmosphere:** No requirement

10.4 **Venting:** Open (flame arrester)

11. HAZARD ASSESSMENT CODE

(See Hazard Assessment Handbook, CG 446-3)

A-T-U

12. HAZARD CLASSIFICATIONS

12.1 **Code of Federal Regulations:** Combustible Liquid

12.2 **NAS Hazard Rating for Bulk Water Transportation:**

Category	Rating
Fire	2
Health	
Vapor Irritant	1
Liquid or Solid Irritant	1
Poisons	1
Water Pollution	
Human Toxicity	1
Aquatic Toxicity	1
Aesthetic Effect	3
Reactivity	
Other Chemicals	0
Water	0
Self-Reaction	0

12.3 **NFPA Hazard Classifications:**

Category	Classification
Health Hazard (Blue)	0
Flammability (Red)	2
Reactivity (Yellow)	0

13. PHYSICAL AND CHEMICAL PROPERTIES

13.1 **Physical State at 15°C and 1 atm:** Liquid

13.2 **Molecular Weight:** Not pertinent

13.3 **Boiling Point at 1 atm:** 380–560°F = 193–293°C = 466–566°K

13.4 **Freezing Point:** −45 to −55° F = −43 to −48°C = 230 to 225°K

13.5 **Critical Temperature:** Not pertinent

13.6 **Critical Pressure:** Not pertinent

13.7 **Specific Gravity:** 0.81–0.85 at 15°C (liquid)

13.8 **Liquid Surface Tension:** 23–32 dynes/cm = 0.023–0.032 N/m at 20°C

13.9 **Liquid-Water Interfacial Tension:** 47–49 dynes/cm = 0.047–0.049 N/m at 20°C

13.10 **Vapor (Gas) Specific Gravity:** Not pertinent

13.11 **Ratio of Specific Heats of Vapor (Gas):** Not pertinent

13.12 **Latent Heat of Vaporization:** 110 Btu/lb = 60 cal/g = 2.5×10^5 J/kg

13.13 **Heat of Combustion:** −18,540 Btu/lb = −10,300 cal/g = -431.24×10^5 J/kg

13.14 **Heat of Decomposition:** Not pertinent

13.15 **Heat of Solution:** Not pertinent

13.16 **Heat of Polymerization:** Not pertinent

NOTES

REVISED 1978

OILS, FUEL: 2

Common Synonyms
Home-heating oil

Oily liquid — Yellow-brown — Lube or fuel oil odor

Floats on water.

Stop discharge if possible.
Call fire department.
Avoid contact with liquid.
Isolate and remove discharged material.
Notify local health and pollution control agencies.

Fire

Combustible.
Extinguish with dry chemical, foam or carbon dioxide.
Water may be ineffective on fire.
Cool exposed containers with water.

Exposure

CALL FOR MEDICAL AID.

LIQUID
Irritating to skin and eyes.
If swallowed, will cause nausea, vomiting.
Remove contaminated clothing and shoes.
Flush affected areas with plenty of water.
IF IN EYES, hold eyelids open and flush with plenty of water.
IF SWALLOWED and victim is CONSCIOUS, have victim drink water or milk.
DO NOT INDUCE VOMITING.

Water Pollution

Dangerous to aquatic life in high concentrations.
Fouling to shoreline.
May be dangerous if it enters water intakes.
Notify local health and wildlife officials.
Notify operators of nearby water intakes.

1. RESPONSE TO DISCHARGE

(See Response Methods Handbook, CG 446-4)

Mechanical containment
Should be removed
Chemical and physical treatment

2. LABELS

No hazard label required by Code of Federal Regulations

3. CHEMICAL DESIGNATIONS

3.1 **Synonyms:** Home heating oil
3.2 **Coast Guard Compatibility Classification:** Miscellaneous hydrocarbon mixtures
3.3 **Chemical Formula:** Not applicable
3.4 **IMCO/United Nations Numerical Designation:** 3.3/1223

4. OBSERVABLE CHARACTERISTICS

4.1 **Physical State (as shipped):** Liquid
4.2 **Color:** Light brown
4.3 **Odor:** Like kerosine; characteristic

5. HEALTH HAZARDS

5.1 **Personal Protective Equipment:** Protective gloves; goggles or face shield.
5.2 **Symptoms Following Exposure:** INHALATION causes headache and slight giddiness. INGESTION causes nausea, vomiting, and cramping; depression of central nervous system ranging from mild headache to anesthesia, coma, and death; pulmonary irritation secondary to exhalation of solvent; signs of kidney and liver damage may be delayed. ASPIRATION causes severe lung irritation with coughing, gagging, dyspnea, substernal distress, and rapidly developing pulmonary edema; later, signs of bronchopneumonia and pneumonitis; acute onset of central nervous system excitement followed by depression.
5.3 **Treatment for Exposure:** INGESTION: do NOT induce vomiting. ASPIRATION: enforce bed rest; administer oxygen; seek medical attention. EYES: wash with copious quantity of water. SKIN: remove solvent by wiping and wash with soap and water.
5.4 **Toxicity by Inhalation (Threshold Limit Value):** No single value applicable.
5.5 **Short-Term Inhalation Limits:** Data not available
5.6 **Toxicity by Ingestion:** Grade 1; LD_{50} 5-15 g/kg
5.7 **Late Toxicity:** Data not available
5.8 **Vapor (Gas) Irritant Characteristics:** Slight smarting of eyes or respiratory system if present in high concentrations. The effect is temporary.
5.9 **Liquid or Solid Irritant Characteristics:** Minimum hazard. If spilled on clothing and allowed to remain, may cause smarting and reddening of skin.
5.10 **Odor Threshold:** Data not available

6. FIRE HAZARDS

6.1 **Flash Point:** 136°F C.C.
6.2 **Flammable Limits in Air:** Data not available
6.3 **Fire Extinguishing Agents:** Dry chemical, foam, or carbon dioxide
6.4 **Fire Extinguishing Agents Not to be Used:** Water may be ineffective
6.5 **Special Hazards of Combustion Products:** Not pertinent
6.6 **Behavior in Fire:** Not pertinent
6.7 **Ignition Temperature:** 494°F
6.8 **Electrical Hazard:** Not pertinent
6.9 **Burning Rate:** 4 mm/min.

7. CHEMICAL REACTIVITY

7.1 **Reactivity with Water:** No reaction
7.2 **Reactivity with Common Materials:** No reaction
7.3 **Stability During Transport:** Stable
7.4 **Neutralizing Agents for Acids and Caustics:** Not pertinent
7.5 **Polymerization:** Not pertinent
7.6 **Inhibitor of Polymerization:** Not pertinent

8. WATER POLLUTION

8.1 **Aquatic Toxicity:**
200 ppm/24 hr/juvenile American shad/TL_m/fresh water
20 ppm/96 hr/rainbow trout eggs/TL_m/salt water
8.2 **Waterfowl Toxicity:** Data not available
8.3 **Biological Oxygen Demand (BOD):** Data not available
8.4 **Food Chain Concentration Potential:** None

9. SELECTED MANUFACTURERS

1. Atlantic Richfield Co.
717 Fifth Ave.
New York, N. Y. 10022
2. Shell Oil Co.
1 Shell Plaza
Houston, Tex. 77001
3. Sun Oil Co.
St. Davids, Pa. 19087

10. SHIPPING INFORMATION

10.1 **Grades or Purity:** Commercial
10.2 **Storage Temperature:** Ambient
10.3 **Inert Atmosphere:** No requirement
10.4 **Venting:** Open (flame arrester)

11. HAZARD ASSESSMENT CODE

(See Hazard Assessment Handbook, CG 446-3)

A-T-U

12. HAZARD CLASSIFICATIONS

12.1 **Code of Federal Regulations:** Combustible Liquid
12.2 **NAS Hazard Rating for Bulk Water Transportation:** Not listed
12.3 **NFPA Hazard Classifications:**

Category	Classification
Health Hazard (Blue)	0
Flammability (Red)	2
Reactivity (Yellow)	0

13. PHYSICAL AND CHEMICAL PROPERTIES

13.1 **Physical State at 15°C and 1 atm:** Liquid
13.2 **Molecular Weight:** Not pertinent
13.3 **Boiling Point at 1 atm:** 540-640°F = 282-338°C = 555-611°K
13.4 **Freezing Point:** −20°F = −29°C = 244°K
13.5 **Critical Temperature:** Not pertinent
13.6 **Critical Pressure:** Not pertinent
13.7 **Specific Gravity:** 0.879 at 20°C (liquid)
13.8 **Liquid Surface Tension:** (est.) 25 dynes/cm = 0.025 N/m at 20°C
13.9 **Liquid-Water Interfacial Tension:** (est.) 50 dynes/cm = 0.05 N/m at 20°C
13.10 **Vapor (Gas) Specific Gravity:** Not pertinent
13.11 **Ratio of Specific Heats of Vapor (Gas):** Not pertinent
13.12 **Latent Heat of Vaporization:** Not pertinent
13.13 **Heat of Combustion:** −19,440 Btu/lb = −10,800 cal/g = -452.17×10^5 J/kg
13.14 **Heat of Decomposition:** Not pertinent
13.15 **Heat of Solution:** Not pertinent
13.16 **Heat of Polymerization:** Not pertinent

NOTES

OFR — OILS, FUEL: 4

Common Synonyms			
Residual fuel oil No. 4	Oily liquid	Dark	Lube or fuel oil odor
	Floats on water.		

Stop discharge if possible.
Call fire department.
Avoid contact with liquid.
Isolate and remove discharged material.
Notify local health and pollution control agencies.

Fire

Combustible.
Extinguish with dry chemical, foam or carbon dioxide.
Water may be ineffective on fire.
Cool exposed containers with water.

Exposure

CALL FOR MEDICAL AID.

LIQUID
Irritating to skin and eyes.
Harmful if swallowed.
Remove contaminated clothing and shoes.
Flush affected areas with plenty of water.
IF IN EYES, hold eyelids open and flush with plenty of water.
IF SWALLOWED and victim is CONSCIOUS, have victim drink water or milk.
DO NOT INDUCE VOMITING.

Water Pollution

Effect of low concentrations on aquatic life is unknown.
Fouling to shoreline.
May be dangerous if it enters water intakes.
Notify local health and wildlife officials.
Notify operators of nearby water intakes.

1. RESPONSE TO DISCHARGE

(See Response Methods Handbook, CG 446-4)

Mechanical containment
Should be removed
Chemical and physical treatment

2. LABELS

No hazard label required by Code of Federal Regulations

3. CHEMICAL DESIGNATIONS

3.1 **Synonyms:** Residual fuel oil, No. 4
3.2 **Coast Guard Compatibility Classification:** Miscellaneous hydrocarbon mixtures
3.3 **Chemical Formula:** Not applicable
3.4 **IMCO/United Nations Numerical Designation:** 3.3/1223

4. OBSERVABLE CHARACTERISTICS

4.1 **Physical State (as shipped):** Liquid
4.2 **Color:** Brown
4.3 **Odor:** Characteristic; like kerosene

5. HEALTH HAZARDS

5.1 **Personal Protective Equipment:** Protective gloves; goggles or face shield.
5.2 **Symptoms Following Exposure:** INGESTION: gastrointestinal irritation. ASPIRATION: pulmonary irritation is normally minimal but may become more severe several hours after exposure.
5.3 **Treatment for Exposure:** INGESTION: do NOT lavage or induce vomiting. ASPIRATION: treatment probably not required; delayed development of pulmonary irritation can be detected by serial chest x-rays; consider prophylactic antibiotic regime if condition warrants. EYES: wash with copious quantity of water. SKIN: wipe off and wash with soap and water.
5.4 **Toxicity by Inhalation (Threshold Limit Value):** Not pertinent
5.5 **Short-Term Inhalation Limits:** Not pertinent
5.6 **Toxicity by Ingestion:** Grade 1; LD_{50} 5 to 15 g/kg
5.7 **Late Toxicity:** Data not available
5.8 **Vapor (Gas) Irritant Characteristics:** None
5.9 **Liquid or Solid Irritant Characteristics:** Minimum hazard. If spilled on clothing and allowed to remain, may cause smarting and reddening of the skin.
5.10 **Odor Threshold:** Data not available

6. FIRE HAZARDS

6.1 **Flash Point:** >130°F C.C.
6.2 **Flammable Limits in Air:** 1.0% – 5%
6.3 **Fire Extinguishing Agents:** Dry chemical, foam, or carbon dioxide
6.4 **Fire Extinguishing Agents Not to be Used:** Water may be ineffective.
6.5 **Special Hazards of Combustion Products:** Not pertinent
6.6 **Behavior in Fire:** Not pertinent
6.7 **Ignition Temperature:** 505°F
6.8 **Electrical Hazard:** Not pertinent
6.9 **Burning Rate:** 4 mm/min.

7. CHEMICAL REACTIVITY

7.1 **Reactivity with Water:** No reaction
7.2 **Reactivity with Common Materials:** No reaction
7.3 **Stability During Transport:** Stable
7.4 **Neutralizing Agents for Acids and Caustics:** Not pertinent
7.5 **Polymerization:** Not pertinent
7.6 **Inhibitor of Polymerization:** Not pertinent

8. WATER POLLUTION

8.1 **Aquatic Toxicity:** Data not available
8.2 **Waterfowl Toxicity:** Data not available
8.3 **Biological Oxygen Demand (BOD):** Data not available
8.4 **Food Chain Concentration Potential:** None

9. SELECTED MANUFACTURERS

1. Atlantic Richfield Co.
717 Fifth Ave.
New York, N. Y. 10022
2. Shell Oil Co.
1 Shell Plaza
Houston, Tex. 77001
3. Sun Oil Co.
St. Davids, Pa. 19087

10. SHIPPING INFORMATION

10.1 **Grades or Purity:** Commercial
10.2 **Storage Temperature:** Ambient
10.3 **Inert Atmosphere:** No requirement
10.4 **Venting:** Open (flame arrester)

11. HAZARD ASSESSMENT CODE

(See Hazard Assessment Handbook, CG 446-3)

A-T-U

12. HAZARD CLASSIFICATIONS

12.1 **Code of Federal Regulations:** Combustible Liquid
12.2 **NAS Hazard Rating for Bulk Water Transportation:** Not listed
12.3 **NFPA Hazard Classifications:**

Category	Classification
Health Hazard (Blue)	0
Flammability (Red)	2
Reactivity (Yellow)	0

13. PHYSICAL AND CHEMICAL PROPERTIES

13.1 **Physical State at 15°C and 1 atm:** Liquid
13.2 **Molecular Weight:** Not pertinent
13.3 **Boiling Point at 1 atm:** 214 to >1092°F = 101 to >588° C = 374 to 861° K
13.4 **Freezing Point:** −20 to +15°F = −29 to −9°C = 244 to 264° K
13.5 **Critical Temperature:** Not pertinent
13.6 **Critical Pressure:** Not pertinent
13.7 **Specific Gravity:** 0.904 at 15°C (liquid)
13.8 **Liquid Surface Tension:** Data not available
13.9 **Liquid-Water Interfacial Tension:** Data not available
13.10 **Vapor (Gas) Specific Gravity:** Not pertinent
13.11 **Ratio of Specific Heats of Vapor (Gas):** Not pertinent
13.12 **Latent Heat of Vaporization:** Not pertinent
13.13 **Heat of Combustion:** −17,460 Btu/lb = −9,700 cal/g = -406.1×10^5 J/kg
13.14 **Heat of Decomposition:** Not pertinent
13.15 **Heat of Solution:** Not pertinent
13.16 **Heat of Polymerization:** Not pertinent

NOTES

REVISED 1978

OFV	OILS, FUEL: 5

Common Synonyms
Residual fuel oil No. 5

Oily liquid — Dark — Strong lube oil odor

Usually floats on water.

Stop discharge if possible.
Call fire department.
Avoid contact with liquid.
Isolate and remove discharged material.
Notify local health and pollution control agencies.

Fire

Combustible.
Extinguish with dry chemical, foam or carbon dioxide.
Water may be ineffective on fire.
Cool exposed containers with water.

Exposure

CALL FOR MEDICAL AID.

LIQUID
Irritating to skin and eyes.
Harmful if swallowed.
Remove contaminated clothing and shoes.
Flush affected areas with plenty of water.
IF IN EYES, hold eyelids open and flush with plenty of water.
IF SWALLOWED and victim is CONSCIOUS, have victim drink water or milk.
DO NOT INDUCE VOMITING.

Water Pollution

Effect of low concentrations on aquatic life is unknown.
Fouling to shoreline.
May be dangerous if it enters water intakes.
Notify local health and wildlife officials.
Notify operators of nearby water intakes.

1. RESPONSE TO DISCHARGE

(See Response Methods Handbook, CG 446-4)

Mechanical containment
Should be removed
Chemical and physical treatment

2. LABELS

No hazard label required by Code of Federal Regulations

3. CHEMICAL DESIGNATIONS

3.1 **Synonyms:** Residual fuel oil, No. 5
3.2 **Coast Guard Compatibility Classification:** Miscellaneous hydrocarbon mixtures
3.3 **Chemical Formula:** Not applicable
3.4 **IMCO/United Nations Numerical Designation:** 3.3/1223

4. OBSERVABLE CHARACTERISTICS

4.1 **Physical State (as shipped):** Liquid
4.2 **Color:** Brown
4.3 **Odor:** Characteristic; like kerosene

5. HEALTH HAZARDS

5.1 **Personal Protective Equipment:** Protective gloves; goggles or face shield.
5.2 **Symptoms Following Exposure:** INGESTION: gastrointestinal irritation. ASPIRATION: pulmonary irritation is normally minimal but may become more severe several hours after exposure.
5.3 **Treatment for Exposure:** INGESTION: do NOT lavage or induce vomiting. ASPIRATION: treatment probably not required; delayed development of pulmonary irritation can be detected by serial chest x-rays; consider prophylactic antibiotic regime if condition warrants. EYES: wash with copious quantity of water. SKIN: wipe off and wash with soap and water.
5.4 **Toxicity by Inhalation (Threshold Limit Value):** Not pertinent
5.5 **Short-Term Inhalation Limits:** Not pertinent
5.6 **Toxicity by Ingestion:** Grade 1; LD_{50} 5 to 15 g/kg
5.7 **Late Toxicity:** Data not available
5.8 **Vapor (Gas) Irritant Characteristics:** None
5.9 **Liquid or Solid Irritant Characteristics:** Minimum hazard. If spilled on clothing and allowed to remain, may cause smarting and reddening of the skin.
5.10 **Odor Threshold:** Data not available

6. FIRE HAZARDS

6.1 **Flash Point:** > 130°F C.C.
6.2 **Flammable Limits in Air:** 1% 5%
6.3 **Fire Extinguishing Agents:** Dry chemical, foam, or carbon dioxide
6.4 **Fire Extinguishing Agents Not to be Used:** Water may be ineffective.
6.5 **Special Hazards of Combustion Products:** Not pertinent
6.6 **Behavior in Fire:** Not pertinent
6.7 **Ignition Temperature:** Data not available
6.8 **Electrical Hazard:** Not pertinent
6.9 **Burning Rate:** 4 mm/min.

7. CHEMICAL REACTIVITY

7.1 **Reactivity with Water:** No reaction
7.2 **Reactivity with Common Materials:** No reaction
7.3 **Stability During Transport:** Stable
7.4 **Neutralizing Agents for Acids and Caustics:** Not pertinent
7.5 **Polymerization:** Not pertinent
7.6 **Inhibitor of Polymerization:** Not pertinent

8. WATER POLLUTION

8.1 **Aquatic Toxicity:** Data not available
8.2 **Waterfowl Toxicity:** Data not available
8.3 **Biological Oxygen Demand (BOD):** Data not available
8.4 **Food Chain Concentration Potential:** None

9. SELECTED MANUFACTURERS

1. Atlantic Richfield Co.
717 Fifth Ave.
New York, N. Y. 10022
2. Shell Oil Co.
1 Shell Plaza
Houston, Tex. 77001
3. Sun Oil Co.
St. Davids, Pa. 19087

10. SHIPPING INFORMATION

10.1 **Grades or Purity:** Fuel oil No. 5 (heavy); Fuel oil No. 5 (light)
10.2 **Storage Temperature:** Ambient
10.3 **Inert Atmosphere:** No requirement
10.4 **Venting:** Open (flame arrester)

11. HAZARD ASSESSMENT CODE

(See Hazard Assessment Handbook, CG 446-3)

A-T-U

12. HAZARD CLASSIFICATIONS

12.1 **Code of Federal Regulations:** Combustible Liquid
12.2 **NAS Hazard Rating for Bulk Water Transportation:** Not listed
12.3 **NFPA Hazard Classifications:**

Category	Classification
Health Hazard (Blue)	0
Flammability (Red)	2
Reactivity (Yellow)	0

13. PHYSICAL AND CHEMICAL PROPERTIES

13.1 **Physical State at 15°C and 1 atm:** Liquid
13.2 **Molecular Weight:** Not pertinent
13.3 **Boiling Point at 1 atm:** 426–> 1062°F = 218–>570°C = 491–>843°K
13.4 **Freezing Point:** 0°F = −18°C = 255°K
13.5 **Critical Temperature:** Not pertinent
13.6 **Critical Pressure:** Not pertinent
13.7 **Specific Gravity:** 0.936 at 16°C (liquid)
13.8 **Liquid Surface Tension:** Data not available
13.9 **Liquid-Water Interfacial Tension:** Data not available
13.10 **Vapor (Gas) Specific Gravity:** Not pertinent
13.11 **Ratio of Specific Heats of Vapor (Gas):** Not pertinent
13.12 **Latent Heat of Vaporization:** Not pertinent
13.13 **Heat of Combustion:** −18,000 Btu/lb = −10,000 cal/g = -418.68×10^5 J/kg
13.14 **Heat of Decomposition:** Not pertinent
13.15 **Heat of Solution:** Not pertinent
13.16 **Heat of Polymerization:** Not pertinent

NOTES

REVISED 1978

OSX — OILS, FUEL: 6

Common Synonyms
Bunker C oil
Residual fuel oil
No. 6

Thick liquid — Black — Tar odor

Usually floats on water.

Stop discharge if possible.
Call fire department.
Avoid contact with liquid.
Isolate and remove discharged material.
Notify local health and pollution control agencies.

Fire

Combustible.
Extinguish with dry chemical, foam or carbon dioxide.
Water may be ineffective on fire.
Cool exposed containers with water.

Exposure

CALL FOR MEDICAL AID.

LIQUID
Irritating to skin and eyes.
Harmful if swallowed.
Remove contaminated clothing and shoes.
Flush affected areas with plenty of water.
IF IN EYES, hold eyelids open and flush with plenty of water.
IF SWALLOWED and victim is CONSCIOUS, have victim drink water or milk.
DO NOT INDUCE VOMITING.

Water Pollution

Dangerous to aquatic life in high concentrations.
Fouling to shoreline.
May be dangerous if it enters water intakes.
Notify local health and wildlife officials.
Notify operators of nearby water intakes.

6. FIRE HAZARDS

6.1 **Flash Point:** > 150°F C.C.
6.2 **Flammable Limits in Air:** 1%—5%
6.3 **Fire Extinguishing Agents:** Dry chemical, foam, or carbon dioxide
6.4 **Fire Extinguishing Agents Not to be Used:** Water may be ineffective
6.5 **Special Hazards of Combustion Products:** Not pertinent
6.6 **Behavior in Fire:** Not pertinent
6.7 **Ignition Temperature:** 765°F
6.8 **Electrical Hazard:** Not pertinent
6.9 **Burning Rate:** 4 mm/min.

7. CHEMICAL REACTIVITY

7.1 **Reactivity with Water:** No reaction
7.2 **Reactivity with Common Materials:** No reaction
7.3 **Stability During Transport:** Stable
7.4 **Neutralizing Agents for Acids and Caustics:** Not pertinent
7.5 **Polymerization:** Not pertinent
7.6 **Inhibitor of Polymerization:** Not pertinent

8. WATER POLLUTION

8.1 **Aquatic Toxicity:**
2400 ppm/48 hr/juvenile American shad/ TL_m/fresh water
2417 mg/l /48 hr/juvenile American shad/ TL_m/salt water
8.2 **Waterfowl Toxicity:** Data not available
8.3 **Biological Oxygen Demand (BOD):** Data not available
8.4 **Food Chain Concentration Potential:** Data not available

9. SELECTED MANUFACTURERS

1. Atlantic Richfield Co.
717 Fifth Ave.
New York, N. Y. 10022
2. Shell Oil Co.
1 Shell Plaza
Houston, Tex. 77001
3. Sun Oil Co.
St. Davids, Pa. 19087

10. SHIPPING INFORMATION

10.1 **Grades or Purity:** Commercial
10.2 **Storage Temperature:** Elevated
10.3 **Inert Atmosphere:** No requirement
10.4 **Venting:** Open (flame arrester)

11. HAZARD ASSESSMENT CODE

(See Hazard Assessment Handbook, CG 446-3)

A-T-U

12. HAZARD CLASSIFICATIONS

12.1 **Code of Federal Regulations:** Combustible Liquid
12.2 **NAS Hazard Rating for Bulk Water Transportation:** Not listed
12.3 **NFPA Hazard Classifications:**

Category	Classification
Health Hazard (Blue)	0
Flammability (Red)	2
Reactivity (Yellow)	0

13. PHYSICAL AND CHEMICAL PROPERTIES

13.1 **Physical State at 15°C and 1 atm:** Liquid
13.2 **Molecular Weight:** Not pertinent
13.3 **Boiling Point at 1 atm:** 415—>> 1093°F = 212—>> 588°C = 485—>> 861°K
13.4 **Freezing Point:** 25 to 55°F = –4 to +13°C = 269 to 286°K
13.5 **Critical Temperature:** Not pertinent
13.6 **Critical Pressure:** Not pertinent
13.7 **Specific Gravity:** 0.967 at 16°C (liquid)
13.8 **Liquid Surface Tension:** Data not available
13.9 **Liquid-Water Interfacial Tension:** Data not available
13.10 **Vapor (Gas) Specific Gravity:** Not pertinent
13.11 **Ratio of Specific Heats of Vapor (Gas):** Not pertinent
13.12 **Latent Heat of Vaporization:** Not pertinent
13.13 **Heat of Combustion:** –18,000 Btu/lb = –10,000 cal/g = -418.68×10^5 J/kg
13.14 **Heat of Decomposition:** Not pertinent
13.15 **Heat of Solution:** Not pertinent
13.16 **Heat of Polymerization:** Not pertinent

NOTES

REVISED 1978

OOD

OILS, FUEL: 1-D

Common Synonyms
Diesel oil (light)

Oily liquid Yellow-brown Lube or fuel oil odor

Floats on water.

Stop discharge if possible.
Call fire department.
Avoid contact with liquid.
Isolate and remove discharged material.
Notify local health and pollution control agencies.

Fire

Combustible.

Extinguish with dry chemical, foam or carbon dioxide.
Water may be ineffective on fire.
Cool exposed containers with water.

Exposure

CALL FOR MEDICAL AID.

LIQUID
Irritating to skin and eyes.
Harmful if swallowed.

Remove contaminated clothing and shoes.
Flush affected areas with plenty of water.
IF IN EYES, hold eyelids open and flush with plenty of water.
IF SWALLOWED and victim is CONSCIOUS, have victim drink water or milk.
DO NOT INDUCE VOMITING.

Water Pollution

Dangerous to aquatic life in high concentrations.
Fouling to shoreline.
May be dangerous if it enters water intakes.

Notify local health and wildlife officials.
Notify operators of nearby water intakes.

1. RESPONSE TO DISCHARGE

(See Response Methods Handbook, CG 446-4)

Mechanical containment
Should be removed
Chemical and physical treatment

2. LABELS

No hazard label required by Code of Federal Regulations

3. CHEMICAL DESIGNATIONS

3.1 **Synonyms:** Diesel oil, light
3.2 **Coast Guard Compatibility Classification:** Miscellaneous hydrocarbon mixtures
3.3 **Chemical Formula:** Not applicable
3.4 **IMCO/United Nations Numerical Designation:** 3.1/1270

4. OBSERVABLE CHARACTERISTICS

4.1 **Physical State (as shipped):** Liquid
4.2 **Color:** Light brown
4.3 **Odor:** Characteristic

5. HEALTH HAZARDS

5.1 **Personal Protective Equipment:** Protective gloves; goggles or face shield.
5.2 **Symptoms Following Exposure:** INHALATION causes headache and slight giddiness. INGESTION causes nausea, vomiting, and cramping; depression of central nervous system ranging from mild headache to anesthesia, coma, and death; pulmonary irritation secondary to exhalation of solvent; signs of kidney and liver damage may be delayed. ASPIRATION causes severe lung irritation with coughing, gagging, dyspnea, substernal distress, and rapidly developing pulmonary edema; later, signs of bronchopneumonia and pneumonitis; acute onset of central nervous system excitement followed by depression.
5.3 **Treatment for Exposure:** INGESTION: do NOT induce vomiting; seek medical attention. ASPIRATION: enforce bed rest; administer oxygen. EYES: wash with copious quantity of water. SKIN: remove solvent by wiping and wash with soap and water.
5.4 **Toxicity by Inhalation (Threshold Limit Value):** No single value applicable.
5.5 **Short-Term Inhalation Limits:** Data not available
5.6 **Toxicity by Ingestion:** Grade 1; LD_{50} 5-15 g/kg
5.7 **Late Toxicity:** Data not available
5.8 **Vapor (Gas) Irritant Characteristics:** Slight smarting of eyes or respiratory system if present in high concentrations. The effect is temporary.
5.9 **Liquid or Solid Irritant Characteristics:** Minimum hazard. If spilled on clothing and allowed to remain, may cause smarting and reddening of skin.
5.10 **Odor Threshold:** 0.7 ppm

6. FIRE HAZARDS

6.1 **Flash Point:** 100°F C.C.
6.2 **Flammable Limits in Air:** 1.3%–6%
6.3 **Fire Extinguishing Agents:** Dry chemical, foam, or carbon dioxide
6.4 **Fire Extinguishing Agents Not to be Used:** Water may be ineffective.
6.5 **Special Hazards of Combustion Products:** Not pertinent
6.6 **Behavior in Fire:** Not pertinent
6.7 **Ignition Temperature:** 350-625°F
6.8 **Electrical Hazard:** Not pertinent
6.9 **Burning Rate:** 4 mm/min.

7. CHEMICAL REACTIVITY

7.1 **Reactivity with Water:** No reaction
7.2 **Reactivity with Common Materials:** No reaction
7.3 **Stability During Transport:** Stable
7.4 **Neutralizing Agents for Acids and Caustics:** Not pertinent
7.5 **Polymerization:** Not pertinent
7.6 **Inhibitor of Polymerization:** Not pertinent

8. WATER POLLUTION

8.1 **Aquatic Toxicity:** 204 mg/l/24 hr/juvenile American shad/ TL_m/salt water
8.2 **Waterfowl Toxicity:** 20 mg/kg LD_{50} (mallard)
8.3 **Biological Oxygen Demand (BOD):** Data not available
8.4 **Food Chain Concentration Potential:** None

9. SELECTED MANUFACTURERS

1. Atlantic Richfield Co.
717 Fifth Ave.
New York, N. Y. 10022
2. Shell Oil Co.
1 Shell Plaza
Houston, Tex. 77001
3. Sun Oil Co.
St. Davids, Pa. 19087

10. SHIPPING INFORMATION

10.1 **Grades or Purity:** Diesel fuel 1-D (ASTM)
10.2 **Storage Temperature:** Ambient
10.3 **Inert Atmosphere:** No requirement
10.4 **Venting:** Open (flame arrester)

11. HAZARD ASSESSMENT CODE

(See Hazard Assessment Handbook, CG 446-3)

A-T-U

12. HAZARD CLASSIFICATIONS

12.1 **Code of Federal Regulations:** Combustible Liquid
12.2 **NAS Hazard Rating for Bulk Water Transportation:** Not listed
12.3 **NFPA Hazard Classifications:**

Category	Classification
Health Hazard (Blue)	0
Flammability (Red)	2
Reactivity (Yellow)	0

13. PHYSICAL AND CHEMICAL PROPERTIES

13.1 **Physical State at 15°C and 1 atm:** Liquid
13.2 **Molecular Weight:** Not pertinent
13.3 **Boiling Point at 1 atm:** 380-560°F = 193-293°C = 466-566°K
13.4 **Freezing Point:** −30°F = −34°C = 240°K
13.5 **Critical Temperature:** Not pertinent
13.6 **Critical Pressure:** Not pertinent
13.7 **Specific Gravity:** 0.81-0.85 at 15°C (liquid)
13.8 **Liquid Surface Tension:** 23-32 dynes/cm = 0.023-0.032 N/m at 20°C
13.9 **Liquid-Water Interfacial Tension:** 47-49 dynes/cm = 0.047-0.049 N/m at 20°C
13.10 **Vapor (Gas) Specific Gravity:** Not pertinent
13.11 **Ratio of Specific Heats of Vapor (Gas):** Not pertinent
13.12 **Latent Heat of Vaporization:** 110 Btu/lb = 60 cal/g = 2.5×10^5 J/kg
13.13 **Heat of Combustion:** −18,540 Btu/lb = −10,300 cal/g = -431.24×10^5 J/kg
13.14 **Heat of Decomposition:** Not pertinent
13.15 **Heat of Solution:** Not pertinent
13.16 **Heat of Polymerization:** Not pertinent

NOTES

REVISED 1978

OTD — OILS, FUEL: 2-D

Common Synonyms
Diesel oil, medium

Oily liquid | Yellow-brown | Lube or fuel oil odor

Floats on water.

Stop discharge if possible.
Call fire department.
Avoid contact with liquid.
Isolate and remove discharged material.
Notify local health and pollution control agencies.

Fire

Combustible.
Extinguish with dry chemical, foam, carbon dioxide.
Water may be ineffective on fire.
Cool exposed containers with water.

Exposure

CALL FOR MEDICAL AID.

LIQUID
Irritating to skin and eyes.
Harmful if swallowed.
Remove contaminated clothing and shoes.
Flush affected areas with plenty of water.
IF IN EYES, hold eyelids open and flush with plenty of water.
IF SWALLOWED, and victim is CONSCIOUS, have victim drink water or milk.
DO NOT INDUCE VOMITING.

Water Pollution

Dangerous to aquatic life in high concentrations.
Fouling to shoreline.
May be dangerous if it enters water intakes.
Notify local health and wildlife officials.
Notify operators of nearby water intakes.

1. RESPONSE TO DISCHARGE

(See Response Methods Handbook, CG 446-4)

Mechanical containment
Should be removed
Chemical and physical treatment

2. LABELS

No hazard label required by Code of Federal Regulations

3. CHEMICAL DESIGNATIONS

3.1 **Synonyms:** Diesel oil, medium
3.2 **Coast Guard Compatibility Classification:** Miscellaneous hydrocarbon mixtures
3.3 **Chemical Formula:** Not applicable
3.4 **IMCO/United Nations Numerical Designation:** 3.1/1270

4. OBSERVABLE CHARACTERISTICS

4.1 **Physical State (as shipped):** Liquid
4.2 **Color:** Light brown
4.3 **Odor:** Characteristic; like kerosene

5. HEALTH HAZARDS

5.1 **Personal Protective Equipment:** Protective gloves; goggles or face shield.
5.2 **Symptoms Following Exposure:** INGESTION causes nausea, vomiting, and cramping; depression of central nervous system ranging from mild headache to anesthesia, coma, and death; pulmonary irritation secondary to exhalation of solvent; signs of kidney and liver damage may be delayed. ASPIRATION causes severe lung irritation with coughing, gagging, dyspnea, substernal distress, and rapidly developing pulmonary edema; later, signs of bronchopneumonia and pneumonitis; acute onset of central nervous system excitement followed by depression.
5.3 **Treatment for Exposure:** INGESTION: do NOT induce vomiting. ASPIRATION: enforce bed rest; administer oxygen; seek medical attention. EYES: wash with copious quantity of water. SKIN: remove solvent by wiping and wash with soap and water.
5.4 **Toxicity by Inhalation (Threshold Limit Value):** No single TLV applicable.
5.5 **Short-Term Inhalation Limits:** Data not available
5.6 **Toxicity by Ingestion:** Grade 1; LD_{50} 5-15 g/kg
5.7 **Late Toxicity:** Data not available
5.8 **Vapor (Gas) Irritant Characteristics:** Slight smarting of eyes or respiratory system if present in high concentrations. The effect is temporary.
5.9 **Liquid or Solid Irritant Characteristics:** Minimum hazard. If spilled on clothing and allowed to remain, may cause smarting and reddening of skin.
5.10 **Odor Threshold:** Data not available

6. FIRE HAZARDS

6.1 **Flash Point:** 125°F C.C.
6.2 **Flammable Limits in Air:** 1.3%–6.0%
6.3 **Fire Extinguishing Agents:** Dry chemical, foam, or carbon dioxide
6.4 **Fire Extinguishing Agents Not to be Used:** Water may be ineffective
6.5 **Special Hazards of Combustion Products:** Not pertinent
6.6 **Behavior in Fire:** Not pertinent
6.7 **Ignition Temperature:** 490-545°F
6.8 **Electrical Hazard:** Not pertinent
6.9 **Burning Rate:** 4 mm/min.

7. CHEMICAL REACTIVITY

7.1 **Reactivity with Water:** No reaction
7.2 **Reactivity with Common Materials:** No reaction
7.3 **Stability During Transport:** Stable
7.4 **Neutralizing Agents for Acids and Caustics:** Not pertinent
7.5 **Polymerization:** Not pertinent
7.6 **Inhibitor of Polymerization:** Not pertinent

8. WATER POLLUTION

8.1 **Aquatic Toxicity:** 204 mg/1/24 hr/juvenile American shad/TL_m/salt water
8.2 **Waterfowl Toxicity:** Data not available
8.3 **Biological Oxygen Demand (BOD):** Data not available
8.4 **Food Chain Concentration Potential:** None

9. SELECTED MANUFACTURERS

1. Atlantic Richfield Co.
717 Fifth Ave.
New York, N. Y. 10022
2. Shell Oil Co.
1 Shell Plaza
Houston, Tex. 77001
3. Sun Oil Co.
St. Davids, Pa. 19087

10. SHIPPING INFORMATION

10.1 **Grades or Purity:** Diesel fuel 2-D (ASTM)
10.2 **Storage Temperature:** Ambient
10.3 **Inert Atmosphere:** No requirement
10.4 **Venting:** Open (flame arrester)

11. HAZARD ASSESSMENT CODE

(See Hazard Assessment Handbook, CG 446-3)

A-T-U

12. HAZARD CLASSIFICATIONS

12.1 **Code of Federal Regulations:** Combustible Liquid
12.2 **NAS Hazard Rating for Bulk Water Transportation:** Not listed
12.3 **NFPA Hazard Classifications:**

Category	Classification
Health Hazard (Blue)	0
Flammability (Red)	2
Reactivity (Yellow)	0

13. PHYSICAL AND CHEMICAL PROPERTIES

13.1 **Physical State at 15°C and 1 atm:** Liquid
13.2 **Molecular Weight:** Not pertinent
13.3 **Boiling Point at 1 atm:** 540-640°F = 282-338°C = 555-611°K
13.4 **Freezing Point:** 0°F = −18°C = 255°K
13.5 **Critical Temperature:** Not pertinent
13.6 **Critical Pressure:** Not pertinent
13.7 **Specific Gravity:** 0.87-0.90 at 20°C (liquid)
13.8 **Liquid Surface Tension:** Data not available
13.9 **Liquid-Water Interfacial Tension:** Data not available
13.10 **Vapor (Gas) Specific Gravity:** Not pertinent
13.11 **Ratio of Specific Heats of Vapor (Gas):** Not pertinent
13.12 **Latent Heat of Vaporization:** Not pertinent
13.13 **Heat of Combustion:** −19,440 Btu/lb = −10,800 cal/g = -452.17×10^5 J/kg
13.14 **Heat of Decomposition:** Not pertinent
13.15 **Heat of Solution:** Not pertinent
13.16 **Heat of Polymerization:** Not pertinent

NOTES

REVISED 1978

OAS	OILS, MISCELLANEOUS: ABSORPTION

Common Synonyms
Absorbent oil

Liquid — Colorless to pale yellow — Fuel oil odor

Floats on water.

Stop discharge if possible.
Call fire department.
Avoid contact with liquid.
Isolate and remove discharged material.
Notify local health and pollution control agencies.

Fire

Combustible.
Extinguish with dry chemical, foam or carbon dioxide.
Water may be ineffective on fire.
Cool exposed containers with water.

Exposure

CALL FOR MEDICAL AID.

LIQUID
Irritating to skin and eyes.
Harmful if swallowed.
Remove contaminated clothing and shoes.
Flush affected areas with plenty of water.
IF IN EYES, hold eyelids open and flush with plenty of water.
IF SWALLOWED and victim is CONSCIOUS, have victim drink water or milk.
DO NOT INDUCE VOMITING.

Water Pollution

Effect of low concentrations on aquatic life is unknown.
Fouling to shoreline.
May be dangerous if it enters water intakes.
Notify local health and wildlife officials.
Notify operators of nearby water intakes.

1. RESPONSE TO DISCHARGE

(See Response Methods Handbook, CG 446-4)

Mechanical containment
Should be removed
Chemical and physical treatment

2. LABELS

No hazard label required by Code of Federal Regulations

3. CHEMICAL DESIGNATIONS

3.1 **Synonyms:** Absorbent oil
3.2 **Coast Guard Compatibility Classification:** Miscellaneous hydrocarbon mixtures
3.3 **Chemical Formula:** Not applicable
3.4 **IMCO/United Nations Numerical Designation:** 3.3/1270

4. OBSERVABLE CHARACTERISTICS

4.1 **Physical State (as shipped):** Liquid
4.2 **Color:** Pale yellow to colorless
4.3 **Odor:** Like fuel oil

5. HEALTH HAZARDS

5.1 **Personal Protective Equipment:** Protective gloves; goggles or face shield.
5.2 **Symptoms Following Exposure:** INGESTION: irritation of stomach. ASPIRATION: pulmonary irritation is normally minimal but may become more severe several hours after exposure. (Delayed development can be detected by serial chest x-rays.)
5.3 **Treatment for Exposure:** INGESTION: have victim drink water or milk; do NOT induce vomiting. EYES: wash with copious amounts of water. SKIN: wipe off and wash with soap and water.
5.4 **Toxicity by Inhalation (Threshold Limit Value):** Not pertinent
5.5 **Short-Term Inhalation Limits:** Not pertinent
5.6 **Toxicity by Ingestion:** Grade 1; LD_{50} 5-15 g/kg
5.7 **Late Toxicity:** Data not available
5.8 **Vapor (Gas) Irritant Characteristics:** None
5.9 **Liquid or Solid Irritant Characteristics:** Minimum hazard. If spilled on clothing and allowed to remain, may cause smarting and reddening of skin.
5.10 **Odor Threshold:** Data not available

6. FIRE HAZARDS

6.1 **Flash Point:** 255°F
6.2 **Flammable Limits in Air:** Data not available
6.3 **Fire Extinguishing Agents:** Dry chemical, foam, or carbon dioxide
6.4 **Fire Extinguishing Agents Not to be Used:** Water may be ineffective
6.5 **Special Hazards of Combustion Products:** Not pertinent
6.6 **Behavior in Fire:** Not pertinent
6.7 **Ignition Temperature:** 300°F
6.8 **Electrical Hazard:** Not pertinent
6.9 **Burning Rate:** 4 mm/min.

7. CHEMICAL REACTIVITY

7.1 **Reactivity with Water:** No reaction
7.2 **Reactivity with Common Materials:** No reaction
7.3 **Stability During Transport:** Stable
7.4 **Neutralizing Agents for Acids and Caustics:** Not pertinent
7.5 **Polymerization:** Not pertinent
7.6 **Inhibitor of Polymerization:** Not pertinent

8. WATER POLLUTION

8.1 **Aquatic Toxicity:** Data not available
8.2 **Waterfowl Toxicity:** Data not available
8.3 **Biological Oxygen Demand (BOD):** Data not available
8.4 **Food Chain Concentration Potential:** None

9. SELECTED MANUFACTURERS

1. Cities Service Co., Inc.
   60 Wall Tower
   New York, N. Y. 10005
2. Exxon Co.
   Houston, Tex. 77001
3. Standard Oil Co. (Indiana)
   910 S. Michigan Ave.
   Chicago, Ill. 60605

10. SHIPPING INFORMATION

10.1 **Grades or Purity:** Commercial
10.2 **Storage Temperature:** Ambient
10.3 **Inert Atmosphere:** No requirement
10.4 **Venting:** Open (flame arrester)

11. HAZARD ASSESSMENT CODE

(See Hazard Assessment Handbook, CG 446-3)

A-T-U

12. HAZARD CLASSIFICATIONS

12.1 **Code of Federal Regulations:** Combustible Liquid
12.2 **NAS Hazard Rating for Bulk Water Transportation:** Not listed
12.3 **NFPA Hazard Classifications:** Not listed

13. PHYSICAL AND CHEMICAL PROPERTIES

13.1 **Physical State at 15°C and 1 atm:** Liquid
13.2 **Molecular Weight:** Not pertinent
13.3 **Boiling Point at 1 atm:** >500°F = >260°C = >533°K
13.4 **Freezing Point:** Not pertinent
13.5 **Critical Temperature:** Not pertinent
13.6 **Critical Pressure:** Not pertinent
13.7 **Specific Gravity:** (est.) 0.85 at 20°C (liquid)
13.8 **Liquid Surface Tension:** (est.) 25 dynes/cm = 0.025 N/m at 20°C
13.9 **Liquid-Water Interfacial Tension:** (est.) 50 dynes/cm = 0.05 N/m at 20°C
13.10 **Vapor (Gas) Specific Gravity:** Not pertinent
13.11 **Ratio of Specific Heats of Vapor (Gas):** Not pertinent
13.12 **Latent Heat of Vaporization:** Not pertinent
13.13 **Heat of Combustion:** (est.) −18,000 Btu/lb = −10,000 cal/g = -420×10^5 J/kg
13.14 **Heat of Decomposition:** Not pertinent
13.15 **Heat of Solution:** Not pertinent
13.16 **Heat of Polymerization:** Not pertinent

NOTES

REVISED 1978

OCT — OILS, MISCELLANEOUS: COAL TAR

Common Synonyms			
Light oil	Liquid	Colorless to yellow	Pleasant odor
	Floats on water. Flammable, irritating vapor is produced.		

Stop discharge if possible. Keep people away.
Shut off ignition sources and call fire department.
Stay upwind and use water spray to "knock down" vapor.
Avoid contact with liquid and vapor.
Isolate and remove discharged material.
Notify local health and pollution control agencies.

Fire

FLAMMABLE.
Flashback along vapor trail may occur.
Vapor may explode if ignited in an enclosed area.
Extinguish with dry chemical, foam or carbon dioxide.
Water may be ineffective on fire.
Cool exposed containers with water.

Exposure

CALL FOR MEDICAL AID.

VAPOR
Irritating to eyes, nose and throat.
Move to fresh air.

LIQUID
Irritating to skin and eyes.
Harmful if swallowed.
Remove contaminated clothing and shoes.
Flush affected areas with plenty of water.
IF IN EYES, hold eyelids open and flush with plenty of water.
IF SWALLOWED and victim is CONSCIOUS, have victim drink water or milk.
DO NOT INDUCE VOMITING.

Water Pollution

Effect of low concentrations on aquatic life is unknown.
Fouling to shoreline.
May be dangerous if it enters water intakes.
Notify local health and wildlife officials.
Notify operators of nearby water intakes.

1. RESPONSE TO DISCHARGE

(See Response Methods Handbook, CG 446-4)

Issue warning—high flammability
Evacuate area

2. LABEL

Red

3. CHEMICAL DESIGNATIONS

3.1 **Synonyms:** Light oil
3.2 **Coast Guard Compatibility Classification:** Miscellaneous hydrocarbon mixtures
3.3 **Chemical Formula:** Not applicable
3.4 **IMCO/United Nations Numerical Designation:** 3.2/1136

4. OBSERVABLE CHARACTERISTICS

4.1 **Physical State (as shipped):** Liquid
4.2 **Color:** Colorless to yellow
4.3 **Odor:** Aromatic; like benzene; pleasant

5. HEALTH HAZARDS

5.1 **Personal Protective Equipment:** Protective gloves; goggles or face shield.
5.2 **Symptoms Following Exposure:** Vapor causes slight irritation of nose and throat, smarting of eyes. Liquid may irritate skin on prolonged contact.
5.3 **Treatment for Exposure:** INGESTION: have victim drink water or milk; do NOT induce vomiting. EYES: flush with water for at least 15 min. SKIN: wipe off and wash with soap and water.
5.4 **Toxicity by Inhalation (Threshold Limit Value):** Data not available
5.5 **Short-Term Inhalation Limits:** Data not available
5.6 **Toxicity by Ingestion:** Data not available
5.7 **Late Toxicity:** Data not available
5.8 **Vapor (Gas) Irritant Characteristics:** Vapors cause a slight smarting of the eyes or respiratory system if present in high concentrations. The effect is temporary.
5.9 **Liquid or Solid Irritant Characteristics:** Minimum hazard. If spilled on clothing and allowed to remain, may cause smarting and reddening of the skin.
5.10 **Odor Threshold:** Data not available

6. FIRE HAZARDS

6.1 **Flash Point:** 60°–77°F C.C.
6.2 **Flammable Limits in Air:** 1.3%–8%
6.3 **Fire Extinguishing Agents:** Dry chemical, foam, or carbon dioxide
6.4 **Fire Extinguishing Agents Not to be Used:** Water may be ineffective.
6.5 **Special Hazards of Combustion Products:** Not pertinent
6.6 **Behavior in Fire:** Not pertinent
6.7 **Ignition Temperature:** Data not available
6.8 **Electrical Hazard:** Not pertinent
6.9 **Burning Rate:** 4 mm/min.

7. CHEMICAL REACTIVITY

7.1 **Reactivity with Water:** No reaction
7.2 **Reactivity with Common Materials:** No reaction
7.3 **Stability During Transport:** Stable
7.4 **Neutralizing Agents for Acids and Caustics:** Not pertinent
7.5 **Polymerization:** Not pertinent
7.6 **Inhibitor of Polymerization:** Not pertinent

8. WATER POLLUTION

8.1 **Aquatic Toxicity:** Data not available
8.2 **Waterfowl Toxicity:** Data not available
8.3 **Biological Oxygen Demand (BOD):** Data not available
8.4 **Food Chain Concentration Potential:** None

9. SELECTED MANUFACTURERS

1. Crowley Tar Products Co., Inc.
   271 Madison Ave.
   New York, N. Y. 10016
2. Koppers Co.
   Organic Materials Division
   Koppers Bldg.
   Pittsburgh, Pa. 15219
3. Neville Chemical Co.
   Neville Island
   Pittsburgh, Pa. 15225

10. SHIPPING INFORMATION

10.1 **Grades or Purity:** Various compositions, depending on type of coal used and boiling range taken.
10.2 **Storage Temperature:** Ambient
10.3 **Inert Atmosphere:** No requirement
10.4 **Venting:** Open (flame arrester) or pressure-vacuum

11. HAZARD ASSESSMENT CODE

(See Hazard Assessment Handbook, CG 446-3)

A-T-U

12. HAZARD CLASSIFICATIONS

12.1 **Code of Federal Regulations:** Flammable liquid
12.2 **NAS Hazard Rating for Bulk Water Transportation:**

Category	Rating
Fire	3
Health	
Vapor Irritant	1
Liquid or Solid Irritant	1
Poisons	2
Water Pollution	
Human Toxicity	2
Aquatic Toxicity	2
Aesthetic Effect	4
Reactivity	
Other Chemicals	2
Water	0
Self-Reaction	0

12.3 **NFPA Hazard Classifications:** Not listed

13. PHYSICAL AND CHEMICAL PROPERTIES

13.1 **Physical State at 15°C and 1 atm:** Liquid
13.2 **Molecular Weight:** Not pertinent
13.3 **Boiling Point at 1 atm:** 223-333°F = 106-167°C = 379-440°K
13.4 **Freezing Point:** Not pertinent
13.5 **Critical Temperature:** Not pertinent
13.6 **Critical Pressure:** Not pertinent
13.7 **Specific Gravity:** (est.) 0.90 at 20°C (liquid)
13.8 **Liquid Surface Tension:** (est.) 25 dynes/cm = 0.025 N/m at 20°C
13.9 **Liquid-Water Interfacial Tension:** Data not available
13.10 **Vapor (Gas) Specific Gravity:** Not pertinent
13.11 **Ratio of Specific Heats of Vapor (Gas):** (est.) 1.071
13.12 **Latent Heat of Vaporization:** (est.) 107 Btu/lb = 59.8 cal/g = 2.5×10^5 J/kg
13.13 **Heat of Combustion:** −17,440 Btu/lb = −9,690 cal/g = -405.7×10^5 J/kg
13.14 **Heat of Decomposition:** Not pertinent
13.15 **Heat of Solution:** Not pertinent
13.16 **Heat of Polymerization:** Not pertinent

NOTES

REVISED 1978

OCR

OILS, MISCELLANEOUS: CROTON

Common Synonyms			
Croton tiglium oil Croton oil	Liquid	Dark	Unpleasant odor
	Floats on water.		

AVOID CONTACT WITH LIQUID. KEEP PEOPLE AWAY.
Stop discharge if possible.
Call fire department.
Isolate and remove discharged material.
Notify local health and pollution control agencies.

Fire

Combustible.
Extinguish with dry chemicals, foam or carbon dioxide.
Water may be ineffective on fire.

Exposure

CALL FOR MEDICAL AID.
LIQUID
POISONOUS IF SWALLOWED OR IF SKIN IS EXPOSED.
Irritating to skin and eyes.
Remove contaminated clothing and shoes.
Flush affected areas with plenty of water.
IF IN EYES, hold eyelids open and flush with plenty of water.
IF SWALLOWED and victim is CONSCIOUS, have victim drink water or milk.
DO NOT INDUCE VOMITING.

Water Pollution

Effect of low concentrations on aquatic life is unknown.
Fouling to shoreline.
May be dangerous if it enters water intakes.
Notify local health and wildlife officials.
Notify operators of nearby water intakes.

1. RESPONSE TO DISCHARGE

(See Response Methods Handbook, CG 446-4)

Mechanical containment
Should be removed
Chemical and physical treatment

2. LABELS

No label required by Code of Federal Regulations

3. CHEMICAL DESIGNATIONS

3.1 **Synonyms:** Crotonoel; *Croton tiglium L.* oil
3.2 **Coast Guard Compatibility Classification:** Esters (13)
3.3 **Chemical Formula:** Not applicable
3.4 **IMCO/United Nations Numerical Designation:** Not listed

4. OBSERVABLE CHARACTERISTICS

4.1 **Physical State (as shipped):** Liquid
4.2 **Color:** Dark
4.3 **Odor:** Unpleasant, acrid

5. HEALTH HAZARDS

5.1 **Personal Protective Equipment:** Goggles or face shield; rubber gloves and any other protective clothing to prevent contact with skin.
5.2 **Symptoms Following Exposure:** Contact of liquid with eyes causes severe irritation. May induce severe skin irritation, inflammation, swelling, and pustule formation. Absorption through the skin may cause purging. Ingestion causes burning of the mouth and stomach and drastic purging, possibly leading to collapse and death. Small doses have a strong laxative effect.
5.3 **Treatment for Exposure:** EYES: flush with water; a 2.5% hydroxycortisone ointment is recommended. SKIN: remove as much liquid as possible from skin by use of a good solvent such as acetone or alcohol; wash with soap and water. INGESTION: for gastrointestinal symptoms, use demulcents; further treatment is symptomatic; do NOT induce vomiting.
5.4 **Toxicity by Inhalation (Threshold Limit Value):** Data not available
5.5 **Short-Term Inhalation Limits:** Data not available
5.6 **Toxicity by Ingestion:** Grade 4; LD_{50} < 50 mg/kg
5.7 **Late Toxicity:** Has been used in cancer research as a promoter for other compounds that cause skin cancer.
5.8 **Vapor (Gas) Irritant Characteristics:** Data not available
5.9 **Liquid or Solid Irritant Characteristics:** Data not available
5.10 **Odor Threshold:** Data not available

6. FIRE HAZARDS

6.1 **Flash Point:** Data not available
6.2 **Flammable Limits in Air:** Not pertinent
6.3 **Fire Extinguishing Agents:** Dry chemical, foam, carbon dioxide
6.4 **Fire Extinguishing Agents Not to be Used:** Water may be ineffective.
6.5 **Special Hazards of Combustion Products:** Not pertinent
6.6 **Behavior in Fire:** Not pertinent
6.7 **Ignition Temperature:** Data not available
6.8 **Electrical Hazard:** Not pertinent
6.9 **Burning Rate:** 4 mm/min.

7. CHEMICAL REACTIVITY

7.1 **Reactivity with Water:** No reaction
7.2 **Reactivity with Common Materials:** No reaction
7.3 **Stability During Transport:** Stable
7.4 **Neutralizing Agents for Acids and Caustics:** Not pertinent
7.5 **Polymerization:** Not pertinent
7.6 **Inhibitor of Polymerization:** Not pertinent

8. WATER POLLUTION

8.1 **Aquatic Toxicity:** Data not available
8.2 **Waterfowl Toxicity:** Data not available
8.3 **Biological Oxygen Demand (BOD):** Data not available
8.4 **Food Chain Concentration Potential:** None

9. SELECTED MANUFACTURERS

Consolidated Midland Corporation
195 East Main Street
Brewster, N. Y. 10509

10. SHIPPING INFORMATION

10.1 **Grades or Purity:** Technical
10.2 **Storage Temperature:** Ambient
10.3 **Inert Atmosphere:** No requirement
10.4 **Venting:** Open (flame arrester)

11. HAZARD ASSESSMENT CODE

(See Hazard Assessment Handbook, CG 446-3)

A-T-U

12. HAZARD CLASSIFICATIONS

12.1 **Code of Federal Regulations:** Not listed
12.2 **NAS Hazard Rating for Bulk Water Transportation:** Not listed
12.3 **NFPA Hazard Classifications:** Not listed

13. PHYSICAL AND CHEMICAL PROPERTIES

13.1 **Physical State at 15°C and 1 atm:** Liquid
13.2 **Molecular Weight:** Not pertinent
13.3 **Boiling Point at 1 atm:** Not pertinent (very high)
13.4 **Freezing Point:** 0 to 18°F = −18 to −8°C = 255 to 265°K
13.5 **Critical Temperature:** Not pertinent
13.6 **Critical Pressure:** Not pertinent
13.7 **Specific Gravity:** (est.) 0.946 at 15°C (liquid)
13.8 **Liquid Surface Tension:** (est.) 25 dynes/cm = 0.025 N/m at 20°C
13.9 **Liquid-Water Interfacial Tension:** (est.) 50 dynes/cm = 0.050 N/m at 20°C
13.10 **Vapor (Gas) Specific Gravity:** Not pertinent
13.11 **Ratio of Specific Heats of Vapor (Gas):** Not pertinent
13.12 **Latent Heat of Vaporization:** Not pertinent
13.13 **Heat of Combustion:** (est.) −16,800 Btu/lb = −9,300 cal/g = -390×10^5 J/kg
13.14 **Heat of Decomposition:** Not pertinent
13.15 **Heat of Solution:** Not pertinent
13.16 **Heat of Polymerization:** Not pertinent

NOTES

OILS, MISCELLANEOUS: LINSEED

Common Synonyms	Liquid	Light yellow to dark yellow	Paint-like odor
Flaxseed oil Raw linseed oil Linseed oil	Floats on water.		

Stop discharge if possible.
Call fire department.
Isolate and remove discharged material.
Notify local health and pollution control agencies.

Fire

Combustible.
Extinguish with dry chemicals, foam or carbon dioxide.
Water may be ineffective on fire.

Exposure

LIQUID
Not harmful.
Do not induce vomiting.

Water Pollution

Effect of low concentrations on aquatic life is unknown.
Fouling to shoreline.
May be dangerous if it enters water intakes.
Notify local health and wildlife officials.
Notify operators of nearby water intakes.

1. RESPONSE TO DISCHARGE

(See Response Methods Handbook, CG 446-4)

Mechanical containment
Should be removed
Chemical and physical treatment

2. LABELS

No label required by Code of Federal Regulations

3. CHEMICAL DESIGNATIONS

3.1 **Synonyms:** Flaxseed oil; Raw linseed oil
3.2 **Coast Guard Compatibility Classification:** Esters (13)
3.3 **Chemical Formula:** Not applicable
3.4 **IMCO/United Nations Numerical Designation:** Not listed

4. OBSERVABLE CHARACTERISTICS

4.1 **Physical State (as shipped):** Liquid
4.2 **Color:** Pale yellow to dark amber
4.3 **Odor:** Like oil-base paint

5. HEALTH HAZARDS

5.1 **Personal Protective Equipment:** Goggles or face shield; rubber gloves
5.2 **Symptoms Following Exposure:** Contact of liquid with eyes causes mild irritation. Prolonged contact with skin can cause dermatitis. Ingestion of large doses (over 1 oz) has laxative effect.
5.3 **Treatment for Exposure:** EYES: flush with water for at least 15 min. SKIN: wipe off; wash with soap and water. INGESTION: do NOT induce vomiting.
5.4 **Toxicity by Inhalation (Threshold Limit Value):** Data not available
5.5 **Short-Term Inhalation Limits:** Not pertinent
5.6 **Toxicity by Ingestion:** Grade 0; LD_{50} > 15 g/kg
5.7 **Late Toxicity:** Liver damage in rats (from addition of oil to diet)
5.8 **Vapor (Gas) Irritant Characteristics:** Data not available
5.9 **Liquid or Solid Irritant Characteristics:** Data not available
5.10 **Odor Threshold:** Data not available

6. FIRE HAZARDS

6.1 **Flash Point:** 535°F O.C.; 403°F C.C.
6.2 **Flammable Limits in Air:** Not pertinent
6.3 **Fire Extinguishing Agents:** Dry chemical, foam, carbon dioxide
6.4 **Fire Extinguishing Agents Not to be Used:** Water or foam may cause frothing; water may be ineffective
6.5 **Special Hazards of Combustion Products:** Not pertinent
6.6 **Behavior in Fire:** Not pertinent
6.7 **Ignition Temperature:** 650°F
6.8 **Electrical Hazard:** Not pertinent
6.9 **Burning Rate:** 4 mm/min.

7. CHEMICAL REACTIVITY

7.1 **Reactivity with Water:** No reaction
7.2 **Reactivity with Common Materials:** No reaction
7.3 **Stability During Transport:** Stable
7.4 **Neutralizing Agents for Acids and Caustics:** Not pertinent
7.5 **Polymerization:** Not pertinent
7.6 **Inhibitor of Polymerization:** Not pertinent

8. WATER POLLUTION

8.1 **Aquatic Toxicity:** Data not available
8.2 **Waterfowl Toxicity:** Data not available
8.3 **Biological Oxygen Demand (BOD):** Data not available
8.4 **Food Chain Concentration Potential:** None

9. SELECTED MANUFACTURERS

1. Cargill, Inc.
   Cargill Building
   Minneapolis, Minn. 55402
2. PVO International, Inc.
   416 Division Street
   Boonton, N. J. 07005
3. Textron, Inc.
   Spencer Kellogg Division
   120 Delaware Ave.
   Buffalo, N. Y. 14240

10. SHIPPING INFORMATION

10.1 **Grades or Purity:** Raw grade, varnish grade; grinding grade; heat-bodied grade; blown grade
10.2 **Storage Temperature:** Ambient
10.3 **Inert Atmosphere:** No requirement
10.4 **Venting:** Open (flame arrester)

11. HAZARD ASSESSMENT CODE

(See Hazard Assessment Handbook, CG 446-3)

A-T-U

12. HAZARD CLASSIFICATIONS

12.1 **Code of Federal Regulations:** Not listed
12.2 **NAS Hazard Rating for Bulk Water Transportation:** Not listed
12.3 **NFPA Hazard Classifications:**

Category	Classification
Health Hazard (Blue)	0
Flammability (Red)	1
Reactivity (Yellow)	0

13. PHYSICAL AND CHEMICAL PROPERTIES

13.1 **Physical State at 15°C and 1 atm:** Liquid
13.2 **Molecular Weight:** Not pertinent
13.3 **Boiling Point at 1 atm:** Not pertinent (very high)
13.4 **Freezing Point:** −2°F = −19°C = 254°K
13.5 **Critical Temperature:** Not pertinent
13.6 **Critical Pressure:** Not pertinent
13.7 **Specific Gravity:** 0.932 at 20°C (liquid)
13.8 **Liquid Surface Tension:** (est.) 25 dynes/cm = 0.025 N/m at 20°C
13.9 **Liquid-Water Interfacial Tension:** (est.) 50 dynes/cm = 0.050 N/m at 20°C
13.10 **Vapor (Gas) Specific Gravity:** Not pertinent
13.11 **Ratio of Specific Heats of Vapor (Gas):** Not pertinent
13.12 **Latent Heat of Vaporization:** Not pertinent
13.13 **Heat of Combustion:** −16,800 Btu/lb = −9,300 cal/g = -390×10^5 J/kg
13.14 **Heat of Decomposition:** Not pertinent
13.15 **Heat of Solution:** Not pertinent
13.16 **Heat of Polymerization:** Not pertinent

NOTES

OLB

OILS, MISCELLANEOUS: LUBRICATING

Common Synonyms			
Crankcase oil Transmission oil Motor oil	Oily liquid	Yellow-brown	Lube oil odor
	Floats on water.		

Stop discharge if possible.
Call fire department.
Avoid contact with liquid.
Isolate and remove discharged material.
Notify local health and pollution control agencies.

Fire

Combustible.
Extinguish with dry chemical, foam or carbon dioxide.
Water may be ineffective on fire.
Cool exposed containers with water.

Exposure

CALL FOR MEDICAL AID.

LIQUID
Irritating to skin and eyes.
Harmful if swallowed.
Remove contaminated clothing and shoes.
Flush affected areas with plenty of water.
IF IN EYES, hold eyelids open and flush with plenty of water.
IF SWALLOWED and victim is CONSCIOUS, have victim drink water or milk.
DO NOT INDUCE VOMITING.

Water Pollution

Effect of low concentrations on aquatic life is unknown.
Fouling to shoreline.
May be dangerous if it enters water intakes.
Notify local health and wildlife officials.
Notify operators of nearby water intakes.

1. RESPONSE TO DISCHARGE

(See Response Methods Handbook, CG 446-4)

Mechanical containment
Should be removed
Chemical and physical treatment

2. LABELS

No hazard label required by Code of Federal Regulations

3. CHEMICAL DESIGNATIONS

3.1 **Synonyms:** Crankcase oil; Motor oil; Transmission oil
3.2 **Coast Guard Compatibility Classification:** Miscellaneous hydrocarbon mixtures
3.3 **Chemical Formula:** Not applicable
3.4 **IMCO/United Nations Numerical Designation:** 3.3/1270

4. OBSERVABLE CHARACTERISTICS

4.1 **Physical State (as shipped):** Liquid
4.2 **Color:** Yellow fluorescent
4.3 **Odor:** Characteristic

5. HEALTH HAZARDS

5.1 **Personal Protective Equipment:** Protective gloves; goggles or face shield.
5.2 **Symptoms Following Exposure:** INGESTION: minimal gastrointestinal tract irritation; increased frequency of bowel passage may occur. ASPIRATION: pulmonary irritation is normally minimal but may become more severe several hours after exposure.
5.3 **Treatment for Exposure:** INGESTION: do NOT lavage or induce vomiting. ASPIRATION: treatment probably not required; delayed development of pulmonary irritation can be detected by serial chest x-rays. EYES: wash with copious quantity of water. SKIN: wipe off and wash with soap and water.
5.4 **Toxicity by Inhalation (Threshold Limit Value):** Data not available
5.5 **Short-Term Inhalation Limits:** Data not available
5.6 **Toxicity by Ingestion:** Grade 1; LD_{50} 5 to 15 g/kg
5.7 **Late Toxicity:** Data not available
5.8 **Vapor (Gas) Irritant Characteristics:** Vapors cause a slight smarting of the eyes or respiratory system if present in high concentrations. The effect is temporary.
5.9 **Liquid or Solid Irritant Characteristics:** Minimum hazard. If spilled on clothing and allowed to remain, may cause smarting and reddening of the skin.
5.10 **Odor Threshold:** Data not available

6. FIRE HAZARDS

6.1 **Flash Point:** 300°F-450°F C.C.
6.2 **Flammable Limits in Air:** Data not available
6.3 **Fire Extinguishing Agents:** Dry chemical, foam, or carbon dioxide
6.4 **Fire Extinguishing Agents Not to be Used:** Water or foam may cause frothing.
6.5 **Special Hazards of Combustion Products:** Not pertinent
6.6 **Behavior in Fire:** Not pertinent
6.7 **Ignition Temperature:** 500°F 700°F
6.8 **Electrical Hazard:** Not pertinent
6.9 **Burning Rate:** 4 mm/min.

7. CHEMICAL REACTIVITY

7.1 **Reactivity with Water:** No reaction
7.2 **Reactivity with Common Materials:** No reaction
7.3 **Stability During Transport:** Stable
7.4 **Neutralizing Agents for Acids and Caustics:** Not pertinent
7.5 **Polymerization:** Not pertinent
7.6 **Inhibitor of Polymerization:** Not pertinent

8. WATER POLLUTION

8.1 **Aquatic Toxicity:** Data not available
8.2 **Waterfowl Toxicity:** Data not available
8.3 **Biological Oxygen Demand (BOD):** Data not available
8.4 **Food Chain Concentration Potential:** None

9. SELECTED MANUFACTURERS

1. Shell Oil Co.
   1 Shell Plaza
   Houston, Tex. 77001
2. Standard Oil Co. (Indiana)
   910 S. Michigan Ave.
   Chicago, Ill. 60605
3. Sun Oil Co.
   St. Davids, Pa. 19087

10. SHIPPING INFORMATION

10.1 **Grades or Purity:** Various viscosities
10.2 **Storage Temperature:** Ambient
10.3 **Inert Atmosphere:** No requirement
10.4 **Venting:** Open (flame arrester)

11. HAZARD ASSESSMENT CODE

(See Hazard Assessment Handbook, CG 446-3)

A-T-U

12. HAZARD CLASSIFICATIONS

12.1 **Code of Federal Regulations:** Not listed
12.2 **NAS Hazard Rating for Bulk Water Transportation:** Not listed
12.3 **NFPA Hazard Classifications:**

Category	Classification
Health Hazard (Blue)	0
Flammability (Red)	1
Reactivity (Yellow)	0

13. PHYSICAL AND CHEMICAL PROPERTIES

13.1 **Physical State at 15°C and 1 atm:** Liquid
13.2 **Molecular Weight:** Not pertinent
13.3 **Boiling Point at 1 atm:** Very high
13.4 **Freezing Point:** Not pertinent
13.5 **Critical Temperature:** Not pertinent
13.6 **Critical Pressure:** Not pertinent
13.7 **Specific Gravity:** (est.) 0.902 at 20°C(liquid)
13.8 **Liquid Surface Tension:** 36-37.5 dynes/cm = 0.036-0.0375 N/m at 20°C
13.9 **Liquid-Water Interfacial Tension:** 33-54 dynes/cm = 0.033-0.054 N/m at 20°C
13.10 **Vapor (Gas) Specific Gravity:** Not pertinent
13.11 **Ratio of Specific Heats of Vapor (Gas):** Not pertinent
13.12 **Latent Heat of Vaporization:** Not pertinent
13.13 **Heat of Combustion:** $-18{,}486$ Btu/lb = $-10{,}270$ cal/g = -429.98×10^5 J/kg
13.14 **Heat of Decomposition:** Not pertinent
13.15 **Heat of Solution:** Not pertinent
13.16 **Heat of Polymerization:** Not pertinent

NOTES

REVISED 1978

OMN	OILS, MISCELLANEOUS: MINERAL

Common Synonyms White oil Liquid Petrolatum	Oily liquid Colorless Odorless Floats on water.

Stop discharge if possible.
Call fire department.
Avoid contact with liquid.
Isolate and remove discharged material.
Notify local health and pollution control agencies.

Fire	Combustible. Extinguish with dry chemical, foam or carbon dioxide. Water may be ineffective on fire. Cool exposed containers with water.
Exposure	CALL FOR MEDICAL AID. LIQUID Irritating to skin and eyes. Remove contaminated clothing and shoes. Flush affected areas with plenty of water. IF IN EYES, hold eyelids open and flush with plenty of water. IF SWALLOWED and victim is CONSCIOUS, have victim drink water or milk. DO NOT INDUCE VOMITING.
Water Pollution	Effect of low concentrations on aquatic life is unknown. Fouling to shoreline. May be dangerous if it enters water intakes. Notify local health and wildlife officials. Notify operators of nearby water intakes.

1. RESPONSE TO DISCHARGE

(See Response Methods Handbook, CG 446-4)

Mechanical containment
Should be removed
Chemical and physical treatment

2. LABELS

No hazard label required by Code of Federal Regulations

3. CHEMICAL DESIGNATIONS

3.1 **Synonyms:** Liquid petrolatum
White oil
3.2 **Coast Guard Compatibility Classification:** Miscellaneous hydrocarbon mixtures
3.3 **Chemical Formula:** Not applicable
3.4 **IMCO/United Nations Numerical Designation:** 3.3/1270

4. OBSERVABLE CHARACTERISTICS

4.1 **Physical State (as shipped):** Liquid
4.2 **Color:** Colorless
4.3 **Odor:** Very faint

5. HEALTH HAZARDS

5.1 **Personal Protective Equipment:** Goggles or face shield.
5.2 **Symptoms Following Exposure:** Ingestion of liquid can cause very loose bowel movements.
5.3 **Treatment for Exposure:** EYES: wash with water.
5.4 **Toxicity by Inhalation (Threshold Limit Value):** Not pertinent
5.5 **Short-Term Inhalation Limits:** Not pertinent
5.6 **Toxicity by Ingestion:** Grade 1; LD_{50} 5 to 15 g/kg
5.7 **Late Toxicity:** None
5.8 **Vapor (Gas) Irritant Characteristics:** None
5.9 **Liquid or Solid Irritant Characteristics:** None
5.10 **Odor Threshold:** Not pertinent

6. FIRE HAZARDS

6.1 **Flash Point:** 380°F O.C.
6.2 **Flammable Limits in Air:** Data not available
6.3 **Fire Extinguishing Agents:** Dry chemical, foam, or carbon dioxide
6.4 **Fire Extinguishing Agents Not to be Used:** Water or foam may cause frothing.
6.5 **Special Hazards of Combustion Products:** Not pertinent
6.6 **Behavior in Fire:** Not pertinent
6.7 **Ignition Temperature:** 500-700°F
6.8 **Electrical Hazard:** Not pertinent
6.9 **Burning Rate:** 4 mm/min.

7. CHEMICAL REACTIVITY

7.1 **Reactivity with Water:** No reaction
7.2 **Reactivity with Common Materials:** No reaction
7.3 **Stability During Transport:** Stable
7.4 **Neutralizing Agents for Acids and Caustics:** Not pertinent
7.5 **Polymerization:** Not pertinent
7.6 **Inhibitor of Polymerization:** Not pertinent

8. WATER POLLUTION

8.1 **Aquatic Toxicity:** Data not available
8.2 **Waterfowl Toxicity:** Data not available
8.3 **Biological Oxygen Demand (BOD):** Data not available
8.4 **Food Chain Concentration Potential:** None

9. SELECTED MANUFACTURERS

1. Shell Oil Co.
1 Shell Plaza
Houston, Tex. 77001
2. Standard Oil Co. (Indiana)
910 S. Michigan Ave.
Chicago, Ill. 60605
3. Sun Oil Co.
St. Davids, Pa. 19087

10. SHIPPING INFORMATION

10.1 **Grades or Purity:** Commercial; refined
10.2 **Storage Temperature:** Ambient
10.3 **Inert Atmosphere:** No requirement
10.4 **Venting:** Open (flame arrester)

11. HAZARD ASSESSMENT CODE

(See Hazard Assessment Handbook, CG 446-3)

A-T-U

12. HAZARD CLASSIFICATIONS

12.1 **Code of Federal Regulations:** Not listed
12.2 **NAS Hazard Rating for Bulk Water Transportation:** Not listed
12.3 **NFPA Hazard Classifications:**

Category	Classification
Health Hazard (Blue)	0
Flammability (Red)	1
Reactivity (Yellow)	0

13. PHYSICAL AND CHEMICAL PROPERTIES

13.1 **Physical State at 15°C and 1 atm:** Liquid
13.2 **Molecular Weight:** Not pertinent
13.3 **Boiling Point at 1 atm:** Very high
13.4 **Freezing Point:** Not pertinent
13.5 **Critical Temperature:** Not pertinent
13.6 **Critical Pressure:** Not pertinent
13.7 **Specific Gravity:** 0.822 at 20°C (liquid)
13.8 **Liquid Surface Tension:** 27 dynes/cm = 0.027 N/m at 20°C
13.9 **Liquid-Water Interfacial Tension:** 47 dynes/cm = 0.047 N/m at 70°C
13.10 **Vapor (Gas) Specific Gravity:** Not pertinent
13.11 **Ratio of Specific Heats of Vapor (Gas):** Not pertinent
13.12 **Latent Heat of Vaporization:** Not pertinent
13.13 **Heat of Combustion:** Data not available
13.14 **Heat of Decomposition:** Not pertinent
13.15 **Heat of Solution:** Not pertinent
13.16 **Heat of Polymerization:** Not pertinent

NOTES

REVISED 1978

OMS

OILS, MISCELLANEOUS: MINERAL SEAL

Common Synonyms			
Long-time burning oil 300° oil Mineral colza oil	Oily liquid	Colorless to yellow	Kerosene odor
	Floats on water.		

Stop discharge if possible.
Call fire department.
Avoid contact with liquid.
Isolate and remove discharged material.
Notify local health and pollution control agencies.

Fire	Combustible. **Extinguish with dry chemical, foam or carbon dioxide.** **Water may be ineffective on fire.** **Cool exposed containers with water.**
Exposure	**CALL FOR MEDICAL AID.** LIQUID Irritating to skin and eyes. Harmful if swallowed. **Remove contaminated clothing and shoes.** **Flush affected areas with plenty of water.** **IF IN EYES, hold eyelids open and flush with plenty of water.** **IF SWALLOWED and victim is CONSCIOUS, have victim drink water or milk.** **DO NOT INDUCE VOMITING.**
Water Pollution	Effect of low concentrations on aquatic life is unknown. Fouling to shoreline. May be dangerous if it enters water intakes. **Notify local health and wildlife officials.** **Notify operators of nearby water intakes.**

1. RESPONSE TO DISCHARGE

(See Response Methods Handbook, CG 446-4)

Mechanical containment
Should be removed
Chemical and physical treatment

2. LABELS

No hazard label required by Code of Federal Regulations

3. CHEMICAL DESIGNATIONS

3.1 **Synonyms:** Long-time burning oil
Mineral colza oil
300° oil
Signal oil

3.2 **Coast Guard Compatibility Classification:** Miscellaneous hydrocarbon mixtures

3.3 **Chemical Formula:** Not applicable

3.4 **IMCO/United Nations Numerical Designation:** 3.3/1270

4. OBSERVABLE CHARACTERISTICS

4.1 **Physical State (as shipped):** Liquid

4.2 **Color:** Colorless to light brown

4.3 **Odor:** Like kerosene

5. HEALTH HAZARDS

5.1 **Personal Protective Equipment:** Protective gloves; goggles or face shield.

5.2 **Symptoms Following Exposure:** Vapors cause slight irritation of eyes and nose. Liquid irritates stomach; if taken into lungs causes coughing, distress, and rapidly developing pulmonary edema.

5.3 **Treatment for Exposure:** ASPIRATION: enforced bed rest; administer oxygen; call a doctor. INGESTION: do NOT induce vomiting; have victim drink water or milk. EYES: wash with copious amounts of water. SKIN: wipe off, wash with soap and water.

5.4 **Toxicity by Inhalation (Threshold Limit Value):** 200 ppm

5.5 **Short-Term Inhalation Limits:** Data not available

5.6 **Toxicity by Ingestion:** Grade 2; LD_{50} 0.5 to 5 g/kg

5.7 **Late Toxicity:** Data not available

5.8 **Vapor (Gas) Irritant Characteristics:** Vapors cause a slight smarting of the eyes or respiratory system if present in high concentrations. The effect is temporary.

5.9 **Liquid or Solid Irritant Characteristics:** Minimum hazard. If spilled on clothing and allowed to remain, may cause smarting and reddening of the skin.

5.10 **Odor Threshold:** 1 ppm

6. FIRE HAZARDS

6.1 **Flash Point:** 170-275°F O.C.

6.2 **Flammable Limits in Air:** Data not available

6.3 **Fire Extinguishing Agents:** Dry chemical, foam, or carbon dioxide

6.4 **Fire Extinguishing Agents Not to be Used:** Water may be ineffective.

6.5 **Special Hazards of Combustion Products:** Not pertinent

6.6 **Behavior in Fire:** Not pertinent

6.7 **Ignition Temperature:** Data not available

6.8 **Electrical Hazard:** Not pertinent

6.9 **Burning Rate:** 4 mm/min.

7. CHEMICAL REACTIVITY

7.1 **Reactivity with Water:** No reaction

7.2 **Reactivity with Common Materials:** No reaction

7.3 **Stability During Transport:** Stable

7.4 **Neutralizing Agents for Acids and Caustics:** Not pertinent

7.5 **Polymerization:** Not pertinent

7.6 **Inhibitor of Polymerization:** Not pertinent

8. WATER POLLUTION

8.1 **Aquatic Toxicity:** Data not available

8.2 **Waterfowl Toxicity:** Data not available

8.3 **Biological Oxygen Demand (BOD):** Data not available

8.4 **Food Chain Concentration Potential:** None

9. SELECTED MANUFACTURERS

1. Shell Oil Co.
1 Shell Plaza
Houston, Tex. 77001
2. Standard Oil Co. (Indiana)
910 S. Michigan Ave.
Chicago, Ill. 60605
3. Sun Oil Co.
St. Davids, Pa. 19087

10. SHIPPING INFORMATION

10.1 **Grades or Purity:** Several grades of varying pour points, all highly refined.

10.2 **Storage Temperature:** Ambient

10.3 **Inert Atmosphere:** No requirement

10.4 **Venting:** Open (flame arrester)

11. HAZARD ASSESSMENT CODE

(See Hazard Assessment Handbook, CG 446-3)

A-T-U

12. HAZARD CLASSIFICATIONS

12.1 **Code of Federal Regulations:** Not listed

12.2 **NAS Hazard Rating for Bulk Water Transportation:** Not listed

12.3 **NFPA Hazard Classifications:** Not listed

13. PHYSICAL AND CHEMICAL PROPERTIES

13.1 **Physical State at 15°C and 1 atm:** Liquid

13.2 **Molecular Weight:** Not pertinent

13.3 **Boiling Point at 1 atm:** >500°F = >260°C = >533°K

13.4 **Freezing Point:** 10.0°F = −12.2°C = 261.0°K

13.5 **Critical Temperature:** Not pertinent

13.6 **Critical Pressure:** Not pertinent

13.7 **Specific Gravity:** 0.811-0.825 at 15°C (liquid)

13.8 **Liquid Surface Tension:** (est.) 25 dynes/cm = 0.025 N/m at 20°C

13.9 **Liquid-Water Interfacial Tension:** 47-50 dynes/cm = 0.047-0.050 N/m at 20°C

13.10 **Vapor (Gas) Specific Gravity:** Not pertinent

13.11 **Ratio of Specific Heats of Vapor (Gas):** Not pertinent

13.12 **Latent Heat of Vaporization:** Not pertinent

13.13 **Heat of Combustion:** (est.) −18,000 Btu/lb = −10,000 cal/g = -420×10^5 J/kg

13.14 **Heat of Decomposition:** Not pertinent

13.15 **Heat of Solution:** Not pertinent

13.16 **Heat of Polymerization:** Not pertinent

NOTES

REVISED 1978

OMT — OILS, MISCELLANEOUS: MOTOR

Common Synonyms
Crankcase oil
Lubricating oil

Oily liquid — Yellow-brown — Lube oil odor

Floats on water.

Stop discharge if possible.
Call fire department.
Avoid contact with liquid.
Isolate and remove discharged material.
Notify local health and pollution control agencies.

Fire

Combustible.
Extinguish with dry chemical, foam or carbon dioxide.
Water may be ineffective on fire.
Cool exposed containers with water.

Exposure

CALL FOR MEDICAL AID.

LIQUID
Irritating to skin and eyes.
Harmful if swallowed.
Remove contaminated clothing and shoes.
Flush affected areas with plenty of water.
IF IN EYES, hold eyelids open and flush with plenty of water.
IF SWALLOWED and victim is CONSCIOUS, have victim drink water or milk.
DO NOT INDUCE VOMITING.

Water Pollution

Effect of low concentrations on aquatic life is unknown.
Fouling to shoreline.
May be dangerous if it enters water intakes.
Notify local health and wildlife officials.
Notify operators of nearby water intakes.

1. RESPONSE TO DISCHARGE

(See Response Methods Handbook, CG 446-4)

Mechanical containment
Should be removed
Chemical and physical treatment

2. LABELS

No hazard label required by Code of Federal Regulations

3. CHEMICAL DESIGNATIONS

3.1 **Synonyms:** Crankcase oil
Lubricating oil
Transmission oil

3.2 **Coast Guard Compatibility Classification:** Miscellaneous hydrocarbon mixtures

3.3 **Chemical Formula:** Not applicable

3.4 **IMCO/United Nations Numerical Designation:** 3.3/1270

4. OBSERVABLE CHARACTERISTICS

4.1 **Physical State (as shipped):** Liquid

4.2 **Color:** Yellow fluorescent

4.3 **Odor:** Characteristic

5. HEALTH HAZARDS

5.1 **Personal Protective Equipment:** Protective gloves; goggles or face shield.

5.2 **Symptoms Following Exposure:** INGESTION: minimal gastrointestinal irritation; increased frequency of bowel passage may occur. ASPIRATION: pulmonary irritation is normally minimal but may become more severe several hours after exposure.

5.3 **Treatment for Exposure:** INGESTION: do NOT lavage or induce vomiting. ASPIRATION: treatment probably not required; delayed development of pulmonary irritation can be detected by serial chest x-rays. EYES: wash with copious amounts of water. SKIN: wipe off oil and wash with soap and water.

5.4 **Toxicity by Inhalation (Threshold Limit Value):** Data not available

5.5 **Short-Term Inhalation Limits:** Data not available

5.6 **Toxicity by Ingestion:** Grade 1; LD_{50} 5 to 15 g/kg

5.7 **Late Toxicity:** Data not available

5.8 **Vapor (Gas) Irritant Characteristics:** Vapors cause a slight smarting of the eyes or respiratory system if present in high concentrations. The effect is temporary.

5.9 **Liquid or Solid Irritant Characteristics:** Minimum hazard. If spilled on clothing and allowed to remain, may cause smarting and reddening of the skin.

5.10 **Odor Threshold:** Data not available

6. FIRE HAZARDS

6.1 **Flash Point:** 275-600°F C.C.

6.2 **Flammable Limits in Air:** Data not available

6.3 **Fire Extinguishing Agents:** Dry chemical, foam, or carbon dioxide

6.4 **Fire Extinguishing Agents Not to be Used:** Water may be ineffective

6.5 **Special Hazards of Combustion Products:** Not pertinent

6.6 **Behavior in Fire:** Not pertinent

6.7 **Ignition Temperature:** 325-625°F

6.8 **Electrical Hazard:** Not pertinent

6.9 **Burning Rate:** 4 mm/min.

7. CHEMICAL REACTIVITY

7.1 **Reactivity with Water:** No reaction

7.2 **Reactivity with Common Materials:** No reaction

7.3 **Stability During Transport:** Stable

7.4 **Neutralizing Agents for Acids and Caustics:** Not pertinent

7.5 **Polymerization:** Not pertinent

7.6 **Inhibitor of Polymerization:** Not pertinent

8. WATER POLLUTION

8.1 **Aquatic Toxicity:** Data not available

8.2 **Waterfowl Toxicity:** Data not available

8.3 **Biological Oxygen Demand (BOD):** Data not available

8.4 **Food Chain Concentration Potential:** None

9. SELECTED MANUFACTURERS

1. Shell Oil Co.
   1 Shell Plaza
   Houston, Tex. 77001
2. Standard Oil Co. (Indiana)
   910 S. Michigan Ave.
   Chicago, Ill. 60605
3. Sun Oil Co.
   St. Davids, Pa. 19087

10. SHIPPING INFORMATION

10.1 **Grades or Purity:** Various viscosities

10.2 **Storage Temperature:** Ambient

10.3 **Inert Atmosphere:** No requirement

10.4 **Venting:** Open (flame arrester)

11. HAZARD ASSESSMENT CODE

(See Hazard Assessment Handbook, CG 446-3)

A-T-U

12. HAZARD CLASSIFICATIONS

12.1 **Code of Federal Regulations:** Not listed

12.2 **NAS Hazard Rating for Bulk Water Transportation:** Not listed

12.3 **NFPA Hazard Classifications:** Not listed

13. PHYSICAL AND CHEMICAL PROPERTIES

13.1 **Physical State at 15°C and 1 atm:** Liquid

13.2 **Molecular Weight:** Not pertinent

13.3 **Boiling Point at 1 atm:** Very high

13.4 **Freezing Point:** −29.9°F = −34.4°C = 238.8°K

13.5 **Critical Temperature:** Not pertinent

13.6 **Critical Pressure:** Not pertinent

13.7 **Specific Gravity:** 0.84-0.96 at 15°C (liquid)

13.8 **Liquid Surface Tension:** 36-37.5 dynes/cm = 0.036-0.0375 N/m at 20°C

13.9 **Liquid-Water Interfacial Tension:** 33-54 dynes/cm = 0.033-0.054 N/m at 20°C

13.10 **Vapor (Gas) Specific Gravity:** Not pertinent

13.11 **Ratio of Specific Heats of Vapor (Gas):** Not pertinent

13.12 **Latent Heat of Vaporization:** Not pertinent

13.13 **Heat of Combustion:** −18,486 Btu/lb = −10,270 cal/g = -429.98×10^5 J/kg

13.14 **Heat of Decomposition:** Not pertinent

13.15 **Heat of Solution:** Not pertinent

13.16 **Heat of Polymerization:** Not pertinent

NOTES

REVISED 1978

ONF	OILS, MISCELLANEOUS: NEATSFOOT

Common Synonyms	Oily liquid — Pale yellow — Peculiar odor Floats on water.

Stop discharge if possible.
Call fire department.
Isolate and remove discharged material.
Notify local health and pollution control agencies.

Fire	Combustible. Extinguish with dry chemical, foam or carbon dioxide. Water may be ineffective on fire.
Exposure	Not harmful.
Water Pollution	Effect of low concentrations on aquatic life is unknown. Fouling to shoreline. May be dangerous if it enters water intakes. Notify local health and wildlife officials. Notify operators of nearby water intakes.

1. RESPONSE TO DISCHARGE
(See Response Methods Handbook, CG 446-4)

Mechanical containment
Should be removed
Chemical and physical treatment

2. LABELS

No hazard label required by Code of Federal Regulations

3. CHEMICAL DESIGNATIONS

3.1 **Synonyms:** No common synonyms
3.2 **Coast Guard Compatibility Classification:** Miscellaneous hydrocarbon mixtures
3.3 **Chemical Formula:** Not applicable
3.4 **IMCO/United Nations Numerical Designation:** Not listed

4. OBSERVABLE CHARACTERISTICS

4.1 **Physical State (as shipped):** Liquid
4.2 **Color:** Pale yellow
4.3 **Odor:** Peculiar

5. HEALTH HAZARDS

5.1 **Personal Protective Equipment:** Data not available
5.2 **Symptoms Following Exposure:** May cause dermatitis in sensitive individuals (humans)
5.3 **Treatment for Exposure:** Data not available
5.4 **Toxicity by Inhalation (Threshold Limit Value):** Data not available
5.5 **Short-Term Inhalation Limits:** Data not available
5.6 **Toxicity by Ingestion:** Grade 0; LD_{50} above 15 g/kg
5.7 **Late Toxicity:** Data not available
5.8 **Vapor (Gas) Irritant Characteristics:** Data not available
5.9 **Liquid or Solid Irritant Characteristics:** Data not available
5.10 **Odor Threshold:** Data not available

6. FIRE HAZARDS

6.1 **Flash Point:** 430°F O.C.; 470°F C.C.
6.2 **Flammable Limits in Air:** Not pertinent
6.3 **Fire Extinguishing Agents:** Foam, dry chemical, or carbon dioxide
6.4 **Fire Extinguishing Agents Not to be Used:** Water or foam may cause frothing.
6.5 **Special Hazards of Combustion Products:** Not pertinent
6.6 **Behavior in Fire:** Not pertinent
6.7 **Ignition Temperature:** 828°F
6.8 **Electrical Hazard:** Not pertinent
6.9 **Burning Rate:** Data not available

7. CHEMICAL REACTIVITY

7.1 **Reactivity with Water:** No reaction
7.2 **Reactivity with Common Materials:** No reaction
7.3 **Stability During Transport:** Stable
7.4 **Neutralizing Agents for Acids and Caustics:** Not pertinent
7.5 **Polymerization:** Not pertinent
7.6 **Inhibitor of Polymerization:** Not pertinent

8. WATER POLLUTION

8.1 **Aquatic Toxicity:** Data not available
8.2 **Waterfowl Toxicity:** Data not available
8.3 **Biological Oxygen Demand (BOD):** Data not available
8.4 **Food Chain Concentration Potential:** None

9. SELECTED MANUFACTURERS

1. Atlas Refinery, Inc.
142 Lockwood St.
Newark, N. J. 07105
2. Neatsfoot Oil Refineries Corp.
East Ontario and Bath Sts.
Philadelphia, Pa. 19134
3. Arthur C. Trask Corp.
7666 W. 63 St.
Argo, Ill. 60501

10. SHIPPING INFORMATION

10.1 **Grades or Purity:** Various grades designated by pour point (20°-40°); also various refined grades
10.2 **Storage Temperature:** Ambient
10.3 **Inert Atmosphere:** No requirement
10.4 **Venting:** Data not available

11. HAZARD ASSESSMENT CODE
(See Hazard Assessment Handbook, CG 446-3)

A-T-U

12. HAZARD CLASSIFICATIONS

12.1 **Code of Federal Regulations:** Not listed
12.2 **NAS Hazard Rating for Bulk Water Transportation:** Not listed
12.3 **NFPA Hazard Classifications:** Not listed

13. PHYSICAL AND CHEMICAL PROPERTIES

13.1 **Physical State at 15°C and 1 atm:** Liquid
13.2 **Molecular Weight:** Not pertinent
13.3 **Boiling Point at 1 atm:** Very high
13.4 **Freezing Point:** 32 to 14°F = 0 to −10°C = 273 to 263°K
13.5 **Critical Temperature:** Not pertinent
13.6 **Critical Pressure:** Not pertinent
13.7 **Specific Gravity:** 0.915 at 16°C (liquid)
13.8 **Liquid Surface Tension:** Data not available
13.9 **Liquid-Water Interfacial Tension:** Data not available
13.10 **Vapor (Gas) Specific Gravity:** Not pertinent
13.11 **Ratio of Specific Heats of Vapor (Gas):** Not pertinent
13.12 **Latent Heat of Vaporization:** Not pertinent
13.13 **Heat of Combustion:** Data not available
13.14 **Heat of Decomposition:** Not pertinent
13.15 **Heat of Solution:** Not pertinent
13.16 **Heat of Polymerization:** Not pertinent

NOTES

REVISED 1978

OPT

OILS, MISCELLANEOUS: PENETRATING

Common Synonyms

Oily liquid — Yellow — Motor oil-like odor

Floats on water.

Stop discharge if possible.
Call fire department.
Avoid contact with liquid.
Isolate and remove discharged material.
Notify local health and pollution control agencies.

Fire

Combustible.
Extinguish with foam, dry chemical, or carbon dioxide.
Water may be ineffective on fire.

Exposure

CALL FOR MEDICAL AID.

LIQUID
Irritating to skin and eyes.
Harmful if swallowed.
Remove contaminated clothing and shoes.
Flush affected areas with plenty of water.
IF IN EYES, hold eyelids open and flush with plenty of water.
IF SWALLOWED and victim is CONSCIOUS, have victim drink water or milk.
DO NOT INDUCE VOMITING.

Water Pollution

Effect of low concentrations on aquatic life is unknown.
Fouling to shoreline.
May be dangerous if it enters water intakes.
Notify local health and wildlife officials.
Notifv operators of nearby water intakes.

1. RESPONSE TO DISCHARGE

(See Response Methods Handbook, CG 446-4)

Mechanical containment
Should be removed
Chemical and physical treatment

2. LABELS

No hazard label required by Code of Federal Regulations

3. CHEMICAL DESIGNATIONS

3.1 **Synonyms:** Preservative oil; Water displacing oil
3.2 **Coast Guard Compatibility Classification:** Miscellaneous hydrocarbon mixtures
3.3 **Chemical Formula:** Not applicable
3.4 **IMCO/United Nations Numerical Designation:** 3.3/1270

4. OBSERVABLE CHARACTERISTICS

4.1 **Physical State (as shipped):** Liquid
4.2 **Color:** Yellowish
4.3 **Odor:** Like motor oil

5. HEALTH HAZARDS

5.1 **Personal Protective Equipment:** Protective gloves; goggles or face shield.
5.2 **Symptoms Following Exposure:** Liquid may irritate stomach and increase frequency of bowel movements.
5.3 **Treatment for Exposure:** INGESTION: have victim drink water or milk; do NOT induce vomiting. ASPIRATION: check for delayed development of pulmonary irritation by serial x-rays. EYES: wash with copious amounts of water. SKIN: wipe off, wash with soap and water.
5.4 **Toxicity by Inhalation (Threshold Limit Value):** Data not available
5.5 **Short-Term Inhalation Limits:** Data not available
5.6 **Toxicity by Ingestion:** Grade 1; LD_{50} 5 to 15 g/kg
5.7 **Late Toxicity:** Data not available
5.8 **Vapor (Gas) Irritant Characteristics:** Vapors cause a slight smarting of the eyes or respiratory system if present in high concentrations. The effect is temporary.
5.9 **Liquid or Solid Irritant Characteristics:** Minimum hazard. If spilled on clothing and allowed to remain, may cause smarting and reddening of skin.
5.10 **Odor Threshold:** Data not available

6. FIRE HAZARDS

6.1 **Flash Point:** 295°F
6.2 **Flammable Limits in Air:** Data not available
6.3 **Fire Extinguishing Agents:** Foam, dry chemical, or carbon dioxide
6.4 **Fire Extinguishing Agents Not to be Used:** Water or foam may cause frothing.
6.5 **Special Hazards of Combustion Products:** Not pertinent
6.6 **Behavior in Fire:** Not pertinent
6.7 **Ignition Temperature:** Data not available
6.8 **Electrical Hazard:** Not pertinent
6.9 **Burning Rate:** Data not available

7. CHEMICAL REACTIVITY

7.1 **Reactivity with Water:** No reaction
7.2 **Reactivity with Common Materials:** No reaction
7.3 **Stability During Transport:** Stable
7.4 **Neutralizing Agents for Acids and Caustics:** Not pertinent
7.5 **Polymerization:** Not pertinent
7.6 **Inhibitor of Polymerization:** Not pertinent

8. WATER POLLUTION

8.1 **Aquatic Toxicity:** Data not available
8.2 **Waterfowl Toxicity:** Data not available
8.3 **Biological Oxygen Demand (BOD):** Data not available
8.4 **Food Chain Concentration Potential:** None

9. SELECTED MANUFACTURERS

1. Ashland Petroleum Co.
   1401 Winchester Ave.
   Ashland, Ky. 41101
2. Penreco, Inc.
   Butler, Pa. 16001
3. Standard Oil Co. (Indiana)
   910 S. Michigan Ave.
   Chicago, Ill. 60605

10. SHIPPING INFORMATION

10.1 **Grades or Purity:** Commercial
10.2 **Storage Temperature:** Ambient
10.3 **Inert Atmosphere:** No requirement
10.4 **Venting:** Open (flame arrester)

11. HAZARD ASSESSMENT CODE

(See Hazard Assessment Handbook, CG 446-3)

A-T-U

12. HAZARD CLASSIFICATIONS

12.1 **Code of Federal Regulations:** Not listed
12.2 **NAS Hazard Rating for Bulk Water Transportation:** Not listed
12.3 **NFPA Hazard Classifications:** Not listed

13. PHYSICAL AND CHEMICAL PROPERTIES

13.1 **Physical State at 15°C and 1 atm:** Liquid
13.2 **Molecular Weight:** Not pertinent
13.3 **Boiling Point at 1 atm:** Very high
13.4 **Freezing Point:** Not pertinent
13.5 **Critical Temperature:** Not pertinent
13.6 **Critical Pressure:** Not pertinent
13.7 **Specific Gravity:** 0.8961 at 20°C (liquid)
13.8 **Liquid Surface Tension:** 29.8 dynes/cm = 0.0298 N/m at 24°C
13.9 **Liquid-Water Interfacial Tension:** 5.5 dynes/cm = 0.0055 N/m at 22°C
13.10 **Vapor (Gas) Specific Gravity:** Not pertinent
13.11 **Ratio of Specific Heats of Vapor (Gas):** Not pertinent
13.12 **Latent Heat of Vaporization:** Not pertinent
13.13 **Heat of Combustion:** (est.) −18,000 Btu/lb = −10,000 cal/g = -420×10^5 J/kg
13.14 **Heat of Decomposition:** Not pertinent
13.15 **Heat of Solution:** Not pertinent
13.16 **Heat of Polymerization:** Not pertinent

NOTES

REVISED 1978

ORG

OILS, MISCELLANEOUS: RANGE

Common Synonyms			
Kerosine Kerosene Fuel oil No. 1 JP-1	Watery liquid	Colorless	Kerosene odor
	Floats on water.		

Stop discharge if possible.
Call fire department.
Avoid contact with liquid.
Isolate and remove discharged material.
Notify local health and pollution control agencies.

Fire	Combustible. **Extinguish with foam, dry chemical, or carbon dioxide.** **Water may be ineffective on fire.** **Cool exposed containers with water.**
Exposure	**CALL FOR MEDICAL AID.** LIQUID Irritating to skin and eyes. Harmful if swallowed. **Remove contaminated clothing and shoes.** **Flush affected areas with plenty of water.** **IF IN EYES, hold eyelids open and flush with plenty of water.** **IF SWALLOWED and victim is CONSCIOUS, have victim drink water or milk.** **DO NOT INDUCE VOMITING.**
Water Pollution	Dangerous to aquatic life in high concentrations. Fouling to shoreline. May be dangerous if it enters water intakes. **Notify local health and wildlife officials.** **Notify operators of nearby water intakes.**

1. RESPONSE TO DISCHARGE

(See Response Methods Handbook, CG 446-4)

Mechanical containment
Should be removed
Chemical and physical treatment

2. LABELS

No hazard label required by Code of Federal Regulations

3. CHEMICAL DESIGNATIONS

3.1 **Synonyms:** No. 1 Fuel Oil
JP-1
Kerosene
Kerosine

3.2 **Coast Guard Compatibility Classification:** Miscellaneous hydrocarbon mixtures

3.3 **Chemical Formula:** Not applicable

3.4 **IMCO/United Nations Numerical Designation:** 3.3/1223

4. OBSERVABLE CHARACTERISTICS

4.1 **Physical State (as shipped):** Liquid

4.2 **Color:** Colorless

4.3 **Odor:** Like kerosene

5. HEALTH HAZARDS

5.1 **Personal Protective Equipment:** Protective gloves; goggles or face shield.

5.2 **Symptoms Following Exposure:** Vapor causes slight irritation of eyes and nose. Liquid irritates stomach; if taken into lungs, causes coughing, distress, and rapidly developing pulmonary edema.

5.3 **Treatment for Exposure:** ASPIRATION: enforce bed rest; administer oxygen; call a doctor. INGESTION: do NOT induce vomiting; call a doctor. EYES: wash with copious amounts of water. SKIN: wipe off and wash with soap and water.

5.4 **Toxicity by Inhalation (Threshold Limit Value):** 200 ppm

5.5 **Short-Term Inhalation Limits:** Data not available

5.6 **Toxicity by Ingestion:** Grade 1; LD_{50} 5 to 15 g/kg

5.7 **Late Toxicity:** Data not available

5.8 **Vapor (Gas) Irritant Characteristics:** Vapors cause slight smarting of eyes and respiratory system if present in high concentrations. The effect is temporary.

5.9 **Liquid or Solid Irritant Characteristics:** Minimum hazard. If spilled on clothing and allowed to remain, may cause smarting and reddening of the skin.

5.10 **Odor Threshold:** 1 ppm

6. FIRE HAZARDS

6.1 **Flash Point:** 100°F C.C.

6.2 **Flammable Limits in Air:** 0.7%–5%

6.3 **Fire Extinguishing Agents:** Foam, dry chemical, or carbon dioxide

6.4 **Fire Extinguishing Agents Not to be Used:** Water may be ineffective

6.5 **Special Hazards of Combustion Products:** Not pertinent

6.6 **Behavior in Fire:** Not pertinent

6.7 **Ignition Temperature:** 444°F

6.8 **Electrical Hazard:** Not pertinent

6.9 **Burning Rate:** 4 mm/min.

7. CHEMICAL REACTIVITY

7.1 **Reactivity with Water:** No reaction

7.2 **Reactivity with Common Materials:** No reaction

7.3 **Stability During Transport:** Stable

7.4 **Neutralizing Agents for Acids and Caustics:** Not pertinent

7.5 **Polymerization:** Not pertinent

7.6 **Inhibitor of Polymerization:** Not pertinent

8. WATER POLLUTION

8.1 **Aquatic Toxicity:** 2990 ppm/24 hr/bluegill/TL_m/fresh water

8.2 **Waterfowl Toxicity:** Data not available

8.3 **Biological Oxygen Demand (BOD):** 53%, 5 days

8.4 **Food Chain Concentration Potential:** None

9. SELECTED MANUFACTURERS

1. Shell Oil Co.
   1 Shell Plaza
   Houston, Tex. 77001
2. Standard Oil Co. (Indiana)
   910 S. Michigan Ave.
   Chicago, Ill. 60605
3. Sun Oil Co.
   St. Davids, Pa. 19087

10. SHIPPING INFORMATION

10.1 **Grades or Purity:** Light hydrocarbon distillate: 100%

10.2 **Storage Temperature:** Ambient

10.3 **Inert Atmosphere:** No requirement

10.4 **Venting:** Open (flame arrester)

11. HAZARD ASSESSMENT CODE

(See Hazard Assessment Handbook, CG 446-3)

A-T-U

12. HAZARD CLASSIFICATIONS

12.1 **Code of Federal Regulations:** Combustible Liquid

12.2 **NAS Hazard Rating for Bulk Water Transportation:**

Category	Rating
Fire	2
Health	
Vapor Irritant	1
Liquid or Solid Irritant	1
Poisons	1
Water Pollution	
Human Toxicity	1
Aquatic Toxicity	1
Aesthetic Effect	3
Reactivity	
Other Chemicals	0
Water	0
Self-Reaction	0

12.3 **NFPA Hazard Classifications:**

Category	Classification
Health Hazard (Blue)	0
Flammability (Red)	2
Reactivity (Yellow)	0

13. PHYSICAL AND CHEMICAL PROPERTIES

13.1 **Physical State at 15°C and 1 atm:** Liquid

13.2 **Molecular Weight:** Not pertinent

13.3 **Boiling Point at 1 atm:** 392-500°F = 200-260°C = 473-533°K

13.4 **Freezing Point:** −45 to −55°F = −43 to −48°C = 230 to 225°K

13.5 **Critical Temperature:** Not pertinent

13.6 **Critical Pressure:** Not pertinent

13.7 **Specific Gravity:** 0.80-0.85 at 20°C (liquid)

13.8 **Liquid Surface Tension:** 23-32 dynes/cm = 0.023-0.032 N/m at 20°C

13.9 **Liquid-Water Interfacial Tension:** 47-49 dynes/cm = 0.047-0.049 N/m at 20°C

13.10 **Vapor (Gas) Specific Gravity:** Not pertinent

13.11 **Ratio of Specific Heats of Vapor (Gas):** Not pertinent

13.12 **Latent Heat of Vaporization:** 108 Btu/lb = 60 cal/g = 2.51×10^5 J/kg

13.13 **Heat of Combustion:** −18,540 Btu/lb = −10,300 cal/g = -431.24×10^5 J/kg

13.14 **Heat of Decomposition:** Not pertinent

13.15 **Heat of Solution:** Not pertinent

13.16 **Heat of Polymerization:** Not pertinent

NOTES

REVISED 1978

ORS — OILS, MISCELLANEOUS: RESIN

Common Synonyms			
Rosin oil Retinol Rosinol Codoil	Liquid	Light amber to red to black	Pinetree pitch odor
	Floats on water.		

Stop discharge if possible.
Call fire department.
Avoid contact with liquid.
Isolate and remove discharged material.
Notify local health and pollution control agencies.

Fire	Combustible. Extinguish with foam, dry chemical, or carbon dioxide. Water may be ineffective on fire.
Exposure	CALL FOR MEDICAL AID. Exposure data not available. Flush affected areas with plenty of water.
Water Pollution	Effect of low concentrations on aquatic life is unknown. Fouling to shoreline. May be dangerous if it enters water intakes. Notify local health and wildlife officials. Notify operators of nearby water intakes.

1. RESPONSE TO DISCHARGE

(See Response Methods Handbook, CG 446-4)

Mechanical containment
Should be removed
Chemical and physical treatment

2. LABELS

No hazard label required by Code of Federal Regulations

3. CHEMICAL DESIGNATIONS

3.1 **Synonyms:** Codoil
Retinol
Rosin oil
Rosinol

3.2 **Coast Guard Compatibility Classification:** Miscellaneous hydrocarbon mixtures

3.3 **Chemical Formula:** Not applicable

3.4 **IMCO/United Nations Numerical Designation:** 3.2/1286

4. OBSERVABLE CHARACTERISTICS

4.1 **Physical State (as shipped):** Liquid

4.2 **Color:** Light amber to red to black, depending on grade

4.3 **Odor:** Characteristic; like pinetree pitch

5. HEALTH HAZARDS

5.1 **Personal Protective Equipment:** Data not available

5.2 **Symptoms Following Exposure:** Data not available

5.3 **Treatment for Exposure:** Data not available

5.4 **Toxicity by Inhalation (Threshold Limit Value):** Data not available

5.5 **Short-Term Inhalation Limits:** Data not available

5.6 **Toxicity by Ingestion:** Data not available, but toxicity is probably low.

5.7 **Late Toxicity:** Data not available

5.8 **Vapor (Gas) Irritant Characteristics:** Data not available

5.9 **Liquid or Solid Irritant Characteristics:** Data not available

5.10 **Odor Threshold:** Data not available

6. FIRE HAZARDS

6.1 **Flash Point:** 255-390°F O.C.

6.2 **Flammable Limits in Air:** Data not available

6.3 **Fire Extinguishing Agents:** Foam, dry chemical, or carbon dioxide

6.4 **Fire Extinguishing Agents Not to be Used:** Water may be ineffective

6.5 **Special Hazards of Combustion Products:** Not pertinent

6.6 **Behavior in Fire:** Not pertinent

6.7 **Ignition Temperature:** 648°F

6.8 **Electrical Hazard:** Not pertinent

6.9 **Burning Rate:** Data not available

7. CHEMICAL REACTIVITY

7.1 **Reactivity with Water:** No reaction

7.2 **Reactivity with Common Materials:** No reaction

7.3 **Stability During Transport:** Stable

7.4 **Neutralizing Agents for Acids and Caustics:** Not pertinent

7.5 **Polymerization:** Not pertinent

7.6 **Inhibitor of Polymerization:** Not pertinent

8. WATER POLLUTION

8.1 **Aquatic Toxicity:** Data not available

8.2 **Waterfowl Toxicity:** Data not available

8.3 **Biological Oxygen Demand (BOD):** Data not available

8.4 **Food Chain Concentration Potential:** None

9. SELECTED MANUFACTURERS

1. Crosby Chemicals, Inc.
   Picayune, Miss. 39466
2. Crowley Hydrocarbon Chemicals, Inc.
   271 Madison Ave.
   New York, N. Y. 10016
3. Natrochem, Inc.
   Savannah, Ga. 31402

10. SHIPPING INFORMATION

10.1 **Grades or Purity:** A variety of grades that differ primarily in color and flash point.

10.2 **Storage Temperature:** Ambient

10.3 **Inert Atmosphere:** No requirement

10.4 **Venting:** Open (flame arrester)

11. HAZARD ASSESSMENT CODE

(See Hazard Assessment Handbook, CG 446-3)

A-T-U

12. HAZARD CLASSIFICATIONS

12.1 **Code of Federal Regulations:** Not listed

12.2 **NAS Hazard Rating for Bulk Water Transportation:**

Category	Rating
Fire	2
Health	
Vapor Irritant	2
Liquid or Solid Irritant	2
Poisons	2
Water Pollution	
Human Toxicity	1
Aquatic Toxicity	3
Aesthetic Effect	2
Reactivity	
Other Chemicals	2
Water	0
Self-Reaction	3

12.3 **NFPA Hazard Classifications:** Not listed

13. PHYSICAL AND CHEMICAL PROPERTIES

13.1 **Physical State at 15°C and 1 atm:** Liquid

13.2 **Molecular Weight:** Not pertinent

13.3 **Boiling Point at 1 atm:** 572-750°F = 300-400°C = 573-673°K

13.4 **Freezing Point:** Not pertinent

13.5 **Critical Temperature:** Not pertinent

13.6 **Critical Pressure:** Not pertinent

13.7 **Specific Gravity:** 0.96-1.02 at 15°C (liquid)

13.8 **Liquid Surface Tension:** (est.) 25 dynes/cm = 0.025 N/m at 20°C

13.9 **Liquid-Water Interfacial Tension:** (est.) 50 dynes/cm = 0.05 N/m at 20°C

13.10 **Vapor (Gas) Specific Gravity:** Not pertinent

13.11 **Ratio of Specific Heats of Vapor (Gas):** Not pertinent

13.12 **Latent Heat of Vaporization:** Not pertinent

13.13 **Heat of Combustion:** (est.) −18,000 Btu/lb = −10,000 cal/g = -420×10^5 J/kg

13.14 **Heat of Decomposition:** Not pertinent

13.15 **Heat of Solution:** Not pertinent

13.16 **Heat of Polymerization:** Not pertinent

NOTES

REVISED 1978

ORD	OILS, MISCELLANEOUS: ROAD

Common Synonyms
Slow curing asphalt
Petroleum asphalt
Liquid asphalt

Oily liquid | Black | Tar odor

Floats on water.

Stop discharge if possible.
Call fire department.
Avoid contact with liquid.
Isolate and remove discharged material.
Notify local health and pollution control agencies.

Fire

Combustible.
Extinguish with foam, dry chemical, or carbon dioxide.
Water may be ineffective on fire.
Cool exposed containers with water.

Exposure

CALL FOR MEDICAL AID.

LIQUID
Will burn skin and eyes.
Harmful if swallowed.
Remove contaminated clothing and shoes.
Flush affected areas with plenty of water.
IF IN EYES, hold eyelids open and flush with plenty of water.
IF SWALLOWED and victim is CONSCIOUS, have victim drink water or milk.
DO NOT INDUCE VOMITING.

Water Pollution

Effect of low concentrations on aquatic life is unknown.
Fouling to shoreline.
May be dangerous if it enters water intakes.
Notify local health and wildlife officials.
Notify operators of nearby water intakes.

1. RESPONSE TO DISCHARGE
(See Response Methods Handbook, CG 446-4)

Mechanical containment
Chemical and physical treatment

2. LABELS

No hazard label required by Code of Federal Regulations

3. CHEMICAL DESIGNATIONS

3.1 **Synonyms:** Liquid asphalt
Petroleum asphalt
Slow-curing asphalt

3.2 **Coast Guard Compatibility Classification:** Miscellaneous hydrocarbon mixtures

3.3 **Chemical Formula:** Not applicable

3.4 **IMCO/United Nations Numerical Designation:** 3.2/1999; 3.3/1999

4. OBSERVABLE CHARACTERISTICS

4.1 **Physical State (as shipped):** Liquid

4.2 **Color:** Dark brown to black

4.3 **Odor:** Tarry

5. HEALTH HAZARDS

5.1 **Personal Protective Equipment:** Protective clothing for hot asphalt; face and eye protection when hot.

5.2 **Symptoms Following Exposure:** Contact with skin may cause dermatitis. Inhalation of vapors may cause moderate irritation of nose and throat. Hot liquid burns skin.

5.3 **Treatment for Exposure:** Severe burns may result from hot liquid. Cool the skin at once with water. Cover burn with sterile dressing and seek medical attention.

5.4 **Toxicity by Inhalation (Threshold Limit Value):** 5 mg/m^3

5.5 **Short-Term Inhalation Limits:** Data not available

5.6 **Toxicity by Ingestion:** Grade 2; LD_{50} 0.5 to 5 g/kg

5.7 **Late Toxicity:** None observed

5.8 **Vapor (Gas) Irritant Characteristics:** Vapors cause moderate irritation such that personnel will find high concentrations unpleasant. The effect is temporary.

5.9 **Liquid or Solid Irritant Characteristics:** Causes smarting of the skin and first-degree burns on short exposure; may cause secondary burns on long exposure.

5.10 **Odor Threshold:** Data not available

6. FIRE HAZARDS

6.1 **Flash Point:** 300-550°F

6.2 **Flammable Limits in Air:** Not pertinent

6.3 **Fire Extinguishing Agents:** Foam, dry chemical, or carbon dioxide

6.4 **Fire Extinguishing Agents Not to be Used:** Water may be ineffective

6.5 **Special Hazards of Combustion Products:** Not pertinent

6.6 **Behavior in Fire:** Not pertinent

6.7 **Ignition Temperature:** 400-700°F

6.8 **Electrical Hazard:** Not pertinent

6.9 **Burning Rate:** Data not available

7. CHEMICAL REACTIVITY

7.1 **Reactivity with Water:** No reaction

7.2 **Reactivity with Common Materials:** No reaction

7.3 **Stability During Transport:** Stable

7.4 **Neutralizing Agents for Acids and Caustics:** Not pertinent

7.5 **Polymerization:** Not pertinent

7.6 **Inhibitor of Polymerization:** Not pertinent

8. WATER POLLUTION

8.1 **Aquatic Toxicity:** Data not available

8.2 **Waterfowl Toxicity:** Data not available

8.3 **Biological Oxygen Demand (BOD):** Data not available

8.4 **Food Chain Concentration Potential:** None

9. SELECTED MANUFACTURERS

1. Johns-Manville Co.
Greenwood Plaza
Denver, Colo. 80217
2. Sun Oil Co.
St. Davids, Pa. 19087
3. Witco Chemical Corp.
Pioneer Division
277 Park Ave.
New York, N. Y. 10017

10. SHIPPING INFORMATION

10.1 **Grades or Purity:** SC-0 to SC-5

10.2 **Storage Temperature:** Ambient

10.3 **Inert Atmosphere:** No requirement

10.4 **Venting:** Open (flame arrester)

11. HAZARD ASSESSMENT CODE
(See Hazard Assessment Handbook, CG 446-3)

A-T-U

12. HAZARD CLASSIFICATIONS

12.1 **Code of Federal Regulations:** Not listed

12.2 **NAS Hazard Rating for Bulk Water Transportation:**

Category	Rating
Fire	1
Health	
Vapor Irritant	1
Liquid or Solid Irritant	2
Poisons	1
Water Pollution	
Human Toxicity	0
Aquatic Toxicity	1
Aesthetic Effect	4
Reactivity	
Other Chemicals	0
Water	0
Self-Reaction	0

12.3 **NFPA Hazard Classifications:**

Category	Classification
Health Hazard (Blue)	0
Flammability (Red)	1
Reactivity (Yellow)	0

13. PHYSICAL AND CHEMICAL PROPERTIES

13.1 **Physical State at 15°C and 1 atm:** Liquid

13.2 **Molecular Weight:** Not pertinent

13.3 **Boiling Point at 1 atm:** Very high

13.4 **Freezing Point:** Not pertinent

13.5 **Critical Temperature:** Not pertinent

13.6 **Critical Pressure:** Not pertinent

13.7 **Specific Gravity:** 1.0-1.2 at 25°C (liquid)

13.8 **Liquid Surface Tension:** (est.) 25 dynes/cm = 0.025 N/m at 20°C

13.9 **Liquid-Water Interfacial Tension:** (est.) 50 dynes/cm = 0.05 N/m at 20°C

13.10 **Vapor (Gas) Specific Gravity:** Not pertinent

13.11 **Ratio of Specific Heats of Vapor (Gas):** Not pertinent

13.12 **Latent Heat of Vaporization:** Not pertinent

13.13 **Heat of Combustion:** (est.) $-18{,}000$ Btu/lb = $-10{,}000$ cal/g = -420×10^5 J/kg

13.14 **Heat of Decomposition:** Not pertinent

13.15 **Heat of Solution:** Not pertinent

13.16 **Heat of Polymerization:** Not pertinent

NOTES

REVISED 1978

ORN — OILS, MISCELLANEOUS: ROSIN

Common Synonyms			
Resin oil Retinol Rosinol Codoil	Liquid	Light amber to red to black	Pinetree pitch odor
	Floats on water.		

Stop discharge if possible.
Call fire department.
Avoid contact with liquid.
Isolate and remove discharged material.
Notify local health and pollution control agencies.

Fire	Combustible. Extinguish with foam, dry chemical or carbon dioxide. Water may be ineffective on fire.
Exposure	CALL FOR MEDICAL AID. Exposure data not available. Flush affected areas with plenty of water.
Water Pollution	Effect of low concentrations on aquatic life is unknown. Fouling to shoreline. May be dangerous if it enters water intakes. Notify local health and wildlife officials. Notify operators of nearby water intakes.

1. RESPONSE TO DISCHARGE

(See Response Methods Handbook, CG 446-4)

Mechanical containment
Should be removed
Chemical and physical treatment

2. LABELS

No hazard label required by Code of Federal Regulations

3. CHEMICAL DESIGNATIONS

3.1 **Synonyms:** Codoil
Resin oil
Retinol
Rosinol

3.2 **Coast Guard Compatibility Classification:** Miscellaneous hydrocarbon mixtures

3.3 **Chemical Formula:** Not applicable

3.4 **IMCO/United Nations Numerical Designation:** 3.2/1286

4. OBSERVABLE CHARACTERISTICS

4.1 **Physical State (as shipped):** Liquid

4.2 **Color:** Light amber to red to black, depending on grade

4.3 **Odor:** Characteristic; like pinetree pitch

5. HEALTH HAZARDS

5.1 **Personal Protective Equipment:** Data not available

5.2 **Symptoms Following Exposure:** Data not available

5.3 **Treatment for Exposure:** Data not available

5.4 **Toxicity by Inhalation (Threshold Limit Value):** Data not available

5.5 **Short-Term Inhalation Limits:** Data not available

5.6 **Toxicity by Ingestion:** No data available, but toxicity is probably low.

5.7 **Late Toxicity:** Data not available

5.8 **Vapor (Gas) Irritant Characteristics:** Data not available

5.9 **Liquid or Solid Irritant Characteristics:** Data not available

5.10 **Odor Threshold:** Data not available

6. FIRE HAZARDS

6.1 **Flash Point:** 255-390°F C.C.

6.2 **Flammable Limits in Air:** Data not available

6.3 **Fire Extinguishing Agents:** Foam, dry chemical, or carbon dioxide

6.4 **Fire Extinguishing Agents Not to be Used:** Water may be ineffective

6.5 **Special Hazards of Combustion Products:** Data not available

6.6 **Behavior in Fire:** Data not available

6.7 **Ignition Temperature:** 648°F

6.8 **Electrical Hazard:** Data not available

6.9 **Burning Rate:** Data not available

7. CHEMICAL REACTIVITY

7.1 **Reactivity with Water:** No reaction

7.2 **Reactivity with Common Materials:** No reaction

7.3 **Stability During Transport:** Stable

7.4 **Neutralizing Agents for Acids and Caustics:** Not pertinent

7.5 **Polymerization:** Not pertinent

7.6 **Inhibitor of Polymerization:** Not pertinent

8. WATER POLLUTION

8.1 **Aquatic Toxicity:** Data not available

8.2 **Waterfowl Toxicity:** Data not available

8.3 **Biological Oxygen Demand (BOD):** Data not available

8.4 **Food Chain Concentration Potential:** None

9. SELECTED MANUFACTURERS

1. Crosby Chemicals, Inc.
   Picayune, Miss. 39466
2. Crowley Hydrocarbon Chemicals, Inc.
   271 Madison Ave.
   New York, N. Y. 10016
3. Natrochem, Inc.
   Savannah, Ga. 31402

10. SHIPPING INFORMATION

10.1 **Grades or Purity:** A variety of grades that differ primarily in color and flash point.

10.2 **Storage Temperature:** Ambient

10.3 **Inert Atmosphere:** No requirement

10.4 **Venting:** Open (flame arrester)

11. HAZARD ASSESSMENT CODE

(See Hazard Assessment Handbook, CG 446-3)

A-T-U

12. HAZARD CLASSIFICATIONS

12.1 **Code of Federal Regulations:** Not listed

12.2 **NAS Hazard Rating for Bulk Water Transportation:**

Category	Rating
Fire	2
Health	
Vapor Irritant	2
Liquid or Solid Irritant	2
Poisons	2
Water Pollution	
Human Toxicity	1
Aquatic Toxicity	3
Aesthetic Effect	2
Reactivity	
Other Chemicals	2
Water	0
Self-Reaction	3

12.3 **NFPA Hazard Classifications:** Not listed

13. PHYSICAL AND CHEMICAL PROPERTIES

13.1 **Physical State at 15°C and 1 atm:** Liquid

13.2 **Molecular Weight:** Not pertinent

13.3 **Boiling Point at 1 atm:** 572-750°F = 300-400°C = 573-673°K

13.4 **Freezing Point:** Not pertinent

13.5 **Critical Temperature:** Not pertinent

13.6 **Critical Pressure:** Not pertinent

13.7 **Specific Gravity:** 0.96-1.02 at 15°C (liquid)

13.8 **Liquid Surface Tension:** (est.) 25 dynes/cm = 0.025 N/m at 20°C

13.9 **Liquid-Water Interfacial Tension:** (est.) 50 dynes/cm = 0.05 N/m at 20°C

13.10 **Vapor (Gas) Specific Gravity:** Not pertinent

13.11 **Ratio of Specific Heats of Vapor (Gas):** Not pertinent

13.12 **Latent Heat of Vaporization:** Not pertinent

13.13 **Heat of Combustion:** (est.) $-18{,}000$ Btu/lb $= -10{,}000$ cal/g $= -420 \times 10^5$ J/kg

13.14 **Heat of Decomposition:** Not pertinent

13.15 **Heat of Solution:** Not pertinent

13.16 **Heat of Polymerization:** Not pertinent

NOTES

REVISED 1978

OSP	OILS, MISCELLANEOUS: SPERM

Common Synonyms	Oily liquid	Pale yellow
	Floats on water.	

Stop discharge if possible.
Call fire department.
Avoid contact with liquid.
Isolate and remove discharged material.
Notify local health and pollution control agencies.

Fire	Combustible. Extinguish with foam, dry chemical, or carbon dioxide. Water may be ineffective on fire.
Exposure	CALL FOR MEDICAL AID. Exposure data not available. Flush affected areas with plenty of water.
Water Pollution	Effect of low concentrations on aquatic life is unknown. Fouling to shoreline. May be dangerous if it enters water intakes. Notify local health and wildlife officials. Notify operators of nearby water intakes.

1. RESPONSE TO DISCHARGE

(See Response Methods Handbook, CG 446-4)

Mechanical containment
Should be removed
Chemical and physical treatment

2. LABELS

No hazard label required by Code of Federal Regulations

3. CHEMICAL DESIGNATIONS

3.1 **Synonyms:** No common synonyms
3.2 **Coast Guard Compatibility Classification:** Miscellaneous hydrocarbon mixtures
3.3 **Chemical Formula:** Not applicable
3.4 **IMCO/United Nations Numerical Designation:** Not listed

4. OBSERVABLE CHARACTERISTICS

4.1 **Physical State (as shipped):** Liquid
4.2 **Color:** Colorless to pale yellow
4.3 **Odor:** Characteristic

5. HEALTH HAZARDS

5.1 **Personal Protective Equipment:** Data not available
5.2 **Symptoms Following Exposure:** Data not available
5.3 **Treatment for Exposure:** Data not available
5.4 **Toxicity by Inhalation (Threshold Limit Value):** Data not available
5.5 **Short-Term Inhalation Limits:** Data not available
5.6 **Toxicity by Ingestion:** Data not available, but toxicity is probably low.
5.7 **Late Toxicity:** Data not available
5.8 **Vapor (Gas) Irritant Characteristics:** Data not available
5.9 **Liquid or Solid Irritant Characteristics:** Data not available
5.10 **Odor Threshold:** Data not available

6. FIRE HAZARDS

6.1 **Flash Point:** 428°F C.C. (No. 1); 460°F C.C. (No. 2); 500-510°F O.C.
6.2 **Flammable Limits in Air:** Data not available
6.3 **Fire Extinguishing Agents:** Dry chemical, foam, or carbon dioxide
6.4 **Fire Extinguishing Agents Not to be Used:** Water or foam may cause frothing.
6.5 **Special Hazards of Combustion Products:** Not pertinent
6.6 **Behavior in Fire:** Not pertinent
6.7 **Ignition Temperature:** 586°F (No. 1)
6.8 **Electrical Hazard:** Not pertinent
6.9 **Burning Rate:** Data not available

7. CHEMICAL REACTIVITY

7.1 **Reactivity with Water:** No reaction
7.2 **Reactivity with Common Materials:** No reaction
7.3 **Stability During Transport:** Stable
7.4 **Neutralizing Agents for Acids and Caustics:** Not pertinent
7.5 **Polymerization:** Not pertinent
7.6 **Inhibitor of Polymerization:** Not pertinent

8. WATER POLLUTION

8.1 **Aquatic Toxicity:** Data not available
8.2 **Waterfowl Toxicity:** Data not available
8.3 **Biological Oxygen Demand (BOD):** Data not available
8.4 **Food Chain Concentration Potential:** None

9. SELECTED MANUFACTURERS

1. Archer Daniels Midland Co.
   4666 Faries Pkwy.
   Decatur, Ill. 62526
2. Neatsfoot Oil Refineries Corp.
   East Ontario and Bath Sts.
   Philadelphia, Pa. 19134
3. Arthur C. Trask Corp.
   7666 W. 63 St.
   Argo, Ill. 60501

10. SHIPPING INFORMATION

10.1 **Grades or Purity:** No. 1, No. 2, Winterized (these differ in purity and flash point)
10.2 **Storage Temperature:** Ambient
10.3 **Inert Atmosphere:** No requirement
10.4 **Venting:** Data not available

11. HAZARD ASSESSMENT CODE

(See Hazard Assessment Handbook, CG 446-3)

A-T-U

12. HAZARD CLASSIFICATIONS

12.1 **Code of Federal Regulations:** Not listed
12.2 **NAS Hazard Rating for Bulk Water Transportation:** Not listed
12.3 **NFPA Hazard Classifications:** Not listed

13. PHYSICAL AND CHEMICAL PROPERTIES

13.1 **Physical State at 15°C and 1 atm:** Liquid
13.2 **Molecular Weight:** Not pertinent
13.3 **Boiling Point at 1 atm:** Very high
13.4 **Freezing Point:** Not pertinent
13.5 **Critical Temperature:** Not pertinent
13.6 **Critical Pressure:** Not pertinent
13.7 **Specific Gravity:** 0.882 at 20°C (liquid)
13.8 **Liquid Surface Tension:** Data not available
13.9 **Liquid-Water Interfacial Tension:** 5.7 dynes/cm = 0.0057 N/m at 30°C
13.10 **Vapor (Gas) Specific Gravity:** Not pertinent
13.11 **Ratio of Specific Heats of Vapor (Gas):** Not pertinent
13.12 **Latent Heat of Vaporization:** Not pertinent
13.13 **Heat of Combustion:** $-17{,}900$ Btu/lb $= -9943$ cal/g $= -416.3 \times 10^5$ J/kg
13.14 **Heat of Decomposition:** Not pertinent
13.15 **Heat of Solution:** Not pertinent
13.16 **Heat of Polymerization:** Not pertinent

NOTES

REVISED 1978

OSD	OILS, MISCELLANEOUS: SPINDLE

Common Synonyms			
High speed bearing oil Bearing oil	Oily liquid	Light brown	Weak kerosene-like odor
	Floats on water.		

Stop discharge if possible.
Call fire department.
Avoid contact with liquid.
Isolate and remove discharged material.
Notify local health and pollution control agencies.

Fire	Combustible. Extinguish with foam, dry chemical, or carbon dioxide. Water may be ineffective on fire.
Exposure	CALL FOR MEDICAL AID. LIQUID Irritating to skin and eyes. Harmful if swallowed. Remove contaminated clothing and shoes. Flush affected areas with plenty of water. IF IN EYES, hold eyelids open and flush with plenty of water. IF SWALLOWED and victim is CONSCIOUS, have victim drink water or milk. DO NOT INDUCE VOMITING.
Water Pollution	Dangerous to aquatic life in high concentrations. Fouling to shoreline. May be dangerous if it enters water intakes. Notify local health and wildlife officials. Notify operators of nearby water intakes.

1. RESPONSE TO DISCHARGE

(See Response Methods Handbook, CG 446-4)

Mechanical containment
Should be removed
Chemical and physical treatment

2. LABELS

No hazard label required by Code of Federal Regulations

3. CHEMICAL DESIGNATIONS

3.1 **Synonyms:** Bearing oil
High-speed bearing oil
3.2 **Coast Guard Compatibility Classification:** Miscellaneous hydrocarbon mixtures
3.3 **Chemical Formula:** Not applicable
3.4 **IMCO/United Nations Numerical Designation:** 3.3/1270

4. OBSERVABLE CHARACTERISTICS

4.1 **Physical State (as shipped):** Liquid
4.2 **Color:** Light brown
4.3 **Odor:** Weak, like kerosene

5. HEALTH HAZARDS

5.1 **Personal Protective Equipment:** Protective gloves; goggles or face shield.
5.2 **Symptoms Following Exposure:** Vapor causes slight irritation of eyes and nose. Liquid irritates stomach; if taken into lungs, causes coughing, distress, and rapidly developing pulmonary edema.
5.3 **Treatment for Exposure:** ASPIRATION: enforce bed rest; administer oxygen; call a doctor. INGESTION: do NOT induce vomiting; call a doctor. EYES: wash with copious amounts of water. SKIN: wipe off and wash with soap and water.
5.4 **Toxicity by Inhalation (Threshold Limit Value):** Data not available
5.5 **Short-Term Inhalation Limits:** Data not available
5.6 **Toxicity by Ingestion:** Grade 1; LD_{50} 5 to 15 g/kg
5.7 **Late Toxicity:** Data not available
5.8 **Vapor (Gas) Irritant Characteristics:** Vapors cause a slight smarting of eyes or respiratory system if present in high concentrations. The effect is temporary.
5.9 **Liquid or Solid Irritant Characteristics:** Minimum hazard. If spilled on clothing and allowed to remain, may cause smarting and reddening of skin.
5.10 **Odor Threshold:** Data not available

6. FIRE HAZARDS

6.1 **Flash Point:** 169°F C.C.
6.2 **Flammable Limits in Air:** Data not available
6.3 **Fire Extinguishing Agents:** Foam, dry chemical, or carbon dioxide
6.4 **Fire Extinguishing Agents Not to be Used:** Water may be ineffective.
6.5 **Special Hazards of Combustion Products:** Not pertinent
6.6 **Behavior in Fire:** Not pertinent
6.7 **Ignition Temperature:** 478°F
6.8 **Electrical Hazard:** Not pertinent
6.9 **Burning Rate:** Data not available

7. CHEMICAL REACTIVITY

7.1 **Reactivity with Water:** No reaction
7.2 **Reactivity with Common Materials:** No reaction
7.3 **Stability During Transport:** Stable
7.4 **Neutralizing Agents for Acids and Caustics:** Not pertinent
7.5 **Polymerization:** Not pertinent
7.6 **Inhibitor of Polymerization:** Not pertinent

8. WATER POLLUTION

8.1 **Aquatic Toxicity:** 2990 ppm/24 hr/bluegill/TL_m/fresh water
8.2 **Waterfowl Toxicity:** Data not available
8.3 **Biological Oxygen Demand (BOD):** 53%, 5 days
8.4 **Food Chain Concentration Potential:** None

9. SELECTED MANUFACTURERS

1. Keystone Lubricating Co.
21 and Lippincott Streets
Philadelphia, Pa. 19132
2. Penreco, Inc.
Butler, Pa. 16001
3. White and Bagley Co.
Worcester, Mass. 01601

10. SHIPPING INFORMATION

10.1 **Grades or Purity:** Several grades, all with same hazard assessment.
10.2 **Storage Temperature:** Ambient
10.3 **Inert Atmosphere:** No requirement
10.4 **Venting:** Open (flame arrester)

11. HAZARD ASSESSMENT CODE

(See Hazard Assessment Handbook, CG 446-3)

A-T-U

12. HAZARD CLASSIFICATIONS

12.1 **Code of Federal Regulations:** Combustible liquid
12.2 **NAS Hazard Rating for Bulk Water Transportation:** Not listed
12.3 **NFPA Hazard Classifications:** Not listed

13. PHYSICAL AND CHEMICAL PROPERTIES

13.1 **Physical State at 15°C and 1 atm:** Liquid
13.2 **Molecular Weight:** Not pertinent
13.3 **Boiling Point at 1 atm:** Very high
13.4 **Freezing Point:** Not pertinent
13.5 **Critical Temperature:** Not pertinent
13.6 **Critical Pressure:** Not pertinent
13.7 **Specific Gravity:** 0.881 at 15°C (liquid)
13.8 **Liquid Surface Tension:** Data not available
13.9 **Liquid-Water Interfacial Tension:** Data not available
13.10 **Vapor (Gas) Specific Gravity:** Not pertinent
13.11 **Ratio of Specific Heats of Vapor (Gas):** Not pertinent
13.12 **Latent Heat of Vaporization:** Not pertinent
13.13 **Heat of Combustion:** Data not available
13.14 **Heat of Decomposition:** Not pertinent
13.15 **Heat of Solution:** Not pertinent
13.16 **Heat of Polymerization:** Not pertinent

NOTES

REVISED 1978

OSY	OILS, MISCELLANEOUS: SPRAY

Common Synonyms			
Plant spray oil Dormant oil Foliage oil	Oily liquid	Light brown	Kerosene-like odor

Stop discharge if possible.
Call fire department.
Avoid contact with liquid.
Isolate and remove discharged material.
Notify local health and pollution control agencies.

Fire	Combustible. Extinguish with foam, dry chemical, or carbon dioxide. Water may be ineffective on fire.
Exposure	CALL FOR MEDICAL AID. LIQUID Irritating to skin and eyes. Harmful if swallowed. Remove contaminated clothing and shoes. Flush affected areas with plenty of water. IF IN EYES, hold eyelids open and flush with plenty of water. IF SWALLOWED and victim is CONSCIOUS, have victim drink water or milk. DO NOT INDUCE VOMITING.
Water Pollution	Dangerous to aquatic life in high concentrations. Fouling to shoreline. May be dangerous if it enters water intakes. Notify local health and wildlife officials. Notify operators of nearby water intakes.

1. RESPONSE TO DISCHARGE
(See Response Methods Handbook, CG 446-4)

Mechanical containment
Should be removed
Chemical and physical treatment

2. LABELS

No hazard label required by Code of Federal Regulations

3. CHEMICAL DESIGNATIONS

3.1 **Synonyms:** Dormant oil
Foliage oil
Kerosene, heavy
Plant spray oil
3.2 **Coast Guard Compatibility Classification:** Miscellaneous hydrocarbon mixtures
3.3 **Chemical Formula:** Not applicable
3.4 **IMCO/United Nations Numerical Designation:** 3.3/1270

4. OBSERVABLE CHARACTERISTICS

4.1 **Physical State (as shipped):** Liquid
4.2 **Color:** Colorless to light brown
4.3 **Odor:** Like kerosene; like fuel oil

5. HEALTH HAZARDS

5.1 **Personal Protective Equipment:** Protective gloves; goggles or face shield.
5.2 **Symptoms Following Exposure:** Vapor causes slight irritation of eyes and nose. Liquid irritates stomach; if taken into lungs, causes coughing, distress, and rapidly developing edema.
5.3 **Treatment for Exposure:** ASPIRATION: enforce bed rest; administer oxygen; call a doctor. INGESTION: do NOT induce vomiting; call a doctor. EYES: wash with copious amounts of water. SKIN: wipe off and wash with soap and water.
5.4 **Toxicity by Inhalation (Threshold Limit Value):** 200 ppm
5.5 **Short-Term Inhalation Limits:** Data not available
5.6 **Toxicity by Ingestion:** Grade 2; LD_{50} 0.5 to 5 g/kg
5.7 **Late Toxicity:** Data not available
5.8 **Vapor (Gas) Irritant Characteristics:** Vapors cause a slight smarting of the eyes or respiratory system if present in high concentrations. The effect is temporary.
5.9 **Liquid or Solid Irritant Characteristics:** Minimum hazard. If spilled on clothing and allowed to remain, may cause smarting and reddening of the skin.
5.10 **Odor Threshold:** 1 ppm

6. FIRE HAZARDS

6.1 **Flash Point:** 140°F (min.) C.C.
6.2 **Flammable Limits in Air:** 0.6% – 4.6%
6.3 **Fire Extinguishing Agents:** Foam, dry chemical, or carbon dioxide
6.4 **Fire Extinguishing Agents Not to be Used:** Water may be ineffective
6.5 **Special Hazards of Combustion Products:** Not pertinent
6.6 **Behavior in Fire:** Not pertinent
6.7 **Ignition Temperature:** 475°F
6.8 **Electrical Hazard:** Not pertinent
6.9 **Burning Rate:** 4 mm/min.

7. CHEMICAL REACTIVITY

7.1 **Reactivity with Water:** No reaction
7.2 **Reactivity with Common Materials:** No reaction
7.3 **Stability During Transport:** Stable
7.4 **Neutralizing Agents for Acids and Caustics:** Not pertinent
7.5 **Polymerization:** Not pertinent
7.6 **Inhibitor of Polymerization:** Not pertinent

8. WATER POLLUTION

8.1 **Aquatic Toxicity:**
500 ppm/*/salmon fingerling/lethal/fresh water
*Time period not specified
8.2 **Waterfowl Toxicity:** Data not available
8.3 **Biological Oxygen Demand (BOD):** 53%, 5 days
8.4 **Food Chain Concentration Potential:** None

9. SELECTED MANUFACTURERS

1. Shell Oil Co.
   1 Shell Plaza
   Houston, Tex. 77001
2. Standard Oil Co. (Indiana)
   910 S. Michigan Ave.
   Chicago, Ill. 60605
3. Sun Oil Co.
   St. Davids, Pa. 19087

10. SHIPPING INFORMATION

10.1 **Grades or Purity:** Commercial
10.2 **Storage Temperature:** Ambient
10.3 **Inert Atmosphere:** No requirement
10.4 **Venting:** Open (flame arrester)

11. HAZARD ASSESSMENT CODE
(See Hazard Assessment Handbook, CG 446-3)

A-T-U

12. HAZARD CLASSIFICATIONS

12.1 **Code of Federal Regulations:** Combustible Liquid
12.2 **NAS Hazard Rating for Bulk Water Transportation:**

Category	Rating
Fire	1-2
Health	
Vapor Irritant	1
Liquid or Solid Irritant	1
Poisons	1
Water Pollution *	
Human Toxicity	1
Aquatic Toxicity	1
Aesthetic Effect	3
Reactivity	
Other Chemicals	0
Water	0
Self-Reaction	0

12.3 **NFPA Hazard Classifications:**

Category	Classification
Health Hazard (Blue)	0
Flammability (Red)	2
Reactivity (Yellow)	0

13. PHYSICAL AND CHEMICAL PROPERTIES

13.1 **Physical State at 15°C and 1 atm:** Liquid
13.2 **Molecular Weight:** Not pertinent
13.3 **Boiling Point at 1 atm:** 590-700°F = 310-371°C = 583-644°K
13.4 **Freezing Point:** Not pertinent
13.5 **Critical Temperature:** Not pertinent
13.6 **Critical Pressure:** Not pertinent
13.7 **Specific Gravity:** 0.82 at 15°C (liquid)
13.8 **Liquid Surface Tension:** (est.) 25 dynes/cm = 0.025 N/m at 20°C
13.9 **Liquid-Water Interfacial Tension:** (est.) 50 dynes/cm = 0.05 N/m at 20°C
13.10 **Vapor (Gas) Specific Gravity:** Not pertinent
13.11 **Ratio of Specific Heats of Vapor (Gas):** Not pertinent
13.12 **Latent Heat of Vaporization:** Not pertinent
13.13 **Heat of Combustion:** −18,540 Btu/lb = −10,300 cal/g = -431.24×10^5 J/kg
13.14 **Heat of Decomposition:** Not pertinent
13.15 **Heat of Solution:** Not pertinent
13.16 **Heat of Polymerization:** Not pertinent

NOTES

REVISED 1978

OTL — OILS, MISCELLANEOUS: TALL

Common Synonyms	Oily liquid	Yellow	Characteristic odor
	Floats on water.		

Stop discharge if possible.
Call fire department.
Avoid contact with liquid.
Isolate and remove discharged material.
Notify local health and pollution control agencies.

Fire	Combustible. Extinguish with foam, dry chemical, or carbon dioxide. Water may be ineffective on fire.
Exposure	CALL FOR MEDICAL AID. Exposure data not available. Flush affected areas with water.
Water Pollution	Effect of low concentrations on aquatic life is unknown. Fouling to shoreline. May be dangerous if it enters water intakes. Notify local health and wildlife officials. Notify operators of nearby water intakes.

1. RESPONSE TO DISCHARGE

(See Response Methods Handbook, CG 446-4)

Mechanical containment
Should be removed
Chemical and physical treatment

2. LABELS

No hazard label required by Code of Federal Regulations

3. CHEMICAL DESIGNATIONS

3.1 **Synonyms:** No common synonyms
3.2 **Coast Guard Compatibility Classification:** Ester
3.3 **Chemical Formula:** Not applicable
3.4 **IMCO/United Nations Numerical Designation:** Not listed

4. OBSERVABLE CHARACTERISTICS

4.1 **Physical State (as shipped):** Liquid
4.2 **Color:** Yellow
4.3 **Odor:** Characteristic

5. HEALTH HAZARDS

5.1 **Personal Protective Equipment:** Data not available
5.2 **Symptoms Following Exposure:** Data not available
5.3 **Treatment for Exposure:** Data not available
5.4 **Toxicity by Inhalation (Threshold Limit Value):** Data not available
5.5 **Short-Term Inhalation Limits:** Data not available
5.6 **Toxicity by Ingestion:** Data not available
5.7 **Late Toxicity:** Data not available
5.8 **Vapor (Gas) Irritant Characteristics:** Data not available
5.9 **Liquid or Solid Irritant Characteristics:** Data not available
5.10 **Odor Threshold:** Data not available

6. FIRE HAZARDS

6.1 **Flash Point:** 380°F O.C.
6.2 **Flammable Limits in Air:** Data not available
6.3 **Fire Extinguishing Agents:** Foam, dry chemical, or carbon dioxide
6.4 **Fire Extinguishing Agents Not to be Used:** Water may be ineffective.
6.5 **Special Hazards of Combustion Products:** Not pertinent
6.6 **Behavior in Fire:** Not pertinent
6.7 **Ignition Temperature:** Data not available
6.8 **Electrical Hazard:** Not pertinent
6.9 **Burning Rate:** Data not available

7. CHEMICAL REACTIVITY

7.1 **Reactivity with Water:** No reaction
7.2 **Reactivity with Common Materials:** No reaction
7.3 **Stability During Transport:** Stable
7.4 **Neutralizing Agents for Acids and Caustics:** Not pertinent
7.5 **Polymerization:** Not pertinent
7.6 **Inhibitor of Polymerization:** Not pertinent

8. WATER POLLUTION

8.1 **Aquatic Toxicity:** Data not available
8.2 **Waterfowl Toxicity:** Data not available
8.3 **Biological Oxygen Demand (BOD):** Data not available
8.4 **Food Chain Concentration Potential:** None

9. SELECTED MANUFACTURERS

1. Georgia-Pacific Corp.
   Commonwealth Bldg.
   Portland, Ore. 97204
2. SCM Corp.
   Glidden-Durkee Division
   900 Union Commerce Bldg.
   Cleveland, Ohio 44115
3. Union Camp Corp.
   Chemical Division
   Jacksonville, Fla. 32205

10. SHIPPING INFORMATION

10.1 **Grades or Purity:** Various grades, which differ primarily in the relative content of fatty acids and rosin acids.
10.2 **Storage Temperature:** Ambient
10.3 **Inert Atmosphere:** No requirement
10.4 **Venting:** Open (flame arrester)

11. HAZARD ASSESSMENT CODE

(See Hazard Assessment Handbook, CG 446-3)

A-T-U

12. HAZARD CLASSIFICATIONS

12.1 **Code of Federal Regulations:** Not listed
12.2 **NAS Hazard Rating for Bulk Water Transportation:** Not listed
12.3 **NFPA Hazard Classifications:** Not listed

13. PHYSICAL AND CHEMICAL PROPERTIES

13.1 **Physical State at 15°C and 1 atm:** Liquid
13.2 **Molecular Weight:** Not pertinent
13.3 **Boiling Point at 1 atm:** Very high
13.4 **Freezing Point:** Not pertinent
13.5 **Critical Temperature:** Not pertinent
13.6 **Critical Pressure:** Not pertinent
13.7 **Specific Gravity:** 0.951 at 16°C (liquid)
13.8 **Liquid Surface Tension:** 34.3 dynes/cm = 0.0343 N/m at 24°C
13.9 **Liquid-Water Interfacial Tension:** 11 dynes/cm = 0.011 N/m at 22.5°C
13.10 **Vapor (Gas) Specific Gravity:** Not pertinent
13.11 **Ratio of Specific Heats of Vapor (Gas):** Not pertinent
13.12 **Latent Heat of Vaporization:** Not pertinent
13.13 **Heat of Combustion:** (est.) $-18{,}000$ Btu/lb = $-10{,}000$ cal/g = -420×10^5 J/kg
13.14 **Heat of Decomposition:** Not pertinent
13.15 **Heat of Solution:** Not pertinent
13.16 **Heat of Polymerization:** Not pertinent

NOTES

REVISED 1978

OTN — OILS, MISCELLANEOUS: TANNER'S

Common Synonyms	
Sulfated Neatsfoot oil	Oily liquid

Stop discharge if possible.
Call fire department.
Avoid contact with liquid.
Isolate and remove discharged material.
Notify local health and pollution control agencies.

Fire	Combustible. Extinguish with foam, dry chemical, carbon dioxide. Water may be ineffective on fire.
Exposure	CALL FOR MEDICAL AID. Exposure data not available. Flush affected areas with plenty of water.
Water Pollution	Effect of low concentrations on aquatic life is unknown. Fouling to shoreline. May be dangerous if it enters water intakes. Notify local health and wildlife officials. Notify operators of nearby water intakes.

1. RESPONSE TO DISCHARGE

(See Response Methods Handbook, CG 446-4)

Data not available

2. LABELS

No hazard label required by Code of Federal Regulations

3. CHEMICAL DESIGNATIONS

3.1 **Synonyms:** Sulfated neatsfoot oil, sodium salt
3.2 **Coast Guard Compatibility Classification:** Miscellaneous hydrocarbon mixtures
3.3 **Chemical Formula:** Not applicable
3.4 **IMCO/United Nations Numerical Designation:** Not listed

4. OBSERVABLE CHARACTERISTICS

4.1 **Physical State (as shipped):** Data not available
4.2 **Color:** Data not available
4.3 **Odor:** Data not available

5. HEALTH HAZARDS

5.1 **Personal Protective Equipment:** Data not available
5.2 **Symptoms Following Exposure:** Data not available
5.3 **Treatment for Exposure:** Data not available
5.4 **Toxicity by Inhalation (Threshold Limit Value):** Data not available
5.5 **Short-Term Inhalation Limits:** Data not available
5.6 **Toxicity by Ingestion:** Data not available
5.7 **Late Toxicity:** Data not available
5.8 **Vapor (Gas) Irritant Characteristics:** Data not available
5.9 **Liquid or Solid Irritant Characteristics:** Data not available
5.10 **Odor Threshold:** Data not available

6. FIRE HAZARDS

6.1 **Flash Point:** Data not available
6.2 **Flammable Limits in Air:** Data not available
6.3 **Fire Extinguishing Agents:** Foam, dry chemical, or carbon dioxide
6.4 **Fire Extinguishing Agents Not to be Used:** Water may be ineffective.
6.5 **Special Hazards of Combustion Products:** Not pertinent
6.6 **Behavior in Fire:** Not pertinent
6.7 **Ignition Temperature:** Data not available
6.8 **Electrical Hazard:** Not pertinent
6.9 **Burning Rate:** Data not available

7. CHEMICAL REACTIVITY

7.1 **Reactivity with Water:** No reaction
7.2 **Reactivity with Common Materials:** No reaction
7.3 **Stability During Transport:** Stable
7.4 **Neutralizing Agents for Acids and Caustics:** Not pertinent
7.5 **Polymerization:** Not pertinent
7.6 **Inhibitor of Polymerization:** Not pertinent

8. WATER POLLUTION

8.1 **Aquatic Toxicity:** Data not available
8.2 **Waterfowl Toxicity:** Data not available
8.3 **Biological Oxygen Demand (BOD):** Data not available
8.4 **Food Chain Concentration Potential:** Data not available

9. SELECTED MANUFACTURERS

1. Diamond Shamrock Corp.
   Nopco Chemicals Division
   Morristown, N. J. 07960
2. Reilly-Whiteman, Inc.
   Conshohocken, Pa. 19428
3. Arthur C. Trask Corp.
   7666 W. 63 St.
   Argo, Ill. 60501

10. SHIPPING INFORMATION

10.1 **Grades or Purity:** Data not available
10.2 **Storage Temperature:** Ambient
10.3 **Inert Atmosphere:** No requirement
10.4 **Venting:** Data not available

11. HAZARD ASSESSMENT CODE

(See Hazard Assessment Handbook, CG 446-3)

A-T-U

12. HAZARD CLASSIFICATIONS

12.1 **Code of Federal Regulations:** Not listed
12.2 **NAS Hazard Rating for Bulk Water Transportation:** Not listed
12.3 **NFPA Hazard Classifications:** Not listed

13. PHYSICAL AND CHEMICAL PROPERTIES

13.1 **Physical State at 15°C and 1 atm:** Liquid
13.2 **Molecular Weight:** Not pertinent
13.3 **Boiling Point at 1 atm:** Very high
13.4 **Freezing Point:** Data not available
13.5 **Critical Temperature:** Not pertinent
13.6 **Critical Pressure:** Not pertinent
13.7 **Specific Gravity:** (est.) 0.85 at 20°C (liquid)
13.8 **Liquid Surface Tension:** (est.) 25 dynes/cm = 0.025 N/m at 20°C
13.9 **Liquid-Water Interfacial Tension:** (est.) 50 dynes/cm = 0.05 N/m at 20°C
13.10 **Vapor (Gas) Specific Gravity:** Not pertinent
13.11 **Ratio of Specific Heats of Vapor (Gas):** Not pertinent
13.12 **Latent Heat of Vaporization:** Not pertinent
13.13 **Heat of Combustion:** (est.) $-18{,}000$ Btu/lb $= -10{,}000$ cal/g $= -420 \times 10^5$ J/kg
13.14 **Heat of Decomposition:** Not pertinent
13.15 **Heat of Solution:** Not pertinent
13.16 **Heat of Polymerization:** Not pertinent

NOTES

REVISED 1978

OTF	OILS, MISCELLANEOUS: TRANSFORMER

Common Synonyms
Insulating oil
Electrical insulating oil
Petroleum insulating oil

Oily liquid — Colorless to light brown — Motor oil-like odor

Floats on water.

Stop discharge if possible.
Call fire department.
Avoid contact with liquid.
Isolate and remove discharged material.
Notify local health and pollution control agencies.

Fire

Combustible.
Extinguish with foam, dry chemical, carbon dioxide.
Water may be ineffective on fire.

Exposure

CALL FOR MEDICAL AID.

LIQUID
Irritating to skin and eyes.
Harmful if swallowed.
Remove contaminated clothing.
Flush affected areas with plenty of water.
IF IN EYES, hold eyelids open and flush with plenty of water.
IF SWALLOWED and victim is CONSCIOUS, have victim drink water or milk.
DO NOT INDUCE VOMITING.

Water Pollution

Effect of low concentrations on aquatic life is unknown.
Fouling to shoreline.
May be dangerous if it enters water intakes.
Notify local health and wildlife officials.
Notify operators of nearby water intakes

1. RESPONSE TO DISCHARGE

(See Response Methods Handbook, CG 446-4)

Mechanical containment
Should be removed
Chemical and physical treatment

2. LABELS

No hazard label required by Code of Federal Regulations

3. CHEMICAL DESIGNATIONS

3.1 **Synonyms:** Electrical insulating oil
Insulating oil
Petroleum insulating oil

3.2 **Coast Guard Compatibility Classification:** Miscellaneous hydrocarbon mixtures

3.3 **Chemical Formula:** Not applicable

3.4 **IMCO/United Nations Numerical Designation:** 3.3/1270

4. OBSERVABLE CHARACTERISTICS

4.1 **Physical State (as shipped):** Liquid

4.2 **Color:** Colorless to light brown

4.3 **Odor:** Like motor oil

5. HEALTH HAZARDS

5.1 **Personal Protective Equipment:** Protective gloves; goggles or face shield.

5.2 **Symptoms Following Exposure:** Ingestion of liquid may irritate stomach and cause increased frequency of bowel movements. If taken into lungs, delayed pulmonary irritation may occur.

5.3 **Treatment for Exposure:** INGESTION: do NOT induce vomiting. ASPIRATION: check for delayed irritation by serial X-rays. EYES: wash with copious amounts of water. SKIN: wipe off and wash with soap and water.

5.4 **Toxicity by Inhalation (Threshold Limit Value):** Data not available

5.5 **Short-Term Inhalation Limits:** Data not available

5.6 **Toxicity by Ingestion:** Grade 1; LD_{50} 5 to 15 g/kg

5.7 **Late Toxicity:** Data not available

5.8 **Vapor (Gas) Irritant Characteristics:** Vapors cause a slight smarting of the eyes or respiratory system if present in high concentrations. The effect is temporary.

5.9 **Liquid or Solid Irritant Characteristics:** Minimum hazard. If spilled on clothing and allowed to remain, may cause smarting and reddening of the skin.

5.10 **Odor Threshold:** Data not available

6. FIRE HAZARDS

6.1 **Flash Point:** 295°F O.C.

6.2 **Flammable Limits in Air:** Data not available

6.3 **Fire Extinguishing Agents:** Foam, dry chemical, or carbon dioxide

6.4 **Fire Extinguishing Agents Not to be Used:** Water may be ineffective.

6.5 **Special Hazards of Combustion Products:** Not pertinent

6.6 **Behavior in Fire:** Not pertinent

6.7 **Ignition Temperature:** Data not available

6.8 **Electrical Hazard:** Not pertinent

6.9 **Burning Rate:** Data not available

7. CHEMICAL REACTIVITY

7.1 **Reactivity with Water:** No reaction

7.2 **Reactivity with Common Materials:** No reaction

7.3 **Stability During Transport:** Stable

7.4 **Neutralizing Agents for Acids and Caustics:** Not pertinent

7.5 **Polymerization:** Not pertinent

7.6 **Inhibitor of Polymerization:** Not pertinent

8. WATER POLLUTION

8.1 **Aquatic Toxicity:** Data not available

8.2 **Waterfowl Toxicity:** Data not available

8.3 **Biological Oxygen Demand (BOD):** Data not available

8.4 **Food Chain Concentration Potential:** None

9. SELECTED MANUFACTURERS

1. Shell Oil Co.
1 Shell Plaza
Houston, Tex. 77001
2. Standard Oil Co. (Indiana)
910 S. Michigan Ave.
Chicago, Ill. 60605
3. Sun Oil Co.
St. Davids, Pa. 19087

10. SHIPPING INFORMATION

10.1 **Grades or Purity:** Data not available

10.2 **Storage Temperature:** Ambient

10.3 **Inert Atmosphere:** No requirement

10.4 **Venting:** Open (flame arrester)

11. HAZARD ASSESSMENT CODE

(See Hazard Assessment Handbook, CG 446-3)

A-T-U

12. HAZARD CLASSIFICATIONS

12.1 **Code of Federal Regulations:** Not listed

12.2 **NAS Hazard Rating for Bulk Water Transportation:** Not listed

12.3 **NFPA Hazard Classifications:**

Category	Classification
Health Hazard (Blue)	0
Flammability (Red)	1
Reactivity (Yellow)	0

13. PHYSICAL AND CHEMICAL PROPERTIES

13.1 **Physical State at 15°C and 1 atm:** Liquid

13.2 **Molecular Weight:** Not pertinent

13.3 **Boiling Point at 1 atm:** Very high

13.4 **Freezing Point:** $-75°F = -59°C = 214°K$

13.5 **Critical Temperature:** Not pertinent

13.6 **Critical Pressure:** Not pertinent

13.7 **Specific Gravity:** 0.891 at 15°C (liquid)

13.8 **Liquid Surface Tension:** Data not available

13.9 **Liquid-Water Interfacial Tension:** 49 dynes/cm = 0.049 N/m at 25°C

13.10 **Vapor (Gas) Specific Gravity:** Not pertinent

13.11 **Ratio of Specific Heats of Vapor (Gas):** Not pertinent

13.12 **Latent Heat of Vaporization:** Not pertinent

13.13 **Heat of Combustion:** Data not available

13.14 **Heat of Decomposition:** Not pertinent

13.15 **Heat of Solution:** Not pertinent

13.16 **Heat of Polymerization:** Not pertinent

NOTES

REVISED 1978

OTB — OILS, MISCELLANEOUS: TURBINE

Common Synonyms			
Steam turbine oil Steam turbine lube oil Turbine oil	Liquid	Colorless to light brown	Kerosene-like odor
	Floats on water.		

Stop discharge if possible. Keep people away.
Call fire department.
Isolate and remove discharged material.
Notify local health and pollution control agencies.

Fire

Combustible.
Extinguish with dry chemicals, foam or carbon dioxide.
Water may be ineffective on fire.

Exposure

Call for medical aid.

VAPOR
Irritating to eyes, nose and throat.
Move victim to fresh air.
If breathing is difficult, give oxygen.

LIQUID
Irritating to skin and eyes.
Harmful if swallowed.
Remove contaminated clothing and shoes.
Flush affected areas with plenty of water.
IF IN EYES, hold eyelids open and flush with plenty of water.
IF SWALLOWED and victim is CONSCIOUS, have victim drink water or milk.
DO NOT INDUCE VOMITING.

Water Pollution

Effect of low concentrations on aquatic life is unknown.
Fouling to shoreline.
May be dangerous if it enters water intakes.
Notify local health and wi'dlife c.ficials.
Notify operators of nearl.y water intakes.

1. RESPONSE TO DISCHARGE
(See Response Methods Handbook, CG 446-4)

Mechanical containment
Should be removed
Chemical and physical treatment

2. LABELS

No label required by Code of Federal Regulations

3. CHEMICAL DESIGNATIONS

3.1 **Synonyms:** Steam turbine oil; Steam turbine lube oil
3.2 **Coast Guard Compatibility Classification:** Petroleum oils (12)
3.3 **Chemical Formula:** Not applicable
3.4 **IMCO/United Nations Numerical Designation:** Not listed

4. OBSERVABLE CHARACTERISTICS

4.1 **Physical State (as shipped):** Liquid
4.2 **Color:** Colorless to pale brown
4.3 **Odor:** Weak, like lube oil

5. HEALTH HAZARDS

5.1 **Personal Protective Equipment:** Goggles or face shield; rubber gloves.
5.2 **Symptoms Following Exposure:** Contact with liquid causes slight irritation of eyes and (on prolonged contact) skin. Ingestion causes slight irritation of stomach and bowel, increased frequency of bowel movement.
5.3 **Treatment for Exposure:** EYES: wash with copious quantity of water for at least 15 min. SKIN: wipe off; wash with soap and water. INGESTION: do NOT induce vomiting; do NOT lavage; 2-4 oz. olive oil and 1-2 oz. activated charcoal may be given.
5.4 **Toxicity by Inhalation (Threshold Limit Value):** 5 mg/m^3
5.5 **Short-Term Inhalation Limits:** Data not available
5.6 **Toxicity by Ingestion:** Grade 0; LD_{50} > 15 g/kg (rat)
5.7 **Late Toxicity:** Data not available
5.8 **Vapor (Gas) Irritant Characteristics:** Data not available
5.9 **Liquid or Solid Irritant Characteristics:** Data not available
5.10 **Odor Threshold:** Data not available

6. FIRE HAZARDS

6.1 **Flash Point:** 390—485°F O.C.
6.2 **Flammable Limits in Air:** Not pertinent
6.3 **Fire Extinguishing Agents:** Dry chemical, foam, carbon dioxide, water fog.
6.4 **Fire Extinguishing Agents Not to be Used:** Water or foam may cause frothing; water may be ineffective.
6.5 **Special Hazards of Combustion Products:** Not pertinent
6.6 **Behavior in Fire:** Not pertinent
6.7 **Ignition Temperature:** 700°F
6.8 **Electrical Hazard:** Not pertinent
6.9 **Burning Rate:** (approx.) 4 mm/min.

7. CHEMICAL REACTIVITY

7.1 **Reactivity with Water:** No reaction
7.2 **Reactivity with Common Materials:** No reaction
7.3 **Stability During Transport:** Stable
7.4 **Neutralizing Agents for Acids and Caustics:** Not pertinent
7.5 **Polymerization:** Not pertinent
7.6 **Inhibitor of Polymerization:** Not pertinent

8. WATER POLLUTION

8.1 **Aquatic Toxicity:** Data not available
8.2 **Waterfowl Toxicity:** Data not available
8.3 **Biological Oxygen Demand (BOD):** Data not available
8.4 **Food Chain Concentration Potential:** None

9. SELECTED MANUFACTURERS

1. Shell Oil Co.
   One Shell Plaza
   Houston, Texas 77002
2. Exxon Co.
   P. O. Box 2180
   Houston, Texas 77001
3. Sun Oil Co.
   P. O. Box 426
   Marcus Hook, Pa. 19061

10. SHIPPING INFORMATION

10.1 **Grades or Purity:** Solvent refined paraffinic oils: 98.5+%. Grades vary in viscosity and flash point.
10.2 **Storage Temperature:** Ambient
10.3 **Inert Atmosphere:** No requirement
10.4 **Venting:** Open (flame arrester)

11. HAZARD ASSESSMENT CODE
(See Hazard Assessment Handbook, CG 446-3)

A-T-U

12. HAZARD CLASSIFICATIONS

12.1 **Code of Federal Regulations:** Not listed
12.2 **NAS Hazard Rating for Bulk Water Transportation:** Not listed
12.3 **NFPA Hazard Classifications:**

Category	Classification
Health Hazard (Blue)	0
Flammability (Red)	1
Reactivity (Yellow)	0

13. PHYSICAL AND CHEMICAL PROPERTIES

13.1 **Physical State at 15°C and 1 atm:** Liquid
13.2 **Molecular Weight:** Not pertinent
13.3 **Boiling Point at 1 atm:** Not pertinent
13.4 **Freezing Point:** Not pertinent
13.5 **Critical Temperature:** Not pertinent
13.6 **Critical Pressure:** Not pertinent
13.7 **Specific Gravity:** 0.87 at 20°C (liquid)
13.8 **Liquid Surface Tension:** (est.) 25 dynes/cm = 0.025 N/m at 20°C
13.9 **Liquid-Water Interfacial Tension:** (est.) 50 dynes/cm = 0.050 N/m at 20°C
13.10 **Vapor (Gas) Specific Gravity:** Not pertinent
13.11 **Ratio of Specific Heats of Vapor (Gas):** Not pertinent
13.12 **Latent Heat of Vaporization:** Not pertinent
13.13 **Heat of Combustion:** (est.) −17,600 Btu/lb = −9,800 cal/g = -410×10^5 J/kg
13.14 **Heat of Decomposition:** Not pertinent
13.15 **Heat of Solution:** Not pertinent
13.16 **Heat of Polymerization:** Not pertinent

NOTES

OLA

OLEIC ACID

Common Synonyms			
cis-8-Heptadecylenecarboxylic acid cis-9-Octadecenoic acid cis-Δ^9-Octadecylenic acid Red oil	Liquid	Colorless to pale yellow	Mild odor
	Floats on water.		

Avoid contact with liquid. Keep people away.
Stop discharge if possible.
Isolate and remove discharged material.
Notify local health and pollution control agencies.

Fire

Combustible.
Extinguish with dry chemicals or carbon dioxide.
Water or foam may be ineffective on fire.
Cool exposed containers with water.

Exposure

CALL FOR MEDICAL AID.

LIQUID
Irritating to skin and eyes.
If swallowed will cause nausea.

Remove contaminated clothing and shoes.
Flush affected areas with plenty of water.
IF IN EYES, hold eyelids open and flush with plenty of water.
IF SWALLOWED and victim is CONSCIOUS, have victim drink water or milk.
IF SWALLOWED and victim is UNCONSCIOUS OR HAVING CONVULSIONS, do nothing except keep victim warm.

Water Pollution

Effect of low concentrations on aquatic life is unknown.
Fouling to shoreline.
May be dangerous if it enters water intakes.
Notify local health and wildlife officials.
Notify operators of nearby water intake

1. RESPONSE TO DISCHARGE

(See Response Methods Handbook, CG 446-4)

Mechanical containment
Should be removed
Chemical and physical treatment

2. LABELS

No hazard label required by Code of Federal Regulations

3. CHEMICAL DESIGNATIONS

3.1 **Synonyms:** *cis*-8-Heptadecylene-carboxylic acid; *cis*-9-Octadecenoic acid; *cis*-Δ^9-Octadecylenic acid; Red oil

3.2 **Coast Guard Compatibility Classification:** Not listed

3.3 **Chemical Formula:** $CH_3(CH_2)_7CH{=}CH(CH_2)_7COOH$

3.4 **IMCO/United Nations Numerical Designation:** Not listed

4. OBSERVABLE CHARACTERISTICS

4.1 **Physical State (as shipped):** Liquid

4.2 **Color:** Pale

4.3 **Odor:** Faint

5. HEALTH HAZARDS

5.1 **Personal Protective Equipment:** Impervious gloves; goggles or face shield; impervious apron

5.2 **Symptoms Following Exposure:** Industrial use of compound involves no known hazards. Ingestion causes mild irritation of mouth and stomach. Contact with eyes or skin causes mild irritation.

5.3 **Treatment for Exposure:** INGESTION: give large amount of water. EYES: if eye irritation occurs, flush with water and get medical attention. SKIN: wash thoroughly with soap and water.

5.4 **Toxicity by Inhalation (Threshold Limit Value):** Data not available

5.5 **Short-Term Inhalation Limits:** Data not available

5.6 **Toxicity by Ingestion:** Grade 1; $LD_{50} > 15$ g/kg

5.7 **Late Toxicity:** Data not available

5.8 **Vapor (Gas) Irritant Characteristics:** Data not available

5.9 **Liquid or Solid Irritant Characteristics:** Data not available

5.10 **Odor Threshold:** Data not available

6. FIRE HAZARDS

6.1 **Flash Point:** 390—425°F O.C.

6.2 **Flammable Limits in Air:** Data not available

6.3 **Fire Extinguishing Agents:** Dry chemical, carbon dioxide

6.4 **Fire Extinguishing Agents Not to be Used:** Water or foam may cause frothing.

6.5 **Special Hazards of Combustion Products:**

6.6 **Behavior in Fire:**

6.7 **Ignition Temperature:** 685°F

6.8 **Electrical Hazard:** Data not available

6.9 **Burning Rate:** Data not available

7. CHEMICAL REACTIVITY

7.1 **Reactivity with Water:** No reaction

7.2 **Reactivity with Common Materials:**

7.3 **Stability During Transport:** Stable

7.4 **Neutralizing Agents for Acids and Caustics:** Not pertinent

7.5 **Polymerization:** Not pertinent

7.6 **Inhibitor of Polymerization:** Not pertinent

8. WATER POLLUTION

8.1 **Aquatic Toxicity:** Data not available

8.2 **Waterfowl Toxicity:** Data not available

8.3 **Biological Oxygen Demand (BOD):** Data not available

8.4 **Food Chain Concentration Potential:** None

9. SELECTED MANUFACTURERS

1. Union Camp Corp.
2051 N. Lane Ave.
Jacksonville, Fla. 32205
2. Hercules, Inc.
Wilmington, Del. 19899
3. Armak Chemicals Div.
Box 1805
Chicago, Ill. 60690

10. SHIPPING INFORMATION

10.1 **Grades or Purity:** Commercial, 79—83%

10.2 **Storage Temperature:** Ambient

10.3 **Inert Atmosphere:** No requirement

10.4 **Venting:** Open (flame arrester)

11. HAZARD ASSESSMENT CODE

(See Hazard Assessment Handbook, CG 446-3)

A-T-U

12. HAZARD CLASSIFICATIONS

12.1 **Code of Federal Regulations:** Not listed

12.2 **NAS Hazard Rating for Bulk Water Transportation:** Not listed

12.3 **NFPA Hazard Classifications:**

Category	Classification
Health Hazard (Blue)	0
Flammability (Red)	1
Reactivity (Yellow)	0

13. PHYSICAL AND CHEMICAL PROPERTIES

13.1 **Physical State at 15°C and 1 atm:** Liquid

13.2 **Molecular Weight:** 277 (avg.)

13.3 **Boiling Point at 1 atm:** 432°F = 222°C = 495°K

13.4 **Freezing Point:** 57°F = 14°C = 287°K

13.5 **Critical Temperature:** Not pertinent

13.6 **Critical Pressure:** Not pertinent

13.7 **Specific Gravity:** 0.89 at 25°C (liquid)

13.8 **Liquid Surface Tension:** 32.8 dynes/cm = 0.0328 N/m at 20°C

13.9 **Liquid-Water Interfacial Tension:** 15.59 dynes/cm = 0.01559 N/m at 20°C

13.10 **Vapor (Gas) Specific Gravity:** Not pertinent

13.11 **Ratio of Specific Heats of Vapor (Gas):** Not pertinent

13.12 **Latent Heat of Vaporization:** 103 Btu/lb = 57 cal/g = 2.4×10^5 J/kg

13.13 **Heat of Combustion:** Data not available

13.14 **Heat of Decomposition:** Not pertinent

13.15 **Heat of Solution:** Not pertinent

13.16 **Heat of Polymerization:** Not pertinent

NOTES

OAP

OLEIC ACID, POTASSIUM SALT

Common Synonyms			
Potassium oleate	Solid or liquid	Brown	Soapy odor
	Sinks and mixes slowly with water.		

Avoid contact with liquid and solid. Keep people away.
Stop discharge if possible.
Isolate and remove discharged material.
Notify local health and pollution control agencies.

Fire	Combustible. **Extinguish with water, dry chemicals, alcohol foam, or carbon dioxide.**
Exposure	**CALL FOR MEDICAL AID.** LIQUID OR SOLID Irritating to skin and eyes. If swallowed will cause nausea and vomiting. **Remove contaminated clothing and shoes.** **Flush affected areas with plenty of water.** **IF IN EYES, hold eyelids open and flush with plenty of water.** **IF SWALLOWED and victim is CONSCIOUS, have victim drink water or milk.** **IF SWALLOWED and victim is UNCONSCIOUS OR HAVING CONVULSIONS, do nothing except keep victim warm.**
Water Pollution	Effect of low concentrations on aquatic life is unknown. May be dangerous if it enters water intakes. **Notify local health and wildlife officials.** **Notify operators of nearby water intakes.**

1. RESPONSE TO DISCHARGE
(See Response Methods Handbook, CG 446-4)

Disperse and flush

2. LABELS

No hazard label required by Code of Federal Regulations

3. CHEMICAL DESIGNATIONS

3.1 **Synonyms:** Potassium oleate
3.2 **Coast Guard Compatibility Classification:** Not listed
3.3 **Chemical Formula:** $C_{17}H_{33}COOK$
3.4 **IMCO/United Nations Numerical Designation:** Not listed

4. OBSERVABLE CHARACTERISTICS

4.1 **Physical State (as shipped):** Solid or liquid
4.2 **Color:** Brown
4.3 **Odor:** Faint soapy

5. HEALTH HAZARDS

5.1 **Personal Protective Equipment:** Chemical goggles and rubber gloves
5.2 **Symptoms Following Exposure:** Inhalation of dust causes irritation of nose and throat, coughing, and sneezing. Ingestion causes mild irritation of mouth and stomach. Contact with eyes causes irritation.
5.3 **Treatment for Exposure:** INHALATION: move to fresh air. INGESTION: give large amount of water. EYES: flush with copious quantities of tap water for 15 min. and seek appropriate medical attention. SKIN: flush with water.
5.4 **Toxicity by Inhalation (Threshold Limit Value):** Data not available
5.5 **Short-Term Inhalation Limits:** Data not available
5.6 **Toxicity by Ingestion:** Data not available
5.7 **Late Toxicity:** Data not available
5.8 **Vapor (Gas) Irritant Characteristics:** Data not available
5.9 **Liquid or Solid Irritant Characteristics:** Data not available
5.10 **Odor Threshold:** Data not available

6. FIRE HAZARDS

6.1 **Flash Point:** 140°F C.C.
6.2 **Flammable Limits in Air:** Data not available
6.3 **Fire Extinguishing Agents:** Dry chemical, alcohol foam, carbon dioxide
6.4 **Fire Extinguishing Agents Not to be Used:** Water may be ineffective.
6.5 **Special Hazards of Combustion Products:**
6.6 **Behavior in Fire:**
6.7 **Ignition Temperature:** Data not available
6.8 **Electrical Hazard:** Not pertinent
6.9 **Burning Rate:** Not pertinent

7. CHEMICAL REACTIVITY

7.1 **Reactivity with Water:** No reaction
7.2 **Reactivity with Common Materials:**
7.3 **Stability During Transport:** Stable
7.4 **Neutralizing Agents for Acids and Caustics:** Not pertinent
7.5 **Polymerization:** Not pertinent
7.6 **Inhibitor of Polymerization:** Not pertinent

8. WATER POLLUTION

8.1 **Aquatic Toxicity:** Data not available
8.2 **Waterfowl Toxicity:** Data not available
8.3 **Biological Oxygen Demand (BOD):** Data not available
8.4 **Food Chain Concentration Potential:** None

9. SELECTED MANUFACTURERS

1. Diamond Shamrock Chemical Co.
   1100 Superior Ave.
   Cleveland, Ohio 44114
2. Alco Chemical Corp.
   Trenton Avenue and William Street
   Philadelphia, Pa. 19134

10. SHIPPING INFORMATION

10.1 **Grades or Purity:** 19% solution in water
10.2 **Storage Temperature:** Ambient
10.3 **Inert Atmosphere:** No requirement
10.4 **Venting:** Open

11. HAZARD ASSESSMENT CODE
(See Hazard Assessment Handbook, CG 446-3)

(liquid) A-P
(solid) SS

12. HAZARD CLASSIFICATIONS

12.1 **Code of Federal Regulations:** Not listed
12.2 **NAS Hazard Rating for Bulk Water Transportation:** Not listed
12.3 **NFPA Hazard Classifications:** Not listed

13. PHYSICAL AND CHEMICAL PROPERTIES

13.1 **Physical State at 15°C and 1 atm:** Solid or liquid
13.2 **Molecular Weight:** 320 (solute only)
13.3 **Boiling Point at 1 atm:** Not pertinent (decomposes)
13.4 **Freezing Point:** 455 – 464°F = 235 – 240°C = 508 – 513°K
13.5 **Critical Temperature:** Not pertinent
13.6 **Critical Pressure:** Not pertinent
13.7 **Specific Gravity:** > 1.1 at 20°C (solid or liquid)
13.8 **Liquid Surface Tension:** Not pertinent
13.9 **Liquid-Water Interfacial Tension:** Not pertinent
13.10 **Vapor (Gas) Specific Gravity:** Not pertinent
13.11 **Ratio of Specific Heats of Vapor (Gas):** Not pertinent
13.12 **Latent Heat of Vaporization:** Not pertinent
13.13 **Heat of Combustion:** Not pertinent
13.14 **Heat of Decomposition:** Not pertinent
13.15 **Heat of Solution:** Not pertinent
13.16 **Heat of Polymerization:** Not pertinent

NOTES

OAC — OLEIC ACID, SODIUM SALT

Common Synonyms
Eunatrol
Sodium oleate

Solid — Light tan — Slight tallow-like odor

Sinks and mixes slowly with water.

Avoid contact with solid and dust. Keep people away.
Stop discharge if possible.
Isolate and remove discharged material.
Notify local health and pollution control agencies.

Fire

Combustible.
Extinguish with water, dry chemicals, foam, or carbon dioxide.

Exposure

CALL FOR MEDICAL AID.

SOLID
Irritating to skin and eyes.
If swallowed will cause nausea and vomiting.
Remove contaminated clothing and shoes.
Flush affected areas with plenty of water.
IF IN EYES, hold eyelids open and flush with plenty of water.
IF SWALLOWED and victim is CONSCIOUS, have victim drink water or milk.
IF SWALLOWED and victim is UNCONSCIOUS OR HAVING CONVULSIONS, do nothing except keep victim warm.

Water Pollution

Effect of low concentrations on aquatic life is unknown.
May be dangerous if it enters water intakes.
Notify local health and wildlife officials.
Notify operators of nearby water intakes.

1. RESPONSE TO DISCHARGE

(See Response Methods Handbook, CG 446-4)

Disperse and flush

2. LABELS

No hazard label required by Code of Federal Regulations

3. CHEMICAL DESIGNATIONS

3.1 **Synonyms:** Eunatrol; Sodium oleate
3.2 **Coast Guard Compatibility Classification:** Not listed
3.3 **Chemical Formula:** $C_{17}H_{33}COONa$
3.4 **IMCO/United Nations Numerical Designation:** Not listed

4. OBSERVABLE CHARACTERISTICS

4.1 **Physical State (as shipped):** Solid
4.2 **Color:** Light tan
4.3 **Odor:** Slight tallow-like

5. HEALTH HAZARDS

5.1 **Personal Protective Equipment:** Dust mask and gloves
5.2 **Symptoms Following Exposure:** Inhalation of dust causes irritation of nose and throat, coughing, and sneezing. Ingestion causes mild irritation of mouth. Contact with eyes causes irritation.
5.3 **Treatment for Exposure:** INHALATION: move to fresh air. INGESTION: give large amount of water. EYES: flush with copious quantities of tap water. SKIN: flush with water.
5.4 **Toxicity by Inhalation (Threshold Limit Value):** Data not available
5.5 **Short-Term Inhalation Limits:** Data not available
5.6 **Toxicity by Ingestion:** Data not available
5.7 **Late Toxicity:** Data not available
5.8 **Vapor (Gas) Irritant Characteristics:** Data not available
5.9 **Liquid or Solid Irritant Characteristics:** Data not available
5.10 **Odor Threshold:** Data not available

6. FIRE HAZARDS

6.1 **Flash Point:** Not pertinent (combustible solid)
6.2 **Flammable Limits in Air:** Not pertinent
6.3 **Fire Extinguishing Agents:** Water, foam, dry chemical, carbon dioxide
6.4 **Fire Extinguishing Agents Not to be Used:**
6.5 **Special Hazards of Combustion Products:**
6.6 **Behavior in Fire:**
6.7 **Ignition Temperature:** Data not available
6.8 **Electrical Hazard:** Not pertinent
6.9 **Burning Rate:** Not pertinent

7. CHEMICAL REACTIVITY

7.1 **Reactivity with Water:** No reaction
7.2 **Reactivity with Common Materials:**
7.3 **Stability During Transport:** Stable
7.4 **Neutralizing Agents for Acids and Caustics:** Not pertinent
7.5 **Polymerization:** Not pertinent
7.6 **Inhibitor of Polymerization:** Not pertinent

8. WATER POLLUTION

8.1 **Aquatic Toxicity:** Data not available
8.2 **Waterfowl Toxicity:** Data not available
8.3 **Biological Oxygen Demand (BOD):** Data not available
8.4 **Food Chain Concentration Potential:** None

9. SELECTED MANUFACTURERS

1. Diamond Shamrock Chemical Co.
   1100 Superior Ave.
   Cleveland, Ohio 44114
2. Witco Chemical Corp.
   Organics Division
   277 Park Avenue
   New York, N. Y. 10017

10. SHIPPING INFORMATION

10.1 **Grades or Purity:** Commercial
10.2 **Storage Temperature:** Ambient
10.3 **Inert Atmosphere:** No requirement
10.4 **Venting:** Open

11. HAZARD ASSESSMENT CODE

(See Hazard Assessment Handbook, CG 446-3)

SS

12. HAZARD CLASSIFICATIONS

12.1 **Code of Federal Regulations:** Not listed
12.2 **NAS Hazard Rating for Bulk Water Transportation:** Not listed
12.3 **NFPA Hazard Classifications:** Not listed

13. PHYSICAL AND CHEMICAL PROPERTIES

13.1 **Physical State at 15°C and 1 atm:** Solid
13.2 **Molecular Weight:** 304 (approx.)
13.3 **Boiling Point at 1 atm:** Not pertinent (decomposes)
13.4 **Freezing Point:** 450—455°F = 232—235°C = 505—508°K
13.5 **Critical Temperature:** Not pertinent
13.6 **Critical Pressure:** Not pertinent
13.7 **Specific Gravity:** >1.1 at 20°C (solid)
13.8 **Liquid Surface Tension:** Not pertinent
13.9 **Liquid-Water Interfacial Tension:** Not pertinent
13.10 **Vapor (Gas) Specific Gravity:** Not pertinent
13.11 **Ratio of Specific Heats of Vapor (Gas):** Not pertinent
13.12 **Latent Heat of Vaporization:** Not pertinent
13.13 **Heat of Combustion:** Not pertinent
13.14 **Heat of Decomposition:** Not pertinent
13.15 **Heat of Solution:** Not pertinent
13.16 **Heat of Polymerization:** Not pertinent

NOTES

OLM	# OLEUM

Common Synonyms			
Fuming sulfuric acid	Oily liquid	Colorless to cloudy	Sharp, choking odor
	Mixes and reacts with water producing heat. Irritating mist is produced.		

AVOID CONTACT WITH LIQUID AND MIST. Keep people away.
Wear chemical protective suit with self-contained breathing apparatus.
Stop discharge if possible.
Stay upwind and use water spray to "knock down" mist.
Isolate and remove discharged material.
Notify local health and pollution control agencies.

Fire

Not flammable.
May cause fire on contact with combustibles.
Flammable gas may be produced on contact with metals.
Wear chemical protective suit with self-contained breathing apparatus.
DO NOT USE WATER ON ADJACENT FIRES.

Exposure

CALL FOR MEDICAL AID.

MIST
Irritating to eyes, nose and throat.
If inhaled, will cause coughing or difficult breathing.
Move to fresh air.
If breathing has stopped, give artificial respiration (but NOT mouth to mouth).
If breathing is difficult, give oxygen.

LIQUID
Will burn skin and eyes.
Harmful if swallowed.
Remove contaminated clothing and shoes.
Flush affected areas with plenty of water.
IF IN EYES, hold eyelids open and flush with plenty of water.
IF SWALLOWED and victim is CONSCIOUS, have victim drink water or milk.
DO NOT INDUCE VOMITING.

Water Pollution

HARMFUL TO AQUATIC LIFE IN VERY LOW CONCENTRATIONS.
Fouling to shoreline.
May be dangerous if it enters water intakes.
Notify local health and wildlife officials.
Notify operators of nearby water intakes.

1. RESPONSE TO DISCHARGE

(See Response Methods Handbook, CG 446-4)

Issue warning—corrosive
Restrict access
Chemical and physical treatment
Disperse and flush

2. LABEL

3. CHEMICAL DESIGNATIONS

3.1 **Synonyms:** Fuming sulfuric acid
3.2 **Coast Guard Compatibility Classification:** Inorganic acid
3.3 **Chemical Formula:** $SO_3-H_2SO_4$
3.4 **IMCO/United Nations Numerical Designation:** 8.0/1831

4. OBSERVABLE CHARACTERISTICS

4.1 **Physical State (as shipped):** Liquid
4.2 **Color:** Colorless to cloudy
4.3 **Odor:** Sharp penetrating; choking

5. HEALTH HAZARDS

5.1 **Personal Protective Equipment:** Respirator approved by U.S. Bureau of Mines for acid mists; rubber gloves; splashproof goggles; eyewash fountain and safety shower; rubber footwear; face shield.
5.2 **Symptoms Following Exposure:** Acid mist is irritating to eyes, nose and throat. Liquid causes severe burns of skin and eyes.
5.3 **Treatment for Exposure:** INGESTION: have victim drink water or milk; do NOT induce vomiting. EYES: flush with plenty of water for at least 15 min.; call a doctor. SKIN: flush with plenty of water.
5.4 **Toxicity by Inhalation (Threshold Limit Value):** 1 mg/m^3
5.5 **Short-Term Inhalation Limits:** 5 mg/m^3 for 5 min.; 3 mg/m^3 for 10 min.; 2 mg/m^3 for 30 min.; 1 mg/m^3 for 60 min.
5.6 **Toxicity by Ingestion:** Severe burns of mouth and stomach.
5.7 **Late Toxicity:** None
5.8 **Vapor (Gas) Irritant Characteristics:** Vapors cause severe irritation of eye and throat and can cause eye and lung injury. They cannot be tolerated even at low concentrations.
5.9 **Liquid or Solid Irritant Characteristics:** Severe skin irritant. Causes second- and third-degree burns on short contact; very injurious to the eyes.
5.10 **Odor Threshold:** 1 mg/m^3

6. FIRE HAZARDS

6.1 **Flash Point:** Not flammable
6.2 **Flammable Limits in Air:** Not flammable
6.3 **Fire Extinguishing Agents:** Not pertinent
6.4 **Fire Extinguishing Agents Not to be Used:** Avoid use of water on adjacent fires.
6.5 **Special Hazards of Combustion Products:** Toxic and irritating vapors are generated.
6.6 **Behavior in Fire:** Not pertinent
6.7 **Ignition Temperature:** Not flammable
6.8 **Electrical Hazard:** Not pertinent
6.9 **Burning Rate:** Not flammable

7. CHEMICAL REACTIVITY

7.1 **Reactivity with Water:** Vigorous reaction with water; spatters.
7.2 **Reactivity with Common Materials:** May react with cast iron with explosive violence. Attacks many metals, releasing flammable hydrogen gas. Capable of igniting finely divided combustible material on contact. Extremely hazardous in contact with many materials.
7.3 **Stability During Transport:** Normally stable
7.4 **Neutralizing Agents for Acids and Caustics:** Cautious dilution with water, with protection against violent spattering. Diluted acid may be neutralized with lime or soda ash.
7.5 **Polymerization:** Not pertinent
7.6 **Inhibitor of Polymerization:** Not pertinent

8. WATER POLLUTION

8.1 **Aquatic Toxicity:**
24 ppm/24 hr/bluegill/lethal/fresh water
42 ppm/48 hr/prawn/LC_{50}/salt water
8.2 **Waterfowl Toxicity:** Data not available
8.3 **Biological Oxygen Demand (BOD):** None
8.4 **Food Chain Concentration Potential:** None

9. SELECTED MANUFACTURERS

1. Allied Chemical Corp.
Industrial Chemicals Division
Morristown, N. J. 07960
2. Monsanto Co.
Monsanto Industrial Chemicals Co.
800 N. Lindbergh Blvd.
St. Louis, Mo. 63166
3. Stauffer Chemical Co.
Industrial Chemicals Division
299 Park Ave.
New York, N. Y. 10017

10. SHIPPING INFORMATION

10.1 **Grades or Purity:** 20% (104.5% sulfuric acid) to 65% (114.6% sulfuric acid)
10.2 **Storage Temperature:** Ambient
10.3 **Inert Atmosphere:** No requirement
10.4 **Venting:** Open

11. HAZARD ASSESSMENT CODE

(See Hazard Assessment Handbook, CG 446-3)

A-O-P

12. HAZARD CLASSIFICATIONS

12.1 **Code of Federal Regulations:** Corrosive material
12.2 **NAS Hazard Rating for Bulk Water Transportation:**

Category	Rating
Fire	0
Health	
Vapor Irritant	4
Liquid or Solid Irritant	4
Poisons	3
Water Pollution	
Human Toxicity	2
Aquatic Toxicity	3
Aesthetic Effect	2
Reactivity	
Other Chemicals	4
Water	3
Self-Reaction	0

12.3 **NFPA Hazard Classifications:**

Category	Classification
Health Hazard (Blue)	3
Flammability (Red)	0
Reactivity (Yellow)	2
	W̶

13. PHYSICAL AND CHEMICAL PROPERTIES

13.1 **Physical State at 15°C and 1 atm:** Liquid
13.2 **Molecular Weight:** Not pertinent
13.3 **Boiling Point at 1 atm:** Decomposes
13.4 **Freezing Point:** Not pertinent
13.5 **Critical Temperature:** Not pertinent
13.6 **Critical Pressure:** Not pertinent
13.7 **Specific Gravity:** 1.91-1.97 at 15°C (liquid)
13.8 **Liquid Surface Tension:** Not pertinent
13.9 **Liquid-Water Interfacial Tension:** Not pertinent
13.10 **Vapor (Gas) Specific Gravity:** Not pertinent
13.11 **Ratio of Specific Heats of Vapor (Gas):** Not pertinent
13.12 **Latent Heat of Vaporization:** Not pertinent
13.13 **Heat of Combustion:** Not pertinent
13.14 **Heat of Decomposition:** Not pertinent
13.15 **Heat of Solution:** Not pertinent
13.16 **Heat of Polymerization:** Not pertinent

NOTES

REVISED 1978

OXA — OXALIC ACID

Common Synonyms			
Ethanedioic acid	Solid crystals	White	Odorless
	Sinks and mixes with water.		

Avoid contact with solid and dust. Keep people away.
Wear goggles, self-contained breathing apparatus and rubber overclothing (including gloves).
Stop discharge if possible.
Isolate and remove discharged material.
Notify local health and pollution control agencies.

Fire

Not flammable.
Poisonous gases are produced in fire.
Wear goggles, self-contained breathing apparatus, and rubber overclothing (including gloves).

Exposure

CALL FOR MEDICAL AID.

DUST
Will burn eyes, nose and throat.
If inhaled, will cause difficult breathing.
Move to fresh air.
If breathing has stopped, give artificial respiration.
If breathing is difficult, give oxygen.

SOLID
Will burn skin and eyes.
If swallowed, will cause nausea or loss of consciousness.
Remove contaminated clothing and shoes.
Flush affected areas with plenty of water.
IF IN EYES, hold eyelids open and flush with plenty of water.
IF SWALLOWED and victim is CONSCIOUS, have victim drink water or milk.
DO NOT INDUCE VOMITING.

Water Pollution

Dangerous to aquatic life in high concentrations.
May be dangerous if it enters water intakes.
Notify local health and wildlife officials.
Notify operators of nearby water intakes.

1. RESPONSE TO DISCHARGE

(See Response Methods Handbook, CG 446-4)

Issue warning - poison
Should be removed
Chemical and physical treatment

2. LABEL

3. CHEMICAL DESIGNATIONS

3.1 **Synonyms:** Ethanedioic acid
3.2 **Coast Guard Compatibility Classification:** Not applicable
3.3 **Chemical Formula:** $C_2H_2O_4$
3.4 **IMCO/United Nations Numerical Designation:** Not listed

4. OBSERVABLE CHARACTERISTICS

4.1 **Physical State (as shipped):** Solid
4.2 **Color:** White
4.3 **Odor:** Odorless

5. HEALTH HAZARDS

5.1 **Personal Protective Equipment:** Respirator for dust or mist protection; rubber, neoprene, or vinyl gloves; chemical safety glasses; rubbers, over leather or rubber safety shoes; apron or impervious clothing for splash protection.
5.2 **Symptoms Following Exposure:** As dust or as a solution, can cause severe burns of eyes, skin, or mucous membranes. Ingestion of 5 grams has caused death with symptoms of nausea, shock, collapse, and convulsions coming on rapidly. Repeated or prolonged skin exposure can cause dermatitis and slow-healing ulcers.
5.3 **Treatment for Exposure:** Get medical attention for all eye exposures and any serious overexposures; treatment is symptomatic. INHALATION: rinse mouth and/or gargle repeatedly with cold water. INGESTION: dilute by drinking large amounts of water; repeat at least once and then administer milk or milk of magnesia as an emollient; do NOT induce vomiting. EYES AND SKIN: flush thoroughly with water.
5.4 **Toxicity by Inhalation (Threshold Limit Value):** Not pertinent
5.5 **Short-Term Inhalation Limits:** Not pertinent
5.6 **Toxicity by Ingestion:** Grade 3; LD_{50} 50 to 500 mg/kg
5.7 **Late Toxicity:** Data not available
5.8 **Vapor (Gas) Irritant Characteristics:** Not pertinent
5.9 **Liquid or Solid Irritant Characteristics:** Severe skin irritant. Causes second- and third-degree burns on short contact and is very injurious to the eyes.
5.10 **Odor Threshold:** Not pertinent

6. FIRE HAZARDS

6.1 **Flash Point:** Not flammable
6.2 **Flammable Limits in Air:** Not flammable
6.3 **Fire Extinguishing Agents:** Not pertinent
6.4 **Fire Extinguishing Agents Not to be Used:** Not pertinent
6.5 **Special Hazards of Combustion Products:** Generates poisonous gases
6.6 **Behavior in Fire:** Not pertinent
6.7 **Ignition Temperature:** Not flammable
6.8 **Electrical Hazard:** Not pertinent
6.9 **Burning Rate:** Not flammable

7. CHEMICAL REACTIVITY

7.1 **Reactivity with Water:** No reaction
7.2 **Reactivity with Common Materials:** No reaction
7.3 **Stability During Transport:** Stable
7.4 **Neutralizing Agents for Acids and Caustics:** Lime or soda ash
7.5 **Polymerization:** Not pertinent
7.6 **Inhibitor of Polymerization:** Not pertinent

8. WATER POLLUTION

8.1 **Aquatic Toxicity:** 4000 mg/l/24 hr/bluegill/TL_m/fresh water
8.2 **Waterfowl Toxicity:** Data not available
8.3 **Biological Oxygen Demand (BOD):** 14%, 5 days
8.4 **Food Chain Concentration Potential:** None

9. SELECTED MANUFACTURERS

1. Allied Chemical Corp.
   Industrial Chemicals Division
   Morristown, N. J. 07960
2. Mallinckrodt Chemical Works
   Industrial Chemicals Division
   2nd and Mallinckrodt St.
   St. Louis, Mo. 63160
3. Pfizer, Inc.
   Chemicals Division
   235 E. 42 St.
   New York, N. Y. 10017

10. SHIPPING INFORMATION

10.1 **Grades or Purity:** Technical: 99.8% min.
10.2 **Storage Temperature:** Ambient
10.3 **Inert Atmosphere:** None required
10.4 **Venting:** Open

11. HAZARD ASSESSMENT CODE

(See Hazard Assessment Handbook, CG 446-3)

SS

12. HAZARD CLASSIFICATIONS

12.1 **Code of Federal Regulations:** Corrosive Material
12.2 **NAS Hazard Rating for Bulk Water Transportation:** Not listed
12.3 **NFPA Hazard Classifications:**

Category	Classification*	
Health Hazard (Blue)	1	2
Flammability (Red)	1	1
Reactivity (Yellow)	0	0

*First column refers to non-fire situation

13. PHYSICAL AND CHEMICAL PROPERTIES

13.1 **Physical State at 15°C and 1 atm:** Solid
13.2 **Molecular Weight:** 126.07
13.3 **Boiling Point at 1 atm:** Decomposes
13.4 **Freezing Point:** 214.7°F = 101.5°C = 374.7°K
13.5 **Critical Temperature:** Not pertinent
13.6 **Critical Pressure:** Not pertinent
13.7 **Specific Gravity:** 1.90 at 15°C (solid)
13.8 **Liquid Surface Tension:** Not pertinent
13.9 **Liquid-Water Interfacial Tension:** Not pertinent
13.10 **Vapor (Gas) Specific Gravity:** Not pertinent
13.11 **Ratio of Specific Heats of Vapor (Gas):** Not pertinent
13.12 **Latent Heat of Vaporization:** Not pertinent
13.13 **Heat of Combustion:** Not pertinent
13.14 **Heat of Decomposition:** Not pertinent
13.15 **Heat of Solution:** Not pertinent
13.16 **Heat of Polymerization:** Not pertinent

NOTES

REVISED 1978

OXY	OXYGEN, LIQUEFIED

Common Synonyms			
Liquid oxygen LOX	Liquefied gas	Light blue	Odorless
	Sinks and boils in water.		

Avoid contact with liquid. Keep people away.
Wear rubber overclothing (including gloves).
Stop discharge if possible.

Fire	Not flammable. Containers may explode in fire. **Cool exposed containers with water.**
Exposure	**Call for medical aid.** VAPOR If inhaled will cause dizziness, or difficult breathing. LIQUID Will cause frostbite. **Flush affected areas with plenty of water.** **DO NOT RUB AFFECTED AREAS.**
Water Pollution	Not harmful to aquatic life.

1. RESPONSE TO DISCHARGE

(See Response Methods Handbook, CG 446-4)

Restrict access

2. LABEL

Yellow

3. CHEMICAL DESIGNATIONS

3.1 **Synonyms:** Liquid oxygen; LOX

3.2 **Coast Guard Compatibility Classification:** Not applicable

3.3 **Chemical Formula:** O_2

3.4 **IMCO/United Nations Numerical Designation:** 2/1073

4. OBSERVABLE CHARACTERISTICS

4.1 **Physical State (as shipped):** Liquefied gas

4.2 **Color:** Light blue

4.3 **Odor:** None

5. HEALTH HAZARDS

5.1 **Personal Protective Equipment:** Safety goggles or face shield; insulated gloves; long sleeves; trousers worn outside boots or over high-top shoes to shed spilled liquid.

5.2 **Symptoms Following Exposure:** Inhalation of 100% oxygen can cause nausea, dizziness, irritation of lungs, pulmonary edema, pneumonia, and collapse. Liquid may cause frostbite of eyes and skin.

5.3 **Treatment for Exposure:** INHALATION: in all but the most severe cases (pneumonia), recovery is rapid after reduction of oxygen pressure; supportive treatment should include immediate sedation, anticonvulsive therapy if needed, and rest. EYES: treat frostbite burns. SKIN: treat frostbite; soak in lukewarm water.

5.4 **Toxicity by Inhalation (Threshold Limit Value):** Not pertinent

5.5 **Short-Term Inhalation Limits:** Not pertinent

5.6 **Toxicity by Ingestion:** Not pertinent

5.7 **Late Toxicity:** Not pertinent

5.8 **Vapor (Gas) Irritant Characteristics:** Data not available

5.9 **Liquid or Solid Irritant Characteristics:** Data not available

5.10 **Odor Threshold:** Not pertinent

6. FIRE HAZARDS

6.1 **Flash Point:** Not flammable but supports combustion

6.2 **Flammable Limits in Air:** Not flammable

6.3 **Fire Extinguishing Agents:** Not pertinent

6.4 **Fire Extinguishing Agents Not to be Used:** Not pertinent

6.5 **Special Hazards of Combustion Products:** Not pertinent

6.6 **Behavior in Fire:** Increases intensity of any fire. Mixtures of liquid oxygen and any fuel are highly explosive.

6.7 **Ignition Temperature:** Not pertinent

6.8 **Electrical Hazard:** Not pertinent

6.9 **Burning Rate:** Not pertinent

7. CHEMICAL REACTIVITY

7.1 **Reactivity with Water:** Heat of water will vigorously vaporize liquid oxygen.

7.2 **Reactivity with Common Materials:** Avoid organic and combustible materials, such as oil, grease, coal dust, etc. If ignited, such mixtures can explode. The low temperature may cause brittleness in some materials.

7.3 **Stability During Transport:** Stable

7.4 **Neutralizing Agents for Acids and Caustics:** Not pertinent

7.5 **Polymerization:** Not pertinent

7.6 **Inhibitor of Polymerization:** Not pertinent

8. WATER POLLUTION

8.1 **Aquatic Toxicity:** None

8.2 **Waterfowl Toxicity:** None

8.3 **Biological Oxygen Demand (BOD):** None

8.4 **Food Chain Concentration Potential:** None

9. SELECTED MANUFACTURERS

1. Union Carbide Corporation
   Linde Division
   Morristown, N. J. 08057
2. Airco, Inc.
   Industrial Gases Division
   575 Mountain Avenue
   Murray Hill, N. J. 07974
3. Liquid Air, Inc.
   1 Embarcadero Center
   San Francisco, Calif. 94111

10. SHIPPING INFORMATION

10.1 **Grades or Purity:** 99.5+%

10.2 **Storage Temperature:** −183°C

10.3 **Inert Atmosphere:** No requirement

10.4 **Venting:** Safety relief

11. HAZARD ASSESSMENT CODE

(See Hazard Assessment Handbook, CG 446-3)

A-I

12. HAZARD CLASSIFICATIONS

12.1 **Code of Federal Regulations:** Nonflammable compressed gas

12.2 **NAS Hazard Rating for Bulk Water Transportation:** Not listed

12.3 **NFPA Hazard Classifications:**

Category	Classification
Health Hazard (Blue)	3
Flammability (Red)	0
Reactivity (Yellow)	0
	oxy

13. PHYSICAL AND CHEMICAL PROPERTIES

13.1 **Physical State at 15°C and 1 atm:** Gas

13.2 **Molecular Weight:** 32.0

13.3 **Boiling Point at 1 atm:** −297.3°F = −182.9°C = 90.3°K

13.4 **Freezing Point:** −361°F = −218°C = 55°K

13.5 **Critical Temperature:** −180°F = −118°C = 155°K

13.6 **Critical Pressure:** 738 psia = 50.1 atm = 5.09 MN/m²

13.7 **Specific Gravity:** 1.14 at −183°C (liquid)

13.8 **Liquid Surface Tension:** 13.47 dynes/cm = 0.01347 N/m at −183°C

13.9 **Liquid-Water Interfacial Tension:** Not pertinent

13.10 **Vapor (Gas) Specific Gravity:** 1.1

13.11 **Ratio of Specific Heats of Vapor (Gas):** 1.3962

13.12 **Latent Heat of Vaporization:** 91.6 Btu/lb = 50.9 cal/g = 2.13×10^5 J/kg

13.13 **Heat of Combustion:** Not pertinent

13.14 **Heat of Decomposition:** Not pertinent

13.15 **Heat of Solution:** Not pertinent

13.16 **Heat of Polymerization:** Not pertinent

NOTES

PFA

PARAFORMALDEHYDE

Common Synonyms
Polyformaldehyde
Polyoxymethylene
Formaldehyde polymer

Solid powder | White | Irritating odor

Sinks and mixes with water.

Stop discharge if possible. Keep people away.
Avoid contact with solid and dust.
Call fire department.
Isolate and remove discharged material.
Notify local health and pollution control agencies.

Fire

Combustible.
Wear goggles and self-contained breathing apparatus.
Extinguish with water, foam, dry chemical or carbon dioxide.

Exposure

CALL FOR MEDICAL AID.

DUST
Irritating to eyes, nose and throat.
Harmful if inhaled.
Move to fresh air.
If breathing has stopped, give artificial respiration.
If breathing is difficult, give oxygen.

SOLID
Irritating to skin and eyes.
If swallowed, will cause nausea, vomiting or loss of consciousness.
Remove contaminated clothing and shoes.
Flush affected areas with plenty of water.
IF IN EYES, hold eyelids open and flush with plenty of water.
IF SWALLOWED and victim is CONSCIOUS, have victim drink water or milk.

Water Pollution

HARMFUL TO AQUATIC LIFE IN VERY LOW CONCENTRATIONS.
May be dangerous if it enters water intakes.
Notify local health and wildlife officials.
Notify operators of nearby water intakes.

1. RESPONSE TO DISCHARGE

(See Response Methods Handbook, CG 446-4)

Disperse and flush

2. LABELS

No hazard label required by Code of Federal Regulations

3. CHEMICAL DESIGNATIONS

3.1 **Synonyms:** Formaldehyde polymer
Polyformaldehyde
Polyoxymethylene
Polyoxymethylene glycol

3.2 **Coast Guard Compatibility Classification:** Aldehyde

3.3 **Chemical Formula:** $HO(CH_2O)_nH$

3.4 **IMCO/United Nations Numerical Designation:** 9.0

4. OBSERVABLE CHARACTERISTICS

4.1 **Physical State (as shipped):** Solid

4.2 **Color:** White

4.3 **Odor:** Pungent and irritating; like formaldehyde

5. HEALTH HAZARDS

5.1 **Personal Protective Equipment:** Goggles or face shield; protective clothing.

5.2 **Symptoms Following Exposure:** Vapor or dust irritates eyes, mucous membranes, and skin; may cause dermatitis. Ingestion of solid or of a solution in water irritates mouth, throat, and stomach and may cause death.

5.3 **Treatment for Exposure:** INGESTION: give milk or white of egg beaten with water; call a doctor. SKIN OR EYES: rinse with copious amounts of water.

5.4 **Toxicity by Inhalation (Threshold Limit Value):** 5 ppm

5.5 **Short-Term Inhalation Limits:** Data not available

5.6 **Toxicity by Ingestion:** Grade 3; LD_{50} 50 to 500 mg/kg

5.7 **Late Toxicity:** Data not available

5.8 **Vapor (Gas) Irritant Characteristics:** Vapor is moderately irritating such that personnel will not usually tolerate moderate or high vapor concentrations.

5.9 **Liquid or Solid Irritant Characteristics:** Minimum hazard. If spilled on clothing and allowed to remain, may cause smarting and reddening of the skin.

5.10 **Odor Threshold:** Data not available

6. FIRE HAZARDS

6.1 **Flash Point:** 199°F (approx.) O.C.; 160°F (approx.) C.C.

6.2 **Flammable Limits in Air:** (formaldehyde gas) 7.0%–73.0%

6.3 **Fire Extinguishing Agents:** Water, foam, dry chemical, or carbon dioxide

6.4 **Fire Extinguishing Agents Not to be Used:** Data not available

6.5 **Special Hazards of Combustion Products:** Not pertinent

6.6 **Behavior in Fire:** Changes to formaldehyde gas, which is highly flammable.

6.7 **Ignition Temperature:** 572°F (approx.)

6.8 **Electrical Hazard:** Not pertinent

6.9 **Burning Rate:** Not pertinent

7. CHEMICAL REACTIVITY

7.1 **Reactivity with Water:** Forms water solution of formaldehyde.

7.2 **Reactivity with Common Materials:** No reaction

7.3 **Stability During Transport:** Slowly decomposes to formaldehyde gas.

7.4 **Neutralizing Agents for Acids and Caustics:** Not pertinent

7.5 **Polymerization:** Not pertinent

7.6 **Inhibitor of Polymerization:** Not pertinent

8. WATER POLLUTION

8.1 **Aquatic Toxicity:**
32 ppm/24 hr/catfish/TL_m/fresh water
100-300 ppm/48 hr/flounder/TL_m/salt water

8.2 **Waterfowl Toxicity:** Data not available

8.3 **Biological Oxygen Demand (BOD):** 37%, 5 days; (theor.) 47%, 5 days

8.4 **Food Chain Concentration Potential:** None

9. SELECTED MANUFACTURERS

1. Celanese Corp.
Celanese Chemical Co. Division
245 Park Ave.
New York, N. Y. 10017

2. Mallinckrodt Chemical Works
2nd and Mallinckrodt St.
St. Louis, Mo. 63160

3. Tenneco Chemicals, Inc.
Tenneco Intermediates Division
5 Turner Place
Piscataway, N. J. 08854

10. SHIPPING INFORMATION

10.1 **Grades or Purity:** 91-99%, powder and flake

10.2 **Storage Temperature:** Ambient

10.3 **Inert Atmosphere:** No requirement

10.4 **Venting:** Open (flame arrester)

11. HAZARD ASSESSMENT CODE

(See Hazard Assessment Handbook, CG 446-3)

RR-C

12. HAZARD CLASSIFICATIONS

12.1 **Code of Federal Regulations:** Combustible Liquid

12.2 **NAS Hazard Rating for Bulk Water Transportation:** Not listed

12.3 **NFPA Hazard Classifications:** Not listed

13. PHYSICAL AND CHEMICAL PROPERTIES

13.1 **Physical State at 15°C and 1 atm:** Solid

13.2 **Molecular Weight:** 600 (approx.)

13.3 **Boiling Point at 1 atm:** Decomposes

13.4 **Freezing Point:** 311-342°F = 155-172°C = 428-445°K

13.5 **Critical Temperature:** Not pertinent

13.6 **Critical Pressure:** Not pertinent

13.7 **Specific Gravity:** 1.46 at 15°C (solid)

13.8 **Liquid Surface Tension:** Not pertinent

13.9 **Liquid-Water Interfacial Tension:** Not pertinent

13.10 **Vapor (Gas) Specific Gravity:** Not pertinent

13.11 **Ratio of Specific Heats of Vapor (Gas):** Not pertinent

13.12 **Latent Heat of Vaporization:** Not pertinent

13.13 **Heat of Combustion:** -6682 Btu/lb $= -3712$ cal/g $= -155.4 \times 10^5$ J/kg

13.14 **Heat of Decomposition:** Not pertinent

13.15 **Heat of Solution:** -150 Btu/lb $= -83.5$ cal/g $= -3.50 \times 10^5$ J/kg

13.16 **Heat of Polymerization:** Not pertinent

NOTES

PTO	PARATHION, LIQUID

Common Synonyms		
Ethyl parathion Phosphorothioic acid, 0,0-diethyl 0-p-nitrophenyl ester 0,0 Diethyl 0-(p-nitrophenyl) phosphorothioate 0,0-diethyl 0-p-nitrophenyl thiophosphate	Liquid Sinks in water. Freezing point is 43° F.	Light to dark brown

AVOID CONTACT WITH LIQUID AND VAPOR. KEEP PEOPLE AWAY.
Wear goggles, self-contained breathing apparatus, and rubber overclothing (including gloves).
Stop discharge if possible.
Isolate and remove discharged material.
Notify local health and pollution control agencies.

Fire

Not flammable.
POISONOUS GASES ARE PRODUCED WHEN HEATED.

Exposure

CALL FOR MEDICAL AID.

LIQUID
POISONOUS IF SWALLOWED OR IF SKIN IS EXPOSED.
Remove contaminated clothing and shoes.
Flush affected areas with plenty of water.
IF IN EYES, hold eyelids open and flush with plenty of water.
IF SWALLOWED and victim is CONSCIOUS, have victim drink water or milk and have victim induce vomiting.
IF SWALLOWED and victim is UNCONSCIOUS OR HAVING CONVULSIONS, do nothing except keep victim warm.

Water Pollution

HARMFUL TO AQUATIC LIFE IN VERY LOW CONCENTRATIONS.
May be dangerous if it enters water intakes.
Notify local health and wildlife officials.
Notify operators of nearby water intakes.

1. RESPONSE TO DISCHARGE

(See Response Methods Handbook, CG 446-4)

Issue warning—poison, water contaminant
Restrict access
Should be removed
Chemical and physical treatment

2. LABEL

3. CHEMICAL DESIGNATIONS

3.1 **Synonyms:** O,O-Diethyl O-(p-nitrophenyl) phosphorothioate; O,O-Diethyl O-(p-nitrophenyl) thiophosphate; Ethyl Parathion; Phosphorothioic acid, O,O-diethyl O-p-nitrophenyl ester

3.2 **Coast Guard Compatibility Classification:** Not applicable

3.3 **Chemical Formula:** $(C_2H_5O)_2PSOC_6H_4NO_2$

3.4 **IMCO/United Nations Numerical Designation:** 6.1/1668

4. OBSERVABLE CHARACTERISTICS

4.1 **Physical State (as shipped):** Liquid

4.2 **Color:** Deep brown to yellow

4.3 **Odor:** Characteristic

5. HEALTH HAZARDS

5.1 **Personal Protective Equipment:** Neoprene-coated gloves; rubber workshoes or overshoes; latex rubber apron; goggles; respirator or mask approved for toxic dusts and organic vapors

5.2 **Symptoms Following Exposure:** Inhalation of mist, dust, or vapor (or ingestion, or absorption through the skin) cause dizziness, usually accompanied by constriction of the pupils, headache, and tightness of the chest. Nausea, vomiting, abdominal cramps, diarrhea, muscular twitchings, convulsions and possibly death may follow. An increase in salivary and bronchial secretions may result which simulate severe pulmonary edema. Contact with eyes causes irritation.

5.3 **Treatment for Exposure:** Call a doctor for all exposures to this compound. INHALATION: remove victim from exposure immediately; have physician treat with atropine injections until full atropinization; 2-PAM may also be administered by physician. EYES: flush with water immediately after contact for at least 15 min. SKIN: remove all clothing and shoes immediately; quickly wipe off the affected area with a clean cloth; follow immediately with a shower, using plenty of soap. If a complete shower is impossible, wash the affected skin repeatedly with soap and water. INGESTION: if victim is conscious, induce vomiting and repeat until vomit fluid is clear; make victim drink plenty of milk or water; have him lie down and keep warm.

5.4 **Toxicity by Inhalation (Threshold Limit Value):** 0.01 mg/m³

5.5 **Short-Term Inhalation Limits:** 0.5 mg/m³ for 30 min.

5.6 **Toxicity by Ingestion:** Grade 4; oral LD_{50} = 2 mg/kg (rat)

5.7 **Late Toxicity:** Birth defects in chick embryos

5.8 **Vapor (Gas) Irritant Characteristics:** Data not available

5.9 **Liquid or Solid Irritant Characteristics:** Data not available

6. FIRE HAZARDS

6.1 **Flash Point:** Not flammable

6.2 **Flammable Limits in Air:** Not flammable

6.3 **Fire Extinguishing Agents:** Water on adjacent fires

6.4 **Fire Extinguishing Agents Not to be Used:** High-pressure water hoses may scatter parathion from broken containers, increasing contamination hazard.

6.5 **Special Hazards of Combustion Products:** Fumes from decomposing material may contain oxides of sulfur and nitrogen.

6.6 **Behavior in Fire:** Containers may explode when heated.

6.7 **Ignition Temperature:** Not pertinent

6.8 **Electrical Hazard:** Not pertinent

6.9 **Burning Rate:** Not pertinent

7. CHEMICAL REACTIVITY

7.1 **Reactivity with Water:** Slow reaction, not considered hazardous

7.2 **Reactivity with Common Materials:** No reaction

7.3 **Stability During Transport:** Stable

7.4 **Neutralizing Agents for Acids and Caustics:** Not pertinent

7.5 **Polymerization:** Not pertinent

7.6 **Inhibitor of Polymerization:** Not pertinent

8. WATER POLLUTION

8.1 **Aquatic Toxicity:**
1.6 ppm/96 hr/minnow/TL_m/fresh water
0.43 ppm/<24 hr/brine shrimp/lethal/salt water

8.2 **Waterfowl Toxicity:** LD_{50}=2.13 mg/kg

8.3 **Biological Oxygen Demand (BOD):** Data not available

8.4 **Food Chain Concentration Potential:** No buildup in food chain

9. SELECTED MANUFACTURERS

1. Monsanto Commercial Products Company
Agricultural Division
800 North Lindbergh Boulevard
St. Louis, Mo. 63166

2. Stauffer Chemical Co.
Agricultural Chemical Division
Westport, Conn. 06880

10. SHIPPING INFORMATION

10.1 **Grades or Purity:** 98.5+%
Sometimes distributed as solutions emulsifiable in water.

10.2 **Storage Temperature:** Ambient

10.3 **Inert Atmosphere:** No requirement

10.4 **Venting:** Pressure-vacuum

11. HAZARD ASSESSMENT CODE

(See Hazard Assessment Handbook, CG 446-3)

A-O-X-Y

12. HAZARD CLASSIFICATIONS

12.1 **Code of Federal Regulations:** Poisonous, Class B

12.2 **NAS Hazard Rating for Bulk Water Transportation:** Not listed

12.3 **NFPA Hazard Classifications:**

Category	Classification
Health Hazard (Blue)	4
Flammability (Red)	1
Reactivity (Yellow)	2

13. PHYSICAL AND CHEMICAL PROPERTIES

13.1 **Physical State at 15°C and 1 atm:** Liquid

13.2 **Molecular Weight:** 291.3

13.3 **Boiling Point at 1 atm:** Very high; decomposes

13.4 **Freezing Point:** 43°F = 6°C = 279°K

13.5 **Critical Temperature:** Not pertinent

13.6 **Critical Pressure:** Not pertinent

13.7 **Specific Gravity:** 1.269 at 25°C (liquid)

13.8 **Liquid Surface Tension:** Data not available

13.9 **Liquid-Water Interfacial Tension:** Data not available

13.10 **Vapor (Gas) Specific Gravity:** Not pertinent

13.11 **Ratio of Specific Heats of Vapor (Gas):** Not pertinent

13.12 **Latent Heat of Vaporization:** Not pertinent

13.13 **Heat of Combustion:** −9,240 Btu/lb = −5,140 cal/g = −215 × 10^5 J/kg

13.14 **Heat of Decomposition:** Not pertinent

13.15 **Heat of Solution:** Not pertinent

13.16 **Heat of Polymerization:** Not pertinent

5. HEALTH HAZARDS (Cont'd.)

5.10 **Odor Threshold:** 0.04 ppm

PTB	PENTABORANE

Common Synonyms	Liquid / Colorless / Strong sour milk odor
(9)-Pentaboron nonahydride	Ignites when exposed to air. Floats on water.

AVOID CONTACT WITH LIQUID AND VAPOR. KEEP PEOPLE AWAY.
Wear goggles, self-contained breathing apparatus, and rubber overclothing (including gloves).
Call fire department.
Stop discharge if possible.
Evacuate area in case of large discharge.
Isolate and remove discharged material.
Notify local health and pollution control agencies.

Fire

IGNITES WHEN EXPOSED TO AIR.
Water may be ineffective on fire.
DO NOT USE WATER OR FOAM ON FIRE.
DO NOT USE WATER ON ADJACENT FIRES.

Exposure

CALL FOR MEDICAL AID.

HAZARD IS FROM PRODUCTS OF COMBUSTION.

VAPOR
POISONOUS IF INHALED.
Move victim to fresh air.
If breathing is difficult, give oxygen.

LIQUID
POISONOUS IF SWALLOWED OR IF SKIN IS EXPOSED.
Irritating to eyes.
Remove contaminated clothing and shoes.
Flush affected areas with plenty of water.
IF IN EYES, hold eyelids open and flush with plenty of water.
IF SWALLOWED and victim is CONSCIOUS, have victim drink water or milk and have victim induce vomiting.
IF SWALLOWED and victim is UNCONSCIOUS OR HAVING CONVULSIONS, do nothing except keep victim warm.

Water Pollution

Effect of low concentrations on aquatic life is unknown.
May be dangerous if it enters water intakes.
Notify local health and wildlife officials.
Notify operators of nearby water intakes.

1. RESPONSE TO DISCHARGE

(See Response Methods Handbook, CG 446-4)

Issue warning—high flammability, water contaminant
Restrict access
Evacuate area
Disperse and flush

2. LABELS

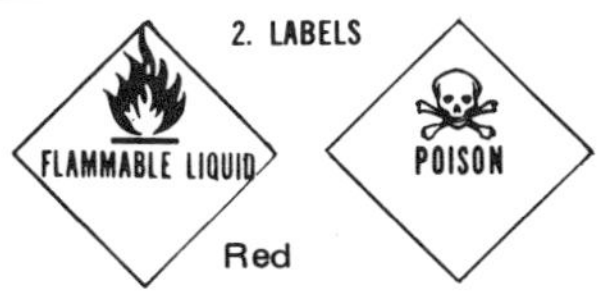

3. CHEMICAL DESIGNATIONS

3.1 **Synonyms:** (9)-Pentaboron nonahydride
3.2 **Coast Guard Compatibility Classification:** Not applicable
3.3 **Chemical Formula:** B_5H_9
3.4 **IMCO/United Nations Numerical Designation:** 4.2/1380

4. OBSERVABLE CHARACTERISTICS

4.1 **Physical State (as shipped):** Liquid
4.2 **Color:** Colorless
4.3 **Odor:** Characteristic; strong, pungent; foul; sour milk

5. HEALTH HAZARDS

5.1 **Personal Protective Equipment:** Self-contained breathing apparatus or air-line mask; goggles or face shield; rubber gloves and protective clothing
5.2 **Symptoms Following Exposure:** Inhalation of low concentrations causes dizziness, blurred vision, nausea, fatigue, light headedness or nervousness; higher concentrations also cause abnormal muscular contractions or twitching of any part of the body, difficult breathing, poor muscular coordination, imperfect articulation of speech, convulsions, and (rarely) coma. Contact with liquid causes severe irritation of eyes and irritation of skin (acute local inflammation with the formation of small blisters, redness and swelling). Can be absorbed through the skin. Compound cannot be swallowed, because it is spontaneously flammable in air.
5.3 **Treatment for Exposure:** Get medical attention following all exposures to this compound. INHALATION: remove victim to fresh air; watch for delayed symptoms for 1-2 days. EYES: wash with copious quantities of water for at least 30 min., holding eyelids apart to insure thorough flushing. SKIN: wash immediately with soap and water; rinse affected area with a 3% ammonia solution followed by additional flushing with water.
5.4 **Toxicity by Inhalation (Threshold Limit Value):** 0.005 ppm
5.5 **Short-Term Inhalation Limits:** 25 ppm for 5 min., 8 ppm for 15 min., 4 ppm for 30 min., 2 ppm for 60 min.
5.6 **Toxicity by Ingestion:** Grade 4; LD_{50} < 50 mg/kg
5.7 **Late Toxicity:** Data not available

6. FIRE HAZARDS

6.1 **Flash Point:** Not pertinent (ignites spontaneously in air)
6.2 **Flammable Limits in Air:** 0.42% – 98%
6.3 **Fire Extinguishing Agents:** Preferably shut off leak and let fire burn; extinguish with dry chemical or carbon dioxide.
6.4 **Fire Extinguishing Agents Not to be Used:** Halogenated hydrocarbons, water
6.5 **Special Hazards of Combustion Products:** Toxic fumes may be formed.
6.6 **Behavior in Fire:** Tends to reignite. Contact with water applied to adjacent fires produces flammable hydrogen gas.
6.7 **Ignition Temperature:** Spontaneously flammable if impure. Approx. 35°C when pure.
6.8 **Electrical Hazard:** Not pertinent
6.9 **Burning Rate:** Not pertinent

7. CHEMICAL REACTIVITY

7.1 **Reactivity with Water:** Reacts slowly to form flammable hydrogen gas. Reaction is not hazardous unless water is hot or unless confined.
7.2 **Reactivity with Common Materials:** Corrosive to natural rubber, some synthetic rubbers, some greases and some lubricants
7.3 **Stability During Transport:** Stable below 302°F
7.4 **Neutralizing Agents for Acids and Caustics:** Not pertinent
7.5 **Polymerization:** Not pertinent
7.6 **Inhibitor of Polymerization:** Not pertinent

8. WATER POLLUTION

8.1 **Aquatic Toxicity:** Data not available
8.2 **Waterfowl Toxicity:** Data not available
8.3 **Biological Oxygen Demand (BOD):** Data not available
8.4 **Food Chain Concentration Potential:** None

9. SELECTED MANUFACTURERS

Callery Chemical Company
Callery, Pa. 16024

10. SHIPPING INFORMATION

10.1 **Grades or Purity:** Technical: 95+%; Hi-Purity: 99+%
10.2 **Storage Temperature:** Cool ambient
10.3 **Inert Atmosphere:** Inerted with dry nitrogen
10.4 **Venting:** Safety relief

11. HAZARD ASSESSMENT CODE

(See Hazard Assessment Handbook, CG 446-3)

A-O-T-U-V-W

12. HAZARD CLASSIFICATIONS

12.1 **Code of Federal Regulations:** Flammable liquid
12.2 **NAS Hazard Rating for Bulk Water Transportation:**

Category	Rating
Fire	4
Health	
Vapor Irritant	2
Liquid or Solid Irritant	
Poisons	4
Water Pollution	
Human Toxicity	4
Aquatic Toxicity	—
Aesthetic Effect	2
Reactivity	
Other Chemicals	4
Water	2
Self-Reaction	4

12.3 **NFPA Hazard Classifications:**

Category	Classification
Health Hazard (Blue)	3
Flammability (Red)	3
Reactivity (Yellow)	2

13. PHYSICAL AND CHEMICAL PROPERTIES

13.1 **Physical State at 15°C and 1 atm:** Liquid
13.2 **Molecular Weight:** 63.2
13.3 **Boiling Point at 1 atm:** 137.1°F = 58.4°C = 331.5°K
13.4 **Freezing Point:** −52.2°F = −46.8°C = 224.6°K
13.5 **Critical Temperature:** 441°F = 227°C = 500°K
13.6 **Critical Pressure:** 570 psia = 38 atm = 3.9 MN/m^2
13.7 **Specific Gravity:** 0.623 at 20°C (liquid)
13.8 **Liquid Surface Tension:** 20.8 dynes/cm = 0.0208 N/m at 25°C
13.9 **Liquid-Water Interfacial Tension:** Not pertinent
13.10 **Vapor (Gas) Specific Gravity:** 2.2
13.11 **Ratio of Specific Heats of Vapor (Gas):** 1.0399
13.12 **Latent Heat of Vaporization:** 219 Btu/lb = 122 cal/g = 5.10×10^5 J/kg
13.13 **Heat of Combustion:** −29,100 Btu/lb = −16,200 cal/g = -677×10^5 J/kg
13.14 **Heat of Decomposition:** Not pertinent
13.15 **Heat of Solution:** Not pertinent
13.16 **Heat of Polymerization:** Not pertinent

5. HEALTH HAZARDS (Cont'd.)

5.8 **Vapor (Gas) Irritant Characteristics:** Vapors cause moderate irritation such that personnel will find high concentrations unpleasant. The effect is temporary.
5.9 **Liquid or Solid Irritant Characteristics:** Data not available
5.10 **Odor Threshold:** 0.8 ppm

PENTACHLOROPHENOL

Common Synonyms
Dowicide 7
Penta
Santophen 20

Solid beads or flakes White to light brown

Sinks in water.

Avoid contact with solid and dust. Keep people away.
Wear goggles and self-contained breathing apparatus.
Stop discharge if possible.
Isolate and remove discharged material.
Notify local health and pollution control agencies.

Fire

Not flammable.

Exposure

CALL FOR MEDICAL AID.

DUST
Irritating to eyes, nose and throat.
If inhaled, will cause coughing or difficult breathing.
Move to fresh air.
If breathing has stopped, give artificial respiration.
If breathing is difficult, give oxygen.

SOLID
POISONOUS IF SWALLOWED.
Will burn skin and eyes.
Remove contaminated clothing and shoes.
Flush affected areas with plenty of water.
IF IN EYES, hold eyelids open and flush with plenty of water.
IF SWALLOWED and victim is CONSCIOUS, have victim drink water or milk and have victim induce vomiting.
IF SWALLOWED and victim is UNCONSCIOUS OR HAVING CONVULSIONS, do nothing except keep victim warm.

Water Pollution

HARMFUL TO AQUATIC LIFE IN VERY LOW CONCENTRATIONS.
May be dangerous if it enters water intakes.
Notify local health and wildlife officials.
Notify operators of nearby water intakes.

1. RESPONSE TO DISCHARGE

(See Response Methods Handbook, CG 446-4)

Issue warning—poison
Restrict access
Should be removed

2. LABELS

No hazard label required by Code of Federal Regulations

3. CHEMICAL DESIGNATIONS

3.1 **Synonyms:** Dowicide 7
Penta
Santophen 20

3.2 **Coast Guard Compatibility Classification:** Not applicable

3.3 **Chemical Formula:** C_6Cl_5OH

3.4 **IMCO/United Nations Numerical Designation:** 6.1/2020

4. OBSERVABLE CHARACTERISTICS

4.1 **Physical State (as shipped):** Solid

4.2 **Color:** Colorless to light brown

4.3 **Odor:** Very weak

5. HEALTH HAZARDS

5.1 **Personal Protective Equipment:** Respirator for dust; goggles; protective clothing.

5.2 **Symptoms Following Exposure:** Dust or vapor irritates skin and mucous membranes, causing coughing and sneezing. Ingestion causes loss of appetite, respiratory difficulties, anesthesia, sweating, coma. Overexposure can cause death.

5.3 **Treatment for Exposure:** Call a doctor! INGESTION: induce vomiting at once. EYES: flush with water for 15-30 min. SKIN: wash well with soap and water.

5.4 **Toxicity by Inhalation (Threshold Limit Value):** 0.5 mg/m^3

5.5 **Short-Term Inhalation Limits:** Data not available

5.6 **Toxicity by Ingestion:** Grade 3; LD_{50} 50 to 500 mg/kg (rat)

5.7 **Late Toxicity:** Data not available

5.8 **Vapor (Gas) Irritant Characteristics:** Vapor is moderately irritating such that personnel will not usually tolerate moderate or high vapor concentrations.

5.9 **Liquid or Solid Irritant Characteristics:** Causes smarting of the skin and first-degree burns on short exposure; may cause secondary burns on long exposure.

5.10 **Odor Threshold:** Data not available

6. FIRE HAZARDS

6.1 **Flash Point:** Not flammable

6.2 **Flammable Limits in Air:** Not flammable

6.3 **Fire Extinguishing Agents:** Not pertinent

6.4 **Fire Extinguishing Agents Not to be Used:** Not pertinent

6.5 **Special Hazards of Combustion Products:** Generates toxic and irritating vapors.

6.6 **Behavior in Fire:** Not pertinent

6.7 **Ignition Temperature:** Not flammable

6.8 **Electrical Hazard:** Not pertinent

6.9 **Burning Rate:** Not flammable

7. CHEMICAL REACTIVITY

7.1 **Reactivity with Water:** No reaction

7.2 **Reactivity with Common Materials:** No reaction

7.3 **Stability During Transport:** Stable

7.4 **Neutralizing Agents for Acids and Caustics:** Not pertinent

7.5 **Polymerization:** Not pertinent

7.6 **Inhibitor of Polymerization:** Not pertinent

8. WATER POLLUTION

8.1 **Aquatic Toxicity:** 5 ppm/3 hr/trout/lethal/fresh water

8.2 **Waterfowl Toxicity:** 4500 ppm/LC_{50}/mallards

8.3 **Biological Oxygen Demand (BOD):** Data not available

8.4 **Food Chain Concentration Potential:** Data not available

9. SELECTED MANUFACTURERS

1. Dow Chemical Co.
Midland, Mich. 48640
2. Reichold Chemicals, Inc.
RCI Bldg.
White Plains, N. Y. 10602
3. Sonford Chemical Co.
Port Neches, Tex. 77651

10. SHIPPING INFORMATION

10.1 **Grades or Purity:** 86-100%

10.2 **Storage Temperature:** Ambient

10.3 **Inert Atmosphere:** No requirement

10.4 **Venting:** Open

11. HAZARD ASSESSMENT CODE

(See Hazard Assessment Handbook, CG 446-3)

II

12. HAZARD CLASSIFICATIONS

12.1 **Code of Federal Regulations:** Not listed

12.2 **NAS Hazard Rating for Bulk Water Transportation:** Not listed

12.3 **NFPA Hazard Classifications:**

Category	Classification
Health Hazard (Blue)	3
Flammability (Red)	0
Reactivity (Yellow)	0

13. PHYSICAL AND CHEMICAL PROPERTIES

13.1 **Physical State at 15°C and 1 atm:** Solid

13.2 **Molecular Weight:** 266.35

13.3 **Boiling Point at 1 atm:** 590°F = 310°C = 583°K

13.4 **Freezing Point:** 370°F = 188°C = 461°K

13.5 **Critical Temperature:** Not pertinent

13.6 **Critical Pressure:** Not pertinent

13.7 **Specific Gravity:** 1.98 at 15°C (solid)

13.8 **Liquid Surface Tension:** Not pertinent

13.9 **Liquid-Water Interfacial Tension:** Not pertinent

13.10 **Vapor (Gas) Specific Gravity:** Not pertinent

13.11 **Ratio of Specific Heats of Vapor (Gas):** Not pertinent

13.12 **Latent Heat of Vaporization:** Not pertinent

13.13 **Heat of Combustion:** Not pertinent

13.14 **Heat of Decomposition:** Not pertinent

13.15 **Heat of Solution:** Not pertinent

13.16 **Heat of Polymerization:** Not pertinent

NOTES

REVISED 1978

PDC — PENTADECANOL

Common Synonyms	Liquid	Colorless	Faint alcohol odor
	Floats on water.		

Stop discharge if possible. Keep people away.
Call fire department.
Avoid contact with liquid.
Isolate and remove discharged material.
Notify local health and pollution control agencies.

Fire

Combustible.
Extinguish with foam, dry chemical, or carbon dioxide.
Water may be ineffective on fire.
Cool exposed containers with water.

Exposure

CALL FOR MEDICAL AID.

LIQUID
Irritating to skin and eyes.
Remove contaminated clothing and shoes.
Flush affected areas with plenty of water.
IF IN EYES, hold eyelids open and flush with plenty of water.

Water Pollution

Harmful to fish and water fowl.
Fouling to shoreline.
May be dangerous if it enters water intakes.
Notify local health and pollution control officials.
Notify operators of nearby water intakes.

1. RESPONSE TO DISCHARGE

(See Response Methods Handbook, CG 446-4)

Mechanical containment
Should be removed

2. LABELS

No hazard label required by Code of Federal Regulations

3. CHEMICAL DESIGNATIONS

3.1 **Synonyms:** 1-Pentadecanol
Pentadecyl alcohol
3.2 **Coast Guard Compatibility Classification:** Alcohol
3.3 **Chemical Formula:** $CH_3(CH_2)_{13}CH_2OH$
3.4 **IMCO/United Nations Numerical Designation:** Not listed

4. OBSERVABLE CHARACTERISTICS

4.1 **Physical State (as shipped):** Liquid
4.2 **Color:** Colorless
4.3 **Odor:** Weak alcoholic

5. HEALTH HAZARDS

5.1 **Personal Protective Equipment:** Goggles or face shield.
5.2 **Symptoms Following Exposure:** Low toxicity. Excessive exposure produces some central nervous system depression. Prolonged contact produces skin irritation.
5.3 **Treatment for Exposure:** INHALATION: if necessary, support respiration. INGESTION: induce vomiting and call a doctor. SKIN OR EYES: wash with copious amounts of water.
5.4 **Toxicity by Inhalation (Threshold Limit Value):** Data not available
5.5 **Short-Term Inhalation Limits:** Data not available
5.6 **Toxicity by Ingestion:** Data not available
5.7 **Late Toxicity:** Data not available
5.8 **Vapor (Gas) Irritant Characteristics:** Data not available
5.9 **Liquid or Solid Irritant Characteristics:** Data not available
5.10 **Odor Threshold:** Data not available

6. FIRE HAZARDS

6.1 **Flash Point:** Data not available
6.2 **Flammable Limits in Air:** Data not available
6.3 **Fire Extinguishing Agents:** Foam, dry chemical or carbon dioxide
6.4 **Fire Extinguishing Agents Not to be Used:** Water or foam may cause frothing.
6.5 **Special Hazards of Combustion Products:** Not pertinent
6.6 **Behavior in Fire:** Not pertinent
6.7 **Ignition Temperature:** Data not available
6.8 **Electrical Hazard:** Not pertinent
6.9 **Burning Rate:** Data not available

7. CHEMICAL REACTIVITY

7.1 **Reactivity with Water:** No reaction
7.2 **Reactivity with Common Materials:** No reaction
7.3 **Stability During Transport:** Stable
7.4 **Neutralizing Agents for Acids and Caustics:** Not pertinent
7.5 **Polymerization:** Not pertinent
7.6 **Inhibitor of Polymerization:** Not pertinent

8. WATER POLLUTION

8.1 **Aquatic Toxicity:** Data not available
8.2 **Waterfowl Toxicity:** Data not available
8.3 **Biological Oxygen Demand (BOD):** Data not available
8.4 **Food Chain Concentration Potential:** None

9. SELECTED MANUFACTURERS

No bulk shipments

10. SHIPPING INFORMATION

10.1 **Grades or Purity:** Data not available
10.2 **Storage Temperature:** Ambient
10.3 **Inert Atmosphere:** No requirement
10.4 **Venting:** Open (flame arrester)

11. HAZARD ASSESSMENT CODE

(See Hazard Assessment Handbook, CG 446-3)

A-T-U

12. HAZARD CLASSIFICATIONS

12.1 **Code of Federal Regulations:** Not listed
12.2 **NAS Hazard Rating for Bulk Water Transportation:** Not listed
12.3 **NFPA Hazard Classifications:** Not listed

13. PHYSICAL AND CHEMICAL PROPERTIES

13.1 **Physical State at 15°C and 1 atm:** Liquid
13.2 **Molecular Weight:** 228.42
13.3 **Boiling Point at 1 atm:** 572°F = 300°C = 573°K
13.4 **Freezing Point:** 111°F = 44°C = 317°K
13.5 **Critical Temperature:** 824°F = 440°C = 713°K
13.6 **Critical Pressure:** Not pertinent
13.7 **Specific Gravity:** 0.829 at 50°C (liquid)
13.8 **Liquid Surface Tension:** (est.) 25 dynes/cm = 0.025 N/m at 50°C
13.9 **Liquid-Water Interfacial Tension:** (est.) 35 dynes/cm = 0.035 N/m at 50°C
13.10 **Vapor (Gas) Specific Gravity:** Not pertinent
13.11 **Ratio of Specific Heats of Vapor (Gas):** 1.024
13.12 **Latent Heat of Vaporization:** Not pertinent
13.13 **Heat of Combustion:** Data not available
13.14 **Heat of Decomposition:** Not pertinent
13.15 **Heat of Solution:** Not pertinent
13.16 **Heat of Polymerization:** Not pertinent

NOTES

REVISED 1978

PET — PENTAERYTHRITOL

Common Synonyms			
2,2-Bis(hydroxymethyl)-1,3-propanediol Pentaerythrite Tetrahydroxymethylmethane PE Mono PE Pentek	Solid	White	Odorless
	Sinks and mixes slowly with water.		

Call fire department.
Isolate and remove discharged material.
Notify local health and pollution control agencies.

Fire	Combustible. Extinguish with water, dry chemicals, foam, or carbon dioxide.
Exposure	VAPOR OR DUST Not harmful. If in eyes, hold eyelids open and flush with plenty of water.
Water Pollution	Effect of low concentrations on aquatic life is unknown. May be dangerous if it enters water intakes. Notify local health and wildlife officials. Notify operators of nearby water intakes.

1. RESPONSE TO DISCHARGE

(See Response Methods Handbook, CG 446-4)

Should be removed
Chemical and physical treatment

2. LABELS

No label required by Code of Federal Regulations

3. CHEMICAL DESIGNATIONS

3.1 **Synonyms:** Mono PE; PE; Pentaerythrite; Pentek; Tetrahydroxymethylmethane; Tetrakis (hydroxymethyl) methane; Tetramethylolmethane

3.2 **Coast Guard Compatibility Classification:** Not applicable

3.3 **Chemical Formula:** $C(CH_2OH)_4$

3.4 **IMCO/United Nations Numerical Designation:** Not listed

4. OBSERVABLE CHARACTERISTICS

4.1 **Physical State (as shipped):** Solid

4.2 **Color:** White

4.3 **Odor:** None

5. HEALTH HAZARDS

5.1 **Personal Protective Equipment:** Dust mask; goggles

5.2 **Symptoms Following Exposure:** Non-toxic; no symptoms likely

5.3 **Treatment for Exposure:** None needed

5.4 **Toxicity by Inhalation (Threshold Limit Value):** Not pertinent ("Inert" particulate)

5.5 **Short-Term Inhalation Limits:** Not pertinent

5.6 **Toxicity by Ingestion:** Grade 0; LD_{50} > 15 g/kg

5.7 **Late Toxicity:** None

5.8 **Vapor (Gas) Irritant Characteristics:** Data not available

5.9 **Liquid or Solid Irritant Characteristics:** Data not available

5.10 **Odor Threshold:** Not pertinent

6. FIRE HAZARDS

6.1 **Flash Point:** Not pertinent (combustible solid)

6.2 **Flammable Limits in Air:** Not pertinent

6.3 **Fire Extinguishing Agents:** Water, dry chemical, carbon dioxide

6.4 **Fire Extinguishing Agents Not to be Used:** Not pertinent

6.5 **Special Hazards of Combustion Products:** Not pertinent

6.6 **Behavior in Fire:** Not pertinent

6.7 **Ignition Temperature:** 842°F (dust cloud)

6.8 **Electrical Hazard:** Not pertinent

6.9 **Burning Rate:** Not pertinent

7. CHEMICAL REACTIVITY

7.1 **Reactivity with Water:** No reaction

7.2 **Reactivity with Common Materials:** No reaction

7.3 **Stability During Transport:** Stable

7.4 **Neutralizing Agents for Acids and Caustics:** Not pertinent

7.5 **Polymerization:** Not pertinent

7.6 **Inhibitor of Polymerization:** Not pertinent

8. WATER POLLUTION

8.1 **Aquatic Toxicity:** Data not available

8.2 **Waterfowl Toxicity:** Data not available

8.3 **Biological Oxygen Demand (BOD):** Data not available

8.4 **Food Chain Concentration Potential:** None

9. SELECTED MANUFACTURERS

1. Celanese Chemical Company
   245 Park Avenue
   New York, N. Y. 10017
2. Commercial Solvents Corporation
   Terre Haute, Ind. 47808
3. Hercules Incorporated
   Coatings & Specialty Products Department
   910 Market Street
   Wilmington, Del. 19899

10. SHIPPING INFORMATION

10.1 **Grades or Purity:** Technical: 86—90% plus 10—14% dimer. Pure: 98+%

10.2 **Storage Temperature:** Ambient

10.3 **Inert Atmosphere:** No requirement

10.4 **Venting:** Open

11. HAZARD ASSESSMENT CODE

(See Hazard Assessment Handbook, CG 446-3)

SS

12. HAZARD CLASSIFICATIONS

12.1 **Code of Federal Regulations:** Not listed

12.2 **NAS Hazard Rating for Bulk Water Transportation:** Not listed

12.3 **NFPA Hazard Classifications:** Not listed

13. PHYSICAL AND CHEMICAL PROPERTIES

13.1 **Physical State at 15°C and 1 atm:** Solid

13.2 **Molecular Weight:** 136.2

13.3 **Boiling Point at 1 atm:** Not pertinent (sublimes)

13.4 **Freezing Point:** 502°F = 261°C = 534°K

13.5 **Critical Temperature:** Not pertinent

13.6 **Critical Pressure:** Not pertinent

13.7 **Specific Gravity:** 1.39 at 25°C (solid)

13.8 **Liquid Surface Tension:** Not pertinent

13.9 **Liquid-Water Interfacial Tension:** Not pertinent

13.10 **Vapor (Gas) Specific Gravity:** Not pertinent

13.11 **Ratio of Specific Heats of Vapor (Gas):** Not pertinent

13.12 **Latent Heat of Vaporization:** Not pertinent

13.13 **Heat of Combustion:** −8,730 Btu/lb = −4,850 cal/g = -203×10^5 J/kg

13.14 **Heat of Decomposition:** Not pertinent

13.15 **Heat of Solution:** Not pertinent

13.16 **Heat of Polymerization:** Not pertinent

NOTES

PTA	**PENTANE**

Common Synonyms	Liquid — Colorless — Gasoline odor Floats on water. Flammable vapor is produced. Boiling point is 97° F.

Stop discharge if possible. Keep people away.
Shut off ignition sources and call fire department.
Stay upwind and use water spray to "knock down" vapor.
Avoid contact with liquid and vapor.
Isolate and remove discharged material.
Notify local health and pollution control agencies.

Fire	FLAMMABLE. Flashback along vapor trail may occur. Containers may explode when heated. Vapor may explode if ignited in an enclosed area. Extinguish with foam, dry chemical or carbon dioxide. Water may be ineffective on fire. Cool exposed containers with water.
Exposure	CALL FOR MEDICAL AID. VAPOR If inhaled, will cause dizziness or difficult breathing. Move to fresh air. If breathing has stopped, give artificial respiration. If breathing is difficult, give oxygen. LIQUID Harmful if swallowed. IF SWALLOWED and victim is CONSCIOUS, have victim drink water or milk. DO NOT INDUCE VOMITING.
Water Pollution	Effect of low concentrations on aquatic life is unknown. May be dangerous if it enters water intakes. Notify local health and wildlife officials. Notify operators of nearby water intakes.

1. RESPONSE TO DISCHARGE

(See Response Methods Handbook, CG 446-4)

Issue warning—high flammability
Restrict access
Evacuate area

2. LABEL

Red

3. CHEMICAL DESIGNATIONS

3.1 **Synonyms:** No common synonyms
3.2 **Coast Guard Compatibility Classification:** Paraffin
3.3 **Chemical Formula:** n-C_5H_{12}
3.4 **IMCO/United Nations Numerical Designation:** 3.1/1265

4. OBSERVABLE CHARACTERISTICS

4.1 **Physical State (as shipped):** Liquid
4.2 **Color:** Colorless
4.3 **Odor:** Like gasoline

5. HEALTH HAZARDS

5.1 **Personal Protective Equipment:** Goggles or face shield (as for gasoline).
5.2 **Symptoms Following Exposure:** Low toxicity. Very high vapor concentrations produce narcosis. Aspiration into lungs can produce chemical pneumonitis and/or pulmonary edema.
5.3 **Treatment for Exposure:** INHALATION: remove from exposure; support respiration if needed. INGESTION: do NOT induce vomiting; call physician.
5.4 **Toxicity by Inhalation (Threshold Limit Value):** 500 ppm
5.5 **Short-Term Inhalation Limits:** Data not available
5.6 **Toxicity by Ingestion:** Data not available
5.7 **Late Toxicity:** None
5.8 **Vapor (Gas) Irritant Characteristics:** Vapors are nonirritating to the eyes and throat.
5.9 **Liquid or Solid Irritant Characteristics:** No appreciable hazard. Practically harmless to the skin.
5.10 **Odor Threshold:** 10 ppm

6. FIRE HAZARDS

6.1 **Flash Point:** −57°F C.C.
6.2 **Flammable Limits in Air:** 1.4%–8.3% (by vol.)
6.3 **Fire Extinguishing Agents:** Foam, dry chemical, carbon dioxide
6.4 **Fire Extinguishing Agents Not to be Used:** Water may be ineffective.
6.5 **Special Hazards of Combustion Products:** Not pertinent
6.6 **Behavior in Fire:** Containers may explode
6.7 **Ignition Temperature:** 544° F
6.8 **Electrical Hazard:** Class I, Group D
6.9 **Burning Rate:** 8.6 mm/min.

7. CHEMICAL REACTIVITY

7.1 **Reactivity with Water:** No reaction
7.2 **Reactivity with Common Materials:** No reaction
7.3 **Stability During Transport:** Stable
7.4 **Neutralizing Agents for Acids and Caustics:** Not pertinent
7.5 **Polymerization:** Not pertinent
7.6 **Inhibitor of Polymerization:** Not pertinent

8. WATER POLLUTION

8.1 **Aquatic Toxicity:**
> 60ppm/*/roach/lethal/fresh water
*Time period not specified
8.2 **Waterfowl Toxicity:** Data not available
8.3 **Biological Oxygen Demand (BOD):** Data not available
8.4 **Food Chain Concentration Potential:** None

9. SELECTED MANUFACTURERS

1. Ashland Oil, Inc.
   Petrochemicals Dept.
   Columbus, Ohio 43216
2. Marathon Oil Co.
   539 South Main St.
   Findlay, Ohio 45840
3. Phillips Petroleum Co.
   Bartlesville, Okla. 74004

10. SHIPPING INFORMATION

10.1 **Grades or Purity:** Pure (99.2%); technical; research (99.98%)
10.2 **Storage Temperature:** Ambient
10.3 **Inert Atmosphere:** No requirement
10.4 **Venting:** Open (flame arrester) or pressure-vacuum

11. HAZARD ASSESSMENT CODE

(See Hazard Assessment Handbook, CG 446-3)

A-T-U-V-W

12. HAZARD CLASSIFICATIONS

12.1 **Code of Federal Regulations:** Flammable liquid
12.2 **NAS Hazard Rating for Bulk Water Transportation:**

Category	Rating
Fire	4
Health	
Vapor Irritant	0
Liquid or Solid Irritant	0
Poisons	1
Water Pollution	
Human Toxicity	1
Aquatic Toxicity	2
Aesthetic Effect	1
Reactivity	
Other Chemicals	0
Water	0
Self-Reaction	0

12.3 **NFPA Hazard Classifications:**

Category	Classification
Health Hazard (Blue)	1
Flammability (Red)	4
Reactivity (Yellow)	0

13. PHYSICAL AND CHEMICAL PROPERTIES

13.1 **Physical State at 15°C and 1 atm:** Liquid
13.2 **Molecular Weight:** 72.15
13.3 **Boiling Point at 1 atm:** 97.0°F = 36.1°C = 309.3°K
13.4 **Freezing Point:** −201.0°F = −129.4°C = 143.8°K
13.5 **Critical Temperature:** 385.7°F = 196.5°C = 469.7°K
13.6 **Critical Pressure:** 490 psia = 33.3 atm = 3.37 MN/m^2
13.7 **Specific Gravity:** 0.626 at 20°C (liquid)
13.8 **Liquid Surface Tension:** 16 dynes/cm = 0.016 N/m at 20°C
13.9 **Liquid-Water Interfacial Tension:** 50.2 dynes/cm = 0.0502 N/m at 20°C
13.10 **Vapor (Gas) Specific Gravity:** 2.5
13.11 **Ratio of Specific Heats of Vapor (Gas):** 1.075
13.12 **Latent Heat of Vaporization:** 153.7 Btu/lb = 85.38 cal/g = 3.575×10^5 J/kg
13.13 **Heat of Combustion:** −19,352 Btu/lb = −10,751 cal/g = -450.12×10^5 J/kg
13.14 **Heat of Decomposition:** Not pertinent
13.15 **Heat of Solution:** Not pertinent
13.16 **Heat of Polymerization:** Not pertinent

NOTES

REVISED 1978

PTE 1-PENTENE

Common Synonyms
alpha-n-Amylene
Propylethylene

Liquid Colorless Gasoline odor

Floats on water. Flammable vapor is produced. Boiling point is 86° F.

Stop discharge if possible. Keep people away.
Shut off ignition sources and call fire department.
Stay upwind and use water spray to "knock down" vapor.
Avoid contact with vapor.
Isolate and remove discharged material.
Notify local health and pollution control agencies.

Fire

FLAMMABLE.
Flashback along vapor trail may occur.
Containers may explode when heated.
Vapor may explode if ignited in an enclosed area.
Extinguish with dry chemical, foam, or carbon dioxide.
Water may be ineffective on fire.
Cool exposed containers with water.

Exposure

CALL FOR MEDICAL AID.

VAPOR
If inhaled, will cause dizziness.
Move to fresh air.
If breathing has stopped, give artificial respiration.
If breathing is difficult, give oxygen.

LIQUID
Harmful if swallowed.
IF SWALLOWED and victim is CONSCIOUS, have victim drink water or milk.
DO NOT INDUCE VOMITING.

Water Pollution

Effect of low concentrations on aquatic life is unknown.
May be dangerous if it enters water intakes.
Notify local health and wildlife officials.
Notify operators of nearby water intakes.

1. RESPONSE TO DISCHARGE

(See Response Methods Handbook, CG 446-4)

Issue warning—high flammability
Restrict access
Evacuate area

2. LABEL

Red

3. CHEMICAL DESIGNATIONS

3.1 **Synonyms:** alpha-n-Amylene
Propylethylene
3.2 **Coast Guard Compatibility Classification:** Olefin
3.3 **Chemical Formula:** $CH_3(CH_2)_2CH{=}CH_2$
3.4 **IMCO/United Nations Numerical Designation:** 3.1/1108

4. OBSERVABLE CHARACTERISTICS

4.1 **Physical State (as shipped):** Liquid
4.2 **Color:** Colorless
4.3 **Odor:** Like gasoline

5. HEALTH HAZARDS

5.1 **Personal Protective Equipment:** Goggles or face shield (as for gasoline).
5.2 **Symptoms Following Exposure:** Acts as a simple asphyxiant or weak anesthetic in high vapor concentrations. Similar to effects caused by gasoline vapors.
5.3 **Treatment for Exposure:** INHALATION: remove victim from exposure. SKIN: wash with soap and water. EYES: flush with water.
5.4 **Toxicity by Inhalation (Threshold Limit Value):** Data not available
5.5 **Short-Term Inhalation Limits:** Data not available
5.6 **Toxicity by Ingestion:** Data not available
5.7 **Late Toxicity:** Data not available
5.8 **Vapor (Gas) Irritant Characteristics:** Not irritating
5.9 **Liquid or Solid Irritant Characteristics:** Not irritating
5.10 **Odor Threshold:** Data not available

6. FIRE HAZARDS

6.1 **Flash Point:** −60°F C.C.; 0°F O.C.
6.2 **Flammable Limits in Air:** 1.4%–8.7%
6.3 **Fire Extinguishing Agents:** Foam, dry chemical, or carbon dioxide. Stop flow of vapor.
6.4 **Fire Extinguishing Agents Not to be Used:** Water may be ineffective.
6.5 **Special Hazards of Combustion Products:** Not pertinent
6.6 **Behavior in Fire:** Containers may explode in fire.
6.7 **Ignition Temperature:** 527°F
6.8 **Electrical Hazard:** Data not available
6.9 **Burning Rate:** 9.1 mm/min.

7. CHEMICAL REACTIVITY

7.1 **Reactivity with Water:** No reaction
7.2 **Reactivity with Common Materials:** No reaction
7.3 **Stability During Transport:** Stable
7.4 **Neutralizing Agents for Acids and Caustics:** Not pertinent
7.5 **Polymerization:** Not pertinent
7.6 **Inhibitor of Polymerization:** Not pertinent

8. WATER POLLUTION

8.1 **Aquatic Toxicity:**
655 ppm/1hr/sunfish/lethal/fresh water
8.2 **Waterfowl Toxicity:** Data not available
8.3 **Biological Oxygen Demand (BOD):**
0.8% (theor.), 0.5 day; 0.5% (theor.), 1 day
8.4 **Food Chain Concentration Potential:** None

9. SELECTED MANUFACTURERS

1. Exxon Chemical Co.
Houston, Tex. 77001
2. Phillips Petroleum
Bartlesville, Okla. 74004
3. Texaco, Inc.
135 E. 42 St.
New York, N. Y. 10017

10. SHIPPING INFORMATION

10.1 **Grades or Purity:** Research: 99.9%; pure: 99.4%; technical: 97.0%
10.2 **Storage Temperature:** Ambient
10.3 **Inert Atmosphere:** No requirement
10.4 **Venting:** Open (flame arrester) or pressure-vacuum

11. HAZARD ASSESSMENT CODE

(See Hazard Assessment Handbook, CG 446-3)

A-T-U-V-W

12. HAZARD CLASSIFICATIONS

12.1 **Code of Federal Regulations:** Flammable liquid
12.2 **NAS Hazard Rating for Bulk Water Transportation:** Not listed
12.3 **NFPA Hazard Classifications:** Not listed

13. PHYSICAL AND CHEMICAL PROPERTIES

13.1 **Physical State at 15°C and 1 atm:** Liquid
13.2 **Molecular Weight:** 70.13
13.3 **Boiling Point at 1 atm:**
85.8°F = 29.9°C = 303.1°K
13.4 **Freezing Point:**
−265°F = −165°C = 108°K
13.5 **Critical Temperature:**
376.9°F = 191.6°C = 464.8°K
13.6 **Critical Pressure:**
588 psia = 40 atm = 4.05 MN/m²
13.7 **Specific Gravity:** 0.641 at 20°C (liquid)
13.8 **Liquid Surface Tension:**
16.5 dynes/cm = 0.0165 N/m at 20°C
13.9 **Liquid-Water Interfacial Tension:**
(est.) 50 dynes/cm = 0.05 N/m at 20°C
13.10 **Vapor (Gas) Specific Gravity:** 2.4
13.11 **Ratio of Specific Heats of Vapor (Gas):** 1.083
13.12 **Latent Heat of Vaporization:**
154.6 Btu/lb = 85.87 cal/g = 3.595×10^5 J/kg
13.13 **Heat of Combustion:** −19,359 Btu/lb = −10,755 cal/g = -450.29×10^5 J/kg
13.14 **Heat of Decomposition:** Not pertinent
13.15 **Heat of Solution:** Not pertinent
13.16 **Heat of Polymerization:** Not pertinent

NOTES

REVISED 1978

PAA — PERACETIC ACID

Common Synonyms			
Peroxyacetic acid Acetyl hydroperoxide	Liquid	Colorless	Strong odor
	Mixes with water. Flammable, irritating vapor is produced.		

Stop discharge if possible. Keep people away.
Call fire department.
Isolate and remove discharged material.
Notify local health and pollution control agencies.

Fire	Combustible. May cause fire on contact with combustibles. Containers may explode in fire. Flood discharge area with water. Cool exposed containers with water.
Exposure	Call for medical aid. VAPOR Irritating to eyes, nose and throat. Move victim to fresh air. If breathing has stopped, give artificial respiration. If breathing is difficult, give oxygen. LIQUID Irritating to skin and eyes. Harmful if swallowed. Remove contaminated clothing and shoes. Flush affected areas with plenty of water. IF IN EYES, hold eyelids open and flush with plenty of water. IF SWALLOWED and victim is CONSCIOUS, have victim drink water or milk.
Water Pollution	Effect of low concentrations on aquatic life is unknown. May be dangerous if it enters water intakes. Notify local health and wildlife officials. Notify operators of nearby water intakes.

1. RESPONSE TO DISCHARGE

(See Response Methods Handbook, CG 446-4)

Issue warning—oxidizing material, water contaminant
Restrict access
Disperse and flush

2. LABEL

Yellow

3. CHEMICAL DESIGNATIONS

3.1 **Synonyms:** Acetyl hydroperoxide; Peroxyacetic acid
3.2 **Coast Guard Compatibility Classification:** Not applicable
3.3 **Chemical Formula:** $CH_3COOOH—CH_3COOH$
3.4 **IMCO/United Nations Numerical Designation:** 5.2/1532

4. OBSERVABLE CHARACTERISTICS

4.1 **Physical State (as shipped):** Liquid
4.2 **Color:** Colorless
4.3 **Odor:** Pungent; strong

5. HEALTH HAZARDS

5.1 **Personal Protective Equipment:** Self-contained breathing apparatus; full protective clothing (goggles, rubber gloves, etc.)
5.2 **Symptoms Following Exposure:** Inhalation causes severe irritation of mucous membrane. Contact with liquid causes severe irritation of eyes and skin. Ingestion causes severe distress, including burns of mouth and stomach.
5.3 **Treatment for Exposure:** INHALATION: remove victim to fresh air; if he is not breathing, apply artificial respiration and oxygen; call a doctor. EYES: flush with water for at least 15 min.; call a doctor. SKIN: flush with water and treat burns. INGESTION: give plenty of warm water; call a doctor.
5.4 **Toxicity by Inhalation (Threshold Limit Value):** Data not available
5.5 **Short-Term Inhalation Limits:** Data not available
5.6 **Toxicity by Ingestion:** Grade 4; oral LD_{50} = 10 mg/kg (guinea pig)
5.7 **Late Toxicity:** Data not available
5.8 **Vapor (Gas) Irritant Characteristics:** Data not available
5.9 **Liquid or Solid Irritant Characteristics:** Data not available
5.10 **Odor Threshold:** Data not available

6. FIRE HAZARDS

6.1 **Flash Point:** 104°F O.C.
6.2 **Flammable Limits in Air:** Data not available
6.3 **Fire Extinguishing Agents:** Water
6.4 **Fire Extinguishing Agents Not to be Used:** Not pertinent
6.5 **Special Hazards of Combustion Products:** Not pertinent
6.6 **Behavior in Fire:** Vapors are very flammable and explosive. Liquid will detonate if concentration rises above 56% because of evaporation of acetic acid.
6.7 **Ignition Temperature:** 392°F
6.8 **Electrical Hazard:** Not pertinent
6.9 **Burning Rate:** Data not available

7. CHEMICAL REACTIVITY

7.1 **Reactivity with Water:** No reaction
7.2 **Reactivity with Common Materials:** May cause fire in contact with organic materials such as wood, cotton or straw. Corrosive to most metals, including aluminum.
7.3 **Stability During Transport:** Stable if kept cool and out of contact with most metals. At 30°C concentration decreases about 0.4% each month.
7.4 **Neutralizing Agents for Acids and Caustics:** Flush with water
7.5 **Polymerization:** Not pertinent
7.6 **Inhibitor of Polymerization:** Not pertinent

8. WATER POLLUTION

8.1 **Aquatic Toxicity:** Data not available
8.2 **Waterfowl Toxicity:** Data not available
8.3 **Biological Oxygen Demand (BOD):** Data not available
8.4 **Food Chain Concentration Potential:** None

9. SELECTED MANUFACTURERS

1. FMC Corporation
Industrial Chemical Division
633 Third Avenue
New York, N. Y. 10017
2. High Point Chemical Corp.
609 Taylor Street
High Point, N. C. 27261
3. Pfaltz and Bauer, Inc.
126–04 Northern Boulevard
Flushing, N. Y. 11368

10. SHIPPING INFORMATION

10.1 **Grades or Purity:** 40% peracetic acid, 40% acetic acid, 5% hydrogen peroxide, 13% water, 500 ppm stabilizer
10.2 **Storage Temperature:** 60°F—122°F
10.3 **Inert Atmosphere:** No requirement
10.4 **Venting:** Safety relief

11. HAZARD ASSESSMENT CODE

(See Hazard Assessment Handbook, CG 446-3)

A-P-Q

12. HAZARD CLASSIFICATIONS

12.1 **Code of Federal Regulations:** Organic Peroxide
12.2 **NAS Hazard Rating for Bulk Water Transportation:** Not listed
12.3 **NFPA Hazard Classifications:** Not listed

13. PHYSICAL AND CHEMICAL PROPERTIES

13.1 **Physical State at 15°C and 1 atm:** Liquid
13.2 **Molecular Weight:** Not pertinent
13.3 **Boiling Point at 1 atm:** Not pertinent (mixture)
13.4 **Freezing Point:** (approx.)* −22°F = −30°C = 243°K
13.5 **Critical Temperature:** Not pertinent
13.6 **Critical Pressure:** Not pertinent
13.7 **Specific Gravity:** (est.) 1.153 at 25°C (liquid)
13.8 **Liquid Surface Tension:** Data not available
13.9 **Liquid-Water Interfacial Tension:** Not pertinent
13.10 **Vapor (Gas) Specific Gravity:** Not pertinent
13.11 **Ratio of Specific Heats of Vapor (Gas):** Not pertinent
13.12 **Latent Heat of Vaporization:** Data not available
13.13 **Heat of Combustion:** Data not available
13.14 **Heat of Decomposition:** Data not available
13.15 **Heat of Solution:** Not pertinent
13.16 **Heat of Polymerization:** Not pertinent

*40% solution in acetic acid/H_2O.

NOTES

PCL	# PERCHLORIC ACID

Common Synonyms			
Dioxonium perchlorate solution	Liquid	Colorless	Odorless
	Sinks and mixes with water.		

Avoid contact with liquid and vapor. Keep people away.
Wear rubber overclothing (including gloves).
Stop discharge if possible.
Isolate and remove discharged material.
Notify local health and pollution control agencies.

Fire

Not flammable.
Will increase the intensity of a fire.
May cause fire on contact with combustibles.
Containers may explode in fire.
POISONOUS GASES MAY BE PRODUCED IN FIRE.
Combat fires from safe distance or protected location.
Flood discharge area with water.
Cool exposed containers with water.

Exposure

CALL FOR MEDICAL AID.

VAPOR
Irritating to eyes, nose and throat.
If inhaled will cause coughing or difficult breathing.
If in eyes, hold eyelids open and flush with plenty of water.
If breathing has stopped, give artificial respiration.
If breathing is difficult, give oxygen.

LIQUID
Will burn skin and eyes.
If swallowed will cause nausea and vomiting.
Remove contaminated clothing and shoes.
Flush affected areas with plenty of water.
IF IN EYES, hold eyelids open and flush with plenty of water.
IF SWALLOWED and victim is CONSCIOUS, have victim drink water or milk.
IF SWALLOWED and victim is UNCONSCIOUS OR HAVING CONVULSIONS, do nothing except keep victim warm.

Water Pollution

Effect of low concentrations on aquatic life is unknown.
May be dangerous if it enters water intakes.
Notify local health and wildlife officials.
Notify operators of nearby water intakes.

1. RESPONSE TO DISCHARGE

(See Response Methods Handbook, CG 446-4)

Issue warning—corrosive, oxidizing material
Restrict access
Disperse and flush

2. LABEL

Yellow

3. CHEMICAL DESIGNATIONS

3.1 **Synonyms:** Dioxonium perchlorate solution; Perchloric acid solution
3.2 **Coast Guard Compatibility Classification:** Not listed
3.3 **Chemical Formula:** $HClO_4-H_2O$
3.4 **IMCO/United Nations Numerical Designation:** 5.1/1873 (50—72% acid); 8/1802 (less than 50% acid)

4. OBSERVABLE CHARACTERISTICS

4.1 **Physical State (as shipped):** Liquid
4.2 **Color:** Colorless
4.3 **Odor:** None

5. HEALTH HAZARDS

5.1 **Personal Protective Equipment:** Rubber gloves; face shield or vapor-tight chemical-type safety goggles; rubber apron; rubber boots or shoes
5.2 **Symptoms Following Exposure:** Inhalation of vapors or mist causes burning sensation of nose and throat, and lung irritation with coughing; prolonged or excessive exposure could cause vomiting and severe coughing. Ingestion causes blistering and burns of mouth and stomach. Contact with eyes or skin causes blistering and burns.
5.3 **Treatment for Exposure:** *Get medical attention following all exposures to this compound.* INHALATION: move to fresh air; give oxygen if necessary. INGESTION: give large amounts of water. EYES: flush with water for at least 15 min. SKIN: flush with water.
5.4 **Toxicity by Inhalation (Threshold Limit Value):** Data not available
5.5 **Short-Term Inhalation Limits:** Data not available
5.6 **Toxicity by Ingestion:** Data not available
5.7 **Late Toxicity:** Data not available
5.8 **Vapor (Gas) Irritant Characteristics:** Data not available
5.9 **Liquid or Solid Irritant Characteristics:** Data not available
5.10 **Odor Threshold:** Odorless

6. FIRE HAZARDS

6.1 **Flash Point:** Not flammable, but may explode in fire
6.2 **Flammable Limits in Air:** Not flammable
6.3 **Fire Extinguishing Agents:** Water from protected area
6.4 **Fire Extinguishing Agents Not to be Used:**
6.5 **Special Hazards of Combustion Products:**
6.6 **Behavior in Fire:** Above 160°C (320°F) will react with combustible material and increase intensity of fire. Containers may explode.
6.7 **Ignition Temperature:** Not pertinent
6.8 **Electrical Hazard:** Not pertinent
6.9 **Burning Rate:** Not pertinent

7. CHEMICAL REACTIVITY

7.1 **Reactivity with Water:** No reaction
7.2 **Reactivity with Common Materials:** Contact with most combustible materials may cause fires and explosions. Corrosive to most metals with formation of flammable hydrogen gas, which may collect in enclosed spaces.
7.3 **Stability During Transport:** Unstable if heated
7.4 **Neutralizing Agents for Acids and Caustics:** Flush with water and rinse with dilute sodium bicarbonate or soda ash solution.
7.5 **Polymerization:** Not pertinent
7.6 **Inhibitor of Polymerization:** Not pertinent

8. WATER POLLUTION

8.1 **Aquatic Toxicity:** Data not available
8.2 **Waterfowl Toxicity:** Data not available
8.3 **Biological Oxygen Demand (BOD):** None
8.4 **Food Chain Concentration Potential:** None

9. SELECTED MANUFACTURERS

1. Allied Chemical Corp.
Specialty Chemicals Div.
P. O. Box 1087R
Morristown, N. J. 07960
2. J. T. Baker Chemical Co.
Phillipsburg, N. J. 08865
3. Hooker Chemical Corporation
Industrial Chemicals Division
Niagara Falls, N. Y. 14302

10. SHIPPING INFORMATION

10.1 **Grades or Purity:** ACS, 60—72% solution in water.
10.2 **Storage Temperature:** Ambient
10.3 **Inert Atmosphere:** No requirement
10.4 **Venting:** Open

11. HAZARD ASSESSMENT CODE

(See Hazard Assessment Handbook, CG 446-3)

A-P

12. HAZARD CLASSIFICATIONS

12.1 **Code of Federal Regulations:** Oxidizer
12.2 **NAS Hazard Rating for Bulk Water Transportation:** Not listed
12.3 **NFPA Hazard Classifications:**

Category	Classification
Health Hazard (Blue)	3
Flammability (Red)	0
Reactivity (Yellow)	3
	oxy

13. PHYSICAL AND CHEMICAL PROPERTIES

13.1 **Physical State at 15°C and 1 atm:** Liquid
13.2 **Molecular Weight:** 100.46 (solute only)
13.3 **Boiling Point at 1 atm:** Not pertinent (decomposes)
13.4 **Freezing Point:** −170°F = −112°C = 161°K
13.5 **Critical Temperature:** Not pertinent
13.6 **Critical Pressure:** Not pertinent
13.7 **Specific Gravity:** 1.6—1.7 at 25°C (liquid)
13.8 **Liquid Surface Tension:** Data not available
13.9 **Liquid-Water Interfacial Tension:** Not pertinent
13.10 **Vapor (Gas) Specific Gravity:** Not pertinent
13.11 **Ratio of Specific Heats of Vapor (Gas):** Not pertinent
13.12 **Latent Heat of Vaporization:** Not pertinent
13.13 **Heat of Combustion:** Not pertinent
13.14 **Heat of Decomposition:** Data not available
13.15 **Heat of Solution:** Not pertinent
13.16 **Heat of Polymerization:** Not pertinent

NOTES

PCM — PERCHLOROMETHYL MERCAPTAN

Common Synonyms			
Trichlormethyl sulfur chloride Thiocarbonyl tetrachloride Trichloromethyl sulfochloride Trichloromethane sulfuryl chloride	Liquid	Yellow to orange-red	Strong unpleasant odor
	Sinks in water. Poisonous vapor is produced.		

AVOID CONTACT WITH LIQUID AND VAPOR. KEEP PEOPLE AWAY.
Wear goggles, self-contained breathing apparatus, and rubber overclothing (including gloves).
Stop discharge if possible.
Evacuate area in case of large discharge.
Stay upwind. Use water spray to "knock down" vapor.
Isolate and remove discharged material.
Notify local health and pollution control agencies.

Fire

Not flammable.
POISONOUS GASES MAY BE PRODUCED WHEN HEATED.

Exposure

CALL FOR MEDICAL AID.

VAPOR
POISONOUS IF INHALED.
Irritating to eyes, nose and throat.
Move victim to fresh air.
If breathing has stopped, give artificial respiration.
If breathing is difficult, give oxygen.

LIQUID
POISONOUS IF SWALLOWED OR IF SKIN IS EXPOSED.
Irritating to eyes.
Remove contaminated clothing and shoes.
Flush affected areas with plenty of water.
IF IN EYES, hold eyelids open and flush with plenty of water.
IF SWALLOWED and victim is CONSCIOUS, have victim drink water or milk and have victim induce vomiting.
IF SWALLOWED and victim is UNCONSCIOUS OR HAVING CONVULSIONS, do nothing except keep victim warm.

Water Pollution

Effect of low concentrations on aquatic life is unknown.
May be dangerous if it enters water intakes.
Notify local health and wildlife officials.
Notify operators of nearby water intakes.

1. RESPONSE TO DISCHARGE

(See Response Methods Handbook, CG 446-4)

Issue warning—poison, water contaminant, air contaminant
Restrict access
Disperse and flush

2. LABEL

3. CHEMICAL DESIGNATIONS

3.1 **Synonyms:** Trichloromethanesulfenyl chloride; Trichloromethane sulphuryl chloride; Trichloromethyl sulphochloride; Trichloromethylsulfur chloride
3.2 **Coast Guard Compatibility Classification:** Not applicable
3.3 **Chemical Formula:** Cl_3CSCl
3.4 **IMCO/United Nations Numerical Designation:** 6.1/1670

4. OBSERVABLE CHARACTERISTICS

4.1 **Physical State (as shipped):** Liquid
4.2 **Color:** Bright yellow; pale yellow; orange-red
4.3 **Odor:** Intensely unpleasant; strong, unbearable; strongly acrid

5. HEALTH HAZARDS

5.1 **Personal Protective Equipment:** Organic- and acid-type canister mask or self-contained breathing apparatus; goggles or face shield; rubber gloves.
5.2 **Symptoms Following Exposure:** Inhalation may cause severe irritation of upper respiratory tract; brief exposure to low concentrations may produce central nervous system depression and lung, liver, and heart congestion. Severe exposures may be fatal. Contract of liquid with eyes causes immediate eye irritation and severe conjunctivitis; if not promptly washed away, severe corneal damage may result. Liquid also causes severe local skin irritation; may be absorbed through skin in quantities sufficient to produce systemic poisoning. Ingestion may cause damage to the membranes of the mouth, throat, and gastrointestinal tract. Pain and burning of the mouth and throat, nausea, vomiting, cramps and diarrhea may occur. In severe cases, tissue ulceration and central nervous system depression may result.
5.3 **Treatment for Exposure:** Get medical attention at once following all exposures to this compound. INHALATION: remove victim to fresh air; give mouth-to-mouth resuscitation if needed. EYES: flush with water for 15 min; if physician is not available, flush for another 15 min. SKIN: flush with water. INGESTION: give large amounts of water, then induce vomiting until vomitus is clear; give milk, eggs, or olive oil to soothe stomach.
5.4 **Toxicity by Inhalation (Threshold Limit Value):** 0.1 ppm
5.5 **Short-Term Inhalation Limits:** Data not available
5.6 **Toxicity by Ingestion:** Grade 3; oral LD_{50} = 83 mg/kg (rat)
5.7 **Late Toxicity:** Data not available

6. FIRE HAZARDS

6.1 **Flash Point:** Not flammable
6.2 **Flammable Limits in Air:** Not flammable
6.3 **Fire Extinguishing Agents:** Not pertinent
6.4 **Fire Extinguishing Agents Not to be Used:** Not pertinent
6.5 **Special Hazards of Combustion Products:** Very irritating vapors formed from hot material. May form toxic phosgene gas, hydrogen chloride, and sulfur dioxide.
6.6 **Behavior in Fire:** At elevated temperatures will decompose to carbon tetrachloride, sulfur chloride, and heavy oily polymers.
6.7 **Ignition Temperature:** Not pertinent
6.8 **Electrical Hazard:** Not pertinent
6.9 **Burning Rate:** Not pertinent

7. CHEMICAL REACTIVITY

7.1 **Reactivity with Water:** Reacts only when hot to give carbon dioxide, hydrochloric acid, and sulfur.
7.2 **Reactivity with Common Materials:** Reacts with iron or steel, evolving carbon tetrachloride. Corrosive to most metals.
7.3 **Stability During Transport:** Stable
7.4 **Neutralizing Agents for Acids and Caustics:** Flood with water, rinse with dilute sodium bicarbonate or lime solution.
7.5 **Polymerization:** Not pertinent
7.6 **Inhibitor of Polymerization:** Not pertinent

8. WATER POLLUTION

8.1 **Aquatic Toxicity:** Data not available
8.2 **Waterfowl Toxicity:** Data not available
8.3 **Biological Oxygen Demand (BOD):** Data not available
8.4 **Food Chain Concentration Potential:** None

9. SELECTED MANUFACTURERS

1. Stauffer Chemical Co.
   1200 South 47th Street
   Richmond, Calif. 94804
2. Aldrich Chemical Co.
   940 West St. Paul Ave.
   Milwaukee, Wis. 53233
3. Pfaltz and Bauer, Inc.
   126-04 Northern Boulevard
   Flushing, N. Y. 11368

10. SHIPPING INFORMATION

10.1 **Grades or Purity:** Technical: 97+%
10.2 **Storage Temperature:** Ambient
10.3 **Inert Atmosphere:** No requirement
10.4 **Venting:** Pressure-vacuum

11. HAZARD ASSESSMENT CODE

(See Hazard Assessment Handbook, CG 446-3)

A-O-X

12. HAZARD CLASSIFICATIONS

12.1 **Code of Federal Regulations:** Poisonous, Class B
12.2 **NAS Hazard Rating for Bulk Water Transportation:** Not listed
12.3 **NFPA Hazard Classifications:** Not listed

13. PHYSICAL AND CHEMICAL PROPERTIES

13.1 **Physical State at 15°C and 1 atm:** Liquid
13.2 **Molecular Weight:** 185.9
13.3 **Boiling Point at 1 atm:** 300°F = 148°C = 421°K
13.4 **Freezing Point:** Not pertinent
13.5 **Critical Temperature:** Not pertinent
13.6 **Critical Pressure:** Not pertinent
13.7 **Specific Gravity:** 1.706 at 11°C (liquid)
13.8 **Liquid Surface Tension:** 35.02 dynes/cm = 0.03502 N/m at 20°C
13.9 **Liquid-Water Interfacial Tension:** Not pertinent
13.10 **Vapor (Gas) Specific Gravity:** Not pertinent
13.11 **Ratio of Specific Heats of Vapor (Gas):** Not pertinent
13.12 **Latent Heat of Vaporization:** (est.) 94 Btu/lb = 52 cal/g = 2.2×10^5 J/kg
13.13 **Heat of Combustion:** Not pertinent
13.14 **Heat of Decomposition:** Not pertinent
13.15 **Heat of Solution:** Not pertinent
13.16 **Heat of Polymerization:** Not pertinent

5. HEALTH HAZARDS (Cont'd.)

5.8 **Vapor (Gas) Irritant Characteristics:** Data not available
5.9 **Liquid or Solid Irritant Characteristics:** Data not available
5.10 **Odor Threshold:** Data not available

PTL	PETROLATUM

Common Synonyms Petroleum jelly	Thick liquid	Dark brown, green, amber or white	Odorless
	Floats on water.		

Stop discharge if possible. Keep people away.
Call fire department.
Avoid contact with liquid.
Isolate and remove discharged material.
Notify local health and pollution control agencies.

Fire	Combustible. Extinguish with foam, dry chemical or carbon dioxide. Cool exposed containers with water.
Exposure	CALL FOR MEDICAL AID. LIQUID Irritating to eyes. Flush affected areas with plenty of water. IF IN EYES, hold eyelids open and flush with plenty of water.
Water Pollution	Effect of low concentrations on aquatic life is unknown. Fouling to shoreline. May be dangerous if it enters water intakes. Notify local health and wildlife officials. Notify operators of nearby water intakes.

1. RESPONSE TO DISCHARGE

(See Response Methods Handbook, CG 446-4)

Mechanical containment
Should be removed
Chemical and physical treatment

2. LABELS

No hazard label required by Code of Federal Regulations

3. CHEMICAL DESIGNATIONS

3.1 **Synonyms:** Petrolatum jelly
Petroleum jelly
Vaseline
Yellow petrolatum

3.2 **Coast Guard Compatibility Classification:** Miscellaneous hydrocarbon mixtures

3.3 **Chemical Formula:** Not applicable

3.4 **IMCO/United Nations Numerical Designation:** Not listed

4. OBSERVABLE CHARACTERISTICS

4.1 **Physical State (as shipped):** Liquid

4.2 **Color:** Colorless, amber, green, dark brown

4.3 **Odor:** None

5. HEALTH HAZARDS

5.1 **Personal Protective Equipment:** Goggles or face shield

5.2 **Symptoms Following Exposure:** None

5.3 **Treatment for Exposure:** EYES: wash with water.

5.4 **Toxicity by Inhalation (Threshold Limit Value):** Not pertinent

5.5 **Short-Term Inhalation Limits:** Not pertinent

5.6 **Toxicity by Ingestion:** Grade 1; LD_{50} 5 to 15 g/kg

5.7 **Late Toxicity:** None

5.8 **Vapor (Gas) Irritant Characteristics:** None

5.9 **Liquid or Solid Irritant Characteristics:** None

5.10 **Odor Threshold:** Not pertinent

6. FIRE HAZARDS

6.1 **Flash Point:** 360-430°F C.C.

6.2 **Flammable Limits in Air:** Data not available

6.3 **Fire Extinguishing Agents:** Water, foam, dry chemical, or carbon dioxide

6.4 **Fire Extinguishing Agents Not to be Used:** Not pertinent

6.5 **Special Hazards of Combustion Products:** Not pertinent

6.6 **Behavior in Fire:** Not pertinent

6.7 **Ignition Temperature:** Data not available

6.8 **Electrical Hazard:** Not pertinent

6.9 **Burning Rate:** Data not available

7. CHEMICAL REACTIVITY

7.1 **Reactivity with Water:** No reaction

7.2 **Reactivity with Common Materials:** No reaction

7.3 **Stability During Transport:** Stable

7.4 **Neutralizing Agents for Acids and Caustics:** Not pertinent

7.5 **Polymerization:** Not pertinent

7.6 **Inhibitor of Polymerization:** Not pertinent

8. WATER POLLUTION

8.1 **Aquatic Toxicity:** Data not available

8.2 **Waterfowl Toxicity:** Data not available

8.3 **Biological Oxygen Demand (BOD):** Data not available

8.4 **Food Chain Concentration Potential:** None

9. SELECTED MANUFACTURERS

1. Mobil Oil Corp.
North American Division
Paulsboro, N. J. 08066
2. Penreco, Inc.
Butler, Pa. 16001
3. Quaker State Oil Refining Co.
Oil City, Pa. 16301

10. SHIPPING INFORMATION

10.1 **Grades or Purity:** USP, NF, technical (these vary in color and in melting point)

10.2 **Storage Temperature:** Ambient

10.3 **Inert Atmosphere:** No requirement

10.4 **Venting:** Open (flame arrester)

11. HAZARD ASSESSMENT CODE

(See Hazard Assessment Handbook, CG 446-3)

A-T-U

12. HAZARD CLASSIFICATIONS

12.1 **Code of Federal Regulations:** Not listed

12.2 **NAS Hazard Rating for Bulk Water Transportation:** Not listed

12.3 **NFPA Hazard Classifications:** Not listed

13. PHYSICAL AND CHEMICAL PROPERTIES

13.1 **Physical State at 15°C and 1 atm:** Grease

13.2 **Molecular Weight:** Not pertinent

13.3 **Boiling Point at 1 atm:** Very high

13.4 **Freezing Point:** 100-135°F = 38-57°C = 311-330°K

13.5 **Critical Temperature:** Not pertinent

13.6 **Critical Pressure:** Not pertinent

13.7 **Specific Gravity:** (est.) 0.865 at 60°C (liquid)

13.8 **Liquid Surface Tension:** Data not available

13.9 **Liquid-Water Interfacial Tension:** 35-50 dynes/cm = 0.035-0.050 N/m

13.10 **Vapor (Gas) Specific Gravity:** Not pertinent

13.11 **Ratio of Specific Heats of Vapor (Gas):** Not pertinent

13.12 **Latent Heat of Vaporization:** 97-100 Btu/lb = 54-63 cal/g = $2.3\text{-}2.6 \times 10^5$ J/kg

13.13 **Heat of Combustion:** Data not available

13.14 **Heat of Decomposition:** Not pertinent

13.15 **Heat of Solution:** Not pertinent

13.16 **Heat of Polymerization:** Not pertinent

NOTES

REVISED 1978

PTN

PETROLEUM NAPHTHA

Common Synonyms			
Petroleum solvent	Liquid	Colorless	Gasoline odor
	Floats on water. Flammable vapor is produced.		

Stop discharge if possible. Keep people away.
Shut off ignition sources and call fire department.
Stay upwind and use water spray to "knock down" vapor.
Avoid contact with liquid.
Isolate and remove discharged material.
Notify local health and pollution control agencies.

Fire	FLAMMABLE. Flashback along vapor trail may occur. Vapor may explode if ignited in an enclosed area. Extinguish with foam, dry chemical, or carbon dioxide. Cool exposed containers with water.
Exposure	CALL FOR MEDICAL AID. VAPOR Not irritating to eyes, nose, or throat. LIQUID Harmful if swallowed. IF SWALLOWED and victim is CONSCIOUS, have victim drink water or milk. DO NOT INDUCE VOMITING.
Water Pollution	Effect of low concentrations on aquatic life is unknown. Fouling to shoreline. May be dangerous if it enters water intakes. Notify local health and wildlife officials. Notify operators of nearby water intakes.

1. RESPONSE TO DISCHARGE

(See Response Methods Handbook, CG 446-4)

Issue warning—high flammability
Restrict access
Evacuate area

2. LABEL

Red

3. CHEMICAL DESIGNATIONS

3.1 **Synonyms:** Petroleum solvent
3.2 **Coast Guard Compatibility Classification:** Miscellaneous hydrocarbon mixtures
3.3 **Chemical Formula:** Not applicable
3.4 **IMCO/United Nations Numerical Designation:** 3.2/1255

4. OBSERVABLE CHARACTERISTICS

4.1 **Physical State (as shipped):** Liquid
4.2 **Color:** Colorless
4.3 **Odor:** Like gasoline and kerosene

5. HEALTH HAZARDS

5.1 **Personal Protective Equipment:** Goggles or face shield (as for gasoline).
5.2 **Symptoms Following Exposure:** Inhalation of concentrated vapor may cause intoxication. Liquid is not very irritating to skin or eyes but may get into lungs by aspiration.
5.3 **Treatment for Exposure:** INHALATION: remove victim to fresh air and treat symptoms. INGESTION: have victim drink water or milk; do NOT induce vomiting. EYES: flush with water for 15 min. SKIN: wipe off and wash with soap and water.
5.4 **Toxicity by Inhalation (Threshold Limit Value):** No single TLV applicable.
5.5 **Short-Term Inhalation Limits:** 500 ppm for 30 min.
5.6 **Toxicity by Ingestion:** Grade 2; LD_{50} 0.5 to 5 g/kg
5.7 **Late Toxicity:** None
5.8 **Vapor (Gas) Irritant Characteristics:** Vapors are non-irritating to the eyes and throat.
5.9 **Liquid or Solid Irritant Characteristics:** No appreciable hazard. Practically harmless to the skin.
5.10 **Odor Threshold:** Data not available

6. FIRE HAZARDS

6.1 **Flash Point:** 20°F (approx.) C.C.
6.2 **Flammable Limits in Air:** 0.9% 6.0%
6.3 **Fire Extinguishing Agents:** Foam, carbon dioxide, or dry chemical
6.4 **Fire Extinguishing Agents Not to be Used:** Water may be ineffective.
6.5 **Special Hazards of Combustion Products:** Not pertinent
6.6 **Behavior in Fire:** Not pertinent
6.7 **Ignition Temperature:** 450°F (approx.)
6.8 **Electrical Hazard:** Not pertinent
6.9 **Burning Rate:** 4 mm/min.

7. CHEMICAL REACTIVITY

7.1 **Reactivity with Water:** No reaction
7.2 **Reactivity with Common Materials:** No reaction
7.3 **Stability During Transport:** Stable
7.4 **Neutralizing Agents for Acids and Caustics:** Not pertinent
7.5 **Polymerization:** Not pertinent
7.6 **Inhibitor of Polymerization:** Not pertinent

8. WATER POLLUTION

8.1 **Aquatic Toxicity:** Data not available
8.2 **Waterfowl Toxicity:** Data not available
8.3 **Biological Oxygen Demand (BOD):** Data not available
8.4 **Food Chain Concentration Potential:** None

9. SELECTED MANUFACTURERS

1. Atlantic Richfield Co.
   ARCO Chemical Co. Division
   260 S. Broad St.
   Philadelphia, Pa. 19101
2. Skelly Oil Co.
   Kansas City, Mo. 64141
3. Union Oil Co. of California
   Amsco Division
   3100 S. Meacham Rd.
   Palatine, Ill. 60067

10. SHIPPING INFORMATION

10.1 **Grades or Purity:** Data not available
10.2 **Storage Temperature:** Ambient
10.3 **Inert Atmosphere:** No requirement
10.4 **Venting:** Open (flame arrester) or pressure-vacuum

11. HAZARD ASSESSMENT CODE

(See Hazard Assessment Handbook, CG 446-3)

A-T-U-V-W

12. HAZARD CLASSIFICATIONS

12.1 **Code of Federal Regulations:** Flammable liquid
12.2 **NAS Hazard Rating for Bulk Water Transportation:** Not listed
12.3 **NFPA Hazard Classifications:** Not listed

13. PHYSICAL AND CHEMICAL PROPERTIES

13.1 **Physical State at 15°C and 1 atm:** Liquid
13.2 **Molecular Weight:** Not pertinent
13.3 **Boiling Point at 1 atm:** 207.0°F = 97.2°C = 370.4°K
13.4 **Freezing Point:** Not pertinent
13.5 **Critical Temperature:** Not pertinent
13.6 **Critical Pressure:** Not pertinent
13.7 **Specific Gravity:** 0.74 at 20°C (liquid)
13.8 **Liquid Surface Tension:** 19-23 dynes/cm = 0.019-0.023 N/m at 20°C
13.9 **Liquid-Water Interfacial Tension:** 39-51 dynes/cm = 0.039-0.051 N/m at 20°C
13.10 **Vapor (Gas) Specific Gravity:** Not pertinent
13.11 **Ratio of Specific Heats of Vapor (Gas):** (est.) 1.030
13.12 **Latent Heat of Vaporization:** 130-150 Btu/lb = 71-81 cal/g = $3.0\text{-}3.4 \times 10^5$ J/kg
13.13 **Heat of Combustion:** Data not available
13.14 **Heat of Decomposition:** Not pertinent
13.15 **Heat of Solution:** Not pertinent
13.16 **Heat of Polymerization:** Not pertinent

NOTES

REVISED 1978

PHN	PHENOL

Common Synonyms			
Carbolic acid Hydroxybenzene	Solid crystals; or watery liquid	White solid, or light pink liquid	Sweet tarry odor
	May float or sink, and mixes slowly with water.		

AVOID CONTACT WITH LIQUID AND SOLID. Keep people away.
Wear goggles, self-contained breathing apparatus, and rubber overclothing (including gloves).
Stop discharge if possible. Call fire department.
Evacuate area in case of large discharge.
Isolate and remove discharged material.
Notify local health and pollution control agencies.

Fire

Combustible.
POISONOUS GASES ARE PRODUCED IN FIRE.
Wear goggles, self-contained breathing apparatus, and rubber overclothing (including gloves).
Extinguish with water, carbon dioxide, dry chemical, or foam.
Cool exposed containers with water.

Exposure

CALL FOR MEDICAL AID.

LIQUID OR SOLID
POISONOUS IF SWALLOWED.
Will burn skin and eyes.
Remove contaminated clothing and shoes.
Flush affected areas with plenty of water.
IF IN EYES, hold eyelids open and flush with plenty of water.
IF SWALLOWED and victim is CONSCIOUS, have victim drink water or milk.
DO NOT INDUCE VOMITING.

Water Pollution

HARMFUL TO AQUATIC LIFE IN VERY LOW CONCENTRATIONS.
May be dangerous if it enters water intakes.
Notify local health and wildlife officials.
Notify operators of nearby water intakes.

1. RESPONSE TO DISCHARGE

(See Response Methods Handbook, CG 446-4)

Issue warning—poison
Restrict access
Should be removed
Chemical and physical treatment

2. LABEL

3. CHEMICAL DESIGNATIONS

3.1 **Synonyms:** Carbolic acid
Hydroxybenzene
Phenic acid
Phenyl hydroxide

3.2 **Coast Guard Compatibility Classification:** Phenol

3.3 **Chemical Formula:** C_6H_5OH

3.4 **IMCO/United Nations Numerical Designation:** 9.0/1671

4. OBSERVABLE CHARACTERISTICS

4.1 **Physical State (as shipped):** Solid or liquid

4.2 **Color:** Colorless to light pink

4.3 **Odor:** Characteristically sweet; sweet, tarry; pungent, distinctive; distinct, aromatic, somewhat sickening sweet and acrid

5. HEALTH HAZARDS

5.1 **Personal Protective Equipment:** Fresh-air mask for confined areas; rubber gloves; protective clothing; full face shield.

5.2 **Symptoms Following Exposure:** Will burn eyes and skin. The analgesic action may cause loss of pain sensation. Readily absorbed through skin, causing increase in heart rate, convulsions, and death.

5.3 **Treatment for Exposure:** INHALATION: if victim shows any ill effects, move him to fresh air, keep him quiet and warm, and call a doctor immediately; if breathing stops, give artificial respiration. INGESTION: do NOT induce vomiting; give milk, egg whites, or large amounts of water and call doctor immediately; no known antidote; treat the symptoms. EYES: immediately flush with plenty of water for at least 15 min.; continue for another 15 min. if doctor has not taken over. SKIN: immediately remove all clothing while in a shower and wash affected area with abundant flowing water or soap and water for at least 15 min.; clean clothing thoroughly or discard.

5.4 **Toxicity by Inhalation (Threshold Limit Value):** 5 ppm (includes skin exposure).

5.5 **Short-Term Inhalation Limits:** Data not available

5.6 **Toxicity by Ingestion:** Grade 2; LD_{50} 0.5 to 5 g/kg (rat)

5.7 **Late Toxicity:** Carcinogenic in laboratory animals

5.8 **Vapor (Gas) Irritant Characteristics:** Vapors cause moderate irritation such that personnel will find high concentrations unpleasant. The effect is temporary.

5.9 **Liquid or Solid Irritant Characteristics:** Fairly severe skin irritant; may cause pain and second-degree burns after a few minutes' contact.

5.10 **Odor Threshold:** 0.05 ppm

6. FIRE HAZARDS

6.1 **Flash Point:** 185°F O.C.; 175°F C.C.

6.2 **Flammable Limits in Air:** 1.7%–8.6%

6.3 **Fire Extinguishing Agents:** Water fog, foam, carbon dioxide, or dry chemical

6.4 **Fire Extinguishing Agents Not to be Used:** Not pertinent

6.5 **Special Hazards of Combustion Products:** Toxic and irritating vapors are generated when heated.

6.6 **Behavior in Fire:** Yields flammable vapors when heated which will form explosive mixtures with air.

6.7 **Ignition Temperature:** 1319°F

6.8 **Electrical Hazard:** Not pertinent

6.9 **Burning Rate:** 3.5 mm/min.

7. CHEMICAL REACTIVITY

7.1 **Reactivity with Water:** No reaction

7.2 **Reactivity with Common Materials:** No reaction

7.3 **Stability During Transport:** Stable

7.4 **Neutralizing Agents for Acids and Caustics:** Not pertinent

7.5 **Polymerization:** Not pertinent

7.6 **Inhibitor of Polymerization:** Not pertinent

8. WATER POLLUTION

8.1 **Aquatic Toxicity:**
11.5-28.5 mg/l /96 hr/bluegill/TL_m/fresh water
1.5 ppm/48 hr/rainbow trout/TL_m/fresh water

8.2 **Waterfowl Toxicity:** Data not available

8.3 **Biological Oxygen Demand (BOD):** 200%, 5 days

8.4 **Food Chain Concentration Potential:** None

9. SELECTED MANUFACTURERS

1. Allied Chemical Corp.
Plastics Division
Morristown, N. J. 07960
2. Dow Chemical Co.
Midland, Mich. 48640
3. Monsanto Company
Monsanto Polymers & Petrochemicals Co.
800 North Lindbergh Blvd.
St. Louis, Mo. 63166

10. SHIPPING INFORMATION

10.1 **Grades or Purity:** 90-99% (solid), 60-85% (liquid). Technical: 82-92% (contains cresols)

10.2 **Storage Temperature:** Ambient

10.3 **Inert Atmosphere:** No requirement

10.4 **Venting:** Pressure-vacuum

11. HAZARD ASSESSMENT CODE

(See Hazard Assessment Handbook, CG 446-3)

A-P-Q

12. HAZARD CLASSIFICATIONS

12.1 **Code of Federal Regulations:** Poisonous liquid or solid, Class B

12.2 **NAS Hazard Rating for Bulk Water Transportation:**

Category	Rating
Fire	1
Health	
Vapor Irritant	2
Liquid or Solid Irritant	3
Poisons	3
Water Pollution	
Human Toxicity	2
Aquatic Toxicity	3
Aesthetic Effect	3
Reactivity	
Other Chemicals	2
Water	0
Self-Reaction	0

12.3 **NFPA Hazard Classifications:**

Category	Classification
Health Hazard (Blue)	3
Flammability (Red)	2
Reactivity (Yellow)	0

13. PHYSICAL AND CHEMICAL PROPERTIES

13.1 **Physical State at 15°C and 1 atm:** Solid or liquid

13.2 **Molecular Weight:** 94.11

13.3 **Boiling Point at 1 atm:** 359.2°F = 181.8°C = 455.0°K

13.4 **Freezing Point:** 105.6°F = 40.9°C = 314.1°K

13.5 **Critical Temperature:** 790.0°F = 421.1°C = 694.3°K

13.6 **Critical Pressure:** 889 psia = 60.5 atm = 6.13 MN/m^2

13.7 **Specific Gravity:** 1.058 at 41°C (liquid)

13.8 **Liquid Surface Tension:** 36.5 dynes/cm = 0.0365 N/m at 55°C

13.9 **Liquid-Water Interfacial Tension:** (est.) 20 dynes/cm = 0.02 N/m at 42°C

13.10 **Vapor (Gas) Specific Gravity:** Not pertinent

13.11 **Ratio of Specific Heats of Vapor (Gas):** 1.089

13.12 **Latent Heat of Vaporization:** 130 Btu/lb = 72 cal/g = 3.0×10^5 J/kg

13.13 **Heat of Combustion:** −13,400 Btu/lb = −7,445 cal/g = -311.7×10^5 J/kg

13.14 **Heat of Decomposition:** Not pertinent

13.15 **Heat of Solution:** Not pertinent

13.16 **Heat of Polymerization:** Not pertinent

NOTES

REVISED 1978

PDL — PHENYLDICHLOROARSINE, LIQUID

Common Synonyms			
Phenylarsenic dichloride	Liquid	Colorless to yellow	Weak unpleasant odor
	Sinks in water.		

AVOID CONTACT WITH LIQUID. KEEP PEOPLE AWAY.
Wear rubber overclothing (including gloves).
Stop discharge if possible.
Isolate and remove discharged material.
Notify local health and pollution control agencies.

Fire

Fire data not available.
POISONOUS GASES ARE PRODUCED WHEN HEATED.

Exposure

CALL FOR MEDICAL AID.

LIQUID
POISONOUS IF SWALLOWED.
Will burn skin and eyes.
Remove contaminated clothing and shoes.
Flush affected areas with plenty of water.
IF IN EYES, hold eyelids open and flush with plenty of water.
IF SWALLOWED and victim is CONSCIOUS, have victim drink water or milk.
DO NOT INDUCE VOMITING.

Water Pollution

Effect of low concentrations on aquatic life is unknown.
May be dangerous if it enters water intakes.
Notify local health and wildlife officials.
Notify operators of nearby water intakes.

1. RESPONSE TO DISCHARGE

(See Response Methods Handbook, CG 446-4)

Issue warning—poison, water contaminant, corrosive
Restrict access
Disperse and flush

2. LABEL

3. CHEMICAL DESIGNATIONS

3.1 **Synonyms:** Phenylarsenic dichloride
3.2 **Coast Guard Compatibility Classification:** Not applicable
3.3 **Chemical Formula:** $C_6H_5AsCl_2$
3.4 **IMCO/United Nations Numerical Designation:** 6.1/1556

4. OBSERVABLE CHARACTERISTICS

4.1 **Physical State (as shipped):** Liquid
4.2 **Color:** Colorless to yellow
4.3 **Odor:** Weak, but very unpleasant

5. HEALTH HAZARDS

5.1 **Personal Protective Equipment:** Full protective clothing; gas mask or self-contained breathing apparatus
5.2 **Symptoms Following Exposure:** Inhalation causes irritation of respiratory system, pulmonary edema, and systemic effects. Vapor irritates eyes. Liquid causes severe burns of eyes and severe irritation or burns of skin. Ingestion causes severe irritation or burns of mouth and stomach.
5.3 **Treatment for Exposure:** *Get medical attention at once following all exposures to this compound.* INHALATION: remove victim from exposure; give artificial respiration if breathing has ceased. EYES: immediately wash with copious amounts of water for at least 15 min. SKIN: flush with water and wash well with soap and water; compound can be absorbed through skin and cause toxic systemic effects. INGESTION: give large amounts of water.
5.4 **Toxicity by Inhalation (Threshold Limit Value):** Data not available
5.5 **Short-Term Inhalation Limits:** Data not available
5.6 **Toxicity by Ingestion:** Data not available
5.7 **Late Toxicity:** Data not available
5.8 **Vapor (Gas) Irritant Characteristics:** Data not available
5.9 **Liquid or Solid Irritant Characteristics:** Data not available
5.10 **Odor Threshold:** Data not available

6. FIRE HAZARDS

6.1 **Flash Point:** Data not available
6.2 **Flammable Limits in Air:** Data not available
6.3 **Fire Extinguishing Agents:** Water
6.4 **Fire Extinguishing Agents Not to be Used:** Not pertinent
6.5 **Special Hazards of Combustion Products:** Highly toxic arsenic fumes are formed when hot.
6.6 **Behavior in Fire:** Data not available
6.7 **Ignition Temperature:** Data not available
6.8 **Electrical Hazard:** Data not available
6.9 **Burning Rate:** 1.8 mm/min.

7. CHEMICAL REACTIVITY

7.1 **Reactivity with Water:** Very slow reaction, considered non-hazardous. Hydrochloric acid is formed.
7.2 **Reactivity with Common Materials:** Corrodes metals because of acid formed.
7.3 **Stability During Transport:** Stable
7.4 **Neutralizing Agents for Acids and Caustics:** Not pertinent
7.5 **Polymerization:** Not pertinent
7.6 **Inhibitor of Polymerization:** Not pertinent

8. WATER POLLUTION

8.1 **Aquatic Toxicity:** Data not available
8.2 **Waterfowl Toxicity:** Data not available
8.3 **Biological Oxygen Demand (BOD):** Data not available
8.4 **Food Chain Concentration Potential:** None

9. SELECTED MANUFACTURERS

1. Strem Chemicals, Inc.
150 Andover St.
Danvers, Mass. 01923
2. Ventron Corporation
Alfa Products Division
8 Congress St.
Beverly, Mass. 01915

10. SHIPPING INFORMATION

10.1 **Grades or Purity:** Commercial
10.2 **Storage Temperature:** Ambient
10.3 **Inert Atmosphere:** No requirement
10.4 **Venting:** Pressure-vacuum

11. HAZARD ASSESSMENT CODE

(See Hazard Assessment Handbook, CG 446-3)

A-O-X

12. HAZARD CLASSIFICATIONS

12.1 **Code of Federal Regulations:** Poisonous, Class B
12.2 **NAS Hazard Rating for Bulk Water Transportation:** Not listed
12.3 **NFPA Hazard Classifications:** Not listed

13. PHYSICAL AND CHEMICAL PROPERTIES

13.1 **Physical State at 15°C and 1 atm:** Liquid
13.2 **Molecular Weight:** 222.9
13.3 **Boiling Point at 1 atm:** 495°F = 257°C = 530°K
13.4 **Freezing Point:** 3.9°F = −15.6°C = 257.6°K
13.5 **Critical Temperature:** Not pertinent
13.6 **Critical Pressure:** Not pertinent
13.7 **Specific Gravity:** 1.657 at 20°C (liquid)
13.8 **Liquid Surface Tension:** 44.64 dynes/cm = 0.04464 N/m at 18°C
13.9 **Liquid-Water Interfacial Tension:** Not pertinent
13.10 **Vapor (Gas) Specific Gravity:** Not pertinent
13.11 **Ratio of Specific Heats of Vapor (Gas):** Not pertinent
13.12 **Latent Heat of Vaporization:** 99 Btu/lb = 55 cal/g = 2.3×10^5 J/kg
13.13 **Heat of Combustion:** (est.) −6,450 Btu/lb = −3,600 cal/g = -150×10^5 J/kg
13.14 **Heat of Decomposition:** Not pertinent
13.15 **Heat of Solution:** Not pertinent
13.16 **Heat of Polymerization:** Not pertinent

NOTES

PHH — PHENYLHYDRAZINE HYDROCHLORIDE

Common Synonyms			
Phenylhydrazinium chloride	Solid	White to tan	Weak odor
	Sinks and mixes with water.		

AVOID CONTACT WITH SOLID AND DUST. KEEP PEOPLE AWAY.
Wear rubber overclothing (including gloves).
Stop discharge if possible.
Isolate and remove discharged material.
Notify local health and pollution control agencies.

Fire

Combustible.
POISONOUS GASES MAY BE PRODUCED IN FIRE.
Wear goggles and self-contained breathing apparatus.
Extinguish with water, dry chemicals, foam, or carbon dioxide.

Exposure

CALL FOR MEDICAL AID.

DUST
Irritating to eyes, nose and throat.
If inhaled will cause coughing or difficult breathing.
If in eyes, hold eyelids open and flush with plenty of water.
If breathing has stopped, give artificial respiration.
If breathing is difficult, give oxygen.

SOLID
POISONOUS IF SWALLOWED.
If swallowed will cause nausea.
Remove contaminated clothing and shoes.
Flush affected areas with plenty of water.
IF IN EYES, hold eyelids open and flush with plenty of water.
IF SWALLOWED and victim is CONSCIOUS, have victim drink water or milk and have victim induce vomiting.
IF SWALLOWED and victim is UNCONSCIOUS OR HAVING CONVULSIONS, do nothing except keep victim warm.

Water Pollution

Effect of low concentrations on aquatic life is unknown.
May be dangerous if it enters water intakes.
Notify local health and wildlife officials.
Notify operators of nearby water intakes.

1. RESPONSE TO DISCHARGE

(See Response Methods Handbook, CG 446-4)

Issue warning—poison, water contaminant
Restrict access
Disperse and flush

2. LABEL

3. CHEMICAL DESIGNATIONS

3.1 **Synonyms:** Phenylhydrazinium chloride
3.2 **Coast Guard Compatibility Classification:** Not listed
3.3 **Chemical Formula:** $C_6H_5NHNH_2{\cdot}HCl$
3.4 **IMCO/United Nations Numerical Designation:** Not listed

4. OBSERVABLE CHARACTERISTICS

4.1 **Physical State (as shipped):** Solid
4.2 **Color:** White to tan
4.3 **Odor:** Weak aromatic

5. HEALTH HAZARDS

5.1 **Personal Protective Equipment:** Dust respirator; rubber gloves; goggles
5.2 **Symptoms Following Exposure:** Inhalation of dust irritates nose and throat; fumes from hot material may cause same symptoms as ingestion. Phenylhydrazine is a chronic poison; ingestion can cause jaundice, anorexia, nausea, and vascular thrombosis; may also cause anemia and liver injury. Contact with eyes causes irritation. Contact with skin causes irritation and dermatitis.
5.3 **Treatment for Exposure:** INHALATION: move to fresh air; get medical attention. INGESTION: give large amount of water; induce vomiting; get medical attention. EYES: flush with water for at least 15 min.; if exposure is prolonged or repeated, get medical attention. SKIN: flush with water.
5.4 **Toxicity by Inhalation (Threshold Limit Value):** Data not available
5.5 **Short-Term Inhalation Limits:** Data not available
5.6 **Toxicity by Ingestion:** Data not available
5.7 **Late Toxicity:** Causes tumors in mice
5.8 **Vapor (Gas) Irritant Characteristics:** Data not available
5.9 **Liquid or Solid Irritant Characteristics:** Data not available
5.10 **Odor Threshold:** Data not available

6. FIRE HAZARDS

6.1 **Flash Point:** Not pertinent (combustible solid)
6.2 **Flammable Limits in Air:** Not pertinent
6.3 **Fire Extinguishing Agents:** Water, foam, dry chemical, carbon dioxide
6.4 **Fire Extinguishing Agents Not to be Used:**
6.5 **Special Hazards of Combustion Products:** Toxic and irritating oxides of nitrogen and hydrogen chloride may form in fire.
6.6 **Behavior in Fire:** The solid may sublime without melting and deposit on cool surfaces.
6.7 **Ignition Temperature:** Data not available
6.8 **Electrical Hazard:** Not pertinent
6.9 **Burning Rate:** Not pertinent

7. CHEMICAL REACTIVITY

7.1 **Reactivity with Water:** No reaction
7.2 **Reactivity with Common Materials:** May be corrosive to metals
7.3 **Stability During Transport:** Stable
7.4 **Neutralizing Agents for Acids and Caustics:** Not pertinent
7.5 **Polymerization:** Not pertinent
7.6 **Inhibitor of Polymerization:** Not pertinent

8. WATER POLLUTION

8.1 **Aquatic Toxicity:** Data not available
8.2 **Waterfowl Toxicity:** Data not available
8.3 **Biological Oxygen Demand (BOD):** Data not available
8.4 **Food Chain Concentration Potential:** None

9. SELECTED MANUFACTURERS

1. Eastman Organic Chemicals
343 State Street
Rochester, N. Y. 14650
2. Aldrich Chemical Co.
940 W. Saint Paul Ave.
Milwaukee, Wis. 53233
3. Gallard-Schlesinger Chemical Mfg. Co.
584 Mineola Ave.
Carle Place, N. Y. 11514

10. SHIPPING INFORMATION

10.1 **Grades or Purity:** Commercial; Pure
10.2 **Storage Temperature:** Ambient
10.3 **Inert Atmosphere:** No requirement
10.4 **Venting:** Open

11. HAZARD ASSESSMENT CODE

(See Hazard Assessment Handbook, CG 446-3)

SS

12. HAZARD CLASSIFICATIONS

12.1 **Code of Federal Regulations:** Poisonous solid, Class B
12.2 **NAS Hazard Rating for Bulk Water Transportation:** Not listed
12.3 **NFPA Hazard Classifications:** Not listed

13. PHYSICAL AND CHEMICAL PROPERTIES

13.1 **Physical State at 15°C and 1 atm:** Solid
13.2 **Molecular Weight:** 144.6
13.3 **Boiling Point at 1 atm:** Not pertinent (decomposes)
13.4 **Freezing Point:** 469°F = 243°C = 516°K
13.5 **Critical Temperature:** Not pertinent
13.6 **Critical Pressure:** Not pertinent
13.7 **Specific Gravity:** > 1 at 20°C (solid)
13.8 **Liquid Surface Tension:** Not pertinent
13.9 **Liquid-Water Interfacial Tension:** Not pertinent
13.10 **Vapor (Gas) Specific Gravity:** Not pertinent
13.11 **Ratio of Specific Heats of Vapor (Gas):** Not pertinent
13.12 **Latent Heat of Vaporization:** Not pertinent
13.13 **Heat of Combustion:** Not pertinent
13.14 **Heat of Decomposition:** Not pertinent
13.15 **Heat of Solution:** Not pertinent
13.16 **Heat of Polymerization:** Not pertinent

NOTES

PHG — PHOSGENE

Common Synonyms			
Carbonyl chloride	Liquefied compressed gas	Colorless gas, or light yellow liquid	Sweet or sharp odor
	Liquid sinks in water. Poisonous vapor is produced. Boiling point is 47° F.		

AVOID CONTACT WITH LIQUID AND VAPOR. Keep people away.
Wear goggles and self-contained breathing apparatus.
Stop discharge if possible.
Evacuate area in case of large discharge.
Stay upwind and use water spray to "knock down" vapor.
Isolate and remove discharged material.
Notify local health and pollution control agencies.

Fire

Not flammable.
POISONOUS GASES ARE PRODUCED WHEN HEATED.
Wear goggles and self-contained breathing apparatus.
Cool exposed containers and protect men effecting shutoff with water.

Exposure

CALL FOR MEDICAL AID.

VAPOR
POISONOUS IF INHALED.
Irritating to eyes, nose, and throat.
Effects may be delayed.
Move to fresh air.
If breathing has stopped, give artificial respiration (but NOT mouth-to-mouth).
If breathing is difficult, give oxygen.
Maintain absolute rest until medical aid arrives.

Water Pollution

Effect of low concentrations on aquatic life is unknown.
May be dangerous if it enters water intakes.
Notify local health and wildlife officials.
Notify operators of nearby water intakes.

1. RESPONSE TO DISCHARGE

(See Response Methods Handbook, CG 446-4)

Issue warning—poison
Restrict access
Evacuate area

2. LABEL

3. CHEMICAL DESIGNATIONS

3.1 **Synonyms:** Carbonyl chloride
Chloroformyl chloride
3.2 **Coast Guard Compatibility Classification:** Not applicable
3.3 **Chemical Formula:** $COCl_2$
3.4 **IMCO/United Nations Numerical Designation:** 2.0/1076

4. OBSERVABLE CHARACTERISTICS

4.1 **Physical State (as shipped):** Compressed gas
4.2 **Color:** Colorless
4.3 **Odor:** Sharp, pungent odor in higher concentrations; like new-mown grass in low concentrations

5. HEALTH HAZARDS

5.1 **Personal Protective Equipment:** Approved U. S. Bureau of Mines respirator; protective clothing.
5.2 **Symptoms Following Exposure:** Irritates lungs, causing delayed pulmonary edema. Slight gassing produces dryness or burning sensation in the throat, numbness, pain in the chest, bronchitis, and shortness of breath.
5.3 **Treatment for Exposure:** INHALATION: remove victim from contaminated area; enforce absolute rest; call a doctor.
5.4 **Toxicity by Inhalation (Threshold Limit Value):** 0.1 ppm
5.5 **Short-Term Inhalation Limits:** 1 ppm for 5 min.
5.6 **Toxicity by Ingestion:** Not pertinent
5.7 **Late Toxicity:** Severe delayed pulmonary edema.
5.8 **Vapor (Gas) Irritant Characteristics:** Vapors cause severe irritation of eyes and throat and can cause eye and lung injury. They cannot be tolerated even at low concentrations.
5.9 **Liquid or Solid Irritant Characteristics:** Severe irritant to all tissues.
5.10 **Odor Threshold:** 0.5 ppm

6. FIRE HAZARDS

6.1 **Flash Point:** Not flammable
6.2 **Flammable Limits in Air:** Not flammable
6.3 **Fire Extinguishing Agents:** Water to cool containers
6.4 **Fire Extinguishing Agents Not to be Used:** Not pertinent
6.5 **Special Hazards of Combustion Products:** Toxic gas is generated when heated.
6.6 **Behavior in Fire:** Not pertinent
6.7 **Ignition Temperature:** Not flammable
6.8 **Electrical Hazard:** Not pertinent
6.9 **Burning Rate:** Not flammable

7. CHEMICAL REACTIVITY

7.1 **Reactivity with Water:** Decomposes, but not vigorously.
7.2 **Reactivity with Common Materials:** No reaction
7.3 **Stability During Transport:** Stable
7.4 **Neutralizing Agents for Acids and Caustics:** Can be absorbed in caustic soda solution. One ton of phosgene requires 2,480 lbs. of caustic soda dissolved in 1000 gal. of water.
7.5 **Polymerization:** Not pertinent
7.6 **Inhibitor of Polymerization:** Not pertinent

8. WATER POLLUTION

8.1 **Aquatic Toxicity:** Data not available
8.2 **Waterfowl Toxicity:** Data not available
8.3 **Biological Oxygen Demand (BOD):** None
8.4 **Food Chain Concentration Potential:** None

9. SELECTED MANUFACTURERS

1. Baychem Corp.
Mobay Chemical Co. Division
Penn Lincoln Pkwy., West
Pittsburg, Pa. 15205
2. E. I. duPont de Nemours & Co., Inc.
Elastomer Chemicals Dept.
Wilmington, Del. 19898
3. The Upjohn Co.
Polymer Chemicals Division
La Porte, Tex. 77571

10. SHIPPING INFORMATION

10.1 **Grades or Purity:** Commercial: 100%
10.2 **Storage Temperature:** Ambient
10.3 **Inert Atmosphere:** No requirement
10.4 **Venting:** Safety relief

11. HAZARD ASSESSMENT CODE

(See Hazard Assessment Handbook, CG 446-3)

A-C-I-J-O

12. HAZARD CLASSIFICATIONS

12.1 **Code of Federal Regulations:** Poisonous gas or liquid, Class A
12.2 **NAS Hazard Rating for Bulk Water Transportation:** Not listed
12.3 **NFPA Hazard Classifications:**

Category	Classification
Health Hazard (Blue)	4
Flammability (Red)	0
Reactivity (Yellow)	0

13. PHYSICAL AND CHEMICAL PROPERTIES

13.1 **Physical State at 15°C and 1 atm:** Gas
13.2 **Molecular Weight:** 98.92
13.3 **Boiling Point at 1 atm:** 46.8°F = 8.2°C = 281.4°K
13.4 **Freezing Point:** −195°F = −126°C = 147°K
13.5 **Critical Temperature:** 360°F = 182°C = 455°K
13.6 **Critical Pressure:** 823 psia = 56.0 atm = 5.67 MN/m^2
13.7 **Specific Gravity:** 1.38 at 20°C (liquid)
13.8 **Liquid Surface Tension:** 22.8 dynes/cm = 0.0228 N/m at 0°C
13.9 **Liquid-Water Interfacial Tension:** Not pertinent
13.10 **Vapor (Gas) Specific Gravity:** 3.4
13.11 **Ratio of Specific Heats of Vapor (Gas):** 1.170
13.12 **Latent Heat of Vaporization:** 110 Btu/lb = 59 cal/g = 2.5×10^5 J/kg
13.13 **Heat of Combustion:** Not pertinent
13.14 **Heat of Decomposition:** Not pertinent
13.15 **Heat of Solution:** Not pertinent
13.16 **Heat of Polymerization:** Not pertinent

NOTES

REVISED 1978

PAC	# PHOSPHORIC ACID

Common Synonyms
Orthophosphoric acid

Thick liquid — Colorless — Odorless

Sinks and mixes with water

AVOID CONTACT WITH LIQUID. Keep people away.
Wear chemical protective suit.
Stop discharge if possible.
Isolate and remove discharged material.
Notify local health and pollution control agencies.

Fire

Not flammable.
Flammable gas is formed on contact with metals.
Wear chemical protective suit with self-contained breathing apparatus.

Exposure

CALL FOR MEDICAL AID.

LIQUID
Will burn skin and eyes.
If swallowed, will cause nausea, vomiting, or loss of consciousness.
Remove contaminated clothing and shoes.
Flush affected areas with plenty of water.
IF IN EYES, hold eyelids open and flush with plenty of water.
IF SWALLOWED and victim is CONSCIOUS, have victim drink water or milk.
DO NOT INDUCE VOMITING.

Water Pollution

Dangerous to aquatic life in high concentrations.
May be dangerous if it enters water intakes.
Notify local health and wildlife officials.
Notify operators of nearby water intakes.

1. RESPONSE TO DISCHARGE

(See Response Methods Handbook, CG 446-4)

Issue warning—corrosive
Restrict access
Disperse and flush

2. LABEL

3. CHEMICAL DESIGNATIONS

3.1 **Synonyms:** Orthophosphoric acid
3.2 **Coast Guard Compatibility Classification:** Non-oxidizing mineral acid
3.3 **Chemical Formula:** H_3PO_4
3.4 **IMCO/United Nations Numerical Designation:** 9.0/1805

4. OBSERVABLE CHARACTERISTICS

4.1 **Physical State (as shipped):** Liquid
4.2 **Color:** Colorless
4.3 **Odor:** Odorless

5. HEALTH HAZARDS

5.1 **Personal Protective Equipment:** Goggles or face shield; rubber gloves and protective clothing.
5.2 **Symptoms Following Exposure:** Burns on mouth and lips, sour acrid taste, severe gastrointestinal irritation, nausea, vomiting, bloody diarrhea, difficult swallowing, severe abdominal pains, thirst, acidemia, difficult breathing, convulsions, collapse, shock, death.
5.3 **Treatment for Exposure:** INGESTION: do NOT induce vomiting; give water, milk, or vegetable oil. SKIN OR EYE CONTACT: flush with water for at least 15 min.
5.4 **Toxicity by Inhalation (Threshold Limit Value):** 1.0 mg/m^3
5.5 **Short-Term Inhalation Limits:** Not pertinent
5.6 **Toxicity by Ingestion:** Grade 3; LD_{50} 50 to 500 mg/kg
5.7 **Late Toxicity:** None
5.8 **Vapor (Gas) Irritant Characteristics:** Not volatile.
5.9 **Liquid or Solid Irritant Characteristics:** Fairly severe skin irritant; may cause pain and second-degree burns after a few minutes' contact.
5.10 **Odor Threshold:** Not pertinent

6. FIRE HAZARDS

6.1 **Flash Point:** Not flammable
6.2 **Flammable Limits in Air:** Not flammable
6.3 **Fire Extinguishing Agents:** Not pertinent
6.4 **Fire Extinguishing Agents Not to be Used:** Not pertinent
6.5 **Special Hazards of Combustion Products:** Not pertinent
6.6 **Behavior in Fire:** Not pertinent
6.7 **Ignition Temperature:** Not flammable
6.8 **Electrical Hazard:** Not pertinent
6.9 **Burning Rate:** Not flammable

7. CHEMICAL REACTIVITY

7.1 **Reactivity with Water:** Mild evolution of heat.
7.2 **Reactivity with Common Materials:** Reacts with metals to liberate flammable hydrogen gas.
7.3 **Stability During Transport:** Stable
7.4 **Neutralizing Agents for Acids and Caustics:** Flush with water, neutralize with lime.
7.5 **Polymerization:** Not pertinent
7.6 **Inhibitor of Polymerization:** Not pertinent

8. WATER POLLUTION

8.1 **Aquatic Toxicity:** 138 ppm/24 hr/mosquito fish/TL_m/fresh water
8.2 **Waterfowl Toxicity:** Data not available
8.3 **Biological Oxygen Demand (BOD):** None
8.4 **Food Chain Concentration Potential:** None

9. SELECTED MANUFACTURERS

1. CF Industries Inc.
   Bartow Phosphate Complex
   Bartow, Fla. 33830
2. Cities Service Co., Inc.
   North American Chemicals & Metals Group
   60 Wall St.
   New York, N. Y. 10005
3. Freeport Minerals Co.
   Freeport Chemical Co. Division
   161 East 42 St.
   New York, N. Y. 10017

10. SHIPPING INFORMATION

10.1 **Grades or Purity:** NF, food, fertilizer, commercial: all 75-85%, the balance being water.
10.2 **Storage Temperature:** Ambient
10.3 **Inert Atmosphere:** No requirement
10.4 **Venting:** Open

11. HAZARD ASSESSMENT CODE

(See Hazard Assessment Handbook, CG 446-3)

A-P

12. HAZARD CLASSIFICATIONS

12.1 **Code of Federal Regulations:** Corrosive liquid
12.2 **NAS Hazard Rating for Bulk Water Transportation:**

Category	Rating
Fire	0
Health	
Vapor Irritant	0
Liquid or Solid Irritant	3
Poisons	1
Water Pollution	
Human Toxicity	2
Aquatic Toxicity	3
Aesthetic Effect	2
Reactivity	
Other Chemicals	3
Water	0
Self-Reaction	0

12.3 **NFPA Hazard Classifications:**

Category	Classification
Health Hazard (Blue)	2
Flammability (Red)	0
Reactivity (Yellow)	0

13. PHYSICAL AND CHEMICAL PROPERTIES

13.1 **Physical State at 15°C and 1 atm:** Liquid
13.2 **Molecular Weight:** 98.00
13.3 **Boiling Point at 1 atm:** >266°F = >130°C = >403°K
13.4 **Freezing Point:** Not pertinent
13.5 **Critical Temperature:** Not pertinent
13.6 **Critical Pressure:** Not pertinent
13.7 **Specific Gravity:** 1.892 at 25°C (liquid)
13.8 **Liquid Surface Tension:** Not pertinent
13.9 **Liquid-Water Interfacial Tension:** Not pertinent
13.10 **Vapor (Gas) Specific Gravity:** Not pertinent
13.11 **Ratio of Specific Heats of Vapor (Gas):** Not pertinent
13.12 **Latent Heat of Vaporization:** Not pertinent
13.13 **Heat of Combustion:** Not pertinent
13.14 **Heat of Decomposition:** Not pertinent
13.15 **Heat of Solution:** -52 Btu/lb = -29 cal/g = -1.2×10^5 J/kg
13.16 **Heat of Polymerization:** Not pertinent

NOTES

REVISED 1978

PPO — PHOSPHORUS OXYCHLORIDE

Common Synonyms	Oily liquid	Colorless to light yellow	Musty odor
	Fumes in air, sinks and reacts with water. Poisonous gas is produced. Freezing point is 34° F.		

Avoid contact with liquid and vapor. Keep people away.
Wear chemical protective suit with self-contained breathing apparatus.
Stop discharge if possible.
Isolate and remove discharged material.
Notify local health and pollution control agencies.

Fire

Not flammable.
Wear chemical protective suit with self-contained breathing apparatus.
DO NOT USE WATER ON ADJACENT FIRES.

Exposure

CALL FOR MEDICAL AID.

VAPOR
Irritating to eyes, nose, and throat.
Harmful if inhaled.
Move to fresh air.
If breathing has stopped, give artificial respiration.
If breathing is difficult, give oxygen.

LIQUID
Will burn skin and eyes.
Harmful if swallowed.
Remove to fresh air.
Flush affected areas with plenty of water.
IF IN EYES, hold eyelids open and flush with plenty of water.
IF SWALLOWED and victim is CONSCIOUS, have victim drink water or milk.
DO NOT INDUCE VOMITING.

Water Pollution

Dangerous to aquatic life in high concentrations.
May be dangerous if it enters water intakes.
Notify local health and wildlife officials.
Notify operators of nearby water intakes.

1. RESPONSE TO DISCHARGE

(See Response Methods Handbook, CG 446-4)

Issue warning—corrosive
Restrict access
Disperse and flush with care

2. LABEL

3. CHEMICAL DESIGNATIONS

3.1 **Synonyms:** Phosphoryl chloride
3.2 **Coast Guard Compatibility Classification:** Not applicable
3.3 **Chemical Formula:** $POCl_3$
3.4 **IMCO/United Nations Numerical Designation:** 8.0/1810

4. OBSERVABLE CHARACTERISTICS

4.1 **Physical State (as shipped):** Liquid
4.2 **Color:** Colorless to pale yellow
4.3 **Odor:** Pungent and musty; disagreeable and lingering

5. HEALTH HAZARDS

5.1 **Personal Protective Equipment:** Chemical safety goggles; face shield; self-contained or air-line respirator; hard hat; foot protection; rubber gloves and clothing.
5.2 **Symptoms Following Exposure:** Vapors burn eyes and respiratory tract. Liquid is very corrosive to body tissues because of reaction with water to form hydrochloric and phosphoric acids.
5.3 **Treatment for Exposure:** *CAUTION: persons doing treatment should protect themselves against exposure.* INHALATION: remove victim from contaminated area at once; if breathing has stopped, start artifical respiration; call a doctor. INGESTION: give water or milk; do NOT induce vomiting. SKIN: remove contaminated clothing and flood exposed skin surfaces with water. EYES: retract eyelids and wash with water for at least 15 min.; call a doctor.
5.4 **Toxicity by Inhalation (Threshold Limit Value):** 0.5 ppm
5.5 **Short-Term Inhalation Limits:** Data not available
5.6 **Toxicity by Ingestion:** Grade 3; oral rat LD_{50} = 380 mg/kg
5.7 **Late Toxicity:** Data not available
5.8 **Vapor (Gas) Irritant Characteristics:** Vapors cause severe irritation of eyes and throat and can cause eye and lung injury. They cannot be tolerated even at low concentrations.
5.9 **Liquid or Solid Irritant Characteristics:** Severe skin irritant. Causes second- and third-degree burns on short contact and is very injurious to the eyes.
5.10 **Odor Threshold:** Data not available

6. FIRE HAZARDS

6.1 **Flash Point:** Not flammable
6.2 **Flammable Limits in Air:** Not flammable
6.3 **Fire Extinguishing Agents:** Sand and carbon dioxide on adjacent fires
6.4 **Fire Extinguishing Agents Not to be Used:** Water
6.5 **Special Hazards of Combustion Products:** Not pertinent
6.6 **Behavior in Fire:** Poisonous, corrosive, irritating gases are generated when heated or when in contact with water.
6.7 **Ignition Temperature:** Not flammable
6.8 **Electrical Hazard:** Not pertinent
6.9 **Burning Rate:** Not flammable

7. CHEMICAL REACTIVITY

7.1 **Reactivity with Water:** Vigorous reaction with evolution of hydrogen chloride fumes.
7.2 **Reactivity with Common Materials:** Corrosive to most metals except nickel and lead. Products of its reaction with water rapidly corrode steel and most metals with formation of flammable hydrogen gas.
7.3 **Stability During Transport:** Stable
7.4 **Neutralizing Agents for Acids and Caustics:** Flush with water, neutralize acids formed with lime or soda ash.
7.5 **Polymerization:** Not pertinent
7.6 **Inhibitor of Polymerization:** Not pertinent

8. WATER POLLUTION

8.1 **Aquatic Toxicity:** Data not available
8.2 **Waterfowl Toxicity:** Data not available
8.3 **Biological Oxygen Demand (BOD):** None
8.4 **Food Chain Concentration Potential:** None

9. SELECTED MANUFACTURERS

1. FMC Corp.
   Organic Chemicals Division
   633 Third Ave.
   New York, N. Y. 10017
2. Monsanto Co.
   Monsanto Industrial Chemicals Co.
   800 North Lindbergh Blvd.
   St. Louis, Mo. 63166
3. Stauffer Chemical Co.
   Specialty Chemical Division
   Cold Creek, Ala. 36512

10. SHIPPING INFORMATION

10.1 **Grades or Purity:** 99-99.9%
10.2 **Storage Temperature:** Above 35°F
10.3 **Inert Atmosphere:** No requirement
10.4 **Venting:** Pressure-vacuum

11. HAZARD ASSESSMENT CODE

(See Hazard Assessment Handbook, CG 446-3)

A-O

12. HAZARD CLASSIFICATIONS

12.1 **Code of Federal Regulations:** Corrosive material
12.2 **NAS Hazard Rating for Bulk Water Transportation:**

Category	Rating
Fire	0
Health	
Vapor Irritant	4
Liquid or Solid Irritant	4
Poisons	4
Water Pollution	
Human Toxicity	4
Aquatic Toxicity	3
Aesthetic Effect	2
Reactivity	
Other Chemicals	4
Water	4
Self-Reaction	0

12.3 **NFPA Hazard Classifications:** Not listed

13. PHYSICAL AND CHEMICAL PROPERTIES

13.1 **Physical State at 15°C and 1 atm:** Liquid
13.2 **Molecular Weight:** 153.33
13.3 **Boiling Point at 1 atm:** 225 °F = 107°C = 380°K
13.4 **Freezing Point:** 34°F = 1°C = 274°K
13.5 **Critical Temperature:** 630°F = 332°C = 605°K
13.6 **Critical Pressure:** Not pertinent
13.7 **Specific Gravity:** 1.675 at 20°C (liquid)
13.8 **Liquid Surface Tension:** Not pertinent
13.9 **Liquid-Water Interfacial Tension:** Not pertinent
13.10 **Vapor (Gas) Specific Gravity:** Not pertinent
13.11 **Ratio of Specific Heats of Vapor (Gas):** (est.) 1.290
13.12 **Latent Heat of Vaporization:** 97 Btu/lb = 54 cal/g = 2.3×10^5 J/kg
13.13 **Heat of Combustion:** Not pertinent
13.14 **Heat of Decomposition:** Not pertinent
13.15 **Heat of Solution:** Not pertinent
13.16 **Heat of Polymerization:** Not pertinent

NOTES

REVISED 1978

PPP	# PHOSPHORUS PENTASULFIDE

Common Synonyms	Solid-flakes or powder — Yellow to green — Odorless or rotten egg odor Sinks and reacts with water. Poisonous gas is produced.

Avoid contact with solid and dust. Keep people away.
Wear goggles and self-contained breathing apparatus.
Stop discharge if possible.
Shut off ignition sources and call fire department.
Stay upwind. Evacuate area in case of large discharge.
Notify local health and pollution control agencies.

Fire	FLAMMABLE. MAY BE IGNITED BY SPARK, FRICTION, OR STATIC DISCHARGE. Containers may explode in fire. POISONOUS, IRRITATING GASES ARE PRODUCED IN FIRE. **Wear goggles and self-contained breathing apparatus.** **DO NOT USE WATER.** **Extinguish with carbon dioxide or sand.**
Exposure	**CALL FOR MEDICAL AID.** DUST Irritating to eyes, nose and throat. **Move to fresh air.** **If breathing has stopped, give artificial respiration.** **If breathing is difficult, give oxygen.** SOLID Irritating to skin and eyes. Harmful if swallowed. **Remove contaminated clothing and shoes.** **Flush affected areas with plenty of water.** **IF IN EYES, hold eyelids open and flush with plenty of water.** **IF SWALLOWED and victim is CONSCIOUS, have victim drink water or milk and have victim induce vomiting.** **IF SWALLOWED and victim is UNCONSCIOUS OR HAVING CONVULSIONS, do nothing except keep victim warm.**
Water Pollution	HARMFUL TO AQUATIC LIFE IN VERY LOW CONCENTRATIONS. May be dangerous if it enters water intakes. **Notify local health and wildlife officials.** **Notify operators of nearby water intakes.**

1. RESPONSE TO DISCHARGE
(See Response Methods Handbook, CG 446-4)

Issue warning—high flammability, poison
Restrict access
Evacuate area
Should be removed

2. LABELS

3. CHEMICAL DESIGNATIONS

3.1 **Synonyms:** Phosphoric sulfide; Phosphorus persulfide; Thiophosphoric anhydride
3.2 **Coast Guard Compatibility Classification:** Not applicable
3.3 **Chemical Formula:** P_2S_5-P_4S_{10}
3.4 **IMCO/United Nations Numerical Designation:** 4.1/1340

4. OBSERVABLE CHARACTERISTICS

4.1 **Physical State (as shipped):** Solid
4.2 **Color:** Light greenish yellow; greenish gray
4.3 **Odor:** Like rotten eggs. High (lethal) concentrations can paralyze the sense of smell.

5. HEALTH HAZARDS

5.1 **Personal Protective Equipment:** Chemical safety goggles; plastic face shield; self-contained or air-line respirator.
5.2 **Symptoms Following Exposure:** Hydrogen sulfide gas formed by reaction with moisture can cause death by respiratory failure. The gas also irritates eyes and respiratory system. The solid irritates skin and eyes; the symptoms may be delayed several hours.
5.3 **Treatment for Exposure:** INHALATION: remove victim from contaminated area; if breathing has stopped, begin artificial respiration. INGESTION: induce vomiting; call physician. SKIN: remove contaminated clothing and wash areas with large amounts of water. EYES: flush with large amounts of water.
5.4 **Toxicity by Inhalation (Threshold Limit Value):** 10 ppm (hydrogen sulfide).
5.5 **Short-Term Inhalation Limits:** 20 ppm for 5 min. (hydrogen sulfide)
5.6 **Toxicity by Ingestion:** Data not available
5.7 **Late Toxicity:** Data not available
5.8 **Vapor (Gas) Irritant Characteristics:** Hydrogen sulfide gas, formed by reaction with moisture, causes severe irritation of eyes and throat and can cause eye and lung injury. It cannot be tolerated even at low concentrations.
5.9 **Liquid or Solid Irritant Characteristics:** Minimum hazard. If spilled on clothing and allowed to remain, may cause smarting and reddening of the skin.
5.10 **Odor Threshold:** 0.0047 ppm (hydrogen sulfide). See also note in par. 4.3.

6. FIRE HAZARDS

6.1 **Flash Point:** Flammable solid
6.2 **Flammable Limits in Air:** Not pertinent
6.3 **Fire Extinguishing Agents:** Sand and carbon dioxide
6.4 **Fire Extinguishing Agents Not to be Used:** Water
6.5 **Special Hazards of Combustion Products:** Products of combustion include sulfur dioxide and phosphorus pentoxide, which are irritating, toxic, and corrosive.
6.6 **Behavior in Fire:** Not pertinent
6.7 **Ignition Temperature:** 527°F (liquid)
6.8 **Electrical Hazard:** Not pertinent
6.9 **Burning Rate:** Not pertinent

7. CHEMICAL REACTIVITY

7.1 **Reactivity with Water:** Reacts with liquid water or atmospheric moisture to liberate toxic hydrogen sulfide gas.
7.2 **Reactivity with Common Materials:** No reaction
7.3 **Stability During Transport:** Can be ignited by friction.
7.4 **Neutralizing Agents for Acids and Caustics:** Not pertinent
7.5 **Polymerization:** Not pertinent
7.6 **Inhibitor of Polymerization:** Not pertinent

8. WATER POLLUTION

8.1 **Aquatic Toxicity:** Data not available
8.2 **Waterfowl Toxicity:** Data not available
8.3 **Biological Oxygen Demand (BOD):** Data not available
8.4 **Food Chain Concentration Potential:** None

9. SELECTED MANUFACTURERS

1. Monsanto Co.
   Monsanto Industrial Chemicals Co.
   800 North Lindbergh Blvd.
   St. Louis, Mo. 63166
2. Occidental Petroleum Corp.
   Hooker Chemical Corp.
   Industrial Chemicals Div.
   Niagara Falls, N. Y. 14302
3. Stauffer Chemical Co.
   Specialty Chemical Division
   Morrisville, Pa. 19067

10. SHIPPING INFORMATION

10.1 **Grades or Purity:** "Regular" (low reactivity), "reactive" (high reactivity), distilled, undistilled: all 99+%
10.2 **Storage Temperature:** Ambient
10.3 **Inert Atmosphere:** No requirement
10.4 **Venting:** Sealed containers must be stored in a well-ventilated area

11. HAZARD ASSESSMENT CODE
(See Hazard Assessment Handbook, CG 446-3)

RR-C

12. HAZARD CLASSIFICATIONS

12.1 **Code of Federal Regulations:** Flammable solid
12.2 **NAS Hazard Rating for Bulk Water Transportation:** Not listed
12.3 **NFPA Hazard Classifications:**

Category	**Classification***	
Health Hazard (Blue)	2	3
Flammability (Red)	1	1
Reactivity (Yellow)	3	2
		~~W~~

*First column refers to non-fire situation

13. PHYSICAL AND CHEMICAL PROPERTIES

13.1 **Physical State at 15°C and 1 atm:** Solid
13.2 **Molecular Weight:** 222.27
13.3 **Boiling Point at 1 atm:** 957°F = 514°C = 787°K
13.4 **Freezing Point:** 527°F = 275°C = 548°K
13.5 **Critical Temperature:** Not pertinent
13.6 **Critical Pressure:** Not pertinent
13.7 **Specific Gravity:** 2.03 at 20°C (solid)
13.8 **Liquid Surface Tension:** Not pertinent
13.9 **Liquid-Water Interfacial Tension:** Not pertinent
13.10 **Vapor (Gas) Specific Gravity:** Not pertinent
13.11 **Ratio of Specific Heats of Vapor (Gas):** Not pertinent
13.12 **Latent Heat of Vaporization:** 184 Btu/lb = 102 cal/g = 4.27×10^5 J/kg
13.13 **Heat of Combustion:** −10,890 Btu/lb = −6,050 cal/g = -253.3×10^5 J/kg
13.14 **Heat of Decomposition:** Not pertinent
13.15 **Heat of Solution:** (est.) −20 Btu/lb = −12 cal/g = -0.5×10^5 J/kg
13.16 **Heat of Polymerization:** Not pertinent

NOTES

REVISED 1978

PPR — PHOSPHORUS, RED

Common Synonyms			
Amorphous phosphorus	Solid powder	Reddish brown	Odorless
	Sinks in water.		

AVOID CONTACT WITH SOLID. Keep people away.
Wear rubber overclothing (including gloves).
Stop discharge if possible.
Call fire department.
Isolate and remove discharged material.
Notify local health and pollution control agencies.

Fire

FLAMMABLE.
POISONOUS, IRRITATING AND FLAMMABLE GASES ARE PRODUCED IN FIRE.
Wear rubber overclothing (including gloves).
Flood discharge area with water.
Cool exposed containers with water.
Continue cooling after the fire has been extinguished.

Exposure

CALL FOR MEDICAL AID.

SOLID
Will burn eyes.
Harmful if swallowed.
Flush affected areas with plenty of water.
IF IN EYES, hold eyelids open and flush with plenty of water.
IF SWALLOWED and victim is CONSCIOUS, have victim drink water or milk.

Water Pollution

HARMFUL TO AQUATIC LIFE IN VERY LOW CONCENTRATIONS.
May be dangerous if it enters water intakes.
Notify local health and wildlife officials.
Notify operators of nearby water intakes.

1. RESPONSE TO DISCHARGE

(See Response Methods Handbook, CG 446-4)

Issue warning—high flammability
Should be removed
Chemical and physical treatment

2. LABEL

Red Stripe

3. CHEMICAL DESIGNATIONS

3.1 **Synonyms:** Amorphous phosphorus
3.2 **Coast Guard Compatibility Classification:** Phosphorus, elemental
3.3 **Chemical Formula:** P
3.4 **IMCO/United Nations Numerical Designation:** 4.1/1338

4. OBSERVABLE CHARACTERISTICS

4.1 **Physical State (as shipped):** Solid
4.2 **Color:** Reddish brown
4.3 **Odor:** Odorless

5. HEALTH HAZARDS

5.1 **Personal Protective Equipment:** Dust mask; gloves of rubber or vinyl; chemical safety glasses; rubber shoes.
5.2 **Symptoms Following Exposure:** Physically irritating to eyes, otherwise essentially harmless and non-toxic unless contaminated by highly toxic yellow phosphorus as an impurity.
5.3 **Treatment for Exposure:** EYES: flush thoroughly with water and get medical attention. SKIN: flush with water and wash with soap and water. *CAUTION: Avoid brushing, as friction may ignite material on skin or clothing.*
5.4 **Toxicity by Inhalation (Threshold Limit Value):** Not pertinent
5.5 **Short-Term Inhalation Limits:** Not pertinent
5.6 **Toxicity by Ingestion:** Data not available
5.7 **Late Toxicity:** Data not available
5.8 **Vapor (Gas) Irritant Characteristics:** Nonvolatile
5.9 **Liquid or Solid Irritant Characteristics:** Data not available
5.10 **Odor Threshold:** Not pertinent

6. FIRE HAZARDS

6.1 **Flash Point:** Flammable solid
6.2 **Flammable Limits in Air:** Not pertinent
6.3 **Fire Extinguishing Agents:** Water
6.4 **Fire Extinguishing Agents Not to be Used:** Not pertinent
6.5 **Special Hazards of Combustion Products:** Heat may cause reversion to yellow phosphorus which is toxic and spontaneously flammable upon contact with air. Burning yields toxic oxides of phosphorus.
6.6 **Behavior in Fire:** Refer to 6.5
6.7 **Ignition Temperature:** 395°F
6.8 **Electrical Hazard:** Not pertinent
6.9 **Burning Rate:** Not pertinent

7. CHEMICAL REACTIVITY

7.1 **Reactivity with Water:** No reaction
7.2 **Reactivity with Common Materials:** Avoid uncontrolled contact with oxidizing agents (chlorates, nitrates, halogens, etc.) or with strong alkaline hydroxides. Can react violently with oxidizing agent in presence of air and moisture, liberating phosphorus acids and toxic, spontaneously flammable phosphine gas.
7.3 **Stability During Transport:** Stable
7.4 **Neutralizing Agents for Acids and Caustics:** Not pertinent
7.5 **Polymerization:** Not pertinent
7.6 **Inhibitor of Polymerization:** Not pertinent

8. WATER POLLUTION

8.1 **Aquatic Toxicity:** 0.105 ppm/48 hr/bluegill/TL_m/fresh water
8.2 **Waterfowl Toxicity:** Data not available
8.3 **Biological Oxygen Demand (BOD):** Data not available
8.4 **Food Chain Concentration Potential:** None

9. SELECTED MANUFACTURERS

1. FMC Corp.
   Inorganic Chemicals Division
   633 Third Ave.
   New York, N. Y. 10017
2. Monsanto Co.
   Monsanto Industrial Chemicals Co.
   800 North Lindbergh Blvd.
   St. Louis, Mo. 63166
3. Stauffer Chemical Co.
   Industrial Chemical Division
   Mt. Pleasant, Tenn. 38474

10. SHIPPING INFORMATION

10.1 **Grades or Purity:** 99.9% Technical
10.2 **Storage Temperature:** Ambient
10.3 **Inert Atmosphere:** No requirement
10.4 **Venting:** Open

11. HAZARD ASSESSMENT CODE

(See Hazard Assessment Handbook, CG 446-3)

II

12. HAZARD CLASSIFICATIONS

12.1 **Code of Federal Regulations:** Flammable solid
12.2 **NAS Hazard Rating for Bulk Water Transportation:** Not listed
12.3 **NFPA Hazard Classifications:**

Category	Classification
Health Hazard (Blue)	0
Flammability (Red)	1
Reactivity (Yellow)	1

13. PHYSICAL AND CHEMICAL PROPERTIES

13.1 **Physical State at 15°C and 1 atm:** Solid
13.2 **Molecular Weight:** 123.89
13.3 **Boiling Point at 1 atm:** Catches fire
13.4 **Freezing Point:** Not pertinent
13.5 **Critical Temperature:** Not pertinent
13.6 **Critical Pressure:** Not pertinent
13.7 **Specific Gravity:** 2.20 at 20°C (solid)
13.8 **Liquid Surface Tension:** Not pertinent
13.9 **Liquid-Water Interfacial Tension:** Not pertinent
13.10 **Vapor (Gas) Specific Gravity:** Not pertinent
13.11 **Ratio of Specific Heats of Vapor (Gas):** Not pertinent
13.12 **Latent Heat of Vaporization:** Not pertinent
13.13 **Heat of Combustion:** Not pertinent
13.14 **Heat of Decomposition:** Not pertinent
13.15 **Heat of Solution:** Not pertinent
13.16 **Heat of Polymerization:** Not pertinent

NOTES

REVISED 1978

PBR	PHOSPHORUS TRIBROMIDE

Common Synonyms Phosphorus bromide	Liquid Colorless to pale yellow Sharp, penetrating odor Sinks and mixes violently with water.

Avoid contact with liquid. Keep people away.
Wear rubber overclothing (including gloves).
Stop discharge if possible.
Isolate and remove discharged material.
Notify local health and pollution control agencies.

Fire	Not flammable. Irritating gases may be produced when heated.
Exposure	CALL FOR MEDICAL AID. LIQUID Will burn skin and eyes. If swallowed will cause nausea. Remove contaminated clothing and shoes. Flush affected areas with plenty of water. IF IN EYES, hold eyelids open and flush with plenty of water. IF SWALLOWED and victim is CONSCIOUS, have victim drink water or milk. IF SWALLOWED and victim is UNCONSCIOUS OR HAVING CONVULSIONS, do nothing except keep victim warm.
Water Pollution	Effect of low concentrations on aquatic life is unknown. May be dangerous if it enters water intakes. Notify local health and wildlife officials. Notify operators of nearby water intakes.

1. RESPONSE TO DISCHARGE

(See Response Methods Handbook, CG 446-4)

Issue warning—corrosive, air contaminant, water contaminant
Restrict access
Disperse and flush

2. LABEL

3. CHEMICAL DESIGNATIONS

3.1 **Synonyms:** Phosphorus bromide
3.2 **Coast Guard Compatibility Classification:** Not listed
3.3 **Chemical Formula:** PBr_3
3.4 **IMCO/United Nations Numerical Designation:** 8/1808

4. OBSERVABLE CHARACTERISTICS

4.1 **Physical State (as shipped):** Liquid
4.2 **Color:** Colorless or slightly yellow
4.3 **Odor:** Pungent; sharp, penetrating

5. HEALTH HAZARDS

5.1 **Personal Protective Equipment:** Acid-gas canister-type mask (full face type for emergencies); chemical safety goggles; apron, gloves, clothing, and safety shoes all made from rubber
5.2 **Symptoms Following Exposure:** Inhalation causes severe irritation of nose, throat, and lungs. Ingestion causes burns of mouth and stomach. Contact with eyes or skin causes severe burns.
5.3 **Treatment for Exposure:** INHALATION: remove victim to clear air; if necessary, apply artificial respiration and/or administer oxygen. INGESTION: dilute by drinking water, then neutralize with milk of magnesia, egg white, etc.; do not use sodium bicarbonate. EYES: immediately flush with large amounts of water for at least 15 min. SKIN: immediately flush with large amounts of water; remove contaminated clothing.
5.4 **Toxicity by Inhalation (Threshold Limit Value):** Data not available
5.5 **Short-Term Inhalation Limits:** Data not available
5.6 **Toxicity by Ingestion:** Data not available
5.7 **Late Toxicity:** Data not available
5.8 **Vapor (Gas) Irritant Characteristics:** Data not available
5.9 **Liquid or Solid Irritant Characteristics:** Data not available
5.10 **Odor Threshold:** Data not available

6. FIRE HAZARDS

6.1 **Flash Point:** Not flammable
6.2 **Flammable Limits in Air:** Not flammable
6.3 **Fire Extinguishing Agents:** Not pertinent
6.4 **Fire Extinguishing Agents Not to be Used:** Do not use water on adjacent fires.
6.5 **Special Hazards of Combustion Products:** Irritating hydrogen bromide and phosphoric acid vapors may form in fire.
6.6 **Behavior in Fire:** Acids formed by reaction with water will attack metals and generate flammable hydrogen gas, which may form explosive mixtures in enclosed spaces.
6.7 **Ignition Temperature:** Not pertinent
6.8 **Electrical Hazard:** Not pertinent
6.9 **Burning Rate:** Not pertinent

7. CHEMICAL REACTIVITY

7.1 **Reactivity with Water:** Reacts violently with water, evolving hydrogen bromide, an irritating and corrosive gas apparent as white fumes
7.2 **Reactivity with Common Materials:** In the presence of moisture, highly corrosive to most metals except lead and nickel
7.3 **Stability During Transport:** Unstable if heated
7.4 **Neutralizing Agents for Acids and Caustics:** Flush with water and rinse with dilute aqueous sodium bicarbonate or soda ash.
7.5 **Polymerization:** Not pertinent
7.6 **Inhibitor of Polymerization:** Not pertinent

8. WATER POLLUTION

8.1 **Aquatic Toxicity:** Data not available
8.2 **Waterfowl Toxicity:** Data not available
8.3 **Biological Oxygen Demand (BOD):** Data not available
8.4 **Food Chain Concentration Potential:** None

9. SELECTED MANUFACTURERS

1. Michigan Chemical Corp.
   351 East Ohio St.
   Chicago, Ill. 60611
2. White Chemical Corp.
   East 22 Street
   Bayonne, N. J. 07002
3. Pfaltz and Bauer, Inc.
   375 Fairfield Ave.
   Stamford, Conn. 06902

10. SHIPPING INFORMATION

10.1 **Grades or Purity:** Purified, 88.55%
10.2 **Storage Temperature:** Ambient
10.3 **Inert Atmosphere:** No requirement
10.4 **Venting:** Open

11. HAZARD ASSESSMENT CODE

(See Hazard Assessment Handbook, CG 446-3)

A-O

12. HAZARD CLASSIFICATIONS

12.1 **Code of Federal Regulations:** Corrosive liquid
12.2 **NAS Hazard Rating for Bulk Water Transportation:** Not listed
12.3 **NFPA Hazard Classifications:** Not listed

13. PHYSICAL AND CHEMICAL PROPERTIES

13.1 **Physical State at 15°C and 1 atm:** Liquid
13.2 **Molecular Weight:** 270.73
13.3 **Boiling Point at 1 atm:** 343°F = 173°C = 446°K
13.4 **Freezing Point:** −42.9°F = −40.5°C = 232.7°K
13.5 **Critical Temperature:** Not pertinent
13.6 **Critical Pressure:** Not pertinent
13.7 **Specific Gravity:** 2.862 at 30°C (liquid)
13.8 **Liquid Surface Tension:** 45.8 dynes/cm = 0.0458 N/m at 24°C
13.9 **Liquid-Water Interfacial Tension:** Not pertinent
13.10 **Vapor (Gas) Specific Gravity:** Not pertinent
13.11 **Ratio of Specific Heats of Vapor (Gas):** Not pertinent
13.12 **Latent Heat of Vaporization:** 64.4 Btu/lb = 35.8 cal/g = 1.50×10^5 J/kg
13.13 **Heat of Combustion:** Not pertinent
13.14 **Heat of Decomposition:** Not pertinent
13.15 **Heat of Solution:** −446 Btu/lb = −248 cal/g = -10.4×10^5 J/kg
13.16 **Heat of Polymerization:** Not pertinent

NOTES

PPT	PHOSPHORUS TRICHLORIDE

Common Synonyms	Liquid	Colorless to slightly yellow	Sharp irritating odor
	Fumes in air, sinks and reacts with water. Harmful vapor is produced.		

AVOID CONTACT WITH LIQUID AND VAPOR. Keep people away.
Wear goggles, self-contained breathing apparatus, and rubber overclothing (including gloves).
Stop discharge if possible.
Isolate and remove discharged material.
Notify local health and pollution control agencies.

Fire

Not flammable.
Wear goggles, self-contained breathing apparatus, and rubber overclothing (including gloves).
DO NOT USE WATER ON ADJACENT FIRES.
Extinguish adjacent fires with carbon dioxide or dry chemical.

Exposure

CALL FOR MEDICAL AID.

VAPOR
Irritating to eyes, nose, and throat.
Harmful if inhaled.
Move to fresh air.
If breathing has stopped, give artificial respiration.
If breathing is difficult, give oxygen.

LIQUID
Will burn skin and eyes.
Poisonous if swallowed.
Remove contaminated clothing and shoes.
Flush affected areas with plenty of water.
IF IN EYES, hold eyelids open and flush with plenty of water.
IF SWALLOWED and victim is CONSCIOUS, have victim drink water or milk.
DO NOT INDUCE VOMITING.

Water Pollution

Dangerous to aquatic life in high concentrations.
May be dangerous if it enters water intakes.
Notify local health and wildlife officials.
Notify operators of nearby water intakes.

1. RESPONSE TO DISCHARGE
(See Response Methods Handbook, CG 446-4)

Issue warning—corrosive
Restrict access
Disperse and flush with care

2. LABEL

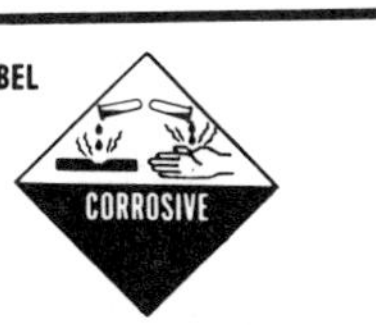

3. CHEMICAL DESIGNATIONS

3.1 **Synonyms:** No common synonyms
3.2 **Coast Guard Compatibility Classification:** Not applicable
3.3 **Chemical Formula:** PCl_3
3.4 **IMCO/United Nations Numerical Designation:** 8.0/1809

4. OBSERVABLE CHARACTERISTICS

4.1 **Physical State (as shipped):** Liquid
4.2 **Color:** Colorless to slightly yellow
4.3 **Odor:** Pungent, irritating, like hydrochloric acid

5. HEALTH HAZARDS

5.1 **Personal Protective Equipment:** Chemical safety goggles; plastic face shield; self-contained or air-line respirator; safety hat; rubber gloves and protective clothing.
5.2 **Symptoms Following Exposure:** Vapors cause severe irritation of eyes and respiratory tract. Liquid burns eyes and skin.
5.3 **Treatment for Exposure:** *CAUTION: Persons doing treatment should protect themselves.* INHALATION: remove victim from contaminated area; if breathing has stopped, start artificial respiration; call a doctor. INGESTION: if victim is conscious, give large quantities of water; do NOT induce vomiting; call a doctor. EYES: retract eyelids and wash eyes with water for at least 15 min.; call a doctor. SKIN: remove contaminated clothing and wash exposed skin with water.
5.4 **Toxicity by Inhalation (Threshold Limit Value):** 0.5 ppm
5.5 **Short-Term Inhalation Limits:** Data not available
5.6 **Toxicity by Ingestion:** Grade 2; oral rat LD_{50} = 550 mg/kg
5.7 **Late Toxicity:** None
5.8 **Vapor (Gas) Irritant Characteristics:** Vapors cause severe irritation of eyes and throat and can cause eye and lung injury. They cannot be tolerated even at low concentrations.
5.9 **Liquid or Solid Irritant Characteristics:** Severe skin irritant. Causes second- and third-degree burns on short contact and is very injurious to the eyes.
5.10 **Odor Threshold:** Data not available

6. FIRE HAZARDS

6.1 **Flash Point:** Not flammable
6.2 **Flammable Limits in Air:** Not flammable
6.3 **Fire Extinguishing Agents:** Sand, carbon dioxide and dry chemicals on adjacent fires
6.4 **Fire Extinguishing Agents Not to be Used:** Water
6.5 **Special Hazards of Combustion Products:** Not pertinent
6.6 **Behavior in Fire:** Generates toxic, irritating gases
6.7 **Ignition Temperature:** Not flammable
6.8 **Electrical Hazard:** Not pertinent
6.9 **Burning Rate:** Not flammable

7. CHEMICAL REACTIVITY

7.1 **Reactivity with Water:** Reacts violently and may cause flashes of fire. Hydrochloric acid fumes are formed in the reaction.
7.2 **Reactivity with Common Materials:** Corrodes most common construction materials. Reacts with water to form hydrochloric acid, which reacts with most metals to form flammable hydrogen gas.
7.3 **Stability During Transport:** Stable
7.4 **Neutralizing Agents for Acids and Caustics:** Flush with water; neutralize acids formed with lime or soda ash.
7.5 **Polymerization:** Not pertinent
7.6 **Inhibitor of Polymerization:** Not pertinent

8. WATER POLLUTION

8.1 **Aquatic Toxicity:** Data not available
8.2 **Waterfowl Toxicity:** Data not available
8.3 **Biological Oxygen Demand (BOD):** None
8.4 **Food Chain Concentration Potential:** None

9. SELECTED MANUFACTURERS

1. FMC Corp.
   Organic Chemicals Division
   633 Third Ave.
   New York, N. Y. 10012
2. Monsanto Co.
   Monsanto Industrial Chemicals Co.
   800 North Lindbergh Blvd.
   St. Louis, Mo. 63166
3. Stauffer Chemical Co.
   Specialty Chemical Division
   Cold Creek, Ala. 36512

10. SHIPPING INFORMATION

10.1 **Grades or Purity:** Pure: 99.5+%; technical: 98.5+%
10.2 **Storage Temperature:** Ambient
10.3 **Inert Atmosphere:** No requirement
10.4 **Venting:** Pressure-vacuum

11. HAZARD ASSESSMENT CODE
(See Hazard Assessment Handbook, CG 446-3)

A-O

12. HAZARD CLASSIFICATIONS

12.1 **Code of Federal Regulations:** Corrosive material
12.2 **NAS Hazard Rating for Bulk Water Transportation:** Not listed
12.3 **NFPA Hazard Classifications:**

Category	Classification
Health Hazard (Blue)	3
Flammability (Red)	0
Reactivity (Yellow)	2
	W

13. PHYSICAL AND CHEMICAL PROPERTIES

13.1 **Physical State at 15°C and 1 atm:** Liquid
13.2 **Molecular Weight:** 137.33
13.3 **Boiling Point at 1 atm:** 169°F = 76°C = 349°K
13.4 **Freezing Point:** −170°F = −112°C = 161°K
13.5 **Critical Temperature:** 547°F = 286°C = 559°K
13.6 **Critical Pressure:** Not pertinent
13.7 **Specific Gravity:** 1.575 at 20°C (liquid)
13.8 **Liquid Surface Tension:** 25.6 dynes/cm = 0.0256 N/m at 20°C
13.9 **Liquid-Water Interfacial Tension:** Not pertinent
13.10 **Vapor (Gas) Specific Gravity:** 4.7
13.11 **Ratio of Specific Heats of Vapor (Gas):** (est.) 1.290
13.12 **Latent Heat of Vaporization:** 95 Btu/lb = 53 cal/g = 2.2×10^5 J/kg
13.13 **Heat of Combustion:** Not pertinent
13.14 **Heat of Decomposition:** Not pertinent
13.15 **Heat of Solution:** Not pertinent
13.16 **Heat of Polymerization:** Not pertinent

NOTES

REVISED 1978

PPW	PHOSPHORUS WHITE

Common Synonyms			
Yellow phosphorus	Waxy solid	Light yellow	Garlic odor
	Fumes and burns in air, sinks in water.		

AVOID CONTACT WITH SOLID. Keep people away.
Wear rubber overclothing (including gloves).
Call fire department.
Isolate and remove discharged material.
Notify local health and pollution control agencies.

Fire	FLAMMABLE. May ignite on contact with air. POISONOUS, IRRITATING GASES ARE PRODUCED IN FIRE. **Wear rubber overclothing (including gloves).** **Flood discharge area with water.** **Cool exposed containers with water.** **Continue cooling after fire has been extinguished.**
Exposure	**CALL FOR MEDICAL AID.** SOLID Will burn skin and eyes. If swallowed, will cause nausea, vomiting or loss of consciousness. **Flush affected areas with plenty of water.** **IF IN EYES, hold eyelids open and flush with plenty of water.** **IF SWALLOWED and victim is CONSCIOUS WITH NO CONVULSIONS, have victim drink water or milk.** **DO NOT INDUCE VOMITING.**
Water Pollution	HARMFUL TO AQUATIC LIFE IN VERY LOW CONCENTRATIONS. May be dangerous if it enters water intakes. **Notify local health and wildlife officials.** **Notify operators of nearby water intakes.**

1. RESPONSE TO DISCHARGE
(See Response Methods Handbook, CG 446-4)

Issue warning—high flammability, poison
Restrict access
Evacuate area
Should be removed
Chemical and physical treatment

2. LABELS

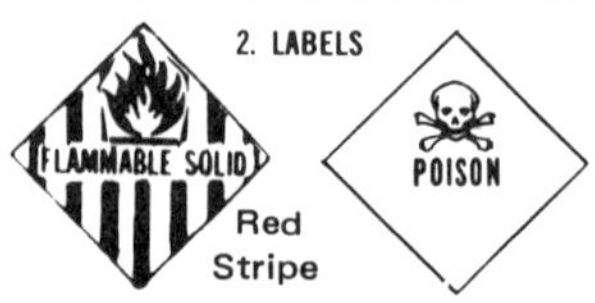

3. CHEMICAL DESIGNATIONS

3.1 **Synonyms:** Yellow phosphorus
3.2 **Coast Guard Compatibility Classification:** Phosphorus, elemental
3.3 **Chemical Formula:** P
3.4 **IMCO/United Nations Numerical Designation:** 4.2/1381

4. OBSERVABLE CHARACTERISTICS

4.1 **Physical State (as shipped):** Liquid, or waxy solid
4.2 **Color:** Pale yellow to deep straw
4.3 **Odor:** Distinctive, disagreeable; pungent, sharp; like garlic

5. HEALTH HAZARDS

5.1 **Personal Protective Equipment:** Heavy rubber gloves and goggles or face shield.
5.2 **Symptoms Following Exposure:** Solid or liquid causes severe burns of skin. If ingested, causes nausea, vomiting, jaundice, low blood pressure, depression, delirium, coma, death. Symptoms after ingestion may be delayed for from a few hours to 3 days.
5.3 **Treatment for Exposure:** INGESTION: if ingested, do NOT induce vomiting; call a doctor at once. SKIN OR EYE CONTACT: immediately flush with plenty of water for at least 15 min.; keep skin area wet until medical attention is obtained.
5.4 **Toxicity by Inhalation (Threshold Limit Value):** 0.1 mg/m^3
5.5 **Short-Term Inhalation Limits:** Not pertinent
5.6 **Toxicity by Ingestion:** Grade 4; LD_{50} below 50 mg/kg
5.7 **Late Toxicity:** Severe attack of liver and bones.
5.8 **Vapor (Gas) Irritant Characteristics:** Nonvolatile
5.9 **Liquid or Solid Irritant Characteristics:** Severe skin irritant. Causes second- and third-degree burns on short contact, and is very injurious to the eyes.
5.10 **Odor Threshold:** Not pertinent

6. FIRE HAZARDS

6.1 **Flash Point:** Ignites spontaneously in air.
6.2 **Flammable Limits in Air:** Not pertinent
6.3 **Fire Extinguishing Agents:** Water
6.4 **Fire Extinguishing Agents Not to be Used:** Not pertinent
6.5 **Special Hazards of Combustion Products:** Fumes from burning phosphorus are highly irritating.
6.6 **Behavior in Fire:** Intense white smoke is formed.
6.7 **Ignition Temperature:** 86°F
6.8 **Electrical Hazard:** Not pertinent
6.9 **Burning Rate:** Not pertinent

7. CHEMICAL REACTIVITY

7.1 **Reactivity with Water:** No reaction
7.2 **Reactivity with Common Materials:** Ignites when exposed to air.
7.3 **Stability During Transport:** Stable
7.4 **Neutralizing Agents for Acids and Caustics:** Not pertinent
7.5 **Polymerization:** Not pertinent
7.6 **Inhibitor of Polymerization:** Not pertinent

8. WATER POLLUTION

8.1 **Aquatic Toxicity:** 0.105 ppm/48 hr/bluegill/TL_m/fresh water
8.2 **Waterfowl Toxicity:** Data not available
8.3 **Biological Oxygen Demand (BOD):** Data not available
8.4 **Food Chain Concentration Potential:** None

9. SELECTED MANUFACTURERS

1. FMC Corp.
   Inorganic Chemicals Division
   633 Third Ave.
   New York, N. Y. 10017
2. Monsanto Co.
   Monsanto Industrial Chemicals Co.
   800 North Lindbergh Blvd.
   St. Louis, Mo. 63166
3. Stauffer Chemical Co.
   Industrial Chemical Division
   Mt. Pleasant, Tenn. 38474

10. SHIPPING INFORMATION

10.1 **Grades or Purity:** 99.8-99.9%
10.2 **Storage Temperature:** Elevated (if liquid)
10.3 **Inert Atmosphere:** Padded
10.4 **Venting:** Pressure-vacuum

11. HAZARD ASSESSMENT CODE
(See Hazard Assessment Handbook, CG 446-3)

II-RR

12. HAZARD CLASSIFICATIONS

12.1 **Code of Federal Regulations:** Flammable solid
12.2 **NAS Hazard Rating for Bulk Water Transportation:**

Category	Rating
Fire	3
Health	
Vapor Irritant	NA
Liquid or Solid Irritant	4
Poisons	4
Water Pollution	
Human Toxicity	3
Aquatic Toxicity	4
Aesthetic Effect	1
Reactivity	
Other Chemicals	4
Water	0
Self-Reaction	0

12.3 **NFPA Hazard Classifications:**

Category	Classification
Health Hazard (Blue)	3
Flammability (Red)	3
Reactivity (Yellow)	1

13. PHYSICAL AND CHEMICAL PROPERTIES

13.1 **Physical State at 15°C and 1 atm:** Solid
13.2 **Molecular Weight:** 123.89
13.3 **Boiling Point at 1 atm:** 535.5°F = 279.7°C = 552.9°K
13.4 **Freezing Point:** 111.4°F = 44.1°C = 317.3°K
13.5 **Critical Temperature:** Not pertinent
13.6 **Critical Pressure:** Not pertinent
13.7 **Specific Gravity:** 1.82 at 20°C (solid)
13.8 **Liquid Surface Tension:** Not pertinent
13.9 **Liquid-Water Interfacial Tension:** Not pertinent
13.10 **Vapor (Gas) Specific Gravity:** Not pertinent
13.11 **Ratio of Specific Heats of Vapor (Gas):** Not pertinent
13.12 **Latent Heat of Vaporization:** Not pertinent
13.13 **Heat of Combustion:** Not pertinent
13.14 **Heat of Decomposition:** Not pertinent
13.15 **Heat of Solution:** Not pertinent
13.16 **Heat of Polymerization:** Not pertinent

NOTES

REVISED 1978

PAN	# PHTHALIC ANHYDRIDE

Common Synonyms			
Phthalic acid anhydride PAN	Solid flakes or liquid (heated)	Colorless or pale yellow	Choking odor
	Solid sinks in water, liquid solidifies and sinks in water.		

Stop discharge if possible. Keep people away.
Avoid contact with liquid.
Wear rubber overclothing (including gloves).
Call fire department.
Isolate and remove discharged material.
Notify local health and pollution control agencies.

Fire

Combustible.
Extinguish with water, dry chemical, or carbon dioxide.

Exposure

CALL FOR MEDICAL AID.

DUST
Irritating to eyes, nose, and throat.
If inhaled, with cause coughing.
Move to fresh air.
If breathing has stopped, give artificial respiration.
If breathing is difficult, give oxygen.

LIQUID OR SOLID
Will burn skin or eyes.
Harmful if swallowed.
Remove contaminated clothing and shoes.
Flush affected areas with plenty of water.
IF IN EYES, hold eyelids open and flush with plenty of water.
IF SWALLOWED and victim is CONSCIOUS, have victim drink water or milk and have victim induce vomiting.
IF SWALLOWED and victim is UNCONSCIOUS OR HAVING CONVULSIONS, do nothing except keep victim warm.

Water Pollution

Effect of low concentrations on aquatic life is unknown.
May be dangerous if it enters water intakes.
Notify local health and wildlife officials.
Notify operators of nearby water intakes.

1. RESPONSE TO DISCHARGE

(See Response Methods Handbook, CG 446-4)

Disperse and flush

2. LABELS

No hazard label required by Code of Federal Regulations

3. CHEMICAL DESIGNATIONS

3.1 **Synonyms:** 1,2-Benzenedicarboxylic acid anhydride; 1,3-Dioxophthalan; PAN; Phthalandione; Phthalic acid anhydride

3.2 **Coast Guard Compatibility Classification:** Organic anhydride

3.3 **Chemical Formula:** $C_6H_4(CO)_2O$

3.4 **IMCO/United Nations Numerical Designation:** Not listed

4. OBSERVABLE CHARACTERISTICS

4.1 **Physical State (as shipped):** Solid or liquid

4.2 **Color:** Colorless or pale yellow

4.3 **Odor:** Characteristic choking odor; choking, acrid

5. HEALTH HAZARDS

5.1 **Personal Protective Equipment:** Coveralls and/or rubber apron; rubber shoes or boots; chemical goggles and/or face shield; Bureau of Mines organic vapor respirator (Type AB); gauntlet-type leather or rubber gloves.

5.2 **Symptoms Following Exposure:** Solid irritates skin and eyes, causing coughing and sneezing. Liquid causes severe thermal burns.

5.3 **Treatment for Exposure:** INHALATION: gargle with water and use a sedative cough mixture. INGESTION: induce vomiting and give water, milk, or vegetable oil. SKIN OR EYE CONTACT: Flush with water for at least 15 min.; if burned by molten material, remove as much solid as possible, soak off the remainder in cold water, and then treat the burn.

5.4 **Toxicity by Inhalation (Threshold Limit Value):** 2 ppm

5.5 **Short-Term Inhalation Limits:** 4 ppm for 5 min.

5.6 **Toxicity by Ingestion:** Grade 2; LD_{50} 0.5 to 5 g/kg (rat)

5.7 **Late Toxicity:** Data not available

5.8 **Vapor (Gas) Irritant Characteristics:** Vapor is moderately irritating such that personnel will not usually tolerate moderate or high vapor concentrations.

5.9 **Liquid or Solid Irritant Characteristics:** Causes smarting of the skin and first-degree burns on short exposure; may cause secondary burns on long exposure.

5.10 **Odor Threshold:** 0.32 - 0.72 mg/m^3

6. FIRE HAZARDS

6.1 **Flash Point:** 329°F O.C.; 305°F C.C.

6.2 **Flammable Limits in Air:** 1.7% – 10.5%

6.3 **Fire Extinguishing Agents:** Water fog, dry chemical, carbon dioxide, or foam

6.4 **Fire Extinguishing Agents Not to be Used:** Water may cause frothing.

6.5 **Special Hazards of Combustion Products:** Not pertinent

6.6 **Behavior in Fire:** Not pertinent

6.7 **Ignition Temperature:** 1058°F

6.8 **Electrical Hazard:** Not pertinent

6.9 **Burning Rate:** Data not available

7. CHEMICAL REACTIVITY

7.1 **Reactivity with Water:** Solid has very slow reaction; no hazard. Liquid spatters when in contact with water.

7.2 **Reactivity with Common Materials:** No reaction

7.3 **Stability During Transport:** Stable

7.4 **Neutralizing Agents for Acids and Caustics:** Water and sodium bicarbonate

7.5 **Polymerization:** Not pertinent

7.6 **Inhibitor of Polymerization:** Not pertinent

8. WATER POLLUTION

8.1 **Aquatic Toxicity:** > 56 ppm/96 hr/fathead minnow/TL_m/fresh water

8.2 **Waterfowl Toxicity:** Data not available

8.3 **Biological Oxygen Demand (BOD):** 106%, 5 days

8.4 **Food Chain Concentration Potential:** None

9. SELECTED MANUFACTURERS

1. Allied Chemical Corp.
Plastics Division
Morristown, N. J. 07960

2. Koppers Co., Inc.
Organic Materials Division
Koppers Bldg.
Pittsburgh, Pa. 15219

3. Monsanto Co.
Monsanto Industrial Chemicals Co.
800 North Lindbergh Blvd.
St. Louis, Mo. 63166

10. SHIPPING INFORMATION

10.1 **Grades or Purity:** Flake; molten; commercial: 99.8%

10.2 **Storage Temperature:** 268-320°F (liquid); Ambient (solid)

10.3 **Inert Atmosphere:** No requirement

10.4 **Venting:** Open (flame arrester)

11. HAZARD ASSESSMENT CODE

(See Hazard Assessment Handbook, CG 446-3)

II or A-X-Y

12. HAZARD CLASSIFICATIONS

12.1 **Code of Federal Regulations:** Not listed

12.2 **NAS Hazard Rating for Bulk Water Transportation:**

Category	Rating
Fire	1
Health	
Vapor Irritant	2
Liquid or Solid Irritant	3
Poisons	1
Water Pollution	
Human Toxicity	2
Aquatic Toxicity	2
Aesthetic Effect	2
Reactivity	
Other Chemicals	3
Water	1
Self-Reaction	0

12.3 **NFPA Hazard Classifications:**

Category	Classification
Health Hazard (Blue)	2
Flammability (Red)	1
Reactivity (Yellow)	0

13. PHYSICAL AND CHEMICAL PROPERTIES

13.1 **Physical State at 15°C and 1 atm:** Solid

13.2 **Molecular Weight:** 148.12

13.3 **Boiling Point at 1 atm:** 544.3°F = 284.6°C = 557.8°K

13.4 **Freezing Point:** 268°F = 131°C = 404°K

13.5 **Critical Temperature:** Not pertinent

13.6 **Critical Pressure:** Not pertinent

13.7 **Specific Gravity:** 1.20 at 135°C (liquid) 1.53 at 20°C (solid)

13.8 **Liquid Surface Tension:** 35.5 dynes/cm = 0.0355 N/m at 155°C

13.9 **Liquid-Water Interfacial Tension:** (est.) 30 dynes/cm = 0.03 N/m at 155°C

13.10 **Vapor (Gas) Specific Gravity:** Not pertinent

13.11 **Ratio of Specific Heats of Vapor (Gas):** 1.080

13.12 **Latent Heat of Vaporization:** 189 Btu/lb = 105 cal/g = 4.40×10^5 J/kg

13.13 **Heat of Combustion:** −9473 Btu/lb = −5263 cal/g = -220.4×10^5 J/kg

13.14 **Heat of Decomposition:** Not pertinent

13.15 **Heat of Solution:** −127 Btu/lb = −70.8 cal/g = -2.96×10^5 J/kg

13.16 **Heat of Polymerization:** Not pertinent

NOTES

REVISED 1978

PPZ — PIPERAZINE

Common Synonyms			
Hexahydro-1, 4-diazine Hexahydropyrazine Diethylenediamine Lumbrical Piperazidine Pyrazine hexahydride	Solid	White	Mild, fishy odor
	Sinks and mixes with water.		

Avoid contact with solid and dust. Keep people away.
Stop discharge if possible.
Isolate and remove discharged material.
Notify local health and pollution control agencies.

Fire	Combustible. POISONOUS GASES MAY BE PRODUCED IN FIRE. Wear goggles and self-contained breathing apparatus. Extinguish with dry chemicals, alcohol foam, or carbon dioxide. Water may be ineffective on fire. Cool exposed containers with water.
Exposure	CALL FOR MEDICAL AID. DUST Irritating to eyes, nose and throat. If inhaled will cause coughing or difficult breathing. If in eyes, hold eyelids open and flush with plenty of water. If breathing has stopped, give artificial respiration. If breathing is difficult, give oxygen. SOLID Will burn eyes. Irritating to eyes. If swallowed will cause nausea and vomiting. Remove contaminated clothing and shoes. Flush affected areas with plenty of water. IF IN EYES, hold eyelids open and flush with plenty of water. IF SWALLOWED and victim is CONSCIOUS, have victim drink water or milk and have victim induce vomiting. IF SWALLOWED and victim is UNCONSCIOUS OR HAVING CONVULSIONS, do nothing except keep victim warm.
Water Pollution	Effect of low concentrations on aquatic life is unknown. May be dangerous if it enters water intakes. Notify local health and wildlife officials. Notify operators of nearby water intakes.

1. RESPONSE TO DISCHARGE

(See Response Methods Handbook, CG 446-4)

Issue warning—water contaminant
Disperse and flush

2. LABELS

No hazard label required by Code of Federal Regulations

3. CHEMICAL DESIGNATIONS

3.1 **Synonyms:** Diethylenediamine; Hexahydro-1,4-diazine; Hexahydropyrazine; Lumbrical; Piperazidine; Pyrazine hexahydride

3.2 **Coast Guard Compatibility Classification:** Not listed

3.3 **Chemical Formula:** $NHCH_2CH_2NHCH_2CH_2$

3.4 **IMCO/United Nations Numerical Designation:** Not listed

4. OBSERVABLE CHARACTERISTICS

4.1 **Physical State (as shipped):** Solid

4.2 **Color:** White

4.3 **Odor:** Mild, amine-like

5. HEALTH HAZARDS

5.1 **Personal Protective Equipment:** Monogoggles or face shield; rubber gloves; dust mask

5.2 **Symptoms Following Exposure:** Inhalation of dust irritates nose and throat. Ingestion causes irritation of mouth and stomach; has been known to cause severe allergic reaction. Contact with eyes causes burns. Repeated contact with skin may cause irritation and sensitization.

5.3 **Treatment for Exposure:** INHALATION: move to fresh air. INGESTION: give large amount of water; induce vomiting; get medical attention. EYES: flush with plenty of water for at least 15 min.; get medical attention. SKIN: wash with soap and water.

5.4 **Toxicity by Inhalation (Threshold Limit Value):** Data not available

5.5 **Short-Term Inhalation Limits:** Data not available

5.6 **Toxicity by Ingestion:** Grade 2; LD_{50} 0.5—5 g/kg

5.7 **Late Toxicity:** Data not available

5.8 **Vapor (Gas) Irritant Characteristics:** Data not available

5.9 **Liquid or Solid Irritant Characteristics:** Data not available

5.10 **Odor Threshold:** Data not available

6. FIRE HAZARDS

6.1 **Flash Point:** 225°F O.C. (molten solid)

6.2 **Flammable Limits in Air:** Not pertinent

6.3 **Fire Extinguishing Agents:** Water, dry chemical, "alcohol" foam, carbon dioxide

6.4 **Fire Extinguishing Agents Not to be Used:** Water may cause frothing.

6.5 **Special Hazards of Combustion Products:** Toxic oxides of nitrogen may form in fires.

6.6 **Behavior in Fire:**

6.7 **Ignition Temperature:** 851°F

6.8 **Electrical Hazard:** Not pertinent

6.9 **Burning Rate:** Not pertinent

7. CHEMICAL REACTIVITY

7.1 **Reactivity with Water:** No reaction

7.2 **Reactivity with Common Materials:** May be corrosive to aluminum, magnesium, and zinc

7.3 **Stability During Transport:** Stable

7.4 **Neutralizing Agents for Acids and Caustics:** Flush with water.

7.5 **Polymerization:** Not pertinent

7.6 **Inhibitor of Polymerization:** Not pertinent

8. WATER POLLUTION

8.1 **Aquatic Toxicity:** Data not available

8.2 **Waterfowl Toxicity:** Data not available

8.3 **Biological Oxygen Demand (BOD):** Data not available

8.4 **Food Chain Concentration Potential:** None

9. SELECTED MANUFACTURERS

1. Jefferson Chemical Co., Inc.
   3336 Richmond Avenue
   P. O. Box 53300
   Houston, Texas 77052
2. Union Carbide Corp.
   Chemicals and Plastics Div.
   270 Park Avenue
   New York, N. Y. 10017
3. Eastman Organic Chemicals
   Rochester, N. Y. 14650

10. SHIPPING INFORMATION

10.1 **Grades or Purity:** Commercial, 99+%; may also be shipped as a solid hexahydrate, whose hazardous properties are similar.

10.2 **Storage Temperature:** Ambient

10.3 **Inert Atmosphere:** No requirement

10.4 **Venting:** Open (flame arrester)

11. HAZARD ASSESSMENT CODE

(See Hazard Assessment Handbook, CG 446-3)

A-P-Q

12. HAZARD CLASSIFICATIONS

12.1 **Code of Federal Regulations:** Not listed

12.2 **NAS Hazard Rating for Bulk Water Transportation:** Not listed

12.3 **NFPA Hazard Classifications:**

Category	Classification
Health Hazard (Blue)	2
Flammability (Red)	2
Reactivity (Yellow)	0

13. PHYSICAL AND CHEMICAL PROPERTIES

13.1 **Physical State at 15°C and 1 atm:** Solid

13.2 **Molecular Weight:** 86

13.3 **Boiling Point at 1 atm:** 299°F = 148°C = 421°K

13.4 **Freezing Point:** 223°F = 106°C = 379°K

13.5 **Critical Temperature:** Not pertinent

13.6 **Critical Pressure:** Not pertinent

13.7 **Specific Gravity:** 1.1 at 20°C (solid)

13.8 **Liquid Surface Tension:** Not pertinent

13.9 **Liquid-Water Interfacial Tension:** Not pertinent

13.10 **Vapor (Gas) Specific Gravity:** Not pertinent

13.11 **Ratio of Specific Heats of Vapor (Gas):** Not pertinent

13.12 **Latent Heat of Vaporization:** Not pertinent

13.13 **Heat of Combustion:** $-14,800$ Btu/lb $= -8,200$ cal/g $= -343 \times 10^5$ J/kg

13.14 **Heat of Decomposition:** Not pertinent

13.15 **Heat of Solution:** -34.9 Btu/lb $= -19.4$ cal/g $= -0.812 \times 10^5$ J/kg

13.16 **Heat of Polymerization:** Not pertinent

NOTES

PLB — POLYBUTENE

Common Synonyms	Oily liquid	Colorless	Odorless
Butene resins Polyisobutylene waxes	Floats on water.		

Stop discharge if possible. Keep people away.
Call fire department.
Isolate and remove discharged material.
Notify local health and pollution control agencies.

Fire	Combustible. Extinguish with dry chemical, foam, or carbon dioxide. Water may be ineffective on fire. Cool exposed containers with water.
Exposure	Not harmful.
Water Pollution	Effect of low concentrations on aquatic life is unknown. Fouling to shoreline. May be dangerous if it enters water intakes. Notify local health and wildlife officials. Notify operators of nearby water intakes.

1. RESPONSE TO DISCHARGE
(See Response Methods Handbook, CG 446-4)

Mechanical containment
Should be removed
Chemical and physical treatment

2. LABELS
No hazard label required by Code of Federal Regulations

3. CHEMICAL DESIGNATIONS
3.1 **Synonyms:** Butene resins
Polyisobutylene plastics
Polyisobutylene resins
Polyisobutylene waxes
3.2 **Coast Guard Compatibility Classification:** Olefin
3.3 **Chemical Formula:** $-(C(CH_3)_2CH_2)-$
3.4 **IMCO/United Nations Numerical Designation:** Not listed

4. OBSERVABLE CHARACTERISTICS
4.1 **Physical State (as shipped):** Liquid
4.2 **Color:** Data not available
4.3 **Odor:** Data not available

5. HEALTH HAZARDS
5.1 **Personal Protective Equipment:** Goggles or face shield.
5.2 **Symptoms Following Exposure:** Low toxicity. Vapor may act as a simple asphyxiant in high concentration.
5.3 **Treatment for Exposure:** INHALATION: remove victim from exposure.
5.4 **Toxicity by Inhalation (Threshold Limit Value):** Data not available
5.5 **Short-Term Inhalation Limits:** Data not available
5.6 **Toxicity by Ingestion:** Grade 0; LD_{50} above 15 g/kg (animals)
5.7 **Late Toxicity:** None
5.8 **Vapor (Gas) Irritant Characteristics:** Vapors are nonirritating to the eyes and throat.
5.9 **Liquid or Solid Irritant Characteristics:** No appreciable hazard. Practically harmless to the skin.
5.10 **Odor Threshold:** Data not available

6. FIRE HAZARDS
6.1 **Flash Point:** 215–470°F O.C.
6.2 **Flammable Limits in Air:** Data not available
6.3 **Fire Extinguishing Agents:** Carbon dioxide, dry chemical, or foam
6.4 **Fire Extinguishing Agents Not to be Used:** Water may be ineffective.
6.5 **Special Hazards of Combustion Products:** Not pertinent
6.6 **Behavior in Fire:** Not pertinent
6.7 **Ignition Temperature:** Data not available
6.8 **Electrical Hazard:** Not pertinent
6.9 **Burning Rate:** Data not available

7. CHEMICAL REACTIVITY
7.1 **Reactivity with Water:** No reaction
7.2 **Reactivity with Common Materials:** No reaction
7.3 **Stability During Transport:** Stable
7.4 **Neutralizing Agents for Acids and Caustics:** Not pertinent
7.5 **Polymerization:** Not pertinent
7.6 **Inhibitor of Polymerization:** Not pertinent

8. WATER POLLUTION
8.1 **Aquatic Toxicity:** Data not available
8.2 **Waterfowl Toxicity:** Data not available
8.3 **Biological Oxygen Demand (BOD):** Data not available
8.4 **Food Chain Concentration Potential:** None

9. SELECTED MANUFACTURERS
1. Amoco Chemicals Corp.
130 East Randolph Drive
Chicago, Ill. 60601
2. Chevron Chemical Co.
Industrial Chemicals Division
200 Bush St.
San Francisco, Calif. 94120
3. Exxon Chemical Co.
Houston, Tex. 77001

10. SHIPPING INFORMATION
10.1 **Grades or Purity:** 85%–98%
10.2 **Storage Temperature:** Ambient
10.3 **Inert Atmosphere:** No requirement
10.4 **Venting:** Open (flame arrester)

11. HAZARD ASSESSMENT CODE
(See Hazard Assessment Handbook, CG 446-3)

A-T-U

12. HAZARD CLASSIFICATIONS
12.1 **Code of Federal Regulations:** Not listed
12.2 **NAS Hazard Rating for Bulk Water Transportation:** Not listed
12.3 **NFPA Hazard Classifications:**

Category	Classification
Health Hazard (Blue)	0
Flammability (Red)	1
Reactivity (Yellow)	0

13. PHYSICAL AND CHEMICAL PROPERTIES
13.1 **Physical State at 15°C and 1 atm:** Liquid
13.2 **Molecular Weight:** 225–2300
13.3 **Boiling Point at 1 atm:** Very high
13.4 **Freezing Point:** Not pertinent
13.5 **Critical Temperature:** Not pertinent
13.6 **Critical Pressure:** Not pertinent
13.7 **Specific Gravity:** 0.81–0.91 at 15°C (liquid)
13.8 **Liquid Surface Tension:** (est.) 25 dynes/cm = 0.025 N/m at 20°C
13.9 **Liquid-Water Interfacial Tension:** (est.) 50 dynes/cm = 0.05 N/m at 20°C
13.10 **Vapor (Gas) Specific Gravity:** Not pertinent
13.11 **Ratio of Specific Heats of Vapor (Gas):** Not pertinent
13.12 **Latent Heat of Vaporization:** Not pertinent
13.13 **Heat of Combustion:** (est.) –20,000 Btu/lb = –11,000 cal/g = -470×10^5 J/kg
13.14 **Heat of Decomposition:** Not pertinent
13.15 **Heat of Solution:** (est.) –9 Btu/lb = –5 cal/g = -0.2×10^5 J/kg
13.16 **Heat of Polymerization:** Not pertinent

NOTES

REVISED 1978

PCB — POLYCHLORINATED BIPHENYL

Common Synonyms			
PCB Chlorinated biphenyl Arochlor	Oily liquid to solid powder	Light yellow liquid, or white powder	Weak odor
	Sinks in water.		

Stop discharge if possible. Keep people away.
Avoid contact with liquid and solid.
Call fire department.
Isolate and remove discharged material.
Notify local health and pollution control agencies.

Fire

Combustible.
Extinguish with water, foam, dry chemical, or carbon dioxide.

Exposure

CALL FOR MEDICAL AID.

LIQUID OR SOLID
Irritating to skin and eyes.
Flush affected areas with plenty of water.
IF IN EYES, hold eyelids open and flush with plenty of water.

Water Pollution

HARMFUL TO AQUATIC LIFE IN VERY LOW CONCENTRATIONS.
May be dangerous if it enters water intakes.
Notify local health and wildlife officials.
Notify operators of nearby water intakes.

1. RESPONSE TO DISCHARGE

(See Response Methods Handbook, CG 446-4)

Issue warning—water contaminant
Should be removed
Chemical and physical treatment

2. LABELS

No hazard label required by Code of Federal Regulations

3. CHEMICAL DESIGNATIONS

3.1 **Synonyms:** Aroclor; Chlorinated biphenyl; Halogenated waxes; PCB; Polychloropolyphenyls
3.2 **Coast Guard Compatibility Classification:** Not applicable
3.3 **Chemical Formula:** $(C_{12}H_{10-x})Cl_x$
3.4 **IMCO/United Nations Numerical Designation:** Not listed

4. OBSERVABLE CHARACTERISTICS

4.1 **Physical State (as shipped):** Liquid or solid
4.2 **Color:** Pale yellow (liquid); colorless (solid)
4.3 **Odor:** Practically odorless

5. HEALTH HAZARDS

5.1 **Personal Protective Equipment:** Gloves and protective garments.
5.2 **Symptoms Following Exposure:** Acne from skin contact.
5.3 **Treatment for Exposure:** SKIN: wash with soap and water.
5.4 **Toxicity by Inhalation (Threshold Limit Value):** 0.5 to 1.0 mg/m^3
5.5 **Short-Term Inhalation Limits:** Data not available
5.6 **Toxicity by Ingestion:** Grade 2; oral rat LD_{50} = 3980 mg/kg
5.7 **Late Toxicity:** Causes chromosomal abnormalities in rats, birth defects in birds
5.8 **Vapor (Gas) Irritant Characteristics:** Vapors cause severe irritation of eyes and throat and cause eye and lung injury. They cannot be tolerated even at low concentrations.
5.9 **Liquid or Solid Irritant Characteristics:** Contact with skin may cause irritation.
5.10 **Odor Threshold:** Data not available

6. FIRE HAZARDS

6.1 **Flash Point:** >286°F
6.2 **Flammable Limits in Air:** Data not available
6.3 **Fire Extinguishing Agents:** Water, foam, dry chemical, or carbon dioxide
6.4 **Fire Extinguishing Agents Not to be Used:** Not pertinent
6.5 **Special Hazards of Combustion Products:** Irritating gases are generated in fires.
6.6 **Behavior in Fire:** Not pertinent
6.7 **Ignition Temperature:** Data not available
6.8 **Electrical Hazard:** Not pertinent
6.9 **Burning Rate:** Data not available

7. CHEMICAL REACTIVITY

7.1 **Reactivity with Water:** No reaction
7.2 **Reactivity with Common Materials:** No reaction
7.3 **Stability During Transport:** Stable
7.4 **Neutralizing Agents for Acids and Caustics:** Not pertinent
7.5 **Polymerization:** Not pertinent
7.6 **Inhibitor of Polymerization:** Not pertinent

8. WATER POLLUTION

8.1 **Aquatic Toxicity:**
0.278 ppm/96 hr/bluegill/TL_m/fresh water
0.005 ppm/336-1080 hr/pinfish/TL_m/salt water
8.2 **Waterfowl Toxicity:** LD_{50} 2000 ppm (mallard duck)
8.3 **Biological Oxygen Demand (BOD):** Very low
8.4 **Food Chain Concentration Potential:** High

9. SELECTED MANUFACTURERS

Monsanto Industrial Chemicals Co.
800 North Lindbergh Blvd.
St. Louis, Mo. 63166

10. SHIPPING INFORMATION

10.1 **Grades or Purity:** 11 grades (some liquid, some solids) which differ primarily in their chlorine content (20-68% by weight)
10.2 **Storage Temperature:** Ambient
10.3 **Inert Atmosphere:** No requirement
10.4 **Venting:** Open

11. HAZARD ASSESSMENT CODE

(See Hazard Assessment Handbook, CG 446-3)

II

12. HAZARD CLASSIFICATIONS

12.1 **Code of Federal Regulations:** Not listed
12.2 **NAS Hazard Rating for Bulk Water Transportation:** Not listed
12.3 **NFPA Hazard Classifications:** Not listed

13. PHYSICAL AND CHEMICAL PROPERTIES

13.1 **Physical State at 15°C and 1 atm:** Solid or liquid
13.2 **Molecular Weight:** Not pertinent
13.3 **Boiling Point at 1 atm:** Very high
13.4 **Freezing Point:** Not pertinent
13.5 **Critical Temperature:** Not pertinent
13.6 **Critical Pressure:** Not pertinent
13.7 **Specific Gravity:** 1.3-1.8 at 20°C (liquid)
13.8 **Liquid Surface Tension:** Not pertinent
13.9 **Liquid-Water Interfacial Tension:** Not pertinent
13.10 **Vapor (Gas) Specific Gravity:** Not pertinent
13.11 **Ratio of Specific Heats of Vapor (Gas):** Not pertinent
13.12 **Latent Heat of Vaporization:** Not pertinent
13.13 **Heat of Combustion:** Not pertinent
13.14 **Heat of Decomposition:** Not pertinent
13.15 **Heat of Solution:** Not pertinent
13.16 **Heat of Polymerization:** Not pertinent

NOTES

REVISED 1978

PPI | POLYMETHYLENE POLYPHENYL ISOCYANATE

Common Synonyms			
PAPI	Liquid	Dark brown	Weak odor
	Sinks in water.		

AVOID CONTACT WITH LIQUID. KEEP PEOPLE AWAY.
Stop discharge if possible.
Call fire department.
Isolate and remove discharged material.
Notify local health and pollution control agencies.

Fire

Combustible.
Containers may explode in fire.
Extinguish with dry chemicals or carbon dioxide.
Cool exposed containers with water.

Exposure

CALL FOR MEDICAL AID.

LIQUID
POISONOUS IF SWALLOWED.
Irritating to skin and eyes.
Remove contaminated clothing and shoes.
Flush affected areas with plenty of water.
IF IN EYES, hold eyelids open and flush with plenty of water.
IF SWALLOWED and victim is CONSCIOUS, have victim drink water or milk and have victim induce vomiting.
IF SWALLOWED and victim is UNCONSCIOUS OR HAVING CONVULSIONS, do nothing except keep victim warm.

Water Pollution

Effect of low concentrations on aquatic life is unknown.
May be dangerous if it enters water intakes.
Notify local health and wildlife officials.
Notify operators of nearby water intakes.

1. RESPONSE TO DISCHARGE

(See Response Methods Handbook, CG 446-4)

Issue warning—poison, water contaminant
Restrict access
Should be removed
Chemical and physical treatment

2. LABEL

3. CHEMICAL DESIGNATIONS

3.1 **Synonyms:** PAPI
3.2 **Coast Guard Compatibility Classification:** Isocyanate
3.3 **Chemical Formula:** $C_6H_4(NCO)CH_2C_6H_4(NCO)$—and polymer
3.4 **IMCO/United Nations Numerical Designation:** 6.1/—

4. OBSERVABLE CHARACTERISTICS

4.1 **Physical State (as shipped):** Liquid
4.2 **Color:** Dark brown
4.3 **Odor:** Very weak

5. HEALTH HAZARDS

5.1 **Personal Protective Equipment:** Air-line or organic canister mask; goggles or face shield; rubber gloves and other protective clothing to prevent contact with skin.
5.2 **Symptoms Following Exposure:** Inhalation causes breathlessness, chest discomfort, and reduced pulmonary function; wheezing, cough, and sputum may also occur. Contact with liquid irritates eyes and skin. Ingestion causes irritation of mouth and stomach.
5.3 **Treatment for Exposure:** *Get medical attention at once following all exposures to this compound.* INHALATION: remove victim to fresh air; give artificial respiration if breathing has stopped; oxygen can be given by qualified personnel. EYES: immediately wash with large amounts of water for at least 15 min. SKIN: flush immediately with water, wipe off, treat with 30% isopropyl alcohol (rubbing alcohol), and wash with soap and water. INGESTION: induce vomiting at least 3 times by giving warm salt water (one tablespoon of salt per cup); follow with a quart of milk and a mild cathartic such as milk of magnesia.
5.4 **Toxicity by Inhalation (Threshold Limit Value):** 0.02 ppm
5.5 **Short-Term Inhalation Limits:** Data not available
5.6 **Toxicity by Ingestion:** Grade 1; LD_{50} 5 to 15 g/kg
5.7 **Late Toxicity:** Data not available
5.8 **Vapor (Gas) Irritant Characteristics:** Vapors are moderately irritating such that personnel will not usually tolerate moderate or high concentrations.
5.9 **Liquid or Solid Irritant Characteristics:** Causes smarting of the skin and first-degree burns on short exposure; may cause second-degree burns on long exposure.
5.10 **Odor Threshold:** Data not available

6. FIRE HAZARDS

6.1 **Flash Point:** 425°F O.C.
6.2 **Flammable Limits in Air:** Not pertinent
6.3 **Fire Extinguishing Agents:** Dry chemical or carbon dioxide
6.4 **Fire Extinguishing Agents Not to be Used:** Not pertinent
6.5 **Special Hazards of Combustion Products:** Not pertinent
6.6 **Behavior in Fire:** Containers may explode.
6.7 **Ignition Temperature:** Data not available
6.8 **Electrical Hazard:** Not pertinent
6.9 **Burning Rate:** Data not available

7. CHEMICAL REACTIVITY

7.1 **Reactivity with Water:** Reacts slowly, forming heavy scum and liberating carbon dioxide gas. Dangerous pressure can build up if container is sealed.
7.2 **Reactivity with Common Materials:** No hazardous reaction unless confined and wet.
7.3 **Stability During Transport:** Stable if kept sealed and dry
7.4 **Neutralizing Agents for Acids and Caustics:** Not pertinent
7.5 **Polymerization:** Not pertinent
7.6 **Inhibitor of Polymerization:** Not pertinent

8. WATER POLLUTION

8.1 **Aquatic Toxicity:** Data not available
8.2 **Waterfowl Toxicity:** Data not available
8.3 **Biological Oxygen Demand (BOD):** Data not available
8.4 **Food Chain Concentration Potential:** None

9. SELECTED MANUFACTURERS

The Upjohn Company
Polymer Chemicals Division
P. O. Box 685
LaPorte, Texas 77571

10. SHIPPING INFORMATION

10.1 **Grades or Purity:** 50% methylenebis-phenylisocyanate plus 50% polymer
10.2 **Storage Temperature:** 35—125°F
10.3 **Inert Atmosphere:** Low-pressure dry nitrogen
10.4 **Venting:** Safety relief

11. HAZARD ASSESSMENT CODE

(See Hazard Assessment Handbook, CG 446-3)

A-O-X-Y

12. HAZARD CLASSIFICATIONS

12.1 **Code of Federal Regulations:** Poisonous liquid, Class B
12.2 **NAS Hazard Rating for Bulk Water Transportation:**

Category	Rating
Fire	1
Health	
Vapor Irritant	3
Liquid or Solid Irritant	2
Poisons	4
Water Pollution	
Human Toxicity	1
Aquatic Toxicity	
Aesthetic Effect	2
Reactivity	
Other Chemicals	1
Water	3
Self-Reaction	1

12.3 **NFPA Hazard Classifications:** Not listed

13. PHYSICAL AND CHEMICAL PROPERTIES

13.1 **Physical State at 15°C and 1 atm:** Liquid
13.2 **Molecular Weight:** 400 (approx.)
13.3 **Boiling Point at 1 atm:** 392°F = 200°C = 473°K
13.4 **Freezing Point:** Not pertinent
13.5 **Critical Temperature:** Not pertinent
13.6 **Critical Pressure:** Not pertinent
13.7 **Specific Gravity:** 1.20 at 20°C (liquid)
13.8 **Liquid Surface Tension:** Data not available
13.9 **Liquid-Water Interfacial Tension:** Not pertinent
13.10 **Vapor (Gas) Specific Gravity:** Not pertinent
13.11 **Ratio of Specific Heats of Vapor (Gas):** Not pertinent
13.12 **Latent Heat of Vaporization:** Not pertinent
13.13 **Heat of Combustion:** (est.) −13,000 Btu/lb = −7,200 cal/g = -300×10^5 J/kg
13.14 **Heat of Decomposition:** Not pertinent
13.15 **Heat of Solution:** Not pertinent
13.16 **Heat of Polymerization:** Not pertinent

NOTES

PPA	POLYPHOSPHORIC ACID

Common Synonyms Condensed phosphoric acid	Liquid Colorless Odorless Sinks and mixes with water.

AVOID CONTACT WITH LIQUID. Keep people away.
Wear chemical protective suit with self-contained breathing apparatus.
Stop discharge if possible.
Isolate and remove discharged material.
Notify local health and pollution control agencies.

Fire	Not flammable. Flammable gas may be produced on contact with metal. Wear chemical protective suit with self-contained breathing apparatus.
Exposure	CALL FOR MEDICAL AID. LIQUID Will burn skin and eyes. Harmful if swallowed. Remove contaminated clothing and shoes. Flush affected areas with plenty of water. IF IN EYES, hold eyelids open and flush with plenty of water. IF SWALLOWED and victim is CONSCIOUS, have victim drink water or milk. DO NOT INDUCE VOMITING.
Water Pollution	Dangerous to aquatic life in high concentrations. May be dangerous if it enters water intakes. Notify local health and wildlife officials. Notify operators of nearby water intakes.

1. RESPONSE TO DISCHARGE
(See Response Methods Handbook, CG 446-4)

Issue warning—corrosive
Restrict access
Disperse and flush

2. LABEL

3. CHEMICAL DESIGNATIONS

3.1 **Synonyms:** Condensed phosphoric acid
3.2 **Coast Guard Compatibility Classification:** Not applicable
3.3 **Chemical Formula:** $(P_2O_5)(H_2O)_{<3}$
3.4 **IMCO/United Nations Numerical Designation:** Not listed

4. OBSERVABLE CHARACTERISTICS

4.1 **Physical State (as shipped):** Liquid
4.2 **Color:** Colorless
4.3 **Odor:** Odorless

5. HEALTH HAZARDS

5.1 **Personal Protective Equipment:** Goggles or face shield; rubber gloves or protective clothing.
5.2 **Symptoms Following Exposure:** Liquid burns skin and eyes unless washed off quickly. If ingested will burn mouth and stomach unless diluted at once.
5.3 **Treatment for Exposure:** INGESTION: give victim water, milk, or vegetable oil; do NOT induce vomiting. SKIN OR EYES: flush with water for at least 15 min.; call doctor for eye exposure.
5.4 **Toxicity by Inhalation (Threshold Limit Value):** Not pertinent
5.5 **Short-Term Inhalation Limits:** Not pertinent
5.6 **Toxicity by Ingestion:** Grade 3; LD_{50} 50 to 500 mg/kg
5.7 **Late Toxicity:** None
5.8 **Vapor (Gas) Irritant Characteristics:** Non-volatile
5.9 **Liquid or Solid Irritant Characteristics:** Fairly severe skin irritant; may cause pain and second-degree burns after a few minutes' contact.
5.10 **Odor Threshold:** Not pertinent

6. FIRE HAZARDS

6.1 **Flash Point:** Not flammable
6.2 **Flammable Limits in Air:** Not flammable
6.3 **Fire Extinguishing Agents:** Not pertinent
6.4 **Fire Extinguishing Agents Not to be Used:** Not pertinent
6.5 **Special Hazards of Combustion Products:** Not pertinent
6.6 **Behavior in Fire:** Not pertinent
6.7 **Ignition Temperature:** Not flammable
6.8 **Electrical Hazard:** Not pertinent
6.9 **Burning Rate:** Not flammable

7. CHEMICAL REACTIVITY

7.1 **Reactivity with Water:** Reacts with water to generate heat and form phosphoric acid. The reaction is not violent.
7.2 **Reactivity with Common Materials:** Reacts with metals to liberate flammable hydrogen gas.
7.3 **Stability During Transport:** Stable
7.4 **Neutralizing Agents for Acids and Caustics:** Flush with water, neutralize acid with lime or soda ash.
7.5 **Polymerization:** Not pertinent
7.6 **Inhibitor of Polymerization:** Not pertinent

8. WATER POLLUTION

8.1 **Aquatic Toxicity:** 138 ppm/24 hr/mosquito fish/TL_m/fresh water
8.2 **Waterfowl Toxicity:** Data not available
8.3 **Biological Oxygen Demand (BOD):** None
8.4 **Food Chain Concentration Potential:** None

9. SELECTED MANUFACTURERS

1. Allied Chemical Corp.
   Agricultural Division
   Morristown, N. J. 07960
2. International Minerals & Chemical Corp.
   Agricultural Operations
   Bonnie, Fla. 33830
3. Texas Gulf Sulphur Co.
   200 Park Ave.
   New York, N. Y. 10017

10. SHIPPING INFORMATION

10.1 **Grades or Purity:** 115% phosphoric acid
10.2 **Storage Temperature:** Ambient
10.3 **Inert Atmosphere:** No requirement
10.4 **Venting:** Open or pressure-vacuum

11. HAZARD ASSESSMENT CODE
(See Hazard Assessment Handbook, CG 446-3)

A-P-O

12. HAZARD CLASSIFICATIONS

12.1 **Code of Federal Regulations:** Corrosive liquid
12.2 **NAS Hazard Rating for Bulk Water Transportation:** Not listed
12.3 **NFPA Hazard Classifications:** Not listed

13. PHYSICAL AND CHEMICAL PROPERTIES

13.1 **Physical State at 15°C and 1 atm:** Liquid
13.2 **Molecular Weight:** Not pertinent
13.3 **Boiling Point at 1 atm:** 1022°F = 550°C = 823°K
13.4 **Freezing Point:** 100°F = 38°C = 311°K
13.5 **Critical Temperature:** Not pertinent
13.6 **Critical Pressure:** Not pertinent
13.7 **Specific Gravity:** 2.05 at 38°C (liquid)
13.8 **Liquid Surface Tension:** Not pertinent
13.9 **Liquid-Water Interfacial Tension:** Not pertinent
13.10 **Vapor (Gas) Specific Gravity:** Not pertinent
13.11 **Ratio of Specific Heats of Vapor (Gas):** Not pertinent
13.12 **Latent Heat of Vaporization:** Not pertinent
13.13 **Heat of Combustion:** Not pertinent
13.14 **Heat of Decomposition:** Not pertinent
13.15 **Heat of Solution:** Not pertinent
13.16 **Heat of Polymerization:** Not pertinent

NOTES

REVISED 1978

PLP	POLYPROPYLENE

Common Synonyms			
Propene polymer	Solid	Tan to white	Odorless
	Floats on water.		

Stop discharge if possible.
Call fire department.
Isolate and remove discharged material.
Notify local health and pollution control agencies.

Fire	Combustible. Extinguish with water, dry chemicals, foam, or carbon dioxide. Water may be ineffective on fire. Cool exposed containers with water.
Exposure	DUST Not harmful. SOLID Not harmful.
Water Pollution	Effect of low concentrations on aquatic life is unknown. May be dangerous if it enters water intakes. Notify local health and wildlife officials. Notify operators of nearby water intakes.

1. RESPONSE TO DISCHARGE
(See Response Methods Handbook, CG 446-4)

Mechanical containment

2. LABELS

No hazard label required by Code of Federal Regulations

3. CHEMICAL DESIGNATIONS

3.1 **Synonyms:** Propene polymer
3.2 **Coast Guard Compatibility Classification:** Olefin
3.3 **Chemical Formula:** $\text{+}CH(CH_3)-CH_2\text{+}_n$ where n is large
3.4 **IMCO/United Nations Numerical Designation:** Not listed

4. OBSERVABLE CHARACTERISTICS

4.1 **Physical State (as shipped):** Solid
4.2 **Color:** Tan or white
4.3 **Odor:** None

5. HEALTH HAZARDS

5.1 **Personal Protective Equipment:** Filter respirator
5.2 **Symptoms Following Exposure:** No apparent toxicity
5.3 **Treatment for Exposure:** None required
5.4 **Toxicity by Inhalation (Threshold Limit Value):** Data not available
5.5 **Short-Term Inhalation Limits:** Data not available
5.6 **Toxicity by Ingestion:** Data not available
5.7 **Late Toxicity:** Causes central bronchitis and peripheral bronchitis in rats and rabbits
5.8 **Vapor (Gas) Irritant Characteristics:** Data not available
5.9 **Liquid or Solid Irritant Characteristics:** Data not available
5.10 **Odor Threshold:** Data not available

6. FIRE HAZARDS

6.1 **Flash Point:** Not pertinent (combustible solid)
6.2 **Flammable Limits in Air:** Not pertinent
6.3 **Fire Extinguishing Agents:** Foam, dry chemical, carbon dioxide
6.4 **Fire Extinguishing Agents Not to be Used:** Water may be ineffective.
6.5 **Special Hazards of Combustion Products:**
6.6 **Behavior in Fire:**
6.7 **Ignition Temperature:** Data not available
6.8 **Electrical Hazard:** Not pertinent
6.9 **Burning Rate:** Not pertinent

7. CHEMICAL REACTIVITY

7.1 **Reactivity with Water:** No reaction
7.2 **Reactivity with Common Materials:**
7.3 **Stability During Transport:** Stable
7.4 **Neutralizing Agents for Acids and Caustics:** Not pertinent
7.5 **Polymerization:** Not pertinent
7.6 **Inhibitor of Polymerization:** Not pertinent

8. WATER POLLUTION

8.1 **Aquatic Toxicity:** None
8.2 **Waterfowl Toxicity:** None
8.3 **Biological Oxygen Demand (BOD):** Data not available
8.4 **Food Chain Concentration Potential:** None

9. SELECTED MANUFACTURERS

1. Diamond Shamrock Corp.
   1100 Superior Avenue
   Cleveland, Ohio 44114
2. Exxon Chemical Co.
   1333 West Loop South
   Houston, Texas 77027
3. Amoco Chemicals Corp.
   130 East Randolph Drive
   Chicago, Ill. 60601

10. SHIPPING INFORMATION

10.1 **Grades or Purity:** Commercial, 100%
10.2 **Storage Temperature:** Ambient
10.3 **Inert Atmosphere:** No requirement
10.4 **Venting:** Open

11. HAZARD ASSESSMENT CODE
(See Hazard Assessment Handbook, CG 446-3)

II

12. HAZARD CLASSIFICATIONS

12.1 **Code of Federal Regulations:** Not listed
12.2 **NAS Hazard Rating for Bulk Water Transportation:** Not listed
12.3 **NFPA Hazard Classifications:** Not listed

13. PHYSICAL AND CHEMICAL PROPERTIES

13.1 **Physical State at 15°C and 1 atm:** Solid
13.2 **Molecular Weight:** Mixture
13.3 **Boiling Point at 1 atm:** Not pertinent (decomposes)
13.4 **Freezing Point:** Not pertinent
13.5 **Critical Temperature:** Not pertinent
13.6 **Critical Pressure:** Not pertinent
13.7 **Specific Gravity:** 0.90 at 20°C (solid)
13.8 **Liquid Surface Tension:** Not pertinent
13.9 **Liquid-Water Interfacial Tension:** Not pertinent
13.10 **Vapor (Gas) Specific Gravity:** Not pertinent
13.11 **Ratio of Specific Heats of Vapor (Gas):** Not pertinent
13.12 **Latent Heat of Vaporization:** Not pertinent
13.13 **Heat of Combustion:** (est.) $-19{,}600$ Btu/lb $= -10{,}900$ cal/g $= -456 \times 10^5$ J/kg
13.14 **Heat of Decomposition:** Not pertinent
13.15 **Heat of Solution:** Not pertinent
13.16 **Heat of Polymerization:** Not pertinent

NOTES

PGC POLYPROPYLENE GLYCOL

Common Synonyms	Liquid	Colorless	Odorless or mild sweet odor
Polyoxypropylene glycol Polyoxpropylene glycol Pluracol polyol	May float or sink in water.		

Stop discharge if possible. Keep people away.
Call fire department.
Isolate and remove discharged material.
Notify local health and pollution control agencies.

Fire

Combustible.
Extinguish with dry chemicals, foam or carbon dioxide.
Water may be ineffective on fire.

Exposure

Call for medical aid.

LIQUID
Irritating to eyes.
Harmful if swallowed.
IF IN EYES, hold eyelids open and flush with plenty of water.
IF SWALLOWED and victim is CONSCIOUS, have victim drink water or milk.

Water Pollution

Effect of low concentrations on aquatic life is unknown.
Fouling to shoreline.
May be dangerous if it enters water intakes.
Notify local health and wildlife officials.
Notify operators of nearby water intakes.

1. RESPONSE TO DISCHARGE
(See Response Methods Handbook, CG 446-4)

Issue warning—water contaminant
Mechanical containment
Should be removed
Disperse and flush

2. LABELS

No label required by Code of Federal Regulations

3. CHEMICAL DESIGNATIONS

3.1 **Synonyms:** Polyoxypropylene glycol; Polyoxpropylene ether, PPG; Pluracol polyol; Polypropylene glycols P400 to P4000; Thanol PPG
3.2 **Coast Guard Compatibility Classification:** Alcohols (6)
3.3 **Chemical Formula:** $HOCH(CH_3)CH_2O[CH_2CH(CH_3)O]_n-H$
n averages 2–34

(Continued on page 4)

4. OBSERVABLE CHARACTERISTICS

4.1 **Physical State (as shipped):** Liquid
4.2 **Color:** Colorless; clear, lightly colored
4.3 **Odor:** None; slight sweet; faint ether-like

5. HEALTH HAZARDS

5.1 **Personal Protective Equipment:** Safety glasses or face shield; rubber gloves
5.2 **Symptoms Following Exposure:** The compound has a very low toxicity; few, if any, symptoms will be observed. Contact of liquid with eyes causes slight transient pain and irritation similar to that caused by a mild soap.
5.3 **Treatment for Exposure:** EYES: flush with water until mild irritation is gone.
5.4 **Toxicity by Inhalation (Threshold Limit Value):** Data not available
5.5 **Short-Term Inhalation Limits:** Not pertinent
5.6 **Toxicity by Ingestion:** (depends on molecular wt.)
Grade 2; oral LD_{50} = 2,150 mg/kg (rat)
Grade 1; LD_{50} 5 to 15 g/kg
Grade 0; LD_{50} > 15 g/kg
5.7 **Late Toxicity:** Data not available
5.8 **Vapor (Gas) Irritant Characteristics:** Data not available
5.9 **Liquid or Solid Irritant Characteristics:** Data not available
5.10 **Odor Threshold:** Not pertinent

6. FIRE HAZARDS

6.1 **Flash Point:** 390—495°F O.C.
6.2 **Flammable Limits in Air:** Not pertinent
6.3 **Fire Extinguishing Agents:** Water, dry chemical, foam, carbon dioxide
6.4 **Fire Extinguishing Agents Not to be Used:** Not pertinent
6.5 **Special Hazards of Combustion Products:** Not pertinent
6.6 **Behavior in Fire:** Not pertinent
6.7 **Ignition Temperature:** Data not available
6.8 **Electrical Hazard:** Not pertinent
6.9 **Burning Rate:** Data not available

7. CHEMICAL REACTIVITY

7.1 **Reactivity with Water:** No reaction
7.2 **Reactivity with Common Materials:** No reaction
7.3 **Stability During Transport:** Stable
7.4 **Neutralizing Agents for Acids and Caustics:** Not pertinent
7.5 **Polymerization:** Not pertinent
7.6 **Inhibitor of Polymerization:** Not pertinent

8. WATER POLLUTION

8.1 **Aquatic Toxicity:** Data not available
8.2 **Waterfowl Toxicity:** Data not available
8.3 **Biological Oxygen Demand (BOD):** Data not available
8.4 **Food Chain Concentration Potential:** Data not available

9. SELECTED MANUFACTURERS

1. Union Carbide Corp.
Chemicals and Plastics Division
270 Park Avenue
New York, N. Y. 10017
2. BASF Wyandotte Corporation
Industrial Chemicals Group
1609 Biddle Avenue
Wyandotte, Mich. 48192
3. Dow Chemical Co.
Midland, Mich. 48640

10. SHIPPING INFORMATION

10.1 **Grades or Purity:**
Low mol. wt. (miscible with water)
Medium mol. wt. (2% soluble in water)
High mol. wt. (insoluble in water)
10.2 **Storage Temperature:** Below 140°F
10.3 **Inert Atmosphere:** No requirement
10.4 **Venting:** Open (flame arrester)

11. HAZARD ASSESSMENT CODE
(See Hazard Assessment Handbook, CG 446-3)

A-T-U-X-Y

12. HAZARD CLASSIFICATIONS

12.1 **Code of Federal Regulations:** Not listed
12.2 **NAS Hazard Rating for Bulk Water Transportation:** Not listed
12.3 **NFPA Hazard Classifications:** Not listed

13. PHYSICAL AND CHEMICAL PROPERTIES

13.1 **Physical State at 15°C and 1 atm:** Liquid
13.2 **Molecular Weight:** Variable—200 to 2000
13.3 **Boiling Point at 1 atm:** Not pertinent (decomposes)
13.4 **Freezing Point:** −22 to −58°F = −30 to −50°C = 243 to 223°K
13.5 **Critical Temperature:** Not pertinent
13.6 **Critical Pressure:** Not pertinent
13.7 **Specific Gravity:** 1.012 at 20°C (liquid)
13.8 **Liquid Surface Tension:** Data not available
13.9 **Liquid-Water Interfacial Tension:** Not pertinent
13.10 **Vapor (Gas) Specific Gravity:** Not pertinent
13.11 **Ratio of Specific Heats of Vapor (Gas):** Not pertinent
13.12 **Latent Heat of Vaporization:** Not pertinent
13.13 **Heat of Combustion:** −14,200 Btu/lb = −7,900 cal/g = -330×10^5 J/kg
13.14 **Heat of Decomposition:** Not pertinent
13.15 **Heat of Solution:** Not pertinent
13.16 **Heat of Polymerization:** Not pertinent

3. CHEMICAL DESIGNATIONS (Cont'd.)

3.4 **IMCO/United Nations Numerical Designation:** Not listed

PGM

POLYPROPYLENE GLYCOL METHYL ETHER

Common Synonyms			
Polyoxypropylene glycol methyl ether Poly (propylene glycol) methyl ether	Liquid	Colorless	Odorless
	Mixes with water.		

Stop discharge if possible. Keep people away.
Shut off ignition sources and call fire department.
Avoid contact with liquid.
Stay upwind and use water spray to "knock down" vapor.
Isolate and remove discharged material.
Notify local health and pollution control agencies.

Fire	Fire data not available.
Exposure	CALL FOR MEDICAL AID. LIQUID Irritating to skin and eyes. Flush affected areas with plenty of water. IF IN EYES, hold eyelids open and flush with plenty of water.
Water Pollution	Effect of low concentrations on aquatic life is unknown. May be dangerous if it enters water intakes. Notify local health and wildlife officials. Notify operators of nearby water intakes.

1. RESPONSE TO DISCHARGE

(See Response Methods Handbook, CG 446-4)

Disperse and flush

2. LABELS

No hazard label required by Code of Federal Regulations

3. CHEMICAL DESIGNATIONS

3.1 **Synonyms:** Polyoxypropylene glycol methyl ether; Poly (propylene glycol) methyl ether

3.2 **Coast Guard Compatibility Classification:** Glycol ether

3.3 **Chemical Formula:** $CH_3O(C_3H_6O)_nH$

3.4 **IMCO/United Nations Numerical Designation:** Not listed

4. OBSERVABLE CHARACTERISTICS

4.1 **Physical State (as shipped):** Liquid

4.2 **Color:** Colorless

4.3 **Odor:** None

5. HEALTH HAZARDS

5.1 **Personal Protective Equipment:** Goggles or face shield.

5.2 **Symptoms Following Exposure:** Low toxicity. In high concentrations may be mildly anesthetic. Direct eye contact produces slight irritation.

5.3 **Treatment for Exposure:** Remove victim from exposure.

5.4 **Toxicity by Inhalation (Threshold Limit Value):** 100 ppm

5.5 **Short-Term Inhalation Limits:** Data not available

5.6 **Toxicity by Ingestion:** Data not available

5.7 **Late Toxicity:** Data not available

5.8 **Vapor (Gas) Irritant Characteristics:** Data not available

5.9 **Liquid or Solid Irritant Characteristics:** Hazard to skin considered minor.

5.10 **Odor Threshold:** Data not available

6. FIRE HAZARDS

6.1 **Flash Point:** Data not available

6.2 **Flammable Limits in Air:** Data not available

6.3 **Fire Extinguishing Agents:** Data not available

6.4 **Fire Extinguishing Agents Not to be Used:** Data not available

6.5 **Special Hazards of Combustion Products:** Not pertinent

6.6 **Behavior in Fire:** Not pertinent

6.7 **Ignition Temperature:** Data not available

6.8 **Electrical Hazard:** Not pertinent

6.9 **Burning Rate:** Data not available

7. CHEMICAL REACTIVITY

7.1 **Reactivity with Water:** No reaction

7.2 **Reactivity with Common Materials:** No reaction

7.3 **Stability During Transport:** Stable

7.4 **Neutralizing Agents for Acids and Caustics:** Not pertinent

7.5 **Polymerization:** Not pertinent

7.6 **Inhibitor of Polymerization:** Not pertinent

8. WATER POLLUTION

8.1 **Aquatic Toxicity:** Data not available

8.2 **Waterfowl Toxicity:** Data not available

8.3 **Biological Oxygen Demand (BOD):** Data not available

8.4 **Food Chain Concentration Potential:** None

9. SELECTED MANUFACTURERS

Union Carbide Corp.
Chemicals and Plastics Division
270 Park Ave.
New York, N. Y. 10017
(No bulk shipment)

10. SHIPPING INFORMATION

10.1 **Grades or Purity:** Technical

10.2 **Storage Temperature:** Ambient

10.3 **Inert Atmosphere:** No requirement

10.4 **Venting:** Open

11. HAZARD ASSESSMENT CODE

(See Hazard Assessment Handbook, CG 446-3)

A-P-Q

12. HAZARD CLASSIFICATIONS

12.1 **Code of Federal Regulations:** Not listed

12.2 **NAS Hazard Rating for Bulk Water Transportation:** Not listed

12.3 **NFPA Hazard Classifications:** Not listed

13. PHYSICAL AND CHEMICAL PROPERTIES

13.1 **Physical State at 15°C and 1 atm:** Liquid

13.2 **Molecular Weight:** Data not available

13.3 **Boiling Point at 1 atm:** Very high

13.4 **Freezing Point:** Not pertinent

13.5 **Critical Temperature:** Not pertinent

13.6 **Critical Pressure:** Not pertinent

13.7 **Specific Gravity:** (est.) 0.90 at 20°C (liquid)

13.8 **Liquid Surface Tension:** Not pertinent

13.9 **Liquid-Water Interfacial Tension:** Not pertinent

13.10 **Vapor (Gas) Specific Gravity:** Not pertinent

13.11 **Ratio of Specific Heats of Vapor (Gas):** (est.) 1.066

13.12 **Latent Heat of Vaporization:** Not pertinent

13.13 **Heat of Combustion:** (est.) $-12,300$ Btu/lb $= -6860$ cal/g $= -287 \times 10^5$ J/kg

13.14 **Heat of Decomposition:** Not pertinent

13.15 **Heat of Solution:** Not pertinent

13.16 **Heat of Polymerization:** Not pertinent

NOTES

REVISED 1978

POTASSIUM ARSENATE

PAS

Common Synonyms			
Macquer's Salt Potassium acid arsenate Potassium dihydrogen arsenate	Solid	White	Odorless
	Mixes with water.		

AVOID CONTACT WITH SOLID AND DUST. KEEP PEOPLE AWAY.
Wear goggles and dust respirator.
Stay upwind. Use water spray to "knock down" dust.
Isolate and remove discharged material.
Notify local health and pollution control agencies.

Fire

Not flammable.

Exposure

CALL FOR MEDICAL AID.

DUST
POISONOUS IF INHALED OR SWALLOWED.
Irritating to eyes, nose and throat.
Move victim to fresh air.
If in eyes, hold eyelids open and flush with plenty of water.
If breathing is difficult, give oxygen.

SOLID
POISONOUS IF SWALLOWED.
Irritating to skin and eyes.
Remove contaminated clothing and shoes.
Flush affected areas with plenty of water.
IF IN EYES, hold eyelids open and flush with plenty of water.
IF SWALLOWED and victim is CONSCIOUS, have victim drink water or milk and have victim induce vomiting.
IF SWALLOWED and victim is UNCONSCIOUS OR HAVING CONVULSIONS, do nothing except keep victim warm.

Water Pollution

HARMFUL TO AQUATIC LIFE IN VERY LOW CONCENTRATIONS.
May be dangerous if it enters water intakes.
Notify local health and wildlife officials.
Notify operators of nearby water intakes.

1. RESPONSE TO DISCHARGE

(See Response Methods Handbook, CG 446-4)

Issue warning—poison, water contaminant
Restrict access
Disperse and flush

2. LABEL

3. CHEMICAL DESIGNATIONS

3.1 **Synonyms:** Macquer's salt; Potassium acid arsenate; Potassium dihydrogen arsenate

3.2 **Coast Guard Compatibility Classification:** Not applicable

3.3 **Chemical Formula:** KH_2AsO_4

3.4 **IMCO/United Nations Numerical Designation:** 6.1/1677

4. OBSERVABLE CHARACTERISTICS

4.1 **Physical State (as shipped):** Solid

4.2 **Color:** White

4.3 **Odor:** None

5. HEALTH HAZARDS

5.1 **Personal Protective Equipment:** Dust respirator; rubber gloves

5.2 **Symptoms Following Exposure:** Dust may irritate eyes. Ingestion or severe exposure by inhalation can cause burning of throat and mouth, abdominal pain, vomiting, diarrhea with hemorrhage, dehydration, jaundice, and collapse.

5.3 **Treatment for Exposure:** EYES: flush with water to remove dust. INGESTION: immediately induce evacuation of intestinal tract by inducing vomiting, giving gastric lavage and saline cathartic; see physician at once; consider possible development of arsenic poisoning.

5.4 **Toxicity by Inhalation (Threshold Limit Value):** 0.5 mg/m³ (as arsenic)

5.5 **Short-Term Inhalation Limits:** Data not available

5.6 **Toxicity by Ingestion:** Data not available

5.7 **Late Toxicity:** May be carcinogenic; arsenic poisoning may develop.

5.8 **Vapor (Gas) Irritant Characteristics:** Data not available

5.9 **Liquid or Solid Irritant Characteristics:** Data not available

5.10 **Odor Threshold:** Not pertinent

6. FIRE HAZARDS

6.1 **Flash Point:** Not flammable

6.2 **Flammable Limits in Air:** Not flammable

6.3 **Fire Extinguishing Agents:** Not pertinent

6.4 **Fire Extinguishing Agents Not to be Used:** Not pertinent

6.5 **Special Hazards of Combustion Products:** Not pertinent

6.6 **Behavior in Fire:** Not pertinent

6.7 **Ignition Temperature:** Not pertinent

6.8 **Electrical Hazard:** Not pertinent

6.9 **Burning Rate:** Not pertinent

7. CHEMICAL REACTIVITY

7.1 **Reactivity with Water:** No reaction

7.2 **Reactivity with Common Materials:** No reaction

7.3 **Stability During Transport:** Stable

7.4 **Neutralizing Agents for Acids and Caustics:** Not pertinent

7.5 **Polymerization:** Not pertinent

7.6 **Inhibitor of Polymerization:** Not pertinent

8. WATER POLLUTION

8.1 **Aquatic Toxicity:** 250 ppm as As/16 hr/minnows/survived

8.2 **Waterfowl Toxicity:** Data not available

8.3 **Biological Oxygen Demand (BOD):** None

8.4 **Food Chain Concentration Potential:** Data not available

9. SELECTED MANUFACTURERS

1. Pfaltz and Bauer, Inc.
   126-04 Northern Boulevard
   Flushing, N. Y. 11368
2. Cerac, Inc.
   13460 W. Silver Spring Rd.
   Menomenee Falls, Wis. 53051
3. Fielding Chemical Co.
   Forrest and Halladay Streets
   Jersey City, N. J. 07305

10. SHIPPING INFORMATION

10.1 **Grades or Purity:** Commercial

10.2 **Storage Temperature:** Ambient

10.3 **Inert Atmosphere:** No requirement

10.4 **Venting:** Open

11. HAZARD ASSESSMENT CODE

(See Hazard Assessment Handbook, CG 446-3)

SS

12. HAZARD CLASSIFICATIONS

12.1 **Code of Federal Regulations:** Poisonous, Class B

12.2 **NAS Hazard Rating for Bulk Water Transportation:** Not listed

12.3 **NFPA Hazard Classifications:** Not listed

13. PHYSICAL AND CHEMICAL PROPERTIES

13.1 **Physical State at 15°C and 1 atm:** Solid

13.2 **Molecular Weight:** 180.0

13.3 **Boiling Point at 1 atm:** Not pertinent (decomposes)

13.4 **Freezing Point:** 550°F = 288°C = 561°K

13.5 **Critical Temperature:** Not pertinent

13.6 **Critical Pressure:** Not pertinent

13.7 **Specific Gravity:** 2.8 at 20°C (solid)

13.8 **Liquid Surface Tension:** Not pertinent

13.9 **Liquid-Water Interfacial Tension:** Not pertinent

13.10 **Vapor (Gas) Specific Gravity:** Not pertinent

13.11 **Ratio of Specific Heats of Vapor (Gas):** Not pertinent

13.12 **Latent Heat of Vaporization:** Not pertinent

13.13 **Heat of Combustion:** Not pertinent

13.14 **Heat of Decomposition:** Not pertinent

13.15 **Heat of Solution:** 49 Btu/lb = 27 cal/g = 1.1×10^5 J/kg

13.16 **Heat of Polymerization:** Not pertinent

NOTES

PBO	POTASSIUM BINOXALATE

Common Synonyms			
Potassium acid oxalate Salt of sorrel Sal acetosella	Solid	White	Odorless
	Sinks in water.		

Stop discharge if possible. Keep people away.
Avoid contact with solid and dust.
Isolate and remove discharged material.
Notify local health and pollution control agencies.

Fire	Not flammable.
Exposure	CALL FOR MEDICAL AID. DUST Irritating to eyes, nose and throat. If inhaled will cause coughing, difficult breathing, or loss of consciousness. If in eyes, hold eyelids open and flush with plenty of water. If breathing has stopped, give artificial respiration. If breathing is difficult, give oxygen. SOLID Irritating to skin and eyes. If swallowed will cause nausea, vomiting or loss of consciousness. Remove contaminated clothing and shoes. Flush affected areas with plenty of water. IF IN EYES, hold eyelids open and flush with plenty of water. IF SWALLOWED and victim is CONSCIOUS, have victim drink water or milk. IF SWALLOWED and victim is UNCONSCIOUS OR HAVING CONVULSIONS, do nothing except keep victim warm.
Water Pollution	Effect of low concentrations on aquatic life is unknown. May be dangerous if it enters water intakes. Notify local health and wildlife officials. Notify operators of nearby water intakes.

1. RESPONSE TO DISCHARGE
(See Response Methods Handbook, CG 446-4)

Issue warning—poison, water contaminant
Restrict access
Disperse and flush

2. LABEL

3. CHEMICAL DESIGNATIONS

3.1 **Synonyms:** Potassium acid oxalate; Sal acetosella; Salt of sorrel
3.2 **Coast Guard Compatibility Classification:** Not listed
3.3 **Chemical Formula:** KHC_2O_4
3.4 **IMCO/United Nations Numerical Designation:** Not listed

4. OBSERVABLE CHARACTERISTICS

4.1 **Physical State (as shipped):** Solid
4.2 **Color:** White
4.3 **Odor:** None

5. HEALTH HAZARDS

5.1 **Personal Protective Equipment:** Dust mask; goggles or face shield; protective gloves
5.2 **Symptoms Following Exposure:** Inhalation of dust causes irritation of nose and throat. Ingestion causes burning pain in throat, esophagus, and stomach; exposed areas of mucous membrane turn white; vomiting, severe purging, weak pulse, and cardiovascular collapse; if death is delayed, neuromuscular symptoms develop. Contact with dust irritates eyes and may cause mild irritation of skin.
5.3 **Treatment for Exposure:** INHALATION: move to fresh air; if exposure to dust is severe, get medical attention. INGESTION: give immediately by mouth a dilute solution of any soluble calcium salt (calcium lactate, lime water, chalk solution, or even milk); large amounts of Ca are required; administer gastric lavage with dilute lime water; consult physician; watch for edema of the glottis and delayed constriction of esophagus. EYES: flush with water for at least 15 min. SKIN: wash well with soap and water.
5.4 **Toxicity by Inhalation (Threshold Limit Value):** Data not available
5.5 **Short-Term Inhalation Limits:** Data not available
5.6 **Toxicity by Ingestion:** Grade 3; LD_{50} 50—500 mg/kg
5.7 **Late Toxicity:** Data not available
5.8 **Vapor (Gas) Irritant Characteristics:** Data not available
5.9 **Liquid or Solid Irritant Characteristics:** Data not available
5.10 **Odor Threshold:** Odorless

6. FIRE HAZARDS

6.1 **Flash Point:** Not flammable
6.2 **Flammable Limits in Air:** Not flammable
6.3 **Fire Extinguishing Agents:** Not pertinent
6.4 **Fire Extinguishing Agents Not to be Used:** Not pertinent
6.5 **Special Hazards of Combustion Products:**
6.6 **Behavior in Fire:**
6.7 **Ignition Temperature:** Not pertinent
6.8 **Electrical Hazard:** Not pertinent
6.9 **Burning Rate:** Not pertinent

7. CHEMICAL REACTIVITY

7.1 **Reactivity with Water:** Below 50°C dissolves in water and reacts to form the much less soluble potassium tetraoxalate, which separates out.
7.2 **Reactivity with Common Materials:**
7.3 **Stability During Transport:** Stable
7.4 **Neutralizing Agents for Acids and Caustics:** Not pertinent
7.5 **Polymerization:** Not pertinent
7.6 **Inhibitor of Polymerization:** Not pertinent

8. WATER POLLUTION

8.1 **Aquatic Toxicity:** Data not available
8.2 **Waterfowl Toxicity:** Data not available
8.3 **Biological Oxygen Demand (BOD):** Data not available
8.4 **Food Chain Concentration Potential:** None

9. SELECTED MANUFACTURERS

1. J. T. Baker Chemical Co.
   Phillipsburg, N. J. 08865
2. Fielding Chemical Co.
   P. O. Box 21
   Jersey City, N. J. 07303

10. SHIPPING INFORMATION

10.1 **Grades or Purity:** Technical
10.2 **Storage Temperature:** Ambient
10.3 **Inert Atmosphere:** No requirement
10.4 **Venting:** Open

11. HAZARD ASSESSMENT CODE
(See Hazard Assessment Handbook, CG 446-3)

II-SS

12. HAZARD CLASSIFICATIONS

12.1 **Code of Federal Regulations:** Poison, Class B
12.2 **NAS Hazard Rating for Bulk Water Transportation:** Not listed
12.3 **NFPA Hazard Classifications:** Not listed

13. PHYSICAL AND CHEMICAL PROPERTIES

13.1 **Physical State at 15°C and 1 atm:** Solid
13.2 **Molecular Weight:** 128.11
13.3 **Boiling Point at 1 atm:** Not pertinent (decomposes)
13.4 **Freezing Point:** Not pertinent
13.5 **Critical Temperature:** Not pertinent
13.6 **Critical Pressure:** Not pertinent
13.7 **Specific Gravity:** 2.0 at 20°C (solid)
13.8 **Liquid Surface Tension:** Not pertinent
13.9 **Liquid-Water Interfacial Tension:** Not pertinent
13.10 **Vapor (Gas) Specific Gravity:** Not pertinent
13.11 **Ratio of Specific Heats of Vapor (Gas):** Not pertinent
13.12 **Latent Heat of Vaporization:** Not pertinent
13.13 **Heat of Combustion:** Not pertinent
13.14 **Heat of Decomposition:** Not pertinent
13.15 **Heat of Solution:** Not pertinent
13.16 **Heat of Polymerization:** Not pertinent

NOTES

PCR POTASSIUM CHLORATE

Common Synonyms			
Potcrate Chlorate of potassium Chlorate of potash	Solid Mixes with water.	White	Odorless

Call fire department.
Keep people away.
Isolate and remove discharged material.
Notify local health and pollution control agencies.

Fire	Not flammable. May cause fire on contact with combustibles. POISONOUS GASES ARE PRODUCED IN FIRE. Flood discharge area with water.
Exposure	Call for medical aid. DUST Irritating to eyes, nose and throat. Move victim to fresh air. If in eyes, hold eyelids open and flush with plenty of water. SOLID Irritating to skin and eyes. If swallowed will cause nausea, vomiting or loss of consciousness. Remove contaminated clothing and shoes. Flush affected areas with plenty of water. IF IN EYES, hold eyelids open and flush with plenty of water. IF SWALLOWED and victim is CONSCIOUS, have victim drink water or milk and have victim induce vomiting. IF SWALLOWED and victim is UNCONSCIOUS OR HAVING CONVULSIONS, do nothing except keep victim warm.
Water Pollution	Effect of low concentrations on aquatic life is unknown. May be dangerous if it enters water intakes. Notify local health and wildlife officials. Notify operators of nearby water intakes.

1. RESPONSE TO DISCHARGE

(See Response Methods Handbook, CG 446-4)

Issue warning—oxidizing material, water contaminant
Restrict access
Disperse and flush

2. LABEL

Yellow

3. CHEMICAL DESIGNATIONS

3.1 **Synonyms:** Chlorate of potash; Chlorate of potassium; potcrate
3.2 **Coast Guard Compatibility Classification:** Not applicable
3.3 **Chemical Formula:** $KClO_3$
3.4 **IMCO/United Nations Numerical Designation:** 5.1/1485

4. OBSERVABLE CHARACTERISTICS

4.1 **Physical State (as shipped):** Solid
4.2 **Color:** White
4.3 **Odor:** None

5. HEALTH HAZARDS

5.1 **Personal Protective Equipment:** Dust mask; rubber gloves; goggles; protective clothing to prevent contact with skin.
5.2 **Symptoms Following Exposure:** Inhalation of dust can irritate nose and throat. Contact with eyes or skin causes irritation. Ingestion causes abdominal pain, nausea, vomiting, cyanosis, collapse.
5.3 **Treatment for Exposure:** EYES: flush with water for at least 15 min. SKIN: wash with soap and water. INGESTION: dilute by drinking soapy or salty water to induce vomiting; call physician.
5.4 **Toxicity by Inhalation (Threshold Limit Value):** Data not available
5.5 **Short-Term Inhalation Limits:** Data not available
5.6 **Toxicity by Ingestion:** Grade 2; LD_{50} 0.5 to 5 g/kg
5.7 **Late Toxicity:** Data not available
5.8 **Vapor (Gas) Irritant Characteristics:** Data not available
5.9 **Liquid or Solid Irritant Characteristics:** Data not available
5.10 **Odor Threshold:** Not pertinent

6. FIRE HAZARDS

6.1 **Flash Point:** Not flammable, but may cause fire upon contact with ordinary combustibles
6.2 **Flammable Limits in Air:** Not pertinent
6.3 **Fire Extinguishing Agents:** Water on adjacent fires
6.4 **Fire Extinguishing Agents Not to be Used:** Not pertinent
6.5 **Special Hazards of Combustion Products:** Toxic fumes are formed in fires.
6.6 **Behavior in Fire:** Decomposes when hot to form oxygen, which increases severity of fire.
6.7 **Ignition Temperature:** Not pertinent
6.8 **Electrical Hazard:** Not pertinent
6.9 **Burning Rate:** Not pertinent

7. CHEMICAL REACTIVITY

7.1 **Reactivity with Water:** No reaction
7.2 **Reactivity with Common Materials:** Contact with combustible material may cause fire.
7.3 **Stability During Transport:** Stable
7.4 **Neutralizing Agents for Acids and Caustics:** Not pertinent
7.5 **Polymerization:** Not pertinent
7.6 **Inhibitor of Polymerization:** Not pertinent

8. WATER POLLUTION

8.1 **Aquatic Toxicity:** Data not available
8.2 **Waterfowl Toxicity:** Data not available
8.3 **Biological Oxygen Demand (BOD):** None
8.4 **Food Chain Concentration Potential:** None

9. SELECTED MANUFACTURERS

1. Pennwalt Corporation
   Chemicals Division
   Three Parkway
   Philadelphia, Pa. 19102
2. Hooker Chemical Corporation
   Specialty Chemicals Division
   P. O. Box 344
   Niagara Falls, N. Y. 14302
3. Mallinckrodt Chemical Works
   Industrial Chemicals Division
   Second and Mallinckrodt Streets
   P. O. Box 5439
   St. Louis, Mo. 63160

10. SHIPPING INFORMATION

10.1 **Grades or Purity:** Commercial: 99.7+%; Reagent; Purified
10.2 **Storage Temperature:** Ambient
10.3 **Inert Atmosphere:** No requirement
10.4 **Venting:** Open

11. HAZARD ASSESSMENT CODE

(See Hazard Assessment Handbook, CG 446-3)

SS

12. HAZARD CLASSIFICATIONS

12.1 **Code of Federal Regulations:** Oxidizing material
12.2 **NAS Hazard Rating for Bulk Water Transportation:** Not listed
12.3 **NFPA Hazard Classifications:**

Category	Classification
Health Hazard (Blue)	1
Flammability (Red)	0
Reactivity (Yellow)	2 oxy

13. PHYSICAL AND CHEMICAL PROPERTIES

13.1 **Physical State at 15°C and 1 atm:** Solid
13.2 **Molecular Weight:** 122.6
13.3 **Boiling Point at 1 atm:** Not pertinent (decomposes)
13.4 **Freezing Point:** 680°F = 360°C = 633°K
13.5 **Critical Temperature:** Not pertinent
13.6 **Critical Pressure:** Not pertinent
13.7 **Specific Gravity:** 2.34 at 20°C (solid)
13.8 **Liquid Surface Tension:** Not pertinent
13.9 **Liquid-Water Interfacial Tension:** Not pertinent
13.10 **Vapor (Gas) Specific Gravity:** Not pertinent
13.11 **Ratio of Specific Heats of Vapor (Gas):** Not pertinent
13.12 **Latent Heat of Vaporization:** Not pertinent
13.13 **Heat of Combustion:** Not pertinent
13.14 **Heat of Decomposition:** -176 Btu/lb $= -98$ cal/g $= -4.1 \times 10^5$ J/kg
13.15 **Heat of Solution:** 147 Btu/lb $= 81.9$ cal/g $= 3.43 \times 10^5$ J/kg
13.16 **Heat of Polymerization:** Not pertinent

NOTES

POTASSIUM CHROMATE

PCH

Common Synonyms			
Potassium chromate (VI) Neutral potassium chromate	Solid	Bright yellow	Odorless
	Sinks and mixes with water.		

Stop discharge if possible. Keep people away.
Avoid contact with solid and dust. Keep people away.
Isolate and remove discharged material.
Notify local health and pollution control agencies.

Fire

Not flammable.
Will increase the intensity of a fire.
May cause fire on contact with combustibles.
Flood discharge area with water.
Cool exposed containers with water.

Exposure

CALL FOR MEDICAL AID.

DUST
Irritating to eyes, nose and throat.
If inhaled will cause coughing or difficult breathing.
If in eyes, hold eyelids open and flush with plenty of water.
If breathing has stopped, give artificial respiration.
If breathing is difficult, give oxygen.

SOLID
POISONOUS IF SWALLOWED.
Irritating to skin and eyes.
If swallowed will cause nausea, vomiting or loss of consciousness.
Remove contaminated clothing and shoes.
Flush affected areas with plenty of water.
IF IN EYES, hold eyelids open and flush with plenty of water.
IF SWALLOWED and victim is CONSCIOUS, have victim drink water or milk and have victim induce vomiting.
IF SWALLOWED and victim is UNCONSCIOUS OR HAVING CONVULSIONS, do nothing except keep victim warm.

Water Pollution

HARMFUL TO AQUATIC LIFE IN VERY LOW CONCENTRATIONS.
May be dangerous if it enters water intakes.
Notify local health and wildlife officials.
Notify operators of nearby water intakes.

1. RESPONSE TO DISCHARGE

(See Response Methods Handbook, CG 446-4)

Disperse and flush

2. LABELS

No hazard label required by Code of Federal Regulations

3. CHEMICAL DESIGNATIONS

3.1 **Synonyms:** Neutral potassium chromate; Potassium chromate (VI)
3.2 **Coast Guard Compatibility Classification:** Not listed
3.3 **Chemical Formula:** K_2CrO_4
3.4 **IMCO/United Nations Numerical Designation:** Not listed

4. OBSERVABLE CHARACTERISTICS

4.1 **Physical State (as shipped):** Solid
4.2 **Color:** Bright yellow
4.3 **Odor:** None

5. HEALTH HAZARDS

5.1 **Personal Protective Equipment:** Bu. of Mines approved filter-type respirator; close-fitting safety goggles; rubber boots and apron; safety hat; face shield
5.2 **Symptoms Following Exposure:** Inhalation causes local irritation of mucous membranes; continuing nose irritation can result in perforation of nasal septum. Ingestion may cause violent gastroenteritis, circulatory collapse, vertigo, coma, and toxic nephritis; ingestion of excessive quantities can be fatal. Contact with eyes causes severe irritation and conjunctivitis. Repeated or prolonged exposure to dust, mist, or solutions may cause dermatitis; contact with breaks in the skin may cause "chrome sores" appearing as slow-healing, hard-rimmed ulcers which leave the area vulnerable to infection.
5.3 **Treatment for Exposure:** INHALATION: move to fresh air. INGESTION: give large amount of water; induce vomiting; treat peripheral vascular shock vigorously; get medical attention. EYES: flush with water for at least 15 min.; get medical attention. SKIN: flush with water; if irritation persists, get medical attention.
5.4 **Toxicity by Inhalation (Threshold Limit Value):** 0.1 mg/m^3 (as chromic acid)
5.5 **Short-Term Inhalation Limits:** Data not available
5.6 **Toxicity by Ingestion:** Grade 3; LD_{50} 50—500 mg/kg
5.7 **Late Toxicity:** Lung cancer may occur.
5.8 **Vapor (Gas) Irritant Characteristics:** Data not available
5.9 **Liquid or Solid Irritant Characteristics:** Data not available
5.10 **Odor Threshold:** Data not available

6. FIRE HAZARDS

6.1 **Flash Point:** Not flammable but may intensify fire.
6.2 **Flammable Limits in Air:** Not flammable
6.3 **Fire Extinguishing Agents:** Not pertinent
6.4 **Fire Extinguishing Agents Not to be Used:** Not pertinent
6.5 **Special Hazards of Combustion Products:** Not pertinent
6.6 **Behavior in Fire:** May increase intensity of fire if in contact with combustible materials. Cool containers and spilled material with plenty of water.
6.7 **Ignition Temperature:** Not pertinent
6.8 **Electrical Hazard:** Not pertinent
6.9 **Burning Rate:** Not pertinent

7. CHEMICAL REACTIVITY

7.1 **Reactivity with Water:** No reaction
7.2 **Reactivity with Common Materials:** In contact with combustible materials may cause fire
7.3 **Stability During Transport:** Stable
7.4 **Neutralizing Agents for Acids and Caustics:** Not pertinent
7.5 **Polymerization:** Not pertinent
7.6 **Inhibitor of Polymerization:** Not pertinent

8. WATER POLLUTION

8.1 **Aquatic Toxicity:**
45.6 ppm/96 hr/fathead minnow/TL_m/soft water
17.8 ppm/*/silver salmon/toxic/salt water
*Time period not specified.
8.2 **Waterfowl Toxicity:** Data not available
8.3 **Biological Oxygen Demand (BOD):** None
8.4 **Food Chain Concentration Potential:** Plants can absorb compound from water and pass it on up the food chain. Bioconcentrative up to 2,000 fold. Not likely to be a problem in a spill situation.

9. SELECTED MANUFACTURERS

1. Allied Chemical Corp.
   P. O. Box 1087R
   Morristown, N. J. 07960
2. J. T. Baker Chemical Co.
   Phillipsburg, N. J. 08865
3. Gallard-Schlesinger Chemical Mfg. Co.
   584 Mineola Ave.
   Carle Place, N. Y. 11514

10. SHIPPING INFORMATION

10.1 **Grades or Purity:** Reagent, 99%; CP; Technical
10.2 **Storage Temperature:** Ambient
10.3 **Inert Atmosphere:** No requirement
10.4 **Venting:** Open

11. HAZARD ASSESSMENT CODE

(See Hazard Assessment Handbook, CG 446-3)

SS

12. HAZARD CLASSIFICATIONS

12.1 **Code of Federal Regulations:** Not listed
12.2 **NAS Hazard Rating for Bulk Water Transportation:** Not listed
12.3 **NFPA Hazard Classifications:** Not listed

13. PHYSICAL AND CHEMICAL PROPERTIES

13.1 **Physical State at 15°C and 1 atm:** Solid
13.2 **Molecular Weight:** 194.20
13.3 **Boiling Point at 1 atm:** Not pertinent (decomposes)
13.4 **Freezing Point:** Not pertinent
13.5 **Critical Temperature:** Not pertinent
13.6 **Critical Pressure:** Not pertinent
13.7 **Specific Gravity:** 2.73 at 18°C (solid)
13.8 **Liquid Surface Tension:** Not pertinent
13.9 **Liquid-Water Interfacial Tension:** Not pertinent
13.10 **Vapor (Gas) Specific Gravity:** Not pertinent
13.11 **Ratio of Specific Heats of Vapor (Gas):** Not pertinent
13.12 **Latent Heat of Vaporization:** Not pertinent
13.13 **Heat of Combustion:** Not pertinent
13.14 **Heat of Decomposition:** Not pertinent
13.15 **Heat of Solution:** Not pertinent
13.16 **Heat of Polymerization:** Not pertinent

NOTES

PTC	POTASSIUM CYANIDE

Common Synonyms	Solid crystals	White	Almond odor
	Sinks and mixes with water.		

AVOID CONTACT WITH SOLID, DUST AND WATER SOLUTION.
Keep people away.
Wear chemical protective suit with self-contained breathing apparatus.
Isolate and remove discharged material.
Notify local health and pollution control agencies.

Fire	Not flammable.
Exposure	CALL FOR MEDICAL AID. DUST POISONOUS IF INHALED OR IF SKIN IS EXPOSED. Move to fresh air. If breathing has stopped, give artificial respiration. If breathing is difficult, give oxygen. SOLID POISONOUS IF SWALLOWED OR IF SKIN IS EXPOSED. Irritating to eyes. Remove contaminated clothing and shoes. Flush affected areas with plenty of water. IF IN EYES, hold eyelids open and flush with plenty of water. IF SWALLOWED and victim is CONSCIOUS, have victim drink water or milk and have victim induce vomiting. IF SWALLOWED and victim is UNCONSCIOUS OR HAVING CONVULSIONS, do nothing except keep victim warm.
Water Pollution	HARMFUL TO AQUATIC LIFE IN VERY LOW CONCENTRATIONS. May be dangerous if it enters water intakes. Notify local health and wildlife officials. Notify operators of nearby water intakes.

1. RESPONSE TO DISCHARGE

(See Response Methods Handbook, CG 446-4)

Issue warning—poison
Restrict access
Evacuate area
Chemical and physical treatment

2. LABEL

3. CHEMICAL DESIGNATIONS

3.1 **Synonyms:** Cyanide
3.2 **Coast Guard Compatibility Classification:** Not applicable
3.3 **Chemical Formula:** KCN
3.4 **IMCO/United Nations Numerical Designation:** 6.1/1680

4. OBSERVABLE CHARACTERISTICS

4.1 **Physical State (as shipped):** Solid
4.2 **Color:** White
4.3 **Odor:** Like hydrogen cyanide; almond-like

5. HEALTH HAZARDS

5.1 **Personal Protective Equipment:** Wear dry cotton gloves and U. S. Bureau of Mines approved dust respirator when handling solid potassium cyanide. Wear rubber gloves and approved chemical safety goggles when handling solutions.

5.2 **Symptoms Following Exposure:** Is a rapidly fatal poison when taken into the digestive system. Dust may cause toxic symptoms when inhaled, and prolonged contact with the skin may cause irritation and possibly poisoning if skin is broken. Strong solutions are corrosive to skin and may cause deep ulcers that heal slowly.

5.3 **Treatment for Exposure:** INGESTION: call physician immediately; have victim lie down and keep him quiet and warm. If he is CONSCIOUS, induce vomiting by having him drink warm salt water (1 tablespoon per cup of water); repeat until vomit fluid is clear; then give orally 1 pint of 1% solution of sodium thiosulfate, to be repeated in 15 min. If victim is NOT BREATHING, give artificial respiration until breathing starts. If victim is UNCONSCIOUS BUT BREATHING, give oxygen from an inhalator if he does not respond to treatment. In all cases, break an amyl nitrite pearl* in a cloth and hold lightly under victim's nose for 15 sec., repeating 5 times at about 15-sec. intervals; if necessary, repeat procedure every 3 min. with fresh pearls until 3 or 4 have been used.

* *Amyl nitrite pearls must not be over 2 years old. Avoid breathing the vapor while administering it to the victim.*

5.4 **Toxicity by Inhalation (Threshold Limit Value):** Not pertinent
5.5 **Short-Term Inhalation Limits:** Not pertinent
5.6 **Toxicity by Ingestion:** Grade 4; below 50 mg/kg (mice)

6. FIRE HAZARDS

6.1 **Flash Point:** Not flammable
6.2 **Flammable Limits in Air:** Not flammable
6.3 **Fire Extinguishing Agents:** Not pertinent
6.4 **Fire Extinguishing Agents Not to be Used:** Not pertinent
6.5 **Special Hazards of Combustion Products:** Not pertinent
6.6 **Behavior in Fire:** Not pertinent
6.7 **Ignition Temperature:** Not flammable
6.8 **Electrical Hazard:** Not pertinent
6.9 **Burning Rate:** Not flammable

7. CHEMICAL REACTIVITY

7.1 **Reactivity with Water:** When potassium cyanide dissolves in water, a mild reaction occurs and some poisonous hydrogen cyanide gas is released. This gas is not hazardous except in an enclosed space. If the water is acidic, however, toxic amounts of the gas will form at once.
7.2 **Reactivity with Common Materials:** Contact with even weak acids causes formation of deadly hydrogen cyanide gas.
7.3 **Stability During Transport:** Stable
7.4 **Neutralizing Agents for Acids and Caustics:** Not pertinent
7.5 **Polymerization:** Not pertinent
7.6 **Inhibitor of Polymerization:** Not pertinent

8. WATER POLLUTION

8.1 **Aquatic Toxicity:**
0.16 ppm/48 hr/bluegill/TL_m/fresh water
0.49 ppm/48 hr/adult zebrafish/TL_m/salt water
8.2 **Waterfowl Toxicity:** Data not available
8.3 **Biological Oxygen Demand (BOD):** 0% of theoretical in 7 days
8.4 **Food Chain Concentration Potential:** None

9. SELECTED MANUFACTURERS

1. Allied Chemical Corp.
   Specialty Chemicals Division
   Baton Rouge, La. 70821
2. E. I. duPont de Nemours & Co., Inc.
   Electrochemicals Dept.
   Wilmington, Del. 19898
3. Harstan Chemical Corp.
   1247 38 St.
   Brooklyn, N. Y. 11218

10. SHIPPING INFORMATION

10.1 **Grades or Purity:** 99.0%
10.2 **Storage Temperature:** Ambient
10.3 **Inert Atmosphere:** No requirement
10.4 **Venting:** Sealed containers must be stored in a well-ventilated area.

11. HAZARD ASSESSMENT CODE

(See Hazard Assessment Handbook, CG 446-3)

SS

12. HAZARD CLASSIFICATIONS

12.1 **Code of Federal Regulations:** Poisonous liquid or solid, Class B
12.2 **NAS Hazard Rating for Bulk Water Transportation:** Not listed
12.3 **NFPA Hazard Classifications:**

Category	Classification
Health Hazard (Blue)	3
Flammability (Red)	0
Reactivity (Yellow)	0

13. PHYSICAL AND CHEMICAL PROPERTIES

13.1 **Physical State at 15°C and 1 atm:** Solid
13.2 **Molecular Weight:** 65.12
13.3 **Boiling Point at 1 atm:** Very high
13.4 **Freezing Point:** 1174.1°F = 634.5°C = 907.7°K
13.5 **Critical Temperature:** Not pertinent
13.6 **Critical Pressure:** Not pertinent
13.7 **Specific Gravity:** 1.52 at 16°C (solid)
13.8 **Liquid Surface Tension:** Not pertinent
13.9 **Liquid-Water Interfacial Tension:** Not pertinent
13.10 **Vapor (Gas) Specific Gravity:** Not pertinent
13.11 **Ratio of Specific Heats of Vapor (Gas):** Not pertinent
13.12 **Latent Heat of Vaporization:** Not pertinent
13.13 **Heat of Combustion:** Not pertinent
13.14 **Heat of Decomposition:** Not pertinent
13.15 **Heat of Solution:** Not pertinent
13.16 **Heat of Polymerization:** Not pertinent

5. HEALTH HAZARDS (Cont'd.)

5.7 **Late Toxicity:** None
5.8 **Vapor (Gas) Irritant Characteristics:** Non-volatile, but moisture in air can liberate some lethal hydrogen cyanide gas.
5.9 **Liquid or Solid Irritant Characteristics:** Moist solid can cause caustic-type irritation of skin and formation of ulcers.
5.10 **Odor Threshold:** Not pertinent

REVISED 1978

PDT | POTASSIUM DICHLORO-s-TRIAZINETRIONE

Common Synonyms			
Potassium dichloroisocyanurate	Solid	White	Chlorine-like odor
	Mixes with water.		

Keep people away.
Isolate and remove discharged material.
Notify local health and pollution control agencies.

Fire

Not flammable.
May cause fire on contact with combustibles.
POISONOUS GASES MAY BE PRODUCED IN FIRE.
Wear goggles and self-contained breathing apparatus.
Flood discharge area with water.

Exposure

Call for medical aid.

DUST
Irritating to eyes, nose and throat.
If inhaled will cause coughing or difficult breathing.
Move victim to fresh air.
If in eyes, hold eyelids open and flush with plenty of water.
If breathing has stopped, give artificial respiration.
If breathing is difficult, give oxygen.

SOLID
Irritating to skin and eyes.
Harmful if swallowed.
Remove contaminated clothing and shoes.
Flush affected areas with plenty of water.
IF IN EYES, hold eyelids open and flush with plenty of water.
IF SWALLOWED and victim is CONSCIOUS, have victim drink water or milk and have victim induce vomiting.
IF SWALLOWED and victim is UNCONSCIOUS OR HAVING CONVULSIONS, do nothing except keep victim warm.

Water Pollution

Effect of low concentrations on aquatic life is unknown.
May be dangerous if it enters water intakes.
Notify local health and wildlife officials.
Notify operators of nearby water intakes.

1. RESPONSE TO DISCHARGE

(See Response Methods Handbook, CG 446-4)

Issue warning—oxidizing material, water contaminant
Restrict access
Disperse and flush

2. LABEL

Yellow

3. CHEMICAL DESIGNATIONS

3.1 **Synonyms:** Potassium dichloroisocyanurate
3.2 **Coast Guard Compatibility Classification:** Not applicable
3.3 **Chemical Formula:** $KCl_2(NCO)_3$
3.4 **IMCO/United Nations Numerical Designation:** 5.1/—

4. OBSERVABLE CHARACTERISTICS

4.1 **Physical State (as shipped):** Solid
4.2 **Color:** White
4.3 **Odor:** Like chlorine

5. HEALTH HAZARDS

5.1 **Personal Protective Equipment:** Dust mask or chlorine canister mask; goggles; rubber gloves and other protective clothing to prevent contact with skin.
5.2 **Symptoms Following Exposure:** Dust causes sneezing and coughing; is moderately irritating to the eyes and causes itching and redness of skin. Ingestion causes burns of mouth and stomach.
5.3 **Treatment for Exposure:** INHALATION: remove victim to fresh air. EYES: irrigate with running water for 15 min.; call a physician. SKIN: flush with water. INGESTION: induce vomiting and call physician.
5.4 **Toxicity by Inhalation (Threshold Limit Value):** Data not available
5.5 **Short-Term Inhalation Limits:** Data not available
5.6 **Toxicity by Ingestion:** Grade 2; LD_{50} 0.5 to 5 g/kg
5.7 **Late Toxicity:** Data not available
5.8 **Vapor (Gas) Irritant Characteristics:** Not pertinent
5.9 **Liquid or Solid Irritant Characteristics:** Data not available
5.10 **Odor Threshold:** Data not available

6. FIRE HAZARDS

6.1 **Flash Point:** Not flammable but may cause fire upon contact with ordinary combustibles
6.2 **Flammable Limits in Air:** Not pertinent
6.3 **Fire Extinguishing Agents:** Water
6.4 **Fire Extinguishing Agents Not to be Used:** Not pertinent
6.5 **Special Hazards of Combustion Products:** May form toxic chlorine and other gases in fire
6.6 **Behavior in Fire:** Decomposition can be initiated with a heat source and can propagate throughout the mass with the evolution of dense fumes. Containers may explode when heated.
6.7 **Ignition Temperature:** Not pertinent
6.8 **Electrical Hazard:** Not pertinent
6.9 **Burning Rate:** Not pertinent

7. CHEMICAL REACTIVITY

7.1 **Reactivity with Water:** Forms a bleach solution; the reaction is not violent.
7.2 **Reactivity with Common Materials:** Contact with most foreign materials, organic matter, or easily chlorinated or oxidized materials may result in fire. Avoid oil, grease, sawdust, floor sweepings, other easily oxidized organic compounds.
7.3 **Stability During Transport:** Stable if dry
7.4 **Neutralizing Agents for Acids and Caustics:** Not pertinent
7.5 **Polymerization:** Not pertinent
7.6 **Inhibitor of Polymerization:** Not pertinent

8. WATER POLLUTION

8.1 **Aquatic Toxicity:** Data not available
8.2 **Waterfowl Toxicity:** Data not available
8.3 **Biological Oxygen Demand (BOD):** None
8.4 **Food Chain Concentration Potential:** None

9. SELECTED MANUFACTURERS

FMC Corporation
Industrial Chemical Division
633 Third Avenue
New York, N. Y. 10017

10. SHIPPING INFORMATION

10.1 **Grades or Purity:** Technical; 39—59% available chlorine.
10.2 **Storage Temperature:** Cool ambient
10.3 **Inert Atmosphere:** No requirement
10.4 **Venting:** Pressure-vacuum

11. HAZARD ASSESSMENT CODE

(See Hazard Assessment Handbook, CG 446-3)

SS

12. HAZARD CLASSIFICATIONS

12.1 **Code of Federal Regulations:** Oxidizing material
12.2 **NAS Hazard Rating for Bulk Water Transportation:** Not listed
12.3 **NFPA Hazard Classifications:**

Category	Classification
Health Hazard (Blue)	3
Flammability (Red)	0
Reactivity (Yellow)	3 oxy

13. PHYSICAL AND CHEMICAL PROPERTIES

13.1 **Physical State at 15°C and 1 atm:** Solid
13.2 **Molecular Weight:** 236.1
13.3 **Boiling Point at 1 atm:** Not pertinent (decomposes)
13.4 **Freezing Point:** Not pertinent
13.5 **Critical Temperature:** Not pertinent
13.6 **Critical Pressure:** Not pertinent
13.7 **Specific Gravity:** 0.96 at 20°C (solid)
13.8 **Liquid Surface Tension:** Not pertinent
13.9 **Liquid-Water Interfacial Tension:** Not pertinent
13.10 **Vapor (Gas) Specific Gravity:** Not pertinent
13.11 **Ratio of Specific Heats of Vapor (Gas):** Not pertinent
13.12 **Latent Heat of Vaporization:** Not pertinent
13.13 **Heat of Combustion:** Not pertinent
13.14 **Heat of Decomposition:** Data not available
13.15 **Heat of Solution:** Not pertinent
13.16 **Heat of Polymerization:** Not pertinent

NOTES

PTD	POTASSIUM DICHROMATE

Common Synonyms Potassium bichromate Bichrome	Solid crystals · Red to orange · Odorless Sinks and mixes with water.

Avoid contact with solid and dust. Keep people away.
Isolate and remove discharged material.
Notify local health and pollution control agencies.

Fire	Not flammable. May cause fire on contact with combustibles. Flood discharge area with water. Cool exposed containers with water.
Exposure	CALL FOR MEDICAL AID. DUST Irritating to eyes, nose, and throat. If inhaled, will cause difficult breathing. Move to fresh air. If breathing has stopped, give artificial respiration. If breathing is difficult, give oxygen. SOLIDS Will burn skin and eyes. If swallowed, will cause nausea, vomiting, or loss of consciousness. Remove contaminated clothing and shoes. Flush affected areas with plenty of water. IF IN EYES, hold eyelids open and flush with plenty of water. IF SWALLOWED and victim is CONSCIOUS, have victim drink water or milk. DO NOT INDUCE VOMITING.
Water Pollution	Dangerous to aquatic life in high concentrations. May be dangerous if it enters water intakes. Notify local health and wildlife officials. Notify operators of nearby water intakes.

1. RESPONSE TO DISCHARGE

(See Response Methods Handbook, CG 446-4)

Issue warning—water contaminant
Disperse and flush

2. LABELS

No hazard label required by Code of Federal Regulations

3. CHEMICAL DESIGNATIONS

3.1 **Synonyms:** Potassium bichromate
Bichrome
3.2 **Coast Guard Compatibility Classification:** Not applicable
3.3 **Chemical Formula:** $K_2Cr_2O_7$
3.4 **IMCO/United Nations Numerical Designation:** 9.0/1874

4. OBSERVABLE CHARACTERISTICS

4.1 **Physical State (as shipped):** Solid
4.2 **Color:** Orange-red
4.3 **Odor:** Odorless

5. HEALTH HAZARDS

5.1 **Personal Protective Equipment:** Approved dust mask; protective gloves; goggles or face shield.
5.2 **Symptoms Following Exposure:** Highly corrosive to skin and mucous membranes. If ingested, causes violent gastroenteritis, peripheral vascular collapse, vertigo, muscle cramps, coma, and (later) toxic nephritis with glycosuria. Allergic reactions may also occur.
5.3 **Treatment for Exposure:** INGESTION: have victim drink water or milk; do NOT induce vomiting. SKIN: treat like acid burns; external lesions may be scrubbed with 2% solution of sodium thiosulfate.
5.4 **Toxicity by Inhalation (Threshold Limit Value):** Not pertinent
5.5 **Short-Term Inhalation Limits:** Not pertinent
5.6 **Toxicity by Ingestion:** Grade 3; 50 to 500 mg/kg (human)
5.7 **Late Toxicity:** Some suggestion of lung cancer.
5.8 **Vapor (Gas) Irritant Characteristics:** Dusts or mists may cause severe irritation of eyes and throat and can cause eye and lung injury. They cannot be tolerated even at low concentrations.
5.9 **Liquid or Solid Irritant Characteristics:** Severe skin irritant; causes second- and third-degree burns on short contact and is very injurious to the eyes.
5.10 **Odor Threshold:** Not pertinent

6. FIRE HAZARDS

6.1 **Flash Point:** Not flammable
6.2 **Flammable Limits in Air:** Not flammable
6.3 **Fire Extinguishing Agents:** Flood spill area with water.
6.4 **Fire Extinguishing Agents Not to be Used:** Not pertinent
6.5 **Special Hazards of Combustion Products:** Not pertinent
6.6 **Behavior in Fire:** May decompose, generating oxygen. Supports the combustion of other materials.
6.7 **Ignition Temperature:** Not flammable
6.8 **Electrical Hazard:** Not pertinent
6.9 **Burning Rate:** Not flammable

7. CHEMICAL REACTIVITY

7.1 **Reactivity with Water:** No reaction
7.2 **Reactivity with Common Materials:** Ignition may occur when in contact with finely divided combustibles, such as sawdust.
7.3 **Stability During Transport:** Stable
7.4 **Neutralizing Agents for Acids and Caustics:** Not pertinent
7.5 **Polymerization:** Not pertinent
7.6 **Inhibitor of Polymerization:** Not pertinent

8. WATER POLLUTION

8.1 **Aquatic Toxicity:**
739 ppm/24 hr/bluegill/TL_m/fresh water
17.8 ppm/*/silver salmon/toxic/salt water
*Time period not specified.
8.2 **Waterfowl Toxicity:** Data not available
8.3 **Biological Oxygen Demand (BOD):** None
8.4 **Food Chain Concentration Potential:** None

9. SELECTED MANUFACTURERS

1. Allied Chemical Corp.
Industrial Chemicals Division
Morristown, N. J. 07960
2. Diamond Shamrock Corp.
Soda Products Division
Belleville Tpk.
Kearny, N. J. 07032
3. Mallinckrodt Chemical Works
Industrial Chemicals Division
2nd and Mallinckrodt Sts.
St. Louis, Mo. 63160

10. SHIPPING INFORMATION

10.1 **Grades or Purity:** 99.9%
10.2 **Storage Temperature:** Ambient
10.3 **Inert Atmosphere:** No requirement
10.4 **Venting:** Open

11. HAZARD ASSESSMENT CODE

(See Hazard Assessment Handbook, CG 446-3)

SS

12. HAZARD CLASSIFICATIONS

12.1 **Code of Federal Regulations:** ORM-A
12.2 **NAS Hazard Rating for Bulk Water Transportation:** Not listed
12.3 **NFPA Hazard Classifications:**

Category	Classification
Health Hazard (Blue)	1
Flammability (Red)	0
Reactivity (Yellow)	1 oxy

13. PHYSICAL AND CHEMICAL PROPERTIES

13.1 **Physical State at 15°C and 1 atm:** Solid
13.2 **Molecular Weight:** 294.19
13.3 **Boiling Point at 1 atm:** Decomposes
13.4 **Freezing Point:** 748°F = 398°C = 671°K
13.5 **Critical Temperature:** Not pertinent
13.6 **Critical Pressure:** Not pertinent
13.7 **Specific Gravity:** 2.676 at 25°C (solid)
13.8 **Liquid Surface Tension:** Not pertinent
13.9 **Liquid-Water Interfacial Tension:** Not pertinent
13.10 **Vapor (Gas) Specific Gravity:** Not pertinent
13.11 **Ratio of Specific Heats of Vapor (Gas):** Not pertinent
13.12 **Latent Heat of Vaporization:** Not pertinent
13.13 **Heat of Combustion:** Not pertinent
13.14 **Heat of Decomposition:** Not pertinent
13.15 **Heat of Solution:** Not pertinent
13.16 **Heat of Polymerization:** Not pertinent

NOTES

REVISED 1978

PTH	# POTASSIUM HYDROXIDE

Common Synonyms			
Caustic potash Lye	Solid crystals, or watery liquid	White solid or colorless liquid	Odorless
	Solid sinks and mixes slowly with water. Liquid mixes with water.		

Avoid contact with liquid, solid, vapor, and dust.
Keep people away.
Wear rubber overclothing (including gloves).
Stop discharge if possible.
Isolate and remove discharged material.
Notify local health and pollution control agencies.

Fire	Not flammable. Flammable gas may be produced on contact with metals. May cause fire on contact with moisture and combustibles. Wear rubber overclothing (including gloves). Flood discharge area with water. Cool exposed containers with water.
Exposure	CALL FOR MEDICAL AID. DUST OR MIST Irritating to eyes, nose, and throat. Harmful if inhaled. Move to fresh air. If breathing has stopped, give artificial respiration. If breathing is difficult, give oxygen. IF IN EYES, hold eyelids open and flush with plenty of water. SOLID OR LIQUID Will burn skin and eyes. Harmful if swallowed. Remove contaminated clothing or shoes. Flush affected areas with plenty of water. IF IN EYES, hold eyelids open and flush with plenty of water. IF SWALLOWED and victim is CONSCIOUS, have victim drink water or milk. DO NOT INDUCE VOMITING.
Water Pollution	HARMFUL TO AQUATIC LIFE IN VERY LOW CONCENTRATIONS. May be dangerous if it enters water intakes. Notify local health and wildlife officials. Notify operators of nearby water intakes.

1. RESPONSE TO DISCHARGE

(See Response Methods Handbook, CG 446-4)

Issue warning—corrosive
Restrict access
Disperse and flush

2. LABEL

3. CHEMICAL DESIGNATIONS

3.1 **Synonyms:** Caustic potash
3.2 **Coast Guard Compatibility Classification:** Not applicable
3.3 **Chemical Formula:** KOH
3.4 **IMCO/United Nations Numerical Designation:** 8.0/1814

4. OBSERVABLE CHARACTERISTICS

4.1 **Physical State (as shipped):** Solid
4.2 **Color:** White
4.3 **Odor:** None

5. HEALTH HAZARDS

5.1 **Personal Protective Equipment:** Wide-brimmed hat and close-fitting safety goggles with rubber side shields; respirator for dust; long-sleeved cotton shirt or jacket with buttoned collar and buttoned sleeves; rubber or rubber-coated canvas gloves (shirt sleeves should be buttoned over the gloves); rubber shoes or boots; cotton coveralls (with trouser cuffs worn over boots); rubber apron.
5.2 **Symptoms Following Exposure:** Causes severe burns of eyes, skin, and mucous membranes.
5.3 **Treatment for Exposure:** (Act quickly!) Call a physician at once, even when injury seems to be slight. INGESTION: give water and milk; do NOT induce vomiting. EYES: flush with water at once for at least 15 min. SKIN: flush with water, then rinse with dilute vinegar.
5.4 **Toxicity by Inhalation (Threshold Limit Value):** Not pertinent
5.5 **Short-Term Inhalation Limits:** Not pertinent
5.6 **Toxicity by Ingestion:** Grade 3; oral rat LD_{50} = 364 mg/kg
5.7 **Late Toxicity:** None
5.8 **Vapor (Gas) Irritant Characteristics:** Not pertinent
5.9 **Liquid or Solid Irritant Characteristics:** Severe skin irritant. Causes second- and third-degree burns on short contact; and is very injurious to the eyes.
5.10 **Odor Threshold:** Not pertinent

6. FIRE HAZARDS

6.1 **Flash Point:** Not flammable
6.2 **Flammable Limits in Air:** Not flammable
6.3 **Fire Extinguishing Agents:** Not pertinent
6.4 **Fire Extinguishing Agents Not to be Used:** Not pertinent
6.5 **Special Hazards of Combustion Products:** Not pertinent
6.6 **Behavior in Fire:** Not pertinent
6.7 **Ignition Temperature:** Not flammable
6.8 **Electrical Hazard:** Not pertinent
6.9 **Burning Rate:** Not flammable

7. CHEMICAL REACTIVITY

7.1 **Reactivity with Water:** Dissolves with liberation of much heat; may steam and spatter.
7.2 **Reactivity with Common Materials:** When wet, attacks metals such as aluminum, tin, lead, and zinc to produce flammable hydrogen gas.
7.3 **Stability During Transport:** Stable
7.4 **Neutralizing Agents for Acids and Caustics:** Flush with water, rinse with dilute acetic acid.
7.5 **Polymerization:** Not pertinent
7.6 **Inhibitor of Polymerization:** Not pertinent

8. WATER POLLUTION

8.1 **Aquatic Toxicity:** 80 ppm/24 hr/mosquito fish/TL_m/fresh water
8.2 **Waterfowl Toxicity:** Data not available
8.3 **Biological Oxygen Demand (BOD):** None
8.4 **Food Chain Concentration Potential:** None

9. SELECTED MANUFACTURERS

1. Diamond Shamrock Corp.
Electro Chemicals Division
300 Union Commerce Bldg.
Cleveland, Ohio 44115
2. Monsanto Co.
Monsanto Industrial Chemicals Co.
800 North Lindbergh Blvd.
St. Louis, Mo. 63166
3. Occidental Petroleum Corp.
Hooker Chemical Corp.
Industrial Chemicals Division
Niagara Falls, N. Y. 14302

10. SHIPPING INFORMATION

10.1 **Grades or Purity:** Technical flake: 85-90%; USP pellets: 85-90%
10.2 **Storage Temperature:** Ambient
10.3 **Inert Atmosphere:** No requirement
10.4 **Venting:** Open

11. HAZARD ASSESSMENT CODE

(See Hazard Assessment Handbook, CG 446-3)

SS

12. HAZARD CLASSIFICATIONS

12.1 **Code of Federal Regulations:** Corrosive Material
12.2 **NAS Hazard Rating for Bulk Water Transportation:** Not listed
12.3 **NFPA Hazard Classifications:**

Category	Classification
Health Hazard (Blue)	3
Flammability (Red)	0
Reactivity (Yellow)	1

13. PHYSICAL AND CHEMICAL PROPERTIES

13.1 **Physical State at 15°C and 1 atm:** Solid
13.2 **Molecular Weight:** 56.11
13.3 **Boiling Point at 1 atm:** Very high
13.4 **Freezing Point:** 716°F = 380°C = 653°K
13.5 **Critical Temperature:** Not pertinent
13.6 **Critical Pressure:** Not pertinent
13.7 **Specific Gravity:** 2.04 at 15°C (solid)
13.8 **Liquid Surface Tension:** Not pertinent
13.9 **Liquid-Water Interfacial Tension:** Not pertinent
13.10 **Vapor (Gas) Specific Gravity:** Not pertinent
13.11 **Ratio of Specific Heats of Vapor (Gas):** Not pertinent
13.12 **Latent Heat of Vaporization:** Not pertinent
13.13 **Heat of Combustion:** Not pertinent
13.14 **Heat of Decomposition:** Not pertinent
13.15 **Heat of Solution:** Not pertinent
13.16 **Heat of Polymerization:** Not pertinent

NOTES

REVISED 1978

PTI	# POTASSIUM IODIDE

Common Synonyms	Solid crystals White Odorless Sinks and mixes with water.
Stop discharge if possible. Keep people away. Isolate and remove discharged material. Notify local health and pollution control agencies.	
Fire	Not flammable.
Exposure	CALL FOR MEDICAL AID. SOLID Harmful if swallowed. Flush affected areas with plenty of water. IF SWALLOWED and victim is CONSCIOUS, have victim drink water or milk.
Water Pollution	Effect of low concentrations on aquatic life is unknown. May be dangerous if it enters water intakes. Notify local health and wildlife officials. Notify operators of nearby water intakes.

1. RESPONSE TO DISCHARGE
(See Response Methods Handbook, CG 446-4)

Issue warning—water contaminant
Disperse and flush

2. LABELS
No hazard label required by Code of Federal Regulations

3. CHEMICAL DESIGNATIONS
3.1 **Synonyms:** No common synonyms
3.2 **Coast Guard Compatibility Classification:** Not applicable
3.3 **Chemical Formula:** KI
3.4 **IMCO/United Nations Numerical Designation:** Not listed

4. OBSERVABLE CHARACTERISTICS
4.1 **Physical State (as shipped):** Solid
4.2 **Color:** White
4.3 **Odor:** Odorless

5. HEALTH HAZARDS
5.1 **Personal Protective Equipment:** Goggles or face shield.
5.2 **Symptoms Following Exposure:** May irritate eyes or open cuts.
5.3 **Treatment for Exposure:** Flush all affected areas with water.
5.4 **Toxicity by Inhalation (Threshold Limit Value):** Not pertinent
5.5 **Short-Term Inhalation Limits:** Not pertinent
5.6 **Toxicity by Ingestion:** Grade 2; 0.5 to 5 g/kg (human)
5.7 **Late Toxicity:** Data not available
5.8 **Vapor (Gas) Irritant Characteristics:** Non-volatile
5.9 **Liquid or Solid Irritant Characteristics:** None
5.10 **Odor Threshold:** Not pertinent

6. FIRE HAZARDS
6.1 **Flash Point:** Not flammable
6.2 **Flammable Limits in Air:** Not flammable
6.3 **Fire Extinguishing Agents:** Not pertinent
6.4 **Fire Extinguishing Agents Not to be Used:** Not pertinent
6.5 **Special Hazards of Combustion Products:** Not pertinent
6.6 **Behavior in Fire:** Not pertinent
6.7 **Ignition Temperature:** Not flammable
6.8 **Electrical Hazard:** Not pertinent
6.9 **Burning Rate:** Not flammable

7. CHEMICAL REACTIVITY
7.1 **Reactivity with Water:** No reaction
7.2 **Reactivity with Common Materials:** Corrosive in all concentrations to most metals, except stainless steel, titanium, and tantalum.
7.3 **Stability During Transport:** Stable
7.4 **Neutralizing Agents for Acids and Caustics:** Not pertinent
7.5 **Polymerization:** Not pertinent
7.6 **Inhibitor of Polymerization:** Not pertinent

8. WATER POLLUTION
8.1 **Aquatic Toxicity:**
7.5 ppm/*/daphnia/toxic/fresh water
*Time period not specified.
8.2 **Waterfowl Toxicity:** Data not available
8.3 **Biological Oxygen Demand (BOD):** None
8.4 **Food Chain Concentration Potential:** None

9. SELECTED MANUFACTURERS
1. Allied Chemical Corp.
Specialty Chemicals Division
Marcus Hook, Pa. 19061
2. Kewanee Oil Co.
Harshaw Chemical Co. Division
1945 E. 97 St.
Cleveland, Ohio 44106
3. Mallinckrodt Chemical Works
Industrial Chemicals Division
2nd and Mallinckrodt Sts.
St. Louis, Mo. 63160

10. SHIPPING INFORMATION
10.1 **Grades or Purity:** USP, ACS, CP (all 99+%)
10.2 **Storage Temperature:** Ambient
10.3 **Inert Atmosphere:** No requirement
10.4 **Venting:** Open

11. HAZARD ASSESSMENT CODE
(See Hazard Assessment Handbook, CG 446-3)

SS

12. HAZARD CLASSIFICATIONS
12.1 **Code of Federal Regulations:** Not listed
12.2 **NAS Hazard Rating for Bulk Water Transportation:** Not listed
12.3 **NFPA Hazard Classifications:** Not listed

13. PHYSICAL AND CHEMICAL PROPERTIES
13.1 **Physical State at 15°C and 1 atm:** Solid
13.2 **Molecular Weight:** 166.01
13.3 **Boiling Point at 1 atm:** Very high
13.4 **Freezing Point:** 1258°F = 681°C = 954°K
13.5 **Critical Temperature:** Not pertinent
13.6 **Critical Pressure:** Not pertinent
13.7 **Specific Gravity:** 3.13 at 15°C (solid)
13.8 **Liquid Surface Tension:** Not pertinent
13.9 **Liquid-Water Interfacial Tension:** Not pertinent
13.10 **Vapor (Gas) Specific Gravity:** Not pertinent
13.11 **Ratio of Specific Heats of Vapor (Gas):** Not pertinent
13.12 **Latent Heat of Vaporization:** Not pertinent
13.13 **Heat of Combustion:** Not pertinent
13.14 **Heat of Decomposition:** Not pertinent
13.15 **Heat of Solution:** Not pertinent
13.16 **Heat of Polymerization:** Not pertinent

NOTES

PTM POTASSIUM, METALLIC

Common Synonyms

Solid Silver-white Odorless

Reacts violently with water.
Flammable gas is produced.

Avoid contact with solid. Keep people away.
Call fire department.
Isolate and remove discharged material.
Notify local health and pollution control agencies.

Fire

Combustible.
IGNITES WHEN EXPOSED TO WATER OR MOISTURE.
Flammable gas is produced on contact with water.
Extinguish with dry graphite, soda ash, or other inert powder.
DO NOT USE WATER OR FOAM ON FIRE.
DO NOT USE WATER OR FOAM ON ADJACENT FIRES.

Exposure

CALL FOR MEDICAL AID.

SOLID
Will burn skin and eyes.
Flush affected areas with plenty of water.
IF IN EYES, hold eyelids open and flush with plenty of water.

Water Pollution

HARMFUL TO AQUATIC LIFE IN VERY LOW CONCENTRATIONS.
May be dangerous if it enters water intakes.
Notify local health and wildlife officials.
Notify operators of nearby water intakes.

1. RESPONSE TO DISCHARGE

(See Response Methods Handbook, CG 446-4)

Issue warning—high flammability, corrosive
Restrict access
Disperse and flush

2. LABELS

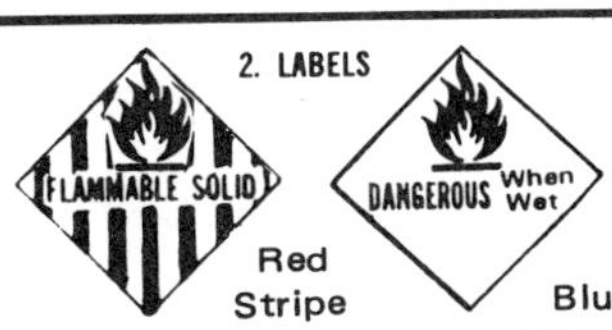

Red Stripe Blue

3. CHEMICAL DESIGNATIONS

3.1 **Synonyms:** No common synonyms
3.2 **Coast Guard Compatibility Classification:** Not listed
3.3 **Chemical Formula:** K
3.4 **IMCO/United Nations Numerical Designation:** Not listed

4. OBSERVABLE CHARACTERISTICS

4.1 **Physical State (as shipped):** Solid
4.2 **Color:** Silvery white
4.3 **Odor:** None

5. HEALTH HAZARDS

5.1 **Personal Protective Equipment:** Goggles or face shield; rubber gloves
5.2 **Symptoms Following Exposure:** Contact with eyes or skin causes severe burns.
5.3 **Treatment for Exposure:** EYES or SKIN: flush with water; treat caustic burns.
5.4 **Toxicity by Inhalation (Threshold Limit Value):** Data not available
5.5 **Short-Term Inhalation Limits:** Not pertinent
5.6 **Toxicity by Ingestion:** Not pertinent
5.7 **Late Toxicity:** Data not available
5.8 **Vapor (Gas) Irritant Characteristics:** Data not available
5.9 **Liquid or Solid Irritant Characteristics:** Data not available
5.10 **Odor Threshold:** Not pertinent

6. FIRE HAZARDS

6.1 **Flash Point:** Not pertinent (combustible solid)
6.2 **Flammable Limits in Air:** Not pertinent
6.3 **Fire Extinguishing Agents:** Graphite, sand, sodium chloride
6.4 **Fire Extinguishing Agents Not to be Used:** Water, foam, carbon dioxide, or halogenated hydrocarbons
6.5 **Special Hazards of Combustion Products:**
6.6 **Behavior in Fire:** Reacts violently with water, forming flammable and explosive hydrogen gas. May ignite spontaneously in air.
6.7 **Ignition Temperature:** Data not available
6.8 **Electrical Hazard:** Not pertinent
6.9 **Burning Rate:** Not pertinent

7. CHEMICAL REACTIVITY

7.1 **Reactivity with Water:** Reacts violently to form flammable hydrogen gas and a strong caustic solution.
7.2 **Reactivity with Common Materials:** May ignite combustible materials if they are damp
7.3 **Stability During Transport:** Stable, if protected from air and moisture
7.4 **Neutralizing Agents for Acids and Caustics:** Caustic formed by reaction with water should be flushed with water, then area can be rinsed with dilute acetic acid.
7.5 **Polymerization:** Not pertinent
7.6 **Inhibitor of Polymerization:** Not pertinent

8. WATER POLLUTION

8.1 **Aquatic Toxicity:** 80 ppm/24 hr/mosquito fish/TL_m/fresh water
8.2 **Waterfowl Toxicity:** Data not available
8.3 **Biological Oxygen Demand (BOD):** None
8.4 **Food Chain Concentration Potential:** None

9. SELECTED MANUFACTURERS

1. Kawecki Berylco Industries, Inc.
220 East 42 St.
New York, N. Y. 10017
2. Gallard-Schlesinger Chemical Mfg. Co.
584 Mineola Ave.
Carle Place, N. Y. 11514
3. Ventron Corp.
P. O. Box 159
Beverly, Mass. 01915

10. SHIPPING INFORMATION

10.1 **Grades or Purity:** Commercial, 99.9+% (Shipped under oil)
10.2 **Storage Temperature:** Ambient
10.3 **Inert Atmosphere:** Inerted
10.4 **Venting:** Pressure-vacuum

11. HAZARD ASSESSMENT CODE

(See Hazard Assessment Handbook, CG 446-3)

RR

12. HAZARD CLASSIFICATIONS

12.1 **Code of Federal Regulations:** Flammable solid
12.2 **NAS Hazard Rating for Bulk Water Transportation:** Not listed
12.3 **NFPA Hazard Classifications:** Not listed

13. PHYSICAL AND CHEMICAL PROPERTIES

13.1 **Physical State at 15°C and 1 atm:** Solid
13.2 **Molecular Weight:** 39
13.3 **Boiling Point at 1 atm:** 1,425°F = 774°C = 1,047°K
13.4 **Freezing Point:** 145°F = 63°C = 336°K
13.5 **Critical Temperature:** Not pertinent
13.6 **Critical Pressure:** Not pertinent
13.7 **Specific Gravity:** 0.86 at 20°C (solid)
13.8 **Liquid Surface Tension:** Not pertinent
13.9 **Liquid-Water Interfacial Tension:** Not pertinent
13.10 **Vapor (Gas) Specific Gravity:** Not pertinent
13.11 **Ratio of Specific Heats of Vapor (Gas):** Not pertinent
13.12 **Latent Heat of Vaporization:** Not pertinent
13.13 **Heat of Combustion:** $-2{,}003$ Btu/lb $= -1{,}113$ cal/g $= -46.57 \times 10^5$ J/kg
13.14 **Heat of Decomposition:** Not pertinent
13.15 **Heat of Solution:** $-2{,}104$ Btu/lb $= -1{,}169$ cal/g $= -48.91 \times 10^5$ J/kg
13.16 **Heat of Polymerization:** Not pertinent

NOTES

PTS	POTASSIUM OXALATE

Common Synonyms			
Potassium oxalate monohydrate	Solid	White	Odorless
	Sinks and mixes with water.		

Avoid contact with solid and dust. Keep people away.
Stop discharge if possible.
Isolate and remove discharged material.
Notify local health and pollution control agencies.

Fire

Not flammable.

Exposure

CALL FOR MEDICAL AID.

DUST
Irritating to eyes, nose and throat.
If inhaled will cause coughing or difficult breathing.
If in eyes, hold eyelids open and flush with plenty of water.
If breathing has stopped, give artificial respiration.
If breathing is difficult, give oxygen.

SOLID
POISONOUS IF SWALLOWED.
Irritating to skin and eyes.
If swallowed will cause nausea, vomiting or loss of consciousness.
Remove contaminated clothing and shoes.
Flush affected areas with plenty of water.
IF IN EYES, hold eyelids open and flush with plenty of water.
IF SWALLOWED and victim is CONSCIOUS, have victim drink water or milk.
IF SWALLOWED and victim is UNCONSCIOUS OR HAVING CONVULSIONS, do nothing except keep victim warm.

Water Pollution

Effect of low concentrations on aquatic life is unknown.
May be dangerous if it enters water intakes.
Notify local health and wildlife officials.
Notify operators of nearby water intakes.

1. RESPONSE TO DISCHARGE
(See Response Methods Handbook, CG 446-4)

Issue warning—water contaminant
Disperse and flush

2. LABEL

POISON

3. CHEMICAL DESIGNATIONS

3.1 **Synonyms:** Potassium oxalate monohydrate
3.2 **Coast Guard Compatibility Classification:** Not listed
3.3 **Chemical Formula:** $K_2C_2O_4 \cdot H_2O$
3.4 **IMCO/United Nations Numerical Designation:** Not listed

4. OBSERVABLE CHARACTERISTICS

4.1 **Physical State (as shipped):** Solid
4.2 **Color:** Colorless; white
4.3 **Odor:** None

5. HEALTH HAZARDS

5.1 **Personal Protective Equipment:** Approved dust respirator; chemical goggles; rubber or plastic-coated gloves
5.2 **Symptoms Following Exposure:** Inhalation of dust can cause systemic poisoning. Ingestion causes burning pain in throat, esophagus, and stomach; exposed areas of mucous membrane turn white; vomiting, severe purging, weak pulse, and cardiovascular collapse may result; if death is delayed, neuromuscular symptoms develop. Contact with eyes or skin causes irritation.
5.3 **Treatment for Exposure:** *Act promptly!* INHALATION: remove victim to fresh air; if exposure to dust is severe, get medical attention. INGESTION: call physician immediately; have victim drink dilute calcium lactate, lime water, chalk soln., or even milk; large amounts of calcium are required; administer gastric lavage with dilute lime water; watch for edema of the glottis and delayed constriction of esophagus. EYES: flush with water and seek medical attention. SKIN: flush with water.
5.4 **Toxicity by Inhalation (Threshold Limit Value):** Data not available
5.5 **Short-Term Inhalation Limits:** Data not available
5.6 **Toxicity by Ingestion:** Grade 3; LD_{50} 50—500 mg/kg
5.7 **Late Toxicity:** Data not available
5.8 **Vapor (Gas) Irritant Characteristics:** Data not available
5.9 **Liquid or Solid Irritant Characteristics:** Data not available
5.10 **Odor Threshold:** Odorless

6. FIRE HAZARDS

6.1 **Flash Point:** Not flammable
6.2 **Flammable Limits in Air:** Not flammable
6.3 **Fire Extinguishing Agents:** Not pertinent
6.4 **Fire Extinguishing Agents Not to be Used:** Not pertinent
6.5 **Special Hazards of Combustion Products:**
6.6 **Behavior in Fire:** Loses water at about 160° and decomposes to carbonate with no charring. The reaction is not hazardous.
6.7 **Ignition Temperature:** Not pertinent
6.8 **Electrical Hazard:** Not pertinent
6.9 **Burning Rate:** Not pertinent

7. CHEMICAL REACTIVITY

7.1 **Reactivity with Water:** No reaction
7.2 **Reactivity with Common Materials:**
7.3 **Stability During Transport:** Stable
7.4 **Neutralizing Agents for Acids and Caustics:** Not pertinent
7.5 **Polymerization:** Not pertinent
7.6 **Inhibitor of Polymerization:** Not pertinent

8. WATER POLLUTION

8.1 **Aquatic Toxicity:** Data not available
8.2 **Waterfowl Toxicity:** Data not available
8.3 **Biological Oxygen Demand (BOD):** Data not available
8.4 **Food Chain Concentration Potential:** None

9. SELECTED MANUFACTURERS

1. Pfizer Chemicals Div.
   215 East 42 St.
   New York, N. Y. 10017
2. J. T. Baker Chemical Co.
   Phillipsburg, N. J. 08865
3. Gallard-Schlesinger Chemical Mfg. Co.
   584 Mineola Ave.
   Carle Place, N. Y. 11514

10. SHIPPING INFORMATION

10.1 **Grades or Purity:** Reagent, 99.0%
10.2 **Storage Temperature:** Ambient
10.3 **Inert Atmosphere:** No requirement
10.4 **Venting:** Open

11. HAZARD ASSESSMENT CODE
(See Hazard Assessment Handbook, CG 446-3)

SS

12. HAZARD CLASSIFICATIONS

12.1 **Code of Federal Regulations:** Poison, Class B
12.2 **NAS Hazard Rating for Bulk Water Transportation:** Not listed
12.3 **NFPA Hazard Classifications:** Not listed

13. PHYSICAL AND CHEMICAL PROPERTIES

13.1 **Physical State at 15°C and 1 atm:** Solid
13.2 **Molecular Weight:** 184.24
13.3 **Boiling Point at 1 atm:** Not pertinent (decomposes)
13.4 **Freezing Point:** Not pertinent
13.5 **Critical Temperature:** Not pertinent
13.6 **Critical Pressure:** Not pertinent
13.7 **Specific Gravity:** 2.13 at 18.5°C (solid)
13.8 **Liquid Surface Tension:** Not pertinent
13.9 **Liquid-Water Interfacial Tension:** Not pertinent
13.10 **Vapor (Gas) Specific Gravity:** Not pertinent
13.11 **Ratio of Specific Heats of Vapor (Gas):** Not pertinent
13.12 **Latent Heat of Vaporization:** Not pertinent
13.13 **Heat of Combustion:** Not pertinent
13.14 **Heat of Decomposition:** Not pertinent
13.15 **Heat of Solution:** Not pertinent
13.16 **Heat of Polymerization:** Not pertinent

NOTES

PTP — POTASSIUM PERMANGANATE

Common Synonyms	Solid crystals	Dark purple	Odorless
	Sinks and mixes slowly with water.		

Stop discharge if possible. Keep people away.
Avoid contact with solid and dust.
Isolate and remove discharged material.
Notify local health and pollution control agencies.

Fire

Not flammable.
Containers may explode in fire.
May cause fire and explode on contact with combustibles.

Flood discharge area with water.
Cool exposed containers with water.

Exposure

CALL FOR MEDICAL AID.

SOLID
Irritating to skin and eyes.
If swallowed, will cause nausea, vomiting, or loss of consciousness.

Flush affected areas with plenty of water.
IF IN EYES, hold eyelids open and flush with plenty of water.
IF SWALLOWED and victim is CONSCIOUS, have victim drink water or milk and have victim induce vomiting.
IF SWALLOWED and victim is UNCONSCIOUS OR HAVING CONVULSIONS, do nothing except keep victim warm.

HARMFUL TO AQUATIC LIFE IN VERY LOW CONCENTRATIONS.
May be dangerous if it enters water intakes.

1. RESPONSE TO DISCHARGE

(See Response Methods Handbook, CG 446-4)

Issue warning—water contaminant
Disperse and flush

2. LABEL

Yellow

3. CHEMICAL DESIGNATIONS

3.1 **Synonyms:** No common synonyms
3.2 **Coast Guard Compatibility Classification:** Not applicable
3.3 **Chemical Formula:** $KMnO_4$
3.4 **IMCO/United Nations Numerical Designation:** 5.1/1490

4. OBSERVABLE CHARACTERISTICS

4.1 **Physical State (as shipped):** Solid
4.2 **Color:** Dark purple or bronze-like
4.3 **Odor:** Odorless

5. HEALTH HAZARDS

5.1 **Personal Protective Equipment:** Goggles or face shield; rubber gloves.
5.2 **Symptoms Following Exposure:** Burns and stains the skin dark brown. If ingested will cause severe distress of gastro-intestinal system. May be fatal if over 4 oz. are consumed.
5.3 **Treatment for Exposure:** INGESTION: induce vomiting and follow with thorough gastric lavage, demulcents, glucose I.V., fluid therapy, and antibiotics. Tracheostomy may be lifesaving.
5.4 **Toxicity by Inhalation (Threshold Limit Value):** Not pertinent
5.5 **Short-Term Inhalation Limits:** Not pertinent
5.6 **Toxicity by Ingestion:** Grade 3; LD_{50} 50 to 500 mg/kg
5.7 **Late Toxicity:** None
5.8 **Vapor (Gas) Irritant Characteristics:** Non-volatile
5.9 **Liquid or Solid Irritant Characteristics:** Can burn skin if not flushed with water.
5.10 **Odor Threshold:** Not pertinent

6. FIRE HAZARDS

6.1 **Flash Point:** Not flammable
6.2 **Flammable Limits in Air:** Not flammable
6.3 **Fire Extinguishing Agents:** Flood spill area with water.
6.4 **Fire Extinguishing Agents Not to be Used:** Not pertinent
6.5 **Special Hazards of Combustion Products:** Not pertinent
6.6 **Behavior in Fire:** May cause fire on contact with combustibles. Containers may explode.
6.7 **Ignition Temperature:** Not flammable
6.8 **Electrical Hazard:** Not pertinent
6.9 **Burning Rate:** Not flammable

7. CHEMICAL REACTIVITY

7.1 **Reactivity with Water:** No reaction
7.2 **Reactivity with Common Materials:** Attacks rubber and most fibers. May cause ignition of wood. Some acids, such as sulfuric acid, may cause an explosion.
7.3 **Stability During Transport:** Stable
7.4 **Neutralizing Agents for Acids and Caustics:** Not pertinent
7.5 **Polymerization:** Not pertinent
7.6 **Inhibitor of Polymerization:** Not pertinent

8. WATER POLLUTION

8.1 **Aquatic Toxicity:** 5.4 ppm/48 hr/bluegill/TL_m/fresh water
8.2 **Waterfowl Toxicity:** Data not available
8.3 **Biological Oxygen Demand (BOD):** None
8.4 **Food Chain Concentration Potential:** None

9. SELECTED MANUFACTURERS

1. Carus Chemical Co., Inc.
   1375 8th St.
   La Salle, Ill. 61301
2. Filo Color and Chemical Corp.
   347 Madison Ave.
   New York, N. Y. 10017
3. Mallinckrodt Chemical Works
   Industrial Chemicals Division
   2nd and Mallinckrodt Sts.
   St. Louis, Mo. 63160

10. SHIPPING INFORMATION

10.1 **Grades or Purity:** USP, Reagent (both 99+%)
10.2 **Storage Temperature:** Ambient
10.3 **Inert Atmosphere:** No requirement
10.4 **Venting:** Open

11. HAZARD ASSESSMENT CODE

(See Hazard Assessment Handbook, CG 446-3)

SS

12. HAZARD CLASSIFICATIONS

12.1 **Code of Federal Regulations:** Oxidizing material
12.2 **NAS Hazard Rating for Bulk Water Transportation:** Not listed
12.3 **NFPA Hazard Classifications:**

Category	Classification
Health Hazard (Blue)	1
Flammability (Red)	0
Reactivity (Yellow)	0 oxy

13. PHYSICAL AND CHEMICAL PROPERTIES

13.1 **Physical State at 15°C and 1 atm:** Solid
13.2 **Molecular Weight:** 158.04
13.3 **Boiling Point at 1 atm:** Decomposes
13.4 **Freezing Point:** >464°F = >240°C = >513°K
13.5 **Critical Temperature:** Not pertinent
13.6 **Critical Pressure:** Not pertinent
13.7 **Specific Gravity:** 2.70 at 15°C (solid)
13.8 **Liquid Surface Tension:** Not pertinent
13.9 **Liquid-Water Interfacial Tension:** Not pertinent
13.10 **Vapor (Gas) Specific Gravity:** Not pertinent
13.11 **Ratio of Specific Heats of Vapor (Gas):** Not pertinent
13.12 **Latent Heat of Vaporization:** Not pertinent
13.13 **Heat of Combustion:** Not pertinent
13.14 **Heat of Decomposition:** Not pertinent
13.15 **Heat of Solution:** Not pertinent
13.16 **Heat of Polymerization:** Not pertinent

NOTES

REVISED 1978

POP	# POTASSIUM PEROXIDE

Common Synonyms			
Potassium superoxide	Solid (powder)	Yellow	Odorless
	Sinks and mixes violently with water.		

Avoid contact with solid and dust. Keep people away.
Stop discharge if possible.
Isolate and remove discharged material.
Notify local health and pollution control agencies.

Fire

Not flammable.
Will increase the intensity of a fire.
May cause fire on contact with combustibles.

Combat fires from safe distance or protected location.
Flood discharge area with water.

Exposure

CALL FOR MEDICAL AID.

DUST
Irritating to eyes, nose and throat.
If inhaled will cause coughing or difficult breathing.

If in eyes, hold eyelids open and flush with plenty of water.
If breathing has stopped, give artificial respiration.
If breathing is difficult, give oxygen.

SOLID
Will burn skin and eyes.
If swallowed will cause nausea.

Remove contaminated clothing and shoes.
Flush affected areas with plenty of water.
IF IN EYES, hold eyelids open and flush with plenty of water.
IF SWALLOWED and victim is CONSCIOUS, have victim drink water or milk.
IF SWALLOWED and victim is UNCONSCIOUS OR HAVING CONVULSIONS, do nothing except keep victim warm.
DO NOT INDUCE VOMITING.

Water Pollution

HARMFUL TO AQUATIC LIFE IN VERY LOW CONCENTRATIONS.
May be dangerous if it enters water intakes.

Notify local health and wildlife officials.
Notify operators of nearby water intakes.

1. RESPONSE TO DISCHARGE

(See Response Methods Handbook, CG 446-4)

Issue warning—oxidizing material
Restrict access
Disperse and flush

2. LABEL

Yellow

3. CHEMICAL DESIGNATIONS

3.1 **Synonyms:** Potassium superoxide
3.2 **Coast Guard Compatibility Classification:** Not listed
3.3 **Chemical Formula:** K_2O_2
3.4 **IMCO/United Nations Numerical Designation:** 5.1/1491

4. OBSERVABLE CHARACTERISTICS

4.1 **Physical State (as shipped):** Powder
4.2 **Color:** Yellow
4.3 **Odor:** None

5. HEALTH HAZARDS

5.1 **Personal Protective Equipment:** Dust mask; goggles or face shield; protective gloves
5.2 **Symptoms Following Exposure:** Inhalation causes respiratory irritation. Ingestion causes severe burns of mouth and stomach. Contact with eyes or skin causes irritation and caustic burns.
5.3 **Treatment for Exposure:** INHALATION: remove from exposure; support respiration. INGESTION: give large amount of water; do NOT induce vomiting; get medical attention. EYES: irrigate with large quantities of water for at least 15 min.; get medical attention for caustic burns. SKIN: flush with water; treat caustic burns.
5.4 **Toxicity by Inhalation (Threshold Limit Value):** Data not available
5.5 **Short-Term Inhalation Limits:** Data not available
5.6 **Toxicity by Ingestion:** Data not available
5.7 **Late Toxicity:** Data not available
5.8 **Vapor (Gas) Irritant Characteristics:** Data not available
5.9 **Liquid or Solid Irritant Characteristics:** Data not available
5.10 **Odor Threshold:** Data not available

6. FIRE HAZARDS

6.1 **Flash Point:** Not pertinent
6.2 **Flammable Limits in Air:** Not pertinent
6.3 **Fire Extinguishing Agents:** Flood with water from a protected area.
6.4 **Fire Extinguishing Agents Not to be Used:** A small amount of water may cause explosions.
6.5 **Special Hazards of Combustion Products:**
6.6 **Behavior in Fire:** Increases intensity of fire and can start fires when in contact with organic combustibles
6.7 **Ignition Temperature:** Not pertinent
6.8 **Electrical Hazard:** Not pertinent
6.9 **Burning Rate:** Not pertinent

7. CHEMICAL REACTIVITY

7.1 **Reactivity with Water:** Reacts violently with liberation of heat and oxygen and the formation of caustic solution
7.2 **Reactivity with Common Materials:** Can form explosive and self-igniting mixtures with wood or other combustible materials
7.3 **Stability During Transport:** Stable if kept dry
7.4 **Neutralizing Agents for Acids and Caustics:** Following reaction with water, caustic formed can be flushed away with water and area rinsed with dilute acetic acid.
7.5 **Polymerization:** Not pertinent
7.6 **Inhibitor of Polymerization:** Not pertinent

8. WATER POLLUTION

8.1 **Aquatic Toxicity:** 80 ppm/24 hr/mosquitofish/TL_m/fresh water
8.2 **Waterfowl Toxicity:** Data not available
8.3 **Biological Oxygen Demand (BOD):** None
8.4 **Food Chain Concentration Potential:** None

9. SELECTED MANUFACTURERS

1. Pfaltz and Bauer, Inc.
375 Fairfield Ave.
Stamford, Conn. 06902
2. Ventron, Inc.
P. O. Box 159
Beverly, Mass. 01915
3. Gallard-Schlesinger Chemical Mfg. Co.
584 Mineola Ave.
Carle Place, N. Y. 11514

10. SHIPPING INFORMATION

10.1 **Grades or Purity:** Commercial; Pure
10.2 **Storage Temperature:** Ambient
10.3 **Inert Atmosphere:** No requirement
10.4 **Venting:** Pressure-vacuum

11. HAZARD ASSESSMENT CODE

(See Hazard Assessment Handbook, CG 446-3)

RR

12. HAZARD CLASSIFICATIONS

12.1 **Code of Federal Regulations:** Oxidizing material, solid
12.2 **NAS Hazard Rating for Bulk Water Transportation:** Not listed
12.3 **NFPA Hazard Classifications:**

Category	Classification
Health Hazard (Blue)	3
Flammability (Red)	0
Reactivity (Yellow)	2
	~~W~~

13. PHYSICAL AND CHEMICAL PROPERTIES

13.1 **Physical State at 15°C and 1 atm:** Solid
13.2 **Molecular Weight:** 110
13.3 **Boiling Point at 1 atm:** Not pertinent (decomposes)
13.4 **Freezing Point:** 914°F = 490°C = 763°K
13.5 **Critical Temperature:** Not pertinent
13.6 **Critical Pressure:** Not pertinent
13.7 **Specific Gravity:** > 1 at 20°C (solid)
13.8 **Liquid Surface Tension:** Not pertinent
13.9 **Liquid-Water Interfacial Tension:** Not pertinent
13.10 **Vapor (Gas) Specific Gravity:** Not pertinent
13.11 **Ratio of Specific Heats of Vapor (Gas):** Not pertinent
13.12 **Latent Heat of Vaporization:** Not pertinent
13.13 **Heat of Combustion:** Not pertinent
13.14 **Heat of Decomposition:** Not pertinent
13.15 **Heat of Solution:** Data not available
13.16 **Heat of Polymerization:** Not pertinent

NOTES

PRP

PROPANE

Common Synonyms

Liquefied compressed gas — Colorless — Odorless—may have skunk odor added

Liquid floats and boils on water. Flammable visible vapor cloud is produced.

Stop discharge if possible. Keep people away.
Shut off ignition sources and call fire department.
Stay upwind and use water spray to "knock down" vapor.
Notify local health and pollution control agencies.

Fire

FLAMMABLE.
Containers may explode in fire.
Flashback along vapor trail may occur.
Vapor may explode if ignited in an enclosed area.

Stop flow of gas if possible.
Cool exposed containers and protect men effecting shut-off with water.
Let fire burn.

Exposure

CALL FOR MEDICAL AID.

VAPOR
Not irritating to eyes, nose or throat.
If inhaled, will cause dizziness, difficult breathing, or loss of consciousness.
Move to fresh air.
If breathing has stopped, give artificial respiration.
If breathing is difficult, give oxygen.

LIQUID
May cause frostbite.
Flush affected areas with plenty of water.
DO NOT RUB AFFECTED AREAS.

Water Pollution

Not harmful to aquatic life.

1. RESPONSE TO DISCHARGE

(See Response Methods Handbook, CG 446-4)

Issue warning—high flammability
Restrict access
Evacuate area

2. LABEL

Red

3. CHEMICAL DESIGNATIONS

3.1 **Synonyms:** Dimethylmethane
3.2 **Coast Guard Compatibility Classification:** Paraffin
3.3 **Chemical Formula:** $CH_3CH_2CH_3$
3.4 **IMCO/United Nations Numerical Designation:** 2.0/1978

4. OBSERVABLE CHARACTERISTICS

4.1 **Physical State (as shipped):** Liquefied compressed gas
4.2 **Color:** Colorless
4.3 **Odor:** Faint gassy

5. HEALTH HAZARDS

5.1 **Personal Protective Equipment:** Self-contained breathing apparatus for high concentrations of gas.
5.2 **Symptoms Following Exposure:** Vaporizing liquid may cause frostbite. Concentrations in air greater than 10% cause dizziness in a few minutes. 1% concentrations give the same effect in 10 min. High concentrations cause asphyxiation.
5.3 **Treatment for Exposure:** Remove to open air. If victim is overcome by gas, apply artificial respiration. Guard against self-injury if confused.
5.4 **Toxicity by Inhalation (Threshold Limit Value):** 1000 ppm
5.5 **Short-Term Inhalation Limits:** Data not available
5.6 **Toxicity by Ingestion:** Not pertinent
5.7 **Late Toxicity:** None
5.8 **Vapor (Gas) Irritant Characteristics:** Vapors are nonirritating to the eyes and throat.
5.9 **Liquid or Solid Irritant Characteristics:** No appreciable hazard Practically harmless to the skin because it evaporates quickly.
5.10 **Odor Threshold:** 5,000—20,000 ppm

6. FIRE HAZARDS

6.1 **Flash Point:** −156°F C.C.
6.2 **Flammable Limits in Air:** 2.1%—9.5%
6.3 **Fire Extinguishing Agents:** Stop flow of gas. For small fires use dry chemicals. Cool adjacent areas with water spray.
6.4 **Fire Extinguishing Agents Not to be Used:** Water
6.5 **Special Hazards of Combustion Products:** Not pertinent
6.6 **Behavior in Fire:** Containers may explode. Vapor is heavier than air and may travel a long distance to a source of ignition and flash back.
6.7 **Ignition Temperature:** 842°F
6.8 **Electrical Hazard:** Class I, Group D
6.9 **Burning Rate:** 8.2 mm/min.

7. CHEMICAL REACTIVITY

7.1 **Reactivity with Water:** No reaction
7.2 **Reactivity with Common Materials:** No reaction
7.3 **Stability During Transport:** Stable
7.4 **Neutralizing Agents for Acids and Caustics:** Not pertinent
7.5 **Polymerization:** Not pertinent
7.6 **Inhibitor of Polymerization:** Not pertinent

8. WATER POLLUTION

8.1 **Aquatic Toxicity:** None
8.2 **Waterfowl Toxicity:** None
8.3 **Biological Oxygen Demand (BOD):** None
8.4 **Food Chain Concentration Potential:** None

9. SELECTED MANUFACTURERS

1. Atlantic Richfield
   ARCO Chemical Co. Division
   260 S. Broad St.
   Philadelphia, Pa. 19101
2. Mobil Chemical Co.
   North American Division
   Beaumont, Texas 77704
3. Union Carbide Corp.
   Linde Division
   270 Park Ave.
   New York, N. Y. 10017

10. SHIPPING INFORMATION

10.1 **Grades or Purity:**

	wt%
Research, Instrument, Pure	99.35+
Technical	97.50

10.2 **Storage Temperature:** Ambient
10.3 **Inert Atmosphere:** No requirement
10.4 **Venting:** Safety relief

11. HAZARD ASSESSMENT CODE

(See Hazard Assessment Handbook, CG 446-3)

A-B-C-D-E-F-G

12. HAZARD CLASSIFICATIONS

12.1 **Code of Federal Regulations:** Flammable compressed gas

12.2 **NAS Hazard Rating for Bulk Water Transportation:**

Category	Rating
Fire	4
Health	
Vapor Irritant	0
Liquid or Solid Irritant	0
Poisons	0
Water Pollution	
Human Toxicity	0
Aquatic Toxicity	0
Aesthetic Effect	0
Reactivity	
Other Chemicals	0
Water	0
Self-Reaction	0

12.3 **NFPA Hazard Classifications:**

Category	Classification
Health Hazard (Blue)	1
Flammability (Red)	4
Reactivity (Yellow)	0

13. PHYSICAL AND CHEMICAL PROPERTIES

13.1 **Physical State at 15°C and 1 atm:** Gas
13.2 **Molecular Weight:** 44.09
13.3 **Boiling Point at 1 atm:** −43.8°F = −42.1°C = 231.1°K
13.4 **Freezing Point:** −305.9°F = −187.7°C = 85.5°K
13.5 **Critical Temperature:** −142.01°F = −96.67°F = 176.53°K
13.6 **Critical Pressure:** 616.5 psia = 41.94 atm = 4.249 MN/m^2
13.7 **Specific Gravity:** 0.590 at −50°C (liquid)
13.8 **Liquid Surface Tension:** 16 dynes/cm = 0.016 N/m at −47°C
13.9 **Liquid-Water Interfacial Tension:** (est.) 50 dynes/cm = 0.05 N/m at −50°C
13.10 **Vapor (Gas) Specific Gravity:** 1.5
13.11 **Ratio of Specific Heats of Vapor (Gas):** 1.130
13.12 **Latent Heat of Vaporization:** 183.2 Btu/lb = 101.8 cal/g = 4.262 × 10^5 J/kg
13.13 **Heat of Combustion:** −19,782 Btu/lb = −10,990 cal/g = −460.13 × 10^5 J/kg
13.14 **Heat of Decomposition:** Not pertinent
13.15 **Heat of Solution:** Not pertinent
13.16 **Heat of Polymerization:** Not pertinent

NOTES

REVISED 1978

PLT	beta-PROPIOLACTONE

Common Synonyms			
beta-Propionolactone Betraprone Propanolide 2-Oxetanone Hydracrylic acid, beta-lactone	Liquid Mixes with water.	Colorless	Irritating odor

AVOID CONTACT WITH LIQUID. KEEP PEOPLE AWAY.
Wear rubber overclothing (including gloves).
Stop discharge if possible.
Call fire department.
Isolate and remove discharged material.
Notify local health and pollution control agencies.

Fire

Combustible.
Containers may explode in fire.
Extinguish with water, dry chemicals, foam, or carbon dioxide.
Cool exposed containers with water.

Exposure

CALL FOR MEDICAL AID.

LIQUID
POISONOUS IF SWALLOWED OR IF SKIN IS EXPOSED.
Irritating to skin and eyes.
Remove contaminated clothing and shoes.
Flush affected areas with plenty of water.
IF IN EYES, hold eyelids open and flush with plenty of water.
IF SWALLOWED and victim is CONSCIOUS, have victim drink water or milk and have victim induce vomiting.
IF SWALLOWED and victim is UNCONSCIOUS OR HAVING CONVULSIONS, do nothing except keep victim warm.

Water Pollution

Effect of low concentrations on aquatic life is unknown.
May be dangerous if it enters water intakes.
Notify local health and wildlife officials.
Notify operators of nearby water intakes.

1. RESPONSE TO DISCHARGE

(See Response Methods Handbook, CG 446-4)

Issue warning—poison, water contaminant
Restrict access
Disperse and flush

2. LABEL

3. CHEMICAL DESIGNATIONS

3.1 **Synonyms:** Betaprone; Hydracrylic acid, beta-lactone; 2-Oxetanone; Propanolide; beta-Propionolactone
3.2 **Coast Guard Compatibility Classification:** Monomers (14)
3.3 **Chemical Formula:** $\underline{OCH_2CH_2CO}$
3.4 **IMCO/United Nations Numerical Designation:** 6.1/—

4. OBSERVABLE CHARACTERISTICS

4.1 **Physical State (as shipped):** Liquid
4.2 **Color:** Colorless
4.3 **Odor:** Pungent, acrylic; irritating

5. HEALTH HAZARDS

5.1 **Personal Protective Equipment:** Air mask or organic canister mask; goggles or face shield; rubber gloves; protective clothing to prevent all contact with skin.
5.2 **Symptoms Following Exposure:** Inhalation causes irritation of nose, throat, and respiratory tract. Contact of liquid with eyes causes irritation and tears. Contact with skin causes irritation and blistering. Ingestion causes burns of mouth and stomach.
5.3 **Treatment for Exposure:** *Get medical attention following all exposures to this compound.* INHALATION: move victim to fresh air; if breathing has stopped, give artificial respiration. EYES: flush continuously with water for at least 15 min. SKIN: flush with water; if blistering occurs, alert physician to fact that fluid from blister will cause additional blistering of adjacent skin. INGESTION: give large amount of water and induce vomiting.
5.4 **Toxicity by Inhalation (Threshold Limit Value):** 0.0 ppm. Because of the high incidence of cancer, either in man or animals, no exposure or contact by any route—respiratory, oral, or skin—should be permitted.
5.5 **Short-Term Inhalation Limits:** Data not available
5.6 **Toxicity by Ingestion:** Grade 3; oral LDL_O = 50 mg/kg (rat)
5.7 **Late Toxicity:** Because of the high incidence of cancer, either in man or animals, no exposure or contact by any route—respiratory, oral, or skin—should be permitted.
5.8 **Vapor (Gas) Irritant Characteristics:** Vapors are moderately irritating such that personnel will not usually tolerate moderate or high concentrations.
5.9 **Liquid or Solid Irritant Characteristics:** Fairly severe skin irritant. May cause pain and second-degree burns after a few minutes' contact.
5.10 **Odor Threshold:** Data not available

6. FIRE HAZARDS

6.1 **Flash Point:** 165°F C.C.
6.2 **Flammable Limits in Air:** 2.9% (LFL)
6.3 **Fire Extinguishing Agents:** Water, dry chemical, foam, carbon dioxide
6.4 **Fire Extinguishing Agents Not to be Used:** Not pertinent
6.5 **Special Hazards of Combustion Products:** Vapors of unburned material are very toxic.
6.6 **Behavior in Fire:** Containers may explode.
6.7 **Ignition Temperature:** Data not available
6.8 **Electrical Hazard:** Data not available
6.9 **Burning Rate:** Data not available

7. CHEMICAL REACTIVITY

7.1 **Reactivity with Water:** Slow, non-hazardous reaction to form beta-hydroxypropionic acid
7.2 **Reactivity with Common Materials:** No reaction
7.3 **Stability During Transport:** Stable
7.4 **Neutralizing Agents for Acids and Caustics:** Not pertinent
7.5 **Polymerization:** Can polymerize and rupture container, especially at elevated temperatures. At 22°C, 0.04% polymerizes each day.
7.6 **Inhibitor of Polymerization:** None used

8. WATER POLLUTION

8.1 **Aquatic Toxicity:** Data not available
8.2 **Waterfowl Toxicity:** Data not available
8.3 **Biological Oxygen Demand (BOD):** Data not available
8.4 **Food Chain Concentration Potential:** None

9. SELECTED MANUFACTURERS

1. Celanese Chemical Company
245 Park Avenue
New York, N. Y. 10017
2. Diamond-Shamrock Chemical Corp.
Nopco Chemical Division
Morristown, N. J. 07960
3. Eastman Kodak Co.
Eastman Organic Chemicals
Rochester, N. Y. 14650

10. SHIPPING INFORMATION

10.1 **Grades or Purity:** 97+%
10.2 **Storage Temperature:** Below 60°F
10.3 **Inert Atmosphere:** No requirement
10.4 **Venting:** Pressure-vacuum

11. HAZARD ASSESSMENT CODE

(See Hazard Assessment Handbook, CG 446-3)

A-P-Q

12. HAZARD CLASSIFICATIONS

12.1 **Code of Federal Regulations:** Poisonous liquid, Class B
12.2 **NAS Hazard Rating for Bulk Water Transportation:**

Category	Rating
Fire	1
Health	
Vapor Irritant	3
Liquid or Solid Irritant	3
Poisons	3
Water Pollution	
Human Toxicity	3
Aquatic Toxicity	2
Aesthetic Effect	1
Reactivity	
Other Chemicals	2
Water	0
Self-Reaction	1

12.3 **NFPA Hazard Classifications:**

Category	Classification
Health Hazard (Blue)	0
Flammability (Red)	2
Reactivity (Yellow)	0

13. PHYSICAL AND CHEMICAL PROPERTIES

13.1 **Physical State at 15°C and 1 atm:** Liquid
13.2 **Molecular Weight:** 72.1
13.3 **Boiling Point at 1 atm:** Not pertinent (decomposes)
13.4 **Freezing Point:** −28.1°F = −33.4°C = 239.8°K
13.5 **Critical Temperature:** Not pertinent
13.6 **Critical Pressure:** Not pertinent
13.7 **Specific Gravity:** 1.148 at 20°C (liquid)
13.8 **Liquid Surface Tension:** (est.) 22 dynes/cm = 0.022 N/m at 20°C
13.9 **Liquid-Water Interfacial Tension:** (est.) 25 dynes/cm = 0.025 N/m at 20°C
13.10 **Vapor (Gas) Specific Gravity:** 2.5
13.11 **Ratio of Specific Heats of Vapor (Gas):** 1.1089
13.12 **Latent Heat of Vaporization:** Not pertinent
13.13 **Heat of Combustion:** −8,510 Btu/lb = −4,730 cal/g = −198 X 10^5 J/kg
13.14 **Heat of Decomposition:** Not pertinent
13.15 **Heat of Solution:** Not pertinent
13.16 **Heat of Polymerization:** Data not available

NOTES

PAD	PROPIONALDEHYDE

Common Synonyms			
Propyl aldehyde Propionic aldehyde Methyl acetaldehyde Propanal Propylic aldehyde	Liquid	Colorless	Suffocating, unpleasant odor
	Floats and mixes slowly with water. Flammable, irritating vapor is produced.		

Avoid contact with liquid and vapor. Keep people away.
Wear goggles, self-contained breathing apparatus, and rubber overclothing (including gloves).
Shut off ignition sources and call fire department.
Stop discharge if possible.
Stay upwind and use water spray to "knock down" vapor.
Isolate and remove discharged material.
Notify local health and pollution control agencies.

Fire

FLAMMABLE.
Flashback along vapor trail may occur.
Vapor may explode if ignited in an enclosed area.
Wear goggles, self-contained breathing apparatus, and rubber overclothing (including gloves).
Extinguish with dry chemical, alcohol foam, carbon dioxide.
Water may be ineffective on fire.
Cool exposed containers with water.

Exposure

CALL FOR MEDICAL AID.

VAPOR
Irritating to eyes, nose and throat.
If inhaled, will cause nausea or vomiting.
Move to fresh air.
If breathing has stopped, give artificial respiration.
If breathing is difficult, give oxygen.

LIQUID
Irritating to skin and eyes.
Harmful if swallowed.
Remove contaminated clothing and shoes.
Flush affected areas with plenty of water.
IF IN EYES, hold eyelids open and flush with plenty of water.
IF SWALLOWED and victim is CONSCIOUS, have victim drink water or milk.

Water Pollution

Effect of low concentrations on aquatic life is unknown.
May be dangerous if it enters water intakes.
Notify local health and wildlife officials.
Notify operators of nearby water intakes.

1. RESPONSE TO DISCHARGE
(See Response Methods Handbook, CG 446-4)

Issue warning—high flammability
Evacuate area

2. LABEL

Red

3. CHEMICAL DESIGNATIONS

3.1 **Synonyms:** Methylacetaldehyde
Propaldehyde
Propanal
Propionic aldehyde
Propyl aldehyde

3.2 **Coast Guard Compatibility Classification:** Aldehyde

3.3 **Chemical Formula:** CH_3CH_2CHO

3.4 **IMCO/United Nations Numerical Designation:** 3.2/1275

4. OBSERVABLE CHARACTERISTICS

4.1 **Physical State (as shipped):** Liquid

4.2 **Color:** Colorless

4.3 **Odor:** Pungent, unpleasant; suffocating

5. HEALTH HAZARDS

5.1 **Personal Protective Equipment:** Air-supplied mask for high vapor concentrations; plastic gloves; goggles.

5.2 **Symptoms Following Exposure:** Vapors will irritate nose and throat, and may cause nausea and vomiting. Liquid causes eye irritation.

5.3 **Treatment for Exposure:** INHALATION: remove victim to fresh air; give oxygen if breathing is difficult; call a physician. EYES: flush with plenty of water for at least 15 min., and call a physician. SKIN: flush with water.

5.4 **Toxicity by Inhalation (Threshold Limit Value):** Data not available

5.5 **Short-Term Inhalation Limits:** Data not available

5.6 **Toxicity by Ingestion:** Grade 2; LD_{50} 0.5 to 5 g/kg (rat)

5.7 **Late Toxicity:** Data not available

5.8 **Vapor (Gas) Irritant Characteristics:** Vapors cause moderate irritation, such that personnel will find high concentrations unpleasant. The effect is temporary.

5.9 **Liquid or Solid Irritant Characteristics:** Minimum hazard. If spilled on clothing and allowed to remain, may cause smarting and reddening of the skin.

5.10 **Odor Threshold:** 1 ppm

6. FIRE HAZARDS

6.1 **Flash Point:** $-22°F$ O.C.

6.2 **Flammable Limits in Air:** 2.6%–16.1%

6.3 **Fire Extinguishing Agents:** Carbon dioxide or dry chemical for small fires, alcohol-type foam for large fires.

6.4 **Fire Extinguishing Agents Not to be Used:** Water may be ineffective.

6.5 **Special Hazards of Combustion Products:** Not pertinent

6.6 **Behavior in Fire:** Vapor is heavier than air and may travel considerable distance to a source of ignition and flash back.

6.7 **Ignition Temperature:** 405°F

6.8 **Electrical Hazard:** Not pertinent

6.9 **Burning Rate:** 4.4 mm/min.

7. CHEMICAL REACTIVITY

7.1 **Reactivity with Water:** No reaction

7.2 **Reactivity with Common Materials:** No reaction

7.3 **Stability During Transport:** Stable

7.4 **Neutralizing Agents for Acids and Caustics:** Not pertinent

7.5 **Polymerization:** May occur in presence of acids or caustics.

7.6 **Inhibitor of Polymerization:** Not pertinent

8. WATER POLLUTION

8.1 **Aquatic Toxicity:** Data not available

8.2 **Waterfowl Toxicity:** Data not available

8.3 **Biological Oxygen Demand (BOD):** 95% of theoretical in 5 days

8.4 **Food Chain Concentration Potential:** None

9. SELECTED MANUFACTURERS

1. Eastman Kodak Co.
   Texas Eastman Co. Division
   Longview, Tex. 75601
2. Union Carbide Corp.
   Chemicals and Plastics Division
   270 Park Ave.
   New York, N. Y. 10017

10. SHIPPING INFORMATION

10.1 **Grades or Purity:** 97–99+%

10.2 **Storage Temperature:** Ambient

10.3 **Inert Atmosphere:** No requirement

10.4 **Venting:** Open (flame arrester) or pressure-vacuum

11. HAZARD ASSESSMENT CODE
(See Hazard Assessment Handbook, CG 446-3)

A-P-Q

12. HAZARD CLASSIFICATIONS

12.1 **Code of Federal Regulations:** Flammable liquid

12.2 **NAS Hazard Rating for Bulk Water Transportation:**

Category	Rating
Fire	3
Health	
Vapor Irritant	2
Liquid or Solid Irritant	1
Poisons	2
Water Pollution	
Human Toxicity	2
Aquatic Toxicity	2
Aesthetic Effect	2
Reactivity	
Other Chemicals	2
Water	0
Self-Reaction	1

12.3 **NFPA Hazard Classifications:**

Category	Classification
Health Hazard (Blue)	2
Flammability (Red)	3
Reactivity (Yellow)	1

13. PHYSICAL AND CHEMICAL PROPERTIES

13.1 **Physical State at 15°C and 1 atm:** Liquid

13.2 **Molecular Weight:** 58.08

13.3 **Boiling Point at 1 atm:** $118.4°F = 48.0°C = 321.2°K$

13.4 **Freezing Point:** $-112°F = -80°C = 193°K$

13.5 **Critical Temperature:** $433°F = 223°C = 496°K$

13.6 **Critical Pressure:** 690 psia = 47 atm = 4.8 MN/m^2

13.7 **Specific Gravity:** 0.805 at 20°C (liquid)

13.8 **Liquid Surface Tension:** 23.4 dynes/cm = 0.0234 N/m at 20°C

13.9 **Liquid-Water Interfacial Tension:** 29 dynes/cm = 0.029 N/m at 22.7°C

13.10 **Vapor (Gas) Specific Gravity:** 2.0

13.11 **Ratio of Specific Heats of Vapor (Gas):** 1.120

13.12 **Latent Heat of Vaporization:** 211 Btu/lb = 117 cal/g = 4.90×10^5 J/kg

13.13 **Heat of Combustion:** $-12{,}470$ Btu/lb $= -6930$ cal/g $= -290.1 \times 10^5$ J/kg

13.14 **Heat of Decomposition:** Not pertinent

13.15 **Heat of Solution:** (est.) -9 Btu/lb $= -5$ cal/g $= -0.2 \times 10^5$ J/kg

13.16 **Heat of Polymerization:** Not pertinent

NOTES

REVISED 1978

PNA	PROPIONIC ACID

Common Synonyms Propanoic acid Ethanecarboxylic acid	Liquid Colorless Sharp rancid odor Mixes with water. Irritating vapor is produced.

AVOID CONTACT WITH LIQUID AND VAPOR. Keep people away.
Wear goggles, self-contained breathing apparatus, and rubber overclothing (including gloves).
Stop discharge if possible.
Call fire department.
Isolate and remove discharged material.
Notify local health and pollution control agencies.

Fire	Combustible. **Wear goggles, self-contained breathing apparatus, and rubber overclothing (including gloves).** **Extinguish with water, dry chemical, alcohol foam, or carbon dioxide.** **Cool exposed containers with water.**
Exposure	**CALL FOR MEDICAL AID.** VAPOR Irritating to eyes, nose and throat. **Move to fresh air.** LIQUID Will burn skin and eyes. Harmful if swallowed. **Remove contaminated clothing and shoes.** **Flush affected areas with plenty of water.** **IF IN EYES, hold eyelids open and flush with plenty of water.** **IF SWALLOWED and victim is CONSCIOUS, have victim drink water or milk.** **DO NOT INDUCE VOMITING.**
Water Pollution	Dangerous to aquatic life in high concentrations. May be dangerous if it enters water intakes. **Notify local health and wildlife officials.** **Notify operators of nearby water intakes.**

1. RESPONSE TO DISCHARGE

(See Response Methods Handbook, CG 446-4)

Disperse and flush

2. LABEL

3. CHEMICAL DESIGNATIONS

3.1 **Synonyms:** Ethanecarboxylic acid
Methylacetic acid
Propanoic acid

3.2 **Coast Guard Compatibility Classification:** Organic acid

3.3 **Chemical Formula:** CH_3CH_2COOH

3.4 **IMCO/United Nations Numerical Designation:** 8.0/1848

4. OBSERVABLE CHARACTERISTICS

4.1 **Physical State (as shipped):** Liquid

4.2 **Color:** Colorless

4.3 **Odor:** Sharp, irritating; slightly pungent, rancid

5. HEALTH HAZARDS

5.1 **Personal Protective Equipment:** Air-supplied mask for high vapor concentrations; plastic gloves; goggles or face shield.

5.2 **Symptoms Following Exposure:** Liquid causes skin and eye burns. Vapors may irritate eyes, nose, and throat, but should not cause systemic illness.

5.3 **Treatment for Exposure:** INHALATION: remove victim to fresh air. INGESTION: have victim drink water or milk; do NOT induce vomiting. SKIN OR EYE CONTACT: immediately flush with plenty of water for at least 15 min.; get medical care for eyes; remove contaminated clothing.

5.4 **Toxicity by Inhalation (Threshold Limit Value):** 10 ppm

5.5 **Short-Term Inhalation Limits:** Data not available

5.6 **Toxicity by Ingestion:** Grade 2; oral rat LD_{50} = 2.6 g/kg

5.7 **Late Toxicity:** None

5.8 **Vapor (Gas) Irritant Characteristics:** Vapors cause moderate irritation such that personnel will find high concentrations unpleasant. The effect is temporary.

5.9 **Liquid or Solid Irritant Characteristics:** Fairly severe skin irritant; may cause pain and second-degree burns after a few minutes' contact.

5.10 **Odor Threshold:** Data not available

6. FIRE HAZARDS

6.1 **Flash Point:** 134°F O.C.; 130°F C.C.

6.2 **Flammable Limits in Air:** 2.9% (calc.)–14.8% (est.)

6.3 **Fire Extinguishing Agents:** Water, carbon dioxide, dry chemical, or alcohol foam.

6.4 **Fire Extinguishing Agents Not to be Used:** Not pertinent

6.5 **Special Hazards of Combustion Products:** Not pertinent

6.6 **Behavior in Fire:** Not pertinent

6.7 **Ignition Temperature:** 1105°F

6.8 **Electrical Hazard:** Not pertinent

6.9 **Burning Rate:** 2.2 mm/min.

7. CHEMICAL REACTIVITY

7.1 **Reactivity with Water:** No reaction

7.2 **Reactivity with Common Materials:** Corrodes ordinary steel and many other metals, but reaction is not hazardous.

7.3 **Stability During Transport:** Stable

7.4 **Neutralizing Agents for Acids and Caustics:** Dilute with water, then neutralize with lime or soda ash.

7.5 **Polymerization:** Not pertinent

7.6 **Inhibitor of Polymerization:** Not pertinent

8. WATER POLLUTION

8.1 **Aquatic Toxicity:** >1000 mg/l /48 hr/culex sp. larvae/TL_m/fresh water

8.2 **Waterfowl Toxicity:** Data not available

8.3 **Biological Oxygen Demand (BOD):** 78%, 5 days

8.4 **Food Chain Concentration Potential:** None

9. SELECTED MANUFACTURERS

1. Celanese Corp.
   Celanese Chemical Co.
   245 Park Ave.
   New York, N. Y. 10017
2. Eastman Kodak Co.
   Tennessee Eastman Co. Division
   Kingsport, Tenn. 37662
3. Union Carbide Corp.
   Chemicals and Plastics Division
   270 Park Ave.
   New York, N. Y. 10017

10. SHIPPING INFORMATION

10.1 **Grades or Purity:** 99+%

10.2 **Storage Temperature:** Ambient

10.3 **Inert Atmosphere:** No requirement

10.4 **Venting:** Open (flame arrester)

11. HAZARD ASSESSMENT CODE

(See Hazard Assessment Handbook, CG 446-3)

A-P-Q

12. HAZARD CLASSIFICATIONS

12.1 **Code of Federal Regulations:** Corrosive Material

12.2 **NAS Hazard Rating for Bulk Water Transportation:**

Category	Rating
Fire	2
Health	
Vapor Irritant	2
Liquid or Solid Irritant	3
Poisons	2
Water Pollution	
Human Toxicity	2
Aquatic Toxicity	2
Aesthetic Effect	2
Reactivity	
Other Chemicals	2
Water	0
Self-Reaction	0

12.3 **NFPA Hazard Classifications:**

Category	Classification
Health Hazard (Blue)	2
Flammability (Red)	2
Reactivity (Yellow)	0

13. PHYSICAL AND CHEMICAL PROPERTIES

13.1 **Physical State at 15°C and 1 atm:** Liquid

13.2 **Molecular Weight:** 74.08

13.3 **Boiling Point at 1 atm:** 285.4°F = 140.8°C = 414.0°K

13.4 **Freezing Point:** −5.3°F = −20.7°C = 252.5°K

13.5 **Critical Temperature:** 642°F = 339°C = 612°K

13.6 **Critical Pressure:** 779 psia = 53 atm = 5.37 MN/m^2

13.7 **Specific Gravity:** 0.995 at 20°C (liquid)

13.8 **Liquid Surface Tension:** 26.2 dynes/cm = 0.0262 N/m at 25°C

13.9 **Liquid-Water Interfacial Tension:** Not pertinent

13.10 **Vapor (Gas) Specific Gravity:** Not pertinent

13.11 **Ratio of Specific Heats of Vapor (Gas):** 1.103

13.12 **Latent Heat of Vaporization:** 248 Btu/lb = 138 cal/g = 5.78×10^5 J/kg

13.13 **Heat of Combustion:** −8,883 Btu/lb = −4,935 cal/g = 206.6×10^5 J/kg

13.14 **Heat of Decomposition:** Not pertinent

13.15 **Heat of Solution:** Not pertinent

13.16 **Heat of Polymerization:** Not pertinent

NOTES

REVISED 1978

PAH	# PROPIONIC ANHYDRIDE

Common Synonyms			
Methylacetic anhydride Propanoic anhydride Propionyl oxide	Liquid	Colorless	Sharp odor
	Sinks and mixes slowly with water.		

Avoid contact with liquid. Keep people away.
Wear rubber overclothing (including gloves).
Stop discharge if possible.
Call fire department.
Isolate and remove discharged material.
Notify local health and pollution control agencies.

Fire	Combustible. Extinguish with water, dry chemicals, alcohol foam, or carbon dioxide.
Exposure	Call for medical aid. LIQUID Will burn skin and eyes. Harmful if swallowed. Remove contaminated clothing and shoes. Flush affected areas with plenty of water. IF IN EYES, hold eyelids open and flush with plenty of water. IF SWALLOWED and victim is CONSCIOUS, have victim drink water or milk. DO NOT INDUCE VOMITING.
Water Pollution	Dangerous to aquatic life in high concentrations. May be dangerous if it enters water intakes. Notify local health and wildlife officials. Notify operators of nearby water intakes.

1. RESPONSE TO DISCHARGE
(See Response Methods Handbook, CG 446-4)

Issue warning—corrosive
Restrict access
Disperse and flush

2. LABEL

3. CHEMICAL DESIGNATIONS

3.1 **Synonyms:** Methylacetic anhydride; Propanoic anhydride; Propionyl oxide
3.2 **Coast Guard Compatibility Classification:** Organic anhydride
3.3 **Chemical Formula:** $(CH_3CH_2CO)_2O$
3.4 **IMCO/United Nations Numerical Designation:** 8/—

4. OBSERVABLE CHARACTERISTICS

4.1 **Physical State (as shipped):** Liquid
4.2 **Color:** Colorless
4.3 **Odor:** Pungent

5. HEALTH HAZARDS

5.1 **Personal Protective Equipment:** Organic canister mask; goggles or face shield; rubber gloves
5.2 **Symptoms Following Exposure:** Inhalation causes irritation of eyes and respiratory tract. Contact with liquid causes burns of eyes and skin. Ingestion causes burns of mouth and stomach.
5.3 **Treatment for Exposure:** INHALATION: move victim to fresh air; if breathing has stopped, give artificial respiration. EYES: immediately flush with plenty of water for at least 15 min.; get medical attention. SKIN: immediately flush with plenty of water for at least 15 min. INGESTION: give large amount of water; do NOT induce vomiting.
5.4 **Toxicity by Inhalation (Threshold Limit Value):** Data not available
5.5 **Short-Term Inhalation Limits:** Data not available
5.6 **Toxicity by Ingestion:** Grade 1; LD_{50} 5 to 15 g/kg; Grade 2; LD_{50} 0.5 to 5 g/kg.
5.7 **Late Toxicity:** Data not available
5.8 **Vapor (Gas) Irritant Characteristics:** Vapors are moderately irritating such that personnel will not usually tolerate moderate or high concentrations.
5.9 **Liquid or Solid Irritant Characteristics:** Causes smarting of the skin and first-degree burns on short exposure; may cause second-degree burns on long exposure.
5.10 **Odor Threshold:** Data not available

6. FIRE HAZARDS

6.1 **Flash Point:** 156°F O.C.; 145°F C.C.
6.2 **Flammable Limits in Air:** 1.48% 11.9%
6.3 **Fire Extinguishing Agents:** Water, dry chemical, alcohol foam, carbon dioxide.
6.4 **Fire Extinguishing Agents Not to be Used:** Not pertinent
6.5 **Special Hazards of Combustion Products:** Not pertinent
6.6 **Behavior in Fire:** Not pertinent
6.7 **Ignition Temperature:** 545°F
6.8 **Electrical Hazard:** Data not available
6.9 **Burning Rate:** 3.0 mm/min.

7. CHEMICAL REACTIVITY

7.1 **Reactivity with Water:** Reacts slowly to form weak propionic acid; the reaction is not hazardous.
7.2 **Reactivity with Common Materials:** Slowly corrosive if wet
7.3 **Stability During Transport:** Stable
7.4 **Neutralizing Agents for Acids and Caustics:** Flush with water, rinse with sodium bicarbonate or lime solution
7.5 **Polymerization:** Not pertinent
7.6 **Inhibitor of Polymerization:** Not pertinent

8. WATER POLLUTION

8.1 **Aquatic Toxicity:**
50 ppm/48 hr/water fleas/TL_m/fresh water
188 ppm/24 hr/bluegill/TL_m/fresh water
8.2 **Waterfowl Toxicity:** Data not available
8.3 **Biological Oxygen Demand (BOD):** 1.3 lb/lb, 5 days
8.4 **Food Chain Concentration Potential:** None

9. SELECTED MANUFACTURERS

1. Union Carbide Corporation
Chemicals and Plastics Division
270 Park Avenue
New York, N. Y. 10017
2. Eastman Chemical Products, Inc.
Kingsport, Tenn. 37662
3. Progil, Inc.
509 Madison Avenue
New York, N. Y. 10022

10. SHIPPING INFORMATION

10.1 **Grades or Purity:** 97+%
10.2 **Storage Temperature:** Ambient
10.3 **Inert Atmosphere:** No requirement
10.4 **Venting:** Pressure-vacuum

11. HAZARD ASSESSMENT CODE
(See Hazard Assessment Handbook, CG 446-3)

A-O-X-Y

12. HAZARD CLASSIFICATIONS

12.1 **Code of Federal Regulations:** Corrosive liquid
12.2 **NAS Hazard Rating for Bulk Water Transportation:**

Category	Rating
Fire	1
Health	
Vapor Irritant	3
Liquid or Solid Irritant	2
Poisons	1
Water Pollution	
Human Toxicity	1
Aquatic Toxicity	3
Aesthetic Effect	2
Reactivity	
Other Chemicals	3
Water	1
Self-Reaction	0

12.3 **NFPA Hazard Classifications:**

Category	Classification
Health Hazard (Blue)	2
Flammability (Red)	2
Reactivity (Yellow)	2 W

13. PHYSICAL AND CHEMICAL PROPERTIES

13.1 **Physical State at 15°C and 1 atm:** Liquid
13.2 **Molecular Weight:** 130.1
13.3 **Boiling Point at 1 atm:** 336°F = 169°C = 442°K
13.4 **Freezing Point:** −45°F = −43°C = 230°K
13.5 **Critical Temperature:** 660°F = 349°C = 622°K
13.6 **Critical Pressure:** 490 psia = 33 atm = 3.3 MN/m^2
13.7 **Specific Gravity:** 1.01 at 20°C (liquid)
13.8 **Liquid Surface Tension:** 30 dynes/cm = 0.030 N/m at 25°C
13.9 **Liquid-Water Interfacial Tension:** Not pertinent
13.10 **Vapor (Gas) Specific Gravity:** 4.5
13.11 **Ratio of Specific Heats of Vapor (Gas):** 1.0543
13.12 **Latent Heat of Vaporization:** 149 Btu/lb = 83 cal/g = 3.5×10^5 J/kg
13.13 **Heat of Combustion:** (at 15°C) −10,320 Btu/lb = −5,740 cal/g = -240×10^5 J/kg
13.14 **Heat of Decomposition:** Not pertinent
13.15 **Heat of Solution:** (est.) −36 Btu/lb = −20 cal/g = -0.84×10^5 J/kg
13.16 **Heat of Polymerization:** Not pertinent

NOTES

PAT	# n-PROPYL ACETATE

Common Synonyms			
Acetic acid, n-propyl ester	Liquid	Colorless	Mild odor
	Floats on water. Flammable, irritating vapor is produced.		

Stop discharge if possible. Keep people away.
Shut off ignition sources and call fire department.
Stay upwind and use water spray to "knock down" vapor.
Isolate and remove discharged material.
Notify local health and pollution control agencies.

Fire

FLAMMABLE.
Flashback along vapor trail may occur.
Vapor may explode if ignited in an enclosed area.
Extinguish with dry chemical, carbon dioxide, or alcohol foam.
Water may be ineffective on fire.
Cool exposed containers with water.

Exposure

CALL FOR MEDICAL AID.

VAPOR
Irritating to eyes, nose and throat.
If inhaled, will cause nausea, vomiting, dizziness, or loss of consciousness.
Move to fresh air.
If breathing has stopped, give artificial respiration.
If breathing is difficult, give oxygen.

LIQUID
Irritating to skin and eyes.
Harmful if swallowed.
Remove contaminated clothing and shoes.
Flush affected areas with plenty of water.
IF IN EYES, hold eyelids open and flush with plenty of water.
IF SWALLOWED and victim is CONSCIOUS, have victim drink water or milk.

Water Pollution

Effect of low concentrations on aquatic life is unknown.
May be dangerous if it enters water intakes.
Notify local health and wildlife officials.
Notify operators of nearby water intakes.

1. RESPONSE TO DISCHARGE

(See Response Methods Handbook, CG 446-4)

Issue warning—high flammability
Evacuate area

2. LABEL

Red

3. CHEMICAL DESIGNATIONS

3.1 **Synonyms:** Acetic acid, propyl ester
3.2 **Coast Guard Compatibility Classification:** Ester
3.3 **Chemical Formula:** $CH_3COOCH_2CH_2CH_3$
3.4 **IMCO/United Nations Numerical Designation:** 3.2/1276

4. OBSERVABLE CHARACTERISTICS

4.1 **Physical State (as shipped):** Liquid
4.2 **Color:** Colorless
4.3 **Odor:** Mild fruity

5. HEALTH HAZARDS

5.1 **Personal Protective Equipment:** Air-supplied mask or chemical canister; goggles or face shield; protective gloves.
5.2 **Symptoms Following Exposure:** Contact with skin and eyes causes no serious injury. High vapor concentrations will be irritating and will cause nausea, vomiting, and dizziness, with final loss of consciousness.
5.3 **Treatment for Exposure:** INHALATION: remove victim to fresh air; give artificial respiration if breathing has stopped; give oxygen if breathing is difficult. SKIN AND EYES: flush with water.
5.4 **Toxicity by Inhalation (Threshold Limit Value):** 200 ppm
5.5 **Short-Term Inhalation Limits:** 200 ppm for 60 min.
5.6 **Toxicity by Ingestion:** Grade 2; LD_{50} 0.5 to 5 g/kg
5.7 **Late Toxicity:** None
5.8 **Vapor (Gas) Irritant Characteristics:** Vapors cause a slight smarting of the eyes or respiratory system if present in high concentrations. The effect is temporary.
5.9 **Liquid or Solid Irritant Characteristics:** Minimum hazard. If spilled on clothing and allowed to remain, may cause smarting and reddening of the skin.
5.10 **Odor Threshold:** 70 mg/m^3

6. FIRE HAZARDS

6.1 **Flash Point:** 58°F C.C.; 65°F O.C.
6.2 **Flammable Limits in Air:** 2.0%–8.0%
6.3 **Fire Extinguishing Agents:** Carbon dioxide or dry chemical for small fires; alcohol foam for large fires.
6.4 **Fire Extinguishing Agents Not to be Used:** Water may be ineffective.
6.5 **Special Hazards of Combustion Products:** Not pertinent
6.6 **Behavior in Fire:** Not pertinent
6.7 **Ignition Temperature:** 842°F
6.8 **Electrical Hazard:** Not pertinent
6.9 **Burning Rate:** Data not available

7. CHEMICAL REACTIVITY

7.1 **Reactivity with Water:** No reaction
7.2 **Reactivity with Common Materials:** No reaction
7.3 **Stability During Transport:** Stable
7.4 **Neutralizing Agents for Acids and Caustics:** Not pertinent
7.5 **Polymerization:** Not pertinent
7.6 **Inhibitor of Polymerization:** Not pertinent

8. WATER POLLUTION

8.1 **Aquatic Toxicity:** 820 ppm/24 hr/brine shrimp/TL_m
8.2 **Waterfowl Toxicity:** Data not available
8.3 **Biological Oxygen Demand (BOD):** 62% of theoretical in 5 days/freshwater
8.4 **Food Chain Concentration Potential:** None

9. SELECTED MANUFACTURERS

1. Celanese Corp.
   Celanese Chemical Co. Division
   245 Park Ave.
   New York, N. Y. 10017
2. Eastman Kodak Co.
   Tennessee Eastman Co. Division
   Kingsport, Tenn. 37662
3. Union Carbide Corp.
   Chemicals and Plastics Division
   270 Park Ave.
   New York, N. Y. 10017

10. SHIPPING INFORMATION

10.1 **Grades or Purity:** 90–100%
10.2 **Storage Temperature:** Ambient
10.3 **Inert Atmosphere:** No requirement
10.4 **Venting:** Open (flame arrester) or pressure-vacuum

11. HAZARD ASSESSMENT CODE

(See Hazard Assessment Handbook, CG 446-3)

A-P-Q-T-U

12. HAZARD CLASSIFICATIONS

12.1 **Code of Federal Regulations:** Flammable liquid
12.2 **NAS Hazard Rating for Bulk Water Transportation:**

Category	Rating
Fire	3
Health	
Vapor Irritant	1
Liquid or Solid Irritant	1
Poisons	2
Water Pollution	
Human Toxicity	1
Aquatic Toxicity	2
Aesthetic Effect	2
Reactivity	
Other Chemicals	1
Water	0
Self-Reaction	0

12.3 **NFPA Hazard Classifications:** Not listed

13. PHYSICAL AND CHEMICAL PROPERTIES

13.1 **Physical State at 15°C and 1 atm:** Liquid
13.2 **Molecular Weight:** 102.13
13.3 **Boiling Point at 1 atm:** 214.9°F = 101.6°C = 374.8°K
13.4 **Freezing Point:** −139°F = −95.0°C = 178.2°K
13.5 **Critical Temperature:** 529°F = 276°C = 549°K
13.6 **Critical Pressure:** 485 psia = 33 atm = 3.3 MN/m^2
13.7 **Specific Gravity:** 0.886 at 20°C (liquid)
13.8 **Liquid Surface Tension:** 24.3 dynes/cm = 0.0243 N/m at 20°C
13.9 **Liquid-Water Interfacial Tension:** Data not available
13.10 **Vapor (Gas) Specific Gravity:** Not pertinent
13.11 **Ratio of Specific Heats of Vapor (Gas):** 1.071
13.12 **Latent Heat of Vaporization:** 145 Btu/lb = 80.3 cal/g = 3.36×10^5 J/kg
13.13 **Heat of Combustion:** Data not available
13.14 **Heat of Decomposition:** Not pertinent
13.15 **Heat of Solution:** Not pertinent
13.16 **Heat of Polymerization:** Not pertinent

NOTES

REVISED 1978

n-PROPYL ALCOHOL

PAL

Common Synonyms
Propyl alcohol
l-Propanol

Liquid | Colorless | Alcohol odor

Mixes with water. Flammable, irritating vapor is produced.

Stop discharge if possible. Keep people away.
Shut off ignition sources and call fire department.
Stay upwind and use water spray to "knock down" vapor.
Avoid contact with liquid and vapor.
Isolate and remove discharged material.
Notify local health and pollution control agencies.

Fire

FLAMMABLE.
Flashback along vapor trail may occur.
Vapor may explode if ignited in an enclosed area.
Extinguish with dry chemical, alcohol foam, or carbon dioxide.
Water may be ineff :tive on fire.
Cool exposed containers with water.

Exposure

CALL FOR MEDICAL AID.

VAPOR
Irritating to eyes, nose, and throat.
If inhaled, will cause nausea, dizziness, or headache.
Move to fresh air.
If breathing has stopped, give artificial respiration.
If breathing is difficult, give oxygen.

LIQUID
Will burn eyes.
Harmful if swallowed.
Flush affected areas with plenty of water.
IF IN EYES, hold eyelids open and flush with plenty of water.
IF SWALLOWED and victim is CONSCIOUS, have victim drink water or milk.

Water Pollution

Dangerous to aquatic life in high concentrations.
May be dangerous if it enters water intakes.
Notify local health and wildlife officials.
Notify operators of nearby water intakes.

1. RESPONSE TO DISCHARGE

(See Response Methods Handbook, CG 446-4)

Issue warning—high flammability
Disperse and flush

2. LABEL

Red

3. CHEMICAL DESIGNATIONS

3.1 **Synonyms:** Ethylcarbinol
l-Propanol
Propyl alcohol

3.2 **Coast Guard Compatibility Classification:** Alcohol

3.3 **Chemical Formula:** $CH_3CH_2CH_2OH$

3.4 **IMCO/United Nations Numerical Designation:** 3.2/1274

4. OBSERVABLE CHARACTERISTICS

4.1 **Physical State (as shipped):** Liquid

4.2 **Color:** Colorless

4.3 **Odor:** Resembles that of ethyl alcohol

5. HEALTH HAZARDS

5.1 **Personal Protective Equipment:** Air-supplied respirator for high concentrations; goggles or face shield; plastic gloves.

5.2 **Symptoms Following Exposure:** Contact with eyes is extremely irritating and may cause burns. Vapors irritate nose and throat. In high concentrations, may cause nausea, dizziness, headache, and stupor.

5.3 **Treatment for Exposure:** INHALATION: remove victim to fresh air; call a physician. SKIN OR EYE CONTACT: flush at once with plenty of water; get medical care for eyes.

5.4 **Toxicity by Inhalation (Threshold Limit Value):** 200 ppm

5.5 **Short-Term Inhalation Limits:** 400 ppm for 30 min

5.6 **Toxicity by Ingestion:** Grade 2; LD_{50} 0.5 to 5 g/kg (rat)

5.7 **Late Toxicity:** None

5.8 **Vapor (Gas) Irritant Characteristics:** Vapors cause a slight smarting of the eyes or respiratory system if present in high concentrations. The effect is temporary.

5.9 **Liquid or Solid Irritant Characteristics:** No appreciable hazard. Practically harmless to the skin.

5.10 **Odor Threshold:** 30 ppm

6. FIRE HAZARDS

6.1 **Flash Point:** 81°F O.C.; 77°F C.C.

6.2 **Flammable Limits in Air:** 2.1%–13.5%

6.3 **Fire Extinguishing Agents:** Carbon dioxide or dry chemical for small fires; alcohol foam for large fires.

6.4 **Fire Extinguishing Agents Not to be Used:** Water may be ineffective

6.5 **Special Hazards of Combustion Products:** Not pertinent

6.6 **Behavior in Fire:** Not pertinent

6.7 **Ignition Temperature:** 700°F

6.8 **Electrical Hazard:** Class I, Group D

6.9 **Burning Rate:** 2.9 mm/min.

7. CHEMICAL REACTIVITY

7.1 **Reactivity with Water:** No reaction

7.2 **Reactivity with Common Materials:** No reaction

7.3 **Stability During Transport:** Stable

7.4 **Neutralizing Agents for Acids and Caustics:** Not pertinent

7.5 **Polymerization:** Not pertinent

7.6 **Inhibitor of Polymerization:** Not pertinent

8. WATER POLLUTION

8.1 **Aquatic Toxicity:** 500 ppm/24 hr/goldfish/died/fresh water

8.2 **Waterfowl Toxicity:** Data not available

8.3 **Biological Oxygen Demand (BOD):** 0.47-1.5 lb/lb, 5 days

8.4 **Food Chain Concentration Potential:** None

9. SELECTED MANUFACTURERS

1. Celanese Chemical Co.
245 Park Ave.
New York, N. Y. 10017
2. Eastman Kodak Co.
Texas Eastman Co. Division
Longview, Tex. 75601
3. Union Carbide Corp.
Chemicals & Plastics Division
270 Park Ave.
New York, N. Y. 10017

10. SHIPPING INFORMATION

10.1 **Grades or Purity:** 99.8+%

10.2 **Storage Temperature:** Ambient

10.3 **Inert Atmosphere:** No requirement

10.4 **Venting:** Open (flame arrester)

11. HAZARD ASSESSMENT CODE

(See Hazard Assessment Handbook, CG 446-3)

A-P-Q

12. HAZARD CLASSIFICATIONS

12.1 **Code of Federal Regulations:** Flammable liquid

12.2 **NAS Hazard Rating for Bulk Water Transportation:**

Category	Rating
Fire	3
Health	
Vapor Irritant	1
Liquid or Solid Irritant	0
Poisons	2
Water Pollution	
Human Toxicity	2
Aquatic Toxicity	2
Aesthetic Effect	1
Reactivity	
Other Chemicals	2
Water	0
Self-Reaction	0

12.3 **NFPA Hazard Classifications:** Not listed

13. PHYSICAL AND CHEMICAL PROPERTIES

13.1 **Physical State at 15°C and 1 atm:** Liquid

13.2 **Molecular Weight:** 60.10

13.3 **Boiling Point at 1 atm:** 207.0°F = 97.2°C = 370.4°K

13.4 **Freezing Point:** −195.2°F = −126.2°C = 147.0°K

13.5 **Critical Temperature:** 506.5°F = 263.6°C = 536.8°K

13.6 **Critical Pressure:** 750 psia = 51 atm = 5.2 MN/m^2

13.7 **Specific Gravity:** 0.803 at 20°C (liquid)

13.8 **Liquid Surface Tension:** Not pertinent

13.9 **Liquid-Water Interfacial Tension:** Not pertinent

13.10 **Vapor (Gas) Specific Gravity:** 2.1

13.11 **Ratio of Specific Heats of Vapor (Gas):** 1.107

13.12 **Latent Heat of Vaporization:** 292.7 Btu/lb = 162.6 cal/g = 6.808×10^5 J/kg

13.13 **Heat of Combustion:** −13,130 Btu/lb = −7296 cal/g = -305.5×10^5 J/kg

13.14 **Heat of Decomposition:** Not pertinent

13.15 **Heat of Solution:** (est.) −9 Btu/lb = −5 cal/g = -0.2×10^5 J/kg

13.16 **Heat of Polymerization:** Not pertinent

NOTES

REVISED 1978

PPL | PROPYLENE

Common Synonyms
Propene
Methylethylene

Liquefied compressed gas — Colorless — Mild odor

Floats and boils on water. Flammable, visible vapor cloud is produced.

Stop discharge if possible. Keep people away.
Shut off ignition sources and call fire department.
Stay upwind and use water spray to "knock down" vapor.
Evacuate area in case of large discharge.
Avoid contact with liquid.
Notify local health and pollution control agencies.

Fire

FLAMMABLE.
Container may explode in fire.
Flashback along vapor trail may occur.
May explode if ignited in an enclosed area.
Stop flow of gas if possible.
Cool exposed containers and protect men effecting shutoff with water.
Let fire burn.

Exposure

CALL FOR MEDICAL AID.

VAPOR
If inhaled, will cause dizziness or loss of consciousness.
Move to fresh air.
If breathing has stopped, give artificial respiration.
If breathing is difficult, give oxygen.

LIQUID
Will cause frostbite.
Flush affected areas with plenty of water.
DO NOT RUB AFFECTED AREAS.

Water Pollution

Not harmful to aquatic life.

1. RESPONSE TO DISCHARGE
(See Response Methods Handbook, CG 446-4)

Issue warning — high flammability
Evacuate area

2. LABEL

Red

3. CHEMICAL DESIGNATIONS

3.1 **Synonyms:** Methylethylene
Propene
3.2 **Coast Guard Compatibility Classification:** Olefin
3.3 **Chemical Formula:** $CH_3CH{=}CH_2$
3.4 **IMCO/United Nations Numerical Designation:** 2.0/1077

4. OBSERVABLE CHARACTERISTICS

4.1 **Physical State (as shipped):** Liquefied gas
4.2 **Color:** Colorless
4.3 **Odor:** Weak gassy

5. HEALTH HAZARDS

5.1 **Personal Protective Equipment:** Organic vapor canister or air-supplied mask; goggles or face shield (for liquid); protective clothing (for liquid).
5.2 **Symptoms Following Exposure:** Moderate concentration in air causes dizziness, drowsiness, and unconsciousness. Contact with liquefied propylene will cause "freezing burn."
5.3 **Treatment for Exposure:** INHALATION: remove victim from exposure; if breathing is irregular or has stopped, start resuscitation; give oxygen; call a doctor.
5.4 **Toxicity by Inhalation (Threshold Limit Value):** 4000 ppm
5.5 **Short-Term Inhalation Limits:** Data not available
5.6 **Toxicity by Ingestion:** Not pertinent
5.7 **Late Toxicity:** None
5.8 **Vapor (Gas) Irritant Characteristics:** Vapors are nonirritating to the eyes and throat.
5.9 **Liquid or Solid Irritant Characteristics:** No appreciable hazard. Practically harmless to the skin because it evaporates quickly.
5.10 **Odor Threshold:** Data not available

6. FIRE HAZARDS

6.1 **Flash Point:** −162°F C.C.
6.2 **Flammable Limits in Air:** 2.0%–11%
6.3 **Fire Extinguishing Agents:** Stop flow of gas.
6.4 **Fire Extinguishing Agents Not to be Used:** Not pertinent
6.5 **Special Hazards of Combustion Products:** Not pertinent
6.6 **Behavior in Fire:** Containers may explode. Vapor is heavier than air and may travel considerable distance to a source of ignition and flash back.
6.7 **Ignition Temperature:** 927°F
6.8 **Electrical Hazard:** Class I, Group D
6.9 **Burning Rate:** 8 mm/min. (liquid)

7. CHEMICAL REACTIVITY

7.1 **Reactivity with Water:** No reaction
7.2 **Reactivity with Common Materials:** No reaction
7.3 **Stability During Transport:** Stable
7.4 **Neutralizing Agents for Acids and Caustics:** Not pertinent
7.5 **Polymerization:** Not pertinent
7.6 **Inhibitor of Polymerization:** Not pertinent

8. WATER POLLUTION

8.1 **Aquatic Toxicity:** None
8.2 **Waterfowl Toxicity:** None
8.3 **Biological Oxygen Demand (BOD):** None
8.4 **Food Chain Concentration Potential:** None

9. SELECTED MANUFACTURERS

1. Dow Chemical Co.
Midland, Mich. 48640
2. Exxon Chemical Co.
Houston, Tex. 77001
3. Union Carbide Corp.
Chemicals and Plastics Division
270 Park Ave.
New York, N. Y. 10017

10. SHIPPING INFORMATION

10.1 **Grades or Purity:** Chemical: 92+%; polymerization: 99+%; research: 99+%; propylene concentrate: 80+%
10.2 **Storage Temperature:** Ambient
10.3 **Inert Atmosphere:** No requirement
10.4 **Venting:** Safety relief

11. HAZARD ASSESSMENT CODE
(See Hazard Assessment Handbook, CG 446-3)

A-B-C-D-E-F-G

12. HAZARD CLASSIFICATIONS

12.1 **Code of Federal Regulations:** Flammable compressed gas
12.2 **NAS Hazard Rating for Bulk Water Transportation:**

Category	Rating
Fire	4
Health	
Vapor Irritant	0
Liquid or Solid Irritant	0
Poisons	1
Water Pollution	
Human Toxicity	0
Aquatic Toxicity	1
Aesthetic Effect	0
Reactivity	
Other Chemicals	1
Water	0
Self-Reaction	1

12.3 **NFPA Hazard Classifications:**

Category	Classification
Health Hazard (Blue)	1
Flammability (Red)	4
Reactivity (Yellow)	1

13. PHYSICAL AND CHEMICAL PROPERTIES

13.1 **Physical State at 15°C and 1 atm:** Gas
13.2 **Molecular Weight:** 42.08
13.3 **Boiling Point at 1 atm:** −53.9°F = −47.7°C = 225.5°K
13.4 **Freezing Point:** −301.4°F = −185.2°C = 88°K
13.5 **Critical Temperature:** 197.2°F = 91.8°C = 365.0°K
13.6 **Critical Pressure:** 670 psia = 45.6 atm = 4.62 MN/m^2
13.7 **Specific Gravity:** 0.609 at −47°C (liquid)
13.8 **Liquid Surface Tension:** 16.7 dynes/cm = 0.0167 N/m at −47°C
13.9 **Liquid-Water Interfacial Tension:** Not pertinent
13.10 **Vapor (Gas) Specific Gravity:** 1.4
13.11 **Ratio of Specific Heats of Vapor (Gas):** 1.152
13.12 **Latent Heat of Vaporization:** 187 Btu/lb = 104 cal/g = 435 × 10^5 J/kg
13.13 **Heat of Combustion:** −19,692 Btu/lb = −10,940 cal/g = −458.04 × 10^5 J/kg
13.14 **Heat of Decomposition:** Not pertinent
13.15 **Heat of Solution:** Not pertinent
13.16 **Heat of Polymerization:** Not pertinent

NOTES

REVISED 1978

PBP

PROPYLENE–BUTYLENE POLYMER

Common Synonyms	Liquid

Stop discharge if possible. Keep people away.
Isolate and remove discharged material.
Notify local health and pollution control agencies.

Fire

Fire data not available.

Exposure

CALL FOR MEDICAL AID.
Exposure data not available.
Flush affected areas with plenty of water.

Water Pollution

Effect of low concentrations on aquatic life is unknown.
Fouling to shoreline.
May be dangerous if it enters water intakes.
Notify local health and wildlife officials.
Notify operators of nearby water intakes.

1. RESPONSE TO DISCHARGE

(See Response Methods Handbook, CG 446-4)

Data not available

2. LABELS

No hazard label required by Code of Federal Regulations

3. CHEMICAL DESIGNATIONS

3.1 **Synonyms:** No common synonyms
3.2 **Coast Guard Compatibility Classification:** Olefin
3.3 **Chemical Formula:** Not applicable
3.4 **IMCO/United Nations Numerical Designation:** Not listed

4. OBSERVABLE CHARACTERISTICS

4.1 **Physical State (as shipped):** Liquid
4.2 **Color:** Data not available
4.3 **Odor:** Data not available

5. HEALTH HAZARDS

5.1 **Personal Protective Equipment:** Data not available
5.2 **Symptoms Following Exposure:** Data not available
5.3 **Treatment for Exposure:** Data not available
5.4 **Toxicity by Inhalation (Threshold Limit Value):** Data not available
5.5 **Short-Term Inhalation Limits:** Data not available
5.6 **Toxicity by Ingestion:** Data not available
5.7 **Late Toxicity:** Data not available
5.8 **Vapor (Gas) Irritant Characteristics:** Data not available
5.9 **Liquid or Solid Irritant Characteristics:** Data not available
5.10 **Odor Threshold:** Data not available

6. FIRE HAZARDS

6.1 **Flash Point:** Data not available
6.2 **Flammable Limits in Air:** Data not available
6.3 **Fire Extinguishing Agents:** Data not available
6.4 **Fire Extinguishing Agents Not to be Used:** Data not available
6.5 **Special Hazards of Combustion Products:** Data not available
6.6 **Behavior in Fire:** Data not available
6.7 **Ignition Temperature:** Data not available
6.8 **Electrical Hazard:** Data not available
6.9 **Burning Rate:** Data not available

7. CHEMICAL REACTIVITY

7.1 **Reactivity with Water:** No reaction
7.2 **Reactivity with Common Materials:** No reaction
7.3 **Stability During Transport:** Stable
7.4 **Neutralizing Agents for Acids and Caustics:** Not pertinent
7.5 **Polymerization:** Not pertinent
7.6 **Inhibitor of Polymerization:** Not pertinent

8. WATER POLLUTION

8.1 **Aquatic Toxicity:** Data not available
8.2 **Waterfowl Toxicity:** Data not available
8.3 **Biological Oxygen Demand (BOD):** Data not available
8.4 **Food Chain Concentration Potential:** None

9. SELECTED MANUFACTURERS

1. American Petrofina, Inc.
   Cosden Oil and Chemical Co. Division
   Big Spring, Tex. 79720
2. Amoco Chemicals Corp.
   910 S. Michigan Ave.
   Chicago, Ill. 60605
3. Exxon Chemical Co. Division
   Houston, Tex. 77001

10. SHIPPING INFORMATION

10.1 **Grades or Purity:** Data not available
10.2 **Storage Temperature:** Ambient
10.3 **Inert Atmosphere:** No requirement
10.4 **Venting:** Open

11. HAZARD ASSESSMENT CODE

(See Hazard Assessment Handbook, CG 446-3)

A-T-U

12. HAZARD CLASSIFICATIONS

12.1 **Code of Federal Regulations:** Not listed
12.2 **NAS Hazard Rating for Bulk Water Transportation:** Not listed
12.3 **NFPA Hazard Classifications:** Not listed

13. PHYSICAL AND CHEMICAL PROPERTIES

13.1 **Physical State at 15°C and 1 atm:** Liquid
13.2 **Molecular Weight:** Not pertinent
13.3 **Boiling Point at 1 atm:** Very high
13.4 **Freezing Point:** Not pertinent
13.5 **Critical Temperature:** Not pertinent
13.6 **Critical Pressure:** Not pertinent
13.7 **Specific Gravity:** Data not available
13.8 **Liquid Surface Tension:** (est.) 25 dynes/cm = 0.025 N/m at 20°C
13.9 **Liquid-Water Interfacial Tension:** (est.) 50 dynes/cm = 0.05 N/m at 20°C
13.10 **Vapor (Gas) Specific Gravity:** Not pertinent
13.11 **Ratio of Specific Heats of Vapor (Gas):** Not pertinent
13.12 **Latent Heat of Vaporization:** Not pertinent
13.13 **Heat of Combustion:** (est.) –17,000 Btu/lb = –9600 cal/g = -400×10^5 J/kg
13.14 **Heat of Decomposition:** Not pertinent
13.15 **Heat of Solution:** Not pertinent
13.16 **Heat of Polymerization:** Not pertinent

NOTES

REVISED 1978

PPG

PROPYLENE GLYCOL

Common Synonyms
1, 2-Propanediol
1, 2-Dihydroxypropane

Thick liquid | Colorless | Odorless

Mixes with water.

Stop discharge if possible. Keep people away.
Call fire department.
Isolate and remove discharged material.
Notify local health and pollution control agencies.

Fire

Combustible.
Extinguish with water, dry chemical, alcohol foam, or carbon dioxide.
Cool exposed containers with water.

Exposure

Not harmful.

Water Pollution

Effect of low concentrations on aquatic life is unknown.
May be dangerous if it enters water intakes.
Notify local health and wildlife officials.
Notify operators of nearby water intakes.

1. RESPONSE TO DISCHARGE

(See Response Methods Handbook, CG 446-4)

Disperse and flush

2. LABELS

No hazard label required by Code of Federal Regulations

3. CHEMICAL DESIGNATIONS

3.1 **Synonyms:** 1,2-Dihydroxypropane
Methylethylene glycol
1,2-Propanediol

3.2 **Coast Guard Compatibility Classification:** Glycol

3.3 **Chemical Formula:** $CH_3CH(OH)CH_2OH$

3.4 **IMCO/United Nations Numerical Designation:** Not listed

4. OBSERVABLE CHARACTERISTICS

4.1 **Physical State (as shipped):** Liquid

4.2 **Color:** Colorless

4.3 **Odor:** Odorless

5. HEALTH HAZARDS

5.1 **Personal Protective Equipment:** Goggles.

5.2 **Symptoms Following Exposure:** Liquid may irritate eyes.

5.3 **Treatment for Exposure:** Flush eyes with plenty of water.

5.4 **Toxicity by Inhalation (Threshold Limit Value):** Not pertinent

5.5 **Short-Term Inhalation Limits:** Not pertinent

5.6 **Toxicity by Ingestion:** Grade 2; LD_{50} 0.5 to 5 g/kg (mouse)

5.7 **Late Toxicity:** None

5.8 **Vapor (Gas) Irritant Characteristics:** Vapors are nonirritating to the eyes and throat.

5.9 **Liquid or Solid Irritant Characteristics:** No appreciable hazard. Practically harmless to the skin.

5.10 **Odor Threshold:** Not pertinent

6. FIRE HAZARDS

6.1 **Flash Point:** 210°F C.C.; 225°F O.C.

6.2 **Flammable Limits in Air:** 2.6% 12.5%

6.3 **Fire Extinguishing Agents:** Water fog, alcohol foam, carbon dioxide, dry chemical.

6.4 **Fire Extinguishing Agents Not to be Used:** Not pertinent

6.5 **Special Hazards of Combustion Products:** Not pertinent

6.6 **Behavior in Fire:** Not pertinent

6.7 **Ignition Temperature:** 790°F

6.8 **Electrical Hazard:** Not pertinent

6.9 **Burning Rate:** 1.5 mm/min.

7. CHEMICAL REACTIVITY

7.1 **Reactivity with Water:** No reaction

7.2 **Reactivity with Common Materials:** None

7.3 **Stability During Transport:** Stable

7.4 **Neutralizing Agents for Acids and Caustics:** Not pertinent

7.5 **Polymerization:** Not pertinent

7.6 **Inhibitor of Polymerization:** Not pertinent

8. WATER POLLUTION

8.1 **Aquatic Toxicity:** Data not available

8.2 **Waterfowl Toxicity:** Data not available

8.3 **Biological Oxygen Demand (BOD):** 2.2% of theoretical in 5 days

8.4 **Food Chain Concentration Potential:** None

9. SELECTED MANUFACTURERS

1. Dow Chemical Co.
Midland, Mich. 48640
2. Oxirane Corp.
120 Alexander St.
Princeton, N. J. 08540
3. Union Carbide Corp.
Chemicals and Plastics Division
270 Park Ave.
New York, N. Y. 10017

10. SHIPPING INFORMATION

10.1 **Grades or Purity:** USP, industrial, food (all 99+%)

10.2 **Storage Temperature:** Ambient

10.3 **Inert Atmosphere:** No requirement

10.4 **Venting:** Open (flame arrester)

11. HAZARD ASSESSMENT CODE

(See Hazard Assessment Handbook, CG 446-3)

A-P-Q

12. HAZARD CLASSIFICATIONS

12.1 **Code of Federal Regulations:** Not listed

12.2 **NAS Hazard Rating for Bulk Water Transportation:**

Category	Rating
Fire	1
Health	
Vapor Irritant	0
Liquid or Solid Irritant	0
Poisons	0
Water Pollution	
Human Toxicity	0
Aquatic Toxicity	1
Aesthetic Effect	1
Reactivity	
Other Chemicals	2
Water	0
Self-Reaction	0

12.3 **NFPA Hazard Classifications:** Not listed

13. PHYSICAL AND CHEMICAL PROPERTIES

13.1 **Physical State at 15°C and 1 atm:** Liquid

13.2 **Molecular Weight:** 76.10

13.3 **Boiling Point at 1 atm:** 369.1°F = 187.3°C = 460.5°K

13.4 **Freezing Point:** <−76°F = < −60°C = < 213°K

13.5 **Critical Temperature:** Not pertinent

13.6 **Critical Pressure:** Not pertinent

13.7 **Specific Gravity:** 1.04 at 20°C (liquid)

13.8 **Liquid Surface Tension:** 36 dynes/cm = 0.036 N/m at 25°C

13.9 **Liquid-Water Interfacial Tension:** Not pertinent

13.10 **Vapor (Gas) Specific Gravity:** Not pertinent

13.11 **Ratio of Specific Heats of Vapor (Gas):** 1.073

13.12 **Latent Heat of Vaporization:** 306 Btu/lb = 170 cal/g = 7.12×10^5 J/kg

13.13 **Heat of Combustion:** −10,310 Btu/lb = −5,728 cal/g = -239.8×10^5 J/kg

13.14 **Heat of Decomposition:** Not pertinent

13.15 **Heat of Solution:** Not pertinent

13.16 **Heat of Polymerization:** Not pertinent

NOTES

REVISED 1978

PME	# PROPYLENE GLYCOL METHYL ETHER

Common Synonyms 1-Methoxy-2-propanol Dowanol PM Dowanol 33B	Liquid Colorless Mild odor Mixes with water. Irritating vapor is produced.

Stop discharge if possible.
Call fire department.
Isolate and remove discharged material.
Notify local health and pollution control agencies.

Fire	FLAMMABLE Flashback along vapor trail may occur. Vapor may explode if ignited in an enclosed area. Extinguish with dry chemical, alcohol foam, or carbon dioxide. Water may be ineffective on fire. Cool exposed containers with water.
Exposure	CALL FOR MEDICAL AID. VAPOR Irritating to eyes, nose, and throat. Move to fresh air. LIQUID Irritating to skin and eyes. Remove contaminated clothing and shoes. Flush affected areas with plenty of water. IF IN EYES, hold eyelids open and flush with plenty of water.
Water Pollution	Effect of low concentrations on aquatic life is unknown. May be dangerous if it enters water intakes. Notify local health and wildlife officials. Notify operators of nearby water intakes.

1. RESPONSE TO DISCHARGE
(See Response Methods Handbook, CG 446-4)

Disperse and flush

2. LABEL

Red

3. CHEMICAL DESIGNATIONS

3.1 **Synonyms:** Dowanol 33B
Dowanol PM
1-Methoxy-2-propanol

3.2 **Coast Guard Compatibility Classification:** Glycol ether

3.3 **Chemical Formula:** $CH_3CH(OH)CH_2OCH_3$

3.4 **IMCO/United Nations Numerical Designation:** Not listed

4. OBSERVABLE CHARACTERISTICS

4.1 **Physical State (as shipped):** Liquid

4.2 **Color:** Colorless

4.3 **Odor:** Mild pleasant; mild ethereal

5. HEALTH HAZARDS

5.1 **Personal Protective Equipment:** Safety goggles, protective clothing.

5.2 **Symptoms Following Exposure:** Liquid irritates eyes and skin.

5.3 **Treatment for Exposure:** EYES: wash with water for 15 min.; call a physician. SKIN: remove contaminated clothing and wash skin with water.

5.4 **Toxicity by Inhalation (Threshold Limit Value):** 100 ppm

5.5 **Short-Term Inhalation Limits:** Data not available

5.6 **Toxicity by Ingestion:** Grade 1; LD_{50} 5 to 15 g/kg (rat)

5.7 **Late Toxicity:** Data not available

5.8 **Vapor (Gas) Irritant Characteristics:** Vapors cause moderate irritation, such that personnel will find high concentrations unpleasant. The effect is temporary.

5.9 **Liquid or Solid Irritant Characteristics:** Minimum hazard. If spilled on clothing and allowed to remain, may cause smarting and reddening of the skin.

5.10 **Odor Threshold:** Data not available

6. FIRE HAZARDS

6.1 **Flash Point:** 97°F O.C.

6.2 **Flammable Limits in Air:** Data not available

6.3 **Fire Extinguishing Agents:** Alcohol foam, dry chemical, or carbon dioxide

6.4 **Fire Extinguishing Agents Not to be Used:** Water may be ineffective.

6.5 **Special Hazards of Combustion Products:** Not pertinent

6.6 **Behavior in Fire:** Not pertinent

6.7 **Ignition Temperature:** Data not available

6.8 **Electrical Hazard:** Not pertinent

6.9 **Burning Rate:** Data not available

7. CHEMICAL REACTIVITY

7.1 **Reactivity with Water:** No reaction

7.2 **Reactivity with Common Materials:** No reaction

7.3 **Stability During Transport:** Stable

7.4 **Neutralizing Agents for Acids and Caustics:** Not pertinent

7.5 **Polymerization:** Not pertinent

7.6 **Inhibitor of Polymerization:** Not pertinent

8. WATER POLLUTION

8.1 **Aquatic Toxicity:** Data not available

8.2 **Waterfowl Toxicity:** Data not available

8.3 **Biological Oxygen Demand (BOD):** Data not available

8.4 **Food Chain Concentration Potential:** None

9. SELECTED MANUFACTURERS

1. Dow Chemical Co.
Midland, Mich. 48640
2. Jefferson Chemical Co., Inc.
7114 N. Lamar Blvd.
Austin, Tex. 78745
3. Olin Corp.
Chemicals Division
120 Long Ridge Rd.
Stamford, Conn. 06904

10. SHIPPING INFORMATION

10.1 **Grades or Purity:** Technical

10.2 **Storage Temperature:** Ambient

10.3 **Inert Atmosphere:** No requirement

10.4 **Venting:** Open (flame arrester)

11. HAZARD ASSESSMENT CODE
(See Hazard Assessment Handbook, CG 446-3)

A-P-Q

12. HAZARD CLASSIFICATIONS

12.1 **Code of Federal Regulations:** Flammable Liquid

12.2 **NAS Hazard Rating for Bulk Water Transportation:** Not listed

12.3 **NFPA Hazard Classifications:** Not listed

13. PHYSICAL AND CHEMICAL PROPERTIES

13.1 **Physical State at 15°C and 1 atm:** Liquid

13.2 **Molecular Weight:** 90.12

13.3 **Boiling Point at 1 atm:** 250°F = 121°C = 394°K

13.4 **Freezing Point:** Not pertinent

13.5 **Critical Temperature:** 538°F = 281°C = 554°K

13.6 **Critical Pressure:** Not pertinent

13.7 **Specific Gravity:** 0.924 at 20°C (liquid)

13.8 **Liquid Surface Tension:** Not pertinent

13.9 **Liquid-Water Interfacial Tension:** Not pertinent

13.10 **Vapor (Gas) Specific Gravity:** Not pertinent

13.11 **Ratio of Specific Heats of Vapor (Gas):** 1.066

13.12 **Latent Heat of Vaporization:** (est.) 166 Btu/lb = 92.3 cal/g = 3.86×10^5 J/kg

13.13 **Heat of Combustion:** (est.) −13,600 Btu/lb = −7580 cal/g = -317×10^5 J/kg

13.14 **Heat of Decomposition:** Not pertinent

13.15 **Heat of Solution:** (est.) −9 Btu/lb = −5 cal/g = -0.2×10^5 J/kg

13.16 **Heat of Polymerization:** Not pertinent

NOTES

REVISED 1978

PROPYLENEIMINE, INHIBITED

Common Synonyms			
Propyleneimine 2-Methylaziridine 2-Methylethyleneimine	Liquid	Colorless	Strong ammonia-like odor
	Mixes with water. Flammable, irritating vapor is produced.		

AVOID CONTACT WITH LIQUID AND VAPOR. KEEP PEOPLE AWAY.
Wear rubber overclothing (including gloves).
Shut off ignition sources. Call fire department.
Stop discharge if possible.
Stay upwind. Use water spray to "knock down" vapor.
Isolate and remove discharged material.
Notify local health and pollution control agencies.

Fire	FLAMMABLE. Irritating gases are produced when heated. Containers may explode in fire. Flashback along vapor trail may occur. Vapor may explode if ignited in an enclosed area. Combat fires from safe distance or protected location. Extinguish with dry chemicals or carbon dioxide. Water may be ineffective on fire. Cool exposed containers with water.
Exposure	CALL FOR MEDICAL AID. VAPOR Irritating to eyes, nose and throat. If inhaled will cause nausea, vomiting or difficult breathing. Move victim to fresh air. If breathing has stopped, give artificial respiration. If breathing is difficult, give oxygen. LIQUID POISONOUS IF SWALLOWED. Will burn skin and eyes. Remove contaminated clothing and shoes. Flush affected areas with plenty of water. IF IN EYES, hold eyelids open and flush with plenty of water. IF SWALLOWED and victim is CONSCIOUS, have victim drink water or milk. DO NOT INDUCE VOMITING.
Water Pollution	Effect of low concentrations on aquatic life is unknown. May be dangerous if it enters water intakes. Notify local health and wildlife officials. Notify operators of nearby water intakes.

1. RESPONSE TO DISCHARGE

(See Response Methods Handbook, CG 446-4)

Issue warning—high flammability, water contaminant, air contaminant
Restrict access
Evacuate area
Disperse and flush

2. LABEL

Red

3. CHEMICAL DESIGNATIONS

3.1 **Synonyms:** 2-Methylaziridine; 2-Methylethylenimine; Propylenimine
3.2 **Coast Guard Compatibility Classification:** Not applicable
3.3 **Chemical Formula:** CH_3CHCH_2NH
3.4 **IMCO/United Nations Numerical Designation:** 3.2/1921

4. OBSERVABLE CHARACTERISTICS

4.1 **Physical State (as shipped):** Liquid
4.2 **Color:** Colorless
4.3 **Odor:** Strong, ammonia-like

5. HEALTH HAZARDS

5.1 **Personal Protective Equipment:** Self-contained breathing apparatus; goggles or face shield; rubber gloves
5.2 **Symptoms Following Exposure:** Inhalation causes vomiting, breathing difficulty, and irritation of eyes, nose, and throat; on prolonged exposure, vapors tend to redden the whites of the eyes. Contact with liquid causes eye irritation, like that caused by strong ammonia. Liquid causes skin burns, which are slow to heal. Ingestion causes burns of mouth and stomach.
5.3 **Treatment for Exposure:** INHALATION: move victim to fresh air; if he is not breathing, apply artificial respiration, oxygen; if breathing is difficult, administer oxygen; call physician. EYES: flush with plenty of water for at least 30 min. and obtain prompt medical attention. SKIN: remove all contaminated clothing and flush with water; rinse with vinegar and water. INGESTION: drink large amounts of milk or water; get prompt medical attention.
5.4 **Toxicity by Inhalation (Threshold Limit Value):** 2 ppm
5.5 **Short-Term Inhalation Limits:** Data not available
5.6 **Toxicity by Ingestion:** Grade 4; oral LD_{50} = 19 mg/kg (rat)
5.7 **Late Toxicity:** Data not available
5.8 **Vapor (Gas) Irritant Characteristics:** Data not available
5.9 **Liquid or Solid Irritant Characteristics:** Data not available
5.10 **Odor Threshold:** Data not available

6. FIRE HAZARDS

6.1 **Flash Point:** 25°F O.C.
6.2 **Flammable Limits in Air:** Data not available
6.3 **Fire Extinguishing Agents:** Dry chemical or carbon dioxide
6.4 **Fire Extinguishing Agents Not to be Used:** Water or foam may be ineffective.
6.5 **Special Hazards of Combustion Products:** Irritating nitrogen oxides are produced.
6.6 **Behavior in Fire:** Containers may explode.
6.7 **Ignition Temperature:** Data not available
6.8 **Electrical Hazard:** Data not available
6.9 **Burning Rate:** 4.1 mm/min.

7. CHEMICAL REACTIVITY

7.1 **Reactivity with Water:** Reacts slowly to form propanolamine. The reaction is not hazardous.
7.2 **Reactivity with Common Materials:** No reaction
7.3 **Stability During Transport:** Stable if kept in contact with solid caustic soda (sodium hydroxide)
7.4 **Neutralizing Agents for Acids and Caustics:** Dilute with water, rinse with vinegar
7.5 **Polymerization:** Polymerizes explosively when in contact with any acid
7.6 **Inhibitor of Polymerization:** Solid sodium hydroxide (caustic soda)

8. WATER POLLUTION

8.1 **Aquatic Toxicity:** Data not available
8.2 **Waterfowl Toxicity:** Data not available
8.3 **Biological Oxygen Demand (BOD):** Data not available
8.4 **Food Chain Concentration Potential:** None

9. SELECTED MANUFACTURERS

1. Arsynco, Incorporated
   P. O Box 8
   Carlstadt, N. J. 07072
2. Polysciences, Inc.
   Paul Valley Industrial Park
   Warrington, Pa. 18976
3. Pfaltz and Bauer, Inc.
   126-04 Northern Boulevard
   Flushing, N. Y. 11368

10. SHIPPING INFORMATION

10.1 **Grades or Purity:** Technical
10.2 **Storage Temperature:** Ambient
10.3 **Inert Atmosphere:** Exclude air
10.4 **Venting:** Pressure-vacuum

11. HAZARD ASSESSMENT CODE

(See Hazard Assessment Handbook, CG 446-3)

A-O-P-Q-R-S-Z

12. HAZARD CLASSIFICATIONS

12.1 **Code of Federal Regulations:** Flammable liquid
12.2 **NAS Hazard Rating for Bulk Water Transportation:** Not listed
12.3 **NFPA Hazard Classifications:** Not listed

13. PHYSICAL AND CHEMICAL PROPERTIES

13.1 **Physical State at 15°C and 1 atm:** Liquid
13.2 **Molecular Weight:** 57.1
13.3 **Boiling Point at 1 atm:** 151°F = 66°C = 339°K
13.4 **Freezing Point:** Not pertinent
13.5 **Critical Temperature:** Not pertinent
13.6 **Critical Pressure:** Not pertinent
13.7 **Specific Gravity:** 0.802 at 25°C (liquid)
13.8 **Liquid Surface Tension:** (est.) 25 dynes/cm = 0.025 N/m at 20°C
13.9 **Liquid-Water Interfacial Tension:** Not pertinent
13.10 **Vapor (Gas) Specific Gravity:** 2
13.11 **Ratio of Specific Heats of Vapor (Gas):** Data not available
13.12 **Latent Heat of Vaporization:** 250 Btu/lb = 139 cal/g = 5.82×10^5 J/kg
13.13 **Heat of Combustion:** (est.) −15,500 Btu/lb = −8,600 cal/g = -360×10^5 J/kg
13.14 **Heat of Decomposition:** Not pertinent
13.15 **Heat of Solution:** −140 Btu/lb = −78 cal/g = -3.3×10^5 J/kg
13.16 **Heat of Polymerization:** (est.) −720 Btu/lb = −400 cal/g = -17×10^5 J/kg

NOTES

PROPYLENE OXIDE

Common Synonyms
Methyloxirane
1, 2-Epoxypropane
Propeneoxide

Liquid | Colorless | Sweet, alcohol odor

Mixes with water. Flammable, irritating vapor is produced.

AVOID CONTACT WITH LIQUID AND VAPOR. Keep people away.
Wear goggles, self-contained breathing apparatus, and rubber overclothing (including gloves).
Shut off ignition sources and call fire department.
Stop discharge if possible.
Stay upwind and use water spray to "knock down" vapor.
Isolate and remove discharged material.
Notify local health and pollution control agencies.

Fire

FLAMMABLE.
Containers may explode in fire.
Flashback along vapor trail may occur.
Vapor may explode if ignited in an enclosed area.
Wear goggles, self-contained breathing apparatus, and rubber overclothing (including gloves).
Combat fires from safe distance or protected location.
Extinguish with dry chemical, alcohol foam, or carbon dioxide.
Water may be ineffective on fire.
Cool exposed containers with water.

Exposure

CALL FOR MEDICAL AID.

VAPOR
Irritating to eyes, nose, and throat.
If inhaled, will cause headache, nausea, vomiting, or loss of consciousness.
Move to fresh air.
If breathing has stopped, give artificial respiration.
If breathing is difficult, give oxygen.

LIQUID
Will burn skin and eyes.
Harmful if swallowed.
Remove contaminated clothing and shoes.
Flush affected areas with plenty of water.
IF IN EYES, hold eyelids open and flush with plenty of water.
IF SWALLOWED and victim is CONSCIOUS, have victim drink water or milk.

Water Pollution

Effect of low concentrations on aquatic life is unknown.
May be dangerous if it enters water intakes.
Notify local health and wildlife officials.
Notify operators of nearby water intakes.

1. RESPONSE TO DISCHARGE

(See Response Methods Handbook, CG 446-4)

Issue warning—high flammability
Restrict access
Evacuate area

2. LABEL

Red

3. CHEMICAL DESIGNATIONS

3.1 **Synonyms:** 1,2-Epoxypropane
Methyl oxirane
Propene oxide

3.2 **Coast Guard Compatibility Classification:** Alkylene oxide

3.3 **Chemical Formula:** CH_3CHCH_2O (ring)

3.4 **IMCO/United Nations Numerical Designation:** 3.1/1280

4. OBSERVABLE CHARACTERISTICS

4.1 **Physical State (as shipped):** Liquid

4.2 **Color:** Colorless

4.3 **Odor:** Ethereal; characteristic; sweet, alcoholic; like natural gas

5. HEALTH HAZARDS

5.1 **Personal Protective Equipment:** Air-supplied mask; rubber or plastic gloves; vapor-proof goggles.

5.2 **Symptoms Following Exposure:** Inhalation may produce headache, nausea, vomiting, and unconsciousness; mild depression of central nervous system; lung irritation. Slightly irritating to skin, but covered contact may cause burn. Very irritating to eyes.

5.3 **Treatment for Exposure:** INHALATION: remove person to fresh air immediately, keep quiet and warm; call a physician; if breathing stops, start artificial respiration. SKIN OR EYE CONTACT: immediately flush with plenty of water for at least 15 min.; immediately remove contaminated clothing, watch bands, rings, etc. to prevent confining product to skin; for eyes get medical attention.

5.4 **Toxicity by Inhalation (Threshold Limit Value):** 100 ppm

5.5 **Short-Term Inhalation Limits:** Data not available

5.6 **Toxicity by Ingestion:** Grade 2; LD_{50} 0.5 to 5 g/kg (rat)

5.7 **Late Toxicity:** Data not available

5.8 **Vapor (Gas) Irritant Characteristics:** Vapor is moderately irritating such that personnel will not usually tolerate moderate or high vapor concentrations.

5.9 **Liquid or Solid Irritant Characteristics:** Causes smarting of the skin and first-degree burns on short exposure; may cause secondary burns on long exposure.

5.10 **Odor Threshold:** 200 ppm

6. FIRE HAZARDS

6.1 **Flash Point:** −35°F C.C.; <−20°F O.C.

6.2 **Flammable Limits in Air:** 2.1%–38.5%

6.3 **Fire Extinguishing Agents:** Carbon dioxide or dry chemical for small fires; alcohol or polymer foam for large fires.

6.4 **Fire Extinguishing Agents Not to be Used:** Water may be ineffective.

6.5 **Special Hazards of Combustion Products:** Not pertinent

6.6 **Behavior in Fire:** Containers may explode. Vapor is heavier than air and may travel considerable distance to a source of ignition and flash back.

6.7 **Ignition Temperature:** 869°F

6.8 **Electrical Hazard:** Class I, Group B

6.9 **Burning Rate:** 3.3 mm/min.

7. CHEMICAL REACTIVITY

7.1 **Reactivity with Water:** No reaction

7.2 **Reactivity with Common Materials:** No reaction

7.3 **Stability During Transport:** Stable

7.4 **Neutralizing Agents for Acids and Caustics:** Not pertinent

7.5 **Polymerization:** May occur due to high temperatures, contamination with alkalies, aqueous acids, amines, and acidic alcohols.

7.6 **Inhibitor of Polymerization:** Not pertinent

8. WATER POLLUTION

8.1 **Aquatic Toxicity:** Data not available

8.2 **Waterfowl Toxicity:** Data not available

8.3 **Biological Oxygen Demand (BOD):** Data not available

8.4 **Food Chain Concentration Potential:** None

9. SELECTED MANUFACTURERS

1. BASF Wyandotte Corp.
Wyandotte, Mich. 48192
2. Dow Chemical Co.
Midland, Mich. 48640
3. Oxirane Corp.
120 Alexander St.
Princeton, N. J. 08540

10. SHIPPING INFORMATION

10.1 **Grades or Purity:** 99.99% (must contain no acetylene)

10.2 **Storage Temperature:** Ambient

10.3 **Inert Atmosphere:** Inerted

10.4 **Venting:** Safety relief

11. HAZARD ASSESSMENT CODE

(See Hazard Assessment Handbook, CG 446-3)

A-P-Q-R-S

12. HAZARD CLASSIFICATIONS

12.1 **Code of Federal Regulations:** Flammable liquid

12.2 **NAS Hazard Rating for Bulk Water Transportation:**

Category	Rating
Fire	4
Health	
Vapor Irritant	3
Liquid or Solid Irritant	2
Poisons	2
Water Pollution	
Human Toxicity	2
Aquatic Toxicity	1
Aesthetic Effect	1
Reactivity	
Other Chemicals	3
Water	1
Self-Reaction	3

12.3 **NFPA Hazard Classifications:**

Category	Classification
Health Hazard (Blue)	2
Flammability (Red)	4
Reactivity (Yellow)	2

13. PHYSICAL AND CHEMICAL PROPERTIES

13.1 **Physical State at 15°C and 1 atm:** Liquid

13.2 **Molecular Weight:** 58.08

13.3 **Boiling Point at 1 atm:** 93.7°F = 34.3°C = 307.5°K

13.4 **Freezing Point:** −169.4°F = −111.9°C = 161.3°K

13.5 **Critical Temperature:** 408.4°F = 209.1°C = 482.3°K

13.6 **Critical Pressure:** 714 psia = 48.6 atm = 4.92 MN/m^2

13.7 **Specific Gravity:** 0.830 at 20°C (liquid)

13.8 **Liquid Surface Tension:** 24.5 dynes/cm = 0.0245 N/m at 15°C

13.9 **Liquid-Water Interfacial Tension:** Not pertinent

13.10 **Vapor (Gas) Specific Gravity:** 2.0

13.11 **Ratio of Specific Heats of Vapor (Gas):** 1.133

13.12 **Latent Heat of Vaporization:** 205 Btu/lb = 114 cal/g = 4.77 × 10^5 J/kg

13.13 **Heat of Combustion:** −13,000 Btu/lb = −7,221 cal/g = −302.3 × 10^5 J/kg

13.14 **Heat of Decomposition:** Not pertinent

13.15 **Heat of Solution:** (est.) −19 Btu/lb = −11 cal/g = −0.45 × 10^5 J/kg

13.16 **Heat of Polymerization:** Not pertinent

NOTES

REVISED 1978

PTT PROPYLENE TETRAMER

Common Synonyms
Dodecene (non-linear)
Tetrapropylene

Liquid — Colorless

Floats on water.

Stop discharge if possible.
Call fire department.
Avoid contact with liquid.
Isolate and remove discharged material.
Notify local health and pollution control agencies.

Fire

Combustible.
Extinguish with foam, water, dry chemical, or carbon dioxide.
Cool exposed containers with water.

Exposure

CALL FOR MEDICAL AID.

LIQUID
Irritating to skin and eyes.
Harmful if swallowed.
Flush affected areas with plenty of water.
IF IN EYES, hold eyelids open and flush with plenty of water.
IF SWALLOWED and victim is CONSCIOUS, have victim drink water or milk.
DO NOT INDUCE VOMITING.

Water Pollution

Effect of low concentrations on aquatic life is unknown.
Fouling to shoreline.
May be dangerous if it enters water intakes.
Notify local health and wildlife officials.
Notify operators of nearby water intakes.

1. RESPONSE TO DISCHARGE

(See Response Methods Handbook, CG 446-4)

Mechanical containment
Should be removed
Chemical and physical treatment

2. LABELS

No hazard label required by Code of Federal Regulations

3. CHEMICAL DESIGNATIONS

3.1 **Synonyms:** Dodecene (nonlinear)
Tetrapropylene

3.2 **Coast Guard Compatibility Classification:** Olefin

3.3 **Chemical Formula:** $C_{12}H_{14}$

3.4 **IMCO/United Nations Numerical Designation:** Not listed

4. OBSERVABLE CHARACTERISTICS

4.1 **Physical State (as shipped):** Liquid

4.2 **Color:** Colorless

4.3 **Odor:** Data not available

5. HEALTH HAZARDS

5.1 **Personal Protective Equipment:** Goggles or face shield.

5.2 **Symptoms Following Exposure:** No inhalation hazard expected. Aspiration hazard if ingested. Low skin and eye irritation.

5.3 **Treatment for Exposure:** INHALATION: remove victim to fresh air. INGESTION: do NOT lavage or induce vomiting; give vegetable oil and demulcents; call physician. EYES: flush with water for 15 min. SKIN: wash with soap and water.

5.4 **Toxicity by Inhalation (Threshold Limit Value):** 200 ppm

5.5 **Short-Term Inhalation Limits:** Data not available

5.6 **Toxicity by Ingestion:** Grade 0; LD_{50} above 15 g/kg

5.7 **Late Toxicity:** Data not available

5.8 **Vapor (Gas) Irritant Characteristics:** Vapors cause a slight smarting of the eyes or respiratory system if present in high concentrations. The effect is temporary.

5.9 **Liquid or Solid Irritant Characteristics:** Minimum hazard. If spilled on clothing and allowed to remain, may cause smarting and reddening of the skin.

5.10 **Odor Threshold:** Data not available

6. FIRE HAZARDS

6.1 **Flash Point:** 120°F C.C.; 134°F O.C.

6.2 **Flammable Limits in Air:** Data not available

6.3 **Fire Extinguishing Agents:** Water fog, foam, carbon dioxide or dry chemical

6.4 **Fire Extinguishing Agents Not to be Used:** Not pertinent

6.5 **Special Hazards of Combustion Products:** Not pertinent

6.6 **Behavior in Fire:** Not pertinent

6.7 **Ignition Temperature:** 400°F (est.)

6.8 **Electrical Hazard:** Not pertinent

6.9 **Burning Rate:** Data not available

7. CHEMICAL REACTIVITY

7.1 **Reactivity with Water:** No reaction

7.2 **Reactivity with Common Materials:** No reaction

7.3 **Stability During Transport:** Stable

7.4 **Neutralizing Agents for Acids and Caustics:** Not pertinent

7.5 **Polymerization:** Not pertinent

7.6 **Inhibitor of Polymerization:** Not pertinent

8. WATER POLLUTION

8.1 **Aquatic Toxicity:** Data not available

8.2 **Waterfowl Toxicity:** Data not available

8.3 **Biological Oxygen Demand (BOD):** Data not available

8.4 **Food Chain Concentration Potential:** None

9. SELECTED MANUFACTURERS

1. Atlantic Richfield Co.
ARCO Chemical Co. Division
260 S. Broad St.
Philadelphia, Pa. 19101
2. Continental Oil Co.
Conoco Chemicals Division
Park 80 Plaza East
Saddle Brook, N. J. 07662
3. Sun Oil Co.
St. Davids, Pa. 19087

10. SHIPPING INFORMATION

10.1 **Grades or Purity:** 98.5+%

10.2 **Storage Temperature:** Ambient

10.3 **Inert Atmosphere:** No requirement

10.4 **Venting:** Open (flame arrester)

11. HAZARD ASSESSMENT CODE

(See Hazard Assessment Handbook, CG 446-3)

A-T-U

12. HAZARD CLASSIFICATIONS

12.1 **Code of Federal Regulations:** Not listed
Combustible Liquid

12.2 **NAS Hazard Rating for Bulk Water Transportation:**

Category	Rating
Fire	2
Health	
Vapor Irritant	1
Liquid or Solid Irritant	1
Poisons	1
Water Pollution	
Human Toxicity	0
Aquatic Toxicity	1
Aesthetic Effect	2
Reactivity	
Other Chemicals	1
Water	0
Self-Reaction	1

12.3 **NFPA Hazard Classifications:**

Category	Classification
Health Hazard (Blue)	0
Flammability (Red)	2
Reactivity (Yellow)	0

13. PHYSICAL AND CHEMICAL PROPERTIES

13.1 **Physical State at 15°C and 1 atm:** Liquid

13.2 **Molecular Weight:** 168.31

13.3 **Boiling Point at 1 atm:** 365 – 385°F = 185 – 196°C = 458 – 469°K

13.4 **Freezing Point:** Not pertinent

13.5 **Critical Temperature:** Not pertinent

13.6 **Critical Pressure:** Not pertinent

13.7 **Specific Gravity:** 0.2937 at 20°C (liquid)

13.8 **Liquid Surface Tension:** 23.9 dynes/cm = 0.0239 N/m at 24°C

13.9 **Liquid-Water Interfacial Tension:** 44.5 dynes/cm = 0.0445 N/m at 22.7°C

13.10 **Vapor (Gas) Specific Gravity:** Not pertinent

13.11 **Ratio of Specific Heats of Vapor (Gas):** Not pertinent

13.12 **Latent Heat of Vaporization:** (est.) 154 Btu/lb = 58.6 cal/g = 2.45×10^5 J/kg

13.13 **Heat Combustion:** -19,100 Btu/lb = -10,600 cal/g = -444×10^5 J/kg

13.14 **Heat of Decomposition:** Not pertinent

13.15 **Heat of Solution:** Not pertinent

13.16 **Heat of Polymerization:** Not pertinent

NOTES

REVISED 1978

PMN	# n-PROPYL MERCAPTAN

Common Synonyms	Liquid	Colorless	Skunk-like odor
1-Propanethiol Propane-1-thiol	Floats on water. Flammable, irritating vapor is produced.		

Stop discharge if possible. Keep people away.
Shut off ignition sources. Call fire department.
Isolate and remove discharged material.
Notify local health and pollution control agencies.

Fire

FLAMMABLE.
POISONOUS GASES ARE PRODUCED IN FIRE.
Containers may explode in fire.
Flashback along vapor trail may occur.
Vapor may explode if ignited in an enclosed area.
Wear goggles and self-contained breathing apparatus.
Extinguish with dry chemicals, foam or carbon dioxide.
Water may be ineffective on fire.
Cool exposed containers with water.

Exposure

Call for medical aid.

VAPOR
If inhaled will cause difficult breathing.
Move victim to fresh air.
If breathing has stopped, give artificial respiration.
If breathing is difficult, give oxygen.

LIQUID
Irritating to skin and eyes.
Harmful if swallowed.
Remove contaminated clothing and shoes.
Flush affected areas with plenty of water.
IF IN EYES, hold eyelids open and flush with plenty of water.
IF SWALLOWED and victim is CONSCIOUS, have victim drink water or milk and have victim induce vomiting.
IF SWALLOWED and victim is UNCONSCIOUS OR HAVING CONVULSIONS, do nothing except keep victim warm.

Water Pollution

Effect of low concentrations on aquatic life is unknown.
Fouling to shoreline.
May be dangerous if it enters water intakes.
Notify local health and wildlife officials.
Notify operators of nearby water intakes.

1. RESPONSE TO DISCHARGE

(See Response Methods Handbook, CG 446-4)

Issue warning—high flammability, water contaminant, air contaminant; restrict access; evacuate area; mechanical containment; should be removed; chemical and physical treatment

2. LABEL

Red

3. CHEMICAL DESIGNATIONS

3.1 **Synonyms:** 1-Propanethiol; Propane-1-thiol
3.2 **Coast Guard Compatibility Classification:** Not applicable
3.3 **Chemical Formula:** $CH_3CH_2CH_2SH$
3.4 **IMCO/United Nations Numerical Designation:** 3.1/1228

4. OBSERVABLE CHARACTERISTICS

4.1 **Physical State (as shipped):** Liquid
4.2 **Color:** Colorless
4.3 **Odor:** Skunky

5. HEALTH HAZARDS

5.1 **Personal Protective Equipment:** Goggles or face shield; rubber gloves; self-contained breathing apparatus or organic canister mask
5.2 **Symptoms Following Exposure:** Inhalation causes muscular weakness, convulsions, and respiratory paralysis; high concentrations may cause pulmonary irritation. Contact with liquid causes irritation of eyes and skin. Ingestion causes irritation of mouth and stomach.
5.3 **Treatment for Exposure:** INHALATION: remove victim from contaminated atmosphere; give artificial respiration and oxygen if needed; observe for premonitory signs of pulmonary edema. EYES: flush with water for 15 min.; if irritation persists, see a physician. SKIN: flush with water; wash with soap and water. INGESTION: induce vomiting and follow with gastric lavage.
5.4 **Toxicity by Inhalation (Threshold Limit Value):** Data not available
5.5 **Short-Term Inhalation Limits:** Data not available
5.6 **Toxicity by Ingestion:** Grade 2; oral LD_{50} = 1,790 mg/kg (rat)
5.7 **Late Toxicity:** Data not available
5.8 **Vapor (Gas) Irritant Characteristics:** Data not available
5.9 **Liquid or Solid Irritant Characteristics:** Data not available
5.10 **Odor Threshold:** 0.00075 ppm

6. FIRE HAZARDS

6.1 **Flash Point:** 5°F O.C.
6.2 **Flammable Limits in Air:** Data not available
6.3 **Fire Extinguishing Agents:** Dry chemical, foam, carbon dioxide
6.4 **Fire Extinguishing Agents Not to be Used:** Water may be ineffective.
6.5 **Special Hazards of Combustion Products:** Toxic sulfur dioxide is generated.
6.6 **Behavior in Fire:** Not pertinent
6.7 **Ignition Temperature:** Data not available
6.8 **Electrical Hazard:** Data not available
6.9 **Burning Rate:** 5.1 mm/min.

7. CHEMICAL REACTIVITY

7.1 **Reactivity with Water:** No reaction
7.2 **Reactivity with Common Materials:** No reaction
7.3 **Stability During Transport:** Stable
7.4 **Neutralizing Agents for Acids and Caustics:** Not pertinent
7.5 **Polymerization:** Not pertinent
7.6 **Inhibitor of Polymerization:** Not pertinent

8. WATER POLLUTION

8.1 **Aquatic Toxicity:** Data not available
8.2 **Waterfowl Toxicity:** Data not available
8.3 **Biological Oxygen Demand (BOD):** Data not available
8.4 **Food Chain Concentration Potential:** None

9. SELECTED MANUFACTURERS

1. Pennwalt Corporation
Chemicals Division
Three Parkway
Philadelphia, Pa. 19102
2. Phillips Petroleum Company
Chemical Department
Special Products Division
Bartlesville, Okla. 74004
3. Aldrich Chemical Co.
940 West St. Paul Avenue
Milwaukee, Wis. 53233

10. SHIPPING INFORMATION

10.1 **Grades or Purity:** 98+%
10.2 **Storage Temperature:** Ambient
10.3 **Inert Atmosphere:** No requirement
10.4 **Venting:** Pressure-vacuum

11. HAZARD ASSESSMENT CODE

(See Hazard Assessment Handbook, CG 446-3)

A-T-U-V-W

12. HAZARD CLASSIFICATIONS

12.1 **Code of Federal Regulations:** Flammable liquid
12.2 **NAS Hazard Rating for Bulk Water Transportation:** Not listed
12.3 **NFPA Hazard Classifications:** Not listed

13. PHYSICAL AND CHEMICAL PROPERTIES

13.1 **Physical State at 15°C and 1 atm:** Liquid
13.2 **Molecular Weight:** 76.2
13.3 **Boiling Point at 1 atm:** 153°F = 67°C = 340°K
13.4 **Freezing Point:** −171°F = −113°C = 160°K
13.5 **Critical Temperature:** (est.) 495°F = 257°C = 530°K
13.6 **Critical Pressure:** (est.) 667 psia = 45.3 atm = 4.60 MN/m^2
13.7 **Specific Gravity:** 0.841 at 20°C (liquid)
13.8 **Liquid Surface Tension:** 24.7 dynes/cm = 0.0247 N/m at 20°C
13.9 **Liquid-Water Interfacial Tension:** (est.) 18 dynes/cm = 0.018 N/m at 20°C
13.10 **Vapor (Gas) Specific Gravity:** 2.6
13.11 **Ratio of Specific Heats of Vapor (Gas):** 1.0984
13.12 **Latent Heat of Vaporization:** 179 Btu/lb = 99 cal/g = 4.16 × 10^5 J/kg
13.13 **Heat of Combustion:** −15,990 Btu/lb = −8,890 cal/g = −372 × 10^5 J/kg
13.14 **Heat of Decomposition:** Not pertinent
13.15 **Heat of Solution:** Not pertinent
13.16 **Heat of Polymerization:** Not pertinent

NOTES

PRD	# PYRIDINE

Common Synonyms	Liquid	Colorless to yellow	Sharp, nauseating odor
	Mixes with water. Poisonous, flammable vapor is produced.		

AVOID CONTACT WITH LIQUID AND VAPOR. Keep people away.
Wear chemical protective suit with self-contained breathing apparatus.
Shut off ignition sources and call fire department.
Stop discharge if possible.
Stay upwind and use water spray to "knock down" vapor.
Isolate and remove discharged material.
Notify local health and pollution control agencies.

Fire	FLAMMABLE. Flashback along vapor trail may occur. Vapor may explode if ignited in an enclosed area. **Wear chemical protective suit with self-contained breathing apparatus.** **Extinguish with dry chemical, alcohol foam, or carbon dioxide.** **Water may be ineffective on fire.** **Cool exposed containers with water.**
Exposure	CALL FOR MEDICAL AID. VAPOR Poisonous if inhaled or if skin is exposed. Irritating to eyes, nose, and throat. **Move to fresh air.** **If breathing has stopped, give artificial respiration.** **If breathing is difficult, give oxygen.** LIQUID Poisonous if swallowed or if skin is exposed. Will burn eyes. **Remove contaminated clothing and shoes.** **Flush affected areas with plenty of water.** **IF IN EYES, hold eyelids open and flush with plenty of water.** **IF SWALLOWED and victim is CONSCIOUS, have victim drink water or milk and have victim induce vomiting.** **IF SWALLOWED and victim is UNCONSCIOUS OR HAVING CONVULSIONS, do nothing except keep victim warm.**
Water Pollution	Dangerous to aquatic life in high concentrations. May be dangerous if it enters water intakes. **Notify local health and wildlife officials.** **Notify operators of nearby water intakes.**

1. RESPONSE TO DISCHARGE
(See Response Methods Handbook, CG 446-4)

Issue warning—high flammability
Restrict access
Evacuate area
Disperse and flush

2. LABEL

Red

3. CHEMICAL DESIGNATIONS

3.1 **Synonyms:** No common synonyms
3.2 **Coast Guard Compatibility Classification:** Aromatic amine
3.3 **Chemical Formula:** C_5H_5N
3.4 **IMCO/United Nations Numerical Designation:** 3.2/1282

4. OBSERVABLE CHARACTERISTICS

4.1 **Physical State (as shipped):** Liquid
4.2 **Color:** Yellow or colorless
4.3 **Odor:** Disagreeable; strong unpleasant; characteristic unpleasant; sharp penetrating, unpleasant

5. HEALTH HAZARDS

5.1 **Personal Protective Equipment:** Air-supplied mask or organic canister; vapor-proof goggles; rubber gloves and protective clothing.
5.2 **Symptoms Following Exposure:** Vapor irritates eyes and nose. Liquid irritates skin and is absorbed through the skin. Overexposure causes nausea, headache, nervous symptoms, increased urinary frequency.
5.3 **Treatment for Exposure:** INHALATION: remove individual promptly from contaminated area; give artificial respiration and oxygen if necessary; treat symptomatically. INGESTION: induce vomiting and follow with gastric lavage. SKIN: wash thoroughly with large amounts of water. EYES: irrigate with water for at least 15 min.
5.4 **Toxicity by Inhalation (Threshold Limit Value):** 5 ppm
5.5 **Short-Term Inhalation Limits:** Data not available
5.6 **Toxicity by Ingestion:** Grade 2; LD_{50} 0.5 to 5 g/kg (rat)
5.7 **Late Toxicity:** Liver and kidney damage after ingestion.
5.8 **Vapor (Gas) Irritant Characteristics:** Vapors cause moderate irritation such that personnel will find high concentrations unpleasant. The effect is temporary.
5.9 **Liquid or Solid Irritant Characteristics:** Causes smarting of the skin and first-degree burns on short exposure; may cause secondary burns on long exposure.
5.10 **Odor Threshold:** 0.021 ppm

6. FIRE HAZARDS

6.1 **Flash Point:** 68°F C.C.
6.2 **Flammable Limits in Air:** 1.8%–12.4%
6.3 **Fire Extinguishing Agents:** Alcohol foam, dry chemical, or carbon dioxide
6.4 **Fire Extinguishing Agents Not to be Used:** Water may be ineffective.
6.5 **Special Hazards of Combustion Products:** Not pertinent
6.6 **Behavior in Fire:** Vapor is heavier than air and may travel considerable distance to source of ignition and flash back.
6.7 **Ignition Temperature:** 900°F
6.8 **Electrical Hazard:** Class I, Group D
6.9 **Burning Rate:** 4.3 mm/min.

7. CHEMICAL REACTIVITY

7.1 **Reactivity with Water:** No reaction
7.2 **Reactivity with Common Materials:** None
7.3 **Stability During Transport:** Stable
7.4 **Neutralizing Agents for Acids and Caustics:** Flush with water.
7.5 **Polymerization:** Not pertinent
7.6 **Inhibitor of Polymerization:** Not pertinent

8. WATER POLLUTION

8.1 **Aquatic Toxicity:** 1350 mg/l/96 hr/fish/TL_m/fresh water
8.2 **Waterfowl Toxicity:** Data not available
8.3 **Biological Oxygen Demand (BOD):** 1.15–1.47 lb/lb, 5 days
8.4 **Food Chain Concentration Potential:** None

9. SELECTED MANUFACTURERS

1. Koppers Co., Inc.
Organic Materials Division
Pittsburgh, Pa. 15219
2. Reilly Tar and Chemical Corp.
1615 Merchants Bank Bldg.
Indianapolis, Ind. 46204
3. Warner-Lambert Co.
Nepera Chemical Co. Division
Harriman, N. Y. 10926

10. SHIPPING INFORMATION

10.1 **Grades or Purity:** Technical; Pure
10.2 **Storage Temperature:** Ambient
10.3 **Inert Atmosphere:** No requirement
10.4 **Venting:** Pressure-vacuum

11. HAZARD ASSESSMENT CODE
(See Hazard Assessment Handbook, CG 446-3)

A-P-Q-R-S

12. HAZARD CLASSIFICATIONS

12.1 **Code of Federal Regulations:** Flammable liquid
12.2 **NAS Hazard Rating for Bulk Water Transportation:**

Category	Rating
Fire	3
Health	
Vapor Irritant	2
Liquid or Solid Irritant	2
Poisons	1
Water Pollution	
Human Toxicity	2
Aquatic Toxicity	2
Aesthetic Effect	3
Reactivity	
Other Chemicals	3
Water	0
Self-Reaction	0

12.3 **NFPA Hazard Classifications:**

Category	Classification
Health Hazard (Blue)	2
Flammability (Red)	3
Reactivity (Yellow)	0

13. PHYSICAL AND CHEMICAL PROPERTIES

13.1 **Physical State at 15°C and 1 atm:** Liquid
13.2 **Molecular Weight:** 79.10
13.3 **Boiling Point at 1 atm:** 239.5°F = 115.3°C = 388.5°K
13.4 **Freezing Point:** −44°F = −42°C = 231°K
13.5 **Critical Temperature:** 656.2°F = 346.8°C = 620°K
13.6 **Critical Pressure:** 817.3 psia = 55.6 atm = 5.63 MN/m²
13.7 **Specific Gravity:** 0.983 at 20°C (liquid)
13.8 **Liquid Surface Tension:** 38.0 dynes/cm = 0.038 N/m at 20°C
13.9 **Liquid-Water Interfacial Tension:** Not pertinent
13.10 **Vapor (Gas) Specific Gravity:** 2.73
13.11 **Ratio of Specific Heats of Vapor (Gas):** 1.123
13.12 **Latent Heat of Vaporization:** 193 Btu/lb = 107 cal/g = 4.48×10^5 J/kg
13.13 **Heat of Combustion:** −14,390 Btu/lb = −7992 cal/g = -334.6×10^5 J/kg
13.14 **Heat of Decomposition:** Not pertinent
13.15 **Heat of Solution:** (est.) −13 Btu/lb = −7 cal/g = -0.3×10^5 J/kg
13.16 **Heat of Polymerization:** Not pertinent

NOTES

REVISED 1978

PGA PYROGALLIC ACID

Common Synonyms	Solid	White to Gray	Odoroless
Pyrogallol 1,2,3-Benzenetriol 1,2,3-Trihydroxybenzene	Sinks and mixes with water.		

Avoid contact with solid and dust. Keep people away.
Stop discharge if possible.
Call fire department.
Isolate and remove discharged material.
Notify local health and pollution control agencies.

Fire

Combustible.
Extinguish with water, dry chemicals, foam, or carbon dioxide.

Exposure

CALL FOR MEDICAL AID.

DUST
Irritating to eyes, nose and thorat.
If inhaled will cause coughing or difficult breathing.
If in eyes, hold eyelids open and flush with plenty of water.
If breathing has stopped, give artificial respiration.
If breathing is difficult, give oxygen.

SOLID
Irritating to skin and eyes.
If swallowed will cause nausea, vomiting or loss of consciousness.
Remove contaminated clothing and shoes.
Flush affected areas with plenty of water.
IF IN EYES, hold eyepids open and flush with plenty of water.
IF SWALLOWED and victim is CONSCIOUS, have victim drink water or milk and have victim induce vomiting.
IF SWALLOWED and victim is UNCONSCIOUS OR HAVING CONVULSIONS, do nothing except keep victim warm.

Water Pollution

HARMFUL TO AQUATIC LIFE IN VERY LOW CONCENTRATIONS.
May be dangerous if it enters water intakes.
Notify local health and wildlife officials.
Notify operators of nearby water intakes.

1. RESPONSE TO DISCHARGE

(See Response Methods Handbook, CG 446-4)

Issue warning—water contaminant
Disperse and flush

2. LABELS

No hazard label required by Code of Federal Regulations

3. CHEMICAL DESIGNATIONS

3.1 **Synonyms:** 1,2,3-Benzenetriol; Pyrogallol; 1,2,3-Trihydroxybenzene
3.2 **Coast Guard Compatibility Classification:** Not listed
3.3 **Chemical Formula:** 1,2,3-$C_6H_3(OH)_3$
3.4 **IMCO/United Nations Numerical Designation:** Not listed

4. OBSERVABLE CHARACTERISTICS

4.1 **Physical State (as shipped):** Solid
4.2 **Color:** White to gray
4.3 **Odor:** None

5. HEALTH HAZARDS

5.1 **Personal Protective Equipment:** Rubber gloves; safety goggles; dust mask
5.2 **Symptoms Following Exposure:** Inhalation of dust causes irritation of nose and throat. Ingestion may cause severe gastrointestinal irritation, convulsions, circulatory collapse, and death. Contact with eyes causes irritation. Skin contact can cause local discoloration, irritation, eczema, and death; repeated contact can cause sensitization.
5.3 **Treatment for Exposure:** INHALATION: remove victim to fresh air. INGESTION: give large amount of water; induce vomiting immediately; consult a physician. EYES: flush with water for at least 15 min.; consult a physician. SKIN: wash immediately with soap and water; consult a physician if exposure has been severe.
5.4 **Toxicity by Inhalation (Threshold Limit Value):** Data not available
5.5 **Short-Term Inhalation Limits:** Data not available
5.6 **Toxicity by Ingestion:** Grade 2; oral LD_{50} = 789 mg/kg (rat)
5.7 **Late Toxicity:** Depresses growth in chicks
5.8 **Vapor (Gas) Irritant Characteristics:** Data not available
5.9 **Liquid or Solid Irritant Characteristics:** Data not available
5.10 **Odor Threshold:** Odorless

6. FIRE HAZARDS

6.1 **Flash Point:** Not pertinent (combustible solid)
6.2 **Flammable Limits in Air:** Not pertinent
6.3 **Fire Extinguishing Agents:** Water, foam, dry chemical, carbon dioxide
6.4 **Fire Extinguishing Agents Not to be Used:**
6.5 **Special Hazards of Combustion Products:**
6.6 **Behavior in Fire:**
6.7 **Ignition Temperature:** Data not available
6.8 **Electrical Hazard:** Not pertinent
6.9 **Burning Rate:** Not pertinent

7. CHEMICAL REACTIVITY

7.1 **Reactivity with Water:** No reaction
7.2 **Reactivity with Common Materials:**
7.3 **Stability During Transport:** Stable
7.4 **Neutralizing Agents for Acids and Caustics:** Not pertinent
7.5 **Polymerization:** Not pertinent
7.6 **Inhibitor of Polymerization:** Not pertinent

8. WATER POLLUTION

8.1 **Aquatic Toxicity:** 18 ppm/48 hr/goldfish/TLm/fresh water
8.2 **Waterfowl Toxicity:** Data not available
8.3 **Biological Oxygen Demand (BOD):** 1.6%, 5 days
8.4 **Food Chain Concentration Potential:** None

9. SELECTED MANUFACTURERS

1. The Harshaw Chemical Co.
   1945 East 97 St.
   Cleveland, Ohio 44106
2. Aldrich Chemical Co.
   940 W. Saint Paul Ave.
   Milwaukee, Wis. 53233
3. Eastman Organic Chemicals
   Rochester, N. Y. 14650

10. SHIPPING INFORMATION

10.1 **Grades or Purity:** N.F.; Reagent
10.2 **Storage Temperature:** Ambient
10.3 **Inert Atmosphere:** No requirement
10.4 **Venting:** Open

11. HAZARD ASSESSMENT CODE

(See Hazard Assessment Handbook, CG 446-3)

SS

12. HAZARD CLASSIFICATIONS

12.1 **Code of Federal Regulations:** Not listed
12.2 **NAS Hazard Rating for Bulk Water Transportation:** Not listed
12.3 **NFPA Hazard Classifications:**

Category	Classification
Health Hazard (Blue)	1
Flammability (Red)	1
Reactivity (Yellow)	0

13. PHYSICAL AND CHEMICAL PROPERTIES

13.1 **Physical State at 15°C and 1 atm:** Solid
13.2 **Molecular Weight:** 126
13.3 **Boiling Point at 1 atm:** 588°F = 309°C = 582°K
13.4 **Freezing Point:** 268°F = 131°C = 404°K
13.5 **Critical Temperature:** Not pertinent
13.6 **Critical Pressure:** Not pertinent
13.7 **Specific Gravity:** 1.45 at 20°C (solid)
13.8 **Liquid Surface Tension:** Not pertinent
13.9 **Liquid-Water Interfacial Tension:** Not pertinent
13.10 **Vapor (Gas) Specific Gravity:** Not pertinent
13.11 **Ratio of Specific Heats of Vapor (Gas):** Not pertinent
13.12 **Latent Heat of Vaporization:** Not pertinent
13.13 **Heat of Combustion:** −9,130 Btu/lb = −5,070 cal/g = -212×10^5 J/kg
13.14 **Heat of Decomposition:** Not pertinent
13.15 **Heat of Solution:** Not pertinent
13.16 **Heat of Polymerization:** Not pertinent

NOTES

QNL QUINOLINE

Common Synonyms
1-Benzazine
Benzo (b) pyridine
Chinoline
1-Azanaphthalene

Liquid | Colorless to brown | Strong unpleasant odor

Sinks in water.

Avoid contact with liquid. Keep people away.
Stop discharge if possible.
Call fire department.
Isolate and remove discharged material.
Notify local health and pollution control agencies.

Fire

Combustible.
POISONOUS GASES MAY BE PRODUCED IN FIRE.
Wear goggles and self-contained breathing apparatus.
Extinguish with water, dry chemicals, foam, or carbon dioxide.

Exposure

CALL FOR MEDICAL AID.

LIQUID
Irritating to skin and eyes.
If swallowed will cause nausea and vomiting.
Remove contaminated clothing and shoes.
Flush affected areas with plenty of water.
IF IN EYES, hold eyelids open and flush with plenty of water.
IF SWALLOWED and victim is CONSCIOUS, have victim drink water or milk and have victim induce vomiting.
IF SWALLOWED and victim is UNCONSCIOUS OR HAVING CONVULSIONS, do nothing except keep victim warm.

Water Pollution

HARMFUL TO AQUATIC LIFE IN VERY LOW CONCENTRATIONS.
Fouling to shoreline.
May be dangerous if it enters water intakes.
Notify local health and wildlife officials.
Notify operators of nearby water intakes.

1. RESPONSE TO DISCHARGE

(See Response Methods Handbook, CG 446-4)

Issue warning—water contaminant
Restrict access
Mechanical containment
Should be removed
Chemical and physical treatment

2. LABELS

No hazard label required by Code of Federal Regulations

3. CHEMICAL DESIGNATIONS

3.1 **Synonyms:** 1-Azanaphthalene; 1-Benzazine; Benzo(b)pyridine; Chinoline; Leucol
3.2 **Coast Guard Compatibility Classification:** Not listed
3.3 **Chemical Formula:** C_9H_7N
3.4 **IMCO/United Nations Numerical Designation:** Not listed

4. OBSERVABLE CHARACTERISTICS

4.1 **Physical State (as shipped):** Liquid
4.2 **Color:** Colorless to brown
4.3 **Odor:** Strong, unpleasant

5. HEALTH HAZARDS

5.1 **Personal Protective Equipment:** U. S. Bu. Mines approved vapor unit; chemical safety goggles; face shield; rubber gloves; coveralls and/or rubber apron; rubber shoes and boots.
5.2 **Symptoms Following Exposure:** Vapors are irritating to nose and throat and may cause headaches, dizziness, and nausea if inhaled. Ingestion causes irritation of mouth and stomach; vomiting may occur. Contact with eyes or skin causes irritation.
5.3 **Treatment for Exposure:** INHALATION: remove victim to fresh air. INGESTION: give large amount of water; induce vomiting; get medical attention. EYES: flush immediately with plenty of water for at least 15 min.; call physician. SKIN: flush with water.
5.4 **Toxicity by Inhalation (Threshold Limit Value):** Data not available
5.5 **Short-Term Inhalation Limits:** Data not available
5.6 **Toxicity by Ingestion:** Grade 3; oral LD_{50} = 460 mg/kg (rat)
5.7 **Late Toxicity:** Data not available
5.8 **Vapor (Gas) Irritant Characteristics:** Data not available
5.9 **Liquid or Solid Irritant Characteristics:** Data not available
5.10 **Odor Threshold:** 71 ppm

6. FIRE HAZARDS

6.1 **Flash Point:** 225°F C.C.
6.2 **Flammable Limits in Air:** Data not available
6.3 **Fire Extinguishing Agents:** Water, dry chemical, foam, carbon dioxide
6.4 **Fire Extinguishing Agents Not to be Used:**
6.5 **Special Hazards of Combustion Products:** Toxic oxides of nitrogen may form in fires.
6.6 **Behavior in Fire:** Heat exposure may cause pressure build-up in closed containers.
6.7 **Ignition Temperature:** 896°F
6.8 **Electrical Hazard:** Data not available
6.9 **Burning Rate:** 4.06 mm/min.

7. CHEMICAL REACTIVITY

7.1 **Reactivity with Water:** No reaction
7.2 **Reactivity with Common Materials:** May attack some forms of plastics
7.3 **Stability During Transport:** Stable
7.4 **Neutralizing Agents for Acids and Caustics:** Not pertinent
7.5 **Polymerization:** Not pertinent
7.6 **Inhibitor of Polymerization:** Not pertinent

8. WATER POLLUTION

8.1 **Aquatic Toxicity:**
52-56 ppm/96 hr/sunfish/TL_m/fresh water
5 ppm/96 hr/trout/TL_m/fresh water
8.2 **Waterfowl Toxicity:** Data not available
8.3 **Biological Oxygen Demand (BOD):** 175%, 5 days
8.4 **Food Chain Concentration Potential:** None

9. SELECTED MANUFACTURERS

1. Koppers Co., Inc.
   Organic Materials Div.
   Koppers Bldg.
   Pittsburgh, Pa. 15219
2. Eastman Organic Chemicals
   Rochester, N. Y. 14650
3. Gallard-Schlesinger Chemical Mfg. Co.
   584 Mineola Ave.
   Carle Place, N. Y. 11514

10. SHIPPING INFORMATION

10.1 **Grades or Purity:** Reagent; Technical
10.2 **Storage Temperature:** Ambient
10.3 **Inert Atmosphere:** No requirement
10.4 **Venting:** Open (flame arrester)

11. HAZARD ASSESSMENT CODE

(See Hazard Assessment Handbook, CG 446-3)

A-T-U-X-Y

12. HAZARD CLASSIFICATIONS

12.1 **Code of Federal Regulations:** Not listed
12.2 **NAS Hazard Rating for Bulk Water Transportation:** Not listed
12.3 **NFPA Hazard Classifications:**

Category	Classification
Health Hazard (Blue)	2
Flammability (Red)	1
Reactivity (Yellow)	0

13. PHYSICAL AND CHEMICAL PROPERTIES

13.1 **Physical State at 15°C and 1 atm:** Liquid
13.2 **Molecular Weight:** 129
13.3 **Boiling Point at 1 atm:** 459°F = 237°C = 510°K
13.4 **Freezing Point:** 5°F = −15°C = 258°K
13.5 **Critical Temperature:** 948°F = 509°C = 782°K
13.6 **Critical Pressure:** Data not available
13.7 **Specific Gravity:** 1.095 at 20°C (liquid)
13.8 **Liquid Surface Tension:** 45.0 dynes/cm = 0.0450 N/m at 20°C
13.9 **Liquid-Water Interfacial Tension:** Not pertinent
13.10 **Vapor (Gas) Specific Gravity:** 4.5
13.11 **Ratio of Specific Heats of Vapor (Gas):** Not pertinent
13.12 **Latent Heat of Vaporization:** (est.) 155 Btu/lb = 86 cal/g = 3.6×10^5 J/kg
13.13 **Heat of Combustion:** −15,700 Btu/lb = −8,710 cal/g = -365×10^5 J/kg
13.14 **Heat of Decomposition:** Not pertinent
13.15 **Heat of Solution:** Not pertinent
13.16 **Heat of Polymerization:** Not pertinent

NOTES

RSC — RESORCINOL

Common Synonyms			
1,3 Benzenediol m-Dihydroxybenzene Resorcin 1,3-Dihydroxybenzene Dihydroxybenzol	Solid	White or off-white	Faint odor
	Sinks and mixes with water.		

Avoid contact with solid and dust. Keep people away.
Stop discharge if possible.
Call fire department.
Isolate and remove discharged material.
Notify local health and pollution control agencies.

Fire	Combustible. CONTAINERS MAY EXPLODE IN FIRE. Extinguish with dry chemicals, foam, or carbon dioxide. Water may be ineffective on fire. Cool exposed containers with water.
Exposure	CALL FOR MEDICAL AID. DUST Irritating to eyes, nose and throat. If inhaled will cause coughing or difficult breathing. If in eyes, hold eyelids open and flush with plenty of water. If breathing has stopped, give artificial respiration. If breathing is difficult, give oxygen. SOLID Irritating to skin and eyes. If swallowed will cause nausea or loss of consciousness. Remove contaminated clothing and shoes. Flush affected areas with plenty of water. IF IN EYES, hold eyelids open and flush with plenty of water. IF SWALLOWED and victim is CONSCIOUS, have victim drink water or milk. IF SWALLOWED and victim is UNCONSCIOUS OR HAVING CONVULSIONS, do nothing except keep victim warm.
Water Pollution	HARMFUL TO AQUATIC LIFE IN VERY LOW CONCENTRATIONS. May be dangerous if it enters water intakes. Notify local health and wildlife officials. Notify operators of nearby water intakes.

1. RESPONSE TO DISCHARGE
(See Response Methods Handbook, CG 446-4)

Issue warning—water contaminant
Disperse and flush

2. LABELS

No hazard label required by Code of Federal Regulations

3. CHEMICAL DESIGNATIONS

3.1 **Synonyms:** 1,3-Benzenediol; m-Dihydroxybenzene; 1,3-Dihydroxybenzene; Dihydroxybenzol; Resorcin
3.2 **Coast Guard Compatibility Classification:** Not listed
3.3 **Chemical Formula:** 1,3-$C_6H_4(OH)_2$
3.4 **IMCO/United Nations Numerical Designation:** Not listed

4. OBSERVABLE CHARACTERISTICS

4.1 **Physical State (as shipped):** Solid
4.2 **Color:** White or nearly white
4.3 **Odor:** Faint, characteristic aromatic

5. HEALTH HAZARDS

5.1 **Personal Protective Equipment:** U. S. Bu. Mines approved respirator; rubber gloves; safety glasses with side shields or chemical goggles; coveralls or rubber apron
5.2 **Symptoms Following Exposure:** Inhalation of vapors or dust causes irritation of respiratory tract. Ingestion causes burns of mucous membranes, severe diarrhea, pallor, sweating, weakness, headache, dizziness, tinnitus, shock, and severe convulsions; may also cause siderosis of the spleen and tubular injury to the kidney. Contact with eyes causes irritation. Can be absorbed from wounds or through unbroken skin, producing severe dermatitis, methemoglobinemia, cyanosis, convulsions, tachycardia, dyspnea, and death.
5.3 **Treatment for Exposure:** INHALATION: remove victim to fresh air; if he is not breathing, give artificial respiration, preferably mouth-to-mouth; if breathing is difficult, give oxygen; call a physician. INGESTION: give activated charcoal; administer gastric lavage with water; consult physician. EYES: flush with water for 15 min. SKIN: flush with water.
5.4 **Toxicity by Inhalation (Threshold Limit Value):** Data not available
5.5 **Short-Term Inhalation Limits:** Data not available
5.6 **Toxicity by Ingestion:** Grade 2; LD_{50} 0.5—5 g/kg
5.7 **Late Toxicity:** Produces goiters in rats
5.8 **Vapor (Gas) Irritant Characteristics:** Data not available
5.9 **Liquid or Solid Irritant Characteristics:** Data not available
5.10 **Odor Threshold:** Data not available

6. FIRE HAZARDS

6.1 **Flash Point:** Not pertinent (combustible solid)
6.2 **Flammable Limits in Air:** Not pertinent
6.3 **Fire Extinguishing Agents:** Water, foam, dry chemical, carbon dioxide
6.4 **Fire Extinguishing Agents Not to be Used:** Water may cause frothing
6.5 **Special Hazards of Combustion Products:**
6.6 **Behavior in Fire:** Containers may explode.
6.7 **Ignition Temperature:** 1,125°F
6.8 **Electrical Hazard:** Not pertinent
6.9 **Burning Rate:** Not pertinent

7. CHEMICAL REACTIVITY

7.1 **Reactivity with Water:** No reaction
7.2 **Reactivity with Common Materials:**
7.3 **Stability During Transport:** Stable
7.4 **Neutralizing Agents for Acids and Caustics:** Not pertinent
7.5 **Polymerization:** Not pertinent
7.6 **Inhibitor of Polymerization:** Not pertinent

8. WATER POLLUTION

8.1 **Aquatic Toxicity:**
35 ppm/*/bleak and carp/toxic threshold/fresh water
56.4 ppm/48 hr/daphnia/TL_m/fresh water
*Time period not specified.
8.2 **Waterfowl Toxicity:** Data not available
8.3 **Biological Oxygen Demand (BOD):** 61%, 5 days
8.4 **Food Chain Concentration Potential:** None

9. SELECTED MANUFACTURERS

1. Koppers Company, Inc.
   Koppers Building
   Pittsburgh, Pa. 15219
2. Eastman Organic Chemicals
   Rochester, N. Y. 14650
3. Aldrich Chemical Co.
   940 W. Saint Paul Ave.
   Milwaukee, Wis. 53233

10. SHIPPING INFORMATION

10.1 **Grades or Purity:** USP, 99.5+%; Technical, 99%
10.2 **Storage Temperature:** Ambient
10.3 **Inert Atmosphere:** No requirement
10.4 **Venting:** Open

11. HAZARD ASSESSMENT CODE
(See Hazard Assessment Handbook, CG 446-3)

SS

12. HAZARD CLASSIFICATIONS

12.1 **Code of Federal Regulations:** Not listed
12.2 **NAS Hazard Rating for Bulk Water Transportation:** Not listed
12.3 **NFPA Hazard Classifications:**

Category	Classification
Health Hazard (Blue)	—
Flammability (Red)	1
Reactivity (Yellow)	0

13. PHYSICAL AND CHEMICAL PROPERTIES

13.1 **Physical State at 15°C and 1 atm:** Solid
13.2 **Molecular Weight:** 110.11
13.3 **Boiling Point at 1 atm:** (sublimes) 531°F = 277°C = 550°K
13.4 **Freezing Point:** 228°F = 109°C = 382°K
13.5 **Critical Temperature:** Not pertinent
13.6 **Critical Pressure:** Not pertinent
13.7 **Specific Gravity:** 1.2 at 20°C (solid)
13.8 **Liquid Surface Tension:** Not pertinent
13.9 **Liquid-Water Interfacial Tension:** Not pertinent
13.10 **Vapor (Gas) Specific Gravity:** Not pertinent
13.11 **Ratio of Specific Heats of Vapor (Gas):** Not pertinent
13.12 **Latent Heat of Vaporization:** Not pertinent
13.13 **Heat of Combustion:** −11,200 Btu/lb = −6,200 cal/g = -259×10^5 J/kg
13.14 **Heat of Decomposition:** Not pertinent
13.15 **Heat of Solution:** Not pertinent
13.16 **Heat of Polymerization:** Not pertinent

NOTES

SLA — SALICYLIC ACID

Common Synonyms
o-Hydroxybenzoic acid
Retarder W

Solid — White to light tan — Odorless

Sinks and mixes slowly with water.

Avoid contact with solid and dust. Keep people away.
Stop discharge if possible.
Call fire department.
Isolate and remove discharged material.
Notify local health and pollution control agencies.

Fire

Combustible
Dust cloud may explode if ignited in an enclosed area.
Irritating gases may be produced when heated.
Extinguish with dry chemicals or carbon dioxide.
Water and foam may be ineffective on fire.
Cool exposed containers with water.

Exposure

CALL FOR MEDICAL AID.

DUST
Irritating to eyes, nose and throat.
If inhaled will cause coughing or difficult breathing.
If in eyes, hold eyelids open and flush with plenty of water.
If breathing has stopped, give artificial respiration.
If breathing is difficult, give oxygen.

SOLID
Irritating to skin and eyes.
If swallowed will cause vomiting.
Remove contaminated clothing and shoes.
Flush affected areas with plenty of water.
IF IN EYES, hold eyelids open and flush with plenty of water.
IF SWALLOWED and victim is CONSCIOUS, have victim drink water or milk and have victim induce vomiting.
IF SWALLOWED and victim is UNCONSCIOUS OR HAVING CONVULSIONS, do nothing except keep victim warm.

Water Pollution

Dangerous to aquatic life in high concentrations.
May be dangerous if it enters water intakes.
Notify local health and wildlife officials.
Notify operators of nearby water intakes.

1. RESPONSE TO DISCHARGE

(See Response Methods Handbook, CG 446-4)

Should be removed
Chemical and physical treatment

2. LABELS

No hazard label required by Code of Federal Regulations

3. CHEMICAL DESIGNATIONS

3.1 **Synonyms:** o-Hydroxybenzoic acid; Retarder W
3.2 **Coast Guard Compatibility Classification:** Not listed
3.3 **Chemical Formula:** 1,2-HOC_6H_4COOH
3.4 **IMCO/United Nations Numerical Designation:** Not listed

4. OBSERVABLE CHARACTERISTICS

4.1 **Physical State (as shipped):** Solid
4.2 **Color:** White to very light tan
4.3 **Odor:** None

5. HEALTH HAZARDS

5.1 **Personal Protective Equipment:** Gloves; goggles; respirator for dust; clean body-covering clothing
5.2 **Symptoms Following Exposure:** Inhalation of dust irritates nose and throat. Vomiting may occur spontaneously if large amounts are swallowed. Contact with eyes causes irritation, marked pain, and corneal injury which should heal. Prolonged or repeated skin contact may cause marked irritation or even a mild burn.
5.3 **Treatment for Exposure:** INHALATION: move to fresh air. INGESTION: induce vomiting and get medical attention promptly. EYES: promptly flush with water for 15 min. and get medical attention. SKIN: wash with soap and water.
5.4 **Toxicity by Inhalation (Threshold Limit Value):** Data not available
5.5 **Short-Term Inhalation Limits:** Data not available
5.6 **Toxicity by Ingestion:** Grade 2; LD_{50} 0.5–5 g/kg
5.7 **Late Toxicity:** Data not available
5.8 **Vapor (Gas) Irritant Characteristics:** Data not available
5.9 **Liquid or Solid Irritant Characteristics:** Data not available
5.10 **Odor Threshold:** Data not available

6. FIRE HAZARDS

6.1 **Flash Point:** Not pertinent (combustible solid)
6.2 **Flammable Limits in Air:** Not pertinent
6.3 **Fire Extinguishing Agents:** Water, foam, dry chemical, carbon dioxide
6.4 **Fire Extinguishing Agents Not to be Used:** Water or foam may cause frothing.
6.5 **Special Hazards of Combustion Products:** Irritating vapors of unburned material and phenol may form in fire.
6.6 **Behavior in Fire:** Sublimes and forms vapor or dust that may explode
6.7 **Ignition Temperature:** Data not available
6.8 **Electrical Hazard:** Not pertinent
6.9 **Burning Rate:** Not pertinent

7. CHEMICAL REACTIVITY

7.1 **Reactivity with Water:** No reaction
7.2 **Reactivity with Common Materials:**
7.3 **Stability During Transport:** Stable
7.4 **Neutralizing Agents for Acids and Caustics:** Not pertinent
7.5 **Polymerization:** Not pertinent
7.6 **Inhibitor of Polymerization:** Not pertinent

8. WATER POLLUTION

8.1 **Aquatic Toxicity:** 1,450 ppm/48 hr/daphnia/threshold for immob./fresh water
8.2 **Waterfowl Toxicity:** Data not available
8.3 **Biological Oxygen Demand (BOD):** 141%, 5 days
8.4 **Food Chain Concentration Potential:** None

9. SELECTED MANUFACTURERS

1. Tenneco Chemicals, Inc.
   Intermediates Div.
   Turner Place
   P. O. Box #2
   Piscataway, N. J. 08854
2. Monsanto Company
   800 North Lindbergh Blvd.
   St. Louis, Mo. 63166
3. Dow Chemical Co.
   Midland, Mich. 48640

10. SHIPPING INFORMATION

10.1 **Grades or Purity:** Technical, 98+%; Pure, 99+%
10.2 **Storage Temperature:** Ambient
10.3 **Inert Atmosphere:** No requirement
10.4 **Venting:** Open

11. HAZARD ASSESSMENT CODE

(See Hazard Assessment Handbook, CG 446-3)

II

12. HAZARD CLASSIFICATIONS

12.1 **Code of Federal Regulations:** Not listed
12.2 **NAS Hazard Rating for Bulk Water Transportation:** Not listed
12.3 **NFPA Hazard Classifications:**

Category	Classification
Health Hazard (Blue)	0
Flammability (Red)	1
Reactivity (Yellow)	0

13. PHYSICAL AND CHEMICAL PROPERTIES

13.1 **Physical State at 15°C and 1 atm:** Solid
13.2 **Molecular Weight:** 138.13
13.3 **Boiling Point at 1 atm:** Not pertinent (decomposes)
13.4 **Freezing Point:** 315°F = 157°C = 430°K
13.5 **Critical Temperature:** Not pertinent
13.6 **Critical Pressure:** Not pertinent
13.7 **Specific Gravity:** 1.44 at 20°C (solid)
13.8 **Liquid Surface Tension:** Not pertinent
13.9 **Liquid-Water Interfacial Tension:** Not pertinent
13.10 **Vapor (Gas) Specific Gravity:** Not pertinent
13.11 **Ratio of Specific Heats of Vapor (Gas):** Not pertinent
13.12 **Latent Heat of Vaporization:** Not pertinent
13.13 **Heat of Combustion:** −9,420 Btu/lb = −5,230 cal/g = -219×10^5 J/kg
13.14 **Heat of Decomposition:** Not pertinent
13.15 **Heat of Solution:** Not pertinent
13.16 **Heat of Polymerization:** Not pertinent

NOTES

SLD | SELENIUM DIOXIDE

Common Synonyms	Solid	White	Sour odor
Selenium oxide Selenious anhydride	Sinks and mixes with water.		

AVOID CONTACT WITH SOLID AND DUST. KEEP PEOPLE AWAY.
Wear goggles, dust respirator, and rubber overclothing (including gloves).
Stop discharge if possible.
Isolate and remove discharged material.
Notify local health and pollution control agencies.

Fire

Not flammable.
POISONOUS GASES MAY BE PRODUCED WHEN HEATED.

Exposure

CALL FOR MEDICAL AID.

DUST
POISONOUS IF INHALED OR IF SKIN IS EXPOSED.
If inhaled will cause coughing or difficult breathing.
If in eyes, hold eyelids open and flush with plenty of water.
If breathing has stopped, give artificial respiration.
If breathing is difficult, give oxygen.

SOLID
Irritating to skin and eyes.
If swallowed will cuase coughing, nausea, or vomiting.
Remove contaminated clothing and shoes.
Flush affected areas with plenty of water.
IF IN EYES, hold eyelids open and flush with plenty of water.
IF SWALLOWED and victim is CONSCIOUS, have victim drink water or milk and have victim induce vomiting.
IF SWALLOWED and victim is UNCONSCIOUS OR HAVING CONVULSIONS, do nothing except keep victim warm.

Water Pollution

HARMFUL TO AQUATIC LIFE IN VERY LOW CONCENTRATIONS.
May be dangerous if it enters water intakes.
Notify local health and wildlife officials.
Notify operators of nearby water intakes.

1. RESPONSE TO DISCHARGE

(See Response Methods Handbook, CG 446-4)

Issue warning—poison, water contaminant
Restrict access
Disperse and flush

2. LABEL

3. CHEMICAL DESIGNATIONS

3.1 **Synonyms:** Selenious anhydride; Selenium oxide
3.2 **Coast Guard Compatibility Classification:** Not listed
3.3 **Chemical Formula:** SeO_2
3.4 **IMCO/United Nations Numerical Designation:** Not listed

4. OBSERVABLE CHARACTERISTICS

4.1 **Physical State (as shipped):** Solid
4.2 **Color:** White
4.3 **Odor:** Pungent; sour

5. HEALTH HAZARDS

This compound is highly toxic if inhaled or ingested.

5.1 **Personal Protective Equipment:** Dust mask; rubber gloves; protective clothing
5.2 **Symptoms Following Exposure:** Absorption of selenium may be demonstrated by presence of the element in the urine and by a garlic-like odor of the breath. Inhalation of dust can cause bronchial spasms, symptoms of asphyxiation, and pneumonitis. Acute symptoms of ingestion include sternal pain, cough, nausea, pallor, coated tongue, gastrointestinal disorders, nervousness, and conjunctivitis. Contact with eyes causes irritation.
5.3 **Treatment for Exposure:** Consult physician after all exposures to this compound. INHALATION: remove victim to fresh air; give oxygen if needed. INGESTION: induce vomiting; follow with gastric lavage and saline cathartics. EYES: flush immediately and thoroughly with water. SKIN: flush with water.
5.4 **Toxicity by Inhalation (Threshold Limit Value):** 0.2 mg/m^3 (as selenium)
5.5 **Short-Term Inhalation Limits:** 0.3 mg/m^3, 30 min. (as selenium)
5.6 **Toxicity by Ingestion:** Data not available
5.7 **Late Toxicity:** Data not available.
5.8 **Vapor (Gas) Irritant Characteristics:** Data not available
5.9 **Liquid or Solid Irritant Characteristics:** Data not available
5.10 **Odor Threshold:** 0.0002 mg/m^3

6. FIRE HAZARDS

6.1 **Flash Point:** Not flammable
6.2 **Flammable Limits in Air:** Not flammable
6.3 **Fire Extinguishing Agents:**
6.4 **Fire Extinguishing Agents Not to be Used:**
6.5 **Special Hazards of Combustion Products:** Sublimes and forms toxic vapors when heated in fire
6.6 **Behavior in Fire:**
6.7 **Ignition Temperature:** Not pertinent
6.8 **Electrical Hazard:** Not pertinent
6.9 **Burning Rate:** Not pertinent

7. CHEMICAL REACTIVITY

7.1 **Reactivity with Water:** No reaction
7.2 **Reactivity with Common Materials:** In presence of water will corrode most metals
7.3 **Stability During Transport:** Stable
7.4 **Neutralizing Agents for Acids and Caustics:** Not pertinent
7.5 **Polymerization:** Not pertinent
7.6 **Inhibitor of Polymerization:** Not pertinent

8. WATER POLLUTION

8.1 **Aquatic Toxicity:** 12 ppm/7 days/goldfish/TL_m/fresh water
8.2 **Waterfowl Toxicity:** Data not available
8.3 **Biological Oxygen Demand (BOD):** None
8.4 **Food Chain Concentration Potential:** None

9. SELECTED MANUFACTURERS

1. American Hoechst Corp. Somerville, N. J. 08876
2. Ventron, Inc. P. O. Box 159 Beverly, Mass. 01915
3. Gallard-Schlesinger Chemical Mfg. Co. 584 Mineola Ave. Carle Place, N. Y. 11514

10. SHIPPING INFORMATION

10.1 **Grades or Purity:** Commercial, 99.5+%
10.2 **Storage Temperature:** Cool ambient
10.3 **Inert Atmosphere:** No requirement
10.4 **Venting:** Open

11. HAZARD ASSESSMENT CODE

(See Hazard Assessment Handbook, CG 446-3)

SS

12. HAZARD CLASSIFICATIONS

12.1 **Code of Federal Regulations:** Poisonous solid, Class B
12.2 **NAS Hazard Rating for Bulk Water Transportation:** Not listed
12.3 **NFPA Hazard Classifications:** Not listed

13. PHYSICAL AND CHEMICAL PROPERTIES

13.1 **Physical State at 15°C and 1 atm:** Solid
13.2 **Molecular Weight:** 111
13.3 **Boiling Point at 1 atm:** (sublimes) 599°F = 315°C = 583°K
13.4 **Freezing Point:** Not pertinent
13.5 **Critical Temperature:** Not pertinent
13.6 **Critical Pressure:** Not pertinent
13.7 **Specific Gravity:** 3.95 at 20°C (solid)
13.8 **Liquid Surface Tension:** Not pertinent
13.9 **Liquid-Water Interfacial Tension:** Not pertinent
13.10 **Vapor (Gas) Specific Gravity:** Not pertinent
13.11 **Ratio of Specific Heats of Vapor (Gas):** Not pertinent
13.12 **Latent Heat of Vaporization:** Not pertinent
13.13 **Heat of Combustion:** Not pertinent
13.14 **Heat of Decomposition:** Not pertinent
13.15 **Heat of Solution:** 12.1 Btu/lb = 6.7 cal/g = 0.28×10^5 J/kg
13.16 **Heat of Polymerization:** Not pertinent

NOTES

STO SELENIUM TRIOXIDE

Common Synonyms
Selenic anhydride

Solid | White

AVOID CONTACT WITH SOLID AND DUST. KEEP PEOPLE AWAY.
Wear goggles, dust respirator, and rubber overclothing (including gloves).
Stop discharge if possible.
Isolate and remove discharged material.
Notify local health and pollution control agencies.

Fire

Not flammable.

Exposure

CALL FOR MEDICAL AID.

DUST
POISONOUS IF INHALED OR IF SKIN IS EXPOSED.
If inhaled will cause coughing or difficult breathing.
If in eyes, hold eyelids open and flush with plenty of water.
If breathing has stopped, give artificial respiration.
If breathing is difficult, give oxygen.

SOLID
Irritating to skin and eyes.
If swallowed will cause coughing, nausea, and vomiting.
Remove contaminated clothing and shoes.
Flush affected areas with plenty of water.
IF IN EYES, hold eyelids open and flush with plenty of water.
IF SWALLOWED and victim is CONSCIOUS, have victim drink water or milk and have victim induce vomiting.
IF SWALLOWED and victim is UNCONSCIOUS OR HAVING CONVULSIONS, do nothing except keep victim warm.

Water Pollution

Effect of low concentrations on aquatic life is unknown.
May be dangerous if it enters water intakes.
Notify local health and wildlife officials.
Notify operators of nearby water intakes.

1. RESPONSE TO DISCHARGE

(See Response Methods Handbook, CG 446-4)

Issue warning—corrosive, water contaminant
Restrict access
Disperse and flush

2. LABEL

3. CHEMICAL DESIGNATIONS

3.1 **Synonyms:** Selenic anhydride
3.2 **Coast Guard Compatibility Classification:** Not listed
3.3 **Chemical Formula:** SeO_3
3.4 **IMCO/United Nations Numerical Designation:** Not listed

4. OBSERVABLE CHARACTERISTICS

4.1 **Physical State (as shipped):** Solid
4.2 **Color:** White
4.3 **Odor:** Data not available

5. HEALTH HAZARDS

This compound is highly toxic if ingested or inhaled.

5.1 **Personal Protective Equipment:** Dust mask; goggles or face shield; rubber gloves
5.2 **Symptoms Following Exposure:** Absorption of selenium may be demonstrated by presence of the element in the urine and by a garlic-like odor of breath. Inhalation can cause bronchial spasms, symptoms of asphyxiation, and pneumonitis. Acute symptoms of ingestion include sternal pain, cough, nausea, pallor, coated tongue, gastrointestinal disorders, nervousness and conjunctivitis. Contact with eyes or skin causes irritation.
5.3 **Treatment for Exposure:** INHALATION: remove victim to fresh air; give oxygen if necessary. INGESTION: induce vomiting; follow with gastric lavage and saline cathartics. EYES: flush immediately and thoroughly with water. SKIN: flush with water.
5.4 **Toxicity by Inhalation (Threshold Limit Value):** 0.2 mg/m^3 (as selenium)
5.5 **Short-Term Inhalation Limits:** 0.3 mg/m^3, 30 min. (as selenium)
5.6 **Toxicity by Ingestion:** Data not available
5.7 **Late Toxicity:** Data not available
5.8 **Vapor (Gas) Irritant Characteristics:** Data not available
5.9 **Liquid or Solid Irritant Characteristics:** Data not available
5.10 **Odor Threshold:** Data not available

6. FIRE HAZARDS

6.1 **Flash Point:** Not flammable
6.2 **Flammable Limits in Air:** Not flammable
6.3 **Fire Extinguishing Agents:** Not pertinent
6.4 **Fire Extinguishing Agents Not to be Used:** Not pertinent
6.5 **Special Hazards of Combustion Products:**
6.6 **Behavior in Fire:**
6.7 **Ignition Temperature:** Not pertinent
6.8 **Electrical Hazard:** Not pertinent
6.9 **Burning Rate:** Not pertinent

7. CHEMICAL REACTIVITY

7.1 **Reactivity with Water:** Reacts vigorously with water to form selenic acid solution
7.2 **Reactivity with Common Materials:** Corrodes all metals when moisture is present
7.3 **Stability During Transport:** Stable
7.4 **Neutralizing Agents for Acids and Caustics:** Flush with water, rinse with dilute solution of sodium bicarbonate or soda ash.
7.5 **Polymerization:** Not pertinent
7.6 **Inhibitor of Polymerization:** Not pertinent

8. WATER POLLUTION

8.1 **Aquatic Toxicity:** Data not available
8.2 **Waterfowl Toxicity:** Data not available
8.3 **Biological Oxygen Demand (BOD):** None
8.4 **Food Chain Concentration Potential:** None

9. SELECTED MANUFACTURERS

1. Gallard-Schlesinger Chemical Mfg. Co.
   584 Mineola Ave.
   Carle Place, N. Y. 11514
2. Pfaltz and Bauer, Inc.
   375 Fairfield Ave.
   Stamford, Conn. 06902

10. SHIPPING INFORMATION

10.1 **Grades or Purity:** Commercial; also shipped as a 40% solution in water (selenic acid)
10.2 **Storage Temperature:** Ambient
10.3 **Inert Atmosphere:** No requirement
10.4 **Venting:** Open

11. HAZARD ASSESSMENT CODE

(See Hazard Assessment Handbook, CG 446-3)

RR

12. HAZARD CLASSIFICATIONS

12.1 **Code of Federal Regulations:** Corrosive solid
12.2 **NAS Hazard Rating for Bulk Water Transportation:** Not listed
12.3 **NFPA Hazard Classifications:** Not listed

13. PHYSICAL AND CHEMICAL PROPERTIES

13.1 **Physical State at 15°C and 1 atm:** Solid
13.2 **Molecular Weight:** 126.9
13.3 **Boiling Point at 1 atm:** Not pertinent (decomposes)
13.4 **Freezing Point:** 244°F = 118°C = 391°K
13.5 **Critical Temperature:** Not pertinent
13.6 **Critical Pressure:** Not pertinent
13.7 **Specific Gravity:** 3.6 at 20°C (solid)
13.8 **Liquid Surface Tension:** Not pertinent
13.9 **Liquid-Water Interfacial Tension:** Not pertinent
13.10 **Vapor (Gas) Specific Gravity:** Not pertinent
13.11 **Ratio of Specific Heats of Vapor (Gas):** Not pertinent
13.12 **Latent Heat of Vaporization:** Not pertinent
13.13 **Heat of Combustion:** Not pertinent
13.14 **Heat of Decomposition:** Not pertinent
13.15 **Heat of Solution:** Data not available
13.16 **Heat of Polymerization:** Not pertinent

NOTES

STC	SILICON TETRACHLORIDE

Common Synonyms			
Silicon chloride	Liquid	Colorless to light yellow	Suffocating odor
	Reacts violently with water. Irritating gas is produced on contact with water.		

AVOID CONTACT WITH LIQUID AND VAPOR. KEEP PEOPLE AWAY.
Wear goggles, self-contained breathing apparatus, and rubber overclothing (including gloves).
Stop discharge if possible.
Isolate and remove discharged material.
Notify local health and pollution control agencies.

Fire

Not flammable.
DO NOT USE WATER OR FOAM ON ADJACENT FIRES.

Exposure

CALL FOR MEDICAL AID.

VAPOR
Irritating to eyes, nose and throat.
If inhaled will cause difficult breathing.
Move victim to fresh air.
If breathing has stopped, give artificial respiration.
If breathing is difficult, give oxygen.

LIQUID
POISONOUS IF SWALLOWED.
Will burn skin and eyes.
Remove contaminated clothing and shoes.
Flush affected areas with plenty of water.
IF IN EYES, hold eyelids open and flush with plenty of water.
IF SWALLOWED and victim is CONSCIOUS, have victim drink water or milk.
DO NOT INDUCE VOMITING.

Water Pollution

Effect of low concentrations on aquatic life is unknown.
May be dangerous if it enters water intakes.
Notify local health and wildlife officials.
Notify operators of nearby water intakes.

1. RESPONSE TO DISCHARGE

(See Response Methods Handbook, CG 446-4)

Issue warning—corrosive, air contaminant
Restrict access
Evacuate area
Disperse and flush with care

2. LABEL

3. CHEMICAL DESIGNATIONS

3.1 **Synonyms:** Silicon chloride
3.2 **Coast Guard Compatibility Classification:** To be developed
3.3 **Chemical Formula:** $SiCl_4$
3.4 **IMCO/United Nations Numerical Designation:** 8/1818

4. OBSERVABLE CHARACTERISTICS

4.1 **Physical State (as shipped):** Liquid
4.2 **Color:** Colorless to pale yellow
4.3 **Odor:** Suffocating; acrid

5. HEALTH HAZARDS

5.1 **Personal Protective Equipment:** Acid-canister-type gas mask or self-contained breathing apparatus; goggles or face shield; rubber gloves; other protective clothing to prevent contact with skin.
5.2 **Symptoms Following Exposure:** Inhalation causes severe irritation of upper respiratory tract resulting in couging, choking, and a feeling of suffocation; continued inhalation may produce ulceration of the nose, throat, and larynx; if inhaled deeply, edema of the lungs may occur. Contact of liquid with eyes causes severe irritation and painful burns; may cause permanent visual impairment. Liquid may cause severe burns of skin. Repeated skin contact with dilute solutions or exposure to concentrated vapors may cause dermatitis. Ingestion causes severe internal injury with pain in the throat and stomach, intense thirst, difficulty in swallowing, nausea, vomiting, and diarrhea; in severe cases, collapse and unconsciousness may result.
5.3 **Treatment for Exposure:** *Get medical attention at once following any exposure to this compound.* INHALATION: remove victim from contaminated atmosphere; if breathing has ceased, start mouth-to-mouth resuscitation; oxygen should only be administered by an experienced person when authorized by a physician; keep patient warm and comfortable. EYES: immediately flush with large quantities of running water for a minimum of 15 min.; continue irrigation for an additional 15 min. if physician is not available. SKIN: immediately flush affected area with water; severe or extensive burns may be caused by silicon tetrachloride, producing shock symptoms (rapid pulse, sweating and collapse); keep patient comfortably warm. INGESTION: if patient is conscious give large amounts of lime water or milk of magnesia; plain water should be given if neither of these is available; do NOT give sodium bicarbonate or make any attempt to induce vomiting; if patient is unconscious, do not give anything but ensure there is no obstruction to breathing (tongue should be kept forward and false teeth removed); he will be less likely to aspirate vomitus if placed in a face-downward position.

6. FIRE HAZARDS

6.1 **Flash Point:** Not flammable
6.2 **Flammable Limits in Air:** Not flammable
6.3 **Fire Extinguishing Agents:** Not pertinent
6.4 **Fire Extinguishing Agents Not to be Used:** Water or foam on adjacent fires
6.5 **Special Hazards of Combustion Products:** Not pertinent
6.6 **Behavior in Fire:** Contact with water in foam applied to adjacent fires will produce irritating fumes of hydrogen chloride.
6.7 **Ignition Temperature:** Not pertinent
6.8 **Electrical Hazard:** Not pertinent
6.9 **Burning Rate:** Not pertinent

7. CHEMICAL REACTIVITY

7.1 **Reactivity with Water:** Reacts vigorously with water to evolve hydrogen chloride (hydrochloric acid)
7.2 **Reactivity with Common Materials:** In presence of moisture will corrode metals; the reaction is not violent.
7.3 **Stability During Transport:** Stable
7.4 **Neutralizing Agents for Acids and Caustics:** Flood with water, rinse with sodium bicarbonate or lime solution.
7.5 **Polymerization:** Not pertinent
7.6 **Inhibitor of Polymerization:** Not pertinent

8. WATER POLLUTION

8.1 **Aquatic Toxicity:** Data not available
8.2 **Waterfowl Toxicity:** Data not available
8.3 **Biological Oxygen Demand (BOD):** None
8.4 **Food Chain Concentration Potential:** None

9. SELECTED MANUFACTURERS

1. Stauffer Chemical Company
   Industrial Chemical Division
   Westport, Conn. 06880
2. Union Carbide Corporation
   Chemicals and Plastics Division
   270 Park Avenue
   New York, N. Y. 10017
3. Dynamit-Nobel of America, Inc.
   105 Stonehurst Court
   Northvale, N. J. 07647

10. SHIPPING INFORMATION

10.1 **Grades or Purity:** Technical: 99.7+%; C.P.: 99.9+%
10.2 **Storage Temperature:** Ambient
10.3 **Inert Atmosphere:** Dry air
10.4 **Venting:** Pressure-vacuum

11. HAZARD ASSESSMENT CODE

(See Hazard Assessment Handbook, CG 446-3)

A-O

12. HAZARD CLASSIFICATIONS

12.1 **Code of Federal Regulations:** Corrosive
12.2 **NAS Hazard Rating for Bulk Water Transportation:**

Category	Rating
Fire	0
Health	
Vapor Irritant	4
Liquid or Solid Irritant	4
Poisons	4
Water Pollution	
Human Toxicity	4
Aquatic Toxicity	2
Aesthetic Effect	2
Reactivity	
Other Chemicals	4
Water	4
Self-Reaction	0

12.3 **NFPA Hazard Classifications:** Not listed

13. PHYSICAL AND CHEMICAL PROPERTIES

13.1 **Physical State at 15°C and 1 atm:** Liquid
13.2 **Molecular Weight:** 169.9
13.3 **Boiling Point at 1 atm:** 135.7°F = 57.6°C = 330.8°K
13.4 **Freezing Point:** −94°F = −70°C = 203°K
13.5 **Critical Temperature:** 472.5°F = 233.6°C = 506.8°K
13.6 **Critical Pressure:** 542 psia = 36.8 atm = 3.74 MN/m^2
13.7 **Specific Gravity:** 1.48 at 20°C (liquid)
13.8 **Liquid Surface Tension:** 19.6 dynes/cm = 0.0196 N/m at 20°C
13.9 **Liquid-Water Interfacial Tension:** Not pertinent
13.10 **Vapor (Gas) Specific Gravity:** 5.86
13.11 **Ratio of Specific Heats of Vapor (Gas):** Data not available
13.12 **Latent Heat of Vaporization:** 74.2 Btu/lb = 41.2 cal/g = 1.73×10^5 J/kg
13.13 **Heat of Combustion:** Not pertinent
13.14 **Heat of Decomposition:** Not pertinent
13.15 **Heat of Solution:** −742 Btu/lb = −412 cal/g = -17.3×10^5 J/kg
13.16 **Heat of Polymerization:** Not pertinent

5. HEALTH HAZARDS (Cont'd.)

5.4 **Toxicity by Inhalation (Threshold Limit Value):** Data not available
5.5 **Short-Term Inhalation Limits:** Data not available
5.6 **Toxicity by Ingestion:** Grade 4; LD_{50} <50 mg/kg
5.7 **Late Toxicity:** Data not available
5.8 **Vapor (Gas) Irritant Characteristics:** Vapors cause severe irritation of eyes and throat and can cause eye and lung injury. They cannot be tolerated even at low concentrations.
5.9 **Liquid or Solid Irritant Characteristics:** Data not available
5.10 **Odor Threshold:** Data not available

SVA | SILVER ACETATE

Common Synonyms	Solid	White to gray	Odorless
	Sinks in water.		

Avoid contact with solid and dust. Keep people away.
Stop discharge if possible.
Isolate and remove discharged material.
Notify local health and pollution control agencies.

Fire

Not flammable.

Exposure

CALL FOR MEDICAL AID.

DUST
Irritating to eyes, nose and throat.
If inhaled will cause coughing or difficult breathing.
If in eyes, hold eyelids open and flush with plenty of water.
If breathing has stopped, give artificial respiration.
If breathing is difficult, give oxygen.

SOLID
Irritating to skin and eyes.
Harmful if swallowed.
Remove contaminated clothing and shoes.
Flush affected areas with plenty of water.
IF IN EYES, hold eyelids open and flush with plenty of water.
IF SWALLOWED and victim is CONSCIOUS, have victim drink water or milk and have victim induce vomiting.
IF SWALLOWED and victim is UNCONSCIOUS OR HAVING CONVULSIONS, do nothing except keep victim warm.

Water Pollution

Effect of low concentrations on aquatic life is unknown.
May be dangerous if it enters water intakes.
Notify local health and wildlife officials.
Notify operators of nearby water intakes.

1. RESPONSE TO DISCHARGE

(See Response Methods Handbook, CG 446-4)

Issue warning—water contaminant
Restrict access
Disperse and flush

2. LABELS

No hazard label required by Code of Federal Regulations

3. CHEMICAL DESIGNATIONS

3.1 **Synonyms:** No common synonyms
3.2 **Coast Guard Compatibility Classification:** Not listed
3.3 **Chemical Formula:** CH_3COOAg
3.4 **IMCO/United Nations Numerical Designation:** Not listed

4. OBSERVABLE CHARACTERISTICS

4.1 **Physical State (as shipped):** Solid
4.2 **Color:** White to gray
4.3 **Odor:** None

5. HEALTH HAZARDS

5.1 **Personal Protective Equipment:** Dust mask; goggles or face shield; protective gloves
5.2 **Symptoms Following Exposure:** Inhalation of dust irritates nose and throat. Contact with eyes or skin causes irritation. If continued for a long period, ingestion or inhalation of silver compounds can cause permanent discoloration of skin (argyria).
5.3 **Treatment for Exposure:** INHALATION: move to fresh air. INGESTION: give large amount of water; induce vomiting. EYES: flush with water for at least 15 min. SKIN: flush with water.
5.4 **Toxicity by Inhalation (Threshold Limit Value):** 0.01 mg/m^3
5.5 **Short-Term Inhalation Limits:** Data not available
5.6 **Toxicity by Ingestion:** Data not available
5.7 **Late Toxicity:** Data not available
5.8 **Vapor (Gas) Irritant Characteristics:** Data not available
5.9 **Liquid or Solid Irritant Characteristics:** Data not available
5.10 **Odor Threshold:** Data not available

6. FIRE HAZARDS

6.1 **Flash Point:** Not flammable
6.2 **Flammable Limits in Air:** Not flammable
6.3 **Fire Extinguishing Agents:** Not pertinent
6.4 **Fire Extinguishing Agents Not to be Used:** Not pertinent
6.5 **Special Hazards of Combustion Products:**
6.6 **Behavior in Fire:**
6.7 **Ignition Temperature:** Not pertinent
6.8 **Electrical Hazard:** Not pertinent
6.9 **Burning Rate:** Not pertinent

7. CHEMICAL REACTIVITY

7.1 **Reactivity with Water:** No reaction
7.2 **Reactivity with Common Materials:**
7.3 **Stability During Transport:** Stable
7.4 **Neutralizing Agents for Acids and Caustics:** Not pertinent
7.5 **Polymerization:** Not pertinent
7.6 **Inhibitor of Polymerization:** Not pertinent

8. WATER POLLUTION

8.1 **Aquatic Toxicity:** Data not available
8.2 **Waterfowl Toxicity:** Data not available
8.3 **Biological Oxygen Demand (BOD):** Data not available
8.4 **Food Chain Concentration Potential:** None

9. SELECTED MANUFACTURERS

1. R.S.A. Corporation
690 Saw Mill River Rd.
Ardsley, N. Y. 10502
2. Ventron Corp.
P. O. Box 159
Beverly, Mass. 01915
3. J. T. Baker Chemical Co.
Phillipsburg, N. J. 08865

10. SHIPPING INFORMATION

10.1 **Grades or Purity:** Commercial; Purified
10.2 **Storage Temperature:** Ambient
10.3 **Inert Atmosphere:** No requirement
10.4 **Venting:** Open

11. HAZARD ASSESSMENT CODE

(See Hazard Assessment Handbook, CG 446-3)

II-SS

12. HAZARD CLASSIFICATIONS

12.1 **Code of Federal Regulations:** Not listed
12.2 **NAS Hazard Rating for Bulk Water Transportation:** Not listed
12.3 **NFPA Hazard Classifications:** Not listed

13. PHYSICAL AND CHEMICAL PROPERTIES

13.1 **Physical State at 15°C and 1 atm:** Solid
13.2 **Molecular Weight:** 166.9
13.3 **Boiling Point at 1 atm:** Not pertinent (decomposes)
13.4 **Freezing Point:** Not pertinent
13.5 **Critical Temperature:** Not pertinent
13.6 **Critical Pressure:** Not pertinent
13.7 **Specific Gravity:** 3.26 at 20°C (solid)
13.8 **Liquid Surface Tension:** Not pertinent
13.9 **Liquid-Water Interfacial Tension:** Not pertinent
13.10 **Vapor (Gas) Specific Gravity:** Not pertinent
13.11 **Ratio of Specific Heats of Vapor (Gas):** Not pertinent
13.12 **Latent Heat of Vaporization:** Not pertinent
13.13 **Heat of Combustion:** Not pertinent
13.14 **Heat of Decomposition:** Not pertinent
13.15 **Heat of Solution:** Not pertinent
13.16 **Heat of Polymerization:** Not pertinent

NOTES

SVC	# SILVER CARBONATE

Common Synonyms	Solid	Yellow to brown	Odorless
	Sinks in water.		

Avoid contact with solid and dust. Keep people away.
Stop discharge if possible.
Isolate and remove discharged material.
Notify local health and pollution control agencies.

Fire	Not flammable.
Exposure	CALL FOR MEDICAL AID. DUST Irritating to eyes, nose and throat. If inhaled will cause coughing or difficult breathing. If in eyes, hold eyelids open and flush with plenty of water. If breathing has stopped, give artificial respiration. If breathing is difficult, give oxygen. SOLID Irritating to skin and eyes. Harmful if swallowed. Remove contaminated clothing and shoes. Flush affected areas with plenty of water. IF IN EYES, hold eyelids open and flush with plenty of water. IF SWALLOWED and victim is CONSCIOUS, have victim drink water or milk and have victim induce vomiting. IF SWALLOWED and victim is UNCONSCIOUS OR HAVING CONVULSIONS, do nothing except keep victim warm.
Water Pollution	Effect of low concentrations on aquatic life is unknown. May be dangerous if it enters water intakes. Notify local health and wildlife officials. Notify operators of nearby water intakes.

1. RESPONSE TO DISCHARGE

(See Response Methods Handbook, CG 446-4)

Issue warning—water contaminant
Should be removed
Chemical and physical treatment

2. LABELS

No hazard label required by Code of Federal Regulations

3. CHEMICAL DESIGNATIONS

3.1 **Synonyms:** No common synonyms
3.2 **Coast Guard Compatibility Classification:** Not listed
3.3 **Chemical Formula:** Ag_2CO_3
3.4 **IMCO/United Nations Numerical Designation:** Not listed

4. OBSERVABLE CHARACTERISTICS

4.1 **Physical State (as shipped):** Solid
4.2 **Color:** Yellow to brown
4.3 **Odor:** None

5. HEALTH HAZARDS

5.1 **Personal Protective Equipment:** Dust mask; goggles or face shield; rubber gloves
5.2 **Symptoms Following Exposure:** Contact with eyes causes irritation. If continued for a long period, ingestion or inhalation of silver compounds can cause permanent discoloration of the skin (argyria).
5.3 **Treatment for Exposure:** INHALATION: move to fresh air. INGESTION: give large amount of water; induce vomiting. EYES: flush with water for at least 15 min. SKIN: flush with water; wash with soap and water.
5.4 **Toxicity by Inhalation (Threshold Limit Value):** 0.01 mg/m^3
5.5 **Short-Term Inhalation Limits:** Data not available
5.6 **Toxicity by Ingestion:** Data not available
5.7 **Late Toxicity:** Data not available
5.8 **Vapor (Gas) Irritant Characteristics:** Data not available
5.9 **Liquid or Solid Irritant Characteristics:** Data not available
5.10 **Odor Threshold:** Data not available

6. FIRE HAZARDS

6.1 **Flash Point:** Not flammable
6.2 **Flammable Limits in Air:** Not flammable
6.3 **Fire Extinguishing Agents:** Not pertinent
6.4 **Fire Extinguishing Agents Not to be Used:** Not pertinent
6.5 **Special Hazards of Combustion Products:** Not pertinent
6.6 **Behavior in Fire:** Decomposes to silver oxide, silver, and carbon dioxide; the reaction is not hazardous.
6.7 **Ignition Temperature:** Not pertinent
6.8 **Electrical Hazard:** Not pertinent
6.9 **Burning Rate:** Not pertinent

7. CHEMICAL REACTIVITY

7.1 **Reactivity with Water:** No reaction
7.2 **Reactivity with Common Materials:**
7.3 **Stability During Transport:** Stable
7.4 **Neutralizing Agents for Acids and Caustics:** Not pertinent
7.5 **Polymerization:** Not pertinent
7.6 **Inhibitor of Polymerization:** Not pertinent

8. WATER POLLUTION

8.1 **Aquatic Toxicity:** Data not available
8.2 **Waterfowl Toxicity:** Data not available
8.3 **Biological Oxygen Demand (BOD):** None
8.4 **Food Chain Concentration Potential:** None

9. SELECTED MANUFACTURERS

1. R.S.A. Corporation
690 Saw Mill River Rd.
Ardsley, N. Y. 10502
2. J. T. Baker Chemical Co.
Phillipsburg, N. J. 08865
3. Gallard-Schlesinger Chemical Mfg. Co.
584 Mineola Ave.
Carle Place, N. Y. 11514

10. SHIPPING INFORMATION

10.1 **Grades or Purity:** Reagent, 98+%
10.2 **Storage Temperature:** Ambient
10.3 **Inert Atmosphere:** No requirement
10.4 **Venting:** Open

11. HAZARD ASSESSMENT CODE

(See Hazard Assessment Handbook, CG 446-3)

II

12. HAZARD CLASSIFICATIONS

12.1 **Code of Federal Regulations:** Not listed
12.2 **NAS Hazard Rating for Bulk Water Transportation:** Not listed
12.3 **NFPA Hazard Classifications:** Not listed

13. PHYSICAL AND CHEMICAL PROPERTIES

13.1 **Physical State at 15°C and 1 atm:** Solid
13.2 **Molecular Weight:** 275.75
13.3 **Boiling Point at 1 atm:** Not pertinent (decomposes)
13.4 **Freezing Point:** Not pertinent
13.5 **Critical Temperature:** Not pertinent
13.6 **Critical Pressure:** Not pertinent
13.7 **Specific Gravity:** 6.1 at 20°C (solid)
13.8 **Liquid Surface Tension:** Not pertinent
13.9 **Liquid-Water Interfacial Tension:** Not pertinent
13.10 **Vapor (Gas) Specific Gravity:** Not pertinent
13.11 **Ratio of Specific Heats of Vapor (Gas):** Not pertinent
13.12 **Latent Heat of Vaporization:** Not pertinent
13.13 **Heat of Combustion:** Not pertinent
13.14 **Heat of Decomposition:** Not pertinent
13.15 **Heat of Solution:** Not pertinent
13.16 **Heat of Polymerization:** Not pertinent

NOTES

SVF — SILVER FLUORIDE

Common Synonyms
Argentous fluoride
Silver monofluoride

Solid — Yellow to gray — Odorless

Sinks and mixes with water.

Avoid contact with solid and dust. Keep people away.
Stop discharge if possible.
Isolate and remove discharged material.
Notify local health and pollution control agencies.

Fire

Not flammable.

Exposure

CALL FOR MEDICAL AID.

DUST
Irritating to eyes, nose and throat.
If inhaled will cause coughing or difficult breathing.

If in eyes, hold eyelids open and flush with plenty of water.
If breathing has stopped, give artificial respiration.
If breathing is difficult, give oxygen.

SOLID
Irritating to skin and eyes.
If swallowed will cause nausea, vomiting or loss of consciousness.

Remove contaminated clothing and shoes.
Flush affected areas with plenty of water.
IF IN EYES, hold eyelids open and flush with plenty of water.
IF SWALLOWED and victim is CONSCIOUS, have victim drink water or milk and have victim induce vomiting.
IF SWALLOWED and victim is UNCONSCIOUS OR HAVING CONVULSIONS, do nothing except keep victim warm.

Water Pollution

Effect of low concentrations on aquatic life is unknown.
May be dangerous if it enters water intakes.

Notify local health and wildlife officials.
Notify operators of nearby water intakes.

1. RESPONSE TO DISCHARGE
(See Response Methods Handbook, CG 446-4)

Issue warning—water contaminant
Disperse and flush

2. LABELS

No hazard label required by Code of Federal Regulations

3. CHEMICAL DESIGNATIONS

3.1 **Synonyms:** Argentous fluoride; Silver monofluoride
3.2 **Coast Guard Compatibility Classification:** Not listed
3.3 **Chemical Formula:** AgF
3.4 **IMCO/United Nations Numerical Designation:** Not listed

4. OBSERVABLE CHARACTERISTICS

4.1 **Physical State (as shipped):** Solid
4.2 **Color:** Yellow to gray
4.3 **Odor:** None

5. HEALTH HAZARDS

5.1 **Personal Protective Equipment:** Dust mask; goggles or face shield; protective gloves.
5.2 **Symptoms Following Exposure:** Inhalation of dust causes irritation of nose and throat. Ingestion may cause vomiting, salty taste, abdominal pain, diarrhea, convulsions, collapse, thirst, disturbed color vision, and acute toxic nephritis. Contact with eyes causes irritation. Skin may be blackened on prolonged exposure.
5.3 **Treatment for Exposure:** INHALATION: move to fresh air. INGESTION: get medical attention at once; give large amount of water and induce vomiting. EYES: flush with water for at least 15 min. SKIN: flush with water.
5.4 **Toxicity by Inhalation (Threshold Limit Value):** 0.01 mg/m^3
5.5 **Short-Term Inhalation Limits:** Data not available
5.6 **Toxicity by Ingestion:** Data not available
5.7 **Late Toxicity:** Data not available
5.8 **Vapor (Gas) Irritant Characteristics:** Data not available
5.9 **Liquid or Solid Irritant Characteristics:** Data not available
5.10 **Odor Threshold:** Data not available

6. FIRE HAZARDS

6.1 **Flash Point:** Not flammable
6.2 **Flammable Limits in Air:** Not flammable
6.3 **Fire Extinguishing Agents:** Not pertinent
6.4 **Fire Extinguishing Agents Not to be Used:** Not pertinent
6.5 **Special Hazards of Combustion Products:** Not pertinent
6.6 **Behavior in Fire:**
6.7 **Ignition Temperature:** Not pertinent
6.8 **Electrical Hazard:** Not pertinent
6.9 **Burning Rate:** Not pertinent

7. CHEMICAL REACTIVITY

7.1 **Reactivity with Water:** No reaction
7.2 **Reactivity with Common Materials:**
7.3 **Stability During Transport:** Stable
7.4 **Neutralizing Agents for Acids and Caustics:** Not pertinent
7.5 **Polymerization:** Not pertinent
7.6 **Inhibitor of Polymerization:** Not pertinent

8. WATER POLLUTION

8.1 **Aquatic Toxicity:** Data not available
8.2 **Waterfowl Toxicity:** Data not available
8.3 **Biological Oxygen Demand (BOD):** None
8.4 **Food Chain Concentration Potential:** None

9. SELECTED MANUFACTURERS

1. Ozark-Mahoning Co.
   1870 South Boulder
   Tulsa, Okla. 74119
2. Ventron, Inc.
   P. O. Box 159
   Beverly, Mass. 01915
3. Gallard-Schlesinger Chemical Mfg. Co.
   584 Mineola Ave.
   Carle Place, N. Y. 11514

10. SHIPPING INFORMATION

10.1 **Grades or Purity:** Commercial; Pure, 99.9+%
10.2 **Storage Temperature:** Ambient
10.3 **Inert Atmosphere:** No requirement
10.4 **Venting:** Open

11. HAZARD ASSESSMENT CODE
(See Hazard Assessment Handbook, CG 446-3)

II-SS

12. HAZARD CLASSIFICATIONS

12.1 **Code of Federal Regulations:** Not listed
12.2 **NAS Hazard Rating for Bulk Water Transportation:** Not listed
12.3 **NFPA Hazard Classifications:** Not listed

13. PHYSICAL AND CHEMICAL PROPERTIES

13.1 **Physical State at 15°C and 1 atm:** Solid
13.2 **Molecular Weight:** 126.9
13.3 **Boiling Point at 1 atm:** 2,118°F = 1,159°C = 1,432°K
13.4 **Freezing Point:** Not pertinent
13.5 **Critical Temperature:** Not pertinent
13.6 **Critical Pressure:** Not pertinent
13.7 **Specific Gravity:** 5.82 at 20°C (solid)
13.8 **Liquid Surface Tension:** Not pertinent
13.9 **Liquid-Water Interfacial Tension:** Not pertinent
13.10 **Vapor (Gas) Specific Gravity:** Not pertinent
13.11 **Ratio of Specific Heats of Vapor (Gas):** Not pertinent
13.12 **Latent Heat of Vaporization:** Not pertinent
13.13 **Heat of Combustion:** Not pertinent
13.14 **Heat of Decomposition:** Not pertinent
13.15 **Heat of Solution:** Not pertinent
13.16 **Heat of Polymerization:** Not pertinent

NOTES

SVI	# SILVER IODATE

Common Synonyms	Solid	White	Odorless
	Sinks in water.		

Avoid contact with solid and dust. Keep people away.
Stop discharge if possible.
Isolate and remove discharged material.
Notify local health and pollution control agencies.

Fire	Not flammable.
Exposure	CALL FOR MEDICAL AID. DUST Irritating to eyes, nose and throat. If inhaled will cause coughing or difficult breathing. If in eyes, hold eyelids open and flush with plenty of water. If breathing has stopped, give artificial respiration. If breathing is difficult, give oxygen. SOLID Irritating to skin and eyes. Harmful if swallowed. Remove contaminated clothing and shoes. Flush affected areas with plenty of water. IF IN EYES, hold eyelids open and flush with plenty of water. IF SWALLOWED and victim is CONSCIOUS, have victim drink water or milk and have victim induce vomiting. IF SWALLOWED and victim is UNCONSCIOUS OR HAVING CONVULSIONS, do nothing except keep victim warm.
Water Pollution	Effect of low concentrations on aquatic life is unknown. May be dangerous if it enters water intakes. Notify local health and wildlife officials. Notify operators of nearby water intakes.

1. RESPONSE TO DISCHARGE
(See Response Methods Handbook, CG 446-4)

Issue warning—water contaminant
Should be removed
Chemical and physical treatment

2. LABELS

No hazard label required by Code of Federal Regulations

3. CHEMICAL DESIGNATIONS

3.1 **Synonyms:** No common synonyms
3.2 **Coast Guard Compatibility Classification:** Not listed
3.3 **Chemical Formula:** $AgIO_3$
3.4 **IMCO/United Nations Numerical Designation:** Not listed

4. OBSERVABLE CHARACTERISTICS

4.1 **Physical State (as shipped):** Solid
4.2 **Color:** White
4.3 **Odor:** None

5. HEALTH HAZARDS

5.1 **Personal Protective Equipment:** Dust mask; goggles or face shield; protective gloves
5.2 **Symptoms Following Exposure:** Contact with eyes causes irritation. If continued for a long period, ingestion or inhalation of silver compounds can cause permanent discoloration of the skin (argyria).
5.3 **Treatment for Exposure:** INHALATION: move to fresh air. INGESTION: give large amount of water; induce vomiting. EYES: flush with water for at least 15 min. SKIN: flush with water, wash with soap and water.
5.4 **Toxicity by Inhalation (Threshold Limit Value):** 0.01 mg/m^3
5.5 **Short-Term Inhalation Limits:** Data not available
5.6 **Toxicity by Ingestion:** Data not available
5.7 **Late Toxicity:** Data not available
5.8 **Vapor (Gas) Irritant Characteristics:** Data not available
5.9 **Liquid or Solid Irritant Characteristics:** Data not available
5.10 **Odor Threshold:** Data not available

6. FIRE HAZARDS

6.1 **Flash Point:** Not flammable
6.2 **Flammable Limits in Air:** Not flammable
6.3 **Fire Extinguishing Agents:** Not pertinent
6.4 **Fire Extinguishing Agents Not to be Used:** Not pertinent
6.5 **Special Hazards of Combustion Products:** Not pertinent
6.6 **Behavior in Fire:**
6.7 **Ignition Temperature:** Not pertinent
6.8 **Electrical Hazard:** Not pertinent
6.9 **Burning Rate:** Not pertinent

7. CHEMICAL REACTIVITY

7.1 **Reactivity with Water:** No reaction
7.2 **Reactivity with Common Materials:**
7.3 **Stability During Transport:** Stable
7.4 **Neutralizing Agents for Acids and Caustics:** Not pertinent
7.5 **Polymerization:** Not pertinent
7.6 **Inhibitor of Polymerization:** Not pertinent

8. WATER POLLUTION

8.1 **Aquatic Toxicity:** Data not available
8.2 **Waterfowl Toxicity:** Data not available
8.3 **Biological Oxygen Demand (BOD):** None
8.4 **Food Chain Concentration Potential:** None

9. SELECTED MANUFACTURERS

1. R.S.A. Corporation
   690 Saw Mill River Rd.
   Ardsley, N. Y. 10502
2. Gallard-Schlesinger Chemical Mfg. Co.
   584 Mineola Ave.
   Carle Place, N. Y. 11514
3. Pfaltz and Bauer, Inc.
   375 Fairfield Ave.
   Stamford, Conn. 06902

10. SHIPPING INFORMATION

10.1 **Grades or Purity:** Commercial, 99+%
10.2 **Storage Temperature:** Ambient
10.3 **Inert Atmosphere:** No requirement
10.4 **Venting:** Open

11. HAZARD ASSESSMENT CODE
(See Hazard Assessment Handbook, CG 446-3)

II

12. HAZARD CLASSIFICATIONS

12.1 **Code of Federal Regulations:** Not listed
12.2 **NAS Hazard Rating for Bulk Water Transportation:** Not listed
12.3 **NFPA Hazard Classifications:** Not listed

13. PHYSICAL AND CHEMICAL PROPERTIES

13.1 **Physical State at 15°C and 1 atm:** Solid
13.2 **Molecular Weight:** 282.1
13.3 **Boiling Point at 1 atm:** Not pertinent (decomposes)
13.4 **Freezing Point:** Not pertinent
13.5 **Critical Temperature:** Not pertinent
13.6 **Critical Pressure:** Not pertinent
13.7 **Specific Gravity:** 5.53 at 20°C (solid)
13.8 **Liquid Surface Tension:** Not pertinent
13.9 **Liquid-Water Interfacial Tension:** Not pertinent
13.10 **Vapor (Gas) Specific Gravity:** Not pertinent
13.11 **Ratio of Specific Heats of Vapor (Gas):** Not pertinent
13.12 **Latent Heat of Vaporization:** Not pertinent
13.13 **Heat of Combustion:** Not pertinent
13.14 **Heat of Decomposition:** Not pertinent
13.15 **Heat of Solution:** Not pertinent
13.16 **Heat of Polymerization:** Not pertinent

NOTES

SVN — SILVER NITRATE

Common Synonyms: Lunar caustic

Solid crystals — Colorless to grayish black — Odorless

Sinks and mixes with water.

Isolate and remove discharged material.
Notify local health and pollution control agencies.

Fire

Not flammable.

Exposure

CALL FOR MEDICAL AID.

SOLIDS
Irritating to skin and eyes.
Harmful if swallowed.
Flush affected areas with plenty of water.
IF IN EYES, hold eyelids open and flush with plenty of water.
IF SWALLOWED and victim is CONSCIOUS, have victim drink water or milk.

Water Pollution

HARMFUL TO AQUATIC LIFE IN VERY LOW CONCENTRATIONS.
May be dangerous if it enters water intakes.
Notify local health and wildlife officials.
Notify operators of nearby water intakes.

1. RESPONSE TO DISCHARGE

(See Response Methods Handbook, CG 446-4)

Issue warning—water contaminant
Should be removed

2. LABEL

Yellow

3. CHEMICAL DESIGNATIONS

3.1 **Synonyms:** Lunar caustic
3.2 **Coast Guard Compatibility Classification:** Not applicable
3.3 **Chemical Formula:** $AgNO_3$
3.4 **IMCO/United Nations Numerical Designation:** 5.1/1493

4. OBSERVABLE CHARACTERISTICS

4.1 **Physical State (as shipped):** Solid
4.2 **Color:** Colorless
4.3 **Odor:** Odorless

5. HEALTH HAZARDS

5.1 **Personal Protective Equipment:** Goggles or face shield; rubber gloves.
5.2 **Symptoms Following Exposure:** Concentrated solutions will produce irritation, ulceration, and discoloration of the skin; also causes severe irritation of the eyes. Ingestion will produce violent abdominal pain and other gastroenteric symptoms.
5.3 **Treatment for Exposure:** INGESTION: gastric lavage with dilute solution of sodium chloride, followed by cathartics and demulcents. Other treatment is symptomatic. SKIN: wash promptly.
5.4 **Toxicity by Inhalation (Threshold Limit Value):** Not pertinent
5.5 **Short-Term Inhalation Limits:** Not pertinent
5.6 **Toxicity by Ingestion:** Grade 3; LD_{50} 50 to 500 mg/kg
5.7 **Late Toxicity:** Data not available
5.8 **Vapor (Gas) Irritant Characteristics:** Non-volatile
5.9 **Liquid or Solid Irritant Characteristics:** Burns skin on prolonged contact.
5.10 **Odor Threshold:** Not pertinent

6. FIRE HAZARDS

6.1 **Flash Point:** Not flammable
6.2 **Flammable Limits in Air:** Not flammable
6.3 **Fire Extinguishing Agents:** Not pertinent
6.4 **Fire Extinguishing Agents Not to be Used:** Not pertinent
6.5 **Special Hazards of Combustion Products:** Not pertinent
6.6 **Behavior in Fire:** Increases flammability of combustibles.
6.7 **Ignition Temperature:** Not flammable
6.8 **Electrical Hazard:** Not pertinent
6.9 **Burning Rate:** Not flammable

7. CHEMICAL REACTIVITY

7.1 **Reactivity with Water:** No reaction
7.2 **Reactivity with Common Materials:** No reaction
7.3 **Stability During Transport:** Stable
7.4 **Neutralizing Agents for Acids and Caustics:** Not pertinent
7.5 **Polymerization:** Not pertinent
7.6 **Inhibitor of Polymerization:** Not pertinent

8. WATER POLLUTION

8.1 **Aquatic Toxicity:**
0.44 mg/l/48 hr/salmon fry/decidedly toxic/fresh water
0.01-0.1 ppm/*/paracentrotus/inhibit egg develop./salt water
*Time period not specified.
8.2 **Waterfowl Toxicity:** Data not available
8.3 **Biological Oxygen Demand (BOD):** None
8.4 **Food Chain Concentration Potential:** Data not available

9. SELECTED MANUFACTURERS

1. Eastman Kodak Co.
   Eastman Organic Chemicals Division
   Rochester, N. Y. 14650
2. Engelhard Minerals and Chemicals Corp.
   Silver Products Dept.
   231 New Jersey Railroad Ave.
   Menlo Park, N. J. 08817
3. Mallinckrodt Chemical Works
   Industrial Chemicals Division
   2nd and Mallinckrodt Sts.
   St. Louis, Mo. 63160

10. SHIPPING INFORMATION

10.1 **Grades or Purity:** Reagent 99.8+%
10.2 **Storage Temperature:** Ambient
10.3 **Inert Atmosphere:** No requirement
10.4 **Venting:** Open

11. HAZARD ASSESSMENT CODE

(See Hazard Assessment Handbook, CG 446-3)

SS

12. HAZARD CLASSIFICATIONS

12.1 **Code of Federal Regulations:** Oxidizing material
12.2 **NAS Hazard Rating for Bulk Water Transportation:** Not listed
12.3 **NFPA Hazard Classifications:**

Category	Classification
Health Hazard (Blue)	1
Flammability (Red)	0
Reactivity (Yellow)	0
	oxy

13. PHYSICAL AND CHEMICAL PROPERTIES

13.1 **Physical State at 15°C and 1 atm:** Solid
13.2 **Molecular Weight:** 169.87
13.3 **Boiling Point at 1 atm:** Decomposes
13.4 **Freezing Point:** 414°F = 212°C = 485°K
13.5 **Critical Temperature:** Not pertinent
13.6 **Critical Pressure:** Not pertinent
13.7 **Specific Gravity:** 4.35 at 19°C (solid)
13.8 **Liquid Surface Tension:** Not pertinent
13.9 **Liquid-Water Interfacial Tension:** Not pertinent
13.10 **Vapor (Gas) Specific Gravity:** Not pertinent
13.11 **Ratio of Specific Heats of Vapor (Gas):** Not pertinent
13.12 **Latent Heat of Vaporization:** Not pertinent
13.13 **Heat of Combustion:** Not pertinent
13.14 **Heat of Decomposition:** Not pertinent
13.15 **Heat of Solution:** Not pertinent
13.16 **Heat of Polymerization:** Not pertinent

NOTES

REVISED 1978

SVO — SILVER OXIDE

Common Synonyms			
Argentous oxide	Solid	Brown-black	Odorless
	Sinks in water.		

Avoid contact with solid and dust. Keep people away.
Stop discharge if possible.
Isolate and remove discharged material.
Notify local health and pollution control agencies.

Fire

Not flammable.

Exposure

CALL FOR MEDICAL AID.

DUST
Irritating to eyes, nose and throat.
If inhaled will cause coughing or difficult breathing.
If in eyes, hold eyelids open and flush with plenty of water.
If breathing has stopped, give artificial respiration.
If breathing is difficult, give oxygen.

SOLID
Irritating to skin and eyes.
Harmful if swallowed.
Remove contaminated clothing and shoes.
Flush affected areas with plenty of water.
IF IN EYES, hold eyelids open and flush with plenty of water.
IF SWALLOWED and victim is CONSCIOUS, have victim drink water or milk and have victim induce vomiting.
IF SWALLOWED and victim is UNCONSCIOUS OR HAVING CONVULSIONS, do nothing except keep victim warm.

Water Pollution

Effect of low concentrations on aquatic life is unknown.
May be dangerous if it enters water intakes.
Notify local health and wildlife officials.
Notify operators of nearby water intakes.

1. RESPONSE TO DISCHARGE

(See Response Methods Handbook, CG 446-4)

Should be removed
Chemical and physical treatment

2. LABELS

No hazard label required by Code of Federal Regulations

3. CHEMICAL DESIGNATIONS

3.1 **Synonyms:** Argentous oxide
3.2 **Coast Guard Compatibility Classification:** Not listed
3.3 **Chemical Formula:** Ag_2O
3.4 **IMCO/United Nations Numerical Designation:** Not listed

4. OBSERVABLE CHARACTERISTICS

4.1 **Physical State (as shipped):** Solid
4.2 **Color:** Brownish-black
4.3 **Odor:** None

5. HEALTH HAZARDS

5.1 **Personal Protective Equipment:** Dust mask; goggles or face shield; protective gloves
5.2 **Symptoms Following Exposure:** Contact with eyes causes mild irritation. If continued for a long period, ingestion or inhalation of silver compounds can cause permanent discoloration of the skin (argyria).
5.3 **Treatment for Exposure:** EYES: flush with water. SKIN: flush with water; wash with soap and water.
5.4 **Toxicity by Inhalation (Threshold Limit Value):** 0.01 mg/m^3
5.5 **Short-Term Inhalation Limits:** Data not available
5.6 **Toxicity by Ingestion:** Grade 2; LD_{50} 0.5—5 g/kg
5.7 **Late Toxicity:** Data not available
5.8 **Vapor (Gas) Irritant Characteristics:** Data not available
5.9 **Liquid or Solid Irritant Characteristics:** Data not available
5.10 **Odor Threshold:** Odorless

6. FIRE HAZARDS

6.1 **Flash Point:** Not flammable
6.2 **Flammable Limits in Air:** Not flammable
6.3 **Fire Extinguishing Agents:** Not pertinent
6.4 **Fire Extinguishing Agents Not to be Used:** Not pertinent
6.5 **Special Hazards of Combustion Products:** Not pertinent
6.6 **Behavior in Fire:** Decomposes into metallic silver and oxygen. If large quantities are involved, the oxygen might increase the intensity of the fire.
6.7 **Ignition Temperature:** Not pertinent
6.8 **Electrical Hazard:** Not pertinent
6.9 **Burning Rate:** Not pertinent

7. CHEMICAL REACTIVITY

7.1 **Reactivity with Water:** No reaction
7.2 **Reactivity with Common Materials:**
7.3 **Stability During Transport:** Stable
7.4 **Neutralizing Agents for Acids and Caustics:** Not pertinent
7.5 **Polymerization:** Not pertinent
7.6 **Inhibitor of Polymerization:** Not pertinent

8. WATER POLLUTION

8.1 **Aquatic Toxicity:** Data not available
8.2 **Waterfowl Toxicity:** Data not available
8.3 **Biological Oxygen Demand (BOD):** None
8.4 **Food Chain Concentration Potential:** None

9. SELECTED MANUFACTURERS

1. Pfaltz and Bauer, Inc.
   375 Fairfield Ave.
   Stamford, Conn. 06902
2. Gallard-Schlesinger Chemical Mfg. Co.
   584 Mineola Ave.
   Carle Place, N. Y. 11514
3. Fisher Scientific Co.
   711 Forbes Ave.
   Pittsburgh, Pa. 15219

10. SHIPPING INFORMATION

10.1 **Grades or Purity:** Commercial, 99+%
10.2 **Storage Temperature:** Ambient
10.3 **Inert Atmosphere:** No requirement
10.4 **Venting:** Open

11. HAZARD ASSESSMENT CODE

(See Hazard Assessment Handbook, CG 446-3)

II

12. HAZARD CLASSIFICATIONS

12.1 **Code of Federal Regulations:** Not listed
12.2 **NAS Hazard Rating for Bulk Water Transportation:** Not listed
12.3 **NFPA Hazard Classifications:** Not listed

13. PHYSICAL AND CHEMICAL PROPERTIES

13.1 **Physical State at 15°C and 1 atm:** Solid
13.2 **Molecular Weight:** 231.8
13.3 **Boiling Point at 1 atm:** Not pertinent (decomposes)
13.4 **Freezing Point:** Not pertinent
13.5 **Critical Temperature:** Not pertinent
13.6 **Critical Pressure:** Not pertinent
13.7 **Specific Gravity:** 7.14 at 20°C (solid)
13.8 **Liquid Surface Tension:** Not pertinent
13.9 **Liquid-Water Interfacial Tension:** Not pertinent
13.10 **Vapor (Gas) Specific Gravity:** Not pertinent
13.11 **Ratio of Specific Heats of Vapor (Gas):** Not pertinent
13.12 **Latent Heat of Vaporization:** Not pertinent
13.13 **Heat of Combustion:** Not pertinent
13.14 **Heat of Decomposition:** Not pertinent
13.15 **Heat of Solution:** Not pertinent
13.16 **Heat of Polymerization:** Not pertinent

NOTES

SVS

SILVER SULFATE

Common Synonyms	Solid	White to gray	Odorless
	Sinks and mixes with water.		

Avoid contact with solid and dust. Keep people away.
Stop discharge if possible.
Isolate and remove discharged material.
Notify local health and pollution control agencies.

Fire

Not flammable.

Exposure

CALL FOR MEDICAL AID.

DUST
Irritating to eyes, nose and throat.
If inhaled will cause coughing or difficult breathing.
If in eyes, hold eyelids open and flush with plenty of water.
If breathing has stopped, give artificial respiration.
If breathing is difficult, give oxygen.

SOLID
Irritating to skin and eyes.
Harmful if swallowed.
Remove contaminated clothing and shoes.
Flush affected areas with plenty of water.
IF IN EYES, hold eyelids open and flush with plenty of water.
IF SWALLOWED and victim is CONSCIOUS, have victim drink water or milk and have victim induce vomiting.
IF SWALLOWED and victim is UNCONSCIOUS OR HAVING CONVULSIONS, do nothing except keep victim warm.

Water Pollution

HARMFUL TO AQUATIC LIFE IN VERY LOW CONCENTRATIONS.
May be dangerous if it enters water intakes.
Notify local health and wildlife officials.
Notify operators of nearby water intakes.

1. RESPONSE TO DISCHARGE

(See Response Methods Handbook, CG 446-4)

Issue warning—water contaminant
Restrict access
Should be removed
Chemical and physical treatment

2. LABELS

No hazard label required by Code of Federal Regulations

3. CHEMICAL DESIGNATIONS

3.1 **Synonyms:** No common synonyms
3.2 **Coast Guard Compatibility Classification:** Not listed
3.3 **Chemical Formula:** Ag_2SO_4
3.4 **IMCO/United Nations Numerical Designation:** Not listed

4. OBSERVABLE CHARACTERISTICS

4.1 **Physical State (as shipped):** Solid
4.2 **Color:** White to gray
4.3 **Odor:** None

5. HEALTH HAZARDS

5.1 **Personal Protective Equipment:** Dust mask; goggles or face shield; protective gloves
5.2 **Symptoms Following Exposure:** Contact with eyes causes irritation. If continued for a long period, ingestion or inhalation of silver compounds can cause permanent discoloration of the skin (argyria).
5.3 **Treatment for Exposure:** INHALATION: move to fresh air. INGESTION: give large amount of water; induce vomiting; EYES: flush with water for at least 15 min. SKIN: flush with water; wash with soap and water.
5.4 **Toxicity by Inhalation (Threshold Limit Value):** 0.01 mg/m^3
5.5 **Short-Term Inhalation Limits:** Data not available
5.6 **Toxicity by Ingestion:** Data not available
5.7 **Late Toxicity:** Data not available
5.8 **Vapor (Gas) Irritant Characteristics:** Data not available
5.9 **Liquid or Solid Irritant Characteristics:** Data not available
5.10 **Odor Threshold:** Data not available

6. FIRE HAZARDS

6.1 **Flash Point:** Not flammable
6.2 **Flammable Limits in Air:** Not flammable
6.3 **Fire Extinguishing Agents:** Not pertinent
6.4 **Fire Extinguishing Agents Not to be Used:** Not pertinent
6.5 **Special Hazards of Combustion Products:** Not pertinent
6.6 **Behavior in Fire:**
6.7 **Ignition Temperature:** Not pertinent
6.8 **Electrical Hazard:** Not pertinent
6.9 **Burning Rate:** Not pertinent

7. CHEMICAL REACTIVITY

7.1 **Reactivity with Water:** No reaction
7.2 **Reactivity with Common Materials:**
7.3 **Stability During Transport:** Stable
7.4 **Neutralizing Agents for Acids and Caustics:** Not pertinent
7.5 **Polymerization:** Not pertinent
7.6 **Inhibitor of Polymerization:** Not pertinent

8. WATER POLLUTION

8.1 **Aquatic Toxicity:**
0.4 ppm/48 hr/barnacles/lethal 90%/salt water
8.2 **Waterfowl Toxicity:** Data not available
8.3 **Biological Oxygen Demand (BOD):** None
8.4 **Food Chain Concentration Potential:** None

9. SELECTED MANUFACTURERS

1. R.S.A. Corporation
690 Saw Mill River Rd.
Ardsley, N. Y. 10502
2. J. T. Baker Chemical Co.
Phillipsburg, N. J. 08865
3. Ventron, Inc.
P. O. Box 159
Beverly, Mass. 01915

10. SHIPPING INFORMATION

10.1 **Grades or Purity:** Reagent; Commercial
10.2 **Storage Temperature:** Ambient
10.3 **Inert Atmosphere:** No requirement
10.4 **Venting:** Open

11. HAZARD ASSESSMENT CODE

(See Hazard Assessment Handbook, CG 446-3)

II-SS

12. HAZARD CLASSIFICATIONS

12.1 **Code of Federal Regulations:** Not listed
12.2 **NAS Hazard Rating for Bulk Water Transportation:** Not listed
12.3 **NFPA Hazard Classifications:** Not listed

13. PHYSICAL AND CHEMICAL PROPERTIES

13.1 **Physical State at 15°C and 1 atm:** Solid
13.2 **Molecular Weight:** 311.80
13.3 **Boiling Point at 1 atm:** Not pertinent
13.4 **Freezing Point:** Not pertinent
13.5 **Critical Temperature:** Not pertinent
13.6 **Critical Pressure:** Not pertinent
13.7 **Specific Gravity:** 5.45 at 20°C (solid)
13.8 **Liquid Surface Tension:** Not pertinent
13.9 **Liquid-Water Interfacial Tension:** Not pertinent
13.10 **Vapor (Gas) Specific Gravity:** Not pertinent
13.11 **Ratio of Specific Heats of Vapor (Gas):** Not pertinent
13.12 **Latent Heat of Vaporization:** Not pertinent
13.13 **Heat of Combustion:** Not pertinent
13.14 **Heat of Decomposition:** Not pertinent
13.15 **Heat of Solution:** Not pertinent
13.16 **Heat of Polymerization:** Not pertinent

NOTES

SDU — SODIUM

Common Synonyms	Soft solid under kerosene — Silver to grayish-white — Odorless Floats and reacts violently with water. Flammable gas is produced.

AVOID CONTACT WITH SOLID. Keep people away.
Call fire department.
Wear goggles, and rubber overclothing (including gloves).
Notify local health and pollution control agencies.

Fire	FLAMMABLE. FIRE MAY START ON CONTACT WITH AIR. Flammable gas formed on contact with water or moisture. **Wear goggles, self-contained breathing apparatus, and rubber overclothing (including gloves).** **DO NOT USE WATER, CARBON DIOXIDE, OR VAPORIZING LIQUIDS.** **Extinguish with dry graphite, soda ash, powdered sodium chloride or other approved dry powder.**
Exposure	**CALL FOR MEDICAL AID.** SOLID Will burn skin and eyes. **Remove contaminated clothing and shoes.** **Flush affected areas with plenty of water.** **IF IN EYES, hold eyelids open and flush with plenty of water.**
Water Pollution	Dangerous to aquatic life in high concentrations. May be dangerous if it enters water intakes. **Notify local health and wildlife officials.** **Notify operators of nearby water intakes.**

1. RESPONSE TO DISCHARGE

(See Response Methods Handbook, CG 446-4)

Issue warning—high flammability
Restrict access
Evacuate area
Chemical and physical treatment

2. LABELS

3. CHEMICAL DESIGNATIONS

3.1 **Synonyms:** No common synonyms
3.2 **Coast Guard Compatibility Classification:** Not applicable
3.3 **Chemical Formula:** Na
3.4 **IMCO/United Nations Numerical Designation:** 4.3/1428

4. OBSERVABLE CHARACTERISTICS

4.1 **Physical State (as shipped):** Soft solid or liquid
4.2 **Color:** Silvery white, changing to gray on exposure to air
4.3 **Odor:** Odorless

5. HEALTH HAZARDS

5.1 **Personal Protective Equipment:** Maximum protective clothing; goggles and face shield.
5.2 **Symptoms Following Exposure:** Severe burns caused by burning metal or by caustic soda formed by reaction with moisture on skin.
5.3 **Treatment for Exposure:** SKIN: brush off any metal, then flood with water for at least 15 min.; treat as heat or caustic burn; call a doctor.
5.4 **Toxicity by Inhalation (Threshold Limit Value):** Not pertinent
5.5 **Short-Term Inhalation Limits:** Not pertinent
5.6 **Toxicity by Ingestion:** Not pertinent
5.7 **Late Toxicity:** None
5.8 **Vapor (Gas) Irritant Characteristics:** Non-volatile
5.9 **Liquid or Solid Irritant Characteristics:** Severe skin irritant. Causes second- and third-degree burns on short contact and is very injurious to the eyes.
5.10 **Odor Threshold:** Not pertinent

6. FIRE HAZARDS

6.1 **Flash Point:** Not pertinent
6.2 **Flammable Limits in Air:** Not pertinent
6.3 **Fire Extinguishing Agents:** Dry soda ash, graphite, salt, or other approved dry powder such as dry limestone.
6.4 **Fire Extinguishing Agents Not to be Used:** Water, carbon dioxide or halogenated extinguishing agents.
6.5 **Special Hazards of Combustion Products:** Fumes of burning Na are highly irritating to skin, eyes, and mucous membranes.
6.6 **Behavior in Fire:** Not pertinent
6.7 **Ignition Temperature:** 250°F
6.8 **Electrical Hazard:** Not pertinent
6.9 **Burning Rate:** Not pertinent

7. CHEMICAL REACTIVITY

7.1 **Reactivity with Water:** Reacts violently, with formation of flammable hydrogen gas and caustic soda solution. A fire often occurs.
7.2 **Reactivity with Common Materials:** No reaction
7.3 **Stability During Transport:** Stable
7.4 **Neutralizing Agents for Acids and Caustics:** After reaction with water, caustic soda formed can be diluted with water and/or neutralized with acetic acid.
7.5 **Polymerization:** Not pertinent
7.6 **Inhibitor of Polymerization:** Not pertinent

8. WATER POLLUTION

8.1 **Aquatic Toxicity:** Not pertinent
8.2 **Waterfowl Toxicity:** Not pertinent
8.3 **Biological Oxygen Demand (BOD):** None
8.4 **Food Chain Concentration Potential:** None

9. SELECTED MANUFACTURERS

1. E. I. duPont de Nemours
   Electrochemicals Dept.
   Wilmington, Del. 19898
2. Ethyl Corp.
   Industrial Chemicals Division
   451 Florida
   Baton Rouge, La. 70801
3. RMI Company
   State Rd.
   Ashtabula, Ohio 44004

10. SHIPPING INFORMATION

10.1 **Grades or Purity:** Commercial grade: 99.95%
10.2 **Storage Temperature:** 230° – 250°F (liquid); ambient (solid)
10.3 **Inert Atmosphere:** Dry nitrogen or argon (for liquid); under kerosene (for solid)
10.4 **Venting:** Pressure-vacuum

11. HAZARD ASSESSMENT CODE

(See Hazard Assessment Handbook, CG 446-3)

RR-C

12. HAZARD CLASSIFICATIONS

12.1 **Code of Federal Regulations:** Flammable solid
12.2 **NAS Hazard Rating for Bulk Water Transportation:** Not listed
12.3 **NFPA Hazard Classifications:**

Category	Classification
Health Hazard (Blue)	3
Flammability (Red)	1
Reactivity (Yellow)	2
	W̶

13. PHYSICAL AND CHEMICAL PROPERTIES

13.1 **Physical State at 15°C and 1 atm:** Solid
13.2 **Molecular Weight:** 22.49
13.3 **Boiling Point at 1 atm:** 1621°F = 883°C = 1156°K
13.4 **Freezing Point:** 207.5°F = 97.5°C = 370.7°K
13.5 **Critical Temperature:** 3632°F = 2000°C = 2273°K
13.6 **Critical Pressure:** 5040 psia = 343 atm = 34.8 MN/m^2
13.7 **Specific Gravity:** 0.971 at 20°C (solid)
13.8 **Liquid Surface Tension:** Not pertinent
13.9 **Liquid-Water Interfacial Tension:** Not pertinent
13.10 **Vapor (Gas) Specific Gravity:** Not pertinent
13.11 **Ratio of Specific Heats of Vapor (Gas):** Not pertinent
13.12 **Latent Heat of Vaporization:** Not pertinent
13.13 **Heat of Combustion:** Not pertinent
13.14 **Heat of Decomposition:** Not pertinent
13.15 **Heat of Solution:** Not pertinent
13.16 **Heat of Polymerization:** Not pertinent

NOTES

REVISED 1978

SAB SODIUM ALKYLBENZENESULFONATES

Common Synonyms
Sulfonated alkylbenzene, sodium salt
Alkylbenzenesulfonic acid, sodium salt

Thick liquid or solid — Pale yellow — Faint detergent odor

Mixes with water. Soap bubbles may be produced.

Stop discharge if possible.
Isolate and remove discharged material.
Notify local health and pollution control agencies.

Fire

Not flammable.

Exposure

CALL FOR MEDICAL AID.

LIQUID
Irritating to skin and eyes.
If swallowed, will cause nausea or vomiting.

Remove contaminated clothing and shoes.
Flush affected areas with plenty of water.
IF IN EYES, hold eyelids open and flush with plenty of water.
IF SWALLOWED and victim is CONSCIOUS, have victim drink water or milk and have victim induce vomiting.
IF SWALLOWED and victim is UNCONSCIOUS OR HAVING CONVULSIONS, do nothing except keep victim warm.

Water Pollution

HARMFUL TO AQUATIC LIFE IN VERY LOW CONCENTRATIONS.
May be dangerous if it enters water intakes.

Notify local health and wildlife officials.
Notify operators of nearby water intakes.

1. RESPONSE TO DISCHARGE

(See Response Methods Handbook, CG 446-4)

Issue warning—water contaminant
Should be removed
Chemical and physical treatment

2. LABELS

No hazard label required by Code of Federal Regulations

3. CHEMICAL DESIGNATIONS

3.1 **Synonyms:** Alkylbenzenesulfonic acid, sodium salt; Sulfonated alkylbenzene, sodium salt
3.2 **Coast Guard Compatibility Classification:** Not applicable
3.3 **Chemical Formula:** $C_nH_{2n+1}C_6H_4SO_3Na$
3.4 **IMCO/United Nations Numerical Designation:** Not listed

4. OBSERVABLE CHARACTERISTICS

4.1 **Physical State (as shipped):** Powder or thick liquid
4.2 **Color:** Pale yellow
4.3 **Odor:** Faint detergent

5. HEALTH HAZARDS

5.1 **Personal Protective Equipment:** Goggles or face shield; rubber gloves.
5.2 **Symptoms Following Exposure:** In general, these chemicals have a moderate order of toxicity. Repeated skin contact with concentrated solutions may cause dermatitis. Ingestion may cause gastrointestinal irritation, vomiting, and diarrhea.
5.3 **Treatment for Exposure:** INGESTION: induce vomiting and call a doctor. EYES OR SKIN: flush with copious amounts of water.
5.4 **Toxicity by Inhalation (Threshold Limit Value):** Not pertinent
5.5 **Short-Term Inhalation Limits:** Not pertinent
5.6 **Toxicity by Ingestion:** Grade 2; LD_{50} 0.5 to 5 g/kg
5.7 **Late Toxicity:** None
5.8 **Vapor (Gas) Irritant Characteristics:** Non-volatile
5.9 **Liquid or Solid Irritant Characteristics:** Minimum hazard. If spilled on clothing and allowed to remain, may cause smarting and reddening of the skin.
5.10 **Odor Threshold:** Not pertinent

6. FIRE HAZARDS

6.1 **Flash Point:** Not flammable
6.2 **Flammable Limits in Air:** Not flammable
6.3 **Fire Extinguishing Agents:** Not pertinent
6.4 **Fire Extinguishing Agents Not to be Used:** Not pertinent
6.5 **Special Hazards of Combustion Products:** Not pertinent
6.6 **Behavior in Fire:** Irritating vapors may be generated.
6.7 **Ignition Temperature:** Not flammable
6.8 **Electrical Hazard:** Not pertinent
6.9 **Burning Rate:** Not flammable

7. CHEMICAL REACTIVITY

7.1 **Reactivity with Water:** No reaction
7.2 **Reactivity with Common Materials:** No reaction
7.3 **Stability During Transport:** Stable
7.4 **Neutralizing Agents for Acids and Caustics:** Not pertinent
7.5 **Polymerization:** Not pertinent
7.6 **Inhibitor of Polymerization:** Not pertinent

8. WATER POLLUTION

8.1 **Aquatic Toxicity:**
5.6 ppm (100% SAB)/96 hr/bluegill/TL_m/fresh water
8.2 ppm/96 hr/winter flounder/TL_m/salt water
8.2 **Waterfowl Toxicity:** Data not available
8.3 **Biological Oxygen Demand (BOD):** 4%, 5 days
8.4 **Food Chain Concentration Potential:** None

9. SELECTED MANUFACTURERS

1. Colgate Palmolive Co.
300 Park Ave.
New York, N. Y. 10022
2. Lever Bros. Co.
390 Park Ave.
New York, N. Y. 10022
3. Witco Chemical Corp.
Organic Division
1914 S. Kilbourn Ave.
Chicago, Ill. 60623

10. SHIPPING INFORMATION

10.1 **Grades or Purity:** Vary with each manufacturer and with intended use. Some is shipped as a thick, concentrated water solution, some as a solid, often mixed with other solids such as sodium phosphate. Ordinary household detergents are good examples of this substance.
10.2 **Storage Temperature:** Ambient
10.3 **Inert Atmosphere:** Not required
10.4 **Venting:** Open

11. HAZARD ASSESSMENT CODE

(See Hazard Assessment Handbook, CG 446-3)

SS

12. HAZARD CLASSIFICATIONS

12.1 **Code of Federal Regulations:** Not listed
12.2 **NAS Hazard Rating for Bulk Water Transportation:** Not listed
12.3 **NFPA Hazard Classifications:** Not listed

13. PHYSICAL AND CHEMICAL PROPERTIES

13.1 **Physical State at 15°C and 1 atm:** Liquid or solid
13.2 **Molecular Weight:** Not pertinent
13.3 **Boiling Point at 1 atm:** Decomposes
13.4 **Freezing Point:** Not pertinent
13.5 **Critical Temperature:** Not pertinent
13.6 **Critical Pressure:** Not pertinent
13.7 **Specific Gravity:** 1.0 at 20°C (liquid)
13.8 **Liquid Surface Tension:** Not pertinent
13.9 **Liquid-Water Interfacial Tension:** Not pertinent
13.10 **Vapor (Gas) Specific Gravity:** Not pertinent
13.11 **Ratio of Specific Heats of Vapor (Gas):** Not pertinent
13.12 **Latent Heat of Vaporization:** Not pertinent
13.13 **Heat of Combustion:** Not pertinent
13.14 **Heat of Decomposition:** Not pertinent
13.15 **Heat of Solution:** Not pertinent
13.16 **Heat of Polymerization:** Not pertinent

NOTES

SAS	SODIUM ALKYL SULFATES

Common Synonyms Sodium hydrogen alkyl sulfate	Solid or thick liquid Pale yellow Faint detergent odor Mixes with water. Soap bubbles may be produced.

Stop discharge if possible.
Isolate and remove discharged material.
Notify local health and pollution control agencies.

Fire	Not flammable.
Exposure	CALL FOR MEDICAL AID. LIQUID OR SOLID Irritating to skin and eyes. If swallowed, will cause nausea or vomiting. Flush affected areas with plenty of water. IF IN EYES, hold eyelids open and flush with plenty of water. IF SWALLOWED and victim is CONSCIOUS, have victim drink water or milk and have victim induce vomiting. IF SWALLOWED and victim is UNCONSCIOUS OR HAVING CONVULSIONS, do nothing except keep victim warm.
Water Pollution	HARMFUL TO AQUATIC LIFE IN VERY LOW CONCENTRATIONS. May be dangerous if it enters water intakes. Notify local health and wildlife officials. Notify operators of nearby water intakes.

1. RESPONSE TO DISCHARGE

(See Response Methods Handbook, CG 446-4)

Issue warning — water contaminant
Should be removed
Chemical and physical treatment

2. LABELS

No hazard label required by Code of Federal Regulations

3. CHEMICAL DESIGNATIONS

3.1 **Synonyms:** Sodium hydrogen alkyl sulfate
3.2 **Coast Guard Compatibility Classification:** Not applicable
3.3 **Chemical Formula:** $C_nH_{2n+1}OSO_2ONa$
3.4 **IMCO/United Nations Numerical Designation:** Not listed

4. OBSERVABLE CHARACTERISTICS

4.1 **Physical State (as shipped):** Liquid or solid
4.2 **Color:** Colorless to pale yellow
4.3 **Odor:** Weak detergent

5. HEALTH HAZARDS

5.1 **Personal Protective Equipment:** Goggles or face shield; rubber gloves.
5.2 **Symptoms Following Exposure:** In general, these chemicals have a moderate order of toxicity. Repeated skin contact with concentrated solutions may cause dermatitis. Ingestion may cause gastrointestinal irritation, vomiting, and diarrhea.
5.3 **Treatment for Exposure:** INGESTION: induce vomiting and follow with gastric lavage. SKIN: wash off with water.
5.4 **Toxicity by Inhalation (Threshold Limit Value):** Not pertinent
5.5 **Short-Term Inhalation Limits:** Not pertinent
5.6 **Toxicity by Ingestion:** Grade 2; LD_{50} 0.5 to 5 g/kg
5.7 **Late Toxicity:** None
5.8 **Vapor (Gas) Irritant Characteristics:** Non-volatile
5.9 **Liquid or Solid Irritant Characteristics:** Minimum hazard. If spilled on clothing and allowed to remain, may cause smarting and reddening of the skin.
5.10 **Odor Threshold:** Not pertinent

6. FIRE HAZARDS

6.1 **Flash Point:** Not flammable
6.2 **Flammable Limits in Air:** Not flammable
6.3 **Fire Extinguishing Agents:** Not pertinent
6.4 **Fire Extinguishing Agents Not to be Used:** Not pertinent
6.5 **Special Hazards of Combustion Products:** Not pertinent
6.6 **Behavior in Fire:** May produce irritating vapors.
6.7 **Ignition Temperature:** Not flammable
6.8 **Electrical Hazard:** Not pertinent
6.9 **Burning Rate:** Not flammable

7. CHEMICAL REACTIVITY

7.1 **Reactivity with Water:** No reaction
7.2 **Reactivity with Common Materials:** No reaction
7.3 **Stability During Transport:** Stable
7.4 **Neutralizing Agents for Acids and Caustics:** Not pertinent
7.5 **Polymerization:** Not pertinent
7.6 **Inhibitor of Polymerization:** Not pertinent

8. WATER POLLUTION

8.1 **Aquatic Toxicity:**
6-7 ppm/*/fish/min lethal/fresh water
*Time period not specified
8.2 **Waterfowl Toxicity:** Data not available
8.3 **Biological Oxygen Demand (BOD):** (theor.) 60%, 5 days
8.4 **Food Chain Concentration Potential:** None

9. SELECTED MANUFACTURERS

1. E. I. duPont de Nemours & Co., Inc.
   Dyes and Chemicals Division
   Wilmington, Del. 19898
2. Millmaster Onyx Corp.
   Onyx Chemical Co. Division
   190 Warren St.
   Jersey City, N. J. 07302
3. Union Carbide Corp.
   Chemicals and Plastics Division
   270 Park Ave.
   New York, N. Y. 10017

10. SHIPPING INFORMATION

10.1 **Grades or Purity:** Vary with each manufacturer and with intended use. Some is shipped as a thick concentrated water solution, some as a solid often mixed with other solids such as sodium phosphate. Ordinary household detergents are good examples of this substance.
10.2 **Storage Temperature:** Ambient
10.3 **Inert Atmosphere:** No requirement
10.4 **Venting:** Open

11. HAZARD ASSESSMENT CODE

(See Hazard Assessment Handbook, CG 446-3)

SS

12. HAZARD CLASSIFICATIONS

12.1 **Code of Federal Regulations:** Not listed
12.2 **NAS Hazard Rating for Bulk Water Transportation:** Not listed
12.3 **NFPA Hazard Classifications:** Not listed

13. PHYSICAL AND CHEMICAL PROPERTIES

13.1 **Physical State at 15°C and 1 atm:** Liquid or solid
13.2 **Molecular Weight:** Not pertinent
13.3 **Boiling Point at 1 atm:** Decomposes
13.4 **Freezing Point:** Not pertinent
13.5 **Critical Temperature:** Not pertinent
13.6 **Critical Pressure:** Not pertinent
13.7 **Specific Gravity:** Data not available
13.8 **Liquid Surface Tension:** Not pertinent
13.9 **Liquid-Water Interfacial Tension:** Not pertinent
13.10 **Vapor (Gas) Specific Gravity:** Not pertinent
13.11 **Ratio of Specific Heats of Vapor (Gas):** Not pertinent
13.12 **Latent Heat of Vaporization:** Not pertinent
13.13 **Heat of Combustion:** Not pertinent
13.14 **Heat of Decomposition:** Not pertinent
13.15 **Heat of Solution:** Not pertinent
13.16 **Heat of Polymerization:** Not pertinent

NOTES

SAM — SODIUM AMIDE

Common Synonyms			
Sodamide	Solid	Colorless	Odorless
	Sinks and reacts violently with water.		

Avoid contact with liquid. Keep people away.
Wear rubber overclothing (including gloves).
Call fire department.
Notify local health and pollution control agencies.

Fire	FLAMMABLE. POISONOUS GAS IS PRODUCED IN FIRE. DO NOT USE WATER ON FIRE. Extinguish with dry graphite, soda ash, powdered limestone, or other approved dry powder.
Exposure	CALL FOR MEDICAL AID. SOLID Will burn skin and eyes. Remove contaminated clothing and shoes. Flush affected areas with plenty of water. IF IN EYES, hold eyelids open and flush with plenty of water.
Water Pollution	Dangerous to aquatic life in high concentrations. May be dangerous if it enters water intakes. Notify local health and wildlife officials. Notify operators of nearby water intakes.

1. RESPONSE TO DISCHARGE

(See Response Methods Handbook, CG 446-4)

Issue warning—corrosive, flammable
Restrict access
Should be removed
Disperse and flush

2. LABELS

3. CHEMICAL DESIGNATIONS

3.1 **Synonyms:** Sodamide
3.2 **Coast Guard Compatibility Classification:** Not applicable
3.3 **Chemical Formula:** $NaNH_2$
3.4 **IMCO/United Nations Numerical Designation:** 4.3/1425

4. OBSERVABLE CHARACTERISTICS

4.1 **Physical State (as shipped):** Solid
4.2 **Color:** Gray
4.3 **Odor:** Like ammonia

5. HEALTH HAZARDS

5.1 **Personal Protective Equipment:** Goggles or face shield; dust respirator; rubber gloves and shoes.
5.2 **Symptoms Following Exposure:** Ammonia gas formed by reaction of solid with moisture irritates eyes and skin. Solid causes caustic burns of eyes and skin. Ingestion burns mouth and stomach in same way as caustic soda and may cause perforation of tissue.
5.3 **Treatment for Exposure:** INGESTION: give water or milk followed by dilute vinegar or fruit juice; do NOT induce vomiting; call a doctor. SKIN OR EYES: flood all affected areas with copious amounts of water.
5.4 **Toxicity by Inhalation (Threshold Limit Value):** Not pertinent
5.5 **Short-Term Inhalation Limits:** Not pertinent
5.6 **Toxicity by Ingestion:** Data not available
5.7 **Late Toxicity:** None
5.8 **Vapor (Gas) Irritant Characteristics:** Only that of ammonia formed by reaction of solid with moisture in air.
5.9 **Liquid or Solid Irritant Characteristics:** Burns skin and eyes just like caustic soda.
5.10 **Odor Threshold:** Not pertinent

6. FIRE HAZARDS

6.1 **Flash Point:** Flammable solid
6.2 **Flammable Limits in Air:** Not pertinent
6.3 **Fire Extinguishing Agents:** Dry soda ash, graphite, salt, or other recommended dry powder, such as dry limestone.
6.4 **Fire Extinguishing Agents Not to be Used:** Water
6.5 **Special Hazards of Combustion Products:** Toxic and irritating ammonia gas may be formed.
6.6 **Behavior in Fire:** Data not available
6.7 **Ignition Temperature:** Not pertinent
6.8 **Electrical Hazard:** Data not available
6.9 **Burning Rate:** Not pertinent

7. CHEMICAL REACTIVITY

7.1 **Reactivity with Water:** Reacts violently and frequently bursts into flames. Forms caustic soda solution.
7.2 **Reactivity with Common Materials:** Data not available
7.3 **Stability During Transport:** Stable
7.4 **Neutralizing Agents for Acids and Caustics:** Caustic solution formed by reaction with water can be diluted with water and/or neutralized by acetic acid.
7.5 **Polymerization:** Not pertinent
7.6 **Inhibitor of Polymerization:** Not pertinent

8. WATER POLLUTION

8.1 **Aquatic Toxicity:** Data not available
8.2 **Waterfowl Toxicity:** Data not available
8.3 **Biological Oxygen Demand (BOD):** None
8.4 **Food Chain Concentration Potential:** None

9. SELECTED MANUFACTURERS

1. Farchan Research Laboratories, Inc.
   4702 East 355 St.
   Willoughby, Ohio
2. Fike Chemicals Inc.
   P. O. Box 546
   Nitro, W. Va. 25143
3. Ventron Corp.
   Alfa Products Division
   Beverly, Mass. 01915

10. SHIPPING INFORMATION

10.1 **Grades or Purity:** Pure; Technical
10.2 **Storage Temperature:** Ambient
10.3 **Inert Atmosphere:** Must be dry
10.4 **Venting:** Sealed containers must be stored in well-ventilated area.

11. HAZARD ASSESSMENT CODE

(See Hazard Assessment Handbook, CG 446-3)

RR

12. HAZARD CLASSIFICATIONS

12.1 **Code of Federal Regulations:** Flammable solid
12.2 **NAS Hazard Rating for Bulk Water Transportation:** Not listed
12.3 **NFPA Hazard Classifications:** Not listed

13. PHYSICAL AND CHEMICAL PROPERTIES

13.1 **Physical State at 15°C and 1 atm:** Solid
13.2 **Molecular Weight:** 39.01
13.3 **Boiling Point at 1 atm:** 752°F = 400°C = 673°K
13.4 **Freezing Point:** 410°F = 210°C = 483°K
13.5 **Critical Temperature:** Not pertinent
13.6 **Critical Pressure:** Not pertinent
13.7 **Specific Gravity:** 1.39 at 20°C (solid)
13.8 **Liquid Surface Tension:** Not pertinent
13.9 **Liquid-Water Interfacial Tension:** Not pertinent
13.10 **Vapor (Gas) Specific Gravity:** Not pertinent
13.11 **Ratio of Specific Heats of Vapor (Gas):** Not pertinent
13.12 **Latent Heat of Vaporization:** Not pertinent
13.13 **Heat of Combustion:** Not pertinent
13.14 **Heat of Decomposition:** Not pertinent
13.15 **Heat of Solution:** Not pertinent
13.16 **Heat of Polymerization:** Not pertinent

NOTES

REVISED 1978

SDA	SODIUM ARSENATE

Common Synonyms Sodium arsenate, dibasic Disodium arsenate heptahydrate	Solid White Odorless Sinks and mixes with water.

Avoid contact with solid and dust. Keep people away.
Stop discharge if possible.
Isolate and remove discharged material.
Notify local health and pollution control agencies.

Fire	Not flammable.
Exposure	CALL FOR MEDICAL AID. DUST Irritating to eyes, nose and throat. If inhaled will cause difficult breathing. If in eyes, hold eyelids open and flush with plenty of water. If breathing has stopped, give artificial respiration. If breathing is difficult, give oxygen. SOLID Irritating to skin and eyes. If swallowed will cause coughing, nausea, vomiting or loss of consciousness. Remove contaminated clothing and shoes. Flush affected areas with plenty of water. IF IN EYES, hold eyelids open and flush with plenty of water. IF SWALLOWED and victim is CONSCIOUS, have victim drink water or milk. IF SWALLOWED and victim is UNCONSCIOUS OR HAVING CONVULSIONS, do nothing except keep victim warm.
Water Pollution	Dangerous to aquatic life in high concentrations. May be dangerous if it enters water intakes. Notify local health and wildlife officials. Notify operators of nearby water intakes.

1. RESPONSE TO DISCHARGE

(See Response Methods Handbook, CG 446-4)

Issue warning—poison, water contaminant
Restrict access
Disperse and flush

2. LABEL

POISON

3. CHEMICAL DESIGNATIONS

3.1 **Synonyms:** Disodium arsenate heptahydrate; Sodium arsenate, dibasic
3.2 **Coast Guard Compatibility Classification:** Not listed
3.3 **Chemical Formula:** $Na_2HAsO_4 \cdot 7H_2O$
3.4 **IMCO/United Nations Numerical Designation:** 6.1/1685

4. OBSERVABLE CHARACTERISTICS

4.1 **Physical State (as shipped):** Solid
4.2 **Color:** White
4.3 **Odor:** None

5. HEALTH HAZARDS

5.1 **Personal Protective Equipment:** Dust mask; goggles or face shield; protective gloves
5.2 **Symptoms Following Exposure:** Inhalation of massive doses can cause laryngitis, bronchitis. Ingestion causes constriction in throat and difficulty in swallowing; also causes burning and pain, vomiting, profuse diarrhea, dehydration, cyanosis, coma, convulsions, and death. Contact with eyes causes irritation. Contact with skin causes various skin eruptions, more often as a late manifestation, or chronic poisoning.
5.3 **Treatment for Exposure:** INHALATION: remove victim from exposure; support respiration. INGESTION: gastric lavage with water, followed by 1 glass of milk; consult physician. EYES: flush with water for at least 15 min. SKIN: flush with water.
5.4 **Toxicity by Inhalation (Threshold Limit Value):** 0.5 mg/m^3 (as arsenic)
5.5 **Short-Term Inhalation Limits:** Data not available
5.6 **Toxicity by Ingestion:** Grade 4; LD_{50} <50 mg/kg
5.7 **Late Toxicity:** Possible carcinogenic effects on skin and lungs.
5.8 **Vapor (Gas) Irritant Characteristics:** Data not available
5.9 **Liquid or Solid Irritant Characteristics:** Data not available
5.10 **Odor Threshold:** Data not available

6. FIRE HAZARDS

6.1 **Flash Point:** Not flammable
6.2 **Flammable Limits in Air:** Not flammable
6.3 **Fire Extinguishing Agents:** Not pertinent
6.4 **Fire Extinguishing Agents Not to be Used:** Not pertinent
6.5 **Special Hazards of Combustion Products:** Not pertinent
6.6 **Behavior in Fire:**
6.7 **Ignition Temperature:** Not pertinent
6.8 **Electrical Hazard:** Not pertinent
6.9 **Burning Rate:** Not pertinent

7. CHEMICAL REACTIVITY

7.1 **Reactivity with Water:** No reaction
7.2 **Reactivity with Common Materials:**
7.3 **Stability During Transport:** Stable
7.4 **Neutralizing Agents for Acids and Caustics:** Not pertinent
7.5 **Polymerization:** Not pertinent
7.6 **Inhibitor of Polymerization:** Not pertinent

8. WATER POLLUTION

8.1 **Aquatic Toxicity:**
234 ppm/*/minnows/lethal conc./fresh water
32 ppm/7 days/goldfish/TL_m/fresh water
*Time period not specified.
8.2 **Waterfowl Toxicity:** Data not available
8.3 **Biological Oxygen Demand (BOD):** None
8.4 **Food Chain Concentration Potential:** Bioconcentrative only 300 fold; not likely to be a problem in spill situation.

9. SELECTED MANUFACTURERS

1. J. T. Baker Chemical Co.
Phillipsburg, N. J. 08865
2. Gallard-Schlesinger Chemical Mfg. Co.
584 Mineola Ave.
Carle Place, N. Y. 11514
3. Fisher Scientific Co.
711 Forbes Ave.
Pittsburgh, Pa. 15219

10. SHIPPING INFORMATION

10.1 **Grades or Purity:** Reagent; Technical, 98+%
10.2 **Storage Temperature:** Ambient
10.3 **Inert Atmosphere:** No requirement
10.4 **Venting:** Open

11. HAZARD ASSESSMENT CODE

(See Hazard Assessment Handbook, CG 446-3)

SS

12. HAZARD CLASSIFICATIONS

12.1 **Code of Federal Regulations:** Poisonous solid, Class B
12.2 **NAS Hazard Rating for Bulk Water Transportation:** Not listed
12.3 **NFPA Hazard Classifications:** Not listed

13. PHYSICAL AND CHEMICAL PROPERTIES

13.1 **Physical State at 15°C and 1 atm:** Solid
13.2 **Molecular Weight:** 312
13.3 **Boiling Point at 1 atm:** (decomposes) 356°F = 180°C = 453°K
13.4 **Freezing Point:** 135°F = 57°C = 330°K
13.5 **Critical Temperature:** Not pertinent
13.6 **Critical Pressure:** Not pertinent
13.7 **Specific Gravity:** 1.87 at 20°C (solid)
13.8 **Liquid Surface Tension:** Not pertinent
13.9 **Liquid-Water Interfacial Tension:** Not pertinent
13.10 **Vapor (Gas) Specific Gravity:** Not pertinent
13.11 **Ratio of Specific Heats of Vapor (Gas):** Not pertinent
13.12 **Latent Heat of Vaporization:** Not pertinent
13.13 **Heat of Combustion:** Not pertinent
13.14 **Heat of Decomposition:** Not pertinent
13.15 **Heat of Solution:** Not pertinent
13.16 **Heat of Polymerization:** Not pertinent

NOTES

SAR	SODIUM ARSENITE

Common Synonyms			
Sodium meta arsenite Sodium ortho arsenite	Solid	White to gray	Odorless
	Mixes with water.		

AVOID CONTACT WITH SOLID AND DUST. KEEP PEOPLE AWAY.
Wear goggles and dust respirator.
Stay upwind. Use water spray to "knock down" dust.
Isolate and remove discharged material.
Notify local health and pollution control agencies.

Fire

Not flammable.
POISONOUS GASES MAY BE PRODUCED WHEN HEATED.

Exposure

CALL FOR MEDICAL AID.

DUST
POISONOUS IF INHALED.
Irritating to eyes, nose and throat.
Move victim to fresh air.
If in eyes, hold eyelids open and flush with plenty of water.
If breathing is difficult, give oxygen.

SOLID
POISONOUS IF SWALLOWED.
Irritating to skin and eyes.
Remove contaminated clothing and shoes.
Flush affected areas with plenty of water.
IF IN EYES, hold eyelids open and flush with plenty of water.
IF SWALLOWED and victim is CONSCIOUS, have victim drink water or milk and have victim induce vomiting.
IF SWALLOWED and victim is UNCONSCIOUS OR HAVING CONVULSIONS, do nothing except keep victim warm.

Water Pollution

HARMFUL TO AQUATIC LIFE IN VERY LOW CONCENTRATIONS.
May be dangerous if it enters water intakes.
Notify local health and wildlife officials.
Notify operators of nearby water intakes.

1. RESPONSE TO DISCHARGE

(See Response Methods Handbook, CG 446-4)

Issue warning — poison, water contaminant
Restrict access
Disperse and flush

2. LABEL

3. CHEMICAL DESIGNATIONS

3.1 **Synonyms:** Sodium metaarsenite; Sodium orthoarsenite
3.2 **Coast Guard Compatibility Classification:** Not applicable
3.3 **Chemical Formula:** $Na_3AsO_3 - NaAsO_2$
3.4 **IMCO/United Nations Numerical Designation:** Solid: 6.1/2027; Aq. soln.: 6.1/1686

4. OBSERVABLE CHARACTERISTICS

4.1 **Physical State (as shipped):** Solid
4.2 **Color:** White to gray
4.3 **Odor:** None

5. HEALTH HAZARDS

5.1 **Personal Protective Equipment:** Dust mask; rubber gloves; goggles or face shield
5.2 **Symptoms Following Exposure:** Dust may irritate eyes. Ingestion or excessive inhalation of dust causes irritation of stomach and intestines with nausea, vomiting, and diarrhea; bloody stools, shock, rapid pulse, coma.
5.3 **Treatment for Exposure:** EYES: flush with water for at least 15 min. SKIN: wash with large amounts of water. INGESTION: immediately induce evacuation of intestinal tract by gastric lavage and saline cathartic; see physician immediately; consider possible development of arsenic poisoning.
5.4 **Toxicity by Inhalation (Threshold Limit Value):** 0.5 mg/m^3 (as arsenic)
5.5 **Short-Term Inhalation Limits:** Data not available
5.6 **Toxicity by Ingestion:** Grade 4; oral LD_{50} = 42 mg/kg (rat)
5.7 **Late Toxicity:** May be carcinogenic. Arsenic poisoning may develop.
5.8 **Vapor (Gas) Irritant Characteristics:** Data not available
5.9 **Liquid or Solid Irritant Characteristics:** Data not available
5.10 **Odor Threshold:** Not pertinent

6. FIRE HAZARDS

6.1 **Flash Point:** Not flammable
6.2 **Flammable Limits in Air:** Not flammable
6.3 **Fire Extinguishing Agents:** Not pertinent
6.4 **Fire Extinguishing Agents Not to be Used:** Not pertinent
6.5 **Special Hazards of Combustion Products:** Toxic arsenic fumes may be formed in fires. Self-contained breathing apparatus should be used.
6.6 **Behavior in Fire:** Not pertinent
6.7 **Ignition Temperature:** Not pertinent
6.8 **Electrical Hazard:** Not pertinent
6.9 **Burning Rate:** Not pertinent

7. CHEMICAL REACTIVITY

7.1 **Reactivity with Water:** No reaction
7.2 **Reactivity with Common Materials:** No reaction
7.3 **Stability During Transport:** Stable
7.4 **Neutralizing Agents for Acids and Caustics:** Not pertinent
7.5 **Polymerization:** Not pertinent
7.6 **Inhibitor of Polymerization:** Not pertinent

8. WATER POLLUTION

8.1 **Aquatic Toxicity:** 36.5 mg/l/48 hr/rainbow trout/TL_m/fresh water
8.2 **Waterfowl Toxicity:** 32 mg/kg LD_{50}
8.3 **Biological Oxygen Demand (BOD):** Data not available
8.4 **Food Chain Concentration Potential:** Data not available

9. SELECTED MANUFACTURERS

1. Mallinckrodt Chemical Works
   Industrial Chemicals Division
   Second and Mallinckrodt Streets
   P. O. Box 5439
   St. Louis, Mo. 63160
2. Chevron Chemical Co.
   940 Hensley St.
   Richmond, Calif. 94804
3. J. T. Baker Chemical Co.
   Phillipsburg, N. J. 08865

10. SHIPPING INFORMATION

10.1 **Grades or Purity:** Pure; Technical (55—98%)
10.2 **Storage Temperature:** Ambient
10.3 **Inert Atmosphere:** No requirement
10.4 **Venting:** Pressure-vacuum

11. HAZARD ASSESSMENT CODE

(See Hazard Assessment Handbook, CG 446-3)

SS

12. HAZARD CLASSIFICATIONS

12.1 **Code of Federal Regulations:** Poisonous, Class B
12.2 **NAS Hazard Rating for Bulk Water Transportation:** Not listed
12.3 **NFPA Hazard Classifications:** Not listed

13. PHYSICAL AND CHEMICAL PROPERTIES

13.1 **Physical State at 15°C and 1 atm:** Solid
13.2 **Molecular Weight:** Not pertinent
13.3 **Boiling Point at 1 atm:** Not pertinent (decomposes)
13.4 **Freezing Point:** 1,139°F = 615°C = 888°K
13.5 **Critical Temperature:** Not pertinent
13.6 **Critical Pessure:** Not pertinent
13.7 **Specific Gravity:** 1.87 at 20°C (solid)
13.8 **Liquid Surface Tension:** Not pertinent
13.9 **Liquid-Water Interfacial Tension:** Not pertinent
13.10 **Vapor (Gas) Specific Gravity:** Not pertinent
13.11 **Ratio of Specfic Heats of Vapor (Gas):** Not pertinent
13.12 **Latent Heat of Vaporization:** Not pertinent
13.13 **Heat of Combustion:** Not pertinent
13.14 **Heat of Decomposition:** Not pertinent
13.15 **Heat of Solution:** Not pertinent
13.16 **Heat of Polymerization:** Not pertinent

NOTES

SAZ	# SODIUM AZIDE

Common Synonyms	Solid	White	Odorless
	Mixes with water.		

AVOID CONTACT WITH SOLID AND DUST. KEEP PEOPLE AWAY.
Wear goggles and dust respirator.
Stay upwind. Use water spray to "knock down" dust.
Isolate and remove discharged material.
Notify local health and pollution control agencies.

Fire

Not flammable.
POISONOUS GASES MAY BE PRODUCED WHEN HEATED.
Containers may explode in fire.
Cool exposed containers with water.

Exposure

CALL FOR MEDICAL AID.

DUST
POISONOUS IF INHALED.
Move victim to fresh air.
If breathing has stopped, give artificial respiration.
If breathing is difficult, give oxygen.

SOLID
POISONOUS IF SWALLOWED.
Remove contaminated clothing and shoes.
Flush affected areas with plenty of water.
IF IN EYES, hold eyelids open and flush with plenty of water.
IF SWALLOWED and victim is CONSCIOUS, have victim drink water or milk and have victim induce vomiting.
IF SWALLOWED and victim is UNCONSCIOUS OR HAVING CONVULSIONS, do nothing except keep victim warm.

Water Pollution

HARMFUL TO AQUATIC LIFE IN VERY LOW CONCENTRATIONS.
May be dangerous if it enters water intakes.
Notify local health and wildlife officials.
Notify operators of nearby water intakes.

1. RESPONSE TO DISCHARGE

(See Response Methods Handbook, CG 446-4)

Issue warning—poison, water contaminant
Restrict access
Disperse and flush

2. LABEL

3. CHEMICAL DESIGNATIONS

3.1 **Synonyms:** Hydrazoic acid, sodium salt
3.2 **Coast Guard Compatibility Classification:** Not applicable
3.3 **Chemical Formula:** NaN_3
3.4 **IMCO/United Nations Numerical Designation:** 6.1/1687

4. OBSERVABLE CHARACTERISTICS

4.1 **Physical State (as shipped):** Solid
4.2 **Color:** White
4.3 **Odor:** None

5. HEALTH HAZARDS

5.1 **Personal Protective Equipment:** Dust mask; protective clothing; goggles.
5.2 **Symptoms Following Exposure:** Inhalation or ingestion causes dizziness, weakness, blurred vision, slight shortness of breath, and feeling of going to faint; moderate reduction of blood pressure and bradycardia. Contact with eyes or skin causes irritation.
5.3 **Treatment for Exposure:** Give oxygen if weakness, pallor, or low blood pressure is observed. INHALATION: remove victim to fresh air; enforce rest; call a doctor. EYES: flush with water for at least 15 min. SKIN: flush with water; wash with soap and water. INGESTION: give large amount of water and induce vomiting at once; get medical attention.
5.4 **Toxicity by Inhalation (Threshold Limit Value):** Data not available
5.5 **Short-Term Inhalation Limits:** Data not available
5.6 **Toxicity by Ingestion:** Grade 4; oral rat LD_{50} = 27 mg/kg (technical)
5.7 **Late Toxicity:** Potent mutagen of salmon-sperm DNA
5.8 **Vapor (Gas) Irritant Characteristics:** Data not available
5.9 **Liquid or Solid Irritant Characteristics:** Data not available
5.10 **Odor Threshold:** Not pertinent

6. FIRE HAZARDS

6.1 **Flash Point:** Not flammable
6.2 **Flammable Limits in Air:** Not flammable
6.3 **Fire Extinguishing Agents:** Not pertinent
6.4 **Fire Extinguishing Agents Not to be Used:** Not pertinent
6.5 **Special Hazards of Combustion Products:** May form toxic hydrazoic acid fumes in fire
6.6 **Behavior in Fire:** Containers may explode.
6.7 **Ignition Temperature:** Not pertinent
6.8 **Electrical Hazard:** Not pertinent
6.9 **Burning Rate:** Not pertinent

7. CHEMICAL REACTIVITY

7.1 **Reactivity with Water:** Dissolves to form an alkaline solution. The reaction is not hazardous.
7.2 **Reactivity with Common Materials:** Forms explosion-sensitive materials with some metals such as lead, silver, mercury, and copper
7.3 **Stability During Transport:** Stable unless in contact with acids
7.4 **Neutralizing Agents for Acids and Caustics:** Not pertinent
7.5 **Polymerization:** Not pertinent
7.6 **Inhibitor of Polymerization:** Not pertinent

8. WATER POLLUTION

8.1 **Aquatic Toxicity:** 1.5 ppm/24 hr/bluegill/TL_m/fresh water
8.2 **Waterfowl Toxicity:** Data not available
8.3 **Biological Oxygen Demand (BOD):** None
8.4 **Food Chain Concentration Potential:** None

9. SELECTED MANUFACTURERS

1. Fairmount Chemical Co., Inc.
   117 Blanchard Street
   Newark, N. J. 07105
2. Eastman Kodak Co.
   Eastman Organic Chemicals
   Rochester, N. Y. 14650
3. PCR, Inc.
   P. O. Box 1466
   Gainesville, Fla. 32602

10. SHIPPING INFORMATION

10.1 **Grades or Purity:** Pure: 99+%; Practical grade
10.2 **Storage Temperature:** Ambient
10.3 **Inert Atmosphere:** No requirement
10.4 **Venting:** Open

11. HAZARD ASSESSMENT CODE

(See Hazard Assessment Handbook, CG 446-3)

SS

12. HAZARD CLASSIFICATIONS

12.1 **Code of Federal Regulations:** Poisonous, Class B
12.2 **NAS Hazard Rating for Bulk Water Transportation:** Not listed
12.3 **NFPA Hazard Classifications:** Not listed

13. PHYSICAL AND CHEMICAL PROPERTIES

13.1 **Physical State at 15°C and 1 atm:** Solid
13.2 **Molecular Weight:** 65
13.3 **Boiling Point at 1 atm:** Not pertinent (decomposes)
13.4 **Freezing Point:** Not pertinent
13.5 **Critical Temperature:** Not pertinent
13.6 **Critical Pressure:** Not pertinent
13.7 **Specific Gravity:** 1.85 at 20°C (solid)
13.8 **Liquid Surface Tension:** Not pertinent
13.9 **Liquid-Water Interfacial Tension:** Not pertinent
13.10 **Vapor (Gas) Specific Gravity:** Not pertinent
13.11 **Ratio of Specific Heats of Vapor (Gas):** Not pertinent
13.12 **Latent Heat of Vaporization:** Not pertinent
13.13 **Heat of Combustion:** Not pertinent
13.14 **Heat of Decomposition:** Not pertinent
13.15 **Heat of Solution:** Not pertinent
13.16 **Heat of Polymerization:** Not pertinent

NOTES

SBS	SODIUM BISULFITE

Common Synonyms Sodium acid sulfite Sodium metabisulfite	Solid powder or granules — White — Odorless to irritating odor Sinks and mixes with water.

Avoid contact with solid and dust. Keep people away.
Isolate and remove discharged material.
Notify local health and pollution control agencies.

Fire	Not flammable.
Exposure	CALL FOR MEDICAL AID. DUST Harmful if inhaled. Move to fresh air. If breathing has stopped, give artificial respiration. If breathing is difficult, give oxygen. SOLID Irritating to skin and eyes. Harmful if swallowed. Flush affected areas with plenty of water. IF IN EYES, hold eyelids open and flush with plenty of water. IF SWALLOWED and victim is CONSCIOUS, have victim drink water or milk.
Water Pollution	Dangerous to aquatic life in high concentrations. May be dangerous if it enters water intakes. Notify local health and wildlife officials. Notify operators of nearby water intakes.

1. RESPONSE TO DISCHARGE

(See Response Methods Handbook, CG 446-4)

Issue warning—water contaminant
Disperse and flush

2. LABELS

No hazard label required by Code of Federal Regulations

3. CHEMICAL DESIGNATIONS

3.1 **Synonyms:** Sodium acid sulfite
Sodium metabisulfite
Sodium pyrosulfite

3.2 **Coast Guard Compatibility Classification:** Not applicable

3.3 **Chemical Formula:** $NaHSO_3$—$Na_2S_2O_5$

3.4 **IMCO/United Nations Numerical Designation:** Not listed

4. OBSERVABLE CHARACTERISTICS

4.1 **Physical State (as shipped):** Solid

4.2 **Color:** White

4.3 **Odor:** Pungent odor of sulfur dioxide when moist

5. HEALTH HAZARDS

5.1 **Personal Protective Equipment:** Dust mask; goggles or face shield.

5.2 **Symptoms Following Exposure:** Powder is irritating to eyes, nose, and throat and can irritate skin. Ingestion may cause irritation of stomach. Very large doses cause violent colic, diarrhea, depression, and death.

5.3 **Treatment for Exposure:** INHALATION OR INGESTION: get medical attention at once. SKIN: wash with plenty of water. EYES: flush with plenty of water for at least 15 min. and get medical attention at once.

5.4 **Toxicity by Inhalation (Threshold Limit Value):** Not pertinent

5.5 **Short-Term Inhalation Limits:** Not pertinent

5.6 **Toxicity by Ingestion:** Grade 2; LD_{50} 0.5 to 5 g/kg

5.7 **Late Toxicity:** None

5.8 **Vapor (Gas) Irritant Characteristics:** Non-volatile

5.9 **Liquid or Solid Irritant Characteristics:** Irritates skin and mucous membranes.

5.10 **Odor Threshold:** Not pertinent

6. FIRE HAZARDS

6.1 **Flash Point:** Not flammable

6.2 **Flammable Limits in Air:** Not flammable

6.3 **Fire Extinguishing Agents:** Not pertinent

6.4 **Fire Extinguishing Agents Not to be Used:** Not pertinent

6.5 **Special Hazards of Combustion Products:** Not pertinent

6.6 **Behavior in Fire:** Not pertinent

6.7 **Ignition Temperature:** Not flammable

6.8 **Electrical Hazard:** Not pertinent

6.9 **Burning Rate:** Not flammable

7. CHEMICAL REACTIVITY

7.1 **Reactivity with Water:** No reaction

7.2 **Reactivity with Common Materials:** No reaction

7.3 **Stability During Transport:** Stable

7.4 **Neutralizing Agents for Acids and Caustics:** Not pertinent

7.5 **Polymerization:** Not pertinent

7.6 **Inhibitor of Polymerization:** Not pertinent

8. WATER POLLUTION

8.1 **Aquatic Toxicity:**
240 ppm/24, 48, & 96 hr/mosquito fish/TLm/fresh water

8.2 **Waterfowl Toxicity:** Data not available

8.3 **Biological Oxygen Demand (BOD):** Reacts chemically with dissolved oxygen even in absence of seeded organisms.

8.4 **Food Chain Concentration Potential:** None

9. SELECTED MANUFACTURERS

1. Allied Chemical Corp.
Industrial Chemicals Division
Morristown, N. J. 07960
2. Monsanto Co.
Monsanto Industrial Chemicals Co.
800 North Lindbergh Blvd.
St. Louis, Mo. 63166
3. Virginia Chemicals, Inc.
3340 West Norfolk Rd.
Portsmouth, Va. 23703

10. SHIPPING INFORMATION

10.1 **Grades or Purity:** Various grades (87-100%)

10.2 **Storage Temperature:** Ambient

10.3 **Inert Atmosphere:** No requirement

10.4 **Venting:** Open

11. HAZARD ASSESSMENT CODE

(See Hazard Assessment Handbook, CG 446-3)

SS

12. HAZARD CLASSIFICATIONS

12.1 **Code of Federal Regulations:** ORM-B

12.2 **NAS Hazard Rating for Bulk Water Transportation:** Not listed

12.3 **NFPA Hazard Classifications:** Not listed

13. PHYSICAL AND CHEMICAL PROPERTIES

13.1 **Physical State at 15°C and 1 atm:** Solid

13.2 **Molecular Weight:** 104.06

13.3 **Boiling Point at 1 atm:** Decomposes

13.4 **Freezing Point:** Not pertinent

13.5 **Critical Temperature:** Not pertinent

13.6 **Critical Pressure:** Not pertinent

13.7 **Specific Gravity:** 1.48 at 20°C (solid)

13.8 **Liquid Surface Tension:** Not pertinent

13.9 **Liquid-Water Interfacial Tension:** Not pertinent

13.10 **Vapor (Gas) Specific Gravity:** Not pertinent

13.11 **Ratio of Specific Heats of Vapor (Gas):** Not pertinent

13.12 **Latent Heat of Vaporization:** Not pertinent

13.13 **Heat of Combustion:** Not pertinent

13.14 **Heat of Decomposition:** Not pertinent

13.15 **Heat of Solution:** Not pertinent

13.16 **Heat of Polymerization:** Not pertinent

NOTES

REVISED 1978

SDB	# SODIUM BORATE

Common Synonyms			
Sodium tetraborate, anhydrous Borax, anhydrous Sodium biborate Sodium pyroborate	Solid	White	Odorless
	Sinks and mixes slowly with water.		

Avoid contact with solid and dust. Keep people away.
Stop discharge if possible.
Isolate and remove discharged material.
Notify local health and pollution control agencies.

Fire

Not flammable.

Exposure

CALL FOR MEDICAL AID.

DUST
Irritating to eyes, nose and throat.
If in eyes, hold eyelids open and flush with plenty of water.

SOLID
Irritating to skin and eyes.
If swallowed will cause headache, dizziness, nausea, or vomiting.

Remove contaminated clothing and shoes.
Flush affected areas with plenty of water.
IF IN EYES, hold eyelids open and flush with plenty of water.
IF SWALLOWED and victim is CONSCIOUS, have victim drink water or milk and have victim induce vomiting.
IF SWALLOWED and victim is UNCONSCIOUS OR HAVING CONVULSIONS, do nothing except keep victim warm.

Water Pollution

Dangerous to aquatic life in high concentrations.
May be dangerous if it enters water intakes.

Notify local health and wildlife officials.
Notify operators of nearby water intakes.

1. RESPONSE TO DISCHARGE

(See Response Methods Handbook, CG 446-4)

Disperse and flush

2. LABELS

No hazard label required by Code of Federal Regulations

3. CHEMICAL DESIGNATIONS

3.1 **Synonyms:** Borax, anhydrous; Sodium biborate; Sodium pyroborate; Sodium tetraborate, anhydrous

3.2 **Coast Guard Compatibility Classification:** Not listed

3.3 **Chemical Formula:** $Na_2B_4O_7$

3.4 **IMCO/United Nations Numerical Designation:** Not listed

4. OBSERVABLE CHARACTERISTICS

4.1 **Physical State (as shipped):** Solid

4.2 **Color:** White

4.3 **Odor:** None

5. HEALTH HAZARDS

5.1 **Personal Protective Equipment:** Dust mask and goggles or face shield

5.2 **Symptoms Following Exposure:** No adverse effects from inhaling borax have been reported. Ingestion may cause acute or chronic effects; initial symptoms are nausea, vomiting, and diarrhea; these may be followed by weakness, depression, headaches, skin rashes, drying skin, cracked lips, and loss of hair; shock may follow ingestion of large doses and may interfere with breathing. Eye contact with powder or solutions may cause irritation; no chronic effects have been recognized, but continued contact should be avoided. Local skin irritation may result from contact with powder or strong solutions; the latter may cause chronic dermatitis on prolonged contact, and if skin is broken, enough boron may be absorbed to cause boron poisoning (symptoms are similar to those for ingestion).

5.3 **Treatment for Exposure:** INHALATION: move to fresh air; call physician immediately; give mouth-to-mouth resuscitation if breathing has ceased; give oxygen if authorized by physician; keep victim warm. INGESTION: get medical attention quickly; if victim is conscious, give warm salty or soapy water to induce vomiting; repeat until vomitus is clear; additional water may be given to wash out stomach. EYES: get medical attention quickly; flush with copious amounts of water for at least 15 min. (30 min. if physician is not available), holding eyelids apart. SKIN: flush with water; remove contaminated clothing under shower; do not use chemical neutralizers; get medical attention unless burn is minor.

5.4 **Toxicity by Inhalation (Threshold Limit Value):** Data not available

5.5 **Short-Term Inhalation Limits:** Data not available

5.6 **Toxicity by Ingestion:** Grade 2; LD_{50} 0.5—5 g/kg

6. FIRE HAZARDS

6.1 **Flash Point:** Not flammable

6.2 **Flammable Limits in Air:** Not flammable

6.3 **Fire Extinguishing Agents:** Not pertinent

6.4 **Fire Extinguishing Agents Not to be Used:** Not pertinent

6.5 **Special Hazards of Combustion Products:** Not pertinent

6.6 **Behavior in Fire:** Compound melts to a glassy material that may flow in large quantities and ignite combustibles elsewhere.

6.7 **Ignition Temperature:** Not pertinent

6.8 **Electrical Hazard:** Not pertinent

6.9 **Burning Rate:** Not pertinent

7. CHEMICAL REACTIVITY

7.1 **Reactivity with Water:** No reaction

7.2 **Reactivity with Common Materials:**

7.3 **Stability During Transport:** Stable

7.4 **Neutralizing Agents for Acids and Caustics:** Not pertinent

7.5 **Polymerization:** Not pertinent

7.6 **Inhibitor of Polymerization:** Not pertinent

8. WATER POLLUTION

8.1 **Aquatic Toxicity:**
3,000—3,300 ppm/*/minnow/min. lethal dose/fresh water
8,200 ppm/48 hr/mosquitofish/TL_m/fresh water
*Time period not specified.

8.2 **Waterfowl Toxicity:** Data not available

8.3 **Biological Oxygen Demand (BOD):** None

8.4 **Food Chain Concentration Potential:** None

9. SELECTED MANUFACTURERS

1. Stauffer Chemical Co.
Industrial Chemicals Div.
Westport, Conn. 06880

2. Kerr-McGee Chemical Corp.
Kerr McGee Bldg.
Oklahoma City, Okla. 73102

3. U. S. Borax and Chemical Corp.
3075 Wilshire Blvd.
Los Angeles, Calif. 90010

10. SHIPPING INFORMATION

10.1 **Grades or Purity:** In addition to anhydrous sodium borate, both pentahydrate and decahydrate are commercially available, all in technical grade. The decahydrate is also available in Radio and U.S.P. grades.

10.2 **Storage Temperature:** Ambient

10.3 **Inert Atmosphere:** No requirement

10.4 **Venting:** Open

11. HAZARD ASSESSMENT CODE

(See Hazard Assessment Handbook, CG 446-3)

SS

12. HAZARD CLASSIFICATIONS

12.1 **Code of Federal Regulations:** Not listed

12.2 **NAS Hazard Rating for Bulk Water Transportation:** Not listed

12.3 **NFPA Hazard Classifications:** Not listed

13. PHYSICAL AND CHEMICAL PROPERTIES

13.1 **Physical State at 15°C and 1 atm:** Solid

13.2 **Molecular Weight:** 201.26

13.3 **Boiling Point at 1 atm:** Not pertinent (decomposes)

13.4 **Freezing Point:** Not pertinent

13.5 **Critical Temperature:** Not pertinent

13.6 **Critical Pressure:** Not pertinent

13.7 **Specific Gravity:** 2.367 at 20°C (solid)

13.8 **Liquid Surface Tension:** Not pertinent

13.9 **Liquid-Water Interfacial Tension:** Not pertinent

13.10 **Vapor (Gas) Specific Gravity:** Not pertinent

13.11 **Ratio of Specific Heats of Vapor (Gas):** Not pertinent

13.12 **Latent Heat of Vaporization:** Not pertinent

13.13 **Heat of Combustion:** Not pertinent

13.14 **Heat of Decomposition:** Not pertinent

13.15 **Heat of Solution:** -92 Btu/lb $= -51$ cal/g $= -2.1 \times 10^5$ J/kg

13.16 **Heat of Polymerization:** Not pertinent

5. HEALTH HAZARDS (Cont'd.)

5.7 **Late Toxicity:** Data not available

5.8 **Vapor (Gas) Irritant Characteristics:** Data not available

5.9 **Liquid or Solid Irritant Characteristics:** Data not available

5.10 **Odor Threshold:** Data not available

SBH	SODIUM BOROHYDRIDE

Common Synonyms			
Borohydride	Solid powder or pellets	White	Odorless
	Sinks and mixes with water. Flammable gas is produced.		

Avoid contact with solid and dust. Keep people away.
Wear rubber overclothing (including gloves) and dust respirator.
Call fire department.
Isolate and remove discharged material.
Notify local health and pollution control agencies.

Fire	Combustible. Flammable, explosive gas may be produced on contact with metals, acids or when heated. **DO NOT USE WATER, FOAM, CARBON DIOXIDE, OR VAPORIZING LIQUIDS.** **Extinguish with dry graphite, powdered limestone, soda ash, or powdered sodium chloride.**
Exposure	**CALL FOR MEDICAL AID.** DUST Harmful if inhaled. **Move to fresh air.** **If breathing has stopped, give artificial respiration.** **If breathing is difficult, give oxygen.** SOLID Irritating to skin and eyes. Harmful if swallowed or if skin is exposed. **Flush affected areas with plenty of water.** **IF IN EYES, hold eyelids open and flush with plenty of water.** **IF SWALLOWED and victim is CONSCIOUS, have victim drink water or milk.** **DO NOT INDUCE VOMITING.**
Water Pollution	Effect of low concentrations on aquatic life is unknown. May be dangerous if it enters water intakes. **Notify local health and wildlife officials.** **Notify operators of nearby water intakes.**

1. RESPONSE TO DISCHARGE

(See Response Methods Handbook, CG 446-4)

Issue warning—flammable, corrosive
Restrict access
Should be removed

2. LABELS

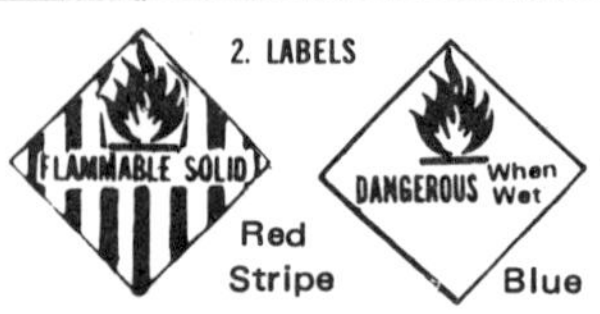

3. CHEMICAL DESIGNATIONS

3.1 **Synonyms:** No common synonyms
3.2 **Coast Guard Compatibility Classification:** Not applicable
3.3 **Chemical Formula:** $NaBH_4$
3.4 **IMCO/United Nations Numerical Designation:** 4.3/1426

4. OBSERVABLE CHARACTERISTICS

4.1 **Physical State (as shipped):** Solid or solution in caustic soda
4.2 **Color:** White
4.3 **Odor:** Odorless

5. HEALTH HAZARDS

5.1 **Personal Protective Equipment:** Goggles, rubber gloves, and protective clothing.
5.2 **Symptoms Following Exposure:** Solid irritates skin. If ingested can form large volume of gas and lead to a gas embolism.
5.3 **Treatment for Exposure:** INGESTION: do NOT induce vomiting; give dilute vinegar, lemon juice, milk, or olive oil; call a doctor. SKIN AND EYES: flood with large amount of water.
5.4 **Toxicity by Inhalation (Threshold Limit Value):** Not pertinent
5.5 **Short-Term Inhalation Limits:** Not pertinent
5.6 **Toxicity by Ingestion:** Violent reaction with acid in stomach. Considered toxic because of boron content.
5.7 **Late Toxicity:** Data not available
5.8 **Vapor (Gas) Irritant Characteristics:** Non-volatile
5.9 **Liquid or Solid Irritant Characteristics:** Irritates skin.
5.10 **Odor Threshold:** Not pertinent

6. FIRE HAZARDS

6.1 **Flash Point:** Flammable solid
6.2 **Flammable Limits in Air:** Not pertinent
6.3 **Fire Extinguishing Agents:** Graphite, limestone, soda ash, sodium chloride powders
6.4 **Fire Extinguishing Agents Not to be Used:** Water, carbon dioxide, or halogenated extinguishing agents.
6.5 **Special Hazards of Combustion Products:** Not pertinent
6.6 **Behavior in Fire:** Decomposes and produces highly flammable hydrogen gas.
6.7 **Ignition Temperature:** Not pertinent
6.8 **Electrical Hazard:** Data not available
6.9 **Burning Rate:** Not pertinent

7. CHEMICAL REACTIVITY

7.1 **Reactivity with Water:** Reacts to form flammable hydrogen gas.
7.2 **Reactivity with Common Materials:** Reacts with acids to form toxic, flammable diborane gas. Slowly corrodes glass.
7.3 **Stability During Transport:** Stable unless mixed with acids or overheated, when flammable hydrogen gas is formed.
7.4 **Neutralizing Agents for Acids and Caustics:** Caustic formed by reaction with water can be diluted and/or neutralized with acetic acid.
7.5 **Polymerization:** Not pertinent
7.6 **Inhibitor of Polymerization:** Not pertinent

8. WATER POLLUTION

8.1 **Aquatic Toxicity:** Data not available
8.2 **Waterfowl Toxicity:** Data not available
8.3 **Biological Oxygen Demand (BOD):** Data not available
8.4 **Food Chain Concentration Potential:** None

9. SELECTED MANUFACTURERS

1. Callery Chemical Co.
   Callery, Pa. 16024
2. Ventron Corp.
   Alfa Products Division
   Beverly, Mass. 01915

10. SHIPPING INFORMATION

10.1 **Grades or Purity:** 95-98% minimum purity; dry powder; pellets; 12% solution in 43% aqueous sodium hydroxide
10.2 **Storage Temperature:** Ambient
10.3 **Inert Atmosphere:** No requirement
10.4 **Venting:** Sealed containers must be stored in well-ventilated area

11. HAZARD ASSESSMENT CODE

(See Hazard Assessment Handbook, CG 446-3)

RR-C

12. HAZARD CLASSIFICATIONS

12.1 **Code of Federal Regulations:** Flammable solid
12.2 **NAS Hazard Rating for Bulk Water Transportation:** Not listed
12.3 **NFPA Hazard Classifications:** Not listed

13. PHYSICAL AND CHEMICAL PROPERTIES

13.1 **Physical State at 15°C and 1 atm:** Solid
13.2 **Molecular Weight:** 37.83
13.3 **Boiling Point at 1 atm:** Decomposes
13.4 **Freezing Point:** Not pertinent
13.5 **Critical Temperature:** Not pertinent
13.6 **Critical Pressure:** Not pertinent
13.7 **Specific Gravity:** 1.074 at 20°C (solid)
13.8 **Liquid Surface Tension:** Not pertinent
13.9 **Liquid-Water Interfacial Tension:** Not pertinent
13.10 **Vapor (Gas) Specific Gravity:** Not pertinent
13.11 **Ratio of Specific Heats of Vapor (Gas):** Not pertinent
13.12 **Latent Heat of Vaporization:** Not pertinent
13.13 **Heat of Combustion:** Not pertinent
13.14 **Heat of Decomposition:** Not pertinent
13.15 **Heat of Solution:** Not pertinent
13.16 **Heat of Polymerization:** Not pertinent

NOTES

REVISED 1978

SCD	SODIUM CACODYLATE

Common Synonyms	Solid or solution	White solid or colorless to light yellow solution	Odorless
Sodium dimethylarsenate Arsecodile Arsicodile Arsycodile Phytar	Mixes with water.		

AVOID CONTACT WITH SOLID AND DUST. KEEP PEOPLE AWAY.
Wear goggles and self-contained breathing apparatus.
Stay upwind. Use water spray to "knock down" dust.
Isolate and remove discharged material.
Notify local health and pollution control agencies.

Fire

Not flammable.
POISONOUS GASES MAY BE PRODUCED WHEN HEATED.

Exposure

CALL FOR MEDICAL AID.

VAPOR OR DUST
POISONOUS IF INHALED.
Irritating to eyes.
Move victim to fresh air.
If in eyes, hold eyelids open and flush with plenty of water.
If breathing is difficult, give oxygen.

LIQUID OR SOLID
POISONOUS IF SWALLOWED.
Irritating to skin and eyes.
Remove contaminated clothing and shoes.
Flush affected areas with plenty of water.
IF IN EYES, hold eyelids open and flush with plenty of water.
IF SWALLOWED and victim is CONSCIOUS, have victim drink water or milk and have victim induce vomiting.
IF SWALLOWED and victim is UNCONSCIOUS OR HAVING CONVULSIONS, do nothing except keep victim warm.

Water Pollution

Effect of low concentrations on aquatic life is unknown.
May be dangerous if it enters water intakes.
Notify local health and wildlife officials.
Notify operators of nearby water intakes.

1. RESPONSE TO DISCHARGE

(See Response Methods Handbook, CG 446-4)

Issue warning—poison, water contaminant
Restrict access
Disperse and flush

2. LABEL

3. CHEMICAL DESIGNATIONS

3.1 **Synonyms:** Arsecodile; Arsicodile; Arsycodile; Phytar 160; Phytar 560; Sodium dimethylarsonate
3.2 **Coast Guard Compatibility Classification:** Not applicable
3.3 **Chemical Formula:** $(CH_3)_2AsOONa$
3.4 **IMCO/United Nations Numerical Designation:** 6.1/1688

4. OBSERVABLE CHARACTERISTICS

4.1 **Physical State (as shipped):** Solid, or water solutions
4.2 **Color:** Colorless to light yellow
4.3 **Odor:** None

5. HEALTH HAZARDS

5.1 **Personal Protective Equipment:** Goggles or face shield; dust mask; rubber gloves
5.2 **Symptoms Following Exposure:** Dust may irritate eyes. Ingestion or excessive inhalation causes irritation of stomach and intestines with nausea, vomiting, diarrhea, shock, rapid pulse, coma.
5.3 **Treatment for Exposure:** INHALATION: remove victim from exposure; call physician. EYES: flush with water. SKIN: flush with water and wash well with soap and water. INGESTION: call physician; induce evacuation of intestinal tract by inducing vomiting, giving gastric lavage and a saline cathartic. Do NOT use BAL as an antidote.
5.4 **Toxicity by Inhalation (Threshold Limit Value):** Data not available
5.5 **Short-Term Inhalation Limits:** Data not available
5.6 **Toxicity by Ingestion:** Grade 2; oral LD_{50} = 2,600 mg/kg (rat)
5.7 **Late Toxicity:** Data not available
5.8 **Vapor (Gas) Irritant Characteristics:** Data not available
5.9 **Liquid or Solid Irritant Characteristics:** Data not available
5.10 **Odor Threshold:** Not pertinent

6. FIRE HAZARDS

6.1 **Flash Point:** Not flammable
6.2 **Flammable Limits in Air:** Not flammable
6.3 **Fire Extinguishing Agents:** Not pertinent
6.4 **Fire Extinguishing Agents Not to be Used:** Not pertinent
6.5 **Special Hazards of Combustion Products:** Arsenic-containing fumes are emitted.
6.6 **Behavior in Fire:** Not pertinent
6.7 **Ignition Temperature:** Not pertinent
6.8 **Electrical Hazard:** Not pertinent
6.9 **Burning Rate:** Not pertinent

7. CHEMICAL REACTIVITY

7.1 **Reactivity with Water:** No reaction
7.2 **Reactivity with Common Materials:** Corrodes common metals, but reaction is not hazardous.
7.3 **Stability During Transport:** Stable
7.4 **Neutralizing Agents for Acids and Caustics:** Not pertinent
7.5 **Polymerization:** Not pertinent
7.6 **Inhibitor of Polymerization:** Not pertinent

8. WATER POLLUTION

8.1 **Aquatic Toxicity:** Data not available
8.2 **Waterfowl Toxicity:** Data not available
8.3 **Biological Oxygen Demand (BOD):** Data not available
8.4 **Food Chain Concentration Potential:** Data not available

9. SELECTED MANUFACTURERS

The Ansul Company
One Stanton Street
Marinette, Wis. 54143

10. SHIPPING INFORMATION

10.1 **Grades or Purity:** 22—28% sodium cacodylate, 3— 5% cacodylic acid, balance inert solid (or water)
10.2 **Storage Temperature:** Ambient
10.3 **Inert Atmosphere:** No requirement
10.4 **Venting:** Open

11. HAZARD ASSESSMENT CODE

(See Hazard Assessment Handbook, CG 446-3)

SS

12. HAZARD CLASSIFICATIONS

12.1 **Code of Federal Regulations:** Poisonous, Class B
12.2 **NAS Hazard Rating for Bulk Water Transportation:** Not listed
12.3 **NFPA Hazard Classifications:** Not listed

13. PHYSICAL AND CHEMICAL PROPERTIES

13.1 **Physical State at 15°C and 1 atm:** Solid
13.2 **Molecular Weight:** 160.0
13.3 **Boiling Point at 1 atm:** Not pertinent (decomposes)
13.4 **Freezing Point:** Not pertinent
13.5 **Critical Temperature:** Not pertinent
13.6 **Critical Pressure:** Not pertinent
13.7 **Specific Gravity:** (est) >1 at 20°C (solid)
13.8 **Liquid Surface Tension:** Not pertinent
13.9 **Liquid-Water Interfacial Tension:** Not pertinent
13.10 **Vapor (Gas) Specific Gravity:** Not pertinent
13.11 **Ratio of Specific Heats of Vapor (Gas):** Not pertinent
13.12 **Latent Heat of Vaporization:** Not pertinent
13.13 **Heat of Combustion:** Not pertinent
13.14 **Heat of Decomposition:** Not pertinent
13.15 **Heat of Solution:** Not pertinent
13.16 **Heat of Polymerization:** Not pertinent

NOTES

SDC	SODIUM CHLORATE

Common Synonyms
Chlorate of soda

Solid crystals or powder — Colorless to pale yellow — Odorless

Sinks and mixes with water.

Stop discharge if possible.
Avoid contact with solid.
Isolate and remove discharged material.
Notify local health and pollution control agencies.

Fire

Not flammable.
CONTAINERS MAY EXPLODE IN FIRE.
May cause fire on contact with combustibles.
Combat adjacent fires from a safe distance or protected location.
Flood discharge area with water.
Cool exposed containers with water.
Continue cooling after fire has been extinguished.

Exposure

CALL FOR MEDICAL AID.

SOLID
Irritating to skin and eyes.
Harmful if swallowed.
Flush affected areas with plenty of water.
IF IN EYES, hold eyelids open and flush with plenty of water.
IF SWALLOWED and victim is CONSCIOUS, have victim drink water or milk and have victim induce vomiting.
IF SWALLOWED and victim is UNCONSCIOUS OR HAVING CONVULSIONS, do nothing except keep victim warm.

Water Pollution

Dangerous to aquatic life in high concentrations.
May be dangerous if it enters water intakes.
Notify local health and wildlife officials.
Notify operators of nearby water intakes.

1. RESPONSE TO DISCHARGE

(See Response Methods Handbook, CG 446-4)

Issue warning—high flammability
Should be removed
Disperse and flush

2. LABEL

Yellow

3. CHEMICAL DESIGNATIONS

3.1 **Synonyms:** Chlorate of soda
3.2 **Coast Guard Compatibility Classification:** Not applicable
3.3 **Chemical Formula:** $NaClO_3$
3.4 **IMCO/United Nations Numerical Designation:** 5.1/1495

4. OBSERVABLE CHARACTERISTICS

4.1 **Physical State (as shipped):** Solid
4.2 **Color:** Pale yellow to white
4.3 **Odor:** Odorless

5. HEALTH HAZARDS

5.1 **Personal Protective Equipment:** Clean work clothing (must be washed well with water after each exposure); rubber gloves and shoes; where dusty, goggles and an approved dust respirator. Do NOT use oils, greases, or protective creams on skin.
5.2 **Symptoms Following Exposure:** Ingestion of a toxic dose (at least 1/2 oz.) leads to severe gastroenteric pain, vomiting, and diarrhea. Possible respiratory difficulties, including failure of respiration. Kidney and liver injury may also be produced. The lethal oral dose for an adult is approximately 15 gm. Contact with eyes causes irritation.
5.3 **Treatment for Exposure:** INGESTION: induce vomiting and follow with gastric lavage, saline cathartics, fluid therapy, and oxygen. EYES: wash thoroughly with water.
5.4 **Toxicity by Inhalation (Threshold Limit Value):** Not pertinent
5.5 **Short-Term Inhalation Limits:** Not pertinent
5.6 **Toxicity by Ingestion:** Grade 3; LD_{50} 50 to 500 mg/kg
5.7 **Late Toxicity:** Data not available
5.8 **Vapor (Gas) Irritant Characteristics:** Non-volatile
5.9 **Liquid or Solid Irritant Characteristics:** Prolonged exposure to solid or dust may irritate skin.
5.10 **Odor Threshold:** Not pertinent

6. FIRE HAZARDS

6.1 **Flash Point:** Not flammable but supports combustion.
6.2 **Flammable Limits in Air:** Not pertinent
6.3 **Fire Extinguishing Agents:** Water
6.4 **Fire Extinguishing Agents Not to be Used:** Fire blankets
6.5 **Special Hazards of Combustion Products:** In fire situations oxygen may be liberated and increase the intensity of the fire.
6.6 **Behavior in Fire:** Melts, then decomposes to give oxygen gas that increases the intensity of fire. Reacts explosively, either as a solid or a liquid, with all organic matter and some metals.
6.7 **Ignition Temperature:** Not pertinent
6.8 **Electrical Hazard:** Not pertinent
6.9 **Burning Rate:** Not pertinent

7. CHEMICAL REACTIVITY

7.1 **Reactivity with Water:** No reaction
7.2 **Reactivity with Common Materials:** Chlorates are powerful oxidizing agents and can cause explosions when heated or rubbed with wood, organic matter, sulfur, and many metals. Even water solutions react in this way if stronger than 30%, especially when warm.
7.3 **Stability During Transport:** Starts at 572°F with evolution of oxygen gas. Decomposition may be self-sustaining. Oxygen increases intensity of fires.
7.4 **Neutralizing Agents for Acids and Caustics:** Not pertinent
7.5 **Polymerization:** Not pertinent
7.6 **Inhibitor of Polymerization:** Not pertinent

8. WATER POLLUTION

8.1 **Aquatic Toxicity:**
11,000 ppm/perch/threshold toxicity/fresh water
3.8 ppm/scenedesmus/threshold toxicity/fresh water
8.2 **Waterfowl Toxicity:** Data not available
8.3 **Biological Oxygen Demand (BOD):** None
8.4 **Food Chain Concentration Potential:** None

9. SELECTED MANUFACTURERS

1. Kerr-McGee Chemical Corp.
   Kerr-McGee Bldg.
   Oklahoma City, Okla. 73102
2. Occidental Petroleum Corp.
   Hooker Chemical Co.
   Industrial Chemicals Division
   Niagara Falls, N. Y. 14302
3. Penn-Olin Chemical Co.
   Calvert City, Ky. 42029

10. SHIPPING INFORMATION

10.1 **Grades or Purity:** Technical (99.5% minimum); treated (99.0% minimum)
10.2 **Storage Temperature:** Ambient
10.3 **Inert Atmosphere:** No requirement
10.4 **Venting:** Open

11. HAZARD ASSESSMENT CODE

(See Hazard Assessment Handbook, CG 446-3)

SS

12. HAZARD CLASSIFICATIONS

12.1 **Code of Federal Regulations:** Oxidizing material
12.2 **NAS Hazard Rating for Bulk Water Transportation:** Not listed
12.3 **NFPA Hazard Classifications:**

Category	Classification*	
Health Hazard (Blue)	0	1
Flammability (Red)	0	0
Reactivity (Yellow)	2	2
	oxy	oxy

*First column refers to non-fire situation

13. PHYSICAL AND CHEMICAL PROPERTIES

13.1 **Physical State at 15°C and 1 atm:** Solid
13.2 **Molecular Weight:** 106.45
13.3 **Boiling Point at 1 atm:** Decomposes
13.4 **Freezing Point:** 478°F = 248°C = 521°K
13.5 **Critical Temperature:** Not pertinent
13.6 **Critical Pressure:** Not pertinent
13.7 **Specific Gravity:** 2.49 at 15°C (solid)
13.8 **Liquid Surface Tension:** Not pertinent
13.9 **Liquid-Water Interfacial Tension:** Not pertinent
13.10 **Vapor (Gas) Specific Gravity:** Not pertinent
13.11 **Ratio of Specific Heats of Vapor (Gas):** Not pertinent
13.12 **Latent Heat of Vaporization:** Not pertinent
13.13 **Heat of Combustion:** Not pertinent
13.14 **Heat of Decomposition:** Not pertinent
13.15 **Heat of Solution:** Not pertinent
13.16 **Heat of Polymerization:** Not pertinent

NOTES

REVISED 1978

SCH

SODIUM CHROMATE

Common Synonyms			
Sodium chromate (VI) Neutral sodium chromate anhydrous	Solid	Yellow	Odorless
	Sinks and mixes with water.		

Avoid contact with solid and dust. Keep people away.
Stop discharge if possible.
Isolate and remove discharged material.
Notify local health and pollution control agencies.

Fire

Not flammable.
Will increase the intensity of a fire.
May cause fire on contact with combustibles.
Flood discharge area with water.
Cool exposed containers with water.

Exposure

CALL FOR MEDICAL AID.

DUST
Irritating to eyes, nose and throat.
If inhaled will cause coughing or difficult breathing.
If in eyes, hold eyelids open and flush with plenty of water.
If breathing has stopped, give artificial respiration.
If breathing is difficult, give oxygen.

SOLID
POISONOUS IF SWALLOWED.
Irritating to skin and eyes.
If swallowed will cause nausea or loss of consciousness.
Remove contaminated clothing and shoes.
Flush affected areas with plenty of water.
IF IN EYES, hold eyelids open and flush with plenty of water.
IF SWALLOWED and victim is CONSCIOUS, have victim drink water or milk and have victim induce vomiting.
IF SWALLOWED and victim is UNCONSCIOUS OR HAVING CONVULSIONS, do nothing except keep victim warm.

Water Pollution

Dangerous to aquatic life in high concentrations.
May be dangerous if it enters water intakes.
Notify local health and wildlife officials.
Notify operators of nearby water intakes.

1. RESPONSE TO DISCHARGE

(See Response Methods Handbook, CG 446-4)

Disperse and flush

2. LABELS

No hazard label required by Code of Federal Regulations

3. CHEMICAL DESIGNATIONS

3.1 **Synonyms:** Neutral sodium chromate, anhydrous; Sodium chromate (VI)

3.2 **Coast Guard Compatibility Classification:** Not listed

3.3 **Chemical Formula:** Na_2CrO_4

3.4 **IMCO/United Nations Numerical Designation:** Not listed

4. OBSERVABLE CHARACTERISTICS

4.1 **Physical State (as shipped):** Solid

4.2 **Color:** Yellow

4.3 **Odor:** None

5. HEALTH HAZARDS

5.1 **Personal Protective Equipment:** U. S. Bu. Mines approved respirator; rubber gloves; chemical safety goggles; rubber apron and sleeves, face shield, rubber shoes, protective clothing.

5.2 **Symptoms Following Exposure:** Inhalation causes irritation and may ulcerate mucous membranes; continued irritation of the nose may lead to perforation of the septum. Ingestion causes severe circulatory collapse and toxic nephritis; may be fatal. Contact with eyes causes severe irritation and possible conjunctivitis. Irritates skin and can cause ulcers; if skin is broken, prolonged contact may cause "chrome sores" (slow-healing, hard-rimmed ulcers), which leave the area vulnerable to infection as a secondary effect.

5.3 **Treatment for Exposure:** INHALATION: remove victim to fresh air; get medical attention. INGESTION: get immediate medical help; if vomiting is not spontaneous, give an emetic such as soapy water followed by copious water intake. EYES: immediately flush with plenty of water for at least 15 min.; consult physician promptly. SKIN: immediately flush with plenty of water for at least 15 min.; persistent dermatitis should be referred to a physician; wash contaminated skin or clothing until chromate color disappears.

5.4 **Toxicity by Inhalation (Threshold Limit Value):** 0.1 mg/m^3 (as chromic acid)

5.5 **Short-Term Inhalation Limits:** Data not available

5.6 **Toxicity by Ingestion:** Grade 3; LD_{50} 50—500 mg/kg

5.7 **Late Toxicity:** Possible lung cancer.

5.8 **Vapor (Gas) Irritant Characteristics:** Data not available

5.9 **Liquid or Solid Irritant Characteristics:** Data not available

5.10 **Odor Threshold:** Data not available

6. FIRE HAZARDS

6.1 **Flash Point:** Not flammable

6.2 **Flammable Limits in Air:** Not flammable

6.3 **Fire Extinguishing Agents:** Not pertinent

6.4 **Fire Extinguishing Agents Not to be Used:** Not pertinent

6.5 **Special Hazards of Combustion Products:** Toxic chromium oxide fumes may form in fire.

6.6 **Behavior in Fire:** May increase intensity of fire when in contact with combustible material

6.7 **Ignition Temperature:** Not pertinent

6.8 **Electrical Hazard:** Not pertinent

6.9 **Burning Rate:** Not pertinent

7. CHEMICAL REACTIVITY

7.1 **Reactivity with Water:** No reaction

7.2 **Reactivity with Common Materials:** In contact with combustible materials may cause fire.

7.3 **Stability During Transport:** Stable

7.4 **Neutralizing Agents for Acids and Caustics:** Not pertinent

7.5 **Polymerization:** Not pertinent

7.6 **Inhibitor of Polymerization:** Not pertinent

8. WATER POLLUTION

8.1 **Aquatic Toxicity:**
300 mg/l /24 hr/bluegill/TL_m/fresh water
40 60 ppm/288 hr/shore crab/toxic/salt water

8.2 **Waterfowl Toxicity:** Data not available

8.3 **Biological Oxygen Demand (BOD):** None

8.4 **Food Chain Concentration Potential:** Bioconcentrative to 2,000 fold but not likely to be a problem in a spill situation.

9. SELECTED MANUFACTURERS

1. Allied Chemical Corp.
Industrial Chemicals Div.
P. O. Box 1139R
Morristown, N. J. 07960

2. PPG Industries, Inc.
One Gateway Center
Pittsburgh, Pa. 15222

3. J. T. Baker Chemical Co.
Phillipsburg, N. J. 08865

10. SHIPPING INFORMATION

10.1 **Grades or Purity:** Reagent; Commercial; Tetrahydrate grade

10.2 **Storage Temperature:** Ambient

10.3 **Inert Atmosphere:** No requirement

10.4 **Venting:** Open

11. HAZARD ASSESSMENT CODE

(See Hazard Assessment Handbook, CG 446-3)

SS

12. HAZARD CLASSIFICATIONS

12.1 **Code of Federal Regulations:** Not listed

12.2 **NAS Hazard Rating for Bulk Water Transportation:** Not listed

12.3 **NFPA Hazard Classifications:** Not listed

13. PHYSICAL AND CHEMICAL PROPERTIES

13.1 **Physical State at 15°C and 1 atm:** Solid

13.2 **Molecular Weight:** 162

13.3 **Boiling Point at 1 atm:** Not pertinent (decomposes)

13.4 **Freezing Point:** Not pertinent

13.5 **Critical Temperature:** Not pertinent

13.6 **Critical Pressure:** Not pertinent

13.7 **Specific Gravity:** 2.723 at 25°C (solid)

13.8 **Liquid Surface Tension:** Not pertinent

13.9 **Liquid-Water Interfacial Tension:** Not pertinent

13.10 **Vapor (Gas) Specific Gravity:** Not pertinent

13.11 **Ratio of Specific Heats of Vapor (Gas):** Not pertinent

13.12 **Latent Heat of Vaporization:** Not pertinent

13.13 **Heat of Combustion:** Not pertinent

13.14 **Heat of Decomposition:** Not pertinent

13.15 **Heat of Solution:** -24.5 Btu/lb $= -13.6$ cal/g $= -0.57 \times 10^5$ J/kg

13.16 **Heat of Polymerization:** Not pertinent

NOTES

SODIUM CYANIDE

Common Synonyms	Solid granules, flakes or lumps	White	Almond odor
	Sinks and mixes with water.		

AVOID CONTACT WITH SOLID, DUST, AND SOLUTION. Keep people away.
Wear chemical protective suit with self-contained breathing apparatus.
Isolate and remove discharged material.
Notify local health and pollution control agencies.

Fire

Not flammable.
Wear chemical protective suit with self-contained breathing apparatus.

Exposure

CALL FOR MEDICAL AID.

DUST
POISONOUS IF INHALED OR IF SKIN IS EXPOSED.
Move to fresh air.
If breathing has stopped, give artificial respiration.
If breathing is difficult, give oxygen.

SOLID
POISONOUS IF SWALLOWED OR IF SKIN IS EXPOSED.
Will burn eyes.
Remove contaminated clothing and shoes.
Flush affected areas with plenty of water.
IF IN EYES, hold eyelids open and flush with plenty of water.
IF SWALLOWED and victim is CONSCIOUS, have victim drink water or milk and have victim induce vomiting.
IF SWALLOWED and victim is UNCONSCIOUS OR HAVING CONVULSIONS, do nothing except keep victim warm.

Water Pollution

HARMFUL TO AQUATIC LIFE IN VERY LOW CONCENTRATIONS.
May be dangerous if it enters water intakes.
Notify local health and wildlife officials.
Notify operators of nearby water intakes.

1. RESPONSE TO DISCHARGE

(See Response Methods Handbook, CG 446-4)

Issue warning—poison
Restrict access
Evacuate area
Chemical and physical treatment

2. LABEL

3. CHEMICAL DESIGNATIONS

3.1 **Synonyms:** Hydrocyanic acid, sodium salt
3.2 **Coast Guard Compatibility Classification:** Not applicable
3.3 **Chemical Formula:** NaCN
3.4 **IMCO/United Nations Numerical Designation:** 6.1/1689

4. OBSERVABLE CHARACTERISTICS

4.1 **Physical State (as shipped):** Solid
4.2 **Color:** White
4.3 **Odor:** Odorless when dry. When moist it has a slight odor of hydrocyanic acid.

5. HEALTH HAZARDS

5.1 **Personal Protective Equipment:** Protective gloves when handling solid sodium cyanide; rubber gloves when handling cyanide solutions (wash hands and rubber gloves thoroughly with running water after handling cyanides); U. S. Bureau of Mines approved dust respirator; approved chemical safety goggles.

5.2 **Symptoms Following Exposure:** As little as 180 milligrams is a rapidly fatal poison if ingested. Non-lethal doses may cause toxic symptoms. Strong water solutions, or the solid itself, can be absorbed by the skin and cause deep ulcers which heal slowly.

5.3 **Treatment for Exposure:** INGESTION: start treatment immediately; call a physician; carry victim to fresh air; have him lie down; keep him quiet and warm until physician arrives.
- If victim is conscious and breathing: induce vomiting by giving emetic of warm salt water (1 tablespoon salt/cup water); repeat until vomit fluid is clear; then have victim drink one pint of 1% solution of sodium thiosulfate, to be repeated in 15 min.
- If victim has stopped breathing: give artificial respiration until breathing starts.
- If victim is unconscious but breathing: give oxygen from an inhalator.

For all of above conditions, have victim breathe amyl nitrite. Break nitrite pearl in a cloth and hold lightly under victim's nose for 15 sec., repeating 5 times at about 15-sec. intervals. If necessary, repeat this procedure every 3 min. with fresh pearls until 3 or 4 have been given. (Pearls must not be over 2 years old. Avoid breathing amyl nitrite while administering it to victim.)

5.4 **Toxicity by Inhalation (Threshold Limit Value):** Not pertinent
5.5 **Short-Term Inhalation Limits:** Not pertinent
5.6 **Toxicity by Ingestion:** Grade 4; below 50 mg/kg

6. FIRE HAZARDS

6.1 **Flash Point:** Not flammable
6.2 **Flammable Limits in Air:** Not flammable
6.3 **Fire Extinguishing Agents:** Not pertinent
6.4 **Fire Extinguishing Agents Not to be Used:** Not pertinent
6.5 **Special Hazards of Combustion Products:** Not pertinent
6.6 **Behavior in Fire:** Not pertinent
6.7 **Ignition Temperature:** Not flammable
6.8 **Electrical Hazard:** Not pertinent
6.9 **Burning Rate:** Not flammable

7. CHEMICAL REACTIVITY

7.1 **Reactivity with Water:** When sodium cyanide dissolves in water, a mild reaction occurs and some poisonous hydrogen cyanide gas is released. This gas is not hazardous except in an enclosed space. If the water is acidic, however, toxic amounts of the gas will form at once.
7.2 **Reactivity with Common Materials:** No reaction
7.3 **Stability During Transport:** Stable
7.4 **Neutralizing Agents for Acids and Caustics:** Not pertinent
7.5 **Polymerization:** Not pertinent
7.6 **Inhibitor of Polymerization:** Not pertinent

8. WATER POLLUTION

8.1 **Aquatic Toxicity:**
0.15 ppm/96 hr/bluegill/TL_m/fresh water
0.25 ppm/48 hr/prawn/LC_{50}/salt water
8.2 **Waterfowl Toxicity:** Data not available
8.3 **Biological Oxygen Demand (BOD):** (theor.) 6%, 7 days
8.4 **Food Chain Concentration Potential:** None

9. SELECTED MANUFACTURERS

1. E. I. duPont de Nemours & Co., Inc.
   Electrochemicals Dept.
   Wilmington, Del. 19898
2. Mallinckrodt Chemical Works
   Industrial Chemicals Division
   2nd and Mallinckrodt Sts.
   St. Louis, Mo. 63160

10. SHIPPING INFORMATION

10.1 **Grades or Purity:** 99+%
10.2 **Storage Temperature:** Ambient
10.3 **Inert Atmosphere:** No requirement
10.4 **Venting:** Sealed containers must be stored in well-ventilated area.

11. HAZARD ASSESSMENT CODE

(See Hazard Assessment Handbook, CG 446-3)

SS

12. HAZARD CLASSIFICATIONS

12.1 **Code of Federal Regulations:** Poisonous liquid or solid, Class B
12.2 **NAS Hazard Rating for Bulk Water Transportation:** Not listed
12.3 **NFPA Hazard Classifications:**

Category	Classification
Health Hazard (Blue)	3
Flammability (Red)	0
Reactivity (Yellow)	0

13. PHYSICAL AND CHEMICAL PROPERTIES

13.1 **Physical State at 15°C and 1 atm:** Solid
13.2 **Molecular Weight:** 49.01
13.3 **Boiling Point at 1 atm:** Very high
13.4 **Freezing Point:** 1047°F = 564°C = 837°K
13.5 **Critical Temperature:** Not pertinent
13.6 **Critical Pressure:** Not pertinent
13.7 **Specific Gravity:** 1.60 at 25°C (solid)
13.8 **Liquid Surface Tension:** Not pertinent
13.9 **Liquid-Water Interfacial Tension:** Not pertinent
13.10 **Vapor (Gas) Specific Gravity:** Not pertinent
13.11 **Ratio of Specific Heats of Vapor (Gas):** Not pertinent
13.12 **Latent Heat of Vaporization:** Not pertinent
13.13 **Heat of Combustion:** Not pertinent
13.14 **Heat of Decomposition:** Not pertinent
13.15 **Heat of Solution:** Not pertinent
13.16 **Heat of Polymerization:** Not pertinent

5. HEALTH HAZARDS (Cont'd.)

5.7 **Late Toxicity:** None
5.8 **Vapor (Gas) Irritant Characteristics:** Non-volatile, but moisture in air can liberate some lethal hydrogen cyanide gas.
5.9 **Liquid or Solid Irritant Characteristics:** Fairly severe skin irritant; may cause pain and second-degree burns after a few minutes' contact.
5.10 **Odor Threshold:** Not pertinent

REVISED 1978

SDT | SODIUM DICHLORO-s-TRIAZINETRIONE

Common Synonyms			
Sodium dichloroisocyanurate	Solid	White	Bleach-like odor
	Mixes with water.		

Keep people away.
Isolate and remove discharged material.
Notify local health and pollution control agencies.

Fire

Not flammable.
May cause fire on contact with combustibles.
POISONOUS GASES MAY BE PRODUCED IN FIRE.
Wear goggles and self-contained breathing apparatus.
Flood discharge area with water.

Exposure

Call for medical aid.

DUST
Irritating to eyes, nose and throat.
If inhaled will cause coughing or difficult breathing.
Move victim to fresh air.
If breathing has stopped, give artificial respiration.
If in eyes, hold eyelids open and flush with plenty of water.
If breathing is difficult, give oxygen.

SOLID
Irritating to skin and eyes.
Harmful if swallowed.
Remove contaminated clothing and shoes.
Flush affected areas with plenty of water.
IF IN EYES, hold eyelids open and flush with plenty of water.
IF SWALLOWED and victim is CONSCIOUS, have victim drink water or milk and have victim induce vomiting.
IF SWALLOWED and victim is UNCONSCIOUS OR HAVING CONVULSIONS, do nothing except keep victim warm.

Water Pollution

Effect of low concentrations on aquatic life is unknown.
May be dangerous if it enters water intakes.
Notify local health and wildlife officials.
Notify operators of nearby water intakes.

1. RESPONSE TO DISCHARGE

(See Response Methods Handbook, CG 446-4)

Issue warning—oxidizing material, water contaminant
Restrict access
Disperse and flush

2. LABEL

Yellow

3. CHEMICAL DESIGNATIONS

3.1 **Synonyms:** Sodium dichloroisocyanurate
3.2 **Coast Guard Compatibility Classification:** Not applicable
3.3 **Chemical Formula:** $NaCl_2(NCO)_3$
3.4 **IMCO/United Nations Numerical Designation:** 5.1/—

4. OBSERVABLE CHARACTERISTICS

4.1 **Physical State (as shipped):** Solid
4.2 **Color:** White
4.3 **Odor:** Like chlorine

5. HEALTH HAZARDS

5.1 **Personal Protective Equipment:** Dust mask or chlorine-canister mask; goggles; rubber gloves and other protective clothing to prevent contact with skin
5.2 **Symptoms Following Exposure:** Dust causes sneezing and coughing, moderate irritation of the eyes, and itchiness and redness of the skin. Ingestion causes burns of mouth and stomach.
5.3 **Treatment for Exposure:** INHALATION: remove victim to fresh air. EYES: flush with water for 15 min.; call a physician. SKIN: flush with water. INGESTION: induce vomiting; call physician.
5.4 **Toxicity by Inhalation (Threshold Limit Value):** Data not available
5.5 **Short-Term Inhalation Limits:** Data not available
5.6 **Toxicity by Ingestion:** Grade 2; oral LD_{50} = 1,670 mg/kg (rat)
5.7 **Late Toxicity:** Effects unknown in experimental animals
5.8 **Vapor (Gas) Irritant Characteristics:** Data not available
5.9 **Liquid or Solid Irritant Characteristics:** Data not available
5.10 **Odor Threshold:** Data not available

6. FIRE HAZARDS

6.1 **Flash Point:** Not flammable, but contact with ordinary combustibles may cause fire.
6.2 **Flammable Limits in Air:** Not flammable
6.3 **Fire Extinguishing Agents:** Water
6.4 **Fire Extinguishing Agents Not to be Used:** Not pertinent
6.5 **Special Hazards of Combustion Products:** May form toxic chlorine and other gases in fire.
6.6 **Behavior in Fire:** Decomposition can be initiated with a heat source and can propagate throughout the mass with the evolution of dense fumes. Containers may explode when heated.
6.7 **Ignition Temperature:** Not pertinent
6.8 **Electrical Hazard:** Not pertinent
6.9 **Burning Rate:** Not pertinent

7. CHEMICAL REACTIVITY

7.1 **Reactivity with Water:** Forms a bleach solution; the reaction is not violent.
7.2 **Reactivity with Common Materials:** Contact with most foreign materials, organic matter, or easily chlorinated or oxidized materials may result in fire. Avoid contact with oil, grease, sawdust, floor sweepings, easily oxidized organics.
7.3 **Stability During Transport:** Stable if dry
7.4 **Neutralizing Agents for Acids and Caustics:** Not pertinent
7.5 **Polymerization:** Not pertinent
7.6 **Inhibitor of Polymerization:** Not pertinent

8. WATER POLLUTION

8.1 **Aquatic Toxicity:** Data not available
8.2 **Waterfowl Toxicity:** Data not available
8.3 **Biological Oxygen Demand (BOD):** None
8.4 **Food Chain Concentration Potential:** None

9. SELECTED MANUFACTURERS

FMC Corporation
Industrial Chemical Division
633 Third Avenue
New York, N. Y. 10017

10. SHIPPING INFORMATION

10.1 **Grades or Purity:** Technical: 39—60% available chlorine
10.2 **Storage Temperature:** Cool ambient
10.3 **Inert Atmosphere:** No requirement
10.4 **Venting:** Pressure-vacuum

11. HAZARD ASSESSMENT CODE

(See Hazard Assessment Handbook, CG 446-3)

SS

12. HAZARD CLASSIFICATIONS

12.1 **Code of Federal Regulations:** Oxidizing material
12.2 **NAS Hazard Rating for Bulk Water Transportation:** Not listed
12.3 **NFPA Hazard Classifications:**

Category	Classification
Health Hazard (Blue)	3
Flammability (Red)	0
Reactivity (Yellow)	2 oxy

13. PHYSICAL AND CHEMICAL PROPERTIES

13.1 **Physical State at 15°C and 1 atm:** Solid
13.2 **Molecular Weight:** 220.0
13.3 **Boiling Point at 1 atm:** Not pertinent (decomposes)
13.4 **Freezing Point:** Not pertinent
13.5 **Critical Temperature:** Not pertinent
13.6 **Critical Pressure:** Not pertinent
13.7 **Specific Gravity:** 0.96 at 20°C (solid)
13.8 **Liquid Surface Tension:** Not pertinent
13.9 **Liquid-Water Interfacial Tension:** Not pertinent
13.10 **Vapor (Gas) Specific Gravity:** Not pertinent
13.11 **Ratio of Specific Heats of Vapor (Gas):** Not pertinent
13.12 **Latent Heat of Vaporization:** Not pertinent
13.13 **Heat of Combustion:** Not pertinent
13.14 **Heat of Decomposition:** Data not available
13.15 **Heat of Solution:** Not pertinent
13.16 **Heat of Polymerization:** Not pertinent

NOTES

SCR

SODIUM DICHROMATE

Common Synonyms
Sodium bichromate

Solid crystals | Red to orange | Odorless

Sinks and mixes with water.

Avoid contact with solid and dust. Keep people away.
Isolate and remove discharged material.
Notify local health and pollution control agencies.

Fire

Not flammable.
May cause fire on contact with combustibles.

Flood spill area with water.

Exposure

CALL FOR MEDICAL AID.
DUST
Irritating to eyes, nose, and throat.
If inhaled, will cause difficult breathing.

Move to fresh air.
If breathing has stopped, give artificial respiration.
If breathing is difficult, give oxygen.
SOLID
Will burn skin and eyes.
If swallowed, will cause nausea and vomiting.

Remove contaminated clothing and shoes.
Flush affected areas with plenty of water.
IF IN EYES, hold eyelids open and flush with plenty of water.
IF SWALLOWED and victim is CONSCIOUS, have victim drink water or milk.
DO NOT INDUCE VOMITING.

Water Pollution

Dangerous to aquatic life in high concentrations.
May be dangerous if it enters water intakes.

Notify local health and wildlife officials.
Notify operators of nearby water intakes.

1. RESPONSE TO DISCHARGE

(See Response Methods Handbook, CG 446-4)

Issue warning—water contaminant
Disperse and flush

2. LABELS

No hazard label required by Code of Federal Regulations

3. CHEMICAL DESIGNATIONS

3.1 **Synonyms:** No common synonyms
3.2 **Coast Guard Compatibility Classification:** Not applicable
3.3 **Chemical Formula:** $Na_2Cr_2O_7$
3.4 **IMCO/United Nations Numerical Designation:** 9.0/1497

4. OBSERVABLE CHARACTERISTICS

4.1 **Physical State (as shipped):** Solid
4.2 **Color:** Bright orange red
4.3 **Odor:** Odorless

5. HEALTH HAZARDS

5.1 **Personal Protective Equipment:** Approved dust mask; protective gloves; goggles or face shield.
5.2 **Symptoms Following Exposure:** Inhalation of dust or mist causes respiratory irritation sometimes resembling asthma; nasal septal perforation may occur. Ingestion causes vomiting, diarrhea, and (rarely) stomach and kidney complications. Contact with eyes or skin produces local irritation; repeated skin exposure causes dermatitis.
5.3 **Treatment for Exposure:** INGESTION: have victim drink water or milk; do NOT induce vomiting; call a doctor. SKIN OR EYE CONTACT: treat like acid burns; flush eyes with water for at least 15 min.; external lesions can be scrubbed with a 2% solution of sodium thiosulfate.
5.4 **Toxicity by Inhalation (Threshold Limit Value):** Not pertinent
5.5 **Short-Term Inhalation Limits:** Data not available
5.6 **Toxicity by Ingestion:** Grade 3; LD_{50} 50 to 500 mg/kg
5.7 **Late Toxicity:** Some suggestion of lung cancer.
5.8 **Vapor (Gas) Irritant Characteristics:** Dusts or mists may cause severe irritation of eye and throat and can cause eye and lung injury. They cannot be tolerated even at low concentrations.
5.9 **Liquid or Solid Irritant Characteristics:** Severe skin irritant. Causes second- and third-degree burns on short contact and is very injurious to the eyes.
5.10 **Odor Threshold:** Odorless

6. FIRE HAZARDS

6.1 **Flash Point:** Not flammable
6.2 **Flammable Limits in Air:** Not flammable
6.3 **Fire Extinguishing Agents:** Flood with water
6.4 **Fire Extinguishing Agents Not to be Used:** Not pertinent
6.5 **Special Hazards of Combustion Products:** Not pertinent
6.6 **Behavior in Fire:** Decomposes to produce oxygen when heated. May ignite other combustibles upon contact.
6.7 **Ignition Temperature:** Not flammable
6.8 **Electrical Hazard:** Not pertinent
6.9 **Burning Rate:** Not flammable

7. CHEMICAL REACTIVITY

7.1 **Reactivity with Water:** No reaction
7.2 **Reactivity with Common Materials:** In contact with finely divided combustibles, such as sawdust, ignition may occur.
7.3 **Stability During Transport:** Stable
7.4 **Neutralizing Agents for Acids and Caustics:** Not pertinent
7.5 **Polymerization:** Not pertinent
7.6 **Inhibitor of Polymerization:** Not pertinent

8. WATER POLLUTION

8.1 **Aquatic Toxicity:** 145 ppm/24 hr/bluegill/TL_m/fresh water
8.2 **Waterfowl Toxicity:** Data not available
8.3 **Biological Oxygen Demand (BOD):** None
8.4 **Food Chain Concentration Potential:** None

9. SELECTED MANUFACTURERS

1. Diamond Shamrock Corp.
Soda Products Division
Belleville Tpke.
Kearny, N. J. 07032
2. Hercules, Inc.
Coatings and Specialty Products Dept.
910 Market St.
Wilmington, Del. 19899
3. PPG Industries, Inc.
Industrial Chemical Division
1 Gateway Center
Pittsburgh, Pa. 15222

10. SHIPPING INFORMATION

10.1 **Grades or Purity:** Technical grades: 98.8% to 99.9%; high-purity grades: 99.3%-99.9%
10.2 **Storage Temperature:** Ambient
10.3 **Inert Atmosphere:** No requirement
10.4 **Venting:** Open

11. HAZARD ASSESSMENT CODE

(See Hazard Assessment Handbook, CG 446-3)

SS

12. HAZARD CLASSIFICATIONS

12.1 **Code of Federal Regulations:** ORM-A
12.2 **NAS Hazard Rating for Bulk Water Transportation:** Not listed
12.3 **NFPA Hazard Classifications:**

Category	Classification
Health Hazard (Blue)	1
Flammability (Red)	0
Reactivity (Yellow)	1 oxy

13. PHYSICAL AND CHEMICAL PROPERTIES

13.1 **Physical State at 15°C and 1 atm:** Solid
13.2 **Molecular Weight:** 262.01
13.3 **Boiling Point at 1 atm:** Decomposes
13.4 **Freezing Point:** 675°F = 357°C = 630°K
13.5 **Critical Temperature:** Not pertinent
13.6 **Critical Pressure:** Not pertinent
13.7 **Specific Gravity:** 2.35 at 25°C (solid)
13.8 **Liquid Surface Tension:** Not pertinent
13.9 **Liquid-Water Interfacial Tension:** Not pertinent
13.10 **Vapor (Gas) Specific Gravity:** Not pertinent
13.11 **Ratio of Specific Heats of Vapor (Gas):** Not pertinent
13.12 **Latent Heat of Vaporizaton:** Not pertinent
13.13 **Heat of Combustion:** Not pertinent
13.14 **Heat of Decomposition:** Not pertinent
13.15 **Heat of Solution:** Not pertinent
13.16 **Heat of Polymerization:** Not pertinent

NOTES

REVISED 1978

SFC	SODIUM FERROCYANIDE

Common Synonyms	Solid	Yellow	Odorless
	Sinks and mixes with water.		

Avoid contact with solid.
Isolate and remove discharged material.
Notify local health and pollution control agencies.

Fire	Not flammable.
Exposure	CALL FOR MEDICAL AID. SOLID Harmful if swallowed. IF SWALLOWED and victim is CONSCIOUS, have victim drink water or milk, have victim induce vomiting. IF SWALLOWED and victim is UNCONSCIOUS OR HAVING CONVULSIONS, do nothing except keep victim warm.
Water Pollution	Effect of low concentrations on aquatic life is unknown. May be dangerous if it enters water intakes. Notify local health and wildlife officials. Notify operators of nearby water intakes.

1. RESPONSE TO DISCHARGE
(See Response Methods Handbook, CG 446-4)

Disperse and flush

2. LABELS

No hazard label required by Code of Federal Regulations

3. CHEMICAL DESIGNATIONS

3.1 **Synonyms:** No common synonyms
3.2 **Coast Guard Compatibility Classification:** Not applicable
3.3 **Chemical Formula:** $Na_4Fe(CN)_6$
3.4 **IMCO/United Nations Numerical Designation:** Not listed

4. OBSERVABLE CHARACTERISTICS

4.1 **Physical State (as shipped):** Solid
4.2 **Color:** Yellow
4.3 **Odor:** Odorless

5. HEALTH HAZARDS

5.1 **Personal Protective Equipment:** U. S. Bureau of Mines respirator for excessive dustiness; safety goggles.
5.2 **Symptoms Following Exposure:** None recorded.
5.3 **Treatment for Exposure:** None is required except for ingestion of a huge dose, in which case the stomach should be emptied by inducing vomiting.
5.4 **Toxicity by Inhalation (Threshold Limit Value):** Not pertinent
5.5 **Short-Term Inhalation Limits:** Not pertinent
5.6 **Toxicity by Ingestion:** Grade 2; LD_{50} 0.5 to 5 g/kg
5.7 **Late Toxicity:** Data not available
5.8 **Vapor (Gas) Irritant Characteristics:** Non-volatile
5.9 **Liquid or Solid Irritant Characteristics:** None
5.10 **Odor Threshold:** Not pertinent

6. FIRE HAZARDS

6.1 **Flash Point:** Not flammable
6.2 **Flammable Limits in Air:** Not flammable
6.3 **Fire Extinguishing Agents:** Not pertinent
6.4 **Fire Extinguishing Agents Not to be Used:** Not pertinent
6.5 **Special Hazards of Combustion Products:** Not pertinent
6.6 **Behavior in Fire:** Not pertinent
6.7 **Ignition Temperature:** Not flammable
6.8 **Electrical Hazard:** Not pertinent
6.9 **Burning Rate:** Not flammable

7. CHEMICAL REACTIVITY

7.1 **Reactivity with Water:** No reaction
7.2 **Reactivity with Common Materials:** No reaction
7.3 **Stability During Transport:** Stable
7.4 **Neutralizing Agents for Acids and Caustics:** Not pertinent
7.5 **Polymerization:** Not pertinent
7.6 **Inhibitor of Polymerization:** Not pertinent

8. WATER POLLUTION

8.1 **Aquatic Toxicity:**
540 mg/l /*/daphnia magna/toxic conc./fresh water
*Time period not specified
8.2 **Waterfowl Toxicity:** Data not available
8.3 **Biological Oxygen Demand (BOD):** 0 lb/lb
8.4 **Food Chain Concentration Potential:** None

9. SELECTED MANUFACTURERS

1. American Cyanamid Co.
Industrial Chemicals & Plastics Division
Wayne, N. J. 07470
2. Filo Color and Chemical Corp.
347 Madison Ave.
New York, N. Y. 10017

10. SHIPPING INFORMATION

10.1 **Grades or Purity:** Commercial
10.2 **Storage Temperature:** Ambient
10.3 **Inert Atmosphere:** No requirement
10.4 **Venting:** Open

11. HAZARD ASSESSMENT CODE
(See Hazard Assessment Handbook, CG 446-3)

SS

12. HAZARD CLASSIFICATIONS

12.1 **Code of Federal Regulations:** Not listed
12.2 **NAS Hazard Rating for Bulk Water Transportation:** Not listed
12.3 **NFPA Hazard Classifications:** Not listed

13. PHYSICAL AND CHEMICAL PROPERTIES

13.1 **Physical State at 15°C and 1 atm:** Solid
13.2 **Molecular Weight:** 484.04
13.3 **Boiling Point at 1 atm:** Decomposes
13.4 **Freezing Point:** Not pertinent
13.5 **Critical Temperature:** Not pertinent
13.6 **Critical Pressure:** Not pertinent
13.7 **Specific Gravity:** 1.458 at 25°C (solid)
13.8 **Liquid Surface Tension:** Not pertinent
13.9 **Liquid-Water Interfacial Tension:** Not pertinent
13.10 **Vapor (Gas) Specific Gravity:** Not pertinent
13.11 **Ratio of Specific Heats of Vapor (Gas):** Not pertinent
13.12 **Latent Heat of Vaporization:** Not pertinent
13.13 **Heat of Combustion:** Not pertinent
13.14 **Heat of Decomposition:** Not pertinent
13.15 **Heat of Solution:** Not pertinent
13.16 **Heat of Polymerization:** Not pertinent

NOTES

REVISED 1978

SDF	# SODIUM FLUORIDE

Common Synonyms	Solid crystals or powder White or tinted blue Odorless Sinks in water.

Isolate and remove discharged material.
Notify local health and pollution control agencies.

Fire	Not flammable.
Exposure	CALL FOR MEDICAL AID. SOLID Poisonous if swallowed. IF SWALLOWED and victim is CONSCIOUS, have victim drink water or milk.
Water Pollution	Dangerous to aquatic life in high concentrations. May be dangerous if it enters water intakes. Notify local health and wildlife officials. Notify operators of nearby water intakes.

1. RESPONSE TO DISCHARGE
(See Response Methods Handbook, CG 446-4)

Disperse and flush

2. LABELS
No hazard label required by Code of Federal Regulations

3. CHEMICAL DESIGNATIONS
3.1 **Synonyms:** No common synonyms
3.2 **Coast Guard Compatibility Classification:** Not applicable
3.3 **Chemical Formula:** NaF
3.4 **IMCO/United Nations Numerical Designation:** 6.1/1690

4. OBSERVABLE CHARACTERISTICS
4.1 **Physical State (as shipped):** Solid (powder)
4.2 **Color:** White (may be tinted Nile blue for identification purposes)
4.3 **Odor:** Odorless

5. HEALTH HAZARDS
5.1 **Personal Protective Equipment:** Goggles or face shield.
5.2 **Symptoms Following Exposure:** Ingestion may cause vomiting, abdominal pain, diarrhea, convulsions, collapse, thirst, disturbed color vision, acute toxic nephritis.
5.3 **Treatment for Exposure:** INGESTION: promptly administer gastric lavage with lime water or 1% calcium chloride solution; support respiration; call a physician.
5.4 **Toxicity by Inhalation (Threshold Limit Value):** Not pertinent
5.5 **Short-Term Inhalation Limits:** Not pertinent
5.6 **Toxicity by Ingestion:** Grade 3; LD_{50} 50 to 500 mg/kg (rabbit, rat)
5.7 **Late Toxicity:** Data not available
5.8 **Vapor (Gas) Irritant Characteristics:** Non-volatile
5.9 **Liquid or Solid Irritant Characteristics:** None
5.10 **Odor Threshold:** Not pertinent

6. FIRE HAZARDS
6.1 **Flash Point:** Not flammable
6.2 **Flammable Limits in Air:** Not flammable
6.3 **Fire Extinguishing Agents:** Not pertinent
6.4 **Fire Extinguishing Agents Not to be Used:** Not pertinent
6.5 **Special Hazards of Combustion Products:** Not pertinent
6.6 **Behavior in Fire:** Not pertinent
6.7 **Ignition Temperature:** Not flammable
6.8 **Electrical Hazard:** Not pertinent
6.9 **Burning Rate:** Not flammable

7. CHEMICAL REACTIVITY
7.1 **Reactivity with Water:** No reaction
7.2 **Reactivity with Common Materials:** No reaction
7.3 **Stability During Transport:** Stable
7.4 **Neutralizing Agents for Acids and Caustics:** Not pertinent
7.5 **Polymerization:** Not pertinent
7.6 **Inhibitor of Polymerization:** Not pertinent

8. WATER POLLUTION
8.1 **Aquatic Toxicity:**
419 ppm/96 hr/mosquito fish/TL_m/fresh water
>300 ppm/48 hr/shrimp/LC_{50}/salt water
8.2 **Waterfowl Toxicity:** Data not available
8.3 **Biological Oxygen Demand (BOD):** None
8.4 **Food Chain Concentration Potential:** None

9. SELECTED MANUFACTURERS
1. Allied Chemical Corp.
Industrial Chemicals Division
Morristown, N. J. 07960
2. Chemtech Corp.
Chemtech Products Inc.
East St. Louis, Ill. 62200
3. Olin Corp.
Chemicals Division
120 Long Ridge Rd.
Stamford, Conn. 06904

10. SHIPPING INFORMATION
10.1 **Grades or Purity:** 97.5+%
10.2 **Storage Temperature:** Ambient
10.3 **Inert Atmosphere:** No requirement
10.4 **Venting:** Open

11. HAZARD ASSESSMENT CODE
(See Hazard Assessment Handbook, CG 446-3)

SS

12. HAZARD CLASSIFICATIONS
12.1 **Code of Federal Regulations:** ORM-B
12.2 **NAS Hazard Rating for Bulk Water Transportation:** Not listed
12.3 **NFPA Hazard Classifications:**

Category	Classification
Health Hazard (Blue)	2
Flammability (Red)	0
Reactivity (Yellow)	0

13. PHYSICAL AND CHEMICAL PROPERTIES
13.1 **Physical State at 15°C and 1 atm:** Solid
13.2 **Molecular Weight:** 41.99
13.3 **Boiling Point at 1 atm:** Very high
13.4 **Freezing Point:** Not pertinent
13.5 **Critical Temperature:** Not pertinent
13.6 **Critical Pressure:** Not pertinent
13.7 **Specific Gravity:** 2.79 at 20°C (solid)
13.8 **Liquid Surface Tension:** Not pertinent
13.9 **Liquid-Water Interfacial Tension:** Not pertinent
13.10 **Vapor (Gas) Specific Gravity:** Not pertinent
13.11 **Ratio of Specific Heats of Vapor (Gas):** Not pertinent
13.12 **Latent Heat of Vaporization:** Not pertinent
13.13 **Heat of Combustion:** Not pertinent
13.14 **Heat of Decomposition:** Not pertinent
13.15 **Heat of Solution:** Not pertinent
13.16 **Heat of Polymerization:** Not pertinent

NOTES

SDH	SODIUM HYDRIDE

Common Synonyms	Powder in oil Gray Kerosene odor Reacts violently with water. Flammable gas is produced.

Avoid contact with discharged material. Keep people away.
Wear rubber overclothing (including gloves).
Shut off ignition sources and call fire department.
Isolate and remove discharged material.
Notify local health and pollution control agencies.

Fire	FLAMMABLE. MAY EXPLODE ON CONTACT WITH WATER. DO NOT USE WATER, FOAM, SODA ACID, DRY CHEMICAL OR CARBON DIOXIDE ON FIRE. Extinguish with powdered limestone, soda ash, or dry graphite.
Exposure	CALL FOR MEDICAL AID. SOLID Will burn skin and eyes. Harmful if swallowed. Remove contaminated clothing and shoes. Flush affected areas with plenty of water. IF IN EYES, hold eyelids open and flush with plenty of water. IF SWALLOWED and victim is CONSCIOUS, have victim drink water or milk. DO NOT INDUCE VOMITING.
Water Pollution	Dangerous to aquatic life in high concentrations. May be dangerous if it enters water intakes. Notify local health and wildlife officials. Notify operators of nearby water intakes.

1. RESPONSE TO DISCHARGE

(See Response Methods Handbook, CG 446-4)

Issue warning—high flammability
Restrict access
Should be removed
Chemical and physical treatment

2. LABELS

3. CHEMICAL DESIGNATIONS

3.1 **Synonyms:** No common synonyms
3.2 **Coast Guard Compatibility Classification:** Not applicable
3.3 **Chemical Formula:** NaH
3.4 **IMCO/United Nations Numerical Designation:** 4.3/1427

4. OBSERVABLE CHARACTERISTICS

4.1 **Physical State (as shipped):** Solid
4.2 **Color:** Gray
4.3 **Odor:** Odorless

5. HEALTH HAZARDS

5.1 **Personal Protective Equipment:** Face shield; rubber gloves.
5.2 **Symptoms Following Exposure:** Moisture of body converts compound to caustic soda, which irritates all tissues.
5.3 **Treatment for Exposure:** INGESTION: do NOT induce vomiting; neutralize alkali in stomach by drinking dilute vinegar, lemon juice, or orange juice; call a physician. SKIN CONTACT: brush off all particles at once and flood the affected area with water.
5.4 **Toxicity by Inhalation (Threshold Limit Value):** Not pertinent
5.5 **Short-Term Inhalation Limits:** Not pertinent
5.6 **Toxicity by Ingestion:** Data not available
5.7 **Late Toxicity:** None
5.8 **Vapor (Gas) Irritant Characteristics:** Non-volatile
5.9 **Liquid or Solid Irritant Characteristics:** Severe skin irritant. Causes second- and third-degree burns on short contact and is very injurious to the eyes.
5.10 **Odor Threshold:** Not pertinent

6. FIRE HAZARDS

6.1 **Flash Point:** Oil is flammable
6.2 **Flammable Limits in Air:** Not pertinent
6.3 **Fire Extinguishing Agents:** Powdered limestone and nitrogen-propelled dry powder.
6.4 **Fire Extinguishing Agents Not to be Used:** Water, soda acid, chemical, or carbon dioxide
6.5 **Special Hazards of Combustion Products:** Not pertinent
6.6 **Behavior in Fire:** Accidental contact with water used to extinguish surrounding fire will result in the release of hydrogen gas and possible explosion
6.7 **Ignition Temperature:** Data not available
6.8 **Electrical Hazard:** Not pertinent
6.9 **Burning Rate:** Not pertinent

7. CHEMICAL REACTIVITY

7.1 **Reactivity with Water:** Vigorous reaction with release of flammable hydrogen gas. Ignition of hydrogen is infrequent.
7.2 **Reactivity with Common Materials:** No reaction
7.3 **Stability During Transport:** Stable below 225°C
7.4 **Neutralizing Agents for Acids and Caustics:** Neutralize only when accidental reaction with water is complete. Do not neutralize the flammable solid with aqueous solutions. Spent reaction solution may be neutralized with dilute solutions of acetic acid.
7.5 **Polymerization:** Not pertinent
7.6 **Inhibitor of Polymerization:** Not pertinent

8. WATER POLLUTION

8.1 **Aquatic Toxicity:**
125 ppm/96 hr/mosquito fish/TL_m/fresh water
180 ppm/23 hr/oysters/lethal/salt water
8.2 **Waterfowl Toxicity:** Data not available
8.3 **Biological Oxygen Demand (BOD):** None
8.4 **Food Chain Concentration Potential:** None

9. SELECTED MANUFACTURERS

1. Callery Chemical Co.
   Callery, Pa. 16024
2. Ventron Corp.
   Alfa Products Division
   Beverly, Mass. 01915

10. SHIPPING INFORMATION

10.1 **Grades or Purity:** 55% plus 45% mineral oil
10.2 **Storage Temperature:** Ambient
10.3 **Inert Atmosphere:** Must be dry
10.4 **Venting:** Pressure-vacuum

11. HAZARD ASSESSMENT CODE

(See Hazard Assessment Handbook, CG 446-3)

RR

12. HAZARD CLASSIFICATIONS

12.1 **Code of Federal Regulations:** Flammable solid
12.2 **NAS Hazard Rating for Bulk Water Transportation:** Not listed
12.3 **NFPA Hazard Classifications:** Not listed

13. PHYSICAL AND CHEMICAL PROPERTIES

13.1 **Physical State at 15°C and 1 atm:** Solid
13.2 **Molecular Weight:** Not applicable
13.3 **Boiling Point at 1 atm:** Very high
13.4 **Freezing Point:** Not pertinent
13.5 **Critical Temperature:** Not pertinent
13.6 **Critical Pressure:** Not pertinent
13.7 **Specific Gravity:** Data not available
13.8 **Liquid Surface Tension:** Not pertinent
13.9 **Liquid-Water Interfacial Tension:** Not pertinent
13.10 **Vapor (Gas) Specific Gravity:** Not pertinent
13.11 **Ratio of Specific Heats of Vapor (Gas):** Not pertinent
13.12 **Latent Heat of Vaporization:** Not pertinent
13.13 **Heat of Combustion:** Not pertinent
13.14 **Heat of Decomposition:** Not pertinent
13.15 **Heat of Solution:** Not pertinent
13.16 **Heat of Polymerization:** Not pertinent

NOTES

REVISED 1978

SHS — SODIUM HYDROSULFIDE SOLUTION

Common Synonyms

Sodium bisulfide
Sodium sulfhydrate
Sodium hydrogen sulfide

Liquid — Light yellow to red — Rotten egg odor

Mixes with water.

Stop discharge if possible. Keep people away.
Isolate and remove discharged material.
Notify local health and pollution control agencies.

Fire

Not flammable.

Exposure

Call for medical aid.

LIQUID
Irritating to skin and eyes.
If swallowed will cause nausea, vomiting, or loss of consciousness.
Remove contaminated clothing and shoes.
Flush affected areas with plenty of water.
IF IN EYES, hold eyelids open and flush with plenty of water.
IF SWALLOWED and victim is CONSCIOUS, have victim drink water or milk and have victim induce vomiting.
IF SWALLOWED and victim is UNCONSCIOUS OR HAVING CONVULSIONS, do nothing except keep victim warm.

Water Pollution

Dangerous to aquatic life in high concentrations.
May be dangerous if it enters water intakes.
Notify local health and wildlife officials.
Notify operators of nearby water intakes.

1. RESPONSE TO DISCHARGE

(See Response Methods Handbook, CG 446-4)

Issue warning—water contaminant, corrosive
Restrict access
Disperse and flush

2. LABELS

No label required by Code of Federal Regulations

3. CHEMICAL DESIGNATIONS

3.1 **Synonyms:** Sodium bisulfide; Sodium hydrogen sulfide; Sodium sulfhydrate

3.2 **Coast Guard Compatibility Classification:** To be developed

3.3 **Chemical Formula:** $NaSH \cdot H_2O$

3.4 **IMCO/United Nations Numerical Designation:** Not listed

4. OBSERVABLE CHARACTERISTICS

4.1 **Physical State (as shipped):** Liquid (water solution)

4.2 **Color:** Light lemon; pale yellow; amber to dark red

4.3 **Odor:** Rotten eggs

5. HEALTH HAZARDS

5.1 **Personal Protective Equipment:** Rubber protective equipment, such as apron, boots, splash-proof goggles, gloves; canister-type respirator or self-contained breathing apparatus.

5.2 **Symptoms Following Exposure:** Inhalation of mist causes irritation of respiratory tract and possible systemic poisoning; hydrogen sulfide gas, which may be given off when acid is present, causes headache, dizziness, nausea, vomiting; continued exposure can lead to loss of consciousness, respiratory failure, and death.
Liquid causes marked eye irritation; itching, lachrymation, swelling, and corneal injury causing blurring of vision are the most common effects; exposure to light may increase the painful effects. Contact of liquid with skin causes irritation and corrosion of tissue; continued exposure may cause dermatitis. Ingestion causes severe burning and corrosion of all portions of the gastro-intestinal tract, pain in the throat and abdomen, nausea, and vomiting, followed by diarrhea. In severe cases, collapse, unconsciousness, and paralysis of respiration may be expected.

5.3 **Treatment for Exposure:** INHALATION: move victim from contaminated atmosphere; call physician; if breathing has ceased, start mouth-to-mouth resuscitation. EYES: immediately flush with large quantities of running water for a minimum of 15 min.; obtain medical attention as soon as possible; while awaiting instructions from physician, patient may be kept in a dark room and ice compresses applied to the eyes and forehead. SKIN: immediately flush affected areas with water; obtain medical attention if irritation persists. INGESTION: obtain medical attention as soon as possible; if patient is conscious, induce vomiting by giving large amounts of water or warm salty water (2 tablespoons of table salt to a pint of water); if this measure is unsuccessful, vomiting may be induced by tickling back of patient's throat with a finger. Vomiting should be encouraged until the vomitus is clear.

6. FIRE HAZARDS

6.1 **Flash Point:** Not flammable

6.2 **Flammable Limits in Air:** Not flammable

6.3 **Fire Extinguishing Agents:** Not pertinent

6.4 **Fire Extinguishing Agents Not to be Used:** Not pertinent

6.5 **Special Hazards of Combustion Products:** Not pertinent

6.6 **Behavior in Fire:** Not pertinent

6.7 **Ignition Temperature:** Not pertinent

6.8 **Electrical Hazard:** Not pertinent

6.9 **Burning Rate:** Not pertinent

7. CHEMICAL REACTIVITY

7.1 **Reactivity with Water:** No reaction

7.2 **Reactivity with Common Materials:** Corrodes most metals, but reaction is not hazardous.

7.3 **Stability During Transport:** Stable

7.4 **Neutralizing Agents for Acids and Caustics:** Flood with water.

7.5 **Polymerization:** Not pertinent

7.6 **Inhibitor of Polymerization:** Not pertinent

8. WATER POLLUTION

8.1 **Aquatic Toxicity:** 206 mg/l /96 hr/mosquito fish/TL_m/fresh water

8.2 **Waterfowl Toxicity:** Data not available

8.3 **Biological Oxygen Demand (BOD):** Data not available

8.4 **Food Chain Concentration Potential:** None

9. SELECTED MANUFACTURERS

1. Stauffer Chemical Company
Industrial Chemical Division
Westport, Conn. 06880

2. PPG Industries, Incorporated
Industrial Chemical Division
One Gateway Center
Pittsburgh, Pa. 15222

3. Chemical Products Corporation
King Philip Rd.
East Providence, R. I. 02914

10. SHIPPING INFORMATION

10.1 **Grades or Purity:** 40—70% solutions in water

10.2 **Storage Temperature:** >63°F

10.3 **Inert Atmosphere:** No requirement

10.4 **Venting:** Pressure-vacuum

11. HAZARD ASSESSMENT CODE

(See Hazard Assessment Handbook, CG 446-3)

A-P

12. HAZARD CLASSIFICATIONS

12.1 **Code of Federal Regulations:** Not listed

12.2 **NAS Hazard Rating for Bulk Water Transportation:**

Category	Rating
Fire	0
Health	
Vapor Irritant	2
Liquid or Solid Irritant	3
Poisons	4
Water Pollution	
Human Toxicity	2
Aquatic Toxicity	3
Aesthetic Effect	3
Reactivity	
Other Chemicals	3
Water	0
Self-Reaction	0

12.3 **NFPA Hazard Classifications:** Not listed

13. PHYSICAL AND CHEMICAL PROPERTIES

13.1 **Physical State at 15°C and 1 atm:** Liquid

13.2 **Molecular Weight:** Not pertinent

13.3 **Boiling Point at 1 atm:** (approx.) 212°F = 100°C = 373°K

13.4 **Freezing Point:** (approx.) 63°F = 17°C = 290°K

13.5 **Critical Temperature:** Not pertinent

13.6 **Critical Pressure:** Not pertinent

13.7 **Specific Gravity:** 1.3 at 15°C (liquid)

13.8 **Liquid Surface Tension:** Data not available

13.9 **Liquid-Water Interfacial Tension:** Not pertinent

13.10 **Vapor (Gas) Specific Gravity:** Not pertinent

13.11 **Ratio of Specific Heats of Vapor (Gas):** Not pertinent

13.12 **Latent Heat of Vaporization:** Not pertinent

13.13 **Heat of Combustion:** Not pertinent

13.14 **Heat of Decomposition:** Not pertinent

13.15 **Heat of Solution:** Not pertinent

13.16 **Heat of Polymerization:** Not pertinent

5. HEALTH HAZARDS (Con't.)

If patient is unconscious, do not give anything but ensure there is no obstruction to breathing (his tongue should be kept forward and false teeth removed). He will be less likely to aspirate vomitus if he is placed in a face-down position.

5.4 **Toxicity by Inhalation (Threshold Limit Value):** Data not available

5.5 **Short-Term Inhalation Limits:** Data not available

5.6 **Toxicity by Ingestion:** Grade 2; LD_{50} 0.5 to 5 g/kg

5.7 **Late Toxicity:** Data not available

5.8 **Vapor (Gas) Irritant Characteristics:** Vapors cause moderate irritation such that personnel will find high concentrations unpleasant. The effect is temporary.

5.9 **Liquid or Solid Irritant Characteristics:** Fairly severe skin irritant. May cause pain and second-degree burns after a few minutes' contact.

5.10 **Odor Threshold:** 0.0047 ppm

SHD	SODIUM HYDROXIDE

Common Synonyms
Caustic soda
Lye

Solid flakes or pellets — White — Odorless

Sinks and mixes with water.

Avoid contact with solid and dust. Keep people away.
Wear rubber overclothing (including gloves).
Stop discharge if possible.
Isolate and remove discharged material.
Notify local health and pollution control agencies.

Fire

Not flammable.
May cause fire on contact with combustibles.
Flammable gas may be produced on contact with metals.
Wear rubber overclothing (including gloves).
Flood discharge area with water.
Cool exposed containers with water.

Exposure

CALL FOR MEDICAL AID.

DUST
Irritating to eyes, nose and throat.
Move to fresh air.
If breathing has stopped, give artificial respiration.
If breathing is difficult, give oxygen.
IF IN EYES, hold eyelids open and flush with plenty of water.

SOLID
Will burn skin and eyes.
Harmful if swallowed.
Remove contaminated clothing and shoes.
Flush affected areas with plenty of water.
IF IN EYES, hold eyelids open and flush with plenty of water.
IF SWALLOWED and victim is CONSCIOUS, have victim drink water or milk.
DO NOT INDUCE VOMITING.

Water Pollution

Dangerous to aquatic life in high concentrations.
May be dangerous if it enters water intakes.
Notify local health and wildlife officials.
Notify operators of nearby water intakes.

1. RESPONSE TO DISCHARGE

(See Response Methods Handbook, CG 446-4)

Issue warning—corrosive
Restrict access
Disperse and flush

2. LABEL

3. CHEMICAL DESIGNATIONS

3.1 **Synonyms:** Caustic soda
3.2 **Coast Guard Compatibility Classification:** Not applicable
3.3 **Chemical Formula:** NaOH
3.4 **IMCO/United Nations Numerical Designation:** 8.0/1823

4. OBSERVABLE CHARACTERISTICS

4.1 **Physical State (as shipped):** Solid
4.2 **Color:** White
4.3 **Odor:** Odorless

5. HEALTH HAZARDS

5.1 **Personal Protective Equipment:** Chemical safety goggles; face shield; filter or dust-type respirator; rubber boots; rubber gloves.
5.2 **Symptoms Following Exposure:** Strong corrosive action on contacted tissues. INHALATION: dust may cause damage to upper respiratory tract and lung itself, producing from mild nose irritation to pneumonitis. INGESTION: severe damage to mucous membranes; severe scar formation or perforation may occur. EYE CONTACT: produces severe damage.
5.3 **Treatment for Exposure:** INHALATION: remove from exposure; support respiration; call physician. INGESTION: give water or milk followed by dilute vinegar or fruit juice; do NOT induce vomiting. SKIN: wash immediately with large quantities of water under emergency safety shower while removing clothing; continue washing until medical help arrives; call physician. EYES: irrigate immediately with copious amounts of water for at least 15 min.; call physician.
5.4 **Toxicity by Inhalation (Threshold Limit Value):** Not pertinent
5.5 **Short-Term Inhalation Limits:** Not pertinent
5.6 **Toxicity by Ingestion:** (10% solution) oral rabbit LD_{Lo} = 500 mg/kg
5.7 **Late Toxicity:** None
5.8 **Vapor (Gas) Irritant Characteristics:** Non-volatile
5.9 **Liquid or Solid Irritant Characteristics:** Severe skin irritant. Causes second- and third-degree burns on short contact and is very injurious to the eyes.
5.10 **Odor Threshold:** Not pertinent

6. FIRE HAZARDS

6.1 **Flash Point:** Not flammable
6.2 **Flammable Limits in Air:** Not flammable
6.3 **Fire Extinguishing Agents:** Not pertinent
6.4 **Fire Extinguishing Agents Not to be Used:** Not pertinent
6.5 **Special Hazards of Combustion Products:** Not pertinent
6.6 **Behavior in Fire:** Not pertinent
6.7 **Ignition Temperature:** Not flammable
6.8 **Electrical Hazard:** Not pertinent
6.9 **Burning Rate:** Not flammable

7. CHEMICAL REACTIVITY

7.1 **Reactivity with Water:** Dissolves with liberation of much heat; may steam and splatter
7.2 **Reactivity with Common Materials:** When wet, attacks metals such as aluminum, tin, lead, and zinc to produce flammable hydrogen gas.
7.3 **Stability During Transport:** Stable
7.4 **Neutralizing Agents for Acids and Caustics:** Flush with water, rinse with dilute acetic acid
7.5 **Polymerization:** Not pertinent
7.6 **Inhibitor of Polymerization:** Not pertinent

8. WATER POLLUTION

8.1 **Aquatic Toxicity:**
125 ppm/96 hr/mosquito fish/TL_m/ fresh water
180 ppm/23 hr/oysters/lethal/salt water
8.2 **Waterfowl Toxicity:** Data not available
8.3 **Biological Oxygen Demand (BOD):** None
8.4 **Food Chain Concentration Potential:** None

9. SELECTED MANUFACTURERS

1. Diamond Shamrock Corp.
   Electro Chemicals Division
   Deer Park, Tex. 77536
2. Dow Chemical Co.
   Midland, Mich. 48640
3. PPG Industries, Inc.
   Industrial Chemical Division
   Barberton, Ohio 44203

10. SHIPPING INFORMATION

10.1 **Grades or Purity:** Technical flakes; USP pellets
10.2 **Storage Temperature:** Ambient
10.3 **Inert Atmosphere:** No requirement
10.4 **Venting:** Open

11. HAZARD ASSESSMENT CODE

(See Hazard Assessment Handbook, CG 446-3)

SS

12. HAZARD CLASSIFICATIONS

12.1 **Code of Federal Regulations:** Corrosive Material
12.2 **NAS Hazard Rating for Bulk Water Transportation:** Not listed
12.3 **NFPA Hazard Classifications:**

Category	Classification
Health Hazard (Blue)	3
Flammability (Red)	0
Reactivity (Yellow)	1

13. PHYSICAL AND CHEMICAL PROPERTIES

13.1 **Physical State at 15°C and 1 atm:** Solid
13.2 **Molecular Weight:** 40.00
13.3 **Boiling Point at 1 atm:** Very high
13.4 **Freezing Point:** 604°F = 318°C = 591°K
13.5 **Critical Temperature:** Not pertinent
13.6 **Critical Pressure:** Not pertinent
13.7 **Specific Gravity:** 2.13 at 20°C (solid)
13.8 **Liquid Surface Tension:** Not pertinent
13.9 **Liquid-Water Interfacial Tension:** Not pertinent
13.10 **Vapor (Gas) Specific Gravity:** Not pertinent
13.11 **Ratio of Specific Heats of Vapor (Gas):** Not pertinent
13.12 **Latent Heat of Vaporization:** Not pertinent
13.13 **Heat of Combustion:** Not pertinent
13.14 **Heat of Decomposition:** Not pertinent
13.15 **Heat of Solution:** Not pertinent
13.16 **Heat of Polymerization:** Not pertinent

NOTES

REVISED 1978

SHC	SODIUM HYPOCHLORITE

Common Synonyms Clorox Liquid bleach	Watery liquid · Green to yellow · Bleaching liquid odor Sinks and mixes with water.

Stop discharge if possible.
Avoid contact with liquid.
Isolate and remove discharged material.
Notify local health and pollution control agencies.

Fire	Not flammable. Cool exposed containers with water.
Exposure	CALL FOR MEDICAL AID. LIQUID Irritating to skin and eyes. Harmful if swallowed. Remove contaminated clothing and shoes. Flush affected areas with plenty of water. IF IN EYES, hold eyelids open and flush with plenty of water. IF SWALLOWED and victim is CONSCIOUS, have victim drink water or milk, and have victim induce vomiting. IF SWALLOWED and victim is UNCONSCIOUS OR HAVING CONVULSIONS, do nothing except keep victim warm.
Water Pollution	Harmful to aquatic life in very low concentrations. May be dangerous if it enters water intakes. Notify local health and wildlife officials. Notify operators of nearby water intakes.

1. RESPONSE TO DISCHARGE
(See Response Methods Handbook, CG 446-4)

Issue warning—corrosive
Disperse and flush

2. LABELS

No hazard label required by Code of Federal Regulations

3. CHEMICAL DESIGNATIONS

3.1 **Synonyms:** Clorox; Liquid bleach
3.2 **Coast Guard Compatibility Classification:** Not applicable
3.3 **Chemical Formula:** $NaOCl-H_2O$
3.4 **IMCO/United Nations Numerical Designation:** 8.0/1791

4. OBSERVABLE CHARACTERISTICS

4.1 **Physical State (as shipped):** Liquid
4.2 **Color:** Green-yellow
4.3 **Odor:** Like bleach solution

5. HEALTH HAZARDS

5.1 **Personal Protective Equipment:** Rubber gloves; goggles.
5.2 **Symptoms Following Exposure:** Liquid can be irritating to skin and eyes if contact is maintained.
5.3 **Treatment for Exposure:** INGESTION: induce vomiting, give water, and repeat. SKIN: wash off contacted skin area. EYES: flush with plenty of water for 15 min. and consult a physician.
5.4 **Toxicity by Inhalation (Threshold Limit Value):** Not pertinent
5.5 **Short-Term Inhalation Limits:** Not pertinent
5.6 **Toxicity by Ingestion:** Grade 1; oral rat LD_{50} = 8.91 g/kg
5.7 **Late Toxicity:** None
5.8 **Vapor (Gas) Irritant Characteristics:** Slight irritation of eyes and mucous membranes.
5.9 **Liquid or Solid Irritant Characteristics:** Irritates eyes and skin on prolonged contact.
5.10 **Odor Threshold:** Data not available

6. FIRE HAZARDS

6.1 **Flash Point:** Not flammable
6.2 **Flammable Limits in Air:** Not flammable
6.3 **Fire Extinguishing Agents:** Not pertinent
6.4 **Fire Extinguishing Agents Not to be Used:** Not pertinent
6.5 **Special Hazards of Combustion Products:** Not pertinent
6.6 **Behavior in Fire:** May decompose, generating irritating chlorine gas.
6.7 **Ignition Temperature:** Not flammable
6.8 **Electrical Hazard:** Not pertinent
6.9 **Burning Rate:** Not flammable

7. CHEMICAL REACTIVITY

7.1 **Reactivity with Water:** No reaction
7.2 **Reactivity with Common Materials:** No reaction
7.3 **Stability During Transport:** Stable
7.4 **Neutralizing Agents for Acids and Caustics:** Destroy with sodium bisulfite or hypo and water, then neutralize with soda ash.
7.5 **Polymerization:** Not pertinent
7.6 **Inhibitor of Polymerization:** Not pertinent

8. WATER POLLUTION

8.1 **Aquatic Toxicity:** Data not available
8.2 **Waterfowl Toxicity:** Data not available
8.3 **Biological Oxygen Demand (BOD):** None
8.4 **Food Chain Concentration Potential:** None

9. SELECTED MANUFACTURERS

1. Clorox Co.
   7901 Oakport St.
   Oakland, Calif. 94621
2. Purex Corp., Ltd.
   5101 Clark Ave.
   Lakewood, Calif. 90712
3. Sobin Chemicals, Inc.
   Sobin Park
   Boston, Mass. 02210

10. SHIPPING INFORMATION

10.1 **Grades or Purity:** Several grades and concentrations, typified by ordinary household bleach.
10.2 **Storage Temperature:** Ambient
10.3 **Inert Atmosphere:** No requirement
10.4 **Venting:** Pressure-vacuum

11. HAZARD ASSESSMENT CODE
(See Hazard Assessment Handbook, CG 446-3)

A-P

12. HAZARD CLASSIFICATIONS

12.1 **Code of Federal Regulations:** Not Listed
12.2 **NAS Hazard Rating for Bulk Water Transportation:** Not listed
12.3 **NFPA Hazard Classifications:** Not listed

13. PHYSICAL AND CHEMICAL PROPERTIES

13.1 **Physical State at 15°C and 1 atm:** Liquid
13.2 **Molecular Weight:** Not applicable
13.3 **Boiling Point at 1 atm:** Decomposes
13.4 **Freezing Point:** Not pertinent
13.5 **Critical Temperature:** Not pertinent
13.6 **Critical Pressure:** Not pertinent
13.7 **Specific Gravity:** 1.06 at 20°C (liquid)
13.8 **Liquid Surface Tension:** Not pertinent
13.9 **Liquid-Water Interfacial Tension:** Not pertinent
13.10 **Vapor (Gas) Specific Gravity:** Not pertinent
13.11 **Ratio of Specific Heats of Vapor (Gas):** Not pertinent
13.12 **Latent Heat of Vaporization:** Not pertinent
13.13 **Heat of Combustion:** Not pertinent
13.14 **Heat of Decomposition:** Not pertinent
13.15 **Heat of Solution:** (est.) −90 Btu/lb = −50 cal/g = -2×10^5 J/kg
13.16 **Heat of Polymerization:** Not pertinent

NOTES

REVISED 1978

SML	SODIUM METHYLATE

Common Synonyms			
Sodium methoxide	Solid	White	Odorless
	Mixes with water.		

Stop discharge if possible. Keep people away.
Call fire department.
Isolate and remove discharged material.
Notify local health and pollution control agencies.

Fire	FLAMMABLE. Flood discharge area with water. Extinguish with dry chemicals or carbon dioxide.
Exposure	Call for medical aid. DUST Irritating to eyes, nose and throat. Move victim to fresh air. If in eyes, hold eyelids open and flush with plenty of water. If breathing has stopped, give artificial respiration. If breathing is difficult, give oxygen. SOLID Irritating to skin and eyes. Harmful if swallowed. Remove contaminated clothing and shoes. Flush affected areas with plenty of water. IF IN EYES, hold eyelids open and flush with plenty of water. IF SWALLOWED and victim is CONSCIOUS, have victim drink water or milk and have victim induce vomiting. IF SWALLOWED and victim is UNCONSCIOUS OR HAVING CONVULSIONS, do nothing except keep victim warm.
Water Pollution	Effect of low concentrations on aquatic life is unknown. May be dangerous if it enters water intakes. Notify local health and wildlife officials. Notify operators of nearby water intakes.

1. RESPONSE TO DISCHARGE

(See Response Methods Handbook, CG 446-4)

Issue warning—high flammability, corrosive
Restrict access
Disperse and flush

2. LABEL

Red Stripe

3. CHEMICAL DESIGNATIONS

3.1 **Synonyms:** Sodium methoxide
3.2 **Coast Guard Compatibility Classification:** Not applicable
3.3 **Chemical Formula:** CH_3ONa
3.4 **IMCO/United Nations Numerical Designation:** 4.3/1431

4. OBSERVABLE CHARACTERISTICS

4.1 **Physical State (as shipped):** Solid
4.2 **Color:** White
4.3 **Odor:** None

5. HEALTH HAZARDS

5.1 **Personal Protective Equipment:** Self-contained breathing apparatus; rubber gloves and apron; goggles or face shield
5.2 **Symptoms Following Exposure:** Inhalation of dust causes severe irritation of nose and throat. Contact with eyes or skin causes severe irritation and burns. Ingestion causes irritation of mouth and stomach.
5.3 **Treatment for Exposure:** *Get medical attention at once following all exposures to this compound.* INHALATION: remove victim from contamination and keep him quiet and warm. Rest is essential. Hot tea or coffee may be given as a stimulant if patient is conscious. If breathing has apparently ceased, give artificial respiration. If available, oxygen should be administered by experienced personnel. EYES: wash well with water, then with 3% boric acid solution and additional water washes. SKIN: wash well with water, then with dilute vinegar. INGESTION: if victim is conscious, induce vomiting by administering a glassful of warm water containing a teaspoon full of salt; repeat until vomit is clear, then give two teaspoons of baking soda every 15 min.; keep victim's eyes covered until all visual and retinal changes have disappeared; alert physician to possibility of methyl alcohol poisoning.
5.4 **Toxicity by Inhalation (Threshold Limit Value):** Data not available
5.5 **Short-Term Inhalation Limits:** Data not available
5.6 **Toxicity by Ingestion:** Data not available
5.7 **Late Toxicity:** Data not available
5.8 **Vapor (Gas) Irritant Characteristics:** Not pertinent
5.9 **Liquid or Solid Irritant Characteristics:** Data not available
5.10 **Odor Threshold:** Not pertinent

6. FIRE HAZARDS

6.1 **Flash Point:** Not pertinent (flammable solid)
6.2 **Flammable Limits in Air:** Not pertinent
6.3 **Fire Extinguishing Agents:** Dry chemical, inert powders such as sand or limestone, carbon dioxide
6.4 **Fire Extinguishing Agents Not to be Used:** Water, foam
6.5 **Special Hazards of Combustion Products:** Not pertinent
6.6 **Behavior in Fire:** Contact with water or foam applied to adjacent fires will produce flammable methanol.
6.7 **Ignition Temperature:** Not pertinent
6.8 **Electrical Hazard:** Not pertinent
6.9 **Burning Rate:** Not pertinent

7. CHEMICAL REACTIVITY

7.1 **Reactivity with Water:** Produces a caustic soda solution and a solution of methyl alcohol. The reaction is not violent.
7.2 **Reactivity with Common Materials:** Attacks certain plastics such as nylon and polyesters
7.3 **Stability During Transport:** Stable if dry
7.4 **Neutralizing Agents for Acids and Caustics:** Water, followed by dilute acetic acid or vinegar
7.5 **Polymerization:** Not pertinent
7.6 **Inhibitor of Polymerization:** Not pertinent

8. WATER POLLUTION

8.1 **Aquatic Toxicity:** Data not available
8.2 **Waterfowl Toxicity:** Data not available
8.3 **Biological Oxygen Demand (BOD):** Data not available
8.4 **Food Chain Concentration Potential:** None

9. SELECTED MANUFACTURERS

1. Harshaw Chemical Company
   Industrial Chemicals Department
   1945 East 97th Street
   Cleveland, Ohio 44106
2. Olin Corporation
   120 Long Ridge Road
   Stamford, Conn. 06904
3. Diamond Shamrock Chemical Co.
   Electro Chemicals Division
   P. O. Box 500
   Deer Park, Texas 77536

10. SHIPPING INFORMATION

10.1 **Grades or Purity:** 97+%
10.2 **Storage Temperature:** Ambient
10.3 **Inert Atmosphere:** Padded, dry nitrogen
10.4 **Venting:** Safety relief

11. HAZARD ASSESSMENT CODE

(See Hazard Assessment Handbook, CG 446-3)

SS

12. HAZARD CLASSIFICATIONS

12.1 **Code of Federal Regulations:** Flammable solid
12.2 **NAS Hazard Rating for Bulk Water Transportation:** Not listed
12.3 **NFPA Hazard Classifications:** Not listed

13. PHYSICAL AND CHEMICAL PROPERTIES

13.1 **Physical State at 15°C and 1 atm:** Solid
13.2 **Molecular Weight:** 54.0
13.3 **Boiling Point at 1 atm:** Not pertinent (decomposes)
13.4 **Freezing Point:** Not pertinent
13.5 **Critical Temperature:** Not pertinent
13.6 **Critical Pressure:** Not pertinent
13.7 **Specific Gravity:** >1 at 20°C (solid)
13.8 **Liquid Surface Tension:** Not pertinent
13.9 **Liquid-Water Interfacial Tension:** Not pertinent
13.10 **Vapor (Gas) Specific Gravity:** Not pertinent
13.11 **Ratio of Specific Heats of Vapor (Gas):** Not pertinent
13.12 **Latent Heat of Vaporization:** Not pertinent
13.13 **Heat of Combustion:** Data not available
13.14 **Heat of Decomposition:** Not pertinent
13.15 **Heat of Solution:** Not pertinent
13.16 **Heat of Polymerization:** Not pertinent

NOTES

SNT	SODIUM NITRITE

Common Synonyms			
Ecrinitrit Filmerine	Solid	White	Odorless
	Sinks and mixes with water.		

Avoid contact with solid and dust. Keep people away.
Stop discharge if possible.
Isolate and remove discharged material.
Notify local health and pollution control agencies.

Fire

Not flammable.
Will increase the intensity of a fire.
May cause fire on contact with combustibles.
POISONOUS GASES MAY BE PRODUCED IN FIRE.
Wear goggles and self contained breathing apparatus.
Flood discharage area with water.

Exposure

CALL FOR MEDICAL AID.

DUST
Irritating to eyes, nose and throat.
If inhaled will cause headache, difficult breathing, or loss of consciousness.
If in eyes, hold eyelids open and flush with plenty of water.
If breathing has stopped, give artificial respiration.
If breathing is difficult, give oxygen.

SOLID
POISONOUS IF SWALLOWED.
Irritating to skin and eyes.
If swallowed will cause headache, nausea, vomiting or loss of consciousness.
Remove contaminated clothing and shoes.
Flush affected areas with plenty of water.
IF IN EYES, hold eyelids open and flush with plenty of water.
IF SWALLOWED and victim is CONSCIOUS, have victim drink water or milk and have victim induce vomiting.
IF SWALLOWED and victim is UNCONSCIOUS OR HAVING CONVULSIONS, do nothing except keep victim warm.

Water Pollution

HARMFUL TO AQUATIC LIFE IN VERY LOW CONCENTRATIONS.
May be dangerous if it enters water intakes.
Notify local health and wildlife officials.
Notify operators of nearby water intakes.

1. RESPONSE TO DISCHARGE

(See Response Methods Handbook, CG 446-4)

Issue warning—oxidizing material
Restrict access
Disperse and flush

2. LABEL

Yellow

3. CHEMICAL DESIGNATIONS

3.1 **Synonyms:** Erinitrit; Filmerine
3.2 **Coast Guard Compatibility Classification:** Not listed
3.3 **Chemical Formula:** $NaNO_2$
3.4 **IMCO/United Nations Numerical Designation:** 5.1/1500

4. OBSERVABLE CHARACTERISTICS

4.1 **Physical State (as shipped):** Solid
4.2 **Color:** White
4.3 **Odor:** None

5. HEALTH HAZARDS

5.1 **Personal Protective Equipment:** Dust mask; goggles or face shield; protective gloves
5.2 **Symptoms Following Exposure:** Ingestion (or inhalation of excessive amounts of dust) causes rapid drop in blood pressure, persistent and throbbing headache, vertigo, palpitations, and visual disturbances; skin becomes flushed and sweaty, later cold and cyanotic; other symptoms include nausea, vomiting, diarrhea (sometimes), fainting, methemoglobinemia. Contact with eyes causes irritation.
5.3 **Treatment for Exposure:** INHALATION: move to fresh air; if exposure is severe, get medical attention. INGESTION: keep patient recumbent in a shock position and comfortably warm; administer gastric lavage; consult a physician. EYES or SKIN: flush with water.
5.4 **Toxicity by Inhalation (Threshold Limit Value):** Data not available
5.5 **Short-Term Inhalation Limits:** Data not available
5.6 **Toxicity by Ingestion:** Grade 3; LD_{50} 50—500 mg/kg
5.7 **Late Toxicity:** Data not available
5.8 **Vapor (Gas) Irritant Characteristics:** Data not available
5.9 **Liquid or Solid Irritant Characteristics:** Data not available
5.10 **Odor Threshold:** Data not available

6. FIRE HAZARDS

6.1 **Flash Point:** Not flammable, but may intensify fire
6.2 **Flammable Limits in Air:** Not flammable
6.3 **Fire Extinguishing Agents:** Apply plenty of water to adjacent fires. Cool exposed containers with water.
6.4 **Fire Extinguishing Agents Not to be Used:**
6.5 **Special Hazards of Combustion Products:** Toxic oxides of nitrogen may form in fires.
6.6 **Behavior in Fire:** May increase intensity of fire if in contact with combustible material. May melt and flow at elevated temperatures.
6.7 **Ignition Temperature:** Not pertinent
6.8 **Electrical Hazard:** Not pertinent
6.9 **Burning Rate:** Not pertinent

7. CHEMICAL REACTIVITY

7.1 **Reactivity with Water:** No reaction
7.2 **Reactivity with Common Materials:**
7.3 **Stability During Transport:** Stable
7.4 **Neutralizing Agents for Acids and Caustics:** Not pertinent
7.5 **Polymerization:** Not pertinent
7.6 **Inhibitor of Polymerization:** Not pertinent

8. WATER POLLUTION

8.1 **Aquatic Toxicity:**
17.1 ppm/24 hr/minnow/no effect/fresh water
7.5 ppm/48 hr/mosquitofish/TL_m/fresh water
8.2 **Waterfowl Toxicity:** Data not available
8.3 **Biological Oxygen Demand (BOD):** Data not available
8.4 **Food Chain Concentration Potential:** None

9. SELECTED MANUFACTURERS

1. Allied Chemical Corp.
   Specialty Chemicals Div.
   P. O. Box 1087R
   Morristown, N. J. 07960
2. J. T. Baker Chemical Co.
   Phillipsburg, N. J. 08865
3. Fisher Scientific Co.
   711 Forbes Ave.
   Pittsburgh, Pa. 15219

10. SHIPPING INFORMATION

10.1 **Grades or Purity:** USP; Reagent
10.2 **Storage Temperature:** Ambient
10.3 **Inert Atmosphere:** No requirement
10.4 **Venting:** Open

11. HAZARD ASSESSMENT CODE

(See Hazard Assessment Handbook, CG 446-3)

SS

12. HAZARD CLASSIFICATIONS

12.1 **Code of Federal Regulations:** Solid oxidizing material
12.2 **NAS Hazard Rating for Bulk Water Transportation:** Not listed
12.3 **NFPA Hazard Classifications:** Not listed

13. PHYSICAL AND CHEMICAL PROPERTIES

13.1 **Physical State at 15°C and 1 atm:** Solid
13.2 **Molecular Weight:** 69
13.3 **Boiling Point at 1 atm:** (decomposes) >608°F = >320°C = >593°K
13.4 **Freezing Point:** 520°F = 271°C = 544°K
13.5 **Critical Temperature:** Not pertinent
13.6 **Critical Pressure:** Not pertinent
13.7 **Specific Gravity:** 2.17 at 20°C (solid)
13.8 **Liquid Surface Tension:** Not pertinent
13.9 **Liquid-Water Interfacial Tension:** Not pertinent
13.10 **Vapor (Gas) Specific Gravity:** Not pertinent
13.11 **Ratio of Specific Heats of Vapor (Gas):** Not pertinent
13.12 **Latent Heat of Vaporization:** Not pertinent
13.13 **Heat of Combustion:** Not pertinent
13.14 **Heat of Decomposition:** Not pertinent
13.15 **Heat of Solution:** Not pertinent
13.16 **Heat of Polymerization:** Not pertinent

NOTES

SOX — SODIUM OXALATE

Common Synonyms			
Ethanedioic acid, disodium salt	Solid	White	Odorless
	Sinks and mixes slowly with water.		

Avoid contact with solid and dust. Keep people away.
Stop discharge if possible.
Isolate and remove discharged material.
Notify local health and pollution control agencies.

Fire

Not flammable.

Exposure

CALL FOR MEDICAL AID.

DUST
Irritating to eyes, nose and throat.
If inhaled will cause difficult breathing or loss of consciousness.
If in eyes, hold eyelids open and flush with plenty of water.
If breathing has stopped, give artificial respiration.
If breathing is difficult, give oxygen.

SOLID
POISONOUS IF SWALLOWED.
Irritating to skin and eyes.
If swallowed will cause nausea, vomiting or loss of consciousness.
Remove contaminated clothing and shoes.
Flush affected areas with plenty of water.
IF IN EYES, hold eyelids open and flush with plenty of water.
IF SWALLOWED and victim is CONSCIOUS, have victim drink water or milk and have victim induce vomiting.
IF SWALLOWED and victim is UNCONSCIOUS OR HAVING CONVULSIONS, do nothing except keep victim warm.

Water Pollution

Dangerous to aquatic life in high concentrations.
May be dangerous if it enters water intakes.
Notify local health and wildlife officials.
Notify operators of nearby water intakes.

1. RESPONSE TO DISCHARGE
(See Response Methods Handbook, CG 446-4)

Issue warning—water contaminant
Disperse and flush

2. LABELS

No hazard label required by Code of Federal Regulations

3. CHEMICAL DESIGNATIONS

3.1 **Synonyms:** Ethanedioic acid, disodium salt
3.2 **Coast Guard Compatibility Classification:** Not listed
3.3 **Chemical Formula:** $Na_2C_2O_4$
3.4 **IMCO/United Nations Numerical Designation:** Not listed

4. OBSERVABLE CHARACTERISTICS

4.1 **Physical State (as shipped):** Solid
4.2 **Color:** White
4.3 **Odor:** None

5. HEALTH HAZARDS

5.1 **Personal Protective Equipment:** Dust mask; goggles or face shield; rubber gloves
5.2 **Symptoms Following Exposure:** Inhalation or ingestion causes pain in throat, esophagus, and stomach; mucous membranes turn white; other symptoms include vomiting, severe purging, weak pulse, cardiovascular collapse, neuromuscular symptoms, and kidney damage. Contact with eyes or skin causes irritation.
5.3 **Treatment for Exposure:** *Act promptly!* INHALATION: move to fresh air; if exposure to dust is severe, get medical attention. INGESTION: give dilute calcium lactate, lime water, or milk; administer gastric lavage; consult physician; watch for edema of the glottis and delayed constriction of esophagus. EYES or SKIN: flush with water.
5.4 **Toxicity by Inhalation (Threshold Limit Value):** Data not available
5.5 **Short-Term Inhalation Limits:** Data not available
5.6 **Toxicity by Ingestion:** Grade 3; LD_{50} 50—500 mg/kg
5.7 **Late Toxicity:** Data not available
5.8 **Vapor (Gas) Irritant Characteristics:** Data not available
5.9 **Liquid or Solid Irritant Characteristics:** Data not available
5.10 **Odor Threshold:** Data not available

6. FIRE HAZARDS

6.1 **Flash Point:** Not flammable
6.2 **Flammable Limits in Air:** Not flammable
6.3 **Fire Extinguishing Agents:** Not pertinent
6.4 **Fire Extinguishing Agents Not to be Used:** Not pertinent
6.5 **Special Hazards of Combustion Products:** Not pertinent
6.6 **Behavior in Fire:**
6.7 **Ignition Temperature:** Not pertinent
6.8 **Electrical Hazard:** Not pertinent
6.9 **Burning Rate:** Not pertinent

7. CHEMICAL REACTIVITY

7.1 **Reactivity with Water:** No reaction
7.2 **Reactivity with Common Materials:**
7.3 **Stability During Transport:** Stable
7.4 **Neutralizing Agents for Acids and Caustics:** Not pertinent
7.5 **Polymerization:** Not pertinent
7.6 **Inhibitor of Polymerization:** Not pertinent

8. WATER POLLUTION

8.1 **Aquatic Toxicity:** 1,350 ppm/48 hr/mosquitofish/TL_m/fresh water
8.2 **Waterfowl Toxicity:** Data not available
8.3 **Biological Oxygen Demand (BOD):** 11%, 5 days
8.4 **Food Chain Concentration Potential:** None

9. SELECTED MANUFACTURERS

1. J. T. Baker Chemical Co.
   Phillipsburg, N. J. 08865
2. Fisher Scientific Co.
   711 Forbes Ave.
   Pittsburgh, Pa. 15219
3. Gallard-Schlesinger Chemical Mfg. Co.
   584 Mineola Ave.
   Carle Place, N. Y. 11514

10. SHIPPING INFORMATION

10.1 **Grades or Purity:** Reagent; Primary standard grade
10.2 **Storage Temperature:** Ambient
10.3 **Inert Atmosphere:** No requirement
10.4 **Venting:** Open

11. HAZARD ASSESSMENT CODE
(See Hazard Assessment Handbook, CG 446-3)

II-SS

12. HAZARD CLASSIFICATIONS

12.1 **Code of Federal Regulations:** Not listed
12.2 **NAS Hazard Rating for Bulk Water Transportation:** Not listed
12.3 **NFPA Hazard Classifications:** Not listed

13. PHYSICAL AND CHEMICAL PROPERTIES

13.1 **Physical State at 15°C and 1 atm:** Solid
13.2 **Molecular Weight:** 134.0
13.3 **Boiling Point at 1 atm:** Not pertinent (decomposes)
13.4 **Freezing Point:** Not pertinent
13.5 **Critical Temperature:** Not pertinent
13.6 **Critical Pressure:** Not pertinent
13.7 **Specific Gravity:** 2.27 at 20°C (solid)
13.8 **Liquid Surface Tension:** Not pertinent
13.9 **Liquid-Water Interfacial Tension:** Not pertinent
13.10 **Vapor (Gas) Specific Gravity:** Not pertinent
13.11 **Ratio of Specific Heats of Vapor (Gas):** Not pertinent
13.12 **Latent Heat of Vaporization:** Not pertinent
13.13 **Heat of Combustion:** Not pertinent
13.14 **Heat of Decomposition:** Not pertinent
13.15 **Heat of Solution:** Not pertinent
13.16 **Heat of Polymerization:** Not pertinent

NOTES

SPP SODIUM PHOSPHATE

Common Synonyms	Solid (powder or granules) White Odorless
	Sinks and mixes with water.

Avoid contact with solid and dust. Keep people away.
Stop discharge if possible.
Isolate and remove discharged material.
Notify local health and pollution control agencies.

Fire	Not flammable.
Exposure	CALL FOR MEDICAL AID. DUST Irritating to eyes, nose and throat. If inhaled will cause coughing or difficult breathing. If in eyes, hold eyelids open and flush with plenty of water. If breathing has stopped, give artificial respiration. If breathing is difficult, give oxygen. SOLID Irritating to skin and eyes. If swallowed will cause nausea and vomiting. Remove contaminated clothing and shoes. Flush affected areas with plenty of water. IF IN EYES, hold eyelids open and flush with plenty of water. IF SWALLOWED and victim is CONSCIOUS, have victim drink water or milk and have victim induce vomiting. IF SWALLOWED and victim is UNCONSCIOUS OR HAVING CONVULSIONS, do nothing except keep victim warm.
Water Pollution	Dangerous to aquatic life in high concentrations. May be dangerous if it enters water intakes. Notify local health and wildlife officials. Notify operators of nearby water intakes.

1. RESPONSE TO DISCHARGE
(See Response Methods Handbook, CG 446-4)

Disperse and flush

2. LABELS

No hazard label required by Code of Federal Regulations

3. CHEMICAL DESIGNATIONS

3.1 **Synonyms:** "Sodium phosphate" is generic term and includes the following: (1) monosodium phosphate (MSP; sodium phosphate, monobasic), (2) disodium phosphate (DSP; sodium phosphate, dibasic), (3) trisodium phosphate (TSP; sodium phosphate, tribasic), (4) sodium acid pyrophosphate (ASPP; SAPP; disodium dihydrogen pyrophosphate), (5) tetrasodium pyrophosphate (TSPP),

(Continued on page 4)

4. OBSERVABLE CHARACTERISTICS

4.1 **Physical State (as shipped):** Granular or powdered solid; some may appear glassy.
4.2 **Color:** White
4.3 **Odor:** None

5. HEALTH HAZARDS

5.1 **Personal Protective Equipment:** U. S. Bu. Mines toxic dust mask; protective gloves; chemical-type goggles; full-cover clothing
5.2 **Symptoms Following Exposure:** Inhalation of heavy dust may irritate nose and throat. Ingestion may injure mouth, throat, and gastrointestinal tract, resulting in nausea, vomiting, cramps and diarrhea; pain and burning in mouth may occur. Contact with eyes produces local irritation; can lead to chronic damage. Contact with skin produces local irritation; repeated or prolonged contact can lead to dermatitis.
5.3 **Treatment for Exposure:** If the following measures do not eliminate the symptoms, see a physician. INHALATION: give large amounts of water or warm salty water to induce vomiting; repeat until vomitus is clear; milk, eggs, or olive oil may then be given to soothe stomach. EYES: immediately flush with large amounts of water for at least 15 min., holding eyelids apart to ensure flushing of entire surface; avoid chemical neutralizers. SKIN: flush with water; avoid chemical neutralizers.
5.4 **Toxicity by Inhalation (Threshold Limit Value):** Data not available
5.5 **Short-Term Inhalation Limits:** Data not available
5.6 **Toxicity by Ingestion:** Data not available
5.7 **Late Toxicity:** Data not available
5.8 **Vapor (Gas) Irritant Characteristics:** Data not available
5.9 **Liquid or Solid Irritant Characteristics:** Data not available
5.10 **Odor Threshold:** Data not available

6. FIRE HAZARDS

6.1 **Flash Point:** Not flammable
6.2 **Flammable Limits in Air:** Not flammable
6.3 **Fire Extinguishing Agents:** Not pertinent
6.4 **Fire Extinguishing Agents Not to be Used:** Not pertinent
6.5 **Special Hazards of Combustion Products:** Not pertinent
6.6 **Behavior in Fire:** May melt with loss of steam
6.7 **Ignition Temperature:** Not pertinent
6.8 **Electrical Hazard:** Not pertinent
6.9 **Burning Rate:** Not pertinent

7. CHEMICAL REACTIVITY

7.1 **Reactivity with Water:** All dissolve readily. MSP and ASPP form weakly acidic solutions; TSP forms strong caustic solution, similar to soda lye; TSPP forms weakly alkaline solution.
7.2 **Reactivity with Common Materials:** When wet, mild steel or brass may be corroded by MSP, ASPP, and TSP. The others are not considered corrosive.
7.3 **Stability During Transport:** All forms of sodium phosphate are stable. TSP tends to pick up moisture from air and form a hard cake.
7.4 **Neutralizing Agents for Acids and Caustics:** For those sodium phosphates that form acidic or basic solutions, dilution with water removes hazard.
7.5 **Polymerization:** Not pertinent
7.6 **Inhibitor of Polymerization:** Not pertinent

8. WATER POLLUTION

8.1 **Aquatic Toxicity:** 126 ppm/72 hr/daphnia magna/TL_m/fresh water
8.2 **Waterfowl Toxicity:** Data not available
8.3 **Biological Oxygen Demand (BOD):** None
8.4 **Food Chain Concentration Potential:** None

9. SELECTED MANUFACTURERS

1. Stauffer Chemical Co. Industrial Chemical Division Westport, Conn. 06880
2. FMC Corporation Inorganic Chemicals Div. 633 Third Avenue New York, N. Y. 10017
3. Monsanto Company 800 North Lindbergh Blvd. St. Louis, Mo. 63166

10. SHIPPING INFORMATION

10.1 **Grades or Purity:** All are available in Technical Grade, some in Food Grade and Reagent Grade. Some are available as hydrates as well as anhydrous forms.
10.2 **Storage Temperature:** Ambient
10.3 **Inert Atmosphere:** No requirement
10.4 **Venting:** Open

11. HAZARD ASSESSMENT CODE
(See Hazard Assessment Handbook, CG 446-3)

SS

12. HAZARD CLASSIFICATIONS

12.1 **Code of Federal Regulations:** Not listed
12.2 **NAS Hazard Rating for Bulk Water Transportation:** Not listed
12.3 **NFPA Hazard Classifications:** Not listed

3. CHEMICAL DESIGNATIONS (Cont'd.)

(6) sodium metaphosphate (insoluble sodium metaphosphate), (7) sodium trimetaphosphate, (8) sodium hexametaphosphate, and (9) sodium tripolyphosphate (STPP; TPP).
3.2 **Coast Guard Compatibility Classification:** Not listed
3.3 **Chemical Formula:** (1) NaH_2PO_4; (2) Na_2HPO_4; (3) Na_3PO_4; (4) $Na_2H_2P_2O_7$; (5) $Na_4P_2O_7$; (6) $(NaPO_3)_n$; (7) $(NaPO_3)_3$; (8) $(NaPO_3)_n \cdot NaO$; (9) $Na_5P_3O_{10}$
3.4 **IMCO/United Nations Numerical Designation:** Not listed

13. PHYSICAL AND CHEMICAL PROPERTIES

13.1 **Physical State at 15°C and 1 atm:** Solid
13.2 **Molecular Weight:** Values for anhydrous salts run from 120 to high polymer values.
13.3 **Boiling Point at 1 atm:** Not pertinent (decomposes)
13.4 **Freezing Point:** Not pertinent
13.5 **Critical Temperature:** Not pertinent
13.6 **Critical Pressure:** Not pertinent
13.7 **Specific Gravity:** 1.8—2.5 at 25°C (solid)
13.8 **Liquid Surface Tension:** Not pertinent
13.9 **Liquid-Water Interfacial Tension:** Not pertinent
13.10 **Vapor (Gas) Specific Gravity:** Not pertinent
13.11 **Ratio of Specific Heats of Vapor (Gas):** Not pertinent
13.12 **Latent Heat of Vaporization:** Not pertinent
13.13 **Heat of Combustion:** Not pertinent
13.14 **Heat of Decomposition:** Not pertinent
13.15 **Heat of Solution:** +83 to −81 Btu/lb = +46 to −45 cal/g = $+1.93$ to -1.88×10^5 J/kg
13.16 **Heat of Polymerization:** Not pertinent

NOTES

SSC SODIUM SILICATE

Common Synonyms	Thick liquid	Colorless	Odorless
Water glass Soluble glass	Sinks and mixes with water.		

Stop discharge if possible.
Isolate and remove discharged material.
Notify local health and pollution control agencies.

Fire	Not flammable.
Exposure	CALL FOR MEDICAL AID. LIQUID Harmful if swallowed. IF SWALLOWED and victim is CONSCIOUS, have victim drink water or milk. DO NOT INDUCE VOMITING.
Water Pollution	Dangerous to aquatic life in high concentrations. May be dangerous if it enters water intakes. Notify local health and wildlife officials. Notify operators of nearby water intakes.

1. RESPONSE TO DISCHARGE

(See Response Methods Handbook, CG 446-4)

Disperse and flush

2. LABELS

No hazard label required by Code of Federal Regulations

3. CHEMICAL DESIGNATIONS

3.1 **Synonyms:** Water glass
Soluble glass

3.2 **Coast Guard Compatibility Classification:** Not applicable

3.3 **Chemical Formula:** $Na_2SiO_3—Na_4SiO_4—H_2O$

3.4 **IMCO/United Nations Numerical Designation:** Not listed

4. OBSERVABLE CHARACTERISTICS

4.1 **Physical State (as shipped):** High-viscosity liquid

4.2 **Color:** Colorless

4.3 **Odor:** Odorless

5. HEALTH HAZARDS

5.1 **Personal Protective Equipment:** Goggles or face shield.

5.2 **Symptoms Following Exposure:** If large doses are ingested, some irritation of mucous membranes may occur, similar to that caused by caustic soda solution.

5.3 **Treatment for Exposure:** INGESTION (large doses): give water or milk; do NOT induce vomiting.

5.4 **Toxicity by Inhalation (Threshold Limit Value):** Not pertinent

5.5 **Short-Term Inhalation Limits:** Not pertinent

5.6 **Toxicity by Ingestion:** Grade 2; 0.5 to 5 g/kg (human)

5.7 **Late Toxicity:** None

5.8 **Vapor (Gas) Irritant Characteristics:** Non-volatile

5.9 **Liquid or Solid Irritant Characteristics:** None

5.10 **Odor Threshold:** Not pertinent

6. FIRE HAZARDS

6.1 **Flash Point:** Not flammable

6.2 **Flammable Limits in Air:** Not flammable

6.3 **Fire Extinguishing Agents:** Not pertinent

6.4 **Fire Extinguishing Agents Not to be Used:** Not pertinent

6.5 **Special Hazards of Combustion Products:** Not pertinent

6.6 **Behavior in Fire:** Not pertinent

6.7 **Ignition Temperature:** Not flammable

6.8 **Electrical Hazard:** Not pertinent

6.9 **Burning Rate:** Not flammable

7. CHEMICAL REACTIVITY

7.1 **Reactivity with Water:** No reaction

7.2 **Reactivity with Common Materials:** No reaction

7.3 **Stability During Transport:** Stable

7.4 **Neutralizing Agents for Acids and Caustics:** Not pertinent

7.5 **Polymerization:** Not pertinent

7.6 **Inhibitor of Polymerization:** Not pertinent

8. WATER POLLUTION

8.1 **Aquatic Toxicity:**
2320 ppm/96 hr/mosquito fish/TL_m/fresh water

8.2 **Waterfowl Toxicity:** Data not available

8.3 **Biological Oxygen Demand (BOD):** None

8.4 **Food Chain Concentration Potential:** None

9. SELECTED MANUFACTURERS

1. Diamond Shamrock Chemical Co.
1100 Superior Ave.
Cleveland, Ohio 44114
2. E. I. duPont de Nemours & Co., Inc.
Industrial and Biochemicals Dept.
Wilmington, Del. 19898
3. Philadelphia Quartz Co.
Public Ledger Bldg.
Philadelphia, Pa. 19106

10. SHIPPING INFORMATION

10.1 **Grades or Purity:** A wide variety of grades, which differ in concentration of sodium silicate in water, in specific gravity, and in viscosity.

10.2 **Storage Temperature:** Ambient

10.3 **Inert Atmosphere:** No requirement

10.4 **Venting:** Open

11. HAZARD ASSESSMENT CODE

(See Hazard Assessment Handbook, CG 446-3)

A-P

12. HAZARD CLASSIFICATIONS

12.1 **Code of Federal Regulations:** Not listed

12.2 **NAS Hazard Rating for Bulk Water Transportation:** Not listed

12.3 **NFPA Hazard Classifications:** Not listed

13. PHYSICAL AND CHEMICAL PROPERTIES

13.1 **Physical State at 15°C and 1 atm:** Liquid

13.2 **Molecular Weight:** Not applicable

13.3 **Boiling Point at 1 atm:** Decomposes

13.4 **Freezing Point:** Not pertinent

13.5 **Critical Temperature:** Not pertinent

13.6 **Critical Pressure:** Not pertinent

13.7 **Specific Gravity:** 1.1—1.7 at 20°C (liquid)

13.8 **Liquid Surface Tension:** Not pertinent

13.9 **Liquid-Water Interfacial Tension:** Not pertinent

13.10 **Vapor (Gas) Specific Gravity:** Not pertinent

13.11 **Ratio of Specific Heats of Vapor (Gas):** Not pertinent

13.12 **Latent Heat of Vaporization:** Not pertinent

13.13 **Heat of Combustion:** Not pertinent

13.14 **Heat of Decomposition:** Not pertinent

13.15 **Heat of Solution:** (est.) −20 Btu/lb = −10 cal/g = -0.4×10^5 J/kg

13.16 **Heat of Polymerization:** Not pertinent

NOTES

SFR	SODIUM SILICOFLUORIDE

Common Synonyms			
Sodium fluosilicate Sodium hexafluorosilicate Salufer	Solid	White	Odorless
	Sinks in water.		

Avoid contact with solid and dust. Keep people away.
Stop discharge if possible.
Isolate and remove discharged material.
Notify local health and pollution control agencies.

Fire	Not flammable.
Exposure	CALL FOR MEDICAL AID. DUST Irritating to eyes, nose and throat. If inhaled will cause coughing, or difficult breathing. If in eyes, hold eyelids open and flush with plenty of water. If breathing has stopped, give artificial respiration. If breathing is difficult, give oxygen. SOLID POISONOUS IF SWALLOWED. Will burn skin and eyes. If swallowed will cause nausea, vomiting or loss of consciousness. Remove contaminated clothing and shoes. Flush affected areas with plenty of water. IF IN EYES, hold eyelids open and flush with plenty of water. IF SWALLOWED and victim is CONSCIOUS, have victim drink water or milk. IF SWALLOWED and victim is UNCONSCIOUS OR HAVING CONVULSIONS, do nothing except keep victim warm.
Water Pollution	Effect of low concentrations on aquatic life is unknown. May be dangerous if it enters water intakes. Notify local health and wildlife officials. Notify operators of nearby water intakes.

1. RESPONSE TO DISCHARGE
(See Response Methods Handbook, CG 446-4)

Issue warning—water contaminant
Disperse and flush

2. LABELS

No hazard label required by Code of Federal Regulations

3. CHEMICAL DESIGNATIONS

3.1 **Synonyms:** Salufer; Sodium fluosilicate; Sodium hexafluorosilicate
3.2 **Coast Guard Compatibility Classification:** Not listed
3.3 **Chemical Formula:** Na_2SiF_6
3.4 **IMCO/United Nations Numerical Designation:** Not listed

4. OBSERVABLE CHARACTERISTICS

4.1 **Physical State (as shipped):** Solid
4.2 **Color:** White
4.3 **Odor:** None

5. HEALTH HAZARDS

5.1 **Personal Protective Equipment:** Dust respirator; goggles or face shield; protective gloves
5.2 **Symptoms Following Exposure:** Inhalation of dust may irritate nose and throat. Ingestion causes symptoms similar to fluoride poisoning; compound is highly toxic; initial symptoms include nausea, cramps, vomiting, diarrhea, and dehydration; in severe cases, convulsions, shock, and cyanosis are followed by death in 2-4 hr. Contact with eyes causes irritation. Contact with skin causes rash, redness, and burning, sometimes followed by ulcer formation.
5.3 **Treatment for Exposure:** INHALATION: move to fresh air. INGESTION: seek medical attention; administer gastric lavage with lime water, then give lime water or milk at frequent intervals. EYES: flush with water for at least 15 min. SKIN: flush with water; wash with soap and water.
5.4 **Toxicity by Inhalation (Threshold Limit Value):** 2.5 mg/m^3 (as fluoride)
5.5 **Short-Term Inhalation Limits:** Data not available
5.6 **Toxicity by Ingestion:** Grade 3; LD_{50} 50—500 mg/kg
5.7 **Late Toxicity:** Data not available
5.8 **Vapor (Gas) Irritant Characteristics:** Data not available
5.9 **Liquid or Solid Irritant Characteristics:** Data not available
5.10 **Odor Threshold:** Data not available

6. FIRE HAZARDS

6.1 **Flash Point:** Not flammable
6.2 **Flammable Limits in Air:** Not flammable
6.3 **Fire Extinguishing Agents:** Not pertinent
6.4 **Fire Extinguishing Agents Not to be Used:** Not pertinent
6.5 **Special Hazards of Combustion Products:** Not pertinent
6.6 **Behavior in Fire:** Decomposes at red heat
6.7 **Ignition Temperature:** Not pertinent
6.8 **Electrical Hazard:** Not pertinent
6.9 **Burning Rate:** Not pertinent

7. CHEMICAL REACTIVITY

7.1 **Reactivity with Water:** No reaction
7.2 **Reactivity with Common Materials:**
7.3 **Stability During Transport:** Stable
7.4 **Neutralizing Agents for Acids and Caustics:** Not pertinent
7.5 **Polymerization:** Not pertinent
7.6 **Inhibitor of Polymerization:** Not pertinent

8. WATER POLLUTION

8.1 **Aquatic Toxicity:** Data not available
8.2 **Waterfowl Toxicity:** Data not available
8.3 **Biological Oxygen Demand (BOD):** None
8.4 **Food Chain Concentration Potential:** None

9. SELECTED MANUFACTURERS

1. Gardinier, Inc.
P. O. Box 3269
Tampa, Fla. 33601
2. Fisher Scientific Co.
711 Forbes Ave.
Pittsburgh, Pa. 15219
3. Pfaltz and Bauer, Inc.
375 Fairfield Ave.
Stamford, Conn. 06902

10. SHIPPING INFORMATION

10.1 **Grades or Purity:** Technical, 98+%
10.2 **Storage Temperature:** Ambient
10.3 **Inert Atmosphere:** No requirement
10.4 **Venting:** Open

11. HAZARD ASSESSMENT CODE
(See Hazard Assessment Handbook, CG 446-3)

II

12. HAZARD CLASSIFICATIONS

12.1 **Code of Federal Regulations:** Not listed
12.2 **NAS Hazard Rating for Bulk Water Transportation:** Not listed
12.3 **NFPA Hazard Classifications:** Not listed

13. PHYSICAL AND CHEMICAL PROPERTIES

13.1 **Physical State at 15°C and 1 atm:** Solid
13.2 **Molecular Weight:** 188
13.3 **Boiling Point at 1 atm:** Not pertinent (decomposes)
13.4 **Freezing Point:** Not pertinent
13.5 **Critical Temperature:** Not pertinent
13.6 **Critical Pressure:** Not pertinent
13.7 **Specific Gravity:** 2.68 at 20°C (solid)
13.8 **Liquid Surface Tension:** Not pertinent
13.9 **Liquid-Water Interfacial Tension:** Not pertinent
13.10 **Vapor (Gas) Specific Gravity:** Not pertinent
13.11 **Ratio of Specific Heats of Vapor (Gas):** Not pertinent
13.12 **Latent Heat of Vaporization:** Not pertinent
13.13 **Heat of Combustion:** Not pertinent
13.14 **Heat of Decomposition:** Not pertinent
13.15 **Heat of Solution:** Not pertinent
13.16 **Heat of Polymerization:** Not pertinent

NOTES

SDS — SODIUM SULFIDE

Common Synonyms			
	Solid flakes	Yellow to red	Rotten eggs odor
	Sinks and mixes with water.		

Avoid contact with solid and dust.
Wear rubber overclothing (including gloves) and dust respirator.
Call fire department.
Isolate and remove discharged material.
Notify local health and pollution control agencies.

Fire	Combustible. POISONOUS GAS MAY BE PRODUCED IN FIRE. Flammable, poisonous gas is formed on contact with acids. **WEAR GOGGLES AND SELF-CONTAINED BREATHING APPARATUS.** **Extinguish with water.**
Exposure	**CALL FOR MEDICAL AID.** DUST Irritating to eyes, nose and throat. **Move to fresh air.** **IF IN EYES, hold eyelids open and flush with plenty of water.** SOLID Harmful if swallowed. **IF SWALLOWED and victim is CONSCIOUS, have victim drink water or milk.**
Water Pollution	HARMFUL TO AQUATIC LIFE IN VERY LOW CONCENTRATIONS. May be dangerous if it enters water intakes. **Notify local health and wildlife officials.** **Notify operators of nearby water intakes.**

1. RESPONSE TO DISCHARGE

(See Response Methods Handbook, CG 446-4)

Issue warning—corrosive
Restrict access
Disperse and flush

2. LABEL

Red Stripe

3. CHEMICAL DESIGNATIONS

3.1 **Synonyms:** No common synonyms
3.2 **Coast Guard Compatibility Classification:** Not applicable
3.3 **Chemical Formula:** Na_2S
3.4 **IMCO/United Nations Numerical Designation:** 4.2/1385

4. OBSERVABLE CHARACTERISTICS

4.1 **Physical State (as shipped):** Solid
4.2 **Color:** Yellow or light buff
4.3 **Odor:** Like rotten eggs

5. HEALTH HAZARDS

5.1 **Personal Protective Equipment:** Goggles or face shield.
5.2 **Symptoms Following Exposure:** Caustic action on skin and eyes. If ingested may liberate hydrogen sulfide in stomach.
5.3 **Treatment for Exposure:** INGESTION: give water; induce vomiting; call a doctor. SKIN OR EYE CONTACT: wash with water for at least 15 min.
5.4 **Toxicity by Inhalation (Threshold Limit Value):** Not pertinent
5.5 **Short-Term Inhalation Limits:** Not pertinent
5.6 **Toxicity by Ingestion:** Grade 3; LD_{50} 50-500 mg/kg (human)
5.7 **Late Toxicity:** None
5.8 **Vapor (Gas) Irritant Characteristics:** Non-volatile
5.9 **Liquid or Solid Irritant Characteristics:** Irritates skin and mucous membranes.
5.10 **Odor Threshold:** Not pertinent

6. FIRE HAZARDS

6.1 **Flash Point:** Moderately flammable solid
6.2 **Flammable Limits in Air:** Not pertinent
6.3 **Fire Extinguishing Agents:** Water
6.4 **Fire Extinguishing Agents Not to be Used:** Not pertinent
6.5 **Special Hazards of Combustion Products:** Irritating sulfur dioxide is produced in fire.
6.6 **Behavior in Fire:** Not pertinent
6.7 **Ignition Temperature:** Not pertinent
6.8 **Electrical Hazard:** Not pertinent
6.9 **Burning Rate:** Not pertinent

7. CHEMICAL REACTIVITY

7.1 **Reactivity with Water:** No reaction
7.2 **Reactivity with Common Materials:** No reaction
7.3 **Stability During Transport:** Stable
7.4 **Neutralizing Agents for Acids and Caustics:** Not pertinent
7.5 **Polymerization:** Not pertinent
7.6 **Inhibitor of Polymerization:** Not pertinent

8. WATER POLLUTION

8.1 **Aquatic Toxicity:** 61 ppm/48 hr/bluegill, sunfish/TL_m/fresh water
8.2 **Waterfowl Toxicity:** Data not available
8.3 **Biological Oxygen Demand (BOD):** Data not available
8.4 **Food Chain Concentration Potential:** None

9. SELECTED MANUFACTURERS

1. Allied Chemical Corp.
   Specialty Chemicals Division
   Marcus Hook, Pa. 19061
2. FMC Corp.
   Inorganic Chemicals Division
   633 Third Ave.
   New York, N. Y. 10017
3. Mobil Oil Corp.
   North American Division
   Paulsboro, N. J. 08066

10. SHIPPING INFORMATION

10.1 **Grades or Purity:** Crystals, 60-62%, plus water; fused chips
10.2 **Storage Temperature:** Ambient
10.3 **Inert Atmosphere:** No requirement
10.4 **Venting:** Open

11. HAZARD ASSESSMENT CODE

(See Hazard Assessment Handbook, CG 446-3)

SS

12. HAZARD CLASSIFICATIONS

12.1 **Code of Federal Regulations:** Flammable solid
12.2 **NAS Hazard Rating for Bulk Water Transportation:** Not listed
12.3 **NFPA Hazard Classifications:**

Category	Classification
Health Hazard (Blue)	2
Flammability (Red)	1
Reactivity (Yellow)	0

13. PHYSICAL AND CHEMICAL PROPERTIES

13.1 **Physical State at 15°C and 1 atm:** Solid
13.2 **Molecular Weight:** 78.4
13.3 **Boiling Point at 1 atm:** Very high
13.4 **Freezing Point:** Not pertinent
13.5 **Critical Temperature:** Not pertinent
13.6 **Critical Pressure:** Not pertinent
13.7 **Specific Gravity:** 1.856 at 20°C (solid)
13.8 **Liquid Surface Tension:** Not pertinent
13.9 **Liquid-Water Interfacial Tension:** Not pertinent
13.10 **Vapor (Gas) Specific Gravity:** Not pertinent
13.11 **Ratio of Specific Heats of Vapor (Gas):** Not pertinent
13.12 **Latent Heat of Vaporization:** Not pertinent
13.13 **Heat of Combustion:** Not pertinent
13.14 **Heat of Decomposition:** Not pertinent
13.15 **Heat of Solution:** Not pertinent
13.16 **Heat of Polymerization:** Not pertinent

NOTES

REVISED 1978

SSF

SODIUM SULFITE

Common Synonyms	Solid	Colorless	Odorless
	Sinks and mixes slowly with water.		

Isolate and remove discharged material.
Notify local health and pollution control agencies.

Fire	Not flammable.
Exposure	**CALL FOR MEDICAL AID.** SOLIDS If swallowed, may cause loss of consciousness. **IF SWALLOWED and victim is CONSCIOUS, have victim drink water or milk.**
Water Pollution	Dangerous to aquatic life in high concentrations. May be dangerous if it enters water intakes. **Notify local health and wildlife officials.** **Notify operators of nearby water intakes.**

1. RESPONSE TO DISCHARGE
(See Response Methods Handbook, CG 446-4)

Issue warning—water contaminant
Disperse and flush

2. LABELS

No hazard label required by Code of Federal Regulations

3. CHEMICAL DESIGNATIONS

3.1 **Synonyms:** No common synonyms
3.2 **Coast Guard Compatibility Classification:** Not applicable
3.3 **Chemical Formula:** Na_2SO_3
3.4 **IMCO/United Nations Numerical Designation:** Not listed

4. OBSERVABLE CHARACTERISTICS

4.1 **Physical State (as shipped):** Solid
4.2 **Color:** White
4.3 **Odor:** Odorless

5. HEALTH HAZARDS

5.1 **Personal Protective Equipment:** Dust mask; goggles or face shield.
5.2 **Symptoms Following Exposure:** When ingested, solutions cause gastric irritation by the liberation of sulfurous acid. Because of rapid oxidation to sulfate, sulfites are well tolerated until large doses are reached; then violent colic and diarrhea, circulatory disturbances, central nervous depression, and death can occur.
5.3 **Treatment for Exposure:** INGESTION: treatment is symptomatic and supportive; call a doctor.
5.4 **Toxicity by Inhalation (Threshold Limit Value):** Not pertinent
5.5 **Short-Term Inhalation Limits:** Not pertinent
5.6 **Toxicity by Ingestion:** Grade 2; LD_{50} 0.5 to 5 g/kg
5.7 **Late Toxicity:** None
5.8 **Vapor (Gas) Irritant Characteristics:** Non-volatile
5.9 **Liquid or Solid Irritant Characteristics:** Not very irritating
5.10 **Odor Threshold:** Not pertinent

6. FIRE HAZARDS

6.1 **Flash Point:** Not flammable
6.2 **Flammable Limits in Air:** Not flammable
6.3 **Fire Extinguishing Agents:** Not pertinent
6.4 **Fire Extinguishing Agents Not to be Used:** Not pertinent
6.5 **Special Hazards of Combustion Products:** Not pertinent
6.6 **Behavior in Fire:** Not pertinent
6.7 **Ignition Temperature:** Not flammable
6.8 **Electrical Hazard:** Not pertinent
6.9 **Burning Rate:** Not flammable

7. CHEMICAL REACTIVITY

7.1 **Reactivity with Water:** No reaction
7.2 **Reactivity with Common Materials:** No reaction
7.3 **Stability During Transport:** Stable
7.4 **Neutralizing Agents for Acids and Caustics:** Not pertinent
7.5 **Polymerization:** Not pertinent
7.6 **Inhibitor of Polymerization:** Not pertinent

8. WATER POLLUTION

8.1 **Aquatic Toxicity:** 2600 ppm/24, 48 & 96 hr/mosquito fish/TL_m/fresh water
8.2 **Waterfowl Toxicity:** Data not available
8.3 **Biological Oxygen Demand (BOD):** 0.12 lb/lb, instantaneous
8.4 **Food Chain Concentration Potential:** None

9. SELECTED MANUFACTURERS

1. Koppers Co., Inc. Organic Materials Division Pittsburgh, Pa. 15219
2. Monsanto Co. Monsanto Industrial Chemicals Co. 800 North Lindbergh Blvd. St. Louis, Mo. 63166
3. Reichhold Chemicals Inc. RCI Bldg. White Plains, N. Y. 10602

10. SHIPPING INFORMATION

10.1 **Grades or Purity:** Technical anhydrous: 90 ± 1%
10.2 **Storage Temperature:** Ambient
10.3 **Inert Atmosphere:** No requirement
10.4 **Venting:** Open

11. HAZARD ASSESSMENT CODE
(See Hazard Assessment Handbook, CG 446-3)

SS

12. HAZARD CLASSIFICATIONS

12.1 **Code of Federal Regulations:** Not listed
12.2 **NAS Hazard Rating for Bulk Water Transportation:** Not listed
12.3 **NFPA Hazard Classifications:** Not listed

13. PHYSICAL AND CHEMICAL PROPERTIES

13.1 **Physical State at 15°C and 1 atm:** Solid
13.2 **Molecular Weight:** 126.04
13.3 **Boiling Point at 1 atm:** Decomposes
13.4 **Freezing Point:** Not pertinent
13.5 **Critical Temperature:** Not pertinent
13.6 **Critical Pressure:** Not pertinent
13.7 **Specific Gravity:** 2.633 at 15°C (solid)
13.8 **Liquid Surface Tension:** Not pertinent
13.9 **Liquid-Water Interfacial Tension:** Not pertinent
13.10 **Vapor (Gas) Specific Gravity:** Not pertinent
13.11 **Ratio of Specific Heats of Vapor (Gas):** Not pertinent
13.12 **Latent Heat of Vaporization:** Not pertinent
13.13 **Heat of Combustion:** Not pertinent
13.14 **Heat of Decomposition:** Not pertinent
13.15 **Heat of Solution:** Not pertinent
13.16 **Heat of Polymerization:** Not pertinent

NOTES

SCY — SODIUM THIOCYANATE

Common Synonyms			
Sodium sulfocyanate Sodium rhodanide Rhodanate	Solid	White	Odorless
	Sinks and mixes with water.		

Avoid contact with solid and dust. Keep people away.
Stop discharge if possible.
Isolate and remove discharged material.
Notify local health and pollution control agencies.

Fire	Not flammable. POISONOUS GASES MAY BE PRODUCED WHEN HEATED.
Exposure	CALL FOR MEDICAL AID. DUST Irritating to eyes, nose and throat. If inhaled will cause coughing or difficult breathing. If in eyes, hold eyelids open and flush with plenty of water. If breathing has stopped, give artificial respiration. If breathing is difficult, give oxygen. SOLID Irritating to skin and eyes. If swallowed will cause nausea, vomiting or loss of consciousness. Remove contaminated clothing and shoes. Flush affected areas with plenty of water. IF IN EYES, hold eyelids open and flush with plenty of water. IF SWALLOWED and victim is CONSCIOUS, have victim drink water or milk. IF SWALLOWED and victim is UNCONSCIOUS OR HAVING CONVULSIONS, do nothing except keep victim warm.
Water Pollution	Dangerous to aquatic life in high concentrations. May be dangerous if it enters water intakes. Notify local health and wildlife officials. Notify operators of nearby water intakes.

1. RESPONSE TO DISCHARGE

(See Response Methods Handbook, CG 446-4)

Issue warning—water contaminant
Disperse and flush

2. LABELS

No hazard label required by Code of Federal Regulations

3. CHEMICAL DESIGNATIONS

3.1 **Synonyms:** Rhodanate; Sodium rhodanide; Sodium sulfocyanate
3.2 **Coast Guard Compatibility Classification:** Not listed
3.3 **Chemical Formula:** NaSCN
3.4 **IMCO/United Nations Numerical Designation:** Not listed

4. OBSERVABLE CHARACTERISTICS

4.1 **Physical State (as shipped):** Solid
4.2 **Color:** White
4.3 **Odor:** None

5. HEALTH HAZARDS

5.1 **Personal Protective Equipment:** Rubber or plastic gloves; standard goggles; rubber or plastic apron
5.2 **Symptoms Following Exposure:** Inhalation of dust causes irritation of nose and throat. Ingestion of large doses causes vomiting, extreme cerebral excitement, convulsions, and death in 10—48 hrs.; chronic poisoning can cause flu-like symptoms, skin rashes, weakness, fatigue, vertigo, nausea, vomiting, diarrhea, confusion. Contact with eyes causes irritation. Prolonged contact with skin may produce various skin eruptions, dizziness, cramps, nausea, and mild to severe disturbance of the nervous system.
5.3 **Treatment for Exposure:** INHALATION: move to fresh air; if exposure has been great, get medical attention. INGESTION: consult physician; hemodialysis is recommended as the treatment of choice. EYES or SKIN: flush with water for 15 min.
5.4 **Toxicity by Inhalation (Threshold Limit Value):** Data not available
5.5 **Short-Term Inhalation Limits:** Data not available
5.6 **Toxicity by Ingestion:** Grade 2; LD_{50} 0.5—5 g/kg
5.7 **Late Toxicity:** Causes birth defects in chick embryos
5.8 **Vapor (Gas) Irritant Characteristics:** Data not available
5.9 **Liquid or Solid Irritant Characteristics:** Data not available
5.10 **Odor Threshold:** Data not available

6. FIRE HAZARDS

6.1 **Flash Point:** Not flammable
6.2 **Flammable Limits in Air:** Not flammable
6.3 **Fire Extinguishing Agents:** Not pertinent
6.4 **Fire Extinguishing Agents Not to be Used:** Not pertinent
6.5 **Special Hazards of Combustion Products:** Irritating oxides of sulfur and nitrogen may form in fire.
6.6 **Behavior in Fire:**
6.7 **Ignition Temperature:** Not pertinent
6.8 **Electrical Hazard:** Not pertinent
6.9 **Burning Rate:** Not pertinent

7. CHEMICAL REACTIVITY

7.1 **Reactivity with Water:** No reaction
7.2 **Reactivity with Common Materials:**
7.3 **Stability During Transport:** Stable
7.4 **Neutralizing Agents for Acids and Caustics:** Not pertinent
7.5 **Polymerization:** Not pertinent
7.6 **Inhibitor of Polymerization:** Not pertinent

8. WATER POLLUTION

8.1 **Aquatic Toxicity:**
3,240 ppm/2 hr/fish/killed/fresh water
> 500 ppm/48 hr/green crab/TL_m/aerated salt water
8.2 **Waterfowl Toxicity:** Data not available
8.3 **Biological Oxygen Demand (BOD):** Data not available
8.4 **Food Chain Concentration Potential:** None

9. SELECTED MANUFACTURERS

1. Argus Chemical Corp.
Halby Div.
600 Terminal Ave.
New Castle, Del. 19720
2. J. T. Baker Chemical Co.
Phillipsburg, N. J. 08865
3. Fisher Scientific Co.
711 Forbes Ave.
Pittsburgh, Pa. 15219

10. SHIPPING INFORMATION

10.1 **Grades or Purity:** Commercial, 98%; Reagent, 99%; 50—60% solutions in water.
10.2 **Storage Temperature:** Ambient
10.3 **Inert Atmosphere:** No requirement
10.4 **Venting:** Open

11. HAZARD ASSESSMENT CODE

(See Hazard Assessment Handbook, CG 446-3)

SS

12. HAZARD CLASSIFICATIONS

12.1 **Code of Federal Regulations:** Not listed
12.2 **NAS Hazard Rating for Bulk Water Transportation:** Not listed
12.3 **NFPA Hazard Classifications:** Not listed

13. PHYSICAL AND CHEMICAL PROPERTIES

13.1 **Physical State at 15°C and 1 atm:** Solid
13.2 **Molecular Weight:** 81.08
13.3 **Boiling Point at 1 atm:** Not pertinent (decomposes)
13.4 **Freezing Point:** 572°F = 300°C = 573°K
13.5 **Critical Temperature:** Not pertinent
13.6 **Critical Pressure:** Not pertinent
13.7 **Specific Gravity:** > 1 at 20°C (solid)
13.8 **Liquid Surface Tension:** Not pertinent
13.9 **Liquid-Water Interfacial Tension:** Not pertinent
13.10 **Vapor (Gas) Specific Gravity:** Not pertinent
13.11 **Ratio of Specific Heats of Vapor (Gas):** Not pertinent
13.12 **Latent Heat of Vaporization:** Not pertinent
13.13 **Heat of Combustion:** Not pertinent
13.14 **Heat of Decomposition:** Not pertinent
13.15 **Heat of Solution:** 34.9 Btu/lb = 19.4 cal/g = 0.812×10^5 J/kg
13.16 **Heat of Polymerization:** Not pertinent

NOTES

SBT — SORBITOL

Common Synonyms			
Sorbol Sorbit 1,2,3,4,5,6,-Hexanehexol D-Glucitol	Liquid	Colorless	Odorless
	Sinks and mixes with water.		

Avoid contact with liquid.
Stop discharge if possible.
Isolate and remove discharged material.
Notify local health and pollution control agencies.

Fire	Not flammable.
Exposure	CALL FOR MEDICAL AID. LIQUID Will burn skin and eyes. Flush affected areas with plenty of water. IF IN EYES, hold eyelids open and flush with plenty of water.
Water Pollution	Dangerous to aquatic life in high concentrations. May be dangerous if it enters water intakes. Notify local health and wildlife officials. Notify operators of nearby water intakes.

1. RESPONSE TO DISCHARGE
(See Response Methods Handbook, CG 446-4)

Disperse and flush

2. LABELS

No hazard label required by Code of Federal Regulations

3. CHEMICAL DESIGNATIONS

3.1 **Synonyms:** D-Glucitol; Hexahydric alcohol; 1,2,3,4,5,6-Hexanehexol; Sorbit; Sorbo; Sorbol

3.2 **Coast Guard Compatibility Classification:** Glycol

3.3 **Chemical Formula:** $CH_2OH(CHOH)_4CH_2OH$

3.4 **IMCO/United Nations Numerical Designation:** Not listed

4. OBSERVABLE CHARACTERISTICS

4.1 **Physical State (as shipped):** Liquid

4.2 **Color:** Colorless

4.3 **Odor:** Odorless

5. HEALTH HAZARDS

5.1 **Personal Protective Equipment:** Goggles or face shield; protective clothing for hot liquid.

5.2 **Symptoms Following Exposure:** Hot liquid will burn skin.

5.3 **Treatment for Exposure:** Only for burns caused by hot liquid.

5.4 **Toxicity by Inhalation (Threshold Limit Value):** Not pertinent

5.5 **Short-Term Inhalation Limits:** Not pertinent

5.6 **Toxicity by Ingestion:** None

5.7 **Late Toxicity:** None

5.8 **Vapor (Gas) Irritant Characteristics:** Non-volatile.

5.9 **Liquid or Solid Irritant Characteristics:** Not pertinent

5.10 **Odor Threshold:** Not pertinent

6. FIRE HAZARDS

6.1 **Flash Point:** 542°F O.C.

6.2 **Flammable Limits in Air:** Not pertinent

6.3 **Fire Extinguishing Agents:** Water

6.4 **Fire Extinguishing Agents Not to be Used:** Data not available

6.5 **Special Hazards of Combustion Products:** Not pertinent

6.6 **Behavior in Fire:** Not pertinent

6.7 **Ignition Temperature:** Data not available

6.8 **Electrical Hazard:** Not pertinent

6.9 **Burning Rate:** Data not available

7. CHEMICAL REACTIVITY

7.1 **Reactivity with Water:** No reaction

7.2 **Reactivity with Common Materials:** No reaction

7.3 **Stability During Transport:** Stable

7.4 **Neutralizing Agents for Acids and Caustics:** Not pertinent

7.5 **Polymerization:** Not pertinent

7.6 **Inhibitor of Polymerization:** Not pertinent

8. WATER POLLUTION

8.1 **Aquatic Toxicity:** Data not available

8.2 **Waterfowl Toxicity:** Data not available

8.3 **Biological Oxygen Demand (BOD):** 43% (theor.), 6 days

8.4 **Food Chain Concentration Potential:** None

9. SELECTED MANUFACTURERS

1. ICI America Inc.
   Atlas Chemicals Division
   New Murphy Road & Concord Pike
   Wilmington, Del. 19899
2. Merck & Co., Inc.
   Merck Chemical Division
   Danville, Pa. 17821
3. Pfizer Inc.
   Chemicals Division
   Groton, Conn. 06340

10. SHIPPING INFORMATION

10.1 **Grades or Purity:** USP crystalline; 93+%

10.2 **Storage Temperature:** Elevated

10.3 **Inert Atmosphere:** No requirement

10.4 **Venting:** Open (flame arrester)

11. HAZARD ASSESSMENT CODE
(See Hazard Assessment Handbook, CG 446-3)

A-P-Q

12. HAZARD CLASSIFICATIONS

12.1 **Code of Federal Regulations:** Not listed

12.2 **NAS Hazard Rating for Bulk Water Transportation:** Not listed

12.3 **NFPA Hazard Classifications:** Not listed

13. PHYSICAL AND CHEMICAL PROPERTIES

13.1 **Physical State at 15°C and 1 atm:** Solid

13.2 **Molecular Weight:** 182.17

13.3 **Boiling Point at 1 atm:** Very high

13.4 **Freezing Point:** 230°F = 110°C = 383°K

13.5 **Critical Temperature:** Not pertinent

13.6 **Critical Pressure:** Not pertinent

13.7 **Specific Gravity:** 1.49 at 15°C (liquid)

13.8 **Liquid Surface Tension:** Not pertinent

13.9 **Liquid-Water Interfacial Tension:** Not pertinent

13.10 **Vapor (Gas) Specific Gravity:** Not pertinent

13.11 **Ratio of Specific Heats of Vapor (Gas):** Not pertinent

13.12 **Latent Heat of Vaporization:** Not pertinent

13.13 **Heat of Combustion:** (est.) −6,750 Btu/lb = −3,750 cal/g = − 157×10^5 J/kg

13.14 **Heat of Decomposition:** Not pertinent

13.15 **Heat of Solution:** (est.) −22 Btu/lb = −12 cal/g = -0.5×10^5 J/kg

13.16 **Heat of Polymerization:** Not pertinent

NOTES

REVISED 1978

SRA	STEARIC ACID

Common Synonyms
Octadecanoic acid
n-Octadecylic acid
Stearophanic acid
1-Heptadecanecarboxylic acid

Solid White Mild odor

Floats on water.

Avoid contact with solid and dust. Keep people away.
Stop discharge if possible.
Call fire department.
Isolate and remove discharged material.
Notify local health and pollution control agencies.

Fire

Combustible.
Extinguish with dry chemicals, foam or carbon dioxide.
Water may be ineffective on fire.
Cool exposed containers with water.

Exposure

CALL FOR MEDICAL AID.

DUST
Irritating to eyes, nose and throat.
If inhaled will cause coughing or difficult breathing.
If in eyes, hold eyelids open and flush with plenty of water.
If breathing has stopped, give artificial respiration.
If breathing is difficult, give oxygen.

SOLID
Irritating to skin and eyes.
Harmful if swallowed.
Remove contaminated clothing and shoes.
Flush affected areas with plenty of water.
IF IN EYES, hold eyelids open and flush with plenty of water.
IF SWALLOWED and victim is CONSCIOUS, have victim drink water or milk and have victim induce vomiting.
IF SWALLOWED and victim is UNCONSCIOUS OR HAVING CONVULSIONS, do nothing except keep victim warm.

Water Pollution

Effect of low concentrations on aquatic life is unknown.
May be dangerous if it enters water intakes.
Notify local health and wildlife officials.
Notify operators of nearby water intakes.

1. RESPONSE TO DISCHARGE

(See Response Methods Handbook, CG 446-4)

Mechanical containment
Should be removed
Chemical and physical treatment

2. LABELS

No hazard label required by Code of Federal Regulations

3. CHEMICAL DESIGNATIONS

3.1 **Synonyms:** 1-Heptadecanecarboxylic acid; Octadecanoic acid; n-Octadecylic acid; Stearophanic acid

3.2 **Coast Guard Compatibility Classification:** Not listed

3.3 **Chemical Formula:** $CH_3(CH_2)_{16}CO_2H$

3.4 **IMCO/United Nations Numerical Designation:** Not listed

4. OBSERVABLE CHARACTERISTICS

4.1 **Physical State (as shipped):** Solid

4.2 **Color:** White

4.3 **Odor:** Fatty

5. HEALTH HAZARDS

5.1 **Personal Protective Equipment:** For prolonged exposure to vapors, use air-supplied mask or chemical cartridge respirator; impervious gloves; goggles; impervious apron

5.2 **Symptoms Following Exposure:** Compound is generally considered nontoxic. Inhalation of dust irritates nose and throat. Dust causes mild irritation of eyes.

5.3 **Treatment for Exposure:** INGESTION: drink large volume of water; induce vomiting; call a physician. EYES: flush with clean water; if irritation persists, get medical attention. SKIN: wash thoroughly with soap and water.

5.4 **Toxicity by Inhalation (Threshold Limit Value):** Data not available

5.5 **Short-Term Inhalation Limits:** Data not available

5.6 **Toxicity by Ingestion:** Grade 0; LD_{50} > 15 g/kg

5.7 **Late Toxicity:** Data not available

5.8 **Vapor (Gas) Irritant Characteristics:** Data not available

5.9 **Liquid or Solid Irritant Characteristics:** Data not available

5.10 **Odor Threshold:** 20 ppm

6. FIRE HAZARDS

6.1 **Flash Point:** (molten solid) 410—435°F O.C.; 365°F C.C.

6.2 **Flammable Limits in Air:** Not pertinent

6.3 **Fire Extinguishing Agents:** Foam, dry chemical, or carbon dioxide

6.4 **Fire Extinguishing Agents Not to be Used:** Water or foam may cause frothing.

6.5 **Special Hazards of Combustion Products:**

6.6 **Behavior in Fire:**

6.7 **Ignition Temperature:** 743°F

6.8 **Electrical Hazard:** Not pertinent

6.9 **Burning Rate:** Not pertinent

7. CHEMICAL REACTIVITY

7.1 **Reactivity with Water:** No reaction

7.2 **Reactivity with Common Materials:**

7.3 **Stability During Transport:** Stable

7.4 **Neutralizing Agents for Acids and Caustics:** Not pertinent

7.5 **Polymerization:** Not pertinent

7.6 **Inhibitor of Polymerization:** Not pertinent

8. WATER POLLUTION

8.1 **Aquatic Toxicity:** Data not available

8.2 **Waterfowl Toxicity:** Data not available

8.3 **Biological Oxygen Demand (BOD):** 144%, 5 days

8.4 **Food Chain Concentration Potential:** None

9. SELECTED MANUFACTURERS

1. Union Camp Corp.
   2051 N. Lane Avenue
   Jacksonville, Fla. 32205
2. Ashland Chemical Co.
   P. O. Box 2219
   Columbus, Ohio 43216
3. Emery Industries, Inc.
   4900 Este Avenue
   Cincinnati, Ohio 45232

10. SHIPPING INFORMATION

10.1 **Grades or Purity:** USP; Commercial; Triple pressed; Double pressed

10.2 **Storage Temperature:** Ambient

10.3 **Inert Atmosphere:** No requirement

10.4 **Venting:** Open

11. HAZARD ASSESSMENT CODE

(See Hazard Assessment Handbook, CG 446-3)

II

12. HAZARD CLASSIFICATIONS

12.1 **Code of Federal Regulations:** Not listed

12.2 **NAS Hazard Rating for Bulk Water Transportation:** Not listed

12.3 **NFPA Hazard Classifications:**

Category	Classification
Health Hazard (Blue)	1
Flammability (Red)	1
Reactivity (Yellow)	0

13. PHYSICAL AND CHEMICAL PROPERTIES

13.1 **Physical State at 15°C and 1 atm:** Solid

13.2 **Molecular Weight:** (avg.) 282

13.3 **Boiling Point at 1 atm:** Not pertinent (decomposes)

13.4 **Freezing Point:** 157°F = 70°C = 343°K

13.5 **Critical Temperature:** Not pertinent

13.6 **Critical Pressure:** Not pertinent

13.7 **Specific Gravity:** 0.86 at 20°C (solid)

13.8 **Liquid Surface Tension:** Not pertinent

13.9 **Liquid-Water Interfacial Tension:** Not pertinent

13.10 **Vapor (Gas) Specific Gravity:** Not pertinent

13.11 **Ratio of Specific Heats of Vapor (Gas):** Not pertinent

13.12 **Latent Heat of Vaporization:** Not pertinent

13.13 **Heat of Combustion:** −17,310 Btu/lb = −9,616 cal/g = -402.3×10^5 J/kg

13.14 **Heat of Decomposition:** Not pertinent

13.15 **Heat of Solution:** Not pertinent

13.16 **Heat of Polymerization:** Not pertinent

NOTES

STYRENE

Common Synonyms		
Styrol Vinylbenzene Phenylethylene	Watery liquid	Colorless to light yellow Sweet pleasant odor
	Floats on water. Flammable, irritating vapor is produced.	

Avoid contact with liquid and vapor. Keep people away.
Wear chemical protective suit with self-contained breathing apparatus.
Stop discharge if possible.
Call fire department.
Isolate and remove discharged material.
Notify local health and pollution control agencies.

Fire	FLAMMABLE CONTAINERS MAY EXPLODE IN FIRE. Flashback along vapor trail may occur. Vapor may explode if ignited in an enclosed area. **Wear chemical protective suit with self-contained breathing apparatus.** **Combat fires from safe distance or protected location.** **Extinguish with dry chemical, foam, or carbon dioxide.** **Water may be ineffective on fire.** **Cool exposed containers with water.**
Exposure	CALL FOR MEDICAL AID. VAPOR Irritating to eyes, nose and throat. If inhaled, will cause dizziness or loss of consciousness. **Move to fresh air.** **If breathing has stopped, give artificial respiration.** **If breathing is difficult, give oxygen.** LIQUID Will burn skin and eyes. Harmful if swallowed. **Remove contaminated clothing and shoes.** **Flush affected areas with plenty of water.** **IF IN EYES, hold eyelids open and flush with plenty of water.** **IF SWALLOWED and victim is CONSCIOUS, have victim drink water or milk.** **DO NOT INDUCE VOMITING.**
Water Pollution	HARMFUL TO AQUATIC LIFE IN VERY LOW CONCENTRATIONS. Fouling to shoreline. May be dangerous if it enters water intakes. **Notify local health and wildlife officials.** **Notify operators of nearby water intakes.**

1. RESPONSE TO DISCHARGE

(See Response Methods Handbook, CG 446-4)

Issue warning - air contaminant
Mechanical containment
Should be removed
Chemical and physical treatment

2. LABEL

Red

3. CHEMICAL DESIGNATIONS

3.1 **Synonyms:** Phenethylene; Phenylethylene; Styrol; Styrolene; Vinylbenzene
3.2 **Coast Guard Compatibility Classification:** Olefin
3.3 **Chemical Formula:** $C_6H_5CH{=}CH_2$
3.4 **IMCO/United Nations Numerical Designation:** 3.3/2055

4. OBSERVABLE CHARACTERISTICS

4.1 **Physical State (as shipped):** Liquid
4.2 **Color:** Colorless
4.3 **Odor:** Sweet at low concentrations; characteristic pungent; sharp; disagreeable

5. HEALTH HAZARDS

5.1 **Personal Protective Equipment:** Air-supplied mask or approved canister; rubber or plastic gloves; boots; goggles or face shield.
5.2 **Symptoms Following Exposure:** Moderate irritation of eyes and skin. High vapor concentrations cause dizziness, drunkenness, and anesthesia.
5.3 **Treatment for Exposure:** INHALATION: remove to fresh air; keep warm and quiet; use artificial respiration if needed. INGESTION: do NOT induce vomiting; call physician; no known antidote. SKIN OR EYE CONTACT: flush with plenty of water; for eyes get medical attention.
5.4 **Toxicity by Inhalation (Threshold Limit Value):** 100 ppm
5.5 **Short-Term Inhalation Limits:** 100 ppm for 30 min.
5.6 **Toxicity by Ingestion:** Grade 2; LD_{50} 0.5 to 5 g/kg
5.7 **Late Toxicity:** Data not available
5.8 **Vapor (Gas) Irritant Characteristics:** Vapors cause moderate irritation such that personnel will find high concentrations unpleasant. The effect is temporary.
5.9 **Liquid or Solid Irritant Characteristics:** Causes smarting of the skin and first-degree burns on short exposure; may cause secondary burns on long exposure.
5.10 **Odor Threshold:** 0.148 ppm

6. FIRE HAZARDS

6.1 **Flash Point:** 93°F O.C.; 88°F C.C.
6.2 **Flammable Limits in Air:** 1.1% - 6.1%
6.3 **Fire Extinguishing Agents:** Water fog, foam, carbon dioxide, or dry chemical
6.4 **Fire Extinguishing Agents Not to be Used:** Water may be ineffective.
6.5 **Special Hazards of Combustion Products:** Not pertinent
6.6 **Behavior in Fire:** Vapor is heavier than air and may travel considerable distance to a source of ignition and flash back. At elevated temperatures such as in fire conditions, polymerization may take place which may lead to container explosion.
6.7 **Ignition Temperature:** 914°F
6.8 **Electrical Hazard:** Class I, Group D
6.9 **Burning Rate:** 5.2 mm/min.

7. CHEMICAL REACTIVITY

7.1 **Reactivity with Water:** No reaction
7.2 **Reactivity with Common Materials:** No reaction
7.3 **Stability During Transport:** Stable
7.4 **Neutralizing Agents for Acids and Caustics:** Not pertinent
7.5 **Polymerization:** May occur if heated above 150°F. Can cause rupture of container. Metal salts, peroxides, and strong acids may also cause polymerization.
7.6 **Inhibitor of Polymerization:** Tertiary-butylcatechol, 10-15 ppm

8. WATER POLLUTION

8.1 **Aquatic Toxicity:** 22 ppm/96 hr/bluegill/TL_m/fresh water
8.2 **Waterfowl Toxicity:** Data not available
8.3 **Biological Oxygen Demand (BOD):** 18% (theor.), 412 days
8.4 **Food Chain Concentration Potential:** None

9. SELECTED MANUFACTURERS

1. Amoco Chemicals Corp.
   910 South Michigan Ave.
   Chicago, Ill. 60605
2. Dow Chemical Co.
   Midland, Mich. 48640
3. Monsanto Co.
   Monsanto Polymers & Petrochemicals Co.
   800 North Lindbergh Blvd.
   St. Louis, Mo. 63166

10. SHIPPING INFORMATION

10.1 **Grades or Purity:** 99.5+%
10.2 **Storage Temperature:** Ambient
10.3 **Inert Atmosphere:** No requirement
10.4 **Venting:** Open (flame arrester)

11. HAZARD ASSESSMENT CODE

(See Hazard Assessment Handbook, CG 446-3)

A-T-U-Z

12. HAZARD CLASSIFICATIONS

12.1 **Code of Federal Regulations:** Flammable Liquid
12.2 **NAS Hazard Rating for Bulk Water Transportation:**

Category	Rating
Fire	3
Health	
Vapor Irritant	2
Liquid or Solid Irritant	2
Poisons	2
Water Pollution	
Human Toxicity	1
Aquatic Toxicity	3
Aesthetic Effect	2
Reactivity	
Other Chemicals	2
Water	0
Self-Reaction	3

12.3 **NFPA Hazard Classifications:**

Category	Classification
Health Hazard (Blue)	2
Flammability (Red)	3
Reactivity (Yellow)	2

13. PHYSICAL AND CHEMICAL PROPERTIES

13.1 **Physical State at 15°C and 1 atm:** Liquid
13.2 **Molecular Weight:** 104.15
13.3 **Boiling Point at 1 atm:** 293.4°F = 145.2°C = 418.4°K
13.4 **Freezing Point:** −23.1°F = −30.6°C = 242.6°K
13.5 **Critical Temperature:** 703°F = 373°C = 646°K
13.6 **Critical Pressure:** 580 psia = 39.46 atm = 4.00 MN/m²
13.7 **Specific Gravity:** 0.906 at 20°C (liquid)
13.8 **Liquid Surface Tension:** 32.14 dynes/cm = 0.03214 N/m at 19°C
13.9 **Liquid-Water Interfacial Tension:** 35.48 dynes/cm = 0.03548 N/m at 19°C
13.10 **Vapor (Gas) Specific Gravity:** Not pertinent
13.11 **Ratio of Specific Heats of Vapor (Gas):** 1.074
13.12 **Latent Heat of Vaporization:** 156 Btu/lb = 86.8 cal/g = 3.63×10^5 J/kg
13.13 **Heat of Combustion:** Not pertinent
13.14 **Heat of Decomposition:** Not pertinent
13.15 **Heat of Solution:** Not pertinent
13.16 **Heat of Polymerization:** −277 Btu/lb = −154 cal/g = -6.45×10^5 J/kg

NOTES

REVISED 1978

SRS	SUCROSE

Common Synonyms			
Beet sugar Cane sugar Saccharose Saccharum Sugar	Solid	White	Odorless
	Sinks in water.		

Stop discharge if possible.
Isolate and remove discharged material.
Notify local health and pollution control agencies.

Fire	Combustible. Extinguish with water.
Exposure	DUST Not harmful. SOLID Not harmful.
Water Pollution	Effect of low concentrations on aquatic life is unknown. May be dangerous if it enters water intakes. Notify local health and wildlife officials. Notify operators of nearby water intakes.

1. RESPONSE TO DISCHARGE

(See Response Methods Handbook, CG 446-4)

Disperse and flush

2. LABELS

No hazard label required by Code of Federal Regulations

3. CHEMICAL DESIGNATIONS

3.1 **Synonyms:** Beet sugar; Cane sugar; Saccharose; Saccharum; Sugar

3.2 **Coast Guard Compatibility Classification:** Not listed

3.3 **Chemical Formula:** $C_{12}H_{22}O_{11}$

3.4 **IMCO/United Nations Numerical Designation:** Not listed

4. OBSERVABLE CHARACTERISTICS

4.1 **Physical State (as shipped):** Solid

4.2 **Color:** White

4.3 **Odor:** None

5. HEALTH HAZARDS

5.1 **Personal Protective Equipment:** Dust mask and goggles or face shield

5.2 **Symptoms Following Exposure:** None

5.3 **Treatment for Exposure:** EYES: flush with water.

5.4 **Toxicity by Inhalation (Threshold Limit Value):** Data not available

5.5 **Short-Term Inhalation Limits:** Data not available

5.6 **Toxicity by Ingestion:** Grade 0; oral LD_{50} (100 days) = 28,500 mg/kg/day (rat)

5.7 **Late Toxicity:** None

5.8 **Vapor (Gas) Irritant Characteristics:** Data not available

5.9 **Liquid or Solid Irritant Characteristics:** Data not available

5.10 **Odor Threshold:** Not pertinent

6. FIRE HAZARDS

6.1 **Flash Point:** Not pertinent (combustible solid)

6.2 **Flammable Limits in Air:** Not pertinent

6.3 **Fire Extinguishing Agents:** Water

6.4 **Fire Extinguishing Agents Not to be Used:** Not pertinent

6.5 **Special Hazards of Combustion Products:** Irritating fumes may form in fires.

6.6 **Behavior in Fire:** Melts and chars

6.7 **Ignition Temperature:** Not pertinent

6.8 **Electrical Hazard:** Not pertinent

6.9 **Burning Rate:** Not pertinent

7. CHEMICAL REACTIVITY

7.1 **Reactivity with Water:** No reaction

7.2 **Reactivity with Common Materials:**

7.3 **Stability During Transport:** Stable

7.4 **Neutralizing Agents for Acids and Caustics:** Not pertinent

7.5 **Polymerization:** Not pertinent

7.6 **Inhibitor of Polymerization:** Not pertinent

8. WATER POLLUTION

8.1 **Aquatic Toxicity:** Indirect effect from high BOD

8.2 **Waterfowl Toxicity:** Data not available

8.3 **Biological Oxygen Demand (BOD):** 69%, 5 days

8.4 **Food Chain Concentration Potential:** None

9. SELECTED MANUFACTURERS

1. Amstar Corporation
   1251 Avenue of the Americas
   New York, N. Y.
2. Great Western Sugar Corp.
   1530 16th Street
   Denver, Colo.
3. Revere Sugar Refining Co.
   Prudential Center
   Boston, Mass.

10. SHIPPING INFORMATION

10.1 **Grades or Purity:** Food grade; Technical

10.2 **Storage Temperature:** Ambient

10.3 **Inert Atmosphere:** No requirement

10.4 **Venting:** Open

11. HAZARD ASSESSMENT CODE

(See Hazard Assessment Handbook, CG 446-3)

SS

12. HAZARD CLASSIFICATIONS

12.1 **Code of Federal Regulations:** Not listed

12.2 **NAS Hazard Rating for Bulk Water Transportation:** Not listed

12.3 **NFPA Hazard Classifications:** Not listed

13. PHYSICAL AND CHEMICAL PROPERTIES

13.1 **Physical State at 15°C and 1 atm:** Solid

13.2 **Molecular Weight:** 342.3

13.3 **Boiling Point at 1 atm:** Not pertinent (decomposes)

13.4 **Freezing Point:** (decomposes) 320—367°F = 160—186°C = 433—459°K

13.5 **Critical Temperature:** Not pertinent

13.6 **Critical Pressure:** Not pertinent

13.7 **Specific Gravity:** 1.59 at 20°C (solid)

13.8 **Liquid Surface Tension:** Not pertinent

13.9 **Liquid-Water Interfacial Tension:** Not pertinent

13.10 **Vapor (Gas) Specific Gravity:** Not pertinent

13.11 **Ratio of Specific Heats of Vapor (Gas):** Not pertinent

13.12 **Latent Heat of Vaporization:** Not pertinent

13.13 **Heat of Combustion:** –6,400 Btu/lb = –3,600 cal/g = -150×10^5 J/kg

13.14 **Heat of Decomposition:** Not pertinent

13.15 **Heat of Solution:** Not pertinent

13.16 **Heat of Polymerization:** Not pertinent

NOTES

SFL	SULFOLANE

Common Synonyms			
Tetrahydrothiophene-1, 1-dioxide Tetramethylene sulfone	Oily liquid	Colorless	Weak oily odor
	Solidifies and sinks and mixes with water. Freezing point is 79° F.		

Stop discharge if possible.
Call fire department.
Avoid contact with liquid.
Isolate and remove discharged material.
Notify local health and pollution control agencies.

Fire	Combustible. POISONOUS GASES ARE PRODUCED IN FIRE. Wear goggles and self-contained breathing apparatus. Extinguish with water, foam, dry chemical, or carbon dioxide.
Exposure	CALL FOR MEDICAL AID. LIQUID Not irritating to skin. Irritating to eyes. Harmful if swallowed. IF IN EYES, hold eyelids open and flush with plenty of water. IF SWALLOWED and victim is CONSCIOUS, have victim drink water or milk, and have victim induce vomiting. IF SWALLOWED and victim is UNCONSCIOUS OR HAVING CONVULSIONS, do nothing except keep victim warm.
Water Pollution	Effect of low concentrations on aquatic life is unknown. May be dangerous if it enters water intakes. Notify local health and wildlife officials. Notify operators of nearby water intakes.

1. RESPONSE TO DISCHARGE
(See Response Methods Handbook, CG 446-4)

Disperse and flush

2. LABELS

No hazard label required by Code of Federal Regulations

3. CHEMICAL DESIGNATIONS

3.1 **Synonyms:** Sulfolane-W
Tetrahydrothiophene-1, 1-dioxide
Tetramethylene sulfone

3.2 **Coast Guard Compatibility Classification:** Sulfolane

3.3 **Chemical Formula:** $\overline{CH_2CH_2CH_2CH_2}SO_2$

3.4 **IMCO/United Nations Numerical Designation:** Not listed

4. OBSERVABLE CHARACTERISTICS

4.1 **Physical State (as shipped):** Liquid
4.2 **Color:** Colorless
4.3 **Odor:** Weak oily

5. HEALTH HAZARDS

5.1 **Personal Protective Equipment:** Goggles or face shield; rubber gloves.
5.2 **Symptoms Following Exposure:** Very mildly irritating to the eyes.
5.3 **Treatment for Exposure:** INGESTION: induce vomiting. SKIN OR EYE CONTACT: flush with water.
5.4 **Toxicity by Inhalation (Threshold Limit Value):** Not pertinent
5.5 **Short-Term Inhalation Limits:** Not pertinent
5.6 **Toxicity by Ingestion:** Grade 2; LD_{50} 0.5 to 5 g/kg (rat, mouse)
5.7 **Late Toxicity:** Data not available
5.8 **Vapor (Gas) Irritant Characteristics:** Vapors are nonirritating to the eyes and throat.
5.9 **Liquid or Solid Irritant Characteristics:** No appreciable hazard. Practically harmless to the skin.
5.10 **Odor Threshold:** Not pertinent

6. FIRE HAZARDS

6.1 **Flash Point:** 330°F C.C.
6.2 **Flammable Limits in Air:** Data not available
6.3 **Fire Extinguishing Agents:** Water, foam, dry chemicals, or carbon dioxide
6.4 **Fire Extinguishing Agents Not to be Used:** Not pertinent
6.5 **Special Hazards of Combustion Products:** Toxic, irritating gases may be generated in fires.
6.6 **Behavior in Fire:** Not pertinent
6.7 **Ignition Temperature:** Data not available
6.8 **Electrical Hazard:** Not pertinent
6.9 **Burning Rate:** Data not available

7. CHEMICAL REACTIVITY

7.1 **Reactivity with Water:** No reaction
7.2 **Reactivity with Common Materials:** No reaction
7.3 **Stability During Transport:** Stable
7.4 **Neutralizing Agents for Acids and Caustics:** Not pertinent
7.5 **Polymerization:** Data not available
7.6 **Inhibitor of Polymerization:** Not pertinent

8. WATER POLLUTION

8.1 **Aquatic Toxicity:** Data not available
8.2 **Waterfowl Toxicity:** Data not available
8.3 **Biological Oxygen Demand (BOD):** Data not available
8.4 **Food Chain Concentration Potential:** None

9. SELECTED MANUFACTURERS

1. Phillips Petroleum Co.
Bartlesville, Okla. 74004
2. Shell Chemical Co.
Industrial Chemicals Division
Houston, Tex. 77001

10. SHIPPING INFORMATION

10.1 **Grades or Purity:** Anhydrous: 99+%; standard water blend: 97% plus 3% water
10.2 **Storage Temperature:** Ambient
10.3 **Inert Atmosphere:** No requirement
10.4 **Venting:** Open (flame arrester)

11. HAZARD ASSESSMENT CODE
(See Hazard Assessment Handbook, CG 446-3)

A-P-Q

12. HAZARD CLASSIFICATIONS

12.1 **Code of Federal Regulations:** Not listed

12.2 **NAS Hazard Rating for Bulk Water Transportation:**

Category	Rating
Fire	1
Health	
Vapor Irritant	0
Liquid or Solid Irritant	0
Poisons	1
Water Pollution	
Human Toxicity	2
Aquatic Toxicity	3
Aesthetic Effect	2
Reactivity	
Other Chemicals	1
Water	0
Self-Reaction	0

12.3 **NFPA Hazard Classifications:** Not listed

13. PHYSICAL AND CHEMICAL PROPERTIES

13.1 **Physical State at 15°C and 1 atm:** Solid
13.2 **Molecular Weight:** 120.17
13.3 **Boiling Point at 1 atm:** 545°F = 285°C = 558°K
13.4 **Freezing Point:** 79°F = 26°C = 299°K
13.5 **Critical Temperature:** Not pertinent
13.6 **Critical Pressure:** Not pertinent
13.7 **Specific Gravity:** 1.26 at 30°C (liquid)
13.8 **Liquid Surface Tension:** Not pertinent
13.9 **Liquid-Water Interfacial Tension:** Not pertinent
13.10 **Vapor (Gas) Specific Gravity:** Not pertinent
13.11 **Ratio of Specific Heats of Vapor (Gas):** Not pertinent
13.12 **Latent Heat of Vaporization:** Not pertinent
13.13 **Heat of Combustion:** (est.) –9,500 Btu/lb = –5,300 cal/g = -220×10^5 J/kg
13.14 **Heat of Decomposition:** Not pertinent
13.15 **Heat of Solution:** (est.) –22 Btu/lb = –12 cal/g = -0.5×10^5 J/kg
13.16 **Heat of Polymerization:** Not pertinent

NOTES

REVISED 1978

SFD — SULFUR DIOXIDE

Common Synonyms	Liquefied compressed gas · Colorless · Sharp irritating odor Liquid sinks and boils in water. Poisonous visible vapor cloud is produced.

Avoid contact with liquid and vapor. Keep people away.
Wear goggles and self-contained breathing apparatus.
Stop discharge if possible.
Notify local health and pollution control agencies.

Fire

Not flammable.
Wear goggles and self-contained breathing apparatus.
Cool exposed containers with water.

Exposure

CALL FOR MEDICAL AID.

VAPOR
POISONOUS IF INHALED.
Move to fresh air.
If breathing has stopped, give artificial respiration.
If breathing is difficult, give oxygen.

LIQUID
Will cause frostbite.
Flush affected areas with plenty of water.
IF IN EYES, hold eyelids open and flush with plenty of water.
DO NOT RUB AFFECTED AREAS.

Water Pollution

HARMFUL TO AQUATIC LIFE IN VERY LOW CONCENTRATIONS.
May be dangerous if it enters water intakes.
Notify local health and wildlife officials.
Notify operators of nearby water intakes.

1. RESPONSE TO DISCHARGE
(See Response Methods Handbook, CG 446-4)

Issue warning-air contaminant
Restrict access
Evacuate area

2. LABEL

Green

3. CHEMICAL DESIGNATIONS
3.1 **Synonyms:** No common synonyms
3.2 **Coast Guard Compatibility Classification:** To be developed
3.3 **Chemical Formula:** SO_2
3.4 **IMCO/United Nations Numerical Designation:** 2.0/1079

4. OBSERVABLE CHARACTERISTICS
4.1 **Physical State (as shipped):** Liquefied gas
4.2 **Color:** Colorless
4.3 **Odor:** Sharp, pungent; characteristic; like burning sulfur

5. HEALTH HAZARDS
5.1 **Personal Protective Equipment:** Air-supplied mask or approved canister; goggles or face shield; rubber gloves; rubber clothing where contact with liquid is possible.
5.2 **Symptoms Following Exposure:** Vapors cause irritation of eyes and lungs, with severe choking.
5.3 **Treatment for Exposure:** INHALATION: remove from exposure; support respiration; administer oxygen; call a doctor. SKIN: flush with water after exposure to liquid. EYES: wash promptly for at least 15 min.; call physician.
5.4 **Toxicity by Inhalation (Threshold Limit Value):** 5 ppm
5.5 **Short-Term Inhalation Limits:** 20 ppm for 5 min.
5.6 **Toxicity by Ingestion:** Not pertinent
5.7 **Late Toxicity:** Data not available
5.8 **Vapor (Gas) Irritant Characteristics:** Vapors cause severe irritation of eyes and throat and can cause eye and lung injury. They cannot be tolerated even at low concentrations.
5.9 **Liquid or Solid Irritant Characteristics:** Liquid can cause frostbite.
5.10 **Odor Threshold:** 3 ppm

6. FIRE HAZARDS
6.1 **Flash Point:** Not flammable
6.2 **Flammable Limits in Air:** Not flammable
6.3 **Fire Extinguishing Agents:** Not pertinent
6.4 **Fire Extinguishing Agents Not to be Used:** Not pertinent
6.5 **Special Hazards of Combustion Products:** Not pertinent
6.6 **Behavior in Fire:** Containers may rupture and release irritating, toxic sulfur dioxide
6.7 **Ignition Temperature:** Not flammable
6.8 **Electrical Hazard:** Not pertinent
6.9 **Burning Rate:** Not flammable

7. CHEMICAL REACTIVITY
7.1 **Reactivity with Water:** Reacts with water to form corrosive acid. The reaction is not hazardous.
7.2 **Reactivity with Common Materials:** Corrodes aluminum
7.3 **Stability During Transport:** Stable
7.4 **Neutralizing Agents for Acids and Caustics:** Mild acidity of water solution may be neutralized by dilute caustic soda
7.5 **Polymerization:** Not pertinent
7.6 **Inhibitor of Polymerization:** Not pertinent

8. WATER POLLUTION
8.1 **Aquatic Toxicity:** 5 ppm/1 hr/trout/lethal/fresh water
8.2 **Waterfowl Toxicity:** Data not available
8.3 **Biological Oxygen Demand (BOD):** Data not available
8.4 **Food Chain Concentration Potential:** None

9. SELECTED MANUFACTURERS
1. Cities Service Co., Inc.
   Industrial Chemicals Division
   P. O. Drawer 50360
   Atlanta, Ga. 30302
2. Stauffer Chemical Co.
   Industrial Chemical Division
   Baton Rouge, La. 70821
3. Virginia Chemicals Inc.
   3340 West Norfolk Rd.
   Portsmouth, Va. 23703

10. SHIPPING INFORMATION
10.1 **Grades or Purity:** Refrigeration grade 99.98%; commercial grade 99.90%
10.2 **Storage Temperature:** Less than 130°F
10.3 **Inert Atmosphere:** No requirement
10.4 **Venting:** Safety relief

11. HAZARD ASSESSMENT CODE
(See Hazard Assessment Handbook, CG 446-3)

A-C-K-M-N

12. HAZARD CLASSIFICATIONS
12.1 **Code of Federal Regulations:** Nonflammable compressed gas
12.2 **NAS Hazard Rating for Bulk Water Transportation:**

Category	Rating
Fire	0
Health	
Vapor Irritant	4
Liquid or Solid Irritant	1
Poisons	4
Water Pollution	
Human Toxicity	0
Aquatic Toxicity	3
Aesthetic Effect	1
Reactivity	
Other Chemicals	1
Water	1
Self-Reaction	0

12.3 **NFPA Hazard Classifications:**

Category	Classification
Health Hazard (Blue)	3
Flammability (Red)	0
Reactivity (Yellow)	0

13. PHYSICAL AND CHEMICAL PROPERTIES
13.1 **Physical State at 15°C and 1 atm:** Gas
13.2 **Molecular Weight:** 64.06
13.3 **Boiling Point at 1 atm:** 14°F = −10.0°C = 263.2°K
13.4 **Freezing Point:** −103.9°F = −75.5°C = 197.7°K
13.5 **Critical Temperature:** 315°F = 157°C = 430°K
13.6 **Critical Pressure:** 1142 psia = 77.69 atm = 7.870 MN/m^2
13.7 **Specific Gravity:** 1.45 at −10°C (liquid)
13.8 **Liquid Surface Tension:** Not pertinent
13.9 **Liquid-Water Interfacial Tension:** Not pertinent
13.10 **Vapor (Gas) Specific Gravity:** 2.2
13.11 **Ratio of Specific Heats of Vapor (Gas):** 1.265
13.12 **Latent Heat of Vaporization:** 171 Btu/lb = 94.8 cal/g = 3.97×10^5 J/kg
13.13 **Heat of Combustion:** Not pertinent
13.14 **Heat of Decomposition:** Not pertinent
13.15 **Heat of Solution:** −94.1 Btu/lb = −52.3 cal/g = -2.19×10^5 J/kg
13.16 **Heat of Polymerization:** Not pertinent

NOTES

REVISED 1978

SULFURIC ACID

Common Synonyms			
Oil of vitriol Battery acid Fertilizer acid	Oily liquid	Colorless	Odorless
	Sinks and mixes violently with water. Irritating mist is produced.		

AVOID CONTACT WITH LIQUID. Keep people away.
Wear goggles, self-contained breathing apparatus, and rubber overclothing.
Stop discharge if possible.
Isolate and remove discharged material.
Notify local health and pollution control agencies.

Fire	Not flammable. May cause fire on contact with combustibles. Flammable gas may be produced on contact with metals. POISONOUS GAS MAY BE PRODUCED IN FIRE. **Wear goggles, self-contained breathing apparatus, and rubber overclothing.** **DO NOT USE WATER ON ADJACENT FIRES.** **Extinguish with dry chemical or carbon dioxide.**
Exposure	CALL FOR MEDICAL AID. MIST Irritating to eyes, nose and throat. If inhaled, will cause coughing, difficult breathing, or loss of consciousness. **Move to fresh air.** **IF IN EYES, hold eyelids open and flush with plenty of water.** **If breathing has stopped, give artificial respiration.** **If breathing is difficult, give oxygen.** LIQUID Will burn skin and eyes. Harmful if swallowed. **Remove contaminated clothing and shoes.** **Flush affected areas with plenty of water.** **IF IN EYES, hold eyelids open and flush with plenty of water.** **IF SWALLOWED and victim is CONSCIOUS, have victim drink water or milk.** **DO NOT INDUCE VOMITING.**
Water Pollution	HARMFUL TO AQUATIC LIFE IN VERY LOW CONCENTRATIONS. May be dangerous if it enters water intakes. **Notify local health and wildlife officials.** **Notify operators of nearby water intakes.**

1. RESPONSE TO DISCHARGE

(See Response Methods Handbook, CG 446-4)

Issue warning—corrosive
Restrict access
Disperse and flush with care

2. LABEL

3. CHEMICAL DESIGNATIONS

3.1 **Synonyms:** Battery acid
Chamber acid
Fertilizer acid
Oil of vitriol

3.2 **Coast Guard Compatibility Classification:** Sulfuric acid

3.3 **Chemical Formula:** H_2SO_4

3.4 **IMCO/United Nations Numerical Designation:** 8.0/1830

4. OBSERVABLE CHARACTERISTICS

4.1 **Physical State (as shipped):** Liquid

4.2 **Color:** Colorless (pure) to dark brown

4.3 **Odor:** Odorless unless hot, then choking

5. HEALTH HAZARDS

5.1 **Personal Protective Equipment:** Safety shower; eyewash fountain; safety goggles; face shield; approved respirator (self-contained or air-line); rubber safety shoes; rubber apron.

5.2 **Symptoms Following Exposure:** Inhalation of vapor from hot, concentrated acid may injure lungs. Swallowing may cause severe injury or death. Contact with skin or eyes causes severe burns.

5.3 **Treatment for Exposure:** Call a doctor. INHALATION: observe victim for delayed pulmonary reaction. INGESTION: have victim drink water if possible; do NOT induce vomiting. EYES AND SKIN: wash with large amounts of water for at least 15 min.; do not use oils or ointments in eyes; treat skin burns.

5.4 **Toxicity by Inhalation (Threshold Limit Value):** 1 mg/m³

5.5 **Short-Term Inhalation Limits:** 10 mg/m³ for 5 min.; 5 mg/m³ for 10 min.; 2 mg/m³ for 30 min; 1 mg/m³ for 60 min.

5.6 **Toxicity by Ingestion:** No effects except those secondary to tissue damage.

5.7 **Late Toxicity:** None

5.8 **Vapor (Gas) Irritant Characteristics:** Vapors from hot acid (77-98%) cause moderate irritation of eyes and respiratory system. Effect is temporary.

5.9 **Liquid or Solid Irritant Characteristics:** 77-98% acid causes severe second- and third-degree burns of skin on short contact and is very injurious to the eyes.

5.10 **Odor Threshold:** Greater than 1 mg/m³

6. FIRE HAZARDS

6.1 **Flash Point:** Not flammable

6.2 **Flammable Limits in Air:** Not flammable

6.3 **Fire Extinguishing Agents:** Not pertinent

6.4 **Fire Extinguishing Agents Not to be Used:** Water used on adjacent fires should be carefully handled.

6.5 **Special Hazards of Combustion Products:** Not pertinent

6.6 **Behavior in Fire:** Not flammable

6.7 **Ignition Temperature:** Not flammable

6.8 **Electrical Hazard:** None

6.9 **Burning Rate:** Not flammable

7. CHEMICAL REACTIVITY

7.1 **Reactivity with Water:** Reacts violently with evolution of heat. Spattering occurs when water is added to the compound.

7.2 **Reactivity with Common Materials:** Extremely hazardous in contact with many materials, particularly metals and combustibles. Dilute acid reacts with most metals, releasing hydrogen which can form explosive mixtures with air in confined spaces.

7.3 **Stability During Transport:** Stable

7.4 **Neutralizing Agents for Acids and Caustics:** Dilute with water, then neutralize with lime, limestone, or soda ash.

7.5 **Polymerization:** Not pertinent

7.6 **Inhibitor of Polymerization:** Not pertinent

8. WATER POLLUTION

8.1 **Aquatic Toxicity:**
24.5 ppm/24 hr/bluegill/lethal/fresh water
42.5 ppm/48 hr/prawn/LC_{50}/salt water

8.2 **Waterfowl Toxicity:** Data not available

8.3 **Biological Oxygen Demand (BOD):** None

8.4 **Food Chain Concentration Potential:** None

9. SELECTED MANUFACTURERS

1. Allied Chemical Corp.
Industrial Chemicals Division
P. O. Box 1139R
Morristown, N. J. 07960

2. Cities Service Co., Inc.
North American Chemicals and Metals Group
60 Wall St.
New York, N. Y. 10005

3. E. I. duPont de Nemours & Co., Inc.
Explosives Dept.
Wilmington, Del. 19898

10. SHIPPING INFORMATION

10.1 **Grades or Purity:** CP; USP; Technical, at 33% to 98% (50° Be to 66° Be).

10.2 **Storage Temperature:** Ambient

10.3 **Inert Atmosphere:** No requirement

10.4 **Venting:** Open

11. HAZARD ASSESSMENT CODE

(See Hazard Assessment Handbook, CG 446-3)

A-P-O

12. HAZARD CLASSIFICATIONS

12.1 **Code of Federal Regulations:** Corrosive Material

12.2 **NAS Hazard Rating for Bulk Water Transportation:**

Category	Rating
Fire	0
Health	
Vapor Irritant	2
Liquid or Solid Irritant	4
Poisons	2
Water Pollution	
Human Toxicity	2
Aquatic Toxicity	3
Aesthetic Effect	2
Reactivity	
Other Chemicals	4
Water	3
Self-Reaction	0

12.3 **NFPA Hazard Classifications:**

Category	Classification
Health Hazard (Blue)	3
Flammability (Red)	0
Reactivity (Yellow)	2
	W

13. PHYSICAL AND CHEMICAL PROPERTIES

13.1 **Physical State at 15°C and 1 atm:** Liquid

13.2 **Molecular Weight:** 98.08

13.3 **Boiling Point at 1 atm:** 644°F = 340°C = 613°K

13.4 **Freezing Point:** Not pertinent

13.5 **Critical Temperature:** Not pertinent

13.6 **Critical Pressure:** Not pertinent

13.7 **Specific Gravity:** 1.84 at 20°C (liquid)

13.8 **Liquid Surface Tension:** Not pertinent

13.9 **Liquid-Water Interfacial Tension:** Not pertinent

13.10 **Vapor (Gas) Specific Gravity:** Not pertinent

13.11 **Ratio of Specific Heats of Vapor (Gas):** Not pertinent

13.12 **Latent Heat of Vaporization:** Not pertinent

13.13 **Heat of Combustion:** Not pertinent

13.14 **Heat of Decomposition:** Not pertinent

13.15 **Heat of Solution:** −418.0 Btu/lb = −232.2 cal/g = -9.715×10^5 J/kg

13.16 **Heat of Polymerization:** Not pertinent

*Physical properties apply to concentrated (98%) acid unless otherwise stated. More dilute acid is more water-like.

NOTES

REVISED 1978

SAC

SULFURIC ACID, SPENT

Common Synonyms	Oily liquid	Colorless to dark brown	Odorless
Dilute sulfuric acid	Sinks and mixes with water.		

AVOID CONTACT WITH LIQUID. Keep people away.
Wear goggles, and rubber overclothing (including gloves).
Stop discharge if possible.
Isolate and remove discharged material.
Notify local health and pollution control agencies.

Fire

Not flammable.
Poisonous gas may be produced in fire.
Flammable gas may be produced on contact with metals.
Wear goggles, self-contained breathing apparatus, and rubber overclothing.
Extinguish adjacent fires with dry chemical or carbon dioxide.

Exposure

CALL FOR MEDICAL AID.

LIQUID
Will burn skin and eyes.
Harmful if swallowed.
Remove contaminated clothing and shoes.
Flush affected areas with plenty of water.
IF IN EYES, hold eyelids open and flush with plenty of water.
IF SWALLOWED and victim is CONSCIOUS, have victim drink water or milk.
DO NOT INDUCE VOMITING.

Water Pollution

HARMFUL TO AQUATIC LIFE IN VERY LOW CONCENTRATIONS.
May be dangerous if it enters water intakes.
Notify local health and wildlife officials.
Notify operators of nearby water intakes.

1. RESPONSE TO DISCHARGE

(See Response Methods Handbook, CG 446-4)

Issue warning—corrosive
Restrict access
Disperse and flush

2. LABEL

3. CHEMICAL DESIGNATIONS

3.1 **Synonyms:** No common synonyms
3.2 **Coast Guard Compatibility Classification:** Sulfuric acid
3.3 **Chemical Formula:** H_2SO_4–H_2O
3.4 **IMCO/United Nations Numerical Designation:** 8.0/1832

4. OBSERVABLE CHARACTERISTICS

4.1 **Physical State (as shipped):** Liquid
4.2 **Color:** Colorless to dark brown
4.3 **Odor:** Odorless

5. HEALTH HAZARDS

5.1 **Personal Protective Equipment:** Chemical safety goggles and face shield; rubber gloves, boots, and apron.
5.2 **Symptoms Following Exposure:** Contact with eyes or skin causes severe burns, the severity depending on the strength of the acid. Ingestion can cause severe irritation of mouth and stomach.
5.3 **Treatment for Exposure:** Call a doctor. INGESTION: do NOT induce vomiting. SKIN OR EYES: flush affected parts with large amounts of water for at least 15 min.; do NOT use oils or ointments in eyes; treat burns.
5.4 **Toxicity by Inhalation (Threshold Limit Value):** Not pertinent
5.5 **Short-Term Inhalation Limits:** Not pertinent
5.6 **Toxicity by Ingestion:** No effects except those stemming from tissue damage.
5.7 **Late Toxicity:** None
5.8 **Vapor (Gas) Irritant Characteristics:** Non-volatile
5.9 **Liquid or Solid Irritant Characteristics:** Severe skin irritant. Causes second- and third-degree burns on short contact and is very injurious to the eyes.
5.10 **Odor Threshold:** Not pertinent

6. FIRE HAZARDS

6.1 **Flash Point:** Not flammable
6.2 **Flammable Limits in Air:** Not flammable
6.3 **Fire Extinguishing Agents:** Not pertinent
6.4 **Fire Extinguishing Agents Not to be Used:** Not pertinent
6.5 **Special Hazards of Combustion Products:** Not pertinent
6.6 **Behavior in Fire:** Not pertinent
6.7 **Ignition Temperature:** Not flammable
6.8 **Electrical Hazard:** Not pertinent
6.9 **Burning Rate:** Not flammable

7. CHEMICAL REACTIVITY

7.1 **Reactivity with Water:** None, unless strength is above 80–90%, in which case heat is liberated.
7.2 **Reactivity with Common Materials:** Attacks many metals, releasing flammable hydrogen gas.
7.3 **Stability During Transport:** Stable
7.4 **Neutralizing Agents for Acids and Caustics:** Limestone, lime, or soda ash.
7.5 **Polymerization:** Not pertinent
7.6 **Inhibitor of Polymerization:** Not pertinent

8. WATER POLLUTION

8.1 **Aquatic Toxicity:**
24.5 ppm/24 hr/bluegill/lethal/fresh water
42.5 ppm/48 hr/prawn/LC_{50}/salt water
8.2 **Waterfowl Toxicity:** Data not available
8.3 **Biological Oxygen Demand (BOD):** None
8.4 **Food Chain Concentration Potential:** None

9. SELECTED MANUFACTURERS

Not manufactured. Dilute, or "spent", sulfuric acid is formed as a result of various uses of the concentrated acid.

10. SHIPPING INFORMATION

10.1 **Grades or Purity:** Purity depends on the process in which the original acid is used. The strength (in water) is probably below 80%, and the solution may contain a wide variety of metals and organic compounds in solution.
10.2 **Storage Temperature:** Ambient
10.3 **Inert Atmosphere:** No requirement
10.4 **Venting:** Open

11. HAZARD ASSESSMENT CODE

(See Hazard Assessment Handbook, CG 446-3)

A-P

12. HAZARD CLASSIFICATIONS

12.1 **Code of Federal Regulations:** Corrosive material
12.2 **NAS Hazard Rating for Bulk Water Transportation:** Not listed
12.3 **NFPA Hazard Classifications:**

Category	Classification
Health Hazard (Blue)	3
Flammability (Red)	0
Reactivity (Yellow)	2

13. PHYSICAL AND CHEMICAL PROPERTIES

13.1 **Physical State at 15°C and 1 atm:** Liquid
13.2 **Molecular Weight:** Not pertinent
13.3 **Boiling Point at 1 atm:** 212°F = 100°C = 373°K
13.4 **Freezing Point:** Not pertinent
13.5 **Critical Temperature:** Not pertinent
13.6 **Critical Pressure:** Not pertinent
13.7 **Specific Gravity:** 1.39 at 20°C (liquid)
13.8 **Liquid Surface Tension:** Not pertinent
13.9 **Liquid-Water Interfacial Tension:** Not pertinent
13.10 **Vapor (Gas) Specific Gravity:** Not pertinent
13.11 **Ratio of Specific Heats of Vapor (Gas):** Not pertinent
13.12 **Latent Heat of Vaporization:** Not pertinent
13.13 **Heat of Combustion:** Not pertinent
13.14 **Heat of Decomposition:** Not pertinent
13.15 **Heat of Solution:** < -418 Btu/lb $= < -232$ cal/g $= < -9.71 \times 10^5$ J/kg
13.16 **Heat of Polymerization:** Not pertinent

NOTES

REVISED 1978

SXX

SULFUR (LIQUID)

Common Synonyms
Brimstone

Liquid (molten solid) | Yellow, orange, tan, brown, or gray | Faint rotten eggs odor

Thickens and sinks in water.

Avoid contact with liquid and solid. Keep people away.
Stop discharge if possible.
Call fire department.
Isolate and remove discharged material.
Notify local health and pollution control agencies.

Fire

Combustible.
POISONOUS GAS IS PRODUCED IN FIRE.

Wear goggles and self-contained breathing apparatus.
Extinguish with water or sand.

Exposure

CALL FOR MEDICAL AID.

LIQUID
Will burn skin and eyes.
Harmful if swallowed.

Flush affected areas with plenty of water.
IF IN EYES, hold eyelids open and flush with plenty of water.
IF SWALLOWED and victim is CONSCIOUS, have victim drink water or milk.

Water Pollution

Dangerous to aquatic life in high concentrations.
May be dangerous if it enters water intakes.

Notify local health and wildlife officials.
Notify operators of nearby water intakes.

1. RESPONSE TO DISCHARGE

(See Response Methods Handbook, CG 446-4)

Should be removed
Chemical and physical treatment

2. LABELS

No hazard label required by Code of Federal Regulations

3. CHEMICAL DESIGNATIONS

3.1 **Synonyms:** Sulphur (liquid)
Brimstone (liquid)

3.2 **Coast Guard Compatibility Classification:** Sulfur, molten

3.3 **Chemical Formula:** S

3.4 **IMCO/United Nations Numerical Designation:** Not listed

4. OBSERVABLE CHARACTERISTICS

4.1 **Physical State (as shipped):** Liquid

4.2 **Color:** Bright yellow, orange, tan, brown, gray, depending upon amount and type of hydrocarbon impurity.

4.3 **Odor:** Pure sulfur is odorless, but traces of hydrocarbon impurity may impart an oily and/or rotten egg odor. Recovered sulfur usually has a strong rotten egg odor

5. HEALTH HAZARDS

5.1 **Personal Protective Equipment:** Safety goggles with side shields; approved respirator; heat-resistant gloves; leather heat-resistant clothing. If recovered sulfur, refer to hydrogen sulfide.*

5.2 **Symptoms Following Exposure:** Can cause eye irritation; may rarely irritate skin. If recovered sulfur, refer to hydrogen sulfide.*

5.3 **Treatment for Exposure:** EYES: wash eyes carefully for at least 15 min. SKIN: Treat molten sulfur burns with petroleum jelly or mineral oil. If recovered sulfur, treat as for hydrogen sulfide.*

5.4 **Toxicity by Inhalation (Threshold Limit Value):** If recovered sulfur, see hydrogen sulfide.*

5.5 **Short-Term Inhalation Limits:** If recovered sulfur, see hydrogen sulfide.*

5.6 **Toxicity by Ingestion:** Grade 2; LD_{50} 0.5 to 5 g/kg

5.7 **Late Toxicity:** None

5.8 **Vapor (Gas) Irritant Characteristics:** Non-volatile

5.9 **Liquid or Solid Irritant Characteristics:** Minimum hazard. If spilled on clothing and allowed to remain, may cause smarting and reddening of the skin.

5.10 **Odor Threshold:** If recovered sulfur, see hydrogen sulfide.*

*Significant amounts of hydrogen sulfide, a very poisonous gas, may collect in poorly ventilated containers of liquid sulfur that has been recovered from hydrogen sulfide.

6. FIRE HAZARDS

6.1 **Flash Point:** 405°F C.C.; for recovered sulfur, see hydrogen sulfide.

6.2 **Flammable Limits in Air:** Not pertinent

6.3 **Fire Extinguishing Agents:** Water

6.4 **Fire Extinguishing Agents Not to be Used:** Not pertinent

6.5 **Special Hazards of Combustion Products:** Produces toxic sulfur dioxide gas.

6.6 **Behavior in Fire:** Burns with a pale blue flame that may be difficult to see in daylight.

6.7 **Ignition Temperature:** 450°F; for recovered sulfur, see hydrogen sulfide.

6.8 **Electrical Hazard:** Not pertinent

6.9 **Burning Rate:** Not pertinent

7. CHEMICAL REACTIVITY

7.1 **Reactivity with Water:** No reaction

7.2 **Reactivity with Common Materials:** No hazardous reaction

7.3 **Stability During Transport:** Stable

7.4 **Neutralizing Agents for Acids and Caustics:** Not pertinent

7.5 **Polymerization:** Not pertinent

7.6 **Inhibitor of Polymerization:** Not pertinent

8. WATER POLLUTION

8.1 **Aquatic Toxicity:** 10,000 ppm/96 hr/mosquito fish/TL_m/fresh water

8.2 **Waterfowl Toxicity:** Data not available

8.3 **Biological Oxygen Demand (BOD):** Data not available

8.4 **Food Chain Concentration Potential:** None

9. SELECTED MANUFACTURERS

1. Freeport Minerals Co.
Freeport Sulphur Co. Division
821 Gravieur St.
New Orleans, La. 70112

2. Pennzoil United, Inc.
Duval Corp.
1906 First City Nat'l Bank Bldg.
Houston, Tex. 77002

3. Texas Gulf Sulfur
200 Park Ave.
New York, N. Y. 10017

10. SHIPPING INFORMATION

10.1 **Grades or Purity:** Frasch liquid sulfur: 99.8+%; solid sulfur is sold in many varieties and grades; these are not presently covered in this manual.

10.2 **Storage Temperature:** 270°F

10.3 **Inert Atmosphere:** Ventilated (natural)

10.4 **Venting:** Open

11. HAZARD ASSESSMENT CODE

(See Hazard Assessment Handbook, CG 446-3)

A-X-Y

12. HAZARD CLASSIFICATIONS

12.1 **Code of Federal Regulations:** Not listed

12.2 **NAS Hazard Rating for Bulk Water Transportation:**

Category	Rating
Fire	1-3
Health	
Vapor Irritant	1
Liquid or Solid Irritant	1
Poisons	1
Water Pollution	
Human Toxicity	0
Aquatic Toxicity	1
Aesthetic Effect	1
Reactivity	
Other Chemicals	4
Water	0
Self-Reaction	0

12.3 **NFPA Hazard Classifications:**

Category	Classification*	
Health Hazard (Blue)	1	2
Flammability (Red)	0	0
Reactivity (Yellow)	1	1

*First column refers to non-fire situations. Ratings pertain to solid sulfur.

13. PHYSICAL AND CHEMICAL PROPERTIES

13.1 **Physical State at 15°C and 1 atm:** Solid

13.2 **Molecular Weight:** 256.51

13.3 **Boiling Point at 1 atm:** 832.3°F = 444.6°C = 717.8°K

13.4 **Freezing Point:** 251°F = 121.7°C = 394.9°K

13.5 **Critical Temperature:** Not pertinent

13.6 **Critical Pressure:** Not pertinent

13.7 **Specific Gravity:** 1.80 at 120°C (liquid)

13.8 **Liquid Surface Tension:** 60.8 dynes/cm = 0.0608 N/m at 120°C

13.9 **Liquid-Water Interfacial Tension:** (est.) 50 dynes/cm = 0.05 N/m at 127°C

13.10 **Vapor (Gas) Specific Gravity:** Not pertinent

13.11 **Ratio of Specific Heats of Vapor (Gas):** 1.582 (est.)

13.12 **Latent Heat of Vaporization:** 120 Btu/lb = 69 cal/g = 2.9×10^5 J/kg

13.13 **Heat of Combustion:** −4,741 Btu/lb = −2,634 cal/g = -110.3×10^5 J/kg

13.14 **Heat of Decomposition:** Not pertinent

13.15 **Heat of Solution:** Not pertinent

13.16 **Heat of Polymerization:** Not pertinent

NOTES

SFM	SULFUR MONOCHLORIDE

Common Synonyms	Oily liquid — Yellow to red — Irritating sharp odor Mixes and reacts with water. Poisonous vapor is produced.

Avoid contact with liquid and vapor. Keep people away.
Wear goggles, self-contained breathing apparatus, and rubber overclothing (including gloves).
Stop discharge if possible.
Call fire department.
Isolate and remove discharged material.
Notify local health and pollution control agencies.

Fire	Combustible. Wear goggles, self-contained breathing apparatus, and rubber overclothing (including gloves). Extinguish with dry chemical, or carbon dioxide. Cool exposed containers with water.
 **Exposure**	CALL FOR MEDICAL AID. VAPOR Irritating to eyes. Poisonous if inhaled. Move to fresh air. If breathing has stopped, give artificial respiration. If breathing is difficult, give oxygen. LIQUID Will burn skin and eyes. Poisonous if swallowed. Remove contaminated clothing and shoes. Flush affected areas with plenty of water. IF IN EYES, hold eyelids open and flush with plenty of water. IF SWALLOWED and victim is CONSCIOUS, have victim drink water or milk. DO NOT INDUCE VOMITING.
Water Pollution	Dangerous to aquatic life in high concentrations. May be dangerous if it enters water intakes. Notify local health and wildlife officials. Notify operators of nearby water intakes.

1. RESPONSE TO DISCHARGE
(See Response Methods Handbook, CG 446-4)

Issue warning—corrosive, air contaminant, water contaminant
Restrict access
Chemical and physical treatment

2. LABEL

3. CHEMICAL DESIGNATIONS

3.1 **Synonyms:** No common synonyms
3.2 **Coast Guard Compatibility Classification:** Not applicable
3.3 **Chemical Formula:** S_2Cl_2
3.4 **IMCO/United Nations Numerical Designation:** 8.0/1828

4. OBSERVABLE CHARACTERISTICS

4.1 **Physical State (as shipped):** Liquid
4.2 **Color:** Yellowish red
4.3 **Odor:** Suffocating; pungent, nauseating, irritating

5. HEALTH HAZARDS

5.1 **Personal Protective Equipment:** Chemical safety goggles and/or face shield; canister-type gas mask (light concentrations) or self-contained breathing apparatus (heavy concentrations); chemically resistant shoes or boots, apron, and long-sleeve gloves.
5.2 **Symptoms Following Exposure:** Vapors irritate eyes and respiratory system; pulmonary edema may result. Liquid burns and damages eyes. Unless removed at once, it burns the skin. Ingestion causes severe damage to mouth and stomach.
5.3 **Treatment for Exposure:** INHALATION: remove to fresh air; use artificial respiration and oxygen if required; call a doctor. INGESTION: give water; do NOT induce vomiting; call a doctor. EYES: flush with water for at least 15 min.; obtain medical attention at once. SKIN: flush with water; remove contaminated clothing under shower.
5.4 **Toxicity by Inhalation (Threshold Limit Value):** 1 ppm
5.5 **Short-Term Inhalation Limits:** Data not available
5.6 **Toxicity by Ingestion:** Data not available
5.7 **Late Toxicity:** None
5.8 **Vapor (Gas) Irritant Characteristics:** Vapors cause severe irritation of eye and throat and can cause eye and lung injury. They cannot be tolerated even at low concentrations.
5.9 **Liquid or Solid Irritant Characteristics:** Severe skin irritant. Causes second- and third degree burns on short contact and is very injurious to the eyes.
5.10 **Odor Threshold:** Data not available

6. FIRE HAZARDS

6.1 **Flash Point:** 245°F C.C.; 266°F O.C.
6.2 **Flammable Limits in Air:** Data not available
6.3 **Fire Extinguishing Agents:** Dry chemical, carbon dioxide, water spray
6.4 **Fire Extinguishing Agents Not to be Used:** Water reacts violently with compound.
6.5 **Special Hazards of Combustion Products:** Toxic and corrosive fumes are evolved when heated.
6.6 **Behavior in Fire:** Not pertinent
6.7 **Ignition Temperature:** 453°F
6.8 **Electrical Hazard:** Not pertinent
6.9 **Burning Rate:** Data not available

7. CHEMICAL REACTIVITY

7.1 **Reactivity with Water:** Reacts violently with water to produce heat and hydrogen chloride fumes. The solution is strongly acid.
7.2 **Reactivity with Common Materials:** The liquid dissolves rubber and plastics. After reaction with water, the strong acid formed attacks metals, generating flammable hydrogen gas.
7.3 **Stability During Transport:** Stable
7.4 **Neutralizing Agents for Acids and Caustics:** After reaction with water, the acid formed can be neutralized with lime or soda ash.
7.5 **Polymerization:** Not pertinent
7.6 **Inhibitor of Polymerization:** Not pertinent

8. WATER POLLUTION

8.1 **Aquatic Toxicity:** Data not available
8.2 **Waterfowl Toxicity:** Data not available
8.3 **Biological Oxygen Demand (BOD):** Data not available
8.4 **Food Chain Concentration Potential:** None

9. SELECTED MANUFACTURERS

1. Hooker Chemical Corp.
   Industrial Chemicals Division
   Niagara Falls, N. Y. 14302
2. Stauffer Chemical Co.
   Industrial Chemical Division
   Niagara Falls, N. Y. 14302

10. SHIPPING INFORMATION

10.1 **Grades or Purity:** Commercial material may contain 0—5% free sulfur.
10.2 **Storage Temperature:** Ambient
10.3 **Inert Atmosphere:** No requirement
10.4 **Venting:** Pressure-vacuum

11. HAZARD ASSESSMENT CODE
(See Hazard Assessment Handbook, CG 446-3)

A-O

12. HAZARD CLASSIFICATIONS

12.1 **Code of Federal Regulations:** Corrosive material
12.2 **NAS Hazard Rating for Bulk Water Transportation:**

Category	Rating
Fire	1
Health	
Vapor Irritant	4
Liquid or Solid Irritant	4
Poisons	4
Water Pollution	
Human Toxicity	3
Aquatic Toxicity	2
Aesthetic Effect	4
Reactivity	
Other Chemicals	3
Water	4
Self-Reaction	0

12.3 **NFPA Hazard Classifications:**

Category	Classification
Health Hazard (Blue)	2
Flammability (Red)	1
Reactivity (Yellow)	1

13. PHYSICAL AND CHEMICAL PROPERTIES

13.1 **Physical State at 15°C and 1 atm:** Liquid
13.2 **Molecular Weight:** 135.03
13.3 **Boiling Point at 1 atm:** 280°F = 138°C = 411°K
13.4 **Freezing Point:** −112°F = −80°C = 193°K
13.5 **Critical Temperature:** Not pertinent
13.6 **Critical Pressure:** Not pertinent
13.7 **Specific Gravity:** 1.68 at 20°C (liquid)
13.8 **Liquid Surface Tension:** Not pertinent
13.9 **Liquid-Water Interfacial Tension:** Not pertinent
13.10 **Vapor (Gas) Specific Gravity:** Not pertinent
13.11 **Ratio of Specific Heats of Vapor (Gas):** 1.129
13.12 **Latent Heat of Vaporization:** 115 Btu/lb = 63.8 cal/g = 2.67×10^5 J/kg
13.13 **Heat of Combustion:** Not pertinent
13.14 **Heat of Decomposition:** Not pertinent
13.15 **Heat of Solution:** −502.2 Btu/lb = −279.0 cal/g = -11.67×10^5 J/kg
13.16 **Heat of Polymerization:** Not pertinent

NOTES

REVISED 1978

SCL — SULFURYL CHLORIDE

Common Synonyms	Watery liquid — Colorless to light yellow — Acrid odor Mixes and reacts violently with water. Poisonous gas is produced.

Avoid contact with liquid and gas. Keep people away.
Wear goggles, self-contained breathing apparatus, and rubber overclothing (including gloves).
Stop discharge if possible.
Isolate and remove discharged material.
Notify local health and pollution control agencies.

Fire	Not flammable. Flammable gas may be produced on contact with metals.
Exposure	CALL FOR MEDICAL AID. VAPOR Irritating to eyes, nose, and throat. If inhaled, will cause coughing, difficult breathing, or loss of consciousness. Move to fresh air. IF IN EYES, hold eyelids open and flush with plenty of water. If breathing has stopped, give artificial respiration. If breathing is difficult, give oxygen. LIQUID Will burn skin and eyes. Harmful if swallowed. Remove contaminated clothing and shoes. Flush affected areas with plenty of water. IF IN EYES, hold eyelids open and flush with plenty of water. IF SWALLOWED and victim is CONSCIOUS, have victim drink water or milk. DO NOT INDUCE VOMITING.
Water Pollution	HARMFUL TO AQUATIC LIFE IN VERY LOW CONCENTRATIONS. May be dangerous if it enters water intakes. Notify local health and wildlife officials. Notify operators of nearby water intakes.

1. RESPONSE TO DISCHARGE
(See Response Methods Handbook, CG 446-4)

Issue warning — corrosive, air contaminant
Restrict access
Evacuate area
Disperse and flush with care

2. LABEL

3. CHEMICAL DESIGNATIONS

3.1 **Synonyms:** No common synonyms
3.2 **Coast Guard Compatibility Classification:** Not applicable
3.3 **Chemical Formula:** SO_2Cl_2
3.4 **IMCO/United Nations Numerical Designation:** 8.0/1834

4. OBSERVABLE CHARACTERISTICS

4.1 **Physical State (as shipped):** Liquid
4.2 **Color:** Colorless to light yellow
4.3 **Odor:** Acrid, choking

5. HEALTH HAZARDS

5.1 **Personal Protective Equipment:** Chemical goggles and face shield; mask with acid-type canister; rubber gloves and boots.
5.2 **Symptoms Following Exposure:** Vapors cause severe irritation of eyes and respiratory system. Liquid burns eyes and skin. If ingested, can cause severe burns of mouth and stomach.
5.3 **Treatment for Exposure:** Call a doctor. INHALATION: remove to fresh air; administer artificial respiration if required. INGESTION: give water or milk; do NOT induce vomiting. EYES: flush with water for at least 15 min. SKIN: wash with large amounts of water.
5.4 **Toxicity by Inhalation (Threshold Limit Value):** Data not available
5.5 **Short-Term Inhalation Limits:** Data not available
5.6 **Toxicity by Ingestion:** Data not available
5.7 **Late Toxicity:** None
5.8 **Vapor (Gas) Irritant Characteristics:** Vapors cause severe irritation of eyes and throat and can cause eye and lung injury. They cannot be tolerated even at low concentrations.
5.9 **Liquid or Solid Irritant Characteristics:** Fairly severe skin irritant; may cause pain and second-degree burns after a few minutes' contact.
5.10 **Odor Threshold:** Data not available

6. FIRE HAZARDS

6.1 **Flash Point:** Not flammable
6.2 **Flammable Limits in Air:** Not flammable
6.3 **Fire Extinguishing Agents:** Not pertinent
6.4 **Fire Extinguishing Agents Not to be Used:** Water applied to adjacent fires should be handled carefully.
6.5 **Special Hazards of Combustion Products:** Not pertinent
6.6 **Behavior in Fire:** Toxic and irritating gases are generated.
6.7 **Ignition Temperature:** Not flammable
6.8 **Electrical Hazard:** Not pertinent
6.9 **Burning Rate:** Not flammable

7. CHEMICAL REACTIVITY

7.1 **Reactivity with Water:** Reacts vigorously with water, releasing hydrogen chloride fumes and forming sulfuric acid.
7.2 **Reactivity with Common Materials:** Acids formed by reaction with moisture attack metals and liberate flammable hydrogen gas.
7.3 **Stability During Transport:** Stable
7.4 **Neutralizing Agents for Acids and Caustics:** Acid formed by reaction with water can be neutralized by limestone, lime, or soda ash.
7.5 **Polymerization:** Not pertinent
7.6 **Inhibitor of Polymerization:** Not pertinent

8. WATER POLLUTION

8.1 **Aquatic Toxicity:** Data not available
8.2 **Waterfowl Toxicity:** Data not available
8.3 **Biological Oxygen Demand (BOD):** None
8.4 **Food Chain Concentration Potential:** None

9. SELECTED MANUFACTURERS

1. Hooker Chemical Corp.
   Industrial Chemicals Division
   Niagara Falls, N. Y. 14302
2. Union Carbide Corp.
   Linde Division
   270 Park Ave.
   New York, N. Y. 10017

10. SHIPPING INFORMATION

10.1 **Grades or Purity:** 99%
10.2 **Storage Temperature:** Ambient
10.3 **Inert Atmosphere:** No requirement
10.4 **Venting:** Pressure-vacuum

11. HAZARD ASSESSMENT CODE
(See Hazard Assessment Handbook, CG 446-3)

A-O

12. HAZARD CLASSIFICATIONS

12.1 **Code of Federal Regulations:** Corrosive material
12.2 **NAS Hazard Rating for Bulk Water Transportation:**

Category	Rating
Fire	0
Health	
Vapor Irritant	4
Liquid or Solid Irritant	3
Poisons	2
Water Pollution	
Human Toxicity	—
Aquatic Toxicity	3
Aesthetic Effect	2
Reactivity	
Other Chemicals	4
Water	4
Self-Reaction	0

12.3 **NFPA Hazard Classifications:** Not listed

13. PHYSICAL AND CHEMICAL PROPERTIES

13.1 **Physical State at 15°C and 1 atm:** Liquid
13.2 **Molecular Weight:** 134.97
13.3 **Boiling Point at 1 atm:** 156.4°F = 69.1°C = 342.3°K
13.4 **Freezing Point:** Not pertinent
13.5 **Critical Temperature:** Not pertinent
13.6 **Critical Pressure:** Not pertinent
13.7 **Specific Gravity:** 1.67 at 20°C (liquid)
13.8 **Liquid Surface Tension:** Not pertinent
13.9 **Liquid-Water Interfacial Tension:** Not pertinent
13.10 **Vapor (Gas) Specific Gravity:** 4.6
13.11 **Ratio of Specific Heats of Vapor (Gas):** 1.122
13.12 **Latent Heat of Vaporization:** 89.1 Btu/lb = 49.5 cal/g = 2.07×10^5 J/kg
13.13 **Heat of Combustion:** Not pertinent
13.14 **Heat of Decomposition:** Not pertinent
13.15 **Heat of Solution:** –885.5 Btu/lb = –491.9 cal/g = -20.58×10^5 J/kg
13.16 **Heat of Polymerization:** Not pertinent

NOTES

REVISED 1978

TLO	# TALLOW

Common Synonyms Edible tallow Inedible tallow	Oily liquid Dark yellow Waxy odor Floats on water. Freezing point is 35° F – 45° F.

Stop discharge if possible.
Call fire department.
Isolate and remove discharged material.
Notify local health and pollution control agencies.

Fire	Combustible. Extinguish with foam, dry chemical, or carbon dioxide. Water may be ineffective on fire. Cool exposed containers with water.
Exposure	Not harmful.
Water Pollution	Effect of low concentrations on aquatic life is unknown. Fouling to shoreline. May be dangerous if it enters water intakes. Notify local health and wildlife officials. Notify operators of nearby water intakes.

1. RESPONSE TO DISCHARGE

(See Response Methods Handbook, CG 446-4)

Mechanical containment
Should be removed
Chemical and physical treatment

2. LABELS

No hazard label required by Code of Federal Regulations

3. CHEMICAL DESIGNATIONS

3.1 **Synonyms:** Edible tallow
Inedible tallow
Tallow oil

3.2 **Coast Guard Compatibility Classification:** Ester

3.3 **Chemical Formula:** Not pertinent

3.4 **IMCO/United Nations Numerical Designation:** Not listed

4. OBSERVABLE CHARACTERISTICS

4.1 **Physical State (as shipped):** Liquid

4.2 **Color:** Dark yellow

4.3 **Odor:** Waxy

5. HEALTH HAZARDS

5.1 **Personal Protective Equipment:** Goggles or face shield; protective clothing, if exposure to hot liquid is possible.

5.2 **Symptoms Following Exposure:** Hot liquid can burn eyes and skin.

5.3 **Treatment for Exposure:** Treat burns caused by hot liquid.

5.4 **Toxicity by Inhalation (Threshold Limit Value):** Not pertinent

5.5 **Short-Term Inhalation Limits:** Not pertinent

5.6 **Toxicity by Ingestion:** Grade 0; LD_{50} above 15 g/kg

5.7 **Late Toxicity:** None

5.8 **Vapor (Gas) Irritant Characteristics:** Non-volatile

5.9 **Liquid or Solid Irritant Characteristics:** None

5.10 **Odor Threshold:** Not pertinent

6. FIRE HAZARDS

6.1 **Flash Point:** 509°F C.C.

6.2 **Flammable Limits in Air:** Not pertinent

6.3 **Fire Extinguishing Agents:** Foam, water, carbon dioxide, or dry chemical

6.4 **Fire Extinguishing Agents Not to be Used:** Water or foam may cause frothing.

6.5 **Special Hazards of Combustion Products:** Not pertinent

6.6 **Behavior in Fire:** Not pertinent

6.7 **Ignition Temperature:** Data not available

6.8 **Electrical Hazard:** Not pertinent

6.9 **Burning Rate:** Data not available

7. CHEMICAL REACTIVITY

7.1 **Reactivity with Water:** No reaction

7.2 **Reactivity with Common Materials:** No reaction

7.3 **Stability During Transport:** Stable

7.4 **Neutralizing Agents for Acids and Caustics:** Not pertinent

7.5 **Polymerization:** Not pertinent

7.6 **Inhibitor of Polymerization:** Not pertinent

8. WATER POLLUTION

8.1 **Aquatic Toxicity:** Data not available

8.2 **Waterfowl Toxicity:** Data not available

8.3 **Biological Oxygen Demand (BOD):** 152%, 5 days

8.4 **Food Chain Concentration Potential:** None

9. SELECTED MANUFACTURERS

1. Atlas Refinery, Inc.
142 Lockwood St.
Newark, N. J. 07105
2. Reilly-Whiteman, Inc.
Conshohocken, Pa. 19428
3. Swift and Co.
115 W. Jackson Blvd.
Chicago, Ill. 60604

10. SHIPPING INFORMATION

10.1 **Grades or Purity:** Acidless; buffing; industrial fancy; edible; inedible

10.2 **Storage Temperature:** Ambient

10.3 **Inert Atmosphere:** No requirement

10.4 **Venting:** Open (flame arrester)

11. HAZARD ASSESSMENT CODE

(See Hazard Assessment Handbook, CG 446-3)

A-T-U

12. HAZARD CLASSIFICATIONS

12.1 **Code of Federal Regulations:** Not listed

12.2 **NAS Hazard Rating for Bulk Water Transportation:** Not listed

12.3 **NFPA Hazard Classifications:**

Category	Classification
Health Hazard (Blue)	0
Flammability (Red)	1
Reactivity (Yellow)	0

13. PHYSICAL AND CHEMICAL PROPERTIES

13.1 **Physical State at 15°C and 1 atm:** Solid

13.2 **Molecular Weight:** Not pertinent

13.3 **Boiling Point at 1 atm:** Very high

13.4 **Freezing Point:** 35—45 °F = 2—7°C = 275—280°K

13.5 **Critical Temperature:** Not pertinent

13.6 **Critical Pressure:** Not pertinent

13.7 **Specific Gravity:** (est.) 0.87 at 80°C (liquid)

13.8 **Liquid Surface Tension:** Data not available

13.9 **Liquid-Water Interfacial Tension:** Data not available

13.10 **Vapor (Gas) Specific Gravity:** Not pertinent

13.11 **Ratio of Specific Heats of Vapor (Gas):** Not pertinent

13.12 **Latent Heat of Vaporization:** Not pertinent

13.13 **Heat of Combustion:** (est.) $-18{,}000$ Btu/lb = $-10{,}000$ cal/g = -420×10^5 J/kg

13.14 **Heat of Decomposition:** Not pertinent

13.15 **Heat of Solution:** Not pertinent

13.16 **Heat of Polymerization:** Not pertinent

NOTES

REVISED 1978

TFA	# TALLOW FATTY ALCOHOL

Common Synonyms			
Higher fatty alcohol Stearyl alcohol, crude	Waxy solid Floats on water.	White	Mild soapy odor

Call fire department.
Isolate and remove discharged material.
Notify local health and pollution control agencies.

Fire	Combustible. Extinguish with dry chemicals, foam or carbon dioxide. Water may be ineffective on fire.
Exposure	SOLID Not harmful.
Water Pollution	Effect of low concentrations on aquatic life is unknown. Fouling to shoreline. May be dangerous if it enters water intakes. Notify local health and wildlife officials. Notify operators of nearby water intakes.

1. RESPONSE TO DISCHARGE

(See Response Methods Handbook, CG 446-4)

Mechanical containment
Should be removed
Chemical and physical treatment

2. LABELS

No label required by Code of Federal Regulations

3. CHEMICAL DESIGNATIONS

3.1 **Synonyms:** Higher fatty alcohol; Stearyl alcohol, crude
3.2 **Coast Guard Compatibility Classification:** Alcohol
3.3 **Chemical Formula:** $C_{18}H_{37}OH$ (approx.)
3.4 **IMCO/United Nations Numerical Designation:** Not listed

4. OBSERVABLE CHARACTERISTICS

4.1 **Physical State (as shipped):** Solid
4.2 **Color:** Waxy white
4.3 **Odor:** Mild soapy

5. HEALTH HAZARDS

5.1 **Personal Protective Equipment:** Goggles or face shield; rubber gloves.
5.2 **Symptoms Following Exposure:** Material is practically non-toxic. Contact with eyes or prolonged contact with skin can cause mild irritation.
5.3 **Treatment for Exposure:** EYES: flush with water; if irritation persists, see a doctor. SKIN: wipe off, wash with soap and water.
5.4 **Toxicity by Inhalation (Threshold Limit Value):** Data not available
5.5 **Short-Term Inhalation Limits:** Not pertinent
5.6 **Toxicity by Ingestion:** Grade 2; oral LD_{50} = 1,900 mg/kg (rat)
5.7 **Late Toxicity:** Data not available
5.8 **Vapor (Gas) Irritant Characteristics:** Data not available
5.9 **Liquid or Solid Irritant Characteristics:** Data not available
5.10 **Odor Threshold:** Data not available

6. FIRE HAZARDS

6.1 **Flash Point:** >270°F C.C.
6.2 **Flammable Limits in Air:** Not pertinent
6.3 **Fire Extinguishing Agents:** Dry chemical, foam, carbon dioxide
6.4 **Fire Extinguishing Agents Not to be Used:** Water may be ineffective.
6.5 **Special Hazards of Combustion Products:** Not pertinent
6.6 **Behavior in Fire:** Not pertinent
6.7 **Ignition Temperature:** Not pertinent
6.8 **Electrical Hazard:** Not pertinent
6.9 **Burning Rate:** Not pertinent

7. CHEMICAL REACTIVITY

7.1 **Reactivity with Water:** No reaction
7.2 **Reactivity with Common Materials:** No reaction
7.3 **Stability During Transport:** Stable
7.4 **Neutralizing Agents for Acids and Caustics:** Not pertinent
7.5 **Polymerization:** Not pertinent
7.6 **Inhibitor of Polymerization:** Not pertinent

8. WATER POLLUTION

8.1 **Aquatic Toxicity:** Data not available
8.2 **Waterfowl Toxicity:** Data not available
8.3 **Biological Oxygen Demand (BOD):** Data not available
8.4 **Food Chain Concentration Potential:** None

9. SELECTED MANUFACTURERS

1. Procter & Gamble Distributing Co.
Industrial Chemicals Division
P. O. Box 599
Cincinnati, Ohio 45201
2. Ashland Chemical Co.
Industrial Chemicals and Solvents Div.
P. O. Box 2219
Columbus, Ohio 43216
3. M. Michel and Co., Inc.
90 Broad Street
New York, N. Y. 10004

10. SHIPPING INFORMATION

10.1 **Grades or Purity:** Technical
10.2 **Storage Temperature:** Ambient
10.3 **Inert Atmosphere:** No requirement
10.4 **Venting:** Open (flame arrester)

11. HAZARD ASSESSMENT CODE

(See Hazard Assessment Handbook, CG 446-3)

II

12. HAZARD CLASSIFICATIONS

12.1 **Code of Federal Regulations:** Not listed
12.2 **NAS Hazard Rating for Bulk Water Transportation:** Not listed
12.3 **NFPA Hazard Classifications:** Not listed

13. PHYSICAL AND CHEMICAL PROPERTIES

13.1 **Physical State at 15°C and 1 atm:** Solid
13.2 **Molecular Weight:** 262 (avg.)
13.3 **Boiling Point at 1 atm:** >480°F = >249°C = >522°K
13.4 **Freezing Point:** 127°F = 53°C = 326°K
13.5 **Critical Temperature:** Not pertinent
13.6 **Critical Pressure:** Not pertinent
13.7 **Specific Gravity:** 0.810 at 25°C (solid)
13.8 **Liquid Surface Tension:** Not pertinent
13.9 **Liquid-Water Interfacial Tension:** Not pertinent
13.10 **Vapor (Gas) Specific Gravity:** Not pertinent
13.11 **Ratio of Specific Heats of Vapor (Gas):** Not pertinent
13.12 **Latent Heat of Vaporization:** Not pertinent
13.13 **Heat of Combustion:** (est.) −18,500 Btu/lb = −10,300 cal/g = -430×10^5 J/kg
13.14 **Heat of Decomposition:** Not pertinent
13.15 **Heat of Solution:** Not pertinent
13.16 **Heat of Polymerization:** Not pertinent

NOTES

TNA — TANNIC ACID

Common Synonyms	Solid	Light yellow to tan	Faint odor
Tannin Chinese tannin Gallotannic acid Gallotannin Glycerite	Sinks and mixes with water.		

Avoid contact with solid and dust. Keep people away.
Stop discharge if possible.
Call fire department.
Isolate and remove discharged material.
Notify local health and pollution control agencies.

Fire

Combustible.
Extinguish with dry chemicals or carbon dioxide.
Water and foam may be ineffective on fire.
Cool exposed containers with water.

Exposure

CALL FOR MEDICAL AID.

DUST
Irritating to eyes, nose and throat.
If inhaled will cause coughing.
If in eyes, hold eyelids open and flush with plenty of water.
If breathing has stopped, give artificial respiration.
If breathing is difficult, give oxygen.

SOLID
Irritating to skin and eyes.
If swallowed will cause nausea and vomiting.
Remove contaminated clothing and shoes.
Flush affected areas with plenty of water.
IF IN EYES, hold eyelids open and flush with plenty of water.
IF SWALLOWED and victim is CONSCIOUS, have victim drink water or milk and have victim induce vomiting.
IF SWALLOWED and victim is UNCONSCIOUS OR HAVING CONVULSIONS, do nothing except keep victim warm.

Water Pollution

HARMFUL TO AQUATIC LIFE IN VERY LOW CONCENTRATIONS.
May be dangerous if it enters water intakes.
Notify local health and wildlife officials.
Notify operators of nearby water intakes.

1. RESPONSE TO DISCHARGE
(See Response Methods Handbook, CG 446-4)

Disperse and flush

2. LABELS

No hazard label required by Code of Federal Regulations

3. CHEMICAL DESIGNATIONS

3.1 **Synonyms:** Chinese tannin; Gallotannic acid; Gallotannin; Glycerite; Tannin
3.2 **Coast Guard Compatibility Classification:** Not listed
3.3 **Chemical Formula:** $C_{76}H_{52}O_{46}$
3.4 **IMCO/United Nations Numerical Designation:** Not listed

4. OBSERVABLE CHARACTERISTICS

4.1 **Physical State (as shipped):** Solid
4.2 **Color:** Light tan to yellow
4.3 **Odor:** Faint characteristic

5. HEALTH HAZARDS

5.1 **Personal Protective Equipment:** Dust mask; goggles or face shield; protective gloves
5.2 **Symptoms Following Exposure:** Inhalation causes irritation of nose and throat, coughing, and sneezing. Ingestion may cause gastric disturbance. Contact with eyes causes irritation.
5.3 **Treatment for Exposure:** INHALATION: move to fresh air. INGESTION: give large amount of water; induce vomiting. EYES or SKIN: flush with water.
5.4 **Toxicity by Inhalation (Threshold Limit Value):** Data not available
5.5 **Short-Term Inhalation Limits:** Data not available
5.6 **Toxicity by Ingestion:** Grade 2; oral LD_{50} = 2,300 mg/kg (rat)
5.7 **Late Toxicity:** Causes cancer of liver in rats
5.8 **Vapor (Gas) Irritant Characteristics:** Data not available
5.9 **Liquid or Solid Irritant Characteristics:** Data not available
5.10 **Odor Threshold:** Data not available

6. FIRE HAZARDS

6.1 **Flash Point:** Not pertinent (combustible solid)
6.2 **Flammable Limits in Air:** Not pertinent
6.3 **Fire Extinguishing Agents:** Water, foam, dry chemical, carbon dioxide
6.4 **Fire Extinguishing Agents Not to be Used:** Water or foam may cause frothing.
6.5 **Special Hazards of Combustion Products:** Decomposes at 210° to carbon dioxide and pyrogallol, which can form irritating vapors.
6.6 **Behavior in Fire:**
6.7 **Ignition Temperature:** 980°F
6.8 **Electrical Hazard:** Not pertinent
6.9 **Burning Rate:** Not pertinent

7. CHEMICAL REACTIVITY

7.1 **Reactivity with Water:** No reaction
7.2 **Reactivity with Common Materials:**
7.3 **Stability During Transport:** Stable
7.4 **Neutralizing Agents for Acids and Caustics:** Not pertinent
7.5 **Polymerization:** Not pertinent
7.6 **Inhibitor of Polymerization:** Not pertinent

8. WATER POLLUTION

8.1 **Aquatic Toxicity:**
37 ppm/96 hr/mosquitofish/TL_m/fresh water
<1.7 mg/l /72 hr/young chinook salmon/crical level/salt water
8.2 **Waterfowl Toxicity:** Data not available
8.3 **Biological Oxygen Demand (BOD):** 38%, 5 days
8.4 **Food Chain Concentration Potential:** None

9. SELECTED MANUFACTURERS

1. The Harshaw Chemical Co.
   1945 East 97 Street
   Cleveland, Ohio 44106
2. J. T. Baker Chemical Co.
   Phillipsburg, N. J. 08865
3. Fisher Scientific Co.
   711 Forbes Ave.
   Pittsburgh, Pa. 15219

10. SHIPPING INFORMATION

10.1 **Grades or Purity:** Commercial, 84%; Reagent
10.2 **Storage Temperature:** Ambient
10.3 **Inert Atmosphere:** No requirement
10.4 **Venting:** Open

11. HAZARD ASSESSMENT CODE
(See Hazard Assessment Handbook, CG 446-3)

SS

12. HAZARD CLASSIFICATIONS

12.1 **Code of Federal Regulations:** Not listed
12.2 **NAS Hazard Rating for Bulk Water Transportation:** Not listed
12.3 **NFPA Hazard Classifications:**

Category	Classification
Health Hazard (Blue)	0
Flammability (Red)	1
Reactivity (Yellow)	0

13. PHYSICAL AND CHEMICAL PROPERTIES

13.1 **Physical State at 15°C and 1 atm:** Solid
13.2 **Molecular Weight:** 1,701
13.3 **Boiling Point at 1 atm:** Not pertinent (decomposes)
13.4 **Freezing Point:** Not pertinent
13.5 **Critical Temperature:** Not pertinent
13.6 **Critical Pressure:** Not pertinent
13.7 **Specific Gravity:** >1 at 20°C (solid)
13.8 **Liquid Surface Tension:** Not pertinent
13.9 **Liquid-Water Interfacial Tension:** Not pertinent
13.10 **Vapor (Gas) Specific Gravity:** Not pertinent
13.11 **Ratio of Specific Heats of Vapor (Gas):** Not pertinent
13.12 **Latent Heat of Vaporization:** Not pertinent
13.13 **Heat of Combustion:** (est.) −9,810 Btu/lb = −5,450 cal/g = -228×10^5 J/kg
13.14 **Heat of Decomposition:** Not pertinent
13.15 **Heat of Solution:** Not pertinent
13.16 **Heat of Polymerization:** Not pertinent

NOTES

TES — 2,4,5-T ESTERS

Common Synonyms	Liquid	Yellowish brown	Mild odor
Butyl 2,4,5-trichlorophenoxyacetate Butoxypropyl trichlorophenoxyacetate Isooctyl trichlorophenoxyacetate 2-Ethylhexyl trichlorophenoxyacetate	Sinks in water.		

Avoid contact with liquid. Keep people away.
Stop discharge if possible.
Isolate and remove discharged material.
Notify local health and pollution control agencies.

Fire

Combustible.
Irritating gases may be produced when heated.
Wear goggles and self-contained breathing apparatus.
Extinguish with dry chemicals or carbon dioxide.
Water and foam may be ineffective on fire.
Cool exposed containers with water.

Exposure

CALL FOR MEDICAL AID.

LIQUID
Irritating to skin and eyes.
If swallowed will cause nausea and vomiting.
Remove contaminated clothing and shoes.
Flush affected areas with plenty of water.
IF IN EYES, hold eyelids open and flush with plenty of water.
IF SWALLOWED and victim is CONSCIOUS, have victim drink water or milk and have victim induce vomiting.
IF SWALLOWED and victim is UNCONSCIOUS OR HAVING CONVULSIONS, do nothing except keep victim warm.

Water Pollution

Effect of low concentrations on aquatic life is unknown.
May be dangerous if it enters water intakes.
Notify local health and wildlife officials.
Notify operators of nearby water intakes.

1. RESPONSE TO DISCHARGE

(See Response Methods Handbook, CG 446-4)

Issue warning—water contaminant
Should be removed
Chemical and physical treatment

2. LABELS

No hazard label required by Code of Federal Regulations

3. CHEMICAL DESIGNATIONS

3.1 **Synonyms:** Butyl 2,4,5-trichlorophenoxyacetate; Butoxypropyl trichlorophenoxyacetate; Isooctyl trichlorophenoxyacetate; 2-Ethylhexyl trichlorophenoxyacetate

3.2 **Coast Guard Compatibility Classification:** Not listed

(Continued on page 4)

4. OBSERVABLE CHARACTERISTICS

4.1 **Physical State (as shipped):** Liquid

4.2 **Color:** Amber; dark amber

4.3 **Odor:** Very weak; mixtures with kerosene or diesel oil have odor of the solvent.

5. HEALTH HAZARDS

5.1 **Personal Protective Equipment:** Goggles or face shield and rubber gloves

5.2 **Symptoms Following Exposure:** Ingestion causes intestinal disturbances. Contact with eyes or skin causes mild irritation; transient corneal injury may occur.

5.3 **Treatment for Exposure:** INGESTION: promptly induce vomiting and get medical attention. EYES: flush with flowing water and get medical attention. SKIN: wash with soap and water.

5.4 **Toxicity by Inhalation (Threshold Limit Value):** Data not available

5.5 **Short-Term Inhalation Limits:** Data not available

5.6 **Toxicity by Ingestion:** Grade 3; LD_{50} 50–500 mg/kg

5.7 **Late Toxicity:** Data not available

5.8 **Vapor (Gas) Irritant Characteristics:** Data not available

5.9 **Liquid or Solid Irritant Characteristics:** Data not available

5.10 **Odor Threshold:** Data not available

6. FIRE HAZARDS

6.1 **Flash Point:** 265–420°F O.C.

6.2 **Flammable Limits in Air:** Data not available

6.3 **Fire Extinguishing Agents:** Water, foam, dry chemical, carbon dioxide

6.4 **Fire Extinguishing Agents Not to be Used:** Water or foam may cause frothing.

6.5 **Special Hazards of Combustion Products:** Hydrogen chloride gas and other irritating fumes may form in fires.

6.6 **Behavior in Fire:**

6.7 **Ignition Temperature:** Data not available

6.8 **Electrical Hazard:** Data not available

6.9 **Burning Rate:** Data not available

7. CHEMICAL REACTIVITY

7.1 **Reactivity with Water:** No reaction

7.2 **Reactivity with Common Materials:** May attack some forms of plastics

7.3 **Stability During Transport:** Stable

7.4 **Neutralizing Agents for Acids and Caustics:** Not pertinent

7.5 **Polymerization:** Not pertinent

7.6 **Inhibitor of Polymerization:** Not pertinent

8. WATER POLLUTION

8.1 **Aquatic Toxicity:**
Isooctyl ester: 26 ppm/48 hr/bluegill/TL_m/fresh water
Butoxypropyl ester: 17 ppm/48 hr/bluegill/TL_m/fresh water

8.2 **Waterfowl Toxicity:** Data not available

8.3 **Biological Oxygen Demand (BOD):** Data not available

8.4 **Food Chain Concentration Potential:** Data not available

9. SELECTED MANUFACTURERS

1. Dow Chemical Co.
Midland, Mich. 48640
2. Transvaal, Inc.
P. O. Box 69
Jacksonville, Ark. 72076
3. Rhodia, Inc.
Chipman Division
23 Belmont Drive
Somerset, N. J. 08873

10. SHIPPING INFORMATION

10.1 **Grades or Purity:** Technical, 96–99%; 55–65% solutions in kerosene or diesel oil, which are combustible.

10.2 **Storage Temperature:** Ambient

10.3 **Inert Atmosphere:** No requirement

10.4 **Venting:** Open

11. HAZARD ASSESSMENT CODE

(See Hazard Assessment Handbook, CG 446-3)

A-X-Y

12. HAZARD CLASSIFICATIONS

12.1 **Code of Federal Regulations:** Not listed

12.2 **NAS Hazard Rating for Bulk Water Transportation:** Not listed

12.3 **NFPA Hazard Classifications:** Not listed

13. PHYSICAL AND CHEMICAL PROPERTIES

13.1 **Physical State at 15°C and 1 atm:** Liquid

13.2 **Molecular Weight:** Mixtures, all greater than 300

13.3 **Boiling Point at 1 atm:**

Butyl	Butoxypropyl	Isooctyl	2-Ethylhexyl	
639	651	770	~770	°F
337	344	410	~410	°C
610	617	683	~683	°K

13.4 **Freezing Point:** Not pertinent

13.5 **Critical Temperature:** Not pertinent

13.6 **Critical Pressure:** Not pertinent

13.7 **Specific Gravity:** 1.2 at 20°C (liquid)

13.8 **Liquid Surface Tension:** Data not available

13.9 **Liquid-Water Interfacial Tension:** Data not available

13.10 **Vapor (Gas) Specific Gravity:** Not pertinent

13.11 **Ratio of Specific Heats of Vapor (Gas):** Not pertinent

13.12 **Latent Heat of Vaporization:** Data not available

13.13 **Heat of Combustion:** Data not available

13.14 **Heat of Decomposition:** Not pertinent

13.15 **Heat of Solution:** Not pertinent

13.16 **Heat of Polymerization:** Not pertinent

3. CHEMICAL DESIGNATIONS (Cont'd.)

3.3 **Chemical Formula:** 2,4,5-$Cl_3C_6H_2OCH_2COOR$ where R = C_4H_9, C_8H_{17}, etc.

3.4 **IMCO/United Nations Numerical Designation:** Not listed

TBT	TETRABUTYL TITANATE

Common Synonyms			
Titanium butoxide Butyl titanate Titanium tetrabutoxide Butyl titanate monomer	Liquid	Colorless to light yellow	Weak alcohol-like odor
	May float or sink in water. Reacts with water.		

Stop discharge if possible. Keep people away.
Call fire department.
Isolate and remove discharged material.
Notify local health and pollution control agencies.

Fire

Combustible.
Containers may explode in fire.
Extinguish with dry chemicals or carbon dioxide.
DO NOT USE WATER ON FIRE.
Cool exposed containers with water.

Exposure

Call for medical aid.

LIQUID
Irritating to skin and eyes.
If swallowed will cause nausea and vomiting.
Remove contaminated clothing and shoes.
Flush affected areas with plenty of water.
IF IN EYES, hold eyelids open and flush with plenty of water.
IF SWALLOWED and victim is CONSCIOUS, have victim drink water or milk and have victim induce vomiting.
IF SWALLOWED and victim is UNCONSCIOUS OR HAVING CONVULSIONS, do nothing except keep victim warm.

Water Pollution

Effect of low concentrations on aquatic life is unknown.
Fouling to shoreline.
May be dangerous if it enters water intakes.
Notify local health and wildlife officials.
Notify operators of nearby water intakes.

1. RESPONSE TO DISCHARGE

(See Response Methods Handbook, CG 446-4)

Issue warning—water contaminant
Restrict access
Mechanical containment
Disperse and flush

2. LABELS

No label required by Code of Federal Regulations.

3. CHEMICAL DESIGNATIONS

3.1 **Synonyms:** Butyl titanate; Butyl titanate monomer; Orthotitanic acid, tetrabutyl ester; Titanium butoxide; Titanium tetrabutoxide

3.2 **Coast Guard Compatibility Classification:** Not applicable

3.3 **Chemical Formula:** $Ti(OC_4H_9)_4$

3.4 **IMCO/United Nations Numerical Designation:** Not listed

4. OBSERVABLE CHARACTERISTICS

4.1 **Physical State (as shipped):** Liquid

4.2 **Color:** Colorless to pale yellow

4.3 **Odor:** Weak alcoholic

5. HEALTH HAZARDS

5.1 **Personal Protective Equipment:** Self-contained breathing apparatus or organic canister mask; goggles or face shield; rubber gloves

5.2 **Symptoms Following Exposure:** Inhalation causes nonspecific irritation of the upper respiratory tract. Contact with liquid may cause corneal damage in eyes and local irritation of skin. Ingestion causes nonspecific irritation of gastrointestinal tract, nausea, vomiting, cramps, and diarrhea; in severe cases, central nervous system depression may result.

5.3 **Treatment for Exposure:** INHALATION: move from contaminated atmosphere; if symptoms of respiratory discomfort persist, see a physician. EYES: immediately flush with large quantities of running water for a minimum of 15 min.; obtain medical attention if irritation persists. SKIN: immediately flush affected areas with water; obtain medical attention if irritation persists. INGESTION: give large amounts of water or warm salty water to induce vomiting; if this measure is unsuccessful, vomiting may be induced by tickling the back of the patient's throat with a finger; vomiting should be encouraged until vomitus is clear; obtain medical attention if abdominal discomfort persists.

5.4 **Toxicity by Inhalation (Threshold Limit Value):** Data not available

5.5 **Short-Term Inhalation Limits:** Data not available

5.6 **Toxicity by Ingestion:** Data not available

5.7 **Late Toxicity:** Data not available

5.8 **Vapor (Gas) Irritant Characteristics:** Data not available

5.9 **Liquid or Solid Irritant Characteristics:** Data not available

5.10 **Odor Threshold:** Data not available

6. FIRE HAZARDS

6.1 **Flash Point:** 170°F C.C.

6.2 **Flammable Limits in Air:** 2%—12%

6.3 **Fire Extinguishing Agents:** Dry chemical, carbon dioxide

6.4 **Fire Extinguishing Agents Not to be Used:** Water

6.5 **Special Hazards of Combustion Products:** Not pertinent

6.6 **Behavior in Fire:** May give off dense white smoke. Containers may explode.

6.7 **Ignition Temperature:** Data not available

6.8 **Electrical Hazard:** Data not available

6.9 **Burning Rate:** 3.4 mm/min.

7. CHEMICAL REACTIVITY

7.1 **Reactivity with Water:** Reacts to form butanol and titanium dioxide; the reaction is not hazardous.

7.2 **Reactivity with Common Materials:** No reaction

7.3 **Stability During Transport:** Stable

7.4 **Neutralizing Agents for Acids and Caustics:** Not pertinent

7.5 **Polymerization:** Not pertinent

7.6 **Inhibitor of Polymerization:** Not pertinent

8. WATER POLLUTION

8.1 **Aquatic Toxicity:** Data not available

8.2 **Waterfowl Toxicity:** Data not available

8.3 **Biological Oxygen Demand (BOD):** Data not available

8.4 **Food Chain Concentration Potential:** Data not available

9. SELECTED MANUFACTURERS

1. Stauffer Chemical Company
   Specialty Chemical Division
   Westport, Conn. 06880
2. Ventron Corporation
   Chemicals Division
   8 Congress Street
   Beverly, Mass. 01915
3. Dynamit Nobel of America, Inc.
   105 Stonehurst Court
   Northvale, N. J. 07647

10. SHIPPING INFORMATION

10.1 **Grades or Purity:** Technical

10.2 **Storage Temperature:** Ambient

10.3 **Inert Atmosphere:** No requirement

10.4 **Venting:** Open (flame arrester)

11. HAZARD ASSESSMENT CODE

(See Hazard Assessment Handbook, CG 446-3)

A-O-T-U-X-Y

12. HAZARD CLASSIFICATIONS

12.1 **Code of Federal Regulations:** Not listed

12.2 **NAS Hazard Rating for Bulk Water Transportation:** Not listed

12.3 **NFPA Hazard Classifications:** Not listed

13. PHYSICAL AND CHEMICAL PROPERTIES

13.1 **Physical State at 15°C and 1 atm:** Liquid

13.2 **Molecular Weight:** 340

13.3 **Boiling Point at 1 atm:** 593°F = 312°C = 585°K

13.4 **Freezing Point:** −67°F = −55°C = 218°K

13.5 **Critical Temperature:** Not pertinent

13.6 **Critical Pressure:** Not pertinent

13.7 **Specific Gravity:** 0.998 at 25°C (liquid)

13.8 **Liquid Surface Tension:** Data not available

13.9 **Liquid-Water Interfacial Tension:** Not pertinent

13.10 **Vapor (Gas) Specific Gravity:** Not pertinent

13.11 **Ratio of Specific Heats of Vapor (Gas):** Not pertinent

13.12 **Latent Heat of Vaporization:** 142 Btu/lb = 79 cal/g = 3.3×10^5 J/kg

13.13 **Heat of Combustion:** (est.) −14,600 Btu/lb = −8,100 cal/g = -340×10^5 J/kg

13.14 **Heat of Decomposition:** Not pertinent

13.15 **Heat of Solution:** Not pertinent

13.16 **Heat of Polymerization:** Not pertinent

NOTES

TEC	**TETRACHLOROETHANE**

Common Synonyms			
1,1,2,2-Tetrachloroethane Acetylene tetrachloride	Liquid	Colorless to pale yellow	Sweet odor
	Sinks in water.		

AVOID CONTACT WITH LIQUID AND VAPOR. KEEP PEOPLE AWAY.
Wear rubber overclothing (including gloves).
Stop discharge if possible.
Isolate and remove discharged material.
Notify local health and pollution control agencies.

Fire	Not flammable. Poisonous gases may be produced when heated.
Exposure	CALL FOR MEDICAL AID. VAPOR Irritating to eyes, nose and throat. Harmful if inhaled. If in eyes, hold eyelids open and flush with plenty of water. If breathing has stopped, give artificial respiration. If breathing is difficult, give oxygen. LIQUID POISONOUS IF SWALLOWED OR IF SKIN IS EXPOSED. Irritating to skin and eyes. If swallowed will cause nausea and vomiting. Remove contaminated clothing and shoes. Flush affected areas with plenty of water. IF IN EYES, hold eyelids open and flush with plenty of water. IF SWALLOWED and victim is CONSCIOUS, have victim drink water or milk and have victim induce vomiting. IF SWALLOWED and victim is UNCONSCIOUS OR HAVING CONVULSIONS, do nothing except keep victim warm.
Water Pollution	Effect of low concentrations on aquatic life is unknown. May be dangerous if it enters water intakes. Notify local health and wildlife officials. Notify operators of nearby water intakes.

1. RESPONSE TO DISCHARGE

(See Response Methods Handbook, CG 446-4)

Issue warning—poison, air contaminant
Restrict access
Should be removed
Chemical and physical treatment

2. LABELS

No hazard label required by Code of Federal Regulations

3. CHEMICAL DESIGNATIONS

3.1 **Synonyms:** Acetylene tetrachloride; 1,1,2,2-Tetrachloroethane
3.2 **Coast Guard Compatibility Classification:** Halogenated hydrocarbon
3.3 **Chemical Formula:** $Cl_2CHCHCl_2$
3.4 **IMCO/United Nations Numerical Designation:** Not listed

4. OBSERVABLE CHARACTERISTICS

4.1 **Physical State (as shipped):** Liquid
4.2 **Color:** Colorless; slightly yellow; light yellowish green
4.3 **Odor:** Chloroform-like, pleasant; like carbon tetrachloride; mild, sweetish, similar to several other chlorinated hydrocarbons.

5. HEALTH HAZARDS

5.1 **Personal Protective Equipment:** Chemical safety goggles; plastic face shield; air- or oxygen-supplied mask; safety hat with brim; solvent-proof apron; synthetic rubber gloves
5.2 **Symptoms Following Exposure:** Compound is a powerful narcotic and liver poison; may also cause changes in blood composition and neurological disturbances. Repeated exposure by inhalation can be fatal. Ingestion causes vomiting, diarrhea, severe mucosal injury, liver necrosis, cyanosis, unconsciousness, loss of reflexes, and death. Contact with eyes causes irritation and lachrymation. Can be absorbed through the skin and may produce severe skin lesions.
5.3 **Treatment for Exposure:** INHALATION: remove victim from exposure; begin artificial respiration if breathing has ceased. INGESTION: induce vomiting; call a physician. EYES: irrigate with water for 15 min. SKIN: remove clothing; wash skin thoroughly with warm water and soap.
5.4 **Toxicity by Inhalation (Threshold Limit Value):** 5 ppm
5.5 **Short-Term Inhalation Limits:** 10 ppm, 30 min.
5.6 **Toxicity by Ingestion:** Grade 3; oral LD_{50} = 200 mg/kg (rat)
5.7 **Late Toxicity:** Liver poisoning, nervous disorders
5.8 **Vapor (Gas) Irritant Characteristics:** Vapor is moderately irritating such that personnel will not usually tolerate moderate or high vapor concentrations.
5.9 **Liquid or Solid Irritant Characteristics:** Minimum hazard. If spilled on clothing and allowed to remain, may cause smarting and reddening of the skin.
5.10 **Odor Threshold:** 0.5 ppm

6. FIRE HAZARDS

6.1 **Flash Point:** Not flammable
6.2 **Flammable Limits in Air:** Not flammable
6.3 **Fire Extinguishing Agents:** Not pertinent
6.4 **Fire Extinguishing Agents Not to be Used:** Not pertinent
6.5 **Special Hazards of Combustion Products:** Irritating hydrogen chloride vapor may form in fire.
6.6 **Behavior in Fire:**
6.7 **Ignition Temperature:** Not pertinent
6.8 **Electrical Hazard:** Not pertinent
6.9 **Burning Rate:** Not pertinent

7. CHEMICAL REACTIVITY

7.1 **Reactivity with Water:** No reaction
7.2 **Reactivity with Common Materials:** May attack some forms of plastics
7.3 **Stability During Transport:** Stable
7.4 **Neutralizing Agents for Acids and Caustics:** Not pertinent
7.5 **Polymerization:** Not pertinent
7.6 **Inhibitor of Polymerization:** Not pertinent

8. WATER POLLUTION

8.1 **Aquatic Toxicity:** Data not available
8.2 **Waterfowl Toxicity:** Data not available
8.3 **Biological Oxygen Demand (BOD):** Data not available
8.4 **Food Chain Concentration Potential:** Data not available

9. SELECTED MANUFACTURERS

1. Hooker Chemical Corp.
   Industrial Chemicals Div.
   Niagara Falls, N. Y. 14302
2. Aldrich Chemical Co.
   940 W. Saint Paul Ave.
   Milwaukee, Wis. 53233
3. Eastman Organic Chemicals
   Rochester, N. Y. 14650

10. SHIPPING INFORMATION

10.1 **Grades or Purity:** Technical, 98%
10.2 **Storage Temperature:** Ambient
10.3 **Inert Atmosphere:** No requirement
10.4 **Venting:** Open

11. HAZARD ASSESSMENT CODE

(See Hazard Assessment Handbook, CG 446-3)

A-X

12. HAZARD CLASSIFICATIONS

12.1 **Code of Federal Regulations:** ORM-A
12.2 **NAS Hazard Rating for Bulk Water Transportation:** Not listed
12.3 **NFPA Hazard Classifications:** Not listed

13. PHYSICAL AND CHEMICAL PROPERTIES

13.1 **Physical State at 15°C and 1 atm:** Liquid
13.2 **Molecular Weight:** 167.85
13.3 **Boiling Point at 1 atm:** 295.3°F = 146.3°C = 419.5°K
13.4 **Freezing Point:** −46.8°F = −43.8°C = 229.4°K
13.5 **Critical Temperature:** Data not available
13.6 **Critical Pressure:** Data not available
13.7 **Specific Gravity:** 1.595 at 20°C (liquid)
13.8 **Liquid Surface Tension:** 37.85 dynes/cm = 0.03785 N/m at 20°C
13.9 **Liquid-Water Interfacial Tension:** Data not available
13.10 **Vapor (Gas) Specific Gravity:** 5.79
13.11 **Ratio of Specific Heats of Vapor (Gas):** 1.090 at 25°C
13.12 **Latent Heat of Vaporization:** 99.2 Btu/lb = 55.1 cal/g = 2.30×10^5 J/kg
13.13 **Heat of Combustion:** Not pertinent
13.14 **Heat of Decomposition:** Not pertinent
13.15 **Heat of Solution:** Not pertinent
13.16 **Heat of Polymerization:** Not pertinent

NOTES

TTE — TETRACHLOROETHYLENE

Common Synonyms	Watery liquid	Colorless	Sweet odor
Tetracap Perclene Perchloroethylene	Sinks in water. Irritating vapor is produced.		

Stop discharge if possible.
Avoid contact with liquid and vapor.
Isolate and remove discharged material.
Notify local health and pollution control agencies.

Fire

Not flammable.
Poisonous gases are produced when heated.

Exposure

CALL FOR MEDICAL AID.

VAPOR
Irritating to eyes, nose, and throat.
If inhaled, will cause difficult breathing, or loss of consciousness.
Move to fresh air.
If breathing has stopped, give artificial respiration.
If breathing is difficult, give oxygen.

LIQUID
Irritating to skin and eyes.
Harmful if swallowed.
Remove contaminated clothing and shoes.
Flush affected areas with plenty of water.
IF IN EYES, hold eyelids open and flush with plenty of water.
IF SWALLOWED and victim is CONSCIOUS, have victim drink water or milk.

Water Pollution

Effect of low concentrations on aquatic life is unknown.
May be dangerous if it enters water intakes.
Notify local health and wildlife officials.
Notify operators of nearby water intakes.

1. RESPONSE TO DISCHARGE

(See Response Methods Handbook, CG 446-4)

Should be removed
Chemical and physical treatment

2. LABELS

No hazard label required by Code of Federal Regulations

3. CHEMICAL DESIGNATIONS

3.1 **Synonyms:** Perchloroethylene
Perclene
Perk
3.2 **Coast Guard Compatibility Classification:** Halogenated hydrocarbon
3.3 **Chemical Formula:** $Cl_2C{=}CCl_2$
3.4 **IMCO/United Nations Numerical Designation:** 9.0/1897

4. OBSERVABLE CHARACTERISTICS

4.1 **Physical State (as shipped):** Liquid
4.2 **Color:** Colorless
4.3 **Odor:** Ethereal; like chloroform; mildly sweet

5. HEALTH HAZARDS

5.1 **Personal Protective Equipment:** For high vapor concentrations use approved canister or air-supplied mask; chemical goggles or face shield; plastic gloves.
5.2 **Symptoms Following Exposure:** Vapor can affect central nervous system and cause anesthesia. Liquid may irritate skin after prolonged contact. May irritate eyes but causes no injury.
5.3 **Treatment for Exposure:** INHALATION: if illness occurs, remove patient to fresh air, keep him warm and quiet, and get medical attention. INGESTION: induce vomiting only on physician's recommendation. EYES AND SKIN: flush with plenty of water and get medical attention if irritation or injury occurs.
5.4 **Toxicity by Inhalation (Threshold Limit Value):** 100 ppm
5.5 **Short-Term Inhalation Limits:** 100 ppm for 60 min.
5.6 **Toxicity by Ingestion:** Grade 2; LD_{50} 0.5 to 5 g/kg
5.7 **Late Toxicity:** None
5.8 **Vapor (Gas) Irritant Characteristics:** Vapors cause a slight smarting of the eyes or throat if present in high concentrations. The effect is temporary.
5.9 **Liquid or Solid Irritant Characteristics:** Minimum hazard. If spilled on clothing and allowed to remain, may cause smarting and reddening of the skin.
5.10 **Odor Threshold:** 5 ppm

6. FIRE HAZARDS

6.1 **Flash Point:** Not flammable
6.2 **Flammable Limits in Air:** Not flammable
6.3 **Fire Extinguishing Agents:** Not pertinent
6.4 **Fire Extinguishing Agents Not to be Used:** Not pertinent
6.5 **Special Hazards of Combustion Products:** Toxic, irritating gases may be generated in fires.
6.6 **Behavior in Fire:** Not pertinent
6.7 **Ignition Temperature:** Not flammable
6.8 **Electrical Hazard:** Not pertinent
6.9 **Burning Rate:** Not flammable

7. CHEMICAL REACTIVITY

7.1 **Reactivity with Water:** No reaction
7.2 **Reactivity with Common Materials:** No reaction
7.3 **Stability During Transport:** Stable
7.4 **Neutralizing Agents for Acids and Caustics:** Not pertinent
7.5 **Polymerization:** Not pertinent
7.6 **Inhibitor of Polymerization:** Not pertinent

8. WATER POLLUTION

8.1 **Aquatic Toxicity:** Data not available
8.2 **Waterfowl Toxicity:** Data not available
8.3 **Biological Oxygen Demand (BOD):** None
8.4 **Food Chain Concentration Potential:** None

9. SELECTED MANUFACTURERS

1. Dow Chemical Co.
Midland, Mich. 48640
2. PPG Industries Inc.
Industrial Chemical Division
Barberton, Ohio 44203
3. Vulcan Materials Co.
Chemicals Division
Wichita, Kan. 67201

10. SHIPPING INFORMATION

10.1 **Grades or Purity:** Dry cleaning and industrial grades: 95+%
10.2 **Storage Temperature:** Ambient
10.3 **Inert Atmosphere:** No requirement
10.4 **Venting:** Pressure-vacuum

11. HAZARD ASSESSMENT CODE

(See Hazard Assessment Handbook, CG 446-3)

A-X

12. HAZARD CLASSIFICATIONS

12.1 **Code of Federal Regulations:** ORM–A
12.2 **NAS Hazard Rating for Bulk Water Transportation:**

Category	Rating
Fire	0
Health	
Vapor Irritant	1
Liquid or Solid Irritant	1
Poisons	2
Water Pollution	
Human Toxicity	1
Aquatic Toxicity	3
Aesthetic Effect	2
Reactivity	
Other Chemicals	1
Water	0
Self-Reaction	1

12.3 **NFPA Hazard Classifications:** Not listed

13. PHYSICAL AND CHEMICAL PROPERTIES

13.1 **Physical State at 15°C and 1 atm:** Liquid
13.2 **Molecular Weight:** 165.83
13.3 **Boiling Point at 1 atm:** 250°F = 121°C = 394°K
13.4 **Freezing Point:** −8.3°F = −22.4°C = 250.8°K
13.5 **Critical Temperature:** 657°F = 347°C = 620°K
13.6 **Critical Pressure:** Not pertinent
13.7 **Specific Gravity:** 1.63 at 20°C (liquid)
13.8 **Liquid Surface Tension:** 31.3 dynes/cm = 0.0313 N/m at 20°C
13.9 **Liquid-Water Interfacial Tension:** 44.4 dynes/cm = 0.0444 N/m at 25°C
13.10 **Vapor (Gas) Specific Gravity:** Not pertinent
13.11 **Ratio of Specific Heats of Vapor (Gas):** 1.116
13.12 **Latent Heat of Vaporization:** 90.2 Btu/lb = 50.1 cal/g = 2.10×10^5 J/kg
13.13 **Heat of Combustion:** Not pertinent
13.14 **Heat of Decomposition:** Not pertinent
13.15 **Heat of Solution:** Not pertinent
13.16 **Heat of Polymerization:** Not pertinent

NOTES

REVISED 1978

TTN TETRADECANOL

Common Synonyms			
1-Tetradecanol Myristic alcohol n-Tetradecyl alcohol	Thick liquid (heated)	Colorless	Faint alcohol odor
	Solidifies and floats on water.		

Stop discharge if possible.
Call fire department.
Avoid contact with liquid.
Isolate and remove discharged material.
Notify local health and pollution control agencies.

Fire

Combustible.

Extinguish with foam, dry chemical, or carbon dioxide.
Water may be ineffective on fire.
Cool exposed containers with water.

Exposure

CALL FOR MEDICAL AID.
LIQUID
Irritating to skin and eyes.

Remove contaminated clothing and shoes.
Flush affected areas with plenty of water.
IF IN EYES, hold eyelids open and flush with plenty of water.

Water Pollution

Effect of low concentrations on aquatic life is unknown.
Fouling to shoreline.
May be dangerous if it enters water intakes.

Notify local health and wildlife officials.
Notify operators of nearby water intakes.

1. RESPONSE TO DISCHARGE

(See Response Methods Handbook, CG 446-4)

Mechanical containment
Should be removed
Chemical and physical treatment

2. LABELS

No hazard label required by Code of Federal Regulations

3. CHEMICAL DESIGNATIONS

3.1 **Synonyms:** Myristic alcohol
Myristyl alcohol
1-Tetradecanol
n-Tetradecyl alcohol

3.2 **Coast Guard Compatibility Classification:** Alcohol

3.3 **Chemical Formula:** $CH_3(CH_2)_{12}CH_2OH$

3.4 **IMCO/United Nations Numerical Designation:** Not listed

4. OBSERVABLE CHARACTERISTICS

4.1 **Physical State (as shipped):** Liquid

4.2 **Color:** Colorless

4.3 **Odor:** Typical fatty alcohol

5. HEALTH HAZARDS

5.1 **Personal Protective Equipment:** Goggles or face shield.

5.2 **Symptoms Following Exposure:** Low toxicity. Overexposure causes some central nervous system depression. Prolonged skin contact causes skin irritation.

5.3 **Treatment for Exposure:** INGESTION: induce vomiting and call a doctor; if necessary, support respiration. EYES AND SKIN: flush with copious amounts of water.

5.4 **Toxicity by Inhalation (Threshold Limit Value):** Not pertinent

5.5 **Short-Term Inhalation Limits:** Not pertinent

5.6 **Toxicity by Ingestion:** Grade 1; LD_{50} 5 to 15 g/kg

5.7 **Late Toxicity:** Data not available

5.8 **Vapor (Gas) Irritant Characteristics:** Non-volatile

5.9 **Liquid or Solid Irritant Characteristics:** Prolonged contact with skin may cause irritation.

5.10 **Odor Threshold:** Not pertinent

6. FIRE HAZARDS

6.1 **Flash Point:** 285°F O.C.

6.2 **Flammable Limits in Air:** Not pertinent

6.3 **Fire Extinguishing Agents:** Foam, dry chemical, or carbon dioxide

6.4 **Fire Extinguishing Agents Not to be Used:** Water or foam may cause frothing.

6.5 **Special Hazards of Combustion Products:** Not pertinent

6.6 **Behavior in Fire:** Not pertinent

6.7 **Ignition Temperature:** Data not available

6.8 **Electrical Hazard:** Not pertinent

6.9 **Burning Rate:** Data not available

7. CHEMICAL REACTIVITY

7.1 **Reactivity with Water:** No reaction

7.2 **Reactivity with Common Materials:** No reaction

7.3 **Stability During Transport:** Stable

7.4 **Neutralizing Agents for Acids and Caustics:** Not pertinent

7.5 **Polymerization:** Not pertinent

7.6 **Inhibitor of Polymerization:** Not pertinent

8. WATER POLLUTION

8.1 **Aquatic Toxicity:** Data not available

8.2 **Waterfowl Toxicity:** Data not available

8.3 **Biological Oxygen Demand (BOD):** Data not available

8.4 **Food Chain Concentration Potential:** None

9. SELECTED MANUFACTURERS

1. Continental Oil Co.
Conoco Chemicals Division
Park 80 Plaza East
Saddle Brook, N. J. 07662

2. Procter and Gamble Co.
301 E. Sixth St.
Cincinnati, Ohio 45201

10. SHIPPING INFORMATION

10.1 **Grades or Purity:** 96—99+%

10.2 **Storage Temperature:** Ambient

10.3 **Inert Atmosphere:** No requirement

10.4 **Venting:** Open (flame arrester)

11. HAZARD ASSESSMENT CODE

(See Hazard Assessment Handbook, CG 446-3)

A-T-U

12. HAZARD CLASSIFICATIONS

12.1 **Code of Federal Regulations:** Not listed

12.2 **NAS Hazard Rating for Bulk Water Transportation:** Not listed

12.3 **NFPA Hazard Classifications:** Not listed

13. PHYSICAL AND CHEMICAL PROPERTIES

13.1 **Physical State at 15°C and 1 atm:** Solid

13.2 **Molecular Weight:** 214.38

13.3 **Boiling Point at 1 atm:** 505.8°F = 263.2°C = 536.4°K

13.4 **Freezing Point:** 99.7°F = 37.6°C = 310.8°K

13.5 **Critical Temperature:** 804°F = 429°C = 702°K

13.6 **Critical Pressure:** Not pertinent

13.7 **Specific Gravity:** 0.824 at 38°C (liquid)

13.8 **Liquid Surface Tension:** 23.7 dynes/cm = 0.0237 N/m at 20°C

13.9 **Liquid-Water Interfacial Tension:** Data not available

13.10 **Vapor (Gas) Specific Gravity:** Not pertinent

13.11 **Ratio of Specific Heats of Vapor (Gas):** 1.026

13.12 **Latent Heat of Vaporization:** Not pertinent

13.13 **Heat of Combustion:** Data not available

13.14 **Heat of Decomposition:** Not pertinent

13.15 **Heat of Solution:** Not pertinent

13.16 **Heat of Polymerization:** Not pertinent

NOTES

REVISED 1978

1-TETRADECENE

TTD

Common Synonyms

Watery liquid | Colorless | Mild pleasant odor

Floats on water.

Stop discharge if possible.
Call fire department.
Isolate and remove discharged material.
Notify local health and pollution control agencies.

Fire

Combustible.
Extinguish with dry chemical, foam, or carbon dioxide.
Water may be ineffective on fire.
Cool exposed containers with water.

Exposure

CALL FOR MEDICAL AID.

LIQUID
Irritating to eyes.
IF IN EYES, hold eyelids open and flush with plenty of water.

Water Pollution

Effect of low concentrations on aquatic life is unknown.
Fouling to shoreline.
May be dangerous if it enters water intakes.
Notify local health and wildlife officials.
Notify operators of nearby water intakes.

1. RESPONSE TO DISCHARGE

(See Response Methods Handbook, CG 446-4)

Mechanical containment
Should be removed
Chemical and physical treatment

2. LABELS

No hazard label required by Code of Federal Regulations

3. CHEMICAL DESIGNATIONS

3.1 **Synonyms:** Dodecylethylene
3.2 **Coast Guard Compatibility Classification:** Olefin
3.3 **Chemical Formula:** $CH_3(CH_2)_{11}CH=CH_2$
3.4 **IMCO/United Nations Numerical Designation:** Not listed

4. OBSERVABLE CHARACTERISTICS

4.1 **Physical State (as shipped):** Liquid
4.2 **Color:** Colorless
4.3 **Odor:** Mild, pleasant

5. HEALTH HAZARDS

5.1 **Personal Protective Equipment:** Goggles or face shield.
5.2 **Symptoms Following Exposure:** Liquid may irritate eyes.
5.3 **Treatment for Exposure:** EYES: flush with water for 15 min.
5.4 **Toxicity by Inhalation (Threshold Limit Value):** Not pertinent
5.5 **Short-Term Inhalation Limits:** Not pertinent
5.6 **Toxicity by Ingestion:** Data not available
5.7 **Late Toxicity:** Data not available
5.8 **Vapor (Gas) Irritant Characteristics:** Non-volatile
5.9 **Liquid or Solid Irritant Characteristics:** Data not available
5.10 **Odor Threshold:** Not pertinent

6. FIRE HAZARDS

6.1 **Flash Point:** 230°F C.C.
6.2 **Flammable Limits in Air:** Data not available
6.3 **Fire Extinguishing Agents:** Dry chemical, foam, or carbon dioxide
6.4 **Fire Extinguishing Agents Not to be Used:** Water or foam may cause frothing.
6.5 **Special Hazards of Combustion Products:** Not pertinent
6.6 **Behavior in Fire:** Not pertinent
6.7 **Ignition Temperature:** 455°F
6.8 **Electrical Hazard:** Not pertinent
6.9 **Burning Rate:** Data not available

7. CHEMICAL REACTIVITY

7.1 **Reactivity with Water:** No reaction
7.2 **Reactivity with Common Materials:** No reaction
7.3 **Stability During Transport:** Stable
7.4 **Neutralizing Agents for Acids and Caustics:** Not pertinent
7.5 **Polymerization:** Not pertinent
7.6 **Inhibitor of Polymerization:** Not pertinent

8. WATER POLLUTION

8.1 **Aquatic Toxicity:** Data not available
8.2 **Waterfowl Toxicity:** Data not available
8.3 **Biological Oxygen Demand (BOD):** Data not available
8.4 **Food Chain Concentration Potential:** None

9. SELECTED MANUFACTURERS

1. Gulf Oil Corp.
   Petrochemicals Division
   Cedar Bayou, Tex. 77520
2. The Humphrey Chemical Co.
   Devine Street
   North Haven, Conn. 06473
3. Phillips Petroleum Co.
   Bartlesville, Okla. 74004

10. SHIPPING INFORMATION

10.1 **Grades or Purity:** Technical: 96—99.6%
10.2 **Storage Temperature:** Ambient
10.3 **Inert Atmosphere:** No requirement
10.4 **Venting:** Open (flame arrester)

11. HAZARD ASSESSMENT CODE

(See Hazard Assessment Handbook, CG 446-3)

A-T-U

12. HAZARD CLASSIFICATIONS

12.1 **Code of Federal Regulations:** Not listed
12.2 **NAS Hazard Rating for Bulk Water Transportation:** Not listed
12.3 **NFPA Hazard Classifications:** Not listed

13. PHYSICAL AND CHEMICAL PROPERTIES

13.1 **Physical State at 15°C and 1 atm:** Liquid
13.2 **Molecular Weight:** 196.38
13.3 **Boiling Point at 1 atm:** 484.0°F = 251.1°C = 524.3°K
13.4 **Freezing Point:** 8.8°F = −12.9°C = 260.3°K
13.5 **Critical Temperature:** Not pertinent
13.6 **Critical Pressure:** Not pertinent
13.7 **Specific Gravity:** 0.771 at 20°C (liquid)
13.8 **Liquid Surface Tension:** 25.0 dynes/cm = 0.025 N/m at 20°C
13.9 **Liquid-Water Interfacial Tension:** 32.8 dynes/cm = 0.0328 N/m at 22.7°C
13.10 **Vapor (Gas) Specific Gravity:** Not pertinent
13.11 **Ratio of Specific Heats of Vapor (Gas):** 1.027
13.12 **Latent Heat of Vaporization:** 103 Btu/lb = 57.1 cal/g = 2.39×10^5 J/kg
13.13 **Heat of Combustion:** −17,600 Btu/lb = −9,779 cal/g = -409.4×10^5 J/kg
13.14 **Heat of Decomposition:** Not pertinent
13.15 **Heat of Solution:** Not pertinent
13.16 **Heat of Polymerization:** Not pertinent

NOTES

REVISED 1978

TDB	TETRADECYLBENZENE

Common Synonyms			
1-Phenyltetradecane	Liquid	Colorless	Mild odor
	Floats on water.		

Avoid contact with liquid and solid. Keep people away.
Stop discharge if possible.
Isolate and remove discharged material.
Notify local health and pollution control agencies.

Fire

Combustible.

Extinguish with dry chemicals, foam or carbon dioxide.
Water may be ineffective on fire.
Cool exposed containers with water.

Exposure

CALL FOR MEDICAL AID.

LIQUID
Irritating to skin and eyes.
If swallowed will cause nausea and vomiting.

Remove contaminated clothing and shoes.
Flush affected areas with plenty of water.
IF IN EYES, hold eyelids open and flush with plenty of water.
IF SWALLOWED and victim is CONSCIOUS, have victim drink water or milk and have victim induce vomiting.
IF SWALLOWED and victim is UNCONSCIOUS OR HAVING CONVULSIONS, do nothing except keep victim warm.

Water Pollution

Effect of low concentrations on aquatic life is unknown.
Fouling to shoreline.
May be dangerous if it enters water intakes.

Notify local health and wildlife officials.
Notify operators of nearby water intakes.

1. RESPONSE TO DISCHARGE

(See Response Methods Handbook, CG 446-4)

Mechanical containment
Should be removed
Chemical and physical treatment

2. LABELS

No hazard label required by Code of Federal Regulations

3. CHEMICAL DESIGNATIONS

3.1 **Synonyms:** 1-Phenyltetradecane
3.2 **Coast Guard Compatibility Classification:** Aromatic hydrocarbon
3.3 **Chemical Formula:** $C_6H_5(CH_2)_{13}CH_3$
3.4 **IMCO/United Nations Numerical Designation:** Not listed

4. OBSERVABLE CHARACTERISTICS

4.1 **Physical State (as shipped):** Liquid
4.2 **Color:** Colorless
4.3 **Odor:** Mild

5. HEALTH HAZARDS

5.1 **Personal Protective Equipment:** Goggles or face shield; rubber gloves
5.2 **Symptoms Following Exposure:** Ingestion may cause intestinal disturbances. Contact with eyes causes mild irritation.
5.3 **Treatment for Exposure:** INGESTION: induce vomiting if large amount has been swallowed. EYES: flush with water. SKIN: wipe off; wash with soap and water.
5.4 **Toxicity by Inhalation (Threshold Limit Value):** Data not available
5.5 **Short-Term Inhalation Limits:** Data not available
5.6 **Toxicity by Ingestion:** Data not available
5.7 **Late Toxicity:** Data not available
5.8 **Vapor (Gas) Irritant Characteristics:** Data not available
5.9 **Liquid or Solid Irritant Characteristics:** Data not available
5.10 **Odor Threshold:** Data not available

6. FIRE HAZARDS

6.1 **Flash Point:** Data not available
6.2 **Flammable Limits in Air:** Data not available
6.3 **Fire Extinguishing Agents:** Foam, dry chemical, carbon dioxide
6.4 **Fire Extinguishing Agents Not to be Used:** Water may be ineffective.
6.5 **Special Hazards of Combustion Products:**
6.6 **Behavior in Fire:**
6.7 **Ignition Temperature:** Data not available
6.8 **Electrical Hazard:** Not pertinent
6.9 **Burning Rate:** 4.42 mm/min.

7. CHEMICAL REACTIVITY

7.1 **Reactivity with Water:** No reaction
7.2 **Reactivity with Common Materials:** May attack some forms of plastics
7.3 **Stability During Transport:** Stable
7.4 **Neutralizing Agents for Acids and Caustics:** Not pertinent
7.5 **Polymerization:** Not pertinent
7.6 **Inhibitor of Polymerization:** Not pertinent

8. WATER POLLUTION

8.1 **Aquatic Toxicity:** Data not available
8.2 **Waterfowl Toxicity:** Data not available
8.3 **Biological Oxygen Demand (BOD):** Data not available
8.4 **Food Chain Concentration Potential:** None

9. SELECTED MANUFACTURERS

1. Conoco Chemicals
   Park-Eighty Plaza West-One
   Saddle Brook, N. J. 07662
2. The Humphrey Chemical Co.
   Devine Street
   North Haven, Conn. 06473
3. Fisher Scientific Co.
   711 Forbes Ave.
   Pittsburgh, Pa. 15219

10. SHIPPING INFORMATION

10.1 **Grades or Purity:** Commercial
10.2 **Storage Temperature:** Ambient
10.3 **Inert Atmosphere:** No requirement
10.4 **Venting:** Open (flame arrester)

11. HAZARD ASSESSMENT CODE

(See Hazard Assessment Handbook, CG 446-3)

A-T-U

12. HAZARD CLASSIFICATIONS

12.1 **Code of Federal Regulations:** Not listed
12.2 **NAS Hazard Rating for Bulk Water Transportation:** Not listed
12.3 **NFPA Hazard Classifications:** Not listed

13. PHYSICAL AND CHEMICAL PROPERTIES

13.1 **Physical State at 15°C and 1 atm:** Solid or liquid
13.2 **Molecular Weight:** 274.47
13.3 **Boiling Point at 1 atm:** 678°F = 359°C = 632°K
13.4 **Freezing Point:** 61°F = 16°C = 289°K
13.5 **Critical Temperature:** Not pertinent
13.6 **Critical Pressure:** Not pertinent
13.7 **Specific Gravity:** 0.855 at 20°C (liquid)
13.8 **Liquid Surface Tension:** 30.27 dynes/cm = 0.03027 N/m at 20°C
13.9 **Liquid-Water Interfacial Tension:** Data not available
13.10 **Vapor (Gas) Specific Gravity:** Not pertinent
13.11 **Ratio of Specific Heats of Vapor (Gas):** Not pertinent
13.12 **Latent Heat of Vaporization:** 95.18 Btu/lb = 52.88 cal/g = 2.212×10^5 J/kg
13.13 **Heat of Combustion:** −18,430 Btu/lb = −10,240 cal/g = -428.4×10^5 J/kg
13.14 **Heat of Decomposition:** Not pertinent
13.15 **Heat of Solution:** Not pertinent
13.16 **Heat of Polymerization:** Not pertinent

NOTES

TED	TETRAETHYL DITHIOPYROPHOSPHATE

Common Synonyms		
Tetraethyl dithionopyrophosphate	Liquid	Colorless
	Sinks in water.	

AVOID CONTACT WITH LIQUID. KEEP PEOPLE AWAY.
Wear goggles and rubber overclothing (including gloves).
Stop discharge if possible.
Isolate and remove discharged material.
Notify local health and pollution control agencies.

Fire	Fire data not available.
Exposure	CALL FOR MEDICAL AID. LIQUID POISONOUS IF SWALLOWED OR IF SKIN IS EXPOSED. Remove contaminated clothing and shoes. Flush affected areas with plenty of water. IF IN EYES, hold eyelids open and flush with plenty of water. IF SWALLOWED and victim is CONSCIOUS, have victim drink water or milk and have victim induce vomiting. IF SWALLOWED and victim is UNCONSCIOUS OR HAVING CONVULSIONS, do nothing except keep victim warm.
Water Pollution	Effect of low concentrations on aquatic life is unknown. May be dangerous if it enters water intakes. Notify local health and wildlife officials. Notify operators of nearby water intakes.

1. RESPONSE TO DISCHARGE
(See Response Methods Handbook, CG 446-4)

Issue warning—poison, water contaminant
Restrict access
Should be removed
Chemical and physical treatment
Disperse and flush (small discharges)

2. LABEL

3. CHEMICAL DESIGNATIONS

3.1 **Synonyms:** Dithiopyrophosphoric acid, O,O,O,O—tetraethyl ester; Tetraethyl dithionopyrophosphate; O,O,O,O—Tetraethyl pyrophosphorodithionate
3.2 **Coast Guard Compatibility Classification:** Not applicable
3.3 **Chemical Formula:** $(C_2H_5O)_2PSOPS(OC_2H_5)_2$
3.4 **IMCO/United Nations Numerical Designation:** 6.1/1704

4. OBSERVABLE CHARACTERISTICS

4.1 **Physical State (as shipped):** Liquid
4.2 **Color:** Colorless
4.3 **Odor:** Data not available

5. HEALTH HAZARDS

5.1 **Personal Protective Equipment:** Mask with canister approved for organic phosphate pesticides; goggles or face shield; rubber gloves and other protective clothing to prevent contact with skin.
5.2 **Symptoms Following Exposure:** Contact with liquid causes irritation of eyes and skin. Compound can be absorbed through skin. Ingestion of liquid or inhalation of mist causes nausea, vomiting, mental confusion, abominal pain, sweating, giddiness, apprehension, and restlessness; later, muscular twitching of the eyelids and tongue begin, then other muscles of face and neck become involved; generalized twitching and muscle weakness may occur; pulmonary edema, ataxia, tremor, and convulsions may advance to coma.
5.3 **Treatment for Exposure:** *Call physician for all exposures to this compound.* INHALATION: support respiration; keep airway clear; use artificial respiration if breathing is difficult or has stopped. EYES: flush with water immediately after contact for at least 15 min. SKIN: remove victim's clothing and shoes immediately using rubber gloves; quickly wipe off affected area with clean cloths; immediately follow with a shower using plenty of soap; if complete shower is impossible, wash affected skin, hair, and fingernails repeatedly with soap and water using clean cloths each time to prevent spreading the contamination. INGESTION: induce vomiting repeatedly until vomit fluid is clear (save fluid for physician's examination); if vomiting cannot be induced within 5 min., have victim drink plenty of milk or water; have him lie down and keep him warm; if increased secretions make breathing difficult, prop patient up; if he stops breathing, apply artificial or mouth-to-mouth respiration, preferably through an airway; wash victim's mouth of contamination; mechanical resuscitator should be used if available; oxygen may be necessary; keep patient under observation for 24 hrs.
5.4 **Toxicity by Inhalation (Threshold Limit Value):** Data not available

6. FIRE HAZARDS

6.1 **Flash Point:** Not flammable
6.2 **Flammable Limits in Air:** Not flammable
6.3 **Fire Extinguishing Agents:** Water, dry chemical, foam, carbon dioxide
6.4 **Fire Extinguishing Agents Not to be Used:** Not pertinent
6.5 **Special Hazards of Combustion Products:** Toxic phosphorus and sulfur oxides are produced.
6.6 **Behavior in Fire:** Not pertinent
6.7 **Ignition Temperature:** Not pertinent
6.8 **Electrical Hazard:** Not pertinent
6.9 **Burning Rate:** Not pertinent

7. CHEMICAL REACTIVITY

7.1 **Reactivity with Water:** Reacts slowly to form non-hazardous products
7.2 **Reactivity with Common Materials:** Corrosive to most metals in the presence of moisture
7.3 **Stability During Transport:** Stable
7.4 **Neutralizing Agents for Acids and Caustics:** Not pertinent
7.5 **Polymerization:** Not pertinent
7.6 **Inhibitor of Polymerization:** Not pertinent

8. WATER POLLUTION

8.1 **Aquatic Toxicity:** Data not available
8.2 **Waterfowl Toxicity:** Data not available
8.3 **Biological Oxygen Demand (BOD):** Data not available
8.4 **Food Chain Concentration Potential:** None

9. SELECTED MANUFACTURERS

Plant Products Corp.
Kennedy Avenue
Blue Point, N. Y. 11715

10. SHIPPING INFORMATION

10.1 **Grades or Purity:** Technical; Emulsifiable concentrate; Dry mixtures with inert solid, > 2%.
10.2 **Storage Temperature:** Ambient
10.3 **Inert Atmosphere:** No requirement
10.4 **Venting:** Pressure-vacuum

11. HAZARD ASSESSMENT CODE
(See Hazard Assessment Handbook, CG 446-3)

A-X-Y

12. HAZARD CLASSIFICATIONS

12.1 **Code of Federal Regulations:** Poisonous, Class B
12.2 **NAS Hazard Rating for Bulk Water Transportation:** Not listed
12.3 **NFPA Hazard Classifications:** Not listed

13. PHYSICAL AND CHEMICAL PROPERTIES

13.1 **Physical State at 15°C and 1 atm:** Liquid
13.2 **Molecular Weight:** 322.3
13.3 **Boiling Point at 1 atm:** Not pertinent (very high)
13.4 **Freezing Point:** Not pertinent
13.5 **Critical Temperature:** Not pertinent
13.6 **Critical Pressure:** Not pertinent
13.7 **Specific Gravity:** 1.19 at 25°C (liquid)
13.8 **Liquid Surface Tension:** Data not available
13.9 **Liquid-Water Interfacial Tension:** Not pertinent
13.10 **Vapor (Gas) Specific Gravity:** Not pertinent
13.11 **Ratio of Specific Heats of Vapor (Gas):** Not pertinent
13.12 **Latent Heat of Vaporization:** Not pertinent
13.13 **Heat of Combustion:** Not pertinent
13.14 **Heat of Decomposition:** Not pertinent
13.15 **Heat of Solution:** Not pertinent
13.16 **Heat of Polymerization:** Not pertinent

5. HEALTH HAZARDS (Cont'd.)

5.5 **Short-Term Inhalation Limits:** Data not available
5.6 **Toxicity by Ingestion:** Grade 4; oral LD_{50} = 5 mg/kg (rat)
5.7 **Late Toxicity:** Causes chromosomal damage in mice
5.8 **Vapor (Gas) Irritant Characteristics:** Data not available
5.9 **Liquid or Solid Irritant Characteristics:** Data not available
5.10 **Odor Threshold:** Data not available

TTG — TETRAETHYLENE GLYCOL

Common Synonyms
Hi-Dry
Bis-[2-(2-hydroxyethoxy) ethyl] ether
3,6,9-Trioxaundecan-1,11-diol

Liquid — Colorless to straw color — Mild odor

Sinks and mixes with water.

Avoid contact with liquid. Keep people away.
Stop discharge if possible.
Isolate and remove discharged material.
Notify local health and pollution control agencies.

Fire

Combustible.
Extinguish with dry chemicals, alcohol foam, or carbon dioxide.
Water may be ineffective on fire.
Cool exposed containers with water.

Exposure

LIQUID
Not harmful.

Water Pollution

Effect of low concentrations on aquatic life is unknown.
May be dangerous if it enters water intakes.
Notify local health and wildlife officials.
Notify operators of nearby water intakes.

1. RESPONSE TO DISCHARGE

(See Response Methods Handbook, CG 446-4)

Disperse and flush

2. LABELS

No hazard label required by Code of Federal Regulations

3. CHEMICAL DESIGNATIONS

3.1 **Synonyms:** Bis-[2-(2-hydroxyethoxy)ethyl] ether; Hi-Dry; 3,6,9-Trioxaundecan-1,11-diol
3.2 **Coast Guard Compatibility Classification:** Glycol ether
3.3 **Chemical Formula:** $HO(C_2H_4O)_4H$
3.4 **IMCO/United Nations Numerical Designation:** Not listed

4. OBSERVABLE CHARACTERISTICS

4.1 **Physical State (as shipped):** Liquid
4.2 **Color:** Light straw; colorless
4.3 **Odor:** Mild

5. HEALTH HAZARDS

5.1 **Personal Protective Equipment:** Goggles or face shield and rubber gloves
5.2 **Symptoms Following Exposure:** Compound is nonirritating. No symptoms observed by any exposure route.
5.3 **Treatment for Exposure:** INGESTION: if large amounts are swallowed, induce vomiting; treat symptomatically. EYES: flush with water; get medical attention if ill effects develop. SKIN: wash with soap and water.
5.4 **Toxicity by Inhalation (Threshold Limit Value):** Data not available
5.5 **Short-Term Inhalation Limits:** Data not available
5.6 **Toxicity by Ingestion:** Grade 0; oral rat LD_{50} = 28-34 g/kg
5,7 **Late Toxicity:** Data not available
5.8 **Vapor (Gas) Irritant Characteristics:** Vapors are nonirritating to eyes and throat.
5.9 **Liquid or Solid Irritant Characteristics:** No appreciable hazard. Practically harmless to the skin.
5.10 **Odor Threshold:** Data not available

6. FIRE HAZARDS

6.1 **Flash Point:** 360°F O.C.
6.2 **Flammable Limits in Air:** Data not available
6.3 **Fire Extinguishing Agents:** "Alcohol" foam, dry chemical, or carbon dioxide
6.4 **Fire Extinguishing Agents Not to be Used:** Water or foam may cause frothing.
6.5 **Special Hazards of Combustion Products:**
6.6 **Behavior in Fire:**
6.7 **Ignition Temperature:** Data not available
6.8 **Electrical Hazard:** Data not available
6.9 **Burning Rate:** Data not available

7. CHEMICAL REACTIVITY

7.1 **Reactivity with Water:** No reaction
7.2 **Reactivity with Common Materials:** May attack some forms of plastics
7.3 **Stability During Transport:** Stable
7.4 **Neutralizing Agents for Acids and Caustics:** Not pertinent
7.5 **Polymerization:** Not pertinent
7.6 **Inhibitor of Polymerization:** Not pertinent

8. WATER POLLUTION

8.1 **Aquatic Toxicity:** Data not available
8.2 **Waterfowl Toxicity:** Data not available
8.3 **Biological Oxygen Demand (BOD):** 50%, 5 days
8.4 **Food Chain Concentration Potential:** None

9. SELECTED MANUFACTURERS

1. Olin Corp. 120 Long Ridge Rd. Stamford, Conn. 06904
2. Union Carbide Corp. Chemicals & Plastics Div. 270 Park Avenue New York, N. Y. 10017
3. The Dow Chemical Co. 2030 Dow Center Midland, Mich. 48640

10. SHIPPING INFORMATION

10.1 **Grades or Purity:** Commercial, 99+%
10.2 **Storage Temperature:** Ambient
10.3 **Inert Atmosphere:** No requirement
10.4 **Venting:** Open (flame arrester)

11. HAZARD ASSESSMENT CODE

(See Hazard Assessment Handbook, CG 446-3)

A-P-Q

12. HAZARD CLASSIFICATIONS

12.1 **Code of Federal Regulations:** Not listed
12.2 **NAS Hazard Rating for Bulk Water Transportation:**

Category	Rating
Fire	1
Health	
Vapor Irritant	0
Liquid or Solid Irritant	0
Poisons	0
Water Pollution	
Human Toxicity	0
Aquatic Toxicity	1
Aesthetic Effect	1
Reactivity	
Other Chemicals	2
Water	0
Self-Reaction	0

12.3 **NFPA Hazard Classifications:**

Category	Classification
Health Hazard (Blue)	1
Flammability (Red)	1
Reactivity (Yellow)	0

13. PHYSICAL AND CHEMICAL PROPERTIES

13.1 **Physical State at 15°C and 1 atm:** Liquid
13.2 **Molecular Weight:** 194.23
13.3 **Boiling Point at 1 atm:** 621°F = 327°C = 600°K
13.4 **Freezing Point:** 24.8°F = −4.0°C = 269.0°K
13.5 **Critical Temperature:** Not pertinent
13.6 **Critical Pressure:** Not pertinent
13.7 **Specific Gravity:** 1.12 at 20°C (liquid)
13.8 **Liquid Surface Tension:** 18.81 dynes/cm = 0.01881 N/m at 327°C
13.9 **Liquid-Water Interfacial Tension:** Not pertinent
13.10 **Vapor (Gas) Specific Gravity:** Not pertinent
13.11 **Ratio of Specific Heats of Vapor (Gas):** Not pertinent
13.12 **Latent Heat of Vaporization:** Data not available
13.13 **Heat of Combustion:** −10,530 Btu/lb = −5,850 cal/g = -245×10^5 J/kg
13.14 **Heat of Decomposition:** Not pertinent
13.15 **Heat of Solution:** Not pertinent
13.16 **Heat of Polymerization:** Not pertinent

NOTES

TTP — TETRAETHYLENEPENTAMINE

Common Synonyms
1,11-Diamino-3,6,9-triazaundecane

Liquid | Yellow | Ammonia odor

May float or sink in water.

Avoid contact with liquid. Keep people away.
Stop discharge if possible.
Isolate and remove discharged material.
Notify local health and pollution control agencies.

Fire

Combustible.
POISONOUS GASES MAY BE PRODUCED IN FIRE.
Wear goggles and self-contained breathing apparatus.
Extinguish with dry chemicals, foam, or carbon dioxide.
Water may be ineffective on fire.
Cool exposed containers with water.

Exposure

CALL FOR MEDICAL AID.
LIQUID
Will burn skin and eyes.
If swallowed will cause nausea.
Remove contaminated clothing and shoes.
Flush affected areas with plenty of water.
IF IN EYES, hold eyelids open and flush with plenty of water.
IF SWALLOWED and victim is CONSCIOUS, have victim drink water or milk.
IF SWALLOWED and victim is UNCONSCIOUS OR HAVING CONVULSIONS, do nothing except keep victim warm.
DO NOT INDUCE VOMITING.

Water Pollution

Effect of low concentrations on aquatic life is unknown.
May be fouling to shoreline.
May be dangerous if it enters water intakes.
Notify local health and wildlife officials.
Notify operators of nearby water intakes.

1. RESPONSE TO DISCHARGE

(See Response Methods Handbook, CG 446-4)

Issue warning—water contaminant
Restrict access
Disperse and flush

2. LABELS

No hazard label required by Code of Federal Regulations

3. CHEMICAL DESIGNATIONS

3.1 **Synonyms:** 1,11-Diamino-3,6,9-triazaundecane
3.2 **Coast Guard Compatibility Classification:** Aliphatic amine
3.3 **Chemical Formula:** $H_2N(C_2H_4NH)_4H$
3.4 **IMCO/United Nations Numerical Designation:** Not listed

4. OBSERVABLE CHARACTERISTICS

4.1 **Physical State (as shipped):** Liquid
4.2 **Color:** Yellow
4.3 **Odor:** Ammoniacal; disagreeable, penetrating

5. HEALTH HAZARDS

5.1 **Personal Protective Equipment:** Air-supplied respirator; rubber gloves; complete eye protection; impervious apron
5.2 **Symptoms Following Exposure:** Inhalation may cause nausea and slight irritation; compound is a sensitizer, and prolonged contact may cause asthma. Ingestion can cause burns of mouth, esophagus, and possibly stomach. Contact with eyes or skin may cause burns. Repeated skin contact may cause dermatitis.
5.3 **Treatment for Exposure:** INHALATION: remove victim to fresh air; give oxygen if breathing is difficult; treat allergic manifestations by usual methods. INGESTION: do NOT induce vomiting; give large quantities of water; give at least one ounce of vinegar in an equal amount of water; get medical attention. EYES: immediately flush with plenty of water; get medical care. SKIN: flush with plenty of water.
5.4 **Toxicity by Inhalation (Threshold Limit Value):** Data not available
5.5 **Short-Term Inhalation Limits:** Data not available
5.6 **Toxicity by Ingestion:** Grade 2; oral LD_{50} = 3,990 mg/kg (rat)
5.7 **Late Toxicity:** Data not available
5.8 **Vapor (Gas) Irritant Characteristics:** Vapors cause a slight smarting of the eyes or respiratory system if present in high concentrations. The effect is temporary.
5.9 **Liquid or Solid Irritant Characteristics:** Causes smarting of the skin and first-degree burns on short exposure; may cause second-degree burns on long exposure.
5.10 **Odor Threshold:** Data not available

6. FIRE HAZARDS

6.1 **Flash Point:** 340°F O.C.
6.2 **Flammable Limits in Air:** (est.) 0.8%—4.6%
6.3 **Fire Extinguishing Agents:** "Alcohol" foam, dry chemical, or carbon dioxide
6.4 **Fire Extinguishing Agents Not to be Used:** Water or foam may cause frothing.
6.5 **Special Hazards of Combustion Products:** Ammonia and toxic oxides of nitrogen may form in fires.
6.6 **Behavior in Fire:**
6.7 **Ignition Temperature:** 572°F
6.8 **Electrical Hazard:** Data not available
6.9 **Burning Rate:** Data not available

7. CHEMICAL REACTIVITY

7.1 **Reactivity with Water:** No reaction
7.2 **Reactivity with Common Materials:** May attack some forms of plastics
7.3 **Stability During Transport:** Stable
7.4 **Neutralizing Agents for Acids and Caustics:** Not pertinent
7.5 **Polymerization:** Not pertinent
7.6 **Inhibitor of Polymerization:** Not pertinent

8. WATER POLLUTION

8.1 **Aquatic Toxicity:** Data not available
8.2 **Waterfowl Toxicity:** Data not available
8.3 **Biological Oxygen Demand (BOD):** Data not available
8.4 **Food Chain Concentration Potential:** None

9. SELECTED MANUFACTURERS

1. Union Carbide Corp.
Chemicals & Plastics Div.
270 Park Avenue
New York, N. Y. 10017
2. Jefferson Chemical Co., Inc.
3336 Richmond Avenue
P. O. Box 53300
Houston, Texas 77052
3. Dow Chemical Co.
Midland, Mich. 48640

10. SHIPPING INFORMATION

10.1 **Grades or Purity:** Commercial
10.2 **Storage Temperature:** Ambient
10.3 **Inert Atmosphere:** No requirement
10.4 **Venting:** Open (flame arrester)

11. HAZARD ASSESSMENT CODE

(See Hazard Assessment Handbook, CG 446-3)

A-P-Q

12. HAZARD CLASSIFICATIONS

12.1 **Code of Federal Regulations:** Not listed
12.2 **NAS Hazard Rating for Bulk Water Transportation:**

Category	Rating
Fire	1
Health	
Vapor Irritant	1
Liquid or Solid Irritant	2
Poisons	2
Water Pollution	
Human Toxicity	2
Aquatic Toxicity	2
Aesthetic Effect	2
Reactivity	
Other Chemicals	3
Water	0
Self-Reaction	0

12.3 **NFPA Hazard Classifications:**

Category	Classification
Health Hazard (Blue)	2
Flammability (Red)	1
Reactivity (Yellow)	0

13. PHYSICAL AND CHEMICAL PROPERTIES

13.1 **Physical State at 15°C and 1 atm:** Liquid
13.2 **Molecular Weight:** 189
13.3 **Boiling Point at 1 atm:** 644°F = 340°C = 613°K
13.4 **Freezing Point:** −22°F = −30°C = 243°K
13.5 **Critical Temperature:** Not pertinent
13.6 **Critical Pressure:** Not pertinent
13.7 **Specific Gravity:** 0.998 at 20°C (liquid)
13.8 **Liquid Surface Tension:** Data not available
13.9 **Liquid-Water Interfacial Tension:** Not pertinent
13.10 **Vapor (Gas) Specific Gravity:** Not pertinent
13.11 **Ratio of Specific Heats of Vapor (Gas):** Not pertinent
13.12 **Latent Heat of Vaporization:** Data not available
13.13 **Heat of Combustion:** Data not available
13.14 **Heat of Decomposition:** Not pertinent
13.15 **Heat of Solution:** Data not available
13.16 **Heat of Polymerization:** Not pertinent

NOTES

TETRAETHYL LEAD

Common Synonyms	
TEL Lead tetraethyl	Oily liquid Colorless, but generally dyed red Fruity odor Sinks in water. Poisonous, flammable vapor is produced.

AVOID CONTACT WITH LIQUID AND VAPOR. Keep people away.
Wear goggles, self-contained breathing apparatus, and rubber overclothing (including gloves).
Stop discharge if possible.
Call fire department.
Stay upwind and use water spray to "knock down" vapor.
Isolate and remove discharged material.
Notify local health and pollution control agencies.

Fire

Combustible.
POISONOUS GASES ARE PRODUCED IN FIRE.
Containers may explode in fire.
Vapor may explode if ignited in an enclosed area.
Wear goggles, self-contained breathing apparatus, and rubber overclothing (including gloves).
Combat fires from behind barrier or protected location.
Flood discharge area with water.
Extinguish with water, dry chemical, foam, or carbon dioxide.
Cool exposed containers with water.

Exposure

CALL FOR MEDICAL AID.

VAPOR
POISONOUS IF INHALED OR IF SKIN IS EXPOSED.
Irritating to eyes.
Move to fresh air.
If breathing has stopped, give artificial respiration.
If breathing is difficult, give oxygen.

LIQUID
POISONOUS IF SWALLOWED OR IF SKIN IS EXPOSED.
Will burn eyes.
Remove contaminated clothing and shoes.
Flush affected areas with plenty of water.
IF IN EYES, hold eyelids open and flush with plenty of water.
IF SWALLOWED and victim is CONSCIOUS, have victim drink water or milk and have victim induce vomiting.
IF SWALLOWED and victim is UNCONSCIOUS OR HAVING CONVULSIONS, do nothing except keep victim warm.

Water Pollution

HARMFUL TO AQUATIC LIFE IN VERY LOW CONCENTRATIONS.
May be dangerous if it enters water intakes.
Notify local health and wildlife officials.
Notify operators of nearby water intakes.

1. RESPONSE TO DISCHARGE

(See Response Methods Handbook, CG 446-4)

Issue warning—poison, water contaminant
Restrict access
Should be removed
Chemical and physical treatment

2. LABEL

3. CHEMICAL DESIGNATIONS

3.1 **Synonyms:** Lead tetraethyl
3.2 **Coast Guard Compatibility Classification:** Special class
3.3 **Chemical Formula:** $Pb(C_2H_5)_4$
3.4 **IMCO/United Nations Numerical Designation:** 6.1/1649

4. OBSERVABLE CHARACTERISTICS

4.1 **Physical State (as shipped):** Liquid
4.2 **Color:** Dyed red or other distinctive color.
4.3 **Odor:** Sweet

5. HEALTH HAZARDS

5.1 **Personal Protective Equipment:** Organic vapor type canister face mask for short periods; air line type for longer periods; neoprene-coated, liquid-proof gloves; protective goggles or face shield; white or light-colored clothing; rubber shoes or boots.
5.2 **Symptoms Following Exposure:** Increased urinary output of lead. If a large degree of absorption from inhalation or skin contact, may cause insomnia, excitability, delirium, coma and death. Do not confuse with inorganic lead.
5.3 **Treatment for Exposure:** Remove victim from contaminated area and consult physician immediately. INGESTION: induce vomiting. SKIN: wash immediately with kerosene or similar petroleum distillate followed by soap and water.
5.4 **Toxicity by Inhalation (Threshold Limit Value):** 0.1 mg/m^3
5.5 **Short-Term Inhalation Limits:** 0.15 mg Pb/m^3 for 30 min.
5.6 **Toxicity by Ingestion:** Oral rat LDL_O = 17 mg/kg
5.7 **Late Toxicity:** Lead poisoning
5.8 **Vapor (Gas) Irritant Characteristics:** Vapors cause a slight smarting of the eyes or respiratory system if present in high concentrations. The effect is temporary.
5.9 **Liquid or Solid Irritant Characteristics:** Causes smarting of the skin and first-degree burns on short exposure; may cause secondary burns on long exposure.
5.10 **Odor Threshold:** Data not available

6. FIRE HAZARDS

6.1 **Flash Point:** 200°F C.C.; 185°F O.C.
6.2 **Flammable Limits in Air:** Data not available
6.3 **Fire Extinguishing Agents:** Water, foam, dry chemical, or carbon dioxide
6.4 **Fire Extinguishing Agents Not to be Used:** Not pertinent
6.5 **Special Hazards of Combustion Products:** Toxic gases are generated in fires.
6.6 **Behavior in Fire:** May explode in fires.
6.7 **Ignition Temperature:** Decomposes above 230°F
6.8 **Electrical Hazard:** Not pertinent
6.9 **Burning Rate:** Data not available

7. CHEMICAL REACTIVITY

7.1 **Reactivity with Water:** No reaction
7.2 **Reactivity with Common Materials:** Rust and some metals cause decomposition.
7.3 **Stability During Transport:** Stable below 230°F. At higher temperatures, may detonate or explode when confined.
7.4 **Neutralizing Agents for Acids and Caustics:** Not pertinent
7.5 **Polymerization:** Not pertinent
7.6 **Inhibitor of Polymerization:** Not pertinent

8. WATER POLLUTION

8.1 **Aquatic Toxicity:** 0.20 mg/l /96 hr/bluegill/TL_m/fresh water
8.2 **Waterfowl Toxicity:** Data not available
8.3 **Biological Oxygen Demand (BOD):** Data not available
8.4 **Food Chain Concentration Potential:** Data not available

9. SELECTED MANUFACTURERS

1. E. I. duPont de Nemours & Co., Inc.
   Petroleum Chemicals Division
   Antioch, Calif. 94509
2. Ethyl Corp.
   Industrial Chemicals Division
   451 Florida St.
   Baton Rouge, La. 70801
3. PPG Industries, Inc.
   Houston Chemical Co. Division
   Beaumont, Tex. 77704

10. SHIPPING INFORMATION

10.1 **Grades or Purity:** Technical
10.2 **Storage Temperature:** Ambient
10.3 **Inert Atmosphere:** No requirement
10.4 **Venting:** Pressure-vacuum

11. HAZARD ASSESSMENT CODE

(See Hazard Assessment Handbook, CG 446-3)

A-X-Y

12. HAZARD CLASSIFICATIONS

12.1 **Code of Federal Regulations:** Poisonous liquid or solid, Class B
12.2 **NAS Hazard Rating for Bulk Water Transportation:** Not listed
12.3 **NFPA Hazard Classifications:**

Category	Classification
Health Hazard (Blue)	3
Flammability (Red)	2
Reactivity (Yellow)	3

13. PHYSICAL AND CHEMICAL PROPERTIES

13.1 **Physical State at 15°C and 1 atm:** Liquid
13.2 **Molecular Weight:** 323.44
13.3 **Boiling Point at 1 atm:** Decomposes
13.4 **Freezing Point:** −215°F = −137°C = 136°K
13.5 **Critical Temperature:** Not pertinent
13.6 **Critical Pressure:** Not pertinent
13.7 **Specific Gravity:** 1.633 at 20°C (liquid)
13.8 **Liquid Surface Tension:** 28.5 dynes/cm = 0.0285 N/m at (est.) 25°C
13.9 **Liquid-Water Interfacial Tension:** (est.) 40 dynes/cm = 0.04 N/m at 20°C
13.10 **Vapor (Gas) Specific Gravity:** Not pertinent
13.11 **Ratio of Specific Heats of Vapor (Gas):** Not pertinent
13.12 **Latent Heat of Vaporization:** Not pertinent
13.13 **Heat of Combustion:** (est.) −7,870 Btu/lb = −4,380 cal/g = -183×10^5 J/kg
13.14 **Heat of Decomposition:** Not pertinent
13.15 **Heat of Solution:** Not pertinent
13.16 **Heat of Polymerization:** Not pertinent

NOTES

REVISED 1978

TEP	# TETRAETHYL PYROPHOSPHATE

Common Synonyms			
Ethylpyrophosphate Nitos T.E.P. T.E.P.P.	Liquid Mixes with water.	Colorless to yellow	Faint fruity odor

AVOID CONTACT WITH LIQUID. KEEP PEOPLE AWAY.
Wear rubber overclothing (including gloves).
Stop discharge if possible.
Isolate and remove discharged material.
Notify local health and pollution control agencies.

Fire	Not flammable. POISONOUS GASES ARE PRODUCED WHEN HEATED.
Exposure	CALL FOR MEDICAL AID. LIQUID POISONOUS IF SWALLOWED OR IF SKIN IS EXPOSED. Remove contaminated clothing and shoes. Flush affected areas with plenty of water. IF IN EYES, hold eyelids open and flush with plenty of water. IF SWALLOWED and victim is CONSCIOUS, have victim drink water or milk and have victim induce vomiting. IF SWALLOWED and victim is UNCONSCIOUS OR HAVING CONVULSIONS, do nothing except keep victim warm.
Water Pollution	HARMFUL TO AQUATIC LIFE IN VERY LOW CONCENTRATIONS. May be dangerous if it enters water intakes. Notify local health and wildlife officials. Notify operators of nearby water intakes.

1. RESPONSE TO DISCHARGE
(See Response Methods Handbook, CG 446-4)

Issue warning—poison, water contaminant
Restrict access
Should be removed
Chemical and physical treatment
Disperse and flush (small discharges)

2. LABEL

3. CHEMICAL DESIGNATIONS

3.1 **Synonyms:** Bladan; Ethyl pyrophosphate; Killax; Mortopal; Nifos; TEP; TEPP; Tetron; Vapotone

3.2 **Coast Guard Compatibility Classification:** Not applicable

3.3 **Chemical Formula:** $(C_2H_5O)_2POOPO(OC_2H_5)_2$ or $(C_2H_5)_4P_2O_7$

3.4 **IMCO/United Nations Numerical Designation:** 6.1/1705

4. OBSERVABLE CHARACTERISTICS

4.1 **Physical State (as shipped):** Liquid

4.2 **Color:** Water-white to amber

4.3 **Odor:** Aromatic; faint fruity

5. HEALTH HAZARDS

5.1 **Personal Protective Equipment:** Mask with canister approved for organic phosphate pesticides; goggles or face shield; rubber gloves and other protective clothing to prevent contact with skin.

5.2 **Symptoms Following Exposure:** Contact with liquid causes irritation of eyes and skin. Compound can be absorbed through skin. Ingestion of liquid or inhalation of mist causes nausea, vomiting, mental confusion, abdominal pain, sweating, giddiness, apprehension, and restlessness; later, muscular twitching of eyelids and tongue begin, then other muscles of face and neck become involved; pulmonary edema, ataxia, tremor, and convulsions may advance to coma.

5.3 **Treatment for Exposure:** *Call physician for all exposures to this compound.* INHALATION: support respiration; keep airway clear; use artificial respiration if breathing is difficult or has stopped. EYES: flush with water immediately after contact for at least 15 min. SKIN: remove victim's clothing and shoes immediately using rubber gloves; quickly wipe off affected area with clean cloths; immediately follow with a shower using plenty of soap; if complete shower is impossible, wash affected skin, hair, and fingernails repeatedly with soap and water using clean cloths each time to prevent spreading the contamination. INGESTION: induce vomiting by putting a finger down the throat or by giving warm salt water (one tablespoon salt per glass). Repeat until vomit fluid is clear (save fluid for physician's examination); if vomiting cannot be induced within five minutes, have victim drink plenty of milk or water; have him lie down and keep him warm; if there is difficulty in breathing due to increased secretions, chest may be cleared by propping patient up; if he stops breathing, use artificial or mouth-to-mouth respiration, preferably through an airway; wash victim's mouth of contamination; mechanical resuscitator should be used if available; oxygen may be necessary; keep patient under observation for 24 hrs.

5.4 **Toxicity by Inhalation (Threshold Limit Value):** 0.05 mg/m³

6. FIRE HAZARDS

6.1 **Flash Point:** Not flammable

6.2 **Flammable Limits in Air:** Not flammable

6.3 **Fire Extinguishing Agents:** Not pertinent

6.4 **Fire Extinguishing Agents Not to be Used:** Not pertinent

6.5 **Special Hazards of Combustion Products:** Highly toxic gases and vapors of unburned material and phosphoric acid are formed in fires.

6.6 **Behavior in Fire:** Water streams applied to adjacent fires will spread contamination of pesticide over wide area.

6.7 **Ignition Temperature:** Not pertinent

6.8 **Electrical Hazard:** Not pertinent

6.9 **Burning Rate:** Not pertinent

7. CHEMICAL REACTIVITY

7.1 **Reactivity with Water:** Reacts slowly to form phosphoric acid

7.2 **Reactivity with Common Materials:** Corrosive to aluminum, slowly corrosive to copper, brass, zinc, and tin

7.3 **Stability During Transport:** Stable

7.4 **Neutralizing Agents for Acids and Caustics:** Flush with water, rinse with sodium bicarbonate or lime solution.

7.5 **Polymerization:** Not pertinent

7.6 **Inhibitor of Polymerization:** Not pertinent

8. WATER POLLUTION

8.1 **Aquatic Toxicity:**
1.0 ppm/96 hr/fathead/TL_m/fresh water
500 ppm/*/marine plankton/no growth or lethal/salt water
*Time period not specified.

8.2 **Waterfowl Toxicity:** LD_{50} = 3.56 mg/kg

8.3 **Biological Oxygen Demand (BOD):** Data not available

8.4 **Food Chain Concentration Potential:** None

9. SELECTED MANUFACTURERS

1. Miller Chemical & Fertilizer Corp.
   P. O. Box 333
   Hanover, Pa. 17331
2. Chevron Chemical Co.
   Ortho Division
   200 Bush Street
   San Francisco, Calif. 94120

10. SHIPPING INFORMATION

10.1 **Grades or Purity:** Technical: 40% plus 60% related ethyl phosphates; Aerosols (5–10%) (Class A poisons); Dusts (0.66–1.2%); Sprays 10–40%.

10.2 **Storage Temperature:** Ambient

10.3 **Inert Atmosphere:** No requirement

10.4 **Venting:** Open

11. HAZARD ASSESSMENT CODE
(See Hazard Assessment Handbook, CG 446-3)

A-P

12. HAZARD CLASSIFICATIONS

12.1 **Code of Federal Regulations:** Poisonous, Class B

12.2 **NAS Hazard Rating for Bulk Water Transportation:** Not listed

12.3 **NFPA Hazard Classifications:** Not listed

13. PHYSICAL AND CHEMICAL PROPERTIES

13.1 **Physical State at 15°C and 1 atm:** Liquid

13.2 **Molecular Weight:** 290.2

13.3 **Boiling Point at 1 atm:** Not pertinent (decomposes)

13.4 **Freezing Point:** Not pertinent

13.5 **Critical Temperature:** Not pertinent

13.6 **Critical Pressure:** Not pertinent

13.7 **Specific Gravity:** 1.18 at 25°C (liquid)

13.8 **Liquid Surface Tension:** Data not available

13.9 **Liquid-Water Interfacial Tension:** Not pertinent

13.10 **Vapor (Gas) Specific Gravity:** Not pertinent

13.11 **Ratio of Specific Heats of Vapor (Gas):** Not pertinent

13.12 **Latent Heat of Vaporization:** Not pertinent

13.13 **Heat of Combustion:** Not pertinent

13.14 **Heat of Decomposition:** Not pertinent

13.15 **Heat of Solution:** Not pertinent

13.16 **Heat of Polymerization:** Not pertinent

5. HEALTH HAZARDS (Cont'd.)

5.5 **Short-Term Inhalation Limits:** 0.25 mg/m³ for 30 min.

5.6 **Toxicity by Ingestion:** Grade 4; LD_{50} < 50 mg/kg

5.7 **Late Toxicity:** Data not available

5.8 **Vapor (Gas) Irritant Characteristics:** Data not available

5.9 **Liquid or Solid Irritant Characteristics:** Data not available

5.10 **Odor Threshold:** Data not available

TFE	TETRAFLUOROETHYLENE, INHIBITED

Common Synonyms	
Teflon monomer	Compressed Gas Colorless Odorless or faint odor Visible vapor cloud is produced.

Shut off ignition sources, call fire department.
Stop discharge if possible. Keep people away.
Stay upwind. Use water spray to "knock down" vapor.
Evacuate area in case of large discharge.
Notify local health and pollution control agencies.

Fire	FLAMMABLE. POISONOUS GASES ARE PRODUCED IN FIRE. Containers may explode in fire. Flashback along vapor trail may occur. Vapor may explode if ignited in an enclosed area. Wear goggles and self-contained breathing apparatus. Let fire burn. Stop flow of gas if possible. Cool exposed containers and protect men effecting shutoff with water.
Exposure	Call for medical aid. VAPOR Irritating to eyes, nose and throat. Move victim to fresh air. If breathing has stopped, give artificial respiration. If breathing is difficult, give oxygen.
Water Pollution	Not harmful to aquatic life.

1. RESPONSE TO DISCHARGE
(See Response Methods Handbook, CG 446-4)

Issue warning—high flammability
Restrict access

2. LABEL

Red

3. CHEMICAL DESIGNATIONS

3.1 **Synonyms:** Teflon monomer
3.2 **Coast Guard Compatibility Classification:** Not applicable
3.3 **Chemical Formula:** $F_2C{=}CF_2$
3.4 **IMCO/United Nations Numerical Designation:** 2/1081

4. OBSERVABLE CHARACTERISTICS

4.1 **Physical State (as shipped):** Compressed gas
4.2 **Color:** Colorless
4.3 **Odor:** None or faint

5. HEALTH HAZARDS

5.1 **Personal Protective Equipment:** Self-contained breathing apparatus for high gas concentrations
5.2 **Symptoms Following Exposure:** Inhalation causes irritation of respiratory system. Contact with eyes causes slight irritation.
5.3 **Treatment for Exposure:** INHALATION: remove victim from exposure; if breathing is difficult, give artificial respiration and call physician.
5.4 **Toxicity by Inhalation (Threshold Limit Value):** Data not available
5.5 **Short-Term Inhalation Limits:** Data not available
5.6 **Toxicity by Ingestion:** Data not available
5.7 **Late Toxicity:** Causes possible impairment of immunological defense system in rats
5.8 **Vapor (Gas) Irritant Characteristics:** Data not available
5.9 **Liquid or Solid Irritant Characteristics:** Data not available
5.10 **Odor Threshold:** Data not available

6. FIRE HAZARDS

6.1 **Flash Point:** Not pertinent (gas)
6.2 **Flammable Limits in Air:** 10%–50%
6.3 **Fire Extinguishing Agents:** Let fire burn; stop flow of gas; cool containers with water.
6.4 **Fire Extinguishing Agents Not to be Used:** Not pertinent
6.5 **Special Hazards of Combustion Products:** When burned in air, gas forms toxic carbonyl fluoride and hydrogen fluoride.
6.6 **Behavior in Fire:** Vapor is heavier than air and may travel considerable distance to a source of ignition and flash back. Containers may explode.
6.7 **Ignition Temperature:** 370°F
6.8 **Electrical Hazard:** C_2F_4—air mixtures produced explosions which propagated through the smallest clearance in the standard test conducted by Underwriters Laboratories. It does not meet any group classification. *(Continued on page 4)*

7. CHEMICAL REACTIVITY

7.1 **Reactivity with Water:** No reaction
7.2 **Reactivity with Common Materials:** No reaction
7.3 **Stability During Transport:** Stable
7.4 **Neutralizing Agents for Acids and Caustics:** Not pertinent
7.5 **Polymerization:** Can polymerize in the absence of inhibitor, especially when heated or in presence of oxygen.
7.6 **Inhibitor of Polymerization:** d-limonene; pinene; tetrahydronaphthalene; 1-octene; methyl methacrylate

8. WATER POLLUTION

8.1 **Aquatic Toxicity:** None
8.2 **Waterfowl Toxicity:** None
8.3 **Biological Oxygen Demand (BOD):** None
8.4 **Food Chain Concentration Potential:** None

9. SELECTED MANUFACTURERS

1. E. I. duPont de Nemours & Co. Plastics Department Wilmington, Del. 19898
2. ICI America Wilmington, Del. 19899
3. Matheson Gas Products Co. East Rutherford, N. J. 07073

10. SHIPPING INFORMATION

10.1 **Grades or Purity:** 98+%
10.2 **Storage Temperature:** Cool ambient
10.3 **Inert Atmosphere:** No requirement
10.4 **Venting:** Safety relief

11. HAZARD ASSESSMENT CODE
(See Hazard Assessment Handbook, CG 446-3)

A-B-C-Z

12. HAZARD CLASSIFICATIONS

12.1 **Code of Federal Regulations:** Flammable compressed gas
12.2 **NAS Hazard Rating for Bulk Water Transportation:** Not listed
12.3 **NFPA Hazard Classifications:**

Category	Classification*	
Health Hazard (Blue)	2	3
Flammability (Red)	4	4
Reactivity (Yellow)	3	3

*First column refers to nonfire situation.

13. PHYSICAL AND CHEMICAL PROPERTIES

13.1 **Physical State at 15°C and 1 atm:** Gas
13.2 **Molecular Weight:** 100.0
13.3 **Boiling Point at 1 atm:** $-105°F = -76°C = 197°K$
13.4 **Freezing Point:** $-224°F = -142°C = 131°K$
13.5 **Critical Temperature:** (est.) $92°F = 33°C = 306°K$
13.6 **Critical Pressure:** (est.) $573\ psia = 38.9\ atm = 3.95\ MN/m^2$
13.7 **Specific Gravity:** Not pertinent
13.8 **Liquid Surface Tension:** Not pertinent
13.9 **Liquid-Water Interfacial Tension:** Not pertinent
13.10 **Vapor (Gas) Specific Gravity:** 3.45
13.11 **Ratio of Specific Heats of Vapor (Gas):** (est.) 1.1261
13.12 **Latent Heat of Vaporization:** Not pertinent
13.13 **Heat of Combustion:** (est.) $-4{,}000\ Btu/lb = -2{,}000\ cal/g = -90 \times 10^5\ J/kg$
13.14 **Heat of Decomposition:** Not pertinent
13.15 **Heat of Solution:** Not pertinent
13.16 **Heat of Polymerization:** $-450\ Btu/lb = -250\ cal/g = -10.5 \times 10^5\ J/kg$

6. FIRE HAZARDS (Cont'd.)

6.9 **Burning Rate:** Not pertinent

THF	TETRAHYDROFURAN

Common Synonyms			
Diethylene oxide Tetramethylene oxide THF	Liquid	Colorless	Faint fruity odor
	Floats and mixes with water. Flammable, irritating vapor is produced.		

Stop discharge if possible. Keep people away.
Shut off ignition sources and call fire department.
Stay upwind and use water spray to "knock down" vapor.
Avoid contact with liquid and vapor.
Isolate and remove discharged material.
Notify local health and pollution control agencies.

Fire

FLAMMABLE.
Flashback along vapor trail may occur.
Vapor may explode if ignited in an enclosed area.
Wear goggles and self-contained breathing apparatus.
Extinguish with dry chemical or carbon dioxide.
Water may be ineffective on fire.
Cool exposed containers with water.

Exposure

CALL FOR MEDICAL AID.

VAPOR
Irritating to eyes, nose and throat.
If inhaled, will cause nausea, headache or loss of consciousness.
Move to fresh air.
If breathing as stopped, give artificial respiration.
If breathing is difficult, give oxygen.

LIQUID
Irritating to skin and eyes.
Harmful if swallowed.
Remove contaminated clothing and shoes.
Flush affected areas with plenty of water.
IF IN EYES, hold eyelids open and flush with plenty of water.
IF SWALLOWED and victim is CONSCIOUS, have victim drink water or milk.

Water Pollution

Effect of low concentrations on aquatic life is unknown.
May be dangerous if it enters water intakes.
Notify local health and wildlife officials.
Notify operators of nearby water intakes.

1. RESPONSE TO DISCHARGE

(See Response Methods Handbook, CG 446-4)

Issue warning—high flammability
Disperse and flush

2. LABEL

Red

3. CHEMICAL DESIGNATIONS

3.1 **Synonyms:** Tetramethylene oxide THF
3.2 **Coast Guard Compatibility Classification:** Ether
3.3 **Chemical Formula:** $\overline{CH_2CH_2CH_2CH_2—O}$
3.4 **IMCO/United Nations Numerical Designation:** 3.1/2056

4. OBSERVABLE CHARACTERISTICS

4.1 **Physical State (as shipped):** Liquid
4.2 **Color:** Colorless
4.3 **Odor:** Ether-like; similar to acetone

5. HEALTH HAZARDS

5.1 **Personal Protective Equipment:** Self-contained breathing apparatus; goggles or face shield; rubber or plastic gloves.
5.2 **Symptoms Following Exposure:** Vapors cause nausea, dizziness, headache, and anesthesia. Liquid can de-fat the skin and cause irritation. Liquid also irritates eyes.
5.3 **Treatment for Exposure:** INHALATION: remove victim from contaminated area; administer artificial respiration and oxygen if necessary. INGESTION: gastric lavage and saline cathartics are usually helpful; subsequent treatment is symptomatic and supportive. SKIN OR EYE CONTACT: wash with copious amounts of water.
5.4 **Toxicity by Inhalation (Threshold Limit Value):** 200 ppm
5.5 **Short-Term Inhalation Limits:** 500 ppm for 30 min.
5.6 **Toxicity by Ingestion:** Grade 3; LD_{50} 50 to 500 mg/kg
5.7 **Late Toxicity:** Data not available
5.8 **Vapor (Gas) Irritant Characteristics:** Vapors cause a slight smarting of the eyes or respiratory system if present in high concentrations. The effect is temporary.
5.9 **Liquid or Solid Irritant Characteristics:** Minimum hazard. If spilled on clothing and allowed to remain, may cause smarting and reddening of the skin.
5.10 **Odor Threshold:** 20—50 ppm

6. FIRE HAZARDS

6.1 **Flash Point:** 6°F C.C.; −4°F O.C.
6.2 **Flammable Limits in Air:** 1.8%—11.8%
6.3 **Fire Extinguishing Agents:** Dry chemical or carbon dioxide
6.4 **Fire Extinguishing Agents Not to be Used:** Water may be ineffective.
6.5 **Special Hazards of Combustion Products:** Irritating vapor is generated when heated.
6.6 **Behavior in Fire:** May explode. Vapor is heavier than air and may travel considerable distance to a source of ignition and flash back.
6.7 **Ignition Temperature:** 610°F
6.8 **Electrical Hazard:** Not pertinent
6.9 **Burning Rate:** 4.7 mm/min.

7. CHEMICAL REACTIVITY

7.1 **Reactivity with Water:** No reaction
7.2 **Reactivity with Common Materials:** No reaction
7.3 **Stability During Transport:** Stable unless 0.1% of peroxides has accumulated because of prolonged storage in presence of air. When concentrated by evaporation of solution, they explode.
7.4 **Neutralizing Agents for Acids and Caustics:** Not pertinent
7.5 **Polymerization:** Not pertinent
7.6 **Inhibitor of Polymerization:** 0.025% butylated hydroxytoluene (BHT) present to prevent peroxide formation.

8. WATER POLLUTION

8.1 **Aquatic Toxicity:** Data not available
8.2 **Waterfowl Toxicity:** Data not available
8.3 **Biological Oxygen Demand (BOD):** Data not available
8.4 **Food Chain Concentration Potential:** None

9. SELECTED MANUFACTURERS

1. E. I. duPont de Nemours & Co., Inc.
   Electrochemicals Dept.
   Wilmington, Del. 19898
2. The Quaker Oats Co.
   Chemical Division
   3324 Chelsea Ave.
   Memphis, Tenn. 38101

10. SHIPPING INFORMATION

10.1 **Grades or Purity:** Data not available
10.2 **Storage Temperature:** Ambient
10.3 **Inert Atmosphere:** Padded
10.4 **Venting:** Pressure-vacuum

11. HAZARD ASSESSMENT CODE

(See Hazard Assessment Handbook, CG 446-3)

A-P-Q-R-S

12. HAZARD CLASSIFICATIONS

12.1 **Code of Federal Regulations:** Flammable liquid
12.2 **NAS Hazard Rating for Bulk Water Transportation:**

Category	Rating
Fire	3
Health	
Vapor Irritant	1
Liquid or Solid Irritant	1
Poisons	2
Water Pollution	
Human Toxicity	2
Aquatic Toxicity	2
Aesthetic Effect	0
Reactivity	
Other Chemicals	1
Water	0
Self-Reaction	3

12.3 **NFPA Hazard Classifications:**

Category	Classification
Health Hazard (Blue)	2
Flammability (Red)	3
Reactivity (Yellow)	1

13. PHYSICAL AND CHEMICAL PROPERTIES

13.1 **Physical State at 15°C and 1 atm:** Liquid
13.2 **Molecular Weight:** 72.10
13.3 **Boiling Point at 1 atm:** 151°F = 66°C = 339°K
13.4 **Freezing Point:** −163.3°F = −108.5°C = 164.7°K
13.5 **Critical Temperature:** 512.6°F = 267.0°C = 540.2°K
13.6 **Critical Pressure:** 753 psia = 51.2 atm = 5.19 MN/m^2
13.7 **Specific Gravity:** 0.888 at 20°C (liquid)
13.8 **Liquid Surface Tension:** 28 dynes/cm = 0.028 N/m at 20°C
13.9 **Liquid-Water Interfacial Tension:** Not pertinent
13.10 **Vapor (Gas) Specific Gravity:** Not pertinent
13.11 **Ratio of Specific Heats of Vapor (Gas):** (est.) 1.083
13.12 **Latent Heat of Vaporization:** 180 Btu/lb = 98 cal/g = 4.1×10^5 J/kg
13.13 **Heat of Combustion:** −14,990 Btu/lb = −8330 cal/g = -348.8×10^5 J/kg
13.14 **Heat of Decomposition:** Not pertinent
13.15 **Heat of Solution:** Not pertinent
13.16 **Heat of Polymerization:** Not pertinent

NOTES

REVISED 1978

THN

TETRAHYDRONAPHTHALENE

Common Synonyms
1,2,3,4-tetrahydronaphthalene
Tetralin
Tetranap

Watery liquid — Colorless — Moldy, turpentine odor

Floats on water.

Stop discharge if possible.
Call fire department.
Avoid contact with liquid.
Isolate and remove discharged material.
Notify local health and pollution control agencies.

Fire

Combustible.
Extinguish with foam, dry chemical, or carbon dioxide.
Water may be ineffective on fire.
Cool exposed containers with water.

Exposure

CALL FOR MEDICAL AID.

LIQUID
Irritating to skin and eyes.
Harmful if swallowed.
Remove contaminated clothing and shoes.
Flush affected areas with plenty of water.
IF IN EYES, hold eyelids open and flush with plenty of water.
IF SWALLOWED and victim is CONSCIOUS, have victim drink water or milk.

Water Pollution

Effect of low concentrations on aquatic life is unknown.
Fouling to shoreline.
May be dangerous if it enters water intakes.
Notify local health and wildlife officials.
Notify operators of nearby water intakes.

1. RESPONSE TO DISCHARGE

(See Response Methods Handbook, CG 446-4)

Mechanical containment
Should be removed
Chemical and physical treatment

2. LABELS

No hazard label required by Code of Federal Regulations

3. CHEMICAL DESIGNATIONS

3.1 **Synonyms:** 1,2,3,4-Tetrahydronaphthalene; Tetralin; Tetranap
3.2 **Coast Guard Compatibility Classification:** Aromatic hydrocarbon
3.3 **Chemical Formula:** $C_{10}H_{12}$
3.4 **IMCO/United Nations Numerical Designation:** Not listed

4. OBSERVABLE CHARACTERISTICS

4.1 **Physical State (as shipped):** Liquid
4.2 **Color:** Colorless
4.3 **Odor:** Mild characteristic; resembling that of naphthalene or turpentine

5. HEALTH HAZARDS

5.1 **Personal Protective Equipment:** Air-supplied mask in closed tanks; goggles or face shield; rubber gloves.
5.2 **Symptoms Following Exposure:** Liquid may cause nervous disturbance, green coloration of urine, and skin and eye irritation.
5.3 **Treatment for Exposure:** INGESTION: induce vomiting; call a doctor; medical treatment should be aimed at conservation of liver and kidney function. EYES: flush with water for at least 15 min; call a doctor. SKIN: wipe off, wash with soap and water.
5.4 **Toxicity by Inhalation (Threshold Limit Value):** 25 ppm (suggested)
5.5 **Short-Term Inhalation Limits:** Data not available
5.6 **Toxicity by Ingestion:** Grade 2; LD_{50} 0.5 to 5 g/kg
5.7 **Late Toxicity:** Liver and kidney damage from high dose.
5.8 **Vapor (Gas) Irritant Characteristics:** Vapors cause a slight smarting of the eyes or respiratory system if present in high concentrations. The effect is temporary.
5.9 **Liquid or Solid Irritant Characteristics:** Minimum hazard. If spilled on clothing and allowed to remain, may cause smarting and reddening of the skin.
5.10 **Odor Threshold:** Data not available

6. FIRE HAZARDS

6.1 **Flash Point:** 176°F C.C.; 190°F O.C.
6.2 **Flammable Limits in Air:** 0.8%–5%
6.3 **Fire Extinguishing Agents:** Foam, dry chemical, or carbon dioxide
6.4 **Fire Extinguishing Agents Not to be Used:** Avoid water
6.5 **Special Hazards of Combustion Products:** Not pertinent
6.6 **Behavior in Fire:** Not pertinent
6.7 **Ignition Temperature:** 725°F
6.8 **Electrical Hazard:** Not pertinent
6.9 **Burning Rate:** Data not available

7. CHEMICAL REACTIVITY

7.1 **Reactivity with Water:** No reaction
7.2 **Reactivity with Common Materials:** No reaction
7.3 **Stability During Transport:** Stable
7.4 **Neutralizing Agents for Acids and Caustics:** Not pertinent
7.5 **Polymerization:** Not pertinent
7.6 **Inhibitor of Polymerization:** Not pertinent

8. WATER POLLUTION

8.1 **Aquatic Toxicity:** 78 ppm/24 hr/brine shrimp/TL_m
8.2 **Waterfowl Toxicity:** Data not available
8.3 **Biological Oxygen Demand (BOD):** 0 lb/lb, 5 days
8.4 **Food Chain Concentration Potential:** None

9. SELECTED MANUFACTURERS

1. E. I. duPont de Nemours & Co., Inc.
   Organic Chemicals Dept.
   Wilmington, Del. 19898
2. Lonza, Inc.
   22-10 Route 208
   Fair Lawn, N. J. 07410
3. Union Carbide Corp.
   Chemicals and Plastics Division
   270 Park Ave.
   New York, N. Y. 10017

10. SHIPPING INFORMATION

10.1 **Grades or Purity:** 90+%
10.2 **Storage Temperature:** Ambient
10.3 **Inert Atmosphere:** No requirement
10.4 **Venting:** Open (flame arrester)

11. HAZARD ASSESSMENT CODE

(See Hazard Assessment Handbook, CG 446-3)

A-T-U

12. HAZARD CLASSIFICATIONS

12.1 **Code of Federal Regulations:** Combustible Liquid
12.2 **NAS Hazard Rating for Bulk Water Transportation:**

Category	Rating
Fire	1
Health	
Vapor Irritant	1
Liquid or Solid Irritant	1
Poisons	2
Water Pollution	
Human Toxicity	1
Aquatic Toxicity	3
Aesthetic Effect	2
Reactivity	
Other Chemicals	0
Water	0
Self-Reaction	0

12.3 **NFPA Hazard Classifications:**

Category	Classification
Health Hazard (Blue)	1
Flammability (Red)	2
Reactivity (Yellow)	0

13. PHYSICAL AND CHEMICAL PROPERTIES

13.1 **Physical State at 15°C and 1 atm:** Liquid
13.2 **Molecular Weight:** 132.21
13.3 **Boiling Point at 1 atm:** 406°F = 208°C = 481°K
13.4 **Freezing Point:** −23.1°F = 30.6°C = 242.6°K
13.5 **Critical Temperature:** Not pertinent
13.6 **Critical Pressure:** Not pertinent
13.7 **Specific Gravity:** 0.974 at 20°C (liquid
13.8 **Liquid Surface Tension:** 35.5 dynes/cm = 0.0355 N/m at 20°C
13.9 **Liquid-Water Interfacial Tension:** Data not available
13.10 **Vapor (Gas) Specific Gravity:** Not pertinent
13.11 **Ratio of Specific Heats of Vapor (Gas):** Not pertinent
13.12 **Latent Heat of Vaporization:** 138 Btu/lb = 76.5 cal/g = 3.20×10^5 J/kg
13.13 **Heat of Combustion:** -18,400 Btu/lb = -10,200 cal/g = -429×10^5 J/kg
13.14 **Heat of Decomposition:** Not pertinent
13.15 **Heat of Solution:** Not pertinent
13.16 **Heat of Polymerization:** Not pertinent

NOTES

REVISED 1978

TETRAMETHYL LEAD

Common Synonyms
Lead tetramethyl

Oily liquid — Colorless — Fruity odor

Sinks in water. Poisonous, flammable vapor is produced.

AVOID CONTACT WITH LIQUID AND VAPOR. Keep people away.
Wear goggles, self-contained breathing apparatus, and rubber overclothing (including gloves).
Stop discharge if possible.
Call fire department.
Stay upwind and use water spray to "knock down" vapor.
Isolate and remove discharged material.
Notify local health and pollution control agencies.

Fire

Combustible.
Containers may explode in fire.
Vapor may explode if ignited in an enclosed area.
Wear goggles, self-contained breathing apparatus, and rubber overclothing (including gloves).
Combat fires from behind barrier or protected location.
Extinguish with water, dry chemicals, foam, or carbon dioxide.
Cool exposed containers with water.

Exposure

CALL FOR MEDICAL AID.

VAPOR
POISONOUS IF INHALED.
Irritating to eyes.
Move to fresh air.
If breathing has stopped, give artifical respiration.
If breathing is difficult, give oxygen.

LIQUID
POISONOUS IF SWALLOWED OR IF SKIN IS EXPOSED.
Will burn eyes.
Remove contaminated clothing and shoes.
Flush affected areas with plenty of water.
IF IN EYES, hold eyelids open and flush with plenty of water.
IF SWALLOWED and victim is CONSCIOUS, have victim drink water or milk.

Water Pollution

Effect of low concentrations on aquatic life is unknown.
May be dangerous if it enters water intakes.
Notify local health and wildlife officials.
Notify operators of nearby water intakes.

1. RESPONSE TO DISCHARGE
(See Response Methods Handbook, CG 446-4)

Issue warning—poison, water contaminant
Restrict access
Should be removed
Chemical and physical treatment

2. LABEL

3. CHEMICAL DESIGNATIONS

3.1 **Synonyms:** Lead tetramethyl
3.2 **Coast Guard Compatibility Classification:** Special class
3.3 **Chemical Formula:** $Pb(CH_3)_4$
3.4 **IMCO/United Nations Numerical Designation:** 6.1/1649

4. OBSERVABLE CHARACTERISTICS

4.1 **Physical State (as shipped):** Liquid
4.2 **Color:** Dyed red or other distinctive color
4.3 **Odor:** Faint fruity odor

5. HEALTH HAZARDS

5.1 **Personal Protective Equipment:** Organic vapor canister face mask for short periods, air line mask for longer periods; protective goggles or face shield; neoprene-coated protective gloves; rubber shoes or boots; white or light-colored clothing.
5.2 **Symptoms Following Exposure:** Increased urinary output of lead. If inhaled or absorbed by skin, may cause insomnia, excitability, delirium, coma, and death. Do not confuse with inorganic lead.
5.3 **Treatment for Exposure:** Remove victim from contaminated area. SKIN: wash with kerosene or similar petroleum distillate, followed by soap and water; consult physician immediately.
5.4 **Toxicity by Inhalation (Threshold Limit Value):** 0.15 mg/m³
5.5 **Short-Term Inhalation Limits:** Data not available
5.6 **Toxicity by Ingestion:** Grade 3; oral rat LD_{50} = 109 mg/kg
5.7 **Late Toxicity:** Lead poisoning
5.8 **Vapor (Gas) Irritant Characteristics:** Vapors cause a slight smarting of the eyes or respiratory system if present in high concentrations. The effect is temporary.
5.9 **Liquid or Solid Irritant Characteristics:** Causes smarting of the skin and first-degree burns on short exposure; may cause secondary burns on long exposure.
5.10 **Odor Threshold:** Data not available

6. FIRE HAZARDS

6.1 **Flash Point:** 100°F O.C.
6.2 **Flammable Limits in Air:** Data not available
6.3 **Fire Extinguishing Agents:** Water, foam, dry chemical, or carbon dioxide
6.4 **Fire Extinguishing Agents Not to be Used:** Not pertinent
6.5 **Special Hazards of Combustion Products:** Toxic gases are generated in fire.
6.6 **Behavior in Fire:** May explode
6.7 **Ignition Temperature:** Data not available
6.8 **Electrical Hazard:** Not pertinent
6.9 **Burning Rate:** Data not available

7. CHEMICAL REACTIVITY

7.1 **Reactivity with Water:** No reaction
7.2 **Reactivity with Common Materials:** No reaction
7.3 **Stability During Transport:** Starts to decompose at about 212°F. If confined, may explode or detonate at high temperatures.
7.4 **Neutralizing Agents for Acids and Caustics:** Not pertinent
7.5 **Polymerization:** Not pertinent
7.6 **Inhibitor of Polymerization:** Not pertinent

8. WATER POLLUTION

8.1 **Aquatic Toxicity:** Data not available
8.2 **Waterfowl Toxicity:** Data not available
8.3 **Biological Oxygen Demand (BOD):** Data not available
8.4 **Food Chain Concentration Potential:** Data not available

9. SELECTED MANUFACTURERS

1. Ethyl Corp.
   Industrial Chemicals Division
   451 Florida St.
   Baton Rouge, La. 70801
2. E. I. duPont de Nemours & Co., Inc.
   Petroleum Chemicals Division
   Antioch, Calif. 94509
3. PPG Industries Inc.
   Houston Chemical Co. Division
   Beaumont, Tex. 77704

10. SHIPPING INFORMATION

10.1 **Grades or Purity:** Technical
10.2 **Storage Temperature:** Ambient
10.3 **Inert Atmosphere:** No requirement
10.4 **Venting:** Pressure-vacuum

11. HAZARD ASSESSMENT CODE
(See Hazard Assessment Handbook, CG 446-3)

A-X-Y

12. HAZARD CLASSIFICATIONS

12.1 **Code of Federal Regulations:** Poisonous liquid or solid, Class B
12.2 **NAS Hazard Rating for Bulk Water Transportation:** Not listed
12.3 **NFPA Hazard Classifications:**

Category	Classification
Health Hazard (Blue)	3
Flammability (Red)	3
Reactivity (Yellow)	3

13. PHYSICAL AND CHEMICAL PROPERTIES

13.1 **Physical State at 15°C and 1 atm:** Liquid
13.2 **Molecular Weight:** 267.33
13.3 **Boiling Point at 1 atm:** 230°F = 110°C = 383°K
13.4 **Freezing Point:** −17.5°F = −27.5°C = 245.7°K
13.5 **Critical Temperature:** Not pertinent
13.6 **Critical Pressure:** Not pertinent
13.7 **Specific Gravity:** 1.999 at 20°C (liquid)
13.8 **Liquid Surface Tension:** Data not available
13.9 **Liquid-Water Interfacial Tension:** Data not available
13.10 **Vapor (Gas) Specific Gravity:** Not pertinent
13.11 **Ratio of Specific Heats of Vapor (Gas):** Not pertinent
13.12 **Latent Heat of Vaporization:** (est.) 55.5 Btu/lb = 30.8 cal/g = 1.29×10^5 J/kg
13.13 **Heat of Combustion:** (est.) −5290 Btu/lb = −2940 cal/g = -123×10^5 J/kg
13.14 **Heat of Decomposition:** Not pertinent
13.15 **Heat of Solution:** Not pertinent
13.16 **Heat of Polymerization:** Not pertinent

NOTES

REVISED 1978

TPG — THIOPHOSGENE

Common Synonyms			
Thiocarbonyl chloride	Liquid	Red	Sharp choking odor
	Sinks in water. Reacts slowly with water and produces poisonous vapor.		

AVOID CONTACT WITH LIQUID AND VAPOR. KEEP PEOPLE AWAY.
Wear goggles and self-contained breathing apparatus.
Stop discharge if possible.
Stay upwind. Use water spray to "knock down" vapor.
Isolate and remove discharged material.
Notify local health and pollution control agencies.

Fire

Combustible.
POISONOUS GASES ARE PRODUCED IN FIRE.
Wear goggles and self-contained breathing apparatus.
Extinguish with dry chemicals or carbon dioxide.
DO NOT USE WATER OR FOAM ON FIRE.
Cool exposed containers with water.

Exposure

CALL FOR MEDICAL AID.

VAPOR
POISONOUS IF INHALED.
Irritating to eyes, nose and throat.
Move victim to fresh air.
If breathing has stopped, give artificial respiration (but not mouth-to-mouth).
If breathing is difficult, give oxygen.

LIQUID
POISONOUS IF SWALLOWED.
Irritating to skin and eyes.
Remove contaminated clothing and shoes.
Flush affected areas with plenty of water.
IF IN EYES, hold eyelids open and flush with plenty of water.
IF SWALLOWED and victim is CONSCIOUS, have victim drink water or milk and have victim induce vomiting.
IF SWALLOWED and victim is UNCONSCIOUS OR HAVING CONVULSIONS, do nothing except keep victim warm.

Water Pollution

Effect of low concentrations on aquatic life is unknown.
May be dangerous if it enters water intakes.
Notify local health and wildlife officials.
Notify operators of nearby water intakes.

1. RESPONSE TO DISCHARGE

(See Response Methods Handbook, CG 446-4)

Issue warning—poison, air contaminant, water contaminant
Restrict access
Evacuate area
Disperse and flush

2. LABEL

3. CHEMICAL DESIGNATIONS

3.1 **Synonyms:** Thiocarbonyl chloride
3.2 **Coast Guard Compatibility Classification:** Not applicable
3.3 **Chemical Formula:** $CSCl_2$
3.4 **IMCO/United Nations Numerical Designation:** 6.1/—

4. OBSERVABLE CHARACTERISTICS

4.1 **Physical State (as shipped):** Liquid
4.2 **Color:** Red
4.3 **Odor:** Choking, very irritating, sharp

5. HEALTH HAZARDS

5.1 **Personal Protective Equipment:** Self-contained breathing apparatus or organic canister mask; goggles or face shield; rubber gloves
5.2 **Symptoms Following Exposure:** Inhalation causes irritation of respiratory system and delayed pulmonary edema. Vapor irritates eyes. Liquid burns skin and eyes. Ingestion causes irritation of mouth and stomach.
5.3 **Treatment for Exposure:** *Get medical attention at once after any exposure to this compound.* INHALATION: remove victim from exposure; support respiration; watch for pulmonary edema. EYES: irrigate with large quantities of water for 15 min. SKIN: flush with water. INGESTION: do NOT induce vomiting; give large amount of water.
5.4 **Toxicity by Inhalation (Threshold Limit Value):** Data not available
5.5 **Short-Term Inhalation Limits:** Data not available
5.6 **Toxicity by Ingestion:** Grade 2; oral LD_{50} = 929 mg/kg (rat)
5.7 **Late Toxicity:** Data not available
5.8 **Vapor (Gas) Irritant Characteristics:** Data not available
5.9 **Liquid or Solid Irritant Characteristics:** Data not available
5.10 **Odor Threshold:** Data not available

6. FIRE HAZARDS

6.1 **Flash Point:** Data not available
6.2 **Flammable Limits in Air:** Data not available
6.3 **Fire Extinguishing Agents:** Dry chemical, carbon dioxide
6.4 **Fire Extinguishing Agents Not to be Used:** Water, foam
6.5 **Special Hazards of Combustion Products:** Toxic phosgene, hydrogen chloride, and sulfur dioxide may be generated in a fire.
6.6 **Behavior in Fire:** Decomposes above 200°C to carbon bisulfide (very flammable) and carbon tetrachloride.
6.7 **Ignition Temperature:** Data not available
6.8 **Electrical Hazard:** Data not available
6.9 **Burning Rate:** Data not available

7. CHEMICAL REACTIVITY

7.1 **Reactivity with Water:** Evolves hydrogen chloride, carbon disulfide, and carbon dioxide. Reaction is slow unless water is hot.
7.2 **Reactivity with Common Materials:** Corrodes metals in presence of moisture.
7.3 **Stability During Transport:** Stable
7.4 **Neutralizing Agents for Acids and Caustics:** Flush with water, rinse with sodium bicarbonate or lime solution
7.5 **Polymerization:** Not pertinent
7.6 **Inhibitor of Polymerization:** Not pertinent

8. WATER POLLUTION

8.1 **Aquatic Toxicity:** Data not available
8.2 **Waterfowl Toxicity:** Data not available
8.3 **Biological Oxygen Demand (BOD):** Data not available
8.4 **Food Chain Concentration Potential:** None

9. SELECTED MANUFACTURERS

1. Aldrich Chemical Co.
   940 West St. Paul Avenue
   Milwaukee, Wis. 53233
2. Pfaltz and Bauer, Inc.
   126-04 Northern Boulevard
   Flushing, N. Y. 11368

10. SHIPPING INFORMATION

10.1 **Grades or Purity:** Commercial
10.2 **Storage Temperature:** Ambient
10.3 **Inert Atmosphere:** No requirement
10.4 **Venting:** Pressure-vacuum

11. HAZARD ASSESSMENT CODE

(See Hazard Assessment Handbook, CG 446-3)

A-O-X-Y

12. HAZARD CLASSIFICATIONS

12.1 **Code of Federal Regulations:** Poisonous, Class B
12.2 **NAS Hazard Rating for Bulk Water Transportation:** Not listed
12.3 **NFPA Hazard Classifications:** Not listed

13. PHYSICAL AND CHEMICAL PROPERTIES

13.1 **Physical State at 15°C and 1 atm:** Liquid
13.2 **Molecular Weight:** 115.0
13.3 **Boiling Point at 1 atm:** 163°F = 73°C = 346°K
13.4 **Freezing Point:** Not pertinent
13.5 **Critical Temperature:** Not pertinent
13.6 **Critical Pressure:** Not pertinent
13.7 **Specific Gravity:** 1.513 at 20°C
13.8 **Liquid Surface Tension:** (est.) 25 dynes/cm = 0.025 N/m at 20°C
13.9 **Liquid-Water Interfacial Tension:** Not pertinent
13.10 **Vapor (Gas) Specific Gravity:** 4
13.11 **Ratio of Specific Heats of Vapor (Gas):** Data not available
13.12 **Latent Heat of Vaporization:** (est.) 128 Btu/lb = 71 cal/g = 3.0×10^5 J/kg
13.13 **Heat of Combustion:** (est.) −3,400 Btu/lb = −1,900 cal/g = -80×10^5 J/kg
13.14 **Heat of Decomposition:** Not pertinent
13.15 **Heat of Solution:** Not pertinent
13.16 **Heat of Polymerization:** Not pertinent

NOTES

THR	THIRAM

Common Synonyms		
Bis (dimethylthiocarbamoyl) disulfide Bis (dimethylthiocarbamyl) disulfide Tetramethyl thiuram disulfide	Solid Sinks in water.	White to light yellow

AVOID CONTACT WITH SOLID AND DUST. KEEP PEOPLE AWAY.
Wear goggles and dust respirator.
Call fire department.
Stay upwind. Use water spray to "knock down" dust.
Isolate and remove discharged material.
Notify local health and pollution control agencies.

Fire	Combustible. POISONOUS GASES ARE PRODUCED IN FIRE. Wear goggles and self-contained breathing apparatus. Flood discharge area with water.
Exposure	CALL FOR MEDICAL AID. DUST POISONOUS IF INHALED. Irritating to eyes, nose and throat. Move victim to fresh air. If in eyes, hold eyelids open and flush with plenty of water. If breathing has stopped, give artificial respiration. If breathing is difficult, give oxygen. SOLID POISONOUS IF SWALLOWED. Irritating to skin and eyes. Remove contaminated clothing and shoes. Flush affected areas with plenty of water. IF IN EYES, hold eyelids open and flush with plenty of water. IF SWALLOWED and victim is CONSCIOUS, have victim drink water or milk and have victim induce vomiting. IF SWALLOWED and victim is UNCONSCIOUS OR HAVING CONVULSIONS, do nothing except keep victim warm.
Water Pollution	HARMFUL TO AQUATIC LIFE IN VERY LOW CONCENTRATIONS. May be dangerous if it enters water intakes. Notify local health and wildlife officials. Notify operators of nearby water intakes.

1. RESPONSE TO DISCHARGE

(See Response Methods Handbook, CG 446-4)

Issue warning—water contaminant
Restrict access
Should be removed
Chemical and physical treatment

2. LABELS

No hazard label required by Code of Federal Regulations

3. CHEMICAL DESIGNATIONS

3.1 **Synonyms:** Arasan; Bis(dimethylthiocarbamoyl) disulfide; Bis(dimethylthiocarbamyl) disulfide; Methyl thiram; Methyl tuads; Tetramethylthiuram disulfide; Thiuram

3.2 **Coast Guard Compatibility Classification:** Not applicable

3.3 **Chemical Formula:** $(CH_3)_2NC(S)SSC(S)N(CH_3)_2$ or $C_6H_{12}N_2S_4$

(Continued on page 4)

4. OBSERVABLE CHARACTERISTICS

4.1 **Physical State (as shipped):** Solid

4.2 **Color:** White to slightly yellow

4.3 **Odor:** Data not available

5. HEALTH HAZARDS

5.1 **Personal Protective Equipment:** Rubber gloves; goggles; dust mask

5.2 **Symptoms Following Exposure:** Inhalation of dust may cause respiratory irritation. Liquid irritates eyes and skin and may cause allergic eczema in sensitive individuals. Ingestion causes nausea, vomiting, and diarrhea, all of which may be persistent; paralysis may develop.

5.3 **Treatment for Exposure:** INHALATION: remove victim from exposure; if breathing has stopped or is difficult, give artificial respiration and call physician. EYES or SKIN: wash with water; if irritation persists, consult a physician. INGESTION: call physician; induce vomiting and follow with gastric lavage; treatment thereafter is symptomatic and supportive; avoid fats, oils, and lipid solvents, which enhance absorption; rigorously prohibit ethyl alcohol in all forms for at least 10 days; inform doctor if patient has used alcohol within 48 hrs.

5.4 **Toxicity by Inhalation (Threshold Limit Value):** 5 mg/m^3

5.5 **Short-Term Inhalation Limits:** Data not available

5.6 **Toxicity by Ingestion:** Grade 2; oral LD_{50} = 560 mg/kg (rat); 375-865 mg/kg (mammal)

5.7 **Late Toxicity:** Causes birth defects in mice and hamsters

5.8 **Vapor (Gas) Irritant Characteristics:** Data not available

5.9 **Liquid or Solid Irritant Characteristics:** Data not available

5.10 **Odor Threshold:** Data not available

6. FIRE HAZARDS

6.1 **Flash Point:** Not pertinent (solid)

6.2 **Flammable Limits in Air:** Not pertinent

6.3 **Fire Extinguishing Agents:** Water, dry chemical, carbon dioxide

6.4 **Fire Extinguishing Agents Not to be Used:** Not pertinent

6.5 **Special Hazards of Combustion Products:** Toxic and irritating oxides of sulfur are formed. Carbon disulfide may be formed from unburned material.

6.6 **Behavior in Fire:** Not pertinent

6.7 **Ignition Temperature:** Data not available

6.8 **Electrical Hazard:** Not pertinent

6.9 **Burning Rate:** Not pertinent

7. CHEMICAL REACTIVITY

7.1 **Reactivity with Water:** No reaction

7.2 **Reactivity with Common Materials:** No reaction

7.3 **Stability During Transport:** Stable

7.4 **Neutralizing Agents for Acids and Caustics:** Not pertinent

7.5 **Polymerization:** Not pertinent

7.6 **Inhibitor of Polymerization:** Not pertinent

8. WATER POLLUTION

8.1 **Aquatic Toxicity:** 0.76 ppm/96 hr/catfish/TL_m/fresh water

8.2 **Waterfowl Toxicity:** LD_{50}=2800 mg/kg

8.3 **Biological Oxygen Demand (BOD):** Data not available

8.4 **Food Chain Concentration Potential:** None

9. SELECTED MANUFACTURERS

1. Pennwalt Corporation
   Chemicals Division
   Three Parkway
   Philadelphia, Pa. 19102
2. E. I. duPont de Nemours & Co.
   Wilmington, Del. 19898
3. R. T. Vanderbilt Co., Inc.
   230 Park Avenue
   New York, N. Y. 10017

10. SHIPPING INFORMATION

10.1 **Grades or Purity:** 98% plus 2% oil

10.2 **Storage Temperature:** Ambient

10.3 **Inert Atmosphere:** No requirement

10.4 **Venting:** Pressure-vacuum

11. HAZARD ASSESSMENT CODE

(See Hazard Assessment Handbook, CG 446-3)

II

12. HAZARD CLASSIFICATIONS

12.1 **Code of Federal Regulations:** ORM–A

12.2 **NAS Hazard Rating for Bulk Water Transportation:** Not listed

12.3 **NFPA Hazard Classifications:** Not listed

13. PHYSICAL AND CHEMICAL PROPERTIES

13.1 **Physical State at 15°C and 1 atm:** Solid

13.2 **Molecular Weight:** 240.4

13.3 **Boiling Point at 1 atm:** Not pertinent (decomposes)

13.4 **Freezing Point:** 288—313°F = 142—156°C = 415—429°K

13.5 **Critical Temperature:** Not pertinent

13.6 **Critical Pressure:** Not pertinent

13.7 **Specific Gravity:** 1.43 at 20°C (solid)

13.8 **Liquid Surface Tension:** Not pertinent

13.9 **Liquid-Water Interfacial Tension:** Not pertinent

13.10 **Vapor (Gas) Specific Gravity:** Not pertinent

13.11 **Ratio of Specific Heats of Vapor (Gas):** Not pertinent

13.12 **Latent Heat of Vaporization:** Not pertinent

13.13 **Heat of Combustion:** Data not available

13.14 **Heat of Decomposition:** Not pertinent

13.15 **Heat of Solution:** Not pertinent

13.16 **Heat of Polymerization:** Not pertinent

3. CHEMICAL DESIGNATIONS (Cont'd.)

3.4 **IMCO/United Nations Numerical Designation:** 6.1/1609

TRN	THORIUM NITRATE

Common Synonyms			
Thorium nitrate tetrahydrate	Solid	White	Odorless
	Mixes with water.		

Keep people away.
Call fire department.
Isolate and remove discharged material.
Notify local health and pollution control agencies.

Fire	Not flammable. May cause fire on contact with combustibles. POISONOUS GASES ARE PRODUCED IN FIRE. Wear goggles and self-contained breathing apparatus. Flood discharge area with water.
Exposure	Call for medical aid. DUST Irritating to eyes, nose and throat. Harmful if inhaled. Move victim to fresh air. If in eyes, hold eyelids open and flush with plenty of water. SOLID Irritating to skin and eyes. Harmful if swallowed. Remove contaminated clothing and shoes. Flush affected areas with plenty of water. IF IN EYES, hold eyelids open and flush with plenty of water. IF SWALLOWED and victim is CONSCIOUS, have victim drink water or milk.
Water Pollution	Dangerous to aquatic life in high concentrations. May be dangerous if it enters water intakes. Notify local health and wildlife officials. Notify operators of nearby water intakes.

1. RESPONSE TO DISCHARGE

(See Response Methods Handbook, CG 446-4)

Issue warning—radioactive, oxidizing material, water contaminant
Restrict access
Should be removed
Chemical and physical treatment

2. LABELS

Yellow

3. CHEMICAL DESIGNATIONS

3.1 **Synonyms:** Thorium nitrate tetrahydrate
3.2 **Coast Guard Compatibility Classification:** Not listed
3.3 **Chemical Formula:** $Th(NO_3)_4 \cdot 4H_2O$ (approx.)
3.4 **IMCO/United Nations Numerical Designation:** 7/Low Specific Activity

4. OBSERVABLE CHARACTERISTICS

4.1 **Physical State (as shipped):** Solid
4.2 **Color:** White
4.3 **Odor:** None

5. HEALTH HAZARDS

5.1 **Personal Protective Equipment:** Dust respirator; gloves; rubber shoes or boots
5.2 **Symptoms Following Exposure:** Compound has low chemical toxicity, but alpha emission is expected to constitute a hazard if a fairly large amount is inhaled or ingested. Dust may irritate eyes and cause diffuse dermatitis. Beta and gamma emission is small.
5.3 **Treatment for Exposure:** EYES: flush with water until solid is removed. SKIN: wash thoroughly with soap and water. INGESTION: get medical attention.
5.4 **Toxicity by Inhalation (Threshold Limit Value):** Data not available
5.5 **Short-Term Inhalation Limits:** Data not available
5.6 **Toxicity by Ingestion:** Data not available
5.7 **Late Toxicity:** Genetic effects of long exposure to low level radiation are suspected to be harmful.
5.8 **Vapor (Gas) Irritant Characteristics:** Data not available
5.9 **Liquid or Solid Irritant Characteristics:** Data not available
5.10 **Odor Threshold:** Not pertinent

6. FIRE HAZARDS

6.1 **Flash Point:** Not flammable, but may cause fire on contact with ordinary combustibles
6.2 **Flammable Limits in Air:** Not flammable
6.3 **Fire Extinguishing Agents:** Water
6.4 **Fire Extinguishing Agents Not to be Used:** Not pertinent
6.5 **Special Hazards of Combustion Products:** Yields toxic gaseous oxides of nitrogen when involved in fire
6.6 **Behavior in Fire:** When large quantities are involved in fire, nitrate may fuse or melt, in which condition application of water may result in extensive scattering of molten material. Will increase the intensity of a fire.
6.7 **Ignition Temperature:** Not pertinent
6.8 **Electrical Hazard:** Not pertinent
6.9 **Burning Rate:** Not pertinent

7. CHEMICAL REACTIVITY

7.1 **Reactivity with Water:** Forms a weak solution of nitric acid; the reaction is not hazardous.
7.2 **Reactivity with Common Materials:** In contact with easily oxidizable substances, may react rapidly enough to cause ignition, violent combustion, or explosion. Solutions in water are acidic and can corrode metals.
7.3 **Stability During Transport:** Stable
7.4 **Neutralizing Agents for Acids and Caustics:** Not pertinent
7.5 **Polymerization:** Not pertinent
7.6 **Inhibitor of Polymerization:** Not pertinent

8. WATER POLLUTION

8.1 **Aquatic Toxicity:** 1,000 ppm/48 hr/water flea/not toxic/fresh water
8.2 **Waterfowl Toxicity:** Data not available
8.3 **Biological Oxygen Demand (BOD):** None
8.4 **Food Chain Concentration Potential:** Data not available

9. SELECTED MANUFACTURERS

1. W. R. Grace & Co.
   Davison Chemical Division
   P. O. Box 188
   Pompton Plains, N. J. 07444
2. American Potash & Chemical Corp.
   Lindsay Rare Earths Division
   258 Ann Street
   West Chicago, Ill. 60185
3. J. T. Baker Chemical Co.
   Phillipsburg, N. J. 08865

10. SHIPPING INFORMATION

10.1 **Grades or Purity:** Reagent
10.2 **Storage Temperature:** Ambient
10.3 **Inert Atmosphere:** No requirement
10.4 **Venting:** Open

11. HAZARD ASSESSMENT CODE

(See Hazard Assessment Handbook, CG 446-3)

SS

12. HAZARD CLASSIFICATIONS

12.1 **Code of Federal Regulations:** Radioactive material
12.2 **NAS Hazard Rating for Bulk Water Transportation:** Not listed
12.3 **NFPA Hazard Classifications:**

Category	Classification*	
Health Hazard (Blue)	0	1
Flammability (Red)	0	0
Reactivity (Yellow)	0	0
	oxy	oxy

*First column refers to nonfire situation.

13. PHYSICAL AND CHEMICAL PROPERTIES

13.1 **Physical State at 15°C and 1 atm:** Solid
13.2 **Molecular Weight:** 555.2
13.3 **Boiling Point at 1 atm:** Not pertinent (decomposes)
13.4 **Freezing Point:** Not pertinent
13.5 **Critical Temperature:** Not pertinent
13.6 **Critical Pressure:** Not pertinent
13.7 **Specific Gravity:** (est.) > 1 at 20°C (solid)
13.8 **Liquid Surface Tension:** Not pertinent
13.9 **Liquid-Water Interfacial Tension:** Not pertinent
13.10 **Vapor (Gas) Specific Gravity:** Not pertinent
13.11 **Ratio of Specific Heats of Vapor (Gas):** Not pertinent
13.12 **Latent Heat of Vaporization:** Not pertinent
13.13 **Heat of Combustion:** Not pertinent
13.14 **Heat of Decomposition:** Not pertinent
13.15 **Heat of Solution:** Not pertinent
13.16 **Heat of Polymerization:** Not pertinent

NOTES

TTT

TITANIUM TETRACHLORIDE

Common Synonyms	Watery liquid	Colorless to light yellow	Irritating odor
	Reacts violently with water. Produces dense fumes in air.		

AVOID CONTACT WITH LIQUID AND VAPOR. Keep people away.
Wear goggles, self-contained breathing apparatus, and rubber overclothing (including gloves).
Stop discharge if possible.
Isolate and remove discharged material.
Notify local health and pollution control agencies.

Fire

Not flammable.
DO NOT USE WATER ON ADJACENT FIRES.

Exposure

CALL FOR MEDICAL AID.

VAPOR
Irritating to eyes, nose and throat.
If inhaled, will cause coughing or headache.
Move to fresh air.
If breathing has stopped, give artificial respiration.
If breathing is difficult, give oxygen.

LIQUID
Will burn skin and eyes.
If swallowed, will cause nausea and vomiting.
Remove contaminated clothing and shoes.
Flush affected areas with plenty of water.
IF IN EYES, hold eyelids open and flush with plenty of water.
IF SWALLOWED and victim is CONSCIOUS, have victim drink water or milk and have victim induce vomiting.
IF SWALLOWED and victim is UNCONSCIOUS OR HAVING CONVULSIONS, do nothing except keep victim warm.

Water Pollution

Dangerous to aquatic life in high concentrations.
May be dangerous if it enters water intakes.
Notify local health and wildlife officials.
Notify operators of nearby water intakes.

1. RESPONSE TO DISCHARGE

(See Response Methods Handbook, CG 446-4)

Issue warning—corrosive, air contaminant
Restrict access
Evacuate area

2. LABEL

3. CHEMICAL DESIGNATIONS

3.1 **Synonyms:** No common synonyms
3.2 **Coast Guard Compatibility Classification:** Not applicable
3.3 **Chemical Formula:** $TiCl_4$
3.4 **IMCO/United Nations Numerical Designation:** 8.0/1838

4. OBSERVABLE CHARACTERISTICS

4.1 **Physical State (as shipped):** Liquid
4.2 **Color:** Colorless
4.3 **Odor:** Acrid; choking

5. HEALTH HAZARDS

5.1 **Personal Protective Equipment:** Goggles and face shield; air-supplied mask or approved canister; rubber gloves; protective clothing.
5.2 **Symptoms Following Exposure:** Vapors can cause severe irritation and damage to eyes, coughing, headache, dizziness, lung damage, bronchiai pneumonia. Liquid causes thermal and acid burns of eyes, skin, throat, and stomach. If ingested, causes nausea, vomiting, cramps, diarrhea, and possible tissue ulceration.
5.3 **Treatment for Exposure:** INHALATION: remove victim to fresh air; if symptoms persist, call a doctor. INGESTION: give large amounts of water, then induce vomiting; give milk, eggs or olive oil; call a doctor. EYES: immediately flush with copious amounts of water for at least 15 min; call a doctor. SKIN: flush with water; obtain medical attention if irritation persists.
5.4 **Toxicity by Inhalation (Threshold Limit Value):** 5 ppm in moist air
5.5 **Short-Term Inhalation Limits:** Data not available
5.6 **Toxicity by Ingestion:** Data not available
5.7 **Late Toxicity:** Disturbances of upper respiratory and nervous system in man.
5.8 **Vapor (Gas) Irritant Characteristics:** Vapor is moderately irritating such that personnel will not tolerate moderate or high vapor concentrations.
5.9 **Liquid or Solid Irritant Characteristics:** Fairly severe skin irritant; may cause pain and second-degree burns after a few minutes' contact.
5.10 **Odor Threshold:** Data not available

6. FIRE HAZARDS

6.1 **Flash Point:** Not flammable
6.2 **Flammable Limits in Air:** Not flammable
6.3 **Fire Extinguishing Agents:** Dry powder or carbon dioxide on adjacent fires.
6.4 **Fire Extinguishing Agents Not to be Used:** Do not use water if it can contact titanium tetrachloride.
6.5 **Special Hazards of Combustion Products:** Not pertinent
6.6 **Behavior in Fire:** If container leaks, a very dense white fume forms and can obscure operations.
6.7 **Ignition Temperature:** Not flammable
6.8 **Electrical Hazard:** Not pertinent
6.9 **Burning Rate:** Not flammable

7. CHEMICAL REACTIVITY

7.1 **Reactivity with Water:** Reacts with moisture in air, forming dense white fume. Reaction with liquid water gives off heat and forms hydrochloric acid.
7.2 **Reactivity with Common Materials:** The acid formed by reaction with moisture attacks metals, forming flammable hydrogen gas.
7.3 **Stability During Transport:** Stable
7.4 **Neutralizing Agents for Acids and Caustics:** Acid formed by reaction with water can be neutralized by limestone, lime, or soda ash.
7.5 **Polymerization:** Not pertinent
7.6 **Inhibitor of Polymerization:** Not pertinent

8. WATER POLLUTION

8.1 **Aquatic Toxicity:** Data not available
8.2 **Waterfowl Toxicity:** Data not available
8.3 **Biological Oxygen Demand (BOD):** None
8.4 **Food Chain Concentration Potential:** None

9. SELECTED MANUFACTURERS

1. Cabot Titania Inc.
   Middle Road
   Ashtabula, Ohio 44004
2. N. L. Industries Inc.
   Titanium Pigment Division
   St. Louis, Mo. 63111
3. Oregon Metallurgical Corp.
   Albany, Ore. 97321

10. SHIPPING INFORMATION

10.1 **Grades or Purity:** Technical
10.2 **Storage Temperature:** Ambient
10.3 **Inert Atmosphere:** No requirement
10.4 **Venting:** Pressure-vacuum

11. HAZARD ASSESSMENT CODE

(See Hazard Assessment Handbook, CG 446-3)

A-O

12. HAZARD CLASSIFICATIONS

12.1 **Code of Federal Regulations:** Corrosive material
12.2 **NAS Hazard Rating for Bulk Water Transportation:**

Category	Rating
Fire	0
Health	
Vapor Irritant	3
Liquid or Solid Irritant	3
Poisons	2
Water Pollution	
Human Toxicity	2
Aquatic Toxicity	2
Aesthetic Effect	2
Reactivity	
Other Chemicals	3
Water	4
Self-Reaction	0

12.3 **NFPA Hazard Classifications:**

Category	Classification
Health Hazard (Blue)	3
Flammability (Red)	0
Reactivity (Yellow)	1

13. PHYSICAL AND CHEMICAL PROPERTIES

13.1 **Physical State at 15°C and 1 atm:** Liquid
13.2 **Molecular Weight:** 189.71
13.3 **Boiling Point at 1 atm:** 277°F = 136°C = 409°K
13.4 **Freezing Point:** −11°F = −24°C = 249°K
13.5 **Critical Temperature:** Not pertinent
13.6 **Critical Pressure:** Not pertinent
13.7 **Specific Gravity:** 1.726 at 20°C (liquid)
13.8 **Liquid Surface Tension:** Not pertinent
13.9 **Liquid-Water Interfacial Tension:** Not pertinent
13.10 **Vapor (Gas) Specific Gravity:** Not pertinent
13.11 **Ratio of Specific Heats of Vapor (Gas):** (est.) 1.221
13.12 **Latent Heat of Vaporization:** 79.7 Btu/lb = 44.3 cal/g = 1.86×10^5 J/kg
13.13 **Heat of Combustion:** Not pertinent
13.14 **Heat of Decomposition:** Not pertinent
13.15 **Heat of Solution:** −482.8 Btu/lb = −268.2 cal/g = -11.22×10^5 J/kg
13.16 **Heat of Polymerization:** Not pertinent

NOTES

REVISED 1978

TOL — TOLUENE

Common Synonyms	Watery liquid	Colorless	Pleasant odor
Toluol Methylbenzene Methylbenzol	Floats on water. Flammable, irritating vapor is produced.		

Stop discharge if possible. Keep people away.
Shut off ignition sources and call fire department.
Stay upwind and use water spray to "knock down" vapor.
Avoid contact with liquid and vapor.
Isolate and remove discharged material.
Notify local health and pollution control agencies.

Fire

FLAMMABLE.
Flashback along vapor trail may occur.
Vapor may explode if ignited in an enclosed area.
Wear goggles and self-contained breathing apparatus.
Extinguish with dry chemical, foam, or carbon dioxide.
Water may be ineffective on fire.
Cool exposed containers with water.

Exposure

CALL FOR MEDICAL AID.

VAPOR
Irritating to eyes, nose and throat.
If inhaled, will cause nausea, vomiting, headache, dizziness, difficult breathing, or loss of consciousness.
Move to fresh air.
If breathing has stopped, give artificial respiration.
If breathing difficult, give oxygen.

LIQUID
Irritating to skin and eyes.
If swallowed, will cause nausea, vomiting or loss of consciousness.
Remove contaminated clothing and shoes.
Flush affected areas with plenty of water.
IF IN EYES, hold eyelids open and flush with plenty of water.
IF SWALLOWED and victim is CONSCIOUS, have victim drink water or milk.
DO NOT INDUCE VOMITING.

Water Pollution

Dangerous to aquatic life in high concentrations.
Fouling to shoreline.
May be dangerous if it enters water intakes.
Notify local health and wildlife officials.
Notify operators of nearby water intakes.

1. RESPONSE TO DISCHARGE

(See Response Methods Handbook, CG 446-4)

Issue warning—high flammability
Evacuate area

2. LABEL

Red

3. CHEMICAL DESIGNATIONS

3.1 **Synonyms:** Methylbenzene
Methylbenzol
Toluol

3.2 **Coast Guard Compatibility Classification:** Aromatic hydrocarbon

3.3 **Chemical Formula:** $C_6H_5CH_3$

3.4 **IMCO/United Nations Numerical Designation:** 3.2/1294

4. OBSERVABLE CHARACTERISTICS

4.1 **Physical State (as shipped):** Liquid

4.2 **Color:** Colorless

4.3 **Odor:** Pungent; aromatic, benzene-like; distinct, pleasant

5. HEALTH HAZARDS

5.1 **Personal Protective Equipment:** Air-supplied mask; goggles or face shield; plastic gloves.

5.2 **Symptoms Following Exposure:** Vapors irritate eyes and upper respiratory tract; cause dizziness, headache, anesthesia, respiratory arrest. Liquid irritates eyes and causes drying of skin. If aspirated, causes coughing, gagging, distress, and rapidly developing pulmonary edema. If ingested causes vomiting, griping, diarrhea, depressed respiration.

5.3 **Treatment for Exposure:** INHALATION: remove to fresh air, give artificial respiration and oxygen if needed; call a doctor. INGESTION: do NOT induce vomiting; call a doctor. EYES: flush with water for at least 15 min. SKIN: wipe off, wash with soap and water.

5.4 **Toxicity by Inhalation (Threshold Limit Value):** 100 ppm

5.5 **Short-Term Inhalation Limits:** 600 ppm for 30 min.

5.6 **Toxicity by Ingestion:** Grade 2; LD_{50} 0.5 to 5 g/kg

5.7 **Late Toxicity:** Kidney and liver damage may follow ingestion.

5.8 **Vapor (Gas) Irritant Characteristics:** Vapors cause a slight smarting of the eyes or respiratory system if present in high concentrations. The effect is temporary.

5.9 **Liquid or Solid Irritant Characteristics:** Minimum hazard. If spilled on clothing and allowed to remain, may cause smarting and reddening of the skin.

5.10 **Odor Threshold:** 0.17 ppm

6. FIRE HAZARDS

6.1 **Flash Point:** 40°F C.C.; 55°F O.C.

6.2 **Flammable Limits in Air:** 1.27%–7%

6.3 **Fire Extinguishing Agents:** Carbon dioxide or dry chemical for small fires, ordinary foam for large fires.

6.4 **Fire Extinguishing Agents Not to be Used:** Water may be ineffective

6.5 **Special Hazards of Combustion Products:** Not pertinent

6.6 **Behavior in Fire:** Vapor is heavier than air and may travel a considerable distance to a source of ignition and flash back.

6.7 **Ignition Temperature:** 997°F

6.8 **Electrical Hazard:** Class I, Group D

6.9 **Burning Rate:** 5.7 mm/min.

7. CHEMICAL REACTIVITY

7.1 **Reactivity with Water:** No reaction

7.2 **Reactivity with Common Materials:** No reaction

7.3 **Stability During Transport:** Stable

7.4 **Neutralizing Agents for Acids and Caustics:** Not pertinent

7.5 **Polymerization:** Not pertinent

7.6 **Inhibitor of Polymerization:** Not pertinent

8. WATER POLLUTION

8.1 **Aquatic Toxicity:** 1180 mg/l/96 hr/sunfish/TL_m/fresh water

8.2 **Waterfowl Toxicity:** Data not available

8.3 **Biological Oxygen Demand (BOD):** 0%, 5 days; 38% (theor.), 8 days

8.4 **Food Chain Concentration Potential:** None

9. SELECTED MANUFACTURERS

1. Exxon Chemical Co.
Houston, Tex. 77001
2. Shell Chemical Co.
Petrochemicals Division
Houston, Tex. 77001
3. Sun Oil Co.
St. Davids, Pa. 19087

10. SHIPPING INFORMATION

10.1 **Grades or Purity:** Research, reagent, nitration—all 99.8+%; industrial: contains 94+%, with 5% xylene and small amounts of benzene and nonaromatic hydrocarbons; 90/120: less pure than industrial.

10.2 **Storage Temperature:** Ambient

10.3 **Inert Atmosphere:** No requirement

10.4 **Venting:** Open (flame arrester) or pressure-vacuum

11. HAZARD ASSESSMENT CODE

(See Hazard Assessment Handbook, CG 446-3)

A-T-U

12. HAZARD CLASSIFICATIONS

12.1 **Code of Federal Regulations:** Flammable liquid

12.2 **NAS Hazard Rating for Bulk Water Transportation:**

Category	Rating
Fire	3
Health	
Vapor Irritant	1
Liquid or Solid Irritant	1
Poisons	2
Water Pollution	
Human Toxicity	1
Aquatic Toxicity	3
Aesthetic Effect	2
Reactivity	
Other Chemicals	1
Water	0
Self-Reaction	0

12.3 **NFPA Hazard Classifications:**

Category	Classification
Health Hazard (Blue)	2
Flammability (Red)	3
Reactivity (Yellow)	0

13. PHYSICAL AND CHEMICAL PROPERTIES

13.1 **Physical State at 15°C and 1 atm:** Liquid

13.2 **Molecular Weight:** 92.14

13.3 **Boiling Point at 1 atm:** 231.1°F = 110.6°C = 383.8°K

13.4 **Freezing Point:** −139°F = −95.0°C = 178.2°K

13.5 **Critical Temperature:** 605.4°F = 318.6°C = 591.8°K

13.6 **Critical Pressure:** 596.1 psia = 40.55 atm = 4.108 MN/m²

13.7 **Specific Gravity:** 0.867 at 20°C (liquid)

13.8 **Liquid Surface Tension:** 29.0 dynes/cm = 0.0290 N/m at 20°C

13.9 **Liquid-Water Interfacial Tension:** 36.1 dynes/cm = 0.0361 N/m at 25°C

13.10 **Vapor (Gas) Specific Gravity:** Not pertinent

13.11 **Ratio of Specific Heats of Vapor (Gas):** 1.089

13.12 **Latent Heat of Vaporization:** 155 Btu/lb = 86.1 cal/g = 3.61×10^5 J/kg

13.13 **Heat of Combustion:** −17,430 Btu/lb = −9686 cal/g = -405.5×10^5 J/kg

13.14 **Heat of Decomposition:** Not pertinent

13.15 **Heat of Solution:** Not pertinent

13.16 **Heat of Polymerization:** Not pertinent

NOTES

REVISED 1978

TDI | TOLUENE 2,4-DIISOCYANATE

Common Synonyms			
TDI Hylene T Mondur TDS Nacconate 100	Liquid	Colorless to light yellow	Sharp, sweet, fruity odor
	Sinks and reacts with water. Freezing point is 68° F – 72° F.		

AVOID CONTACT WITH LIQUID AND VAPOR. Keep people away.
Wear goggles, self-contained breathing apparatus, and rubber overclothing (including gloves).
Stop discharge if possible.
Call fire department.
Isolate and remove discharged material.
Notify local health and pollution control agencies.

Fire

Combustible.
POISONOUS GAS IS PRODUCED IN FIRE.
Wear goggles, self-contained breathing apparatus, and rubber overclothing (including gloves).
Extinguish with dry chemical or carbon dioxide.
Water and foam may be ineffective on fire.
Cool exposed containers with water.

Exposure

CALL FOR MEDICAL AID.

LIQUID
POISONOUS IF SWALLOWED.
Will burn skin and eyes.
Remove contaminated clothing and shoes.
Flush affected areas with plenty of water.
IF IN EYES, hold eyelids open and flush with plenty of water.
IF SWALLOWED and victim is CONSCIOUS, have victim drink water or milk.
DO NOT INDUCE VOMITING.

Water Pollution

Effect of low concentrations on aquatic life is unknown.
May be dangerous if it enters water intakes.
Notify local health and wildlife officials.
Notify operators of nearby water intakes.

1. RESPONSE TO DISCHARGE

(See Response Methods Handbook, CG 446-4)

Issue warning—water contaminant
Restrict access
Should be removed
Chemical and physical treatment

2. LABEL

3. CHEMICAL DESIGNATIONS

3.1 **Synonyms:** Hylene T; Mondur TDS; Nacconate 100; 2,4-Tolylene diisocyanate; TDI
3.2 **Coast Guard Compatibility Classification:** Isocyanate
3.3 **Chemical Formula:** 1-$CH_3C_6H_3(NCO)_2$-2,4
3.4 **IMCO/United Nations Numerical Designation:** Not listed

4. OBSERVABLE CHARACTERISTICS

4.1 **Physical State (as shipped):** Liquid
4.2 **Color:** Colorless to light yellow
4.3 **Odor:** Sweet, fruity, pungent

5. HEALTH HAZARDS

5.1 **Personal Protective Equipment:** Organic vapor canister; goggles or face shield; rubber gloves, boots and apron.
5.2 **Symptoms Following Exposure:** Irritates eyes and skin. Potent sensitizer and lung irritant if inhaled. May produce bronchospasm (asthma), pneumonitis, bronchitis, and pulmonary edema. Nocturnal cough and shortness of breath are common. Repeated low-level exposure may produce chronic lung disease. Oral toxicity is low.
5.3 **Treatment for Exposure:** INHALATION: remove victim to fresh air; administer artificial respiration and oxygen if needed; call a doctor at once. INGESTION: do NOT induce vomiting; call a doctor. EYES: flush with water for at least 15 min.; call doctor at once. SKIN: flush with water; wipe off; wipe with rubbing alcohol; wash with soap and water.
5.4 **Toxicity by Inhalation (Threshold Limit Value):** 0.02 ppm
5.5 **Short-Term Inhalation Limits:** 0.02 ppm for 5 min.
5.6 **Toxicity by Ingestion:** Grade 2; LD_{50} 0.5 to 5 g/kg
5.7 **Late Toxicity:** Data not available
5.8 **Vapor (Gas) Irritant Characteristics:** Vapor is moderately irritating such that personnel will not usually tolerate moderate or high vapor concentrations.
5.9 **Liquid or Solid Irritant Characteristics:** Fairly severe skin irritant; may cause pain and second-degree burns after a few minutes' contact.
5.10 **Odor Threshold:** 0.4—2.14 ppm

6. FIRE HAZARDS

6.1 **Flash Point:** 270°F O.C.
6.2 **Flammable Limits in Air:** 0.9%—9.5%
6.3 **Fire Extinguishing Agents:** Water, foam, dry chemical, or carbon dioxide
6.4 **Fire Extinguishing Agents Not to be Used:** Water or foam may cause frothing.
6.5 **Special Hazards of Combustion Products:** Irritating vapors are generated when heated.
6.6 **Behavior in Fire:** Not pertinent
6.7 **Ignition Temperature:** >300°F
6.8 **Electrical Hazard:** Not pertinent
6.9 **Burning Rate:** Data not available

7. CHEMICAL REACTIVITY

7.1 **Reactivity with Water:** Forms carbon dioxide gas and an organic base; the reaction is not violent.
7.2 **Reactivity with Common Materials:** No reaction
7.3 **Stability During Transport:** Stable
7.4 **Neutralizing Agents for Acids and Caustics:** Not pertinent
7.5 **Polymerization:** Slow, not hazardous, above 113°F
7.6 **Inhibitor of Polymerization:** Not pertinent

8. WATER POLLUTION

8.1 **Aquatic Toxicity:** Data not available
8.2 **Waterfowl Toxicity:** Data not available
8.3 **Biological Oxygen Demand (BOD):** Data not available
8.4 **Food Chain Concentration Potential:** None

9. SELECTED MANUFACTURERS

1. BASF Wyandotte Corp.
   Industrial Chemicals Group
   1609 Biddle Ave.
   Wyandotte, Mich. 48192
2. Baychem Corp.
   Mobay Chemical Co. Division
   Penn Lincoln Pkwy. West
   Pittsburgh, Pa. 15205
3. E. I. duPont de Nemours & Co., Inc.
   Elastomer Chemicals Dept.
   Wilmington, Del. 19898

10. SHIPPING INFORMATION

10.1 **Grades or Purity:** Commercial distilled, 99% total diisocyanate. The following isomer ratios are shipped: (a) 100% 2,4-; (b) 80% 2,4-; 20% 2,6- (most common); (c) 65% 2,4-; 35% 2,6-. All mixtures have about the same hazard characteristics.
10.2 **Storage Temperature:** 75—100°F
10.3 **Inert Atmosphere:** Inerted
10.4 **Venting:** Pressure-vacuum

11. HAZARD ASSESSMENT CODE

(See Hazard Assessment Handbook, CG 446-3)

A-X-Y

12. HAZARD CLASSIFICATIONS

12.1 **Code of Federal Regulations:** Poison, Class B
12.2 **NAS Hazard Rating for Bulk Water Transportation:**

Category	Rating
Fire	1
Health	
Vapor Irritant	3
Liquid or Solid Irritant	3
Poisons	4
Water Pollution	
Human Toxicity	2
Aquatic Toxicity	—
Aesthetic Effect	4
Reactivity	
Other Chemicals	3
Water	3
Self-Reaction	3

12.3 **NFPA Hazard Classifications:**

Category	Classification
Health Hazard (Blue)	3
Flammability (Red)	1
Reactivity (Yellow)	1

13. PHYSICAL AND CHEMICAL PROPERTIES

13.1 **Physical State at 15°C and 1 atm:** Solid
13.2 **Molecular Weight:** 174.16
13.3 **Boiling Point at 1 atm:** 482°F = 250°C = 523°K
13.4 **Freezing Point:** 68—72°F = 20—22°C = 293—295°K
13.5 **Critical Temperature:** Not pertinent
13.6 **Critical Pressure:** Not pertinent
13.7 **Specific Gravity:** 1.22 at 25°C (liquid)
13.8 **Liquid Surface Tension:** (est.) 25 dynes/cm = 0.025 N/m at 25°C
13.9 **Liquid-Water Interfacial Tension:** (est.) 45 dynes/cm = 0.045 N/m at 25°C
13.10 **Vapor (Gas) Specific Gravity:** Not pertinent
13.11 **Ratio of Specific Heats of Vapor (Gas):** Not pertinent
13.12 **Latent Heat of Vaporization:** Not pertinent
13.13 **Heat of Combustion:** (est.) −10,300 Btu/lb = −5720 cal/g = -239×10^5 J/kg
13.14 **Heat of Decomposition:** Not pertinent
13.15 **Heat of Solution:** Not pertinent
13.16 **Heat of Polymerization:** Not pertinent

NOTES

REVISED 1978

TAP

p-TOLUENESULFONIC ACID

Common Synonyms

Methylbenzenesulfonic acid
Tosic acid
p-TSA

Solid | Colorless to black | Odorless or slight odor

Mixes with water.

Keep people away.
Call fire department.
Isolate and remove discharged material.
Notify local health and pollution control agencies.

Fire

Combustible.
Irritating gases may be produced when heated.
Flood discharge area with water.

Exposure

Call for medical aid.

SOLID
Irritating to skin and eyes.
Harmful if swallowed.
Remove contaminated clothing and shoes.
Flush affected areas with plenty of water.
IF IN EYES, hold eyelids open and flush with plenty of water.
IF SWALLOWED and victim is CONSCIOUS, have victim drink water or milk.

Water Pollution

Effect of low concentrations on aquatic life is unknown.
May be dangerous if it enters water intakes.
Notify local health and wildlife officials.
Notify operators of nearby water intakes.

1. RESPONSE TO DISCHARGE

(See Response Methods Handbook, CG 446-4)

Issue warning—corrosive
Disperse and flush

2. LABEL

3. CHEMICAL DESIGNATIONS

3.1 **Synonyms:** Methylbenzenesulfonic acid; Tosic acid; p-TSA
3.2 **Coast Guard Compatibility Classification:** Not applicable
3.3 **Chemical Formula:** $CH_3C_6H_4SO_3H$
3.4 **IMCO/United Nations Numerical Designation:** Not listed

4. OBSERVABLE CHARACTERISTICS

4.1 **Physical State (as shipped):** Solid
4.2 **Color:** White to brown to black; yellow to amber
4.3 **Odor:** None when pure; technical grade has slight aromatic odor.

5. HEALTH HAZARDS

5.1 **Personal Protective Equipment:** Chemical goggles or face shield; rubber gloves
5.2 **Symptoms Following Exposure:** Contact with eyes or skin causes severe irritation. Ingestion causes irritation of mouth and stomach.
5.3 **Treatment for Exposure:** EYES: wash thoroughly with copious amounts of water for at least 15 min.; call physician if irritation persists. SKIN: wash thoroughly with large amounts of water for at least 15 min. INGESTION: give large amount of water.
5.4 **Toxicity by Inhalation (Threshold Limit Value):** Data not available
5.5 **Short-Term Inhalation Limits:** Not pertinent
5.6 **Toxicity by Ingestion:** Grade 3; oral LD_{50} = 400 mg/kg (rat)
5.7 **Late Toxicity:** Data not available
5.8 **Vapor (Gas) Irritant Characteristics:** Data not available
5.9 **Liquid or Solid Irritant Characteristics:** Data not available
5.10 **Odor Threshold:** Not pertinent

6. FIRE HAZARDS

6.1 **Flash Point:** Not pertinent (solid with low flammability)
6.2 **Flammable Limits in Air:** Not pertinent
6.3 **Fire Extinguishing Agents:** Water
6.4 **Fire Extinguishing Agents Not to be Used:** Not pertinent
6.5 **Special Hazards of Combustion Products:** Irritating oxides of sulfur may be formed.
6.6 **Behavior in Fire:** Not pertinent
6.7 **Ignition Temperature:** Data not available
6.8 **Electrical Hazard:** Data not available
6.9 **Burning Rate:** Not pertinent

7. CHEMICAL REACTIVITY

7.1 **Reactivity with Water:** No reaction
7.2 **Reactivity with Common Materials:** A strong acid which can react with common metals
7.3 **Stability During Transport:** Stable
7.4 **Neutralizing Agents for Acids and Caustics:** Flush with water, rinse with dilute sodium bicarbonate or lime solution.
7.5 **Polymerization:** Not pertinent
7.6 **Inhibitor of Polymerization:** Not pertinent

8. WATER POLLUTION

8.1 **Aquatic Toxicity:** Data not available
8.2 **Waterfowl Toxicity:** Data not available
8.3 **Biological Oxygen Demand (BOD):** 0%, 5 days
8.4 **Food Chain Concentration Potential:** None

9. SELECTED MANUFACTURERS

1. Cities Service Company
Industrial Chemicals Division
3445 Peachtree Road, N. E.
P. O. Box 50360
Atlanta, Ga. 30302
2. Nease Chemical Co., Inc.
P. O. Box 221
State College, Pa. 16801
3. Aldrich Chemical Co.
940 West St. Paul Avenue
Milwaukee, Wis. 53233

10. SHIPPING INFORMATION

10.1 **Grades or Purity:** 93+%
10.2 **Storage Temperature:** Ambient
10.3 **Inert Atmosphere:** No requirement
10.4 **Venting:** Open

11. HAZARD ASSESSMENT CODE

(See Hazard Assessment Handbook, CG 446-3)

SS

12. HAZARD CLASSIFICATIONS

12.1 **Code of Federal Regulations:** Corrosive solid
12.2 **NAS Hazard Rating for Bulk Water Transportation:** Not listed
12.3 **NFPA Hazard Classifications:** Not listed

13. PHYSICAL AND CHEMICAL PROPERTIES

13.1 **Physical State at 15°C and 1 atm:** Solid
13.2 **Molecular Weight:** 172.2
13.3 **Boiling Point at 1 atm:** Not pertinent (decomposes)
13.4 **Freezing Point:** 219—221°F = 104—105°C = 377—378°K
13.5 **Critical Temperature:** Not pertinent
13.6 **Critical Pressure:** Not pertinent
13.7 **Specific Gravity:** 1.45 at 25°C (solid)
13.8 **Liquid Surface Tension:** Not pertinent
13.9 **Liquid-Water Interfacial Tension:** Not pertinent
13.10 **Vapor (Gas) Specific Gravity:** Not pertinent
13.11 **Ratio of Specific Heats of Vapor (Gas):** Not pertinent
13.12 **Latent Heat of Vaporization:** Not pertinent
13.13 **Heat of Combustion:** Not pertinent
13.14 **Heat of Decomposition:** Not pertinent
13.15 **Heat of Solution:** −50 Btu/lb = −28 cal/g = -1.2×10^5 J/kg
13.16 **Heat of Polymerization:** Not pertinent

NOTES

TLI	o-TOLUIDINE

Common Synonyms	Liquid	Colorless to yellow-brown	Chemical odor
2-Amino-1-methylbenzene 2-Aminotoluene 2-Methylaniline o-Methylaniline	May float or sink in water.		

Shut off ignition sources. Call fire department.
Avoid contact with liquid. Keep people away.
Stop discharge if possible.
Isolate and remove discharged material.
Notify local health and pollution control agencies.

Fire

Combustible.
POISONOUS GASES MAY BE PRODUCED IN FIRE.
Wear goggles and self-contained breathing apparatus.
Extinguish with dry chemicals, foam or carbon dioxide.
Water may be ineffective on fire.
Cool exposed containers with water.

Exposure

CALL FOR MEDICAL AID.

LIQUID
Irritating to skin and eyes.
If swallowed will cause nausea, vomiting or loss of consciousness.
Remove contaminated clothing and shoes.
Flush affected areas with plenty of water.
IF IN EYES, hold eyelids open and flush with plenty of water.
IF SWALLOWED and victim is CONSCIOUS, have victim drink water or milk and have victim induce vomiting.
IF SWALLOWED and victim is UNCONSCIOUS OR HAVING CONVULSIONS, do nothing except keep victim warm.

Water Pollution

Effect of low concentrations on aquatic life is unknown.
Fouling to shoreline.
May be dangerous if it enters water intakes.
Notify local health and wildlife officials.
Notify operators of nearby water intakes.

1. RESPONSE TO DISCHARGE

(See Response Methods Handbook, CG 446-4)

Issue warning—water contaminant
Restrict access
Mechanical containment
Should be removed
Chemical and physical treatment

2. LABELS

No hazard label required by Code of Federal Regulations

3. CHEMICAL DESIGNATIONS

3.1 **Synonyms:** 2-Amino-1-methylbenzene; 2-Aminotoluene; 2-Methylaniline; o-Methylaniline
3.2 **Coast Guard Compatibility Classification:** Not listed
3.3 **Chemical Formula:** 1,2-$CH_3C_6H_4NH_2$
3.4 **IMCO/United Nations Numerical Designation:** 6.1/1708

4. OBSERVABLE CHARACTERISTICS

4.1 **Physical State (as shipped):** Liquid
4.2 **Color:** Clear to light yellow; turns yellow, brown or deep red on exposure to air and light.
4.3 **Odor:** Aromatic, aniline-like

5. HEALTH HAZARDS

5.1 **Personal Protective Equipment:** Chemical safety goggles; face shield; Bu. Mines approved respirator; leather or rubber safety shoes; butyl rubber gloves
5.2 **Symptoms Following Exposure:** Absorption of toxic quantities by any route causes cyanosis (blue discoloration of lips, nails, skin); nausea, vomiting, and coma may follow. Repeated inhalation of low concentrations may cause pallor, low-grade secondary anemia, fatigability, and loss of appetite. Contact with eyes causes irritation.
5.3 **Treatment for Exposure:** Get medical attention following all exposures to this compound. INHALATION: move to fresh air. INGESTION: if victim is conscious, promptly induce vomiting by giving lukewarm soapy water or mustard and water. EYES: flush with copious amounts of water for at least 15 min., holding lids apart. SKIN: remove all contaminated clothing; wash affected areas immediately and thoroughly with plenty of warm water and soap.
5.4 **Toxicity by Inhalation (Threshold Limit Value):** 5 ppm
5.5 **Short-Term Inhalation Limits:** Data not available
5.6 **Toxicity by Ingestion:** Grade 2; oral LD_{50} = 900 mg/kg (rat)
5.7 **Late Toxicity:** Causes tumors in urinary bladder of rats
5.8 **Vapor (Gas) Irritant Characteristics:** Data not available
5.9 **Liquid or Solid Irritant Characteristics:** Data not available
5.10 **Odor Threshold:** Data not available

6. FIRE HAZARDS

6.1 **Flash Point:** 167°F O.C.; 85°F C.C.
6.2 **Flammable Limits in Air:** Data not available
6.3 **Fire Extinguishing Agents:** Foam, dry chemical, or carbon dioxide
6.4 **Fire Extinguishing Agents Not to be Used:** Water may be ineffective.
6.5 **Special Hazards of Combustion Products:** Toxic oxides of nitrogen and flammable vapors may form in fire.
6.6 **Behavior in Fire:**
6.7 **Ignition Temperature:** 900°F
6.8 **Electrical Hazard:** Data not available
6.9 **Burning Rate:** 3.62 mm/min.

7. CHEMICAL REACTIVITY

7.1 **Reactivity with Water:** No reaction
7.2 **Reactivity with Common Materials:**
7.3 **Stability During Transport:** Stable
7.4 **Neutralizing Agents for Acids and Caustics:** Not pertinent
7.5 **Polymerization:** Not pertinent
7.6 **Inhibitor of Polymerization:** Not pertinent

8. WATER POLLUTION

8.1 **Aquatic Toxicity:** 100 ppm/*/fish/lethal/fresh water
*Time period not specified.
8.2 **Waterfowl Toxicity:** Data not available
8.3 **Biological Oxygen Demand (BOD):** 143%, 5 days
8.4 **Food Chain Concentration Potential:** None

9. SELECTED MANUFACTURERS

1. First Chemical Corp.
656 N. State St.
Jackson, Miss. 39205
2. E. I. du Pont de Nemours & Co., Inc.
Wilmington, Del. 19598
3. Aldrich Chemical Co.
940 W. Saint Paul Ave.
Milwaukee, Wis. 53233

10. SHIPPING INFORMATION

10.1 **Grades or Purity:** Commercial, 99.5+%
10.2 **Storage Temperature:** Ambient
10.3 **Inert Atmosphere:** No requirement
10.4 **Venting:** Open (flame arrester)

11. HAZARD ASSESSMENT CODE

(See Hazard Assessment Handbook, CG 446-3)

A-P-Q-T-U-X-Y

12. HAZARD CLASSIFICATIONS

12.1 **Code of Federal Regulations:** Not listed
12.2 **NAS Hazard Rating for Bulk Water Transportation:** Not listed
12.3 **NFPA Hazard Classifications:**

Category	Classification
Health Hazard (Blue)	3
Flammability (Red)	2
Reactivity (Yellow)	0

13. PHYSICAL AND CHEMICAL PROPERTIES

13.1 **Physical State at 15°C and 1 atm:** Liquid
13.2 **Molecular Weight:** 107.2
13.3 **Boiling Point at 1 atm:** 392°F = 200°C = 473°K
13.4 **Freezing Point:** −11°F = −24°C = 249°K
13.5 **Critical Temperature:** 790°F = 421°C = 694°K
13.6 **Critical Pressure:** 544 psia = 37.0 atm = 3.75 MN/m^2
13.7 **Specific Gravity:** 0.998 at 20°C (liquid)
13.8 **Liquid Surface Tension:** 43.55 dynes/cm = 0.04355 N/m at 20°C
13.9 **Liquid-Water Interfacial Tension:** Data not available
13.10 **Vapor (Gas) Specific Gravity:** Not pertinent
13.11 **Ratio of Specific Heats of Vapor (Gas):** Not pertinent
13.12 **Latent Heat of Vaporization:** 179.1 Btu/lb = 99.5 cal/g = 4.16×10^5 J/kg
13.13 **Heat of Combustion:** −16,180 Btu/lb = −8,990 cal/g = -376×10^5 J/kg
13.14 **Heat of Decomposition:** Not pertinent
13.15 **Heat of Solution:** Not pertinent
13.16 **Heat of Polymerization:** Not pertinent

NOTES

TOXAPHENE

Common Synonyms		
Octachlorocamphene	Waxy solid or in solution	Amber
	Solid sinks in water, solution floats on water.	

Stop discharge if possible. Keep people away.
Call fire department.
Avoid contact with solid and solution.
Isolate and remove discharged material.
Notify local health and pollution control agencies.

Fire

Solid not flammable, but usually dissolved in combustible liquid.
POISONOUS GASES ARE PRODUCED IN FIRE.
Wear goggles, self-contained breathing apparatus, and rubber overclothing (including gloves).
Extinguish with foam, dry chemical or carbon dioxide.
Water may be ineffective on fire.
Cool exposed containers with water.

Exposure

CALL FOR MEDICAL AID.

SOLID OR SOLUTION
POISONOUS IF SWALLOWED.
Irritating to skin and eyes.
Flush affected areas with plenty of water.
IF IN EYES, hold eyelids open and flush with plenty of water.
IF SWALLOWED and victim is CONSCIOUS, have victim drink water or milk and have victim induce vomiting.
IF SWALLOWED and victim is UNCONSCIOUS OR HAVING CONVULSIONS, do nothing except keep victim quiet and warm.

Water Pollution

HARMFUL TO AQUATIC LIFE IN VERY LOW CONCENTRATIONS.
Solution is fouling to shoreline.
May be dangerous if it enters water intakes.
Notify local health and wildlife officials.
Notify operators of nearby water intakes.

1. RESPONSE TO DISCHARGE

(See Response Methods Handbook, CG 446-4)

Issue warning—poison, water contaminant
Should be removed
Chemical and physical treatment

2. LABEL

3. CHEMICAL DESIGNATIONS

3.1 **Synonyms:** Octachlorocamphene
3.2 **Coast Guard Compatibility Classification:** Not applicable
3.3 **Chemical Formula:** $C_{10}H_8Cl_8$
3.4 **IMCO/United Nations Numerical Designation:** Not listed

4. OBSERVABLE CHARACTERISTICS

4.1 **Physical State (as shipped):** Waxy solid
4.2 **Color:** Amber
4.3 **Odor:** Mild turpentine odor

5. HEALTH HAZARDS

5.1 **Personal Protective Equipment:** Chemical-type respirator; rubber gloves; chemical goggles or face shield.
5.2 **Symptoms Following Exposure:** May be absorbed through skin, lungs, or intestinal tract. Symptoms include salivation, leg and back muscle spasms, nausea, vomiting, hyperexcitability, tremors, shivering, clonic convulsions, then tetanic contractions of all skeletal muscles. Lethal doses cause respiratory failure. Respiration, affected as a result of the exertion from vomiting or convulsions, is first arrested because of tetanic muscular contractions, then increased in both amplitude and rate as the muscles relax.
5.3 **Treatment for Exposure:** If symptoms of poisoning appear, promptly remove the unabsorbed pesticide from the stomach by inducing vomiting with warm salty or soapy water (if the patient is conscious) or from the skin with soap and water. Keep patient warm and quiet. Call a physician.
5.4 **Toxicity by Inhalation (Threshold Limit Value):** Not pertinent
5.5 **Short-Term Inhalation Limits:** Not pertinent
5.6 **Toxicity by Ingestion:** Grade 4; LD_{50} below 50 mg/kg (dog)
5.7 **Late Toxicity:** Data not available
5.8 **Vapor (Gas) Irritant Characteristics:** The solid is non-volatile. For solutions, see meta-xylene.
5.9 **Liquid or Solid Irritant Characteristics:** Minimum hazard. If spilled on clothing and allowed to remain, may cause smarting and reddening of the skin.
5.10 **Odor Threshold:** Not pertinent

6. FIRE HAZARDS

6.1 **Flash Point:** 84°F C.C. (solution)
6.2 **Flammable Limits in Air:** 1.1%–6.4% (solvent only)
6.3 **Fire Extinguishing Agents:** Foam, dry chemical, or carbon dioxide
6.4 **Fire Extinguishing Agents Not to be Used:** Water may be ineffective.
6.5 **Special Hazards of Combustion Products:** Toxic vapors are generated when heated.
6.6 **Behavior in Fire:** Solution in xylene may produce corrosive products when heated.
6.7 **Ignition Temperature:** 986°F (solution)
6.8 **Electrical Hazard:** Not pertinent
6.9 **Burning Rate:** 5.8 mm/min.

7. CHEMICAL REACTIVITY

7.1 **Reactivity with Water:** No reaction
7.2 **Reactivity with Common Materials:** No reaction
7.3 **Stability During Transport:** Stable
7.4 **Neutralizing Agents for Acids and Caustics:** Not pertinent
7.5 **Polymerization:** Not pertinent
7.6 **Inhibitor of Polymerization:** Not pertinent

8. WATER POLLUTION

8.1 **Aquatic Toxicity:**
0.05 ppm/20 days/goldfish/100% kill/fresh water
0.02 ppm/*/bluegill/100% kill/fresh water
0.0032 ppm/24 hr/spot/100% kill/salt water
*Time period not specified.
8.2 **Waterfowl Toxicity:** 30.8 mg/kg
8.3 **Biological Oxygen Demand (BOD):** Data not available
8.4 **Food Chain Concentration Potential:** None

9. SELECTED MANUFACTURERS

1. Hercules Inc.
   Synthetics Dept.
   Brunswick, Ga. 31521
2. Sonford Chemical Co.
   Pure-Atlantic Highway
   Port Neches, Tex. 77651

10. SHIPPING INFORMATION

10.1 **Grades or Purity:** Technical; 40% dust concentrate; 90% solution in xylene
10.2 **Storage Temperature:** Ambient
10.3 **Inert Atmosphere:** No requirement
10.4 **Venting:** Sealed containers in well-ventilated area

11. HAZARD ASSESSMENT CODE

(See Hazard Assessment Handbook, CG 446-3)

II-A-T-U

12. HAZARD CLASSIFICATIONS

12.1 **Code of Federal Regulations:** Poisonous liquid or solid, Class B
12.2 **NAS Hazard Rating for Bulk Water Transportation:** Not listed
12.3 **NFPA Hazard Classifications:** Not listed

13. PHYSICAL AND CHEMICAL PROPERTIES

13.1 **Physical State at 15°C and 1 atm:** Waxy solid
13.2 **Molecular Weight:** 414 (avg.)
13.3 **Boiling Point at 1 atm:** Decomposes
13.4 **Freezing Point:** 149–194°F = 65–90°C = 338–363°K
13.5 **Critical Temperature:** Not pertinent
13.6 **Critical Pressure:** Not pertinent
13.7 **Specific Gravity:** 1.6 at 15°C (solid)
13.8 **Liquid Surface Tension:** Not pertinent
13.9 **Liquid-Water Interfacial Tension:** Not pertinent
13.10 **Vapor (Gas) Specific Gravity:** Not pertinent
13.11 **Ratio of Specific Heats of Vapor (Gas):** Not pertinent
13.12 **Latent Heat of Vaporization:** Not pertinent
13.13 **Heat of Combustion:** Not pertinent
13.14 **Heat of Decomposition:** Not pertinent
13.15 **Heat of Solution:** Not pertinent
13.16 **Heat of Polymerization:** Not pertinent

NOTES

REVISED 1978

TCE — TRICHLOROETHANE

Common Synonyms
1,1,1-Trichloroethane
Methylchloroform

Watery liquid — Colorless — Sweet odor

Sinks in water. Irritating vapor is produced.

Stop discharge if possible. Keep people away.
Avoid contact with liquid and vapor.
Call fire department.
Isolate and remove discharged material.
Notify local health and pollution control agencies.

Fire

Combustible.
POISONOUS GASES ARE PRODUCED IN FIRE.
Wear goggles and self-contained breathing apparatus.
Extinguish with dry chemical, carbon dioxide, or foam.

Exposure

CALL FOR MEDICAL AID.

VAPOR
Irritating to eyes, nose and throat.
If inhaled, will cause dizziness or difficult breathing.
Move to fresh air.
If breathing has stopped, give artificial respiration.
If breathing is difficult, give oxygen.

LIQUID
Irritating to skin and eyes.
If swallowed, may produce nausea.
Remove contaminated clothing and shoes.
Flush affected areas with plenty of water.
IF IN EYES, hold eyelids open and flush with plenty of water.
IF SWALLOWED and victim is CONSCIOUS, have victim drink water or milk and have victim induce vomiting.
IF SWALLOWED and victim is UNCONSCIOUS OR HAVING CONVULSIONS, do nothing except keep victim warm.

Water Pollution

Effect of low concentrations on aquatic life is unknown.
May be dangerous if it enters water intakes.
Notify local health and wildlife officials.
Notify operators of nearby water intakes.

1. RESPONSE TO DISCHARGE

(See Response Methods Handbook, CG 446-4)

Should be removed
Chemical and physical treatment

2. LABELS

No hazard label required by Code of Federal Regulations

3. CHEMICAL DESIGNATIONS

3.1 **Synonyms:** Aerothene
Chlorothene
Methyl chloroform
1,1,1-Trichloroethane

3.2 **Coast Guard Compatibility Classification:** Halogenated hydrocarbon

3.3 **Chemical Formula:** CH_3CCl_3

3.4 **IMCU/United Nations Numerical Designation:** Not listed

4. OBSERVABLE CHARACTERISTICS

4.1 **Physical State (as shipped):** Liquid

4.2 **Color:** Colorless

4.3 **Odor:** Chloroform-like; sweetish

5. HEALTH HAZARDS

5.1 **Personal Protective Equipment:** Organic vapor-acid gas canister; self-contained breathing apparatus for emergencies; neoprene or polyvinyl-alcohol-type gloves; chemical safety goggles and face shield; neoprene safety shoes (or leather safety shoes plus neoprene footwear); neoprene or polyvinyl alcohol suit or apron for splash protection.

5.2 **Symptoms Following Exposure:** INHALATION: symptoms range from loss of equilibrium and incoordination to loss of consciousness; high concentration can be fatal due to simple asphyxiation combined with loss of consciousness. INGESTION: produces effects similar to inhalation and may cause some feeling of nausea. EYES: slightly irritating and lachrymatory. SKIN: defatting action may cause dermatitis.

5.3 **Treatment for Exposure:** Get medical attention for all eye exposures and any other serious overexposures. Do NOT administer adrenalin or epinephrine; otherwise, treatment is symptomatic. INHALATION: remove victim to fresh air; if necessary, apply artificial respiration and/or administer oxygen. INGESTION: have victim drink water and induce vomiting. EYES: flush thoroughly with water. SKIN: remove contaminated clothing and wash exposed area thoroughly with soap and warm water.

5.4 **Toxicity by Inhalation (Threshold Limit Value):** 350 ppm

5.5 **Short-Term Inhalation Limits:** 1000 ppm for 60 min. in man

5.6 **Toxicity by Ingestion:** Grade 1; LD_{50} 5 to 15 g/kg (rat, mouse, rabbit, guinea pig)

5.7 **Late Toxicity:** Data not available

5.8 **Vapor (Gas) Irritant Characteristics:** Vapors cause a slight smarting of the eyes or respiratory system if present in high concentrations. The effect is temporary.

6. FIRE HAZARDS

6.1 **Flash Point:** Data not available

6.2 **Flammable Limits in Air:** 7%–16%

6.3 **Fire Extinguishing Agents:** Dry chemical, foam, or carbon dioxide

6.4 **Fire Extinguishing Agents Not to be Used:** Not pertinent

6.5 **Special Hazards of Combustion Products:** Toxic and irritating gases are generated in fires.

6.6 **Behavior in Fire:** Not pertinent

6.7 **Ignition Temperature:** 932°F

6.8 **Electrical Hazard:** Not pertinent

6.9 **Burning Rate:** (est.) 2.9 mm/min.

7. CHEMICAL REACTIVITY

7.1 **Reactivity with Water:** Reacts slowly, releasing corrosive hydrochloric acid.

7.2 **Reactivity with Common Materials:** Corrodes aluminum, but reaction is not hazardous.

7.3 **Stability During Transport:** Stable

7.4 **Neutralizing Agents for Acids and Caustics:** Not pertinent

7.5 **Polymerization:** Not pertinent

7.6 **Inhibitor of Polymerization:** Not pertinent

8. WATER POLLUTION

8.1 **Aquatic Toxicity:**
75–150 ppm/*/pinfish/TL_m/salt water
*Time period not specified.

8.2 **Waterfowl Toxicity:** Data not available

8.3 **Biological Oxygen Demand (BOD):** Data not available

8.4 **Food Chain Concentration Potential:** None

9. SELECTED MANUFACTURERS

1. Dow Chemical Co.
   Midland, Mich. 48640
2. PPG Industries, Inc.
   Industrial Chemicals Division
   1 Gateway Center
   Pittsburgh, Pa. 15222
3. Vulcan Materials Co.
   Chemicals Division
   Wichita, Kan. 67201

10. SHIPPING INFORMATION

10.1 **Grades or Purity:** Uninhibited; inhibited; industrial inhibited; white room; cold cleaning

10.2 **Storage Temperature:** Ambient

10.3 **Inert Atmosphere:** No requirement

10.4 **Venting:** Pressure-vacuum

11. HAZARD ASSESSMENT CODE

(See Hazard Assessment Handbook, CG 446-3)

A-X-Y

12. HAZARD CLASSIFICATIONS

12.1 **Code of Federal Regulations:** Not listed

12.2 **NAS Hazard Rating for Bulk Water Transportation:**

Category	Rating
Fire	1
Health	
Vapor Irritant	1
Liquid or Solid Irritant	1
Poisons	2
Water Pollution	
Human Toxicity	1
Aquatic Toxicity	3
Aesthetic Effect	2
Reactivity	
Other Chemicals	1
Water	0
Self-Reaction	0

12.3 **NFPA Hazard Classifications:** Not listed

13. PHYSICAL AND CHEMICAL PROPERTIES

13.1 **Physical State at 15°C and 1 atm:** Liquid

13.2 **Molecular Weight:** 133.41

13.3 **Boiling Point at 1 atm:** 165°F = 74°C = 347°K

13.4 **Freezing Point:** <−38°F = <−39°C = <234°K

13.5 **Critical Temperature:** Not pertinent

13.6 **Critical Pressure:** Not pertinent

13.7 **Specific Gravity:** 1.31 at 20°C (liquid)

13.8 **Liquid Surface Tension:** 25.4 dynes/cm = 0.0254 N/m at 20°C

13.9 **Liquid-Water Interfacial Tension:** (est.) 45 dynes/cm = 0.045 N/m at 20°C

13.10 **Vapor (Gas) Specific Gravity:** 4.6

13.11 **Ratio of Specific Heats of Vapor (Gas):** 1.104

13.12 **Latent Heat of Vaporization:** 100 Btu/lb = 58 cal/g = 2.4×10^5 J/kg

13.13 **Heat of Combustion:** (est.) 4700 Btu/lb = 2600 cal/g = 110×10^5 J/kg

13.14 **Heat of Decomposition:** Not pertinent

13.15 **Heat of Solution:** Not pertinent

13.16 **Heat of Polymerization:** Not pertinent

5. HEALTH HAZARDS (Cont'd.)

5.9 **Liquid or Solid Irritant Characteristics:** Minimum hazard. If spilled on clothing and allowed to remain, may cause smarting and reddening of the skin.

5.10 **Odor Threshold:** 100 ppm

REVISED 1978

TCL TRICHLOROETHYLENE

Common Synonyms
Trichloroethene
Triclene

Watery liquid | Colorless | Sweet odor

Sinks in water. Irritating vapor is produced.

Stop discharge if possible. Keep people away.
Avoid contact with liquid and vapor.
Call fire department.
Isolate and remove discharged material.
Notify local health and pollution control agencies.

Fire

Combustible.
POISONOUS GASES ARE PRODUCED IN FIRE.
Wear goggles and self-contained breathing apparatus.
Extinguish with dry chemical, carbon dioxide, or foam.

Exposure

CALL FOR MEDICAL AID.

VAPOR
Irritating to eyes, nose and throat.
If inhaled, will cause nausea, vomiting, difficult breathing, or loss of consciousness.
Move to fresh air.
If breathing has stopped, give artificial respiration.
If breathing is difficult, give oxygen.

LIQUID
Irritating to skin and eyes.
If swallowed, will cause nausea, vomiting, difficult breathing, or loss of consciousness.
Remove contaminated clothing and shoes.
Flush affected areas with plenty of water.
IF IN EYES, hold eyelids open and flush with plenty of water.
IF SWALLOWED and victim is CONSCIOUS, have victim drink water or milk and have victim induce vomiting.
IF SWALLOWED and victim is UNCONSCIOUS OR HAVING CONVULSIONS, do nothing except keep victim warm.

Water Pollution

Effect of low concentrations on aquatic life is unknown.
May be dangerous if it enters water intakes.
Notify local health and wildlife officials.
Notify operators of nearby water intakes.

1. RESPONSE TO DISCHARGE

(See Response Methods Handbook, CG 446-4)

Should be removed
Chemical and physical treatment

2. LABELS

No hazard label required by Code of Federal Regulations

3. CHEMICAL DESIGNATIONS

3.1 **Synonyms:** Algylen; Chlorylen; Gemalgene; Threthylene; Trethylene; Tri; Trichloran; Trichloroethene; Tri-Clene; Trielene; Trilene; Triline; Trimar

3.2 **Coast Guard Compatibility Classification:** Halogenated hydrocarbon

3.3 **Chemical Formula:** $CHCl{=}CCl_2$

3.4 **IMCO/United Nations Numerical Designation:** 9.0/1710

4. OBSERVABLE CHARACTERISTICS

4.1 **Physical State (as shipped):** Liquid

4.2 **Color:** Colorless

4.3 **Odor:** Chloroform-like; ethereal

5. HEALTH HAZARDS

5.1 **Personal Protective Equipment:** Organic vapor-acid gas canister; self-contained breathing apparatus for emergencies; neoprene or vinyl gloves; chemical safety goggles; face-shield; neoprene safety shoes; neoprene suit or apron for splash protection.

5.2 **Symptoms Following Exposure:** INHALATION: symptoms range from irritation of the nose and throat to nausea, an attitude of irresponsibility, blurred vision, and finally disturbance of central nervous system resulting in cardiac failure. Chronic exposure may cause organic injury. INGESTION: symptoms similar to inhalation. SKIN: defatting action can cause dermatitis. EYES: slightly irritating sensation and lachrymation.

5.3 **Treatment for Exposure:** Do NOT administer adrenalin or epinephrine; get medical attention for all cases of overexposure. INHALATION: remove victim to fresh air; if necessary, apply artificial respiration and/or administer oxygen. INGESTION: have victim drink water and induce vomiting; repeat three times; then give 1 tablespoon epsom salts in water. EYES; flush thoroughly with water. SKIN: wash thoroughly with soap and warm water.

5.4 **Toxicity by Inhalation (Threshold Limit Value):** 100 ppm

5.5 **Short-Term Inhalation Limits:** 200 ppm for 30 min.

5.6 **Toxicity by Ingestion:** Grade 3; LD_{50} 50 to 500 mg/kg

5.7 **Late Toxicity:** Data not available

5.8 **Vapor (Gas) Irritant Characteristics:** Vapors cause a slight smarting of the eyes or respiratory system if present in high concentrations. The effect is temporary.

5.9 **Liquid or Solid Irritant Characteristics:** Minimum hazard. If spilled on clothing and allowed to remain, may cause smarting and reddening of the skin.

5.10 **Odor Threshold:** 50 ppm

6. FIRE HAZARDS

6.1 **Flash Point:** 90°F C.C.; practically nonflammable

6.2 **Flammable Limits in Air:** 8.0%–10.5%

6.3 **Fire Extinguishing Agents:** Water fog

6.4 **Fire Extinguishing Agents Not to be Used:** Not pertinent

6.5 **Special Hazards of Combustion Products:** Toxic and irritating gases are produced in fire situations.

6.6 **Behavior in Fire:** Not pertinent

6.7 **Ignition Temperature:** 770 °F

6.8 **Electrical Hazard:** Not pertinent

6.9 **Burning Rate:** Not pertinent

7. CHEMICAL REACTIVITY

7.1 **Reactivity with Water:** No reaction

7.2 **Reactivity with Common Materials:** No reaction

7.3 **Stability During Transport:** Stable

7.4 **Neutralizing Agents for Acids and Caustics:** Not pertinent

7.5 **Polymerization:** Not pertinent

7.6 **Inhibitor of Polymerization:** Not pertinent

8. WATER POLLUTION

8.1 **Aquatic Toxicity:** 660 mg/l /40 hr/daphnia/kill/fresh water

8.2 **Waterfowl Toxicity:** Data not available

8.3 **Biological Oxygen Demand (BOD):** Data not available

8.4 **Food Chain Concentration Potential:** None

9. SELECTED MANUFACTURERS

1. Dow Chemical Co. Midland, Mich. 48640
2. E. I. duPont de Nemours & Co., Inc. Electrochemicals Dept. Wilmington, Del. 19898
3. PPG Industries Inc. Industrial Chemical Division Lake Charles, La. 70601

10. SHIPPING INFORMATION

10.1 **Grades or Purity:** Technical; dry cleaning; degreasing; extraction

10.2 **Storage Temperature:** Ambient

10.3 **Inert Atmosphere:** No requirement

10.4 **Venting:** Pressure-vacuum

11. HAZARD ASSESSMENT CODE

(See Hazard Assessment Handbook, CG 446-3)

A-X-Y

12. HAZARD CLASSIFICATIONS

12.1 **Code of Federal Regulations:** ORM–A

12.2 **NAS Hazard Rating for Bulk Water Transportation:**

Category	Rating
Fire	1
Health	
Vapor Irritant	1
Liquid or Solid Irritant	1
Poisons	2
Water Pollution	
Human Toxicity	1
Aquatic Toxicity	2
Aesthetic Effect	2
Reactivity	
Other Chemicals	1
Water	0
Self-Reaction	1

12.3 **NFPA Hazard Classifications:** Not listed

13. PHYSICAL AND CHEMICAL PROPERTIES

13.1 **Physical State at 15°C and 1 atm:** Liquid

13.2 **Molecular Weight:** 131.39

13.3 **Boiling Point at 1 atm:** 189°F = 87°C = 360°K

13.4 **Freezing Point:** −123.5°F = −86.4°C = 186.8°K

13.5 **Critical Temperature:** Not pertinent

13.6 **Critical Pressure:** Not pertinent

13.7 **Specific Gravity:** 1.46 at 20°C (liquid)

13.8 **Liquid Surface Tension:** 29.3 dynes/cm = 0.0293 N/m at 20°C

13.9 **Liquid-Water Interfacial Tension:** 34.5 dynes/cm = 0.0345 N/m at 24°C

13.10 **Vapor (Gas) Specific Gravity:** 4.5

13.11 **Ratio of Specific Heats of Vapor (Gas):** 1.116

13.12 **Latent Heat of Vaporization:** 103 Btu/lb = 57.2 cal/g = 2.40×10^5 J/kg

13.13 **Heat of Combustion:** Not pertinent

13.14 **Heat of Decomposition:** Not pertinent

13.15 **Heat of Solution:** Not pertinent

13.16 **Heat of Polymerization:** Not pertinent

NOTES

REVISED 1978

TCF	TRICHLOROFLUOROMETHANE

Common Synonyms			
F-11 Freon 11 Genetron 11	Liquid	Colorless	Odorless
	Sinks in water. Harmful vapor is produced. Boiling point is 75° F.		

Stop discharge if possible. Keep people away.
Avoid contact with liquid.
Isolate and remove discharged material.
Notify local health and pollution control agencies.

Fire	Not flammable. POISONOUS GASES MAY BE PRODUCED IN FIRE. **Wear goggles and self-contained breathing apparatus.**
Exposure	**CALL FOR MEDICAL AID.** VAPOR If inhaled, will cause dizziness or difficult breathing. **Move to fresh air.** **If breathing has stopped, give artificial respiration.** **If breathing is difficult, give oxygen.** LIQUID Not harmful.
Water Pollution	Not harmful to aquatic life. May be dangerous if it enters water intakes. **Notify local health and wildlife officials.** **Notify operators of nearby water intakes.**

1. RESPONSE TO DISCHARGE

(See Response Methods Handbook, CG 446-4)

Should be removed
Chemical and physical treatment

2. LABELS

No hazard label required by Code of Federal Regulations

3. CHEMICAL DESIGNATIONS

3.1 **Synonyms:** Arcton 9; Freon 11; Isceon 11; Eskimon 11; Frigen 11; Isotron 11; F-11; Genetron 11; Ucon 11

3.2 **Coast Guard Compatibility Classification:** Halogenated hydrocarbon

3.3 **Chemical Formula:** $CFCl_3$

3.4 **IMCO/United Nations Numerical Designation:** Not listed

4. OBSERVABLE CHARACTERISTICS

4.1 **Physical State (as shipped):** Liquid

4.2 **Color:** Colorless

4.3 **Odor:** Odorless; weak chlorinated solvent

5. HEALTH HAZARDS

5.1 **Personal Protective Equipment:** Air line respirator; rubber gloves; monogoggles

5.2 **Symptoms Following Exposure:** Breathing concentrations approaching 10% in air will cause dizziness and drowsiness. Contact with tissues may cause frostbite.

5.3 **Treatment for Exposure:** INHALATION: remove victim to non-contaminated area and apply artificial respiration if breathing has stopped; call a physician immediately; oxygen inhalation may be utilized. SKIN: if frostbite has occurred, flush areas with warm water.

5.4 **Toxicity by Inhalation (Threshold Limit Value):** 1000 ppm

5.5 **Short-Term Inhalation Limits:** Data not available

5.6 **Toxicity by Ingestion:** Data not available

5.7 **Late Toxicity:** Data not available

5.8 **Vapor (Gas) Irritant Characteristics:** Non-irritating

5.9 **Liquid or Solid Irritant Characteristics:** May cause frostbite.

5.10 **Odor Threshold:** Data not available

6. FIRE HAZARDS

6.1 **Flash Point:** Not flammable

6.2 **Flammable Limits in Air:** Not flammable

6.3 **Fire Extinguishing Agents:** Not pertinent

6.4 **Fire Extinguishing Agents Not to be Used:** Not pertinent

6.5 **Special Hazards of Combustion Products:** Produces irritating and toxic products when heated to decomposition temperatures.

6.6 **Behavior in Fire:** Not pertinent

6.7 **Ignition Temperature:** Not-flammable

6.8 **Electrical Hazard:** Not pertinent

6.9 **Burning Rate:** Not flammable

7. CHEMICAL REACTIVITY

7.1 **Reactivity with Water:** No reaction

7.2 **Reactivity with Common Materials:** No reaction

7.3 **Stability During Transport:** Stable

7.4 **Neutralizing Agents for Acids and Caustics:** Not pertinent

7.5 **Polymerization:** Not pertinent

7.6 **Inhibitor of Polymerization:** Not pertinent

8. WATER POLLUTION

8.1 **Aquatic Toxicity:** None

8.2 **Waterfowl Toxicity:** None

8.3 **Biological Oxygen Demand (BOD):** None

8.4 **Food Chain Concentration Potential:** None

9. SELECTED MANUFACTURERS

1. E.I. duPont de Nemours & Co., Inc.
   Freon Products Division
   Wilmington, Del. 19898
2. Pennwalt Corp.
   Chemical Division
   3 Penn Center
   Philadelphia, Pa. 19102
3. Union Carbide Corp.
   Chemicals and Plastics Division
   270 Park Ave.
   New York, N. Y. 10017

10. SHIPPING INFORMATION

10.1 **Grades or Purity:** Technical

10.2 **Storage Temperature:** Ambient

10.3 **Inert Atmosphere:** No requirement

10.4 **Venting:** Safety relief

11. HAZARD ASSESSMENT CODE

(See Hazard Assessment Handbook, CG 446-3)

A-C-I-J

12. HAZARD CLASSIFICATIONS

12.1 **Code of Federal Regulations:** Not listed

12.2 **NAS Hazard Rating for Bulk Water Transportation:** Not listed

12.3 **NFPA Hazard Classifications:** Not listed

13. PHYSICAL AND CHEMICAL PROPERTIES

13.1 **Physical State at 15°C and 1 atm:** Liquid

13.2 **Molecular Weight:** 137.4

13.3 **Boiling Point at 1 atm:** 74.8°F = 23.8°C = 297.0°K

13.4 **Freezing Point:** −168°F = −111°C = 162°K

13.5 **Critical Temperature:** 388°F = 198°C = 471°K

13.6 **Critical Pressure:** 639.4 psia = 43.5 atm = 4.41 MN/m^2

13.7 **Specific Gravity:** 1.49 at 20°C (liquid)

13.8 **Liquid Surface Tension:** 23 dynes/cm = 0.023 N/m at 20°C

13.9 **Liquid-Water Interfacial Tension:** Not pertinent

13.10 **Vapor (Gas) Specific Gravity:** 4.7

13.11 **Ratio of Specific Heats of Vapor (Gas):** (est.) 1.128

13.12 **Latent Heat of Vaporization:** 78.3 Btu/lb = 43.5 cal/g = 1.82×10^5 J/kg

13.13 **Heat of Combustion:** Not pertinent

13.14 **Heat of Decomposition:** Not pertinent

13.15 **Heat of Solution:** Not pertinent

13.16 **Heat of Polymerization:** Not pertinent

NOTES

REVISED 1978

TPH	TRICHLOROPHENOL

Common Synonyms 2,4,5-Trichlorophenol Dowicide 2 Phenachlor	Solid crystals or flakes — Yellow — Strong disinfectant odor Sinks in water.

Isolate and remove discharged material.
Notify local health and pollution control agencies.

Fire	Not flammable.
Exposure	CALL FOR MEDICAL AID. DUST OR SOLID Irritating to eyes, nose, and throat. Move to fresh air. IF IN EYES, hold eyelids open and flush with plenty of water.
Water Pollution	HARMFUL TO AQUATIC LIFE IN VERY LOW CONCENTRATIONS. May be dangerous if it enters water intakes. Notify local health and wildlife officials. Notify operators of nearby water intakes.

1. RESPONSE TO DISCHARGE

(See Response Methods Handbook, CG 446-4)

Issue warning—water contaminant
Should be removed
Chemical and physical treatment

2. LABELS

No hazard label required by Code of Federal Regulations

3. CHEMICAL DESIGNATIONS

3.1 **Synonyms:** Dowicide 2
Omal
Phenachlor
2,4,5-Trichlorophenol

3.2 **Coast Guard Compatibility Classification:** Not applicable

3.3 **Chemical Formula:** 1—$HOC_6H_2Cl_3$—2,4,5

3.4 **IMCO/United Nations Numerical Designation:** Not listed

4. OBSERVABLE CHARACTERISTICS

4.1 **Physical State (as shipped):** Solid

4.2 **Color:** Colorless to gray

4.3 **Odor:** Strong disinfectant

5. HEALTH HAZARDS

5.1 **Personal Protective Equipment:** Approved dust respirator for toxic dusts; goggles; protective clothing to prevent contact with skin.

5.2 **Symptoms Following Exposure:** Dust may cause swelling of eyes and eye injury, irritation of nose and throat. Solid irritates skin on prolonged contact.

5.3 **Treatment for Exposure:** INHALATION: remove to fresh air; get medical attention. EYES: flush with water for at least 15 min; get medical attention. SKIN: wash with soap and water; if irritation occurs, get medical attention.

5.4 **Toxicity by Inhalation (Threshold Limit Value):** Not pertinent

5.5 **Short-Term Inhalation Limits:** Data not available

5.6 **Toxicity by Ingestion:** 20% solution in fuel oil: Grade 2; LD_{50} 0.5 to 5 g/kg (rat)

5.7 **Late Toxicity:** Data not available

5.8 **Vapor (Gas) Irritant Characteristics:** Essentially non-volatile at ordinary temperatures.

5.9 **Liquid or Solid Irritant Characteristics:** May cause injury to eye. Prolonged contact with skin causes a slight burn. Dust irritates nose and throat.

5.10 **Odor Threshold:** Data not available

6. FIRE HAZARDS

6.1 **Flash Point:** Not flammable

6.2 **Flammable Limits in Air:** Not flammable

6.3 **Fire Extinguishing Agents:** Not pertinent

6.4 **Fire Extinguishing Agents Not to be Used:** Not pertinent

6.5 **Special Hazards of Combustion Products:** Not pertinent

6.6 **Behavior in Fire:** Not pertinent

6.7 **Ignition Temperature:** Not flammable

6.8 **Electrical Hazard:** Not pertinent

6.9 **Burning Rate:** Not flammable

7. CHEMICAL REACTIVITY

7.1 **Reactivity with Water:** No reaction

7.2 **Reactivity with Common Materials:** None

7.3 **Stability During Transport:** Stable

7.4 **Neutralizing Agents for Acids and Caustics:** Not pertinent

7.5 **Polymerization:** Not pertinent

7.6 **Inhibitor of Polymerization:** Not pertinent

8. WATER POLLUTION

8.1 **Aquatic Toxicity:** 3.2 ppm/24 hr/fish/TL_m/fresh water

8.2 **Waterfowl Toxicity:** Data not available

8.3 **Biological Oxygen Demand (BOD):** Data not available

8.4 **Food Chain Concentration Potential:** None

9. SELECTED MANUFACTURERS

1. Dow Chemical Co.
Midland, Mich. 48640
2. Fike Chemicals, Inc.
Nitro, W. Va. 25143
3. Hooker Chemical Corp.
Industrial Chemicals Division
Niagara Falls, N. Y. 14302

10. SHIPPING INFORMATION

10.1 **Grades or Purity:** Technical: 95%

10.2 **Storage Temperature:** Ambient

10.3 **Inert Atmosphere:** No requirement

10.4 **Venting:** Open

11. HAZARD ASSESSMENT CODE

(See Hazard Assessment Handbook, CG 446-3)

II

12. HAZARD CLASSIFICATIONS

12.1 **Code of Federal Regulations:** Not listed

12.2 **NAS Hazard Rating for Bulk Water Transportation:** Not listed

12.3 **NFPA Hazard Classifications:** Solid form not listed

13. PHYSICAL AND CHEMICAL PROPERTIES

13.1 **Physical State at 15°C and 1 atm:** Solid

13.2 **Molecular Weight:** 197.5

13.3 **Boiling Point at 1 atm:** 485°F = 252°C = 525°K

13.4 **Freezing Point:** 135°F = 57°C = 330°K

13.5 **Critical Temperature:** Not pertinent

13.6 **Critical Pressure:** Not pertinent

13.7 **Specific Gravity:** 1.7 at 25°C (solid)

13.8 **Liquid Surface Tension:** Not pertinent

13.9 **Liquid-Water Interfacial Tension:** Not pertinent

13.10 **Vapor (Gas) Specific Gravity:** Not pertinent

13.11 **Ratio of Specific Heats of Vapor (Gas):** Not pertinent

13.12 **Latent Heat of Vaporization:** Not pertinent

13.13 **Heat of Combustion:** Not pertinent

13.14 **Heat of Decomposition:** Not pertinent

13.15 **Heat of Solution:** Not pertinent

13.16 **Heat of Polymerization:** Not pertinent

NOTES

TCA 2,4,5-TRICHLOROPHENOXYACETIC ACID

Common Synonyms			
2, 4, 5-T	Solid	White	Odorless
	Sinks in water.		

Keep people away.
Call fire department.
Isolate and remove discharged material.
Notify local health and pollution control agencies.

Fire	Combustible. POISONOUS GASES MAY BE PRODUCED IN FIRE. Extinguish with water, dry chemicals, foam, or carbon dioxide.
Exposure	CALL FOR MEDICAL AID. SOLID POISONOUS IF SWALLOWED. Irritating to skin and eyes. Remove contaminated clothing and shoes. Flush affected areas with plenty of water. IF IN EYES, hold eyelids open and flush with plenty of water. IF SWALLOWED and victim is CONSCIOUS, have victim drink water or milk and have victim induce vomiting. IF SWALLOWED and victim is UNCONSCIOUS OR HAVING CONVULSIONS, do nothing except keep victim warm.
Water Pollution	HARMFUL TO AQUATIC LIFE IN VERY LOW CONCENTRATIONS. May be dangerous if it enters water intakes. Notify local health and wildlife officials. Notify operators of nearby water intakes.

1. RESPONSE TO DISCHARGE
(See Response Methods Handbook, CG 446-4)

Issue warning—poison, water contaminant
Restrict access
Should be removed
Chemical and physical treatment

2. LABEL

3. CHEMICAL DESIGNATIONS

3.1 **Synonyms:** 2,4,5-T
3.2 **Coast Guard Compatibility Classification:** Not applicable
3.3 **Chemical Formula:** 2,4,5-$Cl_3C_6H_2OCH_2COOH$
3.4 **IMCO/United Nations Numerical Designation:** 6.1/1609

4. OBSERVABLE CHARACTERISTICS

4.1 **Physical State (as shipped):** Solid
4.2 **Color:** White
4.3 **Odor:** None

5. HEALTH HAZARDS

5.1 **Personal Protective Equipment:** Dust mask and rubber gloves
5.2 **Symptoms Following Exposure:** Overexposure to dust by inhalation or ingestion may cause fatigue, nausea, vomiting, lowered blood pressure, convulsions, coma. Dust may irritate eyes and skin.
5.3 **Treatment for Exposure:** INHALATION: remove victim to fresh air; if required, give artificial respiration. EYES: flush with water until irritating dust is removed. SKIN: wash with soap and water. INGESTION: call physician at once; induce vomiting and administer gastric lavage.
5.4 **Toxicity by Inhalation (Threshold Limit Value):** 10 mg/m^3
5.5 **Short-Term Inhalation Limits:** Data not available
5.6 **Toxicity by Ingestion:** Grade 3; oral LD_{50} = 500 mg/kg (rat)
5.7 **Late Toxicity:** Birth defects in rats and mice. Causes an acne-like skin eruption among human workers
5.8 **Vapor (Gas) Irritant Characteristics:** Not pertinent
5.9 **Liquid or Solid Irritant Characteristics:** Data not available
5.10 **Odor Threshold:** Not pertinent

6. FIRE HAZARDS

6.1 **Flash Point:** Not pertinent (solid)
6.2 **Flammable Limits in Air:** Not pertinent
6.3 **Fire Extinguishing Agents:** Water, foam, dry chemical, carbon dioxide
6.4 **Fire Extinguishing Agents Not to be Used:** Not pertinent
6.5 **Special Hazards of Combustion Products:** Toxic hydrogen chloride and phosgene gases may be formed.
6.6 **Behavior in Fire:** Not pertinent
6.7 **Ignition Temperature:** Data not available
6.8 **Electrical Hazard:** Not pertinent
6.9 **Burning Rate:** Not pertinent

7. CHEMICAL REACTIVITY

7.1 **Reactivity with Water:** No reaction
7.2 **Reactivity with Common Materials:** Can be corrosive to common metals
7.3 **Stability During Transport:** Stable
7.4 **Neutralizing Agents for Acids and Caustics:** Not pertinent
7.5 **Polymerization:** Not pertinent
7.6 **Inhibitor of Polymerization:** Not pertinent

8. WATER POLLUTION

8.1 **Aquatic Toxicity:**
11.0 ppm/24 hr/bluegill/TL_m/fresh water
0.32 ppm/48 hr/spot/50% kill/salt water
8.2 **Waterfowl Toxicity:** 21,000 ppm LD_{50}
8.3 **Biological Oxygen Demand (BOD):** Data not available
8.4 **Food Chain Concentration Potential:** Will not bioconcentrate

9. SELECTED MANUFACTURERS

1. Dow Chemical Co.
Midland, Mich. 48640
2. Transvaal, Inc.
P.O. Box 69
Jacksonville, Ark. 72076
3. Monsanto Co.
800 N. Lindbergh Boulevard
St. Louis, Mo. 63166

10. SHIPPING INFORMATION

10.1 **Grades or Purity:** Commercial
10.2 **Storage Temperature:** Ambient
10.3 **Inert Atmosphere:** No requirement
10.4 **Venting:** Open (flame arrester)

11. HAZARD ASSESSMENT CODE
(See Hazard Assessment Handbook, CG 446-3)

II

12. HAZARD CLASSIFICATIONS

12.1 **Code of Federal Regulations:** Poisonous, Class B
12.2 **NAS Hazard Rating for Bulk Water Transportation:** Not listed
12.3 **NFPA Hazard Classifications:** Not listed

13. PHYSICAL AND CHEMICAL PROPERTIES

13.1 **Physical State at 15°C and 1 atm:** Solid
13.2 **Molecular Weight:** 255.5
13.3 **Boiling Point at 1 atm:** Not pertinent (decomposes)
13.4 **Freezing Point:** 316°F = 158°C = 431°K
13.5 **Critical Temperature:** Not pertinent
13.6 **Critical Pressure:** Not pertinent
13.7 **Specific Gravity:** 1.803 at 20°C (solid)
13.8 **Liquid Surface Tension:** Not pertinent
13.9 **Liquid-Water Interfacial Tension:** Not pertinent
13.10 **Vapor (Gas) Specific Gravity:** Not pertinent
13.11 **Ratio of Specific Heats of Vapor (Gas):** Not pertinent
13.12 **Latent Heat of Vaporization:** Not pertinent
13.13 **Heat of Combustion:** (est.) −6,500 Btu/lb = −3,600 cal/g = -150×10^5 J/kg
13.14 **Heat of Decomposition:** Not pertinent
13.15 **Heat of Solution:** Not pertinent
13.16 **Heat of Polymerization:** Not pertinent

NOTES

TCS TRICHLOROSILANE

Common Synonyms			
Trichloromonosilane Silicochloroform	Liquid	Colorless	Sharp choking odor
	Reacts violently with water. Irritating gas is produced on contact with water. Boiling point is 90° F.		

Shut off ignition sources. Call fire department.
Avoid contact with liquid. Keep people away.
Stop discharge if possible.
Isolate and remove discharged material.
Notify local health and pollution control agencies.

Fire

FLAMMABLE.
POISONOUS GASES MAY BE PRODUCED IN FIRE.
Containers may explode in fire.
Flashback along vapor trail may occur.
Vapor may explode if ignited in an enclosed area.
Wear goggles and self-contained breathing apparatus.
Extinguish with dry chemicals or carbon dioxide.
DO NOT USE WATER OR FOAM ON FIRE.
DO NOT USE WATER ON ADJACENT FIRES.
Cool exposed containers with water.

Exposure

Call for medical aid.

VAPOR
Irritating to eyes, nose and throat.
Harmful if inhaled.
Move victim to fresh air.
If breathing has stopped, give artificial respiration.
If breathing is difficult, give oxygen.

LIQUID
Will burn skin and eyes.
Harmful if swallowed.
Remove contaminated clothing and shoes.
Flush affected areas with plenty of water.
IF IN EYES, hold eyelids open and flush with plenty of water.
IF SWALLOWED and victim is CONSCIOUS, have victim drink water or milk.
DO NOT INDUCE VOMITING.

Water Pollution

Effect of low concentrations on aquatic life is unknown.
May be dangerous if it enters water intakes.
Notify local health and wildlife officials.
Notify operators of nearby water intakes.

1. RESPONSE TO DISCHARGE

(See Response Methods Handbook, CG 446-4)

Issue warning—high flammability, corrosive, air contaminant
Restrict access
Evacuate area
Disperse and flush

2. LABEL

Red

3. CHEMICAL DESIGNATIONS

3.1 **Synonyms:** Silicochloroform; Trichloromonosilane
3.2 **Coast Guard Compatibility Classification:** Not applicable
3.3 **Chemical Formula:** $SiHCl_3$
3.4 **IMCO/United Nations Numerical Designation:** 4.3/1295

4. OBSERVABLE CHARACTERISTICS

4.1 **Physical State (as shipped):** Liquid
4.2 **Color:** Colorless
4.3 **Odor:** Sharp, choking; like hydrochloric acid

5. HEALTH HAZARDS

5.1 **Personal Protective Equipment:** Acid-vapor-type respiratory protection; rubber gloves; chemical worker's goggles; other protective equipment as necessary to protect skin and eyes.
5.2 **Symptoms Following Exposure:** Inhalation causes severe irritation of respiratory system. Liquid causes severe burns of eyes and skin. Ingestion causes severe burns of mouth and stomach.
5.3 **Treatment for Exposure:** INHALATION: remove victim from exposure; if breathing is difficult or stopped, give artificial respiration; call physician. EYES or SKIN: flush with plenty of water immediately for at least 15 min. and get medical attention. INGESTION: do NOT induce vomiting; give large amount of water; get medical attention.
5.4 **Toxicity by Inhalation (Threshold Limit Value):** Data not available
5.5 **Short-Term Inhalation Limits:** Data not available
5.6 **Toxicity by Ingestion:** Grade 2; oral LD_{50} = 1,000 mg/kg (rat)
5.7 **Late Toxicity:** Data not available
5.8 **Vapor (Gas) Irritant Characteristics:** Vapors cause severe irritation of eyes and throat and can cause eye and lung injury. They cannot be tolerated even at low concentrations.
5.9 **Liquid or Solid Irritant Characteristics:** Severe skin irritant. Causes second- and third-degree burns on short contact and is very injurious to the eyes.
5.10 **Odor Threshold:** Data not available

6. FIRE HAZARDS

6.1 **Flash Point:** −18°F O.C., > −58°F C.C.
6.2 **Flammable Limits in Air:** 1.2%–90.5%
6.3 **Fire Extinguishing Agents:** Dry chemical, carbon dioxide
6.4 **Fire Extinguishing Agents Not to be Used:** Water, foam
6.5 **Special Hazards of Combustion Products:** Toxic hydrogen chloride and phosgene gases may form in fires.
6.6 **Behavior in Fire:** Difficult to extinguish; re-ignition may occur. Vapor is heavier than air and may travel a considerable distance to a source of ignition and flash back.
6.7 **Ignition Temperature:** 220°F
6.8 **Electrical Hazard:** Data not available
6.9 **Burning Rate:** Data not available

7. CHEMICAL REACTIVITY

7.1 **Reactivity with Water:** Reacts violently to form hydrogen chloride fumes (hydrochloric acid).
7.2 **Reactivity with Common Materials:** Reacts with surface moisture to form hydrochloric acid, which corrodes common metals and forms flammable hydrogen gas.
7.3 **Stability During Transport:** Stable
7.4 **Neutralizing Agents for Acids and Caustics:** Flush with water, rinse with sodium bicarbonate or lime solution.
7.5 **Polymerization:** Not pertinent
7.6 **Inhibitor of Polymerization:** Not pertinent

8. WATER POLLUTION

8.1 **Aquatic Toxicity:** Data not available
8.2 **Waterfowl Toxicity:** Data not available
8.3 **Biological Oxygen Demand (BOD):** None
8.4 **Food Chain Concentration Potential:** None

9. SELECTED MANUFACTURERS

1. Ventron Corp.
   Chemicals Division
   8 Congress Street
   Beverly, Mass. 01915
2. Union Carbide Corporation
   Chemicals and Plastics Division
   270 Park Avenue
   New York, N. Y. 10017
3. Dow Corning Corporation
   P. O. Box 592
   Midland, Mich. 48640

10. SHIPPING INFORMATION

10.1 **Grades or Purity:** 99+%
10.2 **Storage Temperature:** Ambient
10.3 **Inert Atmosphere:** No requirement
10.4 **Venting:** Pressure-vacuum

11. HAZARD ASSESSMENT CODE

(See Hazard Assessment Handbook, CG 446-3)

A-O

12. HAZARD CLASSIFICATIONS

12.1 **Code of Federal Regulations:** Flammable liquid
12.2 **NAS Hazard Rating for Bulk Water Transportation:**

Category	Rating
Fire	4
Health	
Vapor Irritant	4
Liquid or Solid Irritant	4
Poisons	3
Water Pollution	
Human Toxicity	3
Aquatic Toxicity	3
Aesthetic Effect	2
Reactivity	
Other Chemicals	3
Water	4
Self-Reaction	1

12.3 **NFPA Hazard Classifications:**

Category	Classification
Health Hazard (Blue)	3
Flammability (Red)	4
Reactivity (Yellow)	1

13. PHYSICAL AND CHEMICAL PROPERTIES

13.1 **Physical State at 15°C and 1 atm:** Liquid
13.2 **Molecular Weight:** 135.5
13.3 **Boiling Point at 1 atm:** 90°F = 32°C = 305°K
13.4 **Freezing Point:** −197°F = −127°C = 146°K
13.5 **Critical Temperature:** Not pertinent
13.6 **Critical Pressure:** Not pertinent
13.7 **Specific Gravity:** 1.344 at 20°C (liquid)
13.8 **Liquid Surface Tension:** (est.) 18.3 dynes/cm = 0.0183 N/m at 20°C
13.9 **Liquid-Water Interfacial Tension:** Not pertinent
13.10 **Vapor (Gas) Specific Gravity:** 4.9
13.11 **Ratio of Specific Heats of Vapor (Gas):** Data not available
13.12 **Latent Heat of Vaporization:** 85 Btu/lb = 47 cal/g = 2.0×10^5 J/kg
13.13 **Heat of Combustion:** Data not available
13.14 **Heat of Decomposition:** Not pertinent
13.15 **Heat of Solution:** Data not available
13.16 **Heat of Polymerization:** Not pertinent

NOTES

TCT | TRICHLORO-s-TRIAZINETRIONE

Common Synonyms			
Trichloro-s-triazine-2, 4, 6-(1H, 3H, 5H)-trione Trichlorotriazinetrione Trichloroisocyanuric acid Trichloroiminoisocyanuric acid	Solid	White	Bleach-like odor
	Sinks and mixes slowly with water.		

Keep people away.
Call fire department.
Isolate and remove discharged material.
Notify local health and pollution control agencies.

Fire

Not flammable.
May cause fire on contact with combustibles.
POISONOUS GASES ARE PRODUCED IN FIRE.
Containers may explode in fire.
Wear goggles and self-contained breathing apparatus.
Flood discharge area with water.
Cool exposed containers with water.

Exposure

Call for medical aid.

DUST
Irritating to eyes, nose and throat.
If inhaled will cause coughing or difficult breathing.
Move victim to fresh air.
If in eyes, hold eyelids open and flush with plenty of water.
If breathing is difficult, give oxygen.

SOLID
Irritating to skin and eyes.
Harmful if swallowed.
Remove contaminated clothing and shoes.
Flush affected areas with plenty of water.
IF IN EYES, hold eyelids open and flush with plenty of water.
IF SWALLOWED and victim is CONSCIOUS, have victim drink water or milk and have victim induce vomiting.
IF SWALLOWED and victim is UNCONSCIOUS OR HAVING CONVULSIONS, do nothing except keep victim warm.

Water Pollution

Effect of low concentrations on aquatic life is unknown.
May be dangerous if it enters water intakes.
Notify local health and wildlife officials.
Notify operators of nearby water intakes.

1. RESPONSE TO DISCHARGE

(See Response Methods Handbook, CG 446-4)

Issue warning—oxidizing material, water contaminant
Restrict access
Disperse and flush

2. LABEL

Yellow

3. CHEMICAL DESIGNATIONS

3.1 **Synonyms:** Trichloroiminoisocyanuric acid; Trichloroisocyanuric acid; Trichloro-s-triazine-2,4,6-(1H,3H,5H)-trione; Trichlorotriazinetrione; 1,3,5-Trichloro-2,4,6-trioxo-1,3,5-triazine

3.2 **Coast Guard Compatibility Classification:** Not applicable

3.3 **Chemical Formula:** $Cl_3(NCO)_3$

3.4 **IMCO/United Nations Numerical Designation:** 5.1/—

4. OBSERVABLE CHARACTERISTICS

4.1 **Physical State (as shipped):** Solid

4.2 **Color:** White

4.3 **Odor:** Like chlorine

5. HEALTH HAZARDS

5.1 **Personal Protective Equipment:** Dust mask or chlorine canister mask; goggles; rubber gloves

5.2 **Symptoms Following Exposure:** Inhalation causes sneezing and coughing. Contact with dust causes moderate irritation of eyes and itching and redness of skin. Ingestion causes burns of mouth and stomach.

5.3 **Treatment for Exposure:** INHALATION: remove victim to fresh air. EYES: irrigate with running water for 15 min.; call physician. SKIN: flush with water. INGESTION: induce vomiting and call physician.

5.4 **Toxicity by Inhalation (Threshold Limit Value):** Data not available

5.5 **Short-Term Inhalation Limits:** Data not available

5.6 **Toxicity by Ingestion:** Grade 2; oral LD_{50} = 750 mg/kg (rat)

5.7 **Late Toxicity:** Data not available

5.8 **Vapor (Gas) Irritant Characteristics:** Data not available

5.9 **Liquid or Solid Irritant Characteristics:** Data not available

5.10 **Odor Threshold:** Data not available

6. FIRE HAZARDS

6.1 **Flash Point:** Not flammable but may cause fire on contact with ordinary combustibles

6.2 **Flammable Limits in Air:** Not pertinent

6.3 **Fire Extinguishing Agents:** Water in large amounts

6.4 **Fire Extinguishing Agents Not to be Used:** Not pertinent

6.5 **Special Hazards of Combustion Products:** Toxic chlorine or nitrogen trichloride may be formed in fires.

6.6 **Behavior in Fire:** Containers may explode when heated.

6.7 **Ignition Temperature:** Not pertinent

6.8 **Electrical Hazard:** Not pertinent

6.9 **Burning Rate:** Not pertinent

7. CHEMICAL REACTIVITY

7.1 **Reactivity with Water:** Reacts to form a bleach solution. The reaction is not hazardous.

7.2 **Reactivity with Common Materials:** Contact with most foreign material, organic matter, or easily chlorinated or oxidized materials may result in fire. Avoid oil, grease, sawdust, floor sweepings, other easily oxidized organic compounds.

7.3 **Stability During Transport:** Stable

7.4 **Neutralizing Agents for Acids and Caustics:** Not pertinent

7.5 **Polymerization:** Not pertinent

7.6 **Inhibitor of Polymerization:** Not pertinent

8. WATER POLLUTION

8.1 **Aquatic Toxicity:** Data not available

8.2 **Waterfowl Toxicity:** Data not available

8.3 **Biological Oxygen Demand (BOD):** Data not available

8.4 **Food Chain Concentration Potential:** None

9. SELECTED MANUFACTURERS

1. FMC Corporation
   Industrial Chemical Division
   633 Third Avenue
   New York, N. Y. 10017

2. Monsanto Co.
   800 N. Lindbergh Boulevard
   St. Louis, Mo. 63166

3. Eastman Kodak Co.
   Eastman Organic Chemicals
   Rochester, N. Y. 14650

10. SHIPPING INFORMATION

10.1 **Grades or Purity:** 39 90% available chlorine

10.2 **Storage Temperature:** Ambient. Avoid elevated temperatures.

10.3 **Inert Atmosphere:** No requirement if dry

10.4 **Venting:** Pressure-vacuum

11. HAZARD ASSESSMENT CODE

(See Hazard Assessment Handbook, CG 446-3)

II-SS

12. HAZARD CLASSIFICATIONS

12.1 **Code of Federal Regulations:** Oxidizing material

12.2 **NAS Hazard Rating for Bulk Water Transportation:** Not listed

12.3 **NFPA Hazard Classifications:**

Category	Classification
Health Hazard (Blue)	3
Flammability (Red)	0
Reactivity (Yellow)	2 oxy

13. PHYSICAL AND CHEMICAL PROPERTIES

13.1 **Physical State at 15°C and 1 atm:** Solid

13.2 **Molecular Weight:** 232.5

13.3 **Boiling Point at 1 atm:** Not pertinent (decomposes)

13.4 **Freezing Point:** Not pertinent

13.5 **Critical Temperature:** Not pertinent

13.6 **Critical Pressure:** Not pertinent

13.7 **Specific Gravity:** (est.) > 1 at 20°C (solid)

13.8 **Liquid Surface Tension:** Not pertinent

13.9 **Liquid-Water Interfacial Tension:** Not pertinent

13.10 **Vapor (Gas) Specific Gravity:** Not pertinent

13.11 **Ratio of Specific Heats of Vapor (Gas):** Not pertinent

13.12 **Latent Heat of Vaporization:** Not pertinent

13.13 **Heat of Combustion:** Not pertinent

13.14 **Heat of Decomposition:** Not pertinent

13.15 **Heat of Solution:** Not pertinent

13.16 **Heat of Polymerization:** Not pertinent

NOTES

TCP	# TRICRESYL PHOSPHATE

Common Synonyms TCP Tri-p-cresyl phosphate	Liquid Colorless Odorless Sinks in water.

Stop discharge if possible.
Call fire department.
Isolate and remove discharged material.
Notify local health and pollution control agencies.

Fire	Combustible. Extinguish with dry chemical, foam, or carbon dioxide. Water may be ineffective on fire. Cool exposed containers with water.
Exposure	LIQUID Harmful if swallowed. IF SWALLOWED and victim is CONSCIOUS, have victim drink water or milk.
Water Pollution	Effect of low concentrations on aquatic life is unknown. May be dangerous if it enters water intakes. Notify local health and wildlife officials. Notify operators of nearby water intakes.

1. RESPONSE TO DISCHARGE

(See Response Methods Handbook, CG 446-4)

Should be removed
Chemical and physical treatment

2. LABELS

No hazard label required by Code of Federal Regulations

3. CHEMICAL DESIGNATIONS

3.1 **Synonyms:** TCP
Tri-p-tolyl phosphate
Tri-p-cresyl phosphate

3.2 **Coast Guard Compatibility Classification:** Not applicable

3.3 **Chemical Formula:** $(p\text{-}CH_3C_6H_4O)_3PO$

3.4 **IMCO/United Nations Numerical Designation:** Not listed

4. OBSERVABLE CHARACTERISTICS

4.1 **Physical State (as shipped):** Liquid

4.2 **Color:** Colorless

4.3 **Odor:** Odorless

5. HEALTH HAZARDS

5.1 **Personal Protective Equipment:** Goggles or face shield.

5.2 **Symptoms Following Exposure:** Vapors may irritate eyes, but only at high temperatures. Ingestion of liquid may cause severe damage to central nervous system and death if significant amounts of the toxic ortho-isomer are present.

5.3 **Treatment for Exposure:** INGESTION: induce vomiting and call a doctor. EYES: flush with water for at least 15 min. SKIN: wipe off, wash with soap and water.

5.4 **Toxicity by Inhalation (Threshold Limit Value):** Not pertinent

5.5 **Short-Term Inhalation Limits:** Not pertinent

5.6 **Toxicity by Ingestion:** Grade 2; LD_{50} 0.5 to 5 g/kg (chicken LD_{50} >2 g/kg)

5.7 **Late Toxicity:** Data not available

5.8 **Vapor (Gas) Irritant Characteristics:** Vapors cause a slight smarting of the eyes or respiratory system if present in high concentrations. The effect is temporary. The compound is non-volatile for all practical purposes.

5.9 **Liquid or Solid Irritant Characteristics:** No appreciable hazard. Practically harmless to the skin.

5.10 **Odor Threshold:** Not pertinent

6. FIRE HAZARDS

6.1 **Flash Point:** 410°F C.C.

6.2 **Flammable Limits in Air:** Data not available

6.3 **Fire Extinguishing Agents:** Foam, dry chemical, or carbon dioxide

6.4 **Fire Extinguishing Agents Not to be Used:** Water or foam may cause frothing.

6.5 **Special Hazards of Combustion Products:** Not pertinent

6.6 **Behavior in Fire:** Not pertinent

6.7 **Ignition Temperature:** Data not available

6.8 **Electrical Hazard:** Not pertinent

6.9 **Burning Rate:** Data not available

7. CHEMICAL REACTIVITY

7.1 **Reactivity with Water:** No reaction

7.2 **Reactivity with Common Materials:** No reaction

7.3 **Stability During Transport:** Stable

7.4 **Neutralizing Agents for Acids and Caustics:** Not pertinent

7.5 **Polymerization:** Not pertinent

7.6 **Inhibitor of Polymerization:** Not pertinent

8. WATER POLLUTION

8.1 **Aquatic Toxicity:** Data not available

8.2 **Waterfowl Toxicity:** Data not available

8.3 **Biological Oxygen Demand (BOD):** Data not available

8.4 **Food Chain Concentration Potential:** None

9. SELECTED MANUFACTURERS

1. FMC Corp.
Organic Chemicals Division
633 Third Ave.
New York, N. Y. 10012
2. Monsanto Co.
Monsanto Industrial Chemicals Co.
800 N. Lindbergh Blvd.
St. Louis, Mo. 63166
3. Stauffer Chemical Co.
Specialty Chemicals Division
Gallipolis Ferry, W. Va. 25515

10. SHIPPING INFORMATION

10.1 **Grades or Purity:** Consists primarily of the para isomer, but several commercial grades may contain a significant proportion of tri-orthocresyl phosphate. Latter is considerably more toxic than the para-isomer if ingested.

10.2 **Storage Temperature:** Ambient

10.3 **Inert Atmosphere:** No requirement

10.4 **Venting:** Open (flame arrester)

11. HAZARD ASSESSMENT CODE

(See Hazard Assessment Handbook, CG 446-3)

A-X

12. HAZARD CLASSIFICATIONS

12.1 **Code of Federal Regulations:** Not listed

12.2 **NAS Hazard Rating for Bulk Water Transportation:** Not listed

12.3 **NFPA Hazard Classifications:** Not listed

13. PHYSICAL AND CHEMICAL PROPERTIES

13.1 **Physical State at 15°C and 1 atm:** Liquid

13.2 **Molecular Weight:** 368

13.3 **Boiling Point at 1 atm:** 770°F = 410°C = 683°K

13.4 **Freezing Point:** −27°F = −33°C = 240°K

13.5 **Critical Temperature:** Not pertinent

13.6 **Critical Pressure:** Not pertinent

13.7 **Specific Gravity:** 1.16 at 20°C (liquid)

13.8 **Liquid Surface Tension:** 44 dynes/cm = 0.044 N/m at 25°C

13.9 **Liquid-Water Interfacial Tension:** Data not available

13.10 **Vapor (Gas) Specific Gravity:** Not pertinent

13.11 **Ratio of Specific Heats of Vapor (Gas):** Not pertinent

13.12 **Latent Heat of Vaporization:** (est.) 80.0 Btu/lb = 44.5 cal/g = 1.86×10^5 J/kg

13.13 **Heat of Combustion:** Not pertinent

13.14 **Heat of Decomposition:** Not pertinent

13.15 **Heat of Solution:** Not pertinent

13.16 **Heat of Polymerization:** Not pertinent

NOTES

REVISED 1978

TDN	TRIDECANOL

Common Synonyms 1-Tridecanol Oxotridecyl alcohol	Oily liquid Colorless Mild, pleasant odor Floats on water.

Stop discharge if possible.
Call fire department.
Isolate and remove discharged material.
Notify local health and pollution control agencies.

Fire	Combustible. Extinguish with dry chemical or alcohol foam. Water may be ineffective on fire. Cool exposed containers with water.
Exposure	Not harmful.
Water Pollution	Effect of low concentrations on aquatic life is unknown. Fouling to shoreline. May be dangerous if it enters water intakes. Notify local health and wildlife officials. Notify operators of nearby water intakes.

1. RESPONSE TO DISCHARGE

(See Response Methods Handbook, CG 446-4)

Mechanical containment
Should be removed
Chemical and physical treatment

2. LABELS

No hazard label required by Code of Federal Regulations

3. CHEMICAL DESIGNATIONS

3.1 **Synonyms:** Isotridecanol
Isotridecyl alcohol
1-Tridecanol

3.2 **Coast Guard Compatibility Classification:** Alcohol

3.3 **Chemical Formula:** $C_{12}H_{25}CH_2OH$

3.4 **IMCO/United Nations Numerical Designation:** Not listed

4. OBSERVABLE CHARACTERISTICS

4.1 **Physical State (as shipped):** Liquid

4.2 **Color:** Colorless

4.3 **Odor:** Mild alcoholic

5. HEALTH HAZARDS

5.1 **Personal Protective Equipment:** Synthetic rubber gloves; chemical goggles.

5.2 **Symptoms Following Exposure:** Inhalation hazard slight. Skin contact results in moderate irritation. Liquid contact with eyes causes severe irritation and possible eye damage.

5.3 **Treatment for Exposure:** EYES: promptly flush with clean water for at least 15 min. and see a physician. SKIN: wash exposed area with soap and water.

5.4 **Toxicity by Inhalation (Threshold Limit Value):** Data not available

5.5 **Short-Term Inhalation Limits:** Data not available

5.6 **Toxicity by Ingestion:** Data not available

5.7 **Late Toxicity:** Data not available

5.8 **Vapor (Gas) Irritant Characteristics:** Vapors are nonirritating to the eyes and throat.

5.9 **Liquid or Solid Irritant Characteristics:** No appreciable hazard. Practically harmless to the skin.

5.10 **Odor Threshold:** Data not available

6. FIRE HAZARDS

6.1 **Flash Point:** 250°F O.C.

6.2 **Flammable Limits in Air:** Data not available

6.3 **Fire Extinguishing Agents:** Alcohol foam, dry chemical, water fog

6.4 **Fire Extinguishing Agents Not to be Used:** Water or foam may cause frothing.

6.5 **Special Hazards of Combustion Products:** Not pertinent

6.6 **Behavior in Fire:** Not pertinent

6.7 **Ignition Temperature:** Data not available

6.8 **Electrical Hazard:** Not pertinent

6.9 **Burning Rate:** Data not available

7. CHEMICAL REACTIVITY

7.1 **Reactivity with Water:** No reaction

7.2 **Reactivity with Common Materials:** No reaction

7.3 **Stability During Transport:** Stable

7.4 **Neutralizing Agents for Acids and Caustics:** Not pertinent

7.5 **Polymerization:** Not pertinent

7.6 **Inhibitor of Polymerization:** Not pertinent

8. WATER POLLUTION

8.1 **Aquatic Toxicity:** Data not available

8.2 **Waterfowl Toxicity:** Data not available

8.3 **Biological Oxygen Demand (BOD):** Data not available

8.4 **Food Chain Concentration Potential:** None

9. SELECTED MANUFACTURERS

1. Exxon Chemical Co.
   Houston, Tex. 77001
2. Union Carbide Corp.
   Chemicals and Plastics Division
   270 Park Ave.
   New York, N. Y. 10017

10. SHIPPING INFORMATION

10.1 **Grades or Purity:** Mixed isomers: 99+%

10.2 **Storage Temperature:** Ambient

10.3 **Inert Atmosphere:** No requirement

10.4 **Venting:** Open (flame arrester)

11. HAZARD ASSESSMENT CODE

(See Hazard Assessment Handbook, CG 446-3)

A-T-U

12. HAZARD CLASSIFICATIONS

12.1 **Code of Federal Regulations:** Not listed

12.2 **NAS Hazard Rating for Bulk Water Transportation:**

Category	Rating
Fire	1
Health	
Vapor Irritant	0
Liquid or Solid Irritant	0
Poisons	0
Water Pollution	
Human Toxicity	0
Aquatic Toxicity	0
Aesthetic Effect	3
Reactivity	
Other Chemicals	2
Water	0
Self-Reaction	0

12.3 **NFPA Hazard Classifications:** Not listed

13. PHYSICAL AND CHEMICAL PROPERTIES

13.1 **Physical State at 15°C and 1 atm:** Liquid

13.2 **Molecular Weight:** 200.37

13.3 **Boiling Point at 1 atm:** 525°F = 274°C = 547°K

13.4 **Freezing Point:** Not pertinent

13.5 **Critical Temperature:** Not pertinent

13.6 **Critical Pressure:** Not pertinent

13.7 **Specific Gravity:** 0.846 at 20°C (liquid)

13.8 **Liquid Surface Tension:** (est.) 30 dynes/cm = 0.03 N/m at 20°C

13.9 **Liquid-Water Interfacial Tension:** (est.) 30 dynes/cm = 0.03 N/m at 20°C

13.10 **Vapor (Gas) Specific Gravity:** Not pertinent

13.11 **Ratio of Specific Heats of Vapor (Gas):** 1.027

13.12 **Latent Heat of Vaporization:** 120 Btu/lb = 64 cal/g = 2.7×10^5 J/kg

13.13 **Heat of Combustion:** (est.) −12,200 Btu/lb = −6,790 cal/g = -284×10^5 J/kg

13.14 **Heat of Decomposition:** Not pertinent

13.15 **Heat of Solution:** Not pertinent

13.16 **Heat of Polymerization:** Not pertinent

NOTES

TDC	# 1-TRIDECENE

Common Synonyms			
Undecylethylene	Watery liquid	Colorless	Mild pleasant odor
	Floats on water.		

Stop discharge if possible.
Call fire department.
Isolate and remove discharged material.
Notify local health and pollution control agencies.

Fire	Combustible. Extinguish with dry chemical, foam, or carbon dioxide. Water may be ineffective on fire. Cool exposed containers with water.
Exposure	LIQUID Irritating to eyes. IF IN EYES, hold eyelids open and flush with plenty of water.
Water Pollution	Effect of low concentrations on aquatic life is unknown. Fouling to shoreline. May be dangerous if it enters water intakes. Notify local health and wildlife officials. Notify operators of nearby water intakes.

1. RESPONSE TO DISCHARGE

(See Response Methods Handbook, CG 446-4)

Mechanical containment
Should be removed
Chemical and physical treatment

2. LABELS

No hazard label required by Code of Federal Regulations

3. CHEMICAL DESIGNATIONS

3.1 **Synonyms:** Undecylethylene
3.2 **Coast Guard Compatibility Classification:** Olefin
3.3 **Chemical Formula:** $CH_3(CH_2)_{10}CH{=}CH_2$
3.4 **IMCO/United Nations Numerical Designation:** Not listed

4. OBSERVABLE CHARACTERISTICS

4.1 **Physical State (as shipped):** Liquid
4.2 **Color:** Colorless
4.3 **Odor:** Mild, pleasant

5. HEALTH HAZARDS

5.1 **Personal Protective Equipment:** Goggles or face shield.
5.2 **Symptoms Following Exposure:** Liquid may irritate eyes.
5.3 **Treatment for Exposure:** EYES: flush with water for 15 min.
5.4 **Toxicity by Inhalation (Threshold Limit Value):** Not pertinent
5.5 **Short-Term Inhalation Limits:** Not pertinent
5.6 **Toxicity by Ingestion:** Data not available
5.7 **Late Toxicity:** Data not available
5.8 **Vapor (Gas) Irritant Characteristics:** Non-volatile
5.9 **Liquid or Solid Irritant Characteristics:** Data not available
5.10 **Odor Threshold:** Not pertinent

6. FIRE HAZARDS

6.1 **Flash Point:** 175°F (approx.)
6.2 **Flammable Limits in Air:** Data not available
6.3 **Fire Extinguishing Agents:** Dry chemical, foam, or carbon dioxide
6.4 **Fire Extinguishing Agents Not to be Used:** Water may be ineffective.
6.5 **Special Hazards of Combustion Products:** Not pertinent
6.6 **Behavior in Fire:** Not pertinent
6.7 **Ignition Temperature:** Data not available
6.8 **Electrical Hazard:** Not pertinent
6.9 **Burning Rate:** Data not available

7. CHEMICAL REACTIVITY

7.1 **Reactivity with Water:** No reaction
7.2 **Reactivity with Common Materials:** No reaction
7.3 **Stability During Transport:** Stable
7.4 **Neutralizing Agents for Acids and Caustics:** Not pertinent
7.5 **Polymerization:** Not pertinent
7.6 **Inhibitor of Polymerization:** Not pertinent

8. WATER POLLUTION

8.1 **Aquatic Toxicity:** Data not available
8.2 **Waterfowl Toxicity:** Data not available
8.3 **Biological Oxygen Demand (BOD):** Data not available
8.4 **Food Chain Concentration Potential:** None

9. SELECTED MANUFACTURERS

1. Exxon Chemical Co.
   Houston, Tex. 77001
2. The Humphrey Chemical Co.
   Devine Street
   North Haven, Conn. 06473
3. Phillips Petroleum Co.
   Bartlesville, Okla. 74004

10. SHIPPING INFORMATION

10.1 **Grades or Purity:** Technical: 95%
10.2 **Storage Temperature:** Ambient
10.3 **Inert Atmosphere:** No requirement
10.4 **Venting:** Open (flame arrester)

11. HAZARD ASSESSMENT CODE

(See Hazard Assessment Handbook, CG 446-3)

A-T-U

12. HAZARD CLASSIFICATIONS

12.1 **Code of Federal Regulations:** Combustible Liquid
12.2 **NAS Hazard Rating for Bulk Water Transportation:** Not listed
12.3 **NFPA Hazard Classifications:** Not listed

13. PHYSICAL AND CHEMICAL PROPERTIES

13.1 **Physical State at 15°C and 1 atm:** Liquid
13.2 **Molecular Weight:** 182.35
13.3 **Boiling Point at 1 atm:** 451°F = 233°C = 506°K
13.4 **Freezing Point:** −11°F = −24°C = 249°K
13.5 **Critical Temperature:** Not pertinent
13.6 **Critical Pressure:** Not pertinent
13.7 **Specific Gravity:** 0.765 at 20°C (liquid)
13.8 **Liquid Surface Tension:** 24.5 dynes/cm = 0.0245 N/m at 20°C
13.9 **Liquid-Water Interfacial Tension:** Data not available
13.10 **Vapor (Gas) Specific Gravity:** Not pertinent
13.11 **Ratio of Specific Heats of Vapor (Gas):** 1.029
13.12 **Latent Heat of Vaporization:** 110 Btu/lb = 59 cal/g = 2.5×10^5 J/kg
13.13 **Heat of Combustion:** −19,048 Btu/lb = −10,582 cal/g = -443.05×10^5 J/kg
13.14 **Heat of Decomposition:** Not pertinent
13.15 **Heat of Solution:** Not pertinent
13.16 **Heat of Polymerization:** Not pertinent

NOTES

TEA	TRIETHANOLAMINE

Common Synonyms			
Triethylolamine Trihydroxytriethylamine Tris(hydroxyethyl)amine	Oily liquid	Colorless	Mild ammonia odor
	Sinks and mixes with water. Freezing point is 71° F.		

Stop discharge if possible.
Call fire department.
Avoid contact with liquid.
Isolate and remove discharged material.
Notify local health and pollution control agencies.

Fire	Combustible. POISONOUS GASES MAY BE PRODUCED IN FIRE. Wear goggles and self-contained breathing apparatus. Extinguish with dry chemical, alcohol foam, or carbon dioxide. Water may be ineffective on fire. Cool exposed containers with water.
Exposure	CALL FOR MEDICAL AID. LIQUID Irritating to skin and eyes. Harmful if swallowed. Remove contaminated clothing and shoes. Flush affected areas with plenty of water. IF IN EYES, hold eyelids open and flush with plenty of water. IF SWALLOWED and victim is CONSCIOUS, have victim drink water or milk.
Water Pollution	Effect of low concentrations on aquatic life is unknown. May be dangerous if it enters water intakes. Notify local health and wildlife officials. Notify operators of nearby water intakes.

1. RESPONSE TO DISCHARGE

(See Response Methods Handbook, CG 446-4)

Disperse and flush

2. LABELS

No hazard label required by Code of Federal Regulations

3. CHEMICAL DESIGNATIONS

3.1 **Synonyms:** 2,2',2"-Nitrilotriethanol
Triethylolamine
Trihydroxytriethylamine
Tris (hydroxyethyl) amine

3.2 **Coast Guard Compatibility Classification:** Alkanolamine

3.3 **Chemical Formula:** $(HOCH_2CH_2)_3N$

3.4 **IMCO/United Nations Numerical Designation:** Not listed

4. OBSERVABLE CHARACTERISTICS

4.1 **Physical State (as shipped):** Liquid

4.2 **Color:** Colorless

4.3 **Odor:** Mild ammoniacal

5. HEALTH HAZARDS

5.1 **Personal Protective Equipment:** Goggles or face shield; rubber gloves and boots.

5.2 **Symptoms Following Exposure:** Liquid may irritate eyes and skin.

5.3 **Treatment for Exposure:** EYES: flush with water for at least 15 min.; call a doctor. SKIN: wipe off, wash with soap and water.

5.4 **Toxicity by Inhalation (Threshold Limit Value):** Not pertinent

5.5 **Short-Term Inhalation Limits:** Not pertinent

5.6 **Toxicity by Ingestion:** Grade 2; LD_{50} 0.5 to 5 g/kg (guinea pig).

5.7 **Late Toxicity:** Data not available

5.8 **Vapor (Gas) Irritant Characteristics:** Non-volatile

5.9 **Liquid or Solid Irritant Characteristics:** Minimum hazard. If spilled on clothing and allowed to remain, may cause smarting and reddening of the skin.

5.10 **Odor Threshold:** Not pertinent

6. FIRE HAZARDS

6.1 **Flash Point:** 355°F C.C.; 375°F O.C.

6.2 **Flammable Limits in Air:** Data not available

6.3 **Fire Extinguishing Agents:** Alcohol foam, dry chemical, or carbon dioxide

6.4 **Fire Extinguishing Agents Not to be Used:** Water or foam may cause frothing.

6.5 **Special Hazards of Combustion Products:** Not pertinent

6.6 **Behavior in Fire:** Not pertinent

6.7 **Ignition Temperature:** Data not available

6.8 **Electrical Hazard:** Not pertinent

6.9 **Burning Rate:** Data not available

7. CHEMICAL REACTIVITY

7.1 **Reactivity with Water:** No reaction

7.2 **Reactivity with Common Materials:** No reaction

7.3 **Stability During Transport:** Stable

7.4 **Neutralizing Agents for Acids and Caustics:** Dilute with water

7.5 **Polymerization:** Not pertinent

7.6 **Inhibitor of Polymerization:** Not pertinent

8. WATER POLLUTION

8.1 **Aquatic Toxicity:** > 100 ppm/48 hr/shrimp/LC_{50}//salt water

8.2 **Waterfowl Toxicity:** Data not available

8.3 **Biological Oxygen Demand (BOD):** 1%, 5 days; 0%, 5 days; 6.2% (theor.), 20 days

8.4 **Food Chain Concentration Potential:** None

9. SELECTED MANUFACTURERS

1. Dow Chemical Co.
   Midland, Mich. 48640
2. Jefferson Chemical Co., Inc.
   3336 Richmond Ave.
   Houston, Tex. 77052
3. Union Carbide Corp.
   Chemicals and Plastics Division
   270 Park Ave.
   New York, N. Y. 10017

10. SHIPPING INFORMATION

10.1 **Grades or Purity:** 85—99%

10.2 **Storage Temperature:** Ambient

10.3 **Inert Atmosphere:** No requirement

10.4 **Venting:** Open

11. HAZARD ASSESSMENT CODE

(See Hazard Assessment Handbook, CG 446-3)

A-P-Q

12. HAZARD CLASSIFICATIONS

12.1 **Code of Federal Regulations:** Not listed

12.2 **NAS Hazard Rating for Bulk Water Transportation:**

Category	Rating
Fire	1
Health	
Vapor Irritant	0
Liquid or Solid Irritant	1
Poisons	1
Water Pollution	
Human Toxicity	1
Aquatic Toxicity	1
Aesthetic Effect	2
Reactivity	
Other Chemicals	3
Water	0
Self-Reaction	0

12.3 **NFPA Hazard Classifications:**

Category	Classification*	
Health Hazard (Blue)	1	2
Flammability (Red)	1	1
Reactivity (Yellow)	1	1

*First column refers to non-fire situations.

13. PHYSICAL AND CHEMICAL PROPERTIES

13.1 **Physical State at 15°C and 1 atm:** Liquid

13.2 **Molecular Weight:** 149.19

13.3 **Boiling Point at 1 atm:** Decomposes

13.4 **Freezing Point:** 70.9°F = 21.6°C = 294.8°K

13.5 **Critical Temperature:** Not pertinent

13.6 **Critical Pressure:** Not pertinent

13.7 **Specific Gravity:** 1.13 at 20°C (liquid)

13.8 **Liquid Surface Tension:** Not pertinent

13.9 **Liquid-Water Interfacial Tension:** Not pertinent

13.10 **Vapor (Gas) Specific Gravity:** Not pertinent

13.11 **Ratio of Specific Heats of Vapor (Gas):** 1.036

13.12 **Latent Heat of Vaporization:** 176 Btu/lb = 97.8 cal/g = 4.10×10^5 J/kg

13.13 **Heat of Combustion:** -11,050 Btu/lb = -6140 cal/g = -257×10^5 J/kg

13.14 **Heat of Decomposition:** Not pertinent

13.15 **Heat of Solution:** (est.) –20 Btu/lb = –12 cal/g = -0.5×10^5 J/kg

13.16 **Heat of Polymerization:** Not pertinent

NOTES

REVISED 1978

TAL — TRIETHYLALUMINUM

Common Synonyms		
TEA Aluminum triethyl ATE	Liquid	Colorless
	IGNITES WHEN EXPOSED TO AIR. Flammable gas is produced on contact with water.	

Stop discharge if possible. Keep people away.
Call fire department.
Wear rubber overclothing (including gloves).
Isolate and remove discharged material.
Notify local health and pollution control agencies.

Fire

IGNITES WHEN EXPOSED TO AIR.
POISONOUS GASES MAY BE PRODUCED IN FIRE.
Wear goggles and self-contained breathing apparatus.
Extinguish with dry graphite, soda ash, or other inert powder.
DO NOT USE WATER, FOAM, CARBON DIOXIDE, DRY CHEMICALS OR VAPORIZING LIQUIDS ON FIRE.
DO NOT USE WATER ON ADJACENT FIRES.

Exposure

Call for medical aid.

LIQUID
Will burn skin and eyes.
Harmful if swallowed.
Remove contaminated clothing and shoes.
Flush affected areas with plenty of water.
IF IN EYES, hold eyelids open and flush with plenty of water.
IF SWALLOWED and victim is CONSCIOUS, have victim drink water or milk.
DO NOT INDUCE VOMITING.

Water Pollution

Effect of low concentrations on aquatic life is unknown.
May be dangerous if it enters water intakes.
Notify local health and wildlife officials.
Notify operators of nearby water intakes.

1. RESPONSE TO DISCHARGE

(See Response Methods Handbook, CG 446-4)

Issue warning—high flammability
Restrict access
Evacuate area
Disperse and flush with care

2. LABELS

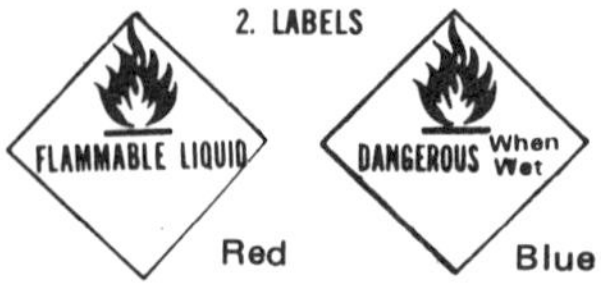

3. CHEMICAL DESIGNATIONS

3.1 **Synonyms:** ATE; Aluminum triethyl; TEA
3.2 **Coast Guard Compatibility Classification:** Not applicable
3.3 **Chemical Formula:** $(C_2H_5)_3Al$
3.4 **IMCO/United Nations Numerical Designation:** 4.2/1102

4. OBSERVABLE CHARACTERISTICS

4.1 **Physical State (as shipped):** Liquid
4.2 **Color:** Colorless
4.3 **Odor:** Not pertinent

5. HEALTH HAZARDS

5.1 **Personal Protective Equipment:** Full protective clothing, preferably of aluminized glass cloth; goggles; face shield; gloves. In case of fire, all-purpose canister or self-contained breathing apparatus.
5.2 **Symptoms Following Exposure:** Exposure to smoke from fire causes metal-fume fever (flu-like symptoms). Since liquid ignites spontaneously, contact with eyes or skin causes severe burns.
5.3 **Treatment for Exposure:** INHALATION: only fumes from fire need be considered; metal-fume fever is not critical, lasting less than 36 hrs. EYES: flush gently with copious quantities of water for 15 min. with lids held open; treat burns if fire occurred; get medical attention. SKIN: wash with water; treat burns caused by fire; get medical attention.
5.4 **Toxicity by Inhalation (Threshold Limit Value):** Not pertinent
5.5 **Short-Term Inhalation Limits:** Not pertinent
5.6 **Toxicity by Ingestion:** Not pertinent (ignites instantly in air, reacts vigorously with water)
5.7 **Late Toxicity:** Data not available
5.8 **Vapor (Gas) Irritant Characteristics:** Not pertinent
5.9 **Liquid or Solid Irritant Characteristics:** Severe skin irritant. Causes second- and third-degree burns on short contact and is very injurious to the eyes.
5.10 **Odor Threshold:** Not pertinent

6. FIRE HAZARDS

6.1 **Flash Point:** Ignites spontaneously in air at all temperatures
6.2 **Flammable Limits in Air:** Not pertinent
6.3 **Fire Extinguishing Agents:** Inert powders (sand, limestone), dry chemical
6.4 **Fire Extinguishing Agents Not to be Used:** Water, foam, halogenated extinguishing agents
6.5 **Special Hazards of Combustion Products:** Intense smoke may cause metal-fume fever.
6.6 **Behavior in Fire:** Dense smoke of aluminum oxide forms. Contact with water applied to adjacent fires causes violent reaction producing toxic and flammable gases.
6.7 **Ignition Temperature:** Not pertinent (self-ignites at ambient temperature)
6.8 **Electrical Hazard:** Not pertinent
6.9 **Burning Rate:** Not pertinent

7. CHEMICAL REACTIVITY

7.1 **Reactivity with Water:** Reacts violently to form flammable ethane gas
7.2 **Reactivity with Common Materials:** No significant reaction
7.3 **Stability During Transport:** Stable
7.4 **Neutralizing Agents for Acids and Caustics:** Not pertinent
7.5 **Polymerization:** Not pertinent
7.6 **Inhibitor of Polymerization:** Not pertinent

8. WATER POLLUTION

8.1 **Aquatic Toxicity:** Not pertinent
8.2 **Waterfowl Toxicity:** Not pertinent
8.3 **Biological Oxygen Demand (BOD):** Not pertinent
8.4 **Food Chain Concentration Potential:** None

9. SELECTED MANUFACTURERS

1. Ethyl Corporation
Industrial Chemicals Division
Ethyl Tower, 451 Florida
Baton Rouge, La. 70801
2. Texas Alkyls, Incorporated
P. O. Box 600
Deer Park, Texas 77536
3. Ventron Corporation
Alfa Products
P. O. Box 159
Beverly, Mass. 01915

10. SHIPPING INFORMATION

10.1 **Grades or Purity:** 92+%. 20% or less by weight in benzene, hexane, or heptane. Solutions are not pyrophoric.
10.2 **Storage Temperature:** Ambient
10.3 **Inert Atmosphere:** Inerted; dry nitrogen at 5 psig
10.4 **Venting:** Safety relief, with rupture disc

11. HAZARD ASSESSMENT CODE

(See Hazard Assessment Handbook, CG 446-3)

A-O-Z

12. HAZARD CLASSIFICATIONS

12.1 **Code of Federal Regulations:** Flammable liquid
12.2 **NAS Hazard Rating for Bulk Water Transportation:**

Category	Rating
Fire	4
Health	
Vapor Irritant	—
Liquid or Solid Irritant*	4
Poisons	3
Water Pollution	
Human Toxicity	0
Aquatic Toxicity	2
Aesthetic Effect	3
Reactivity	
Other Chemicals	4
Water	4
Self-Reaction	0

12.3 **NFPA Hazard Classifications:**

Category	Classification
Health Hazard (Blue)	3
Flammability (Red)	3
Reactivity (Yellow)	3
	W

*Up to 20% by wt. in hydrocarbon solution.

13. PHYSICAL AND CHEMICAL PROPERTIES

13.1 **Physical State at 15°C and 1 atm:** Liquid
13.2 **Molecular Weight:** 114.2
13.3 **Boiling Point at 1 atm:** 367.9°F = 186.6°C = 459.8°K
13.4 **Freezing Point:** −51°F = −46°C = 227°K
13.5 **Critical Temperature:** 761°F = 405°C = 678°K
13.6 **Critical Pressure:** 1,970 psia = 134 atm = 13.6 MN/m^2
13.7 **Specific Gravity:** 0.836 at 20°C (liquid)
13.8 **Liquid Surface Tension:** 26.1 dynes/cm = 0.0261 N/m at 28°C
13.9 **Liquid-Water Interfacial Tension:** Not pertinent
13.10 **Vapor (Gas) Specific Gravity:** Not pertinent
13.11 **Ratio of Specific Heats of Vapor (Gas):** Data not available
13.12 **Latent Heat of Vaporization:** 216 Btu/lb = 120 cal/g = 5.02×10^5 J/kg
13.13 **Heat of Combustion:** −18,364 Btu/lb = −10,202 cal/g = -426.85×10^5 J/kg
13.14 **Heat of Decomposition:** Not pertinent
13.15 **Heat of Solution:** −1,995 Btu/lb = −1,109 cal/g = -46.40×10^5 J/kg
13.16 **Heat of Polymerization:** Not pertinent

NOTES

TEN | TRIETHYLAMINE

Common Synonyms	Watery liquid	Colorless	Fishy odor
TEN	Floats on water. Flammable, irritating vapor is produced.		

Avoid contact with liquid and vapor. Keep people away.
Wear goggles, self-contained breathing apparatus, and rubber overclothing (including gloves).
Shut off ignition sources and call fire department.
Stop discharge if possible.
Stay upwind and use water spray to "knock down" vapor.
Isolate and remove discharged material.
Notify local health and pollution control agencies.

Fire

FLAMMABLE.
Flashback along vapor trail may occur.
Vapor may explode if ignited in an enclosed area.
Wear goggles, self-contained breathing apparatus, and rubber overclothing (including gloves).
Extinguish with dry chemical, alcohol foam, or carbon dioxide.
Water may be ineffective on fire.
Cool exposed containers with water.

Exposure

CALL FOR MEDICAL AID.

VAPOR
Irritating to eyes, nose and throat.
If inhaled, will cause coughing, difficult breathing, or loss of consciousness.
Move to fresh air.
If breathing has stopped, give artificial respiration.
If breathing is difficult, give oxygen.

LIQUID
Will burn skin and eyes.
Harmful if swallowed.
Remove contaminated clothing and shoes.
Flush affected areas with plenty of water.
IF IN EYES, hold eyelids open and flush with plenty of water.
IF SWALLOWED and victim is CONSCIOUS, have victim drink water or milk and have victim induce vomiting.
IF SWALLOWED and victim is UNCONSCIOUS OR HAVING CONVULSIONS, do nothing except keep victim warm.

Water Pollution

HARMFUL TO AQUATIC LIFE IN VERY LOW CONCENTRATIONS.
Fouling to shoreline.
May be dangerous if it enters water intakes.
Notify local health and wildlife officials.
Notify operators of nearby water intakes.

1. RESPONSE TO DISCHARGE

(See Response Methods Handbook, CG 446-4)

Issue warning—high flammability
Evacuate area
Disperse and flush

2. LABEL

Red

3. CHEMICAL DESIGNATIONS

3.1 **Synonyms:** TEN
3.2 **Coast Guard Compatibility Classification:** Aliphatic amine
3.3 **Chemical Formula:** $(C_2H_5)_3N$
3.4 **IMCO/United Nations Numerical Designation:** 3.2/1296

4. OBSERVABLE CHARACTERISTICS

4.1 **Physical State (as shipped):** Liquid
4.2 **Color:** Colorless
4.3 **Odor:** Fishy

5. HEALTH HAZARDS

5.1 **Personal Protective Equipment:** Air-supplied mask; goggles or face shield; rubber gloves.
5.2 **Symptoms Following Exposure:** Vapors irritate nose, throat, and lungs, causing coughing, choking, and difficult breathing. Contact with eyes causes severe burns. Clothing wet with chemical causes skin burns.
5.3 **Treatment for Exposure:** INHALATION: remove victim to fresh air; give artificial respiration if needed; call a doctor. INGESTION: induce vomiting if patient is conscious. EYES: flush with water for at least 30 min.; call a doctor. SKIN: flush with water for at least 30 min.
5.4 **Toxicity by Inhalation (Threshold Limit Value):** 25 ppm
5.5 **Short-Term Inhalation Limits:** 100 ppm for 30 min.
5.6 **Toxicity by Ingestion:** Grade 3; LD_{50} 50 to 500 mg/kg (rat—LD_{50} 460 mg/kg)
5.7 **Late Toxicity:** Data not available
5.8 **Vapor (Gas) Irritant Characteristics:** Vapors cause moderate irritation, such that personnel will find high concentrations unpleasant. The effect is temporary.
5.9 **Liquid or Solid Irritant Characteristics:** Causes smarting of the skin and first-degree burns on short exposure; may cause secondary burns on long exposure.
5.10 **Odor Threshold:** Data not available

6. FIRE HAZARDS

6.1 **Flash Point:** 20°F O.C.
6.2 **Flammable Limits in Air:** 1.2%–8.0%
6.3 **Fire Extinguishing Agents:** Carbon dioxide or dry chemical for small fires; alcohol foam for large fires.
6.4 **Fire Extinguishing Agents Not to be Used:** Water may be ineffective.
6.5 **Special Hazards of Combustion Products:** Not pertinent
6.6 **Behavior in Fire:** Not pertinent
6.7 **Ignition Temperature:** 842°F
6.8 **Electrical Hazard:** Not pertinent
6.9 **Burning Rate:** 6.2 mm/min.

7. CHEMICAL REACTIVITY

7.1 **Reactivity with Water:** No reaction
7.2 **Reactivity with Common Materials:** No reaction
7.3 **Stability During Transport:** Stable
7.4 **Neutralizing Agents for Acids and Caustics:** Dilute with water.
7.5 **Polymerization:** Not pertinent
7.6 **Inhibitor of Polymerization:** Not pertinent

8. WATER POLLUTION

8.1 **Aquatic Toxicity:** 80 ppm/24 hr/fish/lethal/fresh water
8.2 **Waterfowl Toxicity:** Data not available
8.3 **Biological Oxygen Demand (BOD):** Data not available
8.4 **Food Chain Concentration Potential:** None

9. SELECTED MANUFACTURERS

1. Air Products and Chemicals, Inc.
   Allentown, Pa. 18105
2. Pennwalt Corp.
   Chemical Division
   4655 Biddle Ave.
   Wyandotte, Mich. 48193
3. Union Carbide Corp.
   Chemicals and Plastics Division
   270 Park Ave.
   New York, N. Y. 10017

10. SHIPPING INFORMATION

10.1 **Grades or Purity:** 98.5+%
10.2 **Storage Temperature:** Ambient
10.3 **Inert Atmosphere:** No requirement
10.4 **Venting:** Open (flame arrester)

11. HAZARD ASSESSMENT CODE

(See Hazard Assessment Handbook, CG 446-3)

A-P-Q-R-S

12. HAZARD CLASSIFICATIONS

12.1 **Code of Federal Regulations:** Flammable liquid
12.2 **NAS Hazard Rating for Bulk Water Transportation:**

Category	Rating
Fire	3
Health	
Vapor Irritant	2
Liquid or Solid Irritant	2
Poisons	2
Water Pollution	
Human Toxicity	3
Aquatic Toxicity	3
Aesthetic Effect	2
Reactivity	
Other Chemicals	3
Water	0
Self-Reaction	0

12.3 **NFPA Hazard Classifications:** Not listed

13. PHYSICAL AND CHEMICAL PROPERTIES

13.1 **Physical State at 15°C and 1 atm:** Liquid
13.2 **Molecular Weight:** 101.19
13.3 **Boiling Point at 1 atm:** 193.1°F = 89.5°C = 362.7°K
13.4 **Freezing Point:** −174.5°F = −114.7°C = 158.5°K
13.5 **Critical Temperature:** 504°F = 262°C = 535°K
13.6 **Critical Pressure:** 440 psia = 30 atm = 3.0 MN/m²
13.7 **Specific Gravity:** 0.729 at 20°C (liquid)
13.8 **Liquid Surface Tension:** 20.7 dynes/cm = 0.0207 N/m at 20°C
13.9 **Liquid-Water Interfacial Tension:** Not pertinent
13.10 **Vapor (Gas) Specific Gravity:** 3.5
13.11 **Ratio of Specific Heats of Vapor (Gas):** 1.055
13.12 **Latent Heat of Vaporization:** 140 Btu/lb = 80 cal/g = 3.3×10^5 J/kg
13.13 **Heat of Combustion:** −17,040 Btu/lb = −9,466 cal/g = -396.3×10^5 J/kg
13.14 **Heat of Decomposition:** Not pertinent
13.15 **Heat of Solution:** −180 Btu/lb = −99 cal/g = -4.1×10^5 J/kg
13.16 **Heat of Polymerization:** Not pertinent

NOTES

REVISED 1978

TEB	TRIETHYLBENZENE

Common Synonyms 1,3,5-Triethylbenzene sym-Triethylbenzene	Liquid Colorless Weak chemical odor Floats on water.

Stop discharge if possible.
Call fire department.
Avoid contact with liquid.
Isolate and remove discharged material.
Notify local health and pollution control agencies.

Fire	Combustible. Extinguish with dry chemical, foam, or carbon dioxide. Water may be ineffective on fire. Cool exposed containers with water.
Exposure	CALL FOR MEDICAL AID. LIQUID Irritating to skin and eyes. Remove contaminated clothing and shoes. Flush affected areas with plenty of water. IF IN EYES, hold eyelids open and flush with plenty of water.
Water Pollution	Effect of low concentrations on aquatic life is unknown. Fouling to shoreline. May be dangerous if it enters water intakes. Notify local health and wildlife officials. Notify operators of nearby water intakes.

1. RESPONSE TO DISCHARGE
(See Response Methods Handbook, CG 446-4)

Mechanical containment
Should be removed
Chemical and physical treatment

2. LABELS

No hazard label required by Code of Federal Regulations

3. CHEMICAL DESIGNATIONS

3.1 **Synonyms:** 1,3,5-Triethylbenzene sym-Triethylbenzene
3.2 **Coast Guard Compatibility Classification:** Aromatic hydrocarbon
3.3 **Chemical Formula:** $C_6H_3(C_2H_5)_3$—1,3,5
3.4 **IMCO/United Nations Numerical Designation:** Not listed

4. OBSERVABLE CHARACTERISTICS

4.1 **Physical State (as shipped):** Liquid
4.2 **Color:** Colorless
4.3 **Odor:** Weak aromatic

5. HEALTH HAZARDS

5.1 **Personal Protective Equipment:** Goggles or face shield; rubber gloves.
5.2 **Symptoms Following Exposure:** Eye irritation by vapors or liquid. Central nervous system depression. Prolonged skin contact with liquid can cause dermatitis.
5.3 **Treatment for Exposure:** EYES: flush with water for at least 15 min.; call a doctor. SKIN: wipe off, wash with soap and water.
5.4 **Toxicity by Inhalation (Threshold Limit Value):** Data not available
5.5 **Short-Term Inhalation Limits:** Data not available
5.6 **Toxicity by Ingestion:** Data not available
5.7 **Late Toxicity:** Data not available
5.8 **Vapor (Gas) Irritant Characteristics:** Vapors cause a slight smarting of the eyes or respiratory system if present in high concentrations. The effect is temporary.
5.9 **Liquid or Solid Irritant Characteristics:** Minimum hazard. If spilled on clothing and allowed to remain, may cause smarting and reddening of the skin.
5.10 **Odor Threshold:** Data not available

6. FIRE HAZARDS

6.1 **Flash Point:** 181°F O.C.
6.2 **Flammable Limits in Air:** Data not available
6.3 **Fire Extinguishing Agents:** Dry chemical, foam, or carbon dioxide
6.4 **Fire Extinguishing Agents Not to be Used:** Water may be ineffective.
6.5 **Special Hazards of Combustion Products:** Not pertinent
6.6 **Behavior in Fire:** Not pertinent
6.7 **Ignition Temperature:** Data not available
6.8 **Electrical Hazard:** Not pertinent
6.9 **Burning Rate:** Data not available

7. CHEMICAL REACTIVITY

7.1 **Reactivity with Water:** No reaction
7.2 **Reactivity with Common Materials:** No reaction
7.3 **Stability During Transport:** Stable
7.4 **Neutralizing Agents for Acids and Caustics:** Not pertinent
7.5 **Polymerization:** Not pertinent
7.6 **Inhibitor of Polymerization:** Not pertinent

8. WATER POLLUTION

8.1 **Aquatic Toxicity:** Data not available
8.2 **Waterfowl Toxicity:** Data not available
8.3 **Biological Oxygen Demand (BOD):** Data not available
8.4 **Food Chain Concentration Potential:** Data not available

9. SELECTED MANUFACTURERS

No bulk shipments

10. SHIPPING INFORMATION

10.1 **Grades or Purity:** Data not available
10.2 **Storage Temperature:** Ambient
10.3 **Inert Atmosphere:** No requirement
10.4 **Venting:** Open (flame arrester)

11. HAZARD ASSESSMENT CODE
(See Hazard Assessment Handbook, CG 446-3)

A-T-U

12. HAZARD CLASSIFICATIONS

12.1 **Code of Federal Regulations:** Combustible Liquid
12.2 **NAS Hazard Rating for Bulk Water Transportation:**

Category	Rating
Fire	1
Health	
Vapor Irritant	1
Liquid or Solid Irritant	1
Poisons	1
Water Pollution	
Human Toxicity	1
Aquatic Toxicity	2
Aesthetic Effect	2
Reactivity	
Other Chemicals	1
Water	0
Self-Reaction	0

12.3 **NFPA Hazard Classifications:** Not listed

13. PHYSICAL AND CHEMICAL PROPERTIES

13.1 **Physical State at 15°C and 1 atm:** Liquid
13.2 **Molecular Weight:** 162.27
13.3 **Boiling Point at 1 atm:** 421°F = 216°C = 489°K
13.4 **Freezing Point:** Not pertinent
13.5 **Critical Temperature:** Not pertinent
13.6 **Critical Pressure:** Not pertinent
13.7 **Specific Gravity:** 0.861 at 20°C (liquid)
13.8 **Liquid Surface Tension:** Data not available
13.9 **Liquid-Water Interfacial Tension:** Data not available
13.10 **Vapor (Gas) Specific Gravity:** Not pertinent
13.11 **Ratio of Specific Heats of Vapor (Gas):** 1.039
13.12 **Latent Heat of Vaporization:** (est.) 120 Btu/lb = 65 cal/g = 2.7×10^5 J/kg
13.13 **Heat of Combustion:** Data not available
13.14 **Heat of Decomposition:** Not pertinent
13.15 **Heat of Solution:** Not pertinent
13.16 **Heat of Polymerization:** Not pertinent

NOTES

TEG	TRIETHYLENE GLYCOL

Common Synonyms			
Triglycol	Liquid	Colorless	Mild odor
	Sinks and mixes with water.		

Stop discharge if possible.
Call fire department.
Isolate and remove discharged material.
Notify local health and pollution control agencies.

Fire	Combustible. Extinguish with dry chemical, alcohol foam, or carbon dioxide. Water may be ineffective on fire. Cool exposed containers with water.
Exposure	Not harmful.
Water Pollution	Effect of low concentrations on aquatic life is unknown. May be dangerous if it enters water intakes. Notify local health and wildlife officials. Notify operators of nearby water intakes.

1. RESPONSE TO DISCHARGE

(See Response Methods Handbook, CG 446-4)

Disperse and flush

2. LABELS

No hazard label required by Code of Federal Regulations

3. CHEMICAL DESIGNATIONS

3.1 **Synonyms:** Di-beta-hydroxyethoxy-ethane; 2,2'-Ethylenedioxydiethanol; Ethylene glycol dihydroxydiethyl ether; TEG; Triglycol

3.2 **Coast Guard Compatibility Classification:** Glycol

3.3 **Chemical Formula:** $HO(CH_2CH_2O)_3CH$

3.4 **IMCO/United Nations Numerical Designation:** Not listed

4. OBSERVABLE CHARACTERISTICS

4.1 **Physical State (as shipped):** Liquid

4.2 **Color:** Colorless

4.3 **Odor:** Very mild, sweet

5. HEALTH HAZARDS

5.1 **Personal Protective Equipment:** Goggles; plastic gloves.

5.2 **Symptoms Following Exposure:** Vapor and liquid are unlikely to cause harm.

5.3 **Treatment for Exposure:** Flush eyes and skin with water.

5.4 **Toxicity by Inhalation (Threshold Limit Value):** Not pertinent

5.5 **Short-Term Inhalation Limits:** Not pertinent

5.6 **Toxicity by Ingestion:** Grade 1; LD_{50} 5 to 15 g/kg (guinea pig)

5.7 **Late Toxicity:** Data not available

5.8 **Vapor (Gas) Irritant Characteristics:** Vapors are non-irritating to the eyes and throat.

5.9 **Liquid or Solid Irritant Characteristics:** No appreciable hazard. Practically harmless to the skin.

5.10 **Odor Threshold:** Not pertinent

6. FIRE HAZARDS

6.1 **Flash Point:** 350°F C.C.; 330°F O.C.

6.2 **Flammable Limits in Air:** 0.9%—9.2%

6.3 **Fire Extinguishing Agents:** Alcohol foam, dry chemical, or carbon dioxide

6.4 **Fire Extinguishing Agents Not to be Used:** Water or foam may cause frothing.

6.5 **Special Hazards of Combustion Products:** Not pertinent

6.6 **Behavior in Fire:** Not pertinent

6.7 **Ignition Temperature:** 700°F

6.8 **Electrical Hazard:** Not pertinent

6.9 **Burning Rate:** 1.7 mm/min.

7. CHEMICAL REACTIVITY

7.1 **Reactivity with Water:** No reaction

7.2 **Reactivity with Common Materials:** No reaction

7.3 **Stability During Transport:** Stable

7.4 **Neutralizing Agents for Acids and Caustics:** Not pertinent

7.5 **Polymerization:** Not pertinent

7.6 **Inhibitor of Polymerization:** Not pertinent

8. WATER POLLUTION

8.1 **Aquatic Toxicity:** Data not available

8.2 **Waterfowl Toxicity:** Data not available

8.3 **Biological Oxygen Demand (BOD):** 50%, 5 days

8.4 **Food Chain Concentration Potential:** None

9. SELECTED MANUFACTURERS

1. Dow Chemical Co.
   Midland, Mich. 48640
2. Jefferson Chemical Co.
   3336 Richmond Ave.
   Houston, Tex. 77052
3. Union Carbide Corp.
   Chemicals and Plastics Division
   270 Park Ave.
   New York, N. Y. 10017

10. SHIPPING INFORMATION

10.1 **Grades or Purity:** High purity; air treatment; commercial

10.2 **Storage Temperature:** Ambient

10.3 **Inert Atmosphere:** No requirement

10.4 **Venting:** Open (flame arrester)

11. HAZARD ASSESSMENT CODE

(See Hazard Assessment Handbook, CG 446-3)

A-P-Q

12. HAZARD CLASSIFICATIONS

12.1 **Code of Federal Regulations:** Not listed

12.2 **NAS Hazard Rating for Bulk Water Transportation:**

Category	Rating
Fire	1
Health	
Vapor Irritant	0
Liquid or Solid Irritant	0
Poisons	0
Water Pollution	
Human Toxicity	0
Aquatic Toxicity	1
Aesthetic Effect	1
Reactivity	
Other Chemicals	2
Water	0
Self-Reaction	0

12.3 **NFPA Hazard Classifications:** Not listed

13. PHYSICAL AND CHEMICAL PROPERTIES

13.1 **Physical State at 15°C and 1 atm:** Liquid

13.2 **Molecular Weight:** 150.17

13.3 **Boiling Point at 1 atm:** 550°F = 288°C = 561°K

13.4 **Freezing Point:** 24.3°F = −4.3°C = 268.9°K

13.5 **Critical Temperature:** Not pertinent

13.6 **Critical Pressure:** Not pertinent

13.7 **Specific Gravity:** 1.125 at 20°C (liquid)

13.8 **Liquid Surface Tension:** 45.2 dynes/cm = 0.0452 N/m at 20°C

13.9 **Liquid-Water Interfacial Tension:** Not pertinent

13.10 **Vapor (Gas) Specific Gravity:** Not pertinent

13.11 **Ratio of Specific Heats of Vapor (Gas):** 1.039

13.12 **Latent Heat of Vaporization:** 180 Btu/lb = 99 cal/g = 4.1×10^5 J/kg

13.13 **Heat of Combustion:** −10,190 Btu/lb = −5,660 cal/g = -237.0×10^5 J/kg

13.14 **Heat of Decomposition:** Not pertinent

13.15 **Heat of Solution:** (est.) −13 Btu/lb = −7 cal/g = $-.3 \times 10^5$ J/kg

13.16 **Heat of Polymerization:** Not pertinent

NOTES

TET — TRIETHYLENETETRAMINE

Common Synonyms			
TETA N,N'-Bis (2-aminoethyl) ethylenediamine	Oily liquid	Light straw to amber	Ammonia odor
	Floats and mixes with water.		

Avoid contact with liquid. Keep people away.
Wear goggles, self-contained breathing apparatus, and rubber overclothing (including gloves).
Stop discharge if possible.
Call fire department.
Isolate and remove discharged material.
Notify local health and pollution control agencies.

Fire

Combustible.
Extinguish with dry chemical, alcohol foam, or carbon dioxide.
Water may be ineffective on fire.

Exposure

CALL FOR MEDICAL AID.

LIQUID
Will burn skin and eyes.
Harmful if swallowed.
Remove contaminated clothing and shoes.
Flush affected areas with plenty of water.
IF IN EYES, hold eyelids open and flush with plenty of water.
IF SWALLOWED and victim is CONSCIOUS, have victim drink water or milk.
DO NOT INDUCE VOMITING.

Water Pollution

Effect of low concentrations on aquatic life is unknown.
May be dangerous if it enters water intakes.
Notify local health and wildlife officials.
Notify operators of nearby water intakes.

1. RESPONSE TO DISCHARGE

(See Response Methods Handbook, CG 446-4)
Disperse and flush

2. LABELS

No hazard label required by Code of Federal Regulations

3. CHEMICAL DESIGNATIONS

3.1 **Synonyms:** N,N'-Bis(2-aminoethyl)-ethylenediamine
TETA
Trien

3.2 **Coast Guard Compatibility Classification:** Aliphatic amine

3.3 **Chemical Formula:** $NH_2(CH_2)_2NH(CH_2)_2NH(CH_2)_2NH_2$

3.4 **IMCO/United Nations Numerical Designation:** Not listed

4. OBSERVABLE CHARACTERISTICS

4.1 **Physical State (as shipped):** Liquid
4.2 **Color:** Light straw; amber
4.3 **Odor:** Ammoniacal

5. HEALTH HAZARDS

5.1 **Personal Protective Equipment:** Amine-type canister; goggles or face shield; rubber gloves.

5.2 **Symptoms Following Exposure:** Vapors from hot liquid can irritate eyes and upper respiratory system. Liquid burns eyes and skin. May cause sensitization of skin.

5.3 **Treatment for Exposure:** INHALATION: remove victim to fresh air. INGESTION: do NOT induce vomiting; give large quantities of water; give at least one ounce of vinegar in equal amount of water; get medical attention. SKIN: flush with plenty of water. EYES: flush with plenty of water for at least 15 min. and get medical attention.

5.4 **Toxicity by Inhalation (Threshold Limit Value):** Data not available

5.5 **Short-Term Inhalation Limits:** Data not available

5.6 **Toxicity by Ingestion:** Grade 2; LD_{50} 0.5 to 5 g/kg (rat)

5.7 **Late Toxicity:** May cause dermatitis, asthma and other allergic reactions in man

5.8 **Vapor (Gas) Irritant Characteristics:** Vapors cause moderate irritation such that personnel will find high concentrations unpleasant. The effect is temporary.

5.9 **Liquid or Solid Irritant Characteristics:** Causes smarting of the skin and first-degree burns on short exposure; may cause secondary burns on long exposure.

5.10 **Odor Threshold:** Data not available

6. FIRE HAZARDS

6.1 **Flash Point:** 275°F C.C.; 290°F O.C.
6.2 **Flammable Limits in Air:** Data not available
6.3 **Fire Extinguishing Agents:** Dry chemical, alcohol foam, carbon dioxide
6.4 **Fire Extinguishing Agents Not to be Used:** Water or foam may cause frothing.
6.5 **Special Hazards of Combustion Products:** Not pertinent
6.6 **Behavior in Fire:** Not pertinent
6.7 **Ignition Temperature:** 640°F
6.8 **Electrical Hazard:** Not pertinent
6.9 **Burning Rate:** Data not available

7. CHEMICAL REACTIVITY

7.1 **Reactivity with Water:** No reaction
7.2 **Reactivity with Common Materials:** No reaction
7.3 **Stability During Transport:** Stable
7.4 **Neutralizing Agents for Acids and Caustics:** After dilution with water, can be neutralized with acetic acid.
7.5 **Polymerization:** Not pertinent
7.6 **Inhibitor of Polymerization:** Not pertinent

8. WATER POLLUTION

8.1 **Aquatic Toxicity:** Data not available
8.2 **Waterfowl Toxicity:** Data not available
8.3 **Biological Oxygen Demand (BOD):** Data not available
8.4 **Food Chain Concentration Potential:** None

9. SELECTED MANUFACTURERS

1. Dow Chemical Co.
Midland, Mich. 48640
2. Jefferson Chemical Co., Inc.
3336 Richmond Ave.
Houston, Tex. 77052
3. Union Carbide Corp.
Chemicals and Plastics Division
270 Park Ave.
New York, N. Y. 10017

10. SHIPPING INFORMATION

10.1 **Grades or Purity:** 99+%
10.2 **Storage Temperature:** Ambient
10.3 **Inert Atmosphere:** No requirement
10.4 **Venting:** Open

11. HAZARD ASSESSMENT CODE

(See Hazard Assessment Handbook, CG 446-3)
A-P-Q

12. HAZARD CLASSIFICATIONS

12.1 **Code of Federal Regulations:** Not listed

12.2 **NAS Hazard Rating for Bulk Water Transportation:**

Category	Rating
Fire	1
Health	
Vapor Irritant	2
Liquid or Solid Irritant	2
Poisons	1
Water Pollution	
Human Toxicity	1
Aquatic Toxicity	1
Aesthetic Effect	3
Reactivity	
Other Chemicals	3
Water	0
Self-Reaction	0

12.3 **NFPA Hazard Classifications:** Not listed

13. PHYSICAL AND CHEMICAL PROPERTIES

13.1 **Physical State at 15°C and 1 atm:** Liquid
13.2 **Molecular Weight:** 146.24
13.3 **Boiling Point at 1 atm:** 531.3°F = 277.4°C = 550.6°K
13.4 **Freezing Point:** −31°F = −35°C = 238°K
13.5 **Critical Temperature:** 860°F = 460°C = 733°K
13.6 **Critical Pressure:** 470 psia = 32 atm = 3.2 MN/m²
13.7 **Specific Gravity:** 0.982 at 20°C (liquid)
13.8 **Liquid Surface Tension:** Not pertinent
13.9 **Liquid-Water Interfacial Tension:** Not pertinent
13.10 **Vapor (Gas) Specific Gravity:** Not pertinent
13.11 **Ratio of Specific Heats of Vapor (Gas):** 1.037
13.12 **Latent Heat of Vaporization:** Not pertinent
13.13 **Heat of Combustion:** (est.) −13,500 Btu/lb = −7,530 cal/g = -315×10^5 J/kg
13.14 **Heat of Decomposition:** Not pertinent
13.15 **Heat of Solution:** (est.) −13 Btu/lb = −7 cal/g = -0.3×10^5 J/kg
13.16 **Heat of Polymerization:** Not pertinent

NOTES

REVISED 1978

TFC	TRIFLUOROCHLOROETHYLENE

Common Synonyms		
Chlorotrifluoroethylene Trifluorovinyl chloride CTFE Genetron 1113 Trifluoromonochloroethylene Kel F monomer Plaskon monomer	Liquefied compressed gas Colorless Odorless or faint odor Sinks and boils in water. Flammable visible vapor cloud is produced.	

Shut off ignition sources. Call fire department.
Avoid contact with liquid. Keep people away.
Stop discharge if possible.
Stay upwind. Use water spray to "knock down" vapor.
Isolate and remove discharged material.
Notify local health and pollution control agencies.

Fire	FLAMMABLE. POISONOUS GASES ARE PRODUCED IN FIRE. Containers may explode in fire. Flashback along vapor trail may occur. Vapor may explode if ignited in an enclosed area. Wear goggles and self-contained breathing apparatus. Let fire burn. Stop flow of gas if possible. Cool exposed containers and protect men effecting shutoff with water.
Exposure	Call for medical aid. VAPOR If inhaled will cause dizziness, nausea, or vomiting. Move victim to fresh air. If breathing is difficult, give oxygen. LIQUID Will cause frostbite. Flush affected areas with plenty of water. DO NOT RUB AFFECTED AREAS.
Water Pollution	Not harmful to aquatic life.

1. RESPONSE TO DISCHARGE

(See Response Methods Handbook, CG 446-4)

Issue warning—high flammability, air contaminant
Restrict access
Evacuate area

2. LABEL

Red

3. CHEMICAL DESIGNATIONS

3.1 **Synonyms:** Chlorotrifluoroethylene; Kel F monomer; Plaskon monomer; Trifluoromonochloroethylene; Trifluorovinyl chloride

3.2 **Coast Guard Compatibility Classification:** Halogenated compounds (5)

3.3 **Chemical Formula:** $F_2C{=}CFCl$

3.4 **IMCO/United Nations Numerical Designation:** 2/1082

4. OBSERVABLE CHARACTERISTICS

4.1 **Physical State (as shipped):** Compressed liquefied gas

4.2 **Color:** Colorless

4.3 **Odor:** None; faint ethereal

5. HEALTH HAZARDS

5.1 **Personal Protective Equipment:** Self-contained breathing apparatus; goggles; rubber gloves.

5.2 **Symptoms Following Exposure:** Inhalation causes dizziness, nausea, vomiting; liver and kidney injury may develop after several hours and cause jaundice and necrosis of the kidney. Contact with liquid causes frostbite of eyes and possibly of skin.

5.3 **Treatment for Exposure:** *Call a physician after all exposures to this compound; it is more toxic than most of the closely related propellant gases.* INHALATION: remove victim to fresh air; enforce bed rest; administer oxygen for 30 min. of every hour for 6 hours, even if no symptoms appear. SKIN: if frostbite has occurred, apply warm water and treat burn.

5.4 **Toxicity by Inhalation (Threshold Limit Value):** 20 ppm (suggested)

5.5 **Short-Term Inhalation Limits:** Data not available

5.6 **Toxicity by Ingestion:** Not pertinent (TFC is a gas at normal temperatures)

5.7 **Late Toxicity:** Data not available

5.8 **Vapor (Gas) Irritant Characteristics:** Data not available

5.9 **Liquid or Solid Irritant Characteristics:** Data not available

5.10 **Odor Threshold:** Data not available

6. FIRE HAZARDS

6.1 **Flash Point:** Not pertinent (gas)

6.2 **Flammable Limits in Air:** 16%—34%

6.3 **Fire Extinguishing Agents:** Let fire burn; stop gas flow; cool containers with water.

6.4 **Fire Extinguishing Agents Not to be Used:** Not pertinent

6.5 **Special Hazards of Combustion Products:** Toxic hydrogen chloride and hydrogen fluoride gases are formed.

6.6 **Behavior in Fire:** Vapor is heavier than air and may travel considerable distance to a source of ignition and flash back. Containers may explode in a fire.

6.7 **Ignition Temperature:** Data not available

6.8 **Electrical Hazard:** Not pertinent

6.9 **Burning Rate:** Not pertinent

7. CHEMICAL REACTIVITY

7.1 **Reactivity with Water:** No reaction

7.2 **Reactivity with Common Materials:** No reaction

7.3 **Stability During Transport:** Stable

7.4 **Neutralizing Agents for Acids and Caustics:** Not pertinent

7.5 **Polymerization:** Can occur

7.6 **Inhibitor of Polymerization:** Terpenes; Tributylamine (1%)

8. WATER POLLUTION

8.1 **Aquatic Toxicity:** None

8.2 **Waterfowl Toxicity:** None

8.3 **Biological Oxygen Demand (BOD):** None

8.4 **Food Chain Concentration Potential:** None

9. SELECTED MANUFACTURERS

1. 3 M Company
3 M Center
Saint Paul, Minn. 55101

2. Matheson Gas Products Co.
East Rutherford, N. J. 07073

3. Allied Chemical Corporation
Plastics Division
Morristown, N. J. 07960

10. SHIPPING INFORMATION

10.1 **Grades or Purity:** Polymerization grade, 99.0+%

10.2 **Storage Temperature:** Ambient, but less than 150°F

10.3 **Inert Atmosphere:** Air must be excluded.

10.4 **Venting:** Safety relief

11. HAZARD ASSESSMENT CODE

(See Hazard Assessment Handbook, CG 446-3)

A-B-C-H-I-J

12. HAZARD CLASSIFICATIONS

12.1 **Code of Federal Regulations:** Flammable compressed gas

12.2 **NAS Hazard Rating for Bulk Water Transportation:** Not listed

12.3 **NFPA Hazard Classifications:**

Category	Classification
Health Hazard (Blue)	—
Flammability (Red)	4
Reactivity (Yellow)	0

13. PHYSICAL AND CHEMICAL PROPERTIES

13.1 **Physical State at 15°C and 1 atm:** Gas

13.2 **Molecular Weight:** 116.5

13.3 **Boiling Point at 1 atm:** $-18°F = -28°C = 245°K$

13.4 **Freezing Point:** Not pertinent

13.5 **Critical Temperature:** (est.) $223.2°F = 106.2°C = 379.4°K$

13.6 **Critical Pressure:** (est.) 592 psia = 40.2 atm = 4.08 MN/m^2

13.7 **Specific Gravity:** 1.307 at 20°C (liquid)

13.8 **Liquid Surface Tension:** (est.) 12 dynes/cm = 0.012 N/m at 20°C

13.9 **Liquid-Water Interfacial Tension:** Not pertinent

13.10 **Vapor (Gas) Specific Gravity:** 4.02

13.11 **Ratio of Specific Heats of Vapor (Gas):** Data not available

13.12 **Latent Heat of Vaporization:** 83 Btu/lb = 46 cal/g = 1.92×10^5 J/kg

13.13 **Heat of Combustion:** Data not available

13.14 **Heat of Decomposition:** Not pertinent

13.15 **Heat of Solution:** Not pertinent

13.16 **Heat of Polymerization:** Data not available

NOTES

TFR — TRIFLURALIN

Common Synonyms

alpha, alpha, alpha-trifluoro-2, 6-dinitro-N, N-dipropyl-p-toluidine
Treflan

Solid — Yellow-orange

Sinks in water.

AVOID CONTACT WITH SOLID AND DUST. KEEP PEOPLE AWAY.
Wear goggles and dust respirator.
Call fire department.
Isolate and remove discharged material.
Notify local health and pollution control agencies.

Fire

Combustible.
POISONOUS GASES ARE PRODUCED IN FIRE.
Wear goggles and self-contained breathing apparatus.
Extinguish with water, dry chemicals, foam, or carbon dioxide.

Exposure

CALL FOR MEDICAL AID.

DUST
POISONOUS IF INHALED
Move victim to fresh air.
If in eyes, hold eyelids open and flush with plenty of water.
If breathing is difficult, give oxygen.

SOLID
POISONOUS IF SWALLOWED.
Irritating to skin and eyes.
Remove contaminated clothing and shoes.
Flush affected areas with plenty of water.
IF IN EYES, hold eyelids open and flush with plenty of water.
IF SWALLOWED and victim is CONSCIOUS, have victim drink water or milk and have victim induce vomiting.
IF SWALLOWED and victim is UNCONSCIOUS OR HAVING CONVULSIONS, do nothing except keep victim warm.

Water Pollution

HARMFUL TO AQUATIC LIFE IN VERY LOW CONCENTRATIONS.
May be dangerous if it enters water intakes.
Notify local health and wildlife officials.
Notify operators of nearby water intakes.

1. RESPONSE TO DISCHARGE

(See Response Methods Handbook, CG 446-4)

Issue warning—poison, water contaminant
Restrict access
Should be removed
Chemical and physical treatment

2. LABELS

No hazard label required by Code of Federal Regulations

3. CHEMICAL DESIGNATIONS

3.1 **Synonyms:** 2, 6-Dinitro-N,N-dipropyl-4-trifluoromethylaniline; 2, 6-Dinitro-N, N-dipropyl-alpha,alpha,alpha-trifluoro-p-toluidine; N,N-Dipropyl-2, 6-dinitro-4-trifluoro-methylaniline; Treflan; alpha, alpha,alpha-Trifluoro-2, 6-dinitro-N,N-dipropyl-p-toluidine

3.2 **Coast Guard Compatibility Classification:** Not applicable

4. OBSERVABLE CHARACTERISTICS

4.1 **Physical State (as shipped):** Solid
4.2 **Color:** Yellow-orange
4.3 **Odor:** Data not available

5. HEALTH HAZARDS

5.1 **Personal Protective Equipment:** Protective gloves; goggles; dust mask
5.2 **Symptoms Following Exposure:** Dust may irritate eyes. No toxic symptoms have been observed during the manufacture and use of this compound.
5.3 **Treatment for Exposure:** INHALATION: move to fresh air. EYES: wash with running water; call physician if irritation persists. SKIN: wash with soap and running water. INGESTION: induce vomiting; call physician.
5.4 **Toxicity by Inhalation (Threshold Limit Value):** Data not available
5.5 **Short-Term Inhalation Limits:** Data not available
5.6 **Toxicity by Ingestion:** Grade 3; oral LD_{50} = 500 mg/kg (rat)
5.7 **Late Toxicity:** Data not available
5.8 **Vapor (Gas) Irritant Characteristics:** Data not available
5.9 **Liquid or Solid Irritant Characteristics:** Data not available
5.10 **Odor Threshold:** Data not available

6. FIRE HAZARDS

6.1 **Flash Point:** >185°F O.C.
6.2 **Flammable Limits in Air:** Not pertinent
6.3 **Fire Extinguishing Agents:** Water, foam, dry chemical, carbon dioxide
6.4 **Fire Extinguishing Agents Not to be Used:** Not pertinent
6.5 **Special Hazards of Combustion Products:** Toxic and irritating hydrogen fluoride gas may be formed in fires.
6.6 **Behavior in Fire:** Not pertinent
6.7 **Ignition Temperature:** Not pertinent
6.8 **Electrical Hazard:** Not pertinent
6.9 **Burning Rate:** Not pertinent

7. CHEMICAL REACTIVITY

7.1 **Reactivity with Water:** No reaction
7.2 **Reactivity with Common Materials:** No reaction
7.3 **Stability During Transport:** Stable
7.4 **Neutralizing Agents for Acids and Caustics:** Not pertinent
7.5 **Polymerization:** Not pertinent
7.6 **Inhibitor of Polymerization:** Not pertinent

8. WATER POLLUTION

8.1 **Aquatic Toxicity:**
11 μg/l /48 hr/rainbow trout/TL_m/fresh water
0.58 ppm/48 hr/bluegill/LC_{50}/fresh water
8.2 **Waterfowl Toxicity:** Data not available
8.3 **Biological Oxygen Demand (BOD):** Data not available
8.4 **Food Chain Concentration Potential:** Data not available

9. SELECTED MANUFACTURERS

Elanco Products Company
P. O. Box 1750
Indianapolis, Ind. 46206

10. SHIPPING INFORMATION

10.1 **Grades or Purity:** Technical: 95%. Emulsifiable concentrate in flammable solvents.
10.2 **Storage Temperature:** Ambient
10.3 **Inert Atmosphere:** No requirement
10.4 **Venting:** Pressure-vacuum

11. HAZARD ASSESSMENT CODE

(See Hazard Assessment Handbook, CG 446-3)

II

12. HAZARD CLASSIFICATIONS

12.1 **Code of Federal Regulations:** Not Listed
12.2 **NAS Hazard Rating for Bulk Water Transportation:** Not listed
12.3 **NFPA Hazard Classifications:** Not listed

13. PHYSICAL AND CHEMICAL PROPERTIES

13.1 **Physical State at 15°C and 1 atm:** Solid
13.2 **Molecular Weight:** 335.3
13.3 **Boiling Point at 1 atm:** Not pertinent (decomposes)
13.4 **Freezing Point:** 108°F = 42°C = 315°K
13.5 **Critical Temperature:** Not pertinent
13.6 **Critical Pressure:** Not pertinent
13.7 **Specific Gravity:** 1.294 at 25°C (solid)
13.8 **Liquid Surface Tension:** Not pertinent
13.9 **Liquid-Water Interfacial Tension:** Not pertinent
13.10 **Vapor (Gas) Specific Gravity:** Not pertinent
13.11 **Ratio of Specific Heats of Vapor (Gas):** Not pertinent
13.12 **Latent Heat of Vaporization:** Not pertinent
13.13 **Heat of Combustion:** (est.) −9,040 Btu/lb = −5,020 cal/g = -210×10^5 J/kg
13.14 **Heat of Decomposition:** Not pertinent
13.15 **Heat of Solution:** Not pertinent
13.16 **Heat of Polymerization:** Not pertinent

3. CHEMICAL DESIGNATIONS (Cont'd.)

3.3 **Chemical Formula:** $C_{13}H_{16}F_3N_3O_4$
3.4 **IMCO/United Nations Numerical Designation:** 6.1/1609

TIA — TRIISOBUTYLALUMINUM

Common Synonyms	
TIBA TIBAL Aluminum triisobutyl	Liquid Colorless IGNITES WHEN EXPOSED TO AIR. Flammable gas is produced on contact with water.

Call fire department.
Avoid contact with liquid. Keep people away.
Wear rubber overclothing (including gloves).
Stop discharge if possible.
Isolate and remove discharged material.
Notify local health and pollution control agencies.

Fire	IGNITES WHEN EXPOSED TO AIR. POISONOUS GASES MAY BE PRODUCED IN FIRE. Wear goggles and self-contained breathing apparatus. Extinguish with dry graphite, soda ash, or other inert powder. DO NOT USE WATER, FOAM, CARBON DIOXIDE, DRY CHEMICALS OR VAPORIZING LIQUIDS ON FIRE.
Exposure	Call for medical aid. LIQUID Will burn skin and eyes. Harmful if swallowed. Remove contaminated clothing and shoes. Flush affected areas with plenty of water. IF IN EYES, hold eyelids open and flush with plenty of water. IF SWALLOWED and victim is CONSCIOUS, have victim drink water or milk. DO NOT INDUCE VOMITING.
Water Pollution	Effect of low concentrations on aquatic life is unknown. May be dangerous if it enters water intakes. Notify local health and wildlife officials. Notify operators of nearby water intakes.

1. RESPONSE TO DISCHARGE

(See Response Methods Handbook, CG 446-4)

Issue warning—high flammability
Restrict access
Evacuate area
Disperse and flush with care

2. LABELS

Red

Blue

3. CHEMICAL DESIGNATIONS

3.1 **Synonyms:** Aluminum triisobutyl; TIBA; TIBAL
3.2 **Coast Guard Compatibility Classification:** Not applicable
3.3 **Chemical Formula:** $(iso\text{-}C_4H_9)_3Al$
3.4 **IMCO/United Nations Numerical Designation:** 4.2/1930

4. OBSERVABLE CHARACTERISTICS

4.1 **Physical State (as shipped):** Liquid
4.2 **Color:** Colorless
4.3 **Odor:** Not pertinent

5. HEALTH HAZARDS

5.1 **Personal Protective Equipment:** Full protective clothing, preferably of aluminized glass cloth; goggles; face shield; gloves. In case of fire, all-purpose canister or self-contained breathing apparatus.
5.2 **Symptoms Following Exposure:** Inhalation of smoke from fire causes metal-fume fever (flu-like symptoms). Contact with liquid can cause severe burns of eyes and skin because of spontaneous ignition.
5.3 **Treatment for Exposure:** INHALATION: only fumes from fire need be considered; metal-fume fever lasts less than 36 hrs. and is not critical. EYES: flush gently with copious quantities of water for 15 min. with lids open; treat burns, if fire occurred; get medical attention. SKIN: wash with water; treat burns caused by fire; get medical attention. INGESTION: not pertinent
5.4 **Toxicity by Inhalation (Threshold Limit Value):** Not pertinent
5.5 **Short-Term Inhalation Limits:** Not pertinent
5.6 **Toxicity by Ingestion:** Not pertinent
5.7 **Late Toxicity:** Metal fume fever may develop following exposure to smoke from fire.
5.8 **Vapor (Gas) Irritant Characteristics:** Not pertinent
5.9 **Liquid or Solid Irritant Characteristics:** Severe skin irritant. Causes second- and third-degree burns on short contact and is very injurious to the eyes.
5.10 **Odor Threshold:** Not pertinent

6. FIRE HAZARDS

6.1 **Flash Point:** Not pertinent (ignites spontaneously)
6.2 **Flammable Limits in Air:** Not pertinent
6.3 **Fire Extinguishing Agents:** Inert powder (e.g., sand, limestone), dry chemical
6.4 **Fire Extinguishing Agents Not to be Used:** Water, foam, halogenated extinguishing agents
6.5 **Special Hazards of Combustion Products:** Dense smoke may cause metal-fume fever.
6.6 **Behavior in Fire:** Dense smoke of aluminum oxide forms.
6.7 **Ignition Temperature:** Ignites spontaneously under ambient conditions
6.8 **Electrical Hazard:** Not pertinent
6.9 **Burning Rate:** Not pertinent

7. CHEMICAL REACTIVITY

7.1 **Reactivity with Water:** Reacts violently to form flammable hydrocarbon gases
7.2 **Reactivity with Common Materials:** Not compatible with silicone rubber or urethane rubber
7.3 **Stability During Transport:** Stable
7.4 **Neutralizing Agents for Acids and Caustics:** Not pertinent
7.5 **Polymerization:** Not pertinent
7.6 **Inhibitor of Polymerization:** Not pertinent

8. WATER POLLUTION

8.1 **Aquatic Toxicity:** Not pertinent
8.2 **Waterfowl Toxicity:** Not pertinent
8.3 **Biological Oxygen Demand (BOD):** None
8.4 **Food Chain Concentration Potential:** None

9. SELECTED MANUFACTURERS

1. Ethyl Corporation
   Industrial Chemicals Division
   Ethyl Tower, 451 Florida
   Baton Rouge, La. 70801
2. Texas Alkyls, Incorporated
   P. O. Box 600
   Deer Park, Texas 77536
3. Ventron Corporation
   Alfa Products
   P. O. Box 159
   Beverly, Mass. 01915

10. SHIPPING INFORMATION

10.1 **Grades or Purity:** Technical, 95+%; 20% or less by weight in benzene, hexane, or heptane (solutions are not pyrophoric); electronic grade.
10.2 **Storage Temperature:** Ambient
10.3 **Inert Atmosphere:** Inerted; dry nitrogen at 5 psig.
10.4 **Venting:** Safety relief, with rupture disc

11. HAZARD ASSESSMENT CODE

(See Hazard Assessment Handbook, CG 446-3)

A-O-Z

12. HAZARD CLASSIFICATIONS

12.1 **Code of Federal Regulations:** Flammable liquid
12.2 **NAS Hazard Rating for Bulk Water Transportation:** Not listed
12.3 **NFPA Hazard Classifications:**

Category	Classification*
Health Hazard (Blue)	3
Flammability (Red)	3
Reactivity (Yellow)	3
	~~W~~

*Refers to 20% or less by weight in hydrocarbon solution.

13. PHYSICAL AND CHEMICAL PROPERTIES

13.1 **Physical State at 15°C and 1 atm:** Liquid
13.2 **Molecular Weight:** 198.3
13.3 **Boiling Point at 1 atm:** 414°F = 212°C = 485°K
13.4 **Freezing Point:** 33.8°F = 1.0°C = 274.2°K
13.5 **Critical Temperature:** Not pertinent
13.6 **Critical Pressure:** Not pertinent
13.7 **Specific Gravity:** 0.788 at 20°C (liquid)
13.8 **Liquid Surface Tension:** (est.) 24 dynes/cm = 0.024 N/m at 20°C
13.9 **Liquid-Water Interfacial Tension:** Not pertinent
13.10 **Vapor (Gas) Specific Gravity:** Not pertinent
13.11 **Ratio of Specific Heats of Vapor (Gas):** Not pertinent
13.12 **Latent Heat of Vaporization:** 101 Btu/lb = 56 cal/g = 2.3×10^5 J/kg
13.13 **Heat of Combustion:** −18,423 Btu/lb = −10,235 cal/g = -428.23×10^5 J/kg
13.14 **Heat of Decomposition:** Not pertinent
13.15 **Heat of Solution:** Not pertinent
13.16 **Heat of Polymerization:** Not pertinent

NOTES

TMA	TRIMETHYLAMINE

Common Synonyms	Liquefied compressed gas — Colorless — Fish or ammonia odor Floats and mixes and boils on water. Poisonous, flammable visible vapor cloud is produced.

Avoid contact with liquid and vapor. Keep people away.
Wear goggles and self-contained breathing apparatus.
Shut off ignition sources and call fire department.
Stop discharge if possible.
Stay upwind and use water spray to "knock down" vapor.
Notify local health and pollution control agencies.

Fire	FLAMMABLE. Flashback along vapor trail may occur. Vapor may explode if ignited in an enclosed area. Wear goggles and self-contained breathing apparatus. Stop flow of gas if possible. Cool exposed containers and protect men effecting shutoff with water. Let gas fires burn. Extinguish water solution fires with water spray, dry chemical, alcohol foam, or carbon dioxide.
Exposure	CALL FOR MEDICAL AID. VAPOR POISONOUS IF INHALED. Irritating to eyes, nose, and throat. Move to fresh air. IF IN EYES, hold eyelids open and flush with plenty of water. If breathing has stopped, give artificial respiration. If breathing is difficult, give oxygen. LIQUID Will burn skin and eyes. Harmful if swallowed. Remove contaminated clothing and shoes. Flush affected areas with plenty of water. IF IN EYES, hold eyelids open and flush with plenty of water. IF SWALLOWED and victim is CONSCIOUS, have victim drink water or milk.
Water Pollution	Effect of low concentrations on aquatic life is unknown. May be dangerous if it enters water intakes. Notify local health and wildlife officials. Notify operators of nearby water intakes.

1. RESPONSE TO DISCHARGE

(See Response Methods Handbook, CG 446-4)

Issue warning—high flammability, air contaminant
Restrict access
Evacuate area

2. LABEL

Red

3. CHEMICAL DESIGNATIONS

3.1 **Synonyms:** No common synonyms
3.2 **Coast Guard Compatibility Classification:** Aliphatic amine
3.3 **Chemical Formula:** $(CH_3)_3N$
3.4 **IMCO/United Nations Numerical Designation:** 2.0/1083

4. OBSERVABLE CHARACTERISTICS

4.1 **Physical State (as shipped):** Liquefied compressed gas
4.2 **Color:** Colorless
4.3 **Odor:** Ammoniacal

5. HEALTH HAZARDS

5.1 **Personal Protective Equipment:** Vapor-proof goggles and face shield; rubber gloves; air-supplied mask
5.2 **Symptoms Following Exposure:** Vapor irritates eyes, nose, and throat; high concentrations can cause pulmonary edema. Liquid burns eyes and skin.
5.3 **Treatment for Exposure:** INHALATION: remove victim to fresh air and call a doctor; give artificial respiration and oxygen if needed. EYES: flush with water for at least 15 min.; consult an eye doctor. SKIN: flush with water; wash with soap and water.
5.4 **Toxicity by Inhalation (Threshold Limit Value):** Data not available
5.5 **Short-Term Inhalation Limits:** Data not available
5.6 **Toxicity by Ingestion:** Data not available
5.7 **Late Toxicity:** Data not available
5.8 **Vapor (Gas) Irritant Characteristics:** Vapor is moderately irritating such that personnel will not usually tolerate moderate or high concentrations.
5.9 **Liquid or Solid Irritant Characteristics:** Causes smarting of the skin and first-degree burns on short exposure; may cause secondary burns on long exposure.
5.10 **Odor Threshold:** Less than 100 ppm

6. FIRE HAZARDS

6.1 **Flash Point:** Not pertinent (gas)
6.2 **Flammable Limits in Air:** 2.0%–11.6%
6.3 **Fire Extinguishing Agents:** Stop flow of gas. Use water, alcohol foam, dry chemical, or carbon dioxide on water solution fires.
6.4 **Fire Extinguishing Agents Not to be Used:** Not pertinent
6.5 **Special Hazards of Combustion Products:** Not pertinent
6.6 **Behavior in Fire:** Vapor is heavier than air and may travel a considerable distance to a source of ignition and flash back.
6.7 **Ignition Temperature:** 374°F
6.8 **Electrical Hazard:** Not pertinent
6.9 **Burning Rate:** 8 mm/min.

7. CHEMICAL REACTIVITY

7.1 **Reactivity with Water:** No reaction
7.2 **Reactivity with Common Materials:** No reaction
7.3 **Stability During Transport:** Stable
7.4 **Neutralizing Agents for Acids and Caustics:** Although water solutions may be neutralized with acetic acid, simple evaporation will remove all of the compound.
7.5 **Polymerization:** Not pertinent
7.6 **Inhibitor of Polymerization:** Not pertinent

8. WATER POLLUTION

8.1 **Aquatic Toxicity:** Data not available
8.2 **Waterfowl Toxicity:** Data not available
8.3 **Biological Oxygen Demand (BOD):** Data not available
8.4 **Food Chain Concentration Potential:** None

9. SELECTED MANUFACTURERS

1. Air Products and Chemicals Inc. Allentown, Pa. 18105
2. Commercial Solvents Corp. 245 Park Ave. New York, N. Y. 10017
3. Rohm and Haas Co. Independence Mall West Philadelphia, Pa. 19105

10. SHIPPING INFORMATION

10.1 **Grades or Purity:** Anhydrous, 98.5+%; also shipped as 25–30% solution in water.
10.2 **Storage Temperature:** Ambient
10.3 **Inert Atmosphere:** No requirement
10.4 **Venting:** Safety relief

11. HAZARD ASSESSMENT CODE

(See Hazard Assessment Handbook, CG 446-3)

A-B-C-K-L-M-N

12. HAZARD CLASSIFICATIONS

12.1 **Code of Federal Regulations:** Flammable compressed gas

12.2 **NAS Hazard Rating for Bulk Water Transportation:**

Category	Rating
Fire	4
Health	
Vapor Irritant	3
Liquid or Solid Irritant	2
Poisons	3
Water Pollution	
Human Toxicity	2
Aquatic Toxicity	3
Aesthetic Effect	2
Reactivity	
Other Chemicals	3
Water	0
Self-Reaction	0

12.3 **NFPA Hazard Classifications:**

Category	Classification
Health Hazard (Blue)	3
Flammability (Red)	4
Reactivity (Yellow)	0

13. PHYSICAL AND CHEMICAL PROPERTIES*

13.1 **Physical State at 15°C and 1 atm:** Gas
13.2 **Molecular Weight:** 59.11
13.3 **Boiling Point at 1 atm:** 37.2°F = 2.9°C = 276.1°K
13.4 **Freezing Point:** −178.8°F = −117.1°C = 156.1°K
13.5 **Critical Temperature:** 320.2°F = 160.1°C = 433.3°K
13.6 **Critical Pressure:** 591 psia = 40.2 atm = 4.07 MN/m^2
13.7 **Specific Gravity:** 0.633 at 20°C (liquid)
13.8 **Liquid Surface Tension:** 17.4 dynes/cm = 0.0174 N/m at −4°C
13.9 **Liquid-Water Interfacial Tension:** Not pertinent
13.10 **Vapor (Gas) Specific Gravity:** 2.0
13.11 **Ratio of Specific Heats of Vapor (Gas):** 1.139
13.12 **Latent Heat of Vaporization:** 174 Btu/lb = 96.5 cal/g = 4.04 × 10^5 J/kg
13.13 **Heat of Combustion:** −17,660 Btu/lb = −9,810 cal/g = −410.7 × 10^5 J/kg
13.14 **Heat of Decomposition:** Not pertinent
13.15 **Heat of Solution:** −385 Btu/lb = −214 cal/g = −8.96 × 10^5 J/kg
13.16 **Heat of Polymerization:** Not pertinent

*Physical properties apply to anhydrous material.

NOTES

REVISED 1978

TMC — TRIMETHYLCHLOROSILANE

Common Synonyms			
Chlorotrimethylsilane Trimethylsilyl chloride	Liquid	Colorless	Sharp irritating odor

Reacts violently with water. Irritating gas is produced on contact with water.

Shut off ignition sources. Call fire department.
Avoid contact with liquid. Keep people away.
Stop discharge if possible.
Isolate and remove discharged material.
Notify local health and pollution control agencies.

Fire

FLAMMABLE.
POISONOUS GASES MAY BE PRODUCED IN FIRE.
Containers may explode in fire.
Flashback along vapor trail may occur.
Vapor may explode if ignited in an enclosed area.
Extinguish with dry chemicals or carbon dioxide.
DO NOT USE WATER OR FOAM ON FIRE.
Cool exposed containers with water.

Exposure

Call for medical aid.

VAPOR
Irritating to eyes, nose and throat.
Harmful if inhaled.
Move victim to fresh air.
If breathing has stopped, give artificial respiration.
If breathing is difficult, give oxygen.

LIQUID
Will burn skin and eyes.
Harmful if swallowed.
Remove contaminated clothing and shoes.
Flush affected areas with plenty of water.
IF IN EYES, hold eyelids open and flush with plenty of water.
IF SWALLOWED and victim is CONSCIOUS, have victim drink water or milk.
DO NOT INDUCE VOMITING.

Water Pollution

Effect of low concentrations on aquatic life is unknown.
May be dangerous if it enters water intakes.
Notify local health and wildlife officials.
Notify operators of nearby water intakes.

1. RESPONSE TO DISCHARGE

(See Response Methods Handbook, CG 446-4)

Issue warning - high flammability, air contaminant, corrosive
Restrict access
Evacuate area
Disperse and flush with care

2. LABEL

Red

3. CHEMICAL DESIGNATIONS

3.1 **Synonyms:** Chlorotrimethylsilane; Trimethylsilyl chloride
3.2 **Coast Guard Compatibility Classification:** Not applicable
3.3 **Chemical Formula:** $(CH_3)_3SiCl$
3.4 **IMCO/United Nations Numerical Designation:** 3.2/1298

4. OBSERVABLE CHARACTERISTICS

4.1 **Physical State (as shipped):** Liquid
4.2 **Color:** Colorless
4.3 **Odor:** Sharp, hydrochloric acid-like; acrid

5. HEALTH HAZARDS

5.1 **Personal Protective Equipment:** Acid-vapor-type respiratory protection; rubber gloves; chemical worker's goggles; other protective equipment as necessary to protect skin and eyes.
5.2 **Symptoms Following Exposure:** Inhalation of vapor irritates mucous membranes. Contact of liquid with eyes or skin causes severe burns. Ingestion causes severe burns of mouth and stomach.
5.3 **Treatment for Exposure:** *Get medical attention following all exposures to this compound.* INHALATION: remove victim from exposure; if breathing is difficult or stopped, give artificial respiration. EYES: flush with water for 15 min. SKIN: flush with water. INGESTION: do NOT induce vomiting; give large amount of water.
5.4 **Toxicity by Inhalation (Threshold Limit Value):** Data not available
5.5 **Short-Term Inhalation Limits:** Data not available
5.6 **Toxicity by Ingestion:** Grade 3; LD_{50} 0.5 to 5 g/kg
5.7 **Late Toxicity:** Data not available
5.8 **Vapor (Gas) Irritant Characteristics:** Vapors cause severe irritation of eyes and throat and can cause eye and lung injury. They cannot be tolerated even at low concentrations.
5.9 **Liquid or Solid Irritant Characteristics:** Severe skin irritant. Causes second- and third-degree burns on short contact and is very injurious to the eyes.
5.10 **Odor Threshold:** Data not available

6. FIRE HAZARDS

6.1 **Flash Point:** 0°F O.C.
6.2 **Flammable Limits in Air:** 1.8% (LFL)
6.3 **Fire Extinguishing Agents:** Dry chemical
6.4 **Fire Extinguishing Agents Not to be Used:** Water, foam
6.5 **Special Hazards of Combustion Products:** Toxic and irritating hydrogen chloride and phosgene may be formed in fires.
6.6 **Behavior in Fire:** Difficult to extinguish; re-ignition may occur. Contact with water applied to adjacent fires produces irritating hydrogen chloride gas.
6.7 **Ignition Temperature:** 743°F
6.8 **Electrical Hazard:** Data not available
6.9 **Burning Rate:** 5.3 mm/min.

7. CHEMICAL REACTIVITY

7.1 **Reactivity with Water:** Reacts vigorously, evolving hydrogen chloride (hydrochloric acid).
7.2 **Reactivity with Common Materials:** Reacts with surface moisture to evolve hydrogen chloride, which will corrode common metals and form flammable hydrogen gas.
7.3 **Stability During Transport:** Stable
7.4 **Neutralizing Agents for Acids and Caustics:** Flush with water, rinse with sodium bicarbonate or lime solution.
7.5 **Polymerization:** Not pertinent
7.6 **Inhibitor of Polymerization:** Not pertinent

8. WATER POLLUTION

8.1 **Aquatic Toxicity:** Data not available
8.2 **Waterfowl Toxicity:** Data not available
8.3 **Biological Oxygen Demand (BOD):** Data not available
8.4 **Food Chain Concentration Potential:** None

9. SELECTED MANUFACTURERS

1. Dow Corning Corporation
   P. O. Box 592
   Midland, Mich. 48640
2. Aldrich Chemical Co.
   940 West St. Paul Avenue
   Milwaukee, Wis. 53233
3. Pierce Chemical Co.
   P. O. Box 117
   Rockford, Ill. 61105

10. SHIPPING INFORMATION

10.1 **Grades or Purity:** 98+%
10.2 **Storage Temperature:** Ambient
10.3 **Inert Atmosphere:** No requirement
10.4 **Venting:** Pressure-vacuum

11. HAZARD ASSESSMENT CODE

(See Hazard Assessment Handbook, CG 446-3)

A-O

12. HAZARD CLASSIFICATIONS

12.1 **Code of Federal Regulations:** Flammable liquid
12.2 **NAS Hazard Rating for Bulk Water Transportation:**

Category	Rating
Fire	3
Health	
Vapor Irritant	4
Liquid or Solid Irritant	4
Poisons	3
Water Pollution	
Human Toxicity	3
Aquatic Toxicity	3
Aesthetic Effect	2
Reactivity	
Other Chemicals	3
Water	4
Self-Reaction	1

12.3 **NFPA Hazard Classifications:** Not listed

13. PHYSICAL AND CHEMICAL PROPERTIES

13.1 **Physical State at 15°C and 1 atm:** Liquid
13.2 **Molecular Weight:** 108.7
13.3 **Boiling Point at 1 atm:** 135°F = 57°C = 330°K
13.4 **Freezing Point:** Not pertinent
13.5 **Critical Temperature:** Not pertinent
13.6 **Critical Pressure:** Not pertinent
13.7 **Specific Gravity:** 0.846 at 25°C (liquid)
13.8 **Liquid Surface Tension:** (est.) 17.8 dynes/cm = 0.0178 N/m at 20°C
13.9 **Liquid-Water Interfacial Tension:** Not pertinent
13.10 **Vapor (Gas) Specific Gravity:** 3.7
13.11 **Ratio of Specific Heats of Vapor (Gas):** (est.) 1.0683
13.12 **Latent Heat of Vaporization:** 126 Btu/lb = 70 cal/g = 2.9×10^5 J/kg
13.13 **Heat of Combustion:** (est.) −10,300 Btu/lb = −5,700 cal/g = -240×10^5 J/kg
13.14 **Heat of Decomposition:** Not pertinent
13.15 **Heat of Solution:** Data not available
13.16 **Heat of Polymerization:** Not pertinent

NOTES

TGC — TRIPROPYLENE GLYCOL

Common Synonyms	Liquid — Colorless — Characteristic odor May float or sink and mix with water.

Stop discharge if possible.
Call fire department.
Isolate and remove discharged material.
Notify local health and pollution control agencies.

Fire	Combustible. Extinguish with dry chemicals, alcohol foam, or carbon dioxide. Water may be ineffective on fire. Cool exposed containers with water.
Exposure	LIQUID Not harmful.
Water Pollution	Effect of low concentrations on aquatic life is unknown. May be dangerous if it enters water intakes. Notify local health and wildlife officials. Notify operators of nearby water intakes.

1. RESPONSE TO DISCHARGE

(See Response Methods Handbook, CG 446-4)

Disperse and flush

2. LABELS

No hazard label required by Code of Federal Regulations

3. CHEMICAL DESIGNATIONS

3.1 **Synonyms:** No common synonyms
3.2 **Coast Guard Compatibility Classification:** Glycol ether
3.3 **Chemical Formula:** $HO(C_3H_6O)_2C_3H_6OH$
3.4 **IMCO/United Nations Numerical Designation:** Not listed

4. OBSERVABLE CHARACTERISTICS

4.1 **Physical State (as shipped):** Liquid
4.2 **Color:** Colorless
4.3 **Odor:** Characteristic

5. HEALTH HAZARDS

5.1 **Personal Protective Equipment:** Plastic gloves; safety glasses or face shield
5.2 **Symptoms Following Exposure:** Non-irritating; no symptoms observed by any exposure route.
5.3 **Treatment for Exposure:** INGESTION: if large amounts are swallowed, induce vomiting; treat symptomatically. EYES or SKIN: flush with water; get medical attention if ill effects develop.
5.4 **Toxicity by Inhalation (Threshold Limit Value):** Data not available
5.5 **Short-Term Inhalation Limits:** Data not available
5.6 **Toxicity by Ingestion:** Grade 2; oral LD_{50} = 3,000 mg/kg (rat)
5.7 **Late Toxicity:** Data not available
5.8 **Vapor (Gas) Irritant Characteristics:** Vapors are nonirritating to eyes and throat.
5.9 **Liquid or Solid Irritant Characteristics:** No appreciable hazard. Practically harmless to the skin.
5.10 **Odor Threshold:** Odorless

6. FIRE HAZARDS

6.1 **Flash Point:** 285°F O.C.
6.2 **Flammable Limits in Air:** 0.8% – 5.0% (est.)
6.3 **Fire Extinguishing Agents:** "Alcohol" foam, dry chemical, carbon dioxide
6.4 **Fire Extinguishing Agents Not to be Used:** Water may be ineffective
6.5 **Special Hazards of Combustion Products:** Acrid fumes of acids and aldehydes may form in fires.
6.6 **Behavior in Fire:**
6.7 **Ignition Temperature:** Data not available
6.8 **Electrical Hazard:** Data not available
6.9 **Burning Rate:** Data not available

7. CHEMICAL REACTIVITY

7.1 **Reactivity with Water:** No reaction
7.2 **Reactivity with Common Materials:** May attack some forms of plastics
7.3 **Stability During Transport:** Stable
7.4 **Neutralizing Agents for Acids and Caustics:** Not pertinent
7.5 **Polymerization:** Not pertinent
7.6 **Inhibitor of Polymerization:** Not pertinent

8. WATER POLLUTION

8.1 **Aquatic Toxicity:** Data not available
8.2 **Waterfowl Toxicity:** Data not available
8.3 **Biological Oxygen Demand (BOD):** Data not available
8.4 **Food Chain Concentration Potential:** None

9. SELECTED MANUFACTURERS

1. Union Carbide Corp.
   Chemicals and Plastics Div.
   274 Park Avenue
   New York, N. Y. 10017
2. The Dow Chemical Co.
   2030 Dow Center
   Midland, Mich. 48640

10. SHIPPING INFORMATION

10.1 **Grades or Purity:** Commercial, 99%
10.2 **Storage Temperature:** Ambient
10.3 **Inert Atmosphere:** No requirement
10.4 **Venting:** Open (flame arrester)

11. HAZARD ASSESSMENT CODE

(See Hazard Assessment Handbook, CG 446-3)

A-P-Q

12. HAZARD CLASSIFICATIONS

12.1 **Code of Federal Regulations:** Not listed
12.2 **NAS Hazard Rating for Bulk Water Transportation:**

Category	Rating
Fire	1
Health	
Vapor Irritant	0
Liquid or Solid Irritant	0
Poisons	0
Water Pollution	
Human Toxicity	0
Aquatic Toxicity	0
Aesthetic Effect	0
Reactivity	
Other Chemicals	1
Water	0
Self-Reaction	0

12.3 **NFPA Hazard Classifications:** Not listed

13. PHYSICAL AND CHEMICAL PROPERTIES

13.1 **Physical State at 15°C and 1 atm:** Liquid
13.2 **Molecular Weight:** 192.26
13.3 **Boiling Point at 1 atm:** 523°F = 273°C = 546°K
13.4 **Freezing Point:** (sets to glass) −49°F = −45°C = 228°K
13.5 **Critical Temperature:** Not pertinent
13.6 **Critical Pressure:** Not pertinent
13.7 **Specific Gravity:** 1.022 at 20°C (liquid)
13.8 **Liquid Surface Tension:** Data not available
13.9 **Liquid-Water Interfacial Tension:** Not pertinent
13.10 **Vapor (Gas) Specific Gravity:** Not pertinent
13.11 **Ratio of Specific Heats of Vapor (Gas):** Not pertinent
13.12 **Latent Heat of Vaporization:** Data not available
13.13 **Heat of Combustion:** (est.) −13,700 Btu/lb = −7,610 cal/g = -318×10^5 J/kg
13.14 **Heat of Decomposition:** Not pertinent
13.15 **Heat of Solution:** Not pertinent
13.16 **Heat of Polymerization:** Not pertinent

NOTES

TPO — TRIS(AZIRIDINYL)PHOSPHINE OXIDE

Common Synonyms	Solid	White
APO Tris(1-aziridinyl) phosphine oxide Phosphoric acid triethyleneimide Triethylenephosphoramide	Mixes with water.	

Keep people away.
Isolate and remove discharged material.
Notify local health and pollution control agencies.

Fire

Not flammable.
POISONOUS GASES MAY BE PRODUCED WHEN HEATED.

Exposure

Call for medical aid.

SOLID

Irritating to skin and eyes.
Harmful if swallowed.
Remove contaminated clothing and shoes.
Flush affected areas with plenty of water.
IF IN EYES, hold eyelids open and flush with plenty of water.
IF SWALLOWED and victim is CONSCIOUS, have victim drink water or milk.

Water Pollution

Effect of low concentrations on aquatic life is unknown.
May be dangerous if it enters water intakes.
Notify local health and wildlife officials.
Notify operators of nearby water intakes.

1. RESPONSE TO DISCHARGE

(See Response Methods Handbook, CG 446-4)

Issue warning—corrosive, water contaminant
Disperse and flush

2. LABEL

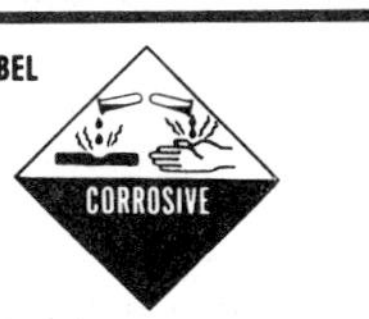

3. CHEMICAL DESIGNATIONS

3.1 **Synonyms:** APO; Phosphoric acid triethyleneimide; Triethylenephosphoramide; Tris (1-aziridinyl) phosphine oxide

3.2 **Coast Guard Compatibility Classification:** Not applicable

3.3 **Chemical Formula:** $(\underline{CH_2CH_2N})_3PO$ or $C_6H_{12}N_3PO$

3.4 **IMCO/United Nations Numerical Designation:** Not listed

4. OBSERVABLE CHARACTERISTICS

4.1 **Physical State (as shipped):** Solid

4.2 **Color:** White

4.3 **Odor:** Data not available

5. HEALTH HAZARDS

5.1 **Personal Protective Equipment:** Protective clothing and gloves to prevent contact with skin; goggles.

5.2 **Symptoms Following Exposure:** Inhalation (unlikely unless a heavy mist is formed) causes symptoms similar to those observed after ingestion. Contact with liquid or powder causes irritation of eyes and (on prolonged contact) irritation and burns of skin. Burns are slow to develop and slow to heal. May sensitize on repeated contact. Ingestion causes depression, anorexia, and diarrhea, appearing 2-3 days before death, followed by terminal dyspnea, incoordination, epistaxis, salivation, prostration and cyanosis.

5.3 **Treatment for Exposure:** INHALATION: remove victim to fresh air. EYES: flush with water at once for at least 15 min.; get medical attention. SKIN: flush with water at once, followed by vinegar and dilute hydrogen peroxide. INGESTION: only symptomatic and supportive measures are available.

5.4 **Toxicity by Inhalation (Threshold Limit Value):** Data not available

5.5 **Short-Term Inhalation Limits:** Data not available

5.6 **Toxicity by Ingestion:** Grade 4; oral rat LD_{50} = 37 mg/kg

5.7 **Late Toxicity:** None observed

5.8 **Vapor (Gas) Irritant Characteristics:** Data not available

5.9 **Liquid or Solid Irritant Characteristics:** Data not available

5.10 **Odor Threshold:** Data not available

6. FIRE HAZARDS

6.1 **Flash Point:** Not flammable

6.2 **Flammable Limits in Air:** Not flammable

6.3 **Fire Extinguishing Agents:** Not pertinent

6.4 **Fire Extinguishing Agents Not to be Used:** Not pertinent

6.5 **Special Hazards of Combustion Products:** Phosphoric acid mist may form in fire. Toxic oxides of nitrogen may form.

6.6 **Behavior in Fire:** Data not available

6.7 **Ignition Temperature:** Not pertinent

6.8 **Electrical Hazard:** Data not available

6.9 **Burning Rate:** Not pertinent

7. CHEMICAL REACTIVITY

7.1 **Reactivity with Water:** No reaction, unless in presence of acids or strong caustics

7.2 **Reactivity with Common Materials:** Slow decomposition, not considered hazardous

7.3 **Stability During Transport:** Stable if cool

7.4 **Neutralizing Agents for Acids and Caustics:** Not pertinent

7.5 **Polymerization:** Violent polymerization occurs at about 255°F. Acid fumes also cause polymerization at ordinary temperatures.

7.6 **Inhibitor of Polymerization:** None used

8. WATER POLLUTION

8.1 **Aquatic Toxicity:** Data not available

8.2 **Waterfowl Toxicity:** 8.5-13 mg/kg LD_{50}

8.3 **Biological Oxygen Demand (BOD):** Data not available

8.4 **Food Chain Concentration Potential:** None

9. SELECTED MANUFACTURERS

Chemirad Corporation
P. O. Box 152
Port Washington, N. Y. 11050

10. SHIPPING INFORMATION

10.1 **Grades or Purity:** 85% solution in acetone - methylene chloride

10.2 **Storage Temperature:** Below 100°F

10.3 **Inert Atmosphere:** No requirement

10.4 **Venting:** Open

11. HAZARD ASSESSMENT CODE

(See Hazard Assessment Handbook, CG 446-3)

SS

12. HAZARD CLASSIFICATIONS

12.1 **Code of Federal Regulations:** Corrosive

12.2 **NAS Hazard Rating for Bulk Water Transportation:** Not listed

12.3 **NFPA Hazard Classifications:** Not listed

13. PHYSICAL AND CHEMICAL PROPERTIES

13.1 **Physical State at 15°C and 1 atm:** Solid

13.2 **Molecular Weight:** 173.16

13.3 **Boiling Point at 1 atm:** Not pertinent (decomposes)

13.4 **Freezing Point:** 106°F = 41°C = 314°K

13.5 **Critical Temperature:** Not pertinent

13.6 **Critical Pressure:** Not pertinent

13.7 **Specific Gravity:** (est.) >1 at 20°C (solid)

13.8 **Liquid Surface Tension:** Not pertinent

13.9 **Liquid-Water Interfacial Tension:** Not pertinent

13.10 **Vapor (Gas) Specific Gravity:** Not pertinent

13.11 **Ratio of Specific Heats of Vapor (Gas):** Not pertinent

13.12 **Latent Heat of Vaporization:** Not pertinent

13.13 **Heat of Combustion:** Not pertinent

13.14 **Heat of Decomposition:** Not pertinent

13.15 **Heat of Solution:** Not pertinent

13.16 **Heat of Polymerization:** Data not available

NOTES

TPT	# TURPENTINE

Common Synonyms		
Spirits of turpentine Turps Gum turpentine Wood turpentine	Watery liquid Colorless Penetrating, unpleasant odor Floats on water. Irritating vapor is produced.	

Stop discharge if possible. Keep people away.
Call fire department.
Avoid contact with liquid and vapor.
Isolate and remove discharged material.
Notify local health and pollution control agencies.

Fire

FLAMMABLE
Flashback along vapor trail may occur.
Vapor may explode if ignited in an enclosed area.

Extinguish with dry chemical, foam, or carbon dioxide.
Water may be ineffective on fire.
Cool exposed containers with water.

Exposure

CALL FOR MEDICAL AID.

VAPOR
Irritating to eyes, nose, and throat.
If inhaled, will cause nausea, vomiting, headache, difficult breathing, or loss of consciousness.
Move to fresh air.
If breathing has stopped, give artificial respiration.
If breathing is difficult, give oxygen.

LIQUID
POISONOUS IF SWALLOWED.
Irritating to skin and eyes.
Remove contaminated clothing and shoes.
Flush affected areas with plenty of water.
IF IN EYES, hold eyelids open and flush with plenty of water.
IF SWALLOWED and victim is CONSCIOUS, have victim drink water or milk.

Water Pollution

Dangerous to aquatic life in high concentrations.
Fouling to shoreline.
May be dangerous if it enters water intakes.
Notify local health and wildlife officials.
Notify operators of nearby water intakes.

1. RESPONSE TO DISCHARGE

(See Response Methods Handbook, CG 446-4)

Mechanical containment
Should be removed
Chemical and physical treatment

2. LABEL

Red

3. CHEMICAL DESIGNATIONS

3.1 **Synonyms:**
D.D. turpentine
Gum turpentine *
Spirits of turpentine
Sulfate turpentine
Turps
Wood turpentine

3.2 **Coast Guard Compatibility Classification:** Olefin

3.3 **Chemical Formula:** $C_{10}H_{16}$

3.4 **IMCO/United Nations Numerical Designation:** 3.3/1299

4. OBSERVABLE CHARACTERISTICS

4.1 **Physical State (as shipped):** Liquid

4.2 **Color:** Colorless

4.3 **Odor:** Aromatic, rather unpleasant, penetrating

5. HEALTH HAZARDS

5.1 **Personal Protective Equipment:** Organic canister or air-supplied mask; goggles or face shield; rubber gloves.

5.2 **Symptoms Following Exposure:** Vapors cause headache, confusion, respiratory distress. Liquid irritates skin. If ingested, can irritate the entire digestive system and may injure kidneys. If liquid is taken into lungs, causes severe pneumonitis.

5.3 **Treatment for Exposure:** INHALATION: remove victim to fresh air; call a doctor; administer artificial respiration and oxygen if required. INGESTION: give water and induce vomiting; call a doctor. EYES: flush with water for at least 15 min. SKIN: wipe off, wash with soap and water.

5.4 **Toxicity by Inhalation (Threshold Limit Value):** 100 ppm

5.5 **Short-Term Inhalation Limits:** 200 ppm for 30 min.

5.6 **Toxicity by Ingestion:** Grade 2; LD_{50} 0.5 to 5 g/kg

5.7 **Late Toxicity:** None

5.8 **Vapor (Gas) Irritant Characteristics:** Vapors cause a slight smarting of the eyes or respiratory system if present in high concentrations. The effect is temporary.

5.9 **Liquid or Solid Irritant Characteristics:** Minimum hazard. If spilled on clothing and allowed to remain, may cause smarting and reddening of the skin.

5.10 **Odor Threshold:** Data not available

6. FIRE HAZARDS

6.1 **Flash Point:** 95°F C.C.

6.2 **Flammable Limits in Air:** 0.8% (LEL)

6.3 **Fire Extinguishing Agents:** Foam, dry chemical, or carbon dioxide

6.4 **Fire Extinguishing Agents Not to be Used:** Water may be ineffective.

6.5 **Special Hazards of Combustion Products:** Not pertinent

6.6 **Behavior in Fire:** Forms heavy black smoke and soot

6.7 **Ignition Temperature:** 488°F

6.8 **Electrical Hazard:** Not pertinent

6.9 **Burning Rate:** 2.4 mm/min.

7. CHEMICAL REACTIVITY

7.1 **Reactivity with Water:** No reaction

7.2 **Reactivity with Common Materials:** No reaction

7.3 **Stability During Transport:** Stable

7.4 **Neutralizing Agents for Acids and Caustics:** Not pertinent

7.5 **Polymerization:** Not pertinent

7.6 **Inhibitor of Polymerization:** Not pertinent

8. WATER POLLUTION

8.1 **Aquatic Toxicity:**
100 ppm/*/fish/toxic/fresh water
*Time period not specified.

8.2 **Waterfowl Toxicity:** Data not available

8.3 **Biological Oxygen Demand (BOD):** Data not available

8.4 **Food Chain Concentration Potential:** None

9. SELECTED MANUFACTURERS

1. Hercules, Inc.
Hattiesburg, Miss. 39401

2. Tenneco Chemicals, Inc.
Newport Division
Pensacola, Fla. 32501

3. Union Camp Corp.
Chemical Division
Jacksonville, Fla. 32205

10. SHIPPING INFORMATION

10.1 **Grades or Purity:** A wide variety of grades and purities are shipped. All have about the same hazardous properties.

10.2 **Storage Temperature:** Ambient

10.3 **Inert Atmosphere:** No requirement

10.4 **Venting:** Open (flame arrester)

11. HAZARD ASSESSMENT CODE

(See Hazard Assessment Handbook, CG 446-3)

A-T-U

12. HAZARD CLASSIFICATIONS

12.1 **Code of Federal Regulations:** Flammable Liquid

12.2 **NAS Hazard Rating for Bulk Water Transportation:**

Category	Rating
Fire	3
Health	
Vapor Irritant	1
Liquid or Solid Irritant	1
Poisons	1
Water Pollution	
Human Toxicity	2
Aquatic Toxicity	3
Aesthetic Effect	2
Reactivity	
Other Chemicals	1
Water	0
Self-Reaction	0

12.3 **NFPA Hazard Classifications:**

Category	Classification
Health Hazard (Blue)	1
Flammability (Red)	3
Reactivity (Yellow)	0

13. PHYSICAL AND CHEMICAL PROPERTIES

13.1 **Physical State at 15°C and 1 atm:** Liquid

13.2 **Molecular Weight:** Not pertinent

13.3 **Boiling Point at 1 atm:** 302 – 320°F = 150 – 160°C = 423 – 433°K

13.4 **Freezing Point:** Not pertinent

13.5 **Critical Temperature:** Not pertinent

13.6 **Critical Pressure:** Not pertinent

13.7 **Specific Gravity:** 0.86 at 15°C (liquid)

13.8 **Liquid Surface Tension:** Data not available

13.9 **Liquid-Water Interfacial Tension:** 14 dynes/cm = 0.014 N/m at 22.7°C

13.10 **Vapor (Gas) Specific Gravity:** Not pertinent

13.11 **Ratio of Specific Heats of Vapor (Gas):** Not pertinent

13.12 **Latent Heat of Vaporization:** Not pertinent

13.13 **Heat of Combustion:** Data not available

13.14 **Heat of Decomposition:** Not pertinent

13.15 **Heat of Solution:** Not pertinent

13.16 **Heat of Polymerization:** Not pertinent

NOTES

REVISED 1978

UND	UNDECANOL

Common Synonyms			
Alcohol C-11 (undecylic) 1-Hendecanol Undecyl alcohol 1-Undecanol	Solid or liquid	Colorless	Mild odor
	Floats on water.		

Stop discharge if possible.
Call fire department.
Isolate and remove discharged material.
Notify local health and pollution control agencies.

Fire	Combustible. Extinguish with foam, dry chemical, or carbon dioxide.
Exposure	CALL FOR MEDICAL AID. LIQUID Irritating to eyes. IF IN EYES, hold eyelids open and flush with plenty of water.
Water Pollution	Effect of low concentrations on aquatic life is unknown. Fouling to shoreline. May be dangerous if it enters water intakes. Notify local health and wildlife officials. Notify operators of nearby water intakes.

1. RESPONSE TO DISCHARGE

(See Response Methods Handbook, CG 446-4)

Mechanical containment
Should be removed
Chemical and physical treatment

2. LABELS

No hazard label required by Code of Federal Regulations

3. CHEMICAL DESIGNATIONS

3.1 **Synonyms:** Hendecanoic alcohol; 1-Hendecanol; n-Hendecylenic alcohol; 1-Undecanol; Undecyl alcohol; Undecylic alcohol

3.2 **Coast Guard Compatibility Classification:** Alcohol

3.3 **Chemical Formula:** $CH_3(CH_2)_9CH_2OH$

3.4 **IMCO/United Nations Numerical Designation:** Not listed

4. OBSERVABLE CHARACTERISTICS

4.1 **Physical State (as shipped):** Liquid

4.2 **Color:** Colorless

4.3 **Odor:** Faint alcoholic

5. HEALTH HAZARDS

5.1 **Personal Protective Equipment:** Goggles or face shield.

5.2 **Symptoms Following Exposure:** Liquid can irritate eyes.

5.3 **Treatment for Exposure:** Wash eyes with water for at least 15 min.

5.4 **Toxicity by Inhalation (Threshold Limit Value):** Not pertinent

5.5 **Short-Term Inhalation Limits:** Not pertinent

5.6 **Toxicity by Ingestion:** Grade 2; LD_{50} 0.5 to 5 g/kg

5.7 **Late Toxicity:** Data not available

5.8 **Vapor (Gas) Irritant Characteristics:** None

5.9 **Liquid or Solid Irritant Characteristics:** No appreciable hazard. Practically harmless to the skin.

5.10 **Odor Threshold:** Not pertinent

6. FIRE HAZARDS

6.1 **Flash Point:** 200°F O.C.

6.2 **Flammable Limits in Air:** Data not available

6.3 **Fire Extinguishing Agents:** Foam, carbon dioxide, or dry chemical

6.4 **Fire Extinguishing Agents Not to be Used:** Water or foam may cause frothing.

6.5 **Special Hazards of Combustion Products:** Not pertinent

6.6 **Behavior in Fire:** Not pertinent

6.7 **Ignition Temperature:** Data not available

6.8 **Electrical Hazard:** Not pertinent

6.9 **Burning Rate:** Data not available

7. CHEMICAL REACTIVITY

7.1 **Reactivity with Water:** No reaction

7.2 **Reactivity with Common Materials:** No reaction

7.3 **Stability During Transport:** Stable

7.4 **Neutralizing Agents for Acids and Caustics:** Not pertinent

7.5 **Polymerization:** Not pertinent

7.6 **Inhibitor of Polymerization:** Not pertinent

8. WATER POLLUTION

8.1 **Aquatic Toxicity:** Data not available

8.2 **Waterfowl Toxicity:** Data not available

8.3 **Biological Oxygen Demand (BOD):** 12.1% (theor.), 5 days; 25.3% (theor.), 1 day

8.4 **Food Chain Concentration Potential:** None

9. SELECTED MANUFACTURERS

Union Carbide Corp.
Chemicals and Plastics Division
270 Park Ave.
New York, N. Y. 10017

10. SHIPPING INFORMATION

10.1 **Grades or Purity:** Technical

10.2 **Storage Temperature:** Ambient

10.3 **Inert Atmosphere:** No requirement

10.4 **Venting:** Open (flame arrester)

11. HAZARD ASSESSMENT CODE

(See Hazard Assessment Handbook, CG 446-3)

A-T-U

12. HAZARD CLASSIFICATIONS

12.1 **Code of Federal Regulations:** Not listed

12.2 **NAS Hazard Rating for Bulk Water Transportation:** Not listed

12.3 **NFPA Hazard Classifications:** Not listed

13. PHYSICAL AND CHEMICAL PROPERTIES

13.1 **Physical State at 15°C and 1 atm:** Liquid

13.2 **Molecular Weight:** 172.30

13.3 **Boiling Point at 1 atm:** 473°F = 245°C = 518°K

13.4 **Freezing Point:** 60.6°F = 15.9°C = 289.1°K

13.5 **Critical Temperature:** 739°F = 393°C = 666°K

13.6 **Critical Pressure:** 308 psia = 21 atm = 2.1 MN/m^2

13.7 **Specific Gravity:** 0.835 at 20°C (liquid)

13.8 **Liquid Surface Tension:** 26.5 dynes/cm = 0.0265 N/m at 20°C

13.9 **Liquid-Water Interfacial Tension:** (est.) 40 dynes/cm = 0.04 N/m at 20°C

13.10 **Vapor (Gas) Specific Gravity:** Not pertinent

13.11 **Ratio of Specific Heats of Vapor (Gas):** 1.032

13.12 **Latent Heat of Vaporization:** Not pertinent

13.13 **Heat of Combustion:** (est.) −18,000 Btu/lb = −10,000 cal/g = -419×10^5 J/kg

13.14 **Heat of Decomposition:** Not pertinent

13.15 **Heat of Solution:** Not pertinent

13.16 **Heat of Polymerization:** Not pertinent

NOTES

REVISED 1978

UDC	# 1-UNDECENE

Common Synonyms n-Nonylethylene	Liquid	Colorless	Mild odor
	Floats on water.		

Stop discharge if possible.
Call fire department.
Avoid contact with liquid.
Isolate and remove discharged material.
Notify local health and pollution control agencies.

Fire	Combustible. Extinguish with foam, dry chemical, or carbon dioxide. Water may be ineffective on fire. Cool exposed containers with water.
Exposure	CALL FOR MEDICAL AID. LIQUID Irritating to skin and eyes. Harmful if swallowed. Remove contaminated clothing and shoes. Flush affected areas with plenty of water. IF IN EYES, hold eyelids open and flush with plenty of water. IF SWALLOWED and victim is CONSCIOUS, have victim drink water or milk. DO NOT INDUCE VOMITING.
Water Pollution	Effect of low concentrations on aquatic life is unknown. Fouling to shoreline. May be dangerous if it enters water intakes. Notify local health and wildlife officials. Notify operators of nearby water intakes.

1. RESPONSE TO DISCHARGE

(See Response Methods Handbook, CG 446-4)

Mechanical containment
Should be removed
Chemical and physical treatment

2. LABELS

No hazard label required by Code of Federal Regulations

3. CHEMICAL DESIGNATIONS

3.1 **Synonyms:** n-Nonylethylene
3.2 **Coast Guard Compatibility Classification:** Olefin
3.3 **Chemical Formula:** $CH_3(CH_2)_8CH{=}CH_2$
3.4 **IMCO/United Nations Numerical Designation:** Not listed

4. OBSERVABLE CHARACTERISTICS

4.1 **Physical State (as shipped):** Liquid
4.2 **Color:** Colorless
4.3 **Odor:** Mild, pleasant

5. HEALTH HAZARDS

5.1 **Personal Protective Equipment:** Goggles or face shield; rubber gloves.
5.2 **Symptoms Following Exposure:** Aspiration hazard if ingested. Slight skin and eye irritation. No inhalation hazard expected.
5.3 **Treatment for Exposure:** INHALATION: remove victim to fresh air. INGESTION: do NOT lavage or induce vomiting; give vegetable oil and demulcents; call a doctor. EYES: flush with water for 15 min. SKIN: wipe off, wash with soap and water.
5.4 **Toxicity by Inhalation (Threshold Limit Value):** Data not available
5.5 **Short-Term Inhalation Limits:** Data not available
5.6 **Toxicity by Ingestion:** Data not available
5.7 **Late Toxicity:** Data not available
5.8 **Vapor (Gas) Irritant Characteristics:** Slight smarting of eyes and respiratory system at high concentrations. The effect is temporary.
5.9 **Liquid or Solid Irritant Characteristics:** Minimum hazard. If spilled on clothing and allowed to remain, may cause smarting and reddening of the skin.
5.10 **Odor Threshold:** Data not available

6. FIRE HAZARDS

6.1 **Flash Point:** 160°F O.C.
6.2 **Flammable Limits in Air:** Data not available
6.3 **Fire Extinguishing Agents:** Foam, dry chemical, or carbon dioxide
6.4 **Fire Extinguishing Agents Not to be Used:** Water may be ineffective.
6.5 **Special Hazards of Combustion Products:** Not pertinent
6.6 **Behavior in Fire:** Not pertinent
6.7 **Ignition Temperature:** Data not available
6.8 **Electrical Hazard:** Not pertinent
6.9 **Burning Rate:** 4.8 mm/min.

7. CHEMICAL REACTIVITY

7.1 **Reactivity with Water:** No reaction
7.2 **Reactivity with Common Materials:** No reaction
7.3 **Stability During Transport:** Stable
7.4 **Neutralizing Agents for Acids and Caustics:** Not pertinent
7.5 **Polymerization:** Not pertinent
7.6 **Inhibitor of Polymerization:** Not pertinent

8. WATER POLLUTION

8.1 **Aquatic Toxicity:** Data not available
8.2 **Waterfowl Toxicity:** Data not available
8.3 **Biological Oxygen Demand (BOD):** Data not available
8.4 **Food Chain Concentration Potential:** None

9. SELECTED MANUFACTURERS

The Humphrey Chemical Co.
Devine Street
North Haven, Conn. 06473

10. SHIPPING INFORMATION

10.1 **Grades or Purity:** Technical: 99%
10.2 **Storage Temperature:** Ambient
10.3 **Inert Atmosphere:** No requirement
10.4 **Venting:** Open (flame arrester)

11. HAZARD ASSESSMENT CODE

(See Hazard Assessment Handbook, CG 446-3)

A-T-U

12. HAZARD CLASSIFICATIONS

12.1 **Code of Federal Regulations:** Combustible Liquid
12.2 **NAS Hazard Rating for Bulk Water Transportation:** Not listed
12.3 **NFPA Hazard Classifications:** Not listed

13. PHYSICAL AND CHEMICAL PROPERTIES

13.1 **Physical State at 15°C and 1 atm:** Liquid
13.2 **Molecular Weight:** 154.2
13.3 **Boiling Point at 1 atm:** 378.9°F = 192.7°C = 465.9°K
13.4 **Freezing Point:** −56°F = −49°C = 224°K
13.5 **Critical Temperature:** Not pertinent
13.6 **Critical Pressure:** Not pertinent
13.7 **Specific Gravity:** 0.750 at 20°C (liquid)
13.8 **Liquid Surface Tension:** 23.4 dynes/cm = 0.0234 N/m at 20°C
13.9 **Liquid-Water Interfacial Tension:** (est.) 50 dynes/cm = 0.050 N/m at 20°C
13.10 **Vapor (Gas) Specific Gravity:** Not pertinent
13.11 **Ratio of Specific Heats of Vapor (Gas):** 1.035
13.12 **Latent Heat of Vaporization:** 154 Btu/lb = 85.8 cal/g = 3.59×10^5 J/kg
13.13 **Heat of Combustion:** −19,084 Btu/lb = −10,602 cal/g = -443.89×10^5 J/kg
13.14 **Heat of Decomposition:** Not pertinent
13.15 **Heat of Solution:** Not pertinent
13.16 **Heat of Polymerization:** Not pertinent

NOTES

UDB n-UNDECYLBENZENE

Common Synonyms: 1-Phenylundecane

Liquid | Colorless | Mild odor

Floats on water.

Avoid contact with liquid. Keep people away.
Stop discharge if possible.
Call fire department.
Isolate and remove discharged material.
Notify local health and pollution control agencies.

Fire

Combustible.

Extinguish with dry chemicals, foam or carbon dioxide.
Water may be ineffective on fire.
Cool exposed containers with water.

Exposure

CALL FOR MEDICAL AID.

LIQUID
Irritating to skin and eyes.
If swallowed will cause nausea and vomiting.

Remove contaminated clothing and shoes.
Flush affected areas with plenty of water.
IF IN EYES, hold eyelids open and flush with plenty of water.
IF SWALLOWED and victim is CONSCIOUS, have victim drink water or milk and have victim induce vomiting.
IF SWALLOWED and victim is UNCONSCIOUS OR HAVING CONVULSIONS, do nothing except keep victim warm.

Water Pollution

Effect of low concentrations on aquatic life is unknown.
Fouling to shoreline.
May be dangerous if it enters water intakes.

Notify local health and wildlife officials.
Notify operators of nearby water intakes.

1. RESPONSE TO DISCHARGE

(See Response Methods Handbook, CG 446-4)

Mechanical containment
Should be removed
Chemical and physical treatment

2. LABELS

No hazard label required by Code of Federal Regulations

3. CHEMICAL DESIGNATIONS

3.1 **Synonyms:** 1-Phenylundecane
3.2 **Coast Guard Compatibility Classification:** Not listed
3.3 **Chemical Formula:** $C_6H_5(CH_2)_{10}CH_3$
3.4 **IMCO/United Nations Numerical Designation:** Not listed

4. OBSERVABLE CHARACTERISTICS

4.1 **Physical State (as shipped):** Liquid
4.2 **Color:** Colorless
4.3 **Odor:** Mild

5. HEALTH HAZARDS

5.1 **Personal Protective Equipment:** Goggles or face shield and rubber gloves.
5.2 **Symptoms Following Exposure:** Ingestion may cause intestinal disturbances. Contact with eyes causes mild irritation.
5.3 **Treatment for Exposure:** INGESTION: induce vomiting if large amount has been swallowed. EYES: flush with water. SKIN: remove spills on skin or clothing by washing with soap and water.
5.4 **Toxicity by Inhalation (Threshold Limit Value):** Data not available
5.5 **Short-Term Inhalation Limits:** Data not available
5.6 **Toxicity by Ingestion:** Data not available
5.7 **Late Toxicity:** Data not available
5.8 **Vapor (Gas) Irritant Characteristics:** Data not available
5.9 **Liquid or Solid Irritant Characteristics:** Data not available
5.10 **Odor Threshold:** Data not available

6. FIRE HAZARDS

6.1 **Flash Point:** 285°F C.C.
6.2 **Flammable Limits in Air:** Data not available
6.3 **Fire Extinguishing Agents:** Foam, dry chemical, carbon dioxide
6.4 **Fire Extinguishing Agents Not to be Used:** Water may be ineffective.
6.5 **Special Hazards of Combustion Products:**
6.6 **Behavior in Fire:**
6.7 **Ignition Temperature:** Data not available
6.8 **Electrical Hazard:** Data not available
6.9 **Burning Rate:** Data not available

7. CHEMICAL REACTIVITY

7.1 **Reactivity with Water:** No reaction
7.2 **Reactivity with Common Materials:** May attack some forms of plastics
7.3 **Stability During Transport:** Stable
7.4 **Neutralizing Agents for Acids and Caustics:** Not pertinent
7.5 **Polymerization:** Not pertinent
7.6 **Inhibitor of Polymerization:** Not pertinent

8. WATER POLLUTION

8.1 **Aquatic Toxicity:** Data not available
8.2 **Waterfowl Toxicity:** Data not available
8.3 **Biological Oxygen Demand (BOD):** Data not available
8.4 **Food Chain Concentration Potential:** None

9. SELECTED MANUFACTURERS

1. The Humphrey Chemical Co.
   Devine Street
   North Haven, Conn. 06473
2. Conoco Chemicals
   Park 80 Plaza East
   Saddle Brook, N. J. 07662
3. Fisher Scientific Co.
   711 Forbes Ave.
   Pittsburgh, Pa. 15219

10. SHIPPING INFORMATION

10.1 **Grades or Purity:** Mixture with decylbenzene and dodecylbenzene, all of which have same general properties
10.2 **Storage Temperature:** Ambient
10.3 **Inert Atmosphere:** No requirement
10.4 **Venting:** Open (flame arrester)

11. HAZARD ASSESSMENT CODE

(See Hazard Assessment Handbook, CG 446-3)

A-T-U

12. HAZARD CLASSIFICATIONS

12.1 **Code of Federal Regulations:** Not listed
12.2 **NAS Hazard Rating for Bulk Water Transportation:** Not listed
12.3 **NFPA Hazard Classifications:** Not listed

13. PHYSICAL AND CHEMICAL PROPERTIES

13.1 **Physical State at 15°C and 1 atm:** Liquid
13.2 **Molecular Weight:** 232.4
13.3 **Boiling Point at 1 atm:** 601°F = 316°C = 589°K
13.4 **Freezing Point:** 23°F = −5°C = 268°K
13.5 **Critical Temperature:** 918.1°F = 492.3°C = 765.5°K
13.6 **Critical Pressure:** 234 psia = 15.9 atm = 1.61 MN/m^2
13.7 **Specific Gravity:** 0.855 at 20°C (liquid)
13.8 **Liquid Surface Tension:** Data not available
13.9 **Liquid-Water Interfacial Tension:** Data not available
13.10 **Vapor (Gas) Specific Gravity:** Not pertinent
13.11 **Ratio of Specific Heats of Vapor (Gas):** Not pertinent
13.12 **Latent Heat of Vaporization:** 101.27 Btu/lb = 56.26 cal/g = 2.354×10^5 J/kg
13.13 **Heat of Combustion:** −19,490 Btu/lb = −10,830 cal/g = -453.1×10^5 J/kg
13.14 **Heat of Decomposition:** Not pertinent
13.15 **Heat of Solution:** Not pertinent
13.16 **Heat of Polymerization:** Not pertinent

NOTES

URA	# URANYL ACETATE

Common Synonyms			
Uranyl acetate dihydrate Uranium acetate Uranium oxyacetate dihydrate Bis (acetate) dioxouranium Uranium acetate dihydrate	Solid	Yellow	Slight vinegar odor
	Sinks and mixes slowly with water.		

Avoid contact with solid and dust. Keep people away.
Stop discharge if possible.
Isolate and remove discharged material.
Notify local health and pollution control agencies.

Fire	Not flammable.
Exposure	CALL FOR MEDICAL AID. DUST Irritating to eyes, nose and throat. If inhaled will cause coughing or difficult breathing. If in eyes, hold eyelids open and flush with plenty of water. If breathing has stopped, give artificial respiration. If breathing is difficult, give oxygen. SOLID Irritating to skin and eyes. Harmful if swallowed. Remove contaminated clothing and shoes. Flush affected areas with plenty of water. IF IN EYES, hold eyelids open and flush with plenty of water. IF SWALLOWED and victim is CONSCIOUS, have victim drink water or milk and have victim induce vomiting. IF SWALLOWED and victim is UNCONSCIOUS OR HAVING CONVULSIONS, do nothing except keep victim warm.
Water Pollution	HARMFUL TO AQUATIC LIFE IN VERY LOW CONCENTRATIONS. May be dangerous if it enters water intakes. Notify local health and wildlife officials. Notify operators of nearby water intakes.

1. RESPONSE TO DISCHARGE
(See Response Methods Handbook, CG 446-4)

Issue warning—water contaminant
Restrict access
Disperse and flush

2. LABEL

3. CHEMICAL DESIGNATIONS

3.1 **Synonyms:** Bis(acetato)dioxouranium; Uranium acetate; Uranium acetate dihydrate; Uranium oxyacetate dihydrate; Uranyl acetate dihydrate
3.2 **Coast Guard Compatibility Classification:** Not listed
3.3 **Chemical Formula:** $UO_2(C_2H_3O_2)_2 \cdot 2H_2O$
3.4 **IMCO/United Nations Numerical Designation:** Not listed

4. OBSERVABLE CHARACTERISTICS

4.1 **Physical State (as shipped):** Solid
4.2 **Color:** Yellow
4.3 **Odor:** Slight vinegar

5. HEALTH HAZARDS

5.1 **Personal Protective Equipment:** Approved dust respirator; goggles or face shield; protective clothing
5.2 **Symptoms Following Exposure:** Inhalation of dust may irritate nose and throat. Contact with eyes causes irritation.
5.3 **Treatment for Exposure:** *Get medical attention after all exposures to this compound.* INHALATION: move to fresh air. INGESTION: give large amount of water; induce vomiting. EYES: flush with water for at least 15 min. SKIN: flush with water.
5.4 **Toxicity by Inhalation (Threshold Limit Value):** 0.2 mg/m^3 (as uranium)
5.5 **Short-Term Inhalation Limits:** Data not available
5.6 **Toxicity by Ingestion:** Grade 1; LD_{50} 5—15 g/kg
5.7 **Late Toxicity:** Data not available
5.8 **Vapor (Gas) Irritant Characteristics:** Data not available
5.9 **Liquid or Solid Irritant Characteristics:** Data not available
5.10 **Odor Threshold:** Data not available

6. FIRE HAZARDS

6.1 **Flash Point:** Not flammable
6.2 **Flammable Limits in Air:** Not flammable
6.3 **Fire Extinguishing Agents:** Not pertinent
6.4 **Fire Extinguishing Agents Not to be Used:** Not pertinent
6.5 **Special Hazards of Combustion Products:**
6.6 **Behavior in Fire:**
6.7 **Ignition Temperature:** Not pertinent
6.8 **Electrical Hazard:** Not pertinent
6.9 **Burning Rate:** Not pertinent

7. CHEMICAL REACTIVITY

7.1 **Reactivity with Water:** Dissolves and reacts to give a milky solution. The reaction is not hazardous.
7.2 **Reactivity with Common Materials:**
7.3 **Stability During Transport:** Stable
7.4 **Neutralizing Agents for Acids and Caustics:** Not pertinent
7.5 **Polymerization:** Not pertinent
7.6 **Inhibitor of Polymerization:** Not pertinent

8. WATER POLLUTION

8.1 **Aquatic Toxicity:** 3.7 ppm/96 hr/fathead minnow/TL_m/soft water
8.2 **Waterfowl Toxicity:** Data not available
8.3 **Biological Oxygen Demand (BOD):** Data not available
8.4 **Food Chain Concentration Potential:** Data not available

9. SELECTED MANUFACTURERS

1. Var-lac-oid Chemical Co.
666 South Front St.
Elizabeth, N. J. 07202
2. J. T. Baker Chemical Co.
Phillipsburg, N. J. 08865
3. Mallinckrodt Chemical Works
2nd and Mallinckrodt St.
St. Louis, Mo. 63160

10. SHIPPING INFORMATION

10.1 **Grades or Purity:** Commercial; Reagent
10.2 **Storage Temperature:** Ambient
10.3 **Inert Atmosphere:** No requirement
10.4 **Venting:** Open

11. HAZARD ASSESSMENT CODE
(See Hazard Assessment Handbook, CG 446-3)

SS

12. HAZARD CLASSIFICATIONS

12.1 **Code of Federal Regulations:** Radioactive material
12.2 **NAS Hazard Rating for Bulk Water Transportation:** Not listed
12.3 **NFPA Hazard Classifications:** Not listed

13. PHYSICAL AND CHEMICAL PROPERTIES

13.1 **Physical State at 15°C and 1 atm:** Solid
13.2 **Molecular Weight:** 424.2
13.3 **Boiling Point at 1 atm:** Not pertinent (decomposes)
13.4 **Freezing Point:** Not pertinent
13.5 **Critical Temperature:** Not pertinent
13.6 **Critical Pressure:** Not pertinent
13.7 **Specific Gravity:** 2.89 at 20°C (solid)
13.8 **Liquid Surface Tension:** Not pertinent
13.9 **Liquid-Water Interfacial Tension:** Not pertinent
13.10 **Vapor (Gas) Specific Gravity:** Not pertinent
13.11 **Ratio of Specific Heats of Vapor (Gas):** Not pertinent
13.12 **Latent Heat of Vaporization:** Not pertinent
13.13 **Heat of Combustion:** Not pertinent
13.14 **Heat of Decomposition:** Not pertinent
13.15 **Heat of Solution:** Not pertinent
13.16 **Heat of Polymerization:** Not pertinent

NOTES

UAN	URANYL NITRATE

Common Synonyms			
Uranium nitrate	Solid	Light yellow	Odorless
	Mixes with water.		

Keep people away.
Call fire department.
Isolate and remove discharged material.
Notify local health and pollution control agencies.

Fire	Not flammable. May cause fire on contact with combustibles. POISONOUS GASES MAY BE PRODUCED IN FIRE. Combat fires from safe distance or protected location. Flood discharge area with water.
Exposure	Call for medical aid. DUST Irritating to eyes, nose and throat. Harmful if inhaled. Move victim to fresh air. If in eyes, hold eyelids open and flush with plenty of water. SOLID Irritating to skin and eyes. Harmful if swallowed. Remove contaminated clothing and shoes. Flush affected areas with plenty of water. IF IN EYES, hold eyelids open and flush with plenty of water. IF SWALLOWED and victim is CONSCIOUS, have victim drink water or milk.
Water Pollution	HARMFUL TO AQUATIC LIFE IN VERY LOW CONCENTRATIONS. May be dangerous if it enters water intakes. Notify local health and wildlife officials. Notify operators of nearby water intakes.

1. RESPONSE TO DISCHARGE
(See Response Methods Handbook, CG 446-4)

Issue warning—radioactive, oxidizing material, water contaminant
Restrict access
Disperse and flush

2. LABELS

Yellow

3. CHEMICAL DESIGNATIONS
3.1 **Synonyms:** Uranium nitrate
3.2 **Coast Guard Compatibility Classification:** Not applicable
3.3 **Chemical Formula:** $UO_2(NO_3)_2 \cdot 6H_2O$
3.4 **IMCO/United Nations Numerical Designation:** 7/Low specific activity

4. OBSERVABLE CHARACTERISTICS
4.1 **Physical State (as shipped):** Solid
4.2 **Color:** Pale yellow
4.3 **Odor:** None

5. HEALTH HAZARDS
5.1 **Personal Protective Equipment:** Dust mask, gloves, goggles
5.2 **Symptoms Following Exposure:** Excessive inhalation of dust may cause irritation of lungs and delayed symptoms similar to those observed after ingestion. Dust irritates eyes and skin and may be absorbed through skin on prolonged exposure. Ingestion causes irritation of mouth and stomach; inflammation of kidney and liver develops 1 to 4 days after exposure.
5.3 **Treatment for Exposure:** INHALATION: remove victim to fresh air. EYES: flush with water for at least 15 min.; see physician if irritation persists. SKIN: wash thoroughly with soap and water. INGESTION: administer large doses of sodium bicarbonate.(This will convert the uranium salt to the bicarbonate, which is much less toxic.) Additional treatment is symptomatic; get medical attention.
5.4 **Toxicity by Inhalation (Threshold Limit Value):** 0.05 mg/m^3
5.5 **Short-Term Inhalation Limits:** Data not available
5.6 **Toxicity by Ingestion:** Grade 3; LD_{50} 50 to 500 mg/kg
5.7 **Late Toxicity:** Delayed inflammation of kidneys. Airborne radioactive particles have apparently been responsible for a significantly increased death rate from lung cancer among long-term uranium miners.
5.8 **Vapor (Gas) Irritant Characteristics:** Data not available
5.9 **Liquid or Solid Irritant Characteristics:** Data not available
5.10 **Odor Threshold:** Not pertinent

6. FIRE HAZARDS
6.1 **Flash Point:** Not flammable, but may cause fire on contact with combustibles
6.2 **Flammable Limits in Air:** Not flammable
6.3 **Fire Extinguishing Agents:** Flooding amounts of water
6.4 **Fire Extinguishing Agents Not to be Used:** Not pertinent
6.5 **Special Hazards of Combustion Products:** Toxic oxides of nitrogen formed in fires.
6.6 **Behavior in Fire:** Intensifies fires. When large quantities are involved, nitrate may fuse or melt; application of water may then cause extensive scattering of molten material.
6.7 **Ignition Temperature:** Not pertinent
6.8 **Electrical Hazard:** Not pertinent
6.9 **Burning Rate:** Not pertinent

7. CHEMICAL REACTIVITY
7.1 **Reactivity with Water:** Dissolves, forming weak solution of nitric acid; the reaction is not hazardous.
7.2 **Reactivity with Common Materials:** In contact with easily oxidizable substances, may react rapidly enough to cause ignition, violent combustion, or explosion. Water solutions are acidic and can corrode metals.
7.3 **Stability During Transport:** Stable
7.4 **Neutralizing Agents for Acids and Caustics:** Wash with water.
7.5 **Polymerization:** Not pertinent
7.6 **Inhibitor of Polymerization:** Not pertinent

8. WATER POLLUTION
8.1 **Aquatic Toxicity:** 3.1 mg/l /96 hr/fathead minnow/TL_m/fresh water
8.2 **Waterfowl Toxicity:** Data not available
8.3 **Biological Oxygen Demand (BOD):** None
8.4 **Food Chain Concentration Potential:** Data not available

9. SELECTED MANUFACTURERS
1. The S. W. Shattuck Chemical Co., Inc.
   1805 South Bannock Street
   Denver, Colo. 80223
2. J. T. Baker Chemical Co.
   222 Red School Lane
   Phillipsburg, N. J. 08865
3. Mallinckrodt Chemical Works
   223 West Side Avenue
   P. O. Box 384
   Jersey City, N. J. 07303

10. SHIPPING INFORMATION
10.1 **Grades or Purity:** Analytical reagent
10.2 **Storage Temperature:** Ambient
10.3 **Inert Atmosphere:** No requirement
10.4 **Venting:** Open

11. HAZARD ASSESSMENT CODE
(See Hazard Assessment Handbook, CG 446-3)

SS

12. HAZARD CLASSIFICATIONS
12.1 **Code of Federal Regulations:** Radioactive material
12.2 **NAS Hazard Rating for Bulk Water Transportation:** Not listed
12.3 **NFPA Hazard Classifications:**

Category	Classification*	
Health Hazard (Blue)	0	1
Flammability (Red)	0	0
Reactivity (Yellow)	0	0
	oxy	oxy

*First column refers to non-fire situation.

13. PHYSICAL AND CHEMICAL PROPERTIES
13.1 **Physical State at 15°C and 1 atm:** Solid
13.2 **Molecular Weight:** 502.13
13.3 **Boiling Point at 1 atm:** Not pertinent (decomposes)
13.4 **Freezing Point:** 140.4°F = 60.2°C = 333.4°K
13.5 **Critical Temperature:** Not pertinent
13.6 **Critical Pressure:** Not pertinent
13.7 **Specific Gravity:** 2.81 at 13°C (solid)
13.8 **Liquid Surface Tension:** Not pertinent
13.9 **Liquid-Water Interfacial Tension:** Not pertinent
13.10 **Vapor (Gas) Specific Gravity:** Not pertinent
13.11 **Ratio of Specific Heats of Vapor (Gas):** Not pertinent
13.12 **Latent Heat of Vaporization:** Not pertinent
13.13 **Heat of Combustion:** Not pertinent
13.14 **Heat of Decomposition:** Not pertinent
13.15 **Heat of Solution:** Not pertinent
13.16 **Heat of Polymerization:** Not pertinent

NOTES

URS

URANYL SULFATE

Common Synonyms			
Uranyl sulfate trihydrate Uranium sulfate Uranium sulfate trihydrate	Solid	Yellow	Odorless
	Sinks and mixes with water.		

Avoid contact with solid and dust. Keep people away.
Stop discharge if possible.
Isolate and remove discharged material.
Notify local health and pollution control agencies.

Fire	Not flammable.
Exposure	CALL FOR MEDICAL AID. DUST Irritating to eyes, nose and throat. If inhaled will cause coughing or difficult breathing. If in eyes, hold eyelids open and flush with plenty of water. If breathing has stopped, give artificial respiration. If breathing is difficult, give oxygen. SOLID Irritating to skin and eyes. Harmful if swallowed. Remove contaminated clothing and shoes. Flush affected areas with plenty of water. IF IN EYES, hold eyelids open and flush with plenty of water. IF SWALLOWED and victim is CONSCIOUS, have victim drink water or milk and have victim induce vomiting. IF SWALLOWED and victim is UNCONSCIOUS OR HAVING CONVULSIONS, do nothing except keep victim warm.
Water Pollution	Effect of low concentrations on aquatic life is unknown. May be dangerous if it enters water intakes. Notify local health and wildlife officials. Notify operators of nearby water intakes.

1. RESPONSE TO DISCHARGE

(See Response Methods Handbook, CG 446-4)

Issue warning—water contaminant
Restrict access
Disperse and flush

2. LABEL

3. CHEMICAL DESIGNATIONS

3.1 **Synonyms:** Uranium sulfate; Uranium sulfate trihydrate; Uranyl sulfate trihydrate
3.2 **Coast Guard Compatibility Classification:** Not listed
3.3 **Chemical Formula:** $UO_2SO_4 \cdot 3H_2O$
3.4 **IMCO/United Nations Numerical Designation:** Not listed

4. OBSERVABLE CHARACTERISTICS

4.1 **Physical State (as shipped):** Solid
4.2 **Color:** Yellow
4.3 **Odor:** None

5. HEALTH HAZARDS

5.1 **Personal Protective Equipment:** Approved dust respirator; goggles or face shield; protective clothing
5.2 **Symptoms Following Exposure:** Inhalation of dust may irritate nose and throat. Contact with eyes causes irritation.
5.3 **Treatment for Exposure:** *Get medical attention after all exposures to this compound.* INGESTION: give large amounts of water; induce vomiting. EYES: flush with water for at least 15 min. SKIN: flush with water.
5.4 **Toxicity by Inhalation (Threshold Limit Value):** 0.2 mg/m³ (as uranium)
5.5 **Short-Term Inhalation Limits:** Data not available
5.6 **Toxicity by Ingestion:** Grade 1; LD_{50} 5–15 g/kg
5.7 **Late Toxicity:** Data not available
5.8 **Vapor (Gas) Irritant Characteristics:** Data not available
5.9 **Liquid or Solid Irritant Characteristics:** Data not available
5.10 **Odor Threshold:** Data not available

6. FIRE HAZARDS

6.1 **Flash Point:** Not flammable
6.2 **Flammable Limits in Air:** Not flammable
6.3 **Fire Extinguishing Agents:** Not pertinent
6.4 **Fire Extinguishing Agents Not to be Used:** Not pertinent
6.5 **Special Hazards of Combustion Products:**
6.6 **Behavior in Fire:**
6.7 **Ignition Temperature:** Not pertinent
6.8 **Electrical Hazard:** Not pertinent
6.9 **Burning Rate:** Not pertinent

7. CHEMICAL REACTIVITY

7.1 **Reactivity with Water:** No reaction
7.2 **Reactivity with Common Materials:**
7.3 **Stability During Transport:** Stable
7.4 **Neutralizing Agents for Acids and Caustics:** Not pertinent
7.5 **Polymerization:** Not pertinent
7.6 **Inhibitor of Polymerization:** Not pertinent

8. WATER POLLUTION

8.1 **Aquatic Toxicity:**
135 ppm/96 hr/fathead minnow/TL_m/hard water
2.8 ppm/96 hr/fathead minnow/TL_m/soft water
8.2 **Waterfowl Toxicity:** Data not available
8.3 **Biological Oxygen Demand (BOD):** None
8.4 **Food Chain Concentration Potential:** Data not available

9. SELECTED MANUFACTURERS

1. Var-lac-oid Chemical Co.
666 South Front St.
Elizabeth, N. J. 07202
2. Gallard-Schlesinger Chemical Mfg. Co.
584 Mineola Ave.
Carle Place, N. Y. 11514
3. Ventron Corp.
P. O. Box 159
Beverly, Mass. 01915

10. SHIPPING INFORMATION

10.1 **Grades or Purity:** Commercial; Pure
10.2 **Storage Temperature:** Ambient
10.3 **Inert Atmosphere:** No requirement
10.4 **Venting:** Open

11. HAZARD ASSESSMENT CODE

(See Hazard Assessment Handbook, CG 446-3)

SS

12. HAZARD CLASSIFICATIONS

12.1 **Code of Federal Regulations:** Radioactive material
12.2 **NAS Hazard Rating for Bulk Water Transportation:** Not listed
12.3 **NFPA Hazard Classifications:** Not listed

13. PHYSICAL AND CHEMICAL PROPERTIES

13.1 **Physical State at 15°C and 1 atm:** Solid
13.2 **Molecular Weight:** 420.2
13.3 **Boiling Point at 1 atm:** Not pertinent (decomposes)
13.4 **Freezing Point:** Not pertinent
13.5 **Critical Temperature:** Not pertinent
13.6 **Critical Pressure:** Not pertinent
13.7 **Specific Gravity:** 3.28 at 20°C (solid)
13.8 **Liquid Surface Tension:** Not pertinent
13.9 **Liquid-Water Interfacial Tension:** Not pertinent
13.10 **Vapor (Gas) Specific Gravity:** Not pertinent
13.11 **Ratio of Specific Heats of Vapor (Gas):** Not pertinent
13.12 **Latent Heat of Vaporization:** Not pertinent
13.13 **Heat of Combustion:** Not pertinent
13.14 **Heat of Decomposition:** Not pertinent
13.15 **Heat of Solution:** Not pertinent
13.16 **Heat of Polymerization:** Not pertinent

NOTES

URE	# UREA

Common Synonyms		
Carbamide Carbonyldiamide	Solid crystals or pellets White Odorless Sinks and mixes with water.	

Call fire department.
Isolate and remove discharged material.
Notify local health and pollution control agencies.

Fire	Combustible. Extinguish with water.
Exposure	Not harmful.
Water Pollution	Effect of low concentrations on aquatic life is unknown. May be dangerous if it enters water intakes. Notify local health and wildlife officials. Notify operators of nearby water intakes.

1. RESPONSE TO DISCHARGE
(See Response Methods Handbook, CG 446-4)

Disperse and flush

2. LABELS

No hazard label required by Code of Federal Regulations

3. CHEMICAL DESIGNATIONS

3.1 **Synonyms:** Carbamide
Carbonyldiamide
3.2 **Coast Guard Compatibility Classification:** Not applicable
3.3 **Chemical Formula:** NH_2CONH_2
3.4 **IMCO/United Nations Numerical Designation:** Not listed

4. OBSERVABLE CHARACTERISTICS

4.1 **Physical State (as shipped):** Solid
4.2 **Color:** White
4.3 **Odor:** Odorless, or slight ammonia odor

5. HEALTH HAZARDS

5.1 **Personal Protective Equipment:** Goggles or face shield; dust mask.
5.2 **Symptoms Following Exposure:** May irritate eyes.
5.3 **Treatment for Exposure:** Wash eyes with water.
5.4 **Toxicity by Inhalation (Threshold Limit Value):** Not pertinent
5.5 **Short-Term Inhalation Limits:** Not pertinent
5.6 **Toxicity by Ingestion:** Data not available
5.7 **Late Toxicity:** None
5.8 **Vapor (Gas) Irritant Characteristics:** Non-volatile
5.9 **Liquid or Solid Irritant Characteristics:** None
5.10 **Odor Threshold:** Not pertinent

6. FIRE HAZARDS

6.1 **Flash Point:** Not flammable
6.2 **Flammable Limits in Air:** Not flammable
6.3 **Fire Extinguishing Agents:** Water
6.4 **Fire Extinguishing Agents Not to be Used:** Not pertinent
6.5 **Special Hazards of Combustion Products:** Not pertinent
6.6 **Behavior in Fire:** Melts and decomposes, generating ammonia.
6.7 **Ignition Temperature:** Not flammable
6.8 **Electrical Hazard:** Not pertinent
6.9 **Burning Rate:** Not flammable

7. CHEMICAL REACTIVITY

7.1 **Reactivity with Water:** No reaction
7.2 **Reactivity with Common Materials:** No reaction
7.3 **Stability During Transport:** Occurs only above melting point (132°C), yielding ammonia and other products. The decomposition is not explosive.
7.4 **Neutralizing Agents for Acids and Caustics:** Not pertinent
7.5 **Polymerization:** Not pertinent
7.6 **Inhibitor of Polymerization:** Not pertinent

8. WATER POLLUTION

8.1 **Aquatic Toxicity:**
30,000 mg/l /24 hr/creek chub/all died fresh water
16,000 mg/l /24 hr/creek chub/all survived/ fresh water
8.2 **Waterfowl Toxicity:** Data not available
8.3 **Biological Oxygen Demand (BOD):** 9%, 5 days
8.4 **Food Chain Concentration Potential:** None

9. SELECTED MANUFACTURERS

1. Allied Chemical Corp.
Agricultural Division
Morristown, N. J. 07960
2. Lone Star Gas Co.
Nipak, Inc. Division
Dallas, Tex. 75221
3. Triad Chemical Co.
Yazoo City, Miss. 39194

10. SHIPPING INFORMATION

10.1 **Grades or Purity:** Various grades and purities, which depend on manufacturing process and intended use. All have essentially the same hazardous properties.
10.2 **Storage Temperature:** Ambient
10.3 **Inert Atmosphere:** No requirement
10.4 **Venting:** Open

11. HAZARD ASSESSMENT CODE
(See Hazard Assessment Handbook, CG 446-3)

SS

12. HAZARD CLASSIFICATIONS

12.1 **Code of Federal Regulations:** Not listed
12.2 **NAS Hazard Rating for Bulk Water Transportation:** Not listed
12.3 **NFPA Hazard Classifications:** Not listed

13. PHYSICAL AND CHEMICAL PROPERTIES

13.1 **Physical State at 15°C and 1 atm:** Solid
13.2 **Molecular Weight:** 60.06
13.3 **Boiling Point at 1 atm:** Decomposes
13.4 **Freezing Point:** 271°F = 133°C = 406°K
13.5 **Critical Temperature:** Not pertinent
13.6 **Critical Pressure:** Not pertinent
13.7 **Specific Gravity:** 1.34 at 20°C (solid)
13.8 **Liquid Surface Tension:** Not pertinent
13.9 **Liquid-Water Interfacial Tension:** Not pertinent
13.10 **Vapor (Gas) Specific Gravity:** Not pertinent
13.11 **Ratio of Specific Heats of Vapor (Gas):** Not pertinent
13.12 **Latent Heat of Vaporization:** Not pertinent
13.13 **Heat of Combustion:** -3913 Btu/lb $= -2174$ cal/g $= -91.02 \times 10^5$ J/kg
13.14 **Heat of Decomposition:** Not pertinent
13.15 **Heat of Solution:** -108 Btu/lb $= -60.1$ cal/g $= -2.52 \times 10^5$ J/kg
13.16 **Heat of Polymerization:** Not pertinent

NOTES

UPO	# UREA PEROXIDE

Common Synonyms			
Urea hydrogen peroxide Hydrogen peroxide carbamide Urea, hydrogen peroxide salt Carbamide peroxide Carbonyl diamine peroxide	Solid Mixes with water.	White	Odorless

Keep people away.
Call fire department.
Isolate and remove discharged material.
Notify local health and pollution control agencies.

Fire	Combustible. May cause fire on contact with combustibles. POISONOUS GASES MAY BE PRODUCED IN FIRE. Containers may explode in fire. Flood discharge area with water. Cool exposed containers with water.
Exposure	Call for medical aid. SOLID Irritating to skin and eyes. Remove contaminated clothing and shoes. Flush affected areas with plenty of water. IF IN EYES, hold eyelids open and flush with plenty of water. IF SWALLOWED and victim is CONSCIOUS, have victim drink water or milk.
Water Pollution	Effect of low concentrations on aquatic life is unknown. May be dangerous if it enters water intakes. Notify local health and wildlife officials. Notify operators of nearby water intakes.

1. RESPONSE TO DISCHARGE

(See Response Methods Handbook, CG 446-4)

Issue warning—oxidizing material
Restrict access
Disperse and flush

2. LABEL

Yellow

3. CHEMICAL DESIGNATIONS

3.1 **Synonyms:** Carbamide peroxide; Carbonyl diamine peroxide; Hydrogen peroxide carbamide; Percarbamide; Perhydrol-Urea; Urea hydrogen peroxide
3.2 **Coast Guard Compatibility Classification:** Not applicable
3.3 **Chemical Formula:** $CO(NH_2)_2 \cdot H_2O_2$
3.4 **IMCO/United Nations Numerical Designation:** 5.1/1511

4. OBSERVABLE CHARACTERISTICS

4.1 **Physical State (as shipped):** Solid
4.2 **Color:** White
4.3 **Odor:** None

5. HEALTH HAZARDS

5.1 **Personal Protective Equipment:** Rubber gloves and protective goggles
5.2 **Symptoms Following Exposure:** Inhalation of dust causes irritation of nose from hydrogen peroxide formed when heated. Contact with eyes causes severe damage. Contact with moist skin causes temporary itching or burning sensation. Ingestion causes irritation of mouth and stomach.
5.3 **Treatment for Exposure:** INHALATION: remove victim from exposure; call physician. EYES: wash thoroughly with large quantities of water for at least 15 min.; call physician. SKIN: flush with water. INGESTION: give large ampunts of water; get medical attention.
5.4 **Toxicity by Inhalation (Threshold Limit Value):** Data not available
5.5 **Short-Term Inhalation Limits:** Data not available
5.6 **Toxicity by Ingestion:** Data not available
5.7 **Late Toxicity:** Data hot available
5.8 **Vapor (Gas) Irritant Characteristics:** Data not available
5.9 **Liquid or Solid Irritant Characteristics:** Data not available
5.10 **Odor Threshold:** Not pertinent

6. FIRE HAZARDS

6.1 **Flash Point:** Not pertinent (combustible solid; may cause fire upon contact with ordinary combustibles)
6.2 **Flammable Limits in Air:** Not pertinent
6.3 **Fire Extinguishing Agents:** Inert powders (e.g., sand, limestone), water
6.4 **Fire Extinguishing Agents Not to be Used:** Not pertinent
6.5 **Special Hazards of Combustion Products:** Irritating ammonia gas may be formed in fire.
6.6 **Behavior in Fire:** Melts and decomposes, giving off oxygen and ammonia. Increases severity of fire. Containers may explode.
6.7 **Ignition Temperature:** >680°F
6.8 **Electrical Hazard:** Data not available
6.9 **Burning Rate:** Not pertinent

7. CHEMICAL REACTIVITY

7.1 **Reactivity with Water:** Forms solution of hydrogen peroxide (non-hazardous reaction)
7.2 **Reactivity with Common Materials:** No significant reaction at ordinary temperatures. At 50°C (122°F) reacts with dust and rubbish.
7.3 **Stability During Transport:** Stable below 60°C (140°F).
7.4 **Neutralizing Agents for Acids and Caustics:** Not pertinent
7.5 **Polymerization:** Not pertinent
7.6 **Inhibitor of Polymerization:** Not pertinent

8. WATER POLLUTION

8.1 **Aquatic Toxicity:** Data not available
8.2 **Waterfowl Toxicity:** Data not available
8.3 **Biological Oxygen Demand (BOD):** None
8.4 **Food Chain Concentration Potential:** None

9. SELECTED MANUFACTURERS

1. The Norac Company, Inc.
   405 South Motor Avenue
   P. O. Box F
   Azusa, Calif. 91702
2. Degussa Incorporated
   Chemicals Division
   2 Pennsylvania Plaza
   New York, N. Y. 10001
3. Gallard-Schlesinger Chemical Mfg. Co.
   584 Mineola Avenue
   Carle Place, N. Y. 11514

10. SHIPPING INFORMATION

10.1 **Grades or Purity:** 98—100%
10.2 **Storage Temperature:** Below 60°C (140°F)
10.3 **Inert Atmosphere:** No requirement
10.4 **Venting:** Open (flame arrester)

11. HAZARD ASSESSMENT CODE

(See Hazard Assessment Handbook, CG 446-3)

SS

12. HAZARD CLASSIFICATIONS

12.1 **Code of Federal Regulations:** Organic Peroxide
12.2 **NAS Hazard Rating for Bulk Water Transportation:** Not listed
12.3 **NFPA Hazard Classifications:** Not listed

13. PHYSICAL AND CHEMICAL PROPERTIES

13.1 **Physical State at 15°C and 1 atm:** Solid
13.2 **Molecular Weight:** 94.1
13.3 **Boiling Point at 1 atm:** Not pertinent (decomposes)
13.4 **Freezing Point:** Not pertinent
13.5 **Critical Temperature:** Not pertinent
13.6 **Critical Pressure:** Not pertinent
13.7 **Specific Gravity:** 0.8 at 20°C (solid)
13.8 **Liquid Surface Tension:** Not pertinent
13.9 **Liquid-Water Interfacial Tension:** Not pertinent
13.10 **Vapor (Gas) Specific Gravity:** Not pertinent
13.11 **Ratio of Specific Heats of Vapor (Gas):** Not pertinent
13.12 **Latent Heat of Vaporization:** Not pertinent
13.13 **Heat of Combustion:** Data not available
13.14 **Heat of Decomposition:** -540 Btu/lb $= -300$ cal/g $= -12.5 \times 10^5$ J/kg
13.15 **Heat of Solution:** Not pertinent
13.16 **Heat of Polymerization:** Not pertinent

NOTES

VALERALDEHYDE

Common Synonyms	Watery liquid	Colorless	Fruity odor
Amyl aldehyde Valeric aldehyde Pentanal Valeral	Floats on water. Flammable, irritating vapor is produced.		

Stop discharge if possible. Keep people away.
Shut off ignition sources and call fire department.
Avoid contact with liquid and vapor.
Stay upwind and use water spray to "knock down" vapor.
Isolate and remove discharged material.
Notify local health and pollution control agencies.

Fire

FLAMMABLE
Flashback along vapor trail may occur.
Vapor may explode if ignited in an enclosed area.
Extinguish with foam, dry chemical, or carbon dioxide.
Water may be ineffective on fires.
Cool exposed containers with water.

Exposure

VAPOR
Irritating to eyes, nose, and throat.
Move to fresh air.
LIQUID
Irritating to skin and eyes.
Flush affected areas with plenty of water.

Water Pollution

Effect of low concentrations on aquatic life is unknown.
Fouling to shoreline.
May be dangerous if it enters water intakes.
Notify local health and wildlife officials.
Notify operators of nearby water intakes.

1. RESPONSE TO DISCHARGE

(See Response Methods Handbook, CG 446-4)

Issue warning—high flammability
Evacuate area
Disperse and flush

2. LABEL

Red

3. CHEMICAL DESIGNATIONS

3.1 **Synonyms:** Amyl aldehyde; Pentanal; Valeral; Valeric aldehyde
3.2 **Coast Guard Compatibility Classification:** Aldehyde
3.3 **Chemical Formula:** $CH_3(CH_2)_3CHO$
3.4 **IMCO/United Nations Numerical Designation:** 3.2/2058

4. OBSERVABLE CHARACTERISTICS

4.1 **Physical State (as shipped):** Liquid
4.2 **Color:** Colorless
4.3 **Odor:** Fruity

5. HEALTH HAZARDS

5.1 **Personal Protective Equipment:** Goggles or face shield; rubber gloves and boots.
5.2 **Symptoms Following Exposure:** Vapor may irritate eyes. Liquid irritates eyes and skin.
5.3 **Treatment for Exposure:** EYES: flush with water for at least 15 min. SKIN: wipe off, wash with soap and water.
5.4 **Toxicity by Inhalation (Threshold Limit Value):** Data not available
5.5 **Short-Term Inhalation Limits:** Data not available
5.6 **Toxicity by Ingestion:** Grade 1; LD_{50} 5 to 15 g/kg (mouse)
5.7 **Late Toxicity:** Data not available
5.8 **Vapor (Gas) Irritant Characteristics:** Vapors cause a slight smarting of the eyes or respiratory system if present in high concentrations. The effect is temporary.
5.9 **Liquid or Solid Irritant Characteristics:** Minimum hazard. If spilled on clothing and allowed to remain, may cause smarting and reddening of the skin.
5.10 **Odor Threshold:** Data not available

6. FIRE HAZARDS

6.1 **Flash Point:** 54°F O.C.
6.2 **Flammable Limits in Air:** Data not available
6.3 **Fire Extinguishing Agents:** Foam, dry chemical, or carbon dioxide
6.4 **Fire Extinguishing Agents Not to be Used:** Water may be ineffective.
6.5 **Special Hazards of Combustion Products:** Not pertinent
6.6 **Behavior in Fire:** Not pertinent
6.7 **Ignition Temperature:** Data not available
6.8 **Electrical Hazard:** Not pertinent
6.9 **Burning Rate:** 1.9 mm/min

7. CHEMICAL REACTIVITY

7.1 **Reactivity with Water:** No reaction
7.2 **Reactivity with Common Materials:** No reaction
7.3 **Stability During Transport:** Stable
7.4 **Neutralizing Agents for Acids and Caustics:** Not pertinent
7.5 **Polymerization:** Not pertinent
7.6 **Inhibitor of Polymerization:** Not pertinent

8. WATER POLLUTION

8.1 **Aquatic Toxicity:** Data not available
8.2 **Waterfowl Toxicity:** Data not available
8.3 **Biological Oxygen Demand (BOD):** 26% (theor.), 5 days
8.4 **Food Chain Concentration Potential:** None

9. SELECTED MANUFACTURERS

Union Carbide Corp.
Chemicals and Plastics Division
270 Park Ave.
New York, N. Y. 10017

10. SHIPPING INFORMATION

10.1 **Grades or Purity:** 98.5+%
10.2 **Storage Temperature:** Ambient
10.3 **Inert Atmosphere:** No requirement
10.4 **Venting:** Open (flame arrester), or pressure-vacuum

11. HAZARD ASSESSMENT CODE

(See Hazard Assessment Handbook, CG 446-3)

A-P-Q-T-U

12. HAZARD CLASSIFICATIONS

12.1 **Code of Federal Regulations:** Flammable liquid
12.2 **NAS Hazard Rating for Bulk Water Transportation:**

Category	Rating
Fire	3
Health	
Vapor Irritant	1
Liquid or Solid Irritant	1
Poisons	2
Water Pollution	
Human Toxicity	1
Aquatic Toxicity	2
Aesthetic Effect	3
Reactivity	
Other Chemicals	2
Water	0
Self-Reaction	1

12.3 **NFPA Hazard Classifications:** Not listed

13. PHYSICAL AND CHEMICAL PROPERTIES

13.1 **Physical State at 15°C and 1 atm:** Liquid
13.2 **Molecular Weight:** 86.13
13.3 **Boiling Point at 1 atm:** 217.4°F = 103.0°C = 376.2°K
13.4 **Freezing Point:** −132°F = −91°C = 182°K
13.5 **Critical Temperature:** 538°F = 281°C = 554°K
13.6 **Critical Pressure:** 514 psia = 35 atm = 3.5 MN/m^2
13.7 **Specific Gravity:** 0.811 at 20°C (liquid)
13.8 **Liquid Surface Tension:** (est.) 30 dynes/cm = 0.03 N/m at 20°C
13.9 **Liquid-Water Interfacial Tension:** (est.) 30 dynes/cm = 0.03 N/m at 20°C
13.10 **Vapor (Gas) Specific Gravity:** Not pertinent
13.11 **Ratio of Specific Heats of Vapor (Gas):** 1.072
13.12 **Latent Heat of Vaporization:** 170 Btu/lb = 93 cal/g = 3.9×10^5 J/kg
13.13 **Heat of Combustion:** −15,500 Btu/lb = −8,610 cal/g = -360.5×10^5 J/kg
13.14 **Heat of Decomposition:** Not pertinent
13.15 **Heat of Solution:** Not pertinent
13.16 **Heat of Polymerization:** Not pertinent

NOTES

REVISED 1978

VOT

VANADIUM OXYTRICHLORIDE

Common Synonyms			
Vanadyl trichloride Trichlorooxovanadium Vanadyl chloride	Liquid	Lemon yellow	Sharp unpleasant odor
	Sinks and mixes violently with water.		

AVOID CONTACT WITH LIQUID AND VAPOR. KEEP PEOPLE AWAY.
Wear rubber overclothing (including gloves).
Stop discharge if possible.
Isolate and remove discharged material.
Notify local health and pollution control agencies.

Fire

Not flammable.
DO NOT USE WATER ON ADJACENT FIRES.

Exposure

CALL FOR MEDICAL AID.

VAPOR
Irritating to eyes, nose and throat.
If inhaled will cause coughing or difficult breathing.

If in eyes, hold eyelids open and flush with plenty of water.
If breathing has stopped, give artificial respiration.
If breathing is difficult, give oxygen.

LIQUID
POISONOUS IF SWALLOWED OR IF SKIN IS EXPOSED.
Irritating to skin and eyes.
Harmful if swallowed.

Remove contaminated clothing and shoes.
Flush affected areas with plenty of water.
IF IN EYES, hold eyelids open and flush with plenty of water.
IF SWALLOWED and victim is CONSCIOUS, have victim drink water or milk.
IF SWALLOWED and victim is UNCONSCIOUS OR HAVING CONVULSIONS, do nothing except keep victim warm.

Water Pollution

Effect of low concentrations on aquatic life is unknown.
May be dangerous if it enters water intakes.

Notify local health and wildlife officials.
Notify operators of nearby water intakes.

1. RESPONSE TO DISCHARGE
(See Response Methods Handbook, CG 446-4)

Issue warning—corrosive, air contaminant
Restrict access
Disperse and flush

2. LABEL

3. CHEMICAL DESIGNATIONS

3.1 **Synonyms:** Trichlorooxo vanadium; Vanadyl chloride; Vanadyl trichloride
3.2 **Coast Guard Compatibility Classification:** Not listed
3.3 **Chemical Formula:** $VOCl_3$
3.4 **IMCO/United Nations Numerical Designation:** Not listed

4. OBSERVABLE CHARACTERISTICS

4.1 **Physical State (as shipped):** Liquid
4.2 **Color:** Lemon yellow
4.3 **Odor:** Acrid

5. HEALTH HAZARDS

5.1 **Personal Protective Equipment:** Acid vapor mask; rubber gloves; face shield; acid-resistant clothing
5.2 **Symptoms Following Exposure:** Inhalation of vapor causes irritation of nose and throat. Ingestion causes irritation of mouth and stomach. Contact with eyes or skin causes severe irritation.
5.3 **Treatment for Exposure:** *Consult a physician after all exposures to this compound.* INHALATION: move to fresh air; give artificial respiration if necessary. INGESTION: give large amount of water. EYES: flush with water for 15 min. SKIN: wipe exposed areas free of the chemical with a dry cloth, then flush thoroughly with water.
5.4 **Toxicity by Inhalation (Threshold Limit Value):** Data not available
5.5 **Short-Term Inhalation Limits:** 5 ppm (HCl) ceiling level, based on fact that compound decomposes in moist air into vanadic acid and HCl
5.6 **Toxicity by Ingestion:** Grade 3; oral rat LD_{50} = 140 mg/kg
5.7 **Late Toxicity:** Repeated exposures may cause discoloration of tongue, loss of appetite, anemia, kidney disorders, and blindness.
5.8 **Vapor (Gas) Irritant Characteristics:** Data not available
5.9 **Liquid or Solid Irritant Characteristics:** Data not available
5.10 **Odor Threshold:** 10 ppm HCl, based on decomposition of compound in moist air

6. FIRE HAZARDS

6.1 **Flash Point:** Not flammable
6.2 **Flammable Limits in Air:** Not flammable
6.3 **Fire Extinguishing Agents:** Not pertinent
6.4 **Fire Extinguishing Agents Not to be Used:** Water, unless in flooding amounts, should not be used on adjacent fires.
6.5 **Special Hazards of Combustion Products:** Irritating fumes of hydrogen chloride may form in fires.
6.6 **Behavior in Fire:**
6.7 **Ignition Temperature:** Not pertinent
6.8 **Electrical Hazard:** Not pertinent
6.9 **Burning Rate:** Not pertinent

7. CHEMICAL REACTIVITY

7.1 **Reactivity with Water:** Reacts to form a solution of hydrochloric acid
7.2 **Reactivity with Common Materials:** In presence of moisture will corrode most metals
7.3 **Stability During Transport:** Stable
7.4 **Neutralizing Agents for Acids and Caustics:** Flush with water and sprinkle with powdered limestone or rinse with dilute solution of sodium bicarbonate or soda ash.
7.5 **Polymerization:** Not pertinent
7.6 **Inhibitor of Polymerization:** Not pertinent

8. WATER POLLUTION

8.1 **Aquatic Toxicity:** See vanadyl sulfate
8.2 **Waterfowl Toxicity:** Data not available
8.3 **Biological Oxygen Demand (BOD):** None
8.4 **Food Chain Concentration Potential:** Data not available

9. SELECTED MANUFACTURERS

1. Foote Mineral Co.
   Route 100
   Exton, Pa. 19341
2. Stauffer Chemical Company
   Specialty Chemical Div.
   Westport, Conn. 06880
3. Ventron, Inc.
   P. O. Box 159
   Beverly, Mass. 01915

10. SHIPPING INFORMATION

10.1 **Grades or Purity:** Technical, 99%
10.2 **Storage Temperature:** Ambient
10.3 **Inert Atmosphere:** No requirement
10.4 **Venting:** Pressure-vacuum

11. HAZARD ASSESSMENT CODE
(See Hazard Assessment Handbook, CG 446-3)

A-O

12. HAZARD CLASSIFICATIONS

12.1 **Code of Federal Regulations:** Corrosive liquid
12.2 **NAS Hazard Rating for Bulk Water Transportation:** Not listed
12.3 **NFPA Hazard Classifications:** Not listed

13. PHYSICAL AND CHEMICAL PROPERTIES

13.1 **Physical State at 15°C and 1 atm:** Liquid
13.2 **Molecular Weight:** 173.3
13.3 **Boiling Point at 1 atm:** 259°F = 126°C = 399°K
13.4 **Freezing Point:** −107°F = −77°C = 196°K
13.5 **Critical Temperature:** Data not available
13.6 **Critical Pressure:** Data not available
13.7 **Specific Gravity:** 1.83 at 20°C (liquid)
13.8 **Liquid Surface Tension:** Data not available
13.9 **Liquid-Water Interfacial Tension:** Not pertinent
13.10 **Vapor (Gas) Specific Gravity:** 5.98
13.11 **Ratio of Specific Heats of Vapor (Gas):** Data not available
13.12 **Latent Heat of Vaporization:** Data not available
13.13 **Heat of Combustion:** Not pertinent
13.14 **Heat of Decomposition:** Not pertinent
13.15 **Heat of Solution:** Data not available
13.16 **Heat of Polymerization:** Not pertinent

NOTES

VOX	VANADIUM PENTOXIDE

Common Synonyms			
Vanadic anhydride Vanadium pentaoxide	Solid	Yellowish-brown	Odorless
	Sinks in water.		

Avoid contact with solid and dust. Keep people away.
Stop discharge if possible.
Call fire department.
Isolate and remove discharged material.
Notify local health and pollution control agencies.

Fire

Not flammable.
Will increase the intensity of a fire.
Flood discharge area with water.

Exposure

CALL FOR MEDICAL AID.

DUST
Irritating to eyes, nose and throat.
If inhaled will cause coughing or difficult breathing.

If in eyes, hold eyelids open and flush with plenty of water.
If breathing has stopped, give artificial respiration.
If breathing is difficult, give oxygen.

SOLID
Irritating to skin and eyes.
If swallowed will cause nausea.

Remove contaminated clothing and shoes.
Flush affected areas with plenty of water.
IF IN EYES, hold eyelids open and flush with plenty of water.
IF SWALLOWED and victim is CONSCIOUS, have victim drink water or milk and have victim induce vomiting.
IF SWALLOWED and victim is UNCONSCIOUS OR HAVING CONVULSIONS, do nothing except keep victim warm.

Water Pollution

HARMFUL TO AQUATIC LIFE IN VERY LOW CONCENTRATIONS.
May be dangerous if it enters water intakes.
Notify local health and wildlife officials.
Notify operators of nearby water intakes.

1. RESPONSE TO DISCHARGE
(See Response Methods Handbook, CG 446-4)

Should be removed
Chemical and physical treatment

2. LABELS

No hazard label required by Code of Federal Regulations

3. CHEMICAL DESIGNATIONS

3.1 **Synonyms:** Vanadic anhydride; Vanadium pentaoxide
3.2 **Coast Guard Compatibility Classification:** Not listed
3.3 **Chemical Formula:** V_2O_5
3.4 **IMCO/United Nations Numerical Designation:** Not listed

4. OBSERVABLE CHARACTERISTICS

4.1 **Physical State (as shipped):** Solid
4.2 **Color:** Yellow-orange (powder), dark gray (flakes); yellowish brown
4.3 **Odor:** None

5. HEALTH HAZARDS

5.1 **Personal Protective Equipment:** Bu. Mines approved respirator; rubber gloves; goggles for prolonged exposure
5.2 **Symptoms Following Exposure:** Inhalation of dust irritates nose and throat. Ingestion causes irritation of mouth and stomach. Contact with eyes or skin causes irritation; eczema may develop.
5.3 **Treatment for Exposure:** INHALATION: move to fresh air; if exposure to dust has been severe, get medical attention. INGESTION: induce vomiting; get medical attention. EYES: flush with water for at least 15 min. SKIN: flush with water; wash with soap and water.
5.4 **Toxicity by Inhalation (Threshold Limit Value):** 0.5 mg/m^3
5.5 **Short-Term Inhalation Limits:** Data not available
5.6 **Toxicity by Ingestion:** Grade 4; oral LD_{50} = 23 mg/kg (mouse)
5.7 **Late Toxicity:** Repeated exposures may cause discoloration of tongue, loss of appetite, anemia, kidney disorders, and blindness.
5.8 **Vapor (Gas) Irritant Characteristics:** Data not available
5.9 **Liquid or Solid Irritant Characteristics:** Data not available
5.10 **Odor Threshold:** Data not available

6. FIRE HAZARDS

6.1 **Flash Point:** Not flammable
6.2 **Flammable Limits in Air:** Not flammable
6.3 **Fire Extinguishing Agents:** Not pertinent
6.4 **Fire Extinguishing Agents Not to be Used:** Not pertinent
6.5 **Special Hazards of Combustion Products:** Not pertinent
6.6 **Behavior in Fire:** May increase intensity of fire when in contact with combustible materials
6.7 **Ignition Temperature:** Not pertinent
6.8 **Electrical Hazard:** Not pertinent
6.9 **Burning Rate:** Not pertinent

7. CHEMICAL REACTIVITY

7.1 **Reactivity with Water:** No reaction
7.2 **Reactivity with Common Materials:**
7.3 **Stability During Transport:** Stable
7.4 **Neutralizing Agents for Acids and Caustics:** Not pertinent
7.5 **Polymerization:** Not pertinent
7.6 **Inhibitor of Polymerization:** Not pertinent

8. WATER POLLUTION

8.1 **Aquatic Toxicity:**
55 ppm/96 hr/fathead minnows/TL_m/hard water
13 ppm/96 hr/fathead minnows/TL_m/soft water
8.2 **Waterfowl Toxicity:** Data not available
8.3 **Biological Oxygen Demand (BOD):** None
8.4 **Food Chain Concentration Potential:** Data not available

9. SELECTED MANUFACTURERS

1. Foote Mineral Co.
Route 100
Exton, Pa. 19341
2. Kerr-McGee Chemical Corp.
Soda Springs
Idaho 83276
3. Stauffer Chemical Co.
Industrial Chemicals Div.
Westport, Conn. 06880

10. SHIPPING INFORMATION

10.1 **Grades or Purity:** Commercial, 98—99%
10.2 **Storage Temperature:** Ambient
10.3 **Inert Atmosphere:** No requirement
10.4 **Venting:** Open

11. HAZARD ASSESSMENT CODE
(See Hazard Assessment Handbook, CG 446-3)

II

12. HAZARD CLASSIFICATIONS

12.1 **Code of Federal Regulations:** Not listed
12.2 **NAS Hazard Rating for Bulk Water Transportation:** Not listed
12.3 **NFPA Hazard Classifications:** Not listed

13. PHYSICAL AND CHEMICAL PROPERTIES

13.1 **Physical State at 15°C and 1 atm:** Solid
13.2 **Molecular Weight:** 181.88
13.3 **Boiling Point at 1 atm:** Not pertinent (decomposes)
13.4 **Freezing Point:** Not pertinent
13.5 **Critical Temperature:** Not pertinent
13.6 **Critical Pressure:** Not pertinent
13.7 **Specific Gravity:** 3.36 at 20°C (solid)
13.8 **Liquid Surface Tension:** Not pertinent
13.9 **Liquid-Water Interfacial Tension:** Not pertinent
13.10 **Vapor (Gas) Specific Gravity:** Not pertinent
13.11 **Ratio of Specific Heats of Vapor (Gas):** Not pertinent
13.12 **Latent Heat of Vaporization:** Not pertinent
13.13 **Heat of Combustion:** Not pertinent
13.14 **Heat of Decomposition:** Not pertinent
13.15 **Heat of Solution:** Not pertinent
13.16 **Heat of Polymerization:** Not pertinent

NOTES

VSF	# VANADYL SULFATE

Common Synonyms Vanadium oxysulfate Vanadyl sulfate dihydrate	Solid — Pale blue — Odorless Sinks and mixes with water.

Avoid contact with solid and dust. Keep people away.
Stop discharge if possible.
Isolate and remove discharged material.
Notify local health and pollution control agencies.

Fire	Not flammable.
Exposure	CALL FOR MEDICAL AID. DUST Irritating to eyes, nose and throat. If inhaled will cause coughing or difficult breathing. If in eyes, hold eyelids open and flush with plenty of water. If breathing has stopped, give artificial respiration. If breathing is difficult, give oxygen. SOLID Irritating to skin and eyes. If swallowed will cause nausea or coughing. Remove contaminated clothing and shoes. Flush affected areas with plenty of water. IF IN EYES, hold eyelids open and flush with plenty of water. IF SWALLOWED and victim is CONSCIOUS, have victim drink water or milk and have victim induce vomiting. IF SWALLOWED and victim is UNCONSCIOUS OR HAVING CONVULSIONS, do nothing except keep victim warm.
Water Pollution	HARMFUL TO AQUATIC LIFE IN VERY LOW CONCENTRATIONS. May be dangerous if it enters water intakes. Notify local health and wildlife officials. Notify operators of nearby water intakes.

1. RESPONSE TO DISCHARGE
(See Response Methods Handbook, CG 446-4)

Issue warning—water contaminant
Disperse and flush

2. LABELS

No hazard label required by Code of Federal Regulations

3. CHEMICAL DESIGNATIONS

3.1 **Synonyms:** Vanadium oxysulfate; Vanadyl sulfate dihydrate
3.2 **Coast Guard Compatibility Classification:** Not listed
3.3 **Chemical Formula:** $VOSO_4 \cdot 2H_2O$
3.4 **IMCO/United Nations Numerical Designation:** Not listed

4. OBSERVABLE CHARACTERISTICS

4.1 **Physical State (as shipped):** Solid
4.2 **Color:** Pale blue
4.3 **Odor:** None

5. HEALTH HAZARDS

5.1 **Personal Protective Equipment:** Dust mask; goggles or face shield; protective gloves
5.2 **Symptoms Following Exposure:** Inhalation of dust causes irritation of nose and throat. Ingestion causes irritation of mouth and stomach. Contact with eyes or skin causes irritation; eczema may develop. (Results of prolonged exposure are described in section 5.7).
5.3 **Treatment for Exposure:** INHALATION: move to fresh air; if exposure to dust has been severe, get medical attention. INGESTION: give large amount of water; induce vomiting; get medical attention. EYES: flush with water for at least 15 min. SKIN: flush with water.
5.4 **Toxicity by Inhalation (Threshold Limit Value):** Data not available
5.5 **Short-Term Inhalation Limits:** Data not available
5.6 **Toxicity by Ingestion:** Grade 3; LD_{50} 50—500 mg/kg
5.7 **Late Toxicity:** Repeated exposures may cause discoloration of tongue, loss of appetite, anemia, kidney disorders, and blindness.
5.8 **Vapor (Gas) Irritant Characteristics:** Data not available
5.9 **Liquid or Solid Irritant Characteristics:** Data not available
5.10 **Odor Threshold:** Data not available

6. FIRE HAZARDS

6.1 **Flash Point:** Not flammable
6.2 **Flammable Limits in Air:** Not flammable
6.3 **Fire Extinguishing Agents:** Not pertinent
6.4 **Fire Extinguishing Agents Not to be Used:** Not pertinent
6.5 **Special Hazards of Combustion Products:** Not pertinent
6.6 **Behavior in Fire:**
6.7 **Ignition Temperature:** Not pertinent
6.8 **Electrical Hazard:** Not pertinent
6.9 **Burning Rate:** Not pertinent

7. CHEMICAL REACTIVITY

7.1 **Reactivity with Water:** No reaction
7.2 **Reactivity with Common Materials:**
7.3 **Stability During Transport:** Stable
7.4 **Neutralizing Agents for Acids and Caustics:** Not pertinent
7.5 **Polymerization:** Not pertinent
7.6 **Inhibitor of Polymerization:** Not pertinent

8. WATER POLLUTION

8.1 **Aquatic Toxicity:**
30 ppm/96 hr/fathead minnow/TL_m/hard water
4.8 ppm/96 hr/fathead minnow/TL_m/soft water
8.2 **Waterfowl Toxicity:** Data not available
8.3 **Biological Oxygen Demand (BOD):** Data not available
8.4 **Food Chain Concentration Potential:** Data not available

9. SELECTED MANUFACTURERS

1. Gallard-Schlesinger Chemical Mfg. Co.
584 Mineola Ave.
Carle Place, N. Y. 11514
2. Eastman Organic Chemicals
Rochester, N. Y. 14650
3. Ventron, Inc.
P. O. Box 159
Beverly, Mass. 01915

10. SHIPPING INFORMATION

10.1 **Grades or Purity:** Commercial; Pure
10.2 **Storage Temperature:** Ambient
10.3 **Inert Atmosphere:** No requirement
10.4 **Venting:** Open

11. HAZARD ASSESSMENT CODE
(See Hazard Assessment Handbook, CG 446-3)

SS

12. HAZARD CLASSIFICATIONS

12.1 **Code of Federal Regulations:** Not listed
12.2 **NAS Hazard Rating for Bulk Water Transportation:** Not listed
12.3 **NFPA Hazard Classifications:** Not listed

13. PHYSICAL AND CHEMICAL PROPERTIES

13.1 **Physical State at 15°C and 1 atm:** Solid
13.2 **Molecular Weight:** 199.1
13.3 **Boiling Point at 1 atm:** Not pertinent (decomposes)
13.4 **Freezing Point:** Not pertinent
13.5 **Critical Temperature:** Not pertinent
13.6 **Critical Pressure:** Not pertinent
13.7 **Specific Gravity:** (approx.) 2.5 at 20°C (solid)
13.8 **Liquid Surface Tension:** Not pertinent
13.9 **Liquid-Water Interfacial Tension:** Not pertinent
13.10 **Vapor (Gas) Specific Gravity:** Not pertinent
13.11 **Ratio of Specific Heats of Vapor (Gas):** Not pertinent
13.12 **Latent Heat of Vaporization:** Not pertinent
13.13 **Heat of Combustion:** Not pertinent
13.14 **Heat of Decomposition:** Not pertinent
13.15 **Heat of Solution:** Data not available
13.16 **Heat of Polymerization:** Not pertinent

NOTES

VINYL ACETATE

Common Synonyms
VyAc
Vinyl A monomer

Watery liquid — Colorless — Pleasant fruity odor

Floats on water. Flammable, irritating vapor is produced.

Stop discharge if possible. Keep people away.
Shut off ignition sources and call fire department.
Avoid contact with liquid and vapor.
Stay upwind and use water spray to "knock down" vapor.
Isolate and remove discharged material.
Notify local health and pollution control agencies.

Fire

FLAMMABLE.
Containers may explode in fire.
Flashback along vapor trail may occur.
Vapor may explode if ignited in an enclosed area.
Wear goggles and self-contained breathing apparatus.
Combat fires from safe distance or protected location.
Extinguish with dry chemical, foam, carbon dioxide.
Cool exposed containers with water.

Exposure

CALL FOR MEDICAL AID.

VAPOR
Irritating to eyes, nose and throat.
If inhaled, will cause dizziness or difficult breathing.
Move to fresh air.
If breathing has stopped, give artificial respiration.
If breathing is difficult, give oxygen.

LIQUID
Irritating to skin and eyes.
Harmful if swallowed or if spilled on skin.
Remove contaminated clothing and shoes.
Flush affected areas with plenty of water.
IF IN EYES, hold eyelids open and flush with plenty of water.
IF SWALLOWED and victim is CONSCIOUS, have victim drink water or milk.

Water Pollution

HARMFUL TO AQUATIC LIFE IN VERY LOW CONCENTRATIONS.
Fouling to shoreline.
May be dangerous if it enters water intakes.
Notify local health and wildlife officials.
Notify operators of nearby water intakes.

1. RESPONSE TO DISCHARGE

(See Response Methods Handbook, CG 446-4)

Issue warning—high flammability, air contaminant
Evacuate area

2. LABEL

Red

3. CHEMICAL DESIGNATIONS

3.1 **Synonyms:** VAM
Vinyl A monomer
Vy Ac
3.2 **Coast Guard Compatibility Classification:** Vinyl acetate
3.3 **Chemical Formula:** $CH_3COOCH{=}CH_2$
3.4 **IMCO/United Nations Numerical Designation:** 3.2/1301

4. OBSERVABLE CHARACTERISTICS

4.1 **Physical State (as shipped):** Liquid
4.2 **Color:** Colorless
4.3 **Odor:** Not unpleasant, sweet smell in small quantities; pleasant fruity; characteristic

5. HEALTH HAZARDS

5.1 **Personal Protective Equipment:** Approved canister or air-supplied mask; goggles or face shield; rubber or plastic gloves.
5.2 **Symptoms Following Exposure:** High vapor concentrations cause narcosis. Liquid irritates eyes and may irritate skin.
5.3 **Treatment for Exposure:** INHALATION: remove victim to fresh air; give artificial respiration if required. EYES: flush with water for at least 15 min.
5.4 **Toxicity by Inhalation (Threshold Limit Value):** 10 ppm
5.5 **Short-Term Inhalation Limits:** Data not available
5.6 **Toxicity by Ingestion:** Grade 2; LD_{50} 0.5 to 5 g/kg (rat)
5.7 **Late Toxicity:** Data not available
5.8 **Vapor (Gas) Irritant Characteristics:** Vapors cause a slight smarting of the eyes or respiratory system if present in high concentrations. The effect is temporary.
5.9 **Liquid or Solid Irritant Characteristics:** Minimum hazard. If spilled on clothing and allowed to remain, may cause smarting and reddening of the skin.
5.10 **Odor Threshold:** 0.12 ppm

6. FIRE HAZARDS

6.1 **Flash Point:** 18°F C.C.; 23°F O.C.
6.2 **Flammable Limits in Air:** 2.6%–13.4%
6.3 **Fire Extinguishing Agents:** Carbon dioxide or dry chemical for small fires; ordinary foam for large fires.
6.4 **Fire Extinguishing Agents Not to be Used:** Water may be ineffective.
6.5 **Special Hazards of Combustion Products:** Not pertinent
6.6 **Behavior in Fire:** Vapor is heavier than air and may travel considerable distance to a source of ignition and flash back. May polymerize when heated in a fire and rupture container.
6.7 **Ignition Temperature:** 800°F
6.8 **Electrical Hazard:** Class I, Group D
6.9 **Burning Rate:** 3.8 mm/min.

7. CHEMICAL REACTIVITY

7.1 **Reactivity with Water:** No reaction
7.2 **Reactivity with Common Materials:** No reaction
7.3 **Stability During Transport:** Stable
7.4 **Neutralizing Agents for Acids and Caustics:** Not pertinent
7.5 **Polymerization:** Can occur when in contact with peroxides and strong acids; but only under extreme conditions.
7.6 **Inhibitor of Polymerization:** 3–5 ppm or 14–17 ppm hydroquinone. Shipments usually also contain 200 ppm of diphenylamine.

8. WATER POLLUTION

8.1 **Aquatic Toxicity:**
18 ppm/96 hr/bluegill/TL_m/fresh water
> 100 ppm/48 hr/flounder/LC_{50}/salt water
8.2 **Waterfowl Toxicity:** Data not available
8.3 **Biological Oxygen Demand (BOD):** 62% of theoretical in 5 days, freshwater, acclimated seed
8.4 **Food Chain Concentration Potential:** None

9. SELECTED MANUFACTURERS

1. Celanese Corp.
Celanese Chemical Co. Division
245 Park Ave.
New York, N. Y. 10017
2. National Distillers & Chemical Corp.
U. S. Industrial Chemicals Co. Division
Houston, Tex. 77000
3. Union Carbide Corp.
Chemicals and Plastics Division
270 Park Ave.
New York, N. Y. 10017

10. SHIPPING INFORMATION

10.1 **Grades or Purity:**
Grade A (Diphenylamine-inhibited): 99.8%
Grade H (Hydroquinone-inhibited): 99.8%
10.2 **Storage Temperature:** Ambient
10.3 **Inert Atmosphere:** No requirement
10.4 **Venting:** Pressure-vacuum

11. HAZARD ASSESSMENT CODE

(See Hazard Assessment Handbook, CG 446-3)

A-P-Q-T-U-V-W-Z

12. HAZARD CLASSIFICATIONS

12.1 **Code of Federal Regulations:** Flammable liquid
12.2 **NAS Hazard Rating for Bulk Water Transportation:**

Category	Rating
Fire	3
Health	
Vapor Irritant	1
Liquid or Solid Irritant	1
Poisons	2
Water Pollution	
Human Toxicity	2
Aquatic Toxicity	1
Aesthetic Effect	2
Reactivity	
Other Chemicals	2
Water	0
Self-Reaction	3

12.3 **NFPA Hazard Classifications:**

Category	Classification
Health Hazard (Blue)	2
Flammability (Red)	3
Reactivity (Yellow)	2

13. PHYSICAL AND CHEMICAL PROPERTIES

13.1 **Physical State at 15°C and 1 atm:** Liquid
13.2 **Molecular Weight:** 86.09
13.3 **Boiling Point at 1 atm:** 163.2°F = 72.9°C = 346.1°K
13.4 **Freezing Point:** −135.0°F = −92.8°C = 180.4°K
13.5 **Critical Temperature:** 486°F = 252°C = 525°K
13.6 **Critical Pressure:** 617 atm = 42 psia = 4.25 MN/m^2
13.7 **Specific Gravity:** 0.934 at 20°C (liquid)
13.8 **Liquid Surface Tension:** 23.95 dynes/cm = 0.02395 N/m at 20°C
13.9 **Liquid-Water Interfacial Tension:** (est.) 30 dynes/cm = 0.03 N/m at 20°C
13.10 **Vapor (Gas) Specific Gravity:** Not pertinent
13.11 **Ratio of Specific Heats of Vapor (Gas):** 1.103
13.12 **Latent Heat of Vaporization:** 163 Btu/lb = 90.6 cal/g = 3.79×10^5 J/kg
13.13 **Heat of Combustion:** −9754 Btu/lb = −5419 cal/g = -226.9×10^5 J/kg
13.14 **Heat of Decomposition:** Not pertinent
13.15 **Heat of Solution:** Not pertinent
13.16 **Heat of Polymerization:** −439 Btu/lb = −244 cal/g = -10.2×10^5 J/kg

NOTES

REVISED 1978

VCM	VINYL CHLORIDE

Common Synonyms
Cloroethylene
VCL
Vinyl C Monomer

Liquefied compressed gas   Colorless   Sweet odor

Liquid floats and boils on water. Flammable, irritating visible vapor cloud is produced.

Stop discharge if possible. Keep people away.
Shut off ignition sources and call fire department.
Stay upwind and use water spray to "knock down" vapor.
Evacuate area in case of large discharge.
Avoid contact with liquid and vapor.
Notify local health and pollution control agencies.

Fire

FLAMMABLE.
POISONOUS GAS IS PRODUCED IN FIRE.
Flashback along vapor trail may occur.
May explode if ignited in an enclosed area.
Wear self-contained breathing apparatus.
Cool exposed containers and protect men effecting shutoff with water.
Stop flow of gas if possible.
Let fire burn.
Extinguish small fires with dry chemical.

Exposure

CALL FOR MEDICAL AID.

VAPOR
Irritating to eyes, nose, and throat.
If inhaled, will cause dizziness or difficult breathing.
Move to fresh air.
If breathing has stopped, give artificial respiration.
If breathing is difficult, give oxygen.

LIQUID
Will cause frostbite.
Flush affected areas with plenty of water.
DO NOT RUB AFFECTED AREAS.

Water Pollution

Not harmful to aquatic life.

1. RESPONSE TO DISCHARGE

(See Response Methods Handbook, CG 446-4)

Issue warning—high flammability
Evacuate area

2. LABEL

Red

3. CHEMICAL DESIGNATIONS

3.1 **Synonyms:** Chloroethene
Chloroethylene
Vinyl C Monomer
VCL
VCM

3.2 **Coast Guard Compatibility Classification:** Vinyl halides

3.3 **Chemical Formula:** $CH_2{=}CHCl$

3.4 **IMCO/United Nations Numerical Designation:** 2.0/1086

4. OBSERVABLE CHARACTERISTICS

4.1 **Physical State (as shipped):** Liquefied gas

4.2 **Color:** Colorless

4.3 **Odor:** Pleasant, sweet

5. HEALTH HAZARDS

5.1 **Personal Protective Equipment:** Rubber gloves and shoes; gas-tight goggles; organic vapor canister or self-contained breathing apparatus.

5.2 **Symptoms Following Exposure:** INHALATION: high concentrations cause dizziness, anesthesia, lung irritation. SKIN: may cause frostbite; phenol inhibitor may be absorbed through skin if large amounts of liquid evaporate.

5.3 **Treatment for Exposure:** INHALATION: remove patient to fresh air and keep him quiet and warm; call a doctor; give artificial respiration if breathing stops. EYES AND SKIN: flush with plenty of water for at least 15 min.; for eyes, get medical attention; remove contaminated clothing.

5.4 **Toxicity by Inhalation (Threshold Limit Value):** 200 ppm

5.5 **Short-Term Inhalation Limits:** 500 ppm for 5 min.

5.6 **Toxicity by Ingestion:** Not pertinent

5.7 **Late Toxicity:** Chronic exposure may cause liver damage.

5.8 **Vapor (Gas) Irritant Characteristics:** Vapors cause moderate irritation such that personnel will find high concentrations unpleasant. The effect is temporary.

5.9 **Liquid or Solid Irritant Characteristics:** Minimum hazard. If spilled on clothing and allowed to remain, may cause smarting and reddening of skin. May cause frostbite.

5.10 **Odor Threshold:** 260 ppm

6. FIRE HAZARDS

6.1 **Flash Point:** −110°F O.C.

6.2 **Flammable Limits in Air:** 4% 26%

6.3 **Fire Extinguishing Agents:** For small fires use dry chemical or carbon dioxide. For large fires stop flow of gas. Cool exposed containers with water.

6.4 **Fire Extinguishing Agents Not to be Used:** Not pertinent

6.5 **Special Hazards of Combustion Products:** Forms highly toxic combustion products such as hydrogen chloride, phosgene, and carbon monoxide.

6.6 **Behavior in Fire:** Container may explode in fire. Gas is heavier than air and may travel considerable distance to a source of ignition and flash back.

6.7 **Ignition Temperature:** 882°F

6.8 **Electrical Hazard:** Class I, Group D

6.9 **Burning Rate:** 4.3 mm/min.

7. CHEMICAL REACTIVITY

7.1 **Reactivity with Water:** No reaction

7.2 **Reactivity with Common Materials:** No reaction

7.3 **Stability During Transport:** Stable

7.4 **Neutralizing Agents for Acids and Caustics:** Not pertinent

7.5 **Polymerization:** Polymerizes in presence of air, sunlight, or heat unless stabilized by inhibitors.

7.6 **Inhibitor of Polymerization:** Not normally used except when high temperatures are expected. Then 40-100 ppm of phenol used.

8. WATER POLLUTION

8.1 **Aquatic Toxicity:** None

8.2 **Waterfowl Toxicity:** None

8.3 **Biological Oxygen Demand (BOD):** None

8.4 **Food Chain Concentration Potential:** None

9. SELECTED MANUFACTURERS

1. Dow Chemical Co.
Midland, Mich. 48640

2. The B. F. Goodrich Co.
6100 Oak Tree Blvd.
Cleveland, Ohio 44131

3. PPG Industries, Inc.
Chemical Division
Guayanilla, Puerto Rico 00656

10. SHIPPING INFORMATION

10.1 **Grades or Purity:** Commercial or technical 99+%

10.2 **Storage Temperature:**
Under pressure: ambient
At atm. pressure: low

10.3 **Inert Atmosphere:** No requirement

10.4 **Venting:**
Under pressure: safety relief
At atm. pressure: pressure-vacuum

11. HAZARD ASSESSMENT CODE

(See Hazard Assessment Handbook, CG 446-3)

A-B-C-D-E-F-G-Z

12. HAZARD CLASSIFICATIONS

12.1 **Code of Federal Regulations:** Flammable compressed gas

12.2 **NAS Hazard Rating for Bulk Water Transportation:**

Category	Rating
Fire	4
Health	
Vapor Irritant	2
Liquid or Solid Irritant	1
Poisons	2
Water Pollution	
Human Toxicity	0
Aquatic Toxicity	0
Aesthetic Effect	0
Reactivity	
Other Chemicals	2
Water	0
Self-Reaction	2

12.3 **NFPA Hazard Classifications:**

Category	Classification
Health Hazard (Blue)	2
Flammability (Red)	4
Reactivity (Yellow)	1

13. PHYSICAL AND CHEMICAL PROPERTIES

13.1 **Physical State at 15°C and 1 atm:** Gas

13.2 **Molecular Weight:** 62.50

13.3 **Boiling Point at 1 atm:** 7.2°F = −13.8°C = 259.4°K

13.4 **Freezing Point:** −244.8°F = −153.8°C = −119.4°K

13.5 **Critical Temperature:** 317.1°F = 158.4°C = 431.6°K

13.6 **Critical Pressure:** 775 psia = 52.7 atm = 5.34 MN/m²

13.7 **Specific Gravity:** 0.969 at −13°C (liquid)

13.8 **Liquid Surface Tension:** 16.0 dynes/cm = 0.0160 N/m at 25°C

13.9 **Liquid-Water Interfacial Tension:** (est.) 30 dynes/cm = 0.03 N/m at 20°C

13.10 **Vapor (Gas) Specific Gravity:** 2.2

13.11 **Ratio of Specific Heats of Vapor (Gas):** 1.186

13.12 **Latent Heat of Vaporization:** 160 Btu/lb = 88 cal/g = 3.7×10^5 J/kg

13.13 **Heat of Combustion:** −8136 Btu/lb = −4520 cal/g = -189.1×10^5 J/kg

13.14 **Heat of Decomposition:** Not pertinent

13.15 **Heat of Solution:** Not pertinent

13.16 **Heat of Polymerization:** −729 Btu/lb = −405 cal/g = -16.9×10^5 J/kg

NOTES

REVISED 1978

VFI	# VINYL FLUORIDE, INHIBITED

Common Synonyms			
Fluoroethylene Monofluoroethylene	Liquefied compressed gas	Colorless	Faint odor
	Floats and boils on water. Flammable visible vapor cloud is produced.		

Shut off ignition sources. Call fire department.
Avoid contact with liquid. Keep people away.
Stop discharge if possible.
Stay upwind. Use water spray to "knock down" vapor.
Isolate and remove discharged material.
Notify local health and pollution control agencies.

Fire	FLAMMABLE. POISONOUS GASES ARE PRODUCED IN FIRE. Containers may explode in fire. Flashback along vapor trail may occur. Vapor may explode if ignited in an enclosed area. Wear goggles and self-contained breathing apparatus. Let fire burn. Stop flow of gas if possible. Cool exposed containers and protect men effecting shutoff with water.
Exposure	Call for medical aid. VAPOR If inhaled will cause headache, or dizziness. Move victim to fresh air. If breathing is difficult, give oxygen. LIQUID Will cause frostbite. Flush affected areas with plenty of water. DO NOT RUB AFFECTED AREAS.
Water Pollution	Not harmful to aquatic life.

1. RESPONSE TO DISCHARGE
(See Response Methods Handbook, CG 446-4)

Issue warning—high flammability, air contaminant
Restrict access
Evacuate area

2. LABEL

Red

3. CHEMICAL DESIGNATIONS

3.1 **Synonyms:** Fluoroethylene; Monofluoro ethylene
3.2 **Coast Guard Compatibility Classification:** Not applicable
3.3 **Chemical Formula:** $CH_2{=}CHF$
3.4 **IMCO/United Nations Numerical Designation:** 2/1860

4. OBSERVABLE CHARACTERISTICS

4.1 **Physical State (as shipped):** Liquefied compressed gas
4.2 **Color:** Colorless
4.3 **Odor:** Faint ethereal

5. HEALTH HAZARDS

5.1 **Personal Protective Equipment:** Protective gloves; safety glasses; self-contained breathing apparatus
5.2 **Symptoms Following Exposure:** Inhalation of vapor causes slight intoxication, some shortness of breath. Liquid may cause frostbite of eyes or skin.
5.3 **Treatment for Exposure:** INHALATION: remove victim to fresh air. SKIN: if frostbite has occurred, immerse in warm water, treat burn.
5.4 **Toxicity by Inhalation (Threshold Limit Value):** Data not available
5.5 **Short-Term Inhalation Limits:** Data not available
5.6 **Toxicity by Ingestion:** Not pertinent (gas at normal temperatures)
5.7 **Late Toxicity:** Data not available
5.8 **Vapor (Gas) Irritant Characteristics:** Data not available
5.9 **Liquid or Solid Irritant Characteristics:** Data not available
5.10 **Odor Threshold:** Data not available

6. FIRE HAZARDS

6.1 **Flash Point:** Not pertinent (flammable compressed liquefied gas)
6.2 **Flammable Limits in Air:** 2.6%–21.7%
6.3 **Fire Extinguishing Agents:** Let fire burn; shut off flow of gas; cool adjacent containers with water.
6.4 **Fire Extinguishing Agents Not to be Used:** Not pertinent
6.5 **Special Hazards of Combustion Products:** Toxic hydrogen fluoride gas is generated in a fire.
6.6 **Behavior in Fire:** Vapor is heavier than air and may travel considerable distance to a source of ignition and flash back. Containers may explode.
6.7 **Ignition Temperature:** ~725°F
6.8 **Electrical Hazard:** Data not available
6.9 **Burning Rate:** Not pertinent

7. CHEMICAL REACTIVITY

7.1 **Reactivity with Water:** No reaction
7.2 **Reactivity with Common Materials:** No reaction
7.3 **Stability During Transport:** Stable
7.4 **Neutralizing Agents for Acids and Caustics:** Not pertinent
7.5 **Polymerization:** In the absence of inhibitor, polymerization can occur.
7.6 **Inhibitor of Polymerization:** Terpene B—0.2%

8. WATER POLLUTION

8.1 **Aquatic Toxicity:** None
8.2 **Waterfowl Toxicity:** None
8.3 **Biological Oxygen Demand (BOD):** None
8.4 **Food Chain Concentration Potential:** None

9. SELECTED MANUFACTURERS

1. E. I. duPont de Nemours & Co., Inc.
   Organic Chemicals Department
   Freon Products Division
   Nemours Building
   Wilmington, Del. 19898
2. Matheson Gas Products Co.
   East Rutherford, N. J. 07073
3. Scientific Gas Products, Inc.
   513 Raritan Center
   Edison, N. J. 08817

10. SHIPPING INFORMATION

10.1 **Grades or Purity:** 99.9+%
10.2 **Storage Temperature:** Ambient
10.3 **Inert Atmosphere:** No requirement
10.4 **Venting:** Safety relief

11. HAZARD ASSESSMENT CODE
(See Hazard Assessment Handbook, CG 446-3)

A-B-C-D-E-F-G-Z

12. HAZARD CLASSIFICATIONS

12.1 **Code of Federal Regulations:** Flammable compressed gas
12.2 **NAS Hazard Rating for Bulk Water Transportation:** Not listed
12.3 **NFPA Hazard Classifications:**

Category	Classification
Health Hazard (Blue)	1
Flammability (Red)	4
Reactivity (Yellow)	2

13. PHYSICAL AND CHEMICAL PROPERTIES

13.1 **Physical State at 15°C and 1 atm:** Gas
13.2 **Molecular Weight:** 46.1
13.3 **Boiling Point at 1 atm:** −98°F = −72°C = 201°K
13.4 **Freezing Point:** −258°F = −161°C = 112°K
13.5 **Critical Temperature:** 130.5°F = 54.7°C = 327.9°K
13.6 **Critical Pressure:** 760 psia = 51.6 atm = 5.24 MN/m²
13.7 **Specific Gravity:** 0.707 at 0°C (liquid)
13.8 **Liquid Surface Tension:** 5 dynes/cm = 0.005 N/m at 15°C
13.9 **Liquid-Water Interfacial Tension:** Not pertinent
13.10 **Vapor (Gas) Specific Gravity:** 1.6
13.11 **Ratio of Specific Heats of Vapor (Gas):** 1.2097
13.12 **Latent Heat of Vaporization:** 156 Btu/lb = 86.5 cal/g = 3.62×10^5 J/kg
13.13 **Heat of Combustion:** (est.) −6,500 Btu/lb = −3,600 cal/g = -150×10^5 J/kg
13.14 **Heat of Decomposition:** Not pertinent
13.15 **Heat of Solution:** Not pertinent
13.16 **Heat of Polymerization:** Data not available

NOTES

VINYLIDENE CHLORIDE, INHIBITED

Common Synonyms			
1, 1-Dichloroethylene unsym-Dichloroethylene	Watery liquid	Colorless	Sweet odor
	Sinks in water. Flammable, irritating vapor is produced. Boiling point is 89° F.		

Avoid contact with liquid and vapor. Keep people away.
Wear goggles, self-contained breathing apparatus, and rubber overclothing (including gloves).
Shut off ignition sources and call fire department.
Stop discharge if possible.
Stay upwind and use water spray to "knock down" vapor.
Evacuate area in case of large discharge.
Isolate and remove discharged material.
Notify local health and pollution control agencies.

Fire

FLAMMABLE.
POISONOUS GAS IS PRODUCED IN FIRE.
Containers may explode in fire.
Flashback along vapor trail may occur.
Vapor may explode if ignited in an enclosed area.
Wear self-contained breathing apparatus.
Combat fires from safe distance or protected location.
Extinguish with dry chemical, foam, or carbon dioxide.
Cool exposed containers with water.

Exposure

CALL FOR MEDICAL AID.

VAPOR
Irritating to eyes, nose, and throat.
If inhaled, will cause dizziness or difficult breathing.
Move to fresh air.
If breathing has stopped, give artificial respiration.
If breathing is difficult, give oxygen.

LIQUID
Will burn skin and eyes.
Harmful if swallowed.
Remove contaminated clothing and shoes.
Flush affected areas with plenty of water.
IF IN EYES, hold eyelids open and flush with plenty of water.
IF SWALLOWED and victim is CONSCIOUS, have victim drink water or milk.

Water Pollution

Effect of low concentrations on aquatic life is unknown.
May be dangerous if it enters water intakes.
Notify local health and wildlife officials.
Notify operators of nearby water intakes.

1. RESPONSE TO DISCHARGE

(See Response Methods Handbook, CG 446-4)

Issue warning—high flammability
Evacuate area.

2. LABEL

Red

3. CHEMICAL DESIGNATIONS

3.1 **Synonyms:** 1,1-Dichloroethylene unsym-Dichloroethylene

3.2 **Coast Guard Compatibility Classification:** Vinyl halides

3.3 **Chemical Formula:** $CH_2{=}CCl_2$

3.4 **IMCO/United Nations Numerical Designation:** 3.1/1303

4. OBSERVABLE CHARACTERISTICS

4.1 **Physical State (as shipped):** Liquid

4.2 **Color:** Colorless

4.3 **Odor:** Sweet; like carbon tetrachloride or chloroform

5. HEALTH HAZARDS

5.1 **Personal Protective Equipment:** Approved canister or air-supplied mask; goggles or face shield; rubber gloves and boots.

5.2 **Symptoms Following Exposure:** Vapor can cause dizziness and drunkenness; high levels cause anesthesia. Liquid irritates eyes and skin.

5.3 **Treatment for Exposure:** INHALATION: if any illness develops, remove person to fresh air promptly, keep warm and quiet, and get medical attention; if breathing stops, start artificial respiration. INGESTION: not likely a problem; no known antidote; treat symptomatically. EYES OR SKIN: flush with plenty of water for at least 15 min; get medical attention for eyes; remove contaminated clothing and wash before reuse.

5.4 **Toxicity by Inhalation (Threshold Limit Value):** 25 ppm (suggested)

5.5 **Short-Term Inhalation Limits:** Data not available

5.6 **Toxicity by Ingestion:** Grade 3; Oral LD_{50} 24 hr = 84 mg/kg (adrenalectomized rat)

5.7 **Late Toxicity:** Data not available

5.8 **Vapor (Gas) Irritant Characteristics:** Vapors cause moderate irritation such that personnel will find high concentrations unpleasant. The effect is temporary.

5.9 **Liquid or Solid Irritant Characteristics:** Causes smarting of the skin and first-degree burns on short exposure; may cause secondary burns on long exposure.

5.10 **Odor Threshold:** Data not available

6. FIRE HAZARDS

6.1 **Flash Point:** 0°F O.C.

6.2 **Flammable Limits in Air:** 7.3%–16.0%

6.3 **Fire Extinguishing Agents:** Foam, carbon dioxide, dry chemical

6.4 **Fire Extinguishing Agents Not to be Used:** Water may be ineffective.

6.5 **Special Hazards of Combustion Products:** Toxic hydrogen chloride and phosgene are generated in fires.

6.6 **Behavior in Fire:** May explode in fire due to polymerization. Vapor is heavier than air and may travel considerable distance to a source of ignition and flash back.

6.7 **Ignition Temperature:** 955–1031°F

6.8 **Electrical Hazard:** Not pertinent

6.9 **Burning Rate:** 2.7 mm/min.

7. CHEMICAL REACTIVITY

7.1 **Reactivity with Water:** No reaction

7.2 **Reactivity with Common Materials:** Copper and aluminum can cause polymerization.

7.3 **Stability During Transport:** Stable

7.4 **Neutralizing Agents for Acids and Caustics:** Not pertinent

7.5 **Polymerization:** Can occur if exposed to sunlight, air, copper, aluminum, heat.

7.6 **Inhibitor of Polymerization:** 200 ppm methyl ether of hydroquinone; 0.6–0.8% phenol

8. WATER POLLUTION

8.1 **Aquatic Toxicity:** Data not available

8.2 **Waterfowl Toxicity:** Data not available

8.3 **Biological Oxygen Demand (BOD):** Data not available

8.4 **Food Chain Concentration Potential:** None

9. SELECTED MANUFACTURERS

1. Dow Chemical Co.
Midland, Mich. 48640
2. P P G Industries, Inc.
Industrial Chemicals Division
1 Gateway Center
Pittsburgh, Pa. 15222
3. Vulcan Materials Co.
Chemicals Division
Wichita, Kans. 67201

10. SHIPPING INFORMATION

10.1 **Grades or Purity:** 99%

10.2 **Storage Temperature:** Ambient

10.3 **Inert Atmosphere:** Padded

10.4 **Venting:** Pressure-vacuum

11. HAZARD ASSESSMENT CODE

(See Hazard Assessment Handbook, CG 446-3)

A-X-Y-Z

12. HAZARD CLASSIFICATIONS

12.1 **Code of Federal Regulations:** Flammable liquid

12.2 **NAS Hazard Rating for Bulk Water Transportation:**

Category	Rating
Fire	3
Health	
Vapor Irritant	2
Liquid or Solid Irritant	2
Poisons	3
Water Pollution	
Human Toxicity	0
Aquatic Toxicity	2
Aesthetic Effect	2
Reactivity	
Other Chemicals	2
Water	0
Self-Reaction	3

12.3 **NFPA Hazard Classifications:**

Category	Classification
Health Hazard (Blue)	2
Flammability (Red)	4
Reactivity (Yellow)	2

13. PHYSICAL AND CHEMICAL PROPERTIES

13.1 **Physical State at 15°C and 1 atm:** Liquid

13.2 **Molecular Weight:** 96.95

13.3 **Boiling Point at 1 atm:** 88.9°F = 31.6°C = 304.8°K

13.4 **Freezing Point:** −187.6°F = −122.0°C = 151.2°K

13.5 **Critical Temperature:** Not pertinent

13.6 **Critical Pressure:** Not pertinent

13.7 **Specific Gravity:** 1.21 at 20°C (liquid)

13.8 **Liquid Surface Tension:** 24 dynes/cm = 0.024 N/m at 15°C

13.9 **Liquid-Water Interfacial Tension:** 37 dynes/cm = 0.037 N/m at 22.7°C

13.10 **Vapor (Gas) Specific Gravity:** 3.3

13.11 **Ratio of Specific Heats of Vapor (Gas):** Data not available

13.12 **Latent Heat of Vaporization:** 130 Btu/lb = 72 cal/g = 3.0×10^5 J/kg

13.13 **Heat of Combustion:** −4860 Btu/lb = −2700 cal/g = -113.0×10^5 J/kg

13.14 **Heat of Decomposition:** Not pertinent

13.15 **Heat of Solution:** Not pertinent

13.16 **Heat of Polymerization:** −333 Btu/lb = −185 cal/g = -7.75×10^5 J/kg

NOTES

REVISED 1978

VME — VINYL METHYL ETHER, INHIBITED

Common Synonyms			
Methyl vinyl ether Methoxyethylene	Liquefied compressed gas	Colorless	Sweet pleasant odor
	Floats and may boil on water. Boiling point is 54° F.		

Shut off ignition sources. Call fire department.
Stop discharge if possible. Keep people away.
Isolate and remove discharged material.
Notify local health and pollution control agencies.

Fire

FLAMMABLE.
Containers may explode in fire.
Flashback along vapor trail may occur.
Vapor may explode if ignited in an enclosed area.
DO NOT USE WATER ON FIRE.
Let fire burn.
Stop flow of gas if possible.
Cool exposed containers and protect men effecting shutoff with water.

Exposure

Call for medical aid.

VAPOR
Irritating to eyes, nose and throat.
If inhaled will cause headache, dizziness, or loss of consciousness.
Move victim to fresh air.
If breathing is difficult, give oxygen.

LIQUID
Irritating to eyes.
Will cause frostbite.
Harmful if swallowed.
DO NOT RUB AFFECTED AREAS.
Flush affected areas with plenty of water.
IF IN EYES, hold eyelids open and flush with plenty of water.
IF SWALLOWED and victim is CONSCIOUS, have victim drink water or milk.
DO NOT INDUCE VOMITING.

Water Pollution

Effect of low concentrations on aquatic life is unknown.
May be dangerous if it enters water intakes.
Notify local health and wildlife officials.
Notify operators of nearby water intakes.

1. RESPONSE TO DISCHARGE

(See Response Methods Handbook, CG 446-4)

Issue warning—high flammability, air contaminant
Restrict access
Evacuate area

2. LABEL

Red

3. CHEMICAL DESIGNATIONS

3.1 **Synonyms:** Methoxyethylene; Methyl vinyl ether
3.2 **Coast Guard Compatibility Classification:** Not applicable
3.3 **Chemical Formula:** $CH_2=CH-O-CH_3$
3.4 **IMCO/United Nations Numerical Designation:** 2/1087

4. OBSERVABLE CHARACTERISTICS

4.1 **Physical State (as shipped):** Liquefied compressed gas
4.2 **Color:** Colorless
4.3 **Odor:** Sweet, pleasant

5. HEALTH HAZARDS

5.1 **Personal Protective Equipment:** Organic-vapor mask; plastic or rubber gloves; safety glasses
5.2 **Symptoms Following Exposure:** Inhalation causes intoxication, blurring of vision, headache, dizziness, excitation, loss of consciousness. Liquid or concentrated vapor irritates eyes and causes frostbite of skin. Aspiration of the liquid will cause chemical pneumonitis.
5.3 **Treatment for Exposure:** INHALATION: remove victim to fresh air; if breathing is difficult, administer oxygen; call physician. EYES: wash with copious quantities of water; consult an eye specialist. SKIN: wash with copious quantities of water; treat frostbite by use of warm water or blankets. INGESTION: do NOT induce vomiting; get medical attention.
5.4 **Toxicity by Inhalation (Threshold Limit Value):** Data not available
5.5 **Short-Term Inhalation Limits:** Data not available
5.6 **Toxicity by Ingestion:** Grade 2; LD_{50} 0.5 to 5 g/kg
5.7 **Late Toxicity:** Data not available
5.8 **Vapor (Gas) Irritant Characteristics:** Data not available
5.9 **Liquid or Solid Irritant Characteristics:** Data not available
5.10 **Odor Threshold:** Data not available

6. FIRE HAZARDS

6.1 **Flash Point:** −69°F O.C.
6.2 **Flammable Limits in Air:** 2.6%–39%
6.3 **Fire Extinguishing Agents:** Let fire burn; shut off gas flow; extinguish small fires with dry chemical or carbon dioxide.
6.4 **Fire Extinguishing Agents Not to be Used:** Water may be ineffective.
6.5 **Special Hazards of Combustion Products:** Not pertinent
6.6 **Behavior in Fire:** Containers may explode. Vapor is heavier than air and may travel a considerable distance to a source of ignition and flash back.
6.7 **Ignition Temperature:** Data not available
6.8 **Electrical Hazard:** Data not available
6.9 **Burning Rate:** Not pertinent

7. CHEMICAL REACTIVITY

7.1 **Reactivity with Water:** Reacts slowly to form acetaldehyde; reaction is not hazardous unless water is hot or acids are present.
7.2 **Reactivity with Common Materials:** Acids will cause polymerization.
7.3 **Stability During Transport:** Stable if kept free from acids
7.4 **Neutralizing Agents for Acids and Caustics:** Not pertinent
7.5 **Polymerization:** Can polymerize in the presence of acids.
7.6 **Inhibitor of Polymerization:** Dioctylamine; Triethanolamine; Solid potassium hydroxide

8. WATER POLLUTION

8.1 **Aquatic Toxicity:** Data not available
8.2 **Waterfowl Toxicity:** Data not available
8.3 **Biological Oxygen Demand (BOD):** Data not available
8.4 **Food Chain Concentration Potential:** None

9. SELECTED MANUFACTURERS

1. GAF Corporation
140 West 51st Street
New York, N. Y. 10020
2. Union Carbide Corporation
Chemicals and Plastics Division
270 Park Avenue
New York, N. Y. 10017
3. Matheson Gas Products Co.
East Rutherford, N. J. 07073

10. SHIPPING INFORMATION

10.1 **Grades or Purity:** 99.7+%
10.2 **Storage Temperature:** Ambient
10.3 **Inert Atmosphere:** No requirement
10.4 **Venting:** Safety relief

11. HAZARD ASSESSMENT CODE

(See Hazard Assessment Handbook, CG 446-3)

A-B-C-D-E-F-G-Z

12. HAZARD CLASSIFICATIONS

12.1 **Code of Federal Regulations:** Flammable compressed gas
12.2 **NAS Hazard Rating for Bulk Water Transportation:** Not listed
12.3 **NFPA Hazard Classifications:**

Category	Classification
Health Hazard (Blue)	2
Flammability (Red)	4
Reactivity (Yellow)	2

13. PHYSICAL AND CHEMICAL PROPERTIES

13.1 **Physical State at 15°C and 1 atm:** Gas
13.2 **Molecular Weight:** 58.1
13.3 **Boiling Point at 1 atm:** 41.9°F = 5.5°C = 278.7°K
13.4 **Freezing Point:** −188°F = −122°C = 151°K
13.5 **Critical Temperature:** Not pertinent
13.6 **Critical Pressure:** Not pertinent
13.7 **Specific Gravity:** 0.777 at 0°C (liquid)
13.8 **Liquid Surface Tension:** (est.) 10 dynes/cm = 0.010 N/m at 0°C
13.9 **Liquid-Water Interfacial Tension:** (est.) 25 dynes/cm = 0.025 N/m at 0°C
13.10 **Vapor (Gas) Specific Gravity:** 2.0
13.11 **Ratio of Specific Heats of Vapor (Gas):** (est.) 1.1473
13.12 **Latent Heat of Vaporization:** (est.) 180 Btu/lb = 100 cal/g = 4.2×10^5 J/kg
13.13 **Heat of Combustion:** (est.) −14,200 Btu/lb = −7,900 cal/g = -330×10^5 J/kg
13.14 **Heat of Decomposition:** Not pertinent
13.15 **Heat of Solution:** Not pertinent
13.16 **Heat of Polymerization:** Data not available

NOTES

VNT	VINYLTOLUENE

Common Synonyms			
p-Methylstyrene	Watery liquid	Colorless	Unpleasant odor
	Floats on water.		

Stop discharge if possible.
Avoid contact with liquid.
Call fire department.
Isolate and remove discharged material.
Notify local health and pollution control agencies.

Fire	Combustible. Containers may explode in fire. Wear goggles and self-contained breathing apparatus. Extinguish with water, dry chemical, foam, or carbon dioxide. Cool exposed containers with water.
Exposure	CALL FOR MEDICAL AID. LIQUID Irritating to skin and eyes. Harmful if swallowed. Remove contaminated clothing and shoes. Flush affected areas with plenty of water. IF IN EYES, hold eyelids open and flush with plenty of water. IF SWALLOWED and victim is CONSCIOUS, have victim drink water or milk.
Water Pollution	Effect of low concentrations on aquatic life is unknown. Fouling to shoreline. May be dangerous if it enters water intakes. Notify local health and wildlife officials. Notify operators of nearby water intakes.

1. RESPONSE TO DISCHARGE

(See Response Methods Handbook, CG 446-4)

Mechanical containment
Should be removed
Chemical and physical treatment

2. LABELS

No hazard label required by Code of Federal Regulations

3. CHEMICAL DESIGNATIONS

3.1 **Synonyms:** Methylstyrene
3.2 **Coast Guard Compatibility Classification:** Olefin
3.3 **Chemical Formula:** $CH_3C_6H_4CH{=}CH_2$
3.4 **IMCO/United Nations Numerical Designation:** Not listed

4. OBSERVABLE CHARACTERISTICS

4.1 **Physical State (as shipped):** Liquid
4.2 **Color:** Colorless
4.3 **Odor:** Disagreeable

5. HEALTH HAZARDS

5.1 **Personal Protective Equipment:** Air-supplied mask; goggles or face shield; plastic gloves.
5.2 **Symptoms Following Exposure:** Vapors irritate eyes and nose; high levels cause dizziness, drunkenness, and anesthesia. Liquid irritates eyes and may irritate skin.
5.3 **Treatment for Exposure:** INHALATION: remove person to fresh air; give artificial respiration and oxygen if needed; call a doctor. INGESTION: do NOT induce vomiting; no known antidote; call a doctor. EYES: flush with water for at least 15 min. SKIN: wipe off, wash with soap and water.
5.4 **Toxicity by Inhalation (Threshold Limit Value):** 100 ppm
5.5 **Short-Term Inhalation Limits:** 400 ppm for 5 min.
5.6 **Toxicity by Ingestion:** Grade 2; LD_{50} 0.5 to 5 g/kg (rat)
5.7 **Late Toxicity:** Data not available
5.8 **Vapor (Gas) Irritant Characteristics:** Vapors cause moderate irritation such that personnel will find high concentrations unpleasant. The effect is temporary.
5.9 **Liquid or Solid Irritant Characteristics:** Minimum hazard. If spilled on clothing and allowed to remain, may cause smarting and reddening of the skin.
5.10 **Odor Threshold:** 50 ppm

6. FIRE HAZARDS

6.1 **Flash Point:** 137°F O.C.; 125°F C.C.
6.2 **Flammable Limits in Air:** 0.8%–11%
6.3 **Fire Extinguishing Agents:** Water fog, foam, carbon dioxide, or dry chemical
6.4 **Fire Extinguishing Agents Not to be Used:** Not pertinent
6.5 **Special Hazards of Combustion Products:** Not pertinent
6.6 **Behavior in Fire:** Containers may explode or rupture in a fire due to polymerization.
6.7 **Ignition Temperature:** 914°F
6.8 **Electrical Hazard:** Not pertinent
6.9 **Burning Rate:** 6.0 mm/min.

7. CHEMICAL REACTIVITY

7.1 **Reactivity with Water:** No reaction
7.2 **Reactivity with Common Materials:** No reaction
7.3 **Stability During Transport:** Stable
7.4 **Neutralizing Agents for Acids and Caustics:** Not pertinent
7.5 **Polymerization:** Slow at ordinary temperatures but when hot may rupture container. Also polymerized by metal salts such as those of iron or aluminum.
7.6 **Inhibitor of Polymerization:** 10–50 ppm tertiary butylcatechol

8. WATER POLLUTION

8.1 **Aquatic Toxicity:** Data not available
8.2 **Waterfowl Toxicity:** Data not available
8.3 **Biological Oxygen Demand (BOD):** Data not available
8.4 **Food Chain Concentration Potential:** None

9. SELECTED MANUFACTURERS

1. Borden, Inc.
   Borden Chemical Division
   50 W. Broad St.
   Columbus, Ohio 43215
2. Dow Chemical Co.
   Midland, Mich. 48640
3. Foster Grant Co., Inc.
   289 N. Main St.
   Leominster, Mass. 01453

10. SHIPPING INFORMATION

10.1 **Grades or Purity:** 99.2+%
10.2 **Storage Temperature:** Ambient
10.3 **Inert Atmosphere:** No requirement
10.4 **Venting:** Open (flame arrester)

11. HAZARD ASSESSMENT CODE

(See Hazard Assessment Handbook, CG 446-3)

A-T-U-Z

12. HAZARD CLASSIFICATIONS

12.1 **Code of Federal Regulations:** Combustible Liquid
12.2 **NAS Hazard Rating for Bulk Water Transportation:**

Category	Rating
Fire	2
Health	
Vapor Irritant	2
Liquid or Solid Irritant	1
Poisons	1
Water Pollution	
Human Toxicity	1
Aquatic Toxicity	3
Aesthetic Effect	2
Reactivity	
Other Chemicals	2
Water	0
Self-Reaction	3

12.3 **NFPA Hazard Classifications:**

Category	Classification
Health Hazard (Blue)	2
Flammability (Red)	2
Reactivity (Yellow)	1

13. PHYSICAL AND CHEMICAL PROPERTIES

13.1 **Physical State at 15°C and 1 atm:** Liquid
13.2 **Molecular Weight:** 118.18
13.3 **Boiling Point at 1 atm:** 333.9°F = 167.7°C = 440.9°K
13.4 **Freezing Point:** −106.6°F = −77.0°C = 196.2°K
13.5 **Critical Temperature:** Not pertinent
13.6 **Critical Pressure:** Not pertinent
13.7 **Specific Gravity:** 0.897 at 20°C (liquid)
13.8 **Liquid Surface Tension:** 31.53 dynes/cm = 0.03153 N/m at 20°C
13.9 **Liquid-Water Interfacial Tension:** (est.) 45 dynes/cm = 0.045 N/m at 20°C
13.10 **Vapor (Gas) Specific Gravity:** Not pertinent
13.11 **Ratio of Specific Heats of Vapor (Gas):** (est.) 1.060
13.12 **Latent Heat of Vaporization:** 150 Btu/lb = 83.5 cal/g = 3.50×10^5 J/kg
13.13 **Heat of Combustion:** −17,710 Btu/lb = −9840 cal/g = -412.0×10^5 J/kg
13.14 **Heat of Decomposition:** Not pertinent
13.15 **Heat of Solution:** Not pertinent
13.16 **Heat of Polymerization:** −243 Btu/lb = −135 cal/g = -5.65×10^5 J/kg

NOTES

REVISED 1978

VTS	# VINYLTRICHLOROSILANE

Common Synonyms			
Trichlorovinylsilane Trichlorovinyl silicane Vinylsilicon trichloride	Liquid	Colorless to light yellow	Sharp choking odor
	Reacts violently with water. Irritating gas is produced on contact with water.		

Shut off ignition sources. Call fire department.
Avoid contact with liquid. Keep people away.
Stop discharge if possible.
Isolate and remove discharged material.
Notify local health and pollution control agencies.

Fire

FLAMMABLE.
POISONOUS GASES MAY BE PRODUCED IN FIRE.
Containers may explode in fire.
Flashback along vapor trail may occur.
Vapor may explode if ignited in an enclosed area.
Extinguish with dry chemicals or carbon dioxide.
DO NOT USE WATER OR FOAM ON FIRE.
Cool exposed containers with water.

Exposure

Call for medical aid.

VAPOR
Irritating to eyes, nose and throat.
Harmful if inhaled.
Move victim to fresh air.
If breathing is difficult, give oxygen.

LIQUID
Will burn skin and eyes.
Harmful if swallowed.
Remove contaminated clothing and shoes.
Flush affected areas with plenty of water.
IF IN EYES, hold eyelids open and flush with plenty of water.
IF SWALLOWED and victim is CONSCIOUS, have victim drink water or milk.
DO NOT INDUCE VOMITING.

Water Pollution

Effect of low concentrations on aquatic life is unknown.
May be dangerous if it enters water intakes.
Notify local health and wildlife officials.
Notify operators of nearby water intakes.

1. RESPONSE TO DISCHARGE

(See Response Methods Handbook, CG 446-4)

Issue warning—high flammability, corrosive, air contaminant
Restrict access
Disperse and flush with care

2. LABEL

Red

3. CHEMICAL DESIGNATIONS

3.1 **Synonyms:** Trichlorovinylsilane; Trichlorovinyl silicane; Vinylsilicon trichloride
3.2 **Coast Guard Compatibility Classification:** Not applicable
3.3 **Chemical Formula:** $CH_2{=}CHSiCl_3$
3.4 **IMCO/United Nations Numerical Designation:** 3.2/1305

4. OBSERVABLE CHARACTERISTICS

4.1 **Physical State (as shipped):** Liquid
4.2 **Color:** Colorless or pale yellow
4.3 **Odor:** Sharp, choking; like hydrochloric acid

5. HEALTH HAZARDS

5.1 **Personal Protective Equipment:** Acid-vapor-type respiratory protection; rubber gloves; chemical worker's goggles; other protective equipment as necessary to protect skin and eyes.
5.2 **Symptoms Following Exposure:** Inhalation causes irritation of mucous membranes. Vapor irritates eyes. Contact with liquid causes severe burns of eyes and skin. Ingestion causes burns of mouth and stomach.
5.3 **Treatment for Exposure:** *Get medical attention following all exposures to this compound.* INHALATION: remove victim from exposure; give artificial respiration if required. EYES: flush with water for 15 min. SKIN: flush with water. INGESTION: do NOT induce vomiting; give large amount of water.
5.4 **Toxicity by Inhalation (Threshold Limit Value):** Data not available
5.5 **Short-Term Inhalation Limits:** Data not available
5.6 **Toxicity by Ingestion:** Grade 2; oral LD_{50} = 1,280 mg/kg (rat)
5.7 **Late Toxicity:** Data not available
5.8 **Vapor (Gas) Irritant Characteristics:** Vapors cause severe irritation of eyes and throat and can cause eye and lung injury. They cannot be tolerated even at low concentrations.
5.9 **Liquid or Solid Irritant Characteristics:** Severe skin irritant. Causes second- and third-degree burns on short contact and is very injurious to the eyes.
5.10 **Odor Threshold:** Data not available

6. FIRE HAZARDS

6.1 **Flash Point:** 60°F O.C.; 52°F C.C.
6.2 **Flammable Limits in Air:** 3% (LFL)
6.3 **Fire Extinguishing Agents:** Dry chemical, carbon dioxide
6.4 **Fire Extinguishing Agents Not to be Used:** Water, foam
6.5 **Special Hazards of Combustion Products:** Toxic chlorine and phosgene gases may be formed in fires.
6.6 **Behavior in Fire:** Difficult to extinguish; re-ignition may occur. Contact with water applied to adjacent fires produces irritating hydrogen chloride gas.
6.7 **Ignition Temperature:** 505°F
6.8 **Electrical Hazard:** Data not available
6.9 **Burning Rate:** 2.9 mm/min.

7. CHEMICAL REACTIVITY

7.1 **Reactivity with Water:** Reacts vigorously, evolving hydrogen chloride (hydrochloric acid)
7.2 **Reactivity with Common Materials:** Reacts with surface moisture to evolve hydrogen chloride, which will corrode common metals and form flammable hydrogen gas.
7.3 **Stability During Transport:** Stable if protected from moisture
7.4 **Neutralizing Agents for Acids and Caustics:** Flush with water, rinse with sodium bicarbonate or lime solution.
7.5 **Polymerization:** May occur in absence of inhibitor
7.6 **Inhibitor of Polymerization:** Diphenylamine; Hydroquinone

8. WATER POLLUTION

8.1 **Aquatic Toxicity:** Data not available
8.2 **Waterfowl Toxicity:** Data not available
8.3 **Biological Oxygen Demand (BOD):** Data not available
8.4 **Food Chain Concentration Potential:** None

9. SELECTED MANUFACTURERS

1. Union Carbide Corp.
   Chemicals and Plastics Division
   270 Park Avenue
   New York, N. Y. 10017
2. Dow Corning Corporation
   P. O. Box 592
   Midland, Mich. 48640
3. Dynamit Nobel of America, Inc.
   105 Stonehurst Court
   Northvale, N. J. 07647

10. SHIPPING INFORMATION

10.1 **Grades or Purity:** 96+%; 98.5+%
10.2 **Storage Temperature:** Ambient
10.3 **Inert Atmosphere:** No requirement
10.4 **Venting:** Pressure-vacuum

11. HAZARD ASSESSMENT CODE

(See Hazard Assessment Handbook, CG 446-3)

A-O

12. HAZARD CLASSIFICATIONS

12.1 **Code of Federal Regulations:** Flammable liquid
12.2 **NAS Hazard Rating for Bulk Water Transportation:**

Category	Rating
Fire	3
Health	
Vapor Irritant	4
Liquid or Solid Irritant	4
Poisons	3
Water Pollution	
Human Toxicity	3
Aquatic Toxicity	3
Aesthetic Effect	3
Reactivity	
Other Chemicals	3
Water	4
Self-Reaction	3

12.3 **NFPA Hazard Classifications:** Not listed

13. PHYSICAL AND CHEMICAL PROPERTIES

13.1 **Physical State at 15°C and 1 atm:** Liquid
13.2 **Molecular Weight:** 161.5
13.3 **Boiling Point at 1 atm:** 195.1°F = 90.6°C = 363.8°K
13.4 **Freezing Point:** −139°F = −95°C = 178°K
13.5 **Critical Temperature:** Not pertinent
13.6 **Critical Pressure:** Not pertinent
13.7 **Specific Gravity:** 1.26 at 20°C (liquid)
13.8 **Liquid Surface Tension:** (est.) 28 dynes/cm = 0.028 N/m at 20°C
13.9 **Liquid-Water Interfacial Tension:** Not pertinent
13.10 **Vapor (Gas) Specific Gravity:** 5.61
13.11 **Ratio of Specific Heats of Vapor (Gas):** Data not available
13.12 **Latent Heat of Vaporization:** 88 Btu/lb = 49 cal/g = 2.0×10^5 J/kg
13.13 **Heat of Combustion:** (est.) −4,300 Btu/lb = −2,400 cal/g = -100×10^5 J/kg
13.14 **Heat of Decomposition:** Not pertinent
13.15 **Heat of Solution:** Data not available
13.16 **Heat of Polymerization:** Not pertinent

NOTES

WCA WAXES: CARNAUBA

Common Synonyms

Thick liquid (heated) Yellow to dark brownish green Odorless

Floats on water and solidifies.

Stop discharge if possible. Keep people away.
Call fire department.
Isolate and remove discharged material.
Notify local health and pollution control agencies.

Fire

Combustible.
Extinguish with foam, water, dry chemical, or carbon dioxide.

Exposure

LIQUID
Will burn skin and eyes.
Remove wax.
Flush affected areas with plenty of water.

Water Pollution

Effect of low concentrations on aquatic life is unknown
Fouling to shoreline.
May be dangerous if it enters water intakes.
Notify local health and wildlife officials.
Notify operators of nearby water intakes.

1. RESPONSE TO DISCHARGE

(See Response Methods Handbook, CG 446-4)

Mechanical containment
Should be removed
Chemical and physical treatment

2. LABELS

No hazard label required by Code of Federal Regulations

3. CHEMICAL DESIGNATIONS

3.1 **Synonyms:** No common synonyms
3.2 **Coast Guard Compatibility Classification:** Ester
3.3 **Chemical Formula:** Not pertinent
3.4 **IMCO/United Nations Numerical Designation:** Not listed

4. OBSERVABLE CHARACTERISTICS

4.1 **Physical State (as shipped):** Liquid
4.2 **Color:** Yellow to dark brownish green
4.3 **Odor:** None

5. HEALTH HAZARDS

5.1 **Personal Protective Equipment:** Goggles or face shield; protective gloves and clothing for hot liquid wax.
5.2 **Symptoms Following Exposure:** Hot wax can burn skin and eyes.
5.3 **Treatment for Exposure:** SKIN OR EYE CONTACT: remove solidified wax from skin, wash with soap and water; if in eyes, or if skin is burned, call a doctor.
5.4 **Toxicity by Inhalation (Threshold Limit Value):** Not pertinent
5.5 **Short-Term Inhalation Limits:** Not pertinent
5.6 **Toxicity by Ingestion:** Data not available
5.7 **Late Toxicity:** Data not available
5.8 **Vapor (Gas) Irritant Characteristics:** Non-volatile
5.9 **Liquid or Solid Irritant Characteristics:** Hot wax can burn skin and eyes.
5.10 **Odor Threshold:** Not pertinent

6. FIRE HAZARDS

6.1 **Flash Point:** 540°F C.C.; 595°F O.C.
6.2 **Flammable Limits in Air:** Not pertinent
6.3 **Fire Extinguishing Agents:** Water, foam, dry chemical, or carbon dioxide
6.4 **Fire Extinguishing Agents Not to be Used:** Water or foam may cause frothing.
6.5 **Special Hazards of Combustion Products:** Not pertinent
6.6 **Behavior in Fire:** Not pertinent
6.7 **Ignition Temperature:** Data not available
6.8 **Electrical Hazard:** Not pertinent
6.9 **Burning Rate:** Not pertinent

7. CHEMICAL REACTIVITY

7.1 **Reactivity with Water:** No reaction
7.2 **Reactivity with Common Materials:** No reaction
7.3 **Stability During Transport:** Stable
7.4 **Neutralizing Agents for Acids and Caustics:** Not pertinent
7.5 **Polymerization:** Not pertinent
7.6 **Inhibitor of Polymerization:** Not pertinent

8. WATER POLLUTION

8.1 **Aquatic Toxicity:** Data not available
8.2 **Waterfowl Toxicity:** Data not available
8.3 **Biological Oxygen Demand (BOD):** Data not available
8.4 **Food Chain Concentration Potential:** None

9. SELECTED MANUFACTURERS

1. Dura Commodities Corp.
   111 Calvert St.
   Harrison, N. Y. 10528
2. W. R. Grace & Co.
   Cornelius Wax Refining Corp. Division
   50 National Rd.
   Edison, N. J. 08817
3. Frank B. Ross Co.
   6-10 Ash St.
   Jersey City, N. J. 07304

10. SHIPPING INFORMATION

10.1 **Grades or Purity:** Shipped in a variety of grades, depending on source of wax and intended use. All have about the same hazardous properties.
10.2 **Storage Temperature:** Elevated
10.3 **Inert Atmosphere:** No requirement
10.4 **Venting:** Open (flame arrester)

11. HAZARD ASSESSMENT CODE

(See Hazard Assessment Handbook, CG 446-3)

A-T-U

12. HAZARD CLASSIFICATIONS

12.1 **Code of Federal Regulations:** Not listed
12.2 **NAS Hazard Rating for Bulk Water Transportation:** Not listed
12.3 **NFPA Hazard Classifications:** Not listed

13. PHYSICAL AND CHEMICAL PROPERTIES

13.1 **Physical State at 15°C and 1 atm:** Solid
13.2 **Molecular Weight:** Not pertinent
13.3 **Boiling Point at 1 atm:** Very high
13.4 **Freezing Point:** 176—187°F = 80—86°C = 353—359°K
13.5 **Critical Temperature:** Not pertinent
13.6 **Critical Pressure:** Not pertinent
13.7 **Specific Gravity:** 0.998 at 25°C (solid)
13.8 **Liquid Surface Tension:** 32 dynes/cm = 0.032 N/m at 100°C
13.9 **Liquid-Water Interfacial Tension:** Data not available
13.10 **Vapor (Gas) Specific Gravity:** Not pertinent
13.11 **Ratio of Specific Heats of Vapor (Gas):** Not pertinent
13.12 **Latent Heat of Vaporization:** Not pertinent
13.13 **Heat of Combustion:** Data not available
13.14 **Heat of Decomposition:** Not pertinent
13.15 **Heat of Solution:** Not pertinent
13.16 **Heat of Polymerization:** Not pertinent

NOTES

REVISED 1978

WPF	# WAXES: PARAFFIN

Common Synonyms Petroleum wax	Thick liquid (heated) Yellow to white Waxy odor Floats on water and solidifies.

Stop discharge if possible.
Isolate and remove discharged material.
Call fire department.
Notify local health and pollution control agencies.

Fire	Combustible. Extinguish with water, foam, dry chemical, or carbon dioxide.
Exposure	LIQUID Will burn skin and eyes. Remove wax. Flush affected areas with plenty of water.
Water Pollution	Effect of low concentrations on aquatic life is unknown. Fouling to shoreline. May be dangerous if it enters water intakes. Notify local health and wildlife officials. Notify operators of nearby water intakes.

1. RESPONSE TO DISCHARGE

(See Response Methods Handbook, CG 446-4)

Mechanical containment
Should be removed
Chemical and physical treatment

2. LABELS

No hazard label required by Code of Federal Regulations

3. CHEMICAL DESIGNATIONS

3.1 **Synonyms:** Petroleum wax
3.2 **Coast Guard Compatibility Classification:** Saturated hydrocarbon
3.3 **Chemical Formula:** Not pertinent
3.4 **IMCO/United Nations Numerical Designation:** Not listed

4. OBSERVABLE CHARACTERISTICS

4.1 **Physical State (as shipped):** Liquid to hard solid
4.2 **Color:** Yellow to white
4.3 **Odor:** Very weak

5. HEALTH HAZARDS

5.1 **Personal Protective Equipment:** Goggles or face shield; protective gloves and clothing for hot liquid wax.
5.2 **Symptoms Following Exposure:** Hot wax can burn eyes and skin.
5.3 **Treatment for Exposure:** SKIN OR EYE CONTACT: remove solidified wax; wash skin with soap and water; if in eyes, call a doctor.
5.4 **Toxicity by Inhalation (Threshold Limit Value):** Not pertinent
5.5 **Short-Term Inhalation Limits:** Not pertinent
5.6 **Toxicity by Ingestion:** Grade 1; LD_{50} 5 to 15 g/kg
5.7 **Late Toxicity:** None
5.8 **Vapor (Gas) Irritant Characteristics:** Non-volatile
5.9 **Liquid or Solid Irritant Characteristics:** None
5.10 **Odor Threshold:** Not pertinent

6. FIRE HAZARDS

6.1 **Flash Point:** 390°F C.C.; 380 465°F O.C.
6.2 **Flammable Limits in Air:** Not pertinent
6.3 **Fire Extinguishing Agents:** Water, foam, dry chemical, or carbon dioxide
6.4 **Fire Extinguishing Agents Not to be Used:** Water or foam may cause frothing.
6.5 **Special Hazards of Combustion Products:** Not pertinent
6.6 **Behavior in Fire:** Not pertinent
6.7 **Ignition Temperature:** 473°F
6.8 **Electrical Hazard:** Not pertinent
6.9 **Burning Rate:** Not pertinent

7. CHEMICAL REACTIVITY

7.1 **Reactivity with Water:** No reaction
7.2 **Reactivity with Common Materials:** No reaction
7.3 **Stability During Transport:** Stable
7.4 **Neutralizing Agents for Acids and Caustics:** Not pertinent
7.5 **Polymerization:** Not pertinent
7.6 **Inhibitor of Polymerization:** Not pertinent

8. WATER POLLUTION

8.1 **Aquatic Toxicity:** Data not available
8.2 **Waterfowl Toxicity:** Data not available
8.3 **Biological Oxygen Demand (BOD):** Data not available
8.4 **Food Chain Concentration Potential:** Data not available

9. SELECTED MANUFACTURERS

1. W. R. Grace & Co.
   Cornelius Wax Refining Co. Division
   50 National Rd.
   Edison, N. J. 08817
2. Exxon Chemical Co.
   Houston, Tex. 77001
3. Sun Oil Co.
   St. Davids, Pa. 19087

10. SHIPPING INFORMATION

10.1 **Grades or Purity:** Crude-scale; refined
10.2 **Storage Temperature:** Ambient
10.3 **Inert Atmosphere:** No requirement
10.4 **Venting:** Open (flame arrester)

11. HAZARD ASSESSMENT CODE

(See Hazard Assessment Handbook, CG 446-3)

A-T-U

12. HAZARD CLASSIFICATIONS

12.1 **Code of Federal Regulations:** Not listed
12.2 **NAS Hazard Rating for Bulk Water Transportation:** Not listed
12.3 **NFPA Hazard Classifications:** Not listed

13. PHYSICAL AND CHEMICAL PROPERTIES

13.1 **Physical State at 15°C and 1 atm:** Solid
13.2 **Molecular Weight:** Not pertinent
13.3 **Boiling Point at 1 atm:** Very high
13.4 **Freezing Point:** 118–149°F = 48–65°C = 321–338°K
13.5 **Critical Temperature:** Not pertinent
13.6 **Critical Pressure:** Not pertinent
13.7 **Specific Gravity:** 0.78–0.79 at 20°C (liquid)
13.8 **Liquid Surface Tension:** 30.6 dynes/cm = 0.0306 N/m at 54°C
13.9 **Liquid-Water Interfacial Tension:** 35–50 dynes/cm = 0.035–0.050 N/m at 54°C
13.10 **Vapor (Gas) Specific Gravity:** Not pertinent
13.11 **Ratio of Specific Heats of Vapor (Gas):** Not pertinent
13.12 **Latent Heat of Vaporization:** Not pertinent
13.13 **Heat of Combustion:** -18,000 Btu/lb = -10,000 cal/g = -430×10^5 J/kg
13.14 **Heat of Decomposition:** Not pertinent
13.15 **Heat of Solution:** Not pertinent
13.16 **Heat of Polymerization:** Not pertinent

NOTES

REVISED 1978

XLM

m-XYLENE

Common Synonyms			
1, 3-Dimethylbenzene Xylol	Watery liquid	Colorless	Sweet odor
	Floats on water. Flammable, irritating vapor is produced.		

Stop discharge if possible. Keep people away.
Call fire department.
Avoid contact with liquid and vapor.
Isolate and remove discharged material.
Notify local health and pollution control agencies.

Fire

FLAMMABLE
Flashback along vapor trail may occur.
Vapor may explode if ignited in an enclosed area.
Wear self-contained breathing apparatus.
Extinguish with foam, dry chemical, or carbon dioxide.
Water may be ineffective on fire.
Cool exposed containers with water.

Exposure

CALL FOR MEDICAL AID.

VAPOR
Irritating to eyes, nose, and throat.
If inhaled, will cause headache, difficult breathing, or loss of consciousness.
Move to fresh air.
If breathing has stopped, give artificial respiration.
If breathing is difficult, give oxygen.

LIQUID
Irritating to skin and eyes.
If swallowed, will cause nausea, vomiting, or loss of consciousness
Remove contaminated clothing and shoes.
Flush affected areas with plenty of water.
IF IN EYES, hold eyelids open and flush with plenty of water.
IF SWALLOWED and victim is CONSCIOUS, have victim drink water or milk.
DO NOT INDUCE VOMITING.

Water Pollution

HARMFUL TO AQUATIC LIFE IN VERY LOW CONCENTRATIONS.
Fouling to shoreline.
May be dangerous if it enters water intakes.
Notify local health and wildlife officials.
Notify operators of nearby water intakes.

1. RESPONSE TO DISCHARGE

(See Response Methods Handbook, CG 446-4)

Issue warning - high flammability
Evacuate area
Should be removed
Chemical and physical treatment

2. LABEL

Red

3. CHEMICAL DESIGNATIONS

3.1 **Synonyms:** 1,3-Dimethylbenzene Xylol
3.2 **Coast Guard Compatibility Classification:** Aromatic hydrocarbon
3.3 **Chemical Formula:** $m\text{-}C_6H_4(CH_3)_2$
3.4 **IMCO/United Nations Numerical Designation:** 3.2/1307

4. OBSERVABLE CHARACTERISTICS

4.1 **Physical State (as shipped):** Liquid
4.2 **Color:** Colorless
4.3 **Odor:** Like benzene; characteristic aromatic

5. HEALTH HAZARDS

5.1 **Personal Protective Equipment:** Approved canister or air-supplied mask; goggles or face shield; plastic gloves and boots.
5.2 **Symptoms Following Exposure:** Vapors cause headache and dizziness. Liquid irritates eyes and skin. If taken into lungs, causes severe coughing, distress, and rapidly developing pulmonary edema. If ingested, causes nausea, vomiting, cramps, headache, and coma; can be fatal. Kidney and liver damage can occur.
5.3 **Treatment for Exposure:** INHALATION: remove to fresh air; administer artificial respiration and oxygen if required; call a doctor. INGESTION: do NOT induce vomiting; call a doctor. EYES: flush with water for at least 15 min. SKIN: wipe off, wash with soap and water.
5.4 **Toxicity by Inhalation (Threshold Limit Value):** 100 ppm
5.5 **Short-Term Inhalation Limits:** 300 ppm for 30 min.
5.6 **Toxicity by Ingestion:** Grade 3; LD_{50} 50 to 500 g/kg
5.7 **Late Toxicity:** Kidney and liver damage.
5.8 **Vapor (Gas) Irritant Characteristics:** Vapors cause a slight smarting of the eyes or respiratory system if present in high concentrations. The effect is temporary.
5.9 **Liquid or Solid Irritant Characteristics:** Minimum hazard. If spilled on clothing and allowed to remain, may cause smarting and reddening of the skin.
5.10 **Odor Threshold:** 0.05 ppm

6. FIRE HAZARDS

6.1 **Flash Point:** 84°F C.C.
6.2 **Flammable Limits in Air:** 1.1% - 6.4%
6.3 **Fire Extinguishing Agents:** Foam, dry chemical, or carbon dioxide
6.4 **Fire Extinguishing Agents Not to be Used:** Water may be ineffective.
6.5 **Special Hazards of Combustion Products:** Not pertinent
6.6 **Behavior in Fire:** Vapor is heavier than air and may travel considerable distance to a source of ignition and flash back.
6.7 **Ignition Temperature:** 986°F
6.8 **Electrical Hazard:** Class I, Group D
6.9 **Burning Rate:** 5.8 mm/min.

7. CHEMICAL REACTIVITY

7.1 **Reactivity with Water:** No reaction
7.2 **Reactivity with Common Materials:** No reaction
7.3 **Stability During Transport:** Stable
7.4 **Neutralizing Agents for Acids and Caustics:** Not pertinent
7.5 **Polymerization:** Not pertinent
7.6 **Inhibitor of Polymerization:** Not pertinent

8. WATER POLLUTION

8.1 **Aquatic Toxicity:** 22 ppm/96 hr/bluegill/TL_m/fresh water
8.2 **Waterfowl Toxicity:** Data not available
8.3 **Biological Oxygen Demand (BOD):** 0 lb/lb, 5 days; 0% (theor.), 8 days
8.4 **Food Chain Concentration Potential:** Data not available

9. SELECTED MANUFACTURERS

1. ARCO Chemical Co.
   260 South Broad St.
   Philadelphia, Pa. 19101
2. Chevron Chemical Co.
   Industrial Chemicals Division
   200 Bush St.
   San Francisco, Calif. 94120
3. Fallek Chemical Corp.
   460 Park Ave.
   New York, N. Y. 10022

10. SHIPPING INFORMATION

10.1 **Grades or Purity:** Research: 99.99%; Pure: 99.9%; Technical: 99.2%
10.2 **Storage Temperature:** Ambient
10.3 **Inert Atmosphere:** No requirement
10.4 **Venting:** Open (flame arrester) or pressure-vacuum

11. HAZARD ASSESSMENT CODE

(See Hazard Assessment Handbook, CG 446-3)

A-T-U

12. HAZARD CLASSIFICATIONS

12.1 **Code of Federal Regulations:** Flammable liquid
12.2 **NAS Hazard Rating for Bulk Water Transportation:**

Category	Rating
Fire	3
Health	
Vapor Irritant	1
Liquid or Solid Irritant	1
Poisons	2
Water Pollution	
Human Toxicity	1
Aquatic Toxicity	3
Aesthetic Effect	2
Reactivity	
Other Chemicals	1
Water	0
Self-Reaction	0

12.3 **NFPA Hazard Classifications:**

Category	Classification
Health Hazard (Blue)	2
Flammability (Red)	3
Reactivity (Yellow)	0

13. PHYSICAL AND CHEMICAL PROPERTIES

13.1 **Physical State at 15°C and 1 atm:** Liquid
13.2 **Molecular Weight:** 106.16
13.3 **Boiling Point at 1 atm:** 269.4°F = 131.9°C = 405.1°K
13.4 **Freezing Point:** −54.2°F = −47.9°C = 225.3°K
13.5 **Critical Temperature:** 650.8°F = 343.8°C = 617.0°K
13.6 **Critical Pressure:** 513.8 atm = 34.95 psia = 3.540 MN/m²
13.7 **Specific Gravity:** 0.864 at 20°C (liquid)
13.8 **Liquid Surface Tension:** 28.6 dynes/cm = 0.0286 N/m at 20°C
13.9 **Liquid-Water Interfacial Tension:** 36.4 dynes/cm = 0.0364 N/m at 30°C
13.10 **Vapor (Gas) Specific Gravity:** Not pertinent
13.11 **Ratio of Specific Heats of Vapor (Gas):** 1.071
13.12 **Latent Heat of Vaporization:** 147 Btu/lb = 81.9 cal/g = 3.43×10^5 J/kg
13.13 **Heat of Combustion:** −17,554 Btu/lb = −9752.4 cal/g = -408.31×10^5 J/kg
13.14 **Heat of Decomposition:** Not pertinent
13.15 **Heat of Solution:** Not pertinent
13.16 **Heat of Polymerization:** Not pertinent

NOTES

REVISED 1978

XLO	# o-XYLENE

Common Synonyms 1, 2-Dimethylbenzene Xylol	Watery liquid Colorless Sweet odor Floats on water. Flammable, irritating vapor is produced.

Stop discharge if possible. Keep people away.
Call fire department.
Avoid contact with liquid and vapor.
Isolate and remove discharged material.
Notify local health and pollution control agencies.

Fire	FLAMMABLE Flashback along vapor trail may occur. Vapor may explode if ignited in an enclosed area. Wear self-contained breathing apparatus. Extinguish with foam, dry chemical, or carbon dioxide. Water may be ineffective on fire. Cool exposed containers with water.
Exposure	CALL FOR MEDICAL AID. VAPOR Irritating to eyes, nose and throat. If inhaled, will cause headache, difficult breathing, or loss of consciousness. Move to fresh air. If breathing has stopped, give artificial respiration. If breathing is difficult, give oxygen. LIQUID Irritating to skin and eyes. If swallowed, will cause nausea, vomiting, or loss of consciousness. Remove contaminated clothing and shoes. Flush affected areas with plenty of water. IF IN EYES, hold eyelids open and flush with plenty of water. IF SWALLOWED and victim is CONSCIOUS, have victim drink water or milk. DO NOT INDUCE VOMITING.
Water Pollution	Dangerous to aquatic life in high concentrations. Fouling to shoreline. May be dangerous if it enters water intakes. Notify local health and wildlife officials. Notify operators of nearby water intakes.

1. RESPONSE TO DISCHARGE

(See Response Methods Handbook, CG 446-4)

Issue warning — high flammability
Evacuate area
Should be removed
Chemical and physical treatment

2. LABEL

Red

3. CHEMICAL DESIGNATIONS

3.1 **Synonyms:** 1,2-Dimethylbenzene
3.2 **Coast Guard Compatibility Classification:** Aromatic hydrocarbon
3.3 **Chemical Formula:** o-$C_6H_4(CH_3)_2$
3.4 **IMCO/United Nations Numerical Designation:** 3.2/1307

4. OBSERVABLE CHARACTERISTICS

4.1 **Physical State (as shipped):** Liquid
4.2 **Color:** Colorless
4.3 **Odor:** Benzene-like; characteristic aromatic

5. HEALTH HAZARDS

5.1 **Personal Protective Equipment:** Approved canister or air-supplied mask; goggles or face shield; plastic gloves and boots.
5.2 **Symptoms Following Exposure:** Vapors cause headache and dizziness. Liquid irritates eyes and skin. If taken into lungs, causes severe coughing, distress, and rapidly developing pulmonary edema. If ingested, causes nausea, vomiting, cramps, headache, and coma. Can be fatal. Kidney and liver damage can occur.
5.3 **Treatment for Exposure:** INHALATION: remove to fresh air; administer artificial respiration and oxygen if required; call a doctor. INGESTION: do NOT induce vomiting; call a doctor. EYES: flush with water for at least 15 min. SKIN: wipe off, wash with soap and water.
5.4 **Toxicity by Inhalation (Threshold Limit Value):** 100 ppm
5.5 **Short-Term Inhalation Limits:** 300 ppm for 30 min.
5.6 **Toxicity by Ingestion:** Grade 3; LD_{50} 50 to 500 mg/kg
5.7 **Late Toxicity:** Kidney and liver damage.
5.8 **Vapor (Gas) Irritant Characteristics:** Vapors cause a slight smarting of the eyes or respiratory system if present in high concentrations. The effect is temporary.
5.9 **Liquid or Solid Irritant Characteristics:** Minimum hazard. If spilled on clothing and allowed to remain, may cause smarting and reddening of the skin.
5.10 **Odor Threshold:** 0.05 ppm

6. FIRE HAZARDS

6.1 **Flash Point:** 63°F C.C.; 75°F O.C.
6.2 **Flammable Limits in Air:** 1.1% 7.0%
6.3 **Fire Extinguishing Agents:** Foam, dry chemical, or carbon dioxide
6.4 **Fire Extinguishing Agents Not to be Used:** Water may be ineffective.
6.5 **Special Hazards of Combustion Products:** Not pertinent
6.6 **Behavior in Fire:** Vapor is heavier than air and may travel considerable distance to a source of ignition and flash back.
6.7 **Ignition Temperature:** 869°F
6.8 **Electrical Hazard:** Class I, Group D
6.9 **Burning Rate:** 5.8 mm/min.

7. CHEMICAL REACTIVITY

7.1 **Reactivity with Water:** No reaction
7.2 **Reactivity with Common Materials:** No reaction
7.3 **Stability During Transport:** Stable
7.4 **Neutralizing Agents for Acids and Caustics:** Not pertinent
7.5 **Polymerization:** Not pertinent
7.6 **Inhibitor of Polymerization:** Not pertinent

8. WATER POLLUTION

8.1 **Aquatic Toxicity:** > 100 mg/l /96 hr/D. magna/TL_m/fresh water
8.2 **Waterfowl Toxicity:** Data not available
8.3 **Biological Oxygen Demand (BOD):** 0 lb/lb, 5 days; 2.5% (theor.), 8 days
8.4 **Food Chain Concentration Potential:** Data not available

9. SELECTED MANUFACTURERS

1. Atlantic Richfield Co.
   ARCO Chemical Co. Division
   260 South Broad St.
   Philadelphia, Pa. 19101
2. Cities Service Co., Inc.
   Petrochemicals Division
   60 Wall St.
   New York, N. Y. 10005
3. Shell Chemical Co.
   Petrochemicals Division
   Houston, Tex. 77001

10. SHIPPING INFORMATION

10.1 **Grades or Purity:** Research: 99.99%; Pure: 99.7%; Commercial: 95+%
10.2 **Storage Temperature:** Ambient
10.3 **Inert Atmosphere:** No requirement
10.4 **Venting:** Open (flame arrester), or pressure-vacuum

11. HAZARD ASSESSMENT CODE

(See Hazard Assessment Handbook, CG 446-3)

A-T-U

12. HAZARD CLASSIFICATIONS

12.1 **Code of Federal Regulations:** Flammable liquid
12.2 **NAS Hazard Rating for Bulk Water Transportation:**

Category	Rating
Fire	3
Health	
Vapor Irritant	1
Liquid or Solid Irritant	1
Poisons	2
Water Pollution	
Human Toxicity	1
Aquatic Toxicity	3
Aesthetic Effect	2
Reactivity	
Other Chemicals	1
Water	0
Self-Reaction	0

12.3 **NFPA Hazard Classifications:**

Category	Classification
Health Hazard (Blue)	2
Flammability (Red)	3
Reactivity (Yellow)	0

13. PHYSICAL AND CHEMICAL PROPERTIES

13.1 **Physical State at 15°C and 1 atm:** Liquid
13.2 **Molecular Weight:** 106.16
13.3 **Boiling Point at 1 atm:** 291.9°F = 144.4°C = 417.6°K
13.4 **Freezing Point:** −13.3°F = −25.2°C = 248.0°K
13.5 **Critical Temperature:** 674.8°F = 357.1°C = 630.3°K
13.6 **Critical Pressure:** 541.5 atm = 36.84 psia = 3.732 MN/m²
13.7 **Specific Gravity:** 0.880 at 20°C (liquid)
13.8 **Liquid Surface Tension:** 30.53 dynes/cm = 0.03053 N/m at 15.5°C
13.9 **Liquid-Water Interfacial Tension:** 36.06 dynes/cm = 0.03606 N/m at 20°C
13.10 **Vapor (Gas) Specific Gravity:** Not pertinent
13.11 **Ratio of Specific Heats of Vapor (Gas):** 1.068
13.12 **Latent Heat of Vaporization:** 149 Btu/lb = 82.9 cal/g = 3.47×10^5 J/kg
13.13 **Heat of Combustion:** −17,558 Btu/lb = −9754.7 cal/g = -408.41×10^5 J/kg
13.14 **Heat of Decomposition:** Not pertinent
13.15 **Heat of Solution:** Not pertinent
13.16 **Heat of Polymerization:** Not pertinent

NOTES

REVISED 1978

XLP	# p-XYLENE

Common Synonyms			
1, 4-Dimethylbenzene Xylol	Watery liquid	Colorless	Sweet odor
	Floats on water. Flammable, irritating vapor is produced. Freezing point is 56° F.		

Stop discharge if possible. Keep people away.
Call fire department.
Avoid contact with liquid and vapor.
Isolate and remove discharged material.
Notify local health and pollution control agencies.

Fire

FLAMMABLE
Flashback along vapor trail may occur.
Vapor may explode if ignited in an enclosed area.
Wear self-contained breathing apparatus.
Extinguish with foam, dry chemical, or carbon dioxide.
Water may be ineffective on fire.
Cool exposed containers with water.

Exposure

CALL FOR MEDICAL AID.

VAPOR
Irritating to eyes, nose and throat.
If inhaled, will cause dizziness, difficult breathing, or loss of consciousness.
Move to fresh air.
If breathing has stopped, give artificial respiration.
If breathing is difficult, give oxygen.

LIQUID
Irritating to skin and eyes.
If swallowed, will cause nausea, vomiting, loss of consciousness.
Remove contaminated clothing and shoes.
Flush affected areas with plenty of water.
IF IN EYES, hold eyelids open and flush with plenty of water.
IF SWALLOWED and victim is CONSCIOUS, have victim drink water or milk.
DO NOT INDUCE VOMITING.

Water Pollution

HARMFUL TO AQUATIC LIFE IN VERY LOW CONCENTRATIONS.
Fouling to shoreline.
May be dangerous if it enters water intakes.
Notify local health and wildlife officials.
Notify operators of nearby water intakes.

1. RESPONSE TO DISCHARGE

(See Response Methods Handbook, CG 446-4)

Issue warning—high flammability
Evacuate area
Should be removed
Chemical and physical treatment

2. LABEL

Red

3. CHEMICAL DESIGNATIONS

3.1 **Synonyms:** 1,4-Dimethylbenzene
3.2 **Coast Guard Compatibility Classification:** Aromatic hydrocarbon
3.3 **Chemical Formula:** p—$C_6H_4(CH_3)_2$
3.4 **IMCO/United Nations Numerical Designation:** 3.2/1307

4. OBSERVABLE CHARACTERISTICS

4.1 **Physical State (as shipped):** Liquid
4.2 **Color:** Colorless
4.3 **Odor:** Like benzene; characteristic aromatic

5. HEALTH HAZARDS

5.1 **Personal Protective Equipment:** Approved canister or air-supplied mask; goggles or face shield; plastic gloves and boots.
5.2 **Symptoms Following Exposure:** Vapors cause headache and dizziness. Liquid irritates eyes and skin. If taken into lungs, causes severe coughing, distress, and rapidly developing pulmonary edema. If ingested, causes nausea, vomiting, cramps, headache, and coma. Can be fatal. Kidney and liver damage can occur.
5.3 **Treatment for Exposure:** INHALATION: remove to fresh air; administer artificial respiration and oxygen if required; call a doctor. INGESTION: do NOT induce vomiting; call a doctor. EYES: flush with water for at least 15 min. SKIN: wipe off, wash with soap and water.
5.4 **Toxicity by Inhalation (Threshold Limit Value):** 100 ppm
5.5 **Short-Term Inhalation Limits:** 300 ppm for 30 min.
5.6 **Toxicity by Ingestion:** Grade 3; LD_{50} 50 to 500 mg/kg
5.7 **Late Toxicity:** Kidney and liver damage.
5.8 **Vapor (Gas) Irritant Characteristics:** Vapors cause a slight smarting of the eyes or respiratory system if present in high concentrations. The effect is temporary.
5.9 **Liquid or Solid Irritant Characteristics:** Minimum hazard. If spilled on clothing and allowed to remain, may cause smarting and reddening of the skin.
5.10 **Odor Threshold:** 0.05 ppm

6. FIRE HAZARDS

6.1 **Flash Point:** 81°F C.C.
6.2 **Flammable Limits in Air:** 1.1%–6.6%
6.3 **Fire Extinguishing Agents:** Foam, dry chemical, or carbon dioxide
6.4 **Fire Extinguishing Agents Not to be Used:** Water may be ineffective.
6.5 **Special Hazards of Combustion Products:** Not pertinent
6.6 **Behavior in Fire:** Vapor is heavier than air and may travel considerable distance to a source of ignition and flash back.
6.7 **Ignition Temperature:** 870°F
6.8 **Electrical Hazard:** Class I, Group D
6.9 **Burning Rate:** 5.8 mm/min.

7. CHEMICAL REACTIVITY

7.1 **Reactivity with Water:** No reaction
7.2 **Reactivity with Common Materials:** No reaction
7.3 **Stability During Transport:** Stable
7.4 **Neutralizing Agents for Acids and Caustics:** Not pertinent
7.5 **Polymerization:** Not pertinent
7.6 **Inhibitor of Polymerization:** Not pertinent

8. WATER POLLUTION

8.1 **Aquatic Toxicity:** 22 ppm/96 hr/bluegill/TL_m/fresh water
8.2 **Waterfowl Toxicity:** Data not available
8.3 **Biological Oxygen Demand (BOD):** 0 lb/lb in 5 days
8.4 **Food Chain Concentration Potential:** Data not available

9. SELECTED MANUFACTURERS

1. Amoco Chemicals Corp.
130 East Randolph Drive
Chicago, Ill. 60601
2. ARCO Chemical Co.
260 South Broad St.
Philadelphia, Pa. 19101
3. Chevron Chemical Co.
Industrial Chemicals Division
200 Bush St.
San Francisco, Calif. 94120

10. SHIPPING INFORMATION

10.1 **Grades or Purity:** Research: 99.99%; Pure: 99.8%; Technical: 99.0%
10.2 **Storage Temperature:** Ambient
10.3 **Inert Atmosphere:** No requirement
10.4 **Venting:** Open (flame arrester), or pressure-vacuum

11. HAZARD ASSESSMENT CODE

(See Hazard Assessment Handbook, CG 446-3)

A-T-U

12. HAZARD CLASSIFICATIONS

12.1 **Code of Federal Regulations:** Flammable liquid
12.2 **NAS Hazard Rating for Bulk Water Transportation:**

Category	Rating
Fire	3
Health	
Vapor Irritant	1
Liquid or Solid Irritant	1
Poisons	2
Water Pollution	
Human Toxicity	1
Aquatic Toxicity	3
Aesthetic Effect	2
Reactivity	
Other Chemicals	1
Water	0
Self-Reaction	0

12.3 **NFPA Hazard Classifications:**

Category	Classification
Health Hazard (Blue)	2
Flammability (Red)	3
Reactivity (Yellow)	0

13. PHYSICAL AND CHEMICAL PROPERTIES

13.1 **Physical State at 15°C and 1 atm:** Liquid
13.2 **Molecular Weight:** 106.16
13.3 **Boiling Point at 1 atm:** 280.9°F = 138.3°C = 411.5°K
13.4 **Freezing Point:** 55.9°F = 13.3°C = 286.5°K
13.5 **Critical Temperature:** 649.4°F = 343.0°C = 616.2°K
13.6 **Critical Pressure:** 509.4 atm = 34.65 psia = 3.510 MN/m^2
13.7 **Specific Gravity:** 0.861 at 20°C (liquid)
13.8 **Liquid Surface Tension:** 28.3 dynes/cm = 0.0283 N/m at 20°C
13.9 **Liquid-Water Interfacial Tension:** 37.8 dynes/cm = 0.0378 N/m at 20°C
13.10 **Vapor (Gas) Specific Gravity:** Not pertinent
13.11 **Ratio of Specific Heats of Vapor (Gas):** 1.071
13.12 **Latent Heat of Vaporization:** 150 Btu/lb = 81 cal/g = 3.4×10^5 J/kg
13.13 **Heat of Combustion:** −17,559 Btu/lb = −9754.7 cal/g = -408.41×10^5 J/kg
13.14 **Heat of Decomposition:** Not pertinent
13.15 **Heat of Solution:** Not pertinent
13.16 **Heat of Polymerization:** Not pertinent

NOTES

REVISED 1978

XYL XYLENOL

Common Synonyms			
Dimethylphenol 2, 6-Xylenol Cresylic acid 2-Hydroxy-m-xylene	Solid or liquid	Light yellowish brown	Sweet tarry odor
	May float or sink in water.		

Stop discharge if possible. Keep people away.
Call fire department.
Isolate and remove discharged material.
Notify local health and pollution control agencies.

Fire

Combustible.
POISONOUS GASES ARE PRODUCED IN FIRE.
Wear goggles and self-contained breathing apparatus.
Extinguish with dry chemicals, foam or carbon dioxide.
Water may be ineffective on fire.

Exposure

Call for medical aid.

DUST
Irritating to eyes, nose and throat.
Harmful if inhaled.
Move victim to fresh air.
If in eyes, hold eyelids open and flush with plenty of water.
If breathing has stopped, give artificial respiration.
If breathing is difficult, give oxygen.

LIQUID OR SOLID
Irritating to skin and eyes.
If swallowed or skin is exposed will cause nausea and vomiting.
Remove contaminated clothing and shoes.
Flush affected areas with plenty of water.
IF IN EYES, hold eyelids open and flush with plenty of water.
IF SWALLOWED and victim is CONSCIOUS, have victim drink water or milk and have victim induce vomiting.
IF SWALLOWED and victim is UNCONSCIOUS OR HAVING CONVULSIONS, do nothing except keep victim warm.

Water Pollution

HARMFUL TO AQUATIC LIFE IN VERY LOW CONCENTRATIONS.
Fouling to shoreline.
May be dangerous if it enters water intakes.
Notify local health and wildlife officials.
Notify operators of nearby water intakes.

1. RESPONSE TO DISCHARGE

(See Response Methods Handbook, CG 446-4)

Issue warning water contaminant
Restrict access
Mechanical containment
Should be removed
Chemical and physical treatment

2. LABELS

No label required by Code of Federal Regulations

3. CHEMICAL DESIGNATIONS

3.1 **Synonyms:** Cresylic acid; 2,6-Dimethylphenol; 2-Hydroxy-m-xylene; 2,6-Xylenol; *vic*-m-Xylenol
3.2 **Coast Guard Compatibility Classification:** Phenols (15)
3.3 **Chemical Formula:** $2,6\text{-}(CH_3)_2C_6H_3OH$
3.4 **IMCO/United Nations Numerical Designation:** Not listed

4. OBSERVABLE CHARACTERISTICS

4.1 **Physical State (as shipped):** Solid or liquid
4.2 **Color:** Light yellow-brown
4.3 **Odor:** Sweet tarry

5. HEALTH HAZARDS

5.1 **Personal Protective Equipment:** Organic canister mask; goggles or face shield; rubber gloves; other protective clothing to prevent contact with skin.
5.2 **Symptoms Following Exposure:** Vapor irritates eyes, nose, and throat and is readily absorbed through mucous membranes and lungs, producing general toxic symptoms (weakness, dizziness, headache, difficult breathing, twitching).
Contact with skin causes temporary prickling and intense burning, then local anesthesia. Affected areas initially show white discoloration, wrinkling, and softening, then become red, then brown or black (signs of gangrene). Extensive burns may permit absorption of chemical to produce toxic symptoms described above.
Ingestion causes irritation of mouth and stomach, nausea, abdominal pain, weakness, dizziness, headache, difficult breathing, and twitching.
5.3 **Treatment for Exposure:** *Get medical attention at once following exposure to this compound.* INHALATION: remove patient immediately to fresh air; irritation of nose or throat may be somewhat relieved by spraying or gargling with water until all odor is gone; 100% oxygen inhalation is indicated for cyanosis or respiratory distress; keep patient warm, but not hot. EYES: flood with running water for 15 min.; if physician is not immediately available, continue irrigation for another 15 min.; 2-3 drops of 0.5% pontocaine or equivalent may be instilled after first 15 min.; do not use oils or oily ointments unless ordered by physician. SKIN: wash affected areas with large quantities of water or soapy water until all odor is gone; then wash with alcohol or 20% glycerin solution and more water; keep patient warm, but not hot; cover chemical burns continuously with compresses wet with saturated solution of sodium thiosulfate; apply no salves or ointments for 24 hrs after injury. INGESTION: give large quantities of liquid (salt water, weak sodium bicarbonate solution, milk, or gruel) followed by demulcent such as raw egg white or corn starch paste; if profuse vomiting does not follow immediately, give a mild emetic (such as 1 tbsp. mustard in glass of water), or tickle back of throat. Repeat procedure until vomitus is free of the odor. Some demulcent should be left in stomach after vomiting. Keep patient comfortably warm.

6. FIRE HAZARDS

6.1 **Flash Point:** 186°F C.C.
6.2 **Flammable Limits in Air:** 1.4% (LFL)
6.3 **Fire Extinguishing Agents:** Water, dry chemical, foam, carbon dioxide
6.4 **Fire Extinguishing Agents Not to be Used:** Not pertinent
6.5 **Special Hazards of Combustion Products:** Toxic vapors of unburned material may form in fire.
6.6 **Behavior in Fire:** Not pertinent
6.7 **Ignition Temperature:** 1110°F
6.8 **Electrical Hazard:** Data not available
6.9 **Burning Rate:** Data not available

7. CHEMICAL REACTIVITY

7.1 **Reactivity with Water:** No reaction
7.2 **Reactivity with Common Materials:** No reaction
7.3 **Stability During Transport:** Stable
7.4 **Neutralizing Agents for Acids and Caustics:** Not pertinent
7.5 **Polymerization:** Not pertinent
7.6 **Inhibitor of Polymerization:** Not pertinent

8. WATER POLLUTION

8.1 **Aquatic Toxicity:** (2,6 isomer)
7-9 ppm/*/trout/lethal/fresh water
*Time period not specified.
8.2 **Waterfowl Toxicity:** Data not available
8.3 **Biological Oxygen Demand (BOD):** 31% of theoretical in 5 days
8.4 **Food Chain Concentration Potential:** None

9. SELECTED MANUFACTURERS

1. Continental Oil Company
Pitt-Consol Chemicals
Park-Eighty Plaza East
Saddle Brook, N. J. 07662
2. Koppers Company, Incorporated
Organic Materials Division
1350 Koppers Building
Pittsburgh, Pa. 15219
3. Crowley Tar Products Co.
267 Madison Avenue
New York, N. Y. 10016

10. SHIPPING INFORMATION

10.1 **Grades or Purity:** 99% 2,6-Xylenol. Other commercial Xylenols include 2,3-; 2,4-; 2,5-; 3,4-; 3,5-; and various mixtures of these. Properties are similar to those of the 2,6- compound.
10.2 **Storage Temperature:** Ambient
10.3 **Inert Atmosphere:** No requirement
10.4 **Venting:** Open (flame arrester)

11. HAZARD ASSESSMENT CODE

(See Hazard Assessment Handbook, CG 446-3)

A-T-U-X-Y

12. HAZARD CLASSIFICATIONS

12.1 **Code of Federal Regulations:** Not listed
12.2 **NAS Hazard Rating for Bulk Water Transportation:** Not listed
12.3 **NFPA Hazard Classifications:**

Category	Classification
Health Hazard (Blue)	2
Flammability (Red)	1
Reactivity (Yellow)	0

13. PHYSICAL AND CHEMICAL PROPERTIES

13.1 **Physical State at 15°C and 1 atm:** Solid or liquid
13.2 **Molecular Weight:** 122.2
13.3 **Boiling Point at 1 atm:** 413°F = 212°C = 485°K
13.4 **Freezing Point:** −40 to +106°F = −40 to +45°C = 233 to 318°K
13.5 **Critical Temperature:** Not pertinent
13.6 **Critical Pressure:** Not pertinent
13.7 **Specific Gravity:** 1.01 at 20°C (liquid)
13.8 **Liquid Surface Tension:** (est.) 30 dynes/cm = 0.030 N/m at 30°C
13.9 **Liquid-Water Interfacial Tension:** (est.) 25 dynes/cm = 0.025 N/m at 25°C
13.10 **Vapor (Gas) Specific Gravity:** Not pertinent
13.11 **Ratio of Specific Heats of Vapor (Gas):** Not pertinent
13.12 **Latent Heat of Vaporization:** 212.74 Btu/lb = 118.19 cal/g at 25°C = 4.9451×10^5 J/kg
13.13 **Heat of Combustion:** −15,310 Btu/lb = −8,500 cal/g = -356×10^5 J/kg
13.14 **Heat of Decomposition:** Not pertinent
13.15 **Heat of Solution:** Not pertinent
13.16 **Heat of Polymerization:** Not pertinent

5. HEALTH HAZARDS (Cont'd.)

5.4 **Toxicity by Inhalation (Threshold Limit Value):** 45 ppm
5.5 **Short-Term Inhalation Limits:** Data not available
5.6 **Toxicity by Ingestion:** Grade 2; oral LD_{50} = 1,070 mg/kg (mouse)
5.7 **Late Toxicity:** Damage to heart muscle, and changes in liver, kidney, and spleen in rats
5.8 **Vapor (Gas) Irritant Characteristics:** Data not available
5.9 **Liquid or Solid Irritant Characteristics:** Data not available
5.10 **Odor Threshold:** Data not available

ZNA	ZINC ACETATE

Common Synonyms			
Zinc acetate dihydrate Dicarbomethoxyzinc Zinc diacetate Acetic acid, zinc salt	Solid	White	Faint vinegar odor
	Sinks and mixes with water.		

Avoid contact with solid and dust. Keep people away.
Stop discharge if possible.
Isolate and remove discharged material.
Notify local health and pollution control agencies.

Fire	Not flammable.
Exposure	CALL FOR MEDICAL AID. DUST Irritating to eyes, nose and throat. If inhaled will cause coughing or difficult breathing. If in eyes, hold eyelids open and flush with plenty of water. If breathing has stopped, give artificial respiration. If breathing is difficult, give oxygen. SOLID Irritating to skin and eyes. If swallowed will cause nausea and vomiting. Remove contaminated clothing and shoes. Flush affected areas with plenty of water. IF IN EYES, hold eyelids open and flush with plenty of water. IF SWALLOWED and victim is CONSCIOUS, have victim drink water or milk and have victim induce vomiting. IF SWALLOWED and victim is UNCONSCIOUS OR HAVING CONVULSIONS, do nothing except keep victim warm.
Water Pollution	HARMFUL TO AQUATIC LIFE IN VERY LOW CONCENTRATIONS. May be dangerous if it enters water intakes. Notify local health and wildlife officials. Notify operators of nearby water intakes.

1. RESPONSE TO DISCHARGE
(See Response Methods Handbook, CG 446-4)

Disperse and flush

2. LABELS

No hazard label required by Code of Federal Regulations

3. CHEMICAL DESIGNATIONS

3.1 **Synonyms:** Acetic acid, zinc salt; Dicarbomethoxyzinc; Zinc acetate dihydrate; Zinc diacetate

3.2 **Coast Guard Compatibility Classification:** Not listed

3.3 **Chemical Formula:** $Zn(C_2H_3O_2)_2$ or $Zn(C_2H_3O_2)_2 \cdot 2H_2O$

3.4 **IMCO/United Nations Numerical Designation:** Not listed

4. OBSERVABLE CHARACTERISTICS

4.1 **Physical State (as shipped):** Solid

4.2 **Color:** White

4.3 **Odor:** Faint acetic acid

5. HEALTH HAZARDS

5.1 **Personal Protective Equipment:** Bu. Mines approved respirator; rubber gloves; chemical goggles

5.2 **Symptoms Following Exposure:** Inhalation causes mild irritation of nose and throat, coughing, and sneezing. Ingestion can cause irritation or corrosion of the alimentary tract, resulting in vomiting. Contact with dust causes irritation of eyes and mild irritation of skin.

5.3 **Treatment for Exposure:** INHALATION: move to fresh air; if exposure is severe, get medical attention. INGESTION: induce vomiting, followed by prompt and complete gastric lavage, cathartics, and demulcents. EYES: flush with water for at least 10 min.; consult physician if irritation persists. SKIN: wash with soap and water.

5.4 **Toxicity by Inhalation (Threshold Limit Value):** Data not available

5.5 **Short-Term Inhalation Limits:** Data not available

5.6 **Toxicity by Ingestion:** Grade 2; LD_{50} 0.5—5 g/kg

5.7 **Late Toxicity:** Data not available

5.8 **Vapor (Gas) Irritant Characteristics:** Data not available

5.9 **Liquid or Solid Irritant Characteristics:** Data not available

5.10 **Odor Threshold:** Data not available

6. FIRE HAZARDS

6.1 **Flash Point:** Not flammable

6.2 **Flammable Limits in Air:** Not flammable

6.3 **Fire Extinguishing Agents:** Not pertinent

6.4 **Fire Extinguishing Agents Not to be Used:** Not pertinent

6.5 **Special Hazards of Combustion Products:** Not pertinent

6.6 **Behavior in Fire:**

6.7 **Ignition Temperature:** Not pertinent

6.8 **Electrical Hazard:** Not pertinent

6.9 **Burning Rate:** Not pertinent

7. CHEMICAL REACTIVITY

7.1 **Reactivity with Water:** No reaction

7.2 **Reactivity with Common Materials:**

7.3 **Stability During Transport:** Stable

7.4 **Neutralizing Agents for Acids and Caustics:** Not pertinent

7.5 **Polymerization:** Not pertinent

7.6 **Inhibitor of Polymerization:** Not pertinent

8. WATER POLLUTION

8.1 **Aquatic Toxicity:** 0.88 ppm/96 hr/fathead minnow/TL_m/soft water

8.2 **Waterfowl Toxicity:** Data not available

8.3 **Biological Oxygen Demand (BOD):** Data not available

8.4 **Food Chain Concentration Potential:** Zinc is not considered to be bioconcentrative even though it accumulates in some organisms.

9. SELECTED MANUFACTURERS

1. The Harshaw Chemical Co.
   1945 E. 97 St.
   Cleveland, Ohio 44106
2. Allied Chemical Corp.
   Specialty Chemicals Div.
   P. O. Box 1087R
   Morristown, N. J. 07960
3. Union Carbide Corp.
   Chemicals and Plastics Div.
   270 Park Avenue
   New York, N. Y. 10017

10. SHIPPING INFORMATION

10.1 **Grades or Purity:** Reagent, 99%; Commercial, 98.4%

10.2 **Storage Temperature:** Ambient

10.3 **Inert Atmosphere:** No requirement

10.4 **Venting:** Open

11. HAZARD ASSESSMENT CODE
(See Hazard Assessment Handbook, CG 446-3)

SS

12. HAZARD CLASSIFICATIONS

12.1 **Code of Federal Regulations:** Not listed

12.2 **NAS Hazard Rating for Bulk Water Transportation:** Not listed

12.3 **NFPA Hazard Classifications:**

Category	Classification
Health Hazard (Blue)	1
Flammability (Red)	1
Reactivity (Yellow)	0

13. PHYSICAL AND CHEMICAL PROPERTIES

13.1 **Physical State at 15°C and 1 atm:** Solid

13.2 **Molecular Weight:** 219.49

13.3 **Boiling Point at 1 atm:** Not pertinent (decomposes)

13.4 **Freezing Point:** Not pertinent

13.5 **Critical Temperature:** Not pertinent

13.6 **Critical Pressure:** Not pertinent

13.7 **Specific Gravity:** 1.74 at 20°C (solid)

13.8 **Liquid Surface Tension:** Not pertinent

13.9 **Liquid-Water Interfacial Tension:** Not pertinent

13.10 **Vapor (Gas) Specific Gravity:** Not pertinent

13.11 **Ratio of Specific Heats of Vapor (Gas):** Not pertinent

13.12 **Latent Heat of Vaporization:** Not pertinent

13.13 **Heat of Combustion:** Not pertinent

13.14 **Heat of Decomposition:** Not pertinent

13.15 **Heat of Solution:** (approx.) −0.5 Btu/lb = −0.3 cal/g = -0.01×10^5 J/kg

13.16 **Heat of Polymerization:** Not pertinent

NOTES

ZAC ZINC AMMONIUM CHLORIDE

Common Synonyms
Ammonium pentachlorozincate
Ammonium zinc chloride

Solid White Odorless

Sinks and mixes with water.

Avoid contact with solid and dust. Keep people away.
Stop discharge if possible.
Isolate and remove discharged material.
Notify local health and pollution control agencies.

Fire

Not flammable.

Exposure

CALL FOR MEDICAL AID.

DUST
Irritating to eyes, nose and throat.
If inhaled will cause coughing or difficult breathing.

If in eyes, hold eyelids open and flush with plenty of water.
If breathing has stopped, give artificial respiration.
If breathing is difficult, give oxygen.

SOLID
Irritating to skin and eyes.
If swallowed will cause nausea and vomiting.

Remove contaminated clothing and shoes.
Flush affected areas with plenty of water.
IF IN EYES, hold eyelids open and flush with plenty of water.
IF SWALLOWED and victim is CONSCIOUS, have victim drink water or milk and have victim induce vomiting.
IF SWALLOWED and victim is UNCONSCIOUS OR HAVING CONVULSIONS, do nothing except keep victim warm.

Water Pollution

Effect of low concentrations on aquatic life is unknown.
May be dangerous if it enters water intakes.
Notify local health and wildlife officials.
Notify operators of nearby water intakes.

1. RESPONSE TO DISCHARGE

(See Response Methods Handbook, CG 446-4)

Disperse and flush

2. LABELS

No hazard label required by Code of Federal Regulations

3. CHEMICAL DESIGNATIONS

3.1 **Synonyms:** Ammonium pentachlorozincate; Ammonium zinc chloride
3.2 **Coast Guard Compatibility Classification:** Not listed
3.3 **Chemical Formula:** $ZnCl_2 \cdot 3NH_4Cl$
3.4 **IMCO/United Nations Numerical Designation:** Not listed

4. OBSERVABLE CHARACTERISTICS

4.1 **Physical State (as shipped):** Solid
4.2 **Color:** White
4.3 **Odor:** None

5. HEALTH HAZARDS

5.1 **Personal Protective Equipment:** Dust mask; goggles or face shield; protective gloves
5.2 **Symptoms Following Exposure:** Inhalation of dust irritates nose and throat. Ingestion can cause irritation or corrosion of the alimentary tract. Contact with eyes or skin causes irritation.
5.3 **Treatment for Exposure:** INHALATION: remove from exposure; begin artificial respiration if breathing has ceased. INGESTION: induce vomiting, followed by prompt and complete gastric lavage, catharsis, and demulcents. EYES: flush with water for at least 15 min. SKIN: flush with water.
5.4 **Toxicity by Inhalation (Threshold Limit Value):** Data not available
5.5 **Short-Term Inhalation Limits:** Data not available
5.6 **Toxicity by Ingestion:** Grade 2; LD_{50} 0.5—5 g/kg
5.7 **Late Toxicity:** Data not available
5.8 **Vapor (Gas) Irritant Characteristics:** Data not available
5.9 **Liquid or Solid Irritant Characteristics:** Data not available
5.10 **Odor Threshold:** Data not available

6. FIRE HAZARDS

6.1 **Flash Point:** Not flammable
6.2 **Flammable Limits in Air:** Not flammable
6.3 **Fire Extinguishing Agents:** Not pertinent
6.4 **Fire Extinguishing Agents Not to be Used:** Not pertinent
6.5 **Special Hazards of Combustion Products:** Not pertinent
6.6 **Behavior in Fire:**
6.7 **Ignition Temperature:** Not pertinent
6.8 **Electrical Hazard:** Not pertinent
6.9 **Burning Rate:** Not pertinent

7. CHEMICAL REACTIVITY

7.1 **Reactivity with Water:** No reaction
7.2 **Reactivity with Common Materials:**
7.3 **Stability During Transport:** Stable
7.4 **Neutralizing Agents for Acids and Caustics:** Not pertinent
7.5 **Polymerization:** Not pertinent
7.6 **Inhibitor of Polymerization:** Not pertinent

8. WATER POLLUTION

8.1 **Aquatic Toxicity:** Data not available
8.2 **Waterfowl Toxicity:** Data not available
8.3 **Biological Oxygen Demand (BOD):** Data not available
8.4 **Food Chain Concentration Potential:** Zinc accumulates in some organisms but is not considered to be bioconcentrative.

9. SELECTED MANUFACTURERS

1. Chemical and Pigment Co.
   600 Nichols Road
   Pittsburg, Calif. 94565
2. Pfaltz and Bauer, Inc.
   375 Fairfield Ave.
   Stamford, Conn. 06902

10. SHIPPING INFORMATION

10.1 **Grades or Purity:** Commercial
10.2 **Storage Temperature:** Ambient
10.3 **Inert Atmosphere:** No requirement
10.4 **Venting:** Open

11. HAZARD ASSESSMENT CODE

(See Hazard Assessment Handbook, CG 446-3)

SS

12. HAZARD CLASSIFICATIONS

12.1 **Code of Federal Regulations:** Not listed
12.2 **NAS Hazard Rating for Bulk Water Transportation:** Not listed
12.3 **NFPA Hazard Classifications:** Not listed

13. PHYSICAL AND CHEMICAL PROPERTIES

13.1 **Physical State at 15°C and 1 atm:** Solid
13.2 **Molecular Weight:** 296.8
13.3 **Boiling Point at 1 atm:** (sublimes) 644°F = 340°C = 613°K
13.4 **Freezing Point:** Not pertinent
13.5 **Critical Temperature:** Not pertinent
13.6 **Critical Pressure:** Not pertinent
13.7 **Specific Gravity:** 1.81 at 20°C (solid)
13.8 **Liquid Surface Tension:** Not pertinent
13.9 **Liquid-Water Interfacial Tension:** Not pertinent
13.10 **Vapor (Gas) Specific Gravity:** Not pertinent
13.11 **Ratio of Specific Heats of Vapor (Gas):** Not pertinent
13.12 **Latent Heat of Vaporization:** Not pertinent
13.13 **Heat of Combustion:** Not pertinent
13.14 **Heat of Decomposition:** Not pertinent
13.15 **Heat of Solution:** Data not available
13.16 **Heat of Polymerization:** Not pertinent

NOTES

ZAR ZINC ARSENATE

Common Synonyms	Solid	Colorless	Odorless
	Sinks in water.		

AVOID CONTACT WITH SOLID AND DUST. KEEP PEOPLE AWAY.
Wear goggles and dust respirator.
Isolate and remove discharged material.
Notify local health and pollution control agencies.

Fire

Not flammable.

Exposure

CALL FOR MEDICAL AID.

DUST
POISONOUS IF INHALED.
Irritating to eyes, nose and throat.
Move victim to fresh air.
If in eyes, hold eyelids open and flush with plenty of water.
If breathing is difficult, give oxygen.

SOLID
POISONOUS IF SWALLOWED.
Irritating to skin and eyes.
Remove contaminated clothing and shoes.
Flush affected areas with plenty of water.
IF IN EYES, hold eyelids open and flush with plenty of water.
IF SWALLOWED and victim is CONSCIOUS, have victim drink water or milk and have victim induce vomiting.
IF SWALLOWED and victim is UNCONSCIOUS OR HAVING CONVULSIONS, do nothing except keep victim warm.

Water Pollution

Effect of low concentrations on aquatic life is unknown.
May be dangerous if it enters water intakes.
Notify local health and wildlife officials.
Notify operators of nearby water intakes.

1. RESPONSE TO DISCHARGE

(See Response Methods Handbook, CG 446-4)

Issue warning – poison, water contaminant
Restrict access
Should be removed
Chemical and physical treatment

2. LABEL

POISON

3. CHEMICAL DESIGNATIONS

3.1 **Synonyms:** No common synonyms
3.2 **Coast Guard Compatibility Classification:** Not applicable
3.3 **Chemical Formula:** (approx.) $(ZnO)_5(As_2O_5)_2$
3.4 **IMCO/United Nations Numerical Designation:** 6.1/1712

4. OBSERVABLE CHARACTERISTICS

4.1 **Physical State (as shipped):** Solid
4.2 **Color:** Colorless
4.3 **Odor:** None

5. HEALTH HAZARDS

5.1 **Personal Protective Equipment:** Dust mask; rubber gloves
5.2 **Symptoms Following Exposure:** Dust may irritate eyes. Ingestion or excessive inhalation of dust causes burning of mouth, abdominal pain, vomiting, diarrhea with hemorrhage, dehydration, jaundice, and collapse.
5.3 **Treatment for Exposure:** EYES: flush with water to remove dust. INGESTION: immediately induce evacuation of intestinal tract by inducing vomiting and giving gastric lavage and saline cathartic; see physician at once; consider development of arsenic poisoning.
5.4 **Toxicity by Inhalation (Threshold Limit Value):** 0.5 mg/m^3
5.5 **Short-Term Inhalation Limits:** Data not available
5.6 **Toxicity by Ingestion:** Data not available
5.7 **Late Toxicity:** May be carcinogenic. Arsenic poisoning may develop.
5.8 **Vapor (Gas) Irritant Characteristics:** Data not available
5.9 **Liquid or Solid Irritant Characteristics:** Data not available
5.10 **Odor Threshold:** Not pertinent

6. FIRE HAZARDS

6.1 **Flash Point:** Not flammable
6.2 **Flammable Limits in Air:** Not flammable
6.3 **Fire Extinguishing Agents:** Not pertinent
6.4 **Fire Extinguishing Agents Not to be Used:** Not pertinent
6.5 **Special Hazards of Combustion Products:** Not pertinent
6.6 **Behavior in Fire:** Not pertinent
6.7 **Ignition Temperature:** Not pertinent
6.8 **Electrical Hazard:** Not pertinent
6.9 **Burning Rate:** Not pertinent

7. CHEMICAL REACTIVITY

7.1 **Reactivity with Water:** No reaction
7.2 **Reactivity with Common Materials:** No reaction
7.3 **Stability During Transport:** Stable
7.4 **Neutralizing Agents for Acids and Caustics:** Not pertinent
7.5 **Polymerization:** Not pertinent
7.6 **Inhibitor of Polymerization:** Not pertinent

8. WATER POLLUTION

8.1 **Aquatic Toxicity:** Data not available
8.2 **Waterfowl Toxicity:** Data not available
8.3 **Biological Oxygen Demand (BOD):** None
8.4 **Food Chain Concentration Potential:** Data not available

9. SELECTED MANUFACTURERS

Pfaltz and Bauer, Inc.
126-04 Northern Boulevard
Flushing, N. Y. 11368

10. SHIPPING INFORMATION

10.1 **Grades or Purity:** Technical
10.2 **Storage Temperature:** Ambient
10.3 **Inert Atmosphere:** No requirement
10.4 **Venting:** Open

11. HAZARD ASSESSMENT CODE

(See Hazard Assessment Handbook, CG 446-3)

II

12. HAZARD CLASSIFICATIONS

12.1 **Code of Federal Regulations:** Poisonous, Class B
12.2 **NAS Hazard Rating for Bulk Water Transportation:** Not listed
12.3 **NFPA Hazard Classifications:** Not listed

13. PHYSICAL AND CHEMICAL PROPERTIES

13.1 **Physical State at 15°C and 1 atm:** Solid
13.2 **Molecular Weight:** 866 (approx.)
13.3 **Boiling Point at 1 atm:** Not pertinent (decomposes)
13.4 **Freezing Point:** Not pertinent
13.5 **Critical Temperature:** Not pertinent
13.6 **Critical Pressure:** Not pertinent
13.7 **Specific Gravity:** 3.31 at 15°C (solid)
13.8 **Liquid Surface Tension:** Not pertinent
13.9 **Liquid-Water Interfacial Tension:** Not pertinent
13.10 **Vapor (Gas) Specific Gravity:** Not pertinent
13.11 **Ratio of Specific Heats of Vapor (Gas):** Not pertinent
13.12 **Latent Heat of Vaporization:** Not pertinent
13.13 **Heat of Combustion:** Not pertinent
13.14 **Heat of Decomposition:** Not pertinent
13.15 **Heat of Solution:** Not pertinent
13.16 **Heat of Polymerization:** Not pertinent

NOTES

ZBO	# ZINC BORATE

Common Synonyms	Solid	White	Odorless
	Sinks in water.		

Avoid contact with solid and dust. Keep people away.
Stop discharge if possible.
Isolate and remove discharged material.
Notify local health and pollution control agencies.

Fire	Not flammable.
Exposure	CALL FOR MEDICAL AID. DUST Irritating to eyes, nose and throat. If inhaled will cause coughing or difficult breathing. If in eyes, hold eyelids open and flush with plenty of water. If breathing has stopped, give artificial respiration. If breathing is difficult, give oxygen. SOLID Irritating to skin and eyes. If swallowed will cause nausea and vomiting. Remove contaminated clothing and shoes. Flush affected areas with plenty of water. IF IN EYES, hold eyelids open and flush with plenty of water. IF SWALLOWED and victim is CONSCIOUS, have victim drink water or milk. IF SWALLOWED and victim is UNCONSCIOUS OR HAVING CONVULSIONS, do nothing except keep victim warm.
Water Pollution	Effect of low concentrations on aquatic life is unknown. May be dangerous if it enters water intakes. Notify local health and wildlife officials. Notify operators of nearby water intakes.

1. RESPONSE TO DISCHARGE
(See Response Methods Handbook, CG 446-4)

Should be removed
Chemical and physical treatment

2. LABELS

No hazard label required by Code of Federal Regulations

3. CHEMICAL DESIGNATIONS

3.1 **Synonyms:** No common synonyms
3.2 **Coast Guard Compatibility Classification:** Not listed
3.3 **Chemical Formula:** (approx.) $2ZnO{\cdot}3B_2O_3{\cdot}3.5H_2O$
3.4 **IMCO/United Nations Numerical Designation:** Not listed

4. OBSERVABLE CHARACTERISTICS

4.1 **Physical State (as shipped):** Solid
4.2 **Color:** White
4.3 **Odor:** None

5. HEALTH HAZARDS

5.1 **Personal Protective Equipment:** Dust mask; goggles or face shield; protective gloves
5.2 **Symptoms Following Exposure:** Inhalation of dust may irritate nose and throat. Ingestion can cause gastrointestinal disturbances, convulsions, central nervous depressions, skin eruptions, shock, and death. Contact with eyes or skin causes irritation.
5.3 **Treatment for Exposure:** INHALATION: move to fresh air. INGESTION: administer gastric lavage with warm tap water; saline catharsis; consult physician. EYES or SKIN: flush with water.
5.4 **Toxicity by Inhalation (Threshold Limit Value):** 10 mg/m³ (as boric oxide)
5.5 **Short-Term Inhalation Limits:** Data not available
5.6 **Toxicity by Ingestion:** Grade 1; LD_{50} 5—15 g/kg
5.7 **Late Toxicity:** Data not available
5.8 **Vapor (Gas) Irritant Characteristics:** Data not available
5.9 **Liquid or Solid Irritant Characteristics:** Data not available
5.10 **Odor Threshold:** Not pertinent

6. FIRE HAZARDS

6.1 **Flash Point:** Not flammable
6.2 **Flammable Limits in Air:** Not flammable
6.3 **Fire Extinguishing Agents:** Not pertinent
6.4 **Fire Extinguishing Agents Not to be Used:** Not pertinent
6.5 **Special Hazards of Combustion Products:** Not pertinent
6.6 **Behavior in Fire:**
6.7 **Ignition Temperature:** Not pertinent
6.8 **Electrical Hazard:** Not pertinent
6.9 **Burning Rate:** Not pertinent

7. CHEMICAL REACTIVITY

7.1 **Reactivity with Water:** No reaction
7.2 **Reactivity with Common Materials:**
7.3 **Stability During Transport:** Stable
7.4 **Neutralizing Agents for Acids and Caustics:** Not pertinent
7.5 **Polymerization:** Not pertinent
7.6 **Inhibitor of Polymerization:** Not pertinent

8. WATER POLLUTION

8.1 **Aquatic Toxicity:** Data not available
8.2 **Waterfowl Toxicity:** Data not available
8.3 **Biological Oxygen Demand (BOD):** None
8.4 **Food Chain Concentration Potential:** Zinc accumulates in some organisms but is not considered to be bioconcentrative.

9. SELECTED MANUFACTURERS

1. U. S. Borax & Chemical Co.
   3075 Wilshire Blvd.
   Los Angeles, Calif. 90010
2. Humphrey Chemical Corp.
   P. O. Box 21
   Edgewood Arsenal, Md. 21010
3. American Hoechst Corp.
   Chemicals and Plastics Div.
   Rt. 202—206 North
   Somerville, N. J. 08876

10. SHIPPING INFORMATION

10.1 **Grades or Purity:** Commercial grades contain 45—52% zinc oxide, 29—35% boric anhydride
10.2 **Storage Temperature:** Ambient
10.3 **Inert Atmosphere:** No requirement
10.4 **Venting:** Open

11. HAZARD ASSESSMENT CODE
(See Hazard Assessment Handbook, CG 446-3)

II

12. HAZARD CLASSIFICATIONS

12.1 **Code of Federal Regulations:** Not listed
12.2 **NAS Hazard Rating for Bulk Water Transportation:** Not listed
12.3 **NFPA Hazard Classifications:** Not listed

13. PHYSICAL AND CHEMICAL PROPERTIES

13.1 **Physical State at 15°C and 1 atm:** Solid
13.2 **Molecular Weight:** 434.75
13.3 **Boiling Point at 1 atm:** Not pertinent (decomposes)
13.4 **Freezing Point:** Not pertinent
13.5 **Critical Temperature:** Not pertinent
13.6 **Critical Pressure:** Not pertinent
13.7 **Specific Gravity:** 2.7 at 20°C (solid)
13.8 **Liquid Surface Tension:** Not pertinent
13.9 **Liquid-Water Interfacial Tension:** Not pertinent
13.10 **Vapor (Gas) Specific Gravity:** Not pertinent
13.11 **Ratio of Specific Heats of Vapor (Gas):** Not pertinent
13.12 **Latent Heat of Vaporization:** Not pertinent
13.13 **Heat of Combustion:** Not pertinent
13.14 **Heat of Decomposition:** Not pertinent
13.15 **Heat of Solution:** Not pertinent
13.16 **Heat of Polymerization:** Not pertinent

NOTES

ZBR — ZINC BROMIDE

Common Synonyms

Solid White Odorless

Sinks and mixes with water.

Avoid contact with solid and dust. Keep people away.
Stop discharge if possible.
Isolate and remove discharged material.
Notify local health and pollution control agencies.

Fire

Not flammable.

Exposure

CALL FOR MEDICAL AID.

DUST
Irritating to eyes, nose and throat.
If inhaled will cause coughing or difficult breathing.
If in eyes, hold eyelids open and flush with plenty of water.
If breathing has stopped, give artificial respiration.
If breathing is difficult, give oxygen.

SOLID
Irritating to skin and eyes.
If swallowed will cause nausea and vomiting.

Remove contaminated clothing and shoes.
Flush affected areas with plenty of water.
IF IN EYES, hold eyelids open and flush with plenty of water.
IF SWALLOWED and victim is CONSCIOUS, have victim drink water or milk and have victim induce vomiting.
IF SWALLOWED and victim is UNCONSCIOUS OR HAVING CONVULSIONS, do nothing except keep victim warm.

Water Pollution

Effect of low concentrations on aquatic life is unknown.
May be dangerous if it enters water intakes.
Notify local health and wildlife officials.
Notify operators of nearby water intakes.

1. RESPONSE TO DISCHARGE
(See Response Methods Handbook, CG 446-4)

Issue warning—water contaminant
Disperse and flush

2. LABELS

No hazard label required by Code of Federal Regulations

3. CHEMICAL DESIGNATIONS

3.1 **Synonyms:** No common synonyms
3.2 **Coast Guard Compatibility Classification:** Not listed
3.3 **Chemical Formula:** $ZnBr_2$
3.4 **IMCO/United Nations Numerical Designation:** Not listed

4. OBSERVABLE CHARACTERISTICS

4.1 **Physical State (as shipped):** Solid
4.2 **Color:** White
4.3 **Odor:** None

5. HEALTH HAZARDS

5.1 **Personal Protective Equipment:** Chemical goggles or face shield; rubber gloves; dust mask
5.2 **Symptoms Following Exposure:** Inhalation of dust may irritate nose and throat. Ingestion can cause irritation or corrosion of the alimentary tract; if large amount is swallowed and not thrown up, drowsiness and other symptoms of bromide poisoning may occur. Contact with eyes or skin causes irritation.
5.3 **Treatment for Exposure:** INHALATION: move to fresh air. INGESTION: give large amount of water; induce vomiting, followed by prompt and complete gastric lavage, catharsis, and demulcents. EYES or SKIN: wash immediately with large volumes of water.
5.4 **Toxicity by Inhalation (Threshold Limit Value):** Data not available
5.5 **Short-Term Inhalation Limits:** Data not available
5.6 **Toxicity by Ingestion:** Grade 2; LD_{50} 0.5—5 g/kg
5.7 **Late Toxicity:** Data not available
5.8 **Vapor (Gas) Irritant Characteristics:** Data not available
5.9 **Liquid or Solid Irritant Characteristics:** Data not available
5.10 **Odor Threshold:** Data not available

6. FIRE HAZARDS

6.1 **Flash Point:** Not flammable
6.2 **Flammable Limits in Air:** Not flammable
6.3 **Fire Extinguishing Agents:** Not pertinent
6.4 **Fire Extinguishing Agents Not to be Used:** Not pertinent
6.5 **Special Hazards of Combustion Products:** Not pertinent
6.6 **Behavior in Fire:**
6.7 **Ignition Temperature:** Not pertinent
6.8 **Electrical Hazard:** Not pertinent
6.9 **Burning Rate:** Not pertinent

7. CHEMICAL REACTIVITY

7.1 **Reactivity with Water:** No reaction
7.2 **Reactivity with Common Materials:**
7.3 **Stability During Transport:** Stable
7.4 **Neutralizing Agents for Acids and Caustics:** Not pertinent
7.5 **Polymerization:** Not pertinent
7.6 **Inhibitor of Polymerization:** Not pertinent

8. WATER POLLUTION

8.1 **Aquatic Toxicity:** Data not available
8.2 **Waterfowl Toxicity:** Data not available
8.3 **Biological Oxygen Demand (BOD):** None
8.4 **Food Chain Concentration Potential:** Zinc accumulates in some organisms, but is not considered to be bioconcentrative.

9. SELECTED MANUFACTURERS

1. American Hoechst Corp.
   Chemicals and Plastics Div.
   Rt. 202—206 North
   Somerville, N. J. 08876
2. Michigan Chemical Corp.
   351 E. Ohio St.
   Chicago, Ill. 60611
3. Fisher Scientific Co.
   711 Forbes Ave.
   Pittsburgh, Pa. 15219

10. SHIPPING INFORMATION

10.1 **Grades or Purity:** Technical, 98+%
10.2 **Storage Temperature:** Ambient
10.3 **Inert Atmosphere:** No requirement
10.4 **Venting:** Open

11. HAZARD ASSESSMENT CODE
(See Hazard Assessment Handbook, CG 446-3)

SS

12. HAZARD CLASSIFICATIONS

12.1 **Code of Federal Regulations:** Not listed
12.2 **NAS Hazard Rating for Bulk Water Transportation:** Not listed
12.3 **NFPA Hazard Classifications:** Not listed

13. PHYSICAL AND CHEMICAL PROPERTIES

13.1 **Physical State at 15°C and 1 atm:** Solid
13.2 **Molecular Weight:** 225.18
13.3 **Boiling Point at 1 atm:** Not pertinent (decomposes)
13.4 **Freezing Point:** Not pertinent
13.5 **Critical Temperature:** Not pertinent
13.6 **Critical Pressure:** Not pertinent
13.7 **Specific Gravity:** 4.22 at 20°C (solid)
13.8 **Liquid Surface Tension:** Not pertinent
13.9 **Liquid-Water Interfacial Tension:** Not pertinent
13.10 **Vapor (Gas) Specific Gravity:** Not pertinent
13.11 **Ratio of Specific Heats of Vapor (Gas):** Not pertinent
13.12 **Latent Heat of Vaporization:** Not pertinent
13.13 **Heat of Combustion:** Not pertinent
13.14 **Heat of Decomposition:** Not pertinent
13.15 **Heat of Solution:** Not pertinent
13.16 **Heat of Polymerization:** Not pertinent

NOTES

ZCL — ZINC CHLORIDE

Common Synonyms

Solid | White solid | Odorless

Solid sinks and mixes with water.

Keep people away.
Avoid contact with solid and solution.
Isolate and remove discharged material.
Notify local health and pollution control agencies.

Fire

Not flammable.

Exposure

CALL FOR MEDICAL AID.

SOLID OR SOLUTION
Irritating to skin and eyes.
If swallowed, will cause nausea or vomiting.

Flush affected areas with plenty of water.
IF IN EYES, hold eyelids open and flush with plenty of water.
IF SWALLOWED and victim is CONSCIOUS, have victim drink water or milk and have victim induce vomiting.
IF SWALLOWED and victim is UNCONSCIOUS OR HAVING CONVULSIONS, do nothing except keep victim warm.

Water Pollution

HARMFUL TO AQUATIC LIFE IN VERY LOW CONCENTRATIONS.
May be dangerous if it enters water intakes.
Notify local health and wildlife officials.
Notify operators of nearby water intakes.

1. RESPONSE TO DISCHARGE
(See Response Methods Handbook, CG 446-4)

Issue warning – water contaminant
Disperse and flush

2. LABEL

3. CHEMICAL DESIGNATIONS

3.1 **Synonyms:** No common synonyms
3.2 **Coast Guard Compatibility Classification:** Not applicable
3.3 **Chemical Formula:** $ZnCl_2$
3.4 **IMCO/United Nations Numerical Designation:** 8.0/1840

4. OBSERVABLE CHARACTERISTICS

4.1 **Physical State (as shipped):** Solid
4.2 **Color:** White
4.3 **Odor:** Odorless

5. HEALTH HAZARDS

5.1 **Personal Protective Equipment:** Goggles or face shield.
5.2 **Symptoms Following Exposure:** Solid or water solution is astringent and can irritate the eyes. When ingested, can cause intoxication, severe irritation of stomach, nausea, vomiting, and diarrhea.
5.3 **Treatment for Exposure:** INGESTION: give large volumes of water and induce vomiting; repeat process; call a doctor. EYES: wash with water for at least 15 min.
5.4 **Toxicity by Inhalation (Threshold Limit Value):** 1 mg/m^3 (dust).
5.5 **Short-Term Inhalation Limits:** Not pertinent
5.6 **Toxicity by Ingestion:** Grade 3; LD_{50} 50 to 500 mg/kg
5.7 **Late Toxicity:** Data not available
5.8 **Vapor (Gas) Irritant Characteristics:** Non-volatile
5.9 **Liquid or Solid Irritant Characteristics:** Solid irritates skin on prolonged contact.
5.10 **Odor Threshold:** Not pertinent

6. FIRE HAZARDS

6.1 **Flash Point:** Not flammable
6.2 **Flammable Limits in Air:** Not flammable
6.3 **Fire Extinguishing Agents:** Not pertinent
6.4 **Fire Extinguishing Agents Not to be Used:** Not pertinent
6.5 **Special Hazards of Combustion Products:** Not pertinent
6.6 **Behavior in Fire:** Not pertinent
6.7 **Ignition Temperature:** Not flammable
6.8 **Electrical Hazard:** Not pertinent
6.9 **Burning Rate:** Not flammable

7. CHEMICAL REACTIVITY

7.1 **Reactivity with Water:** No reaction
7.2 **Reactivity with Common Materials:** No reaction
7.3 **Stability During Transport:** Stable
7.4 **Neutralizing Agents for Acids and Caustics:** Not pertinent
7.5 **Polymerization:** Not pertinent
7.6 **Inhibitor of Polymerization:** Not pertinent

8. WATER POLLUTION

8.1 **Aquatic Toxicity:**
7.2 ppm/96 hr/medium bluegill/TL_m/fresh water
28 ppm/48 hr/zebrafish/TL_m/salt water
8.2 **Waterfowl Toxicity:** Data not available
8.3 **Biological Oxygen Demand (BOD):** None
8.4 **Food Chain Concentration Potential:** None

9. SELECTED MANUFACTURERS

1. Allied Chemical Corp.
   Industrial Chemicals Division
   Morristown, N. J. 07960
2. Gould Marathon Battery Co.
   Wausau, Wis. 54402
3. Mallinckrodt Chemical Works
   Industrial Chemicals Division
   2nd and Mallinckrodt Sts.
   St. Louis, Mo. 63160

10. SHIPPING INFORMATION

10.1 **Grades or Purity:** Reagent; USP; technical; 50% solution in water
10.2 **Storage Temperature:** Ambient
10.3 **Inert Atmosphere:** Not required
10.4 **Venting:** Data not available

11. HAZARD ASSESSMENT CODE
(See Hazard Assessment Handbook, CG 446-3)

SS

12. HAZARD CLASSIFICATIONS

12.1 **Code of Federal Regulations:** Corrosive Material
12.2 **NAS Hazard Rating for Bulk Water Transportation:** Not listed
12.3 **NFPA Hazard Classifications:** Not listed

13. PHYSICAL AND CHEMICAL PROPERTIES

13.1 **Physical State at 15°C and 1 atm:** Solid
13.2 **Molecular Weight:** 136.28
13.3 **Boiling Point at 1 atm:** Very high
13.4 **Freezing Point:** 541°F = 283°C = 556°K
13.5 **Critical Temperature:** Not pertinent
13.6 **Critical Pressure:** Not pertinent
13.7 **Specific Gravity:** 2.91 at 25°C (solid)
13.8 **Liquid Surface Tension:** Not pertinent
13.9 **Liquid-Water Interfacial Tension:** Not pertinent
13.10 **Vapor (Gas) Specific Gravity:** Not pertinent
13.11 **Ratio of Specific Heats of Vapor (Gas):** Not pertinent
13.12 **Latent Heat of Vaporization:** Not pertinent
13.13 **Heat of Combustion:** Not pertinent
13.14 **Heat of Decomposition:** Not pertinent
13.15 **Heat of Solution:** Not pertinent
13.16 **Heat of Polymerization:** Not pertinent

NOTES

REVISED 1978

ZCR — ZINC CHROMATE

Common Synonyms			
Zinc chromate(VI) hydroxide Zinc yellow Buttercup yellow	Solid	Yellow	Odorless
	Sinks in water.		

Avoid contact with solid and dust. Keep people away.
Stop discharge if possible.
Isolate and remove discharged material.
Notify local health and pollution control agencies.

Fire

Not flammable.

Exposure

CALL FOR MEDICAL AID.

DUST
Irritating to eyes, nose and throat.
If inhaled will cause coughing or difficult breathing.
If in eyes, hold eyelids open and flush with plenty of water.
If breathing has stopped, give artificial respiration.
If breathing is difficult, give oxygen.

SOLID
Irritating to skin and eyes.
If swallowed will cause nausea and vomiting.
Remove contaminated clothing and shoes.
Flush affected areas with plenty of water.
IF IN EYES, hold eyelids open and flush with plenty of water.
IF SWALLOWED and victim is CONSCIOUS, have victim drink water or milk and have victim induce vomiting.
IF SWALLOWED and victim is UNCONSCIOUS OR HAVING CONVULSIONS, do nothing except keep victim warm.

Water Pollution

Effect of low concentrations on aquatic life is unknown.
May be dangerous if it enters water intakes.
Notify local health and wildlife officials.
Notify operators of nearby water intakes.

1. RESPONSE TO DISCHARGE

(See Response Methods Handbook, CG 446-4)

Should be removed
Chemical and physical treatment

2. LABELS

No hazard label required by Code of Federal Regulations

3. CHEMICAL DESIGNATIONS

3.1 **Synonyms:** Buttercup yellow; Zinc chromate (VI) hydroxide; Zinc yellow
3.2 **Coast Guard Compatibility Classification:** Not listed
3.3 **Chemical Formula:** (approx.) $4ZnO \cdot K_2O \cdot 4CrO_3 \cdot 3H_2O$
3.4 **IMCO/United Nations Numerical Designation:** Not listed

4. OBSERVABLE CHARACTERISTICS

4.1 **Physical State (as shipped):** Solid
4.2 **Color:** Yellow
4.3 **Odor:** None

5. HEALTH HAZARDS

5.1 **Personal Protective Equipment:** Suitable respirator (for dust); rubber gloves; chemical goggles or face shield
5.2 **Symptoms Following Exposure:** Inhalation of dust causes irritation of nose and throat. Ingestion can cause irritation or corrosion of the alimentary tract, circulatory collapse, and toxic nephritis. Contact with eyes or skin causes irritation.
5.3 **Treatment for Exposure:** INHALATION: move to fresh air; if exposure has been severe, get medical attention. INGESTION: induce vomiting, followed by prompt and complete gastric lavage, catharsis, and demulcents. EYES: flush with water. SKIN: wash thoroughly with soap and water.
5.4 **Toxicity by Inhalation (Threshold Limit Value):** 0.1 mg/m^3
5.5 **Short-Term Inhalation Limits:** Data not available
5.6 **Toxicity by Ingestion:** Grade 2; LD_{50} 0.5—5 g/kg
5.7 **Late Toxicity:** Possible lung cancer.
5.8 **Vapor (Gas) Irritant Characteristics:** Data not available
5.9 **Liquid or Solid Irritant Characteristics:** Data not available
5.10 **Odor Threshold:** Odorless

6. FIRE HAZARDS

6.1 **Flash Point:** Not flammable
6.2 **Flammable Limits in Air:** Not flammable
6.3 **Fire Extinguishing Agents:** Not pertinent
6.4 **Fire Extinguishing Agents Not to be Used:** Not pertinent
6.5 **Special Hazards of Combustion Products:** Not pertinent
6.6 **Behavior in Fire:**
6.7 **Ignition Temperature:** Not pertinent
6.8 **Electrical Hazard:** Not pertinent
6.9 **Burning Rate:** Not pertinent

7. CHEMICAL REACTIVITY

7.1 **Reactivity with Water:** No reaction
7.2 **Reactivity with Common Materials:**
7.3 **Stability During Transport:** Stable
7.4 **Neutralizing Agents for Acids and Caustics:** Not pertinent
7.5 **Polymerization:** Not pertinent
7.6 **Inhibitor of Polymerization:** Not pertinent

8. WATER POLLUTION

8.1 **Aquatic Toxicity:** Data not available
8.2 **Waterfowl Toxicity:** Data not available
8.3 **Biological Oxygen Demand (BOD):** None
8.4 **Food Chain Concentration Potential:** Both chromium and zinc are concentrated by some organisms but are not considered to be bioconcentrative in a spill situation.

9. SELECTED MANUFACTURERS

1. Mineral Pigments Corp.
7011 Muirkirk Road
Beltsville, Md. 20705
2. Fisher Scientific Co.
711 Forbes Ave.
Pittsburgh, Pa. 15219
3. Pfaltz and Bauer, Inc.
375 Fairfield Ave.
Stamford, Conn. 06902

10. SHIPPING INFORMATION

10.1 **Grades or Purity:** Commercial, 100%
10.2 **Storage Temperature:** Ambient
10.3 **Inert Atmosphere:** No requirement
10.4 **Venting:** Open

11. HAZARD ASSESSMENT CODE

(See Hazard Assessment Handbook, CG 446-3)

SS

12. HAZARD CLASSIFICATIONS

12.1 **Code of Federal Regulations:** Not listed
12.2 **NAS Hazard Rating for Bulk Water Transportation:** Not listed
12.3 **NFPA Hazard Classifications:**

Category	Classification
Health Hazard (Blue)	1
Flammability (Red)	0
Reactivity (Yellow)	0

13. PHYSICAL AND CHEMICAL PROPERTIES

13.1 **Physical State at 15°C and 1 atm:** Solid
13.2 **Molecular Weight:** 874 (approx.)
13.3 **Boiling Point at 1 atm:** Not pertinent (decomposes)
13.4 **Freezing Point:** Not pertinent
13.5 **Critical Temperature:** Not pertinent
13.6 **Critical Pressure:** Not pertinent
13.7 **Specific Gravity:** 3.43 at 20°C (solid)
13.8 **Liquid Surface Tension:** Not pertinent
13.9 **Liquid-Water Interfacial Tension:** Not pertinent
13.10 **Vapor (Gas) Specific Gravity:** Not pertinent
13.11 **Ratio of Specific Heats of Vapor (Gas):** Not pertinent
13.12 **Latent Heat of Vaporization:** Not pertinent
13.13 **Heat of Combustion:** Not pertinent
13.14 **Heat of Decomposition:** Not pertinent
13.15 **Heat of Solution:** Not pertinent
13.16 **Heat of Polymerization:** Not pertinent

NOTES

ZDP	# ZINC DIALKYLDITHIOPHOSPHATE

Common Synonyms		
Zinc dihexyldithiophosphate Zinc dihexylphosphorodithioate	Solid or liquid Straw yellow to green Sweet, alcohol-like odor Sinks in water.	

Avoid contact with liquid and solid. Keep people away.
Stop discharge if possible.
Isolate and remove discharged material.
Notify local health and pollution control agencies.

Fire	Combustible. Irritating gases may be produced when heated. Extinguish with water, dry chemicals, foam, or carbon dioxide.
Exposure	CALL FOR MEDICAL AID. DUST Irritating to eyes, nose and throat. If inhaled will cause coughing or difficult breathing. If in eyes, hold eyelids open and flush with plenty of water. If breathing has stopped, give artificial respiration. If breathing is difficult, give oxygen. LIQUID OR SOLID Irritating to skin and eyes. If swallowed will cause nausea and vomiting. Remove contaminated clothing and shoes. Flush affected areas with plenty of water. IF IN EYES, hold eyelids open and flush with plenty of water. IF SWALLOWED and victim is CONSCIOUS, have victim drink water or milk and have victim induce vomiting. IF SWALLOWED and victim is UNCONSCIOUS OR HAVING CONVULSIONS, do nothing except keep victim warm.
Water Pollution	Effect of low concentrations on aquatic life is unknown. May be dangerous if it enters water intakes. Notify local health and wildlife officials. Notify operators of nearby water intakes.

1. RESPONSE TO DISCHARGE
(See Response Methods Handbook, CG 446-4)

Issue warning—water contaminant
Restrict access
Should be removed
Chemical and physical treatment

2. LABELS

No hazard label required by Code of Federal Regulations

3. CHEMICAL DESIGNATIONS

3.1 **Synonyms:** Zinc O,O-di-n-butylphosphorodithioate; Zinc dihexyldithiophosphate; Zinc dihexylphosphorodithioate
3.2 **Coast Guard Compatibility Classification:** Not listed
3.3 **Chemical Formula:** $[(RO)_2PSS]_2Zn$ where $R=C_4H_9$ etc.
3.4 **IMCO/United Nations Numerical Designation:** 6.1/1893

4. OBSERVABLE CHARACTERISTICS

4.1 **Physical State (as shipped):** Solid or liquid
4.2 **Color:** Straw yellow; yellow-green
4.3 **Odor:** Sweet, alcohol-like

5. HEALTH HAZARDS

5.1 **Personal Protective Equipment:** Rubber gloves; safety glasses or face shield; dust respirator for solid form
5.2 **Symptoms Following Exposure:** (All commercially available members of this class have about the same health hazards.) Inhalation of dust can cause respiratory discomfort. Ingestion causes irritation of mouth and stomach. Contact with eyes causes moderately severe irritation. Contact with skin causes mild irritation.
5.3 **Treatment for Exposure:** INHALATION: move from exposure. INGESTION: if large amounts have been ingested, induce vomiting. EYES: flush with copious amounts of water; if irritation persists, consult a physician. SKIN: wash affected areas with soap and water.
5.4 **Toxicity by Inhalation (Threshold Limit Value):** Data not available
5.5 **Short-Term Inhalation Limits:** Data not available
5.6 **Toxicity by Ingestion:** Grade 2; LD_{50} 0.5—5 g/kg
5.7 **Late Toxicity:** Data not available
5.8 **Vapor (Gas) Irritant Characteristics:** Data not available
5.9 **Liquid or Solid Irritant Characteristics:** Data not available
5.10 **Odor Threshold:** Data not available

6. FIRE HAZARDS

6.1 **Flash Point:** 360°F C.C.
6.2 **Flammable Limits in Air:** Data not available
6.3 **Fire Extinguishing Agents:** Water, dry chemical, foam, carbon dioxide
6.4 **Fire Extinguishing Agents Not to be Used:**
6.5 **Special Hazards of Combustion Products:** Irritating oxides of sulfur and phosphorus may form in fires.
6.6 **Behavior in Fire:**
6.7 **Ignition Temperature:** Data not available
6.8 **Electrical Hazard:** Not pertinent
6.9 **Burning Rate:** Not pertinent

7. CHEMICAL REACTIVITY

7.1 **Reactivity with Water:** No reaction at ordinary temperatures
7.2 **Reactivity with Common Materials:**
7.3 **Stability During Transport:** Stable
7.4 **Neutralizing Agents for Acids and Caustics:** Not pertinent
7.5 **Polymerization:** Not pertinent
7.6 **Inhibitor of Polymerization:** Not pertinent

8. WATER POLLUTION

8.1 **Aquatic Toxicity:** Data not available
8.2 **Waterfowl Toxicity:** Data not available
8.3 **Biological Oxygen Demand (BOD):** Data not available
8.4 **Food Chain Concentration Potential:** Zinc accumulates in some organisms but is not considered to be bioconcentrative.

9. SELECTED MANUFACTURERS

1. Elco Corporation
P. O. Box 09168
Cleveland, Ohio 44109
2. Monsanto Company
800 N. Lindbergh Blvd.
St. Louis, Mo. 63166

10. SHIPPING INFORMATION

10.1 **Grades or Purity:** Technical; 62% on inert filler.
10.2 **Storage Temperature:** Below 66°C (150°F)
10.3 **Inert Atmosphere:** No requirement
10.4 **Venting:** Open (flame arrester)

11. HAZARD ASSESSMENT CODE
(See Hazard Assessment Handbook, CG 446-3)

(Liquid) A-X-Y
(Solid) II

12. HAZARD CLASSIFICATIONS

12.1 **Code of Federal Regulations:** Not listed
12.2 **NAS Hazard Rating for Bulk Water Transportation:** Not listed
12.3 **NFPA Hazard Classifications:** Not listed

13. PHYSICAL AND CHEMICAL PROPERTIES

13.1 **Physical State at 15°C and 1 atm:** Solid or liquid
13.2 **Molecular Weight:** 548 (approx.)
13.3 **Boiling Point at 1 atm:** Not pertinent (decomposes)
13.4 **Freezing Point:** Not pertinent
13.5 **Critical Temperature:** Not pertinent
13.6 **Critical Pressure:** Not pertinent
13.7 **Specific Gravity:** 1.12—1.26 at 20°C (liquid) 1.6 at 20°C (solid)
13.8 **Liquid Surface Tension:** Data not available
13.9 **Liquid-Water Interfacial Tension:** Data not available
13.10 **Vapor (Gas) Specific Gravity:** Not pertinent
13.11 **Ratio of Specific Heats of Vapor (Gas):** Not pertinent
13.12 **Latent Heat of Vaporization:** Not pertinent
13.13 **Heat of Combustion:** Not pertinent
13.14 **Heat of Decomposition:** Not pertinent
13.15 **Heat of Solution:** Not pertinent
13.16 **Heat of Polymerization:** Not pertinent

NOTES

ZFB

ZINC FLUOROBORATE

Common Synonyms
Zinc fluoborate solution

Liquid Colorless Odorless

Sinks and mixes with water.

Avoid contact with liquid. Keep people away.
Stop discharge if possible.
Isolate and remove discharged material.
Notify local health and pollution control agencies.

Fire

Not flammable.

Exposure

CALL FOR MEDICAL AID.

LIQUID
Irritating to skin and eyes.
If swallowed will cause nausea and vomiting.
Remove contaminated clothing and shoes.
Flush affected areas with plenty of water.
IF IN EYES, hold eyelids open and flush with plenty of water.
IF SWALLOWED and victim is CONSCIOUS, have victim drink water or milk and have victim induce vomiting.
IF SWALLOWED and victim is UNCONSCIOUS OR HAVING CONVULSIONS, do nothing except keep victim warm.

Water Pollution

Effect of low concentrations on aquatic life is unknown.
May be dangerous if it enters water intakes.

Notify local health and wildlife officials.
Notify operators of nearby water intakes.

1. RESPONSE TO DISCHARGE

(See Response Methods Handbook, CG 446-4)

Issue warning—water contaminant
Disperse and flush

2. LABELS

No hazard label required by Code of Federal Regulations

3. CHEMICAL DESIGNATIONS

3.1 **Synonyms:** Zinc fluoborate solution
3.2 **Coast Guard Compatibility Classification:** Not listed
3.3 **Chemical Formula:** $Zn(BF_4)_2-H_2O$
3.4 **IMCO/United Nations Numerical Designation:** Not listed

4. OBSERVABLE CHARACTERISTICS

4.1 **Physical State (as shipped):** Liquid
4.2 **Color:** Colorless
4.3 **Odor:** None

5. HEALTH HAZARDS

5.1 **Personal Protective Equipment:** Rubber gloves; safety glasses or face shield
5.2 **Symptoms Following Exposure:** Ingestion may cause irritation or corrosion of the alimentary tract. Contact with eyes or skin causes irritation.
5.3 **Treatment for Exposure:** INGESTION: give gastric lavage, cathartics, and demulcents. EYES: flush with plenty of water; get medical attention. SKIN: flush with plenty of water.
5.4 **Toxicity by Inhalation (Threshold Limit Value):** Data not available
5.5 **Short-Term Inhalation Limits:** Data not available
5.6 **Toxicity by Ingestion:** Grade 2; LD_{50} 0.5—5 g/kg
5.7 **Late Toxicity:** Data not available
5.8 **Vapor (Gas) Irritant Characteristics:** Data not available
5.9 **Liquid or Solid Irritant Characteristics:** Data not available
5.10 **Odor Threshold:** Data not available

6. FIRE HAZARDS

6.1 **Flash Point:** Not flammable
6.2 **Flammable Limits in Air:** Not flammable
6.3 **Fire Extinguishing Agents:** Not pertinent
6.4 **Fire Extinguishing Agents Not to be Used:** Not pertinent
6.5 **Special Hazards of Combustion Products:** Not pertinent
6.6 **Behavior in Fire:**
6.7 **Ignition Temperature:** Not pertinent
6.8 **Electrical Hazard:** Not pertinent
6.9 **Burning Rate:** Not pertinent

7. CHEMICAL REACTIVITY

7.1 **Reactivity with Water:** No reaction
7.2 **Reactivity with Common Materials:**
7.3 **Stability During Transport:** Stable
7.4 **Neutralizing Agents for Acids and Caustics:** Not pertinent
7.5 **Polymerization:** Not pertinent
7.6 **Inhibitor of Polymerization:** Not pertinent

8. WATER POLLUTION

8.1 **Aquatic Toxicity:** Data not available
8.2 **Waterfowl Toxicity:** Data not available
8.3 **Biological Oxygen Demand (BOD):** None
8.4 **Food Chain Concentration Potential:** Zinc accumulates in some organisms but is not considered to be bioconcentrative.

9. SELECTED MANUFACTURERS

1. Allied Chemical Corp.
P. O. Box 1087R
Morristown, N. J. 07960
2. Pfaltz and Bauer, Inc.
375 Fairfield Ave.
Stamford, Conn. 06902

10. SHIPPING INFORMATION

10.1 **Grades or Purity:** Purified, 41% solution in water.
10.2 **Storage Temperature:** Ambient
10.3 **Inert Atmosphere:** No requirement
10.4 **Venting:** Open

11. HAZARD ASSESSMENT CODE

(See Hazard Assessment Handbook, CG 446-3)

A-P

12. HAZARD CLASSIFICATIONS

12.1 **Code of Federal Regulations:** Not listed
12.2 **NAS Hazard Rating for Bulk Water Transportation:** Not listed
12.3 **NFPA Hazard Classifications:** Not listed

13. PHYSICAL AND CHEMICAL PROPERTIES

13.1 **Physical State at 15°C and 1 atm:** Liquid
13.2 **Molecular Weight:** 238.98 (solute only)
13.3 **Boiling Point at 1 atm:** (approx.) 212°F = 100°C = 373°K
13.4 **Freezing Point:** Data not available
13.5 **Critical Temperature:** Not pertinent
13.6 **Critical Pressure:** Not pertinent
13.7 **Specific Gravity:** 1.45 at 20°C (liquid)
13.8 **Liquid Surface Tension:** Data not available
13.9 **Liquid-Water Interfacial Tension:** Not pertinent
13.10 **Vapor (Gas) Specific Gravity:** Not pertinent
13.11 **Ratio of Specific Heats of Vapor (Gas):** Not pertinent
13.12 **Latent Heat of Vaporization:** Not pertinent
13.13 **Heat of Combustion:** Not pertinent
13.14 **Heat of Decomposition:** Not pertinent
13.15 **Heat of Solution:** Not pertinent
13.16 **Heat of Polymerization:** Not pertinent

NOTES

ZNT	# ZINC NITRATE

Common Synonyms			
Zinc nitrate hexahydrate	Solid	White	Odorless
	Sinks and mixes with water.		

Avoid contact with solid and dust. Keep people away.
Stop discharge if possible.
Call fire department.
Isolate and remove discharged material.
Notify local health and pollution control agencies.

Fire

Not flammable.
Will increase the intensity of a fire.
POISONOUS GASES MAY BE PRODUCED IN FIRE.
Wear goggles and self-contained breathing apparatus.
Flood discharge area with water.

Exposure

CALL FOR MEDICAL AID.

DUST
Irritating to eyes, nose and throat.
If inhaled will cause coughing or difficult breathing.
If in eyes, hold eyelids open and flush with plenty of water.
If breathing has stopped, give artificial respiration.
If breathing is difficult, give oxygen.

SOLID
Irritating to skin and eyes.
If swallowed will cause nausea and vomiting.
Remove contaminated clothing and shoes.
Flush affected areas with plenty of water.
IF IN EYES, hold eyelids open and flush with plenty of water.
IF SWALLOWED and victim is CONSCIOUS, have victim drink water or milk and have victim induce vomiting.
IF SWALLOWED and victim is UNCONSCIOUS OR HAVING CONVULSIONS, do nothing except keep victim warm.

Water Pollution

HARMFUL TO AQUATIC LIFE IN VERY LOW CONCENTRATIONS.
May be dangerous if it enters water intakes.
Notify local health and wildlife officials.
Notify bperators of nearby water intakes.

1. RESPONSE TO DISCHARGE

(See Response Methods Handbook, CG 446-4)

Disperse and flush

2. LABEL

Yellow

3. CHEMICAL DESIGNATIONS

3.1 **Synonyms:** Zinc nitrate hexahydrate
3.2 **Coast Guard Compatibility Classification:** Not listed
3.3 **Chemical Formula:** $Zn(NO_3)_2 \cdot 6H_2O$
3.4 **IMCO/United Nations Numerical Designation:** 5.1/1514

4. OBSERVABLE CHARACTERISTICS

4.1 **Physical State (as shipped):** Solid
4.2 **Color:** White; colorless
4.3 **Odor:** None

5. HEALTH HAZARDS

5.1 **Personal Protective Equipment:** Dust mask; goggles or face shield; protective gloves.
5.2 **Symptoms Following Exposure:** Inhalation of dust may irritate nose and throat. Ingestion can cause irritation or corrosion of the alimentary tract. Contact with eyes causes irritation, which may be delayed. Contact with skin causes irritation.
5.3 **Treatment for Exposure:** INHALATION: move to fresh air. INGESTION: induce vomiting, followed by prompt and complete gastric lavage, cathartics, and demulcents. EYES: flush with water; consult a physician. SKIN: wash with soap and water.
5.4 **Toxicity by Inhalation (Threshold Limit Value):** Data not available
5.5 **Short-Term Inhalation Limits:** Data not available
5.6 **Toxicity by Ingestion:** Grade 2; oral LD_{50} = 2,500 mg/kg (rat)
5.7 **Late Toxicity:** Causes enlarged liver, spleen, and bone marrow in rabbits
5.8 **Vapor (Gas) Irritant Characteristics:** Data not available
5.9 **Liquid or Solid Irritant Characteristics:** Data not available
5.10 **Odor Threshold:** Odorless

6. FIRE HAZARDS

6.1 **Flash Point:** Not flammable
6.2 **Flammable Limits in Air:** Not flammable
6.3 **Fire Extinguishing Agents:** Not pertinent
6.4 **Fire Extinguishing Agents Not to be Used:** Not pertinent
6.5 **Special Hazards of Combustion Products:** Toxic oxides of nitrogen may form in fire.
6.6 **Behavior in Fire:** May increase intensity of fire when in contact with combustible material
6.7 **Ignition Temperature:** Not pertinent
6.8 **Electrical Hazard:** Not pertinent
6.9 **Burning Rate:** Not pertinent

7. CHEMICAL REACTIVITY

7.1 **Reactivity with Water:** No reaction
7.2 **Reactivity with Common Materials:**
7.3 **Stability During Transport:** Stable
7.4 **Neutralizing Agents for Acids and Caustics:** Not pertinent
7.5 **Polymerization:** Not pertinent
7.6 **Inhibitor of Polymerization:** Not pertinent

8. WATER POLLUTION

8.1 **Aquatic Toxicity:**
1.89 ppm/3 mo./tadpoles/survived but no limb buds/fresh water
32 ppm/48 hr/barnacles/90% lethal/salt water
8.2 **Waterfowl Toxicity:** Data not available
8.3 **Biological Oxygen Demand (BOD):** None
8.4 **Food Chain Concentration Potential:** Zinc is accumulated in some organisms but is not considered to be bioconcentrative.

9. SELECTED MANUFACTURERS

1. The Harshaw Chemical Co.
   1945 E. 97 St.
   Cleveland, Ohio 44106
2. J. T. Baker Chemical Co.
   Phillipsburg, N. J. 08865
3. Mallinckrodt Chemical Works
   2nd and Mallinckrodt Streets
   St. Louis, Mo. 63160

10. SHIPPING INFORMATION

10.1 **Grades or Purity:** Reagent; Technical
10.2 **Storage Temperature:** Ambient
10.3 **Inert Atmosphere:** No requirement
10.4 **Venting:** Open

11. HAZARD ASSESSMENT CODE

(See Hazard Assessment Handbook, CG 446-3)

SS

12. HAZARD CLASSIFICATIONS

12.1 **Code of Federal Regulations:** Solid oxidizing material
12.2 **NAS Hazard Rating for Bulk Water Transportation:** Not listed
12.3 **NFPA Hazard Classifications:**

Category	Classification
Health Hazard (Blue)	1
Flammability (Red)	0
Reactivity (Yellow)	0
	oxy

13. PHYSICAL AND CHEMICAL PROPERTIES

13.1 **Physical State at 15°C and 1 atm:** Solid
13.2 **Molecular Weight:** 297.47
13.3 **Boiling Point at 1 atm:** Not pertinent (decomposes)
13.4 **Freezing Point:** 97°F = 36°C = 309°K
13.5 **Critical Temperature:** Not pertinent
13.6 **Critical Pressure:** Not pertinent
13.7 **Specific Gravity:** 2.07 at 20°C (solid)
13.8 **Liquid Surface Tension:** Not pertinent
13.9 **Liquid-Water Interfacial Tension:** Not pertinent
13.10 **Vapor (Gas) Specific Gravity:** Not pertinent
13.11 **Ratio of Specific Heats of Vapor (Gas):** Not pertinent
13.12 **Latent Heat of Vaporization:** Not pertinent
13.13 **Heat of Combustion:** Not pertinent
13.14 **Heat of Decomposition:** Not pertinent
13.15 **Heat of Solution:** Not pertinent
13.16 **Heat of Polymerization:** Not pertinent

NOTES

ZPS	# ZINC PHENOLSULFONATE

Common Synonyms			
Zinc p-phenolsulfonate Zinc sulfocarbolate Zinc sulfophenate Zinc phenolsulfonate octahydrate	Solid	White	Odorless
	Sinks and mixes with water.		

Avoid contact with solid and dust. Keep people away.
Stop discharge if possible.
Isolate and remove discharged material.
Notify local health and pollution control agencies.

Fire	Not flammable. Irritating gases may be produced when heated.
Exposure	CALL FOR MEDICAL AID. DUST Irritating to eyes, nose and throat. If inhaled will cause coughing or difficult breathing. If in eyes, hold eyelids open and flush with plenty of water. If breathing has stopped, give artificial respiration. If breathing is difficult, give oxygen. SOLID Irritating to skin and eyes. If swallowed will cause nausea and vomiting. Remove contaminated clothing and shoes. Flush affected areas with plenty of water. IF IN EYES, hold eyelids open and flush with plenty of water. IF SWALLOWED and victim is CONSCIOUS, have victim drink water or milk and have victim induce vomiting. IF SWALLOWED and victim is UNCONSCIOUS OR HAVING CONVULSIONS, do nothing except keep victim warm.
Water Pollution	Effect of low concentrations on aquatic life is unknown. May be dangerous if it enters water intakes. Notify local health and wildlife officials. Notify operators of nearby water intakes.

1. RESPONSE TO DISCHARGE

(See Response Methods Handbook, CG 446-4)

Issue warning—water contaminant
Disperse and flush

2. LABELS

No hazard label required by Code of Federal Regulations

3. CHEMICAL DESIGNATIONS

3.1 **Synonyms:** p-Hydroxybenzenesulfonic acid, zinc salt; Zinc p-phenolsulfonate; Zinc phenolsulfonate octahydrate; Zinc sulfocarbolate; Zinc sulfophenate

3.2 **Coast Guard Compatibility Classification:** Not listed

3.3 **Chemical Formula:** $(1,4\text{-}HOC_6H_4SO_3)_2Zn \cdot 8H_2O$

3.4 **IMCO/United Nations Numerical Designation:** Not listed

4. OBSERVABLE CHARACTERISTICS

4.1 **Physical State (as shipped):** Solid

4.2 **Color:** White

4.3 **Odor:** None

5. HEALTH HAZARDS

5.1 **Personal Protective Equipment:** Dust mask; goggles or face shield; protective gloves

5.2 **Symptoms Following Exposure:** Inhalation of dust may irritate nose and throat. Ingestion of large doses has emetic and astringent effects; can cause irritation or corrosion of the alimentary tract. Contact with eyes causes irritation. Contact with skin causes mild irritation.

5.3 **Treatment for Exposure:** INHALATION: move to fresh air. INGESTION: if large amount has been swallowed, induce vomiting, followed by prompt and complete gastric lavage, cathartics, and demulcents. EYES or SKIN: flush with water.

5.4 **Toxicity by Inhalation (Threshold Limit Value):** Data not available

5.5 **Short-Term Inhalation Limits:** Data not available

5.6 **Toxicity by Ingestion:** Grade 2; LD_{50} 0.5—5 g/kg

5.7 **Late Toxicity:** Data not available

5.8 **Vapor (Gas) Irritant Characteristics:** Data not available

5.9 **Liquid or Solid Irritant Characteristics:** Data not available

5.10 **Odor Threshold:** Odorless

6. FIRE HAZARDS

6.1 **Flash Point:** Not flammable

6.2 **Flammable Limits in Air:** Not flammable

6.3 **Fire Extinguishing Agents:** Not pertinent

6.4 **Fire Extinguishing Agents Not to be Used:** Not pertinent

6.5 **Special Hazards of Combustion Products:** Irritating oxides of sulfur may form in fires.

6.6 **Behavior in Fire:**

6.7 **Ignition Temperature:** Not pertinent

6.8 **Electrical Hazard:** Not pertinent

6.9 **Burning Rate:** Not pertinent

7. CHEMICAL REACTIVITY

7.1 **Reactivity with Water:** No reaction

7.2 **Reactivity with Common Materials:**

7.3 **Stability During Transport:** Stable

7.4 **Neutralizing Agents for Acids and Caustics:** Not pertinent

7.5 **Polymerization:** Not pertinent

7.6 **Inhibitor of Polymerization:** Not pertinent

8. WATER POLLUTION

8.1 **Aquatic Toxicity:** Data not available

8.2 **Waterfowl Toxicity:** Data not available

8.3 **Biological Oxygen Demand (BOD):** Data not available

8.4 **Food Chain Concentration Potential:** Zinc is accumulated by some organisms but is not considered to be bioconcentrative.

9. SELECTED MANUFACTURERS

1. Allied Chemical Corp.
   Specialty Chemicals Div.
   P. O. Box 1087R
   Morristown, N. J. 07960
2. Mallinckrodt Chemical Works
   2nd and Mallinckrodt Streets
   St. Louis, Mo. 63160
3. Pfaltz and Bauer, Inc.
   375 Fairfield Ave.
   Stamford, Conn. 06902

10. SHIPPING INFORMATION

10.1 **Grades or Purity:** Purified

10.2 **Storage Temperature:** Ambient

10.3 **Inert Atmosphere:** No requirement

10.4 **Venting:** Open

11. HAZARD ASSESSMENT CODE

(See Hazard Assessment Handbook, CG 446-3)

SS

12. HAZARD CLASSIFICATIONS

12.1 **Code of Federal Regulations:** Not listed

12.2 **NAS Hazard Rating for Bulk Water Transportation:** Not listed

12.3 **NFPA Hazard Classifications:** Not listed

13. PHYSICAL AND CHEMICAL PROPERTIES

13.1 **Physical State at 15°C and 1 atm:** Solid

13.2 **Molecular Weight:** 555.8

13.3 **Boiling Point at 1 atm:** (decomposes) 248°F = 120°C = 393°K

13.4 **Freezing Point:** Not pertinent

13.5 **Critical Temperature:** Not pertinent

13.6 **Critical Pressure:** Not pertinent

13.7 **Specific Gravity:** > 1 at 20°C (solid)

13.8 **Liquid Surface Tension:** Not pertinent

13.9 **Liquid-Water Interfacial Tension:** Not pertinent

13.10 **Vapor (Gas) Specific Gravity:** Not pertinent

13.11 **Ratio of Specific Heats of Vapor (Gas):** Not pertinent

13.12 **Latent Heat of Vaporization:** Not pertinent

13.13 **Heat of Combustion:** Not pertinent

13.14 **Heat of Decomposition:** Not pertinent

13.15 **Heat of Solution:** Not pertinent

13.16 **Heat of Polymerization:** Not pertinent

NOTES

ZPP ZINC PHOSPHIDE

Common Synonyms	Solid	Grey to black	Faint odor
	Sinks in water.		

Avoid contact with solid and dust. Keep people away.
Stop discharge if possible.
Isolate and remove discharged material.
Notify local health and pollution control agencies.

Fire	Not flammable. Irritating gases may be produced when heated.
Exposure	CALL FOR MEDICAL AID. DUST Irritating to eyes, nose and throat. If inhaled will cause dizziness, difficult breathing, or loss of consciousness If in eyes, hold eyelids open and flush with plenty of water. If breathing has stopped, give artificial respiration. If breathing is difficult, give oxygen. SOLID POISONOUS IF SWALLOWED. Irritating to skin and eyes. If swallowed will cause dizziness, nausea, vomiting or loss of consciousness. Remove contaminated clothing and shoes. Flush affected areas with plenty of water. IF IN EYES, hold eyelids open and flush with plenty of water. IF SWALLOWED and victim is CONSCIOUS, have victim drink water or milk and have victim induce vomiting. IF SWALLOWED and victim is UNCONSCIOUS, OR HAVING CONVULSIONS, do nothing except keep victim warm.
Water Pollution	Effect of low concentrations on aquatic life is unknown. May be dangerous if it enters water intakes. Notify local health and wildlife officials. Notify operators of nearby water intakes.

1. RESPONSE TO DISCHARGE
(See Response Methods Handbook, CG 446-4)

Issue warning—poison, water contaminant, air contaminant
Restrict access
Should be removed
Chemical and physical treatment

2. LABEL

POISON

3. CHEMICAL DESIGNATIONS

3.1 **Synonyms:** No common synonyms
3.2 **Coast Guard Compatibility Classification:** Not listed
3.3 **Chemical Formula:** Zn_3P_2
3.4 **IMCO/United Nations Numerical Designation:** 6.1/1714

4. OBSERVABLE CHARACTERISTICS

4.1 **Physical State (as shipped):** Solid
4.2 **Color:** Gray or gray-black
4.3 **Odor:** Faint phosphorus

5. HEALTH HAZARDS

5.1 **Personal Protective Equipment:** Dust mask or self-contained breathing apparatus; goggles or face shield; protective gloves
5.2 **Symptoms Following Exposure:** When inhaled or ingested, compound releases phosphine, which causes faintness, weakness, nausea, vomiting, dyspnea, fall in blood pressure, change in pulse rate, diarrhea, intense thirst, convulsions, paralysis, and coma. Contact with eyes or skin causes irritation.
5.3 **Treatment for Exposure:** INHALATION: move to fresh air; give artificial respiration if required; get medical attention for phosphine poisoning. INGESTION: give one tablespoonful of mustard in a glass of warm water; repeat until vomit fluid is clear; avoid use of all oils; call a physician immediately; have patient lie down and keep warm. EYES: flush with water for at least 15 min. SKIN: flush with water, wash with soap and water.
5.4 **Toxicity by Inhalation (Threshold Limit Value):** Data not available
5.5 **Short-Term Inhalation Limits:** Data not available
5.6 **Toxicity by Ingestion:** Grade 4; oral LD_{50} = 40 mg/kg (rat)
5.7 **Late Toxicity:** Data not available
5.8 **Vapor (Gas) Irritant Characteristics:** Data not available
5.9 **Liquid or Solid Irritant Characteristics:** Data not available
5.10 **Odor Threshold:** Data not available

6. FIRE HAZARDS

6.1 **Flash Point:** Not flammable
6.2 **Flammable Limits in Air:** Not flammable
6.3 **Fire Extinguishing Agents:** Use water, foam, or dry chemical on adjacent fires.
6.4 **Fire Extinguishing Agents Not to be Used:** Any agent with an acid reaction (e.g., carbon dioxide or halogenated agents) will liberate phosphine, a toxic and spontaneously flammable gas.
6.5 **Special Hazards of Combustion Products:** Irritating oxides of phosphorus may be formed in fires.
6.6 **Behavior in Fire:**
6.7 **Ignition Temperature:** Not pertinent
6.8 **Electrical Hazard:** Not pertinent
6.9 **Burning Rate:** Not pertinent

7. CHEMICAL REACTIVITY

7.1 **Reactivity with Water:** Reacts slowly with water, more rapidly with dilute acid, to form phosphine gas, which is toxic and spontaneously flammable.
7.2 **Reactivity with Common Materials:**
7.3 **Stability During Transport:** Stable unless exposed to moisture; toxic phosphine gas may then be released and collect in closed spaces.
7.4 **Neutralizing Agents for Acids and Caustics:** Not pertinent
7.5 **Polymerization:** Not pertinent
7.6 **Inhibitor of Polymerization:** Not pertinent

8. WATER POLLUTION

8.1 **Aquatic Toxicity:** Data not available
8.2 **Waterfowl Toxicity:** 1,285 ppm LC_{50}
8.3 **Biological Oxygen Demand (BOD):** Data not available
8.4 **Food Chain Concentration Potential:** Zinc is accumulated by some organisms but is not considered to be bioconcentrative.

9. SELECTED MANUFACTURERS

1. Hooker Chemical Corp.
   Industrial Chemicals Div.
   Niagara Falls, N. Y. 14302
2. American Hoechst Corp.
   Chemicals and Plastics Div.
   Rt. 202—206 North
   Somerville, N. J. 08876
3. Fisher Scientific Co.
   711 Forbes Ave.
   Pittsburgh, Pa. 15219

10. SHIPPING INFORMATION

10.1 **Grades or Purity:** Technical, 94+%
10.2 **Storage Temperature:** Ambient
10.3 **Inert Atmosphere:** No requirement
10.4 **Venting:** Pressure-vacuum

11. HAZARD ASSESSMENT CODE
(See Hazard Assessment Handbook, CG 446-3)

II

12. HAZARD CLASSIFICATIONS

12.1 **Code of Federal Regulations:** Poisonous solid, Class B
12.2 **NAS Hazard Rating for Bulk Water Transportation:** Not listed
12.3 **NFPA Hazard Classifications:** Not listed

13. PHYSICAL AND CHEMICAL PROPERTIES

13.1 **Physical State at 15°C and 1 atm:** Solid
13.2 **Molecular Weight:** 258.10
13.3 **Boiling Point at 1 atm:** 2,012°F = 1,110°C = 1,373°K
13.4 **Freezing Point:** (sublimes) 788°F = 420°C = 693°K
13.5 **Critical Temperature:** Not pertinent
13.6 **Critical Pressure:** Not pertinent
13.7 **Specific Gravity:** 4.55 at 13°C (solid)
13.8 **Liquid Surface Tension:** Not pertinent
13.9 **Liquid-Water Interfacial Tension:** Not pertinent
13.10 **Vapor (Gas) Specific Gravity:** Not pertinent
13.11 **Ratio of Specific Heats of Vapor (Gas):** Not pertinent
13.12 **Latent Heat of Vaporization:** Not pertinent
13.13 **Heat of Combustion:** –4,100 Btu/lb = –2,270 cal/g = -95×10^5 J/kg
13.14 **Heat of Decomposition:** Not pertinent
13.15 **Heat of Solution:** Not pertinent
13.16 **Heat of Polymerization:** Not pertinent

NOTES

ZSL	ZINC SILICOFLUORIDE

Common Synonyms			
Zinc hexafluorosilicate Zinc fluosilicate Zinc silicofluoride hexahydrate	Solid	White	Odorless
	Sinks and mixes with water.		

Avoid contact with solid and dust. Keep people away.
Stop discharge if possible.
Isolate and remove discharged material.
Notify local health and pollution control agencies.

Fire	Not flammable. Irritating gases may be produced when heated.
Exposure	CALL FOR MEDICAL AID DUST Irritating to eyes, nose and throat. If inhaled will cause coughing or difficult breathing. If in eyes, hold eyelids open and flush with plenty of water. If breathing has stopped, give artificial respiration. If breathing is difficult, give oxygen. SOLID POISONOUS IF SWALLOWED. Irritating to skin and eyes. If swallowed will cause nausea, vomiting or loss of consciousness. Remove contaminated clothing and shoes. Flush affected areas with plenty of water. IF IN EYES, hold eyelids open and flush with plenty of water. IF SWALLOWED and victim is CONSCIOUS, have victim drink water or milk and have victim induce vomiting. IF SWALLOWED and victim is UNCONSCIOUS OR HAVING CONVULSIONS, do nothing except keep victim warm.
Water Pollution	Effect of low concentrations on aquatic life is unknown. May be dangerous if it enters water intakes. Notify local health and wildlife officials. Notify operators of nearby water intakes.

1. RESPONSE TO DISCHARGE

(See Response Methods Handbook, CG 446-4)

Issue warning—water contaminant
Disperse and flush

2. LABELS

No hazard label required by Code of Federal Regulations

3. CHEMICAL DESIGNATIONS

3.1 **Synonyms:** Zinc fluosilicate; Zinc hexafluorosilicate; Zinc silicofluoride hexahydrate
3.2 **Coast Guard Compatibility Classification:** Not listed
3.3 **Chemical Formula:** $ZnSiF_6 \cdot 6H_2O$
3.4 **IMCO/United Nations Numerical Designation:** Not listed

4. OBSERVABLE CHARACTERISTICS

4.1 **Physical State (as shipped):** Solid
4.2 **Color:** White; transparent
4.3 **Odor:** None

5. HEALTH HAZARDS

5.1 **Personal Protective Equipment:** Dust respirator; chemical goggles or face shield; protective gloves
5.2 **Symptoms Following Exposure:** Inhalation of dust irritates nose and throat; excessive inhalation may cause severe pulmonary inflammation. Ingestion causes nausea, cramps, vomiting, shock, convulsions, cyanosis, and other symptoms of fluoride poisoning. Contact with eyes or skin causes irritation; skin ulcers may develop.
5.3 **Treatment for Exposure:** INHALATION: move to fresh air. INGESTION: cause vomiting by giving soapy water or mustard water; have patient drink large quantities of lime water; if necessary, give stimulant such as strong coffee. EYES: flush with water; call physician as necessary. SKIN: wash with soap and water.
5.4 **Toxicity by Inhalation (Threshold Limit Value):** 2.5 mg/m^3 (as fluoride)
5.5 **Short-Term Inhalation Limits:** Data not available
5.6 **Toxicity by Ingestion:** Oral LD_{LO} = 100 mg/kg (rat)
5.7 **Late Toxicity:** Data not available
5.8 **Vapor (Gas) Irritant Characteristics:** Data not available
5.9 **Liquid or Solid Irritant Characteristics:** Data not available
5.10 **Odor Threshold:** Data not available

6. FIRE HAZARDS

6.1 **Flash Point:** Not flammable
6.2 **Flammable Limits in Air:** Not flammable
6.3 **Fire Extinguishing Agents:** Not pertinent
6.4 **Fire Extinguishing Agents Not to be Used:** Not pertinent
6.5 **Special Hazards of Combustion Products:** Toxic and irritating hydrogen fluoride and silicon tetrafluoride are formed in fires.
6.6 **Behavior in Fire:**
6.7 **Ignition Temperature:** Not pertinent
6.8 **Electrical Hazard:** Not pertinent
6.9 **Burning Rate:** Not pertinent

7. CHEMICAL REACTIVITY

7.1 **Reactivity with Water:** No reaction
7.2 **Reactivity with Common Materials:**
7.3 **Stability During Transport:** Stable
7.4 **Neutralizing Agents for Acids and Caustics:** Not pertinent
7.5 **Polymerization:** Not pertinent
7.6 **Inhibitor of Polymerization:** Not pertinent

8. WATER POLLUTION

8.1 **Aquatic Toxicity:** Data not available
8.2 **Waterfowl Toxicity:** Data not available
8.3 **Biological Oxygen Demand (BOD):** None
8.4 **Food Chain Concentration Potential:** Zinc is accumulated by some organisms but is not considered to be bioconcentrative.

9. SELECTED MANUFACTURERS

1. W. R. Grace & Co.
   Agricultural Chemicals Group
   Baltimore, Md. 21203
2. Agrico Chemical Co.
   P. O. Box 67
   Pierce (Polk County), Fla. 33867
3. American Hoechst Corp.
   Chemicals and Plastics Div.
   Rt. 202—206 North
   Somerville, N. J. 08876

10. SHIPPING INFORMATION

10.1 **Grades or Purity:** Technical, 98—99%
10.2 **Storage Temperature:** Ambient
10.3 **Inert Atmosphere:** No requirement
10.4 **Venting:** Open

11. HAZARD ASSESSMENT CODE

(See Hazard Assessment Handbook, CG 446-3)

SS

12. HAZARD CLASSIFICATIONS

12.1 **Code of Federal Regulations:** Not listed
12.2 **NAS Hazard Rating for Bulk Water Transportation:** Not listed
12.3 **NFPA Hazard Classifications:** Not listed

13. PHYSICAL AND CHEMICAL PROPERTIES

13.1 **Physical State at 15°C and 1 atm:** Solid
13.2 **Molecular Weight:** 315.5
13.3 **Boiling Point at 1 atm:** (decomposes) 122—158°F = 50—70°C = 323—343°K
13.4 **Freezing Point:** Not pertinent
13.5 **Critical Temperature:** Not pertinent
13.6 **Critical Pressure:** Not pertinent
13.7 **Specific Gravity:** 2.10 at 20°C (solid)
13.8 **Liquid Surface Tension:** Not pertinent
13.9 **Liquid-Water Interfacial Tension:** Not pertinent
13.10 **Vapor (Gas) Specific Gravity:** Not pertinent
13.11 **Ratio of Specific Heats of Vapor (Gas):** Not pertinent
13.12 **Latent Heat of Vaporization:** Not pertinent
13.13 **Heat of Combustion:** Not pertinent
13.14 **Heat of Decomposition:** Not pertinent
13.15 **Heat of Solution:** Data not available
13.16 **Heat of Polymerization:** Not pertinent

NOTES

ZSF ZINC SULFATE

Common Synonyms			
White vitriol Zinc vitriol Zinc suflate heptahydrate	Solid	White	Odorless
	Sinks and mixes with water.		

Avoid contact with solid and dust. Keep people away.
Stop discharge if possible.
Isolate and remove discharged material.
Notify local health and pollution control agencies.

Fire	Not flammable.
Exposure	CALL FOR MEDICAL AID. DUST Irritating to eyes, nose and throat. If inhaled will cause coughing or difficult breathing. If in eyes, hold eyelids open and flush with plenty of water. If breathing has stopped, give artificial respiration. If breathing is difficult, give oxygen. SOLID Irritating to skin and eyes. If swallowed will cause nausea and vomiting. Remove contaminated clothing and shoes. Flush affected areas with plenty of water. IF IN EYES, hold eyelids open and flush with plenty of water. IF SWALLOWED and victim is CONSCIOUS, have victim drink water or milk and have victim induce vomiting. IF SWALLOWED and victim is UNCONSCIOUS OR HAVING CONVULSIONS, do nothing except keep victim warm.
Water Pollution	HARMFUL TO AQUATIC LIFE IN VERY LOW CONCENTRATIONS. May be dangerous if it enters water intakes. Notify local health and wildlife officials. Notify operators of nearby water intakes.

1. RESPONSE TO DISCHARGE
(See Response Methods Handbook, CG 446-4)

Disperse and flush

2. LABELS

No hazard label required by Code of Federal Regulations

3. CHEMICAL DESIGNATIONS

3.1 **Synonyms:** White vitriol; Zinc sulfate heptahydrate; Zinc vitriol

3.2 **Coast Guard Compatibility Classification:** Not listed

3.3 **Chemical Formula:** $ZnSO_4 \cdot 7H_2O$

3.4 **IMCO/United Nations Numerical Designation:** Not listed

4. OBSERVABLE CHARACTERISTICS

4.1 **Physical State (as shipped):** Solid

4.2 **Color:** Colorless

4.3 **Odor:** None

5. HEALTH HAZARDS

5.1 **Personal Protective Equipment:** Dust mask; goggles or face shield; protective gloves

5.2 **Symptoms Following Exposure:** Inhalation of dust causes irritation of nose and throat. Ingestion can cause irritation or corrosion of the alimentary tract. Contact with eyes or skin causes irritation.

5.3 **Treatment for Exposure:** INHALATION: move to fresh air. INGESTION: induce vomiting, followed by prompt and complete gastric lavage, cathartics, and demulcents. EYES or SKIN: flush with water.

5.4 **Toxicity by Inhalation (Threshold Limit Value):** Data not available

5.5 **Short-Term Inhalation Limits:** Data not available

5.6 **Toxicity by Ingestion:** Grade 2; LD_{50} 0.5—5 g/kg

5.7 **Late Toxicity:** Data not available

5.8 **Vapor (Gas) Irritant Characteristics:** Data not available

5.9 **Liquid or Solid Irritant Characteristics:** Data not available

5.10 **Odor Threshold:** Data not available

6. FIRE HAZARDS

6.1 **Flash Point:** Not flammable

6.2 **Flammable Limits in Air:** Not flammable

6.3 **Fire Extinguishing Agents:** Not pertinent

6.4 **Fire Extinguishing Agents Not to be Used:** Not pertinent

6.5 **Special Hazards of Combustion Products:** Not pertinent

6.6 **Behavior in Fire:**

6.7 **Ignition Temperature:** Not pertinent

6.8 **Electrical Hazard:** Not pertinent

6.9 **Burning Rate:** Not pertinent

7. CHEMICAL REACTIVITY

7.1 **Reactivity with Water:** No reaction

7.2 **Reactivity with Common Materials:**

7.3 **Stability During Transport:** Stable

7.4 **Neutralizing Agents for Acids and Caustics:** Not pertinent

7.5 **Polymerization:** Not pertinent

7.6 **Inhibitor of Polymerization:** Not pertinent

8. WATER POLLUTION

8.1 **Aquatic Toxicity:**
0.3 mg/l as Zn/120 hr/stickleback/lethal/*
4.6 ppm/96 hr/rainbow trout/LC_{50}/fresh water
*Water type not specified

8.2 **Waterfowl Toxicity:** Data not available

8.3 **Biological Oxygen Demand (BOD):** None

8.4 **Food Chain Concentration Potential:** Zinc is accumulated by some organisms but is not considered to be bioconcentrative.

9. SELECTED MANUFACTURERS

1. Chemical and Pigment Co.
600 Nichols Road
Pittsburg, Calif. 94565

2. Eagle-Picher Industries, Inc.
Agricultural Chemicals Division
P. O. Box 187
Galena, Kan. 66739

3. J. T. Baker Chemical Co.
Phillipsburg, N. J. 08865

10. SHIPPING INFORMATION

10.1 **Grades or Purity:** Reagent; Technical

10.2 **Storage Temperature:** Ambient

10.3 **Inert Atmosphere:** No requirement

10.4 **Venting:** Open

11. HAZARD ASSESSMENT CODE
(See Hazard Assessment Handbook, CG 446-3)

SS

12. HAZARD CLASSIFICATIONS

12.1 **Code of Federal Regulations:** Not listed

12.2 **NAS Hazard Rating for Bulk Water Transportation:** Not listed

12.3 **NFPA Hazard Classifications:** Not listed

13. PHYSICAL AND CHEMICAL PROPERTIES

13.1 **Physical State at 15°C and 1 atm:** Solid

13.2 **Molecular Weight:** 287.54

13.3 **Boiling Point at 1 atm:** Not pertinent (decomposes)

13.4 **Freezing Point:** (decomposes) 122—212°F = 50—100°C = 323—373°K

13.5 **Critical Temperature:** Not pertinent

13.6 **Critical Pressure:** Not pertinent

13.7 **Specific Gravity:** 1.96 at 20°C (solid)

13.8 **Liquid Surface Tension:** Not pertinent

13.9 **Liquid-Water Interfacial Tension:** Not pertinent

13.10 **Vapor (Gas) Specific Gravity:** Not pertinent

13.11 **Ratio of Specific Heats of Vapor (Gas):** Not pertinent

13.12 **Latent Heat of Vaporization:** Not pertinent

13.13 **Heat of Combustion:** Not pertinent

13.14 **Heat of Decomposition:** Not pertinent

13.15 **Heat of Solution:** Not pertinent

13.16 **Heat of Polymerization:** Not pertinent

NOTES

ZCA ZIRCONIUM ACETATE

Common Synonyms	Liquid	White	Weak vinegar odor
	Sinks and mixes with water.		

Avoid contact with liquid. Keep people away.
Stop discharge if possible.
Isolate and remove discharged material.
Notify local health and pollution control agencies.

Fire

Not flammable.

Exposure

CALL FOR MEDICAL AID.

LIQUID
Irritating to skin and eyes.
Harmful if swallowed.
Remove contaminated clothing and shoes.
Flush affected areas with plenty of water.
IF IN EYES, hold eyelids open and flush with plenty of water.
IF SWALLOWED and victim is CONSCIOUS, have victim drink water or milk.
IF SWALLOWED and victim is UNCONSCIOUS OR HAVING CONVULSIONS, do nothing except keep victim warm.

Water Pollution

Effect of low concentrations on aquatic life is unknown.
May be dangerous if it enters water intakes.
Notify local health and wildlife officials.
Notify operators of nearby water intakes.

1. RESPONSE TO DISCHARGE
(See Response Methods Handbook, CG 446-4)

Issue warning—water contaminant
Disperse and flush

2. LABELS

No hazard label required by Code of Federal Regulations

3. CHEMICAL DESIGNATIONS

3.1 **Synonyms:** Zirconium acetate solution
3.2 **Coast Guard Compatibility Classification:** Not listed
3.3 **Chemical Formula:** $Zr(C_2H_3O_2)_4-H_2O$
3.4 **IMCO/United Nations Numerical Designation:** Not listed

4. OBSERVABLE CHARACTERISTICS

4.1 **Physical State (as shipped):** Liquid
4.2 **Color:** Colorless
4.3 **Odor:** Weak vinegar

5. HEALTH HAZARDS

5.1 **Personal Protective Equipment:** Rubber gloves; chemical goggles or face shield
5.2 **Symptoms Following Exposure:** Has only a mild pharmacological action. Contact with eyes or skin may cause irritation.
5.3 **Treatment for Exposure:** INGESTION: give large amount of water. EYES: flush with water for at least 15 min.; consult a physician if irritation persists. SKIN: flush with water.
5.4 **Toxicity by Inhalation (Threshold Limit Value):** 5 mg/m^3 (as zirconium)
5.5 **Short-Term Inhalation Limits:** Data not available
5.6 **Toxicity by Ingestion:** Grade 2; LD_{50} 0.5—5 g/kg (rat)
5.7 **Late Toxicity:** Data not available
5.8 **Vapor (Gas) Irritant Characteristics:** Data not available
5.9 **Liquid or Solid Irritant Characteristics:** Data not available
5.10 **Odor Threshold:** Data not available

6. FIRE HAZARDS

6.1 **Flash Point:** Not flammable
6.2 **Flammable Limits in Air:** Not flammable
6.3 **Fire Extinguishing Agents:** Not pertinent
6.4 **Fire Extinguishing Agents Not to be Used:** Not pertinent
6.5 **Special Hazards of Combustion Products:** Not pertinent
6.6 **Behavior in Fire:**
6.7 **Ignition Temperature:** Not pertinent
6.8 **Electrical Hazard:** Not pertinent
6.9 **Burning Rate:** Not pertinent

7. CHEMICAL REACTIVITY

7.1 **Reactivity with Water:** No reaction
7.2 **Reactivity with Common Materials:**
7.3 **Stability During Transport:** Stable
7.4 **Neutralizing Agents for Acids and Caustics:** Not pertinent
7.5 **Polymerization:** Not pertinent
7.6 **Inhibitor of Polymerization:** Not pertinent

8. WATER POLLUTION

8.1 **Aquatic Toxicity:** Data not available
8.2 **Waterfowl Toxicity:** Data not available
8.3 **Biological Oxygen Demand (BOD):** Data not available
8.4 **Food Chain Concentration Potential:** Data not available

9. SELECTED MANUFACTURERS

1. The Harshaw Chemical Co.
   1945 E. 97 St.
   Cleveland, Ohio 44106
2. Gallard-Schlesinger Chemical Mfg. Co.
   584 Mineola Ave.
   Carle Place, N. Y. 11514
3. Pfaltz and Bauer, Inc.
   375 Fairfield Ave.
   Stamford, Conn. 06902

10. SHIPPING INFORMATION

10.1 **Grades or Purity:** 25% solution in water
10.2 **Storage Temperature:** Ambient
10.3 **Inert Atmosphere:** No requirement
10.4 **Venting:** Open

11. HAZARD ASSESSMENT CODE
(See Hazard Assessment Handbook, CG 446-3)

A-P

12. HAZARD CLASSIFICATIONS

12.1 **Code of Federal Regulations:** Not listed
12.2 **NAS Hazard Rating for Bulk Water Transportation:** Not listed
12.3 **NFPA Hazard Classifications:**

Category	Classification
Health Hazard (Blue)	0
Flammability (Red)	0
Reactivity (Yellow)	0

13. PHYSICAL AND CHEMICAL PROPERTIES

13.1 **Physical State at 15°C and 1 atm:** Liquid
13.2 **Molecular Weight:** 327 (solute only)
13.3 **Boiling Point at 1 atm:** Not pertinent
13.4 **Freezing Point:** Not pertinent
13.5 **Critical Temperature:** Not pertinent
13.6 **Critical Pressure:** Not pertinent
13.7 **Specific Gravity:** 1.37 at 20°C (liquid)
13.8 **Liquid Surface Tension:** Data not available
13.9 **Liquid-Water Interfacial Tension:** Not pertinent
13.10 **Vapor (Gas) Specific Gravity:** Not pertinent
13.11 **Ratio of Specific Heats of Vapor (Gas):** Not pertinent
13.12 **Latent Heat of Vaporization:** Not pertinent
13.13 **Heat of Combustion:** Not pertinent
13.14 **Heat of Decomposition:** Not pertinent
13.15 **Heat of Solution:** Not pertinent
13.16 **Heat of Polymerization:** Not pertinent

NOTES

ZIR — ZIRCONIUM NITRATE

Common Synonyms			
Zirconium nitrate pentahydrate	Solid	White	Odorless
	Sinks and mixes with water.		

Avoid contact with solid and dust. Keep people away.
Stop discharge if possible.
Isolate and remove discharged material.
Notify local health and pollution control agencies.

Fire

Not flammable.
Will increase the intensity of a fire.
POISONOUS GASES MAY BE PRODUCED IN FIRE.
Wear goggles and self-contained breathing apparatus.
Flood discharge area with water.

Exposure

CALL FOR MEDICAL AID.

DUST
Irritating to eyes, nose and throat.
If inhaled will cause coughing or difficult breathing.
If in eyes, hold eyelids open and flush with plenty of water.
If breathing has stopped, give artificial respiration.
If breathing is difficult, give oxygen.

SOLID
Irritating to skin and eyes.
If swallowed will cause nausea and vomiting.
Remove contaminated clothing and shoes.
Flush affected areas with plenty of water.
IF IN EYES, hold eyelids open and flush with plenty of water.
IF SWALLOWED and victim is CONSCIOUS, have victim drink water or milk.
IF SWALLOWED and victim is UNCONSCIOUS OR HAVING CONVULSIONS, do nothing except keep victim warm.

Water Pollution

Effect of low concentrations on aquatic life is unknown.
May be dangerous if it enters water intakes.
Notify local health and wildlife officials.
Notify operators of nearby water intakes.

1. RESPONSE TO DISCHARGE

(See Response Methods Handbook, CG 446-4)

Issue warning—water contaminant
Disperse and flush

2. LABELS

No hazard label required by Code of Federal Regulations

3. CHEMICAL DESIGNATIONS

3.1 **Synonyms:** Zirconium nitrate pentahydrate
3.2 **Coast Guard Compatibility Classification:** Not listed
3.3 **Chemical Formula:** $Zr(NO_3)_4 \cdot 5H_2O$
3.4 **IMCO/United Nations Numerical Designation:** Not listed

4. OBSERVABLE CHARACTERISTICS

4.1 **Physical State (as shipped):** Solid
4.2 **Color:** White
4.3 **Odor:** None

5. HEALTH HAZARDS

5.1 **Personal Protective Equipment:** Dust mask; goggles or face shield; protective gloves
5.2 **Symptoms Following Exposure:** Has only a mild pharmacological action. Inhalation of dust may irritate nose and throat. Contact with eyes or skin causes irritation.
5.3 **Treatment for Exposure:** INHALATION: move to fresh air. INGESTION: give large amount of water. EYES or SKIN: flush with water.
5.4 **Toxicity by Inhalation (Threshold Limit Value):** 5 mg/m³ (as zirconium)
5.5 **Short-Term Inhalation Limits:** Data not available
5.6 **Toxicity by Ingestion:** Grade 2; oral LD_{50} = 2.5 g/kg (rat)
5.7 **Late Toxicity:** Data not available
5.8 **Vapor (Gas) Irritant Characteristics:** Data not available
5.9 **Liquid or Solid Irritant Characteristics:** Data not available
5.10 **Odor Threshold:** Data not available

6. FIRE HAZARDS

6.1 **Flash Point:** Not flammable but may intensify fire
6.2 **Flammable Limits in Air:** Not flammable
6.3 **Fire Extinguishing Agents:** Not pertinent
6.4 **Fire Extinguishing Agents Not to be Used:** Not pertinent
6.5 **Special Hazards of Combustion Products:** Toxic oxides of nitrogen may form in fire.
6.6 **Behavior in Fire:** May increase intensity of fire when in contact with combustible materials.
6.7 **Ignition Temperature:** Not pertinent
6.8 **Electrical Hazard:** Not pertinent
6.9 **Burning Rate:** Not pertinent

7. CHEMICAL REACTIVITY

7.1 **Reactivity with Water:** Dissolves to give an acid solution
7.2 **Reactivity with Common Materials:** Will corrode most metals
7.3 **Stability During Transport:** Stable
7.4 **Neutralizing Agents for Acids and Caustics:** Flush with water
7.5 **Polymerization:** Not pertinent
7.6 **Inhibitor of Polymerization:** Not pertinent

8. WATER POLLUTION

8.1 **Aquatic Toxicity:** Data not available
8.2 **Waterfowl Toxicity:** Data not available
8.3 **Biological Oxygen Demand (BOD):** None
8.4 **Food Chain Concentration Potential:** Data not available

9. SELECTED MANUFACTURERS

1. Var-lac-oid Chemical Co.
   666 South Front Street
   Elizabeth, N. J. 07202
2. Gallard-Schlesinger Chemical Mfg. Co.
   584 Mineola Ave.
   Carle Place, N. Y. 11514
3. Pfaltz and Bauer, Inc.
   375 Fairfield Ave.
   Stamford, Conn. 06902

10. SHIPPING INFORMATION

10.1 **Grades or Purity:** Commercial; 99+%
10.2 **Storage Temperature:** Ambient
10.3 **Inert Atmosphere:** No requirement
10.4 **Venting:** Open

11. HAZARD ASSESSMENT CODE

(See Hazard Assessment Handbook, CG 446-3)

SS

12. HAZARD CLASSIFICATIONS

12.1 **Code of Federal Regulations:** Not listed
12.2 **NAS Hazard Rating for Bulk Water Transportation:** Not listed
12.3 **NFPA Hazard Classifications:** Not listed

13. PHYSICAL AND CHEMICAL PROPERTIES

13.1 **Physical State at 15°C and 1 atm:** Solid
13.2 **Molecular Weight:** 429.3
13.3 **Boiling Point at 1 atm:** Not pertinent (decomposes)
13.4 **Freezing Point:** Not pertinent
13.5 **Critical Temperature:** Not pertinent
13.6 **Critical Pressure:** Not pertinent
13.7 **Specific Gravity:** > 1 at 20°C (solid)
13.8 **Liquid Surface Tension:** Not pertinent
13.9 **Liquid-Water Interfacial Tension:** Not pertinent
13.10 **Vapor (Gas) Specific Gravity:** Not pertinent
13.11 **Ratio of Specific Heats of Vapor (Gas):** Not pertinent
13.12 **Latent Heat of Vaporization:** Not pertinent
13.13 **Heat of Combustion:** Not pertinent
13.14 **Heat of Decomposition:** Not pertinent
13.15 **Heat of Solution:** Not pertinent
13.16 **Heat of Polymerization:** Not pertinent

NOTES

ZCO	# ZIRCONIUM OXYCHLORIDE

Common Synonyms			
Zirconyl chloride Basic zirconium chloride Zirconium oxychloride hydrate Zirconium oxide chloride	Solid Sinks and mixes with water.	White to yellow	Odorless

Avoid contact with solid and dust. Keep people away.
Stop discharge if possible.
Isolate and remove discharged material.
Notify local health and pollution control agencies.

Fire	Not flammable.
Exposure	CALL FOR MEDICAL AID. DUST Irritating to eyes, nose and throat. If inhaled will cause coughing or difficult breathing. If in eyes, hold eyelids open and flush with plenty of water. If breathing has stopped, give artificial respiration. If breathing is difficult, give oxygen. SOLID Irritating to skin and eyes. If swallowed will cause nausea and vomiting. Remove contaminated clothing and shoes. Flush affected areas with plenty of water. IF IN EYES, hold eyelids open and flush with plenty of water. IF SWALLOWED and victim is CONSCIOUS, have victim drink water or milk and have victim induce vomiting. IF SWALLOWED and victim is UNCONSCIOUS OR HAVING CONVULSIONS, do nothing except keep victim warm.
Water Pollution	Effect of low concentrations on aquatic life is unknown. May be dangerous if it enters water intakes. Notify local health and wildlife officials. Notify operators of nearby water intakes.

1. RESPONSE TO DISCHARGE
(See Response Methods Handbook, CG 446-4)

Issue warning—water contaminant
Disperse and flush

2. LABELS
No hazard label required by Code of Federal Regulations

3. CHEMICAL DESIGNATIONS
3.1 **Synonyms:** Basic zirconium chloride; Zirconium oxide chloride; Zirconium oxychloride hydrate; Zirconyl chloride
3.2 **Coast Guard Compatibility Classification:** Not listed
3.3 **Chemical Formula:** $ZrOCl_2 \cdot 8H_2O$
3.4 **IMCO/United Nations Numerical Designation:** Not listed

4. OBSERVABLE CHARACTERISTICS
4.1 **Physical State (as shipped):** Solid
4.2 **Color:** Yellow to white
4.3 **Odor:** None

5. HEALTH HAZARDS
5.1 **Personal Protective Equipment:** Safety glasses or face shield; protective gloves; dust mask
5.2 **Symptoms Following Exposure:** Has only a mild pharmacological action. Inhalation of dust may irritate nose and throat. Contact with eyes or skin causes irritation.
5.3 **Treatment for Exposure:** INHALATION: move to fresh air. INGESTION: give large amount of water. EYES or SKIN: flush with water.
5.4 **Toxicity by Inhalation (Threshold Limit Value):** 5 mg/m^3 (as zirconium)
5.5 **Short-Term Inhalation Limits:** Data not available
5.6 **Toxicity by Ingestion:** Grade 2; oral LD_{50} = 3.5 g/kg (rat)
5.7 **Late Toxicity:** Data not available
5.8 **Vapor (Gas) Irritant Characteristics:** Data not available
5.9 **Liquid or Solid Irritant Characteristics:** Data not available
5.10 **Odor Threshold:** Data not available

6. FIRE HAZARDS
6.1 **Flash Point:** Not flammable
6.2 **Flammable Limits in Air:** Not flammable
6.3 **Fire Extinguishing Agents:** Not pertinent
6.4 **Fire Extinguishing Agents Not to be Used:** Not pertinent
6.5 **Special Hazards of Combustion Products:** Not pertinent
6.6 **Behavior in Fire:**
6.7 **Ignition Temperature:** Not pertinent
6.8 **Electrical Hazard:** Not pertinent
6.9 **Burning Rate:** Not pertinent

7. CHEMICAL REACTIVITY
7.1 **Reactivity with Water:** No reaction
7.2 **Reactivity with Common Materials:**
7.3 **Stability During Transport:** Stable
7.4 **Neutralizing Agents for Acids and Caustics:** Not pertinent
7.5 **Polymerization:** Not pertinent
7.6 **Inhibitor of Polymerization:** Not pertinent

8. WATER POLLUTION
8.1 **Aquatic Toxicity:**
240 ppm/96 hr/fathead minnow/TL_m/hard water
18 ppm/96 hr/fathead minnow/TL_m/soft water
8.2 **Waterfowl Toxicity:** Data not available
8.3 **Biological Oxygen Demand (BOD):** None
8.4 **Food Chain Concentration Potential:** Data not available

9. SELECTED MANUFACTURERS
1. NL Industries, Inc.
TAM Division
4511 Hyde Park Blvd.
Niagara Falls, N. Y. 14305
2. Ventron, Inc.
P. O. Box 159
Beverly, Mass. 01915
3. Gallard-Schlesinger Chemical Mfg. Co.
584 Mineola Ave.
Carle Place, N. Y. 11514

10. SHIPPING INFORMATION
10.1 **Grades or Purity:** Technical; Pure
10.2 **Storage Temperature:** Ambient
10.3 **Inert Atmosphere:** No requirement
10.4 **Venting:** Open

11. HAZARD ASSESSMENT CODE
(See Hazard Assessment Handbook, CG 446-3)

SS

12. HAZARD CLASSIFICATIONS
12.1 **Code of Federal Regulations:** Not listed
12.2 **NAS Hazard Rating for Bulk Water Transportation:** Not listed
12.3 **NFPA Hazard Classifications:** Not listed

13. PHYSICAL AND CHEMICAL PROPERTIES
13.1 **Physical State at 15°C and 1 atm:** Solid
13.2 **Molecular Weight:** 322.3
13.3 **Boiling Point at 1 atm:** Not pertinent (decomposes)
13.4 **Freezing Point:** Not pertinent
13.5 **Critical Temperature:** Not pertinent
13.6 **Critical Pressure:** Not pertinent
13.7 **Specific Gravity:** >1 at 20°C (solid)
13.8 **Liquid Surface Tension:** Not pertinent
13.9 **Liquid-Water Interfacial Tension:** Not pertinent
13.10 **Vapor (Gas) Specific Gravity:** Not pertinent
13.11 **Ratio of Specific Heats of Vapor (Gas):** Not pertinent
13.12 **Latent Heat of Vaporization:** Not pertinent
13.13 **Heat of Combustion:** Not pertinent
13.14 **Heat of Decomposition:** Not pertinent
13.15 **Heat of Solution:** Not pertinent
13.16 **Heat of Polymerization:** Not pertinent

NOTES

ZCS	ZIRCONIUM SULFATE

Common Synonyms			
Disulfatozirconic acid Zirconium sulfate tetrahydrate	Solid	White	Odorless
	Sinks and mixes with water.		

Avoid contact with solid and dust. Keep people away.
Stop discharge if possible.
Isolate and remove discharged material.
Notify local health and pollution control agencies.

Fire	Not flammable.
Exposure	CALL FOR MEDICAL AID DUST Irritating to eyes, nose and throat. If inhaled will cause coughing or difficult breathing. **If in eyes, hold eyelids open and flush with plenty of water.** **If breathing has stopped, give artificial respiration.** **If breathing is difficult, give oxygen.** SOLID Irritating to skin and eyes. If swallowed will cause nausea and vomiting. **Remove contaminated clothing and shoes.** **Flush affected areas with plenty of water.** **IF IN EYES, hold eyelids open and flush with plenty of water.** **IF SWALLOWED and victim is CONSCIOUS, have victim drink water or milk.** **IF SWALLOWED and victim is UNCONSCIOUS OR HAVING CONVULSIONS, do nothing except keep victim warm.**
Water Pollution	Effect of low concentrations on aquatic life is unknown. May be dangerous to water intakes. **Notify local health and wildlife officials.** **Notify opeartors of nearby water intakes.**

1. RESPONSE TO DISCHARGE
(See Response Methods Handbook, CG 446-4)

Issue warning—water contaminant
Disperse and flush

2. LABELS

No hazard label required by Code of Federal Regulations

3. CHEMICAL DESIGNATIONS

3.1 **Synonyms:** Disulfatozirconic acid; Zirconium sulfate tetrahydrate
3.2 **Coast Guard Compatibility Classification:** Not listed
3.3 **Chemical Formula:** $Zr(SO_4)_2{\cdot}4H_2O$
3.4 **IMCO/Uited Nations Numerical Designation:** Not listed

4. OBSERVABLE CHARACTERISTICS

4.1 **Physical State (as shipped):** Solid
4.2 **Color:** White
4.3 **Odor:** None

5. HEALTH HAZARDS

5.1 **Personal Protective Equipment:** Dust mask; goggles or face shield; protective gloves.
5.2 **Symptoms Following Exposure:** Has only a mild pharmacological action. Inhalation of dust may irritate nose and throat. Contact with eyes or skin causes irritation.
5.3 **Treatment for Exposure:** INHALATION: move to fresh air. INGESTION: give large amount of water. EYES or SKIN: flush with water.
5.4 **Toxicity by Inhalation (Threshold Limit Value):** 5 mg/m^3 (as zirconium)
5.5 **Short-Term Inhalation Limits:** Data not available
5.6 **Toxicity by Ingestion:** Grade 2; oral LD_{50} = 3.5 g/kg (rat)
5.7 **Late Toxicity:** Data not available
5.8 **Vapor (Gas) Irritant Characteristics:** Data not available
5.9 **Liquid or Solid Irritant Characteristics:** Data not available
5.10 **Odor Threshold:** Data not available

6. FIRE HAZARDS

6.1 **Flash Point:** Not flammable
6.2 **Flammable Limits in Air:** Not flammable
6.3 **Fire Extinguishing Agents:** Not pertinent
6.4 **Fire Extinguishing Agents Not to be Used:** Not pertinent
6.5 **Special Hazards of Combustion Products:** Not pertinent
6.6 **Behavior in Fire:**
6.7 **Ignition Temperature:** Not pertinent
6.8 **Electrical Hazard:** Not pertinent
6.9 **Burning Rate:** Not pertinent

7. CHEMICAL REACTIVITY

7.1 **Reactivity with Water:** No reaction
7.2 **Reactivity with Common Materials:**
7.3 **Stability During Transport:** Stable
7.4 **Neutralizing Agents for Acids and Caustics:** Not pertinent
7.5 **Polymerization:** Not pertinent
7.6 **Inhibitor of Polymerization:** Not pertinent

8. WATER POLLUTION

8.1 **Aquatic Toxicity:**
14 ppm as Zr/96 hr/fathead minnow/TL_m/soft water
145 ppm/96 hr/fathead minnow/TL_m/hard water
8.2 **Waterfowl Toxicity:** Data not available
8.3 **Biological Oxygen Demand (BOD):** None
8.4 **Food Chain Concentration Potential:** Data not available

9. SELECTED MANUFACTURERS

1. Var-lac-oid Chemical Co.
666 South Front Street
Elizabeth, N. J. 07202
2. Gallard-Schlesinger Chemical Mfg. Co.
584 Mineola Ave.
Carle Place, N. Y. 11514
3. Pfaltz and Bauer, Inc.
375 Fairfield Ave.
Stamford, Conn. 06902

10. SHIPPING INFORMATION

10.1 **Grades or Purity:** Technical
10.2 **Storage Temperature:** Ambient
10.3 **Inert Atmosphere:** No requirement
10.4 **Venting:** Open

11. HAZARD ASSESSMENT CODE
(See Hazard Assessment Handbook, CG 446-3)

SS

12. HAZARD CLASSIFICATIONS

12.1 **Code of Federal Regulations:** Not listed
12.2 **NAS Hazard Rating for Bulk Water Transportation:** Not listed
12.3 **NFPA Hazard Classifications:** Not listed

13. PHYSICAL AND CHEMICAL PROPERTIES

13.1 **Physical State at 15°C and 1 atm:** Solid
13.2 **Molecular Weight:** 355.4
13.3 **Boiling Point at 1 atm:** Not pertinent (decomposes)
13.4 **Freezing Point:** Not pertinent
13.5 **Critical Temperature:** Not pertinent
13.6 **Critical Pressure:** Not pertinent
13.7 **Specific Gravity:** (approx.) 3.0 at 20°C (solid)
13.8 **Liquid Surface Tension:** Not pertinent
13.9 **Liquid-Water Interfacial Tension:** Not pertinent
13.10 **Vapor (Gas) Specific Gravity:** Not pertinent
13.11 **Ratio of Specific Heats of Vapor (Gas):** Not pertinent
13.12 **Latent Heat of Vaporization:** Not pertinent
13.13 **Heat of Combustion:** Not pertinent
13.14 **Heat of Decomposition:** Not pertinent
13.15 **Heat of Solution:** Not pertinent
13.16 **Heat of Polymerization:** Not pertinent

NOTES

Part 2

Oak Ridge National Laboratory Chemicals

This part was prepared from *Material Safety Data Sheets—The Basis for Control of Toxic Chemicals, Volumes 1-3* (ORNL/TM-6981/V1, V2, V3), by E.E. Ketchen and W.E. Porter of Oak Ridge National Laboratory for the U.S. Department of Energy, September 1979.

INTRODUCTION

The material safety data sheets contained in this part are the basis for the Toxic Chemical Control Program developed by the Industrial Hygiene Department, Health Division, Oak Ridge National Laboratory (ORNL). They are designed to give the reader comprehensive information on a particular material.

The data sheets were developed and compiled to aid in apprising employees of hazards peculiar to the handling and/or use of specific toxic chemicals stored at ORNL. Space limitations necessitate the use of descriptive medical terms and toxicological abbreviations. A glossary and an abbreviation list were developed to define some of those sometimes unfamiliar terms and abbreviations. These are included as separate sections of this Introduction.

The chemicals are classified as to toxicity, flammability, reactivity, and special properties using a 0 through 4 index (1)(2). At this point those materials which are extremely toxic or highly toxic (Class 3 or 4) have been classified and put into the system.

The toxicity rating appears on the diamond-shaped portion of the page, in the left quadrant. Descriptive terminology is given in Table 1. The materials are placed in a 0 through 4 toxicity class based on the range of threshold limit values (TLV's) (1)(2)(3), also shown in this table. No claim is made as to the validity of the TLV ranges shown in Table 1. They were selected somewhat arbitrarily by mutual agreement of representatives of the Industrial Hygiene Departments of the Nuclear Division plants.

In cases where the TLV of a material is given both in parts per million parts air (ppm) and in milligrams per cubic meter of air (mg/m^3), the one which results in the more restrictive (higher toxicity) rating is preferentially used. This is, almost without exception, the ppm limit.

In cases where no TLV's have been set, the oral LD_{50} for rats, the inhalation LC_{50} for rats, or the skin LD_{50} for rabbits is used as shown in Table 2 (1)(2). Many of the chemicals stocked at ORNL cannot be classified using Table 1 or 2. These materials are classified on the basis of other toxicity data or by comparison with similar materials.

The health (left quadrant), flammability (upper quadrant), and reactivity (right quadrant) hazard ratings are shown on the diamond as 0-4 designation, as well as a special hazard notice (if any). The current TLV's (if available) and the routes of absorption into the body are listed. The reader is informed of the acute symptoms of exposure and the effects of chronic exposure. Brief emergency procedures, which are to be used in the event of an exposure, and unusual hazards are also given. In the space designated as "carcinogenic," a "yes," "no," or "neoplastic" is entered, followed by "mutagen" and/or "teratogen" if applicable. Also included are useful physical and chemical characteristics, fire and explosion data.

TABLE 1

Relative Toxicity Criteria Based on TLV's

Rating	Commonly Used Term	TLV ppm	TLV mg/c.m.
4	Extremely Toxic	≤10	≤.25
3	Highly Toxic	11 to 250	.26 to 2.5
2	Moderately Toxic	251 to 500	2.51 to 5.0
1	Slightly Toxic or Practically Non-Toxic	501 to 1000	5.1 to 10
0	Relatively Harmless	>1000	>10

TABLE 2

Relative Toxicity Criteria Based on LD_{50} and LC_{50} Data

Commonly Used Term	LD_{50} Single Oral Dose Rats mg/kg	LC_{50} Inhalation 4-hr. Vapor Exposure Rats ppm	LD_{50}-Skin Rabbits mg/kg
Extremely Toxic	≤1	≤10	≤5
Highly Toxic	1.1 to 50	11 to 100	5.1 to 43
Moderately Toxic	50.1 to 500	101 to 1,000	44 to 340
Slightly Toxic or Practically Non-Toxic	501 to 15,000	1,001 to 100,000	350 to 22,600
Relatively Harmless	>15,000	>100,000	>22,600

REFERENCES

1. Porter, W. E., C. L. Hunt and N. E. Bolton: A System for Labeling and Control of Toxic Materials in a Large Research Facility. Am. Ind. Hyg. Assoc. J. 36: 51-56 (1977).

2. U. S. Department of Health, Education and Welfare: A Recommended Standard .... An Identification System for Occupationally Hazardous Materials. HEW Publication No. (NIOSH) 75-126:13-23 (1974).

3. Threshold Limit Values for Chemical Substances in Work Room Air Adopted by ACGIH for 1978. Am. Conf. of Gov. Ind. Hyg. Cincinnati (1978).

GLOSSARY

Albumin	A water-soluble protein excreted in urine
Anemia	A condition in which the blood is deficient in hemoglobin and/or numbers of red cells compared with normal blood
Anesthesia	Loss of feeling or sensation, especially the loss of tactile sensibility
Anorexia	Lack or loss of the appetite for food
Anosmia	Absence of the sense of smell
Anuria	Total suppression of urinary output by the kidney
Apathy	Lack of feeling or emotion, indifference
Arthralgia	Neuralgia or pain in a joint
Asbestosis	Lung disease caused by prolonged inhaling of asbestos
Asphyxia	Suffocation
Bronchiolitis	Capillary bronchitis
Bronchitis	Inflammation of the bronchial tubes
Bronchopneumonia	Inflammation of the lungs following infection of the upper respiratory tract, infectious fevers, or debilitating diseases
Bronchospasms	Bronchial spasm
Carcinoma	A malignant new growth made up of epithelial cells, which give rise to metastases
Cataract	An opacity of the crystalline eye lens or its capsule
Cirrhosis	A disease of the liver marked by the progressive destruction of liver cells
Conjunctivitis	Inflammation of the conjunctiva, the delicate membrane that lines the eyelids and covers the eyeballs
Convulsion	A violent contraction of the voluntary muscles
Cornea	Transparent structure of the external layer of the eyeball
Cyanosis	Blueness of the skin caused by insufficient oxygenation of the blood
Delirium	A mental disturbance marked by illusion, hallucinations, physical restlessness, and incoherence

Dermatitis	Inflammation of the skin
Dyspnea	Difficult or labored breathing
Emphysema	A swelling or inflation due to presence of air in the interstices of the connective tissues of the lungs
Epigastric	Pertaining to the belly or (epigastrium)
Erosion	An eating or gnawing away
Euphoria	Exaggerated sense of well-being
Gangrene	Anemic necrosis of tissue
Gastroenteritis	Inflammation of the stomach and intestines
Goiters	Enlargement of the thyroid gland
Hallucinations	A sense perception not founded upon objective reality
Hematemesis	Vomiting of blood
Hematuria	Discharge of blood in the urine
Hemoglobinuria	Presence of hemoglobin in the urine
Hemolytic anemia	Anemia due to active destruction of red cells
Hepatitis	Inflammation of the liver
Hyperactivity	Overactivity
Hyperemia	Excess of blood in any part of the body
Hyperpnea	Abnormal exaggeration of the respiratory movements
Hypotension	Abnormally low blood pressure
Hypothermia	Abnormally low temperature
Inebriation	The condition of being drunk
Ingestion	The act of taking food, medicines, etc. into the body by mouth
Insomnia	Inability to sleep
Intoxication	The condition produced by excessive use of alcohol; also, the state of being poisoned
Lacrimation	The secretion and discharge of tears
Laryngitis	Sore throat, hoarseness, and cough due to inflammation of the larynx
Lassitude	Weakness, exhaustion
Lethargy	A condition of drowsiness of mental origin
Leukopenia	Reduction of white cells in the blood
Malaise	Uneasiness, discomfort, and distress

Malignant tumors	Tumors that tend to become progressively worse and result in death
Mottling	A condition of a tissue spotted with patches of color
Myalgia	Pain in a muscle or muscles
Narcosis	A state of profound unconsciousness produced by a drug
Narcotic	Having the effect of producing sleep or stupor and at the same time relieving pain (adj.)
Nausea	Tending to vomit, feeling of sickness at the stomach
Necrosis	Death of a circumscribed portion of tissue
Nephritis	Inflammation of the kidney (hemorrhagic)
Neuritis	Inflammation of a nerve
Neurotoxic	Poisonous or destructive to nerve tissue
Oliguria	Reduced daily output of urine
Opacity	The condition of being opaque
Opaque	Impervious to light rays
Pallor	Paleness
Palpitation	Unduly rapid beating of the heart
Papilledema	Edema of the optic papilla
Paralysis	Loss or impairment due to lesion of muscular mechanism
Petechia	A small area of discoloration formed by the effusion of blood
Pharyngitis	Inflammation of the pharynx
Pneumonitis	Localized inflammation of the lung
Polyneuritis	Inflammation of many nerves at once
Prostration	Extreme exhaustion or powerlessness
Psychic disturbances	Mental disturbances
Pulmonary edema	Fluid in the lungs
Pulmonary fibrosis	Fibrous tissues in the lungs
Pyorrhea	A discharge of pus, particularly from the gums
Retching	A strong involuntary effort to vomit
Salivation	An excessive discharge of saliva
Somnolence	Sleepiness
Stomatitis	Inflammation of the mouth
Stupor	Partial or nearly complete unconsciousness

Thiocyanate intoxication	Characterized by central nervous system stimulation and depression, and muscular incoordination
Tinnitus	Ringing in the ears
Ulceration	The formation of an ulcer
Vertigo	A sensation as if the external world were revolving around the patient or if he were revolving in space; the term is sometimes erroneously used as a synonym for dizziness

KEY TO ABBREVIATIONS

cc	cubic centimeter
CNS	central nervous system effects
gm	gram
H	hour
IRR	irritant effects
kg	kilogram (one thousand grams)
LC_{50}	lethal concentration 50 percent kill
LCLO	lowest published lethal concentration
LD_{50}	lethal dose 50 percent kill
LDLO	lowest published lethal dose
C.M.	cubic meter
mg	milligram (one thousandth of a gram; 10^{-3} gm)
ml	milliliter
ppm	parts per million (v/v)
TCLO	lowest published toxic concentration
TDLO	lowest published toxic dose
TLV	threshold limit value
TWA	time weighted average
W	week
Y	year

ACETAL

Hazard diamond: 3 (top), 2 (left), 4 (right)

SECTION I SOURCE AND NOMENCLATURE

MANUFACTURER'S NAME	EMERGENCY TELEPHONE NO.

ADDRESS (NUMBER, STREET, CITY, STATE, ZIP CODE)

TRADE NAME AND SYNONYMS	CHEMICAL FAMILY
Acetaldehydediethylacetal	
CHEMICAL NAME AND SYNONYMS	**FORMULA**
Acetal	$(C_2H_5O)_2CHCH_3$

SECTION II HAZARDOUS INGREDIENTS

BASIC MATERIAL	APPROXIMATE OR MAXIMUM % WT. OR VOL.	ESTABLISHED OSHA STANDARD	LD_{50} mg/kg ORAL	LD_{50} mg/kg PERCUT.	SPECIES	LC_{50} CONC.
			4570		Rat	
			2582		Rabbit	

Lowest published lethal concentration in air - inhalation - rat 4000 PPM/4H

Lowest published lethal dose oral - human - 500 mg/kg

SECTION III PHYSICAL DATA MW - 118.17

BOILING POINT (°F.): 102°F.	VAPOR PRESSURE (mm Hg.):
MELTING POINT (°F.):	VAPOR DENSITY (Air = 1): 4.1
SPECIFIC GRAVITY (H_2O = 1): 0.82	EVAPORATION RATE (____ = 1):
SOLUBILITY IN WATER Pts/100 pts H_2O: 6	VOLATILE % Vol. % Wt.

APPEARANCE AND ODOR: Liquid

SECTION IV FIRE AND EXPLOSION HAZARD DATA

FLASH POINT	FLAMMABLE (EXPLOSIVE) LIMITS	UPPER: 10.4%
-50 F.		
METHOD USED		LOWER: 1.6%

EXTINGUISHING MEDIA: Carbon dioxide, dry chemical, alcohol foam.

SPECIAL FIRE FIGHTING PROCEDURES: Water may be ineffective.

UNUSUAL FIRE AND EXPLOSION HAZARDS: Former explosive peroxides

SECTION V HEALTH HAZARD DATA

TOXIC LEVEL: None available — CARCINOGENIC: No

PRINCIPLE ROUTES OF ABSORBTION: Inhalation, ingestion — SKIN AND EYE IRRITATION: Slight

REVELANT SYMPTOMS OF EXPOSURE: Central nervous system depression.

EFFECTS OF CHRONIC EXPOSURE: Dermatitis

EMERGENCY AND FIRST AID PROCEDURES: Inhalation - rest; if swallowed - give emetic; get medical aid; wash skin & eyes.

SECTION VI REACTIVITY DATA

CONDITIONS CONTRIBUTING TO INSTABILITY: Reacts with oxygen to produce explosive peroxides shelf life - 12 months.

CONDITIONS CONTRIBUTING TO HAZARDOUS POLYMERIZATION:

INCOMPATABILITY (MATERIALS TO AVOID):

HAZARDOUS DECOMPOSITION PRODUCTS:

SECTION VII SPILL OR LEAK PROCEDURES

STEPS TO BE TAKEN IN CASE MATERIAL IS RELEASED OR SPILLED:

WASTE DISPOSAL METHOD:

SECTION VIII SPECIAL PROTECTION INFORMATION

VENTILATION REQUIREMENTS LOCAL EXHAUST	PROTECTIVE EQUIPMENT (SPECIFY TYPES) EYE: Safety glasses
MECHANICAL (GENERAL)	GLOVES
SPECIAL	RESPIRATOR: Canister type

OTHER PROTECTIVE EQUIPMENT:

SECTION IX SPECIAL PRECAUTIONS

PRECAUTIONS TO BE TAKEN IN HANDLING AND STORAGE:

OTHER PRECAUTIONS:

SIGNATURE	DATE

ACETAMIDE

Hazard diamond: 1 (top), 4 (left), 0 (right)

SECTION I SOURCE AND NOMENCLATURE

MANUFACTURER'S NAME	EMERGENCY TELEPHONE NO.

ADDRESS (NUMBER, STREET, CITY, STATE, ZIP CODE)

TRADE NAME AND SYNONYMS	CHEMICAL FAMILY
Ethanamide	
CHEMICAL NAME AND SYNONYMS	**FORMULA**
Acetic Acid Amide	CH_3CONH_2

SECTION II HAZARDOUS INGREDIENTS

BASIC MATERIAL	APPROXIMATE OR MAXIMUM % WT. OR VOL.	ESTABLISHED OSHA STANDARD	LD_{50} mg/kg ORAL	LD_{50} mg/kg PERCUT.	SPECIES	LC_{50} CONC.
				10,000	Rat	
				8,300	Mouse	

Note: Lowest published toxic dose, oral - rat 360 mg/kg, 152 weeks, carcinogen

SECTION III PHYSICAL DATA Mol. Wt. 59.07

BOILING POINT (°F.): 222°C	VAPOR PRESSURE (mm Hg.): at 65°C 1
MELTING POINT (°F.): 81°C	VAPOR DENSITY (Air = 1):
SPECIFIC GRAVITY (H_2O = 1): 1.159	EVAPORATION RATE (____ = 1):
SOLUBILITY IN WATER Pts/100 pts H_2O: 200	VOLATILE % Vol. % Wt.

APPEARANCE AND ODOR: White solid, odor, mouse nest

SECTION IV FIRE AND EXPLOSION HAZARD DATA

FLASH POINT	FLAMMABLE (EXPLOSIVE) LIMITS	UPPER
None available		
METHOD USED		LOWER

EXTINGUISHING MEDIA: Carbon dioxide, dry chemical

SPECIAL FIRE FIGHTING PROCEDURES:

UNUSUAL FIRE AND EXPLOSION HAZARDS:

SECTION V HEALTH HAZARD DATA

TOXIC LEVEL: Not available — CARCINOGENIC: Yes

PRINCIPLE ROUTES OF ABSORBTION: Ingestion, inhalation — SKIN AND EYE IRRITATION: No

REVELANT SYMPTOMS OF EXPOSURE: Drowsiness, fatigue, nausea, acidosis, skin eruptions.

EFFECTS OF CHRONIC EXPOSURE:

EMERGENCY AND FIRST AID PROCEDURES: If swallowed - wash mouth, induce vomiting; if inhaled - rest.

SECTION VI REACTIVITY DATA

CONDITIONS CONTRIBUTING TO INSTABILITY:

CONDITIONS CONTRIBUTING TO HAZARDOUS POLYMERIZATION:

INCOMPATABILITY (MATERIALS TO AVOID):

HAZARDOUS DECOMPOSITION PRODUCTS:

SECTION VII SPILL OR LEAK PROCEDURES

STEPS TO BE TAKEN IN CASE MATERIAL IS RELEASED OR SPILLED:

WASTE DISPOSAL METHOD:

SECTION VIII SPECIAL PROTECTION INFORMATION

VENTILATION REQUIREMENTS LOCAL EXHAUST	PROTECTIVE EQUIPMENT (SPECIFY TYPES) EYE: Goggles
MECHANICAL (GENERAL)	GLOVES: Rubber
SPECIAL	RESPIRATOR: With proper filter

OTHER PROTECTIVE EQUIPMENT:

SECTION IX SPECIAL PRECAUTIONS

PRECAUTIONS TO BE TAKEN IN HANDLING AND STORAGE:

OTHER PRECAUTIONS:

SIGNATURE	DATE

ACETONYLACETONE

Hazard diamond: 1 (top), 3 (left), 0 (right)

SECTION I SOURCE AND NOMENCLATURE

MANUFACTURER'S NAME | EMERGENCY TELEPHONE NO.

ADDRESS (NUMBER, STREET, CITY, STATE, ZIP CODE)

TRADE NAME AND SYNONYMS | CHEMICAL FAMILY

CHEMICAL NAME AND SYNONYMS: 2,5 Hexenedione

FORMULA: $(CH_2COCH_3)_2$

SECTION II HAZARDOUS INGREDIENTS

BASIC MATERIAL	APPROXIMATE OR MAXIMUM % WT. OR VOL.	ESTABLISHED OSHA STANDARD	LD_{50} mg/kg ORAL	PERCUT.	SPECIES	LC_{50} CONC.
			2700		Rat	

SECTION III PHYSICAL DATA MW-114.16

BOILING POINT: 192.4°C. °F.

VAPOR PRESSURE: at 25°C. 1.6 mm Hg.

MELTING POINT: -9°C. °F.

VAPOR DENSITY (Air = 1): 3.9

SPECIFIC GRAVITY (H_2O = 1): 0.97

EVAPORATION RATE (______ = 1)

SOLUBILITY IN WATER Pts/100 pts H_2O: Very soluble

VOLATILE % Vol. % Wt.

APPEARANCE AND ODOR: Colorless liquid

SECTION IV FIRE AND EXPLOSION HAZARD DATA

FLASH POINT: 174°F.

FLAMMABLE (EXPLOSIVE) LIMITS: UPPER / LOWER

METHOD USED: Closed cup

EXTINGUISHING MEDIA: Alcohol foam, carbon dioxide, dry chemical.

SPECIAL FIRE FIGHTING PROCEDURES

UNUSUAL FIRE AND EXPLOSION HAZARDS

SECTION V HEALTH HAZARD DATA

TOXIC LEVEL: MAC-75 PPM

CARCINOGENIC: No

PRINCIPLE ROUTES OF ABSORBTION: Inhalation, ingestion, skin

SKIN AND EYE IRRITATION: Yes

REVELANT SYMPTOMS OF EXPOSURE: Irritation of eyes and mucous membranes. Narcosis at higher concentrations.

EFFECTS OF CHRONIC EXPOSURE: --

EMERGENCY AND FIRST AID PROCEDURES: Inhalation - rest; if swallowed - induce vomiting, get medical aid; wash skin & eyes.

SECTION VI REACTIVITY DATA

CONDITIONS CONTRIBUTING TO INSTABILITY

CONDITIONS CONTRIBUTING TO HAZARDOUS POLYMERIZATION

INCOMPATABILITY (MATERIALS TO AVOID): can react with oxidizing materials.

HAZARDOUS DECOMPOSITION PRODUCTS

SECTION VII SPILL OR LEAK PROCEDURES

STEPS TO BE TAKEN IN CASE MATERIAL IS RELEASED OR SPILLED

WASTE DISPOSAL METHOD

SECTION VIII SPECIAL PROTECTION INFORMATION

VENTILATION REQUIREMENTS LOCAL EXHAUST

MECHANICAL (GENERAL)

SPECIAL

OTHER PROTECTIVE EQUIPMENT

PROTECTIVE EQUIPMENT (SPECIFY TYPES) EYE: Goggles

GLOVES: Rubber

RESPIRATOR: With chemical cartridge filter

SECTION IX SPECIAL PRECAUTIONS

PRECAUTIONS TO BE TAKEN IN HANDLING AND STORAGE: Store in well-ventilated area.

OTHER PRECAUTIONS

SIGNATURE | DATE

ACETYL METHYL AMINE

Hazard diamond: 1 (top), 4 (left), 0 (right)

SECTION I SOURCE AND NOMENCLATURE

MANUFACTURER'S NAME | EMERGENCY TELEPHONE NO.

ADDRESS (NUMBER, STREET, CITY, STATE, ZIP CODE)

TRADE NAME AND SYNONYMS: Methyl Acetamide

CHEMICAL FAMILY

CHEMICAL NAME AND SYNONYMS: Acetyl-Methyl amine

FORMULA: $CH_3CONHCH_3$

SECTION II HAZARDOUS INGREDIENTS

BASIC MATERIAL	APPROXIMATE OR MAXIMUM % WT. OR VOL.	ESTABLISHED OSHA STANDARD	LD_{50} ORAL	PERCUT.	LC_{50} SPECIES	CONC.
			None available			

SECTION III PHYSICAL DATA MW - 73.1

BOILING POINT: 206°C °F.

VAPOR PRESSURE: mm Hg.

MELTING POINT: 31°C °F.

VAPOR DENSITY (Air = 1)

SPECIFIC GRAVITY (H_2O = 1): Not available

EVAPORATION RATE (______ = 1)

SOLUBILITY IN WATER Pts/100 pts H_2O: Very soluble

VOLATILE % Vol. % Wt.

APPEARANCE AND ODOR: Colorless, needle-like crystals

SECTION IV FIRE AND EXPLOSION HAZARD DATA

FLASH POINT: None available

FLAMMABLE (EXPLOSIVE) LIMITS: UPPER / LOWER

METHOD USED: Not available

EXTINGUISHING MEDIA

SPECIAL FIRE FIGHTING PROCEDURES

UNUSUAL FIRE AND EXPLOSION HAZARDS

SECTION V HEALTH HAZARD DATA

TOXIC LEVEL: None available

CARCINOGENIC: No

PRINCIPLE ROUTES OF ABSORBTION: Inhalation, ingestion, skin

SKIN AND EYE IRRITATION: moderate

REVELANT SYMPTOMS OF EXPOSURE: Jaundice, drowsiness, fatigue, nausea

EFFECTS OF CHRONIC EXPOSURE: Liver damage

EMERGENCY AND FIRST AID PROCEDURES: If swallowed - induce vomiting, get medical aid; inhalation - rest; wash skin & eyes.

SECTION VI REACTIVITY DATA

CONDITIONS CONTRIBUTING TO INSTABILITY

CONDITIONS CONTRIBUTING TO HAZARDOUS POLYMERIZATION

INCOMPATABILITY (MATERIALS TO AVOID)

HAZARDOUS DECOMPOSITION PRODUCTS

SECTION VII SPILL OR LEAK PROCEDURES

STEPS TO BE TAKEN IN CASE MATERIAL IS RELEASED OR SPILLED

WASTE DISPOSAL METHOD

SECTION VIII SPECIAL PROTECTION INFORMATION

VENTILATION REQUIREMENTS LOCAL EXHAUST

MECHANICAL (GENERAL)

SPECIAL

OTHER PROTECTIVE EQUIPMENT

PROTECTIVE EQUIPMENT (SPECIFY TYPES) EYE: Goggles

GLOVES: Rubber gloves

RESPIRATOR: With chemical filter

SECTION IX SPECIAL PRECAUTIONS

PRECAUTIONS TO BE TAKEN IN HANDLING AND STORAGE

OTHER PRECAUTIONS

SIGNATURE | DATE

ACETYL METHYLUREA

1 / 3 / 0

SECTION I SOURCE AND NOMENCLATURE

MANUFACTURER'S NAME

EMERGENCY TELEPHONE NO.

ADDRESS (NUMBER, STREET, CITY, STATE, ZIP CODE)

TRADE NAME AND SYNONYMS

CHEMICAL FAMILY

CHEMICAL NAME AND SYNONYMS: Acetyl methylurea

FORMULA: $CH_3CONHCONHCH_3$

SECTION II HAZARDOUS INGREDIENTS

BASIC MATERIAL	APPROXIMATE OR MAXIMUM % WT. OR VOL.	ESTABLISHED OSHA STANDARD	LD_{50} ORAL	LD_{50} PERCUT.	LC_{50} SPECIES	LC_{50} CONC.
None available						

SECTION III PHYSICAL DATA MW - 116.12

BOILING POINT °F.: decomposes -

VAPOR PRESSURE mm Hg.

MELTING POINT °F.: 180°C.

VAPOR DENSITY (Air = 1)

SPECIFIC GRAVITY (H_2O = 1)

EVAPORATION RATE (______ = 1)

SOLUBILITY IN WATER Pts/100 pts H_2O: Soluble

VOLATILE % Vol. % Wt.

APPEARANCE AND ODOR: Crystal

SECTION IV FIRE AND EXPLOSION HAZARD DATA

FLASH POINT: None available

FLAMMABLE (EXPLOSIVE) LIMITS: UPPER / LOWER

METHOD USED

EXTINGUISHING MEDIA: Alcohol foam, carbon dioxide, dry chemical.

SPECIAL FIRE FIGHTING PROCEDURES

UNUSUAL FIRE AND EXPLOSION HAZARDS

SECTION V HEALTH HAZARD DATA

TOXIC LEVEL: None available

CARCINOGENIC: No

PRINCIPLE ROUTES OF ABSORBTION: Inhalation, ingestion

SKIN AND EYE IRRITATION: Slight

REVELANT SYMPTOMS OF EXPOSURE: Dermatitis

EFFECTS OF CHRONIC EXPOSURE: Depression of bone marrow, anemia, allergic skin eruptions.

EMERGENCY AND FIRST AID PROCEDURES: If swallowed - induce vomiting, get medical aid; inhalation - rest; wash from eyes & skin.

SECTION VI REACTIVITY DATA

CONDITIONS CONTRIBUTING TO INSTABILITY

CONDITIONS CONTRIBUTING TO HAZARDOUS POLYMERIZATION

INCOMPATABILITY (MATERIALS TO AVOID)

HAZARDOUS DECOMPOSITION PRODUCTS

SECTION VII SPILL OR LEAK PROCEDURES

STEPS TO BE TAKEN IN CASE MATERIAL IS RELEASED OR SPILLED: Cover with weak calcium hypochlorite solution, neutralize with dilute HCl after 24 hours and wash into drain.

WASTE DISPOSAL METHOD

SECTION VIII SPECIAL PROTECTION INFORMATION

VENTILATION REQUIREMENTS LOCAL EXHAUST

PROTECTIVE EQUIPMENT (SPECIFY TYPES) EYE: Goggles

MECHANICAL (GENERAL)

GLOVES: Rubber

SPECIAL

RESPIRATOR: With proper filter

OTHER PROTECTIVE EQUIPMENT

SECTION IX SPECIAL PRECAUTIONS

PRECAUTIONS TO BE TAKEN IN HANDLING AND STORAGE

OTHER PRECAUTIONS

SIGNATURE | DATE

ACETYL THIOUREA

1 / 3 / 0

SECTION I SOURCE AND NOMENCLATURE

MANUFACTURER'S NAME

EMERGENCY TELEPHONE NO.

ADDRESS (NUMBER, STREET, CITY, STATE, ZIP CODE)

TRADE NAME AND SYNONYMS

CHEMICAL FAMILY

CHEMICAL NAME AND SYNONYMS: 1,acetyl-2-thiourea

FORMULA: $CH_3CONHCSNH_2$

SECTION II HAZARDOUS INGREDIENTS

BASIC MATERIAL	APPROXIMATE OR MAXIMUM % WT. OR VOL.	ESTABLISHED OSHA STANDARD	LD_{50} Mg/kg ORAL	LD_{50} Mg/kg PERCUT.	LC_{50} SPECIES	LC_{50} CONC.
			50		Rat	

SECTION III PHYSICAL DATA MW-118.17

BOILING POINT °F.

VAPOR PRESSURE mm Hg.

MELTING POINT °F.: 167°C.

VAPOR DENSITY (Air = 1)

SPECIFIC GRAVITY (H_2O = 1)

EVAPORATION RATE (______ = 1)

SOLUBILITY IN WATER Pts/100 pts H_2O: Soluble

VOLATILE % Vol. % Wt.

APPEARANCE AND ODOR: Powder

SECTION IV FIRE AND EXPLOSION HAZARD DATA

FLASH POINT: None available

FLAMMABLE (EXPLOSIVE) LIMITS: UPPER / LOWER

METHOD USED

EXTINGUISHING MEDIA: Alcohol foam, carbon dioxide, dry chemical.

SPECIAL FIRE FIGHTING PROCEDURES

UNUSUAL FIRE AND EXPLOSION HAZARDS

SECTION V HEALTH HAZARD DATA

TOXIC LEVEL: None available

CARCINOGENIC: No

PRINCIPLE ROUTES OF ABSORBTION: Inhalation, ingestion

SKIN AND EYE IRRITATION: Slight

REVELANT SYMPTOMS OF EXPOSURE: Dermatitis

EFFECTS OF CHRONIC EXPOSURE: Depression of bone marrow, anemia, allergic skin eruptions.

EMERGENCY AND FIRST AID PROCEDURES: If swallowed - induce vomiting, get medical aid; wash from eyes.

SECTION VI REACTIVITY DATA

CONDITIONS CONTRIBUTING TO INSTABILITY

CONDITIONS CONTRIBUTING TO HAZARDOUS POLYMERIZATION

INCOMPATABILITY (MATERIALS TO AVOID)

HAZARDOUS DECOMPOSITION PRODUCTS

SECTION VII SPILL OR LEAK PROCEDURES

STEPS TO BE TAKEN IN CASE MATERIAL IS RELEASED OR SPILLED: Cover with weak solution calcium hypochlorite. Neutralize with dilute HCl and wash into drain.

WASTE DISPOSAL METHOD

SECTION VIII SPECIAL PROTECTION INFORMATION

VENTILATION REQUIREMENTS LOCAL EXHAUST

PROTECTIVE EQUIPMENT (SPECIFY TYPES) EYE: Goggles

MECHANICAL (GENERAL)

GLOVES: Rubber

SPECIAL

RESPIRATOR: With proper filter

OTHER PROTECTIVE EQUIPMENT

SECTION IX SPECIAL PRECAUTIONS

PRECAUTIONS TO BE TAKEN IN HANDLING AND STORAGE

OTHER PRECAUTIONS

SIGNATURE | DATE

ACETYL UREA

Hazard diamond: 1 (top), 3 (left), 0 (right)

SECTION I SOURCE AND NOMENCLATURE

MANUFACTURER'S NAME	EMERGENCY TELEPHONE NO.
ADDRESS (NUMBER, STREET, CITY, STATE, ZIP CODE)	
TRADE NAME AND SYNONYMS	CHEMICAL FAMILY
CHEMICAL NAME AND SYNONYMS Acetyl urea	FORMULA $CH_3CONHCONH_2$

SECTION II HAZARDOUS INGREDIENTS

BASIC MATERIAL	APPROXIMATE OR MAXIMUM % WT. OR VOL.	ESTABLISHED OSHA STANDARD	LD_{50} ORAL	LD_{50} PERCUT.	LC_{50} SPECIES	LC_{50} CONC.
			See	note		
Lowest published lethal dose						
Oral - rat 10 mg/Kg.						
Oral - mouse 40 mg/kg						

SECTION III PHYSICAL DATA MW - 102.09

BOILING POINT °F. Sublimes at 190°C	VAPOR PRESSURE mm Hg.
MELTING POINT °F. 218°C	VAPOR DENSITY (Air = 1)
SPECIFIC GRAVITY (H_2O = 1)	EVAPORATION RATE (______ = 1)
SOLUBILITY IN WATER Pts/100 pts H_2O 2	VOLATILE % Vol. % Wt.

APPEARANCE AND ODOR
Needle-like crystals

SECTION IV FIRE AND EXPLOSION HAZARD DATA

FLASH POINT None available	FLAMMABLE (EXPLOSIVE) LIMITS	UPPER
METHOD USED		LOWER

EXTINGUISHING MEDIA
Carbon dioxide, dry chemical

SPECIAL FIRE FIGHTING PROCEDURES

UNUSUAL FIRE AND EXPLOSION HAZARDS

SECTION V HEALTH HAZARD DATA

TOXIC LEVEL
None available — CARCINOGENIC No

PRINCIPLE ROUTES OF ABSORBTION
Inhalation, ingestion — SKIN AND EYE IRRITATION Slight

REVELANT SYMPTOMS OF EXPOSURE
Dermatitis

EFFECTS OF CHRONIC EXPOSURE Depression of bone marrow, anemia, allergic skin eruptions.

EMERGENCY AND FIRST AID PROCEDURES If swallowed - induce vomiting, get medical aid; inhalation - rest; wash from eyes.

SECTION VI REACTIVITY DATA

CONDITIONS CONTRIBUTING TO INSTABILITY

CONDITIONS CONTRIBUTING TO HAZARDOUS POLYMERIZATION

INCOMPATABILITY (MATERIALS TO AVOID)

HAZARDOUS DECOMPOSITION PRODUCTS

SECTION VII SPILL OR LEAK PROCEDURES

STEPS TO BE TAKEN IN CASE MATERIAL IS RELEASED OR SPILLED

WASTE DISPOSAL METHOD

SECTION VIII SPECIAL PROTECTION INFORMATION

VENTILATION REQUIREMENTS LOCAL EXHAUST	PROTECTIVE EQUIPMENT (SPECIFY TYPES) EYE Goggles
MECHANICAL (GENERAL)	GLOVES
SPECIAL	RESPIRATOR With proper filter
OTHER PROTECTIVE EQUIPMENT	

SECTION IX SPECIAL PRECAUTIONS

PRECAUTIONS TO BE TAKEN IN HANDLING AND STORAGE

OTHER PRECAUTIONS

SIGNATURE	DATE

ALLYLAMINE

Hazard diamond: 3 (top), 3 (left), 1 (right)

SECTION I SOURCE AND NOMENCLATURE

MANUFACTURER'S NAME	EMERGENCY TELEPHONE NO.
ADDRESS (NUMBER, STREET, CITY, STATE, ZIP CODE)	
TRADE NAME AND SYNONYMS	CHEMICAL FAMILY
CHEMICAL NAME AND SYNONYMS 2 Propenylamine, Allylamine	FORMULA $CH_2{=}CHCH_2NH_2$

SECTION II HAZARDOUS INGREDIENTS

BASIC MATERIAL	APPROXIMATE OR MAXIMUM % WT. OR VOL.	ESTABLISHED OSHA STANDARD	LD_{50} mg/kg ORAL	LD_{50} mg/kg PERCUT.	LC_{50} PPM/4H SPECIES	LC_{50} PPM/4H CONC.
TCLO - inhalation - man 5 ppm/5 min.					Rat	286
			106		Rat	
				35	Rabbit	
			57		Mouse	

SECTION III PHYSICAL DATA MW - 57.11

BOILING POINT °F. 53.2°C	VAPOR PRESSURE mm Hg.
MELTING POINT °F. -58°C.	VAPOR DENSITY (Air = 1) 2.00
SPECIFIC GRAVITY (H_2O = 1) 0.716	EVAPORATION RATE (______ = 1)
SOLUBILITY IN WATER Pts/100 pts H_2O ∞	VOLATILE % Vol. % Wt.

APPEARANCE AND ODOR
Colorless liquid, with strong ammonical odor.

SECTION IV FIRE AND EXPLOSION HAZARD DATA

FLASH POINT -20°F	FLAMMABLE (EXPLOSIVE) LIMITS	UPPER 22%
METHOD USED		LOWER 2.2%

EXTINGUISHING MEDIA
Alcohol foam, carbon dioxide, dry chemical, water spray

SPECIAL FIRE FIGHTING PROCEDURES
Water may be ineffective - wear full protective clothing.

UNUSUAL FIRE AND EXPLOSION HAZARDS
When heated to decomposition temperature, emits fumes.

SECTION V HEALTH HAZARD DATA

TOXIC LEVEL
Not available — CARCINOGENIC No

PRINCIPLE ROUTES OF ABSORBTION
Inhalation, ingestion, skin — SKIN AND EYE IRRITATION Severe

REVELANT SYMPTOMS OF EXPOSURE
Irritation of mucous membranes of eyes, nose and mouth

EFFECTS OF CHRONIC EXPOSURE
Liver and kidney damage

EMERGENCY AND FIRST AID PROCEDURES Wash from eyes and skin; if swallowed - wash mouth, drink water, get medical aid; inhalation - rest.

SECTION VI REACTIVITY DATA

CONDITIONS CONTRIBUTING TO INSTABILITY

CONDITIONS CONTRIBUTING TO HAZARDOUS POLYMERIZATION

INCOMPATABILITY (MATERIALS TO AVOID)
Reacts with acids.

HAZARDOUS DECOMPOSITION PRODUCTS

SECTION VII SPILL OR LEAK PROCEDURES

STEPS TO BE TAKEN IN CASE MATERIAL IS RELEASED OR SPILLED
Spread sodium bisulfate over area and sprinkle with water.

WASTE DISPOSAL METHOD
Dissolve in combustible solvent such as alcohol and burn in open furnace.

SECTION VIII SPECIAL PROTECTION INFORMATION

VENTILATION REQUIREMENTS LOCAL EXHAUST	PROTECTIVE EQUIPMENT (SPECIFY TYPES) EYE Safety goggles
MECHANICAL (GENERAL)	GLOVES Rubber
SPECIAL	RESPIRATOR With proper filter
OTHER PROTECTIVE EQUIPMENT Overalls	

SECTION IX SPECIAL PRECAUTIONS

PRECAUTIONS TO BE TAKEN IN HANDLING AND STORAGE
Protect against physical damage outside or detach storage preferred; inside storage in standard liquids storage cabinet.

OTHER PRECAUTIONS

SIGNATURE	DATE

p-AMINOAZOBENZENE

3 / 4 / 0

SECTION I SOURCE AND NOMENCLATURE

MANUFACTURER'S NAME | EMERGENCY TELEPHONE NO.

ADDRESS (NUMBER, STREET, CITY, STATE, ZIP CODE)

TRADE NAME AND SYNONYMS: P-Phenylazoaniline | CHEMICAL FAMILY

CHEMICAL NAME AND SYNONYMS: P-aminoazobenzene | FORMULA: $C_6H_5N_2C_6H_4.NH_2$

SECTION II HAZARDOUS INGREDIENTS

BASIC MATERIAL	APPROXIMATE OR MAXIMUM % WT. OR VOL.	ESTABLISHED OSHA STANDARD	LD_{50}		LC_{50}	
			ORAL	PERCUT.	SPECIES	CONC.
Lowest published toxic dose						
Oral - rat 128 g/Kg over 2 years carcinogen						
Skin - rat - 1550 mg/kg/97 weeks						

SECTION III PHYSICAL DATA MW 197.2

BOILING POINT °F: > 360°C. | VAPOR PRESSURE mm Hg: 225 120

MELTING POINT °F: 124-126°C. | VAPOR DENSITY (Air = 1)

SPECIFIC GRAVITY (H_2O = 1) | EVAPORATION RATE (____ = 1)

SOLUBILITY IN WATER Pts/100 pts H_2O: Slightly soluble | VOLATILE % Vol. % Wt.

APPEARANCE AND ODOR: Brownish-yellow needles

SECTION IV FIRE AND EXPLOSION HAZARD DATA

FLASH POINT: None available | FLAMMABLE (EXPLOSIVE) LIMITS | UPPER | LOWER

METHOD USED

EXTINGUISHING MEDIA: Water spray, foam, carbon dioxide.

SPECIAL FIRE FIGHTING PROCEDURES

UNUSUAL FIRE AND EXPLOSION HAZARDS: Vapor forms explosive mixture in air.

SECTION V HEALTH HAZARD DATA

TOXIC LEVEL: None available | CARCINOGENIC: Yes

PRINCIPLE ROUTES OF ABSORBTION: Inhalation, ingestion, skin | SKIN AND EYE IRRITATION: Slight

RELEVANT SYMPTOMS OF EXPOSURE: Cyanosis, headache, blood pressure drop, lethargy, jaundice and anemia.

EFFECTS OF CHRONIC EXPOSURE: Weight loss, anemia, weakness, irritability, liver damage, kidney irritation, liver cancer, kidney cancer.

EMERGENCY AND FIRST AID PROCEDURES: Inhalation - rest; wash eyes and skin; ingestion - remove by gastric lavage.

SECTION VI REACTIVITY DATA

CONDITIONS CONTRIBUTING TO INSTABILITY

CONDITIONS CONTRIBUTING TO HAZARDOUS POLYMERIZATION

INCOMPATABILITY (MATERIALS TO AVOID)

HAZARDOUS DECOMPOSITION PRODUCTS

SECTION VII SPILL OR LEAK PROCEDURES

STEPS TO BE TAKEN IN CASE MATERIAL IS RELEASED OR SPILLED

WASTE DISPOSAL METHOD

SECTION VIII SPECIAL PROTECTION INFORMATION

VENTILATION REQUIREMENTS LOCAL EXHAUST | PROTECTIVE EQUIPMENT (SPECIFY TYPES) EYE: Goggles

MECHANICAL (GENERAL) | GLOVES: Rubber

SPECIAL | RESPIRATOR: Chemical respirator

OTHER PROTECTIVE EQUIPMENT

SECTION IX SPECIAL PRECAUTIONS

PRECAUTIONS TO BE TAKEN IN HANDLING AND STORAGE

OTHER PRECAUTIONS

SIGNATURE | DATE

p-AMINOAZOBENZENE HYDROCHLORIDE

2 / 4 / 0

SECTION I SOURCE AND NOMENCLATURE

MANUFACTURER'S NAME | EMERGENCY TELEPHONE NO.

ADDRESS (NUMBER, STREET, CITY, STATE, ZIP CODE)

TRADE NAME AND SYNONYMS | CHEMICAL FAMILY

CHEMICAL NAME AND SYNONYMS: P-Aminoazobenzene Hydrochloride | FORMULA: $C_6H_5N{=}NC_6H_4NH_2HCl$

SECTION II HAZARDOUS INGREDIENTS

BASIC MATERIAL	APPROXIMATE OR MAXIMUM % WT. OR VOL.	ESTABLISHED OSHA STANDARD	LD_{50} mg/kg		LC_{50}	
			ORAL	PERCUT.	SPECIES	CONC.
			1250		Rat	
	None available					

SECTION III PHYSICAL DATA MW-233.7

BOILING POINT °F | VAPOR PRESSURE mm Hg

MELTING POINT °F: 228°C. | VAPOR DENSITY (Air = 1)

SPECIFIC GRAVITY (H_2O = 1) | EVAPORATION RATE (____ = 1)

SOLUBILITY IN WATER Pts/100 pts H_2O: Soluble | VOLATILE % Vol. % Wt.

APPEARANCE AND ODOR: Steel blue crystals

SECTION IV FIRE AND EXPLOSION HAZARD DATA

FLASH POINT: None available | FLAMMABLE (EXPLOSIVE) LIMITS | UPPER | LOWER

METHOD USED

EXTINGUISHING MEDIA: Water spray, foam, carbon dioxide

SPECIAL FIRE FIGHTING PROCEDURES

UNUSUAL FIRE AND EXPLOSION HAZARDS: Vapor forms explosive mixture in air.

SECTION V HEALTH HAZARD DATA

TOXIC LEVEL: None available | CARCINOGENIC: Yes

PRINCIPLE ROUTES OF ABSORBTION: Inhalation, ingestion, skin | SKIN AND EYE IRRITATION: Slight

RELEVANT SYMPTOMS OF EXPOSURE: Cyanosis, headache, blood pressure drop, lethargy, jaundice and anemia.

EFFECTS OF CHRONIC EXPOSURE: Liver cancer.

EMERGENCY AND FIRST AID PROCEDURES: Inhalation - rest; if swallowed - induce vomiting, get medical aid; wash skin & eyes.

SECTION VI REACTIVITY DATA

CONDITIONS CONTRIBUTING TO INSTABILITY

CONDITIONS CONTRIBUTING TO HAZARDOUS POLYMERIZATION

INCOMPATABILITY (MATERIALS TO AVOID)

HAZARDOUS DECOMPOSITION PRODUCTS

SECTION VII SPILL OR LEAK PROCEDURES

STEPS TO BE TAKEN IN CASE MATERIAL IS RELEASED OR SPILLED

WASTE DISPOSAL METHOD

SECTION VIII SPECIAL PROTECTION INFORMATION

VENTILATION REQUIREMENTS LOCAL EXHAUST | PROTECTIVE EQUIPMENT (SPECIFY TYPES) EYE: Safety glasses

MECHANICAL (GENERAL) | GLOVES: Rubber

SPECIAL | RESPIRATOR

OTHER PROTECTIVE EQUIPMENT

SECTION IX SPECIAL PRECAUTIONS

PRECAUTIONS TO BE TAKEN IN HANDLING AND STORAGE

OTHER PRECAUTIONS

SIGNATURE | DATE

2-AMINO-1,4-DIMETHYLBENZENE

Health 1 / Flammability 4 / Reactivity 0

SECTION I SOURCE AND NOMENCLATURE

MANUFACTURER'S NAME

EMERGENCY TELEPHONE NO.

ADDRESS (NUMBER, STREET, CITY, STATE, ZIP CODE)

TRADE NAME AND SYNONYMS: 2,5 dimethyl aniline

CHEMICAL FAMILY

CHEMICAL NAME AND SYNONYMS: 2,5 xylidine, 2-amino-1, 4-Dimethylbenzene

FORMULA: $(CH_3)_2C_6H_3NH_2$

SECTION II HAZARDOUS INGREDIENTS

BASIC MATERIAL	APPROXIMATE OR MAXIMUM % WT. OR VOL.	ESTABLISHED OSHA STANDARD	LD_{50} mg/Kg ORAL	PERCUT.	SPECIES	LC_{50} CONC.
			1297		Rat	
			841		Mouse	

SECTION III PHYSICAL DATA MW-121.18

BOILING POINT: 215°C

VAPOR PRESSURE (mm Hg.)

MELTING POINT: 16°C

VAPOR DENSITY (Air = 1)

SPECIFIC GRAVITY (H_2O = 1): 0.98

EVAPORATION RATE (= 1)

SOLUBILITY IN WATER (Pts/100 pts H_2O): Very slightly soluble

VOLATILE % Vol. % Wt.

APPEARANCE AND ODOR: Red to dark brown liquid

SECTION IV FIRE AND EXPLOSION HAZARD DATA

FLASH POINT: 206°F

FLAMMABLE (EXPLOSIVE) LIMITS: UPPER / LOWER

METHOD USED: C.C.

EXTINGUISHING MEDIA: Foam, carbon dioxide, dry chemical

SPECIAL FIRE FIGHTING PROCEDURES

UNUSUAL FIRE AND EXPLOSION HAZARDS

SECTION V HEALTH HAZARD DATA

TOXIC LEVEL: TLV 5 PPM, 25 mg/C.M.

CARCINOGENIC: No

PRINCIPLE ROUTES OF ABSORBTION: Inhalation, ingestion, skin

SKIN AND EYE IRRITATION: slight

REVELANT SYMPTOMS OF EXPOSURE: Headache, drowsiness, cyanosis, mental confusion, convulsions, central nervous system depression

EFFECTS OF CHRONIC EXPOSURE: Liver and kidney damage, fatigue, loss of appetite

EMERGENCY AND FIRST AID PROCEDURES: If swallowed - give emetic, get medical aid; inhalation - rest; wash from skin & eyes.

SECTION VI REACTIVITY DATA

CONDITIONS CONTRIBUTING TO INSTABILITY

CONDITIONS CONTRIBUTING TO HAZARDOUS POLYMERIZATION

INCOMPATABILITY (MATERIALS TO AVOID)

HAZARDOUS DECOMPOSITION PRODUCTS

SECTION VII SPILL OR LEAK PROCEDURES

STEPS TO BE TAKEN IN CASE MATERIAL IS RELEASED OR SPILLED

WASTE DISPOSAL METHOD

SECTION VIII SPECIAL PROTECTION INFORMATION

VENTILATION REQUIREMENTS LOCAL EXHAUST

PROTECTIVE EQUIPMENT (SPECIFY TYPES) EYE: Goggles

MECHANICAL (GENERAL)

GLOVES: Rubber

SPECIAL

RESPIRATOR: With proper filter

OTHER PROTECTIVE EQUIPMENT

SECTION IX SPECIAL PRECAUTIONS

PRECAUTIONS TO BE TAKEN IN HANDLING AND STORAGE

OTHER PRECAUTIONS

SIGNATURE

DATE

4-AMINO-1,3-DIMETHYLBENZENE

Health 1 / Flammability 4 / Reactivity 0

SECTION I SOURCE AND NOMENCLATURE

MANUFACTURER'S NAME

EMERGENCY TELEPHONE NO.

ADDRESS (NUMBER, STREET, CITY, STATE, ZIP CODE)

TRADE NAME AND SYNONYMS: 2,4 Xylidine

CHEMICAL FAMILY

CHEMICAL NAME AND SYNONYMS: 2,4 Dimethyl aniline, 4 amino-1, 3 Dimethylbenzene

FORMULA: $(CH_3)_2C_6H_3NH_2$

SECTION II HAZARDOUS INGREDIENTS

BASIC MATERIAL	APPROXIMATE OR MAXIMUM % WT. OR VOL.	ESTABLISHED OSHA STANDARD	LD_{50} mg/Kg ORAL	PERCUT.	SPECIES	LC_{50} CONC.
			467		Rat	
			250		Mouse	

SECTION III PHYSICAL DATA MW-121.18

BOILING POINT: 214°C

VAPOR PRESSURE (mm Hg.): 12 at 104°C

MELTING POINT

VAPOR DENSITY (Air = 1)

SPECIFIC GRAVITY (H_2O = 1): 0.98

EVAPORATION RATE (= 1)

SOLUBILITY IN WATER (Pts/100 pts H_2O): Very slightly soluble

VOLATILE % Vol. % Wt.

APPEARANCE AND ODOR: Red to dark brown liquid

SECTION IV FIRE AND EXPLOSION HAZARD DATA

FLASH POINT: 206°F

FLAMMABLE (EXPLOSIVE) LIMITS: UPPER / LOWER

METHOD USED: C.C.

EXTINGUISHING MEDIA: Foam, carbon dioxide, dry chemical

SPECIAL FIRE FIGHTING PROCEDURES

UNUSUAL FIRE AND EXPLOSION HAZARDS

SECTION V HEALTH HAZARD DATA

TOXIC LEVEL: TLV 5 PPM, 25 mg/C.M.

CARCINOGENIC: no

PRINCIPLE ROUTES OF ABSORBTION: Inhalation, ingestion, skin

SKIN AND EYE IRRITATION: slight

REVELANT SYMPTOMS OF EXPOSURE: Headache, drowsiness, cyanosis, mental confusion, convulsions, central nervous system depression

EFFECTS OF CHRONIC EXPOSURE: Liver and kidney damage, fatigue, loss of appetite

EMERGENCY AND FIRST AID PROCEDURES: If swallowed - give emetic, get medical aid; inhalation - rest; wash from skin and eyes.

SECTION VI REACTIVITY DATA

CONDITIONS CONTRIBUTING TO INSTABILITY

CONDITIONS CONTRIBUTING TO HAZARDOUS POLYMERIZATION

INCOMPATABILITY (MATERIALS TO AVOID): Forms explosive chloroamines on exposure to hypochlorite.

HAZARDOUS DECOMPOSITION PRODUCTS

SECTION VII SPILL OR LEAK PROCEDURES

STEPS TO BE TAKEN IN CASE MATERIAL IS RELEASED OR SPILLED

WASTE DISPOSAL METHOD

SECTION VIII SPECIAL PROTECTION INFORMATION

VENTILATION REQUIREMENTS LOCAL EXHAUST

PROTECTIVE EQUIPMENT (SPECIFY TYPES) EYE: Goggles

MECHANICAL (GENERAL)

GLOVES: Rubber

SPECIAL

RESPIRATOR: With proper filter

OTHER PROTECTIVE EQUIPMENT: Full protective clothing

SECTION IX SPECIAL PRECAUTIONS

PRECAUTIONS TO BE TAKEN IN HANDLING AND STORAGE: Protect against physical damage, store in cool, dry place.

OTHER PRECAUTIONS

SIGNATURE

DATE

5-AMINO-1,3-DIMETHYLBENZENE

1 4 0

SECTION I SOURCE AND NOMENCLATURE

MANUFACTURER'S NAME | EMERGENCY TELEPHONE NO.

ADDRESS (NUMBER, STREET, CITY, STATE, ZIP CODE)

TRADE NAME AND SYNONYMS: 3,5 Dimethylaniline | CHEMICAL FAMILY

CHEMICAL NAME AND SYNONYMS: 3,5 Xylidine, 5 amino-1, 3-Dimethylbenzene | FORMULA: $(CH_3)_2C_6H_3NH_2$

SECTION II HAZARDOUS INGREDIENTS

BASIC MATERIAL	APPROXIMATE OR MAXIMUM % WT. OR VOL.	ESTABLISHED OSHA STANDARD	LD_{50} mg/kg ORAL	PERCUT.	SPECIES	LC_{50} CONC.
			707		Rat	
			421		Mouse	

SECTION III PHYSICAL DATA MW-121.18

BOILING POINT °F.: 220°C | VAPOR PRESSURE mm Hg.: at 105°C 14

MELTING POINT °F. | VAPOR DENSITY (Air = 1)

SPECIFIC GRAVITY (H_2O = 1): 0.97 | EVAPORATION RATE (____ = 1)

SOLUBILITY IN WATER Pts/100 pts H_2O: Very slightly soluble | VOLATILE % Vol. % Wt.

APPEARANCE AND ODOR: Pale brown crystals

SECTION IV FIRE AND EXPLOSION HAZARD DATA

FLASH POINT: 206°F | FLAMMABLE (EXPLOSIVE) LIMITS UPPER / LOWER

METHOD USED: CC

EXTINGUISHING MEDIA: Foam, Carbon dioxide, dry chemical

SPECIAL FIRE FIGHTING PROCEDURES

UNUSUAL FIRE AND EXPLOSION HAZARDS

SECTION V HEALTH HAZARD DATA

TOXIC LEVEL: TLV 5 PPM, 25 mg/C.M. | CARCINOGENIC: no

PRINCIPLE ROUTES OF ABSORBTION: Inhalation, ingestion, skin | SKIN AND EYE IRRITATION: slight

REVELANT SYMPTOMS OF EXPOSURE: Headache, drowsiness, cyanosis, mental confusion, convulsions, central nervous system depression

EFFECTS OF CHRONIC EXPOSURE: Liver and kidney damage, fatigue, loss of appetite.

EMERGENCY AND FIRST AID PROCEDURES: If swallowed, give emetic, get medical aid; inhalation - rest; wash from skin & eyes.

SECTION VI REACTIVITY DATA

CONDITIONS CONTRIBUTING TO INSTABILITY

CONDITIONS CONTRIBUTING TO HAZARDOUS POLYMERIZATION

INCOMPATABILITY (MATERIALS TO AVOID): Forms explosive chloroamines on exposure to hypochlorites.

HAZARDOUS DECOMPOSITION PRODUCTS

SECTION VII SPILL OR LEAK PROCEDURES

STEPS TO BE TAKEN IN CASE MATERIAL IS RELEASED OR SPILLED

WASTE DISPOSAL METHOD

SECTION VIII SPECIAL PROTECTION INFORMATION

VENTILATION REQUIREMENTS LOCAL EXHAUST | PROTECTIVE EQUIPMENT (SPECIFY TYPES) EYE: Goggles

MECHANICAL (GENERAL) | GLOVES: Rubber

SPECIAL | RESPIRATOR: With proper filter

OTHER PROTECTIVE EQUIPMENT: Full protective clothing

SECTION IX SPECIAL PRECAUTIONS

PRECAUTIONS TO BE TAKEN IN HANDLING AND STORAGE: Protect against physical damage. Store in cool, dry place.

OTHER PRECAUTIONS

SIGNATURE | DATE

o-AMINODIPHENYL

1 4 0

SECTION I SOURCE AND NOMENCLATURE

MANUFACTURER'S NAME | EMERGENCY TELEPHONE NO.

ADDRESS (NUMBER, STREET, CITY, STATE, ZIP CODE)

TRADE NAME AND SYNONYMS: O-aminobiphenyl | CHEMICAL FAMILY

CHEMICAL NAME AND SYNONYMS: O-aminodiphenyl | FORMULA: $NH_2C_6H_4 - C_6H_5$

SECTION II HAZARDOUS INGREDIENTS

BASIC MATERIAL	APPROXIMATE OR MAXIMUM % WT. OR VOL.	ESTABLISHED OSHA STANDARD	LD_{50} mg/kg ORAL	PERCUT.	SPECIES	LC_{50} CONC.
			2340		Rat	
			1020		Rabbit	

SECTION III PHYSICAL DATA MW - 169.23

BOILING POINT °F.: 299°C | VAPOR PRESSURE mm Hg.

MELTING POINT °F.: 52°C | VAPOR DENSITY (Air = 1): 5.8

SPECIFIC GRAVITY (H_2O = 1) | EVAPORATION RATE (____ = 1)

SOLUBILITY IN WATER Pts/100 pts H_2O: Insoluble | VOLATILE % Vol. % Wt.

APPEARANCE AND ODOR: Leaf-like crystals

SECTION IV FIRE AND EXPLOSION HAZARD DATA

FLASH POINT: None available | FLAMMABLE (EXPLOSIVE) LIMITS UPPER / LOWER

METHOD USED

EXTINGUISHING MEDIA: Carbon-dioxide, dry chemical

SPECIAL FIRE FIGHTING PROCEDURES

UNUSUAL FIRE AND EXPLOSION HAZARDS

SECTION V HEALTH HAZARD DATA

TOXIC LEVEL: None available | CARCINOGENIC: Yes

PRINCIPLE ROUTES OF ABSORBTION: Inhalation, ingestion, skin | SKIN AND EYE IRRITATION: Severe

REVELANT SYMPTOMS OF EXPOSURE: Mucous membrane & skin irritation, cyanosis & jaundice.

EFFECTS OF CHRONIC EXPOSURE: Bladder cancer

EMERGENCY AND FIRST AID PROCEDURES: If swallowed - give emetic, get medical aid; inhalation - oxygen as needed; wash skin & eyes.

SECTION VI REACTIVITY DATA

CONDITIONS CONTRIBUTING TO INSTABILITY

CONDITIONS CONTRIBUTING TO HAZARDOUS POLYMERIZATION

INCOMPATABILITY (MATERIALS TO AVOID)

HAZARDOUS DECOMPOSITION PRODUCTS

SECTION VII SPILL OR LEAK PROCEDURES

STEPS TO BE TAKEN IN CASE MATERIAL IS RELEASED OR SPILLED

WASTE DISPOSAL METHOD

SECTION VIII SPECIAL PROTECTION INFORMATION

VENTILATION REQUIREMENTS LOCAL EXHAUST | PROTECTIVE EQUIPMENT (SPECIFY TYPES) EYE: Goggles

MECHANICAL (GENERAL) | GLOVES

SPECIAL | RESPIRATOR: With proper filter

OTHER PROTECTIVE EQUIPMENT

SECTION IX SPECIAL PRECAUTIONS

PRECAUTIONS TO BE TAKEN IN HANDLING AND STORAGE

OTHER PRECAUTIONS

SIGNATURE | DATE

2-AMINOPYRIDINE

4 1 0

SECTION I SOURCE AND NOMENCLATURE

MANUFACTURER'S NAME	EMERGENCY TELEPHONE NO.
ADDRESS (NUMBER, STREET, CITY, STATE, ZIP CODE)	
TRADE NAME AND SYNONYMS	CHEMICAL FAMILY
CHEMICAL NAME AND SYNONYMS: α aminopyridine	FORMULA: $C_5H_6N_2$

SECTION II HAZARDOUS INGREDIENTS

BASIC MATERIAL	APPROXIMATE OR MAXIMUM % WT. OR VOL.	ESTABLISHED OSHA STANDARD	LD_{50} mg/kg ORAL	LD_{50} PERCUT.	SPECIES	LC_{50} CONC.
TCLO inhalation - human	5 ppm/5 hrs.		200		Rat	

SECTION III PHYSICAL DATA MW - 94.11

BOILING POINT °F.: 210.6°C	VAPOR PRESSURE mm Hg.
MELTING POINT °F.: 58°C	VAPOR DENSITY (Air = 1)
SPECIFIC GRAVITY (H_2O = 1)	EVAPORATION RATE (= 1)
SOLUBILITY IN WATER Pts/100 pts. H_2O: 5	VOLATILE % Vol. % Wt.

APPEARANCE AND ODOR: White powder or crystals.

SECTION IV FIRE AND EXPLOSION HAZARD DATA

	FLAMMABLE (EXPLOSIVE) LIMITS	
FLASH POINT: N/A		UPPER
METHOD USED		LOWER

EXTINGUISHING MEDIA: Alcohol foam, carbon dioxide, dry chemical water may be ineffective.

SPECIAL FIRE FIGHTING PROCEDURES: Wear goggles and self-contained breathing apparatus.

UNUSUAL FIRE AND EXPLOSION HAZARDS: When heated to decomposition, it emits highly toxic fumes.

SECTION V HEALTH HAZARD DATA

TOXIC LEVEL: TLV - 0.5 PPM, 2 mg/C.M. CARCINOGENIC: No

PRINCIPLE ROUTES OF ABSORBTION: Skin, inhalation, ingestion SKIN AND EYE IRRITATION: Moderate

REVELANT SYMPTOMS OF EXPOSURE: Headache, increased blood pressure, nausea, convulsions, respiratory failure.

EFFECTS OF CHRONIC EXPOSURE: --

EMERGENCY AND FIRST AID PROCEDURES: Inhalation - remove from exposure, rest, artificial respiration as needed, get medical aid; ingestion - induce vomiting, get medical aid; wash from skin & eyes.

SECTION VI REACTIVITY DATA

CONDITIONS CONTRIBUTING TO INSTABILITY

CONDITIONS CONTRIBUTING TO HAZARDOUS POLYMERIZATION

INCOMPATABILITY (MATERIALS TO AVOID)

HAZARDOUS DECOMPOSITION PRODUCTS

SECTION VII SPILL OR LEAK PROCEDURES

STEPS TO BE TAKEN IN CASE MATERIAL IS RELEASED OR SPILLED: Cover with a 9/1 mixture of sand and soda ash; transfer into paper carton, burn in furnace.

WASTE DISPOSAL METHOD

SECTION VIII SPECIAL PROTECTION INFORMATION

VENTILATION REQUIREMENTS LOCAL EXHAUST	PROTECTIVE EQUIPMENT (SPECIFY TYPES) EYE: Goggles
MECHANICAL (GENERAL)	GLOVES: Rubber (butyl)
SPECIAL	RESPIRATOR: Self-contained respirator

OTHER PROTECTIVE EQUIPMENT: Plastic Coveralls

SECTION IX SPECIAL PRECAUTIONS

PRECAUTIONS TO BE TAKEN IN HANDLING AND STORAGE

OTHER PRECAUTIONS

SIGNATURE	DATE

AMMONIUM SULFIDE SOLUTION

3 2 0

SECTION I SOURCE AND NOMENCLATURE

MANUFACTURER'S NAME	EMERGENCY TELEPHONE NO.
ADDRESS (NUMBER, STREET, CITY, STATE, ZIP CODE)	
TRADE NAME AND SYNONYMS	CHEMICAL FAMILY
CHEMICAL NAME AND SYNONYMS: Ammonium Sulfide	FORMULA: $(NH_4)_2S$

SECTION II HAZARDOUS INGREDIENTS

BASIC MATERIAL	APPROXIMATE OR MAXIMUM % WT. OR VOL.	ESTABLISHED OSHA STANDARD	LD_{50} mg/kg ORAL	LD_{50} PERCUT.	SPECIES	LC_{50} CONC.
				132	mouse	
Lowest published lethal dose						
oral - mouse 80 mg/kg						
skin - mouse 245 mg/c.m.						

SECTION III PHYSICAL DATA MW-51.12

BOILING POINT °F.: N/A	VAPOR PRESSURE mm Hg.
MELTING POINT °F.: -18°C decomposes	VAPOR DENSITY (Air = 1): 1.2
SPECIFIC GRAVITY (H_2O = 1): N/A	EVAPORATION RATE (= 1)
SOLUBILITY IN WATER Pts/100 pts H_2O: Very soluble	VOLATILE % Vol. % Wt.

APPEARANCE AND ODOR: Clear solution with H_2S odor.

SECTION IV FIRE AND EXPLOSION HAZARD DATA

	FLAMMABLE (EXPLOSIVE) LIMITS	
FLASH POINT: 500° F. ignition temperature		UPPER 45%
METHOD USED		LOWER 4.3%

EXTINGUISHING MEDIA: Use water spray alcohol foam, carbon dioxide or dry chemical.

SPECIAL FIRE FIGHTING PROCEDURES: Wear goggles and self-contained breathing apparatus.

UNUSUAL FIRE AND EXPLOSION HAZARDS: Produces inflammable gas upon heating.

SECTION V HEALTH HAZARD DATA

TOXIC LEVEL: TLV - (10 PPM H_2S value) CARCINOGENIC: no

PRINCIPLE ROUTES OF ABSORBTION: Inhalation of vapor, ingestion of liquid. SKIN AND EYE IRRITATION: Yes

REVELANT SYMPTOMS OF EXPOSURE: Irritation of respiratory tract; headache, dizziness, pulmonary edema.

EFFECTS OF CHRONIC EXPOSURE: Low blood pressure, nausea, loss of appetite, weight loss, chronic cough.

EMERGENCY AND FIRST AID PROCEDURES: Remove from exposure, rest, keep warm, get medical aid, if swallowed - wash mouth, give emetic, get medical aid; wash from skin and eyes.

SECTION VI REACTIVITY DATA

CONDITIONS CONTRIBUTING TO INSTABILITY

CONDITIONS CONTRIBUTING TO HAZARDOUS POLYMERIZATION

INCOMPATABILITY (MATERIALS TO AVOID)

HAZARDOUS DECOMPOSITION PRODUCTS

SECTION VII SPILL OR LEAK PROCEDURES

STEPS TO BE TAKEN IN CASE MATERIAL IS RELEASED OR SPILLED

WASTE DISPOSAL METHOD

SECTION VIII SPECIAL PROTECTION INFORMATION

VENTILATION REQUIREMENTS LOCAL EXHAUST	PROTECTIVE EQUIPMENT (SPECIFY TYPES) EYE: Safety glasses
MECHANICAL (GENERAL)	GLOVES: Rubber gloves
SPECIAL	RESPIRATOR: Self-contained breathing apparatus.

OTHER PROTECTIVE EQUIPMENT

SECTION IX SPECIAL PRECAUTIONS

PRECAUTIONS TO BE TAKEN IN HANDLING AND STORAGE

OTHER PRECAUTIONS

SIGNATURE	DATE

AMMONIUM VANADATE

(Hazard diamond: 0 / 3 / 0)

SECTION I SOURCE AND NOMENCLATURE

MANUFACTURER'S NAME | EMERGENCY TELEPHONE NO.

ADDRESS (NUMBER, STREET, CITY, STATE, ZIP CODE)

TRADE NAME AND SYNONYMS | CHEMICAL FAMILY

CHEMICAL NAME AND SYNONYMS: Ammonium Vanadate | FORMULA: NH_4VO_3

SECTION II HAZARDOUS INGREDIENTS

BASIC MATERIAL	APPROXIMATE OR MAXIMUM % WT. OR VOL.	ESTABLISHED OSHA STANDARD	LD50 mg/kg ORAL	LD50 mg/kg PERCUT.	LD50 mg/kg SPECIES	LC50 CONC.
			18		Rat	

SECTION III PHYSICAL DATA MW - 116.99

BOILING POINT °F. | VAPOR PRESSURE mm Hg.

MELTING POINT °F.: d | VAPOR DENSITY (Air = 1)

SPECIFIC GRAVITY (H_2O = 1): 2.3 | EVAPORATION RATE (= 1)

SOLUBILITY IN WATER 44 Pts/100 pts H_2O at 18°C. | VOLATILE % Vol. % Wt.

APPEARANCE AND ODOR: Colored crystal

SECTION IV FIRE AND EXPLOSION HAZARD DATA

FLASH POINT: Not flammable | FLAMMABLE (EXPLOSIVE) LIMITS: UPPER / LOWER

METHOD USED

EXTINGUISHING MEDIA

SPECIAL FIRE FIGHTING PROCEDURES

UNUSUAL FIRE AND EXPLOSION HAZARDS

SECTION V HEALTH HAZARD DATA

TOXIC LEVEL: TLV - 0.5 mg/m³ (V_2O_5) | CARCINOGENIC: No

PRINCIPLE ROUTES OF ABSORBTION: Inhalation, ingestion | SKIN AND EYE IRRITATION: Yes

REVELANT SYMPTOMS OF EXPOSURE: Irritation of the respiratory system, vomiting, excessive salivation, diarrhea, nervous system damage.

EFFECTS OF CHRONIC EXPOSURE: Bronchial asthma

EMERGENCY AND FIRST AID PROCEDURES: Inhalation - rest, keep warm; if swallowed - wash mouth, give emetic, get medical attention; wash eyes with water.

SECTION VI REACTIVITY DATA

CONDITIONS CONTRIBUTING TO INSTABILITY

CONDITIONS CONTRIBUTING TO HAZARDOUS POLYMERIZATION

INCOMPATABILITY (MATERIALS TO AVOID)

HAZARDOUS DECOMPOSITION PRODUCTS

SECTION VII SPILL OR LEAK PROCEDURES

STEPS TO BE TAKEN IN CASE MATERIAL IS RELEASED OR SPILLED: Vacuum or wet clean; do not dry sweep.

WASTE DISPOSAL METHOD: In seated plastic bag.

SECTION VIII SPECIAL PROTECTION INFORMATION

VENTILATION REQUIREMENTS LOCAL EXHAUST | PROTECTIVE EQUIPMENT (SPECIFY TYPES) EYE: Safety glasses

MECHANICAL (GENERAL) | GLOVES

SPECIAL | RESPIRATOR: With appropriate filter

OTHER PROTECTIVE EQUIPMENT

SECTION IX SPECIAL PRECAUTIONS

PRECAUTIONS TO BE TAKEN IN HANDLING AND STORAGE

OTHER PRECAUTIONS

SIGNATURE | DATE

ANILINE HYDROCHLORIDE

(Hazard diamond: 1 / 4 / 0)

SECTION I SOURCE AND NOMENCLATURE

MANUFACTURER'S NAME | EMERGENCY TELEPHONE NO.

ADDRESS (NUMBER, STREET, CITY, STATE, ZIP CODE)

TRADE NAME AND SYNONYMS: Aniline Chloride | CHEMICAL FAMILY

CHEMICAL NAME AND SYNONYMS: Aniline Hydrochloride | FORMULA: $C_6H_5NH_2.H\ Cl$

SECTION II HAZARDOUS INGREDIENTS

BASIC MATERIAL	APPROXIMATE OR MAXIMUM % WT. OR VOL.	ESTABLISHED OSHA STANDARD	LD50 mg/kg ORAL	LD50 mg/kg PERCUT.	LD50 mg/kg SPECIES	LC50 CONC.
			1072		Rat	
			841		Mouse	

SECTION III PHYSICAL DATA MW-129.60

BOILING POINT °F.: 245°C. | VAPOR PRESSURE mm Hg.

MELTING POINT °F.: 198°C. | VAPOR DENSITY (Air = 1): 4.46

SPECIFIC GRAVITY (H_2O = 1): 1.22 | EVAPORATION RATE (= 1)

SOLUBILITY IN WATER 88 Pts/100 pts H_2O at 15°C. | VOLATILE % Vol. % Wt.

APPEARANCE AND ODOR: Clear crystal, turns brown when exposed to light.

SECTION IV FIRE AND EXPLOSION HAZARD DATA

FLASH POINT: 380°F. | FLAMMABLE (EXPLOSIVE) LIMITS: UPPER / LOWER

METHOD USED: O.C.

EXTINGUISHING MEDIA: Alcohol foam, dry chemical, carbon dioxide.

SPECIAL FIRE FIGHTING PROCEDURES: Self-contained gas mask, full protective clothing.

UNUSUAL FIRE AND EXPLOSION HAZARDS: Vapors of aniline and H Cl.

SECTION V HEALTH HAZARD DATA

TOXIC LEVEL: None available (skin) | CARCINOGENIC: No

PRINCIPLE ROUTES OF ABSORBTION: Ingestion, skin, inhalation | SKIN AND EYE IRRITATION: Severe

REVELANT SYMPTOMS OF EXPOSURE: Cyanosis, headache, blood pressure drop, lethargy, jaundice and anemia, pain in joints, fainting.

EFFECTS OF CHRONIC EXPOSURE: Weight loss, anemia, weakness & irritability, possible liver damage, inflammation of kidneys, ringing in ear.

EMERGENCY AND FIRST AID PROCEDURES: Remove from exposure; wash from eyes and skin; if ingested - remove by gastric lavage.

SECTION VI REACTIVITY DATA

CONDITIONS CONTRIBUTING TO INSTABILITY

CONDITIONS CONTRIBUTING TO HAZARDOUS POLYMERIZATION

INCOMPATABILITY (MATERIALS TO AVOID): Ozone, perchloric acid, sodium peroxide, potassium peroxide.

HAZARDOUS DECOMPOSITION PRODUCTS

SECTION VII SPILL OR LEAK PROCEDURES

STEPS TO BE TAKEN IN CASE MATERIAL IS RELEASED OR SPILLED: Wear protective clothing during clean ups.

WASTE DISPOSAL METHOD

SECTION VIII SPECIAL PROTECTION INFORMATION

VENTILATION REQUIREMENTS LOCAL EXHAUST | PROTECTIVE EQUIPMENT (SPECIFY TYPES) EYE: Safety goggles

MECHANICAL (GENERAL) | GLOVES: Rubber

SPECIAL | RESPIRATOR: Self-contained

OTHER PROTECTIVE EQUIPMENT: Full protective clothing.

SECTION IX SPECIAL PRECAUTIONS

PRECAUTIONS TO BE TAKEN IN HANDLING AND STORAGE: Preferred outside storage in solvent storage cabinet.

OTHER PRECAUTIONS

SIGNATURE | DATE

ANILINE SULFATE

Hazard diamond: 1 (top), 4 (left), 0 (right)

SECTION I SOURCE AND NOMENCLATURE

MANUFACTURER'S NAME	EMERGENCY TELEPHONE NO.
ADDRESS (NUMBER, STREET, CITY, STATE, ZIP CODE)	
TRADE NAME AND SYNONYMS	CHEMICAL FAMILY
CHEMICAL NAME AND SYNONYMS Aniline Sulfate	FORMULA $(C_6H_5NH_2)_2H_2SO_4$

SECTION II HAZARDOUS INGREDIENTS

BASIC MATERIAL	APPROXIMATE OR MAXIMUM % WT. OR VOL.	ESTABLISHED OSHA STANDARD	LD_{50} ORAL	LD_{50} PERCUT.	LC_{50} SPECIES	LC_{50} CONC.
			None	available		

SECTION III PHYSICAL DATA MW - 284.32

BOILING POINT °F.	VAPOR PRESSURE mm Hg.
MELTING POINT °F. decomposes	VAPOR DENSITY (Air = 1)
SPECIFIC GRAVITY (H_2O = 1) 1.38	EVAPORATION RATE (= 1)
SOLUBILITY IN WATER 5 Pts/100 pts H_2O at 14°C.	VOLATILE % Vol. % Wt.
APPEARANCE AND ODOR White crystals	

SECTION IV FIRE AND EXPLOSION HAZARD DATA

FLASH POINT None available	FLAMMABLE (EXPLOSIVE) LIMITS	UPPER
METHOD USED		LOWER

EXTINGUISHING MEDIA
Alcohol foam, carbon dioxide, dry chemical.

SPECIAL FIRE FIGHTING PROCEDURES
Self-contained mask, full protective clothing.

UNUSUAL FIRE AND EXPLOSION HAZARDS
Vapors of aniline & SO_2

SECTION V HEALTH HAZARD DATA

TOXIC LEVEL
None available (skin) — CARCINOGENIC No

PRINCIPLE ROUTES OF ABSORPTION
Inhalation, ingestion, skin — SKIN AND EYE IRRITATION Severe

REVELANT SYMPTOMS OF EXPOSURE Cyanosis, headache, blood pressure drop, lethargy, jaundice and anemia, pain in joints, fainting.

EFFECTS OF CHRONIC EXPOSURE Weight loss, anemia, weakness and irritability, possible liver damage, inflammation of kidneys, ringing in ears.

EMERGENCY AND FIRST AID PROCEDURES Remove from exposure; wash from skin & eyes; if swallowed - remove by gastric lavage.

SECTION VI REACTIVITY DATA

CONDITIONS CONTRIBUTING TO INSTABILITY

CONDITIONS CONTRIBUTING TO HAZARDOUS POLYMERIZATION

INCOMPATABILITY (MATERIALS TO AVOID)
Ozone, perchloric acid, sodium peroxide, potassium peroxide.

HAZARDOUS DECOMPOSITION PRODUCTS

SECTION VII SPILL OR LEAK PROCEDURES

STEPS TO BE TAKEN IN CASE MATERIAL IS RELEASED OR SPILLED
Wear full, protective clothing during cleanup.

WASTE DISPOSAL METHOD

SECTION VIII SPECIAL PROTECTION INFORMATION

VENTILATION REQUIREMENTS LOCAL EXHAUST	PROTECTIVE EQUIPMENT (SPECIFY TYPES) EYE Safety goggles
MECHANICAL (GENERAL)	GLOVES Rubber
SPECIAL	RESPIRATOR Self-contained

OTHER PROTECTIVE EQUIPMENT
Wear protective clothing.

SECTION IX SPECIAL PRECAUTIONS

PRECAUTIONS TO BE TAKEN IN HANDLING AND STORAGE
Preferred outside storage in solvent storage cabinet.

OTHER PRECAUTIONS

SIGNATURE	DATE

o-ANISIDINE

Hazard diamond: 2 (top), 4 (left), 0 (right)

SECTION I SOURCE AND NOMENCLATURE

MANUFACTURER'S NAME	EMERGENCY TELEPHONE NO.
ADDRESS (NUMBER, STREET, CITY, STATE, ZIP CODE)	
TRADE NAME AND SYNONYMS	CHEMICAL FAMILY
CHEMICAL NAME AND SYNONYMS Amino-Anisole, 2 Methoxyaniline	FORMULA $NH_2C_6H_4OCH_3$

SECTION II HAZARDOUS INGREDIENTS

BASIC MATERIAL	APPROXIMATE OR MAXIMUM % WT. OR VOL.	ESTABLISHED OSHA STANDARD	LD_{50} ORAL	LD_{50} PERCUT.	LC_{50} SPECIES	LC_{50} CONC.

Note: Lowest published toxic concentration in air, inhalation - human 1.9 mg/C.M., 24 weeks; central nervous system effects.

SECTION III PHYSICAL DATA MW - 123.17

BOILING POINT °F. 225°C	VAPOR PRESSURE mm Hg.
MELTING POINT °F. 5.2°C	VAPOR DENSITY (Air = 1)
SPECIFIC GRAVITY (H_2O = 1) 1.098	EVAPORATION RATE (= 1)
SOLUBILITY IN WATER Pts/100 pts H_2O Slightly soluble	VOLATILE % Vol. % Wt.
APPEARANCE AND ODOR Pale yellow to orange liquid	

SECTION IV FIRE AND EXPLOSION HAZARD DATA

FLASH POINT N/A	FLAMMABLE (EXPLOSIVE) LIMITS	UPPER
METHOD USED		LOWER

EXTINGUISHING MEDIA
Similar to aniline; water spray, alcohol foam, dry chemical

SPECIAL FIRE FIGHTING PROCEDURES

UNUSUAL FIRE AND EXPLOSION HAZARDS
Emits toxic fumes, when heated to the decomposition point.

SECTION V HEALTH HAZARD DATA

TOXIC LEVEL
TLV 0.1 ppm, 0.5 mg/C.M. — CARCINOGENIC No

PRINCIPLE ROUTES OF ABSORPTION
Inhalation, skin absorption — SKIN AND EYE IRRITATION Yes

REVELANT SYMPTOMS OF EXPOSURE
Headache, drowsiness, cyanosis, skin irritant

EFFECTS OF CHRONIC EXPOSURE

EMERGENCY AND FIRST AID PROCEDURES
Wash skin with soap and water, flush eyes with water; if swallowed, wash mouth with water and get medical attention.

SECTION VI REACTIVITY DATA

CONDITIONS CONTRIBUTING TO INSTABILITY

CONDITIONS CONTRIBUTING TO HAZARDOUS POLYMERIZATION

INCOMPATABILITY (MATERIALS TO AVOID)

HAZARDOUS DECOMPOSITION PRODUCTS

SECTION VII SPILL OR LEAK PROCEDURES

STEPS TO BE TAKEN IN CASE MATERIAL IS RELEASED OR SPILLED

WASTE DISPOSAL METHOD

SECTION VIII SPECIAL PROTECTION INFORMATION

VENTILATION REQUIREMENTS LOCAL EXHAUST Use in hood	PROTECTIVE EQUIPMENT (SPECIFY TYPES) EYE Safety goggles
MECHANICAL (GENERAL)	GLOVES Rubber gloves (butyl)
SPECIAL	RESPIRATOR Self-contained breathing apparatus

OTHER PROTECTIVE EQUIPMENT
Neoprene protective clothing

SECTION IX SPECIAL PRECAUTIONS

PRECAUTIONS TO BE TAKEN IN HANDLING AND STORAGE

OTHER PRECAUTIONS

SIGNATURE	DATE

p-ANISIDINE

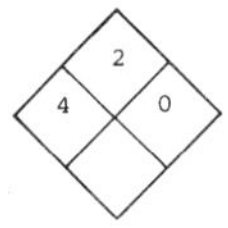

SECTION I SOURCE AND NOMENCLATURE

MANUFACTURER'S NAME | EMERGENCY TELEPHONE NO.

ADDRESS (NUMBER, STREET, CITY, STATE, ZIP CODE)

TRADE NAME AND SYNONYMS | CHEMICAL FAMILY

CHEMICAL NAME AND SYNONYMS: Amino-Anisoles | FORMULA: $NH_2C_6H_4OCH_3$

SECTION II HAZARDOUS INGREDIENTS

BASIC MATERIAL	APPROXIMATE OR MAXIMUM % WT. OR VOL.	ESTABLISHED OSHA STANDARD	LD_{50} mg/kg ORAL	LD_{50} mg/kg PERCUT.	LD_{50} mg/kg SPECIES	LC_{50} CONC.
			~~1400~~	~~3200~~	~~Rat~~	

Note: Lowest published toxic concentration in air, inhalation - human 1.9 mg/C.M., 24 weeks; central nervous system effects.

SECTION III PHYSICAL DATA

BOILING POINT °F.: 243

VAPOR PRESSURE mm Hg.:

MELTING POINT °F.: 57.2

VAPOR DENSITY (Air = 1):

SPECIFIC GRAVITY (H_2O = 1): 1.089

EVAPORATION RATE (______ = 1):

SOLUBILITY IN WATER Pts/100 pts H_2O: Slightly soluble

VOLATILE % Vol. % Wt.:

APPEARANCE AND ODOR: Pale yellow to brown solid

SECTION IV FIRE AND EXPLOSION HAZARD DATA

FLASH POINT: Not available

METHOD USED:

FLAMMABLE (EXPLOSIVE) LIMITS: UPPER / LOWER

EXTINGUISHING MEDIA: Similar to aniline; water spray, alcohol foam, dry chemical

SPECIAL FIRE FIGHTING PROCEDURES:

UNUSUAL FIRE AND EXPLOSION HAZARDS: Emits toxic fumes when heated to decomposition point.

SECTION V HEALTH HAZARD DATA

TOXIC LEVEL: TLV 0.1 ppm, 0.5 mg/C.M. — CARCINOGENIC: No

PRINCIPLE ROUTES OF ABSORBTION: Inhalation, skin absorption — SKIN AND EYE IRRITATION: Yes

REVELANT SYMPTOMS OF EXPOSURE: Headache, drowsiness, cyanosis, skin irritant

EFFECTS OF CHRONIC EXPOSURE:

EMERGENCY AND FIRST AID PROCEDURES: Wash skin with soap and water, flush eyes with water, if swallowed, wash mouth with water and get medical attention.

SECTION VI REACTIVITY DATA

CONDITIONS CONTRIBUTING TO INSTABILITY:

CONDITIONS CONTRIBUTING TO HAZARDOUS POLYMERIZATION:

INCOMPATABILITY (MATERIALS TO AVOID):

HAZARDOUS DECOMPOSITION PRODUCTS:

SECTION VII SPILL OR LEAK PROCEDURES

STEPS TO BE TAKEN IN CASE MATERIAL IS RELEASED OR SPILLED:

WASTE DISPOSAL METHOD:

SECTION VIII SPECIAL PROTECTION INFORMATION

VENTILATION REQUIREMENTS LOCAL EXHAUST: Use in hood

PROTECTIVE EQUIPMENT (SPECIFY TYPES) EYE: Safety goggles

MECHANICAL (GENERAL):

GLOVES: Rubber gloves

SPECIAL:

RESPIRATOR: Self-contained breathing apparatus

OTHER PROTECTIVE EQUIPMENT: Neoprene protective clothing

SECTION IX SPECIAL PRECAUTIONS

PRECAUTIONS TO BE TAKEN IN HANDLING AND STORAGE:

OTHER PRECAUTIONS:

SIGNATURE | DATE

ANTIMONY

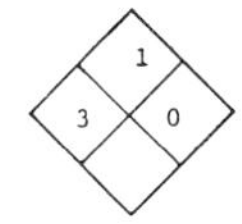

SECTION I SOURCE AND NOMENCLATURE

MANUFACTURER'S NAME | EMERGENCY TELEPHONE NO.

ADDRESS (NUMBER, STREET, CITY, STATE, ZIP CODE)

TRADE NAME AND SYNONYMS | CHEMICAL FAMILY

CHEMICAL NAME AND SYNONYMS: Antimony | FORMULA: Sb

SECTION II HAZARDOUS INGREDIENTS

BASIC MATERIAL	APPROXIMATE OR MAXIMUM % WT. OR VOL.	ESTABLISHED OSHA STANDARD	LD_{50} mg/kg ORAL	LD_{50} mg/kg PERCUT.	LC_{50} SPECIES	LC_{50} CONC.
#1 Lowest published toxic concentration in air - inhalation - human 4.7 mg/C.M.			100		Rat	
			See Note 1			

SECTION III PHYSICAL DATA Aw. 121.76

BOILING POINT °F.: 1380°C.

VAPOR PRESSURE mm Hg.: 1 at 886°C.

MELTING POINT °F.: 630°

VAPOR DENSITY (Air = 1):

SPECIFIC GRAVITY (H_2O = 1): 6.684 at 25°C.

EVAPORATION RATE (______ = 1):

SOLUBILITY IN WATER Pts/100 pts H_2O: Insoluble

VOLATILE % Vol. % Wt.:

APPEARANCE AND ODOR: Grey metal (powder).

SECTION IV FIRE AND EXPLOSION HAZARD DATA

FLASH POINT: None available

METHOD USED:

FLAMMABLE (EXPLOSIVE) LIMITS: UPPER / LOWER

EXTINGUISHING MEDIA: Carbon dioxide, dry chemicals.

SPECIAL FIRE FIGHTING PROCEDURES:

UNUSUAL FIRE AND EXPLOSION HAZARDS:

SECTION V HEALTH HAZARD DATA

TOXIC LEVEL: TLV-TWA 0.5 mg/m³ — CARCINOGENIC: No

PRINCIPLE ROUTES OF ABSORBTION: Inhalation — SKIN AND EYE IRRITATION: Skin

REVELANT SYMPTOMS OF EXPOSURE: Acute pneumonitis, nausea, vomiting

EFFECTS OF CHRONIC EXPOSURE: Bleeding gums, conjunctivitis, laryngitis, headache, weight loss and anemia

EMERGENCY AND FIRST AID PROCEDURES: Remove from exposure - if ingested give gastric lavage, emesis and catharsis. (Give dimercaprol). Administer oxygen if necessary.

SECTION VI REACTIVITY DATA

CONDITIONS CONTRIBUTING TO INSTABILITY:

CONDITIONS CONTRIBUTING TO HAZARDOUS POLYMERIZATION:

INCOMPATABILITY (MATERIALS TO AVOID): Oxidizing materials

HAZARDOUS DECOMPOSITION PRODUCTS:

SECTION VII SPILL OR LEAK PROCEDURES

STEPS TO BE TAKEN IN CASE MATERIAL IS RELEASED OR SPILLED:

WASTE DISPOSAL METHOD:

SECTION VIII SPECIAL PROTECTION INFORMATION

VENTILATION REQUIREMENTS LOCAL EXHAUST:

PROTECTIVE EQUIPMENT (SPECIFY TYPES) EYE: Goggles or safety glasses

MECHANICAL (GENERAL):

GLOVES: Rubber

SPECIAL:

RESPIRATOR: With appropriate filter

OTHER PROTECTIVE EQUIPMENT: Rubber apron

SECTION IX SPECIAL PRECAUTIONS

PRECAUTIONS TO BE TAKEN IN HANDLING AND STORAGE:

OTHER PRECAUTIONS:

SIGNATURE | DATE

p-ARSANILIC ACID

Hazard diamond: 2 (top), 3 (left), 0 (right)

SECTION I SOURCE AND NOMENCLATURE

MANUFACTURER'S NAME | EMERGENCY TELEPHONE NO.

ADDRESS (NUMBER, STREET, CITY, STATE, ZIP CODE)

TRADE NAME AND SYNONYMS | CHEMICAL FAMILY

CHEMICAL NAME AND SYNONYMS: P-Aminobenzenearsonic acid

FORMULA: $NH_2C_6H_4AsO(OH)_2$

SECTION II HAZARDOUS INGREDIENTS

BASIC MATERIAL	APPROXIMATE OR MAXIMUM % WT. OR VOL.	ESTABLISHED OSHA STANDARD	LD_{50} mg/kg ORAL	PERCUT.	SPECIES	LC_{50} CONC.
			216		Rat	

SECTION III PHYSICAL DATA Mol. Wt. 217.0

BOILING POINT °F.: N/A

VAPOR PRESSURE mm Hg.

MELTING POINT °F.: 232°C

VAPOR DENSITY (Air = 1)

SPECIFIC GRAVITY (H_2O = 1): N/A

EVAPORATION RATE (_____ = 1)

SOLUBILITY IN WATER Pts/100 pts H_2O

VOLATILE % Vol. % Wt.

APPEARANCE AND ODOR: White crystaline powder

SECTION IV FIRE AND EXPLOSION HAZARD DATA

FLASH POINT: N/A

FLAMMABLE (EXPLOSIVE) LIMITS: UPPER / LOWER

METHOD USED: N/A

EXTINGUISHING MEDIA: N/A

SPECIAL FIRE FIGHTING PROCEDURES: N/A

UNUSUAL FIRE AND EXPLOSION HAZARDS: Decomposed by heat to yield flammable vapors

SECTION V HEALTH HAZARD DATA

TOXIC LEVEL: TLV 0.72 mg/C.M.

CARCINOGENIC: No

PRINCIPLE ROUTES OF ABSORBTION: Inhalation, ingestion, skin absorption

SKIN AND EYE IRRITATION: Moderate

REVELANT SYMPTOMS OF EXPOSURE: Burning of mouth and throat, gastroenteric pain, vomiting, diarrhea, hematuris, dehydration, oliguria and collapse

EFFECTS OF CHRONIC EXPOSURE: Arsenic poisoning and skin ulceration, falling out of hair, loss of appetite

EMERGENCY AND FIRST AID PROCEDURES: Inhalation - rest, get medical attention; if swallowed - wash mouth, gastric lavage with 1% sodium thiosulfate; wash skin and eyes, get medical attention.

SECTION VI REACTIVITY DATA

CONDITIONS CONTRIBUTING TO INSTABILITY

CONDITIONS CONTRIBUTING TO HAZARDOUS POLYMERIZATION

INCOMPATABILITY (MATERIALS TO AVOID)

HAZARDOUS DECOMPOSITION PRODUCTS

SECTION VII SPILL OR LEAK PROCEDURES

STEPS TO BE TAKEN IN CASE MATERIAL IS RELEASED OR SPILLED

WASTE DISPOSAL METHOD

SECTION VIII SPECIAL PROTECTION INFORMATION

VENTILATION REQUIREMENTS LOCAL EXHAUST

PROTECTIVE EQUIPMENT (SPECIFY TYPES) EYE: Goggles

MECHANICAL (GENERAL)

GLOVES: Rubber

SPECIAL

RESPIRATOR: with appropriate filter

OTHER PROTECTIVE EQUIPMENT

SECTION IX SPECIAL PRECAUTIONS

PRECAUTIONS TO BE TAKEN IN HANDLING AND STORAGE

OTHER PRECAUTIONS

SIGNATURE | DATE

ARSENIC METAL

Hazard diamond: 0 (top), 4 (left), 0 (right)

SECTION I SOURCE AND NOMENCLATURE

MANUFACTURER'S NAME | EMERGENCY TELEPHONE NO.

ADDRESS (NUMBER, STREET, CITY, STATE, ZIP CODE)

TRADE NAME AND SYNONYMS | CHEMICAL FAMILY

CHEMICAL NAME AND SYNONYMS

FORMULA: As

SECTION II HAZARDOUS INGREDIENTS

BASIC MATERIAL	APPROXIMATE OR MAXIMUM % WT. OR VOL.	ESTABLISHED OSHA STANDARD	LD_{50} ORAL	PERCUT.	SPECIES	LC_{50} CONC.

Lowest published lethal dose oral-human 5 mg/kg

intramuscular - rat - 25 mg/kg.

subcutaneous - rabbit 300 mg/kg

subcutaneous - guinea pig 300 mg/kg

SECTION III PHYSICAL DATA MW 74.92

BOILING POINT °F.: subl. 613°C.

VAPOR PRESSURE mm Hg.

MELTING POINT °F.: 817°C.

VAPOR DENSITY (Air = 1)

SPECIFIC GRAVITY (H_2O = 1): 5.727

EVAPORATION RATE (_____ = 1)

SOLUBILITY IN WATER Pts/100 pts H_2O

VOLATILE % Vol. % Wt.

APPEARANCE AND ODOR: Grey metal

SECTION IV FIRE AND EXPLOSION HAZARD DATA

FLASH POINT: None available

FLAMMABLE (EXPLOSIVE) LIMITS: UPPER / LOWER

METHOD USED

EXTINGUISHING MEDIA: Carbon dioxide

SPECIAL FIRE FIGHTING PROCEDURES

UNUSUAL FIRE AND EXPLOSION HAZARDS

SECTION V HEALTH HAZARD DATA

TOXIC LEVEL: TLV 0.5 mg/m^3 Teratogen

CARCINOGENIC: Yes

PRINCIPLE ROUTES OF ABSORBTION: Ingestion, inhalation, skin

SKIN AND EYE IRRITATION: Moderate

REVELANT SYMPTOMS OF EXPOSURE: Ingestion: vomiting, headache, dizziness. Inhalation: pulmonary edema, cyanosis, cough.

EFFECTS OF CHRONIC EXPOSURE: Anemia, weight loss, lung cancer, lymphatic cancer.

EMERGENCY AND FIRST AID PROCEDURES: Inhalation - rest; if swallowed - give emetic, get medical aid; wash from skin and eyes.

SECTION VI REACTIVITY DATA

CONDITIONS CONTRIBUTING TO INSTABILITY

CONDITIONS CONTRIBUTING TO HAZARDOUS POLYMERIZATION

INCOMPATABILITY (MATERIALS TO AVOID): Avoid bromates, chlorates and iodates, also fluorides, nitrates and peroxides.

HAZARDOUS DECOMPOSITION PRODUCTS

SECTION VII SPILL OR LEAK PROCEDURES

STEPS TO BE TAKEN IN CASE MATERIAL IS RELEASED OR SPILLED

WASTE DISPOSAL METHOD

SECTION VIII SPECIAL PROTECTION INFORMATION

VENTILATION REQUIREMENTS LOCAL EXHAUST

PROTECTIVE EQUIPMENT (SPECIFY TYPES) EYE: Goggles

MECHANICAL (GENERAL)

GLOVES: Rubber

SPECIAL

RESPIRATOR: With proper filter

OTHER PROTECTIVE EQUIPMENT: Long sleeve coveralls.

SECTION IX SPECIAL PRECAUTIONS

PRECAUTIONS TO BE TAKEN IN HANDLING AND STORAGE: Store in well-ventilated area.

OTHER PRECAUTIONS

SIGNATURE | DATE

ASBESTOS FIBER

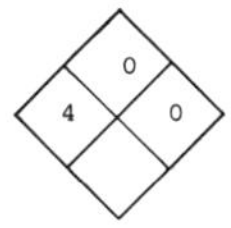

SECTION I SOURCE AND NOMENCLATURE

MANUFACTURER'S NAME	EMERGENCY TELEPHONE NO.
ADDRESS (NUMBER, STREET, CITY, STATE, ZIP CODE)	
TRADE NAME AND SYNONYMS	CHEMICAL FAMILY
CHEMICAL NAME AND SYNONYMS	FORMULA

SECTION II HAZARDOUS INGREDIENTS

BASIC MATERIAL	APPROXIMATE OR MAXIMUM % WT. OR VOL.	ESTABLISHED OSHA STANDARD	LD_{50} ORAL	LD_{50} PERCUT.	LC_{50} SPECIES	LC_{50} CONC.
#1 lowest published toxic dose inhalation - human 1.? fibers 1 CC						

SECTION III PHYSICAL DATA

BOILING POINT °F. N/A	VAPOR PRESSURE mm Hg.
MELTING POINT °F. N/A	VAPOR DENSITY (Air = 1)
SPECIFIC GRAVITY (H_2O = 1)	EVAPORATION RATE (= 1)
SOLUBILITY IN WATER Pts/100 pts H_2O	VOLATILE % Vol. % Wt.

APPEARANCE AND ODOR
Grey-white fibers

SECTION IV FIRE AND EXPLOSION HAZARD DATA

FLASH POINT Not flammable	FLAMMABLE (EXPLOSIVE) LIMITS	UPPER
METHOD USED		LOWER

EXTINGUISHING MEDIA
N/A

SPECIAL FIRE FIGHTING PROCEDURES
N/A

UNUSUAL FIRE AND EXPLOSION HAZARDS
N/A

SECTION V HEALTH HAZARD DATA

TOXIC LEVEL 5 fibers/CC >5μ long — CARCINOGENIC yes

PRINCIPLE ROUTES OF ABSORBTION Inhalation — SKIN AND EYE IRRITATION No

REVELANT SYMPTOMS OF EXPOSURE None

EFFECTS OF CHRONIC EXPOSURE Asbestosis, lung cancer

EMERGENCY AND FIRST AID PROCEDURES Not available

SECTION VI REACTIVITY DATA

CONDITIONS CONTRIBUTING TO INSTABILITY N/A

CONDITIONS CONTRIBUTING TO HAZARDOUS POLYMERIZATION N/A

INCOMPATABILITY (MATERIALS TO AVOID) N/A

HAZARDOUS DECOMPOSITION PRODUCTS N/A

SECTION VII SPILL OR LEAK PROCEDURES

STEPS TO BE TAKEN IN CASE MATERIAL IS RELEASED OR SPILLED
Vacuum or wet clean only. No dry sweeping allowed.

WASTE DISPOSAL METHOD
In sealed plastic bag.

SECTION VIII SPECIAL PROTECTION INFORMATION

VENTILATION REQUIREMENTS LOCAL EXHAUST Advise of Ind. Hygiene Dept.	PROTECTIVE EQUIPMENT (SPECIFY TYPES) EYE None
MECHANICAL (GENERAL) " "	GLOVES None
SPECIAL " "	RESPIRATOR > 5 fiber/CC >5 μ long

OTHER PROTECTIVE EQUIPMENT

SECTION IX SPECIAL PRECAUTIONS

PRECAUTIONS TO BE TAKEN IN HANDLING AND STORAGE

OTHER PRECAUTIONS

SIGNATURE	DATE

AZOBENZENE

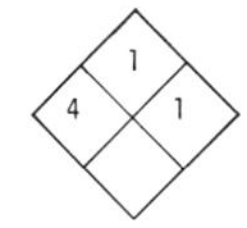

SECTION I SOURCE AND NOMENCLATURE

MANUFACTURER'S NAME	EMERGENCY TELEPHONE NO.
ADDRESS (NUMBER, STREET, CITY, STATE, ZIP CODE)	
TRADE NAME AND SYNONYMS Diphenyldiazene	CHEMICAL FAMILY
CHEMICAL NAME AND SYNONYMS Azobenzene	FORMULA $C_6H_5N_2C_6H_5$

SECTION II HAZARDOUS INGREDIENTS

BASIC MATERIAL	APPROXIMATE OR MAXIMUM % WT. OR VOL.	ESTABLISHED OSHA STANDARD	LD_{50} mg/kg ORAL	LD_{50} PERCUT.	LC_{50} SPECIES	LC_{50} CONC.
LDLO oral - human 50 mg/kg			1000		Rat	

SECTION III PHYSICAL DATA MW - 182.22

BOILING POINT °F. 293°C.	VAPOR PRESSURE at 103°C. 1 mm Hg.
MELTING POINT °F. 68°C.	VAPOR DENSITY (Air = 1)
SPECIFIC GRAVITY (H_2O = 1) 1.20	EVAPORATION RATE (= 1)
SOLUBILITY IN WATER Pts/100 pts H_2O Insoluble in water	VOLATILE % Vol. % Wt.

APPEARANCE AND ODOR
Orange-red crystals.

SECTION IV FIRE AND EXPLOSION HAZARD DATA

FLASH POINT Not available	FLAMMABLE (EXPLOSIVE) LIMITS	UPPER
METHOD USED		LOWER

EXTINGUISHING MEDIA
Not specified.

SPECIAL FIRE FIGHTING PROCEDURES

UNUSUAL FIRE AND EXPLOSION HAZARDS

SECTION V HEALTH HAZARD DATA

TOXIC LEVEL None available — CARCINOGENIC Yes

PRINCIPLE ROUTES OF ABSORBTION Inhalation, ingestion — SKIN AND EYE IRRITATION Slight

REVELANT SYMPTOMS OF EXPOSURE Slight respiratory irritation.

EFFECTS OF CHRONIC EXPOSURE Liver and lung cancer.

EMERGENCY AND FIRST AID PROCEDURES Inhalation - rest; if swallowed - emetic, get medical aid; wash skin & eyes.

SECTION VI REACTIVITY DATA

CONDITIONS CONTRIBUTING TO INSTABILITY

CONDITIONS CONTRIBUTING TO HAZARDOUS POLYMERIZATION

INCOMPATABILITY (MATERIALS TO AVOID)

HAZARDOUS DECOMPOSITION PRODUCTS

SECTION VII SPILL OR LEAK PROCEDURES

STEPS TO BE TAKEN IN CASE MATERIAL IS RELEASED OR SPILLED

WASTE DISPOSAL METHOD

SECTION VIII SPECIAL PROTECTION INFORMATION

VENTILATION REQUIREMENTS LOCAL EXHAUST	PROTECTIVE EQUIPMENT (SPECIFY TYPES) EYE Safety goggles
MECHANICAL (GENERAL)	GLOVES Rubber
SPECIAL	RESPIRATOR With chem. filter

OTHER PROTECTIVE EQUIPMENT

SECTION IX SPECIAL PRECAUTIONS

PRECAUTIONS TO BE TAKEN IN HANDLING AND STORAGE

OTHER PRECAUTIONS

SIGNATURE	DATE

AZOXYBENZENE

1
4 0

SECTION I SOURCE AND NOMENCLATURE

MANUFACTURER'S NAME

EMERGENCY TELEPHONE NO.

ADDRESS (NUMBER, STREET, CITY, STATE, ZIP CODE)

TRADE NAME AND SYNONYMS

CHEMICAL FAMILY

CHEMICAL NAME AND SYNONYMS: Azoxybenzene

FORMULA: $C_{12}H_{10}N_2O$

SECTION II HAZARDOUS INGREDIENTS

BASIC MATERIAL	APPROXIMATE OR MAXIMUM % WT. OR VOL.	ESTABLISHED OSHA STANDARD	LD_{50} Mg/Kg. ORAL	LD_{50} Mg/Kg. PERCUT.	LC_{50} SPECIES	LC_{50} CONC.
			620		Rat	
				1090	Rabbit	

SECTION III PHYSICAL DATA MW - 198.24

BOILING POINT °F.: Decomposes

VAPOR PRESSURE mm Hg.

MELTING POINT °F.: 36°C

VAPOR DENSITY (Air = 1)

SPECIFIC GRAVITY (H_2O = 1): 1.16

EVAPORATION RATE (= 1)

SOLUBILITY IN WATER Pts/100 pts H_2O: Insoluble

VOLATILE % Vol. % Wt.

APPEARANCE AND ODOR: Pale yellow crystals

SECTION IV FIRE AND EXPLOSION HAZARD DATA

FLASH POINT: None available

FLAMMABLE (EXPLOSIVE) LIMITS: UPPER / LOWER

METHOD USED

EXTINGUISHING MEDIA: Carbon dioxide, dry chemical

SPECIAL FIRE FIGHTING PROCEDURES

UNUSUAL FIRE AND EXPLOSION HAZARDS: On decomposition emits toxic fumes

SECTION V HEALTH HAZARD DATA

TOXIC LEVEL: None available

CARCINOGENIC: Yes

PRINCIPLE ROUTES OF ABSORBTION: Inhalation, ingestion, skin

SKIN AND EYE IRRITATION: Slight

REVELANT SYMPTOMS OF EXPOSURE: Skin irritation, jaundice, cyanosis, dermatitis

EFFECTS OF CHRONIC EXPOSURE: Liver & kidney damage, malignant tumors

EMERGENCY AND FIRST AID PROCEDURES: If swallowed - give emetic; get medical aid; inhalation - rest; wash skin & eyes.

SECTION VI REACTIVITY DATA

CONDITIONS CONTRIBUTING TO INSTABILITY

CONDITIONS CONTRIBUTING TO HAZARDOUS POLYMERIZATION

INCOMPATABILITY (MATERIALS TO AVOID)

HAZARDOUS DECOMPOSITION PRODUCTS

SECTION VII SPILL OR LEAK PROCEDURES

STEPS TO BE TAKEN IN CASE MATERIAL IS RELEASED OR SPILLED

WASTE DISPOSAL METHOD

SECTION VIII SPECIAL PROTECTION INFORMATION

VENTILATION REQUIREMENTS LOCAL EXHAUST

PROTECTIVE EQUIPMENT (SPECIFY TYPES) EYE: Goggles

MECHANICAL (GENERAL)

GLOVES

SPECIAL

RESPIRATOR: With proper filter

OTHER PROTECTIVE EQUIPMENT

SECTION IX SPECIAL PRECAUTIONS

PRECAUTIONS TO BE TAKEN IN HANDLING AND STORAGE

OTHER PRECAUTIONS

SIGNATURE

DATE

BARIUM ACETATE

0
3 1

SECTION I SOURCE AND NOMENCLATURE

MANUFACTURER'S NAME

EMERGENCY TELEPHONE NO.

ADDRESS (NUMBER, STREET, CITY, STATE, ZIP CODE)

TRADE NAME AND SYNONYMS

CHEMICAL FAMILY

CHEMICAL NAME AND SYNONYMS: Barium Acetate

FORMULA: $Ba(C_2H_3O_2)_2 \cdot 2H_2O$

SECTION II HAZARDOUS INGREDIENTS

BASIC MATERIAL	APPROXIMATE OR MAXIMUM % WT. OR VOL.	ESTABLISHED OSHA STANDARD	LD_{50} mg/kg ORAL	LD_{50} mg/kg PERCUT.	LC_{50} SPECIES	LC_{50} CONC.
#1 lowest published lethal dose oral - rabbit 236 mg/kg			921		rat	

SECTION III PHYSICAL DATA MW - 255.44

BOILING POINT °F.

VAPOR PRESSURE mm Hg.

MELTING POINT °F.: Looses water at 41°C.

VAPOR DENSITY (Air = 1)

SPECIFIC GRAVITY (H_2O = 1): 2.19

EVAPORATION RATE (= 1)

SOLUBILITY IN WATER Pts/100 pts H_2O

VOLATILE % Vol. % Wt.

APPEARANCE AND ODOR: White crystal

SECTION IV FIRE AND EXPLOSION HAZARD DATA

FLASH POINT: N/A

FLAMMABLE (EXPLOSIVE) LIMITS: UPPER / LOWER

METHOD USED

EXTINGUISHING MEDIA: N/A

SPECIAL FIRE FIGHTING PROCEDURES: N/A

UNUSUAL FIRE AND EXPLOSION HAZARDS: N/A

SECTION V HEALTH HAZARD DATA

TOXIC LEVEL: TLV - 0.93 mg/C.M.

CARCINOGENIC: No

PRINCIPLE ROUTES OF ABSORBTION: Ingestion, inhalation of dust

SKIN AND EYE IRRITATION: Moderate

REVELANT SYMPTOMS OF EXPOSURE: Nausea, vomiting, stomach pains, diarrhea, headache, weakness, paralysis of central nervous system, digitalis like action on the heart.

EFFECTS OF CHRONIC EXPOSURE

EMERGENCY AND FIRST AID PROCEDURES: If swallowed - wash mouth, drink 2 tablespoons of epsom salts in water, then emetic; rest, keep warm, get medical aid.

SECTION VI REACTIVITY DATA

CONDITIONS CONTRIBUTING TO INSTABILITY

CONDITIONS CONTRIBUTING TO HAZARDOUS POLYMERIZATION

INCOMPATABILITY (MATERIALS TO AVOID)

HAZARDOUS DECOMPOSITION PRODUCTS

SECTION VII SPILL OR LEAK PROCEDURES

STEPS TO BE TAKEN IN CASE MATERIAL IS RELEASED OR SPILLED

WASTE DISPOSAL METHOD

SECTION VIII SPECIAL PROTECTION INFORMATION

VENTILATION REQUIREMENTS LOCAL EXHAUST

PROTECTIVE EQUIPMENT (SPECIFY TYPES) EYE: Goggles

MECHANICAL (GENERAL)

GLOVES: Rubber gloves

SPECIAL

RESPIRATOR: With proper filter

OTHER PROTECTIVE EQUIPMENT

SECTION IX SPECIAL PRECAUTIONS

PRECAUTIONS TO BE TAKEN IN HANDLING AND STORAGE

OTHER PRECAUTIONS

SIGNATURE

DATE

BARIUM CHLORIDE

Hazard diamond: 0 (top), 3 (left), 0 (right)

SECTION I SOURCE AND NOMENCLATURE

MANUFACTURER'S NAME	EMERGENCY TELEPHONE NO.
ADDRESS (NUMBER, STREET, CITY, STATE, ZIP CODE)	
TRADE NAME AND SYNONYMS	CHEMICAL FAMILY
CHEMICAL NAME AND SYNONYMS: Barium Chloride	FORMULA: $BaCl_2 . 2H_2O$

SECTION II HAZARDOUS INGREDIENTS

BASIC MATERIAL	APPROXIMATE OR MAXIMUM % WT. OR VOL.	ESTABLISHED OSHA STANDARD	LD_{50} mg/kg. ORAL	LD_{50} PERCUT.	SPECIES	LC_{50} CONC.
#1 Lowest published lethal dose oral - human 11.4 mg/kg.			118		Rat	
			76		guinea pig	

SECTION III PHYSICAL DATA MW - 244.28

BOILING POINT °F.: 35.7^{20}	VAPOR PRESSURE mm Hg.
MELTING POINT °F.: $-2H_2O$ 113°C.	VAPOR DENSITY (Air = 1)
SPECIFIC GRAVITY (H_2O = 1): 3.097 ($24^\circ/4^\circ$)	EVAPORATION RATE (= 1)
SOLUBILITY IN WATER Pts/100 pts H_2O: 39.3 gr	VOLATILE % Vol. % Wt.

APPEARANCE AND ODOR: Colorless crystals

SECTION IV FIRE AND EXPLOSION HAZARD DATA

FLASH POINT: Not flammable	FLAMMABLE (EXPLOSIVE) LIMITS	UPPER
METHOD USED		LOWER

EXTINGUISHING MEDIA

SPECIAL FIRE FIGHTING PROCEDURES

UNUSUAL FIRE AND EXPLOSION HAZARDS

SECTION V HEALTH HAZARD DATA

TOXIC LEVEL: TLV - 0.806 mg/m^3 CARCINOGENIC: no

PRINCIPLE ROUTES OF ABSORBTION: Ingestion or inhalation of dust SKIN AND EYE IRRITATION: Moderate

REVELANT SYMPTOMS OF EXPOSURE: Dermatitis, conjunctivitis and bronchial irritation, arterial hypertension, muscle twitching.

EFFECTS OF CHRONIC EXPOSURE

EMERGENCY AND FIRST AID PROCEDURES: If swallowed - give 2 tablespoons of epsom salts in water, then emetic, get medical aid.

SECTION VI REACTIVITY DATA

CONDITIONS CONTRIBUTING TO INSTABILITY

CONDITIONS CONTRIBUTING TO HAZARDOUS POLYMERIZATION

INCOMPATABILITY (MATERIALS TO AVOID)

HAZARDOUS DECOMPOSITION PRODUCTS

SECTION VII SPILL OR LEAK PROCEDURES

STEPS TO BE TAKEN IN CASE MATERIAL IS RELEASED OR SPILLED

WASTE DISPOSAL METHOD

SECTION VIII SPECIAL PROTECTION INFORMATION

VENTILATION REQUIREMENTS LOCAL EXHAUST	PROTECTIVE EQUIPMENT (SPECIFY TYPES) EYE: Goggles
MECHANICAL (GENERAL)	GLOVES: Rubber
SPECIAL	RESPIRATOR: W/proper filter

OTHER PROTECTIVE EQUIPMENT

SECTION IX SPECIAL PRECAUTIONS

PRECAUTIONS TO BE TAKEN IN HANDLING AND STORAGE

OTHER PRECAUTIONS

SIGNATURE	DATE

BARIUM HYDROXIDE

Hazard diamond: 0 (top), 3 (left), 1 (right)

SECTION I SOURCE AND NOMENCLATURE

MANUFACTURER'S NAME	EMERGENCY TELEPHONE NO.
ADDRESS (NUMBER, STREET, CITY, STATE, ZIP CODE)	
TRADE NAME AND SYNONYMS	CHEMICAL FAMILY
CHEMICAL NAME AND SYNONYMS: Barium Hydroxide	FORMULA: $Ba(OH)_2 . 8H_2O$

SECTION II HAZARDOUS INGREDIENTS

BASIC MATERIAL	APPROXIMATE OR MAXIMUM % WT. OR VOL.	ESTABLISHED OSHA STANDARD	LD_{50} ORAL	LD_{50} PERCUT.	SPECIES	LC_{50} CONC.
			Not available			

SECTION III PHYSICAL DATA MW 315.48

BOILING POINT °F.: $-8 H_2O$. 780° C.	VAPOR PRESSURE mm Hg.
MELTING POINT °F.: 78° C.	VAPOR DENSITY (Air = 1)
SPECIFIC GRAVITY (H_2O = 1): 2.18	EVAPORATION RATE (= 1)
SOLUBILITY IN WATER Pts/100 pts H_2O: Slightly soluble	VOLATILE % Vol. % Wt.

APPEARANCE AND ODOR: Colorless crystals

SECTION IV FIRE AND EXPLOSION HAZARD DATA

FLASH POINT: Not flammable	FLAMMABLE (EXPLOSIVE) LIMITS	UPPER
METHOD USED		LOWER

EXTINGUISHING MEDIA

SPECIAL FIRE FIGHTING PROCEDURES

UNUSUAL FIRE AND EXPLOSION HAZARDS

SECTION V HEALTH HAZARD DATA

TOXIC LEVEL: TLV 1.15 mg/m^3 CARCINOGENIC: No

PRINCIPLE ROUTES OF ABSORBTION: Ingestion or inhalation of dust SKIN AND EYE IRRITATION: Moderate

REVELANT SYMPTOMS OF EXPOSURE: Dermatitis, conjunctivis and bronchial irritation, tightness of face and neck muscles, vomiting, diarrhea, weakness, difficulty in breathing.

EFFECTS OF CHRONIC EXPOSURE

EMERGENCY AND FIRST AID PROCEDURES: Give artificial respiration if necessary, specifically with oxygen if available. Antidote to be administered as soon as possible 10% sodium sulfate slowly I.V. - repeat every 15 min. until symtom subsides. 30 Gm of sodium sulfate in 250 ml of water orally.

SECTION VI REACTIVITY DATA

CONDITIONS CONTRIBUTING TO INSTABILITY

CONDITIONS CONTRIBUTING TO HAZARDOUS POLYMERIZATION

INCOMPATABILITY (MATERIALS TO AVOID)

HAZARDOUS DECOMPOSITION PRODUCTS

SECTION VII SPILL OR LEAK PROCEDURES

STEPS TO BE TAKEN IN CASE MATERIAL IS RELEASED OR SPILLED

WASTE DISPOSAL METHOD

SECTION VIII SPECIAL PROTECTION INFORMATION

VENTILATION REQUIREMENTS LOCAL EXHAUST	PROTECTIVE EQUIPMENT (SPECIFY TYPES) EYE: Goggles
MECHANICAL (GENERAL)	GLOVES: Rubber
SPECIAL	RESPIRATOR: With appropriate filter

OTHER PROTECTIVE EQUIPMENT

SECTION IX SPECIAL PRECAUTIONS

PRECAUTIONS TO BE TAKEN IN HANDLING AND STORAGE: Store in closed containers.

OTHER PRECAUTIONS

SIGNATURE	DATE

BARIUM METAL

1
3 2
W

SECTION I SOURCE AND NOMENCLATURE

MANUFACTURER'S NAME | EMERGENCY TELEPHONE NO.

ADDRESS (NUMBER, STREET, CITY, STATE, ZIP CODE)

TRADE NAME AND SYNONYMS | CHEMICAL FAMILY

CHEMICAL NAME AND SYNONYMS: Barium | FORMULA: Ba

SECTION II HAZARDOUS INGREDIENTS

BASIC MATERIAL	APPROXIMATE OR MAXIMUM % WT. OR VOL.	ESTABLISHED OSHA STANDARD	LD_{50} ORAL	LD_{50} PERCUT.	LC_{50} SPECIES	LC_{50} CONC.
			Not available			

SECTION III PHYSICAL DATA AW - 137.34

BOILING POINT: 1140°C. | VAPOR PRESSURE (mm Hg.)

MELTING POINT: 850° C. | VAPOR DENSITY (Air = 1)

SPECIFIC GRAVITY (H_2O = 1): 3.5 | EVAPORATION RATE (____ = 1)

SOLUBILITY IN WATER (Pts/100 pts H_2O): d | VOLATILE % Vol. % Wt.

APPEARANCE AND ODOR: Silver metal

SECTION IV FIRE AND EXPLOSION HAZARD DATA

FLASH POINT: None available | FLAMMABLE (EXPLOSIVE) LIMITS: UPPER / LOWER

METHOD USED

EXTINGUISHING MEDIA: Carbon dioxide and dry chemical, dry graphite soda ash, powdered NaCl.

SPECIAL FIRE FIGHTING PROCEDURES: Wear goggles and self-contained breathing apparatus.

UNUSUAL FIRE AND EXPLOSION HAZARDS

SECTION V HEALTH HAZARD DATA

TOXIC LEVEL: TLV - 0.5 mg/m^3 | CARCINOGENIC

PRINCIPLE ROUTES OF ABSORBTION | SKIN AND EYE IRRITATION: Severe

REVELANT SYMPTOMS OF EXPOSURE: Nausea, vomiting, stomach pain, diarrhea, headache, weakness, paralysis of central nervous system - attack on eye and skin.

EFFECTS OF CHRONIC EXPOSURE: --

EMERGENCY AND FIRST AID PROCEDURES: If swallowed - wash mouth, drink glass of water with 2 tablespoons of epsom salt; inhalation - rest, keep warm; get medical attention; wash from eyes and skin with water.

SECTION VI REACTIVITY DATA

CONDITIONS CONTRIBUTING TO INSTABILITY

CONDITIONS CONTRIBUTING TO HAZARDOUS POLYMERIZATION

INCOMPATABILITY (MATERIALS TO AVOID): Water, acids

HAZARDOUS DECOMPOSITION PRODUCTS

SECTION VII SPILL OR LEAK PROCEDURES

STEPS TO BE TAKEN IN CASE MATERIAL IS RELEASED OR SPILLED

WASTE DISPOSAL METHOD

SECTION VIII SPECIAL PROTECTION INFORMATION

VENTILATION REQUIREMENTS LOCAL EXHAUST | PROTECTIVE EQUIPMENT (SPECIFY TYPES) EYE: Goggles

MECHANICAL (GENERAL) | GLOVES: Rubber gloves

SPECIAL | RESPIRATOR: Self-contained breathing mask

OTHER PROTECTIVE EQUIPMENT: Rubber protective clothing, apron

SECTION IX SPECIAL PRECAUTIONS

PRECAUTIONS TO BE TAKEN IN HANDLING AND STORAGE: Store in tightly closed container under inert gas.

OTHER PRECAUTIONS

SIGNATURE | DATE

1,2-BENZANTHRACENE

1
4 0

SECTION I SOURCE AND NOMENCLATURE

MANUFACTURER'S NAME | EMERGENCY TELEPHONE NO.

ADDRESS (NUMBER, STREET, CITY, STATE, ZIP CODE)

TRADE NAME AND SYNONYMS: Naphthanthracene | CHEMICAL FAMILY

CHEMICAL NAME AND SYNONYMS: 1,2-Benzanthracene | FORMULA: $C_{18}H_{12}$

SECTION II HAZARDOUS INGREDIENTS

BASIC MATERIAL	APPROXIMATE OR MAXIMUM % WT. OR VOL.	ESTABLISHED OSHA STANDARD	LD_{50} ORAL	LD_{50} PERCUT.	LC_{50} SPECIES	LC_{50} CONC.
Notes: Lowest publsihed toxic dose						
subcutaneous - mouse 2 mg/Kg.						
Oral - Mouse 4 mg/Kg/5 weeks						
Skin - Mouse 240 mg/Kg. /5 weeks						

SECTION III PHYSICAL DATA MW - 228.29

BOILING POINT: Sublimes | VAPOR PRESSURE (mm Hg.)

MELTING POINT: 160°C | VAPOR DENSITY (Air = 1)

SPECIFIC GRAVITY (H_2O = 1) | EVAPORATION RATE (____ = 1)

SOLUBILITY IN WATER (Pts/100 pts H_2O): Insoluble | VOLATILE % Vol. % Wt.

APPEARANCE AND ODOR

SECTION IV FIRE AND EXPLOSION HAZARD DATA

FLASH POINT: N/A | FLAMMABLE (EXPLOSIVE) LIMITS: UPPER / LOWER

METHOD USED

EXTINGUISHING MEDIA: Water, carbon dioxide, foam or dry chemical

SPECIAL FIRE FIGHTING PROCEDURES: Wear goggles and self contained breathing apparatus.

UNUSUAL FIRE AND EXPLOSION HAZARDS

SECTION V HEALTH HAZARD DATA

TOXIC LEVEL: Not available | CARCINOGENIC: Yes

PRINCIPLE ROUTES OF ABSORBTION: Inhalation, ingestion | SKIN AND EYE IRRITATION: Slight

REVELANT SYMPTOMS OF EXPOSURE: Nausea, vomiting

EFFECTS OF CHRONIC EXPOSURE: --

EMERGENCY AND FIRST AID PROCEDURES: Inhalation - rest; if swallowed - remove by gastric lavage or emetic; wash from skin & eyes.

SECTION VI REACTIVITY DATA

CONDITIONS CONTRIBUTING TO INSTABILITY

CONDITIONS CONTRIBUTING TO HAZARDOUS POLYMERIZATION

INCOMPATABILITY (MATERIALS TO AVOID)

HAZARDOUS DECOMPOSITION PRODUCTS

SECTION VII SPILL OR LEAK PROCEDURES

STEPS TO BE TAKEN IN CASE MATERIAL IS RELEASED OR SPILLED

WASTE DISPOSAL METHOD

SECTION VIII SPECIAL PROTECTION INFORMATION

VENTILATION REQUIREMENTS LOCAL EXHAUST | PROTECTIVE EQUIPMENT (SPECIFY TYPES) EYE: Safety goggles

MECHANICAL (GENERAL) | GLOVES: Rubber

SPECIAL | RESPIRATOR: With proper filter

OTHER PROTECTIVE EQUIPMENT

SECTION IX SPECIAL PRECAUTIONS

PRECAUTIONS TO BE TAKEN IN HANDLING AND STORAGE

OTHER PRECAUTIONS

SIGNATURE | DATE

BENZIDINE DIHYDROCHLORIDE

1 4 0

SECTION I SOURCE AND NOMENCLATURE

MANUFACTURER'S NAME	EMERGENCY TELEPHONE NO.
ADDRESS (NUMBER, STREET, CITY, STATE, ZIP CODE)	
TRADE NAME AND SYNONYMS	CHEMICAL FAMILY
CHEMICAL NAME AND SYNONYMS Benzidine Dihydrochloride	FORMULA $H_2N(C_6H_4)_2NH_2 . 2HCl$

SECTION II HAZARDOUS INGREDIENTS

BASIC MATERIAL	APPROXIMATE OR MAXIMUM % WT. OR VOL.	ESTABLISHED OSHA STANDARD	LD_{50}		LC_{50}	
			ORAL	PERCUT.	SPECIES	CONC.
Lowest published lethal dose						
Oral - dog 200 mg/kg						

SECTION III PHYSICAL DATA MW - 257.17

BOILING POINT °F. -	VAPOR PRESSURE mm Hg.
MELTING POINT °F. -	VAPOR DENSITY (Air = 1)
SPECIFIC GRAVITY (H_2O = 1)	EVAPORATION RATE (_____ = 1)
SOLUBILITY IN WATER Pts/100 pts H_2O Soluble	VOLATILE % Vol. % Wt.

APPEARANCE AND ODOR

SECTION IV FIRE AND EXPLOSION HAZARD DATA

FLASH POINT None available	FLAMMABLE (EXPLOSIVE) LIMITS	UPPER
METHOD USED		LOWER

EXTINGUISHING MEDIA
Carbon dioxide, dry chemical.

SPECIAL FIRE FIGHTING PROCEDURES

UNUSUAL FIRE AND EXPLOSION HAZARDS

SECTION V HEALTH HAZARD DATA

TOXIC LEVEL None available — CARCINOGENIC Yes

PRINCIPLE ROUTES OF ABSORBTION Inhalation, ingestion, skin — SKIN AND EYE IRRITATION Severe

REVELANT SYMPTOMS OF EXPOSURE Eye irritation

EFFECTS OF CHRONIC EXPOSURE Bladder tumors.

EMERGENCY AND FIRST AID PROCEDURES If swallowed - give emetic, get medical aid; inhalation - rest; wash skin & eyes.

SECTION VI REACTIVITY DATA

CONDITIONS CONTRIBUTING TO INSTABILITY

CONDITIONS CONTRIBUTING TO HAZARDOUS POLYMERIZATION

INCOMPATABILITY (MATERIALS TO AVOID)

HAZARDOUS DECOMPOSITION PRODUCTS

SECTION VII SPILL OR LEAK PROCEDURES

STEPS TO BE TAKEN IN CASE MATERIAL IS RELEASED OR SPILLED

WASTE DISPOSAL METHOD

SECTION VIII SPECIAL PROTECTION INFORMATION

VENTILATION REQUIREMENTS LOCAL EXHAUST	PROTECTIVE EQUIPMENT (SPECIFY TYPES) EYE Goggles
MECHANICAL (GENERAL)	GLOVES Rubber
SPECIAL	RESPIRATOR Filter type for dust mist and fumes

OTHER PROTECTIVE EQUIPMENT
Full protective clothing.

SECTION IX SPECIAL PRECAUTIONS

PRECAUTIONS TO BE TAKEN IN HANDLING AND STORAGE
Store in sealed container.

OTHER PRECAUTIONS

SIGNATURE	DATE

BENZOYL BROMIDE

2 3 1

SECTION I SOURCE AND NOMENCLATURE

MANUFACTURER'S NAME	EMERGENCY TELEPHONE NO.
ADDRESS (NUMBER, STREET, CITY, STATE, ZIP CODE)	
TRADE NAME AND SYNONYMS	CHEMICAL FAMILY
CHEMICAL NAME AND SYNONYMS Benzoyl Bromide	FORMULA C_6H_5COBr

SECTION II HAZARDOUS INGREDIENTS

BASIC MATERIAL	APPROXIMATE OR MAXIMUM % WT. OR VOL.	ESTABLISHED OSHA STANDARD	LD_{50}		LC_{50}	
			ORAL	PERCUT.	SPECIES	CONC.
			Not available			

SECTION III PHYSICAL DATA

BOILING POINT °F. 218-19°C	VAPOR PRESSURE mm Hg.
MELTING POINT °F. 5.5 - 7.5°C	VAPOR DENSITY (Air = 1)
SPECIFIC GRAVITY (H_2O = 1) 1.57	EVAPORATION RATE (_____ = 1)
SOLUBILITY IN WATER Pts/100 pts H_2O Decomposes	VOLATILE % Vol. % Wt.

APPEARANCE AND ODOR
Colorless liquid

SECTION IV FIRE AND EXPLOSION HAZARD DATA

FLASH POINT N/A	FLAMMABLE (EXPLOSIVE) LIMITS	UPPER
METHOD USED		LOWER

EXTINGUISHING MEDIA Use water spray sparingly, foam, carbon dioxide or dry chemical.

SPECIAL FIRE FIGHTING PROCEDURES
Wear full protective clothing.

UNUSUAL FIRE AND EXPLOSION HAZARDS

SECTION V HEALTH HAZARD DATA

TOXIC LEVEL Not available — CARCINOGENIC No

PRINCIPLE ROUTES OF ABSORBTION Inhalation, ingestion — SKIN AND EYE IRRITATION Severe

REVELANT SYMPTOMS OF EXPOSURE Irritates respiratory system; burns skin, eyes, mouth and digestive system.

EFFECTS OF CHRONIC EXPOSURE --

EMERGENCY AND FIRST AID PROCEDURES Inhalation - rest; get medical aid; if swallowed - wash mouth, drink water, milk of magnesia; get medical aid; wash skin & eyes.

SECTION VI REACTIVITY DATA

CONDITIONS CONTRIBUTING TO INSTABILITY
Reacts with oxidizing materials

CONDITIONS CONTRIBUTING TO HAZARDOUS POLYMERIZATION

INCOMPATABILITY (MATERIALS TO AVOID)

HAZARDOUS DECOMPOSITION PRODUCTS

SECTION VII SPILL OR LEAK PROCEDURES

STEPS TO BE TAKEN IN CASE MATERIAL IS RELEASED OR SPILLED

WASTE DISPOSAL METHOD

SECTION VIII SPECIAL PROTECTION INFORMATION

VENTILATION REQUIREMENTS LOCAL EXHAUST	PROTECTIVE EQUIPMENT (SPECIFY TYPES) EYE Safety goggles
MECHANICAL (GENERAL)	GLOVES Rubber
SPECIAL	RESPIRATOR Self contained

OTHER PROTECTIVE EQUIPMENT

SECTION IX SPECIAL PRECAUTIONS

PRECAUTIONS TO BE TAKEN IN HANDLING AND STORAGE

OTHER PRECAUTIONS

SIGNATURE	DATE

BENZYL-m-TOLUIDINE

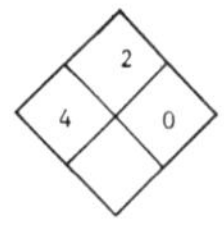

SECTION I SOURCE AND NOMENCLATURE

MANUFACTURER'S NAME | EMERGENCY TELEPHONE NO.

ADDRESS (NUMBER, STREET, CITY, STATE, ZIP CODE)

TRADE NAME AND SYNONYMS | CHEMICAL FAMILY

CHEMICAL NAME AND SYNONYMS: Phenyl-tolyl-methane | FORMULA: $C_6H_5CH_2NHC_6H_4CH_3$

SECTION II HAZARDOUS INGREDIENTS

BASIC MATERIAL	APPROXIMATE OR MAXIMUM % WT. OR VOL.	ESTABLISHED OSHA STANDARD	LD_{50} ORAL	LD_{50} PERCUT.	LC_{50} SPECIES	LC_{50} CONC.
			None available			

SECTION III PHYSICAL DATA MW - 182.25

BOILING POINT °F.: 275°C. | VAPOR PRESSURE: at 193°C. 15 mm Hg.

MELTING POINT °F.: 0.997 | VAPOR DENSITY (Air = 1)

SPECIFIC GRAVITY (H_2O = 1) | EVAPORATION RATE (= 1)

SOLUBILITY IN WATER Pts/100 pts H_2O | VOLATILE % Vol. % Wt.

APPEARANCE AND ODOR: Colorless liquid

SECTION IV FIRE AND EXPLOSION HAZARD DATA

FLASH POINT: None available | FLAMMABLE (EXPLOSIVE) LIMITS: UPPER / LOWER

METHOD USED

EXTINGUISHING MEDIA: Carbon dioxide, dry chemical, foam or water spray.

SPECIAL FIRE FIGHTING PROCEDURES: Wear full protective clothing.

UNUSUAL FIRE AND EXPLOSION HAZARDS: When heated emits highly toxic fumes.

SECTION V HEALTH HAZARD DATA

TOXIC LEVEL: None available | CARCINOGENIC: No

PRINCIPLE ROUTES OF ABSORBTION: Inhalation, ingestion, skin | SKIN AND EYE IRRITATION: Moderate

REVELANT SYMPTOMS OF EXPOSURE: Cyanosis, air hunger, nausea, vomiting, low blood pressure, convulsions.

EFFECTS OF CHRONIC EXPOSURE: Weight loss, anemia, weakness.

EMERGENCY AND FIRST AID PROCEDURES: Inhalation - rest; if swallowed - give emetic, get medical aid; wash from skin & eyes.

SECTION VI REACTIVITY DATA

CONDITIONS CONTRIBUTING TO INSTABILITY

CONDITIONS CONTRIBUTING TO HAZARDOUS POLYMERIZATION

INCOMPATABILITY (MATERIALS TO AVOID)

HAZARDOUS DECOMPOSITION PRODUCTS

SECTION VII SPILL OR LEAK PROCEDURES

STEPS TO BE TAKEN IN CASE MATERIAL IS RELEASED OR SPILLED

WASTE DISPOSAL METHOD

SECTION VIII SPECIAL PROTECTION INFORMATION

VENTILATION REQUIREMENTS LOCAL EXHAUST | PROTECTIVE EQUIPMENT (SPECIFY TYPES) EYE: Safety goggles

MECHANICAL (GENERAL) | GLOVES: Rubber gloves

SPECIAL | RESPIRATOR: With proper filter

OTHER PROTECTIVE EQUIPMENT

SECTION IX SPECIAL PRECAUTIONS

PRECAUTIONS TO BE TAKEN IN HANDLING AND STORAGE

OTHER PRECAUTIONS

SIGNATURE | DATE

BENZYL-o-TOLUIDINE

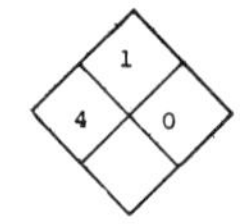

SECTION I SOURCE AND NOMENCLATURE

MANUFACTURER'S NAME | EMERGENCY TELEPHONE NO.

ADDRESS (NUMBER, STREET, CITY, STATE, ZIP CODE)

TRADE NAME AND SYNONYMS | CHEMICAL FAMILY

CHEMICAL NAME AND SYNONYMS: Benzyl-O-Toluidine | FORMULA: $C_6H_5\ CH_2NHC_6H_4CH_3$

SECTION II HAZARDOUS INGREDIENTS

BASIC MATERIAL	APPROXIMATE OR MAXIMUM % WT. OR VOL.	ESTABLISHED OSHA STANDARD	LD_{50} ORAL	LD_{50} PERCUT.	LC_{50} SPECIES	LC_{50} CONC.
			None available			

SECTION III PHYSICAL DATA MW - 182.25

BOILING POINT °F. | VAPOR PRESSURE mm Hg.

MELTING POINT °F.: 60°C | VAPOR DENSITY (Air = 1)

SPECIFIC GRAVITY (H_2O = 1) | EVAPORATION RATE (= 1)

SOLUBILITY IN WATER Pts/100 pts H_2O | VOLATILE % Vol. % Wt.

APPEARANCE AND ODOR: Crystaline solid

SECTION IV FIRE AND EXPLOSION HAZARD DATA

FLASH POINT: None available | FLAMMABLE (EXPLOSIVE) LIMITS: UPPER / LOWER

METHOD USED

EXTINGUISHING MEDIA: Carbon dioxide, dry chemical, foam or water spray.

SPECIAL FIRE FIGHTING PROCEDURES: --

UNUSUAL FIRE AND EXPLOSION HAZARDS: --

SECTION V HEALTH HAZARD DATA

TOXIC LEVEL: None available | CARCINOGENIC: No

PRINCIPLE ROUTES OF ABSORBTION: Inhalation, ingestion, skin | SKIN AND EYE IRRITATION: Moderate

REVELANT SYMPTOMS OF EXPOSURE: Cyanosis, air hunger, nausea, vomiting, low blood pressure, convulsions.

EFFECTS OF CHRONIC EXPOSURE: Weight loss, anemia, weakness.

EMERGENCY AND FIRST AID PROCEDURES: Inhalation - rest; if swallowed, give emetic, get medical aid; wash from skin & hands.

SECTION VI REACTIVITY DATA

CONDITIONS CONTRIBUTING TO INSTABILITY

CONDITIONS CONTRIBUTING TO HAZARDOUS POLYMERIZATION

INCOMPATABILITY (MATERIALS TO AVOID)

HAZARDOUS DECOMPOSITION PRODUCTS

SECTION VII SPILL OR LEAK PROCEDURES

STEPS TO BE TAKEN IN CASE MATERIAL IS RELEASED OR SPILLED

WASTE DISPOSAL METHOD

SECTION VIII SPECIAL PROTECTION INFORMATION

VENTILATION REQUIREMENTS LOCAL EXHAUST | PROTECTIVE EQUIPMENT (SPECIFY TYPES) EYE: Safety goggles

MECHANICAL (GENERAL) | GLOVES: Rubber gloves

SPECIAL | RESPIRATOR

OTHER PROTECTIVE EQUIPMENT

SECTION IX SPECIAL PRECAUTIONS

PRECAUTIONS TO BE TAKEN IN HANDLING AND STORAGE

OTHER PRECAUTIONS

SIGNATURE | DATE

BENZYL-p-TOLUIDINE

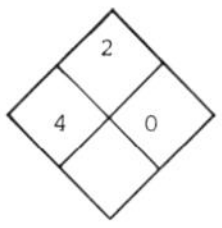

SECTION I SOURCE AND NOMENCLATURE

MANUFACTURER'S NAME	EMERGENCY TELEPHONE NO.
ADDRESS (NUMBER, STREET, CITY, STATE, ZIP CODE)	
TRADE NAME AND SYNONYMS	CHEMICAL FAMILY
CHEMICAL NAME AND SYNONYMS Benzyl-P-Toluidine	FORMULA $C_6H_5CH_2NHC_6H_4CH_3$

SECTION II HAZARDOUS INGREDIENTS

BASIC MATERIAL	APPROXIMATE OR MAXIMUM % WT. OR VOL.	ESTABLISHED OSHA STANDARD	LD_{50} ORAL	LD_{50} PERCUT.	LC_{50} SPECIES	LC_{50} CONC.
			None	available		

SECTION III PHYSICAL DATA MW - 182.25

BOILING POINT °F. 286°C.	VAPOR PRESSURE mm Hg. at 180°C. 10
MELTING POINT °F. -30°C.	VAPOR DENSITY (Air = 1)
SPECIFIC GRAVITY (H_2O = 1) 0.994	EVAPORATION RATE (______ = 1)
SOLUBILITY IN WATER Pts/100 pts H_2O Very soluble	VOLATILE % Vol. % Wt.

APPEARANCE AND ODOR
Liquid

SECTION IV FIRE AND EXPLOSION HAZARD DATA

FLASH POINT None available	FLAMMABLE (EXPLOSIVE) LIMITS	UPPER
METHOD USED		LOWER

EXTINGUISHING MEDIA
Carbon dioxide, dry chemical, foam or water spray.

SPECIAL FIRE FIGHTING PROCEDURES
Wear full protective clothing.

UNUSUAL FIRE AND EXPLOSION HAZARDS
When heated emits highly toxic fumes.

SECTION V HEALTH HAZARD DATA

TOXIC LEVEL
None available — CARCINOGENIC No

PRINCIPLE ROUTES OF ABSORBTION
Inhalation, ingestion, skin — SKIN AND EYE IRRITATION Moderate

REVELANT SYMPTOMS OF EXPOSURE Cyanosis, air hunger, nausea, vomiting, low blood pressure, convulsions.

EFFECTS OF CHRONIC EXPOSURE
Weight loss, anemia, weakness.

EMERGENCY AND FIRST AID PROCEDURES Inhalation - rest; if swallowed - give emetic, get medical aid; wash from skin & hands.

SECTION VI REACTIVITY DATA

CONDITIONS CONTRIBUTING TO INSTABILITY

CONDITIONS CONTRIBUTING TO HAZARDOUS POLYMERIZATION

INCOMPATABILITY (MATERIALS TO AVOID)

HAZARDOUS DECOMPOSITION PRODUCTS

SECTION VII SPILL OR LEAK PROCEDURES

STEPS TO BE TAKEN IN CASE MATERIAL IS RELEASED OR SPILLED

WASTE DISPOSAL METHOD

SECTION VIII SPECIAL PROTECTION INFORMATION

VENTILATION REQUIREMENTS LOCAL EXHAUST	PROTECTIVE EQUIPMENT (SPECIFY TYPES) EYE Safety goggles
MECHANICAL (GENERAL)	GLOVES Rubber gloves
SPECIAL	RESPIRATOR

OTHER PROTECTIVE EQUIPMENT

SECTION IX SPECIAL PRECAUTIONS

PRECAUTIONS TO BE TAKEN IN HANDLING AND STORAGE

OTHER PRECAUTIONS

SIGNATURE	DATE

BROMAL

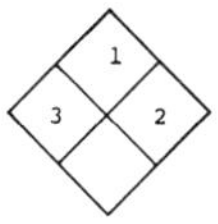

SECTION I SOURCE AND NOMENCLATURE

MANUFACTURER'S NAME	EMERGENCY TELEPHONE NO.
ADDRESS (NUMBER, STREET, CITY, STATE, ZIP CODE)	
TRADE NAME AND SYNONYMS Tribromoacetaldehyde	CHEMICAL FAMILY
CHEMICAL NAME AND SYNONYMS Bromal	FORMULA CBr_3CHO

SECTION II HAZARDOUS INGREDIENTS

BASIC MATERIAL	APPROXIMATE OR MAXIMUM % WT. OR VOL.	ESTABLISHED OSHA STANDARD	LD_{50} ORAL	LD_{50} PERCUT.	LC_{50} SPECIES	LC_{50} CONC.
For Bromal Hydrate						
Lowest published lethal dose oral - rat 40 mg/kg.						

SECTION III PHYSICAL DATA MW - 280.76

BOILING POINT °F. 174°C-	VAPOR PRESSURE mm Hg. at 73°C. 18
MELTING POINT °F.	VAPOR DENSITY (Air = 1)
SPECIFIC GRAVITY (H_2O = 1) 2.67	EVAPORATION RATE (______ = 1)
SOLUBILITY IN WATER Pts/100 pts H_2O Decomposes	VOLATILE % Vol. % Wt.

APPEARANCE AND ODOR
Yellow liquid, irritating odor.

SECTION IV FIRE AND EXPLOSION HAZARD DATA

FLASH POINT None available	FLAMMABLE (EXPLOSIVE) LIMITS	UPPER
METHOD USED		LOWER

EXTINGUISHING MEDIA
Alcohol foam, carbon dioxide, dry chemical.

SPECIAL FIRE FIGHTING PROCEDURES

UNUSUAL FIRE AND EXPLOSION HAZARDS
When heated to decomposition emits toxic fumes.

SECTION V HEALTH HAZARD DATA

TOXIC LEVEL
None available — CARCINOGENIC No

PRINCIPLE ROUTES OF ABSORBTION
Inhalation, ingestion — SKIN AND EYE IRRITATION Severe

REVELANT SYMPTOMS OF EXPOSURE Ingestion - nausea, vomiting, coldness of extremities, unconsciousness, skin & eye irritation, hypnotic effects.

EFFECTS OF CHRONIC EXPOSURE Dermatitis, eye irritation.

EMERGENCY AND FIRST AID PROCEDURES
Ingestion - give emetic, get medical aid, inhalation - rest; wash skin and eyes with water.

SECTION VI REACTIVITY DATA

CONDITIONS CONTRIBUTING TO INSTABILITY

CONDITIONS CONTRIBUTING TO HAZARDOUS POLYMERIZATION

INCOMPATABILITY (MATERIALS TO AVOID)

HAZARDOUS DECOMPOSITION PRODUCTS

SECTION VII SPILL OR LEAK PROCEDURES

STEPS TO BE TAKEN IN CASE MATERIAL IS RELEASED OR SPILLED

WASTE DISPOSAL METHOD

SECTION VIII SPECIAL PROTECTION INFORMATION

VENTILATION REQUIREMENTS LOCAL EXHAUST	PROTECTIVE EQUIPMENT (SPECIFY TYPES) EYE Goggles
MECHANICAL (GENERAL)	GLOVES
SPECIAL	RESPIRATOR Self-contained breathing apparatus

OTHER PROTECTIVE EQUIPMENT

SECTION IX SPECIAL PRECAUTIONS

PRECAUTIONS TO BE TAKEN IN HANDLING AND STORAGE
Protect against physical damage, detached storage preferred. Inside storage in flammable liquids cabinet.

OTHER PRECAUTIONS

SIGNATURE	DATE

BROMOACETIC ACID

Hazard diamond: 2 (top), 3 (left), 1 (right)

SECTION I SOURCE AND NOMENCLATURE

MANUFACTURER'S NAME	EMERGENCY TELEPHONE NO.
ADDRESS (NUMBER, STREET, CITY, STATE, ZIP CODE)	
TRADE NAME AND SYNONYMS	CHEMICAL FAMILY
CHEMICAL NAME AND SYNONYMS Bromacetic Acid	FORMULA $BrCH_2COOH$

SECTION II HAZARDOUS INGREDIENTS

BASIC MATERIAL	APPROXIMATE OR MAXIMUM % WT. OR VOL.	ESTABLISHED OSHA STANDARD	LD_{50} mg/kg ORAL	PERCUT.	SPECIES	LC_{50} CONC.
			100		Mouse	

SECTION III PHYSICAL DATA MW - 138.96

BOILING POINT °F. 208°C	VAPOR PRESSURE mm Hg.
MELTING POINT °F. 50°C	VAPOR DENSITY (Air = 1)
SPECIFIC GRAVITY (H_2O = 1) 1.94	EVAPORATION RATE (_____ = 1)
SOLUBILITY IN WATER Pts/100 pts H_2O Very soluble	VOLATILE % Vol. % Wt.

APPEARANCE AND ODOR
Colorless crystals

SECTION IV FIRE AND EXPLOSION HAZARD DATA

FLASH POINT None available	FLAMMABLE (EXPLOSIVE) LIMITS	UPPER
METHOD USED		LOWER

EXTINGUISHING MEDIA
Use water spray, dry chemical, carbon dioxide alcohol foam.

SPECIAL FIRE FIGHTING PROCEDURES
Wear goggles & self-contained breathing apparatus.

UNUSUAL FIRE AND EXPLOSION HAZARDS

SECTION V HEALTH HAZARD DATA

TOXIC LEVEL
None available — CARCINOGENIC: No

PRINCIPLE ROUTES OF ABSORBTION
Inhalation, ingestion — SKIN AND EYE IRRITATION: Severe

REVELANT SYMPTOMS OF EXPOSURE
Severe eye & skin burns; severe respiratory and digestive system irritation

EFFECTS OF CHRONIC EXPOSURE
-

EMERGENCY AND FIRST AID PROCEDURES
If swallowed - drink water, milk of magnesia, get medical aid; inhalation - rest; wash eyes & skin.

SECTION VI REACTIVITY DATA

CONDITIONS CONTRIBUTING TO INSTABILITY

CONDITIONS CONTRIBUTING TO HAZARDOUS POLYMERIZATION

INCOMPATABILITY (MATERIALS TO AVOID)
Reacts with alkalies.

HAZARDOUS DECOMPOSITION PRODUCTS

SECTION VII SPILL OR LEAK PROCEDURES

STEPS TO BE TAKEN IN CASE MATERIAL IS RELEASED OR SPILLED
Pour sodium bicarbonate on spill.

WASTE DISPOSAL METHOD
Dissolve in combustible solvent and burn.

SECTION VIII SPECIAL PROTECTION INFORMATION

VENTILATION REQUIREMENTS LOCAL EXHAUST	PROTECTIVE EQUIPMENT (SPECIFY TYPES) EYE Goggles
MECHANICAL (GENERAL)	GLOVES Rubber
SPECIAL	RESPIRATOR w/appropriate filter

OTHER PROTECTIVE EQUIPMENT

SECTION IX SPECIAL PRECAUTIONS

PRECAUTIONS TO BE TAKEN IN HANDLING AND STORAGE
Store in dry, air-excluded place.

OTHER PRECAUTIONS

SIGNATURE	DATE

α-BROMOACETOPHENONE

Hazard diamond: 1 (top), 4 (left), 0 (right)

SECTION I SOURCE AND NOMENCLATURE

MANUFACTURER'S NAME	EMERGENCY TELEPHONE NO.
ADDRESS (NUMBER, STREET, CITY, STATE, ZIP CODE)	
TRADE NAME AND SYNONYMS	CHEMICAL FAMILY
CHEMICAL NAME AND SYNONYMS Phenacyl Bromide, α -Bromo-acetophenone	FORMULA $BrCH_2COC_6H_5$

SECTION II HAZARDOUS INGREDIENTS

BASIC MATERIAL	APPROXIMATE OR MAXIMUM % WT. OR VOL.	ESTABLISHED OSHA STANDARD	LD_{50} ORAL	PERCUT.	SPECIES	LC_{50} CONC.
			None available			

SECTION III PHYSICAL DATA MW - 199.05

BOILING POINT °F. 119°C	VAPOR PRESSURE at 140°C 12 mm Hg.
MELTING POINT °F. 50°C	VAPOR DENSITY (Air = 1)
SPECIFIC GRAVITY (H_2O = 1) 1.647	EVAPORATION RATE (_____ = 1)
SOLUBILITY IN WATER Pts/100 pts H_2O Insoluble	VOLATILE % Vol. % Wt.

APPEARANCE AND ODOR
White to greenish crystals

SECTION IV FIRE AND EXPLOSION HAZARD DATA

FLASH POINT Not available	FLAMMABLE (EXPLOSIVE) LIMITS	UPPER
METHOD USED Carbon dioxide, dry chemical		LOWER

EXTINGUISHING MEDIA

SPECIAL FIRE FIGHTING PROCEDURES

UNUSUAL FIRE AND EXPLOSION HAZARDS

SECTION V HEALTH HAZARD DATA

TOXIC LEVEL
Not available — CARCINOGENIC: No

PRINCIPLE ROUTES OF ABSORBTION
Inhalation, ingestion — SKIN AND EYE IRRITATION: Severe

REVELANT SYMPTOMS OF EXPOSURE Eye irritation, corneal haziness and pain; burning of throat and eyes; irritation of skin and eyes.

EFFECTS OF CHRONIC EXPOSURE
--

EMERGENCY AND FIRST AID PROCEDURES Inhalation - remove from exposure, rest. If swallowed - wash mouth, give emetic, get medical aid; wash from skin & eyes.

SECTION VI REACTIVITY DATA

CONDITIONS CONTRIBUTING TO INSTABILITY

CONDITIONS CONTRIBUTING TO HAZARDOUS POLYMERIZATION

INCOMPATABILITY (MATERIALS TO AVOID)

HAZARDOUS DECOMPOSITION PRODUCTS

SECTION VII SPILL OR LEAK PROCEDURES

STEPS TO BE TAKEN IN CASE MATERIAL IS RELEASED OR SPILLED

WASTE DISPOSAL METHOD

SECTION VIII SPECIAL PROTECTION INFORMATION

VENTILATION REQUIREMENTS LOCAL EXHAUST	PROTECTIVE EQUIPMENT (SPECIFY TYPES) EYE Goggles
MECHANICAL (GENERAL)	GLOVES Rubber
SPECIAL	RESPIRATOR With proper filter

OTHER PROTECTIVE EQUIPMENT

SECTION IX SPECIAL PRECAUTIONS

PRECAUTIONS TO BE TAKEN IN HANDLING AND STORAGE

OTHER PRECAUTIONS

SIGNATURE	DATE

BROMOACETYL BROMIDE

Hazard rating: 2 (top), 4 (left), 2 (right), -W- (bottom)

SECTION I SOURCE AND NOMENCLATURE

MANUFACTURER'S NAME	EMERGENCY TELEPHONE NO.
ADDRESS (NUMBER, STREET, CITY, STATE, ZIP CODE)	
TRADE NAME AND SYNONYMS	CHEMICAL FAMILY
CHEMICAL NAME AND SYNONYMS: Bromoacetyl Bromide	FORMULA: $BrCH_2COBr$

SECTION II HAZARDOUS INGREDIENTS

BASIC MATERIAL	APPROXIMATE OR MAXIMUM % WT. OR VOL.	ESTABLISHED OSHA STANDARD	LD_{50} ORAL	LD_{50} PERCUT.	LC_{50} SPECIES	LC_{50} CONC.
			None	available		

SECTION III PHYSICAL DATA MW-201.87

BOILING POINT °F.: 150°C	VAPOR PRESSURE mm Hg.
MELTING POINT °F.: --	VAPOR DENSITY (Air = 1)
SPECIFIC GRAVITY (H_2O = 1): 2.32	EVAPORATION RATE (= 1)
SOLUBILITY IN WATER Pts/100 pts H_2O: Decomposes	VOLATILE % Vol. % Wt.
APPEARANCE AND ODOR: Fuming liquid	

SECTION IV FIRE AND EXPLOSION HAZARD DATA

FLASH POINT: Not available

METHOD USED

FLAMMABLE (EXPLOSIVE) LIMITS: UPPER / LOWER

EXTINGUISHING MEDIA: Do not use water or foam, use carbon dioxide or dry chemical.

SPECIAL FIRE FIGHTING PROCEDURES

UNUSUAL FIRE AND EXPLOSION HAZARDS: When heated to decomposition point gives off toxic HBr.

SECTION V HEALTH HAZARD DATA

TOXIC LEVEL: None available — CARCINOGENIC: No

PRINCIPLE ROUTES OF ABSORBTION: Inhalation, ingestion — SKIN AND EYE IRRITATION: Severe

REVELANT SYMPTOMS OF EXPOSURE: Irritation of eyes & respiratory system; severe internal damage if swallowed; eye & skin burns.

EFFECTS OF CHRONIC EXPOSURE: Bronchial irritation, errosion of teeth.

EMERGENCY AND FIRST AID PROCEDURES: Inhalation - rest; get medical aid; if swallowed - drink water, milk of magnesia, get medical aid, wash eyes & skin.

SECTION VI REACTIVITY DATA

CONDITIONS CONTRIBUTING TO INSTABILITY

CONDITIONS CONTRIBUTING TO HAZARDOUS POLYMERIZATION

INCOMPATABILITY (MATERIALS TO AVOID): Reacts violently with water or alcohol.

HAZARDOUS DECOMPOSITION PRODUCTS

SECTION VII SPILL OR LEAK PROCEDURES

STEPS TO BE TAKEN IN CASE MATERIAL IS RELEASED OR SPILLED

WASTE DISPOSAL METHOD

SECTION VIII SPECIAL PROTECTION INFORMATION

VENTILATION REQUIREMENTS LOCAL EXHAUST	PROTECTIVE EQUIPMENT (SPECIFY TYPES) EYE: Goggles
MECHANICAL (GENERAL)	GLOVES: Rubber
SPECIAL	RESPIRATOR: Self contained

OTHER PROTECTIVE EQUIPMENT: Rubber apron

SECTION IX SPECIAL PRECAUTIONS

PRECAUTIONS TO BE TAKEN IN HANDLING AND STORAGE

OTHER PRECAUTIONS

SIGNATURE — DATE

3-BROMO-4-AMINOTOLUENE

Hazard rating: 2 (top), 4 (left), 0 (right)

SECTION I SOURCE AND NOMENCLATURE

MANUFACTURER'S NAME	EMERGENCY TELEPHONE NO.
ADDRESS (NUMBER, STREET, CITY, STATE, ZIP CODE)	
TRADE NAME AND SYNONYMS: 3 Bromo P Toluidine	CHEMICAL FAMILY
CHEMICAL NAME AND SYNONYMS: 2 Bromo-4 Methyl Aniline	FORMULA: $CH_3C_6H_3BrNH_2$

SECTION II HAZARDOUS INGREDIENTS

BASIC MATERIAL	APPROXIMATE OR MAXIMUM % WT. OR VOL.	ESTABLISHED OSHA STANDARD	LD_{50} mg/Kg. ORAL	LD_{50} PERCUT.	LC_{50} SPECIES	LC_{50} CONC.
			1.9		Bird	

SECTION III PHYSICAL DATA MW-186.07

BOILING POINT °F.: 240	VAPOR PRESSURE mm Hg.
MELTING POINT °F.: 16-18	VAPOR DENSITY (Air = 1)
SPECIFIC GRAVITY (H_2O = 1): 1.5	EVAPORATION RATE (= 1)
SOLUBILITY IN WATER Pts/100 pts H_2O: Insoluble	VOLATILE % Vol. % Wt.
APPEARANCE AND ODOR: Liquid	

SECTION IV FIRE AND EXPLOSION HAZARD DATA

FLASH POINT: N/A

METHOD USED

FLAMMABLE (EXPLOSIVE) LIMITS: UPPER / LOWER

EXTINGUISHING MEDIA: Carbon dioxide, dry chemical, foam or water spray

SPECIAL FIRE FIGHTING PROCEDURES: Wear full protective clothing

UNUSUAL FIRE AND EXPLOSION HAZARDS: When heated emits highly toxic fumes.

SECTION V HEALTH HAZARD DATA

TOXIC LEVEL: Not available — CARCINOGENIC: No

PRINCIPLE ROUTES OF ABSORBTION: Inhalation, ingestion, skin — SKIN AND EYE IRRITATION: Yes

REVELANT SYMPTOMS OF EXPOSURE: Cyanosis, air hunger, nausea, vomiting, low blood pressure, convulsions

EFFECTS OF CHRONIC EXPOSURE: Weight loss, anemia, weakness

EMERGENCY AND FIRST AID PROCEDURES: Inhalation - rest. If swallowed - wash mouth, give emetic, get medical aid; wash from skin & eyes.

SECTION VI REACTIVITY DATA

CONDITIONS CONTRIBUTING TO INSTABILITY

CONDITIONS CONTRIBUTING TO HAZARDOUS POLYMERIZATION

INCOMPATABILITY (MATERIALS TO AVOID)

HAZARDOUS DECOMPOSITION PRODUCTS

SECTION VII SPILL OR LEAK PROCEDURES

STEPS TO BE TAKEN IN CASE MATERIAL IS RELEASED OR SPILLED

WASTE DISPOSAL METHOD

SECTION VIII SPECIAL PROTECTION INFORMATION

VENTILATION REQUIREMENTS LOCAL EXHAUST	PROTECTIVE EQUIPMENT (SPECIFY TYPES) EYE: Safety goggles
MECHANICAL (GENERAL)	GLOVES: Rubber
SPECIAL	RESPIRATOR: With proper filter

OTHER PROTECTIVE EQUIPMENT

SECTION IX SPECIAL PRECAUTIONS

PRECAUTIONS TO BE TAKEN IN HANDLING AND STORAGE

OTHER PRECAUTIONS

SIGNATURE — DATE

5-BROMO-2-AMINOTOLUENE

(Hazard diamond: 2 / 4 / 0)

SECTION I SOURCE AND NOMENCLATURE

MANUFACTURER'S NAME | EMERGENCY TELEPHONE NO.

ADDRESS (NUMBER, STREET, CITY, STATE, ZIP CODE)

TRADE NAME AND SYNONYMS: 4 Bromo-2-toluidine

CHEMICAL FAMILY:

CHEMICAL NAME AND SYNONYMS: 4 Bromo 2 Methyl Analine

FORMULA: $CH_3C_6H_3BrNH_2$

SECTION II HAZARDOUS INGREDIENTS

BASIC MATERIAL	APPROXIMATE OR MAXIMUM % WT. OR VOL.	ESTABLISHED OSHA STANDARD	LD_{50} ORAL	LD_{50} PERCUT.	LC_{50} SPECIES	LC_{50} CONC.
			None available			

SECTION III PHYSICAL DATA MW-686.07

BOILING POINT: 240°C

VAPOR PRESSURE: mm Hg.

MELTING POINT: 58-59°C

VAPOR DENSITY (Air = 1)

SPECIFIC GRAVITY (H_2O = 1)

EVAPORATION RATE (= 1)

SOLUBILITY IN WATER Pts/100 pts H_2O

VOLATILE % Vol. % Wt.

APPEARANCE AND ODOR: Crystaline

SECTION IV FIRE AND EXPLOSION HAZARD DATA

FLASH POINT: N/A

FLAMMABLE (EXPLOSIVE) LIMITS: UPPER / LOWER

METHOD USED

EXTINGUISHING MEDIA: Carbon dioxide, dry chemical, foam or water spray

SPECIAL FIRE FIGHTING PROCEDURES: Wear full protective clothing

UNUSUAL FIRE AND EXPLOSION HAZARDS: When heated emits highly toxic fumes.

SECTION V HEALTH HAZARD DATA

TOXIC LEVEL: Not available

CARCINOGENIC: No

PRINCIPLE ROUTES OF ABSORBTION: Inhalation, ingestion, skin

SKIN AND EYE IRRITATION: Eyes

REVELANT SYMPTOMS OF EXPOSURE: Cyanosis, air hunger, nausea, vomiting, low blood pressure, convulsions.

EFFECTS OF CHRONIC EXPOSURE: Weight loss, anemia, weakness

EMERGENCY AND FIRST AID PROCEDURES: Inhalation - rest. If swallowed - wash mouth, give emetic, get medial aid; wash from skin & eyes.

SECTION VI REACTIVITY DATA

CONDITIONS CONTRIBUTING TO INSTABILITY

CONDITIONS CONTRIBUTING TO HAZARDOUS POLYMERIZATION

INCOMPATABILITY (MATERIALS TO AVOID)

HAZARDOUS DECOMPOSITION PRODUCTS

SECTION VII SPILL OR LEAK PROCEDURES

STEPS TO BE TAKEN IN CASE MATERIAL IS RELEASED OR SPILLED

WASTE DISPOSAL METHOD

SECTION VIII SPECIAL PROTECTION INFORMATION

VENTILATION REQUIREMENTS LOCAL EXHAUST

PROTECTIVE EQUIPMENT (SPECIFY TYPES) EYE: Safety goggles

MECHANICAL (GENERAL)

GLOVES: Rubber

SPECIAL

RESPIRATOR: With proper filter

OTHER PROTECTIVE EQUIPMENT

SECTION IX SPECIAL PRECAUTIONS

PRECAUTIONS TO BE TAKEN IN HANDLING AND STORAGE

OTHER PRECAUTIONS

SIGNATURE | DATE

m-BROMOANILINE

(Hazard diamond: 1 / 4 / 0)

SECTION I SOURCE AND NOMENCLATURE

MANUFACTURER'S NAME | EMERGENCY TELEPHONE NO.

ADDRESS (NUMBER, STREET, CITY, STATE, ZIP CODE)

TRADE NAME AND SYNONYMS

CHEMICAL FAMILY

CHEMICAL NAME AND SYNONYMS: M-Bromoaniline

FORMULA: $Br(C_6H_4)NH_2$

SECTION II HAZARDOUS INGREDIENTS

BASIC MATERIAL	APPROXIMATE OR MAXIMUM % WT. OR VOL.	ESTABLISHED OSHA STANDARD	LD_{50} ORAL	LD_{50} PERCUT.	LC_{50} SPECIES	LC_{50} CONC.
		None available				

SECTION III PHYSICAL DATA MW - 172.03

BOILING POINT: 251°C.

VAPOR PRESSURE: at 16°C. 16 mm Hg.

MELTING POINT: 19°C.

VAPOR DENSITY (Air = 1)

SPECIFIC GRAVITY (H_2O = 1): 1.58

EVAPORATION RATE (= 1)

SOLUBILITY IN WATER Pts/100 pts H_2O: Insoluble

VOLATILE % Vol. % Wt.

APPEARANCE AND ODOR: Liquid

SECTION IV FIRE AND EXPLOSION HAZARD DATA

FLASH POINT: None available

FLAMMABLE (EXPLOSIVE) LIMITS: UPPER / LOWER

METHOD USED

EXTINGUISHING MEDIA: Water spray, carbon dioxide, alcohol foam.

SPECIAL FIRE FIGHTING PROCEDURES: Self-contained breathing apparatus.

UNUSUAL FIRE AND EXPLOSION HAZARDS: Vapor may form explosive mixture in air.

SECTION V HEALTH HAZARD DATA

TOXIC LEVEL: None available

CARCINOGENIC: No

PRINCIPLE ROUTES OF ABSORBTION: Inhalation, skin, ingestion

SKIN AND EYE IRRITATION: Low

REVELANT SYMPTOMS OF EXPOSURE: Cyanosis, headache, blood pressure drop, lethargy, jaundice and anemia, pain in joints, fainting.

EFFECTS OF CHRONIC EXPOSURE: Weight loss, anemia, weakness and irritability, possible liver damage, inflammation of kidney, ringing in ear.

EMERGENCY AND FIRST AID PROCEDURES: Remove from exposure; wash from skin & eyes. If swallowed - remove by gastric lavage and cartharsis; give oxygen as needed.

SECTION VI REACTIVITY DATA

CONDITIONS CONTRIBUTING TO INSTABILITY

CONDITIONS CONTRIBUTING TO HAZARDOUS POLYMERIZATION

INCOMPATABILITY (MATERIALS TO AVOID): Sodium and potassium peroxide ozone, perchloric acid.

HAZARDOUS DECOMPOSITION PRODUCTS

SECTION VII SPILL OR LEAK PROCEDURES

STEPS TO BE TAKEN IN CASE MATERIAL IS RELEASED OR SPILLED: Wear protective clothing during clean up.

WASTE DISPOSAL METHOD

SECTION VIII SPECIAL PROTECTION INFORMATION

VENTILATION REQUIREMENTS LOCAL EXHAUST

PROTECTIVE EQUIPMENT (SPECIFY TYPES) EYE: Goggles

MECHANICAL (GENERAL)

GLOVES: Rubber

SPECIAL

RESPIRATOR: Self-contained

OTHER PROTECTIVE EQUIPMENT: Protective clothing

SECTION IX SPECIAL PRECAUTIONS

PRECAUTIONS TO BE TAKEN IN HANDLING AND STORAGE: Store in flammable liquids cabinet.

OTHER PRECAUTIONS

SIGNATURE | DATE

o-BROMOANILINE

Hazard rating diamond: 1 (top), 3 (left), 0 (right)

SECTION I SOURCE AND NOMENCLATURE

MANUFACTURER'S NAME | EMERGENCY TELEPHONE NO.

ADDRESS (NUMBER, STREET, CITY, STATE, ZIP CODE) —

TRADE NAME AND SYNONYMS | CHEMICAL FAMILY

CHEMICAL NAME AND SYNONYMS: O-Bromoaniline | FORMULA: $Br(C_6H_4)NH_2$

SECTION II HAZARDOUS INGREDIENTS

BASIC MATERIAL	APPROXIMATE OR MAXIMUM % WT. OR VOL.	ESTABLISHED OSHA STANDARD	LD_{50} ORAL	LD_{50} PERCUT.	LC_{50} SPECIES	LC_{50} CONC.
			None available			

SECTION III PHYSICAL DATA MW - 172.03

BOILING POINT: 229°C. °F. | VAPOR PRESSURE: Low mm Hg.

MELTING POINT: 32°C. °F. | VAPOR DENSITY (Air = 1)

SPECIFIC GRAVITY (H_2O = 1): -- | EVAPORATION RATE (= 1)

SOLUBILITY IN WATER Pts/100 pts H_2O: Insoluble | VOLATILE % Vol. % Wt.

APPEARANCE AND ODOR: Crystals

SECTION IV FIRE AND EXPLOSION HAZARD DATA

FLASH POINT: None available | FLAMMABLE (EXPLOSIVE) LIMITS: UPPER / LOWER

METHOD USED

EXTINGUISHING MEDIA: Water spray, carbon dioxide, alcohol foam.

SPECIAL FIRE FIGHTING PROCEDURES: Self-contained breathing apparatus.

UNUSUAL FIRE AND EXPLOSION HAZARDS: Vapor may form explosive mixture in air.

SECTION V HEALTH HAZARD DATA

TOXIC LEVEL: None available | CARCINOGENIC: No

PRINCIPLE ROUTES OF ABSORBTION: Inhalation, skin, ingestion | SKIN AND EYE IRRITATION: Low

REVELANT SYMPTOMS OF EXPOSURE: Headache, nausea, vomiting; jaundice, convulsions.

EFFECTS OF CHRONIC EXPOSURE: Severe kidney damage, some liver damage, dermatitis.

EMERGENCY AND FIRST AID PROCEDURES: Remove from exposure; wash skin & eyes; if swallowed - remove by gastric lavage and cathartics, oxygen as needed

SECTION VI REACTIVITY DATA

CONDITIONS CONTRIBUTING TO INSTABILITY

CONDITIONS CONTRIBUTING TO HAZARDOUS POLYMERIZATION

INCOMPATABILITY (MATERIALS TO AVOID): Sodium & potassium peroxide, ozone, perchloric acid.

HAZARDOUS DECOMPOSITION PRODUCTS

SECTION VII SPILL OR LEAK PROCEDURES

STEPS TO BE TAKEN IN CASE MATERIAL IS RELEASED OR SPILLED: Wear protective clothing during cleanup.

WASTE DISPOSAL METHOD

SECTION VIII SPECIAL PROTECTION INFORMATION

VENTILATION REQUIREMENTS LOCAL EXHAUST | PROTECTIVE EQUIPMENT (SPECIFY TYPES) EYE: Goggles

MECHANICAL (GENERAL) | GLOVES: Rubber

SPECIAL | RESPIRATOR: Self-contained

OTHER PROTECTIVE EQUIPMENT: Protective clothing.

SECTION IX SPECIAL PRECAUTIONS

PRECAUTIONS TO BE TAKEN IN HANDLING AND STORAGE: Store in flammable liquids cabinet.

OTHER PRECAUTIONS

SIGNATURE | DATE

p-BROMOANILINE

Hazard rating diamond: 1 (top), 4 (left), 0 (right)

SECTION I SOURCE AND NOMENCLATURE

MANUFACTURER'S NAME | EMERGENCY TELEPHONE NO.

ADDRESS (NUMBER, STREET, CITY, STATE, ZIP CODE)

TRADE NAME AND SYNONYMS | CHEMICAL FAMILY

CHEMICAL NAME AND SYNONYMS: P-Bromoaniline | FORMULA: $Br(C_6H_4)NH_2$

SECTION II HAZARDOUS INGREDIENTS

BASIC MATERIAL	APPROXIMATE OR MAXIMUM % WT. OR VOL.	ESTABLISHED OSHA STANDARD	LD_{50} ORAL	LD_{50} PERCUT.	LC_{50} SPECIES	LC_{50} CONC.
			None Available			

SECTION III PHYSICAL DATA MW - 172.03

BOILING POINT °F. | VAPOR PRESSURE: Low mm Hg.

MELTING POINT: 62-64°C. °F. | VAPOR DENSITY (Air = 1)

SPECIFIC GRAVITY (H_2O = 1): 1.80 | EVAPORATION RATE (= 1)

SOLUBILITY IN WATER Pts/100 pts H_2O: Insoluble | VOLATILE % Vol. % Wt.

APPEARANCE AND ODOR: Crystalline, sweet odor.

SECTION IV FIRE AND EXPLOSION HAZARD DATA

FLASH POINT: None available | FLAMMABLE (EXPLOSIVE) LIMITS: UPPER / LOWER

METHOD USED

EXTINGUISHING MEDIA: Water spray, carbon dioxide, alcohol foam.

SPECIAL FIRE FIGHTING PROCEDURES: Self-contained breathing apparatus.

UNUSUAL FIRE AND EXPLOSION HAZARDS: Vapor may form explosive mixtures in air.

SECTION V HEALTH HAZARD DATA

TOXIC LEVEL: None available | CARCINOGENIC: No

PRINCIPLE ROUTES OF ABSORBTION: Inhalation, skin, ingestion | SKIN AND EYE IRRITATION: Low

REVELANT SYMPTOMS OF EXPOSURE: Cyanosis, headache, blood pressure drop, lethargy, jaundice and anemia, pain in joints, fainting.

EFFECTS OF CHRONIC EXPOSURE: Weight loss, anemia, weakness and irritability, possible liver damage, inflammation of kidneys, ringing in ear.

EMERGENCY AND FIRST AID PROCEDURES: Remove from exposure; wash from eyes and skin; if ingested - remove by gastric lavage and catharsis; give oxygen as needed.

SECTION VI REACTIVITY DATA

CONDITIONS CONTRIBUTING TO INSTABILITY

CONDITIONS CONTRIBUTING TO HAZARDOUS POLYMERIZATION

INCOMPATABILITY (MATERIALS TO AVOID): Sodium or potassium peroxide, ozone, perchloric acid.

HAZARDOUS DECOMPOSITION PRODUCTS

SECTION VII SPILL OR LEAK PROCEDURES

STEPS TO BE TAKEN IN CASE MATERIAL IS RELEASED OR SPILLED: Wear protective clothing during clean up.

WASTE DISPOSAL METHOD

SECTION VIII SPECIAL PROTECTION INFORMATION

VENTILATION REQUIREMENTS LOCAL EXHAUST | PROTECTIVE EQUIPMENT (SPECIFY TYPES) EYE: Safety goggles

MECHANICAL (GENERAL) | GLOVES: Rubber

SPECIAL | RESPIRATOR: Self-contained

OTHER PROTECTIVE EQUIPMENT: Protective clothing

SECTION IX SPECIAL PRECAUTIONS

PRECAUTIONS TO BE TAKEN IN HANDLING AND STORAGE: Store in flammable liquids cabinet.

OTHER PRECAUTIONS

SIGNATURE | DATE

m-BROMOBENZOYL BROMIDE

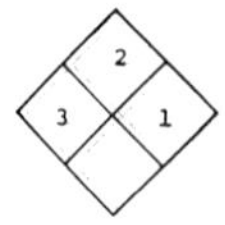

SECTION I SOURCE AND NOMENCLATURE

MANUFACTURER'S NAME | EMERGENCY TELEPHONE NO.

ADDRESS (NUMBER, STREET, CITY, STATE, ZIP CODE)

TRADE NAME AND SYNONYMS | CHEMICAL FAMILY

CHEMICAL NAME AND SYNONYMS: M-Bromobenzoyl Bromide

FORMULA: BrC_6H_4COBr

SECTION II HAZARDOUS INGREDIENTS

BASIC MATERIAL	APPROXIMATE OR MAXIMUM % WT. OR VOL.	ESTABLISHED OSHA STANDARD	LD_{50} ORAL	LD_{50} PERCUT.	LC_{50} SPECIES	LC_{50} CONC.
				Not available		

SECTION III PHYSICAL DATA

BOILING POINT °F.: 124-126°C

VAPOR PRESSURE mm Hg.

MELTING POINT °F.: N/A

VAPOR DENSITY (Air = 1)

SPECIFIC GRAVITY (H_2O = 1)

EVAPORATION RATE (______ = 1)

SOLUBILITY IN WATER Pts/100 pts H_2O

VOLATILE % Vol. % Wt.

APPEARANCE AND ODOR: Colorless liquid

SECTION IV FIRE AND EXPLOSION HAZARD DATA

FLASH POINT: N/A

METHOD USED

FLAMMABLE (EXPLOSIVE) LIMITS: UPPER / LOWER

EXTINGUISHING MEDIA: Use water spray sparingly, foam, carbon dioxide or dry chemical.

SPECIAL FIRE FIGHTING PROCEDURES: Wear full protective clothing.

UNUSUAL FIRE AND EXPLOSION HAZARDS

SECTION V HEALTH HAZARD DATA

TOXIC LEVEL: Not available

CARCINOGENIC: No

PRINCIPLE ROUTES OF ABSORBTION: Inhalation, ingestion

SKIN AND EYE IRRITATION: Severe

REVELANT SYMPTOMS OF EXPOSURE: Irritation of respiratory system, burns skin, eyes, mouth and digestive tract

EFFECTS OF CHRONIC EXPOSURE: --

EMERGENCY AND FIRST AID PROCEDURES: Inhalation - rest; if swallowed - wash mouth, drink water, milk of magnesia, get medical aid; wash skin & eyes.

SECTION VI REACTIVITY DATA

CONDITIONS CONTRIBUTING TO INSTABILITY: Reacts with oxidizing materials

CONDITIONS CONTRIBUTING TO HAZARDOUS POLYMERIZATION

INCOMPATABILITY (MATERIALS TO AVOID)

HAZARDOUS DECOMPOSITION PRODUCTS

SECTION VII SPILL OR LEAK PROCEDURES

STEPS TO BE TAKEN IN CASE MATERIAL IS RELEASED OR SPILLED

WASTE DISPOSAL METHOD

SECTION VIII SPECIAL PROTECTION INFORMATION

VENTILATION REQUIREMENTS LOCAL EXHAUST

MECHANICAL (GENERAL)

SPECIAL

OTHER PROTECTIVE EQUIPMENT

PROTECTIVE EQUIPMENT (SPECIFY TYPES) EYE: Goggles

GLOVES: Rubber

RESPIRATOR: Self-contained

SECTION IX SPECIAL PRECAUTIONS

PRECAUTIONS TO BE TAKEN IN HANDLING AND STORAGE

OTHER PRECAUTIONS

SIGNATURE | DATE

BROMOCHLOROMETHANE

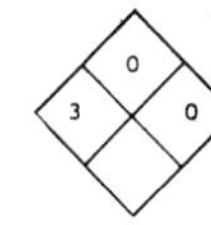

SECTION I SOURCE AND NOMENCLATURE

MANUFACTURER'S NAME | EMERGENCY TELEPHONE NO.

ADDRESS (NUMBER, STREET, CITY, STATE, ZIP CODE)

TRADE NAME AND SYNONYMS: Methylene chlorobromide

CHEMICAL FAMILY

CHEMICAL NAME AND SYNONYMS: Bromochloromethane

FORMULA: $ClCH_2Br$

SECTION II HAZARDOUS INGREDIENTS

BASIC MATERIAL	APPROXIMATE OR MAXIMUM % WT. OR VOL.	ESTABLISHED OSHA STANDARD	LD_{50} mg/kg ORAL	LD_{50} mg/kg PERCUT.	LC_{50} PPM SPECIES	LC_{50} PPM CONC.
LDLO - oral - human 50	mg/kg		5000		Rat	465-15 M.
			4300		Mouse	2300-72H.
Lowest published lethal	dose in air					
inhalation - mouse 15	00 PPM					

SECTION III PHYSICAL DATA MW 129.4

BOILING POINT °F.: 68°C

VAPOR PRESSURE mm Hg.

MELTING POINT °F.: -55°C

VAPOR DENSITY (Air = 1): 4.46

SPECIFIC GRAVITY (H_2O = 1): 1.99

EVAPORATION RATE (______ = 1)

SOLUBILITY IN WATER Pts/100 pts H_2O: Sparingly ~~soluble in water~~

VOLATILE % Vol. % Wt.

APPEARANCE AND ODOR: Colorless liquid, sweet chloroform like odor

SECTION IV FIRE AND EXPLOSION HAZARD DATA

FLASH POINT: None

METHOD USED

FLAMMABLE (EXPLOSIVE) LIMITS: UPPER / LOWER

EXTINGUISHING MEDIA

SPECIAL FIRE FIGHTING PROCEDURES

UNUSUAL FIRE AND EXPLOSION HAZARDS: May emit toxic fumes when heated.

SECTION V HEALTH HAZARD DATA

TOXIC LEVEL: TLV - 200 PPM, 1050 mg/C.M.

CARCINOGENIC: No

PRINCIPLE ROUTES OF ABSORBTION: Inhalation, ingestion, skin

SKIN AND EYE IRRITATION: Severe

REVELANT SYMPTOMS OF EXPOSURE: Inhalation - pulmonary edema; irritation of eyes or skin; central nervous system depression

EFFECTS OF CHRONIC EXPOSURE: Liver and kidney damage

EMERGENCY AND FIRST AID PROCEDURES: Inhalation - rest; if swallowed, wash mouth, give emetic, get medical aid; wash from eyes & skin.

SECTION VI REACTIVITY DATA

CONDITIONS CONTRIBUTING TO INSTABILITY

CONDITIONS CONTRIBUTING TO HAZARDOUS POLYMERIZATION

INCOMPATABILITY (MATERIALS TO AVOID)

HAZARDOUS DECOMPOSITION PRODUCTS

SECTION VII SPILL OR LEAK PROCEDURES

STEPS TO BE TAKEN IN CASE MATERIAL IS RELEASED OR SPILLED: Absorb spill with paper towels. Let liquid evaporate in hood.

WASTE DISPOSAL METHOD: Dissolve in combustible solvent and burn.

SECTION VIII SPECIAL PROTECTION INFORMATION

VENTILATION REQUIREMENTS LOCAL EXHAUST

MECHANICAL (GENERAL)

SPECIAL

OTHER PROTECTIVE EQUIPMENT: Rubber coveralls

PROTECTIVE EQUIPMENT (SPECIFY TYPES) EYE: Goggles

GLOVES: Rubber

RESPIRATOR: With proper filter

SECTION IX SPECIAL PRECAUTIONS

PRECAUTIONS TO BE TAKEN IN HANDLING AND STORAGE

OTHER PRECAUTIONS

SIGNATURE | DATE

p-BROMODIETHYLANILINE

Hazard diamond: 1 (top), 4 (left), 0 (right)

SECTION I SOURCE AND NOMENCLATURE

MANUFACTURER'S NAME | EMERGENCY TELEPHONE NO.

ADDRESS (NUMBER, STREET, CITY, STATE, ZIP CODE)

TRADE NAME AND SYNONYMS | CHEMICAL FAMILY

CHEMICAL NAME AND SYNONYMS: P-Bromo diethylaniline | FORMULA: $BrC_6H_4N(C_2H_5)_2$

SECTION II HAZARDOUS INGREDIENTS

BASIC MATERIAL	APPROXIMATE OR MAXIMUM % WT. OR VOL.	ESTABLISHED OSHA STANDARD	LD_{50} ORAL	LD_{50} PERCUT.	LC_{50} SPECIES	LC_{50} CONC.
			Not available			

SECTION III PHYSICAL DATA

BOILING POINT °F.: 270 °C | VAPOR PRESSURE mm Hg.

MELTING POINT °F.: 33°C | VAPOR DENSITY (Air = 1)

SPECIFIC GRAVITY (H_2O = 1) | EVAPORATION RATE (= 1)

SOLUBILITY IN WATER Pts/100 pts H_2O: Insoluble | VOLATILE % Vol. % Wt.

APPEARANCE AND ODOR: Yellow crystal

SECTION IV FIRE AND EXPLOSION HAZARD DATA

FLASH POINT: Probably above 200°F estimated | FLAMMABLE (EXPLOSIVE) LIMITS: UPPER / LOWER

METHOD USED

EXTINGUISHING MEDIA: Alcohol foam, water spray, carbon dioxide, dry chemical

SPECIAL FIRE FIGHTING PROCEDURES

UNUSUAL FIRE AND EXPLOSION HAZARDS

SECTION V HEALTH HAZARD DATA

TOXIC LEVEL: Not available | CARCINOGENIC: No

PRINCIPLE ROUTES OF ABSORBTION: Inhalation, ingestion, skin | SKIN AND EYE IRRITATION: yes

REVELANT SYMPTOMS OF EXPOSURE: Weakness, tremors, cyanosis, drowsiness, central nervous system depressant

EFFECTS OF CHRONIC EXPOSURE: Fatigue, loss of appetite, headache, dizziness

EMERGENCY AND FIRST AID PROCEDURES: Inhalation - rest, get medical aid; if swallowed - wash mouth, give emetic; get medical aid; wash from skin & eyes.

SECTION VI REACTIVITY DATA

CONDITIONS CONTRIBUTING TO INSTABILITY

CONDITIONS CONTRIBUTING TO HAZARDOUS POLYMERIZATION

INCOMPATABILITY (MATERIALS TO AVOID)

HAZARDOUS DECOMPOSITION PRODUCTS

SECTION VII SPILL OR LEAK PROCEDURES

STEPS TO BE TAKEN IN CASE MATERIAL IS RELEASED OR SPILLED

WASTE DISPOSAL METHOD

SECTION VIII SPECIAL PROTECTION INFORMATION

VENTILATION REQUIREMENTS LOCAL EXHAUST | PROTECTIVE EQUIPMENT (SPECIFY TYPES) EYE: Goggles

MECHANICAL (GENERAL) | GLOVES: Gloves

SPECIAL | RESPIRATOR: With proper filter

OTHER PROTECTIVE EQUIPMENT

SECTION IX SPECIAL PRECAUTIONS

PRECAUTIONS TO BE TAKEN IN HANDLING AND STORAGE

OTHER PRECAUTIONS

SIGNATURE | DATE

p-BROMODIMETHYLANILINE

Hazard diamond: 1 (top), 4 (left), 0 (right)

SECTION I SOURCE AND NOMENCLATURE

MANUFACTURER'S NAME | EMERGENCY TELEPHONE NO.

ADDRESS (NUMBER, STREET, CITY, STATE, ZIP CODE)

TRADE NAME AND SYNONYMS: P-bromo, N,N'dimethylaniline | CHEMICAL FAMILY

CHEMICAL NAME AND SYNONYMS: P-bromodimethylaniline | FORMULA: $BrC_6H_4N(CH_3)_2$

SECTION II HAZARDOUS INGREDIENTS

BASIC MATERIAL	APPROXIMATE OR MAXIMUM % WT. OR VOL.	ESTABLISHED OSHA STANDARD	LD_{50} ORAL	LD_{50} PERCUT.	LC_{50} SPECIES	LC_{50} CONC.
LDLO oral - rat 500 mg/kg			None available			

SECTION III PHYSICAL DATA

BOILING POINT °F.: 264 °C | VAPOR PRESSURE mm Hg.

MELTING POINT °F.: 55 °C | VAPOR DENSITY (Air = 1)

SPECIFIC GRAVITY (H_2O = 1) | EVAPORATION RATE (= 1)

SOLUBILITY IN WATER Pts/100 pts H_2O: Insoluble | VOLATILE % Vol. % Wt.

APPEARANCE AND ODOR: Crystaline

SECTION IV FIRE AND EXPLOSION HAZARD DATA

FLASH POINT: Probably above 200°F estimated | FLAMMABLE (EXPLOSIVE) LIMITS: UPPER / LOWER

METHOD USED

EXTINGUISHING MEDIA: Alcohol foam, water spray, carbon dioxide, dry chemical

SPECIAL FIRE FIGHTING PROCEDURES

UNUSUAL FIRE AND EXPLOSION HAZARDS

SECTION V HEALTH HAZARD DATA

TOXIC LEVEL: Not available | CARCINOGENIC: No

PRINCIPLE ROUTES OF ABSORBTION: Inhalation, ingestion, skin | SKIN AND EYE IRRITATION: Yes

REVELANT SYMPTOMS OF EXPOSURE: Weakness, tremors, cyanosis, drowsiness, central nervous system depressant

EFFECTS OF CHRONIC EXPOSURE: Fatigue, loss of appetite, headache, dizziness

EMERGENCY AND FIRST AID PROCEDURES: Inhalation - rest, get medical aid; if swallowed - wash mouth, give emetic, get medical aid; wash from eyes & skin.

SECTION VI REACTIVITY DATA

CONDITIONS CONTRIBUTING TO INSTABILITY

CONDITIONS CONTRIBUTING TO HAZARDOUS POLYMERIZATION

INCOMPATABILITY (MATERIALS TO AVOID)

HAZARDOUS DECOMPOSITION PRODUCTS

SECTION VII SPILL OR LEAK PROCEDURES

STEPS TO BE TAKEN IN CASE MATERIAL IS RELEASED OR SPILLED

WASTE DISPOSAL METHOD

SECTION VIII SPECIAL PROTECTION INFORMATION

VENTILATION REQUIREMENTS LOCAL EXHAUST | PROTECTIVE EQUIPMENT (SPECIFY TYPES) EYE: Goggles

MECHANICAL (GENERAL) | GLOVES: Rubber

SPECIAL | RESPIRATOR: With proper filter

OTHER PROTECTIVE EQUIPMENT

SECTION IX SPECIAL PRECAUTIONS

PRECAUTIONS TO BE TAKEN IN HANDLING AND STORAGE

OTHER PRECAUTIONS

SIGNATURE | DATE

2-BROMO-1,4-DIMETHYLBENZENE

2 / 3 / 0

Similar to xylene

SECTION I SOURCE AND NOMENCLATURE

MANUFACTURER'S NAME

EMERGENCY TELEPHONE NO.

ADDRESS (NUMBER, STREET, CITY, STATE, ZIP CODE)

TRADE NAME AND SYNONYMS

CHEMICAL FAMILY

CHEMICAL NAME AND SYNONYMS: 2-Bromo-P-Xylene

FORMULA: $BrC_6H_3(CH_3)_2$

SECTION II HAZARDOUS INGREDIENTS

BASIC MATERIAL	APPROXIMATE OR MAXIMUM % WT. OR VOL.	ESTABLISHED OSHA STANDARD	LD_{50} ORAL	LD_{50} PERCUT.	LC_{50} SPECIES	LC_{50} CONC.
			See Note			
Note:						
Lowest published toxic concentration in air						
inhalation - human 75 PPM 110 months						

SECTION III PHYSICAL DATA

BOILING POINT °F: 205°C at 755 mm

VAPOR PRESSURE mm Hg.

MELTING POINT °F: 9-10°C

VAPOR DENSITY (Air = 1)

SPECIFIC GRAVITY (H_2O = 1): 1.356

EVAPORATION RATE (____ = 1)

SOLUBILITY IN WATER Pts/100 pts H_2O

VOLATILE % Vol. % Wt.

APPEARANCE AND ODOR: Colorless liquid

SECTION IV FIRE AND EXPLOSION HAZARD DATA

FLASH POINT: Flash point estimated 100-200°F

METHOD USED

FLAMMABLE (EXPLOSIVE) LIMITS: UPPER / LOWER

EXTINGUISHING MEDIA: Foam, carbon dioxide, dry chemical water spray may be used to reduce rate of burning

SPECIAL FIRE FIGHTING PROCEDURES: Wear self contained breathing apparatus

UNUSUAL FIRE AND EXPLOSION HAZARDS

SECTION V HEALTH HAZARD DATA

TOXIC LEVEL: Not available

CARCINOGENIC: No

PRINCIPLE ROUTES OF ABSORBTION: Inhalation, ingestion

SKIN AND EYE IRRITATION: Severe

REVELANT SYMPTOMS OF EXPOSURE: Irritation of respiratory tract, disturbance of vision, dizziness, nausea, collapse, coma; burning of skin or eyes.

EFFECTS OF CHRONIC EXPOSURE: Dermatitis, blood & liver damage

EMERGENCY AND FIRST AID PROCEDURES: Inhalation - rest, get medical aid; if swallowed - give saline cathartics and demulcents, get medical aid; wash skin & eyes.

SECTION VI REACTIVITY DATA

CONDITIONS CONTRIBUTING TO INSTABILITY

CONDITIONS CONTRIBUTING TO HAZARDOUS POLYMERIZATION

INCOMPATABILITY (MATERIALS TO AVOID)

HAZARDOUS DECOMPOSITION PRODUCTS

SECTION VII SPILL OR LEAK PROCEDURES

STEPS TO BE TAKEN IN CASE MATERIAL IS RELEASED OR SPILLED

WASTE DISPOSAL METHOD

SECTION VIII SPECIAL PROTECTION INFORMATION

VENTILATION REQUIREMENTS LOCAL EXHAUST

MECHANICAL (GENERAL)

SPECIAL

OTHER PROTECTIVE EQUIPMENT

PROTECTIVE EQUIPMENT (SPECIFY TYPES) EYE: Goggles

GLOVES: Rubber

RESPIRATOR: Self contained breathing apparatus

SECTION IX SPECIAL PRECAUTIONS

PRECAUTIONS TO BE TAKEN IN HANDLING AND STORAGE

OTHER PRECAUTIONS

SIGNATURE

DATE

o-BROMODIPHENYL

1 / 3 / 0

SECTION I SOURCE AND NOMENCLATURE

MANUFACTURER'S NAME

EMERGENCY TELEPHONE NO.

ADDRESS (NUMBER, STREET, CITY, STATE, ZIP CODE)

TRADE NAME AND SYNONYMS

CHEMICAL FAMILY

CHEMICAL NAME AND SYNONYMS: O-bromodiphenyl

FORMULA: $BrC_6H_4C_6H_5$

SECTION II HAZARDOUS INGREDIENTS

BASIC MATERIAL	APPROXIMATE OR MAXIMUM % WT. OR VOL.	ESTABLISHED OSHA STANDARD	LD_{50} ORAL	LD_{50} PERCUT.	LC_{50} SPECIES	LC_{50} CONC.
			None	available		

SECTION III PHYSICAL DATA MW-233.11

BOILING POINT °F: 296-8°C

VAPOR PRESSURE mm Hg.

MELTING POINT °F: <-20°C

VAPOR DENSITY (Air = 1)

SPECIFIC GRAVITY (H_2O = 1)

EVAPORATION RATE (____ = 1)

SOLUBILITY IN WATER Pts/100 pts H_2O: Insoluble

VOLATILE % Vol. % Wt.

APPEARANCE AND ODOR: Liquid

SECTION IV FIRE AND EXPLOSION HAZARD DATA

FLASH POINT: N/A

METHOD USED

FLAMMABLE (EXPLOSIVE) LIMITS: UPPER / LOWER

EXTINGUISHING MEDIA: Water, foam, carbon dioxide, dry chemical

SPECIAL FIRE FIGHTING PROCEDURES

UNUSUAL FIRE AND EXPLOSION HAZARDS

SECTION V HEALTH HAZARD DATA

TOXIC LEVEL: Not available

CARCINOGENIC: No

PRINCIPLE ROUTES OF ABSORBTION: Inhalation, ingestion skin

SKIN AND EYE IRRITATION: Yes

REVELANT SYMPTOMS OF EXPOSURE: Acne, irritation of respiratory passages

EFFECTS OF CHRONIC EXPOSURE: Liver damage

EMERGENCY AND FIRST AID PROCEDURES: Inhalation - rest, get medical aid; if swallowed, wash mouth, give emetic, get medical aid, wash skin & eyes.

SECTION VI REACTIVITY DATA

CONDITIONS CONTRIBUTING TO INSTABILITY

CONDITIONS CONTRIBUTING TO HAZARDOUS POLYMERIZATION

INCOMPATABILITY (MATERIALS TO AVOID)

HAZARDOUS DECOMPOSITION PRODUCTS

SECTION VII SPILL OR LEAK PROCEDURES

STEPS TO BE TAKEN IN CASE MATERIAL IS RELEASED OR SPILLED

WASTE DISPOSAL METHOD

SECTION VIII SPECIAL PROTECTION INFORMATION

VENTILATION REQUIREMENTS LOCAL EXHAUST

MECHANICAL (GENERAL)

SPECIAL

OTHER PROTECTIVE EQUIPMENT

PROTECTIVE EQUIPMENT (SPECIFY TYPES) EYE: Goggles

GLOVES: Rubber

RESPIRATOR: Self contained breathing apparatus

SECTION IX SPECIAL PRECAUTIONS

PRECAUTIONS TO BE TAKEN IN HANDLING AND STORAGE

OTHER PRECAUTIONS

SIGNATURE

DATE

p-BROMODIPHENYL

Hazard diamond: 1 (top), 3 (left), 0 (right)

SECTION I SOURCE AND NOMENCLATURE

MANUFACTURER'S NAME	EMERGENCY TELEPHONE NO.
ADDRESS (NUMBER, STREET, CITY, STATE, ZIP CODE)	
TRADE NAME AND SYNONYMS	CHEMICAL FAMILY
CHEMICAL NAME AND SYNONYMS: P-Bromodiphenyl	FORMULA: $BrC_6H_4C_6H_5$

SECTION II HAZARDOUS INGREDIENTS

BASIC MATERIAL	APPROXIMATE OR MAXIMUM % WT. OR VOL.	ESTABLISHED OSHA STANDARD	LD_{50} ORAL	LD_{50} PERCUT.	LC_{50} SPECIES	LC_{50} CONC.
			None	available		

SECTION III PHYSICAL DATA MW-233.11

BOILING POINT °F: 310.8°C	VAPOR PRESSURE mm Hg.
MELTING POINT °F: 89.8-90.3°C	VAPOR DENSITY (Air = 1)
SPECIFIC GRAVITY (H_2O = 1)	EVAPORATION RATE (______ = 1)
SOLUBILITY IN WATER Pts/100 pts H_2O: Insoluble	VOLATILE % Vol. % Wt.

APPEARANCE AND ODOR: White crystals

SECTION IV FIRE AND EXPLOSION HAZARD DATA

	FLAMMABLE (EXPLOSIVE) LIMITS	
FLASH POINT: 290°F		UPPER
METHOD USED		LOWER

EXTINGUISHING MEDIA: Water, foam, carbon dioxide, dry chemical

SPECIAL FIRE FIGHTING PROCEDURES

UNUSUAL FIRE AND EXPLOSION HAZARDS

SECTION V HEALTH HAZARD DATA

TOXIC LEVEL: Not available — CARCINOGENIC: No

PRINCIPLE ROUTES OF ABSORBTION: Inhalation, ingestion skin — SKIN AND EYE IRRITATION: Yes

REVELANT SYMPTOMS OF EXPOSURE: Acne, irritation of respiratory passages

EFFECTS OF CHRONIC EXPOSURE: Liver damage

EMERGENCY AND FIRST AID PROCEDURES: Inhalation - rest, get medical aid; if swallowed - wash mouth, give emetic, get medical aid, wash skin & eyes.

SECTION VI REACTIVITY DATA

CONDITIONS CONTRIBUTING TO INSTABILITY

CONDITIONS CONTRIBUTING TO HAZARDOUS POLYMERIZATION

INCOMPATABILITY (MATERIALS TO AVOID)

HAZARDOUS DECOMPOSITION PRODUCTS

SECTION VII SPILL OR LEAK PROCEDURES

STEPS TO BE TAKEN IN CASE MATERIAL IS RELEASED OR SPILLED

WASTE DISPOSAL METHOD

SECTION VIII SPECIAL PROTECTION INFORMATION

VENTILATION REQUIREMENTS LOCAL EXHAUST	PROTECTIVE EQUIPMENT (SPECIFY TYPES) EYE: Goggles
MECHANICAL (GENERAL)	GLOVES
SPECIAL	RESPIRATOR: With proper filter
OTHER PROTECTIVE EQUIPMENT	

SECTION IX SPECIAL PRECAUTIONS

PRECAUTIONS TO BE TAKEN IN HANDLING AND STORAGE

OTHER PRECAUTIONS

SIGNATURE	DATE

α-BROMOETHYLBENZENE

Hazard diamond: 1 (top), 3 (left), 0 (right)

SECTION I SOURCE AND NOMENCLATURE

MANUFACTURER'S NAME	EMERGENCY TELEPHONE NO.
ADDRESS (NUMBER, STREET, CITY, STATE, ZIP CODE)	
TRADE NAME AND SYNONYMS: α Phenylethylbromide	CHEMICAL FAMILY
CHEMICAL NAME AND SYNONYMS: (1 Bromoethyl) benzene	FORMULA: $CH_3CHBrC_6H_5$

SECTION II HAZARDOUS INGREDIENTS

BASIC MATERIAL	APPROXIMATE OR MAXIMUM % WT. OR VOL.	ESTABLISHED OSHA STANDARD	LD_{50} ORAL	LD_{50} PERCUT.	LC_{50} SPECIES	LC_{50} CONC.
			None	available		

SECTION III PHYSICAL DATA

BOILING POINT °F: 201°C	VAPOR PRESSURE: at 105-107 30 mm Hg.
MELTING POINT °F: -65°C	VAPOR DENSITY (Air = 1): 6.4
SPECIFIC GRAVITY (H_2O = 1): 1.35	EVAPORATION RATE (______ = 1)
SOLUBILITY IN WATER Pts/100 pts H_2O: Insoluble	VOLATILE % Vol. % Wt.

APPEARANCE AND ODOR: Colorless to pale yellow liquid

SECTION IV FIRE AND EXPLOSION HAZARD DATA

	FLAMMABLE (EXPLOSIVE) LIMITS	
FLASH POINT: 205°F		UPPER
METHOD USED		LOWER

EXTINGUISHING MEDIA: Water, foam, carbon dioxide, dry chemical

SPECIAL FIRE FIGHTING PROCEDURES: Wear goggles and self contained breathing apparatus

UNUSUAL FIRE AND EXPLOSION HAZARDS

SECTION V HEALTH HAZARD DATA

TOXIC LEVEL: Not available — CARCINOGENIC: No

PRINCIPLE ROUTES OF ABSORBTION: Inhalation, ingestion, skin — SKIN AND EYE IRRITATION: Yes

REVELANT SYMPTOMS OF EXPOSURE: Cyanosis, deep & rapid respiration, irregular pulse, drowsiness, unconsciousness

EFFECTS OF CHRONIC EXPOSURE: Lung, liver & kidney damage

EMERGENCY AND FIRST AID PROCEDURES: Inhalation - remove from exposure, rest; if swallowed - wash mouth, give emetic, get medical aid; wash eyes & skin.

SECTION VI REACTIVITY DATA

CONDITIONS CONTRIBUTING TO INSTABILITY

CONDITIONS CONTRIBUTING TO HAZARDOUS POLYMERIZATION

INCOMPATABILITY (MATERIALS TO AVOID): May react with oxidizing materials

HAZARDOUS DECOMPOSITION PRODUCTS

SECTION VII SPILL OR LEAK PROCEDURES

STEPS TO BE TAKEN IN CASE MATERIAL IS RELEASED OR SPILLED

WASTE DISPOSAL METHOD

SECTION VIII SPECIAL PROTECTION INFORMATION

VENTILATION REQUIREMENTS LOCAL EXHAUST	PROTECTIVE EQUIPMENT (SPECIFY TYPES) EYE: Goggles
MECHANICAL (GENERAL)	GLOVES: Rubber
SPECIAL	RESPIRATOR: Self contained breathing apparatus
OTHER PROTECTIVE EQUIPMENT	

SECTION IX SPECIAL PRECAUTIONS

PRECAUTIONS TO BE TAKEN IN HANDLING AND STORAGE

OTHER PRECAUTIONS

SIGNATURE	DATE

BROMOFORM

Hazard rating: top 0, left 4, right 0

SECTION I SOURCE AND NOMENCLATURE

- MANUFACTURER'S NAME:
- EMERGENCY TELEPHONE NO.:
- ADDRESS (NUMBER, STREET, CITY, STATE, ZIP CODE):
- TRADE NAME AND SYNONYMS:
- CHEMICAL FAMILY:
- CHEMICAL NAME AND SYNONYMS: Tribromomethane, Bromoform
- FORMULA: Br_3CH

SECTION II HAZARDOUS INGREDIENTS

BASIC MATERIAL	APPROXIMATE OR MAXIMUM % WT. OR VOL.	ESTABLISHED OSHA STANDARD	LD_{50} ORAL	LD_{50} PERCUT.	LD_{50} SPECIES	LC_{50} CONC.
#1 lowest published lethal dose subcutaneous - rabbit	410 mg/kg.			1820	mouse	

SECTION III PHYSICAL DATA MW-252.77

- BOILING POINT °F.: 149-150°C
- VAPOR PRESSURE mm Hg.:
- MELTING POINT °F.: 8.3°C
- VAPOR DENSITY (Air = 1):
- SPECIFIC GRAVITY (H_2O = 1): 2.9
- EVAPORATION RATE (= 1):
- SOLUBILITY IN WATER Pts/100 pts H_2O: 0.1
- VOLATILE % Vol. % Wt.:
- APPEARANCE AND ODOR: Colorless liquid, chloroform odor

SECTION IV FIRE AND EXPLOSION HAZARD DATA

- FLASH POINT: None
- METHOD USED:
- FLAMMABLE (EXPLOSIVE) LIMITS: UPPER / LOWER
- EXTINGUISHING MEDIA:
- SPECIAL FIRE FIGHTING PROCEDURES:
- UNUSUAL FIRE AND EXPLOSION HAZARDS: Emits toxic fumes when heated

SECTION V HEALTH HAZARD DATA

- TOXIC LEVEL: TLV - 0.5 PPM, 5 mg/C.M.
- CARCINOGENIC: No
- PRINCIPLE ROUTES OF ABSORBTION: Inhalation, ingestion, skin
- SKIN AND EYE IRRITATION: Yes
- REVELANT SYMPTOMS OF EXPOSURE: Narcosis, irritation of respiratory tract, liver damage, tremors, unconsciousness.
- EFFECTS OF CHRONIC EXPOSURE: Liver damage
- EMERGENCY AND FIRST AID PROCEDURES: Inhalation - rest, give oxygen as needed; ingestion - gastric lavage, and oxygen; wash eyes and skin.

SECTION VI REACTIVITY DATA

- CONDITIONS CONTRIBUTING TO INSTABILITY:
- CONDITIONS CONTRIBUTING TO HAZARDOUS POLYMERIZATION:
- INCOMPATABILITY (MATERIALS TO AVOID):
- HAZARDOUS DECOMPOSITION PRODUCTS:

SECTION VII SPILL OR LEAK PROCEDURES

- STEPS TO BE TAKEN IN CASE MATERIAL IS RELEASED OR SPILLED:
- WASTE DISPOSAL METHOD:

SECTION VIII SPECIAL PROTECTION INFORMATION

- VENTILATION REQUIREMENTS LOCAL EXHAUST:
- MECHANICAL (GENERAL):
- SPECIAL:
- PROTECTIVE EQUIPMENT (SPECIFY TYPES) EYE: Goggles
- GLOVES: Rubber
- RESPIRATOR: Self contained breathing apparatus
- OTHER PROTECTIVE EQUIPMENT:

SECTION IX SPECIAL PRECAUTIONS

- PRECAUTIONS TO BE TAKEN IN HANDLING AND STORAGE:
- OTHER PRECAUTIONS:
- SIGNATURE:
- DATE:

BROMOHYDROQUINONE

Hazard rating: top 1, left 3, right 0

SECTION I SOURCE AND NOMENCLATURE

- MANUFACTURER'S NAME:
- EMERGENCY TELEPHONE NO.:
- ADDRESS (NUMBER, STREET, CITY, STATE, ZIP CODE):
- TRADE NAME AND SYNONYMS:
- CHEMICAL FAMILY:
- CHEMICAL NAME AND SYNONYMS: Bromohydroquinone
- FORMULA: $BrC_6H_3(OH)_2$

SECTION II HAZARDOUS INGREDIENTS

BASIC MATERIAL	APPROXIMATE OR MAXIMUM % WT. OR VOL.	ESTABLISHED OSHA STANDARD	LD_{50} ORAL	LD_{50} PERCUT.	LC_{50} SPECIES	LC_{50} CONC.
			None	available		

SECTION III PHYSICAL DATA MW - 189.02

- BOILING POINT °F.: sublimes
- VAPOR PRESSURE mm Hg.:
- MELTING POINT °F.: 113°C.
- VAPOR DENSITY (Air = 1):
- SPECIFIC GRAVITY (H_2O = 1): --
- EVAPORATION RATE (= 1):
- SOLUBILITY IN WATER Pts/100 pts H_2O: Very soluble
- VOLATILE % Vol. % Wt.:
- APPEARANCE AND ODOR: Leaflets

SECTION IV FIRE AND EXPLOSION HAZARD DATA

- FLASH POINT: None available
- METHOD USED:
- FLAMMABLE (EXPLOSIVE) LIMITS: UPPER / LOWER
- EXTINGUISHING MEDIA: Water, carbon dioxide, dry chemical
- SPECIAL FIRE FIGHTING PROCEDURES:
- UNUSUAL FIRE AND EXPLOSION HAZARDS:

SECTION V HEALTH HAZARD DATA

- TOXIC LEVEL: None available
- CARCINOGENIC: No
- PRINCIPLE ROUTES OF ABSORBTION: Inhalation of dust, ingestion, skin
- SKIN AND EYE IRRITATION: Severe
- REVELANT SYMPTOMS OF EXPOSURE: Nausea, dizziness, increased respiration, headache, cyanosis, delirium and collpase, depigmentation of eyes and skin.
- EFFECTS OF CHRONIC EXPOSURE: Eye damage, dermatitis
- EMERGENCY AND FIRST AID PROCEDURES: Inhalation - rest; if swallowed - induce vomiting, gastric lavage, medical aid; wash from skin & eyes.

SECTION VI REACTIVITY DATA

- CONDITIONS CONTRIBUTING TO INSTABILITY:
- CONDITIONS CONTRIBUTING TO HAZARDOUS POLYMERIZATION:
- INCOMPATABILITY (MATERIALS TO AVOID):
- HAZARDOUS DECOMPOSITION PRODUCTS:

SECTION VII SPILL OR LEAK PROCEDURES

- STEPS TO BE TAKEN IN CASE MATERIAL IS RELEASED OR SPILLED:
- WASTE DISPOSAL METHOD:

SECTION VIII SPECIAL PROTECTION INFORMATION

- VENTILATION REQUIREMENTS LOCAL EXHAUST:
- MECHANICAL (GENERAL):
- SPECIAL:
- PROTECTIVE EQUIPMENT (SPECIFY TYPES) EYE: Safety goggles
- GLOVES: Rubber
- RESPIRATOR:
- OTHER PROTECTIVE EQUIPMENT:

SECTION IX SPECIAL PRECAUTIONS

- PRECAUTIONS TO BE TAKEN IN HANDLING AND STORAGE:
- OTHER PRECAUTIONS:
- SIGNATURE:
- DATE:

BROMOMESITYLENE

Hazard diamond: 1 (top), 3 (left), 0 (right)

SECTION I SOURCE AND NOMENCLATURE

MANUFACTURER'S NAME | EMERGENCY TELEPHONE NO.

ADDRESS (NUMBER, STREET, CITY, STATE, ZIP CODE)

TRADE NAME AND SYNONYMS | CHEMICAL FAMILY

CHEMICAL NAME AND SYNONYMS: Bromomesitylene | FORMULA: $Br(C_6H_2)(CH_3)_3$

SECTION II HAZARDOUS INGREDIENTS

BASIC MATERIAL	APPROXIMATE OR MAXIMUM % WT. OR VOL.	ESTABLISHED OSHA STANDARD	LD_{50} ORAL	LD_{50} PERCUT.	LC_{50} SPECIES	LC_{50} CONC.
			None Available			

SECTION III PHYSICAL DATA MW - 199.09

BOILING POINT °F.: 230°C.	VAPOR PRESSURE mm Hg.
MELTING POINT °F.: -1°C.	VAPOR DENSITY (Air = 1)
SPECIFIC GRAVITY (H_2O = 1): 1.32	EVAPORATION RATE (= 1)
SOLUBILITY IN WATER Pts/100 pts H_2O: --	VOLATILE % Vol. % Wt.

APPEARANCE AND ODOR: Liquid

SECTION IV FIRE AND EXPLOSION HAZARD DATA

FLASH POINT: None available

METHOD USED

FLAMMABLE (EXPLOSIVE) LIMITS: UPPER / LOWER

EXTINGUISHING MEDIA: Carbon dioxide, dry chemical, alcohol foam.

SPECIAL FIRE FIGHTING PROCEDURES

UNUSUAL FIRE AND EXPLOSION HAZARDS

SECTION V HEALTH HAZARD DATA

TOXIC LEVEL: None available | CARCINOGENIC: No

PRINCIPLE ROUTES OF ABSORBTION: Inhalation, ingestion, skin | SKIN AND EYE IRRITATION: Severe

REVELANT SYMPTOMS OF EXPOSURE: Chemical pneumonia, central nervous system depression, frequent bleeding.

EFFECTS OF CHRONIC EXPOSURE: Central nervous system depression, nervousness, tension, anxiety, asthmatic bronchitis.

EMERGENCY AND FIRST AID PROCEDURES: Inhalation - oxygen as needed; if swallowed - get medical aid, gastric lavage; wash skin & eyes.

SECTION VI REACTIVITY DATA

CONDITIONS CONTRIBUTING TO INSTABILITY

CONDITIONS CONTRIBUTING TO HAZARDOUS POLYMERIZATION

INCOMPATABILITY (MATERIALS TO AVOID)

HAZARDOUS DECOMPOSITION PRODUCTS

SECTION VII SPILL OR LEAK PROCEDURES

STEPS TO BE TAKEN IN CASE MATERIAL IS RELEASED OR SPILLED

WASTE DISPOSAL METHOD

SECTION VIII SPECIAL PROTECTION INFORMATION

VENTILATION REQUIREMENTS LOCAL EXHAUST | PROTECTIVE EQUIPMENT (SPECIFY TYPES) EYE: Goggles

MECHANICAL (GENERAL) | GLOVES: Rubber

SPECIAL | RESPIRATOR: With proper filter

OTHER PROTECTIVE EQUIPMENT

SECTION IX SPECIAL PRECAUTIONS

PRECAUTIONS TO BE TAKEN IN HANDLING AND STORAGE

OTHER PRECAUTIONS

SIGNATURE | DATE

BROMOMETHYL ACETATE

Hazard diamond: 2 (top), 3 (left), 0 (right)

SECTION I SOURCE AND NOMENCLATURE

MANUFACTURER'S NAME | EMERGENCY TELEPHONE NO.

ADDRESS (NUMBER, STREET, CITY, STATE, ZIP CODE)

TRADE NAME AND SYNONYMS | CHEMICAL FAMILY

CHEMICAL NAME AND SYNONYMS: Bromomethyl Acetate | FORMULA: CH_3COOCH_2Br

SECTION II HAZARDOUS INGREDIENTS

BASIC MATERIAL	APPROXIMATE OR MAXIMUM % WT. OR VOL.	ESTABLISHED OSHA STANDARD	LD_{50} ORAL	LD_{50} PERCUT.	LC_{50} SPECIES	LC_{50} CONC.
Note: Lowest published lethal dose Intravenous - mouse 10 mg/kg						

SECTION III PHYSICAL DATA MW - 153.00

BOILING POINT °F.: 133°C.	VAPOR PRESSURE mm Hg.
MELTING POINT °F.	VAPOR DENSITY (Air = 1)
SPECIFIC GRAVITY (H_2O = 1): 1.20	EVAPORATION RATE (= 1)
SOLUBILITY IN WATER Pts/100 pts H_2O: Insoluble	VOLATILE % Vol. % Wt.

APPEARANCE AND ODOR: Colorless liquid

SECTION IV FIRE AND EXPLOSION HAZARD DATA

FLASH POINT: None Available

METHOD USED

FLAMMABLE (EXPLOSIVE) LIMITS: UPPER / LOWER

EXTINGUISHING MEDIA: Carbon dioxide, dry chemical, alcohol foam.

SPECIAL FIRE FIGHTING PROCEDURES

UNUSUAL FIRE AND EXPLOSION HAZARDS

SECTION V HEALTH HAZARD DATA

TOXIC LEVEL: None available | CARCINOGENIC: No

PRINCIPLE ROUTES OF ABSORBTION: Inhalation, ingestion, skin | SKIN AND EYE IRRITATION: Severe

REVELANT SYMPTOMS OF EXPOSURE: Irritation of eyes and upper respiratory system, dizziness and depression, palpitation of heart.

EFFECTS OF CHRONIC EXPOSURE: --

EMERGENCY AND FIRST AID PROCEDURES: If swallowed - give emetic, get medical aid; inhalation - rest; wash skin & eyes.

SECTION VI REACTIVITY DATA

CONDITIONS CONTRIBUTING TO INSTABILITY

CONDITIONS CONTRIBUTING TO HAZARDOUS POLYMERIZATION

INCOMPATABILITY (MATERIALS TO AVOID)

HAZARDOUS DECOMPOSITION PRODUCTS

SECTION VII SPILL OR LEAK PROCEDURES

STEPS TO BE TAKEN IN CASE MATERIAL IS RELEASED OR SPILLED

WASTE DISPOSAL METHOD

SECTION VIII SPECIAL PROTECTION INFORMATION

VENTILATION REQUIREMENTS LOCAL EXHAUST | PROTECTIVE EQUIPMENT (SPECIFY TYPES) EYE: Goggles

MECHANICAL (GENERAL) | GLOVES: Neoprene

SPECIAL | RESPIRATOR: Self-contained

OTHER PROTECTIVE EQUIPMENT

SECTION IX SPECIAL PRECAUTIONS

PRECAUTIONS TO BE TAKEN IN HANDLING AND STORAGE

OTHER PRECAUTIONS

SIGNATURE | DATE

α-BROMONAPHTHALENE

1
3 0

SECTION I SOURCE AND NOMENCLATURE

MANUFACTURER'S NAME | EMERGENCY TELEPHONE NO.

ADDRESS (NUMBER, STREET, CITY, STATE, ZIP CODE)

TRADE NAME AND SYNONYMS | CHEMICAL FAMILY

CHEMICAL NAME AND SYNONYMS: α Naphthyl Bromide

FORMULA: $C_{10}H_7Br$

SECTION II HAZARDOUS INGREDIENTS

BASIC MATERIAL	APPROXIMATE OR MAXIMUM % WT. OR VOL.	ESTABLISHED OSHA STANDARD	LD_{50} ORAL	LD_{50} PERCUT.	LC_{50} SPECIES	LC_{50} CONC.
			None Available			

SECTION III PHYSICAL DATA MW - 207.07

BOILING POINT °F.: 281°C.

VAPOR PRESSURE mm Hg.: at 84° 1

MELTING POINT °F.: 6°C.

VAPOR DENSITY (Air = 1)

SPECIFIC GRAVITY (H_2O = 1): 1.48

EVAPORATION RATE (____ = 1)

SOLUBILITY IN WATER Pts/100 pts H_2O: Slightly soluble

VOLATILE % Vol. % Wt.

APPEARANCE AND ODOR: Colorless oil

SECTION IV FIRE AND EXPLOSION HAZARD DATA

FLASH POINT: None available

METHOD USED

FLAMMABLE (EXPLOSIVE) LIMITS: UPPER / LOWER

EXTINGUISHING MEDIA: Carbon dioxide, dry chemical, or foam.

SPECIAL FIRE FIGHTING PROCEDURES

UNUSUAL FIRE AND EXPLOSION HAZARDS

SECTION V HEALTH HAZARD DATA

TOXIC LEVEL: None available

CARCINOGENIC: No

PRINCIPLE ROUTES OF ABSORBTION: Inhalation, ingestion

SKIN AND EYE IRRITATION: Yes

REVELANT SYMPTOMS OF EXPOSURE: Gastroenteric distress, tremors, convulsions, fever, respiratory failure, headache, nausea, vomiting.

EFFECTS OF CHRONIC EXPOSURE: Clouding of lens.

EMERGENCY AND FIRST AID PROCEDURES: Inhalation - rest; if swallowed - gastric lavage, blood transfusion, oxygen.

SECTION VI REACTIVITY DATA

CONDITIONS CONTRIBUTING TO INSTABILITY

CONDITIONS CONTRIBUTING TO HAZARDOUS POLYMERIZATION

INCOMPATABILITY (MATERIALS TO AVOID)

HAZARDOUS DECOMPOSITION PRODUCTS

SECTION VII SPILL OR LEAK PROCEDURES

STEPS TO BE TAKEN IN CASE MATERIAL IS RELEASED OR SPILLED

WASTE DISPOSAL METHOD

SECTION VIII SPECIAL PROTECTION INFORMATION

VENTILATION REQUIREMENTS LOCAL EXHAUST | PROTECTIVE EQUIPMENT (SPECIFY TYPES) EYE

MECHANICAL (GENERAL) | GLOVES

SPECIAL | RESPIRATOR

OTHER PROTECTIVE EQUIPMENT

SECTION IX SPECIAL PRECAUTIONS

PRECAUTIONS TO BE TAKEN IN HANDLING AND STORAGE

OTHER PRECAUTIONS

SIGNATURE | DATE

β-BROMONAPHTHALENE

1
3 0

SECTION I SOURCE AND NOMENCLATURE

MANUFACTURER'S NAME | EMERGENCY TELEPHONE NO.

ADDRESS (NUMBER, STREET, CITY, STATE, ZIP CODE)

TRADE NAME AND SYNONYMS | CHEMICAL FAMILY

CHEMICAL NAME AND SYNONYMS: B-Naphthyl bromide

FORMULA: $C_{10}H_7Br$

SECTION II HAZARDOUS INGREDIENTS

BASIC MATERIAL	APPROXIMATE OR MAXIMUM % WT. OR VOL.	ESTABLISHED OSHA STANDARD	LD_{50} ORAL	LD_{50} PERCUT.	LC_{50} SPECIES	LC_{50} CONC.
			None available			

SECTION III PHYSICAL DATA MW - 207.07

BOILING POINT °F.: 282°C.

VAPOR PRESSURE mm Hg.

MELTING POINT °F.: 59°C.

VAPOR DENSITY (Air = 1)

SPECIFIC GRAVITY (H_2O = 1): 1.61

EVAPORATION RATE (____ = 1)

SOLUBILITY IN WATER Pts/100 pts H_2O: Slightly soluble

VOLATILE % Vol. % Wt.

APPEARANCE AND ODOR: Leaflets

SECTION IV FIRE AND EXPLOSION HAZARD DATA

FLASH POINT: None available

METHOD USED

FLAMMABLE (EXPLOSIVE) LIMITS: UPPER / LOWER

EXTINGUISHING MEDIA: Carbon dioxide, dry chemical, or foam.

SPECIAL FIRE FIGHTING PROCEDURES

UNUSUAL FIRE AND EXPLOSION HAZARDS

SECTION V HEALTH HAZARD DATA

TOXIC LEVEL: None available

CARCINOGENIC: No

PRINCIPLE ROUTES OF ABSORBTION: Inhalation, ingestion

SKIN AND EYE IRRITATION: Yes

REVELANT SYMPTOMS OF EXPOSURE: Gastroenteric distress, tremors, convulsions, fever, respiratory failure, headache, nausea, vomiting.

EFFECTS OF CHRONIC EXPOSURE: Clouding of lens.

EMERGENCY AND FIRST AID PROCEDURES: Inhalation - rest; if swallowed - gastric lavage, blood transfusion, oxygen.

SECTION VI REACTIVITY DATA

CONDITIONS CONTRIBUTING TO INSTABILITY

CONDITIONS CONTRIBUTING TO HAZARDOUS POLYMERIZATION

INCOMPATABILITY (MATERIALS TO AVOID)

HAZARDOUS DECOMPOSITION PRODUCTS

SECTION VII SPILL OR LEAK PROCEDURES

STEPS TO BE TAKEN IN CASE MATERIAL IS RELEASED OR SPILLED

WASTE DISPOSAL METHOD

SECTION VIII SPECIAL PROTECTION INFORMATION

VENTILATION REQUIREMENTS LOCAL EXHAUST | PROTECTIVE EQUIPMENT (SPECIFY TYPES) EYE

MECHANICAL (GENERAL) | GLOVES

SPECIAL | RESPIRATOR

OTHER PROTECTIVE EQUIPMENT

SECTION IX SPECIAL PRECAUTIONS

PRECAUTIONS TO BE TAKEN IN HANDLING AND STORAGE

OTHER PRECAUTIONS

SIGNATURE | DATE

p-BROMOPHENACYL BROMIDE

NFPA: 1 / 4 / 0

SECTION I SOURCE AND NOMENCLATURE

MANUFACTURER'S NAME | EMERGENCY TELEPHONE NO.

ADDRESS (NUMBER, STREET, CITY, STATE, ZIP CODE)

TRADE NAME AND SYNONYMS | CHEMICAL FAMILY

CHEMICAL NAME AND SYNONYMS: P-Bromophenacyl Bromide

FORMULA: $BrC_6H_4COCH_2Br$

SECTION II HAZARDOUS INGREDIENTS

BASIC MATERIAL	APPROXIMATE OR MAXIMUM % WT. OR VOL.	ESTABLISHED OSHA STANDARD	LD_{50} ORAL	LD_{50} PERCUT.	LC_{50} SPECIES	LC_{50} CONC.

SECTION III PHYSICAL DATA MW - 277.96

BOILING POINT °F. | VAPOR PRESSURE mm Hg.

MELTING POINT °F.: 110 | VAPOR DENSITY (Air = 1)

SPECIFIC GRAVITY (H_2O = 1) | EVAPORATION RATE (= 1)

SOLUBILITY IN WATER Pts/100 pts H_2O: Insoluble | VOLATILE % Vol. % Wt.

APPEARANCE AND ODOR: Needle-like crystals

SECTION IV FIRE AND EXPLOSION HAZARD DATA

FLASH POINT: N/A | FLAMMABLE (EXPLOSIVE) LIMITS: UPPER / LOWER

METHOD USED

EXTINGUISHING MEDIA: N/A

SPECIAL FIRE FIGHTING PROCEDURES

UNUSUAL FIRE AND EXPLOSION HAZARDS

SECTION V HEALTH HAZARD DATA

TOXIC LEVEL: Not available | CARCINOGENIC: No

PRINCIPLE ROUTES OF ABSORBTION: Inhalation, ingestion | SKIN AND EYE IRRITATION: Yes

REVELANT SYMPTOMS OF EXPOSURE: Eye irritation: corneal haziness and pain: burning of throat & eyes; irritation of skin & eyes.

EFFECTS OF CHRONIC EXPOSURE: --

EMERGENCY AND FIRST AID PROCEDURES: Inhalation - rest; if swallowed - wash mouth, give emetic, get medical aid; wash from eyes and skin.

SECTION VI REACTIVITY DATA

CONDITIONS CONTRIBUTING TO INSTABILITY

CONDITIONS CONTRIBUTING TO HAZARDOUS POLYMERIZATION

INCOMPATABILITY (MATERIALS TO AVOID)

HAZARDOUS DECOMPOSITION PRODUCTS

SECTION VII SPILL OR LEAK PROCEDURES

STEPS TO BE TAKEN IN CASE MATERIAL IS RELEASED OR SPILLED

WASTE DISPOSAL METHOD

SECTION VIII SPECIAL PROTECTION INFORMATION

VENTILATION REQUIREMENTS LOCAL EXHAUST | PROTECTIVE EQUIPMENT (SPECIFY TYPES) EYE: Goggles

MECHANICAL (GENERAL) | GLOVES

SPECIAL | RESPIRATOR: With chemical filter

OTHER PROTECTIVE EQUIPMENT

SECTION IX SPECIAL PRECAUTIONS

PRECAUTIONS TO BE TAKEN IN HANDLING AND STORAGE

OTHER PRECAUTIONS

SIGNATURE | DATE

o-BROMOPHENOL

NFPA: 1 / 3 / 0

SECTION I SOURCE AND NOMENCLATURE

MANUFACTURER'S NAME | EMERGENCY TELEPHONE NO.

ADDRESS (NUMBER, STREET, CITY, STATE, ZIP CODE)

TRADE NAME AND SYNONYMS | CHEMICAL FAMILY

CHEMICAL NAME AND SYNONYMS: O-Bromophenol

FORMULA: $Br\ C_6H_4OH$

SECTION II HAZARDOUS INGREDIENTS

BASIC MATERIAL	APPROXIMATE OR MAXIMUM % WT. OR VOL.	ESTABLISHED OSHA STANDARD	LD_{50} ORAL	LD_{50} PERCUT.	LC_{50} SPECIES	LC_{50} CONC.
LD_{50}-1500 mg/kg subcutaneous - rat						

SECTION III PHYSICAL DATA MW - 178.02

BOILING POINT °F.: 195°C. | VAPOR PRESSURE mm Hg.

MELTING POINT °F.: 6°C. | VAPOR DENSITY (Air = 1)

SPECIFIC GRAVITY (H_2O = 1): 1.55 | EVAPORATION RATE (= 1)

SOLUBILITY IN WATER Pts/100 pts H_2O: Soluble | VOLATILE % Vol. % Wt.

APPEARANCE AND ODOR: Colorless liquid

SECTION IV FIRE AND EXPLOSION HAZARD DATA

FLASH POINT: None available | FLAMMABLE (EXPLOSIVE) LIMITS: UPPER / LOWER

METHOD USED

EXTINGUISHING MEDIA: Alcohol foam, carbon dioxide, dry chemical.

SPECIAL FIRE FIGHTING PROCEDURES

UNUSUAL FIRE AND EXPLOSION HAZARDS

SECTION V HEALTH HAZARD DATA

TOXIC LEVEL: None available | CARCINOGENIC: No

PRINCIPLE ROUTES OF ABSORBTION: Inhalation, ingestion, skin | SKIN AND EYE IRRITATION: Severe

REVELANT SYMPTOMS OF EXPOSURE: Eye, skin and mucous membrane irritation, dermatitis, cyanosis.

EFFECTS OF CHRONIC EXPOSURE: Liver damage

EMERGENCY AND FIRST AID PROCEDURES: Inhalation - rest; wash skin & eyes, if swallowed - give emetic, get medical aid.

SECTION VI REACTIVITY DATA

CONDITIONS CONTRIBUTING TO INSTABILITY

CONDITIONS CONTRIBUTING TO HAZARDOUS POLYMERIZATION

INCOMPATABILITY (MATERIALS TO AVOID)

HAZARDOUS DECOMPOSITION PRODUCTS

SECTION VII SPILL OR LEAK PROCEDURES

STEPS TO BE TAKEN IN CASE MATERIAL IS RELEASED OR SPILLED: Put sand and soda ash (9:1) mixture on spill and sweep into box for disposal.

WASTE DISPOSAL METHOD

SECTION VIII SPECIAL PROTECTION INFORMATION

VENTILATION REQUIREMENTS LOCAL EXHAUST | PROTECTIVE EQUIPMENT (SPECIFY TYPES) EYE: Goggles

MECHANICAL (GENERAL) | GLOVES: Butyl Rubber

SPECIAL | RESPIRATOR: Self-contained

OTHER PROTECTIVE EQUIPMENT

SECTION IX SPECIAL PRECAUTIONS

PRECAUTIONS TO BE TAKEN IN HANDLING AND STORAGE: Store in cool, dry place.

OTHER PRECAUTIONS

SIGNATURE | DATE

p-BROMOPHENOL

Hazard rating: 1 (top), 3 (left), 0 (right)

SECTION I SOURCE AND NOMENCLATURE

MANUFACTURER'S NAME | EMERGENCY TELEPHONE NO.

ADDRESS (NUMBER, STREET, CITY, STATE, ZIP CODE)

TRADE NAME AND SYNONYMS | CHEMICAL FAMILY

CHEMICAL NAME AND SYNONYMS: P-Bromophenol | FORMULA: Br C_6H_4OH

SECTION II HAZARDOUS INGREDIENTS

BASIC MATERIAL	APPROXIMATE OR MAXIMUM % WT. OR VOL.	ESTABLISHED OSHA STANDARD	LD_{50} ORAL	LD_{50} PERCUT.	LC_{50} SPECIES	LC_{50} CONC.
TDLO 7200 mg/kg skin mouse						

SECTION III PHYSICAL DATA MW - 173.02

BOILING POINT (°F): 238°C. | VAPOR PRESSURE (mm Hg.):

MELTING POINT (°F): 64°C. | VAPOR DENSITY (Air = 1):

SPECIFIC GRAVITY (H_2O = 1): 1.59 | EVAPORATION RATE (____ = 1)

SOLUBILITY IN WATER (Pts. 100 pts. H_2O): 1.4 | VOLATILE % Vol. % Wt.

APPEARANCE AND ODOR: Crystals

SECTION IV FIRE AND EXPLOSION HAZARD DATA

FLASH POINT: None available | FLAMMABLE (EXPLOSIVE) LIMITS: UPPER / LOWER

METHOD USED

EXTINGUISHING MEDIA: Alcohol foam, carbon dioxide, dry chemical.

SPECIAL FIRE FIGHTING PROCEDURES

UNUSUAL FIRE AND EXPLOSION HAZARDS

SECTION V HEALTH HAZARD DATA

TOXIC LEVEL: None available | CARCINOGENIC: No

PRINCIPLE ROUTES OF ABSORBTION: Inhalation, ingestion, skin | SKIN AND EYE IRRITATION: Severe

REVELANT SYMPTOMS OF EXPOSURE: Eyes, skin and mucous membrane irritation; dermatitis, cyanosis.

EFFECTS OF CHRONIC EXPOSURE: Liver damage.

EMERGENCY AND FIRST AID PROCEDURES: Inhalation - rest; wash from skin, eyes; if swallowed - give emetic, get medical aid.

SECTION VI REACTIVITY DATA

CONDITIONS CONTRIBUTING TO INSTABILITY

CONDITIONS CONTRIBUTING TO HAZARDOUS POLYMERIZATION

INCOMPATABILITY (MATERIALS TO AVOID)

HAZARDOUS DECOMPOSITION PRODUCTS

SECTION VII SPILL OR LEAK PROCEDURES

STEPS TO BE TAKEN IN CASE MATERIAL IS RELEASED OR SPILLED: Pour sand to soda ash (9:1) mixture on spill and sweep into disposal box.

WASTE DISPOSAL METHOD

SECTION VIII SPECIAL PROTECTION INFORMATION

VENTILATION REQUIREMENTS LOCAL EXHAUST | PROTECTIVE EQUIPMENT (SPECIFY TYPES) EYE: Goggles

MECHANICAL (GENERAL) | GLOVES: Butyl rubber

SPECIAL | RESPIRATOR: Self-contained

OTHER PROTECTIVE EQUIPMENT: Protective clothing.

SECTION IX SPECIAL PRECAUTIONS

PRECAUTIONS TO BE TAKEN IN HANDLING AND STORAGE: Store in cool, dry place.

OTHER PRECAUTIONS

SIGNATURE | DATE

BROMOPICRIN

Hazard rating: 0 (top), 4 (left), 3 (right)

SECTION I SOURCE AND NOMENCLATURE

MANUFACTURER'S NAME | EMERGENCY TELEPHONE NO.

ADDRESS (NUMBER, STREET, CITY, STATE, ZIP CODE)

TRADE NAME AND SYNONYMS | CHEMICAL FAMILY

CHEMICAL NAME AND SYNONYMS: Bromopicrin | FORMULA: Br_3CNO_2

SECTION II HAZARDOUS INGREDIENTS

BASIC MATERIAL	APPROXIMATE OR MAXIMUM % WT. OR VOL.	ESTABLISHED OSHA STANDARD	LD_{50} ORAL	LD_{50} PERCUT.	LC_{50} SPECIES	LC_{50} CONC.
			None available			

SECTION III PHYSICAL DATA MW - 297.77

BOILING POINT (°F): 127°C | VAPOR PRESSURE: at 86°C 15 mm Hg.

MELTING POINT (°F): 103°C. | VAPOR DENSITY (Air = 1):

SPECIFIC GRAVITY (H_2O = 1): 2.79 | EVAPORATION RATE (____ = 1)

SOLUBILITY IN WATER (Pts/100 pts H_2O): Insoluble | VOLATILE % Vol. % Wt.

APPEARANCE AND ODOR: Oily colorless liquid

SECTION IV FIRE AND EXPLOSION HAZARD DATA

FLASH POINT: None available | FLAMMABLE (EXPLOSIVE) LIMITS: UPPER / LOWER

METHOD USED

EXTINGUISHING MEDIA: Water, carbon dioxide, dry chemical.

SPECIAL FIRE FIGHTING PROCEDURES

UNUSUAL FIRE AND EXPLOSION HAZARDS: May explode on heating.

SECTION V HEALTH HAZARD DATA

TOXIC LEVEL: None available | CARCINOGENIC: No

PRINCIPLE ROUTES OF ABSORBTION: Inhalation, ingestion | SKIN AND EYE IRRITATION: Severe

REVELANT SYMPTOMS OF EXPOSURE: Skin, eye, lung irritatnt, vomiting, bronchitis and pulmonary edema.

EFFECTS OF CHRONIC EXPOSURE: --

EMERGENCY AND FIRST AID PROCEDURES: If swallowed - give emetic, get medical aid; inhalation - rest; wash from skin & eyes.

SECTION VI REACTIVITY DATA

CONDITIONS CONTRIBUTING TO INSTABILITY

CONDITIONS CONTRIBUTING TO HAZARDOUS POLYMERIZATION

INCOMPATABILITY (MATERIALS TO AVOID): Oxidizing agents

HAZARDOUS DECOMPOSITION PRODUCTS

SECTION VII SPILL OR LEAK PROCEDURES

STEPS TO BE TAKEN IN CASE MATERIAL IS RELEASED OR SPILLED

WASTE DISPOSAL METHOD

SECTION VIII SPECIAL PROTECTION INFORMATION

VENTILATION REQUIREMENTS LOCAL EXHAUST | PROTECTIVE EQUIPMENT (SPECIFY TYPES) EYE: Safety goggles

MECHANICAL (GENERAL) | GLOVES: Rubber

SPECIAL | RESPIRATOR: Self-contained

OTHER PROTECTIVE EQUIPMENT: Protective clothing

SECTION IX SPECIAL PRECAUTIONS

PRECAUTIONS TO BE TAKEN IN HANDLING AND STORAGE: Protect against physical damage, outside storage recommended.

OTHER PRECAUTIONS

SIGNATURE | DATE

β-BROMOPROPIONIC ACID

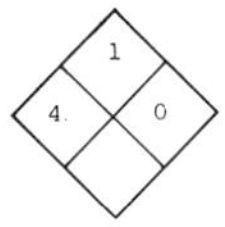

SECTION I SOURCE AND NOMENCLATURE

MANUFACTURER'S NAME | EMERGENCY TELEPHONE NO.

ADDRESS (NUMBER, STREET, CITY, STATE, ZIP CODE)

TRADE NAME AND SYNONYMS | CHEMICAL FAMILY

CHEMICAL NAME AND SYNONYMS: 3 Bromopropionic Acid

FORMULA: $BrCH_2CH_2COOH$

SECTION II HAZARDOUS INGREDIENTS

BASIC MATERIAL	APPROXIMATE OR MAXIMUM % WT. OR VOL.	ESTABLISHED OSHA STANDARD	LD_{50} ORAL	LD_{50} PERCUT.	LC_{50} SPECIES	LC_{50} CONC.
Note:						
Lowest published lethal	dose -					
intraperitoneal - mouse	500 mg/kg.					

SECTION III PHYSICAL DATA MW - 152.99

BOILING POINT (°F.): --

VAPOR PRESSURE (mm Hg.): 45 at 142°C.

MELTING POINT (°F.): 63°C.

VAPOR DENSITY (Air = 1)

SPECIFIC GRAVITY (H_2O = 1): 1.48

EVAPORATION RATE (______ = 1)

SOLUBILITY IN WATER Pts/100 pts H_2O: very soluble

VOLATILE % Vol. % Wt.

APPEARANCE AND ODOR: white solid

SECTION IV FIRE AND EXPLOSION HAZARD DATA

FLASH POINT: None available

FLAMMABLE (EXPLOSIVE) LIMITS: UPPER / LOWER

METHOD USED

EXTINGUISHING MEDIA: Water, foam, carbon dioxide or dry chemical.

SPECIAL FIRE FIGHTING PROCEDURES

UNUSUAL FIRE AND EXPLOSION HAZARDS

SECTION V HEALTH HAZARD DATA

TOXIC LEVEL: None available

CARCINOGENIC: yes

PRINCIPLE ROUTES OF ABSORBTION: Inhalation, ingestion

SKIN AND EYE IRRITATION: Severe

REVELANT SYMPTOMS OF EXPOSURE: Severe eye or skin burns, irritation of respiratory and digestive systems.

EFFECTS OF CHRONIC EXPOSURE: --

EMERGENCY AND FIRST AID PROCEDURES: If swallowed - drink water, milk of magnesia, get medical aid; inhalation - rest; wash skin & eyes.

SECTION VI REACTIVITY DATA

CONDITIONS CONTRIBUTING TO INSTABILITY

CONDITIONS CONTRIBUTING TO HAZARDOUS POLYMERIZATION

INCOMPATABILITY (MATERIALS TO AVOID)

HAZARDOUS DECOMPOSITION PRODUCTS

SECTION VII SPILL OR LEAK PROCEDURES

STEPS TO BE TAKEN IN CASE MATERIAL IS RELEASED OR SPILLED

WASTE DISPOSAL METHOD

SECTION VIII SPECIAL PROTECTION INFORMATION

VENTILATION REQUIREMENTS LOCAL EXHAUST

PROTECTIVE EQUIPMENT (SPECIFY TYPES) EYE: Goggles

MECHANICAL (GENERAL)

GLOVES: Rubber

SPECIAL

RESPIRATOR: With proper filter

OTHER PROTECTIVE EQUIPMENT

SECTION IX SPECIAL PRECAUTIONS

PRECAUTIONS TO BE TAKEN IN HANDLING AND STORAGE

OTHER PRECAUTIONS

SIGNATURE | DATE

α-BROMOPROPIONYL BROMIDE

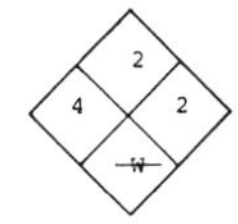

SECTION I SOURCE AND NOMENCLATURE

MANUFACTURER'S NAME | EMERGENCY TELEPHONE NO.

ADDRESS (NUMBER, STREET, CITY, STATE, ZIP CODE)

TRADE NAME AND SYNONYMS | CHEMICAL FAMILY

CHEMICAL NAME AND SYNONYMS: 2 Bromopropionyl Bromide

FORMULA: $CH_3CHBrCOBr$

SECTION II HAZARDOUS INGREDIENTS

BASIC MATERIAL	APPROXIMATE OR MAXIMUM % WT. OR VOL.	ESTABLISHED OSHA STANDARD	LD_{50} ORAL	LD_{50} PERCUT.	LC_{50} SPECIES	LC_{50} CONC.
			None	available		

SECTION III PHYSICAL DATA MW - 215.89

BOILING POINT (°F.): 154°C.

VAPOR PRESSURE (mm Hg.)

MELTING POINT (°F.): --

VAPOR DENSITY (Air = 1)

SPECIFIC GRAVITY (H_2O = 1): 2.06

EVAPORATION RATE (______ = 1)

SOLUBILITY IN WATER Pts/100 pts H_2O: -

VOLATILE % Vol. % Wt.

APPEARANCE AND ODOR: Liquid

SECTION IV FIRE AND EXPLOSION HAZARD DATA

FLASH POINT: None available

FLAMMABLE (EXPLOSIVE) LIMITS: UPPER / LOWER

METHOD USED

EXTINGUISHING MEDIA: Do not use water or foam, use carbon dioxide or dry chemical.

SPECIAL FIRE FIGHTING PROCEDURES: --

UNUSUAL FIRE AND EXPLOSION HAZARDS: When heated to decomposition point gives off toxic bromide fumes.

SECTION V HEALTH HAZARD DATA

TOXIC LEVEL: None available

CARCINOGENIC: No

PRINCIPLE ROUTES OF ABSORBTION: Inhalation, ingestion

SKIN AND EYE IRRITATION: Severe

REVELANT SYMPTOMS OF EXPOSURE: Irritation of eyes and respiratory system; severe internal damage if swallowed; eye and skin burns.

EFFECTS OF CHRONIC EXPOSURE: Bronchial irritation, errosion of teeth.

EMERGENCY AND FIRST AID PROCEDURES: Inhalation - rest; if swallowed - drink water, milk of magnesia, get medical aid; wash skin & eyes.

SECTION VI REACTIVITY DATA

CONDITIONS CONTRIBUTING TO INSTABILITY

CONDITIONS CONTRIBUTING TO HAZARDOUS POLYMERIZATION

INCOMPATABILITY (MATERIALS TO AVOID): Reacts with water to produce H Br

HAZARDOUS DECOMPOSITION PRODUCTS

SECTION VII SPILL OR LEAK PROCEDURES

STEPS TO BE TAKEN IN CASE MATERIAL IS RELEASED OR SPILLED

WASTE DISPOSAL METHOD

SECTION VIII SPECIAL PROTECTION INFORMATION

VENTILATION REQUIREMENTS LOCAL EXHAUST

PROTECTIVE EQUIPMENT (SPECIFY TYPES) EYE: Goggles

MECHANICAL (GENERAL)

GLOVES: Rubber

SPECIAL

RESPIRATOR: Self-contained

OTHER PROTECTIVE EQUIPMENT: Rubber apron

SECTION IX SPECIAL PRECAUTIONS

PRECAUTIONS TO BE TAKEN IN HANDLING AND STORAGE

OTHER PRECAUTIONS

SIGNATURE | DATE

2-BROMOPYRIDINE

Hazard rating: 3 (top), 4 (left), 0 (right)

SECTION I SOURCE AND NOMENCLATURE

MANUFACTURER'S NAME | EMERGENCY TELEPHONE NO.

ADDRESS (NUMBER, STREET, CITY, STATE, ZIP CODE)

TRADE NAME AND SYNONYMS | CHEMICAL FAMILY

CHEMICAL NAME AND SYNONYMS: α Bromo-pyridine

FORMULA: BrC_5H_4N

SECTION II HAZARDOUS INGREDIENTS

BASIC MATERIAL	APPROXIMATE OR MAXIMUM % WT. OR VOL.	ESTABLISHED OSHA STANDARD	LD_{50} ORAL	LD_{50} PERCUT.	LC_{50} SPECIES	LC_{50} CONC.
#1 lowest published lethal dose intraperitoneal - mouse	31.3 mg/kg.					

SECTION III PHYSICAL DATA MW - 158.01

BOILING POINT °F.: 193°C

VAPOR PRESSURE mm Hg.:

MELTING POINT °F.: --

VAPOR DENSITY (Air = 1):

SPECIFIC GRAVITY (H_2O = 1): 1.657

EVAPORATION RATE (______ = 1):

SOLUBILITY IN WATER Pts/100 pts H_2O: Slightly soluble

VOLATILE % Vol. % Wt.:

APPEARANCE AND ODOR: Liquid with fish-like odor

SECTION IV FIRE AND EXPLOSION HAZARD DATA

FLASH POINT: N/A

METHOD USED:

FLAMMABLE (EXPLOSIVE) LIMITS: UPPER / LOWER

EXTINGUISHING MEDIA: Alcohol foam, carbon dioxide, dry chemical

SPECIAL FIRE FIGHTING PROCEDURES: Wear goggles and self-contained breathing apparatus.

UNUSUAL FIRE AND EXPLOSION HAZARDS:

SECTION V HEALTH HAZARD DATA

TOXIC LEVEL: None available

CARCINOGENIC: No

PRINCIPLE ROUTES OF ABSORBTION: Inhalation, ingestion, skin

SKIN AND EYE IRRITATION: Severe

REVELANT SYMPTOMS OF EXPOSURE: Narcosis, headache, nausea, giddiness, vomiting; oral - liver & kidney damage.

EFFECTS OF CHRONIC EXPOSURE: Occasional vomiting and diarrhea; weight loss and anemia; ocular and facial paralysis.

EMERGENCY AND FIRST AID PROCEDURES: Inhalation - rest; if swallowed - induce vomiting, then gastric lavage; wash skin & eyes.

SECTION VI REACTIVITY DATA

CONDITIONS CONTRIBUTING TO INSTABILITY:

CONDITIONS CONTRIBUTING TO HAZARDOUS POLYMERIZATION:

INCOMPATABILITY (MATERIALS TO AVOID):

HAZARDOUS DECOMPOSITION PRODUCTS:

SECTION VII SPILL OR LEAK PROCEDURES

STEPS TO BE TAKEN IN CASE MATERIAL IS RELEASED OR SPILLED:

WASTE DISPOSAL METHOD:

SECTION VIII SPECIAL PROTECTION INFORMATION

VENTILATION REQUIREMENTS LOCAL EXHAUST: Use in hood

MECHANICAL (GENERAL):

SPECIAL:

OTHER PROTECTIVE EQUIPMENT:

PROTECTIVE EQUIPMENT (SPECIFY TYPES) EYE: Goggles

GLOVES: Rubber

RESPIRATOR: Self-contained breathing apparatus

SECTION IX SPECIAL PRECAUTIONS

PRECAUTIONS TO BE TAKEN IN HANDLING AND STORAGE:

OTHER PRECAUTIONS:

SIGNATURE | DATE

m-BROMOTOLUENE

Hazard rating: 2 (top), 3 (left), 0 (right)

SECTION I SOURCE AND NOMENCLATURE

MANUFACTURER'S NAME | EMERGENCY TELEPHONE NO.

ADDRESS (NUMBER, STREET, CITY, STATE, ZIP CODE)

TRADE NAME AND SYNONYMS | CHEMICAL FAMILY

CHEMICAL NAME AND SYNONYMS: 3-Bromotoluene

FORMULA: $CH_3C_6H_4Br$

SECTION II HAZARDOUS INGREDIENTS

BASIC MATERIAL	APPROXIMATE OR MAXIMUM % WT. OR VOL.	ESTABLISHED OSHA STANDARD	LD_{50} ORAL	LD_{50} PERCUT.	LC_{50} SPECIES	LC_{50} CONC.
			None	available		

SECTION III PHYSICAL DATA MW - 171.04

BOILING POINT °F.: 183.7

VAPOR PRESSURE mm Hg.:

MELTING POINT °F.: -39.8

VAPOR DENSITY (Air = 1):

SPECIFIC GRAVITY (H_2O = 1): 1.41

EVAPORATION RATE (______ = 1):

SOLUBILITY IN WATER Pts/100 pts H_2O: Insoluble

VOLATILE % Vol. % Wt.:

APPEARANCE AND ODOR: Colorless liquid

SECTION IV FIRE AND EXPLOSION HAZARD DATA

FLASH POINT: N/A

METHOD USED:

FLAMMABLE (EXPLOSIVE) LIMITS: UPPER / LOWER

EXTINGUISHING MEDIA: Alcohol foam, carbon dioxide, dry chemical

SPECIAL FIRE FIGHTING PROCEDURES:

UNUSUAL FIRE AND EXPLOSION HAZARDS:

SECTION V HEALTH HAZARD DATA

TOXIC LEVEL: None available

CARCINOGENIC: No

PRINCIPLE ROUTES OF ABSORBTION: Inhalation, ingestion

SKIN AND EYE IRRITATION: Severe

REVELANT SYMPTOMS OF EXPOSURE: Eye & respiratory system irritation, severe internal irritation of digestive tract.

EFFECTS OF CHRONIC EXPOSURE: --

EMERGENCY AND FIRST AID PROCEDURES: Inhalation - rest; if swallowed - give emetic, get medical aid; wash eyes & skin.

SECTION VI REACTIVITY DATA

CONDITIONS CONTRIBUTING TO INSTABILITY:

CONDITIONS CONTRIBUTING TO HAZARDOUS POLYMERIZATION:

INCOMPATABILITY (MATERIALS TO AVOID):

HAZARDOUS DECOMPOSITION PRODUCTS:

SECTION VII SPILL OR LEAK PROCEDURES

STEPS TO BE TAKEN IN CASE MATERIAL IS RELEASED OR SPILLED:

WASTE DISPOSAL METHOD:

SECTION VIII SPECIAL PROTECTION INFORMATION

VENTILATION REQUIREMENTS LOCAL EXHAUST:

MECHANICAL (GENERAL):

SPECIAL:

OTHER PROTECTIVE EQUIPMENT:

PROTECTIVE EQUIPMENT (SPECIFY TYPES) EYE: Goggles

GLOVES: Rubber

RESPIRATOR: With proper filter

SECTION IX SPECIAL PRECAUTIONS

PRECAUTIONS TO BE TAKEN IN HANDLING AND STORAGE:

OTHER PRECAUTIONS:

SIGNATURE | DATE

o-BROMOTOLUENE

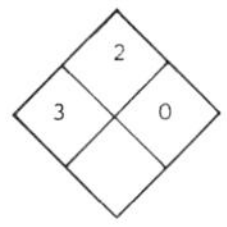

SECTION I SOURCE AND NOMENCLATURE

MANUFACTURER'S NAME	EMERGENCY TELEPHONE NO.
ADDRESS (NUMBER, STREET, CITY, STATE, ZIP CODE)	
TRADE NAME AND SYNONYMS	CHEMICAL FAMILY
CHEMICAL NAME AND SYNONYMS 2-Bromotoluene	FORMULA $CH_3C_6H_4Br$

SECTION II HAZARDOUS INGREDIENTS

BASIC MATERIAL	APPROXIMATE OR MAXIMUM % WT. OR VOL.	ESTABLISHED OSHA STANDARD	LD_{50} ORAL	LD_{50} PERCUT.	LC_{50} SPECIES	LC_{50} CONC.
			None	available		

SECTION III PHYSICAL DATA MW - 171.04

BOILING POINT °F. 181°C	VAPOR PRESSURE mm Hg.
MELTING POINT °F. -28°C	VAPOR DENSITY (Air = 1) 5.9
SPECIFIC GRAVITY (H_2O = 1) 1.42	EVAPORATION RATE (= 1)
SOLUBILITY IN WATER Pts/100 pts H_2O Insoluble	VOLATILE % Vol. % Wt.

APPEARANCE AND ODOR
Colorless liquid

SECTION IV FIRE AND EXPLOSION HAZARD DATA

FLASH POINT 175°F	FLAMMABLE (EXPLOSIVE) LIMITS	UPPER
METHOD USED		LOWER

EXTINGUISHING MEDIA
Alcohol foam, carbon dioxide, dry chemical

SPECIAL FIRE FIGHTING PROCEDURES

UNUSUAL FIRE AND EXPLOSION HAZARDS

SECTION V HEALTH HAZARD DATA

TOXIC LEVEL
None available — CARCINOGENIC No

PRINCIPLE ROUTES OF ABSORBTION
Inhalation, ingestion — SKIN AND EYE IRRITATION Severe

REVELANT SYMPTOMS OF EXPOSURE Eye & respiratory system irritation, severe internal irritation of digestive tract.

EFFECTS OF CHRONIC EXPOSURE
--

EMERGENCY AND FIRST AID PROCEDURES Inhalation - rest; if swallowed - give emetic, get medical aid; wash eyes & skin.

SECTION VI REACTIVITY DATA

CONDITIONS CONTRIBUTING TO INSTABILITY

CONDITIONS CONTRIBUTING TO HAZARDOUS POLYMERIZATION

INCOMPATABILITY (MATERIALS TO AVOID)

HAZARDOUS DECOMPOSITION PRODUCTS

SECTION VII SPILL OR LEAK PROCEDURES

STEPS TO BE TAKEN IN CASE MATERIAL IS RELEASED OR SPILLED

WASTE DISPOSAL METHOD

SECTION VIII SPECIAL PROTECTION INFORMATION

VENTILATION REQUIREMENTS LOCAL EXHAUST	PROTECTIVE EQUIPMENT (SPECIFY TYPES) EYE Goggles
MECHANICAL (GENERAL)	GLOVES Rubber
SPECIAL	RESPIRATOR With proper filter

OTHER PROTECTIVE EQUIPMENT

SECTION IX SPECIAL PRECAUTIONS

PRECAUTIONS TO BE TAKEN IN HANDLING AND STORAGE

OTHER PRECAUTIONS

SIGNATURE	DATE

p-BROMOTOLUENE

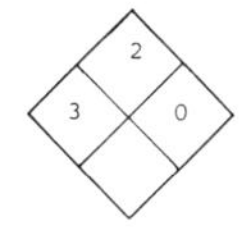

SECTION I SOURCE AND NOMENCLATURE

MANUFACTURER'S NAME	EMERGENCY TELEPHONE NO.
ADDRESS (NUMBER, STREET, CITY, STATE, ZIP CODE)	
TRADE NAME AND SYNONYMS	CHEMICAL FAMILY
CHEMICAL NAME AND SYNONYMS P-Tolyl Bromide, P-Bromotoluene	FORMULA $CH_3C_6H_4Br$

SECTION II HAZARDOUS INGREDIENTS

BASIC MATERIAL	APPROXIMATE OR MAXIMUM % WT. OR VOL.	ESTABLISHED OSHA STANDARD	LD_{50} ORAL	LD_{50} PERCUT.	LC_{50} SPECIES	LC_{50} CONC.
			None	available		

SECTION III PHYSICAL DATA MW-171.04

BOILING POINT °F. 185°C	VAPOR PRESSURE mm Hg.
MELTING POINT °F. 27°C	VAPOR DENSITY (Air = 1) 5.9
SPECIFIC GRAVITY (H_2O = 1) 1.39	EVAPORATION RATE (= 1)
SOLUBILITY IN WATER Pts/100 pts H_2O Insoluble	VOLATILE % Vol.

APPEARANCE AND ODOR
Liquid or crystal

SECTION IV FIRE AND EXPLOSION HAZARD DATA

FLASH POINT 185°F	FLAMMABLE (EXPLOSIVE) LIMITS	UPPER
METHOD USED		LOWER

EXTINGUISHING MEDIA
Alcohol foam, carbon dioxide, dry chemical

SPECIAL FIRE FIGHTING PROCEDURES

UNUSUAL FIRE AND EXPLOSION HAZARDS

SECTION V HEALTH HAZARD DATA

TOXIC LEVEL
None available — CARCINOGENIC No

PRINCIPLE ROUTES OF ABSORBTION
Inhalation, ingestion — SKIN AND EYE IRRITATION Severe

REVELANT SYMPTOMS OF EXPOSURE Eye & respiratory system irritation, severe internal irritation of digestive tract.

EFFECTS OF CHRONIC EXPOSURE
--

EMERGENCY AND FIRST AID PROCEDURES Inhalation - rest; if swallowed - give emetic, get medical aid; wash eyes & skin.

SECTION VI REACTIVITY DATA

CONDITIONS CONTRIBUTING TO INSTABILITY

CONDITIONS CONTRIBUTING TO HAZARDOUS POLYMERIZATION

INCOMPATABILITY (MATERIALS TO AVOID)

HAZARDOUS DECOMPOSITION PRODUCTS

SECTION VII SPILL OR LEAK PROCEDURES

STEPS TO BE TAKEN IN CASE MATERIAL IS RELEASED OR SPILLED

WASTE DISPOSAL METHOD

SECTION VIII SPECIAL PROTECTION INFORMATION

VENTILATION REQUIREMENTS LOCAL EXHAUST	PROTECTIVE EQUIPMENT (SPECIFY TYPES) EYE Goggles
MECHANICAL (GENERAL)	GLOVES Rubber
SPECIAL	RESPIRATOR With proper filter

OTHER PROTECTIVE EQUIPMENT

SECTION IX SPECIAL PRECAUTIONS

PRECAUTIONS TO BE TAKEN IN HANDLING AND STORAGE

OTHER PRECAUTIONS

SIGNATURE	DATE

n-BUTYLBENZENE

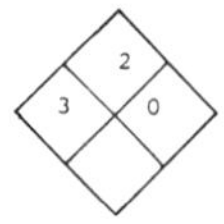

SECTION I SOURCE AND NOMENCLATURE

MANUFACTURER'S NAME	EMERGENCY TELEPHONE NO.
ADDRESS (NUMBER, STREET, CITY, STATE, ZIP CODE)	
TRADE NAME AND SYNONYMS	CHEMICAL FAMILY
CHEMICAL NAME AND SYNONYMS 1-Phenylbutane	FORMULA $C_6H_5(CH_2)_3CH_3$

SECTION II HAZARDOUS INGREDIENTS

BASIC MATERIAL	APPROXIMATE OR MAXIMUM % WT. OR VOL.	ESTABLISHED OSHA STANDARD	LD_{50} ORAL	LD_{50} PERCUT.	LC_{50} SPECIES	LC_{50} CONC.
Lowest published lethal dose						
Oral - rat 5000 mg/kg.						

SECTION III PHYSICAL DATA MW - 124.31

BOILING POINT °F. 182°C.	VAPOR PRESSURE mm Hg. 1 at 23°C.
MELTING POINT °F. -82°C.	VAPOR DENSITY (Air = 1) 4.6
SPECIFIC GRAVITY (H_2O = 1) 0.86	EVAPORATION RATE (________ = 1)
SOLUBILITY IN WATER Pts/100 pts H_2O Insoluble	VOLATILE % Vol. % Wt.
APPEARANCE AND ODOR Colorless liquid	

SECTION IV FIRE AND EXPLOSION HAZARD DATA

FLASH POINT 160°F.	FLAMMABLE (EXPLOSIVE) LIMITS	UPPER 5.8%
METHOD USED OC		LOWER 0.8%

EXTINGUISHING MEDIA
Foam, carbon dioxide, dry chemical

SPECIAL FIRE FIGHTING PROCEDURES

UNUSUAL FIRE AND EXPLOSION HAZARDS

SECTION V HEALTH HAZARD DATA

TOXIC LEVEL None available CARCINOGENIC No

PRINCIPLE ROUTES OF ABSORBTION Inhalation (mainly) ingestion SKIN AND EYE IRRITATION Moderate

REVELANT SYMPTOMS OF EXPOSURE Depression of central nervous system, eye and respiratory system irritation.

EFFECTS OF CHRONIC EXPOSURE Possible liver and kidney damage.

EMERGENCY AND FIRST AID PROCEDURES Inhalation - rest; if swallowed - wash mouth, give emetic, get medical aid; wash skin & eyes.

SECTION VI REACTIVITY DATA

CONDITIONS CONTRIBUTING TO INSTABILITY

CONDITIONS CONTRIBUTING TO HAZARDOUS POLYMERIZATION

INCOMPATABILITY (MATERIALS TO AVOID)

HAZARDOUS DECOMPOSITION PRODUCTS

SECTION VII SPILL OR LEAK PROCEDURES

STEPS TO BE TAKEN IN CASE MATERIAL IS RELEASED OR SPILLED

WASTE DISPOSAL METHOD

SECTION VIII SPECIAL PROTECTION INFORMATION

VENTILATION REQUIREMENTS LOCAL EXHAUST	PROTECTIVE EQUIPMENT (SPECIFY TYPES) EYE Goggles
MECHANICAL (GENERAL)	GLOVES Rubber
SPECIAL	RESPIRATOR Self-contained
OTHER PROTECTIVE EQUIPMENT	

SECTION IX SPECIAL PRECAUTIONS

PRECAUTIONS TO BE TAKEN IN HANDLING AND STORAGE

OTHER PRECAUTIONS

SIGNATURE	DATE

sec-BUTYLBENZENE

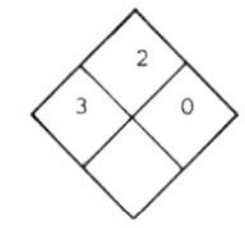

SECTION I SOURCE AND NOMENCLATURE

MANUFACTURER'S NAME	EMERGENCY TELEPHONE NO.
ADDRESS (NUMBER, STREET, CITY, STATE, ZIP CODE)	
TRADE NAME AND SYNONYMS	CHEMICAL FAMILY
CHEMICAL NAME AND SYNONYMS 2 Phenylbutane	FORMULA $C_6H_5CH(CH_3)C_2H_5$

SECTION II HAZARDOUS INGREDIENTS

BASIC MATERIAL	APPROXIMATE OR MAXIMUM % WT. OR VOL.	ESTABLISHED OSHA STANDARD	LD_{50}mg/kg ORAL	LD_{50}mg/kg PERCUT.	LC_{50} SPECIES	LC_{50} CONC.
			2240		Rat	

SECTION III PHYSICAL DATA MW-134.25

BOILING POINT °F. 174°C.	VAPOR PRESSURE mm Hg. 1 at 19°C.
MELTING POINT °F. -83°C.	VAPOR DENSITY (Air = 1) 4.62
SPECIFIC GRAVITY (H_2O = 1) 0.86	EVAPORATION RATE (________ = 1)
SOLUBILITY IN WATER Pts/100 pts H_2O Insoluble	VOLATILE % Vol. % Wt.
APPEARANCE AND ODOR Colorless liquid	

SECTION IV FIRE AND EXPLOSION HAZARD DATA

FLASH POINT 126°F.	FLAMMABLE (EXPLOSIVE) LIMITS	UPPER 6.9%
METHOD USED O.C.		LOWER 0.8%

EXTINGUISHING MEDIA
Foam, carbon dioxide, dry chemical.

SPECIAL FIRE FIGHTING PROCEDURES

UNUSUAL FIRE AND EXPLOSION HAZARDS

SECTION V HEALTH HAZARD DATA

TOXIC LEVEL None Available CARCINOGENIC No

PRINCIPLE ROUTES OF ABSORBTION Inhalation (mainly) ingestion SKIN AND EYE IRRITATION Moderate

REVELANT SYMPTOMS OF EXPOSURE Depression of central nervous system, eye and respiratory system irritation.

EFFECTS OF CHRONIC EXPOSURE Possible liver and kidney damage.

EMERGENCY AND FIRST AID PROCEDURES Inhalation - rest; if swallowed - give emetic, get medical aid; wash from skin and eyes.

SECTION VI REACTIVITY DATA

CONDITIONS CONTRIBUTING TO INSTABILITY

CONDITIONS CONTRIBUTING TO HAZARDOUS POLYMERIZATION

INCOMPATABILITY (MATERIALS TO AVOID)

HAZARDOUS DECOMPOSITION PRODUCTS

SECTION VII SPILL OR LEAK PROCEDURES

STEPS TO BE TAKEN IN CASE MATERIAL IS RELEASED OR SPILLED

WASTE DISPOSAL METHOD

SECTION VIII SPECIAL PROTECTION INFORMATION

VENTILATION REQUIREMENTS LOCAL EXHAUST	PROTECTIVE EQUIPMENT (SPECIFY TYPES) EYE Goggles
MECHANICAL (GENERAL)	GLOVES Rubber
SPECIAL	RESPIRATOR Self-contained
OTHER PROTECTIVE EQUIPMENT	

SECTION IX SPECIAL PRECAUTIONS

PRECAUTIONS TO BE TAKEN IN HANDLING AND STORAGE

OTHER PRECAUTIONS

SIGNATURE	DATE

tert-BUTYLBENZENE

Hazard diamond: 2 (top), 3 (left), 0 (right)

SECTION I SOURCE AND NOMENCLATURE

- MANUFACTURER'S NAME:
- EMERGENCY TELEPHONE NO.:
- ADDRESS (NUMBER, STREET, CITY, STATE, ZIP CODE):
- TRADE NAME AND SYNONYMS:
- CHEMICAL FAMILY:
- CHEMICAL NAME AND SYNONYMS: Tert-butylbenzene
- FORMULA: $C_6H_5C(CH_3)_3$

SECTION II HAZARDOUS INGREDIENTS

BASIC MATERIAL	APPROXIMATE OR MAXIMUM % WT. OR VOL.	ESTABLISHED OSHA STANDARD	LD_{50} ORAL	LD_{50} PERCUT.	LC_{50} SPECIES	LC_{50} CONC.
LDLO - oral - rat 500 mg/kg						

SECTION III PHYSICAL DATA MW-134.24

- BOILING POINT °F.: 169°C
- VAPOR PRESSURE mm Hg.: 1 at 13°C
- MELTING POINT °F.: -58°C
- VAPOR DENSITY (Air = 1): 4.62
- SPECIFIC GRAVITY (H_2O = 1): 0.867
- EVAPORATION RATE (= 1):
- SOLUBILITY IN WATER Pts/100 pts H_2O: Insoluble
- VOLATILE % Vol. % Wt.:
- APPEARANCE AND ODOR: Colorless liquid

SECTION IV FIRE AND EXPLOSION HAZARD DATA

- FLASH POINT: 160°F
- METHOD USED: C.C.
- FLAMMABLE (EXPLOSIVE) LIMITS: UPPER 5.8; LOWER 0.8
- EXTINGUISHING MEDIA: Foam, carbon dioxide, dry chemical
- SPECIAL FIRE FIGHTING PROCEDURES: Wear goggles and self contained breathing apparatus.
- UNUSUAL FIRE AND EXPLOSION HAZARDS:

SECTION V HEALTH HAZARD DATA

- TOXIC LEVEL: Not available
- CARCINOGENIC: No
- PRINCIPLE ROUTES OF ABSORBTION: Inhalation (mainly) ingestion
- SKIN AND EYE IRRITATION: Moderate
- REVELANT SYMPTOMS OF EXPOSURE: Depression of central nervous system, eye & respiratory system irritation.
- EFFECTS OF CHRONIC EXPOSURE: Possible liver & kidney damage.
- EMERGENCY AND FIRST AID PROCEDURES: Inhalation - rest; if swallowed - wash mouth, give emetic, get medical aid; wash skin & eyes.

SECTION VI REACTIVITY DATA

- CONDITIONS CONTRIBUTING TO INSTABILITY:
- CONDITIONS CONTRIBUTING TO HAZARDOUS POLYMERIZATION:
- INCOMPATABILITY (MATERIALS TO AVOID):
- HAZARDOUS DECOMPOSITION PRODUCTS:

SECTION VII SPILL OR LEAK PROCEDURES

- STEPS TO BE TAKEN IN CASE MATERIAL IS RELEASED OR SPILLED:
- WASTE DISPOSAL METHOD:

SECTION VIII SPECIAL PROTECTION INFORMATION

- VENTILATION REQUIREMENTS LOCAL EXHAUST:
- PROTECTIVE EQUIPMENT (SPECIFY TYPES) EYE: Goggles
- MECHANICAL (GENERAL):
- GLOVES: Rubber
- SPECIAL:
- RESPIRATOR: Self contained
- OTHER PROTECTIVE EQUIPMENT:

SECTION IX SPECIAL PRECAUTIONS

- PRECAUTIONS TO BE TAKEN IN HANDLING AND STORAGE:
- OTHER PRECAUTIONS:
- SIGNATURE:
- DATE:

n-BUTYL CHLOROACETATE

Hazard diamond: 2 (top), 3 (left), 0 (right)

SECTION I SOURCE AND NOMENCLATURE

- MANUFACTURER'S NAME:
- EMERGENCY TELEPHONE NO.:
- ADDRESS (NUMBER, STREET, CITY, STATE, ZIP CODE):
- TRADE NAME AND SYNONYMS:
- CHEMICAL FAMILY:
- CHEMICAL NAME AND SYNONYMS: N-Butyl Chloroacetate
- FORMULA: $C_4H_9O_2CCH_2Cl$

SECTION II HAZARDOUS INGREDIENTS

BASIC MATERIAL	APPROXIMATE OR MAXIMUM % WT. OR VOL.	ESTABLISHED OSHA STANDARD	LD_{50} ORAL	LD_{50} PERCUT.	LC_{50} SPECIES	LC_{50} CONC.
			None available			

SECTION III PHYSICAL DATA MM - 150.61

- BOILING POINT °F.: 183°C.
- VAPOR PRESSURE mm Hg.:
- MELTING POINT °F.: --
- VAPOR DENSITY (Air = 1):
- SPECIFIC GRAVITY (H_2O = 1): 1.08
- EVAPORATION RATE (= 1):
- SOLUBILITY IN WATER Pts/100 pts H_2O:
- VOLATILE % Vol. % Wt.:
- APPEARANCE AND ODOR: Colorless liquid

SECTION IV FIRE AND EXPLOSION HAZARD DATA

- FLASH POINT: None available
- METHOD USED:
- FLAMMABLE (EXPLOSIVE) LIMITS: UPPER; LOWER
- EXTINGUISHING MEDIA: Alcohol foam, carbon dioxide, dry chemical.
- SPECIAL FIRE FIGHTING PROCEDURES: Water may be ineffective.
- UNUSUAL FIRE AND EXPLOSION HAZARDS:

SECTION V HEALTH HAZARD DATA

- TOXIC LEVEL: None available
- CARCINOGENIC: No
- PRINCIPLE ROUTES OF ABSORBTION: Inhalation, ingestion, skin
- SKIN AND EYE IRRITATION: Moderate to severe
- REVELANT SYMPTOMS OF EXPOSURE: Eye and respiratory passage irritation, salivation.
- EFFECTS OF CHRONIC EXPOSURE: Fatigue, loss of weight, blood changes.
- EMERGENCY AND FIRST AID PROCEDURES: Inhalation - give oxygen as needed; if swallowed - give emetic, get medical aid; wash skin & eyes.

SECTION VI REACTIVITY DATA

- CONDITIONS CONTRIBUTING TO INSTABILITY:
- CONDITIONS CONTRIBUTING TO HAZARDOUS POLYMERIZATION:
- INCOMPATABILITY (MATERIALS TO AVOID): May react with oxidizing materials.
- HAZARDOUS DECOMPOSITION PRODUCTS:

SECTION VII SPILL OR LEAK PROCEDURES

- STEPS TO BE TAKEN IN CASE MATERIAL IS RELEASED OR SPILLED:
- WASTE DISPOSAL METHOD:

SECTION VIII SPECIAL PROTECTION INFORMATION

- VENTILATION REQUIREMENTS LOCAL EXHAUST:
- PROTECTIVE EQUIPMENT (SPECIFY TYPES) EYE: Goggles
- MECHANICAL (GENERAL):
- GLOVES: Rubber
- SPECIAL:
- RESPIRATOR: With proper filter
- OTHER PROTECTIVE EQUIPMENT:

SECTION IX SPECIAL PRECAUTIONS

- PRECAUTIONS TO BE TAKEN IN HANDLING AND STORAGE:
- OTHER PRECAUTIONS:
- SIGNATURE:
- DATE:

α-BUTYLENE BROMIDE

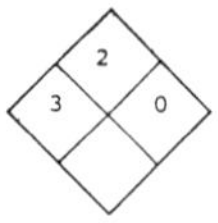

SECTION I SOURCE AND NOMENCLATURE

MANUFACTURER'S NAME | EMERGENCY TELEPHONE NO.

ADDRESS (NUMBER, STREET, CITY, STATE, ZIP CODE)

TRADE NAME AND SYNONYMS | CHEMICAL FAMILY

CHEMICAL NAME AND SYNONYMS: 1,2 dibromobutane | FORMULA: $CH_3CH_2CHBrCH_2Br$

SECTION II HAZARDOUS INGREDIENTS

BASIC MATERIAL	APPROXIMATE OR MAXIMUM % WT. OR VOL.	ESTABLISHED OSHA STANDARD	LD_{50} ORAL	LD_{50} PERCUT.	LC_{50} SPECIES	LC_{50} CONC.
			None available			

SECTION III PHYSICAL DATA

BOILING POINT °F.: 166°C	VAPOR PRESSURE mm Hg.
MELTING POINT °F.: -65°C	VAPOR DENSITY (Air = 1)
SPECIFIC GRAVITY (H_2O = 1): 1.82	EVAPORATION RATE (= 1)
SOLUBILITY IN WATER Pts/100 pts H_2O: Insoluble	VOLATILE % Vol. % Wt.

APPEARANCE AND ODOR: Yellowish liquid

SECTION IV FIRE AND EXPLOSION HAZARD DATA

FLASH POINT: N/A | FLAMMABLE (EXPLOSIVE) LIMITS: UPPER / LOWER

METHOD USED

EXTINGUISHING MEDIA: Water, foam, carbon dioxide, dry chemical

SPECIAL FIRE FIGHTING PROCEDURES: Water may be ineffective.

UNUSUAL FIRE AND EXPLOSION HAZARDS

SECTION V HEALTH HAZARD DATA

TOXIC LEVEL: None available | CARCINOGENIC: No

PRINCIPLE ROUTES OF ABSORBTION: Inhalation, ingestion | SKIN AND EYE IRRITATION: Moderate

REVELANT SYMPTOMS OF EXPOSURE: Eye irritation, headache, vomiting, drowsiness, narcosis

EFFECTS OF CHRONIC EXPOSURE: Dermatitis, liver & kidney damage

EMERGENCY AND FIRST AID PROCEDURES: Inhalation - rest; if swallowed - wash mouth, give emetic, get medical aid; wash from skin & eyes.

SECTION VI REACTIVITY DATA

CONDITIONS CONTRIBUTING TO INSTABILITY

CONDITIONS CONTRIBUTING TO HAZARDOUS POLYMERIZATION

INCOMPATABILITY (MATERIALS TO AVOID)

HAZARDOUS DECOMPOSITION PRODUCTS

SECTION VII SPILL OR LEAK PROCEDURES

STEPS TO BE TAKEN IN CASE MATERIAL IS RELEASED OR SPILLED: Wear self contained breathing apparatus

WASTE DISPOSAL METHOD

SECTION VIII SPECIAL PROTECTION INFORMATION

VENTILATION REQUIREMENTS LOCAL EXHAUST | PROTECTIVE EQUIPMENT (SPECIFY TYPES) EYE: Goggles

MECHANICAL (GENERAL) | GLOVES: Rubber

SPECIAL | RESPIRATOR: With organic vapor resistant canister

OTHER PROTECTIVE EQUIPMENT

SECTION IX SPECIAL PRECAUTIONS

PRECAUTIONS TO BE TAKEN IN HANDLING AND STORAGE

OTHER PRECAUTIONS

SIGNATURE | DATE

β-BUTYLENE BROMIDE

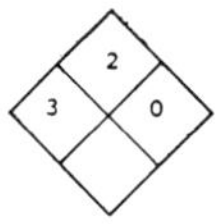

SECTION I SOURCE AND NOMENCLATURE

MANUFACTURER'S NAME | EMERGENCY TELEPHONE NO.

ADDRESS (NUMBER, STREET, CITY, STATE, ZIP CODE)

TRADE NAME AND SYNONYMS | CHEMICAL FAMILY

CHEMICAL NAME AND SYNONYMS: 2,3 dibromobutane | FORMULA: $CH_3CHBrCHBrCH_3$

SECTION II HAZARDOUS INGREDIENTS

BASIC MATERIAL	APPROXIMATE OR MAXIMUM % WT. OR VOL.	ESTABLISHED OSHA STANDARD	LD_{50} ORAL	LD_{50} PERCUT.	LC_{50} SPECIES	LC_{50} CONC.
			None available			

SECTION III PHYSICAL DATA MW - 215.94

BOILING POINT °F.: 158°C	VAPOR PRESSURE mm Hg.
MELTING POINT °F.: -34.5°C	VAPOR DENSITY (Air = 1)
SPECIFIC GRAVITY (H_2O = 1): 1.78	EVAPORATION RATE (= 1)
SOLUBILITY IN WATER Pts/100 pts H_2O: Insoluble	VOLATILE % Vol. % Wt.

APPEARANCE AND ODOR: Colorless liquid

SECTION IV FIRE AND EXPLOSION HAZARD DATA

FLASH POINT: N/A | FLAMMABLE (EXPLOSIVE) LIMITS: UPPER / LOWER

METHOD USED

EXTINGUISHING MEDIA: Water, foam, carbon dioxide, dry chemical

SPECIAL FIRE FIGHTING PROCEDURES: Water may be ineffective

UNUSUAL FIRE AND EXPLOSION HAZARDS

SECTION V HEALTH HAZARD DATA

TOXIC LEVEL: None available | CARCINOGENIC: No

PRINCIPLE ROUTES OF ABSORBTION: Inhalation, ingestion | SKIN AND EYE IRRITATION: Moderate

REVELANT SYMPTOMS OF EXPOSURE: Eye irritation, headache, vomiting, drowsiness, narcosis, liver & kidney damage.

EFFECTS OF CHRONIC EXPOSURE: Dermatitis, liver & kidney damage

EMERGENCY AND FIRST AID PROCEDURES: Inhalation - rest; if swallowed - wash mouth, give emetic; get medical aid; wash from skin & eyes.

SECTION VI REACTIVITY DATA

CONDITIONS CONTRIBUTING TO INSTABILITY

CONDITIONS CONTRIBUTING TO HAZARDOUS POLYMERIZATION

INCOMPATABILITY (MATERIALS TO AVOID)

HAZARDOUS DECOMPOSITION PRODUCTS

SECTION VII SPILL OR LEAK PROCEDURES

STEPS TO BE TAKEN IN CASE MATERIAL IS RELEASED OR SPILLED: Wear self contained breathing apparatus.

WASTE DISPOSAL METHOD

SECTION VIII SPECIAL PROTECTION INFORMATION

VENTILATION REQUIREMENTS LOCAL EXHAUST | PROTECTIVE EQUIPMENT (SPECIFY TYPES) EYE: Goggles

MECHANICAL (GENERAL) | GLOVES: Rubber

SPECIAL | RESPIRATOR: With organic vapor resistant canister

OTHER PROTECTIVE EQUIPMENT

SECTION IX SPECIAL PRECAUTIONS

PRECAUTIONS TO BE TAKEN IN HANDLING AND STORAGE

OTHER PRECAUTIONS

SIGNATURE | DATE

n-BUTYL FORMATE

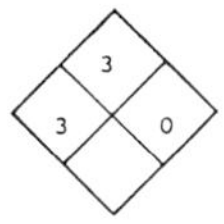

SECTION I SOURCE AND NOMENCLATURE

MANUFACTURER'S NAME	EMERGENCY TELEPHONE NO.
ADDRESS (NUMBER, STREET, CITY, STATE, ZIP CODE)	
TRADE NAME AND SYNONYMS	CHEMICAL FAMILY
CHEMICAL NAME AND SYNONYMS: Butyl Methanoate	FORMULA: $C_4H_9OOCH_3$

SECTION II HAZARDOUS INGREDIENTS

BASIC MATERIAL	APPROXIMATE OR MAXIMUM % WT. OR VOL.	ESTABLISHED OSHA STANDARD	LD_{50} mg/kg ORAL	LD_{50} PERCUT.	LC_{50} SPECIES	LC_{50} CONC.
TCLO - inhalation - human	10418 ppm		2660		rabbit	

SECTION III PHYSICAL DATA MW - 101.12

BOILING POINT °F.: 106°C	VAPOR PRESSURE mm Hg.: at 31.6°C 40
MELTING POINT °F.: -90°C	VAPOR DENSITY (Air = 1): 3.52
SPECIFIC GRAVITY (H_2O = 1): 0.911	EVAPORATION RATE (= 1)
SOLUBILITY IN WATER Pts/100 pts H_2O: slightly	VOLATILE % Vol. % Wt.
APPEARANCE AND ODOR: colorless liquid	

SECTION IV FIRE AND EXPLOSION HAZARD DATA

FLASH POINT: 64°F	FLAMMABLE (EXPLOSIVE) LIMITS	UPPER: 8.0%
METHOD USED: CC		LOWER: 1.7%

EXTINGUISHING MEDIA: Alcohol foam, carbon dioxide, dry chemical

SPECIAL FIRE FIGHTING PROCEDURES: Water may be ineffective

UNUSUAL FIRE AND EXPLOSION HAZARDS

SECTION V HEALTH HAZARD DATA

TOXIC LEVEL: None available — CARCINOGENIC: No

PRINCIPLE ROUTES OF ABSORBTION: Inhalation, ingestion — SKIN AND EYE IRRITATION: Severe

REVELANT SYMPTOMS OF EXPOSURE: Irritant of mucous membranes & respiratory tract, eye irritation, salivation, vomiting.

EFFECTS OF CHRONIC EXPOSURE: --

EMERGENCY AND FIRST AID PROCEDURES: Inhalation - rest; if swallowed - give emetic; get medical aid; wash from eyes & skin.

SECTION VI REACTIVITY DATA

CONDITIONS CONTRIBUTING TO INSTABILITY

CONDITIONS CONTRIBUTING TO HAZARDOUS POLYMERIZATION

INCOMPATABILITY (MATERIALS TO AVOID)

HAZARDOUS DECOMPOSITION PRODUCTS

SECTION VII SPILL OR LEAK PROCEDURES

STEPS TO BE TAKEN IN CASE MATERIAL IS RELEASED OR SPILLED: Absorb on paper, evaporate in hood.

WASTE DISPOSAL METHOD: Mix with more flammable solvent and burn.

SECTION VIII SPECIAL PROTECTION INFORMATION

VENTILATION REQUIREMENTS LOCAL EXHAUST	PROTECTIVE EQUIPMENT (SPECIFY TYPES) EYE: Goggles
MECHANICAL (GENERAL)	GLOVES: Rubber
SPECIAL	RESPIRATOR: Chemical cartridge respirator
OTHER PROTECTIVE EQUIPMENT: Overalls	

SECTION IX SPECIAL PRECAUTIONS

PRECAUTIONS TO BE TAKEN IN HANDLING AND STORAGE: Keep away from sources of ignition.

OTHER PRECAUTIONS

SIGNATURE	DATE

n-BUTYL IODIDE

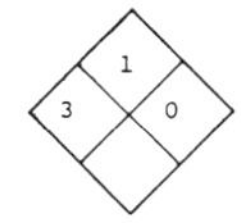

SECTION I SOURCE AND NOMENCLATURE

MANUFACTURER'S NAME	EMERGENCY TELEPHONE NO.
ADDRESS (NUMBER, STREET, CITY, STATE, ZIP CODE)	
TRADE NAME AND SYNONYMS	CHEMICAL FAMILY
CHEMICAL NAME AND SYNONYMS: 1-Iodobutane	FORMULA: $CH_3CH_2CH_2CH_2I$

SECTION II HAZARDOUS INGREDIENTS

BASIC MATERIAL	APPROXIMATE OR MAXIMUM % WT. OR VOL.	ESTABLISHED OSHA STANDARD	LD_{50} ORAL	LD_{50} PERCUT.	LC_{50} SPECIES	LC_{50} CONC.
			None	available		

SECTION III PHYSICAL DATA MW - 184.03

BOILING POINT °F.: 131°C	VAPOR PRESSURE mm Hg.
MELTING POINT °F.: -103°C	VAPOR DENSITY (Air = 1)
SPECIFIC GRAVITY (H_2O = 1): 1.62	EVAPORATION RATE (= 1)
SOLUBILITY IN WATER Pts/100 pts H_2O: Insoluble	VOLATILE % Vol. % Wt.
APPEARANCE AND ODOR: Colorless to light brown liquid	

SECTION IV FIRE AND EXPLOSION HAZARD DATA

FLASH POINT: Not available	FLAMMABLE (EXPLOSIVE) LIMITS	UPPER
METHOD USED		LOWER

EXTINGUISHING MEDIA: Water, carbon dioxide, dry chemical

SPECIAL FIRE FIGHTING PROCEDURES: --

UNUSUAL FIRE AND EXPLOSION HAZARDS: When heated to decomposition point emits toxic fumes of iodates.

SECTION V HEALTH HAZARD DATA

TOXIC LEVEL: Not available — CARCINOGENIC: No

PRINCIPLE ROUTES OF ABSORBTION: Inhalation, ingestion, skin — SKIN AND EYE IRRITATION: Slight

REVELANT SYMPTOMS OF EXPOSURE: Nausea, vomiting, diarrhea, dizziness, visual disturbances, irritability, drowsiness, central nervous system depressant.

EFFECTS OF CHRONIC EXPOSURE: Dermatitis

EMERGENCY AND FIRST AID PROCEDURES: Inhalation - rest; if swallowed - wash mouth, give emetic, get medical aid; wash skin & eyes.

SECTION VI REACTIVITY DATA

CONDITIONS CONTRIBUTING TO INSTABILITY

CONDITIONS CONTRIBUTING TO HAZARDOUS POLYMERIZATION

INCOMPATABILITY (MATERIALS TO AVOID)

HAZARDOUS DECOMPOSITION PRODUCTS

SECTION VII SPILL OR LEAK PROCEDURES

STEPS TO BE TAKEN IN CASE MATERIAL IS RELEASED OR SPILLED

WASTE DISPOSAL METHOD

SECTION VIII SPECIAL PROTECTION INFORMATION

VENTILATION REQUIREMENTS LOCAL EXHAUST	PROTECTIVE EQUIPMENT (SPECIFY TYPES) EYE: Safety goggles
MECHANICAL (GENERAL)	GLOVES: Rubber
SPECIAL	RESPIRATOR: With proper filter
OTHER PROTECTIVE EQUIPMENT	

SECTION IX SPECIAL PRECAUTIONS

PRECAUTIONS TO BE TAKEN IN HANDLING AND STORAGE

OTHER PRECAUTIONS

SIGNATURE	DATE

sec-BUTYL IODIDE

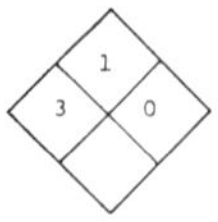

SECTION I SOURCE AND NOMENCLATURE

MANUFACTURER'S NAME | EMERGENCY TELEPHONE NO.

ADDRESS (NUMBER, STREET, CITY, STATE, ZIP CODE)

TRADE NAME AND SYNONYMS | CHEMICAL FAMILY

CHEMICAL NAME AND SYNONYMS: 2 Iodobutane | FORMULA: $C_2H_5CHICH_3$

SECTION II HAZARDOUS INGREDIENTS

BASIC MATERIAL	APPROXIMATE OR MAXIMUM % WT. OR VOL.	ESTABLISHED OSHA STANDARD	LD_{50}		LC_{50}	
			ORAL	PERCUT.	SPECIES	CONC.
			None	available		

SECTION III PHYSICAL DATA MW 184.03

BOILING POINT °F.: $120^{0}C$	VAPOR PRESSURE mm Hg.
MELTING POINT °F.: -104	VAPOR DENSITY (Air = 1)
SPECIFIC GRAVITY (H_2O = 1): 1.60	EVAPORATION RATE (= 1)
SOLUBILITY IN WATER Pts 100 pts H_2O: Insoluble	VOLATILE % Vol. % Wt.

APPEARANCE AND ODOR: Colorless to light brown liquid

SECTION IV FIRE AND EXPLOSION HAZARD DATA

FLASH POINT: Not available | FLAMMABLE (EXPLOSIVE) LIMITS: UPPER / LOWER

METHOD USED

EXTINGUISHING MEDIA: Water, carbon dioxide, dry chemical

SPECIAL FIRE FIGHTING PROCEDURES: --

UNUSUAL FIRE AND EXPLOSION HAZARDS: When heated to decomposition point emits toxic fumes of iodates.

SECTION V HEALTH HAZARD DATA

TOXIC LEVEL: Not available | CARCINOGENIC: No

PRINCIPLE ROUTES OF ABSORBTION: Inhalation, ingestion, skin | SKIN AND EYE IRRITATION: Slight

REVELANT SYMPTOMS OF EXPOSURE: Nausea, vomiting, diarrhea, dizziness, visual disturbances, irritability, drowsiness, central nervous system depressant

EFFECTS OF CHRONIC EXPOSURE: Dermatitis

EMERGENCY AND FIRST AID PROCEDURES: Inhalation - rest; if swallowed - wash mouth, give emetic; get medical aid; wash skin & eyes.

SECTION VI REACTIVITY DATA

CONDITIONS CONTRIBUTING TO INSTABILITY

CONDITIONS CONTRIBUTING TO HAZARDOUS POLYMERIZATION

INCOMPATABILITY (MATERIALS TO AVOID)

HAZARDOUS DECOMPOSITION PRODUCTS

SECTION VII SPILL OR LEAK PROCEDURES

STEPS TO BE TAKEN IN CASE MATERIAL IS RELEASED OR SPILLED

WASTE DISPOSAL METHOD

SECTION VIII SPECIAL PROTECTION INFORMATION

VENTILATION REQUIREMENTS LOCAL EXHAUST | PROTECTIVE EQUIPMENT (SPECIFY TYPES) EYE: Goggles

MECHANICAL (GENERAL) | GLOVES: Rubber

SPECIAL | RESPIRATOR: With proper filter

OTHER PROTECTIVE EQUIPMENT

SECTION IX SPECIAL PRECAUTIONS

PRECAUTIONS TO BE TAKEN IN HANDLING AND STORAGE

OTHER PRECAUTIONS

SIGNATURE | DATE

tert-BUTYL IODIDE

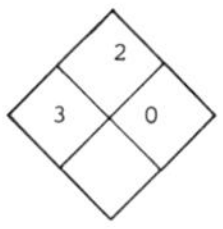

SECTION I SOURCE AND NOMENCLATURE

MANUFACTURER'S NAME | EMERGENCY TELEPHONE NO.

ADDRESS (NUMBER, STREET, CITY, STATE, ZIP CODE)

TRADE NAME AND SYNONYMS | CHEMICAL FAMILY

CHEMICAL NAME AND SYNONYMS: 2 iodo-2 methylpropane | FORMULA: $(CH_3)_3CI$

SECTION II HAZARDOUS INGREDIENTS MW - 184.03

BASIC MATERIAL	APPROXIMATE OR MAXIMUM % WT. OR VOL.	ESTABLISHED OSHA STANDARD	LD_{50}		LC_{50}	
			ORAL	PERCUT.	SPECIES	CONC.
			None	available		

SECTION III PHYSICAL DATA

BOILING POINT °F.: $99^{0}C$	VAPOR PRESSURE mm Hg.
MELTING POINT °F.: $-34^{0}C$	VAPOR DENSITY (Air = 1)
SPECIFIC GRAVITY (H_2O = 1): 1.57	EVAPORATION RATE (= 1)
SOLUBILITY IN WATER Pts 100 pts H_2O: Insoluble	VOLATILE % Vol. % Wt.

APPEARANCE AND ODOR: Colorless liquid

SECTION IV FIRE AND EXPLOSION HAZARD DATA

FLASH POINT: Not available | FLAMMABLE (EXPLOSIVE) LIMITS: UPPER / LOWER

METHOD USED

EXTINGUISHING MEDIA: Water, carbon dioxide, dry chemical

SPECIAL FIRE FIGHTING PROCEDURES: --

UNUSUAL FIRE AND EXPLOSION HAZARDS: When heated to decomposition point emits toxic fumes of iodates.

SECTION V HEALTH HAZARD DATA

TOXIC LEVEL: Not available | CARCINOGENIC: No

PRINCIPLE ROUTES OF ABSORBTION: Inhalation, ingestion, skin | SKIN AND EYE IRRITATION: slight

REVELANT SYMPTOMS OF EXPOSURE: Nausea, vomiting, diarrhea, dizziness, visual disturbances, irritability, drowsiness, central nervous system depression.

EFFECTS OF CHRONIC EXPOSURE: Dermatitis

EMERGENCY AND FIRST AID PROCEDURES: Inhalation - rest; if swallowed - wash mouth, give emetic, get medical aid; wash from skin & eyes.

SECTION VI REACTIVITY DATA

CONDITIONS CONTRIBUTING TO INSTABILITY

CONDITIONS CONTRIBUTING TO HAZARDOUS POLYMERIZATION

INCOMPATABILITY (MATERIALS TO AVOID)

HAZARDOUS DECOMPOSITION PRODUCTS

SECTION VII SPILL OR LEAK PROCEDURES

STEPS TO BE TAKEN IN CASE MATERIAL IS RELEASED OR SPILLED

WASTE DISPOSAL METHOD

SECTION VIII SPECIAL PROTECTION INFORMATION

VENTILATION REQUIREMENTS LOCAL EXHAUST | PROTECTIVE EQUIPMENT (SPECIFY TYPES) EYE: Goggles

MECHANICAL (GENERAL) | GLOVES: Rubber

SPECIAL | RESPIRATOR: With proper filter

OTHER PROTECTIVE EQUIPMENT

SECTION IX SPECIAL PRECAUTIONS

PRECAUTIONS TO BE TAKEN IN HANDLING AND STORAGE

OTHER PRECAUTIONS

SIGNATURE | DATE

n-BUTYRONITRILE

SECTION I SOURCE AND NOMENCLATURE

MANUFACTURER'S NAME	EMERGENCY TELEPHONE NO.
ADDRESS (NUMBER, STREET, CITY, STATE, ZIP CODE)	
TRADE NAME AND SYNONYMS	CHEMICAL FAMILY
CHEMICAL NAME AND SYNONYMS: Propyl Cyanide	FORMULA: C_3H_7CN

SECTION II HAZARDOUS INGREDIENTS

BASIC MATERIAL	APPROXIMATE OR MAXIMUM % WT. OR VOL.	ESTABLISHED OSHA STANDARD	LD_{50} mg/Kg ORAL	LD_{50} PERCUT.	LC_{50} SPECIES	LC_{50} CONC.
Lowest published lethal dose				400	Rabbit	
oral - rat 50 mg/kg.			200		Rat	
Lowest published lethal concentration in air						
inhalation - rat 400 PPM.						

SECTION III PHYSICAL DATA MW - 69.10

BOILING POINT °F.: 118°C.	VAPOR PRESSURE mm Hg.: 10 at 15°C.
MELTING POINT °F.: -113°C.	VAPOR DENSITY (Air = 1): 1.07
SPECIFIC GRAVITY (H_2O = 1): 0.796	EVAPORATION RATE (= 1)
SOLUBILITY IN WATER Pts/100 pts H_2O: Slightly soluble	VOLATILE % Vol. % Wt.

APPEARANCE AND ODOR: Colorless liquid, sharp suffocating odor.

SECTION IV FIRE AND EXPLOSION HAZARD DATA

FLASH POINT: 79°F.	FLAMMABLE (EXPLOSIVE) LIMITS	UPPER
METHOD USED: O.C.		LOWER: 1.65%

EXTINGUISHING MEDIA: Dry chemical, alcohol foam or carbon dioxide.

SPECIAL FIRE FIGHTING PROCEDURES: Protective clothing, self-contained breathing apparatus.

UNUSUAL FIRE AND EXPLOSION HAZARDS

SECTION V HEALTH HAZARD DATA

TOXIC LEVEL: TWA-TLV-8 PPM, 22 mg/C.M. (NIOSH Value) CARCINOGENIC: No

PRINCIPLE ROUTES OF ABSORBTION: Inhalation, ingestion, skin SKIN AND EYE IRRITATION: Moderate

REVELANT SYMPTOMS OF EXPOSURE: Dizziness, rapid respiration, headache, drowsiness, drop in blood pressure and pulse, delayed symptoms.

EFFECTS OF CHRONIC EXPOSURE: Eye irritation, loss of appetite, weight loss, mental deterioration.

EMERGENCY AND FIRST AID PROCEDURES: If swallowed - inhalation of amyl nitrite, induce vomiting, or gastric lavage, wash skin & eyes; inhalation, rest.

SECTION VI REACTIVITY DATA

CONDITIONS CONTRIBUTING TO INSTABILITY

CONDITIONS CONTRIBUTING TO HAZARDOUS POLYMERIZATION

INCOMPATABILITY (MATERIALS TO AVOID)

HAZARDOUS DECOMPOSITION PRODUCTS: When heated to decomposition omits highly toxic cyanide fumes.

SECTION VII SPILL OR LEAK PROCEDURES

STEPS TO BE TAKEN IN CASE MATERIAL IS RELEASED OR SPILLED: Absorb with paper towels - evaporate in hood.

WASTE DISPOSAL METHOD: Add excess alcoholic sodium hydroxide, after one hour, evaporate alcohol, then add calcium hypochlorite.

SECTION VIII SPECIAL PROTECTION INFORMATION

VENTILATION REQUIREMENTS LOCAL EXHAUST	PROTECTIVE EQUIPMENT (SPECIFY TYPES) EYE: Goggles
MECHANICAL (GENERAL)	GLOVES: Rubber
SPECIAL	RESPIRATOR: Self-contained

OTHER PROTECTIVE EQUIPMENT: Protective apron, full protective clothing.

SECTION IX SPECIAL PRECAUTIONS

PRECAUTIONS TO BE TAKEN IN HANDLING AND STORAGE

OTHER PRECAUTIONS

SIGNATURE	DATE

CADMIUM METAL

SECTION I SOURCE AND NOMENCLATURE

MANUFACTURER'S NAME	EMERGENCY TELEPHONE NO.
ADDRESS (NUMBER, STREET, CITY, STATE, ZIP CODE)	
TRADE NAME AND SYNONYMS	CHEMICAL FAMILY
CHEMICAL NAME AND SYNONYMS: Cadmium	FORMULA: Cd

SECTION II HAZARDOUS INGREDIENTS

BASIC MATERIAL	APPROXIMATE OR MAXIMUM % WT. OR VOL.	ESTABLISHED OSHA STANDARD	LD_{50} ORAL	LD_{50} PERCUT.	LC_{50} SPECIES	LC_{50} CONC.
			See Note 1			
#1 Lowest published toxic concentration in air						
Inhalation - Man 0.088 mg/c.m. - 6 years						

SECTION III PHYSICAL DATA MW - 112.41

BOILING POINT °F.: 767±2° C.	VAPOR PRESSURE mm Hg.: 1 at 394° C.
MELTING POINT °F.: 320.9° C.	VAPOR DENSITY (Air = 1)
SPECIFIC GRAVITY (H_2O = 1)	EVAPORATION RATE (= 1)
SOLUBILITY IN WATER Pts/100 pts H_2O	VOLATILE % Vol. % Wt.

APPEARANCE AND ODOR: Hexagonal crystals or silver-white malleable metal.

SECTION IV FIRE AND EXPLOSION HAZARD DATA

FLASH POINT: None available	FLAMMABLE (EXPLOSIVE) LIMITS	UPPER
METHOD USED		LOWER

EXTINGUISHING MEDIA: None available

SPECIAL FIRE FIGHTING PROCEDURES: Use self-contained breathing apparatus.

UNUSUAL FIRE AND EXPLOSION HAZARDS

SECTION V HEALTH HAZARD DATA

TOXIC LEVEL: TLV 0.05 mg/m³ CARCINOGENIC: Yes

PRINCIPLE ROUTES OF ABSORBTION: Inhalation or ingestion of dust or fume SKIN AND EYE IRRITATION

REVELANT SYMPTOMS OF EXPOSURE: Inhalation causes a metallic taste in mouth, shortness of breath, chest pains, ingestion: nausea, vomiting, diarrhea, headache, muscle aches, salivation, abdominal pain, shock, liver damage and renal failure.

EFFECTS OF CHRONIC EXPOSURE: Inhalation - loss of sense of smell, cough, dyspnea, weight loss, anemia, irritability and yellowstained teeth. Probable liver and kidney damage.

EMERGENCY AND FIRST AID PROCEDURES: Remove from exposure; treat pulmonary edema, liver and kidney damage if any. Calcium edetate orally is effective for inhalation. Obtain medical treatment.

SECTION VI REACTIVITY DATA

CONDITIONS CONTRIBUTING TO INSTABILITY

CONDITIONS CONTRIBUTING TO HAZARDOUS POLYMERIZATION

INCOMPATABILITY (MATERIALS TO AVOID): Ammonium Nitrate, Tellurium, Zinc

HAZARDOUS DECOMPOSITION PRODUCTS

SECTION VII SPILL OR LEAK PROCEDURES

STEPS TO BE TAKEN IN CASE MATERIAL IS RELEASED OR SPILLED

WASTE DISPOSAL METHOD

SECTION VIII SPECIAL PROTECTION INFORMATION

VENTILATION REQUIREMENTS LOCAL EXHAUST	PROTECTIVE EQUIPMENT (SPECIFY TYPES) EYE: Goggles or safety glasses
MECHANICAL (GENERAL)	GLOVES: Rubber
SPECIAL	RESPIRATOR: Mechanical filter respirator

OTHER PROTECTIVE EQUIPMENT: Protective coveralls and face shield

SECTION IX SPECIAL PRECAUTIONS

PRECAUTIONS TO BE TAKEN IN HANDLING AND STORAGE: Store in closed containers, protect containers against physical damage.

OTHER PRECAUTIONS

SIGNATURE	DATE

CAMPHOR GUM

(Hazard diamond: 4, 2, 0)

SECTION I SOURCE AND NOMENCLATURE

MANUFACTURER'S NAME | EMERGENCY TELEPHONE NO.

ADDRESS (NUMBER, STREET, CITY, STATE, ZIP CODE)

TRADE NAME AND SYNONYMS: 2 Camphanone

CHEMICAL FAMILY

CHEMICAL NAME AND SYNONYMS: Camphor

FORMULA: $C_{10}H_{16}O$

SECTION II HAZARDOUS INGREDIENTS

BASIC MATERIAL	APPROXIMATE OR MAXIMUM % WT. OR VOL.	ESTABLISHED OSHA STANDARD	LD_{50} ORAL	LD_{50} PERCUT.	LC_{50} SPECIES	LC_{50} CONC.
Oral - human LDLO 50 mg/kg						
(1) Intraperitoneal - rat LD50 900 mg/kg						
(2) Lowest published toxic dose						
skin - mouse 84 mg/C.M.						

SECTION III PHYSICAL DATA MW - 152.23

BOILING POINT °F.: 204°C.

VAPOR PRESSURE mm Hg.:

MELTING POINT °F.: 177°C.

VAPOR DENSITY (Air = 1): 5.24

SPECIFIC GRAVITY (H_2O = 1): 0.992

EVAPORATION RATE (= 1)

SOLUBILITY IN WATER Pts/100 pts H_2O: Insoluble

VOLATILE % Vol. % Wt.

APPEARANCE AND ODOR: White transparent crystaline mass, aromatic odor.

SECTION IV FIRE AND EXPLOSION HAZARD DATA

FLASH POINT: 150°F.

METHOD USED: C.C.

FLAMMABLE (EXPLOSIVE) LIMITS: UPPER 3.5%; LOWER 0.6%

EXTINGUISHING MEDIA: Foam, carbon dioxide, dry chemical

SPECIAL FIRE FIGHTING PROCEDURES

UNUSUAL FIRE AND EXPLOSION HAZARDS

SECTION V HEALTH HAZARD DATA

TOXIC LEVEL: TLV - 2 PPM, 12 mg/C.M.

CARCINOGENIC: neoplastic

PRINCIPLE ROUTES OF ABSORBTION: Inhalation, ingestion

SKIN AND EYE IRRITATION: Moderate

REVELANT SYMPTOMS OF EXPOSURE: Rigidity of jaw muscles, headache, convulsive seizures, coma.

EFFECTS OF CHRONIC EXPOSURE: --

EMERGENCY AND FIRST AID PROCEDURES: If swallowed - medical aid, gastric lavage, cathartics, inhalation - rest; wash from eyes and skin.

SECTION VI REACTIVITY DATA

CONDITIONS CONTRIBUTING TO INSTABILITY

CONDITIONS CONTRIBUTING TO HAZARDOUS POLYMERIZATION

INCOMPATABILITY (MATERIALS TO AVOID)

HAZARDOUS DECOMPOSITION PRODUCTS

SECTION VII SPILL OR LEAK PROCEDURES

STEPS TO BE TAKEN IN CASE MATERIAL IS RELEASED OR SPILLED

WASTE DISPOSAL METHOD

SECTION VIII SPECIAL PROTECTION INFORMATION

VENTILATION REQUIREMENTS LOCAL EXHAUST

PROTECTIVE EQUIPMENT (SPECIFY TYPES) EYE

MECHANICAL (GENERAL)

GLOVES

SPECIAL

RESPIRATOR: With proper filter

OTHER PROTECTIVE EQUIPMENT

SECTION IX SPECIAL PRECAUTIONS

PRECAUTIONS TO BE TAKEN IN HANDLING AND STORAGE

OTHER PRECAUTIONS

SIGNATURE | DATE

CARBON TETRABROMIDE

(Hazard diamond: 4, 0, 0)

SECTION I SOURCE AND NOMENCLATURE

MANUFACTURER'S NAME | EMERGENCY TELEPHONE NO.

ADDRESS (NUMBER, STREET, CITY, STATE, ZIP CODE)

TRADE NAME AND SYNONYMS

CHEMICAL FAMILY

CHEMICAL NAME AND SYNONYMS: Tetrabromomethane

FORMULA: $C Br_4$

SECTION II HAZARDOUS INGREDIENTS

BASIC MATERIAL	APPROXIMATE OR MAXIMUM % WT. OR VOL.	ESTABLISHED OSHA STANDARD	LD_{50} ORAL	LD_{50} PERCUT.	LC_{50} SPECIES	LC_{50} CONC.
Lowest Published lethaldose						
oral - rat 1000 mg/kg.						
LD50 - subcutaneous - mouse 298 mg/kg						

SECTION III PHYSICAL DATA MW - 331.67

BOILING POINT °F.: 190°C

VAPOR PRESSURE mm Hg.:

MELTING POINT °F.: α-48°C, β90°C

VAPOR DENSITY (Air = 1): 11.4

SPECIFIC GRAVITY (H_2O = 1): 3.42

EVAPORATION RATE (= 1)

SOLUBILITY IN WATER Pts/100 pts H_2O: 0.024

VOLATILE % Vol. % Wt.

APPEARANCE AND ODOR: Colorless solid

SECTION IV FIRE AND EXPLOSION HAZARD DATA

FLASH POINT: Nonflammable

METHOD USED

FLAMMABLE (EXPLOSIVE) LIMITS: UPPER; LOWER

EXTINGUISHING MEDIA: --

SPECIAL FIRE FIGHTING PROCEDURES: Self contained respirator if heated in a fire.

UNUSUAL FIRE AND EXPLOSION HAZARDS

SECTION V HEALTH HAZARD DATA

TOXIC LEVEL: TLV 0.1 PPM, 1.4 mg/C.M.

CARCINOGENIC: No

PRINCIPLE ROUTES OF ABSORBTION: Inhalation, ingestion, skin

SKIN AND EYE IRRITATION: Severe

REVELANT SYMPTOMS OF EXPOSURE: Eye and upper respiratory irritation, lung, liver & kidney damage, damage to cornea.

EFFECTS OF CHRONIC EXPOSURE: Liver damage

EMERGENCY AND FIRST AID PROCEDURES: Inhalation & rest, get medical aid; if swallowed - give emetic, get medical aid; wash from eyes and skin.

SECTION VI REACTIVITY DATA

CONDITIONS CONTRIBUTING TO INSTABILITY

CONDITIONS CONTRIBUTING TO HAZARDOUS POLYMERIZATION

INCOMPATABILITY (MATERIALS TO AVOID)

HAZARDOUS DECOMPOSITION PRODUCTS

SECTION VII SPILL OR LEAK PROCEDURES

STEPS TO BE TAKEN IN CASE MATERIAL IS RELEASED OR SPILLED: Collect on paper, evaporate on dish in hood.

WASTE DISPOSAL METHOD

SECTION VIII SPECIAL PROTECTION INFORMATION

VENTILATION REQUIREMENTS LOCAL EXHAUST

PROTECTIVE EQUIPMENT (SPECIFY TYPES) EYE: Safety goggles

MECHANICAL (GENERAL)

GLOVES: Rubber

SPECIAL

RESPIRATOR: Self contained

OTHER PROTECTIVE EQUIPMENT

SECTION IX SPECIAL PRECAUTIONS

PRECAUTIONS TO BE TAKEN IN HANDLING AND STORAGE

OTHER PRECAUTIONS

SIGNATURE | DATE

CHLORANIL

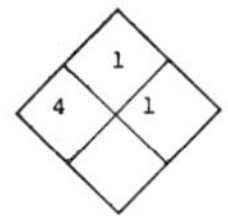

SECTION I SOURCE AND NOMENCLATURE

MANUFACTURER'S NAME

EMERGENCY TELEPHONE NO.

ADDRESS (NUMBER, STREET, CITY, STATE, ZIP CODE)

TRADE NAME AND SYNONYMS: Chloranil

CHEMICAL FAMILY

CHEMICAL NAME AND SYNONYMS: 2,3,5,6 Tetrachloric P-Benzoquinone

FORMULA: OC_6Cl_4O

SECTION II HAZARDOUS INGREDIENTS

BASIC MATERIAL	APPROXIMATE OR MAXIMUM % WT. OR VOL.	ESTABLISHED OSHA STANDARD	LD_{50} mg/kg ORAL	LD_{50} mg/kg PERCUT.	LC_{50} SPECIES	LC_{50} CONC.
			4000		Rat	
LDLO - oral - human 500	mg/kg					

SECTION III PHYSICAL DATA MW 245.86

BOILING POINT °F.

VAPOR PRESSURE mm Hg.

MELTING POINT °F.: 290°C.

VAPOR DENSITY (Air = 1)

SPECIFIC GRAVITY (H_2O = 1)

EVAPORATION RATE (______ = 1)

SOLUBILITY IN WATER Pts/100 pts H_2O

VOLATILE % Vol. % Wt.

APPEARANCE AND ODOR: Yellow crystals

SECTION IV FIRE AND EXPLOSION HAZARD DATA

FLASH POINT: None available

FLAMMABLE (EXPLOSIVE) LIMITS — UPPER / LOWER

METHOD USED

EXTINGUISHING MEDIA: Dry chemical, carbon dioxide, alcohol foam.

SPECIAL FIRE FIGHTING PROCEDURES

UNUSUAL FIRE AND EXPLOSION HAZARDS

SECTION V HEALTH HAZARD DATA

TOXIC LEVEL: None available

CARCINOGENIC: Yes

PRINCIPLE ROUTES OF ABSORBTION: Inhalation, ingestion

SKIN AND EYE IRRITATION: Severe

REVELANT SYMPTOMS OF EXPOSURE: Irritation to skin, eye and mucous membrane, headache, dizziness.

EFFECTS OF CHRONIC EXPOSURE: May cause cancer of liver or kidneys.

EMERGENCY AND FIRST AID PROCEDURES: Wash skin & eyes; if swallowed - gastric lavage, get medical aid; inhalation - rest.

SECTION VI REACTIVITY DATA

CONDITIONS CONTRIBUTING TO INSTABILITY

CONDITIONS CONTRIBUTING TO HAZARDOUS POLYMERIZATION

INCOMPATABILITY (MATERIALS TO AVOID)

HAZARDOUS DECOMPOSITION PRODUCTS

SECTION VII SPILL OR LEAK PROCEDURES

STEPS TO BE TAKEN IN CASE MATERIAL IS RELEASED OR SPILLED

WASTE DISPOSAL METHOD

SECTION VIII SPECIAL PROTECTION INFORMATION

VENTILATION REQUIREMENTS LOCAL EXHAUST

PROTECTIVE EQUIPMENT (SPECIFY TYPES) EYE: Safety goggles

MECHANICAL (GENERAL)

GLOVES: Rubber

SPECIAL

RESPIRATOR: With mechanical filter

OTHER PROTECTIVE EQUIPMENT

SECTION IX SPECIAL PRECAUTIONS

PRECAUTIONS TO BE TAKEN IN HANDLING AND STORAGE

OTHER PRECAUTIONS

SIGNATURE

DATE

CHLOROACETIC ACID

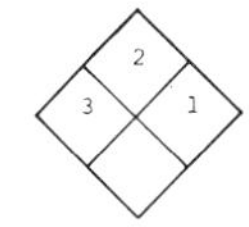

SECTION I SOURCE AND NOMENCLATURE

MANUFACTURER'S NAME

EMERGENCY TELEPHONE NO.

ADDRESS (NUMBER, STREET, CITY, STATE, ZIP CODE)

TRADE NAME AND SYNONYMS

CHEMICAL FAMILY

CHEMICAL NAME AND SYNONYMS: Monochloroacetic acid

FORMULA: $ClCH_2COOH$

SECTION II HAZARDOUS INGREDIENTS

BASIC MATERIAL	APPROXIMATE OR MAXIMUM % WT. OR VOL.	ESTABLISHED OSHA STANDARD	LD_{50} mg/kg ORAL	LD_{50} mg/kg PERCUT.	LC_{50} SPECIES	LC_{50} CONC.
			76		Rat	
Lowest published lethal dose - human - oral 50 mg/kg				5	Rat	
			165		Mouse	
Note: Subcutaneous - rat	LD50 - 5 mg/kg.					

SECTION III PHYSICAL DATA Mol. Wt. 94.50

BOILING POINT °F.: 189°C

VAPOR PRESSURE mm Hg.

MELTING POINT °F.: 63°C

VAPOR DENSITY (Air = 1): 3.26

SPECIFIC GRAVITY (H_2O = 1): 1.58

EVAPORATION RATE (______ = 1)

SOLUBILITY IN WATER Pts/100 pts H_2O: Very soluble

VOLATILE % Vol. % Wt.

APPEARANCE AND ODOR: Colorless crystals

SECTION IV FIRE AND EXPLOSION HAZARD DATA

FLASH POINT: None available

FLAMMABLE (EXPLOSIVE) LIMITS — UPPER / LOWER

METHOD USED

EXTINGUISHING MEDIA: Use water spray, dry chemical, CO_2, or alcohol foam.

SPECIAL FIRE FIGHTING PROCEDURES: Wear goggles & self-contained breathing apparatus.

UNUSUAL FIRE AND EXPLOSION HAZARDS

SECTION V HEALTH HAZARD DATA

TOXIC LEVEL: None available

CARCINOGENIC: No

PRINCIPLE ROUTES OF ABSORBTION: Inhalation, ingestion

SKIN AND EYE IRRITATION: Severe

REVELANT SYMPTOMS OF EXPOSURE: Severe eye & skin burns, severe respiratory and digestive system irritation.

EFFECTS OF CHRONIC EXPOSURE: --

EMERGENCY AND FIRST AID PROCEDURES: If swallowed - drink water, milk of magnesia, get medical aid; inhalation - rest; wash eyes & skin.

SECTION VI REACTIVITY DATA

CONDITIONS CONTRIBUTING TO INSTABILITY

CONDITIONS CONTRIBUTING TO HAZARDOUS POLYMERIZATION

INCOMPATABILITY (MATERIALS TO AVOID): Reacts with alkalies.

HAZARDOUS DECOMPOSITION PRODUCTS

SECTION VII SPILL OR LEAK PROCEDURES

STEPS TO BE TAKEN IN CASE MATERIAL IS RELEASED OR SPILLED

WASTE DISPOSAL METHOD: Dissolve in combustible liquid and burn.

SECTION VIII SPECIAL PROTECTION INFORMATION

VENTILATION REQUIREMENTS LOCAL EXHAUST

PROTECTIVE EQUIPMENT (SPECIFY TYPES) EYE: Goggles

MECHANICAL (GENERAL)

GLOVES: Rubber

SPECIAL

RESPIRATOR: w/proper filter

OTHER PROTECTIVE EQUIPMENT

SECTION IX SPECIAL PRECAUTIONS

PRECAUTIONS TO BE TAKEN IN HANDLING AND STORAGE: Keep containers closed, store in a cool place.

OTHER PRECAUTIONS

SIGNATURE

DATE

CHLOROACETONE

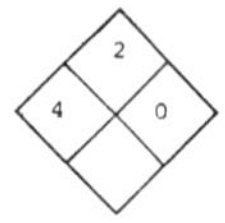

SECTION I SOURCE AND NOMENCLATURE

MANUFACTURER'S NAME

EMERGENCY TELEPHONE NO.

ADDRESS (NUMBER, STREET, CITY, STATE, ZIP CODE)

TRADE NAME AND SYNONYMS

CHEMICAL FAMILY

CHEMICAL NAME AND SYNONYMS: 1 Chloro 2 Propanone

FORMULA: CH_3COCH_2Cl

SECTION II HAZARDOUS INGREDIENTS

BASIC MATERIAL	APPROXIMATE OR MAXIMUM % WT. OR VOL.	ESTABLISHED OSHA STANDARD	LD_{50} ORAL	LD_{50} PERCUT.	LC_{50} SPECIES	LC_{50} CONC.
Lowest published lethal concentration in air						
Inhalation - human - 605 PPM/10 minutes						

SECTION III PHYSICAL DATA MW - 95.53

BOILING POINT: 121°C.

VAPOR PRESSURE (mm Hg.)

MELTING POINT: -44.5°

VAPOR DENSITY (Air = 1)

SPECIFIC GRAVITY (H_2O = 1): 1.16

EVAPORATION RATE (=1)

SOLUBILITY IN WATER Pts/100 pts H_2O: ∞

VOLATILE % Vol. % Wt.

APPEARANCE AND ODOR: Colorless liquid, pungent odor.

SECTION IV FIRE AND EXPLOSION HAZARD DATA

FLASH POINT: None available

FLAMMABLE (EXPLOSIVE) LIMITS: UPPER / LOWER

METHOD USED

EXTINGUISHING MEDIA: Alcohol foam. Water may be ineffective.

SPECIAL FIRE FIGHTING PROCEDURES

UNUSUAL FIRE AND EXPLOSION HAZARDS: Emits phosgene on heating to decomposition point.

SECTION V HEALTH HAZARD DATA

TOXIC LEVEL: None available

CARCINOGENIC: Neoplastic

PRINCIPLE ROUTES OF ABSORBTION: Inhalation, ingestion

SKIN AND EYE IRRITATION: Severe

RELEVANT SYMPTOMS OF EXPOSURE: Irritation of respiratory system and eyes; ingestion - severe internal irritation.

EFFECTS OF CHRONIC EXPOSURE: --

EMERGENCY AND FIRST AID PROCEDURES: Inhalation - rest; if swallowed - give emetic, get medical aid; wash eyes and skin.

SECTION VI REACTIVITY DATA

CONDITIONS CONTRIBUTING TO INSTABILITY

CONDITIONS CONTRIBUTING TO HAZARDOUS POLYMERIZATION

INCOMPATABILITY (MATERIALS TO AVOID)

HAZARDOUS DECOMPOSITION PRODUCTS

SECTION VII SPILL OR LEAK PROCEDURES

STEPS TO BE TAKEN IN CASE MATERIAL IS RELEASED OR SPILLED

WASTE DISPOSAL METHOD

SECTION VIII SPECIAL PROTECTION INFORMATION

VENTILATION REQUIREMENTS LOCAL EXHAUST

PROTECTIVE EQUIPMENT (SPECIFY TYPES) EYE: Goggles

MECHANICAL (GENERAL)

GLOVES: Rubber

SPECIAL

RESPIRATOR: Self-contained

OTHER PROTECTIVE EQUIPMENT

SECTION IX SPECIAL PRECAUTIONS

PRECAUTIONS TO BE TAKEN IN HANDLING AND STORAGE

OTHER PRECAUTIONS

SIGNATURE

DATE

CHLOROACETONITRILE

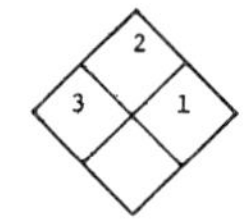

SECTION I SOURCE AND NOMENCLATURE

MANUFACTURER'S NAME

EMERGENCY TELEPHONE NO.

ADDRESS (NUMBER, STREET, CITY, STATE, ZIP CODE)

TRADE NAME AND SYNONYMS

CHEMICAL FAMILY

CHEMICAL NAME AND SYNONYMS: Chloromethyl Cyanide

FORMULA: $ClCH_2CN$

SECTION II HAZARDOUS INGREDIENTS

BASIC MATERIAL	APPROXIMATE OR MAXIMUM % WT. OR VOL.	ESTABLISHED OSHA STANDARD	LD_{50} mg/kg ORAL	LD_{50} PERCUT.	LC_{50} SPECIES	LC_{50} CONC.
			220		Rat	
LCLO inhalation - rat 250 ppm/4 hrs.						
LDLO - skin - rabbit 71 mg/kg						

SECTION III PHYSICAL DATA MW-75.50

BOILING POINT: 126°C.

VAPOR PRESSURE (mm Hg.)

MELTING POINT: --

VAPOR DENSITY (Air = 1)

SPECIFIC GRAVITY (H_2O = 1): 1.19

EVAPORATION RATE (=1)

SOLUBILITY IN WATER Pts/100 pts H_2O: --

VOLATILE % Vol. % Wt.

APPEARANCE AND ODOR: Crystals

SECTION IV FIRE AND EXPLOSION HAZARD DATA

FLASH POINT: None available

FLAMMABLE (EXPLOSIVE) LIMITS: UPPER / LOWER

METHOD USED

EXTINGUISHING MEDIA: Alcohol foam, dry chemical, carbon dioxide, water may be ineffective.

SPECIAL FIRE FIGHTING PROCEDURES: Wear self-contained breathing apparatus.

UNUSUAL FIRE AND EXPLOSION HAZARDS

SECTION V HEALTH HAZARD DATA

TOXIC LEVEL: None available

CARCINOGENIC: No

PRINCIPLE ROUTES OF ABSORBTION: Inhalation, ingestion, skin

SKIN AND EYE IRRITATION: Mild to moderate

RELEVANT SYMPTOMS OF EXPOSURE: Dizziness, rapid respiration, headache, drowsiness, drop in blood pressure, rapid pulse, unconsciousness.

EFFECTS OF CHRONIC EXPOSURE: Dizziness, weakness, lung congestion, eye irritation, loss of appetite and weight, mental deterioration.

EMERGENCY AND FIRST AID PROCEDURES: Remove from area, give amyl nitrite inhalation every 5 minutes, artificial respiration with oxygen, wash skin & eyes, get medical aid.

SECTION VI REACTIVITY DATA

CONDITIONS CONTRIBUTING TO INSTABILITY

CONDITIONS CONTRIBUTING TO HAZARDOUS POLYMERIZATION

INCOMPATABILITY (MATERIALS TO AVOID)

HAZARDOUS DECOMPOSITION PRODUCTS: Cyanides

SECTION VII SPILL OR LEAK PROCEDURES

STEPS TO BE TAKEN IN CASE MATERIAL IS RELEASED OR SPILLED

WASTE DISPOSAL METHOD

SECTION VIII SPECIAL PROTECTION INFORMATION

VENTILATION REQUIREMENTS LOCAL EXHAUST

PROTECTIVE EQUIPMENT (SPECIFY TYPES) EYE: Goggles

MECHANICAL (GENERAL)

GLOVES: Rubber

SPECIAL

RESPIRATOR: Self-contained

OTHER PROTECTIVE EQUIPMENT

SECTION IX SPECIAL PRECAUTIONS

PRECAUTIONS TO BE TAKEN IN HANDLING AND STORAGE: Outside or detached storage preferred inside storage in standard liquids storage room.

OTHER PRECAUTIONS

SIGNATURE

DATE

o-CHLOROANILINE

Hazard diamond: 1 (top), 3 (left), 0 (right)

SECTION I SOURCE AND NOMENCLATURE

MANUFACTURER'S NAME | EMERGENCY TELEPHONE NO.

ADDRESS (NUMBER, STREET, CITY, STATE, ZIP CODE)

TRADE NAME AND SYNONYMS: Fast Yellow GC Base

CHEMICAL FAMILY

CHEMICAL NAME AND SYNONYMS: O-Chloroaniline

FORMULA: $Cl(C_6H_4)NH_2$

SECTION II HAZARDOUS INGREDIENTS

BASIC MATERIAL	APPROXIMATE OR MAXIMUM % WT. OR VOL.	ESTABLISHED OSHA STANDARD	LD_{50} mg/kg ORAL	PERCUT.	LC_{50} SPECIES	CONC.
			256		Mouse	
				222	Cat	

SECTION III PHYSICAL DATA MW - 127.58

BOILING POINT °F: 211°C.

VAPOR PRESSURE mm Hg: 1 at 46°C.

MELTING POINT °F: 0°C.

VAPOR DENSITY (Air = 1)

SPECIFIC GRAVITY (H_2O = 1): 1.21

EVAPORATION RATE (____ = 1)

SOLUBILITY IN WATER Pts/100 pts H_2O: Insoluble

VOLATILE % Vol. % Wt.

APPEARANCE AND ODOR: Liquid, sweet odor

SECTION IV FIRE AND EXPLOSION HAZARD DATA

FLASH POINT: None available

METHOD USED

FLAMMABLE (EXPLOSIVE) LIMITS: UPPER / LOWER

EXTINGUISHING MEDIA: Water spray, carbon dioxide, alcohol foam.

SPECIAL FIRE FIGHTING PROCEDURES: Self-contained breathing apparatus.

UNUSUAL FIRE AND EXPLOSION HAZARDS: Vapor may form explosive mixtures in air.

SECTION V HEALTH HAZARD DATA

TOXIC LEVEL: None available

CARCINOGENIC: No

PRINCIPLE ROUTES OF ABSORBTION: Inhalation, skin, ingestion

SKIN AND EYE IRRITATION: Low

REVELANT SYMPTOMS OF EXPOSURE: Headache, nausea, vomiting, jaundic, convulsions

EFFECTS OF CHRONIC EXPOSURE: Severe kidney damage, some liver damage, dermatitis.

EMERGENCY AND FIRST AID PROCEDURES: Remove from exposure; wash skin & eyes; if swallowed - remove by gastric lavage and cathartics; oxygen as needed.

SECTION VI REACTIVITY DATA

CONDITIONS CONTRIBUTING TO INSTABILITY

CONDITIONS CONTRIBUTING TO HAZARDOUS POLYMERIZATION

INCOMPATABILITY (MATERIALS TO AVOID): Sodium and potassium peroxide, ozone, perchloric acid.

HAZARDOUS DECOMPOSITION PRODUCTS

SECTION VII SPILL OR LEAK PROCEDURES

STEPS TO BE TAKEN IN CASE MATERIAL IS RELEASED OR SPILLED: Wear protective clothing during cleanup.

WASTE DISPOSAL METHOD

SECTION VIII SPECIAL PROTECTION INFORMATION

VENTILATION REQUIREMENTS LOCAL EXHAUST

PROTECTIVE EQUIPMENT (SPECIFY TYPES) EYE: Goggles

MECHANICAL (GENERAL)

GLOVES: Rubber

SPECIAL

RESPIRATOR: Self-contained

OTHER PROTECTIVE EQUIPMENT: Protective clothing

SECTION IX SPECIAL PRECAUTIONS

PRECAUTIONS TO BE TAKEN IN HANDLING AND STORAGE: Store in flammable liquids cabinets.

OTHER PRECAUTIONS

SIGNATURE | DATE

o-CHLOROBENZALDEHYDE

Hazard diamond: 2 (top), 3 (left), 0 (right)

SECTION I SOURCE AND NOMENCLATURE

MANUFACTURER'S NAME | EMERGENCY TELEPHONE NO.

ADDRESS (NUMBER, STREET, CITY, STATE, ZIP CODE)

TRADE NAME AND SYNONYMS

CHEMICAL FAMILY

CHEMICAL NAME AND SYNONYMS: 2-chlorobenzaldehyde

FORMULA: $Cl\ C_6H_4CHO$

SECTION II HAZARDOUS INGREDIENTS

BASIC MATERIAL	APPROXIMATE OR MAXIMUM % WT. OR VOL.	ESTABLISHED OSHA STANDARD	LD_{50} ORAL	PERCUT.	LC_{50} SPECIES	CONC.
#1 lowest published lethal dose intraperitoneal - mouse	10 mg/kg					

SECTION III PHYSICAL DATA MW - 140.57

BOILING POINT °F: 208

VAPOR PRESSURE mm Hg

MELTING POINT °F: 11°C

VAPOR DENSITY (Air = 1)

SPECIFIC GRAVITY (H_2O = 1): 1.29

EVAPORATION RATE (____ = 1)

SOLUBILITY IN WATER Pts/100 pts H_2O: Very slightly soluble

VOLATILE % Vol. % Wt.

APPEARANCE AND ODOR: White crystal

SECTION IV FIRE AND EXPLOSION HAZARD DATA

FLASH POINT: Not available

METHOD USED

FLAMMABLE (EXPLOSIVE) LIMITS: UPPER / LOWER

EXTINGUISHING MEDIA: Foam, carbon dioxide, dry chemical

SPECIAL FIRE FIGHTING PROCEDURES

UNUSUAL FIRE AND EXPLOSION HAZARDS

SECTION V HEALTH HAZARD DATA

TOXIC LEVEL: None available

CARCINOGENIC: No

PRINCIPLE ROUTES OF ABSORBTION: Ingestion, skin, inhalation

SKIN AND EYE IRRITATION: Moderate

REVELANT SYMPTOMS OF EXPOSURE: Eye irritation, headache, drowsiness, gastroentric distress, pulmonary edema

EFFECTS OF CHRONIC EXPOSURE: Anemia, dermatitis

EMERGENCY AND FIRST AID PROCEDURES: If swallowed - induce vomiting, get medical aid; wash eyes & skin; inhalation - rest.

SECTION VI REACTIVITY DATA

CONDITIONS CONTRIBUTING TO INSTABILITY

CONDITIONS CONTRIBUTING TO HAZARDOUS POLYMERIZATION

INCOMPATABILITY (MATERIALS TO AVOID)

HAZARDOUS DECOMPOSITION PRODUCTS

SECTION VII SPILL OR LEAK PROCEDURES

STEPS TO BE TAKEN IN CASE MATERIAL IS RELEASED OR SPILLED

WASTE DISPOSAL METHOD

SECTION VIII SPECIAL PROTECTION INFORMATION

VENTILATION REQUIREMENTS LOCAL EXHAUST

PROTECTIVE EQUIPMENT (SPECIFY TYPES) EYE: Goggles

MECHANICAL (GENERAL)

GLOVES: Rubber

SPECIAL

RESPIRATOR: With proper filter

OTHER PROTECTIVE EQUIPMENT

SECTION IX SPECIAL PRECAUTIONS

PRECAUTIONS TO BE TAKEN IN HANDLING AND STORAGE

OTHER PRECAUTIONS

SIGNATURE | DATE

p-CHLOROBENZALDEHYDE

2 / 3 / 0

SECTION I SOURCE AND NOMENCLATURE

MANUFACTURER'S NAME | EMERGENCY TELEPHONE NO.

ADDRESS (NUMBER, STREET, CITY, STATE, ZIP CODE)

TRADE NAME AND SYNONYMS | CHEMICAL FAMILY

CHEMICAL NAME AND SYNONYMS: P-chlorobenzaldehyde

FORMULA: $Cl\ C_6H_4CHO$

SECTION II HAZARDOUS INGREDIENTS

BASIC MATERIAL	APPROXIMATE OR MAXIMUM % WT. OR VOL.	ESTABLISHED OSHA STANDARD	LD_{50} ORAL	LD_{50} PERCUT.	LC_{50} SPECIES	LC_{50} CONC.
			None available			

SECTION III PHYSICAL DATA MW – 140.57

BOILING POINT °F.: 213°C

VAPOR PRESSURE mm Hg.:

MELTING POINT °F.: 48°C

VAPOR DENSITY (Air = 1):

SPECIFIC GRAVITY (H_2O = 1): 1.12

EVAPORATION RATE (= 1):

SOLUBILITY IN WATER Pts/100 pts H_2O: Soluble in hot water

VOLATILE % Vol. % Wt.:

APPEARANCE AND ODOR: White needle-like crystals

SECTION IV FIRE AND EXPLOSION HAZARD DATA

FLASH POINT: Not available

METHOD USED:

FLAMMABLE (EXPLOSIVE) LIMITS: UPPER / LOWER

EXTINGUISHING MEDIA: Foam, carbon dioxide, dry chemical

SPECIAL FIRE FIGHTING PROCEDURES:

UNUSUAL FIRE AND EXPLOSION HAZARDS:

SECTION V HEALTH HAZARD DATA

TOXIC LEVEL: Not available

CARCINOGENIC: No

PRINCIPLE ROUTES OF ABSORBTION: Ingestion, skin, inhalation of dust

SKIN AND EYE IRRITATION: moderate

REVELANT SYMPTOMS OF EXPOSURE: Eye irritation, headache, drowsiness, gastroentric distress, pulmonary edema

EFFECTS OF CHRONIC EXPOSURE: Anemia, dermatitis

EMERGENCY AND FIRST AID PROCEDURES: If swallowed - induce vomiting, get medical aid; wash eyes & skin; inhalation - rest.

SECTION VI REACTIVITY DATA

CONDITIONS CONTRIBUTING TO INSTABILITY:

CONDITIONS CONTRIBUTING TO HAZARDOUS POLYMERIZATION:

INCOMPATABILITY (MATERIALS TO AVOID):

HAZARDOUS DECOMPOSITION PRODUCTS:

SECTION VII SPILL OR LEAK PROCEDURES

STEPS TO BE TAKEN IN CASE MATERIAL IS RELEASED OR SPILLED:

WASTE DISPOSAL METHOD:

SECTION VIII SPECIAL PROTECTION INFORMATION

VENTILATION REQUIREMENTS LOCAL EXHAUST:

MECHANICAL (GENERAL):

SPECIAL:

PROTECTIVE EQUIPMENT (SPECIFY TYPES) EYE: Goggles

GLOVES: Rubber

RESPIRATOR: With proper filter

OTHER PROTECTIVE EQUIPMENT:

SECTION IX SPECIAL PRECAUTIONS

PRECAUTIONS TO BE TAKEN IN HANDLING AND STORAGE:

OTHER PRECAUTIONS:

SIGNATURE | DATE

p-CHLOROBENZONITRILE

1 / 3 / 0

SECTION I SOURCE AND NOMENCLATURE

MANUFACTURER'S NAME | EMERGENCY TELEPHONE NO.

ADDRESS (NUMBER, STREET, CITY, STATE, ZIP CODE)

TRADE NAME AND SYNONYMS | CHEMICAL FAMILY

CHEMICAL NAME AND SYNONYMS: P-Chlorobenzonitrile

FORMULA: ClC_6H_4CN

SECTION II HAZARDOUS INGREDIENTS

BASIC MATERIAL	APPROXIMATE OR MAXIMUM % WT. OR VOL.	ESTABLISHED OSHA STANDARD	LD_{50} mg/kg ORAL	LD_{50} mg/kg PERCUT.	LC_{50} SPECIES	LC_{50} CONC.
			887		Rat	

SECTION III PHYSICAL DATA MW – 137.57

BOILING POINT °F.: 223°C.

VAPOR PRESSURE mm Hg.:

MELTING POINT °F.: 94°C.

VAPOR DENSITY (Air = 1):

SPECIFIC GRAVITY (H_2O = 1):

EVAPORATION RATE (= 1):

SOLUBILITY IN WATER Pts/100 pts H_2O: Insoluble

VOLATILE % Vol. % Wt.:

APPEARANCE AND ODOR: White crystals

SECTION IV FIRE AND EXPLOSION HAZARD DATA

FLASH POINT: None available

METHOD USED:

FLAMMABLE (EXPLOSIVE) LIMITS: UPPER / LOWER

EXTINGUISHING MEDIA: Alcohol foam, carbon dioxide, dry chemical.

SPECIAL FIRE FIGHTING PROCEDURES: Wear self-contained breathing apparatus.

UNUSUAL FIRE AND EXPLOSION HAZARDS:

SECTION V HEALTH HAZARD DATA

TOXIC LEVEL: None available

CARCINOGENIC: No

PRINCIPLE ROUTES OF ABSORBTION: Inhalation, ingestion, skin

SKIN AND EYE IRRITATION: Mild to moderate

REVELANT SYMPTOMS OF EXPOSURE: Dizziness, rapid respiration, headache, drowsiness, drop in blood pressure, rapid pulse.

EFFECTS OF CHRONIC EXPOSURE: Dizziness, weakness, eye irritation, lung congestion, weigh and appetite loss, mental degradation.

EMERGENCY AND FIRST AID PROCEDURES: If swallowed - amyl nitrite inhalation, induce vomiting, get medical aid; wash skin & eyes; inhalation - rest.

SECTION VI REACTIVITY DATA

CONDITIONS CONTRIBUTING TO INSTABILITY:

CONDITIONS CONTRIBUTING TO HAZARDOUS POLYMERIZATION:

INCOMPATABILITY (MATERIALS TO AVOID):

HAZARDOUS DECOMPOSITION PRODUCTS: Cyanides

SECTION VII SPILL OR LEAK PROCEDURES

STEPS TO BE TAKEN IN CASE MATERIAL IS RELEASED OR SPILLED:

WASTE DISPOSAL METHOD:

SECTION VIII SPECIAL PROTECTION INFORMATION

VENTILATION REQUIREMENTS LOCAL EXHAUST:

MECHANICAL (GENERAL):

SPECIAL:

PROTECTIVE EQUIPMENT (SPECIFY TYPES) EYE: Goggles

GLOVES: Rubber

RESPIRATOR: With proper filter

OTHER PROTECTIVE EQUIPMENT: Protective apron

SECTION IX SPECIAL PRECAUTIONS

PRECAUTIONS TO BE TAKEN IN HANDLING AND STORAGE: Outside or detached storage preferred, inside storage in standard flammable liquids storage cabinet.

OTHER PRECAUTIONS:

SIGNATURE | DATE

o-CHLOROBENZOYL CHLORIDE

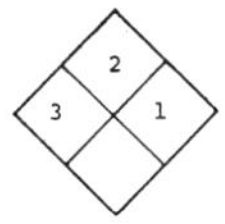

SECTION I SOURCE AND NOMENCLATURE

MANUFACTURER'S NAME | EMERGENCY TELEPHONE NO.

ADDRESS (NUMBER, STREET, CITY, STATE, ZIP CODE)

TRADE NAME AND SYNONYMS | CHEMICAL FAMILY

CHEMICAL NAME AND SYNONYMS: 1,2 chlorobenzoyl chloride | FORMULA: Cl C_6H_4COCl

SECTION II HAZARDOUS INGREDIENTS

BASIC MATERIAL	APPROXIMATE OR MAXIMUM % WT. OR VOL.	ESTABLISHED OSHA STANDARD	LD_{50} ORAL	LD_{50} PERCUT.	LC_{50} SPECIES	LC_{50} CONC.
			None	available		

SECTION III PHYSICAL DATA MW - 175.02

BOILING POINT °F.: 230 | VAPOR PRESSURE mm Hg.

MELTING POINT °F.: $-4^{\circ}C$ | VAPOR DENSITY (Air = 1)

SPECIFIC GRAVITY (H_2O = 1) | EVAPORATION RATE (______ = 1)

SOLUBILITY IN WATER Pts/100 pts H_2O: Decomposes | VOLATILE % Vol. % Wt.

APPEARANCE AND ODOR: Colorless liquid - fumes

SECTION IV FIRE AND EXPLOSION HAZARD DATA

FLASH POINT: N/A | FLAMMABLE (EXPLOSIVE) LIMITS: UPPER / LOWER

METHOD USED

EXTINGUISHING MEDIA: Water spray, foam, carbon dioxide or dry chemical

SPECIAL FIRE FIGHTING PROCEDURES: Full protective clothing

UNUSUAL FIRE AND EXPLOSION HAZARDS

SECTION V HEALTH HAZARD DATA

TOXIC LEVEL: None available | CARCINOGENIC: No

PRINCIPLE ROUTES OF ABSORBTION: Inhalation, ingestion | SKIN AND EYE IRRITATION: Severe

REVELANT SYMPTOMS OF EXPOSURE: Irritates respiratory system, skin & eye burns, mouth and digestive tract burns.

EFFECTS OF CHRONIC EXPOSURE: --

EMERGENCY AND FIRST AID PROCEDURES: Inhalation - rest, artificial respiration if needed; if swallowed - drink water, milk of magnesia, get medical aid; wash skin & eyes.

SECTION VI REACTIVITY DATA

CONDITIONS CONTRIBUTING TO INSTABILITY

CONDITIONS CONTRIBUTING TO HAZARDOUS POLYMERIZATION

INCOMPATABILITY (MATERIALS TO AVOID): dimethyl sulfoxide

HAZARDOUS DECOMPOSITION PRODUCTS: When heated, phosgene in produced

SECTION VII SPILL OR LEAK PROCEDURES

STEPS TO BE TAKEN IN CASE MATERIAL IS RELEASED OR SPILLED

WASTE DISPOSAL METHOD

SECTION VIII SPECIAL PROTECTION INFORMATION

VENTILATION REQUIREMENTS LOCAL EXHAUST | PROTECTIVE EQUIPMENT (SPECIFY TYPES) EYE: Goggles

MECHANICAL (GENERAL) | GLOVES: Rubber

SPECIAL | RESPIRATOR: With proper filter

OTHER PROTECTIVE EQUIPMENT

SECTION IX SPECIAL PRECAUTIONS

PRECAUTIONS TO BE TAKEN IN HANDLING AND STORAGE: Store in coll, dry, well-ventilated room.

OTHER PRECAUTIONS

SIGNATURE | DATE

p-CHLOROBENZOYL CHLORIDE

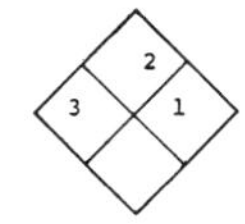

SECTION I SOURCE AND NOMENCLATURE

MANUFACTURER'S NAME | EMERGENCY TELEPHONE NO.

ADDRESS (NUMBER, STREET, CITY, STATE, ZIP CODE)

TRADE NAME AND SYNONYMS | CHEMICAL FAMILY

CHEMICAL NAME AND SYNONYMS: 1,4 Chlorobenzoyl Chloride | FORMULA: ClC_6H_4COCl

SECTION II HAZARDOUS INGREDIENTS

BASIC MATERIAL	APPROXIMATE OR MAXIMUM % WT. OR VOL.	ESTABLISHED OSHA STANDARD	LD_{50} ORAL	LD_{50} PERCUT.	LC_{50} SPECIES	LC_{50} CONC.
			None	available		

SECTION III PHYSICAL DATA MW-175.02

BOILING POINT °F.: $221^{\circ}C$ | VAPOR PRESSURE mm Hg.

MELTING POINT °F.: $16^{\circ}C$ | VAPOR DENSITY (Air = 1)

SPECIFIC GRAVITY (H_2O = 1): 1.38 | EVAPORATION RATE (______ = 1)

SOLUBILITY IN WATER Pts/100 pts H_2O: Decomposes | VOLATILE % Vol. % Wt.

APPEARANCE AND ODOR: Colorless liquid - fumes

SECTION IV FIRE AND EXPLOSION HAZARD DATA

FLASH POINT: N/A | FLAMMABLE (EXPLOSIVE) LIMITS: UPPER / LOWER

METHOD USED

EXTINGUISHING MEDIA: Water spray, foam, carbon dioxide or dry chemical.

SPECIAL FIRE FIGHTING PROCEDURES: Full protective clothing.

UNUSUAL FIRE AND EXPLOSION HAZARDS

SECTION V HEALTH HAZARD DATA

TOXIC LEVEL: None available | CARCINOGENIC: No

PRINCIPLE ROUTES OF ABSORBTION: Inhalation, ingestion | SKIN AND EYE IRRITATION: Severe

REVELANT SYMPTOMS OF EXPOSURE: Irritates respiratory system, skin & eye burns, mouth and digestive tract burns.

EFFECTS OF CHRONIC EXPOSURE: --

EMERGENCY AND FIRST AID PROCEDURES: Inhalation - rest, artificial respiration as needed; ingestion-drink water, milk of magnesia, get medical aid; wash eyes & skin.

SECTION VI REACTIVITY DATA

CONDITIONS CONTRIBUTING TO INSTABILITY

CONDITIONS CONTRIBUTING TO HAZARDOUS POLYMERIZATION

INCOMPATABILITY (MATERIALS TO AVOID): Dimethyl Sulfoxide

HAZARDOUS DECOMPOSITION PRODUCTS: When heated, phosgene is produced

SECTION VII SPILL OR LEAK PROCEDURES

STEPS TO BE TAKEN IN CASE MATERIAL IS RELEASED OR SPILLED

WASTE DISPOSAL METHOD

SECTION VIII SPECIAL PROTECTION INFORMATION

VENTILATION REQUIREMENTS LOCAL EXHAUST | PROTECTIVE EQUIPMENT (SPECIFY TYPES) EYE: Goggles

MECHANICAL (GENERAL) | GLOVES: Rubber

SPECIAL | RESPIRATOR: With proper filter

OTHER PROTECTIVE EQUIPMENT

SECTION IX SPECIAL PRECAUTIONS

PRECAUTIONS TO BE TAKEN IN HANDLING AND STORAGE: Store in cool, dry, well-ventilated location.

OTHER PRECAUTIONS

SIGNATURE | DATE

o-CHLOROBENZYL CHLORIDE

2
4 1

SECTION I SOURCE AND NOMENCLATURE

MANUFACTURER'S NAME | EMERGENCY TELEPHONE NO.

ADDRESS (NUMBER, STREET, CITY, STATE, ZIP CODE)

TRADE NAME AND SYNONYMS | CHEMICAL FAMILY

CHEMICAL NAME AND SYNONYMS: 1,2 Chlorobenzyl Chloride | FORMULA: $ClC_6H_4CH_2Cl$

SECTION II HAZARDOUS INGREDIENTS

BASIC MATERIAL	APPROXIMATE OR MAXIMUM % WT. OR VOL.	ESTABLISHED OSHA STANDARD	LD_{50} ORAL	LD_{50} PERCUT.	LC_{50} SPECIES	LC_{50} CONC.
			None available			

SECTION III PHYSICAL DATA MW-161.08

BOILING POINT: 219°C °F. | VAPOR PRESSURE mm Hg.

MELTING POINT: -30°C °F. | VAPOR DENSITY (Air = 1): 5.55

SPECIFIC GRAVITY (H_2O = 1): 1.27 | EVAPORATION RATE (____ = 1)

SOLUBILITY IN WATER Pts. 100 pts H_2O: Insoluble | VOLATILE % Vol. % Wt.

APPEARANCE AND ODOR: Clear liquid

SECTION IV FIRE AND EXPLOSION HAZARD DATA

FLASH POINT: N/A | FLAMMABLE (EXPLOSIVE) LIMITS: UPPER / LOWER

METHOD USED

EXTINGUISHING MEDIA: Water spray, dry chemical, carbon dioxide

SPECIAL FIRE FIGHTING PROCEDURES

UNUSUAL FIRE AND EXPLOSION HAZARDS

SECTION V HEALTH HAZARD DATA

TOXIC LEVEL: None available | CARCINOGENIC: No

PRINCIPLE ROUTES OF ABSORBTION: Inhalation, ingestion | SKIN AND EYE IRRITATION: High

REVELANT SYMPTOMS OF EXPOSURE: Eye, nose & throat irritation; severe burns to mouth and digestive tract if swallowed.

EFFECTS OF CHRONIC EXPOSURE: --

EMERGENCY AND FIRST AID PROCEDURES: If inhaled - rest; if swallowed - give emetic, get medical aid; wash skin & eyes.

SECTION VI REACTIVITY DATA

CONDITIONS CONTRIBUTING TO INSTABILITY

CONDITIONS CONTRIBUTING TO HAZARDOUS POLYMERIZATION: May undergo exothermic condensation in presence of copper aluminum, iron, zinc, magnesium, tin and other metals.

INCOMPATABILITY (MATERIALS TO AVOID)

HAZARDOUS DECOMPOSITION PRODUCTS

SECTION VII SPILL OR LEAK PROCEDURES

STEPS TO BE TAKEN IN CASE MATERIAL IS RELEASED OR SPILLED: Flush spills with large quantities of water and ventilate area to remove vapors.

WASTE DISPOSAL METHOD

SECTION VIII SPECIAL PROTECTION INFORMATION

VENTILATION REQUIREMENTS LOCAL EXHAUST | PROTECTIVE EQUIPMENT (SPECIFY TYPES) EYE: Goggles

MECHANICAL (GENERAL) | GLOVES: Rubber

SPECIAL | RESPIRATOR: With proper filter

OTHER PROTECTIVE EQUIPMENT

SECTION IX SPECIAL PRECAUTIONS

PRECAUTIONS TO BE TAKEN IN HANDLING AND STORAGE

OTHER PRECAUTIONS

SIGNATURE | DATE

m-CHLOROBROMOBENZENE

2
3 0

SECTION I SOURCE AND NOMENCLATURE

MANUFACTURER'S NAME | EMERGENCY TELEPHONE NO.

ADDRESS (NUMBER, STREET, CITY, STATE, ZIP CODE)

TRADE NAME AND SYNONYMS | CHEMICAL FAMILY

CHEMICAL NAME AND SYNONYMS: 1,3 Chlorobromobenzene | FORMULA: ClC_6H_4Br

SECTION II HAZARDOUS INGREDIENTS

BASIC MATERIAL	APPROXIMATE OR MAXIMUM % WT. OR VOL.	ESTABLISHED OSHA STANDARD	LD_{50} ORAL	LD_{50} PERCUT.	LC_{50} SPECIES	LC_{50} CONC.
			None available			

SECTION III PHYSICAL DATA MW 191.47

BOILING POINT: 196°C °F. | VAPOR PRESSURE mm Hg.

MELTING POINT: -21°C °F. | VAPOR DENSITY (Air = 1)

SPECIFIC GRAVITY (H_2O = 1): 1.63 | EVAPORATION RATE (____ = 1)

SOLUBILITY IN WATER Pts. 100 pts H_2O: Insoluble | VOLATILE % Vol. % Wt.

APPEARANCE AND ODOR: White crystal

SECTION IV FIRE AND EXPLOSION HAZARD DATA

FLASH POINT: N/A | FLAMMABLE (EXPLOSIVE) LIMITS: UPPER / LOWER

METHOD USED

EXTINGUISHING MEDIA: Water spray, carbon dioxide, foam, dry chemical

SPECIAL FIRE FIGHTING PROCEDURES: Wear goggles and self-contained breathing apparatus.

UNUSUAL FIRE AND EXPLOSION HAZARDS

SECTION V HEALTH HAZARD DATA

TOXIC LEVEL: Not available | CARCINOGENIC: No

PRINCIPLE ROUTES OF ABSORBTION: Inhalation, ingestion, skin | SKIN AND EYE IRRITATION: Severe

REVELANT SYMPTOMS OF EXPOSURE: Cyanosis, deep rapid respiration, irregular pulse, drowsiness, unconsciousness.

EFFECTS OF CHRONIC EXPOSURE: Lung, liver & kidney damage

EMERGENCY AND FIRST AID PROCEDURES: Inhalation - rest; if swallowed - give emetic; get medical aid; wash eyes and skin.

SECTION VI REACTIVITY DATA

CONDITIONS CONTRIBUTING TO INSTABILITY

CONDITIONS CONTRIBUTING TO HAZARDOUS POLYMERIZATION

INCOMPATABILITY (MATERIALS TO AVOID)

HAZARDOUS DECOMPOSITION PRODUCTS

SECTION VII SPILL OR LEAK PROCEDURES

STEPS TO BE TAKEN IN CASE MATERIAL IS RELEASED OR SPILLED

WASTE DISPOSAL METHOD

SECTION VIII SPECIAL PROTECTION INFORMATION

VENTILATION REQUIREMENTS LOCAL EXHAUST | PROTECTIVE EQUIPMENT (SPECIFY TYPES) EYE: Goggles

MECHANICAL (GENERAL) | GLOVES: Polyvinyl alcohol

SPECIAL | RESPIRATOR: With proper filter

OTHER PROTECTIVE EQUIPMENT

SECTION IX SPECIAL PRECAUTIONS

PRECAUTIONS TO BE TAKEN IN HANDLING AND STORAGE

OTHER PRECAUTIONS

SIGNATURE | DATE

p-CHLOROBROMOBENZENE

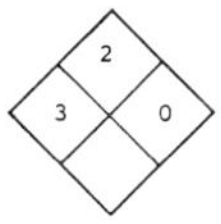

SECTION I SOURCE AND NOMENCLATURE

MANUFACTURER'S NAME | EMERGENCY TELEPHONE NO.

ADDRESS (NUMBER, STREET, CITY, STATE, ZIP CODE)

TRADE NAME AND SYNONYMS: 1, 4 Chlorobromobenzene

CHEMICAL FAMILY:

CHEMICAL NAME AND SYNONYMS: P Chlorobromobenzene

FORMULA: ClC_6H_4Br

SECTION II HAZARDOUS INGREDIENTS

BASIC MATERIAL	APPROXIMATE OR MAXIMUM % WT. OR VOL.	ESTABLISHED OSHA STANDARD	LD_{50} ORAL	LD_{50} PERCUT.	LC_{50} SPECIES	LC_{50} CONC.
			none	available		

SECTION III PHYSICAL DATA MW - 191.47

BOILING POINT °F.: 196°C.

VAPOR PRESSURE mm Hg.:

MELTING POINT °F.: 67°C.

VAPOR DENSITY (Air = 1):

SPECIFIC GRAVITY (H_2O = 1): 1.58

EVAPORATION RATE (____ = 1):

SOLUBILITY IN WATER Pts/100 pts H_2O: Insoluble

VOLATILE % Vol. % Wt.:

APPEARANCE AND ODOR: White crystal

SECTION IV FIRE AND EXPLOSION HAZARD DATA

FLASH POINT: N/A

FLAMMABLE (EXPLOSIVE) LIMITS: UPPER / LOWER

METHOD USED:

EXTINGUISHING MEDIA: Water spray, carbon dioxide, foam, dry chemical.

SPECIAL FIRE FIGHTING PROCEDURES: Wear goggles and self-contained breathing apparatus.

UNUSUAL FIRE AND EXPLOSION HAZARDS:

SECTION V HEALTH HAZARD DATA

TOXIC LEVEL: Not available

CARCINOGENIC: No

PRINCIPLE ROUTES OF ABSORBTION: Inhalation, ingestion, skin

SKIN AND EYE IRRITATION: Severe

REVELANT SYMPTOMS OF EXPOSURE: Cyanosis, deep and rapid respiration, irregualr pulse, drowsiness, unconsciousness.

EFFECTS OF CHRONIC EXPOSURE: Lung, liver and kidney damage.

EMERGENCY AND FIRST AID PROCEDURES: Inhalation - rest; if swallowed - give emetic, get medical aid; wash eyes and skin.

SECTION VI REACTIVITY DATA

CONDITIONS CONTRIBUTING TO INSTABILITY:

CONDITIONS CONTRIBUTING TO HAZARDOUS POLYMERIZATION:

INCOMPATABILITY (MATERIALS TO AVOID):

HAZARDOUS DECOMPOSITION PRODUCTS:

SECTION VII SPILL OR LEAK PROCEDURES

STEPS TO BE TAKEN IN CASE MATERIAL IS RELEASED OR SPILLED:

WASTE DISPOSAL METHOD:

SECTION VIII SPECIAL PROTECTION INFORMATION

VENTILATION REQUIREMENTS LOCAL EXHAUST:

PROTECTIVE EQUIPMENT (SPECIFY TYPES) EYE: Goggles

MECHANICAL (GENERAL):

GLOVES: Polyvinyl Alcohol

SPECIAL:

RESPIRATOR: Self-contained

OTHER PROTECTIVE EQUIPMENT:

SECTION IX SPECIAL PRECAUTIONS

PRECAUTIONS TO BE TAKEN IN HANDLING AND STORAGE:

OTHER PRECAUTIONS:

SIGNATURE | DATE

γ-CHLOROBUTYRONITRILE

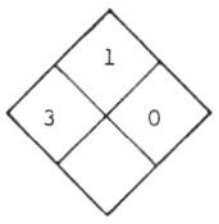

SECTION I SOURCE AND NOMENCLATURE

MANUFACTURER'S NAME | EMERGENCY TELEPHONE NO.

ADDRESS (NUMBER, STREET, CITY, STATE, ZIP CODE)

TRADE NAME AND SYNONYMS:

CHEMICAL FAMILY:

CHEMICAL NAME AND SYNONYMS: G Chlorobutyronitrile

FORMULA: $Cl\ CH_2CH_2CH_2CN$

SECTION II HAZARDOUS INGREDIENTS

BASIC MATERIAL	APPROXIMATE OR MAXIMUM % WT. OR VOL.	ESTABLISHED OSHA STANDARD	LD_{50} ORAL	LD_{50} PERCUT.	LC_{50} SPECIES	LC_{50} CONC.
			None	available		

SECTION III PHYSICAL DATA MW - 103.55

BOILING POINT °F.: 198° C.

VAPOR PRESSURE mm Hg.: at 96° C. 25

MELTING POINT °F.: -

VAPOR DENSITY (Air = 1):

SPECIFIC GRAVITY (H_2O = 1): 1.16

EVAPORATION RATE (____ = 1):

SOLUBILITY IN WATER Pts/100 pts H_2O: Insoluble

VOLATILE % Vol. % Wt.:

APPEARANCE AND ODOR: Liquid

SECTION IV FIRE AND EXPLOSION HAZARD DATA

FLASH POINT: None available

FLAMMABLE (EXPLOSIVE) LIMITS: UPPER / LOWER

METHOD USED:

EXTINGUISHING MEDIA: Alcohol foam, carbon dioxide, dry chemical.

SPECIAL FIRE FIGHTING PROCEDURES: Self-contained breathing apparatus.

UNUSUAL FIRE AND EXPLOSION HAZARDS:

SECTION V HEALTH HAZARD DATA

TOXIC LEVEL: None available

CARCINOGENIC: No

PRINCIPLE ROUTES OF ABSORBTION: Inhalation, ingestion, skin

SKIN AND EYE IRRITATION: Moderate

REVELANT SYMPTOMS OF EXPOSURE: Dizziness, rapid respiration, headache, drowsiness, drop in blood pressure, pulse, delayed symptoms.

EFFECTS OF CHRONIC EXPOSURE: Eye irritation, loss of appetite and weight, mental deterioration.

EMERGENCY AND FIRST AID PROCEDURES: If swallowed - inhalation of amyl nitrite, induce vomiting or gastric lavage; wash skin & eyes; inhalation - rest.

SECTION VI REACTIVITY DATA

CONDITIONS CONTRIBUTING TO INSTABILITY:

CONDITIONS CONTRIBUTING TO HAZARDOUS POLYMERIZATION:

INCOMPATABILITY (MATERIALS TO AVOID):

HAZARDOUS DECOMPOSITION PRODUCTS: When heated to decomposition emits highly toxic cyanide fumes.

SECTION VII SPILL OR LEAK PROCEDURES

STEPS TO BE TAKEN IN CASE MATERIAL IS RELEASED OR SPILLED: Absorb with paper towels - evaporate in hood.

WASTE DISPOSAL METHOD: Add excess alcoholic sodium hydroxide, after one hour, evaporate alcohol, then add calcium hypochloride.

SECTION VIII SPECIAL PROTECTION INFORMATION

VENTILATION REQUIREMENTS LOCAL EXHAUST:

PROTECTIVE EQUIPMENT (SPECIFY TYPES) EYE: Goggles

MECHANICAL (GENERAL):

GLOVES: Rubber

SPECIAL:

RESPIRATOR: Self-contained

OTHER PROTECTIVE EQUIPMENT: Protective apron.

SECTION IX SPECIAL PRECAUTIONS

PRECAUTIONS TO BE TAKEN IN HANDLING AND STORAGE:

OTHER PRECAUTIONS:

SIGNATURE | DATE

2-CHLORO-5-HYDROXYTOLUENE

1
3 0

SECTION I SOURCE AND NOMENCLATURE

MANUFACTURER'S NAME

EMERGENCY TELEPHONE NO.

ADDRESS (NUMBER, STREET, CITY, STATE, ZIP CODE)

TRADE NAME AND SYNONYMS

CHEMICAL FAMILY

CHEMICAL NAME AND SYNONYMS: 6 Chloro-M-Cresol

FORMULA: $Cl(C_6H_3)(OH)CH_3$

SECTION II HAZARDOUS INGREDIENTS

BASIC MATERIAL	APPROXIMATE OR MAXIMUM % WT. OR VOL.	ESTABLISHED OSHA STANDARD	LD_{50} ORAL	LD_{50} PERCUT.	LC_{50} SPECIES	LC_{50} CONC.
			None Available			

SECTION III PHYSICAL DATA MW - 142.59

BOILING POINT °F.: 235°C.

VAPOR PRESSURE mm Hg.

MELTING POINT °F.: 66°C.

VAPOR DENSITY (Air = 1)

SPECIFIC GRAVITY (H_2O = 1)

EVAPORATION RATE (= 1)

SOLUBILITY IN WATER Pts/100 pts H_2O: .38

VOLATILE % Vol. % Wt.

APPEARANCE AND ODOR: White crystals

SECTION IV FIRE AND EXPLOSION HAZARD DATA

FLASH POINT: None available

METHOD USED

FLAMMABLE (EXPLOSIVE) LIMITS: UPPER / LOWER

EXTINGUISHING MEDIA: Alcohol foam, carbon dioxide, dry chemical.

SPECIAL FIRE FIGHTING PROCEDURES

UNUSUAL FIRE AND EXPLOSION HAZARDS

SECTION V HEALTH HAZARD DATA

TOXIC LEVEL: None available

CARCINOGENIC: No

PRINCIPLE ROUTES OF ABSORBTION: Inhalation, ingestion

SKIN AND EYE IRRITATION: Severe

REVELANT SYMPTOMS OF EXPOSURE: Eye, skin irritation, sensitizes, irritation of digestive tract, C.N.S. depression, convulsions.

EFFECTS OF CHRONIC EXPOSURE: Kidney damage

EMERGENCY AND FIRST AID PROCEDURES: Wash eyes and skin; if swallowed - gastric lavage with olive oil, get medical aid.

SECTION VI REACTIVITY DATA

CONDITIONS CONTRIBUTING TO INSTABILITY

CONDITIONS CONTRIBUTING TO HAZARDOUS POLYMERIZATION

INCOMPATABILITY (MATERIALS TO AVOID)

HAZARDOUS DECOMPOSITION PRODUCTS

SECTION VII SPILL OR LEAK PROCEDURES

STEPS TO BE TAKEN IN CASE MATERIAL IS RELEASED OR SPILLED

WASTE DISPOSAL METHOD

SECTION VIII SPECIAL PROTECTION INFORMATION

VENTILATION REQUIREMENTS LOCAL EXHAUST

MECHANICAL (GENERAL)

SPECIAL

PROTECTIVE EQUIPMENT (SPECIFY TYPES) EYE: Safety goggles

GLOVES: Rubber

RESPIRATOR

OTHER PROTECTIVE EQUIPMENT

SECTION IX SPECIAL PRECAUTIONS

PRECAUTIONS TO BE TAKEN IN HANDLING AND STORAGE

OTHER PRECAUTIONS

SIGNATURE

DATE

m-CHLOROIODOBENZENE

2
3 0

SECTION I SOURCE AND NOMENCLATURE

MANUFACTURER'S NAME

EMERGENCY TELEPHONE NO.

ADDRESS (NUMBER, STREET, CITY, STATE, ZIP CODE)

TRADE NAME AND SYNONYMS

CHEMICAL FAMILY

CHEMICAL NAME AND SYNONYMS: 3 Chloroiodobenzene

FORMULA: ClC_6H_4I

SECTION II HAZARDOUS INGREDIENTS

BASIC MATERIAL	APPROXIMATE OR MAXIMUM % WT. OR VOL.	ESTABLISHED OSHA STANDARD	LD_{50} ORAL	LD_{50} PERCUT.	LC_{50} SPECIES	LC_{50} CONC.
			None available			

SECTION III PHYSICAL DATA MW - 238.47

BOILING POINT °F.: 230°C

VAPOR PRESSURE mm Hg.: at 114°C 14

MELTING POINT °F.

VAPOR DENSITY (Air = 1)

SPECIFIC GRAVITY (H_2O = 1)

EVAPORATION RATE (= 1)

SOLUBILITY IN WATER Pts/100 pts H_2O

VOLATILE % Vol. % Wt.

APPEARANCE AND ODOR: Yellow liquid

SECTION IV FIRE AND EXPLOSION HAZARD DATA

FLASH POINT: N/A

METHOD USED

FLAMMABLE (EXPLOSIVE) LIMITS: UPPER / LOWER

EXTINGUISHING MEDIA: Water spray, carbon dioxide, foam dry chemical

SPECIAL FIRE FIGHTING PROCEDURES: Wear goggles and self-contained breathing apparatus.

UNUSUAL FIRE AND EXPLOSION HAZARDS

SECTION V HEALTH HAZARD DATA

TOXIC LEVEL: None available

CARCINOGENIC: No

PRINCIPLE ROUTES OF ABSORBTION: Inhalation, ingestion, skin

SKIN AND EYE IRRITATION: Severe

REVELANT SYMPTOMS OF EXPOSURE: Cyanosis, deep & rapid respiration; irregular pulse, drowsiness, unconsciousness.

EFFECTS OF CHRONIC EXPOSURE: Lung, liver & kidney damage.

EMERGENCY AND FIRST AID PROCEDURES: Inhalation - rest; if swallowed - give emetic, get medical aid; wash skin & eyes.

SECTION VI REACTIVITY DATA

CONDITIONS CONTRIBUTING TO INSTABILITY

CONDITIONS CONTRIBUTING TO HAZARDOUS POLYMERIZATION

INCOMPATABILITY (MATERIALS TO AVOID)

HAZARDOUS DECOMPOSITION PRODUCTS

SECTION VII SPILL OR LEAK PROCEDURES

STEPS TO BE TAKEN IN CASE MATERIAL IS RELEASED OR SPILLED

WASTE DISPOSAL METHOD

SECTION VIII SPECIAL PROTECTION INFORMATION

VENTILATION REQUIREMENTS LOCAL EXHAUST

MECHANICAL (GENERAL)

SPECIAL

PROTECTIVE EQUIPMENT (SPECIFY TYPES) EYE: Goggles

GLOVES: Polyvinyl gloves

RESPIRATOR: With proper filter

OTHER PROTECTIVE EQUIPMENT

SECTION IX SPECIAL PRECAUTIONS

PRECAUTIONS TO BE TAKEN IN HANDLING AND STORAGE

OTHER PRECAUTIONS

SIGNATURE

DATE

o-CHLOROIODOBENZENE

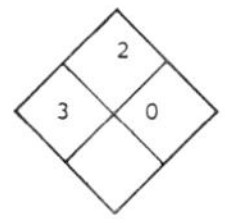

SECTION I SOURCE AND NOMENCLATURE

MANUFACTURER'S NAME	EMERGENCY TELEPHONE NO.
ADDRESS (NUMBER, STREET, CITY, STATE, ZIP CODE)	
TRADE NAME AND SYNONYMS	CHEMICAL FAMILY
CHEMICAL NAME AND SYNONYMS 2 Chloro Iodobenzene	FORMULA ClC_6H_4I

SECTION II HAZARDOUS INGREDIENTS

BASIC MATERIAL	APPROXIMATE OR MAXIMUM % WT. OR VOL.	ESTABLISHED OSHA STANDARD	LC_{50} ORAL	LC_{50} PERCUT.	LC_{50} SPECIES	LC_{50} CONC.
			None available			

SECTION III PHYSICAL DATA MW - 238.47

BOILING POINT °F. 234	VAPOR PRESSURE mm Hg.
MELTING POINT °F. 0.7	VAPOR DENSITY (Air = 1)
SPECIFIC GRAVITY (H_2O = 1) 1.95	EVAPORATION RATE (= 1)
SOLUBILITY IN WATER Pts 100 pts H_2O	VOLATILE % Vol. % Wt.

APPEARANCE AND ODOR
Colorless oil

SECTION IV FIRE AND EXPLOSION HAZARD DATA

FLASH POINT N/A	FLAMMABLE (EXPLOSIVE) LIMITS	UPPER
METHOD USED		LOWER

EXTINGUISHING MEDIA
Water spray, carbon dioxide, foam, dry chemical

SPECIAL FIRE FIGHTING PROCEDURES
Wear goggles and self contained breathing apparatus.

UNUSUAL FIRE AND EXPLOSION HAZARDS

SECTION V HEALTH HAZARD DATA

TOXIC LEVEL None available — CARCINOGENIC No

PRINCIPLE ROUTES OF ABSORBTION Inhalation, ingestion, skin — SKIN AND EYE IRRITATION Severe

REVELANT SYMPTOMS OF EXPOSURE Cyanosis, deep & rapid respiration; irregular pulse, drowsiness, unconsciousness.

EFFECTS OF CHRONIC EXPOSURE Liver, lung & kidney damage.

EMERGENCY AND FIRST AID PROCEDURES Inhalation - rest; if swallowed - give emetic; get medical aid; wash eyes & skin.

SECTION VI REACTIVITY DATA

CONDITIONS CONTRIBUTING TO INSTABILITY

CONDITIONS CONTRIBUTING TO HAZARDOUS POLYMERIZATION

INCOMPATABILITY (MATERIALS TO AVOID)

HAZARDOUS DECOMPOSITION PRODUCTS

SECTION VII SPILL OR LEAK PROCEDURES

STEPS TO BE TAKEN IN CASE MATERIAL IS RELEASED OR SPILLED

WASTE DISPOSAL METHOD

SECTION VIII SPECIAL PROTECTION INFORMATION

VENTILATION REQUIREMENTS LOCAL EXHAUST	PROTECTIVE EQUIPMENT (SPECIFY TYPES) EYE Goggles
MECHANICAL (GENERAL)	GLOVES Polyvinyl alcohol
SPECIAL	RESPIRATOR With proper filter
OTHER PROTECTIVE EQUIPMENT	

SECTION IX SPECIAL PRECAUTIONS

PRECAUTIONS TO BE TAKEN IN HANDLING AND STORAGE

OTHER PRECAUTIONS

SIGNATURE	DATE

p-CHLOROIODOBENZENE

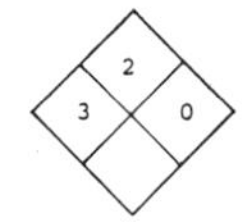

SECTION I SOURCE AND NOMENCLATURE

MANUFACTURER'S NAME	EMERGENCY TELEPHONE NO.
ADDRESS (NUMBER, STREET, CITY, STATE, ZIP CODE)	
TRADE NAME AND SYNONYMS	CHEMICAL FAMILY
CHEMICAL NAME AND SYNONYMS 4 Chloro Iodobenzene	FORMULA ClC_6H_4I

SECTION II HAZARDOUS INGREDIENTS

BASIC MATERIAL	APPROXIMATE OR MAXIMUM % WT. OR VOL.	ESTABLISHED OSHA STANDARD	LD_{50} ORAL	LD_{50} PERCUT.	LC_{50} SPECIES	LC_{50} CONC.
			None available			

SECTION III PHYSICAL DATA MW - 238.47

BOILING POINT °F. 228	VAPOR PRESSURE mm Hg.
MELTING POINT °F. 56°C	VAPOR DENSITY (Air = 1)
SPECIFIC GRAVITY (H_2O = 1) 1.89	EVAPORATION RATE (= 1)
SOLUBILITY IN WATER Pts 100 pts H_2O Insoluble	VOLATILE % Vol. % Wt.

APPEARANCE AND ODOR
White solid

SECTION IV FIRE AND EXPLOSION HAZARD DATA

FLASH POINT N/A	FLAMMABLE (EXPLOSIVE) LIMITS	UPPER
METHOD USED		LOWER

EXTINGUISHING MEDIA
Water spray, carbon dioxide, foam, dry chemical

SPECIAL FIRE FIGHTING PROCEDURES
Wear goggles and self-contained breathing apparatus.

UNUSUAL FIRE AND EXPLOSION HAZARDS

SECTION V HEALTH HAZARD DATA

TOXIC LEVEL None available — CARCINOGENIC No

PRINCIPLE ROUTES OF ABSORBTION Inhalation, ingestion, skin — SKIN AND EYE IRRITATION Severe

REVELANT SYMPTOMS OF EXPOSURE Cyanosis, deep and rapid respiration, irregular pulse, drowsiness, unconsciousness.

EFFECTS OF CHRONIC EXPOSURE Lung, liver & kidney damage.

EMERGENCY AND FIRST AID PROCEDURES Inhalation - rest; if swallowed - give emetic, get medical aid; wash eyes & skin.

SECTION VI REACTIVITY DATA

CONDITIONS CONTRIBUTING TO INSTABILITY

CONDITIONS CONTRIBUTING TO HAZARDOUS POLYMERIZATION

INCOMPATABILITY (MATERIALS TO AVOID)

HAZARDOUS DECOMPOSITION PRODUCTS

SECTION VII SPILL OR LEAK PROCEDURES

STEPS TO BE TAKEN IN CASE MATERIAL IS RELEASED OR SPILLED

WASTE DISPOSAL METHOD

SECTION VIII SPECIAL PROTECTION INFORMATION

VENTILATION REQUIREMENTS LOCAL EXHAUST	PROTECTIVE EQUIPMENT (SPECIFY TYPES) EYE Goggles
MECHANICAL (GENERAL)	GLOVES Polyvinyl gloves
SPECIAL	RESPIRATOR With proper filter
OTHER PROTECTIVE EQUIPMENT	

SECTION IX SPECIAL PRECAUTIONS

PRECAUTIONS TO BE TAKEN IN HANDLING AND STORAGE

OTHER PRECAUTIONS

SIGNATURE	DATE

CHLOROMALEIC ANHYDRIDE

NFPA: 1 (top), 4 (left), 1 (right)

SECTION I SOURCE AND NOMENCLATURE

MANUFACTURER'S NAME | EMERGENCY TELEPHONE NO.

ADDRESS (NUMBER, STREET, CITY, STATE, ZIP CODE)

TRADE NAME AND SYNONYMS | CHEMICAL FAMILY

CHEMICAL NAME AND SYNONYMS: Chloromaleic Anhydride | FORMULA: C_4HClO_3

SECTION II HAZARDOUS INGREDIENTS

BASIC MATERIAL	APPROXIMATE OR MAXIMUM % WT. OR VOL.	ESTABLISHED OSHA STANDARD	LD_{50} ORAL	LD_{50} PERCUT.	LC_{50} SPECIES	LC_{50} CONC.
~~None available~~						

SECTION III PHYSICAL DATA MW - 132.5

BOILING POINT °F. | VAPOR PRESSURE at 95°C. 25 mm Hg.

MELTING POINT °F. | VAPOR DENSITY (Air = 1)

SPECIFIC GRAVITY (H_2O = 1) | EVAPORATION RATE (= 1)

SOLUBILITY IN WATER Pts/100 pts H_2O: Hydrolyzes in water | VOLATILE % Vol. % Wt.

APPEARANCE AND ODOR: Corrosive fumes, liquid.

SECTION IV FIRE AND EXPLOSION HAZARD DATA

FLASH POINT: None available | FLAMMABLE (EXPLOSIVE) LIMITS: UPPER / LOWER

METHOD USED

EXTINGUISHING MEDIA: Alcohol foam, dry chemical, carbon dioxide.

SPECIAL FIRE FIGHTING PROCEDURES

UNUSUAL FIRE AND EXPLOSION HAZARDS

SECTION V HEALTH HAZARD DATA

TOXIC LEVEL: None available | CARCINOGENIC: No

PRINCIPLE ROUTES OF ABSORBTION: Inhalation, ingestion | SKIN AND EYE IRRITATION: Severe

REVELANT SYMPTOMS OF EXPOSURE: Coughing, choking, headache, dizziness, cyanosis, pulmonary edema.

EFFECTS OF CHRONIC EXPOSURE: Gastric pain, erosion of teeth, jaw necrosis, bronchial irritation.

EMERGENCY AND FIRST AID PROCEDURES: Use respiratory resuscitation measures; treat shock; get medical aid.

SECTION VI REACTIVITY DATA

CONDITIONS CONTRIBUTING TO INSTABILITY

CONDITIONS CONTRIBUTING TO HAZARDOUS POLYMERIZATION

INCOMPATABILITY (MATERIALS TO AVOID): Alkali metals, amines, calcium hydroxide, potassium hydroxide, sodium hydroxide, pyridine, tertiary amines.

HAZARDOUS DECOMPOSITION PRODUCTS

SECTION VII SPILL OR LEAK PROCEDURES

STEPS TO BE TAKEN IN CASE MATERIAL IS RELEASED OR SPILLED

WASTE DISPOSAL METHOD

SECTION VIII SPECIAL PROTECTION INFORMATION

VENTILATION REQUIREMENTS LOCAL EXHAUST | PROTECTIVE EQUIPMENT (SPECIFY TYPES) EYE: Goggles

MECHANICAL (GENERAL) | GLOVES: Neoprene

SPECIAL | RESPIRATOR: W/chemical filter

OTHER PROTECTIVE EQUIPMENT: Rubber apron, full-face shield

SECTION IX SPECIAL PRECAUTIONS

PRECAUTIONS TO BE TAKEN IN HANDLING AND STORAGE: Protect from moisture and oxidizing materials. Store in cool, dry, well-ventilated area where fire hazard is not acute.

OTHER PRECAUTIONS

SIGNATURE | DATE

1-CHLORO-1-NITROPROPANE

NFPA: 2 (top), 3 (left), 3 (right)

SECTION I SOURCE AND NOMENCLATURE

MANUFACTURER'S NAME | EMERGENCY TELEPHONE NO.

ADDRESS (NUMBER, STREET, CITY, STATE, ZIP CODE)

TRADE NAME AND SYNONYMS | CHEMICAL FAMILY

CHEMICAL NAME AND SYNONYMS: 1,1 Chloronitropropane | FORMULA: $C_3H_6ClNO_2$

SECTION II HAZARDOUS INGREDIENTS

BASIC MATERIAL	APPROXIMATE OR MAXIMUM % WT. OR VOL.	ESTABLISHED OSHA STANDARD	LD_{50} ORAL	LD_{50} PERCUT.	LC_{50} SPECIES	LC_{50} CONC.
LDLO - oral - human - 50 mg/kg			510		Mouse	
			50		Rabbit	

SECTION III PHYSICAL DATA MW - 123.54

BOILING POINT: 140°C °F. | VAPOR PRESSURE at 20°C 5.8 mm Hg.

MELTING POINT °F. | VAPOR DENSITY (Air = 1): 4.26

SPECIFIC GRAVITY (H_2O = 1): 1.21 | EVAPORATION RATE (= 1)

SOLUBILITY IN WATER Pts/100 pts H_2O: 0.8 | VOLATILE % Vol. % Wt.

APPEARANCE AND ODOR: Liquid

SECTION IV FIRE AND EXPLOSION HAZARD DATA

FLASH POINT: 144 | FLAMMABLE (EXPLOSIVE) LIMITS: UPPER / LOWER

METHOD USED: O.C.

EXTINGUISHING MEDIA: Alcohol foam, water, carbon dioxide, dry chemical

SPECIAL FIRE FIGHTING PROCEDURES

UNUSUAL FIRE AND EXPLOSION HAZARDS: May explode when heated

SECTION V HEALTH HAZARD DATA

TOXIC LEVEL: TLV - 20 PPM, 100 mg/C.M. | CARCINOGENIC: No

PRINCIPLE ROUTES OF ABSORBTION: Inhalation, ingestion | SKIN AND EYE IRRITATION: Moderate

REVELANT SYMPTOMS OF EXPOSURE: Irritation of respiratory system, lung, liver, heart muscle and kidney damage.

EFFECTS OF CHRONIC EXPOSURE

EMERGENCY AND FIRST AID PROCEDURES: If swallowed - induce vomiting, get medical aid; wash skin & eyes; inhalation - give oxygen as needed.

SECTION VI REACTIVITY DATA

CONDITIONS CONTRIBUTING TO INSTABILITY: May explode when heated

CONDITIONS CONTRIBUTING TO HAZARDOUS POLYMERIZATION

INCOMPATABILITY (MATERIALS TO AVOID): Oxidizing material

HAZARDOUS DECOMPOSITION PRODUCTS

SECTION VII SPILL OR LEAK PROCEDURES

STEPS TO BE TAKEN IN CASE MATERIAL IS RELEASED OR SPILLED: Cover spill with soda ash, spray and mix with water.

WASTE DISPOSAL METHOD

SECTION VIII SPECIAL PROTECTION INFORMATION

VENTILATION REQUIREMENTS LOCAL EXHAUST | PROTECTIVE EQUIPMENT (SPECIFY TYPES) EYE: Goggles

MECHANICAL (GENERAL) | GLOVES

SPECIAL | RESPIRATOR: With proper filter

OTHER PROTECTIVE EQUIPMENT: Protective clothing made of plastic.

SECTION IX SPECIAL PRECAUTIONS

PRECAUTIONS TO BE TAKEN IN HANDLING AND STORAGE

OTHER PRECAUTIONS

SIGNATURE | DATE

m-CHLOROPHENOL

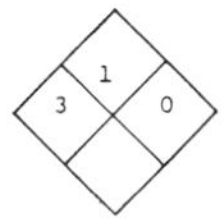

SECTION I SOURCE AND NOMENCLATURE

MANUFACTURER'S NAME | EMERGENCY TELEPHONE NO.

ADDRESS (NUMBER, STREET, CITY, STATE, ZIP CODE)

TRADE NAME AND SYNONYMS: 3 Chloro-1 hydroxybenzene

CHEMICAL FAMILY

CHEMICAL NAME AND SYNONYMS: M chlorophenol

FORMULA: Cl C_6H_4OH

SECTION II HAZARDOUS INGREDIENTS

BASIC MATERIAL	APPROXIMATE OR MAXIMUM % WT. OR VOL.	ESTABLISHED OSHA STANDARD	LD_{50} mg/kg			LC_{50}
			ORAL	PERCUT.	SPECIES	CONC.
			570		rat	

SECTION III PHYSICAL DATA MW - 128.56

BOILING POINT °F.: 214°C.

VAPOR PRESSURE mm Hg.: at 44°C. 1

MELTING POINT °F.: 33°C.

VAPOR DENSITY (Air = 1)

SPECIFIC GRAVITY (H_2O = 1): 1.27

EVAPORATION RATE (______ = 1)

SOLUBILITY IN WATER Pts/100 pts H_2O: 2.6

VOLATILE % Vol. % Wt.

APPEARANCE AND ODOR: Needle-like crystals

SECTION IV FIRE AND EXPLOSION HAZARD DATA

FLASH POINT: None available

FLAMMABLE (EXPLOSIVE) LIMITS: UPPER / LOWER

METHOD USED

EXTINGUISHING MEDIA: Alcohol foam, carbon dioxide, dry chemical.

SPECIAL FIRE FIGHTING PROCEDURES

UNUSUAL FIRE AND EXPLOSION HAZARDS

SECTION V HEALTH HAZARD DATA

TOXIC LEVEL: None available

CARCINOGENIC: Neoplastic

PRINCIPLE ROUTES OF ABSORBTION: Inhalation, ingestion, skin

SKIN AND EYE IRRITATION: Severe

REVELANT SYMPTOMS OF EXPOSURE: Eye skin and mucous membrane irritation, dermatitis, cyanosis.

EFFECTS OF CHRONIC EXPOSURE: Liver damage.

EMERGENCY AND FIRST AID PROCEDURES: Inhalation - rest; wash from skin & eyes; if swallowed - give emetic, get medical aid.

SECTION VI REACTIVITY DATA

CONDITIONS CONTRIBUTING TO INSTABILITY

CONDITIONS CONTRIBUTING TO HAZARDOUS POLYMERIZATION

INCOMPATABILITY (MATERIALS TO AVOID)

HAZARDOUS DECOMPOSITION PRODUCTS

SECTION VII SPILL OR LEAK PROCEDURES

STEPS TO BE TAKEN IN CASE MATERIAL IS RELEASED OR SPILLED: Put a mixture of sand 9 and 1 soda ash. Put intc box for disposal.

WASTE DISPOSAL METHOD

SECTION VIII SPECIAL PROTECTION INFORMATION

VENTILATION REQUIREMENTS LOCAL EXHAUST

PROTECTIVE EQUIPMENT (SPECIFY TYPES) EYE: Goggles

MECHANICAL (GENERAL)

GLOVES: Butyl Rubber

SPECIAL

RESPIRATOR: Self-contained

OTHER PROTECTIVE EQUIPMENT: Protective clothing

SECTION IX SPECIAL PRECAUTIONS

PRECAUTIONS TO BE TAKEN IN HANDLING AND STORAGE: Store in cool, dry place.

OTHER PRECAUTIONS

SIGNATURE | DATE

o-CHLOROPHENOL

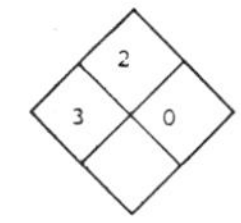

SECTION I SOURCE AND NOMENCLATURE

MANUFACTURER'S NAME | EMERGENCY TELEPHONE NO.

ADDRESS (NUMBER, STREET, CITY, STATE, ZIP CODE)

TRADE NAME AND SYNONYMS: 2 Chloro-1 Hydroxybenzene

CHEMICAL FAMILY

CHEMICAL NAME AND SYNONYMS: O-chlorophenol

FORMULA: Cl C_6H_4OH

SECTION II HAZARDOUS INGREDIENTS

BASIC MATERIAL	APPROXIMATE OR MAXIMUM % WT. OR VOL.	ESTABLISHED OSHA STANDARD	LD_{50} mg/kg			LC_{50}
			ORAL	PERCUT.	SPECIES	CONC.
			670		Rat	
			670		Mouse	
			440		Mammal	

SECTION III PHYSICAL DATA MW-128.56

BOILING POINT °F.: 176°C.

VAPOR PRESSURE mm Hg.: at 12°C. 1

MELTING POINT °F.: 7°C

VAPOR DENSITY (Air = 1)

SPECIFIC GRAVITY (H_2O = 1): 1.24

EVAPORATION RATE (______ = 1)

SOLUBILITY IN WATER Pts/100 pts H_2O: 2.8

VOLATILE % Vol. % Wt.

APPEARANCE AND ODOR: Colorless liquid

SECTION IV FIRE AND EXPLOSION HAZARD DATA

FLASH POINT: 147°F.

FLAMMABLE (EXPLOSIVE) LIMITS: UPPER / LOWER

METHOD USED

EXTINGUISHING MEDIA: Alcohol foam, carbon dioxide - dry chemical.

SPECIAL FIRE FIGHTING PROCEDURES

UNUSUAL FIRE AND EXPLOSION HAZARDS

SECTION V HEALTH HAZARD DATA

TOXIC LEVEL: None available

CARCINOGENIC: Neoplastic

PRINCIPLE ROUTES OF ABSORBTION: Inhalation, ingestion, skin

SKIN AND EYE IRRITATION: Severe

REVELANT SYMPTOMS OF EXPOSURE: Eye, skin and mucose membrane irritation dermatitis, cyanosis.

EFFECTS OF CHRONIC EXPOSURE: Liver damage

EMERGENCY AND FIRST AID PROCEDURES: Inhalation - rest; wash from skin & eyes; if swallowed - give emetic, get medical aid.

SECTION VI REACTIVITY DATA

CONDITIONS CONTRIBUTING TO INSTABILITY

CONDITIONS CONTRIBUTING TO HAZARDOUS POLYMERIZATION

INCOMPATABILITY (MATERIALS TO AVOID)

HAZARDOUS DECOMPOSITION PRODUCTS

SECTION VII SPILL OR LEAK PROCEDURES

STEPS TO BE TAKEN IN CASE MATERIAL IS RELEASED OR SPILLED: Put sand-soda ash (9:1) mixture on spill and sweep into box for disposal.

WASTE DISPOSAL METHOD

SECTION VIII SPECIAL PROTECTION INFORMATION

VENTILATION REQUIREMENTS LOCAL EXHAUST

PROTECTIVE EQUIPMENT (SPECIFY TYPES) EYE: Goggles

MECHANICAL (GENERAL)

GLOVES: Butyl rubber

SPECIAL

RESPIRATOR: Self-contained

OTHER PROTECTIVE EQUIPMENT: Protective clothing

SECTION IX SPECIAL PRECAUTIONS

PRECAUTIONS TO BE TAKEN IN HANDLING AND STORAGE: Store in cool, dry place.

OTHER PRECAUTIONS

SIGNATURE | DATE

m-CHLOROPHENYL ISOCYANATE

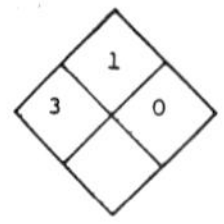

SECTION I SOURCE AND NOMENCLATURE

MANUFACTURER'S NAME | EMERGENCY TELEPHONE NO.

ADDRESS (NUMBER, STREET, CITY, STATE, ZIP CODE)

TRADE NAME AND SYNONYMS | CHEMICAL FAMILY

CHEMICAL NAME AND SYNONYMS: M-Chlorophenyl Isocyanate

FORMULA: ClC_6H_4NCO

SECTION II HAZARDOUS INGREDIENTS

BASIC MATERIAL	APPROXIMATE OR MAXIMUM % WT. OR VOL.	ESTABLISHED OSHA STANDARD	LD_{50} mg/CM ORAL	LD_{50} mg/CM PERCUT.	LC_{50} SPECIES	LC_{50} CONC.
TCLO - ocular-human 1 ppm/1 minute eye			4760		Rat	
			530		Mouse	

SECTION III PHYSICAL DATA MW - 153.6

BOILING POINT °F.: 101^0 C. at 30 min.

VAPOR PRESSURE mm Hg.: at 77^0 C. 10

MELTING POINT °F.: -4^0 C.

VAPOR DENSITY (Air = 1)

SPECIFIC GRAVITY (H_2O = 1)

EVAPORATION RATE (______ = 1)

SOLUBILITY IN WATER Pts/100 pts H_2O

VOLATILE % Vol. % Wt.

APPEARANCE AND ODOR: Water white liquid.

SECTION IV FIRE AND EXPLOSION HAZARD DATA

FLASH POINT: one available

METHOD USED

FLAMMABLE (EXPLOSIVE) LIMITS: UPPER / LOWER

EXTINGUISHING MEDIA: Alcohol foam, carbon dioxide, dry chemical.

SPECIAL FIRE FIGHTING PROCEDURES: Wear self-contained breathing apparatus.

UNUSUAL FIRE AND EXPLOSION HAZARDS: Powerful lachrymater.

SECTION V HEALTH HAZARD DATA

TOXIC LEVEL: None available

CARCINOGENIC: No

PRINCIPLE ROUTES OF ABSORBTION: Inhalation, ingestion

SKIN AND EYE IRRITATION: Severe

REVELANT SYMPTOMS OF EXPOSURE: Skin, eyes and mucous membrane irritation.

EFFECTS OF CHRONIC EXPOSURE: --

EMERGENCY AND FIRST AID PROCEDURES: Inhalation - oxygen as needed; if swallowed - induce vomiting, get medical aid; wash skin & eyes.

SECTION VI REACTIVITY DATA

CONDITIONS CONTRIBUTING TO INSTABILITY

CONDITIONS CONTRIBUTING TO HAZARDOUS POLYMERIZATION

INCOMPATABILITY (MATERIALS TO AVOID)

HAZARDOUS DECOMPOSITION PRODUCTS

SECTION VII SPILL OR LEAK PROCEDURES

STEPS TO BE TAKEN IN CASE MATERIAL IS RELEASED OR SPILLED

WASTE DISPOSAL METHOD

SECTION VIII SPECIAL PROTECTION INFORMATION

VENTILATION REQUIREMENTS LOCAL EXHAUST

PROTECTIVE EQUIPMENT (SPECIFY TYPES) EYE: Goggles

MECHANICAL (GENERAL)

GLOVES: Rubber

SPECIAL

RESPIRATOR: Self-contained

OTHER PROTECTIVE EQUIPMENT

SECTION IX SPECIAL PRECAUTIONS

PRECAUTIONS TO BE TAKEN IN HANDLING AND STORAGE

OTHER PRECAUTIONS

SIGNATURE | DATE

o-CHLOROPHENYL ISOCYANATE

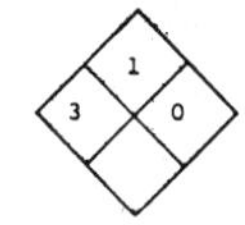

SECTION I SOURCE AND NOMENCLATURE

MANUFACTURER'S NAME | EMERGENCY TELEPHONE NO.

ADDRESS (NUMBER, STREET, CITY, STATE, ZIP CODE)

TRADE NAME AND SYNONYMS | CHEMICAL FAMILY

CHEMICAL NAME AND SYNONYMS: O-Chlorophenyl Isocyanate

FORMULA: $Cl\ C_6H_4NCO$

SECTION II HAZARDOUS INGREDIENTS

BASIC MATERIAL	APPROXIMATE OR MAXIMUM % WT. OR VOL.	ESTABLISHED OSHA STANDARD	LD_{50} ORAL	LD_{50} PERCUT.	LC_{50} SPECIES	LC_{50} CONC.
			None	available		

SECTION III PHYSICAL DATA MW - 153.6

BOILING POINT °F.: 106^0 C. at 30 min.

VAPOR PRESSURE mm Hg.: at 77^0C. 10

MELTING POINT °F

VAPOR DENSITY (Air = 1)

SPECIFIC GRAVITY (H_2O = 1)

EVAPORATION RATE (______ = 1)

SOLUBILITY IN WATER Pts/100 pts H_2O

VOLATILE % Vol. % Wt.

APPEARANCE AND ODOR: Colorless liquid

SECTION IV FIRE AND EXPLOSION HAZARD DATA

FLASH POINT: None available

METHOD USED

FLAMMABLE (EXPLOSIVE) LIMITS: UPPER / LOWER

EXTINGUISHING MEDIA: Alcohol foam, carbon dioxide, dry chemical.

SPECIAL FIRE FIGHTING PROCEDURES: Wear self-contained breathing apparatus.

UNUSUAL FIRE AND EXPLOSION HAZARDS: Powerful lachrymater

SECTION V HEALTH HAZARD DATA

TOXIC LEVEL: None available

CARCINOGENIC: No

PRINCIPLE ROUTES OF ABSORBTION: Inhalation, ingestion

SKIN AND EYE IRRITATION: Severe

REVELANT SYMPTOMS OF EXPOSURE: Irritation of skin, eyes and mucous membrane.

EFFECTS OF CHRONIC EXPOSURE: --

EMERGENCY AND FIRST AID PROCEDURES: Inhaltion - oxygen as needed; if swallowed - induce vomiting, get medical aid; wash skin & eyes.

SECTION VI REACTIVITY DATA

CONDITIONS CONTRIBUTING TO INSTABILITY

CONDITIONS CONTRIBUTING TO HAZARDOUS POLYMERIZATION

INCOMPATABILITY (MATERIALS TO AVOID)

HAZARDOUS DECOMPOSITION PRODUCTS

SECTION VII SPILL OR LEAK PROCEDURES

STEPS TO BE TAKEN IN CASE MATERIAL IS RELEASED OR SPILLED

WASTE DISPOSAL METHOD

SECTION VIII SPECIAL PROTECTION INFORMATION

VENTILATION REQUIREMENTS LOCAL EXHAUST

PROTECTIVE EQUIPMENT (SPECIFY TYPES) EYE: Goggles

MECHANICAL (GENERAL)

GLOVES: Rubber

SPECIAL

RESPIRATOR: Self-contained

OTHER PROTECTIVE EQUIPMENT

SECTION IX SPECIAL PRECAUTIONS

PRECAUTIONS TO BE TAKEN IN HANDLING AND STORAGE

OTHER PRECAUTIONS

SIGNATURE | DATE

CHLOROPLATINIC ACID

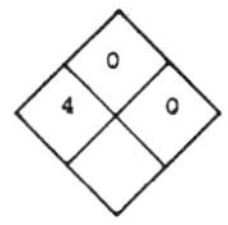

SECTION I SOURCE AND NOMENCLATURE

MANUFACTURER'S NAME	EMERGENCY TELEPHONE NO.
ADDRESS (NUMBER, STREET, CITY, STATE, ZIP CODE)	
TRADE NAME AND SYNONYMS	CHEMICAL FAMILY
CHEMICAL NAME AND SYNONYMS Chloroplatinic Acid	FORMULA $H_2PtCl_6 \cdot 6H_2O$

SECTION II HAZARDOUS INGREDIENTS

BASIC MATERIAL	APPROXIMATE OR MAXIMUM % WT. OR VOL.	ESTABLISHED OSHA STANDARD	LD_{50} ORAL	LD_{50} PERCUT.	LD_{50} SPECIES	LC_{50} CONC.
None available						

SECTION III PHYSICAL DATA MW - 517.92

BOILING POINT °F.	VAPOR PRESSURE mm Hg.
MELTING POINT °F. 60°C.	VAPOR DENSITY (Air = 1)
SPECIFIC GRAVITY (H_2O = 1) 2.41	EVAPORATION RATE (______ = 1)
SOLUBILITY IN WATER Pts/100 pts H_2O Soluble	VOLATILE % Vol. % Wt.
APPEARANCE AND ODOR Brownish-yellow	

SECTION IV FIRE AND EXPLOSION HAZARD DATA

FLASH POINT Not flammable	FLAMMABLE (EXPLOSIVE) LIMITS	UPPER
METHOD USED		LOWER

EXTINGUISHING MEDIA

SPECIAL FIRE FIGHTING PROCEDURES

UNUSUAL FIRE AND EXPLOSION HAZARDS

SECTION V HEALTH HAZARD DATA

TOXIC LEVEL TLV - 0.002 Mg/CM — CARCINOGENIC No

PRINCIPLE ROUTES OF ABSORBTION Inhalation, ingestion — SKIN AND EYE IRRITATION Severe

REVELANT SYMPTOMS OF EXPOSURE Inflammation of mucous membranes in the nose, lacramation and burning in the eyes.

EFFECTS OF CHRONIC EXPOSURE Running of the nose, sneezing, tightness of chest, cyanosis, wheezing, cough, dermatitis.

EMERGENCY AND FIRST AID PROCEDURES Inhalation - remove to fresh air, rest, treat asthmatic attack; ingestion - drink water, get medical aid.

SECTION VI REACTIVITY DATA

CONDITIONS CONTRIBUTING TO INSTABILITY

CONDITIONS CONTRIBUTING TO HAZARDOUS POLYMERIZATION

INCOMPATABILITY (MATERIALS TO AVOID)

HAZARDOUS DECOMPOSITION PRODUCTS

SECTION VII SPILL OR LEAK PROCEDURES

STEPS TO BE TAKEN IN CASE MATERIAL IS RELEASED OR SPILLED

WASTE DISPOSAL METHOD

SECTION VIII SPECIAL PROTECTION INFORMATION

VENTILATION REQUIREMENTS LOCAL EXHAUST	PROTECTIVE EQUIPMENT (SPECIFY TYPES) EYE Goggles
MECHANICAL (GENERAL)	GLOVES
SPECIAL	RESPIRATOR
OTHER PROTECTIVE EQUIPMENT	

SECTION IX SPECIAL PRECAUTIONS

PRECAUTIONS TO BE TAKEN IN HANDLING AND STORAGE

OTHER PRECAUTIONS

SIGNATURE	DATE

3-CHLORO-1,2-PROPANEDIOL

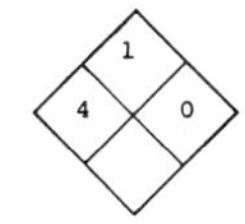

SECTION I SOURCE AND NOMENCLATURE

MANUFACTURER'S NAME	EMERGENCY TELEPHONE NO.
ADDRESS (NUMBER, STREET, CITY, STATE, ZIP CODE)	
TRADE NAME AND SYNONYMS	CHEMICAL FAMILY
CHEMICAL NAME AND SYNONYMS α Chlorohydrin	FORMULA $ClCH_2CHOHCH_2OH$

SECTION II HAZARDOUS INGREDIENTS

BASIC MATERIAL	APPROXIMATE OR MAXIMUM % WT. OR VOL.	ESTABLISHED OSHA STANDARD	LD_{50} mg/kg ORAL	LD_{50} mg/kg PERCUT.	SPECIES	LC_{50} CONC.
			150		Rat	
				5	Mouse	
			160		Mouse	

SECTION III PHYSICAL DATA MW 110.54

BOILING POINT °F. 213°C decomposes	VAPOR PRESSURE mm Hg.
MELTING POINT °F.	VAPOR DENSITY (Air = 1)
SPECIFIC GRAVITY (H_2O = 1) 1.33	EVAPORATION RATE (______ = 1)
SOLUBILITY IN WATER Pts/100 pts H_2O	VOLATILE % Vol. % Wt.
APPEARANCE AND ODOR Colorless liquid	

SECTION IV FIRE AND EXPLOSION HAZARD DATA

FLASH POINT N/A	FLAMMABLE (EXPLOSIVE) LIMITS	UPPER
METHOD USED		LOWER

EXTINGUISHING MEDIA Alcohol foam, carbon dioxide, dry chemical

SPECIAL FIRE FIGHTING PROCEDURES

UNUSUAL FIRE AND EXPLOSION HAZARDS

SECTION V HEALTH HAZARD DATA

TOXIC LEVEL None available — CARCINOGENIC No

PRINCIPLE ROUTES OF ABSORBTION Inhalation, ingestion, skin — SKIN AND EYE IRRITATION Moderate

REVELANT SYMPTOMS OF EXPOSURE eye and nasal irritation, nausea, vomiting, giddiness, dizziness, incoordination, coma

EFFECTS OF CHRONIC EXPOSURE Nausea, vomiting, abdominal pain

EMERGENCY AND FIRST AID PROCEDURES Wash eyes and skin; if swallowed - give emetic; get medical aid; inhalation - rest, give oxygen as needed .

SECTION VI REACTIVITY DATA

CONDITIONS CONTRIBUTING TO INSTABILITY

CONDITIONS CONTRIBUTING TO HAZARDOUS POLYMERIZATION

INCOMPATABILITY (MATERIALS TO AVOID)

HAZARDOUS DECOMPOSITION PRODUCTS

SECTION VII SPILL OR LEAK PROCEDURES

STEPS TO BE TAKEN IN CASE MATERIAL IS RELEASED OR SPILLED

WASTE DISPOSAL METHOD

SECTION VIII SPECIAL PROTECTION INFORMATION

VENTILATION REQUIREMENTS LOCAL EXHAUST	PROTECTIVE EQUIPMENT (SPECIFY TYPES) EYE Goggles
MECHANICAL (GENERAL)	GLOVES Rubber
SPECIAL	RESPIRATOR With proper filter
OTHER PROTECTIVE EQUIPMENT	

SECTION IX SPECIAL PRECAUTIONS

PRECAUTIONS TO BE TAKEN IN HANDLING AND STORAGE

OTHER PRECAUTIONS

SIGNATURE	DATE

β-CHLOROPROPIONYL CHLORIDE

SECTION I SOURCE AND NOMENCLATURE

MANUFACTURER'S NAME	EMERGENCY TELEPHONE NO.
ADDRESS (NUMBER, STREET, CITY, STATE, ZIP CODE)	
TRADE NAME AND SYNONYMS	CHEMICAL FAMILY
CHEMICAL NAME AND SYNONYMS: 3 Chloropropionyl chloride	FORMULA: CH_2ClCH_2COCl

SECTION II HAZARDOUS INGREDIENTS

BASIC MATERIAL	APPROXIMATE OR MAXIMUM % WT. OR VOL.	ESTABLISHED OSHA STANDARD	LD_{50} ORAL	LD_{50} PERCUT.	LC_{50} SPECIES	LC_{50} CONC.
			None available			

SECTION III PHYSICAL DATA MW - 126.98

BOILING POINT °F.: 144°C	VAPOR PRESSURE mm Hg.: 18 at 44°C
MELTING POINT °F.: --	VAPOR DENSITY (Air = 1)
SPECIFIC GRAVITY (H_2O = 1): 1.33	EVAPORATION RATE (____ = 1)
SOLUBILITY IN WATER Pts/100 pts H_2O: Insoluble	VOLATILE % Vol. % Wt.
APPEARANCE AND ODOR: Colorless liquid	

SECTION IV FIRE AND EXPLOSION HAZARD DATA

FLASH POINT: Not available

FLAMMABLE (EXPLOSIVE) LIMITS: UPPER / LOWER

METHOD USED

EXTINGUISHING MEDIA: Carbon dioxide, dry chemical

SPECIAL FIRE FIGHTING PROCEDURES

UNUSUAL FIRE AND EXPLOSION HAZARDS

SECTION V HEALTH HAZARD DATA

TOXIC LEVEL: None available

CARCINOGENIC: No

PRINCIPLE ROUTES OF ABSORBTION: Inhalation, ingestion

SKIN AND EYE IRRITATION: Severe

REVELANT SYMPTOMS OF EXPOSURE: Cyanosis, coughing, headache, pulmonary edema, ulceration of skin, dizziness.

EFFECTS OF CHRONIC EXPOSURE: Erosion of teeth, bronchial irritation and chronic cough.

EMERGENCY AND FIRST AID PROCEDURES: Ingestion - drink water, milk of magnesia, get medical aid; inhalation - artificial respiration as needed; wash eyes & skin.

SECTION VI REACTIVITY DATA

CONDITIONS CONTRIBUTING TO INSTABILITY

CONDITIONS CONTRIBUTING TO HAZARDOUS POLYMERIZATION

INCOMPATABILITY (MATERIALS TO AVOID): May react violently with water, alcohol, and dimethyl sulfoxide.

HAZARDOUS DECOMPOSITION PRODUCTS

SECTION VII SPILL OR LEAK PROCEDURES

STEPS TO BE TAKEN IN CASE MATERIAL IS RELEASED OR SPILLED

WASTE DISPOSAL METHOD

SECTION VIII SPECIAL PROTECTION INFORMATION

VENTILATION REQUIREMENTS LOCAL EXHAUST	PROTECTIVE EQUIPMENT (SPECIFY TYPES) EYE: Goggles
MECHANICAL (GENERAL)	GLOVES: Rubber
SPECIAL	RESPIRATOR: Self-contained
OTHER PROTECTIVE EQUIPMENT	

SECTION IX SPECIAL PRECAUTIONS

PRECAUTIONS TO BE TAKEN IN HANDLING AND STORAGE

OTHER PRECAUTIONS

SIGNATURE | DATE

2-CHLOROPYRIDINE

SECTION I SOURCE AND NOMENCLATURE

MANUFACTURER'S NAME	EMERGENCY TELEPHONE NO.
ADDRESS (NUMBER, STREET, CITY, STATE, ZIP CODE)	
TRADE NAME AND SYNONYMS: α chloropyridine	CHEMICAL FAMILY
CHEMICAL NAME AND SYNONYMS: 2- chloropyridine	FORMULA: C_5H_4N Cl

SECTION II HAZARDOUS INGREDIENTS

BASIC MATERIAL	APPROXIMATE OR MAXIMUM % WT. OR VOL.	ESTABLISHED OSHA STANDARD	LD_{50} mg/kg ORAL	LD_{50} mg/kg PERCUT.	LC_{50} SPECIES	LC_{50} CONC.
			107		Mouse	
LCLO - inhalation - rat	250 ppm/7 hrs.			64	Rabbit	

SECTION III PHYSICAL DATA MW 113.55

BOILING POINT °F.: 170°C	VAPOR PRESSURE mm Hg.: at 13°C
MELTING POINT °F.: --	VAPOR DENSITY (Air = 1): 3.93
SPECIFIC GRAVITY (H_2O = 1): 1.21	EVAPORATION RATE (____ = 1)
SOLUBILITY IN WATER Pts/100 pts H_2O: Soluble	VOLATILE % Vol. % Wt.
APPEARANCE AND ODOR: Oil with fish-like odor.	

SECTION IV FIRE AND EXPLOSION HAZARD DATA

FLASH POINT: N/A

FLAMMABLE (EXPLOSIVE) LIMITS: UPPER / LOWER

METHOD USED

EXTINGUISHING MEDIA: Alcohol foam, carbon dioxide, dry chemical.

SPECIAL FIRE FIGHTING PROCEDURES: Wear goggles and self-contained breathing apparatus.

UNUSUAL FIRE AND EXPLOSION HAZARDS

SECTION V HEALTH HAZARD DATA

TOXIC LEVEL: None available

CARCINOGENIC: No

PRINCIPLE ROUTES OF ABSORBTION: Inhalation, ingestion, skin

SKIN AND EYE IRRITATION: Severe

REVELANT SYMPTOMS OF EXPOSURE: Narcosis, headache, nausea, giddiness, vomiting; oral - liver & kidney damage.

EFFECTS OF CHRONIC EXPOSURE: Occasional vomiting and diarrhea; weight loss & anemia; ocular and facial paralysis.

EMERGENCY AND FIRST AID PROCEDURES: Inhalation - rest; if swallowed - induce vomiting, then gastric lavage; wash skin & eyes.

SECTION VI REACTIVITY DATA

CONDITIONS CONTRIBUTING TO INSTABILITY

CONDITIONS CONTRIBUTING TO HAZARDOUS POLYMERIZATION

INCOMPATABILITY (MATERIALS TO AVOID): May react with oxidizing materials

HAZARDOUS DECOMPOSITION PRODUCTS

SECTION VII SPILL OR LEAK PROCEDURES

STEPS TO BE TAKEN IN CASE MATERIAL IS RELEASED OR SPILLED

WASTE DISPOSAL METHOD

SECTION VIII SPECIAL PROTECTION INFORMATION

VENTILATION REQUIREMENTS LOCAL EXHAUST: Use in hood	PROTECTIVE EQUIPMENT (SPECIFY TYPES) EYE: Goggles
MECHANICAL (GENERAL)	GLOVES: Rubber
SPECIAL	RESPIRATOR: Self-contained breathing apparatus
OTHER PROTECTIVE EQUIPMENT	

SECTION IX SPECIAL PRECAUTIONS

PRECAUTIONS TO BE TAKEN IN HANDLING AND STORAGE

OTHER PRECAUTIONS

SIGNATURE | DATE

m-CHLOROTOLUENE

Hazard diamond: 2 (top), 3 (left), 0 (right)

SECTION I SOURCE AND NOMENCLATURE

MANUFACTURER'S NAME | EMERGENCY TELEPHONE NO.

ADDRESS (NUMBER, STREET, CITY, STATE, ZIP CODE)

TRADE NAME AND SYNONYMS | CHEMICAL FAMILY

CHEMICAL NAME AND SYNONYMS: 3 Chlorotoluene | FORMULA: $CH_3C_6H_4Cl$

SECTION II HAZARDOUS INGREDIENTS

BASIC MATERIAL	APPROXIMATE OR MAXIMUM % WT. OR VOL.	ESTABLISHED OSHA STANDARD	LD_{50} ORAL	LD_{50} PERCUT.	LD_{50} SPECIES	LC_{50} CONC.
			None available			

SECTION III PHYSICAL DATA MW = 126.58

BOILING POINT °F.: 162°C | VAPOR PRESSURE mm Hg.: at 43°C 10

MELTING POINT °F.: -48°C | VAPOR DENSITY (Air = 1)

SPECIFIC GRAVITY (H_2O = 1): 1.08 | EVAPORATION RATE (_____ = 1)

SOLUBILITY IN WATER Pts/100 pts H_2O | VOLATILE % Vol. % Wt.

APPEARANCE AND ODOR: Colorless liquid

SECTION IV FIRE AND EXPLOSION HAZARD DATA

FLASH POINT: Not available | FLAMMABLE (EXPLOSIVE) LIMITS: UPPER / LOWER

METHOD USED

EXTINGUISHING MEDIA: Alcohol foam, carbon dioxide, dry chemical

SPECIAL FIRE FIGHTING PROCEDURES

UNUSUAL FIRE AND EXPLOSION HAZARDS

SECTION V HEALTH HAZARD DATA

TOXIC LEVEL: None available | CARCINOGENIC: No

PRINCIPLE ROUTES OF ABSORBTION: Inhalation, ingestion | SKIN AND EYE IRRITATION: Severe

REVELANT SYMPTOMS OF EXPOSURE: Eye & respiratory system irritation, severe internal damage, narcotic effect

EFFECTS OF CHRONIC EXPOSURE: --

EMERGENCY AND FIRST AID PROCEDURES: Inhalation - rest; if swallowed - give emetic, get medical aid; wash skin & eyes.

SECTION VI REACTIVITY DATA

CONDITIONS CONTRIBUTING TO INSTABILITY

CONDITIONS CONTRIBUTING TO HAZARDOUS POLYMERIZATION

INCOMPATABILITY (MATERIALS TO AVOID)

HAZARDOUS DECOMPOSITION PRODUCTS

SECTION VII SPILL OR LEAK PROCEDURES

STEPS TO BE TAKEN IN CASE MATERIAL IS RELEASED OR SPILLED

WASTE DISPOSAL METHOD

SECTION VIII SPECIAL PROTECTION INFORMATION

VENTILATION REQUIREMENTS LOCAL EXHAUST | PROTECTIVE EQUIPMENT (SPECIFY TYPES) EYE: Goggles

MECHANICAL (GENERAL) | GLOVES

SPECIAL | RESPIRATOR: With proper filter

OTHER PROTECTIVE EQUIPMENT

SECTION IX SPECIAL PRECAUTIONS

PRECAUTIONS TO BE TAKEN IN HANDLING AND STORAGE

OTHER PRECAUTIONS

SIGNATURE | DATE

o-CHLOROTOLUENE

Hazard diamond: 2 (top), 3 (left), 0 (right)

SECTION I SOURCE AND NOMENCLATURE

MANUFACTURER'S NAME | EMERGENCY TELEPHONE NO.

ADDRESS (NUMBER, STREET, CITY, STATE, ZIP CODE)

TRADE NAME AND SYNONYMS | CHEMICAL FAMILY

CHEMICAL NAME AND SYNONYMS: 2 Chlorotoluene | FORMULA: $CH_3C_6H_4Cl$

SECTION II HAZARDOUS INGREDIENTS

BASIC MATERIAL	APPROXIMATE OR MAXIMUM % WT. OR VOL.	ESTABLISHED OSHA STANDARD	LD_{50} ORAL	LD_{50} PERCUT.	LD_{50} SPECIES	LC_{50} CONC.
			None available			

SECTION III PHYSICAL DATA MW 126.58

BOILING POINT °F.: 159°C | VAPOR PRESSURE mm Hg.: at 43°C 10

MELTING POINT °F.: -37°C | VAPOR DENSITY (Air = 1)

SPECIFIC GRAVITY (H_2O = 1): 1.08 | EVAPORATION RATE (_____ = 1)

SOLUBILITY IN WATER Pts/100 pts H_2O: Insoluble | VOLATILE % Vol. % Wt.

APPEARANCE AND ODOR: Colorless liquid

SECTION IV FIRE AND EXPLOSION HAZARD DATA

FLASH POINT: None available | FLAMMABLE (EXPLOSIVE) LIMITS: UPPER / LOWER

METHOD USED

EXTINGUISHING MEDIA: Alcohol foam, carbon dioxide, dry chemical

SPECIAL FIRE FIGHTING PROCEDURES: --

UNUSUAL FIRE AND EXPLOSION HAZARDS: --

SECTION V HEALTH HAZARD DATA

TOXIC LEVEL: TLV - 50 PPM, 250 mg/C.M. | CARCINOGENIC: Neoplastic

PRINCIPLE ROUTES OF ABSORBTION: Inhalation, ingestion | SKIN AND EYE IRRITATION: Severe

REVELANT SYMPTOMS OF EXPOSURE: Eye and respiratory system irritation, severe internal damage.

EFFECTS OF CHRONIC EXPOSURE: --

EMERGENCY AND FIRST AID PROCEDURES: Inhalation - rest; if swallowed - give emetic, get medical aid; wash eyes and skin.

SECTION VI REACTIVITY DATA

CONDITIONS CONTRIBUTING TO INSTABILITY

CONDITIONS CONTRIBUTING TO HAZARDOUS POLYMERIZATION

INCOMPATABILITY (MATERIALS TO AVOID)

HAZARDOUS DECOMPOSITION PRODUCTS

SECTION VII SPILL OR LEAK PROCEDURES

STEPS TO BE TAKEN IN CASE MATERIAL IS RELEASED OR SPILLED

WASTE DISPOSAL METHOD

SECTION VIII SPECIAL PROTECTION INFORMATION

VENTILATION REQUIREMENTS LOCAL EXHAUST | PROTECTIVE EQUIPMENT (SPECIFY TYPES) EYE: Goggles

MECHANICAL (GENERAL) | GLOVES: Rubber

SPECIAL | RESPIRATOR: With proper filter

OTHER PROTECTIVE EQUIPMENT

SECTION IX SPECIAL PRECAUTIONS

PRECAUTIONS TO BE TAKEN IN HANDLING AND STORAGE

OTHER PRECAUTIONS

SIGNATURE | DATE

p-CHLOROTOLUENE

(Hazard diamond: 2 / 3 / 0)

SECTION I SOURCE AND NOMENCLATURE

MANUFACTURER'S NAME | EMERGENCY TELEPHONE NO.

ADDRESS (NUMBER, STREET, CITY, STATE, ZIP CODE)

TRADE NAME AND SYNONYMS | CHEMICAL FAMILY

CHEMICAL NAME AND SYNONYMS: 4 Chlorotoluene | FORMULA: $CH_3C_6H_4Cl$

SECTION II HAZARDOUS INGREDIENTS

BASIC MATERIAL	APPROXIMATE OR MAXIMUM % WT. OR VOL.	ESTABLISHED OSHA STANDARD	LD_{50} ORAL	LD_{50} PERCUT.	LC_{50} SPECIES	LC_{50} CONC.
			None available			

SECTION III PHYSICAL DATA MW - 126.58

BOILING POINT °F.: 162°C | VAPOR PRESSURE mm Hg.: at 44°C 10

MELTING POINT °F.: 7°C | VAPOR DENSITY (Air = 1)

SPECIFIC GRAVITY (H_2O = 1): 1.07 | EVAPORATION RATE (____ = 1)

SOLUBILITY IN WATER Pts/100 pts H_2O | VOLATILE % Vol. % Wt.

APPEARANCE AND ODOR: Colorless liquid

SECTION IV FIRE AND EXPLOSION HAZARD DATA

FLASH POINT: N/A | FLAMMABLE (EXPLOSIVE) LIMITS | UPPER | LOWER

METHOD USED

EXTINGUISHING MEDIA: Alcohol foam, carbon dioxide, dry chemical

SPECIAL FIRE FIGHTING PROCEDURES

UNUSUAL FIRE AND EXPLOSION HAZARDS

SECTION V HEALTH HAZARD DATA

TOXIC LEVEL: None available | CARCINOGENIC: No

PRINCIPLE ROUTES OF ABSORBTION: Inhalation, ingestion | SKIN AND EYE IRRITATION: Severe

REVELANT SYMPTOMS OF EXPOSURE: Eye & respiratory system irritation, severe internal damage if swallowed.

EFFECTS OF CHRONIC EXPOSURE: --

EMERGENCY AND FIRST AID PROCEDURES: Inhalation - rest; if swallowed - give emetic, get medical aid; wash skin & eyes.

SECTION VI REACTIVITY DATA

CONDITIONS CONTRIBUTING TO INSTABILITY

CONDITIONS CONTRIBUTING TO HAZARDOUS POLYMERIZATION

INCOMPATABILITY (MATERIALS TO AVOID)

HAZARDOUS DECOMPOSITION PRODUCTS

SECTION VII SPILL OR LEAK PROCEDURES

STEPS TO BE TAKEN IN CASE MATERIAL IS RELEASED OR SPILLED

WASTE DISPOSAL METHOD

SECTION VIII SPECIAL PROTECTION INFORMATION

VENTILATION REQUIREMENTS LOCAL EXHAUST | PROTECTIVE EQUIPMENT (SPECIFY TYPES) EYE: Goggles

MECHANICAL (GENERAL) | GLOVES

SPECIAL | RESPIRATOR: With proper filter

OTHER PROTECTIVE EQUIPMENT

SECTION IX SPECIAL PRECAUTIONS

PRECAUTIONS TO BE TAKEN IN HANDLING AND STORAGE

OTHER PRECAUTIONS

SIGNATURE | DATE

CHROMIC NITRATE

(Hazard diamond: 0 / 2 / 0)

SECTION I SOURCE AND NOMENCLATURE

MANUFACTURER'S NAME | EMERGENCY TELEPHONE NO.

ADDRESS (NUMBER, STREET, CITY, STATE, ZIP CODE)

TRADE NAME AND SYNONYMS | CHEMICAL FAMILY

CHEMICAL NAME AND SYNONYMS: Chromic Nitrate | FORMULA: $Cr(NO_3)_3 \cdot 9H_2O$

SECTION II HAZARDOUS INGREDIENTS

BASIC MATERIAL	APPROXIMATE OR MAXIMUM % WT. OR VOL.	ESTABLISHED OSHA STANDARD	LD_{50} mg/kg ORAL	LD_{50} mg/kg PERCUT.	LC_{50} SPECIES	LC_{50} CONC.
			3250		Rat	

SECTION III PHYSICAL DATA MW - 400.15

BOILING POINT °F.: d 100° C. | VAPOR PRESSURE mm Hg.

MELTING POINT °F.: 60°C. | VAPOR DENSITY (Air = 1)

SPECIFIC GRAVITY (H_2O = 1) | EVAPORATION RATE (____ = 1)

SOLUBILITY IN WATER Pts/100 pts H_2O: Soluble | VOLATILE % Vol. % Wt.

APPEARANCE AND ODOR: Purple crystals

SECTION IV FIRE AND EXPLOSION HAZARD DATA

FLASH POINT: N/A | FLAMMABLE (EXPLOSIVE) LIMITS | UPPER | LOWER

METHOD USED

EXTINGUISHING MEDIA: N/A

SPECIAL FIRE FIGHTING PROCEDURES: N/A

UNUSUAL FIRE AND EXPLOSION HAZARDS: N/A

SECTION V HEALTH HAZARD DATA

TOXIC LEVEL: TLV - 3.84 mg/m³ | CARCINOGENIC: No

PRINCIPLE ROUTES OF ABSORBTION: Inhalation, ingestion | SKIN AND EYE IRRITATION: Severe

REVELANT SYMPTOMS OF EXPOSURE: Lesions of the skin and mucous membrane.

EFFECTS OF CHRONIC EXPOSURE: Possible hepatitis, jaundice and kidney lesions.

EMERGENCY AND FIRST AID PROCEDURES: After ingestion; induce vomiting or give gastric lavage; if inhaled - rest, keep warm, get medical attention.

SECTION VI REACTIVITY DATA

CONDITIONS CONTRIBUTING TO INSTABILITY

CONDITIONS CONTRIBUTING TO HAZARDOUS POLYMERIZATION: --

INCOMPATABILITY (MATERIALS TO AVOID): Avoid contact with strong reducing agents.

HAZARDOUS DECOMPOSITION PRODUCTS: Oxides of nitrogen.

SECTION VII SPILL OR LEAK PROCEDURES

STEPS TO BE TAKEN IN CASE MATERIAL IS RELEASED OR SPILLED

WASTE DISPOSAL METHOD

SECTION VIII SPECIAL PROTECTION INFORMATION

VENTILATION REQUIREMENTS LOCAL EXHAUST | PROTECTIVE EQUIPMENT (SPECIFY TYPES) EYE: Safety glasses

MECHANICAL (GENERAL) | GLOVES

SPECIAL | RESPIRATOR: With appropriate filter

OTHER PROTECTIVE EQUIPMENT

SECTION IX SPECIAL PRECAUTIONS

PRECAUTIONS TO BE TAKEN IN HANDLING AND STORAGE

OTHER PRECAUTIONS

SIGNATURE | DATE

CHROMIUM METAL

(Hazard diamond: 0 top, 4 left, 0 right)

SECTION I SOURCE AND NOMENCLATURE

MANUFACTURER'S NAME	EMERGENCY TELEPHONE NO.
ADDRESS (NUMBER, STREET, CITY, STATE, ZIP CODE)	
TRADE NAME AND SYNONYMS	CHEMICAL FAMILY
CHEMICAL NAME AND SYNONYMS Chromium	FORMULA Cr

SECTION II HAZARDOUS INGREDIENTS

BASIC MATERIAL	APPROXIMATE OR MAXIMUM % WT. OR VOL.	ESTABLISHED OSHA STANDARD	LD_{50} ORAL	LD_{50} PERCUT.	LC_{50} SPECIES	LC_{50} CONC.
~~#1 Lowest published toxic dose~~						
~~Intraperitoneal - rat~~	~~504 mg/kg.~~					
~~TCLO inhalation - human~~	~~4.5 mg/c.m.~~	~~5 years.~~				

SECTION III PHYSICAL DATA AW-52.00

BOILING POINT °F. 2200° C.	VAPOR PRESSURE mm Hg.
MELTING POINT °F. 1615° C.	VAPOR DENSITY (Air = 1)
SPECIFIC GRAVITY (H_2O = 1) N/A	EVAPORATION RATE (______ = 1)
SOLUBILITY IN WATER Pts/100 pts H_2O	VOLATILE % Vol. % Wt.

APPEARANCE AND ODOR
Silver metal powder.

SECTION IV FIRE AND EXPLOSION HAZARD DATA

FLASH POINT N/A	FLAMMABLE (EXPLOSIVE) LIMITS	UPPER
METHOD USED		LOWER

EXTINGUISHING MEDIA
N/A

SPECIAL FIRE FIGHTING PROCEDURES
N/A

UNUSUAL FIRE AND EXPLOSION HAZARDS
N/A

SECTION V HEALTH HAZARD DATA

TOXIC LEVEL
TLV - 1 mg/CM — CARCINOGENIC Yes

PRINCIPLE ROUTES OF ABSORBTION
Inhalation of dust, ingestion — SKIN AND EYE IRRITATION Mild

REVELANT SYMPTOMS OF EXPOSURE
--

EFFECTS OF CHRONIC EXPOSURE
May produce tumors, dermatitis.

EMERGENCY AND FIRST AID PROCEDURES Inhalation - rest, keep warm, get medical aid; if swallowed - wash mouth, give emetic - get medical aid; wash from skin and eyes.

SECTION VI REACTIVITY DATA

CONDITIONS CONTRIBUTING TO INSTABILITY

CONDITIONS CONTRIBUTING TO HAZARDOUS POLYMERIZATION

INCOMPATABILITY (MATERIALS TO AVOID)

HAZARDOUS DECOMPOSITION PRODUCTS

SECTION VII SPILL OR LEAK PROCEDURES

STEPS TO BE TAKEN IN CASE MATERIAL IS RELEASED OR SPILLED

WASTE DISPOSAL METHOD

SECTION VIII SPECIAL PROTECTION INFORMATION

VENTILATION REQUIREMENTS LOCAL EXHAUST	PROTECTIVE EQUIPMENT (SPECIFY TYPES) EYE Safety glasses
MECHANICAL (GENERAL)	GLOVES
SPECIAL	RESPIRATOR With proper filter
OTHER PROTECTIVE EQUIPMENT	

SECTION IX SPECIAL PRECAUTIONS

PRECAUTIONS TO BE TAKEN IN HANDLING AND STORAGE

OTHER PRECAUTIONS

SIGNATURE	DATE

CHROMIUM POTASSIUM SULFATE

(Hazard diamond: 0 top, 2 left, 0 right)

SECTION I SOURCE AND NOMENCLATURE

MANUFACTURER'S NAME	EMERGENCY TELEPHONE NO.
ADDRESS (NUMBER, STREET, CITY, STATE, ZIP CODE)	
TRADE NAME AND SYNONYMS	CHEMICAL FAMILY
CHEMICAL NAME AND SYNONYMS Chromium Potassium Sulfate	FORMULA $CrK(SO_4)_2 \cdot 12H_2O$

SECTION II HAZARDOUS INGREDIENTS

BASIC MATERIAL	APPROXIMATE OR MAXIMUM % WT. OR VOL.	ESTABLISHED OSHA STANDARD	LD_{50} ORAL	LD_{50} PERCUT.	LC_{50} SPECIES	LC_{50} CONC.
LD50 - intervenous - rat	112 mg/kg					

SECTION III PHYSICAL DATA MW 499.13

BOILING POINT °F. N/A	VAPOR PRESSURE mm Hg.
MELTING POINT °F. N/A	VAPOR DENSITY (Air = 1)
SPECIFIC GRAVITY (H_2O = 1) 1.81	EVAPORATION RATE (______ = 1)
SOLUBILITY IN WATER Pts/100 pts H_2O	VOLATILE % Vol. % Wt.

APPEARANCE AND ODOR
Dark violet-red crystals.

SECTION IV FIRE AND EXPLOSION HAZARD DATA

FLASH POINT N/A	FLAMMABLE (EXPLOSIVE) LIMITS	UPPER
METHOD USED		LOWER

EXTINGUISHING MEDIA
N/A

SPECIAL FIRE FIGHTING PROCEDURES
N/A

UNUSUAL FIRE AND EXPLOSION HAZARDS
N/A

SECTION V HEALTH HAZARD DATA

TOXIC LEVEL
TLV - 4.78 mg/C.M. — CARCINOGENIC No

PRINCIPLE ROUTES OF ABSORBTION
Inhalation of dust, ingestion — SKIN AND EYE IRRITATION Mild

REVELANT SYMPTOMS OF EXPOSURE
--

EFFECTS OF CHRONIC EXPOSURE
Skin ulcers and dermatitis.

EMERGENCY AND FIRST AID PROCEDURES Inhalation - rest, keep warm; if swallowed - wash mouth, give emetic, get medical aid; wash from skin and eyes.

SECTION VI REACTIVITY DATA

CONDITIONS CONTRIBUTING TO INSTABILITY

CONDITIONS CONTRIBUTING TO HAZARDOUS POLYMERIZATION

INCOMPATABILITY (MATERIALS TO AVOID)

HAZARDOUS DECOMPOSITION PRODUCTS

SECTION VII SPILL OR LEAK PROCEDURES

STEPS TO BE TAKEN IN CASE MATERIAL IS RELEASED OR SPILLED

WASTE DISPOSAL METHOD

SECTION VIII SPECIAL PROTECTION INFORMATION

VENTILATION REQUIREMENTS LOCAL EXHAUST	PROTECTIVE EQUIPMENT (SPECIFY TYPES) EYE Safety glasses
MECHANICAL (GENERAL)	GLOVES
SPECIAL	RESPIRATOR With proper filter
OTHER PROTECTIVE EQUIPMENT	

SECTION IX SPECIAL PRECAUTIONS

PRECAUTIONS TO BE TAKEN IN HANDLING AND STORAGE

OTHER PRECAUTIONS

SIGNATURE	DATE

COBALT METAL

Hazard diamond: 0 (top), 4 (left), 1 (right)

SECTION I SOURCE AND NOMENCLATURE

MANUFACTURER'S NAME	EMERGENCY TELEPHONE NO.
ADDRESS (NUMBER, STREET, CITY, STATE, ZIP CODE)	
TRADE NAME AND SYNONYMS	CHEMICAL FAMILY
CHEMICAL NAME AND SYNONYMS: Cobalt	FORMULA: Co

SECTION II HAZARDOUS INGREDIENTS

BASIC MATERIAL	APPROXIMATE OR MAXIMUM % WT. OR VOL.	ESTABLISHED OSHA STANDARD	LD_{50} ORAL	LD_{50} PERCUT.	LC_{50} SPECIES	LC_{50} CONC.
			See Note 1			
#1 lowest published lethal dose oral - rat 1500 mg/kg.						

SECTION III PHYSICAL DATA MW 58.93

BOILING POINT °F.: 2900° C.	VAPOR PRESSURE mm Hg.
MELTING POINT °F.: 1495° C.	VAPOR DENSITY (Air = 1)
SPECIFIC GRAVITY (H_2O = 1): 8.9 20°/4° C.	EVAPORATION RATE (= 1)
SOLUBILITY IN WATER Pts/100 pts H_2O	VOLATILE % Vol. % Wt.

APPEARANCE AND ODOR: Silver-gray metal

SECTION IV FIRE AND EXPLOSION HAZARD DATA

FLASH POINT: N/A

METHOD USED

FLAMMABLE (EXPLOSIVE) LIMITS: UPPER / LOWER

EXTINGUISHING MEDIA: N/A

SPECIAL FIRE FIGHTING PROCEDURES: N/A

UNUSUAL FIRE AND EXPLOSION HAZARDS: N/A

SECTION V HEALTH HAZARD DATA

TOXIC LEVEL: TLV 0.1 mg/m^3 — CARCINOGENIC: Yes

PRINCIPLE ROUTES OF ABSORBTION: Inhalation of dust — SKIN AND EYE IRRITATION: mild

REVELANT SYMPTOMS OF EXPOSURE: Inhalation - shortness of breath, lung densities, dermatitis with hyperemia and vesiculation. Ingestion - hypotension, pain, vomiting, nerve deafness,

EFFECTS OF CHRONIC EXPOSURE: Possible pneumoconiosis

EMERGENCY AND FIRST AID PROCEDURES: Remove from exposure, rest. Keep warm - eye contact - wash with water. Obtain medical attention.

SECTION VI REACTIVITY DATA

CONDITIONS CONTRIBUTING TO INSTABILITY

CONDITIONS CONTRIBUTING TO HAZARDOUS POLYMERIZATION

INCOMPATABILITY (MATERIALS TO AVOID): In the powdered state of cobalt - avoid ammonium nitrate, acetylene.

HAZARDOUS DECOMPOSITION PRODUCTS

SECTION VII SPILL OR LEAK PROCEDURES

STEPS TO BE TAKEN IN CASE MATERIAL IS RELEASED OR SPILLED

WASTE DISPOSAL METHOD

SECTION VIII SPECIAL PROTECTION INFORMATION

VENTILATION REQUIREMENTS LOCAL EXHAUST	PROTECTIVE EQUIPMENT (SPECIFY TYPES) EYE: Safety glasses
MECHANICAL (GENERAL)	GLOVES: Rubber
SPECIAL	RESPIRATOR: W/appropriate filter
OTHER PROTECTIVE EQUIPMENT	

SECTION IX SPECIAL PRECAUTIONS

PRECAUTIONS TO BE TAKEN IN HANDLING AND STORAGE

OTHER PRECAUTIONS

SIGNATURE — DATE

COUMARIN

Hazard diamond: 2 (top), 4 (left), 0 (right)

SECTION I SOURCE AND NOMENCLATURE

MANUFACTURER'S NAME	EMERGENCY TELEPHONE NO.
ADDRESS (NUMBER, STREET, CITY, STATE, ZIP CODE)	
TRADE NAME AND SYNONYMS: 1,2 Benzopyrene	CHEMICAL FAMILY
CHEMICAL NAME AND SYNONYMS: Coumarin	FORMULA: $C_9H_6O_2$

SECTION II HAZARDOUS INGREDIENTS

BASIC MATERIAL	APPROXIMATE OR MAXIMUM % WT. OR VOL.	ESTABLISHED OSHA STANDARD	LD_{50} mg/kg ORAL	LD_{50} mg/kg PERCUT.	LC_{50} SPECIES	LC_{50} CONC.
			293		Rat	
LDLO - oral - human 50 mg/kg			196	242	Mouse	
			202		Guinea Pig	

SECTION III PHYSICAL DATA MW-146.15

BOILING POINT °F.: 299°C.	VAPOR PRESSURE at 106°C. .1 mm Hg.
MELTING POINT °F.: 70°C.	VAPOR DENSITY (Air = 1)
SPECIFIC GRAVITY (H_2O = 1)	EVAPORATION RATE (= 1)
SOLUBILITY IN WATER Pts/100 pts H_2O: 0.25	VOLATILE % Vol. % Wt.

APPEARANCE AND ODOR: Orthorhombic crystals with fragrant odor.

SECTION IV FIRE AND EXPLOSION HAZARD DATA

FLASH POINT: None available

METHOD USED

FLAMMABLE (EXPLOSIVE) LIMITS: UPPER / LOWER

EXTINGUISHING MEDIA: Dry chemical, carbon dioxide

SPECIAL FIRE FIGHTING PROCEDURES

UNUSUAL FIRE AND EXPLOSION HAZARDS

SECTION V HEALTH HAZARD DATA

TOXIC LEVEL: None available — CARCINOGENIC: Yes

PRINCIPLE ROUTES OF ABSORBTION: Inhalation, ingestion — SKIN AND EYE IRRITATION: Slight

REVELANT SYMPTOMS OF EXPOSURE: May cause excessive bleeding.

EFFECTS OF CHRONIC EXPOSURE: Disturbs clothing mechanism of the blood.

EMERGENCY AND FIRST AID PROCEDURES: Wash skin & eyes; if swallowed - gastric lavage, get medical aid.

SECTION VI REACTIVITY DATA

CONDITIONS CONTRIBUTING TO INSTABILITY

CONDITIONS CONTRIBUTING TO HAZARDOUS POLYMERIZATION

INCOMPATABILITY (MATERIALS TO AVOID)

HAZARDOUS DECOMPOSITION PRODUCTS

SECTION VII SPILL OR LEAK PROCEDURES

STEPS TO BE TAKEN IN CASE MATERIAL IS RELEASED OR SPILLED

WASTE DISPOSAL METHOD

SECTION VIII SPECIAL PROTECTION INFORMATION

VENTILATION REQUIREMENTS LOCAL EXHAUST	PROTECTIVE EQUIPMENT (SPECIFY TYPES) EYE: Goggles
MECHANICAL (GENERAL)	GLOVES: Rubber
SPECIAL	RESPIRATOR
OTHER PROTECTIVE EQUIPMENT	

SECTION IX SPECIAL PRECAUTIONS

PRECAUTIONS TO BE TAKEN IN HANDLING AND STORAGE

OTHER PRECAUTIONS

SIGNATURE — DATE

CUPRIC CARBONATE

(hazard diamond: 0 top, 2 left, 0 right)

SECTION I SOURCE AND NOMENCLATURE

MANUFACTURER'S NAME	EMERGENCY TELEPHONE NO.
ADDRESS (NUMBER, STREET, CITY, STATE, ZIP CODE)	
TRADE NAME AND SYNONYMS	CHEMICAL FAMILY
CHEMICAL NAME AND SYNONYMS Cupric Carbonate	FORMULA $CuCO_3 \cdot Cu(OH)_2$

SECTION II HAZARDOUS INGREDIENTS

BASIC MATERIAL	APPROXIMATE OR MAXIMUM % WT. OR VOL.	ESTABLISHED OSHA STANDARD	LD_{50} mg/kg ORAL	PERCUT.	SPECIES	LC_{50} CONC.
			159		Rat	
			810		bird	

SECTION III PHYSICAL DATA MW - 221.17

BOILING POINT °F.	VAPOR PRESSURE mm Hg.
MELTING POINT °F. d 200° C.	VAPOR DENSITY (Air = 1)
SPECIFIC GRAVITY (H_2O = 1) 4.0	EVAPORATION RATE (____ = 1)
SOLUBILITY IN WATER Pts/100 pts H_2O	VOLATILE % Vol. % Wt.

APPEARANCE AND ODOR
Green powder

SECTION IV FIRE AND EXPLOSION HAZARD DATA

FLASH POINT N/A	FLAMMABLE (EXPLOSIVE) LIMITS	UPPER
METHOD USED		LOWER

EXTINGUISHING MEDIA
N/A

SPECIAL FIRE FIGHTING PROCEDURES
N/A

UNUSUAL FIRE AND EXPLOSION HAZARDS
N/A

SECTION V HEALTH HAZARD DATA

TOXIC LEVEL
TLV 3.35 mg/C.M. CARCINOGENIC No

PRINCIPLE ROUTES OF ABSORBTION
Inhalation of dust (mainly),ingestion SKIN AND EYE IRRITATION Moderate

REVELANT SYMPTOMS OF EXPOSURE Irritation of eyes and mucous membranes, vomiting, diarrhea, abdominal pain, collapse.

EFFECTS OF CHRONIC EXPOSURE
Liver and kidney damage.

EMERGENCY AND FIRST AID PROCEDURES Inhalation - rest, keep warm; if swallowed - wash mouth, give emetic, get medical aid; wash from skin and eyes.

SECTION VI REACTIVITY DATA

CONDITIONS CONTRIBUTING TO INSTABILITY

CONDITIONS CONTRIBUTING TO HAZARDOUS POLYMERIZATION

INCOMPATABILITY (MATERIALS TO AVOID)
With acids

HAZARDOUS DECOMPOSITION PRODUCTS

SECTION VII SPILL OR LEAK PROCEDURES

STEPS TO BE TAKEN IN CASE MATERIAL IS RELEASED OR SPILLED

WASTE DISPOSAL METHOD

SECTION VIII SPECIAL PROTECTION INFORMATION

VENTILATION REQUIREMENTS LOCAL EXHAUST	PROTECTIVE EQUIPMENT (SPECIFY TYPES) EYE Safety glasses
MECHANICAL (GENERAL)	GLOVES
SPECIAL	RESPIRATOR

OTHER PROTECTIVE EQUIPMENT

SECTION IX SPECIAL PRECAUTIONS

PRECAUTIONS TO BE TAKEN IN HANDLING AND STORAGE

OTHER PRECAUTIONS

SIGNATURE	DATE

CUPRIC OXIDE

(hazard diamond: 0 top, 2 left, 0 right)

SECTION I SOURCE AND NOMENCLATURE

MANUFACTURER'S NAME	EMERGENCY TELEPHONE NO.
ADDRESS (NUMBER, STREET, CITY, STATE, ZIP CODE)	
TRADE NAME AND SYNONYMS	CHEMICAL FAMILY
CHEMICAL NAME AND SYNONYMS Cupric Oxide	FORMULA CuO

SECTION II HAZARDOUS INGREDIENTS

BASIC MATERIAL	APPROXIMATE OR MAXIMUM % WT. OR VOL.	ESTABLISHED OSHA STANDARD	LD_{50} Mg/Kg ORAL	PERCUT.	SPECIES	LC_{50} CONC.
			Not available			

SECTION III PHYSICAL DATA MW - 79.5

BOILING POINT °F.	VAPOR PRESSURE mm Hg.
MELTING POINT °F. d at 1026	VAPOR DENSITY (Air = 1)
SPECIFIC GRAVITY (H_2O = 1)	EVAPORATION RATE (____ = 1)
SOLUBILITY IN WATER Pts/100 pts H_2O I	VOLATILE % Vol. % Wt.

APPEARANCE AND ODOR
Black Crystals

SECTION IV FIRE AND EXPLOSION HAZARD DATA

FLASH POINT Not flammable	FLAMMABLE (EXPLOSIVE) LIMITS	UPPER
METHOD USED N/A		LOWER

EXTINGUISHING MEDIA
N/A

SPECIAL FIRE FIGHTING PROCEDURES
N/A

UNUSUAL FIRE AND EXPLOSION HAZARDS

SECTION V HEALTH HAZARD DATA

TOXIC LEVEL
TLV 1.25 Mg/M^3 CARCINOGENIC No

PRINCIPLE ROUTES OF ABSORBTION
Inhalation of dust (mainly), ingestion SKIN AND EYE IRRITATION Moderate

REVELANT SYMPTOMS OF EXPOSURE Irritation of mucous membranes, irritation of eyes, vomiting, diarrhea, abdominal pain, collapse.

EFFECTS OF CHRONIC EXPOSURE
Liver and kidney damage.

EMERGENCY AND FIRST AID PROCEDURES Inhalation - rest, keep warm; if swallowed - wash mouth, give emetic, get medical attention, wash from skin and eyes.

SECTION VI REACTIVITY DATA

CONDITIONS CONTRIBUTING TO INSTABILITY

CONDITIONS CONTRIBUTING TO HAZARDOUS POLYMERIZATION

INCOMPATABILITY (MATERIALS TO AVOID)

HAZARDOUS DECOMPOSITION PRODUCTS

SECTION VII SPILL OR LEAK PROCEDURES

STEPS TO BE TAKEN IN CASE MATERIAL IS RELEASED OR SPILLED

WASTE DISPOSAL METHOD

SECTION VIII SPECIAL PROTECTION INFORMATION

VENTILATION REQUIREMENTS LOCAL EXHAUST	PROTECTIVE EQUIPMENT (SPECIFY TYPES) EYE Safety glasses
MECHANICAL (GENERAL)	GLOVES
SPECIAL	RESPIRATOR

OTHER PROTECTIVE EQUIPMENT

SECTION IX SPECIAL PRECAUTIONS

PRECAUTIONS TO BE TAKEN IN HANDLING AND STORAGE

OTHER PRECAUTIONS

SIGNATURE	DATE

CUPROUS CHLORIDE

Hazard diamond: 0 (top), 3 (left), 0 (right)

SECTION I SOURCE AND NOMENCLATURE

MANUFACTURER'S NAME | EMERGENCY TELEPHONE NO.

ADDRESS (NUMBER, STREET, CITY, STATE, ZIP CODE)

TRADE NAME AND SYNONYMS | CHEMICAL FAMILY

CHEMICAL NAME AND SYNONYMS: Cuprous Chloride | FORMULA: Cu_2Cl_2

SECTION II HAZARDOUS INGREDIENTS

BASIC MATERIAL	APPROXIMATE OR MAXIMUM % WT. OR VOL.	ESTABLISHED OSHA STANDARD	LD_{50} Mg/Kg ORAL	LD_{50} Mg/Kg PERCUT.	LC_{50} SPECIES	LC_{50} CONC.
Oral - human LDLO 50 mg/kg						
#1 more toxic than $Cu\ Cl_2$						
LD_{50} for $Cu\ Cl_2$ oral - rat 140 mg/kg						

SECTION III PHYSICAL DATA Mw - 99.00

BOILING POINT °F: 1366°C. | VAPOR PRESSURE mm Hg.

MELTING POINT °F: 422°C. | VAPOR DENSITY (Air = 1)

SPECIFIC GRAVITY (H_2O = 1): 3.53 | EVAPORATION RATE (_____ = 1)

SOLUBILITY IN WATER Pts/100 pts H_2O: 1.52 | VOLATILE % Vol. % Wt.

APPEARANCE AND ODOR: White crystals

SECTION IV FIRE AND EXPLOSION HAZARD DATA

FLASH POINT: Not flammable | FLAMMABLE (EXPLOSIVE) LIMITS: UPPER / LOWER

METHOD USED

EXTINGUISHING MEDIA: N/A

SPECIAL FIRE FIGHTING PROCEDURES: N/A

UNUSUAL FIRE AND EXPLOSION HAZARDS: N/A

SECTION V HEALTH HAZARD DATA

TOXIC LEVEL: TLV 1.56 Mg/C.M. | CARCINOGENIC: No

PRINCIPLE ROUTES OF ABSORBTION: Inhalation of dust, ingestion | SKIN AND EYE IRRITATION: Moderate

REVELANT SYMPTOMS OF EXPOSURE: Irritation of mucous membranes and eyes, gastroenteric pain, vomiting, diarrhea, abdominal pain, collapse.

EFFECTS OF CHRONIC EXPOSURE: Liver and kidney damage.

EMERGENCY AND FIRST AID PROCEDURES: Inhalation - rest, keep warm; if swallowed - wash mouth, give emetic, get medical aid, wash from skin and eyes.

SECTION VI REACTIVITY DATA

CONDITIONS CONTRIBUTING TO INSTABILITY

CONDITIONS CONTRIBUTING TO HAZARDOUS POLYMERIZATION

INCOMPATABILITY (MATERIALS TO AVOID)

HAZARDOUS DECOMPOSITION PRODUCTS

SECTION VII SPILL OR LEAK PROCEDURES

STEPS TO BE TAKEN IN CASE MATERIAL IS RELEASED OR SPILLED

WASTE DISPOSAL METHOD

SECTION VIII SPECIAL PROTECTION INFORMATION

VENTILATION REQUIREMENTS LOCAL EXHAUST | PROTECTIVE EQUIPMENT (SPECIFY TYPES) EYE: Safety glasses

MECHANICAL (GENERAL) | GLOVES

SPECIAL | RESPIRATOR: W/appropriate filter

OTHER PROTECTIVE EQUIPMENT

SECTION IX SPECIAL PRECAUTIONS

PRECAUTIONS TO BE TAKEN IN HANDLING AND STORAGE

OTHER PRECAUTIONS

SIGNATURE | DATE

CUPROUS OXIDE

Hazard diamond: 0 (top), 2 (left), 0 (right)

SECTION I SOURCE AND NOMENCLATURE

MANUFACTURER'S NAME | EMERGENCY TELEPHONE NO.

ADDRESS (NUMBER, STREET, CITY, STATE, ZIP CODE)

TRADE NAME AND SYNONYMS | CHEMICAL FAMILY

CHEMICAL NAME AND SYNONYMS | FORMULA: Cu_2O

SECTION II HAZARDOUS INGREDIENTS

BASIC MATERIAL	APPROXIMATE OR MAXIMUM % WT. OR VOL.	ESTABLISHED OSHA STANDARD	LD_{50} Mg/Kg ORAL	LD_{50} Mg/Kg PERCUT.	LC_{50} SPECIES	LC_{50} CONC.
			470		Rat	

SECTION III PHYSICAL DATA MW - 143.13

BOILING POINT °F: decomposes 1800° C. | VAPOR PRESSURE mm Hg.

MELTING POINT °F: 1235° C. | VAPOR DENSITY (Air = 1)

SPECIFIC GRAVITY (H_2O = 1): 6.0 | EVAPORATION RATE (_____ = 1)

SOLUBILITY IN WATER Pts/100 pts H_2O: Insoluble | VOLATILE % Vol. % Wt.

APPEARANCE AND ODOR: Dark red crystals.

SECTION IV FIRE AND EXPLOSION HAZARD DATA

FLASH POINT: Not flammable | FLAMMABLE (EXPLOSIVE) LIMITS: UPPER / LOWER

METHOD USED

EXTINGUISHING MEDIA: N/A

SPECIAL FIRE FIGHTING PROCEDURES: N/A

UNUSUAL FIRE AND EXPLOSION HAZARDS: N/A

SECTION V HEALTH HAZARD DATA

TOXIC LEVEL: TLV 2.26 Mg/M^3 | CARCINOGENIC: No

PRINCIPLE ROUTES OF ABSORBTION: Inhalation of dust (mainly) ingestion | SKIN AND EYE IRRITATION: Moderate

REVELANT SYMPTOMS OF EXPOSURE: Irritation of mucous membranes, irritation of eyes, vomiting, diarrhea, abdominal pain, collapse.

EFFECTS OF CHRONIC EXPOSURE: Liver and kidney damage.

EMERGENCY AND FIRST AID PROCEDURES: Inhalation - rest, keep warm; if swallowed - wash mouth, give emetic, get medical attention; wash from skin and eyes.

SECTION VI REACTIVITY DATA

CONDITIONS CONTRIBUTING TO INSTABILITY

CONDITIONS CONTRIBUTING TO HAZARDOUS POLYMERIZATION

INCOMPATABILITY (MATERIALS TO AVOID): May react violently with alkali and alkaline earth metals.

HAZARDOUS DECOMPOSITION PRODUCTS

SECTION VII SPILL OR LEAK PROCEDURES

STEPS TO BE TAKEN IN CASE MATERIAL IS RELEASED OR SPILLED

WASTE DISPOSAL METHOD

SECTION VIII SPECIAL PROTECTION INFORMATION

VENTILATION REQUIREMENTS LOCAL EXHAUST | PROTECTIVE EQUIPMENT (SPECIFY TYPES) EYE: Safety glasses

MECHANICAL (GENERAL) | GLOVES

SPECIAL | RESPIRATOR: With appropriate filter

OTHER PROTECTIVE EQUIPMENT

SECTION IX SPECIAL PRECAUTIONS

PRECAUTIONS TO BE TAKEN IN HANDLING AND STORAGE

OTHER PRECAUTIONS

SIGNATURE | DATE

CYANAMIDE

NFPA diamond: Health 3, Flammability 1, Reactivity 3

SECTION I SOURCE AND NOMENCLATURE

MANUFACTURER'S NAME | EMERGENCY TELEPHONE NO.

ADDRESS (NUMBER, STREET, CITY, STATE, ZIP CODE)

TRADE NAME AND SYNONYMS: Hydrogen Cyanamide | CHEMICAL FAMILY

CHEMICAL NAME AND SYNONYMS: Cyanamide | FORMULA: H_2NCN

SECTION II HAZARDOUS INGREDIENTS

BASIC MATERIAL	APPROXIMATE OR MAXIMUM % WT. OR VOL.	ESTABLISHED OSHA STANDARD	LD_{50} mg/kg ORAL	LD_{50} PERCUT.	SPECIES	LC_{50} CONC.
			125		Rat	

SECTION III PHYSICAL DATA MW - 42.05

BOILING POINT °F.: 260°C.

VAPOR PRESSURE mm Hg.: 19 at 140°C.

MELTING POINT °F.: 45°C.

VAPOR DENSITY (Air = 1): 1.45

SPECIFIC GRAVITY (H_2O = 1): 1.07

EVAPORATION RATE (= 1)

SOLUBILITY IN WATER Pts/100 pts H_2O: Very soluble

VOLATILE % Vol. % Wt.

APPEARANCE AND ODOR: Deliquescent Crystals

SECTION IV FIRE AND EXPLOSION HAZARD DATA

FLASH POINT: 285°F.

METHOD USED

FLAMMABLE (EXPLOSIVE) LIMITS: UPPER / LOWER

EXTINGUISHING MEDIA: Water spray, carbon dioxide, dry chemical.

SPECIAL FIRE FIGHTING PROCEDURES: Self-contained breathing apparatus.

UNUSUAL FIRE AND EXPLOSION HAZARDS

SECTION V HEALTH HAZARD DATA

TOXIC LEVEL: TLV 2 mg/C.M.

CARCINOGENIC: No

PRINCIPLE ROUTES OF ABSORBTION: Inhalation, ingestion, skin

SKIN AND EYE IRRITATION: Severe

REVELANT SYMPTOMS OF EXPOSURE: Irritation of mucous membranes, dermatitis, flushing of face, headache, dizziness.

EFFECTS OF CHRONIC EXPOSURE: Dermatitis

EMERGENCY AND FIRST AID PROCEDURES: If swallowed - induce vomiting, get medical aid; wash skin & eyes.

SECTION VI REACTIVITY DATA

CONDITIONS CONTRIBUTING TO INSTABILITY

CONDITIONS CONTRIBUTING TO HAZARDOUS POLYMERIZATION: Polymerizes at 122°C.

INCOMPATABILITY (MATERIALS TO AVOID)

HAZARDOUS DECOMPOSITION PRODUCTS

SECTION VII SPILL OR LEAK PROCEDURES

STEPS TO BE TAKEN IN CASE MATERIAL IS RELEASED OR SPILLED: Collect by sweeping with a broom onto a sheet of paper.

WASTE DISPOSAL METHOD: Dispose by burning the paper.

SECTION VIII SPECIAL PROTECTION INFORMATION

VENTILATION REQUIREMENTS LOCAL EXHAUST

PROTECTIVE EQUIPMENT (SPECIFY TYPES) EYE: Goggles

MECHANICAL (GENERAL)

GLOVES: Long rubber gloves

SPECIAL

RESPIRATOR: Self-contained

OTHER PROTECTIVE EQUIPMENT: Protective clothing

SECTION IX SPECIAL PRECAUTIONS

PRECAUTIONS TO BE TAKEN IN HANDLING AND STORAGE: Store in cool, dry place.

OTHER PRECAUTIONS

SIGNATURE | DATE

CYCLOHEXENE

NFPA diamond: Health 2, Flammability 3, Reactivity 4

SECTION I SOURCE AND NOMENCLATURE

MANUFACTURER'S NAME | EMERGENCY TELEPHONE NO.

ADDRESS (NUMBER, STREET, CITY, STATE, ZIP CODE)

TRADE NAME AND SYNONYMS: Tetrahydrobenzene | CHEMICAL FAMILY

CHEMICAL NAME AND SYNONYMS: Cyclohexene | FORMULA: C_6H_{10}

SECTION II HAZARDOUS INGREDIENTS

BASIC MATERIAL	APPROXIMATE OR MAXIMUM % WT. OR VOL.	ESTABLISHED OSHA STANDARD	LD_{50} ORAL	LD_{50} PERCUT.	SPECIES	LC_{50} CONC.
			2300		Mouse	

SECTION III PHYSICAL DATA MW - 82.14

BOILING POINT °F: 83°C.

VAPOR PRESSURE mm Hg.

MELTING POINT °F: -104°C.

VAPOR DENSITY (Air = 1): 2.8

SPECIFIC GRAVITY (H_2O = 1): 0.81

EVAPORATION RATE (= 1)

SOLUBILITY IN WATER Pts/100 pts H_2O: Very slightly soluble

VOLATILE % Vol. % Wt.

APPEARANCE AND ODOR: Colorless liquid

SECTION IV FIRE AND EXPLOSION HAZARD DATA

FLASH POINT: <20°F.

METHOD USED

FLAMMABLE (EXPLOSIVE) LIMITS: UPPER / LOWER

EXTINGUISHING MEDIA: Dry chemical, carbon dioxide.

SPECIAL FIRE FIGHTING PROCEDURES: Water may be ineffective.

UNUSUAL FIRE AND EXPLOSION HAZARDS: Forms explosive peroxides with oxygen.

SECTION V HEALTH HAZARD DATA

TOXIC LEVEL: TLV - 300 PPM, 1015 mg/C.M.

CARCINOGENIC: No

PRINCIPLE ROUTES OF ABSORBTION: Inhalation, ingestion

SKIN AND EYE IRRITATION: Yes

REVELANT SYMPTOMS OF EXPOSURE: Irritation of eyes and mucous membranes.

EFFECTS OF CHRONIC EXPOSURE: --

EMERGENCY AND FIRST AID PROCEDURES: Inhalation - rest; if swallowed - induce vomiting, get medical aid; wash skin & eyes.

SECTION VI REACTIVITY DATA

CONDITIONS CONTRIBUTING TO INSTABILITY: Reacts with oxygen to produce explosive peroxides shelf life - 12 months.

CONDITIONS CONTRIBUTING TO HAZARDOUS POLYMERIZATION

INCOMPATABILITY (MATERIALS TO AVOID)

HAZARDOUS DECOMPOSITION PRODUCTS

SECTION VII SPILL OR LEAK PROCEDURES

STEPS TO BE TAKEN IN CASE MATERIAL IS RELEASED OR SPILLED

WASTE DISPOSAL METHOD

SECTION VIII SPECIAL PROTECTION INFORMATION

VENTILATION REQUIREMENTS LOCAL EXHAUST

PROTECTIVE EQUIPMENT (SPECIFY TYPES) EYE: Safety glasses

MECHANICAL (GENERAL)

GLOVES: Rubber gloves

SPECIAL

RESPIRATOR: Industrial canister type

OTHER PROTECTIVE EQUIPMENT

SECTION IX SPECIAL PRECAUTIONS

PRECAUTIONS TO BE TAKEN IN HANDLING AND STORAGE

OTHER PRECAUTIONS

SIGNATURE | DATE

3,4-DIAMINOANISOLE DIHYDROCHLORIDE

Hazard rating: 1 / 4 / 0

SECTION I SOURCE AND NOMENCLATURE

MANUFACTURER'S NAME | EMERGENCY TELEPHONE NO.

ADDRESS (NUMBER, STREET, CITY, STATE, ZIP CODE)

TRADE NAME AND SYNONYMS | CHEMICAL FAMILY

CHEMICAL NAME AND SYNONYMS: 2,4 diaminoanisole dihydrochloride

FORMULA: $CH_3OC_6H_3(NH_2)_2 2HCl$

SECTION II HAZARDOUS INGREDIENTS

BASIC MATERIAL	APPROXIMATE OR MAXIMUM % WT. OR VOL.	ESTABLISHED OSHA STANDARD	LD_{50} ORAL	LD_{50} PERCUT.	LC_{50} SPECIES	LC_{50} CONC.
			None available			

SECTION III PHYSICAL DATA MW-211.10

BOILING POINT °F. | VAPOR PRESSURE mm Hg.

MELTING POINT °F. | VAPOR DENSITY (Air = 1)

SPECIFIC GRAVITY (H_2O = 1) | EVAPORATION RATE (= 1)

SOLUBILITY IN WATER Pts/100 pts H_2O: Very soluble | VOLATILE % Vol. % Wt.

APPEARANCE AND ODOR: Tan Crystals

SECTION IV FIRE AND EXPLOSION HAZARD DATA

FLASH POINT: None available | FLAMMABLE (EXPLOSIVE) LIMITS: UPPER / LOWER

METHOD USED

EXTINGUISHING MEDIA: Water spray, alcohol foam, dry chemical.

SPECIAL FIRE FIGHTING PROCEDURES

UNUSUAL FIRE AND EXPLOSION HAZARDS: Emits toxic fumes when heated to decomposition point.

SECTION V HEALTH HAZARD DATA

TOXIC LEVEL: None available | CARCINOGENIC: Yes

PRINCIPLE ROUTES OF ABSORBTION: Inhalation, ingestion, skin | SKIN AND EYE IRRITATION: Severe

RELEVANT SYMPTOMS OF EXPOSURE: Headache, drowsiness, cyanosis, skin irritant.

EFFECTS OF CHRONIC EXPOSURE: Increase of sulphemoglobin, methemoblobin

EMERGENCY AND FIRST AID PROCEDURES: Wash skin and eyes; if swallowed - give emetic, get medical aid; inhalation - rest.

SECTION VI REACTIVITY DATA

CONDITIONS CONTRIBUTING TO INSTABILITY

CONDITIONS CONTRIBUTING TO HAZARDOUS POLYMERIZATION

INCOMPATABILITY (MATERIALS TO AVOID)

HAZARDOUS DECOMPOSITION PRODUCTS

SECTION VII SPILL OR LEAK PROCEDURES

STEPS TO BE TAKEN IN CASE MATERIAL IS RELEASED OR SPILLED

WASTE DISPOSAL METHOD

SECTION VIII SPECIAL PROTECTION INFORMATION

VENTILATION REQUIREMENTS LOCAL EXHAUST | PROTECTIVE EQUIPMENT (SPECIFY TYPES) EYE: Safety goggles

MECHANICAL (GENERAL) | GLOVES: Butyl

SPECIAL | RESPIRATOR: Self-contained

OTHER PROTECTIVE EQUIPMENT: Neoprene protective clothing.

SECTION IX SPECIAL PRECAUTIONS

PRECAUTIONS TO BE TAKEN IN HANDLING AND STORAGE

OTHER PRECAUTIONS

SIGNATURE | DATE

1,4-DIAMINOBUTANE DIHYDROCHLORIDE

Hazard rating: 2 / 4 / 0

SECTION I SOURCE AND NOMENCLATURE

MANUFACTURER'S NAME | EMERGENCY TELEPHONE NO.

ADDRESS (NUMBER, STREET, CITY, STATE, ZIP CODE)

TRADE NAME AND SYNONYMS | CHEMICAL FAMILY

CHEMICAL NAME AND SYNONYMS: Putrescine hydrochloride

FORMULA: $H_2N(CH_2)_4NH_2$ 2HCl

SECTION II HAZARDOUS INGREDIENTS

BASIC MATERIAL	APPROXIMATE OR MAXIMUM % WT. OR VOL.	ESTABLISHED OSHA STANDARD	LD_{50} ORAL	LD_{50} PERCUT.	LC_{50} SPECIES	LC_{50} CONC.
			None available			

SECTION III PHYSICAL DATA MW - 161.08

BOILING POINT °F.: -- | VAPOR PRESSURE mm Hg.

MELTING POINT °F.: >290°C | VAPOR DENSITY (Air = 1)

SPECIFIC GRAVITY (H_2O = 1) | EVAPORATION RATE (= 1)

SOLUBILITY IN WATER Pts/100 pts H_2O: Very | VOLATILE % Vol. % Wt.

APPEARANCE AND ODOR: White crystal

SECTION IV FIRE AND EXPLOSION HAZARD DATA

FLASH POINT: Not available | FLAMMABLE (EXPLOSIVE) LIMITS: UPPER / LOWER

METHOD USED

EXTINGUISHING MEDIA: Alcohol foam, carbon dioxide, dry chemical

SPECIAL FIRE FIGHTING PROCEDURES: Wear goggles and self-contained breathing apparatus.

UNUSUAL FIRE AND EXPLOSION HAZARDS

SECTION V HEALTH HAZARD DATA

TOXIC LEVEL: Not available | CARCINOGENIC: No

PRINCIPLE ROUTES OF ABSORBTION: Ingestion, inhalation, skin | SKIN AND EYE IRRITATION: Moderate

RELEVANT SYMPTOMS OF EXPOSURE: Nose, throat & eye irritation, headache, skin & eye burns, depression of blood pressure, convulsions.

EFFECTS OF CHRONIC EXPOSURE: Dermatitis

EMERGENCY AND FIRST AID PROCEDURES: Inhalation - rest; if swallowed - drink water, get medical aid; wash skin & eyes.

SECTION VI REACTIVITY DATA

CONDITIONS CONTRIBUTING TO INSTABILITY

CONDITIONS CONTRIBUTING TO HAZARDOUS POLYMERIZATION

INCOMPATABILITY (MATERIALS TO AVOID)

HAZARDOUS DECOMPOSITION PRODUCTS

SECTION VII SPILL OR LEAK PROCEDURES

STEPS TO BE TAKEN IN CASE MATERIAL IS RELEASED OR SPILLED

WASTE DISPOSAL METHOD

SECTION VIII SPECIAL PROTECTION INFORMATION

VENTILATION REQUIREMENTS LOCAL EXHAUST | PROTECTIVE EQUIPMENT (SPECIFY TYPES) EYE: Goggles

MECHANICAL (GENERAL) | GLOVES: Rubber

SPECIAL | RESPIRATOR: Self contained breathing apparatus

OTHER PROTECTIVE EQUIPMENT

SECTION IX SPECIAL PRECAUTIONS

PRECAUTIONS TO BE TAKEN IN HANDLING AND STORAGE

OTHER PRECAUTIONS

SIGNATURE | DATE

2,5-DIAMINOCHLOROBENZENE DIHYDROCHLORIDE

Health 4 · Flammability 1 · Reactivity 0

SECTION I SOURCE AND NOMENCLATURE

MANUFACTURER'S NAME	EMERGENCY TELEPHONE NO.
ADDRESS (NUMBER, STREET, CITY, STATE, ZIP CODE)	
TRADE NAME AND SYNONYMS 3 chloro-4 aminoaniline dihydrochloride	CHEMICAL FAMILY
CHEMICAL NAME AND SYNONYMS 2,5 diaminochloro-benzene dihydrochloride	FORMULA $C_6H_3Cl(NH_2)_2 . 2\ HCl$

SECTION II HAZARDOUS INGREDIENTS

BASIC MATERIAL	APPROXIMATE OR MAXIMUM % WT. OR VOL.	ESTABLISHED OSHA STANDARD	LD_{50} ORAL	LD_{50} PERCUT.	LC_{50} SPECIES	LC_{50} CONC.
			None	available		

SECTION III PHYSICAL DATA MW - 215.51

BOILING POINT °F.	VAPOR PRESSURE mm Hg.
MELTING POINT °F.	VAPOR DENSITY (Air = 1)
SPECIFIC GRAVITY (H_2O = 1)	EVAPORATION RATE (____ = 1)
SOLUBILITY IN WATER Pts/100 pts H_2O Soluble	VOLATILE % Vol. % Wt.

APPEARANCE AND ODOR Crystals

SECTION IV FIRE AND EXPLOSION HAZARD DATA

FLASH POINT None available	FLAMMABLE (EXPLOSIVE) LIMITS	UPPER
METHOD USED		LOWER

EXTINGUISHING MEDIA Dry chemical, carbon dioxide, alcohol foam.

SPECIAL FIRE FIGHTING PROCEDURES

UNUSUAL FIRE AND EXPLOSION HAZARDS

SECTION V HEALTH HAZARD DATA

TOXIC LEVEL None available — CARCINOGENIC No

PRINCIPLE ROUTES OF ABSORBTION Inhalation, ingestion, skin — SKIN AND EYE IRRITATION Yes

REVELANT SYMPTOMS OF EXPOSURE Headache, drowsiness, cyanosis, skin & eye irritant.

EFFECTS OF CHRONIC EXPOSURE Dermatitis, allergenic bronchial asthma.

EMERGENCY AND FIRST AID PROCEDURES Wash skin with soap & water, flush eyes with water; if swallowed wash mouth, get medical aid.

SECTION VI REACTIVITY DATA

CONDITIONS CONTRIBUTING TO INSTABILITY

CONDITIONS CONTRIBUTING TO HAZARDOUS POLYMERIZATION

INCOMPATABILITY (MATERIALS TO AVOID)

HAZARDOUS DECOMPOSITION PRODUCTS

SECTION VII SPILL OR LEAK PROCEDURES

STEPS TO BE TAKEN IN CASE MATERIAL IS RELEASED OR SPILLED

WASTE DISPOSAL METHOD

SECTION VIII SPECIAL PROTECTION INFORMATION

VENTILATION REQUIREMENTS LOCAL EXHAUST	PROTECTIVE EQUIPMENT (SPECIFY TYPES) EYE Safety goggles
MECHANICAL (GENERAL)	GLOVES Butyl Rubber
SPECIAL	RESPIRATOR Self-contained

OTHER PROTECTIVE EQUIPMENT

SECTION IX SPECIAL PRECAUTIONS

PRECAUTIONS TO BE TAKEN IN HANDLING AND STORAGE

OTHER PRECAUTIONS

SIGNATURE	DATE

p,p'-DIAMINODIPHENYLMETHANE

Health 4 · Flammability 1 · Reactivity 0

SECTION I SOURCE AND NOMENCLATURE

MANUFACTURER'S NAME	EMERGENCY TELEPHONE NO.
ADDRESS (NUMBER, STREET, CITY, STATE, ZIP CODE)	
TRADE NAME AND SYNONYMS P,P, Methylenedianiline	CHEMICAL FAMILY
CHEMICAL NAME AND SYNONYMS P,P-Diaminodiphenylmethane	FORMULA $(H_2NC_6H_4)_2CH_2$

SECTION II HAZARDOUS INGREDIENTS

BASIC MATERIAL	APPROXIMATE OR MAXIMUM % WT. OR VOL.	ESTABLISHED OSHA STANDARD	LD_{50} mg/kg ORAL	LD_{50} PERCUT.	LD_{50} SPECIES	LC_{50} CONC.
			347		Rat	

Note: Lowest published toxic concentration in air - inhalation - human 169 mg/kg.

Lowest published toxic dose Oral - man 8.42 mg/kg

SECTION III PHYSICAL DATA MW - 198.29

BOILING POINT °F. 399° C.	VAPOR PRESSURE 253 mm Hg. 15
MELTING POINT °F. 93° C.	VAPOR DENSITY (Air = 1)
SPECIFIC GRAVITY (H_2O = 1)	EVAPORATION RATE (____ = 1)
SOLUBILITY IN WATER Pts/100 pts H_2O Slightly soluble	VOLATILE % Vol. % Wt.

APPEARANCE AND ODOR Tan flakes, faint amine-like odor.

SECTION IV FIRE AND EXPLOSION HAZARD DATA

FLASH POINT 440° F.	FLAMMABLE (EXPLOSIVE) LIMITS	UPPER
METHOD USED		LOWER

EXTINGUISHING MEDIA Carbon dioxide, dry chemical, alcohol foam.

SPECIAL FIRE FIGHTING PROCEDURES

UNUSUAL FIRE AND EXPLOSION HAZARDS

SECTION V HEALTH HAZARD DATA

TOXIC LEVEL None available — CARCINOGENIC Yes

PRINCIPLE ROUTES OF ABSORBTION Inhalation, ingestion — SKIN AND EYE IRRITATION Severe

REVELANT SYMPTOMS OF EXPOSURE Mucous membrane and skin irritation, cyanosis and jaundice, digestive tract irritation.

EFFECTS OF CHRONIC EXPOSURE Liver damage, malignant tumors.

EMERGENCY AND FIRST AID PROCEDURES If swallowed - give emetic, get medical aid; inhalation - rest; wash skin and eyes.

SECTION VI REACTIVITY DATA

CONDITIONS CONTRIBUTING TO INSTABILITY

CONDITIONS CONTRIBUTING TO HAZARDOUS POLYMERIZATION

INCOMPATABILITY (MATERIALS TO AVOID)

HAZARDOUS DECOMPOSITION PRODUCTS When heated to decomposition emits toxic aniline fumes.

SECTION VII SPILL OR LEAK PROCEDURES

STEPS TO BE TAKEN IN CASE MATERIAL IS RELEASED OR SPILLED

WASTE DISPOSAL METHOD

SECTION VIII SPECIAL PROTECTION INFORMATION

VENTILATION REQUIREMENTS LOCAL EXHAUST	PROTECTIVE EQUIPMENT (SPECIFY TYPES) EYE Goggles
MECHANICAL (GENERAL)	GLOVES Rubber
SPECIAL	RESPIRATOR With proper filter

OTHER PROTECTIVE EQUIPMENT

SECTION IX SPECIAL PRECAUTIONS

PRECAUTIONS TO BE TAKEN IN HANDLING AND STORAGE

OTHER PRECAUTIONS

SIGNATURE	DATE

DIAMINO PLATINUM NITRITE

Hazard rating: 1 (top), 4 (left), 0 (right)

SECTION I SOURCE AND NOMENCLATURE

MANUFACTURER'S NAME	EMERGENCY TELEPHONE NO.
ADDRESS (NUMBER, STREET, CITY, STATE, ZIP CODE)	
TRADE NAME AND SYNONYMS	CHEMICAL FAMILY
CHEMICAL NAME AND SYNONYMS: Diamino Platinum Nitrite	FORMULA: $(NH_4)_2\ Pt\ (NO_2)_6$

SECTION II HAZARDOUS INGREDIENTS

BASIC MATERIAL	APPROXIMATE OR MAXIMUM % WT. OR VOL.	ESTABLISHED OSHA STANDARD	LD_{50} ORAL	LD_{50} PERCUT.	LC_{50} SPECIES	LC_{50} CONC.
None available						

SECTION III PHYSICAL DATA

BOILING POINT °F.	VAPOR PRESSURE mm Hg.
MELTING POINT °F.	VAPOR DENSITY (Air = 1)
SPECIFIC GRAVITY (H_2O = 1)	EVAPORATION RATE (______ = 1)
SOLUBILITY IN WATER Pts/100 pts H_2O: Yes	VOLATILE % Vol. % Wt.
APPEARANCE AND ODOR	

SECTION IV FIRE AND EXPLOSION HAZARD DATA

FLASH POINT: None available

METHOD USED

FLAMMABLE (EXPLOSIVE) LIMITS: UPPER / LOWER

EXTINGUISHING MEDIA

SPECIAL FIRE FIGHTING PROCEDURES

UNUSUAL FIRE AND EXPLOSION HAZARDS

SECTION V HEALTH HAZARD DATA

TOXIC LEVEL: TLV - 0.002 mg/c.m. CARCINOGENIC: No

PRINCIPLE ROUTES OF ABSORBTION: Inhalation, ingestion SKIN AND EYE IRRITATION: Severe

REVELANT SYMPTOMS OF EXPOSURE: Inflammation of mucous membranes in nose, lacramation and burning in the eyes.

EFFECTS OF CHRONIC EXPOSURE: Running of the nose, sneezing, tightness of chest, cyanosis, wheezing, cough, dermatitis.

EMERGENCY AND FIRST AID PROCEDURES: If swallowed - drink water, get medical aid; inhalation - rest, treat asthmatic symptoms.

SECTION VI REACTIVITY DATA

CONDITIONS CONTRIBUTING TO INSTABILITY

CONDITIONS CONTRIBUTING TO HAZARDOUS POLYMERIZATION

INCOMPATABILITY (MATERIALS TO AVOID)

HAZARDOUS DECOMPOSITION PRODUCTS

SECTION VII SPILL OR LEAK PROCEDURES

STEPS TO BE TAKEN IN CASE MATERIAL IS RELEASED OR SPILLED

WASTE DISPOSAL METHOD

SECTION VIII SPECIAL PROTECTION INFORMATION

VENTILATION REQUIREMENTS LOCAL EXHAUST	PROTECTIVE EQUIPMENT (SPECIFY TYPES) EYE: Goggles
MECHANICAL (GENERAL)	GLOVES
SPECIAL	RESPIRATOR
OTHER PROTECTIVE EQUIPMENT	

SECTION IX SPECIAL PRECAUTIONS

PRECAUTIONS TO BE TAKEN IN HANDLING AND STORAGE

OTHER PRECAUTIONS

SIGNATURE	DATE

2,4-DIAMINOTOLUENE

Hazard rating: 1 (top), 4 (left), 0 (right)

SECTION I SOURCE AND NOMENCLATURE

MANUFACTURER'S NAME	EMERGENCY TELEPHONE NO.
ADDRESS (NUMBER, STREET, CITY, STATE, ZIP CODE)	
TRADE NAME AND SYNONYMS: m-Toluenediamine	CHEMICAL FAMILY
CHEMICAL NAME AND SYNONYMS: 2,4-Diaminotoluene	FORMULA: $CH_3C_6H_3(NH_2)_2$

SECTION II HAZARDOUS INGREDIENTS

BASIC MATERIAL	APPROXIMATE OR MAXIMUM % WT. OR VOL.	ESTABLISHED OSHA STANDARD	LD_{50} ORAL	LD_{50} PERCUT.	LC_{50} SPECIES	LC_{50} CONC.
LDLO - oral - human 50 mg/kg						
Lowest published lethal dose						
Oral - rat 500 mg/kg						

SECTION III PHYSICAL DATA MW - 122.19

BOILING POINT °F.: 292°C.	VAPOR PRESSURE mm Hg.: at 107°C. 1
MELTING POINT °F.: 99°C.	VAPOR DENSITY (Air = 1)
SPECIFIC GRAVITY (H_2O = 1)	EVAPORATION RATE (______ = 1)
SOLUBILITY IN WATER Pts/100 pts H_2O: Very soluble	VOLATILE % Vol. % Wt.
APPEARANCE AND ODOR: Leaflets	

SECTION IV FIRE AND EXPLOSION HAZARD DATA

FLASH POINT: None available

METHOD USED

FLAMMABLE (EXPLOSIVE) LIMITS: UPPER / LOWER

EXTINGUISHING MEDIA: Foam, carbon dioxide, dry chemical

SPECIAL FIRE FIGHTING PROCEDURES

UNUSUAL FIRE AND EXPLOSION HAZARDS

SECTION V HEALTH HAZARD DATA

TOXIC LEVEL: None available CARCINOGENIC: Yes

PRINCIPLE ROUTES OF ABSORBTION: Ingestion, inhalation, skin SKIN AND EYE IRRITATION: Yes

REVELANT SYMPTOMS OF EXPOSURE: Fatty degeneration of liver, diarrhea, vomiting cyanosis, methemoglobinemia.

EFFECTS OF CHRONIC EXPOSURE: Sensitizes, methemoglobinemia, cancer.

EMERGENCY AND FIRST AID PROCEDURES: Wash skin & eyes; if swallowed - induce vomiting, get medical aid.

SECTION VI REACTIVITY DATA

CONDITIONS CONTRIBUTING TO INSTABILITY

CONDITIONS CONTRIBUTING TO HAZARDOUS POLYMERIZATION

INCOMPATABILITY (MATERIALS TO AVOID)

HAZARDOUS DECOMPOSITION PRODUCTS

SECTION VII SPILL OR LEAK PROCEDURES

STEPS TO BE TAKEN IN CASE MATERIAL IS RELEASED OR SPILLED

WASTE DISPOSAL METHOD

SECTION VIII SPECIAL PROTECTION INFORMATION

VENTILATION REQUIREMENTS LOCAL EXHAUST	PROTECTIVE EQUIPMENT (SPECIFY TYPES) EYE: Goggles
MECHANICAL (GENERAL)	GLOVES: Rubber
SPECIAL	RESPIRATOR: Self-contained
OTHER PROTECTIVE EQUIPMENT	

SECTION IX SPECIAL PRECAUTIONS

PRECAUTIONS TO BE TAKEN IN HANDLING AND STORAGE: Protect against physical damage to container; store in cool, dry area.

OTHER PRECAUTIONS

SIGNATURE	DATE

2,4-DIAMINOTOLUENE MONOHYDROCHLORIDE

Hazard diamond: 1 (top), 4 (left), 0 (right)

SECTION I SOURCE AND NOMENCLATURE

MANUFACTURER'S NAME | EMERGENCY TELEPHONE NO.

ADDRESS (NUMBER, STREET, CITY, STATE, ZIP CODE)

TRADE NAME AND SYNONYMS | CHEMICAL FAMILY

CHEMICAL NAME AND SYNONYMS 2,4-Diaminotoluene Monohydrochloride | FORMULA $CH_3C_6H_3(NH_2)_2HCl$

SECTION II HAZARDOUS INGREDIENTS

BASIC MATERIAL	APPROXIMATE OR MAXIMUM % WT. OR VOL.	ESTABLISHED OSHA STANDARD	LD_{50} ORAL	LD_{50} PERCUT.	LC_{50} SPECIES	LC_{50} CONC.

SECTION III PHYSICAL DATA MW - 158.65

BOILING POINT °F. | VAPOR PRESSURE mm Hg.

MELTING POINT °F. | VAPOR DENSITY (Air = 1)

SPECIFIC GRAVITY (H_2O = 1) | EVAPORATION RATE (= 1)

SOLUBILITY IN WATER Pts/100 pts H_2O | VOLATILE % Vol. % Wt.

APPEARANCE AND ODOR

SECTION IV FIRE AND EXPLOSION HAZARD DATA

FLASH POINT None available | FLAMMABLE (EXPLOSIVE) LIMITS UPPER / LOWER

METHOD USED

EXTINGUISHING MEDIA Foam, carbon dioxide, dry chemical

SPECIAL FIRE FIGHTING PROCEDURES

UNUSUAL FIRE AND EXPLOSION HAZARDS

SECTION V HEALTH HAZARD DATA

TOXIC LEVEL None available | CARCINOGENIC Yes

PRINCIPLE ROUTES OF ABSORBTION Inhalation, ingestion, skin | SKIN AND EYE IRRITATION Moderate

REVELANT SYMPTOMS OF EXPOSURE Fatty degeneration of liver, diarrhea, vomiting cyanosis, methemoglobinemia.

EFFECTS OF CHRONIC EXPOSURE Sensitizes, methemoglobinemia, cancer.

EMERGENCY AND FIRST AID PROCEDURES Wash skin & eyes; if swallowed - induce vomiting, get medical aid

SECTION VI REACTIVITY DATA

CONDITIONS CONTRIBUTING TO INSTABILITY

CONDITIONS CONTRIBUTING TO HAZARDOUS POLYMERIZATION

INCOMPATABILITY (MATERIALS TO AVOID)

HAZARDOUS DECOMPOSITION PRODUCTS

SECTION VII SPILL OR LEAK PROCEDURES

STEPS TO BE TAKEN IN CASE MATERIAL IS RELEASED OR SPILLED

WASTE DISPOSAL METHOD

SECTION VIII SPECIAL PROTECTION INFORMATION

VENTILATION REQUIREMENTS LOCAL EXHAUST | PROTECTIVE EQUIPMENT (SPECIFY TYPES) EYE Goggles

MECHANICAL (GENERAL) | GLOVES Rubber

SPECIAL | RESPIRATOR Self-contained

OTHER PROTECTIVE EQUIPMENT

SECTION IX SPECIAL PRECAUTIONS

PRECAUTIONS TO BE TAKEN IN HANDLING AND STORAGE

OTHER PRECAUTIONS

SIGNATURE | DATE

DI-n-AMYLAMINE

Hazard diamond: 2 (top), 3 (left), 0 (right)

SECTION I SOURCE AND NOMENCLATURE

MANUFACTURER'S NAME | EMERGENCY TELEPHONE NO.

ADDRESS (NUMBER, STREET, CITY, STATE, ZIP CODE)

TRADE NAME AND SYNONYMS | CHEMICAL FAMILY

CHEMICAL NAME AND SYNONYMS Dipentylamine | FORMULA $(C_5H_{11})_2NH$

SECTION II HAZARDOUS INGREDIENTS

BASIC MATERIAL	APPROXIMATE OR MAXIMUM % WT. OR VOL.	ESTABLISHED OSHA STANDARD	LD_{50} mg/kg ORAL	LD_{50} mg/kg PERCUT.	LD_{50} mg/kg SPECIES	LC_{50} CONC.
			270		Rat	

Note: Lowest published lethal concentration in air inhalation - rat 63 PPM / 4 hours

Lowest published lethal dose 35 mg/kg skin - rabbit

SECTION III PHYSICAL DATA MW - 157.26

BOILING POINT °F. 202°C | VAPOR PRESSURE mm Hg.

MELTING POINT °F. | VAPOR DENSITY (Air = 1) 5.42

SPECIFIC GRAVITY (H_2O = 1) 0.777 | EVAPORATION RATE (= 1)

SOLUBILITY IN WATER Pts/100 pts H_2O Very slightly soluble | VOLATILE % Vol. % Wt.

APPEARANCE AND ODOR Water white liquid, ammonical odor

SECTION IV FIRE AND EXPLOSION HAZARD DATA

FLASH POINT 124°F | FLAMMABLE (EXPLOSIVE) LIMITS UPPER / LOWER –

METHOD USED Closed Cup

EXTINGUISHING MEDIA Alcohol foam, carbon dioxide, dry chemical

SPECIAL FIRE FIGHTING PROCEDURES

UNUSUAL FIRE AND EXPLOSION HAZARDS

SECTION V HEALTH HAZARD DATA

TOXIC LEVEL Not available | CARCINOGENIC No

PRINCIPLE ROUTES OF ABSORBTION Inhalation, ingestion | SKIN AND EYE IRRITATION Severe

REVELANT SYMPTOMS OF EXPOSURE Irritation of eyes and respiratory system, severe irritation of digestive tract.

EFFECTS OF CHRONIC EXPOSURE Dermatitis

EMERGENCY AND FIRST AID PROCEDURES Inhalation - rest; if swallowed - drink water, get medical aid; wash eyes & skin.

SECTION VI REACTIVITY DATA

CONDITIONS CONTRIBUTING TO INSTABILITY

CONDITIONS CONTRIBUTING TO HAZARDOUS POLYMERIZATION

INCOMPATABILITY (MATERIALS TO AVOID)

HAZARDOUS DECOMPOSITION PRODUCTS

SECTION VII SPILL OR LEAK PROCEDURES

STEPS TO BE TAKEN IN CASE MATERIAL IS RELEASED OR SPILLED Put sodium bisulfate on spills, wash with water.

WASTE DISPOSAL METHOD

SECTION VIII SPECIAL PROTECTION INFORMATION

VENTILATION REQUIREMENTS LOCAL EXHAUST | PROTECTIVE EQUIPMENT (SPECIFY TYPES) EYE Goggles

MECHANICAL (GENERAL) | GLOVES Rubber

SPECIAL | RESPIRATOR With proper filter

OTHER PROTECTIVE EQUIPMENT Wear full protective clothing

SECTION IX SPECIAL PRECAUTIONS

PRECAUTIONS TO BE TAKEN IN HANDLING AND STORAGE Protect against physical damage; store in cool, well-ventilated place.

OTHER PRECAUTIONS

SIGNATURE | DATE

o-DIANISIDINE

1
4 0

SECTION I SOURCE AND NOMENCLATURE

MANUFACTURER'S NAME	EMERGENCY TELEPHONE NO.
ADDRESS (NUMBER, STREET, CITY, STATE, ZIP CODE)	
TRADE NAME AND SYNONYMS	CHEMICAL FAMILY
CHEMICAL NAME AND SYNONYMS: 3,3, Dimethoxybenzidine	FORMULA: $(NH_2C_6H_3OCH_3)_2$

SECTION II HAZARDOUS INGREDIENTS

BASIC MATERIAL	APPROXIMATE OR MAXIMUM % WT. OR VOL.	ESTABLISHED OSHA STANDARD	LD_{50} mg/kg ORAL	PERCUT.	SPECIES	LC_{50} CONC.
Lowest published toxic concentration in air inhalation - human 1.9 mg/kg.			1920	—	Rat	

SECTION III PHYSICAL DATA MW - 244.28

BOILING POINT °F.: -	VAPOR PRESSURE mm Hg.
MELTING POINT °F.: 138°C	VAPOR DENSITY (Air = 1)
SPECIFIC GRAVITY (H_2O = 1)	EVAPORATION RATE (______ = 1)
SOLUBILITY IN WATER Pts/100 pts H_2O: Insoluble	VOLATILE % Vol. % Wt.
APPEARANCE AND ODOR: Colorless crystals	

SECTION IV FIRE AND EXPLOSION HAZARD DATA

FLASH POINT: not available	FLAMMABLE (EXPLOSIVE) LIMITS	UPPER
METHOD USED		LOWER

EXTINGUISHING MEDIA: Alcohol foam, carbon dioxide, dry chemical, water spray

SPECIAL FIRE FIGHTING PROCEDURES

UNUSUAL FIRE AND EXPLOSION HAZARDS: Emits toxic fumes when heated to decomposition point.

SECTION V HEALTH HAZARD DATA

TOXIC LEVEL: Not available — CARCINOGENIC: Neoplastic

PRINCIPLE ROUTES OF ABSORBTION: Inhalation, skin, ingestion — SKIN AND EYE IRRITATION: yes

REVELANT SYMPTOMS OF EXPOSURE: Headache, drowsiness, cyanosis

EFFECTS OF CHRONIC EXPOSURE: -

EMERGENCY AND FIRST AID PROCEDURES: Wash skin with soap and water; if swallowed - give emetic, get medical aid, inhalation - rest.

SECTION VI REACTIVITY DATA

CONDITIONS CONTRIBUTING TO INSTABILITY

CONDITIONS CONTRIBUTING TO HAZARDOUS POLYMERIZATION

INCOMPATABILITY (MATERIALS TO AVOID)

HAZARDOUS DECOMPOSITION PRODUCTS

SECTION VII SPILL OR LEAK PROCEDURES

STEPS TO BE TAKEN IN CASE MATERIAL IS RELEASED OR SPILLED

WASTE DISPOSAL METHOD

SECTION VIII SPECIAL PROTECTION INFORMATION

VENTILATION REQUIREMENTS LOCAL EXHAUST	PROTECTIVE EQUIPMENT (SPECIFY TYPES) EYE: Goggles
MECHANICAL (GENERAL)	GLOVES
SPECIAL	RESPIRATOR
OTHER PROTECTIVE EQUIPMENT	

SECTION IX SPECIAL PRECAUTIONS

PRECAUTIONS TO BE TAKEN IN HANDLING AND STORAGE

OTHER PRECAUTIONS

SIGNATURE	DATE

DIAZOAMINOBENZENE

2
4 3

SECTION I SOURCE AND NOMENCLATURE

MANUFACTURER'S NAME	EMERGENCY TELEPHONE NO.
ADDRESS (NUMBER, STREET, CITY, STATE, ZIP CODE)	
TRADE NAME AND SYNONYMS	CHEMICAL FAMILY
CHEMICAL NAME AND SYNONYMS: 1,3 Diphenyl Triazene	FORMULA: $C_6H_5NHNNC_6H_5$

SECTION II HAZARDOUS INGREDIENTS

BASIC MATERIAL	APPROXIMATE OR MAXIMUM % WT. OR VOL.	ESTABLISHED OSHA STANDARD	LD_{50} Mg/kg ORAL	PERCUT.	SPECIES	LC_{50} CONC.
			224		Rat	

SECTION III PHYSICAL DATA MW - 197.26

BOILING POINT °F.: explodes	VAPOR PRESSURE mm Hg.
MELTING POINT °F.: 98°C.	VAPOR DENSITY (Air = 1): 6.8
SPECIFIC GRAVITY (H_2O = 1)	EVAPORATION RATE (______ = 1)
SOLUBILITY IN WATER Pts/100 pts H_2O: Insoluble	VOLATILE % Vol. % Wt.
APPEARANCE AND ODOR: Small, golden-yellow crystals.	

SECTION IV FIRE AND EXPLOSION HAZARD DATA

FLASH POINT: None available	FLAMMABLE (EXPLOSIVE) LIMITS	UPPER
METHOD USED		LOWER

EXTINGUISHING MEDIA: Explosive do not fight fire except by remote control.

SPECIAL FIRE FIGHTING PROCEDURES

UNUSUAL FIRE AND EXPLOSION HAZARDS: Explodes when heated to 150°C.

SECTION V HEALTH HAZARD DATA

TOXIC LEVEL: None available — CARCINOGENIC: Yes

PRINCIPLE ROUTES OF ABSORBTION: Ingestion, inhalation of dust, skin — SKIN AND EYE IRRITATION: Moderate

REVELANT SYMPTOMS OF EXPOSURE: Irritation of eyes and skin.

EFFECTS OF CHRONIC EXPOSURE: Liver and kidney damage, tumors.

EMERGENCY AND FIRST AID PROCEDURES: Inhalation - rest; if swallowed - give emetic, get medical aid; wash skin & eyes.

SECTION VI REACTIVITY DATA

CONDITIONS CONTRIBUTING TO INSTABILITY

CONDITIONS CONTRIBUTING TO HAZARDOUS POLYMERIZATION

INCOMPATABILITY (MATERIALS TO AVOID)

HAZARDOUS DECOMPOSITION PRODUCTS

SECTION VII SPILL OR LEAK PROCEDURES

STEPS TO BE TAKEN IN CASE MATERIAL IS RELEASED OR SPILLED

WASTE DISPOSAL METHOD

SECTION VIII SPECIAL PROTECTION INFORMATION

VENTILATION REQUIREMENTS LOCAL EXHAUST	PROTECTIVE EQUIPMENT (SPECIFY TYPES) EYE: Goggles
MECHANICAL (GENERAL)	GLOVES: Rubber
SPECIAL	RESPIRATOR: Dust mask
OTHER PROTECTIVE EQUIPMENT	

SECTION IX SPECIAL PRECAUTIONS

PRECAUTIONS TO BE TAKEN IN HANDLING AND STORAGE

OTHER PRECAUTIONS

SIGNATURE	DATE

1,2,5,6-DIBENZANTHRACENE

(Hazard diamond: 1 / 4 / 0)

SECTION I SOURCE AND NOMENCLATURE

MANUFACTURER'S NAME | EMERGENCY TELEPHONE NO.

ADDRESS (NUMBER, STREET, CITY, STATE, ZIP CODE)

TRADE NAME AND SYNONYMS | CHEMICAL FAMILY

CHEMICAL NAME AND SYNONYMS: Dibenz(a,h)anthracene | FORMULA: $C_{22}H_{14}$

SECTION II HAZARDOUS INGREDIENTS

BASIC MATERIAL	APPROXIMATE OR MAXIMUM % WT. OR VOL.	ESTABLISHED OSHA STANDARD	LD_{50} ORAL	LD_{50} PERCUT.	LC_{50} SPECIES	LC_{50} CONC.
Lowest published toxic dose subcutaneous - rat 0.5 mg/kg.						
Lowest published toxic dose oral - mouse - 0.36 mg/kg						
Lowest published lethal dose skin - mouse - 0.6 micrograms per kg.						

SECTION III PHYSICAL DATA MW - 278.33

BOILING POINT (°F.): sublimes | VAPOR PRESSURE (mm Hg.)

MELTING POINT (°F.): 262°C | VAPOR DENSITY (Air = 1)

SPECIFIC GRAVITY (H_2O = 1) | EVAPORATION RATE (= 1)

SOLUBILITY IN WATER Pts/100 pts H_2O: Slightly soluble | VOLATILE % Vol. % Wt.

APPEARANCE AND ODOR: Crystal

SECTION IV FIRE AND EXPLOSION HAZARD DATA

FLASH POINT: not available | FLAMMABLE (EXPLOSIVE) LIMITS: UPPER / LOWER

METHOD USED

EXTINGUISHING MEDIA: Alcohol foam, carbon dioxide, dry chemical

SPECIAL FIRE FIGHTING PROCEDURES

UNUSUAL FIRE AND EXPLOSION HAZARDS

SECTION V HEALTH HAZARD DATA

TOXIC LEVEL: None available | CARCINOGENIC: yes

PRINCIPLE ROUTES OF ABSORPTION: Ingestion, skin | SKIN AND EYE IRRITATION: moderate

REVELANT SYMPTOMS OF EXPOSURE: Gastroenteric distress, tremors, convulsions, fever, dermatitis

EFFECTS OF CHRONIC EXPOSURE: Tumors, cancer, dermatitis

EMERGENCY AND FIRST AID PROCEDURES: If swallowed - induce vomiting, get medical aid; wash skin & eyes.

SECTION VI REACTIVITY DATA

CONDITIONS CONTRIBUTING TO INSTABILITY

CONDITIONS CONTRIBUTING TO HAZARDOUS POLYMERIZATION

INCOMPATABILITY (MATERIALS TO AVOID)

HAZARDOUS DECOMPOSITION PRODUCTS

SECTION VII SPILL OR LEAK PROCEDURES

STEPS TO BE TAKEN IN CASE MATERIAL IS RELEASED OR SPILLED

WASTE DISPOSAL METHOD

SECTION VIII SPECIAL PROTECTION INFORMATION

VENTILATION REQUIREMENTS LOCAL EXHAUST | PROTECTIVE EQUIPMENT (SPECIFY TYPES) EYE: Goggles

MECHANICAL (GENERAL) | GLOVES

SPECIAL | RESPIRATOR

OTHER PROTECTIVE EQUIPMENT

SECTION IX SPECIAL PRECAUTIONS

PRECAUTIONS TO BE TAKEN IN HANDLING AND STORAGE

OTHER PRECAUTIONS

SIGNATURE | DATE

DIBENZOTHIOPHENE

(Hazard diamond: 1 / 4 / 0)

SECTION I SOURCE AND NOMENCLATURE

MANUFACTURER'S NAME | EMERGENCY TELEPHONE NO.

ADDRESS (NUMBER, STREET, CITY, STATE, ZIP CODE)

TRADE NAME AND SYNONYMS | CHEMICAL FAMILY

CHEMICAL NAME AND SYNONYMS: Dibenzothiophene | FORMULA: $C_6H_4SC_6H_4$

SECTION II HAZARDOUS INGREDIENTS

BASIC MATERIAL	APPROXIMATE OR MAXIMUM % WT. OR VOL.	ESTABLISHED OSHA STANDARD	LD_{50} ORAL	LD_{50} PERCUT.	LC_{50} SPECIES	LC_{50} CONC.
			None	available		

SECTION III PHYSICAL DATA MW - 184.26

BOILING POINT (°F.): 333° C. | VAPOR PRESSURE (mm Hg.)

MELTING POINT (°F.): 100° C. | VAPOR DENSITY (Air = 1)

SPECIFIC GRAVITY (H_2O = 1) | EVAPORATION RATE (= 1)

SOLUBILITY IN WATER Pts/100 pts H_2O | VOLATILE % Vol. % Wt.

APPEARANCE AND ODOR

SECTION IV FIRE AND EXPLOSION HAZARD DATA

FLASH POINT: None available | FLAMMABLE (EXPLOSIVE) LIMITS: UPPER / LOWER

METHOD USED

EXTINGUISHING MEDIA: Carbon dioxide, dry chemical.

SPECIAL FIRE FIGHTING PROCEDURES

UNUSUAL FIRE AND EXPLOSION HAZARDS

SECTION V HEALTH HAZARD DATA

TOXIC LEVEL: None available | CARCINOGENIC: Yes

PRINCIPLE ROUTES OF ABSORPTION: Inhalation, ingestion, skin | SKIN AND EYE IRRITATION: Severe

REVELANT SYMPTOMS OF EXPOSURE: Corrosive action on mouth and digestive system, bloody saliva flow, dizziness, rapid heart action, shortness of breath.

EFFECTS OF CHRONIC EXPOSURE: --

EMERGENCY AND FIRST AID PROCEDURES: Inhalation - oxygen as needed; if swallowed - induce vomiting, get medical aid; wash eyes and skin.

SECTION VI REACTIVITY DATA

CONDITIONS CONTRIBUTING TO INSTABILITY

CONDITIONS CONTRIBUTING TO HAZARDOUS POLYMERIZATION

INCOMPATABILITY (MATERIALS TO AVOID)

HAZARDOUS DECOMPOSITION PRODUCTS

SECTION VII SPILL OR LEAK PROCEDURES

STEPS TO BE TAKEN IN CASE MATERIAL IS RELEASED OR SPILLED

WASTE DISPOSAL METHOD

SECTION VIII SPECIAL PROTECTION INFORMATION

VENTILATION REQUIREMENTS LOCAL EXHAUST | PROTECTIVE EQUIPMENT (SPECIFY TYPES) EYE: Goggles

MECHANICAL (GENERAL) | GLOVES: Rubber

SPECIAL | RESPIRATOR

OTHER PROTECTIVE EQUIPMENT

SECTION IX SPECIAL PRECAUTIONS

PRECAUTIONS TO BE TAKEN IN HANDLING AND STORAGE

OTHER PRECAUTIONS

SIGNATURE | DATE

p-DIBROMOBENZENE

NFPA diamond: 1 (top), 3 (left), 0 (right)

SECTION I SOURCE AND NOMENCLATURE

MANUFACTURER'S NAME

EMERGENCY TELEPHONE NO.

ADDRESS (NUMBER, STREET, CITY, STATE, ZIP CODE)

TRADE NAME AND SYNONYMS

CHEMICAL FAMILY

CHEMICAL NAME AND SYNONYMS: 4 dibromobenzene

FORMULA: $Br_2C_6H_4$

SECTION II HAZARDOUS INGREDIENTS

BASIC MATERIAL	APPROXIMATE OR MAXIMUM % WT. OR VOL.	ESTABLISHED OSHA STANDARD	LD_{50} ORAL	LD_{50} PERCUT.	LC_{50} SPECIES	LC_{50} CONC.
			None available			

SECTION III PHYSICAL DATA MW - 235.92

BOILING POINT °F: 219°C

VAPOR PRESSURE mm Hg: at 61°C

MELTING POINT °F: 87°C

VAPOR DENSITY (Air = 1): 8.1

SPECIFIC GRAVITY (H_2O = 1): 2.26

EVAPORATION RATE (= 1)

SOLUBILITY IN WATER Pts/100 pts H_2O: Insoluble

VOLATILE % Vol.

APPEARANCE AND ODOR: Clear crystal, odor of xylene

SECTION IV FIRE AND EXPLOSION HAZARD DATA

FLASH POINT: N/A

METHOD USED

FLAMMABLE (EXPLOSIVE) LIMITS: UPPER / LOWER

EXTINGUISHING MEDIA: Use water spray, foam, carbon dioxide or dry chemical

SPECIAL FIRE FIGHTING PROCEDURES: Wear self contained breathing apparatus.

UNUSUAL FIRE AND EXPLOSION HAZARDS

SECTION V HEALTH HAZARD DATA

TOXIC LEVEL: None available

CARCINOGENIC: no

PRINCIPLE ROUTES OF ABSORBTION: Inhalation, ingestion, skin

SKIN AND EYE IRRITATION: Severe

REVELANT SYMPTOMS OF EXPOSURE: Cyanosis, deep & rapid respiration, irregular pulse, drowsiness, unconsciousness

EFFECTS OF CHRONIC EXPOSURE: Lung, liver & kidney damage

EMERGENCY AND FIRST AID PROCEDURES: Inhalation - rest; if swallowed - give emetic, get medical aid; wash from skin & eyes.

SECTION VI REACTIVITY DATA

CONDITIONS CONTRIBUTING TO INSTABILITY

CONDITIONS CONTRIBUTING TO HAZARDOUS POLYMERIZATION

INCOMPATABILITY (MATERIALS TO AVOID)

HAZARDOUS DECOMPOSITION PRODUCTS

SECTION VII SPILL OR LEAK PROCEDURES

STEPS TO BE TAKEN IN CASE MATERIAL IS RELEASED OR SPILLED

WASTE DISPOSAL METHOD

SECTION VIII SPECIAL PROTECTION INFORMATION

VENTILATION REQUIREMENTS LOCAL EXHAUST

PROTECTIVE EQUIPMENT (SPECIFY TYPES) EYE: Goggles

MECHANICAL (GENERAL)

GLOVES: Polyvinyl alcohol

SPECIAL

RESPIRATOR: Self contained

OTHER PROTECTIVE EQUIPMENT

SECTION IX SPECIAL PRECAUTIONS

PRECAUTIONS TO BE TAKEN IN HANDLING AND STORAGE

OTHER PRECAUTIONS

SIGNATURE

DATE

4,4′-DIBROMOBIPHENYL

NFPA diamond: 1 (top), 3 (left), 0 (right)

SECTION I SOURCE AND NOMENCLATURE

MANUFACTURER'S NAME

EMERGENCY TELEPHONE NO.

ADDRESS (NUMBER, STREET, CITY, STATE, ZIP CODE)

TRADE NAME AND SYNONYMS

CHEMICAL FAMILY

CHEMICAL NAME AND SYNONYMS: P,P-Dibromobiphenyl

FORMULA: $(Br\ C_6H_4)_2$

SECTION II HAZARDOUS INGREDIENTS

BASIC MATERIAL	APPROXIMATE OR MAXIMUM % WT. OR VOL.	ESTABLISHED OSHA STANDARD	LD_{50} ORAL	LD_{50} PERCUT.	LC_{50} SPECIES	LC_{50} CONC.
			None available			

SECTION III PHYSICAL DATA MW - 312.02

BOILING POINT °F: 357°C

VAPOR PRESSURE mm Hg

MELTING POINT °F: 166°C

VAPOR DENSITY (Air = 1)

SPECIFIC GRAVITY (H_2O = 1): 1.90

EVAPORATION RATE (= 1)

SOLUBILITY IN WATER Pts/100 pts H_2O: Insoluble

VOLATILE % Vol. % Wt.

APPEARANCE AND ODOR: Crystal

SECTION IV FIRE AND EXPLOSION HAZARD DATA

FLASH POINT: Not available

METHOD USED

FLAMMABLE (EXPLOSIVE) LIMITS: UPPER / LOWER

EXTINGUISHING MEDIA: Water, foam, carbon dioxide or dry chemical

SPECIAL FIRE FIGHTING PROCEDURES

UNUSUAL FIRE AND EXPLOSION HAZARDS

SECTION V HEALTH HAZARD DATA

TOXIC LEVEL: None available

CARCINOGENIC: No

PRINCIPLE ROUTES OF ABSORBTION: Inhalation, ingestion, skin

SKIN AND EYE IRRITATION: moderate

REVELANT SYMPTOMS OF EXPOSURE: Acne, irritation of respiratory passages

EFFECTS OF CHRONIC EXPOSURE: Liver damage

EMERGENCY AND FIRST AID PROCEDURES: Inhalation - rest, get medical aid; if swallowed - wash mouth, give emetic, get medical aid, wash skin & eyes.

SECTION VI REACTIVITY DATA

CONDITIONS CONTRIBUTING TO INSTABILITY

CONDITIONS CONTRIBUTING TO HAZARDOUS POLYMERIZATION

INCOMPATABILITY (MATERIALS TO AVOID)

HAZARDOUS DECOMPOSITION PRODUCTS

SECTION VII SPILL OR LEAK PROCEDURES

STEPS TO BE TAKEN IN CASE MATERIAL IS RELEASED OR SPILLED

WASTE DISPOSAL METHOD

SECTION VIII SPECIAL PROTECTION INFORMATION

VENTILATION REQUIREMENTS LOCAL EXHAUST

PROTECTIVE EQUIPMENT (SPECIFY TYPES) EYE: Goggles

MECHANICAL (GENERAL)

GLOVES

SPECIAL

RESPIRATOR: With proper filter

OTHER PROTECTIVE EQUIPMENT

SECTION IX SPECIAL PRECAUTIONS

PRECAUTIONS TO BE TAKEN IN HANDLING AND STORAGE

OTHER PRECAUTIONS

SIGNATURE

DATE

4,6-DIBROMO-o-CRESOL

NFPA diamond: Health 3, Flammability 1, Reactivity 0

SECTION I SOURCE AND NOMENCLATURE

MANUFACTURER'S NAME | EMERGENCY TELEPHONE NO.

ADDRESS (NUMBER, STREET, CITY, STATE, ZIP CODE)

TRADE NAME AND SYNONYMS | CHEMICAL FAMILY

CHEMICAL NAME AND SYNONYMS: 3,5 dibromo-2 hydroxytoluene | FORMULA: $Br_2CH_3C_6H_2OH$

SECTION II HAZARDOUS INGREDIENTS

BASIC MATERIAL	APPROXIMATE OR MAXIMUM % WT. OR VOL.	ESTABLISHED OSHA STANDARD	LD50 ORAL	LD50 PERCUT.	LC50 SPECIES	LC50 CONC.
			None available			

SECTION III PHYSICAL DATA MW - 265.95

BOILING POINT (°F): decomposes at 265°C | VAPOR PRESSURE (mm Hg)

MELTING POINT (°F): 56° | VAPOR DENSITY (Air = 1)

SPECIFIC GRAVITY (H2O = 1) | EVAPORATION RATE (= 1)

SOLUBILITY IN WATER (Pts 100 pts H2O): very slightly soluble | VOLATILE % Vol. % Wt.

APPEARANCE AND ODOR: crystal with phenolic odor

SECTION IV FIRE AND EXPLOSION HAZARD DATA

FLASH POINT: none available | FLAMMABLE (EXPLOSIVE) LIMITS: UPPER / LOWER

METHOD USED

EXTINGUISHING MEDIA: Water spray, dry chemical, foam or carbon dioxide

SPECIAL FIRE FIGHTING PROCEDURES: Wear self-contained breathing apparatus

UNUSUAL FIRE AND EXPLOSION HAZARDS

SECTION V HEALTH HAZARD DATA

TOXIC LEVEL: None available | CARCINOGENIC: no

PRINCIPLE ROUTES OF ABSORBTION: Inhalation, ingestion, skin | SKIN AND EYE IRRITATION: Severe

REVELANT SYMPTOMS OF EXPOSURE: Cyanosis, vomiting, nausea, sweating, intense thirst, diarrhea, convulsions, collapse.

EFFECTS OF CHRONIC EXPOSURE: Liver or kidney damage, dermatitis

EMERGENCY AND FIRST AID PROCEDURES: Inhalation - rest, if swallowed - give emetic, get medical aid; wash from skin & eyes.

SECTION VI REACTIVITY DATA

CONDITIONS CONTRIBUTING TO INSTABILITY

CONDITIONS CONTRIBUTING TO HAZARDOUS POLYMERIZATION

INCOMPATABILITY (MATERIALS TO AVOID)

HAZARDOUS DECOMPOSITION PRODUCTS

SECTION VII SPILL OR LEAK PROCEDURES

STEPS TO BE TAKEN IN CASE MATERIAL IS RELEASED OR SPILLED

WASTE DISPOSAL METHOD

SECTION VIII SPECIAL PROTECTION INFORMATION

VENTILATION REQUIREMENTS LOCAL EXHAUST | PROTECTIVE EQUIPMENT (SPECIFY TYPES) EYE: Goggles

MECHANICAL (GENERAL) | GLOVES: Rubber

SPECIAL | RESPIRATOR: Self-contained

OTHER PROTECTIVE EQUIPMENT

SECTION IX SPECIAL PRECAUTIONS

PRECAUTIONS TO BE TAKEN IN HANDLING AND STORAGE

OTHER PRECAUTIONS

SIGNATURE | DATE

4,4′-DIBROMODIPHENYL ETHER

NFPA diamond: Health 3, Flammability 1, Reactivity 0

SECTION I SOURCE AND NOMENCLATURE

MANUFACTURER'S NAME | EMERGENCY TELEPHONE NO.

ADDRESS (NUMBER, STREET, CITY, STATE, ZIP CODE)

TRADE NAME AND SYNONYMS | CHEMICAL FAMILY

CHEMICAL NAME AND SYNONYMS: P,P-Dibromodiphenyl ether | FORMULA: $(BrC_6H_4)_2O$

SECTION II HAZARDOUS INGREDIENTS

BASIC MATERIAL	APPROXIMATE OR MAXIMUM % WT. OR VOL.	ESTABLISHED OSHA STANDARD	LD50 ORAL	LD50 PERCUT.	LC50 SPECIES	LC50 CONC.
Lowest published lethal dose intraperitoneal - mouse 125 mg/kg.						

SECTION III PHYSICAL DATA MW-328.02

BOILING POINT (°F): 340°C | VAPOR PRESSURE (mm Hg)

MELTING POINT (°F): 60°C | VAPOR DENSITY (Air = 1)

SPECIFIC GRAVITY (H2O = 1) | EVAPORATION RATE (= 1)

SOLUBILITY IN WATER (Pts 100 pts H2O): very soluble | VOLATILE % Vol. % Wt.

APPEARANCE AND ODOR: crystal

SECTION IV FIRE AND EXPLOSION HAZARD DATA

FLASH POINT: none available | FLAMMABLE (EXPLOSIVE) LIMITS: UPPER / LOWER

METHOD USED

EXTINGUISHING MEDIA: water alcohol foam, carbon dioxide, dry chemical

SPECIAL FIRE FIGHTING PROCEDURES

UNUSUAL FIRE AND EXPLOSION HAZARDS

SECTION V HEALTH HAZARD DATA

TOXIC LEVEL: None available | CARCINOGENIC: no

PRINCIPLE ROUTES OF ABSORBTION: Inhalation, ingestion | SKIN AND EYE IRRITATION: yes

REVELANT SYMPTOMS OF EXPOSURE: loss of appetite, nausea

EFFECTS OF CHRONIC EXPOSURE: liver, spleen, kidney and thyroid damage

EMERGENCY AND FIRST AID PROCEDURES: If swallowed - give emetic, get medical aid; wash skin & eyes; inhalation - give oxygen as needed.

SECTION VI REACTIVITY DATA

CONDITIONS CONTRIBUTING TO INSTABILITY

CONDITIONS CONTRIBUTING TO HAZARDOUS POLYMERIZATION

INCOMPATABILITY (MATERIALS TO AVOID): May react with oxidizing materials

HAZARDOUS DECOMPOSITION PRODUCTS

SECTION VII SPILL OR LEAK PROCEDURES

STEPS TO BE TAKEN IN CASE MATERIAL IS RELEASED OR SPILLED

WASTE DISPOSAL METHOD

SECTION VIII SPECIAL PROTECTION INFORMATION

VENTILATION REQUIREMENTS LOCAL EXHAUST | PROTECTIVE EQUIPMENT (SPECIFY TYPES) EYE: goggles

MECHANICAL (GENERAL) | GLOVES: rubber

SPECIAL | RESPIRATOR

OTHER PROTECTIVE EQUIPMENT

SECTION IX SPECIAL PRECAUTIONS

PRECAUTIONS TO BE TAKEN IN HANDLING AND STORAGE

OTHER PRECAUTIONS

SIGNATURE | DATE

2,6-DIBROMO-4-NITROPHENOL

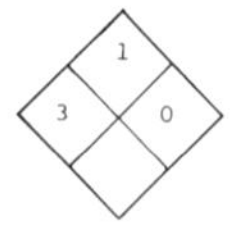

SECTION I SOURCE AND NOMENCLATURE

MANUFACTURER'S NAME | EMERGENCY TELEPHONE NO.

ADDRESS (NUMBER, STREET, CITY, STATE, ZIP CODE)

TRADE NAME AND SYNONYMS | CHEMICAL FAMILY

CHEMICAL NAME AND SYNONYMS: O,O' Dibromo-P Nitrophenol

FORMULA: $Br_2C_6H_2(OH)NO_2$

SECTION II HAZARDOUS INGREDIENTS

BASIC MATERIAL	APPROXIMATE OR MAXIMUM % WT. OR VOL.	ESTABLISHED OSHA STANDARD	LD_{50} ORAL	LD_{50} PERCUT.	LC_{50} SPECIES	LC_{50} CONC.
			None available			

SECTION III PHYSICAL DATA MW - 296.92

BOILING POINT °F: d > 145°C

VAPOR PRESSURE mm Hg:

MELTING POINT °F: 143°C

VAPOR DENSITY (Air = 1):

SPECIFIC GRAVITY (H_2O = 1):

EVAPORATION RATE (= 1):

SOLUBILITY IN WATER Pts/100 pts H_2O: very soluble

VOLATILE % Vol. % Wt.:

APPEARANCE AND ODOR: colorless crystals

SECTION IV FIRE AND EXPLOSION HAZARD DATA

FLASH POINT: none available

METHOD USED:

FLAMMABLE (EXPLOSIVE) LIMITS: UPPER | LOWER

EXTINGUISHING MEDIA: Water, dry chemical, carbon dioxide, alcohol foam

SPECIAL FIRE FIGHTING PROCEDURES: Wear self-contained gas mask

UNUSUAL FIRE AND EXPLOSION HAZARDS:

SECTION V HEALTH HAZARD DATA

TOXIC LEVEL: None available

CARCINOGENIC: No

PRINCIPLE ROUTES OF ABSORBTION: Inhalation, ingestion, skin

SKIN AND EYE IRRITATION: Severe

REVELANT SYMPTOMS OF EXPOSURE: Cyanosis, deep & rapid respiration, irregular pulse, severe damage to digestive tract, drowsiness.

EFFECTS OF CHRONIC EXPOSURE: Lung, liver & kidney damage.

EMERGENCY AND FIRST AID PROCEDURES: Inhalation - rest; if swallowed - drink water, epsom salts, get medical aid; wash from eyes & skin.

SECTION VI REACTIVITY DATA

CONDITIONS CONTRIBUTING TO INSTABILITY: Decomposes on heating to give toxic bromides.

CONDITIONS CONTRIBUTING TO HAZARDOUS POLYMERIZATION:

INCOMPATABILITY (MATERIALS TO AVOID):

HAZARDOUS DECOMPOSITION PRODUCTS:

SECTION VII SPILL OR LEAK PROCEDURES

STEPS TO BE TAKEN IN CASE MATERIAL IS RELEASED OR SPILLED:

WASTE DISPOSAL METHOD:

SECTION VIII SPECIAL PROTECTION INFORMATION

VENTILATION REQUIREMENTS LOCAL EXHAUST:

PROTECTIVE EQUIPMENT (SPECIFY TYPES) EYE: Goggles

MECHANICAL (GENERAL):

GLOVES: Rubber

SPECIAL:

RESPIRATOR: Self-contained

OTHER PROTECTIVE EQUIPMENT:

SECTION IX SPECIAL PRECAUTIONS

PRECAUTIONS TO BE TAKEN IN HANDLING AND STORAGE:

OTHER PRECAUTIONS:

SIGNATURE | DATE

2,4-DIBROMOPHENOL

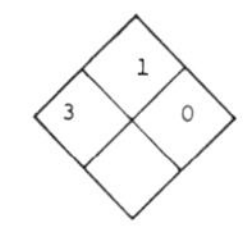

SECTION I SOURCE AND NOMENCLATURE

MANUFACTURER'S NAME | EMERGENCY TELEPHONE NO.

ADDRESS (NUMBER, STREET, CITY, STATE, ZIP CODE)

TRADE NAME AND SYNONYMS | CHEMICAL FAMILY

CHEMICAL NAME AND SYNONYMS: 2,4-Dibromophenol

FORMULA: $Br_2C_6H_3OH$

SECTION II HAZARDOUS INGREDIENTS

BASIC MATERIAL	APPROXIMATE OR MAXIMUM % WT. OR VOL.	ESTABLISHED OSHA STANDARD	LD_{50} ORAL	LD_{50} PERCUT.	LC_{50} SPECIES	LC_{50} CONC.
			None available			

SECTION III PHYSICAL DATA MW - 251.92

BOILING POINT °F: 239°C.

VAPOR PRESSURE mm Hg:

MELTING POINT °F: 40°C.

VAPOR DENSITY (Air = 1):

SPECIFIC GRAVITY (H_2O = 1):

EVAPORATION RATE (= 1):

SOLUBILITY IN WATER Pts/100 pts H_2O: 0.2

VOLATILE % Vol. % Wt.:

APPEARANCE AND ODOR: Crystaline

SECTION IV FIRE AND EXPLOSION HAZARD DATA

FLASH POINT: None available

METHOD USED:

FLAMMABLE (EXPLOSIVE) LIMITS: UPPER | LOWER

EXTINGUISHING MEDIA: Dry chemical, alcohol foam, carbon dioxide.

SPECIAL FIRE FIGHTING PROCEDURES:

UNUSUAL FIRE AND EXPLOSION HAZARDS:

SECTION V HEALTH HAZARD DATA

TOXIC LEVEL: None available

CARCINOGENIC: NO

PRINCIPLE ROUTES OF ABSORBTION: Inhalation, ingestion, skin

SKIN AND EYE IRRITATION: Severe

REVELANT SYMPTOMS OF EXPOSURE: Eye, skin & mucous membrane irritation dermatitis; cyanosis.

EFFECTS OF CHRONIC EXPOSURE: Liver damage

EMERGENCY AND FIRST AID PROCEDURES: Inhalation - rest; wash from skin & eyes; if swallowed - give emetic, get medical aid.

SECTION VI REACTIVITY DATA

CONDITIONS CONTRIBUTING TO INSTABILITY:

CONDITIONS CONTRIBUTING TO HAZARDOUS POLYMERIZATION:

INCOMPATABILITY (MATERIALS TO AVOID):

HAZARDOUS DECOMPOSITION PRODUCTS:

SECTION VII SPILL OR LEAK PROCEDURES

STEPS TO BE TAKEN IN CASE MATERIAL IS RELEASED OR SPILLED: Pour sand to soda ash (9:1) mixture on spill and sweep into box for disposal.

WASTE DISPOSAL METHOD:

SECTION VIII SPECIAL PROTECTION INFORMATION

VENTILATION REQUIREMENTS LOCAL EXHAUST:

PROTECTIVE EQUIPMENT (SPECIFY TYPES) EYE: Goggles

MECHANICAL (GENERAL):

GLOVES: Butyl Rubber

SPECIAL:

RESPIRATOR: Self-contained

OTHER PROTECTIVE EQUIPMENT: Protective clothing

SECTION IX SPECIAL PRECAUTIONS

PRECAUTIONS TO BE TAKEN IN HANDLING AND STORAGE: Store in cool, dry place.

OTHER PRECAUTIONS:

SIGNATURE | DATE

2,6-DIBROMOPHENOL

Health: 3 · Flammability: 1 · Reactivity: 0

SECTION I SOURCE AND NOMENCLATURE

MANUFACTURER'S NAME	EMERGENCY TELEPHONE NO.
ADDRESS (NUMBER, STREET, CITY, STATE, ZIP CODE)	
TRADE NAME AND SYNONYMS	CHEMICAL FAMILY
CHEMICAL NAME AND SYNONYMS: 2,6 Dibromophenol	FORMULA: $Br_2C_6H_3OH$

SECTION II HAZARDOUS INGREDIENTS

BASIC MATERIAL	APPROXIMATE OR MAXIMUM % WT. OR VOL.	ESTABLISHED OSHA STANDARD	LD_{50} ORAL	LD_{50} PERCUT.	LC_{50} SPECIES	LC_{50} CONC.
		None available				

SECTION III PHYSICAL DATA MW - 251.92

BOILING POINT °F.	VAPOR PRESSURE mm Hg.: at 162°C. 21
MELTING POINT °F.: 56°C.	VAPOR DENSITY (Air = 1)
SPECIFIC GRAVITY (H_2O = 1)	EVAPORATION RATE (____ = 1)
SOLUBILITY IN WATER Pts/100 pts H_2O	VOLATILE % Vol. % Wt.

APPEARANCE AND ODOR: Needle-like crystals

SECTION IV FIRE AND EXPLOSION HAZARD DATA

FLASH POINT: None available	FLAMMABLE (EXPLOSIVE) LIMITS	UPPER
METHOD USED		LOWER

EXTINGUISHING MEDIA: Dry chemical, alcohol foam, carbon dioxide.

SPECIAL FIRE FIGHTING PROCEDURES

UNUSUAL FIRE AND EXPLOSION HAZARDS

SECTION V HEALTH HAZARD DATA

TOXIC LEVEL: None available — CARCINOGENIC: No

PRINCIPLE ROUTES OF ABSORBTION: Inhalation, ingestion, skin — SKIN AND EYE IRRITATION: Severe

REVELANT SYMPTOMS OF EXPOSURE: Eye, skin and mucous membrane irritation, dermatitis, cyanosis.

EFFECTS OF CHRONIC EXPOSURE: Liver damage

EMERGENCY AND FIRST AID PROCEDURES: Inhalation - rest; wash skin & eyes; if swallowed - give emetic, get medical aid.

SECTION VI REACTIVITY DATA

CONDITIONS CONTRIBUTING TO INSTABILITY

CONDITIONS CONTRIBUTING TO HAZARDOUS POLYMERIZATION

INCOMPATABILITY (MATERIALS TO AVOID)

HAZARDOUS DECOMPOSITION PRODUCTS

SECTION VII SPILL OR LEAK PROCEDURES

STEPS TO BE TAKEN IN CASE MATERIAL IS RELEASED OR SPILLED: Pour sand to soda ash (9:1) mixture on spill and sweep into box for disposal.

WASTE DISPOSAL METHOD

SECTION VIII SPECIAL PROTECTION INFORMATION

VENTILATION REQUIREMENTS LOCAL EXHAUST	PROTECTIVE EQUIPMENT (SPECIFY TYPES) EYE: Goggles
MECHANICAL (GENERAL)	GLOVES: Butyl Rubber
SPECIAL	RESPIRATOR: Self-contained

OTHER PROTECTIVE EQUIPMENT: Protective clothing

SECTION IX SPECIAL PRECAUTIONS

PRECAUTIONS TO BE TAKEN IN HANDLING AND STORAGE: Store in cool, dry place.

OTHER PRECAUTIONS

SIGNATURE	DATE

3,5-DIBROMOPYRIDINE

Health: 4 · Flammability: 3 · Reactivity: 0

SECTION I SOURCE AND NOMENCLATURE

MANUFACTURER'S NAME	EMERGENCY TELEPHONE NO.
ADDRESS (NUMBER, STREET, CITY, STATE, ZIP CODE)	
TRADE NAME AND SYNONYMS	CHEMICAL FAMILY
CHEMICAL NAME AND SYNONYMS: 3,5 Dibromopyridine	FORMULA: $C_5H_3N\ Br_2$

SECTION II HAZARDOUS INGREDIENTS

BASIC MATERIAL	APPROXIMATE OR MAXIMUM % WT. OR VOL.	ESTABLISHED OSHA STANDARD	LD_{50} ORAL	LD_{50} PERCUT.	LC_{50} SPECIES	LC_{50} CONC.
			none	available		

SECTION III PHYSICAL DATA MW - 236.91

BOILING POINT °F.: 222°C	VAPOR PRESSURE mm Hg.
MELTING POINT °F.: 111°C	VAPOR DENSITY (Air = 1)
SPECIFIC GRAVITY (H_2O = 1): -	EVAPORATION RATE (____ = 1)
SOLUBILITY IN WATER Pts/100 pts H_2O: Slightly soluble	VOLATILE % Vol. % Wt.

APPEARANCE AND ODOR: Solid - fish-like odor.

SECTION IV FIRE AND EXPLOSION HAZARD DATA

FLASH POINT: none available	FLAMMABLE (EXPLOSIVE) LIMITS	UPPER
METHOD USED		LOWER

EXTINGUISHING MEDIA: Alcohol foam, carbon dioxide, dry chemical

SPECIAL FIRE FIGHTING PROCEDURES: Wear goggles and self-contained breathing apparatus

UNUSUAL FIRE AND EXPLOSION HAZARDS

SECTION V HEALTH HAZARD DATA

TOXIC LEVEL: Not available — CARCINOGENIC: no

PRINCIPLE ROUTES OF ABSORBTION: Inhalation, ingestion, skin — SKIN AND EYE IRRITATION: severe

REVELANT SYMPTOMS OF EXPOSURE: Narcosis, headache, nausea, giddiness, vomiting

EFFECTS OF CHRONIC EXPOSURE: Occasional vomiting and diarrhea; weight loss and anemia, ocular and facial paralysis.

EMERGENCY AND FIRST AID PROCEDURES: Inhalation - rest; if swallowed - induce vomiting, get medical aid; wash skin & eyes.

SECTION VI REACTIVITY DATA

CONDITIONS CONTRIBUTING TO INSTABILITY

CONDITIONS CONTRIBUTING TO HAZARDOUS POLYMERIZATION

INCOMPATABILITY (MATERIALS TO AVOID): May react with oxidizing materials

HAZARDOUS DECOMPOSITION PRODUCTS

SECTION VII SPILL OR LEAK PROCEDURES

STEPS TO BE TAKEN IN CASE MATERIAL IS RELEASED OR SPILLED

WASTE DISPOSAL METHOD

SECTION VIII SPECIAL PROTECTION INFORMATION

VENTILATION REQUIREMENTS LOCAL EXHAUST: Use in hood	PROTECTIVE EQUIPMENT (SPECIFY TYPES) EYE: Goggles
MECHANICAL (GENERAL)	GLOVES: Rubber
SPECIAL	RESPIRATOR: Self-contained

OTHER PROTECTIVE EQUIPMENT

SECTION IX SPECIAL PRECAUTIONS

PRECAUTIONS TO BE TAKEN IN HANDLING AND STORAGE

OTHER PRECAUTIONS

SIGNATURE	DATE

2,5-DIBROMOTOLUENE

Hazard diamond: 2 (top), 3 (left), 0 (right)

SECTION I SOURCE AND NOMENCLATURE

MANUFACTURER'S NAME | EMERGENCY TELEPHONE NO.

ADDRESS (NUMBER, STREET, CITY, STATE, ZIP CODE)

TRADE NAME AND SYNONYMS | CHEMICAL FAMILY

CHEMICAL NAME AND SYNONYMS: 2,5 - Dibromotoluene | FORMULA: $Br_2C_6H_3CH_3$

SECTION II HAZARDOUS INGREDIENTS

BASIC MATERIAL	APPROXIMATE OR MAXIMUM % WT. OR VOL.	ESTABLISHED OSHA STANDARD	LD_{50} ORAL	LD_{50} PERCUT.	LC_{50} SPECIES	LC_{50} CONC.
			None	available		

SECTION III PHYSICAL DATA MW - 249.95

BOILING POINT °F.: 236°C | VAPOR PRESSURE mm Hg.

MELTING POINT °F.: <-20°C | VAPOR DENSITY (Air = 1)

SPECIFIC GRAVITY (H_2O = 1): 1.81 | EVAPORATION RATE (= 1)

SOLUBILITY IN WATER Pts/100 pts H_2O: Insoluble | VOLATILE % Vol. % Wt.

APPEARANCE AND ODOR: Colorless liquid

SECTION IV FIRE AND EXPLOSION HAZARD DATA

FLASH POINT: Not available | FLAMMABLE (EXPLOSIVE) LIMITS: UPPER / LOWER

METHOD USED

EXTINGUISHING MEDIA: Alcohol foam, carbon dioxide, dry chemical

SPECIAL FIRE FIGHTING PROCEDURES

UNUSUAL FIRE AND EXPLOSION HAZARDS

SECTION V HEALTH HAZARD DATA

TOXIC LEVEL: None available | CARCINOGENIC: no

PRINCIPLE ROUTES OF ABSORBTION: Inhalation, ingestion | SKIN AND EYE IRRITATION: severe

REVELANT SYMPTOMS OF EXPOSURE: Eye & respiratory system irritation, severe internal irritation of digestive tract.

EFFECTS OF CHRONIC EXPOSURE: --

EMERGENCY AND FIRST AID PROCEDURES: Inhalation - rest; if swallowed - give emetic, get medical aid; wash skin & eyes.

SECTION VI REACTIVITY DATA

CONDITIONS CONTRIBUTING TO INSTABILITY

CONDITIONS CONTRIBUTING TO HAZARDOUS POLYMERIZATION

INCOMPATABILITY (MATERIALS TO AVOID)

HAZARDOUS DECOMPOSITION PRODUCTS

SECTION VII SPILL OR LEAK PROCEDURES

STEPS TO BE TAKEN IN CASE MATERIAL IS RELEASED OR SPILLED

WASTE DISPOSAL METHOD

SECTION VIII SPECIAL PROTECTION INFORMATION

VENTILATION REQUIREMENTS LOCAL EXHAUST | PROTECTIVE EQUIPMENT (SPECIFY TYPES) EYE: Goggles

MECHANICAL (GENERAL) | GLOVES

SPECIAL | RESPIRATOR: With proper filter

OTHER PROTECTIVE EQUIPMENT

SECTION IX SPECIAL PRECAUTIONS

PRECAUTIONS TO BE TAKEN IN HANDLING AND STORAGE

OTHER PRECAUTIONS

SIGNATURE | DATE

DI-sec-BUTYLAMINE

Hazard diamond: 3 (top), 3 (left), 0 (right)

SECTION I SOURCE AND NOMENCLATURE

MANUFACTURER'S NAME | EMERGENCY TELEPHONE NO.

ADDRESS (NUMBER, STREET, CITY, STATE, ZIP CODE)

TRADE NAME AND SYNONYMS | CHEMICAL FAMILY

CHEMICAL NAME AND SYNONYMS: Di-Sec Butylamine | FORMULA: $(C_4H_9)_2NH$

SECTION II HAZARDOUS INGREDIENTS

BASIC MATERIAL	APPROXIMATE OR MAXIMUM % WT. OR VOL.	ESTABLISHED OSHA STANDARD	LD_{50} ORAL	LD_{50} PERCUT.	LC_{50} SPECIES	LC_{50} CONC.
			None	available		

SECTION III PHYSICAL DATA MW - 12.92

BOILING POINT °F.: 132°C. | VAPOR PRESSURE mm Hg.

MELTING POINT °F.: -104° C. | VAPOR DENSITY (Air = 1): 4.5

SPECIFIC GRAVITY (H_2O = 1): 0.78 | EVAPORATION RATE (= 1)

SOLUBILITY IN WATER Pts/100 pts H_2O: Very soluble | VOLATILE % Vol. % Wt.

APPEARANCE AND ODOR: Colorless liquid

SECTION IV FIRE AND EXPLOSION HAZARD DATA

FLASH POINT: 75° F. | FLAMMABLE (EXPLOSIVE) LIMITS: UPPER / LOWER

METHOD USED: O.C.

EXTINGUISHING MEDIA: Alcohol foam, carbon dioxide, dry chemical.

SPECIAL FIRE FIGHTING PROCEDURES

UNUSUAL FIRE AND EXPLOSION HAZARDS

SECTION V HEALTH HAZARD DATA

TOXIC LEVEL: None available | CARCINOGENIC: No

PRINCIPLE ROUTES OF ABSORBTION: Inhalation, ingestion | SKIN AND EYE IRRITATION: Severe

REVELANT SYMPTOMS OF EXPOSURE: Restlessness, excitability, increased pulse and respiration rates, irritates eyes, skin and digestive systems.

EFFECTS OF CHRONIC EXPOSURE: --

EMERGENCY AND FIRST AID PROCEDURES: Inhalation - rest; if swallowed - drink water, get medical aid; wash eyes and skin.

SECTION VI REACTIVITY DATA

CONDITIONS CONTRIBUTING TO INSTABILITY

CONDITIONS CONTRIBUTING TO HAZARDOUS POLYMERIZATION

INCOMPATABILITY (MATERIALS TO AVOID)

HAZARDOUS DECOMPOSITION PRODUCTS

SECTION VII SPILL OR LEAK PROCEDURES

STEPS TO BE TAKEN IN CASE MATERIAL IS RELEASED OR SPILLED

WASTE DISPOSAL METHOD

SECTION VIII SPECIAL PROTECTION INFORMATION

VENTILATION REQUIREMENTS LOCAL EXHAUST | PROTECTIVE EQUIPMENT (SPECIFY TYPES) EYE: Goggles

MECHANICAL (GENERAL) | GLOVES: Rubber

SPECIAL | RESPIRATOR: Self-contained

OTHER PROTECTIVE EQUIPMENT

SECTION IX SPECIAL PRECAUTIONS

PRECAUTIONS TO BE TAKEN IN HANDLING AND STORAGE

OTHER PRECAUTIONS

SIGNATURE | DATE

2-DI-n-BUTYLAMINOETHYL ALCOHOL

Hazard diamond: 2 (top), 4 (left), 0 (right)

SECTION I SOURCE AND NOMENCLATURE

MANUFACTURER'S NAME | EMERGENCY TELEPHONE NO.

ADDRESS (NUMBER, STREET, CITY, STATE, ZIP CODE)

TRADE NAME AND SYNONYMS | CHEMICAL FAMILY

CHEMICAL NAME AND SYNONYMS: 2 Dibutylaminoethanol | FORMULA: $[CH_3(CH_2)_3]_2NCH_2CH_2OH$

SECTION II HAZARDOUS INGREDIENTS

BASIC MATERIAL	APPROXIMATE OR MAXIMUM % WT. OR VOL.	ESTABLISHED OSHA STANDARD	LD_{50} ORAL	LD_{50} PERCUT.	LC_{50} SPECIES	LC_{50} CONC.
			1070		Rat	
				1680	Rabbit	

SECTION III PHYSICAL DATA MW - 173.3

BOILING POINT: 222°C | VAPOR PRESSURE: 91-97°C 7 mm Hg.

MELTING POINT: -- | VAPOR DENSITY (Air = 1): 6.0

SPECIFIC GRAVITY (H_2O = 1): 0.85 | EVAPORATION RATE (____ = 1)

SOLUBILITY IN WATER Pts/100 pts H_2O: Insoluble | VOLATILE % Vol. % Wt.

APPEARANCE AND ODOR: Colorless liquid

SECTION IV FIRE AND EXPLOSION HAZARD DATA

FLASH POINT: 200°F | FLAMMABLE (EXPLOSIVE) LIMITS: UPPER / LOWER

METHOD USED: Open cup

EXTINGUISHING MEDIA: Carbon dioxide, dry chemical

SPECIAL FIRE FIGHTING PROCEDURES

UNUSUAL FIRE AND EXPLOSION HAZARDS

SECTION V HEALTH HAZARD DATA

TOXIC LEVEL: TLV - 2 PPM, 14 mg/C.M. | CARCINOGENIC: No

PRINCIPLE ROUTES OF ABSORBTION: Skin, ingestion, inhalation | SKIN AND EYE IRRITATION: moderate

REVELANT SYMPTOMS OF EXPOSURE: Dermatitis, weight loss, slight increase in clotting time

EFFECTS OF CHRONIC EXPOSURE: Dermatitis

EMERGENCY AND FIRST AID PROCEDURES: If swallowed - drink fruit juice, get medical aid; inhalation rest. Wash from skin & eyes.

SECTION VI REACTIVITY DATA

CONDITIONS CONTRIBUTING TO INSTABILITY

CONDITIONS CONTRIBUTING TO HAZARDOUS POLYMERIZATION

INCOMPATABILITY (MATERIALS TO AVOID)

HAZARDOUS DECOMPOSITION PRODUCTS

SECTION VII SPILL OR LEAK PROCEDURES

STEPS TO BE TAKEN IN CASE MATERIAL IS RELEASED OR SPILLED

WASTE DISPOSAL METHOD

SECTION VIII SPECIAL PROTECTION INFORMATION

VENTILATION REQUIREMENTS LOCAL EXHAUST | PROTECTIVE EQUIPMENT (SPECIFY TYPES) EYE: Safety goggles

MECHANICAL (GENERAL) | GLOVES: Rubber

SPECIAL | RESPIRATOR

OTHER PROTECTIVE EQUIPMENT

SECTION IX SPECIAL PRECAUTIONS

PRECAUTIONS TO BE TAKEN IN HANDLING AND STORAGE

OTHER PRECAUTIONS

SIGNATURE | DATE

DI-n-BUTYLANILINE

Hazard diamond: 1 (top), 3 (left), 0 (right)

SECTION I SOURCE AND NOMENCLATURE

MANUFACTURER'S NAME | EMERGENCY TELEPHONE NO.

ADDRESS (NUMBER, STREET, CITY, STATE, ZIP CODE)

TRADE NAME AND SYNONYMS | CHEMICAL FAMILY

CHEMICAL NAME AND SYNONYMS: N N'-Dibutylaniline | FORMULA: $C_6H_5N(C_4H_9)_2$

SECTION II HAZARDOUS INGREDIENTS

BASIC MATERIAL	APPROXIMATE OR MAXIMUM % WT. OR VOL.	ESTABLISHED OSHA STANDARD	LD_{50} ORAL	LD_{50} PERCUT.	LC_{50} SPECIES	LC_{50} CONC.
			none	available		

SECTION III PHYSICAL DATA MW - 205.34

BOILING POINT: 263°C | VAPOR PRESSURE mm Hg.

MELTING POINT | VAPOR DENSITY (Air = 1): 7.1

SPECIFIC GRAVITY (H_2O = 1): 0.94 | EVAPORATION RATE (____ = 1)

SOLUBILITY IN WATER Pts/100 pts H_2O: Insoluble | VOLATILE % Vol. % Wt.

APPEARANCE AND ODOR: Light yellow liquid

SECTION IV FIRE AND EXPLOSION HAZARD DATA

FLASH POINT: 230°F | FLAMMABLE (EXPLOSIVE) LIMITS: UPPER / LOWER

METHOD USED

EXTINGUISHING MEDIA: Water, alcohol foam, carbon dioxide, dry chemical

SPECIAL FIRE FIGHTING PROCEDURES

UNUSUAL FIRE AND EXPLOSION HAZARDS

SECTION V HEALTH HAZARD DATA

TOXIC LEVEL: None available | CARCINOGENIC: no

PRINCIPLE ROUTES OF ABSORBTION: Inhalation, ingestion, skin | SKIN AND EYE IRRITATION: moderate

REVELANT SYMPTOMS OF EXPOSURE: Headache, drowsiness, cyanosis, mental confusion, convulsions

EFFECTS OF CHRONIC EXPOSURE: Fatigue, loss of appetite, headache, dizziness

EMERGENCY AND FIRST AID PROCEDURES: If swallowed - induce vomiting, get medical aid; inhalation - rest, get medical aid; wash from skin & eyes.

SECTION VI REACTIVITY DATA

CONDITIONS CONTRIBUTING TO INSTABILITY

CONDITIONS CONTRIBUTING TO HAZARDOUS POLYMERIZATION

INCOMPATABILITY (MATERIALS TO AVOID)

HAZARDOUS DECOMPOSITION PRODUCTS

SECTION VII SPILL OR LEAK PROCEDURES

STEPS TO BE TAKEN IN CASE MATERIAL IS RELEASED OR SPILLED

WASTE DISPOSAL METHOD

SECTION VIII SPECIAL PROTECTION INFORMATION

VENTILATION REQUIREMENTS LOCAL EXHAUST | PROTECTIVE EQUIPMENT (SPECIFY TYPES) EYE: Goggles

MECHANICAL (GENERAL) | GLOVES: Polyvinyl alcohol gloves

SPECIAL | RESPIRATOR: Self-contained

OTHER PROTECTIVE EQUIPMENT

SECTION IX SPECIAL PRECAUTIONS

PRECAUTIONS TO BE TAKEN IN HANDLING AND STORAGE

OTHER PRECAUTIONS

SIGNATURE | DATE

DICHLOROACETYL CHLORIDE

Hazard rating: 2 (top), 4 (left), 2 (right), W (bottom)

SECTION I SOURCE AND NOMENCLATURE

MANUFACTURER'S NAME	EMERGENCY TELEPHONE NO.
ADDRESS (NUMBER, STREET, CITY, STATE, ZIP CODE)	
TRADE NAME AND SYNONYMS	CHEMICAL FAMILY
CHEMICAL NAME AND SYNONYMS: Dichloroethanoyl Chloride	FORMULA: $Cl_2CHCOCl$

SECTION II HAZARDOUS INGREDIENTS

BASIC MATERIAL	APPROXIMATE OR MAXIMUM % WT. OR VOL.	ESTABLISHED OSHA STANDARD	LD_{50} mg/kg ORAL	LD_{50} PERCUT.	SPECIES	LC_{50} CONC.
LCLO inh-rat 2000 ppm/4H			2460		Rat	
				650	Rabbit	

SECTION III PHYSICAL DATA MW 147.4

BOILING POINT °F: 108°C	VAPOR PRESSURE mm Hg.
MELTING POINT °F.	VAPOR DENSITY (Air = 1): 5.1
SPECIFIC GRAVITY (H_2O = 1): 1.53	EVAPORATION RATE (= 1)
SOLUBILITY IN WATER Pts/100 pts H_2O: reacts with water	VOLATILE % Vol. % Wt.
APPEARANCE AND ODOR: Clear liquid, acrid odor	

SECTION IV FIRE AND EXPLOSION HAZARD DATA

FLASH POINT: 151°	FLAMMABLE (EXPLOSIVE) LIMITS	UPPER
METHOD USED		LOWER

EXTINGUISHING MEDIA: Do not use water or foam, use carbon dioxide or dry chemical.

SPECIAL FIRE FIGHTING PROCEDURES

UNUSUAL FIRE AND EXPLOSION HAZARDS: When heated to decomposition point, gives off toxic chloride fumes

SECTION V HEALTH HAZARD DATA

TOXIC LEVEL: None available — CARCINOGENIC: no

PRINCIPLE ROUTES OF ABSORBTION: Inhalation, ingestion — SKIN AND EYE IRRITATION: severe

REVELANT SYMPTOMS OF EXPOSURE: Irritation of eyes and respiratory system; severe internal damage if swallowed; eye & skin burns.

EFFECTS OF CHRONIC EXPOSURE: Bronchial irritation, errosion of teeth.

EMERGENCY AND FIRST AID PROCEDURES: Inhalation - rest, get medical aid; if swallowed - drink water, milk of magnesia, get medical aid; wash eyes & skin.

SECTION VI REACTIVITY DATA

CONDITIONS CONTRIBUTING TO INSTABILITY

CONDITIONS CONTRIBUTING TO HAZARDOUS POLYMERIZATION

INCOMPATABILITY (MATERIALS TO AVOID): Reacts with water to produce H Cl, alcohol and dimethyl sulfoxide.

HAZARDOUS DECOMPOSITION PRODUCTS

SECTION VII SPILL OR LEAK PROCEDURES

STEPS TO BE TAKEN IN CASE MATERIAL IS RELEASED OR SPILLED

WASTE DISPOSAL METHOD

SECTION VIII SPECIAL PROTECTION INFORMATION

VENTILATION REQUIREMENTS LOCAL EXHAUST	PROTECTIVE EQUIPMENT (SPECIFY TYPES) EYE: Goggles
MECHANICAL (GENERAL)	GLOVES: Rubber
SPECIAL	RESPIRATOR: Self-contained

OTHER PROTECTIVE EQUIPMENT: Rubber apron

SECTION IX SPECIAL PRECAUTIONS

PRECAUTIONS TO BE TAKEN IN HANDLING AND STORAGE

OTHER PRECAUTIONS

SIGNATURE	DATE

2,4-DICHLOROBENZALDEHYDE

Hazard rating: 2 (top), 3 (left), 0 (right)

SECTION I SOURCE AND NOMENCLATURE

MANUFACTURER'S NAME	EMERGENCY TELEPHONE NO.
ADDRESS (NUMBER, STREET, CITY, STATE, ZIP CODE)	
TRADE NAME AND SYNONYMS	CHEMICAL FAMILY
CHEMICAL NAME AND SYNONYMS: 2,4, Dichlorobenzaldehye	FORMULA: $C_7H_4Cl_2O$

SECTION II HAZARDOUS INGREDIENTS

BASIC MATERIAL	APPROXIMATE OR MAXIMUM % WT. OR VOL.	ESTABLISHED OSHA STANDARD	LD_{50} ORAL	LD_{50} PERCUT.	LC_{50} SPECIES	LC_{50} CONC
			none	available		

SECTION III PHYSICAL DATA MW - 175.02

BOILING POINT °F: --	VAPOR PRESSURE mm Hg.
MELTING POINT °F: 71°C	VAPOR DENSITY (Air = 1): --
SPECIFIC GRAVITY (H_2O = 1): --	EVAPORATION RATE (= 1)
SOLUBILITY IN WATER Pts/100 pts H_2O: Insoluble	VOLATILE % Vol. % Wt.
APPEARANCE AND ODOR: White crystal	

SECTION IV FIRE AND EXPLOSION HAZARD DATA

FLASH POINT: Not available	FLAMMABLE (EXPLOSIVE) LIMITS	UPPER
METHOD USED		LOWER

EXTINGUISHING MEDIA: Foam, carbon dioxide, dry chemical

SPECIAL FIRE FIGHTING PROCEDURES

UNUSUAL FIRE AND EXPLOSION HAZARDS

SECTION V HEALTH HAZARD DATA

TOXIC LEVEL: None available — CARCINOGENIC: no

PRINCIPLE ROUTES OF ABSORBTION: Ingestion, skin, inhalation — SKIN AND EYE IRRITATION: moderate

REVELANT SYMPTOMS OF EXPOSURE: Eye irritation, headache, drowsiness, gastroenteric distress, pulmonary edema

EFFECTS OF CHRONIC EXPOSURE: Anemia, dermatitis

EMERGENCY AND FIRST AID PROCEDURES: If swallowed - induce vomiting, get medical aid; wash skin & eyes; inhalation - rest.

SECTION VI REACTIVITY DATA

CONDITIONS CONTRIBUTING TO INSTABILITY

CONDITIONS CONTRIBUTING TO HAZARDOUS POLYMERIZATION

INCOMPATABILITY (MATERIALS TO AVOID)

HAZARDOUS DECOMPOSITION PRODUCTS

SECTION VII SPILL OR LEAK PROCEDURES

STEPS TO BE TAKEN IN CASE MATERIAL IS RELEASED OR SPILLED

WASTE DISPOSAL METHOD

SECTION VIII SPECIAL PROTECTION INFORMATION

VENTILATION REQUIREMENTS LOCAL EXHAUST	PROTECTIVE EQUIPMENT (SPECIFY TYPES) EYE: Goggles
MECHANICAL (GENERAL)	GLOVES
SPECIAL	RESPIRATOR

OTHER PROTECTIVE EQUIPMENT

SECTION IX SPECIAL PRECAUTIONS

PRECAUTIONS TO BE TAKEN IN HANDLING AND STORAGE

OTHER PRECAUTIONS

SIGNATURE	DATE

m-DICHLOROBENZENE

(Hazard diamond: 2 top, 3 left, 0 right)

SECTION I SOURCE AND NOMENCLATURE

MANUFACTURER'S NAME

EMERGENCY TELEPHONE NO.

ADDRESS (NUMBER, STREET, CITY, STATE, ZIP CODE)

TRADE NAME AND SYNONYMS

CHEMICAL FAMILY

CHEMICAL NAME AND SYNONYMS: M-dichlorobenzene

FORMULA: $C_6H_4Cl_2$

SECTION II HAZARDOUS INGREDIENTS

BASIC MATERIAL	APPROXIMATE OR MAXIMUM % WT. OR VOL.	ESTABLISHED OSHA STANDARD	LD_{50} ORAL	LD_{50} PERCUT.	SPECIES	LC_{50} CONC.
			none available			

SECTION III PHYSICAL DATA

BOILING POINT °F.: 173°C

VAPOR PRESSURE mm Hg.: 1 at 12.1°C

MELTING POINT °F.: 24.8°C

VAPOR DENSITY (Air = 1): 5.08

SPECIFIC GRAVITY (H_2O = 1): 1.288

EVAPORATION RATE (= 1)

SOLUBILITY IN WATER Pts/100 pts H_2O: Insoluble

VOLATILE % Vol. % Wt.

APPEARANCE AND ODOR: Colorless liquid

SECTION IV FIRE AND EXPLOSION HAZARD DATA

FLASH POINT: Not available

FLAMMABLE (EXPLOSIVE) LIMITS: UPPER / LOWER

METHOD USED

EXTINGUISHING MEDIA: Water spray, foam, CO_2 or dry chemical

SPECIAL FIRE FIGHTING PROCEDURES: Wear self-contained breathing apparatus

UNUSUAL FIRE AND EXPLOSION HAZARDS

SECTION V HEALTH HAZARD DATA

TOXIC LEVEL: Unknown

CARCINOGENIC: no

PRINCIPLE ROUTES OF ABSORBTION: Inhalation, ingestion

SKIN AND EYE IRRITATION: severe

REVELANT SYMPTOMS OF EXPOSURE: Eye, skin and mucous membrane irritation, headache; drowsiness, unsteadiness

EFFECTS OF CHRONIC EXPOSURE: Liver damage

EMERGENCY AND FIRST AID PROCEDURES: Inhalation - rest, keep warm; if swallowed - wash mouth, give emetic, get medical attention; wash from skin & eyes.

SECTION VI REACTIVITY DATA

CONDITIONS CONTRIBUTING TO INSTABILITY

CONDITIONS CONTRIBUTING TO HAZARDOUS POLYMERIZATION

INCOMPATABILITY (MATERIALS TO AVOID)

HAZARDOUS DECOMPOSITION PRODUCTS

SECTION VII SPILL OR LEAK PROCEDURES

STEPS TO BE TAKEN IN CASE MATERIAL IS RELEASED OR SPILLED

WASTE DISPOSAL METHOD

SECTION VIII SPECIAL PROTECTION INFORMATION

VENTILATION REQUIREMENTS LOCAL EXHAUST

PROTECTIVE EQUIPMENT (SPECIFY TYPES) EYE: Goggles

MECHANICAL (GENERAL)

GLOVES: Rubber

SPECIAL

RESPIRATOR: Self-contained breathing apparatus

OTHER PROTECTIVE EQUIPMENT: Protective clothing

SECTION IX SPECIAL PRECAUTIONS

PRECAUTIONS TO BE TAKEN IN HANDLING AND STORAGE: Store in well-ventilated, dark, cool area.

OTHER PRECAUTIONS

SIGNATURE

DATE

s-DICHLOROMETHYL ETHER

(Hazard diamond: 2 top, 4 left, 0 right)

SECTION I SOURCE AND NOMENCLATURE

MANUFACTURER'S NAME

EMERGENCY TELEPHONE NO.

ADDRESS (NUMBER, STREET, CITY, STATE, ZIP CODE)

TRADE NAME AND SYNONYMS

CHEMICAL FAMILY

CHEMICAL NAME AND SYNONYMS: Bis(chloromethyl)ether

FORMULA: $(ClCH_2)_2O$

SECTION II HAZARDOUS INGREDIENTS

BASIC MATERIAL	APPROXIMATE OR MAXIMUM % WT. OR VOL.	ESTABLISHED OSHA STANDARD	LD_{50} mg/kg ORAL	LD_{50} mg/kg PERCUT.	SPECIES	LC_{50} mg/C.M. CONC.
			210		Rat	7 PPM/7 hr.
					Mouse	25/6 hr.
				280	Rabbit	
					Hamster	7 PPM/7 hr.

SECTION III PHYSICAL DATA MW - 114.96

BOILING POINT °F.: 104°C.

VAPOR PRESSURE mm Hg.

MELTING POINT °F.: --

VAPOR DENSITY (Air = 1): 4.0

SPECIFIC GRAVITY (H_2O = 1): 1.33

EVAPORATION RATE (= 1)

SOLUBILITY IN WATER Pts/100 pts H_2O: Decomposes

VOLATILE % Vol. % Wt.

APPEARANCE AND ODOR: Colorless liquid

SECTION IV FIRE AND EXPLOSION HAZARD DATA

FLASH POINT: None available

FLAMMABLE (EXPLOSIVE) LIMITS: UPPER / LOWER

METHOD USED

EXTINGUISHING MEDIA: Water may be used to blanket fire.

SPECIAL FIRE FIGHTING PROCEDURES

UNUSUAL FIRE AND EXPLOSION HAZARDS

SECTION V HEALTH HAZARD DATA

TOXIC LEVEL: 1 PPB

CARCINOGENIC: Yes

PRINCIPLE ROUTES OF ABSORBTION: Inhalation, ingestion, skin

SKIN AND EYE IRRITATION: Severe

REVELANT SYMPTOMS OF EXPOSURE: Respiratory and eye irritation, skin cancer.

EFFECTS OF CHRONIC EXPOSURE: Lung cancer.

EMERGENCY AND FIRST AID PROCEDURES: If swallowed - give emetic, get medical aid; inhalation - rest; wash skin & eyes.

SECTION VI REACTIVITY DATA

CONDITIONS CONTRIBUTING TO INSTABILITY

CONDITIONS CONTRIBUTING TO HAZARDOUS POLYMERIZATION

INCOMPATABILITY (MATERIALS TO AVOID)

HAZARDOUS DECOMPOSITION PRODUCTS

SECTION VII SPILL OR LEAK PROCEDURES

STEPS TO BE TAKEN IN CASE MATERIAL IS RELEASED OR SPILLED

WASTE DISPOSAL METHOD

SECTION VIII SPECIAL PROTECTION INFORMATION

VENTILATION REQUIREMENTS LOCAL EXHAUST

PROTECTIVE EQUIPMENT (SPECIFY TYPES) EYE: Goggles

MECHANICAL (GENERAL)

GLOVES: Rubber

SPECIAL

RESPIRATOR: Full face self-contained

OTHER PROTECTIVE EQUIPMENT: Full protective clothing.

SECTION IX SPECIAL PRECAUTIONS

PRECAUTIONS TO BE TAKEN IN HANDLING AND STORAGE

OTHER PRECAUTIONS

SIGNATURE

DATE

1,2-DICHLORO-4-NITROBENZENE

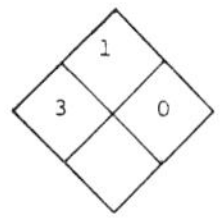

SECTION I SOURCE AND NOMENCLATURE

MANUFACTURER'S NAME | EMERGENCY TELEPHONE NO.

ADDRESS (NUMBER, STREET, CITY, STATE, ZIP CODE)

TRADE NAME AND SYNONYMS | CHEMICAL FAMILY

CHEMICAL NAME AND SYNONYMS: 3,4 dichloronitrobenzene

FORMULA: $Cl_2C_6H_3NO_2$

SECTION II HAZARDOUS INGREDIENTS

BASIC MATERIAL	APPROXIMATE OR MAXIMUM % WT. OR VOL.	ESTABLISHED OSHA STANDARD	LD_{50} mg/kg ORAL	LD_{50} PERCUT.	SPECIES	LC_{50} CONC.
			643		Rat	

SECTION III PHYSICAL DATA MW - 192.00

BOILING POINT °F.: 256°C

VAPOR PRESSURE mm Hg.

MELTING POINT °F.: 43°C

VAPOR DENSITY (Air = 1)

SPECIFIC GRAVITY (H_2O = 1): 1.43

EVAPORATION RATE (______ = 1)

SOLUBILITY IN WATER Pts/100 pts H_2O: Insoluble

VOLATILE % Vol. % Wt.

APPEARANCE AND ODOR: Colorless crystal

SECTION IV FIRE AND EXPLOSION HAZARD DATA

FLASH POINT: Not available

METHOD USED

FLAMMABLE (EXPLOSIVE) LIMITS: UPPER / LOWER

EXTINGUISHING MEDIA: Water spray, foam, carbon dioxide, dry chemical

SPECIAL FIRE FIGHTING PROCEDURES: Self-contained breathing apparatus.

UNUSUAL FIRE AND EXPLOSION HAZARDS

SECTION V HEALTH HAZARD DATA

TOXIC LEVEL: None available

CARCINOGENIC: no

PRINCIPLE ROUTES OF ABSORBTION: Ingestion, inhalation of dust

SKIN AND EYE IRRITATION: Severe

REVELANT SYMPTOMS OF EXPOSURE: Eye, skin & mucous membrane irritation, headache; drowsiness, unsteadiness.

EFFECTS OF CHRONIC EXPOSURE: Liver damage

EMERGENCY AND FIRST AID PROCEDURES: Inhalation - rest, keep warm; if swallowed - wash mouth, give emetic, get medical aid; wash from skin & eyes.

SECTION VI REACTIVITY DATA

CONDITIONS CONTRIBUTING TO INSTABILITY

CONDITIONS CONTRIBUTING TO HAZARDOUS POLYMERIZATION

INCOMPATABILITY (MATERIALS TO AVOID)

HAZARDOUS DECOMPOSITION PRODUCTS

SECTION VII SPILL OR LEAK PROCEDURES

STEPS TO BE TAKEN IN CASE MATERIAL IS RELEASED OR SPILLED

WASTE DISPOSAL METHOD

SECTION VIII SPECIAL PROTECTION INFORMATION

VENTILATION REQUIREMENTS LOCAL EXHAUST

MECHANICAL (GENERAL)

SPECIAL

OTHER PROTECTIVE EQUIPMENT

PROTECTIVE EQUIPMENT (SPECIFY TYPES) EYE: Goggles

GLOVES: Rubber

RESPIRATOR: Self-contained

SECTION IX SPECIAL PRECAUTIONS

PRECAUTIONS TO BE TAKEN IN HANDLING AND STORAGE

OTHER PRECAUTIONS

SIGNATURE | DATE

2,5-DICHLORONITROBENZENE

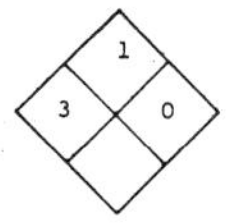

SECTION I SOURCE AND NOMENCLATURE

MANUFACTURER'S NAME | EMERGENCY TELEPHONE NO.

ADDRESS (NUMBER, STREET, CITY, STATE, ZIP CODE)

TRADE NAME AND SYNONYMS | CHEMICAL FAMILY

CHEMICAL NAME AND SYNONYMS: 1,4 dichloro-2-nitrobenzene

FORMULA: $Cl_2C_6H_3NO_2$

SECTION II HAZARDOUS INGREDIENTS

BASIC MATERIAL	APPROXIMATE OR MAXIMUM % WT. OR VOL.	ESTABLISHED OSHA STANDARD	LD_{50} ORAL	LD_{50} PERCUT.	SPECIES	LC_{50} CONC.
			1210		Rat	

SECTION III PHYSICAL DATA MW-192.00

BOILING POINT °F.: 266°C

VAPOR PRESSURE mm Hg.

MELTING POINT °F.: 55°C

VAPOR DENSITY (Air = 1)

SPECIFIC GRAVITY (H_2O = 1): 1.67

EVAPORATION RATE (______ = 1)

SOLUBILITY IN WATER Pts/100 pts H_2O: Insoluble

VOLATILE % Vol. % Wt.

APPEARANCE AND ODOR: Colorless crystal

SECTION IV FIRE AND EXPLOSION HAZARD DATA

FLASH POINT: Not available

METHOD USED

FLAMMABLE (EXPLOSIVE) LIMITS: UPPER / LOWER

EXTINGUISHING MEDIA: Water spray, foam, carbon dioxide, dry chemical

SPECIAL FIRE FIGHTING PROCEDURES: Self-contained breathing apparatus

UNUSUAL FIRE AND EXPLOSION HAZARDS

SECTION V HEALTH HAZARD DATA

TOXIC LEVEL: None available

CARCINOGENIC: No

PRINCIPLE ROUTES OF ABSORBTION: Ingestion, inhalation of dust

SKIN AND EYE IRRITATION: severe

REVELANT SYMPTOMS OF EXPOSURE: Eye, skin & mucous membrane irritation, headache; drowsiness, unsteadiness.

EFFECTS OF CHRONIC EXPOSURE: Liver damage

EMERGENCY AND FIRST AID PROCEDURES: Inhalation - rest; keep warm; if swallowed - wash mouth, give emetic; get medical aid; wash from skin & eyes.

SECTION VI REACTIVITY DATA

CONDITIONS CONTRIBUTING TO INSTABILITY

CONDITIONS CONTRIBUTING TO HAZARDOUS POLYMERIZATION

INCOMPATABILITY (MATERIALS TO AVOID)

HAZARDOUS DECOMPOSITION PRODUCTS

SECTION VII SPILL OR LEAK PROCEDURES

STEPS TO BE TAKEN IN CASE MATERIAL IS RELEASED OR SPILLED

WASTE DISPOSAL METHOD

SECTION VIII SPECIAL PROTECTION INFORMATION

VENTILATION REQUIREMENTS LOCAL EXHAUST

MECHANICAL (GENERAL)

SPECIAL

OTHER PROTECTIVE EQUIPMENT

PROTECTIVE EQUIPMENT (SPECIFY TYPES) EYE: Goggles

GLOVES: Rubber

RESPIRATOR: Self-contained

SECTION IX SPECIAL PRECAUTIONS

PRECAUTIONS TO BE TAKEN IN HANDLING AND STORAGE

OTHER PRECAUTIONS

SIGNATURE | DATE

2,5-DICHLOROPHENYLHYDRAZINE

1
4 0

SECTION I SOURCE AND NOMENCLATURE

MANUFACTURER'S NAME	EMERGENCY TELEPHONE NO.
ADDRESS (NUMBER, STREET, CITY, STATE, ZIP CODE)	
TRADE NAME AND SYNONYMS	CHEMICAL FAMILY
CHEMICAL NAME AND SYNONYMS 2,5 - Dichlorophenylhydrazine	FORMULA $Cl_2C_6H_3NHNH_2$

SECTION II HAZARDOUS INGREDIENTS

BASIC MATERIAL	APPROXIMATE OR MAXIMUM % WT. OR VOL.	ESTABLISHED OSHA STANDARD	LD_{50} ORAL	LD_{50} PERCUT.	LC_{50} SPECIES	LC_{50} CONC.
Lowest published lethal dose						
Oral - rat 500 mg/kg.						

SECTION III PHYSICAL DATA MW - 177.04

BOILING POINT °F.	VAPOR PRESSURE mm Hg.
MELTING POINT °F. 105°C	VAPOR DENSITY (Air = 1)
SPECIFIC GRAVITY (H_2O = 1)	EVAPORATION RATE (____ = 1)
SOLUBILITY IN WATER Pts/100 pts H_2O Slightly soluble	VOLATILE % Vol. % Wt.

APPEARANCE AND ODOR
Yellow needle-like crystal

SECTION IV FIRE AND EXPLOSION HAZARD DATA

	FLAMMABLE (EXPLOSIVE) LIMITS	
FLASH POINT Not available		UPPER
METHOD USED		LOWER

EXTINGUISHING MEDIA
Alcohol foam

SPECIAL FIRE FIGHTING PROCEDURES

UNUSUAL FIRE AND EXPLOSION HAZARDS

SECTION V HEALTH HAZARD DATA

TOXIC LEVEL None available — CARCINOGENIC No

PRINCIPLE ROUTES OF ABSORBTION Inhalation, ingestion, skin — SKIN AND EYE IRRITATION Severe

REVELANT SYMPTOMS OF EXPOSURE Irritation of skin, eyes & mucous membrane, gastroenteric disturbances, anemia

EFFECTS OF CHRONIC EXPOSURE
Blood & liver damage, loss of appetite

EMERGENCY AND FIRST AID PROCEDURES Inhalation - rest; if swallowed - induce vomiting, get medical aid; wash from skin & eyes.

SECTION VI REACTIVITY DATA

CONDITIONS CONTRIBUTING TO INSTABILITY

CONDITIONS CONTRIBUTING TO HAZARDOUS POLYMERIZATION

INCOMPATABILITY (MATERIALS TO AVOID)
Avoid lead dioxide

HAZARDOUS DECOMPOSITION PRODUCTS

SECTION VII SPILL OR LEAK PROCEDURES

STEPS TO BE TAKEN IN CASE MATERIAL IS RELEASED OR SPILLED

WASTE DISPOSAL METHOD

SECTION VIII SPECIAL PROTECTION INFORMATION

VENTILATION REQUIREMENTS LOCAL EXHAUST	PROTECTIVE EQUIPMENT (SPECIFY TYPES) EYE Goggles
MECHANICAL (GENERAL)	GLOVES Rubber
SPECIAL	RESPIRATOR Self-contained
OTHER PROTECTIVE EQUIPMENT	

SECTION IX SPECIAL PRECAUTIONS

PRECAUTIONS TO BE TAKEN IN HANDLING AND STORAGE

OTHER PRECAUTIONS

SIGNATURE	DATE

DICYCLOHEXYLAMINE

1
3 0

SECTION I SOURCE AND NOMENCLATURE

MANUFACTURER'S NAME	EMERGENCY TELEPHONE NO.
ADDRESS (NUMBER, STREET, CITY, STATE, ZIP CODE)	
TRADE NAME AND SYNONYMS	CHEMICAL FAMILY
CHEMICAL NAME AND SYNONYMS Dicyclohexylamine	FORMULA $(C_6H_{11})_2NH$

SECTION II HAZARDOUS INGREDIENTS

BASIC MATERIAL	APPROXIMATE OR MAXIMUM % WT. OR VOL.	ESTABLISHED OSHA STANDARD	LD_{50} ORAL	LD_{50} PERCUT.	LC_{50} SPECIES	LC_{50} CONC.
			373	-	rat	
Note:						
Lowest published lethal dose						
oral - mouse 500 mg/kg.						

SECTION III PHYSICAL DATA MW - 181.31

BOILING POINT °F. 255°C	VAPOR PRESSURE mm Hg.
MELTING POINT °F. 20°C	VAPOR DENSITY (Air = 1)
SPECIFIC GRAVITY (H_2O = 1) 0.93	EVAPORATION RATE (____ = 1)
SOLUBILITY IN WATER Pts/100 pts H_2O 0.16	VOLATILE % Vol. % Wt.

APPEARANCE AND ODOR
Colorless liquid, fishy odor

SECTION IV FIRE AND EXPLOSION HAZARD DATA

	FLAMMABLE (EXPLOSIVE) LIMITS	
FLASH POINT >210°F		UPPER
METHOD USED O.C.		LOWER

EXTINGUISHING MEDIA
Alcohol foam, carbon dioxide, dry chemical

SPECIAL FIRE FIGHTING PROCEDURES
Wear goggles & self-contined breathing apparatus.

UNUSUAL FIRE AND EXPLOSION HAZARDS

SECTION V HEALTH HAZARD DATA

TOXIC LEVEL 20 PPM (suggested) — CARCINOGENIC No

PRINCIPLE ROUTES OF ABSORBTION Inhalation, ingestion — SKIN AND EYE IRRITATION Severe

REVELANT SYMPTOMS OF EXPOSURE Irritation of eyes and respiratory system, if swallowed; irritation of digestive tract, drowsiness, nausea.

EFFECTS OF CHRONIC EXPOSURE
Weight loss

EMERGENCY AND FIRST AID PROCEDURES If swallowed - induce vomiting, get medical aid; inhalation - give oxygen as needed; wash skin & eyes.

SECTION VI REACTIVITY DATA

CONDITIONS CONTRIBUTING TO INSTABILITY

CONDITIONS CONTRIBUTING TO HAZARDOUS POLYMERIZATION

INCOMPATABILITY (MATERIALS TO AVOID)
May react with oxidizing materials

HAZARDOUS DECOMPOSITION PRODUCTS

SECTION VII SPILL OR LEAK PROCEDURES

STEPS TO BE TAKEN IN CASE MATERIAL IS RELEASED OR SPILLED

WASTE DISPOSAL METHOD

SECTION VIII SPECIAL PROTECTION INFORMATION

VENTILATION REQUIREMENTS LOCAL EXHAUST	PROTECTIVE EQUIPMENT (SPECIFY TYPES) EYE Goggles
MECHANICAL (GENERAL)	GLOVES Rubber
SPECIAL	RESPIRATOR Self-contained
OTHER PROTECTIVE EQUIPMENT	

SECTION IX SPECIAL PRECAUTIONS

PRECAUTIONS TO BE TAKEN IN HANDLING AND STORAGE

OTHER PRECAUTIONS

SIGNATURE	DATE

DIETHYLAMINE HYDROCHLORIDE

(Hazard diamond: 1, 3, 0)

SECTION I SOURCE AND NOMENCLATURE

MANUFACTURER'S NAME

EMERGENCY TELEPHONE NO.

ADDRESS (NUMBER, STREET, CITY, STATE, ZIP CODE)

TRADE NAME AND SYNONYMS

CHEMICAL FAMILY

CHEMICAL NAME AND SYNONYMS: Diethylamine Hydrochloride

FORMULA: $(C_2H_5)_2NH_2Cl$

SECTION II HAZARDOUS INGREDIENTS

BASIC MATERIAL	APPROXIMATE OR MAXIMUM % WT. OR VOL.	ESTABLISHED OSHA STANDARD	LD_{50} ORAL	LD_{50} PERCUT.	LC_{50} SPECIES	LC_{50} CONC.
			none	available		

SECTION III PHYSICAL DATA MW - 109.60

BOILING POINT °F: 330°C

VAPOR PRESSURE mm Hg

MELTING POINT °F: 229°C

VAPOR DENSITY (Air = 1)

SPECIFIC GRAVITY (H_2O = 1): 1.05

EVAPORATION RATE (______ = 1)

SOLUBILITY IN WATER Pts/100 pts H_2O: 232

VOLATILE % Vol. % Wt.

APPEARANCE AND ODOR: White crystal

SECTION IV FIRE AND EXPLOSION HAZARD DATA

FLASH POINT: Not available

FLAMMABLE (EXPLOSIVE) LIMITS: UPPER / LOWER

METHOD USED

EXTINGUISHING MEDIA: Alcohol foam, carbon dioxide, dry chemical

SPECIAL FIRE FIGHTING PROCEDURES: Wear self-contained breathing apparatus

UNUSUAL FIRE AND EXPLOSION HAZARDS

SECTION V HEALTH HAZARD DATA

TOXIC LEVEL: Not available

CARCINOGENIC: No

PRINCIPLE ROUTES OF ABSORBTION: Ingestion, inhalation of dust

SKIN AND EYE IRRITATION: moderate

REVELANT SYMPTOMS OF EXPOSURE: Eye and mucous membrane irritation, irritation of gastroenteric tract.

EFFECTS OF CHRONIC EXPOSURE: Irritation of cornea and lungs.

EMERGENCY AND FIRST AID PROCEDURES: If swallowed - drink water, get medical aid; inhalation - rest; wash from skin & eyes.

SECTION VI REACTIVITY DATA

CONDITIONS CONTRIBUTING TO INSTABILITY

CONDITIONS CONTRIBUTING TO HAZARDOUS POLYMERIZATION

INCOMPATABILITY (MATERIALS TO AVOID)

HAZARDOUS DECOMPOSITION PRODUCTS

SECTION VII SPILL OR LEAK PROCEDURES

STEPS TO BE TAKEN IN CASE MATERIAL IS RELEASED OR SPILLED

WASTE DISPOSAL METHOD

SECTION VIII SPECIAL PROTECTION INFORMATION

VENTILATION REQUIREMENTS LOCAL EXHAUST

PROTECTIVE EQUIPMENT (SPECIFY TYPES) EYE: Goggles

MECHANICAL (GENERAL)

GLOVES: Rubber

SPECIAL

RESPIRATOR: Self-contained

OTHER PROTECTIVE EQUIPMENT

SECTION IX SPECIAL PRECAUTIONS

PRECAUTIONS TO BE TAKEN IN HANDLING AND STORAGE

OTHER PRECAUTIONS

SIGNATURE

DATE

DIETHYLAMINOCHLOROBENZENE

(Hazard diamond: 1, 4, 0)

SECTION I SOURCE AND NOMENCLATURE

MANUFACTURER'S NAME

EMERGENCY TELEPHONE NO.

ADDRESS (NUMBER, STREET, CITY, STATE, ZIP CODE)

TRADE NAME AND SYNONYMS

CHEMICAL FAMILY

CHEMICAL NAME AND SYNONYMS: N,N Diethyl chloroaniline

FORMULA: $(C_2H_5)_2NC_6H_4Cl$

SECTION II HAZARDOUS INGREDIENTS

BASIC MATERIAL	APPROXIMATE OR MAXIMUM % WT. OR VOL.	ESTABLISHED OSHA STANDARD	LD_{50} ORAL	LD_{50} PERCUT.	LC_{50} SPECIES	LC_{50} CONC.
			None	available		

SECTION III PHYSICAL DATA MW - 183.69

BOILING POINT °F: --

VAPOR PRESSURE mm Hg: 1.5 at 96°C

MELTING POINT °F: 39°C.

VAPOR DENSITY (Air = 1)

SPECIFIC GRAVITY (H_2O = 1)

EVAPORATION RATE (______ = 1)

SOLUBILITY IN WATER Pts/100 pts H_2O

VOLATILE % Vol. % Wt.

APPEARANCE AND ODOR: Needle-like crystals.

SECTION IV FIRE AND EXPLOSION HAZARD DATA

FLASH POINT: None available

FLAMMABLE (EXPLOSIVE) LIMITS: UPPER / LOWER

METHOD USED

EXTINGUISHING MEDIA: Alcohol foam, carbon dioxide and dry chemical.

SPECIAL FIRE FIGHTING PROCEDURES

UNUSUAL FIRE AND EXPLOSION HAZARDS

SECTION V HEALTH HAZARD DATA

TOXIC LEVEL: None available

CARCINOGENIC: No

PRINCIPLE ROUTES OF ABSORBTION: Inhalation, ingestion skin

SKIN AND EYE IRRITATION: Moderate

REVELANT SYMPTOMS OF EXPOSURE: Weakness, tremors, cyanosis, drowsiness, central nervous system depressed.

EFFECTS OF CHRONIC EXPOSURE: Fatigue, loss of appetite, headache, dizziness.

EMERGENCY AND FIRST AID PROCEDURES: Inhalation - rest, get medical aid; if swallowed - give emetic; get medical aid, wash skin & eyes.

SECTION VI REACTIVITY DATA

CONDITIONS CONTRIBUTING TO INSTABILITY

CONDITIONS CONTRIBUTING TO HAZARDOUS POLYMERIZATION

INCOMPATABILITY (MATERIALS TO AVOID)

HAZARDOUS DECOMPOSITION PRODUCTS

SECTION VII SPILL OR LEAK PROCEDURES

STEPS TO BE TAKEN IN CASE MATERIAL IS RELEASED OR SPILLED

WASTE DISPOSAL METHOD

SECTION VIII SPECIAL PROTECTION INFORMATION

VENTILATION REQUIREMENTS LOCAL EXHAUST

PROTECTIVE EQUIPMENT (SPECIFY TYPES) EYE: Goggles

MECHANICAL (GENERAL)

GLOVES: Polyvinyl alcohol

SPECIAL

RESPIRATOR: With proper filter

OTHER PROTECTIVE EQUIPMENT

SECTION IX SPECIAL PRECAUTIONS

PRECAUTIONS TO BE TAKEN IN HANDLING AND STORAGE

OTHER PRECAUTIONS

SIGNATURE

DATE

2-DIETHYLAMINO-1,4-DIMETHYLBENZENE

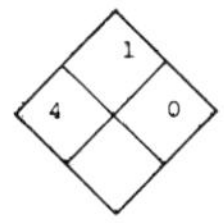

SECTION I SOURCE AND NOMENCLATURE

MANUFACTURER'S NAME

EMERGENCY TELEPHONE NO.

ADDRESS (NUMBER, STREET, CITY, STATE, ZIP CODE)

TRADE NAME AND SYNONYMS

CHEMICAL FAMILY

CHEMICAL NAME AND SYNONYMS
N,N Diethyl 3,6 Dimethylaniline

FORMULA
$(C_2H_5)_2NC_6H_3(CH_3)_2$

SECTION II HAZARDOUS INGREDIENTS

BASIC MATERIAL	APPROXIMATE OR MAXIMUM % WT. OR VOL.	ESTABLISHED OSHA STANDARD	LD_{50}		LC_{50}	
			ORAL	PERCUT.	SPECIES	CONC.
			None	available		

SECTION III PHYSICAL DATA

BOILING POINT °F. --

VAPOR PRESSURE at 96° C. 11 mm Hg.

MELTING POINT °F. --

VAPOR DENSITY (Air = 1)

SPECIFIC GRAVITY (H_2O = 1)

EVAPORATION RATE (= 1)

SOLUBILITY IN WATER Pts/100 pts H_2O

VOLATILE % Vol. % Wt.

APPEARANCE AND ODOR
Colorless to yellow liquid.

SECTION IV FIRE AND EXPLOSION HAZARD DATA

FLASH POINT
None available

METHOD USED

FLAMMABLE (EXPLOSIVE) LIMITS: UPPER / LOWER

EXTINGUISHING MEDIA
Alcohol foam, dry chemical, carbon dioxide.

SPECIAL FIRE FIGHTING PROCEDURES

UNUSUAL FIRE AND EXPLOSION HAZARDS

SECTION V HEALTH HAZARD DATA

TOXIC LEVEL
None available

CARCINOGENIC No

PRINCIPLE ROUTES OF ABSORBTION
Inhalation, ingestion, skin

SKIN AND EYE IRRITATION Moderate

REVELANT SYMPTOMS OF EXPOSURE Weakness, tremors, cyanosis, drowsiness, central nervous system depressant.

EFFECTS OF CHRONIC EXPOSURE
Fatigue, loss of appetite, headache, dizziness.

EMERGENCY AND FIRST AID PROCEDURES
Inhalation - rest; if swallowed - give emetic, get medical aid; wash skin & eyes.

SECTION VI REACTIVITY DATA

CONDITIONS CONTRIBUTING TO INSTABILITY

CONDITIONS CONTRIBUTING TO HAZARDOUS POLYMERIZATION

INCOMPATABILITY (MATERIALS TO AVOID)

HAZARDOUS DECOMPOSITION PRODUCTS

SECTION VII SPILL OR LEAK PROCEDURES

STEPS TO BE TAKEN IN CASE MATERIAL IS RELEASED OR SPILLED

WASTE DISPOSAL METHOD

SECTION VIII SPECIAL PROTECTION INFORMATION

VENTILATION REQUIREMENTS LOCAL EXHAUST

PROTECTIVE EQUIPMENT (SPECIFY TYPES) EYE
Goggles

MECHANICAL (GENERAL)

GLOVES
Polyvinyl Alcohol

SPECIAL

RESPIRATOR
With proper filter

OTHER PROTECTIVE EQUIPMENT

SECTION IX SPECIAL PRECAUTIONS

PRECAUTIONS TO BE TAKEN IN HANDLING AND STORAGE

OTHER PRECAUTIONS

SIGNATURE

DATE

4-DIETHYLAMINO-1,3-DIMETHYLBENZENE

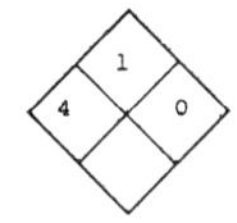

SECTION I SOURCE AND NOMENCLATURE

MANUFACTURER'S NAME

EMERGENCY TELEPHONE NO.

ADDRESS (NUMBER, STREET, CITY, STATE, ZIP CODE)

TRADE NAME AND SYNONYMS

CHEMICAL FAMILY

CHEMICAL NAME AND SYNONYMS
N,N-Diethyl 2,4-Dimethylaniline

FORMULA
$(C_2H_5)_2NC_6H_3(CH_3)_2$

SECTION II HAZARDOUS INGREDIENTS

BASIC MATERIAL	APPROXIMATE OR MAXIMUM % WT. OR VOL.	ESTABLISHED OSHA STANDARD	LD_{50}		LC_{50}	
			ORAL	PERCUT.	SPECIES	CONC.
			None	available		

SECTION III PHYSICAL DATA

BOILING POINT °F. --

VAPOR PRESSURE at 102° C. 14 mm Hg.

MELTING POINT °F. --

VAPOR DENSITY (Air = 1)

SPECIFIC GRAVITY (H_2O = 1)

EVAPORATION RATE (= 1)

SOLUBILITY IN WATER Pts/100 pts H_2O

VOLATILE % Vol. % Wt.

APPEARANCE AND ODOR
Colorelss to yellow liquid.

SECTION IV FIRE AND EXPLOSION HAZARD DATA

FLASH POINT
None available

METHOD USED

FLAMMABLE (EXPLOSIVE) LIMITS: UPPER / LOWER

EXTINGUISHING MEDIA
Alcohol foam, dry chemical, carbon dioxide.

SPECIAL FIRE FIGHTING PROCEDURES

UNUSUAL FIRE AND EXPLOSION HAZARDS

SECTION V HEALTH HAZARD DATA

TOXIC LEVEL
None available

CARCINOGENIC No

PRINCIPLE ROUTES OF ABSORBTION
Inhalation, ingestion, skin

SKIN AND EYE IRRITATION Moderate

REVELANT SYMPTOMS OF EXPOSURE Weakness, tremors, cyanosis, drowsiness, central nervous system depressant.

EFFECTS OF CHRONIC EXPOSURE
Fatigue, loss of appetite, headache, dizziness.

EMERGENCY AND FIRST AID PROCEDURES
Inhalation - rest; if swallowed - give emetic, get medical aid; wash skin and eyes.

SECTION VI REACTIVITY DATA

CONDITIONS CONTRIBUTING TO INSTABILITY

CONDITIONS CONTRIBUTING TO HAZARDOUS POLYMERIZATION

INCOMPATABILITY (MATERIALS TO AVOID)

HAZARDOUS DECOMPOSITION PRODUCTS

SECTION VII SPILL OR LEAK PROCEDURES

STEPS TO BE TAKEN IN CASE MATERIAL IS RELEASED OR SPILLED

WASTE DISPOSAL METHOD

SECTION VIII SPECIAL PROTECTION INFORMATION

VENTILATION REQUIREMENTS LOCAL EXHAUST

PROTECTIVE EQUIPMENT (SPECIFY TYPES) EYE
Goggles

MECHANICAL (GENERAL)

GLOVES
Polyvinyl alcohol

SPECIAL

RESPIRATOR
With proper filter

OTHER PROTECTIVE EQUIPMENT

SECTION IX SPECIAL PRECAUTIONS

PRECAUTIONS TO BE TAKEN IN HANDLING AND STORAGE

OTHER PRECAUTIONS

SIGNATURE

DATE

β-DIETHYLAMINOETHYL ALCOHOL

Hazard diamond: 2 (top), 4 (left), 0 (right)

SECTION I SOURCE AND NOMENCLATURE

MANUFACTURER'S NAME | EMERGENCY TELEPHONE NO.

ADDRESS (NUMBER, STREET, CITY, STATE, ZIP CODE)

TRADE NAME AND SYNONYMS | CHEMICAL FAMILY

CHEMICAL NAME AND SYNONYMS: 2 diethylaminoethanol | FORMULA: $(C_2H_5)_2NCH_2CH_2OH$

SECTION II HAZARDOUS INGREDIENTS

BASIC MATERIAL	APPROXIMATE OR MAXIMUM % WT. OR VOL.	ESTABLISHED OSHA STANDARD	LD_{50} mg/kg ORAL	LD_{50} mg/kg PERCUT.	LD_{50} mg/kg SPECIES	LC_{50} CONC.
			1300		Rat	
TCLO inhalation - human	100 PPM - CNS			1260	Rabbit	

SECTION III PHYSICAL DATA

BOILING POINT (°F): 322°F

VAPOR PRESSURE (mm Hg): at 20°C 1.4

MELTING POINT (°F): --

VAPOR DENSITY (Air = 1): 4.0

SPECIFIC GRAVITY (H_2O = 1): 0.9

EVAPORATION RATE (______ = 1)

SOLUBILITY IN WATER Pts/100 pts H_2O: Yes

VOLATILE % Vol. % Wt.

APPEARANCE AND ODOR: Colorless liquid

SECTION IV FIRE AND EXPLOSION HAZARD DATA

FLASH POINT: 140°C

METHOD USED

FLAMMABLE (EXPLOSIVE) LIMITS: UPPER | LOWER

EXTINGUISHING MEDIA: Alcohol foam, carbon dioxide, dry chemical

SPECIAL FIRE FIGHTING PROCEDURES

UNUSUAL FIRE AND EXPLOSION HAZARDS

SECTION V HEALTH HAZARD DATA

TOXIC LEVEL: TLV - 10 PPM, 50 mg/C.M.

CARCINOGENIC: No

PRINCIPLE ROUTES OF ABSORBTION: Skin, inhalation, ingestion

SKIN AND EYE IRRITATION: Severe

REVELANT SYMPTOMS OF EXPOSURE: Dermatitis, weight loss, slight increase in clotting time

EFFECTS OF CHRONIC EXPOSURE: Dermatitis

EMERGENCY AND FIRST AID PROCEDURES: Wash from skin and eyes; if swallowed - drink fruit juice, get medical aid; inhalation - rest.

SECTION VI REACTIVITY DATA

CONDITIONS CONTRIBUTING TO INSTABILITY

CONDITIONS CONTRIBUTING TO HAZARDOUS POLYMERIZATION

INCOMPATABILITY (MATERIALS TO AVOID)

HAZARDOUS DECOMPOSITION PRODUCTS

SECTION VII SPILL OR LEAK PROCEDURES

STEPS TO BE TAKEN IN CASE MATERIAL IS RELEASED OR SPILLED

WASTE DISPOSAL METHOD

SECTION VIII SPECIAL PROTECTION INFORMATION

VENTILATION REQUIREMENTS LOCAL EXHAUST

MECHANICAL (GENERAL)

SPECIAL

OTHER PROTECTIVE EQUIPMENT

PROTECTIVE EQUIPMENT (SPECIFY TYPES) EYE: Safety goggles

GLOVES: Rubber

RESPIRATOR: With proper filter

SECTION IX SPECIAL PRECAUTIONS

PRECAUTIONS TO BE TAKEN IN HANDLING AND STORAGE

OTHER PRECAUTIONS

SIGNATURE | DATE

DIETHYLANILINE

Hazard diamond: 2 (top), 4 (left), 0 (right)

SECTION I SOURCE AND NOMENCLATURE

MANUFACTURER'S NAME | EMERGENCY TELEPHONE NO.

ADDRESS (NUMBER, STREET, CITY, STATE, ZIP CODE)

TRADE NAME AND SYNONYMS: diethylphenylamine | CHEMICAL FAMILY

CHEMICAL NAME AND SYNONYMS: NN diethylaniline | FORMULA: $C_6H_5N(C_2H)_5$

SECTION II HAZARDOUS INGREDIENTS

BASIC MATERIAL	APPROXIMATE OR MAXIMUM % WT. OR VOL.	ESTABLISHED OSHA STANDARD	LD_{50} mg/kg ORAL	LD_{50} mg/kg PERCUT.	LD_{50} mg/kg SPECIES	LC_{50} CONC.
			782		Rat	

SECTION III PHYSICAL DATA MW = 149.23

BOILING POINT (°F): 215.5°C

VAPOR PRESSURE (mm Hg): at 49.7°C 1

MELTING POINT (°F): -38°C

VAPOR DENSITY (Air = 1): 5.15

SPECIFIC GRAVITY (H_2O = 1): 0.9351

EVAPORATION RATE (______ = 1)

SOLUBILITY IN WATER Pts/100 pts H_2O: 1.44

VOLATILE % Vol. % Wt.

APPEARANCE AND ODOR: Colorless to yellow liquid

SECTION IV FIRE AND EXPLOSION HAZARD DATA

FLASH POINT: 185°F

METHOD USED

FLAMMABLE (EXPLOSIVE) LIMITS: UPPER | LOWER

EXTINGUISHING MEDIA: Alcohol foam, water spray, carbon dioxide, dry chemical

SPECIAL FIRE FIGHTING PROCEDURES

UNUSUAL FIRE AND EXPLOSION HAZARDS

SECTION V HEALTH HAZARD DATA

TOXIC LEVEL: not available

CARCINOGENIC: no

PRINCIPLE ROUTES OF ABSORBTION: Inhalation, ingestion, skin

SKIN AND EYE IRRITATION: yes

REVELANT SYMPTOMS OF EXPOSURE: Weakness, tremors, cyanosis, drowsiness, central nervous system depressant

EFFECTS OF CHRONIC EXPOSURE: Fatigue, loss of appetite, headache, dizziness

EMERGENCY AND FIRST AID PROCEDURES: Inhalation - rest, get medical aid; if swallowed - wash mouth, give emetic, get medical aid; wash from eyes & skin.

SECTION VI REACTIVITY DATA

CONDITIONS CONTRIBUTING TO INSTABILITY

CONDITIONS CONTRIBUTING TO HAZARDOUS POLYMERIZATION

INCOMPATABILITY (MATERIALS TO AVOID)

HAZARDOUS DECOMPOSITION PRODUCTS

SECTION VII SPILL OR LEAK PROCEDURES

STEPS TO BE TAKEN IN CASE MATERIAL IS RELEASED OR SPILLED

WASTE DISPOSAL METHOD

SECTION VIII SPECIAL PROTECTION INFORMATION

VENTILATION REQUIREMENTS LOCAL EXHAUST

MECHANICAL (GENERAL)

SPECIAL

OTHER PROTECTIVE EQUIPMENT

PROTECTIVE EQUIPMENT (SPECIFY TYPES) EYE: Goggles

GLOVES: Polyvinyl alcohol

RESPIRATOR: With proper filter

SECTION IX SPECIAL PRECAUTIONS

PRECAUTIONS TO BE TAKEN IN HANDLING AND STORAGE

OTHER PRECAUTIONS

SIGNATURE | DATE

DIETHYL-m-TOLUIDINE

2
4 0

SECTION I SOURCE AND NOMENCLATURE

MANUFACTURER'S NAME | EMERGENCY TELEPHONE NO.

ADDRESS (NUMBER, STREET, CITY, STATE, ZIP CODE)

TRADE NAME AND SYNONYMS | CHEMICAL FAMILY

CHEMICAL NAME AND SYNONYMS: N,N,-Diethyl-M-Toluidine

FORMULA: $(C_2H_5)_2NC_6H_4CH_3$

SECTION II HAZARDOUS INGREDIENTS

BASIC MATERIAL	APPROXIMATE OR MAXIMUM % WT. OR VOL.	ESTABLISHED OSHA STANDARD	LD_{50} ORAL	LD_{50} PERCUT.	LC_{50} SPECIES	LC_{50} CONC.
			None	available		

SECTION III PHYSICAL DATA

BOILING POINT °F.: --

VAPOR PRESSURE mm Hg.: 97°C at 7

MELTING POINT °F.: --

VAPOR DENSITY (Air = 1)

SPECIFIC GRAVITY (H_2O = 1)

EVAPORATION RATE (______ = 1)

SOLUBILITY IN WATER Pts/100 pts H_2O

VOLATILE % Vol. % Wt.

APPEARANCE AND ODOR: Dark colored liquid

SECTION IV FIRE AND EXPLOSION HAZARD DATA

FLASH POINT: None available

METHOD USED

FLAMMABLE (EXPLOSIVE) LIMITS: UPPER | LOWER

EXTINGUISHING MEDIA: Carbon dioxide, dry chemical, foam or water spray

SPECIAL FIRE FIGHTING PROCEDURES: Wear full protective clothing

UNUSUAL FIRE AND EXPLOSION HAZARDS: Emits toxic fumes when heated

SECTION V HEALTH HAZARD DATA

TOXIC LEVEL: Not available

CARCINOGENIC: no

PRINCIPLE ROUTES OF ABSORBTION: Inhalation, ingestion, skin

SKIN AND EYE IRRITATION: yes

REVELANT SYMPTOMS OF EXPOSURE: Cyanosis, air hunger, nausea, vomiting, low blood pressure, convulsions

EFFECTS OF CHRONIC EXPOSURE: Weight loss, anemia, weakness

EMERGENCY AND FIRST AID PROCEDURES: Inhalation - rest; if swallowed - give emetic, get medical aid; wash skin & eyes, skin may be washed with a 5% solution of acetic acid to aid removal.

SECTION VI REACTIVITY DATA

CONDITIONS CONTRIBUTING TO INSTABILITY

CONDITIONS CONTRIBUTING TO HAZARDOUS POLYMERIZATION

INCOMPATABILITY (MATERIALS TO AVOID)

HAZARDOUS DECOMPOSITION PRODUCTS

SECTION VII SPILL OR LEAK PROCEDURES

STEPS TO BE TAKEN IN CASE MATERIAL IS RELEASED OR SPILLED

WASTE DISPOSAL METHOD

SECTION VIII SPECIAL PROTECTION INFORMATION

VENTILATION REQUIREMENTS LOCAL EXHAUST

PROTECTIVE EQUIPMENT (SPECIFY TYPES) EYE: Goggles

MECHANICAL (GENERAL)

GLOVES: Rubber

SPECIAL

RESPIRATOR: With proper filter

OTHER PROTECTIVE EQUIPMENT

SECTION IX SPECIAL PRECAUTIONS

PRECAUTIONS TO BE TAKEN IN HANDLING AND STORAGE

OTHER PRECAUTIONS

SIGNATURE | DATE

DIETHYL-o-TOLUIDINE

2
4 0

SECTION I SOURCE AND NOMENCLATURE

MANUFACTURER'S NAME | EMERGENCY TELEPHONE NO.

ADDRESS (NUMBER, STREET, CITY, STATE, ZIP CODE)

TRADE NAME AND SYNONYMS | CHEMICAL FAMILY

CHEMICAL NAME AND SYNONYMS: N,N-Diethyl-O -Toluidine

FORMULA: $(C_2H_5)_2NC_6H_4CH_3$

SECTION II HAZARDOUS INGREDIENTS

BASIC MATERIAL	APPROXIMATE OR MAXIMUM % WT. OR VOL.	ESTABLISHED OSHA STANDARD	LD_{50} ORAL	LD_{50} PERCUT.	LC_{50} SPECIES	LC_{50} CONC.
			none	available		

SECTION III PHYSICAL DATA MW135.23

BOILING POINT °F.: 209°C

VAPOR PRESSURE mm Hg.: 10 at 86°C

MELTING POINT °F.: --

VAPOR DENSITY (Air = 1)

SPECIFIC GRAVITY (H_2O = 1): --

EVAPORATION RATE (______ = 1)

SOLUBILITY IN WATER Pts/100 pts H_2O

VOLATILE % Vol. % Wt.

APPEARANCE AND ODOR: Dark colored liquid

SECTION IV FIRE AND EXPLOSION HAZARD DATA

FLASH POINT: None available

METHOD USED

FLAMMABLE (EXPLOSIVE) LIMITS: UPPER | LOWER

EXTINGUISHING MEDIA: Carbon dioxide, dry chemical, foam or water spray

SPECIAL FIRE FIGHTING PROCEDURES: Wear full protective clothing

UNUSUAL FIRE AND EXPLOSION HAZARDS: Emits toxic fumes when heated

SECTION V HEALTH HAZARD DATA

TOXIC LEVEL: None available

CARCINOGENIC: No

PRINCIPLE ROUTES OF ABSORBTION: Inhalation, ingestion, skin

SKIN AND EYE IRRITATION: yes

REVELANT SYMPTOMS OF EXPOSURE: Cyanosis, air hunger, nausea, vomiting, low blood pressure, convulsions

EFFECTS OF CHRONIC EXPOSURE: Weight loss, anemia, weakness

EMERGENCY AND FIRST AID PROCEDURES: Inhalation - rest; if swallowed - give emetic; get medical aid; wash skin & eyes, skin may be washed with a 5% acetic acid to aid removal.

SECTION VI REACTIVITY DATA

CONDITIONS CONTRIBUTING TO INSTABILITY

CONDITIONS CONTRIBUTING TO HAZARDOUS POLYMERIZATION

INCOMPATABILITY (MATERIALS TO AVOID): Can react with oxidizing materials

HAZARDOUS DECOMPOSITION PRODUCTS

SECTION VII SPILL OR LEAK PROCEDURES

STEPS TO BE TAKEN IN CASE MATERIAL IS RELEASED OR SPILLED

WASTE DISPOSAL METHOD

SECTION VIII SPECIAL PROTECTION INFORMATION

VENTILATION REQUIREMENTS LOCAL EXHAUST

PROTECTIVE EQUIPMENT (SPECIFY TYPES) EYE: Goggles

MECHANICAL (GENERAL)

GLOVES: Rubber

SPECIAL

RESPIRATOR: With proper filter

OTHER PROTECTIVE EQUIPMENT

SECTION IX SPECIAL PRECAUTIONS

PRECAUTIONS TO BE TAKEN IN HANDLING AND STORAGE

OTHER PRECAUTIONS

SIGNATURE | DATE

DIETHYL-p-TOLUIDINE

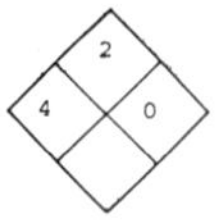

SECTION I SOURCE AND NOMENCLATURE

MANUFACTURER'S NAME | EMERGENCY TELEPHONE NO.

ADDRESS (NUMBER, STREET, CITY, STATE, ZIP CODE)

TRADE NAME AND SYNONYMS | CHEMICAL FAMILY

CHEMICAL NAME AND SYNONYMS: N,N-Diethyl-P-Toluidine

FORMULA: $(C_2H_5)_2NC_6H_4CH_3$

SECTION II HAZARDOUS INGREDIENTS

BASIC MATERIAL	APPROXIMATE OR MAXIMUM % WT. OR VOL.	ESTABLISHED OSHA STANDARD	LD_{50} ORAL	LD_{50} PERCUT.	LC_{50} SPECIES	LC_{50} CONC.
			None available			

SECTION III PHYSICAL DATA MW - 163.26

BOILING POINT °F.: 229°C

VAPOR PRESSURE mm Hg.: 10 at 103°C

MELTING POINT °F.:

VAPOR DENSITY (Air = 1):

SPECIFIC GRAVITY (H_2O = 1): 0.92

EVAPORATION RATE (____ = 1):

SOLUBILITY IN WATER Pts/100 pts H_2O:

VOLATILE % Vol. % Wt.:

APPEARANCE AND ODOR: Dark colored liquid

SECTION IV FIRE AND EXPLOSION HAZARD DATA

FLASH POINT: Not available

FLAMMABLE (EXPLOSIVE) LIMITS: UPPER / LOWER

METHOD USED:

EXTINGUISHING MEDIA: Carbon dioxide, dry chemical, foam or water spray

SPECIAL FIRE FIGHTING PROCEDURES: Wear full protective clothing

UNUSUAL FIRE AND EXPLOSION HAZARDS: Emits toxic fumes when heated

SECTION V HEALTH HAZARD DATA

TOXIC LEVEL: Not available

CARCINOGENIC: no

PRINCIPLE ROUTES OF ABSORBTION: Inhalation, ingestion, skin

SKIN AND EYE IRRITATION: yes

REVELANT SYMPTOMS OF EXPOSURE: Cyanosis, air hunger, nausea, vomiting, low blood pressure, convulsions

EFFECTS OF CHRONIC EXPOSURE: Weight loss, anemia, weakness

EMERGENCY AND FIRST AID PROCEDURES: Inhalation - rest; if swallowed - give emetic, get medical aid; wash from skin & eyes, skin may be washed with a 5% acetic acid solution to aid removal.

SECTION VI REACTIVITY DATA

CONDITIONS CONTRIBUTING TO INSTABILITY

CONDITIONS CONTRIBUTING TO HAZARDOUS POLYMERIZATION

INCOMPATABILITY (MATERIALS TO AVOID): May react vigorously with oxidizing materials.

HAZARDOUS DECOMPOSITION PRODUCTS

SECTION VII SPILL OR LEAK PROCEDURES

STEPS TO BE TAKEN IN CASE MATERIAL IS RELEASED OR SPILLED

WASTE DISPOSAL METHOD

SECTION VIII SPECIAL PROTECTION INFORMATION

VENTILATION REQUIREMENTS LOCAL EXHAUST

PROTECTIVE EQUIPMENT (SPECIFY TYPES) EYE: Goggles

MECHANICAL (GENERAL)

GLOVES: Rubber

SPECIAL

RESPIRATOR: With proper filter

OTHER PROTECTIVE EQUIPMENT

SECTION IX SPECIAL PRECAUTIONS

PRECAUTIONS TO BE TAKEN IN HANDLING AND STORAGE

OTHER PRECAUTIONS

SIGNATURE | DATE

p,p'-DIFLUORODIPHENYL

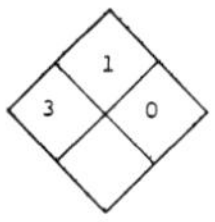

SECTION I SOURCE AND NOMENCLATURE

MANUFACTURER'S NAME | EMERGENCY TELEPHONE NO.

ADDRESS (NUMBER, STREET, CITY, STATE, ZIP CODE)

TRADE NAME AND SYNONYMS | CHEMICAL FAMILY

CHEMICAL NAME AND SYNONYMS: 4,4 Difluorobiphenyl

FORMULA: $(F\ C_6H_4)_2$

SECTION II HAZARDOUS INGREDIENTS

BASIC MATERIAL	APPROXIMATE OR MAXIMUM % WT. OR VOL.	ESTABLISHED OSHA STANDARD	LD_{50} ORAL	LD_{50} PERCUT.	LC_{50} SPECIES	LC_{50} CONC.
			None available			

SECTION III PHYSICAL DATA MW - 190.18

BOILING POINT °F.: 255°C

VAPOR PRESSURE mm Hg.:

MELTING POINT °F.: 95°C

VAPOR DENSITY (Air = 1):

SPECIFIC GRAVITY (H_2O = 1): 1.04

EVAPORATION RATE (____ = 1):

SOLUBILITY IN WATER Pts/100 pts H_2O: Insoluble

VOLATILE % Vol. % Wt.:

APPEARANCE AND ODOR: Crystal power, aromatic odor

SECTION IV FIRE AND EXPLOSION HAZARD DATA

FLASH POINT: None available

FLAMMABLE (EXPLOSIVE) LIMITS: UPPER / LOWER

METHOD USED:

EXTINGUISHING MEDIA: Water, foam, carbon dioxide or dry chemical

SPECIAL FIRE FIGHTING PROCEDURES

UNUSUAL FIRE AND EXPLOSION HAZARDS

SECTION V HEALTH HAZARD DATA

TOXIC LEVEL: None available

CARCINOGENIC: no

PRINCIPLE ROUTES OF ABSORBTION: Inhalation, ingestion, skin

SKIN AND EYE IRRITATION: moderate

REVELANT SYMPTOMS OF EXPOSURE: Serious liver damage, acne, irritation of respiratory passages

EFFECTS OF CHRONIC EXPOSURE: Liver damage

EMERGENCY AND FIRST AID PROCEDURES: Inhalation - rest, get medical aid; if swallowed - give emetic, get medical aid; wash skin & eyes.

SECTION VI REACTIVITY DATA

CONDITIONS CONTRIBUTING TO INSTABILITY

CONDITIONS CONTRIBUTING TO HAZARDOUS POLYMERIZATION

INCOMPATABILITY (MATERIALS TO AVOID)

HAZARDOUS DECOMPOSITION PRODUCTS

SECTION VII SPILL OR LEAK PROCEDURES

STEPS TO BE TAKEN IN CASE MATERIAL IS RELEASED OR SPILLED

WASTE DISPOSAL METHOD

SECTION VIII SPECIAL PROTECTION INFORMATION

VENTILATION REQUIREMENTS LOCAL EXHAUST

PROTECTIVE EQUIPMENT (SPECIFY TYPES) EYE: Goggles

MECHANICAL (GENERAL)

GLOVES

SPECIAL

RESPIRATOR: With proper filter

OTHER PROTECTIVE EQUIPMENT

SECTION IX SPECIAL PRECAUTIONS

PRECAUTIONS TO BE TAKEN IN HANDLING AND STORAGE

OTHER PRECAUTIONS

SIGNATURE | DATE

1,3-DIHYDROXY-4-CHLOROBENZENE

Hazard rating: 1 (top), 4 (left), 0 (right)

SECTION I SOURCE AND NOMENCLATURE

MANUFACTURER'S NAME	EMERGENCY TELEPHONE NO.
ADDRESS (NUMBER, STREET, CITY, STATE, ZIP CODE)	
TRADE NAME AND SYNONYMS	CHEMICAL FAMILY
CHEMICAL NAME AND SYNONYMS: 4 Chloro Resorcinol	FORMULA: $Cl(C_6H_3)(OH)_2$

SECTION II HAZARDOUS INGREDIENTS

BASIC MATERIAL	APPROXIMATE OR MAXIMUM % WT. OR VOL.	ESTABLISHED OSHA STANDARD	LD_{50} ORAL	LD_{50} PERCUT.	LC_{50} SPECIES	LC_{50} CONC.

SECTION III PHYSICAL DATA MW - 144.56

BOILING POINT °F.: 256° C.	VAPOR PRESSURE mm Hg.
MELTING POINT °F.: 107°C.	VAPOR DENSITY (Air = 1)
SPECIFIC GRAVITY (H_2O = 1)	EVAPORATION RATE (______ = 1)
SOLUBILITY IN WATER Pts/100 pts H_2O: Soluble	VOLATILE % Vol. % Wt.

APPEARANCE AND ODOR: White amorphous

SECTION IV FIRE AND EXPLOSION HAZARD DATA

FLASH POINT: None available	FLAMMABLE (EXPLOSIVE) LIMITS	UPPER
METHOD USED		LOWER

EXTINGUISHING MEDIA: Water, carbon dioxide, dry chemical.

SPECIAL FIRE FIGHTING PROCEDURES

UNUSUAL FIRE AND EXPLOSION HAZARDS

SECTION V HEALTH HAZARD DATA

TOXIC LEVEL: None available — CARCINOGENIC: No

PRINCIPLE ROUTES OF ABSORBTION: Inhalation (dust) skin, ingestion — SKIN AND EYE IRRITATION: Moderate

REVELANT SYMPTOMS OF EXPOSURE: Hyperemia, itching, dermatitis, restlessness, cyanosis, dizziness, tremors, convulsions and death.

EFFECTS OF CHRONIC EXPOSURE: Possible liver, kidney and lung damage.

EMERGENCY AND FIRST AID PROCEDURES: Inhalation - rest; if swallowed - induce vomiting, get medical aid; wash skin and eyes.

SECTION VI REACTIVITY DATA

CONDITIONS CONTRIBUTING TO INSTABILITY

CONDITIONS CONTRIBUTING TO HAZARDOUS POLYMERIZATION

INCOMPATABILITY (MATERIALS TO AVOID)

HAZARDOUS DECOMPOSITION PRODUCTS

SECTION VII SPILL OR LEAK PROCEDURES

STEPS TO BE TAKEN IN CASE MATERIAL IS RELEASED OR SPILLED

WASTE DISPOSAL METHOD

SECTION VIII SPECIAL PROTECTION INFORMATION

VENTILATION REQUIREMENTS LOCAL EXHAUST	PROTECTIVE EQUIPMENT (SPECIFY TYPES) EYE: Safety goggles
MECHANICAL (GENERAL)	GLOVES: Rubber
SPECIAL	RESPIRATOR

OTHER PROTECTIVE EQUIPMENT

SECTION IX SPECIAL PRECAUTIONS

PRECAUTIONS TO BE TAKEN IN HANDLING AND STORAGE

OTHER PRECAUTIONS

SIGNATURE	DATE

p-DIIODOBENZENE

Hazard rating: 1 (top), 3 (left), 0 (right)

SECTION I SOURCE AND NOMENCLATURE

MANUFACTURER'S NAME	EMERGENCY TELEPHONE NO.
ADDRESS (NUMBER, STREET, CITY, STATE, ZIP CODE)	
TRADE NAME AND SYNONYMS	CHEMICAL FAMILY
CHEMICAL NAME AND SYNONYMS: 4-Diiodobenzene	FORMULA: $C_6H_4I_2$

SECTION II HAZARDOUS INGREDIENTS

BASIC MATERIAL	APPROXIMATE OR MAXIMUM % WT. OR VOL.	ESTABLISHED OSHA STANDARD	LD_{50} ORAL	LD_{50} PERCUT.	LC_{50} SPECIES	LC_{50} CONC.
			none available			

SECTION III PHYSICAL DATA MW - 329.93

BOILING POINT °F.: 285°C	VAPOR PRESSURE mm Hg.
MELTING POINT °F.: 130°C	VAPOR DENSITY (Air = 1)
SPECIFIC GRAVITY (H_2O = 1)	EVAPORATION RATE (______ = 1)
SOLUBILITY IN WATER Pts/100 pts H_2O: insoluble	VOLATILE % Vol. % Wt.

APPEARANCE AND ODOR: crystaline

SECTION IV FIRE AND EXPLOSION HAZARD DATA

FLASH POINT: none available	FLAMMABLE (EXPLOSIVE) LIMITS	UPPER
METHOD USED		LOWER

EXTINGUISHING MEDIA: water, foam, carbon dioxide, dry chemical

SPECIAL FIRE FIGHTING PROCEDURES: wear goggles and self-contained breathing apparatus

UNUSUAL FIRE AND EXPLOSION HAZARDS

SECTION V HEALTH HAZARD DATA

TOXIC LEVEL: none available — CARCINOGENIC: no

PRINCIPLE ROUTES OF ABSORBTION: inhalation, ingestion, skin — SKIN AND EYE IRRITATION: severe

REVELANT SYMPTOMS OF EXPOSURE: cyanosis, nausea, deep and rapid respiration, irregular pulse, drowsiness, unconsciousness

EFFECTS OF CHRONIC EXPOSURE: Lung, liver and kidney damage

EMERGENCY AND FIRST AID PROCEDURES: Inhalation - rest; if swallowed - give emetic, get medical aid; wash from skin & eyes.

SECTION VI REACTIVITY DATA

CONDITIONS CONTRIBUTING TO INSTABILITY

CONDITIONS CONTRIBUTING TO HAZARDOUS POLYMERIZATION

INCOMPATABILITY (MATERIALS TO AVOID)

HAZARDOUS DECOMPOSITION PRODUCTS

SECTION VII SPILL OR LEAK PROCEDURES

STEPS TO BE TAKEN IN CASE MATERIAL IS RELEASED OR SPILLED

WASTE DISPOSAL METHOD

SECTION VIII SPECIAL PROTECTION INFORMATION

VENTILATION REQUIREMENTS LOCAL EXHAUST	PROTECTIVE EQUIPMENT (SPECIFY TYPES) EYE: Goggles
MECHANICAL (GENERAL)	GLOVES: Polyvinyl alcohol
SPECIAL	RESPIRATOR: Self-contained breathing apparatus

OTHER PROTECTIVE EQUIPMENT

SECTION IX SPECIAL PRECAUTIONS

PRECAUTIONS TO BE TAKEN IN HANDLING AND STORAGE

OTHER PRECAUTIONS

SIGNATURE	DATE

DIISOBUTYLAMINE

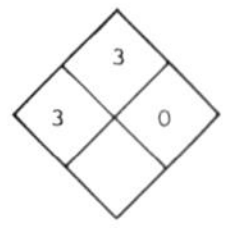

SECTION I SOURCE AND NOMENCLATURE

MANUFACTURER'S NAME	EMERGENCY TELEPHONE NO.
ADDRESS (NUMBER, STREET, CITY, STATE, ZIP CODE)	
TRADE NAME AND SYNONYMS	CHEMICAL FAMILY
CHEMICAL NAME AND SYNONYMS	FORMULA $[(CH_3)_2CHCH_2]_2NH$

SECTION II HAZARDOUS INGREDIENTS

BASIC MATERIAL	APPROXIMATE OR MAXIMUM % WT. OR VOL.	ESTABLISHED OSHA STANDARD	LD_{50} (mg/kg) ORAL	PERCUT.	SPECIES	LC_{50} CONC.
			258		Rat	
			629		Mouse	
			620		Guinea Pig	

SECTION III PHYSICAL DATA MW-129.24

BOILING POINT °F. 140^0 C.	VAPOR PRESSURE mm Hg. At 31^0C. 10
MELTING POINT °F. -70^0	VAPOR DENSITY (Air = 1) 4.46
SPECIFIC GRAVITY (H_2O = 1) 0.75	EVAPORATION RATE (______ = 1)
SOLUBILITY IN WATER Pts/100 pts H_2O	VOLATILE % Vol. % Wt.

APPEARANCE AND ODOR
Water-white amine-like odor.

SECTION IV FIRE AND EXPLOSION HAZARD DATA

	FLAMMABLE (EXPLOSIVE) LIMITS	
FLASH POINT 85^0F.		UPPER
METHOD USED		LOWER

EXTINGUISHING MEDIA
Foam, carbon dioxide, dry chemical.

SPECIAL FIRE FIGHTING PROCEDURES
Wear goggles and self-contained breathing apparatus.

UNUSUAL FIRE AND EXPLOSION HAZARDS

SECTION V HEALTH HAZARD DATA

TOXIC LEVEL
None available CARCINOGENIC No

PRINCIPLE ROUTES OF ABSORPTION
Inhalation, ingestion SKIN AND EYE IRRITATION Severe

REVELANT SYMPTOMS OF EXPOSURE Stimulation of sympathetic nervous system, cardiac depressant, may cause convulsions.

EFFECTS OF CHRONIC EXPOSURE
--

EMERGENCY AND FIRST AID PROCEDURES Inhalation - rest; if swallowed - drink water, get medical aid; wash skin and eyes.

SECTION VI REACTIVITY DATA

CONDITIONS CONTRIBUTING TO INSTABILITY

CONDITIONS CONTRIBUTING TO HAZARDOUS POLYMERIZATION

INCOMPATABILITY (MATERIALS TO AVOID)
May react vigorously with oxidizing materials.

HAZARDOUS DECOMPOSITION PRODUCTS

SECTION VII SPILL OR LEAK PROCEDURES

STEPS TO BE TAKEN IN CASE MATERIAL IS RELEASED OR SPILLED

WASTE DISPOSAL METHOD

SECTION VIII SPECIAL PROTECTION INFORMATION

VENTILATION REQUIREMENTS LOCAL EXHAUST	PROTECTIVE EQUIPMENT (SPECIFY TYPES) EYE
MECHANICAL (GENERAL)	GLOVES
SPECIAL	RESPIRATOR
OTHER PROTECTIVE EQUIPMENT	

SECTION IX SPECIAL PRECAUTIONS

PRECAUTIONS TO BE TAKEN IN HANDLING AND STORAGE

OTHER PRECAUTIONS

SIGNATURE	DATE

DIMETHYLAMINE HYDROCHLORIDE

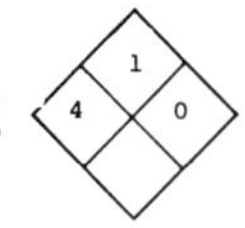

SECTION I SOURCE AND NOMENCLATURE

MANUFACTURER'S NAME	EMERGENCY TELEPHONE NO.
ADDRESS (NUMBER, STREET, CITY, STATE, ZIP CODE)	
TRADE NAME AND SYNONYMS	CHEMICAL FAMILY
CHEMICAL NAME AND SYNONYMS Dimethylamine Hydrochloride	FORMULA $(CH_3)_2NH$ HCl

SECTION II HAZARDOUS INGREDIENTS

BASIC MATERIAL	APPROXIMATE OR MAXIMUM % WT. OR VOL.	ESTABLISHED OSHA STANDARD	LD_{50} ORAL	PERCUT.	SPECIES	LC_{50} CONC.

Note: Lowest published toxic dose oral - mouse 140 mg/kg.

SECTION III PHYSICAL DATA MW - 81.58

BOILING POINT °F. --	VAPOR PRESSURE mm Hg.
MELTING POINT °F. 171^0C	VAPOR DENSITY (Air = 1)
SPECIFIC GRAVITY (H_2O = 1)	EVAPORATION RATE (______ = 1)
SOLUBILITY IN WATER Pts/100 pts H_2O 369	VOLATILE % Vol. % Wt.

APPEARANCE AND ODOR
Colorless crystal

SECTION IV FIRE AND EXPLOSION HAZARD DATA

	FLAMMABLE (EXPLOSIVE) LIMITS	
FLASH POINT None available		UPPER
METHOD USED		LOWER

EXTINGUISHING MEDIA
Foam, dry chemical, carbon dioxide

SPECIAL FIRE FIGHTING PROCEDURES

UNUSUAL FIRE AND EXPLOSION HAZARDS

SECTION V HEALTH HAZARD DATA

TOXIC LEVEL
None available CARCINOGENIC no

PRINCIPLE ROUTES OF ABSORPTION
Ingestion, inhalation of dust SKIN AND EYE IRRITATION moderate

REVELANT SYMPTOMS OF EXPOSURE Headache, faintness, anxiety, cardiac stimulation, pulmonary edema, skin & eye irritation.

EFFECTS OF CHRONIC EXPOSURE
Cardiac depression

EMERGENCY AND FIRST AID PROCEDURES Inhalation - rest; if swallowed - drink water, get medical aid; wash skin & eyes.

SECTION VI REACTIVITY DATA

CONDITIONS CONTRIBUTING TO INSTABILITY

CONDITIONS CONTRIBUTING TO HAZARDOUS POLYMERIZATION

INCOMPATABILITY (MATERIALS TO AVOID)

HAZARDOUS DECOMPOSITION PRODUCTS

SECTION VII SPILL OR LEAK PROCEDURES

STEPS TO BE TAKEN IN CASE MATERIAL IS RELEASED OR SPILLED

WASTE DISPOSAL METHOD

SECTION VIII SPECIAL PROTECTION INFORMATION

VENTILATION REQUIREMENTS LOCAL EXHAUST	PROTECTIVE EQUIPMENT (SPECIFY TYPES) EYE Goggles
MECHANICAL (GENERAL)	GLOVES Rubber
SPECIAL	RESPIRATOR
OTHER PROTECTIVE EQUIPMENT	

SECTION IX SPECIAL PRECAUTIONS

PRECAUTIONS TO BE TAKEN IN HANDLING AND STORAGE

OTHER PRECAUTIONS

SIGNATURE	DATE

p-DIMETHYLAMINOAZOBENZENE

Hazard rating: 2 (top), 4 (left), 0 (right)

SECTION I SOURCE AND NOMENCLATURE

MANUFACTURER'S NAME | EMERGENCY TELEPHONE NO.

ADDRESS (NUMBER, STREET, CITY, STATE, ZIP CODE)

TRADE NAME AND SYNONYMS: N,N-Dimethyl-4-aminoazobenzene

CHEMICAL FAMILY

CHEMICAL NAME AND SYNONYMS: 4 dimethylaminoazobenzene

FORMULA: $(CH_3)_2NC_6H_4NNC_6H_5$

SECTION II HAZARDOUS INGREDIENTS

BASIC MATERIAL	APPROXIMATE OR MAXIMUM % WT. OR VOL.	ESTABLISHED OSHA STANDARD	LD_{50} mg/kg ORAL	LD_{50} mg/kg PERCUT.	LC_{50} SPECIES	LC_{50} CONC.
Notes:			200		Rat	
1. Intraperitoneal - mouse LD_{50} 500 mg/kg			300		Mouse	
2. Lowest published toxic dose						
Oral - rat 800 mg/kg - 4 days carcinogen						

SECTION III PHYSICAL DATA MW - 225.18

BOILING POINT (°F.): d

VAPOR PRESSURE (mm Hg.)

MELTING POINT (°F.): 116

VAPOR DENSITY (Air = 1)

SPECIFIC GRAVITY (H_2O = 1)

EVAPORATION RATE (____ = 1)

SOLUBILITY IN WATER (Pts/100 pts H_2O): Insoluble

VOLATILE (% Vol. / % Wt.)

APPEARANCE AND ODOR: Yellow crystal

SECTION IV FIRE AND EXPLOSION HAZARD DATA

FLASH POINT: None available

METHOD USED

FLAMMABLE (EXPLOSIVE) LIMITS: UPPER / LOWER

EXTINGUISHING MEDIA: Alcohol foam, carbon dioxide, dry chemical.

SPECIAL FIRE FIGHTING PROCEDURES: Self-contained breathing apparatus.

UNUSUAL FIRE AND EXPLOSION HAZARDS

SECTION V HEALTH HAZARD DATA

TOXIC LEVEL: None available — Teratogen

CARCINOGENIC: Yes

PRINCIPLE ROUTES OF ABSORBTION: Inhalation of dust, ingestion, skin

SKIN AND EYE IRRITATION: Moderate

RELEVANT SYMPTOMS OF EXPOSURE: Eye, skin and respiratory system irritant.

EFFECTS OF CHRONIC EXPOSURE: May produce liver cancer, bladder cancer.

EMERGENCY AND FIRST AID PROCEDURES: Inhalation - rest; if swallowed - get medical aid; wash from skin & eyes.

SECTION VI REACTIVITY DATA

CONDITIONS CONTRIBUTING TO INSTABILITY

CONDITIONS CONTRIBUTING TO HAZARDOUS POLYMERIZATION

INCOMPATABILITY (MATERIALS TO AVOID)

HAZARDOUS DECOMPOSITION PRODUCTS

SECTION VII SPILL OR LEAK PROCEDURES

STEPS TO BE TAKEN IN CASE MATERIAL IS RELEASED OR SPILLED

WASTE DISPOSAL METHOD

SECTION VIII SPECIAL PROTECTION INFORMATION

VENTILATION REQUIREMENTS LOCAL EXHAUST

MECHANICAL (GENERAL)

SPECIAL

PROTECTIVE EQUIPMENT (SPECIFY TYPES) EYE: Goggles

GLOVES: Rubber

RESPIRATOR: Half mask filter type for dust, mist and fumes.

OTHER PROTECTIVE EQUIPMENT: Full body protective clothing.

SECTION IX SPECIAL PRECAUTIONS

PRECAUTIONS TO BE TAKEN IN HANDLING AND STORAGE

OTHER PRECAUTIONS

SIGNATURE | DATE

β-DIMETHYLAMINOETHYL ALCOHOL

Hazard rating: 2 (top), 4 (left), 0 (right)

SECTION I SOURCE AND NOMENCLATURE

MANUFACTURER'S NAME | EMERGENCY TELEPHONE NO.

ADDRESS (NUMBER, STREET, CITY, STATE, ZIP CODE)

TRADE NAME AND SYNONYMS

CHEMICAL FAMILY

CHEMICAL NAME AND SYNONYMS: Dimethylaminoethanol

FORMULA: $(CH_3)_2NCH_2CH_2OH$

SECTION II HAZARDOUS INGREDIENTS

BASIC MATERIAL	APPROXIMATE OR MAXIMUM % WT. OR VOL.	ESTABLISHED OSHA STANDARD	LD_{50} mg/kg ORAL	LD_{50} mg/kg PERCUT.	LC_{50} SPECIES	LC_{50} CONC.
			2340		Rat	
				1370	Rabbit	

SECTION III PHYSICAL DATA MW - 89.16

BOILING POINT (°F.): 272°F

VAPOR PRESSURE (mm Hg.)

MELTING POINT (°F.): --

VAPOR DENSITY (Air = 1): 3.1

SPECIFIC GRAVITY (H_2O = 1): 0.9

EVAPORATION RATE (____ = 1)

SOLUBILITY IN WATER (Pts/100 pts H_2O): Yes

VOLATILE (% Vol. / % Wt.)

APPEARANCE AND ODOR: Colorless liquid

SECTION IV FIRE AND EXPLOSION HAZARD DATA

FLASH POINT: 105

METHOD USED: O.C.

FLAMMABLE (EXPLOSIVE) LIMITS: UPPER / LOWER

EXTINGUISHING MEDIA: Alcohol foam, carbon dioxide, dry chemical

SPECIAL FIRE FIGHTING PROCEDURES: Water may be ineffective

UNUSUAL FIRE AND EXPLOSION HAZARDS

SECTION V HEALTH HAZARD DATA

TOXIC LEVEL: Not available

CARCINOGENIC: No

PRINCIPLE ROUTES OF ABSORBTION: Skin, inhalation, ingestion

SKIN AND EYE IRRITATION: Severe

RELEVANT SYMPTOMS OF EXPOSURE: Dermatitis, weight loss, slight increase in clotting time.

EFFECTS OF CHRONIC EXPOSURE: Dermatitis

EMERGENCY AND FIRST AID PROCEDURES: Wash from skin and eyes; if swallowed - drink fruit juice, get medical aid; inhalation - rest.

SECTION VI REACTIVITY DATA

CONDITIONS CONTRIBUTING TO INSTABILITY

CONDITIONS CONTRIBUTING TO HAZARDOUS POLYMERIZATION

INCOMPATABILITY (MATERIALS TO AVOID)

HAZARDOUS DECOMPOSITION PRODUCTS

SECTION VII SPILL OR LEAK PROCEDURES

STEPS TO BE TAKEN IN CASE MATERIAL IS RELEASED OR SPILLED

WASTE DISPOSAL METHOD

SECTION VIII SPECIAL PROTECTION INFORMATION

VENTILATION REQUIREMENTS LOCAL EXHAUST

MECHANICAL (GENERAL)

SPECIAL

PROTECTIVE EQUIPMENT (SPECIFY TYPES) EYE: Safety goggles

GLOVES: Rubber

RESPIRATOR: With proper filter

OTHER PROTECTIVE EQUIPMENT

SECTION IX SPECIAL PRECAUTIONS

PRECAUTIONS TO BE TAKEN IN HANDLING AND STORAGE

OTHER PRECAUTIONS

SIGNATURE | DATE

DIMETHYLANILINE

2
4 0

SECTION I SOURCE AND NOMENCLATURE

MANUFACTURER'S NAME | EMERGENCY TELEPHONE NO.

ADDRESS (NUMBER, STREET, CITY, STATE, ZIP CODE)

TRADE NAME AND SYNONYMS: dimethylphenylamine | CHEMICAL FAMILY

CHEMICAL NAME AND SYNONYMS: NN dimethylaniline | FORMULA: $C_6H_5N(CH_3)_2$

SECTION II HAZARDOUS INGREDIENTS

BASIC MATERIAL	APPROXIMATE OR MAXIMUM % WT. OR VOL.	ESTABLISHED OSHA STANDARD	LD_{50} mg/kg ORAL	PERCUT.	SPECIES	LC_{50} CONC.
LDLO, oral human 50 mg/kg.			1410		Rat	
				1770	Rabbit	

SECTION III PHYSICAL DATA MW – 121.18

BOILING POINT (°F.): 193.1°C | VAPOR PRESSURE (mm Hg.): at 29.5°C 1

MELTING POINT (°F.): 2.5°C | VAPOR DENSITY (AIR = 1): 4.17

SPECIFIC GRAVITY (H_2O = 1): 0.9557 | EVAPORATION RATE (____ = 1)

SOLUBILITY IN WATER Pts/100 pts H_2O: Insoluble | VOLATILE % Vol. % Wt.

APPEARANCE AND ODOR: Colorless to brown liquid, irritation odor

SECTION IV FIRE AND EXPLOSION HAZARD DATA

FLASH POINT: 145°F | FLAMMABLE (EXPLOSIVE) LIMITS | UPPER | LOWER

METHOD USED: CC

EXTINGUISHING MEDIA: "Alcohol" foam, water spray, carbon dioxide, dry chemical

SPECIAL FIRE FIGHTING PROCEDURES

UNUSUAL FIRE AND EXPLOSION HAZARDS

SECTION V HEALTH HAZARD DATA

TOXIC LEVEL: TLV – 5PPM; 25 mg/C.M. | CARCINOGENIC: no

PRINCIPLE ROUTES OF ABSORBTION: Inhalation, ingestion, skin | SKIN AND EYE IRRITATION: Yes

REVELANT SYMPTOMS OF EXPOSURE: Weakness, tremors, central nervous system depressant, cyanosis, drowsiness

EFFECTS OF CHRONIC EXPOSURE: Fatigue, loss of appetite, headache, dizziness

EMERGENCY AND FIRST AID PROCEDURES: Inhalation – rest; if swallowed – wash mouth, give emetic, get medical aid; wash eyes & skin.

SECTION VI REACTIVITY DATA

CONDITIONS CONTRIBUTING TO INSTABILITY

CONDITIONS CONTRIBUTING TO HAZARDOUS POLYMERIZATION

INCOMPATABILITY (MATERIALS TO AVOID)

HAZARDOUS DECOMPOSITION PRODUCTS

SECTION VII SPILL OR LEAK PROCEDURES

STEPS TO BE TAKEN IN CASE MATERIAL IS RELEASED OR SPILLED: Spread a mixture of sand and soda ash (9/1); Transfer to cardboard carton for disposal.

WASTE DISPOSAL METHOD

SECTION VIII SPECIAL PROTECTION INFORMATION

VENTILATION REQUIREMENTS LOCAL EXHAUST | PROTECTIVE EQUIPMENT (SPECIFY TYPES) EYE: Goggles

MECHANICAL (GENERAL) | GLOVES: Rubber

SPECIAL | RESPIRATOR: With filter

OTHER PROTECTIVE EQUIPMENT: Coveralls

SECTION IX SPECIAL PRECAUTIONS

PRECAUTIONS TO BE TAKEN IN HANDLING AND STORAGE

OTHER PRECAUTIONS

SIGNATURE | DATE

9,10-DIMETHYL-1,2-BENZANTHRACENE

1
4 0

SECTION I SOURCE AND NOMENCLATURE

MANUFACTURER'S NAME | EMERGENCY TELEPHONE NO.

ADDRESS (NUMBER, STREET, CITY, STATE, ZIP CODE)

TRADE NAME AND SYNONYMS | CHEMICAL FAMILY

CHEMICAL NAME AND SYNONYMS: 7,12 – Dimethyl, benz(a) anthracene | FORMULA: $(CH_3)_2C_{18}H_{10}$

SECTION II HAZARDOUS INGREDIENTS

BASIC MATERIAL	APPROXIMATE OR MAXIMUM % WT. OR VOL.	ESTABLISHED OSHA STANDARD	LD_{50} ORAL	PERCUT.	SPECIES	LC_{50} CONC.
			340		Mouse	

Notes:

1. Lowest published toxic dose oral – rat 15 mg/kg. car
2. " " " " skin mouse 0.014 mg/kg/17W car
3. " " " " skin rabbit 12 mg/kg/10W1 car
4. LD50 intravenous – rat 54 mg/kg

SECTION III PHYSICAL DATA MW –256.33

BOILING POINT (°F.): – | VAPOR PRESSURE (mm Hg.)

MELTING POINT (°F.): 123°C | VAPOR DENSITY (AIR = 1)

SPECIFIC GRAVITY (H_2O = 1) | EVAPORATION RATE (____ = 1)

SOLUBILITY IN WATER Pts/100 pts H_2O: Insoluble | VOLATILE % Vol. % Wt.

APPEARANCE AND ODOR: Greenish yellow crystals

SECTION IV FIRE AND EXPLOSION HAZARD DATA

FLASH POINT: none available | FLAMMABLE (EXPLOSIVE) LIMITS | UPPER | LOWER

METHOD USED

EXTINGUISHING MEDIA: alcohol foam, carbon dioxide, dry chemical

SPECIAL FIRE FIGHTING PROCEDURES: self-contained breathing apparatus

UNUSUAL FIRE AND EXPLOSION HAZARDS

SECTION V HEALTH HAZARD DATA

TOXIC LEVEL: None available | Teratogen | CARCINOGENIC: yes

PRINCIPLE ROUTES OF ABSORBTION: Inhalation of dust, ingestion, skin | SKIN AND EYE IRRITATION: yes

REVELANT SYMPTOMS OF EXPOSURE: Irritation of gastroenteric tract, eyes & skin

EFFECTS OF CHRONIC EXPOSURE: Cancer formation in many parts of the body

EMERGENCY AND FIRST AID PROCEDURES: Inhalation – rest; if swallowed – get medical aid (gastric lavage); wash from skin & eyes.

SECTION VI REACTIVITY DATA

CONDITIONS CONTRIBUTING TO INSTABILITY

CONDITIONS CONTRIBUTING TO HAZARDOUS POLYMERIZATION

INCOMPATABILITY (MATERIALS TO AVOID)

HAZARDOUS DECOMPOSITION PRODUCTS

SECTION VII SPILL OR LEAK PROCEDURES

STEPS TO BE TAKEN IN CASE MATERIAL IS RELEASED OR SPILLED

WASTE DISPOSAL METHOD

SECTION VIII SPECIAL PROTECTION INFORMATION

VENTILATION REQUIREMENTS LOCAL EXHAUST | PROTECTIVE EQUIPMENT (SPECIFY TYPES) EYE: Goggles

MECHANICAL (GENERAL) | GLOVES: Rubber

SPECIAL | RESPIRATOR: With proper filter

OTHER PROTECTIVE EQUIPMENT

SECTION IX SPECIAL PRECAUTIONS

PRECAUTIONS TO BE TAKEN IN HANDLING AND STORAGE

OTHER PRECAUTIONS

SIGNATURE | DATE

DIMETHYLBENZYLAMINE

(NFPA diamond: 2, 4, 0)

SECTION I SOURCE AND NOMENCLATURE

MANUFACTURER'S NAME | EMERGENCY TELEPHONE NO.

ADDRESS (NUMBER, STREET, CITY, STATE, ZIP CODE)

TRADE NAME AND SYNONYMS | CHEMICAL FAMILY

CHEMICAL NAME AND SYNONYMS: N, N Dimethylbenzylamine | FORMULA: $(CH_3)_2NCH_2C_6H_5$

SECTION II HAZARDOUS INGREDIENTS

BASIC MATERIAL	APPROXIMATE OR MAXIMUM % WT. OR VOL.	ESTABLISHED OSHA STANDARD	LD_{50} mg/kg ORAL	PERCUT.	SPECIES	LC_{50} CONC.
			265		Rat	

SECTION III PHYSICAL DATA MW - 135.20

BOILING POINT °F: 184°C	VAPOR PRESSURE mm Hg: 8 at 55°C
MELTING POINT °F	VAPOR DENSITY (Air = 1)
SPECIFIC GRAVITY (H_2O = 1): 0.920	EVAPORATION RATE (= 1)
SOLUBILITY IN WATER Pts/100 pts H_2O: Slightly soluble	VOLATILE % Vol. % Wt.

APPEARANCE AND ODOR: Liquid

SECTION IV FIRE AND EXPLOSION HAZARD DATA

FLASH POINT: None available | FLAMMABLE (EXPLOSIVE) LIMITS: UPPER / LOWER

METHOD USED

EXTINGUISHING MEDIA: Alcohol foam, carbon dioxide, dry chemical

SPECIAL FIRE FIGHTING PROCEDURES: self-contained breathing apparatus

UNUSUAL FIRE AND EXPLOSION HAZARDS

SECTION V HEALTH HAZARD DATA

TOXIC LEVEL: None available | CARCINOGENIC: No

PRINCIPLE ROUTES OF ABSORBTION: Inhalation, ingestion, skin | SKIN AND EYE IRRITATION: severe

REVELANT SYMPTOMS OF EXPOSURE: Headache, cyanosis, drowsiness, mental confusion, convulsions

EFFECTS OF CHRONIC EXPOSURE: Loss of appetite, headache, dizziness, may effect the nervous system.

EMERGENCY AND FIRST AID PROCEDURES: Inhalation - rest; if swallowed - give emetic; get medical aid; wash from skin & eyes.

SECTION VI REACTIVITY DATA

CONDITIONS CONTRIBUTING TO INSTABILITY

CONDITIONS CONTRIBUTING TO HAZARDOUS POLYMERIZATION

INCOMPATABILITY (MATERIALS TO AVOID)

HAZARDOUS DECOMPOSITION PRODUCTS

SECTION VII SPILL OR LEAK PROCEDURES

STEPS TO BE TAKEN IN CASE MATERIAL IS RELEASED OR SPILLED

WASTE DISPOSAL METHOD

SECTION VIII SPECIAL PROTECTION INFORMATION

VENTILATION REQUIREMENTS LOCAL EXHAUST | PROTECTIVE EQUIPMENT (SPECIFY TYPES) EYE: Goggles

MECHANICAL (GENERAL) | GLOVES: Rubber

SPECIAL | RESPIRATOR: With proper filter

OTHER PROTECTIVE EQUIPMENT

SECTION IX SPECIAL PRECAUTIONS

PRECAUTIONS TO BE TAKEN IN HANDLING AND STORAGE

OTHER PRECAUTIONS

SIGNATURE | DATE

DIMETHYLDIHYDRORESORCINOL

(NFPA diamond: 1, 3, 0)

SECTION I SOURCE AND NOMENCLATURE

MANUFACTURER'S NAME | EMERGENCY TELEPHONE NO.

ADDRESS (NUMBER, STREET, CITY, STATE, ZIP CODE)

TRADE NAME AND SYNONYMS | CHEMICAL FAMILY

CHEMICAL NAME AND SYNONYMS: 5,5 Dimethyl - 1, 3 Cyclohexanedione | FORMULA: $(CH_3)_2C_6H_6O_2$

SECTION II HAZARDOUS INGREDIENTS

BASIC MATERIAL	APPROXIMATE OR MAXIMUM % WT. OR VOL.	ESTABLISHED OSHA STANDARD	LD_{50} ORAL	PERCUT.	SPECIES	LC_{50} CONC.
			none	available		

SECTION III PHYSICAL DATA MW - 140.18

BOILING POINT °F: --	VAPOR PRESSURE mm Hg
MELTING POINT °F: 149°C	VAPOR DENSITY (Air = 1)
SPECIFIC GRAVITY (H_2O = 1)	EVAPORATION RATE (= 1)
SOLUBILITY IN WATER Pts/100 pts H_2O: 0.42	VOLATILE % Vol. % Wt.

APPEARANCE AND ODOR: Yellow crystals

SECTION IV FIRE AND EXPLOSION HAZARD DATA

FLASH POINT: None available | FLAMMABLE (EXPLOSIVE) LIMITS: UPPER / LOWER

METHOD USED

EXTINGUISHING MEDIA: Alcohol foam, carbon dioxide, dry chemical

SPECIAL FIRE FIGHTING PROCEDURES

UNUSUAL FIRE AND EXPLOSION HAZARDS

SECTION V HEALTH HAZARD DATA

TOXIC LEVEL: none available | CARCINOGENIC: no

PRINCIPLE ROUTES OF ABSORBTION: Inhalation of dust, ingestion, skin | SKIN AND EYE IRRITATION: moderate

REVELANT SYMPTOMS OF EXPOSURE: Headache, eye & upper respiratory irritation;

EFFECTS OF CHRONIC EXPOSURE: Liver and kidney damage

EMERGENCY AND FIRST AID PROCEDURES: Inhalation - rest; if swallowed - induce vomiting, get medical aid; wash skin & eyes.

SECTION VI REACTIVITY DATA

CONDITIONS CONTRIBUTING TO INSTABILITY

CONDITIONS CONTRIBUTING TO HAZARDOUS POLYMERIZATION

INCOMPATABILITY (MATERIALS TO AVOID)

HAZARDOUS DECOMPOSITION PRODUCTS

SECTION VII SPILL OR LEAK PROCEDURES

STEPS TO BE TAKEN IN CASE MATERIAL IS RELEASED OR SPILLED

WASTE DISPOSAL METHOD

SECTION VIII SPECIAL PROTECTION INFORMATION

VENTILATION REQUIREMENTS LOCAL EXHAUST | PROTECTIVE EQUIPMENT (SPECIFY TYPES) EYE: Goggles

MECHANICAL (GENERAL) | GLOVES: Rubber

SPECIAL | RESPIRATOR: With proper filter

OTHER PROTECTIVE EQUIPMENT

SECTION IX SPECIAL PRECAUTIONS

PRECAUTIONS TO BE TAKEN IN HANDLING AND STORAGE

OTHER PRECAUTIONS

SIGNATURE | DATE

2,4-DIMETHYLPHENACYL CHLORIDE

Hazard rating: 1 (top), 4 (left), 0 (right)

SECTION I SOURCE AND NOMENCLATURE

MANUFACTURER'S NAME

EMERGENCY TELEPHONE NO.

ADDRESS (NUMBER, STREET, CITY, STATE, ZIP CODE)

TRADE NAME AND SYNONYMS

CHEMICAL FAMILY

CHEMICAL NAME AND SYNONYMS: α chloro-2,4 dimethyl acetophenone

FORMULA: $ClCH_2COC_6H_3(CH_3)_2$

SECTION II HAZARDOUS INGREDIENTS

BASIC MATERIAL	APPROXIMATE OR MAXIMUM % WT. OR VOL.	ESTABLISHED OSHA STANDARD	LD_{50} ORAL	LD_{50} PERCUT.	LC_{50} SPECIES	LC_{50} CONC.
			none	available		

SECTION III PHYSICAL DATA MW - 182.65

BOILING POINT °F.: --

VAPOR PRESSURE mm Hg.

MELTING POINT °F.: $62^{0}C$

VAPOR DENSITY (Air = 1)

SPECIFIC GRAVITY (H_2O = 1)

EVAPORATION RATE (____ = 1)

SOLUBILITY IN WATER Pts/100 pts H_2O

VOLATILE % Vol. % Wt.

APPEARANCE AND ODOR: Needle-like crystals

SECTION IV FIRE AND EXPLOSION HAZARD DATA

FLASH POINT: Not available

METHOD USED

FLAMMABLE (EXPLOSIVE) LIMITS: UPPER / LOWER

EXTINGUISHING MEDIA: Not available

SPECIAL FIRE FIGHTING PROCEDURES

UNUSUAL FIRE AND EXPLOSION HAZARDS: May react with water or steam to produce toxic and corrosive fumes.

SECTION V HEALTH HAZARD DATA

TOXIC LEVEL: None available

CARCINOGENIC: no

PRINCIPLE ROUTES OF ABSORBTION: Inhalation of dust, ingestion

SKIN AND EYE IRRITATION: moderate

REVELANT SYMPTOMS OF EXPOSURE: Eye irritation, corneal haziness & pain; burning of throat and eyes; skin irritation.

EFFECTS OF CHRONIC EXPOSURE: --

EMERGENCY AND FIRST AID PROCEDURES: Inhalation - rest; treat for pulmonary edema; if swallowed - give emetic, get medical aid; wash from skin & eyes.

SECTION VI REACTIVITY DATA

CONDITIONS CONTRIBUTING TO INSTABILITY

CONDITIONS CONTRIBUTING TO HAZARDOUS POLYMERIZATION

INCOMPATABILITY (MATERIALS TO AVOID)

HAZARDOUS DECOMPOSITION PRODUCTS

SECTION VII SPILL OR LEAK PROCEDURES

STEPS TO BE TAKEN IN CASE MATERIAL IS RELEASED OR SPILLED

WASTE DISPOSAL METHOD

SECTION VIII SPECIAL PROTECTION INFORMATION

VENTILATION REQUIREMENTS LOCAL EXHAUST

PROTECTIVE EQUIPMENT (SPECIFY TYPES) EYE: Goggles

MECHANICAL (GENERAL)

GLOVES: Rubber

SPECIAL

RESPIRATOR: Self-contained

OTHER PROTECTIVE EQUIPMENT

SECTION IX SPECIAL PRECAUTIONS

PRECAUTIONS TO BE TAKEN IN HANDLING AND STORAGE

OTHER PRECAUTIONS

SIGNATURE

DATE

2,4-DIMETHYLPIPERIDINE

Hazard rating: 2 (top), 4 (left), 0 (right)

SECTION I SOURCE AND NOMENCLATURE

MANUFACTURER'S NAME

EMERGENCY TELEPHONE NO.

ADDRESS (NUMBER, STREET, CITY, STATE, ZIP CODE)

TRADE NAME AND SYNONYMS

CHEMICAL FAMILY

CHEMICAL NAME AND SYNONYMS: α, γ - Lupetidine

FORMULA: $HNC_6H_9(CH_3)_2$

SECTION II HAZARDOUS INGREDIENTS

BASIC MATERIAL	APPROXIMATE OR MAXIMUM % WT. OR VOL.	ESTABLISHED OSHA STANDARD	LD_{50} ORAL	LD_{50} PERCUT.	LC_{50} SPECIES	LC_{50} CONC.
			none	available		

SECTION III PHYSICAL DATA MW = 113.20

BOILING POINT °F.: $139^{0}C$

VAPOR PRESSURE mm Hg.

MELTING POINT °F.

VAPOR DENSITY (Air = 1)

SPECIFIC GRAVITY (H_2O = 1): 0.845

EVAPORATION RATE (____ = 1)

SOLUBILITY IN WATER Pts/100 pts H_2O

VOLATILE % Vol. % Wt.

APPEARANCE AND ODOR: Liquid

SECTION IV FIRE AND EXPLOSION HAZARD DATA

FLASH POINT: None available

METHOD USED

FLAMMABLE (EXPLOSIVE) LIMITS: UPPER / LOWER

EXTINGUISHING MEDIA: Alcohol foam, carbon dioxide, dry chemical

SPECIAL FIRE FIGHTING PROCEDURES: Wear goggles and self-contained breathing apparatus.

UNUSUAL FIRE AND EXPLOSION HAZARDS

SECTION V HEALTH HAZARD DATA

TOXIC LEVEL: None available

CARCINOGENIC: no

PRINCIPLE ROUTES OF ABSORBTION: Inhalation, ingestion, skin

SKIN AND EYE IRRITATION: severe

REVELANT SYMPTOMS OF EXPOSURE: Increased blood pressure, weakness, nausea, vomiting, salivation, asphyxia, corneal damage.

EFFECTS OF CHRONIC EXPOSURE: Weight loss, anemia, ocular and facial paralysis -

EMERGENCY AND FIRST AID PROCEDURES: Inhalation - rest; if swallowed - induce vomiting then gastric lavage; wash skin & eyes.

SECTION VI REACTIVITY DATA

CONDITIONS CONTRIBUTING TO INSTABILITY

CONDITIONS CONTRIBUTING TO HAZARDOUS POLYMERIZATION

INCOMPATABILITY (MATERIALS TO AVOID)

HAZARDOUS DECOMPOSITION PRODUCTS

SECTION VII SPILL OR LEAK PROCEDURES

STEPS TO BE TAKEN IN CASE MATERIAL IS RELEASED OR SPILLED

WASTE DISPOSAL METHOD

SECTION VIII SPECIAL PROTECTION INFORMATION

VENTILATION REQUIREMENTS LOCAL EXHAUST

PROTECTIVE EQUIPMENT (SPECIFY TYPES) EYE: Goggles

MECHANICAL (GENERAL)

GLOVES: Rubber

SPECIAL

RESPIRATOR: Self-contained

OTHER PROTECTIVE EQUIPMENT

SECTION IX SPECIAL PRECAUTIONS

PRECAUTIONS TO BE TAKEN IN HANDLING AND STORAGE

OTHER PRECAUTIONS

SIGNATURE

DATE

DIMETHYL-n-PROPYLCARBINOL

(Hazard diamond: 2, 3, 0)

SECTION I SOURCE AND NOMENCLATURE

MANUFACTURER'S NAME | EMERGENCY TELEPHONE NO.

ADDRESS (NUMBER, STREET, CITY, STATE, ZIP CODE)

TRADE NAME AND SYNONYMS | CHEMICAL FAMILY

CHEMICAL NAME AND SYNONYMS: 2 Methyl-2-Pentanol

FORMULA: $(CH_3)_2COH(CH_2)_2CH_3$

SECTION II HAZARDOUS INGREDIENTS

BASIC MATERIAL	APPROXIMATE OR MAXIMUM % WT. OR VOL.	ESTABLISHED OSHA STANDARD	LD_{50} ORAL	LD_{50} PERCUT.	LC_{50} SPECIES	LC_{50} CONC.
			None	available		

SECTION III PHYSICAL DATA MW = 102.18

BOILING POINT (°F.): 123°C

VAPOR PRESSURE (mm Hg.):

MELTING POINT (°F.): -108°C

VAPOR DENSITY (Air = 1):

SPECIFIC GRAVITY (H_2O = 1): 0.81

EVAPORATION RATE (___ = 1):

SOLUBILITY IN WATER Pts/100 pts H_2O: Slightly soluble

VOLATILE % Vol.:

APPEARANCE AND ODOR: Clear liquid

SECTION IV FIRE AND EXPLOSION HAZARD DATA

FLASH POINT: None available

METHOD USED:

FLAMMABLE (EXPLOSIVE) LIMITS: UPPER / LOWER

EXTINGUISHING MEDIA: Alcohol foam, carbon dioxide, dry chemical

SPECIAL FIRE FIGHTING PROCEDURES:

UNUSUAL FIRE AND EXPLOSION HAZARDS:

SECTION V HEALTH HAZARD DATA

TOXIC LEVEL: None available

CARCINOGENIC: no

PRINCIPLE ROUTES OF ABSORBTION: Inhalation, ingestion

SKIN AND EYE IRRITATION: moderate

REVELANT SYMPTOMS OF EXPOSURE: Eye, nose & throat irritation; narcosis, ingestion, headache, dizziness, nausea, vomiting, diarrhea, delirium, coma

EFFECTS OF CHRONIC EXPOSURE: Possible liver damage

EMERGENCY AND FIRST AID PROCEDURES: Inhalation - give oxygen as needed; ingestion - induce vomiting, get medical aid; wash from skin & eyes.

SECTION VI REACTIVITY DATA

CONDITIONS CONTRIBUTING TO INSTABILITY

CONDITIONS CONTRIBUTING TO HAZARDOUS POLYMERIZATION

INCOMPATABILITY (MATERIALS TO AVOID)

HAZARDOUS DECOMPOSITION PRODUCTS

SECTION VII SPILL OR LEAK PROCEDURES

STEPS TO BE TAKEN IN CASE MATERIAL IS RELEASED OR SPILLED

WASTE DISPOSAL METHOD

SECTION VIII SPECIAL PROTECTION INFORMATION

VENTILATION REQUIREMENTS LOCAL EXHAUST

PROTECTIVE EQUIPMENT (SPECIFY TYPES) EYE: Goggles

MECHANICAL (GENERAL)

GLOVES: Rubber

SPECIAL

RESPIRATOR: With proper filter

OTHER PROTECTIVE EQUIPMENT

SECTION IX SPECIAL PRECAUTIONS

PRECAUTIONS TO BE TAKEN IN HANDLING AND STORAGE

OTHER PRECAUTIONS

SIGNATURE | DATE

DIMETHYL-m-TOLUIDINE

(Hazard diamond: 2, 4, 0)

SECTION I SOURCE AND NOMENCLATURE

MANUFACTURER'S NAME | EMERGENCY TELEPHONE NO.

ADDRESS (NUMBER, STREET, CITY, STATE, ZIP CODE)

TRADE NAME AND SYNONYMS | CHEMICAL FAMILY

CHEMICAL NAME AND SYNONYMS: N, N Dimethyl-M-aminotoluene

FORMULA: $(CH_3)_2NC_6H_4CH_3$

SECTION II HAZARDOUS INGREDIENTS

BASIC MATERIAL	APPROXIMATE OR MAXIMUM % WT. OR VOL.	ESTABLISHED OSHA STANDARD	LD_{50} ORAL	LD_{50} PERCUT.	LC_{50} SPECIES	LC_{50} CONC.
			none	available		

SECTION III PHYSICAL DATA MW - 135.20

BOILING POINT (°F.): 186°C

VAPOR PRESSURE (mm Hg.): 20 at 98°C

MELTING POINT (°F.): -61°C

VAPOR DENSITY (Air = 1):

SPECIFIC GRAVITY (H_2O = 1): 0.93

EVAPORATION RATE (___ = 1):

SOLUBILITY IN WATER Pts/100 pts H_2O: Insoluble

VOLATILE % Vol. % Wt.:

APPEARANCE AND ODOR: Liquid

SECTION IV FIRE AND EXPLOSION HAZARD DATA

FLASH POINT: Not available

METHOD USED:

FLAMMABLE (EXPLOSIVE) LIMITS: UPPER / LOWER

EXTINGUISHING MEDIA: Dry chemical, carbon dioxide, foam or water spray

SPECIAL FIRE FIGHTING PROCEDURES: Wear full protective clothing

UNUSUAL FIRE AND EXPLOSION HAZARDS: Emits toxic fumes when heated

SECTION V HEALTH HAZARD DATA

TOXIC LEVEL: None available

CARCINOGENIC: No

PRINCIPLE ROUTES OF ABSORBTION: Inhalation, ingestion, skin

SKIN AND EYE IRRITATION: moderate

REVELANT SYMPTOMS OF EXPOSURE: Cyanosis, air hunger, nausea, vomiting, low blood pressure, convulsions.

EFFECTS OF CHRONIC EXPOSURE: Weight loss, anemia, weakness

EMERGENCY AND FIRST AID PROCEDURES: Inhalation - rest; if swallowed - give emetic, get medical aid; wash skin & eyes, skin may be washed with 5% acetic acid to aid removal.

SECTION VI REACTIVITY DATA

CONDITIONS CONTRIBUTING TO INSTABILITY

CONDITIONS CONTRIBUTING TO HAZARDOUS POLYMERIZATION

INCOMPATABILITY (MATERIALS TO AVOID)

HAZARDOUS DECOMPOSITION PRODUCTS

SECTION VII SPILL OR LEAK PROCEDURES

STEPS TO BE TAKEN IN CASE MATERIAL IS RELEASED OR SPILLED

WASTE DISPOSAL METHOD

SECTION VIII SPECIAL PROTECTION INFORMATION

VENTILATION REQUIREMENTS LOCAL EXHAUST

PROTECTIVE EQUIPMENT (SPECIFY TYPES) EYE: Goggles

MECHANICAL (GENERAL)

GLOVES: Rubber

SPECIAL

RESPIRATOR: With proper filter

OTHER PROTECTIVE EQUIPMENT

SECTION IX SPECIAL PRECAUTIONS

PRECAUTIONS TO BE TAKEN IN HANDLING AND STORAGE

OTHER PRECAUTIONS

SIGNATURE | DATE

DIMETHYL-o-TOLUIDINE

2 / 4 / 0

SECTION I SOURCE AND NOMENCLATURE

MANUFACTURER'S NAME | EMERGENCY TELEPHONE NO.

ADDRESS (NUMBER, STREET, CITY, STATE, ZIP CODE)

TRADE NAME AND SYNONYMS | CHEMICAL FAMILY

CHEMICAL NAME AND SYNONYMS: N,N Dimethy-O-Aminotoluene

FORMULA: $(CH_3)_2NC_6H_4CH_3$

SECTION II HAZARDOUS INGREDIENTS

BASIC MATERIAL	APPROXIMATE OR MAXIMUM % WT. OR VOL.	ESTABLISHED OSHA STANDARD	LD_{50} ORAL	LD_{50} PERCUT.	LC_{50} SPECIES	LC_{50} CONC.
LDLO oral - rat 500 mg/kg			none available			

SECTION III PHYSICAL DATA MW - 135.20

BOILING POINT (°F.): 215°C

VAPOR PRESSURE (mm Hg.): at 70°C 10

MELTING POINT (°F.)

VAPOR DENSITY (Air = 1)

SPECIFIC GRAVITY (H_2O = 1): 0.94

EVAPORATION RATE (______ = 1)

SOLUBILITY IN WATER Pts/100 pts H_2O: Insoluble

VOLATILE % Vol. % Wt.

APPEARANCE AND ODOR: Yellow oil

SECTION IV FIRE AND EXPLOSION HAZARD DATA

FLASH POINT: Not available

METHOD USED

FLAMMABLE (EXPLOSIVE) LIMITS: UPPER / LOWER

EXTINGUISHING MEDIA: Dry chemical, carbon dioxide, foam or water spray

SPECIAL FIRE FIGHTING PROCEDURES: Wear full protective clothing

UNUSUAL FIRE AND EXPLOSION HAZARDS: Emits toxic fumes when heated

SECTION V HEALTH HAZARD DATA

TOXIC LEVEL: None available

CARCINOGENIC: no

PRINCIPLE ROUTES OF ABSORBTION: Inhalation, ingestion, skin

SKIN AND EYE IRRITATION: moderate

REVELANT SYMPTOMS OF EXPOSURE: Cyanosis, air hunger, nausea, vomiting, low blood pressure, convulsions.

EFFECTS OF CHRONIC EXPOSURE: Weight loss, anemia, weakness

EMERGENCY AND FIRST AID PROCEDURES: Inhalation - rest; if swallowed - give emetic, get medical aid; wash skin and eyes, skin may be washed with 5% acetic acid to aid removal.

SECTION VI REACTIVITY DATA

CONDITIONS CONTRIBUTING TO INSTABILITY

CONDITIONS CONTRIBUTING TO HAZARDOUS POLYMERIZATION

INCOMPATABILITY (MATERIALS TO AVOID)

HAZARDOUS DECOMPOSITION PRODUCTS

SECTION VII SPILL OR LEAK PROCEDURES

STEPS TO BE TAKEN IN CASE MATERIAL IS RELEASED OR SPILLED

WASTE DISPOSAL METHOD

SECTION VIII SPECIAL PROTECTION INFORMATION

VENTILATION REQUIREMENTS LOCAL EXHAUST

MECHANICAL (GENERAL)

SPECIAL

PROTECTIVE EQUIPMENT (SPECIFY TYPES) EYE: Goggles

GLOVES: Rubber

RESPIRATOR: With proper filter

OTHER PROTECTIVE EQUIPMENT

SECTION IX SPECIAL PRECAUTIONS

PRECAUTIONS TO BE TAKEN IN HANDLING AND STORAGE

OTHER PRECAUTIONS

SIGNATURE | DATE

DIMETHYL-p-TOLUIDINE

2 / 4 / 0

SECTION I SOURCE AND NOMENCLATURE

MANUFACTURER'S NAME | EMERGENCY TELEPHONE NO.

ADDRESS (NUMBER, STREET, CITY, STATE, ZIP CODE)

TRADE NAME AND SYNONYMS | CHEMICAL FAMILY

CHEMICAL NAME AND SYNONYMS: N,N-Dimethyl-P-Aminotoluene

FORMULA: $(CH_3)_2NC_6H_4CH_3$

SECTION II HAZARDOUS INGREDIENTS

BASIC MATERIAL	APPROXIMATE OR MAXIMUM % WT. OR VOL.	ESTABLISHED OSHA STANDARD	LD_{50} ORAL	LD_{50} PERCUT.	LC_{50} SPECIES	LC_{50} CONC.
			None available			

SECTION III PHYSICAL DATA

BOILING POINT (°F.): 211°C

VAPOR PRESSURE (mm Hg.): at 72°C 5

MELTING POINT (°F.)

VAPOR DENSITY (Air = 1)

SPECIFIC GRAVITY (H_2O = 1): 0.94

EVAPORATION RATE (______ = 1)

SOLUBILITY IN WATER Pts/100 pts H_2O: insoluble

VOLATILE % Vol. % Wt.

APPEARANCE AND ODOR: liquid

SECTION IV FIRE AND EXPLOSION HAZARD DATA

FLASH POINT: not available

METHOD USED

FLAMMABLE (EXPLOSIVE) LIMITS: UPPER / LOWER

EXTINGUISHING MEDIA: dry chemical, carbon dioxide, foam or water spray

SPECIAL FIRE FIGHTING PROCEDURES: wear full protective clothing

UNUSUAL FIRE AND EXPLOSION HAZARDS: emits toxic fumes whem heated.

SECTION V HEALTH HAZARD DATA

TOXIC LEVEL: None available

CARCINOGENIC: no

PRINCIPLE ROUTES OF ABSORBTION: Inhalation, ingestion, skin

SKIN AND EYE IRRITATION: moderate

REVELANT SYMPTOMS OF EXPOSURE: Cyanosis, air hunger, nausea vomiting, low blood pressure, convulsions

EFFECTS OF CHRONIC EXPOSURE: weight loss, anemia, weakness

EMERGENCY AND FIRST AID PROCEDURES: inhalation - rest; if swallowed - give emetic, get medical aid; wash skin & eyes, skin may be washed with 5% acetic acid to aid removal.

SECTION VI REACTIVITY DATA

CONDITIONS CONTRIBUTING TO INSTABILITY

CONDITIONS CONTRIBUTING TO HAZARDOUS POLYMERIZATION

INCOMPATABILITY (MATERIALS TO AVOID)

HAZARDOUS DECOMPOSITION PRODUCTS

SECTION VII SPILL OR LEAK PROCEDURES

STEPS TO BE TAKEN IN CASE MATERIAL IS RELEASED OR SPILLED

WASTE DISPOSAL METHOD

SECTION VIII SPECIAL PROTECTION INFORMATION

VENTILATION REQUIREMENTS LOCAL EXHAUST

MECHANICAL (GENERAL)

SPECIAL

PROTECTIVE EQUIPMENT (SPECIFY TYPES) EYE: goggles

GLOVES: rubber

RESPIRATOR: with proper filter

OTHER PROTECTIVE EQUIPMENT

SECTION IX SPECIAL PRECAUTIONS

PRECAUTIONS TO BE TAKEN IN HANDLING AND STORAGE

OTHER PRECAUTIONS

SIGNATURE | DATE

2,4-DINITROANISOLE

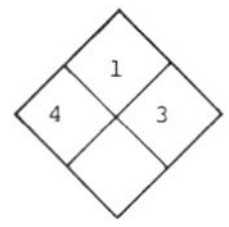

SECTION I SOURCE AND NOMENCLATURE

MANUFACTURER'S NAME | EMERGENCY TELEPHONE NO.

ADDRESS (NUMBER, STREET, CITY, STATE, ZIP CODE)

TRADE NAME AND SYNONYMS: 1 Methoxy-2,4 Dinitrobenzene

CHEMICAL FAMILY:

CHEMICAL NAME AND SYNONYMS: 2,4 - Dinitroanisole

FORMULA: $(NO_2)_2C_6H_3OCH_3$

SECTION II HAZARDOUS INGREDIENTS

BASIC MATERIAL	APPROXIMATE OR MAXIMUM % WT. OR VOL.	ESTABLISHED OSHA STANDARD	LD_{50} ORAL	LD_{50} PERCUT.	LC_{50} SPECIES	LC_{50} CONC.
Note:						
Lowest published lethal	dose					
oral - rat 100 mg/kg						

SECTION III PHYSICAL DATA MW - 198.13

BOILING POINT °F.: Sublimes	VAPOR PRESSURE mm Hg.:
MELTING POINT °F.: 95°C	VAPOR DENSITY (Air = 1): 6.83
SPECIFIC GRAVITY (H_2O = 1): 1.34	EVAPORATION RATE (____ = 1):
SOLUBILITY IN WATER Pts/100 pts H_2O: Slightly soluble	VOLATILE % Vol. % Wt.:

APPEARANCE AND ODOR: Colorless crystal

SECTION IV FIRE AND EXPLOSION HAZARD DATA

FLASH POINT: none available

METHOD USED:

FLAMMABLE (EXPLOSIVE) LIMITS: UPPER / LOWER

EXTINGUISHING MEDIA: explosive - do not fight fire except for unmanned equipment - use water.

SPECIAL FIRE FIGHTING PROCEDURES: Wear full protective clothing for rescue operations.

UNUSUAL FIRE AND EXPLOSION HAZARDS: Severe explosion hazard.

SECTION V HEALTH HAZARD DATA

TOXIC LEVEL: none available

CARCINOGENIC: no

PRINCIPLE ROUTES OF ABSORBTION: Inhalation, ingestion, skin

SKIN AND EYE IRRITATION: moderate

REVELANT SYMPTOMS OF EXPOSURE: Blood damage, cyanosis, dizziness, headache, increased pulse rate, unconsciousness.

EFFECTS OF CHRONIC EXPOSURE: Weight loss, anemia, possible liver damage.

EMERGENCY AND FIRST AID PROCEDURES: Inhalation - rest, give oxygen as needed. If swallowed - give emetic, get medical aid; wash skin & eyes.

SECTION VI REACTIVITY DATA

CONDITIONS CONTRIBUTING TO INSTABILITY:

CONDITIONS CONTRIBUTING TO HAZARDOUS POLYMERIZATION:

INCOMPATABILITY (MATERIALS TO AVOID): Keep away oxidizing materials (explosive) inorganic nitrates, chlorides or ammonium nitrate.

HAZARDOUS DECOMPOSITION PRODUCTS: Oxides of nitrogen

SECTION VII SPILL OR LEAK PROCEDURES

STEPS TO BE TAKEN IN CASE MATERIAL IS RELEASED OR SPILLED:

WASTE DISPOSAL METHOD:

SECTION VIII SPECIAL PROTECTION INFORMATION

VENTILATION REQUIREMENTS LOCAL EXHAUST	PROTECTIVE EQUIPMENT (SPECIFY TYPES) EYE: Safety goggles
MECHANICAL (GENERAL)	GLOVES: Rubber
SPECIAL	RESPIRATOR: Self-contained

OTHER PROTECTIVE EQUIPMENT:

SECTION IX SPECIAL PRECAUTIONS

PRECAUTIONS TO BE TAKEN IN HANDLING AND STORAGE: Keep from oxidizing materials, combustibles and source of heat.

OTHER PRECAUTIONS:

SIGNATURE | DATE

3,5-DINITROBENZOYL CHLORIDE

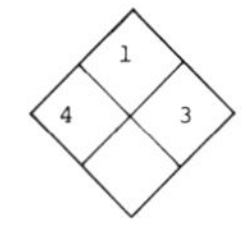

SECTION I SOURCE AND NOMENCLATURE

MANUFACTURER'S NAME | EMERGENCY TELEPHONE NO.

ADDRESS (NUMBER, STREET, CITY, STATE, ZIP CODE)

TRADE NAME AND SYNONYMS:

CHEMICAL FAMILY:

CHEMICAL NAME AND SYNONYMS: 3,5 - Dinitrobenzoyl chloride

FORMULA: $(NO_2)_2C_6H_3COCl$

SECTION II HAZARDOUS INGREDIENTS

BASIC MATERIAL	APPROXIMATE OR MAXIMUM % WT. OR VOL.	ESTABLISHED OSHA STANDARD	LD_{50} ORAL	LD_{50} PERCUT.	LC_{50} SPECIES	LC_{50} CONC.
			none	available		

SECTION III PHYSICAL DATA MW - 230.57

BOILING POINT °F.:	VAPOR PRESSURE mm Hg.: at 196°C 12
MELTING POINT °F.: 69°C	VAPOR DENSITY (Air = 1):
SPECIFIC GRAVITY (H_2O = 1):	EVAPORATION RATE (____ = 1):
SOLUBILITY IN WATER Pts/100 pts H_2O: decomposes	VOLATILE % Vol. % Wt.:

APPEARANCE AND ODOR: yellow green crystals

SECTION IV FIRE AND EXPLOSION HAZARD DATA

FLASH POINT: Not available

METHOD USED:

FLAMMABLE (EXPLOSIVE) LIMITS: UPPER / LOWER

EXTINGUISHING MEDIA: Explosive - do not fight fire except for unmanned equpment - use water.

SPECIAL FIRE FIGHTING PROCEDURES: Wear full protective clothing for rescue operations

UNUSUAL FIRE AND EXPLOSION HAZARDS: Severe explosion hazard.

SECTION V HEALTH HAZARD DATA

TOXIC LEVEL: none available

CARCINOGENIC: no

PRINCIPLE ROUTES OF ABSORBTION: inhalation, ingestion, skin

SKIN AND EYE IRRITATION: severe

REVELANT SYMPTOMS OF EXPOSURE: Headache, cyanosis, dizziness, irritation of respiratory system, skin, eye, mouth and digestive tract burns.

EFFECTS OF CHRONIC EXPOSURE: Weight loss, anemia, possible liver damage.

EMERGENCY AND FIRST AID PROCEDURES: Inhalation - rest; give oxygen as needed; if swallowed - drink water, milk of magnesia, get medical aid; wash skin & eyes.

SECTION VI REACTIVITY DATA

CONDITIONS CONTRIBUTING TO INSTABILITY:

CONDITIONS CONTRIBUTING TO HAZARDOUS POLYMERIZATION:

INCOMPATABILITY (MATERIALS TO AVOID): Keep from oxidizing materials (explosive) inorganic nitrates, chlorides or ammonium nitrate.

HAZARDOUS DECOMPOSITION PRODUCTS:

SECTION VII SPILL OR LEAK PROCEDURES

STEPS TO BE TAKEN IN CASE MATERIAL IS RELEASED OR SPILLED:

WASTE DISPOSAL METHOD:

SECTION VIII SPECIAL PROTECTION INFORMATION

VENTILATION REQUIREMENTS LOCAL EXHAUST	PROTECTIVE EQUIPMENT (SPECIFY TYPES) EYE: Safety goggles
MECHANICAL (GENERAL)	GLOVES: Rubber
SPECIAL	RESPIRATOR: Self-contained

OTHER PROTECTIVE EQUIPMENT:

SECTION IX SPECIAL PRECAUTIONS

PRECAUTIONS TO BE TAKEN IN HANDLING AND STORAGE: Keep from oxidizing materials, combustibles and source of heat.

OTHER PRECAUTIONS:

SIGNATURE | DATE

2,4-DINITROBROMOBENZENE

(Hazard diamond: 1 top, 4 left, 4 right)

SECTION I SOURCE AND NOMENCLATURE

MANUFACTURER'S NAME | EMERGENCY TELEPHONE NO.

ADDRESS (NUMBER, STREET, CITY, STATE, ZIP CODE)

TRADE NAME AND SYNONYMS | CHEMICAL FAMILY

CHEMICAL NAME AND SYNONYMS: 1 bromo-2,4, Dinitrobenzene

FORMULA: $(NO_2)_2C_6H_3Br$

SECTION II HAZARDOUS INGREDIENTS

BASIC MATERIAL	APPROXIMATE OR MAXIMUM % WT. OR VOL.	ESTABLISHED OSHA STANDARD	LD_{50} ORAL	LD_{50} PERCUT.	LC_{50} SPECIES	LC_{50} CONC.
			none available			

SECTION III PHYSICAL DATA MW - 247.02

BOILING POINT °F. | VAPOR PRESSURE mm Hg.

MELTING POINT °F.: 74°C | VAPOR DENSITY (Air = 1)

SPECIFIC GRAVITY (H_2O = 1) | EVAPORATION RATE (= 1)

SOLUBILITY IN WATER Pts/100 pts H_2O | VOLATILE % Vol. % Wt.

APPEARANCE AND ODOR: Yellow crystals

SECTION IV FIRE AND EXPLOSION HAZARD DATA

FLASH POINT: None available | FLAMMABLE (EXPLOSIVE) LIMITS UPPER | LOWER

METHOD USED

EXTINGUISHING MEDIA: Explosive - do not fight fire except for unmanned equipment - use water.

SPECIAL FIRE FIGHTING PROCEDURES: Wear full protective clothing for rescue operations

UNUSUAL FIRE AND EXPLOSION HAZARDS: Severe explosion hazard.

SECTION V HEALTH HAZARD DATA

TOXIC LEVEL: None available | CARCINOGENIC: no

PRINCIPLE ROUTES OF ABSORBTION: Inhalation, ingestion | SKIN AND EYE IRRITATION: moderate

REVELANT SYMPTOMS OF EXPOSURE: Dermatitis, irritation of respiratory and digestive systems.

EFFECTS OF CHRONIC EXPOSURE: Dermatitis

EMERGENCY AND FIRST AID PROCEDURES: Inhalation - rest; if swallowed - give emetic, get medical aid; wash skin & eyes.

SECTION VI REACTIVITY DATA

CONDITIONS CONTRIBUTING TO INSTABILITY

CONDITIONS CONTRIBUTING TO HAZARDOUS POLYMERIZATION

INCOMPATABILITY (MATERIALS TO AVOID): Keep from oxidizing materials (explosive) inorganic nitrates, chlorates or ammonium nitrate.

HAZARDOUS DECOMPOSITION PRODUCTS: Oxides of nitrogen

SECTION VII SPILL OR LEAK PROCEDURES

STEPS TO BE TAKEN IN CASE MATERIAL IS RELEASED OR SPILLED

WASTE DISPOSAL METHOD

SECTION VIII SPECIAL PROTECTION INFORMATION

VENTILATION REQUIREMENTS LOCAL EXHAUST | PROTECTIVE EQUIPMENT (SPECIFY TYPES) EYE: Safety goggles

MECHANICAL (GENERAL) | GLOVES: Rubber

SPECIAL | RESPIRATOR: Self-contained

OTHER PROTECTIVE EQUIPMENT

SECTION IX SPECIAL PRECAUTIONS

PRECAUTIONS TO BE TAKEN IN HANDLING AND STORAGE: Keep from oxidizing materials, combustibles and source of heat.

OTHER PRECAUTIONS

SIGNATURE | DATE

2,4-DINITROCHLOROBENZENE

(Hazard diamond: 1 top, 4 left, 4 right)

SECTION I SOURCE AND NOMENCLATURE

MANUFACTURER'S NAME | EMERGENCY TELEPHONE NO.

ADDRESS (NUMBER, STREET, CITY, STATE, ZIP CODE)

TRADE NAME AND SYNONYMS | CHEMICAL FAMILY

CHEMICAL NAME AND SYNONYMS: 1 chloro 2,4, - dinitrobenzene

FORMULA: $(NO_2)_2C_6H_3Cl$

SECTION II HAZARDOUS INGREDIENTS

BASIC MATERIAL	APPROXIMATE OR MAXIMUM % WT. OR VOL.	ESTABLISHED OSHA STANDARD	LD_{50} mg/kg ORAL	LD_{50} mg/kg PERCUT.	LC_{50} SPECIES	LC_{50} CONC.
			1070		Rat	
				130	Rabbit	

SECTION III PHYSICAL DATA MW - 202.56

BOILING POINT °F.: 315°C | VAPOR PRESSURE mm Hg.

MELTING POINT °F.: 52°C | VAPOR DENSITY (Air = 1): 6.98

SPECIFIC GRAVITY (H_2O = 1): 1.68 | EVAPORATION RATE (= 1)

SOLUBILITY IN WATER Pts/100 pts H_2O: Insoluble | VOLATILE % Vol. % Wt.

APPEARANCE AND ODOR: Crystalline (yellow) almond-like odor

SECTION IV FIRE AND EXPLOSION HAZARD DATA

FLASH POINT: 382°F | FLAMMABLE (EXPLOSIVE) LIMITS UPPER | LOWER

METHOD USED: CC

EXTINGUISHING MEDIA: Explosive - do not fight fire except for unmanned equipment - use water.

SPECIAL FIRE FIGHTING PROCEDURES: Wear full protective clothing for rescue operations.

UNUSUAL FIRE AND EXPLOSION HAZARDS: Severe explosion hazard

SECTION V HEALTH HAZARD DATA

TOXIC LEVEL: None available | CARCINOGENIC: no

PRINCIPLE ROUTES OF ABSORBTION: Inhalation, ingestion, skin | SKIN AND EYE IRRITATION: moderate

REVELANT SYMPTOMS OF EXPOSURE: Dermatitis, irritation of respiratory and digestive systems, skin sensitized.

EFFECTS OF CHRONIC EXPOSURE: Dermatitis

EMERGENCY AND FIRST AID PROCEDURES: Inhalation - rest; if swallowed - give emetic, get medical aid; wash skin & eyes.

SECTION VI REACTIVITY DATA

CONDITIONS CONTRIBUTING TO INSTABILITY

CONDITIONS CONTRIBUTING TO HAZARDOUS POLYMERIZATION

INCOMPATABILITY (MATERIALS TO AVOID): Keep from oxidizing materials (explosive) inorganic nitrates, chlorates and ammonium nitrate

HAZARDOUS DECOMPOSITION PRODUCTS: oxides of nitrogen

SECTION VII SPILL OR LEAK PROCEDURES

STEPS TO BE TAKEN IN CASE MATERIAL IS RELEASED OR SPILLED

WASTE DISPOSAL METHOD

SECTION VIII SPECIAL PROTECTION INFORMATION

VENTILATION REQUIREMENTS LOCAL EXHAUST | PROTECTIVE EQUIPMENT (SPECIFY TYPES) EYE: goggles

MECHANICAL (GENERAL) | GLOVES: rubber

SPECIAL | RESPIRATOR: self-contained

OTHER PROTECTIVE EQUIPMENT

SECTION IX SPECIAL PRECAUTIONS

PRECAUTIONS TO BE TAKEN IN HANDLING AND STORAGE: Keep from oxidizing materials, combustibles, and source of heat.

OTHER PRECAUTIONS

SIGNATURE | DATE

4,4′-DINITRODIPHENYL

Hazard diamond: 1 (top), 4 (left), 0 (right)

SECTION I SOURCE AND NOMENCLATURE

MANUFACTURER'S NAME: | EMERGENCY TELEPHONE NO.:

ADDRESS (NUMBER, STREET, CITY, STATE, ZIP CODE):

TRADE NAME AND SYNONYMS: | CHEMICAL FAMILY:

CHEMICAL NAME AND SYNONYMS: 4,4 - Dinitrobiphenyl | FORMULA: $NO_2C_6H_4C_6H_4NO_2$

SECTION II HAZARDOUS INGREDIENTS

BASIC MATERIAL	APPROXIMATE OR MAXIMUM % WT. OR VOL.	ESTABLISHED OSHA STANDARD	LD_{50} ORAL	LD_{50} PERCUT.	LC_{50} SPECIES	LC_{50} CONC.
Lowest published toxic dose oral - rat 860 mg/kg.	- neoplastic					

SECTION III PHYSICAL DATA MW - 244.20

BOILING POINT: °F. | VAPOR PRESSURE: mm Hg.

MELTING POINT: 233°C | VAPOR DENSITY (Air = 1):

SPECIFIC GRAVITY (H_2O = 1): 1.45 | EVAPORATION RATE (= 1):

SOLUBILITY IN WATER Pts/100 pts H_2O: Insoluble | VOLATILE: % Vol. % Wt.

APPEARANCE AND ODOR: Needle-like crystals

SECTION IV FIRE AND EXPLOSION HAZARD DATA

FLASH POINT: None available | FLAMMABLE (EXPLOSIVE) LIMITS: UPPER / LOWER

METHOD USED:

EXTINGUISHING MEDIA: Alcohol foam, carbon dioxide or dry chemical

SPECIAL FIRE FIGHTING PROCEDURES: Water may cause frothing.

UNUSUAL FIRE AND EXPLOSION HAZARDS:

SECTION V HEALTH HAZARD DATA

TOXIC LEVEL: none available | CARCINOGENIC: neoplastic

PRINCIPLE ROUTES OF ABSORBTION: inhalation (mainly) ingestion | SKIN AND EYE IRRITATION: moderate

REVELANT SYMPTOMS OF EXPOSURE: Nausea, vomiting, central nervous system depression, headache, anemia

EFFECTS OF CHRONIC EXPOSURE: Irritation and lesions of respiratory passages.

EMERGENCY AND FIRST AID PROCEDURES: Inhalation - rest; oxygen as needed; if swallowed - get medical aid, gastric lavage; wash from skin & eyes.

SECTION VI REACTIVITY DATA

CONDITIONS CONTRIBUTING TO INSTABILITY:

CONDITIONS CONTRIBUTING TO HAZARDOUS POLYMERIZATION:

INCOMPATABILITY (MATERIALS TO AVOID):

HAZARDOUS DECOMPOSITION PRODUCTS:

SECTION VII SPILL OR LEAK PROCEDURES

STEPS TO BE TAKEN IN CASE MATERIAL IS RELEASED OR SPILLED:

WASTE DISPOSAL METHOD:

SECTION VIII SPECIAL PROTECTION INFORMATION

VENTILATION REQUIREMENTS LOCAL EXHAUST: | PROTECTIVE EQUIPMENT (SPECIFY TYPES) EYE: Goggles

MECHANICAL (GENERAL): | GLOVES: Rubber

SPECIAL: | RESPIRATOR: With proper filter

OTHER PROTECTIVE EQUIPMENT:

SECTION IX SPECIAL PRECAUTIONS

PRECAUTIONS TO BE TAKEN IN HANDLING AND STORAGE:

OTHER PRECAUTIONS:

SIGNATURE: | DATE:

4,4′-DINITRODIPHENYLMETHANE

Hazard diamond: 1 (top), 4 (left), 3 (right)

SECTION I SOURCE AND NOMENCLATURE

MANUFACTURER'S NAME: | EMERGENCY TELEPHONE NO.:

ADDRESS (NUMBER, STREET, CITY, STATE, ZIP CODE):

TRADE NAME AND SYNONYMS: | CHEMICAL FAMILY:

CHEMICAL NAME AND SYNONYMS: Bis (P-Nitrophenyl) Methane | FORMULA: $(NO_2C_6H_4)_2CH_2$

SECTION II HAZARDOUS INGREDIENTS

BASIC MATERIAL	APPROXIMATE OR MAXIMUM % WT. OR VOL.	ESTABLISHED OSHA STANDARD	LD_{50} ORAL	LD_{50} PERCUT.	LC_{50} SPECIES	LC_{50} CONC.
			None	available		

SECTION III PHYSICAL DATA MW - 258.23

BOILING POINT: °F. | VAPOR PRESSURE: mm Hg.

MELTING POINT: 184°C. | VAPOR DENSITY (Air = 1):

SPECIFIC GRAVITY (H_2O = 1): | EVAPORATION RATE (= 1):

SOLUBILITY IN WATER Pts/100 pts H_2O: Soluble | VOLATILE: % Vol. % Wt.

APPEARANCE AND ODOR: Needle-like crystals.

SECTION IV FIRE AND EXPLOSION HAZARD DATA

FLASH POINT: None available | FLAMMABLE (EXPLOSIVE) LIMITS: UPPER / LOWER

METHOD USED:

EXTINGUISHING MEDIA: Water, carbon dioxide, dry chemicals.

SPECIAL FIRE FIGHTING PROCEDURES:

UNUSUAL FIRE AND EXPLOSION HAZARDS:

SECTION V HEALTH HAZARD DATA

TOXIC LEVEL: None available | CARCINOGENIC: No

PRINCIPLE ROUTES OF ABSORBTION: Inhalation, ingestion, skin | SKIN AND EYE IRRITATION: Slight

REVELANT SYMPTOMS OF EXPOSURE: Difficulty in breathing, cyanosis, vomiting, headache.

EFFECTS OF CHRONIC EXPOSURE: Methemoglobin - former.

EMERGENCY AND FIRST AID PROCEDURES: If swallowed - give emetic, get medical aid; inhalation - rest; wash from skin and eyes.

SECTION VI REACTIVITY DATA

CONDITIONS CONTRIBUTING TO INSTABILITY:

CONDITIONS CONTRIBUTING TO HAZARDOUS POLYMERIZATION:

INCOMPATABILITY (MATERIALS TO AVOID): May explode when heated with sulfuric acid.

HAZARDOUS DECOMPOSITION PRODUCTS:

SECTION VII SPILL OR LEAK PROCEDURES

STEPS TO BE TAKEN IN CASE MATERIAL IS RELEASED OR SPILLED:

WASTE DISPOSAL METHOD:

SECTION VIII SPECIAL PROTECTION INFORMATION

VENTILATION REQUIREMENTS LOCAL EXHAUST: | PROTECTIVE EQUIPMENT (SPECIFY TYPES) EYE: Goggles

MECHANICAL (GENERAL): | GLOVES: Rubber

SPECIAL: | RESPIRATOR: Self-contained

OTHER PROTECTIVE EQUIPMENT:

SECTION IX SPECIAL PRECAUTIONS

PRECAUTIONS TO BE TAKEN IN HANDLING AND STORAGE: Outside or detached storage preferred, inside storage in standard flammable liquids cabinet.

OTHER PRECAUTIONS:

SIGNATURE: | DATE:

2,4-DINITROPHENETOLE

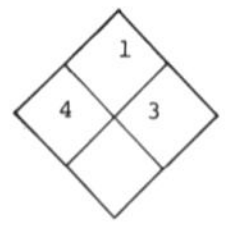

SECTION I SOURCE AND NOMENCLATURE

MANUFACTURER'S NAME

EMERGENCY TELEPHONE NO.

ADDRESS (NUMBER, STREET, CITY, STATE, ZIP CODE)

TRADE NAME AND SYNONYMS

CHEMICAL FAMILY

CHEMICAL NAME AND SYNONYMS: α dinitrophenetole

FORMULA: $(NO_2)_2C_6H_3OC_2H_5$

SECTION II HAZARDOUS INGREDIENTS

BASIC MATERIAL	APPROXIMATE OR MAXIMUM % WT. OR VOL.	ESTABLISHED OSHA STANDARD	LD_{50} ORAL	LD_{50} PERCUT.	LC_{50} SPECIES	LC_{50} CONC.
Lowest published lethal dose						
Oral - rat 250 mg/kg.						

SECTION III PHYSICAL DATA MW - 212.16

BOILING POINT: decomposes °F.

VAPOR PRESSURE: mm Hg.

MELTING POINT: 86°C °F.

VAPOR DENSITY (Air = 1)

SPECIFIC GRAVITY (H_2O = 1)

EVAPORATION RATE (= 1)

SOLUBILITY IN WATER Pts/100 pts H_2O: very slightly soluble

VOLATILE % Vol. % Wt.

APPEARANCE AND ODOR: colorless crystal

SECTION IV FIRE AND EXPLOSION HAZARD DATA

FLASH POINT: none available

METHOD USED

FLAMMABLE (EXPLOSIVE) LIMITS: UPPER / LOWER

EXTINGUISHING MEDIA: Explosive - do not fight fire except for unmanned equipment - use water.

SPECIAL FIRE FIGHTING PROCEDURES: Wear full protective clothing for rescue operations.

UNUSUAL FIRE AND EXPLOSION HAZARDS: Severe explosion hazard

SECTION V HEALTH HAZARD DATA

TOXIC LEVEL: None available (skin)

CARCINOGENIC: no

PRINCIPLE ROUTES OF ABSORBTION: Inhalation of dust, ingestion, skin

SKIN AND EYE IRRITATION: moderate

REVELANT SYMPTOMS OF EXPOSURE: Cyanosis, dizziness, increased pulse rate, headache, unconsciousness.

EFFECTS OF CHRONIC EXPOSURE: Weight loss, anemia, possible liver damage.

EMERGENCY AND FIRST AID PROCEDURES: Inhalation - rest, give oxygen as needed; if swallowed - give emetic, get medical aid; wash from skin & eyes.

SECTION VI REACTIVITY DATA

CONDITIONS CONTRIBUTING TO INSTABILITY

CONDITIONS CONTRIBUTING TO HAZARDOUS POLYMERIZATION

INCOMPATABILITY (MATERIALS TO AVOID): Explosive - keep from oxidizing materials, inorganic nitrates chlorates or ammonium nitrate.

HAZARDOUS DECOMPOSITION PRODUCTS

SECTION VII SPILL OR LEAK PROCEDURES

STEPS TO BE TAKEN IN CASE MATERIAL IS RELEASED OR SPILLED

WASTE DISPOSAL METHOD

SECTION VIII SPECIAL PROTECTION INFORMATION

VENTILATION REQUIREMENTS LOCAL EXHAUST

MECHANICAL (GENERAL)

SPECIAL

OTHER PROTECTIVE EQUIPMENT

PROTECTIVE EQUIPMENT (SPECIFY TYPES) EYE: Goggles

GLOVES: Rubber

RESPIRATOR: Self-contained

SECTION IX SPECIAL PRECAUTIONS

PRECAUTIONS TO BE TAKEN IN HANDLING AND STORAGE: Keep from oxidizing materials, combustibles and source of heat.

OTHER PRECAUTIONS

SIGNATURE | DATE

2,4-DINITROPHENYL HYDRAZINE

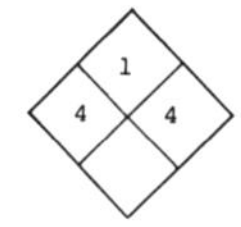

SECTION I SOURCE AND NOMENCLATURE

MANUFACTURER'S NAME

EMERGENCY TELEPHONE NO.

ADDRESS (NUMBER, STREET, CITY, STATE, ZIP CODE)

TRADE NAME AND SYNONYMS

CHEMICAL FAMILY

CHEMICAL NAME AND SYNONYMS: 2,4 - Dinitrophenylhydrazine

FORMULA: $(NO_2)_2C_6H_3NHNH_2$

SECTION II HAZARDOUS INGREDIENTS

BASIC MATERIAL	APPROXIMATE OR MAXIMUM % WT. OR VOL.	ESTABLISHED OSHA STANDARD	LD_{50} mg/kg ORAL	LD_{50} mg/kg PERCUT.	LC_{50} SPECIES	LC_{50} CONC.
			none	available		
			654		Rat	

SECTION III PHYSICAL DATA MW - 198.14

BOILING POINT: -- °F.

VAPOR PRESSURE: mm Hg.

MELTING POINT: 194°C °F.

VAPOR DENSITY (Air = 1)

SPECIFIC GRAVITY (H_2O = 1)

EVAPORATION RATE (= 1)

SOLUBILITY IN WATER Pts/100 pts H_2O: insoluble

VOLATILE % Vol. % Wt.

APPEARANCE AND ODOR: red crystaline powder

SECTION IV FIRE AND EXPLOSION HAZARD DATA

FLASH POINT: none available

METHOD USED

FLAMMABLE (EXPLOSIVE) LIMITS: UPPER / LOWER

EXTINGUISHING MEDIA: explosive - do not fight fire except for unmanned equipment - use water

SPECIAL FIRE FIGHTING PROCEDURES: wear full protective clothing for rescue operation

UNUSUAL FIRE AND EXPLOSION HAZARDS: dangerous explosive

SECTION V HEALTH HAZARD DATA

TOXIC LEVEL: None available (skin)

CARCINOGENIC: no

PRINCIPLE ROUTES OF ABSORBTION: Inhalation, ingestion, skin

SKIN AND EYE IRRITATION: severe

REVELANT SYMPTOMS OF EXPOSURE: Irritation of skin, eyes and mucous membranes, gastroentric disturbances, anemia

EFFECTS OF CHRONIC EXPOSURE: Blood and liver damage, loss of appetite

EMERGENCY AND FIRST AID PROCEDURES: Inhalation - rest; if swallowed - induce vomiting, get medical aid; wash from skin & eyes.

SECTION VI REACTIVITY DATA

CONDITIONS CONTRIBUTING TO INSTABILITY

CONDITIONS CONTRIBUTING TO HAZARDOUS POLYMERIZATION

INCOMPATABILITY (MATERIALS TO AVOID): Avoid strong oxidizing agents.

HAZARDOUS DECOMPOSITION PRODUCTS

SECTION VII SPILL OR LEAK PROCEDURES

STEPS TO BE TAKEN IN CASE MATERIAL IS RELEASED OR SPILLED

WASTE DISPOSAL METHOD

SECTION VIII SPECIAL PROTECTION INFORMATION

VENTILATION REQUIREMENTS LOCAL EXHAUST

MECHANICAL (GENERAL)

SPECIAL

OTHER PROTECTIVE EQUIPMENT

PROTECTIVE EQUIPMENT (SPECIFY TYPES) EYE: Goggles

GLOVES: Rubber

RESPIRATOR: With proper filter

SECTION IX SPECIAL PRECAUTIONS

PRECAUTIONS TO BE TAKEN IN HANDLING AND STORAGE: Keep away from heat and sources of ignition or acute fire hazard.

OTHER PRECAUTIONS

SIGNATURE | DATE

2,6-DINITROTHYMOL

1 / 4 / 3

SECTION I SOURCE AND NOMENCLATURE

MANUFACTURER'S NAME | EMERGENCY TELEPHONE NO.

ADDRESS (NUMBER, STREET, CITY, STATE, ZIP CODE)

TRADE NAME AND SYNONYMS | CHEMICAL FAMILY

CHEMICAL NAME AND SYNONYMS 1,3 dinitro-5 isopropyl - 4 hydroxy - 6 methylbenzene

FORMULA $(NO_2)_2C_6H(CH_3)(C_3H_8)OH$

SECTION II HAZARDOUS INGREDIENTS

BASIC MATERIAL	APPROXIMATE OR MAXIMUM % WT. OR VOL.	ESTABLISHED OSHA STANDARD	LD_{50} ORAL	LD_{50} PERCUT.	LC_{50} SPECIES	LC_{50} CONC.
Lowest published lethal dose oral - rat 100 mg/kg						

SECTION III PHYSICAL DATA MW-240.22

BOILING POINT °F. | VAPOR PRESSURE mm Hg.

MELTING POINT °F. 55°C | VAPOR DENSITY (Air = 1)

SPECIFIC GRAVITY (H_2O = 1) | EVAPORATION RATE (= 1)

SOLUBILITY IN WATER Pts/100 pts H_2O insoluble | VOLATILE % Vol. % Wt.

APPEARANCE AND ODOR prismatic crystals

SECTION IV FIRE AND EXPLOSION HAZARD DATA

FLASH POINT none available

METHOD USED

FLAMMABLE (EXPLOSIVE) LIMITS UPPER LOWER

EXTINGUISHING MEDIA explosive - do not fight fire except for unmanned equipment - use water.

SPECIAL FIRE FIGHTING PROCEDURES Wear full protective clothing for rescue operations.

UNUSUAL FIRE AND EXPLOSION HAZARDS Explosion hazard

SECTION V HEALTH HAZARD DATA

TOXIC LEVEL none available

CARCINOGENIC no

PRINCIPLE ROUTES OF ABSORBTION Inhalation of dust, ingestion, skin

SKIN AND EYE IRRITATION moderate

REVELANT SYMPTOMS OF EXPOSURE temperature elevation, dermatitis, headache

EFFECTS OF CHRONIC EXPOSURE dermatitis, headache

EMERGENCY AND FIRST AID PROCEDURES If swallowed - drink water, epsom salts solution, get medical aid; inhalation - rest; wash skin & eyes.

SECTION VI REACTIVITY DATA

CONDITIONS CONTRIBUTING TO INSTABILITY

CONDITIONS CONTRIBUTING TO HAZARDOUS POLYMERIZATION

INCOMPATABILITY (MATERIALS TO AVOID) Explosive - keep from oxidizing materials

HAZARDOUS DECOMPOSITION PRODUCTS Oxides of nitrogen

SECTION VII SPILL OR LEAK PROCEDURES

STEPS TO BE TAKEN IN CASE MATERIAL IS RELEASED OR SPILLED

WASTE DISPOSAL METHOD

SECTION VIII SPECIAL PROTECTION INFORMATION

VENTILATION REQUIREMENTS LOCAL EXHAUST

PROTECTIVE EQUIPMENT (SPECIFY TYPES) EYE Goggles

MECHANICAL (GENERAL) | GLOVES

SPECIAL | RESPIRATOR

OTHER PROTECTIVE EQUIPMENT

SECTION IX SPECIAL PRECAUTIONS

PRECAUTIONS TO BE TAKEN IN HANDLING AND STORAGE Keep from oxidizing materials, combustibles and source of heat.

OTHER PRECAUTIONS

SIGNATURE | DATE

m-DIPHENYLBENZENE

1 / 4 / 0

SECTION I SOURCE AND NOMENCLATURE

MANUFACTURER'S NAME | EMERGENCY TELEPHONE NO.

ADDRESS (NUMBER, STREET, CITY, STATE, ZIP CODE)

TRADE NAME AND SYNONYMS | CHEMICAL FAMILY

CHEMICAL NAME AND SYNONYMS M Terphenyl

FORMULA $C_6H_5C_6H_4C_6H_5$

SECTION II HAZARDOUS INGREDIENTS

BASIC MATERIAL	APPROXIMATE OR MAXIMUM % WT. OR VOL.	ESTABLISHED OSHA STANDARD	LD_{50} ORAL	LD_{50} PERCUT.	LC_{50} SPECIES	LC_{50} CONC.
			none	available		

SECTION III PHYSICAL DATA MW - 230.29

BOILING POINT °F. 363°C | VAPOR PRESSURE mm Hg.

MELTING POINT °F. 87°C | VAPOR DENSITY (Air = 1) 7.95

SPECIFIC GRAVITY (H_2O = 1) 1.16 | EVAPORATION RATE (= 1)

SOLUBILITY IN WATER Pts/100 pts H_2O insoluble | VOLATILE % Vol. % Wt.

APPEARANCE AND ODOR Crystals

SECTION IV FIRE AND EXPLOSION HAZARD DATA

FLASH POINT 375°F

METHOD USED O.C.

FLAMMABLE (EXPLOSIVE) LIMITS UPPER LOWER

EXTINGUISHING MEDIA Water, carbon dioxide, dry chemical

SPECIAL FIRE FIGHTING PROCEDURES Water or foam may cause frothing.

UNUSUAL FIRE AND EXPLOSION HAZARDS

SECTION V HEALTH HAZARD DATA

TOXIC LEVEL 1 PPM, 9 mg/C.M. (ceiling values)

CARCINOGENIC no

PRINCIPLE ROUTES OF ABSORBTION Inhalation, ingestion

SKIN AND EYE IRRITATION moderate

REVELANT SYMPTOMS OF EXPOSURE Eye, skin & respiratory irritation, nausea, vomiting

EFFECTS OF CHRONIC EXPOSURE Liver and kidney damage, pulmonary edema, bronchopneumonia

EMERGENCY AND FIRST AID PROCEDURES Inhalation - rest; if swallowed - get medical aid, gastric lavage; wash from skin and eyes

SECTION VI REACTIVITY DATA

CONDITIONS CONTRIBUTING TO INSTABILITY

CONDITIONS CONTRIBUTING TO HAZARDOUS POLYMERIZATION

INCOMPATABILITY (MATERIALS TO AVOID)

HAZARDOUS DECOMPOSITION PRODUCTS

SECTION VII SPILL OR LEAK PROCEDURES

STEPS TO BE TAKEN IN CASE MATERIAL IS RELEASED OR SPILLED

WASTE DISPOSAL METHOD

SECTION VIII SPECIAL PROTECTION INFORMATION

VENTILATION REQUIREMENTS LOCAL EXHAUST

PROTECTIVE EQUIPMENT (SPECIFY TYPES) EYE Goggles

MECHANICAL (GENERAL) | GLOVES Polyvinyl alcohol

SPECIAL | RESPIRATOR With proper filter

OTHER PROTECTIVE EQUIPMENT

SECTION IX SPECIAL PRECAUTIONS

PRECAUTIONS TO BE TAKEN IN HANDLING AND STORAGE

OTHER PRECAUTIONS

SIGNATURE | DATE

o-DIPHENYLBENZENE

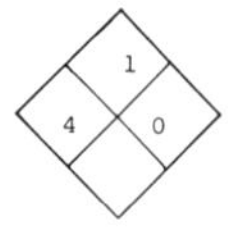

SECTION I SOURCE AND NOMENCLATURE

MANUFACTURER'S NAME | EMERGENCY TELEPHONE NO.

ADDRESS (NUMBER, STREET, CITY, STATE, ZIP CODE)

TRADE NAME AND SYNONYMS | CHEMICAL FAMILY

CHEMICAL NAME AND SYNONYMS: O-Terphenyl | FORMULA: $C_6H_5C_6H_4C_6H_5$

SECTION II HAZARDOUS INGREDIENTS

BASIC MATERIAL	APPROXIMATE OR MAXIMUM % WT. OR VOL.	ESTABLISHED OSHA STANDARD	LD_{50} ORAL	LD_{50} PERCUT.	LC_{50} SPECIES	LC_{50} CONC.
			none available			

SECTION III PHYSICAL DATA MW - 230.29

BOILING POINT °F.: 332°C | VAPOR PRESSURE mm Hg.:

MELTING POINT °F.: 57°C | VAPOR DENSITY (Air = 1): 7.95

SPECIFIC GRAVITY (H_2O = 1): 1.14 | EVAPORATION RATE (______ = 1)

SOLUBILITY IN WATER Pts/100 pts H_2O: Insoluble | VOLATILE % Vol. % Wt.

APPEARANCE AND ODOR: Crystaline

SECTION IV FIRE AND EXPLOSION HAZARD DATA

FLASH POINT: 325°F | FLAMMABLE (EXPLOSIVE) LIMITS: UPPER / LOWER

METHOD USED: O.C.

EXTINGUISHING MEDIA: water, carbon dioxide, dry chemical

SPECIAL FIRE FIGHTING PROCEDURES: water or foam may cause frothing

UNUSUAL FIRE AND EXPLOSION HAZARDS

SECTION V HEALTH HAZARD DATA

TOXIC LEVEL: 1 PPM, 9 mg/C.M. (Ceiling values) | CARCINOGENIC: no

PRINCIPLE ROUTES OF ABSORBTION: Inhalation, ingestion | SKIN AND EYE IRRITATION: moderate

REVELANT SYMPTOMS OF EXPOSURE: Eye, skin and respiratory irritation; nausea, vomiting

EFFECTS OF CHRONIC EXPOSURE: Liver and kidney damage, pulmonary edema, bronchopneumonia

EMERGENCY AND FIRST AID PROCEDURES: Inhalation - rest; if swallowed - get medical aid, gastric lavage; wash from skin and eyes.

SECTION VI REACTIVITY DATA

CONDITIONS CONTRIBUTING TO INSTABILITY

CONDITIONS CONTRIBUTING TO HAZARDOUS POLYMERIZATION

INCOMPATABILITY (MATERIALS TO AVOID)

HAZARDOUS DECOMPOSITION PRODUCTS

SECTION VII SPILL OR LEAK PROCEDURES

STEPS TO BE TAKEN IN CASE MATERIAL IS RELEASED OR SPILLED

WASTE DISPOSAL METHOD

SECTION VIII SPECIAL PROTECTION INFORMATION

VENTILATION REQUIREMENTS LOCAL EXHAUST | PROTECTIVE EQUIPMENT (SPECIFY TYPES) EYE: Goggles

MECHANICAL (GENERAL) | GLOVES: Polyvinyl alcohol

SPECIAL | RESPIRATOR: With proper filter

OTHER PROTECTIVE EQUIPMENT

SECTION IX SPECIAL PRECAUTIONS

PRECAUTIONS TO BE TAKEN IN HANDLING AND STORAGE

OTHER PRECAUTIONS

SIGNATURE | DATE

p-DIPHENYLBENZENE

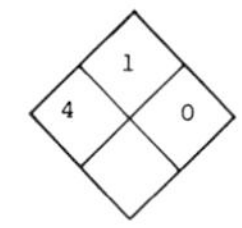

SECTION I SOURCE AND NOMENCLATURE

MANUFACTURER'S NAME | EMERGENCY TELEPHONE NO.

ADDRESS (NUMBER, STREET, CITY, STATE, ZIP CODE)

TRADE NAME AND SYNONYMS | CHEMICAL FAMILY

CHEMICAL NAME AND SYNONYMS: P Terphenyl | FORMULA: $C_6H_5C_6H_4C_6H_5$

SECTION II HAZARDOUS INGREDIENTS

BASIC MATERIAL	APPROXIMATE OR MAXIMUM % WT. OR VOL.	ESTABLISHED OSHA STANDARD	LD_{50} ORAL	LD_{50} PERCUT.	LC_{50} SPECIES	LC_{50} CONC.
Lowest published lethal dose Oral - rat 500 mg/kg.						

SECTION III PHYSICAL DATA MW - 230.29

BOILING POINT °F.: 376°C | VAPOR PRESSURE mm Hg.:

MELTING POINT °F.: 213°C | VAPOR DENSITY (Air = 1): 7.95

SPECIFIC GRAVITY (H_2O = 1): 1.24 | EVAPORATION RATE (______ = 1)

SOLUBILITY IN WATER Pts/100 pts H_2O: Soluble in hot water | VOLATILE % Vol. % Wt.

APPEARANCE AND ODOR: crystaline

SECTION IV FIRE AND EXPLOSION HAZARD DATA

FLASH POINT: 405°C | FLAMMABLE (EXPLOSIVE) LIMITS: UPPER / LOWER

METHOD USED: O.C.

EXTINGUISHING MEDIA: Water, carbon dioxide, dry chemical

SPECIAL FIRE FIGHTING PROCEDURES: Water or foam may cause frothing.

UNUSUAL FIRE AND EXPLOSION HAZARDS

SECTION V HEALTH HAZARD DATA

TOXIC LEVEL: 1 PPM, 9 mg/C.M. (Ceiling Value) | CARCINOGENIC: no

PRINCIPLE ROUTES OF ABSORBTION: Inhalation, ingestion | SKIN AND EYE IRRITATION: Moderate

REVELANT SYMPTOMS OF EXPOSURE: Eye, skin & respiratory irritation; nausea, vomiting

EFFECTS OF CHRONIC EXPOSURE: Liver and kidney damage, pulmonary edema, bronchopneumonia

EMERGENCY AND FIRST AID PROCEDURES: Inhalation - rest; if swallowed - get medical aid, gastric lavage; wash from skin & eyes.

SECTION VI REACTIVITY DATA

CONDITIONS CONTRIBUTING TO INSTABILITY

CONDITIONS CONTRIBUTING TO HAZARDOUS POLYMERIZATION

INCOMPATABILITY (MATERIALS TO AVOID)

HAZARDOUS DECOMPOSITION PRODUCTS

SECTION VII SPILL OR LEAK PROCEDURES

STEPS TO BE TAKEN IN CASE MATERIAL IS RELEASED OR SPILLED

WASTE DISPOSAL METHOD

SECTION VIII SPECIAL PROTECTION INFORMATION

VENTILATION REQUIREMENTS LOCAL EXHAUST | PROTECTIVE EQUIPMENT (SPECIFY TYPES) EYE: Goggles

MECHANICAL (GENERAL) | GLOVES: Polyvinyl alcohol

SPECIAL | RESPIRATOR: With proper filter

OTHER PROTECTIVE EQUIPMENT

SECTION IX SPECIAL PRECAUTIONS

PRECAUTIONS TO BE TAKEN IN HANDLING AND STORAGE

OTHER PRECAUTIONS

SIGNATURE | DATE

as-DIPHENYLHYDRAZINE HYDROCHLORIDE

1

4 0

SECTION I SOURCE AND NOMENCLATURE

MANUFACTURER'S NAME | EMERGENCY TELEPHONE NO.

ADDRESS (NUMBER, STREET, CITY, STATE, ZIP CODE)

TRADE NAME AND SYNONYMS | CHEMICAL FAMILY

CHEMICAL NAME AND SYNONYMS: 1,1-Diphenylhydrazine hydrochloride | FORMULA: $(C_6H_5)_2NNH_2HCl$

SECTION II HAZARDOUS INGREDIENTS

BASIC MATERIAL	APPROXIMATE OR MAXIMUM % WT. OR VOL.	ESTABLISHED OSHA STANDARD	LD_{50}		LC_{50}	
			ORAL	PERCUT.	SPECIES	CONC.
			none available			

SECTION III PHYSICAL DATA MW - 220.70

BOILING POINT °F.: not available | VAPOR PRESSURE mm Hg.

MELTING POINT °F.: not available | VAPOR DENSITY (Air = 1)

SPECIFIC GRAVITY (H_2O = 1) | EVAPORATION RATE (______ = 1)

SOLUBILITY IN WATER Pts/100 pts H_2O: slightly soluble | VOLATILE % Vol. % Wt.

APPEARANCE AND ODOR: white to grey white crystals

SECTION IV FIRE AND EXPLOSION HAZARD DATA

FLASH POINT: none available | FLAMMABLE (EXPLOSIVE) LIMITS: UPPER / LOWER

METHOD USED

EXTINGUISHING MEDIA: alcohol foam, water, carbon dioxide

SPECIAL FIRE FIGHTING PROCEDURES

UNUSUAL FIRE AND EXPLOSION HAZARDS

SECTION V HEALTH HAZARD DATA

TOXIC LEVEL: none available | CARCINOGENIC: no

PRINCIPLE ROUTES OF ABSORBTION: inhalation of dust, ingestion, skin | SKIN AND EYE IRRITATION: severe

REVELANT SYMPTOMS OF EXPOSURE: Irritation of skin, eyes and mucous membranes, gastroentric disturbances, anemia

EFFECTS OF CHRONIC EXPOSURE: Blood and liver damage, loss of appetite.

EMERGENCY AND FIRST AID PROCEDURES: Inhalation - rest; if swallowed - induce vomiting, get medical aid; wash from skin & eyes.

SECTION VI REACTIVITY DATA

CONDITIONS CONTRIBUTING TO INSTABILITY

CONDITIONS CONTRIBUTING TO HAZARDOUS POLYMERIZATION

INCOMPATABILITY (MATERIALS TO AVOID): Avoid lead dioxide

HAZARDOUS DECOMPOSITION PRODUCTS

SECTION VII SPILL OR LEAK PROCEDURES

STEPS TO BE TAKEN IN CASE MATERIAL IS RELEASED OR SPILLED

WASTE DISPOSAL METHOD

SECTION VIII SPECIAL PROTECTION INFORMATION

VENTILATION REQUIREMENTS LOCAL EXHAUST | PROTECTIVE EQUIPMENT (SPECIFY TYPES) EYE: Goggles

MECHANICAL (GENERAL) | GLOVES: Rubber

SPECIAL | RESPIRATOR

OTHER PROTECTIVE EQUIPMENT

SECTION IX SPECIAL PRECAUTIONS

PRECAUTIONS TO BE TAKEN IN HANDLING AND STORAGE

OTHER PRECAUTIONS

SIGNATURE | DATE

1,4-DIPHENYLSEMICARBAZIDE

1

4 0

SECTION I SOURCE AND NOMENCLATURE

MANUFACTURER'S NAME | EMERGENCY TELEPHONE NO.

ADDRESS (NUMBER, STREET, CITY, STATE, ZIP CODE)

TRADE NAME AND SYNONYMS | CHEMICAL FAMILY

CHEMICAL NAME AND SYNONYMS: 1,4 - Diphenylsemicarbazide | FORMULA: $C_6H_5NHNHCONHC_6H_5$

SECTION II HAZARDOUS INGREDIENTS

BASIC MATERIAL	APPROXIMATE OR MAXIMUM % WT. OR VOL.	ESTABLISHED OSHA STANDARD	LD_{50}		LC_{50}	
			ORAL	PERCUT.	SPECIES	CONC.
			none available			

SECTION III PHYSICAL DATA MW - 227.26

BOILING POINT °F. | VAPOR PRESSURE mm Hg.

MELTING POINT °F.: 180°C | VAPOR DENSITY (Air = 1): 7.8

SPECIFIC GRAVITY (H_2O = 1) | EVAPORATION RATE (______ = 1)

SOLUBILITY IN WATER Pts/100 pts H_2O: very slightly soluble | VOLATILE % Vol. % Wt.

APPEARANCE AND ODOR: crystals

SECTION IV FIRE AND EXPLOSION HAZARD DATA

FLASH POINT: none available | FLAMMABLE (EXPLOSIVE) LIMITS: UPPER / LOWER

METHOD USED

EXTINGUISHING MEDIA: alcohol foam, carbon dioxide, dry chemical

SPECIAL FIRE FIGHTING PROCEDURES

UNUSUAL FIRE AND EXPLOSION HAZARDS: when heated to decomposition emits toxic fumes

SECTION V HEALTH HAZARD DATA

TOXIC LEVEL: none available | CARCINOGENIC: no

PRINCIPLE ROUTES OF ABSORBTION: inhalation, ingestion, skin | SKIN AND EYE IRRITATION: severe

REVELANT SYMPTOMS OF EXPOSURE: irritation of skin, eyes and mucous membranes, gastroentric disturbances, anemia

EFFECTS OF CHRONIC EXPOSURE: Blood and liver damage, loss of appetite

EMERGENCY AND FIRST AID PROCEDURES: Inhalacion - rest; if swallowed - induce vomiting, get medical aid; wash from eyes & skin.

SECTION VI REACTIVITY DATA

CONDITIONS CONTRIBUTING TO INSTABILITY

CONDITIONS CONTRIBUTING TO HAZARDOUS POLYMERIZATION

INCOMPATABILITY (MATERIALS TO AVOID)

HAZARDOUS DECOMPOSITION PRODUCTS

SECTION VII SPILL OR LEAK PROCEDURES

STEPS TO BE TAKEN IN CASE MATERIAL IS RELEASED OR SPILLED

WASTE DISPOSAL METHOD

SECTION VIII SPECIAL PROTECTION INFORMATION

VENTILATION REQUIREMENTS LOCAL EXHAUST | PROTECTIVE EQUIPMENT (SPECIFY TYPES) EYE: goggles

MECHANICAL (GENERAL) | GLOVES: rubber

SPECIAL | RESPIRATOR: self-contained

OTHER PROTECTIVE EQUIPMENT

SECTION IX SPECIAL PRECAUTIONS

PRECAUTIONS TO BE TAKEN IN HANDLING AND STORAGE

OTHER PRECAUTIONS

SIGNATURE | DATE

N,N-DI-n-PROPYLANILINE

Hazard diamond: 1 (top), 3 (left), 0 (right)

SECTION I SOURCE AND NOMENCLATURE

MANUFACTURER'S NAME | EMERGENCY TELEPHONE NO.

ADDRESS (NUMBER, STREET, CITY, STATE, ZIP CODE)

TRADE NAME AND SYNONYMS | CHEMICAL FAMILY

CHEMICAL NAME AND SYNONYMS: N,N-Dipropylaniline | FORMULA: $C_6H_5N(CH_2CH_2CH_3)_2$

SECTION II HAZARDOUS INGREDIENTS

BASIC MATERIAL	APPROXIMATE OR MAXIMUM % WT. OR VOL.	ESTABLISHED OSHA STANDARD	LD_{50} ORAL	LD_{50} PERCUT.	LC_{50} SPECIES	LC_{50} CONC.
			none	available		

SECTION III PHYSICAL DATA MW - 177.28

BOILING POINT °F.: 245°C | VAPOR PRESSURE mm Hg.: 12 at 177°C

MELTING POINT °F. | VAPOR DENSITY (Air = 1)

SPECIFIC GRAVITY (H_2O = 1): 0.91 | EVAPORATION RATE (= 1)

SOLUBILITY IN WATER Pts/100 pts H_2O: insoluble | VOLATILE % Vol. % Wt.

APPEARANCE AND ODOR: yellow oil

SECTION IV FIRE AND EXPLOSION HAZARD DATA

FLASH POINT: none available | FLAMMABLE (EXPLOSIVE) LIMITS: UPPER / LOWER

METHOD USED

EXTINGUISHING MEDIA: alcohol foam, carbon dioxide, water spray, dry chemical

SPECIAL FIRE FIGHTING PROCEDURES

UNUSUAL FIRE AND EXPLOSION HAZARDS

SECTION V HEALTH HAZARD DATA

TOXIC LEVEL: none available | CARCINOGENIC: no

PRINCIPLE ROUTES OF ABSORBTION: inhalation, ingestion, skin | SKIN AND EYE IRRITATION: severe

REVELANT SYMPTOMS OF EXPOSURE: headache, drowsiness, cyanosis, mental confusion, convulsions.

EFFECTS OF CHRONIC EXPOSURE: Fatigue, loss of appetite, headache, dizziness

EMERGENCY AND FIRST AID PROCEDURES: If swallowed - induce vomiting, get medical aid; inhalation - rest; wash from skin & eyes.

SECTION VI REACTIVITY DATA

CONDITIONS CONTRIBUTING TO INSTABILITY

CONDITIONS CONTRIBUTING TO HAZARDOUS POLYMERIZATION

INCOMPATABILITY (MATERIALS TO AVOID)

HAZARDOUS DECOMPOSITION PRODUCTS

SECTION VII SPILL OR LEAK PROCEDURES

STEPS TO BE TAKEN IN CASE MATERIAL IS RELEASED OR SPILLED

WASTE DISPOSAL METHOD

SECTION VIII SPECIAL PROTECTION INFORMATION

VENTILATION REQUIREMENTS LOCAL EXHAUST | PROTECTIVE EQUIPMENT (SPECIFY TYPES) EYE: goggles

MECHANICAL (GENERAL) | GLOVES: polyvinyl

SPECIAL | RESPIRATOR: self-contained

OTHER PROTECTIVE EQUIPMENT

SECTION IX SPECIAL PRECAUTIONS

PRECAUTIONS TO BE TAKEN IN HANDLING AND STORAGE

OTHER PRECAUTIONS

SIGNATURE | DATE

DI-n-PROPYLCARBINOL

Hazard diamond: 2 (top), 3 (left), 0 (right)

SECTION I SOURCE AND NOMENCLATURE

MANUFACTURER'S NAME | EMERGENCY TELEPHONE NO.

ADDRESS (NUMBER, STREET, CITY, STATE, ZIP CODE)

TRADE NAME AND SYNONYMS | CHEMICAL FAMILY

CHEMICAL NAME AND SYNONYMS: Heptanol 4 | FORMULA: $(C_3H_7)_2CHOH$

SECTION II HAZARDOUS INGREDIENTS

BASIC MATERIAL	APPROXIMATE OR MAXIMUM % WT. OR VOL.	ESTABLISHED OSHA STANDARD	LD_{50} ORAL	LD_{50} PERCUT.	LC_{50} SPECIES	LC_{50} CONC.
			none	available		

SECTION III PHYSICAL DATA MW - 116.2

BOILING POINT °F.: 156°C | VAPOR PRESSURE mm Hg.

MELTING POINT °F.: -37°C | VAPOR DENSITY (Air = 1)

SPECIFIC GRAVITY (H_2O = 1): 0.82 | EVAPORATION RATE (= 1)

SOLUBILITY IN WATER Pts/100 pts H_2O: insoluble | VOLATILE % Vol. % Wt.

APPEARANCE AND ODOR: liquid

SECTION IV FIRE AND EXPLOSION HAZARD DATA

FLASH POINT: none available | FLAMMABLE (EXPLOSIVE) LIMITS: UPPER / LOWER

METHOD USED

EXTINGUISHING MEDIA: alcohol foam, carbon dioxide, dry chemical

SPECIAL FIRE FIGHTING PROCEDURES: water may be ineffective

UNUSUAL FIRE AND EXPLOSION HAZARDS

SECTION V HEALTH HAZARD DATA

TOXIC LEVEL: not available | CARCINOGENIC: no

PRINCIPLE ROUTES OF ABSORBTION: inhalation, ingestion, skin | SKIN AND EYE IRRITATION: severe

REVELANT SYMPTOMS OF EXPOSURE: Twitching, narcosis, weakness, tremors, convulsions, eye irritation.

EFFECTS OF CHRONIC EXPOSURE: liver damage

EMERGENCY AND FIRST AID PROCEDURES: inhalation - rest; if swallowed - induce vomiting, get medical aid; wash from skin & eyes.

SECTION VI REACTIVITY DATA

CONDITIONS CONTRIBUTING TO INSTABILITY

CONDITIONS CONTRIBUTING TO HAZARDOUS POLYMERIZATION

INCOMPATABILITY (MATERIALS TO AVOID)

HAZARDOUS DECOMPOSITION PRODUCTS

SECTION VII SPILL OR LEAK PROCEDURES

STEPS TO BE TAKEN IN CASE MATERIAL IS RELEASED OR SPILLED

WASTE DISPOSAL METHOD

SECTION VIII SPECIAL PROTECTION INFORMATION

VENTILATION REQUIREMENTS LOCAL EXHAUST | PROTECTIVE EQUIPMENT (SPECIFY TYPES) EYE: goggles

MECHANICAL (GENERAL) | GLOVES: rubber

SPECIAL | RESPIRATOR

OTHER PROTECTIVE EQUIPMENT

SECTION IX SPECIAL PRECAUTIONS

PRECAUTIONS TO BE TAKEN IN HANDLING AND STORAGE

OTHER PRECAUTIONS

SIGNATURE | DATE

ETHANOLAMINE

Hazard diamond: 2 (top), 4 (left), 0 (right)

SECTION I SOURCE AND NOMENCLATURE

MANUFACTURER'S NAME

EMERGENCY TELEPHONE NO.

ADDRESS (NUMBER, STREET, CITY, STATE, ZIP CODE)

TRADE NAME AND SYNONYMS

CHEMICAL FAMILY

CHEMICAL NAME AND SYNONYMS: 2 anino ethanol

FORMULA: $NH_2CH_2CH_2OH$

SECTION II HAZARDOUS INGREDIENTS

BASIC MATERIAL	APPROXIMATE OR MAXIMUM % WT. OR VOL.	ESTABLISHED OSHA STANDARD	LD_{50} mg/kg. ORAL	PERCUT.	SPECIES	LC_{50} CONC.
			2100		rat	
				1000	rabbit	

SECTION III PHYSICAL DATA MW - 61.08

BOILING POINT °F.: 171°C

VAPOR PRESSURE mm Hg.: at 60°C 6

MELTING POINT °F.: 11°C

VAPOR DENSITY (Air = 1): 2.1

SPECIFIC GRAVITY (H_2O = 1): 1.02

EVAPORATION RATE (_______ = 1)

SOLUBILITY IN WATER Pts/100 pts H_2O: very soluble

VOLATILE % Vol. % Wt.

APPEARANCE AND ODOR: colorless oil, ammoniacal odor.

SECTION IV FIRE AND EXPLOSION HAZARD DATA

FLASH POINT: 185°F

METHOD USED

FLAMMABLE (EXPLOSIVE) LIMITS: UPPER / LOWER

EXTINGUISHING MEDIA: alcohol foam, water, carbon dioxide, dry chemical

SPECIAL FIRE FIGHTING PROCEDURES

UNUSUAL FIRE AND EXPLOSION HAZARDS

SECTION V HEALTH HAZARD DATA

TOXIC LEVEL: 3 PPM, 6 mg/C.M.

CARCINOGENIC: no

PRINCIPLE ROUTES OF ABSORBTION: ingestion, inhalation

SKIN AND EYE IRRITATION: severe

REVELANT SYMPTOMS OF EXPOSURE: Affects central nervous system, eye and respiratory tract irritation.

EFFECTS OF CHRONIC EXPOSURE: Weight loss, decrease in alertness, skin irritation, liver and kidney damage.

EMERGENCY AND FIRST AID PROCEDURES: If swallowed - drink water, fruit juice, get medical aid; inhalation - rest; wash eyes and skin.

SECTION VI REACTIVITY DATA

CONDITIONS CONTRIBUTING TO INSTABILITY

CONDITIONS CONTRIBUTING TO HAZARDOUS POLYMERIZATION

INCOMPATABILITY (MATERIALS TO AVOID)

HAZARDOUS DECOMPOSITION PRODUCTS

SECTION VII SPILL OR LEAK PROCEDURES

STEPS TO BE TAKEN IN CASE MATERIAL IS RELEASED OR SPILLED: Take up liquid in sodium bisulfite, then add water.

WASTE DISPOSAL METHOD

SECTION VIII SPECIAL PROTECTION INFORMATION

VENTILATION REQUIREMENTS LOCAL EXHAUST

PROTECTIVE EQUIPMENT (SPECIFY TYPES) EYE: goggles

MECHANICAL (GENERAL)

GLOVES: rubber (butyl)

SPECIAL

RESPIRATOR: with proper filter

OTHER PROTECTIVE EQUIPMENT

SECTION IX SPECIAL PRECAUTIONS

PRECAUTIONS TO BE TAKEN IN HANDLING AND STORAGE: Store in well-ventilated area, away from flammable material.

OTHER PRECAUTIONS

SIGNATURE

DATE

β-ETHOXYETHYL ACETATE

Hazard diamond: 2 (top), 3 (left), 0 (right)

SECTION I SOURCE AND NOMENCLATURE

MANUFACTURER'S NAME

EMERGENCY TELEPHONE NO.

ADDRESS (NUMBER, STREET, CITY, STATE, ZIP CODE)

TRADE NAME AND SYNONYMS: Cellosolve acetate

CHEMICAL FAMILY

CHEMICAL NAME AND SYNONYMS: 2 Ethoxyethyl Acetate

FORMULA: $CH_3CO_2CH_2CH_2OC_2H_5$

SECTION II HAZARDOUS INGREDIENTS

BASIC MATERIAL	APPROXIMATE OR MAXIMUM % WT. OR VOL.	ESTABLISHED OSHA STANDARD	LD_{50} mg/kg ORAL	PERCUT.	SPECIES	LC_{50} CONC.
			1910		guinea pig	
LDLO - oral - human 500 mg/kg			5100		rat	
lowest published lethal concentration in air				15,500	rabbit	
inhalation - rat 1500 PPM			1910		guinea pig	

SECTION III PHYSICAL DATA MW - 132.17

BOILING POINT °F.: 156°C

VAPOR PRESSURE mm Hg.: at 20°C 1.2

MELTING POINT °F.: -62°C

VAPOR DENSITY (Air = 1): 4.72

SPECIFIC GRAVITY (H_2O = 1): 0.97

EVAPORATION RATE (_______ = 1)

SOLUBILITY IN WATER Pts/100 pts H_2O: soluble

VOLATILE % Vol. % Wt.

APPEARANCE AND ODOR: colorless liquid, pleasant odor, bitter taste.

SECTION IV FIRE AND EXPLOSION HAZARD DATA

FLASH POINT: 120°F

METHOD USED: COC

FLAMMABLE (EXPLOSIVE) LIMITS: UPPER / LOWER 1.7%

EXTINGUISHING MEDIA: Alcohol foam, carbon dioxide, dry chemical

SPECIAL FIRE FIGHTING PROCEDURES

UNUSUAL FIRE AND EXPLOSION HAZARDS

SECTION V HEALTH HAZARD DATA

TOXIC LEVEL: TLV - 100 PPM, 540 mg/C.M. (skin)

CARCINOGENIC: no

PRINCIPLE ROUTES OF ABSORBTION: inhalation, ingestion, skin

SKIN AND EYE IRRITATION: slightly

REVELANT SYMPTOMS OF EXPOSURE: restlessness, drowsiness, coma, irritation of eyes & nose.

EFFECTS OF CHRONIC EXPOSURE: kidney injury

EMERGENCY AND FIRST AID PROCEDURES: If swallowed - give emetic, get medical aid; inhalation - rest; wash from eyes and skin.

SECTION VI REACTIVITY DATA

CONDITIONS CONTRIBUTING TO INSTABILITY

CONDITIONS CONTRIBUTING TO HAZARDOUS POLYMERIZATION

INCOMPATABILITY (MATERIALS TO AVOID)

HAZARDOUS DECOMPOSITION PRODUCTS

SECTION VII SPILL OR LEAK PROCEDURES

STEPS TO BE TAKEN IN CASE MATERIAL IS RELEASED OR SPILLED

WASTE DISPOSAL METHOD

SECTION VIII SPECIAL PROTECTION INFORMATION

VENTILATION REQUIREMENTS LOCAL EXHAUST

PROTECTIVE EQUIPMENT (SPECIFY TYPES) EYE: goggles

MECHANICAL (GENERAL)

GLOVES: rubber

SPECIAL

RESPIRATOR

OTHER PROTECTIVE EQUIPMENT

SECTION IX SPECIAL PRECAUTIONS

PRECAUTIONS TO BE TAKEN IN HANDLING AND STORAGE

OTHER PRECAUTIONS

SIGNATURE

DATE

ETHYLAMINE HYDROCHLORIDE

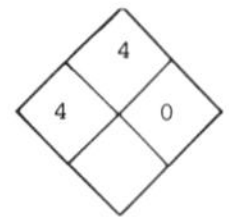

SECTION I SOURCE AND NOMENCLATURE

MANUFACTURER'S NAME | EMERGENCY TELEPHONE NO.

ADDRESS (NUMBER, STREET, CITY, STATE, ZIP CODE)

TRADE NAME AND SYNONYMS: Ethyl ammonium chloride | CHEMICAL FAMILY

CHEMICAL NAME AND SYNONYMS: aminoethane hydrochloride | FORMULA: $C_2H_5NH_2HCl$

SECTION II HAZARDOUS INGREDIENTS

BASIC MATERIAL	APPROXIMATE OR MAXIMUM % WT. OR VOL.	ESTABLISHED OSHA STANDARD	LD_{50} ORAL	LD_{50} PERCUT.	LC_{50} SPECIES	LC_{50} CONC.
Unclassified - Rat LDLO	2500 mg/kg					

SECTION III PHYSICAL DATA MW - 81.56

BOILING POINT °F.: 16.6°C | VAPOR PRESSURE mm Hg.: at 2°C 400

MELTING POINT °F.: -80.6°C | VAPOR DENSITY (Air = 1): 1.56

SPECIFIC GRAVITY (H_2O = 1): 0.8 | EVAPORATION RATE (= 1)

SOLUBILITY IN WATER ∞ Pts/100 pts H_2O: Infinitely soluble | VOLATILE % Vol. % Wt.

APPEARANCE AND ODOR: colorless liquid, strong ammonical odor

SECTION IV FIRE AND EXPLOSION HAZARD DATA

FLASH POINT: <0°F | FLAMMABLE (EXPLOSIVE) LIMITS: UPPER 14.0%, LOWER 3.5%

METHOD USED

EXTINGUISHING MEDIA: alcohol foam, carbon dioxide, dry chemical

SPECIAL FIRE FIGHTING PROCEDURES: water may be ineffective

UNUSUAL FIRE AND EXPLOSION HAZARDS

SECTION V HEALTH HAZARD DATA

TOXIC LEVEL: none available | CARCINOGENIC: no

PRINCIPLE ROUTES OF ABSORBTION: inhalation, ingestion | SKIN AND EYE IRRITATION: severe

REVELANT SYMPTOMS OF EXPOSURE: eye irritation, mucous membrane and respiratory system irritation

EFFECTS OF CHRONIC EXPOSURE: Lung, liver and kidney damage

EMERGENCY AND FIRST AID PROCEDURES: Inhalation - rest, get medical aid; if swallowed - drink water, get medical aid; wash eyes and skin.

SECTION VI REACTIVITY DATA

CONDITIONS CONTRIBUTING TO INSTABILITY

CONDITIONS CONTRIBUTING TO HAZARDOUS POLYMERIZATION

INCOMPATABILITY (MATERIALS TO AVOID)

HAZARDOUS DECOMPOSITION PRODUCTS

SECTION VII SPILL OR LEAK PROCEDURES

STEPS TO BE TAKEN IN CASE MATERIAL IS RELEASED OR SPILLED

WASTE DISPOSAL METHOD

SECTION VIII SPECIAL PROTECTION INFORMATION

VENTILATION REQUIREMENTS LOCAL EXHAUST | PROTECTIVE EQUIPMENT (SPECIFY TYPES) EYE: Safety goggles

MECHANICAL (GENERAL) | GLOVES: Rubber

SPECIAL | RESPIRATOR: With proper filter

OTHER PROTECTIVE EQUIPMENT

SECTION IX SPECIAL PRECAUTIONS

PRECAUTIONS TO BE TAKEN IN HANDLING AND STORAGE

OTHER PRECAUTIONS

SIGNATURE | DATE

N-ETHYLANILINE

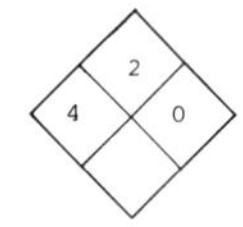

SECTION I SOURCE AND NOMENCLATURE

MANUFACTURER'S NAME | EMERGENCY TELEPHONE NO.

ADDRESS (NUMBER, STREET, CITY, STATE, ZIP CODE)

TRADE NAME AND SYNONYMS | CHEMICAL FAMILY

CHEMICAL NAME AND SYNONYMS: Ethylphenylamine | FORMULA: $C_2H_5NHC_6H_5$

SECTION II HAZARDOUS INGREDIENTS

BASIC MATERIAL	APPROXIMATE OR MAXIMUM % WT. OR VOL.	ESTABLISHED OSHA STANDARD	LD_{50} mg/kg ORAL	LD_{50} mg/kg PERCUT.	SPECIES	LC_{50} CONC.
			334		rat	
				4700	rabbit	

SECTION III PHYSICAL DATA MW - 121.20

BOILING POINT °F.: 204°C | VAPOR PRESSURE mm Hg.: at 39°C 1

MELTING POINT °F.: -64°C | VAPOR DENSITY (Air = 1): 4.18

SPECIFIC GRAVITY (H_2O = 1): 0.96 | EVAPORATION RATE (= 1)

SOLUBILITY IN WATER Pts/100 pts H_2O | VOLATILE % Vol. % Wt.

APPEARANCE AND ODOR: clear, yellow brown oil

SECTION IV FIRE AND EXPLOSION HAZARD DATA

FLASH POINT: 185°F | FLAMMABLE (EXPLOSIVE) LIMITS: UPPER, LOWER

METHOD USED: O.C.

EXTINGUISHING MEDIA: Foam, carbon dioxide, dry chemical

SPECIAL FIRE FIGHTING PROCEDURES

UNUSUAL FIRE AND EXPLOSION HAZARDS: When heated may decompose to give toxic aniline fumes.

SECTION V HEALTH HAZARD DATA

TOXIC LEVEL: none available | CARCINOGENIC: no

PRINCIPLE ROUTES OF ABSORBTION: inhalation, ingestion, skin | SKIN AND EYE IRRITATION: moderate

REVELANT SYMPTOMS OF EXPOSURE: weakness, tremors, cyanosis, drowsiness, central nervous system depression.

EFFECTS OF CHRONIC EXPOSURE: Fatigue, loss of appetite, headache, dizziness

EMERGENCY AND FIRST AID PROCEDURES: Inhalation - rest; if swallowed - give emetic, get medical aid; wash from skin & eyes.

SECTION VI REACTIVITY DATA

CONDITIONS CONTRIBUTING TO INSTABILITY

CONDITIONS CONTRIBUTING TO HAZARDOUS POLYMERIZATION

INCOMPATABILITY (MATERIALS TO AVOID)

HAZARDOUS DECOMPOSITION PRODUCTS

SECTION VII SPILL OR LEAK PROCEDURES

STEPS TO BE TAKEN IN CASE MATERIAL IS RELEASED OR SPILLED

WASTE DISPOSAL METHOD

SECTION VIII SPECIAL PROTECTION INFORMATION

VENTILATION REQUIREMENTS LOCAL EXHAUST | PROTECTIVE EQUIPMENT (SPECIFY TYPES) EYE: goggles

MECHANICAL (GENERAL) | GLOVES: polyvinyl alcohol

SPECIAL | RESPIRATOR: with proper filter

OTHER PROTECTIVE EQUIPMENT

SECTION IX SPECIAL PRECAUTIONS

PRECAUTIONS TO BE TAKEN IN HANDLING AND STORAGE

OTHER PRECAUTIONS

SIGNATURE | DATE

ETHYLBENZYLANILINE

1
4 0

SECTION I SOURCE AND NOMENCLATURE

MANUFACTURER'S NAME | EMERGENCY TELEPHONE NO.

ADDRESS (NUMBER, STREET, CITY, STATE, ZIP CODE)

TRADE NAME AND SYNONYMS | CHEMICAL FAMILY

CHEMICAL NAME AND SYNONYMS: Ethylbenzylphenylamine

FORMULA: $C_6H_5N(C_2H_5)CH_2C_6H_5$

SECTION II HAZARDOUS INGREDIENTS

BASIC MATERIAL	APPROXIMATE OR MAXIMUM % WT. OR VOL.	ESTABLISHED OSHA STANDARD	LD_{50} ORAL	LD_{50} PERCUT.	LC_{50} SPECIES	LC_{50} CONC.
			none	available		

SECTION III PHYSICAL DATA MW=211.3

BOILING POINT °F.: 286°C

VAPOR PRESSURE mm Hg.:

MELTING POINT °F.:

VAPOR DENSITY (Air = 1): 7.28

SPECIFIC GRAVITY (H_2O = 1): 1.03

EVAPORATION RATE (= 1):

SOLUBILITY IN WATER Pts/100 pts H_2O:

VOLATILE % Vol. % Wt.:

APPEARANCE AND ODOR: clear colorless oil

SECTION IV FIRE AND EXPLOSION HAZARD DATA

FLASH POINT: none available

METHOD USED:

FLAMMABLE (EXPLOSIVE) LIMITS: UPPER / LOWER

EXTINGUISHING MEDIA: foam, carbon dioxide, dry chemical

SPECIAL FIRE FIGHTING PROCEDURES:

UNUSUAL FIRE AND EXPLOSION HAZARDS: when heated may decompose to give off toxic aniline fumes.

SECTION V HEALTH HAZARD DATA

TOXIC LEVEL: none available

CARCINOGENIC: no

PRINCIPLE ROUTES OF ABSORBTION: inhalation, ingestion, skin

SKIN AND EYE IRRITATION: moderate

REVELANT SYMPTOMS OF EXPOSURE: weakness, trmors, cyanosis, drowsiness, central nervous system depression

EFFECTS OF CHRONIC EXPOSURE: Fatigue, loss of appetite, headache, dizziness

EMERGENCY AND FIRST AID PROCEDURES: Inhalation - rest; if swallowed - give emetic, get medical aid; wash from skin & eyes.

SECTION VI REACTIVITY DATA

CONDITIONS CONTRIBUTING TO INSTABILITY

CONDITIONS CONTRIBUTING TO HAZARDOUS POLYMERIZATION

INCOMPATABILITY (MATERIALS TO AVOID)

HAZARDOUS DECOMPOSITION PRODUCTS

SECTION VII SPILL OR LEAK PROCEDURES

STEPS TO BE TAKEN IN CASE MATERIAL IS RELEASED OR SPILLED

WASTE DISPOSAL METHOD

SECTION VIII SPECIAL PROTECTION INFORMATION

VENTILATION REQUIREMENTS LOCAL EXHAUST

PROTECTIVE EQUIPMENT (SPECIFY TYPES) EYE: goggles

MECHANICAL (GENERAL)

GLOVES: polyvinyl alcohol

SPECIAL

RESPIRATOR: with proper filter

OTHER PROTECTIVE EQUIPMENT

SECTION IX SPECIAL PRECAUTIONS

PRECAUTIONS TO BE TAKEN IN HANDLING AND STORAGE

OTHER PRECAUTIONS

SIGNATURE | DATE

ETHYL BROMIDE

3
3 0

SECTION I SOURCE AND NOMENCLATURE

MANUFACTURER'S NAME | EMERGENCY TELEPHONE NO.

ADDRESS (NUMBER, STREET, CITY, STATE, ZIP CODE)

TRADE NAME AND SYNONYMS | CHEMICAL FAMILY

CHEMICAL NAME AND SYNONYMS: Bromoethane

FORMULA: C_2H_5Br

SECTION II HAZARDOUS INGREDIENTS

BASIC MATERIAL	APPROXIMATE OR MAXIMUM % WT. OR VOL.	ESTABLISHED OSHA STANDARD	LD_{50} ORAL	LD_{50} PERCUT.	LC_{50} PPM SPECIES	LC_{50} PPM CONC.
LCLO inhalation - rat	149,000 PPM - 15 Min.					
LCLO - inhalation - guinea pig	3,200 PPM/9 hrs.					

SECTION III PHYSICAL DATA MW - 108.98

BOILING POINT °F.: 38.4°C

VAPOR PRESSURE mm Hg.: at 21°C 400

MELTING POINT °F.: -119°C

VAPOR DENSITY (Air = 1): 3.76

SPECIFIC GRAVITY (H_2O = 1): 1.45

EVAPORATION RATE (= 1):

SOLUBILITY IN WATER Pts/100 pts H_2O: 0.91

VOLATILE % Vol. % Wt.:

APPEARANCE AND ODOR: colorless, volatile liquid, ether-like odor.

SECTION IV FIRE AND EXPLOSION HAZARD DATA

FLASH POINT:

METHOD USED:

FLAMMABLE (EXPLOSIVE) LIMITS: UPPER 11.3 / LOWER 6.7

EXTINGUISHING MEDIA: water, carbon dioxide, dry chemical

SPECIAL FIRE FIGHTING PROCEDURES: <-4°F.

UNUSUAL FIRE AND EXPLOSION HAZARDS:

SECTION V HEALTH HAZARD DATA

TOXIC LEVEL: TLV - 200 PPM, 890 mg/C.M.

CARCINOGENIC: no

PRINCIPLE ROUTES OF ABSORBTION: Inhalation, ingestion

SKIN AND EYE IRRITATION: moderate

REVELANT SYMPTOMS OF EXPOSURE: Central nervous system depression, irritation of respiratory system.

EFFECTS OF CHRONIC EXPOSURE: Possible liver and kidney damage.

EMERGENCY AND FIRST AID PROCEDURES: Inhalation - rest; if swallowed - induce vomiting, get medical aid; wash from eyes & skin.

SECTION VI REACTIVITY DATA

CONDITIONS CONTRIBUTING TO INSTABILITY

CONDITIONS CONTRIBUTING TO HAZARDOUS POLYMERIZATION

INCOMPATABILITY (MATERIALS TO AVOID)

HAZARDOUS DECOMPOSITION PRODUCTS

SECTION VII SPILL OR LEAK PROCEDURES

STEPS TO BE TAKEN IN CASE MATERIAL IS RELEASED OR SPILLED: Absorb on paper towels, evaporate in hood.

WASTE DISPOSAL METHOD

SECTION VIII SPECIAL PROTECTION INFORMATION

VENTILATION REQUIREMENTS LOCAL EXHAUST

PROTECTIVE EQUIPMENT (SPECIFY TYPES) EYE: Goggles

MECHANICAL (GENERAL)

GLOVES: Rubber

SPECIAL

RESPIRATOR: Self-contained breathing apparatus

OTHER PROTECTIVE EQUIPMENT: Overalls

SECTION IX SPECIAL PRECAUTIONS

PRECAUTIONS TO BE TAKEN IN HANDLING AND STORAGE: Outside or detached storage preferred, inside storage in flammable liquids storage cabinet.

OTHER PRECAUTIONS

SIGNATURE | DATE

ETHYL BROMOACETATE

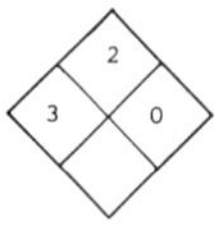

SECTION I SOURCE AND NOMENCLATURE

MANUFACTURER'S NAME | EMERGENCY TELEPHONE NO.

ADDRESS (NUMBER, STREET, CITY, STATE, ZIP CODE)

TRADE NAME AND SYNONYMS | CHEMICAL FAMILY

CHEMICAL NAME AND SYNONYMS: Ethyl bromoethanoate | FORMULA: $BrCH_2COOC_2H_5$

SECTION II HAZARDOUS INGREDIENTS

BASIC MATERIAL	APPROXIMATE OR MAXIMUM % WT. OR VOL.	ESTABLISHED OSHA STANDARD	LD_{50} ORAL	LD_{50} PERCUT.	LC_{50} SPECIES	LC_{50} CONC.
TDLO subcutaneous - mouse	332 mg/kg	83W neoplastic				

SECTION III PHYSICAL DATA MW - 167.01

BOILING POINT °F.: 160°C | VAPOR PRESSURE mm Hg.: at 58°C 15

MELTING POINT °F.: -20°C | VAPOR DENSITY (Air = 1): 5.8

SPECIFIC GRAVITY (H_2O = 1): 1.51 | EVAPORATION RATE (_____ = 1)

SOLUBILITY IN WATER Pts/100 pts H_2O: insoluble | VOLATILE % Vol. % Wt.

APPEARANCE AND ODOR: colorless to straw colored liquid

SECTION IV FIRE AND EXPLOSION HAZARD DATA

FLASH POINT: none available | FLAMMABLE (EXPLOSIVE) LIMITS UPPER / LOWER

METHOD USED

EXTINGUISHING MEDIA: water, alcohol foam, carbon dioxide, dry chemical

SPECIAL FIRE FIGHTING PROCEDURES

UNUSUAL FIRE AND EXPLOSION HAZARDS

SECTION V HEALTH HAZARD DATA

TOXIC LEVEL: none available | CARCINOGENIC: neoplastic

PRINCIPLE ROUTES OF ABSORBTION: inhalation, ingestion | SKIN AND EYE IRRITATION: severe

REVELANT SYMPTOMS OF EXPOSURE: skin and eye irritation, pulmonary edema, dermatitis

EFFECTS OF CHRONIC EXPOSURE

EMERGENCY AND FIRST AID PROCEDURES: If swallowed - give emetic, get medical aid; wash skin and eyes; inhalation - rest.

SECTION VI REACTIVITY DATA

CONDITIONS CONTRIBUTING TO INSTABILITY

CONDITIONS CONTRIBUTING TO HAZARDOUS POLYMERIZATION

INCOMPATABILITY (MATERIALS TO AVOID)

HAZARDOUS DECOMPOSITION PRODUCTS

SECTION VII SPILL OR LEAK PROCEDURES

STEPS TO BE TAKEN IN CASE MATERIAL IS RELEASED OR SPILLED

WASTE DISPOSAL METHOD

SECTION VIII SPECIAL PROTECTION INFORMATION

VENTILATION REQUIREMENTS LOCAL EXHAUST | PROTECTIVE EQUIPMENT (SPECIFY TYPES) EYE: goggles

MECHANICAL (GENERAL) | GLOVES: rubber

SPECIAL | RESPIRATOR: self-contained

OTHER PROTECTIVE EQUIPMENT

SECTION IX SPECIAL PRECAUTIONS

PRECAUTIONS TO BE TAKEN IN HANDLING AND STORAGE

OTHER PRECAUTIONS

SIGNATURE | DATE

ETHYL α-BROMOPROPIONATE

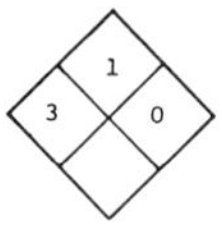

SECTION I SOURCE AND NOMENCLATURE

MANUFACTURER'S NAME | EMERGENCY TELEPHONE NO.

ADDRESS (NUMBER, STREET, CITY, STATE, ZIP CODE)

TRADE NAME AND SYNONYMS | CHEMICAL FAMILY

CHEMICAL NAME AND SYNONYMS: Ethyl-2 Bromopropionate | FORMULA: $CH_3CHBrCOOC_2H_5$

SECTION II HAZARDOUS INGREDIENTS

BASIC MATERIAL	APPROXIMATE OR MAXIMUM % WT. OR VOL.	ESTABLISHED OSHA STANDARD	LD_{50} ORAL	LD_{50} PERCUT.	LC_{50} SPECIES	LC_{50} CONC.
			none available			

SECTION III PHYSICAL DATA MW - 181.04

BOILING POINT °F.: 160-5°C | VAPOR PRESSURE mm Hg.: at 55°C 10

MELTING POINT °F. | VAPOR DENSITY (Air = 1)

SPECIFIC GRAVITY (H_2O = 1): 1.35 | EVAPORATION RATE (_____ = 1)

SOLUBILITY IN WATER Pts/100 pts H_2O: insoluble | VOLATILE % Vol. % Wt.

APPEARANCE AND ODOR: colorless liquid

SECTION IV FIRE AND EXPLOSION HAZARD DATA

FLASH POINT: none available | FLAMMABLE (EXPLOSIVE) LIMITS UPPER / LOWER

METHOD USED

EXTINGUISHING MEDIA: water, alcohol foam, carbon dioxide, dry chemical

SPECIAL FIRE FIGHTING PROCEDURES

UNUSUAL FIRE AND EXPLOSION HAZARDS

SECTION V HEALTH HAZARD DATA

TOXIC LEVEL: none available | CARCINOGENIC: no

PRINCIPLE ROUTES OF ABSORBTION: inhalation, ingestion | SKIN AND EYE IRRITATION: severe

REVELANT SYMPTOMS OF EXPOSURE: skin and eye irritation, pulmonary edema, dermatitis

EFFECTS OF CHRONIC EXPOSURE: --

EMERGENCY AND FIRST AID PROCEDURES: If swallowed - give emetic, get medical aid; wash skin & eyes; inhalation - rest.

SECTION VI REACTIVITY DATA

CONDITIONS CONTRIBUTING TO INSTABILITY

CONDITIONS CONTRIBUTING TO HAZARDOUS POLYMERIZATION

INCOMPATABILITY (MATERIALS TO AVOID)

HAZARDOUS DECOMPOSITION PRODUCTS

SECTION VII SPILL OR LEAK PROCEDURES

STEPS TO BE TAKEN IN CASE MATERIAL IS RELEASED OR SPILLED

WASTE DISPOSAL METHOD

SECTION VIII SPECIAL PROTECTION INFORMATION

VENTILATION REQUIREMENTS LOCAL EXHAUST | PROTECTIVE EQUIPMENT (SPECIFY TYPES) EYE: goggles

MECHANICAL (GENERAL) | GLOVES: rubber

SPECIAL | RESPIRATOR: with proper filter

OTHER PROTECTIVE EQUIPMENT

SECTION IX SPECIAL PRECAUTIONS

PRECAUTIONS TO BE TAKEN IN HANDLING AND STORAGE

OTHER PRECAUTIONS

SIGNATURE | DATE

ETHYL β-BROMOPROPIONATE

NFPA diamond: 1 (top), 3 (left), 0 (right)

SECTION I SOURCE AND NOMENCLATURE

MANUFACTURER'S NAME	EMERGENCY TELEPHONE NO.
ADDRESS (NUMBER, STREET, CITY, STATE, ZIP CODE)	
TRADE NAME AND SYNONYMS	CHEMICAL FAMILY
CHEMICAL NAME AND SYNONYMS: Ethyl - 3 Bromopropionate	FORMULA: $BrCH_2CH_2COOC_2H_5$

SECTION II HAZARDOUS INGREDIENTS

BASIC MATERIAL	APPROXIMATE OR MAXIMUM % WT. OR VOL.	ESTABLISHED OSHA STANDARD	LD_{50} ORAL	LD_{50} PERCUT.	LC_{50} SPECIES	LC_{50} CONC.
			none available			

SECTION III PHYSICAL DATA MW - 181.04

BOILING POINT °F.: 112°C at 44 mm	VAPOR PRESSURE mm Hg.: at 73°C 15
MELTING POINT °F.: --	VAPOR DENSITY (Air = 1)
SPECIFIC GRAVITY (H_2O = 1): 1.41	EVAPORATION RATE (______ = 1)
SOLUBILITY IN WATER Pts/100 pts H_2O	VOLATILE % Vol. % Wt.

APPEARANCE AND ODOR: yellow liquid

SECTION IV FIRE AND EXPLOSION HAZARD DATA

FLASH POINT: none available	FLAMMABLE (EXPLOSIVE) LIMITS	UPPER
METHOD USED		LOWER

EXTINGUISHING MEDIA: water, alcohol foam, carbon dioxide, dry chemical

SPECIAL FIRE FIGHTING PROCEDURES

UNUSUAL FIRE AND EXPLOSION HAZARDS

SECTION V HEALTH HAZARD DATA

TOXIC LEVEL: none available — CARCINOGENIC: no

PRINCIPLE ROUTES OF ABSORBTION: inhalation, ingestion — SKIN AND EYE IRRITATION: severe

REVELANT SYMPTOMS OF EXPOSURE: skin & eye irritation, pulmonary edema, dermatitis

EFFECTS OF CHRONIC EXPOSURE: --

EMERGENCY AND FIRST AID PROCEDURES: If swallowed - give emetic, get medical aid; wash skin & eyes; inhalation - rest.

SECTION VI REACTIVITY DATA

CONDITIONS CONTRIBUTING TO INSTABILITY

CONDITIONS CONTRIBUTING TO HAZARDOUS POLYMERIZATION

INCOMPATABILITY (MATERIALS TO AVOID)

HAZARDOUS DECOMPOSITION PRODUCTS

SECTION VII SPILL OR LEAK PROCEDURES

STEPS TO BE TAKEN IN CASE MATERIAL IS RELEASED OR SPILLED

WASTE DISPOSAL METHOD

SECTION VIII SPECIAL PROTECTION INFORMATION

VENTILATION REQUIREMENTS LOCAL EXHAUST	PROTECTIVE EQUIPMENT (SPECIFY TYPES) EYE: goggles
MECHANICAL (GENERAL)	GLOVES: rubber
SPECIAL	RESPIRATOR: with proper filter

OTHER PROTECTIVE EQUIPMENT

SECTION IX SPECIAL PRECAUTIONS

PRECAUTIONS TO BE TAKEN IN HANDLING AND STORAGE

OTHER PRECAUTIONS

SIGNATURE	DATE

ETHYLENE BROMOHYDRIN

NFPA diamond: 2 (top), 4 (left), 0 (right)

SECTION I SOURCE AND NOMENCLATURE

MANUFACTURER'S NAME	EMERGENCY TELEPHONE NO.
ADDRESS (NUMBER, STREET, CITY, STATE, ZIP CODE)	
TRADE NAME AND SYNONYMS	CHEMICAL FAMILY
CHEMICAL NAME AND SYNONYMS: 2 Bromoethanol	FORMULA: BrC_2H_4OH

SECTION II HAZARDOUS INGREDIENTS

BASIC MATERIAL	APPROXIMATE OR MAXIMUM % WT. OR VOL.	ESTABLISHED OSHA STANDARD	LD_{50} ORAL	LD_{50} PERCUT.	LC_{50} SPECIES	LC_{50} CONC.
Lowest published lethal dose intraperitoneal - mouse	90 mg/kg					

SECTION III PHYSICAL DATA MW - 124.98

BOILING POINT °F.: 149-150°C (Dec)	VAPOR PRESSURE mm Hg.: 50°C 12
MELTING POINT °F.	VAPOR DENSITY (Air = 1)
SPECIFIC GRAVITY (H_2O = 1): 1.79	EVAPORATION RATE (______ = 1)
SOLUBILITY IN WATER Pts/100 pts H_2O: soluble	VOLATILE % Vol. % Wt.

APPEARANCE AND ODOR: colorless liquid

SECTION IV FIRE AND EXPLOSION HAZARD DATA

FLASH POINT: none available	FLAMMABLE (EXPLOSIVE) LIMITS	UPPER
METHOD USED		LOWER

EXTINGUISHING MEDIA: alcohol foam, carbon dioxide, dry chemical

SPECIAL FIRE FIGHTING PROCEDURES: Wear protective clothing and self-contained breathing apparatus.

UNUSUAL FIRE AND EXPLOSION HAZARDS

SECTION V HEALTH HAZARD DATA

TOXIC LEVEL: none available — CARCINOGENIC: no

PRINCIPLE ROUTES OF ABSORBTION: inhalation, ingestion, skin — SKIN AND EYE IRRITATION: severe

REVELANT SYMPTOMS OF EXPOSURE: Eye and mucous membrane irritation, nausea, headache, vomiting, unconsciousness.

EFFECTS OF CHRONIC EXPOSURE: --

EMERGENCY AND FIRST AID PROCEDURES: If swallowed - induce vomiting, get medical aid; inhalation - rest; wash from skin & eyes.

SECTION VI REACTIVITY DATA

CONDITIONS CONTRIBUTING TO INSTABILITY

CONDITIONS CONTRIBUTING TO HAZARDOUS POLYMERIZATION

INCOMPATABILITY (MATERIALS TO AVOID)

HAZARDOUS DECOMPOSITION PRODUCTS

SECTION VII SPILL OR LEAK PROCEDURES

STEPS TO BE TAKEN IN CASE MATERIAL IS RELEASED OR SPILLED

WASTE DISPOSAL METHOD

SECTION VIII SPECIAL PROTECTION INFORMATION

VENTILATION REQUIREMENTS LOCAL EXHAUST	PROTECTIVE EQUIPMENT (SPECIFY TYPES) EYE: goggles
MECHANICAL (GENERAL)	GLOVES: rubber
SPECIAL	RESPIRATOR: with proper filter

OTHER PROTECTIVE EQUIPMENT

SECTION IX SPECIAL PRECAUTIONS

PRECAUTIONS TO BE TAKEN IN HANDLING AND STORAGE

OTHER PRECAUTIONS

SIGNATURE	DATE

ETHYLENE CHLOROBROMIDE

NFPA diamond: 0 (top), 3 (left), 0 (right)

SECTION I SOURCE AND NOMENCLATURE

MANUFACTURER'S NAME | EMERGENCY TELEPHONE NO.

ADDRESS (NUMBER, STREET, CITY, STATE, ZIP CODE)

TRADE NAME AND SYNONYMS | CHEMICAL FAMILY

CHEMICAL NAME AND SYNONYMS: 1 Bromo 2 chloroethane | FORMULA: BrC_2H_4Cl

SECTION II HAZARDOUS INGREDIENTS

BASIC MATERIAL	APPROXIMATE OR MAXIMUM % WT. OR VOL.	ESTABLISHED OSHA STANDARD	LD_{50} mg/kg ORAL	LD_{50} PERCUT.	SPECIES	LC_{50} CONC.
Oral - human LDLO 50 mg/kg			64		rat	

SECTION III PHYSICAL DATA MW - 143.4

BOILING POINT °F: 106°C | VAPOR PRESSURE mm Hg: at 30°C 40

MELTING POINT °F: -18°C | VAPOR DENSITY (Air = 1): 4.94

SPECIFIC GRAVITY (H_2O = 1): 1.72 | EVAPORATION RATE (_____ = 1)

SOLUBILITY IN WATER Pts/100 pts H_2O | VOLATILE % Vol. % Wt.

APPEARANCE AND ODOR: colorless, volatile liquid, sweet chloroform-like odor.

SECTION IV FIRE AND EXPLOSION HAZARD DATA

FLASH POINT: not flammable | FLAMMABLE (EXPLOSIVE) LIMITS: UPPER | LOWER

METHOD USED

EXTINGUISHING MEDIA

SPECIAL FIRE FIGHTING PROCEDURES

UNUSUAL FIRE AND EXPLOSION HAZARDS

SECTION V HEALTH HAZARD DATA

TOXIC LEVEL: none available | CARCINOGENIC: no

PRINCIPLE ROUTES OF ABSORBTION: inhalation, ingestion, skin | SKIN AND EYE IRRITATION: severe

REVELANT SYMPTOMS OF EXPOSURE: Pneumonia, skin & eye irritation, central nervous system depression.

EFFECTS OF CHRONIC EXPOSURE: Liver and kidney damage

EMERGENCY AND FIRST AID PROCEDURES: If swallowed - give emetic, get medical aid; inhalation - rest; wash eyes and skin.

SECTION VI REACTIVITY DATA

CONDITIONS CONTRIBUTING TO INSTABILITY

CONDITIONS CONTRIBUTING TO HAZARDOUS POLYMERIZATION

INCOMPATABILITY (MATERIALS TO AVOID)

HAZARDOUS DECOMPOSITION PRODUCTS

SECTION VII SPILL OR LEAK PROCEDURES

STEPS TO BE TAKEN IN CASE MATERIAL IS RELEASED OR SPILLED

WASTE DISPOSAL METHOD

SECTION VIII SPECIAL PROTECTION INFORMATION

VENTILATION REQUIREMENTS LOCAL EXHAUST | PROTECTIVE EQUIPMENT (SPECIFY TYPES) EYE: goggles

MECHANICAL (GENERAL) | GLOVES: rubber

SPECIAL | RESPIRATOR: with proper filter

OTHER PROTECTIVE EQUIPMENT

SECTION IX SPECIAL PRECAUTIONS

PRECAUTIONS TO BE TAKEN IN HANDLING AND STORAGE

OTHER PRECAUTIONS

SIGNATURE | DATE

ETHYLIDENE BROMIDE

NFPA diamond: 0 (top), 3 (left), 0 (right)

SECTION I SOURCE AND NOMENCLATURE

MANUFACTURER'S NAME | EMERGENCY TELEPHONE NO.

ADDRESS (NUMBER, STREET, CITY, STATE, ZIP CODE)

TRADE NAME AND SYNONYMS | CHEMICAL FAMILY

CHEMICAL NAME AND SYNONYMS: 1,1 dibromoethane | FORMULA: Br_2HCCH_3

SECTION II HAZARDOUS INGREDIENTS

BASIC MATERIAL	APPROXIMATE OR MAXIMUM % WT. OR VOL.	ESTABLISHED OSHA STANDARD	LD_{50} ORAL	LD_{50} PERCUT.	SPECIES	LC_{50} CONC.
LDLO rectal - rabbit	1250 mg/kg		none	available		

SECTION III PHYSICAL DATA MW -187.8°

BOILING POINT °F: 109°C | VAPOR PRESSURE mm Hg

MELTING POINT °F: - | VAPOR DENSITY (Air = 1)

SPECIFIC GRAVITY (H_2O = 1): 2.06 | EVAPORATION RATE (_____ = 1)

SOLUBILITY IN WATER Pts/100 pts H_2O: insoluble | VOLATILE % Vol. % Wt.

APPEARANCE AND ODOR: liquid

SECTION IV FIRE AND EXPLOSION HAZARD DATA

FLASH POINT: not available | FLAMMABLE (EXPLOSIVE) LIMITS: UPPER | LOWER

METHOD USED

EXTINGUISHING MEDIA: not available

SPECIAL FIRE FIGHTING PROCEDURES

UNUSUAL FIRE AND EXPLOSION HAZARDS

SECTION V HEALTH HAZARD DATA

TOXIC LEVEL: none available | CARCINOGENIC: no

PRINCIPLE ROUTES OF ABSORBTION: Inhalation, ingestion, skin | SKIN AND EYE IRRITATION: moderate

REVELANT SYMPTOMS OF EXPOSURE: Central nervous system depression, eye irritation, dermatitis

EFFECTS OF CHRONIC EXPOSURE: Liver damage, dermatitis

EMERGENCY AND FIRST AID PROCEDURES: If swallowed - give emetic, get medical aid; inhalation - rest; wash eyes and skin.

SECTION VI REACTIVITY DATA

CONDITIONS CONTRIBUTING TO INSTABILITY

CONDITIONS CONTRIBUTING TO HAZARDOUS POLYMERIZATION

INCOMPATABILITY (MATERIALS TO AVOID)

HAZARDOUS DECOMPOSITION PRODUCTS

SECTION VII SPILL OR LEAK PROCEDURES

STEPS TO BE TAKEN IN CASE MATERIAL IS RELEASED OR SPILLED

WASTE DISPOSAL METHOD

SECTION VIII SPECIAL PROTECTION INFORMATION

VENTILATION REQUIREMENTS LOCAL EXHAUST | PROTECTIVE EQUIPMENT (SPECIFY TYPES) EYE: goggles

MECHANICAL (GENERAL) | GLOVES: rubber

SPECIAL | RESPIRATOR: with proper filter

OTHER PROTECTIVE EQUIPMENT

SECTION IX SPECIAL PRECAUTIONS

PRECAUTIONS TO BE TAKEN IN HANDLING AND STORAGE

OTHER PRECAUTIONS

SIGNATURE | DATE

ETHYLIDENE CHLORIDE

NFPA diamond: 3 (top), 3 (left), 0 (right)

SECTION I SOURCE AND NOMENCLATURE

MANUFACTURER'S NAME	EMERGENCY TELEPHONE NO.
ADDRESS (NUMBER, STREET, CITY, STATE, ZIP CODE)	
TRADE NAME AND SYNONYMS	CHEMICAL FAMILY
CHEMICAL NAME AND SYNONYMS: 1,1, Dichloroethane	FORMULA: Cl_2HCCH_3

SECTION II HAZARDOUS INGREDIENTS

BASIC MATERIAL	APPROXIMATE OR MAXIMUM % WT. OR VOL.	ESTABLISHED OSHA STANDARD	LD_{50} ORAL	LD_{50} PERCUT.	LC_{50} SPECIES	LC_{50} CONC.
			725		rat	
LDLO oral human 500 mg/kg						

SECTION III PHYSICAL DATA MW - 98.97

BOILING POINT °F.: 58°C	VAPOR PRESSURE mm Hg.: at 25°C 230
MELTING POINT °F.: -97°C	VAPOR DENSITY (Air = 1): 3.44
SPECIFIC GRAVITY (H_2O = 1): 1.17	EVAPORATION RATE (= 1)
SOLUBILITY IN WATER Pts/100 pts H_2O: 0.5	VOLATILE % Vol. % Wt.

APPEARANCE AND ODOR: Colorless liquid, ethereal odor

SECTION IV FIRE AND EXPLOSION HAZARD DATA

FLASH POINT: 22°F	FLAMMABLE (EXPLOSIVE) LIMITS	UPPER: 11.4%
METHOD USED: C.C.		LOWER: 5.6%

EXTINGUISHING MEDIA: Alcohol foam, water, carbon dioxide, dry chemical

SPECIAL FIRE FIGHTING PROCEDURES: Water may be ineffective

UNUSUAL FIRE AND EXPLOSION HAZARDS: Wear full protective clothing

SECTION V HEALTH HAZARD DATA

TOXIC LEVEL: TLV - 200 PPM, 320 mg/C.M. Teratogen CARCINOGENIC: no

PRINCIPLE ROUTES OF ABSORBTION: Inhalation, ingestion, skin SKIN AND EYE IRRITATION: moderate

REVELANT SYMPTOMS OF EXPOSURE: Central nervous system depression, eye irritation, dermatitis.

EFFECTS OF CHRONIC EXPOSURE: Liver damage, dermatitis.

EMERGENCY AND FIRST AID PROCEDURES: If swallowed - get medical aid; inhalation-rest; wash eyes & skin.

SECTION VI REACTIVITY DATA

CONDITIONS CONTRIBUTING TO INSTABILITY

CONDITIONS CONTRIBUTING TO HAZARDOUS POLYMERIZATION

INCOMPATABILITY (MATERIALS TO AVOID)

HAZARDOUS DECOMPOSITION PRODUCTS

SECTION VII SPILL OR LEAK PROCEDURES

STEPS TO BE TAKEN IN CASE MATERIAL IS RELEASED OR SPILLED: Absorb spill with paper towels, place in hood for evaporation.

WASTE DISPOSAL METHOD

SECTION VIII SPECIAL PROTECTION INFORMATION

VENTILATION REQUIREMENTS LOCAL EXHAUST	PROTECTIVE EQUIPMENT (SPECIFY TYPES) EYE: goggles
MECHANICAL (GENERAL)	GLOVES: rubber
SPECIAL	RESPIRATOR: with proper filter

OTHER PROTECTIVE EQUIPMENT: Working clothing

SECTION IX SPECIAL PRECAUTIONS

PRECAUTIONS TO BE TAKEN IN HANDLING AND STORAGE

OTHER PRECAUTIONS

SIGNATURE	DATE

ETHYL IODIDE

NFPA diamond: 1 (top), 3 (left), 0 (right)

SECTION I SOURCE AND NOMENCLATURE

MANUFACTURER'S NAME	EMERGENCY TELEPHONE NO.
ADDRESS (NUMBER, STREET, CITY, STATE, ZIP CODE)	
TRADE NAME AND SYNONYMS	CHEMICAL FAMILY
CHEMICAL NAME AND SYNONYMS: monoiodoethane	FORMULA: C_2H_5I

SECTION II HAZARDOUS INGREDIENTS

BASIC MATERIAL	APPROXIMATE OR MAXIMUM % WT. OR VOL.	ESTABLISHED OSHA STANDARD	LD_{50} ORAL	LD_{50} PERCUT.	LC_{50} SPECIES	LC_{50} CONC.
			none	available		

SECTION III PHYSICAL DATA MW - 155.98

BOILING POINT °F.: 72.2°C	VAPOR PRESSURE mm Hg.: at 25°C 137
MELTING POINT °F.: -108.5°C	VAPOR DENSITY (Air = 1): 5.4
SPECIFIC GRAVITY (H_2O = 1): 1.92	EVAPORATION RATE (= 1)
SOLUBILITY IN WATER Pts/100 pts H_2O: 0.4	VOLATILE % Vol. % Wt.

APPEARANCE AND ODOR: Colorless liquid, ethereal odor

SECTION IV FIRE AND EXPLOSION HAZARD DATA

FLASH POINT: Not available	FLAMMABLE (EXPLOSIVE) LIMITS	UPPER
METHOD USED		LOWER

EXTINGUISHING MEDIA: Water, carbon dioxide, dry chemical

SPECIAL FIRE FIGHTING PROCEDURES: --

UNUSUAL FIRE AND EXPLOSION HAZARDS: When heated to decomposition points, emits toxic fumes of iodates.

SECTION V HEALTH HAZARD DATA

TOXIC LEVEL: Not available CARCINOGENIC: no

PRINCIPLE ROUTES OF ABSORBTION: Inhalation, ingestion, skin SKIN AND EYE IRRITATION: slight

REVELANT SYMPTOMS OF EXPOSURE: Nausea, vomiting, diarrhea, dizziness, visual disturbances, irritability, drowsiness, central nervous ystem depressant.

EFFECTS OF CHRONIC EXPOSURE: dermatitis

EMERGENCY AND FIRST AID PROCEDURES: Inhalation -rest; if swallowed - wash mouth, give emetic, get medical aid; wash from skin & eyes.

SECTION VI REACTIVITY DATA

CONDITIONS CONTRIBUTING TO INSTABILITY

CONDITIONS CONTRIBUTING TO HAZARDOUS POLYMERIZATION

INCOMPATABILITY (MATERIALS TO AVOID)

HAZARDOUS DECOMPOSITION PRODUCTS

SECTION VII SPILL OR LEAK PROCEDURES

STEPS TO BE TAKEN IN CASE MATERIAL IS RELEASED OR SPILLED

WASTE DISPOSAL METHOD

SECTION VIII SPECIAL PROTECTION INFORMATION

VENTILATION REQUIREMENTS LOCAL EXHAUST	PROTECTIVE EQUIPMENT (SPECIFY TYPES) EYE: safety goggles
MECHANICAL (GENERAL)	GLOVES: rubber
SPECIAL	RESPIRATOR: with proper filter

OTHER PROTECTIVE EQUIPMENT

SECTION IX SPECIAL PRECAUTIONS

PRECAUTIONS TO BE TAKEN IN HANDLING AND STORAGE

OTHER PRECAUTIONS

SIGNATURE	DATE

ETHYL ISOTHIOCYANATE

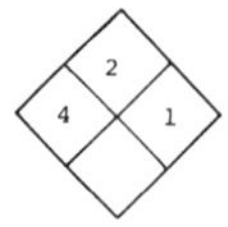

SECTION I SOURCE AND NOMENCLATURE

MANUFACTURER'S NAME | EMERGENCY TELEPHONE NO.

ADDRESS (NUMBER, STREET, CITY, STATE, ZIP CODE)

TRADE NAME AND SYNONYMS | CHEMICAL FAMILY

CHEMICAL NAME AND SYNONYMS: Ethyl Thiocarbimide

FORMULA: C_2H_5NCS

SECTION II HAZARDOUS INGREDIENTS

BASIC MATERIAL	APPROXIMATE OR MAXIMUM % WT. OR VOL.	ESTABLISHED OSHA STANDARD	LD_{50} ORAL	LD_{50} PERCUT.	LC_{50} SPECIES	LC_{50} CONC.
		none available				

SECTION III PHYSICAL DATA MW - 87.2

BOILING POINT °F.: 131°C

VAPOR PRESSURE mm Hg.: at 23°C 10

MELTING POINT °F.: -6°C

VAPOR DENSITY (Air = 1): 3.01

SPECIFIC GRAVITY (H_2O = 1): 1.00

EVAPORATION RATE (____ = 1)

SOLUBILITY IN WATER Pts/100 pts H_2O

VOLATILE % Vol. % Wt.

APPEARANCE AND ODOR: colorless liquid, pungent odor

SECTION IV FIRE AND EXPLOSION HAZARD DATA

FLASH POINT: none available

METHOD USED

FLAMMABLE (EXPLOSIVE) LIMITS: UPPER / LOWER

EXTINGUISHING MEDIA: carbon dioxide, dry chemical

SPECIAL FIRE FIGHTING PROCEDURES: wear full protective clothing, self-contained breathing apparatus.

UNUSUAL FIRE AND EXPLOSION HAZARDS

SECTION V HEALTH HAZARD DATA

TOXIC LEVEL: none available

CARCINOGENIC: no

PRINCIPLE ROUTES OF ABSORBTION: ingestion, inhalation

SKIN AND EYE IRRITATION: severe

REVELANT SYMPTOMS OF EXPOSURE: Irritation of eyes, respiratory system, irritation of digestive tract.

EFFECTS OF CHRONIC EXPOSURE: --

EMERGENCY AND FIRST AID PROCEDURES: If swallowed - induce vomiting, get medical aid; wash skin & eyes; inhalation - rest.

SECTION VI REACTIVITY DATA

CONDITIONS CONTRIBUTING TO INSTABILITY

CONDITIONS CONTRIBUTING TO HAZARDOUS POLYMERIZATION

INCOMPATABILITY (MATERIALS TO AVOID): May react with oxidizing materials and acids.

HAZARDOUS DECOMPOSITION PRODUCTS

SECTION VII SPILL OR LEAK PROCEDURES

STEPS TO BE TAKEN IN CASE MATERIAL IS RELEASED OR SPILLED

WASTE DISPOSAL METHOD

SECTION VIII SPECIAL PROTECTION INFORMATION

VENTILATION REQUIREMENTS LOCAL EXHAUST

MECHANICAL (GENERAL)

SPECIAL

PROTECTIVE EQUIPMENT (SPECIFY TYPES) EYE: goggles

GLOVES: rubber

RESPIRATOR: self-contained

OTHER PROTECTIVE EQUIPMENT

SECTION IX SPECIAL PRECAUTIONS

PRECAUTIONS TO BE TAKEN IN HANDLING AND STORAGE

OTHER PRECAUTIONS

SIGNATURE | DATE

ETHYL m-TOLUIDINE

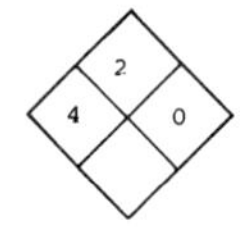

SECTION I SOURCE AND NOMENCLATURE

MANUFACTURER'S NAME | EMERGENCY TELEPHONE NO.

ADDRESS (NUMBER, STREET, CITY, STATE, ZIP CODE)

TRADE NAME AND SYNONYMS | CHEMICAL FAMILY

CHEMICAL NAME AND SYNONYMS: Ethyl-M-Toluidine

FORMULA: $CH_3C_6H_4NHC_2H_5$

SECTION II HAZARDOUS INGREDIENTS

BASIC MATERIAL	APPROXIMATE OR MAXIMUM % WT. OR VOL.	ESTABLISHED OSHA STANDARD	LD_{50} ORAL	LD_{50} PERCUT.	LC_{50} SPECIES	LC_{50} CONC.
			none available			

SECTION III PHYSICAL DATA MW - 135.20

BOILING POINT °F.: 221°C

VAPOR PRESSURE mm Hg.: at 102°C 10

MELTING POINT °F.: --

VAPOR DENSITY (Air = 1)

SPECIFIC GRAVITY (H_2O = 1): -

EVAPORATION RATE (____ = 1)

SOLUBILITY IN WATER Pts/100 pts H_2O: insoluble

VOLATILE % Vol. % Wt.

APPEARANCE AND ODOR: liquid

SECTION IV FIRE AND EXPLOSION HAZARD DATA

FLASH POINT: none available

METHOD USED

FLAMMABLE (EXPLOSIVE) LIMITS: UPPER / LOWER

EXTINGUISHING MEDIA: alcohol foam, carbon dioxide, dry chemical

SPECIAL FIRE FIGHTING PROCEDURES: wear full protective clothing.

UNUSUAL FIRE AND EXPLOSION HAZARDS

SECTION V HEALTH HAZARD DATA

TOXIC LEVEL: none available

CARCINOGENIC: no

PRINCIPLE ROUTES OF ABSORBTION: inhalation, ingestion, skin

SKIN AND EYE IRRITATION: moderate

REVELANT SYMPTOMS OF EXPOSURE: Cyanosis, air hunger, nausea, vomiting, low blood pressure, convulsions.

EFFECTS OF CHRONIC EXPOSURE: Weight loss, anemia, weakness

EMERGENCY AND FIRST AID PROCEDURES: If swallowed - give emetic; get medical aid; inhalation - rest; wash from skin & eyes.

SECTION VI REACTIVITY DATA

CONDITIONS CONTRIBUTING TO INSTABILITY

CONDITIONS CONTRIBUTING TO HAZARDOUS POLYMERIZATION

INCOMPATABILITY (MATERIALS TO AVOID)

HAZARDOUS DECOMPOSITION PRODUCTS

SECTION VII SPILL OR LEAK PROCEDURES

STEPS TO BE TAKEN IN CASE MATERIAL IS RELEASED OR SPILLED

WASTE DISPOSAL METHOD

SECTION VIII SPECIAL PROTECTION INFORMATION

VENTILATION REQUIREMENTS LOCAL EXHAUST

MECHANICAL (GENERAL)

SPECIAL

PROTECTIVE EQUIPMENT (SPECIFY TYPES) EYE: goggles

GLOVES: rubber

RESPIRATOR: with proper filter

OTHER PROTECTIVE EQUIPMENT

SECTION IX SPECIAL PRECAUTIONS

PRECAUTIONS TO BE TAKEN IN HANDLING AND STORAGE: Outside storage preferred. Inside storage in a standard flammable liquids storage cabinet.

OTHER PRECAUTIONS

SIGNATURE | DATE

ETHYL o-TOLUIDINE

Hazard diamond: 2 (top), 4 (left), 0 (right)

SECTION I SOURCE AND NOMENCLATURE

MANUFACTURER'S NAME	EMERGENCY TELEPHONE NO.
ADDRESS (NUMBER, STREET, CITY, STATE, ZIP CODE)	
TRADE NAME AND SYNONYMS	CHEMICAL FAMILY
CHEMICAL NAME AND SYNONYMS: Ethyl-O -Toluidine	FORMULA: $CH_3C_6H_4NHC_2H_5$

SECTION II HAZARDOUS INGREDIENTS

BASIC MATERIAL	APPROXIMATE OR MAXIMUM % WT. OR VOL.	ESTABLISHED OSHA STANDARD	LD_{50} ORAL	LD_{50} PERCUT.	LC_{50} SPECIES	LC_{50} CONC.
			none	available		

SECTION III PHYSICAL DATA MW - 135.20

BOILING POINT °F.: 216°C	VAPOR PRESSURE mm Hg.: at 85°C 6
MELTING POINT °F.: <-15°C	VAPOR DENSITY (Air = 1)
SPECIFIC GRAVITY (H_2O = 1): 0.95	EVAPORATION RATE (______ = 1)
SOLUBILITY IN WATER Pts/100 pts H_2O: insoluble	VOLATILE % Vol. % Wt.
APPEARANCE AND ODOR: liquid	

SECTION IV FIRE AND EXPLOSION HAZARD DATA

FLASH POINT: none available

METHOD USED

FLAMMABLE (EXPLOSIVE) LIMITS: UPPER / LOWER

EXTINGUISHING MEDIA: alcohol foam, carbon dioxide, dry chemical

SPECIAL FIRE FIGHTING PROCEDURES: wear full protective clothing

UNUSUAL FIRE AND EXPLOSION HAZARDS

SECTION V HEALTH HAZARD DATA

TOXIC LEVEL: none available — CARCINOGENIC: no

PRINCIPLE ROUTES OF ABSORBTION: Inhalation, ingestion, skin — SKIN AND EYE IRRITATION: moderate

REVELANT SYMPTOMS OF EXPOSURE: cyanosis, air hunger, nausea, vomiting, low blood pressure, convulsions.

EFFECTS OF CHRONIC EXPOSURE: Weight loss, anemia, weakness.

EMERGENCY AND FIRST AID PROCEDURES: If swallowed - give emetic, get medical aid; inhalation - rest; wash from skin & eyes.

SECTION VI REACTIVITY DATA

CONDITIONS CONTRIBUTING TO INSTABILITY

CONDITIONS CONTRIBUTING TO HAZARDOUS POLYMERIZATION

INCOMPATABILITY (MATERIALS TO AVOID)

HAZARDOUS DECOMPOSITION PRODUCTS

SECTION VII SPILL OR LEAK PROCEDURES

STEPS TO BE TAKEN IN CASE MATERIAL IS RELEASED OR SPILLED

WASTE DISPOSAL METHOD

SECTION VIII SPECIAL PROTECTION INFORMATION

VENTILATION REQUIREMENTS LOCAL EXHAUST	PROTECTIVE EQUIPMENT (SPECIFY TYPES) EYE: goggles
MECHANICAL (GENERAL)	GLOVES: rubber
SPECIAL	RESPIRATOR: with proper filter
OTHER PROTECTIVE EQUIPMENT	

SECTION IX SPECIAL PRECAUTIONS

PRECAUTIONS TO BE TAKEN IN HANDLING AND STORAGE: Outside storage preferred. Inside storage in a standard flammable liquid storage cabinet.

OTHER PRECAUTIONS

SIGNATURE | DATE

ETHYL p-TOLUIDINE

Hazard diamond: 2 (top), 4 (left), 0 (right)

SECTION I SOURCE AND NOMENCLATURE

MANUFACTURER'S NAME	EMERGENCY TELEPHONE NO.
ADDRESS (NUMBER, STREET, CITY, STATE, ZIP CODE)	
TRADE NAME AND SYNONYMS	CHEMICAL FAMILY
CHEMICAL NAME AND SYNONYMS: Ethyl-P-Toluidine	FORMULA: $CH_3C_6H_4NHC_2H_5$

SECTION II HAZARDOUS INGREDIENTS

BASIC MATERIAL	APPROXIMATE OR MAXIMUM % WT. OR VOL.	ESTABLISHED OSHA STANDARD	LD_{50} ORAL	LD_{50} PERCUT.	LC_{50} SPECIES	LC_{50} CONC.
			none	available		

SECTION III PHYSICAL DATA MW - 135.20

BOILING POINT °F.: 217°C	VAPOR PRESSURE mm Hg.: at 101°C 10
MELTING POINT °F.: --	VAPOR DENSITY (Air = 1)
SPECIFIC GRAVITY (H_2O = 1): 0.94	EVAPORATION RATE (______ = 1)
SOLUBILITY IN WATER Pts/100 pts H_2O: insoluble	VOLATILE % Vol. % Wt.
APPEARANCE AND ODOR: liquid	

SECTION IV FIRE AND EXPLOSION HAZARD DATA

FLASH POINT: none available

METHOD USED

FLAMMABLE (EXPLOSIVE) LIMITS: UPPER / LOWER

EXTINGUISHING MEDIA: alcohol foam, carbon dioxide, dry chemical

SPECIAL FIRE FIGHTING PROCEDURES: wear full protective clothing.

UNUSUAL FIRE AND EXPLOSION HAZARDS

SECTION V HEALTH HAZARD DATA

TOXIC LEVEL: none available — CARCINOGENIC: no

PRINCIPLE ROUTES OF ABSORBTION: inhalation, ingestion, skin — SKIN AND EYE IRRITATION: moderate

REVELANT SYMPTOMS OF EXPOSURE: cyanosis, air hunger, nausea, vomiting, low blood pressure, convulsions.

EFFECTS OF CHRONIC EXPOSURE: weight loss, anemia, weakness

EMERGENCY AND FIRST AID PROCEDURES: If swallowed - give emetic, get medical aid; inhalation - rest; wash from skin & eyes.

SECTION VI REACTIVITY DATA

CONDITIONS CONTRIBUTING TO INSTABILITY

CONDITIONS CONTRIBUTING TO HAZARDOUS POLYMERIZATION

INCOMPATABILITY (MATERIALS TO AVOID)

HAZARDOUS DECOMPOSITION PRODUCTS

SECTION VII SPILL OR LEAK PROCEDURES

STEPS TO BE TAKEN IN CASE MATERIAL IS RELEASED OR SPILLED

WASTE DISPOSAL METHOD

SECTION VIII SPECIAL PROTECTION INFORMATION

VENTILATION REQUIREMENTS LOCAL EXHAUST	PROTECTIVE EQUIPMENT (SPECIFY TYPES) EYE: goggles
MECHANICAL (GENERAL)	GLOVES: rubber
SPECIAL	RESPIRATOR: with proper filter
OTHER PROTECTIVE EQUIPMENT	

SECTION IX SPECIAL PRECAUTIONS

PRECAUTIONS TO BE TAKEN IN HANDLING AND STORAGE: Outside storage preferred. Inside storage in a standard flammable liquid storage cabinet.

OTHER PRECAUTIONS

SIGNATURE | DATE

p-FLUOROBROMOBENZENE

Hazard rating: 2 (top), 3 (left), 0 (right)

SECTION I SOURCE AND NOMENCLATURE

MANUFACTURER'S NAME | EMERGENCY TELEPHONE NO.

ADDRESS (NUMBER, STREET, CITY, STATE, ZIP CODE)

TRADE NAME AND SYNONYMS | CHEMICAL FAMILY

CHEMICAL NAME AND SYNONYMS: P-Fluorobromobenzene | FORMULA: FC_6H_4Br

SECTION II HAZARDOUS INGREDIENTS

BASIC MATERIAL	APPROXIMATE OR MAXIMUM % WT. OR VOL.	ESTABLISHED OSHA STANDARD	LD_{50} ORAL	LD_{50} PERCUT.	LC_{50} SPECIES	LC_{50} CONC.
			none	available		

SECTION III PHYSICAL DATA MW - 175.01

BOILING POINT (°F.): $155^{\circ}C$	VAPOR PRESSURE (mm Hg.)
MELTING POINT (°F.): $-8^{\circ}C$	VAPOR DENSITY (Air = 1)
SPECIFIC GRAVITY (H_2O = 1): 1.60	EVAPORATION RATE (____ = 1)
SOLUBILITY IN WATER Pts/100 pts H_2O: insoluble	VOLATILE % Vol. % Wt.

APPEARANCE AND ODOR: colorless liquid

SECTION IV FIRE AND EXPLOSION HAZARD DATA

FLASH POINT: none available | FLAMMABLE (EXPLOSIVE) LIMITS: UPPER / LOWER

METHOD USED

EXTINGUISHING MEDIA: use water spray, foam, carbon dioxide or dry chemical

SPECIAL FIRE FIGHTING PROCEDURES: wear self-contained breathing apparatus.

UNUSUAL FIRE AND EXPLOSION HAZARDS

SECTION V HEALTH HAZARD DATA

TOXIC LEVEL: none available | CARCINOGENIC: no

PRINCIPLE ROUTES OF ABSORBTION: inhalation, ingestion, skin | SKIN AND EYE IRRITATION: severe

REVELANT SYMPTOMS OF EXPOSURE: cyanosis, deep and rapid respiration, irregular pulse, drowsiness, unconsciousness.

EFFECTS OF CHRONIC EXPOSURE: Lung, liver and kidney damage

EMERGENCY AND FIRST AID PROCEDURES: If swallowed - give emetic, get medical aid; inhalation - rest; wash from skin & eyes.

SECTION VI REACTIVITY DATA

CONDITIONS CONTRIBUTING TO INSTABILITY

CONDITIONS CONTRIBUTING TO HAZARDOUS POLYMERIZATION

INCOMPATABILITY (MATERIALS TO AVOID)

HAZARDOUS DECOMPOSITION PRODUCTS

SECTION VII SPILL OR LEAK PROCEDURES

STEPS TO BE TAKEN IN CASE MATERIAL IS RELEASED OR SPILLED

WASTE DISPOSAL METHOD

SECTION VIII SPECIAL PROTECTION INFORMATION

VENTILATION REQUIREMENTS LOCAL EXHAUST | PROTECTIVE EQUIPMENT (SPECIFY TYPES) EYE: goggles

MECHANICAL (GENERAL) | GLOVES: polyvinyl alcohol gloves

SPECIAL | RESPIRATOR: self-contained

OTHER PROTECTIVE EQUIPMENT

SECTION IX SPECIAL PRECAUTIONS

PRECAUTIONS TO BE TAKEN IN HANDLING AND STORAGE

OTHER PRECAUTIONS

SIGNATURE | DATE

m-FLUOROCHLOROBENZENE

Hazard rating: 2 (top), 3 (left), 0 (right)

SECTION I SOURCE AND NOMENCLATURE

MANUFACTURER'S NAME | EMERGENCY TELEPHONE NO.

ADDRESS (NUMBER, STREET, CITY, STATE, ZIP CODE)

TRADE NAME AND SYNONYMS | CHEMICAL FAMILY

CHEMICAL NAME AND SYNONYMS: M-Fluorochlorobenzene | FORMULA: FC_6H_4Cl

SECTION II HAZARDOUS INGREDIENTS

BASIC MATERIAL	APPROXIMATE OR MAXIMUM % WT. OR VOL.	ESTABLISHED OSHA STANDARD	LD_{50} ORAL	LD_{50} PERCUT.	LC_{50} SPECIES	LC_{50} CONC.
			none	available		

SECTION III PHYSICAL DATA MW - 130.55

BOILING POINT (°F.): $125^{\circ}C$	VAPOR PRESSURE (mm Hg.)
MELTING POINT (°F.)	VAPOR DENSITY (Air = 1)
SPECIFIC GRAVITY (H_2O = 1)	EVAPORATION RATE (____ = 1)
SOLUBILITY IN WATER Pts/100 pts H_2O	VOLATILE % Vol. % Wt.

APPEARANCE AND ODOR: colorless liquid

SECTION IV FIRE AND EXPLOSION HAZARD DATA

FLASH POINT: not available | FLAMMABLE (EXPLOSIVE) LIMITS: UPPER / LOWER

METHOD USED

EXTINGUISHING MEDIA: use water spray, foam, carbon dioxide or dry chemical

SPECIAL FIRE FIGHTING PROCEDURES: wear self-contained breathing apparatus.

UNUSUAL FIRE AND EXPLOSION HAZARDS

SECTION V HEALTH HAZARD DATA

TOXIC LEVEL: none available | CARCINOGENIC: no

PRINCIPLE ROUTES OF ABSORBTION: inhalation, ingestion, skin | SKIN AND EYE IRRITATION: severe

REVELANT SYMPTOMS OF EXPOSURE: Cyanosis, deep and rapid respiration, irregular pulse, drowsiness, unconsciousness.

EFFECTS OF CHRONIC EXPOSURE: Lung, liver and kidney damage.

EMERGENCY AND FIRST AID PROCEDURES: If swallowed - give emetic, get medical aid; inhalation - rest; wash from skin & eyes.

SECTION VI REACTIVITY DATA

CONDITIONS CONTRIBUTING TO INSTABILITY

CONDITIONS CONTRIBUTING TO HAZARDOUS POLYMERIZATION

INCOMPATABILITY (MATERIALS TO AVOID)

HAZARDOUS DECOMPOSITION PRODUCTS

SECTION VII SPILL OR LEAK PROCEDURES

STEPS TO BE TAKEN IN CASE MATERIAL IS RELEASED OR SPILLED

WASTE DISPOSAL METHOD

SECTION VIII SPECIAL PROTECTION INFORMATION

VENTILATION REQUIREMENTS LOCAL EXHAUST | PROTECTIVE EQUIPMENT (SPECIFY TYPES) EYE: goggles

MECHANICAL (GENERAL) | GLOVES: polyvinyl alcohol gloves

SPECIAL | RESPIRATOR: self-contained

OTHER PROTECTIVE EQUIPMENT

SECTION IX SPECIAL PRECAUTIONS

PRECAUTIONS TO BE TAKEN IN HANDLING AND STORAGE

OTHER PRECAUTIONS

SIGNATURE | DATE

o-FLUOROCHLOROBENZENE

Hazard diamond: 2 (top), 3 (left), 0 (right)

SECTION I SOURCE AND NOMENCLATURE

MANUFACTURER'S NAME

EMERGENCY TELEPHONE NO.

ADDRESS (NUMBER, STREET, CITY, STATE, ZIP CODE)

TRADE NAME AND SYNONYMS

CHEMICAL FAMILY

CHEMICAL NAME AND SYNONYMS: O-Fluorochlorobenzene

FORMULA: FC_6H_4Cl

SECTION II HAZARDOUS INGREDIENTS

BASIC MATERIAL	APPROXIMATE OR MAXIMUM % WT. OR VOL.	ESTABLISHED OSHA STANDARD	LD_{50} ORAL	LD_{50} PERCUT.	LC_{50} SPECIES	LC_{50} CONC.
			none	available		

SECTION III PHYSICAL DATA MW -130.55

BOILING POINT °F.: 136°C

VAPOR PRESSURE mm Hg.

MELTING POINT °F.: -43°C

VAPOR DENSITY (Air = 1)

SPECIFIC GRAVITY (H_2O = 1)

EVAPORATION RATE (______ = 1)

SOLUBILITY IN WATER Pts/100 pts H_2O

VOLATILE % Vol. % Wt.

APPEARANCE AND ODOR: liquid

SECTION IV FIRE AND EXPLOSION HAZARD DATA

FLASH POINT: not available

METHOD USED

FLAMMABLE (EXPLOSIVE) LIMITS: UPPER / LOWER

EXTINGUISHING MEDIA: Use water spray, foam, carbon dioxide or dry chemical

SPECIAL FIRE FIGHTING PROCEDURES: Wear self-contained breathing apparatus.

UNUSUAL FIRE AND EXPLOSION HAZARDS

SECTION V HEALTH HAZARD DATA

TOXIC LEVEL: none available

CARCINOGENIC: no

PRINCIPLE ROUTES OF ABSORBTION: inhalation, ingestion, skin

SKIN AND EYE IRRITATION: severe

REVELANT SYMPTOMS OF EXPOSURE: Cyanosis, deep and rapid respiration, irregular pulse, drowsiness, unconsciousness.

EFFECTS OF CHRONIC EXPOSURE: Lung, liver and kidney damage.

EMERGENCY AND FIRST AID PROCEDURES: If swallowed - give emetic, get medical aid; inhalation - rest; wash from skin & eyes.

SECTION VI REACTIVITY DATA

CONDITIONS CONTRIBUTING TO INSTABILITY

CONDITIONS CONTRIBUTING TO HAZARDOUS POLYMERIZATION

INCOMPATABILITY (MATERIALS TO AVOID)

HAZARDOUS DECOMPOSITION PRODUCTS

SECTION VII SPILL OR LEAK PROCEDURES

STEPS TO BE TAKEN IN CASE MATERIAL IS RELEASED OR SPILLED

WASTE DISPOSAL METHOD

SECTION VIII SPECIAL PROTECTION INFORMATION

VENTILATION REQUIREMENTS LOCAL EXHAUST

PROTECTIVE EQUIPMENT (SPECIFY TYPES) EYE: goggles

MECHANICAL (GENERAL)

GLOVES: polyvinyl alcohol gloves

SPECIAL

RESPIRATOR: self-contained

OTHER PROTECTIVE EQUIPMENT

SECTION IX SPECIAL PRECAUTIONS

PRECAUTIONS TO BE TAKEN IN HANDLING AND STORAGE

OTHER PRECAUTIONS

SIGNATURE | DATE

m-FLUOROTOLUENE

Hazard diamond: 2 (top), 3 (left), 0 (right)

SECTION I SOURCE AND NOMENCLATURE

MANUFACTURER'S NAME

EMERGENCY TELEPHONE NO.

ADDRESS (NUMBER, STREET, CITY, STATE, ZIP CODE)

TRADE NAME AND SYNONYMS

CHEMICAL FAMILY

CHEMICAL NAME AND SYNONYMS: M-Fluoromethyl benzene

FORMULA: $CH_3C_6H_4F$

SECTION II HAZARDOUS INGREDIENTS

BASIC MATERIAL	APPROXIMATE OR MAXIMUM % WT. OR VOL.	ESTABLISHED OSHA STANDARD	LD_{50} ORAL	LD_{50} PERCUT.	LC_{50} SPECIES	LC_{50} CONC.
			none	available		

SECTION III PHYSICAL DATA MW - 110.13

BOILING POINT °F.: 116°C

VAPOR PRESSURE mm Hg.

MELTING POINT °F.: -111°C

VAPOR DENSITY (Air = 1)

SPECIFIC GRAVITY (H_2O = 1): 0.997

EVAPORATION RATE (______ = 1)

SOLUBILITY IN WATER Pts/100 pts H_2O: insoluble

VOLATILE % Vol. % Wt.

APPEARANCE AND ODOR: colorless liquid

SECTION IV FIRE AND EXPLOSION HAZARD DATA

FLASH POINT: none available

METHOD USED

FLAMMABLE (EXPLOSIVE) LIMITS: UPPER / LOWER

EXTINGUISHING MEDIA: Alcohol foam, carbon dioxide, dry chemical

SPECIAL FIRE FIGHTING PROCEDURES

UNUSUAL FIRE AND EXPLOSION HAZARDS

SECTION V HEALTH HAZARD DATA

TOXIC LEVEL: none available

CARCINOGENIC: no

PRINCIPLE ROUTES OF ABSORBTION: inhalation, ingestion

SKIN AND EYE IRRITATION: severe

REVELANT SYMPTOMS OF EXPOSURE: Eye and respiratory system irritation, severe digestive system attack, narcotic effect.

EFFECTS OF CHRONIC EXPOSURE: --

EMERGENCY AND FIRST AID PROCEDURES: If swallowed - give emetic, get medical aid; inhalation - rest; wash from skin & eyes.

SECTION VI REACTIVITY DATA

CONDITIONS CONTRIBUTING TO INSTABILITY

CONDITIONS CONTRIBUTING TO HAZARDOUS POLYMERIZATION

INCOMPATABILITY (MATERIALS TO AVOID)

HAZARDOUS DECOMPOSITION PRODUCTS

SECTION VII SPILL OR LEAK PROCEDURES

STEPS TO BE TAKEN IN CASE MATERIAL IS RELEASED OR SPILLED

WASTE DISPOSAL METHOD

SECTION VIII SPECIAL PROTECTION INFORMATION

VENTILATION REQUIREMENTS LOCAL EXHAUST

PROTECTIVE EQUIPMENT (SPECIFY TYPES) EYE: goggles

MECHANICAL (GENERAL)

GLOVES: rubber

SPECIAL

RESPIRATOR: with proper filter

OTHER PROTECTIVE EQUIPMENT

SECTION IX SPECIAL PRECAUTIONS

PRECAUTIONS TO BE TAKEN IN HANDLING AND STORAGE

OTHER PRECAUTIONS

SIGNATURE | DATE

o-FLUOROTOLUENE

Hazard diamond: 2 (top), 3 (left), 0 (right)

SECTION I SOURCE AND NOMENCLATURE

MANUFACTURER'S NAME | EMERGENCY TELEPHONE NO.

ADDRESS (NUMBER, STREET, CITY, STATE, ZIP CODE)

TRADE NAME AND SYNONYMS | CHEMICAL FAMILY

CHEMICAL NAME AND SYNONYMS: O-Fluoromethylbenzene | FORMULA: $CH_3C_6H_4F$

SECTION II HAZARDOUS INGREDIENTS

BASIC MATERIAL	APPROXIMATE OR MAXIMUM % WT. OR VOL.	ESTABLISHED OSHA STANDARD	LD_{50} ORAL	LD_{50} PERCUT.	LC_{50} SPECIES	LC_{50} CONC.
			none	available		

SECTION III PHYSICAL DATA MW - 110.13

BOILING POINT °F.: 114°C | VAPOR PRESSURE mm Hg.

MELTING POINT °F.: -80°C | VAPOR DENSITY (Air = 1)

SPECIFIC GRAVITY (H_2O = 1): 1.004 | EVAPORATION RATE (_____ = 1)

SOLUBILITY IN WATER Pts/100 pts H_2O: insoluble | VOLATILE % Vol. % Wt.

APPEARANCE AND ODOR: colorless liquid

SECTION IV FIRE AND EXPLOSION HAZARD DATA

FLASH POINT: none available | FLAMMABLE (EXPLOSIVE) LIMITS | UPPER | LOWER

METHOD USED

EXTINGUISHING MEDIA: alcohol foam, carbon dioxide, dry chemical.

SPECIAL FIRE FIGHTING PROCEDURES

UNUSUAL FIRE AND EXPLOSION HAZARDS

SECTION V HEALTH HAZARD DATA

TOXIC LEVEL: none available | CARCINOGENIC: no

PRINCIPLE ROUTES OF ABSORBTION: inhalation, ingestion | SKIN AND EYE IRRITATION: severe

REVELANT SYMPTOMS OF EXPOSURE: Eye and respiratory system irritation, severe digestive system irritation, narcotic effect.

EFFECTS OF CHRONIC EXPOSURE: --

EMERGENCY AND FIRST AID PROCEDURES: If swallowed - give emetic, get medical aid; inhalation - rest; wash from skin & eyes.

SECTION VI REACTIVITY DATA

CONDITIONS CONTRIBUTING TO INSTABILITY

CONDITIONS CONTRIBUTING TO HAZARDOUS POLYMERIZATION

INCOMPATABILITY (MATERIALS TO AVOID)

HAZARDOUS DECOMPOSITION PRODUCTS

SECTION VII SPILL OR LEAK PROCEDURES

STEPS TO BE TAKEN IN CASE MATERIAL IS RELEASED OR SPILLED

WASTE DISPOSAL METHOD

SECTION VIII SPECIAL PROTECTION INFORMATION

VENTILATION REQUIREMENTS LOCAL EXHAUST | PROTECTIVE EQUIPMENT (SPECIFY TYPES) EYE: goggles

MECHANICAL (GENERAL) | GLOVES: rubber

SPECIAL | RESPIRATOR: with proper filter

OTHER PROTECTIVE EQUIPMENT

SECTION IX SPECIAL PRECAUTIONS

PRECAUTIONS TO BE TAKEN IN HANDLING AND STORAGE

OTHER PRECAUTIONS

SIGNATURE | DATE

p-FLUOROTOLUENE

Hazard diamond: 2 (top), 3 (left), 0 (right)

SECTION I SOURCE AND NOMENCLATURE

MANUFACTURER'S NAME | EMERGENCY TELEPHONE NO.

ADDRESS (NUMBER, STREET, CITY, STATE, ZIP CODE)

TRADE NAME AND SYNONYMS | CHEMICAL FAMILY

CHEMICAL NAME AND SYNONYMS: P-Fluoromethyl benzene | FORMULA: $CH_3C_6H_4F$

SECTION II HAZARDOUS INGREDIENTS

BASIC MATERIAL	APPROXIMATE OR MAXIMUM % WT. OR VOL.	ESTABLISHED OSHA STANDARD	LD_{50} ORAL	LD_{50} PERCUT.	LC_{50} SPECIES	LC_{50} CONC.
LDLO parenteral - mouse	500 mg/kg		none	available		

SECTION III PHYSICAL DATA MW - 110.13

BOILING POINT °F.: 117°C | VAPOR PRESSURE mm Hg.

MELTING POINT °F.: -- | VAPOR DENSITY (Air = 1)

SPECIFIC GRAVITY (H_2O = 1): 1.001 | EVAPORATION RATE (_____ = 1)

SOLUBILITY IN WATER Pts/100 pts H_2O: insoluble | VOLATILE % Vol. % Wt.

APPEARANCE AND ODOR: colorless liquid, odor of oil of bitter almonds.

SECTION IV FIRE AND EXPLOSION HAZARD DATA

FLASH POINT: none available | FLAMMABLE (EXPLOSIVE) LIMITS | UPPER | LOWER

METHOD USED

EXTINGUISHING MEDIA: Dry chemical, carbon dioxide, alcohol foam.

SPECIAL FIRE FIGHTING PROCEDURES

UNUSUAL FIRE AND EXPLOSION HAZARDS

SECTION V HEALTH HAZARD DATA

TOXIC LEVEL: none available | CARCINOGENIC: no

PRINCIPLE ROUTES OF ABSORBTION: inhalation, ingestion | SKIN AND EYE IRRITATION: severe

REVELANT SYMPTOMS OF EXPOSURE: Eye and respiratory system irritation, severe digestive system irritation, narcotic effect.

EFFECTS OF CHRONIC EXPOSURE: --

EMERGENCY AND FIRST AID PROCEDURES: If swallowed - give emetic, get medical aid; inhalation - rest; wash from skin & eyes.

SECTION VI REACTIVITY DATA

CONDITIONS CONTRIBUTING TO INSTABILITY

CONDITIONS CONTRIBUTING TO HAZARDOUS POLYMERIZATION

INCOMPATABILITY (MATERIALS TO AVOID)

HAZARDOUS DECOMPOSITION PRODUCTS

SECTION VII SPILL OR LEAK PROCEDURES

STEPS TO BE TAKEN IN CASE MATERIAL IS RELEASED OR SPILLED

WASTE DISPOSAL METHOD

SECTION VIII SPECIAL PROTECTION INFORMATION

VENTILATION REQUIREMENTS LOCAL EXHAUST | PROTECTIVE EQUIPMENT (SPECIFY TYPES) EYE: goggles

MECHANICAL (GENERAL) | GLOVES: gloved

SPECIAL | RESPIRATOR: with proper filter

OTHER PROTECTIVE EQUIPMENT

SECTION IX SPECIAL PRECAUTIONS

PRECAUTIONS TO BE TAKEN IN HANDLING AND STORAGE

OTHER PRECAUTIONS

SIGNATURE | DATE

FORMAMIDE

Hazard diamond: 1 (top), 3 (left), 0 (right)

SECTION I SOURCE AND NOMENCLATURE

MANUFACTURER'S NAME | EMERGENCY TELEPHONE NO.

ADDRESS (NUMBER, STREET, CITY, STATE, ZIP CODE)

TRADE NAME AND SYNONYMS | CHEMICAL FAMILY

CHEMICAL NAME AND SYNONYMS: Methanamide | FORMULA: $HCONH_2$

SECTION II HAZARDOUS INGREDIENTS

BASIC MATERIAL	APPROXIMATE OR MAXIMUM % WT. OR VOL.	ESTABLISHED OSHA STANDARD	LD_{50} mg/kg ORAL	PERCUT.	SPECIES	LC_{50} CONC.
			7500		Rat	
Intramuscular - guinea pig						
LD50 2539 mg/kg.						

SECTION III PHYSICAL DATA MW - 45.04

BOILING POINT °F.: 211°C | VAPOR PRESSURE mm Hg.: at 130°C 30

MELTING POINT °F.: 2.6°C | VAPOR DENSITY (Air = 1)

SPECIFIC GRAVITY (H_2O = 1): 1.13 | EVAPORATION RATE (= 1)

SOLUBILITY IN WATER Pts/100 pts H_2O: very soluble | VOLATILE % Vol. % Wt.

APPEARANCE AND ODOR: Colorless, hygroscopic, oily, liquid, odorless

SECTION IV FIRE AND EXPLOSION HAZARD DATA

FLASH POINT: 310°F | FLAMMABLE (EXPLOSIVE) LIMITS | UPPER | LOWER

METHOD USED: (COC)

EXTINGUISHING MEDIA: Water, alcohol foam, carbon dioxide, dry chemical

SPECIAL FIRE FIGHTING PROCEDURES

UNUSUAL FIRE AND EXPLOSION HAZARDS

SECTION V HEALTH HAZARD DATA

TOXIC LEVEL: TLV - 20 PPM, 30 mg/kg. | Teratogen | CARCINOGENIC: no

PRINCIPLE ROUTES OF ABSORBTION: Ingestion, inhalation | SKIN AND EYE IRRITATION: moderate

REVELANT SYMPTOMS OF EXPOSURE: Drowsiness, fatigue, nausea, acidosis, skin eruptions.

EFFECTS OF CHRONIC EXPOSURE: --

EMERGENCY AND FIRST AID PROCEDURES: If swallowed - induce vomiting, get medical aid; inhalation - rest; wash from skin & eyes.

SECTION VI REACTIVITY DATA

CONDITIONS CONTRIBUTING TO INSTABILITY

CONDITIONS CONTRIBUTING TO HAZARDOUS POLYMERIZATION

INCOMPATABILITY (MATERIALS TO AVOID)

HAZARDOUS DECOMPOSITION PRODUCTS

SECTION VII SPILL OR LEAK PROCEDURES

STEPS TO BE TAKEN IN CASE MATERIAL IS RELEASED OR SPILLED

WASTE DISPOSAL METHOD

SECTION VIII SPECIAL PROTECTION INFORMATION

VENTILATION REQUIREMENTS LOCAL EXHAUST | PROTECTIVE EQUIPMENT (SPECIFY TYPES) EYE: goggles

MECHANICAL (GENERAL) | GLOVES: rubber

SPECIAL | RESPIRATOR

OTHER PROTECTIVE EQUIPMENT

SECTION IX SPECIAL PRECAUTIONS

PRECAUTIONS TO BE TAKEN IN HANDLING AND STORAGE: Stored in closed container away from moisture.

OTHER PRECAUTIONS

SIGNATURE | DATE

FORMANILIDE

Hazard diamond: 1 (top), 3 (left), 0 (right)

SECTION I SOURCE AND NOMENCLATURE

MANUFACTURER'S NAME | EMERGENCY TELEPHONE NO.

ADDRESS (NUMBER, STREET, CITY, STATE, ZIP CODE)

TRADE NAME AND SYNONYMS | CHEMICAL FAMILY

CHEMICAL NAME AND SYNONYMS: Phenylformamide | FORMULA: $HCONHC_6H_5$

SECTION II HAZARDOUS INGREDIENTS

BASIC MATERIAL	APPROXIMATE OR MAXIMUM % WT. OR VOL.	ESTABLISHED OSHA STANDARD	LD_{50} ORAL	PERCUT.	SPECIES	LC_{50} CONC.
LDLO - oral - dog 400 mg/kg						

SECTION III PHYSICAL DATA MW - 121.13

BOILING POINT °F.: 271°C | VAPOR PRESSURE mm Hg.: 216 120

MELTING POINT °F.: 50°C | VAPOR DENSITY (Air = 1)

SPECIFIC GRAVITY (H_2O = 1): 1.15 | EVAPORATION RATE (= 1)

SOLUBILITY IN WATER Pts/100 pts H_2O: slightly soluble | VOLATILE % Vol. % Wt.

APPEARANCE AND ODOR: crystals

SECTION IV FIRE AND EXPLOSION HAZARD DATA

FLASH POINT: none available | FLAMMABLE (EXPLOSIVE) LIMITS | UPPER | LOWER

METHOD USED

EXTINGUISHING MEDIA: Foam, carbon dioxide, dry chemical

SPECIAL FIRE FIGHTING PROCEDURES

UNUSUAL FIRE AND EXPLOSION HAZARDS

SECTION V HEALTH HAZARD DATA

TOXIC LEVEL: None available | CARCINOGENIC: no

PRINCIPLE ROUTES OF ABSORBTION: Inhalation of dust, ingestion | SKIN AND EYE IRRITATION: moderate

REVELANT SYMPTOMS OF EXPOSURE: Cyanosis, headache, dizziness, confusion, blood pressure fall, convulsions and coma.

EFFECTS OF CHRONIC EXPOSURE: Weight loss, anemia, weakness and irritability

EMERGENCY AND FIRST AID PROCEDURES: If swallowed - induce vomiting, get medical aid; inhalation - give oxygen as needed; wash skin & eyes.

SECTION VI REACTIVITY DATA

CONDITIONS CONTRIBUTING TO INSTABILITY

CONDITIONS CONTRIBUTING TO HAZARDOUS POLYMERIZATION

INCOMPATABILITY (MATERIALS TO AVOID)

HAZARDOUS DECOMPOSITION PRODUCTS

SECTION VII SPILL OR LEAK PROCEDURES

STEPS TO BE TAKEN IN CASE MATERIAL IS RELEASED OR SPILLED

WASTE DISPOSAL METHOD

SECTION VIII SPECIAL PROTECTION INFORMATION

VENTILATION REQUIREMENTS LOCAL EXHAUST | PROTECTIVE EQUIPMENT (SPECIFY TYPES) EYE: goggles

MECHANICAL (GENERAL) | GLOVES: rubber

SPECIAL | RESPIRATOR: with proper filter

OTHER PROTECTIVE EQUIPMENT

SECTION IX SPECIAL PRECAUTIONS

PRECAUTIONS TO BE TAKEN IN HANDLING AND STORAGE

OTHER PRECAUTIONS

SIGNATURE | DATE

FUMARYL CHLORIDE

Hazard diamond: 2 (top), 4 (left), 2 (right), W (bottom)

SECTION I SOURCE AND NOMENCLATURE

MANUFACTURER'S NAME

EMERGENCY TELEPHONE NO.

ADDRESS (NUMBER, STREET, CITY, STATE, ZIP CODE)

TRADE NAME AND SYNONYMS

CHEMICAL FAMILY

CHEMICAL NAME AND SYNONYMS: Trans-Butenedioyl chloride

FORMULA: $(ClOCCH)_2$

SECTION II HAZARDOUS INGREDIENTS

BASIC MATERIAL	APPROXIMATE OR MAXIMUM % WT. OR VOL.	ESTABLISHED OSHA STANDARD	LD_{50} ORAL	LD_{50} PERCUT.	LC_{50} SPECIES	LC_{50} CONC.
Lowest published lethal concentration in air						
inhalation - rat 3000 mg/C.M.						
LCLO - inhalation mouse 1000 mg/C.M./10 minutes						

SECTION III PHYSICAL DATA MW - 153

BOILING POINT: 160°C

VAPOR PRESSURE: mm Hg.

MELTING POINT: °F

VAPOR DENSITY (Air = 1)

SPECIFIC GRAVITY (H_2O = 1): 1.408

EVAPORATION RATE (= 1)

SOLUBILITY IN WATER Pts/100 pts H_2O: decomposes in water

VOLATILE % Vol. % Wt.

APPEARANCE AND ODOR: Clear, straw-colored liquid, pungent odor, fumes in moist air.

SECTION IV FIRE AND EXPLOSION HAZARD DATA

FLASH POINT: Unknown

FLAMMABLE (EXPLOSIVE) LIMITS: UPPER / LOWER

METHOD USED

EXTINGUISHING MEDIA: Alcohol foam, carbon dioxide, dry chemical. Keep away from water.

SPECIAL FIRE FIGHTING PROCEDURES

UNUSUAL FIRE AND EXPLOSION HAZARDS: Reacts with water or steam to produce HCl.

SECTION V HEALTH HAZARD DATA

TOXIC LEVEL: none available

CARCINOGENIC: no

PRINCIPLE ROUTES OF ABSORBTION: Inhalation, ingestion

SKIN AND EYE IRRITATION: severe

REVELANT SYMPTOMS OF EXPOSURE: Cyanosis, coughing, headache, pulmonary edema, ulceration of skin, dizziness.

EFFECTS OF CHRONIC EXPOSURE: Erosion of teeth, jaw necrosis, lung irritation and cough.

EMERGENCY AND FIRST AID PROCEDURES: Inhalation - rest; if swallowed - drink water, milk of magnesia, get medical aid; wash skin & eyes.

SECTION VI REACTIVITY DATA

CONDITIONS CONTRIBUTING TO INSTABILITY

CONDITIONS CONTRIBUTING TO HAZARDOUS POLYMERIZATION

INCOMPATABILITY (MATERIALS TO AVOID): Reacts violently with water, ethal alcohol, dimethyl sulfoxide.

HAZARDOUS DECOMPOSITION PRODUCTS: Evolves phosgene when heated.

SECTION VII SPILL OR LEAK PROCEDURES

STEPS TO BE TAKEN IN CASE MATERIAL IS RELEASED OR SPILLED

WASTE DISPOSAL METHOD

SECTION VIII SPECIAL PROTECTION INFORMATION

VENTILATION REQUIREMENTS LOCAL EXHAUST

MECHANICAL (GENERAL)

SPECIAL

PROTECTIVE EQUIPMENT (SPECIFY TYPES) EYE: goggles

GLOVES: rubber

RESPIRATOR: with proper filter

OTHER PROTECTIVE EQUIPMENT

SECTION IX SPECIAL PRECAUTIONS

PRECAUTIONS TO BE TAKEN IN HANDLING AND STORAGE

OTHER PRECAUTIONS

SIGNATURE

DATE

FUROYL CHLORIDE

Hazard diamond: 2 (top), 4 (left), 2 (right), W (bottom)

SECTION I SOURCE AND NOMENCLATURE

MANUFACTURER'S NAME

EMERGENCY TELEPHONE NO.

ADDRESS (NUMBER, STREET, CITY, STATE, ZIP CODE)

TRADE NAME AND SYNONYMS

CHEMICAL FAMILY

CHEMICAL NAME AND SYNONYMS: Furoyl chloride

FORMULA: C_4H_3OCOCl

SECTION II HAZARDOUS INGREDIENTS

BASIC MATERIAL	APPROXIMATE OR MAXIMUM % WT. OR VOL.	ESTABLISHED OSHA STANDARD	LD_{50} ORAL	LD_{50} PERCUT.	LC_{50} SPECIES	LC_{50} CONC.
			none available			

SECTION III PHYSICAL DATA MW - 130.53

BOILING POINT: 176°C

VAPOR PRESSURE: mm Hg.

MELTING POINT: -2°C

VAPOR DENSITY (Air = 1)

SPECIFIC GRAVITY (H_2O = 1)

EVAPORATION RATE (= 1)

SOLUBILITY IN WATER Pts/100 pts H_2O: decomposes

VOLATILE % Vol. % Wt.

APPEARANCE AND ODOR: colorless liquid

SECTION IV FIRE AND EXPLOSION HAZARD DATA

FLASH POINT: none available

FLAMMABLE (EXPLOSIVE) LIMITS: UPPER / LOWER

METHOD USED

EXTINGUISHING MEDIA: alcohol foam, carbon dioxide, dry chemical

SPECIAL FIRE FIGHTING PROCEDURES

UNUSUAL FIRE AND EXPLOSION HAZARDS: reacts with water or steam.

SECTION V HEALTH HAZARD DATA

TOXIC LEVEL: none available

CARCINOGENIC: no

PRINCIPLE ROUTES OF ABSORBTION: inhalation, ingestion

SKIN AND EYE IRRITATION: severe

REVELANT SYMPTOMS OF EXPOSURE: Irritation of eyes and respiratory system; if swallowed - severe digestive tract attack; eye and skin burns.

EFFECTS OF CHRONIC EXPOSURE: Bronchial irritation, erosion of teeth

EMERGENCY AND FIRST AID PROCEDURES: If swallowed - drink water, milk of magnesia, get medical aid; inhalation - rest; wash skin & eyes.

SECTION VI REACTIVITY DATA

CONDITIONS CONTRIBUTING TO INSTABILITY

CONDITIONS CONTRIBUTING TO HAZARDOUS POLYMERIZATION

INCOMPATABILITY (MATERIALS TO AVOID): May react violently with water or alcohol

HAZARDOUS DECOMPOSITION PRODUCTS

SECTION VII SPILL OR LEAK PROCEDURES

STEPS TO BE TAKEN IN CASE MATERIAL IS RELEASED OR SPILLED

WASTE DISPOSAL METHOD

SECTION VIII SPECIAL PROTECTION INFORMATION

VENTILATION REQUIREMENTS LOCAL EXHAUST

MECHANICAL (GENERAL)

SPECIAL

PROTECTIVE EQUIPMENT (SPECIFY TYPES) EYE: Goggles

GLOVES: Rubber

RESPIRATOR: Self-contained

OTHER PROTECTIVE EQUIPMENT: rubber apron

SECTION IX SPECIAL PRECAUTIONS

PRECAUTIONS TO BE TAKEN IN HANDLING AND STORAGE

OTHER PRECAUTIONS

SIGNATURE

DATE

GOLD CYANIDE

Hazard diamond: 0 (top), 3 (left), 0 (right)

SECTION I SOURCE AND NOMENCLATURE

MANUFACTURER'S NAME | EMERGENCY TELEPHONE NO.

ADDRESS (NUMBER, STREET, CITY, STATE, ZIP CODE)

TRADE NAME AND SYNONYMS | CHEMICAL FAMILY

CHEMICAL NAME AND SYNONYMS: Gold Cyanide | FORMULA: $HAu(CN)_4 \cdot 3H_2O$

SECTION II HAZARDOUS INGREDIENTS

BASIC MATERIAL	APPROXIMATE OR MAXIMUM % WT. OR VOL.	ESTABLISHED OSHA STANDARD	LD_{50} ORAL	LD_{50} PERCUT.	LC_{50} SPECIES	LC_{50} CONC.

SECTION III PHYSICAL DATA MW - 356.33

BOILING POINT °F. | VAPOR PRESSURE mm Hg.

MELTING POINT °F.: D 50°C. | VAPOR DENSITY (Air = 1)

SPECIFIC GRAVITY (H_2O = 1): 7.12 | EVAPORATION RATE (=1)

SOLUBILITY IN WATER Pts/100 pts H_2O: Very soluble | VOLATILE % Vol. % Wt.

APPEARANCE AND ODOR: Light yellow, crystalline powder.

SECTION IV FIRE AND EXPLOSION HAZARD DATA

FLASH POINT: Not flammable | FLAMMABLE (EXPLOSIVE) LIMITS: UPPER / LOWER

METHOD USED

EXTINGUISHING MEDIA

SPECIAL FIRE FIGHTING PROCEDURES

UNUSUAL FIRE AND EXPLOSION HAZARDS

SECTION V HEALTH HAZARD DATA

TOXIC LEVEL: TLV 5 mg/C.M. (CN) | CARCINOGENIC: No

PRINCIPLE ROUTES OF ABSORBTION: Inhalation, skin and ingestion | SKIN AND EYE IRRITATION: Moderate

REVELANT SYMPTOMS OF EXPOSURE: Giddiness, dizziness, rapid heart action followed by unconsciousness; if swallowed - rapid respiration becoming slow.

EFFECTS OF CHRONIC EXPOSURE: Dizziness, weakness, lung congestion, appetite loss, weight loss.

EMERGENCY AND FIRST AID PROCEDURES: If inhaled - breathe amyl nitrite, artificial respiration and oxygen as needed; if swallowed same; wash from skin & eyes.

SECTION VI REACTIVITY DATA

CONDITIONS CONTRIBUTING TO INSTABILITY

CONDITIONS CONTRIBUTING TO HAZARDOUS POLYMERIZATION

INCOMPATABILITY (MATERIALS TO AVOID): Avoid acids.

HAZARDOUS DECOMPOSITION PRODUCTS

SECTION VII SPILL OR LEAK PROCEDURES

STEPS TO BE TAKEN IN CASE MATERIAL IS RELEASED OR SPILLED

WASTE DISPOSAL METHOD

SECTION VIII SPECIAL PROTECTION INFORMATION

VENTILATION REQUIREMENTS LOCAL EXHAUST | PROTECTIVE EQUIPMENT (SPECIFY TYPES) EYE: Face shield

MECHANICAL (GENERAL) | GLOVES: Rubber

SPECIAL | RESPIRATOR: With proper filter

OTHER PROTECTIVE EQUIPMENT: Overalls and aprons

SECTION IX SPECIAL PRECAUTIONS

PRECAUTIONS TO BE TAKEN IN HANDLING AND STORAGE: Store in cool, dry place away from acids,

OTHER PRECAUTIONS

SIGNATURE | DATE

GOLD POTASSIUM CYANIDE

Hazard diamond: 0 (top), 3 (left), 0 (right)

SECTION I SOURCE AND NOMENCLATURE

MANUFACTURER'S NAME | EMERGENCY TELEPHONE NO.

ADDRESS (NUMBER, STREET, CITY, STATE, ZIP CODE)

TRADE NAME AND SYNONYMS | CHEMICAL FAMILY

CHEMICAL NAME AND SYNONYMS: Gold Potassium Cyanide | FORMULA: $KAu(CN)_2 . 2HO$

SECTION II HAZARDOUS INGREDIENTS

BASIC MATERIAL	APPROXIMATE OR MAXIMUM % WT. OR VOL.	ESTABLISHED OSHA STANDARD	LD_{50} ORAL	LD_{50} PERCUT.	LC_{50} SPECIES	LC_{50} CONC.

SECTION III PHYSICAL DATA MW - 324.36

BOILING POINT °F. | VAPOR PRESSURE mm Hg.

MELTING POINT °F. | VAPOR DENSITY (Air = 1)

SPECIFIC GRAVITY (H_2O = 1) | EVAPORATION RATE (=1)

SOLUBILITY IN WATER Pts/100 pts H_2O | VOLATILE % Vol. % Wt.

APPEARANCE AND ODOR: White crystalline powder

SECTION IV FIRE AND EXPLOSION HAZARD DATA

FLASH POINT: Not flammable | FLAMMABLE (EXPLOSIVE) LIMITS: UPPER / LOWER

METHOD USED

EXTINGUISHING MEDIA

SPECIAL FIRE FIGHTING PROCEDURES

UNUSUAL FIRE AND EXPLOSION HAZARDS

SECTION V HEALTH HAZARD DATA

TOXIC LEVEL: TLV - 5 mg/C.M. (CN) | CARCINOGENIC: No

PRINCIPLE ROUTES OF ABSORBTION: Inhalation, ingestion, skin | SKIN AND EYE IRRITATION: Moderate

REVELANT SYMPTOMS OF EXPOSURE: Giddiness, dizziness, rapid heart action followed by unconsciousness; if swallowed - rapid respiration - becoming slow, coma.

EFFECTS OF CHRONIC EXPOSURE: Dizziness, weakness, lung congestion, appetite loss, weight loss.

EMERGENCY AND FIRST AID PROCEDURES: If inhaled or ingested - breathe amyl nitrite - artificial respiration and oxygen as needed.

SECTION VI REACTIVITY DATA

CONDITIONS CONTRIBUTING TO INSTABILITY

CONDITIONS CONTRIBUTING TO HAZARDOUS POLYMERIZATION

INCOMPATABILITY (MATERIALS TO AVOID)

HAZARDOUS DECOMPOSITION PRODUCTS: Avoid contact with acids, produces HCN.

SECTION VII SPILL OR LEAK PROCEDURES

STEPS TO BE TAKEN IN CASE MATERIAL IS RELEASED OR SPILLED

WASTE DISPOSAL METHOD

SECTION VIII SPECIAL PROTECTION INFORMATION

VENTILATION REQUIREMENTS LOCAL EXHAUST | PROTECTIVE EQUIPMENT (SPECIFY TYPES) EYE: Face shield

MECHANICAL (GENERAL) | GLOVES: Rubber

SPECIAL | RESPIRATOR: With proper filter

OTHER PROTECTIVE EQUIPMENT: Overalls and aprons.

SECTION IX SPECIAL PRECAUTIONS

PRECAUTIONS TO BE TAKEN IN HANDLING AND STORAGE: Store in cool, dry place away from acids.

OTHER PRECAUTIONS

SIGNATURE | DATE

HEXACHLOROETHANE

2
4 0

SECTION I SOURCE AND NOMENCLATURE

MANUFACTURER'S NAME | EMERGENCY TELEPHONE NO.

ADDRESS (NUMBER, STREET, CITY, STATE, ZIP CODE)

TRADE NAME AND SYNONYMS | CHEMICAL FAMILY

CHEMICAL NAME AND SYNONYMS: Carbon Hexachloride | FORMULA: C_2Cl_6

SECTION II HAZARDOUS INGREDIENTS

BASIC MATERIAL	APPROXIMATE OR MAXIMUM % WT. OR VOL.	ESTABLISHED OSHA STANDARD	LD_{50} ORAL	LD_{50} PERCUT.	LC_{50} SPECIES	LC_{50} CONC.
Oral - human LDLO - 50 mg/kg						
LDLO intravenous - dog 325 mg/kg						
LDLO subcutaneous - rabbit - 4000 mg/kg.						

SECTION III PHYSICAL DATA MW - 236.76

BOILING POINT °F.: 186°C | VAPOR PRESSURE mm Hg.: 1 at 33°C

MELTING POINT °F. | VAPOR DENSITY (Air = 1)

SPECIFIC GRAVITY (H_2O = 1): 2.09 | EVAPORATION RATE (= 1)

SOLUBILITY IN WATER Pts/100 pts H_2O: insoluble | VOLATILE % Vol. % Wt.

APPEARANCE AND ODOR: colorless crystals with camphor-like odor.

SECTION IV FIRE AND EXPLOSION HAZARD DATA

FLASH POINT: none available | FLAMMABLE (EXPLOSIVE) LIMITS: UPPER / LOWER

METHOD USED

EXTINGUISHING MEDIA: Alcohol foam, carbon dioxide, dry chemical.

SPECIAL FIRE FIGHTING PROCEDURES

UNUSUAL FIRE AND EXPLOSION HAZARDS

SECTION V HEALTH HAZARD DATA

TOXIC LEVEL: TLV - 1 PPM, 10 mg/C.M. skin | CARCINOGENIC: no

PRINCIPLE ROUTES OF ABSORBTION: Inhalation, ingestion | SKIN AND EYE IRRITATION: moderate

REVELANT SYMPTOMS OF EXPOSURE: Eyes, skin, mucous membrane and lung irritation; narcotic in high concentrations.

EFFECTS OF CHRONIC EXPOSURE: Liver injury

EMERGENCY AND FIRST AID PROCEDURES: If swallowed - give emetic, get medical aid; inhalation - rest; wash from skin & eyes.

SECTION VI REACTIVITY DATA

CONDITIONS CONTRIBUTING TO INSTABILITY

CONDITIONS CONTRIBUTING TO HAZARDOUS POLYMERIZATION

INCOMPATABILITY (MATERIALS TO AVOID): Avoid alkalies and metals.

HAZARDOUS DECOMPOSITION PRODUCTS

SECTION VII SPILL OR LEAK PROCEDURES

STEPS TO BE TAKEN IN CASE MATERIAL IS RELEASED OR SPILLED

WASTE DISPOSAL METHOD

SECTION VIII SPECIAL PROTECTION INFORMATION

VENTILATION REQUIREMENTS LOCAL EXHAUST | PROTECTIVE EQUIPMENT (SPECIFY TYPES) EYE: goggles

MECHANICAL (GENERAL) | GLOVES: rubber

SPECIAL | RESPIRATOR: with proper filter

OTHER PROTECTIVE EQUIPMENT: rubber apron

SECTION IX SPECIAL PRECAUTIONS

PRECAUTIONS TO BE TAKEN IN HANDLING AND STORAGE

OTHER PRECAUTIONS

SIGNATURE | DATE

HYDRAZINE DIHYDROCHLORIDE

2
4 2

SECTION I SOURCE AND NOMENCLATURE

MANUFACTURER'S NAME | EMERGENCY TELEPHONE NO.

ADDRESS (NUMBER, STREET, CITY, STATE, ZIP CODE)

TRADE NAME AND SYNONYMS | CHEMICAL FAMILY

CHEMICAL NAME AND SYNONYMS: Hydrazine Dihydrochloride | FORMULA: N_2H_4 2HCL

SECTION II HAZARDOUS INGREDIENTS

BASIC MATERIAL	APPROXIMATE OR MAXIMUM % WT. OR VOL.	ESTABLISHED OSHA STANDARD	LD_{50} ORAL	LD_{50} PERCUT.	LC_{50} SPECIES	LC_{50} CONC.
			none	available		

SECTION III PHYSICAL DATA MW - 104.98

BOILING POINT °F.: -- | VAPOR PRESSURE mm Hg.

MELTING POINT °F.: 198°C | VAPOR DENSITY (Air = 1)

SPECIFIC GRAVITY (H_2O = 1): 1.42 | EVAPORATION RATE (= 1)

SOLUBILITY IN WATER Pts/100 pts H_2O: soluble | VOLATILE % Vol. % Wt.

APPEARANCE AND ODOR: white crystals

SECTION IV FIRE AND EXPLOSION HAZARD DATA

FLASH POINT: none available | FLAMMABLE (EXPLOSIVE) LIMITS: UPPER / LOWER

METHOD USED

EXTINGUISHING MEDIA: Flood with water; dry chemical and carbon dioxide may be used to extinguish fire.

SPECIAL FIRE FIGHTING PROCEDURES: Wear full protective clothing.

UNUSUAL FIRE AND EXPLOSION HAZARDS

SECTION V HEALTH HAZARD DATA

TOXIC LEVEL: None available | CARCINOGENIC: no

PRINCIPLE ROUTES OF ABSORBTION: Inhalation, ingestion, skin | SKIN AND EYE IRRITATION: moderate

REVELANT SYMPTOMS OF EXPOSURE: Ingestion - vomiting and convulsions; inhalation - vomiting, respiratory distress, convulsions, eye damage.

EFFECTS OF CHRONIC EXPOSURE: Liver damage, hemolytic anemia and convulsive seizures, lung and kidney damage, allergic reaction.

EMERGENCY AND FIRST AID PROCEDURES: If swallowed - drink water, get medical aid; inhalation - artificial respiration, oxygen; wash skin & eyes.

SECTION VI REACTIVITY DATA

CONDITIONS CONTRIBUTING TO INSTABILITY

CONDITIONS CONTRIBUTING TO HAZARDOUS POLYMERIZATION

INCOMPATABILITY (MATERIALS TO AVOID): avoid oxidizing agents.

HAZARDOUS DECOMPOSITION PRODUCTS

SECTION VII SPILL OR LEAK PROCEDURES

STEPS TO BE TAKEN IN CASE MATERIAL IS RELEASED OR SPILLED

WASTE DISPOSAL METHOD

SECTION VIII SPECIAL PROTECTION INFORMATION

VENTILATION REQUIREMENTS LOCAL EXHAUST | PROTECTIVE EQUIPMENT (SPECIFY TYPES) EYE: goggles

MECHANICAL (GENERAL) | GLOVES: rubber

SPECIAL | RESPIRATOR: with proper filter

OTHER PROTECTIVE EQUIPMENT

SECTION IX SPECIAL PRECAUTIONS

PRECAUTIONS TO BE TAKEN IN HANDLING AND STORAGE: Outside storage preferred; store in standard flammable liquids storage cabinet. Keep separate from oxidizing materials.

OTHER PRECAUTIONS

SIGNATURE | DATE

HYDRAZINE HYDRATE

Hazard diamond: 2 (top), 4 (left), 2 (right)

SECTION I SOURCE AND NOMENCLATURE

MANUFACTURER'S NAME | EMERGENCY TELEPHONE NO.

ADDRESS (NUMBER, STREET, CITY, STATE, ZIP CODE)

TRADE NAME AND SYNONYMS | CHEMICAL FAMILY

CHEMICAL NAME AND SYNONYMS: Hydrazine hydrate | FORMULA: $N_2H_4H_2O$

SECTION II HAZARDOUS INGREDIENTS

BASIC MATERIAL	APPROXIMATE OR MAXIMUM % WT. OR VOL.	ESTABLISHED OSHA STANDARD	LD_{50} mg/kg ORAL	PERCUT.	SPECIES	LC_{50} CONC.
			129		Rat	
#1 LD_{50} intravenous - rat 25 mg/kg.			83		Mouse	
			55		Rabbit	
			40		Guinea Pig	

SECTION III PHYSICAL DATA MW 50.88

BOILING POINT (°F): 113.5°C

VAPOR PRESSURE (mm Hg): 39.5 at 118.5°C 39.5

MELTING POINT (°F): -51.7°C

VAPOR DENSITY (Air = 1): 1 at 100°F

SPECIFIC GRAVITY (H_2O = 1): 1.03

EVAPORATION RATE (______ = 1)

SOLUBILITY IN WATER ∞ Pts/100 pts H_2O

VOLATILE % Vol. % Wt.

APPEARANCE AND ODOR: colorless - ammonical and fishy odor.

SECTION IV FIRE AND EXPLOSION HAZARD DATA

FLASH POINT: 126°C

METHOD USED: O.C.

FLAMMABLE (EXPLOSIVE) LIMITS: UPPER 100%; LOWER 4.7%

EXTINGUISHING MEDIA: Water, dry chemicals and CO_2, alcohol foam -

SPECIAL FIRE FIGHTING PROCEDURES: Water spray to disperse vapors. Wear full protective clothing.

UNUSUAL FIRE AND EXPLOSION HAZARDS: Spontaneous ignition with oxidants such as nitric acid and hydrogen peroxide.

SECTION V HEALTH HAZARD DATA

TOXIC LEVEL: TLV - 1 PPM, 1.3 mg/C.M.

CARCINOGENIC: yes

PRINCIPLE ROUTES OF ABSORPTION: Ingestion, inhalation, skin absorption

SKIN AND EYE IRRITATION: Severe on liquid contact

REVELANT SYMPTOMS OF EXPOSURE: Inhalation - vomiting, respiratory distress and convulsions; direct skin contact - burns; ingestion - vomiting and convulsions, eye damage.

EFFECTS OF CHRONIC EXPOSURE: Lung and kidney damage, allergic reaction, Liver damage, hemolytic anemia and convulsive seizures.

EMERGENCY AND FIRST AID PROCEDURES

SECTION VI REACTIVITY DATA

CONDITIONS CONTRIBUTING TO INSTABILITY: Heat and strong oxidizing agents.

CONDITIONS CONTRIBUTING TO HAZARDOUS POLYMERIZATION

INCOMPATABILITY (MATERIALS TO AVOID): Spontaneous ignition with strong oxidants such as nitric acid or H_2O_2.

HAZARDOUS DECOMPOSITION PRODUCTS

SECTION VII SPILL OR LEAK PROCEDURES

STEPS TO BE TAKEN IN CASE MATERIAL IS RELEASED OR SPILLED: Use water spray to disperse vapors.

WASTE DISPOSAL METHOD: Dilute to non-inflammable mixture.

SECTION VIII SPECIAL PROTECTION INFORMATION

VENTILATION REQUIREMENTS LOCAL EXHAUST: Use in hood

PROTECTIVE EQUIPMENT (SPECIFY TYPES) EYE: Face shield

MECHANICAL (GENERAL)

GLOVES: Rubber gloves

SPECIAL

RESPIRATOR: With appropriate filter

OTHER PROTECTIVE EQUIPMENT: Full protective clothing recommended.

SECTION IX SPECIAL PRECAUTIONS

PRECAUTIONS TO BE TAKEN IN HANDLING AND STORAGE: Maintain under nitrogen atmosphere.

OTHER PRECAUTIONS: Keep away from strong oxidizing agents.

SIGNATURE | DATE

HYDRAZINE SULFATE

Hazard diamond: 2 (top), 4 (left), 2 (right)

SECTION I SOURCE AND NOMENCLATURE

MANUFACTURER'S NAME | EMERGENCY TELEPHONE NO.

ADDRESS (NUMBER, STREET, CITY, STATE, ZIP CODE)

TRADE NAME AND SYNONYMS | CHEMICAL FAMILY

CHEMICAL NAME AND SYNONYMS: Hydrazine sulfate | FORMULA: $N_2H_4H_2SO_4$

SECTION II HAZARDOUS INGREDIENTS

BASIC MATERIAL	APPROXIMATE OR MAXIMUM % WT. OR VOL.	ESTABLISHED OSHA STANDARD	LD_{50} mg/kg ORAL	PERCUT.	SPECIES	LC_{50} CONC.
Oral - human TDLO 20 mg/C.M.			601		rat	
lowest published - lethal dose						
subcutaneous - rat 250 mg/kg						
intraperitoneal - mouse 250 mg/kg.						

SECTION III PHYSICAL DATA

BOILING POINT (°F): -

VAPOR PRESSURE (mm Hg)

MELTING POINT (°F): 254°C

VAPOR DENSITY (Air = 1)

SPECIFIC GRAVITY (H_2O = 1): 1.378

EVAPORATION RATE (______ = 1)

SOLUBILITY IN WATER Pts/100 pts H_2O: very soluble

VOLATILE % Vol. % Wt.

APPEARANCE AND ODOR: colorless crystal

SECTION IV FIRE AND EXPLOSION HAZARD DATA

FLASH POINT: N/A

METHOD USED

FLAMMABLE (EXPLOSIVE) LIMITS: UPPER; LOWER

EXTINGUISHING MEDIA: Water, dry chemical and CO_2

SPECIAL FIRE FIGHTING PROCEDURES: Water spray to disperse hydrazine vapor - wear full protective clothing.

UNUSUAL FIRE AND EXPLOSION HAZARDS: May react vigorously with strong oxidizing agents.

SECTION V HEALTH HAZARD DATA

TOXIC LEVEL: none available

CARCINOGENIC: carcinogenic

PRINCIPLE ROUTES OF ABSORPTION: ingestion, skin

SKIN AND EYE IRRITATION: severe on direct contact

REVELANT SYMPTOMS OF EXPOSURE: Burns - from direct skin contact. Ingestion - vomiting and convulsions, eye damage, respiratory distress.

EFFECTS OF CHRONIC EXPOSURE: Lung and kidney damage, allergic reaction Liver damage, hemolytic, anemia and convulsive seizures.

EMERGENCY AND FIRST AID PROCEDURES: Skin contact - drench with water; if swallowed - wash mouth, give water to drink, get medical aid; inhalation - oxygen or artificial respiration as necessary.

SECTION VI REACTIVITY DATA

CONDITIONS CONTRIBUTING TO INSTABILITY: Heat and strong oxidizing agents.

CONDITIONS CONTRIBUTING TO HAZARDOUS POLYMERIZATION

INCOMPATABILITY (MATERIALS TO AVOID): Reacts vigorously with strong oxidizing agents.

HAZARDOUS DECOMPOSITION PRODUCTS

SECTION VII SPILL OR LEAK PROCEDURES

STEPS TO BE TAKEN IN CASE MATERIAL IS RELEASED OR SPILLED: flush with water

WASTE DISPOSAL METHOD

SECTION VIII SPECIAL PROTECTION INFORMATION

VENTILATION REQUIREMENTS LOCAL EXHAUST

PROTECTIVE EQUIPMENT (SPECIFY TYPES) EYE: Face shield

MECHANICAL (GENERAL)

GLOVES: Rubber

SPECIAL

RESPIRATOR: With proper filter

OTHER PROTECTIVE EQUIPMENT: Rubber apron, full face shield.

SECTION IX SPECIAL PRECAUTIONS

PRECAUTIONS TO BE TAKEN IN HANDLING AND STORAGE

OTHER PRECAUTIONS: Keep away from strong oxidizing agents

SIGNATURE | DATE

HYDRIODIC ACID

Hazard diamond: 0 (top), 4 (left), 0 (right)

SECTION I SOURCE AND NOMENCLATURE

MANUFACTURER'S NAME	EMERGENCY TELEPHONE NO.
ADDRESS (NUMBER, STREET, CITY, STATE, ZIP CODE)	
TRADE NAME AND SYNONYMS	CHEMICAL FAMILY
CHEMICAL NAME AND SYNONYMS: Hydriotic Acid	FORMULA: HI

SECTION II HAZARDOUS INGREDIENTS

BASIC MATERIAL	APPROXIMATE OR MAXIMUM % WT. OR VOL.	ESTABLISHED OSHA STANDARD	LD_{50} ORAL	LD_{50} PERCUT.	LC_{50} SPECIES	LC_{50} CONC.
			Not	available		

SECTION III PHYSICAL DATA MW - 127.91

BOILING POINT °F: -35.38°C at m	VAPOR PRESSURE mm Hg.
MELTING POINT °F: -50.8°C.	VAPOR DENSITY (Air = 1): 4.4
SPECIFIC GRAVITY (H_2O = 1): 2.85	EVAPORATION RATE (= 1)
SOLUBILITY IN WATER Pts/100 pts H_2O	VOLATILE % Vol. % Wt.

APPEARANCE AND ODOR: Pale yellow liquid.

SECTION IV FIRE AND EXPLOSION HAZARD DATA

FLASH POINT: None available	FLAMMABLE (EXPLOSIVE) LIMITS	UPPER
METHOD USED		LOWER

EXTINGUISHING MEDIA: Use water, neutralize with chemically basic substances.

SPECIAL FIRE FIGHTING PROCEDURES: Wear full protective clothing.

UNUSUAL FIRE AND EXPLOSION HAZARDS

SECTION V HEALTH HAZARD DATA

TOXIC LEVEL: Suggest value 10 PPM — CARCINOGENIC: No

PRINCIPLE ROUTES OF ABSORBTION: Inhalation of vapor — SKIN AND EYE IRRITATION: Severe

REVELANT SYMPTOMS OF EXPOSURE: Conjunctivitis, pulmonary edema, coughing, choking, headache, dizziness, cyanosis.

EFFECTS OF CHRONIC EXPOSURE: Possible jaw necrosis, erosion of teeth, bronchial irritation with chronic cough.

EMERGENCY AND FIRST AID PROCEDURES: Ingestion - give milk of magnesia; get medical aid; inhalation - respiratory resuscitation as needed, wash skin & eyes - get medical aid.

SECTION VI REACTIVITY DATA

CONDITIONS CONTRIBUTING TO INSTABILITY

CONDITIONS CONTRIBUTING TO HAZARDOUS POLYMERIZATION

INCOMPATABILITY (MATERIALS TO AVOID): Fluorine, Perchloric Acid

HAZARDOUS DECOMPOSITION PRODUCTS

SECTION VII SPILL OR LEAK PROCEDURES

STEPS TO BE TAKEN IN CASE MATERIAL IS RELEASED OR SPILLED

WASTE DISPOSAL METHOD

SECTION VIII SPECIAL PROTECTION INFORMATION

VENTILATION REQUIREMENTS LOCAL EXHAUST	PROTECTIVE EQUIPMENT (SPECIFY TYPES) EYE: Goggles
MECHANICAL (GENERAL)	GLOVES: Neoprene rubber
SPECIAL	RESPIRATOR: W/appropriate filter

OTHER PROTECTIVE EQUIPMENT: Rubber apron, full face shield.

SECTION IX SPECIAL PRECAUTIONS

PRECAUTIONS TO BE TAKEN IN HANDLING AND STORAGE: Avoid physical damage.

OTHER PRECAUTIONS

SIGNATURE	DATE

HYDROFLUOSILICIC ACID

Hazard diamond: 0 (top), 4 (left), 0 (right)

SECTION I SOURCE AND NOMENCLATURE

MANUFACTURER'S NAME	EMERGENCY TELEPHONE NO.
ADDRESS (NUMBER, STREET, CITY, STATE, ZIP CODE)	
TRADE NAME AND SYNONYMS	CHEMICAL FAMILY
CHEMICAL NAME AND SYNONYMS: Fluorosilicic Acid	FORMULA: H_2SiF_6

SECTION II HAZARDOUS INGREDIENTS

BASIC MATERIAL	APPROXIMATE OR MAXIMUM % WT. OR VOL.	ESTABLISHED OSHA STANDARD	LD_{50} ORAL	LD_{50} PERCUT.	LC_{50} SPECIES	LC_{50} CONC.

#1 lowest published lethal dose - oral - guinea pig 200 mg/kg.

#2 lowest published lethal dose subcutaneous - guinea pig - 250 mg/kg.

SECTION III PHYSICAL DATA MW 144.11

BOILING POINT °F: N/A	VAPOR PRESSURE mm Hg.
MELTING POINT °F: N/A	VAPOR DENSITY (Air = 1)
SPECIFIC GRAVITY (H_2O = 1): N/A	EVAPORATION RATE (= 1)
SOLUBILITY IN WATER Pts/100 pts H_2O: S	VOLATILE % Vol. % Wt.

APPEARANCE AND ODOR: Clear liquid hydrolyzes when exposed to moist air.

SECTION IV FIRE AND EXPLOSION HAZARD DATA

FLASH POINT: N/A	FLAMMABLE (EXPLOSIVE) LIMITS	UPPER
METHOD USED		LOWER

EXTINGUISHING MEDIA: Water

SPECIAL FIRE FIGHTING PROCEDURES: Wear full protective clothing.

UNUSUAL FIRE AND EXPLOSION HAZARDS: Hydrogen may be formed when in contact with certain metals.

SECTION V HEALTH HAZARD DATA

TOXIC LEVEL: TLV 3 PPM (for HF), 2 Mg/CM — CARCINOGENIC: No

PRINCIPLE ROUTES OF ABSORBTION: Inhalation, skin absorption — SKIN AND EYE IRRITATION: Yes

REVELANT SYMPTOMS OF EXPOSURE: Irritation of respiratory system, severe skin or eye burns.

EFFECTS OF CHRONIC EXPOSURE: Weight loss - brittleness of bones, anemia, weakening, discoloration of teeth.

EMERGENCY AND FIRST AID PROCEDURES: Inhalation - rest - give oxygen if necessary, get medical attention; wash from eyes or skin, get medical attention; if swallowed, wash mouth - drink water, milk of magnesia, get medical attention.

SECTION VI REACTIVITY DATA

CONDITIONS CONTRIBUTING TO INSTABILITY

CONDITIONS CONTRIBUTING TO HAZARDOUS POLYMERIZATION

INCOMPATABILITY (MATERIALS TO AVOID): May react exothermally with As_2O_3 cr P_2O5

HAZARDOUS DECOMPOSITION PRODUCTS

SECTION VII SPILL OR LEAK PROCEDURES

STEPS TO BE TAKEN IN CASE MATERIAL IS RELEASED OR SPILLED

WASTE DISPOSAL METHOD

SECTION VIII SPECIAL PROTECTION INFORMATION

VENTILATION REQUIREMENTS LOCAL EXHAUST	PROTECTIVE EQUIPMENT (SPECIFY TYPES) EYE: Goggles
MECHANICAL (GENERAL)	GLOVES: Rubber gloves
SPECIAL	RESPIRATOR: W/appropriate filter

OTHER PROTECTIVE EQUIPMENT: Full face shield

SECTION IX SPECIAL PRECAUTIONS

PRECAUTIONS TO BE TAKEN IN HANDLING AND STORAGE

OTHER PRECAUTIONS

SIGNATURE	DATE

2-HYDROXY-1,4-DIMETHYLBENZENE

1 / 4 / 0

SECTION I SOURCE AND NOMENCLATURE

MANUFACTURER'S NAME | EMERGENCY TELEPHONE NO.

ADDRESS (NUMBER, STREET, CITY, STATE, ZIP CODE)

TRADE NAME AND SYNONYMS | CHEMICAL FAMILY

CHEMICAL NAME AND SYNONYMS: 2,5 Xylenol | FORMULA: $HO(C_6H_3)(CH_3)_2$

SECTION II HAZARDOUS INGREDIENTS

BASIC MATERIAL	APPROXIMATE OR MAXIMUM % WT. OR VOL.	ESTABLISHED OSHA STANDARD	LD_{50} ORAL	LD_{50} PERCUT.	LC_{50} SPECIES	LC_{50} CONC.
			444		rat	
			383		mouse	
			938		rabbit	

SECTION III PHYSICAL DATA MW - 122.16

BOILING POINT (°F.): $212^{\circ}C$ | VAPOR PRESSURE (mm Hg.)

MELTING POINT (°F.): $76^{\circ}C$ | VAPOR DENSITY (Air = 1)

SPECIFIC GRAVITY (H_2O = 1) | EVAPORATION RATE (______ = 1)

SOLUBILITY IN WATER Pts/100 pts H_2O: slightly soluble | VOLATILE % Vol. % Wt.

APPEARANCE AND ODOR: white crystal

SECTION IV FIRE AND EXPLOSION HAZARD DATA

FLASH POINT: none available | FLAMMABLE (EXPLOSIVE) LIMITS: UPPER / LOWER

METHOD USED

EXTINGUISHING MEDIA: Foam, carbon dioxide, dry chemical

SPECIAL FIRE FIGHTING PROCEDURES: Wear self-contained breathing apparatus.

UNUSUAL FIRE AND EXPLOSION HAZARDS

SECTION V HEALTH HAZARD DATA

TOXIC LEVEL: none available | CARCINOGENIC: yes

PRINCIPLE ROUTES OF ABSORBTION: inhalation, ingestion | SKIN AND EYE IRRITATION: severe

REVELANT SYMPTOMS OF EXPOSURE: Headache, dizziness, nausea, vomiting, stomach pain, exhaustion; skin & eye irritation.

EFFECTS OF CHRONIC EXPOSURE: Liver and kidney damage.

EMERGENCY AND FIRST AID PROCEDURES: If swallowed - drink epsom salt solution, get medical aid; inhalation - rest; wash skin & eyes.

SECTION VI REACTIVITY DATA

CONDITIONS CONTRIBUTING TO INSTABILITY

CONDITIONS CONTRIBUTING TO HAZARDOUS POLYMERIZATION

INCOMPATABILITY (MATERIALS TO AVOID)

HAZARDOUS DECOMPOSITION PRODUCTS

SECTION VII SPILL OR LEAK PROCEDURES

STEPS TO BE TAKEN IN CASE MATERIAL IS RELEASED OR SPILLED

WASTE DISPOSAL METHOD

SECTION VIII SPECIAL PROTECTION INFORMATION

VENTILATION REQUIREMENTS LOCAL EXHAUST | PROTECTIVE EQUIPMENT (SPECIFY TYPES) EYE: goggles

MECHANICAL (GENERAL) | GLOVES: rubber

SPECIAL | RESPIRATOR: self-contained

OTHER PROTECTIVE EQUIPMENT

SECTION IX SPECIAL PRECAUTIONS

PRECAUTIONS TO BE TAKEN IN HANDLING AND STORAGE

OTHER PRECAUTIONS

SIGNATURE | DATE

4-HYDROXY-1,2-DIMETHYLBENZENE

1 / 4 / 0

SECTION I SOURCE AND NOMENCLATURE

MANUFACTURER'S NAME | EMERGENCY TELEPHONE NO.

ADDRESS (NUMBER, STREET, CITY, STATE, ZIP CODE)

TRADE NAME AND SYNONYMS | CHEMICAL FAMILY

CHEMICAL NAME AND SYNONYMS: 3,4 - Xylenol | FORMULA: $HO(C_6H_3)(CH_3)_2$

SECTION II HAZARDOUS INGREDIENTS

BASIC MATERIAL	APPROXIMATE OR MAXIMUM % WT. OR VOL.	ESTABLISHED OSHA STANDARD	LD_{50} mg/kg ORAL	LD_{50} mg/kg PERCUT.	LC_{50} SPECIES	LC_{50} CONC.
			400		mouse	
			800		rabbit	
Lowest published toxic dose						
skin - mouse 90 mg/kg						

SECTION III PHYSICAL DATA MW - 122.16

BOILING POINT (°F.): 225 | VAPOR PRESSURE (mm Hg.)

MELTING POINT (°F.): $65^{\circ}C$ | VAPOR DENSITY (Air = 1)

SPECIFIC GRAVITY (H_2O = 1): 1.02 | EVAPORATION RATE (______ = 1)

SOLUBILITY IN WATER Pts/100 pts H_2O: slightly soluble | VOLATILE % Vol. % Wt.

APPEARANCE AND ODOR: white crystal

SECTION IV FIRE AND EXPLOSION HAZARD DATA

FLASH POINT: none available | FLAMMABLE (EXPLOSIVE) LIMITS: UPPER / LOWER

METHOD USED

EXTINGUISHING MEDIA: foam, carbon dioxide, dry chemical

SPECIAL FIRE FIGHTING PROCEDURES: wear self-contained breathing apparatus.

UNUSUAL FIRE AND EXPLOSION HAZARDS

SECTION V HEALTH HAZARD DATA

TOXIC LEVEL: none available | CARCINOGENIC: yes

PRINCIPLE ROUTES OF ABSORBTION: inhalation, ingestion | SKIN AND EYE IRRITATION: severe

REVELANT SYMPTOMS OF EXPOSURE: Headache, dizziness, nausea, vomiting, stomach pain, exhaustion; skin & eye irritation.

EFFECTS OF CHRONIC EXPOSURE: Liver and kidney damage.

EMERGENCY AND FIRST AID PROCEDURES: If swallowed - drink epsom salt solution, get medical aid; inhalation - rest; wash skin & eyes.

SECTION VI REACTIVITY DATA

CONDITIONS CONTRIBUTING TO INSTABILITY

CONDITIONS CONTRIBUTING TO HAZARDOUS POLYMERIZATION

INCOMPATABILITY (MATERIALS TO AVOID)

HAZARDOUS DECOMPOSITION PRODUCTS

SECTION VII SPILL OR LEAK PROCEDURES

STEPS TO BE TAKEN IN CASE MATERIAL IS RELEASED OR SPILLED

WASTE DISPOSAL METHOD

SECTION VIII SPECIAL PROTECTION INFORMATION

VENTILATION REQUIREMENTS LOCAL EXHAUST | PROTECTIVE EQUIPMENT (SPECIFY TYPES) EYE: goggles

MECHANICAL (GENERAL) | GLOVES: rubber

SPECIAL | RESPIRATOR: self-contained

OTHER PROTECTIVE EQUIPMENT

SECTION IX SPECIAL PRECAUTIONS

PRECAUTIONS TO BE TAKEN IN HANDLING AND STORAGE

OTHER PRECAUTIONS

SIGNATURE | DATE

4-HYDROXY-1,3-DIMETHYLBENZENE

2
4 0

SECTION I SOURCE AND NOMENCLATURE

MANUFACTURER'S NAME | EMERGENCY TELEPHONE NO.

ADDRESS (NUMBER, STREET, CITY, STATE, ZIP CODE)

TRADE NAME AND SYNONYMS | CHEMICAL FAMILY

CHEMICAL NAME AND SYNONYMS: 2,4 - Xylenol

FORMULA: $HO(C_6H_3)(CH_3)_2$

SECTION II HAZARDOUS INGREDIENTS

BASIC MATERIAL	APPROXIMATE OR MAXIMUM % WT. OR VOL.	ESTABLISHED OSHA STANDARD	LD50 mg/kg ORAL	LD50 mg/kg PERCUT.	LD50 mg/kg SPECIES	LC50 CONC.
Lowest published lethal dose			3200	1040	Rat	
Intraperitoneal - mouse 150 mg/kg			809		Mouse	

SECTION III PHYSICAL DATA MW - 122.16

BOILING POINT: 212°C

VAPOR PRESSURE mm Hg: at 120°C 20

MELTING POINT: 26°C

VAPOR DENSITY (Air = 1)

SPECIFIC GRAVITY (H_2O = 1): 1.04

EVAPORATION RATE (= 1)

SOLUBILITY IN WATER Pts/100 pts H_2O: slightly soluble

VOLATILE % Vol. % Wt.

APPEARANCE AND ODOR: yellow-brown liquid

SECTION IV FIRE AND EXPLOSION HAZARD DATA

FLASH POINT: none available

FLAMMABLE (EXPLOSIVE) LIMITS: UPPER / LOWER

METHOD USED

EXTINGUISHING MEDIA: foam, carbon dioxide, dry chemical

SPECIAL FIRE FIGHTING PROCEDURES: Wear self-contained breathing apparatus.

UNUSUAL FIRE AND EXPLOSION HAZARDS

SECTION V HEALTH HAZARD DATA

TOXIC LEVEL: none available

CARCINOGENIC: yes

PRINCIPLE ROUTES OF ABSORBTION: inhalation, ingestion

SKIN AND EYE IRRITATION: severe

REVELANT SYMPTOMS OF EXPOSURE: Headache, dizziness, nausea, vomiting, stomach pain, exhaustion; skin & eye irritation.

EFFECTS OF CHRONIC EXPOSURE: Liver and kidney damage.

EMERGENCY AND FIRST AID PROCEDURES: If swallowed - drink epsom salt solution, get medical aid; inhalation - rest; wash eyes and skin.

SECTION VI REACTIVITY DATA

CONDITIONS CONTRIBUTING TO INSTABILITY

CONDITIONS CONTRIBUTING TO HAZARDOUS POLYMERIZATION

INCOMPATABILITY (MATERIALS TO AVOID)

HAZARDOUS DECOMPOSITION PRODUCTS

SECTION VII SPILL OR LEAK PROCEDURES

STEPS TO BE TAKEN IN CASE MATERIAL IS RELEASED OR SPILLED

WASTE DISPOSAL METHOD

SECTION VIII SPECIAL PROTECTION INFORMATION

VENTILATION REQUIREMENTS LOCAL EXHAUST

PROTECTIVE EQUIPMENT (SPECIFY TYPES) EYE: goggles

MECHANICAL (GENERAL)

GLOVES: rubber

SPECIAL

RESPIRATOR: self-contained

OTHER PROTECTIVE EQUIPMENT

SECTION IX SPECIAL PRECAUTIONS

PRECAUTIONS TO BE TAKEN IN HANDLING AND STORAGE

OTHER PRECAUTIONS

SIGNATURE | DATE

o-HYDROXYDIPHENYL

1
4 0

SECTION I SOURCE AND NOMENCLATURE

MANUFACTURER'S NAME | EMERGENCY TELEPHONE NO.

ADDRESS (NUMBER, STREET, CITY, STATE, ZIP CODE)

TRADE NAME AND SYNONYMS | CHEMICAL FAMILY

CHEMICAL NAME AND SYNONYMS: 2 hydroxybiphenyl

FORMULA: $C_6H_5(C_6H_4)OH$

SECTION II HAZARDOUS INGREDIENTS

BASIC MATERIAL	APPROXIMATE OR MAXIMUM % WT. OR VOL.	ESTABLISHED OSHA STANDARD	LD50 mg/kg ORAL	LD50 mg/kg PERCUT.	LD50 mg/kg SPECIES	LC50 CONC.
			2700		Rat	
LDLO oral - human 500 mg/kg						

SECTION III PHYSICAL DATA MW - 170.20

BOILING POINT: 286°C

VAPOR PRESSURE mm Hg.

MELTING POINT: 57°C

VAPOR DENSITY (Air = 1)

SPECIFIC GRAVITY (H_2O = 1)

EVAPORATION RATE (= 1)

SOLUBILITY IN WATER Pts/100 pts H_2O: insoluble

VOLATILE % Vol. % Wt.

APPEARANCE AND ODOR: white solid

SECTION IV FIRE AND EXPLOSION HAZARD DATA

FLASH POINT: none available

FLAMMABLE (EXPLOSIVE) LIMITS: UPPER / LOWER

METHOD USED

EXTINGUISHING MEDIA: carbon dioxide or dry chemical

SPECIAL FIRE FIGHTING PROCEDURES: water may cause frothing.

UNUSUAL FIRE AND EXPLOSION HAZARDS

SECTION V HEALTH HAZARD DATA

TOXIC LEVEL: none available

CARCINOGENIC: no

PRINCIPLE ROUTES OF ABSORBTION: inhalation, ingestion

SKIN AND EYE IRRITATION: moderate

REVELANT SYMPTOMS OF EXPOSURE: Nausea, vomiting, central nervous system depression.

EFFECTS OF CHRONIC EXPOSURE: Irritation and lesions of respiratory system.

EMERGENCY AND FIRST AID PROCEDURES: If swallowed - get medical aid, gastric lavage; inhalation - rest; wash from skin & eyes.

SECTION VI REACTIVITY DATA

CONDITIONS CONTRIBUTING TO INSTABILITY

CONDITIONS CONTRIBUTING TO HAZARDOUS POLYMERIZATION

INCOMPATABILITY (MATERIALS TO AVOID)

HAZARDOUS DECOMPOSITION PRODUCTS

SECTION VII SPILL OR LEAK PROCEDURES

STEPS TO BE TAKEN IN CASE MATERIAL IS RELEASED OR SPILLED

WASTE DISPOSAL METHOD

SECTION VIII SPECIAL PROTECTION INFORMATION

VENTILATION REQUIREMENTS LOCAL EXHAUST

PROTECTIVE EQUIPMENT (SPECIFY TYPES) EYE: Goggles

MECHANICAL (GENERAL)

GLOVES: Polyvinyl alcohol

SPECIAL

RESPIRATOR: With proper filter

OTHER PROTECTIVE EQUIPMENT

SECTION IX SPECIAL PRECAUTIONS

PRECAUTIONS TO BE TAKEN IN HANDLING AND STORAGE

OTHER PRECAUTIONS

SIGNATURE | DATE

p-HYDROXYDIPHENYL

Hazard rating: 1 / 4 / 0

SECTION I SOURCE AND NOMENCLATURE

MANUFACTURER'S NAME | EMERGENCY TELEPHONE NO.

ADDRESS (NUMBER, STREET, CITY, STATE, ZIP CODE)

TRADE NAME AND SYNONYMS | CHEMICAL FAMILY

CHEMICAL NAME AND SYNONYMS: 4 hydroxybiphenyl | FORMULA: $C_6H_5(C_6H_4)OH$

SECTION II HAZARDOUS INGREDIENTS

BASIC MATERIAL	APPROXIMATE OR MAXIMUM % WT. OR VOL.	ESTABLISHED OSHA STANDARD	LD_{50} ORAL	LD_{50} PERCUT.	LC_{50} SPECIES	LC_{50} CONC.
			none	available		
TCLO subcutaneous - mouse	1000 mg/kg					

SECTION III PHYSICAL DATA MW - 170.20

BOILING POINT °F.: 306°C | VAPOR PRESSURE mm Hg.

MELTING POINT °F.: 165°C | VAPOR DENSITY (Air = 1)

SPECIFIC GRAVITY (H_2O = 1) | EVAPORATION RATE (______ = 1)

SOLUBILITY IN WATER Pts/100 pts H_2O: slightly soluble | VOLATILE % Vol. % Wt.

APPEARANCE AND ODOR: white solid

SECTION IV FIRE AND EXPLOSION HAZARD DATA

FLASH POINT: none available | FLAMMABLE (EXPLOSIVE) LIMITS UPPER / LOWER

METHOD USED

EXTINGUISHING MEDIA: carbon dioxide or dry chemical

SPECIAL FIRE FIGHTING PROCEDURES: water may cause frothing.

UNUSUAL FIRE AND EXPLOSION HAZARDS

SECTION V HEALTH HAZARD DATA

TOXIC LEVEL: none available | CARCINOGENIC: neoplastic

PRINCIPLE ROUTES OF ABSORBTION: inhalation, ingestion | SKIN AND EYE IRRITATION: moderate

REVELANT SYMPTOMS OF EXPOSURE: Nausea, vomiting, central nervous system depression.

EFFECTS OF CHRONIC EXPOSURE: Irritation and lesions of respiratory system.

EMERGENCY AND FIRST AID PROCEDURES: If swallowed - get medical aid, gastric lavage; inhalation - rest; wash from skin & eyes.

SECTION VI REACTIVITY DATA

CONDITIONS CONTRIBUTING TO INSTABILITY

CONDITIONS CONTRIBUTING TO HAZARDOUS POLYMERIZATION

INCOMPATABILITY (MATERIALS TO AVOID)

HAZARDOUS DECOMPOSITION PRODUCTS

SECTION VII SPILL OR LEAK PROCEDURES

STEPS TO BE TAKEN IN CASE MATERIAL IS RELEASED OR SPILLED

WASTE DISPOSAL METHOD

SECTION VIII SPECIAL PROTECTION INFORMATION

VENTILATION REQUIREMENTS LOCAL EXHAUST | PROTECTIVE EQUIPMENT (SPECIFY TYPES) EYE: Goggles

MECHANICAL (GENERAL) | GLOVES: polyvinyl alcohol

SPECIAL | RESPIRATOR: With proper filter

OTHER PROTECTIVE EQUIPMENT

SECTION IX SPECIAL PRECAUTIONS

PRECAUTIONS TO BE TAKEN IN HANDLING AND STORAGE

OTHER PRECAUTIONS

SIGNATURE | DATE

8-HYDROXYQUINOLINE

Hazard rating: 1 / 4 / 0

SECTION I SOURCE AND NOMENCLATURE

MANUFACTURER'S NAME | EMERGENCY TELEPHONE NO.

ADDRESS (NUMBER, STREET, CITY, STATE, ZIP CODE)

TRADE NAME AND SYNONYMS: 8 Quinolinol | CHEMICAL FAMILY

CHEMICAL NAME AND SYNONYMS: 8-Hydroxyquinoline | FORMULA: C_9H_7ON

SECTION II HAZARDOUS INGREDIENTS

BASIC MATERIAL	APPROXIMATE OR MAXIMUM % WT. OR VOL.	ESTABLISHED OSHA STANDARD	LD_{50} mg/kg ORAL	LD_{50} PERCUT.	LC_{50} SPECIES	LC_{50} CONC.
Oral - human LDLO	500 mg/kg		1200		Rat	

SECTION III PHYSICAL DATA MW - 145.10

BOILING POINT °F.: 267° C. | VAPOR PRESSURE mm Hg.

MELTING POINT °F.: 76° C. | VAPOR DENSITY (Air = 1)

SPECIFIC GRAVITY (H_2O = 1) | EVAPORATION RATE (______ = 1)

SOLUBILITY IN WATER Pts/100 pts H_2O: Nearly insoluble | VOLATILE % Vol. % Wt.

APPEARANCE AND ODOR: White crystal or powder.

SECTION IV FIRE AND EXPLOSION HAZARD DATA

FLASH POINT: None available | FLAMMABLE (EXPLOSIVE) LIMITS UPPER / LOWER

METHOD USED

EXTINGUISHING MEDIA: Alcohol foam, carbon dioxide, dry chemical.

SPECIAL FIRE FIGHTING PROCEDURES

UNUSUAL FIRE AND EXPLOSION HAZARDS: When heated to decomposition, it emits highly toxic fumes of oxides of nitrogen.

SECTION V HEALTH HAZARD DATA

TOXIC LEVEL: None available | CARCINOGENIC: Yes

PRINCIPLE ROUTES OF ABSORBTION: Inhalation, ingestion, skin | SKIN AND EYE IRRITATION: Severe

REVELANT SYMPTOMS OF EXPOSURE: Stimulation of central nervous system, digestive system irritation, eye irritation.

EFFECTS OF CHRONIC EXPOSURE: May form cancerous tumors.

EMERGENCY AND FIRST AID PROCEDURES: If swallowed - give emetic, get medical aid; inhalation - rest; wash skin & eyes.

SECTION VI REACTIVITY DATA

CONDITIONS CONTRIBUTING TO INSTABILITY

CONDITIONS CONTRIBUTING TO HAZARDOUS POLYMERIZATION

INCOMPATABILITY (MATERIALS TO AVOID)

HAZARDOUS DECOMPOSITION PRODUCTS

SECTION VII SPILL OR LEAK PROCEDURES

STEPS TO BE TAKEN IN CASE MATERIAL IS RELEASED OR SPILLED

WASTE DISPOSAL METHOD

SECTION VIII SPECIAL PROTECTION INFORMATION

VENTILATION REQUIREMENTS LOCAL EXHAUST | PROTECTIVE EQUIPMENT (SPECIFY TYPES) EYE: Goggles

MECHANICAL (GENERAL) | GLOVES: Rubber

SPECIAL | RESPIRATOR: With organic filter

OTHER PROTECTIVE EQUIPMENT

SECTION IX SPECIAL PRECAUTIONS

PRECAUTIONS TO BE TAKEN IN HANDLING AND STORAGE

OTHER PRECAUTIONS

SIGNATURE | DATE

8-HYDROXY-5-QUINOLINE-SULFONIC ACID

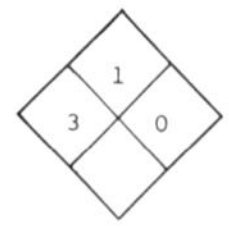

SECTION I SOURCE AND NOMENCLATURE

MANUFACTURER'S NAME

EMERGENCY TELEPHONE NO.

ADDRESS (NUMBER, STREET, CITY, STATE, ZIP CODE)

TRADE NAME AND SYNONYMS: Oxyquinoline Sulfate

CHEMICAL FAMILY

CHEMICAL NAME AND SYNONYMS: 8 hydroxy -5 - quinoline sulfonic acid

FORMULA: $C_9H_7NO_4S$

SECTION II HAZARDOUS INGREDIENTS

BASIC MATERIAL	APPROXIMATE OR MAXIMUM % WT. OR VOL.	ESTABLISHED OSHA STANDARD	LD_{50}		LC_{50}	
			ORAL	PERCUT.	SPECIES	CONC.
			none available			

SECTION III PHYSICAL DATA MW - 225.22

BOILING POINT: - °F.

VAPOR PRESSURE: mm Hg.

MELTING POINT: 275°C °F.

VAPOR DENSITY (Air = 1)

SPECIFIC GRAVITY (H_2O = 1): -

EVAPORATION RATE (= 1)

SOLUBILITY IN WATER Pts/100 pts H_2O: Slightly soluble

VOLATILE % Vol. % Wt.

APPEARANCE AND ODOR: Pale yellow needle-like crystals.

SECTION IV FIRE AND EXPLOSION HAZARD DATA

FLASH POINT: None available

FLAMMABLE (EXPLOSIVE) LIMITS: UPPER / LOWER

METHOD USED

EXTINGUISHING MEDIA: Carbon dioxide, dry chemical

SPECIAL FIRE FIGHTING PROCEDURES

UNUSUAL FIRE AND EXPLOSION HAZARDS

SECTION V HEALTH HAZARD DATA

TOXIC LEVEL: None available

CARCINOGENIC: no

PRINCIPLE ROUTES OF ABSORBTION: Inhalation, ingestion

SKIN AND EYE IRRITATION: Severe

REVELANT SYMPTOMS OF EXPOSURE: Retina damage, eye & mucous membrane irritation, cyanosis, headache, dizziness.

EFFECTS OF CHRONIC EXPOSURE: Possible liver damage

EMERGENCY AND FIRST AID PROCEDURES: If swallowed - induce vomiting, get medical aid; inhalation - rest; wash from skin and eyes.

SECTION VI REACTIVITY DATA

CONDITIONS CONTRIBUTING TO INSTABILITY

CONDITIONS CONTRIBUTING TO HAZARDOUS POLYMERIZATION

INCOMPATABILITY (MATERIALS TO AVOID)

HAZARDOUS DECOMPOSITION PRODUCTS

SECTION VII SPILL OR LEAK PROCEDURES

STEPS TO BE TAKEN IN CASE MATERIAL IS RELEASED OR SPILLED

WASTE DISPOSAL METHOD

SECTION VIII SPECIAL PROTECTION INFORMATION

VENTILATION REQUIREMENTS LOCAL EXHAUST

PROTECTIVE EQUIPMENT (SPECIFY TYPES) EYE: Goggles

MECHANICAL (GENERAL)

GLOVES

SPECIAL

RESPIRATOR: With proper filter

OTHER PROTECTIVE EQUIPMENT

SECTION IX SPECIAL PRECAUTIONS

PRECAUTIONS TO BE TAKEN IN HANDLING AND STORAGE

OTHER PRECAUTIONS

SIGNATURE

DATE

HYPOPHOSPHOROUS ACID (30%)

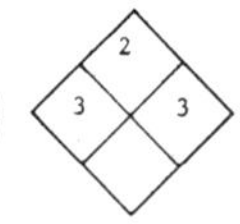

SECTION I SOURCE AND NOMENCLATURE

MANUFACTURER'S NAME

EMERGENCY TELEPHONE NO.

ADDRESS (NUMBER, STREET, CITY, STATE, ZIP CODE)

TRADE NAME AND SYNONYMS

CHEMICAL FAMILY

CHEMICAL NAME AND SYNONYMS: Hypophosphorous Acid

FORMULA: $H(H_2PO_2)$

SECTION II HAZARDOUS INGREDIENTS

BASIC MATERIAL	APPROXIMATE OR MAXIMUM % WT. OR VOL.	ESTABLISHED OSHA STANDARD	LD_{50}		LC_{50}	
			ORAL	PERCUT.	SPECIES	CONC.
			None available			

SECTION III PHYSICAL DATA

BOILING POINT: d at 140°C. °F

VAPOR PRESSURE: mm Hg.

MELTING POINT: 26.5°C. °F

VAPOR DENSITY (Air = 1)

SPECIFIC GRAVITY (H_2O = 1): 1.493

EVAPORATION RATE (= 1)

SOLUBILITY IN WATER Pts/100 pts H_2O: V.S.

VOLATILE % Vol. % Wt.

APPEARANCE AND ODOR: Colorless oily liquid

SECTION IV FIRE AND EXPLOSION HAZARD DATA

FLASH POINT: N/A

FLAMMABLE (EXPLOSIVE) LIMITS: UPPER / LOWER

METHOD USED

EXTINGUISHING MEDIA: Water

SPECIAL FIRE FIGHTING PROCEDURES: N/A

UNUSUAL FIRE AND EXPLOSION HAZARDS: Emits PH_3 and H_2 vapors on heating above 140° C. May explode on heating.

SECTION V HEALTH HAZARD DATA

TOXIC LEVEL: Not available

CARCINOGENIC: No

PRINCIPLE ROUTES OF ABSORBTION: Ingestion

SKIN AND EYE IRRITATION: Yes

REVELANT SYMPTOMS OF EXPOSURE: Slow attack on skin, attack on mucous membranes and digestive tract.

EFFECTS OF CHRONIC EXPOSURE: Erosion of teeth.

EMERGENCY AND FIRST AID PROCEDURES: Wash from skin and eyes; if swallowed - wash mouth, drink water, then milk of magnesia, get medical aid.

SECTION VI REACTIVITY DATA

CONDITIONS CONTRIBUTING TO INSTABILITY

CONDITIONS CONTRIBUTING TO HAZARDOUS POLYMERIZATION

INCOMPATABILITY (MATERIALS TO AVOID): Reacts violently with oxidizing agents.

HAZARDOUS DECOMPOSITION PRODUCTS

SECTION VII SPILL OR LEAK PROCEDURES

STEPS TO BE TAKEN IN CASE MATERIAL IS RELEASED OR SPILLED

WASTE DISPOSAL METHOD

SECTION VIII SPECIAL PROTECTION INFORMATION

VENTILATION REQUIREMENTS LOCAL EXHAUST

PROTECTIVE EQUIPMENT (SPECIFY TYPES) EYE: Face shield

MECHANICAL (GENERAL)

GLOVES: Rubber

SPECIAL

RESPIRATOR

OTHER PROTECTIVE EQUIPMENT: Face shield and rubber apron.

SECTION IX SPECIAL PRECAUTIONS

PRECAUTIONS TO BE TAKEN IN HANDLING AND STORAGE

OTHER PRECAUTIONS

SIGNATURE

DATE

INDENE

3 / 4 / 0

SECTION I SOURCE AND NOMENCLATURE

MANUFACTURER'S NAME | EMERGENCY TELEPHONE NO.

ADDRESS (NUMBER, STREET, CITY, STATE, ZIP CODE)

TRADE NAME AND SYNONYMS | CHEMICAL FAMILY

CHEMICAL NAME AND SYNONYMS: Indonaphthene | FORMULA: $C_6H_4CH_2CH{:}CH$

SECTION II HAZARDOUS INGREDIENTS

BASIC MATERIAL	APPROXIMATE OR MAXIMUM % WT. OR VOL.	ESTABLISHED OSHA STANDARD	LD_{50} ORAL	LD_{50} PERCUT.	LC_{50} SPECIES	LC_{50} CONC.
			See note			
Note: Lowest published lethal dose subcutaneous - rat 100 mg/kg.						

SECTION III PHYSICAL DATA MW - 116.15

BOILING POINT (°F.): 182°C | VAPOR PRESSURE (mm Hg.)

MELTING POINT (°F.): -2°C | VAPOR DENSITY (Air = 1)

SPECIFIC GRAVITY (H_2O = 1): 0.99 | EVAPORATION RATE (= 1)

SOLUBILITY IN WATER Pts/100 pts H_2O: insoluble | VOLATILE % Vol. % Wt.

APPEARANCE AND ODOR: Colorless liquid

SECTION IV FIRE AND EXPLOSION HAZARD DATA

FLASH POINT: none available | FLAMMABLE (EXPLOSIVE) LIMITS UPPER / LOWER

METHOD USED

EXTINGUISHING MEDIA: Water spray carbon dioxide, dry chemical or alcohol foam.

SPECIAL FIRE FIGHTING PROCEDURES: Wear self-contained breathing apparatus.

UNUSUAL FIRE AND EXPLOSION HAZARDS

SECTION V HEALTH HAZARD DATA

TOXIC LEVEL: TLV - 10 PPM, 45 mg/C.M. | CARCINOGENIC: no

PRINCIPLE ROUTES OF ABSORBTION: Inhalation, ingestion, skin | SKIN AND EYE IRRITATION: slight

REVELANT SYMPTOMS OF EXPOSURE: Irritation of mucous membranes and lungs, pulmonary edema.

EFFECTS OF CHRONIC EXPOSURE: Dermatitis, liver and kidney damage.

EMERGENCY AND FIRST AID PROCEDURES: If swallowed - give emetic, get medical aid; inhalation - rest; wash from skin & eyes.

SECTION VI REACTIVITY DATA

CONDITIONS CONTRIBUTING TO INSTABILITY

CONDITIONS CONTRIBUTING TO HAZARDOUS POLYMERIZATION

INCOMPATABILITY (MATERIALS TO AVOID)

HAZARDOUS DECOMPOSITION PRODUCTS

SECTION VII SPILL OR LEAK PROCEDURES

STEPS TO BE TAKEN IN CASE MATERIAL IS RELEASED OR SPILLED

WASTE DISPOSAL METHOD

SECTION VIII SPECIAL PROTECTION INFORMATION

VENTILATION REQUIREMENTS LOCAL EXHAUST | PROTECTIVE EQUIPMENT (SPECIFY TYPES) EYE: Goggles

MECHANICAL (GENERAL) | GLOVES: Rubber

SPECIAL | RESPIRATOR: With proper filter

OTHER PROTECTIVE EQUIPMENT

SECTION IX SPECIAL PRECAUTIONS

PRECAUTIONS TO BE TAKEN IN HANDLING AND STORAGE

OTHER PRECAUTIONS

SIGNATURE | DATE

IODIC ACID

0 / 3 / 1

SECTION I SOURCE AND NOMENCLATURE

MANUFACTURER'S NAME | EMERGENCY TELEPHONE NO.

ADDRESS (NUMBER, STREET, CITY, STATE, ZIP CODE)

TRADE NAME AND SYNONYMS | CHEMICAL FAMILY

CHEMICAL NAME AND SYNONYMS | FORMULA: HIO_3

SECTION II HAZARDOUS INGREDIENTS

BASIC MATERIAL	APPROXIMATE OR MAXIMUM % WT. OR VOL.	ESTABLISHED OSHA STANDARD	LD_{50} ORAL	LD_{50} PERCUT.	LC_{50} SPECIES	LC_{50} CONC.
			None available			

SECTION III PHYSICAL DATA

BOILING POINT (°F.): $-H_2O$ at 170°C. | VAPOR PRESSURE (mm Hg.)

MELTING POINT (°F.): 110° C. | VAPOR DENSITY (Air = 1)

SPECIFIC GRAVITY (H_2O = 1): 4.629 | EVAPORATION RATE (= 1)

SOLUBILITY IN WATER Pts/100 pts H_2O: 2.69 | VOLATILE % Vol. % Wt.

APPEARANCE AND ODOR: White crystal

SECTION IV FIRE AND EXPLOSION HAZARD DATA

FLASH POINT: None available | FLAMMABLE (EXPLOSIVE) LIMITS UPPER / LOWER

METHOD USED

EXTINGUISHING MEDIA: Water

SPECIAL FIRE FIGHTING PROCEDURES

UNUSUAL FIRE AND EXPLOSION HAZARDS: Strong oxidizing agent

SECTION V HEALTH HAZARD DATA

TOXIC LEVEL: Not available | CARCINOGENIC: No

PRINCIPLE ROUTES OF ABSORBTION: Inhalation of dust, ingestion | SKIN AND EYE IRRITATION: Yes

REVELANT SYMPTOMS OF EXPOSURE: Respiratory system irritates; skin and eye irritation; severe internal irritation and damage.

EFFECTS OF CHRONIC EXPOSURE: --

EMERGENCY AND FIRST AID PROCEDURES: Inhalation - rest, get medical attention; wash from skin and eyes; if swallowed - wash mouth, drink water, then milk of magnesia, get medical attention.

SECTION VI REACTIVITY DATA

CONDITIONS CONTRIBUTING TO INSTABILITY

CONDITIONS CONTRIBUTING TO HAZARDOUS POLYMERIZATION

INCOMPATABILITY (MATERIALS TO AVOID): Reducing agents.

HAZARDOUS DECOMPOSITION PRODUCTS: At 300° C. decomposes to give I_2 vapor.

SECTION VII SPILL OR LEAK PROCEDURES

STEPS TO BE TAKEN IN CASE MATERIAL IS RELEASED OR SPILLED

WASTE DISPOSAL METHOD

SECTION VIII SPECIAL PROTECTION INFORMATION

VENTILATION REQUIREMENTS LOCAL EXHAUST | PROTECTIVE EQUIPMENT (SPECIFY TYPES) EYE: Face shield

MECHANICAL (GENERAL) | GLOVES: Rubber gloves

SPECIAL | RESPIRATOR: With proper filter

OTHER PROTECTIVE EQUIPMENT: Face shield and rubber apron.

SECTION IX SPECIAL PRECAUTIONS

PRECAUTIONS TO BE TAKEN IN HANDLING AND STORAGE

OTHER PRECAUTIONS

SIGNATURE | DATE

IODINE

Hazard diamond: 0 (top), 4 (left), 0 (right)

SECTION I SOURCE AND NOMENCLATURE

MANUFACTURER'S NAME | EMERGENCY TELEPHONE NO.

ADDRESS (NUMBER, STREET, CITY, STATE, ZIP CODE)

TRADE NAME AND SYNONYMS | CHEMICAL FAMILY

CHEMICAL NAME AND SYNONYMS: Iodine | FORMULA: I_2

SECTION II HAZARDOUS INGREDIENTS

BASIC MATERIAL	APPROXIMATE OR MAXIMUM % WT. OR VOL.	ESTABLISHED OSHA STANDARD	LD50 ORAL	LD50 PERCUT.	LC50 SPECIES	LC50 CONC.
LDLO oral - human 5 mg/kg.						

SECTION III PHYSICAL DATA MW 253.84

BOILING POINT °F: 183°C. | VAPOR PRESSURE mm Hg: at 38.7°

MELTING POINT °F: 112.9°C. | VAPOR DENSITY (Air = 1)

SPECIFIC GRAVITY (H_2O = 1): 4.93 | EVAPORATION RATE (______ = 1)

SOLUBILITY IN WATER Pts/100 pts H_2O | VOLATILE % Vol. % Wt.

APPEARANCE AND ODOR: Rhombic, violet-black crystals, metallic luster - characteristic odor.

SECTION IV FIRE AND EXPLOSION HAZARD DATA

FLASH POINT: Non-flammable | FLAMMABLE (EXPLOSIVE) LIMITS: UPPER / LOWER

METHOD USED

EXTINGUISHING MEDIA: Water

SPECIAL FIRE FIGHTING PROCEDURES: Wear full protective clothing.

UNUSUAL FIRE AND EXPLOSION HAZARDS: Emits toxic fumes.

SECTION V HEALTH HAZARD DATA

TOXIC LEVEL: TLV - 0.1 PPM, 1 mg/CM. (Ceiling value) | CARCINOGENIC: no

PRINCIPLE ROUTES OF ABSORBTION: Inhalation | SKIN AND EYE IRRITATION: Possible skin sensitizer

REVELANT SYMPTOMS OF EXPOSURE: Lachrymation, blephoritis, rhinitis, catarrhal stomatitis, and pharnygitis, eye irritation.

EFFECTS OF CHRONIC EXPOSURE: Irritation and nervousness.

EMERGENCY AND FIRST AID PROCEDURES: Remove from exposure; rest and keep warm; obtain medical aid.

SECTION VI REACTIVITY DATA

CONDITIONS CONTRIBUTING TO INSTABILITY

CONDITIONS CONTRIBUTING TO HAZARDOUS POLYMERIZATION

INCOMPATABILITY (MATERIALS TO AVOID): Certain Carbides, Acetaldehyde, Acetylene, Aluminum, Ammonia, Antimony, Bromine Penta fluoride, Cesium Monoxide, Cuprous Acetylide.

HAZARDOUS DECOMPOSITION PRODUCTS: When heated releases highly toxic fumes of Iodine.

SECTION VII SPILL OR LEAK PROCEDURES

STEPS TO BE TAKEN IN CASE MATERIAL IS RELEASED OR SPILLED: Cover with weak reducing agent such as hypobisulfites or ferrous salts. Transfer sludge into large container of water and neutralize with soda ash.

WASTE DISPOSAL METHOD

SECTION VIII SPECIAL PROTECTION INFORMATION

VENTILATION REQUIREMENTS LOCAL EXHAUST | PROTECTIVE EQUIPMENT (SPECIFY TYPES) EYE: Goggles

MECHANICAL (GENERAL) | GLOVES: Rubber

SPECIAL | RESPIRATOR: W/appropriate filter

OTHER PROTECTIVE EQUIPMENT: Self-contained breathing apparatus.

SECTION IX SPECIAL PRECAUTIONS

PRECAUTIONS TO BE TAKEN IN HANDLING AND STORAGE: Protect container against physical damage; store in cool, dry area away from sunlight.

OTHER PRECAUTIONS

SIGNATURE | DATE

p-IODOANILINE

Hazard diamond: 1 (top), 4 (left), 0 (right)

SECTION I SOURCE AND NOMENCLATURE

MANUFACTURER'S NAME | EMERGENCY TELEPHONE NO.

ADDRESS (NUMBER, STREET, CITY, STATE, ZIP CODE)

TRADE NAME AND SYNONYMS | CHEMICAL FAMILY

CHEMICAL NAME AND SYNONYMS: P-Iodoaniline | FORMULA: $IC_6H_4NH_2$

SECTION II HAZARDOUS INGREDIENTS

BASIC MATERIAL	APPROXIMATE OR MAXIMUM % WT. OR VOL.	ESTABLISHED OSHA STANDARD	LD50 mg/kg ORAL	LD50 PERCUT.	LC50 SPECIES	LC50 CONC.
			100		Bird	

SECTION III PHYSICAL DATA MW - 219.03

BOILING POINT °F | VAPOR PRESSURE mm Hg

MELTING POINT °F: 68°C. | VAPOR DENSITY (Air = 1)

SPECIFIC GRAVITY (H_2O = 1) | EVAPORATION RATE (______ = 1)

SOLUBILITY IN WATER Pts/100 pts H_2O: Slightly soluble | VOLATILE % Vol. % Wt.

APPEARANCE AND ODOR: Yellow crystal

SECTION IV FIRE AND EXPLOSION HAZARD DATA

FLASH POINT: None available | FLAMMABLE (EXPLOSIVE) LIMITS: UPPER / LOWER

METHOD USED

EXTINGUISHING MEDIA: Water spray, carbon dioxide, alcohol foam.

SPECIAL FIRE FIGHTING PROCEDURES: Self-contained breathing apparatus.

UNUSUAL FIRE AND EXPLOSION HAZARDS: Vapor may form explosive mixture with air.

SECTION V HEALTH HAZARD DATA

TOXIC LEVEL: None available | CARCINOGENIC: No

PRINCIPLE ROUTES OF ABSORBTION: Inhalation, ingestion, skin | SKIN AND EYE IRRITATION: Low

REVELANT SYMPTOMS OF EXPOSURE: Cyanosis, headache, blood pressure drop, lethargy, jaundice, anemia, pain in joints, fainting.

EFFECTS OF CHRONIC EXPOSURE: Weight loss, anemia, weakness, irritability, possible liver damage, inflammation of kidneys, ringing in ear.

EMERGENCY AND FIRST AID PROCEDURES: Remove from exposure; wash skin and eyes

SECTION VI REACTIVITY DATA

CONDITIONS CONTRIBUTING TO INSTABILITY

CONDITIONS CONTRIBUTING TO HAZARDOUS POLYMERIZATION

INCOMPATABILITY (MATERIALS TO AVOID): Sodium and potassium peroxides, ozone, perchloric acid.

HAZARDOUS DECOMPOSITION PRODUCTS

SECTION VII SPILL OR LEAK PROCEDURES

STEPS TO BE TAKEN IN CASE MATERIAL IS RELEASED OR SPILLED: Wear protective clothing during clean-up.

WASTE DISPOSAL METHOD

SECTION VIII SPECIAL PROTECTION INFORMATION

VENTILATION REQUIREMENTS LOCAL EXHAUST | PROTECTIVE EQUIPMENT (SPECIFY TYPES) EYE: Goggles

MECHANICAL (GENERAL) | GLOVES: Rubber

SPECIAL | RESPIRATOR: Self-contained

OTHER PROTECTIVE EQUIPMENT: Protective clothing

SECTION IX SPECIAL PRECAUTIONS

PRECAUTIONS TO BE TAKEN IN HANDLING AND STORAGE: Store in flammable liquids cabinet.

OTHER PRECAUTIONS

SIGNATURE | DATE

IODOBENZENE

SECTION I SOURCE AND NOMENCLATURE

MANUFACTURER'S NAME	EMERGENCY TELEPHONE NO.
ADDRESS (NUMBER, STREET, CITY, STATE, ZIP CODE)	
TRADE NAME AND SYNONYMS	CHEMICAL FAMILY
CHEMICAL NAME AND SYNONYMS: Phenyliodide	FORMULA: C_6H_5I

SECTION II HAZARDOUS INGREDIENTS

BASIC MATERIAL	APPROXIMATE OR MAXIMUM % WT. OR VOL.	ESTABLISHED OSHA STANDARD	LD_{50} ORAL	LD_{50} PERCUT.	LD_{50} SPECIES	LC_{50} CONC.
			none available			

SECTION III PHYSICAL DATA MW - 204.02

BOILING POINT °F.: 188.5	VAPOR PRESSURE mm Hg.: at 71°C 12
MELTING POINT °F.: -31.3	VAPOR DENSITY (Air = 1)
SPECIFIC GRAVITY (H_2O = 1): 1.82	EVAPORATION RATE (= 1)
SOLUBILITY IN WATER Pts/100 pts H_2O: insoluble	VOLATILE % Vol. % Wt.
APPEARANCE AND ODOR: colorless liquid	

SECTION IV FIRE AND EXPLOSION HAZARD DATA

FLASH POINT: none available

METHOD USED

FLAMMABLE (EXPLOSIVE) LIMITS: UPPER / LOWER

EXTINGUISHING MEDIA: water, foam, carbon dioxide, dry chemical.

SPECIAL FIRE FIGHTING PROCEDURES: Wear goggles and self-contained breathing apparatus.

UNUSUAL FIRE AND EXPLOSION HAZARDS

SECTION V HEALTH HAZARD DATA

TOXIC LEVEL: none available

CARCINOGENIC: no

PRINCIPLE ROUTES OF ABSORBTION: inhalation, ingestion, skin

SKIN AND EYE IRRITATION: severe

REVELANT SYMPTOMS OF EXPOSURE: Cyanosis, deep and rapid respiration, irregular pulse, drowsiness, unconsciousness.

EFFECTS OF CHRONIC EXPOSURE: Lung, liver and kidney damage.

EMERGENCY AND FIRST AID PROCEDURES: Inhalation - rest; if swallowed - give emetic, get medical aid; wash from skin & eyes.

SECTION VI REACTIVITY DATA

CONDITIONS CONTRIBUTING TO INSTABILITY

CONDITIONS CONTRIBUTING TO HAZARDOUS POLYMERIZATION

INCOMPATABILITY (MATERIALS TO AVOID)

HAZARDOUS DECOMPOSITION PRODUCTS

SECTION VII SPILL OR LEAK PROCEDURES

STEPS TO BE TAKEN IN CASE MATERIAL IS RELEASED OR SPILLED

WASTE DISPOSAL METHOD

SECTION VIII SPECIAL PROTECTION INFORMATION

VENTILATION REQUIREMENTS LOCAL EXHAUST	PROTECTIVE EQUIPMENT (SPECIFY TYPES) EYE: Goggles
MECHANICAL (GENERAL)	GLOVES: Polyvinyl alcohol
SPECIAL	RESPIRATOR: Self-contained breathing apparatus.
OTHER PROTECTIVE EQUIPMENT	

SECTION IX SPECIAL PRECAUTIONS

PRECAUTIONS TO BE TAKEN IN HANDLING AND STORAGE

OTHER PRECAUTIONS

SIGNATURE	DATE

o-IODOBENZOYL CHLORIDE

SECTION I SOURCE AND NOMENCLATURE

MANUFACTURER'S NAME	EMERGENCY TELEPHONE NO.
ADDRESS (NUMBER, STREET, CITY, STATE, ZIP CODE)	
TRADE NAME AND SYNONYMS	CHEMICAL FAMILY
CHEMICAL NAME AND SYNONYMS: 2 Iodobenzoyl Chloride	FORMULA: IC_6H_4COCl

SECTION II HAZARDOUS INGREDIENTS

BASIC MATERIAL	APPROXIMATE OR MAXIMUM % WT. OR VOL.	ESTABLISHED OSHA STANDARD	LD_{50} ORAL	LD_{50} PERCUT.	LD_{50} SPECIES	LC_{50} CONC.
			none available			

SECTION III PHYSICAL DATA MW - 266.98

BOILING POINT °F.	VAPOR PRESSURE mm Hg.: at 159°C 29
MELTING POINT °F.: 31°C	VAPOR DENSITY (Air = 1)
SPECIFIC GRAVITY (H_2O = 1)	EVAPORATION RATE (= 1)
SOLUBILITY IN WATER Pts/100 pts H_2O: Decomposes	VOLATILE % Vol. % Wt.
APPEARANCE AND ODOR: Crystaline	

SECTION IV FIRE AND EXPLOSION HAZARD DATA

FLASH POINT: None available

METHOD USED

FLAMMABLE (EXPLOSIVE) LIMITS: UPPER / LOWER

EXTINGUISHING MEDIA: Water spray, foam, carbon dioxide or dry chemical.

SPECIAL FIRE FIGHTING PROCEDURES: Full protective clothing.

UNUSUAL FIRE AND EXPLOSION HAZARDS

SECTION V HEALTH HAZARD DATA

TOXIC LEVEL: None available

CARCINOGENIC: no

PRINCIPLE ROUTES OF ABSORBTION: Inhalation, ingestion

SKIN AND EYE IRRITATION: severe

REVELANT SYMPTOMS OF EXPOSURE: Irritation of respiratory system, skin & eye burns, mouth and digestive tract burns.

EFFECTS OF CHRONIC EXPOSURE

EMERGENCY AND FIRST AID PROCEDURES: If swallowed - drink water, milk of magnesia, get medical aid; inhalation - rest, oxygen, artificial respiration; wash skin & eyes.

SECTION VI REACTIVITY DATA

CONDITIONS CONTRIBUTING TO INSTABILITY

CONDITIONS CONTRIBUTING TO HAZARDOUS POLYMERIZATION

INCOMPATABILITY (MATERIALS TO AVOID): May react with oxidizing materials.

HAZARDOUS DECOMPOSITION PRODUCTS: When heated, phosgene is produced.

SECTION VII SPILL OR LEAK PROCEDURES

STEPS TO BE TAKEN IN CASE MATERIAL IS RELEASED OR SPILLED

WASTE DISPOSAL METHOD

SECTION VIII SPECIAL PROTECTION INFORMATION

VENTILATION REQUIREMENTS LOCAL EXHAUST	PROTECTIVE EQUIPMENT (SPECIFY TYPES) EYE: Goggles
MECHANICAL (GENERAL)	GLOVES: Rubber
SPECIAL	RESPIRATOR: With proper filter
OTHER PROTECTIVE EQUIPMENT	

SECTION IX SPECIAL PRECAUTIONS

PRECAUTIONS TO BE TAKEN IN HANDLING AND STORAGE: Store in cool, dry, well-ventilated location.

OTHER PRECAUTIONS

SIGNATURE	DATE

2-IODO-1,3-DIMETHYLBENZENE

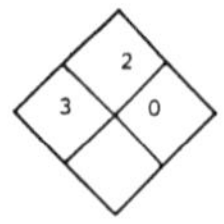

SECTION I SOURCE AND NOMENCLATURE

MANUFACTURER'S NAME

EMERGENCY TELEPHONE NO.

ADDRESS (NUMBER, STREET, CITY, STATE, ZIP CODE)

TRADE NAME AND SYNONYMS

CHEMICAL FAMILY

CHEMICAL NAME AND SYNONYMS: Iodo-M-Xylene 2

FORMULA: $I\ C_6H_3(CH_3)_2$

SECTION II HAZARDOUS INGREDIENTS

BASIC MATERIAL	APPROXIMATE OR MAXIMUM % WT. OR VOL.	ESTABLISHED OSHA STANDARD	LD_{50} ORAL	LD_{50} PERCUT.	LC_{50} SPECIES	LC_{50} CONC.
			none available			

SECTION III PHYSICAL DATA MW - 232.07

BOILING POINT °F.: 230°C

VAPOR PRESSURE mm Hg.

MELTING POINT °F.: --

VAPOR DENSITY (Air = 1)

SPECIFIC GRAVITY (H_2O = 1)

EVAPORATION RATE (______ = 1)

SOLUBILITY IN WATER Pts/100 pts H_2O: insoluble

VOLATILE % Vol. % Wt.

APPEARANCE AND ODOR: clear liquid

SECTION IV FIRE AND EXPLOSION HAZARD DATA

FLASH POINT: none available

FLAMMABLE (EXPLOSIVE) LIMITS: UPPER / LOWER

METHOD USED

EXTINGUISHING MEDIA: Foam, carbon dioxide, dry chemical. Water spray may be used to reduce rate of burning.

SPECIAL FIRE FIGHTING PROCEDURES: Wear self-contained breathing apparatus.

UNUSUAL FIRE AND EXPLOSION HAZARDS

SECTION V HEALTH HAZARD DATA

TOXIC LEVEL: none available

CARCINOGENIC: no

PRINCIPLE ROUTES OF ABSORBTION: inhalation, ingestion

SKIN AND EYE IRRITATION: severe

REVELANT SYMPTOMS OF EXPOSURE: Irritation of respiratory tract, disturbance of vision, dizziness, nausea, collapse, coma; burning of skin and eyes.

EFFECTS OF CHRONIC EXPOSURE: Dermatitis, blood and liver damage.

EMERGENCY AND FIRST AID PROCEDURES: Inhalation - rest, get medical aid; ingestion - give saline cathartics and demulcents, get medical aid; wash skin & eyes.

SECTION VI REACTIVITY DATA

CONDITIONS CONTRIBUTING TO INSTABILITY

CONDITIONS CONTRIBUTING TO HAZARDOUS POLYMERIZATION

INCOMPATABILITY (MATERIALS TO AVOID)

HAZARDOUS DECOMPOSITION PRODUCTS

SECTION VII SPILL OR LEAK PROCEDURES

STEPS TO BE TAKEN IN CASE MATERIAL IS RELEASED OR SPILLED

WASTE DISPOSAL METHOD

SECTION VIII SPECIAL PROTECTION INFORMATION

VENTILATION REQUIREMENTS LOCAL EXHAUST

PROTECTIVE EQUIPMENT (SPECIFY TYPES) EYE: Goggles

MECHANICAL (GENERAL)

GLOVES: Rubber

SPECIAL

RESPIRATOR: Self-contained

OTHER PROTECTIVE EQUIPMENT

SECTION IX SPECIAL PRECAUTIONS

PRECAUTIONS TO BE TAKEN IN HANDLING AND STORAGE: Outside storage preferred; inside storage flammable liquids storage cabinet.

OTHER PRECAUTIONS

SIGNATURE

DATE

2-IODO-1,4-DIMETHYLBENZENE

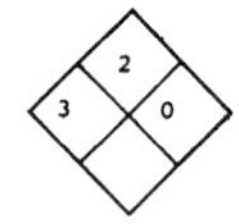

SECTION I SOURCE AND NOMENCLATURE

MANUFACTURER'S NAME

EMERGENCY TELEPHONE NO.

ADDRESS (NUMBER, STREET, CITY, STATE, ZIP CODE)

TRADE NAME AND SYNONYMS

CHEMICAL FAMILY

CHEMICAL NAME AND SYNONYMS: Iodo - P - Xylene 2

FORMULA: $IC_6H_3(CH_3)_2$

SECTION II HAZARDOUS INGREDIENTS

BASIC MATERIAL	APPROXIMATE OR MAXIMUM % WT. OR VOL.	ESTABLISHED OSHA STANDARD	LD_{50} ORAL	LD_{50} PERCUT.	LC_{50} SPECIES	LC_{50} CONC.
			none available			

SECTION III PHYSICAL DATA MW - 232.07

BOILING POINT °F.: 230°C

VAPOR PRESSURE mm Hg.: 20 at 114°C

MELTING POINT °F.

VAPOR DENSITY (Air = 1)

SPECIFIC GRAVITY (H_2O = 1): 1.62

EVAPORATION RATE (______ = 1)

SOLUBILITY IN WATER Pts/100 pts H_2O: insoluble

VOLATILE % Vol. % Wt.

APPEARANCE AND ODOR: clear liquid

SECTION IV FIRE AND EXPLOSION HAZARD DATA

FLASH POINT: none available

FLAMMABLE (EXPLOSIVE) LIMITS: UPPER / LOWER

METHOD USED

EXTINGUISHING MEDIA: Foam, carbon dioxide, dry chemical. Water spray may be used to reduce rate of burning.

SPECIAL FIRE FIGHTING PROCEDURES: Wear self-contained breathing apparatus.

UNUSUAL FIRE AND EXPLOSION HAZARDS

SECTION V HEALTH HAZARD DATA

TOXIC LEVEL: None available

CARCINOGENIC: no

PRINCIPLE ROUTES OF ABSORBTION: Inhalation, ingestion

SKIN AND EYE IRRITATION: severe

REVELANT SYMPTOMS OF EXPOSURE: Irritation of respiratory tract, disturbance of vision, dizziness, nausea, collapse, coma; burning of skin & eyes.

EFFECTS OF CHRONIC EXPOSURE: Dermatitis, blood and liver damage.

EMERGENCY AND FIRST AID PROCEDURES: Inhalation - rest, get medical aid; ingestion - give saline water cathartics and demulcents, get medical aid; wash skin & eyes.

SECTION VI REACTIVITY DATA

CONDITIONS CONTRIBUTING TO INSTABILITY

CONDITIONS CONTRIBUTING TO HAZARDOUS POLYMERIZATION

INCOMPATABILITY (MATERIALS TO AVOID)

HAZARDOUS DECOMPOSITION PRODUCTS

SECTION VII SPILL OR LEAK PROCEDURES

STEPS TO BE TAKEN IN CASE MATERIAL IS RELEASED OR SPILLED

WASTE DISPOSAL METHOD

SECTION VIII SPECIAL PROTECTION INFORMATION

VENTILATION REQUIREMENTS LOCAL EXHAUST

PROTECTIVE EQUIPMENT (SPECIFY TYPES) EYE: Goggles

MECHANICAL (GENERAL)

GLOVES: Rubber

SPECIAL

RESPIRATOR: Self-contained

OTHER PROTECTIVE EQUIPMENT

SECTION IX SPECIAL PRECAUTIONS

PRECAUTIONS TO BE TAKEN IN HANDLING AND STORAGE: Outside storage preferred; inside storage flammable liquids storage cabinet.

OTHER PRECAUTIONS

SIGNATURE

DATE

4-IODO-1,3-DIMETHYLBENZENE

Hazard rating: 2 (top), 3 (left), 0 (right)

SECTION I SOURCE AND NOMENCLATURE

MANUFACTURER'S NAME	EMERGENCY TELEPHONE NO.
ADDRESS (NUMBER, STREET, CITY, STATE, ZIP CODE)	
TRADE NAME AND SYNONYMS	CHEMICAL FAMILY
CHEMICAL NAME AND SYNONYMS Iodo-M-Xylene 4	FORMULA $IC_6H_3(CH_3)_2$

SECTION II HAZARDOUS INGREDIENTS

BASIC MATERIAL	APPROXIMATE OR MAXIMUM % WT. OR VOL.	ESTABLISHED OSHA STANDARD	LD50 ORAL	LD50 PERCUT.	LD50 SPECIES	LC50 CONC.
			none available			

SECTION III PHYSICAL DATA MW - 232.07

BOILING POINT °F. not available	VAPOR PRESSURE mm Hg. at 113°C 20
MELTING POINT °F. --	VAPOR DENSITY (Air = 1)
SPECIFIC GRAVITY (H_2O = 1)	EVAPORATION RATE (=1)
SOLUBILITY IN WATER Pts/100 pts H_2O insoluble	VOLATILE % Vol. % Wt.
APPEARANCE AND ODOR clear liquid	

SECTION IV FIRE AND EXPLOSION HAZARD DATA

FLASH POINT: none available

METHOD USED:

FLAMMABLE (EXPLOSIVE) LIMITS: UPPER / LOWER

EXTINGUISHING MEDIA: Foam, carbon dioxide, dry chemical. Water spray may be used to reduce rate of burning.

SPECIAL FIRE FIGHTING PROCEDURES: Wear self-contained breathing apparatus.

UNUSUAL FIRE AND EXPLOSION HAZARDS:

SECTION V HEALTH HAZARD DATA

TOXIC LEVEL: None available

CARCINOGENIC: no

PRINCIPLE ROUTES OF ABSORBTION: Inhalation, ingestion

SKIN AND EYE IRRITATION: severe

REVELANT SYMPTOMS OF EXPOSURE: Irritation of respiratory tract; disturbance of vision, dizziness, nausea, collapse, coma; burning of skin & eyes.

EFFECTS OF CHRONIC EXPOSURE: Dermatitis, blood & liver damage.

EMERGENCY AND FIRST AID PROCEDURES: Inhalation - rest, get medical aid; if swallowed - give saline water cathartics and demulcents, get medical aid; wash skin and eyes.

SECTION VI REACTIVITY DATA

CONDITIONS CONTRIBUTING TO INSTABILITY:

CONDITIONS CONTRIBUTING TO HAZARDOUS POLYMERIZATION:

INCOMPATABILITY (MATERIALS TO AVOID):

HAZARDOUS DECOMPOSITION PRODUCTS:

SECTION VII SPILL OR LEAK PROCEDURES

STEPS TO BE TAKEN IN CASE MATERIAL IS RELEASED OR SPILLED:

WASTE DISPOSAL METHOD:

SECTION VIII SPECIAL PROTECTION INFORMATION

VENTILATION REQUIREMENTS LOCAL EXHAUST	PROTECTIVE EQUIPMENT (SPECIFY TYPES) EYE Goggles
MECHANICAL (GENERAL)	GLOVES Rubber
SPECIAL	RESPIRATOR Self-contained
OTHER PROTECTIVE EQUIPMENT	

SECTION IX SPECIAL PRECAUTIONS

PRECAUTIONS TO BE TAKEN IN HANDLING AND STORAGE: Outside storage preferred; inside storage flammable liquids storage cabinet.

OTHER PRECAUTIONS:

SIGNATURE	DATE

o-IODODIPHENYL

Hazard rating: 1 (top), 3 (left), 0 (right)

SECTION I SOURCE AND NOMENCLATURE

MANUFACTURER'S NAME	EMERGENCY TELEPHONE NO.
ADDRESS (NUMBER, STREET, CITY, STATE, ZIP CODE)	
TRADE NAME AND SYNONYMS	CHEMICAL FAMILY
CHEMICAL NAME AND SYNONYMS O-Iododiphenyl	FORMULA $IC_6H_4C_6H_5$

SECTION II HAZARDOUS INGREDIENTS

BASIC MATERIAL	APPROXIMATE OR MAXIMUM % WT. OR VOL.	ESTABLISHED OSHA STANDARD	LD50 ORAL	LD50 PERCUT.	LD50 SPECIES	LC50 CONC.
			none available			

SECTION III PHYSICAL DATA MW - 280.11

BOILING POINT °F. 192°C at 36 mm	VAPOR PRESSURE mm Hg. at 175°C 16
MELTING POINT °F. --	VAPOR DENSITY (Air = 1)
SPECIFIC GRAVITY (H_2O = 1) 1.60	EVAPORATION RATE (=1)
SOLUBILITY IN WATER Pts/100 pts H_2O --	VOLATILE % Vol. % Wt.
APPEARANCE AND ODOR colorless liquid	

SECTION IV FIRE AND EXPLOSION HAZARD DATA

FLASH POINT: none available

METHOD USED:

FLAMMABLE (EXPLOSIVE) LIMITS: UPPER / LOWER

EXTINGUISHING MEDIA: water, foam, carbon dioxide, dry chemical

SPECIAL FIRE FIGHTING PROCEDURES:

UNUSUAL FIRE AND EXPLOSION HAZARDS:

SECTION V HEALTH HAZARD DATA

TOXIC LEVEL: none available

CARCINOGENIC: no

PRINCIPLE ROUTES OF ABSORBTION: inhalation, ingestion, skin

SKIN AND EYE IRRITATION: moderate

REVELANT SYMPTOMS OF EXPOSURE: Acne, irritation of respiratory passages.

EFFECTS OF CHRONIC EXPOSURE: Liver damage

EMERGENCY AND FIRST AID PROCEDURES: If swallowed - give emetic, get medical aid; inhalation - rest; wash from skin & eyes.

SECTION VI REACTIVITY DATA

CONDITIONS CONTRIBUTING TO INSTABILITY:

CONDITIONS CONTRIBUTING TO HAZARDOUS POLYMERIZATION:

INCOMPATABILITY (MATERIALS TO AVOID):

HAZARDOUS DECOMPOSITION PRODUCTS:

SECTION VII SPILL OR LEAK PROCEDURES

STEPS TO BE TAKEN IN CASE MATERIAL IS RELEASED OR SPILLED:

WASTE DISPOSAL METHOD:

SECTION VIII SPECIAL PROTECTION INFORMATION

VENTILATION REQUIREMENTS LOCAL EXHAUST	PROTECTIVE EQUIPMENT (SPECIFY TYPES) EYE Goggles
MECHANICAL (GENERAL)	GLOVES Rubber
SPECIAL	RESPIRATOR Self-contained
OTHER PROTECTIVE EQUIPMENT	

SECTION IX SPECIAL PRECAUTIONS

PRECAUTIONS TO BE TAKEN IN HANDLING AND STORAGE:

OTHER PRECAUTIONS:

SIGNATURE	DATE

p-IODODIPHENYL

1
3 0

SECTION I SOURCE AND NOMENCLATURE

MANUFACTURER'S NAME | EMERGENCY TELEPHONE NO.

ADDRESS (NUMBER, STREET, CITY, STATE, ZIP CODE)

TRADE NAME AND SYNONYMS | CHEMICAL FAMILY

CHEMICAL NAME AND SYNONYMS: P-Iododiphenyl

FORMULA: $IC_6H_4C_6H_5$

SECTION II HAZARDOUS INGREDIENTS

BASIC MATERIAL	APPROXIMATE OR MAXIMUM % WT. OR VOL.	ESTABLISHED OSHA STANDARD	LD_{50} ORAL	LD_{50} PERCUT.	LC_{50} SPECIES	LC_{50} CONC.
			none available			

SECTION III PHYSICAL DATA MW - 280.11

BOILING POINT °F: 320°C

VAPOR PRESSURE mm Hg: at 183 — 11

MELTING POINT °F: 113°C

VAPOR DENSITY (Air = 1)

SPECIFIC GRAVITY (H_2O = 1)

EVAPORATION RATE (= 1)

SOLUBILITY IN WATER Pts/100 pts H_2O: insoluble

VOLATILE % Vol. % Wt.

APPEARANCE AND ODOR: colorless crystal

SECTION IV FIRE AND EXPLOSION HAZARD DATA

FLASH POINT: none available

FLAMMABLE (EXPLOSIVE) LIMITS: UPPER / LOWER

METHOD USED

EXTINGUISHING MEDIA: water, foam, carbon dioxide, dry chemical.

SPECIAL FIRE FIGHTING PROCEDURES

UNUSUAL FIRE AND EXPLOSION HAZARDS

SECTION V HEALTH HAZARD DATA

TOXIC LEVEL: none available

CARCINOGENIC: no

PRINCIPLE ROUTES OF ABSORBTION: inhalation, ingestion, skin

SKIN AND EYE IRRITATION: moderate

REVELANT SYMPTOMS OF EXPOSURE: Acne, irritation of respiratory passages.

EFFECTS OF CHRONIC EXPOSURE: Liver damage

EMERGENCY AND FIRST AID PROCEDURES: If swallowed - give emetic, get medical aid; inhalation - rest; wash skin & eyes.

SECTION VI REACTIVITY DATA

CONDITIONS CONTRIBUTING TO INSTABILITY

CONDITIONS CONTRIBUTING TO HAZARDOUS POLYMERIZATION

INCOMPATABILITY (MATERIALS TO AVOID)

HAZARDOUS DECOMPOSITION PRODUCTS

SECTION VII SPILL OR LEAK PROCEDURES

STEPS TO BE TAKEN IN CASE MATERIAL IS RELEASED OR SPILLED

WASTE DISPOSAL METHOD

SECTION VIII SPECIAL PROTECTION INFORMATION

VENTILATION REQUIREMENTS LOCAL EXHAUST

PROTECTIVE EQUIPMENT (SPECIFY TYPES) EYE: goggles

MECHANICAL (GENERAL)

GLOVES: rubber

SPECIAL

RESPIRATOR: self-contained

OTHER PROTECTIVE EQUIPMENT

SECTION IX SPECIAL PRECAUTIONS

PRECAUTIONS TO BE TAKEN IN HANDLING AND STORAGE

OTHER PRECAUTIONS

SIGNATURE | DATE

o-IODOPHENOL

1
3 0

SECTION I SOURCE AND NOMENCLATURE

MANUFACTURER'S NAME | EMERGENCY TELEPHONE NO.

ADDRESS (NUMBER, STREET, CITY, STATE, ZIP CODE)

TRADE NAME AND SYNONYMS | CHEMICAL FAMILY

CHEMICAL NAME AND SYNONYMS: O-Iodophenol

FORMULA: $I\ C_6H_4OH$

SECTION II HAZARDOUS INGREDIENTS

BASIC MATERIAL	APPROXIMATE OR MAXIMUM % WT. OR VOL.	ESTABLISHED OSHA STANDARD	LD_{50} ORAL	LD_{50} PERCUT.	LC_{50} SPECIES	LC_{50} CONC.
Lowest published lethal	dose					
subcutaneous - rat - 4	000 mg/kg					

SECTION III PHYSICAL DATA MW - 220.01

BOILING POINT °F

VAPOR PRESSURE mm Hg: 186°C. — 160

MELTING POINT °F: 40°C.

VAPOR DENSITY (Air = 1)

SPECIFIC GRAVITY (H_2O = 1): 1.88

EVAPORATION RATE (= 1)

SOLUBILITY IN WATER Pts/100 pts H_2O: Soluble in hot water

VOLATILE % Vol. % Wt.

APPEARANCE AND ODOR: Needle-like crystals

SECTION IV FIRE AND EXPLOSION HAZARD DATA

FLASH POINT: None available

FLAMMABLE (EXPLOSIVE) LIMITS: UPPER / LOWER

METHOD USED

EXTINGUISHING MEDIA: Dry chemical, alcohol foam, carbon dioxide

SPECIAL FIRE FIGHTING PROCEDURES: Self-contained

UNUSUAL FIRE AND EXPLOSION HAZARDS: May liberate iodine fumes on heating.

SECTION V HEALTH HAZARD DATA

TOXIC LEVEL: None available

CARCINOGENIC: NO

PRINCIPLE ROUTES OF ABSORBTION: Inhalation, ingestion, skin

SKIN AND EYE IRRITATION: Severe

REVELANT SYMPTOMS OF EXPOSURE: Eye, skin and mucous membrane irritation, dermatitis, cyanosis.

EFFECTS OF CHRONIC EXPOSURE: Liver damage

EMERGENCY AND FIRST AID PROCEDURES: Inhalation - rest; wash skin & eyes; if swallowed - give emetic, get medical aid.

SECTION VI REACTIVITY DATA

CONDITIONS CONTRIBUTING TO INSTABILITY

CONDITIONS CONTRIBUTING TO HAZARDOUS POLYMERIZATION

INCOMPATABILITY (MATERIALS TO AVOID)

HAZARDOUS DECOMPOSITION PRODUCTS

SECTION VII SPILL OR LEAK PROCEDURES

STEPS TO BE TAKEN IN CASE MATERIAL IS RELEASED OR SPILLED: Pour sand to soda ash 9:1 mixture on spill and sweep into box for disposal.

WASTE DISPOSAL METHOD

SECTION VIII SPECIAL PROTECTION INFORMATION

VENTILATION REQUIREMENTS LOCAL EXHAUST

PROTECTIVE EQUIPMENT (SPECIFY TYPES) EYE: Goggles

MECHANICAL (GENERAL)

GLOVES: Butyl Rubber

SPECIAL

RESPIRATOR: Self-contained

OTHER PROTECTIVE EQUIPMENT: Protective clothing

SECTION IX SPECIAL PRECAUTIONS

PRECAUTIONS TO BE TAKEN IN HANDLING AND STORAGE: Store in cool, dry place.

OTHER PRECAUTIONS

SIGNATURE | DATE

m-IODOTOLUENE

2 / 3 / 0

SECTION I SOURCE AND NOMENCLATURE

MANUFACTURER'S NAME | EMERGENCY TELEPHONE NO.

ADDRESS (NUMBER, STREET, CITY, STATE, ZIP CODE)

TRADE NAME AND SYNONYMS | CHEMICAL FAMILY

CHEMICAL NAME AND SYNONYMS: M-Tolyl Iodide | FORMULA: $IC_6H_4CH_3$

SECTION II HAZARDOUS INGREDIENTS

BASIC MATERIAL	APPROXIMATE OR MAXIMUM % WT. OR VOL.	ESTABLISHED OSHA STANDARD	LD_{50} ORAL	LD_{50} PERCUT.	LC_{50} SPECIES	LC_{50} CONC.
			none available			

SECTION III PHYSICAL DATA MW - 218.05

BOILING POINT °F.: 213°C | VAPOR PRESSURE mm Hg.

MELTING POINT °F.: -- | VAPOR DENSITY (Air = 1)

SPECIFIC GRAVITY (H_2O = 1): 1.70 | EVAPORATION RATE (______ = 1)

SOLUBILITY IN WATER Pts/100 pts H_2O: insoluble | VOLATILE % Vol. % Wt.

APPEARANCE AND ODOR: clear liquid

SECTION IV FIRE AND EXPLOSION HAZARD DATA

FLASH POINT: none available | FLAMMABLE (EXPLOSIVE) LIMITS: UPPER / LOWER

METHOD USED

EXTINGUISHING MEDIA: alcohol foam, carbon dioxide, dry chemical.

SPECIAL FIRE FIGHTING PROCEDURES

UNUSUAL FIRE AND EXPLOSION HAZARDS

SECTION V HEALTH HAZARD DATA

TOXIC LEVEL: none available | CARCINOGENIC: no

PRINCIPLE ROUTES OF ABSORBTION: Inhalation, ingestion | SKIN AND EYE IRRITATION: severe

REVELANT SYMPTOMS OF EXPOSURE: Eye and respiratory system irritation; severe irritation of digestive tract if swallowed.

EFFECTS OF CHRONIC EXPOSURE: --

EMERGENCY AND FIRST AID PROCEDURES: If swallowed - give emetic, get medical aid; inhalation - rest; wash from skin & eyes.

SECTION VI REACTIVITY DATA

CONDITIONS CONTRIBUTING TO INSTABILITY

CONDITIONS CONTRIBUTING TO HAZARDOUS POLYMERIZATION

INCOMPATABILITY (MATERIALS TO AVOID)

HAZARDOUS DECOMPOSITION PRODUCTS

SECTION VII SPILL OR LEAK PROCEDURES

STEPS TO BE TAKEN IN CASE MATERIAL IS RELEASED OR SPILLED

WASTE DISPOSAL METHOD

SECTION VIII SPECIAL PROTECTION INFORMATION

VENTILATION REQUIREMENTS LOCAL EXHAUST | PROTECTIVE EQUIPMENT (SPECIFY TYPES) EYE: goggles

MECHANICAL (GENERAL) | GLOVES: rubber

SPECIAL | RESPIRATOR: with proper filter

OTHER PROTECTIVE EQUIPMENT

SECTION IX SPECIAL PRECAUTIONS

PRECAUTIONS TO BE TAKEN IN HANDLING AND STORAGE

OTHER PRECAUTIONS

SIGNATURE | DATE

p-IODOTOLUENE

2 / 3 / 0

SECTION I SOURCE AND NOMENCLATURE

MANUFACTURER'S NAME | EMERGENCY TELEPHONE NO.

ADDRESS (NUMBER, STREET, CITY, STATE, ZIP CODE)

TRADE NAME AND SYNONYMS | CHEMICAL FAMILY

CHEMICAL NAME AND SYNONYMS: P-Tolyl Iodide | FORMULA: $IC_6H_4CH_3$

SECTION II HAZARDOUS INGREDIENTS

BASIC MATERIAL	APPROXIMATE OR MAXIMUM % WT. OR VOL.	ESTABLISHED OSHA STANDARD	LD_{50} ORAL	LD_{50} PERCUT.	LC_{50} SPECIES	LC_{50} CONC.
			none available			

SECTION III PHYSICAL DATA

BOILING POINT °F.: 212°C | VAPOR PRESSURE mm Hg.

MELTING POINT °F.: 36°C | VAPOR DENSITY (Air = 1)

SPECIFIC GRAVITY (H_2O = 1): 1.68 | EVAPORATION RATE (______ = 1)

SOLUBILITY IN WATER Pts/100 pts H_2O: insoluble | VOLATILE % Vol. % Wt.

APPEARANCE AND ODOR: white crystals

SECTION IV FIRE AND EXPLOSION HAZARD DATA

FLASH POINT: none available | FLAMMABLE (EXPLOSIVE) LIMITS: UPPER / LOWER

METHOD USED

EXTINGUISHING MEDIA: alcohol foam, carbon dioxide, dry chemical

SPECIAL FIRE FIGHTING PROCEDURES

UNUSUAL FIRE AND EXPLOSION HAZARDS

SECTION V HEALTH HAZARD DATA

TOXIC LEVEL: None available | CARCINOGENIC: no

PRINCIPLE ROUTES OF ABSORBTION: Inhalation, ingestion | SKIN AND EYE IRRITATION: Severe

REVELANT SYMPTOMS OF EXPOSURE: Eye and respiratory system irritation, severe irritation of digestive tract if swallowed.

EFFECTS OF CHRONIC EXPOSURE: --

EMERGENCY AND FIRST AID PROCEDURES: If swallowed - give emetic, get medical aid; inhalation - rest; wash from skin & eyes.

SECTION VI REACTIVITY DATA

CONDITIONS CONTRIBUTING TO INSTABILITY

CONDITIONS CONTRIBUTING TO HAZARDOUS POLYMERIZATION

INCOMPATABILITY (MATERIALS TO AVOID)

HAZARDOUS DECOMPOSITION PRODUCTS

SECTION VII SPILL OR LEAK PROCEDURES

STEPS TO BE TAKEN IN CASE MATERIAL IS RELEASED OR SPILLED

WASTE DISPOSAL METHOD

SECTION VIII SPECIAL PROTECTION INFORMATION

VENTILATION REQUIREMENTS LOCAL EXHAUST | PROTECTIVE EQUIPMENT (SPECIFY TYPES) EYE: Goggles

MECHANICAL (GENERAL) | GLOVES: Rubber

SPECIAL | RESPIRATOR: With proper filter

OTHER PROTECTIVE EQUIPMENT

SECTION IX SPECIAL PRECAUTIONS

PRECAUTIONS TO BE TAKEN IN HANDLING AND STORAGE

OTHER PRECAUTIONS

SIGNATURE | DATE

ISOAMYL ACETATE

Hazard diamond: 3 (top), 3 (left), 0 (right)

SECTION I SOURCE AND NOMENCLATURE

MANUFACTURER'S NAME

EMERGENCY TELEPHONE NO.

ADDRESS (NUMBER, STREET, CITY, STATE, ZIP CODE)

TRADE NAME AND SYNONYMS: Banana oil

CHEMICAL FAMILY

CHEMICAL NAME AND SYNONYMS: 3 methyl butanol acetate

FORMULA: $CH_3COOCH_2CH(CH_3)C_2H_5$

SECTION II HAZARDOUS INGREDIENTS

BASIC MATERIAL	APPROXIMATE OR MAXIMUM % WT. OR VOL.	ESTABLISHED OSHA STANDARD	LD_{50} mg/kg ORAL	LD_{50} PERCUT.	LD_{50} SPECIES	LC_{50} CONC.
LDLO - oral - human 500 mg/kg			7422		Rabbit	
TCLO inhalation - human 200PPM irritant						

SECTION III PHYSICAL DATA MW - 130.2

BOILING POINT °F: 142°C

VAPOR PRESSURE mm Hg.

MELTING POINT °F

VAPOR DENSITY (Air = 1): 4.5

SPECIFIC GRAVITY (H_2O = 1): 0.9

EVAPORATION RATE (______ = 1)

SOLUBILITY IN WATER Pts/100 pts H_2O: Slight

VOLATILE % Vol. % Wt.

APPEARANCE AND ODOR: Liquid - banana-like odor

SECTION IV FIRE AND EXPLOSION HAZARD DATA

FLASH POINT: 77°F

METHOD USED

FLAMMABLE (EXPLOSIVE) LIMITS: UPPER 7.5; LOWER 1.0

EXTINGUISHING MEDIA: Alcohol foam, carbon dioxide, dry chemical water may be ineffective.

SPECIAL FIRE FIGHTING PROCEDURES

UNUSUAL FIRE AND EXPLOSION HAZARDS

SECTION V HEALTH HAZARD DATA

TOXIC LEVEL: TLV 100 PPM, 525 mg/C.M.

CARCINOGENIC: no

PRINCIPLE ROUTES OF ABSORBTION: Inhalation, ingestion

SKIN AND EYE IRRITATION: Severe

REVELANT SYMPTOMS OF EXPOSURE: Irritation of respiratory tract, intoxication, eye irritation, throat dryness.

EFFECTS OF CHRONIC EXPOSURE: Kidney damage

EMERGENCY AND FIRST AID PROCEDURES: Inhalation - remove from exposure, artificial respiration as necessary; wash from eyes or skin. Ingestion - gastric lavage.

SECTION VI REACTIVITY DATA

CONDITIONS CONTRIBUTING TO INSTABILITY

CONDITIONS CONTRIBUTING TO HAZARDOUS POLYMERIZATION

INCOMPATABILITY (MATERIALS TO AVOID)

HAZARDOUS DECOMPOSITION PRODUCTS

SECTION VII SPILL OR LEAK PROCEDURES

STEPS TO BE TAKEN IN CASE MATERIAL IS RELEASED OR SPILLED: Absorb on paper towels, evaporate in hood.

WASTE DISPOSAL METHOD

SECTION VIII SPECIAL PROTECTION INFORMATION

VENTILATION REQUIREMENTS LOCAL EXHAUST

MECHANICAL (GENERAL)

SPECIAL

PROTECTIVE EQUIPMENT (SPECIFY TYPES) EYE: Goggles

GLOVES: Rubber

RESPIRATOR: With proper filter

OTHER PROTECTIVE EQUIPMENT: Rubber clothing

SECTION IX SPECIAL PRECAUTIONS

PRECAUTIONS TO BE TAKEN IN HANDLING AND STORAGE: Tightly closed, in detached area.

OTHER PRECAUTIONS

SIGNATURE

DATE

ISOBUTYLENE BROMIDE

Hazard diamond: 2 (top), 3 (left), 0 (right)

SECTION I SOURCE AND NOMENCLATURE

MANUFACTURER'S NAME

EMERGENCY TELEPHONE NO.

ADDRESS (NUMBER, STREET, CITY, STATE, ZIP CODE)

TRADE NAME AND SYNONYMS

CHEMICAL FAMILY

CHEMICAL NAME AND SYNONYMS: 1,2 dibromoisobutane

FORMULA: $(CH_3)_2CBrCH_2Br$

SECTION II HAZARDOUS INGREDIENTS

BASIC MATERIAL	APPROXIMATE OR MAXIMUM % WT. OR VOL.	ESTABLISHED OSHA STANDARD	LD_{50} ORAL	LD_{50} PERCUT.	LD_{50} SPECIES	LC_{50} CONC.
			None available			

SECTION III PHYSICAL DATA MW - 215.94

BOILING POINT °F: 149°C at 737 ml

VAPOR PRESSURE mm Hg.

MELTING POINT °F: -70.3°C

VAPOR DENSITY (Air = 1)

SPECIFIC GRAVITY (H_2O = 1): 1.76

EVAPORATION RATE (______ = 1)

SOLUBILITY IN WATER Pts/100 pts H_2O: Insoluble

VOLATILE % Vol. % Wt.

APPEARANCE AND ODOR: Colorless liquid

SECTION IV FIRE AND EXPLOSION HAZARD DATA

FLASH POINT: N/A

METHOD USED

FLAMMABLE (EXPLOSIVE) LIMITS: UPPER; LOWER

EXTINGUISHING MEDIA: Water, foam, carbon dioxide, dry chemical

SPECIAL FIRE FIGHTING PROCEDURES: Water may be ineffective

UNUSUAL FIRE AND EXPLOSION HAZARDS

SECTION V HEALTH HAZARD DATA

TOXIC LEVEL: None available

CARCINOGENIC: No

PRINCIPLE ROUTES OF ABSORBTION: Inhalation, ingestion

SKIN AND EYE IRRITATION: Moderate

REVELANT SYMPTOMS OF EXPOSURE: Eye irritation, headache, vomiting, drowsiness, narcosis, liver & kidney damage.

EFFECTS OF CHRONIC EXPOSURE: Dermatitis, liver & kidney damage

EMERGENCY AND FIRST AID PROCEDURES: Inhalation - rest; if swallowed - wash mouth give emetic, get medical aid; wash from eyes & skin.

SECTION VI REACTIVITY DATA

CONDITIONS CONTRIBUTING TO INSTABILITY

CONDITIONS CONTRIBUTING TO HAZARDOUS POLYMERIZATION

INCOMPATABILITY (MATERIALS TO AVOID)

HAZARDOUS DECOMPOSITION PRODUCTS

SECTION VII SPILL OR LEAK PROCEDURES

STEPS TO BE TAKEN IN CASE MATERIAL IS RELEASED OR SPILLED: Wear self contained breathing apparatus.

WASTE DISPOSAL METHOD

SECTION VIII SPECIAL PROTECTION INFORMATION

VENTILATION REQUIREMENTS LOCAL EXHAUST

MECHANICAL (GENERAL)

SPECIAL

PROTECTIVE EQUIPMENT (SPECIFY TYPES) EYE: Goggles

GLOVES: Rubber

RESPIRATOR: With organic vapor resistant canister

OTHER PROTECTIVE EQUIPMENT

SECTION IX SPECIAL PRECAUTIONS

PRECAUTIONS TO BE TAKEN IN HANDLING AND STORAGE

OTHER PRECAUTIONS

SIGNATURE

DATE

ISOBUTYL IODIDE

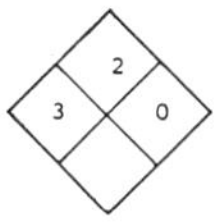

SECTION I SOURCE AND NOMENCLATURE

MANUFACTURER'S NAME | EMERGENCY TELEPHONE NO.

ADDRESS (NUMBER, STREET, CITY, STATE, ZIP CODE)

TRADE NAME AND SYNONYMS | CHEMICAL FAMILY

CHEMICAL NAME AND SYNONYMS: 1 Iodo-2 Methyl Propane

FORMULA: $(CH_3)_2CHCH_2I$

SECTION II HAZARDOUS INGREDIENTS

BASIC MATERIAL	APPROXIMATE OR MAXIMUM % WT. OR VOL.	ESTABLISHED OSHA STANDARD	LD_{50} ORAL	LD_{50} PERCUT.	LC_{50} SPECIES	LC_{50} CONC.
			None available			

SECTION III PHYSICAL DATA MW - 184.03

BOILING POINT °F.: 121°C

VAPOR PRESSURE mm Hg.:

MELTING POINT °F.: -94°C

VAPOR DENSITY (Air = 1):

SPECIFIC GRAVITY (H_2O = 1): 1.60

EVAPORATION RATE (= 1):

SOLUBILITY IN WATER Pts/100 pts H_2O: Insoluble

VOLATILE % Vol. % Wt.

APPEARANCE AND ODOR: Colorless liquid

SECTION IV FIRE AND EXPLOSION HAZARD DATA

FLASH POINT: Not available

METHOD USED:

FLAMMABLE (EXPLOSIVE) LIMITS: UPPER / LOWER

EXTINGUISHING MEDIA: Water, carbon dioxide, dry chemical

SPECIAL FIRE FIGHTING PROCEDURES:

UNUSUAL FIRE AND EXPLOSION HAZARDS: When heated to decomposition point emits toxic fumes of iodates

SECTION V HEALTH HAZARD DATA

TOXIC LEVEL: Not available

CARCINOGENIC: No

PRINCIPLE ROUTES OF ABSORBTION: Inhalation, ingestion, skin

SKIN AND EYE IRRITATION: slight

REVELANT SYMPTOMS OF EXPOSURE: Nausea, vomiting, diarrhea, dizziness, visual disturbances, irritability, drowsiness, central nervous system depression.

EFFECTS OF CHRONIC EXPOSURE: Dermatitis

EMERGENCY AND FIRST AID PROCEDURES: Inhalation - rest; if swallowed - wash mouth, give emetic, get medical aid; wash skin & eyes.

SECTION VI REACTIVITY DATA

CONDITIONS CONTRIBUTING TO INSTABILITY

CONDITIONS CONTRIBUTING TO HAZARDOUS POLYMERIZATION

INCOMPATABILITY (MATERIALS TO AVOID)

HAZARDOUS DECOMPOSITION PRODUCTS

SECTION VII SPILL OR LEAK PROCEDURES

STEPS TO BE TAKEN IN CASE MATERIAL IS RELEASED OR SPILLED

WASTE DISPOSAL METHOD

SECTION VIII SPECIAL PROTECTION INFORMATION

VENTILATION REQUIREMENTS LOCAL EXHAUST

MECHANICAL (GENERAL)

SPECIAL

PROTECTIVE EQUIPMENT (SPECIFY TYPES) EYE: Goggles

GLOVES: Rubber

RESPIRATOR: With proper filter

OTHER PROTECTIVE EQUIPMENT

SECTION IX SPECIAL PRECAUTIONS

PRECAUTIONS TO BE TAKEN IN HANDLING AND STORAGE

OTHER PRECAUTIONS

SIGNATURE | DATE

ISOPROPYL IODIDE

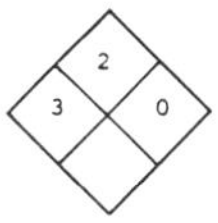

SECTION I SOURCE AND NOMENCLATURE

MANUFACTURER'S NAME | EMERGENCY TELEPHONE NO.

ADDRESS (NUMBER, STREET, CITY, STATE, ZIP CODE)

TRADE NAME AND SYNONYMS | CHEMICAL FAMILY

CHEMICAL NAME AND SYNONYMS: 2-Iodopropane

FORMULA: CH_3CHICH_3

SECTION II HAZARDOUS INGREDIENTS

BASIC MATERIAL	APPROXIMATE OR MAXIMUM % WT. OR VOL.	ESTABLISHED OSHA STANDARD	LD_{50} ORAL	LD_{50} PERCUT.	LC_{50} SPECIES	LC_{50} CONC.
			none	available		

SECTION III PHYSICAL DATA MW - 169.99

BOILING POINT °F.: 90°C

VAPOR PRESSURE mm Hg.:

MELTING POINT °F.: -90°C

VAPOR DENSITY (Air = 1):

SPECIFIC GRAVITY (H_2O = 1): 1.74

EVAPORATION RATE (= 1):

SOLUBILITY IN WATER Pts/100 pts H_2O: 0.14

VOLATILE % Vol. % Wt.

APPEARANCE AND ODOR: Colorless liquid

SECTION IV FIRE AND EXPLOSION HAZARD DATA

FLASH POINT: Not available

METHOD USED:

FLAMMABLE (EXPLOSIVE) LIMITS: UPPER / LOWER

EXTINGUISHING MEDIA: Water, carbon dioxide, dry chemical.

SPECIAL FIRE FIGHTING PROCEDURES: --

UNUSUAL FIRE AND EXPLOSION HAZARDS: When heated to decomposition point emits toxic fumes of iodates.

SECTION V HEALTH HAZARD DATA

TOXIC LEVEL: Not available

CARCINOGENIC: No

PRINCIPLE ROUTES OF ABSORBTION: Inhalation, ingestion, skin

SKIN AND EYE IRRITATION: Slight

REVELANT SYMPTOMS OF EXPOSURE: Nausea, vomiting, diarrhea, dizziness, visual disturbances, irritability, drowsiness, central nervous system depressant.

EFFECTS OF CHRONIC EXPOSURE: Dermatitis

EMERGENCY AND FIRST AID PROCEDURES: Inhalation - rest; if swallowed - wash mouth, give emetic, get medical aid; wash skin & eyes.

SECTION VI REACTIVITY DATA

CONDITIONS CONTRIBUTING TO INSTABILITY

CONDITIONS CONTRIBUTING TO HAZARDOUS POLYMERIZATION

INCOMPATABILITY (MATERIALS TO AVOID)

HAZARDOUS DECOMPOSITION PRODUCTS

SECTION VII SPILL OR LEAK PROCEDURES

STEPS TO BE TAKEN IN CASE MATERIAL IS RELEASED OR SPILLED

WASTE DISPOSAL METHOD

SECTION VIII SPECIAL PROTECTION INFORMATION

VENTILATION REQUIREMENTS LOCAL EXHAUST

MECHANICAL (GENERAL)

SPECIAL

PROTECTIVE EQUIPMENT (SPECIFY TYPES) EYE: Safety goggles

GLOVES: Rubber

RESPIRATOR: With proper filter

OTHER PROTECTIVE EQUIPMENT

SECTION IX SPECIAL PRECAUTIONS

PRECAUTIONS TO BE TAKEN IN HANDLING AND STORAGE

OTHER PRECAUTIONS

SIGNATURE | DATE

LEAD CHLORIDE

Hazard diamond: 0 / 4 / 0

SECTION I SOURCE AND NOMENCLATURE

MANUFACTURER'S NAME | EMERGENCY TELEPHONE NO.

ADDRESS (NUMBER, STREET, CITY, STATE, ZIP CODE)

TRADE NAME AND SYNONYMS | CHEMICAL FAMILY

CHEMICAL NAME AND SYNONYMS: Lead chloride | FORMULA: $PbCl_2$

SECTION II HAZARDOUS INGREDIENTS

BASIC MATERIAL	APPROXIMATE OR MAXIMUM % WT. OR VOL.	ESTABLISHED OSHA STANDARD	LD_{50} ORAL	LD_{50} PERCUT.	LC_{50} SPECIES	LC_{50} CONC.
#1 Lowest published lethal dose oral - guinea pig 2000 Mg/kg.						

SECTION III PHYSICAL DATA MW 278.1

BOILING POINT: 950° C. °F. | VAPOR PRESSURE mm Hg.

MELTING POINT: 501° C. °F. | VAPOR DENSITY (Air = 1)

SPECIFIC GRAVITY (H_2O = 1): 5.85 | EVAPORATION RATE (= 1)

SOLUBILITY IN WATER Pts/100 pts H_2O | VOLATILE % Vol. % Wt.

APPEARANCE AND ODOR: White crystal

SECTION IV FIRE AND EXPLOSION HAZARD DATA

FLASH POINT: Not flammable | FLAMMABLE (EXPLOSIVE) LIMITS: UPPER / LOWER

METHOD USED

EXTINGUISHING MEDIA: N/A

SPECIAL FIRE FIGHTING PROCEDURES: N/A

UNUSUAL FIRE AND EXPLOSION HAZARDS: N/A

SECTION V HEALTH HAZARD DATA

TOXIC LEVEL: TLV 0.067 mg/C.M. | CARCINOGENIC: No

PRINCIPLE ROUTES OF ABSORBTION: Inhalation, ingestion | SKIN AND EYE IRRITATION: No

REVELANT SYMPTOMS OF EXPOSURE: Abdominal pain, diarrhea, constipation, loss of appetite, muscular weakness, headache, blue line on gums, vomiting.

EFFECTS OF CHRONIC EXPOSURE: Loss of appetite, constipation, blue line on gums, loss of weight.

EMERGENCY AND FIRST AID PROCEDURES: If swallowed - give egg white, milk or tannin, also gastric lavage with solution of magnesium sulfate; if inhaled - rest, keep warm, get medical aid.

SECTION VI REACTIVITY DATA

CONDITIONS CONTRIBUTING TO INSTABILITY

CONDITIONS CONTRIBUTING TO HAZARDOUS POLYMERIZATION

INCOMPATABILITY (MATERIALS TO AVOID)

HAZARDOUS DECOMPOSITION PRODUCTS

SECTION VII SPILL OR LEAK PROCEDURES

STEPS TO BE TAKEN IN CASE MATERIAL IS RELEASED OR SPILLED

WASTE DISPOSAL METHOD

SECTION VIII SPECIAL PROTECTION INFORMATION

VENTILATION REQUIREMENTS LOCAL EXHAUST | PROTECTIVE EQUIPMENT (SPECIFY TYPES) EYE

MECHANICAL (GENERAL) | GLOVES

SPECIAL | RESPIRATOR: With proper filter

OTHER PROTECTIVE EQUIPMENT

SECTION IX SPECIAL PRECAUTIONS

PRECAUTIONS TO BE TAKEN IN HANDLING AND STORAGE

OTHER PRECAUTIONS

SIGNATURE | DATE

LEAD DIOXIDE

Hazard diamond: 0 / 4 / 0

SECTION I SOURCE AND NOMENCLATURE

MANUFACTURER'S NAME | EMERGENCY TELEPHONE NO.

ADDRESS (NUMBER, STREET, CITY, STATE, ZIP CODE)

TRADE NAME AND SYNONYMS | CHEMICAL FAMILY

CHEMICAL NAME AND SYNONYMS: Lead Dioxide | FORMULA: PbO_2

SECTION II HAZARDOUS INGREDIENTS

BASIC MATERIAL	APPROXIMATE OR MAXIMUM % WT. OR VOL.	ESTABLISHED OSHA STANDARD	LD_{50} ORAL	LD_{50} PERCUT.	LC_{50} SPECIES	LC_{50} CONC.
Oral - human LDLO 50 mg/kg						
Lowest published lethal dose intraperitoneal - guinea pig 115 mg/kg.						

SECTION III PHYSICAL DATA MW 239.19

BOILING POINT °F. | VAPOR PRESSURE mm Hg.

MELTING POINT: d 290° C. °F. | VAPOR DENSITY (Air = 1)

SPECIFIC GRAVITY (H_2O = 1): 9.375 | EVAPORATION RATE (= 1)

SOLUBILITY IN WATER Pts/100 pts H_2O | VOLATILE % Vol. % Wt.

APPEARANCE AND ODOR: Brown crystals

SECTION IV FIRE AND EXPLOSION HAZARD DATA

FLASH POINT: Not flammable | FLAMMABLE (EXPLOSIVE) LIMITS: UPPER / LOWER

METHOD USED

EXTINGUISHING MEDIA: N/A

SPECIAL FIRE FIGHTING PROCEDURES: N/A

UNUSUAL FIRE AND EXPLOSION HAZARDS: N/A

SECTION V HEALTH HAZARD DATA

TOXIC LEVEL: TLV - 0.058 mg/cm | CARCINOGENIC: No

PRINCIPLE ROUTES OF ABSORBTION: Inhalation, ingestion | SKIN AND EYE IRRITATION: No

REVELANT SYMPTOMS OF EXPOSURE: Abdominal pain, diarrhea, constipation, loss of appetite, muscular weakness, headache, blue line on gums, pyorrhea, vomiting.

EFFECTS OF CHRONIC EXPOSURE: Loss of appetite, constipation, blue line on gums, loss of weight.

EMERGENCY AND FIRST AID PROCEDURES: If swallowed - give egg white, milk or tannin, also gastric lavage with solution of magnesium sulfate if inhaled - rest, keep warm, get medical aid.

SECTION VI REACTIVITY DATA

CONDITIONS CONTRIBUTING TO INSTABILITY: Oxidizing agent

CONDITIONS CONTRIBUTING TO HAZARDOUS POLYMERIZATION

INCOMPATABILITY (MATERIALS TO AVOID): Aluminum carbide, barium sulfide, boron, calcium sulfide, hydrogen sulfide, hydroxylamine, molybdenum, Phenyl hydrazine, phosphorous.

HAZARDOUS DECOMPOSITION PRODUCTS

SECTION VII SPILL OR LEAK PROCEDURES

STEPS TO BE TAKEN IN CASE MATERIAL IS RELEASED OR SPILLED

WASTE DISPOSAL METHOD

SECTION VIII SPECIAL PROTECTION INFORMATION

VENTILATION REQUIREMENTS LOCAL EXHAUST | PROTECTIVE EQUIPMENT (SPECIFY TYPES) EYE

MECHANICAL (GENERAL) | GLOVES

SPECIAL | RESPIRATOR: With proper filter

OTHER PROTECTIVE EQUIPMENT

SECTION IX SPECIAL PRECAUTIONS

PRECAUTIONS TO BE TAKEN IN HANDLING AND STORAGE

OTHER PRECAUTIONS

SIGNATURE | DATE

LEAD METAL

1
4 0

SECTION I SOURCE AND NOMENCLATURE

MANUFACTURER'S NAME | EMERGENCY TELEPHONE NO.

ADDRESS (NUMBER, STREET, CITY, STATE, ZIP CODE)

TRADE NAME AND SYNONYMS | CHEMICAL FAMILY

CHEMICAL NAME AND SYNONYMS: Lead | FORMULA: Pb

SECTION II HAZARDOUS INGREDIENTS

BASIC MATERIAL	APPROXIMATE OR MAXIMUM % WT. OR VOL.	ESTABLISHED OSHA STANDARD	LD_{50} ORAL	LD_{50} PERCUT.	LC_{50} SPECIES	LC_{50} CONC.
Oral - woman - TDLO 450 mg/kg - 6Y						
~~#1 Lowest published lethal dose~~						
~~intraperitoneal - guinea pig 100 mg/kg.~~						

SECTION III PHYSICAL DATA MW 207.19

BOILING POINT (°F.)	1744^0 C.	VAPOR PRESSURE (mm Hg.)	
MELTING POINT (°F.)	327.5	VAPOR DENSITY (Air = 1)	
SPECIFIC GRAVITY (H_2O = 1)	11.3436	EVAPORATION RATE (= 1)	
SOLUBILITY IN WATER Pts/100 pts H_2O	Insoluble	VOLATILE % Vol. % Wt.	

APPEARANCE AND ODOR: Silver blue soft metal.

SECTION IV FIRE AND EXPLOSION HAZARD DATA

FLASH POINT: None available | FLAMMABLE (EXPLOSIVE) LIMITS: UPPER / LOWER

METHOD USED

EXTINGUISHING MEDIA: Not available

SPECIAL FIRE FIGHTING PROCEDURES

UNUSUAL FIRE AND EXPLOSION HAZARDS

SECTION V HEALTH HAZARD DATA

TOXIC LEVEL: TLV - 0.05 mg/C.M. | CARCINOGENIC: No

PRINCIPLE ROUTES OF ABSORBTION: Inhalation of dust or fume, ingestion | SKIN AND EYE IRRITATION: No

REVELANT SYMPTOMS OF EXPOSURE: Abdominal pain, constipation, diarrhea, loss of appetite, muscular weakness, headache, line on gumes, pyorrhea.

EFFECTS OF CHRONIC EXPOSURE: Loss of appetite, muscular weakness, loss of weight, blue line on gums.

EMERGENCY AND FIRST AID PROCEDURES: If swallowed - give egg white, milk or tannin, also gastric lavage with 1% solution of magnesium sulfate if inhaled - rest, keep warm, get medical aid.

SECTION VI REACTIVITY DATA

CONDITIONS CONTRIBUTING TO INSTABILITY

CONDITIONS CONTRIBUTING TO HAZARDOUS POLYMERIZATION

INCOMPATABILITY (MATERIALS TO AVOID): Ammonium nitrate, hydrogen pweroxide, zirconium.

HAZARDOUS DECOMPOSITION PRODUCTS

SECTION VII SPILL OR LEAK PROCEDURES

STEPS TO BE TAKEN IN CASE MATERIAL IS RELEASED OR SPILLED

WASTE DISPOSAL METHOD

SECTION VIII SPECIAL PROTECTION INFORMATION

VENTILATION REQUIREMENTS LOCAL EXHAUST | PROTECTIVE EQUIPMENT (SPECIFY TYPES) EYE: Goggles

MECHANICAL (GENERAL) | GLOVES: Rubber

SPECIAL | RESPIRATOR: With proper filter

OTHER PROTECTIVE EQUIPMENT

SECTION IX SPECIAL PRECAUTIONS

PRECAUTIONS TO BE TAKEN IN HANDLING AND STORAGE

OTHER PRECAUTIONS

SIGNATURE | DATE

LEAD OXIDE

0
4 0

SECTION I SOURCE AND NOMENCLATURE

MANUFACTURER'S NAME | EMERGENCY TELEPHONE NO.

ADDRESS (NUMBER, STREET, CITY, STATE, ZIP CODE)

TRADE NAME AND SYNONYMS: C.I. 77577, C.l. Pigment yellow 46 | CHEMICAL FAMILY

CHEMICAL NAME AND SYNONYMS: Litharge | FORMULA: PbO

SECTION II HAZARDOUS INGREDIENTS

BASIC MATERIAL	APPROXIMATE OR MAXIMUM % WT. OR VOL.	ESTABLISHED OSHA STANDARD	LD_{50} ORAL	LD_{50} PERCUT.	LC_{50} SPECIES	LC_{50} CONC.
			See Note			
Intraperitoneal - rat	LD50 450 mg/kg.					

SECTION III PHYSICAL DATA MW-223.19

BOILING POINT (°F.)		VAPOR PRESSURE (mm Hg.)	
MELTING POINT (°F.)	888^0 C.	VAPOR DENSITY (Air = 1)	
SPECIFIC GRAVITY (H_2O = 1)	9.53	EVAPORATION RATE (= 1)	
SOLUBILITY IN WATER Pts/100 pts H_2O	0.0068	VOLATILE % Vol. % Wt.	

APPEARANCE AND ODOR: Yellow powder

SECTION IV FIRE AND EXPLOSION HAZARD DATA

FLASH POINT: Will not burn | FLAMMABLE (EXPLOSIVE) LIMITS: UPPER / LOWER

METHOD USED

EXTINGUISHING MEDIA: Flood with water

SPECIAL FIRE FIGHTING PROCEDURES

UNUSUAL FIRE AND EXPLOSION HAZARDS

SECTION V HEALTH HAZARD DATA

TOXIC LEVEL: TLV - TWA 0.054 mg/m^3 | CARCINOGENIC: no

PRINCIPLE ROUTES OF ABSORBTION: Inhalation of dust or ingestion | SKIN AND EYE IRRITATION: No

REVELANT SYMPTOMS OF EXPOSURE: Headache, muscular weakness, constipation, collapse, gastro-enteric symptoms, abdominal pain, vomiting.

EFFECTS OF CHRONIC EXPOSURE: Loss of appetite, anemia, loss of weight, blue line bn gums.

EMERGENCY AND FIRST AID PROCEDURES: If inhaled - rest, get medical attention; if swallowed - wash mouth, give emetic, then epsom salts 30 g/250 ml. of hot water, get medical attention; wash from skin and eyes.

SECTION VI REACTIVITY DATA

CONDITIONS CONTRIBUTING TO INSTABILITY

CONDITIONS CONTRIBUTING TO HAZARDOUS POLYMERIZATION

INCOMPATABILITY (MATERIALS TO AVOID)

HAZARDOUS DECOMPOSITION PRODUCTS

SECTION VII SPILL OR LEAK PROCEDURES

STEPS TO BE TAKEN IN CASE MATERIAL IS RELEASED OR SPILLED

WASTE DISPOSAL METHOD

SECTION VIII SPECIAL PROTECTION INFORMATION

VENTILATION REQUIREMENTS LOCAL EXHAUST | PROTECTIVE EQUIPMENT (SPECIFY TYPES) EYE: Face shield

MECHANICAL (GENERAL) | GLOVES

SPECIAL | RESPIRATOR: With appropriate filter

OTHER PROTECTIVE EQUIPMENT: Face shield

SECTION IX SPECIAL PRECAUTIONS

PRECAUTIONS TO BE TAKEN IN HANDLING AND STORAGE

OTHER PRECAUTIONS

SIGNATURE | DATE

LEAD SULFATE

Hazard diamond: 0 / 4 / 0

SECTION I SOURCE AND NOMENCLATURE

MANUFACTURER'S NAME | EMERGENCY TELEPHONE NO.

ADDRESS (NUMBER, STREET, CITY, STATE, ZIP CODE)

TRADE NAME AND SYNONYMS: C I pigment white 3 | CHEMICAL FAMILY

CHEMICAL NAME AND SYNONYMS: White lead | FORMULA: $PbSO_4$

SECTION II HAZARDOUS INGREDIENTS

BASIC MATERIAL	APPROXIMATE OR MAXIMUM % WT. OR VOL.	ESTABLISHED OSHA STANDARD	LD_{50} ORAL	LD_{50} PERCUT.	LC_{50} SPECIES	LC_{50} CONC.
LD_{50} intraperitoneal - guinea pig 300 mg/kg.						
LDLO - oral - guinea pig 30 mg/kg.						

SECTION III PHYSICAL DATA 303.27 - MW

BOILING POINT °F: N/A | VAPOR PRESSURE mm Hg.

MELTING POINT °F: 1170 | VAPOR DENSITY (Air = 1)

SPECIFIC GRAVITY (H_2O = 1): 6.2 | EVAPORATION RATE (____ = 1)

SOLUBILITY IN WATER Pts/100 pts H_2O: 0.0028 | VOLATILE % Vol. % Wt.

APPEARANCE AND ODOR: White powder

SECTION IV FIRE AND EXPLOSION HAZARD DATA

FLASH POINT: Will not burn | FLAMMABLE (EXPLOSIVE) LIMITS: UPPER / LOWER

METHOD USED

EXTINGUISHING MEDIA: Water

SPECIAL FIRE FIGHTING PROCEDURES: N/A

UNUSUAL FIRE AND EXPLOSION HAZARDS: N/A

SECTION V HEALTH HAZARD DATA

TOXIC LEVEL: TLV - 0.073 mg/m^3 | CARCINOGENIC: no

PRINCIPLE ROUTES OF ABSORBTION: Inhalation of dust or ingestion | SKIN AND EYE IRRITATION: No

REVELANT SYMPTOMS OF EXPOSURE: Headache, muscular weakness, constipation, collapse, gastro-enteric symptoms, abdominal pain, vomiting.

EFFECTS OF CHRONIC EXPOSURE: Loss of appetite, anemia, loss of weight, blue line on gums.

EMERGENCY AND FIRST AID PROCEDURES: If inhaled - rest, get medical attention. If swallowed - wash mouth, give emetic, then epsom salts 30 g/250 ml. hot water, get medical attention; wash from skin and eyes.

SECTION VI REACTIVITY DATA

CONDITIONS CONTRIBUTING TO INSTABILITY

CONDITIONS CONTRIBUTING TO HAZARDOUS POLYMERIZATION

INCOMPATABILITY (MATERIALS TO AVOID): Reacts with lithium carbide exothermally.

HAZARDOUS DECOMPOSITION PRODUCTS

SECTION VII SPILL OR LEAK PROCEDURES

STEPS TO BE TAKEN IN CASE MATERIAL IS RELEASED OR SPILLED

WASTE DISPOSAL METHOD

SECTION VIII SPECIAL PROTECTION INFORMATION

VENTILATION REQUIREMENTS LOCAL EXHAUST | PROTECTIVE EQUIPMENT (SPECIFY TYPES) EYE: Face shield

MECHANICAL (GENERAL) | GLOVES

SPECIAL | RESPIRATOR: With appropriate filter

OTHER PROTECTIVE EQUIPMENT

SECTION IX SPECIAL PRECAUTIONS

PRECAUTIONS TO BE TAKEN IN HANDLING AND STORAGE

OTHER PRECAUTIONS

SIGNATURE | DATE

LITHIUM FLUORIDE

Hazard diamond: 0 / 2 / 0

SECTION I SOURCE AND NOMENCLATURE

MANUFACTURER'S NAME | EMERGENCY TELEPHONE NO.

ADDRESS (NUMBER, STREET, CITY, STATE, ZIP CODE)

TRADE NAME AND SYNONYMS | CHEMICAL FAMILY

CHEMICAL NAME AND SYNONYMS: Lithium Fluoride | FORMULA: LiF

SECTION II HAZARDOUS INGREDIENTS

BASIC MATERIAL	APPROXIMATE OR MAXIMUM % WT. OR VOL.	ESTABLISHED OSHA STANDARD	LD_{50} mg/kg ORAL	LD_{50} mg/kg PERCUT.	LC_{50} SPECIES	LC_{50} CONC.
Lowest published lethal dose						
oral guinea pig 200 mg/kg						
subcutaneous - guinea pig 2000 mg/kg.						

SECTION III PHYSICAL DATA MW-25.94

BOILING POINT °F: 1670° C. | VAPOR PRESSURE mm Hg.

MELTING POINT °F: 870° C. | VAPOR DENSITY (Air = 1)

SPECIFIC GRAVITY (H_2O = 1): 2.295 | EVAPORATION RATE (____ = 1)

SOLUBILITY IN WATER Pts/100 pts H_2O: 0.13 at 0° C. | VOLATILE % Vol. % Wt.

APPEARANCE AND ODOR: White powder

SECTION IV FIRE AND EXPLOSION HAZARD DATA

FLASH POINT: Does not burn | FLAMMABLE (EXPLOSIVE) LIMITS: UPPER / LOWER

METHOD USED

EXTINGUISHING MEDIA

SPECIAL FIRE FIGHTING PROCEDURES: May produce HF with water.

UNUSUAL FIRE AND EXPLOSION HAZARDS: Wear self-contained mask.

SECTION V HEALTH HAZARD DATA

TOXIC LEVEL: TLV - 2.4 mg/M^3 | CARCINOGENIC: no

PRINCIPLE ROUTES OF ABSORBTION: Inhalation of dust (mainly) ingestion | SKIN AND EYE IRRITATION: Moderate

REVELANT SYMPTOMS OF EXPOSURE: Vomiting, abdominal pain, diarrhea, convulsions, muscular weakness, salivation, nausea, irritation of respiratory system.

EFFECTS OF CHRONIC EXPOSURE: Increase in fluoride content of teeth.

EMERGENCY AND FIRST AID PROCEDURES: Inhalation - rest, keep warm, get medical attention; if swallowed - wash mouth, drink water, then milk of magnesia, get medical aid; wash from skin and eyes.

SECTION VI REACTIVITY DATA

CONDITIONS CONTRIBUTING TO INSTABILITY

CONDITIONS CONTRIBUTING TO HAZARDOUS POLYMERIZATION

INCOMPATABILITY (MATERIALS TO AVOID): May produce HF with water.

HAZARDOUS DECOMPOSITION PRODUCTS

SECTION VII SPILL OR LEAK PROCEDURES

STEPS TO BE TAKEN IN CASE MATERIAL IS RELEASED OR SPILLED

WASTE DISPOSAL METHOD

SECTION VIII SPECIAL PROTECTION INFORMATION

VENTILATION REQUIREMENTS LOCAL EXHAUST | PROTECTIVE EQUIPMENT (SPECIFY TYPES) EYE: Face shield

MECHANICAL (GENERAL) | GLOVES

SPECIAL | RESPIRATOR: With proper filter

OTHER PROTECTIVE EQUIPMENT: Face shield and protective clothing.

SECTION IX SPECIAL PRECAUTIONS

PRECAUTIONS TO BE TAKEN IN HANDLING AND STORAGE

OTHER PRECAUTIONS

SIGNATURE | DATE

LITHIUM HYDROXIDE

(Health 3, Flammability 0, Reactivity 1)

SECTION I SOURCE AND NOMENCLATURE

MANUFACTURER'S NAME | EMERGENCY TELEPHONE NO.

ADDRESS (NUMBER, STREET, CITY, STATE, ZIP CODE)

TRADE NAME AND SYNONYMS | CHEMICAL FAMILY

CHEMICAL NAME AND SYNONYMS: Lithium Hydroxide | FORMULA: $LiOH.H_2O$

SECTION II HAZARDOUS INGREDIENTS

BASIC MATERIAL	APPROXIMATE OR MAXIMUM % WT. OR VOL.	ESTABLISHED OSHA STANDARD	LD_{50} ORAL	LD_{50} PERCUT.	LC_{50} SPECIES	LC_{50} CONC.
			none available			

SECTION III PHYSICAL DATA

BOILING POINT °F: 925° C.

VAPOR PRESSURE mm Hg.

MELTING POINT °F: 445° C.

VAPOR DENSITY (Air = 1)

SPECIFIC GRAVITY (H_2O = 1): 2.54

EVAPORATION RATE (= 1)

SOLUBILITY IN WATER Pts/100 pts H_2O: 22.3

VOLATILE % Vol. % Wt.

APPEARANCE AND ODOR: White crystals

SECTION IV FIRE AND EXPLOSION HAZARD DATA

FLASH POINT: Not flammable

FLAMMABLE (EXPLOSIVE) LIMITS: UPPER | LOWER

METHOD USED

EXTINGUISHING MEDIA: Water

SPECIAL FIRE FIGHTING PROCEDURES: N/A

UNUSUAL FIRE AND EXPLOSION HAZARDS: N/A

SECTION V HEALTH HAZARD DATA

TOXIC LEVEL: Not available 2 mg/C.M. (NaOH)

CARCINOGENIC: no

PRINCIPLE ROUTES OF ABSORBTION: Inhalation, ingestion

SKIN AND EYE IRRITATION: Severe

REVELANT SYMPTOMS OF EXPOSURE: Severe attack of membranes in digestive tract and stomach, anorexia, nausea, vomiting, diarrhea, salivation, irritation of skin and eyes.

EFFECTS OF CHRONIC EXPOSURE: Kidney damage.

EMERGENCY AND FIRST AID PROCEDURES: Wash from skin and eyes - if swallowed - wash mouth, drink water, then 1% vinegar, get medical aid, if inhaled - rest, keep warm, get medical aid.

SECTION VI REACTIVITY DATA

CONDITIONS CONTRIBUTING TO INSTABILITY

CONDITIONS CONTRIBUTING TO HAZARDOUS POLYMERIZATION

INCOMPATABILITY (MATERIALS TO AVOID): Reacts with acids

HAZARDOUS DECOMPOSITION PRODUCTS

SECTION VII SPILL OR LEAK PROCEDURES

STEPS TO BE TAKEN IN CASE MATERIAL IS RELEASED OR SPILLED

WASTE DISPOSAL METHOD

SECTION VIII SPECIAL PROTECTION INFORMATION

VENTILATION REQUIREMENTS LOCAL EXHAUST

PROTECTIVE EQUIPMENT (SPECIFY TYPES) EYE: Face shield

MECHANICAL (GENERAL)

GLOVES: Rubber

SPECIAL

RESPIRATOR: With proper filter

OTHER PROTECTIVE EQUIPMENT: Protective clothing

SECTION IX SPECIAL PRECAUTIONS

PRECAUTIONS TO BE TAKEN IN HANDLING AND STORAGE

OTHER PRECAUTIONS

SIGNATURE | DATE

2,4-LUTIDINE

(Health 4, Flammability 2, Reactivity 0)

SECTION I SOURCE AND NOMENCLATURE

MANUFACTURER'S NAME | EMERGENCY TELEPHONE NO.

ADDRESS (NUMBER, STREET, CITY, STATE, ZIP CODE)

TRADE NAME AND SYNONYMS | CHEMICAL FAMILY

CHEMICAL NAME AND SYNONYMS: 2,4 Dimethyl Pyridine | FORMULA: $C_5NH_3(CH_3)_2$

SECTION II HAZARDOUS INGREDIENTS

BASIC MATERIAL	APPROXIMATE OR MAXIMUM % WT. OR VOL.	ESTABLISHED OSHA STANDARD	LD_{50} ORAL	LD_{50} PERCUT.	LC_{50} SPECIES	LC_{50} CONC.
			200		rat	

SECTION III PHYSICAL DATA MW - 107.15

BOILING POINT °F: 157°C

VAPOR PRESSURE mm Hg.

MELTING POINT °F

VAPOR DENSITY (Air = 1)

SPECIFIC GRAVITY (H_2O = 1): 0.95

EVAPORATION RATE (= 1)

SOLUBILITY IN WATER Pts/100 pts H_2O: 20

VOLATILE % Vol. % Wt.

APPEARANCE AND ODOR: clear liquid - fish-like odor

SECTION IV FIRE AND EXPLOSION HAZARD DATA

FLASH POINT: none available

FLAMMABLE (EXPLOSIVE) LIMITS: UPPER | LOWER

METHOD USED

EXTINGUISHING MEDIA: alcohol foam, carbon dioxide, dry chemical

SPECIAL FIRE FIGHTING PROCEDURES: Wear goggles and self-contained breathing apparatus.

UNUSUAL FIRE AND EXPLOSION HAZARDS

SECTION V HEALTH HAZARD DATA

TOXIC LEVEL: none available

CARCINOGENIC: no

PRINCIPLE ROUTES OF ABSORBTION: inhalation, ingestion, skin

SKIN AND EYE IRRITATION: severe

REVELANT SYMPTOMS OF EXPOSURE: Narcosis, headache, nausea, giddiness, vomiting.

EFFECTS OF CHRONIC EXPOSURE: Occasional vomiting and diarrhea; weight loss and anemia; ocular and facial paralysis.

EMERGENCY AND FIRST AID PROCEDURES: If swallowed - induce vomiting, get medical aid; inhalation - artificial respiration and oxygen as necessary; wash from skin and eyes.

SECTION VI REACTIVITY DATA

CONDITIONS CONTRIBUTING TO INSTABILITY

CONDITIONS CONTRIBUTING TO HAZARDOUS POLYMERIZATION

INCOMPATABILITY (MATERIALS TO AVOID): May react with oxidizing materials.

HAZARDOUS DECOMPOSITION PRODUCTS

SECTION VII SPILL OR LEAK PROCEDURES

STEPS TO BE TAKEN IN CASE MATERIAL IS RELEASED OR SPILLED

WASTE DISPOSAL METHOD

SECTION VIII SPECIAL PROTECTION INFORMATION

VENTILATION REQUIREMENTS LOCAL EXHAUST: Use in hood

PROTECTIVE EQUIPMENT (SPECIFY TYPES) EYE: goggles

MECHANICAL (GENERAL)

GLOVES: rubber

SPECIAL

RESPIRATOR: self-contained

OTHER PROTECTIVE EQUIPMENT

SECTION IX SPECIAL PRECAUTIONS

PRECAUTIONS TO BE TAKEN IN HANDLING AND STORAGE: Outside storage preferred, store inside in a flammable liquids storage cabinet.

OTHER PRECAUTIONS

SIGNATURE | DATE

MALONIC ACID

1
3 0

SECTION I SOURCE AND NOMENCLATURE

MANUFACTURER'S NAME | EMERGENCY TELEPHONE NO.

ADDRESS (NUMBER, STREET, CITY, STATE, ZIP CODE)

TRADE NAME AND SYNONYMS | CHEMICAL FAMILY

CHEMICAL NAME AND SYNONYMS: Propanedioic acid

FORMULA: $HOOCCH_2COOH$

SECTION II HAZARDOUS INGREDIENTS

BASIC MATERIAL	APPROXIMATE OR MAXIMUM % WT. OR VOL.	ESTABLISHED OSHA STANDARD	LD_{50} mg/kg ORAL	PERCUT.	LC_{50} SPECIES	CONC.
			1310		Rat	
			4000		Mouse	

Note: LD_{50} intraperitoneal - rat 1540 mg/kg
LD_{50} intraperitoneal - mouse 300 mg/kg

SECTION III PHYSICAL DATA Mol. Wt. 104.06

BOILING POINT °F.: Decomposes

VAPOR PRESSURE mm Hg.:

MELTING POINT °F.: 132-134°C

VAPOR DENSITY (Air = 1):

SPECIFIC GRAVITY (H_2O = 1): 1.67

EVAPORATION RATE (________ = 1):

SOLUBILITY IN WATER Pts/100 pts H_2O: V.S.

VOLATILE % Vol. % Wt.

APPEARANCE AND ODOR: White solid

SECTION IV FIRE AND EXPLOSION HAZARD DATA

FLASH POINT: Not available

METHOD USED

FLAMMABLE (EXPLOSIVE) LIMITS: UPPER / LOWER

EXTINGUISHING MEDIA: Alcohol foam, carbon dioxide, dry chemical.

SPECIAL FIRE FIGHTING PROCEDURES: Wear full protective clothing.

UNUSUAL FIRE AND EXPLOSION HAZARDS

SECTION V HEALTH HAZARD DATA

TOXIC LEVEL: Not available

CARCINOGENIC: No

PRINCIPLE ROUTES OF ABSORBTION: Inhalation, ingestion

SKIN AND EYE IRRITATION: Severe

REVELANT SYMPTOMS OF EXPOSURE: Gastric pain, headache, nausea and vomiting, cough, tearing and respiratory oppression.

EFFECTS OF CHRONIC EXPOSURE: Errosion of teeth, bronchial irritation, gastrointestinal disturbance.

EMERGENCY AND FIRST AID PROCEDURES: Inhalation - rest, get medical aid if needed; ingestion - wash mouth, drink water, get medical aid; wash from skin and eyes.

SECTION VI REACTIVITY DATA

CONDITIONS CONTRIBUTING TO INSTABILITY

CONDITIONS CONTRIBUTING TO HAZARDOUS POLYMERIZATION

INCOMPATABILITY (MATERIALS TO AVOID)

HAZARDOUS DECOMPOSITION PRODUCTS: Gives off flammable vapor when heated.

SECTION VII SPILL OR LEAK PROCEDURES

STEPS TO BE TAKEN IN CASE MATERIAL IS RELEASED OR SPILLED

WASTE DISPOSAL METHOD

SECTION VIII SPECIAL PROTECTION INFORMATION

VENTILATION REQUIREMENTS LOCAL EXHAUST

MECHANICAL (GENERAL)

SPECIAL

PROTECTIVE EQUIPMENT (SPECIFY TYPES) EYE: Safety glasses

GLOVES: Rubber

RESPIRATOR: w/proper filter

OTHER PROTECTIVE EQUIPMENT

SECTION IX SPECIAL PRECAUTIONS

PRECAUTIONS TO BE TAKEN IN HANDLING AND STORAGE

OTHER PRECAUTIONS

SIGNATURE | DATE

MESITYLENE

2
3 0

SECTION I SOURCE AND NOMENCLATURE

MANUFACTURER'S NAME | EMERGENCY TELEPHONE NO.

ADDRESS (NUMBER, STREET, CITY, STATE, ZIP CODE)

TRADE NAME AND SYNONYMS | CHEMICAL FAMILY

CHEMICAL NAME AND SYNONYMS: 1,3,5 Trimethylbenzene

FORMULA: $(CH_3)_3C_6H_3$

SECTION II HAZARDOUS INGREDIENTS

BASIC MATERIAL	APPROXIMATE OR MAXIMUM % WT. OR VOL.	ESTABLISHED OSHA STANDARD	LD_{50} ORAL	PERCUT.	LC_{50} SPECIES	CONC.

TCLO inhalation - human 10 PPM - CNS
LDLO intraperitoneal - rat 1500 mg/kg
LCLO inhalation - rat 2400 ppm/24 hr.

SECTION III PHYSICAL DATA MW - 120.19

BOILING POINT °F.: 165°C

VAPOR PRESSURE mm Hg.: at 20°C 1.86

MELTING POINT °F.: -53°C

VAPOR DENSITY (Air = 1): 1.01

SPECIFIC GRAVITY (H_2O = 1): 0.86

EVAPORATION RATE (________ = 1):

SOLUBILITY IN WATER Pts/100 pts H_2O: slghtly soluble

VOLATILE % Vol. % Wt.

APPEARANCE AND ODOR: liquid, peculiar odor

SECTION IV FIRE AND EXPLOSION HAZARD DATA

FLASH POINT: 122°F

METHOD USED

FLAMMABLE (EXPLOSIVE) LIMITS: UPPER / LOWER

EXTINGUISHING MEDIA: Alcohol foam, carbon dioxide, dry chemical

SPECIAL FIRE FIGHTING PROCEDURES: Wear full protective clothing.

UNUSUAL FIRE AND EXPLOSION HAZARDS

SECTION V HEALTH HAZARD DATA

TOXIC LEVEL: TLV 25 ppm - 120 mg/C.M.

CARCINOGENIC: no

PRINCIPLE ROUTES OF ABSORBTION: Inhalation, ingestion, skin

SKIN AND EYE IRRITATION: severe

REVELANT SYMPTOMS OF EXPOSURE: Chemical pneumonia, central nervous system depression, frequent bleeding.

EFFECTS OF CHRONIC EXPOSURE: Central nervous system depression, nervousness, tension, anxiety, asthmatic bronchitis.

EMERGENCY AND FIRST AID PROCEDURES: Inhalation - oxygen as needed; if swallowed - get medical aid, gastric lavage; wash from skin & eyes.

SECTION VI REACTIVITY DATA

CONDITIONS CONTRIBUTING TO INSTABILITY

CONDITIONS CONTRIBUTING TO HAZARDOUS POLYMERIZATION

INCOMPATABILITY (MATERIALS TO AVOID)

HAZARDOUS DECOMPOSITION PRODUCTS

SECTION VII SPILL OR LEAK PROCEDURES

STEPS TO BE TAKEN IN CASE MATERIAL IS RELEASED OR SPILLED: Absorb on paper towels, evaporate in hood.

WASTE DISPOSAL METHOD

SECTION VIII SPECIAL PROTECTION INFORMATION

VENTILATION REQUIREMENTS LOCAL EXHAUST

MECHANICAL (GENERAL)

SPECIAL

PROTECTIVE EQUIPMENT (SPECIFY TYPES) EYE: Goggles

GLOVES: Rubber

RESPIRATOR: With proper filter

OTHER PROTECTIVE EQUIPMENT: Protective clothing

SECTION IX SPECIAL PRECAUTIONS

PRECAUTIONS TO BE TAKEN IN HANDLING AND STORAGE: Store in flammable liquids storage cabinet.

OTHER PRECAUTIONS

SIGNATURE | DATE

METHYLAMINE HYDROCHLORIDE

Hazard diamond: 1 (top), 3 (left), 0 (right)

SECTION I SOURCE AND NOMENCLATURE

MANUFACTURER'S NAME	EMERGENCY TELEPHONE NO.
ADDRESS (NUMBER, STREET, CITY, STATE, ZIP CODE)	
TRADE NAME AND SYNONYMS	CHEMICAL FAMILY
CHEMICAL NAME AND SYNONYMS Methyl ammonium chloride	FORMULA CH_3NH_2HCl

SECTION II HAZARDOUS INGREDIENTS

BASIC MATERIAL	APPROXIMATE OR MAXIMUM % WT. OR VOL.	ESTABLISHED OSHA STANDARD	LD_{50} ORAL	LD_{50} PERCUT.	LC_{50} SPECIES	LC_{50} CONC.
Note: Lowest published lethal dose; skin - rat 800 mg/kg						

SECTION III PHYSICAL DATA Mol. Wt. 67.52

BOILING POINT °F.	VAPOR PRESSURE at 230°C 15 mm Hg.
MELTING POINT °F. 228°C	VAPOR DENSITY (Air = 1)
SPECIFIC GRAVITY (H_2O = 1) 1.23	EVAPORATION RATE (____ = 1)
SOLUBILITY IN WATER Pts/100 pts H_2O Very soluble	VOLATILE % Vol. % Wt.
APPEARANCE AND ODOR	

SECTION IV FIRE AND EXPLOSION HAZARD DATA

FLASH POINT: Estimated 200°C

METHOD USED:

FLAMMABLE (EXPLOSIVE) LIMITS: UPPER / LOWER

EXTINGUISHING MEDIA: Water spray, alcohol foam, CO_2, dry chemical.

SPECIAL FIRE FIGHTING PROCEDURES: Wear full protective clothing

UNUSUAL FIRE AND EXPLOSION HAZARDS:

SECTION V HEALTH HAZARD DATA

TOXIC LEVEL: None available — CARCINOGENIC: No

PRINCIPLE ROUTES OF ABSORBTION: Ingestion, inhalation of dust — SKIN AND EYE IRRITATION: Mild

REVELANT SYMPTOMS OF EXPOSURE: Cardiac stimulation, eye irritation, pulmonary edema, headache, faintness and anxiety.

EFFECTS OF CHRONIC EXPOSURE:

EMERGENCY AND FIRST AID PROCEDURES: Inhalation - rest, keep warm, get medical attention; if swallowed - wash mouth with water, give plenty of water to drink; wash skin and eyes with water.

SECTION VI REACTIVITY DATA

CONDITIONS CONTRIBUTING TO INSTABILITY:

CONDITIONS CONTRIBUTING TO HAZARDOUS POLYMERIZATION:

INCOMPATABILITY (MATERIALS TO AVOID):

HAZARDOUS DECOMPOSITION PRODUCTS:

SECTION VII SPILL OR LEAK PROCEDURES

STEPS TO BE TAKEN IN CASE MATERIAL IS RELEASED OR SPILLED:

WASTE DISPOSAL METHOD:

SECTION VIII SPECIAL PROTECTION INFORMATION

VENTILATION REQUIREMENTS LOCAL EXHAUST	PROTECTIVE EQUIPMENT (SPECIFY TYPES) EYE Safety glasses
MECHANICAL (GENERAL)	GLOVES Rubber
SPECIAL	RESPIRATOR w/appropriate filter

OTHER PROTECTIVE EQUIPMENT: Face shield

SECTION IX SPECIAL PRECAUTIONS

PRECAUTIONS TO BE TAKEN IN HANDLING AND STORAGE:

OTHER PRECAUTIONS:

SIGNATURE: DATE:

METHYL n-AMYL KETONE

Hazard diamond: 2 (top), 3 (left), 0 (right)

SECTION I SOURCE AND NOMENCLATURE

MANUFACTURER'S NAME	EMERGENCY TELEPHONE NO.
ADDRESS (NUMBER, STREET, CITY, STATE, ZIP CODE)	
TRADE NAME AND SYNONYMS	CHEMICAL FAMILY
CHEMICAL NAME AND SYNONYMS 2 Heptanone	FORMULA $CH_3(CH_2)_4COCH_3$

SECTION II HAZARDOUS INGREDIENTS

BASIC MATERIAL	APPROXIMATE OR MAXIMUM % WT. OR VOL.	ESTABLISHED OSHA STANDARD	LD_{50} mg/kg ORAL	LD_{50} mg/kg PERCUT.	LC_{50} SPECIES	LC_{50} CONC.
			1670		rat	
			730		mouse	
				13,000	rabbit	
Lowest published lethal concentration in air inhalation - rat 4000 PPM						

SECTION III PHYSICAL DATA MW - 114.18

BOILING POINT °F. 151°C	VAPOR PRESSURE at 20°C 2.6 mm Hg.
MELTING POINT °F. -36°C	VAPOR DENSITY (Air = 1) 3.9
SPECIFIC GRAVITY (H_2O = 1) 0.82	EVAPORATION RATE (____ = 1)
SOLUBILITY IN WATER Pts/100 pts H_2O 0.4	VOLATILE % Vol. % Wt.
APPEARANCE AND ODOR Water white liquid	

SECTION IV FIRE AND EXPLOSION HAZARD DATA

FLASH POINT: 120°F

METHOD USED: O.C.

FLAMMABLE (EXPLOSIVE) LIMITS: UPPER / LOWER

EXTINGUISHING MEDIA: Foam, carbon dioxide, dry chemical

SPECIAL FIRE FIGHTING PROCEDURES:

UNUSUAL FIRE AND EXPLOSION HAZARDS:

SECTION V HEALTH HAZARD DATA

TOXIC LEVEL: 100 PPM, 465 mg/kg. — CARCINOGENIC: no

PRINCIPLE ROUTES OF ABSORBTION: Inhalation, ingestion — SKIN AND EYE IRRITATION: moderate

REVELANT SYMPTOMS OF EXPOSURE: Irritation of mucous membranes, narcosis, eye irritation.

EFFECTS OF CHRONIC EXPOSURE: --

EMERGENCY AND FIRST AID PROCEDURES: If swallowed - give emetic, get medical aid; inhalation - rest; wash eyes and skin.

SECTION VI REACTIVITY DATA

CONDITIONS CONTRIBUTING TO INSTABILITY:

CONDITIONS CONTRIBUTING TO HAZARDOUS POLYMERIZATION:

INCOMPATABILITY (MATERIALS TO AVOID):

HAZARDOUS DECOMPOSITION PRODUCTS:

SECTION VII SPILL OR LEAK PROCEDURES

STEPS TO BE TAKEN IN CASE MATERIAL IS RELEASED OR SPILLED:

WASTE DISPOSAL METHOD:

SECTION VIII SPECIAL PROTECTION INFORMATION

VENTILATION REQUIREMENTS LOCAL EXHAUST	PROTECTIVE EQUIPMENT (SPECIFY TYPES) EYE Goggles
MECHANICAL (GENERAL)	GLOVES Neoprene
SPECIAL	RESPIRATOR With proper filter

OTHER PROTECTIVE EQUIPMENT:

SECTION IX SPECIAL PRECAUTIONS

PRECAUTIONS TO BE TAKEN IN HANDLING AND STORAGE:

OTHER PRECAUTIONS:

SIGNATURE: DATE:

METHYL BROMOACETATE

2 / 3 / 0

SECTION I SOURCE AND NOMENCLATURE

MANUFACTURER'S NAME	EMERGENCY TELEPHONE NO.
ADDRESS (NUMBER, STREET, CITY, STATE, ZIP CODE)	
TRADE NAME AND SYNONYMS	CHEMICAL FAMILY
CHEMICAL NAME AND SYNONYMS Methyl Bromoacetate	FORMULA $BrCH_2CO_2CH_3$

SECTION II HAZARDOUS INGREDIENTS

BASIC MATERIAL	APPROXIMATE OR MAXIMUM % WT. OR VOL.	ESTABLISHED OSHA STANDARD	LD_{50} ORAL	LD_{50} PERCUT.	LC_{50} SPECIES	LC_{50} CONC.
			See Note			
Note:						
Lowest published lethal dose						
intravenous - mouse	10 mg/kg					

SECTION III PHYSICAL DATA MW - 152.99

BOILING POINT °F. 144 d	VAPOR PRESSURE mm Hg. 52°C 15
MELTING POINT °F.	VAPOR DENSITY (Air = 1)
SPECIFIC GRAVITY (H_2O = 1) > 1	EVAPORATION RATE (____ = 1)
SOLUBILITY IN WATER Pts/100 pts H_2O low solubility	VOLATILE % Vol. % Wt.

APPEARANCE AND ODOR
colorless liquid

SECTION IV FIRE AND EXPLOSION HAZARD DATA

FLASH POINT none available	FLAMMABLE (EXPLOSIVE) LIMITS	UPPER
METHOD USED		LOWER

EXTINGUISHING MEDIA
Alcohol foam, carbon dioxide, dry chemical and water.

SPECIAL FIRE FIGHTING PROCEDURES

UNUSUAL FIRE AND EXPLOSION HAZARDS

SECTION V HEALTH HAZARD DATA

TOXIC LEVEL
None available

CARCINOGENIC

PRINCIPLE ROUTES OF ABSORBTION
Inhalation, ingestion, skin

SKIN AND EYE IRRITATION
severe

REVELANT SYMPTOMS OF EXPOSURE Irritation of eyes and upper respiratory system, dizziness and depression, palpitation of heart.

EFFECTS OF CHRONIC EXPOSURE
--

EMERGENCY AND FIRST AID PROCEDURES If swallowed - give emetic, get medical aid; inhalation - rest; wash from skin & eyes.

SECTION VI REACTIVITY DATA

CONDITIONS CONTRIBUTING TO INSTABILITY

CONDITIONS CONTRIBUTING TO HAZARDOUS POLYMERIZATION

INCOMPATABILITY (MATERIALS TO AVOID)

HAZARDOUS DECOMPOSITION PRODUCTS

SECTION VII SPILL OR LEAK PROCEDURES

STEPS TO BE TAKEN IN CASE MATERIAL IS RELEASED OR SPILLED

WASTE DISPOSAL METHOD

SECTION VIII SPECIAL PROTECTION INFORMATION

VENTILATION REQUIREMENTS LOCAL EXHAUST	PROTECTIVE EQUIPMENT (SPECIFY TYPES) EYE Goggles
MECHANICAL (GENERAL)	GLOVES Rubber
SPECIAL	RESPIRATOR With proper filter
OTHER PROTECTIVE EQUIPMENT	

SECTION IX SPECIAL PRECAUTIONS

PRECAUTIONS TO BE TAKEN IN HANDLING AND STORAGE
Store in standard flammable liquids cabinet.

OTHER PRECAUTIONS

SIGNATURE	DATE

METHYL β-BROMOPROPIONATE

2 / 3 / 0

SECTION I SOURCE AND NOMENCLATURE

MANUFACTURER'S NAME	EMERGENCY TELEPHONE NO.
ADDRESS (NUMBER, STREET, CITY, STATE, ZIP CODE)	
TRADE NAME AND SYNONYMS	CHEMICAL FAMILY
CHEMICAL NAME AND SYNONYMS Methyl 3-Bromopropionate	FORMULA $BrCH_2CH_2COOCH_3$

SECTION II HAZARDOUS INGREDIENTS

BASIC MATERIAL	APPROXIMATE OR MAXIMUM % WT. OR VOL.	ESTABLISHED OSHA STANDARD	LD_{50} ORAL	LD_{50} PERCUT.	LC_{50} SPECIES	LC_{50} CONC.
			none	available		

SECTION III PHYSICAL DATA MW - 167.01

BOILING POINT °F. at 60 mm - 106°C	VAPOR PRESSURE mm Hg. at 60°C 12
MELTING POINT °F. --	VAPOR DENSITY (Air = 1)
SPECIFIC GRAVITY (H_2O = 1) 1.5122	EVAPORATION RATE (____ = 1)
SOLUBILITY IN WATER Pts/100 pts H_2O --	VOLATILE % Vol. % Wt.

APPEARANCE AND ODOR
colorless liquid

SECTION IV FIRE AND EXPLOSION HAZARD DATA

FLASH POINT none available	FLAMMABLE (EXPLOSIVE) LIMITS	UPPER
METHOD USED		LOWER

EXTINGUISHING MEDIA
Foam, carbon dioxide, dry chemical.

SPECIAL FIRE FIGHTING PROCEDURES

UNUSUAL FIRE AND EXPLOSION HAZARDS

SECTION V HEALTH HAZARD DATA

TOXIC LEVEL
none available

CARCINOGENIC
no

PRINCIPLE ROUTES OF ABSORBTION
inhalation, ingestion, skin

SKIN AND EYE IRRITATION
severe

REVELANT SYMPTOMS OF EXPOSURE Irritation of eyes and upper respiratory syste, dizziness and depression, palpitation of heart.

EFFECTS OF CHRONIC EXPOSURE
--

EMERGENCY AND FIRST AID PROCEDURES If swallowed - give emetic, get medical aid; inhalation - rest; wash from skin & eyes.

SECTION VI REACTIVITY DATA

CONDITIONS CONTRIBUTING TO INSTABILITY

CONDITIONS CONTRIBUTING TO HAZARDOUS POLYMERIZATION

INCOMPATABILITY (MATERIALS TO AVOID)

HAZARDOUS DECOMPOSITION PRODUCTS

SECTION VII SPILL OR LEAK PROCEDURES

STEPS TO BE TAKEN IN CASE MATERIAL IS RELEASED OR SPILLED

WASTE DISPOSAL METHOD

SECTION VIII SPECIAL PROTECTION INFORMATION

VENTILATION REQUIREMENTS LOCAL EXHAUST	PROTECTIVE EQUIPMENT (SPECIFY TYPES) EYE Goggles
MECHANICAL (GENERAL)	GLOVES Rubber
SPECIAL	RESPIRATOR With proper filter
OTHER PROTECTIVE EQUIPMENT	

SECTION IX SPECIAL PRECAUTIONS

PRECAUTIONS TO BE TAKEN IN HANDLING AND STORAGE
Store in standard flammable liquids cabinet.

OTHER PRECAUTIONS

SIGNATURE	DATE

METHYL CARBONATE

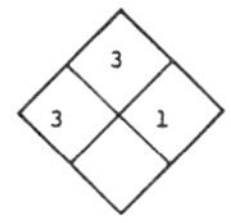

SECTION I SOURCE AND NOMENCLATURE

MANUFACTURER'S NAME	EMERGENCY TELEPHONE NO.
ADDRESS (NUMBER, STREET, CITY, STATE, ZIP CODE)	
TRADE NAME AND SYNONYMS	CHEMICAL FAMILY
CHEMICAL NAME AND SYNONYMS: dimethyl carbonate	FORMULA: $CO(OCH_3)_2$

SECTION II HAZARDOUS INGREDIENTS

BASIC MATERIAL	APPROXIMATE OR MAXIMUM % WT. OR VOL.	ESTABLISHED OSHA STANDARD	LD_{50} ORAL	LD_{50} PERCUT.	LC_{50} SPECIES	LC_{50} CONC.
LD_{50} - 8500 mg/kg subcutaneous rat						

SECTION III PHYSICAL DATA MW - 90.08

BOILING POINT °F.: 90°C	VAPOR PRESSURE mm Hg.:
MELTING POINT °F.: 1°C	VAPOR DENSITY (Air = 1): 3.1
SPECIFIC GRAVITY (H_2O = 1): 1.07	EVAPORATION RATE (______ = 1):
SOLUBILITY IN WATER Pts/100 pts H_2O: insoluble	VOLATILE % Vol. % Wt.

APPEARANCE AND ODOR: colorless liquid, pleasant odor.

SECTION IV FIRE AND EXPLOSION HAZARD DATA

	FLAMMABLE (EXPLOSIVE) LIMITS	
FLASH POINT: 66°F		UPPER
METHOD USED: O.C.		LOWER

EXTINGUISHING MEDIA: Alcohol foam, carbon dioxide, dry chemical. Water may be ineffective.

SPECIAL FIRE FIGHTING PROCEDURES: Wear goggles and self-contained breathing apparatus.

UNUSUAL FIRE AND EXPLOSION HAZARDS

SECTION V HEALTH HAZARD DATA

TOXIC LEVEL: none available — CARCINOGENIC: no

PRINCIPLE ROUTES OF ABSORBTION: inhalation (mainly), ingestion — SKIN AND EYE IRRITATION: moderate

REVELANT SYMPTOMS OF EXPOSURE: Weakness, loss of coordination, gasping, frothing at mouth, pulmonary edema.

EFFECTS OF CHRONIC EXPOSURE: --

EMERGENCY AND FIRST AID PROCEDURES: If swallowed - drink water, milk of magnesia, get medical aid; inhalation - artificial respiration & oxygen as needed - wash from skin and eyes.

SECTION VI REACTIVITY DATA

CONDITIONS CONTRIBUTING TO INSTABILITY

CONDITIONS CONTRIBUTING TO HAZARDOUS POLYMERIZATION

INCOMPATABILITY (MATERIALS TO AVOID)

HAZARDOUS DECOMPOSITION PRODUCTS

SECTION VII SPILL OR LEAK PROCEDURES

STEPS TO BE TAKEN IN CASE MATERIAL IS RELEASED OR SPILLED

WASTE DISPOSAL METHOD

SECTION VIII SPECIAL PROTECTION INFORMATION

	PROTECTIVE EQUIPMENT (SPECIFY TYPES)
VENTILATION REQUIREMENTS LOCAL EXHAUST	EYE: Goggles
MECHANICAL (GENERAL)	GLOVES: Rubber
SPECIAL	RESPIRATOR: Self-contained

OTHER PROTECTIVE EQUIPMENT

SECTION IX SPECIAL PRECAUTIONS

PRECAUTIONS TO BE TAKEN IN HANDLING AND STORAGE: Outside storage preferred, store inside flammable liquids storage cabinet.

OTHER PRECAUTIONS

SIGNATURE	DATE

p-METHYLCYCLOHEXANOL

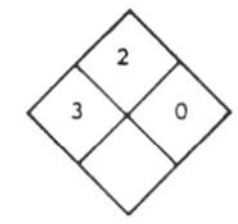

SECTION I SOURCE AND NOMENCLATURE

MANUFACTURER'S NAME	EMERGENCY TELEPHONE NO.
ADDRESS (NUMBER, STREET, CITY, STATE, ZIP CODE)	
TRADE NAME AND SYNONYMS: Hexahydromethyl Phenol	CHEMICAL FAMILY
CHEMICAL NAME AND SYNONYMS: Hexahydrocresol	FORMULA: $CH_3C_6H_{10}OH$

SECTION II HAZARDOUS INGREDIENTS

BASIC MATERIAL	APPROXIMATE OR MAXIMUM % WT. OR VOL.	ESTABLISHED OSHA STANDARD	LD_{50} mg/kg ORAL	LD_{50} mg/kg PERCUT.	LC_{50} SPECIES	LC_{50} CONC.
Inhalation - human	TCLO 500 PPM		1660		Rat	
Lowest published lethal dose oral - rabbit 1750 mg/kg						

SECTION III PHYSICAL DATA MW - 114.18

BOILING POINT °F.: 174°C	VAPOR PRESSURE mm Hg.: at 30°C 1.5
MELTING POINT °F.: -50°C	VAPOR DENSITY (Air = 1): 3.93
SPECIFIC GRAVITY (H_2O = 1): 0.92	EVAPORATION RATE (______ = 1):
SOLUBILITY IN WATER Pts/100 pts H_2O: very slightly soluble	VOLATILE % Vol. % Wt.

APPEARANCE AND ODOR: colorless, viscous liquid, aromatic, menthol odor.

SECTION IV FIRE AND EXPLOSION HAZARD DATA

	FLAMMABLE (EXPLOSIVE) LIMITS	
FLASH POINT: 154°F		UPPER
METHOD USED: C.C.		LOWER

EXTINGUISHING MEDIA: Foam, carbon dioxide or dry chemicals.

SPECIAL FIRE FIGHTING PROCEDURES

UNUSUAL FIRE AND EXPLOSION HAZARDS

SECTION V HEALTH HAZARD DATA

TOXIC LEVEL: 50 PPM, 235 mg/C.M. — CARCINOGENIC: no

PRINCIPLE ROUTES OF ABSORBTION: Inhalation, ingestion, skin — SKIN AND EYE IRRITATION: severe

REVELANT SYMPTOMS OF EXPOSURE: Salivation, eye and mucous membrane irritation, tremors, narcosis, headache, gross hemorrhage.

EFFECTS OF CHRONIC EXPOSURE: Liver and kidney damage

EMERGENCY AND FIRST AID PROCEDURES: If swallowed - give emetic, get medical aid, gastric lavage; inhalation - rest, oxygen as needed; wash from eyes & skin.

SECTION VI REACTIVITY DATA

CONDITIONS CONTRIBUTING TO INSTABILITY

CONDITIONS CONTRIBUTING TO HAZARDOUS POLYMERIZATION

INCOMPATABILITY (MATERIALS TO AVOID): can react with oxidizing materials.

HAZARDOUS DECOMPOSITION PRODUCTS

SECTION VII SPILL OR LEAK PROCEDURES

STEPS TO BE TAKEN IN CASE MATERIAL IS RELEASED OR SPILLED

WASTE DISPOSAL METHOD

SECTION VIII SPECIAL PROTECTION INFORMATION

	PROTECTIVE EQUIPMENT (SPECIFY TYPES)
VENTILATION REQUIREMENTS LOCAL EXHAUST	EYE: goggles
MECHANICAL (GENERAL)	GLOVES: rubber
SPECIAL	RESPIRATOR: with proper filter

OTHER PROTECTIVE EQUIPMENT

SECTION IX SPECIAL PRECAUTIONS

PRECAUTIONS TO BE TAKEN IN HANDLING AND STORAGE

OTHER PRECAUTIONS

SIGNATURE	DATE

m-METHYLCYCLOHEXANONE

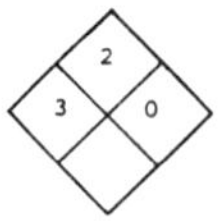

SECTION I SOURCE AND NOMENCLATURE

MANUFACTURER'S NAME

EMERGENCY TELEPHONE NO.

ADDRESS (NUMBER, STREET, CITY, STATE, ZIP CODE)

TRADE NAME AND SYNONYMS

CHEMICAL FAMILY

CHEMICAL NAME AND SYNONYMS: 3 Methylcyclohexanone

FORMULA: $CH_3C_5H_9CO$

SECTION II HAZARDOUS INGREDIENTS

BASIC MATERIAL	APPROXIMATE OR MAXIMUM % WT. OR VOL.	ESTABLISHED OSHA STANDARD	LD_{50}		LC_{50}	
			ORAL	PERCUT.	SPECIES	CONC.
			none	available		

SECTION III PHYSICAL DATA

BOILING POINT: 170°C °F.

VAPOR PRESSURE mm Hg.

MELTING POINT: -74°C °F.

VAPOR DENSITY (Air = 1): 3.86

SPECIFIC GRAVITY (H_2O = 1): 0.92

EVAPORATION RATE (_______ = 1)

SOLUBILITY IN WATER Pts/100 pts H_2O: no

VOLATILE % Vol. % Wt.

APPEARANCE AND ODOR: water white to pale yellow, acetone like odor.

SECTION IV FIRE AND EXPLOSION HAZARD DATA

FLASH POINT: 118°F

METHOD USED: C.C.

FLAMMABLE (EXPLOSIVE) LIMITS: UPPER / LOWER

EXTINGUISHING MEDIA: Foam, carbon dioxide, dry chemical

SPECIAL FIRE FIGHTING PROCEDURES

UNUSUAL FIRE AND EXPLOSION HAZARDS

SECTION V HEALTH HAZARD DATA

TOXIC LEVEL: none available

CARCINOGENIC: no

PRINCIPLE ROUTES OF ABSORBTION: inhalation, ingestion

SKIN AND EYE IRRITATION: moderate

REVELANT SYMPTOMS OF EXPOSURE: eye, skin and mucous membrane irritation, narcosis

EFFECTS OF CHRONIC EXPOSURE: Liver and kidney damage

EMERGENCY AND FIRST AID PROCEDURES: If swallowed - give emetic, get medical aid, gastric lavage; inhalation - give oxygen as needed. Wash from skin & eyes.

SECTION VI REACTIVITY DATA

CONDITIONS CONTRIBUTING TO INSTABILITY

CONDITIONS CONTRIBUTING TO HAZARDOUS POLYMERIZATION

INCOMPATABILITY (MATERIALS TO AVOID)

HAZARDOUS DECOMPOSITION PRODUCTS

SECTION VII SPILL OR LEAK PROCEDURES

STEPS TO BE TAKEN IN CASE MATERIAL IS RELEASED OR SPILLED

WASTE DISPOSAL METHOD

SECTION VIII SPECIAL PROTECTION INFORMATION

VENTILATION REQUIREMENTS LOCAL EXHAUST

PROTECTIVE EQUIPMENT (SPECIFY TYPES) EYE: Safety goggles

MECHANICAL (GENERAL)

GLOVES: Rubber

SPECIAL

RESPIRATOR: With proper filter

OTHER PROTECTIVE EQUIPMENT

SECTION IX SPECIAL PRECAUTIONS

PRECAUTIONS TO BE TAKEN IN HANDLING AND STORAGE

OTHER PRECAUTIONS

SIGNATURE

DATE

p-METHYLCYCLOHEXANONE

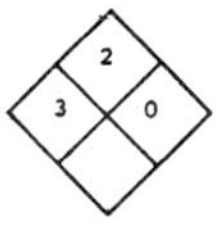

SECTION I SOURCE AND NOMENCLATURE

MANUFACTURER'S NAME

EMERGENCY TELEPHONE NO.

ADDRESS (NUMBER, STREET, CITY, STATE, ZIP CODE)

TRADE NAME AND SYNONYMS

CHEMICAL FAMILY

CHEMICAL NAME AND SYNONYMS: 4 Methylcyclohexanone

FORMULA: $CH_3C_5H_9CO$

SECTION II HAZARDOUS INGREDIENTS

BASIC MATERIAL	APPROXIMATE OR MAXIMUM % WT. OR VOL.	ESTABLISHED OSHA STANDARD	LD_{50}		LC_{50}	
			ORAL	PERCUT.	SPECIES	CONC.
			none	available		

SECTION III PHYSICAL DATA MW = 112.17

BOILING POINT: 171°C °F.

VAPOR PRESSURE mm Hg.

MELTING POINT: -41°C °F.

VAPOR DENSITY (Air = 1): 3.86

SPECIFIC GRAVITY (H_2O = 1): 0.93

EVAPORATION RATE (_______ = 1)

SOLUBILITY IN WATER Pts/100 pts H_2O: no

VOLATILE % Vol. % Wt.

APPEARANCE AND ODOR: water white to pale yellow - acetone like odor.

SECTION IV FIRE AND EXPLOSION HAZARD DATA

FLASH POINT: 118°F

METHOD USED: C.C.

FLAMMABLE (EXPLOSIVE) LIMITS: UPPER / LOWER

EXTINGUISHING MEDIA: Foam, carbon dioxide, dry chemical

SPECIAL FIRE FIGHTING PROCEDURES

UNUSUAL FIRE AND EXPLOSION HAZARDS

SECTION V HEALTH HAZARD DATA

TOXIC LEVEL: none available

CARCINOGENIC: no

PRINCIPLE ROUTES OF ABSORBTION: Inhalation, ingestion

SKIN AND EYE IRRITATION: moderate

REVELANT SYMPTOMS OF EXPOSURE: Eye, skin and mucous membrane irritation, narcosis.

EFFECTS OF CHRONIC EXPOSURE: Liver and kidney damage.

EMERGENCY AND FIRST AID PROCEDURES: If swallowed - give emetic, get medical aid, gastric lavage; inhalation - give oxygen as needed. Wash from skin & eyes.

SECTION VI REACTIVITY DATA

CONDITIONS CONTRIBUTING TO INSTABILITY

CONDITIONS CONTRIBUTING TO HAZARDOUS POLYMERIZATION

INCOMPATABILITY (MATERIALS TO AVOID)

HAZARDOUS DECOMPOSITION PRODUCTS

SECTION VII SPILL OR LEAK PROCEDURES

STEPS TO BE TAKEN IN CASE MATERIAL IS RELEASED OR SPILLED

WASTE DISPOSAL METHOD

SECTION VIII SPECIAL PROTECTION INFORMATION

VENTILATION REQUIREMENTS LOCAL EXHAUST

PROTECTIVE EQUIPMENT (SPECIFY TYPES) EYE: Safety goggles

MECHANICAL (GENERAL)

GLOVES: Rubber

SPECIAL

RESPIRATOR: With proper filter

OTHER PROTECTIVE EQUIPMENT

SECTION IX SPECIAL PRECAUTIONS

PRECAUTIONS TO BE TAKEN IN HANDLING AND STORAGE

OTHER PRECAUTIONS

SIGNATURE

DATE

METHYL α,β-DIBROMOPROPIONATE

Hazard diamond: 1 (top), 3 (left), 0 (right)

SECTION I SOURCE AND NOMENCLATURE

MANUFACTURER'S NAME	EMERGENCY TELEPHONE NO.
ADDRESS (NUMBER, STREET, CITY, STATE, ZIP CODE)	
TRADE NAME AND SYNONYMS	CHEMICAL FAMILY
CHEMICAL NAME AND SYNONYMS: Methyl 2,3 - Dibromopropionate	FORMULA: $BrCH_2CHBrCOOCH_3$

SECTION II HAZARDOUS INGREDIENTS

BASIC MATERIAL	APPROXIMATE OR MAXIMUM % WT. OR VOL.	ESTABLISHED OSHA STANDARD	LD_{50} ORAL	LD_{50} PERCUT.	LD_{50} SPECIES	LC_{50} CONC.
			none available			

SECTION III PHYSICAL DATA MW - 245.92

BOILING POINT °F.: 203 at 115 mm	VAPOR PRESSURE mm Hg.: at 93°C 17
MELTING POINT °F.: --	VAPOR DENSITY (Air = 1)
SPECIFIC GRAVITY (H_2O = 1): 1.96	EVAPORATION RATE (____ = 1)
SOLUBILITY IN WATER Pts/100 pts H_2O	VOLATILE % Vol. % Wt.

APPEARANCE AND ODOR: clear liquid

SECTION IV FIRE AND EXPLOSION HAZARD DATA

FLASH POINT: none available

METHOD USED

FLAMMABLE (EXPLOSIVE) LIMITS: UPPER / LOWER

EXTINGUISHING MEDIA: Foam, carbon dioxide, dry chemical, water.

SPECIAL FIRE FIGHTING PROCEDURES

UNUSUAL FIRE AND EXPLOSION HAZARDS

SECTION V HEALTH HAZARD DATA

TOXIC LEVEL: none available — CARCINOGENIC: no

PRINCIPLE ROUTES OF ABSORBTION: inhalation, ingestion, skin — SKIN AND EYE IRRITATION: severe

REVELANT SYMPTOMS OF EXPOSURE: Irritation of eyes and upper respiratory system, dizziness and depression, palpitation of heart.

EFFECTS OF CHRONIC EXPOSURE: --

EMERGENCY AND FIRST AID PROCEDURES: If swallowed - give emetic, get medical aid; inhalation - rest; wash from skin & eyes.

SECTION VI REACTIVITY DATA

CONDITIONS CONTRIBUTING TO INSTABILITY

CONDITIONS CONTRIBUTING TO HAZARDOUS POLYMERIZATION

INCOMPATABILITY (MATERIALS TO AVOID)

HAZARDOUS DECOMPOSITION PRODUCTS

SECTION VII SPILL OR LEAK PROCEDURES

STEPS TO BE TAKEN IN CASE MATERIAL IS RELEASED OR SPILLED

WASTE DISPOSAL METHOD

SECTION VIII SPECIAL PROTECTION INFORMATION

VENTILATION REQUIREMENTS LOCAL EXHAUST	PROTECTIVE EQUIPMENT (SPECIFY TYPES) EYE: goggles
MECHANICAL (GENERAL)	GLOVES: rubber
SPECIAL	RESPIRATOR: with proper filter
OTHER PROTECTIVE EQUIPMENT	

SECTION IX SPECIAL PRECAUTIONS

PRECAUTIONS TO BE TAKEN IN HANDLING AND STORAGE: Store in standard flammable liquids cabinet.

OTHER PRECAUTIONS

SIGNATURE — DATE

METHYLENE BROMIDE

Hazard diamond: 0 (top), 3 (left), 0 (right)

SECTION I SOURCE AND NOMENCLATURE

MANUFACTURER'S NAME	EMERGENCY TELEPHONE NO.
ADDRESS (NUMBER, STREET, CITY, STATE, ZIP CODE)	
TRADE NAME AND SYNONYMS	CHEMICAL FAMILY
CHEMICAL NAME AND SYNONYMS: Dibromomethane	FORMULA: CH_2Br_2

SECTION II HAZARDOUS INGREDIENTS

BASIC MATERIAL	APPROXIMATE OR MAXIMUM % WT. OR VOL.	ESTABLISHED OSHA STANDARD	LD_{50} mg/kg ORAL	LD_{50} PERCUT.	LD_{50} SPECIES	LC_{50} CONC.
				3738	mouse	

SECTION III PHYSICAL DATA MW - 173.86

BOILING POINT °F.: 97°C	VAPOR PRESSURE mm Hg.
MELTING POINT °F.: -52.7°C	VAPOR DENSITY (Air = 1)
SPECIFIC GRAVITY (H_2O = 1): 2.50	EVAPORATION RATE (____ = 1)
SOLUBILITY IN WATER Pts/100 pts H_2O: 1.17	VOLATILE % Vol. % Wt.

APPEARANCE AND ODOR: colorless liquid

SECTION IV FIRE AND EXPLOSION HAZARD DATA

FLASH POINT: not available

METHOD USED

FLAMMABLE (EXPLOSIVE) LIMITS: UPPER / LOWER

EXTINGUISHING MEDIA: Probably inflammable like CH_2Cl_2

SPECIAL FIRE FIGHTING PROCEDURES

UNUSUAL FIRE AND EXPLOSION HAZARDS: May generate $COBr_2$ when in contact with flame or very hot metal surfaces.

SECTION V HEALTH HAZARD DATA

TOXIC LEVEL: not available — CARCINOGENIC: no

PRINCIPLE ROUTES OF ABSORBTION: inhalation, skin, ingestion — SKIN AND EYE IRRITATION: moderate

REVELANT SYMPTOMS OF EXPOSURE: Dizziness, nausea, numbness of extremities, sense of stupor or dullness, drunkenness.

EFFECTS OF CHRONIC EXPOSURE: Dermatitis

EMERGENCY AND FIRST AID PROCEDURES: Inhalation - rest; if swallowed - induce vomiting, get medical aid; wash from eyes and skin.

SECTION VI REACTIVITY DATA

CONDITIONS CONTRIBUTING TO INSTABILITY

CONDITIONS CONTRIBUTING TO HAZARDOUS POLYMERIZATION

INCOMPATABILITY (MATERIALS TO AVOID): Alkali metals

HAZARDOUS DECOMPOSITION PRODUCTS: May generate toxic $COBr_2$ on contact with flame or very hot metal surfaces.

SECTION VII SPILL OR LEAK PROCEDURES

STEPS TO BE TAKEN IN CASE MATERIAL IS RELEASED OR SPILLED

WASTE DISPOSAL METHOD

SECTION VIII SPECIAL PROTECTION INFORMATION

VENTILATION REQUIREMENTS LOCAL EXHAUST	PROTECTIVE EQUIPMENT (SPECIFY TYPES) EYE: goggles
MECHANICAL (GENERAL)	GLOVES
SPECIAL	RESPIRATOR
OTHER PROTECTIVE EQUIPMENT	

SECTION IX SPECIAL PRECAUTIONS

PRECAUTIONS TO BE TAKEN IN HANDLING AND STORAGE

OTHER PRECAUTIONS

SIGNATURE — DATE

METHYLENE IODIDE

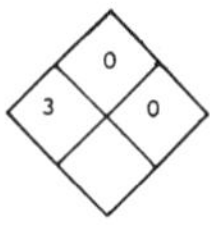

SECTION I SOURCE AND NOMENCLATURE

MANUFACTURER'S NAME | EMERGENCY TELEPHONE NO.

ADDRESS (NUMBER, STREET, CITY, STATE, ZIP CODE)

TRADE NAME AND SYNONYMS | CHEMICAL FAMILY

CHEMICAL NAME AND SYNONYMS: diiodomethane | FORMULA: CH_2I_2

SECTION II HAZARDOUS INGREDIENTS

BASIC MATERIAL	APPROXIMATE OR MAXIMUM % WT. OR VOL.	ESTABLISHED OSHA STANDARD	LD_{50} mg/kg. ORAL	PERCUT.	SPECIES	LC_{50} CONC.
				830	mouse	

SECTION III PHYSICAL DATA MW - 267.83

BOILING POINT °F.: 181°C	VAPOR PRESSURE mm Hg.
MELTING POINT °F.: 5°C	VAPOR DENSITY (Air = 1): 9.25
SPECIFIC GRAVITY (H_2O = 1): 3.33	EVAPORATION RATE (____ = 1)
SOLUBILITY IN WATER Pts/100 pts H_2O: 1.6	VOLATILE % Vol. % Wt.

APPEARANCE AND ODOR: light straw to clear colored liquid.

SECTION IV FIRE AND EXPLOSION HAZARD DATA

FLASH POINT: none available | FLAMMABLE (EXPLOSIVE) LIMITS: UPPER | LOWER

METHOD USED

EXTINGUISHING MEDIA: not available

SPECIAL FIRE FIGHTING PROCEDURES: --

UNUSUAL FIRE AND EXPLOSION HAZARDS: When heated to decomposition point - emits toxic fumes of iodides.

SECTION V HEALTH HAZARD DATA

TOXIC LEVEL: not available | CARCINOGENIC: no

PRINCIPLE ROUTES OF ABSORBTION: inhalation, ingestion, skin | SKIN AND EYE IRRITATION: moderate

REVELANT SYMPTOMS OF EXPOSURE: Nausea, vomiting, diarrhea, dizziness, visual disturbances, irritability, drowsiness, central nervous system depressant

EFFECTS OF CHRONIC EXPOSURE: Dermatitis

EMERGENCY AND FIRST AID PROCEDURES: Inhalation - rest; if swallowed - wash mouth, give emetic; get medical aid; wash from eyes & skin.

SECTION VI REACTIVITY DATA

CONDITIONS CONTRIBUTING TO INSTABILITY

CONDITIONS CONTRIBUTING TO HAZARDOUS POLYMERIZATION

INCOMPATABILITY (MATERIALS TO AVOID)

HAZARDOUS DECOMPOSITION PRODUCTS

SECTION VII SPILL OR LEAK PROCEDURES

STEPS TO BE TAKEN IN CASE MATERIAL IS RELEASED OR SPILLED

WASTE DISPOSAL METHOD

SECTION VIII SPECIAL PROTECTION INFORMATION

VENTILATION REQUIREMENTS LOCAL EXHAUST	PROTECTIVE EQUIPMENT (SPECIFY TYPES) EYE: Safety goggles
MECHANICAL (GENERAL)	GLOVES: Rubber
SPECIAL	RESPIRATOR: With proper filter
OTHER PROTECTIVE EQUIPMENT	

SECTION IX SPECIAL PRECAUTIONS

PRECAUTIONS TO BE TAKEN IN HANDLING AND STORAGE

OTHER PRECAUTIONS

SIGNATURE | DATE

METHYLETHYLAMINE HYDROCHLORIDE

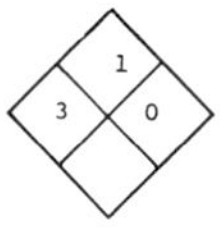

SECTION I SOURCE AND NOMENCLATURE

MANUFACTURER'S NAME | EMERGENCY TELEPHONE NO.

ADDRESS (NUMBER, STREET, CITY, STATE, ZIP CODE)

TRADE NAME AND SYNONYMS | CHEMICAL FAMILY

CHEMICAL NAME AND SYNONYMS: N-Methylethylamine Hydrochloride | FORMULA: $CH_3NHC_2H_5HCl$

SECTION II HAZARDOUS INGREDIENTS

BASIC MATERIAL	APPROXIMATE OR MAXIMUM % WT. OR VOL.	ESTABLISHED OSHA STANDARD	LD_{50} ORAL	PERCUT.	LC_{50} SPECIES	CONC.
			none	available		

SECTION III PHYSICAL DATA MW - 95.58

BOILING POINT °F.: --	VAPOR PRESSURE mm Hg.
MELTING POINT °F.: 133°C	VAPOR DENSITY (Air = 1)
SPECIFIC GRAVITY (H_2O = 1)	EVAPORATION RATE (____ = 1)
SOLUBILITY IN WATER Pts/100 pts H_2O: very soluble	VOLATILE % Vol. % Wt.

APPEARANCE AND ODOR: white crystals

SECTION IV FIRE AND EXPLOSION HAZARD DATA

FLASH POINT: none available | FLAMMABLE (EXPLOSIVE) LIMITS: UPPER | LOWER

METHOD USED

EXTINGUISHING MEDIA: Alcohol foam, carbon dioxide, dry chemical

SPECIAL FIRE FIGHTING PROCEDURES: Wear self-contained breathing apparatus.

UNUSUAL FIRE AND EXPLOSION HAZARDS

SECTION V HEALTH HAZARD DATA

TOXIC LEVEL: none available | CARCINOGENIC: no

PRINCIPLE ROUTES OF ABSORBTION: Ingestion, inhalation of dust | SKIN AND EYE IRRITATION: moderate

REVELANT SYMPTOMS OF EXPOSURE: Eye and mucous membrane irritation, irritation of gastroenteric tract.

EFFECTS OF CHRONIC EXPOSURE: Irritation of cornea and lungs.

EMERGENCY AND FIRST AID PROCEDURES: If swallowed - drink water, get medical aid; inhalation - rest; wash from skin & eyes.

SECTION VI REACTIVITY DATA

CONDITIONS CONTRIBUTING TO INSTABILITY

CONDITIONS CONTRIBUTING TO HAZARDOUS POLYMERIZATION

INCOMPATABILITY (MATERIALS TO AVOID)

HAZARDOUS DECOMPOSITION PRODUCTS

SECTION VII SPILL OR LEAK PROCEDURES

STEPS TO BE TAKEN IN CASE MATERIAL IS RELEASED OR SPILLED

WASTE DISPOSAL METHOD

SECTION VIII SPECIAL PROTECTION INFORMATION

VENTILATION REQUIREMENTS LOCAL EXHAUST	PROTECTIVE EQUIPMENT (SPECIFY TYPES) EYE: Goggles
MECHANICAL (GENERAL)	GLOVES: Rubber
SPECIAL	RESPIRATOR: Self-contained
OTHER PROTECTIVE EQUIPMENT	

SECTION IX SPECIAL PRECAUTIONS

PRECAUTIONS TO BE TAKEN IN HANDLING AND STORAGE

OTHER PRECAUTIONS

SIGNATURE | DATE

METHYLETHYLANILINE

Health 2, Flammability 4, Reactivity 0

SECTION I SOURCE AND NOMENCLATURE

MANUFACTURER'S NAME | EMERGENCY TELEPHONE NO.

ADDRESS (NUMBER, STREET, CITY, STATE, ZIP CODE)

TRADE NAME AND SYNONYMS | CHEMICAL FAMILY

CHEMICAL NAME AND SYNONYMS: Methylethylaniline | FORMULA: $C_6H_5N(CH_3)C_2H_5$

SECTION II HAZARDOUS INGREDIENTS

BASIC MATERIAL	APPROXIMATE OR MAXIMUM % WT. OR VOL.	ESTABLISHED OSHA STANDARD	LD_{50} ORAL	LD_{50} PERCUT.	LC_{50} SPECIES	LC_{50} CONC.
			none	available		

SECTION III PHYSICAL DATA MW - 135.20

BOILING POINT °F: 201°C

VAPOR PRESSURE mm Hg: at 94°C 12

MELTING POINT °F: --

VAPOR DENSITY (Air = 1)

SPECIFIC GRAVITY (H_2O = 1): 0.92

EVAPORATION RATE (________ = 1)

SOLUBILITY IN WATER Pts/100 pts H_2O: insoluble

VOLATILE % Vol. % Wt.

APPEARANCE AND ODOR: colorless to yellow liquid

SECTION IV FIRE AND EXPLOSION HAZARD DATA

FLASH POINT: none available

FLAMMABLE (EXPLOSIVE) LIMITS: UPPER / LOWER

METHOD USED

EXTINGUISHING MEDIA: Alcohol foam, water spray, carbon dioxide, dry chemical

SPECIAL FIRE FIGHTING PROCEDURES

UNUSUAL FIRE AND EXPLOSION HAZARDS

SECTION V HEALTH HAZARD DATA

TOXIC LEVEL: none available | CARCINOGENIC: no

PRINCIPLE ROUTES OF ABSORBTION: inhalation, ingestion, skin | SKIN AND EYE IRRITATION: severe

REVELANT SYMPTOMS OF EXPOSURE: Weakness, tremors, cyanosis, drowsiness, central nervous system depression.

EFFECTS OF CHRONIC EXPOSURE: Fatigue, loss of appetite, headache, dizziness

EMERGENCY AND FIRST AID PROCEDURES: If swallowed - give emetic, get medical aid; inhalation - rest; wash from skin & eyes.

SECTION VI REACTIVITY DATA

CONDITIONS CONTRIBUTING TO INSTABILITY

CONDITIONS CONTRIBUTING TO HAZARDOUS POLYMERIZATION

INCOMPATABILITY (MATERIALS TO AVOID)

HAZARDOUS DECOMPOSITION PRODUCTS

SECTION VII SPILL OR LEAK PROCEDURES

STEPS TO BE TAKEN IN CASE MATERIAL IS RELEASED OR SPILLED

WASTE DISPOSAL METHOD

SECTION VIII SPECIAL PROTECTION INFORMATION

VENTILATION REQUIREMENTS LOCAL EXHAUST | PROTECTIVE EQUIPMENT (SPECIFY TYPES) EYE: Goggles

MECHANICAL (GENERAL) | GLOVES: Polyvinyl gloves

SPECIAL | RESPIRATOR: With proper filter

OTHER PROTECTIVE EQUIPMENT

SECTION IX SPECIAL PRECAUTIONS

PRECAUTIONS TO BE TAKEN IN HANDLING AND STORAGE

OTHER PRECAUTIONS

SIGNATURE | DATE

METHYLHYDRAZINE SULFATE

Health 1, Flammability 4, Reactivity 0

SECTION I SOURCE AND NOMENCLATURE

MANUFACTURER'S NAME | EMERGENCY TELEPHONE NO.

ADDRESS (NUMBER, STREET, CITY, STATE, ZIP CODE)

TRADE NAME AND SYNONYMS | CHEMICAL FAMILY

CHEMICAL NAME AND SYNONYMS: Methylhydrazine Sulfate | FORMULA: $CH_3NHNH_2H_2SO_4$

SECTION II HAZARDOUS INGREDIENTS

BASIC MATERIAL	APPROXIMATE OR MAXIMUM % WT. OR VOL.	ESTABLISHED OSHA STANDARD	LD_{50} ORAL	LD_{50} PERCUT.	LC_{50} SPECIES	LC_{50} CONC.
TCLO - oral - mouse	1551 mg/kg	75 weeks carcinogen	none	available		

SECTION III PHYSICAL DATA MW - 144.15

BOILING POINT °F: --

VAPOR PRESSURE mm Hg

MELTING POINT °F: 142°C

VAPOR DENSITY (Air = 1)

SPECIFIC GRAVITY (H_2O = 1): --

EVAPORATION RATE (________ = 1)

SOLUBILITY IN WATER Pts/100 pts H_2O: very soluble

VOLATILE % Vol. % Wt.

APPEARANCE AND ODOR: white crystal

SECTION IV FIRE AND EXPLOSION HAZARD DATA

FLASH POINT: none available

FLAMMABLE (EXPLOSIVE) LIMITS: UPPER / LOWER

METHOD USED

EXTINGUISHING MEDIA: Alcohol foam, carbon dioxide, dry chemical.

SPECIAL FIRE FIGHTING PROCEDURES

UNUSUAL FIRE AND EXPLOSION HAZARDS

SECTION V HEALTH HAZARD DATA

TOXIC LEVEL: none available | CARCINOGENIC: yes

PRINCIPLE ROUTES OF ABSORBTION: inhalation of dust, ingestion, skin | SKIN AND EYE IRRITATION: severe

REVELANT SYMPTOMS OF EXPOSURE: Respiratory irritation and convulsions, eye and skin irritations, gastroentric disturbances.

EFFECTS OF CHRONIC EXPOSURE: Blood and liver damage.

EMERGENCY AND FIRST AID PROCEDURES: If swallowed - induce vomiting, get medical aid; inhalation - rest; wash from skin & eyes.

SECTION VI REACTIVITY DATA

CONDITIONS CONTRIBUTING TO INSTABILITY

CONDITIONS CONTRIBUTING TO HAZARDOUS POLYMERIZATION

INCOMPATABILITY (MATERIALS TO AVOID)

HAZARDOUS DECOMPOSITION PRODUCTS

SECTION VII SPILL OR LEAK PROCEDURES

STEPS TO BE TAKEN IN CASE MATERIAL IS RELEASED OR SPILLED

WASTE DISPOSAL METHOD

SECTION VIII SPECIAL PROTECTION INFORMATION

VENTILATION REQUIREMENTS LOCAL EXHAUST | PROTECTIVE EQUIPMENT (SPECIFY TYPES) EYE: Goggles

MECHANICAL (GENERAL) | GLOVES: Rubber

SPECIAL | RESPIRATOR: With proper filter

OTHER PROTECTIVE EQUIPMENT

SECTION IX SPECIAL PRECAUTIONS

PRECAUTIONS TO BE TAKEN IN HANDLING AND STORAGE

OTHER PRECAUTIONS

SIGNATURE | DATE

METHYL IODIDE

(Hazard diamond: top 0, left 4, right 0)

SECTION I SOURCE AND NOMENCLATURE

MANUFACTURER'S NAME	EMERGENCY TELEPHONE NO.
ADDRESS (NUMBER, STREET, CITY, STATE, ZIP CODE)	
TRADE NAME AND SYNONYMS	CHEMICAL FAMILY
CHEMICAL NAME AND SYNONYMS: Iodomethane	FORMULA: CH_3I

SECTION II HAZARDOUS INGREDIENTS

BASIC MATERIAL	APPROXIMATE OR MAXIMUM % WT. OR VOL.	ESTABLISHED OSHA STANDARD	LD_{50} mg/kg ORAL	PERCUT.	SPECIES	LC_{50} CONC.
Oral - human LDLO 50 mg/kg			220	110	mouse rat	

SECTION III PHYSICAL DATA MW - 141.95

BOILING POINT °F.: 42.5°C	VAPOR PRESSURE mm Hg.: at 25.3°C 400
MELTING POINT °F.: -64.4°C	VAPOR DENSITY (Air = 1): 4.89
SPECIFIC GRAVITY (H_2O = 1): 2.28	EVAPORATION RATE (= 1)
SOLUBILITY IN WATER Pts/100 pts H_2O: 1.8	VOLATILE % Vol. % Wt.

APPEARANCE AND ODOR: Colorless liquid - turns brown on exposure to light sweetish odor.

SECTION IV FIRE AND EXPLOSION HAZARD DATA

FLASH POINT: Not flammable

METHOD USED

FLAMMABLE (EXPLOSIVE) LIMITS: UPPER / LOWER

EXTINGUISHING MEDIA: not available

SPECIAL FIRE FIGHTING PROCEDURES: --

UNUSUAL FIRE AND EXPLOSION HAZARDS: When heated to decomposition point - emits toxic fumes of iodides.

SECTION V HEALTH HAZARD DATA

TOXIC LEVEL: TLV - 5 PPM, 28 mg/CM. CARCINOGENIC: yes

PRINCIPLE ROUTES OF ABSORBTION: inhalation, ingestion, skin. SKIN AND EYE IRRITATION: severe

REVELANT SYMPTOMS OF EXPOSURE: Nausea, vomiting, diarrhea, dizziness, visual disturbances, irritability, drowsiness, central nervous system depressant.

EFFECTS OF CHRONIC EXPOSURE: Dermatitis

EMERGENCY AND FIRST AID PROCEDURES: Inhalation - rest; if swallowed - wash mouth, give emetic, get medical aid; wash from eyes & skin.

SECTION VI REACTIVITY DATA

CONDITIONS CONTRIBUTING TO INSTABILITY

CONDITIONS CONTRIBUTING TO HAZARDOUS POLYMERIZATION

INCOMPATABILITY (MATERIALS TO AVOID)

HAZARDOUS DECOMPOSITION PRODUCTS

SECTION VII SPILL OR LEAK PROCEDURES

STEPS TO BE TAKEN IN CASE MATERIAL IS RELEASED OR SPILLED: Absorb with paper towels; evaporate in hood.

WASTE DISPOSAL METHOD

SECTION VIII SPECIAL PROTECTION INFORMATION

VENTILATION REQUIREMENTS LOCAL EXHAUST	PROTECTIVE EQUIPMENT (SPECIFY TYPES) EYE: safety goggles
MECHANICAL (GENERAL)	GLOVES: rubber
SPECIAL	RESPIRATOR: with proper filter
OTHER PROTECTIVE EQUIPMENT	

SECTION IX SPECIAL PRECAUTIONS

PRECAUTIONS TO BE TAKEN IN HANDLING AND STORAGE: Store in tightly closed containers in dark place.

OTHER PRECAUTIONS

SIGNATURE | DATE

METHYL ISOPROPYL CARBINOL

(Hazard diamond: top 2, left 3, right 0)

SECTION I SOURCE AND NOMENCLATURE

MANUFACTURER'S NAME	EMERGENCY TELEPHONE NO.
ADDRESS (NUMBER, STREET, CITY, STATE, ZIP CODE)	
TRADE NAME AND SYNONYMS	CHEMICAL FAMILY
CHEMICAL NAME AND SYNONYMS: 2 Methyl butanol 3	FORMULA: $(CH_3)_2CHCHOHCH_3$

SECTION II HAZARDOUS INGREDIENTS

BASIC MATERIAL	APPROXIMATE OR MAXIMUM % WT. OR VOL.	ESTABLISHED OSHA STANDARD	LD_{50} ORAL	PERCUT.	SPECIES	LC_{50} CONC.
			none available			

SECTION III PHYSICAL DATA

BOILING POINT °F.: 112°C	VAPOR PRESSURE mm Hg.
MELTING POINT °F.	VAPOR DENSITY (Air = 1): 3.0
SPECIFIC GRAVITY (H_2O = 1): 0.82	EVAPORATION RATE (= 1)
SOLUBILITY IN WATER Pts/100 pts H_2O: 2.8	VOLATILE % Vol. % Wt.

APPEARANCE AND ODOR: colorless to yellow liquid

SECTION IV FIRE AND EXPLOSION HAZARD DATA

FLASH POINT: 103°F

METHOD USED

FLAMMABLE (EXPLOSIVE) LIMITS: UPPER 9.0% / LOWER 1.2%

EXTINGUISHING MEDIA: alcohol foam, carbon dioxide, dry chemical.

SPECIAL FIRE FIGHTING PROCEDURES

UNUSUAL FIRE AND EXPLOSION HAZARDS

SECTION V HEALTH HAZARD DATA

TOXIC LEVEL: none available. CARCINOGENIC: no

PRINCIPLE ROUTES OF ABSORBTION: inhalation, ingestion. SKIN AND EYE IRRITATION: severe

REVELANT SYMPTOMS OF EXPOSURE: Eye, nose and throat irritation, narcosis; headache, dizziness, nausea, vomiting, diarrhea.

EFFECTS OF CHRONIC EXPOSURE: Liver damage

EMERGENCY AND FIRST AID PROCEDURES: If swallowed - induce vomiting, get medical aid; inhalation - rest, artificial respiration; wash skin & eyes.

SECTION VI REACTIVITY DATA

CONDITIONS CONTRIBUTING TO INSTABILITY

CONDITIONS CONTRIBUTING TO HAZARDOUS POLYMERIZATION

INCOMPATABILITY (MATERIALS TO AVOID)

HAZARDOUS DECOMPOSITION PRODUCTS

SECTION VII SPILL OR LEAK PROCEDURES

STEPS TO BE TAKEN IN CASE MATERIAL IS RELEASED OR SPILLED

WASTE DISPOSAL METHOD

SECTION VIII SPECIAL PROTECTION INFORMATION

VENTILATION REQUIREMENTS LOCAL EXHAUST	PROTECTIVE EQUIPMENT (SPECIFY TYPES) EYE: Goggles
MECHANICAL (GENERAL)	GLOVES: Rubber
SPECIAL	RESPIRATOR: With proper filter
OTHER PROTECTIVE EQUIPMENT	

SECTION IX SPECIAL PRECAUTIONS

PRECAUTIONS TO BE TAKEN IN HANDLING AND STORAGE

OTHER PRECAUTIONS

SIGNATURE | DATE

METHYL ISOPROPYL KETONE

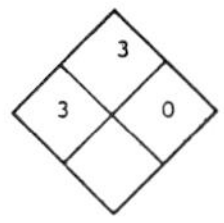

SECTION I SOURCE AND NOMENCLATURE

MANUFACTURER'S NAME | EMERGENCY TELEPHONE NO.

ADDRESS (NUMBER, STREET, CITY, STATE, ZIP CODE)

TRADE NAME AND SYNONYMS | CHEMICAL FAMILY

CHEMICAL NAME AND SYNONYMS: 3 Methyl 2 Butanone | FORMULA: $CH_3COCH(CH_3)_2$

SECTION II HAZARDOUS INGREDIENTS

BASIC MATERIAL	APPROXIMATE OR MAXIMUM % WT. OR VOL.	ESTABLISHED OSHA STANDARD	LD_{50} mg/kg ORAL	LD_{50} mg/kg PERCUT.	LD_{50} mg/kg SPECIES	LC_{50} CONC.
		•	148		rat	
Note: Lowest published lethal concentration in air inhalation rat 5700 PPM/4 hr.						

SECTION III PHYSICAL DATA MW - 86.13

BOILING POINT °F.: 95°C | VAPOR PRESSURE mm Hg.

MELTING POINT °F.: -92°C | VAPOR DENSITY (Air = 1)

SPECIFIC GRAVITY (H_2O = 1): 0.80 | EVAPORATION RATE (______ = 1)

SOLUBILITY IN WATER Pts/100 pts H_2O: very slightly soluble | VOLATILE % Vol. % Wt.

APPEARANCE AND ODOR: colorless liquid, acetone like odor.

SECTION IV FIRE AND EXPLOSION HAZARD DATA

FLASH POINT: none available | FLAMMABLE (EXPLOSIVE) LIMITS: UPPER / LOWER

METHOD USED

EXTINGUISHING MEDIA: alcohol foam, carbon dioxide, dry chemical.

SPECIAL FIRE FIGHTING PROCEDURES

UNUSUAL FIRE AND EXPLOSION HAZARDS

SECTION V HEALTH HAZARD DATA

TOXIC LEVEL: None available | CARCINOGENIC: no

PRINCIPLE ROUTES OF ABSORBTION: Inhalation, ingestion | SKIN AND EYE IRRITATION: severe

REVELANT SYMPTOMS OF EXPOSURE: Irritation to eyes and mucous membranes, narcotic effect.

EFFECTS OF CHRONIC EXPOSURE: --

EMERGENCY AND FIRST AID PROCEDURES: If swallowed - give emetic, get medical aid; inhalation - rest; wash skin & eyes.

SECTION VI REACTIVITY DATA

CONDITIONS CONTRIBUTING TO INSTABILITY

CONDITIONS CONTRIBUTING TO HAZARDOUS POLYMERIZATION

INCOMPATABILITY (MATERIALS TO AVOID)

HAZARDOUS DECOMPOSITION PRODUCTS

SECTION VII SPILL OR LEAK PROCEDURES

STEPS TO BE TAKEN IN CASE MATERIAL IS RELEASED OR SPILLED

WASTE DISPOSAL METHOD

SECTION VIII SPECIAL PROTECTION INFORMATION

VENTILATION REQUIREMENTS LOCAL EXHAUST | PROTECTIVE EQUIPMENT (SPECIFY TYPES) EYE: Goggles

MECHANICAL (GENERAL) | GLOVES: Rubber

SPECIAL | RESPIRATOR: With proper filter

OTHER PROTECTIVE EQUIPMENT

SECTION IX SPECIAL PRECAUTIONS

PRECAUTIONS TO BE TAKEN IN HANDLING AND STORAGE

OTHER PRECAUTIONS

SIGNATURE | DATE

METHYL ISOTHIOCYANATE

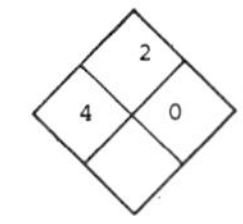

SECTION I SOURCE AND NOMENCLATURE

MANUFACTURER'S NAME | EMERGENCY TELEPHONE NO.

ADDRESS (NUMBER, STREET, CITY, STATE, ZIP CODE)

TRADE NAME AND SYNONYMS: Methyl Mustard Oil | CHEMICAL FAMILY

CHEMICAL NAME AND SYNONYMS: Methyl Isothiocyanate | FORMULA: CH_3NCS

SECTION II HAZARDOUS INGREDIENTS

BASIC MATERIAL	APPROXIMATE OR MAXIMUM % WT. OR VOL.	ESTABLISHED OSHA STANDARD	LD_{50} mg/kg ORAL	LD_{50} mg/kg PERCUT.	LD_{50} mg/kg SPECIES	LC_{50} CONC.
Oral - human - LDLO 50 mg/kg			97		rat	
			97		mouse	

SECTION III PHYSICAL DATA MW - 73.11

BOILING POINT °F.: 119°C | VAPOR PRESSURE mm Hg.

MELTING POINT °F.: 36°C | VAPOR DENSITY (Air = 1)

SPECIFIC GRAVITY (H_2O = 1): 1.07 | EVAPORATION RATE (______ = 1)

SOLUBILITY IN WATER Pts/100 pts H_2O: very slightly soluble | VOLATILE % Vol. % Wt.

APPEARANCE AND ODOR: crystals

SECTION IV FIRE AND EXPLOSION HAZARD DATA

FLASH POINT: none available | FLAMMABLE (EXPLOSIVE) LIMITS: UPPER / LOWER

METHOD USED

EXTINGUISHING MEDIA: Water, alcohol foam, carbon dioxide, dry chemical.

SPECIAL FIRE FIGHTING PROCEDURES: Wear self-contained breathing apparatus and goggles.

UNUSUAL FIRE AND EXPLOSION HAZARDS

SECTION V HEALTH HAZARD DATA

TOXIC LEVEL: none available | CARCINOGENIC: no

PRINCIPLE ROUTES OF ABSORBTION: inhalation, ingestion | SKIN AND EYE IRRITATION: severe

REVELANT SYMPTOMS OF EXPOSURE: Eye and mucous membrane irritation, skin irritation.

EFFECTS OF CHRONIC EXPOSURE: Liver and kidney damage.

EMERGENCY AND FIRST AID PROCEDURES: If swallowed - give emetic, medical aid, gastric lavage; inhalation - rest, give oxygen as needed; wash from eyes & skin.

SECTION VI REACTIVITY DATA

CONDITIONS CONTRIBUTING TO INSTABILITY

CONDITIONS CONTRIBUTING TO HAZARDOUS POLYMERIZATION

INCOMPATABILITY (MATERIALS TO AVOID)

HAZARDOUS DECOMPOSITION PRODUCTS

SECTION VII SPILL OR LEAK PROCEDURES

STEPS TO BE TAKEN IN CASE MATERIAL IS RELEASED OR SPILLED: Absorb on paper towels; evaporate in hood.

WASTE DISPOSAL METHOD

SECTION VIII SPECIAL PROTECTION INFORMATION

VENTILATION REQUIREMENTS LOCAL EXHAUST | PROTECTIVE EQUIPMENT (SPECIFY TYPES) EYE: Goggles

MECHANICAL (GENERAL) | GLOVES: Rubber (butyl)

SPECIAL | RESPIRATOR: Self-contained

OTHER PROTECTIVE EQUIPMENT: Coveralls

SECTION IX SPECIAL PRECAUTIONS

PRECAUTIONS TO BE TAKEN IN HANDLING AND STORAGE

OTHER PRECAUTIONS

SIGNATURE | DATE

α-METHYLNAPHTHALENE

Hazard diamond: 2 (top), 3 (left), 0 (right)

SECTION I SOURCE AND NOMENCLATURE

MANUFACTURER'S NAME | EMERGENCY TELEPHONE NO.

ADDRESS (NUMBER, STREET, CITY, STATE, ZIP CODE)

TRADE NAME AND SYNONYMS | CHEMICAL FAMILY

CHEMICAL NAME AND SYNONYMS: 1 Methylnaphthalene

FORMULA: $C_{10}H_7CH_3$

SECTION II HAZARDOUS INGREDIENTS

BASIC MATERIAL	APPROXIMATE OR MAXIMUM % WT. OR VOL.	ESTABLISHED OSHA STANDARD	LD_{50} mg/kg ORAL	PERCUT.	SPECIES	LC_{50} CONC.
			4360		rat	

SECTION III PHYSICAL DATA MW - 142.21

BOILING POINT °F.: 243^0C

VAPOR PRESSURE mm Hg.:

MELTING POINT °F.: -32^0C

VAPOR DENSITY (Air = 1):

SPECIFIC GRAVITY (H_2O = 1): 1.03

EVAPORATION RATE (____ = 1):

SOLUBILITY IN WATER Pts/100 pts H_2O: insoluble

VOLATILE % Vol. % Wt.:

APPEARANCE AND ODOR: colorless liquid

SECTION IV FIRE AND EXPLOSION HAZARD DATA

FLASH POINT: none available

FLAMMABLE (EXPLOSIVE) LIMITS: UPPER / LOWER

METHOD USED

EXTINGUISHING MEDIA: Alcohol foam, carbon dioxide, dry chemical

SPECIAL FIRE FIGHTING PROCEDURES: Self-contained breathing apparatus.

UNUSUAL FIRE AND EXPLOSION HAZARDS

SECTION V HEALTH HAZARD DATA

TOXIC LEVEL: none available

CARCINOGENIC: no

PRINCIPLE ROUTES OF ABSORBTION: inhalation, ingestion

SKIN AND EYE IRRITATION: moderate

REVELANT SYMPTOMS OF EXPOSURE: Nausea, vomiting, diarrhea, anemia, jaundice, excitement, convulsions, coma, headache, visual disturbances.

EFFECTS OF CHRONIC EXPOSURE: Nausea, vomiting, diarrhea, anemia, jaundice, dermatitis.

EMERGENCY AND FIRST AID PROCEDURES: If swallowed - give emetic, get medical aid; inhalation - rest; wash from skin & eyes.

SECTION VI REACTIVITY DATA

CONDITIONS CONTRIBUTING TO INSTABILITY

CONDITIONS CONTRIBUTING TO HAZARDOUS POLYMERIZATION

INCOMPATABILITY (MATERIALS TO AVOID)

HAZARDOUS DECOMPOSITION PRODUCTS

SECTION VII SPILL OR LEAK PROCEDURES

STEPS TO BE TAKEN IN CASE MATERIAL IS RELEASED OR SPILLED

WASTE DISPOSAL METHOD

SECTION VIII SPECIAL PROTECTION INFORMATION

VENTILATION REQUIREMENTS LOCAL EXHAUST

MECHANICAL (GENERAL)

SPECIAL

PROTECTIVE EQUIPMENT (SPECIFY TYPES) EYE: Goggles

GLOVES: Rubber

RESPIRATOR: Self-contained

OTHER PROTECTIVE EQUIPMENT

SECTION IX SPECIAL PRECAUTIONS

PRECAUTIONS TO BE TAKEN IN HANDLING AND STORAGE

OTHER PRECAUTIONS

SIGNATURE | DATE

β-METHYLNAPHTHALENE

Hazard diamond: 2 (top), 3 (left), 0 (right)

SECTION I SOURCE AND NOMENCLATURE

MANUFACTURER'S NAME | EMERGENCY TELEPHONE NO.

ADDRESS (NUMBER, STREET, CITY, STATE, ZIP CODE)

TRADE NAME AND SYNONYMS | CHEMICAL FAMILY

CHEMICAL NAME AND SYNONYMS: 2 Methylnaphthalene

FORMULA: $C_{10}H_7CH_3$

SECTION II HAZARDOUS INGREDIENTS

BASIC MATERIAL	APPROXIMATE OR MAXIMUM % WT. OR VOL.	ESTABLISHED OSHA STANDARD	LD_{50} mg/kg ORAL	PERCUT.	SPECIES	LC_{50} CONC.
			4360		rat	

SECTION III PHYSICAL DATA MW - 142.21

BOILING POINT °F.: 242^0C

VAPOR PRESSURE mm Hg.:

MELTING POINT °F.: 34^0C

VAPOR DENSITY (Air = 1):

SPECIFIC GRAVITY (H_2O = 1): 0.944

EVAPORATION RATE (____ = 1):

SOLUBILITY IN WATER Pts/100 pts H_2O: insoluble

VOLATILE % Vol. % Wt.:

APPEARANCE AND ODOR: solid

SECTION IV FIRE AND EXPLOSION HAZARD DATA

FLASH POINT: none available

FLAMMABLE (EXPLOSIVE) LIMITS: UPPER / LOWER

METHOD USED

EXTINGUISHING MEDIA: alcohol foam, carbon dioxide, dry chemical

SPECIAL FIRE FIGHTING PROCEDURES: self-contained breathing apparatus.

UNUSUAL FIRE AND EXPLOSION HAZARDS

SECTION V HEALTH HAZARD DATA

TOXIC LEVEL: none available

CARCINOGENIC: no

PRINCIPLE ROUTES OF ABSORBTION: Inhalation, ingestion

SKIN AND EYE IRRITATION: moderate

REVELANT SYMPTOMS OF EXPOSURE: Nausea, vomiting, diarrhea, anemia, jaundice, excitement, convulsions, coma, headache, visual disturbances.

EFFECTS OF CHRONIC EXPOSURE: Nausea, vomiting, diarrhea, anemia, jaundice, dermatitis.

EMERGENCY AND FIRST AID PROCEDURES: Ingestion - give emetic, get medical aid; inhalation - rest; wash skin & eyes.

SECTION VI REACTIVITY DATA

CONDITIONS CONTRIBUTING TO INSTABILITY

CONDITIONS CONTRIBUTING TO HAZARDOUS POLYMERIZATION

INCOMPATABILITY (MATERIALS TO AVOID)

HAZARDOUS DECOMPOSITION PRODUCTS

SECTION VII SPILL OR LEAK PROCEDURES

STEPS TO BE TAKEN IN CASE MATERIAL IS RELEASED OR SPILLED

WASTE DISPOSAL METHOD

SECTION VIII SPECIAL PROTECTION INFORMATION

VENTILATION REQUIREMENTS LOCAL EXHAUST

MECHANICAL (GENERAL)

SPECIAL

PROTECTIVE EQUIPMENT (SPECIFY TYPES) EYE: Goggles

GLOVES: Rubber

RESPIRATOR: Self-contained

OTHER PROTECTIVE EQUIPMENT

SECTION IX SPECIAL PRECAUTIONS

PRECAUTIONS TO BE TAKEN IN HANDLING AND STORAGE

OTHER PRECAUTIONS

SIGNATURE | DATE

METHYL-p-PHENYLENEDIAMINE DIHYDROCHLORIDE

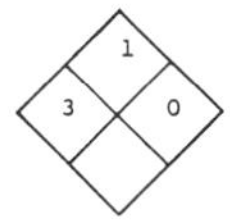

SECTION I SOURCE AND NOMENCLATURE

MANUFACTURER'S NAME | EMERGENCY TELEPHONE NO.

ADDRESS (NUMBER, STREET, CITY, STATE, ZIP CODE)

TRADE NAME AND SYNONYMS | CHEMICAL FAMILY

CHEMICAL NAME AND SYNONYMS: N Methyl - P - Phenylenediamine Hydrochloride

FORMULA: $CH_3NHC_6H_4NH_2 2HCl$

SECTION II HAZARDOUS INGREDIENTS

BASIC MATERIAL	APPROXIMATE OR MAXIMUM % WT. OR VOL.	ESTABLISHED OSHA STANDARD	LD_{50}		LC_{50}	
			ORAL	PERCUT.	SPECIES	CONC.
			none	available		

SECTION III PHYSICAL DATA MW - 195.10

BOILING POINT °F.: --

VAPOR PRESSURE mm Hg.

MELTING POINT °F.: --

VAPOR DENSITY (Air = 1)

SPECIFIC GRAVITY (H_2O = 1): --

EVAPORATION RATE (________ = 1)

SOLUBILITY IN WATER Pts/100 pts H_2O: very soluble

VOLATILE % Vol. % Wt.

APPEARANCE AND ODOR: yellow crystals

SECTION IV FIRE AND EXPLOSION HAZARD DATA

FLASH POINT: none available

METHOD USED

FLAMMABLE (EXPLOSIVE) LIMITS: UPPER / LOWER

EXTINGUISHING MEDIA: Alcohol foam, carbon dioxide, dry chemical

SPECIAL FIRE FIGHTING PROCEDURES

UNUSUAL FIRE AND EXPLOSION HAZARDS

SECTION V HEALTH HAZARD DATA

TOXIC LEVEL: none available

CARCINOGENIC: no

PRINCIPLE ROUTES OF ABSORBTION: Inhalation of dust, ingestion

SKIN AND EYE IRRITATION: moderate

REVELANT SYMPTOMS OF EXPOSURE: Headache, drowsiness, cyanosis, mental confusion, convulsions.

EFFECTS OF CHRONIC EXPOSURE: --

EMERGENCY AND FIRST AID PROCEDURES: If swallowed - give emetic get medical aid; inhalation - rest; wash from skin & eyes.

SECTION VI REACTIVITY DATA

CONDITIONS CONTRIBUTING TO INSTABILITY

CONDITIONS CONTRIBUTING TO HAZARDOUS POLYMERIZATION

INCOMPATABILITY (MATERIALS TO AVOID)

HAZARDOUS DECOMPOSITION PRODUCTS

SECTION VII SPILL OR LEAK PROCEDURES

STEPS TO BE TAKEN IN CASE MATERIAL IS RELEASED OR SPILLED

WASTE DISPOSAL METHOD

SECTION VIII SPECIAL PROTECTION INFORMATION

VENTILATION REQUIREMENTS LOCAL EXHAUST

MECHANICAL (GENERAL)

SPECIAL

PROTECTIVE EQUIPMENT (SPECIFY TYPES) EYE: goggles

GLOVES: Rubber

RESPIRATOR: With proper filter

OTHER PROTECTIVE EQUIPMENT

SECTION IX SPECIAL PRECAUTIONS

PRECAUTIONS TO BE TAKEN IN HANDLING AND STORAGE

OTHER PRECAUTIONS

SIGNATURE | DATE

α-METHYL-α-PHENYLHYDRAZINE

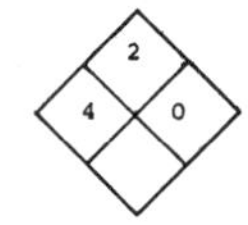

SECTION I SOURCE AND NOMENCLATURE

MANUFACTURER'S NAME | EMERGENCY TELEPHONE NO.

ADDRESS (NUMBER, STREET, CITY, STATE, ZIP CODE)

TRADE NAME AND SYNONYMS | CHEMICAL FAMILY

CHEMICAL NAME AND SYNONYMS: 1 Methyl, 1 Phenylhydrazine

FORMULA: $NH_2N(C_6H_5)CH_3$

SECTION II HAZARDOUS INGREDIENTS

BASIC MATERIAL	APPROXIMATE OR MAXIMUM % WT. OR VOL.	ESTABLISHED OSHA STANDARD	LD_{50}		LC_{50}	
			ORAL	PERCUT.	SPECIES	CONC.
			none available			

SECTION III PHYSICAL DATA MW - 122.17

BOILING POINT °F.: 227°C at 755 mm

VAPOR PRESSURE mm Hg.: at 118°C 21

MELTING POINT °F.: --

VAPOR DENSITY (Air = 1)

SPECIFIC GRAVITY (H_2O = 1): 1.04

EVAPORATION RATE (________ = 1)

SOLUBILITY IN WATER Pts/100 pts H_2O: slightly soluble

VOLATILE % Vol. % Wt.

APPEARANCE AND ODOR: yellow liquid

SECTION IV FIRE AND EXPLOSION HAZARD DATA

FLASH POINT: none available

METHOD USED

FLAMMABLE (EXPLOSIVE) LIMITS: UPPER / LOWER

EXTINGUISHING MEDIA: alcohol foam, water, carbon dioxide

SPECIAL FIRE FIGHTING PROCEDURES

UNUSUAL FIRE AND EXPLOSION HAZARDS: When heated to decomposition point, it gives off fumes of nitrogen compounds.

SECTION V HEALTH HAZARD DATA

TOXIC LEVEL: none available

CARCINOGENIC: no

PRINCIPLE ROUTES OF ABSORBTION: Inhalation, ingestion, skin

SKIN AND EYE IRRITATION: severe

REVELANT SYMPTOMS OF EXPOSURE: Irritation of skin, eyes and mucous membranes, gastroentric disturbances.

EFFECTS OF CHRONIC EXPOSURE: Blood and liver damage, loss of appetite.

EMERGENCY AND FIRST AID PROCEDURES: Inhalation - rest; if swallowed - induce vomiting, get medical aid; wash from skin & eyes.

SECTION VI REACTIVITY DATA

CONDITIONS CONTRIBUTING TO INSTABILITY

CONDITIONS CONTRIBUTING TO HAZARDOUS POLYMERIZATION

INCOMPATABILITY (MATERIALS TO AVOID): Avoid lead dioxide

HAZARDOUS DECOMPOSITION PRODUCTS

SECTION VII SPILL OR LEAK PROCEDURES

STEPS TO BE TAKEN IN CASE MATERIAL IS RELEASED OR SPILLED

WASTE DISPOSAL METHOD

SECTION VIII SPECIAL PROTECTION INFORMATION

VENTILATION REQUIREMENTS LOCAL EXHAUST

MECHANICAL (GENERAL)

SPECIAL

PROTECTIVE EQUIPMENT (SPECIFY TYPES) EYE: Goggles

GLOVES: Rubber

RESPIRATOR: Self-contained

OTHER PROTECTIVE EQUIPMENT

SECTION IX SPECIAL PRECAUTIONS

PRECAUTIONS TO BE TAKEN IN HANDLING AND STORAGE

OTHER PRECAUTIONS

SIGNATURE | DATE

METHYL SULFATE

Hazard rating: 2 (top), 4 (left), 0 (right)

SECTION I SOURCE AND NOMENCLATURE

MANUFACTURER'S NAME

EMERGENCY TELEPHONE NO.

ADDRESS (NUMBER, STREET, CITY, STATE, ZIP CODE)

TRADE NAME AND SYNONYMS

CHEMICAL FAMILY

CHEMICAL NAME AND SYNONYMS: Dimethyl Sulfate

FORMULA: $(CH_3)_2SO_4$

SECTION II HAZARDOUS INGREDIENTS

BASIC MATERIAL	APPROXIMATE OR MAXIMUM % WT. OR VOL.	ESTABLISHED OSHA STANDARD	LD_{50} mg/kg ORAL	PERCUT.	SPECIES	LC_{50} CONC.
Inhalation - human LCLO	97 PPM 110 Min.					
			440		Rat	

Note: Lowest published lethal concentration in air; inhalation - rat 32 ppm.

SECTION III PHYSICAL DATA Mol. Wt. 126.13

BOILING POINT (°F): 188°C

VAPOR PRESSURE (mm Hg.)

MELTING POINT (°F): -38°C

VAPOR DENSITY (Air = 1): 4.35

SPECIFIC GRAVITY (H_2O = 1): 1.3322 at 20°/4°C

EVAPORATION RATE (_____ = 1)

SOLUBILITY IN WATER (Pts/100 pts H_2O): Very slightly soluble

VOLATILE (% Vol.) (% Wt.)

APPEARANCE AND ODOR: Colorless liquid

SECTION IV FIRE AND EXPLOSION HAZARD DATA

FLASH POINT: 182°F

METHOD USED: Open Cup

FLAMMABLE (EXPLOSIVE) LIMITS: UPPER / LOWER

EXTINGUISHING MEDIA: Water spray, dry chemical, foam, or carbon dioxide.

SPECIAL FIRE FIGHTING PROCEDURES: Wear full protective clothing.

UNUSUAL FIRE AND EXPLOSION HAZARDS

SECTION V HEALTH HAZARD DATA (Skin)

TOXIC LEVEL: TLV 0.01 ppm

CARCINOGENIC: Experimental ~~Carcinogen~~

PRINCIPLE ROUTES OF ABSORBTION: Inhalation of vapor; percutaneous absorption of liquid

SKIN AND EYE IRRITATION: Yes

REVELANT SYMPTOMS OF EXPOSURE: Eye irritation, bronchial irritation, lung irritation, kidney damage, dimming of vision (symptoms may be delayed).

EFFECTS OF CHRONIC EXPOSURE: --

EMERGENCY AND FIRST AID PROCEDURES: Inhalation - rest, get medical aid; wash skin & eyes; if swallowed - drink water, then milk of magnesia, get medical aid.

SECTION VI REACTIVITY DATA

CONDITIONS CONTRIBUTING TO INSTABILITY

CONDITIONS CONTRIBUTING TO HAZARDOUS POLYMERIZATION

INCOMPATABILITY (MATERIALS TO AVOID)

HAZARDOUS DECOMPOSITION PRODUCTS

SECTION VII SPILL OR LEAK PROCEDURES

STEPS TO BE TAKEN IN CASE MATERIAL IS RELEASED OR SPILLED: Water spray

WASTE DISPOSAL METHOD

SECTION VIII SPECIAL PROTECTION INFORMATION Full Protection

VENTILATION REQUIREMENTS LOCAL EXHAUST

PROTECTIVE EQUIPMENT (SPECIFY TYPES) EYE: Goggles

MECHANICAL (GENERAL)

GLOVES: Rubber

SPECIAL

RESPIRATOR: w/appropriate filter

OTHER PROTECTIVE EQUIPMENT: Self-contained breathing apparatus, if contaminated area must be entered.

SECTION IX SPECIAL PRECAUTIONS

PRECAUTIONS TO BE TAKEN IN HANDLING AND STORAGE: Store in cool, dry, well-ventilated location away from fire hazards.

OTHER PRECAUTIONS

SIGNATURE

DATE

METHYL THIOCYANATE

Hazard rating: 2 (top), 3 (left), 0 (right)

SECTION I SOURCE AND NOMENCLATURE

MANUFACTURER'S NAME

EMERGENCY TELEPHONE NO.

ADDRESS (NUMBER, STREET, CITY, STATE, ZIP CODE)

TRADE NAME AND SYNONYMS

CHEMICAL FAMILY

CHEMICAL NAME AND SYNONYMS: Methyl Sulfocyanate

FORMULA: CH_3SCN

SECTION II HAZARDOUS INGREDIENTS

BASIC MATERIAL	APPROXIMATE OR MAXIMUM % WT. OR VOL.	ESTABLISHED OSHA STANDARD	LD_{50} ORAL	PERCUT.	SPECIES	LC_{50} CONC.

Notes:

1. Lowest published lethal dose oral - cat 8.5 mg/kg
2. Subcutaneous - rabbit LDLO - 20 mg/kg

SECTION III PHYSICAL DATA MW - 73.11

BOILING POINT (°F): 133°C

VAPOR PRESSURE (mm Hg.)

MELTING POINT (°F): -51°C

VAPOR DENSITY (Air = 1)

SPECIFIC GRAVITY (H_2O = 1): 1.07

EVAPORATION RATE (_____ = 1)

SOLUBILITY IN WATER (Pts/100 pts H_2O): very slightly soluble

VOLATILE (% Vol.) (% Wt.)

APPEARANCE AND ODOR: colorless liquid, onion odor.

SECTION IV FIRE AND EXPLOSION HAZARD DATA

FLASH POINT: none available

METHOD USED

FLAMMABLE (EXPLOSIVE) LIMITS: UPPER / LOWER

EXTINGUISHING MEDIA: water, alcohol foam, carbon dioxide, dry chemical.

SPECIAL FIRE FIGHTING PROCEDURES: Wear self-contained breathing apparatus and goggles.

UNUSUAL FIRE AND EXPLOSION HAZARDS

SECTION V HEALTH HAZARD DATA

TOXIC LEVEL: none available

CARCINOGENIC: no

PRINCIPLE ROUTES OF ABSORBTION: ingestion, inhalation

SKIN AND EYE IRRITATION: moderate

REVELANT SYMPTOMS OF EXPOSURE: eye and mucous membrane irritation. Thiocyanate intoxication.

EFFECTS OF CHRONIC EXPOSURE: Goiters

EMERGENCY AND FIRST AID PROCEDURES: If swallowed - give emetic, get medical aid, gastric lavage; inhalation - rest, give oxygen as needed. Wash from skin & eyes.

SECTION VI REACTIVITY DATA

CONDITIONS CONTRIBUTING TO INSTABILITY

CONDITIONS CONTRIBUTING TO HAZARDOUS POLYMERIZATION

INCOMPATABILITY (MATERIALS TO AVOID)

HAZARDOUS DECOMPOSITION PRODUCTS

SECTION VII SPILL OR LEAK PROCEDURES

STEPS TO BE TAKEN IN CASE MATERIAL IS RELEASED OR SPILLED

WASTE DISPOSAL METHOD

SECTION VIII SPECIAL PROTECTION INFORMATION

VENTILATION REQUIREMENTS LOCAL EXHAUST

PROTECTIVE EQUIPMENT (SPECIFY TYPES) EYE: Goggles

MECHANICAL (GENERAL)

GLOVES: Rubber

SPECIAL

RESPIRATOR: Self-contained

OTHER PROTECTIVE EQUIPMENT

SECTION IX SPECIAL PRECAUTIONS

PRECAUTIONS TO BE TAKEN IN HANDLING AND STORAGE

OTHER PRECAUTIONS

SIGNATURE

DATE

METHYL-m-TOLUIDINE

NFPA hazard diamond: 2 (top), 4 (left), 0 (right)

SECTION I SOURCE AND NOMENCLATURE

Field	Value
MANUFACTURER'S NAME	
EMERGENCY TELEPHONE NO.	
ADDRESS (NUMBER, STREET, CITY, STATE, ZIP CODE)	
TRADE NAME AND SYNONYMS	
CHEMICAL FAMILY	
CHEMICAL NAME AND SYNONYMS	N Methyl-M-Toluidine
FORMULA	$CH_3NHC_6H_4CH_3$

SECTION II HAZARDOUS INGREDIENTS

BASIC MATERIAL	APPROXIMATE OR MAXIMUM % WT. OR VOL.	ESTABLISHED OSHA STANDARD	LD_{50} ORAL	LD_{50} PERCUT.	LC_{50} SPECIES	LC_{50} CONC.
			none available			

SECTION III PHYSICAL DATA MW - 121.18

Field	Value
BOILING POINT °F.	207°C
VAPOR PRESSURE mm Hg.	
MELTING POINT °F.	--
VAPOR DENSITY (Air = 1)	
SPECIFIC GRAVITY (H_2O = 1)	0.95
EVAPORATION RATE (___ = 1)	
SOLUBILITY IN WATER Pts/100 pts H_2O	insoluble
VOLATILE % Vol. % Wt.	
APPEARANCE AND ODOR	liquid

SECTION IV FIRE AND EXPLOSION HAZARD DATA

Field	Value
FLASH POINT	none available
METHOD USED	
FLAMMABLE (EXPLOSIVE) LIMITS UPPER	
FLAMMABLE (EXPLOSIVE) LIMITS LOWER	
EXTINGUISHING MEDIA	Alcohol foam, carbon dioxide, dry chemical.
SPECIAL FIRE FIGHTING PROCEDURES	Wear full protective clothing.
UNUSUAL FIRE AND EXPLOSION HAZARDS	

SECTION V HEALTH HAZARD DATA

Field	Value
TOXIC LEVEL	none available
CARCINOGENIC	no
PRINCIPLE ROUTES OF ABSORBTION	Inhalation, ingestion, skin
SKIN AND EYE IRRITATION	moderate
REVELANT SYMPTOMS OF EXPOSURE	Cyanosis, air hunger, nausea, vomiting, low blood pressure, convulsions.
EFFECTS OF CHRONIC EXPOSURE	Weight loss, anemia, weakness.
EMERGENCY AND FIRST AID PROCEDURES	If swallowed - give emetic, get medical aid; inhalation - rest; wash from skin & eyes.

SECTION VI REACTIVITY DATA

Field	Value
CONDITIONS CONTRIBUTING TO INSTABILITY	
CONDITIONS CONTRIBUTING TO HAZARDOUS POLYMERIZATION	
INCOMPATABILITY (MATERIALS TO AVOID)	
HAZARDOUS DECOMPOSITION PRODUCTS	

SECTION VII SPILL OR LEAK PROCEDURES

Field	Value
STEPS TO BE TAKEN IN CASE MATERIAL IS RELEASED OR SPILLED	
WASTE DISPOSAL METHOD	

SECTION VIII SPECIAL PROTECTION INFORMATION

Field	Value
VENTILATION REQUIREMENTS LOCAL EXHAUST	
PROTECTIVE EQUIPMENT (SPECIFY TYPES) EYE	Goggles
MECHANICAL (GENERAL)	
GLOVES	Rubber
SPECIAL	
RESPIRATOR	With proper filter
OTHER PROTECTIVE EQUIPMENT	

SECTION IX SPECIAL PRECAUTIONS

Field	Value
PRECAUTIONS TO BE TAKEN IN HANDLING AND STORAGE	Outside storage preferred. Inside storage in a standard flammable liquids storage cabinet.
OTHER PRECAUTIONS	
SIGNATURE	
DATE	

METHYL-o-TOLUIDINE

NFPA hazard diamond: 2 (top), 4 (left), 0 (right)

SECTION I SOURCE AND NOMENCLATURE

Field	Value
MANUFACTURER'S NAME	
EMERGENCY TELEPHONE NO.	
ADDRESS (NUMBER, STREET, CITY, STATE, ZIP CODE)	
TRADE NAME AND SYNONYMS	
CHEMICAL FAMILY	
CHEMICAL NAME AND SYNONYMS	N Methyl-O-Toluidine
FORMULA	$CH_3NHC_6H_4CH_3$

SECTION II HAZARDOUS INGREDIENTS

BASIC MATERIAL	APPROXIMATE OR MAXIMUM % WT. OR VOL.	ESTABLISHED OSHA STANDARD	LD_{50} ORAL	LD_{50} PERCUT.	LC_{50} SPECIES	LC_{50} CONC.
			none available			

SECTION III PHYSICAL DATA MW - 121.18

Field	Value
BOILING POINT °F.	207°C
VAPOR PRESSURE mm Hg.	
MELTING POINT °F.	--
VAPOR DENSITY (Air = 1)	
SPECIFIC GRAVITY (H_2O = 1)	0.97
EVAPORATION RATE (___ = 1)	
SOLUBILITY IN WATER Pts/100 pts H_2O	insoluble
VOLATILE % Vol. % Wt.	
APPEARANCE AND ODOR	liquid

SECTION IV FIRE AND EXPLOSION HAZARD DATA

Field	Value
FLASH POINT	none available
METHOD USED	
FLAMMABLE (EXPLOSIVE) LIMITS UPPER	
FLAMMABLE (EXPLOSIVE) LIMITS LOWER	
EXTINGUISHING MEDIA	Alcohol foam, carbon dioxide, dry chemical
SPECIAL FIRE FIGHTING PROCEDURES	Wear full protective clothing.
UNUSUAL FIRE AND EXPLOSION HAZARDS	

SECTION V HEALTH HAZARD DATA

Field	Value
TOXIC LEVEL	none available
CARCINOGENIC	no
PRINCIPLE ROUTES OF ABSORBTION	Inhalation, ingestion, skin
SKIN AND EYE IRRITATION	moderate
REVELANT SYMPTOMS OF EXPOSURE	Cyanosis, air hunger, nausea, vomiting, low blood pressure, convulsions.
EFFECTS OF CHRONIC EXPOSURE	Weight loss, anemia, weakness
EMERGENCY AND FIRST AID PROCEDURES	If swallowed - give emetic, get medical aid; inhalation - rest; wash from skin & eyes.

SECTION VI REACTIVITY DATA

Field	Value
CONDITIONS CONTRIBUTING TO INSTABILITY	
CONDITIONS CONTRIBUTING TO HAZARDOUS POLYMERIZATION	
INCOMPATABILITY (MATERIALS TO AVOID)	
HAZARDOUS DECOMPOSITION PRODUCTS	

SECTION VII SPILL OR LEAK PROCEDURES

Field	Value
STEPS TO BE TAKEN IN CASE MATERIAL IS RELEASED OR SPILLED	
WASTE DISPOSAL METHOD	

SECTION VIII SPECIAL PROTECTION INFORMATION

Field	Value
VENTILATION REQUIREMENTS LOCAL EXHAUST	
PROTECTIVE EQUIPMENT (SPECIFY TYPES) EYE	Goggles
MECHANICAL (GENERAL)	
GLOVES	Rubber
SPECIAL	
RESPIRATOR	With proper filer
OTHER PROTECTIVE EQUIPMENT	

SECTION IX SPECIAL PRECAUTIONS

Field	Value
PRECAUTIONS TO BE TAKEN IN HANDLING AND STORAGE	Outside storage preferred. Inside storage in a standard flammable liquids storage cabinet.
OTHER PRECAUTIONS	
SIGNATURE	
DATE	

METHYL-p-TOLUIDINE

Hazard diamond: 2 (top), 4 (left), 0 (right)

SECTION I SOURCE AND NOMENCLATURE

MANUFACTURER'S NAME | EMERGENCY TELEPHONE NO.

ADDRESS (NUMBER, STREET, CITY, STATE, ZIP CODE)

TRADE NAME AND SYNONYMS | CHEMICAL FAMILY

CHEMICAL NAME AND SYNONYMS: N Methyl P Toluidine

FORMULA: $CH_3NHC_6H_4CH_3$

SECTION II HAZARDOUS INGREDIENTS

BASIC MATERIAL	APPROXIMATE OR MAXIMUM % WT. OR VOL.	ESTABLISHED OSHA STANDARD	LD_{50} ORAL	LD_{50} PERCUT.	LC_{50} SPECIES	LC_{50} CONC.
			none	available		

SECTION III PHYSICAL DATA MW - 121.18

BOILING POINT °F.: 210°C

VAPOR PRESSURE mm Hg.:

MELTING POINT °F.: --

VAPOR DENSITY (Air = 1):

SPECIFIC GRAVITY (H_2O = 1): 0.94

EVAPORATION RATE (____ = 1):

SOLUBILITY IN WATER Pts/100 pts H_2O: insoluble

VOLATILE % Vol. % Wt.:

APPEARANCE AND ODOR: liquid

SECTION IV FIRE AND EXPLOSION HAZARD DATA

FLASH POINT: none available

METHOD USED:

FLAMMABLE (EXPLOSIVE) LIMITS: UPPER / LOWER

EXTINGUISHING MEDIA: alcohol foam, carbon dioxide, dry chemical

SPECIAL FIRE FIGHTING PROCEDURES: wear full protective clothing

UNUSUAL FIRE AND EXPLOSION HAZARDS:

SECTION V HEALTH HAZARD DATA

TOXIC LEVEL: none available

CARCINOGENIC: no

PRINCIPLE ROUTES OF ABSORBTION: inhalation, ingestion, skin

SKIN AND EYE IRRITATION: moderate

REVELANT SYMPTOMS OF EXPOSURE: cyanosis, air hunger, nausea, vomiting, high blood pressure, convulsions.

EFFECTS OF CHRONIC EXPOSURE: weight loss, anemia, weakness

EMERGENCY AND FIRST AID PROCEDURES: If swallowed - give emetic, get medical aid; inhalation - rest; wash from skin & eyes.

SECTION VI REACTIVITY DATA

CONDITIONS CONTRIBUTING TO INSTABILITY:

CONDITIONS CONTRIBUTING TO HAZARDOUS POLYMERIZATION:

INCOMPATABILITY (MATERIALS TO AVOID):

HAZARDOUS DECOMPOSITION PRODUCTS:

SECTION VII SPILL OR LEAK PROCEDURES

STEPS TO BE TAKEN IN CASE MATERIAL IS RELEASED OR SPILLED:

WASTE DISPOSAL METHOD:

SECTION VIII SPECIAL PROTECTION INFORMATION

VENTILATION REQUIREMENTS LOCAL EXHAUST:

PROTECTIVE EQUIPMENT (SPECIFY TYPES) EYE: goggles

MECHANICAL (GENERAL):

GLOVES: rubber

SPECIAL:

RESPIRATOR: with proper filter

OTHER PROTECTIVE EQUIPMENT:

SECTION IX SPECIAL PRECAUTIONS

PRECAUTIONS TO BE TAKEN IN HANDLING AND STORAGE: Outside storage preferred. Inside storage in a standard flammable liquids storage cabinet.

OTHER PRECAUTIONS:

SIGNATURE | DATE

MYRISTOYL CHLORIDE

Hazard diamond: 1 (top), 3 (left), 1 (right), W (bottom)

SECTION I SOURCE AND NOMENCLATURE

MANUFACTURER'S NAME | EMERGENCY TELEPHONE NO.

ADDRESS (NUMBER, STREET, CITY, STATE, ZIP CODE)

TRADE NAME AND SYNONYMS | CHEMICAL FAMILY

CHEMICAL NAME AND SYNONYMS: Myristoyl Chloride

FORMULA: $C_{13}H_{27}COCl$

SECTION II HAZARDOUS INGREDIENTS

BASIC MATERIAL	APPROXIMATE OR MAXIMUM % WT. OR VOL.	ESTABLISHED OSHA STANDARD	LD_{50} ORAL	LD_{50} PERCUT.	LC_{50} SPECIES	LC_{50} CONC.
			none	available		

SECTION III PHYSICAL DATA MW-246.81

BOILING POINT °F.: 168°C 15 mm

VAPOR PRESSURE mm Hg.:

MELTING POINT °F.: 3°C

VAPOR DENSITY (Air = 1):

SPECIFIC GRAVITY (H_2O = 1):

EVAPORATION RATE (____ = 1):

SOLUBILITY IN WATER Pts/100 pts H_2O: decomposes in water

VOLATILE % Vol. % Wt.:

APPEARANCE AND ODOR: colorless liquid

SECTION IV FIRE AND EXPLOSION HAZARD DATA

FLASH POINT: none available

METHOD USED:

FLAMMABLE (EXPLOSIVE) LIMITS: UPPER / LOWER

EXTINGUISHING MEDIA: Carbon dioxide, dry chemical. Do not use water.

SPECIAL FIRE FIGHTING PROCEDURES: Full protective clothing, self-contained breathing apparatus.

UNUSUAL FIRE AND EXPLOSION HAZARDS: When heated to decomposition point, gives off toxic HCl.

SECTION V HEALTH HAZARD DATA

TOXIC LEVEL: none available

CARCINOGENIC: no

PRINCIPLE ROUTES OF ABSORBTION: Inhalation, ingestion

SKIN AND EYE IRRITATION: severe

REVELANT SYMPTOMS OF EXPOSURE: Irritation of eyes and respiratory system; severe internal damage if swallowed; eye and skin burns.

EFFECTS OF CHRONIC EXPOSURE: Bronchial irritation, errosion of teeth.

EMERGENCY AND FIRST AID PROCEDURES: Inhalation - rest; if swallowed - drink water, milk of magnesia, get medical aid, wash eyes and skin.

SECTION VI REACTIVITY DATA

CONDITIONS CONTRIBUTING TO INSTABILITY:

CONDITIONS CONTRIBUTING TO HAZARDOUS POLYMERIZATION:

INCOMPATABILITY (MATERIALS TO AVOID): Reacts with water or alcohol

HAZARDOUS DECOMPOSITION PRODUCTS:

SECTION VII SPILL OR LEAK PROCEDURES

STEPS TO BE TAKEN IN CASE MATERIAL IS RELEASED OR SPILLED: Flush spills with large quantities of water and ventilate area to remove vapor.

WASTE DISPOSAL METHOD:

SECTION VIII SPECIAL PROTECTION INFORMATION

VENTILATION REQUIREMENTS LOCAL EXHAUST:

PROTECTIVE EQUIPMENT (SPECIFY TYPES) EYE: Goggles

MECHANICAL (GENERAL):

GLOVES: Rubber

SPECIAL:

RESPIRATOR: Self-contained

OTHER PROTECTIVE EQUIPMENT: Rubber apron

SECTION IX SPECIAL PRECAUTIONS

PRECAUTIONS TO BE TAKEN IN HANDLING AND STORAGE:

OTHER PRECAUTIONS:

SIGNATURE | DATE

α-NAPHTHALENEACETONITRILE

Hazard diamond: 2 (top), 3 (left), 0 (right)

SECTION I SOURCE AND NOMENCLATURE

MANUFACTURER'S NAME | EMERGENCY TELEPHONE NO.

ADDRESS (NUMBER, STREET, CITY, STATE, ZIP CODE)

TRADE NAME AND SYNONYMS: 1 Naphthaleneacetonitrile

CHEMICAL FAMILY

CHEMICAL NAME AND SYNONYMS: α naphthaleneacetonitrile

FORMULA: $C_{10}H_7CH_2CN$

SECTION II HAZARDOUS INGREDIENTS

BASIC MATERIAL	APPROXIMATE OR MAXIMUM % WT. OR VOL.	ESTABLISHED OSHA STANDARD	LD_{50} ORAL	LD_{50} PERCUT.	LC_{50} SPECIES	LC_{50} CONC.
			none available			

SECTION III PHYSICAL DATA MW - 167.21

BOILING POINT °F.: 194°C at 18 mm

VAPOR PRESSURE mm Hg.: 185°C 10

MELTING POINT °F.: 35°C

VAPOR DENSITY (Air = 1)

SPECIFIC GRAVITY (H_2O = 1)

EVAPORATION RATE (______ = 1)

SOLUBILITY IN WATER Pts/100 pts H_2O: --

VOLATILE % Vol. % Wt.

APPEARANCE AND ODOR: crystal or liquid

SECTION IV FIRE AND EXPLOSION HAZARD DATA

FLASH POINT: none available

FLAMMABLE (EXPLOSIVE) LIMITS: UPPER / LOWER

METHOD USED

EXTINGUISHING MEDIA: alcohol foam, carbon dioxide, dry chemical

SPECIAL FIRE FIGHTING PROCEDURES

UNUSUAL FIRE AND EXPLOSION HAZARDS

SECTION V HEALTH HAZARD DATA

TOXIC LEVEL: none available

CARCINOGENIC: no

PRINCIPLE ROUTES OF ABSORBTION: inhalation, ingestion (mainly)

SKIN AND EYE IRRITATION: moderate

REVELANT SYMPTOMS OF EXPOSURE: Dizziness, rapid respiration, headache, drowsiness, drop in blood pressure, rapid pulse.

EFFECTS OF CHRONIC EXPOSURE: Dizziness, weakness, hoarseness, lung congestion, eye irritation, loss of weight and appetite.

EMERGENCY AND FIRST AID PROCEDURES: If swallowed - cyanide antidote; inhalation - inhale amyl nitrite artificial respiration; get medical aid; wash skin & eyes.

SECTION VI REACTIVITY DATA

CONDITIONS CONTRIBUTING TO INSTABILITY

CONDITIONS CONTRIBUTING TO HAZARDOUS POLYMERIZATION

INCOMPATABILITY (MATERIALS TO AVOID)

HAZARDOUS DECOMPOSITION PRODUCTS

SECTION VII SPILL OR LEAK PROCEDURES

STEPS TO BE TAKEN IN CASE MATERIAL IS RELEASED OR SPILLED

WASTE DISPOSAL METHOD

SECTION VIII SPECIAL PROTECTION INFORMATION

VENTILATION REQUIREMENTS LOCAL EXHAUST

PROTECTIVE EQUIPMENT (SPECIFY TYPES) EYE: Safety goggles

MECHANICAL (GENERAL)

GLOVES: Rubber

SPECIAL

RESPIRATOR: With proper filter

OTHER PROTECTIVE EQUIPMENT

SECTION IX SPECIAL PRECAUTIONS

PRECAUTIONS TO BE TAKEN IN HANDLING AND STORAGE

OTHER PRECAUTIONS

SIGNATURE | DATE

α-NAPHTHOL

Hazard diamond: 1 (top), 3 (left), 0 (right)

SECTION I SOURCE AND NOMENCLATURE

MANUFACTURER'S NAME | EMERGENCY TELEPHONE NO.

ADDRESS (NUMBER, STREET, CITY, STATE, ZIP CODE)

TRADE NAME AND SYNONYMS: A-Hydroxynaphthalene

CHEMICAL FAMILY

CHEMICAL NAME AND SYNONYMS: A-Naphthol

FORMULA: $C_{10}H_7OH$

SECTION II HAZARDOUS INGREDIENTS

BASIC MATERIAL	APPROXIMATE OR MAXIMUM % WT. OR VOL.	ESTABLISHED OSHA STANDARD	LD_{50} Mg/Kg ORAL	LD_{50} PERCUT.	SPECIES	LC_{50} CONC.
			2590		Rat	
				880	Rabbit	

SECTION III PHYSICAL DATA MW - 144.18

BOILING POINT °F.: 283° C.

VAPOR PRESSURE mm Hg.: at 94° C. 1

MELTING POINT °F.: 96° C.

VAPOR DENSITY (Air = 1)

SPECIFIC GRAVITY (H_2O = 1): 1.10

EVAPORATION RATE (______ = 1)

SOLUBILITY IN WATER Pts/100 pts H_2O

VOLATILE % Vol. % Wt.

APPEARANCE AND ODOR: Colorless crystal, disagreeable odor.

SECTION IV FIRE AND EXPLOSION HAZARD DATA

FLASH POINT: None available

FLAMMABLE (EXPLOSIVE) LIMITS: UPPER / LOWER

METHOD USED

EXTINGUISHING MEDIA: Carbon dioxide, dry chemical.

SPECIAL FIRE FIGHTING PROCEDURES

UNUSUAL FIRE AND EXPLOSION HAZARDS

SECTION V HEALTH HAZARD DATA

TOXIC LEVEL: None available

CARCINOGENIC: No

PRINCIPLE ROUTES OF ABSORBTION: Inhalation, ingestion & skin

SKIN AND EYE IRRITATION: Severe

REVELANT SYMPTOMS OF EXPOSURE: Skin and mucous membrane irritation, eye damage.

EFFECTS OF CHRONIC EXPOSURE: Kidney damage, dermatitis.

EMERGENCY AND FIRST AID PROCEDURES: Inhalation - rest; if swallowed - drink water, epsom salt solution, get medical aid; wash skin & eyes.

SECTION VI REACTIVITY DATA

CONDITIONS CONTRIBUTING TO INSTABILITY

CONDITIONS CONTRIBUTING TO HAZARDOUS POLYMERIZATION

INCOMPATABILITY (MATERIALS TO AVOID)

HAZARDOUS DECOMPOSITION PRODUCTS

SECTION VII SPILL OR LEAK PROCEDURES

STEPS TO BE TAKEN IN CASE MATERIAL IS RELEASED OR SPILLED

WASTE DISPOSAL METHOD

SECTION VIII SPECIAL PROTECTION INFORMATION

VENTILATION REQUIREMENTS LOCAL EXHAUST

PROTECTIVE EQUIPMENT (SPECIFY TYPES) EYE: Safety goggles

MECHANICAL (GENERAL)

GLOVES: Rubber

SPECIAL

RESPIRATOR: With proper filter

OTHER PROTECTIVE EQUIPMENT

SECTION IX SPECIAL PRECAUTIONS

PRECAUTIONS TO BE TAKEN IN HANDLING AND STORAGE

OTHER PRECAUTIONS

SIGNATURE | DATE

β-NAPHTHOL

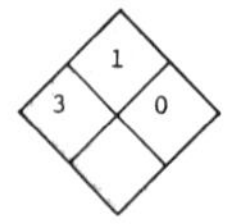

SECTION I SOURCE AND NOMENCLATURE

MANUFACTURER'S NAME

EMERGENCY TELEPHONE NO.

ADDRESS (NUMBER, STREET, CITY, STATE, ZIP CODE)

TRADE NAME AND SYNONYMS: B-Hydroxynaphthalene

CHEMICAL FAMILY

CHEMICAL NAME AND SYNONYMS: B-Naphthol

FORMULA: $C_{10}H_7OH$

SECTION II HAZARDOUS INGREDIENTS

BASIC MATERIAL	APPROXIMATE OR MAXIMUM % WT. OR VOL.	ESTABLISHED OSHA STANDARD	LD_{50} mg/kg ORAL	PERCUT.	SPECIES	LC_{50} CONC.
Oral - human - LDLO 50 mg/kg			2420		Rat	

SECTION III PHYSICAL DATA MW-144.18

BOILING POINT °F.: 280°C.

VAPOR PRESSURE mm Hg.: 10 at 146°C

MELTING POINT °F.: 122.5°C.

VAPOR DENSITY (Air = 1): 4.97

SPECIFIC GRAVITY (H_2O = 1): 1.22

EVAPORATION RATE (= 1)

SOLUBILITY IN WATER Pts/100 pts H_2O

VOLATILE % Vol. % Wt.

APPEARANCE AND ODOR: White to yellow white crystal, p henolic odor.

SECTION IV FIRE AND EXPLOSION HAZARD DATA

FLASH POINT: 322°F.

FLAMMABLE (EXPLOSIVE) LIMITS: UPPER / LOWER

METHOD USED

EXTINGUISHING MEDIA: Carbon dioxide, dry chemical.

SPECIAL FIRE FIGHTING PROCEDURES

UNUSUAL FIRE AND EXPLOSION HAZARDS

SECTION V HEALTH HAZARD DATA

TOXIC LEVEL: None available

CARCINOGENIC: No

PRINCIPLE ROUTES OF ABSORBTION: Inhalation, ingestion & skin

SKIN AND EYE IRRITATION: Severe

REVELANT SYMPTOMS OF EXPOSURE: Skin and mucous membrane irritation, eye damage.

EFFECTS OF CHRONIC EXPOSURE: Kidney damage, dermatitis.

EMERGENCY AND FIRST AID PROCEDURES: Inhalation - rest; if swallowed - drink water, epsom salt solution, get medical aid; wash skin & eyes.

SECTION VI REACTIVITY DATA

CONDITIONS CONTRIBUTING TO INSTABILITY

CONDITIONS CONTRIBUTING TO HAZARDOUS POLYMERIZATION

INCOMPATABILITY (MATERIALS TO AVOID)

HAZARDOUS DECOMPOSITION PRODUCTS

SECTION VII SPILL OR LEAK PROCEDURES

STEPS TO BE TAKEN IN CASE MATERIAL IS RELEASED OR SPILLED

WASTE DISPOSAL METHOD

SECTION VIII SPECIAL PROTECTION INFORMATION

VENTILATION REQUIREMENTS LOCAL EXHAUST

PROTECTIVE EQUIPMENT (SPECIFY TYPES) EYE: Safety goggles

MECHANICAL (GENERAL)

GLOVES: Rubber

SPECIAL

RESPIRATOR: With proper filter

OTHER PROTECTIVE EQUIPMENT

SECTION IX SPECIAL PRECAUTIONS

PRECAUTIONS TO BE TAKEN IN HANDLING AND STORAGE

OTHER PRECAUTIONS

SIGNATURE

DATE

α-NAPHTHONITRILE

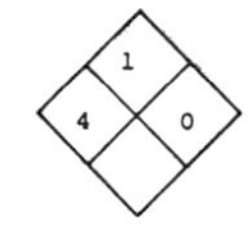

SECTION I SOURCE AND NOMENCLATURE

MANUFACTURER'S NAME

EMERGENCY TELEPHONE NO.

ADDRESS (NUMBER, STREET, CITY, STATE, ZIP CODE)

TRADE NAME AND SYNONYMS

CHEMICAL FAMILY

CHEMICAL NAME AND SYNONYMS: A-Naphthonitrile

FORMULA: $C_{10}H_7CN$

SECTION II HAZARDOUS INGREDIENTS

BASIC MATERIAL	APPROXIMATE OR MAXIMUM % WT. OR VOL.	ESTABLISHED OSHA STANDARD	LD_{50} ORAL	PERCUT.	SPECIES	LC_{50} CONC.
			None available			

SECTION III PHYSICAL DATA MW - 153.19

BOILING POINT °F.: 307° C.

VAPOR PRESSURE mm Hg.: 12 at 158° C.

MELTING POINT °F.: 35° C.

VAPOR DENSITY (Air = 1)

SPECIFIC GRAVITY (H_2O = 1): 1.09

EVAPORATION RATE (= 1)

SOLUBILITY IN WATER Pts/100 pts H_2O: Insoluble

VOLATILE % Vol. % Wt.

APPEARANCE AND ODOR: Leaf-like crystals

SECTION IV FIRE AND EXPLOSION HAZARD DATA

FLASH POINT: None available

FLAMMABLE (EXPLOSIVE) LIMITS: UPPER / LOWER

METHOD USED

EXTINGUISHING MEDIA: Alcohol foam, carbon dioxide, dry chemical.

SPECIAL FIRE FIGHTING PROCEDURES

UNUSUAL FIRE AND EXPLOSION HAZARDS

SECTION V HEALTH HAZARD DATA

TOXIC LEVEL: None available

CARCINOGENIC: No

PRINCIPLE ROUTES OF ABSORBTION: Inhalation, ingestion, skin

SKIN AND EYE IRRITATION: Moderate

REVELANT SYMPTOMS OF EXPOSURE: Headache, nausea, vomiting, abdominal cramping, unconsciousness, convulsions.

EFFECTS OF CHRONIC EXPOSURE: Flushing of face, nausea, jaundice.

EMERGENCY AND FIRST AID PROCEDURES: Wash skin & eyes; inhalation - inhale amyl nitrite vapor; if swallowed - give emetic, get medical aid.

SECTION VI REACTIVITY DATA

CONDITIONS CONTRIBUTING TO INSTABILITY

CONDITIONS CONTRIBUTING TO HAZARDOUS POLYMERIZATION

INCOMPATABILITY (MATERIALS TO AVOID)

HAZARDOUS DECOMPOSITION PRODUCTS

SECTION VII SPILL OR LEAK PROCEDURES

STEPS TO BE TAKEN IN CASE MATERIAL IS RELEASED OR SPILLED

WASTE DISPOSAL METHOD

SECTION VIII SPECIAL PROTECTION INFORMATION

VENTILATION REQUIREMENTS LOCAL EXHAUST

PROTECTIVE EQUIPMENT (SPECIFY TYPES) EYE: Goggles

MECHANICAL (GENERAL)

GLOVES: Rubber

SPECIAL

RESPIRATOR: With organic filter

OTHER PROTECTIVE EQUIPMENT

SECTION IX SPECIAL PRECAUTIONS

PRECAUTIONS TO BE TAKEN IN HANDLING AND STORAGE

OTHER PRECAUTIONS

SIGNATURE

DATE

β-NAPHTHYLAMINE HYDROCHLORIDE

Hazard diamond: 1 (top), 4 (left), 0 (right)

SECTION I SOURCE AND NOMENCLATURE

MANUFACTURER'S NAME	EMERGENCY TELEPHONE NO.

ADDRESS (NUMBER, STREET, CITY, STATE, ZIP CODE)

TRADE NAME AND SYNONYMS	CHEMICAL FAMILY
CHEMICAL NAME AND SYNONYMS 2-Naphthylamine Hydrochloride	FORMULA $C_{10}H_7NH_2HCl$

SECTION II HAZARDOUS INGREDIENTS

BASIC MATERIAL	APPROXIMATE OR MAXIMUM % WT. OR VOL.	ESTABLISHED OSHA STANDARD	LD_{50}		LC_{50}	
			ORAL	PERCUT.	SPECIES	CONC.

SECTION III PHYSICAL DATA MW - 179.65

BOILING POINT °F. --	VAPOR PRESSURE mm Hg.
MELTING POINT °F. --	VAPOR DENSITY (Air = 1)
SPECIFIC GRAVITY (H_2O = 1) --	EVAPORATION RATE (______ = 1)
SOLUBILITY IN WATER Pts/100 pts H_2O Very soluble	VOLATILE % Vol. % Wt.

APPEARANCE AND ODOR
Leaflet crystals

SECTION IV FIRE AND EXPLOSION HAZARD DATA

FLASH POINT None available	FLAMMABLE (EXPLOSIVE) LIMITS	UPPER
METHOD USED		LOWER

EXTINGUISHING MEDIA
Carbon dioxide, dry chemical.

SPECIAL FIRE FIGHTING PROCEDURES

UNUSUAL FIRE AND EXPLOSION HAZARDS

SECTION V HEALTH HAZARD DATA

TOXIC LEVEL
None available — CARCINOGENIC Yes

PRINCIPLE ROUTES OF ABSORBTION
Inhalation, ingestion, skin — SKIN AND EYE IRRITATION Moderate

REVELANT SYMPTOMS OF EXPOSURE
Eye irritation

EFFECTS OF CHRONIC EXPOSURE
Bladder tumors

EMERGENCY AND FIRST AID PROCEDURES If swallowed - give emetic, get medical aid; inhalation - rest; wash skin & hands.

SECTION VI REACTIVITY DATA

CONDITIONS CONTRIBUTING TO INSTABILITY

CONDITIONS CONTRIBUTING TO HAZARDOUS POLYMERIZATION

INCOMPATABILITY (MATERIALS TO AVOID)

HAZARDOUS DECOMPOSITION PRODUCTS

SECTION VII SPILL OR LEAK PROCEDURES

STEPS TO BE TAKEN IN CASE MATERIAL IS RELEASED OR SPILLED

WASTE DISPOSAL METHOD

SECTION VIII SPECIAL PROTECTION INFORMATION

VENTILATION REQUIREMENTS LOCAL EXHAUST	PROTECTIVE EQUIPMENT (SPECIFY TYPES) EYE Safety goggles
MECHANICAL (GENERAL)	GLOVES Rubber
SPECIAL	RESPIRATOR Filter type for dust, mist and fumes

OTHER PROTECTIVE EQUIPMENT
Full protective clothing.

SECTION IX SPECIAL PRECAUTIONS

PRECAUTIONS TO BE TAKEN IN HANDLING AND STORAGE

OTHER PRECAUTIONS

SIGNATURE	DATE

α-NAPHTHYLHYDRAZINE HYDROCHLORIDE

Hazard diamond: 1 (top), 4 (left), 0 (right)

SECTION I SOURCE AND NOMENCLATURE

MANUFACTURER'S NAME	EMERGENCY TELEPHONE NO.

ADDRESS (NUMBER, STREET, CITY, STATE, ZIP CODE)

TRADE NAME AND SYNONYMS	CHEMICAL FAMILY
CHEMICAL NAME AND SYNONYMS α Naphthylhydrazine hydrochloride	FORMULA $C_{10}H_7NHNH_2HCl$

SECTION II HAZARDOUS INGREDIENTS

BASIC MATERIAL	APPROXIMATE OR MAXIMUM % WT. OR VOL.	ESTABLISHED OSHA STANDARD	LD_{50}		LC_{50}	
			ORAL	PERCUT.	SPECIES	CONC.
			none	available		

SECTION III PHYSICAL DATA MW - 194.66

BOILING POINT °F. --	VAPOR PRESSURE mm Hg.
MELTING POINT °F. --	VAPOR DENSITY (Air = 1)
SPECIFIC GRAVITY (H_2O = 1)	EVAPORATION RATE (______ = 1)
SOLUBILITY IN WATER Pts/100 pts H_2O soluble	VOLATILE % Vol. % Wt.

APPEARANCE AND ODOR
crystaline

SECTION IV FIRE AND EXPLOSION HAZARD DATA

FLASH POINT none available	FLAMMABLE (EXPLOSIVE) LIMITS	UPPER
METHOD USED		LOWER

EXTINGUISHING MEDIA

SPECIAL FIRE FIGHTING PROCEDURES

UNUSUAL FIRE AND EXPLOSION HAZARDS

SECTION V HEALTH HAZARD DATA

TOXIC LEVEL
none available — CARCINOGENIC no

PRINCIPLE ROUTES OF ABSORBTION
inhalation of dust, ingestion, skin — SKIN AND EYE IRRITATION severe

REVELANT SYMPTOMS OF EXPOSURE Irritation of skin, eyes and mucous membranes - gastroentric disturbances, anemia.

EFFECTS OF CHRONIC EXPOSURE
Blood and liver damage, loss of appetite.

EMERGENCY AND FIRST AID PROCEDURES Inhalation - rest; if swallowed - induce vomiting, get medical aid; wash from skin & eyes.

SECTION VI REACTIVITY DATA

CONDITIONS CONTRIBUTING TO INSTABILITY

CONDITIONS CONTRIBUTING TO HAZARDOUS POLYMERIZATION

INCOMPATABILITY (MATERIALS TO AVOID)

HAZARDOUS DECOMPOSITION PRODUCTS

SECTION VII SPILL OR LEAK PROCEDURES

STEPS TO BE TAKEN IN CASE MATERIAL IS RELEASED OR SPILLED

WASTE DISPOSAL METHOD

SECTION VIII SPECIAL PROTECTION INFORMATION

VENTILATION REQUIREMENTS LOCAL EXHAUST	PROTECTIVE EQUIPMENT (SPECIFY TYPES) EYE Goggles
MECHANICAL (GENERAL)	GLOVES Rubber
SPECIAL	RESPIRATOR

OTHER PROTECTIVE EQUIPMENT

SECTION IX SPECIAL PRECAUTIONS

PRECAUTIONS TO BE TAKEN IN HANDLING AND STORAGE

OTHER PRECAUTIONS

SIGNATURE	DATE

α-NAPHTHYL ISOCYANATE

Hazard rating: 1 (top), 4 (left), 0 (right)

SECTION I SOURCE AND NOMENCLATURE

MANUFACTURER'S NAME

EMERGENCY TELEPHONE NO.

ADDRESS (NUMBER, STREET, CITY, STATE, ZIP CODE)

TRADE NAME AND SYNONYMS: α- Naphthl Isocyanate

CHEMICAL FAMILY

CHEMICAL NAME AND SYNONYMS: 1 - Naphthyl Isocyanate

FORMULA: $C_{10}H_7NCO$

SECTION II HAZARDOUS INGREDIENTS

BASIC MATERIAL	APPROXIMATE OR MAXIMUM % WT. OR VOL.	ESTABLISHED OSHA STANDARD	LD_{50} mg/kg ORAL	LD_{50} PERCUT.	LD_{50} SPECIES	LC_{50} CONC.
			245		mouse	

SECTION III PHYSICAL DATA MW - 185.25

BOILING POINT °F.: 270°C

VAPOR PRESSURE mm Hg.

MELTING POINT °F.: 4°C

VAPOR DENSITY (Air = 1)

SPECIFIC GRAVITY (H_2O = 1): 1.18

EVAPORATION RATE (= 1)

SOLUBILITY IN WATER Pts/100 pts H_2O: decomposes

VOLATILE % Vol. % Wt.

APPEARANCE AND ODOR: colorless liquid

SECTION IV FIRE AND EXPLOSION HAZARD DATA

FLASH POINT: none available

FLAMMABLE (EXPLOSIVE) LIMITS: UPPER / LOWER

METHOD USED

EXTINGUISHING MEDIA: alcohol foam, carbon dioxide, dry chemical

SPECIAL FIRE FIGHTING PROCEDURES

UNUSUAL FIRE AND EXPLOSION HAZARDS

SECTION V HEALTH HAZARD DATA

TOXIC LEVEL: none available

CARCINOGENIC: no

PRINCIPLE ROUTES OF ABSORBTION: inhalation, ingestion, skin

SKIN AND EYE IRRITATION: severe

REVELANT SYMPTOMS OF EXPOSURE: skin, eye and respiratory tract irritation.

EFFECTS OF CHRONIC EXPOSURE: allergic sensitization of respiratory tract.

EMERGENCY AND FIRST AID PROCEDURES: if swallowed - induce vomiting, get medical aid; inhalation - rest; wash from skin & eyes.

SECTION VI REACTIVITY DATA

CONDITIONS CONTRIBUTING TO INSTABILITY

CONDITIONS CONTRIBUTING TO HAZARDOUS POLYMERIZATION

INCOMPATABILITY (MATERIALS TO AVOID)

HAZARDOUS DECOMPOSITION PRODUCTS

SECTION VII SPILL OR LEAK PROCEDURES

STEPS TO BE TAKEN IN CASE MATERIAL IS RELEASED OR SPILLED

WASTE DISPOSAL METHOD

SECTION VIII SPECIAL PROTECTION INFORMATION

VENTILATION REQUIREMENTS LOCAL EXHAUST

PROTECTIVE EQUIPMENT (SPECIFY TYPES) EYE: goggles

MECHANICAL (GENERAL)

GLOVES: rubber

SPECIAL

RESPIRATOR: with proper filter

OTHER PROTECTIVE EQUIPMENT

SECTION IX SPECIAL PRECAUTIONS

PRECAUTIONS TO BE TAKEN IN HANDLING AND STORAGE

OTHER PRECAUTIONS

SIGNATURE

DATE

β-NAPHTHYL ISOCYANATE

Hazard rating: 1 (top), 4 (left), 0 (right)

SECTION I SOURCE AND NOMENCLATURE

MANUFACTURER'S NAME

EMERGENCY TELEPHONE NO.

ADDRESS (NUMBER, STREET, CITY, STATE, ZIP CODE)

TRADE NAME AND SYNONYMS: 2-Naphthyl Isocyanate

CHEMICAL FAMILY

CHEMICAL NAME AND SYNONYMS: B-Naphthyl Isocyanate,

FORMULA: $C_{10}H_7NCO$

SECTION II HAZARDOUS INGREDIENTS

BASIC MATERIAL	APPROXIMATE OR MAXIMUM % WT. OR VOL.	ESTABLISHED OSHA STANDARD	LD_{50} ORAL	LD_{50} PERCUT.	LC_{50} SPECIES	LC_{50} CONC.
			none	available		

SECTION III PHYSICAL DATA MW - 185.25

BOILING POINT °F.: --

VAPOR PRESSURE mm Hg.

MELTING POINT °F.: 56°C

VAPOR DENSITY (Air = 1)

SPECIFIC GRAVITY (H_2O = 1)

EVAPORATION RATE (= 1)

SOLUBILITY IN WATER Pts/100 pts H_2O: decomposes

VOLATILE % Vol. % Wt.

APPEARANCE AND ODOR: crystals

SECTION IV FIRE AND EXPLOSION HAZARD DATA

FLASH POINT: none available

FLAMMABLE (EXPLOSIVE) LIMITS: UPPER / LOWER

METHOD USED

EXTINGUISHING MEDIA: alcohol foam, carbon dioxide, dry chemical.

SPECIAL FIRE FIGHTING PROCEDURES

UNUSUAL FIRE AND EXPLOSION HAZARDS

SECTION V HEALTH HAZARD DATA

TOXIC LEVEL: none available

CARCINOGENIC: no

PRINCIPLE ROUTES OF ABSORBTION: Inhalation, ingestion, skin

SKIN AND EYE IRRITATION: Severe

REVELANT SYMPTOMS OF EXPOSURE: Skin, eye and respiratory tract irritation.

EFFECTS OF CHRONIC EXPOSURE: Sensitization of respiratory tract.

EMERGENCY AND FIRST AID PROCEDURES: If swallowed - induce vomiting, get medical aid; inhalation - rest; wash from skin & eyes.

SECTION VI REACTIVITY DATA

CONDITIONS CONTRIBUTING TO INSTABILITY

CONDITIONS CONTRIBUTING TO HAZARDOUS POLYMERIZATION

INCOMPATABILITY (MATERIALS TO AVOID)

HAZARDOUS DECOMPOSITION PRODUCTS

SECTION VII SPILL OR LEAK PROCEDURES

STEPS TO BE TAKEN IN CASE MATERIAL IS RELEASED OR SPILLED

WASTE DISPOSAL METHOD

SECTION VIII SPECIAL PROTECTION INFORMATION

VENTILATION REQUIREMENTS LOCAL EXHAUST

PROTECTIVE EQUIPMENT (SPECIFY TYPES) EYE: goggles

MECHANICAL (GENERAL)

GLOVES: rubber

SPECIAL

RESPIRATOR: with proper filter

OTHER PROTECTIVE EQUIPMENT

SECTION IX SPECIAL PRECAUTIONS

PRECAUTIONS TO BE TAKEN IN HANDLING AND STORAGE

OTHER PRECAUTIONS

SIGNATURE

DATE

NICKEL METAL

Hazard diamond: 1 (top), 4 (left), 1 (right)

SECTION I SOURCE AND NOMENCLATURE

MANUFACTURER'S NAME | EMERGENCY TELEPHONE NO.

ADDRESS (NUMBER, STREET, CITY, STATE, ZIP CODE)

TRADE NAME AND SYNONYMS | CHEMICAL FAMILY

CHEMICAL NAME AND SYNONYMS: Nickel | FORMULA: Ni

SECTION II HAZARDOUS INGREDIENTS

BASIC MATERIAL	APPROXIMATE OR MAXIMUM % WT. OR VOL.	ESTABLISHED OSHA STANDARD	LD_{50} ORAL	LD_{50} PERCUT.	LC_{50} SPECIES	LC_{50} CONC.

Notes:
Lowest published toxic concentration in air
Inhalation - guinea pig 15 mg/c.m./91 weeks

Lowest published lethal dose
Oral - guinea pig 5 mg/kg.

SECTION III PHYSICAL DATA 58.71

BOILING POINT °F.: 2835°C.
VAPOR PRESSURE mm Hg.:
MELTING POINT °F.: 1445°C.
VAPOR DENSITY (Air = 1):
SPECIFIC GRAVITY (H_2O = 1): 8.9
EVAPORATION RATE (= 1):
SOLUBILITY IN WATER Pts/100 pts H_2O: 1
VOLATILE % Vol. % Wt.:
APPEARANCE AND ODOR: Silver colored metal.

SECTION IV FIRE AND EXPLOSION HAZARD DATA

FLASH POINT: N/A
FLAMMABLE (EXPLOSIVE) LIMITS: UPPER / LOWER
METHOD USED: N/A
EXTINGUISHING MEDIA: N/A
SPECIAL FIRE FIGHTING PROCEDURES:
UNUSUAL FIRE AND EXPLOSION HAZARDS: Powder may burn.

SECTION V HEALTH HAZARD DATA

TOXIC LEVEL: TLV - 1 mg/m^3
CARCINOGENIC: yes
PRINCIPLE ROUTES OF ABSORBTION: Inhalation (mainly) ingestion
SKIN AND EYE IRRITATION: slight
REVELANT SYMPTOMS OF EXPOSURE: Nausea, vomiting, diarrhea, central nervous system depression.
EFFECTS OF CHRONIC EXPOSURE: Cancer of the lungs, dermatitis.
EMERGENCY AND FIRST AID PROCEDURES: Inhalation - rest; if swallowed - give emetic, get medical attention; wash from eyes.

SECTION VI REACTIVITY DATA

CONDITIONS CONTRIBUTING TO INSTABILITY:
CONDITIONS CONTRIBUTING TO HAZARDOUS POLYMERIZATION:
INCOMPATABILITY (MATERIALS TO AVOID): Vigorous reaction of Ni powder with S or Se.
HAZARDOUS DECOMPOSITION PRODUCTS:

SECTION VII SPILL OR LEAK PROCEDURES

STEPS TO BE TAKEN IN CASE MATERIAL IS RELEASED OR SPILLED:
WASTE DISPOSAL METHOD:

SECTION VIII SPECIAL PROTECTION INFORMATION

VENTILATION REQUIREMENTS LOCAL EXHAUST:
PROTECTIVE EQUIPMENT (SPECIFY TYPES) EYE: Safety glasses
MECHANICAL (GENERAL):
GLOVES: Rubber
SPECIAL:
RESPIRATOR: With appropriate filter
OTHER PROTECTIVE EQUIPMENT:

SECTION IX SPECIAL PRECAUTIONS

PRECAUTIONS TO BE TAKEN IN HANDLING AND STORAGE:
OTHER PRECAUTIONS:
SIGNATURE | DATE

3-NITRO-4-ACETYLAMINOTOLUENE

Hazard diamond: 1 (top), 4 (left), 3 (right)

SECTION I SOURCE AND NOMENCLATURE

MANUFACTURER'S NAME | EMERGENCY TELEPHONE NO.

ADDRESS (NUMBER, STREET, CITY, STATE, ZIP CODE)

TRADE NAME AND SYNONYMS: 2 Nitro-P-Acetotoluidide | CHEMICAL FAMILY

CHEMICAL NAME AND SYNONYMS: 3 Nitro 4 Acetylamino Toluene | FORMULA: $NO_2C_6H_3(CH_3)(NHCOCH_3)$

SECTION II HAZARDOUS INGREDIENTS

BASIC MATERIAL	APPROXIMATE OR MAXIMUM % WT. OR VOL.	ESTABLISHED OSHA STANDARD	LD_{50} ORAL	LD_{50} PERCUT.	LC_{50} SPECIES	LC_{50} CONC.
			none	available		

SECTION III PHYSICAL DATA MW - 194.19

BOILING POINT °F.: --
VAPOR PRESSURE mm Hg.:
MELTING POINT °F.: 93°C
VAPOR DENSITY (Air = 1):
SPECIFIC GRAVITY (H_2O = 1): --
EVAPORATION RATE (= 1):
SOLUBILITY IN WATER Pts/100 pts H_2O: soluble
VOLATILE % Vol. % Wt.:
APPEARANCE AND ODOR: yellow crystals

SECTION IV FIRE AND EXPLOSION HAZARD DATA

FLASH POINT: none available
FLAMMABLE (EXPLOSIVE) LIMITS: UPPER / LOWER
METHOD USED:
EXTINGUISHING MEDIA: carbon dioxide, dry chemical
SPECIAL FIRE FIGHTING PROCEDURES:
UNUSUAL FIRE AND EXPLOSION HAZARDS:

SECTION V HEALTH HAZARD DATA

TOXIC LEVEL: none available
CARCINOGENIC: no
PRINCIPLE ROUTES OF ABSORBTION: inhalation, ingestion, skin
SKIN AND EYE IRRITATION: slight
REVELANT SYMPTOMS OF EXPOSURE: difficulty in breathing, cyanosis, vomiting, headache
EFFECTS OF CHRONIC EXPOSURE: blood damage
EMERGENCY AND FIRST AID PROCEDURES: If swallowed - give emetic, get medical aid; inhalation - rest; wash from skin & eyes.

SECTION VI REACTIVITY DATA

CONDITIONS CONTRIBUTING TO INSTABILITY:
CONDITIONS CONTRIBUTING TO HAZARDOUS POLYMERIZATION:
INCOMPATABILITY (MATERIALS TO AVOID): May explode when heated with sulfuric acid.
HAZARDOUS DECOMPOSITION PRODUCTS:

SECTION VII SPILL OR LEAK PROCEDURES

STEPS TO BE TAKEN IN CASE MATERIAL IS RELEASED OR SPILLED:
WASTE DISPOSAL METHOD:

SECTION VIII SPECIAL PROTECTION INFORMATION

VENTILATION REQUIREMENTS LOCAL EXHAUST:
PROTECTIVE EQUIPMENT (SPECIFY TYPES) EYE: goggles
MECHANICAL (GENERAL):
GLOVES: rubber
SPECIAL:
RESPIRATOR: self-contained
OTHER PROTECTIVE EQUIPMENT:

SECTION IX SPECIAL PRECAUTIONS

PRECAUTIONS TO BE TAKEN IN HANDLING AND STORAGE:
OTHER PRECAUTIONS:
SIGNATURE | DATE

5-NITRO-2-AMINOANISOLE

Hazard diamond: 1 (top), 4 (left), 1 (right)

SECTION I SOURCE AND NOMENCLATURE

MANUFACTURER'S NAME

EMERGENCY TELEPHONE NO.

ADDRESS (NUMBER, STREET, CITY, STATE, ZIP CODE)

TRADE NAME AND SYNONYMS: 5 nitro-o-anisidine

CHEMICAL FAMILY

CHEMICAL NAME AND SYNONYMS: 1-methoxy-4 nitroaniline

FORMULA: $NO_2C_6H_3(NH_2)OCH_3$

SECTION II HAZARDOUS INGREDIENTS

BASIC MATERIAL	APPROXIMATE OR MAXIMUM % WT. OR VOL.	ESTABLISHED OSHA STANDARD	LD_{50} mg/kg ORAL	PERCUT.	SPECIES	LC_{50} CONC.
			704		rat	

SECTION III PHYSICAL DATA MW - 168.15

BOILING POINT °F.: --

VAPOR PRESSURE mm Hg.

MELTING POINT °F.: 141°C

VAPOR DENSITY (Air = 1)

SPECIFIC GRAVITY (H_2O = 1): 1.21

EVAPORATION RATE (________ = 1)

SOLUBILITY IN WATER Pts/100 pts H_2O: --

VOLATILE % Vol. % Wt.

APPEARANCE AND ODOR: yellow crystals

SECTION IV FIRE AND EXPLOSION HAZARD DATA

FLASH POINT: 390°F

METHOD USED

FLAMMABLE (EXPLOSIVE) LIMITS: UPPER / LOWER

EXTINGUISHING MEDIA: water spray, carbon dioxide, dry chemical, foam

SPECIAL FIRE FIGHTING PROCEDURES: wear full protective clothing.

UNUSUAL FIRE AND EXPLOSION HAZARDS

SECTION V HEALTH HAZARD DATA

TOXIC LEVEL: none available

CARCINOGENIC: no

PRINCIPLE ROUTES OF ABSORBTION: inhalation, ingestion, skin

SKIN AND EYE IRRITATION: yes

REVELANT SYMPTOMS OF EXPOSURE: Headache, nausea, vomiting, weakness, stupor, cyanosis.

EFFECTS OF CHRONIC EXPOSURE: Liver damage

EMERGENCY AND FIRST AID PROCEDURES: If swallowed - give emetic, get medical aid; inhalation - rest; wash from skin & eyes.

SECTION VI REACTIVITY DATA

CONDITIONS CONTRIBUTING TO INSTABILITY

CONDITIONS CONTRIBUTING TO HAZARDOUS POLYMERIZATION

INCOMPATABILITY (MATERIALS TO AVOID)

HAZARDOUS DECOMPOSITION PRODUCTS

SECTION VII SPILL OR LEAK PROCEDURES

STEPS TO BE TAKEN IN CASE MATERIAL IS RELEASED OR SPILLED

WASTE DISPOSAL METHOD

SECTION VIII SPECIAL PROTECTION INFORMATION

VENTILATION REQUIREMENTS LOCAL EXHAUST

PROTECTIVE EQUIPMENT (SPECIFY TYPES) EYE: goggles

MECHANICAL (GENERAL)

GLOVES: rubber

SPECIAL

RESPIRATOR: with proper filter

OTHER PROTECTIVE EQUIPMENT

SECTION IX SPECIAL PRECAUTIONS

PRECAUTIONS TO BE TAKEN IN HANDLING AND STORAGE

OTHER PRECAUTIONS

SIGNATURE

DATE

5-NITRO-4-AMINO-1,3-DIMETHYLBENZENE

Hazard diamond: 1 (top), 4 (left), 1 (right)

SECTION I SOURCE AND NOMENCLATURE

MANUFACTURER'S NAME

EMERGENCY TELEPHONE NO.

ADDRESS (NUMBER, STREET, CITY, STATE, ZIP CODE)

TRADE NAME AND SYNONYMS: 5 nitro-4-amino-1,3 dimethyl-benzene

CHEMICAL FAMILY

CHEMICAL NAME AND SYNONYMS: 6 nitro-2,4,xylidine

FORMULA: $(CH_3)_2C_6H_2(NH_2)NO_2$

SECTION II HAZARDOUS INGREDIENTS

BASIC MATERIAL	APPROXIMATE OR MAXIMUM % WT. OR VOL.	ESTABLISHED OSHA STANDARD	LD_{50} ORAL	PERCUT.	LC_{50} SPECIES	CONC.
			none	available		

SECTION III PHYSICAL DATA MW - 166.18

BOILING POINT °F.: --

VAPOR PRESSURE mm Hg.

MELTING POINT °F.: 69°C

VAPOR DENSITY (Air = 1)

SPECIFIC GRAVITY (H_2O = 1): --

EVAPORATION RATE (________ = 1)

SOLUBILITY IN WATER Pts/100 pts H_2O: --

VOLATILE % Vol. % Wt.

APPEARANCE AND ODOR: crystaline

SECTION IV FIRE AND EXPLOSION HAZARD DATA

FLASH POINT: none available

METHOD USED

FLAMMABLE (EXPLOSIVE) LIMITS: UPPER / LOWER

EXTINGUISHING MEDIA: water spray, carbon dioxide, dry chemical

SPECIAL FIRE FIGHTING PROCEDURES

UNUSUAL FIRE AND EXPLOSION HAZARDS

SECTION V HEALTH HAZARD DATA

TOXIC LEVEL: none available

CARCINOGENIC: no

PRINCIPLE ROUTES OF ABSORBTION: inhalation, ingestion, skin

SKIN AND EYE IRRITATION: yes

REVELANT SYMPTOMS OF EXPOSURE: Cyanosis, headache, nausea, vomiting, weakness, stupor.

EFFECTS OF CHRONIC EXPOSURE: Liver damage

EMERGENCY AND FIRST AID PROCEDURES: If swallowed - give emetic, get medical aid; inhalation - rest; wash from skin & eyes.

SECTION VI REACTIVITY DATA

CONDITIONS CONTRIBUTING TO INSTABILITY

CONDITIONS CONTRIBUTING TO HAZARDOUS POLYMERIZATION

INCOMPATABILITY (MATERIALS TO AVOID)

HAZARDOUS DECOMPOSITION PRODUCTS

SECTION VII SPILL OR LEAK PROCEDURES

STEPS TO BE TAKEN IN CASE MATERIAL IS RELEASED OR SPILLED

WASTE DISPOSAL METHOD

SECTION VIII SPECIAL PROTECTION INFORMATION

VENTILATION REQUIREMENTS LOCAL EXHAUST

PROTECTIVE EQUIPMENT (SPECIFY TYPES) EYE: goggles

MECHANICAL (GENERAL)

GLOVES: rubber

SPECIAL

RESPIRATOR: with proper filter

OTHER PROTECTIVE EQUIPMENT

SECTION IX SPECIAL PRECAUTIONS

PRECAUTIONS TO BE TAKEN IN HANDLING AND STORAGE

OTHER PRECAUTIONS

SIGNATURE

DATE

3-NITRO-4-AMINOPHENETOLE

(Hazard diamond: 1 / 4 / 1)

SECTION I SOURCE AND NOMENCLATURE

MANUFACTURER'S NAME	EMERGENCY TELEPHONE NO.
ADDRESS (NUMBER, STREET, CITY, STATE, ZIP CODE)	
TRADE NAME AND SYNONYMS: 3-Nitro-4 aminophenetole	CHEMICAL FAMILY
CHEMICAL NAME AND SYNONYMS: 3 Nitro-P-Phenetidine	FORMULA: $NO_2C_6H_3(NH_2)OC_2H_5$

SECTION II HAZARDOUS INGREDIENTS

BASIC MATERIAL	APPROXIMATE OR MAXIMUM % WT. OR VOL.	ESTABLISHED OSHA STANDARD	LD_{50} ORAL	LD_{50} PERCUT.	LC_{50} SPECIES	LC_{50} CONC.
			none available			

SECTION III PHYSICAL DATA MW - 182.18

BOILING POINT °F.: --	VAPOR PRESSURE mm Hg.
MELTING POINT °F.: 113°C	VAPOR DENSITY (Air = 1)
SPECIFIC GRAVITY (H_2O = 1): --	EVAPORATION RATE (____ = 1)
SOLUBILITY IN WATER Pts/100 pts H_2O: --	VOLATILE % Vol. % Wt.

APPEARANCE AND ODOR: red crystals

SECTION IV FIRE AND EXPLOSION HAZARD DATA

FLASH POINT: none available	FLAMMABLE (EXPLOSIVE) LIMITS	UPPER
METHOD USED		LOWER

EXTINGUISHING MEDIA: water spray, carbon dioxide, dry chemical

SPECIAL FIRE FIGHTING PROCEDURES

UNUSUAL FIRE AND EXPLOSION HAZARDS

SECTION V HEALTH HAZARD DATA

TOXIC LEVEL: none available — CARCINOGENIC: no

PRINCIPLE ROUTES OF ABSORBTION: inhalation, ingestion, skin — SKIN AND EYE IRRITATION: yes

REVELANT SYMPTOMS OF EXPOSURE: Cyanosis, headache, nausea, vomiting, weakness, stupor.

EFFECTS OF CHRONIC EXPOSURE: Liver damage

EMERGENCY AND FIRST AID PROCEDURES: If swallowed - give emetic, get medical aid; inhalation - rest; wash from skin & eyes.

SECTION VI REACTIVITY DATA

CONDITIONS CONTRIBUTING TO INSTABILITY

CONDITIONS CONTRIBUTING TO HAZARDOUS POLYMERIZATION

INCOMPATABILITY (MATERIALS TO AVOID)

HAZARDOUS DECOMPOSITION PRODUCTS

SECTION VII SPILL OR LEAK PROCEDURES

STEPS TO BE TAKEN IN CASE MATERIAL IS RELEASED OR SPILLED

WASTE DISPOSAL METHOD

SECTION VIII SPECIAL PROTECTION INFORMATION

VENTILATION REQUIREMENTS LOCAL EXHAUST	PROTECTIVE EQUIPMENT (SPECIFY TYPES) EYE: goggles
MECHANICAL (GENERAL)	GLOVES: rubber
SPECIAL	RESPIRATOR: with proper filter

OTHER PROTECTIVE EQUIPMENT

SECTION IX SPECIAL PRECAUTIONS

PRECAUTIONS TO BE TAKEN IN HANDLING AND STORAGE

OTHER PRECAUTIONS

SIGNATURE	DATE

3-NITRO-4-AMINOTOLUENE

(Hazard diamond: 1 / 4 / 1)

SECTION I SOURCE AND NOMENCLATURE

MANUFACTURER'S NAME	EMERGENCY TELEPHONE NO.
ADDRESS (NUMBER, STREET, CITY, STATE, ZIP CODE)	
TRADE NAME AND SYNONYMS: 3 Nitro-4-Aminotoluene	CHEMICAL FAMILY
CHEMICAL NAME AND SYNONYMS: 3 Nitro-P-Toluidine	FORMULA: $NO_2C_6H_3(CH_3)NH_2$

SECTION II HAZARDOUS INGREDIENTS

BASIC MATERIAL	APPROXIMATE OR MAXIMUM % WT. OR VOL.	ESTABLISHED OSHA STANDARD	LD_{50} mg/kg ORAL	LD_{50} PERCUT.	LC_{50} SPECIES	LC_{50} CONC.
			3		bird	

SECTION III PHYSICAL DATA MW - 152.15

BOILING POINT °F.: --	VAPOR PRESSURE mm Hg.
MELTING POINT °F.: 117°C	VAPOR DENSITY (Air = 1): 5.2
SPECIFIC GRAVITY (H_2O = 1): 1.31	EVAPORATION RATE (____ = 1)
SOLUBILITY IN WATER Pts/100 pts H_2O: slightly soluble	VOLATILE % Vol. % Wt.

APPEARANCE AND ODOR: red crystals

SECTION IV FIRE AND EXPLOSION HAZARD DATA

FLASH POINT: 315°F	FLAMMABLE (EXPLOSIVE) LIMITS	UPPER
METHOD USED: C.C.		LOWER

EXTINGUISHING MEDIA: Water spray, carbon dioxide, dry chemical

SPECIAL FIRE FIGHTING PROCEDURES: Wear full protective clothing.

UNUSUAL FIRE AND EXPLOSION HAZARDS

SECTION V HEALTH HAZARD DATA

TOXIC LEVEL: none available — CARCINOGENIC: no

PRINCIPLE ROUTES OF ABSORBTION: inhalation, ingestion, skin — SKIN AND EYE IRRITATION: yes

REVELANT SYMPTOMS OF EXPOSURE: cyanosis, headache, nausea, vomiting, weakness, stupor.

EFFECTS OF CHRONIC EXPOSURE: Liver damage.

EMERGENCY AND FIRST AID PROCEDURES: If swallowed - give emetic, get medical aid; inhalation - rest; wash from skin & eyes.

SECTION VI REACTIVITY DATA

CONDITIONS CONTRIBUTING TO INSTABILITY

CONDITIONS CONTRIBUTING TO HAZARDOUS POLYMERIZATION

INCOMPATABILITY (MATERIALS TO AVOID)

HAZARDOUS DECOMPOSITION PRODUCTS

SECTION VII SPILL OR LEAK PROCEDURES

STEPS TO BE TAKEN IN CASE MATERIAL IS RELEASED OR SPILLED

WASTE DISPOSAL METHOD

SECTION VIII SPECIAL PROTECTION INFORMATION

VENTILATION REQUIREMENTS LOCAL EXHAUST	PROTECTIVE EQUIPMENT (SPECIFY TYPES) EYE: goggles
MECHANICAL (GENERAL)	GLOVES: rubber
SPECIAL	RESPIRATOR: with proper filter

OTHER PROTECTIVE EQUIPMENT

SECTION IX SPECIAL PRECAUTIONS

PRECAUTIONS TO BE TAKEN IN HANDLING AND STORAGE

OTHER PRECAUTIONS

SIGNATURE	DATE

4-NITRO-2-AMINOTOLUENE

Hazard rating diamond: 1, 4, 1

SECTION I SOURCE AND NOMENCLATURE

MANUFACTURER'S NAME | EMERGENCY TELEPHONE NO.

ADDRESS (NUMBER, STREET, CITY, STATE, ZIP CODE)

TRADE NAME AND SYNONYMS: 4 nitro-2-aminotoluene

CHEMICAL FAMILY

CHEMICAL NAME AND SYNONYMS: 4 nitro-o-toluidine

FORMULA: $NO_2(C_6H_3)(CH_3)NH_2$

SECTION II HAZARDOUS INGREDIENTS

BASIC MATERIAL	APPROXIMATE OR MAXIMUM % WT. OR VOL.	ESTABLISHED OSHA STANDARD	LD_{50} ORAL	LD_{50} PERCUT.	LC_{50} SPECIES	LC_{50} CONC.
			none available			

SECTION III PHYSICAL DATA MW - 152.15

BOILING POINT °F.: --

VAPOR PRESSURE mm Hg.

MELTING POINT °F.: 107°C

VAPOR DENSITY (Air = 1)

SPECIFIC GRAVITY (H_2O = 1): 1.37

EVAPORATION RATE (= 1)

SOLUBILITY IN WATER Pts/100 pts H_2O: very slightly soluble

VOLATILE % Vol. % Wt.

APPEARANCE AND ODOR: yellow crystals

SECTION IV FIRE AND EXPLOSION HAZARD DATA

FLASH POINT: none available

METHOD USED

FLAMMABLE (EXPLOSIVE) LIMITS: UPPER / LOWER

EXTINGUISHING MEDIA: water spray, carbon dioxide, dry chemical

SPECIAL FIRE FIGHTING PROCEDURES: wear full protective clothing.

UNUSUAL FIRE AND EXPLOSION HAZARDS

SECTION V HEALTH HAZARD DATA

TOXIC LEVEL: None available

CARCINOGENIC: no

PRINCIPLE ROUTES OF ABSORBTION: Inhalation, ingestion, skin

SKIN AND EYE IRRITATION: yes

REVELANT SYMPTOMS OF EXPOSURE: Cyanosis, headache, nausea, vomiting, weakness, stupor.

EFFECTS OF CHRONIC EXPOSURE: Liver damage

EMERGENCY AND FIRST AID PROCEDURES: If swallowed - give emetic, get medical aid; inhalation - rest; wash from skin & eyes.

SECTION VI REACTIVITY DATA

CONDITIONS CONTRIBUTING TO INSTABILITY

CONDITIONS CONTRIBUTING TO HAZARDOUS POLYMERIZATION

INCOMPATABILITY (MATERIALS TO AVOID)

HAZARDOUS DECOMPOSITION PRODUCTS

SECTION VII SPILL OR LEAK PROCEDURES

STEPS TO BE TAKEN IN CASE MATERIAL IS RELEASED OR SPILLED

WASTE DISPOSAL METHOD

SECTION VIII SPECIAL PROTECTION INFORMATION

VENTILATION REQUIREMENTS LOCAL EXHAUST

MECHANICAL (GENERAL)

SPECIAL

OTHER PROTECTIVE EQUIPMENT

PROTECTIVE EQUIPMENT (SPECIFY TYPES) EYE: Goggles

GLOVES: Rubber

RESPIRATOR: With proper filter

SECTION IX SPECIAL PRECAUTIONS

PRECAUTIONS TO BE TAKEN IN HANDLING AND STORAGE

OTHER PRECAUTIONS

SIGNATURE | DATE

m-NITROANILINE

Hazard rating diamond: 1, 4, 1

SECTION I SOURCE AND NOMENCLATURE

MANUFACTURER'S NAME | EMERGENCY TELEPHONE NO.

ADDRESS (NUMBER, STREET, CITY, STATE, ZIP CODE)

TRADE NAME AND SYNONYMS

CHEMICAL FAMILY

CHEMICAL NAME AND SYNONYMS: 3-Nitroaniline

FORMULA: $NO_2C_6H_4NH_2$

SECTION II HAZARDOUS INGREDIENTS

BASIC MATERIAL	APPROXIMATE OR MAXIMUM % WT. OR VOL.	ESTABLISHED OSHA STANDARD	LD_{50} ORAL	LD_{50} PERCUT.	LC_{50} SPECIES	LC_{50} CONC.
			535		rat	
			308		mouse	

SECTION III PHYSICAL DATA MW - 138.12

BOILING POINT °F.: 306°C

VAPOR PRESSURE mm Hg.: 1 at 119°C

MELTING POINT °F.: 114°C

VAPOR DENSITY (Air = 1)

SPECIFIC GRAVITY (H_2O = 1): 1.43

EVAPORATION RATE (= 1)

SOLUBILITY IN WATER Pts/100 pts H_2O: 0.09

VOLATILE % Vol. % Wt.

APPEARANCE AND ODOR: Yellow rhombic crystals, burning sweet taste.

SECTION IV FIRE AND EXPLOSION HAZARD DATA

FLASH POINT: 390°F

METHOD USED

FLAMMABLE (EXPLOSIVE) LIMITS: UPPER / LOWER

EXTINGUISHING MEDIA: Water spray, carbon dioxide, dry chemical

SPECIAL FIRE FIGHTING PROCEDURES: Wear full protective clothing.

UNUSUAL FIRE AND EXPLOSION HAZARDS

SECTION V HEALTH HAZARD DATA

TOXIC LEVEL: none available

CARCINOGENIC: no

PRINCIPLE ROUTES OF ABSORBTION: inhalation, ingestion skin

SKIN AND EYE IRRITATION: yes

REVELANT SYMPTOMS OF EXPOSURE: cyanosis, headache, nausea, vomiting, weakness, stupor.

EFFECTS OF CHRONIC EXPOSURE: Liver damage

EMERGENCY AND FIRST AID PROCEDURES: If swallowed - give emetic, get medical aid; inhalation - rest; wash from skin & eyes.

SECTION VI REACTIVITY DATA

CONDITIONS CONTRIBUTING TO INSTABILITY

CONDITIONS CONTRIBUTING TO HAZARDOUS POLYMERIZATION

INCOMPATABILITY (MATERIALS TO AVOID)

HAZARDOUS DECOMPOSITION PRODUCTS

SECTION VII SPILL OR LEAK PROCEDURES

STEPS TO BE TAKEN IN CASE MATERIAL IS RELEASED OR SPILLED

WASTE DISPOSAL METHOD

SECTION VIII SPECIAL PROTECTION INFORMATION

VENTILATION REQUIREMENTS LOCAL EXHAUST

MECHANICAL (GENERAL)

SPECIAL

OTHER PROTECTIVE EQUIPMENT: Protective work clothes

PROTECTIVE EQUIPMENT (SPECIFY TYPES) EYE: goggles

GLOVES: rubber (butyl)

RESPIRATOR: with proper filter

SECTION IX SPECIAL PRECAUTIONS

PRECAUTIONS TO BE TAKEN IN HANDLING AND STORAGE: Protect against moisture and physical damage.

OTHER PRECAUTIONS

SIGNATURE | DATE

p-NITROBENZYL BROMIDE

Hazard diamond: 1 (top), 4 (left), 3 (right)

SECTION I SOURCE AND NOMENCLATURE

MANUFACTURER'S NAME | EMERGENCY TELEPHONE NO.

ADDRESS (NUMBER, STREET, CITY, STATE, ZIP CODE)

TRADE NAME AND SYNONYMS | CHEMICAL FAMILY

CHEMICAL NAME AND SYNONYMS: P Nitro-a-bromotoluene

FORMULA: $NO_2C_6H_4CH_2Br$

SECTION II HAZARDOUS INGREDIENTS

BASIC MATERIAL	APPROXIMATE OR MAXIMUM % WT. OR VOL.	ESTABLISHED OSHA STANDARD	LD_{50} ORAL	LD_{50} PERCUT.	LC_{50} SPECIES	LC_{50} CONC.
			none available			

SECTION III PHYSICAL DATA MW - 216.04

BOILING POINT °F.: --

VAPOR PRESSURE mm Hg.:

MELTING POINT °F.: 100°C

VAPOR DENSITY (Air = 1):

SPECIFIC GRAVITY (H_2O = 1): --

EVAPORATION RATE (______ = 1):

SOLUBILITY IN WATER Pts/100 pts H_2O: insoluble

VOLATILE % Vol. % Wt.:

APPEARANCE AND ODOR: needle-like crystals.

SECTION IV FIRE AND EXPLOSION HAZARD DATA

FLASH POINT: none available

METHOD USED:

FLAMMABLE (EXPLOSIVE) LIMITS: UPPER / LOWER

EXTINGUISHING MEDIA: water, carbon dioxide, dry chemicals.

SPECIAL FIRE FIGHTING PROCEDURES:

UNUSUAL FIRE AND EXPLOSION HAZARDS:

SECTION V HEALTH HAZARD DATA

TOXIC LEVEL: none available

CARCINOGENIC: no

PRINCIPLE ROUTES OF ABSORBTION: inhalation, ingestion, skin

SKIN AND EYE IRRITATION: slight

REVELANT SYMPTOMS OF EXPOSURE: Difficulty in breathing, cyanosis, vomiting, headache.

EFFECTS OF CHRONIC EXPOSURE: Blood damage.

EMERGENCY AND FIRST AID PROCEDURES: If swallowed - give emetic, get medical aid; inhalation - rest; wash from skin & eyes.

SECTION VI REACTIVITY DATA

CONDITIONS CONTRIBUTING TO INSTABILITY:

CONDITIONS CONTRIBUTING TO HAZARDOUS POLYMERIZATION:

INCOMPATABILITY (MATERIALS TO AVOID): May explode when heated with sulfuric acid.

HAZARDOUS DECOMPOSITION PRODUCTS:

SECTION VII SPILL OR LEAK PROCEDURES

STEPS TO BE TAKEN IN CASE MATERIAL IS RELEASED OR SPILLED:

WASTE DISPOSAL METHOD:

SECTION VIII SPECIAL PROTECTION INFORMATION

VENTILATION REQUIREMENTS LOCAL EXHAUST:

PROTECTIVE EQUIPMENT (SPECIFY TYPES) EYE: goggles

MECHANICAL (GENERAL):

GLOVES: rubber

SPECIAL:

RESPIRATOR: self-contained

OTHER PROTECTIVE EQUIPMENT:

SECTION IX SPECIAL PRECAUTIONS

PRECAUTIONS TO BE TAKEN IN HANDLING AND STORAGE:

OTHER PRECAUTIONS:

SIGNATURE | DATE

p-NITROBENZYL CHLORIDE

Hazard diamond: 1 (top), 4 (left), 3 (right)

SECTION I SOURCE AND NOMENCLATURE

MANUFACTURER'S NAME | EMERGENCY TELEPHONE NO.

ADDRESS (NUMBER, STREET, CITY, STATE, ZIP CODE)

TRADE NAME AND SYNONYMS | CHEMICAL FAMILY

CHEMICAL NAME AND SYNONYMS: P-Nitro- α Chlorotoluene

FORMULA: $NO_2C_6H_4CH_2Cl$

SECTION II HAZARDOUS INGREDIENTS

BASIC MATERIAL	APPROXIMATE OR MAXIMUM % WT. OR VOL.	ESTABLISHED OSHA STANDARD	LD_{50} ORAL	LD_{50} PERCUT.	LC_{50} SPECIES	LC_{50} CONC.
			none available			

SECTION III PHYSICAL DATA MW - 171.58

BOILING POINT °F.: --

VAPOR PRESSURE mm Hg.:

MELTING POINT °F.: 71°C

VAPOR DENSITY (Air = 1):

SPECIFIC GRAVITY (H_2O = 1):

EVAPORATION RATE (______ = 1):

SOLUBILITY IN WATER Pts/100 pts H_2O: insoluble

VOLATILE % Vol. % Wt.:

APPEARANCE AND ODOR: needle-like crystals.

SECTION IV FIRE AND EXPLOSION HAZARD DATA

FLASH POINT: none available

METHOD USED:

FLAMMABLE (EXPLOSIVE) LIMITS: UPPER / LOWER

EXTINGUISHING MEDIA: water, carbon dioxide, dry chemical.

SPECIAL FIRE FIGHTING PROCEDURES:

UNUSUAL FIRE AND EXPLOSION HAZARDS:

SECTION V HEALTH HAZARD DATA

TOXIC LEVEL: none available

CARCINOGENIC: no

PRINCIPLE ROUTES OF ABSORBTION: Inhalation, ingestion, skin

SKIN AND EYE IRRITATION:

REVELANT SYMPTOMS OF EXPOSURE: Difficulty in breathing, cyanosis, vomiting, headache.

EFFECTS OF CHRONIC EXPOSURE: Blood damage.

EMERGENCY AND FIRST AID PROCEDURES: If swallowed - give emetic, get medical aid; inhalation - rest; wash from skin & eyes.

SECTION VI REACTIVITY DATA

CONDITIONS CONTRIBUTING TO INSTABILITY:

CONDITIONS CONTRIBUTING TO HAZARDOUS POLYMERIZATION:

INCOMPATABILITY (MATERIALS TO AVOID): May explode when heated with sulfuric acid.

HAZARDOUS DECOMPOSITION PRODUCTS:

SECTION VII SPILL OR LEAK PROCEDURES

STEPS TO BE TAKEN IN CASE MATERIAL IS RELEASED OR SPILLED:

WASTE DISPOSAL METHOD:

SECTION VIII SPECIAL PROTECTION INFORMATION

VENTILATION REQUIREMENTS LOCAL EXHAUST:

PROTECTIVE EQUIPMENT (SPECIFY TYPES) EYE: Goggles

MECHANICAL (GENERAL):

GLOVES: Rubber

SPECIAL:

RESPIRATOR: Self-contained

OTHER PROTECTIVE EQUIPMENT:

SECTION IX SPECIAL PRECAUTIONS

PRECAUTIONS TO BE TAKEN IN HANDLING AND STORAGE:

OTHER PRECAUTIONS:

SIGNATURE | DATE

m-NITROBROMOBENZENE

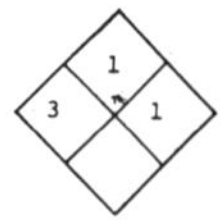

SECTION I SOURCE AND NOMENCLATURE

MANUFACTURER'S NAME | EMERGENCY TELEPHONE NO.

ADDRESS (NUMBER, STREET, CITY, STATE, ZIP CODE)

TRADE NAME AND SYNONYMS | CHEMICAL FAMILY

CHEMICAL NAME AND SYNONYMS: M-Bromonitrobenzene | FORMULA: $BrC_6H_4NO_2$

SECTION II HAZARDOUS INGREDIENTS

BASIC MATERIAL	APPROXIMATE OR MAXIMUM % WT. OR VOL.	ESTABLISHED OSHA STANDARD	LD_{50} ORAL	LD_{50} PERCUT.	LC_{50} SPECIES	LC_{50} CONC.
			none available			

SECTION III PHYSICAL DATA MW - 202.02

BOILING POINT °F.: 257°C | VAPOR PRESSURE mm Hg.

MELTING POINT °F.: 56°C | VAPOR DENSITY (Air = 1)

SPECIFIC GRAVITY (H_2O = 1): 1.70 | EVAPORATION RATE (______ = 1)

SOLUBILITY IN WATER Pts/100 pts H_2O: insoluble | VOLATILE % Vol. % Wt.

APPEARANCE AND ODOR: yellow crystals.

SECTION IV FIRE AND EXPLOSION HAZARD DATA

FLASH POINT: none available | FLAMMABLE (EXPLOSIVE) LIMITS UPPER / LOWER

METHOD USED

EXTINGUISHING MEDIA: water spray, carbon dioxide, dry chemical, foam.

SPECIAL FIRE FIGHTING PROCEDURES: Wear self-contained breathing apparatus.

UNUSUAL FIRE AND EXPLOSION HAZARDS

SECTION V HEALTH HAZARD DATA

TOXIC LEVEL: none available | CARCINOGENIC: no

PRINCIPLE ROUTES OF ABSORBTION: inhalation, ingestion, skin | SKIN AND EYE IRRITATION: moderate

REVELANT SYMPTOMS OF EXPOSURE: cyanosis, headache, weakness, anemia

EFFECTS OF CHRONIC EXPOSURE: hemoglobinuria

EMERGENCY AND FIRST AID PROCEDURES: If swallowed - give emetic, get medical aid; inhalation - oxygen as needed; wash from skin & eyes.

SECTION VI REACTIVITY DATA

CONDITIONS CONTRIBUTING TO INSTABILITY

CONDITIONS CONTRIBUTING TO HAZARDOUS POLYMERIZATION

INCOMPATABILITY (MATERIALS TO AVOID)

HAZARDOUS DECOMPOSITION PRODUCTS

SECTION VII SPILL OR LEAK PROCEDURES

STEPS TO BE TAKEN IN CASE MATERIAL IS RELEASED OR SPILLED

WASTE DISPOSAL METHOD

SECTION VIII SPECIAL PROTECTION INFORMATION

VENTILATION REQUIREMENTS LOCAL EXHAUST | PROTECTIVE EQUIPMENT (SPECIFY TYPES) EYE: Goggles

MECHANICAL (GENERAL) | GLOVES: Rubber

SPECIAL | RESPIRATOR: Self-contained

OTHER PROTECTIVE EQUIPMENT

SECTION IX SPECIAL PRECAUTIONS

PRECAUTIONS TO BE TAKEN IN HANDLING AND STORAGE: Store away from oxidizing materials and sources of heat.

OTHER PRECAUTIONS

SIGNATURE | DATE

o-NITROBROMOBENZENE

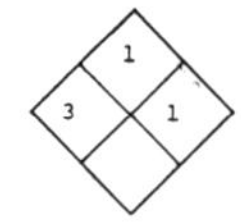

SECTION I SOURCE AND NOMENCLATURE

MANUFACTURER'S NAME | EMERGENCY TELEPHONE NO.

ADDRESS (NUMBER, STREET, CITY, STATE, ZIP CODE)

TRADE NAME AND SYNONYMS | CHEMICAL FAMILY

CHEMICAL NAME AND SYNONYMS: O-Bromonitrobenzene | FORMULA: $BrC_6H_4NO_2$

SECTION II HAZARDOUS INGREDIENTS

BASIC MATERIAL	APPROXIMATE OR MAXIMUM % WT. OR VOL.	ESTABLISHED OSHA STANDARD	LD_{50} ORAL	LD_{50} PERCUT.	LC_{50} SPECIES	LC_{50} CONC.
			none available			

SECTION III PHYSICAL DATA MW - 202.02

BOILING POINT °F.: 261 | VAPOR PRESSURE mm Hg.

MELTING POINT °F.: 43°C | VAPOR DENSITY (Air = 1)

SPECIFIC GRAVITY (H_2O = 1): 1.62 | EVAPORATION RATE (______ = 1)

SOLUBILITY IN WATER Pts/100 pts H_2O: insoluble | VOLATILE % Vol. % Wt.

APPEARANCE AND ODOR: yellow crystals

SECTION IV FIRE AND EXPLOSION HAZARD DATA

FLASH POINT: none available | FLAMMABLE (EXPLOSIVE) LIMITS UPPER / LOWER

METHOD USED

EXTINGUISHING MEDIA: Water spray, carbon dioxide, dry chemical, foam.

SPECIAL FIRE FIGHTING PROCEDURES: Wear self-contained breathing apparatus.

UNUSUAL FIRE AND EXPLOSION HAZARDS

SECTION V HEALTH HAZARD DATA

TOXIC LEVEL: none available | CARCINOGENIC: no

PRINCIPLE ROUTES OF ABSORBTION: inhalation, ingestion, skin | SKIN AND EYE IRRITATION: moderate

REVELANT SYMPTOMS OF EXPOSURE: cyanosis, headache, weakness, anemia.

EFFECTS OF CHRONIC EXPOSURE: hemoglobinuria

EMERGENCY AND FIRST AID PROCEDURES: If swallowed - give emetic, get medical aid; inhalation - oxygen if needed; wash from skin & eyes.

SECTION VI REACTIVITY DATA

CONDITIONS CONTRIBUTING TO INSTABILITY

CONDITIONS CONTRIBUTING TO HAZARDOUS POLYMERIZATION

INCOMPATABILITY (MATERIALS TO AVOID)

HAZARDOUS DECOMPOSITION PRODUCTS

SECTION VII SPILL OR LEAK PROCEDURES

STEPS TO BE TAKEN IN CASE MATERIAL IS RELEASED OR SPILLED

WASTE DISPOSAL METHOD

SECTION VIII SPECIAL PROTECTION INFORMATION

VENTILATION REQUIREMENTS LOCAL EXHAUST | PROTECTIVE EQUIPMENT (SPECIFY TYPES) EYE: Goggles

MECHANICAL (GENERAL) | GLOVES: Rubber

SPECIAL | RESPIRATOR: Self-contained

OTHER PROTECTIVE EQUIPMENT

SECTION IX SPECIAL PRECAUTIONS

PRECAUTIONS TO BE TAKEN IN HANDLING AND STORAGE: Store away from oxidizing materials and sources of heat.

OTHER PRECAUTIONS

SIGNATURE | DATE

p-NITROBROMOBENZENE

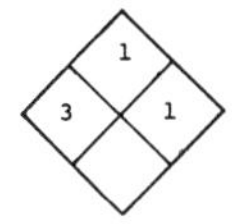

SECTION I SOURCE AND NOMENCLATURE

MANUFACTURER'S NAME | EMERGENCY TELEPHONE NO.

ADDRESS (NUMBER, STREET, CITY, STATE, ZIP CODE)

TRADE NAME AND SYNONYMS | CHEMICAL FAMILY

CHEMICAL NAME AND SYNONYMS: P Bromonitrobenzene | FORMULA: $BrC_6H_4NO_2$

SECTION II HAZARDOUS INGREDIENTS

BASIC MATERIAL	APPROXIMATE OR MAXIMUM % WT. OR VOL.	ESTABLISHED OSHA STANDARD	LD_{50} ORAL	LD_{50} PERCUT.	LC_{50} SPECIES	LC_{50} CONC.
			none available			

SECTION III PHYSICAL DATA

BOILING POINT °F.: 256°C | VAPOR PRESSURE mm Hg.

MELTING POINT °F.: 127°C | VAPOR DENSITY (Air = 1)

SPECIFIC GRAVITY (H_2O = 1): 1.94 | EVAPORATION RATE (________ = 1)

SOLUBILITY IN WATER Pts/100 pts H_2O: insoluble | VOLATILE % Vol. % Wt.

APPEARANCE AND ODOR: yellow crystals

SECTION IV FIRE AND EXPLOSION HAZARD DATA

FLASH POINT: none available | FLAMMABLE (EXPLOSIVE) LIMITS: UPPER / LOWER

METHOD USED

EXTINGUISHING MEDIA: Water spray, carbon dioxide, dry chemical, foam.

SPECIAL FIRE FIGHTING PROCEDURES: Wear self-contained breathing apparatus.

UNUSUAL FIRE AND EXPLOSION HAZARDS

SECTION V HEALTH HAZARD DATA

TOXIC LEVEL: none available | CARCINOGENIC: no

PRINCIPLE ROUTES OF ABSORBTION: inhalation, ingestion, skin | SKIN AND EYE IRRITATION: moderate

REVELANT SYMPTOMS OF EXPOSURE: cyanosis, headache, weakness, anemia

EFFECTS OF CHRONIC EXPOSURE: hemoglobinuria

EMERGENCY AND FIRST AID PROCEDURES: If swallowed - give emetic, get medical aid; inhalation - oxygen as needed; wash from skin & eyes.

SECTION VI REACTIVITY DATA

CONDITIONS CONTRIBUTING TO INSTABILITY

CONDITIONS CONTRIBUTING TO HAZARDOUS POLYMERIZATION

INCOMPATABILITY (MATERIALS TO AVOID)

HAZARDOUS DECOMPOSITION PRODUCTS

SECTION VII SPILL OR LEAK PROCEDURES

STEPS TO BE TAKEN IN CASE MATERIAL IS RELEASED OR SPILLED

WASTE DISPOSAL METHOD

SECTION VIII SPECIAL PROTECTION INFORMATION

VENTILATION REQUIREMENTS LOCAL EXHAUST | PROTECTIVE EQUIPMENT (SPECIFY TYPES) EYE: Goggles

MECHANICAL (GENERAL) | GLOVES: Rubber

SPECIAL | RESPIRATOR: Self-contained

OTHER PROTECTIVE EQUIPMENT

SECTION IX SPECIAL PRECAUTIONS

PRECAUTIONS TO BE TAKEN IN HANDLING AND STORAGE: Store away from oxidizing materials and sources of heat.

OTHER PRECAUTIONS

SIGNATURE | DATE

m-NITROCHLOROBENZENE

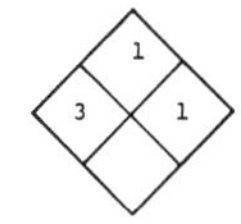

SECTION I SOURCE AND NOMENCLATURE

MANUFACTURER'S NAME | EMERGENCY TELEPHONE NO.

ADDRESS (NUMBER, STREET, CITY, STATE, ZIP CODE)

TRADE NAME AND SYNONYMS | CHEMICAL FAMILY

CHEMICAL NAME AND SYNONYMS: M-Chloronitrobenzene | FORMULA: $ClC_6H_4NO_2$

SECTION II HAZARDOUS INGREDIENTS

BASIC MATERIAL	APPROXIMATE OR MAXIMUM % WT. OR VOL.	ESTABLISHED OSHA STANDARD	LD_{50} mg/kg ORAL	LD_{50} PERCUT.	LC_{50} SPECIES	LC_{50} CONC.
TCLO inhalation - human 0.012 mg/C.M.			390		Mouse	

SECTION III PHYSICAL DATA MW - 157.56

BOILING POINT °F.: 236° | VAPOR PRESSURE mm Hg.

MELTING POINT °F.: 44°C | VAPOR DENSITY (Air = 1)

SPECIFIC GRAVITY (H_2O = 1): 1.34 | EVAPORATION RATE (________ = 1)

SOLUBILITY IN WATER Pts/100 pts H_2O: insoluble | VOLATILE % Vol. % Wt.

APPEARANCE AND ODOR: yellow crystals, aromatic odor.

SECTION IV FIRE AND EXPLOSION HAZARD DATA

FLASH POINT: 261°F | FLAMMABLE (EXPLOSIVE) LIMITS: UPPER / LOWER

METHOD USED

EXTINGUISHING MEDIA: water spray, carbon dioxide, dry chemical, foam

SPECIAL FIRE FIGHTING PROCEDURES: wear self-contained breathing apparatus.

UNUSUAL FIRE AND EXPLOSION HAZARDS

SECTION V HEALTH HAZARD DATA

TOXIC LEVEL: none available | CARCINOGENIC: no

PRINCIPLE ROUTES OF ABSORBTION: inhalation, ingestion, skin | SKIN AND EYE IRRITATION: moderate

REVELANT SYMPTOMS OF EXPOSURE: cyanosis, headache, weakness, anemia.

EFFECTS OF CHRONIC EXPOSURE: Hemoglobinuria

EMERGENCY AND FIRST AID PROCEDURES: If swallowed - give emetic, get medical aid; inhalation - oxygen if needed; wash from skin & eyes.

SECTION VI REACTIVITY DATA

CONDITIONS CONTRIBUTING TO INSTABILITY

CONDITIONS CONTRIBUTING TO HAZARDOUS POLYMERIZATION

INCOMPATABILITY (MATERIALS TO AVOID)

HAZARDOUS DECOMPOSITION PRODUCTS

SECTION VII SPILL OR LEAK PROCEDURES

STEPS TO BE TAKEN IN CASE MATERIAL IS RELEASED OR SPILLED

WASTE DISPOSAL METHOD

SECTION VIII SPECIAL PROTECTION INFORMATION

VENTILATION REQUIREMENTS LOCAL EXHAUST | PROTECTIVE EQUIPMENT (SPECIFY TYPES) EYE: Goggles

MECHANICAL (GENERAL) | GLOVES: Rubber

SPECIAL | RESPIRATOR: Self-contained

OTHER PROTECTIVE EQUIPMENT

SECTION IX SPECIAL PRECAUTIONS

PRECAUTIONS TO BE TAKEN IN HANDLING AND STORAGE: Store away from oxidizing materials and sources of heat.

OTHER PRECAUTIONS

SIGNATURE | DATE

o-NITROCHLOROBENZENE

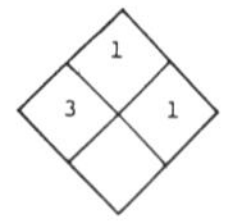

SECTION I SOURCE AND NOMENCLATURE

MANUFACTURER'S NAME | EMERGENCY TELEPHONE NO.

ADDRESS (NUMBER, STREET, CITY, STATE, ZIP CODE)

TRADE NAME AND SYNONYMS | CHEMICAL FAMILY

CHEMICAL NAME AND SYNONYMS: O-chloronitrobenzene | FORMULA: $CLC_6H_4NO_2$

SECTION II HAZARDOUS INGREDIENTS

BASIC MATERIAL	APPROXIMATE OR MAXIMUM % WT. OR VOL.	ESTABLISHED OSHA STANDARD	LD_{50} mg/kg ORAL	PERCUT.	SPECIES	LC_{50} CONC.
			288		rat	
			135		mouse	

SECTION III PHYSICAL DATA MW - 157.56

BOILING POINT °F. 246°C	VAPOR PRESSURE mm Hg.
MELTING POINT °F. 33°C	VAPOR DENSITY (Air = 1)
SPECIFIC GRAVITY (H_2O = 1) 1.31	EVAPORATION RATE (______ = 1)
SOLUBILITY IN WATER Pts/100 pts H_2O insoluble	VOLATILE % Vol. % Wt.

APPEARANCE AND ODOR: yellow crystals, aromatic odor.

SECTION IV FIRE AND EXPLOSION HAZARD DATA

FLASH POINT: 261°F | FLAMMABLE (EXPLOSIVE) LIMITS: UPPER / LOWER

METHOD USED

EXTINGUISHING MEDIA: Water spray, carbon dioxide, dry chemical, foam.

SPECIAL FIRE FIGHTING PROCEDURES: Wear self-contained breathing apparatus.

UNUSUAL FIRE AND EXPLOSION HAZARDS

SECTION V HEALTH HAZARD DATA

TOXIC LEVEL: none available | CARCINOGENIC: no

PRINCIPLE ROUTES OF ABSORBTION: inhalation, ingestion, skin | SKIN AND EYE IRRITATION: moderate

REVELANT SYMPTOMS OF EXPOSURE: cyanosis, headache, weakness, anemia

EFFECTS OF CHRONIC EXPOSURE: hemoglobinuria

EMERGENCY AND FIRST AID PROCEDURES: If swallowed - give emetic, get medical aid; inhalation - oxygen if needed; wash from skin & eyes.

SECTION VI REACTIVITY DATA

CONDITIONS CONTRIBUTING TO INSTABILITY

CONDITIONS CONTRIBUTING TO HAZARDOUS POLYMERIZATION

INCOMPATABILITY (MATERIALS TO AVOID)

HAZARDOUS DECOMPOSITION PRODUCTS

SECTION VII SPILL OR LEAK PROCEDURES

STEPS TO BE TAKEN IN CASE MATERIAL IS RELEASED OR SPILLED

WASTE DISPOSAL METHOD

SECTION VIII SPECIAL PROTECTION INFORMATION.

VENTILATION REQUIREMENTS LOCAL EXHAUST | PROTECTIVE EQUIPMENT (SPECIFY TYPES) EYE: goggles

MECHANICAL (GENERAL) | GLOVES: rubber

SPECIAL | RESPIRATOR: self-contained

OTHER PROTECTIVE EQUIPMENT

SECTION IX SPECIAL PRECAUTIONS

PRECAUTIONS TO BE TAKEN IN HANDLING AND STORAGE: Store away from oxidizing materials and sources of heat.

OTHER PRECAUTIONS

SIGNATURE | DATE

p-NITROCHLOROBENZENE

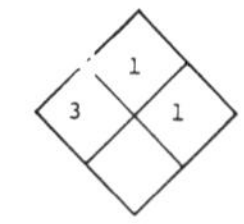

SECTION I SOURCE AND NOMENCLATURE

MANUFACTURER'S NAME | EMERGENCY TELEPHONE NO.

ADDRESS (NUMBER, STREET, CITY, STATE, ZIP CODE)

TRADE NAME AND SYNONYMS | CHEMICAL FAMILY

CHEMICAL NAME AND SYNONYMS: P Chloronitrobenzene | FORMULA: $ClC_6H_4NO_2$

SECTION II HAZARDOUS INGREDIENTS

BASIC MATERIAL	APPROXIMATE OR MAXIMUM % WT. OR VOL.	ESTABLISHED OSHA STANDARD	LD_{50} mg/kg ORAL	PERCUT.	SPECIES	LC_{50} CONC.
			420		rat	
			650		mouse	

SECTION III PHYSICAL DATA MW - 157.56

BOILING POINT °F. 242°C	VAPOR PRESSURE mm Hg.
MELTING POINT °F. 84°C	VAPOR DENSITY (Air = 1) 5.44
SPECIFIC GRAVITY (H_2O = 1) 1.29	EVAPORATION RATE (______ = 1)
SOLUBILITY IN WATER Pts/100 pts H_2O insoluble	VOLATILE % Vol. % Wt.

APPEARANCE AND ODOR: yellow crystals, aromatic odor.

SECTION IV FIRE AND EXPLOSION HAZARD DATA

FLASH POINT: 261°F | FLAMMABLE (EXPLOSIVE) LIMITS: UPPER / LOWER

METHOD USED

EXTINGUISHING MEDIA: water spray, carbon dioxide, dry chemical, foam

SPECIAL FIRE FIGHTING PROCEDURES: wear self-contained breathing apparatus.

UNUSUAL FIRE AND EXPLOSION HAZARDS

SECTION V HEALTH HAZARD DATA

TOXIC LEVEL: TLV - 1 mg/C.M. | CARCINOGENIC: no

PRINCIPLE ROUTES OF ABSORBTION: inhalation, ingestion, skin | SKIN AND EYE IRRITATION: moderate

REVELANT SYMPTOMS OF EXPOSURE: cyanosis, headache, weakness, anemia

EFFECTS OF CHRONIC EXPOSURE: hemoglobinuria

EMERGENCY AND FIRST AID PROCEDURES: If swallowed - give emetic, get medical aid; inhalation - oxygen if needed; wash from skin & eyes.

SECTION VI REACTIVITY DATA

CONDITIONS CONTRIBUTING TO INSTABILITY

CONDITIONS CONTRIBUTING TO HAZARDOUS POLYMERIZATION

INCOMPATABILITY (MATERIALS TO AVOID)

HAZARDOUS DECOMPOSITION PRODUCTS

SECTION VII SPILL OR LEAK PROCEDURES

STEPS TO BE TAKEN IN CASE MATERIAL IS RELEASED OR SPILLED

WASTE DISPOSAL METHOD

SECTION VIII SPECIAL PROTECTION INFORMATION

VENTILATION REQUIREMENTS LOCAL EXHAUST | PROTECTIVE EQUIPMENT (SPECIFY TYPES) EYE: goggles

MECHANICAL (GENERAL) | GLOVES: rubber

SPECIAL | RESPIRATOR: self-contained

OTHER PROTECTIVE EQUIPMENT

SECTION IX SPECIAL PRECAUTIONS

PRECAUTIONS TO BE TAKEN IN HANDLING AND STORAGE: Store away from oxidizing materials and sources of heat.

OTHER PRECAUTIONS

SIGNATURE | DATE

2-NITRO-p-CYMENE

Top: 1, Left: 4, Right: 3

SECTION I SOURCE AND NOMENCLATURE

MANUFACTURER'S NAME	EMERGENCY TELEPHONE NO.
ADDRESS (NUMBER, STREET, CITY, STATE, ZIP CODE)	
TRADE NAME AND SYNONYMS	CHEMICAL FAMILY
CHEMICAL NAME AND SYNONYMS: 2 Nitro-P-Cymene	FORMULA: $NO_2C_6H_3(CH_3)C_3H_7$

SECTION II HAZARDOUS INGREDIENTS

BASIC MATERIAL	APPROXIMATE OR MAXIMUM % WT. OR VOL.	ESTABLISHED OSHA STANDARD	LD_{50} ORAL	LD_{50} PERCUT.	LC_{50} SPECIES	LC_{50} CONC.
			None	available		

SECTION III PHYSICAL DATA MW - 179.21

BOILING POINT °F.: 152° C. - 15 mm	VAPOR PRESSURE mm Hg.: at 128° C. 10
MELTING POINT °F.	VAPOR DENSITY (Air = 1)
SPECIFIC GRAVITY (H_2O = 1): 1.067	EVAPORATION RATE (______ = 1)
SOLUBILITY IN WATER Pts/100 pts H_2O: Insoluble	VOLATILE % Vol. % Wt.
APPEARANCE AND ODOR: Oil	

SECTION IV FIRE AND EXPLOSION HAZARD DATA

FLASH POINT: None available	FLAMMABLE (EXPLOSIVE) LIMITS	UPPER
METHOD USED		LOWER

EXTINGUISHING MEDIA: Foam, carbon dioxide, dry chemical.

SPECIAL FIRE FIGHTING PROCEDURES

UNUSUAL FIRE AND EXPLOSION HAZARDS

SECTION V HEALTH HAZARD DATA

TOXIC LEVEL: None available — CARCINOGENIC: No

PRINCIPLE ROUTES OF ABSORBTION: Inhalation, ingestion, skin — SKIN AND EYE IRRITATION: Slight

REVELANT SYMPTOMS OF EXPOSURE: Difficulty in breathing, cyanosis, vomiting, headache.

EFFECTS OF CHRONIC EXPOSURE: --

EMERGENCY AND FIRST AID PROCEDURES: If swallowed - give emetic, get medical aid; inhalation - rest; wash skin and eyes.

SECTION VI REACTIVITY DATA

CONDITIONS CONTRIBUTING TO INSTABILITY

CONDITIONS CONTRIBUTING TO HAZARDOUS POLYMERIZATION

INCOMPATABILITY (MATERIALS TO AVOID): May explode when heated with sulfuric acid.

HAZARDOUS DECOMPOSITION PRODUCTS

SECTION VII SPILL OR LEAK PROCEDURES

STEPS TO BE TAKEN IN CASE MATERIAL IS RELEASED OR SPILLED

WASTE DISPOSAL METHOD

SECTION VIII SPECIAL PROTECTION INFORMATION

VENTILATION REQUIREMENTS LOCAL EXHAUST	PROTECTIVE EQUIPMENT (SPECIFY TYPES) EYE: Goggles
MECHANICAL (GENERAL)	GLOVES: Rubber
SPECIAL	RESPIRATOR: Self contained
OTHER PROTECTIVE EQUIPMENT	

SECTION IX SPECIAL PRECAUTIONS

PRECAUTIONS TO BE TAKEN IN HANDLING AND STORAGE: Outside or detached storage preferred; inside storage in standard flammable liquids cabinet.

OTHER PRECAUTIONS

SIGNATURE	DATE

2-NITRO-1,4-DIAMINOBENZENE

Top: 1, Left: 4, Right: 1

SECTION I SOURCE AND NOMENCLATURE

MANUFACTURER'S NAME	EMERGENCY TELEPHONE NO.
ADDRESS (NUMBER, STREET, CITY, STATE, ZIP CODE)	
TRADE NAME AND SYNONYMS: 2 Nitro - 1,4 Diaminobenzene	CHEMICAL FAMILY
CHEMICAL NAME AND SYNONYMS: 2 Nitro-P-Phenylenediamine	FORMULA: $NO_2C_6H_3(NH_2)_2$

SECTION II HAZARDOUS INGREDIENTS

BASIC MATERIAL	APPROXIMATE OR MAXIMUM % WT. OR VOL.	ESTABLISHED OSHA STANDARD	LD_{50} ORAL	LD_{50} PERCUT.	LC_{50} SPECIES	LC_{50} CONC.
			none	available		

SECTION III PHYSICAL DATA

BOILING POINT °F.	VAPOR PRESSURE mm Hg.
MELTING POINT °F.: 140°C	VAPOR DENSITY (Air = 1)
SPECIFIC GRAVITY (H_2O = 1): --	EVAPORATION RATE (______ = 1)
SOLUBILITY IN WATER Pts/100 pts H_2O: --	VOLATILE % Vol. % Wt.
APPEARANCE AND ODOR: --	

SECTION IV FIRE AND EXPLOSION HAZARD DATA

FLASH POINT: not available	FLAMMABLE (EXPLOSIVE) LIMITS	UPPER
METHOD USED		LOWER

EXTINGUISHING MEDIA: information not available

SPECIAL FIRE FIGHTING PROCEDURES

UNUSUAL FIRE AND EXPLOSION HAZARDS

SECTION V HEALTH HAZARD DATA

TOXIC LEVEL: None available — CARCINOGENIC: no

PRINCIPLE ROUTES OF ABSORBTION: inhalation, ingestion, skin — SKIN AND EYE IRRITATION: yes

REVELANT SYMPTOMS OF EXPOSURE: cyanosis, headache, nausea, vomiting, weakness, stupor

EFFECTS OF CHRONIC EXPOSURE: Liver damage

EMERGENCY AND FIRST AID PROCEDURES: If swallowed - give emetic, get medical aid; inhalation - rest; wash from skin & eyes.

SECTION VI REACTIVITY DATA

CONDITIONS CONTRIBUTING TO INSTABILITY

CONDITIONS CONTRIBUTING TO HAZARDOUS POLYMERIZATION

INCOMPATABILITY (MATERIALS TO AVOID)

HAZARDOUS DECOMPOSITION PRODUCTS

SECTION VII SPILL OR LEAK PROCEDURES

STEPS TO BE TAKEN IN CASE MATERIAL IS RELEASED OR SPILLED

WASTE DISPOSAL METHOD

SECTION VIII SPECIAL PROTECTION INFORMATION

VENTILATION REQUIREMENTS LOCAL EXHAUST	PROTECTIVE EQUIPMENT (SPECIFY TYPES) EYE: goggles
MECHANICAL (GENERAL)	GLOVES: rubber
SPECIAL	RESPIRATOR: with proper filter
OTHER PROTECTIVE EQUIPMENT	

SECTION IX SPECIAL PRECAUTIONS

PRECAUTIONS TO BE TAKEN IN HANDLING AND STORAGE: information not available

OTHER PRECAUTIONS

SIGNATURE	DATE

4-NITRO-1,2-DIAMINOBENZENE

Hazard rating: 1 (top), 4 (left), 1 (right)

SECTION I SOURCE AND NOMENCLATURE

MANUFACTURER'S NAME | EMERGENCY TELEPHONE NO.

ADDRESS (NUMBER, STREET, CITY, STATE, ZIP CODE)

TRADE NAME AND SYNONYMS: 4 Nitro-1,2-Diaminobenzene | CHEMICAL FAMILY

CHEMICAL NAME AND SYNONYMS: 4 Nitro-P-Phenylenediamine | FORMULA: $NO_2C_6H_3(NH_2)_2$

SECTION II HAZARDOUS INGREDIENTS

BASIC MATERIAL	APPROXIMATE OR MAXIMUM % WT. OR VOL.	ESTABLISHED OSHA STANDARD	LD_{50} mg/kg ORAL	LD_{50} PERCUT.	SPECIES	LC_{50} CONC.
			681		rat	
			681		mouse	

SECTION III PHYSICAL DATA

BOILING POINT °F: -- | VAPOR PRESSURE mm Hg.

MELTING POINT °F: 202°C | VAPOR DENSITY (Air = 1)

SPECIFIC GRAVITY (H_2O = 1): -- | EVAPORATION RATE (= 1)

SOLUBILITY IN WATER Pts/100 pts H_2O: -- | VOLATILE % Vol. % Wt.

APPEARANCE AND ODOR: dark red crystals

SECTION IV FIRE AND EXPLOSION HAZARD DATA

FLASH POINT: not available | FLAMMABLE (EXPLOSIVE) LIMITS: UPPER / LOWER

METHOD USED

EXTINGUISHING MEDIA: information not available

SPECIAL FIRE FIGHTING PROCEDURES

UNUSUAL FIRE AND EXPLOSION HAZARDS

SECTION V HEALTH HAZARD DATA

TOXIC LEVEL: none available | CARCINOGENIC: no

PRINCIPLE ROUTES OF ABSORPTION: inhalation, ingestion, skin | SKIN AND EYE IRRITATION: yes

REVELANT SYMPTOMS OF EXPOSURE: cyanosis, headache, nausea, vomiting, weakness, stupor.

EFFECTS OF CHRONIC EXPOSURE: liver damage

EMERGENCY AND FIRST AID PROCEDURES: if swallowed - give emetic, get medical aid; inhalation - rest; wash from skin & eyes.

SECTION VI REACTIVITY DATA

CONDITIONS CONTRIBUTING TO INSTABILITY

CONDITIONS CONTRIBUTING TO HAZARDOUS POLYMERIZATION

INCOMPATABILITY (MATERIALS TO AVOID)

HAZARDOUS DECOMPOSITION PRODUCTS

SECTION VII SPILL OR LEAK PROCEDURES

STEPS TO BE TAKEN IN CASE MATERIAL IS RELEASED OR SPILLED

WASTE DISPOSAL METHOD

SECTION VIII SPECIAL PROTECTION INFORMATION

VENTILATION REQUIREMENTS LOCAL EXHAUST | PROTECTIVE EQUIPMENT (SPECIFY TYPES) EYE: Goggles

MECHANICAL (GENERAL) | GLOVES: Rubber

SPECIAL | RESPIRATOR: With proper filter

OTHER PROTECTIVE EQUIPMENT

SECTION IX SPECIAL PRECAUTIONS

PRECAUTIONS TO BE TAKEN IN HANDLING AND STORAGE: Information not available.

OTHER PRECAUTIONS

SIGNATURE | DATE

o-NITRODIETHYLANILINE

Hazard rating: 1 (top), 4 (left), 0 (right)

SECTION I SOURCE AND NOMENCLATURE

MANUFACTURER'S NAME | EMERGENCY TELEPHONE NO.

ADDRESS (NUMBER, STREET, CITY, STATE, ZIP CODE)

TRADE NAME AND SYNONYMS | CHEMICAL FAMILY

CHEMICAL NAME AND SYNONYMS: 2-Nitro-N, N-diethylaniline | FORMULA: $NO_2C_6H_4N(C_2H_5)_2$

SECTION II HAZARDOUS INGREDIENTS

BASIC MATERIAL	APPROXIMATE OR MAXIMUM % WT. OR VOL.	ESTABLISHED OSHA STANDARD	LD_{50} ORAL	LD_{50} PERCUT.	SPECIES	LC_{50} CONC.
			none	available		

SECTION III PHYSICAL DATA MW - 194.24

BOILING POINT °F: -- | VAPOR PRESSURE: at 155°C 20 mm Hg.

MELTING POINT °F: -- | VAPOR DENSITY (Air = 1)

SPECIFIC GRAVITY (H_2O = 1): -- | EVAPORATION RATE (= 1)

SOLUBILITY IN WATER Pts/100 pts H_2O: slightly soluble | VOLATILE % Vol. % Wt.

APPEARANCE AND ODOR: yellow liquid

SECTION IV FIRE AND EXPLOSION HAZARD DATA

FLASH POINT: none available | FLAMMABLE (EXPLOSIVE) LIMITS: UPPER / LOWER

METHOD USED

EXTINGUISHING MEDIA: information not available

SPECIAL FIRE FIGHTING PROCEDURES

UNUSUAL FIRE AND EXPLOSION HAZARDS

SECTION V HEALTH HAZARD DATA

TOXIC LEVEL: none available | CARCINOGENIC: no

PRINCIPLE ROUTES OF ABSORPTION: inhalation, ingestion, skin | SKIN AND EYE IRRITATION: yes

REVELANT SYMPTOMS OF EXPOSURE: Weakness, tremors, cyanosis, drowsiness.

EFFECTS OF CHRONIC EXPOSURE: Fatigue, loss of appetite, headache, dizziness.

EMERGENCY AND FIRST AID PROCEDURES: If swallowed - give emetic, get medical aid; inhalation - rest; wash from skin & eyes.

SECTION VI REACTIVITY DATA

CONDITIONS CONTRIBUTING TO INSTABILITY

CONDITIONS CONTRIBUTING TO HAZARDOUS POLYMERIZATION

INCOMPATABILITY (MATERIALS TO AVOID)

HAZARDOUS DECOMPOSITION PRODUCTS

SECTION VII SPILL OR LEAK PROCEDURES

STEPS TO BE TAKEN IN CASE MATERIAL IS RELEASED OR SPILLED

WASTE DISPOSAL METHOD

SECTION VIII SPECIAL PROTECTION INFORMATION

VENTILATION REQUIREMENTS LOCAL EXHAUST | PROTECTIVE EQUIPMENT (SPECIFY TYPES) EYE: Goggles

MECHANICAL (GENERAL) | GLOVES: Rubber

SPECIAL | RESPIRATOR: With proper filter

OTHER PROTECTIVE EQUIPMENT

SECTION IX SPECIAL PRECAUTIONS

PRECAUTIONS TO BE TAKEN IN HANDLING AND STORAGE

OTHER PRECAUTIONS

SIGNATURE | DATE

m-NITRODIMETHYLANILINE

1 4 0

SECTION I SOURCE AND NOMENCLATURE

MANUFACTURER'S NAME	EMERGENCY TELEPHONE NO.
ADDRESS (NUMBER, STREET, CITY, STATE, ZIP CODE)	
TRADE NAME AND SYNONYMS	CHEMICAL FAMILY
CHEMICAL NAME AND SYNONYMS 3-Nitro-N,N-dimethylaniline	FORMULA $NO_2C_6H_4N(CH_3)_2$

SECTION II HAZARDOUS INGREDIENTS

BASIC MATERIAL	APPROXIMATE OR MAXIMUM % WT. OR VOL.	ESTABLISHED OSHA STANDARD	LD_{50} ORAL	LD_{50} PERCUT.	LC_{50} SPECIES	LC_{50} CONC.
			none available			

SECTION III PHYSICAL DATA MW - 166.18

BOILING POINT °F. 280°C	VAPOR PRESSURE mm Hg.
MELTING POINT °F. 60°C	VAPOR DENSITY (Air = 1)
SPECIFIC GRAVITY (H_2O = 1) 1.313	EVAPORATION RATE (_______ = 1)
SOLUBILITY IN WATER Pts/100 pts H_2O insoluble	VOLATILE % Vol. % Wt.

APPEARANCE AND ODOR
yellow to red crystals.

SECTION IV FIRE AND EXPLOSION HAZARD DATA

FLASH POINT none available	FLAMMABLE (EXPLOSIVE) LIMITS	UPPER
METHOD USED		LOWER

EXTINGUISHING MEDIA
Foam, carbon dioxide, dry chemical.

SPECIAL FIRE FIGHTING PROCEDURES

UNUSUAL FIRE AND EXPLOSION HAZARDS

SECTION V HEALTH HAZARD DATA

TOXIC LEVEL
none available — CARCINOGENIC no

PRINCIPLE ROUTES OF ABSORBTION
inhalation, ingestion, skin — SKIN AND EYE IRRITATION yes

REVELANT SYMPTOMS OF EXPOSURE
weakness, tremors, cyanosis, drowsiness

EFFECTS OF CHRONIC EXPOSURE
Fatigue, loss of appetite, headache, dizziness

EMERGENCY AND FIRST AID PROCEDURES
If swallowed - give emetic, get medical aid; inhalation - rest; wash from skin & eyes.

SECTION VI REACTIVITY DATA

CONDITIONS CONTRIBUTING TO INSTABILITY

CONDITIONS CONTRIBUTING TO HAZARDOUS POLYMERIZATION

INCOMPATABILITY (MATERIALS TO AVOID)

HAZARDOUS DECOMPOSITION PRODUCTS

SECTION VII SPILL OR LEAK PROCEDURES

STEPS TO BE TAKEN IN CASE MATERIAL IS RELEASED OR SPILLED

WASTE DISPOSAL METHOD

SECTION VIII SPECIAL PROTECTION INFORMATION

VENTILATION REQUIREMENTS LOCAL EXHAUST	PROTECTIVE EQUIPMENT (SPECIFY TYPES) EYE goggles
MECHANICAL (GENERAL)	GLOVES rubber
SPECIAL	RESPIRATOR with proper filter
OTHER PROTECTIVE EQUIPMENT	

SECTION IX SPECIAL PRECAUTIONS

PRECAUTIONS TO BE TAKEN IN HANDLING AND STORAGE

OTHER PRECAUTIONS

SIGNATURE	DATE

2-NITRO-1,4-DIMETHYLBENZENE

1 4 3

SECTION I SOURCE AND NOMENCLATURE

MANUFACTURER'S NAME	EMERGENCY TELEPHONE NO.
ADDRESS (NUMBER, STREET, CITY, STATE, ZIP CODE)	
TRADE NAME AND SYNONYMS Nitro-P-Xylene	CHEMICAL FAMILY
CHEMICAL NAME AND SYNONYMS 2-Nitro-1,4-Dimethylbenzene	FORMULA $(NO_2)C_6H_3(CH_3)_2$

SECTION II HAZARDOUS INGREDIENTS

BASIC MATERIAL	APPROXIMATE OR MAXIMUM % WT. OR VOL.	ESTABLISHED OSHA STANDARD	LD_{50} mg/kg ORAL	LD_{50} mg/kg PERCUT.	LC_{50} SPECIES	LC_{50} CONC.
			2440		Rat	

SECTION III PHYSICAL DATA MW - 151.16

BOILING POINT °F. 239°C	VAPOR PRESSURE at 102°C 10 mm Hg.
MELTING POINT °F. --	VAPOR DENSITY (Air = 1)
SPECIFIC GRAVITY (H_2O = 1) 1.13	EVAPORATION RATE (_______ = 1)
SOLUBILITY IN WATER Pts/100 pts H_2O	VOLATILE % Vol. % Wt.

APPEARANCE AND ODOR
yellow liquid

SECTION IV FIRE AND EXPLOSION HAZARD DATA

FLASH POINT none available	FLAMMABLE (EXPLOSIVE) LIMITS	UPPER
METHOD USED		LOWER

EXTINGUISHING MEDIA
Information - not available.

SPECIAL FIRE FIGHTING PROCEDURES

UNUSUAL FIRE AND EXPLOSION HAZARDS

SECTION V HEALTH HAZARD DATA

TOXIC LEVEL
none available — CARCINOGENIC no

PRINCIPLE ROUTES OF ABSORBTION
inhalation, ingestion, skin — SKIN AND EYE IRRITATION slight

REVELANT SYMPTOMS OF EXPOSURE
difficulty in breathing, cyanosis, vomiting, headache.

EFFECTS OF CHRONIC EXPOSURE
blood damage.

EMERGENCY AND FIRST AID PROCEDURES
If swallowed - give emetic, get medical aid; inhalation - rest; wash from skin & eyes.

SECTION VI REACTIVITY DATA

CONDITIONS CONTRIBUTING TO INSTABILITY

CONDITIONS CONTRIBUTING TO HAZARDOUS POLYMERIZATION

INCOMPATABILITY (MATERIALS TO AVOID)
May explode when heated with sulfuric acid

HAZARDOUS DECOMPOSITION PRODUCTS

SECTION VII SPILL OR LEAK PROCEDURES

STEPS TO BE TAKEN IN CASE MATERIAL IS RELEASED OR SPILLED

WASTE DISPOSAL METHOD

SECTION VIII SPECIAL PROTECTION INFORMATION

VENTILATION REQUIREMENTS LOCAL EXHAUST	PROTECTIVE EQUIPMENT (SPECIFY TYPES) EYE goggles
MECHANICAL (GENERAL)	GLOVES rubber
SPECIAL	RESPIRATOR self-contained
OTHER PROTECTIVE EQUIPMENT	

SECTION IX SPECIAL PRECAUTIONS

PRECAUTIONS TO BE TAKEN IN HANDLING AND STORAGE
Information not available.

OTHER PRECAUTIONS

SIGNATURE	DATE

3-NITRO-1,2-DIMETHYLBENZENE

Hazard diamond: 1 (top), 4 (left), 3 (right)

SECTION I SOURCE AND NOMENCLATURE

MANUFACTURER'S NAME

EMERGENCY TELEPHONE NO.

ADDRESS (NUMBER, STREET, CITY, STATE, ZIP CODE)

TRADE NAME AND SYNONYMS: 3 Nitro-O-Xylene

CHEMICAL FAMILY

CHEMICAL NAME AND SYNONYMS: 3-Nitro-1,2-Dimethylbenzene

FORMULA: $NO_2C_6H_3(CH_3)_2$

SECTION II HAZARDOUS INGREDIENTS

BASIC MATERIAL	APPROXIMATE OR MAXIMUM % WT. OR VOL.	ESTABLISHED OSHA STANDARD	LD_{50} ORAL	LD_{50} PERCUT.	LC_{50} SPECIES	LC_{50} CONC.
			none	available		

SECTION III PHYSICAL DATA MW - 151.16

BOILING POINT °F.: 245°C

VAPOR PRESSURE mm Hg.

MELTING POINT °F.: 15°C

VAPOR DENSITY (Air = 1)

SPECIFIC GRAVITY (H_2O = 1): 1.147

EVAPORATION RATE (_____ = 1)

SOLUBILITY IN WATER Pts/100 pts H_2O: insoluble

VOLATILE % Vol. % Wt.

APPEARANCE AND ODOR: yellow liquid

SECTION IV FIRE AND EXPLOSION HAZARD DATA

FLASH POINT: none available

FLAMMABLE (EXPLOSIVE) LIMITS — UPPER / LOWER

METHOD USED

EXTINGUISHING MEDIA: information not available

SPECIAL FIRE FIGHTING PROCEDURES

UNUSUAL FIRE AND EXPLOSION HAZARDS

SECTION V HEALTH HAZARD DATA

TOXIC LEVEL: none available

CARCINOGENIC: no

PRINCIPLE ROUTES OF ABSORBTION: inhalation, ingestion, skin

SKIN AND EYE IRRITATION: slight

REVELANT SYMPTOMS OF EXPOSURE: Difficulty in breathing, cyanosis, vomiting, headache.

EFFECTS OF CHRONIC EXPOSURE: Blood damage.

EMERGENCY AND FIRST AID PROCEDURES: If swallowed - give emetic, get medical aid; inhalation - rest; wash from skin & eyes.

SECTION VI REACTIVITY DATA

CONDITIONS CONTRIBUTING TO INSTABILITY

CONDITIONS CONTRIBUTING TO HAZARDOUS POLYMERIZATION

INCOMPATABILITY (MATERIALS TO AVOID): May explode when heated with sulfuric acid.

HAZARDOUS DECOMPOSITION PRODUCTS

SECTION VII SPILL OR LEAK PROCEDURES

STEPS TO BE TAKEN IN CASE MATERIAL IS RELEASED OR SPILLED

WASTE DISPOSAL METHOD

SECTION VIII SPECIAL PROTECTION INFORMATION

VENTILATION REQUIREMENTS LOCAL EXHAUST

PROTECTIVE EQUIPMENT (SPECIFY TYPES) EYE: Goggles

MECHANICAL (GENERAL)

GLOVES: Rubber

SPECIAL

RESPIRATOR: Self-contained

OTHER PROTECTIVE EQUIPMENT

SECTION IX SPECIAL PRECAUTIONS

PRECAUTIONS TO BE TAKEN IN HANDLING AND STORAGE: Information not available.

OTHER PRECAUTIONS

SIGNATURE | DATE

p-NITROFLUOROBENZENE

Hazard diamond: 1 (top), 3 (left), 1 (right)

SECTION I SOURCE AND NOMENCLATURE

MANUFACTURER'S NAME

EMERGENCY TELEPHONE NO.

ADDRESS (NUMBER, STREET, CITY, STATE, ZIP CODE)

TRADE NAME AND SYNONYMS

CHEMICAL FAMILY

CHEMICAL NAME AND SYNONYMS: P-Fluoronitrobenzene

FORMULA: $FC_6H_4NO_2$

SECTION II HAZARDOUS INGREDIENTS

BASIC MATERIAL	APPROXIMATE OR MAXIMUM % WT. OR VOL.	ESTABLISHED OSHA STANDARD	LD_{50} ORAL	LD_{50} PERCUT.	LC_{50} SPECIES	LC_{50} CONC.
LDLO oral - rat 250 mg/kg						

SECTION III PHYSICAL DATA MW - 141.10

BOILING POINT °F.: 207°C

VAPOR PRESSURE mm Hg.

MELTING POINT °F.: 27°C

VAPOR DENSITY (Air = 1)

SPECIFIC GRAVITY (H_2O = 1): 1.32

EVAPORATION RATE (_____ = 1)

SOLUBILITY IN WATER Pts/100 pts H_2O

VOLATILE % Vol. % Wt.

APPEARANCE AND ODOR: yellow crystal

SECTION IV FIRE AND EXPLOSION HAZARD DATA

FLASH POINT: none available

FLAMMABLE (EXPLOSIVE) LIMITS — UPPER / LOWER

METHOD USED

EXTINGUISHING MEDIA: water spray, carbon dioxide, dry chemical, foam.

SPECIAL FIRE FIGHTING PROCEDURES: wear self-contained breathing apparatus.

UNUSUAL FIRE AND EXPLOSION HAZARDS

SECTION V HEALTH HAZARD DATA

TOXIC LEVEL: none available

CARCINOGENIC: no

PRINCIPLE ROUTES OF ABSORBTION: inhalation, ingestion, skin

SKIN AND EYE IRRITATION: moderate

REVELANT SYMPTOMS OF EXPOSURE: cyanosis, headache, weakness, anemia

EFFECTS OF CHRONIC EXPOSURE: hemoglobinuria

EMERGENCY AND FIRST AID PROCEDURES: If swallowed - give emetic, get medical aid, inhalation - oxygen if needed; wash from skin & eyes.

SECTION VI REACTIVITY DATA

CONDITIONS CONTRIBUTING TO INSTABILITY

CONDITIONS CONTRIBUTING TO HAZARDOUS POLYMERIZATION

INCOMPATABILITY (MATERIALS TO AVOID)

HAZARDOUS DECOMPOSITION PRODUCTS

SECTION VII SPILL OR LEAK PROCEDURES

STEPS TO BE TAKEN IN CASE MATERIAL IS RELEASED OR SPILLED

WASTE DISPOSAL METHOD

SECTION VIII SPECIAL PROTECTION INFORMATION

VENTILATION REQUIREMENTS LOCAL EXHAUST

PROTECTIVE EQUIPMENT (SPECIFY TYPES) EYE: Goggles

MECHANICAL (GENERAL)

GLOVES: Rubber

SPECIAL

RESPIRATOR: Self-contained

OTHER PROTECTIVE EQUIPMENT

SECTION IX SPECIAL PRECAUTIONS

PRECAUTIONS TO BE TAKEN IN HANDLING AND STORAGE: Store away from oxidizing materials and sources of heat.

OTHER PRECAUTIONS

SIGNATURE | DATE

m-NITROIODOBENZENE

1
3 1

SECTION I SOURCE AND NOMENCLATURE

MANUFACTURER'S NAME | EMERGENCY TELEPHONE NO.

ADDRESS (NUMBER, STREET, CITY, STATE, ZIP CODE)

TRADE NAME AND SYNONYMS | CHEMICAL FAMILY

CHEMICAL NAME AND SYNONYMS: M-Iodonitrobenzene | FORMULA: $IC_6H_4NO_2$

SECTION II HAZARDOUS INGREDIENTS

BASIC MATERIAL	APPROXIMATE OR MAXIMUM % WT. OR VOL.	ESTABLISHED OSHA STANDARD	LD_{50} ORAL	LD_{50} PERCUT.	LC_{50} SPECIES	LC_{50} CONC.
			none available			

SECTION III PHYSICAL DATA MW - 249.02

BOILING POINT °F.: 280°C | VAPOR PRESSURE mm Hg.

MELTING POINT °F.: 38°C | VAPOR DENSITY (Air = 1)

SPECIFIC GRAVITY (H_2O = 1): 1.88 | EVAPORATION RATE (_____ = 1)

SOLUBILITY IN WATER Pts/100 pts H_2O | VOLATILE % Vol. % Wt.

APPEARANCE AND ODOR: crystaline

SECTION IV FIRE AND EXPLOSION HAZARD DATA

FLASH POINT: none available | FLAMMABLE (EXPLOSIVE) LIMITS | UPPER | LOWER

METHOD USED

EXTINGUISHING MEDIA: water spray, carbon dioxide, dry chemical, foam

SPECIAL FIRE FIGHTING PROCEDURES: Wear self-contained breathing apparatus.

UNUSUAL FIRE AND EXPLOSION HAZARDS

SECTION V HEALTH HAZARD DATA

TOXIC LEVEL: none available | CARCINOGENIC: no

PRINCIPLE ROUTES OF ABSORBTION: inhalation, ingestion, skin | SKIN AND EYE IRRITATION: moderate

REVELANT SYMPTOMS OF EXPOSURE: cyanosis, headache, weakness, anemia

EFFECTS OF CHRONIC EXPOSURE: hemoglobinuria

EMERGENCY AND FIRST AID PROCEDURES: If swallowed - give emetic, get medical aid; inhalation - oxygen if needed; wash from skin & eyes.

SECTION VI REACTIVITY DATA

CONDITIONS CONTRIBUTING TO INSTABILITY

CONDITIONS CONTRIBUTING TO HAZARDOUS POLYMERIZATION

INCOMPATABILITY (MATERIALS TO AVOID)

HAZARDOUS DECOMPOSITION PRODUCTS

SECTION VII SPILL OR LEAK PROCEDURES

STEPS TO BE TAKEN IN CASE MATERIAL IS RELEASED OR SPILLED

WASTE DISPOSAL METHOD

SECTION VIII SPECIAL PROTECTION INFORMATION

VENTILATION REQUIREMENTS LOCAL EXHAUST | PROTECTIVE EQUIPMENT (SPECIFY TYPES) EYE: Goggles

MECHANICAL (GENERAL) | GLOVES: Rubber

SPECIAL | RESPIRATOR: Self-contained

OTHER PROTECTIVE EQUIPMENT

SECTION IX SPECIAL PRECAUTIONS

PRECAUTIONS TO BE TAKEN IN HANDLING AND STORAGE: Store away from oxidizing materials and sources of heat.

OTHER PRECAUTIONS

SIGNATURE | DATE

o-NITROIODOBENZENE

1
3 1

SECTION I SOURCE AND NOMENCLATURE

MANUFACTURER'S NAME | EMERGENCY TELEPHONE NO.

ADDRESS (NUMBER, STREET, CITY, STATE, ZIP CODE)

TRADE NAME AND SYNONYMS | CHEMICAL FAMILY

CHEMICAL NAME AND SYNONYMS: O-Iodonitrobenzene | FORMULA: $IC_6H_4NO_2$

SECTION II HAZARDOUS INGREDIENTS

BASIC MATERIAL	APPROXIMATE OR MAXIMUM % WT. OR VOL.	ESTABLISHED OSHA STANDARD	LD_{50} ORAL	LD_{50} PERCUT.	LC_{50} SPECIES	LC_{50} CONC.
			none available			

SECTION III PHYSICAL DATA MW - 249.02

BOILING POINT °F.: 289°C | VAPOR PRESSURE mm Hg.

MELTING POINT °F.: 52°C | VAPOR DENSITY (Air = 1)

SPECIFIC GRAVITY (H_2O = 1): 1.88 | EVAPORATION RATE (_____ = 1)

SOLUBILITY IN WATER Pts/100 pts H_2O: insoluble | VOLATILE % Vol. % Wt.

APPEARANCE AND ODOR: yellow crystals

SECTION IV FIRE AND EXPLOSION HAZARD DATA

FLASH POINT: none available | FLAMMABLE (EXPLOSIVE) LIMITS | UPPER | LOWER

METHOD USED

EXTINGUISHING MEDIA: water spray, carbon dioxide, dry chemical, foam.

SPECIAL FIRE FIGHTING PROCEDURES: wear self-contained breathing apparatus.

UNUSUAL FIRE AND EXPLOSION HAZARDS

SECTION V HEALTH HAZARD DATA

TOXIC LEVEL: none available | CARCINOGENIC: no

PRINCIPLE ROUTES OF ABSORBTION: inhalation, ingestion, skin | SKIN AND EYE IRRITATION: moderate

REVELANT SYMPTOMS OF EXPOSURE: cyanosis, headache, weakness, anemia.

EFFECTS OF CHRONIC EXPOSURE: hemoglobinuria

EMERGENCY AND FIRST AID PROCEDURES: If swallowed - give emetic, get medical aid; inhalation - oxygen if needed; wash from skin & eyes.

SECTION VI REACTIVITY DATA

CONDITIONS CONTRIBUTING TO INSTABILITY

CONDITIONS CONTRIBUTING TO HAZARDOUS POLYMERIZATION

INCOMPATABILITY (MATERIALS TO AVOID)

HAZARDOUS DECOMPOSITION PRODUCTS

SECTION VII SPILL OR LEAK PROCEDURES

STEPS TO BE TAKEN IN CASE MATERIAL IS RELEASED OR SPILLED

WASTE DISPOSAL METHOD

SECTION VIII SPECIAL PROTECTION INFORMATION

VENTILATION REQUIREMENTS LOCAL EXHAUST | PROTECTIVE EQUIPMENT (SPECIFY TYPES) EYE: Goggles

MECHANICAL (GENERAL) | GLOVES: Rubber

SPECIAL | RESPIRATOR: Self-contained

OTHER PROTECTIVE EQUIPMENT

SECTION IX SPECIAL PRECAUTIONS

PRECAUTIONS TO BE TAKEN IN HANDLING AND STORAGE: Store away from oxidizing materials and sources of heat.

OTHER PRECAUTIONS

SIGNATURE | DATE

p-NITROIODOBENZENE

1 3 1

SECTION I SOURCE AND NOMENCLATURE

MANUFACTURER'S NAME | EMERGENCY TELEPHONE NO.

ADDRESS (NUMBER, STREET, CITY, STATE, ZIP CODE)

TRADE NAME AND SYNONYMS | CHEMICAL FAMILY

CHEMICAL NAME AND SYNONYMS: P-Iodonitrobenzene | FORMULA: $IC_6H_4NO_2$

SECTION II HAZARDOUS INGREDIENTS

BASIC MATERIAL	APPROXIMATE OR MAXIMUM % WT. OR VOL.	ESTABLISHED OSHA STANDARD	LD_{50} ORAL	LD_{50} PERCUT.	LC_{50} SPECIES	LC_{50} CONC.
			none	available		

SECTION III PHYSICAL DATA MW - 249.02

BOILING POINT °F.: 287°C | VAPOR PRESSURE mm Hg.

MELTING POINT °F.: 171°C | VAPOR DENSITY (Air = 1)

SPECIFIC GRAVITY (H_2O = 1): 2.27 | EVAPORATION RATE (_____ = 1)

SOLUBILITY IN WATER Pts/100 pts H_2O | VOLATILE % Vol. % Wt.

APPEARANCE AND ODOR: yellow crystals

SECTION IV FIRE AND EXPLOSION HAZARD DATA

FLASH POINT: none available | FLAMMABLE (EXPLOSIVE) LIMITS: UPPER / LOWER

METHOD USED

EXTINGUISHING MEDIA: water spray, carbon dioxide, dry chemical, foam.

SPECIAL FIRE FIGHTING PROCEDURES: Wear self-contained breathing apparatus.

UNUSUAL FIRE AND EXPLOSION HAZARDS

SECTION V HEALTH HAZARD DATA

TOXIC LEVEL: none available | CARCINOGENIC: no

PRINCIPLE ROUTES OF ABSORBTION: inhalation, ingestion, skin | SKIN AND EYE IRRITATION: moderate

REVELANT SYMPTOMS OF EXPOSURE: cyanosis, headache, weakness, anemia.

EFFECTS OF CHRONIC EXPOSURE: hemoglobinuria

EMERGENCY AND FIRST AID PROCEDURES: If swallowed - give emetic, get medical aid; inhalation - oxygen if needed; wash from skin & eyes.

SECTION VI REACTIVITY DATA

CONDITIONS CONTRIBUTING TO INSTABILITY

CONDITIONS CONTRIBUTING TO HAZARDOUS POLYMERIZATION

INCOMPATABILITY (MATERIALS TO AVOID)

HAZARDOUS DECOMPOSITION PRODUCTS

SECTION VII SPILL OR LEAK PROCEDURES

STEPS TO BE TAKEN IN CASE MATERIAL IS RELEASED OR SPILLED

WASTE DISPOSAL METHOD

SECTION VIII SPECIAL PROTECTION INFORMATION

VENTILATION REQUIREMENTS LOCAL EXHAUST | PROTECTIVE EQUIPMENT (SPECIFY TYPES) EYE: Goggles

MECHANICAL (GENERAL) | GLOVES: Rubber

SPECIAL | RESPIRATOR: With proper filter

OTHER PROTECTIVE EQUIPMENT

SECTION IX SPECIAL PRECAUTIONS

PRECAUTIONS TO BE TAKEN IN HANDLING AND STORAGE: Store away from oxidizing materials and sources of heat.

OTHER PRECAUTIONS

SIGNATURE | DATE

NITROMESITYLENE

1 3 0

SECTION I SOURCE AND NOMENCLATURE

MANUFACTURER'S NAME | EMERGENCY TELEPHONE NO.

ADDRESS (NUMBER, STREET, CITY, STATE, ZIP CODE)

TRADE NAME AND SYNONYMS | CHEMICAL FAMILY

CHEMICAL NAME AND SYNONYMS: 2 Nitromesitylene | FORMULA: $NO_2C_6H_2(CH_3)_3$

SECTION II HAZARDOUS INGREDIENTS

BASIC MATERIAL	APPROXIMATE OR MAXIMUM % WT. OR VOL.	ESTABLISHED OSHA STANDARD	LD_{50} ORAL	LD_{50} PERCUT.	LC_{50} SPECIES	LC_{50} CONC.
			none	available		

SECTION III PHYSICAL DATA MW - 165.19

BOILING POINT °F.: 255°C | VAPOR PRESSURE mm Hg.

MELTING POINT °F.: 44°C | VAPOR DENSITY (Air = 1)

SPECIFIC GRAVITY (H_2O = 1) | EVAPORATION RATE (_____ = 1)

SOLUBILITY IN WATER Pts/100 pts H_2O | VOLATILE % Vol. % Wt.

APPEARANCE AND ODOR: crystaline

SECTION IV FIRE AND EXPLOSION HAZARD DATA

FLASH POINT: none available | FLAMMABLE (EXPLOSIVE) LIMITS: UPPER / LOWER

METHOD USED

EXTINGUISHING MEDIA: information not available.

SPECIAL FIRE FIGHTING PROCEDURES

UNUSUAL FIRE AND EXPLOSION HAZARDS

SECTION V HEALTH HAZARD DATA

TOXIC LEVEL: none available | CARCINOGENIC: no

PRINCIPLE ROUTES OF ABSORBTION: inhalation, ingestion, skin | SKIN AND EYE IRRITATION: severe

REVELANT SYMPTOMS OF EXPOSURE: chemical pneumonia, central nervous system depression, frequent bleeding.

EFFECTS OF CHRONIC EXPOSURE: Nervous tension, anxiety, asthmatic bronchitis.

EMERGENCY AND FIRST AID PROCEDURES: Inhalation - oxygen as needed; if swallowed - get medical aid, gastric lavage; wash from skin & eyes.

SECTION VI REACTIVITY DATA

CONDITIONS CONTRIBUTING TO INSTABILITY

CONDITIONS CONTRIBUTING TO HAZARDOUS POLYMERIZATION

INCOMPATABILITY (MATERIALS TO AVOID)

HAZARDOUS DECOMPOSITION PRODUCTS

SECTION VII SPILL OR LEAK PROCEDURES

STEPS TO BE TAKEN IN CASE MATERIAL IS RELEASED OR SPILLED

WASTE DISPOSAL METHOD

SECTION VIII SPECIAL PROTECTION INFORMATION

VENTILATION REQUIREMENTS LOCAL EXHAUST | PROTECTIVE EQUIPMENT (SPECIFY TYPES) EYE: Goggles

MECHANICAL (GENERAL) | GLOVES: Rubber

SPECIAL | RESPIRATOR: With proper filter

OTHER PROTECTIVE EQUIPMENT

SECTION IX SPECIAL PRECAUTIONS

PRECAUTIONS TO BE TAKEN IN HANDLING AND STORAGE

OTHER PRECAUTIONS

SIGNATURE | DATE

p-NITROMETHYLANILINE

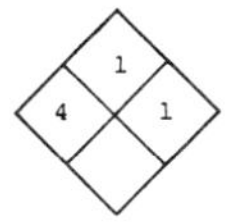

SECTION I SOURCE AND NOMENCLATURE

MANUFACTURER'S NAME	EMERGENCY TELEPHONE NO.
ADDRESS (NUMBER, STREET, CITY, STATE, ZIP CODE)	
TRADE NAME AND SYNONYMS	CHEMICAL FAMILY
CHEMICAL NAME AND SYNONYMS P-Nitro-N-Methylaniline	FORMULA $NO_2C_6H_4NHCH_3$

SECTION II HAZARDOUS INGREDIENTS

BASIC MATERIAL	APPROXIMATE OR MAXIMUM % WT. OR VOL.	ESTABLISHED OSHA STANDARD	LD_{50} ORAL	LD_{50} PERCUT.	LC_{50} SPECIES	LC_{50} CONC.
			none	available		

SECTION III PHYSICAL DATA MW - 152.15

BOILING POINT °F. --	VAPOR PRESSURE mm Hg.
MELTING POINT °F. 152°C	VAPOR DENSITY (Air = 1)
SPECIFIC GRAVITY (H_2O = 1) --	EVAPORATION RATE (____ = 1)
SOLUBILITY IN WATER Pts/100 pts H_2O soluble	VOLATILE % Vol. % Wt.

APPEARANCE AND ODOR
brown-yellow crystals.

SECTION IV FIRE AND EXPLOSION HAZARD DATA

FLASH POINT	FLAMMABLE (EXPLOSIVE) LIMITS	UPPER
none available METHOD USED		LOWER

EXTINGUISHING MEDIA
information not available.

SPECIAL FIRE FIGHTING PROCEDURES

UNUSUAL FIRE AND EXPLOSION HAZARDS

SECTION V HEALTH HAZARD DATA

TOXIC LEVEL none available — CARCINOGENIC no

PRINCIPLE ROUTES OF ABSORBTION inhalation, ingestion, skin — SKIN AND EYE IRRITATION moderate

REVELANT SYMPTOMS OF EXPOSURE Headache, drowsiness, cyanosis, mental confusion, convulsions.

EFFECTS OF CHRONIC EXPOSURE Fatigue, loss of appetite, headache, dizzinesss, blood damage, central nervous system damage.

EMERGENCY AND FIRST AID PROCEDURES If swallowed - give emetic, get medical aid; inhalation - rest; wash from skin & eyes.

SECTION VI REACTIVITY DATA

CONDITIONS CONTRIBUTING TO INSTABILITY

CONDITIONS CONTRIBUTING TO HAZARDOUS POLYMERIZATION

INCOMPATABILITY (MATERIALS TO AVOID)

HAZARDOUS DECOMPOSITION PRODUCTS

SECTION VII SPILL OR LEAK PROCEDURES

STEPS TO BE TAKEN IN CASE MATERIAL IS RELEASED OR SPILLED

WASTE DISPOSAL METHOD

SECTION VIII SPECIAL PROTECTION INFORMATION

VENTILATION REQUIREMENTS LOCAL EXHAUST	PROTECTIVE EQUIPMENT (SPECIFY TYPES) EYE goggles
MECHANICAL (GENERAL)	GLOVES neoprene rubber
SPECIAL	RESPIRATOR with proper filter
OTHER PROTECTIVE EQUIPMENT	

SECTION IX SPECIAL PRECAUTIONS

PRECAUTIONS TO BE TAKEN IN HANDLING AND STORAGE

OTHER PRECAUTIONS

SIGNATURE	DATE

5-NITRO-1-NAPHTHYLAMINE

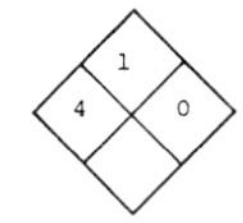

SECTION I SOURCE AND NOMENCLATURE

MANUFACTURER'S NAME	EMERGENCY TELEPHONE NO.
ADDRESS (NUMBER, STREET, CITY, STATE, ZIP CODE)	
TRADE NAME AND SYNONYMS	CHEMICAL FAMILY
CHEMICAL NAME AND SYNONYMS 5 Nitro α -Naphthylamine	FORMULA $NO_2C_{10}H_6NH_2$

SECTION II HAZARDOUS INGREDIENTS

BASIC MATERIAL	APPROXIMATE OR MAXIMUM % WT. OR VOL.	ESTABLISHED OSHA STANDARD	LD_{50} ORAL	LD_{50} PERCUT.	LC_{50} SPECIES	LC_{50} CONC.
			none	available		

SECTION III PHYSICAL DATA MW - 188.19

BOILING POINT °F. --	VAPOR PRESSURE mm Hg.
MELTING POINT °F. 119°C	VAPOR DENSITY (Air = 1)
SPECIFIC GRAVITY (H_2O = 1)	EVAPORATION RATE (____ = 1)
SOLUBILITY IN WATER Pts/100 pts H_2O	VOLATILE % Vol. % Wt.

APPEARANCE AND ODOR
red needle-like crystals.

SECTION IV FIRE AND EXPLOSION HAZARD DATA

FLASH POINT	FLAMMABLE (EXPLOSIVE) LIMITS	UPPER
none available METHOD USED		LOWER

EXTINGUISHING MEDIA
information not available.

SPECIAL FIRE FIGHTING PROCEDURES

UNUSUAL FIRE AND EXPLOSION HAZARDS

SECTION V HEALTH HAZARD DATA

TOXIC LEVEL none available — CARCINOGENIC suspected

PRINCIPLE ROUTES OF ABSORBTION inhalation, ingestion, skin — SKIN AND EYE IRRITATION moderate

REVELANT SYMPTOMS OF EXPOSURE Cyanosis, headache, nausea, vomiting, weakness, stupor.

EFFECTS OF CHRONIC EXPOSURE Liver damage, possible bladder tumors.

EMERGENCY AND FIRST AID PROCEDURES If swallowed - give emetic, get medical aid; inhalation - rest; wash from skin & eyes.

SECTION VI REACTIVITY DATA

CONDITIONS CONTRIBUTING TO INSTABILITY

CONDITIONS CONTRIBUTING TO HAZARDOUS POLYMERIZATION

INCOMPATABILITY (MATERIALS TO AVOID)

HAZARDOUS DECOMPOSITION PRODUCTS

SECTION VII SPILL OR LEAK PROCEDURES

STEPS TO BE TAKEN IN CASE MATERIAL IS RELEASED OR SPILLED

WASTE DISPOSAL METHOD

SECTION VIII SPECIAL PROTECTION INFORMATION

VENTILATION REQUIREMENTS LOCAL EXHAUST	PROTECTIVE EQUIPMENT (SPECIFY TYPES) EYE goggles
MECHANICAL (GENERAL)	GLOVES rubber
SPECIAL	RESPIRATOR with proper filter
OTHER PROTECTIVE EQUIPMENT	

SECTION IX SPECIAL PRECAUTIONS

PRECAUTIONS TO BE TAKEN IN HANDLING AND STORAGE

OTHER PRECAUTIONS

SIGNATURE	DATE

m-NITROPHENOL

Hazard diamond: 1 (top), 3 (left), 0 (right)

SECTION I SOURCE AND NOMENCLATURE

MANUFACTURER'S NAME | EMERGENCY TELEPHONE NO.

ADDRESS (NUMBER, STREET, CITY, STATE, ZIP CODE)

TRADE NAME AND SYNONYMS | CHEMICAL FAMILY

CHEMICAL NAME AND SYNONYMS: M-Nitrophenol | FORMULA: $NO_2C_6H_4OH$

SECTION II HAZARDOUS INGREDIENTS

BASIC MATERIAL	APPROXIMATE OR MAXIMUM % WT. OR VOL.	ESTABLISHED OSHA STANDARD	LD_{50} mg/kg ORAL	PERCUT.	SPECIES	LC_{50} CONC.
			447		Rat	
			1414		Mouse	

SECTION III PHYSICAL DATA MW 139.12

BOILING POINT °F. | VAPOR PRESSURE at 194° C. 70 mm Hg.

MELTING POINT °F.: 97°C. | VAPOR DENSITY (Air = 1)

SPECIFIC GRAVITY (H_2O = 1): 1.49 | EVAPORATION RATE (____ = 1)

SOLUBILITY IN WATER Pts/100 pts H_2O: 1.35 | VOLATILE % Vol. % Wt.

APPEARANCE AND ODOR: Monoclinic prism crystals

SECTION IV FIRE AND EXPLOSION HAZARD DATA

FLASH POINT: None available | FLAMMABLE (EXPLOSIVE) LIMITS: UPPER / LOWER

METHOD USED

EXTINGUISHING MEDIA: Use water to cool down containers.

SPECIAL FIRE FIGHTING PROCEDURES: Wear full protective clothing, self-contained breathing apparatus.

UNUSUAL FIRE AND EXPLOSION HAZARDS: When heated to decomposition gives off toxic nitrogen oxide fumes.

SECTION V HEALTH HAZARD DATA

TOXIC LEVEL: None Available | CARCINOGENIC: No

PRINCIPLE ROUTES OF ABSORBTION: Inhalation, ingestion, skin | SKIN AND EYE IRRITATION: Moderate

REVELANT SYMPTOMS OF EXPOSURE: Eye and mucous membrane irritation, C.N.S. depressant.

EFFECTS OF CHRONIC EXPOSURE: Forms methemoglobin, dermatitis.

EMERGENCY AND FIRST AID PROCEDURES: Inhalation - rest; wash from skin & eyes. If swallowed - remove by gastric lavage or emesis.

SECTION VI REACTIVITY DATA

CONDITIONS CONTRIBUTING TO INSTABILITY

CONDITIONS CONTRIBUTING TO HAZARDOUS POLYMERIZATION

INCOMPATABILITY (MATERIALS TO AVOID): Keep from oxidizing materials.

HAZARDOUS DECOMPOSITION PRODUCTS

SECTION VII SPILL OR LEAK PROCEDURES

STEPS TO BE TAKEN IN CASE MATERIAL IS RELEASED OR SPILLED

WASTE DISPOSAL METHOD

SECTION VIII SPECIAL PROTECTION INFORMATION

VENTILATION REQUIREMENTS LOCAL EXHAUST | PROTECTIVE EQUIPMENT (SPECIFY TYPES) EYE: Goggles

MECHANICAL (GENERAL) | GLOVES: Butyl Rubber

SPECIAL | RESPIRATOR: Self-contained

OTHER PROTECTIVE EQUIPMENT: Protective clothing and shoes.

SECTION IX SPECIAL PRECAUTIONS

PRECAUTIONS TO BE TAKEN IN HANDLING AND STORAGE: Protect from physical damage. Separate from combustibles and other organic materials.

OTHER PRECAUTIONS

SIGNATURE | DATE

o-NITROPHENOL SODIUM SALT

Hazard diamond: 1 (top), 3 (left), 0 (right)

SECTION I SOURCE AND NOMENCLATURE

MANUFACTURER'S NAME | EMERGENCY TELEPHONE NO.

ADDRESS (NUMBER, STREET, CITY, STATE, ZIP CODE)

TRADE NAME AND SYNONYMS | CHEMICAL FAMILY

CHEMICAL NAME AND SYNONYMS: Sodium-O-Nitrophenol | FORMULA: $NO_2C_6H_4O$ Na

SECTION II HAZARDOUS INGREDIENTS

BASIC MATERIAL	APPROXIMATE OR MAXIMUM % WT. OR VOL.	ESTABLISHED OSHA STANDARD	LD_{50} ORAL	PERCUT.	SPECIES	LC_{50} CONC.
			None available			

SECTION III PHYSICAL DATA MW-161.10

BOILING POINT °F.: - | VAPOR PRESSURE mm Hg.

MELTING POINT °F.: - | VAPOR DENSITY (Air = 1)

SPECIFIC GRAVITY (H_2O = 1): - | EVAPORATION RATE (____ = 1)

SOLUBILITY IN WATER Pts/100 pts H_2O: Very soluble | VOLATILE % Vol. % Wt.

APPEARANCE AND ODOR: Reddish crystals.

SECTION IV FIRE AND EXPLOSION HAZARD DATA

FLASH POINT: None available | FLAMMABLE (EXPLOSIVE) LIMITS: UPPER / LOWER

METHOD USED

EXTINGUISHING MEDIA: Use water to wet down containers.

SPECIAL FIRE FIGHTING PROCEDURES: Wear full protective clothing; self-contained breathing apparatus.

UNUSUAL FIRE AND EXPLOSION HAZARDS: When heated to decomposition give off toxic nitrogen oxide fumes.

SECTION V HEALTH HAZARD DATA

TOXIC LEVEL: None available | CARCINOGENIC: No

PRINCIPLE ROUTES OF ABSORBTION: Inhalation, ingestion, skin | SKIN AND EYE IRRITATION: Slight

REVELANT SYMPTOMS OF EXPOSURE: Eye and mucous membrane irritation, CNS depressant.

EFFECTS OF CHRONIC EXPOSURE: Forms methemoglobin, dermatitis

EMERGENCY AND FIRST AID PROCEDURES: Inhalation - rest; wash from eyes & skin; if swallowed - remove by gastric lavage or emesis.

SECTION VI REACTIVITY DATA

CONDITIONS CONTRIBUTING TO INSTABILITY

CONDITIONS CONTRIBUTING TO HAZARDOUS POLYMERIZATION

INCOMPATABILITY (MATERIALS TO AVOID): Keep from oxidizing materials.

HAZARDOUS DECOMPOSITION PRODUCTS

SECTION VII SPILL OR LEAK PROCEDURES

STEPS TO BE TAKEN IN CASE MATERIAL IS RELEASED OR SPILLED

WASTE DISPOSAL METHOD

SECTION VIII SPECIAL PROTECTION INFORMATION

VENTILATION REQUIREMENTS LOCAL EXHAUST | PROTECTIVE EQUIPMENT (SPECIFY TYPES) EYE: Goggles

MECHANICAL (GENERAL) | GLOVES: Butyl Rubber

SPECIAL | RESPIRATOR: Self-contained

OTHER PROTECTIVE EQUIPMENT: Protective clothing

SECTION IX SPECIAL PRECAUTIONS

PRECAUTIONS TO BE TAKEN IN HANDLING AND STORAGE: Keep from combustible materials.

OTHER PRECAUTIONS

SIGNATURE | DATE

p-NITROPHENYLHYDRAZINE

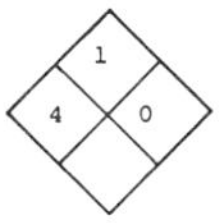

SECTION I SOURCE AND NOMENCLATURE

MANUFACTURER'S NAME	EMERGENCY TELEPHONE NO.
ADDRESS (NUMBER, STREET, CITY, STATE, ZIP CODE)	
TRADE NAME AND SYNONYMS	CHEMICAL FAMILY
CHEMICAL NAME AND SYNONYMS P-Nitrophenylhydrazine	FORMULA $NO_2C_6H_4NHNH_2$

SECTION II HAZARDOUS INGREDIENTS

BASIC MATERIAL	APPROXIMATE OR MAXIMUM % WT. OR VOL.	ESTABLISHED OSHA STANDARD	LD_{50} ORAL	LD_{50} PERCUT.	LC_{50} SPECIES	LC_{50} CONC.
Note:						
Lowest published lethal dose						
Interperitoneal - mouse	250/mg/kg					

SECTION III PHYSICAL DATA MW - 153.14

BOILING POINT °F.	VAPOR PRESSURE mm Hg.
MELTING POINT °F. 157°C	VAPOR DENSITY (Air = 1)
SPECIFIC GRAVITY (H_2O = 1)	EVAPORATION RATE (____ = 1)
SOLUBILITY IN WATER Pts/100 pts H_2O soluble in hot water	VOLATILE % Vol. % Wt.

APPEARANCE AND ODOR
orange-red crystals

SECTION IV FIRE AND EXPLOSION HAZARD DATA

FLASH POINT none available	FLAMMABLE (EXPLOSIVE) LIMITS	UPPER
METHOD USED		LOWER

EXTINGUISHING MEDIA
information not available.

SPECIAL FIRE FIGHTING PROCEDURES

UNUSUAL FIRE AND EXPLOSION HAZARDS

SECTION V HEALTH HAZARD DATA

TOXIC LEVEL none available — CARCINOGENIC no

PRINCIPLE ROUTES OF ABSORBTION Inhalation, ingestion, skin — SKIN AND EYE IRRITATION Severe

REVELANT SYMPTOMS OF EXPOSURE Irritation of skin, eyes and mucous membranes, gastroentric disturbances, anemia.

EFFECTS OF CHRONIC EXPOSURE Blood and liver damage, loss of appetite.

EMERGENCY AND FIRST AID PROCEDURES If swallowed - induce vomiting, get medical aid; inhalation - rest; wash from skin & eyes.

SECTION VI REACTIVITY DATA

CONDITIONS CONTRIBUTING TO INSTABILITY

CONDITIONS CONTRIBUTING TO HAZARDOUS POLYMERIZATION

INCOMPATABILITY (MATERIALS TO AVOID)

HAZARDOUS DECOMPOSITION PRODUCTS

SECTION VII SPILL OR LEAK PROCEDURES

STEPS TO BE TAKEN IN CASE MATERIAL IS RELEASED OR SPILLED

WASTE DISPOSAL METHOD

SECTION VIII SPECIAL PROTECTION INFORMATION

VENTILATION REQUIREMENTS LOCAL EXHAUST	PROTECTIVE EQUIPMENT (SPECIFY TYPES) EYE Goggles
MECHANICAL (GENERAL)	GLOVES Rubber
SPECIAL	RESPIRATOR Self-contained
OTHER PROTECTIVE EQUIPMENT	

SECTION IX SPECIAL PRECAUTIONS

PRECAUTIONS TO BE TAKEN IN HANDLING AND STORAGE

OTHER PRECAUTIONS

SIGNATURE	DATE

p-NITROPHENYLHYDRAZINE HYDROCHLORIDE

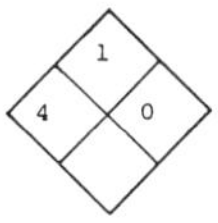

SECTION I SOURCE AND NOMENCLATURE

MANUFACTURER'S NAME	EMERGENCY TELEPHONE NO.
ADDRESS (NUMBER, STREET, CITY, STATE, ZIP CODE)	
TRADE NAME AND SYNONYMS	CHEMICAL FAMILY
CHEMICAL NAME AND SYNONYMS P-Nitrophenylhydrazine Hydrochloride	FORMULA $NO_2C_6H_4NHNH_2HCl$

SECTION II HAZARDOUS INGREDIENTS

BASIC MATERIAL	APPROXIMATE OR MAXIMUM % WT. OR VOL.	ESTABLISHED OSHA STANDARD	LD_{50} ORAL	LD_{50} PERCUT.	LC_{50} SPECIES	LC_{50} CONC.
			none	available		

SECTION III PHYSICAL DATA MW - 189.61

BOILING POINT °F.	VAPOR PRESSURE mm Hg.
MELTING POINT °F. 215°C	VAPOR DENSITY (Air = 1)
SPECIFIC GRAVITY (H_2O = 1)	EVAPORATION RATE (____ = 1)
SOLUBILITY IN WATER Pts/100 pts H_2O	VOLATILE % Vol. % Wt.

APPEARANCE AND ODOR
orange-red crystals.

SECTION IV FIRE AND EXPLOSION HAZARD DATA

FLASH POINT none available	FLAMMABLE (EXPLOSIVE) LIMITS	UPPER
METHOD USED		LOWER

EXTINGUISHING MEDIA
information not available.

SPECIAL FIRE FIGHTING PROCEDURES

UNUSUAL FIRE AND EXPLOSION HAZARDS

SECTION V HEALTH HAZARD DATA

TOXIC LEVEL none available — CARCINOGENIC no

PRINCIPLE ROUTES OF ABSORBTION inhalation, ingestion, skin — SKIN AND EYE IRRITATION severe

REVELANT SYMPTOMS OF EXPOSURE Irritation of eyes, skin & mucous membranes, gastroentric disturbances, anemia.

EFFECTS OF CHRONIC EXPOSURE Blood and liver damage, loss of appetite.

EMERGENCY AND FIRST AID PROCEDURES Inhalation - rest; if swallowed - induce vomiting, get medical aid; wash from skin & eyes.

SECTION VI REACTIVITY DATA

CONDITIONS CONTRIBUTING TO INSTABILITY

CONDITIONS CONTRIBUTING TO HAZARDOUS POLYMERIZATION

INCOMPATABILITY (MATERIALS TO AVOID)

HAZARDOUS DECOMPOSITION PRODUCTS

SECTION VII SPILL OR LEAK PROCEDURES

STEPS TO BE TAKEN IN CASE MATERIAL IS RELEASED OR SPILLED

WASTE DISPOSAL METHOD

SECTION VIII SPECIAL PROTECTION INFORMATION

VENTILATION REQUIREMENTS LOCAL EXHAUST	PROTECTIVE EQUIPMENT (SPECIFY TYPES) EYE Goggles
MECHANICAL (GENERAL)	GLOVES Rubber
SPECIAL	RESPIRATOR Self-contained
OTHER PROTECTIVE EQUIPMENT	

SECTION IX SPECIAL PRECAUTIONS

PRECAUTIONS TO BE TAKEN IN HANDLING AND STORAGE

OTHER PRECAUTIONS

SIGNATURE	DATE

NITROPHTHALIC ANHYDRIDE

Hazard diamond: 1 (top), 4 (left), 0 (right)

SECTION I SOURCE AND NOMENCLATURE

MANUFACTURER'S NAME | EMERGENCY TELEPHONE NO.

ADDRESS (NUMBER, STREET, CITY, STATE, ZIP CODE)

TRADE NAME AND SYNONYMS | CHEMICAL FAMILY

CHEMICAL NAME AND SYNONYMS: Nitro-3-phthalic Anhydride

FORMULA: $NO_2C_6H_3(CO)_2O$

SECTION II HAZARDOUS INGREDIENTS

BASIC MATERIAL	APPROXIMATE OR MAXIMUM % WT. OR VOL.	ESTABLISHED OSHA STANDARD	LD_{50} ORAL	LD_{50} PERCUT.	LC_{50} SPECIES	LC_{50} CONC.
			None available			

SECTION III PHYSICAL DATA MW - 193.11

BOILING POINT °F. | VAPOR PRESSURE mm Hg.

MELTING POINT: 163°C. | VAPOR DENSITY (Air = 1)

SPECIFIC GRAVITY (H_2O = 1) | EVAPORATION RATE (____ = 1)

SOLUBILITY IN WATER Pts/100 pts H_2O: Soluble in hot water | VOLATILE % Vol. % Wt.

APPEARANCE AND ODOR: Needle-like crystals.

SECTION IV FIRE AND EXPLOSION HAZARD DATA

FLASH POINT: None available | FLAMMABLE (EXPLOSIVE) LIMITS: UPPER / LOWER

METHOD USED

EXTINGUISHING MEDIA: Water, carbon dioxide, dry chemical

SPECIAL FIRE FIGHTING PROCEDURES: Foam or direct water spray on molten nitrophthalic anhydride may cause foaming or spattering.

UNUSUAL FIRE AND EXPLOSION HAZARDS

SECTION V HEALTH HAZARD DATA

TOXIC LEVEL: None available | CARCINOGENIC: No

PRINCIPLE ROUTES OF ABSORBTION: Inhalation, ingestion | SKIN AND EYE IRRITATION: Severe

REVELANT SYMPTOMS OF EXPOSURE: Coughing, sneezing, burning in respiratory system, dermatitis, skin irritation, if swallowed - digestive tract irritation.

EFFECTS OF CHRONIC EXPOSURE: Bronchitis, eye irritation, inflammation of respiratory tract; bloody nasal discharge.

EMERGENCY AND FIRST AID PROCEDURES: Inhalation - rest; if swallowed - give milk of magnesia, get medical aid; wash skin & eyes.

SECTION VI REACTIVITY DATA

CONDITIONS CONTRIBUTING TO INSTABILITY

CONDITIONS CONTRIBUTING TO HAZARDOUS POLYMERIZATION

INCOMPATABILITY (MATERIALS TO AVOID)

HAZARDOUS DECOMPOSITION PRODUCTS

SECTION VII SPILL OR LEAK PROCEDURES

STEPS TO BE TAKEN IN CASE MATERIAL IS RELEASED OR SPILLED: Cover with soda ash, mix and add water.

WASTE DISPOSAL METHOD

SECTION VIII SPECIAL PROTECTION INFORMATION

VENTILATION REQUIREMENTS LOCAL EXHAUST | PROTECTIVE EQUIPMENT (SPECIFY TYPES) EYE: Goggles

MECHANICAL (GENERAL) | GLOVES: Rubber

SPECIAL | RESPIRATOR: Self-contained breathing apparatus

OTHER PROTECTIVE EQUIPMENT

SECTION IX SPECIAL PRECAUTIONS

PRECAUTIONS TO BE TAKEN IN HANDLING AND STORAGE: Protect against physical damage, store in cool, well-ventilated location.

OTHER PRECAUTIONS

SIGNATURE | DATE

1-NITROPROPANE

Hazard diamond: 2 (top), 3 (left), 3 (right)

SECTION I SOURCE AND NOMENCLATURE

MANUFACTURER'S NAME | EMERGENCY TELEPHONE NO.

ADDRESS (NUMBER, STREET, CITY, STATE, ZIP CODE)

TRADE NAME AND SYNONYMS | CHEMICAL FAMILY

CHEMICAL NAME AND SYNONYMS: 1-Nitropropane

FORMULA: $C_3H_7NO_2$

SECTION II HAZARDOUS INGREDIENTS

BASIC MATERIAL	APPROXIMATE OR MAXIMUM % WT. OR VOL.	ESTABLISHED OSHA STANDARD	LD_{50} mg/kg ORAL	LD_{50} PERCUT.	LC_{50} SPECIES	LC_{50} CONC.
Oral - human LDLO 500 mg/kg			800		mouse	
LDLO - oral - rat 1000 mg/kg.						
Inhalation - human - TCLO 150 ppm		irritant				

SECTION III PHYSICAL DATA MW - 89.09

BOILING POINT: 132°C | VAPOR PRESSURE at 25°C: 13 mm Hg.

MELTING POINT: -108°C | VAPOR DENSITY (Air = 1): 3.06

SPECIFIC GRAVITY (H_2O = 1): 1.00 | EVAPORATION RATE (____ = 1)

SOLUBILITY IN WATER Pts/100 pts H_2O: 9.5 | VOLATILE % Vol. % Wt.

APPEARANCE AND ODOR: colorless liquid, with disagreeable odor.

SECTION IV FIRE AND EXPLOSION HAZARD DATA

FLASH POINT: 120°F | FLAMMABLE (EXPLOSIVE) LIMITS: UPPER / LOWER

METHOD USED: T.O.C.

EXTINGUISHING MEDIA: Alcohol foam, carbon dioxide, dry chemical or water spray.

SPECIAL FIRE FIGHTING PROCEDURES: Wear protective clothing and self-contained breathing apparatus.

UNUSUAL FIRE AND EXPLOSION HAZARDS

SECTION V HEALTH HAZARD DATA

TOXIC LEVEL: TLV - 25 PPM, 90 mg/C.M. | CARCINOGENIC: no

PRINCIPLE ROUTES OF ABSORBTION: Inhalation, ingestion | SKIN AND EYE IRRITATION: moderate

REVELANT SYMPTOMS OF EXPOSURE: Headache, dizziness, nausea, vomiting, diarrhea, restlessness, muscular uncoordination.

EFFECTS OF CHRONIC EXPOSURE: Liver, kidney and heart damage.

EMERGENCY AND FIRST AID PROCEDURES: If swallowed - induce vomiting, get medical aid, gastric lavage; inhalation - artificial respiration, oxygen as needed, wash skin & eyes.

SECTION VI REACTIVITY DATA

CONDITIONS CONTRIBUTING TO INSTABILITY

CONDITIONS CONTRIBUTING TO HAZARDOUS POLYMERIZATION

INCOMPATABILITY (MATERIALS TO AVOID): Unstable with amines, acids, alkalis.

HAZARDOUS DECOMPOSITION PRODUCTS

SECTION VII SPILL OR LEAK PROCEDURES

STEPS TO BE TAKEN IN CASE MATERIAL IS RELEASED OR SPILLED: After mixing spill with soda ash, spray with water, neutralize with

WASTE DISPOSAL METHOD

SECTION VIII SPECIAL PROTECTION INFORMATION

VENTILATION REQUIREMENTS LOCAL EXHAUST | PROTECTIVE EQUIPMENT (SPECIFY TYPES) EYE: Goggles

MECHANICAL (GENERAL) | GLOVES: Rubber

SPECIAL | RESPIRATOR: Self-contained

OTHER PROTECTIVE EQUIPMENT: Plastic working clothing

SECTION IX SPECIAL PRECAUTIONS

PRECAUTIONS TO BE TAKEN IN HANDLING AND STORAGE: Out side or detached storage preferred, store inside in standard flammable liquids cabinet, isolate from other flammable liquids.

OTHER PRECAUTIONS: Avoid contact with $Ca(OH)_2$, NaOH, KOH, hydrocarbons.

SIGNATURE | DATE

2-NITRORESORCINOL

1 / 4 / 0

SECTION I SOURCE AND NOMENCLATURE

MANUFACTURER'S NAME	EMERGENCY TELEPHONE NO.
ADDRESS (NUMBER, STREET, CITY, STATE, ZIP CODE)	
TRADE NAME AND SYNONYMS	CHEMICAL FAMILY
CHEMICAL NAME AND SYNONYMS 2-Nitroresorcinol	FORMULA $NO_2C_6H_3(OH)_2$

SECTION II HAZARDOUS INGREDIENTS

BASIC MATERIAL	APPROXIMATE OR MAXIMUM % WT. OR VOL.	ESTABLISHED OSHA STANDARD	LD_{50} ORAL	LD_{50} PERCUT.	LC_{50} SPECIES	LC_{50} CONC.
			none available			

SECTION III PHYSICAL DATA MW - 155.11

BOILING POINT °F. d>180°C	VAPOR PRESSURE mm Hg.
MELTING POINT °F. 83°C	VAPOR DENSITY (Air = 1)
SPECIFIC GRAVITY (H_2O = 1) --	EVAPORATION RATE (____ = 1)
SOLUBILITY IN WATER Pts/100 pts H_2O --	VOLATILE % Vol. % Wt.

APPEARANCE AND ODOR crystals

SECTION IV FIRE AND EXPLOSION HAZARD DATA

FLASH POINT none available	FLAMMABLE (EXPLOSIVE) LIMITS	UPPER
METHOD USED		LOWER

EXTINGUISHING MEDIA information not available

SPECIAL FIRE FIGHTING PROCEDURES

UNUSUAL FIRE AND EXPLOSION HAZARDS

SECTION V HEALTH HAZARD DATA

TOXIC LEVEL none available — CARCINOGENIC no

PRINCIPLE ROUTES OF ABSORPTION inhalation, ingestion, skin — SKIN AND EYE IRRITATION moderate

REVELANT SYMPTOMS OF EXPOSURE Itching, dermatitis, restlessness, cyanosis, dizziness, tremors, convulsion and death.

EFFECTS OF CHRONIC EXPOSURE Liver, kidney and lung damage.

EMERGENCY AND FIRST AID PROCEDURES If swallowed - induce vomiting, get medical aid; inhalation - rest; wash skin & eyes.

SECTION VI REACTIVITY DATA

CONDITIONS CONTRIBUTING TO INSTABILITY

CONDITIONS CONTRIBUTING TO HAZARDOUS POLYMERIZATION

INCOMPATABILITY (MATERIALS TO AVOID)

HAZARDOUS DECOMPOSITION PRODUCTS

SECTION VII SPILL OR LEAK PROCEDURES

STEPS TO BE TAKEN IN CASE MATERIAL IS RELEASED OR SPILLED

WASTE DISPOSAL METHOD

SECTION VIII SPECIAL PROTECTION INFORMATION

VENTILATION REQUIREMENTS LOCAL EXHAUST	PROTECTIVE EQUIPMENT (SPECIFY TYPES) EYE Goggles
MECHANICAL (GENERAL)	GLOVES Rubber
SPECIAL	RESPIRATOR

OTHER PROTECTIVE EQUIPMENT

SECTION IX SPECIAL PRECAUTIONS

PRECAUTIONS TO BE TAKEN IN HANDLING AND STORAGE

OTHER PRECAUTIONS

SIGNATURE	DATE

p-NITROSODIETHYLANILINE

2 / 4 / 1

SECTION I SOURCE AND NOMENCLATURE

MANUFACTURER'S NAME	EMERGENCY TELEPHONE NO.
ADDRESS (NUMBER, STREET, CITY, STATE, ZIP CODE)	
TRADE NAME AND SYNONYMS N,N-Diethyl-P-Nitrosoaniline	CHEMICAL FAMILY
CHEMICAL NAME AND SYNONYMS P-Nitrosodiethyaniline	FORMULA $(C_2H_5)_2NC_6H_4NO$

SECTION II HAZARDOUS INGREDIENTS

BASIC MATERIAL	APPROXIMATE OR MAXIMUM % WT. OR VOL.	ESTABLISHED OSHA STANDARD	LD_{50} ORAL	LD_{50} PERCUT.	LC_{50} SPECIES	LC_{50} CONC.
			none available			

SECTION III PHYSICAL DATA MW - 178.23

BOILING POINT °F.	VAPOR PRESSURE mm Hg.
MELTING POINT °F. 84°C	VAPOR DENSITY (Air = 1)
SPECIFIC GRAVITY (H_2O = 1) 1.24	EVAPORATION RATE (____ = 1)
SOLUBILITY IN WATER Pts/100 pts H_2O very slightly soluble	VOLATILE % Vol. % Wt.

APPEARANCE AND ODOR green crystals

SECTION IV FIRE AND EXPLOSION HAZARD DATA

FLASH POINT none available	FLAMMABLE (EXPLOSIVE) LIMITS	UPPER
METHOD USED		LOWER

EXTINGUISHING MEDIA Alcohol foam, water spray, carbon dioxide and dry chemical.

SPECIAL FIRE FIGHTING PROCEDURES

UNUSUAL FIRE AND EXPLOSION HAZARDS

SECTION V HEALTH HAZARD DATA

TOXIC LEVEL none available — CARCINOGENIC no

PRINCIPLE ROUTES OF ABSORPTION inhalation, ingestion, skin — SKIN AND EYE IRRITATION moderate

REVELANT SYMPTOMS OF EXPOSURE Weakness, tremors, cyanosis, drowsiness, dermatitis.

EFFECTS OF CHRONIC EXPOSURE Fatigue, loss of appetite, headache, dizziness, dermatitis.

EMERGENCY AND FIRST AID PROCEDURES If swallowed - give emetic, get medical aid; inhalation - rest; wash skin & eyes.

SECTION VI REACTIVITY DATA

CONDITIONS CONTRIBUTING TO INSTABILITY

CONDITIONS CONTRIBUTING TO HAZARDOUS POLYMERIZATION

INCOMPATABILITY (MATERIALS TO AVOID)

HAZARDOUS DECOMPOSITION PRODUCTS

SECTION VII SPILL OR LEAK PROCEDURES

STEPS TO BE TAKEN IN CASE MATERIAL IS RELEASED OR SPILLED

WASTE DISPOSAL METHOD

SECTION VIII SPECIAL PROTECTION INFORMATION

VENTILATION REQUIREMENTS LOCAL EXHAUST	PROTECTIVE EQUIPMENT (SPECIFY TYPES) EYE Goggles
MECHANICAL (GENERAL)	GLOVES Rubber
SPECIAL	RESPIRATOR With proper filter

OTHER PROTECTIVE EQUIPMENT

SECTION IX SPECIAL PRECAUTIONS

PRECAUTIONS TO BE TAKEN IN HANDLING AND STORAGE

OTHER PRECAUTIONS

SIGNATURE	DATE

p-NITROSODIMETHYLANILINE

(Hazard diamond: 2, 4, 1)

SECTION I SOURCE AND NOMENCLATURE

MANUFACTURER'S NAME:
EMERGENCY TELEPHONE NO.:
ADDRESS (NUMBER, STREET, CITY, STATE, ZIP CODE):
TRADE NAME AND SYNONYMS:
CHEMICAL FAMILY:
CHEMICAL NAME AND SYNONYMS: P-Nitrosodimethylaniline
FORMULA: $(CH_3)_2NC_6H_4NO$

SECTION II HAZARDOUS INGREDIENTS

BASIC MATERIAL	APPROXIMATE OR MAXIMUM % WT. OR VOL.	ESTABLISHED OSHA STANDARD	LD_{50} mg/kg. ORAL	PERCUT.	SPECIES	LC_{50} CONC.
Note:			65		rat	
Lowest published lethal dose						
Oral - Mammal 650 mg/kg.						

SECTION III PHYSICAL DATA MW - 150.14

BOILING POINT: °F.
VAPOR PRESSURE: mm Hg.
MELTING POINT: 87°C
VAPOR DENSITY (Air = 1):
SPECIFIC GRAVITY (H_2O = 1):
EVAPORATION RATE (= 1):
SOLUBILITY IN WATER Pts/100 pts H_2O: insoluble
VOLATILE % Vol. % Wt.:
APPEARANCE AND ODOR: green leaflets

SECTION IV FIRE AND EXPLOSION HAZARD DATA

FLASH POINT: none available
METHOD USED:
FLAMMABLE (EXPLOSIVE) LIMITS: UPPER / LOWER
EXTINGUISHING MEDIA: Alcohol foam, dry chemical, carbon dioxide and water spray.
SPECIAL FIRE FIGHTING PROCEDURES:
UNUSUAL FIRE AND EXPLOSION HAZARDS:

SECTION V HEALTH HAZARD DATA

TOXIC LEVEL: none available
CARCINOGENIC: neoplastic
PRINCIPLE ROUTES OF ABSORBTION: Inhalation, ingestion, skin
SKIN AND EYE IRRITATION: moderate
REVELANT SYMPTOMS OF EXPOSURE: Weakness, tremors, cyanosis, drowsiness, dermatitis.
EFFECTS OF CHRONIC EXPOSURE: Fatigue, loss of appetite, headache, dizziness, dermatitis.
EMERGENCY AND FIRST AID PROCEDURES: Inhalation - rest; if swallowed - give emetic, get medical aid; wash from skin & eyes.

SECTION VI REACTIVITY DATA

CONDITIONS CONTRIBUTING TO INSTABILITY:
CONDITIONS CONTRIBUTING TO HAZARDOUS POLYMERIZATION:
INCOMPATABILITY (MATERIALS TO AVOID):
HAZARDOUS DECOMPOSITION PRODUCTS:

SECTION VII SPILL OR LEAK PROCEDURES

STEPS TO BE TAKEN IN CASE MATERIAL IS RELEASED OR SPILLED:
WASTE DISPOSAL METHOD:

SECTION VIII SPECIAL PROTECTION INFORMATION

VENTILATION REQUIREMENTS LOCAL EXHAUST:
PROTECTIVE EQUIPMENT (SPECIFY TYPES) EYE: Goggles
MECHANICAL (GENERAL):
GLOVES: Rubber
SPECIAL:
RESPIRATOR: With proper filter
OTHER PROTECTIVE EQUIPMENT:

SECTION IX SPECIAL PRECAUTIONS

PRECAUTIONS TO BE TAKEN IN HANDLING AND STORAGE:
OTHER PRECAUTIONS:
SIGNATURE:
DATE:

2-NITROSO-1-NAPHTHOL

(Hazard diamond: 1, 4, 0)

SECTION I SOURCE AND NOMENCLATURE

MANUFACTURER'S NAME:
EMERGENCY TELEPHONE NO.:
ADDRESS (NUMBER, STREET, CITY, STATE, ZIP CODE):
TRADE NAME AND SYNONYMS:
CHEMICAL FAMILY:
CHEMICAL NAME AND SYNONYMS: 2-Nitroso-1-Naphthol
FORMULA: $HOC_{10}H_6NO$

SECTION II HAZARDOUS INGREDIENTS

BASIC MATERIAL	APPROXIMATE OR MAXIMUM % WT. OR VOL.	ESTABLISHED OSHA STANDARD	LD_{50} ORAL	PERCUT.	SPECIES	LC_{50} CONC.
Lowest published toxic dose						
oral - rat 11 g/Kg/Year						

SECTION III PHYSICAL DATA MW - 173.18

BOILING POINT: °F.
VAPOR PRESSURE: mm Hg.
MELTING POINT: Decomp. 140°C.
VAPOR DENSITY (Air = 1):
SPECIFIC GRAVITY (H_2O = 1):
EVAPORATION RATE (= 1):
SOLUBILITY IN WATER Pts/100 pts H_2O: Soluble in hot water
VOLATILE % Vol. % Wt.:
APPEARANCE AND ODOR: Yellow needles

SECTION IV FIRE AND EXPLOSION HAZARD DATA

FLASH POINT: None available
METHOD USED:
FLAMMABLE (EXPLOSIVE) LIMITS: UPPER / LOWER
EXTINGUISHING MEDIA: Dry chemical, carbon dioxide, alcohol foam.
SPECIAL FIRE FIGHTING PROCEDURES:
UNUSUAL FIRE AND EXPLOSION HAZARDS:

SECTION V HEALTH HAZARD DATA

TOXIC LEVEL: None available
CARCINOGENIC: Yes
PRINCIPLE ROUTES OF ABSORBTION: Inhalation, ingestion, skin
SKIN AND EYE IRRITATION: Severe
REVELANT SYMPTOMS OF EXPOSURE: Eye irritation and burns, dermatitis, nausea, vomiting, convulsions.
EFFECTS OF CHRONIC EXPOSURE: Kidney and liver damage, hematuria.
EMERGENCY AND FIRST AID PROCEDURES: Wash eyes and skin; if ingested - give emetic, get medical aid; inhalation - rest.

SECTION VI REACTIVITY DATA

CONDITIONS CONTRIBUTING TO INSTABILITY:
CONDITIONS CONTRIBUTING TO HAZARDOUS POLYMERIZATION:
INCOMPATABILITY (MATERIALS TO AVOID):
HAZARDOUS DECOMPOSITION PRODUCTS:

SECTION VII SPILL OR LEAK PROCEDURES

STEPS TO BE TAKEN IN CASE MATERIAL IS RELEASED OR SPILLED:
WASTE DISPOSAL METHOD:

SECTION VIII SPECIAL PROTECTION INFORMATION

VENTILATION REQUIREMENTS LOCAL EXHAUST:
PROTECTIVE EQUIPMENT (SPECIFY TYPES) EYE: Goggles
MECHANICAL (GENERAL):
GLOVES: Rubber
SPECIAL:
RESPIRATOR: With particulate filter
OTHER PROTECTIVE EQUIPMENT:

SECTION IX SPECIAL PRECAUTIONS

PRECAUTIONS TO BE TAKEN IN HANDLING AND STORAGE:
OTHER PRECAUTIONS:
SIGNATURE:
DATE:

NITROSORESORCINOL SODIUM SALT

Hazard diamond: 1 (top), 4 (left), 0 (right)

SECTION I SOURCE AND NOMENCLATURE

MANUFACTURER'S NAME | EMERGENCY TELEPHONE NO.

ADDRESS (NUMBER, STREET, CITY, STATE, ZIP CODE)

TRADE NAME AND SYNONYMS | CHEMICAL FAMILY

CHEMICAL NAME AND SYNONYMS: Nitrosoresorcinol Sodium Salt | FORMULA: $NOC_6H_3(OH)ONa$

SECTION II HAZARDOUS INGREDIENTS

BASIC MATERIAL	APPROXIMATE OR MAXIMUM % WT. OR VOL.	ESTABLISHED OSHA STANDARD	LD_{50} ORAL	LD_{50} PERCUT.	LC_{50} SPECIES	LC_{50} CONC.
			none	available		

SECTION III PHYSICAL DATA For P-Nitrosoresorcinol

BOILING POINT °F.: --	VAPOR PRESSURE mm Hg.
MELTING POINT °F.: 148 d°C.	VAPOR DENSITY (Air = 1)
SPECIFIC GRAVITY (H_2O = 1): --	EVAPORATION RATE (=1)
SOLUBILITY IN WATER Pts/100 pts H_2O: soluble	VOLATILE % Vol. % Wt.

APPEARANCE AND ODOR: yellow crystals

SECTION IV FIRE AND EXPLOSION HAZARD DATA

FLASH POINT: not available | FLAMMABLE (EXPLOSIVE) LIMITS: UPPER / LOWER

METHOD USED

EXTINGUISHING MEDIA: information not available.

SPECIAL FIRE FIGHTING PROCEDURES

UNUSUAL FIRE AND EXPLOSION HAZARDS

SECTION V HEALTH HAZARD DATA

TOXIC LEVEL: none available | CARCINOGENIC: no

PRINCIPLE ROUTES OF ABSORBTION: inhalation, ingestion, skin | SKIN AND EYE IRRITATION: moderate

REVELANT SYMPTOMS OF EXPOSURE: Itching, dermaitits, restlessness, cyanosis, dizziness, tremors, convulsions and death.

EFFECTS OF CHRONIC EXPOSURE: Liver, kidney and lung damage.

EMERGENCY AND FIRST AID PROCEDURES: If swallowed - induce vomiting, get medical aid; inhalation - rest; wash skin & eyes.

SECTION VI REACTIVITY DATA

CONDITIONS CONTRIBUTING TO INSTABILITY

CONDITIONS CONTRIBUTING TO HAZARDOUS POLYMERIZATION

INCOMPATABILITY (MATERIALS TO AVOID)

HAZARDOUS DECOMPOSITION PRODUCTS

SECTION VII SPILL OR LEAK PROCEDURES

STEPS TO BE TAKEN IN CASE MATERIAL IS RELEASED OR SPILLED

WASTE DISPOSAL METHOD

SECTION VIII SPECIAL PROTECTION INFORMATION

VENTILATION REQUIREMENTS LOCAL EXHAUST | PROTECTIVE EQUIPMENT (SPECIFY TYPES) EYE: Goggles

MECHANICAL (GENERAL) | GLOVES: Rubber

SPECIAL | RESPIRATOR

OTHER PROTECTIVE EQUIPMENT

SECTION IX SPECIAL PRECAUTIONS

PRECAUTIONS TO BE TAKEN IN HANDLING AND STORAGE

OTHER PRECAUTIONS

SIGNATURE | DATE

m-NITROTOLUENE

Hazard diamond: 1 (top), 4 (left), 3 (right)

SECTION I SOURCE AND NOMENCLATURE

MANUFACTURER'S NAME | EMERGENCY TELEPHONE NO.

ADDRESS (NUMBER, STREET, CITY, STATE, ZIP CODE)

TRADE NAME AND SYNONYMS: 3-Methylnitrobenzene | CHEMICAL FAMILY

CHEMICAL NAME AND SYNONYMS: M-Nitrotoluene | FORMULA: $CH_3C_6H_4NO_2$

SECTION II HAZARDOUS INGREDIENTS

BASIC MATERIAL	APPROXIMATE OR MAXIMUM % WT. OR VOL.	ESTABLISHED OSHA STANDARD	LD_{50} mg/kg ORAL	PERCUT.	SPECIES	LC_{50} CONC.
			1072		rat	
			330		mouse	
			2400		rabbit	
			3600		guinea pig	

SECTION III PHYSICAL DATA MW - 137.15

BOILING POINT °F.: 232°C	VAPOR PRESSURE mm Hg.: at 50°C 1
MELTING POINT °F.: 15°C	VAPOR DENSITY (Air = 1): 4.72
SPECIFIC GRAVITY (H_2O = 1): 1.16	EVAPORATION RATE (=1)
SOLUBILITY IN WATER Pts/100 pts H_2O: 0.05	VOLATILE % Vol. % Wt.

APPEARANCE AND ODOR: liquid

SECTION IV FIRE AND EXPLOSION HAZARD DATA

FLASH POINT: 233°F | FLAMMABLE (EXPLOSIVE) LIMITS: UPPER / LOWER

METHOD USED: C.C.

EXTINGUISHING MEDIA: Water, carbon dioxide, dry chemical.

SPECIAL FIRE FIGHTING PROCEDURES

UNUSUAL FIRE AND EXPLOSION HAZARDS

SECTION V HEALTH HAZARD DATA

TOXIC LEVEL: TLV - 5 PPM, 30 mg/CM | CARCINOGENIC: no

PRINCIPLE ROUTES OF ABSORBTION: Inhalation, ingestion, skin | SKIN AND EYE IRRITATION: Slight

REVELANT SYMPTOMS OF EXPOSURE: Difficulty in breathing, cyanosis, vomiting, headache.

EFFECTS OF CHRONIC EXPOSURE: Blood damage.

EMERGENCY AND FIRST AID PROCEDURES: If swallowed - give emetic, get medical aid; inhalation - rest; wash from skin & eyes.

SECTION VI REACTIVITY DATA

CONDITIONS CONTRIBUTING TO INSTABILITY

CONDITIONS CONTRIBUTING TO HAZARDOUS POLYMERIZATION

INCOMPATABILITY (MATERIALS TO AVOID): May explode when heated with sulfuric acid.

HAZARDOUS DECOMPOSITION PRODUCTS

SECTION VII SPILL OR LEAK PROCEDURES

STEPS TO BE TAKEN IN CASE MATERIAL IS RELEASED OR SPILLED: Add sodium bicarbonate or sand-soda ash mixture (9:1). Transfer into cardboard carton for disposal.

WASTE DISPOSAL METHOD

SECTION VIII SPECIAL PROTECTION INFORMATION

VENTILATION REQUIREMENTS LOCAL EXHAUST | PROTECTIVE EQUIPMENT (SPECIFY TYPES) EYE: Goggles

MECHANICAL (GENERAL) | GLOVES: Rubber (butyl)

SPECIAL | RESPIRATOR: Self-contained

OTHER PROTECTIVE EQUIPMENT

SECTION IX SPECIAL PRECAUTIONS

PRECAUTIONS TO BE TAKEN IN HANDLING AND STORAGE: Outside or detached storage preferred inside storage in standard flammable liquids cabinet.

OTHER PRECAUTIONS

SIGNATURE | DATE

o-NITROTOLUENE

Hazard rating diamond: 1, 4, 3

SECTION I SOURCE AND NOMENCLATURE

MANUFACTURER'S NAME | EMERGENCY TELEPHONE NO.

ADDRESS (NUMBER, STREET, CITY, STATE, ZIP CODE)

TRADE NAME AND SYNONYMS: 2 Methyl Nitrobenzene

CHEMICAL FAMILY:

CHEMICAL NAME AND SYNONYMS: O-Nitrotoluene

FORMULA: $CH_3C_6H_4NO_2$

SECTION II HAZARDOUS INGREDIENTS

BASIC MATERIAL	APPROXIMATE OR MAXIMUM % WT. OR VOL.	ESTABLISHED OSHA STANDARD	LD_{50} mg/kg ORAL	PERCUT.	SPECIES	LC_{50} CONC.
			891		rat	
			2462		mouse	

SECTION III PHYSICAL DATA MW - 137.1

BOILING POINT °F.: 222°C

VAPOR PRESSURE mm Hg.: at 50°C 1

MELTING POINT °F.: -4°C

VAPOR DENSITY (Air = 1): 4.72

SPECIFIC GRAVITY (H_2O = 1): 1.16

EVAPORATION RATE (= 1)

SOLUBILITY IN WATER Pts/100 pts H_2O: 0.07

VOLATILE % Vol. % Wt.

APPEARANCE AND ODOR: yellowish liquid

SECTION IV FIRE AND EXPLOSION HAZARD DATA

FLASH POINT: 223°F

FLAMMABLE (EXPLOSIVE) LIMITS: UPPER / LOWER

METHOD USED: C.C.

EXTINGUISHING MEDIA: Water, carbon dioxide, dry chemical

SPECIAL FIRE FIGHTING PROCEDURES

UNUSUAL FIRE AND EXPLOSION HAZARDS

SECTION V HEALTH HAZARD DATA

TOXIC LEVEL: TLV - 5 PPM, 30 mg/C.M.

CARCINOGENIC: no

PRINCIPLE ROUTES OF ABSORBTION: inhalation, ingestion, skin

SKIN AND EYE IRRITATION: slight

RELEVANT SYMPTOMS OF EXPOSURE: Difficulty in breathing, cyanosis, vomiting, headache.

EFFECTS OF CHRONIC EXPOSURE: Blood damage

EMERGENCY AND FIRST AID PROCEDURES: If swallowed - give emetic, get medical aid; inhalation - rest; wash from skin & eyes.

SECTION VI REACTIVITY DATA

CONDITIONS CONTRIBUTING TO INSTABILITY

CONDITIONS CONTRIBUTING TO HAZARDOUS POLYMERIZATION

INCOMPATABILITY (MATERIALS TO AVOID): May explode when heated with sulfuric acid.

HAZARDOUS DECOMPOSITION PRODUCTS

SECTION VII SPILL OR LEAK PROCEDURES

STEPS TO BE TAKEN IN CASE MATERIAL IS RELEASED OR SPILLED: Pour sand soda ash mixture (9:1) on to spill and sweep up into cardboard container for disposal.

WASTE DISPOSAL METHOD

SECTION VIII SPECIAL PROTECTION INFORMATION

VENTILATION REQUIREMENTS LOCAL EXHAUST

MECHANICAL (GENERAL)

SPECIAL

PROTECTIVE EQUIPMENT (SPECIFY TYPES) EYE: Goggles

GLOVES: Rubber (butyl)

RESPIRATOR: Self-contained

OTHER PROTECTIVE EQUIPMENT

SECTION IX SPECIAL PRECAUTIONS

PRECAUTIONS TO BE TAKEN IN HANDLING AND STORAGE: Outside or detached storage preferred; inside storage in standard flammable liquids cabinet.

OTHER PRECAUTIONS

SIGNATURE | DATE

p-NITROTOLUENE

Hazard rating diamond: 1, 4, 3

SECTION I SOURCE AND NOMENCLATURE

MANUFACTURER'S NAME | EMERGENCY TELEPHONE NO.

ADDRESS (NUMBER, STREET, CITY, STATE, ZIP CODE)

TRADE NAME AND SYNONYMS: 4 Methyl Nitrobenzene

CHEMICAL FAMILY:

CHEMICAL NAME AND SYNONYMS: P-Nitrotoluene

FORMULA: $CH_3C_6H_4NO_2$

SECTION II HAZARDOUS INGREDIENTS

BASIC MATERIAL	APPROXIMATE OR MAXIMUM % WT. OR VOL.	ESTABLISHED OSHA STANDARD	LD_{50} mg/kg ORAL	PERCUT.	SPECIES	LC_{50} CONC.
			2144	16,000	rat	
			1231		mouse	

SECTION III PHYSICAL DATA MW - 137.15

BOILING POINT °F.: 238°C

VAPOR PRESSURE mm Hg.: at 54°C 1

MELTING POINT °F.: 52°C

VAPOR DENSITY (Air = 1): 4.72

SPECIFIC GRAVITY (H_2O = 1): 1.12

EVAPORATION RATE (= 1)

SOLUBILITY IN WATER Pts/100 pts H_2O: insoluble

VOLATILE % Vol. % Wt.

APPEARANCE AND ODOR: yellowish crystals

SECTION IV FIRE AND EXPLOSION HAZARD DATA

FLASH POINT: 223°F

FLAMMABLE (EXPLOSIVE) LIMITS: UPPER / LOWER

METHOD USED: C.C.

EXTINGUISHING MEDIA: Water, carbon dioxide, dry chemicals.

SPECIAL FIRE FIGHTING PROCEDURES

UNUSUAL FIRE AND EXPLOSION HAZARDS

SECTION V HEALTH HAZARD DATA

TOXIC LEVEL: TLV - 5 PPM, 30 mg/C.M.

CARCINOGENIC: no

PRINCIPLE ROUTES OF ABSORBTION: inhalation, ingestion, skin

SKIN AND EYE IRRITATION: slight

RELEVANT SYMPTOMS OF EXPOSURE: Difficulty in breathing, cyanosis, vomiting, headache.

EFFECTS OF CHRONIC EXPOSURE: Blood damage

EMERGENCY AND FIRST AID PROCEDURES: If swallowed - give emetic, get medical aid; inhalation - rest; wash from skin & eyes.

SECTION VI REACTIVITY DATA

CONDITIONS CONTRIBUTING TO INSTABILITY

CONDITIONS CONTRIBUTING TO HAZARDOUS POLYMERIZATION

INCOMPATABILITY (MATERIALS TO AVOID): May explode when heated with sulfuric acid.

HAZARDOUS DECOMPOSITION PRODUCTS

SECTION VII SPILL OR LEAK PROCEDURES

STEPS TO BE TAKEN IN CASE MATERIAL IS RELEASED OR SPILLED: Add (9:1) sand to soda ash to spill and transfer to cardboard container for disposal.

WASTE DISPOSAL METHOD

SECTION VIII SPECIAL PROTECTION INFORMATION

VENTILATION REQUIREMENTS LOCAL EXHAUST

MECHANICAL (GENERAL)

SPECIAL

PROTECTIVE EQUIPMENT (SPECIFY TYPES) EYE: Goggles

GLOVES: Rubber (butyl)

RESPIRATOR: Self-contained

OTHER PROTECTIVE EQUIPMENT

SECTION IX SPECIAL PRECAUTIONS

PRECAUTIONS TO BE TAKEN IN HANDLING AND STORAGE: Outside or detached storage preferred inside storage in standard flammable liquids cabinet.

OTHER PRECAUTIONS

SIGNATURE | DATE

OSMIC ACID ANHYDRIDE

(Hazard diamond: 0 / 4 / 0)

SECTION I SOURCE AND NOMENCLATURE

MANUFACTURER'S NAME	EMERGENCY TELEPHONE NO.
ADDRESS (NUMBER, STREET, CITY, STATE, ZIP CODE)	
TRADE NAME AND SYNONYMS: Osmium Tetraoxide	CHEMICAL FAMILY
CHEMICAL NAME AND SYNONYMS: Osmic Acid Anhydride	FORMULA: OsO_4

SECTION II HAZARDOUS INGREDIENTS

BASIC MATERIAL	APPROXIMATE OR MAXIMUM % WT. OR VOL.	ESTABLISHED OSHA STANDARD	LD_{50} mg/kg ORAL	PERCUT.	SPECIES	LC_{50} CONC.
Lowest published toxic concentration in air			14		Rat	
inhalation - human 0.1 mg/C.M.			162		Mouse	

SECTION III PHYSICAL DATA MW - 254.26

BOILING POINT °F.: 130°C.	VAPOR PRESSURE mm Hg.: at 26°C. 10
MELTING POINT °F.: 40°C.	VAPOR DENSITY (Air = 1)
SPECIFIC GRAVITY (H_2O = 1): 4.91	EVAPORATION RATE (______ = 1)
SOLUBILITY IN WATER Pts/100 pts H_2O	VOLATILE % Vol. % Wt.

APPEARANCE AND ODOR: Monoclinic colorless crystals, chlorine-like odor.

SECTION IV FIRE AND EXPLOSION HAZARD DATA

FLASH POINT: Not flammable	FLAMMABLE (EXPLOSIVE) LIMITS	UPPER
METHOD USED		LOWER

EXTINGUISHING MEDIA

SPECIAL FIRE FIGHTING PROCEDURES

UNUSUAL FIRE AND EXPLOSION HAZARDS

SECTION V HEALTH HAZARD DATA

TOXIC LEVEL: TLV - 0.002 mg/C.M. CARCINOGENIC: No

PRINCIPLE ROUTES OF ABSORBTION: Inhalation, ingestion SKIN AND EYE IRRITATION: Severe

REVELANT SYMPTOMS OF EXPOSURE: Irritation of respiratory system, eye and skin irritation; if wallowed severe internal irritation.

EFFECTS OF CHRONIC EXPOSURE: Disturbances of vision, dermatitis and skin ulceration.

EMERGENCY AND FIRST AID PROCEDURES: Inhalation - rest; if swallowed - drink water, milk of magnesia, get medical aid; wash skin & eyes.

SECTION VI REACTIVITY DATA

CONDITIONS CONTRIBUTING TO INSTABILITY

CONDITIONS CONTRIBUTING TO HAZARDOUS POLYMERIZATION

INCOMPATABILITY (MATERIALS TO AVOID)

HAZARDOUS DECOMPOSITION PRODUCTS

SECTION VII SPILL OR LEAK PROCEDURES

STEPS TO BE TAKEN IN CASE MATERIAL IS RELEASED OR SPILLED

WASTE DISPOSAL METHOD

SECTION VIII SPECIAL PROTECTION INFORMATION

VENTILATION REQUIREMENTS LOCAL EXHAUST	PROTECTIVE EQUIPMENT (SPECIFY TYPES) EYE: Safety goggles
MECHANICAL (GENERAL)	GLOVES
SPECIAL	RESPIRATOR: With chemical filter

OTHER PROTECTIVE EQUIPMENT

SECTION IX SPECIAL PRECAUTIONS

PRECAUTIONS TO BE TAKEN IN HANDLING AND STORAGE

OTHER PRECAUTIONS

SIGNATURE	DATE

OXALYL CHLORIDE

(Hazard diamond: 3 / 4 / 2 / W)

SECTION I SOURCE AND NOMENCLATURE

MANUFACTURER'S NAME	EMERGENCY TELEPHONE NO.
ADDRESS (NUMBER, STREET, CITY, STATE, ZIP CODE)	
TRADE NAME AND SYNONYMS	CHEMICAL FAMILY
CHEMICAL NAME AND SYNONYMS: Oxalyl Chloride	FORMULA: ClOCCOCl

SECTION II HAZARDOUS INGREDIENTS

BASIC MATERIAL	APPROXIMATE OR MAXIMUM % WT. OR VOL.	ESTABLISHED OSHA STANDARD	LD_{50} ORAL	PERCUT.	SPECIES	LC_{50} CONC.
			none available			

SECTION III PHYSICAL DATA MW - 126.93

BOILING POINT °F.: 64°C	VAPOR PRESSURE mm Hg.
MELTING POINT °F.: -12°C	VAPOR DENSITY (Air = 1)
SPECIFIC GRAVITY (H_2O = 1): 1.49	EVAPORATION RATE (______ = 1)
SOLUBILITY IN WATER Pts/100 pts H_2O	VOLATILE % Vol. % Wt.

APPEARANCE AND ODOR

SECTION IV FIRE AND EXPLOSION HAZARD DATA

FLASH POINT: none available	FLAMMABLE (EXPLOSIVE) LIMITS	UPPER
METHOD USED		LOWER

EXTINGUISHING MEDIA: carbon dioxide, dry chemical - avoid water.

SPECIAL FIRE FIGHTING PROCEDURES

UNUSUAL FIRE AND EXPLOSION HAZARDS

SECTION V HEALTH HAZARD DATA

TOXIC LEVEL: none available CARCINOGENIC: no

PRINCIPLE ROUTES OF ABSORBTION: Inhalation, ingestion SKIN AND EYE IRRITATION: severe

REVELANT SYMPTOMS OF EXPOSURE: Cyanosis, coughing, headache, pulmonary edema, ulceration of skin, dizziness.

EFFECTS OF CHRONIC EXPOSURE: Erosion of teeth, jaw necrosis, bronchial irritation and chronic cough.

EMERGENCY AND FIRST AID PROCEDURES: If swallowed - drink water, milk of magnesia, get medical aid; inhalation - rest; wash skin & eyes.

SECTION VI REACTIVITY DATA

CONDITIONS CONTRIBUTING TO INSTABILITY

CONDITIONS CONTRIBUTING TO HAZARDOUS POLYMERIZATION

INCOMPATABILITY (MATERIALS TO AVOID): water and alcohol.

HAZARDOUS DECOMPOSITION PRODUCTS

SECTION VII SPILL OR LEAK PROCEDURES

STEPS TO BE TAKEN IN CASE MATERIAL IS RELEASED OR SPILLED: Cover spill with sodium bicarbonate, transfer to plastic bag for disposal.

WASTE DISPOSAL METHOD

SECTION VIII SPECIAL PROTECTION INFORMATION

VENTILATION REQUIREMENTS LOCAL EXHAUST	PROTECTIVE EQUIPMENT (SPECIFY TYPES) EYE: Goggles
MECHANICAL (GENERAL)	GLOVES: Rubber
SPECIAL	RESPIRATOR: With proper filter

OTHER PROTECTIVE EQUIPMENT: Coveralls

SECTION IX SPECIAL PRECAUTIONS

PRECAUTIONS TO BE TAKEN IN HANDLING AND STORAGE

OTHER PRECAUTIONS

SIGNATURE	DATE

PENTAMETHYLENE BROMIDE

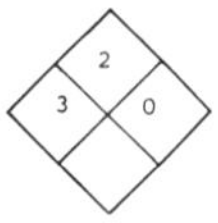

SECTION I SOURCE AND NOMENCLATURE

MANUFACTURER'S NAME | EMERGENCY TELEPHONE NO.

ADDRESS (NUMBER, STREET, CITY, STATE, ZIP CODE)

TRADE NAME AND SYNONYMS | CHEMICAL FAMILY

CHEMICAL NAME AND SYNONYMS: 1,5 Dibromopentane

FORMULA: $Br(CH_2)_5Br$

SECTION II HAZARDOUS INGREDIENTS

BASIC MATERIAL	APPROXIMATE OR MAXIMUM % WT. OR VOL.	ESTABLISHED OSHA STANDARD	LD_{50} ORAL	LD_{50} PERCUT.	LC_{50} SPECIES	LC_{50} CONC.
			none	available		

SECTION III PHYSICAL DATA MW - 229.96

BOILING POINT °F.: 222°C

VAPOR PRESSURE mm Hg.: at 108°C 20

MELTING POINT °F.: -40°C

VAPOR DENSITY (Air = 1):

SPECIFIC GRAVITY (H_2O = 1): 1.69

EVAPORATION RATE (______ = 1):

SOLUBILITY IN WATER Pts/100 pts H_2O: --

VOLATILE % Vol. % Wt.:

APPEARANCE AND ODOR: colorless liquid

SECTION IV FIRE AND EXPLOSION HAZARD DATA

FLASH POINT: none available

METHOD USED:

FLAMMABLE (EXPLOSIVE) LIMITS: UPPER / LOWER

EXTINGUISHING MEDIA: Water, foam, carbon dioxide and dry chemical.

SPECIAL FIRE FIGHTING PROCEDURES: Water may be ineffective.

UNUSUAL FIRE AND EXPLOSION HAZARDS:

SECTION V HEALTH HAZARD DATA

TOXIC LEVEL: none — CARCINOGENIC: no

PRINCIPLE ROUTES OF ABSORBTION: inhalation, ingestion — SKIN AND EYE IRRITATION: moderate

REVELANT SYMPTOMS OF EXPOSURE: eye irritation, narcosis, drowsiness, headache, vomiting.

EFFECTS OF CHRONIC EXPOSURE: dermatitis, liver & kidney damage.

EMERGENCY AND FIRST AID PROCEDURES: Inhalation - rest; if swallowed - give emetic, get medical aid; wash from skin & eyes.

SECTION VI REACTIVITY DATA

CONDITIONS CONTRIBUTING TO INSTABILITY:

CONDITIONS CONTRIBUTING TO HAZARDOUS POLYMERIZATION:

INCOMPATABILITY (MATERIALS TO AVOID):

HAZARDOUS DECOMPOSITION PRODUCTS:

SECTION VII SPILL OR LEAK PROCEDURES

STEPS TO BE TAKEN IN CASE MATERIAL IS RELEASED OR SPILLED:

WASTE DISPOSAL METHOD:

SECTION VIII SPECIAL PROTECTION INFORMATION

VENTILATION REQUIREMENTS LOCAL EXHAUST:

MECHANICAL (GENERAL):

SPECIAL:

PROTECTIVE EQUIPMENT (SPECIFY TYPES) EYE: Goggles

GLOVES: Rubber

RESPIRATOR: With organic vapor resistant canister.

OTHER PROTECTIVE EQUIPMENT:

SECTION IX SPECIAL PRECAUTIONS

PRECAUTIONS TO BE TAKEN IN HANDLING AND STORAGE:

OTHER PRECAUTIONS:

SIGNATURE | DATE

PERIODIC ACID

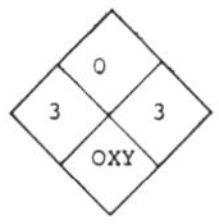

SECTION I SOURCE AND NOMENCLATURE

MANUFACTURER'S NAME | EMERGENCY TELEPHONE NO.

ADDRESS (NUMBER, STREET, CITY, STATE, ZIP CODE)

TRADE NAME AND SYNONYMS | CHEMICAL FAMILY

CHEMICAL NAME AND SYNONYMS: Periodic Acid

FORMULA: $HIO_4{\cdot}2H_2O$

SECTION II HAZARDOUS INGREDIENTS

BASIC MATERIAL	APPROXIMATE OR MAXIMUM % WT. OR VOL.	ESTABLISHED OSHA STANDARD	LD_{50} ORAL	LD_{50} PERCUT.	LC_{50} SPECIES	LC_{50} CONC.
			Not	available		

SECTION III PHYSICAL DATA

BOILING POINT °F.: N/A

VAPOR PRESSURE mm Hg.:

MELTING POINT °F.: Decomp to HIO_4 above 100°C.

VAPOR DENSITY (Air = 1):

SPECIFIC GRAVITY (H_2O = 1): 1.39

EVAPORATION RATE (______ = 1):

SOLUBILITY IN WATER Pts/100 pts H_2O: 7.82

VOLATILE % Vol. % Wt.:

APPEARANCE AND ODOR: White crystal

SECTION IV FIRE AND EXPLOSION HAZARD DATA

FLASH POINT: N/A

METHOD USED:

FLAMMABLE (EXPLOSIVE) LIMITS: UPPER / LOWER

EXTINGUISHING MEDIA: Water spray

SPECIAL FIRE FIGHTING PROCEDURES: N/A

UNUSUAL FIRE AND EXPLOSION HAZARDS: Strong oxidizing agent - explosive properties unknown but assumed to be like perchloric acid.

SECTION V HEALTH HAZARD DATA

TOXIC LEVEL: Not available — CARCINOGENIC: No

PRINCIPLE ROUTES OF ABSORBTION: Inhalation of dust, ingestion — SKIN AND EYE IRRITATION: Severe

REVELANT SYMPTOMS OF EXPOSURE: Respiratory system irritated; skin and eye irritation; severe internal irritation and damage.

EFFECTS OF CHRONIC EXPOSURE: --

EMERGENCY AND FIRST AID PROCEDURES: Inhalation - rest, get medical attention; wash from skin and eyes; if swallowed - wash from skin and eyes; if swallowed - wash mouth, drink water, then milk of magnesia, get medical attention.

SECTION VI REACTIVITY DATA

CONDITIONS CONTRIBUTING TO INSTABILITY: It is a stronger oxidizing agent than $KMnO_4$

CONDITIONS CONTRIBUTING TO HAZARDOUS POLYMERIZATION:

INCOMPATABILITY (MATERIALS TO AVOID): Avoid contact with strong reducing agents.

HAZARDOUS DECOMPOSITION PRODUCTS: Decomposed to oxide of iodine.

SECTION VII SPILL OR LEAK PROCEDURES

STEPS TO BE TAKEN IN CASE MATERIAL IS RELEASED OR SPILLED:

WASTE DISPOSAL METHOD:

SECTION VIII SPECIAL PROTECTION INFORMATION

VENTILATION REQUIREMENTS LOCAL EXHAUST:

MECHANICAL (GENERAL):

SPECIAL:

PROTECTIVE EQUIPMENT (SPECIFY TYPES) EYE: Face shield

GLOVES:

RESPIRATOR: With appropriate filter

OTHER PROTECTIVE EQUIPMENT:

SECTION IX SPECIAL PRECAUTIONS

PRECAUTIONS TO BE TAKEN IN HANDLING AND STORAGE:

OTHER PRECAUTIONS:

SIGNATURE | DATE

m-PHENETIDINE

Hazard rating: 1 (top), 4 (left), 0 (right)

SECTION I SOURCE AND NOMENCLATURE

MANUFACTURER'S NAME | EMERGENCY TELEPHONE NO.

ADDRESS (NUMBER, STREET, CITY, STATE, ZIP CODE)

TRADE NAME AND SYNONYMS: 1 amino - 3 Ethoxybenzene | CHEMICAL FAMILY

CHEMICAL NAME AND SYNONYMS: M Aminophenetol | FORMULA: $NH_2C_6H_4NH_2$

SECTION II HAZARDOUS INGREDIENTS

BASIC MATERIAL	APPROXIMATE OR MAXIMUM % WT. OR VOL.	ESTABLISHED OSHA STANDARD	LD_{50} ORAL	LD_{50} PERCUT.	LC_{50} SPECIES	LC_{50} CONC.
			none available			

SECTION III PHYSICAL DATA MW - 137.18

BOILING POINT °F.: 248°C | VAPOR PRESSURE mm Hg.: at 125°C 7

MELTING POINT °F. | VAPOR DENSITY (Air = 1)

SPECIFIC GRAVITY (H_2O = 1) | EVAPORATION RATE (____ = 1)

SOLUBILITY IN WATER Pts/100 pts H_2O: insoluble | VOLATILE % Vol. % Wt.

APPEARANCE AND ODOR: liquid

SECTION IV FIRE AND EXPLOSION HAZARD DATA

FLASH POINT: none available | FLAMMABLE (EXPLOSIVE) LIMITS: UPPER / LOWER

METHOD USED

EXTINGUISHING MEDIA: carbon dioxide, dry chemical

SPECIAL FIRE FIGHTING PROCEDURES: water or foam may cause frothing.

UNUSUAL FIRE AND EXPLOSION HAZARDS

SECTION V HEALTH HAZARD DATA

TOXIC LEVEL: none available | CARCINOGENIC: no

PRINCIPLE ROUTES OF ABSORBTION: inhalation, ingestion, skin | SKIN AND EYE IRRITATION: moderate

REVELANT SYMPTOMS OF EXPOSURE: cyanosis, hypothermia, headache, drowsiness, vomiting, irritation of alimentary tract, nephritis.

EFFECTS OF CHRONIC EXPOSURE: Weight loss, anemia, weakness, irritability, possible liver damage.

EMERGENCY AND FIRST AID PROCEDURES: Inhalation - rest; if swallowed - emetic or gastric lavage, get medical attention; wash skin & eyes.

SECTION VI REACTIVITY DATA

CONDITIONS CONTRIBUTING TO INSTABILITY

CONDITIONS CONTRIBUTING TO HAZARDOUS POLYMERIZATION

INCOMPATABILITY (MATERIALS TO AVOID)

HAZARDOUS DECOMPOSITION PRODUCTS

SECTION VII SPILL OR LEAK PROCEDURES

STEPS TO BE TAKEN IN CASE MATERIAL IS RELEASED OR SPILLED: Pour (9:1) sand to soda ash mixture on spill, put in cardboard carton for disposal.

WASTE DISPOSAL METHOD

SECTION VIII SPECIAL PROTECTION INFORMATION

VENTILATION REQUIREMENTS LOCAL EXHAUST | PROTECTIVE EQUIPMENT (SPECIFY TYPES) EYE: Goggles

MECHANICAL (GENERAL) | GLOVES: Rubber (butyl)

SPECIAL | RESPIRATOR: With proper filter

OTHER PROTECTIVE EQUIPMENT

SECTION IX SPECIAL PRECAUTIONS

PRECAUTIONS TO BE TAKEN IN HANDLING AND STORAGE: Store in dark, cool area.

OTHER PRECAUTIONS

SIGNATURE | DATE

o-PHENETIDINE

Hazard rating: 1 (top), 4 (left), 0 (right)

SECTION I SOURCE AND NOMENCLATURE

MANUFACTURER'S NAME | EMERGENCY TELEPHONE NO.

ADDRESS (NUMBER, STREET, CITY, STATE, ZIP CODE)

TRADE NAME AND SYNONYMS: 1 amino - 2 Ethoxybenzene | CHEMICAL FAMILY

CHEMICAL NAME AND SYNONYMS: O-Aminophenetol | FORMULA: $C_2H_5OC_6H_4NH_2$

SECTION II HAZARDOUS INGREDIENTS

BASIC MATERIAL	APPROXIMATE OR MAXIMUM % WT. OR VOL.	ESTABLISHED OSHA STANDARD	LD_{50} ORAL	LD_{50} PERCUT.	LC_{50} SPECIES	LC_{50} CONC.
			none available			

SECTION III PHYSICAL DATA MW - 137.18

BOILING POINT °F.: 229°C | VAPOR PRESSURE: 133°C 25 mm Hg.

MELTING POINT °F.: < -21°C | VAPOR DENSITY (Air = 1)

SPECIFIC GRAVITY (H_2O = 1): -- | EVAPORATION RATE (____ = 1)

SOLUBILITY IN WATER Pts/100 pts H_2O: insoluble | VOLATILE % Vol. % Wt.

APPEARANCE AND ODOR: Oil, colorless to brownish

SECTION IV FIRE AND EXPLOSION HAZARD DATA

FLASH POINT: none available | FLAMMABLE (EXPLOSIVE) LIMITS: UPPER / LOWER

METHOD USED

EXTINGUISHING MEDIA: carbon dioxide, dry chemical

SPECIAL FIRE FIGHTING PROCEDURES: water or foam may cause frothing.

UNUSUAL FIRE AND EXPLOSION HAZARDS

SECTION V HEALTH HAZARD DATA

TOXIC LEVEL: none available | CARCINOGENIC: no

PRINCIPLE ROUTES OF ABSORBTION: inhalation, ingestion, skin | SKIN AND EYE IRRITATION: moderate

REVELANT SYMPTOMS OF EXPOSURE: Cyanosis, hypothermia, headache, drowsiness, vomiting, irritation of alimentary tract, nephritis.

EFFECTS OF CHRONIC EXPOSURE: Weight loss, anemia, weakness, irritability, possible liver damage.

EMERGENCY AND FIRST AID PROCEDURES: Inhalation - rest; if swallowed - emetic or gastric lavage, get medical aid; wash skin & eyes.

SECTION VI REACTIVITY DATA

CONDITIONS CONTRIBUTING TO INSTABILITY

CONDITIONS CONTRIBUTING TO HAZARDOUS POLYMERIZATION

INCOMPATABILITY (MATERIALS TO AVOID)

HAZARDOUS DECOMPOSITION PRODUCTS

SECTION VII SPILL OR LEAK PROCEDURES

STEPS TO BE TAKEN IN CASE MATERIAL IS RELEASED OR SPILLED: Pour (9:1) sand to soda ash mixture on spill, put in cardboard carton for disposal.

WASTE DISPOSAL METHOD

SECTION VIII SPECIAL PROTECTION INFORMATION

VENTILATION REQUIREMENTS LOCAL EXHAUST | PROTECTIVE EQUIPMENT (SPECIFY TYPES) EYE: Goggles

MECHANICAL (GENERAL) | GLOVES: Rubber (butyl)

SPECIAL | RESPIRATOR: With proper filter

OTHER PROTECTIVE EQUIPMENT: Plastic coveralls, self-contained breathing apparatus.

SECTION IX SPECIAL PRECAUTIONS

PRECAUTIONS TO BE TAKEN IN HANDLING AND STORAGE: Store in dark, cool area.

OTHER PRECAUTIONS

SIGNATURE | DATE

p-PHENETIDINE

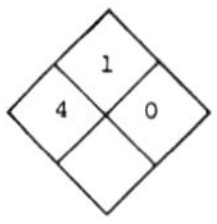

SECTION I SOURCE AND NOMENCLATURE

MANUFACTURER'S NAME | EMERGENCY TELEPHONE NO.

ADDRESS (NUMBER, STREET, CITY, STATE, ZIP CODE)

TRADE NAME AND SYNONYMS: 1 Amino - 4 Ethoxybenzene | CHEMICAL FAMILY

CHEMICAL NAME AND SYNONYMS: P-Aminophenetol, 4 Ethoxy Aniline | FORMULA: $C_2H_5OC_6H_4NH_2$

SECTION II HAZARDOUS INGREDIENTS

BASIC MATERIAL	APPROXIMATE OR MAXIMUM % WT. OR VOL.	ESTABLISHED OSHA STANDARD	LD_{50} ORAL	LD_{50} PERCUT.	LC_{50} SPECIES	LC_{50} CONC.
Unclassified - mouse LD50 83 mg/kg			none available			

SECTION III PHYSICAL DATA MW - 137.18

BOILING POINT °F.: 255°C	VAPOR PRESSURE mm Hg.
MELTING POINT °F.: 3°C	VAPOR DENSITY (Air = 1)
SPECIFIC GRAVITY (H_2O = 1): 1.06	EVAPORATION RATE (____ = 1)
SOLUBILITY IN WATER Pts/100 pts H_2O: very slightly	VOLATILE % Vol. % Wt.

APPEARANCE AND ODOR: liquid

SECTION IV FIRE AND EXPLOSION HAZARD DATA

FLASH POINT: 241°F | FLAMMABLE (EXPLOSIVE) LIMITS: UPPER / LOWER

METHOD USED

EXTINGUISHING MEDIA: carbon dioxide, dry chemical

SPECIAL FIRE FIGHTING PROCEDURES: water or foam may cause frothing.

UNUSUAL FIRE AND EXPLOSION HAZARDS

SECTION V HEALTH HAZARD DATA

TOXIC LEVEL: none available | CARCINOGENIC: no

PRINCIPLE ROUTES OF ABSORBTION: inhalation, ingestion, skin | SKIN AND EYE IRRITATION: moderate

REVELANT SYMPTOMS OF EXPOSURE: Cyanosis, hypothermia, headache, drowsiness, vomiting, ~~irritation of alimentary tract, nephritis.~~

EFFECTS OF CHRONIC EXPOSURE: Weight loss, anemia, weakness, irritability, possible liver damage.

EMERGENCY AND FIRST AID PROCEDURES: Inhalation - rest; if swallowed - emetic or gastric lavage, get medical aid; wash skin & eyes.

SECTION VI REACTIVITY DATA

CONDITIONS CONTRIBUTING TO INSTABILITY

CONDITIONS CONTRIBUTING TO HAZARDOUS POLYMERIZATION

INCOMPATABILITY (MATERIALS TO AVOID)

HAZARDOUS DECOMPOSITION PRODUCTS

SECTION VII SPILL OR LEAK PROCEDURES

STEPS TO BE TAKEN IN CASE MATERIAL IS RELEASED OR SPILLED: Pour (9:1) sand to soda ash mixture on spill, put in cardboard container for disposal.

WASTE DISPOSAL METHOD

SECTION VIII SPECIAL PROTECTION INFORMATION

VENTILATION REQUIREMENTS LOCAL EXHAUST | PROTECTIVE EQUIPMENT (SPECIFY TYPES) EYE: Goggles

MECHANICAL (GENERAL) | GLOVES: Rubber (butyl)

SPECIAL | RESPIRATOR: With proper filter

OTHER PROTECTIVE EQUIPMENT: Store in dark, cool area.

SECTION IX SPECIAL PRECAUTIONS

PRECAUTIONS TO BE TAKEN IN HANDLING AND STORAGE

OTHER PRECAUTIONS

SIGNATURE | DATE

γ-PHENOXYBUTYRONITRILE

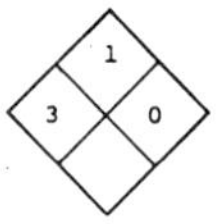

SECTION I SOURCE AND NOMENCLATURE

MANUFACTURER'S NAME | EMERGENCY TELEPHONE NO.

ADDRESS (NUMBER, STREET, CITY, STATE, ZIP CODE)

TRADE NAME AND SYNONYMS | CHEMICAL FAMILY

CHEMICAL NAME AND SYNONYMS: G-Phenoxybutyronitrile | FORMULA: C_6H_5-O-C_3H_6CN

SECTION II HAZARDOUS INGREDIENTS

BASIC MATERIAL	APPROXIMATE OR MAXIMUM % WT. OR VOL.	ESTABLISHED OSHA STANDARD	LD_{50} ORAL	LD_{50} PERCUT.	LC_{50} SPECIES	LC_{50} CONC.
			None	available		

SECTION III PHYSICAL DATA MW - 161.20

BOILING POINT °F.: 285° C. at 765 min.	VAPOR PRESSURE mm Hg.
MELTING POINT °F.: 46° C.	VAPOR DENSITY (Air = 1)
SPECIFIC GRAVITY (H_2O = 1)	EVAPORATION RATE (____ = 1)
SOLUBILITY IN WATER Pts/100 pts H_2O	VOLATILE % Vol. % Wt.

APPEARANCE AND ODOR: Needle-like crystals.

SECTION IV FIRE AND EXPLOSION HAZARD DATA

FLASH POINT: None available | FLAMMABLE (EXPLOSIVE) LIMITS: UPPER / LOWER

METHOD USED

EXTINGUISHING MEDIA: Dry chemical, alcohol foam or carbon dioxide.

SPECIAL FIRE FIGHTING PROCEDURES

UNUSUAL FIRE AND EXPLOSION HAZARDS

SECTION V HEALTH HAZARD DATA

TOXIC LEVEL: None available | CARCINOGENIC: No

PRINCIPLE ROUTES OF ABSORBTION: Inhalation, ingestion, skin | SKIN AND EYE IRRITATION: Moderate

REVELANT SYMPTOMS OF EXPOSURE: Dizziness, rapid respiration, headache, drowsiness, drop in blood pressure and pulse, delayed symptoms.

EFFECTS OF CHRONIC EXPOSURE: Eye irritation, loss of appetite, weight loss, mental deterioration.

EMERGENCY AND FIRST AID PROCEDURES: If swallowed - inhalation of amyl nitrite, induce vomiting or gastric lavage; wash skin and eyes; inhalation - rest.

SECTION VI REACTIVITY DATA

CONDITIONS CONTRIBUTING TO INSTABILITY

CONDITIONS CONTRIBUTING TO HAZARDOUS POLYMERIZATION

INCOMPATABILITY (MATERIALS TO AVOID)

HAZARDOUS DECOMPOSITION PRODUCTS: When heated to decomposition point, emits highly toxic cyanide fumes.

SECTION VII SPILL OR LEAK PROCEDURES

STEPS TO BE TAKEN IN CASE MATERIAL IS RELEASED OR SPILLED

WASTE DISPOSAL METHOD

SECTION VIII SPECIAL PROTECTION INFORMATION

VENTILATION REQUIREMENTS LOCAL EXHAUST | PROTECTIVE EQUIPMENT (SPECIFY TYPES) EYE: Goggles

MECHANICAL (GENERAL) | GLOVES: Rubber

SPECIAL | RESPIRATOR: With proper filter

OTHER PROTECTIVE EQUIPMENT

SECTION IX SPECIAL PRECAUTIONS

PRECAUTIONS TO BE TAKEN IN HANDLING AND STORAGE

OTHER PRECAUTIONS

SIGNATURE | DATE

PHENYLACETYL CHLORIDE

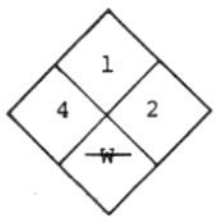

SECTION I SOURCE AND NOMENCLATURE

MANUFACTURER'S NAME	EMERGENCY TELEPHONE NO.
ADDRESS (NUMBER, STREET, CITY, STATE, ZIP CODE)	
TRADE NAME AND SYNONYMS	CHEMICAL FAMILY
CHEMICAL NAME AND SYNONYMS Phenylacetyl Chloride	FORMULA $C_6H_5CH_2COCl$

SECTION II HAZARDOUS INGREDIENTS

BASIC MATERIAL	APPROXIMATE OR MAXIMUM % WT. OR VOL.	ESTABLISHED OSHA STANDARD	LD_{50} ORAL	LD_{50} PERCUT.	SPECIES	LC_{50} CONC.
			none available			

SECTION III PHYSICAL DATA MW - 154.59

BOILING POINT °F. 170°C at 250 mm	VAPOR PRESSURE mm Hg. at 95°C 12
MELTING POINT °F.	VAPOR DENSITY (Air = 1)
SPECIFIC GRAVITY (H_2O = 1) 1.17	EVAPORATION RATE (_____ = 1)
SOLUBILITY IN WATER Pts/100 pts H_2O decomposes	VOLATILE % Vol. % Wt.
APPEARANCE AND ODOR colorless liquid	

SECTION IV FIRE AND EXPLOSION HAZARD DATA

FLASH POINT none available	FLAMMABLE (EXPLOSIVE) LIMITS	UPPER
METHOD USED		LOWER

EXTINGUISHING MEDIA
carbon dioxide, dry chemical

SPECIAL FIRE FIGHTING PROCEDURES
wear full protective clothing.

UNUSUAL FIRE AND EXPLOSION HAZARDS
reacts with water

SECTION V HEALTH HAZARD DATA

TOXIC LEVEL none CARCINOGENIC no

PRINCIPLE ROUTES OF ABSORBTION inhalation, ingestion SKIN AND EYE IRRITATION severe

REVELANT SYMPTOMS OF EXPOSURE Cyanosis, coughing, headache, pulmonary edema, ulceration of skin, dizziness.

EFFECTS OF CHRONIC EXPOSURE Erosion of teeth, jaw necrosis, bronchial irritation and chronic cough.

EMERGENCY AND FIRST AID PROCEDURES Inhalation - artificial respiration if necessary; ingestion - drink water, milk of magnesia, get medical aid; wash from skin & eyes.

SECTION VI REACTIVITY DATA

CONDITIONS CONTRIBUTING TO INSTABILITY

CONDITIONS CONTRIBUTING TO HAZARDOUS POLYMERIZATION

INCOMPATABILITY (MATERIALS TO AVOID)
Avoid water.

HAZARDOUS DECOMPOSITION PRODUCTS
Produces HCl with water.

SECTION VII SPILL OR LEAK PROCEDURES

STEPS TO BE TAKEN IN CASE MATERIAL IS RELEASED OR SPILLED

WASTE DISPOSAL METHOD

SECTION VIII SPECIAL PROTECTION INFORMATION

VENTILATION REQUIREMENTS LOCAL EXHAUST	PROTECTIVE EQUIPMENT (SPECIFY TYPES) EYE Goggles
MECHANICAL (GENERAL)	GLOVES Rubber
SPECIAL	RESPIRATOR With proper filter
OTHER PROTECTIVE EQUIPMENT	

SECTION IX SPECIAL PRECAUTIONS

PRECAUTIONS TO BE TAKEN IN HANDLING AND STORAGE
Store in cool, dry well-ventilated room; inside storage in standard flammable liquids storage cabinet.

OTHER PRECAUTIONS

SIGNATURE	DATE

PHENYLARSONIC ACID

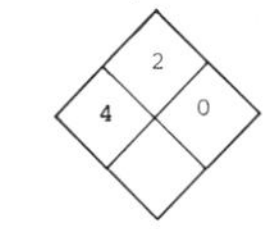

SECTION I SOURCE AND NOMENCLATURE

MANUFACTURER'S NAME	EMERGENCY TELEPHONE NO.
ADDRESS (NUMBER, STREET, CITY, STATE, ZIP CODE)	
TRADE NAME AND SYNONYMS	CHEMICAL FAMILY
CHEMICAL NAME AND SYNONYMS Benzenearsonic Acid	FORMULA $C_6H_5AsO(OH)_2$

SECTION II HAZARDOUS INGREDIENTS

BASIC MATERIAL	APPROXIMATE OR MAXIMUM % WT. OR VOL.	ESTABLISHED OSHA STANDARD	LD_{50} mg/C.M. ORAL	LD_{50} PERCUT.	SPECIES	LC_{50} CONC.
			.270		Mouse	

Note: Lowest published lethal dose, oral - rat 50 mg/kg.

SECTION III PHYSICAL DATA Mol. Wt. 202

BOILING POINT °F.	VAPOR PRESSURE mm Hg.
MELTING POINT °F. 160°C decomposes	VAPOR DENSITY (Air = 1)
SPECIFIC GRAVITY (H_2O = 1) 1.760	EVAPORATION RATE (_____ = 1)
SOLUBILITY IN WATER Pts/100 pts H_2O 3.3 at 28°C	VOLATILE % Vol. % Wt.
APPEARANCE AND ODOR Colorless crystal	

SECTION IV FIRE AND EXPLOSION HAZARD DATA

FLASH POINT None available	FLAMMABLE (EXPLOSIVE) LIMITS	UPPER
METHOD USED		LOWER

EXTINGUISHING MEDIA
Carbon dioxide, dry chemical, foam

SPECIAL FIRE FIGHTING PROCEDURES

UNUSUAL FIRE AND EXPLOSION HAZARDS
Decomposes to yield flammable vapor.

SECTION V HEALTH HAZARD DATA

TOXIC LEVEL TLV 0.68 mg/C.M. Teratogen CARCINOGENIC Yes

PRINCIPLE ROUTES OF ABSORBTION Inhalation, ingestion SKIN AND EYE IRRITATION Yes

REVELANT SYMPTOMS OF EXPOSURE Burning of mouth and throat, gastroentric pain, vomiting, diarrhea, hematuria, dehydration, oliguria and collapse.

EFFECTS OF CHRONIC EXPOSURE Arsenic poisoning and skin ulceration, falling out of hair, loss of appetite.

EMERGENCY AND FIRST AID PROCEDURES Inhalation - rest, keep warm, get medical aid; if swallowed - wash mouth, gastric lavage with 1% sodium thiosulfate; wash skin and eyes.

SECTION VI REACTIVITY DATA

CONDITIONS CONTRIBUTING TO INSTABILITY

CONDITIONS CONTRIBUTING TO HAZARDOUS POLYMERIZATION

INCOMPATABILITY (MATERIALS TO AVOID)

HAZARDOUS DECOMPOSITION PRODUCTS

SECTION VII SPILL OR LEAK PROCEDURES

STEPS TO BE TAKEN IN CASE MATERIAL IS RELEASED OR SPILLED

WASTE DISPOSAL METHOD

SECTION VIII SPECIAL PROTECTION INFORMATION

VENTILATION REQUIREMENTS LOCAL EXHAUST	PROTECTIVE EQUIPMENT (SPECIFY TYPES) EYE Goggles
MECHANICAL (GENERAL)	GLOVES
SPECIAL	RESPIRATOR w/proper filter
OTHER PROTECTIVE EQUIPMENT	

SECTION IX SPECIAL PRECAUTIONS

PRECAUTIONS TO BE TAKEN IN HANDLING AND STORAGE

OTHER PRECAUTIONS

SIGNATURE	DATE

m-PHENYLENEDIAMINE

1
3 0

SECTION I SOURCE AND NOMENCLATURE

MANUFACTURER'S NAME | EMERGENCY TELEPHONE NO.

ADDRESS (NUMBER, STREET, CITY, STATE, ZIP CODE)

TRADE NAME AND SYNONYMS: 3 Aminoaniline | CHEMICAL FAMILY

CHEMICAL NAME AND SYNONYMS: 1,3 Aminobenzene | FORMULA: $C_6H_6(NH_2)_2$

SECTION II HAZARDOUS INGREDIENTS

BASIC MATERIAL	APPROXIMATE OR MAXIMUM % WT. OR VOL.	ESTABLISHED OSHA STANDARD	LD_{50} ORAL	LD_{50} PERCUT.	LC_{50} SPECIES	LC_{50} CONC.
			See Notes			
Notes:						
Lowest published lethal dose						
Oral - rat 80 mg/kg						
Subcutaneous - rat 80 mg/kg						
Intravenous - dog 17 mg/kg						
Oral - cat 300 mg/kg						
Subcutaneous - rabbit 200 mg/kg						

SECTION III PHYSICAL DATA MW - 108.16

BOILING POINT °F: 283°C | VAPOR PRESSURE mm Hg.

MELTING POINT °F: 63°C | VAPOR DENSITY (Air = 1)

SPECIFIC GRAVITY (H_2O = 1): 1.14 | EVAPORATION RATE (______ = 1)

SOLUBILITY IN WATER Pts/100 pts H_2O: 84 | VOLATILE % Vol. % Wt.

APPEARANCE AND ODOR: colorless crystal

SECTION IV FIRE AND EXPLOSION HAZARD DATA

FLASH POINT: none available | FLAMMABLE (EXPLOSIVE) LIMITS | UPPER | LOWER

METHOD USED

EXTINGUISHING MEDIA: water, carbon dioxide, dry chemical.

SPECIAL FIRE FIGHTING PROCEDURES

UNUSUAL FIRE AND EXPLOSION HAZARDS

SECTION V HEALTH HAZARD DATA

TOXIC LEVEL: none available | CARCINOGENIC: no

PRINCIPLE ROUTES OF ABSORBTION: ingestion (mainly) inhalation | SKIN AND EYE IRRITATION: severe

REVELANT SYMPTOMS OF EXPOSURE: Eye irritation, headache, cyanosis, drowsiness, mental confusion, dermatitis.

EFFECTS OF CHRONIC EXPOSURE: Liver damage

EMERGENCY AND FIRST AID PROCEDURES: If swallowed - give emetic, get medical aid; inhalation - rest; wash from skin & eyes.

SECTION VI REACTIVITY DATA

CONDITIONS CONTRIBUTING TO INSTABILITY

CONDITIONS CONTRIBUTING TO HAZARDOUS POLYMERIZATION

INCOMPATABILITY (MATERIALS TO AVOID)

HAZARDOUS DECOMPOSITION PRODUCTS

SECTION VII SPILL OR LEAK PROCEDURES

STEPS TO BE TAKEN IN CASE MATERIAL IS RELEASED OR SPILLED: Cover with (9:1) sand to soda ash mixture; put in cardboard carton for disposal.

WASTE DISPOSAL METHOD

SECTION VIII SPECIAL PROTECTION INFORMATION

VENTILATION REQUIREMENTS LOCAL EXHAUST | PROTECTIVE EQUIPMENT (SPECIFY TYPES) EYE: Goggles

MECHANICAL (GENERAL) | GLOVES: Rubber (butyl)

SPECIAL | RESPIRATOR: With proper filter

OTHER PROTECTIVE EQUIPMENT: Plastic coveralls, self-contained breathing apparatus.

SECTION IX SPECIAL PRECAUTIONS

PRECAUTIONS TO BE TAKEN IN HANDLING AND STORAGE

OTHER PRECAUTIONS

SIGNATURE | DATE

p-PHENYLENEDIAMINE

1
4 0

SECTION I SOURCE AND NOMENCLATURE

MANUFACTURER'S NAME | EMERGENCY TELEPHONE NO.

ADDRESS (NUMBER, STREET, CITY, STATE, ZIP CODE)

TRADE NAME AND SYNONYMS: 4 Aminoaniline | CHEMICAL FAMILY

CHEMICAL NAME AND SYNONYMS: 1,4 Diaminobenzene | FORMULA: $C_6H_4(NH_2)_2$

SECTION II HAZARDOUS INGREDIENTS

BASIC MATERIAL	APPROXIMATE OR MAXIMUM % WT. OR VOL.	ESTABLISHED OSHA STANDARD	LD_{50} ORAL	LD_{50} PERCUT.	LC_{50} SPECIES	LC_{50} CONC.
Oral - human LDLO 50 mg/kg			See Notes			
Notes:						
Lowest published lethal dose						
Oral - rat 100 mg/kg						
Subcutaneous - rat 170 mg/kg						
Intravenous - dog 17 mg/kg						
Intraperitoneal - rat 50 mg/kg						
Oral - cat - 100 mg/kg						
Subcutaneous - rabbit 200 mg/kg.						

SECTION III PHYSICAL DATA MW - 108.16

BOILING POINT °F: 267°C | VAPOR PRESSURE mm Hg.

MELTING POINT °F: 140°C | VAPOR DENSITY (Air = 1): 3.72

SPECIFIC GRAVITY (H_2O = 1) | EVAPORATION RATE (______ = 1)

SOLUBILITY IN WATER Pts/100 pts H_2O: 3.8 | VOLATILE % Vol. % Wt.

APPEARANCE AND ODOR: colorless monoclinic crystals

SECTION IV FIRE AND EXPLOSION HAZARD DATA

FLASH POINT: 312 | FLAMMABLE (EXPLOSIVE) LIMITS | UPPER | LOWER

METHOD USED: C.C.

EXTINGUISHING MEDIA: water, carbon dioxide, dry chemical.

SPECIAL FIRE FIGHTING PROCEDURES

UNUSUAL FIRE AND EXPLOSION HAZARDS

SECTION V HEALTH HAZARD DATA

TOXIC LEVEL: 0.1 mg/C.M. (skin) | CARCINOGENIC: no

PRINCIPLE ROUTES OF ABSORBTION: inhalation, ingestion, skin | SKIN AND EYE IRRITATION: severe

REVELANT SYMPTOMS OF EXPOSURE: dermatitis, eye irritation.

EFFECTS OF CHRONIC EXPOSURE: Allergenic bronchial asthma, dermatitis, liver damage.

EMERGENCY AND FIRST AID PROCEDURES: If swallowed - give emetic, get medical aid; inhalation - rest; wash from skin & eyes.

SECTION VI REACTIVITY DATA

CONDITIONS CONTRIBUTING TO INSTABILITY

CONDITIONS CONTRIBUTING TO HAZARDOUS POLYMERIZATION

INCOMPATABILITY (MATERIALS TO AVOID)

HAZARDOUS DECOMPOSITION PRODUCTS

SECTION VII SPILL OR LEAK PROCEDURES

STEPS TO BE TAKEN IN CASE MATERIAL IS RELEASED OR SPILLED: Cover with (9:1) sand to soda ash mixture; put into cardboard carton for disposal.

WASTE DISPOSAL METHOD

SECTION VIII SPECIAL PROTECTION INFORMATION

VENTILATION REQUIREMENTS LOCAL EXHAUST | PROTECTIVE EQUIPMENT (SPECIFY TYPES) EYE: Goggles

MECHANICAL (GENERAL) | GLOVES: Rubber (butyl)

SPECIAL | RESPIRATOR: With proper filter

OTHER PROTECTIVE EQUIPMENT: Plastic coveralls, self-contained breathing apparatus.

SECTION IX SPECIAL PRECAUTIONS

PRECAUTIONS TO BE TAKEN IN HANDLING AND STORAGE

OTHER PRECAUTIONS

SIGNATURE | DATE

m-PHENYLENEDIAMINE DIHYDROCHLORIDE

(Hazard diamond: 1, 3, 0)

SECTION I SOURCE AND NOMENCLATURE

MANUFACTURER'S NAME | EMERGENCY TELEPHONE NO.

ADDRESS (NUMBER, STREET, CITY, STATE, ZIP CODE)

TRADE NAME AND SYNONYMS | CHEMICAL FAMILY

CHEMICAL NAME AND SYNONYMS: 1,4 Diaminobenzene dihydrochloride | FORMULA: $C_6H_4(NH_2)_2 2HCl$

SECTION II HAZARDOUS INGREDIENTS

BASIC MATERIAL	APPROXIMATE OR MAXIMUM % WT. OR VOL.	ESTABLISHED OSHA STANDARD	LD_{50} ORAL	LD_{50} PERCUT.	LC_{50} SPECIES	LC_{50} CONC.
			see note			

Notes:
Intraperitoneal - mouse LD50 100 mg/kg
Intraperitoneal - rat LD 50 325 mg/kg

SECTION III PHYSICAL DATA MW - 181.08

BOILING POINT °F.: -- | VAPOR PRESSURE mm Hg.

MELTING POINT °F. | VAPOR DENSITY (Air = 1)

SPECIFIC GRAVITY (H_2O = 1) | EVAPORATION RATE (____ = 1)

SOLUBILITY IN WATER Pts/100 pts H_2O: very soluble | VOLATILE % Vol. % Wt.

APPEARANCE AND ODOR: colorless crystals

SECTION IV FIRE AND EXPLOSION HAZARD DATA

FLASH POINT: none available | FLAMMABLE (EXPLOSIVE) LIMITS UPPER / LOWER

METHOD USED

EXTINGUISHING MEDIA: water, carbon dioxide, dry chemical.

SPECIAL FIRE FIGHTING PROCEDURES

UNUSUAL FIRE AND EXPLOSION HAZARDS

SECTION V HEALTH HAZARD DATA

TOXIC LEVEL: none available | CARCINOGENIC: no

PRINCIPLE ROUTES OF ABSORBTION: ingestion (mainly) | SKIN AND EYE IRRITATION: slight

REVELANT SYMPTOMS OF EXPOSURE: Dermatitis, headache, cyanosis, drowsiness, mental confusion.

EFFECTS OF CHRONIC EXPOSURE: Liver damage

EMERGENCY AND FIRST AID PROCEDURES: If swallowed - give emetic, get medical aid; wash from skin & eyes.

SECTION VI REACTIVITY DATA

CONDITIONS CONTRIBUTING TO INSTABILITY

CONDITIONS CONTRIBUTING TO HAZARDOUS POLYMERIZATION

INCOMPATABILITY (MATERIALS TO AVOID)

HAZARDOUS DECOMPOSITION PRODUCTS

SECTION VII SPILL OR LEAK PROCEDURES

STEPS TO BE TAKEN IN CASE MATERIAL IS RELEASED OR SPILLED

WASTE DISPOSAL METHOD

SECTION VIII SPECIAL PROTECTION INFORMATION

VENTILATION REQUIREMENTS LOCAL EXHAUST | PROTECTIVE EQUIPMENT (SPECIFY TYPES) EYE: Goggles

MECHANICAL (GENERAL) | GLOVES

SPECIAL | RESPIRATOR

OTHER PROTECTIVE EQUIPMENT

SECTION IX SPECIAL PRECAUTIONS

PRECAUTIONS TO BE TAKEN IN HANDLING AND STORAGE

OTHER PRECAUTIONS

SIGNATURE | DATE

β-PHENYLETHYL BROMIDE

(Hazard diamond: 2, 4, 1)

SECTION I SOURCE AND NOMENCLATURE

MANUFACTURER'S NAME | EMERGENCY TELEPHONE NO.

ADDRESS (NUMBER, STREET, CITY, STATE, ZIP CODE)

TRADE NAME AND SYNONYMS | CHEMICAL FAMILY

CHEMICAL NAME AND SYNONYMS: 2 Bromoethyl Benzene | FORMULA: $C_6H_5C_2H_4Br$

SECTION II HAZARDOUS INGREDIENTS

BASIC MATERIAL	APPROXIMATE OR MAXIMUM % WT. OR VOL.	ESTABLISHED OSHA STANDARD	LD_{50} ORAL	LD_{50} PERCUT.	LC_{50} SPECIES	LC_{50} CONC.
			none	available		

SECTION III PHYSICAL DATA MW - 185.07

BOILING POINT °F.: See vapor pressure | VAPOR PRESSURE: 102°C 20 mm Hg.

MELTING POINT °F. | VAPOR DENSITY (Air = 1)

SPECIFIC GRAVITY (H_2O = 1) | EVAPORATION RATE (____ = 1)

SOLUBILITY IN WATER Pts/100 pts H_2O: insoluble | VOLATILE % Vol. % Wt.

APPEARANCE AND ODOR: liquid

SECTION IV FIRE AND EXPLOSION HAZARD DATA

FLASH POINT: none available | FLAMMABLE (EXPLOSIVE) LIMITS UPPER / LOWER

METHOD USED

EXTINGUISHING MEDIA: water spray, dry chemical, carbon dioxide.

SPECIAL FIRE FIGHTING PROCEDURES

UNUSUAL FIRE AND EXPLOSION HAZARDS

SECTION V HEALTH HAZARD DATA

TOXIC LEVEL: none available | CARCINOGENIC: no

PRINCIPLE ROUTES OF ABSORBTION: Inhalation, ingestion | SKIN AND EYE IRRITATION: high

REVELANT SYMPTOMS OF EXPOSURE: Irritation of eyes, nose and throat; severe burns to digestive tract if swallowed.

EFFECTS OF CHRONIC EXPOSURE: --

EMERGENCY AND FIRST AID PROCEDURES: If inhaled - rest; if swallowed - give emetic, get medical aid; wash from skin & eyes.

SECTION VI REACTIVITY DATA

CONDITIONS CONTRIBUTING TO INSTABILITY

CONDITIONS CONTRIBUTING TO HAZARDOUS POLYMERIZATION

INCOMPATABILITY (MATERIALS TO AVOID)

HAZARDOUS DECOMPOSITION PRODUCTS

SECTION VII SPILL OR LEAK PROCEDURES

STEPS TO BE TAKEN IN CASE MATERIAL IS RELEASED OR SPILLED

WASTE DISPOSAL METHOD

SECTION VIII SPECIAL PROTECTION INFORMATION

VENTILATION REQUIREMENTS LOCAL EXHAUST | PROTECTIVE EQUIPMENT (SPECIFY TYPES) EYE: Goggles

MECHANICAL (GENERAL) | GLOVES: Rubber

SPECIAL | RESPIRATOR: Self-contained breathing apparatus

OTHER PROTECTIVE EQUIPMENT

SECTION IX SPECIAL PRECAUTIONS

PRECAUTIONS TO BE TAKEN IN HANDLING AND STORAGE

OTHER PRECAUTIONS

SIGNATURE | DATE

PHENYLHYDRAZINE

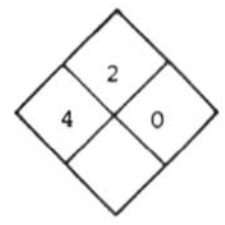

SECTION I SOURCE AND NOMENCLATURE

MANUFACTURER'S NAME	EMERGENCY TELEPHONE NO.
ADDRESS (NUMBER, STREET, CITY, STATE, ZIP CODE)	
TRADE NAME AND SYNONYMS	CHEMICAL FAMILY
CHEMICAL NAME AND SYNONYMS: Phenylhydrazine	FORMULA: $C_6H_5NHNH_2$

SECTION II HAZARDOUS INGREDIENTS

BASIC MATERIAL	APPROXIMATE OR MAXIMUM % WT. OR VOL.	ESTABLISHED OSHA STANDARD	LD_{50} mg/kg ORAL	PERCUT.	SPECIES	LC_{50} CONC.
			188		rat	
			80		rabbit	
			80		guinea pig	

SECTION III PHYSICAL DATA MW - 108.14

BOILING POINT °F.: 243°C	VAPOR PRESSURE mm Hg.: at 72°C 1
MELTING POINT °F.: 19.6°C	VAPOR DENSITY (Air = 1): 3.7
SPECIFIC GRAVITY (H_2O = 1): 1.10	EVAPORATION RATE (= 1)
SOLUBILITY IN WATER Pts/100 pts H_2O: slightly soluble	VOLATILE % Vol. % Wt.

APPEARANCE AND ODOR: yellow, monoclinic crystals or oil.

SECTION IV FIRE AND EXPLOSION HAZARD DATA

FLASH POINT: 192°F.	FLAMMABLE (EXPLOSIVE) LIMITS	UPPER
METHOD USED: C.C.		LOWER

EXTINGUISHING MEDIA: Alcohol foam, water, carbon dioxide.

SPECIAL FIRE FIGHTING PROCEDURES

UNUSUAL FIRE AND EXPLOSION HAZARDS: When heated to decomposition point, it gives off fumes of nitrogen compounds.

SECTION V HEALTH HAZARD DATA

TOXIC LEVEL: TLV 5 PPM, 22 mg/C.M. CARCINOGENIC: no

PRINCIPLE ROUTES OF ABSORBTION: Inhalation, ingestion, skin. SKIN AND EYE IRRITATION: severe

REVELANT SYMPTOMS OF EXPOSURE: Irritation of skin, eyes and mucous membranes, gastroentric disturbances, anemia.

EFFECTS OF CHRONIC EXPOSURE: Blood and liver damage, loss of appetite.

EMERGENCY AND FIRST AID PROCEDURES: Inhalation - rest; if swallowed - induce vomiting, get medical aid; wash from skin & eyes.

SECTION VI REACTIVITY DATA

CONDITIONS CONTRIBUTING TO INSTABILITY

CONDITIONS CONTRIBUTING TO HAZARDOUS POLYMERIZATION

INCOMPATABILITY (MATERIALS TO AVOID): Avoid lead dioxide.

HAZARDOUS DECOMPOSITION PRODUCTS

SECTION VII SPILL OR LEAK PROCEDURES

STEPS TO BE TAKEN IN CASE MATERIAL IS RELEASED OR SPILLED: Dilute with water, neutralize with sulfuric acid.

WASTE DISPOSAL METHOD

SECTION VIII SPECIAL PROTECTION INFORMATION

VENTILATION REQUIREMENTS LOCAL EXHAUST	PROTECTIVE EQUIPMENT (SPECIFY TYPES) EYE: Goggles
MECHANICAL (GENERAL)	GLOVES: Rubber
SPECIAL	RESPIRATOR: Self-contained

OTHER PROTECTIVE EQUIPMENT: Penetration - resistant coveralls, full-body shield.

SECTION IX SPECIAL PRECAUTIONS

PRECAUTIONS TO BE TAKEN IN HANDLING AND STORAGE: Store in flammable liquids storage cabinet; detached storage preferred.

OTHER PRECAUTIONS

SIGNATURE	DATE

PHENYL ISOCYANATE

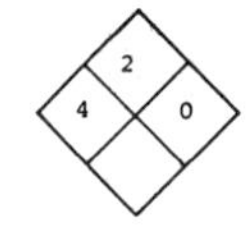

SECTION I SOURCE AND NOMENCLATURE

MANUFACTURER'S NAME	EMERGENCY TELEPHONE NO.
ADDRESS (NUMBER, STREET, CITY, STATE, ZIP CODE)	
TRADE NAME AND SYNONYMS: Benzene Isocyanate	CHEMICAL FAMILY
CHEMICAL NAME AND SYNONYMS: Phenyl Isocyanate	FORMULA: C_6H_5NCO

SECTION II HAZARDOUS INGREDIENTS

BASIC MATERIAL	APPROXIMATE OR MAXIMUM % WT. OR VOL.	ESTABLISHED OSHA STANDARD	LD_{50} mg/kg ORAL	PERCUT.	SPECIES	LC_{50} CONC.
			940		rat	

SECTION III PHYSICAL DATA MW - 119.12

BOILING POINT °F.: 163°C	VAPOR PRESSURE mm Hg.: at 62°C 13
MELTING POINT °F.: -30°C	VAPOR DENSITY (Air = 1)
SPECIFIC GRAVITY (H_2O = 1): 1.10	EVAPORATION RATE (= 1)
SOLUBILITY IN WATER Pts/100 pts H_2O: insoluble	VOLATILE % Vol. % Wt.

APPEARANCE AND ODOR: liquid with acrid odor.

SECTION IV FIRE AND EXPLOSION HAZARD DATA

FLASH POINT: none available	FLAMMABLE (EXPLOSIVE) LIMITS	UPPER
METHOD USED		LOWER

EXTINGUISHING MEDIA: Water, alcohol foam, carbon dioxide or dry chemical.

SPECIAL FIRE FIGHTING PROCEDURES: Wear self-contained breathing apparatus and goggles.

UNUSUAL FIRE AND EXPLOSION HAZARDS

SECTION V HEALTH HAZARD DATA

TOXIC LEVEL: None available. CARCINOGENIC: no

PRINCIPLE ROUTES OF ABSORBTION: Inhalation, ingestion. SKIN AND EYE IRRITATION: severe

REVELANT SYMPTOMS OF EXPOSURE: Eye and mucous membrane irritation; skin irritation. Sensitizer

EFFECTS OF CHRONIC EXPOSURE: Liver and kidney damage.

EMERGENCY AND FIRST AID PROCEDURES: If swallowed - give emetic, medical aid, gastric lavage; inhalation - rest, give oxygen as needed; wash from skin & eyes.

SECTION VI REACTIVITY DATA

CONDITIONS CONTRIBUTING TO INSTABILITY

CONDITIONS CONTRIBUTING TO HAZARDOUS POLYMERIZATION

INCOMPATABILITY (MATERIALS TO AVOID)

HAZARDOUS DECOMPOSITION PRODUCTS

SECTION VII SPILL OR LEAK PROCEDURES

STEPS TO BE TAKEN IN CASE MATERIAL IS RELEASED OR SPILLED

WASTE DISPOSAL METHOD

SECTION VIII SPECIAL PROTECTION INFORMATION

VENTILATION REQUIREMENTS LOCAL EXHAUST	PROTECTIVE EQUIPMENT (SPECIFY TYPES) EYE: Goggles
MECHANICAL (GENERAL)	GLOVES: Rubber
SPECIAL	RESPIRATOR: Self-contained

OTHER PROTECTIVE EQUIPMENT

SECTION IX SPECIAL PRECAUTIONS

PRECAUTIONS TO BE TAKEN IN HANDLING AND STORAGE

OTHER PRECAUTIONS

SIGNATURE	DATE

PHENYL ISOTHIOCYANATE

NFPA diamond: Health 4, Flammability 2, Reactivity 0

SECTION I SOURCE AND NOMENCLATURE

MANUFACTURER'S NAME	EMERGENCY TELEPHONE NO.
ADDRESS (NUMBER, STREET, CITY, STATE, ZIP CODE)	
TRADE NAME AND SYNONYMS: Benzene 1 Isothiocyanate	CHEMICAL FAMILY
CHEMICAL NAME AND SYNONYMS: Phenyl Isothiocyanate	FORMULA: C_6H_5NCS

SECTION II HAZARDOUS INGREDIENTS

BASIC MATERIAL	APPROXIMATE OR MAXIMUM % WT. OR VOL.	ESTABLISHED OSHA STANDARD	LD_{50} mg/kg ORAL	LD_{50} PERCUT.	LC_{50} SPECIES	LC_{50} CONC.
			400		mouse	

SECTION III PHYSICAL DATA MW - 135.19

BOILING POINT °F.: 221°C	VAPOR PRESSURE mm Hg.: at 90°C 15
MELTING POINT °F.: -21°C	VAPOR DENSITY (Air = 1)
SPECIFIC GRAVITY (H_2O = 1): 1.14	EVAPORATION RATE (= 1)
SOLUBILITY IN WATER Pts/100 pts H_2O: insoluble	VOLATILE % Vol. % Wt.

APPEARANCE AND ODOR: Pale yellow liquid

SECTION IV FIRE AND EXPLOSION HAZARD DATA

FLASH POINT	FLAMMABLE (EXPLOSIVE) LIMITS	UPPER
None available		
METHOD USED		LOWER

EXTINGUISHING MEDIA: Water, alcohol foam, carbon dioxide, or dry chemical.

SPECIAL FIRE FIGHTING PROCEDURES: Wear self-contained breathing apparatus and goggles.

UNUSUAL FIRE AND EXPLOSION HAZARDS

SECTION V HEALTH HAZARD DATA

TOXIC LEVEL: None available — CARCINOGENIC: no

PRINCIPLE ROUTES OF ABSORBTION: Inhalation, ingestion — SKIN AND EYE IRRITATION: Severe

REVELANT SYMPTOMS OF EXPOSURE: Eye and mucous membrane irritation; skin irritation, sensitizer.

EFFECTS OF CHRONIC EXPOSURE: Liver and kidney damage.

EMERGENCY AND FIRST AID PROCEDURES: If swallowed - give emetic, medical aid, gastric lavage; inhalation - rest, give oxygen as needed; wash from skin & eyes.

SECTION VI REACTIVITY DATA

CONDITIONS CONTRIBUTING TO INSTABILITY

CONDITIONS CONTRIBUTING TO HAZARDOUS POLYMERIZATION

INCOMPATABILITY (MATERIALS TO AVOID)

HAZARDOUS DECOMPOSITION PRODUCTS

SECTION VII SPILL OR LEAK PROCEDURES

STEPS TO BE TAKEN IN CASE MATERIAL IS RELEASED OR SPILLED

WASTE DISPOSAL METHOD

SECTION VIII SPECIAL PROTECTION INFORMATION

VENTILATION REQUIREMENTS LOCAL EXHAUST	PROTECTIVE EQUIPMENT (SPECIFY TYPES) EYE: Goggles
MECHANICAL (GENERAL)	GLOVES: Rubber
SPECIAL	RESPIRATOR: Self-contained

OTHER PROTECTIVE EQUIPMENT

SECTION IX SPECIAL PRECAUTIONS

PRECAUTIONS TO BE TAKEN IN HANDLING AND STORAGE

OTHER PRECAUTIONS

SIGNATURE	DATE

p-PHENYLPHENACYL BROMIDE

NFPA diamond: Health 4, Flammability 1, Reactivity 0

SECTION I SOURCE AND NOMENCLATURE

MANUFACTURER'S NAME	EMERGENCY TELEPHONE NO.
ADDRESS (NUMBER, STREET, CITY, STATE, ZIP CODE)	
TRADE NAME AND SYNONYMS	CHEMICAL FAMILY
CHEMICAL NAME AND SYNONYMS: Bromo-P-Phenylacetophenone	FORMULA: $C_6H_5C_6H_4COCH_2Br$

SECTION II HAZARDOUS INGREDIENTS

BASIC MATERIAL	APPROXIMATE OR MAXIMUM % WT. OR VOL.	ESTABLISHED OSHA STANDARD	LD_{50} ORAL	LD_{50} PERCUT.	LC_{50} SPECIES	LC_{50} CONC.
			none	available		

SECTION III PHYSICAL DATA MW - 275.15

BOILING POINT °F.: --	VAPOR PRESSURE mm Hg.
MELTING POINT °F.: 126°C	VAPOR DENSITY (Air = 1)
SPECIFIC GRAVITY (H_2O = 1)	EVAPORATION RATE (= 1)
SOLUBILITY IN WATER Pts/100 pts H_2O	VOLATILE % Vol. % Wt.

APPEARANCE AND ODOR: crystals

SECTION IV FIRE AND EXPLOSION HAZARD DATA

FLASH POINT	FLAMMABLE (EXPLOSIVE) LIMITS	UPPER
none available		
METHOD USED		LOWER

EXTINGUISHING MEDIA: carbon dioxide, dry chemical.

SPECIAL FIRE FIGHTING PROCEDURES

UNUSUAL FIRE AND EXPLOSION HAZARDS

SECTION V HEALTH HAZARD DATA

TOXIC LEVEL: none available — CARCINOGENIC: no

PRINCIPLE ROUTES OF ABSORBTION: inhalation, ingestion — SKIN AND EYE IRRITATION: severe

REVELANT SYMPTOMS OF EXPOSURE: Eye irritation, corneal haziness and pain; burning of throat; skin irritation.

EFFECTS OF CHRONIC EXPOSURE: --

EMERGENCY AND FIRST AID PROCEDURES: Inhalation - rest; if swallowed - wash mouth, give emetic, get medical aid; wash from skin & eyes.

SECTION VI REACTIVITY DATA

CONDITIONS CONTRIBUTING TO INSTABILITY

CONDITIONS CONTRIBUTING TO HAZARDOUS POLYMERIZATION

INCOMPATABILITY (MATERIALS TO AVOID)

HAZARDOUS DECOMPOSITION PRODUCTS

SECTION VII SPILL OR LEAK PROCEDURES

STEPS TO BE TAKEN IN CASE MATERIAL IS RELEASED OR SPILLED

WASTE DISPOSAL METHOD

SECTION VIII SPECIAL PROTECTION INFORMATION

VENTILATION REQUIREMENTS LOCAL EXHAUST	PROTECTIVE EQUIPMENT (SPECIFY TYPES) EYE: Goggles
MECHANICAL (GENERAL)	GLOVES: Rubber
SPECIAL	RESPIRATOR: With proper filter

OTHER PROTECTIVE EQUIPMENT

SECTION IX SPECIAL PRECAUTIONS

PRECAUTIONS TO BE TAKEN IN HANDLING AND STORAGE

OTHER PRECAUTIONS

SIGNATURE	DATE

β-PHENYLPROPIONYL CHLORIDE

Hazard rating: 1 (top), 4 (left), 2 (right), W (bottom, struck through)

SECTION I SOURCE AND NOMENCLATURE

MANUFACTURER'S NAME

EMERGENCY TELEPHONE NO.

ADDRESS (NUMBER, STREET, CITY, STATE, ZIP CODE)

TRADE NAME AND SYNONYMS

CHEMICAL FAMILY

CHEMICAL NAME AND SYNONYMS: Hydrocinnamoyl Chloride

FORMULA: $C_6H_5CH_2CH_2COCl$

SECTION II HAZARDOUS INGREDIENTS

BASIC MATERIAL	APPROXIMATE OR MAXIMUM % WT. OR VOL.	ESTABLISHED OSHA STANDARD	LD_{50} ORAL	LD_{50} PERCUT.	LC_{50} SPECIES	LC_{50} CONC.
			none	available		

SECTION III PHYSICAL DATA MW = 168.82

BOILING POINT °F.: see vapor pressure	VAPOR PRESSURE mm Hg.: at 120°C 20
MELTING POINT °F.: −5°C	VAPOR DENSITY (Air = 1)
SPECIFIC GRAVITY (H_2O = 1)	EVAPORATION RATE (____ = 1)
SOLUBILITY IN WATER Pts/100 pts H_2O: decomposes	VOLATILE % Vol. % Wt.

APPEARANCE AND ODOR: colorless liquid

SECTION IV FIRE AND EXPLOSION HAZARD DATA

FLASH POINT: none available

METHOD USED

FLAMMABLE (EXPLOSIVE) LIMITS: UPPER / LOWER

EXTINGUISHING MEDIA: carbon dioxide, dry chemical.

SPECIAL FIRE FIGHTING PROCEDURES: Wear full protective clothing.

UNUSUAL FIRE AND EXPLOSION HAZARDS: Reacts with water.

SECTION V HEALTH HAZARD DATA

TOXIC LEVEL: None available

CARCINOGENIC: no

PRINCIPLE ROUTES OF ABSORBTION: Inhalation, ingestion

SKIN AND EYE IRRITATION: severe

REVELANT SYMPTOMS OF EXPOSURE: Cyanosis, coughing, headache, pulmonary edema, ulceration of skin, dizziness.

EFFECTS OF CHRONIC EXPOSURE: Erosion of teeth, jaw necrosis, bronchial irritation and chronic cough.

EMERGENCY AND FIRST AID PROCEDURES: Inhalation - artificial respiration if necessary; ingestion - drink water, milk of magnesia, get medical aid; wash from skin & eyes.

SECTION VI REACTIVITY DATA

CONDITIONS CONTRIBUTING TO INSTABILITY

CONDITIONS CONTRIBUTING TO HAZARDOUS POLYMERIZATION

INCOMPATABILITY (MATERIALS TO AVOID): Avoid water

HAZARDOUS DECOMPOSITION PRODUCTS: Produces HCl with water

SECTION VII SPILL OR LEAK PROCEDURES

STEPS TO BE TAKEN IN CASE MATERIAL IS RELEASED OR SPILLED

WASTE DISPOSAL METHOD

SECTION VIII SPECIAL PROTECTION INFORMATION

VENTILATION REQUIREMENTS LOCAL EXHAUST

MECHANICAL (GENERAL)

SPECIAL

OTHER PROTECTIVE EQUIPMENT

PROTECTIVE EQUIPMENT (SPECIFY TYPES) EYE: Goggles

GLOVES: Rubber

RESPIRATOR: With proper filter

SECTION IX SPECIAL PRECAUTIONS

PRECAUTIONS TO BE TAKEN IN HANDLING AND STORAGE: Store in cool, dry, well-ventilated room; inside storage in standard flammable liquids storage cabinet.

OTHER PRECAUTIONS

SIGNATURE

DATE

1-PHENYLSEMICARBAZIDE

Hazard rating: 1 (top), 4 (left), 0 (right)

SECTION I SOURCE AND NOMENCLATURE

MANUFACTURER'S NAME

EMERGENCY TELEPHONE NO.

ADDRESS (NUMBER, STREET, CITY, STATE, ZIP CODE)

TRADE NAME AND SYNONYMS: Cryogenine

CHEMICAL FAMILY

CHEMICAL NAME AND SYNONYMS: 1-Phenylsemicarbazide

FORMULA: $C_6H_5NHNHCONH_2$

SECTION II HAZARDOUS INGREDIENTS

BASIC MATERIAL	APPROXIMATE OR MAXIMUM % WT. OR VOL.	ESTABLISHED OSHA STANDARD	LD_{50} ORAL	LD_{50} PERCUT.	LC_{50} SPECIES	LC_{50} CONC.
Intraperitoneal - rat LD50 55 mg/kg						

SECTION III PHYSICAL DATA MW = 151.17

BOILING POINT °F.: --	VAPOR PRESSURE mm Hg.
MELTING POINT °F.: 176	VAPOR DENSITY (Air = 1)
SPECIFIC GRAVITY (H_2O = 1)	EVAPORATION RATE (____ = 1)
SOLUBILITY IN WATER Pts/100 pts H_2O: soluble in hot water	VOLATILE % Vol. % Wt.

APPEARANCE AND ODOR

SECTION IV FIRE AND EXPLOSION HAZARD DATA

FLASH POINT: none available

METHOD USED

FLAMMABLE (EXPLOSIVE) LIMITS: UPPER / LOWER

EXTINGUISHING MEDIA: Alcohol foam, carbon dioxide, dry chemical.

SPECIAL FIRE FIGHTING PROCEDURES

UNUSUAL FIRE AND EXPLOSION HAZARDS

SECTION V HEALTH HAZARD DATA

TOXIC LEVEL: none available

CARCINOGENIC: no

PRINCIPLE ROUTES OF ABSORBTION: inhalation, ingestion, skin

SKIN AND EYE IRRITATION: severe

REVELANT SYMPTOMS OF EXPOSURE: Irritation of skin, eyes and mucous membranes, gastroentric disturbances, anemia.

EFFECTS OF CHRONIC EXPOSURE: Blood and liver damage, loss of appetite.

EMERGENCY AND FIRST AID PROCEDURES: Inhalation - rest; if swallowed - induce vomiting, get medical aid; wash from eyes and skin.

SECTION VI REACTIVITY DATA

CONDITIONS CONTRIBUTING TO INSTABILITY

CONDITIONS CONTRIBUTING TO HAZARDOUS POLYMERIZATION

INCOMPATABILITY (MATERIALS TO AVOID)

HAZARDOUS DECOMPOSITION PRODUCTS

SECTION VII SPILL OR LEAK PROCEDURES

STEPS TO BE TAKEN IN CASE MATERIAL IS RELEASED OR SPILLED

WASTE DISPOSAL METHOD

SECTION VIII SPECIAL PROTECTION INFORMATION

VENTILATION REQUIREMENTS LOCAL EXHAUST

MECHANICAL (GENERAL)

SPECIAL

OTHER PROTECTIVE EQUIPMENT

PROTECTIVE EQUIPMENT (SPECIFY TYPES) EYE: Goggles

GLOVES: Rubber

RESPIRATOR: Self-contained

SECTION IX SPECIAL PRECAUTIONS

PRECAUTIONS TO BE TAKEN IN HANDLING AND STORAGE

OTHER PRECAUTIONS

SIGNATURE

DATE

4-PHENYLTHIOSEMICARBAZIDE

Hazard diamond: 1 (top), 4 (left), 0 (right)

SECTION I SOURCE AND NOMENCLATURE

MANUFACTURER'S NAME | EMERGENCY TELEPHONE NO.

ADDRESS (NUMBER, STREET, CITY, STATE, ZIP CODE)

TRADE NAME AND SYNONYMS | CHEMICAL FAMILY

CHEMICAL NAME AND SYNONYMS: 4 Phenylthiosemicarbazide | FORMULA: $C_6H_5NHNHCSNH_2$

SECTION II HAZARDOUS INGREDIENTS

BASIC MATERIAL	APPROXIMATE OR MAXIMUM % WT. OR VOL.	ESTABLISHED OSHA STANDARD	LD_{50} ORAL	LD_{50} PERCUT.	LC_{50} SPECIES	LC_{50} CONC.

Notes:
1. Lowest published lethal dose oral - rat 50 mg/kg
2. Lowest published lethal dose intraperitoneal - mouse 15 mg/kg.

SECTION III PHYSICAL DATA MW - 167.23

BOILING POINT °F. | VAPOR PRESSURE mm Hg.

MELTING POINT: 140°C | VAPOR DENSITY (Air = 1)

SPECIFIC GRAVITY (H_2O = 1) | EVAPORATION RATE (____ = 1)

SOLUBILITY IN WATER Pts/100 pts H_2O | VOLATILE % Vol. % Wt.

APPEARANCE AND ODOR: crystals

SECTION IV FIRE AND EXPLOSION HAZARD DATA

FLASH POINT: none available | FLAMMABLE (EXPLOSIVE) LIMITS: UPPER / LOWER

METHOD USED

EXTINGUISHING MEDIA: alcohol foam, dry chemical, carbon dioxide.

SPECIAL FIRE FIGHTING PROCEDURES

UNUSUAL FIRE AND EXPLOSION HAZARDS

SECTION V HEALTH HAZARD DATA

TOXIC LEVEL: none available | CARCINOGENIC: no

PRINCIPLE ROUTES OF ABSORBTION: inhalation, ingestion, skin | SKIN AND EYE IRRITATION: severe

REVELANT SYMPTOMS OF EXPOSURE: Irritation of skin, eyes and mucous membranes, gastroentric disturbances, anemia.

EFFECTS OF CHRONIC EXPOSURE: Blood and liver damage, loss of appetite.

EMERGENCY AND FIRST AID PROCEDURES: Inhalation - rest; if swallowed - induce vomiting, get medical aid; wash from skin & eyes.

SECTION VI REACTIVITY DATA

CONDITIONS CONTRIBUTING TO INSTABILITY

CONDITIONS CONTRIBUTING TO HAZARDOUS POLYMERIZATION

INCOMPATABILITY (MATERIALS TO AVOID)

HAZARDOUS DECOMPOSITION PRODUCTS

SECTION VII SPILL OR LEAK PROCEDURES

STEPS TO BE TAKEN IN CASE MATERIAL IS RELEASED OR SPILLED

WASTE DISPOSAL METHOD

SECTION VIII SPECIAL PROTECTION INFORMATION

VENTILATION REQUIREMENTS LOCAL EXHAUST | PROTECTIVE EQUIPMENT (SPECIFY TYPES) EYE: Goggles

MECHANICAL (GENERAL) | GLOVES: Rubber

SPECIAL | RESPIRATOR: Self-contained

OTHER PROTECTIVE EQUIPMENT

SECTION IX SPECIAL PRECAUTIONS

PRECAUTIONS TO BE TAKEN IN HANDLING AND STORAGE

OTHER PRECAUTIONS

SIGNATURE | DATE

PHENYLTHIOUREA

Hazard diamond: 1 (top), 3 (left), 1 (right)

SECTION I SOURCE AND NOMENCLATURE

MANUFACTURER'S NAME | EMERGENCY TELEPHONE NO.

ADDRESS (NUMBER, STREET, CITY, STATE, ZIP CODE)

TRADE NAME AND SYNONYMS | CHEMICAL FAMILY

CHEMICAL NAME AND SYNONYMS: Phenylcarbamide | FORMULA: $C_6H_5HN-CS-NH_2$

SECTION II HAZARDOUS INGREDIENTS

BASIC MATERIAL	APPROXIMATE OR MAXIMUM % WT. OR VOL.	ESTABLISHED OSHA STANDARD	LD_{50} mg/kg ORAL	LD_{50} mg/kg PERCUT.	SPECIES	LC_{50} CONC.
			3		rat	
			10		mouse	
			40		rabbit	

SECTION III PHYSICAL DATA MW - 152.22

BOILING POINT °F. | VAPOR PRESSURE mm Hg.

MELTING POINT: 154°C | VAPOR DENSITY (Air = 1)

SPECIFIC GRAVITY (H_2O = 1): 1.3 | EVAPORATION RATE (____ = 1)

SOLUBILITY IN WATER Pts/100 pts H_2O: slightly soluble | VOLATILE % Vol. % Wt.

APPEARANCE AND ODOR: needle-like crystals - bitter taste.

SECTION IV FIRE AND EXPLOSION HAZARD DATA

FLASH POINT: none available | FLAMMABLE (EXPLOSIVE) LIMITS: UPPER / LOWER

METHOD USED

EXTINGUISHING MEDIA: carbon dioxide, water.

SPECIAL FIRE FIGHTING PROCEDURES

UNUSUAL FIRE AND EXPLOSION HAZARDS: Burns to form SO_2 fumes.

SECTION V HEALTH HAZARD DATA

TOXIC LEVEL: none available | CARCINOGENIC: no

PRINCIPLE ROUTES OF ABSORBTION: inhalation, ingestion | SKIN AND EYE IRRITATION: slight

REVELANT SYMPTOMS OF EXPOSURE: dermatitis

EFFECTS OF CHRONIC EXPOSURE: Ingestion - depression of bone marrow, anemia, leukopenia, allergic skin eruptions.

EMERGENCY AND FIRST AID PROCEDURES: Inhalation - rest; if swallowed - induce vomiting, get medical aid; wash from skin & eyes.

SECTION VI REACTIVITY DATA

CONDITIONS CONTRIBUTING TO INSTABILITY

CONDITIONS CONTRIBUTING TO HAZARDOUS POLYMERIZATION

INCOMPATABILITY (MATERIALS TO AVOID)

HAZARDOUS DECOMPOSITION PRODUCTS: On decomposition emits toxic fumes of sulfur and nitrogen.

SECTION VII SPILL OR LEAK PROCEDURES

STEPS TO BE TAKEN IN CASE MATERIAL IS RELEASED OR SPILLED

WASTE DISPOSAL METHOD

SECTION VIII SPECIAL PROTECTION INFORMATION

VENTILATION REQUIREMENTS LOCAL EXHAUST | PROTECTIVE EQUIPMENT (SPECIFY TYPES) EYE: Goggles

MECHANICAL (GENERAL) | GLOVES

SPECIAL | RESPIRATOR

OTHER PROTECTIVE EQUIPMENT

SECTION IX SPECIAL PRECAUTIONS

PRECAUTIONS TO BE TAKEN IN HANDLING AND STORAGE

OTHER PRECAUTIONS

SIGNATURE | DATE

PHOSPHORIC ANHYDRIDE

0 3 2

SECTION I SOURCE AND NOMENCLATURE

MANUFACTURER'S NAME	EMERGENCY TELEPHONE NO.
ADDRESS (NUMBER, STREET, CITY, STATE, ZIP CODE)	
TRADE NAME AND SYNONYMS	CHEMICAL FAMILY
CHEMICAL NAME AND SYNONYMS Phosphorus Pentoxide	FORMULA P_2O_5

SECTION II HAZARDOUS INGREDIENTS

BASIC MATERIAL	APPROXIMATE OR MAXIMUM % WT. OR VOL.	ESTABLISHED OSHA STANDARD	LD_{50} ORAL	LD_{50} PERCUT.	LC_{50} SPECIES	LC_{50} CONC.
			Not available			

SECTION III PHYSICAL DATA MW - 141.94

BOILING POINT °F. sublimes easily at 300°C.	VAPOR PRESSURE mm Hg. at 384°C. 1
MELTING POINT °F. 657	VAPOR DENSITY (Air = 1)
SPECIFIC GRAVITY (H_2O = 1) 2.39	EVAPORATION RATE (= 1)
SOLUBILITY IN WATER Pts/100 pts H_2O	VOLATILE % Vol. % Wt.

APPEARANCE AND ODOR
White, fluffy, crystalline, deliquescent powder.

SECTION IV FIRE AND EXPLOSION HAZARD DATA

FLASH POINT Not flammable	FLAMMABLE (EXPLOSIVE) LIMITS	UPPER
METHOD USED		LOWER

EXTINGUISHING MEDIA
Avoid water

SPECIAL FIRE FIGHTING PROCEDURES
Wear self-contained respirator.

UNUSUAL FIRE AND EXPLOSION HAZARDS

SECTION V HEALTH HAZARD DATA

TOXIC LEVEL
TLV - 1 mg/m³ CARCINOGENIC No

PRINCIPLE ROUTES OF ABSORBTION
Inhalation of powder, ingestion SKIN AND EYE IRRITATION Severe

REVELANT SYMPTOMS OF EXPOSURE Coughing, choking, headache, dizziness, pulmonary edema, corrosion of skin, possible cyanosis.

EFFECTS OF CHRONIC EXPOSURE Bronchial irritation, possible gastrointestinal disturbances, erosion of teeth and jaw necrosis.

EMERGENCY AND FIRST AID PROCEDURES Inhalation - rest; if swallowed - give milk of magnesia, get medical aid; wash from skin & eyes.

SECTION VI REACTIVITY DATA

CONDITIONS CONTRIBUTING TO INSTABILITY

CONDITIONS CONTRIBUTING TO HAZARDOUS POLYMERIZATION

INCOMPATABILITY (MATERIALS TO AVOID) Hydrogen Fluoride, Perchloric Acid, (Perchloric Acid and Chloroform), Potassium, Propargyl Alcohol, Sodium, Water (Water and Organic Matter).

HAZARDOUS DECOMPOSITION PRODUCTS

SECTION VII SPILL OR LEAK PROCEDURES

STEPS TO BE TAKEN IN CASE MATERIAL IS RELEASED OR SPILLED
Add soda ash to spill. Add Water.

WASTE DISPOSAL METHOD

SECTION VIII SPECIAL PROTECTION INFORMATION Full protection

VENTILATION REQUIREMENTS LOCAL EXHAUST	PROTECTIVE EQUIPMENT (SPECIFY TYPES) EYE Goggles
MECHANICAL (GENERAL)	GLOVES Neoprene
SPECIAL	RESPIRATOR With appropriate filter

OTHER PROTECTIVE EQUIPMENT
Rubber apron, full face-shield.

SECTION IX SPECIAL PRECAUTIONS

PRECAUTIONS TO BE TAKEN IN HANDLING AND STORAGE

OTHER PRECAUTIONS

SIGNATURE	DATE

PHOSPHORUS PENTACHLORIDE

0 3 2 ₩

SECTION I SOURCE AND NOMENCLATURE

MANUFACTURER'S NAME	EMERGENCY TELEPHONE NO.
ADDRESS (NUMBER, STREET, CITY, STATE, ZIP CODE)	
TRADE NAME AND SYNONYMS	CHEMICAL FAMILY
CHEMICAL NAME AND SYNONYMS Phosphorus Pentachloride	FORMULA PCl_5

SECTION II HAZARDOUS INGREDIENTS

BASIC MATERIAL	APPROXIMATE OR MAXIMUM % WT. OR VOL.	ESTABLISHED OSHA STANDARD	LD_{50} mg/kg ORAL	LD_{50} mg/kg PERCUT.	LC_{50} mg/cm SPECIES	LC_{50} mg/cm CONC.
#1 lowest published lethal concentration in air						
inhalation - mouse - 120 ppm.			660		rat	205

SECTION III PHYSICAL DATA MW 208.31

BOILING POINT °F. 162° C. sublimes	VAPOR PRESSURE mm Hg. at 55.5°C. 1
MELTING POINT °F. 148°C. under pressure	VAPOR DENSITY (Air = 1)
SPECIFIC GRAVITY (H_2O = 1) 1.6 g/l	EVAPORATION RATE (= 1)
SOLUBILITY IN WATER Pts/100 pts H_2O Decomposes in water	VOLATILE % Vol. % Wt.

APPEARANCE AND ODOR
Yellowish-white, fuming, crystalline.

SECTION IV FIRE AND EXPLOSION HAZARD DATA

FLASH POINT Will not burn	FLAMMABLE (EXPLOSIVE) LIMITS	UPPER
METHOD USED		LOWER

EXTINGUISHING MEDIA
Avoid water

SPECIAL FIRE FIGHTING PROCEDURES
Wear self-contained respirator.

UNUSUAL FIRE AND EXPLOSION HAZARDS

SECTION V HEALTH HAZARD DATA

TOXIC LEVEL
TLV 1 mg/m³ CARCINOGENIC No

PRINCIPLE ROUTES OF ABSORBTION
Inhalation of dust, ingestion, skin. SKIN AND EYE IRRITATION Severe

REVELANT SYMPTOMS OF EXPOSURE Irritant to mucous membranes and all parts of respiratory system. Coughing, choking, etc.

EFFECTS OF CHRONIC EXPOSURE
Possible damage to lungs, pulmonary edema.

EMERGENCY AND FIRST AID PROCEDURES
Inhalaed - rest, wash skin and eyes.

SECTION VI REACTIVITY DATA

CONDITIONS CONTRIBUTING TO INSTABILITY

CONDITIONS CONTRIBUTING TO HAZARDOUS POLYMERIZATION

INCOMPATABILITY (MATERIALS TO AVOID) Chlorine Trioxide, Flourine, Hydroxylamine, Magnesium Oxide, Phosphorous

HAZARDOUS DECOMPOSITION PRODUCTS
Heated to decomposition, emits highly toxic fumes of chlorides.

SECTION VII SPILL OR LEAK PROCEDURES

STEPS TO BE TAKEN IN CASE MATERIAL IS RELEASED OR SPILLED
Cover with sodium bicarbonate, then water spray, transfer into large container of water for disposal.

WASTE DISPOSAL METHOD

SECTION VIII SPECIAL PROTECTION INFORMATION

VENTILATION REQUIREMENTS LOCAL EXHAUST	PROTECTIVE EQUIPMENT (SPECIFY TYPES) EYE Goggles
MECHANICAL (GENERAL)	GLOVES Rubber (Neoprene)
SPECIAL	RESPIRATOR W/proper filter

OTHER PROTECTIVE EQUIPMENT
Rubber apron

SECTION IX SPECIAL PRECAUTIONS

PRECAUTIONS TO BE TAKEN IN HANDLING AND STORAGE
Protect the sealed glass bottles in which material is stored.

OTHER PRECAUTIONS

SIGNATURE	DATE

PHOSPHOTUNGSTIC ACID

Hazard diamond: 0 (top), 3 (left), 0 (right)

SECTION I SOURCE AND NOMENCLATURE

MANUFACTURER'S NAME	EMERGENCY TELEPHONE NO.
ADDRESS (NUMBER, STREET, CITY, STATE, ZIP CODE)	
TRADE NAME AND SYNONYMS	CHEMICAL FAMILY
CHEMICAL NAME AND SYNONYMS: Phosphotungstic Acid	FORMULA: $P_2O_5 \cdot 12WO_3 \cdot 42H_2O$

SECTION II HAZARDOUS INGREDIENTS

BASIC MATERIAL	APPROXIMATE OR MAXIMUM % WT. OR VOL.	ESTABLISHED OSHA STANDARD	LD_{50} ORAL	LD_{50} PERCUT.	LC_{50} SPECIES	LC_{50} CONC.
			None available			

SECTION III PHYSICAL DATA MW-3680.94

BOILING POINT °F: --	VAPOR PRESSURE mm Hg:
MELTING POINT °F: --	VAPOR DENSITY (Air = 1):
SPECIFIC GRAVITY (H_2O = 1): --	EVAPORATION RATE (= 1):
SOLUBILITY IN WATER Pts/100 pts H_2O: Soluble	VOLATILE % Vol. % Wt.:

APPEARANCE AND ODOR: Yellow-green crystals

SECTION IV FIRE AND EXPLOSION HAZARD DATA

	FLAMMABLE (EXPLOSIVE) LIMITS	
FLASH POINT: Not flammable		UPPER
METHOD USED		LOWER

EXTINGUISHING MEDIA

SPECIAL FIRE FIGHTING PROCEDURES

UNUSUAL FIRE AND EXPLOSION HAZARDS

SECTION V HEALTH HAZARD DATA

TOXIC LEVEL: 1.7 mg/C.M. CARCINOGENIC: No

PRINCIPLE ROUTES OF ABSORBTION: Ingestion (mainly), inhalation of dust SKIN AND EYE IRRITATION: Severe

REVELANT SYMPTOMS OF EXPOSURE: Loss of appetite, colic, loss of coordination, trembling, difficulty in breathing, weight loss.

EFFECTS OF CHRONIC EXPOSURE: --

EMERGENCY AND FIRST AID PROCEDURES: If swallowed - wash mouth, induce vomiting, get medical aid; inhalation - rest; wash skin and eyes.

SECTION VI REACTIVITY DATA

CONDITIONS CONTRIBUTING TO INSTABILITY

CONDITIONS CONTRIBUTING TO HAZARDOUS POLYMERIZATION

INCOMPATABILITY (MATERIALS TO AVOID)

HAZARDOUS DECOMPOSITION PRODUCTS

SECTION VII SPILL OR LEAK PROCEDURES

STEPS TO BE TAKEN IN CASE MATERIAL IS RELEASED OR SPILLED

WASTE DISPOSAL METHOD

SECTION VIII SPECIAL PROTECTION INFORMATION

VENTILATION REQUIREMENTS LOCAL EXHAUST	PROTECTIVE EQUIPMENT (SPECIFY TYPES) EYE: Goggles
MECHANICAL (GENERAL)	GLOVES
SPECIAL	RESPIRATOR

OTHER PROTECTIVE EQUIPMENT

SECTION IX SPECIAL PRECAUTIONS

PRECAUTIONS TO BE TAKEN IN HANDLING AND STORAGE

OTHER PRECAUTIONS

SIGNATURE	DATE

PICOLINE

Hazard diamond: 2 (top), 4 (left), 0 (right)

SECTION I SOURCE AND NOMENCLATURE

MANUFACTURER'S NAME	EMERGENCY TELEPHONE NO.
ADDRESS (NUMBER, STREET, CITY, STATE, ZIP CODE)	
TRADE NAME AND SYNONYMS: Methyl Pyridine	CHEMICAL FAMILY
CHEMICAL NAME AND SYNONYMS: α Picoline	FORMULA: $C_6H_4NCH_3$

SECTION II HAZARDOUS INGREDIENTS

BASIC MATERIAL	APPROXIMATE OR MAXIMUM % WT. OR VOL.	ESTABLISHED OSHA STANDARD	LD_{50} ORAL	LD_{50} PERCUT.	LC_{50} SPECIES	LC_{50} CONC.
			790		rat	
			674		mouse	
Lowest published lethal concentration in air				410	rabbit	
inhalation - rat 4000 PPM			900		guinea pig	

SECTION III PHYSICAL DATA MW - 92.13

BOILING POINT °F: 129°C	VAPOR PRESSURE mm Hg: at 24°C 10
MELTING POINT °F: -70°C	VAPOR DENSITY (Air = 1): 3.2
SPECIFIC GRAVITY (H_2O = 1): 0.95	EVAPORATION RATE (= 1):
SOLUBILITY IN WATER Pts/100 pts H_2O: very soluble	VOLATILE % Vol. % Wt.:

APPEARANCE AND ODOR: colorless liquid with unpleasant odor

SECTION IV FIRE AND EXPLOSION HAZARD DATA

	FLAMMABLE (EXPLOSIVE) LIMITS	
FLASH POINT: 102		UPPER
METHOD USED: O.C.		LOWER

EXTINGUISHING MEDIA: Alcohol foam, carbon dioxide, dry chemical

SPECIAL FIRE FIGHTING PROCEDURES: Wear goggles and self-contained breathing apparatus.

UNUSUAL FIRE AND EXPLOSION HAZARDS

SECTION V HEALTH HAZARD DATA

TOXIC LEVEL: none available CARCINOGENIC: no

PRINCIPLE ROUTES OF ABSORBTION: inhalation, ingestion, skin SKIN AND EYE IRRITATION: severe

REVELANT SYMPTOMS OF EXPOSURE: Narcosis, headache, nausea, giddiness, vomiting

EFFECTS OF CHRONIC EXPOSURE: Occasional vomiting and diarrhea; weight loss and anemia; ocular and facial paralysis.

EMERGENCY AND FIRST AID PROCEDURES: If swallowed - induce vomiting, get medical aid; inhalation - artificial respiration and oxygen as necessary; wash from skin & eyes.

SECTION VI REACTIVITY DATA

CONDITIONS CONTRIBUTING TO INSTABILITY

CONDITIONS CONTRIBUTING TO HAZARDOUS POLYMERIZATION

INCOMPATABILITY (MATERIALS TO AVOID): May react with oxidizing materials.

HAZARDOUS DECOMPOSITION PRODUCTS

SECTION VII SPILL OR LEAK PROCEDURES

STEPS TO BE TAKEN IN CASE MATERIAL IS RELEASED OR SPILLED

WASTE DISPOSAL METHOD

SECTION VIII SPECIAL PROTECTION INFORMATION

VENTILATION REQUIREMENTS LOCAL EXHAUST: Use in hood	PROTECTIVE EQUIPMENT (SPECIFY TYPES) EYE: goggles
MECHANICAL (GENERAL)	GLOVES: rubber
SPECIAL	RESPIRATOR: self-contained

OTHER PROTECTIVE EQUIPMENT

SECTION IX SPECIAL PRECAUTIONS

PRECAUTIONS TO BE TAKEN IN HANDLING AND STORAGE: Outside storage preferred, store inside in a flammable liquids storage cabinet.

OTHER PRECAUTIONS

SIGNATURE	DATE

β-PICOLINE

Hazard rating: 2 / 4 / 0

SECTION I SOURCE AND NOMENCLATURE

MANUFACTURER'S NAME | EMERGENCY TELEPHONE NO.

ADDRESS (NUMBER, STREET, CITY, STATE, ZIP CODE)

TRADE NAME AND SYNONYMS: 3 Methylpyridine

CHEMICAL FAMILY

CHEMICAL NAME AND SYNONYMS: 3 - Picoline

FORMULA: $C_5H_4NCH_3$

SECTION II HAZARDOUS INGREDIENTS

BASIC MATERIAL	APPROXIMATE OR MAXIMUM % WT. OR VOL.	ESTABLISHED OSHA STANDARD	LD_{50} ORAL	LD_{50} PERCUT.	LC_{50} SPECIES	LC_{50} CONC.
			none	available		

SECTION III PHYSICAL DATA MW - 93.12

BOILING POINT °F.: 143°C.

VAPOR PRESSURE mm Hg.

MELTING POINT °F.

VAPOR DENSITY (Air = 1): 3.21

SPECIFIC GRAVITY (H_2O = 1): 0.96

EVAPORATION RATE (= 1)

SOLUBILITY IN WATER Pts/100 pts H_2O: ∞

VOLATILE % Vol. % Wt.

APPEARANCE AND ODOR: Colorless liquid, with unpleasant odor.

SECTION IV FIRE AND EXPLOSION HAZARD DATA

FLASH POINT: None available

FLAMMABLE (EXPLOSIVE) LIMITS: UPPER / LOWER

METHOD USED

EXTINGUISHING MEDIA: Alcohol foam, carbon dioxide, dry chemical.

SPECIAL FIRE FIGHTING PROCEDURES: Wear goggles and self-contained breathing apparatus.

UNUSUAL FIRE AND EXPLOSION HAZARDS

SECTION V HEALTH HAZARD DATA

TOXIC LEVEL: none available

CARCINOGENIC: no

PRINCIPLE ROUTES OF ABSORBTION: Inhalation, ingestion, skin

SKIN AND EYE IRRITATION: Severe

REVELANT SYMPTOMS OF EXPOSURE: Narcosis, headache, nausea, giddiness, vomiting.

EFFECTS OF CHRONIC EXPOSURE: Occasional vomiting and diarrhea; weight loss and anemia; ocular and facial paralysis.

EMERGENCY AND FIRST AID PROCEDURES: If swallowed - induce vomiting, get medical aid; inhalation - artificial respiration and oxygen as needed; wash from skin & eyes.

SECTION VI REACTIVITY DATA

CONDITIONS CONTRIBUTING TO INSTABILITY

CONDITIONS CONTRIBUTING TO HAZARDOUS POLYMERIZATION

INCOMPATABILITY (MATERIALS TO AVOID): May react with oxidizing materials.

HAZARDOUS DECOMPOSITION PRODUCTS

SECTION VII SPILL OR LEAK PROCEDURES

STEPS TO BE TAKEN IN CASE MATERIAL IS RELEASED OR SPILLED

WASTE DISPOSAL METHOD

SECTION VIII SPECIAL PROTECTION INFORMATION

VENTILATION REQUIREMENTS LOCAL EXHAUST: Used in hood

PROTECTIVE EQUIPMENT (SPECIFY TYPES) EYE: Goggles

MECHANICAL (GENERAL)

GLOVES: Rubber

SPECIAL

RESPIRATOR: Self-contained

OTHER PROTECTIVE EQUIPMENT

SECTION IX SPECIAL PRECAUTIONS

PRECAUTIONS TO BE TAKEN IN HANDLING AND STORAGE: Outside storage preferred, store inside in a flammable liquids storage cabinet.

OTHER PRECAUTIONS

SIGNATURE | DATE

γ-PICOLINE

Hazard rating: 2 / 4 / 0

SECTION I SOURCE AND NOMENCLATURE

MANUFACTURER'S NAME | EMERGENCY TELEPHONE NO.

ADDRESS (NUMBER, STREET, CITY, STATE, ZIP CODE)

TRADE NAME AND SYNONYMS: 4 Methyl pyridine

CHEMICAL FAMILY

CHEMICAL NAME AND SYNONYMS: 4 Picoline

FORMULA: $C_5H_4NCH_3$

SECTION II HAZARDOUS INGREDIENTS

BASIC MATERIAL	APPROXIMATE OR MAXIMUM % WT. OR VOL.	ESTABLISHED OSHA STANDARD	LD_{50} mg/kg ORAL	LD_{50} mg/kg PERCUT.	SPECIES	LC_{50} CONC.
			1290		rat	
				270	rabbit	
Lowest published lethal concentration in air inhalation - rat 1000 PPM - 4 hrs.						

SECTION III PHYSICAL DATA MW - 93.12

BOILING POINT °F.: 143

VAPOR PRESSURE mm Hg.

MELTING POINT °F.: 3.7°C

VAPOR DENSITY (Air = 1): 3.21

SPECIFIC GRAVITY (H_2O = 1): 0.96

EVAPORATION RATE (= 1)

SOLUBILITY IN WATER Pts/100 pts H_2O: ∞

VOLATILE % Vol. % Wt.

APPEARANCE AND ODOR: colorless liquid with unpleasant odor.

SECTION IV FIRE AND EXPLOSION HAZARD DATA

FLASH POINT: 134°F

FLAMMABLE (EXPLOSIVE) LIMITS: UPPER / LOWER

METHOD USED: O.C.

EXTINGUISHING MEDIA: Alcohol foam, carbon dioxide, dry chemical

SPECIAL FIRE FIGHTING PROCEDURES: Wear goggles and self-contained breathing apparatus.

UNUSUAL FIRE AND EXPLOSION HAZARDS

SECTION V HEALTH HAZARD DATA

TOXIC LEVEL: none available

CARCINOGENIC: no

PRINCIPLE ROUTES OF ABSORBTION: inhalation, ingestion, skin

SKIN AND EYE IRRITATION: severe

REVELANT SYMPTOMS OF EXPOSURE: Narcosis, headache, nausea giddiness, vomiting.

EFFECTS OF CHRONIC EXPOSURE: Occasional vomiting and diarrhea weight loss and anemia; ocular and facial paralysis.

EMERGENCY AND FIRST AID PROCEDURES: If swallowed - induce vomiting, get medical aid; inhalation - artificial respiration and oxygen as needed; wash from skin & eyes.

SECTION VI REACTIVITY DATA

CONDITIONS CONTRIBUTING TO INSTABILITY

CONDITIONS CONTRIBUTING TO HAZARDOUS POLYMERIZATION

INCOMPATABILITY (MATERIALS TO AVOID): May react with oxidizing materials.

HAZARDOUS DECOMPOSITION PRODUCTS

SECTION VII SPILL OR LEAK PROCEDURES

STEPS TO BE TAKEN IN CASE MATERIAL IS RELEASED OR SPILLED

WASTE DISPOSAL METHOD

SECTION VIII SPECIAL PROTECTION INFORMATION

VENTILATION REQUIREMENTS LOCAL EXHAUST: Use in hood

PROTECTIVE EQUIPMENT (SPECIFY TYPES) EYE: Goggles

MECHANICAL (GENERAL)

GLOVES: Rubber

SPECIAL

RESPIRATOR: Self-contained

OTHER PROTECTIVE EQUIPMENT

SECTION IX SPECIAL PRECAUTIONS

PRECAUTIONS TO BE TAKEN IN HANDLING AND STORAGE: Outside storage preferred, store inside in a flammable liquids storage cabinet.

OTHER PRECAUTIONS

SIGNATURE | DATE

PICRAMIC ACID

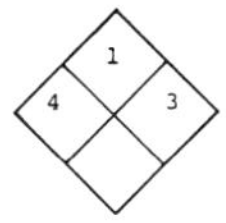

SECTION I SOURCE AND NOMENCLATURE

MANUFACTURER'S NAME | EMERGENCY TELEPHONE NO.

ADDRESS (NUMBER, STREET, CITY, STATE, ZIP CODE)

TRADE NAME AND SYNONYMS | CHEMICAL FAMILY

CHEMICAL NAME AND SYNONYMS: 2 amino-4,6 Dinitro Phenol | FORMULA: $(NH_2)\ (OH)\ C_6H_2(NO_2)_2$

SECTION II HAZARDOUS INGREDIENTS

BASIC MATERIAL	APPROXIMATE OR MAXIMUM % WT. OR VOL.	ESTABLISHED OSHA STANDARD	LD_{50} ORAL	LD_{50} PERCUT.	LC_{50} SPECIES	LC_{50} CONC.
			None available			

SECTION III PHYSICAL DATA MW - 199.13

BOILING POINT °F. | VAPOR PRESSURE mm Hg.

MELTING POINT °F.: 169°C | VAPOR DENSITY (Air = 1)

SPECIFIC GRAVITY (H_2O = 1) | EVAPORATION RATE (______ = 1)

SOLUBILITY IN WATER Pts/100 pts H_2O: Soluble | VOLATILE % Vol. % Wt.

APPEARANCE AND ODOR: Red crystals

SECTION IV FIRE AND EXPLOSION HAZARD DATA

FLASH POINT: None available | FLAMMABLE (EXPLOSIVE) LIMITS: UPPER / LOWER

METHOD USED

EXTINGUISHING MEDIA: Explosive - do not fight fire except for unmanned equipment - use water.

SPECIAL FIRE FIGHTING PROCEDURES: Wear full protective clothing for rescue operation.

UNUSUAL FIRE AND EXPLOSION HAZARDS: Severe explosion hazard

SECTION V HEALTH HAZARD DATA

TOXIC LEVEL: None available | CARCINOGENIC: No

PRINCIPLE ROUTES OF ABSORBTION: Inhalation, ingestion, skin | SKIN AND EYE IRRITATION: Moderate

REVELANT SYMPTOMS OF EXPOSURE: Headache, drowsiness, weakness, cyanosis, respiratory distress.

EFFECTS OF CHRONIC EXPOSURE: Weight loss, anemia, possible liver damage, dermatitis

EMERGENCY AND FIRST AID PROCEDURES: If swallowed - give epsom salt solution, get medical aid, inhalation - rest; wash from skin & eyes.

SECTION VI REACTIVITY DATA

CONDITIONS CONTRIBUTING TO INSTABILITY

CONDITIONS CONTRIBUTING TO HAZARDOUS POLYMERIZATION

INCOMPATABILITY (MATERIALS TO AVOID): Explosive (keep from oxidizing materials)

HAZARDOUS DECOMPOSITION PRODUCTS

SECTION VII SPILL OR LEAK PROCEDURES

STEPS TO BE TAKEN IN CASE MATERIAL IS RELEASED OR SPILLED

WASTE DISPOSAL METHOD

SECTION VIII SPECIAL PROTECTION INFORMATION

VENTILATION REQUIREMENTS LOCAL EXHAUST | PROTECTIVE EQUIPMENT (SPECIFY TYPES) EYE: Goggles

MECHANICAL (GENERAL) | GLOVES: Rubber

SPECIAL | RESPIRATOR: Self contained

OTHER PROTECTIVE EQUIPMENT

SECTION IX SPECIAL PRECAUTIONS

PRECAUTIONS TO BE TAKEN IN HANDLING AND STORAGE: Keep away from oxidizing materials, combustables and sources of heat.

OTHER PRECAUTIONS

SIGNATURE | DATE

PICRAMIDE

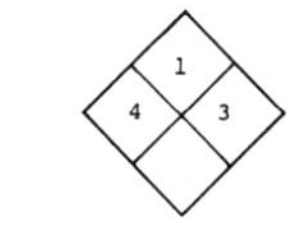

SECTION I SOURCE AND NOMENCLATURE

MANUFACTURER'S NAME | EMERGENCY TELEPHONE NO.

ADDRESS (NUMBER, STREET, CITY, STATE, ZIP CODE)

TRADE NAME AND SYNONYMS | CHEMICAL FAMILY

CHEMICAL NAME AND SYNONYMS: 2,4,6 Trinitroaniline | FORMULA: $NH_2C_6H_2(NO_2)_3$

SECTION II HAZARDOUS INGREDIENTS

BASIC MATERIAL	APPROXIMATE OR MAXIMUM % WT. OR VOL.	ESTABLISHED OSHA STANDARD	LD_{50} ORAL	LD_{50} PERCUT.	LC_{50} SPECIES	LC_{50} CONC.
			none available			

SECTION III PHYSICAL DATA MW - 228.12

BOILING POINT °F.: explodes | VAPOR PRESSURE mm Hg.

MELTING POINT °F.: 190°C | VAPOR DENSITY (Air = 1)

SPECIFIC GRAVITY (H_2O = 1) | EVAPORATION RATE (______ = 1)

SOLUBILITY IN WATER Pts/100 pts H_2O: insoluble | VOLATILE % Vol. % Wt.

APPEARANCE AND ODOR: yellow monoclinic crystals

SECTION IV FIRE AND EXPLOSION HAZARD DATA

FLASH POINT: none available | FLAMMABLE (EXPLOSIVE) LIMITS: UPPER / LOWER

METHOD USED

EXTINGUISHING MEDIA: Explosive - do not fight fire except for unmanned equipment - use water.

SPECIAL FIRE FIGHTING PROCEDURES: Wear full protective clothing for rescue operations.

UNUSUAL FIRE AND EXPLOSION HAZARDS: Severe explosion hazard.

SECTION V HEALTH HAZARD DATA

TOXIC LEVEL: none available | CARCINOGENIC: no

PRINCIPLE ROUTES OF ABSORBTION: Inhalation of dust, ingestion, skin | SKIN AND EYE IRRITATION: moderate

REVELANT SYMPTOMS OF EXPOSURE: Headache, drowsiness, weakness, cyanosis, respiratory distress.

EFFECTS OF CHRONIC EXPOSURE: Weight loss, anemia, possible liver damage.

EMERGENCY AND FIRST AID PROCEDURES: Inhalation - rest; if swallowed - give emetic, get medical aid; wash skin & eyes.

SECTION VI REACTIVITY DATA

CONDITIONS CONTRIBUTING TO INSTABILITY

CONDITIONS CONTRIBUTING TO HAZARDOUS POLYMERIZATION

INCOMPATABILITY (MATERIALS TO AVOID): Reacts with oxidizing materials.

HAZARDOUS DECOMPOSITION PRODUCTS

SECTION VII SPILL OR LEAK PROCEDURES

STEPS TO BE TAKEN IN CASE MATERIAL IS RELEASED OR SPILLED

WASTE DISPOSAL METHOD

SECTION VIII SPECIAL PROTECTION INFORMATION

VENTILATION REQUIREMENTS LOCAL EXHAUST | PROTECTIVE EQUIPMENT (SPECIFY TYPES) EYE: Goggles

MECHANICAL (GENERAL) | GLOVES: Rubber

SPECIAL | RESPIRATOR: With proper filter

OTHER PROTECTIVE EQUIPMENT

SECTION IX SPECIAL PRECAUTIONS

PRECAUTIONS TO BE TAKEN IN HANDLING AND STORAGE: Store away from oxidizing materials and heat.

OTHER PRECAUTIONS

SIGNATURE | DATE

PICRIC ACID

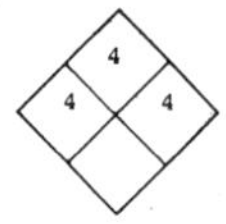

SECTION I SOURCE AND NOMENCLATURE

MANUFACTURER'S NAME	EMERGENCY TELEPHONE NO.
ADDRESS (NUMBER, STREET, CITY, STATE, ZIP CODE)	
TRADE NAME AND SYNONYMS Nitroxanthic acid	CHEMICAL FAMILY
CHEMICAL NAME AND SYNONYMS Carbazotic acid	FORMULA $HOC_6H_2(NO_2)_3$

SECTION II HAZARDOUS INGREDIENTS

BASIC MATERIAL	APPROXIMATE OR MAXIMUM % WT. OR VOL.	ESTABLISHED OSHA STANDARD	LD_{50} ORAL	LD_{50} PERCUT.	LC_{50} SPECIES	LC_{50} CONC.
			See Note			

Note: Lowest published lethal concentration subcutaneous frog - 200 mg/kg.
" " " " oral dog - 120 mg/kg.
" " " " dose-oral-human - 5 mg/kg.

SECTION III PHYSICAL DATA Mol. Wt. 229.11

BOILING POINT (°F.) Explodes >300°C	VAPOR PRESSURE (mm Hg.)
MELTING POINT (°F.) 121.8°C	VAPOR DENSITY (Air = 1) 7.9
SPECIFIC GRAVITY (H_2O = 1) 1.763	EVAPORATION RATE (___ = 1)
SOLUBILITY IN WATER Pts/100 pts H_2O Soluble	VOLATILE % Vol. % Wt.

APPEARANCE AND ODOR
Yellow crystals or yellow liquid, intensely bitter taste.

SECTION IV FIRE AND EXPLOSION HAZARD DATA

FLASH POINT 302°F	FLAMMABLE (EXPLOSIVE) LIMITS	UPPER
METHOD USED		LOWER

EXTINGUISHING MEDIA
Water

SPECIAL FIRE FIGHTING PROCEDURES
Fight from explosion-resistant location

UNUSUAL FIRE AND EXPLOSION HAZARDS
Highly explosive

SECTION V HEALTH HAZARD DATA

TOXIC LEVEL
TLV 0.1 mg/m^3 CARCINOGENIC No

PRINCIPLE ROUTES OF ABSORBTION Percutaneous absorption of dust or inhalation of dust SKIN AND EYE IRRITATION Severe

REVELANT SYMPTOMS OF EXPOSURE
Dermatitis, headache, shivering, fever, insomnia

EFFECTS OF CHRONIC EXPOSURE Diarrhea, ulceration of eyes, headache, nausea, vomiting, skin eruptions, jaundice and severe inflammation of kidneys.

EMERGENCY AND FIRST AID PROCEDURES Remove from skin, by washing thoroughly with water, remove contaminated clothing. If ingested - induce vomiting; if necessary give artificial respiration; obtain medical aid.

SECTION VI REACTIVITY DATA

CONDITIONS CONTRIBUTING TO INSTABILITY

CONDITIONS CONTRIBUTING TO HAZARDOUS POLYMERIZATION

INCOMPATABILITY (MATERIALS TO AVOID)
Copper, lead, zinc, ammonia, bases, concrete

HAZARDOUS DECOMPOSITION PRODUCTS

SECTION VII SPILL OR LEAK PROCEDURES

STEPS TO BE TAKEN IN CASE MATERIAL IS RELEASED OR SPILLED
Mix sodium bicarbonate or a mixture of 9:1 sand - soda ash and transfer to cardboard carton for disposal.

WASTE DISPOSAL METHOD

SECTION VIII SPECIAL PROTECTION INFORMATION

VENTILATION REQUIREMENTS LOCAL EXHAUST	PROTECTIVE EQUIPMENT (SPECIFY TYPES) EYE Goggles
MECHANICAL (GENERAL)	GLOVES Rubber (butyl)
SPECIAL	RESPIRATOR w/appropriate filter

OTHER PROTECTIVE EQUIPMENT
Apron, face shield, self-contained breathing apparatus

SECTION IX SPECIAL PRECAUTIONS

PRECAUTIONS TO BE TAKEN IN HANDLING AND STORAGE
Protect against physical damage

OTHER PRECAUTIONS

SIGNATURE	DATE

PICRYL CHLORIDE

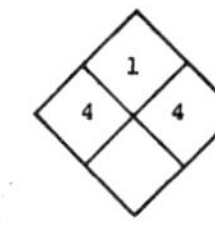

SECTION I SOURCE AND NOMENCLATURE

MANUFACTURER'S NAME	EMERGENCY TELEPHONE NO.
ADDRESS (NUMBER, STREET, CITY, STATE, ZIP CODE)	
TRADE NAME AND SYNONYMS	CHEMICAL FAMILY
CHEMICAL NAME AND SYNONYMS 2,4,6 Trinitrochlorobenzene	FORMULA $(NO_2)_3C_6H_2Cl$

SECTION II HAZARDOUS INGREDIENTS

BASIC MATERIAL	APPROXIMATE OR MAXIMUM % WT. OR VOL.	ESTABLISHED OSHA STANDARD	LD_{50} ORAL	LD_{50} PERCUT.	LC_{50} SPECIES	LC_{50} CONC.
		none available				

SECTION III PHYSICAL DATA MW - 247.56

BOILING POINT (°F.) --	VAPOR PRESSURE (mm Hg.)
MELTING POINT (°F.) 83°C	VAPOR DENSITY (Air = 1)
SPECIFIC GRAVITY (H_2O = 1) 1.80	EVAPORATION RATE (___ = 1)
SOLUBILITY IN WATER Pts/100 pts H_2O insoluble	VOLATILE % Vol. % Wt.

APPEARANCE AND ODOR
yellow monoclinic

SECTION IV FIRE AND EXPLOSION HAZARD DATA

FLASH POINT none available	FLAMMABLE (EXPLOSIVE) LIMITS	UPPER
METHOD USED		LOWER

EXTINGUISHING MEDIA Explosive - do not fight fire except for unmanned equipment - use water.

SPECIAL FIRE FIGHTING PROCEDURES
Wear full protective clothing for rescue operations.

UNUSUAL FIRE AND EXPLOSION HAZARDS
Severe explosion hazard.

SECTION V HEALTH HAZARD DATA

TOXIC LEVEL
none available CARCINOGENIC no

PRINCIPLE ROUTES OF ABSORBTION
inhalation, ingestion, skin SKIN AND EYE IRRITATION moderate

REVELANT SYMPTOMS OF EXPOSURE
Dermatitis, irritation of respiratory and digestive systems.

EFFECTS OF CHRONIC EXPOSURE
Dermatitis.

EMERGENCY AND FIRST AID PROCEDURES Inhalation - rest; if swallowed - give emetic, get medical aid; wash skin & eyes.

SECTION VI REACTIVITY DATA

CONDITIONS CONTRIBUTING TO INSTABILITY

CONDITIONS CONTRIBUTING TO HAZARDOUS POLYMERIZATION

INCOMPATABILITY (MATERIALS TO AVOID)
Keep from oxidizing materials (explosive) inorganic nitrates, chlorates and ammonium nitrate.

HAZARDOUS DECOMPOSITION PRODUCTS

SECTION VII SPILL OR LEAK PROCEDURES

STEPS TO BE TAKEN IN CASE MATERIAL IS RELEASED OR SPILLED

WASTE DISPOSAL METHOD

SECTION VIII SPECIAL PROTECTION INFORMATION

VENTILATION REQUIREMENTS LOCAL EXHAUST	PROTECTIVE EQUIPMENT (SPECIFY TYPES) EYE Goggles
MECHANICAL (GENERAL)	GLOVES Rubber
SPECIAL	RESPIRATOR Self-contained

OTHER PROTECTIVE EQUIPMENT

SECTION IX SPECIAL PRECAUTIONS

PRECAUTIONS TO BE TAKEN IN HANDLING AND STORAGE
Keep from oxidizing materials, combustibles, and source of heat.

OTHER PRECAUTIONS

SIGNATURE	DATE

PLATINUM CHLORIDE

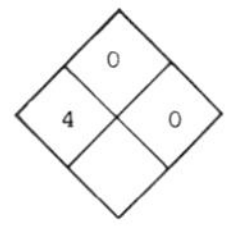

SECTION I SOURCE AND NOMENCLATURE

MANUFACTURER'S NAME	EMERGENCY TELEPHONE NO.

ADDRESS (NUMBER, STREET, CITY, STATE, ZIP CODE)

TRADE NAME AND SYNONYMS	CHEMICAL FAMILY
CHEMICAL NAME AND SYNONYMS	**FORMULA**
Chloroplatinic Acid	$H_2PtCl_6 \cdot 6H_2O$

SECTION II HAZARDOUS INGREDIENTS

BASIC MATERIAL	APPROXIMATE OR MAXIMUM % WT. OR VOL.	ESTABLISHED OSHA STANDARD	LD_{50}		LC_{50}	
			ORAL	PERCUT.	SPECIES	CONC.
LDLO intravenous - rabbit	23 mg/Kg.					

SECTION III PHYSICAL DATA MW - 517.92

BOILING POINT °F.	VAPOR PRESSURE mm Hg.
MELTING POINT °F. 60°C.	**VAPOR DENSITY (Air = 1)**
SPECIFIC GRAVITY (H_2O = 1) 2.41	**EVAPORATION RATE (= 1)**
SOLUBILITY IN WATER Pts/100 pts H_2O Soluble	**VOLATILE % Vol. % Wt.**

APPEARANCE AND ODOR
Brownish-yellow

SECTION IV FIRE AND EXPLOSION HAZARD DATA

FLASH POINT	FLAMMABLE (EXPLOSIVE) LIMITS	UPPER
Not flammable		
METHOD USED		**LOWER**

EXTINGUISHING MEDIA

SPECIAL FIRE FIGHTING PROCEDURES

UNUSUAL FIRE AND EXPLOSION HAZARDS

SECTION V HEALTH HAZARD DATA

TOXIC LEVEL
TLV 0.002 mg/CM — CARCINOGENIC No

PRINCIPLE ROUTES OF ABSORBTION
Inhalation (dust) — SKIN AND EYE IRRITATION Severe

REVELANT SYMPTOMS OF EXPOSURE Inflammation of mucous membranes in nose, lacramation and burning in the eyes.

EFFECTS OF CHRONIC EXPOSURE Running of the nose, sneezing, tightness of chest, cyanosis, wheezing, cough, dermatitis.

EMERGENCY AND FIRST AID PROCEDURES Inhalation - remove to fresh air, rest, treat asthmatic symptoms; drink water, get medical aid.

SECTION VI REACTIVITY DATA

CONDITIONS CONTRIBUTING TO INSTABILITY

CONDITIONS CONTRIBUTING TO HAZARDOUS POLYMERIZATION

INCOMPATABILITY (MATERIALS TO AVOID)

HAZARDOUS DECOMPOSITION PRODUCTS

SECTION VII SPILL OR LEAK PROCEDURES

STEPS TO BE TAKEN IN CASE MATERIAL IS RELEASED OR SPILLED

WASTE DISPOSAL METHOD

SECTION VIII SPECIAL PROTECTION INFORMATION

VENTILATION REQUIREMENTS LOCAL EXHAUST	PROTECTIVE EQUIPMENT (SPECIFY TYPES) EYE Goggles
MECHANICAL (GENERAL)	GLOVES
SPECIAL	RESPIRATOR

OTHER PROTECTIVE EQUIPMENT

SECTION IX SPECIAL PRECAUTIONS

PRECAUTIONS TO BE TAKEN IN HANDLING AND STORAGE

OTHER PRECAUTIONS

SIGNATURE	DATE

POTASSIUM BIFLUORIDE

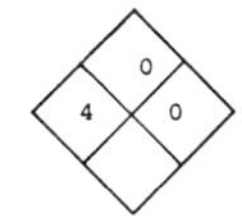

SECTION I SOURCE AND NOMENCLATURE

MANUFACTURER'S NAME	EMERGENCY TELEPHONE NO.

ADDRESS (NUMBER, STREET, CITY, STATE, ZIP CODE)

TRADE NAME AND SYNONYMS	CHEMICAL FAMILY
Potassium acid fluoride	
CHEMICAL NAME AND SYNONYMS	**FORMULA**
Potassium difluoride	KHF_2

SECTION II HAZARDOUS INGREDIENTS

BASIC MATERIAL	APPROXIMATE OR MAXIMUM % WT. OR VOL.	ESTABLISHED OSHA STANDARD	LD_{50} mg/kg		LC_{50}	
			ORAL	PERCUT.	SPECIES	CONC.
#1 Lowest published lethal dose						
Oral - guinea pig	150 mg/kg.					
#2 Lowest published lethal dose						
Subcutaneous - guinea pig	250 mg/kg.					

SECTION III PHYSICAL DATA MW - 78.11

BOILING POINT °F.	VAPOR PRESSURE mm Hg.
MELTING POINT °F. d	**VAPOR DENSITY (Air = 1)**
SPECIFIC GRAVITY (H_2O = 1) --	**EVAPORATION RATE (= 1)**
SOLUBILITY IN WATER Pts/100 pts H_2O 41 at 21°C.	**VOLATILE % Vol. % Wt.**

APPEARANCE AND ODOR
Colorless crystal

SECTION IV FIRE AND EXPLOSION HAZARD DATA

FLASH POINT	FLAMMABLE (EXPLOSIVE) LIMITS	UPPER
N/A		
METHOD USED		**LOWER**

EXTINGUISHING MEDIA
Water

SPECIAL FIRE FIGHTING PROCEDURES
Wear full protective clothing and respirator.

UNUSUAL FIRE AND EXPLOSION HAZARDS
N/A

SECTION V HEALTH HAZARD DATA

TOXIC LEVEL
TLV - 3 PPM (for HF) — CARCINOGENIC no

PRINCIPLE ROUTES OF ABSORBTION
Inhalation of HF, skin absorption — SKIN AND EYE IRRITATION Severe

REVELANT SYMPTOMS OF EXPOSURE
Irritation of respiratory system, skin absorption and eye burns.

EFFECTS OF CHRONIC EXPOSURE
Weight loss, brittleness of bones, anemia, weakness, discoloration of teeth.

EMERGENCY AND FIRST AID PROCEDURES Inhalation - rest, give oxygen if necessary, get medical attention; wash from skin and eyes, get medical attention; if swallowed - wash mouth, drink water, then milk of magnesia, get medical attention.

SECTION VI REACTIVITY DATA

CONDITIONS CONTRIBUTING TO INSTABILITY

CONDITIONS CONTRIBUTING TO HAZARDOUS POLYMERIZATION

INCOMPATABILITY (MATERIALS TO AVOID)

HAZARDOUS DECOMPOSITION PRODUCTS

SECTION VII SPILL OR LEAK PROCEDURES

STEPS TO BE TAKEN IN CASE MATERIAL IS RELEASED OR SPILLED

WASTE DISPOSAL METHOD

SECTION VIII SPECIAL PROTECTION INFORMATION

VENTILATION REQUIREMENTS LOCAL EXHAUST	PROTECTIVE EQUIPMENT (SPECIFY TYPES) EYE Goggles
MECHANICAL (GENERAL)	GLOVES Rubber gloves
SPECIAL	RESPIRATOR W/appropriate filter

OTHER PROTECTIVE EQUIPMENT

SECTION IX SPECIAL PRECAUTIONS

PRECAUTIONS TO BE TAKEN IN HANDLING AND STORAGE

OTHER PRECAUTIONS

SIGNATURE	DATE

POTASSIUM FLUORIDE

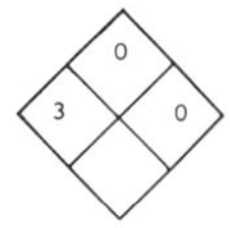

SECTION I SOURCE AND NOMENCLATURE

CHEMICAL NAME AND SYNONYMS: Potassium Fluoride

FORMULA: KF

SECTION II HAZARDOUS INGREDIENTS

BASIC MATERIAL	APPROXIMATE OR MAXIMUM % WT. OR VOL.	ESTABLISHED OSHA STANDARD	LD_{50} mg/kg ORAL	LD_{50} PERCUT.	SPECIES	LC_{50} CONC.
			245		Rat	

SECTION III PHYSICAL DATA MW = 58.10

BOILING POINT: 880°C.

VAPOR PRESSURE: 1 mm at 885°C.

MELTING POINT: 1500°C.

SPECIFIC GRAVITY (H_2O = 1): 2.48

SOLUBILITY IN WATER Pts/100 pts H_2O: 92.3 at 18°C.

APPEARANCE AND ODOR: White solid

SECTION IV FIRE AND EXPLOSION HAZARD DATA

FLASH POINT: Not flammable

EXTINGUISHING MEDIA: Water

SECTION V HEALTH HAZARD DATA

TOXIC LEVEL: 2.5 mg/m³

CARCINOGENIC: No

PRINCIPLE ROUTES OF ABSORBTION: Inhalation of dust, ingestion skin.

SKIN AND EYE IRRITATION: Moderate

REVELANT SYMPTOMS OF EXPOSURE: Irritation of respiratory system; severe internal irritation and damage; skin and eye burns.

EFFECTS OF CHRONIC EXPOSURE: Increase in fluoride content of teeth.

EMERGENCY AND FIRST AID PROCEDURES: Inhalation - rest, keep warm, get medical aid; if swallowed - wash mouth, drink water, then milk of magnesia, get medial aid, wash from skin and eyes, get medical aid.

SECTION VI REACTIVITY DATA

SECTION VII SPILL OR LEAK PROCEDURES

STEPS TO BE TAKEN IN CASE MATERIAL IS RELEASED OR SPILLED: Sweep into container for disposal.

SECTION VIII SPECIAL PROTECTION INFORMATION

PROTECTIVE EQUIPMENT (SPECIFY TYPES) EYE: Face shield

GLOVES: Rubber

RESPIRATOR: With proper filter

SECTION IX SPECIAL PRECAUTIONS

PRECAUTIONS TO BE TAKEN IN HANDLING AND STORAGE: Store away from contacts with acids.

POTASSIUM PERIODATE

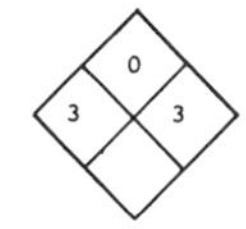

SECTION I SOURCE AND NOMENCLATURE

CHEMICAL NAME AND SYNONYMS: Potassium Periodate

FORMULA: KIO_4

SECTION II HAZARDOUS INGREDIENTS

BASIC MATERIAL	APPROXIMATE OR MAXIMUM % WT. OR VOL.	ESTABLISHED OSHA STANDARD	LD_{50} ORAL	LD_{50} PERCUT.	SPECIES	LC_{50} CONC.
			not	available		

SECTION III PHYSICAL DATA

VAPOR PRESSURE: at 300°C. 0 mm Hg.

MELTING POINT: 582°C.

SPECIFIC GRAVITY (H_2O = 1): 3.618 at 15°C.

SOLUBILITY IN WATER Pts/100 pts H_2O: 0.66 at 13°C.

APPEARANCE AND ODOR: White crystal

SECTION IV FIRE AND EXPLOSION HAZARD DATA

FLASH POINT: N/A

METHOD USED: N/A

EXTINGUISHING MEDIA: N/A

SPECIAL FIRE FIGHTING PROCEDURES: N/A

UNUSUAL FIRE AND EXPLOSION HAZARDS: N/A

SECTION V HEALTH HAZARD DATA

TOXIC LEVEL: Not available

CARCINOGENIC: No

PRINCIPLE ROUTES OF ABSORBTION: Inhalation of dust, ingestion

SKIN AND EYE IRRITATION: Moderate

REVELANT SYMPTOMS OF EXPOSURE: Irritation of respiratory system; eye irritation, internal irritation.

EFFECTS OF CHRONIC EXPOSURE: --

EMERGENCY AND FIRST AID PROCEDURES: Inhalation - rest, get medical attention; wash from eyes; if swallowed - wash mouth, induce vomiting, get medical aid.

SECTION VI REACTIVITY DATA

CONDITIONS CONTRIBUTING TO INSTABILITY: Strong oxidizing agent.

INCOMPATABILITY (MATERIALS TO AVOID): Reducing materials.

SECTION VII SPILL OR LEAK PROCEDURES

SECTION VIII SPECIAL PROTECTION INFORMATION

PROTECTIVE EQUIPMENT (SPECIFY TYPES) EYE: Safety glasses

RESPIRATOR: With proper filter

SECTION IX SPECIAL PRECAUTIONS

PROPIONITRILE

3 / 4 / 1

SECTION I SOURCE AND NOMENCLATURE

MANUFACTURER'S NAME | EMERGENCY TELEPHONE NO.

ADDRESS (NUMBER, STREET, CITY, STATE, ZIP CODE)

TRADE NAME AND SYNONYMS: Hydrocyanic ether | CHEMICAL FAMILY:

CHEMICAL NAME AND SYNONYMS: Ethyl Cyanide | FORMULA: CH_3CH_2CN

SECTION II HAZARDOUS INGREDIENTS

BASIC MATERIAL	APPROXIMATE OR MAXIMUM % WT. OR VOL.	ESTABLISHED OSHA STANDARD	LD_{50} mg/kg ORAL	PERCUT.	SPECIES	LC_{50} CONC.
			39		Rat	
				210	Rabbit	
LCLO - inhalation - rat	500 PPM/4H					

SECTION III PHYSICAL DATA Mol. Wt. 5508

BOILING POINT °F.: 97.1°C | VAPOR PRESSURE: at 22°C 40 mm Hg.

MELTING POINT °F.: -103.5°C | VAPOR DENSITY (Air = 1): 1.9

SPECIFIC GRAVITY (H_2O = 1): 0.783 | EVAPORATION RATE (_____ = 1)

SOLUBILITY IN WATER Pts/100 pts H_2O: Soluble | VOLATILE % Vol. % Wt.

APPEARANCE AND ODOR: Colorless liquid; ethereal odor.

SECTION IV FIRE AND EXPLOSION HAZARD DATA

FLASH POINT: N/A | FLAMMABLE (EXPLOSIVE) LIMITS UPPER:

METHOD USED: Closed Cup | LOWER: 3.1%

EXTINGUISHING MEDIA: Use dry chemicals, alcohol foam or carbon dioxide.

SPECIAL FIRE FIGHTING PROCEDURES: Wear protective clothing; use self-contained breathing apparatus.

UNUSUAL FIRE AND EXPLOSION HAZARDS

SECTION V HEALTH HAZARD DATA

TOXIC LEVEL: TWA-TLV-6 PPM, 14 MG/C.M. (NIOSH Value) | CARCINOGENIC: No

PRINCIPLE ROUTES OF ABSORBTION: Inhalation, ingestion, skin | SKIN AND EYE IRRITATION: Mild to moderate

REVELANT SYMPTOMS OF EXPOSURE: Dizziness, rapid respiration, headache, drowsiness, drop in blood pressure and pulse, and unconsciousness, delayed symptoms.

EFFECTS OF CHRONIC EXPOSURE: Eye irritation, loss of appetite, weight loss, mental deterioration.

EMERGENCY AND FIRST AID PROCEDURES: If swallowed - inhalation of amyl nitrite, immediate vomiting, or gastric lavage, wash from skin and eyes; inhalation - rest, get medical aid.

SECTION VI REACTIVITY DATA

CONDITIONS CONTRIBUTING TO INSTABILITY

CONDITIONS CONTRIBUTING TO HAZARDOUS POLYMERIZATION

INCOMPATABILITY (MATERIALS TO AVOID)

HAZARDOUS DECOMPOSITION PRODUCTS: Heated to decomposition emits highly toxic cyanide fumes.

SECTION VII SPILL OR LEAK PROCEDURES

STEPS TO BE TAKEN IN CASE MATERIAL IS RELEASED OR SPILLED: Absorb with paper towels; evaporate in hood.

WASTE DISPOSAL METHOD

SECTION VIII SPECIAL PROTECTION INFORMATION

VENTILATION REQUIREMENTS LOCAL EXHAUST | PROTECTIVE EQUIPMENT (SPECIFY TYPES) EYE: Goggles - Safety Glasses

MECHANICAL (GENERAL) | GLOVES: Rubber

SPECIAL | RESPIRATOR: w/proper filter

OTHER PROTECTIVE EQUIPMENT: Protective clothing

SECTION IX SPECIAL PRECAUTIONS

PRECAUTIONS TO BE TAKEN IN HANDLING AND STORAGE: Store in flammable liquids storage cabinet.

OTHER PRECAUTIONS

SIGNATURE | DATE

PROPIONYL CHLORIDE

3 / 4 / 1 / ~~W~~

SECTION I SOURCE AND NOMENCLATURE

MANUFACTURER'S NAME | EMERGENCY TELEPHONE NO.

ADDRESS (NUMBER, STREET, CITY, STATE, ZIP CODE)

TRADE NAME AND SYNONYMS | CHEMICAL FAMILY

CHEMICAL NAME AND SYNONYMS: Propionyl Chloride | FORMULA: CH_2CH_2COCl

SECTION II HAZARDOUS INGREDIENTS

BASIC MATERIAL	APPROXIMATE OR MAXIMUM % WT. OR VOL.	ESTABLISHED OSHA STANDARD	LD_{50} ORAL	PERCUT.	SPECIES	LC_{50} CONC.
		none available				

SECTION III PHYSICAL DATA MW - 92.5

BOILING POINT °F.: 80°C | VAPOR PRESSURE mm Hg.

MELTING POINT °F.: -94°C | VAPOR DENSITY (Air = 1): 3.2

SPECIFIC GRAVITY (H_2O = 1): 1.07 | EVAPORATION RATE (_____ = 1)

SOLUBILITY IN WATER Pts/100 pts H_2O: decomposes | VOLATILE % Vol. % Wt.

APPEARANCE AND ODOR: colorless liquid

SECTION IV FIRE AND EXPLOSION HAZARD DATA

FLASH POINT: 54°F | FLAMMABLE (EXPLOSIVE) LIMITS UPPER:

METHOD USED | LOWER:

EXTINGUISHING MEDIA: carbon dioxide, dry chemical.

SPECIAL FIRE FIGHTING PROCEDURES

UNUSUAL FIRE AND EXPLOSION HAZARDS

SECTION V HEALTH HAZARD DATA

TOXIC LEVEL: none available | CARCINOGENIC: no

PRINCIPLE ROUTES OF ABSORBTION: inhalation, ingestion | SKIN AND EYE IRRITATION: severe

REVELANT SYMPTOMS OF EXPOSURE: Cyanosis, coughing, headache, pulmonary edema, ulceration of skin dizziness.

EFFECTS OF CHRONIC EXPOSURE: Erosion of teeth, bronchial irritation and chronic cough.

EMERGENCY AND FIRST AID PROCEDURES: Ingestion - drink water, milk of magnesia, get medical aid; inhalation - artificial respiration as needed; wash eyes and skin.

SECTION VI REACTIVITY DATA

CONDITIONS CONTRIBUTING TO INSTABILITY

CONDITIONS CONTRIBUTING TO HAZARDOUS POLYMERIZATION

INCOMPATABILITY (MATERIALS TO AVOID): May react violently with water, alcohol, and dimethyl sulfoxide.

HAZARDOUS DECOMPOSITION PRODUCTS

SECTION VII SPILL OR LEAK PROCEDURES

STEPS TO BE TAKEN IN CASE MATERIAL IS RELEASED OR SPILLED

WASTE DISPOSAL METHOD

SECTION VIII SPECIAL PROTECTION INFORMATION

VENTILATION REQUIREMENTS LOCAL EXHAUST | PROTECTIVE EQUIPMENT (SPECIFY TYPES) EYE: Goggles

MECHANICAL (GENERAL) | GLOVES: Rubber

SPECIAL | RESPIRATOR: Self-contained

OTHER PROTECTIVE EQUIPMENT

SECTION IX SPECIAL PRECAUTIONS

PRECAUTIONS TO BE TAKEN IN HANDLING AND STORAGE

OTHER PRECAUTIONS

SIGNATURE | DATE

n-PROPYLAMINE

3
3 0

SECTION I SOURCE AND NOMENCLATURE

MANUFACTURER'S NAME | EMERGENCY TELEPHONE NO.

ADDRESS (NUMBER, STREET, CITY, STATE, ZIP CODE)

TRADE NAME AND SYNONYMS | CHEMICAL FAMILY

CHEMICAL NAME AND SYNONYMS: 1 Aminopropane | FORMULA: $C_3H_7NH_2$

SECTION II HAZARDOUS INGREDIENTS

BASIC MATERIAL	APPROXIMATE OR MAXIMUM % WT. OR VOL.	ESTABLISHED OSHA STANDARD	LD_{50} mg/kg		LC_{50} PPM	
			ORAL	PERCUT.	SPECIES	CONC.
Lowest published lethal dose					rat	2310
Oral - rat 570/mg/kg				560	rabbit	

SECTION III PHYSICAL DATA MW - 59.13

BOILING POINT °F.: 49°C | VAPOR PRESSURE mm Hg.: at 20°C 248

MELTING POINT °F.: -83°C | VAPOR DENSITY (Air = 1): 1.7

SPECIFIC GRAVITY (H_2O = 1): 0.72 | EVAPORATION RATE (________ = 1)

SOLUBILITY IN WATER Pts/100 pts H_2O: infinite | VOLATILE % Vol. % Wt.

APPEARANCE AND ODOR: colorless liquid, strong ammonia odor.

SECTION IV FIRE AND EXPLOSION HAZARD DATA

FLASH POINT: -35°F | FLAMMABLE (EXPLOSIVE) LIMITS — UPPER: 10.4% LOWER: 2.0%

METHOD USED

EXTINGUISHING MEDIA: carbon dioxide, dry chemical, alcohol foam. Water may be ineffective.

SPECIAL FIRE FIGHTING PROCEDURES: Wear full protective clothing.

UNUSUAL FIRE AND EXPLOSION HAZARDS

SECTION V HEALTH HAZARD DATA

TOXIC LEVEL: none available | CARCINOGENIC: no

PRINCIPLE ROUTES OF ABSORBTION: Inhalation, ingestion | SKIN AND EYE IRRITATION: severe

REVELANT SYMPTOMS OF EXPOSURE: nose and throat irritation, lung edema, skin & eye burns.

EFFECTS OF CHRONIC EXPOSURE: --

EMERGENCY AND FIRST AID PROCEDURES: Inhalation - rest, keep warm, get medical aid; if swallowed - wash mouth, drink water, get medical aid; wash from skin & eyes.

SECTION VI REACTIVITY DATA

CONDITIONS CONTRIBUTING TO INSTABILITY

CONDITIONS CONTRIBUTING TO HAZARDOUS POLYMERIZATION

INCOMPATABILITY (MATERIALS TO AVOID)

HAZARDOUS DECOMPOSITION PRODUCTS

SECTION VII SPILL OR LEAK PROCEDURES

STEPS TO BE TAKEN IN CASE MATERIAL IS RELEASED OR SPILLED

WASTE DISPOSAL METHOD

SECTION VIII SPECIAL PROTECTION INFORMATION

VENTILATION REQUIREMENTS LOCAL EXHAUST | PROTECTIVE EQUIPMENT (SPECIFY TYPES) EYE: goggles

MECHANICAL (GENERAL) | GLOVES

SPECIAL | RESPIRATOR: with proper filter

OTHER PROTECTIVE EQUIPMENT: Face shield, rubber apron

SECTION IX SPECIAL PRECAUTIONS

PRECAUTIONS TO BE TAKEN IN HANDLING AND STORAGE: Protect against physical damage - outside or detached storage preferred. Inside storage in a standard flammable liquids storage cabinet.

OTHER PRECAUTIONS

SIGNATURE | DATE

n-PROPYLANILINE

2
4 0

SECTION I SOURCE AND NOMENCLATURE

MANUFACTURER'S NAME | EMERGENCY TELEPHONE NO.

ADDRESS (NUMBER, STREET, CITY, STATE, ZIP CODE)

TRADE NAME AND SYNONYMS | CHEMICAL FAMILY

CHEMICAL NAME AND SYNONYMS: N Propylaniline | FORMULA: $C_3H_7NHC_6H_5$

SECTION II HAZARDOUS INGREDIENTS

BASIC MATERIAL	APPROXIMATE OR MAXIMUM % WT. OR VOL.	ESTABLISHED OSHA STANDARD	LD_{50}		LC_{50}	
			ORAL	PERCUT.	SPECIES	CONC.
			none	available		
Unclassified - Mouse LD50	250 mg/kg					

SECTION III PHYSICAL DATA MW - 135.20

BOILING POINT °F.: 222°C. | VAPOR PRESSURE mm Hg.: at 87°C 5

MELTING POINT °F. | VAPOR DENSITY (Air = 1)

SPECIFIC GRAVITY (H_2O = 1): 0.95 | EVAPORATION RATE (________ = 1)

SOLUBILITY IN WATER Pts/100 pts H_2O: insoluble | VOLATILE % Vol. % Wt.

APPEARANCE AND ODOR: liquid

SECTION IV FIRE AND EXPLOSION HAZARD DATA

FLASH POINT: none available | FLAMMABLE (EXPLOSIVE) LIMITS — UPPER | LOWER

METHOD USED

EXTINGUISHING MEDIA: Foam, carbon dioxide, dry chemical.

SPECIAL FIRE FIGHTING PROCEDURES

UNUSUAL FIRE AND EXPLOSION HAZARDS: When heated may decompose to give toxic aniline fumes.

SECTION V HEALTH HAZARD DATA

TOXIC LEVEL: none available | CARCINOGENIC: no

PRINCIPLE ROUTES OF ABSORBTION: inhalation, ingestion, skin | SKIN AND EYE IRRITATION: moderate

REVELANT SYMPTOMS OF EXPOSURE: Weakness, tremors, cyanosis, drowsiness, central nervous system depression.

EFFECTS OF CHRONIC EXPOSURE: Fatigue, loss of appetite, headache, dizziness.

EMERGENCY AND FIRST AID PROCEDURES: Inhalation - rest; if swallowed - give emetic, get medical aid; wash from skin & eyes.

SECTION VI REACTIVITY DATA

CONDITIONS CONTRIBUTING TO INSTABILITY

CONDITIONS CONTRIBUTING TO HAZARDOUS POLYMERIZATION

INCOMPATABILITY (MATERIALS TO AVOID)

HAZARDOUS DECOMPOSITION PRODUCTS

SECTION VII SPILL OR LEAK PROCEDURES

STEPS TO BE TAKEN IN CASE MATERIAL IS RELEASED OR SPILLED

WASTE DISPOSAL METHOD

SECTION VIII SPECIAL PROTECTION INFORMATION

VENTILATION REQUIREMENTS LOCAL EXHAUST | PROTECTIVE EQUIPMENT (SPECIFY TYPES) EYE: Goggles

MECHANICAL (GENERAL) | GLOVES: Polyvinyl alcohol

SPECIAL | RESPIRATOR: With proper filter

OTHER PROTECTIVE EQUIPMENT

SECTION IX SPECIAL PRECAUTIONS

PRECAUTIONS TO BE TAKEN IN HANDLING AND STORAGE

OTHER PRECAUTIONS

SIGNATURE | DATE

n-PROPYLBENZENE

NFPA: Health 3, Flammability 3, Reactivity 0

SECTION I SOURCE AND NOMENCLATURE

MANUFACTURER'S NAME	EMERGENCY TELEPHONE NO.
ADDRESS (NUMBER, STREET, CITY, STATE, ZIP CODE)	
TRADE NAME AND SYNONYMS	CHEMICAL FAMILY
CHEMICAL NAME AND SYNONYMS Phenyl propane	FORMULA $C_3H_7C_6H_5$

SECTION II HAZARDOUS INGREDIENTS

BASIC MATERIAL	APPROXIMATE OR MAXIMUM % WT. OR VOL.	ESTABLISHED OSHA STANDARD	LD_{50} mg/kg ORAL	PERCUT.	LC_{50} SPECIES	CONC.
					rat	
LCLO inhalation - mouse 4100 PPM						
LDLO - oral - rat 4830 mg/kg						

SECTION III PHYSICAL DATA MW - 120.19

BOILING POINT °F. 159.2	VAPOR PRESSURE mm Hg. at 43.4°C 10
MELTING POINT °F. -99.5°C	VAPOR DENSITY (Air = 1) 4.14
SPECIFIC GRAVITY (H_2O = 1) 0.862	EVAPORATION RATE (______ = 1)
SOLUBILITY IN WATER Pts/100 pts H_2O insoluble	VOLATILE % Vol. % Wt.
APPEARANCE AND ODOR colorless liquid	

SECTION IV FIRE AND EXPLOSION HAZARD DATA

FLASH POINT 86°F	FLAMMABLE (EXPLOSIVE) LIMITS	UPPER
METHOD USED C.C.		LOWER

EXTINGUISHING MEDIA
Water spray, carbon dioxide, dry chemical or foam.

SPECIAL FIRE FIGHTING PROCEDURES
Water may be ineffective. Wear goggles and self-contained breathing apparatus.

UNUSUAL FIRE AND EXPLOSION HAZARDS

SECTION V HEALTH HAZARD DATA

TOXIC LEVEL: not available — CARCINOGENIC: no

PRINCIPLE ROUTES OF ABSORBTION: inhalation, ingestion, skin — SKIN AND EYE IRRITATION: moderate

REVELANT SYMPTOMS OF EXPOSURE: Skin and eye irritant, dizziness

EFFECTS OF CHRONIC EXPOSURE: Possible kidney damage.

EMERGENCY AND FIRST AID PROCEDURES: Inhalation - rest; if swallowed - wash mouth, give emetic, get medical aid; wash skin & eyes.

SECTION VI REACTIVITY DATA

CONDITIONS CONTRIBUTING TO INSTABILITY

CONDITIONS CONTRIBUTING TO HAZARDOUS POLYMERIZATION

INCOMPATABILITY (MATERIALS TO AVOID)

HAZARDOUS DECOMPOSITION PRODUCTS

SECTION VII SPILL OR LEAK PROCEDURES

STEPS TO BE TAKEN IN CASE MATERIAL IS RELEASED OR SPILLED

WASTE DISPOSAL METHOD

SECTION VIII SPECIAL PROTECTION INFORMATION

VENTILATION REQUIREMENTS LOCAL EXHAUST	PROTECTIVE EQUIPMENT (SPECIFY TYPES) EYE Goggles
MECHANICAL (GENERAL)	GLOVES Rubber
SPECIAL	RESPIRATOR Self-contained
OTHER PROTECTIVE EQUIPMENT	

SECTION IX SPECIAL PRECAUTIONS

PRECAUTIONS TO BE TAKEN IN HANDLING AND STORAGE

OTHER PRECAUTIONS

SIGNATURE	DATE

n-PROPYL CHLOROCARBONATE

NFPA: Health 3, Flammability 2, Reactivity 0

SECTION I SOURCE AND NOMENCLATURE

MANUFACTURER'S NAME	EMERGENCY TELEPHONE NO.
ADDRESS (NUMBER, STREET, CITY, STATE, ZIP CODE)	
TRADE NAME AND SYNONYMS	CHEMICAL FAMILY
CHEMICAL NAME AND SYNONYMS Propyl Chloroformate	FORMULA $ClCOOC_3H_7$

SECTION II HAZARDOUS INGREDIENTS

BASIC MATERIAL	APPROXIMATE OR MAXIMUM % WT. OR VOL.	ESTABLISHED OSHA STANDARD	LD_{50} ORAL	PERCUT.	LC_{50} SPECIES	CONC.
			none available			

SECTION III PHYSICAL DATA MW - 122.55

BOILING POINT °F. 115°C	VAPOR PRESSURE mm Hg.
MELTING POINT °F. --	VAPOR DENSITY (Air = 1)
SPECIFIC GRAVITY (H_2O = 1) 1.10	EVAPORATION RATE (______ = 1)
SOLUBILITY IN WATER Pts/100 pts H_2O decomposes	VOLATILE % Vol. % Wt.
APPEARANCE AND ODOR liquid	

SECTION IV FIRE AND EXPLOSION HAZARD DATA

FLASH POINT none available	FLAMMABLE (EXPLOSIVE) LIMITS	UPPER
METHOD USED		LOWER

EXTINGUISHING MEDIA
carbon dioxide, dry chemical.

SPECIAL FIRE FIGHTING PROCEDURES

UNUSUAL FIRE AND EXPLOSION HAZARDS

SECTION V HEALTH HAZARD DATA

TOXIC LEVEL: None available — CARCINOGENIC: no

PRINCIPLE ROUTES OF ABSORBTION: Inhalation, ingestion — SKIN AND EYE IRRITATION: severe

REVELANT SYMPTOMS OF EXPOSURE: Eye irritation, irritation of upper respiratory tract, pulmonary edema.

EFFECTS OF CHRONIC EXPOSURE: --

EMERGENCY AND FIRST AID PROCEDURES: If swallowed - give emetic, get medical aid; inhalation - rest; wash from skin & eyes.

SECTION VI REACTIVITY DATA

CONDITIONS CONTRIBUTING TO INSTABILITY

CONDITIONS CONTRIBUTING TO HAZARDOUS POLYMERIZATION

INCOMPATABILITY (MATERIALS TO AVOID)

HAZARDOUS DECOMPOSITION PRODUCTS

SECTION VII SPILL OR LEAK PROCEDURES

STEPS TO BE TAKEN IN CASE MATERIAL IS RELEASED OR SPILLED

WASTE DISPOSAL METHOD

SECTION VIII SPECIAL PROTECTION INFORMATION

VENTILATION REQUIREMENTS LOCAL EXHAUST	PROTECTIVE EQUIPMENT (SPECIFY TYPES) EYE Goggles
MECHANICAL (GENERAL)	GLOVES Rubber
SPECIAL	RESPIRATOR Self-contained
OTHER PROTECTIVE EQUIPMENT	

SECTION IX SPECIAL PRECAUTIONS

PRECAUTIONS TO BE TAKEN IN HANDLING AND STORAGE
Keep away from water.

OTHER PRECAUTIONS

SIGNATURE	DATE

PROPYLENE BROMIDE

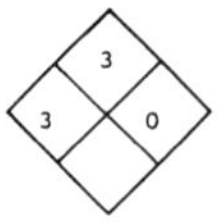

SECTION I SOURCE AND NOMENCLATURE

MANUFACTURER'S NAME

EMERGENCY TELEPHONE NO.

ADDRESS (NUMBER, STREET, CITY, STATE, ZIP CODE)

TRADE NAME AND SYNONYMS: Propylene dibromide

CHEMICAL FAMILY

CHEMICAL NAME AND SYNONYMS: 1,2 dibromopropane

FORMULA: $CH_2BrCHBrCH_3$

SECTION II HAZARDOUS INGREDIENTS

BASIC MATERIAL	APPROXIMATE OR MAXIMUM % WT. OR VOL.	ESTABLISHED OSHA STANDARD	LD_{50} ORAL	LD_{50} PERCUT.	LC_{50} SPECIES	LC_{50} CONC.
			none available			
Intraperitoneal - mouse	LDLO 75 mg/kg					

SECTION III PHYSICAL DATA MW - 201.9

BOILING POINT °F.: 140°C

VAPOR PRESSURE mm Hg.

MELTING POINT °F.: -55°C

VAPOR DENSITY (Air = 1): 7.0

SPECIFIC GRAVITY (H_2O = 1): 1.94

EVAPORATION RATE (______ = 1)

SOLUBILITY IN WATER Pts/100 pts H_2O

VOLATILE % Vol. % Wt.

APPEARANCE AND ODOR: colorless liquid

SECTION IV FIRE AND EXPLOSION HAZARD DATA

FLASH POINT: N/A

FLAMMABLE (EXPLOSIVE) LIMITS: UPPER / LOWER

METHOD USED

EXTINGUISHING MEDIA: Water, foam, carbon dioxide, dry chemical

SPECIAL FIRE FIGHTING PROCEDURES: Water may be ineffective.

UNUSUAL FIRE AND EXPLOSION HAZARDS

SECTION V HEALTH HAZARD DATA

TOXIC LEVEL: not available

CARCINOGENIC: no

PRINCIPLE ROUTES OF ABSORBTION: Inhalation, ingestion

SKIN AND EYE IRRITATION: moderate

REVELANT SYMPTOMS OF EXPOSURE: Eye irritation, narcosis, drowsiness, headache, vomiting.

EFFECTS OF CHRONIC EXPOSURE: Dermatitis, liver and kidney damage.

EMERGENCY AND FIRST AID PROCEDURES: Inhalation - rest; if swallowed - give emetic, get medical aid; wash from skin & eyes.

SECTION VI REACTIVITY DATA

CONDITIONS CONTRIBUTING TO INSTABILITY

CONDITIONS CONTRIBUTING TO HAZARDOUS POLYMERIZATION

INCOMPATABILITY (MATERIALS TO AVOID)

HAZARDOUS DECOMPOSITION PRODUCTS

SECTION VII SPILL OR LEAK PROCEDURES

STEPS TO BE TAKEN IN CASE MATERIAL IS RELEASED OR SPILLED: Wear self-contained breathing apparatus.

WASTE DISPOSAL METHOD

SECTION VIII SPECIAL PROTECTION INFORMATION

VENTILATION REQUIREMENTS LOCAL EXHAUST

MECHANICAL (GENERAL)

SPECIAL

PROTECTIVE EQUIPMENT (SPECIFY TYPES) EYE: Goggles

GLOVES: Rubber

RESPIRATOR: With organic vapor resistant canister

OTHER PROTECTIVE EQUIPMENT

SECTION IX SPECIAL PRECAUTIONS

PRECAUTIONS TO BE TAKEN IN HANDLING AND STORAGE

OTHER PRECAUTIONS

SIGNATURE

DATE

PROPYLENE BROMOHYDRIN

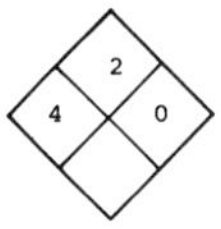

SECTION I SOURCE AND NOMENCLATURE

MANUFACTURER'S NAME

EMERGENCY TELEPHONE NO.

ADDRESS (NUMBER, STREET, CITY, STATE, ZIP CODE)

TRADE NAME AND SYNONYMS

CHEMICAL FAMILY

CHEMICAL NAME AND SYNONYMS: 1 Bromo 2 Propanol

FORMULA: $BrCH_2CHOHCH_3$

SECTION II HAZARDOUS INGREDIENTS

BASIC MATERIAL	APPROXIMATE OR MAXIMUM % WT. OR VOL.	ESTABLISHED OSHA STANDARD	LD_{50} ORAL	LD_{50} PERCUT.	LC_{50} SPECIES	LC_{50} CONC.

SECTION III PHYSICAL DATA MW - 139.0

BOILING POINT °F.: 148°C.

VAPOR PRESSURE mm Hg.: at 52°C. 16

MELTING POINT °F.

VAPOR DENSITY (Air = 1)

SPECIFIC GRAVITY (H_2O = 1): 1.60

EVAPORATION RATE (______ = 1)

SOLUBILITY IN WATER Pts/100 pts H_2O: soluble

VOLATILE % Vol. % Wt.

APPEARANCE AND ODOR: colorless liquid

SECTION IV FIRE AND EXPLOSION HAZARD DATA

FLASH POINT: none available

FLAMMABLE (EXPLOSIVE) LIMITS: UPPER / LOWER

METHOD USED

EXTINGUISHING MEDIA: alcohol foam, carbon dioxide, dry chemical.

SPECIAL FIRE FIGHTING PROCEDURES

UNUSUAL FIRE AND EXPLOSION HAZARDS

SECTION V HEALTH HAZARD DATA

TOXIC LEVEL: None available

CARCINOGENIC: no

PRINCIPLE ROUTES OF ABSORBTION: Inhalation, ingestion, skin

SKIN AND EYE IRRITATION: severe

REVELANT SYMPTOMS OF EXPOSURE: Eye and nasal irritation, nausea, vomiting, giddiness, dizziness, incoordination, visual disturbances, coma.

EFFECTS OF CHRONIC EXPOSURE: --

EMERGENCY AND FIRST AID PROCEDURES: Wash skin & eyes; if swallowed - give emetic, get medical aid; inhalation - rest, give oxygen as needed.

SECTION VI REACTIVITY DATA

CONDITIONS CONTRIBUTING TO INSTABILITY

CONDITIONS CONTRIBUTING TO HAZARDOUS POLYMERIZATION

INCOMPATABILITY (MATERIALS TO AVOID)

HAZARDOUS DECOMPOSITION PRODUCTS

SECTION VII SPILL OR LEAK PROCEDURES

STEPS TO BE TAKEN IN CASE MATERIAL IS RELEASED OR SPILLED

WASTE DISPOSAL METHOD

SECTION VIII SPECIAL PROTECTION INFORMATION

VENTILATION REQUIREMENTS LOCAL EXHAUST

MECHANICAL (GENERAL)

SPECIAL

PROTECTIVE EQUIPMENT (SPECIFY TYPES) EYE: Goggles

GLOVES: Rubber

RESPIRATOR: With proper filter

OTHER PROTECTIVE EQUIPMENT

SECTION IX SPECIAL PRECAUTIONS

PRECAUTIONS TO BE TAKEN IN HANDLING AND STORAGE

OTHER PRECAUTIONS

SIGNATURE

DATE

PROPYLENE CHLORIDE

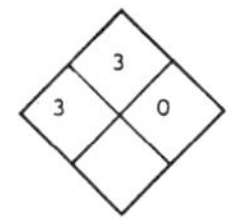

SECTION I SOURCE AND NOMENCLATURE

MANUFACTURER'S NAME

EMERGENCY TELEPHONE NO.

ADDRESS (NUMBER, STREET, CITY, STATE, ZIP CODE)

TRADE NAME AND SYNONYMS: Propylene dichloride

CHEMICAL FAMILY

CHEMICAL NAME AND SYNONYMS: 1,2 dichloropropane

FORMULA: $CH_2ClCHClCH_3$

SECTION II HAZARDOUS INGREDIENTS

BASIC MATERIAL	APPROXIMATE OR MAXIMUM % WT. OR VOL.	ESTABLISHED OSHA STANDARD	LD_{50}			LC_{50}
			ORAL	PERCUT.	SPECIES	CONC.
Oral - human LDLO 50 mg/kg			1900		rat	
LCLO - inhalation - rat 2000 ppm/4hrs.			860		mouse	
				8750	rabbit	
			2000		guinea pig	

SECTION III PHYSICAL DATA MW - 112.99

BOILING POINT °F.: 96.8°C

VAPOR PRESSURE mm Hg.: at 25°C 50

MELTING POINT °F.: -100°C

VAPOR DENSITY (Air = 1): 3.9

SPECIFIC GRAVITY (H_2O = 1): 1.159

EVAPORATION RATE (______ = 1)

SOLUBILITY IN WATER Pts/100 pts H_2O: 0.27

VOLATILE % Vol. % Wt.

APPEARANCE AND ODOR: Colorless liquid, chloroform-like odor.

SECTION IV FIRE AND EXPLOSION HAZARD DATA

FLASH POINT: 60°F

FLAMMABLE (EXPLOSIVE) LIMITS: UPPER / LOWER

METHOD USED: O.C.

EXTINGUISHING MEDIA: Foam, carbon dioxide, dry chemical

SPECIAL FIRE FIGHTING PROCEDURES: Water may be ineffective.

UNUSUAL FIRE AND EXPLOSION HAZARDS

SECTION V HEALTH HAZARD DATA

TOXIC LEVEL: TLV 75 PPM, 350 mg/C.M.

CARCINOGENIC: no

PRINCIPLE ROUTES OF ABSORBTION: Inhalation, ingestion

SKIN AND EYE IRRITATION: moderate

REVELANT SYMPTOMS OF EXPOSURE: Eye irritation, narcosis, sleepiness, headache, vomiting.

EFFECTS OF CHRONIC EXPOSURE: Dermatitis, liver and kidney damage.

EMERGENCY AND FIRST AID PROCEDURES: Inhalation - rest; if swallowed - give emetic, get medical aid; wash from skin & eyes.

SECTION VI REACTIVITY DATA

CONDITIONS CONTRIBUTING TO INSTABILITY

CONDITIONS CONTRIBUTING TO HAZARDOUS POLYMERIZATION

INCOMPATABILITY (MATERIALS TO AVOID)

HAZARDOUS DECOMPOSITION PRODUCTS

SECTION VII SPILL OR LEAK PROCEDURES

STEPS TO BE TAKEN IN CASE MATERIAL IS RELEASED OR SPILLED: Wear self-contained breathing apparatus.

WASTE DISPOSAL METHOD

SECTION VIII SPECIAL PROTECTION INFORMATION

VENTILATION REQUIREMENTS LOCAL EXHAUST

PROTECTIVE EQUIPMENT (SPECIFY TYPES) EYE: Goggles

MECHANICAL (GENERAL)

GLOVES: Rubber

SPECIAL

RESPIRATOR: With organic vapor resistant canister

OTHER PROTECTIVE EQUIPMENT

SECTION IX SPECIAL PRECAUTIONS

PRECAUTIONS TO BE TAKEN IN HANDLING AND STORAGE

OTHER PRECAUTIONS

SIGNATURE | DATE

PROPYLENE CHLOROHYDRIN

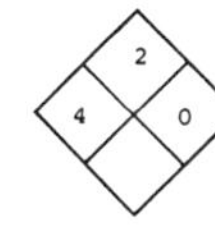

SECTION I SOURCE AND NOMENCLATURE

MANUFACTURER'S NAME

EMERGENCY TELEPHONE NO.

ADDRESS (NUMBER, STREET, CITY, STATE, ZIP CODE)

TRADE NAME AND SYNONYMS

CHEMICAL FAMILY

CHEMICAL NAME AND SYNONYMS: 1 Chloro 2 Propanol

FORMULA: $ClCH_2CHOHCH_3$

SECTION II HAZARDOUS INGREDIENTS

BASIC MATERIAL	APPROXIMATE OR MAXIMUM % WT. OR VOL.	ESTABLISHED OSHA STANDARD	LD_{50} mg/kg			LC_{50}
			ORAL	PERCUT.	SPECIES	CONC.
				480	rabbit	

SECTION III PHYSICAL DATA MW-94.55

BOILING POINT °F.: 127°C.

VAPOR PRESSURE mm Hg.

MELTING POINT °F.

VAPOR DENSITY (Air = 1): 3.26

SPECIFIC GRAVITY (H_2O = 1): 1.10

EVAPORATION RATE (______ = 1)

SOLUBILITY IN WATER Pts/100 pts H_2O: infinitely soluble

VOLATILE % Vol. % Wt.

APPEARANCE AND ODOR: Colorless liquid, mild nonresidual odor.

SECTION IV FIRE AND EXPLOSION HAZARD DATA

FLASH POINT: 125°F.

FLAMMABLE (EXPLOSIVE) LIMITS: UPPER / LOWER

METHOD USED: C.C.

EXTINGUISHING MEDIA: Alcohol foam, carbon dioxide, dry chemical.

SPECIAL FIRE FIGHTING PROCEDURES

UNUSUAL FIRE AND EXPLOSION HAZARDS

SECTION V HEALTH HAZARD DATA

TOXIC LEVEL: None available

CARCINOGENIC: no

PRINCIPLE ROUTES OF ABSORBTION: Inhalation, ingestion, skin

SKIN AND EYE IRRITATION: Severe

REVELANT SYMPTOMS OF EXPOSURE: Eye and nasal irritation, nausea, vomiting, giddiness, dizziness, incoordination, visual disturbances, coma.

EFFECTS OF CHRONIC EXPOSURE: --

EMERGENCY AND FIRST AID PROCEDURES: Wash skin & eyes; if swallowed - give emetic, get medical aid; inhalation - rest, give oxygen as needed.

SECTION VI REACTIVITY DATA

CONDITIONS CONTRIBUTING TO INSTABILITY

CONDITIONS CONTRIBUTING TO HAZARDOUS POLYMERIZATION

INCOMPATABILITY (MATERIALS TO AVOID)

HAZARDOUS DECOMPOSITION PRODUCTS

SECTION VII SPILL OR LEAK PROCEDURES

STEPS TO BE TAKEN IN CASE MATERIAL IS RELEASED OR SPILLED

WASTE DISPOSAL METHOD

SECTION VIII SPECIAL PROTECTION INFORMATION

VENTILATION REQUIREMENTS LOCAL EXHAUST

PROTECTIVE EQUIPMENT (SPECIFY TYPES) EYE: Goggles

MECHANICAL (GENERAL)

GLOVES: Rubber

SPECIAL

RESPIRATOR: With proper filter

OTHER PROTECTIVE EQUIPMENT

SECTION IX SPECIAL PRECAUTIONS

PRECAUTIONS TO BE TAKEN IN HANDLING AND STORAGE

OTHER PRECAUTIONS

SIGNATURE | DATE

QUINONE

Hazard diamond: top 1, left 4, right 0

SECTION I SOURCE AND NOMENCLATURE

MANUFACTURER'S NAME | EMERGENCY TELEPHONE NO.

ADDRESS (NUMBER, STREET, CITY, STATE, ZIP CODE)

TRADE NAME AND SYNONYMS: Chinone | CHEMICAL FAMILY

CHEMICAL NAME AND SYNONYMS: P-Benzoquinone | FORMULA: OC_6H_4O

SECTION II HAZARDOUS INGREDIENTS

BASIC MATERIAL	APPROXIMATE OR MAXIMUM % WT. OR VOL.	ESTABLISHED OSHA STANDARD	LD_{50} mg/kg ORAL	PERCUT.	LC_{50} SPECIES	CONC.
Oral-human LDLO 40 mg/kg			130		rat	
Lowest published lethal concentration in air						
inhalation - mouse 320 mg/C.M.						

SECTION III PHYSICAL DATA MW - 108.09

BOILING POINT °F.: sublimes | VAPOR PRESSURE mm Hg.: 0.1 at 25°C.

MELTING POINT °F.: 116°C | VAPOR DENSITY (Air = 1)

SPECIFIC GRAVITY (H_2O = 1): 1.318 | EVAPORATION RATE (____ = 1)

SOLUBILITY IN WATER Pts/100 pts H_2O: Very slightly soluble | VOLATILE % Vol. % Wt.

APPEARANCE AND ODOR: Yellow crystals, pungent odor.

SECTION IV FIRE AND EXPLOSION HAZARD DATA

FLASH POINT: none available | FLAMMABLE (EXPLOSIVE) LIMITS | UPPER | LOWER

METHOD USED

EXTINGUISHING MEDIA: Water, carbon dioxide, dry chemical.

SPECIAL FIRE FIGHTING PROCEDURES

UNUSUAL FIRE AND EXPLOSION HAZARDS

SECTION V HEALTH HAZARD DATA

TOXIC LEVEL: 0.1 PPM, 0.4 mg/C.M. | CARCINOGENIC: neoplastic

PRINCIPLE ROUTES OF ABSORBTION: Inhalation, ingestion | SKIN AND EYE IRRITATION: Severe

REVELANT SYMPTOMS OF EXPOSURE: Irritation and damage to skin & eyes, loss of vision, vomiting, collapse, coma.

EFFECTS OF CHRONIC EXPOSURE: Eye damage, discoloration, severe irritation, swelling.

EMERGENCY AND FIRST AID PROCEDURES: If swallowed - give milk, emetic, get medical aid; inhalation - rest; wash from skin & eyes.

SECTION VI REACTIVITY DATA

CONDITIONS CONTRIBUTING TO INSTABILITY

CONDITIONS CONTRIBUTING TO HAZARDOUS POLYMERIZATION

INCOMPATABILITY (MATERIALS TO AVOID)

HAZARDOUS DECOMPOSITION PRODUCTS

SECTION VII SPILL OR LEAK PROCEDURES

STEPS TO BE TAKEN IN CASE MATERIAL IS RELEASED OR SPILLED: Collect with brush onto a paper, burn on iron pan in hood.

WASTE DISPOSAL METHOD

SECTION VIII SPECIAL PROTECTION INFORMATION

VENTILATION REQUIREMENTS LOCAL EXHAUST | PROTECTIVE EQUIPMENT (SPECIFY TYPES) EYE: Goggles

MECHANICAL (GENERAL) | GLOVES: Rubber

SPECIAL | RESPIRATOR: With proper filter

OTHER PROTECTIVE EQUIPMENT: Coveralls

SECTION IX SPECIAL PRECAUTIONS

PRECAUTIONS TO BE TAKEN IN HANDLING AND STORAGE

OTHER PRECAUTIONS

SIGNATURE | DATE

RESORCINOL MONOMETHYL ETHER

Hazard diamond: top 1, left 4, right 0

SECTION I SOURCE AND NOMENCLATURE

MANUFACTURER'S NAME | EMERGENCY TELEPHONE NO.

ADDRESS (NUMBER, STREET, CITY, STATE, ZIP CODE)

TRADE NAME AND SYNONYMS | CHEMICAL FAMILY

CHEMICAL NAME AND SYNONYMS: M-Methoxyphenol | FORMULA: $CH_3OC_6H_4OH$

SECTION II HAZARDOUS INGREDIENTS

BASIC MATERIAL	APPROXIMATE OR MAXIMUM % WT. OR VOL.	ESTABLISHED OSHA STANDARD	LD_{50} ORAL	PERCUT.	LC_{50} SPECIES	CONC.
			none	available		

SECTION III PHYSICAL DATA MW - 124.13

BOILING POINT °F.: 244°C | VAPOR PRESSURE mm Hg.: 5 at 115°C

MELTING POINT °F.: -17°C | VAPOR DENSITY (Air = 1)

SPECIFIC GRAVITY (H_2O = 1): > 1 | EVAPORATION RATE (____ = 1)

SOLUBILITY IN WATER Pts/100 pts H_2O: slightly soluble | VOLATILE % Vol. % Wt.

APPEARANCE AND ODOR: liquid

SECTION IV FIRE AND EXPLOSION HAZARD DATA

FLASH POINT: none available | FLAMMABLE (EXPLOSIVE) LIMITS | UPPER | LOWER

METHOD USED

EXTINGUISHING MEDIA: carbon dioxide, alcohol foam, dry chemical

SPECIAL FIRE FIGHTING PROCEDURES

UNUSUAL FIRE AND EXPLOSION HAZARDS

SECTION V HEALTH HAZARD DATA

TOXIC LEVEL: none available | CARCINOGENIC: no

PRINCIPLE ROUTES OF ABSORBTION: inhalation, ingestion | SKIN AND EYE IRRITATION: moderate

REVELANT SYMPTOMS OF EXPOSURE: hyperemia, itching, dermatitis, restlessness, cyanosis, dizziness, tremors, convulsions, death.

EFFECTS OF CHRONIC EXPOSURE: Possible liver, kidney and lung damage

EMERGENCY AND FIRST AID PROCEDURES: If swallowed - induce vomiting, get medical aid; wash eyes & skin; inhalation - rest.

SECTION VI REACTIVITY DATA

CONDITIONS CONTRIBUTING TO INSTABILITY

CONDITIONS CONTRIBUTING TO HAZARDOUS POLYMERIZATION

INCOMPATABILITY (MATERIALS TO AVOID)

HAZARDOUS DECOMPOSITION PRODUCTS

SECTION VII SPILL OR LEAK PROCEDURES

STEPS TO BE TAKEN IN CASE MATERIAL IS RELEASED OR SPILLED

WASTE DISPOSAL METHOD

SECTION VIII SPECIAL PROTECTION INFORMATION

VENTILATION REQUIREMENTS LOCAL EXHAUST | PROTECTIVE EQUIPMENT (SPECIFY TYPES) EYE: Safety goggles

MECHANICAL (GENERAL) | GLOVES: rubber

SPECIAL | RESPIRATOR

OTHER PROTECTIVE EQUIPMENT

SECTION IX SPECIAL PRECAUTIONS

PRECAUTIONS TO BE TAKEN IN HANDLING AND STORAGE

OTHER PRECAUTIONS

SIGNATURE | DATE

RHODIUM TRICHLORIDE

NFPA: 0 / 4 / 0

SECTION I SOURCE AND NOMENCLATURE

MANUFACTURER'S NAME | EMERGENCY TELEPHONE NO.

ADDRESS (NUMBER, STREET, CITY, STATE, ZIP CODE)

TRADE NAME AND SYNONYMS | CHEMICAL FAMILY

CHEMICAL NAME AND SYNONYMS: Rhodium Trichloride | FORMULA: $Rh\ Cl_3$

SECTION II HAZARDOUS INGREDIENTS

BASIC MATERIAL	APPROXIMATE OR MAXIMUM % WT. OR VOL.	ESTABLISHED OSHA STANDARD	LD_{50} ORAL	LD_{50} PERCUT.	LC_{50} SPECIES	LC_{50} CONC.
LD50-intravenous-rat	198 mg/kg					
LD50-intravenous-rabbit	215 mg/kg					

SECTION III PHYSICAL DATA MW - 209.26

BOILING POINT (°F.): sublimes (800°C)

VAPOR PRESSURE (mm Hg.):

MELTING POINT (°F.): 450°C (decomp.)

VAPOR DENSITY (Air = 1):

SPECIFIC GRAVITY (H_2O = 1):

EVAPORATION RATE (______ = 1):

SOLUBILITY IN WATER Pts/100 pts H_2O: Insoluble

VOLATILE % Vol. % Wt.

APPEARANCE AND ODOR: Dark red powder

SECTION IV FIRE AND EXPLOSION HAZARD DATA

FLASH POINT: Not flammable

FLAMMABLE (EXPLOSIVE) LIMITS: UPPER / LOWER

METHOD USED

EXTINGUISHING MEDIA: Use water on fire

SPECIAL FIRE FIGHTING PROCEDURES

UNUSUAL FIRE AND EXPLOSION HAZARDS

SECTION V HEALTH HAZARD DATA

TOXIC LEVEL: TLV - 0.001 mg/C.M. | CARCINOGENIC: Yes

PRINCIPLE ROUTES OF ABSORBTION: Inhalation, ingestion | SKIN AND EYE IRRITATION: Slight

REVELANT SYMPTOMS OF EXPOSURE: Possible allergic asthma

EFFECTS OF CHRONIC EXPOSURE: Possible allergic asthma

EMERGENCY AND FIRST AID PROCEDURES: If swallowed - induce vomiting, get medical aid; inhalation - rest; wash skin and eyes.

SECTION VI REACTIVITY DATA

CONDITIONS CONTRIBUTING TO INSTABILITY

CONDITIONS CONTRIBUTING TO HAZARDOUS POLYMERIZATION

INCOMPATABILITY (MATERIALS TO AVOID)

HAZARDOUS DECOMPOSITION PRODUCTS

SECTION VII SPILL OR LEAK PROCEDURES

STEPS TO BE TAKEN IN CASE MATERIAL IS RELEASED OR SPILLED

WASTE DISPOSAL METHOD

SECTION VIII SPECIAL PROTECTION INFORMATION

VENTILATION REQUIREMENTS LOCAL EXHAUST

PROTECTIVE EQUIPMENT (SPECIFY TYPES) EYE: Safety goggles

MECHANICAL (GENERAL)

GLOVES

SPECIAL

RESPIRATOR: With particulate filter

OTHER PROTECTIVE EQUIPMENT

SECTION IX SPECIAL PRECAUTIONS

PRECAUTIONS TO BE TAKEN IN HANDLING AND STORAGE

OTHER PRECAUTIONS

SIGNATURE | DATE

RUTHENIUM CHLORIDE

NFPA: 0 / 4 / 0

SECTION I SOURCE AND NOMENCLATURE

MANUFACTURER'S NAME | EMERGENCY TELEPHONE NO.

ADDRESS (NUMBER, STREET, CITY, STATE, ZIP CODE)

TRADE NAME AND SYNONYMS | CHEMICAL FAMILY

CHEMICAL NAME AND SYNONYMS: Ruthenium Chloride | FORMULA: $RuCl_4 \cdot 2H_2O$

SECTION II HAZARDOUS INGREDIENTS

BASIC MATERIAL	APPROXIMATE OR MAXIMUM % WT. OR VOL.	ESTABLISHED OSHA STANDARD	LD_{50} ORAL	LD_{50} PERCUT.	LC_{50} SPECIES	LC_{50} CONC.
LD50 intraperitoneal - rat	360 mg/kg					

SECTION III PHYSICAL DATA MW 207.42

BOILING POINT (°F.): --

VAPOR PRESSURE (mm Hg.): --

MELTING POINT (°F.): --

VAPOR DENSITY (Air = 1):

SPECIFIC GRAVITY (H_2O = 1):

EVAPORATION RATE (______ = 1):

SOLUBILITY IN WATER Pts/100 pts H_2O: Soluble

VOLATILE % Vol. % Wt.

APPEARANCE AND ODOR: Dark colored powder

SECTION IV FIRE AND EXPLOSION HAZARD DATA

FLASH POINT: Will not burn.

FLAMMABLE (EXPLOSIVE) LIMITS: UPPER / LOWER

METHOD USED

EXTINGUISHING MEDIA

SPECIAL FIRE FIGHTING PROCEDURES

UNUSUAL FIRE AND EXPLOSION HAZARDS

SECTION V HEALTH HAZARD DATA

TOXIC LEVEL: None available | CARCINOGENIC: no

PRINCIPLE ROUTES OF ABSORBTION: Inhalation, ingestion | SKIN AND EYE IRRITATION: Severe

REVELANT SYMPTOMS OF EXPOSURE: Lacramation, disturbance of vision, headache, eye irritation, cough.

EFFECTS OF CHRONIC EXPOSURE: Pulmonary edema

EMERGENCY AND FIRST AID PROCEDURES: Inhalation - rest; if swallowed - induce vomiting, get medical aid.

SECTION VI REACTIVITY DATA

CONDITIONS CONTRIBUTING TO INSTABILITY

CONDITIONS CONTRIBUTING TO HAZARDOUS POLYMERIZATION

INCOMPATABILITY (MATERIALS TO AVOID)

HAZARDOUS DECOMPOSITION PRODUCTS

SECTION VII SPILL OR LEAK PROCEDURES

STEPS TO BE TAKEN IN CASE MATERIAL IS RELEASED OR SPILLED

WASTE DISPOSAL METHOD

SECTION VIII SPECIAL PROTECTION INFORMATION

VENTILATION REQUIREMENTS LOCAL EXHAUST

PROTECTIVE EQUIPMENT (SPECIFY TYPES) EYE: Goggles

MECHANICAL (GENERAL)

GLOVES

SPECIAL

RESPIRATOR: With proper filter

OTHER PROTECTIVE EQUIPMENT

SECTION IX SPECIAL PRECAUTIONS

PRECAUTIONS TO BE TAKEN IN HANDLING AND STORAGE

OTHER PRECAUTIONS

SIGNATURE | DATE

SELENIUM METAL

Hazard diamond: 1 (top), 4 (left), 0 (right)

SECTION I SOURCE AND NOMENCLATURE

MANUFACTURER'S NAME	EMERGENCY TELEPHONE NO.
ADDRESS (NUMBER, STREET, CITY, STATE, ZIP CODE)	
TRADE NAME AND SYNONYMS	CHEMICAL FAMILY
CHEMICAL NAME AND SYNONYMS Selenium	FORMULA Se

SECTION II HAZARDOUS INGREDIENTS

BASIC MATERIAL	APPROXIMATE OR MAXIMUM % WT. OR VOL.	ESTABLISHED OSHA STANDARD	LD_{50} ORAL	LD_{50} PERCUT.	LC_{50} SPECIES	LC_{50} CONC.
			See Note			
Note:						
Lowest published lethal concentration in air						
inhalation - rat 33 mg/C.M.						

SECTION III PHYSICAL DATA An 78.96 Se_8 631.68

BOILING POINT °F. 688°C.	VAPOR PRESSURE at 356°C. 1 mm Hg.
MELTING POINT °F. 170 - 217°C.	VAPOR DENSITY (Air = 1)
SPECIFIC GRAVITY (H_2O = 1) 4.26 - 4.79	EVAPORATION RATE (______ = 1)
SOLUBILITY IN WATER Pts/100 pts H_2O insoluble	VOLATILE % Vol. % Wt.

APPEARANCE AND ODOR
steel grey, non-metallic element.

SECTION IV FIRE AND EXPLOSION HAZARD DATA

FLASH POINT none available	FLAMMABLE (EXPLOSIVE) LIMITS	UPPER
METHOD USED		LOWER

EXTINGUISHING MEDIA
Use water spray.

SPECIAL FIRE FIGHTING PROCEDURES
Small fires extinguished with sand. Self-contained breathing apparatus.

UNUSUAL FIRE AND EXPLOSION HAZARDS
Same as sulfur. Burning Se produces SeO_2.

SECTION V HEALTH HAZARD DATA

TOXIC LEVEL
TLV 0.2 mg/m^3 — CARCINOGENIC: neoplastic

PRINCIPLE ROUTES OF ABSORBTION
Inhalation of dust, ingestion — SKIN AND EYE IRRITATION: Slight

REVELANT SYMPTOMS OF EXPOSURE
Nervousness, fear, vomiting, difficult respiration, central nervous system depression.

EFFECTS OF CHRONIC EXPOSURE
Possible liver damage, nervousness, tremor, metallic taste, garlic-like odor of breath.

EMERGENCY AND FIRST AID PROCEDURES
Inhalation - rest, oxygen as needed; if swallowed - give emetic, get medical aid.

SECTION VI REACTIVITY DATA

CONDITIONS CONTRIBUTING TO INSTABILITY

CONDITIONS CONTRIBUTING TO HAZARDOUS POLYMERIZATION

INCOMPATABILITY (MATERIALS TO AVOID) Barium Carbide, Bromine Pentafluoride, Calcium Carbide, Chromium Trioxide, Fluorene, Lithium, Silicide, Nickel, Nitric Acid, Nitrogen Trichloride.

HAZARDOUS DECOMPOSITION PRODUCTS

SECTION VII SPILL OR LEAK PROCEDURES

STEPS TO BE TAKEN IN CASE MATERIAL IS RELEASED OR SPILLED

WASTE DISPOSAL METHOD

SECTION VIII SPECIAL PROTECTION INFORMATION

VENTILATION REQUIREMENTS LOCAL EXHAUST	PROTECTIVE EQUIPMENT (SPECIFY TYPES) EYE Goggles
MECHANICAL (GENERAL)	GLOVES Plastic or rubber
SPECIAL	RESPIRATOR With proper filter
OTHER PROTECTIVE EQUIPMENT	

SECTION IX SPECIAL PRECAUTIONS

PRECAUTIONS TO BE TAKEN IN HANDLING AND STORAGE
Store in place free from water.

OTHER PRECAUTIONS

SIGNATURE	DATE

SEMICARBAZIDE HYDROCHLORIDE

Hazard diamond: 1 (top), 4 (left), 0 (right)

SECTION I SOURCE AND NOMENCLATURE

MANUFACTURER'S NAME	EMERGENCY TELEPHONE NO.
ADDRESS (NUMBER, STREET, CITY, STATE, ZIP CODE)	
TRADE NAME AND SYNONYMS	CHEMICAL FAMILY
CHEMICAL NAME AND SYNONYMS Semicarbazide Hydrochloride	FORMULA $H_2NCONHNH_2 \cdot HCl$

SECTION II HAZARDOUS INGREDIENTS

BASIC MATERIAL	APPROXIMATE OR MAXIMUM % WT. OR VOL.	ESTABLISHED OSHA STANDARD	LD_{50} ORAL	LD_{50} PERCUT.	LC_{50} SPECIES	LC_{50} CONC.
Lowest published lethal dose -						
oral - rat 10 mg/kg.						

SECTION III PHYSICAL DATA MW - 111.54

BOILING POINT °F.	VAPOR PRESSURE mm Hg.
MELTING POINT °F. 173°C.D.	VAPOR DENSITY (Air = 1)
SPECIFIC GRAVITY (H_2O = 1)	EVAPORATION RATE (______ = 1)
SOLUBILITY IN WATER Pts/100 pts H_2O very soluble	VOLATILE % Vol. % Wt.

APPEARANCE AND ODOR
Crystals

SECTION IV FIRE AND EXPLOSION HAZARD DATA

FLASH POINT none available	FLAMMABLE (EXPLOSIVE) LIMITS	UPPER
METHOD USED		LOWER

EXTINGUISHING MEDIA
Alcohol foam, carbon dioxide, dry chemical.

SPECIAL FIRE FIGHTING PROCEDURES

UNUSUAL FIRE AND EXPLOSION HAZARDS

SECTION V HEALTH HAZARD DATA

TOXIC LEVEL
none available — Teratogen — CARCINOGENIC: no

PRINCIPLE ROUTES OF ABSORBTION
Inhalation, ingestion, skin — SKIN AND EYE IRRITATION: Severe

REVELANT SYMPTOMS OF EXPOSURE Irritation of skin, eyes and mucous membranes, gastroentric disturbances, anemia.

EFFECTS OF CHRONIC EXPOSURE
Blood and liver damage, loss of appetite.

EMERGENCY AND FIRST AID PROCEDURES Inhalation - rest; if swallowed - induce vomiting, get medical aid; wash from skin & eyes.

SECTION VI REACTIVITY DATA

CONDITIONS CONTRIBUTING TO INSTABILITY

CONDITIONS CONTRIBUTING TO HAZARDOUS POLYMERIZATION

INCOMPATABILITY (MATERIALS TO AVOID)

HAZARDOUS DECOMPOSITION PRODUCTS

SECTION VII SPILL OR LEAK PROCEDURES

STEPS TO BE TAKEN IN CASE MATERIAL IS RELEASED OR SPILLED

WASTE DISPOSAL METHOD

SECTION VIII SPECIAL PROTECTION INFORMATION

VENTILATION REQUIREMENTS LOCAL EXHAUST	PROTECTIVE EQUIPMENT (SPECIFY TYPES) EYE Goggles
MECHANICAL (GENERAL)	GLOVES Rubber
SPECIAL	RESPIRATOR Self-contained
OTHER PROTECTIVE EQUIPMENT	

SECTION IX SPECIAL PRECAUTIONS

PRECAUTIONS TO BE TAKEN IN HANDLING AND STORAGE

OTHER PRECAUTIONS

SIGNATURE	DATE

SILVER CYANIDE

Hazard diamond: 0 (top), 4 (left), 0 (right)

SECTION I SOURCE AND NOMENCLATURE

MANUFACTURER'S NAME	EMERGENCY TELEPHONE NO.
ADDRESS (NUMBER, STREET, CITY, STATE, ZIP CODE)	
TRADE NAME AND SYNONYMS	CHEMICAL FAMILY
CHEMICAL NAME AND SYNONYMS Silver Cyanide	FORMULA AgCN

SECTION II HAZARDOUS INGREDIENTS

BASIC MATERIAL	APPROXIMATE OR MAXIMUM % WT. OR VOL.	ESTABLISHED OSHA STANDARD	LD_{50} mg/CM ORAL	PERCUT.	SPECIES	LC_{50} CONC.
			123		Rat	

SECTION III PHYSICAL DATA MW - 133.89

BOILING POINT °F.	VAPOR PRESSURE mm Hg.
MELTING POINT °F. 320°C.	VAPOR DENSITY (Air = 1)
SPECIFIC GRAVITY (H_2O = 1) 3.95	EVAPORATION RATE (______ = 1)
SOLUBILITY IN WATER Pts/100 pts H_2O Slightly soluble	VOLATILE % Vol. % Wt.

APPEARANCE AND ODOR

SECTION IV FIRE AND EXPLOSION HAZARD DATA

FLASH POINT Not flammable	FLAMMABLE (EXPLOSIVE) LIMITS	UPPER
METHOD USED		LOWER

EXTINGUISHING MEDIA
Water

SPECIAL FIRE FIGHTING PROCEDURES

UNUSUAL FIRE AND EXPLOSION HAZARDS

SECTION V HEALTH HAZARD DATA

TOXIC LEVEL TLV - 0.016 — CARCINOGENIC No

PRINCIPLE ROUTES OF ABSORBTION Inhalation, ingestion, skin — SKIN AND EYE IRRITATION Moderate

REVELANT SYMPTOMS OF EXPOSURE Violent abdominal pain on ingestion, rapid respiration, then gasping; dizziness, shortness of breath.

EFFECTS OF CHRONIC EXPOSURE Blue-grey discoloration of skin, mucous membranes and eyes.

EMERGENCY AND FIRST AID PROCEDURES If swallowed - amyl nitrite inhalation, followed by immediate vomiting or gastric lavage.

SECTION VI REACTIVITY DATA

CONDITIONS CONTRIBUTING TO INSTABILITY

CONDITIONS CONTRIBUTING TO HAZARDOUS POLYMERIZATION

INCOMPATABILITY (MATERIALS TO AVOID)
Fluorine and silver cyanide - may result in violent explosion.

HAZARDOUS DECOMPOSITION PRODUCTS
H Cn with acids.

SECTION VII SPILL OR LEAK PROCEDURES

STEPS TO BE TAKEN IN CASE MATERIAL IS RELEASED OR SPILLED

WASTE DISPOSAL METHOD

SECTION VIII SPECIAL PROTECTION INFORMATION

VENTILATION REQUIREMENTS LOCAL EXHAUST	PROTECTIVE EQUIPMENT (SPECIFY TYPES) EYE Safety glasses
MECHANICAL (GENERAL)	GLOVES Rubber
SPECIAL	RESPIRATOR With particulate filter

OTHER PROTECTIVE EQUIPMENT

SECTION IX SPECIAL PRECAUTIONS

PRECAUTIONS TO BE TAKEN IN HANDLING AND STORAGE
Store away from acids.

OTHER PRECAUTIONS

SIGNATURE	DATE

SODIUM CALCIUM HYDRATE

Hazard diamond: 0 (top), 3 (left), 1 (right)

SECTION I SOURCE AND NOMENCLATURE

MANUFACTURER'S NAME	EMERGENCY TELEPHONE NO.
ADDRESS (NUMBER, STREET, CITY, STATE, ZIP CODE)	
TRADE NAME AND SYNONYMS	CHEMICAL FAMILY
CHEMICAL NAME AND SYNONYMS Soda lime	FORMULA --

SECTION II HAZARDOUS INGREDIENTS

BASIC MATERIAL	APPROXIMATE OR MAXIMUM % WT. OR VOL.	ESTABLISHED OSHA STANDARD	LD_{50} ORAL	PERCUT.	SPECIES	LC_{50} CONC.
CaO+(5-20% NaOH)			not available			

SECTION III PHYSICAL DATA MW - 22.90

BOILING POINT °F. N/A	VAPOR PRESSURE mm Hg.
MELTING POINT °F. N/A	VAPOR DENSITY (Air = 1)
SPECIFIC GRAVITY (H_2O = 1)	EVAPORATION RATE (______ = 1)
SOLUBILITY IN WATER Pts/100 pts H_2O	VOLATILE % Vol. % Wt.

APPEARANCE AND ODOR
White granules

SECTION IV FIRE AND EXPLOSION HAZARD DATA

FLASH POINT Not flammable	FLAMMABLE (EXPLOSIVE) LIMITS	UPPER
METHOD USED N/A		LOWER

EXTINGUISHING MEDIA
N/A

SPECIAL FIRE FIGHTING PROCEDURES
N/A

UNUSUAL FIRE AND EXPLOSION HAZARDS
N/A

SECTION V HEALTH HAZARD DATA

TOXIC LEVEL TLV - 2 mg/C.M. (from NaOH) — CARCINOGENIC No

PRINCIPLE ROUTES OF ABSORBTION Ingestion — SKIN AND EYE IRRITATION Yes

REVELANT SYMPTOMS OF EXPOSURE Severe irritation of skin and eyes, severe internal irritation and damage, vomiting, diarrhea and collapse.

EFFECTS OF CHRONIC EXPOSURE
--

EMERGENCY AND FIRST AID PROCEDURES Flush with water and vinegar; if swallowed, wash mouth and give water to drink, then vinegar, get medical attention.

SECTION VI REACTIVITY DATA

CONDITIONS CONTRIBUTING TO INSTABILITY

CONDITIONS CONTRIBUTING TO HAZARDOUS POLYMERIZATION

INCOMPATABILITY (MATERIALS TO AVOID)
Acids

HAZARDOUS DECOMPOSITION PRODUCTS

SECTION VII SPILL OR LEAK PROCEDURES

STEPS TO BE TAKEN IN CASE MATERIAL IS RELEASED OR SPILLED

WASTE DISPOSAL METHOD

SECTION VIII SPECIAL PROTECTION INFORMATION

VENTILATION REQUIREMENTS LOCAL EXHAUST	PROTECTIVE EQUIPMENT (SPECIFY TYPES) EYE Safety glasses
MECHANICAL (GENERAL)	GLOVES
SPECIAL	RESPIRATOR

OTHER PROTECTIVE EQUIPMENT

SECTION IX SPECIAL PRECAUTIONS

PRECAUTIONS TO BE TAKEN IN HANDLING AND STORAGE

OTHER PRECAUTIONS

SIGNATURE	DATE

SODIUM PEROXIDE

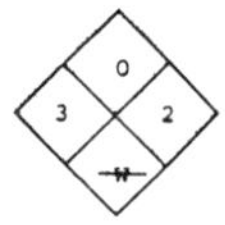

SECTION I SOURCE AND NOMENCLATURE

MANUFACTURER'S NAME | EMERGENCY TELEPHONE NO.

ADDRESS (NUMBER, STREET, CITY, STATE, ZIP CODE)

TRADE NAME AND SYNONYMS | CHEMICAL FAMILY

CHEMICAL NAME AND SYNONYMS: Sodium Peroxide | FORMULA: Na_2O_2

SECTION II HAZARDOUS INGREDIENTS

BASIC MATERIAL	APPROXIMATE OR MAXIMUM % WT. OR VOL.	ESTABLISHED OSHA STANDARD	LD_{50} ORAL	LD_{50} PERCUT.	LC_{50} SPECIES	LC_{50} CONC.
			Not available			

SECTION III PHYSICAL DATA MW - 77.99

BOILING POINT °F: --

VAPOR PRESSURE mm Hg:

MELTING POINT °F: d (460°C.)

VAPOR DENSITY (Air = 1):

SPECIFIC GRAVITY (H_2O = 1): 2.805

EVAPORATION RATE (=1):

SOLUBILITY IN WATER Pts/100 pts H_2O: 5

VOLATILE % Vol. % Wt.

APPEARANCE AND ODOR: White powder

SECTION IV FIRE AND EXPLOSION HAZARD DATA

FLASH POINT: Not flammable

FLAMMABLE (EXPLOSIVE) LIMITS: UPPER / LOWER

METHOD USED

EXTINGUISHING MEDIA

SPECIAL FIRE FIGHTING PROCEDURES: Wear full protective clothing.

UNUSUAL FIRE AND EXPLOSION HAZARDS: Reacts with water to produce oxygen.

SECTION V HEALTH HAZARD DATA

TOXIC LEVEL: TLV 2 Mg/C.M. (NaOH) (ceiling value)

CARCINOGENIC: no

PRINCIPLE ROUTES OF ABSORBTION: Inhalation, ingestion (mainly)

SKIN AND EYE IRRITATION: Severe

REVELANT SYMPTOMS OF EXPOSURE: Extremely corrosive to skin, eyes and mucous membranes, pulmonary irritation.

EFFECTS OF CHRONIC EXPOSURE: Chronic dermatitis.

EMERGENCY AND FIRST AID PROCEDURES: Wash from skin and eyes; if swallowed - wash mouth, drink 1% vinegar, orange juice or lemon juice, get medical aid; if inhaled - rest, keep warm, get medical aid.

SECTION VI REACTIVITY DATA

CONDITIONS CONTRIBUTING TO INSTABILITY: Strong oxidizing agent.

CONDITIONS CONTRIBUTING TO HAZARDOUS POLYMERIZATION

INCOMPATABILITY (MATERIALS TO AVOID): Reducing agents, water, strong acids.

HAZARDOUS DECOMPOSITION PRODUCTS

SECTION VII SPILL OR LEAK PROCEDURES

STEPS TO BE TAKEN IN CASE MATERIAL IS RELEASED OR SPILLED: Cover with (9.1) sand to soda ash

WASTE DISPOSAL METHOD

SECTION VIII SPECIAL PROTECTION INFORMATION

VENTILATION REQUIREMENTS LOCAL EXHAUST

PROTECTIVE EQUIPMENT (SPECIFY TYPES) EYE: Safety glasses

MECHANICAL (GENERAL)

GLOVES: Rubber

SPECIAL

RESPIRATOR

OTHER PROTECTIVE EQUIPMENT

SECTION IX SPECIAL PRECAUTIONS

PRECAUTIONS TO BE TAKEN IN HANDLING AND STORAGE: Store away from sources of water.

OTHER PRECAUTIONS

SIGNATURE | DATE

SODIUM TELLURITE

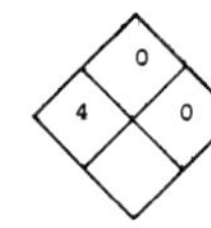

SECTION I SOURCE AND NOMENCLATURE

MANUFACTURER'S NAME | EMERGENCY TELEPHONE NO.

ADDRESS (NUMBER, STREET, CITY, STATE, ZIP CODE)

TRADE NAME AND SYNONYMS: Disodium Tellorus Acid | CHEMICAL FAMILY

CHEMICAL NAME AND SYNONYMS: Sodium tellurite | FORMULA: Na_2TeO_3

SECTION II HAZARDOUS INGREDIENTS

BASIC MATERIAL	APPROXIMATE OR MAXIMUM % WT. OR VOL.	ESTABLISHED OSHA STANDARD	LD_{50} mg/kg ORAL	LD_{50} mg/kg PERCUT.	LC_{50} SPECIES	LC_{50} CONC.
			20		Mouse	
Note:			83		Rat	
Lowest published lethal dose			67		Rabbit	
Oral - human 29 mg/kg.			45		Guinea Pig	

SECTION III PHYSICAL DATA MW-221.58

BOILING POINT °F: N/A

VAPOR PRESSURE mm Hg:

MELTING POINT °F: N/A

VAPOR DENSITY (Air = 1):

SPECIFIC GRAVITY (H_2O = 1): N/A

EVAPORATION RATE (=1):

SOLUBILITY IN WATER Pts/100 pts H_2O

VOLATILE % Vol. % Wt.

APPEARANCE AND ODOR: White powder

SECTION IV FIRE AND EXPLOSION HAZARD DATA

FLASH POINT: Not flammable

FLAMMABLE (EXPLOSIVE) LIMITS: UPPER / LOWER

METHOD USED

EXTINGUISHING MEDIA: Water

SPECIAL FIRE FIGHTING PROCEDURES: N/A

UNUSUAL FIRE AND EXPLOSION HAZARDS: N/A

SECTION V HEALTH HAZARD DATA

TOXIC LEVEL: 0.1 Mg/C.M.

CARCINOGENIC: No

PRINCIPLE ROUTES OF ABSORBTION: Inhalation, ingestion

SKIN AND EYE IRRITATION: Moderate

REVELANT SYMPTOMS OF EXPOSURE: Sleepiness, loss of appetite, nausea, metallic taste, garlic odor to breath.

EFFECTS OF CHRONIC EXPOSURE: Fatigue, loss of appetite, bronchitis, dermatitis.

EMERGENCY AND FIRST AID PROCEDURES: If inhaled - rest, keep warm, get medical aid. If swallowed, wash mouth, give emetic, get medical aid; wash from skin and eyes.

SECTION VI REACTIVITY DATA

CONDITIONS CONTRIBUTING TO INSTABILITY

CONDITIONS CONTRIBUTING TO HAZARDOUS POLYMERIZATION

INCOMPATABILITY (MATERIALS TO AVOID)

HAZARDOUS DECOMPOSITION PRODUCTS

SECTION VII SPILL OR LEAK PROCEDURES

STEPS TO BE TAKEN IN CASE MATERIAL IS RELEASED OR SPILLED

WASTE DISPOSAL METHOD

SECTION VIII SPECIAL PROTECTION INFORMATION

VENTILATION REQUIREMENTS LOCAL EXHAUST

PROTECTIVE EQUIPMENT (SPECIFY TYPES) EYE: Safety glasses

MECHANICAL (GENERAL)

GLOVES: Rubber

SPECIAL

RESPIRATOR: With proper filter

OTHER PROTECTIVE EQUIPMENT

SECTION IX SPECIAL PRECAUTIONS

PRECAUTIONS TO BE TAKEN IN HANDLING AND STORAGE

OTHER PRECAUTIONS

SIGNATURE | DATE

STANNIC CHLORIDE

Hazard diamond: 0 (top), 2 (left), 1 (right)

SECTION I SOURCE AND NOMENCLATURE

MANUFACTURER'S NAME | EMERGENCY TELEPHONE NO.

ADDRESS (NUMBER, STREET, CITY, STATE, ZIP CODE)

TRADE NAME AND SYNONYMS | CHEMICAL FAMILY

CHEMICAL NAME AND SYNONYMS: Tin Tetrachloride | FORMULA: $SnCl_4 \cdot 5H_2O$

SECTION II HAZARDOUS INGREDIENTS

BASIC MATERIAL	APPROXIMATE OR MAXIMUM % WT. OR VOL.	ESTABLISHED OSHA STANDARD	LD_{50} mg/kg ORAL	PERCUT.	SPECIES	LC_{50} CONC.
				21	mouse	

Note: Intraperitoneal - mouse LD50 46 mg/kg.

SECTION III PHYSICAL DATA MW - 350.58

BOILING POINT (°F.): 114.1°C.

VAPOR PRESSURE (mm Hg.): 10 at 10°C.

MELTING POINT (°F.): -33°C.

VAPOR DENSITY (Air = 1)

SPECIFIC GRAVITY (H_2O = 1): 2.232

EVAPORATION RATE (_____ = 1)

SOLUBILITY IN WATER Pts/100 pts H_2O

VOLATILE % Vol. % Wt.

APPEARANCE AND ODOR: Colorless, fuming liquid or crystals.

SECTION IV FIRE AND EXPLOSION HAZARD DATA

FLASH POINT: Not flammable

FLAMMABLE (EXPLOSIVE) LIMITS: UPPER / LOWER

METHOD USED

EXTINGUISHING MEDIA: Do not use hand-held water-base extinguishers.

SPECIAL FIRE FIGHTING PROCEDURES: Full protective clothing.

UNUSUAL FIRE AND EXPLOSION HAZARDS: --

SECTION V HEALTH HAZARD DATA

TOXIC LEVEL: TLV - 4 mg/C.M.

CARCINOGENIC: no

PRINCIPLE ROUTES OF ABSORBTION: Inhalation of dust, ingestion.

SKIN AND EYE IRRITATION: moderately severe skin & eyes

REVELANT SYMPTOMS OF EXPOSURE: Respiratory and mucous membrane irritant, may cause skin burns.

EFFECTS OF CHRONIC EXPOSURE: --

EMERGENCY AND FIRST AID PROCEDURES: Inhaled - wash skin & eyes; if swallowed - induce vomiting, get medical aid.

SECTION VI REACTIVITY DATA

CONDITIONS CONTRIBUTING TO INSTABILITY

CONDITIONS CONTRIBUTING TO HAZARDOUS POLYMERIZATION

INCOMPATABILITY (MATERIALS TO AVOID): Turpentine

HAZARDOUS DECOMPOSITION PRODUCTS

SECTION VII SPILL OR LEAK PROCEDURES

STEPS TO BE TAKEN IN CASE MATERIAL IS RELEASED OR SPILLED

WASTE DISPOSAL METHOD

SECTION VIII SPECIAL PROTECTION INFORMATION

VENTILATION REQUIREMENTS LOCAL EXHAUST

PROTECTIVE EQUIPMENT (SPECIFY TYPES) EYE: Goggles

MECHANICAL (GENERAL)

GLOVES: Rubber

SPECIAL

RESPIRATOR: W/appropriate filter

OTHER PROTECTIVE EQUIPMENT: Rubber apron

SECTION IX SPECIAL PRECAUTIONS

PRECAUTIONS TO BE TAKEN IN HANDLING AND STORAGE: Store in dry, well-ventilated area - Protect against water and direct sunlight.

OTHER PRECAUTIONS: Do not store with combustible materials.

SIGNATURE | DATE

STANNOUS CHLORIDE

Hazard diamond: 0 (top), 2 (left), 0 (right)

SECTION I SOURCE AND NOMENCLATURE

MANUFACTURER'S NAME | EMERGENCY TELEPHONE NO.

ADDRESS (NUMBER, STREET, CITY, STATE, ZIP CODE)

TRADE NAME AND SYNONYMS | CHEMICAL FAMILY

CHEMICAL NAME AND SYNONYMS: Tin dichloride | FORMULA: $SnCl_2 \cdot 2H_2O$

SECTION II HAZARDOUS INGREDIENTS

BASIC MATERIAL	APPROXIMATE OR MAXIMUM % WT. OR VOL.	ESTABLISHED OSHA STANDARD	LD_{50} mg/kg ORAL	PERCUT.	SPECIES	LC_{50} CONC.
			1200	41	mouse	
			700		rat	

Note: Intraperitoneal - mouse 66 mg/kg.

SECTION III PHYSICAL DATA MW 225.65

BOILING POINT (°F.): Decomposes

VAPOR PRESSURE (mm Hg.)

MELTING POINT (°F.): 37.7°C.

VAPOR DENSITY (Air = 1)

SPECIFIC GRAVITY (H_2O = 1): 2.71 at 15.5°C.

EVAPORATION RATE (_____ = 1)

SOLUBILITY IN WATER Pts/100 pts H_2O

VOLATILE % Vol. % Wt.

APPEARANCE AND ODOR: Colorless crystals

SECTION IV FIRE AND EXPLOSION HAZARD DATA

FLASH POINT: None available

FLAMMABLE (EXPLOSIVE) LIMITS: UPPER / LOWER

METHOD USED

EXTINGUISHING MEDIA: None available

SPECIAL FIRE FIGHTING PROCEDURES

UNUSUAL FIRE AND EXPLOSION HAZARDS

SECTION V HEALTH HAZARD DATA

TOXIC LEVEL: TLV - 3.2 mg/CM.

CARCINOGENIC: no

PRINCIPLE ROUTES OF ABSORBTION: Inhalation of dust or ingestion.

SKIN AND EYE IRRITATION: Mild

REVELANT SYMPTOMS OF EXPOSURE: Pain of mouth and throat, vomiting, diarrhea, tenesmus, retching, anuria, liver damage.

EFFECTS OF CHRONIC EXPOSURE: Possible erythematous, papular and granulomatous reactions.

EMERGENCY AND FIRST AID PROCEDURES: Inhalation - rest, keep warm, get medical aid; if swallowed - wash mouth, drink water, milk of magnesia, get medical aid; wash from skin & eyes.

SECTION VI REACTIVITY DATA

CONDITIONS CONTRIBUTING TO INSTABILITY

CONDITIONS CONTRIBUTING TO HAZARDOUS POLYMERIZATION

INCOMPATABILITY (MATERIALS TO AVOID): Calcium Carbide, Ethylene Oxide, Hydrazine Hydrate, Nitrates.

HAZARDOUS DECOMPOSITION PRODUCTS

SECTION VII SPILL OR LEAK PROCEDURES

STEPS TO BE TAKEN IN CASE MATERIAL IS RELEASED OR SPILLED

WASTE DISPOSAL METHOD

SECTION VIII SPECIAL PROTECTION INFORMATION

VENTILATION REQUIREMENTS LOCAL EXHAUST

PROTECTIVE EQUIPMENT (SPECIFY TYPES) EYE: Safety glasses or goggles

MECHANICAL (GENERAL)

GLOVES: Rubber

SPECIAL

RESPIRATOR: W/appropriate filter

OTHER PROTECTIVE EQUIPMENT

SECTION IX SPECIAL PRECAUTIONS

PRECAUTIONS TO BE TAKEN IN HANDLING AND STORAGE

OTHER PRECAUTIONS

SIGNATURE | DATE

STRYCHNINE SULFATE

1
4 0

SECTION I SOURCE AND NOMENCLATURE

MANUFACTURER'S NAME

EMERGENCY TELEPHONE NO.

ADDRESS (NUMBER, STREET, CITY, STATE, ZIP CODE)

TRADE NAME AND SYNONYMS

CHEMICAL FAMILY

CHEMICAL NAME AND SYNONYMS: Strychnine Sulfate

FORMULA: $(C_{21}H_{22}N_2O_2)_2H_2SO_4 \cdot 5H_2O$

SECTION II HAZARDOUS INGREDIENTS

BASIC MATERIAL	APPROXIMATE OR MAXIMUM % WT. OR VOL.	ESTABLISHED OSHA STANDARD	LD_{50} mg/kg ORAL	LD_{50} mg/kg PERCUT.	LD_{50} mg/kg SPECIES	LC_{50} CONC.
			5		rat	
Lowest published lethal dose				1.7	rat	
Oral-human 30 mg/kg			5		bird	

SECTION III PHYSICAL DATA MW - 856.96

BOILING POINT °F.

VAPOR PRESSURE mm Hg.

MELTING POINT °F.: 200°C.

VAPOR DENSITY (Air = 1)

SPECIFIC GRAVITY (H_2O = 1)

EVAPORATION RATE (= 1)

SOLUBILITY IN WATER Pts/100 pts H_2O: 3

VOLATILE % Vol. % Wt.

APPEARANCE AND ODOR: Colorless, odorless, bitter crystal.

SECTION IV FIRE AND EXPLOSION HAZARD DATA

FLASH POINT: not available

FLAMMABLE (EXPLOSIVE) LIMITS: UPPER / LOWER

METHOD USED

EXTINGUISHING MEDIA: Carbon dioxide, dry chemical.

SPECIAL FIRE FIGHTING PROCEDURES

UNUSUAL FIRE AND EXPLOSION HAZARDS

SECTION V HEALTH HAZARD DATA

TOXIC LEVEL: 0.15 mg/C.M.

CARCINOGENIC: No

PRINCIPLE ROUTES OF ABSORBTION: Inhalation, ingestion

SKIN AND EYE IRRITATION: slight

REVELANT SYMPTOMS OF EXPOSURE: Increase in deep reflexes, stiffening of knees, extension spasms in arms and legs, death by respiratory failure.

EFFECTS OF CHRONIC EXPOSURE: --

EMERGENCY AND FIRST AID PROCEDURES: If symptoms have begun, avoid trying to make vomit; keep quiet, artificial respiration as needed.

SECTION VI REACTIVITY DATA

CONDITIONS CONTRIBUTING TO INSTABILITY

CONDITIONS CONTRIBUTING TO HAZARDOUS POLYMERIZATION

INCOMPATABILITY (MATERIALS TO AVOID)

HAZARDOUS DECOMPOSITION PRODUCTS

SECTION VII SPILL OR LEAK PROCEDURES

STEPS TO BE TAKEN IN CASE MATERIAL IS RELEASED OR SPILLED

WASTE DISPOSAL METHOD

SECTION VIII SPECIAL PROTECTION INFORMATION

VENTILATION REQUIREMENTS LOCAL EXHAUST

PROTECTIVE EQUIPMENT (SPECIFY TYPES) EYE

MECHANICAL (GENERAL)

GLOVES

SPECIAL

RESPIRATOR

OTHER PROTECTIVE EQUIPMENT

SECTION IX SPECIAL PRECAUTIONS

PRECAUTIONS TO BE TAKEN IN HANDLING AND STORAGE

OTHER PRECAUTIONS

SIGNATURE

DATE

SUCCINONITRILE

1
4 0

SECTION I SOURCE AND NOMENCLATURE

MANUFACTURER'S NAME

EMERGENCY TELEPHONE NO.

ADDRESS (NUMBER, STREET, CITY, STATE, ZIP CODE)

TRADE NAME AND SYNONYMS: Ethylene dicyanide

CHEMICAL FAMILY

CHEMICAL NAME AND SYNONYMS: Succinonitrile

FORMULA: $CN(CH_2)_2CN$

SECTION II HAZARDOUS INGREDIENTS

BASIC MATERIAL	APPROXIMATE OR MAXIMUM % WT. OR VOL.	ESTABLISHED OSHA STANDARD	LD_{50} mg/kg ORAL	LD_{50} mg/kg PERCUT.	LD_{50} mg/kg SPECIES	LC_{50} CONC.
					rat	

SECTION III PHYSICAL DATA MW - 80.09

BOILING POINT °F.: 267°C.

VAPOR PRESSURE: at 100°C 2 mm Hg.

MELTING POINT °F.: 58°C

VAPOR DENSITY (Air = 1): 2.1

SPECIFIC GRAVITY (H_2O = 1): 1.02

EVAPORATION RATE (= 1)

SOLUBILITY IN WATER Pts/100 pts H_2O: 12.8

VOLATILE % Vol. % Wt.

APPEARANCE AND ODOR: Colorless, odorless, waxy material

SECTION IV FIRE AND EXPLOSION HAZARD DATA

FLASH POINT: 270°F

FLAMMABLE (EXPLOSIVE) LIMITS: UPPER / LOWER

METHOD USED: Closed Cup

EXTINGUISHING MEDIA: Carbon dioxide, dry chemical - water or foam may cause frothing.

SPECIAL FIRE FIGHTING PROCEDURES: Wear self-contained breathing apparatus.

UNUSUAL FIRE AND EXPLOSION HAZARDS

SECTION V HEALTH HAZARD DATA

TOXIC LEVEL: TWA-TLV-6 PPM, 20 mg/c.m. (NIOSH Value)

CARCINOGENIC: no

PRINCIPLE ROUTES OF ABSORBTION: inhalation, ingestion, skin

SKIN AND EYE IRRITATION: moderate

REVELANT SYMPTOMS OF EXPOSURE: Dizziness, rapid respiration, headache, drowsiness, drop in blood pressure, rapid pulse, unconsciousness.

EFFECTS OF CHRONIC EXPOSURE: Dizziness, weakness, hoarseness, lung congestion, eye irritation, loss of weight and appetite.

EMERGENCY AND FIRST AID PROCEDURES: Inhalation - amyl nitrite inhalation, get medical aid; if swallowed - give emetic, get medical aid; wash from skin & eyes.

SECTION VI REACTIVITY DATA

CONDITIONS CONTRIBUTING TO INSTABILITY

CONDITIONS CONTRIBUTING TO HAZARDOUS POLYMERIZATION

INCOMPATABILITY (MATERIALS TO AVOID): With acid may yield HCN

HAZARDOUS DECOMPOSITION PRODUCTS

SECTION VII SPILL OR LEAK PROCEDURES

STEPS TO BE TAKEN IN CASE MATERIAL IS RELEASED OR SPILLED

WASTE DISPOSAL METHOD

SECTION VIII SPECIAL PROTECTION INFORMATION

VENTILATION REQUIREMENTS LOCAL EXHAUST

PROTECTIVE EQUIPMENT (SPECIFY TYPES) EYE: Goggles

MECHANICAL (GENERAL)

GLOVES: Rubber

SPECIAL

RESPIRATOR: Self-contained

OTHER PROTECTIVE EQUIPMENT

SECTION IX SPECIAL PRECAUTIONS

PRECAUTIONS TO BE TAKEN IN HANDLING AND STORAGE

OTHER PRECAUTIONS

SIGNATURE

DATE

SUCCINYL CHLORIDE

Hazard diamond: 2 (top), 4 (left), 2 (right), W (bottom, crossed through)

SECTION I SOURCE AND NOMENCLATURE

MANUFACTURER'S NAME	EMERGENCY TELEPHONE NO.
ADDRESS (NUMBER, STREET, CITY, STATE, ZIP CODE)	
TRADE NAME AND SYNONYMS	CHEMICAL FAMILY
CHEMICAL NAME AND SYNONYMS: Succinyl Chloride	FORMULA: $ClOC(CH_2)_2COCl$

SECTION II HAZARDOUS INGREDIENTS

BASIC MATERIAL	APPROXIMATE OR MAXIMUM % WT. OR VOL.	ESTABLISHED OSHA STANDARD	LD_{50} ORAL	LD_{50} PERCUT.	LC_{50} SPECIES	LC_{50} CONC.
Intraperitoneal - mouse	LD50 - 62 mg/kg					

SECTION III PHYSICAL DATA MW - 154.99

BOILING POINT °F.: 193°C	VAPOR PRESSURE mm Hg.
MELTING POINT °F.: 17°C	VAPOR DENSITY (Air = 1)
SPECIFIC GRAVITY (H_2O = 1): 1.38	EVAPORATION RATE (= 1)
SOLUBILITY IN WATER Pts/100 pts H_2O: decomposes	VOLATILE % Vol. % Wt.

APPEARANCE AND ODOR: Fuming, highly refractive liquid.

SECTION IV FIRE AND EXPLOSION HAZARD DATA

FLASH POINT: none available	FLAMMABLE (EXPLOSIVE) LIMITS	UPPER
METHOD USED		LOWER

EXTINGUISHING MEDIA: Carbon dioxide, dry chemical - do not use foam or water.

SPECIAL FIRE FIGHTING PROCEDURES: Wear full protective clothing.

UNUSUAL FIRE AND EXPLOSION HAZARDS: HCl is produced with water addition.

SECTION V HEALTH HAZARD DATA

TOXIC LEVEL: None available — CARCINOGENIC: no

PRINCIPLE ROUTES OF ABSORBTION: Inhalation, ingestion — SKIN AND EYE IRRITATION: Severe

REVELANT SYMPTOMS OF EXPOSURE: Cyanosis, headache, coughing, dizziness, weakness, pulmonary edema; on skin contact - severe pain, burns.

EFFECTS OF CHRONIC EXPOSURE: Erosion of teeth, jaw necrosis, bronchial irritation and chronic cough.

EMERGENCY AND FIRST AID PROCEDURES: Inhalation - artificial respiration as needed; if swallowed - drink water, milk of magnesia, get medical aid; wash skin & eyes.

SECTION VI REACTIVITY DATA

CONDITIONS CONTRIBUTING TO INSTABILITY

CONDITIONS CONTRIBUTING TO HAZARDOUS POLYMERIZATION

INCOMPATABILITY (MATERIALS TO AVOID): Avoid water

HAZARDOUS DECOMPOSITION PRODUCTS: Produces HCl.

SECTION VII SPILL OR LEAK PROCEDURES

STEPS TO BE TAKEN IN CASE MATERIAL IS RELEASED OR SPILLED

WASTE DISPOSAL METHOD

SECTION VIII SPECIAL PROTECTION INFORMATION

VENTILATION REQUIREMENTS LOCAL EXHAUST	PROTECTIVE EQUIPMENT (SPECIFY TYPES) EYE: Goggles
MECHANICAL (GENERAL)	GLOVES: Rubber
SPECIAL	RESPIRATOR: Self-contained

OTHER PROTECTIVE EQUIPMENT

SECTION IX SPECIAL PRECAUTIONS

PRECAUTIONS TO BE TAKEN IN HANDLING AND STORAGE: Store in cool, dry well-ventilated place inside storage in flammable liquids storage cabinet.

OTHER PRECAUTIONS

SIGNATURE	DATE

SULFANILAMIDE

Hazard diamond: 1 (top), 4 (left), 0 (right)

SECTION I SOURCE AND NOMENCLATURE

MANUFACTURER'S NAME	EMERGENCY TELEPHONE NO.
ADDRESS (NUMBER, STREET, CITY, STATE, ZIP CODE)	
TRADE NAME AND SYNONYMS: P-Aminobenzenesulfonamide	CHEMICAL FAMILY
CHEMICAL NAME AND SYNONYMS: Sulfanilamide	FORMULA: $4\text{-}(NH_2)C_6H_4SO_2NH_2$

SECTION II HAZARDOUS INGREDIENTS

BASIC MATERIAL	APPROXIMATE OR MAXIMUM % WT. OR VOL.	ESTABLISHED OSHA STANDARD	LD_{50} mg/kg ORAL	LD_{50} PERCUT.	LC_{50} SPECIES	LC_{50} CONC.
Oral - human LDLO 500 mg/kg			3900		Rat	
			2000		Dog	
			3700		Mouse	
			2000		Rabbit	

SECTION III PHYSICAL DATA MW - 172.22

BOILING POINT °F.	VAPOR PRESSURE mm Hg.
MELTING POINT °F.: 166°C.	VAPOR DENSITY (Air = 1)
SPECIFIC GRAVITY (H_2O = 1): 1.08	EVAPORATION RATE (= 1)
SOLUBILITY IN WATER Pts/100 pts H_2O: .75	VOLATILE % Vol. % Wt.

APPEARANCE AND ODOR: White to yellow powder.

SECTION IV FIRE AND EXPLOSION HAZARD DATA

FLASH POINT: None available	FLAMMABLE (EXPLOSIVE) LIMITS	UPPER
METHOD USED		LOWER

EXTINGUISHING MEDIA: Foam, carbon dioxide, dry chemical.

SPECIAL FIRE FIGHTING PROCEDURES

UNUSUAL FIRE AND EXPLOSION HAZARDS

SECTION V HEALTH HAZARD DATA

TOXIC LEVEL: None available — CARCINOGENIC: Yes

PRINCIPLE ROUTES OF ABSORBTION: Ingestion — SKIN AND EYE IRRITATION: No

REVELANT SYMPTOMS OF EXPOSURE: Diarrhea, jaundice, cyanosis, headache, dizziness, crystalluria, nausea, vomiting.

EFFECTS OF CHRONIC EXPOSURE

EMERGENCY AND FIRST AID PROCEDURES: Ingestion - large quantities - induce vomiting, drink large quantities of water, get medical aid.

SECTION VI REACTIVITY DATA

CONDITIONS CONTRIBUTING TO INSTABILITY

CONDITIONS CONTRIBUTING TO HAZARDOUS POLYMERIZATION

INCOMPATABILITY (MATERIALS TO AVOID)

HAZARDOUS DECOMPOSITION PRODUCTS

SECTION VII SPILL OR LEAK PROCEDURES

STEPS TO BE TAKEN IN CASE MATERIAL IS RELEASED OR SPILLED

WASTE DISPOSAL METHOD

SECTION VIII SPECIAL PROTECTION INFORMATION

VENTILATION REQUIREMENTS LOCAL EXHAUST	PROTECTIVE EQUIPMENT (SPECIFY TYPES) EYE
MECHANICAL (GENERAL)	GLOVES
SPECIAL	RESPIRATOR

OTHER PROTECTIVE EQUIPMENT

SECTION IX SPECIAL PRECAUTIONS

PRECAUTIONS TO BE TAKEN IN HANDLING AND STORAGE

OTHER PRECAUTIONS

SIGNATURE	DATE

SULFUROUS ACID

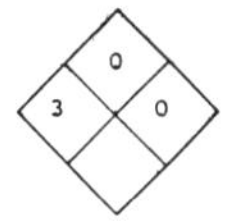

SECTION I SOURCE AND NOMENCLATURE

MANUFACTURER'S NAME | EMERGENCY TELEPHONE NO.

ADDRESS (NUMBER, STREET, CITY, STATE, ZIP CODE)

TRADE NAME AND SYNONYMS | CHEMICAL FAMILY

CHEMICAL NAME AND SYNONYMS: Sulfurous Acid | FORMULA: H_2SO_3

SECTION II HAZARDOUS INGREDIENTS

BASIC MATERIAL	APPROXIMATE OR MAXIMUM % WT. OR VOL.	ESTABLISHED OSHA STANDARD	LD_{50} ORAL	LD_{50} PERCUT.	LC_{50} SPECIES	LC_{50} CONC.

SECTION III PHYSICAL DATA MW-82.02

BOILING POINT °F. | VAPOR PRESSURE mm Hg.

MELTING POINT °F. | VAPOR DENSITY (Air = 1)

SPECIFIC GRAVITY (H_2O = 1) | EVAPORATION RATE (= 1)

SOLUBILITY IN WATER Pts/100 pts H_2O: yes | VOLATILE % Vol. % Wt.

APPEARANCE AND ODOR: Irritating odor, clear

SECTION IV FIRE AND EXPLOSION HAZARD DATA

FLASH POINT: Not flammable | FLAMMABLE (EXPLOSIVE) LIMITS: UPPER / LOWER

METHOD USED

EXTINGUISHING MEDIA: Dry chemical

SPECIAL FIRE FIGHTING PROCEDURES

UNUSUAL FIRE AND EXPLOSION HAZARDS

SECTION V HEALTH HAZARD DATA

TOXIC LEVEL: TLV - 2 ppm, 5 mg/c.m. (SO_2) | CARCINOGENIC: no

PRINCIPLE ROUTES OF ABSORBTION: Ingestion, inhalation | SKIN AND EYE IRRITATION: Moderate

REVELANT SYMPTOMS OF EXPOSURE: Dermatitis, skin burn, irritation of nose and throat, bronchitis, nausea, vomiting

EFFECTS OF CHRONIC EXPOSURE: Pneumonitis, dental erosion, vomiting of blood and eroded tissue, perforation of GI tract.

EMERGENCY AND FIRST AID PROCEDURES: Irrigate eyes with water, wash skin, gastric lavage, oxygen, decogestants

SECTION VI REACTIVITY DATA

CONDITIONS CONTRIBUTING TO INSTABILITY

CONDITIONS CONTRIBUTING TO HAZARDOUS POLYMERIZATION

INCOMPATABILITY (MATERIALS TO AVOID): Reacts with alkaline materials

HAZARDOUS DECOMPOSITION PRODUCTS: Liberates SO2 when heated

SECTION VII SPILL OR LEAK PROCEDURES

STEPS TO BE TAKEN IN CASE MATERIAL IS RELEASED OR SPILLED

WASTE DISPOSAL METHOD

SECTION VIII SPECIAL PROTECTION INFORMATION

VENTILATION REQUIREMENTS LOCAL EXHAUST | PROTECTIVE EQUIPMENT (SPECIFY TYPES) EYE: Safety goggles

MECHANICAL (GENERAL) | GLOVES: Rubber

SPECIAL | RESPIRATOR

OTHER PROTECTIVE EQUIPMENT

SECTION IX SPECIAL PRECAUTIONS

PRECAUTIONS TO BE TAKEN IN HANDLING AND STORAGE: Store in well-ventilated area .

OTHER PRECAUTIONS

SIGNATURE | DATE

TELLURIUM METAL

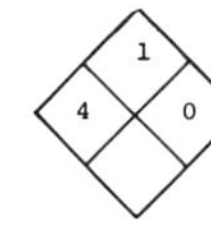

SECTION I SOURCE AND NOMENCLATURE

MANUFACTURER'S NAME | EMERGENCY TELEPHONE NO.

ADDRESS (NUMBER, STREET, CITY, STATE, ZIP CODE)

TRADE NAME AND SYNONYMS | CHEMICAL FAMILY

CHEMICAL NAME AND SYNONYMS: Tellurium | FORMULA: Te

SECTION II HAZARDOUS INGREDIENTS

BASIC MATERIAL	APPROXIMATE OR MAXIMUM % WT. OR VOL.	ESTABLISHED OSHA STANDARD	LD_{50} ORAL	LD_{50} PERCUT.	LC_{50} SPECIES	LC_{50} CONC.
			See Note			

Note: Lowest published lethal dose subcutaneous - dog 290 mg/kg.

SECTION III PHYSICAL DATA A wt. - 127.61

BOILING POINT: 987°C. | VAPOR PRESSURE mm Hg.

MELTING POINT: 450°C. | VAPOR DENSITY (Air = 1)

SPECIFIC GRAVITY (H_2O = 1): 6.24 | EVAPORATION RATE (= 1)

SOLUBILITY IN WATER Pts/100 pts H_2O: Insoluble | VOLATILE % Vol. % Wt.

APPEARANCE AND ODOR: Grey-white powder

SECTION IV FIRE AND EXPLOSION HAZARD DATA

FLASH POINT: Not available | FLAMMABLE (EXPLOSIVE) LIMITS: UPPER / LOWER

METHOD USED

EXTINGUISHING MEDIA: Use water spray

SPECIAL FIRE FIGHTING PROCEDURES: Straight water streams will scatter molten Te.

UNUSUAL FIRE AND EXPLOSION HAZARDS: Burning produces toxic TeO_2 gas.

SECTION V HEALTH HAZARD DATA

TOXIC LEVEL: TLV - 0.1 mg/CM. | CARCINOGENIC: No

PRINCIPLE ROUTES OF ABSORBTION: Inhalation, ingestion | SKIN AND EYE IRRITATION: Moderate

REVELANT SYMPTOMS OF EXPOSURE: Sleepiness, loss of appetite, nausea, metallic taste and garlic odor to breath.

EFFECTS OF CHRONIC EXPOSURE: Fatigue, loss of appetite, bronchitis, dermatitis.

EMERGENCY AND FIRST AID PROCEDURES: If inhaled - keep warm; if swallowed - wash mouth, give emetic, get medical aid; wash from skin and eyes.

SECTION VI REACTIVITY DATA

CONDITIONS CONTRIBUTING TO INSTABILITY

CONDITIONS CONTRIBUTING TO HAZARDOUS POLYMERIZATION

INCOMPATABILITY (MATERIALS TO AVOID)

HAZARDOUS DECOMPOSITION PRODUCTS

SECTION VII SPILL OR LEAK PROCEDURES

STEPS TO BE TAKEN IN CASE MATERIAL IS RELEASED OR SPILLED

WASTE DISPOSAL METHOD

SECTION VIII SPECIAL PROTECTION INFORMATION

VENTILATION REQUIREMENTS LOCAL EXHAUST | PROTECTIVE EQUIPMENT (SPECIFY TYPES) EYE: Goggles

MECHANICAL (GENERAL) | GLOVES: Rubber

SPECIAL | RESPIRATOR: With proper filter

OTHER PROTECTIVE EQUIPMENT

SECTION IX SPECIAL PRECAUTIONS

PRECAUTIONS TO BE TAKEN IN HANDLING AND STORAGE

OTHER PRECAUTIONS

SIGNATURE | DATE

TETRABROMOETHANE (ACETYLENE TETRABROMIDE)

Hazard diamond: 0 (top), 4 (left), 1 (right)

SECTION I SOURCE AND NOMENCLATURE

MANUFACTURER'S NAME | EMERGENCY TELEPHONE NO.

ADDRESS (NUMBER, STREET, CITY, STATE, ZIP CODE)

TRADE NAME AND SYNONYMS | CHEMICAL FAMILY

CHEMICAL NAME AND SYNONYMS: 1,1,2,2 Tetrabromo Ethane | FORMULA: $CHBr_2CHBr_2$

SECTION II HAZARDOUS INGREDIENTS

BASIC MATERIAL	APPROXIMATE OR MAXIMUM % WT. OR VOL.	ESTABLISHED OSHA STANDARD	LD_{50} mg/kg ORAL	PERCUT.	SPECIES	LC_{50} CONC.
			400		rabbit	
			400		guinea pig	

SECTION III PHYSICAL DATA MW 345.7

BOILING POINT °F.: decomposition 239° to 242°C | VAPOR PRESSURE mm Hg.

MELTING POINT °F. | VAPOR DENSITY (Air = 1): 11.92

SPECIFIC GRAVITY (H_2O = 1): 2.9638 (20°/4°C) | EVAPORATION RATE (____ = 1)

SOLUBILITY IN WATER Pts/100 pts H_2O | VOLATILE % Vol. % Wt.

APPEARANCE AND ODOR: Colorless to yellow liquid with chloroform-like odor.

SECTION IV FIRE AND EXPLOSION HAZARD DATA

FLASH POINT: -- | FLAMMABLE (EXPLOSIVE) LIMITS: UPPER | LOWER

METHOD USED

EXTINGUISHING MEDIA: Use water for cooling containers.

SPECIAL FIRE FIGHTING PROCEDURES: Full protective clothing

UNUSUAL FIRE AND EXPLOSION HAZARDS

SECTION V HEALTH HAZARD DATA

TOXIC LEVEL: TLV - 1 ppm, 14 mg/.M. | CARCINOGENIC: No

PRINCIPLE ROUTES OF ABSORBTION: Inhalation of vapor | SKIN AND EYE IRRITATION: yes

REVELANT SYMPTOMS OF EXPOSURE: Narcotic properties; is a c.n.s. and respiratory depressant; drowsiness, etc. Possibility of liver and kidney damage.

EFFECTS OF CHRONIC EXPOSURE: --

EMERGENCY AND FIRST AID PROCEDURES: Remove from exposure, give artificial respiration (if necessary) rest and keep warm. Do not give stimulants.

SECTION VI REACTIVITY DATA

CONDITIONS CONTRIBUTING TO INSTABILITY

CONDITIONS CONTRIBUTING TO HAZARDOUS POLYMERIZATION

INCOMPATABILITY (MATERIALS TO AVOID)

HAZARDOUS DECOMPOSITION PRODUCTS: Toxic fumes of carbonyl bromide.

SECTION VII SPILL OR LEAK PROCEDURES

STEPS TO BE TAKEN IN CASE MATERIAL IS RELEASED OR SPILLED

WASTE DISPOSAL METHOD

SECTION VIII SPECIAL PROTECTION INFORMATION Full protection

VENTILATION REQUIREMENTS LOCAL EXHAUST | PROTECTIVE EQUIPMENT (SPECIFY TYPES) EYE: Goggles

MECHANICAL (GENERAL) | GLOVES: Rubber

SPECIAL | RESPIRATOR: W/Appropriate filter

OTHER PROTECTIVE EQUIPMENT: (face shield), rubber apron.

SECTION IX SPECIAL PRECAUTIONS

PRECAUTIONS TO BE TAKEN IN HANDLING AND STORAGE: Be sure there is adequate ventilation.

OTHER PRECAUTIONS

SIGNATURE | DATE

TETRABROMOPHTHALIC ANHYDRIDE

Hazard diamond: 1 (top), 4 (left), 0 (right)

SECTION I SOURCE AND NOMENCLATURE

MANUFACTURER'S NAME | EMERGENCY TELEPHONE NO.

ADDRESS (NUMBER, STREET, CITY, STATE, ZIP CODE)

TRADE NAME AND SYNONYMS | CHEMICAL FAMILY

CHEMICAL NAME AND SYNONYMS: Tetrabromophthalic Anhydride | FORMULA: C_6Br_4COOCO

SECTION II HAZARDOUS INGREDIENTS

BASIC MATERIAL	APPROXIMATE OR MAXIMUM % WT. OR VOL.	ESTABLISHED OSHA STANDARD	LD_{50} ORAL	PERCUT.	SPECIES	LC_{50} CONC.
			none	available		

SECTION III PHYSICAL DATA MW - 463.74

BOILING POINT °F.: -- | VAPOR PRESSURE mm Hg.

MELTING POINT °F.: 280°C | VAPOR DENSITY (Air = 1)

SPECIFIC GRAVITY (H_2O = 1) | EVAPORATION RATE (____ = 1)

SOLUBILITY IN WATER Pts/100 pts H_2O: very slightly soluble | VOLATILE % Vol. % Wt.

APPEARANCE AND ODOR: crystaline

SECTION IV FIRE AND EXPLOSION HAZARD DATA

FLASH POINT: none available | FLAMMABLE (EXPLOSIVE) LIMITS: UPPER | LOWER

METHOD USED

EXTINGUISHING MEDIA: water, carbon dioxide, dry chemical.

SPECIAL FIRE FIGHTING PROCEDURES

UNUSUAL FIRE AND EXPLOSION HAZARDS

SECTION V HEALTH HAZARD DATA

TOXIC LEVEL: None available | CARCINOGENIC: no

PRINCIPLE ROUTES OF ABSORBTION: Inhalation (mainly) | SKIN AND EYE IRRITATION: moderate

REVELANT SYMPTOMS OF EXPOSURE: Coughing, sneezing, respiratory system irritation, dermatitis, eye irritation.

EFFECTS OF CHRONIC EXPOSURE: Bronchitis, eye irritation, inflammation of respiratory tract; bloody nasal discharge. Sensitization

EMERGENCY AND FIRST AID PROCEDURES: Inhalation - rest, oxygen to relieve respiratory distress; wash skin & eyes; get medical aid.

SECTION VI REACTIVITY DATA

CONDITIONS CONTRIBUTING TO INSTABILITY

CONDITIONS CONTRIBUTING TO HAZARDOUS POLYMERIZATION

INCOMPATABILITY (MATERIALS TO AVOID)

HAZARDOUS DECOMPOSITION PRODUCTS

SECTION VII SPILL OR LEAK PROCEDURES

STEPS TO BE TAKEN IN CASE MATERIAL IS RELEASED OR SPILLED

WASTE DISPOSAL METHOD

SECTION VIII SPECIAL PROTECTION INFORMATION

VENTILATION REQUIREMENTS LOCAL EXHAUST | PROTECTIVE EQUIPMENT (SPECIFY TYPES) EYE: Safety goggles

MECHANICAL (GENERAL) | GLOVES

SPECIAL | RESPIRATOR: With proper filter

OTHER PROTECTIVE EQUIPMENT

SECTION IX SPECIAL PRECAUTIONS

PRECAUTIONS TO BE TAKEN IN HANDLING AND STORAGE

OTHER PRECAUTIONS

SIGNATURE | DATE

TETRA-n-BUTYLAMMONIUM IODIDE

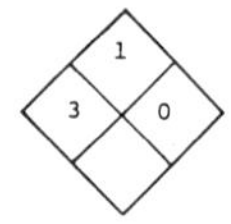

SECTION I SOURCE AND NOMENCLATURE

MANUFACTURER'S NAME | EMERGENCY TELEPHONE NO.

ADDRESS (NUMBER, STREET, CITY, STATE, ZIP CODE)

TRADE NAME AND SYNONYMS | CHEMICAL FAMILY

CHEMICAL NAME AND SYNONYMS: Tetra-N-Butylammonium Iodide | FORMULA: $(C_4H_9)_4NI$

SECTION II HAZARDOUS INGREDIENTS

BASIC MATERIAL	APPROXIMATE OR MAXIMUM % WT. OR VOL.	ESTABLISHED OSHA STANDARD	LD_{50} mg/kg ORAL	PERCUT.	SPECIES	LC_{50} CONC.
			1990		Rat	

SECTION III PHYSICAL DATA MW - 369.38

BOILING POINT °F. | VAPOR PRESSURE mm Hg.

MELTING POINT 148°C | VAPOR DENSITY (Air = 1)

SPECIFIC GRAVITY (H_2O = 1) | EVAPORATION RATE (______ = 1)

SOLUBILITY IN WATER Pts/100 pts H_2O: slightly soluble | VOLATILE % Vol. % Wt.

APPEARANCE AND ODOR: crystal leaves

SECTION IV FIRE AND EXPLOSION HAZARD DATA

FLASH POINT: none available | FLAMMABLE (EXPLOSIVE) LIMITS: UPPER / LOWER

METHOD USED

EXTINGUISHING MEDIA: Water, carbon dioxide, dry chemical.

SPECIAL FIRE FIGHTING PROCEDURES

UNUSUAL FIRE AND EXPLOSION HAZARDS

SECTION V HEALTH HAZARD DATA

TOXIC LEVEL: None available | CARCINOGENIC: No

PRINCIPLE ROUTES OF ABSORBTION: Inhalation, ingestion | SKIN AND EYE IRRITATION: Moderate

REVELANT SYMPTOMS OF EXPOSURE: Irritates eyes & skin, nausea, vomiting, dizziness, central nervous system depression.

EFFECTS OF CHRONIC EXPOSURE: Dermatitis

EMERGENCY AND FIRST AID PROCEDURES: Inhalation - rest; if swallowed - induce vomiting, get medical aid; wash skin & eyes.

SECTION VI REACTIVITY DATA

CONDITIONS CONTRIBUTING TO INSTABILITY

CONDITIONS CONTRIBUTING TO HAZARDOUS POLYMERIZATION

INCOMPATABILITY (MATERIALS TO AVOID)

HAZARDOUS DECOMPOSITION PRODUCTS

SECTION VII SPILL OR LEAK PROCEDURES

STEPS TO BE TAKEN IN CASE MATERIAL IS RELEASED OR SPILLED

WASTE DISPOSAL METHOD

SECTION VIII SPECIAL PROTECTION INFORMATION

VENTILATION REQUIREMENTS LOCAL EXHAUST | PROTECTIVE EQUIPMENT (SPECIFY TYPES) EYE: Safety goggles

MECHANICAL (GENERAL) | GLOVES

SPECIAL | RESPIRATOR

OTHER PROTECTIVE EQUIPMENT

SECTION IX SPECIAL PRECAUTIONS

PRECAUTIONS TO BE TAKEN IN HANDLING AND STORAGE

OTHER PRECAUTIONS

SIGNATURE | DATE

TETRACHLOROETHYLENE (PERCHLOROETHYLENE)

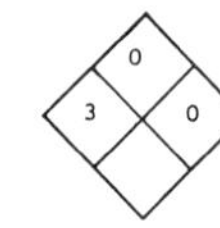

SECTION I SOURCE AND NOMENCLATURE

MANUFACTURER'S NAME | EMERGENCY TELEPHONE NO.

ADDRESS (NUMBER, STREET, CITY, STATE, ZIP CODE)

TRADE NAME AND SYNONYMS: Ethylene Tetrachloride | CHEMICAL FAMILY

CHEMICAL NAME AND SYNONYMS: Tetrachloroethylene Perchloroethylene | FORMULA: $Cl_2C{=}C\ Cl_2$

SECTION II HAZARDOUS INGREDIENTS

BASIC MATERIAL	APPROXIMATE OR MAXIMUM % WT. OR VOL.	ESTABLISHED OSHA STANDARD	LD_{50} mg/kg ORAL	PERCUT.	SPECIES	LC_{50} CONC.
			200		rat	
LDLO oral - human 500 mg/kg			8890		mouse	
Lowest published toxic concentration in air						
inhalation - human 260 PPM Systemic						
" " 280 " Eye						
" " 600 " CNS						

SECTION III PHYSICAL DATA MW - 165.85

BOILING POINT 121.2 °F. | VAPOR PRESSURE 22°C 15.8 mm Hg.

MELTING POINT -22.4°C °F. | VAPOR DENSITY (Air = 1) 5.83

SPECIFIC GRAVITY (H_2O = 1) 1.631 | EVAPORATION RATE (______ = 1)

SOLUBILITY IN WATER Pts/100 pts H_2O: 1 | VOLATILE % Vol. % Wt.

APPEARANCE AND ODOR: colorless liquid, chloroform-like odor.

SECTION IV FIRE AND EXPLOSION HAZARD DATA

FLASH POINT: nonflammable | FLAMMABLE (EXPLOSIVE) LIMITS: UPPER / LOWER

METHOD USED

EXTINGUISHING MEDIA: water spray, CO_2 or dry chemical -

SPECIAL FIRE FIGHTING PROCEDURES: Wear self-contained breathing apparatus.

UNUSUAL FIRE AND EXPLOSION HAZARDS: When heated to decomposition point (above 140°C) emits highly toxic fumes of chlorides.

SECTION V HEALTH HAZARD DATA

TOXIC LEVEL: TLV - 100 PPM, 670 mg/C.M. teratogen | CARCINOGENIC: no

PRINCIPLE ROUTES OF ABSORBTION: inhalation, ingestion, skin | SKIN AND EYE IRRITATION: moderate

REVELANT SYMPTOMS OF EXPOSURE: Dizziness, nausea, vomiting, loss of weight, muscular weakness, acute gastrointestinal distress.

EFFECTS OF CHRONIC EXPOSURE: Liver and kidney damage, weight loss, dermatitis.

EMERGENCY AND FIRST AID PROCEDURES: Inhalation - rest, remove from exposure; if swallowed - wash mouth, induce vomiting, get medical aid; wash from skin & eyes.

SECTION VI REACTIVITY DATA

CONDITIONS CONTRIBUTING TO INSTABILITY

CONDITIONS CONTRIBUTING TO HAZARDOUS POLYMERIZATION

INCOMPATABILITY (MATERIALS TO AVOID)

HAZARDOUS DECOMPOSITION PRODUCTS: Upon decomposition, gives off toxic chloride fumes.

SECTION VII SPILL OR LEAK PROCEDURES

STEPS TO BE TAKEN IN CASE MATERIAL IS RELEASED OR SPILLED: Absorb on paper towels; evaporate in hood.

WASTE DISPOSAL METHOD

SECTION VIII SPECIAL PROTECTION INFORMATION

VENTILATION REQUIREMENTS LOCAL EXHAUST: In hood | PROTECTIVE EQUIPMENT (SPECIFY TYPES) EYE: Goggles

MECHANICAL (GENERAL) | GLOVES: Rubber

SPECIAL | RESPIRATOR: With proper filter

OTHER PROTECTIVE EQUIPMENT

SECTION IX SPECIAL PRECAUTIONS

PRECAUTIONS TO BE TAKEN IN HANDLING AND STORAGE: Store in closed container in cool place.

OTHER PRECAUTIONS

SIGNATURE | DATE

TETRACHLOROHYDROQUINONE

1 3 0

SECTION I SOURCE AND NOMENCLATURE

MANUFACTURER'S NAME	EMERGENCY TELEPHONE NO.
ADDRESS (NUMBER, STREET, CITY, STATE, ZIP CODE)	
TRADE NAME AND SYNONYMS	CHEMICAL FAMILY
CHEMICAL NAME AND SYNONYMS: Tetrachlorohydroquinone	FORMULA: $C_6Cl_4(OH)_2$

SECTION II HAZARDOUS INGREDIENTS

BASIC MATERIAL	APPROXIMATE OR MAXIMUM % WT. OR VOL.	ESTABLISHED OSHA STANDARD	LD_{50} ORAL	LD_{50} PERCUT.	LC_{50} SPECIES	LC_{50} CONC.
Lowest published lethal dose						
intraperitoneal - mouse	25 mg/kg.					

SECTION III PHYSICAL DATA MW - 245.88

BOILING POINT °F.	VAPOR PRESSURE mm Hg.
MELTING POINT °F.: 240°C	VAPOR DENSITY (Air = 1)
SPECIFIC GRAVITY (H_2O = 1)	EVAPORATION RATE (= 1)
SOLUBILITY IN WATER Pts/100 pts H_2O: insoluble	VOLATILE % Vol. % Wt.
APPEARANCE AND ODOR: yellow crystals	

SECTION IV FIRE AND EXPLOSION HAZARD DATA

FLASH POINT: none available

METHOD USED

FLAMMABLE (EXPLOSIVE) LIMITS: UPPER / LOWER

EXTINGUISHING MEDIA: Water, carbon dioxide, dry chemical.

SPECIAL FIRE FIGHTING PROCEDURES

UNUSUAL FIRE AND EXPLOSION HAZARDS

SECTION V HEALTH HAZARD DATA

TOXIC LEVEL: none available

CARCINOGENIC: no

PRINCIPLE ROUTES OF ABSORBTION: Inhalation, ingestion

SKIN AND EYE IRRITATION: Severe

REVELANT SYMPTOMS OF EXPOSURE: Tremors, convulsions, shortness of breath, respiratory system irritation.

EFFECTS OF CHRONIC EXPOSURE: --

EMERGENCY AND FIRST AID PROCEDURES: Inhalation - rest; if swallowed - drink water, epsom salts solution, get medical aid; wash eyes and skin.

SECTION VI REACTIVITY DATA

CONDITIONS CONTRIBUTING TO INSTABILITY

CONDITIONS CONTRIBUTING TO HAZARDOUS POLYMERIZATION

INCOMPATABILITY (MATERIALS TO AVOID)

HAZARDOUS DECOMPOSITION PRODUCTS

SECTION VII SPILL OR LEAK PROCEDURES

STEPS TO BE TAKEN IN CASE MATERIAL IS RELEASED OR SPILLED

WASTE DISPOSAL METHOD

SECTION VIII SPECIAL PROTECTION INFORMATION

VENTILATION REQUIREMENTS LOCAL EXHAUST	PROTECTIVE EQUIPMENT (SPECIFY TYPES) EYE: Goggles
MECHANICAL (GENERAL)	GLOVES: Rubber gloves
SPECIAL	RESPIRATOR: Self-contained
OTHER PROTECTIVE EQUIPMENT	

SECTION IX SPECIAL PRECAUTIONS

PRECAUTIONS TO BE TAKEN IN HANDLING AND STORAGE

OTHER PRECAUTIONS

SIGNATURE	DATE

TETRACHLOROPHTHALIC ANHYDRIDE

1 4 0

SECTION I SOURCE AND NOMENCLATURE

MANUFACTURER'S NAME	EMERGENCY TELEPHONE NO.
ADDRESS (NUMBER, STREET, CITY, STATE, ZIP CODE)	
TRADE NAME AND SYNONYMS	CHEMICAL FAMILY
CHEMICAL NAME AND SYNONYMS: Tetrachlorophthalic Anhydride	FORMULA: C_6Cl_4COOCO

SECTION II HAZARDOUS INGREDIENTS

BASIC MATERIAL	APPROXIMATE OR MAXIMUM % WT. OR VOL.	ESTABLISHED OSHA STANDARD	LD_{50} ORAL	LD_{50} PERCUT.	LC_{50} SPECIES	LC_{50} CONC.
			none	available		

SECTION III PHYSICAL DATA MW - 285.92

BOILING POINT °F.: 371°C	VAPOR PRESSURE mm Hg.
MELTING POINT °F.: 256°C	VAPOR DENSITY (Air = 1)
SPECIFIC GRAVITY (H_2O = 1)	EVAPORATION RATE (= 1)
SOLUBILITY IN WATER Pts/100 pts H_2O: decomposes	VOLATILE % Vol. % Wt.
APPEARANCE AND ODOR: needle-like crystals.	

SECTION IV FIRE AND EXPLOSION HAZARD DATA

FLASH POINT: none available

METHOD USED

FLAMMABLE (EXPLOSIVE) LIMITS: UPPER / LOWER

EXTINGUISHING MEDIA: Water, carbon dioxide, dry chemical.

SPECIAL FIRE FIGHTING PROCEDURES

UNUSUAL FIRE AND EXPLOSION HAZARDS

SECTION V HEALTH HAZARD DATA

TOXIC LEVEL: none available

CARCINOGENIC: no

PRINCIPLE ROUTES OF ABSORBTION: Ingestion, inhalation (mainly)

SKIN AND EYE IRRITATION: Moderate

REVELANT SYMPTOMS OF EXPOSURE: Coughing, sneezing, respiratory irritation, dermatitis, eye irritation.

EFFECTS OF CHRONIC EXPOSURE: Bronchitis, eye irritations, inflammation of respiratory tract; bloody nasal discharge.

EMERGENCY AND FIRST AID PROCEDURES: Inhalation - rest, oxygen to relieve respiratory distress; wash skin & eyes; get medical aid.

SECTION VI REACTIVITY DATA

CONDITIONS CONTRIBUTING TO INSTABILITY

CONDITIONS CONTRIBUTING TO HAZARDOUS POLYMERIZATION

INCOMPATABILITY (MATERIALS TO AVOID)

HAZARDOUS DECOMPOSITION PRODUCTS

SECTION VII SPILL OR LEAK PROCEDURES

STEPS TO BE TAKEN IN CASE MATERIAL IS RELEASED OR SPILLED

WASTE DISPOSAL METHOD

SECTION VIII SPECIAL PROTECTION INFORMATION

VENTILATION REQUIREMENTS LOCAL EXHAUST	PROTECTIVE EQUIPMENT (SPECIFY TYPES) EYE: Safety goggles
MECHANICAL (GENERAL)	GLOVES
SPECIAL	RESPIRATOR: With proper filter
OTHER PROTECTIVE EQUIPMENT	

SECTION IX SPECIAL PRECAUTIONS

PRECAUTIONS TO BE TAKEN IN HANDLING AND STORAGE

OTHER PRECAUTIONS

SIGNATURE	DATE

TETRAMETHYLAMMONIUM BROMIDE

Hazard diamond: 1 (top), 3 (left), 0 (right)

SECTION I SOURCE AND NOMENCLATURE

MANUFACTURER'S NAME | EMERGENCY TELEPHONE NO.

ADDRESS (NUMBER, STREET, CITY, STATE, ZIP CODE)

TRADE NAME AND SYNONYMS | CHEMICAL FAMILY

CHEMICAL NAME AND SYNONYMS: Tetramethylammonium Bromide | FORMULA: $(CH_3)_4NBr$

SECTION II HAZARDOUS INGREDIENTS

BASIC MATERIAL	APPROXIMATE OR MAXIMUM % WT. OR VOL.	ESTABLISHED OSHA STANDARD	LD_{50} ORAL	LD_{50} PERCUT.	LC_{50} SPECIES	LC_{50} CONC.
			none available			

SECTION III PHYSICAL DATA MW - 154.06

BOILING POINT °F: subl. > 360°C. | VAPOR PRESSURE mm Hg:

MELTING POINT °F: d > 230°C. | VAPOR DENSITY (Air = 1):

SPECIFIC GRAVITY (H_2O = 1): 1.56 | EVAPORATION RATE (= 1):

SOLUBILITY IN WATER Pts/100 pts H_2O: 55 | VOLATILE % Vol. % Wt.

APPEARANCE AND ODOR: Crystaline

SECTION IV FIRE AND EXPLOSION HAZARD DATA

FLASH POINT: None available | FLAMMABLE (EXPLOSIVE) LIMITS: UPPER / LOWER

METHOD USED

EXTINGUISHING MEDIA: Carbon dioxide, dry chemical.

SPECIAL FIRE FIGHTING PROCEDURES: Wear full protective clothing.

UNUSUAL FIRE AND EXPLOSION HAZARDS

SECTION V HEALTH HAZARD DATA

TOXIC LEVEL: None available | CARCINOGENIC: No

PRINCIPLE ROUTES OF ABSORPTION: Inhalation, ingestion | SKIN AND EYE IRRITATION: Moderate

REVELANT SYMPTOMS OF EXPOSURE: Central nervous system depression, irritation of respiratory system.

EFFECTS OF CHRONIC EXPOSURE: Possible liver and kidney damage.

EMERGENCY AND FIRST AID PROCEDURES: Inhalation - rest; if swallowed - induce vomiting, get medical aid; wash from skin & eyes.

SECTION VI REACTIVITY DATA

CONDITIONS CONTRIBUTING TO INSTABILITY

CONDITIONS CONTRIBUTING TO HAZARDOUS POLYMERIZATION

INCOMPATABILITY (MATERIALS TO AVOID)

HAZARDOUS DECOMPOSITION PRODUCTS

SECTION VII SPILL OR LEAK PROCEDURES

STEPS TO BE TAKEN IN CASE MATERIAL IS RELEASED OR SPILLED

WASTE DISPOSAL METHOD

SECTION VIII SPECIAL PROTECTION INFORMATION

VENTILATION REQUIREMENTS LOCAL EXHAUST | PROTECTIVE EQUIPMENT (SPECIFY TYPES) EYE: Goggles

MECHANICAL (GENERAL) | GLOVES

SPECIAL | RESPIRATOR: Self-contained breathing apparatus.

OTHER PROTECTIVE EQUIPMENT

SECTION IX SPECIAL PRECAUTIONS

PRECAUTIONS TO BE TAKEN IN HANDLING AND STORAGE

OTHER PRECAUTIONS

SIGNATURE | DATE

TETRAMETHYLAMMONIUM CHLORIDE

Hazard diamond: 1 (top), 3 (left), 0 (right)

SECTION I SOURCE AND NOMENCLATURE

MANUFACTURER'S NAME | EMERGENCY TELEPHONE NO.

ADDRESS (NUMBER, STREET, CITY, STATE, ZIP CODE)

TRADE NAME AND SYNONYMS | CHEMICAL FAMILY

CHEMICAL NAME AND SYNONYMS: Tetramethylammonium Chloride | FORMULA: $(CH_3)_4N\ Cl$

SECTION II HAZARDOUS INGREDIENTS

BASIC MATERIAL	APPROXIMATE OR MAXIMUM % WT. OR VOL.	ESTABLISHED OSHA STANDARD	LD_{50} mg/kg ORAL	LD_{50} mg/kg PERCUT.	LC_{50} SPECIES	LC_{50} CONC.
				40	mouse	
Lowest published lethal dose - Subcutaneous - rabbit - 6 mg/kg.				See Note		

SECTION III PHYSICAL DATA MW - 109.62

BOILING POINT °F: > 300°C. sublimes | VAPOR PRESSURE mm Hg:

MELTING POINT °F: 425°C. | VAPOR DENSITY (Air = 1):

SPECIFIC GRAVITY (H_2O = 1): 1.17 | EVAPORATION RATE (= 1):

SOLUBILITY IN WATER Pts/100 pts H_2O: soluble | VOLATILE % Vol. % Wt.

APPEARANCE AND ODOR: colorless crystal.

SECTION IV FIRE AND EXPLOSION HAZARD DATA

FLASH POINT: none available | FLAMMABLE (EXPLOSIVE) LIMITS: UPPER / LOWER

METHOD USED

EXTINGUISHING MEDIA: Carbon dioxide, dry chemical.

SPECIAL FIRE FIGHTING PROCEDURES: Wear full protective clothing.

UNUSUAL FIRE AND EXPLOSION HAZARDS

SECTION V HEALTH HAZARD DATA

TOXIC LEVEL: None available | CARCINOGENIC: no

PRINCIPLE ROUTES OF ABSORPTION: Inhalation, ingestion | SKIN AND EYE IRRITATION: moderate

REVELANT SYMPTOMS OF EXPOSURE: Central nervous system depression, irritation of respiratory system.

EFFECTS OF CHRONIC EXPOSURE: possible liver and kidney damage.

EMERGENCY AND FIRST AID PROCEDURES: Inhalation - rest; if swallowed - induce vomiting, get medical aid; wash from skin & eyes.

SECTION VI REACTIVITY DATA

CONDITIONS CONTRIBUTING TO INSTABILITY

CONDITIONS CONTRIBUTING TO HAZARDOUS POLYMERIZATION

INCOMPATABILITY (MATERIALS TO AVOID)

HAZARDOUS DECOMPOSITION PRODUCTS

SECTION VII SPILL OR LEAK PROCEDURES

STEPS TO BE TAKEN IN CASE MATERIAL IS RELEASED OR SPILLED

WASTE DISPOSAL METHOD

SECTION VIII SPECIAL PROTECTION INFORMATION

VENTILATION REQUIREMENTS LOCAL EXHAUST | PROTECTIVE EQUIPMENT (SPECIFY TYPES) EYE: Goggles

MECHANICAL (GENERAL) | GLOVES

SPECIAL | RESPIRATOR: Self-contained

OTHER PROTECTIVE EQUIPMENT

SECTION IX SPECIAL PRECAUTIONS

PRECAUTIONS TO BE TAKEN IN HANDLING AND STORAGE

OTHER PRECAUTIONS

SIGNATURE | DATE

TETRAMETHYLAMMONIUM HYDROXIDE

Hazard rating diamond: 3, 3, 0

SECTION I SOURCE AND NOMENCLATURE

MANUFACTURER'S NAME | EMERGENCY TELEPHONE NO.

ADDRESS (NUMBER, STREET, CITY, STATE, ZIP CODE)

TRADE NAME AND SYNONYMS | CHEMICAL FAMILY

CHEMICAL NAME AND SYNONYMS: Tetramethylammonium Hydroxide | FORMULA: $(CH_3)_4NOH$

SECTION II HAZARDOUS INGREDIENTS

BASIC MATERIAL	APPROXIMATE OR MAXIMUM % WT. OR VOL.	ESTABLISHED OSHA STANDARD	LD_{50} ORAL	LD_{50} PERCUT.	LC_{50} SPECIES	LC_{50} CONC.
Lowest published lethal dose						
subcutaneous - mouse 20 mg/kg						

SECTION III PHYSICAL DATA MW - 91.15

BOILING POINT °F.: decomposes | VAPOR PRESSURE mm Hg.

MELTING POINT °F.: 25°C | VAPOR DENSITY (Air = 1)

SPECIFIC GRAVITY (H_2O = 1): ∿ 1.0 | EVAPORATION RATE (____ = 1)

SOLUBILITY IN WATER Pts/100 pts H_2O: 151 | VOLATILE % Vol. % Wt.

APPEARANCE AND ODOR: a liquid

SECTION IV FIRE AND EXPLOSION HAZARD DATA

FLASH POINT: none available | FLAMMABLE (EXPLOSIVE) LIMITS: UPPER / LOWER

METHOD USED

EXTINGUISHING MEDIA: Water, carbon dioxide, dry chemical.

SPECIAL FIRE FIGHTING PROCEDURES: Wear full protective clothing.

UNUSUAL FIRE AND EXPLOSION HAZARDS

SECTION V HEALTH HAZARD DATA

TOXIC LEVEL: none available | CARCINOGENIC: no

PRINCIPLE ROUTES OF ABSORBTION: inhalation, ingestion | SKIN AND EYE IRRITATION: severe

REVELANT SYMPTOMS OF EXPOSURE: Severe irritation of eyes and respiratory system; if swallowed - alkali attack on digestive system.

EFFECTS OF CHRONIC EXPOSURE

EMERGENCY AND FIRST AID PROCEDURES: Inhalation - rest; if swallowed - drink milk, get medical aid, gastric lavage; wash from skin & eyes.

SECTION VI REACTIVITY DATA

CONDITIONS CONTRIBUTING TO INSTABILITY

CONDITIONS CONTRIBUTING TO HAZARDOUS POLYMERIZATION

INCOMPATABILITY (MATERIALS TO AVOID)

HAZARDOUS DECOMPOSITION PRODUCTS

SECTION VII SPILL OR LEAK PROCEDURES

STEPS TO BE TAKEN IN CASE MATERIAL IS RELEASED OR SPILLED

WASTE DISPOSAL METHOD

SECTION VIII SPECIAL PROTECTION INFORMATION

VENTILATION REQUIREMENTS LOCAL EXHAUST | PROTECTIVE EQUIPMENT (SPECIFY TYPES) EYE: Goggles

MECHANICAL (GENERAL) | GLOVES

SPECIAL | RESPIRATOR: Self-contained

OTHER PROTECTIVE EQUIPMENT

SECTION IX SPECIAL PRECAUTIONS

PRECAUTIONS TO BE TAKEN IN HANDLING AND STORAGE

OTHER PRECAUTIONS

SIGNATURE | DATE

TETRAMETHYLAMMONIUM IODIDE

Hazard rating diamond: 1, 3, 0

SECTION I SOURCE AND NOMENCLATURE

MANUFACTURER'S NAME | EMERGENCY TELEPHONE NO.

ADDRESS (NUMBER, STREET, CITY, STATE, ZIP CODE)

TRADE NAME AND SYNONYMS | CHEMICAL FAMILY

CHEMICAL NAME AND SYNONYMS: Tetramethylaminonium Iodide | FORMULA: $(CH_3)_4NI$

SECTION II HAZARDOUS INGREDIENTS

BASIC MATERIAL	APPROXIMATE OR MAXIMUM % WT. OR VOL.	ESTABLISHED OSHA STANDARD	LD_{50} ORAL	LD_{50} PERCUT.	LC_{50} SPECIES	LC_{50} CONC.
Lowest published lethal dose -						
1. Intraperitoneal - mouse - 40 mg/kg						
2. Subcutaneous - mouse 33 mg/kg						
3. Intravenous - mouse 3 mg/kg						

SECTION III PHYSICAL DATA MW - 201.06

BOILING POINT °F. | VAPOR PRESSURE mm Hg.

MELTING POINT °F.: Decomposes 230°C | VAPOR DENSITY (Air = 1)

SPECIFIC GRAVITY (H_2O = 1): 1.84 | EVAPORATION RATE (____ = 1)

SOLUBILITY IN WATER Pts/100 pts H_2O: slightly soluble | VOLATILE % Vol. % Wt.

APPEARANCE AND ODOR: pale yellow crystals

SECTION IV FIRE AND EXPLOSION HAZARD DATA

FLASH POINT: none available | FLAMMABLE (EXPLOSIVE) LIMITS: UPPER / LOWER

METHOD USED

EXTINGUISHING MEDIA: water, carbon dioxide, dry chemical

SPECIAL FIRE FIGHTING PROCEDURES: Wear full protective clothing

UNUSUAL FIRE AND EXPLOSION HAZARDS

SECTION V HEALTH HAZARD DATA

TOXIC LEVEL: none available | CARCINOGENIC: no

PRINCIPLE ROUTES OF ABSORBTION: inhalation, ingestion | SKIN AND EYE IRRITATION: moderate

REVELANT SYMPTOMS OF EXPOSURE: Nausea, vomiting, diarrhea, slurred speech, visual disturbances, drowsiness, coma.

EFFECTS OF CHRONIC EXPOSURE: Neurotoxic effects.

EMERGENCY AND FIRST AID PROCEDURES: Inhalation - rest; if swallowed - induce vomiting, get medical aid; wash from skin & eyes.

SECTION VI REACTIVITY DATA

CONDITIONS CONTRIBUTING TO INSTABILITY

CONDITIONS CONTRIBUTING TO HAZARDOUS POLYMERIZATION

INCOMPATABILITY (MATERIALS TO AVOID)

HAZARDOUS DECOMPOSITION PRODUCTS

SECTION VII SPILL OR LEAK PROCEDURES

STEPS TO BE TAKEN IN CASE MATERIAL IS RELEASED OR SPILLED

WASTE DISPOSAL METHOD

SECTION VIII SPECIAL PROTECTION INFORMATION

VENTILATION REQUIREMENTS LOCAL EXHAUST | PROTECTIVE EQUIPMENT (SPECIFY TYPES) EYE: Goggles

MECHANICAL (GENERAL) | GLOVES

SPECIAL | RESPIRATOR: With proper filter

OTHER PROTECTIVE EQUIPMENT

SECTION IX SPECIAL PRECAUTIONS

PRECAUTIONS TO BE TAKEN IN HANDLING AND STORAGE

OTHER PRECAUTIONS

SIGNATURE | DATE

TETRAMETHYLENE CHLORIDE

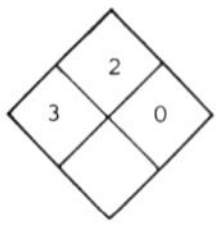

SECTION I SOURCE AND NOMENCLATURE

MANUFACTURER'S NAME | EMERGENCY TELEPHONE NO.

ADDRESS (NUMBER, STREET, CITY, STATE, ZIP CODE)

TRADE NAME AND SYNONYMS | CHEMICAL FAMILY

CHEMICAL NAME AND SYNONYMS: 1,4 Dichlorobutane

FORMULA: ClC_4H_8Cl

SECTION II HAZARDOUS INGREDIENTS

BASIC MATERIAL	APPROXIMATE OR MAXIMUM % WT. OR VOL.	ESTABLISHED OSHA STANDARD	LD_{50} ORAL	LD_{50} PERCUT.	LC_{50} SPECIES	LC_{50} CONC.
			none available			

SECTION III PHYSICAL DATA MW - 127.02

BOILING POINT °F.: 155°C.

VAPOR PRESSURE at 20°C: 4 mm Hg.

MELTING POINT °F.: -39°C

VAPOR DENSITY (Air = 1): 4.4

SPECIFIC GRAVITY (H_2O = 1): 1.14

EVAPORATION RATE (= 1)

SOLUBILITY IN WATER Pts/100 pts H_2O

VOLATILE % Vol. % Wt.

APPEARANCE AND ODOR: Colorless, mobile liquid with mild, pleasant odor.

SECTION IV FIRE AND EXPLOSION HAZARD DATA

FLASH POINT: 126°F

FLAMMABLE (EXPLOSIVE) LIMITS: UPPER / LOWER

METHOD USED: closed cup

EXTINGUISHING MEDIA: carbon dioxide, dry chemical

SPECIAL FIRE FIGHTING PROCEDURES: Water may be ineffective.

UNUSUAL FIRE AND EXPLOSION HAZARDS

SECTION V HEALTH HAZARD DATA

TOXIC LEVEL: none available

CARCINOGENIC: no

PRINCIPLE ROUTES OF ABSORBTION: Inhalation, ingestion

SKIN AND EYE IRRITATION: moderate

REVELANT SYMPTOMS OF EXPOSURE: Eye irritation, headache, vomiting, drowsiness, narcosis.

EFFECTS OF CHRONIC EXPOSURE: Dermatitis, liver and kidney damage.

EMERGENCY AND FIRST AID PROCEDURES: Inhalation - rest; if swallowed - give emetic, get medical aid; wash from skin & eyes.

SECTION VI REACTIVITY DATA

CONDITIONS CONTRIBUTING TO INSTABILITY

CONDITIONS CONTRIBUTING TO HAZARDOUS POLYMERIZATION

INCOMPATABILITY (MATERIALS TO AVOID)

HAZARDOUS DECOMPOSITION PRODUCTS

SECTION VII SPILL OR LEAK PROCEDURES

STEPS TO BE TAKEN IN CASE MATERIAL IS RELEASED OR SPILLED: Wear self-contained breathing apparatus.

WASTE DISPOSAL METHOD

SECTION VIII SPECIAL PROTECTION INFORMATION

VENTILATION REQUIREMENTS LOCAL EXHAUST

PROTECTIVE EQUIPMENT (SPECIFY TYPES) EYE: Goggles

MECHANICAL (GENERAL)

GLOVES: Rubber

SPECIAL

RESPIRATOR: With proper filter

OTHER PROTECTIVE EQUIPMENT

SECTION IX SPECIAL PRECAUTIONS

PRECAUTIONS TO BE TAKEN IN HANDLING AND STORAGE

OTHER PRECAUTIONS

SIGNATURE | DATE

THIOACETAMIDE

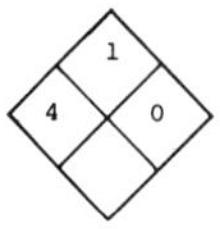

SECTION I SOURCE AND NOMENCLATURE

MANUFACTURER'S NAME | EMERGENCY TELEPHONE NO.

ADDRESS (NUMBER, STREET, CITY, STATE, ZIP CODE)

TRADE NAME AND SYNONYMS | CHEMICAL FAMILY

CHEMICAL NAME AND SYNONYMS: Thioacetamide

FORMULA: CH_3CSNH_2

SECTION II HAZARDOUS INGREDIENTS

BASIC MATERIAL	APPROXIMATE OR MAXIMUM % WT. OR VOL.	ESTABLISHED OSHA STANDARD	LD_{50} ORAL	LD_{50} PERCUT.	LC_{50} SPECIES	LC_{50} CONC.
Lowest published lethal dose - oral - rat 200 mg/kg.						

SECTION III PHYSICAL DATA MW - 75.13

BOILING POINT °F.

VAPOR PRESSURE mm Hg.

MELTING POINT °F.: 115°C.

VAPOR DENSITY (Air = 1)

SPECIFIC GRAVITY (H_2O = 1)

EVAPORATION RATE (= 1)

SOLUBILITY IN WATER Pts/100 pts H_2O: 16.3

VOLATILE % Vol. % Wt.

APPEARANCE AND ODOR: Colorless leaflets, slight odor of mercaptans.

SECTION IV FIRE AND EXPLOSION HAZARD DATA

FLASH POINT: none available

FLAMMABLE (EXPLOSIVE) LIMITS: UPPER / LOWER

METHOD USED

EXTINGUISHING MEDIA: Carbon dioxide, dry chemical.

SPECIAL FIRE FIGHTING PROCEDURES

UNUSUAL FIRE AND EXPLOSION HAZARDS

SECTION V HEALTH HAZARD DATA

TOXIC LEVEL: None available

CARCINOGENIC: Yes

PRINCIPLE ROUTES OF ABSORBTION: Inhalation, ingestion

SKIN AND EYE IRRITATION: Slight

REVELANT SYMPTOMS OF EXPOSURE: Drowsiness, fatigue, nausea, acidosis, skin eruptions.

EFFECTS OF CHRONIC EXPOSURE: Liver damage

EMERGENCY AND FIRST AID PROCEDURES: If swallowed - induce vomiting, get medical aid; inhalation - rest.

SECTION VI REACTIVITY DATA

CONDITIONS CONTRIBUTING TO INSTABILITY

CONDITIONS CONTRIBUTING TO HAZARDOUS POLYMERIZATION

INCOMPATABILITY (MATERIALS TO AVOID)

HAZARDOUS DECOMPOSITION PRODUCTS

SECTION VII SPILL OR LEAK PROCEDURES

STEPS TO BE TAKEN IN CASE MATERIAL IS RELEASED OR SPILLED

WASTE DISPOSAL METHOD

SECTION VIII SPECIAL PROTECTION INFORMATION

VENTILATION REQUIREMENTS LOCAL EXHAUST

PROTECTIVE EQUIPMENT (SPECIFY TYPES) EYE: Goggles

MECHANICAL (GENERAL)

GLOVES: Rubber

SPECIAL

RESPIRATOR

OTHER PROTECTIVE EQUIPMENT

SECTION IX SPECIAL PRECAUTIONS

PRECAUTIONS TO BE TAKEN IN HANDLING AND STORAGE

OTHER PRECAUTIONS

SIGNATURE | DATE

THIOGLYCOLIC ACID

NFPA: Health 3, Flammability 1, Reactivity 0

SECTION I SOURCE AND NOMENCLATURE

MANUFACTURER'S NAME:

EMERGENCY TELEPHONE NO.:

ADDRESS (NUMBER, STREET, CITY, STATE, ZIP CODE):

TRADE NAME AND SYNONYMS: Mercaptoacetic Acid

CHEMICAL FAMILY:

CHEMICAL NAME AND SYNONYMS: Thioglycolic Acid

FORMULA: $HSCH_2COOH$

SECTION II HAZARDOUS INGREDIENTS

BASIC MATERIAL	APPROXIMATE OR MAXIMUM % WT. OR VOL.	ESTABLISHED OSHA STANDARD	LD_{50} mg/kg ORAL	LD_{50} PERCUT.	LC_{50} SPECIES	LC_{50} CONC.
			250		Rat	
			126		Rabbit	
			126		Guinea pig	

SECTION III PHYSICAL DATA MW - 92.11

BOILING POINT °F.: 123°C at 29 mm.

VAPOR PRESSURE mm Hg.: 106°C 11

MELTING POINT °F.: -16.5°C

VAPOR DENSITY (Air = 1):

SPECIFIC GRAVITY (H_2O = 1): 1.33

EVAPORATION RATE (______ = 1):

SOLUBILITY IN WATER Pts/100 pts H_2O: Soluble

VOLATILE % Vol. % Wt.:

APPEARANCE AND ODOR: Colorless liquid, disagreeable odor

SECTION IV FIRE AND EXPLOSION HAZARD DATA

FLASH POINT: None available

METHOD USED:

FLAMMABLE (EXPLOSIVE) LIMITS: UPPER / LOWER

EXTINGUISHING MEDIA: Water

SPECIAL FIRE FIGHTING PROCEDURES:

UNUSUAL FIRE AND EXPLOSION HAZARDS:

SECTION V HEALTH HAZARD DATA

TOXIC LEVEL: None available

CARCINOGENIC: No

PRINCIPLE ROUTES OF ABSORBTION: Inhalation, ingestion, skin

SKIN AND EYE IRRITATION: Severe

REVELANT SYMPTOMS OF EXPOSURE: Weakness, gasping, respiration, convulsions, severe eye and skin irritant.

EFFECTS OF CHRONIC EXPOSURE: --

EMERGENCY AND FIRST AID PROCEDURES: If swallowed - drink water, milk of magnesia, get medical aid; inhalation - rest; wash eyes and skin.

SECTION VI REACTIVITY DATA

CONDITIONS CONTRIBUTING TO INSTABILITY:

CONDITIONS CONTRIBUTING TO HAZARDOUS POLYMERIZATION:

INCOMPATABILITY (MATERIALS TO AVOID):

HAZARDOUS DECOMPOSITION PRODUCTS:

SECTION VII SPILL OR LEAK PROCEDURES

STEPS TO BE TAKEN IN CASE MATERIAL IS RELEASED OR SPILLED:

WASTE DISPOSAL METHOD:

SECTION VIII SPECIAL PROTECTION INFORMATION

VENTILATION REQUIREMENTS LOCAL EXHAUST:

PROTECTIVE EQUIPMENT (SPECIFY TYPES) EYE: Goggles

MECHANICAL (GENERAL):

GLOVES:

SPECIAL:

RESPIRATOR: With proper filter.

OTHER PROTECTIVE EQUIPMENT: Coveralls

SECTION IX SPECIAL PRECAUTIONS

PRECAUTIONS TO BE TAKEN IN HANDLING AND STORAGE:

OTHER PRECAUTIONS:

SIGNATURE: DATE:

THIONYL CHLORIDE

NFPA: Health 4, Flammability 0, Reactivity 1

SECTION I SOURCE AND NOMENCLATURE

MANUFACTURER'S NAME:

EMERGENCY TELEPHONE NO.:

ADDRESS (NUMBER, STREET, CITY, STATE, ZIP CODE):

TRADE NAME AND SYNONYMS: Sulfurous Oxychloride

CHEMICAL FAMILY:

CHEMICAL NAME AND SYNONYMS: Thionyl Chloride

FORMULA: $SOCl_2$

SECTION II HAZARDOUS INGREDIENTS

BASIC MATERIAL	APPROXIMATE OR MAXIMUM % WT. OR VOL.	ESTABLISHED OSHA STANDARD	LD_{50} ORAL	LD_{50} PERCUT.	LC_{50} SPECIES	LC_{50} CONC.
			none available			

SECTION III PHYSICAL DATA MW - 118.97

BOILING POINT °F.: 79°C

VAPOR PRESSURE mm Hg.: at 21.4°C 100

MELTING POINT °F.: -105°C

VAPOR DENSITY (Air = 1): 4.1

SPECIFIC GRAVITY (H_2O = 1): 1.64

EVAPORATION RATE (______ = 1):

SOLUBILITY IN WATER Pts/100 pts H_2O: decomposes

VOLATILE % Vol. % Wt.:

APPEARANCE AND ODOR: Colorless to yellow to rec liquid.

SECTION IV FIRE AND EXPLOSION HAZARD DATA

FLASH POINT: Not flammable

METHOD USED:

FLAMMABLE (EXPLOSIVE) LIMITS: UPPER / LOWER

EXTINGUISHING MEDIA:

SPECIAL FIRE FIGHTING PROCEDURES:

UNUSUAL FIRE AND EXPLOSION HAZARDS:

SECTION V HEALTH HAZARD DATA

TOXIC LEVEL: none available

CARCINOGENIC: no

PRINCIPLE ROUTES OF ABSORBTION: inhalation, ingestion

SKIN AND EYE IRRITATION: severe

REVELANT SYMPTOMS OF EXPOSURE: Irritation of skin, eyes, mucous membranes and respiratory tract.

EFFECTS OF CHRONIC EXPOSURE: --

EMERGENCY AND FIRST AID PROCEDURES: Inhalation - rest; if swallowed - drink water, milk of magnesia, get medical aid; wash skin & eyes.

SECTION VI REACTIVITY DATA

CONDITIONS CONTRIBUTING TO INSTABILITY:

CONDITIONS CONTRIBUTING TO HAZARDOUS POLYMERIZATION:

INCOMPATABILITY (MATERIALS TO AVOID): Water

HAZARDOUS DECOMPOSITION PRODUCTS: Reacts with water to produce HCl and SO_2.

SECTION VII SPILL OR LEAK PROCEDURES

STEPS TO BE TAKEN IN CASE MATERIAL IS RELEASED OR SPILLED: Cover with bicarbonate

WASTE DISPOSAL METHOD:

SECTION VIII SPECIAL PROTECTION INFORMATION

VENTILATION REQUIREMENTS LOCAL EXHAUST:

PROTECTIVE EQUIPMENT (SPECIFY TYPES) EYE: Goggles

MECHANICAL (GENERAL):

GLOVES: Rubber

SPECIAL:

RESPIRATOR: With proper filter

OTHER PROTECTIVE EQUIPMENT: Protective clothing

SECTION IX SPECIAL PRECAUTIONS

PRECAUTIONS TO BE TAKEN IN HANDLING AND STORAGE:

OTHER PRECAUTIONS:

SIGNATURE: DATE:

THIOPHENOL

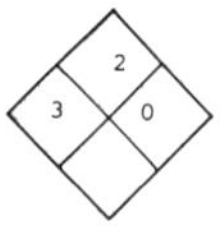

SECTION I SOURCE AND NOMENCLATURE

MANUFACTURER'S NAME:

EMERGENCY TELEPHONE NO.:

ADDRESS (NUMBER, STREET, CITY, STATE, ZIP CODE):

TRADE NAME AND SYNONYMS: Phenyl Mercaptan

CHEMICAL FAMILY:

CHEMICAL NAME AND SYNONYMS: Thiophenol

FORMULA: C_6H_5SH

SECTION II HAZARDOUS INGREDIENTS

BASIC MATERIAL	APPROXIMATE OR MAXIMUM % WT. OR VOL.	ESTABLISHED OSHA STANDARD	LD_{50} mg/kg ORAL	LD_{50} mg/kg PERCUT.	SPECIES	LC_{50} PPM CONC.
			24		Bird	
			46		Rat	
					Mouse	28
					Rat	33
				300	Rat	
				134	Rabbit	

SECTION III PHYSICAL DATA MW - 110.2

BOILING POINT °F.: 168°C

VAPOR PRESSURE mm Hg.:

MELTING POINT °F.: -15°C

VAPOR DENSITY (Air = 1):

SPECIFIC GRAVITY (H_2O = 1): 1.07

EVAPORATION RATE (____ = 1):

SOLUBILITY IN WATER Pts/100 pts H_2O: very slightly soluble

VOLATILE % Vol. % Wt.:

APPEARANCE AND ODOR: liquids, penetrating odor.

SECTION IV FIRE AND EXPLOSION HAZARD DATA

FLASH POINT: none available

FLAMMABLE (EXPLOSIVE) LIMITS: UPPER / LOWER

METHOD USED:

EXTINGUISHING MEDIA: carbon dioxide, dry chemical.

SPECIAL FIRE FIGHTING PROCEDURES:

UNUSUAL FIRE AND EXPLOSION HAZARDS:

SECTION V HEALTH HAZARD DATA

TOXIC LEVEL: none available

CARCINOGENIC: no

PRINCIPLE ROUTES OF ABSORBTION: inhalation, ingestion

SKIN AND EYE IRRITATION: severe

REVELANT SYMPTOMS OF EXPOSURE: Skin and mucous membrane irritation, headache, dizziness, dermatitis.

EFFECTS OF CHRONIC EXPOSURE: --

EMERGENCY AND FIRST AID PROCEDURES: Inhalation - artificial respiration as necessary; if swallowed - induce vomiting, get medical aid, gastric lavage; wash skin & eyes.

SECTION VI REACTIVITY DATA

CONDITIONS CONTRIBUTING TO INSTABILITY:

CONDITIONS CONTRIBUTING TO HAZARDOUS POLYMERIZATION:

INCOMPATABILITY (MATERIALS TO AVOID):

HAZARDOUS DECOMPOSITION PRODUCTS:

SECTION VII SPILL OR LEAK PROCEDURES

STEPS TO BE TAKEN IN CASE MATERIAL IS RELEASED OR SPILLED:

WASTE DISPOSAL METHOD:

SECTION VIII SPECIAL PROTECTION INFORMATION

VENTILATION REQUIREMENTS LOCAL EXHAUST:

PROTECTIVE EQUIPMENT (SPECIFY TYPES) EYE: Goggles

MECHANICAL (GENERAL):

GLOVES: Rubber

SPECIAL:

RESPIRATOR: With proper filter

OTHER PROTECTIVE EQUIPMENT:

SECTION IX SPECIAL PRECAUTIONS

PRECAUTIONS TO BE TAKEN IN HANDLING AND STORAGE:

OTHER PRECAUTIONS:

SIGNATURE: DATE:

THIOSEMICARBAZIDE

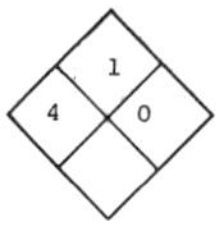

SECTION I SOURCE AND NOMENCLATURE

MANUFACTURER'S NAME:

EMERGENCY TELEPHONE NO.:

ADDRESS (NUMBER, STREET, CITY, STATE, ZIP CODE):

TRADE NAME AND SYNONYMS:

CHEMICAL FAMILY:

CHEMICAL NAME AND SYNONYMS: 3 Thiosemicarbazide

FORMULA: $NH_2CSNHNH_2$

SECTION II HAZARDOUS INGREDIENTS

BASIC MATERIAL	APPROXIMATE OR MAXIMUM % WT. OR VOL.	ESTABLISHED OSHA STANDARD	LD_{50} mg/kg ORAL	LD_{50} mg/kg PERCUT.	SPECIES	LC_{50} CONC.
			9		bird	
			9		rat	
			10		dog	
			20		cat	

Intraperitoneal - mouse LD50 1 mg/kg

Intraperitoneal - guinea pig - LD_{50} 24 mg/kg

SECTION III PHYSICAL DATA MW - 91.13

BOILING POINT °F.: sublimes

VAPOR PRESSURE mm Hg.:

MELTING POINT °F.: 183°C

VAPOR DENSITY (Air = 1):

SPECIFIC GRAVITY (H_2O = 1):

EVAPORATION RATE (____ = 1):

SOLUBILITY IN WATER Pts/100 pts H_2O: S

VOLATILE % Vol. % Wt.:

APPEARANCE AND ODOR:

SECTION IV FIRE AND EXPLOSION HAZARD DATA

FLASH POINT: none available

FLAMMABLE (EXPLOSIVE) LIMITS: UPPER / LOWER

METHOD USED:

EXTINGUISHING MEDIA: Alcohol foam, carbon dioxide, dry chemical.

SPECIAL FIRE FIGHTING PROCEDURES:

UNUSUAL FIRE AND EXPLOSION HAZARDS:

SECTION V HEALTH HAZARD DATA

TOXIC LEVEL: None available

CARCINOGENIC: NO

PRINCIPLE ROUTES OF ABSORBTION: Inhalation, ingestion, skin

SKIN AND EYE IRRITATION: Severe

REVELANT SYMPTOMS OF EXPOSURE: Irritation of skin, eyes and mucous membranes, gastroentric disturbances, anemia.

EFFECTS OF CHRONIC EXPOSURE: Blood and liver damage, loss of appetite.

EMERGENCY AND FIRST AID PROCEDURES: Inhalation - rest; if swallowed - induce vomiting, get medical aid; wash skin & eyes.

SECTION VI REACTIVITY DATA

CONDITIONS CONTRIBUTING TO INSTABILITY:

CONDITIONS CONTRIBUTING TO HAZARDOUS POLYMERIZATION:

INCOMPATABILITY (MATERIALS TO AVOID):

HAZARDOUS DECOMPOSITION PRODUCTS:

SECTION VII SPILL OR LEAK PROCEDURES

STEPS TO BE TAKEN IN CASE MATERIAL IS RELEASED OR SPILLED:

WASTE DISPOSAL METHOD:

SECTION VIII SPECIAL PROTECTION INFORMATION

VENTILATION REQUIREMENTS LOCAL EXHAUST:

PROTECTIVE EQUIPMENT (SPECIFY TYPES) EYE: Goggles

MECHANICAL (GENERAL):

GLOVES: Rubber

SPECIAL:

RESPIRATOR: Self-contained

OTHER PROTECTIVE EQUIPMENT:

SECTION IX SPECIAL PRECAUTIONS

PRECAUTIONS TO BE TAKEN IN HANDLING AND STORAGE:

OTHER PRECAUTIONS:

SIGNATURE: DATE:

THIOUREA

Hazard diamond: 1 (top), 3 (left), 1 (right)

SECTION I SOURCE AND NOMENCLATURE

MANUFACTURER'S NAME | EMERGENCY TELEPHONE NO.

ADDRESS (NUMBER, STREET, CITY, STATE, ZIP CODE)

TRADE NAME AND SYNONYMS | CHEMICAL FAMILY

CHEMICAL NAME AND SYNONYMS: Thiocarbamide | FORMULA: NH_2-CS-NH_2

SECTION II HAZARDOUS INGREDIENTS

BASIC MATERIAL	APPROXIMATE OR MAXIMUM % WT. OR VOL.	ESTABLISHED OSHA STANDARD	LD_{50} mg/kg ORAL	PERCUT.	SPECIES	LC_{50} CONC.
			125		rat	
			11,500		mouse	
TCLO oral - woman 1600 mg/kg - 5 weeks			6,985		rabbit	

SECTION III PHYSICAL DATA MW - 76.12

BOILING POINT °F.: d | VAPOR PRESSURE mm Hg.

MELTING POINT °F.: 180-2°C. | VAPOR DENSITY (Air = 1)

SPECIFIC GRAVITY (H_2O = 1): 1.41 | EVAPORATION RATE (= 1)

SOLUBILITY IN WATER Pts/100 pts H_2O: 9.2 | VOLATILE % Vol. % Wt.

APPEARANCE AND ODOR: Rhombic crystals.

SECTION IV FIRE AND EXPLOSION HAZARD DATA

FLASH POINT: None available | FLAMMABLE (EXPLOSIVE) LIMITS: UPPER / LOWER

METHOD USED

EXTINGUISHING MEDIA: Carbon dioxide, dry chemical.

SPECIAL FIRE FIGHTING PROCEDURES

UNUSUAL FIRE AND EXPLOSION HAZARDS: Burns to form SO_2 fumes.

SECTION V HEALTH HAZARD DATA

TOXIC LEVEL: None available | CARCINOGENIC: neoplastic effects

PRINCIPLE ROUTES OF ABSORBTION: Ingestion, inhalation of dust | SKIN AND EYE IRRITATION: slight

REVELANT SYMPTOMS OF EXPOSURE: Dermatitis

EFFECTS OF CHRONIC EXPOSURE: Ingestion - depression of bone marrow, anemia, leukopenia, allergic skin eruptions.

EMERGENCY AND FIRST AID PROCEDURES: Inhalation - rest; if swallowed - induce vomiting, get medical aid; wash from skin & eyes.

SECTION VI REACTIVITY DATA

CONDITIONS CONTRIBUTING TO INSTABILITY

CONDITIONS CONTRIBUTING TO HAZARDOUS POLYMERIZATION

INCOMPATABILITY (MATERIALS TO AVOID)

HAZARDOUS DECOMPOSITION PRODUCTS: Oxides of sulfur and nitrogen.

SECTION VII SPILL OR LEAK PROCEDURES

STEPS TO BE TAKEN IN CASE MATERIAL IS RELEASED OR SPILLED

WASTE DISPOSAL METHOD

SECTION VIII SPECIAL PROTECTION INFORMATION

VENTILATION REQUIREMENTS LOCAL EXHAUST | PROTECTIVE EQUIPMENT (SPECIFY TYPES) EYE: Goggles

MECHANICAL (GENERAL) | GLOVES

SPECIAL | RESPIRATOR

OTHER PROTECTIVE EQUIPMENT

SECTION IX SPECIAL PRECAUTIONS

PRECAUTIONS TO BE TAKEN IN HANDLING AND STORAGE

OTHER PRECAUTIONS

SIGNATURE | DATE

o-TOLIDINE DIHYDROCHLORIDE

Hazard diamond: 1 (top), 4 (left), 0 (right)

SECTION I SOURCE AND NOMENCLATURE

MANUFACTURER'S NAME | EMERGENCY TELEPHONE NO.

ADDRESS (NUMBER, STREET, CITY, STATE, ZIP CODE)

TRADE NAME AND SYNONYMS | CHEMICAL FAMILY

CHEMICAL NAME AND SYNONYMS | FORMULA: $(HClNH_2(CH_3)C_6H_3)_2$

SECTION II HAZARDOUS INGREDIENTS

BASIC MATERIAL	APPROXIMATE OR MAXIMUM % WT. OR VOL.	ESTABLISHED OSHA STANDARD	LD_{50} ORAL	PERCUT.	SPECIES	LC_{50} CONC.

SECTION III PHYSICAL DATA MW - 285.22

BOILING POINT °F. | VAPOR PRESSURE mm Hg.

MELTING POINT °F.: >340°C | VAPOR DENSITY (Air = 1)

SPECIFIC GRAVITY (H_2O = 1) | EVAPORATION RATE (= 1)

SOLUBILITY IN WATER Pts/100 pts H_2O: 0.9 | VOLATILE % Vol. % Wt.

APPEARANCE AND ODOR: Scales

SECTION IV FIRE AND EXPLOSION HAZARD DATA

FLASH POINT: None available | FLAMMABLE (EXPLOSIVE) LIMITS: UPPER / LOWER

METHOD USED

EXTINGUISHING MEDIA: Carbon dioxide, dry chemical.

SPECIAL FIRE FIGHTING PROCEDURES: Wear full protective clothing.

UNUSUAL FIRE AND EXPLOSION HAZARDS

SECTION V HEALTH HAZARD DATA

TOXIC LEVEL: None available | CARCINOGENIC: Yes

PRINCIPLE ROUTES OF ABSORBTION: Inhalation, ingestion, skin | SKIN AND EYE IRRITATION: Slight

REVELANT SYMPTOMS OF EXPOSURE: Cyanosis, air hunger, nausea, vomiting, low blood pressure, convulsions.

EFFECTS OF CHRONIC EXPOSURE: Weight loss, anemia, weakness.

EMERGENCY AND FIRST AID PROCEDURES: Inhalation - rest; if swallowed - give emetic, get medical aid; wash skin & eyes.

SECTION VI REACTIVITY DATA

CONDITIONS CONTRIBUTING TO INSTABILITY

CONDITIONS CONTRIBUTING TO HAZARDOUS POLYMERIZATION

INCOMPATABILITY (MATERIALS TO AVOID)

HAZARDOUS DECOMPOSITION PRODUCTS

SECTION VII SPILL OR LEAK PROCEDURES

STEPS TO BE TAKEN IN CASE MATERIAL IS RELEASED OR SPILLED

WASTE DISPOSAL METHOD

SECTION VIII SPECIAL PROTECTION INFORMATION

VENTILATION REQUIREMENTS LOCAL EXHAUST | PROTECTIVE EQUIPMENT (SPECIFY TYPES) EYE: Goggles

MECHANICAL (GENERAL) | GLOVES: Rubber

SPECIAL | RESPIRATOR

OTHER PROTECTIVE EQUIPMENT

SECTION IX SPECIAL PRECAUTIONS

PRECAUTIONS TO BE TAKEN IN HANDLING AND STORAGE

OTHER PRECAUTIONS

SIGNATURE | DATE

TOLUHYDROQUINONE

Hazard diamond: 1 (top), 3 (left), 0 (right)

SECTION I SOURCE AND NOMENCLATURE

MANUFACTURER'S NAME

EMERGENCY TELEPHONE NO.

ADDRESS (NUMBER, STREET, CITY, STATE, ZIP CODE)

TRADE NAME AND SYNONYMS

CHEMICAL FAMILY

CHEMICAL NAME AND SYNONYMS: Tolyl Hydroquinone

FORMULA: $CH_3C_6H_3(OH)_2$

SECTION II HAZARDOUS INGREDIENTS

BASIC MATERIAL	APPROXIMATE OR MAXIMUM % WT. OR VOL.	ESTABLISHED OSHA STANDARD	LD_{50} ORAL	LD_{50} PERCUT.	LC_{50} SPECIES	LC_{50} CONC.
LDLO - oral - rat 200 mg/kg						

SECTION III PHYSICAL DATA MW - 124.13

BOILING POINT °F.: --

VAPOR PRESSURE mm Hg.

MELTING POINT °F.: 129°C.

VAPOR DENSITY (Air = 1)

SPECIFIC GRAVITY (H_2O = 1)

EVAPORATION RATE (= 1)

SOLUBILITY IN WATER Pts/100 pts H_2O: soluble

VOLATILE % Vol. % Wt.

APPEARANCE AND ODOR: leaflets

SECTION IV FIRE AND EXPLOSION HAZARD DATA

FLASH POINT: None available

FLAMMABLE (EXPLOSIVE) LIMITS UPPER / LOWER

METHOD USED

EXTINGUISHING MEDIA: Carbon dioxide, dry chemical, water.

SPECIAL FIRE FIGHTING PROCEDURES

UNUSUAL FIRE AND EXPLOSION HAZARDS

SECTION V HEALTH HAZARD DATA

TOXIC LEVEL: ~~None available~~

CARCINOGENIC: no

PRINCIPLE ROUTES OF ABSORBTION: Inhalation, ingestion, skin

SKIN AND EYE IRRITATION: Severe

REVELANT SYMPTOMS OF EXPOSURE: Nausea, dizziness, increased respiration, headache, cyanosis, delirium and collapse, depigmentation of eyes and skin.

EFFECTS OF CHRONIC EXPOSURE: Eye damage, dermatitis.

EMERGENCY AND FIRST AID PROCEDURES: Inhalation - rest; if swallowed - induce vomiting, gastric lavage, medical aid; wash from skin & eyes.

SECTION VI REACTIVITY DATA

CONDITIONS CONTRIBUTING TO INSTABILITY

CONDITIONS CONTRIBUTING TO HAZARDOUS POLYMERIZATION

INCOMPATABILITY (MATERIALS TO AVOID)

HAZARDOUS DECOMPOSITION PRODUCTS

SECTION VII SPILL OR LEAK PROCEDURES

STEPS TO BE TAKEN IN CASE MATERIAL IS RELEASED OR SPILLED

WASTE DISPOSAL METHOD

SECTION VIII SPECIAL PROTECTION INFORMATION

VENTILATION REQUIREMENTS LOCAL EXHAUST

PROTECTIVE EQUIPMENT (SPECIFY TYPES) EYE: Safety goggles

MECHANICAL (GENERAL)

GLOVES: Rubber

SPECIAL

RESPIRATOR

OTHER PROTECTIVE EQUIPMENT

SECTION IX SPECIAL PRECAUTIONS

PRECAUTIONS TO BE TAKEN IN HANDLING AND STORAGE

OTHER PRECAUTIONS

SIGNATURE

DATE

m-TOLUIDINE

Hazard diamond: 2 (top), 4 (left), 0 (right)

SECTION I SOURCE AND NOMENCLATURE

MANUFACTURER'S NAME

EMERGENCY TELEPHONE NO.

ADDRESS (NUMBER, STREET, CITY, STATE, ZIP CODE)

TRADE NAME AND SYNONYMS: 3-Amino Toluene

CHEMICAL FAMILY

CHEMICAL NAME AND SYNONYMS: 3-Methyl Aniline

FORMULA: $CH_3C_6H_4NH_2$

SECTION II HAZARDOUS INGREDIENTS

BASIC MATERIAL	APPROXIMATE OR MAXIMUM % WT. OR VOL.	ESTABLISHED OSHA STANDARD	LD_{50} mg/kg ORAL	LD_{50} PERCUT.	SPECIES	LC_{50} CONC.
			974		rat	
1. Lowest published lethal dose intraperitoneal - mouse 150 mg/kg						

SECTION III PHYSICAL DATA MW - 107.2

BOILING POINT °F.: 203.4°C.

VAPOR PRESSURE mm Hg.: 1 at 41°C.

MELTING POINT °F.: -31.3°C.

VAPOR DENSITY (Air = 1): 3.90

SPECIFIC GRAVITY (H_2O = 1): 0.989

EVAPORATION RATE (= 1)

SOLUBILITY IN WATER Pts/100 pts H_2O

VOLATILE % Vol. % Wt.

APPEARANCE AND ODOR: Colorless liquid

SECTION IV FIRE AND EXPLOSION HAZARD DATA

FLASH POINT: ~~N/A~~

FLAMMABLE (EXPLOSIVE) LIMITS UPPER / LOWER

METHOD USED

EXTINGUISHING MEDIA: Carbon dioxide, dry chemical, foam or water spray.

SPECIAL FIRE FIGHTING PROCEDURES: Wear full protective clothing.

UNUSUAL FIRE AND EXPLOSION HAZARDS: When heated emits highly toxic fumes.

SECTION V HEALTH HAZARD DATA

TOXIC LEVEL: Not available

CARCINOGENIC: Neoplastic

PRINCIPLE ROUTES OF ABSORBTION: Inhalation, ingestion, skin

SKIN AND EYE IRRITATION: Yes

REVELANT SYMPTOMS OF EXPOSURE: Cyanosis, air hunger, nausea, vomiting, low blood pressure, convulsions.

EFFECTS OF CHRONIC EXPOSURE: Weight loss, anemia, weakness.

EMERGENCY AND FIRST AID PROCEDURES: Inhalation - rest; if swallowed - qash mouth, give emetic, get medical aid; wash from skin & eyes, skin may be washed with 5% acetic acid to aid removal.

SECTION VI REACTIVITY DATA

CONDITIONS CONTRIBUTING TO INSTABILITY

CONDITIONS CONTRIBUTING TO HAZARDOUS POLYMERIZATION: May react vigorously with oxidizing materials.

INCOMPATABILITY (MATERIALS TO AVOID)

HAZARDOUS DECOMPOSITION PRODUCTS

SECTION VII SPILL OR LEAK PROCEDURES

STEPS TO BE TAKEN IN CASE MATERIAL IS RELEASED OR SPILLED: Cover with a 9:1 mixture of sand and soda ash, put in cardboard box for disposal.

WASTE DISPOSAL METHOD

SECTION VIII SPECIAL PROTECTION INFORMATION

VENTILATION REQUIREMENTS LOCAL EXHAUST

PROTECTIVE EQUIPMENT (SPECIFY TYPES) EYE: Safety goggles

MECHANICAL (GENERAL)

GLOVES: Rubber

SPECIAL

RESPIRATOR: ~~With proper filter~~

OTHER PROTECTIVE EQUIPMENT

SECTION IX SPECIAL PRECAUTIONS

PRECAUTIONS TO BE TAKEN IN HANDLING AND STORAGE: Outside or detached storage is preferred; store in dry, well-ventilated, cool location away from fire hazards.

OTHER PRECAUTIONS

SIGNATURE

DATE

p-TOLUIDINE

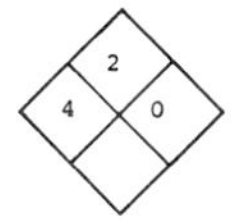

SECTION I SOURCE AND NOMENCLATURE

MANUFACTURER'S NAME | EMERGENCY TELEPHONE NO.

ADDRESS (NUMBER, STREET, CITY, STATE, ZIP CODE)

TRADE NAME AND SYNONYMS: P-Methyl Aniline | CHEMICAL FAMILY

CHEMICAL NAME AND SYNONYMS: P-Amino Toluene | FORMULA: $CH_3C_6H_4NH_2$

SECTION II HAZARDOUS INGREDIENTS

BASIC MATERIAL	APPROXIMATE OR MAXIMUM % WT. OR VOL.	ESTABLISHED OSHA STANDARD	LD_{50} mg/kg ORAL	PERCUT.	SPECIES	LC_{50} CONC.
LDLO oral - human 50 mg/kg			656		Rat	
			42		Bird	

SECTION III PHYSICAL DATA MW - 107.2

BOILING POINT °F.: 200.6°C.

VAPOR PRESSURE mm Hg.: at 42°C. 1

MELTING POINT °F.: 43.8°C.

VAPOR DENSITY (Air = 1): 3.90

SPECIFIC GRAVITY (H_2O = 1): 1.046

EVAPORATION RATE (______ = 1)

SOLUBILITY IN WATER Pts/100 pts H_2O: 0.74

VOLATILE % Vol. % Wt.

APPEARANCE AND ODOR: crystals

SECTION IV FIRE AND EXPLOSION HAZARD DATA

FLASH POINT: 188°F.

METHOD USED: C.C.

FLAMMABLE (EXPLOSIVE) LIMITS: UPPER | LOWER

EXTINGUISHING MEDIA: Carbon dioxide, dry chemical, foam and water spray.

SPECIAL FIRE FIGHTING PROCEDURES: Wear full protective clothing.

UNUSUAL FIRE AND EXPLOSION HAZARDS: Emits toxic fumes when heated.

SECTION V HEALTH HAZARD DATA

TOXIC LEVEL: Not available | CARCINOGENIC: yes

PRINCIPLE ROUTES OF ABSORBTION: Inhalation, ingestion, skin | SKIN AND EYE IRRITATION: Yes

REVELANT SYMPTOMS OF EXPOSURE: Cyanosis, air hunger, nausea, vomiting, low blood pressure, convulsions.

EFFECTS OF CHRONIC EXPOSURE: Weight loss, anemia, weakness.

EMERGENCY AND FIRST AID PROCEDURES: Inhalation - rest; if swallowed - wash mouth, give emetic, get medical aid; wash from skin & eyes, skin may be washed with 5% acetic acid to aid removal.

SECTION VI REACTIVITY DATA

CONDITIONS CONTRIBUTING TO INSTABILITY

CONDITIONS CONTRIBUTING TO HAZARDOUS POLYMERIZATION

INCOMPATABILITY (MATERIALS TO AVOID): May react vigorously with oxidizing materials.

HAZARDOUS DECOMPOSITION PRODUCTS

SECTION VII SPILL OR LEAK PROCEDURES

STEPS TO BE TAKEN IN CASE MATERIAL IS RELEASED OR SPILLED: Cover with a (9:1) sand to soda ash mixture. Transfer to box for disposal.

WASTE DISPOSAL METHOD

SECTION VIII SPECIAL PROTECTION INFORMATION

VENTILATION REQUIREMENTS LOCAL EXHAUST | PROTECTIVE EQUIPMENT (SPECIFY TYPES) EYE: Safety goggles

MECHANICAL (GENERAL) | GLOVES: Rubber

SPECIAL | RESPIRATOR: With proper filter

OTHER PROTECTIVE EQUIPMENT: Protective clothing with self-contained breathing apparatus.

SECTION IX SPECIAL PRECAUTIONS

PRECAUTIONS TO BE TAKEN IN HANDLING AND STORAGE: Protect container from physical damage, outside or detached storage preferred. Store in cool, dry, well-ventilated area.

OTHER PRECAUTIONS

SIGNATURE | DATE

p-TOLUIDINE HYDROCHLORIDE

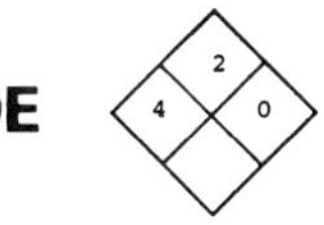

SECTION I SOURCE AND NOMENCLATURE

MANUFACTURER'S NAME | EMERGENCY TELEPHONE NO.

ADDRESS (NUMBER, STREET, CITY, STATE, ZIP CODE)

TRADE NAME AND SYNONYMS: P-Amino toluene hydrochloride | CHEMICAL FAMILY

CHEMICAL NAME AND SYNONYMS: P-Methyl analine hydrochloride | FORMULA: $CH_3C_6H_4NH_2 \cdot HCl$

SECTION II HAZARDOUS INGREDIENTS

BASIC MATERIAL	APPROXIMATE OR MAXIMUM % WT. OR VOL.	ESTABLISHED OSHA STANDARD	LD_{50} mg/kg ORAL	PERCUT.	SPECIES	LC_{50} CONC.
Note: LD50 intraperitoneal - rat 399 mg/kg			1285		rat	

SECTION III PHYSICAL DATA MW - 143.63

BOILING POINT °F.: 257.5°C.

VAPOR PRESSURE mm Hg.

MELTING POINT °F.: 243°C.

VAPOR DENSITY (Air = 1)

SPECIFIC GRAVITY (H_2O = 1)

EVAPORATION RATE (______ = 1)

SOLUBILITY IN WATER Pts/100 pts H_2O: 22.9

VOLATILE % Vol. % Wt.

APPEARANCE AND ODOR: Grey white crystals.

SECTION IV FIRE AND EXPLOSION HAZARD DATA

FLASH POINT: N/A

METHOD USED

FLAMMABLE (EXPLOSIVE) LIMITS: UPPER | LOWER

EXTINGUISHING MEDIA: Carbon dioxide, dry chemical, foam or water spray.

SPECIAL FIRE FIGHTING PROCEDURES: Wear full protective clothing.

UNUSUAL FIRE AND EXPLOSION HAZARDS: Emits toxic fumes when heated.

SECTION V HEALTH HAZARD DATA

TOXIC LEVEL: Not available | CARCINOGENIC: yes

PRINCIPLE ROUTES OF ABSORBTION: Inhalation, ingestion, skin | SKIN AND EYE IRRITATION: Slight

REVELANT SYMPTOMS OF EXPOSURE: Cyanosis, air hunger, nausea, vomiting, low blood pressure, convulsions.

EFFECTS OF CHRONIC EXPOSURE: Weight loss, anemia, weakness.

EMERGENCY AND FIRST AID PROCEDURES: Inhalation - rest; if swallowed - wash mouth, give emetic, get medical aid; wash from skin & eyes, skin may be washed with 5% acetic acid to aid removal.

SECTION VI REACTIVITY DATA

CONDITIONS CONTRIBUTING TO INSTABILITY

CONDITIONS CONTRIBUTING TO HAZARDOUS POLYMERIZATION

INCOMPATABILITY (MATERIALS TO AVOID): May react vigorously with oxidizing materials.

HAZARDOUS DECOMPOSITION PRODUCTS

SECTION VII SPILL OR LEAK PROCEDURES

STEPS TO BE TAKEN IN CASE MATERIAL IS RELEASED OR SPILLED: Cover with a (9:1) sand to soda ash mixture. Transfer to box for disposal.

WASTE DISPOSAL METHOD

SECTION VIII SPECIAL PROTECTION INFORMATION

VENTILATION REQUIREMENTS LOCAL EXHAUST | PROTECTIVE EQUIPMENT (SPECIFY TYPES) EYE: Safety goggles

MECHANICAL (GENERAL) | GLOVES: Rubber

SPECIAL | RESPIRATOR: With proper filter

OTHER PROTECTIVE EQUIPMENT: Protective clothing with self-contained breathing apparatus.

SECTION IX SPECIAL PRECAUTIONS

PRECAUTIONS TO BE TAKEN IN HANDLING AND STORAGE: Protect container from physical damage, outside or detached storage preferred. Store in cool, dry, well-ventilated area.

OTHER PRECAUTIONS

SIGNATURE | DATE

2,4,6-TRIBROMOANILINE

Hazard rating: 1 (top), 3 (left), 0 (right)

SECTION I SOURCE AND NOMENCLATURE

MANUFACTURER'S NAME | EMERGENCY TELEPHONE NO.

ADDRESS (NUMBER, STREET, CITY, STATE, ZIP CODE)

TRADE NAME AND SYNONYMS | CHEMICAL FAMILY

CHEMICAL NAME AND SYNONYMS: 2,4,6-Tribromoaniline | FORMULA: $Br_3C_6H_2NH_2$

SECTION II HAZARDOUS INGREDIENTS

BASIC MATERIAL	APPROXIMATE OR MAXIMUM % WT. OR VOL.	ESTABLISHED OSHA STANDARD	LD_{50} ORAL	LD_{50} PERCUT.	LC_{50} SPECIES	LC_{50} CONC.
Lowest published lethal dose						
intraperitoneal - mouse	500 mg/kg.					

SECTION III PHYSICAL DATA MW - 329.84

BOILING POINT: 300°C | VAPOR PRESSURE (mm Hg.)

MELTING POINT: 122°C. | VAPOR DENSITY (Air = 1)

SPECIFIC GRAVITY (H_2O = 1) | EVAPORATION RATE

SOLUBILITY IN WATER Pts/100 pts H_2O: Insoluble | VOLATILE % Vol. % Wt.

APPEARANCE AND ODOR: Needle-like crystals.

SECTION IV FIRE AND EXPLOSION HAZARD DATA

FLASH POINT: None available | FLAMMABLE (EXPLOSIVE) LIMITS UPPER / LOWER

METHOD USED

EXTINGUISHING MEDIA: Dry chemical or carbon dioxide. Water or foam may cause frothing.

SPECIAL FIRE FIGHTING PROCEDURES

UNUSUAL FIRE AND EXPLOSION HAZARDS

SECTION V HEALTH HAZARD DATA

TOXIC LEVEL: None available | CARCINOGENIC: No

PRINCIPLE ROUTES OF ABSORBTION: Inhalation, ingestion, skin | SKIN AND EYE IRRITATION: Low

REVELANT SYMPTOMS OF EXPOSURE: Headache, nausea, vomiting, jaundice, convulsions.

EFFECTS OF CHRONIC EXPOSURE: Severe kidney damage, liver damage, dermatitis.

EMERGENCY AND FIRST AID PROCEDURES: Inhalation - rest; wash skin & eyes; if swallowed - remove by gastric lavage; give oxygen as needed.

SECTION VI REACTIVITY DATA

CONDITIONS CONTRIBUTING TO INSTABILITY

CONDITIONS CONTRIBUTING TO HAZARDOUS POLYMERIZATION

INCOMPATABILITY (MATERIALS TO AVOID)

HAZARDOUS DECOMPOSITION PRODUCTS

SECTION VII SPILL OR LEAK PROCEDURES

STEPS TO BE TAKEN IN CASE MATERIAL IS RELEASED OR SPILLED: Mix with (9:1) sand to soda ash mixture, sweep into cardboard carton for disposal.

WASTE DISPOSAL METHOD

SECTION VIII SPECIAL PROTECTION INFORMATION

VENTILATION REQUIREMENTS LOCAL EXHAUST | PROTECTIVE EQUIPMENT (SPECIFY TYPES) EYE: Goggles

MECHANICAL (GENERAL) | GLOVES: Rubber

SPECIAL | RESPIRATOR: Self-contained

OTHER PROTECTIVE EQUIPMENT: Protective clothing

SECTION IX SPECIAL PRECAUTIONS

PRECAUTIONS TO BE TAKEN IN HANDLING AND STORAGE: Protect container against physical damage, separate from oxidizing materials.

OTHER PRECAUTIONS

SIGNATURE | DATE

1,3,5-TRIBROMOBENZENE

Hazard rating: 1 (top), 4 (left), 0 (right)

SECTION I SOURCE AND NOMENCLATURE

MANUFACTURER'S NAME | EMERGENCY TELEPHONE NO.

ADDRESS (NUMBER, STREET, CITY, STATE, ZIP CODE)

TRADE NAME AND SYNONYMS | CHEMICAL FAMILY

CHEMICAL NAME AND SYNONYMS: 1,3,5 Tribromobenzene | FORMULA: $C_6H_3Br_3$

SECTION II HAZARDOUS INGREDIENTS

BASIC MATERIAL	APPROXIMATE OR MAXIMUM % WT. OR VOL.	ESTABLISHED OSHA STANDARD	LD_{50} ORAL	LD_{50} PERCUT.	LC_{50} SPECIES	LC_{50} CONC.
			None	Available		

SECTION III PHYSICAL DATA MW - 31.483

BOILING POINT: 271°C at 765 mn | VAPOR PRESSURE (mm Hg.)

MELTING POINT: 120°C. | VAPOR DENSITY (Air = 1)

SPECIFIC GRAVITY (H_2O = 1) | EVAPORATION RATE

SOLUBILITY IN WATER Pts/100 pts H_2O: Insoluble | VOLATILE % Vol. % Wt.

APPEARANCE AND ODOR: Crystalline, aromatic odor.

SECTION IV FIRE AND EXPLOSION HAZARD DATA

FLASH POINT: None available | FLAMMABLE (EXPLOSIVE) LIMITS UPPER / LOWER

METHOD USED

EXTINGUISHING MEDIA: Water may be used to blanket fire, alcohol foam, carbon dioxide.

SPECIAL FIRE FIGHTING PROCEDURES: Self-contained mask.

UNUSUAL FIRE AND EXPLOSION HAZARDS

SECTION V HEALTH HAZARD DATA

TOXIC LEVEL: None available | CARCINOGENIC: No

PRINCIPLE ROUTES OF ABSORBTION: Inhalation, ingestion, skin | SKIN AND EYE IRRITATION: Severe

REVELANT SYMPTOMS OF EXPOSURE: Skin, eye and throat irritation, CNS stimulation, headache, dermatitis.

EFFECTS OF CHRONIC EXPOSURE: Liver and kidney damage.

EMERGENCY AND FIRST AID PROCEDURES: Inhalation - rest; if swallowed - remove by gastric lavage or emesis; wash skin & eyes.

SECTION VI REACTIVITY DATA

CONDITIONS CONTRIBUTING TO INSTABILITY

CONDITIONS CONTRIBUTING TO HAZARDOUS POLYMERIZATION

INCOMPATABILITY (MATERIALS TO AVOID): May react with oxidizing materials.

HAZARDOUS DECOMPOSITION PRODUCTS

SECTION VII SPILL OR LEAK PROCEDURES

STEPS TO BE TAKEN IN CASE MATERIAL IS RELEASED OR SPILLED: Absorb the spill with paper towels; evaporate in hood; then burn towels.

WASTE DISPOSAL METHOD

SECTION VIII SPECIAL PROTECTION INFORMATION

VENTILATION REQUIREMENTS LOCAL EXHAUST | PROTECTIVE EQUIPMENT (SPECIFY TYPES) EYE: Goggles

MECHANICAL (GENERAL) | GLOVES: Rubber

SPECIAL | RESPIRATOR: Self-contained

OTHER PROTECTIVE EQUIPMENT

SECTION IX SPECIAL PRECAUTIONS

PRECAUTIONS TO BE TAKEN IN HANDLING AND STORAGE: Store in cool, dark and well-ventilated area.

OTHER PRECAUTIONS

SIGNATURE | DATE

TRIBROMOETHYLENE

Hazard diamond: 0 (top), 3 (left), 0 (right)

SECTION I SOURCE AND NOMENCLATURE

MANUFACTURER'S NAME | EMERGENCY TELEPHONE NO.

ADDRESS (NUMBER, STREET, CITY, STATE, ZIP CODE)

TRADE NAME AND SYNONYMS | CHEMICAL FAMILY

CHEMICAL NAME AND SYNONYMS: Acetylene Bromide | FORMULA: $CHBr = CBr_2$

SECTION II HAZARDOUS INGREDIENTS

BASIC MATERIAL	APPROXIMATE OR MAXIMUM % WT. OR VOL.	ESTABLISHED OSHA STANDARD	LD_{50} ORAL	LD_{50} PERCUT.	LC_{50} SPECIES	LC_{50} CONC.
			none	available		

SECTION III PHYSICAL DATA MW - 264.78

BOILING POINT °F.: 164°C. | VAPOR PRESSURE mm Hg.

MELTING POINT °F.: - | VAPOR DENSITY (Air = 1)

SPECIFIC GRAVITY (H_2O = 1): 2.71 | EVAPORATION RATE (______ = 1)

SOLUBILITY IN WATER Pts/100 pts H_2O | VOLATILE % Vol. % Wt.

APPEARANCE AND ODOR: liquid

SECTION IV FIRE AND EXPLOSION HAZARD DATA

FLASH POINT: not flammable | FLAMMABLE (EXPLOSIVE) LIMITS: UPPER / LOWER

METHOD USED

EXTINGUISHING MEDIA

SPECIAL FIRE FIGHTING PROCEDURES

UNUSUAL FIRE AND EXPLOSION HAZARDS

SECTION V HEALTH HAZARD DATA

TOXIC LEVEL: none available | CARCINOGENIC: no

PRINCIPLE ROUTES OF ABSORBTION: inhalation, ingestion | SKIN AND EYE IRRITATION: yes

REVELANT SYMPTOMS OF EXPOSURE: inhalation - headache, dizziness, nausea, unconsciousness; ingestion - same as inhalation; eye irritation.

EFFECTS OF CHRONIC EXPOSURE: Anemia, liver damage, respiratory trouble.

EMERGENCY AND FIRST AID PROCEDURES: Inhalation - rest, keep warm; ingestion - wash mouth, give emetic, get medical aid; wash from eyes.

SECTION VI REACTIVITY DATA

CONDITIONS CONTRIBUTING TO INSTABILITY

CONDITIONS CONTRIBUTING TO HAZARDOUS POLYMERIZATION

INCOMPATABILITY (MATERIALS TO AVOID): Reacts explosively with aluminum powder.

HAZARDOUS DECOMPOSITION PRODUCTS

SECTION VII SPILL OR LEAK PROCEDURES

STEPS TO BE TAKEN IN CASE MATERIAL IS RELEASED OR SPILLED: Mop up

WASTE DISPOSAL METHOD: Burial - call safety department

SECTION VIII SPECIAL PROTECTION INFORMATION

VENTILATION REQUIREMENTS LOCAL EXHAUST | PROTECTIVE EQUIPMENT (SPECIFY TYPES) EYE: safety glasses

MECHANICAL (GENERAL) | GLOVES: rubber gloves

SPECIAL | RESPIRATOR: with proper filter

OTHER PROTECTIVE EQUIPMENT

SECTION IX SPECIAL PRECAUTIONS

PRECAUTIONS TO BE TAKEN IN HANDLING AND STORAGE

OTHER PRECAUTIONS

SIGNATURE | DATE

1,2,3-TRIBROMOPROPANE

Hazard diamond: 2 (top), 3 (left), 0 (right)

SECTION I SOURCE AND NOMENCLATURE

MANUFACTURER'S NAME | EMERGENCY TELEPHONE NO.

ADDRESS (NUMBER, STREET, CITY, STATE, ZIP CODE)

TRADE NAME AND SYNONYMS: Glycerol Tribromohydrin | CHEMICAL FAMILY

CHEMICAL NAME AND SYNONYMS: 1,2,3-Tribromopropane | FORMULA: $CH_2BrCHBrCH_2Br$

SECTION II HAZARDOUS INGREDIENTS

BASIC MATERIAL	APPROXIMATE OR MAXIMUM % WT. OR VOL.	ESTABLISHED OSHA STANDARD	LD_{50} ORAL	LD_{50} PERCUT.	LC_{50} SPECIES	LC_{50} CONC.
			None	available		

SECTION III PHYSICAL DATA MW - 280.82

BOILING POINT °F.: 221 | VAPOR PRESSURE mm Hg.

MELTING POINT °F.: 17°C. | VAPOR DENSITY (Air = 1)

SPECIFIC GRAVITY (H_2O = 1): 2.44 | EVAPORATION RATE (______ = 1)

SOLUBILITY IN WATER Pts/100 pts H_2O: insoluble | VOLATILE % Vol. % Wt.

APPEARANCE AND ODOR: Crystaline

SECTION IV FIRE AND EXPLOSION HAZARD DATA

FLASH POINT: None available | FLAMMABLE (EXPLOSIVE) LIMITS: UPPER / LOWER

METHOD USED

EXTINGUISHING MEDIA: Foam, carbon dioxide, dry chemical.

SPECIAL FIRE FIGHTING PROCEDURES: Wear self-contained breathing apparatus.

UNUSUAL FIRE AND EXPLOSION HAZARDS

SECTION V HEALTH HAZARD DATA

TOXIC LEVEL: None available | CARCINOGENIC: no

PRINCIPLE ROUTES OF ABSORBTION: Inhalation, ingestion, skin | SKIN AND EYE IRRITATION: Severe

REVELANT SYMPTOMS OF EXPOSURE: Eye and throat irritation, edema and hemorrhaging in lungs, narcosis.

EFFECTS OF CHRONIC EXPOSURE: Heart and liver damage.

EMERGENCY AND FIRST AID PROCEDURES: If swallowed - give emetic, get medical aid; inhalation - artificial respiration if needed; wash from skin & eyes.

SECTION VI REACTIVITY DATA

CONDITIONS CONTRIBUTING TO INSTABILITY

CONDITIONS CONTRIBUTING TO HAZARDOUS POLYMERIZATION

INCOMPATABILITY (MATERIALS TO AVOID)

HAZARDOUS DECOMPOSITION PRODUCTS

SECTION VII SPILL OR LEAK PROCEDURES

STEPS TO BE TAKEN IN CASE MATERIAL IS RELEASED OR SPILLED

WASTE DISPOSAL METHOD

SECTION VIII SPECIAL PROTECTION INFORMATION

VENTILATION REQUIREMENTS LOCAL EXHAUST | PROTECTIVE EQUIPMENT (SPECIFY TYPES) EYE: Goggles

MECHANICAL (GENERAL) | GLOVES: Rubber

SPECIAL | RESPIRATOR: With proper filter

OTHER PROTECTIVE EQUIPMENT

SECTION IX SPECIAL PRECAUTIONS

PRECAUTIONS TO BE TAKEN IN HANDLING AND STORAGE

OTHER PRECAUTIONS

SIGNATURE | DATE

TRICHLOROACETIC ACID

Hazard rating: 2 (top), 3 (left), 1 (right)

SECTION I SOURCE AND NOMENCLATURE

MANUFACTURER'S NAME | EMERGENCY TELEPHONE NO.

ADDRESS (NUMBER, STREET, CITY, STATE, ZIP CODE)

TRADE NAME AND SYNONYMS | CHEMICAL FAMILY

CHEMICAL NAME AND SYNONYMS: Trichloroacetic

FORMULA: CCl_3COOH

SECTION II HAZARDOUS INGREDIENTS

BASIC MATERIAL	APPROXIMATE OR MAXIMUM % WT. OR VOL.	ESTABLISHED OSHA STANDARD	LD_{50} mg/kg ORAL	PERCUT.	SPECIES	LC_{50} CONC.
			3320		Rat	
			5640		Mouse	

SECTION III PHYSICAL DATA Mol. Wt. 163.4

BOILING POINT (°F.): 197.5°C

VAPOR PRESSURE (mm Hg.): 1 at 51°C

MELTING POINT (°F.): 57.5°C

VAPOR DENSITY (Air = 1): 5.65

SPECIFIC GRAVITY (H_2O = 1): 1.6298

EVAPORATION RATE (= 1)

SOLUBILITY IN WATER Pts/100 pts H_2O: Very soluble

VOLATILE % Vol. % Wt.

APPEARANCE AND ODOR: Colorless, rhombic deliquescent crystals

SECTION IV FIRE AND EXPLOSION HAZARD DATA

FLASH POINT: Non-Combustible

FLAMMABLE (EXPLOSIVE) LIMITS: UPPER / LOWER

METHOD USED

EXTINGUISHING MEDIA: Use water spray, dry chemical, CO_2 or alcohol foam.

SPECIAL FIRE FIGHTING PROCEDURES: Wear goggles and self-contained breathing apparatus.

UNUSUAL FIRE AND EXPLOSION HAZARDS

SECTION V HEALTH HAZARD DATA

TOXIC LEVEL: 1 mg/C.M.

CARCINOGENIC: No

PRINCIPLE ROUTES OF ABSORBTION: Inhalation of vapor, ingestion

SKIN AND EYE IRRITATION: Severe damage

REVELANT SYMPTOMS OF EXPOSURE: Coughing, choking, headache, dizziness, pulmonary edema, very severe damage to tissue.

EFFECTS OF CHRONIC EXPOSURE: Erosion of teeth, possible jaw necrosis, bronchial irritation with chronic cough, gastrointestinal disturbances are possible.

EMERGENCY AND FIRST AID PROCEDURES: Skin - drench with water, remove & wash contaminated clothing; inhalation - respiratory resuscitation, treat for shock, and treat pulmonary edema, receive further medical attention.

SECTION VI REACTIVITY DATA

CONDITIONS CONTRIBUTING TO INSTABILITY

CONDITIONS CONTRIBUTING TO HAZARDOUS POLYMERIZATION

INCOMPATABILITY (MATERIALS TO AVOID)

HAZARDOUS DECOMPOSITION PRODUCTS

SECTION VII SPILL OR LEAK PROCEDURES

STEPS TO BE TAKEN IN CASE MATERIAL IS RELEASED OR SPILLED: Put sodium bicarbonate on spill. Transfer to container of water for 24 hrs. before draining.

WASTE DISPOSAL METHOD

SECTION VIII SPECIAL PROTECTION INFORMATION

VENTILATION REQUIREMENTS LOCAL EXHAUST

PROTECTIVE EQUIPMENT (SPECIFY TYPES) EYE: Goggles

MECHANICAL (GENERAL)

GLOVES: Rubber

SPECIAL: Water eye bubbler, fountain & showers available.

RESPIRATOR: with appropriate filter

OTHER PROTECTIVE EQUIPMENT: Apron

SECTION IX SPECIAL PRECAUTIONS

PRECAUTIONS TO BE TAKEN IN HANDLING AND STORAGE: Keep in tightly closed bottles. Store in cool, dry place.

OTHER PRECAUTIONS

SIGNATURE | DATE

2,4,6-TRICHLOROANILINE

Hazard rating: 1 (top), 3 (left), 0 (right)

SECTION I SOURCE AND NOMENCLATURE

MANUFACTURER'S NAME | EMERGENCY TELEPHONE NO.

ADDRESS (NUMBER, STREET, CITY, STATE, ZIP CODE)

TRADE NAME AND SYNONYMS | CHEMICAL FAMILY

CHEMICAL NAME AND SYNONYMS: 2,4,6 - Trichloroaniline

FORMULA: $Cl_3C_6H_2NH_2$

SECTION II HAZARDOUS INGREDIENTS

BASIC MATERIAL	APPROXIMATE OR MAXIMUM % WT. OR VOL.	ESTABLISHED OSHA STANDARD	LD_{50} ORAL	PERCUT.	SPECIES	LC_{50} CONC.
		None available				

SECTION III PHYSICAL DATA MW - 196.47

BOILING POINT (°F.): 262°C at 746 mn

VAPOR PRESSURE (mm Hg.)

MELTING POINT (°F.): 78°C.

VAPOR DENSITY (Air = 1)

SPECIFIC GRAVITY (H_2O = 1)

EVAPORATION RATE (= 1)

SOLUBILITY IN WATER Pts/100 pts H_2O: Insoluble

VOLATILE % Vol. % Wt.

APPEARANCE AND ODOR: Needle-like crystals

SECTION IV FIRE AND EXPLOSION HAZARD DATA

FLASH POINT: None available

FLAMMABLE (EXPLOSIVE) LIMITS: UPPER / LOWER

METHOD USED

EXTINGUISHING MEDIA: Dry chemical or carbon dioxide. Water or foam may cause frothing.

SPECIAL FIRE FIGHTING PROCEDURES

UNUSUAL FIRE AND EXPLOSION HAZARDS

SECTION V HEALTH HAZARD DATA

TOXIC LEVEL: None available

CARCINOGENIC: No

PRINCIPLE ROUTES OF ABSORBTION: Inhalation, ingestion, skin

SKIN AND EYE IRRITATION: Low

REVELANT SYMPTOMS OF EXPOSURE: Headache, nausea, vomiting, jaundice, convulsions.

EFFECTS OF CHRONIC EXPOSURE: Severe kidney damage, liver damage, dermatitis.

EMERGENCY AND FIRST AID PROCEDURES: Inhalation - rest; wash skin & eyes; if swallowed - remove by gastric lavage; give oxygen as needed.

SECTION VI REACTIVITY DATA

CONDITIONS CONTRIBUTING TO INSTABILITY

CONDITIONS CONTRIBUTING TO HAZARDOUS POLYMERIZATION

INCOMPATABILITY (MATERIALS TO AVOID)

HAZARDOUS DECOMPOSITION PRODUCTS

SECTION VII SPILL OR LEAK PROCEDURES

STEPS TO BE TAKEN IN CASE MATERIAL IS RELEASED OR SPILLED: Mix with (9:1) sand to soda ash mixture, sweep into cardboard carton for disposal.

WASTE DISPOSAL METHOD

SECTION VIII SPECIAL PROTECTION INFORMATION

VENTILATION REQUIREMENTS LOCAL EXHAUST

PROTECTIVE EQUIPMENT (SPECIFY TYPES) EYE: Goggles

MECHANICAL (GENERAL)

GLOVES: Rubber

SPECIAL

RESPIRATOR: Self-contained

OTHER PROTECTIVE EQUIPMENT: Protective clothing

SECTION IX SPECIAL PRECAUTIONS

PRECAUTIONS TO BE TAKEN IN HANDLING AND STORAGE: Protect container against physical damage. Separate from oxidizing materials.

OTHER PRECAUTIONS

SIGNATURE | DATE

1,2,4-TRICHLOROBENZENE

NFPA hazard diamond: 1 (top), 4 (left), 0 (right)

SECTION I SOURCE AND NOMENCLATURE

MANUFACTURER'S NAME | EMERGENCY TELEPHONE NO.

ADDRESS (NUMBER, STREET, CITY, STATE, ZIP CODE)

TRADE NAME AND SYNONYMS | CHEMICAL FAMILY

CHEMICAL NAME AND SYNONYMS: 1,2,4 Trichlorobenzene | FORMULA: $C_6H_3Cl_3$

SECTION II HAZARDOUS INGREDIENTS

BASIC MATERIAL	APPROXIMATE OR MAXIMUM % WT. OR VOL.	ESTABLISHED OSHA STANDARD	LD_{50} mg/kg ORAL	PERCUT.	LC_{50} SPECIES	CONC.
			756		Rat	
			766		Mouse	

SECTION III PHYSICAL DATA MW - 181.44

BOILING POINT °F.: 213°C. | VAPOR PRESSURE mm Hg.: at 38°C. 1

MELTING POINT °F.: 17°C. | VAPOR DENSITY (Air = 1): 6.26

SPECIFIC GRAVITY (H_2O = 1): 1.46 | EVAPORATION RATE (____ = 1)

SOLUBILITY IN WATER Pts/100 pts H_2O: Insoluble | VOLATILE % Vol. % Wt.

APPEARANCE AND ODOR: Colorless liquid, aromatic odor.

SECTION IV FIRE AND EXPLOSION HAZARD DATA

FLASH POINT: 210 | FLAMMABLE (EXPLOSIVE) LIMITS UPPER

METHOD USED: OC | LOWER

EXTINGUISHING MEDIA: Water may be used to blanket fire.

SPECIAL FIRE FIGHTING PROCEDURES: Self-contained mask.

UNUSUAL FIRE AND EXPLOSION HAZARDS

SECTION V HEALTH HAZARD DATA

TOXIC LEVEL: TLV-5 PPM, 40 mg/C.M. ceiling value | CARCINOGENIC: No

PRINCIPLE ROUTES OF ABSORBTION: Inhalation, ingestion, skin | SKIN AND EYE IRRITATION: Severe

REVELANT SYMPTOMS OF EXPOSURE: Skin irritation; eye and throat irritation, CNS stimulation, headache, dermatitis.

EFFECTS OF CHRONIC EXPOSURE: Liver damage, kidney damage.

EMERGENCY AND FIRST AID PROCEDURES: Inhalation - rest; if swallowed - remove by gastric lavage or emesis; wash skin & eyes.

SECTION VI REACTIVITY DATA

CONDITIONS CONTRIBUTING TO INSTABILITY

CONDITIONS CONTRIBUTING TO HAZARDOUS POLYMERIZATION

INCOMPATABILITY (MATERIALS TO AVOID): Reacts vigorously with oxidizing materials.

HAZARDOUS DECOMPOSITION PRODUCTS

SECTION VII SPILL OR LEAK PROCEDURES

STEPS TO BE TAKEN IN CASE MATERIAL IS RELEASED OR SPILLED: Absorb the spill with paper towels; evaporate in hood; then burn towels

WASTE DISPOSAL METHOD

SECTION VIII SPECIAL PROTECTION INFORMATION

VENTILATION REQUIREMENTS LOCAL EXHAUST | PROTECTIVE EQUIPMENT (SPECIFY TYPES) EYE: Goggles

MECHANICAL (GENERAL) | GLOVES: Rubber

SPECIAL | RESPIRATOR: Self-contained

OTHER PROTECTIVE EQUIPMENT

SECTION IX SPECIAL PRECAUTIONS

PRECAUTIONS TO BE TAKEN IN HANDLING AND STORAGE: Store in well-ventilated, dark, cool area.

OTHER PRECAUTIONS

SIGNATURE | DATE

1,1,2-TRICHLOROETHANE

NFPA hazard diamond: 0 (top), 4 (left), 0 (right)

SECTION I SOURCE AND NOMENCLATURE

MANUFACTURER'S NAME | EMERGENCY TELEPHONE NO.

ADDRESS (NUMBER, STREET, CITY, STATE, ZIP CODE)

TRADE NAME AND SYNONYMS | CHEMICAL FAMILY

CHEMICAL NAME AND SYNONYMS: Vinyl Trichloride | FORMULA: $CH_2ClCHCl_2$

SECTION II HAZARDOUS INGREDIENTS

BASIC MATERIAL	APPROXIMATE OR MAXIMUM % WT. OR VOL.	ESTABLISHED OSHA STANDARD	LD_{50} mg/kg ORAL	PERCUT.	LC_{50} SPECIES	CONC.
			1140		Rat	
				3730	Rabbit	

Oral - human LDLO 50 mg/kg

Lowest published lethal concentration in air

Inhalation - rat 500 PPM/8 hours.

SECTION III PHYSICAL DATA MW - 133.4

BOILING POINT °F.: 114°C. | VAPOR PRESSURE mm Hg.: at 35.2°C. 40

MELTING POINT °F.: -35°C. | VAPOR DENSITY (Air = 1): 4.6

SPECIFIC GRAVITY (H_2O = 1): 1.4416 | EVAPORATION RATE (____ = 1)

SOLUBILITY IN WATER Pts/100 pts H_2O: 0.44 | VOLATILE % Vol. % Wt.

APPEARANCE AND ODOR: Liquid, pleasant odor.

SECTION IV FIRE AND EXPLOSION HAZARD DATA

FLASH POINT: None available | FLAMMABLE (EXPLOSIVE) LIMITS UPPER

METHOD USED | LOWER

EXTINGUISHING MEDIA: Information not available.

SPECIAL FIRE FIGHTING PROCEDURES

UNUSUAL FIRE AND EXPLOSION HAZARDS

SECTION V HEALTH HAZARD DATA

TOXIC LEVEL: TLV - 10 PPM, 45 mg/C.M. | CARCINOGENIC: no

PRINCIPLE ROUTES OF ABSORBTION: Inhalation or skin absorption. | SKIN AND EYE IRRITATION: Moderate

REVELANT SYMPTOMS OF EXPOSURE: Irritation of eyes and nose, headache, nausea, cyanosis and C.N.S. depression.

EFFECTS OF CHRONIC EXPOSURE: Headache, tremor, dizziness, peripheral parethesia, hypesthesia, or anesthesia.

EMERGENCY AND FIRST AID PROCEDURES: Artificial respiration, remove contaminated clothing. (Do not give stimulants), 5% glucose I.V. (to maintain blood pressure)

SECTION VI REACTIVITY DATA

CONDITIONS CONTRIBUTING TO INSTABILITY

CONDITIONS CONTRIBUTING TO HAZARDOUS POLYMERIZATION

INCOMPATABILITY (MATERIALS TO AVOID)

HAZARDOUS DECOMPOSITION PRODUCTS

SECTION VII SPILL OR LEAK PROCEDURES

STEPS TO BE TAKEN IN CASE MATERIAL IS RELEASED OR SPILLED

WASTE DISPOSAL METHOD

SECTION VIII SPECIAL PROTECTION INFORMATION

VENTILATION REQUIREMENTS LOCAL EXHAUST | PROTECTIVE EQUIPMENT (SPECIFY TYPES) EYE: Goggles

MECHANICAL (GENERAL) | GLOVES: Rubber

SPECIAL | RESPIRATOR: If MAC < $2X10^4$ PPM

OTHER PROTECTIVE EQUIPMENT: Apron | > 20000 PPM self-contained

SECTION IX SPECIAL PRECAUTIONS

PRECAUTIONS TO BE TAKEN IN HANDLING AND STORAGE

OTHER PRECAUTIONS

SIGNATURE | DATE

2,4,6-TRICHLOROPHENOL

Hazard diamond: 1 (top), 4 (left), 0 (right)

SECTION I SOURCE AND NOMENCLATURE

MANUFACTURER'S NAME | EMERGENCY TELEPHONE NO.

ADDRESS (NUMBER, STREET, CITY, STATE, ZIP CODE)

TRADE NAME AND SYNONYMS | CHEMICAL FAMILY

CHEMICAL NAME AND SYNONYMS: 2,4,6 Trichlorophenol | FORMULA: $Cl_3C_6H_2OH$

SECTION II HAZARDOUS INGREDIENTS

BASIC MATERIAL	APPROXIMATE OR MAXIMUM % WT. OR VOL.	ESTABLISHED OSHA STANDARD	LD_{50} mg/kg			LC_{50}
			ORAL	PERCUT.	SPECIES	CONC.
			820		Rat	

SECTION III PHYSICAL DATA MW - 197.44

BOILING POINT °F.: 246°C.

VAPOR PRESSURE mm Hg.: at 77°C. 1

MELTING POINT °F.: 69°C.

VAPOR DENSITY (Air = 1)

SPECIFIC GRAVITY (H_2O = 1): 1.49

EVAPORATION RATE (______ = 1)

SOLUBILITY IN WATER Pts/100 pts H_2O: 0.09

VOLATILE % Vol. % Wt.

APPEARANCE AND ODOR: Needle-like crystal, colorless to yellow, phenol like odor.

SECTION IV FIRE AND EXPLOSION HAZARD DATA

FLASH POINT: None available

METHOD USED

FLAMMABLE (EXPLOSIVE) LIMITS: UPPER / LOWER

EXTINGUISHING MEDIA: Carbon dioxide, alcohol foam.

SPECIAL FIRE FIGHTING PROCEDURES: Self-contained breathing apparatus.

UNUSUAL FIRE AND EXPLOSION HAZARDS

SECTION V HEALTH HAZARD DATA

TOXIC LEVEL: None available

CARCINOGENIC: Yes

PRINCIPLE ROUTES OF ABSORBTION: Inhalation, ingestion, skin

SKIN AND EYE IRRITATION: Severe

REVELANT SYMPTOMS OF EXPOSURE: Irritation or severe burns on skin & eyes; ingestion - irritation of digestive system.

EFFECTS OF CHRONIC EXPOSURE: --

EMERGENCY AND FIRST AID PROCEDURES: Inhalation - rest; if swallowed - give epsom salts, get medical aid; wash eyes and skin.

SECTION VI REACTIVITY DATA

CONDITIONS CONTRIBUTING TO INSTABILITY

CONDITIONS CONTRIBUTING TO HAZARDOUS POLYMERIZATION

INCOMPATABILITY (MATERIALS TO AVOID)

HAZARDOUS DECOMPOSITION PRODUCTS

SECTION VII SPILL OR LEAK PROCEDURES

STEPS TO BE TAKEN IN CASE MATERIAL IS RELEASED OR SPILLED: Neutralize with sodium bicarbonate or mixture of sand and soda ash. Put into carton for disposal.

WASTE DISPOSAL METHOD

SECTION VIII SPECIAL PROTECTION INFORMATION

VENTILATION REQUIREMENTS LOCAL EXHAUST

PROTECTIVE EQUIPMENT (SPECIFY TYPES) EYE: Goggles

MECHANICAL (GENERAL)

GLOVES: Rubber

SPECIAL

RESPIRATOR: Self-contained

OTHER PROTECTIVE EQUIPMENT: Protective clothing.

SECTION IX SPECIAL PRECAUTIONS

PRECAUTIONS TO BE TAKEN IN HANDLING AND STORAGE

OTHER PRECAUTIONS

SIGNATURE | DATE

1,2,3-TRICHLOROPROPANE

Hazard diamond: 2 (top), 3 (left), 0 (right)

SECTION I SOURCE AND NOMENCLATURE

MANUFACTURER'S NAME | EMERGENCY TELEPHONE NO.

ADDRESS (NUMBER, STREET, CITY, STATE, ZIP CODE)

TRADE NAME AND SYNONYMS: Glycerol Trichlorohydrin | CHEMICAL FAMILY

CHEMICAL NAME AND SYNONYMS: 1,2,3 Trichloropropane | FORMULA: $ClCH_2CHClCH_2Cl$

SECTION II HAZARDOUS INGREDIENTS

BASIC MATERIAL	APPROXIMATE OR MAXIMUM % WT. OR VOL.	ESTABLISHED OSHA STANDARD	LD_{50} mg/kg			LC_{50}
			ORAL	PERCUT.	SPECIES	CONC.
-			320		rat	
Lowest published lethal concentration in air				1770	rabbit	
inhalation - rat 1000 PPM /4 hr.						

SECTION III PHYSICAL DATA MW - 147.44

BOILING POINT °F.: 157°C.

VAPOR PRESSURE mm Hg.: at 46°C. 10

MELTING POINT °F.: -15°C.

VAPOR DENSITY (Air = 1): 5.1

SPECIFIC GRAVITY (H_2O = 1): 1.39

EVAPORATION RATE (______ = 1)

SOLUBILITY IN WATER Pts/100 pts H_2O: <0.1

VOLATILE % Vol. % Wt.

APPEARANCE AND ODOR: colorless liquid

SECTION IV FIRE AND EXPLOSION HAZARD DATA

FLASH POINT: 180°C.

METHOD USED: O.C.

FLAMMABLE (EXPLOSIVE) LIMITS: UPPER 12.6% / LOWER 3.2%

EXTINGUISHING MEDIA: Water may be used to blanket the fire, foam, carbon dioxide, dry chemical.

SPECIAL FIRE FIGHTING PROCEDURES: Wear self-contained breathing apparatus.

UNUSUAL FIRE AND EXPLOSION HAZARDS

SECTION V HEALTH HAZARD DATA

TOXIC LEVEL: 50 PPM, 300 mg/C.M.

CARCINOGENIC: no

PRINCIPLE ROUTES OF ABSORBTION: Inhalation, ingestion, skin

SKIN AND EYE IRRITATION: Severe

REVELANT SYMPTOMS OF EXPOSURE: Eye and throat irritation, edema and hemorrhaging in lungs, narcosis.

EFFECTS OF CHRONIC EXPOSURE: Heart and liver damage.

EMERGENCY AND FIRST AID PROCEDURES: If swallowed - give emetic, get medical aid; inhalation - rest, artificial respiration if needed; wash from skin & eyes.

SECTION VI REACTIVITY DATA

CONDITIONS CONTRIBUTING TO INSTABILITY

CONDITIONS CONTRIBUTING TO HAZARDOUS POLYMERIZATION

INCOMPATABILITY (MATERIALS TO AVOID)

HAZARDOUS DECOMPOSITION PRODUCTS

SECTION VII SPILL OR LEAK PROCEDURES

STEPS TO BE TAKEN IN CASE MATERIAL IS RELEASED OR SPILLED

WASTE DISPOSAL METHOD

SECTION VIII SPECIAL PROTECTION INFORMATION

VENTILATION REQUIREMENTS LOCAL EXHAUST

PROTECTIVE EQUIPMENT (SPECIFY TYPES) EYE: Goggles

MECHANICAL (GENERAL)

GLOVES: Rubber

SPECIAL

RESPIRATOR: With proper filter

OTHER PROTECTIVE EQUIPMENT

SECTION IX SPECIAL PRECAUTIONS

PRECAUTIONS TO BE TAKEN IN HANDLING AND STORAGE

OTHER PRECAUTIONS

SIGNATURE | DATE

1,2,4-TRICHLOROBENZENE

Hazard diamond: 1 (top), 4 (left), 0 (right)

SECTION I SOURCE AND NOMENCLATURE

MANUFACTURER'S NAME | EMERGENCY TELEPHONE NO.

ADDRESS (NUMBER, STREET, CITY, STATE, ZIP CODE)

TRADE NAME AND SYNONYMS | CHEMICAL FAMILY

CHEMICAL NAME AND SYNONYMS: 1,2,4 Trichlorobenzene | FORMULA: $C_6H_3Cl_3$

SECTION II HAZARDOUS INGREDIENTS

BASIC MATERIAL	APPROXIMATE OR MAXIMUM % WT. OR VOL.	ESTABLISHED OSHA STANDARD	LD_{50} mg/kg ORAL	PERCUT.	SPECIES	LC_{50} CONC.
			756		Rat	
			766		Mouse	

SECTION III PHYSICAL DATA MW - 181.44

- BOILING POINT °F.: 213°C.
- VAPOR PRESSURE mm Hg.: at 38°C. 1
- MELTING POINT °F.: 17°C.
- VAPOR DENSITY (Air = 1): 6.26
- SPECIFIC GRAVITY (H_2O = 1): 1.46
- EVAPORATION RATE (____ = 1)
- SOLUBILITY IN WATER Pts/100 pts H_2O: Insoluble
- VOLATILE % Vol. % Wt.
- APPEARANCE AND ODOR: Colorless liquid, aromatic odor.

SECTION IV FIRE AND EXPLOSION HAZARD DATA

- FLASH POINT: 210
- METHOD USED: OC
- FLAMMABLE (EXPLOSIVE) LIMITS: UPPER / LOWER
- EXTINGUISHING MEDIA: Water may be used to blanket fire.
- SPECIAL FIRE FIGHTING PROCEDURES: Self-contained mask.
- UNUSUAL FIRE AND EXPLOSION HAZARDS

SECTION V HEALTH HAZARD DATA

- TOXIC LEVEL: TLV-5 PPM, 40 mg/C.M. ceiling value
- CARCINOGENIC: No
- PRINCIPLE ROUTES OF ABSORBTION: Inhalation, ingestion, skin
- SKIN AND EYE IRRITATION: Severe
- REVELANT SYMPTOMS OF EXPOSURE: Skin irritation; eye and throat irritation, CNS stimulation, headache, dermatitis.
- EFFECTS OF CHRONIC EXPOSURE: Liver damage, kidney damage.
- EMERGENCY AND FIRST AID PROCEDURES: Inhalation - rest; if swallowed - remove by gastric lavage or emesis; wash skin & eyes.

SECTION VI REACTIVITY DATA

- CONDITIONS CONTRIBUTING TO INSTABILITY
- CONDITIONS CONTRIBUTING TO HAZARDOUS POLYMERIZATION
- INCOMPATABILITY (MATERIALS TO AVOID): Reacts vigorously with oxidizing materials.
- HAZARDOUS DECOMPOSITION PRODUCTS

SECTION VII SPILL OR LEAK PROCEDURES

- STEPS TO BE TAKEN IN CASE MATERIAL IS RELEASED OR SPILLED: Absorb the spill with paper towels; evaporate in hood; then burn towels
- WASTE DISPOSAL METHOD

SECTION VIII SPECIAL PROTECTION INFORMATION

- VENTILATION REQUIREMENTS LOCAL EXHAUST
- PROTECTIVE EQUIPMENT (SPECIFY TYPES) EYE: Goggles
- MECHANICAL (GENERAL)
- GLOVES: Rubber
- SPECIAL
- RESPIRATOR: Self-contained
- OTHER PROTECTIVE EQUIPMENT

SECTION IX SPECIAL PRECAUTIONS

- PRECAUTIONS TO BE TAKEN IN HANDLING AND STORAGE: Store in well-ventilated, dark, cool area.
- OTHER PRECAUTIONS

SIGNATURE | DATE

1,1,2-TRICHLOROETHANE

Hazard diamond: 0 (top), 4 (left), 0 (right)

SECTION I SOURCE AND NOMENCLATURE

MANUFACTURER'S NAME | EMERGENCY TELEPHONE NO.

ADDRESS (NUMBER, STREET, CITY, STATE, ZIP CODE)

TRADE NAME AND SYNONYMS | CHEMICAL FAMILY

CHEMICAL NAME AND SYNONYMS: Vinyl Trichloride | FORMULA: $CH_2ClCHCl_2$

SECTION II HAZARDOUS INGREDIENTS

BASIC MATERIAL	APPROXIMATE OR MAXIMUM % WT. OR VOL.	ESTABLISHED OSHA STANDARD	LD_{50} mg/kg ORAL	PERCUT.	SPECIES	LC_{50} CONC.
			1140		Rat	
				3730	Rabbit	

Oral - human LDLO 50 mg/kg

Lowest published lethal concentration in air

Inhalation - rat 500 PPM/8 hours.

SECTION III PHYSICAL DATA MW = 133.4

- BOILING POINT °F.: 114°C.
- VAPOR PRESSURE mm Hg.: at 35.2°C. 40
- MELTING POINT °F.: -35°C.
- VAPOR DENSITY (Air = 1): 4.6
- SPECIFIC GRAVITY (H_2O = 1): 1.4416
- EVAPORATION RATE (____ = 1)
- SOLUBILITY IN WATER Pts/100 pts H_2O: 0.44
- VOLATILE % Vol. % Wt.
- APPEARANCE AND ODOR: Liquid, pleasant odor.

SECTION IV FIRE AND EXPLOSION HAZARD DATA

- FLASH POINT: None available
- METHOD USED
- FLAMMABLE (EXPLOSIVE) LIMITS: UPPER / LOWER
- EXTINGUISHING MEDIA: Information not available.
- SPECIAL FIRE FIGHTING PROCEDURES
- UNUSUAL FIRE AND EXPLOSION HAZARDS

SECTION V HEALTH HAZARD DATA

- TOXIC LEVEL: TLV - 10 PPM, 45 mg/C.M.
- CARCINOGENIC: no
- PRINCIPLE ROUTES OF ABSORBTION: Inhalation or skin absorption.
- SKIN AND EYE IRRITATION: Moderate
- REVELANT SYMPTOMS OF EXPOSURE: Irritation of eyes and nose, headache, nausea, cyanosis and C.N.S. depression.
- EFFECTS OF CHRONIC EXPOSURE: Headache, tremor, dizziness, peripheral parethesia, hypesthesia, or anesthesia.
- EMERGENCY AND FIRST AID PROCEDURES: Artificial respiration, remove contaminated clothing. (Do not give stimulants), 5% glucose I.V. (to maintain blood pressure)

SECTION VI REACTIVITY DATA

- CONDITIONS CONTRIBUTING TO INSTABILITY
- CONDITIONS CONTRIBUTING TO HAZARDOUS POLYMERIZATION
- INCOMPATABILITY (MATERIALS TO AVOID)
- HAZARDOUS DECOMPOSITION PRODUCTS

SECTION VII SPILL OR LEAK PROCEDURES

- STEPS TO BE TAKEN IN CASE MATERIAL IS RELEASED OR SPILLED
- WASTE DISPOSAL METHOD

SECTION VIII SPECIAL PROTECTION INFORMATION

- VENTILATION REQUIREMENTS LOCAL EXHAUST
- PROTECTIVE EQUIPMENT (SPECIFY TYPES) EYE: Goggles
- MECHANICAL (GENERAL)
- GLOVES: Rubber
- SPECIAL
- RESPIRATOR: If MAC < $2X10^4$ PPM; > 20000 PPM self-contained
- OTHER PROTECTIVE EQUIPMENT: Apron

SECTION IX SPECIAL PRECAUTIONS

- PRECAUTIONS TO BE TAKEN IN HANDLING AND STORAGE
- OTHER PRECAUTIONS

SIGNATURE | DATE

2,4,6-TRICHLOROPHENOL

NFPA: 1 (top), 4 (left), 0 (right)

SECTION I SOURCE AND NOMENCLATURE

MANUFACTURER'S NAME | EMERGENCY TELEPHONE NO.

ADDRESS (NUMBER, STREET, CITY, STATE, ZIP CODE)

TRADE NAME AND SYNONYMS | CHEMICAL FAMILY

CHEMICAL NAME AND SYNONYMS: 2,4,6 Trichlorophenol

FORMULA: $Cl_3C_6H_2OH$

SECTION II HAZARDOUS INGREDIENTS

BASIC MATERIAL	APPROXIMATE OR MAXIMUM % WT. OR VOL.	ESTABLISHED OSHA STANDARD	LD_{50} mg/kg ORAL	PERCUT.	LC_{50} SPECIES	CONC.
			820		Rat	

SECTION III PHYSICAL DATA MW - 197.44

BOILING POINT °F.: 246°C.

VAPOR PRESSURE mm Hg.: at 77°C. 1

MELTING POINT °F.: 69°C.

VAPOR DENSITY (Air = 1)

SPECIFIC GRAVITY (H_2O = 1): 1.49

EVAPORATION RATE (________ = 1)

SOLUBILITY IN WATER Pts/100 pts H_2O: 0.09

VOLATILE % Vol. % Wt.

APPEARANCE AND ODOR: Needle-like crystal, colorless to yellow, phenol like odor.

SECTION IV FIRE AND EXPLOSION HAZARD DATA

FLASH POINT: None available

FLAMMABLE (EXPLOSIVE) LIMITS: UPPER / LOWER

METHOD USED

EXTINGUISHING MEDIA: Carbon dioxide, alcohol foam.

SPECIAL FIRE FIGHTING PROCEDURES: Self-contained breathing apparatus.

UNUSUAL FIRE AND EXPLOSION HAZARDS

SECTION V HEALTH HAZARD DATA

TOXIC LEVEL: None available

CARCINOGENIC: Yes

PRINCIPLE ROUTES OF ABSORBTION: Inhalation, ingestion, skin

SKIN AND EYE IRRITATION: Severe

REVELANT SYMPTOMS OF EXPOSURE: Irritation or severe burns on skin & eyes; ingestion - irritation of digestive system.

EFFECTS OF CHRONIC EXPOSURE: --

EMERGENCY AND FIRST AID PROCEDURES: Inhalation - rest; if swallowed - give epsom salts, get medical aid; wash eyes and skin.

SECTION VI REACTIVITY DATA

CONDITIONS CONTRIBUTING TO INSTABILITY

CONDITIONS CONTRIBUTING TO HAZARDOUS POLYMERIZATION

INCOMPATABILITY (MATERIALS TO AVOID)

HAZARDOUS DECOMPOSITION PRODUCTS

SECTION VII SPILL OR LEAK PROCEDURES

STEPS TO BE TAKEN IN CASE MATERIAL IS RELEASED OR SPILLED: Neutralize with sodium bicarbonate or mixture of sand and soda ash. Put into carton for disposal.

WASTE DISPOSAL METHOD

SECTION VIII SPECIAL PROTECTION INFORMATION

VENTILATION REQUIREMENTS LOCAL EXHAUST

PROTECTIVE EQUIPMENT (SPECIFY TYPES) EYE: Goggles

MECHANICAL (GENERAL)

GLOVES: Rubber

SPECIAL

RESPIRATOR: Self-contained

OTHER PROTECTIVE EQUIPMENT: Protective clothing.

SECTION IX SPECIAL PRECAUTIONS

PRECAUTIONS TO BE TAKEN IN HANDLING AND STORAGE

OTHER PRECAUTIONS

SIGNATURE | DATE

1,2,3-TRICHLOROPROPANE

NFPA: 2 (top), 3 (left), 0 (right)

SECTION I SOURCE AND NOMENCLATURE

MANUFACTURER'S NAME | EMERGENCY TELEPHONE NO.

ADDRESS (NUMBER, STREET, CITY, STATE, ZIP CODE)

TRADE NAME AND SYNONYMS: Glycerol Trichlorohydrin

CHEMICAL FAMILY

CHEMICAL NAME AND SYNONYMS: 1,2,3 Trichloropropane

FORMULA: $ClCH_2CHClClCH_2$

SECTION II HAZARDOUS INGREDIENTS

BASIC MATERIAL	APPROXIMATE OR MAXIMUM % WT. OR VOL.	ESTABLISHED OSHA STANDARD	LD_{50} mg/kg ORAL	PERCUT.	LC_{50} SPECIES	CONC.
–			320		rat	
Lowest published lethal concentration in air inhalation - rat 1000 PPM /4 hr.				1770	rabbit	

SECTION III PHYSICAL DATA MW - 147.44

BOILING POINT °F.: 157°C.

VAPOR PRESSURE mm Hg.: at 46°C. 10

MELTING POINT °F.: -15°C.

VAPOR DENSITY (Air = 1): 5.1

SPECIFIC GRAVITY (H_2O = 1): 1.39

EVAPORATION RATE (________ = 1)

SOLUBILITY IN WATER Pts/100 pts H_2O: <0.1

VOLATILE % Vol. % Wt.

APPEARANCE AND ODOR: colorless liquid

SECTION IV FIRE AND EXPLOSION HAZARD DATA

FLASH POINT: 180°C.

FLAMMABLE (EXPLOSIVE) LIMITS: UPPER 12.6% / LOWER 3.2%

METHOD USED: O.C.

EXTINGUISHING MEDIA: Water may be used to blanket the fire,foam, carbon dioxide, dry chemical.

SPECIAL FIRE FIGHTING PROCEDURES: Wear self-contained breathing apparatus.

UNUSUAL FIRE AND EXPLOSION HAZARDS

SECTION V HEALTH HAZARD DATA

TOXIC LEVEL: 50 PPM, 300 mg/C.M.

CARCINOGENIC: no

PRINCIPLE ROUTES OF ABSORBTION: Inhalation, ingestion ,skin

SKIN AND EYE IRRITATION: Severe

REVELANT SYMPTOMS OF EXPOSURE: Eye and throat irritation, edema and hemorrhaging in lungs, narcosis.

EFFECTS OF CHRONIC EXPOSURE: Heart and liver damage.

EMERGENCY AND FIRST AID PROCEDURES: If swallowed - give emetic, get medical aid; inhalation - rest, artificial respiration if needed; wash from skin & eyes.

SECTION VI REACTIVITY DATA

CONDITIONS CONTRIBUTING TO INSTABILITY

CONDITIONS CONTRIBUTING TO HAZARDOUS POLYMERIZATION

INCOMPATABILITY (MATERIALS TO AVOID)

HAZARDOUS DECOMPOSITION PRODUCTS

SECTION VII SPILL OR LEAK PROCEDURES

STEPS TO BE TAKEN IN CASE MATERIAL IS RELEASED OR SPILLED

WASTE DISPOSAL METHOD

SECTION VIII SPECIAL PROTECTION INFORMATION

VENTILATION REQUIREMENTS LOCAL EXHAUST

PROTECTIVE EQUIPMENT (SPECIFY TYPES) EYE: Goggles

MECHANICAL (GENERAL)

GLOVES: Rubber

SPECIAL

RESPIRATOR: With proper filter

OTHER PROTECTIVE EQUIPMENT

SECTION IX SPECIAL PRECAUTIONS

PRECAUTIONS TO BE TAKEN IN HANDLING AND STORAGE

OTHER PRECAUTIONS

SIGNATURE | DATE

TRIMETHYLENE BROMIDE

(Hazard rating: 2 / 3 / 0)

SECTION I SOURCE AND NOMENCLATURE

MANUFACTURER'S NAME | EMERGENCY TELEPHONE NO.

ADDRESS (NUMBER, STREET, CITY, STATE, ZIP CODE)

TRADE NAME AND SYNONYMS | CHEMICAL FAMILY

CHEMICAL NAME AND SYNONYMS: 1,3 - Dibromopropane

FORMULA: $CH_2BrCH_2CH_2Br$

SECTION II HAZARDOUS INGREDIENTS

BASIC MATERIAL	APPROXIMATE OR MAXIMUM % WT. OR VOL.	ESTABLISHED OSHA STANDARD	LD_{50} ORAL	LD_{50} PERCUT.	LC_{50} SPECIES	LC_{50} CONC.
			See Note			
Lowest published lethal dose Intraperitoneal - mouse	750 mg/kg					

SECTION III PHYSICAL DATA MW - 201.91

BOILING POINT (°F.): 167°C.	VAPOR PRESSURE (mm Hg.):
MELTING POINT (°F.): -36°C.	VAPOR DENSITY (Air = 1):
SPECIFIC GRAVITY (H_2O = 1): 1.97	EVAPORATION RATE (= 1):
SOLUBILITY IN WATER Pts/100 pts H_2O: 0.02	VOLATILE % Vol. % Wt.

APPEARANCE AND ODOR: colorless liquid

SECTION IV FIRE AND EXPLOSION HAZARD DATA

FLASH POINT: none available

METHOD USED

FLAMMABLE (EXPLOSIVE) LIMITS: UPPER | LOWER

EXTINGUISHING MEDIA: Foam, carbon dioxide, dry chemical.

SPECIAL FIRE FIGHTING PROCEDURES

UNUSUAL FIRE AND EXPLOSION HAZARDS

SECTION V HEALTH HAZARD DATA

TOXIC LEVEL: none available — CARCINOGENIC: no

PRINCIPLE ROUTES OF ABSORBTION: inhalation, ingestion — SKIN AND EYE IRRITATION: moderate

REVELANT SYMPTOMS OF EXPOSURE: Eye irritation, narcosis, sleepiness, headache, vomiting.

EFFECTS OF CHRONIC EXPOSURE: Dermatitis, liver and kidney damage.

EMERGENCY AND FIRST AID PROCEDURES: Inhalation - rest; if swallowed - give emetic, get medical aid; wash from skin and eyes.

SECTION VI REACTIVITY DATA

CONDITIONS CONTRIBUTING TO INSTABILITY

CONDITIONS CONTRIBUTING TO HAZARDOUS POLYMERIZATION

INCOMPATABILITY (MATERIALS TO AVOID)

HAZARDOUS DECOMPOSITION PRODUCTS

SECTION VII SPILL OR LEAK PROCEDURES

STEPS TO BE TAKEN IN CASE MATERIAL IS RELEASED OR SPILLED

WASTE DISPOSAL METHOD

SECTION VIII SPECIAL PROTECTION INFORMATION

VENTILATION REQUIREMENTS LOCAL EXHAUST — PROTECTIVE EQUIPMENT (SPECIFY TYPES) EYE: Goggles

MECHANICAL (GENERAL) — GLOVES: Rubber

SPECIAL — RESPIRATOR: With proper filter

OTHER PROTECTIVE EQUIPMENT

SECTION IX SPECIAL PRECAUTIONS

PRECAUTIONS TO BE TAKEN IN HANDLING AND STORAGE

OTHER PRECAUTIONS

SIGNATURE | DATE

TRIMETHYLENE GLYCOL

(Hazard rating: 1 / 3 / 0)

SECTION I SOURCE AND NOMENCLATURE

MANUFACTURER'S NAME | EMERGENCY TELEPHONE NO.

ADDRESS (NUMBER, STREET, CITY, STATE, ZIP CODE)

TRADE NAME AND SYNONYMS: 1,3 dihydroxy propane | CHEMICAL FAMILY

CHEMICAL NAME AND SYNONYMS: Trimethylene Glycol

FORMULA: $HOCH_2CH_2CH_2OH$

SECTION II HAZARDOUS INGREDIENTS

BASIC MATERIAL	APPROXIMATE OR MAXIMUM % WT. OR VOL.	ESTABLISHED OSHA STANDARD	LD_{50} ORAL	LD_{50} PERCUT.	LC_{50} SPECIES	LC_{50} CONC.
			none	available		

SECTION III PHYSICAL DATA MW - 76.11

BOILING POINT (°F.): 211°C.	VAPOR PRESSURE (mm Hg.): at 125°C. 30
MELTING POINT (°F.):	VAPOR DENSITY (Air = 1): 2.6
SPECIFIC GRAVITY (H_2O = 1): 1.05	EVAPORATION RATE (= 1):
SOLUBILITY IN WATER Pts/100 pts H_2O:	VOLATILE % Vol. % Wt.

APPEARANCE AND ODOR: Colorless, odorless liquid.

SECTION IV FIRE AND EXPLOSION HAZARD DATA

FLASH POINT: None available

METHOD USED

FLAMMABLE (EXPLOSIVE) LIMITS: UPPER | LOWER

EXTINGUISHING MEDIA: Water, carbon dioxide, dry chemical.

SPECIAL FIRE FIGHTING PROCEDURES

UNUSUAL FIRE AND EXPLOSION HAZARDS

SECTION V HEALTH HAZARD DATA

TOXIC LEVEL: None available — CARCINOGENIC: no

PRINCIPLE ROUTES OF ABSORBTION: Inhalation, ingestion — SKIN AND EYE IRRITATION: Slight

REVELANT SYMPTOMS OF EXPOSURE: Coldness, spastic contraction of muscles, loss of reflexes, coma, death; pulmonary edema, eye irritation.

EFFECTS OF CHRONIC EXPOSURE: Kidney damage, rapid weight loss, weakened reflexes.

EMERGENCY AND FIRST AID PROCEDURES: If swallowed - give emetic; get medical aid; inhalation - rest; wash from eyes.

SECTION VI REACTIVITY DATA

CONDITIONS CONTRIBUTING TO INSTABILITY

CONDITIONS CONTRIBUTING TO HAZARDOUS POLYMERIZATION

INCOMPATABILITY (MATERIALS TO AVOID)

HAZARDOUS DECOMPOSITION PRODUCTS

SECTION VII SPILL OR LEAK PROCEDURES

STEPS TO BE TAKEN IN CASE MATERIAL IS RELEASED OR SPILLED

WASTE DISPOSAL METHOD

SECTION VIII SPECIAL PROTECTION INFORMATION

VENTILATION REQUIREMENTS LOCAL EXHAUST — PROTECTIVE EQUIPMENT (SPECIFY TYPES) EYE: Goggles

MECHANICAL (GENERAL) — GLOVES

SPECIAL — RESPIRATOR

OTHER PROTECTIVE EQUIPMENT

SECTION IX SPECIAL PRECAUTIONS

PRECAUTIONS TO BE TAKEN IN HANDLING AND STORAGE

OTHER PRECAUTIONS

SIGNATURE | DATE

2,4,6-TRINITRORESORCINOL

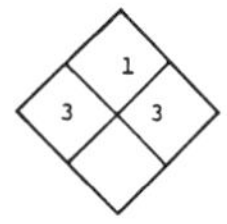

SECTION I SOURCE AND NOMENCLATURE

MANUFACTURER'S NAME | EMERGENCY TELEPHONE NO.

ADDRESS (NUMBER, STREET, CITY, STATE, ZIP CODE)

TRADE NAME AND SYNONYMS: Styphnic acid, 4 di hydroxy-1,3,5 trinitrobenzene

CHEMICAL FAMILY

CHEMICAL NAME AND SYNONYMS: 2,4,6 - Trinitroresorcinol

FORMULA: $(NO_2)_3C_6H(OH)_2$

SECTION II HAZARDOUS INGREDIENTS

BASIC MATERIAL	APPROXIMATE OR MAXIMUM % WT. OR VOL.	ESTABLISHED OSHA STANDARD	LD_{50} ORAL	LD_{50} PERCUT.	SPECIES	LC_{50} CONC.
			None available			

SECTION III PHYSICAL DATA MW 245.12

BOILING POINT °F. | VAPOR PRESSURE mm Hg.

MELTING POINT °F.: 176°C. | VAPOR DENSITY (Air = 1)

SPECIFIC GRAVITY (H_2O = 1) | EVAPORATION RATE (________ = 1)

SOLUBILITY IN WATER Pts/100 pts H_2O: 0.6 | VOLATILE % Vol. % Wt.

APPEARANCE AND ODOR: Yellow crystals

SECTION IV FIRE AND EXPLOSION HAZARD DATA

FLASH POINT: None available

METHOD USED

FLAMMABLE (EXPLOSIVE) LIMITS: UPPER | LOWER

EXTINGUISHING MEDIA: Explosive - do not fight fire except for unmanned equipment - use water.

SPECIAL FIRE FIGHTING PROCEDURES: Full protective clothing for rescue operations.

UNUSUAL FIRE AND EXPLOSION HAZARDS: Severe explosion hazard. Dot class A explosive

SECTION V HEALTH HAZARD DATA

TOXIC LEVEL: None available | CARCINOGENIC: no

PRINCIPLE ROUTES OF ABSORBTION: Inhalation, ingestion, skin | SKIN AND EYE IRRITATION: Moderate

REVELANT SYMPTOMS OF EXPOSURE: Temperature elevation, dermatitis.

EFFECTS OF CHRONIC EXPOSURE: Dermatitis, headache.

EMERGENCY AND FIRST AID PROCEDURES: If swallowed - drink water, epsom salts solution, get medical aid; inhalation - rest; wash skin & eyes.

SECTION VI REACTIVITY DATA

CONDITIONS CONTRIBUTING TO INSTABILITY

CONDITIONS CONTRIBUTING TO HAZARDOUS POLYMERIZATION

INCOMPATABILITY (MATERIALS TO AVOID): Explosive (keep from oxidizing materials) inorganic nitrates and chlorates, ammonium nitrate.

HAZARDOUS DECOMPOSITION PRODUCTS: Oxides of nitrogen.

SECTION VII SPILL OR LEAK PROCEDURES

STEPS TO BE TAKEN IN CASE MATERIAL IS RELEASED OR SPILLED

WASTE DISPOSAL METHOD

SECTION VIII SPECIAL PROTECTION INFORMATION

VENTILATION REQUIREMENTS LOCAL EXHAUST | PROTECTIVE EQUIPMENT (SPECIFY TYPES) EYE: Goggles

MECHANICAL (GENERAL) | GLOVES: Rubber

SPECIAL | RESPIRATOR: Self-contained

OTHER PROTECTIVE EQUIPMENT

SECTION IX SPECIAL PRECAUTIONS

PRECAUTIONS TO BE TAKEN IN HANDLING AND STORAGE: Keep from oxidizing materials, combustibles and source of heat.

OTHER PRECAUTIONS

SIGNATURE | DATE

TRI-n-PROPYLAMINE

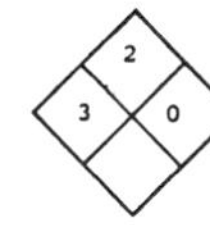

SECTION I SOURCE AND NOMENCLATURE

MANUFACTURER'S NAME | EMERGENCY TELEPHONE NO.

ADDRESS (NUMBER, STREET, CITY, STATE, ZIP CODE)

TRADE NAME AND SYNONYMS

CHEMICAL FAMILY

CHEMICAL NAME AND SYNONYMS: Tri-N-Propylamine

FORMULA: $(C_3H_7)_3N$

SECTION II HAZARDOUS INGREDIENTS

BASIC MATERIAL	APPROXIMATE OR MAXIMUM % WT. OR VOL.	ESTABLISHED OSHA STANDARD	LD_{50} ORAL	LD_{50} PERCUT.	SPECIES	LC_{50} CONC.
			96		rat	
				570	rabbit	
LDLO inhalation - rat 250 PPM/4 hr.						

SECTION III PHYSICAL DATA MW - 143.31

BOILING POINT °F.: 157°C. | VAPOR PRESSURE mm Hg.

MELTING POINT °F.: -94°C. | VAPOR DENSITY (Air = 1): 4.9

SPECIFIC GRAVITY (H_2O = 1): 0.76 | EVAPORATION RATE (________ = 1)

SOLUBILITY IN WATER Pts/100 pts H_2O: very slightly soluble | VOLATILE % Vol. % Wt.

APPEARANCE AND ODOR: colorless liquid

SECTION IV FIRE AND EXPLOSION HAZARD DATA

FLASH POINT: 105°F.

METHOD USED: open cup

FLAMMABLE (EXPLOSIVE) LIMITS: UPPER | LOWER

EXTINGUISHING MEDIA: Carbon dioxide, dry chemical

SPECIAL FIRE FIGHTING PROCEDURES

UNUSUAL FIRE AND EXPLOSION HAZARDS

SECTION V HEALTH HAZARD DATA

TOXIC LEVEL: None available | CARCINOGENIC: no

PRINCIPLE ROUTES OF ABSORBTION: Inhalation, ingestion | SKIN AND EYE IRRITATION: Severe

REVELANT SYMPTOMS OF EXPOSURE: Eye and skin burns; eye irritation and corneal edema, mucous membrane irritation.

EFFECTS OF CHRONIC EXPOSURE: Lung irritation, liver and kidney damage, heart damage and edema.

EMERGENCY AND FIRST AID PROCEDURES: If swallowed - induce vomiting, get medical aid; inhalation - rest; wash eyes and skin.

SECTION VI REACTIVITY DATA

CONDITIONS CONTRIBUTING TO INSTABILITY

CONDITIONS CONTRIBUTING TO HAZARDOUS POLYMERIZATION

INCOMPATABILITY (MATERIALS TO AVOID)

HAZARDOUS DECOMPOSITION PRODUCTS

SECTION VII SPILL OR LEAK PROCEDURES

STEPS TO BE TAKEN IN CASE MATERIAL IS RELEASED OR SPILLED

WASTE DISPOSAL METHOD

SECTION VIII SPECIAL PROTECTION INFORMATION

VENTILATION REQUIREMENTS LOCAL EXHAUST | PROTECTIVE EQUIPMENT (SPECIFY TYPES) EYE: Goggles

MECHANICAL (GENERAL) | GLOVES: Rubber

SPECIAL | RESPIRATOR: With proper filter

OTHER PROTECTIVE EQUIPMENT

SECTION IX SPECIAL PRECAUTIONS

PRECAUTIONS TO BE TAKEN IN HANDLING AND STORAGE

OTHER PRECAUTIONS

SIGNATURE | DATE

m-XYLYL BROMIDE

1 / 3 / 0

SECTION I SOURCE AND NOMENCLATURE

MANUFACTURER'S NAME | EMERGENCY TELEPHONE NO.

ADDRESS (NUMBER, STREET, CITY, STATE, ZIP CODE)

TRADE NAME AND SYNONYMS | CHEMICAL FAMILY

CHEMICAL NAME AND SYNONYMS: M-Xylyl Bromide | FORMULA: $CH_3C_6H_4CH_2Br$

SECTION II HAZARDOUS INGREDIENTS

BASIC MATERIAL	APPROXIMATE OR MAXIMUM % WT. OR VOL.	ESTABLISHED OSHA STANDARD	LD50 ORAL	LD50 PERCUT.	LC50 SPECIES	LC50 CONC.
Lowest published lethal dose						
inhalation - human 75 PPM/10 minutes						

SECTION III PHYSICAL DATA MW - 185.08

BOILING POINT °F.: 215°C.

VAPOR PRESSURE mm Hg.: at 110°C. 25

MELTING POINT °F.: -

VAPOR DENSITY (Air = 1):

SPECIFIC GRAVITY (H_2O = 1): 1.37

EVAPORATION RATE (= 1):

SOLUBILITY IN WATER Pts/100 pts H_2O: insoluble

VOLATILE % Vol. % Wt.

APPEARANCE AND ODOR: Colorless liquid.

SECTION IV FIRE AND EXPLOSION HAZARD DATA

FLASH POINT: None availalbe

FLAMMABLE (EXPLOSIVE) LIMITS: UPPER / LOWER

METHOD USED

EXTINGUISHING MEDIA: Foam, carbon dioxide, dry chemical.

SPECIAL FIRE FIGHTING PROCEDURES: Self-contained breathing apparatus.

UNUSUAL FIRE AND EXPLOSION HAZARDS

SECTION V HEALTH HAZARD DATA

TOXIC LEVEL: None available — CARCINOGENIC: No

PRINCIPLE ROUTES OF ABSORBTION: Inhalation, ingestion — SKIN AND EYE IRRITATION: Severe

REVELANT SYMPTOMS OF EXPOSURE: Eye and respiratory system irritation; severe irritation of digestive tract.

EFFECTS OF CHRONIC EXPOSURE

EMERGENCY AND FIRST AID PROCEDURES: Inhalation - rest; if swallowed - give emetic, get medical aid; wash skin & eyes.

SECTION VI REACTIVITY DATA

CONDITIONS CONTRIBUTING TO INSTABILITY

CONDITIONS CONTRIBUTING TO HAZARDOUS POLYMERIZATION

INCOMPATABILITY (MATERIALS TO AVOID)

HAZARDOUS DECOMPOSITION PRODUCTS

SECTION VII SPILL OR LEAK PROCEDURES

STEPS TO BE TAKEN IN CASE MATERIAL IS RELEASED OR SPILLED

WASTE DISPOSAL METHOD

SECTION VIII SPECIAL PROTECTION INFORMATION

VENTILATION REQUIREMENTS LOCAL EXHAUST | PROTECTIVE EQUIPMENT (SPECIFY TYPES) EYE: Goggles

MECHANICAL (GENERAL) | GLOVES: Rubber

SPECIAL | RESPIRATOR: Self-contained

OTHER PROTECTIVE EQUIPMENT

SECTION IX SPECIAL PRECAUTIONS

PRECAUTIONS TO BE TAKEN IN HANDLING AND STORAGE

OTHER PRECAUTIONS

SIGNATURE | DATE

o-XYLYL BROMIDE

1 / 3 / 0

SECTION I SOURCE AND NOMENCLATURE

MANUFACTURER'S NAME | EMERGENCY TELEPHONE NO.

ADDRESS (NUMBER, STREET, CITY, STATE, ZIP CODE)

TRADE NAME AND SYNONYMS | CHEMICAL FAMILY

CHEMICAL NAME AND SYNONYMS: 2 - Methylbenzyl bromide | FORMULA: $CH_3C_6H_4CH_2Br$

SECTION II HAZARDOUS INGREDIENTS

BASIC MATERIAL	APPROXIMATE OR MAXIMUM % WT. OR VOL.	ESTABLISHED OSHA STANDARD	LD50 ORAL	LD50 PERCUT.	LC50 SPECIES	LC50 CONC.
Lowest published lethal dose -						
Inhalation - human 75 PPM/10 minutes.						

SECTION III PHYSICAL DATA MW - 185.07

BOILING POINT °F.: 224°C.

VAPOR PRESSURE mm Hg.: at 103°C. 25

MELTING POINT °F.: 21°C.

VAPOR DENSITY (Air = 1):

SPECIFIC GRAVITY (H_2O = 1): 1.38

EVAPORATION RATE (= 1):

SOLUBILITY IN WATER Pts/100 pts H_2O: insoluble

VOLATILE % Vol. % Wt.

APPEARANCE AND ODOR: colorless liquid

SECTION IV FIRE AND EXPLOSION HAZARD DATA

FLASH POINT: none available

FLAMMABLE (EXPLOSIVE) LIMITS: UPPER / LOWER

METHOD USED

EXTINGUISHING MEDIA: Foam, carbon dioxide, dry chemical.

SPECIAL FIRE FIGHTING PROCEDURES: Self-contained breathing apparatus.

UNUSUAL FIRE AND EXPLOSION HAZARDS

SECTION V HEALTH HAZARD DATA

TOXIC LEVEL: None available — CARCINOGENIC:

PRINCIPLE ROUTES OF ABSORBTION: Inhalation, ingestion — SKIN AND EYE IRRITATION: Severe

REVELANT SYMPTOMS OF EXPOSURE: Eye and respiratory system irritation; severe irritation of digestive tract.

EFFECTS OF CHRONIC EXPOSURE: --

EMERGENCY AND FIRST AID PROCEDURES: Inhalation - rest; if swallowed - give emetic, get medical aid; wash skin & eyes.

SECTION VI REACTIVITY DATA

CONDITIONS CONTRIBUTING TO INSTABILITY

CONDITIONS CONTRIBUTING TO HAZARDOUS POLYMERIZATION

INCOMPATABILITY (MATERIALS TO AVOID)

HAZARDOUS DECOMPOSITION PRODUCTS

SECTION VII SPILL OR LEAK PROCEDURES

STEPS TO BE TAKEN IN CASE MATERIAL IS RELEASED OR SPILLED

WASTE DISPOSAL METHOD

SECTION VIII SPECIAL PROTECTION INFORMATION

VENTILATION REQUIREMENTS LOCAL EXHAUST | PROTECTIVE EQUIPMENT (SPECIFY TYPES) EYE: Goggles

MECHANICAL (GENERAL) | GLOVES: Rubber gloves

SPECIAL | RESPIRATOR: Self-contained

OTHER PROTECTIVE EQUIPMENT

SECTION IX SPECIAL PRECAUTIONS

PRECAUTIONS TO BE TAKEN IN HANDLING AND STORAGE

OTHER PRECAUTIONS

SIGNATURE | DATE

YTTRIUM METAL

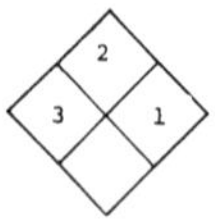

SECTION I SOURCE AND NOMENCLATURE

MANUFACTURER'S NAME | EMERGENCY TELEPHONE NO.

ADDRESS (NUMBER, STREET, CITY, STATE, ZIP CODE)

TRADE NAME AND SYNONYMS | CHEMICAL FAMILY

CHEMICAL NAME AND SYNONYMS: Yttrium | FORMULA: Y

SECTION II HAZARDOUS INGREDIENTS

BASIC MATERIAL	APPROXIMATE OR MAXIMUM % WT. OR VOL.	ESTABLISHED OSHA STANDARD	LD_{50} ORAL	LD_{50} PERCUT.	LC_{50} SPECIES	LC_{50} CONC.
			None available			

SECTION III PHYSICAL DATA AW - 88.91

BOILING POINT °F.: 3200° C.

VAPOR PRESSURE mm Hg.:

MELTING POINT °F.: 1500° C.

VAPOR DENSITY (Air = 1):

SPECIFIC GRAVITY (H_2O = 1): 5.51

EVAPORATION RATE (______ = 1):

SOLUBILITY IN WATER Pts/100 pts H_2O: Slow reaction

VOLATILE % Vol. % Wt.

APPEARANCE AND ODOR: Grey to black metal.

SECTION IV FIRE AND EXPLOSION HAZARD DATA

FLASH POINT: N/A

METHOD USED

FLAMMABLE (EXPLOSIVE) LIMITS: UPPER / LOWER

EXTINGUISHING MEDIA: Sand, dry powder.

SPECIAL FIRE FIGHTING PROCEDURES: Smother with suitable dry powder, no water.

UNUSUAL FIRE AND EXPLOSION HAZARDS: Moderate fire hazard.

SECTION V HEALTH HAZARD DATA

TOXIC LEVEL: TLV - 1 Mg/C.M. CARCINOGENIC: No

PRINCIPLE ROUTES OF ABSORBTION: Inhalation, ingestion

SKIN AND EYE IRRITATION: Moderate

REVELANT SYMPTOMS OF EXPOSURE: Irritation to respiratory system and eyes; kidney damage, internal damage if swallowed.

EFFECTS OF CHRONIC EXPOSURE: --

EMERGENCY AND FIRST AID PROCEDURES: If inhaled - rest, get medical aid; if swallowed - wash mouth, give emetic, get medical aid; wash eyes as needed.

SECTION VI REACTIVITY DATA

CONDITIONS CONTRIBUTING TO INSTABILITY

CONDITIONS CONTRIBUTING TO HAZARDOUS POLYMERIZATION

INCOMPATABILITY (MATERIALS TO AVOID): Oxidizing materials

HAZARDOUS DECOMPOSITION PRODUCTS

SECTION VII SPILL OR LEAK PROCEDURES

STEPS TO BE TAKEN IN CASE MATERIAL IS RELEASED OR SPILLED

WASTE DISPOSAL METHOD

SECTION VIII SPECIAL PROTECTION INFORMATION

VENTILATION REQUIREMENTS LOCAL EXHAUST

PROTECTIVE EQUIPMENT (SPECIFY TYPES) EYE: Safety glasses

MECHANICAL (GENERAL)

GLOVES: Rubber

SPECIAL

RESPIRATOR: With proper filter

OTHER PROTECTIVE EQUIPMENT

SECTION IX SPECIAL PRECAUTIONS

PRECAUTIONS TO BE TAKEN IN HANDLING AND STORAGE

OTHER PRECAUTIONS

SIGNATURE | DATE

YTTRIUM OXIDE

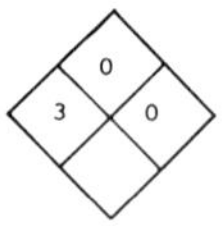

SECTION I SOURCE AND NOMENCLATURE

MANUFACTURER'S NAME | EMERGENCY TELEPHONE NO.

ADDRESS (NUMBER, STREET, CITY, STATE, ZIP CODE)

TRADE NAME AND SYNONYMS | CHEMICAL FAMILY

CHEMICAL NAME AND SYNONYMS: Yttrium Oxide | FORMULA: Y_2O_3

SECTION II HAZARDOUS INGREDIENTS

BASIC MATERIAL	APPROXIMATE OR MAXIMUM % WT. OR VOL.	ESTABLISHED OSHA STANDARD	LD_{50} ORAL	LD_{50} PERCUT.	LC_{50} SPECIES	LC_{50} CONC.
			See Note 1			
#1 LD_{50} intraperitoneal - rat 500 mg/kg.						

SECTION III PHYSICAL DATA

BOILING POINT °F.: 4300

VAPOR PRESSURE mm Hg.:

MELTING POINT °F.: 2410

VAPOR DENSITY (Air = 1):

SPECIFIC GRAVITY (H_2O = 1): 4.84

EVAPORATION RATE (______ = 1):

SOLUBILITY IN WATER Pts/100 pts H_2O: 1.

VOLATILE % Vol. % Wt.

APPEARANCE AND ODOR: Yellow to white powder

SECTION IV FIRE AND EXPLOSION HAZARD DATA

FLASH POINT: Not flammable

METHOD USED

FLAMMABLE (EXPLOSIVE) LIMITS: UPPER / LOWER

EXTINGUISHING MEDIA: Water

SPECIAL FIRE FIGHTING PROCEDURES: N/A

UNUSUAL FIRE AND EXPLOSION HAZARDS: N/A

SECTION V HEALTH HAZARD DATA

TOXIC LEVEL: TLV - 1 Mg/C.M. CARCINOGENIC: No

PRINCIPLE ROUTES OF ABSORBTION: Inhalation, ingestion

SKIN AND EYE IRRITATION: Moderate

REVELANT SYMPTOMS OF EXPOSURE: Irritation of eyes and respiratory tract; kidney damage; internal damage if swallowed.

EFFECTS OF CHRONIC EXPOSURE: --

EMERGENCY AND FIRST AID PROCEDURES: If inhaled - rest; wash from skin and eyes; if swallowed - wash mouth, give emetic, get medical aid.

SECTION VI REACTIVITY DATA

CONDITIONS CONTRIBUTING TO INSTABILITY

CONDITIONS CONTRIBUTING TO HAZARDOUS POLYMERIZATION

INCOMPATABILITY (MATERIALS TO AVOID)

HAZARDOUS DECOMPOSITION PRODUCTS

SECTION VII SPILL OR LEAK PROCEDURES

STEPS TO BE TAKEN IN CASE MATERIAL IS RELEASED OR SPILLED

WASTE DISPOSAL METHOD

SECTION VIII SPECIAL PROTECTION INFORMATION

VENTILATION REQUIREMENTS LOCAL EXHAUST

PROTECTIVE EQUIPMENT (SPECIFY TYPES) EYE: Safety glasses

MECHANICAL (GENERAL)

GLOVES: Rubber

SPECIAL

RESPIRATOR

OTHER PROTECTIVE EQUIPMENT

SECTION IX SPECIAL PRECAUTIONS

PRECAUTIONS TO BE TAKEN IN HANDLING AND STORAGE

OTHER PRECAUTIONS

SIGNATURE | DATE

Index

PRIORITY TOXIC POLLUTANTS
Health Effects and Allowable Limits 1980

Edited by Marshall Sittig

Enivronmental Health Review No. 1

This book is a practical manual with its contents arranged alphabetically in encyclopedic form. It is intended to provide specific information on the 65 priority toxic pollutants (actually reflecting 129 individual compounds), their derivatives or degradation products, or intermedia transfers.

Compelled by a consent decree obtained in a federal court by public interest groups, the Environmental Protection Agency has promulgated criteria on allowable limits and profuse guidelines on how to interpret them and also how to determine further such criteria.

Here, in ready reference form, are the essentials of the criteria and guidelines which so far have been published for the control of priority toxic pollutants. Each typical pollutant is characterized with descriptions of its:

Occurrence
Physical Properties
Chemical Properties
Uses
Toxic Effects
Current Levels of Exposure
Special Groups at Risk
Existing Guidelines & Standards
Summary of Proposed Criteria
Bases for the Human Health Criteria
Pertinent References

A list of the 65 original priority pollutants follows here, however all of the 129 compounds are covered in this book.

1. **Acenaphthene**
2. **Acrolein**
3. **Acrylonitrile**
4. **Aldrin/Dieldrin**
5. **Antimony & Compounds**
6. **Arsenic & Compounds**
7. **Asbestos**
8. **Benzene**
9. **Benzidine**
10. **Beryllium & Compounds**
11. **Cadmium & Compounds**
12. **Carbon Tetrachloride**
13. **Chlordane**
14. **Chlorobenzenes**
15. **Chlorinated Ethanes**
16. **Chloroalkyl Ethers**
17. **Chlorinated Naphthalenes**
18. **Chlorinated Phenols**
19. **Chloroform**
20. **2-Chlorophenol**
21. **Chromium & Compounds**
22. **Copper & Compounds**
23. **Cyanides**
24. **DDT & Metabolites**
25. **Dichlorobenzenes**
26. **Dichlorobenzidine**
27. **Dichloroethylene**
28. **2,4-Dichlorophenol**
29. **Dichloropropane/propene**
30. **2,4-Dimethylphenol**
31. **Dinitrotoluenes**
32. **Diphenylhydrazines**
33. **Endosulfan**
34. **Endrin**
35. **Ethylbenzene**
36. **Fluoranthene**
37. **Haloethers**
38. **Halomethanes**
39. **Heptachlor**
40. **Hexachlorobutadiene**
41. **Hexachlorocyclohexane**
42. **Hexachlorocyclopentadiene**
43. **Isophorone**
44. **Lead & Compounds**
45. **Mercury & Compounds**
46. **Naphthalene**
47. **Nickel & Compounds**
48. **Nitrobenzene**
49. **Nitrophenols**
50. **Nitrosamines**
51. **Pentachlorophenol**
52. **Phenol**
53. **Phthalate Esters**
54. **Polychlorinated Biphenyls**
55. **Polynuclear Aromatics**
56. **Selenium & Compounds**
57. **Silver & Compounds**
58. **Tetrachlorodibenzo-p-dioxin**
59. **Tetrachloroethylene**
60. **Thallium & Compounds**
61. **Toluene**
62. **Toxaphene**
63. **Trichloroethylene**
64. **Vinyl Chloride**
65. **Zinc & Compounds**

ISBN 0-8155-0797-6 **370 pages**

METAL AND INORGANIC WASTE RECLAIMING ENCYCLOPEDIA 1980

by Marshall Sittig

Pollution Technology Review No. 70
Chemical Technology Review No. 175

This book describes processes for reclaiming metal and inorganic wastes arranged in encyclopedic form. It contains approximately 325 recent processes for metal and inorganic waste recycling and recovery.

The recovery and recycling of industrial wastes is assuming ever-increasing importance due to ore and mineral conservation and pollution regulations. The technology of a specific recovery and recycling process is outlined as well as background as to the origin of the waste material, some of the problems posed by the waste and possible alternative processes for recovery and recycling.

The partial table of contents below lists selected examples of *69* recoverable industrial waste processes. In addition there are *259* other recent processes and developments for numerous other products.

ISBN 0-8155-0823-9

591 pages

HAZARDOUS CHEMICAL SPILL CLEANUP 1979

Edited by J. S. Robinson

Pollution Technology Review No. 59

There is a delicate balance in nature involving a perfect blending of chemical, physical and biological elements. Once this balance is upset by a spillage of chemicals, it may take years to restore the original environment. Meanwhile, untold damage results. The deleterious effects, however, can be lessened if there is a rapid response reinforced by expertise in proper procedure. It is the intent of this book to provide the specific know-how needed to cope with chemical spillage quickly by correct containment, countermeasures, and cleanup.

After an introductory overview of legislation, Robinson offers practical advice and response techniques for different kinds of substances, including preferred methods of disposal and restoration. Managers in all fields with problems of potential spills should avail themselves of this information now.

The partial table of contents below lists chapter headings and **examples of some** subtitles.

ISBN 0-8155-0767-4

406 pages